7728
3-90

OCEANOGRAPHIC INDEX

Cumulation 1946-1973:
Marine Organisms, Chiefly Planktonic

Woods Hole Oceanographic Institution

Compiled by
Dr. Mary Sears

Volume 3
OSTRACODS (Hadacypridina) – FUNGI (Zignoella)

G. K. HALL & CO., 70 LINCOLN STREET, BOSTON, MASSACHUSETTS
1974

This publication is printed on permanent/durable acid-free paper.

Copyright 1974 by G. K. Hall & Co.

LIBRARY
LOS ANGELES COUNTY MUSEUM OF NATURAL HISTORY

ISBN 0-8161-0933-8

Oceanographic Index: Marine Organisms Cumulation, 1946-1973

Hadacypridina bruuni n.gen., n.sp.

Poulsen, Erik M., 1962
Ostracoda-Myodocopa. Part 1.
Cyprindiniformes-Cyprindinidae.
Dana-Report No. 57:414 pp.

Halocypris brevirostris

Deevey, Georgiana B. 1970.
Pelagic ostracods (Myodocopa Halocypri-didae) from the North Atlantic off Barbados.
Proc. Biol. Soc. Wash. 82(62): 799-824.

Halocypris brevirostris

Deevey, Georgiana B., 1968.
Pelagic ostracods of the Sargasso Sea off Bermuda.
Bull. Peabody Mus. nat. Hist., Yale Univ., 26:125 pp

Halocypris brevirostris

Granata, L., and L. di Caporiacco, 1949
Ostracodes marins recueillis pendant les croiserieres du Prince Albert, 1. Res. Camp.
Sci. Monaco, 109: 48 pp., 4 pls.

Halocypris brevirostris

Iles, E.J., 1953.
A preliminary report on the Ostracoda of the Benguela Current. Discovery Repts. 26:259-280, 5 textfigs.

Halocypris brevirostris

Ramirez, Fernando C., y Alicia Moguilevesky 1971.
Ostracodos planctonicos hallados en aguas oceanicas frente a la Provincia de Buenos Aires (resultados de la XLI Comissao Oceanografica Costa Sul).
Physis, 30 (81): 637-666.
Buenos Aires

Halocypris brevirostris

Skogsberg, T., (1931) 1944.
Ostracods. Rept. Sci. Res. "Michael Sars" N. Atlantic Deep-sea Exped., 1910, 5(1):1-26, 5 text-figs.

Halocypris concha

Vavra, V., 1906
Die Ostracoden (Halocypriden und Cypridiniden) der Plankton Expedition.
Ergeb. Plankton-Exped. Humboldt-Stiftung.
Vol.2 G.g., 1-76, 8 pls. (of 142 figs.)

Halocypris globosa

Cannon, H. G., 1940
Ostracoda. John Murray Exped., 1933-34,
Sci. Repts. 6(8):319-325, 3 text figs.

Halocypris globosa

Deevey, Georgiana B., 1968.
Pelagic ostracods of the Sargasso Sea off Bermuda.
Bull. Peabody Mus. nat. Hist., Yale Univ., 26:125 pp.

Halocypris globosa

Granata, L., and L. di Caporiacco, 1949
Ostracodes marins recueillis pendant les croiserieres du Prince Albert, 1. Res. Camp.
Sci. Monaco, 109: 48 pp., 4 pls.

Halocypris globosa?

Kielhorn, W.V., 1952
The biology of the surface zone zooplankton of a Boreo-Arctic Atlantic Ocean area. J. Fish Res. Bd., Canada 9 (5): 223-264, 13 text figs.

Halocypris globosa

Klie, W., 1944.
Ostracoda II. Family Conchoeciidae. Fiches d'Ident. Zooplancton, Cons. Perm. Int. Expl. Mer, 6:4 pp., 7 textfigs.

Halocypris globosa

Ramirez, Fernando C., y Alicia Moguilevesky 1971.
Ostracodos planctonicos hallados en aguas oceanicas frente a la Provincia de Buenos Aires (resultados de la XLI Comissao Oceanografica Costa Sul).
Physis, 30 (81): 637-666.
Buenos Aires

Halocypris globosa

Skogsberg, T., (1931) 1944.
Ostracods. Rept. Sci. Res. "Michael Sars" N. Atlantic Deep-sea Exped., 1910, 5(1):1-26, 5 text figs.

Halocypris globosa

Vavra, V., 1906
Die Ostracoden (Halocypriden und Cypridiniden) der Plankton Expedition.
Ergeb. Plankton-Exped. Humboldt-Stiftung.
Vol.2 G.g., 1-76, 8 pls. (of 142 figs.)

Halocypris inflata

Leveau, Michel, 1967.
Ostracodes pélagiques du sud-ouest de l'Océan Indien (Région de Tuléar).
Recl. Trav. Stn. Mar. Endoume, hors Série, Suppl., 6: 63-70.

Halocypris pelagica

Juday, C., 1906.
IX. Ostracoda of the San Diego Region. 1. Halocypridae. Contributions from the Laboratory of the Marine Biological Association of San Diego.
Univ. Calif. Publ. Zool. 3(2):13-38, Pls. 3-7.

Halocypris pelagica

Vavra, V., 1906
Die Ostracoden (Halocypriden und Cypridiniden) der Plankton Expedition.
Ergeb. Plankton-Exped. Humboldt-Stiftung.
Vol.2 G.g., 1-76, 8 pls. (of 142 figs.)

? Halocypris striata

Granata, L., and L. di Caporiacco, 1949
Ostracodes marins recueillis pendant les croiserieres du Prince Albert, 1. Res. Camp.
Sci. Monaco, 109: 48 pp., 4 pls.

Halocypris taurina n.sp.

Vavra, V., 1906
Die Ostracoden (Halocypriden und Cypridiniden) der Plankton Expedition.
Ergeb. Plankton-Exped. Humboldt-Stiftung.
Vol.2 G.g., 1-76, 8 pls. (of 142 figs.)

Haplocytheridea bradyi

King, Charles E. and Louis S. Kornicker, 1970.
Ostracoda in Texas bays and lagoons: an ecological study.
Smithson. Contrib. Zool., 24: 92 pp.

Haplocytheridea maia n.sp.

Benson, Richard H., 1959.
Arthropoda 1. Ecology of recent ostracodes of the Todos Santos Bay region, Baja California, Mexico. Univ. Kansas, Paleontol. Contr., :1-80, 11 pls., 20 figs.

Haplocytheridea palda n.sp.

Benson, Richard H., 1959.
Arthropoda 1. Ecology of recent ostracodes of the Todos Santos Bay region, Baja California, Mexico. Univ. Kansas, Paleontol. Contr., :1-80, 11 pls., 20 figs.

Haplocytheridea setipunctata

King, Charles E. and Louis S. Kornicker, 1970.
Ostracoda in Texas bays and lagoons: an ecological study.
Smithson. Contrib. Zool., 24: 92 pp.

Haplocytheridea setipunctata

Williams, Roger B., 1966.
Recent marine podocopid Ostracoda of Narregansett Bay, Rhode Island.
Univ. Kansas, Paleont. Contrib., Paper 11:36 pp.

Hemicythere (?) spp.

Hartmann, Gerd, 1964.
Zur Kenntnis der Ostracoden des Roten Meeres. Wissenschaftliche Ergebnisse einer Forschungs-reise von A. Remane und E. Schulz nach dem Roten Meer. E. 11.
Kieler Meeresf., 20:35-127.

Hemicythere sp.

Swain, Frederick M., 1967.
Ostracoda from the Gulf of California.
Mem. Geol. Soc. Am., 101: 139 pp.

Hemicythere angulata

Stephensen, K., 1938.
Marine Ostracoda and Cladocera. Zool. Iceland 3(32):19 pp., 1 textfig.

Hemicythere bicarina n.sp.

Smith, Verna Z., 1952.
Further Ostracoda of the Vancouver Island region.
J. Fish. Res. Bd., Canada, 9(1):16-41, 11 pls.

Hemicythere(?) borchersi n.sp.

Hartmann, Gerd, 1964.
Zur Kenntnis der Ostracoden des Roten Meeres. Wissenschaftliche Ergebnisse einer Forschungs-reise von A. Remane und E. Schulz nach dem Roten Meer. E. 11.
Kieler Meeresf., 20:35-127.

Hemicythere californiensis

Benson, Richard H., 1959.
Arthropoda 1. Ecology of recent ostracodes of the Todos Santos Bay region, Baja California, Mexico. Univ. Kansas, Paleontol. Contr., :1-80, 11 pls., 20 figs.

Hemicythere californiensis

Swain, Frederick M., 1967.
Ostracoda from the Gulf of California.
Mem. Geol. Soc. Am., 101: 139 pp.

Hemicythere concinna

Stephensen, K., 1938.
Marine Ostracoda and Cladocera. Zool. Iceland 3(32):19 pp., 1 textfig.

Hemicythere emarginata

Stephensen, K., 1938.
Marine Ostracoda and Cladocera. Zool. Iceland 3(32):19 pp.

Hemicythere emarginata

Williams, Roger B., 1966.
Recent marine podocopid Ostracoda of Narregansett Bay, Rhode Island.
Univ. Kansas, Paleont. Contrib., Paper 11:36 pp.

Hemicythere jollaensis

Benson, Richard H., 1959.
Arthropoda 1. Ecology of recent ostracodes of the Todos Santos Bay region, Baja California, Mexico. Univ. Kansas, Paleontol. Contr., :1-80, 11 pls., 20 figs.

Hemicythere latimarginata

Stephensen, K., 1938.
Marine Ostracoda and Cladocera. Zool. Iceland 3(32):19 pp., 1 textfig.

Hemicythere obesa

Smith, Verna Z., 1952.
Further Ostracoda of the Vancouver Island region.
J. Fish. Res. Bd., Canada, 9(1):16-41, 11 pls.

Hemicythere palosensis

Benson, Richard H., 1959.
Arthropoda 1. Ecology of recent ostracodes of the Todos Santos Bay region, Baja California, Mexico. Univ. Kansas, Paleontol. Contr., :1-80, 11 pls., 20 figs.

Hemicythere quadridentata

Stephensen, K., 1938.
Marine Ostracoda and Cladocera. Zool. Iceland 3(32):19 pp., 1 textfig.

Hemicytherura videns aegyptica n.subsp.

Hartmann, Gerd, 1964.
Zur Kenntnis der Ostracoden des Roten Meeres. Wissenschaftliche Ergebnisse einer Forschungsreise von A. Remane und E. Schulz nach dem Roten Meer. E. 11.
Kieler Meeresf., 20:35-127.

Hemicythere villosa

Stephensen, K., 1938.
Marine Ostracoda and Cladocera. Zool. Iceland 3(32):19 pp., 1 textfig.

Hemicytheridea sp.

Benson, Richard H., 1959.
Arthropoda 1. Ecology of recent ostracodes of the Todos Santos Bay region, Baja California, Mexico. Univ. Kansas, Paleontol. Contr., :1-80, 11 pls., 20 figs.

Hemicytheridea portjacksonensis n.sp.

McKenzie, K.G. 1967.
Recent Ostracoda from Port Phillip Bay, Victoria.
Proc. R. Soc. Victoria, n.s., 80(1): 61-106.

Hemicytherura anomala

Neale, J. W., 1967.
An ostracod fauna from Halley Bay, Coats Land, British Antarctic Territory.
Brit. Antarct. Surv. Sci. Rep. 58: 50 pp.

Hemicytherura sp.cf. H. clathrata

Benson, Richard H., 1959.
Arthropoda 1. Ecology of recent ostracodes of the Todos Santos Bay region, Baja California, Mexico. Univ. Kansas, Paleontol. Contr., :1-80, 11 pls., 20 figs.

Hemicytherura cuneata n.sp.

Hanai, T., 1957.
Studies on the Ostracoda from Japan. III. Subfamilies Cytherurinae G.W. Müller (emend. G.O. Sars 1925) and Cytheropterinae n.subfam.
J. Fac. Sci., Univ. Tokyo, (II), 11(1):11-34.

Hemicythura irregularis

Neale, J. W., 1967.
An ostracod fauna from Halley Bay, Coats Land, British Antarctic Territory.
Brit. Antarct. Surv. Sci. Rep. 58: 50 pp.

Hemicytherura kajiyamai n.sp.

Hanai, T., 1957.
Studies on the Ostracoda from Japan. III. Subfamilies Cytherurinae G.W. Müller (emend. G.O. Sars 1925) and Cytheropterinae n.subfam.
J. Fac. Sci., Univ. Tokyo, (II), 11(1):11-34.

Hemicytherura seaholmensis n.sp.

McKenzie, K.G. 1967.
Recent Ostracoda from Port Phillip Bay, Victoria.
Proc. R. Soc. Victoria n.s. 80(1): 61-106.

Hemicytherura tricarinata n.sp.

Hanai, T., 1957.
Studies on the Ostracoda from Japan. III. Subfamilies Cytherurinae G.W. Müller (emend. G.O. Sars 1925) and Cytheropterinae n.subfam.
J. Fac. Sci., Univ. Tokyo, (II), 11(1):11-34.

Hemikrithe orientalis n.sp.

van den Bold, W.A., 1950.
Hemikrithe, a new genus of Ostracoda from the Indopacific. Ann. Mag. Nat. Hist., 12th ser., 3(34):900-904, 1 fig.

Hemikrithe n.gen.

van den Bold, W.A., 1950.
Hemikrithe, a new genus of Ostracoda from the Indopacific. Ann. Mag. Nat. Hist., 12th ser., 3(34):900-904, 1 fig.

Henryhowella sarsi

Ascoli, P., 1965.
Crociera Talassografia Afriatica 1955. VI. Ricerche ecologiche sugli ostracodi contenuti in 16 carote prelevate sul fondo del Mare Adriatico.
Arch. Oceanogr. e Limnol., 14(1):69-138.

Heptonema keiensis n. sp.

Poulson, Erik M., 1965.
Ostracoda-Myodocopa. II. Cypridiniformes-Rutidermatidae, Sarsiellidae and Asteropidae.
Dana Report, No. 65:484 pp.

Heptonema serrata n. gen., n. sp.

Poulson, Erik M., 1965.
Ostracoda-Myodocopa. II. Cypridiniformes-Rutidermatidae, Sarsiellidae and Asteropidae.
Dana Report, No. 65:484 pp.

Hermanites costata

Hulings, Neil C., 1967.
Marine Ostracoda from the western North Atlantic Ocean: Labrador Sea, Gulf of St. Lawrence, and off Nova Scotia.
Crustaceana, 13(3):310-328.

Herpetocypris agilis n.sp.

Rome, R., 1954.
Contribution à l'étude des ostracodes de Belgique II. Espèces rares et espèces nouvelles.
Bull. Inst. Sci. Nat. Belg. 30(33):1-32, 7 textfigs.

Herpetocypris caerulea n.sp.

Rome, R., 1954.
Contribution à l'étude des ostracodes de Belgique II. Espèces rares et espèces nouvelles.
Bull. Inst. Sci. Nat. Belg. 30(33):1-32, 7 textfigs.

Heterocyprideis sorbyana n.gen.

Elofson, O. 1941
Zur Kenntnis der marinen Ostracodes Schwedens mit besonderer Berücksichtigung des Skageraks. Zool. Bidrag. Uppsala 19: 215-534; 42 maps, 52 figs.

Heterocyprideis sorbyana

Hulings, Neil C., 1967.
Marine Ostracoda from the western North Atlantic Ocean: Labrador Sea, Gulf of St. Lawrence, and off Nova Scotia.
Crustaceana, 13(3):310-328.

Heterocypris incongruens

Dulk, A. den, 1951.
Zeldzame Ostracoden in de Biesbos.
Levende Natuur 54(9):176-178.

Heterocypris salinus

Ganning, B., 1971.
On the ecology of Heterocypris salinus, H. incongruens and Cypridopsis aculeata (Crustacea: Ostracoda) from Baltic brackish-water rockpools.
Mar. Biol., 8(4): 271-279.

Hirschmannia viridis n.gen.

Elofson, O. 1941
Zur Kenntnis der marinen Ostracodes Schwedens mit besonderer Berücksichtigung des Skageraks. Zool. Bidrag. Uppsala 19: 215-534; 42 maps, 52 figs.

Howeina camptocytheroidea n.sp.

Hanai, T., 1957.
Studies on the Ostracoda from Japan. III. Subfamilies Cytherurinae G.W. Müller (emend. G.O. Sars 1925) and Cytheropterinae n.subfam.
J. Fac. Sci., Univ. Tokyo, (II), 11(1):11-34.

Hulingsina americana

Williams, Roger B., 1966.
Recent marine podocopid Ostracoda of Narragansett Bay, Rhode Island.
Univ. Kansas, Paleont. Contrib., Paper 11:36 pp.

Hulingsina sandersi

King, Charles E. and Louis S. Kornicker, 1970.
Ostracoda in Texas bays and lagoons: an ecological study.
Smithson. Contrib. Zool., 24: 92 pp.

Iliocythere sp.

Hartmann, G., 1952.
Zur Kenntnis des Mangrove-Estero-Gebeites von El Salvador und seiner Ostracoden-Fauna. II. Systematischer Teil. Kieler Meeresf. 13(1): 134-159.

Iliocythere dentatomarginata

Hartmann, Gerd, 1959.
Zur Kenntnis der lotischen Lebensbereiche der Pazifischen Küste von El Salvador unter besonderer Berücksichtigung seiner Ostracodenfauna.
Kieler Meeresf., 15(2):187-241.

Iliocythere dentata-marginata n. sp.

Hartmann, G., 1952
Zur Kenntnis des Mangrove-Estero-Gebeites von El Salvador und seiner Ostracoden-Fauna. II. Systematischer Teil. Kieler Meeresf. 13(1): 134-159.

Iliocythere meyer-abichi

Hartmann, G., 1952.
Zur Kenntnis des Mangrove-Estero-Gebeites von El Salvador und seiner Ostracoden-Fauna. II. Systematischer Teil. Kieler Meeresf. 13(1): 134-159.

Iliocythere punctata n. sp.

Hartmann, G., 1952
Zur Kenntnis des Mangrove-Estero-Gebeites von El Salvador und seiner Ostracoden-Fauna. II. Systematischer Teil. Kieler Meeresf. 13(1): 134-159.

Isocypris quadrisetosa

Rome, R., 1954.
Contribution à l'étude des ostracodes de Belgique II. Espèces rares et espèces nouvelles.
Bull. Inst. Sci. Nat. Belg. 30(33):1-32, 7 textfigs.

Jonesia simplex

Hulings, Neil C., 1967.
Marine Ostracoda from the western North Atlantic Ocean: Labrador Sea, Gulf of St. Lawrence, and off Nova Scotia.
Crustaceana, 13(3):310-328.

Jugosocythereis pannosa

Van den Bold, W.A. 1966.
Ostracoda from Colon Harbour, Panama
Carib. J. Sci. 6 (1½):43-64.

Kangarina cf. quellita

Swain, Frederick M., 1967.
Ostracoda from the Gulf of California.
Mem. Geol. Soc. Am., 101: 139 pp.

Kobayashiina hyalinosa n.gen., n.sp.

Hanai, T., 1957.
Studies on the Ostracoda from Japan. III.
Subfamilies Cytherurinae G.W. Müller (emend. G.O. Sars 1925) and Cytheropterinae n.subfam.
J. Fac. Sci., Univ. Tokyo, (II), 11(1):11-34.

Krithe bartonensis

Elofson, O. 1941
Zur Kenntnis der marinen Ostracodes Schwedens mit besonderer Berücksichtigung des Skageraks. Zool. Bidrag. Uppsala 19: 215-534; 42 maps, 52 figs.

Krithe aff. bartonensis

Hanai, T., 1959.
Studies on the Ostracoda from Japan. IV. Family Cytherideidae.
J. Fac. Sci., Univ. Tokyo, (II) 11(3):291-308.

Kritha bartonensis

Stephensen, K., 1938.
Marine Ostracoda and Cladocera. Zool. Iceland 3(32):19 pp., 1 textfig.

Krithe bartonensis levantina n. subsp.

Lerner-Seggev, Ruth, 1965.
Preliminary notes on the Ostracoda of the Mediterranean coast of Israel.
Israel J. Zool., 13(4):145-176.

Krithe droogeri n.sp.

Key, A.J., 1953.
Preliminary note on the recent Ostracoda of the Snellius Expedition. Proc. K. Nederl. Akad. Wetensch., B, 56(2):155-168, 1 textfig., 2 pls.

Krithe monosteraensis

Ascoli, P., 1965.
Crociere Talassografia Afriatica 1955. VI.
Ricerche ecologiche sugli ostracodi contenuti in 16 carote prelevate sul fondo del Mare Adriatico.
Arch. Oceanogr. e Limnol., 14(1):69-138.

Krithe producta

Swain, Frederick M., 1967.
Ostracoda from the Gulf of California.
Mem. Geol. Soc. Am., 101: 139 pp.

Krithe sawanensis n.sp.

Hanai, T., 1959.
Studies on the Ostracoda from Japan. IV. Family Cytherideidae.
J. Fac. Sci., Univ. Tokyo, (II), 11(3):291-308.

Krithe setosa n.sp. (ostracod)

Rudjakov, J.A., 1961.
[A new species of the family Cytheridae from the ultraabyssal depths of the Java Trench.]
Trudy Inst. Okeanol., 51: 116-120.

Leguminocythereis corrugata

Benson, Richard H., 1959.
Arthropoda 1. Ecology of recent ostracodes of the Todos Santos Bay region, Baja California, Mexico. Univ. Kansas, Paleontol. Contr., :1-80, 11 pls., 20 figs.

?Leguminocythereis corrugata

Swain, Frederick M., 1967.
Ostracoda from the Gulf of California.
Mem. Geol. Soc. Am., 101: 139 pp.

Leptocythere sp.

King, Charles E. and Louis S. Kornicker, 1970.
Ostracoda in Texas bays and lagoons: an ecological study.
Smithson. Contrib. Zool., 24: 92 pp.

Leptocythere arenicola n.sp

Hartmann, Gerd, 1964.
Zur Kenntnis der Ostracoden des Roten Meeres. Wissenschaftliche Ergebnisse einer Forschungsreise von A. Remane und E. Schulz nach dem Roten Meer. E. 11.
Kieler Meeresf., 20:35-127.

Leptpoythere bacescoi

Ascoli, P., 1965.
Crociere Talassografia Afriatica 1955. VI.
Ricerche ecologiche sugli ostracodi contenuti in 16 carote prelevate sul fondo del Mare Adriatico.
Arch. Oceanogr. e Limnol., 14(1):69-138.

Leptocythere baltica

Elofson, O. 1941
Zur Kenntnis der marinen Ostracodes Schwedens mit besonderer Berücksichtigung des Skageraks. Zool. Bidrag. Uppsala 19: 215-534; 42 maps, 52 figs.

Leptocythere castanea

Elofson, O. 1941
Zur Kenntnis der marinen Ostracodes Schwedens mit besonderer Berücksichtigung des Skageraks. Zool. Bidrag. Uppsala 19: 215-534; 42 maps, 52 figs.

Leptocythere castranea

Stephensen, K., 1938.
Marine Ostracoda and Cladocera. Zool. Iceland 3(32):19 pp., 1 textfig.

Leptocythere crispata

Elofson, O. 1941
Zur Kenntnis der marinen Ostracodes Schwedens mit besonderer Berücksichtigung des Skageraks. Zool. Bidrag. Uppsala 19: 215-534; 42 maps, 52 figs.

Leptocythere (Callistocythere) diffusa

Marinov, Tenio, 1967.
Le specie del genere Leptocythere (Ostracoda Crustacea) del litorale Bulgaro del Mar Nero.
Pubbl. Staz. Zool., Napoli, 35(3):274-285.

Leptocythere diffusa

Schornikov, E.I., 1966.
Leptocythere (Crustacea Ostracoda) in the Azov-Black Sea Basin. (In Russian; English abstract).
Zool. Zhurn., Akad. Nauk. SSSR, 45(1):32-49.

Leptocythere histriana

Schornikov, E.I., 1966.
Leptocythere (Crustacea Ostracoda) in the Azov-Black Sea Basin. (In Russian; English abstract).
Zool. Zhurn., Akad. Nauk. SSSR, 45(1):32-49.

Leptocythere ilyophila

Elofson, O. 1941
Zur Kenntnis der marinen Ostracodes Schwedens mit besonderer Berücksichtigung des Skageraks. Zool. Bidrag. Uppsala 19: 215-534; 42 maps, 52 figs.

Leptocythere lacertosa

Elofson, O. 1941
Zur Kenntnis der marinen Ostracodes Schwedens mit besonderer Berücksichtigung des Skageraks. Zool. Bidrag. Uppsala 19: 215-534; 42 maps, 52 figs.

Leptocythere lacertosa

Theisen, Bent Friis, 1967.
The life history of seven species of ostracods from a Danish brackish-water locality.
Meddelelser Danmarks Fiskeri- og Havundersøgelser, n.s. 4(8):215-

Leptocythere (L.) lagunae

Hartmann, Gerd, 1960.
Ostracoden von Banyuls-sur-Mer.
Vie et Milieu, 11(3):413-424.

Leptocythere cf. litoralis

Hartmann, Gerd, 1964.
Zur Kenntnis der Ostracoden des Roten Meeres. Wissenschaftliche Ergebnisse einer Forschungsreise von A. Remane und E. Schulz nach dem Roten Meer. E. 11.
Kieler Meeresf., 20:35-127.

Leptocythere (Leptocythere) macallana

Marinov, Tenio, 1967.
Le specie del genere Leptocythere (Ostracoda Crustacea) del litorale Bulgaro del Mar Nero.
Pubbl. Staz. Zool., Napoli, 35(3):274-285.

Leptocythere (Callistocythere) mediterranea

Marinov, Tenio, 1967.
Le specie del genere Leptocythere (Ostracoda Crustacea) del litorale Bulgaro del Mar Nero.
Pubbl. Staz. Zool., Napoli, 35(3):274-285.

Leptocythere mediterranea

Schornikov, E.I., 1966.
Leptocythere (Crustacea Ostracoda) in the Azov-Black Sea Basin. (In Russian; English abstract).
Zool. Zhurn., Akad. Nauk. SSSR, 45(1):32-49.

Leptocythere (Leptocythere) multipunctata

Marinov, Tenio, 1967.
Le specie del genere Leptocythere (Ostracoda Crustacea) del litorale Bulgaro del Mar Nero.
Pubbl. Staz. Zool., Napoli, 35(3):274-285.

Leptocythere (Leptocythere) nitida

Marinov, Tenio, 1967.
Le specie del genere Leptocythere (Ostracoda Crustacea) del litorale Bulgaro del Mar Nero.
Pubbl. Staz. Zool., Napoli, 35(3):274-285.

Leptocythere pellucida

Elofson, O. 1941
Zur Kenntnis der marinen Ostracodes Schwedens mit besonderer Berücksichtigung des Skageraks. Zool. Bidrag. Uppsala 19: 215-534; 42 maps, 52 figs.

Leptocythere pellucida

Stephensen, K., 1938.
Marine Ostracoda and Cladocera. Zool. Iceland 3(32):19 pp., 1 textfig.

Leptocythere petit n.

Hartmann, G., 1953.
Ostracodes des étangs méditerranéens.
Vie et Milieu, Bull. Lab. Arago, 4(4):707-712, 1 textfig.

Leptocythere quinquetuberculata

Marinov, Tenio, 1967.
Le specie del genere Leptocythere (Ostracoda Crustacea) del litorale Bulgaro del Mar Nero.
Pubbl. Staz. Zool., Napoli, 35(3):274-285.

Leptocythere rara

Hartmann, G., 1953.
Ostracodes des étangs méditerranéens.
Vie et Milieu, Bull. Lab. Arago, 4(4):707-712, 1 textfig.

Leptocythere rarepunctata

Ascoli, P., 1965.
Crociere Talassografia Afriatica 1955. VI.
Ricerche ecologiche sugli ostracodi contenuti in 16 carote prelevate sul fondo del Mare Adriatico.
Arch. Oceanogr. e Limnol., 14(1):69-138.

Leptocythere relicta

Marinov, Tenio, 1967.
Le specie del genere Leptocythere (Ostracoda Crustacea) del litorale Bulgaro del Mar Nero.
Pubbl. Staz. Zool., Napoli, 35(3):274-285.

Leptocythere cf. vellicata

McKenzie, K.G. 1967.
Recent Ostracoda from Port Phillip Bay, Victoria.
Proc. R. Soc. Victoria n.s. 80(1): 61-106.

Limnocythere sanctipatricii
King, Charles E. and Louis S. Kornicker, 1970.
Ostracoda in Texas bays and lagoons: an ecological study.
Smithson. Contrib. Zool., 24: 92 pp.

Limnocythere sanctipatricii
Swain, Frederick M., 1967.
Ostracoda from the Gulf of California.
Mem. Geol. Soc. Am., 101: 139 pp.

Linocheles vagans n.gen., n.sp.
Brady, G.S., 1907.
Crustacea. V. Ostracoda. Nat. Antarctic Exped., 1901-1904, 3:9 pp., 3 pls.

Loxoconcha sp.
McKenzie, K.G. 1967.
Recent Ostracoda from Port Phillip Bay, Victoria.
Proc. R. Soc. Victoria n.s. 80(1): 61-106.

Loxoconcha sp.
Van den Bold, W.A. 1966.
Ostracoda from Colon Harbour, Panama.
Carib. J. Sci. 6(1/2): 43-64.

Loxoconcha alata longispina n.var.
Key, A.J., 1953.
Preliminary note on the recent Ostracoda of the Snellius Expedition. Proc. K. Nederl. Akad. Wetensch., B, 56(2):155-168, 1 textfigs., 2 pls.

Loxoconcha australis
McKenzie, K.G. 1967.
Recent Ostracoda from Port Phillip Bay, Victoria.
Proc. R. Soc. Victoria n.s. 80(1): 61-106.

Loxoconcha avellana
Van den Bold, W.A. 1966.
Ostracoda from Colon Harbour, Panama.
Carib. J. Sci. 6(1/2): 43-64.

Loxoconcha baltica
Elofson, O. 1941
Zur Kenntnis der marinen Ostracodes Schwedens mit besonderer Berücksichtigung des Skageraks. Zool. Bidrag. Uppsala 19: 215-534; 42 maps, 52 figs.

Loxoconcha bulgarica n.sp.
Caraion, F.E., 1962.
Loxoconcha bulgarica n.sp. un ostracod nov, colectat in Apele Bulgaresti ale marii negre (Sozopol).
Comun. Acad. Bucuresti, 11(1):21-27.

Loxoconcha dentarticula n.sp.
Smith, Verna Z., 1952.
Further Ostracoda of the Vancouver Island region
J. Fish. Res. Bd., Canada, 9(1):16-41, 11 pls.

Loxoconcha dimorpha, n.sp.
Hartmann, G., 1959.
Neue Ostracoden von Teneriffa.
Zool. Anz., 162(5/6):160-171.

Loxoconcha dorsotuberculata
Van den Bold, W.A. 1966.
Ostracoda from Colon Harbour, Panama.
Carib. J. Sci. 6(1/2): 43-64.

Loxoconcha elliptica
Elofson, O. 1941
Zur Kenntnis der marinen Ostracodes Schwedens mit besonderer Berücksichtigung des Skageraks. Zool. Bidrag. Uppsala 19: 215-534; 42 maps, 52 figs.

Loxoconcha elliptica
Hartmann, G., 1953.
Ostracodes des étangs mediterraneens.
Vie et Milieu, Bull. Lab. Arago, 4(4):707-712, 1 textfig.

Loxoconcha elliptica
Theisen, Bent Friis, 1967.
The life history of seven species of ostracods from a Danish brackish-water locality.
Meddelelser Danmarks Fiskeri- og Havundersøgelser, n.s. 4(8):215-

Loxoconcha? emaciata n.sp.
Swain, Frederick M., 1967.
Ostracoda from the Gulf of California.
Mem. Geol. Soc. Am., 101: 139 pp.

Loxoconcha fischeri
Van den Bold, W.A. 1966.
Ostracoda from Colon Harbour, Panama.
Carib. J. Sci., 6(1/2): 43-64.

Loxoconcha fragilis
Elofson, O. 1941
Zur Kenntnis der marinen Ostracodes Schwedens mit besonderer Berücksichtigung des Skageraks. Zool. Bidrag. Uppsala 19: 215-534; 42 maps, 52 figs.

Loxoconcha fragilis
Lucas, V.Z., 1931.
Some Ostracoda from the Vancouver island region.
Contr. Canadian Biol. Fish., n.s., 6(17):397-416.

Loxoconcha ghardaqensis n.sp.
Hartmann, Gerd, 1964.
Zur Kenntnis der Ostracoden des Roten Meeres. Wissenschaftliche Ergebnisse einer Forschungsreise von A. Remane und E. Schulz nach dem Roten Meer. E. 11.
Kieler Meeresf., 20:35-127.

Loxoconcha gilli n.sp.
McKenzie, K.G. 1967.
Recent Ostracoda from Port Phillip Bay, Victoria.
Proc. R. Soc. Victoria n.s. 80(1): 61-106.

Loxoconcha guttata
Stephensen, K., 1938.
Marine Ostracoda and Cladocera. Zool. Iceland 3(32):19 pp., 1 textfig.

Loxoconcha granulata
Elofson, O. 1941
Zur Kenntnis der marinen Ostracodes Schwedens mit besonderer Berücksichtigung des Skageraks. Zool. Bidrag. Uppsala 19: 215-534; 42 maps, 52 figs.

Loxoconcha granulata
Williams, Roger B., 1966.
Recent marine podocopid Ostracoda of Narregansett Bay, Rhode Island.
Univ. Kansas, Paleont. Contrib., Paper 11:36pp.

Loxoconcha idkeri n.sp.
Hartmann, Gerd, 1964.
Zur Kenntnis der Ostracoden des Roten Meeres. Wissenschaftliche Ergebnisse einer Forschungsreise von A. Remane und E. Schulz nach dem Roten Meer. E. 11.
Kieler Meeresf., 20:35-127.

Loxoconcha impressa
Elofson, O. 1941
Zur Kenntnis der marinen Ostracodes Schwedens mit besonderer Berücksichtigung des Skageraks. Zool. Bidrag. Uppsala 19: 215-534; 42 maps, 52 figs.

Loxoconcha impressa
Hartmann, G., 1953.
Ostracodes des étangs mediterraneens.
Vie et Milieu, Bull. Lab. Arago, 4(4):707-712, 1 textfig.

Loxoconcha lapidiscola n.sp.
Hartmann, Gerd, 1959.
Zur Kenntnis der lotischen Lebensbereiche der Pazifischen Küste von El Salvador unter besonderer Berücksichtigung seiner Ostracodenfauna.
Kieler Meeresf., 15(2):187-241.

Loxoconcha lapidiscola
Van den Bold, W.A. 1966.
Ostracoda from Colon Harbour, Panama.
Carib. J. Sci. 6(1/2): 43-64.

Loxoconcha lenticulata
Benson, Richard H., 1959.
Arthropoda 1. Ecology of recent ostracodes of the Todos Santos Bay region, Baja California, Mexico. Univ. Kansas, Paleontol. Contr., :1-80, 11 pls., 20 figs.

Loxoconcha lenticulata
Swain, Frederick M., 1967.
Ostracoda from the Gulf of California.
Mem. Geol. Soc. Am., 101: 139 pp.

Loxoconcha littoralis
Ruggieri, Giuliano, 1964.
Ecological remarks on the present and past distribution of four species of Loxoconcha in the Mediterranean.
Pubbl., Staz. Zool., Napoli, 33(Suppl.):515-528.

Loxoconcha aff. mediterranea
Ascoli, P., 1965.
Crociera Talassografia Afriatica 1955. VI. Ricerche ecologiche sugli ostracodi contenuti in 16 carote prelevate sul fondo del Mare Adriatico.
Arch. Oceanogr. e Limnol., 14(1):69-138.

Loxoconcha minima
Hartmann, G., 1953.
Ostracodes des étangs mediterraneens.
Vie et Milieu, Bull. Lab. Arago, 4(4): 707-712, 1 textfig.

Loxoconcha multiflora
Elofson, O. 1941
Zur Kenntnis der marinen Ostracodes Schwedens mit besonderer Berücksichtigung des Skageraks. Zool. Bidrag. Uppsala 19: 215-534; 42 maps, 52 figs.

Loxoconcha napoliana
Ruggieri, Giuliano, 1964.
Ecological remarks on the present and past distribution of four species of Loxoconcha in the Mediterranean.
Pubbl., Staz. Zool., Napoli, 33(Suppl.):515-528.

Loxoconcha ornatovalvae n.sp.
Hartmann, Gerd, 1964.
Zur Kenntnis der Ostracoden des Roten Meeres. Wissenschaftliche Ergebnisse einer Forschungsreise von A. Remane und E. Schulz nach dem Roten Meer. E. 11.
Kieler Meeresf., 20:35-127.

Loxoconcha ovulata
Ascoli, P., 1965.
Crociera Talassografia Afriatica 1955. VI. Ricerche ecologiche sugli ostracodi contenuti in 16 carote prelevate sul fondo del Mare Adriatico.
Arch. Oceanogr. e Limnol., 14(1):69-138.

Loxoconcha purisubrhomboidea
Grossman, Stuart, 1965.
Morphology and ecology of two podocopid ostracodes from Redfish Bay, Texas.
Micropaleontology, 11(2):141-150.

Loxoconcha purisubrhomboidea
King, Charles E. and Louis S. Kornicker, 1970.
Ostracoda in Texas bays and lagoons: an ecological study.
Smithson. Contrib. Zool., 24: 92 pp.

Loxoconcha pusilla
Elofson, O. 1941
Zur Kenntnis der marinen Ostracodes Schwedens mit besonderer Berücksichtigung des Skageraks. Zool. Bidrag. Uppsala 19: 215-534; 42 maps, 52 figs.

Loxoconcha rhomboidea
Williams, Roger B., 1966.
Recent marine podocopid Ostracoda of Narragansett Bay, Rhode Island.
Univ. Kansas, Paleont. Contrib., Paper 11:36 pp.

Loxoconcha rubritincta n.sp.
Ruggieri, Giuliano, 1964.
Ecological remarks on the present and past distribution of four species of Loxoconcha in the Mediterranean.
Pubbl., Staz. Zool., Napoli, 33(Suppl.):515-528.

Loxoconcha schusterae n.sp.
Hartmann, Gerd, 1959.
Zur Kenntnis der letischen Lebensbereiche der Pazifischen Küste von El Salvador unter besonderer Berücksichtigung seiner Ostracodenfauna.
Kieler Meeresf., 15(2):187-241.

Loxoconcha sperata n.sp.
Williams, Roger B., 1966.
Recent marine podocopid Ostracoda of Narragansett Bay, Rhode Island.
Univ. Kansas, Paleont. Contrib., Paper 11:36 pp.

Loxoconcha stellifera
Lerner-Seggev, Ruth, 1965.
Preliminary notes on the Ostracoda of the Mediterranean coast of Israel.
Israel J. Zool., 13(4):145-176.

Loxoconcha tamarinda
Ascoli, P., 1965.
Crociere Talassografia Afriatica 1955. VI.
Ricerche ecologiche sugli ostracodi contenuti in 16 carote prelevate sul fondo del Mare Adriatico.
Arch. Oceanogr. e Limnol., 14(1):69-138.

Loxoconcha tamarindus
Elofson, O. 1941
Zur Kenntnis der marinen Ostracodes Schwedens mit besonderer Berücksichtigung des Skageraks. Zool. Bidrag. Uppsala 19: 215-534; 42 maps, 52 figs.

Loxoconcha tamarindus
Stephensen, K., 1938.
Marine Ostracoda and Cladocera. Zool. Iceland 3(32):19 pp., 1 textfig.

Loxoconcha tamarinoidea n. sp.
Swain, Frederick M., 1967.
Ostracoda from the Gulf of California.
Mem. Geol. Soc. Am., 101: 139 pp.

Loxoconcha tenuiungula n.sp.
Smith, Verna Z., 1952.
Further Ostracoda of the Vancouver Island region.
J. Fish. Res. Bd., Canada, 9(1):16-41, 11 pls.

Loxoconcha trita n.sp.
McKenzie, K.G. 1967.
Recent Ostracoda from Port Phillip Bay, Victoria.
Proc. R. Soc. Victoria n.s. 80(1):61-106.

Loxoconcha cf. variolata
McKenzie, K.G. 1967.
Recent Ostracoda from Port Phillip Bay, Victoria.
Proc. R. Soc. Victoria n.s. 80(1):61-106.

Loxoconcha versicolor
Ruggieri, Giuliano, 1964.
Ecological remarks on the present and past distribution of four species of Loxoconcha in the Mediterranean.
Pubbl., Staz. Zool., Napoli, 33(Suppl.):515-528.

Loxoconcha viridis
Stephensen, K., 1938.
Marine Ostracoda and Cladocera. Zool. Iceland 3(32):19 pp., 1 textfig.

Loxoconchella dorsobulleta n.sp.
Hartmann, Gerd, 1964.
Zur Kenntnis der Ostracoden des Roten Meeres.
Wissenschaftliche Ergebnisse einer Forschungsreise von A. Remane und E. Sculz nach dem Roten Meer. E. 11.
Kieler Meeresf., 20:35-127.

Loxoconchella honoluliensis n. gen.
Triebel, E., 1954.
Loxoconchella n.ge. (Crust., Ostr.). Senck. leth 35(1/2):17-21., 2 pls.

Loxoconchella pulchra n.sp.
McKenzie, K.G. 1967.
Recent Ostracoda from Port Phillip Bay, Victoria.
Proc. R. Soc. Victoria n.s. 80(1):61-106.

Loxocorniculum sculptoides n.sp.
Swain, Frederick M., 1967.
Ostracoda from the Gulf of California.
Mem. Geol. Soc. Am., 101: 139 pp.

Loxocythere angulosa
Ascoli, P., 1965.
Crociere Talassografia Afriatica 1955. VI.
Ricerche ecologiche sugli ostracodi contenuti in 16 carote prelevate sul fondo del Mare Adriatico.
Arch. Oceanogr. e Limnol., 14(1):69-138.

Loxocythere crassa
Hanai, T., 1959.
Studies on the Ostracoda from Japan. II. Subfamily Cytherinae Dana, 1852(emend.).
J. Fac. Sci., Univ. Tokyo, (II), 11(4):409-418.

Loxocythere frigida n.sp.
Neale, J. W., 1967.
An ostracod fauna from Halley Bay, Coats Land, British Antarctic Territory.
Brit. Antarct. Surv. Sci. Rep. 58: 50 pp.

Loxocythere hornibrooki n.sp.
McKenzie, K.G. 1967.
Recent Ostracoda from Port Phillip Bay, Victoria.
Proc. R. Soc. Victoria n.s. 80(1):61-106.

Loxocythere inflata n.sp.
Hanai, T., 1959.
Studies on the Ostracoda from Japan. V. Subfamily Cytherinae Dana, 1852(emend.).
J. Fac. Sci., Univ. Tokyo, (II), 11(4):409-418.

Loxoreticulatum fallax
Neale, J. W., 1967.
An ostracod fauna from Halley Bay, Coats Land, British Antarctic Territory.
Brit. Antarct. Surv. Sci. Rep. 58: 50 pp.

Macroconchoecia reticulata
Granata, L., and L. di Caporiacco, 1949
Ostracodes marins recueillis pendant les croiserieres du Prince Albert, 1. Res. Camp. Sci., Monaco, 109; 48 pp., 4 pls.

Macrocypridina castanea
Deevey, Georgiana B., 1968.
Pelagic ostracods of the Sargasso Sea off Bermuda.
Bull. Peabody Mus. nat. Hist., Yale Univ., 26:125 pp.

Macrocypridina costanea
Poulsen, Erik M., 1962
Ostracoda-Myodocopa. Part 1. Cypridiniformes-Cypridinidae.
Dana-Report No. 57:414 pp.

Macrocyprina sp.
McKenzie, K.G. 1967.
Recent Ostracoda from Port Phillip Bay, Victoria.
Proc. R. Soc. Victoria n.s. 80(1):61-106.

Macrocyprina pacifica
Swain, Frederick M., 1967.
Ostracoda from the Gulf of California.
Mem. Geol. Soc. Am., 101: 139 pp.

Macrocypris sp.
Ascoli, P., 1965.
Crociere Talassografia Afriatica 1955. VI.
Ricerche ecologiche sugli ostracodi contenuti in 16 carote prelevate sul fondo del Mare Adriatico.
Arch. Oceanogr. e Limnol., 14(1):69-138.

Macrocypris africana
Tressler, W.L., 1949
Marine ostracoda from Tortugas, Florida.
J. Wash. Acad. Sci. 39(10):334-343, 25 text figs.

Macrocypris angusta
Elofson, O. 1941
Zur Kenntnis der marinen Ostracodes Schwedens mit besonderer Berücksichtigung des Skageraks. Zool. Bidrag. Uppsala 19: 215-534; 42 maps, 52 figs.

Macrocypris bathyalensis
Hulings, Neil C., 1967.
Marine Ostracoda from the western North Atlantic Ocean: Labrador Sea, Gulf of St. Lawrence, and off Nova Scotia.
Crustaceana, 13(3):310-328.

Macrocypris bathyalensis n. sp.
Hulings, Neil C., 1967.
Marine Ostracoda from the western North Atlantic between Cape Hatteras, North Carolina, and Jupiter Inlet, Florida.
Bull. mar. Sci., Miami, 17(3):629-659.

Macrocypris (Macrocyprina) Marchilensis nsp.
Hartmann-Schröder, Gesa, und Gerd Hartmann, 1965.
Zur Kenntnis des Sublitorels der chilenischen Küste unter besonderer Berücksichtigung der Polychaeten und Ostracoden (mit Bermerkungen über den Einfluss saverstoffarmer Strömungen auf die Besiedlung von marinen Sedimenten).
Mitt. Hamburg. Zool. Mus. u Inst., Ergänzungsband zu 62:384 pp.

Macrocypris minna
Elofson, O. 1941
Zur Kenntnis der marinen Ostracodes Schwedens mit besonderer Berücksichtigung des Skageraks. Zool. Bidrag. Uppsala 19: 215-534; 42 maps, 52 figs.

Macrocypris minna

Sars, G.O., 1886
Crustacea JI. Zool. Norwegian North-Atlantic Exped., 1876-1878: 96 pp., 1 map.

Macrocypris schmitti n.sp.

Tressler, W.L., 1949
Marine ostracoda from Tortugas, Florida.
J. Wash. Acad. Sci. 39(10):334-343, 25 text figs.

Neale, J. W., 1967. Macrocypris cf. M.simili
An ostracod fauna from Halley Bay, Coats Land, British Antarctic Territory.
Brit. Antarct. Surv. Sci. Rep. 58: 50 pp.

Macrocythere rostrata n.sp.

Lucas, V.Z., 1931.
Some Ostracoda of the Vancouver island region.
Contr. Canadian Biol. Fish., n.s., 6(17):397-416.

Macrocythere simplex

Elofson, O. 1941
Zur Kenntnis der marinen Ostracodes Schwedens mit besonderer Berücksichtigung des Skageraks. Zool. Bidrag. Uppsala 19: 215-534; 42 maps, 52 figs.

Macrocythere simplex

Stephensen, K., 1938.
Marine Ostracoda and Cladocera. Zool. Iceland 3(32):19 pp., 1 textfig.

Macrodentina calcarata n.sp.

Triebel, E., 1954.
Malm-Ostracoden mit amphidontem Schloss.
Senck. Lett. 35(1/2):3-16, 4 pls.

Macrodentina lineata

Triebel, E., 1954.
Malm-Ostracoden mit amphidontem Schloss.
Senck. Lett. 35(1.2):3-16, 4 pls.

Manawa tryphena n.gen., n.sp.

Hornibrook, N. de B., 1949.
A new family of living Ostracoda with resemblance to some Paleozoic Beyrichiidae.
Trans. R. Soc., New Zealand, 77(4):469-471.

Megacythere johnsoni

King, Charles E. and Louis S. Kornicker, 1970.
Ostracoda in Texas bays and lagoons: an ecological study.
Smithson. Contrib. Zool., 24: 92 pp.

Megacythere punctocostata n.sp.

Swain, Frederick M., 1967.
Ostracoda from the Gulf of California
Mem. Geol. Soc. Am., 101: 139 pp.

Melavargula japonica /n.gen., n.sp.

Poulsen, Erik M., 1962
Ostracoda-Myodocopa. Part 1.
Cypridiniformes-Cyprididae.
Dana-Report No. 57:414 pp.

Melavargula nana n.sp.

Poulsen, Erik M., 1962
Ostracoda-Myodocopa. Part 1.
Cypridiniformes-Cyprididae.
Dana-Report No. 57:414 pp.

Mesocythere elongata

Hartmann, Gerd, 1959.
Zur Kenntnis der lotischen Lebensbereiche der Pazifischen Küste von El Salvador unter Besonderer Berücksichtigung seiner Ostracoden-fauna.
Kieler Meeresf., 15(2):187-241.

Metaconchoecia rotundata

Granata, L., and L.di Caporiacco, 1949
Ostracodes marins recueillis pendant les croisieres du Prince Albert, 1. Res.Camp.
Sei, Monaco, 109: 48 pp., 4 pls.

Microasteropteron parvum n. sp.

Poulson, Erik M., 1965.
Ostracoda-Myodocopa. II. Cypridiniformes-Rutidermatidae, Sarsiellidae and Asteropidae.
Dana Report, No. 65:484 pp.

Microconchoecia curta

Granata, L., and L.di Caporiacco, 1949
Ostracodes marins recueillis pendant les croiseriers du Prince Albert, 1. Res.Camp.
Sei, Monaco, 109: 48 pp., 4 pls.

Microconchoecia echinulata

Granata, L., and L.di Caporiacco, 1949
Ostracodes marins recueillis pendant les croiseriers du Prince Albert, 1. Res.Camp.
Sei, Monaco, 109: 48 pp., 4 pls.

Microcythere macphersoni n. sp.

McKenzie, K.G. 1967.
Recent Ostracoda from Port Phillip Bay, Victoria.
Proc. R. Soc. Victoria n.s. 80(1): 61-106.

Microcythere subterranea

Rao, G. Chandrasekhara, 1970.
On some interstitial fauna in the marine sands on the Indian coast.
Current Sci., 39(22): 504-507.

Microcytherides labiata n. gen., nsp.

Hartmann-Schröder, Gesa, und Gerd Hartmann, 1965.
Zur Kenntnis des Sublitorels der chilenischen Küste unter besonderer Berücksichtigung der Polychaeten und Ostracoden (mit Bermerkungen über den Einfluss saverstoffarmer Strömungen auf die Besiedlung von marinen Sedimenten).
Mitt. Hamburg, Zool. Mus. u Inst., Ergänzungsband zu 62:384 pp.

Microcytherura australis n.sp.

McKenzie, K.G. 1967.
Recent Ostracoda from Port Phillip Bay, Victoria.
Proc. R. Soc. Victoria n.s. 80(1): 61-106.

Microcytherura gawemuelleri n. sp.

McKenzie, K.G. 1967.
Recent Ostracoda from Port Phillip Bay, Victoria.
Proc. R. Soc. Victoria n.s. 80(1): 61-106.

Microcytherura triebeli n. sp.

McKenzie, K.G. 1967.
Recent Ostracoda from Port Phillip Bay, Victoria.
Proc. R. Soc. Victoria n.s. 80(1): 61-106.

Microloxoconcha (?) barralesi nsp.

Hartmann-Schröder, Gesa, und Gerd Hartmann, 1965.
Zur Kenntnis des Sublitorels der chilenischen Küste unter besonderer Berücksichtigung der Polychaeten und Ostracoden (mit Bermerkungen über den Einfluss saverstoffarmer Strömungen auf die Besiedlung von marinen Sedimenten).
Mitt. Hamburg, Zool. Mus. u Inst., Ergänzungsband zu 62:384 pp.

Microloxoconcha compressa

Rao, G. Chandrasekhara, 1970.
On some interstitial fauna in the marine sands on the Indian coast.
Current Sci., 39(22): 504-507.

Miracythere? sp.

Swain, Frederick M., 1967.
Ostracoda from the Gulf of California.
Mem. Geol. Soc. Am., 101: 139 pp.

Monoceratina 2 sp.

Key, A.J., 1953.
Preliminary note on the recent Ostracoda of the Snellius Expedition. Proc. K. Nederl. Akad. Wetensch., B, 56(2):155-168, 1 textfigs., 2 pls.

Monoceratina bifurcata

Swain, Frederick M., 1967.
Ostracoda from the Gulf of California.
Mem. Geol. Soc. Am., 101: 139 pp.

Monopia flaveola

Poulsen, Erik M., 1962
Ostracoda-Myodocopa. Part 1.
Cypridiniformes-Cyprididae.
Dana-Report No. 57:414 pp.

Moosella spp.

Hartmann, Gerd, 1964.
Zur Kenntnis der Ostracoden des Roten Meeres. Wissenschaftliche Ergebnisse einer Forschungsreise von A. Remane und E. Schulz nach dem Roten Meer. E. 11.
Kieler Meeresf., 20:35-127.

Moosella striata n.gen., n.sp.

Hartmann, Gerd, 1964.
Zur Kenntnis der Ostracoden des Roten Meeres. Wissenschaftliche Ergebnisse einer Forschungsreise von A. Remane und E. Schulz nach dem Roten Meer. E. 11.
Kieler Meeresf., 20:35-127.

Muelleriella setifera n. sp.

Poulson, Erik M., 1965.
Ostracoda-Myodocopa. II. Cypridiniformes-Rutidermitidae, Sarsiellidae and Asteropidae.
Dana Report, No. 65:484 pp.

Muelleriella sealandica n. gen., n. sp.

Poulson, Erik M., 1965.
Ostracoda-Myodocopa. II. Cypridiniformes-Rutidermitidae, Sarsiellidae and Asteropidae.
Dana Report, No. 65:484 pp.

Murrayina canadensis

Williams, Roger B., 1966.
Recent marine podocopid Ostracoda of Narragansett Bay, Rhode Island.
Univ. Kansas, Paleont. Contrib., Paper 11:36 pp.

Murrayina micula

Williams, Roger B., 1966.
Recent marine podocopid Ostracoda of Narragansett Bay, Rhode Island.
Univ. Kansas, Paleont. Contrib., Paper 11:36 pp.

Oceanographic Index: Marine Organisms Cumulation, 1946-1973

Mutilas confragosa
Swain, Frederick M., 1967.
Ostracoda from the Gulf of California.
Mem. Geol. Soc. Am., 101: 139 pp.

Myrena n. gen.
Neale, J. W., 1967.
An ostracod fauna from Halley Bay, Coats Land, British Antarctic Territory.
Brit. Antarct. Surv. Sci. Rep. 58: 50 pp.

Myrena meridionalis
Neale, J. W., 1967.
An ostracod fauna from Halley Bay, Coats Land, British Antarctic Territory.
Brit. Antarct. Surv. Sci. Rep. 58: 50 pp.

Nannocythere remanei n.gen., n.sp.
Schäfer, H.W., 1953.
Über Meeres- und Brackwasser-Ostracoden aus dem Deutschen Küstengebiet mit: 2. Mitteilung über die Ostracodenfauna Griechenlands. Hydrobiologia 5(4):351-389.

Neocyprideis pseudodonta n.gen., n.s
Hanai, T., 1959.
Studies on the Ostracoda from Japan. IV. Family Cythrideidae.
J. Fac. Sci., Univ. Tokyo, (II), 11(3):291-308.

Neomonoceratina columbiformis
Key, A.J., 1953.
Preliminary note on the recent Ostracoda of the Snellius Expedition. Proc. K. Nederl. Akad. Wetensch., B, 56(2):155-168, 1 textfig., 2 pls.

Nesidea cushmani n.sp.
Tressler, W.L., 1949
Marine ostracoda from Tortugas, Florida.
J. Wash. Acad. Sci. 39(10):334-343, 25 text figs.

Normanicythere lioderma
Hulings, Neil C., 1967.
Marine Ostracoda from the western North Atlantic Ocean: Labrador Sea, Gulf of St. Lawrence, and off Nova Scotia.
Crustaceana, 13(3):310-328.

Orionina pseudovaughani n.sp.
Swain, Frederick M., 1967.
Ostracoda from the Gulf of California.
Mem. Geol. Soc. Am., 101: 139 pp.

Orionina reticulata
Hartman, Gerd, 1959.
Zur Kenntnis der lotischen Lebensbereiche der Pazifischen Küste von El Salvador unter besonderer Berücksichtigung seiner Ostracodenfauna.
Kieler Meeresf., 15(2):187-241.

Orionina serrulata
Van den Bold, W.A. 1966.
Ostracoda from Colon Harbour, Panama.
Carib. J. Sci. 6(1/2): 43-64.

Orthoconchoecia atlantica
Granata, L., and L.di Caporiacco, 1949
Ostracodes marins recueillis pendant les croiserieres du Prince Albert,1. Res.Camp. Sei. Monaco, 109: 48 pp., 4 pls.

Orthoconchoecia bispinosa
Granata, L., and L.di Caporiacco, 1949
Ostracodes marins recueillis pendant les croiserieres du Prince Albert,1. Res.Camp. Sei. Monaco, 109: 48 pp., 4 pls.

Palaciosavandenboldi n.gen., n.sp.
Hartmann, Gerd, 1959.
Zur Kenntnis der lotischen Lebensbereiche der Pazifischen Küste von El Salvador unter besonderer Berücksichtigung seiner Ostracodenfauna.
Kieler Meeresf., 15(2):187-241.

Palmenella carida n.sp. [a]
Benson, Richard H., 1959.
Arthropoda 1. Ecology of recent ostracodes of the Todos Santos Bay region, Baja California, Mexico. Univ. Kansas, Paleontol. Contr., :1-80, 11 pls., 20 figs.

Palmenella limicola
Elofson, O. 1941
Zur Kenntnis der marinen Ostracodes Schwedens mit besonderer Berücksichtigung des Skageraks. Zool. Bidrag. Uppsala 19: 215-534; 42 maps, 52 figs.

Paraconchoecia gracilis
Brady, G.S., 1907.
Crustacea. V. Ostracoda. Nat. Antarctic Exped., 1901-1904, 3:9 pp., 3 pls.

Paracyprideis fennica
Elofson, O. 1941
Zur Kenntnis der marinen Ostracodes Schwedens mit besonderer Berücksichtigung des Skageraks. Zool. Bidrag. Uppsala 19: 215-534; 42 maps, 52 figs.

Paracypridina aberrata n.gen., n.sp.
Poulsen, Erik M., 1962
Ostracoda-Myodocopa. Part 1. Cypridiniformes-Cypridinidae.
Dana-Report No. 57:414 pp.

Paracypris bradyi n.sp.
McKenzie K. G. 1967.
Recent Ostracoda from Port Phillip Bay, Victoria.
Proc. R. Soc. Victoria, n.s. 80(1): 61-106.

Paracypris franquesoides n.sp.
Swain, Frederick M., 1967.
Ostracoda from the Gulf of California.
Mem. Geol. Soc. Am., 101: 139 pp.

Paracypris pacifica [a]
Benson, Richard H., 1959.
Arthropoda 1. Ecology of recent ostracodes of the Todos Santos Bay region, Baja California, Mexico. Univ. Kansas, Paleontol. Contr., :1-80, 11 pls., 20 figs.

Paracypris polita
Elofson, O. 1941
Zur Kenntnis der marinen Ostracodes Schwedens mit besonderer Berücksichtigung des Skageraks. Zool. Bidrag. Uppsala 19: 215-534; 42 maps, 52 figs.

Paracypris politella n.sp.
Swain, Frederick M., 1967.
Ostracoda from the Gulf of California.
Mem. Geol. Soc. Am., 101: 139 pp.

Paracythereis flexuosa
Elofson, O. 1941
Zur Kenntnis der marinen Ostracodes Schwedens mit besonderer Berücksichtigung des Skageraks. Zool. Bidrag. Uppsala 19: 215-534; 42 maps, 52 figs.

Paracythereis arcuata
Elofson, O. 1941
Zur Kenntnis der marinen Ostracodes Schwedens mit besonderer Berücksichtigung des Skageraks. Zool. Bidrag. Uppsala 19: 215-534; 42 maps, 52 figs.

Paracytheridea granti [a]
Benson, Richard H., 1959.
Arthropoda 1. Ecology of recent ostracodes of the Todos Santos Bay region, Baja California, Mexico. Univ. Kansas, Paleontol. Contr., :1-80, 11 pls., 20 figs.

Paracytheridea granti
Swain, Frederick M., 1967.
Ostracoda from the Gulf of California.
Mem. Geol. Soc. Am., 101: 139 pp.

Paracytheridea? pichelinguensis
Swain, Frederick M., 1967.
Ostracoda from the Gulf of California.
Mem. Geol. Soc. Am., 101: 139 pp.

Paracytheridea remanei n.sp.
Hartmann, Gerd, 1964.
Zur Kenntnis der Ostracoden des Roten Meeres. Wissenschaftliche Ergebnisse einer Forschungsreise von A. Remane und E. Schulz nach dem Roten Meer. E. 11.
Kieler Meeresf., 20:35-127.

"Paracytheridea" simplex n.sp.
Swain, Frederick M., 1967.
Ostracoda from the Gulf of California.
Mem. Geol. Soc. Am., 101: 139 pp.

Paracytheridea troglodyta
King, Charles E. and Louis S. Kornicker, 1970.
Ostracoda in Texas bays and lagoons: an ecological study.
Smithson. Contrib. Zool., 24: 92 pp.

Paracytherma n.gen.
Juday, C., 1907
XIV. Ostracoda of the San Diego region. II. Littoral forms. Contributions from the Laboratory of the Marine Biological Association of San Diego. Univ. Calif. Publ. Zool. 3(9):135-156, Pls. 18-20.

Paracytherois? perspicilli
Swain, Frederick M., 1967.
Ostracoda from the Gulf of California.
Mem. Geol. Soc. Am., 101: 139 pp.

Paracytherois portphillipensis n.sp.
McKenzie K. G. 1967.
Recent Ostracoda from Port Phillip Bay, Victoria.
Proc. R. Soc. Victoria, n.s. 80(1): 61-106.

Paracytheroma costata n.sp.
Hartmann, G., 1952
Zur Kenntnis des Mangrove-Estero-Gebietes von El Salvador und seiner Ostracoden-Fauna. II. Systematischer Teil. Kieler Meeresf. 13(1): 134-159.

Paracytheroma levis n.sp.
Hartmann, G., 1952
Zur Kenntnis des Mangrove-Estero-Gebietes von El Salvador und seiner Ostracoden-Fauna. II. Systematischer Teil. Kieler Meeresf. 13(1): 134-159.

Paracytheroma magna n.sp.
Hartmann, Gerd, 1959.
Zur Kenntnis der lotischen Lebensbereiche der Pazifischen Küste von El Salvador unter besonderer Berücksichtigung seiner Ostracodenfauna.
Kieler Meeresf., 15(2):187-241.

Paracytheroma pedrensis n.sp.
Juday, C., 1907
XIV. Ostracoda of the San Diego region. II. Littoral forms. Contributions from the Laboratory of the Marine Biological Association of San Diego. Univ. Calif. Publ. Zool. 3(9):135-156, Pls. 18-20.

Paracytheroma pedrensis
Skogsberg, T., 1950.
Two new species of marine ostracod from California. Proc. Calif. Acad. Sci. 26(4):483-505, figs.

Paracytheroma unduli-marginata n. sp.
Hartmann, G., 1952
Zur Kenntnis des Mangrove-Estero-Gebietes von El Salvador und seiner Ostracoden-Fauna. II. Systematischer Teil. Kieler Meeresf. 13(1): 134-159.

Paradoloria n.gen.
Poulsen, Erik M., 1962
Ostracoda-Myodocopa. Part 1. Cypridiniformes-Cypridinidae.
Dana-Report No. 57:414 pp.

Paradoloria angulata n.sp.
Poulsen, Erik M., 1962
Ostracoda-Myodocopa. Part 1. Cypridiniformes-Cypridinidae.
Dana-Report No. 57:414 pp.

Paradoloria australis n.sp.
Poulsen, Erik M., 1962
Ostracoda-Myodocopa. Part 1. Cypridiniformes-Cypridinidae.
Dana-Report No. 57:414 pp.

Paradoloria dorsoserrata
Poulsen, Erik M., 1962
Ostracoda-Myodocopa. Part 1. Cypridiniformes-Cypridinidae.
Dana-Report No. 57:414 pp.

Paradoloria nuda n.sp.
Poulsen, Erik M., 1962
Ostracoda-Myodocopa. Part 1. Cypridiniformes-Cypridinidae.
Dana-Report No. 57:414 pp.

Paradoloria vanhöffeni
Poulsen, Erik M., 1962
Ostracoda-Myodocopa. Part 1. Cypridiniformes-Cypridinidae.
Dana-Report No. 57:414 pp.

Paradoxastoma? sp.
King, Charles E. and Louis S. Kornicker, 1970.
Ostracoda in Texas bays and lagoons: an ecological study.
Smithson. Contrib. Zool., 24: 92 pp.

Paradoxostoma abbreviatum
Elofson, O. 1941
Zur Kenntnis der marinen Ostracodes Schwedens mit besonderer Berücksichtigung des Skageraks. Zool. Bidrag. Uppsala 19: 215-534; 42 maps, 52 figs.

Paradoxostoma abbreviatum
Schäfer, H.W., 1953.
Über Meeres- und Brackwasser-Ostracoden aus dem Deutschen Küstengebiet mit 2. Mitteilung über die Ostracodenfauna Griechenlands. Hydrobiologia 5(4):351-389.

Paradoxostoma arcuatum n.sp.
Hartmann, Gerd, 1964.
Zur Kenntnis der Ostracoden des Roten Meeres. Wissenschaftliche Ergebnisse einer Forschungsreise von A. Remane und E. Schulz nach dem Roten Meer. E. 11.
Kieler Meeresf., 20:35-127.

Paradoxostoma bradyi
Elofson, O. 1941
Zur Kenntnis der marinen Ostracodes Schwedens mit besonderer Berücksichtigung des Skageraks. Zool. Bidrag. Uppsala 19: 215-534; 42 maps, 52 figs.

Paradoxostoma breve
Hartmann, Gerd, 1964.
Zur Kenntnis der Ostracoden des Roten Meeres. Wissenschaftliche Ergebnisse einer Forschungsreise von A. Remane und E. Schulz nach dem Roten Meer. E. 11.
Kieler Meeresf., 20:35-127.

Paradoxostoma commune n.sp.
McKenzie, K.G. 1967.
Recent Ostracoda from Port Phillip Bay, Victoria.
Proc. R. Soc. Victoria n.s. 80(1): 61-106.

Paradoxostoma cuneata n.sp.
Lucas, V.Z., 1931.
Some Ostracoda from the Vancouver island region.
Contr. Canadian Biol. Fish., n.s., 6(17):397-416.

Paradoxostoma curtum n.sp.
Hartmann, G., 1959.
Neue Ostracoden von Teneriffa.
Zool. Anz., 162(5/6):160-171.

Paradoxostoma ensiforme
Elofson, O. 1941
Zur Kenntnis der marinen Ostracodes Schwedens mit besonderer Berücksichtigung des Skageraks. Zool. Bidrag. Uppsala 19: 215-534; 42 maps, 52 figs.

Paradoxostoma fraseri n.sp.
Smith, Verna Z., 1952.
Further Ostracoda of the Vancouver Island region. J. Fish. Res. Bd., Canada, 9(1):16-41, 11 pls.

Paradoxostoma cf. hodgei
Swain, Frederick M., 1967.
Ostracoda from the Gulf of California.
Mem. Geol. Soc. Am., 101: 139 pp.

Paradoxostoma hypselum
Neale, J. W., 1967.
An ostracod fauna from Halley Bay, Coats Land, British Antarctic Territory.
Brit. Antarct. Surv. Sci. Rep. 58: 50 pp.

Paradoxostoma insigne n.sp.
Hartmann, G., 1959.
Neue Ostracoden von Teneriffa.
Zool. Anz., 162(5/6):160-171.

Paradoxostoma intermedium
Lerner-Seggev, Ruth, 1965.
Preliminary notes on the Ostracoda of the Mediterranean coast of Israel.
Israel J. Zool., 13(4):145-176.

Paradoxostoma intermedium
Makkaveeva, E.B., 1961.
The biocoenotic history of small worms, Crustacea and marine ostracods.
Trudy Sevastopol Biol. Sta., (14):147-162.

Paradoxostoma longum n.sp.
Hartmann, Gerd, 1964.
Zur Kenntnis der Ostracoden des Roten Meeres. Wissenschaftliche Ergebnisse einer Forschungsreise von A. Remane und E. Schulz nach dem Roten Meer. E. 11.
Kieler Meeresf., 20:35-127.

Paradoxostoma micropunctata n.sp.
Swain, Frederick M., 1967.
Ostracoda from the Gulf of California.
Mem. Geol. Soc. Am., 101: 139 pp.

Paradoxostoma normani
Elofson, O. 1941
Zur Kenntnis der marinen Ostracodes Schwedens mit besonderer Berücksichtigung des Skageraks. Zool. Bidrag. Uppsala 19: 215-534; 42 maps, 52 figs.

Paradoxostoma parabreve n.sp.
Hartmann, Gerd, 1964.
Zur Kenntnis der Ostracoden des Roten Meeres. Wissenschaftliche Ergebnisse einer Forschungsreise von A. Remane und E. Schulz nach dem Roten Meer. E. 11.
Kieler Meeresf., 20:35-127.

Paradoxostoma pequegnati n.sp.
McKenzie, K.G., 1971.
Paradoxostoma pequegnati n.sp. (Ostracoda, Podocopina) from the Gulf of Mexico.
Crustaceana 20(1):46-50.

Paradoxostoma pulchellum
Elofson, O. 1941
Zur Kenntnis der marinen Ostracodes Schwedens mit besonderer Berücksichtigung des Skageraks. Zool. Bidrag. Uppsala 19: 215-534; 42 maps, 52 figs.

Paradoxostoma punctatum n.sp.
Hartmann, Gerd, 1964.
Zur Kenntnis der Ostracoden des Roten Meeres. Wissenschaftliche Ergebnisse einer Forschungsreise von A. Remane und E. Schulz nach dem Roten Meer. E. 11.
Kieler Meeresf., 20:35-127.

Paradoxostoma romei n.sp.
McKenzie, K.G. 1967.
Recent Ostracoda from Port Phillip Bay, Victoria.
Proc. R. Soc. Victoria n.s. 80(1): 61-106.

Paradoxostoma salvadorianus n.sp.
Hartmann, Gerd, 1959.
Zur Kenntnis der lotischen Lebensbereiche der Pazifischen Küste von El Salvador unter besonderer Berücksichtigung seiner Ostracodenfauna.
Kieler Meeresf., 15(2):187-241.

Paradoxostoma striungulum
McHardy, R.A., 1964.
Marine ostracods from the plankton of Indian Arm, British Columbia, including a diminutive subspecies resembling Conchoecia alata, major Rudjakov.
J. Fish. Res. Bd., Canada, 21(3):555-576.

Paradoxostoma striungulum n.sp.
Smith, Verna Z., 1952.
Further Ostracoda of the Vancouver Island region. J. Fish. Res. Bd., Canada, 9(1):16-41, 11 pls.

Paradoxostoma trapezoideum n.sp.
McKenzie, K.G. 1967.
Recent Ostracoda from Port Phillip Bay, Victoria.
Proc. R. Soc. Victoria n.s. 80(1): 61-106.

Paradoxostoma variabile
Elofson, O. 1941
Zur Kenntnis der marinen Ostracodes Schwedens mit besonderer Berücksichtigung des Skageraks. Zool. Bidrag. Uppsala 19: 215-534; 42 maps, 52 figs.

Paradoxostoma variabili
Stephensen, K., 1938.
Marine Ostracoda and Cladocera. Zool. Iceland 3(32):19 pp., 1 textfig.

Parakrithe ? sp.

Swain, Frederick M., 1967.

Ostracoda from the Gulf of California.

Mem. Geol. Soc. Am., 101: 139 pp.

Parakrithella australis n.sp.

McKenzie, K.G. 1967.
Recent Ostracoda from Port Phillip Bay, Victoria.
Proc. R. Soc. Victoria n.s. 80(1): 61-106.

Parakrithella oblonga n.sp.

Swain, Frederick M., 1967.

Ostracoda from the Gulf of California.

Mem. Geol. Soc. Am., 101: 139 pp.

Paraphilomedes tricornuta n.sp.

Poulsen, Erik M., 1962
Ostracoda-Myodocopa. Part 1.
Cypridiniformes-Cypridinidae.
Dana-Report No. 57:414 pp.

Paraphilomedes unicornuta n.sp.

Poulsen, Erik M., 1962
Ostracoda-Myodocopa. Part 1.
Cypridiniformes-Cypridinidae.
Dana-Report No. 57:414 pp.

Parapontoparta arcuata n.gen., n.sp.

Hartmann, G., 1953.
Neue marine Ostracoden der Familie Cypridae und der Subfamilie Cytherideinae der Familie Cytharidae aus Brasiliens. Zool. Anz., 154:109-127.

Parasarsiella globulus n. gen.

Poulson, Erik.M., 1965.
Ostracoda-Myodocopa. II. Cypridiniformes-Rutidermatidae, Sarsiellidae and Asteropidae.
Dana Report, No. 65:484 pp.

Parasterope corrugata n. sp.

Poulson, Erik M., 1965.
Ostracoda-Myodocopa. II. Cypridiniformes-Rutidermatidae, Sarsiellidae and Asteropidae.
Dana Report, No. 65:484 pp.

Parasterope jenseni n. sp.

Poulson, Erik M., 1965.
Ostracoda-Myodocopa. II. Cypridiniformes-Rutidermatidae, Sarsiellidae and Asteropidae.
Dana Report, No. 65:484 pp.

Parasterope longungues n. sp.

Poulson, Erik M., 1965.
Ostracoda-Myodocopa. II. Cypridiniformes-Rutidermatidae, Sarsiellidae and Asteropidae.
Dana Report, No. 65:484 pp.

Parasterope muelleri

Poulson, Erik.M., 1965.
Ostracoda-Myodocopa. II. Cypridiniformes-Rutidermatidae, Sarsiellidae and Asteropidae.
Dana Report, No. 65:484 pp.

Parasterope nana n. sp.

Poulson, Erik M., 1965.
Ostracoda-Myodocopa. II. Cypridiniformes-Rutidermitidae, Sarsiellidae and Asteropidae.
Dana Report, No. 65:484 pp.

Parasterope obesa n. gen, n.sp.

Poulson, Erik.M., 1965.
Ostracoda-Myodocopa. II. Cypridiniformes-Rutidermatidae, Sarsiellidae and Asteropidae.
Dana Report, No. 65:484 pp.

Parasterope pectinata n. sp.

Poulson, Erik.M., 1965.
Ostracoda-Myodocopa. II. Cypridiniformes-Rutidermatidae, Sarsiellidae and Asteropidae.
Dana Report, No. 65:484 pp.

Paravargula n.gen.

Poulsen, Erik M., 1962
Ostracoda-Myodocopa. Part 1.
Cypridiniformes-Cypridinidae.
Dana-Report No. 57:414 pp.

Paravargula arborea

Poulsen, Erik M., 1962
Ostracoda-Myodocopa. Part 1.
Cypridiniformes-Cypridinidae.
Dana-Report No. 57:414 pp.

Paravargula ensifera n.sp.

Poulsen, Erik M., 1962
Ostracoda-Myodocopa. Part 1.
Cypridiniformes-Cypridinidae.
Dana-Report No. 57:414 pp.

Paravargula hirsuta

Poulsen, Erik M., 1962
Ostracoda-Myodocopa. Part 1.
Cypridiniformes-Cypridinidae.
Dana-Report No. 57:414 pp.

Parvocythere dentata n.gen., n.sp.

Hartmann, Gerd, 1959.
Zur Kenntnis der lotischen Lebensbereiche der Pazifischen Küste von El Salvador unter besonderer Berücksichtigung seiner Ostracodenfauna.
Kieler Meeresf., 15(2):187-241.

Parvocytherinae n.subfam.

Hartmann, Gerd, 1959.
Zur Kenntnis der lotischen Lebensbereiche der Pazifischen Küste von El Salvador unter besonderer Berücksichtigung seiner Ostracodenfauna.
Kieler Meeresf., 15(2):187-241.

Patagonacythere devexa

Neale, J. W., 1967.
An ostracod fauna from Halley Bay, Coats Land, British Antarctic Territory.

Brit. Antarct. Surv. Sci. Rep. 58: 50 pp.

Payenborchella malaiensis

Key, A.J., 1953.
Preliminary note on the recent Ostracoda of the Snellius Expedition. Proc. K. Nederl. Akad. Wetensch., B, 56(2):155-168, 1 textfig., 2 pls.

Payenborchella rocosa

Key, A.J., 1953.
Preliminary note on the recent Ostracoda of the Snellius Expedition. Proc. K. Nederl. Akad., Wetensch., B, 56(2):155-168, 1 textfig., 2 pls.

Pectocythere tomalensis n. sp.

Watling, Les, 1970.
Two new species of Cytherinae (Ostracoda) from central California.
Crustaceana 19(3): 251-263

Pellucistoma sp.

King, Charles E. and Louis S. Kornicker, 1970.

Ostracoda in Texas bays and lagoons: an ecological study.
Smithson. Contrib. Zool., 24: 92 pp.

Perissocytheridea brachyforma

King, Charles E. and Louis S. Kornicker, 1970.

Ostracoda in Texas bays and lagoons: an ecological study.
Smithson. Contrib. Zool., 24: 92 pp.

Pellucistoma scrippsi n.sp.

Benson, Richard H., 1959.
Arthropoda 1. Ecology of recent ostracodes of the Todos Santos Bay region, Baja California, Mexico. Univ. Kansas, Paleontol. Contr., :1-80, 11 pls., 20 figs.

Pellucistoma scrippsi

Swain, Frederick M., 1967.

Ostracoda from the Gulf of California

Mem. Geol. Soc. Am., 101: 139 pp.

Pericythere foveata n. gen. n. sp.

Hartmann, G., 1952
Zur Kenntnis des Mangrove-Estero-Gebietes von El Salvador und seiner Ostracoden-Fauna. II. Systematischer Teil. Kieler Meeresf. 13(1): 134-159.

Perissacytheridea meyerabichi

Swain, Frederick M., 1967.

Ostracoda from the Gulf of California.

Mem. Geol. Soc. Am., 101: 139 pp.

Perissocytheridea rugata

King, Charles E. and Louis S. Kornicker, 1970.

Ostracoda in Texas bays and lagoons: an ecological study.
Smithson. Contrib. Zool., 24: 92 pp.

Perissocytheridea swaini

King, Charles E. and Louis S. Kornicker, 1970.

Ostracoda in Texas bays and lagoons: an ecological study.
Smithson. Contrib. Zool., 24: 92 pp.

Philippiella n. gen.

Poulson, Erik M., 1965.
Ostracoda-Myodocopa. II. Cypridiniformes-Rutidermitidae, Sarsiellidae and Asteropidae.
Dana Report, No. 65:484 pp.

Philomedes sp.

McHardy, R.A., 1964.
Marine ostracods from the plankton of Indian Arm, British Columbia, including a diminutive subspecies resembling Conchoecia alata, major Rudjakov.
J. Fish. Res. Bd., Canada, 21(3):555-576.

Philomedes antarctica n.sp.

Brady, G.S., 1907.
Crustacea. V. Ostracoda. Nat. Antarctic Exped., 1901-1904, 3: 9 pp., 3 pls.

Philomedes assimilis n.sp.

Brady, G.S., 1907.
Crustacea. V. Ostracoda. Nat. Antarctic Exped., 1901-1904, 3:9 pp., 3 pls.

Philomedes brenda
Sars, G.O., 1886
Crustacea II. Zool. Norwegian North-Atlantic Exped., 1876-1878: 96 pp., 1 map.

Philomedes brenda
Stephensen, K., 1917
Zoogeographical investigations of certain fjords in southern Greenland with special reference to Crustacea, Pycnogonida and Echinodermata including a list of Alcyonaria and Pisces, Medd. om Grønland, 53(3):229-378.

Philomedes brenda
Sylvester-Bradley, P.C., 1950.
The identity of the ostracod, Philomedes brenda (Baird). Ann. Mag. Nat. Hist., 12th ser., 3(33): 777-778.

Philomedes (Schleroconcha) chilensis n.sp.
Hartmann-Schröder, Gesa, und Gerd Hartmann, 1965.
Zur Kenntnis des Sublitorals der chilenischen Küste unter besonderer Berücksichtigung der Polychaeten und Ostracoden (mit Bermerkungen über den Einfluss saverstoffarmer Strömungen auf die Besiedlung von marinen Sedimenten). Mitt. Hamburg. Zool. Mus. u Inst., Ergärzungsband zu 62:384 pp.

Philomedes carcharodonta n.sp.
Smith, Verna Z., 1952.
Further Ostracoda of the Vancouver Island region J. Fish. Res. Bd., Canada, 9(1):16-41, 11 pls.

Philomedes curvata n.sp.
Poulsen, Erik M., 1962
Ostracoda-Myodocopa. Part 1. Cypridiniformes-Cypridinidae. Dana-Report No. 57:414 pp.

Philomedes dentata n.sp.
Poulsen, Erik M., 1962
Ostracoda-Myodocopa. Part 1. Cypridiniformes-Cypridinidae. Dana-Report No. 57:414 pp.

Philomedes fonsecensis n.sp.
Hartmann, Gerd, 1959.
Zur Kenntnis der lotischen Lebensbereiche der Pazifischen Küste von El Salvador unter besonderer Berücksichtigung seiner Ostracodenfauna.
Kieler Meeresf., 15(2):187-241.

Philomedes globosa
Granata, L., and L. di Caporiacco, 1949
Ostracodes marins recueillis pendant les croiserieres du Prince Albert, 1. Res. Camp. Sci. Monaco, 109: 48 pp., 4 pls.

Philomedes globosa
Poulsen, Erik M., 1962
Ostracoda-Myodocopa. Part 1. Cypridiniformes-Cypridinidae. Dana-Report No. 57:414 pp.

Philomedes globosus
Stephensen, K., 1912
Report on the Malacostraca, Pycnogonida and some Entomostraca collected by the Danmark-Expedition to North East Greenland. Medd. om Grønland, 45(11):501-630, Pls. 39-43.

Philomedes globosa
Klie, W., 1944.
Ostracoda I. Family: Cypridinidae. Fiches d'Ident. Zooplancton, Cons. Perm. Int. Expl. Mer, 5:4 pp., 6 textfigs.

Philomedes globosus
Stephensen, K., 1938.
Marine Ostracoda and Cladocera. Zool. Iceland 3(32):19 pp., 1 textfig.

Philomedes (Philomedes) globulosus
Elofson, O. 1941
Zur Kenntnis der marinen Ostracodes Schwedens mit besonderer Berücksichtigung des Skageraks. Zool. Bidrag. Uppsala 19: 215-534; 42 maps, 52 figs.

Philomedes grafi n.sp.
Hartmann, Gerd, 1964.
Zur Kenntnis der Ostracoden des Roten Meeres. Wissenschaftliche Ergebnisse einer Forschungsreise von A. Remane und E. Schulz nach dem Roten Meer. E. 11.
Kieler Meeresf., 20:35-127.

Philomedes interpuncta
Granata, L., and L. di Caporiacco, 1949
Ostracodes marins recueillis pendant les croiserieres du Prince Albert, 1. Res. Camp. Sci. Monaco, 109: 48 pp., 4 pls.

Philomedes interpuncta
Klie, W., 1944.
Ostracoda I. Family: Cypridinidae. Fiches d'Ident. Zooplancton, Cons. Perm. Int. Expl. Mer, 5:4 pp., 6 textfigs.

Philomedes (Philomedes) lilljeborgi
Elofson, O. 1941
Zur Kenntnis der marinen Ostracodes Schwedens mit besonderer Berücksichtigung des Skageraks. Zool. Bidrag. Uppsala 19: 215-534; 42 maps, 52 figs.

Philomedes lilljeborgi
Klie, W., 1944.
Ostracoda I. Family: Cypridinidae. Fiches d'Ident. Zooplancton, Cons. Perm. Int. Expl. Mer, 5:4 pp., 6 textfigs.

Philomedes lilljeborgi
Poulsen, Erik M., 1962
Ostracoda-Myodocopa. Part 1. Cypridiniformes-Cypridinidae. Dana-Report No. 57:414 pp.

Philomedes lilljeborgii
Sars, G.O., 1886
Crustacea II. Zool. Norwegian North-Atlantic Exped., 1876-1878: 96 pp., 1 map.

Philomedes lilljeborgi
Stephensen, K., 1938.
Marine Ostracoda and Cladocera. Zool. Iceland 3(32):19 pp., 1 textfig.

Philomedes lomae
Hartmann, G., 1959.
Zur Kenntnis des Mangrove-Estero-Gebietes von El Salvador und seiner Ostracoden-Fauna. II. Systematischer Teil. Kieler Meeresf. 13(1): 134-159.

Philomedes longiseta n. sp.
Juday, C., 1907
XIV. Ostracoda of the San Diego region. II. Littoral forms. Contributions from the Laboratory of the Marine Biological Association of San Diego. Univ. Calif. Publ. Zool. 3(9):135-156, Pls. 18-20.

Philomedes longiseta
Lucas, V.Z., 1931.
Some Ostracoda from the Vancouver island region. Contr. Canadian Biol. Fish., n.s., 6(17):397-416.

Philomedes longiseta
Smith, Verna Z., 1952.
Further Ostracoda of the Vancouver Island region J. Fish. Res. Bd., Canada, 9(1):16-41

Philomedes macandrei
Klie, W., 1944.
Ostracoda I. Family: Cypridinidae. Fiches d'Ident. Zooplancton, Cons. Perm. Int. Expl. Mer, 5:4 pp., 6 textfigs.

Philomedes orbicularis n.sp.
Brady, G.S., 1907.
Crustacea. V. Ostracoda. Nat. Antarctic Exped., 1901-1904, 3:9 pp., 3 pls.

Philomedes (Schleroconcha) reticulate nsp.
Hartmann-Schröder, Gesa, und Gerd Hartmann, 1965.
Zur Kenntnis des Sublitorels der chilenischen Küste unter besonderer Berücksichtigung der Polychaeten und Ostracoden (mit Bermerkungen über den Einfluss saverstoffarmer Strömungen auf die Besiedlung von marinen Sedimenten). Mitt. Hamburg, Zool. Mus. u Inst., Ergärzungsband zu 62:384 pp.

Philomedes trituberculatus n.sp.
Lucas, V.Z., 1931.
Some Ostracoda from the Vancouver island region. Contr. Canadian Biol. Fish., n.s., 6(17):397-416.

Phyctenophora sp.
McKenzie K.G. 1967.
Recent Ostracoda from Port Phillip Bay, Victoria.
Proc. R. Soc. Victoria n.s. 80(1):61-106

Physocypria perlata
Rome, R., 1954.
Contribution à l'étude des ostracodes de Belgique II. Espèces rares et espèces nouvelles.
Bull. Inst. Sci. Nat. Belg. 30(33):1-32, 7 textfigs.

Pleoschisma oblonga n.sp.
Juday, C., 1907
XIV. Ostracoda of the San Diego region. II. Littoral forms. Contributions from the Laboratory of the Marine Biological Association of San Diego. Univ. Calif. Publ. Zool. 3(9):135-156, Pls. 18-20.

Polycope spp.
Bonaduce, Gioacchino, 1964.
Contributo alla conoscenza e correlazione sistematica nel l'ambito della Famiglia Polycopida (Ostracoda, Cladocopa) con particolare referimento alle relazioni tra parte molle e carapace. Pubbl. Staz. Zool., Napoli, 34:160-184.

Polycope spp.
Hartmann, Gerd, 1964.
Zur Kenntniss der Ostracoden des Roten Meeres. Wissenschaftliche Ergebnisse einer Forschungsreise von A. Remane und E. Schulz nach dem Roten Meer. E. 11.
Kieler Meeresf., 20:35-127.

Polycope aida n.sp.
Hartmann, Gerd, 1959.
Zur Kenntnis der lostischen Lebensbereiche der Pazifischen Küste von El Salvador unter besonderer Berücksichtigung seiner Ostracodenfauna.
Kieler Meeresf., 15(2):187-241.

Polycope arenicola n.sp.
Hartman, G., 1954.
Neue Polycopidae (Ostracoda: Cladocopa) von europäischen Küsten. Kieler Meeresf. 10(1):84-99, Pls. 24-34.

Polycope areolata
Elofson, O. 1941
Zur Kenntnis der marinen Ostracodes Schwedens mit besonderer Berücksichtigung des Skageraks. Zool. Bidrag. Uppsala 19: 215-534; 42 maps, 52 figs.

Polycope cancellea n.sp.
Hartman, G., 1954.
Neue Polycopidae (Ostracoda:Cladocopa) von europäischen Küsten. Kieler Meeresf. 10(1):84-99, Pls. 24-34.

Polycope clathrata
Elofson, O. 1941
Zur Kenntnis der marinen Ostracodes Schwedens mit besonderer Berücksichtigung des Skageraks. Zool. Bidrag. Uppsala 19: 215-534; 42 maps, 52 figs.

Polycope difficilis
Schäfer, H.W., 1953.
Über Meeres- und Brackwasser-Ostracoden aus dem Deutschen Küstengebiet mit: 2. Mitteilung über die Ostracodenfauna Griechenlands. Hydrobiologia 5(4):351-389.

Polycope dimorpha n.sp.
Hartman, G., 1954.
Neue Polycopidae (Ostracoda: Cladocopa) von europäischen Küsten. Kieler Meeresf. 10(1):84-99, Pls. 24-34.

Polycope duplidentata n.sp.
Hartmann, Gerd, 1959.
Zur Kenntnis der lotischen Lebensbereiche der Pazifischen Küste von El Salvador unter besonderer Berücksichtigung seiner Ostracodenfauna.
Kieler Meeresf., 15(2):187-241.

Polycope elongata n.sp.
Hartman, G., 1954.
Neue Polycopidae (Ostracoda:Cladocopa) von europäischen Küsten. Kieler Meeresf. 10(1):84-99, Pls. 24-34.

Polycope laevis n.sp.
Hartman, G., 1954.
Zwei neue Polycope-Arten (Ostracoda, Cladocopa) aus Brasilien. Zool. Anz., 153(7/8):175-182.

Polycope limbata n.sp.
Hartman, G., 1954.
Neue Policopidae (Ostracoda: Cladocopa) von europäischen Küsten. Kieler Meeresf. 10(1):84-99, Pls. 24-34.

Polycope longipes n.sp.
Hartman, G., 1954.
Neue Polycopidae (Ostracoda:Cladocopa) von europäischen Küsten. Kieler Meeresf. 10(1):84-99, Pls. 24-34.

Polycope loscobanosi n.sp.
Hartmann, Gerd, 1959.
Zur Kenntnis der lotischen Lebensbereiche der Pazifischen Küste von El Salvador unter besonderer Berücksichtigung seiner Ostracodenfauna.
Kieler Meeresf., 15(1):187-241.

Polycope microdispar, n.sp.
Hartman, G., 1954.
Neue Polycopidae (Ostracoda:Cladocopa) von europäischen Küsten. Kieler Meeresf. 10(1):84-99, Pls. 24-34.

Polycope noodti n.sp.
Hartmann, Gerd, 1959.
Zur Kenntnis der lotischen Lebensbereiche der Pazifischen Küste von El Salvador unter besonderer Berücksichtigung seiner Ostracodenfauna.
Kieler Meeresf., 15(2):187-241.

Polycope obtusa n.sp.
Hartman, G., 1954.
Neue Polycopidae (Ostracoda:Cladocopa) vone euro-päischen Küsten. Kieler Meeresf. 10(1):84-99, Pls. 24-34.

Polycope onkophora n.sp.
Hartman, G., 1954.
Neue Polycopidae (Ostracoda: Cladocopa) von europäischen Küsten. Kieler Meeresf. 10(1):84-99, Pls 24-34.

Polycope orbicularis
Elofson,O. 1941
Zur Kenntnis der marinen Ostracodes Schwedens mit besonderer Berücksichtigung des Skageraks. Zool. Bidrag. Uppsala 19: 215-534; 42 maps, 52 figs.

Polycope punctata
Elofson,O. 1941
Zur Kenntnis der marinen Ostracodes Schwedens mit besonderer Berücksichtigung des Skageraks. Zool. Bidrag. Uppsala 19: 215-534; 42 maps, 52 figs.

Polycope ramosa n.sp.
Hartman, G., 1954.
Neue Polycopidae (Ostracoda:Cladocopa) von europäischen Küsten. Kieler Meeresf. 10(1):84-99, Pls. 24-34.

Polycope schulzi n.sp.
Klie, W., 1950.
Eine neue Polycope (Ostr.) aus der Kieler Bucht. Kieler Meeresf. 7(1):129-132, 8 textfigs.

Polycope seridentata n.sp.
Hartmann, Gerd, 1959.
Zur Kenntnis der lotischen Lebensbereiche der Pazifischen Küste von El Salvador unter besonderer Berücksichtigung seiner Ostracodenfauna.
Kieler Meeresf., 15(2):187-241.

Polycope sublaevis
Elofson,O. 1941
Zur Kenntnis der marinen Ostracodes Schwedens mit besonderer Berücksichtigung des Skageraks. Zool. Bidrag. Uppsala 19: 215-534; 42 maps, 52 figs.

Polycope teneriffa n.sp.
Hartmann, G., 1959.
Neue Ostracoden von Teneriffa. Zool. Anz., 162(5/6):160-171.

Ponticocythereis militaris
McKenzie, K.G. 1967.
Recent Ostracoda from Port Phillip Bay, Victoria.
Proc. R. Soc. Victoria n.s. 80(1):61-106.

Pontocypris 6 spp.
Maddocks, Rosalie F., 1969.
Recent Ostracodes of the family Pontocyprididae Chiefly from the Indian Ocean.
Smithson. Contrib. Zool., 7:1-56.

Pontocypris clemensi n.sp.
Smith, Verna Z., 1952.
Further Ostracoda of the Vancouver Island region.
J. Fish. Res. Bd., Canada, 9(1):16-41, 11 pls.

Pontocypris dispar
Hartmann, G., 1953.
Ostracodes des étangs méditerraneens.
Vie et Milieu, Bull. Lab. Arago, 4(4):707-712, 1 textfig.

Pontocypris helenae n.sp.commensal
Maddocks, Rosalie F., 1968.
Commensal and free-living species of Pontocypris Müller, 1894 (Ostrocoda, Pontocyprididae) from the Indian and Southern oceans.
Crustaceana, 15(2):121-136.

Pontocypris humesi n. sp.
Maddocks, Rosalie F., 1968.
Commensal and free-living species of Pontocypris Müller, 1894 (Ostrocoda, Pontocyprididae) from the Indian and Southern Oceans.
Crustaceana, 15(2):121-136.

Pontocypris intermedia
Tressler, W.L., 1949
Marine ostracoda from Tortugas, Florida.
J. Wash. Acad. Sci. 39(10):334-343, 25 text figs.

Pontocypris meridionalis n. comb.
Maddocks, Rosalie F., 1968.
Commensal and free-living species of Pontocypria Müller 1894 (Ostrocoda, Pontocyprididae) from the Indian and Southern Oceans.
Crustaceana, 15(2):121-136.

Pontocypris trigonella
Elofson,O. 1941
Zur Kenntnis der marinen Ostracodes Schwedens mit besonderer Berücksichtigung des Skageraks. Zool. Bidrag. Uppsala 19: 215-534; 42 maps, 52 figs.

Pontocythere ashermani
Hulings, Neil C., 1966.
Marine Ostracoda from western North Atlantic Ocean off the Virginia coast.
Chesapeake Science, 7(1):40-56.

Pontocythere cf. turbida
Hulings, Neil C., 1966.
Marine Ostracoda from western North Atlantic Ocean off the Virginia coast.
Chesapeake Science, 7(1):40-56.

Ponticocythereis n.gen.
McKenzie, K.G. 1967.
Recent Ostracoda from Port Phillip Bay, Victoria.
Proc. R. Soc. Victoria n.s. 80(1):61-106.

Potamocypris sp.
King, Charles E. and Louis S. Kornicker, 1970.
Ostracoda in Texas bays and lagoons: an ecological study.
Smithson. Contrib. Zool., 24: 92 pp.

Potamocypris hambergi n.sp.
Alm, G., 1914.
Beschreibung einiger neue Ostracoden aus Schweden Zool. Anz. 43:468-475, 4 textfigs.

Potamocypris maculata n.sp.
Alm, G., 1914.
Beschreibung einiger neue Ostracoden aus Schweden Zool. Anz. 43:468-475, 4 textfigs.

Potamocypris mazatlanensis n. sp.
Swain, Frederick M., 1967.
Ostracoda from the Gulf of California.
Mem. Geol. Soc. Am., 101: 139 pp.

Pomatocypris pallida n.sp.
Alm, G., 1914.
Beschreibung einiger neue Ostracoden aus Schweden. Zool. Anz. 43:468-475, 4 textfigs.

Potamocypris smeragdina
King, Charles E. and Louis S. Kornicker, 1970.
Ostracoda in Texas bays and lagoons: an ecological study.
Smithson. Contrib. Zool., 24: 92 pp.

Propontocypris (Ekpontocypris 2 spp.
Maddocks, Rosalie F., 1969.
Recent Ostracodes of the family Pontocyprididae Chiefly from the Indian Ocean.
Smithson. Contrib. Zool., 7:1-56.

Propontocypris (Schedopontocypris) 6 spp.
Maddocks, Rosalie F., 1969.
Recent Ostracodes of the family Pontocyprididae Chiefly from the Indian Ocean.
Smithson. Contrib. Zool., 7:1-56.

Propontocypris (Propontocypris) 8 spp
Maddocks, Rosalie F., 1969.
Recent Ostracodes of the family Pontocyprididae Chiefly from the Indian Ocean.
Smithson. Contrib. Zool., 7:1-56.

Propontocypris sp.
McKenzie, K.G. 1967.
Recent Ostracoda from Port Phillip Bay, Victoria.
Proc. R. Soc. Victoria, n.s. 80(1):61-106.

Propontocypris (Schedopontocypris) bengalensis n.sp.
Maddocks, Rosalie F., 1969.
Recent Ostracodes of the family Pontocyprididae
chiefly from the Indian Ocean.
Smithson. Contrib. Zool., 7:1-56.

Propontocypris (Propontocypris) corcata n.sp.
Maddocks, Rosalie F., 1969.
Recent Ostracodes of the family Pontocyprididae
chiefly from the Indian Ocean.
Smithson. Contrib. Zool., 7:1-56.

Propontocypris edwardsi
Williams, Roger B., 1966.
Recent marine podocopid Ostracoda of
Narragansett Bay, Rhode Island.
Univ. Kansas, Paleont. Contrib., Paper 11:36pp.

Propontocypris (Ekpontocypris) epicyrta n.sp.
Maddocks, Rosalie F., 1969.
Recent Ostracodes of the family Pontocyprididae
chiefly from the Indian Ocean.
Smithson. Contrib. Zool., 7:1-56.

Propontocypris (Propontocypris) herdmani
Maddocks, Rosalie F., 1969.
Recent Ostracodes of the family Pontocyprididae
chiefly from the Indian Ocean.
Smithson. Contrib. Zool., 7:1-56.

Propontocypris howei
Hulings, Neil C., 1966.
Marine Ostracods from western North Atlantic
Ocean off the Virginia coast.
Chesapeake Science, 7(1):40-56.

Propontocypris (Ekpontocypris) litoricola n.sp. & n.subsp.
Maddocks, Rosalie F., 1969.
Recent Ostracodes of the family Pontocyprididae
chiefly from the Indian Ocean.
Smithson. Contrib. Zool., 7:1-56.

Propontocypris (Propontocypris?) lobodonta n.sp.
Maddocks, Rosalie F., 1969.
Recent Ostracodes of the family Pontocyprididae
chiefly from the Indian Ocean.
Smithson. Contrib. Zool., 7:1-56.

Propontocypris (Ekpontocypris) McMurdoensis n.sp
Maddocks, Rosalie F., 1969.
Recent Ostracodes of the family Pontocyprididae
chiefly from the Indian Ocean.
Smithson. Contrib. Zool., 7:1-56.

Propontocypris (Propontocypris) paradispar n.sp.
Maddocks, Rosalie F., 1969.
Recent Ostracodes of the family Pontocyprididae
chiefly from the Indian Ocean.
Smithson. Contrib. Zool., 7:1-56.

Propontocypris (Propontocypris) quasicrocata n.sp.
Maddocks, Rosalie F., 1969.
Recent Ostracodes of the family Pontocyprididae
chiefly from the Indian Ocean.
Smithson. Contrib. Zool., 7:1-56.

Propontocypris (schedopontocypris?) simplex
Maddocks, Rosalie F., 1969.
Recent Ostracodes of the family Pontocyprididae
chiefly from the Indian Ocean.
Smithson. Contrib. Zool., 7:1-56.

Propontocypris (Propontocypris) subreniformis
Maddocks, Rosalie F., 1969.
Recent Ostracodes of the family Pontocyprididae
chiefly from the Indian Ocean.
Smithson. Contrib. Zool., 7:1-56.

Pseudoconchoecia serrulata var. laevis var. nov.
Brady, G.S., 1907.
Crustacea. V. Ostracoda. Nat. Antarctic Exped., 1901-1904, 3:9 pp., 3 pls.

Pseudocythere caudata
Ascoli, P., 1965.
Crociere Talassografia Afriatica 1955. VI.
Ricerche ecologiche sugli ostracodi contenuti
in 16 carote prelevate sul fondo del Mare Adriatico.
Arch. Oceanogr. e Limnol., 14(1):69-138.

Pseudocythere caudata
Elofson, O. 1941
Zur Kenntnis der marinen Ostracodes
Schwedens mit besonderer Berücksichtigung
des Skageraks. Zool. Bidrag. Uppsala 19:
215-534; 42 maps, 52 figs.

Pseudocythere of P. caudata
Neale, J. W., 1967.
An ostracod fauna from Halley Bay, Coats Land,
British Antarctic Territory.
Brit. Antarct. Surv. Sci. Rep. 58: 50 pp.

Pseudophilomedes ferulanus
Kornicker, Louis S. 1967.
The myodocopid ostracod families
Philomedidae and Pseudophilomedidae
(new family).
Proc. U.S. natn. Mus. 121(3580):1-35.

Pseudophilomedes foveolatus
Kornicker, Louis S. 1967.
The myodocopid ostracod families
Philomedidae and Pseudophilomedidae
(new family).
Proc. U.S. natn. Mus. 121(3580):1-35.

Pterygocythereis sp.
Van den Bold, W.A. 1966
Ostracoda from Colon Harbour, Panama
Carib. J. Sci., 6(1/2):43-64.

Pterygocythereis sp.
Ascoli, P., 1965.
Crociere Talassografia Afriatica 1955. VI.
Ricerche ecologiche sugli ostracodi contenuti
in 16 carote prelevate sul fondo del Mare Adriatico.
Arch. Oceanogr. e Limnol., 14(1):69-138.

Pterygocythereis ceratoptera
Ascoli, P., 1965.
Crociere Talassografia Afriatica 1955. VI.
Ricerche ecologiche sugli ostracodi contenuti
in 16 carote prelevate sul fondo del mare Adriatico.
Arch. Oceanogr. e Limnol., 14(1):69-138.

Pterygocythereis? cuevasensis n.sp.
Swain, Frederick M., 1967.
Ostracoda from the Gulf of California.
Mem. Geol. Soc. Am., 101: 139 pp.

Pterygocythereis delicata
Swain, Frederick M., 1967.
Ostracoda from the Gulf of California.
Mem. Geol. Soc. Am., 101: 139 pp.

Pterygocythereis jonesii
Ascoli, P., 1965.
Crociere Talassografia Afriatica 1955. VI.
Ricerche ecologiche sugli ostracodi contenuti
in 16 carote prelevate sul fondo del Mare Adriatico.
Arch. Oceanogr. e Limnol., 14(1):69-138.

Pterygocythereis semitranslucens
Benson, Richard H., 1959
Arthropoda 1. Ecology of recent ostracodes
of the Todos Santos Bay region, Baja
California, Mexico.
Univ. Kansas, Paleontol. Contr., :1-80,
11 pls., 20 figs.

Pterocypridina alata n.gen., n.sp
Poulsen, Erik M., 1962
Ostracoda-Myodocopa. Part 1.
Cypridiniformes-Cypridinidae.
Dana-Report No. 57:414 pp.

Pterocypridina birostrata n.sp.
Poulsen, Erik M., 1962
Ostracoda-Myodocopa. Part 1.
Cypridiniformes-Cypridinidae.
Dana-Report No. 57:414 pp.

Pterocypridina excreta n.sp.
Poulsen, Erik M., 1962
Ostracoda-Myodocopa. Part 1.
Cypridiniformes-Cypridinidae.
Dana-Report No. 57:414 pp.

Pteroloxa guaymasensis n.sp.
Swain, Frederick M., 1967.
Ostracoda from the Gulf of California.
Mem. Geol. Soc. Am., 101: 139 pp.

Pumilocytheridea sp.
Swain, Frederick M., 1967.
Ostracoda from the Gulf of California.
Mem. Geol. Soc. Am., 101: 139 pp.

Pumilocytheridea vermiculoides n.sp.
Swain, Frederick M., 1967.
Ostracoda from the Gulf of California.
Mem. Geol. Soc. Am., 101: 139 pp.

Puncia novozealandica n.gen., n.sp.
Hornibrook, N. de B., 1949.
A new family of living Ostracoda with resemblance
to some Paleozoic Beyrichiidae.
Trans. R. Soc., New Zealand, 77(4):469-471.

Puriana sp.
Van den Bold, W.A. 1966
Ostracoda from Colon Harbour, Panama
Carib. J. Sci., 6(1/2):43-64.

Puriana pacifica n.sp.
Benson, Richard H., 1959.
Arthropoda 1. Ecology of recent ostracodes
of the Todos Santos Bay region, Baja
California, Mexico. Univ. Kansas, Paleontol.
Contr., :1-80, 11 pls., 20 figs.

Puriana pacifica
Swain, Frederick M., 1967.
Ostracoda from the Gulf of California.
Mem. Geol. Soc. Am., 101: 139 pp.

Puriana rugipunctata
Williams, Roger B., 1966.
Recent marine podocopid Ostracoda of
Narragansett Bay, Rhode Island.
Univ. Kansas, Paleont. Contrib., Paper 11: 36 pp.

Pyrocypris natans
Delsman, H.C., 1936.
Preliminary plankton investigations in the Java
Sea. Treubia 17:139-181, 8 maps, 41 figs.

Quadracythere compacta
Van den Bold, W.A. 1966
Ostracoda from Colon Harbour, Panama
Carib. J. Sci., 6(1/2):43-64.

Reticulocythereis multicarinata
King, Charles E. and Louis S. Kornicker, 1970.
Ostracoda in Texas bays and lagoons: an
ecological study.
Smithson. Contrib. Zool., 24: 92 pp.

Quadracythere regalia n.sp.
Benson, Richard H., 1959.
Arthropoda 1. Ecology of recent ostracodes
of the Todos Santos Bay region, Baja
California, Mexico. Univ. Kansas, Paleontol.
Contr., :1-80, 11 pls., 20 figs.

Reussicythere n. gen.
Van den Bold, W.A. 1966
Ostracoda from Colon Harbour, Panama.
Carib. J. Sci. 6(1/2):43-64.

Reussicythere reussi
Van den Bold, W.A. 1966.
Ostracoda from Colon Harbour, Panama.
Carib. J. Sci. 6(1/2):43-64.

Robertsonites antarcticus n. sp.
Neale, J.W., 1967.
An ostracod fauna from Halley Bay, Coats Land,
British Antarctic Territory.
Brit. Antarct. Surv. Sci. Rep. 58: 50 pp.

Robertsonites tuberculata
Hulings, Neil C., 1967.
Marine Ostracoda from the western North Atlantic
Ocean: Labrador Sea, Gulf of St. Lawrence, and
off Nova Scotia.
Crustaceana, 13(3):310-328.

Ruggieria dictyon n.sp.
Van den Bold, W.A. 1966
Ostracoda from Colon Harbour, Panama
Carib. J. Sci., 6(1/2):43-64.

Rutiderma californica n.sp.
McKenzie, K.G., 1965.
Myodocopid Ostracoda (Cypridinacea) from Scammon
Lagoon, Baja California, Mexico, and their ecologic associations.
Crustaceana, 9(1):57-69.

Rutiderma (Rutiderma) chacaoi nsp.
Hartmann-Schröder, Gese, und Gerd Hartmann, 1965.
Zur Kenntnis des Sublitorels der chilenischen
Küste unter besonderer Berücksichtigung der
Polychaeten und Ostracoden (mit Bermerkungen
über den Einfluss saverstoffarmer Strömungen
auf die Besiedlung von marinen Sedimenten).
Mitt. Hamburg. Zool. Mus. u Inst.,
Ergerzungsband zu 62:384 pp.

Rutiderma compressa
Hartman, Gerd, 1964.
Zur Kenntnis der Ostracoden des Roten Meeres.
Wissenschaftliche Ergebnisse einer Forschungs-
reise von A. Remane und E. Schulz nach dem Roten
Meer. E. 11.
Kieler Meeresf., 20:35-127.

Rutiderma (Rutiderma) compressa
Hartmann-Schröder, Gese, und Gerd Hartmann, 1965.
Zur Kenntnis des Sublitorels der chilenischen
Küste unter besonderer Berücksichtigung der
Polychaeten und Ostracoden (mit Bermerkungen
über den Einfluss saverstoffarmer Strömungen
auf die Besiedlung von marinen Sedimenten).
Mitt. Hamburg. Zool. Mus. u Inst.,
Ergerzungsband zu 62:384 pp.

Rutiderma fusca n. sp.
Poulson, Erik M., 1965.
Ostracoda-Myodocopa. II. Cypridiniformes-
Rutidermatidae, Sarsiellidae and Asteropidae.
Dana Report, No. 65:484 pp.

Rutiderma hartmanni n. sp.
Poulson, Erik M., 1965.
Ostracoda-Myodocopa. II. Cypridiniformes-
Rutidermatidae, Sarsiellidae and Asteropidae.
Dana Report, No. 65:484 pp.

Rutiderma judayi n.sp.
McKenzie, K.G., 1965.
Myodocopid Ostracoda (Cypridinacea) from Scammon
Lagoon, Baja California, Mexico, and their ecologic associations.
Crustaceana, 9(1):57-69.

Rutiderma mortenseni nsp.
Poulson, Erik M., 1965.
Ostracoda-Myodocopa. II. Cypridiniformes-
Rutidermatidae, Sarsiellidae and Asteropidae.
Dana Report, No. 65:484 pp.

Rutiderma normani n. sp.
Poulson, Erik M., 1965.
Ostracoda-Myodocopa. II. Cypridiniformes-
Rutidermatidae, Sarsiellidae and Asteropidae.
Dana Report, No. 65:484 pp.

Rutiderma rostrata
Hartmann, Gerd, 1959.
Zur Kenntnis der lotischen Lebensbereiche
der Pazifischen Küste von El Salvador unter
besonderer Berücksichtigung seiner Ostracodenfauna.
Kieler Meeresf., 15(2):187-241.

Rutiderma rostrata n. sp.
Juday, C., 1907
XIV. Ostracoda of the San Diego region.
II. Littoral forms. Contributions from the
Laboratory of the Marine Biological Association of San Diego. Univ. Calif. Publ. Zool.
3(9):135-156, Pls. 18-20.

Rutiderma rostrata
McKenzie, K.G., 1965.
Myodocopid Ostracoda (Cypridinacea) from Scammon
Lagoon, Baja California, Mexico, and their ecologic associations.
Crustaceana, 9(1):57-69.

Rutiderma rostrata
Poulson, Erik M., 1965.
Ostracoda-Myodocopa. II. Cypridiniformes-
Rutidermatidae, Sarsiellidae and Asteropidae.
Dana Report, No. 65:484 pp.

Rutiderma rotunda n. sp.
Poulson, Erik M., 1965.
Ostracoda-Myodocopa. II. Cypridiniformes-
Rutidermatidae, Sarsiellidae and Asteropidae.
Dana Report, No. 65:484 pp.

Saipanetta bensoni n.sp.
Maddocks, Rosalie F. 1972
Two new living species of
Saipanetta (Ostracoda Podocopida).
Crustaceana 23(1):28-42.

Saipanetta brooksi n.sp.
Maddocks, Rosalie F. 1973.
Zenker's organ and a new species
of Saipanetta (Ostracoda)
Micropaleontology 19(2):193-208

Saipanetta kelloughae
Maddocks, Rosalie F. 1973.
Zenker's organ and a new species
of Saipanetta (Ostracoda)
Micropaleontology 19(2):193-208

Saipanetta kelloughae n.sp.
Maddocks, Rosalie F. 1972
Two new living species of
Saipanetta (Ostracoda Podocopida).
Crustaceana 23(1):28-42.

Saipanetta tumida
Maddocks, Rosalie F. 1973.
Zenker's organ and a new species
of Saipanetta (Ostracoda)
Micropaleontology 19(2):193-208

Sarsiella sp.
McKenzie, K.G., 1965.
Myodocopid Ostracoda (Cypridinacea) from Scammon
Lagoon, Baja California, Mexico, and their ecologic associations.
Crustaceana, 9(1):57-69.

Sarsiella crispata
Hartmann, Gerd, 1964.
Zur Kenntnis der Ostracoden des Roten Meeres.
Wissenschaftliche Ergebnisse einer Forschungs-
reise von A. Remane und E. Schulz nach dem Roten
Meer. E. 11.
Kieler Meeresf., 20:35-127.

Sarsiella globulus
Kornicker, L.S., and C.D. Wise, 1962.
Sarsiella (Ostracoda) in Texas bays and lagoons.
Crustaceana, 4(1):57-74.

Sarsiella rotunda n.sp.
Hartmann, Gerd, 1959.
Zur Kenntnis der lotischen Lebensbereiche der
Pazifischen Küste von El Salvador unter
besonderer Berücksichtigung seiner Ostracodenfauna.
Kieler Meeresf., 15(2):187-241.

Sarsiella spinosa n.sp.
Kornicker, L.S., and C.D. Wise, 1962.
Sarsiella (Ostracoda) in Texas bays and lagoons.
Crustaceana, 4(1):57-74.

Sarsiella texana n.sp.
Kornicker, L.S., and C.D. Wise, 1962.
Sarsiella (Ostracoda) in Texas bays and lagoons.
Crustaceana, 4(1):57-74.

Sarsiella zostericola
Kornicker, L.S., and C.D. Wise, 1962.
Sarsiella (Ostracoda) in Texas bays and lagoons.
Crustaceana, 4(1):57-74.

Schedopontocypris n. subgen.
Maddocks, Rosalie F., 1969.
Recent ostracodes of the family Pontocyprididae
chiefly from the Indian Ocean.
Smithson. Contrib. Zool., 7:1-56.

Schlerochilus antarcticus
Neale, J.W., 1967.
An ostracod fauna from Halley Bay, Coats Land,
British Antarctic Territory.
Brit. Antarct. Surv. Sci. Rep. 58: 50 pp.

Schlerochilus centroamericanus n.sp.
Hartmann, Gerd, 1959.
Zur Kenntnis der lotischen Lebensbereiche der Pazifischen Küste von El Salvador unter besonderer Berücksichtigung seiner Ostracodenfauna.
Kieler Meeresf., 15(2):187-241.

Schlerochilus? contortellus n.sp.
Swain, Frederick M., 1967.
Ostracoda from the Gulf of California.
Mem. Geol. Soc. Am., 101: 139 pp.

Schlerochilus contortus
Elofson, O. 1941
Zur Kenntnis der marinen Ostracodes Schwedens mit besonderer Berücksichtigung des Skageraks. Zool. Bidrag. Uppsala 19: 215-534; 42 maps, 52 figs.

Sclerochilus contortus
Williams, Rober B., 1966.
Recent marine podocopid Ostracoda of Narragensett Bay, Rhode Island.
Univ. Kansas, Paleont. Contrib., Paper 11:36pp.

Sclerochillus gewemülleri
Makkaveeva, E.B., 1961.
The biocoenotic history of small worms, Crustacea and marine ostracods.
Trudy Sevastopol Biol. Sta., (14):147-162.

Schlerochilus meridionalis
Neale, J. W., 1967.
An ostracod fauna from Halley Bay, Coats Land, British Antarctic Territory.
Brit. Antarct. Surv. Sci. Rep. 58: 50 pp.

Schlerochilus nasus n.sp.
Benson, Richard H., 1959.
Arthropoda 1. Ecology of recent ostracodes of the Todos Santos Bay region, Baja California, Mexico. Univ. Kansas, Paleontol. Contr., :1-80, 11 pls., 20 figs.

Sclerochilus rectomarginatus n.sp.
Hartmann, Gerd, 1964.
Zur Kenntnis der Ostracoden des Roten Meeres. Wissenschaftliche Ergebnisse einer Forschungsreise von A. Remane und E. Schulz nach dem Roten Meer. E. 11.
Kieler Meeresf., 20:35-127.

Schlerochilus reniformis
Neale, J. W., 1967.
An ostracod fauna from Halley Bay, Coats Land, British Antarctic Territory.
Brit. Antarct. Surv. Sci. Rep. 58: 50 pp.

Schleroconcha arcuata n.sp.
Poulsen, Erik M., 1962
Ostracoda-Myodocopa. Part 1. Cypridiniformes-Cypridinidae.
Dana-Report No. 57:414 pp.

Schleroconcha flexilis
Poulsen, Erik M., 1962
Ostracoda-Myodocopa. Part 1. Cypridiniformes-Cypridinidae.
Dana-Report No. 57:414 pp.

Scottiella crispata n. gen., n.sp
Poulson, Erik M., 1965.
Ostracoda-Myodocopa. II. Cypridiniformes-Rutidermatidae, Sarsiellidae and Asteropidae.
Dana Report, No. 65:484 pp.

Semicytherura sp.
McKenzie, K.G. 1967.
Recent Ostracoda from Port Phillip Bay, Victoria.
Proc. R. Soc. Victoria n.s. 80(1):61-106.

Semicytherura angusta n.sp.
McKenzie, K.G. 1967.
Recent Ostracoda from Port Phillip Bay, Victoria.
Proc. R. Soc. Victoria n.s. 80(1):61-106.

Semicytherura cryptifera
McKenzie, K.G. 1967.
Recent Ostracoda from Port Phillip Bay, Victoria.
Proc. R. Soc. Victoria n.s. 80(1):61-106.

Semicytherura incongruens
Ascoli, P., 1965.
Crociera Talassografica Adriatica 1955. VI. Ricerche ecologiche sugli ostracodi contenuti in 16 carote prelevate sul fondo del Mare Adriatico.
Arch Oceanogr. e Limnol., 14(1):69-138.

Semicytherura paenenuda
McKenzie, K.G. 1967.
Recent Ostracoda from Port Phillips Bay, Victoria.
Proc. R. Soc. Victoria n.s. 80(1):61-106.

Semicytherura similis
Williams, Roger B., 1966.
Recent marine podocopid Ostracoda of Narragensett Bay, Rhode Island.
Univ. Kansas, Paleont. Contrib., Paper 11:36 pp.

Semicytherura tenuireticulata n.sp.
McKenzie, K.G. 1967.
Recent Ostracoda from Port Phillip Bay, Victoria.
Proc. R. Soc. Victoria n.s. 80(1):61-106.

Skogsbergia crenulata n.s
Poulsen, Erik M., 1962
Ostracoda-Myodocopa. Part 1. Cypridiniformes-Cypridinidae.
Dana-Report No. 57:414 pp.

Skogsbergia curvata n.sp.
Poulsen, Erik M., 1962
Ostracoda-Myodocopa. Part 1. Cypridiniformes-Cypridinidae.
Dana-Report No. 57:414 pp.

Skogsbergia hesperidea
Poulsen, Erik M., 1962
Ostracoda-Myodocopa. Part 1. Cypridiniformes-Cypridinidae.
Dana-Report No. 57:414 pp.

Skogsbergia minuta n.gen., n.sp.
Poulsen, Erik M., 1962
Ostracoda-Myodocopa. Part 1. Cypridiniformes-Cypridinidae.
Dana-Report No. 57:414 pp.

Sphaeromicola dudichi
Roelofs, H.M.A. 1968.
Etude du développement de l'ostracode marin Sphaeromicola dudichi Klie 1938.
Bull. zool. Mus univ. Amsterdam 1 (5): 39-51.

Spinacopia bisetula n.sp
Kornicker, Louis S. 1969.
Morphology, ontogeny and intraspecific variation of Spinacopia, a new genus of myodocopid ostracod (Sarsiellidae).
Smithson. Contrib. Zool. (8): 1-50.

Spinacopia menziesi n.sp.
Kornicker, Louis S. 1969.
Morphology, ontogeny, and intraspecific variation of Spinacopia, a new genus of myodocopid ostracod (Sarsiellidae).
Smithson. Contrib. Zool. (8): 1-50.

Spinacopia sandersi n.gen.n.sp.
Kornicker, Louis S. 1969.
Morphology, ontogeny, and intraspecific variation of Spinacopia, a new genus of myodocopid ostracod (Sarsiellidae).
Smithson. Contrib. Zool. (8): 1-50.

Spinacopia variabilis n.sp.
Kornicker, Louis S. 1969.
Morphology, ontogeny, and intraspecific variation of Spinacopia, a new genus of myodocopid ostracod (Sarsiellidae).
Smithson. Contrib. Zool. (8): 1-50.

Spinileberis hyalinus n. sp.
Watling, Les, 1970.
Two new species of Cytherinae (Ostracoda) from central California.
Crustaceana 19(3): 251-263.

Stenocypria australis nsp.
Swain, Frederick M., 1967.
Ostracoda from the Gulf of California.
Mem. Geol. Soc. Am., 101: 139 pp.

Stenocypris junodi n.sp.
Delachaux, T., 1920.
Description d'un Ostracode nouveau de l'Afrique portugaise. Bull. Soc. Port. Sci. Nat. 8:145-147, Pl. 7.

Strandesia marina nsp.
Hartmann-Schröder, Gesa, und Gerd Hartmann, 1965.
Zur Kenntnis des Sublitorals der chilenischen Küste unter besonderer Berücksichtigung der Polychaeten und Ostracoden (mit Bermerkungen über den Einfluss saverstoffarmer Strömungen auf die Besiedlung von marinen Sedimenten).
Mitt. Hamburg, Zool. Mus. u Inst., Ergarzungsband zu 62:384 pp.

Synasterope bassana n. gen., n. sp.
Poulson, Erik M., 1965.
Ostracoda-Myodocopa. II. Cypridiniformes-Rutidermatidae, Sarsiellidae and Asteropidae.
Dana Report, No. 65:484 pp.

Synasterope implumis n. sp.
Poulson, Erik M., 1965.
Ostracoda-Myodocopa. II. Cypridiniformes-Rutidermatidae, Sarsiellidae and Asteropidae.
Dana Report, No. 65:484 pp.

Synasterope knudseni n. sp.
Poulson, Erik M., 1965.
Ostracoda-Myodocopa. II. Cypridiniformes-Rutidermatidae, Sarsiellidae and Asteropidae.
Dana Report, No. 65:484 pp.

Synasterope longiseta n. sp.
Poulson, Erik M., 1965.
Ostracoda-Myodocopa. II. Cypridiniformes-Rutidermatidae, Sarsiellidae and Asteropidae.
Dana Report, No. 65:484 pp.

Synasterope oculata
Poulson, Erik M., 1965.
Ostracoda-Myodocopa. II. Cypridiniformes-Rutidermatidae, Sarsiellidae and Asteropidae.
Dana Report, No. 65:484 pp.

Synasterope quadrata

Poulson, Erik M., 1965.
Ostracoda-Myodocopa. II. Cypridiniformes-Rutidermatidae, Sarsiellidae and Asteropidae.
Dana Report, No. 65:484 pp.

Synasterope quatrisetosa

Poulson, Erik M., 1965.
Ostracoda-Myodocopa. II. Cypridiniformes-Rutidermatidae, Sarsiellidae and Asteropidae.
Dana Report, No. 65:484 pp.

Synasterope serrata n. sp.

Poulson, Erik M., 1965.
Ostracoda-Myodocopa. II. Cypridiniformes-Rutidermitidae, Sarsiellidae and Asteropidae.
Dana Report, No. 65:484 pp.

Tetragonodon ctenorhyncus

Van den Bold, W.A. 1966
Ostracoda from Colon Harbour, Panama.
Carib. J. Sci., 6(1/2):43-64.

Thalassocypris sp.

Hartmann, G., 1952.
Zur Kenntnis des Mangrove-Estero-Gebeites von El Salvador und seiner Ostracoden-Fauna. II. Systematischer Teil. Kieler Meeresf. 13(1): 134-159.

Thalassocypris aestuarina n. gen. n. sp.

Hartmann, G., 1952.
Zur Kenntnis des Mangrove-Estero-Gebeites von El Salvador und seiner Ostracoden-Fauna. II. Systematischer Teil. Kieler Meeresf. 13(1): 134-159.

Thalassocypris elongata n.gen., n.sp.

Hartmann, G., 1953.
Neue marine Ostracoden der Familie Cypridae und der Subfamilie Cytherideinae der Familie Cytheridae aus Brasiliens. Zool. Anz., 154:190-127.

Thalassocypris elongata n. sp.

Hartmann, G., 1952
Zur Kenntnis des Mangrove-Estero-Gebeites von El Salvador und seiner Ostracoden-Fauna. II. Systematischer Teil. Kieler Meeresf. 13(1):134-159.

Thaumatocypris bettenstaetti n.sp.

Bartenstein, H., 1949.
Thaumatocypris bettenstaetti n.sp. aus dem nordwestdeutschen Liss Z. (Ostrac.). Senckenbergiana 30:95-98, figs.

Trachyleberis

Harding, J.P., and P.C. Sylvester-Bradley, 1953.
The ostracod genus Trachyleberis.
Bull. Brit. Mus. (N.H.), Zool., 2(1):1-16, 25 textfigs.

Trachyleberis dunelmensis

Hulings, Neil C., 1967.
Marine Ostracoda from the western North Atlantic Ocean: Labrador Sea, Gulf of St. Lawrence, and off Nova Scotia.
Crustaceana, 13(3):310-328.

Trachyleberis lytteltonensis n.sp.

Harding, J.P., and P.C. Sylvester-Bradley, 1953.
The ostracod genus Trachyleberis.
Bull. Brit. Mus. (N.H.), Zool., 2(1):1-16, 25 text-figs.

Trachyleberis (Actinocythereis ?) marchilensis n. sp.

Hartmann-Schröder, Gesa, und Gerd Hartmann, 1965.
Zur Kenntnis des Sublitorels der chilenischen Küste unter besonderer Berücksichtigung der Polychaeten und Ostracoden (mit Bermerkungen über den Einfluss saverstoffarmer Strömungen auf die Besiedlung von marinen Sedimenten).
Mitt. Hamburg. Zool. Mus. u Inst., Ergänzungsband zu 62:384 pp.

Trachyleberis scabrocuneata

Harding, J.P., and P.C. Sylvester-Bradley, 1953.
The ostracod genus Trachyleberis.
Bull. Brit. Mus. (N.H.), Zool., 2(1):1-16, 25 textfigs.

Trachyleberidea tricornis n.sp.

Swain, Frederick M., 1967.
Ostracoda from the Gulf of California.
Mem. Geol. Soc. Am., 101? 139 pp.

Triebelina bradyi

Key, A.J., 1953.
Preliminary note on the recent Ostracoda of the Snellius Expedition. Proc. K. Nederl. Akad. Wetensch., B, 56(2):155-168, 1 textfig., 2 pls.

Triebelina giorloffi n.sp.

Hartmann, Gerd, 1959.
Zur Kenntnis der lotischen Lebensbereiche der Pazifischen Küste von El Salvador unter besonderer Berücksichtigung seiner Ostracodenfauna.
Kieler Meeresf., 15(2):187-241.

Triebelina reticulopunctata n.sp.

Benson, Richard H., 1959.
Arthropoda 1. Ecology of recent ostracodes of the Todos Santos Bay region, Baja California, Mexico. Univ. Kansas, Paleontol. Contr., :1-80, 11 pls., 20 figs.

Triabelina schulzi n.sp.

Hartman, Gerd, 1964.
Zur Kenntnis der Ostracoden des Roten Meeres. Wissenschaftliche Ergebnisse einer Forschungsreise von A. Remane und E. Schulz nach dem Roten Meer., E. 11.
Kieler Meeresf., 20:35-127.

Triebelina sertata

Key, A.J., 1953.
Preliminary note on the recent Ostracoda of the Snellius Expedition. Proc. K. Nederl. Akad. Wetensch., B, 56(2):155-168, 1 textfig., 2 pls.

Trigingbymus arenicola

Williams, Roger B., 1966.
Recent marine podocopid Ostracoda of Narragansett Bay, Rhode Island.
Univ. Kansas, Paleont. Contrib., Paper 11:36 pp.

Trigingbymus denticulata n.sp

Hulings, Neil C., 1966.
Marine Ostracoda from western North Atlantic Ocean off the Virginia coast.
Chesapeake Science, 7(1):40-56.

Urocythereis schulzi

Hartman, Gerd, 1960.
Ostracodes von Banyuls-sur-Mer.
Vie et Milieu, 11(3):413-424.

Vargula antarctica

Poulsen, Erik M., 1962
Ostracoda-Myodocopa. Part 1.
Cypridiniformes-Cypridinidae.
Dana-Report No. 57:414 pp.

Vargula bullae n.sp.

Poulsen, Erik M., 1962
Ostracoda-Myodocopa. Part 1.
Cypridiniformes-Cypridinidae.
Dana-Report No. 57:414 pp.

Vargula harveyi n.sp.

Kornicker, Louis S., and Charles E. King, 1965.
A new species of luminescent Ostracoda from Jamaica, West Indies.
Micropaleontology, 11(1):105-110.

Vargula hilgendorfi

Poulsen, Erik M., 1962
Ostracoda-Myodocopa. Part 1.
Cypridiniformes-Cypridinidae.
Dana-Report No. 57:414 pp.

Vargula plicata n.sp.

Poulsen, Erik M., 1962
Ostracoda-Myodocopa. Part 1.
Cypridiniformes-Cypridinidae.
Dana-Report No. 57:414 pp.

Vargula puppis n.sp.

Poulsen, Erik M., 1962
Ostracoda-Myodocopa. Part 1.
Cypridiniformes-Cypridinidae.
Dana-Report No. 57:414 pp.

Vargula spinosa n.sp.

Poulsen, Erik M., 1962
Ostracoda-Myodocopa. Part 1.
Cypridiniformes-Cypridinidae.
Dana-Report No. 57:414 pp.

Vargula spinulosa n. sp.

Poulsen, Erik M., 1962
Ostracoda-Myodocopa. Part 1.
Cypridiniformes-Cypridinidae.
Dana-Report No. 57:414 pp.

Vargula tubulata n.sp.

Poulsen, Erik M., 1962
Ostracoda-Myodocopa. Part 1.
Cypridiniformes-Cypridinidae.
Dana-Report No. 57:414 pp.

Xenocythere cuneiformis

Elofson, O. 1941
Zur Kenntnis der marinen Ostracodes Schwedens mit besonderer Berücksichtigung des Skageraks. Zool. Bidrag. Uppsala 19: 215-534; 42 maps, 52 figs.

Xestoleberis spp.

Ascoli, P., 1965.
Crociers Telassografia Afriatica 1955. VI. Ricerche ecologiche sugli ostracodi contenuti in 16 carote prelevate sul fondo del Mare Adriatico.
Arch. Oceanogr. e Limnol., 14(1):69-138.

Xestoleberis sp.

King, Charles E. and Louis S. Kornicker, 1970.
Ostracoda in Texas bays and lagoons: an ecological study.
Smithson. Contrib. Zool., 24: 92 pp.

Xestoleberis sp.

Van den Bold, W.A. 1966
Ostracoda from Colon Harbour, Panama.
Carib. J. Sci., 6(1/2):43-64.

Xestoleberis aurantia

Benson, Richard H., 1959.
Arthropoda 1. Ecology of recent ostracodes of the Todos Santos Bay region, Baja California, Mexico. Univ. Kansas, Paleontol. Contr., :1-80, 11 pls., 20 figs.

Xestoliberis aurantia
Elofson, O. 1941
Zur Kenntnis der marinen Ostracodes Schwedens mit besonderer Berücksichtigung des Skageraks. Zool. Bidrag. Uppsala 19: 215-534; 42 maps, 52 figs.

Xestoleberis aurantia
Hartmann, G., 1953.
Ostracodes des étangs méditerranéens. Vie et Milieu, Bull. Lab. Arago, 4(4):707-712, 1 textfig.

Xestoleberis aurantia
Schodduyn, M., 1926
Observations faites dans la baie d'Ambleteuse (Pas de Calais). Bull. Inst. Ocean., Monaco, No. 482: 64 pp.

Xestoleberis banda n.sp.
Benson, Richard H., 1959.
Arthropoda 1. Ecology of recent ostracodes of the Todos Santos Bay region, Baja California, Mexico. Univ. Kansas, Paleontol. Contr., :1-80, 11 pls., 20 figs.

Xestoleberis briggsi n.sp.
McKenzie, K. G. 1967.
Recent Ostracoda from Port Phillip Bay, Victoria.
Proc. R. Soc. Victoria n.s. 80(1): 61-106.

Xestoleberis capillata n.sp
Hartmann, Gerd, 1964.
Zur Kenntnis der Ostracoden des Roten Meeres. Wissenschaftliche Ergebnisse einer Forschungsreise von A. Remane und E. Schulz nach dem Roten Meer. E. 11.
Kieler Meeresf., 20:35-127.

Xestoleberis chipanae nsp.
Hartmann-Schröder, Gesa, und Gerd Hartmann, 1965.
Zur Kenntnis des Sublitorals der chilenischen Küste unter besonderer Berücksichtigung der Polychaeten und Ostracoden (mit Bermerkungen über den Einfluss sauerstoffarmer Strömungen auf die Besiedlung von marinen Sedimenten). Mitt. Hamburg. Zool. Mus. u Inst., Ergärzungsband zu 62:384 pp.

Xestoleberis decipiens
Makkaveeva, E.B., 1961.
[The biocoenotic history of small worms, Crustacea and marine ostracods]
Trudy Sevastopol Biol. Sta., (14):147-162.

Xestoliberis depressa
Elofson, O. 1941
Zur Kenntnis der marinen Ostracodes Schwedens mit besonderer Berücksichtigung des Skageraks. Zool. Bidrag. Uppsala 19: 215-534; 42 maps, 52 figs.

Xestoliberis depressa
Lucas, V.Z., 1931.
Some Ostracoda of Vancouver island region. Contr. Canadian Biol. Fish., n.s., 6(17):397-416.

Xestoleberis depressa
Stephensen, K., 1938.
Marine Ostracoda and Cladocera. Zool. Iceland 3(32):19 pp., 1 textfig.

Xestoleberis dispar
Juday, C., 1907
XIV. Ostracoda of the San Diego region. II. Littoral forms. Contributions from the Marine Biological Association of San Diego. Univ. Calif. Publ. Zool. 3(9):135-156, Pls. 18-20.

Xestoliberis dispar
Lucas, V.Z., 1931.
Some Ostracoda of the Vancouver island region. Contr. Canadian Biol. Fish., n.s., 6(17):397-416.

Xestoleberis dispar
Skogsberg, T., 1950.
Two new species of marine ostracods from California. Proc. Calif. Acad. Sci. 26(4):483-505, figs.

Xestoleberis eulitoralis n.sp.
Hartmann, Gerd, 1959.
Zur Kenntnis der lotischen Lebensbereiche der Pazifischen Küste von El Salvador unter besonderer Berücksichtigung seiner Ostracodenfauna.
Kieler Meeresf., 15(2):187-241.

Xestoleberis flavescens
Skogsberg, T., 1950.
Two new species of marine ostracod from California. Proc. Calif. Acad. Sci. 26(4):483-505, figs.

Xestoleberis ghardaqae n.sp
Hartmann, Gerd, 1964.
Zur Kenntnis der Ostracoden des Roten Meeres. Wissenschaftliche Ergebnisse einer Forschungsreise von A. Remane und E. Schulz nach dem Roten Meer. E. 11.
Kieler Meeresf., 20:35-127.

Xestoleberis hopkinsi
Skogsberg, T., 1950.
Two new species of marine ostracods from California. Proc. Calif. Acad. Sci. 26(4):483-505, figs.

Xestoleberis hopkinsi
Swain, Frederick M., 1967.
Ostracoda from the Gulf of California.
Mem. Geol. Soc. Am., 101: 139. pp.

Xestoleberis multiporosa n.sp.
Hartmann, Gerd, 1964.
Zur Kenntnis der Ostracoden des Roten Meeres. Wissenschaftliche Ergebnisse einer Forschungsreise von A. Remane und E. Schulz nach dem Roten Meer. E. 11.
Kieler Meeresf., 20:35-127.

Xestoleberis cf. nana
Swain, Frederick M., 1967.
Ostracoda from the Gulf of California.
Mem. Geol. Soc. Am., 101: 139 pp.

Xestoleberis parahowei n.sp.
Swain, Frederick M., 1967.
Ostracoda from the Gulf of California.
Mem. Geol. Soc. Am., 101: 139 pp.

Xestoleberis punctata n.sp.
Tressler, W.L., 1949
Marine ostracoda from Tortugas, Florida. J. Wash. Acad. Sci. 39(10):334-343, 25 text figs.

Xestoliberis pusilla n.s]
Elofson, O. 1941
Zur Kenntnis der marinen Ostracodes Schwedens mit besonderer Berücksichtigung des Skageraks. Zool. Bidrag. Uppsala 19: 215-534; 42 maps, 52 figs.

Xestoleberis reniformis n.sp.
Brady, G.S., 1907.
Crustacea. V. Ostracoda. Nat. Antarctic Exped., 1901-1904, 3:9 pp., 3 pls.

Xestoleberis rhomboidea n.s]
Hartmann, Gerd, 1964.
Zur Kenntnis der Ostracoden des Roten Meeres. Wissenschaftliche Ergebnisse einer Forschungsreise von A. Remane und E. Schulz nach dem Roten Meer. E. 11.
Kieler Meeresf., 20:35-127.

Xestoleberis rigusa
Neale, J. W., 1967.
An ostracod fauna from Halley Bay, Coats Land, British Antarctic Territory.
Brit. Antarct. Surv. Sci. Rep. 58: 50 pp.

Xestoleberis rotunda n.sp.
Hartmann, Gerd, 1964.
Zur Kenntnis der Ostracoden des Roten Meeres. Wissenschaftliche Ergebnisse einer Forschungsreise von A. Remane und E. Schulz nach dem Roten Meer. E. 11.
Kieler Meeresf., 20:35-127.

Xestoleberis rubrimaris n.sp.
Hartmann, Gerd, 1964.
Zur Kenntnis der Ostracoden des Roten Meeres. Wissenschaftliche Ergebnisse einer Forschungsreise von A. Remane und E. Schulz nach dem Roten Meer. E. 11.
Kieler Meeresf., 20:35-127.

Xestoleberis setigera
Neale, J. W., 1967.
An ostracod fauna from Halley Bay, Coats Land, British Antarctic Territory.
Brit. Antarct. Surv. Sci. Rep. 58: 50 pp.

Xestoleberis simplex n.sp.
Hartmann, Gerd, 1964.
Zur Kenntnis der Ostracoden des Roten Meeres. Wissenschaftliche Ergebnisse einer Forschungsreise von A. Remane und E. Schulz nach dem Roten Meer. E. 11.
Kieler Meeresf., 20:35-127.

Xestoleberis tigrina
McKenzie, K. G. 1967.
Recent Ostracoda from Port Phillip Bay, Victoria.
Proc. R. Soc. Victoria n.s. 80(1): 61-106.

Xestoleberis transversalis
Skogsberg, T., 1950.
Two new species of marine ostracods from California. Proc. Calif. Acad. Sci. 26(4):483-505, figs.

Xiphichilus gracilis
Neale, J. W., 1967.
An ostracod fauna from Halley Bay, Coats Land, British Antarctic Territory.
Brit. Antarct. Surv. Sci. Rep. 58: 50 pp.

Xiphichilus tenuissimoides n.sp.
Swain, Frederick M., 1967.
Ostracoda from the Gulf of California.
Mem. Geol. Soc. Am., 101: 139 pp.

Zybythocaris helicina
Maddocks, Rosalie F. 1973.
Bythocypris promoza n.sp. and males of Zybythocaris helicina and Bairdoppilata hirsuta (Ostracoda, Podocopida)
Crustaceana 24(1): 33-42.

priapulida
*Kirsteuer, E. and J. van der Land, 1970.
Some notes on Tubiluchus corallicola (Priapulida) from Barbados, West Indies. Marine Biol., 7(3): 230-238.

priapulids
Land, J. van der 1972.
Priapulus from the deepsea (Vermes, Priapulida).
Zool. Mededel. Leiden, 47: 358-368.

Oceanographic Index: Marine Organisms Cumulation, 1946-1973

priapulids
Murina, V.V., 1971.
On the occurrence of deep-sea sipunculids and priapulids in the Kurile-Kamchatka Trench. (In Russian; English abstract). Trudy Inst. Okeanol. P.P. Shirshov 92: 41-45.

priapulids
Murina, V.V., 1964
On the problem of the bipolar distribution of Priapulidea. (In Russian). Okeanologiia. Akad. Nauk. SSSR, 4(5):873-875.

priapulids
Salvini-Plawen, L.v., 1973
Ein Priapulide mit Kleptocniden aus den Adriatischen Meer. Mar. Biol. 20(2): 165-169.

priapulids
van der Land J. 1970
Systematics, zoogeography, and ecology of the Priapulida.
Zool. Verhandel. Leiden 112:1-118.

priapulids, ecol.
Hammond, R.A., 1970.
The burrowing of Priapulus caudatus.
J. Zool., Lond, 162(4):469-480.

pycnogonids
Barnard, K.H., 1955.
Additions to the fauna - list of South African Crustacea and Pycnogonida. Ann. S. African Mus., 13(1):1-107.

Pycnogonida
Barnard, K. H., 1955.
Additions to the fauna-list of South African Crustacea and Pycnogonida. Ann. S. African Mus., 43(1): 1-107.

pycnogonids
Barnard, K.H., 1946.
Diagnoses of new species and a new genus of Pycnogonida in the South African Museum. Ann. Mag. Nat. Hist., ser. 11, 13:60-63.

pycnogonides
Bourdillon, A., 1955.
Les Pycnogonides de la croisière 1951 du "President Theodore Tissier". Rev. Trav. Inst. Pêches Marit., 19(4):581-609.

pycnogonides.
Bourdillon, A., 1954.
Les pycnogonides de Marseille et ses environs. Recueil Trav. Stat. Mar. Endoume, 12:145-158.

Pycnogonida
Calman, W. T., 1938
Pycnogonida. John Murray Exped. 1933-34, Sci. Repts. 5(6):147-166, 10 text figs.

Pycnogonida
Calman, W.T., 1927.
Zoological results of the Cambridge Expedition to the Suez Canal. XXVIII. Report on the Pycnogonida Trans. Zool. Soc., London, 22:403-410, textfigs. 102-104.

pycnogonids
Clark, W.C., 1958.
Some Pycnogonida from Cook Strait, New Zealand, with descriptions of two new species. Victoria Univ., Wellington, Zool. Publ., 23:7 pp.

pycnogonids
de Haro, Andrés, 1967.
Les picnogónidos y su posición sistemática dentro de los Artrópodos. Bol. R. Soc. esp. Hist. nat. (Biol.), 65(3/4): 367-375

pycnogonids
de Haro, Andrés 1967.
Relaciones entre picnogónidos e hidroideos en el medio posidonícola. Bol. R. Soc. esp. Hist. nat (Biol) 65(3/4): 301-303.

pycnogonides
Fage, L., 1956.
Les pycnogonides (excl. le genre Nymphon). Galathea Rept., 2:167-183.

pycnogonids
Fage, L., 1952.
Pycnogonides de la Terre Adélie. Echantillons rapportés par le Docteur Sapin-Jaloustre, Medecin-Biologiste de la Première Expédition en Terre Adélie (1949-1951). (Expeditions Polaires Françaises, Missions Paul-Emile Victor). Bull. Mus. Nat. Hist. Nat., 2nd ser., 24:263-273, 2 textfigs.

pycnogonids
Fage, L., 1952.
II. Pycnogonides. Missions du Bâtiment Polaire "Commandant Charcot", récoltes faites en Terre Adélie (1950) par Paul Tchernia. Bull. Mus. Nat. Hist. Nat., 2 ser., 24:180-186, 2 textfigs.

Pycnogonids
Fage, Louis (posthumous) et Jan H. Stock, 1966.
Pycnogonides. Campagne de la Calypso aux Iles du Cap Vert (1959). Ann. Inst. Oceanogr., Monaco, n.s. 44: 315-328.

Pycnogonids
Fry, W.G., 1962.
Feeding preferences and the ecological role of Antarctic Pycnogonida. (Abstract for Symposium Antarctic Biol., Paris, 2-8 Sept., 1962). Polar Record, 11(72):334.
Also: SCAR Bull. #12.

pycnogonids
Fry, William G., and Joel W. Hedgpeth 1969.
The fauna of the Ross Sea. 7. Pycnogonida I. Colossendeidae, Pycnogonidae, Endeidae, Ammotheidae. Bull. N.Z.D.S.I.R. 198 (Mem. N.Z. Oceanogr. Inst. 49). 139 pp.

pycnogonids
Giltray, L., 1942.
New records of Pycnogonida from the Canadian Atlantic coast. J. Fish. Res. Bd., Canada, 5(5): 459-460.

pycnogonids
Gordon, I., 1932.
Pycnogonida. Discovery Repts., 6:1-138.

pycnogonids
Haswell, W.A., 1885.
On the Pycnogonida of the Australian coast. Proc. Linn. Soc., N.S. Wales, 9:1021-1034, Pls. 54-57.

Pycnogonida
Hedgpeth, J.W., 1954.
On the phylogeny of the Pycnogonida. Acta Zool., Int. Tidskrift för Zool., 35(3): 193-213, 9 textfigs.

Pycnogonida
Hedgpeth, J. W., 1950
Pycnogonida of the United States Navy Antarctic Expedition, 1947-48. Proc. U. S. Nat. Mus. 100 (3260):147-160, 19 text figs.

Pycnogonida
Hedgpeth, J.W., 1948
The Pycnogonida of the western North Atlantic and the Caribbean. Proc. U.S. Nat. Mus. 97:157-342, figs. 4-53, charts 1-3.

pycnogonids
Hodgson, T.V., 1908.
Pycnogonida. Nat. Antarctic Exped., 1901-1904, 3:72 pp., 10 pls.

pycnogonida
Just Jean 1972.
Pycnogonida from Jørgen Brønlund Fjord North Greenland. Meddr Grønland, 184 (7): 23-27

pycnogonids
King P.E. 1973.
Pycnogonids Hutchinson, London (£3.50), 144 pp (not seen)

pycnogonids
Lebour, M.V., 1949.
Two new Pycnogonids from Bermuda. Proc. Zool. Soc., London, 118(4):929-932, 3 textfigs.

Pycnogonida
Lebour, M.V., 1947
Notes on the Pycnogonida of Plymouth. JMBA (2):139-165, 7 text figs.

Pycnogonida
Meinert, Fr., 1899
Pycnogonida. Danish Ingolf Exped. 3(1): 71 pp., 5 pls., 2 text figs., 1 charts, list of stations.

pycnogonids
Nesis, K.N., 1960.
Littoral Pantopoda of the eastern Murmansk coast. Trudy Murmansk. Morsk. Biol. Inst., 2(6):137-161

pycnogonids
Soot-Ryen, T., 1927.
The Folden Fiord. Crustacea III. Isopoda, Cumacea, Ostracoda and Pycnogonida. Tromsø Mus. Skr. 1(5):15-20.

Pycnogonida
Stephensen, K., 1933
Crustacea and Pycnogonida. 2nd East Greenland Expedition. Medd. om Grønland 104 (15):12 pp.

Pycnogonida
Stephensen, K., 1933
The Godthaab expedition 1928. Pycnogonida. Medd. om Grønland 79(6):46 pp.

Pycnogonida
Stephensen, K., 1917
Zoogeographical investigations of certain fjords in southern Greenland with special reference to Crustacea, Pycnogonida and Echinodermata including a list of Alcyonaria and Pisces. Medd. om Grønland, 53(3):229-378.

Pycnogonida
Stephensen, K., 1912
Report on the Malacostraca, Pycnogonida and some Entomostraca collected by the Danmark-Expedition to North East Greenland. Medd. om Grønland, 45(11):501-630, Pls. 39-43.

pycnogonids
Stock, Jan H., 1968.
Pycnogonida collected by the Galathea and Anton Bruun in the Indian and Pacific oceans. Vidensk. Meddr. dansk naturh. Foren., 131: 7-65.

pycnogonids
Stock, Jan H., 1966.
Pycnogonida from West Africa. Atlantide Rep., 9:45-57.

Pycnogonida

Stock, J.H., 1958.
Pycnogonida from the Mediterranean coast of Israel.
Bull. Res. Counc., Israel, (B) 7(3/4):137-142.

Reprinted in:
Bull. Sea Fish. Res. Sta., Haifa, No. 19:

pycnogonids

Stock, J.H., 1957.
Pantopoden aus dem Zoologischen Museum Hamburg.
Mitt. Hamburgischen Zool. Mus. u. Inst., 55:81-106.

pycnogonids

Stock, J.H., 1957.
The pycnogonid family Austrodecidae. Beaufortia 6(68):1-81.

pycnogonids

Turpaeva, E.P., 1971.
The deep-water Pantopoda collected in the Kurile-Kamchatka Trench. (In Russian; English abstract). Trudy Inst. Okeanol. P.P. Shirshov 92: 274-291.

pycnogonids

Turpajeva, E.P., 1955.
New Pantopoda sp of the Kurile-Kamchatka Trench.
Trudy Inst. Oceanol., 12:322-327.

Pycnogonids

Utinomi, Huzio, 1959
Pycnogonida of the Japanese Antarctic Research Expeditions 1956-1958. Biol. Res. Japan Antarctic Res. Exped., 8(Spec. Publ. Seto Mar. Biol. Lab.): 12 pp.

Pycnogonida

Utinomi, H., 1955.
Report on the Pycnogonida collected by the Soya-maru expeditions made on the continental shelf bordering Japan during 1926-1930.
Publ. Seto Mar. Biol. Lab., 5(1):1-42.

pycnogonids

Utinomi, H., 1954
The fauna of Akkeshi Bay. XIX. Littoral Pycnogonida. Publ. Akkeshi Mar. Biol. Sta. No. 3:1-28, 1 pl., 11 textfigs.

pycnogonids

Utinomi, H., 1951.
On some pycnogonids from the sea around the Kii Peninsula.
Publ. Seto Mar. Biol. Sta., Kyoto, Univ., 1(4):159-168.

pycnogonids, anat.-phys.

Fry, William G., 1965.
The feeding mechanism and preferred foods of three species of Pycnogonida.
Bull. Brit. Mus.(N.H.), Zool., 12(6):197-223.

pycnogonids, anat.-physiol.

Hanström, Bertil, 1965.
Indication of neurosecretion and the structure of the Sokolow's organ in pycnogonids.
Sarsia, 18:25-36.

pycnogonids, anat.-physiol.

Henry, L.M., 1953.
The nervous system of the Pycnogonida.
Microentomology 18(1):16-36, figs.

Pycnogonids, anat.-phys.

Krishnan, G., 1955.
Nature of the cuticle of Pycnogonida. Nature 175 (4464):904.

pycnogonids, anat.physiol.

Morgan, Elfed, A. Nelson-Smith and E.W. Knight-Jones, 1964.
Responses of Nymphon gracile (Pycnogonida) to pressure cycles of tidal frequency.
J. Exp. Biol., 41(4):825-836.

pycnogonids, anat. physiol.

Redmond, James R., and Charles D. Swanson 1968.
Preliminary studies of the physiology of the Pycnogonida.
Antarct. Jl USA. 3(4):130-131.

pycnogonids, ecology of

de Haro, Andrés, 1966.
Distribución ecológica de los picnogónidos entre algas y posidonias mediterráneas.
Inv. Pesq., Barcelona, 30:661-667.

pycnogonides, lists of spp.

Arnaud, Françoise, 1972.
Invertébrés marins des XIIème et XVème expéditions antarctiques françaises en Terre Adélie. 9. Pycnogonides.
Téthys, Suppl. 4:135-156

pycnogonids, lists of spp

de Haro, Andrés, 1966.
Distribución ecológica de los picnogónidos entre algas y posidonias mediterráneas.
Inv. Pesq., Barcelona, 30:661-667.

pycnogonids, lists of spp.

de Haro, Andres, 1966.
Picnogónidos del alga parda Halopteris Scoperia (L.) de las islas Medas (Gerona).
Bol. R. Soc. Española Hist. Nat. (Biol).64:5-14.

Pycnogonida, lists of spp.

de Haro, Andrés, 1965.
Picnogónidos de la fauna espanola. Picnogónidos posidonícolas de la isles Medas (Gerona).
Publ. Inst. Biol. Aplic., 39:137-145.

pycnogonids, lists of spp

Hedgpeth, Joel W. 1968.
Pycnogonid studies.
Antarct. Jl USA. 3(4):129-130.

pycnogonids, lists of spp.

King, P.E. 1972.
The marine flora and fauna of the Isles of Scilly: Pycnogonida.
J. nat. Hist. 6(6):621-624.

pycnogonids, lists of spp.

Nesis, K.N. 1970
Feeding habits, water masses and depth as factors in pantopod distribution in the regions off Labrador and Newfoundland. (In Russian; English abstract)
Trudy Inst. Okeanol. Akad. Nauk SSSR 88: 150-173.

pycnogonids, lists of spp.

Stock, Jan H., 1970.
A new species of Endeis and other pycnogonid records from the Gulf of Aqaba.
Bull. Zool. Mus. Amsterdam, 2(1):1-4

pycnogonids, lists of spp.

Stock, Jan H. 1968.
Pycnogonides. Faune marine des Pyrénées-orientales.
Vie Milieu 19 (Suppl. 1A):5-38.

pycnogonids, lists of spp.

Stock, J.H., 1962.
Second list of Pycnogonida of the University of Cape Town ecological survey.
Trans. R. Soc., S. African, 36(4):273-286.

Pycnogonida, lists of spp

Stock, J.H., 1958.
The Pycnogonida of the Erythraean and of the Mediterranean coasts of Israel.
Sea Fish. Res. Sta., Haifa, Bull., No. 16:3-5.

pycnogonids, lists of spp.

Stock, J.H. 1954
Four new Tanystylum species, and other Pycnogonida from the West Indies. Studies Fauna Curaçao and other Carib. Is. 5(24):115-129, 7 pls.

pycnogonids, lists of spp.

Zavodnik, Dušan 1968.
Beitrag zur Kenntnis der Asselspinnen (Pantopoda) der Umgebung von Rovinj (Nördl. Adria).
Thalassia Jugoslavica 4:45-53.

pycnogonid synonymies

Stock, Jan H. 1968.
Pycnogonides. Faune marine des Pyrénées-orientales.
Vie Milieu 19 (Suppl. 1A):5-38.

pycnogonids

Turpaeva, E.P., 1971.
An addition to the pantopod fauna of deep-sea trenches in the northwestern part of the Pacific Ocean. (In Russian; English abstract). Trudy Inst. Okeanol. P.P. Shirshov 92: 292-297.

pycnogonids, lists of spp.

Utinomi, Huzio, 1952
Pycnogonida of Sagami Bay.
Publ. Seto Mar. Biol. Lab., 7(2):197-222.

Achelia sp.

Stock, Jan H. 1973.
Pycnogonida from south-eastern Australia.
Beaufortia 20 (266):99-127

Achelia sp.

Stock, Jan H. 1972.
The Pycnogonida collected off northwestern Africa during the cruise of the Meteor.
Meteor-Forsch. Ergebn. (D)5:6-10.

Achelia adelpha, n.sp.

Child, C. Allan, 1970.
Pycnogonida of the Smithsonian-Bredin Pacific Expedition, 1957.
Proc. Biol. Soc. Wash. 83(27): 287-308.

Achelia alaskensis

Utinomi, H., 1954.
The fauna of Akkeshi Bay. XIX. Littoral Pycnogonida. Publ. Akkeshi Mar. Biol. Sta. No. 3:1-28, 1 pl., 11 textfigs.

Achelia assimilis

Stock, Jan H. 1973.
Pycnogonida from south-eastern Australia.
Beaufortia 20 (266):99-127.

Achelia assimilis

Stock, Jan H., 1968.
Pycnogonida collected by the Galathea and Anton Bruun in the Indian and Pacific oceans.
Vidensk. Meddr dansk naturh. Foren. 131: 7-65.

Achelia assimilis
Stock, J. H. 1954.
Pycnogonida from Indo-West Pacific, Australian and New Zealand waters. Papers from Dr. Th. Mortensen's Pacific Expedition, 1914-1916. LXXVII.
Vidensk. Medd. Dansk Naturhist. Foren., København, 116: 1-168.

Achelia (Ignavogriphus) australiensis [a]
Fry, William G. and Joel W. Hedgpeth, 1969.
The fauna of the Ross Sea. 7. Pychogonida, I. Colossendeidae, Pycnogonidae, Endeidae, Ammotheidae. Bull. N.Z.D.S.I.R. 198 (Mem. N.Z. Oceanogr. Inst. 49): 139 pp.

Achelia australiensis nov. comb.
Stock, J. H. 1954.
Pycnogonida from Indo-West Pacific, Australian and New Zealand waters. Papers from Dr. Th. Mortensen's Pacific Expedition, 1914-1916. LXXVII.
Vidensk. Medd. Dansk Naturhist. Foren., København, 116: 1-168.

Achelia (Pigrolavatus) besnardi [a]
Fry, William G. and Joel W. Hedgpeth, 1969.
The fauna of the Ross Sea. 7. Pychogonida, I. Colossendeidae, Pycnogonidae, Endeidae, Ammotheidae. Bull. N.Z.D.S.I.R. 198 (Mem. N.Z. Oceanogr. Inst. 49): 139 pp.

Achelia bituberculata
Stock, J. H. 1954.
Pycnogonida from Indo-West Pacific, Australian and New Zealand waters. Papers from Dr. Th. Mortensen's Pacific Expedition, 1914-1916. LXXVII.
Vidensk. Medd. Dansk Naturhist. Foren., København, 116: 1-168.

Achelia bituberculata
Utinomi, Huzio, 1962
Pycnogonida of Sagami Bay - Supplement.
Publ. Seto Mar. Biol. Lab., 10(1) (Article 5) 91-104.

Achelia brevichelifera n.sp.
Hedgpeth, J.W., 1948
The Pycnogonida of the western North Atlantic and the Caribbean. Proc. U.S. Nat. Mus. 97:157-342, figs.4-53, charts 1-3.

Achelia brucei
Fage, L., 1952.
II. Pycnogonides. Missions du Bâtiment Polaire "Commandant Charcot", récoltés faîtes en Terre Adélie (1950) par Paul Tchernia. Bull. Mus. Nat. Hist. Nat., 2 ser., 24:180-186, 2 textfigs.

Achelia brucei
Fage, L., 1952.
Pycnogonides de la Terre Adélie. Échantillons rapportés par le Docteur Sapin-Jaloustre, Medecin-Biologiste de la Première Expédition en Terre Adélie (1949-1951). (Expéditions Polaires Françaises, Mission Paul-Émile Victor).
Bull. Mus. Nat. Hist. Nat., 2 ser., 24:263-273, 2 textfigs.

Achelia brucei
Gordon, I., 1932.
Pycnogonida. Discovery Repts., 6:1-138.

Achelia (Ignavogriphus) communis [a]
Fry, William G. and Joel W. Hedgpeth, 1969.
The fauna of the Ross Sea. 7. Pychogonida, I. Colossendeidae, Pycnogonidae, Endeidae, Ammotheidae. Bull. N.Z.D.S.I.R. 198 (Mem. N.Z. Oceanogr. Inst. 49): 139 pp.

Achelia communis
Hedgpeth, J. W., 1950
Pycnogonida of the United States Navy Antarctic Expedition, 1947-48. Proc. U.S. Nat. Mus. 100 (3260):147-160, 19 text figs.

Achelia communis
Stephensen, K., 1947
Tanaidacea, Isopoda, Amphipoda, and Pycnogonida. Sci. Res. Norweg. Antarctic Exp. 1927-1928 et seq. Date Norske Videnskaps-Akademi i Oslo, II, No.27:1-90, 28 text figs.

Achelia dohrni
Clark, W.C. 1971.
Pycnogonida of the Antipodes Islands.
N.Z. Jl mar. Freshwat. Res. 5(3/4):427-452.

Achelia dohrni [a]
Clark, W.C. 1971.
Pycnogonida of the Snares Islands.
N.Z. Jl mar. Freshwat. Res. 5(2): 329-341.

Achelia (Ingnavogriphus) dohrni [a]
Fry, William G. and Joel W. Hedgpeth, 1969.
The fauna of the Ross Sea. 7. Pycnogonida, I. Colossendeidae, Pycnogonidae, Endeidae, Ammotheidae. Bull. N.Z.D.S.I.R. 198 (Mem. N.Z. Oceanogr. Inst. 49): 139 pp.

Achelia dohrni
Stock, Jan H., 1968.
Pycnogonida collected by the Galathea and Anton Bruun in the Indian and Pacific oceans.
Vidensk. Meddr. dansk naturh.Foren., 131: 7-65.

Achelia dohrni
Stock, J. H. 1954.
Pycnogonida from Indo-West Pacific, Australian and New Zealand waters. Papers from Dr. Th. Mortensen's Pacific Expedition, 1914-1916. LXXVII.
Vidensk. Medd. Dansk Naturhist. Foren., København, 116: 1-168.

Achelia echinata
Bourdillon, A., 1954.
Contribution à l'étude des Pycnogonides de Tunisie. Sta. Oceanogr., Salammbo, Notes, 35: 8 pp., 1 pl.

Achelia echinata
Bourdillon, A., 1954.
Les pycnogonides de Marseille et ses environs.
Recueil Trav. Sta. Mar., Endoume, 12:145-158.

Achelia echinata
Calman, W. T., 1938
Pycnogonida. John Murray Exped. 1933-34.
Sci. Repts. 5(6):147-166, 10 text figs.

Achelia echinata
de Haro, Andrés, 1965.
Picnogónidos de la fauna español. Comunidad de picnogonidos en el alga parda Halopteris Scoperia (L.).
Bol. R. Soc. Española Hist. Nat. (Biol.), 63: 213-218.

Achelia echinata
Fage, L., 1952.
Sur quelques Pycnogonides de Dakar. Bull. Mus. Nat. Hist. Nat., 2 ser., 24(6):530-533, 1 textfig.

Achelia echinata [a]
Fage, Louis (posthumous) et Jan H. Stock, 1966.
Pycnogonides. Campagne de la Calypso aux îles du Cap Vert (1959). Ann. Inst. Océanogr., Monaco, n.s., 44: 315-328.

Achelia echinata [a]
Knapp, Franz 1973.
Pycnogonida from Pantelleria and Catánia, Sicily.
Beaufortia 21 (277): 55-74.

Achelia echinata
Stock, Jan H., 1966.
Pycnogonida from West Africa.
Atlantide Rep., 9:45-57.

Achelia echinata
Stock, J.H., 1958
Pycnogonida from the Mediterranean coast of Israel.
Bull. Res. Counc., Israel, (B):7(3/4):137-142.
Reprinted in:
Bull. Sea Fish. Res. Sta., Haifa, No. 19:

Achelia echinata
Utinomi, Huzio 1962
Pycnogonida of Sagami Bay - Supplement.
Publ. Seto Mar. Biol. Lab., 10(1) (Article 5) 91-104.

Achelia echinata
Utinomi, H., 1954.
The fauna of Akkeshi Bay. XIX. Littoral Pycnogonida. Publ. Akkeshi Mar. Biol. Sta. No. 3:1-28, 1 pl., 11 textfigs.
Contr. 221 Seto Mar. Biol. Lab.

Achelia echinata
Utinomi, Huzio, 1952
Pycnogonida of Sagami Bay.
Publ. Seto Mar. Biol. Lab., 7(2):197-222.

Achelia echinata tuberculata n.subsp. [a]
Lozina-Lozinsky, L.K. and L.M. Kopaneva 1973.
New Pantopoda of the Guinean Republic coast. (In Russian, English abstract).
Zool. Zh. 52(7): 1083-1084.

Achelia (Pigrolavatus) fernandeziana
Fry, William G. and Joel W. Hedgpeth, 1969.
The fauna of the Ross Sea. 7. Pycnogonida, I. Colossendeidae, Pycnogonidae, Endeidae, Ammotheidae. Bull. N.Z.D.S.I.R. 198 (Mem. N.Z. Oceanogr. Inst 49): 139 pp.

Achelia gracilis
Bourdillon, A., 1955.
Les Pycnogonides de la croisière 1951 du "President Theodore Tissier".
Rev. Trav. Inst. Pêches Marit., 19(4):581-609.

Achelia (Pigrolavatus) gracilis [a]
Fry, William G. and Joel W. Hedgpeth, 1969.
The fauna of the Ross Sea. 7. Pycnogonida, I. Colossendeidae, Pycnogonidae, Endeidae, Ammotheidae. Bull. N.Z.D.S.I.R. 198 (Mem. N.Z. Oceanogr. Inst. 49): 139 pp.

Achelia gracilis
Hedgpeth, J.W., 1948
The Pycnogonida of the western North Atlantic and the Caribbean. Proc. U.S. Nat. Mus. 97:157-342, figs.4-53, charts 1-3.

Achelia (Ignavogriphus) hoekii [a]
Fry, William G. and Joel W. Hedgpeth, 1969.
The fauna of the Ross Sea. 7. Pycnogonida, I. Colossendeidae, Pycnogonidae, Endeidae, Ammotheidae. Bull. N.Z.D.S.I.R. 198 (Mem. N.Z. Oceanogr. Inst. 49): 139 pp.

Achelia hoeki
Gordon, I., 1932.
Pycnogonida. Discovery Repts., 6:1-138.

Achelia intermedia
Fage, L., 1952.
II. Pycnogonides. Missions du Bâtiment Polaire "Commandant Charcot", récoltés faîtes en Terre Adélie (1950) par Paul Tchernia. Bull. Mus. Nat. Hist. Nat., 2 ser., 24:180-186, 2 textfigs.

Achelia intermedia
Fage, L., 1952.
Pycnogonides de la Terre Adélie. Échantillons rapportés par le Docteur Sapin-Jaloustre, Medecin-Biologiste de la Première Expédition en Terre Adélie (1949-1951). (Expéditions Polaires Françaises, Missions Paul-Émile Victor). Bull. Mus. Nat. Hist. Nat., 2 ser., 24:263-273, 2 textfigs.

Achelia intermedia
Gordon, I., 1932.
Pycnogonida. Discovery Repts., 6:1-138.

Achelia intermedia
Hedgpeth, J. W., 1950
Pycnogonida of the United States Navy Antarctic Expedition, 1947-48. Proc. U. S. Nat. Mus. 100 (3260):147-160, 19 text figs.

Achelia intermedia ?
Stephensen, K., 1947
Tanaidacea, Isopoda, Amphipoda, and Pycnogonida. Sci. Res. Norweg. Antarctic Exp. 1927-1928 et seq. Date Norske Videnskaps-Akademi i Oslo, II, No.27:1-90, 26 text figs.

Achelia kiiensis n.sp.
Utinomi, H., 1951.
On some pycnogonids from the sea around the Kii Peninsula. Publ. Seto Mar. Biol. Sta., Kyoto Univ., 1(4): 159-168.

Achelia latifrons
Stock, J. H. 1954.
Pycnogonida from Indo-West Pacific, Australian and New Zealand waters. Papers from Dr. Th. Mortensen's Pacific Expedition, 1914-1916. LXXVII.
Vidensk. Medd. Dansk Naturhist. Foren., København, 116: 1-168.

Achelia laevis australiensis
Haswell, W.A., 1885.
On the Pycnogonida of the Australian coast. Proc. Linn. Soc., N.S. Wales, 9:1021-1034, Pls. 54-57.

Achelia longipes
Bourdillon, A., 1954.
Les pycnogonides de Marseille et ses environs. Recueil Trav. Sta. Mar., Endoume, 12:145-158.

Achelia nana
Stock, Jan H., 1968.
Pycnogonida collected by the Galathea and Anton Bruun in the Indian and Pacific oceans. Vidensk. Meddr dansk naturh. Foren. 131: 7-65.

Achelia nana
Stock, J. H. 1954.
Pycnogonida from Indo-West Pacific, Australian and New Zealand waters. Papers from Dr. Th. Mortensen's Pacific Expedition, 1914-1916. LXXVII.
Vidensk. Medd. Dansk Naturhist. Foren., København, 116: 1-168.

Achelia nana
Stock, Jan H., 1965.
Pycnogonida from the southwest Indian Ocean. Beaufortia, 13(151):13-33.

Achelia spinosa
Stock, J. H. 1954.
Pycnogonida from Indo-West Pacific, Australian and New Zealand waters. Papers from Dr. Th. Mortensen's Pacific Expedition, 1914-1916. LXXVII.
Vidensk. Medd. Dansk Naturhist. Foren., København, 116: 1-168.

Achelia ohshimai
Utinomi, H., 1954.
The fauna of Akkeshi Bay. XIX. Littoral Pycnogonida. Publ. Akkeshi Mar. Biol. Sta. No. 3:1-28, 1 pl., 11 textfigs.

Contr. 221 Seto Mar. Biol. Lab.

Achelia ohshimai n.sp.
Utinomi, H., 1951.
On some pycnogonids from the sea around the Kii Peninsula. Publ. Seto Mar. Biol. Sta., Kyoto Univ., 1(4): 159-169.

Achelia (Pigrolavatus) parvula [a]
Fry, William G. and Joel W. Hedgpeth, 1969.
The fauna of the Ross Sea. 7. Pycnogonida, I. Colossendeidae, Pycnogonidae, Endeidae, Ammotheidae. Bull. N.Z.D.S.I.R. 198 (Mem. N.Z. Oceanogr. Inst. 49): 139 pp.

Achelia parvula
Gordon, I., 1932.
Pycnogonida. Discovery Repts., 6:1-138.

Achelia parvula
Hedgpeth, J. W., 1950
Pycnogonida of the United States Navy Antarctic Expedition, 1947-48. Proc. U. S. Nat. Mus. 100 (3260):147-160, 19 text figs.

Achelia quadridentata
Stock, J.H., 1962.
Second list of Pycnogonida of the University of Cape Town ecological survey. Trans. R. Soc., S. Africa, 36(4):273-286.

Achelia sawayai [a]
Fage, Louis (posthumous) et Jan H. Stock, 1966.
Pycnogonides. Campagne de la Calypso aux Îles du Cap Vert (1959). Ann. Inst. Océanogr. Monaco, n.s., 44: 315-328.

Achelia (Pigrolavatus) sawayai [a]
Fry, William G. and Joel W. Hedgpeth, 1969.
The fauna of the Ross Sea. 7. Pycnogonida, I. Colossendeidae, Pycnogonidae, Endeidae, Ammotheidae. Bull. N.Z.D.S.I.R. 198 (Mem. N.Z. Oceanogr. Inst. 49): 139 pp.

Achelia sawayai
Stock, Jan H., 1966.
Pycnogonide from West Africa. Atlantide Rep., 9:45-57.

Achelia scabra
Hedgpeth, J.W., 1948
The Pycnogonida of the western North Atlantic and the Caribbean. Proc. U.S. Nat. Mus. 97:157-342, figs.4-53, charts 1-3.

Achelia segmentata n.sp.
Utinomi, H., 1954.
The fauna of Akkeshi Bay. XIX. Littoral Pycnogonida. Publ. Akkeshi Mar. Biol. Sta. No. 3:1-28, 1 pl., 11 textfigs.

Contr. 221 Seto Mar. Biol. Lab.

Achelia (Ignavogriphus) serratipalpis [a]
Fry, William G. and Joel W. Hedgpeth, 1969.
The fauna of the Ross Sea. 7. Pycnogonida, I. Colossendeidae, Pycnogonidae, Endeidae, Ammotheidae. Bull. N.Z.D.S.I.R. 198 (Mem. N.Z. Oceanogr. Inst. 49): 139 pp.

Achelia serratipalpis
Gordon, I., 1932.
Pycnogonida. Discovery Repts., 6:1-138.

Achelia simplex
Wyer, D., P.E. King and J. Jarvis 1971
Achelia simplex (Giltay) a pycnogonid new to the Irish fauna. Ir. nat. J. 17(3): 92-95

Achelia spicata
Fage, L., 1952.
II. Pycnogonides. Missions du Bâtiment Polaire "Commandant Charcot", récoltes faites en Terre Adélie (1950) par Paul Tchernia. Bull. Mus. Nat. Hist. Nat., 2 ser., 24:180-186, 2 textfigs.

Achelia spicata
Fage, L., 1952.
Pycnogonides de la Terre Adélie. Échantillons rapportés par le Docteur Sapin-Jaloustre, Medecin-Biologiste de la Première Expédition en Terre Adélie (1949-1951). (Expéditions Polaires Françaises, Missions Paul-Émile Victor). Bull. Mus. Nat. Hist. Nat., 2 ser., 24:263-273, 2 textfigs.

Achelia (Ignavogriphus) spicata [a]
Fry, William G. and Joel W. Hedgpeth, 1969.
The fauna of the Ross Sea. 7. Pycnogonida, I. Colossendeidae, Pycnogonidae, Endeidae, Ammotheidae. Bull. N.Z.D.S.I.R. 198 (Mem. N.Z. Oceanogr. Inst. 49): 139 pp.

Achelia spicata
Hedgpeth, J. W., 1950
Pycnogonida of the United States Navy Antarctic Expedition, 1947-48. Proc. U. S. Nat. Mus. 100 (3260):147-160, 19 text figs.

Achelia spinosa
Hedgpeth, J.W., 1948
The Pycnogonida of the western North Atlantic and the Caribbean. Proc. U.S. Nat. Mus. 97:157-342, figs.4-53, charts 1-3.

Achelia (Pigrolavatus) sufflata [a]
Fry, William G. and Joel W. Hedgpeth, 1969.
The fauna of the Ross Sea. 7. Pycnogonida, I. Colossendeidae, Pycnogonidae, Endeidae, Ammotheidae. Bull. N.Z.D.S.I.R. 198 (Mem. N.Z. Oceanogr. Inst. 49): 139 pp.

Achelia superba
Stock, J. H. 1954.
Pycnogonida from Indo-West Pacific, Australian and New Zealand waters. Papers from Dr. Th. Mortensen's Pacific Expedition, 1914-1916. LXXVII.
Vidensk. Medd. Dansk Naturhist. Foren., København, 116: 1-168.

Achelia superba
Utinomi, Huzio, 1962
Pycnogonida of Sagami Bay - Supplement. Publ. Seto Mar. Biol. Lab., 10(1) (Article 5) 91-104.

Achelia superba
Utinomi, H., 1955.
Report on the Pycnogonida collected by the Soyo-maru expeditions made on the continental shelf bordering Japan during the years 1926-1930. Publ. Seto Mar. Biol. Lab., 5(1):1-42.

Achelia superba
Utinomi, Huzio, 1952
Pycnogonida of Sagami Bay. Publ. Seto Mar. Biol. Lab., 7(2):197-222.

Achelia superba
Utinomi, H., 1951.
On some pycnogonids from the sea around the Kii Peninsula. Publ. Seto Mar. Biol. Sta., Kyoto Univ., 1(4): 159-168.

Achelia transfuga n.sp.
Stock, J. H. 1954.
Pycnogonida from Indo-West Pacific, Australian and New Zealand waters. Papers from Dr. Th. Mortensen's Pacific Expedition, 1914-1916. LXXVII.
Vidensk. Medd. Dansk Naturhist. Foren., København, 116: 1-168.

Achelia transfugoides n.sp.

Stock, Jan H. 1973.
Pycnogonida from south-eastern Australia.
Beaufortia 20 (266): 99-127

Achelia uni-unguiculata
Bourdillon, A., 1954.
Les pycnogonides de Marseille et ses environs.
Recueil Trav. Sta. Mar. Endoume, 12:145-158.

Achelia uni-unguiculata
de Haro, Andrés, 1965.
Picnogónidos de la fauna español. Comunidad de picnogónidos en el alga parda Halopteris Scoparia (L.).
Bol. R. Soc. Española Hist. Nat. (Biol.), 63: 213-218.

Achelia variabilis
Stock, Jan H., 1968.
Pycnogonida collected by the Galathea and Anton Brunn in the Indian and Pacific oceans.
Vidensk. Meddr dansk naturh. Foren. 131: 7-65.

Achelia variabilis n.sp.
Stock, J. H. 1954.
Pycnogonida from Indo-West Pacific, Australian and New Zealand waters. Papers from Dr. Th. Mortensen's Pacific Expedition, 1914-1916. LXXVII.
Vidensk. Medd. Dansk Naturhist. Foren., København, 116: 1-168.

Achelia vulgaris
de Haro, Andrés, 1965.
Picnogónidos de la fauna español. Comunidad de picnogónidos en el alga parda Halopteris Scoparia (L.).
Bol. R. Soc. Española Hist. Nat. (Biol.), 63: 213-218.

Achelia vulgaris
Fage, Louis (posthumous) et Jan H. Stock, 1966.
Pycnogonides. Campagne de la Calypso aux Îles du Cap Vert (1959). Ann. Inst. Océanogr., Monaco, n.s., 44: 315-328.

Achelia vulgaris
Krapp, Franz 1973.
Pycnogonida from Pantelleria and Catania, Sicily.
Beaufortia 21 (277): 55-74.

Achelia vulgaris
Stock, Jan H., 1966.
Pycnogonida from West Africa.
Atlantide Rep. 9:45-57.

Achelia wilsoni
Stock, J.H., 1957.
Pantopoden aus dem Zoologischen Museum Hamburg.
Mitt. Hamburgischen Zool. Mus. u. Inst. 55:81-106.

Aduncorostris transfuga n.gen.
Fry, William G. and Joel W. Hedgpeth, 1969.
The fauna of the Ross Sea. 7. Pycnogonida, I. Colossendeidae, Pycnogonidae, Endeidae, Ammotheidae. Bull. N.Z.D.S.I.R. 198 (Mem. N.Z. Oceanogr. Inst. 49): 139 pp.

Ammothea sp.?
Gordon, I., 1932.
Pycnogonida. Discovery Repts., 6:1-138.

Ammothea antipodensis n.sp.
Clark, W.C. 1971.
Pycnogonida of the Antipodes Islands.
N.Z. Jl mar. Freshwat. Res. 5 (3/4): 427-452

Ammothea assimilis n.sp.
Haswell, W.A., 1885.
On the Pycnogonida of the Australian coast.
Proc. Linn. Soc., N.S. Wales, 9:1-21-1034, Pls. 54-57.

Ammothea australiensis
Stock, Jan H. 1973.
Pycnogonida from south-eastern Australia.
Beaufortia 20 (266): 99-127

Ammothea (Mathoma) calmani
Fry, William G. and Joel W. Hedgpeth, 1969.
The fauna of the Ross Sea. 7. Pycnogonida, I. Colossendeidae, Pycnogonidae, Endeidae, Ammotheidae. Bull. N.Z.D.S.I.R. 198 (Mem. N.Z. Oceanogr. Inst. 49): 139 pp.

Ammothea calmani n.sp.
Gordon, I., 1932.
Pycnogonida. Discovery Repts., 6: 1-138.

Ammothea (Ammothea) carolinensis
Fry, William G. and Joel W. Hedgpeth, 1969.
The fauna of the Ross Sea. 7. Pycnogonida, I. Colossendeidae, Pycnogonidae, Endeidae, Ammotheidae. Bull. N.Z.D.S.I.R. 198 (Mem. N.Z. Oceanogr. Inst. 49): 139 pp.

Ammothea carolinensis
Gordon, I., 1932.
Pycnogonida. Discovery Repts., 6:1-138.

Ammothea carolinensis
Hedgpeth, J. W., 1950
Pycnogonida of the United States Navy Antarctic Expedition, 1947-48. Proc. U. S. Nat. Mus. 100 (3260):147-160, 19 text figs.

Ammothea carolinensis
Stephensen, K., 1947
Tanaidacea, Isopoda, Amphipoda, and Pycnogonida. Sci. Res. Norweg. Antarctic Exp. 1927-1928 et seq. Date Norske Videnskaps-Akademi i Oslo, II, No.27:1-90, 26 text figs.

Ammothea carolinensis
Stock, J.H., 1957.
Pantopoden aus dem Zoologischen Museum Hamburg.
Mitt. Hamburgischen Zool. Mus. u. Inst., 55:81-106.

Ammothea (Theammoa) clausi
Fry, William G. and Joel W. Hedgpeth, 1969.
The fauna of the Ross Sea. 7. Pycnogonida, I. Colossendeidae, Pycnogonidae, Endeidae, Ammotheidae. Bull. N.Z.D.S.I.R. 198 (Mem. N.Z. Oceanogr. Inst. 49): 139 pp.

Ammothea clausi
Gordon, I., 1932.
Pycnogonida. Discovery Repts., 6:1-138.

Ammothea clausi
Hedgpeth, J. W., 1950
Pycnogonida of the United States Navy Antarctic Expedition, 1947-48. Proc. U. S. Nat. Mus. 100 (3260):147-160, 19 text figs.

Ammothea clausi
Stephensen, K., 1947
Tanaidacea, Isopoda, Amphipoda, and Pycnogonida. Sci. Res. Norweg. Antarctic Exp. 1927-1928 et seq. Date Norske Videnskaps-Akademi i Oslo, II, No.27:1-90, 26 text figs.

Ammothea depolaris n.sp.
Stock, J.H., 1966.
Pycnogonida. Campagne de la Calypso au large des côtes Atlantiques de l'Amerique du Sud (1961-1962). Ann. Inst. Océanogr., Monaco, n.s. 44: 385-406.

Ammothea echinata
Lebour, M.V., 1945.
Notes on the Pycnogonida of Plymouth. J.M.B.A. 26:139-165, 7 textfigs.

Ammothea gibbosa
Hedgpeth, J. W., 1950
Pycnogonida of the United States Navy Antarctic Expedition, 1947-48. Proc. U. S. Nat. Mus. 100 (3260):147-160, 19 text figs.

Ammothea gibbosa
Stephensen, K., 1947
Tanaidacea, Isopoda, Amphipoda, and Pycnogonida. Sci. Res. Norweg. Antarctic Exp. 1927-1928 et seq. Date Norske Videnskaps-Akademi i Oslo, II, No.27:1-90, 26 text figs.

Ammothea gigantea n.sp.
Gordon, I., 1932.
Pycnogonida. Discovery Repts., 6:1-138.

Ammothea glacialis
Fage, L., 1952.
Pycnogonides de la Terre Adélie. Echantillons rapportés par le Docteur Sapin-Jeloustre, Médecin-Biologiste de la Première Expédition en Terre Adélie (1949-1951). (Expédition Polaires Françaises, Missions Paul-Emile Victor).
Bull. Mus. Nat. Hist. Nat., 2 ser., 24:263-273, 2 textfigs.

Ammothea (Ammothea) glacialis
Fry, William G. and Joel W. Hedgpeth, 1969.
The fauna of the Ross Sea. 7. Pycnogonida, I. Colossendeidae, Pycnogonidae, Endeidae, Ammotheidae. Bull. N.Z.D.S.I.R. 198 (Mem. N.Z. Oceanogr. Inst. 49): 139 pp.

Ammothea glacialis
Hedgpeth, J. W., 1950
Pycnogonida of the United States Navy Antarctic Expedition, 1947-48. Proc. U. S. Nat. Mus. 100 (3260):147-160, 19 text figs.

Ammothea hilgendorfi
Child, C. Allan, 1970.
Pycnogonida of the Smithsonian-Bredin Pacific Expedition, 1957.
Proc. Biol. Soc. Wash. 83(27): 287-308.

Ammothea indica
Child, C. Allan, 1970.
Pycnogonida of the Smithsonian-Bredin Pacific Expedition, 1957.
Proc. Biol. Soc. Wash. 83(27): 287-308.

Ammothea laevis
Lebour, M.V., 1945.
Notes on the Pycnogonida of Plymouth. J.M.B.A. 26:139-165, 7 textfigs.

Ammothea longicollis n.sp.
Haswell, W.A., 1885.
On the Pycnogonida of the Australian coast.
Proc. Linn. Soc., N.S. Wales, 9:1021-1034, Pls. 54-57.

Ammothea longipes
Lebour, M.V., 1945.
Notes on the Pycnogonida of Plymouth. J.M.B.A. 26:139-165, 7 textfigs.

Ammothea (Homathea) longispina

Fry, William G. and Joel W. Hedgpeth, 1969.
The fauna of the Ross Sea. 7. Pychogonida,
I. Colossendeidae, Pycnogonidae, Endeidae,
Ammotheidae. Bull. N.Z.D.S.I.R. 198 (Mem.
N.Z. Oceanogr. Inst. 49): 139 pp.

Ammothea longispina n.sp.

Gordon, I., 1932.
Pycnogonida. Discovery Repts., 6:1-138.

Ammothea magniceps

Clark, W.C. 1971
Pycnogonida of the Antipodes Islands.
N.Z. Jl mar. Freshwat. Res. 5(3/4): 427-452

Ammothea magniceps

Clark, W.C. 1971
Pycnogonida of the Snares Islands.
N.Z. Jl mar. Freshwat. Res. 5(2): 329-341

Ammothea (Theammoa) magniceps

Fry, William G. and Joel W. Hedgpeth, 1969.
The fauna of the Ross Sea. 7. Pycnogonida,
I. Colossendeidae, Pycnogonidae, Endeidae,
Ammotheidae. Bull. N.Z.D.S.I.R. 198 (Mem.
N.Z. Oceanogr. Inst. 49): 139 pp.

Ammothea minor

Fage, L., 1952.
Pycnogonides de la Terre Adélie. Échantillons
rapportés par le Docteur Sapin-Jaloustre,
Medecin-Biologiste de la Premiere Expédition en
Terre Adélie (1949-1951). (Expéditions Polaires
Françaises, Paul-Emile Victor).
Bull. Mus. Nat. Hist. Nat., 2 ser., 24:263-273,
2 textfigs. (Missions)

Ammothea (Theammoa) minor

Fry, William G. and Joel W. Hedgpeth, 1969.
The fauna of the Ross Sea. 7. Pycnogonida,
I. Colossendeidae, Pycnogonidae, Endeidae,
Ammotheidae. Bull. N.Z.D.S.I.R. 198 (Mem.
N.Z. Oceanogr. Inst. 49): 139 pp.

Ammothea minor

Gordon, I., 1932.
Pycnogonida. Discovery Repts., 6:1-138.

Ammothea (Lecythorhynchus) ovatoides n.sp.

Stock, Jan H. 1973.
Pycnogonida from south-eastern
Australia.
Beaufortia 20 (266): 99-127

Ammothea scabra

Giltray, L., 1942.
New records of Pycnogonida from the Canadian
Atlantic coast. J. Fish. Res. Bd., Canada, 5(5):
459-460.

Ammothea spinosa

Gordon, I., 1932.
Pycnogonida. Discovery Repts., 6:1-138.

Ammothea striata

Gordon, I., 1932.
Pycnogonida. Discovery Repts., 6:1-138.

Ammothea (Thammota) stylirostris

Fry, William G. and Joel W. Hedgpeth, 1969.
The fauna of the Ross Sea. 7. Pycnogonida,
I. Colossendeidae, Pycnogonidae, Endeidae,
Ammotheidae. Bull. N.Z.D.S.I.R. 198 (Mem.
N.Z. Oceanogr. Inst. 49): 139 pp.

Ammothea stylirostris n.sp.

Gordon, I., 1932.
Pycnogonida. Discovery Repts., 6:1-138.

Ammothea tetrapora n.sp.

Gordon, I., 1932.
Pycnogonida. Discovery Repts., 6:1-138.

Ammothella sp.

Stock, Jan H., 1968.
Pycnogonida collected by the Galathea and Anton
Brunn in the Indian and Pacific oceans.
Vidensk. Meddr dansk naturh. Foren. 131: 7-65.

Ammothella appendiculata

Bourdillon, A., 1954.
Les pycnogonides de Marseille et ses environs.
Recueil Trav. Sta. Mar., Endoume, 12:145-158.

Ammothella appendiculata

Fage, Louis (posthumous) et Jan H. Stock, 1966.
Pycnogonides. Campagne de la Calypso aux Îles
du Cap Vert (1959). Ann. Inst. Océanogr.,
Monaco, n.s., 44: 315-328.

Ammothella appendiculata

Stock, J.H., 1964.
Report on the Pycnogonida of the Israel South
Red Sea Expedition.
Sea Fish Res.Inst.,Israel,Bull.No.35:27-34.

Ammothella appendiculata

Stock, J.H., 1957.
Pantopoden aus dem Zoologischen Museum Hamburg.
Mitt. Hamburgischen Zool. Mus. u. Inst., 55:81-106.

Ammothella appendiculata

Stock, J.H., 1957.
Pycnogonida from the Gulf of Aqaba. Contr. Knowl.
Red Sea, 2. Bull. Sea Fish. Res. Sta., Israel, No
13:13-14.

Ammothella bi-unguiculata

Bourdillon, A., 1954.
Les pycnogonides de Marseille et ses environs.
Recueil Trav. Sta. Mar., Endoume, 12:145-158.

Ammothella dawsoni n. sp.

Child, C. Allan, and Joel W. Hedgpeth, 1971.
Pycnogonida of the Galápagos Islands. J. nat.
Hist. 5(6): 609-634.

Ammothella elegantula n. sp.

Stock, Jan H., 1968.
Pycnogonida collected by the Galathea and Anton
Brunn in the Indian and Pacific oceans.
Vidensk. Meddr dansk naturh. Foren. 131: 7-65.

Ammothella gigas n.sp.

Fage, L., 1956.
Les pycnogonides (excl. le genre Nymphon).
Galathea Rept., 2:167-183.

Ammothella hedgpethi n.sp.

Fage, L., 1953.
Deux pycnogonides nouveaux de la cote occidentale
d'Afrique. Bull. Mus. Nat. Hist. Nat., 2e ser.,
25:376-382, 4 textfigs.

Ammothella indica

Stock, Jan H., 1968.
Pycnogonida collected by the Galathea and Anton
Brunn in the Indian and Pacific oceans.
Vidensk. Meddr dansk naturh. Foren. 131: 7-65.

Ammothella indica n.sp.

Stock, J. H. 1954.
Pycnogonida from Indo-West Pacific,
Australian and New Zealand waters. Papers
from Dr. Th. Mortensen's Pacific Expedition,
1914-1916. LXXVII.
Vidensk. Medd. Dansk Naturhist. Foren.,
København, 116: 1-168.

Ammothella indica

Utinomi, Huzio, 1952
Pycnogonida of Sagami Bay.
Publ. Seto Mar. Biol. Lab., 7(2):197-222.

Ammothella longioculata

Stock, J.H., 1958
Pycnogonida from the Mediterranean coast
of Israel.
Bull. Res. Counc., Israel, (B): 7(3/4):137-142.
Reprinted in:
Bull. Sea Fish. Res. Sta., Haifa, No. 19:

Ammothella longipes

de Haro, Andrés, 1965.
Picnogonidos de la fauna española. Comunidad
de picnogonidos en el alga parda Halopteris
Scoparia (L.).
Bol. R. Soc. Española Hist. Nat. (Biol.), 63:
213-218.

Ammothella marcusi n.sp

Hedgpeth, J.W., 1948
The Pycnogonida of the western North
Atlantic and the Caribbean. Proc. U.S. Nat.
Mus. 97:157-342, figs.4-53, charts 1-3.

Ammothella pacifica

Stock, Jan H., 1968.
Pycnogonida collected by the Galathea and Anton
Brunn in the Indian and Pacific oceans.
Vidensk. Meddr dansk naturh. Foren. 131: 7-65.

Ammothella rugulosa

Bourdillon, A., 1955.
Les Pycnogonides de la croisière 1951 du
"President Theodore Tissier".
Rev. Trav. Inst. Pêches Marit., 19(4):581-609.

Ammothella rugulosa

Hedgpeth, J.W., 1948
The Pycnogonida of the western North
Atlantic and the Caribbean. Proc. U.S. Nat.
Mus. 97:157-342, figs.4-53, charts 1-3.

Ammothella schmitti n.sp.

Child, C. Allan, 1970.
Pycnogonida of the Smithsonian-Bredin
Pacific Expedition, 1957.
Proc. Biol. Soc. Wash. 83(27): 287-308.

Ammothella uniunguiculata

Krapp, Franz 1973.
Pycnogonida from Pantellaria
and Catania, Sicily.
Beaufortia 21 (277): 55-74.

Anoplodactylus sp.

Stock, Jan H. 1973.
Pycnogonida from south-eastern
Australia.
Beaufortia 20 (266): 99-127

Anoplodactylus sp.

Stock, Jan H., 1968.
Pycnogonida collected by the Galathea and Anton
Brunn in the Indian and Pacific oceans.
Vidensk. Meddr dansk naturh. Foren., 131: 7-65.

Anoplodactylis spp (2)

Stock, J. H. 1954.
Pycnogonida from Indo-West Pacific,
Australian and New Zealand waters. Papers
from Dr. Th. Mortensen's Pacific Expedition,
1914-1916. LXXVII.
Vidensk. Medd. Dansk Naturhist. Foren.,
København, 116: 1-168.

Oceanographic Index: Marine Organisms Cumulation, 1946-1973

Anoplodactylus angulatus
Bourdillon, A., 1954.
Les pycnogonides de Marseille et ses environs.
Recueil Trav. Sta. Mar., Endoume, 12:145-158.

Anoplodactylus angulatus
Knapp, Franz 1973.
Pycnogonida from Pantellaria and Catania, Sicily.
Beaufortia 21 (277): 55-74.

Anoplodactylus angulatus
Lebour, M.V., 1945.
Notes on the Pycnogonida of Plymouth. J.M.B.A. 26:139-165, 7 textfigs.

Anoplodactylus aculeatus
Barnard, K.H., 1954.
III(3): South African Pycnogonida.
Ann. S. African Mus., 41:81-158, 34 textfigs.

Anoplodactylus batangensis
Arnaud, Françoise 1972 (1973).
Pycnogonides des récifs coralliens de Madagascar. 4. Colossendeidae, Phoxichilidiidae et Endeidae.
Tethys 4(4): 953-960.

Anoplodactylus batangense
Bourdillon, A., 1955.
Les Pycnogonides de la croisière du "President Theodore Tissier".
Rev. Trav. Inst. Pêches Marit., 19(4):581-609.

Anoplodactylus batangensis n.sp.
Stock, Jan H., 1968.
Pycnogonida collected by the Galathea and Anton Bruun in the Indian and Pacific oceans.
Vidensk. Meddr dansk naturh. Foren., 131: 7-65.

Anoplodactylus carvalhoi
Bourdillon, A., 1955.
Les Pycnogonides de la croisière 1951 du "President Theodore Tissier".
Rev. Trav. Inst. Pêches Marit., 19(4):581-609.

Anoplodactylus carvalhoi
Hedgpeth, J.W., 1948
The Pycnogonida of the western North Atlantic and the Caribbean. Proc. U.S. Nat. Mus. 97:157-342, figs.4-53, charts 1-3.

Anoplodactylus coxalis n.sp.
Stock, Jan H., 1968.
Pycnogonida collected by the Galathea and Anton Bruun in the Indian and Pacific oceans.
Vidensk. Meddr dansk naturh. Foren., 131: 7-65.

Anoplodactylus digitatus
Arnaud, Françoise 1972 (1973).
Pycnogonides des récifs coralliens de Madagascar. 4. Colossendeidae, Phoxichilidiidae et Endeidae.
Tethys 4(4): 953-960.

Anoplodactylus digitatus
Stock, Jan H., 1968.
Pycnogonida collected by the Galathea and Anton Bruun in the Indian and Pacific oceans.
Vidensk. Meddr dansk naturh. Foren., 131: 7-65.

Anoplodactylus digitatus
Stock, Jan H., 1965.
Pycnogonida from the southwest Indian Ocean.
Beaufortia, 13(151):13-33.

Anoplodactylus erectus
Child, C. Allan, 1970.
Pycnogonida of the Smithsonian-Bredin Pacific Expedition, 1957.
Proc. Biol. Soc. Wash. 83(27): 287-308.

Anoplodactylus erectus
Henry, L.M., 1953.
The nervous system of the Pycnogonida.
Microentomology 18(1):16-36, figs.

Anoplodactylus eroticus n.sp
Stock, Jan H., 1968.
Pycnogonida collected by the Galathea and Anton Brunn in the Indian and Pacific oceans.
Vidensk. Meddr dansk naturh. Foren. 131: 7-65.

Anoplodactylus evansi
Stock, Jan H. 1973.
Pycnogonida from south-eastern Australia.
Beaufortia 20 (266), 99-127

Anoplodactylus evelinae
Hedgpeth, J.W., 1948
The Pycnogonida of the western North Atlantic and the Caribbean. Proc. U.S. Nat. Mus. 97:157-342, figs.4-53, charts 1-3.

Anoplodactylus gestiens
Utinomi, Huzio, 1962
Pycnogonida of Sagami Bay - Supplement.
Publ. Seto Mar. Biol. Lab., 10(1) (Article 5) 91-104.

Anoplodactylus glandulifer
Arnaud, Françoise 1972 (1973).
Pycnogonides des récifs coralliens de Madagascar. 4. Colossendeidae, Phoxichilidiidae et Endeidae.
Tethys 4(4): 953-960.

Anoplodactylus glandulifer
Stock, Jan H., 1968.
Pycnogonida collected by the Galathea and Anton Brunn in the Indian and Pacific oceans.
Vidensk. Meddr dansk naturh. Foren. 131: 7-65.

Anoplodactylus glandulifer n.sp.
Stock, J. H. 1954.
Pycnogonida from Indo-West Pacific, Australian and New Zealand waters. Papers from Dr. Th. Mortensen's Pacific Expedition, 1914-1916. LXXVII.
Vidensk. Medd. Dansk Naturhist. Foren., København, 116: 1-168.

Anoplodactylis haswelli
Stock, J. H. 1954.
Pycnogonida from Indo-West Pacific, Australian and New Zealand waters. Papers from Dr. Th. Mortensen's Pacific Expedition, 1914-1916. LXXVII.
Vidensk. Medd. Dansk Naturhist. Foren., København, 116: 1-168.

Anoplodactylus insignis
Hedgpeth, J.W., 1948
The Pycnogonida of the western North Atlantic and the Caribbean. Proc. U.S. Nat. Mus. 97:157-342, figs.4-53, charts 1-3.

Anoplodactylus investigatoris
Bourdillon, A., 1955.
Les Pycnogonides de la croisière 1951 du "President Theodore Tissier".
Rev. Trav. Inst. Pêches Marit., 19(4):581-609.

Anoplodactylus investigatoris
Stock, J. H. 1954.
Pycnogonida from Indo-West Pacific, Australian and New Zealand waters. Papers from Dr. Th. Mortensen's Pacific Expedition, 1914-1916. LXXVII.
Vidensk. Medd. Dansk Naturhist. Foren., København, 116: 1-168.

Anoplodactylus lentus
Hedgpeth, J.W., 1948
The Pycnogonida of the western North Atlantic and the Caribbean. Proc. U.S. Nat. Mus. 97:157-342, figs.4-53, charts 1-3.

Anoplodactylis mamillosus n.sp.
Stock, J. H. 1954.
Pycnogonida from Indo-West Pacific, Australian and New Zealand waters. Papers from Dr. Th. Mortensen's Pacific Expedition, 1914-1916. LXXVII.
Vidensk. Medd. Dansk Naturhist. Foren., København, 116: 1-168.

? Anoplodactylus maritimus
Hedgpeth, J.W., 1948
The Pycnogonida of the western North Atlantic and the Caribbean. Proc. U.S. Nat. Mus. 97:157-342, figs.4-53, charts 1-3.

Anoplodactylus massiliensis
Stock, Jan H. 1970.
The Pycnogonida collected off northwestern Africa during the cruise of the Meteor.
Meteor-Forsch. Ergebn. (D) 5: 6-10.

Anoplodactylus massiliensis
Stock, Jan H., 1966.
Pycnogonida from West Africa.
Atlantide Rep., 9:45-57.

Anoplodactylus micros n.sp.
Bourdillon, A., 1955.
Les Pycnogonides de la croisière 1951 du "President Theodore Tissier".
Rev. Trav. Inst. Pêches Marit., 19(4):581-609.

Anoplodactylus micros
Stock, Jan H. 1973.
Pycnogonida from south-eastern Australia.
Beaufortia 20 (266), 99-127

Anoplodactylus minutissimus n.sp.
Stock, J. H. 1954.
Pycnogonida from Indo-West Pacific, Australian and New Zealand waters. Papers from Dr. Th. Mortensen's Pacific Expedition, 1914-1916. LXXVII.
Vidensk. Medd. Dansk Naturhist. Foren., København, 116: 1-168.

Anoplodactylus parvus
Bourdillon, A., 1955.
Les Pycnogonides de la croisière 1951 du "President Theodore Tissier".
Rev. Trav. Inst. Pêches Marit., 19(4):581-609.

Anoplodactylus parvus
Fage, Louis (posthumous) et Jan H. Stock, 1966.
Pycnogonides. Campagne de la Calypso aux Iles du Cap Vert (1959). Ann. Inst. Océanogr., Monaco, n.s., 44: 315-328.

Anoplodactylus parvus
Hedgpeth, J.W., 1948
The Pycnogonida of the western North Atlantic and the Caribbean. Proc. U.S. Nat. Mus. 97:157-342, figs.4-53, charts 1-3.

Anoplodactylus parvus
Stock, J.H., 1957.
Pantopoden aus dem Zoologischen Museum Hamburg.
Mitt. Hamburgischen Zool. Mus. u. Inst., 55:81-106.

Anoplodactylus pectinus

Arnaud, Françoise 1972 (1973).
Pycnogonides des récifs coralliens
de Madagascar. 4. Colossendeidae,
Rhynchothoracidae et Endeidae.
Téthys 4(4): 953-960.

Anoplodactylus pectinus n.sp.

Hedgpeth, J.W., 1948
The Pycnogonida of the western North
Atlantic and the Caribbean. Proc. U.S. Nat.
Mus. 97:157-342, figs.4-53, charts 1-3.

Anoplodactylus pelagicus

Stock, J.H., 1963
South African deep-sea Pycnogonida, with
descriptions of five new species.
Annals, South African Mus., 46(12):321-340.

Anoplodactylus petiolatus

Bourdillon, A., 1955.
Les Pycnogonides de la croisière 1951 du
"President Theodore Tissier."
Rev. Trav. Inst. Pêches Marit., 19(4):581-609.

Anoplodactylus petiolatus

Bourdillon, A., 1954.
Les pycnogonides de Marseille et ses environs.
Receuil Trav. Sta. Mar. Endoume, 12:145-158.

Anoplodactylus petiolatus

Hedgpeth, J.W., 1948
The Pycnogonida of the western North
Atlantic and the Caribbean. Proc. U.S. Nat.
Mus. 97:157-342, figs.4-53, charts 1-3.

Anoplodactylus petiolatus

Lebour, M.V., 1945.
Notes on the Pycnogonida of Plymouth. J.M.B.A.
26:139-165, 7 textfigs.

Anoplodactylus petiolatus

Stock, Jan H., 1966.
Pycnogonida from West Africa.
Atlantide Rep., 9:45-57.

Anoplodactylus petiolatus

Stock, J.H., 1957.
Pantopoden aus dem Zoologischen Museum Hamburg.
Mitt. Hamburgischen Zool. Mus. u. Inst., 55:81-106.

Anoplodactylus petiolatus

Whitten, H. L., H. F. Rosene, and J. W. Hedgpeth, 1950
The invertebrate fauna of Texas coast
jetties; a preliminary survey. (Systematic
Appendix). Publ. Inst. Mar. Sci. 1(2):53-86,
1 pl., 4 text figs.

Anoplodactylus polignaci

Hedgpeth, J.W., 1948
The Pycnogonida of the western North
Atlantic and the Caribbean. Proc. U.S. Nat.
Mus. 97:157-342, figs.4-53, charts 1-3.

Anoplodactylus polignaci

Stock, Jan H., 1966.
Pycnogonida from West Africa.
Atlantide Rep., 9:45-57.

Anoplodactylus portus n.sp.

Calman, W.T., 1927.
Zoological results of the Cambridge Expedition to
the Suez Canal. XXVIII. Report on the Pycnogonida
Trans. Zool. Soc., London, 22:403-410, textfigs.
102-104.

Anoplodactylus portus

Stock, J.H., 1958
Pycnogonida from the Mediterranean coast of
Israel.
Bull. Res. Counc., Israel, (B) 7(3/4):137-142.
Reprinted in:
Bull. Sea Fish. Res. Sta., Haifa, No. 19:

Anoplodactylus pulcher

Arnaud, Françoise 1972 (1973).
Pycnogonides des récifs coralliens
de Madagascar. 4. Colossendeidae,
Rhynchothoracidae et Endeidae.
Téthys 4(4): 953-960.

Anoplodactylus pulcher

Stock, Jan H., 1968.
Pycnogonida collected by the Galathea and Anton
Bruun in the Indian and Pacific oceans.
Vidensk. Meddr dansk naturh. Foren. 131: 7-65.

Anoplodactylus pulcher

Stock, Jan H., 1965.
Pycnogonida from the southwest Indian Ocean.
Beaufortia, 13(151):13-33.

Anoplodactylus pulcher

Stock, J. H. 1954.
Pycnogonida from Indo-West Pacific,
Australian and New Zealand waters. Papers
from Dr. Th. Mortensen's Pacific Expedition,
1914-1916. LXXVII.
Vidensk. Medd. Dansk Naturhist. Foren.,
København, 116: 1-168.

Anoplodactylis pycnosoma

Stock, J. H. 1954.
Pycnogonida from Indo-West Pacific,
Australian and New Zealand waters. Papers
from Dr. Th. Mortensen's Pacific Expedition,
1914-1916. LXXVII.
Vidensk. Medd. Dansk Naturhist. Foren.,
København, 116: 1-168.

Anoplodactylus pygmaeus

Bourdillon, A., 1954.
Les pycnogonides de Marseille et ses environs.
Recueil Trav. Sta. Mar., Endoume, 12:145-158.

Anoplodactylus pygmaeus

Hedgpeth, J.W., 1948
The Pycnogonida of the western North
Atlantic and the Caribbean. Proc. U.S. Nat.
Mus. 97:157-342, figs.4-53, charts 1-3.

Anoplydactylus pygmaeus

Lebour, M.V., 1945.
Notes on the Pycnogonida of Plymouth. J.M.B.A.
26:139-165, 7 textfigs.

Anoplodactylus pygmaeus

Whitten, H. L., H. F. Rosene, and J. W. Hedgpeth, 1950
The invertebrate fauna of Texas coast
jetties; a preliminary survey. (Systematic
Appendix). Publ. Inst. Mar. Sci. 1(2):53-86,
1 pl., 4 text figs.

Anoplodactylus quadratispinosus

Hedgpeth, J.W., 1948
The Pycnogonida of the western North
Atlantic and the Caribbean. Proc. U.S. Nat.
Mus. 97:157-342, figs.4-53, charts 1-3.

Anoplodactylus robustus

Child, C. Allan, and Joel W. Hedgpeth, 1971.
Pycnogonida of the Galápagos Islands. J. nat.
Hist. 5(6): 609-634.

Anopladactylus robustus

Zilberberg, F., 1963
Notes of Pantopoda.
Bol. Inst. Oceanogr., Sao Paulo, 13(2):21-32.

Anoplodactylus saxatilis

Calman, W.T., 1927.
Zoological results of the Cambridge Expedition to
the Suez Canal. XXVIII. Report on the Pycnogonida
Trans. Zool. Soc., London, 22:403-410, textfigs.
102-104.

Anoplodactylus saxatilis

Stock, J.H., 1958
Pycnogonida from the Mediterranean coast of
Israel.
Bull. Res. Counc., Israel, (B) 7(3/4):137-142.
Reprinted in:
Bull. Sea Fish. Res. Sta., Haifa, No. 19:

Anoplodactylus squalida n.sp.

Clark W.C. 1973.
New species of Pycnogonida from
New Britain and Tonga.
Pacific Sci. 27(1): 28-33.

Anoplodactylus spinirostrum n.sp.

Stock, Jan H. 1973.
Pycnogonida from south-eastern
Australia.
Beaufortia 20 (266): 99-127

Anoplodactylus stictus

Stock, J.H., 1957.
Pantopoden aus dem Zoologischen Museum Hamburg.
Mitt. Hamburgischen Zool. Mus. u. Inst., 55:81-106.

Anoplodactylus stylirostris n.sp.

Hedgpeth, J.W., 1948
The Pycnogonida of the western North
Atlantic and the Caribbean. Proc. U.S. Nat.
Mus. 97:157-342, figs.4-53, charts 1-3.

Anoplodactylus tarsalis

Arnaud, Françoise 1972 (1973).
Pycnogonides des récifs coralliens
de Madagascar. 4. Colossendeidae,
Rhynchothoracidae et Endeidae.
Téthys 4(4): 953-960.

Anoplodactylus tarsalis n.sp

Stock, Jan H., 1968.
Pycnogonida collected by the Galathea and Anton
Bruun in the Indian and Pacific oceans.
Vidensk. Meddr dansk naturh. Foren. 131: 7-65.

Anoplodactylus tenuirostris n. sp.

Lebour, M. V., 1949.
Two new Pycnogonids from Bermuda. Proc. Zool.
Soc., London, 118(4):929-932, 3 textfigs.

Anoplodactylus torus n. sp.

Child, C. Allan, and Joel W. Hedgpeth, 1971.
Pycnogonida of the Galápagos Islands. J. nat.
Hist. 5(6): 609-634.

Anoplodactylus trispinosus

Stock, J.H., 1964.
Report on the Pycnogida of the Israel South
Red Sea Expedition.
Sea Fish Res. Inst., Israel, Bull., No. 35:27-34.

Anoplodactylus typhlops

Hedgpeth, J.W., 1948
The Pycnogonida of the western North
Atlantic and the Caribbean. Proc. U.S. Nat.
Mus. 97:157-342, figs.4-53, charts 1-3.

Anoplodactylus versluysi

Utinomi, Huzio, 1952
Pycnogonida of Sagami Bay.
Publ. Seto Mar. Biol. Lab., 7(2):197-222.

Anoplodactylus versluysi

Stock, Jan H., 1965.
Pycnogonida from the southwest Indian Ocean.
Beaufortia, 13(151):13-33.

Anoplodactylus versluysi
Stock, J. H. 1954.
Pycnogonida from Indo-West Pacific,
Australian and New Zealand waters. Papers
from Dr. Th. Mortensen's Pacific Expedition,
1914-1916. LXXVII.
Vidensk. Medd. Dansk Naturhist. Foren.,
København, 116: 1-168.

Anoplodactylus virescens
Knapp, Franz 1973.
Pycnogonida from Pantelleria
and Catania, Sicily.
Beaufortia 21 (277): 55-74.

Anoplodactylus zeatiens
Stock, J. H. 1954.
Pycnogonida from Indo-West Pacific,
Australian and New Zealand waters. Papers
from Dr. Th. Mortensen's Pacific Expedition,
1914-1916. LXXVII.
Vidensk. Medd. Dansk Naturhist. Foren.,
København, 116: 1-168.

Ascorhynchus sp.
Stock, Jan H. 1966.
Pycnogonida from West Africa.
Atlantide Rep. 9:45-57.

Asorhynchus sp.
Stock, J. H. 1954.
Pycnogonida from Indo-West Pacific,
Australian and New Zealand waters. Papers
from Dr. Th. Mortensen's Pacific Expedition,
1914-1916. LXXVII.
Vidensk. Medd. Dansk Naturhist. Foren.,
København, 116: 1-168.

Ascorhynchus agassizi
Henry, L.M., 1953.
The nervous system of the Pycnogonida.
Microentomology 18(1):16-36, figs.

Ascorhynchus sp.
Bourdillon, A., 1955.
Les Pycnogonides de la croisière 1951 du
"President Theodore Tissier."
Rev. Trav. Inst. Pêches Marit., 19(4):581-609.

Ascorhynchus armatus
Hedgpeth, J.W., 1948
The Pycnogonida of the western North
Atlantic and the Caribbean. Proc. U.S. Nat.
Mus. 97:157-342, figs.4-53, charts 1-3.

Asorhynchus auchenicum
Stock, J. H. 1954.
Pycnogonida from Indo-West Pacific,
Australian and New Zealand waters. Papers
from Dr. Th. Mortensen's Pacific Expedition,
1914-1916. LXXVII.
Vidensk. Medd. Dansk Naturhist. Foren.,
København, 116: 1-168.

Ascorhynchus auchenicum
Utinomi, Huzio, 1952
Pycnogonida of Sagami Bay.
Publ. Seto Mar. Biol. Lab., 7(2):197-222.

Ascorhynchus brevicapus n. sp.
Stock, Jan H., 1968.
Pycnogonida collected by the Galathea and Anton
Bruun in the Indian and Pacific oceans.
Vidensk. Meddr dansk naturh. Foren. 131: 7-65.

Asorhynchus cactoides n.sp.
Stock, J. H. 1954.
Pycnogonida from Indo-West Pacific,
Australian and New Zealand waters. Papers
from Dr. Th. Mortensen's Pacific Expedition,
1914-1916. LXXVII.
Vidensk. Medd. Dansk Naturhist. Foren.,
København, 116: 1-168.

Ascorhynchus castellioides n.sp.
Stock, J.H., 1957.
Pantopoden aus dem Zoologischen Museum Hamburg.
Mitt. Hamburgischen Zool. Mus. u. Inst., 55:81-106.

Ascorhynchus colei
Hedgpeth, J.W., 1948
The Pycnogonida of the western North
Atlantic and the Caribbean. Proc. U.S. Nat.
Mus. 97:157-342, figs.4-53, charts 1-3.

Ascorhynchus corderoi
Stock, Jan H., 1965.
Pycnogonida from the southwest Indian Ocean.
Beaufortia, 13(151):13-33.

Asorhynchus cryptopygium
Stock, J. H. 1954.
Pycnogonida from Indo-West Pacific,
Australian and New Zealand waters. Papers
from Dr. Th. Mortensen's Pacific Expedition,
1914-1916. LXXVII.
Vidensk. Medd. Dansk Naturhist. Foren.,
København, 116: 1-168.

Ascorhynchus cryptoptgium
Utinomi, H., 1955.
Report on the Pycnogonida collected by the
Soyo-maru expeditions made on the continental
shelf bordering Japan during the years 1926-1930.
Publ. Seto Mar. Biol. Lab., 5(1):1-42.

Ascorhynchus cuculus n.sp.
Fry, William G. and Joel W. Hedgpeth, 1969.
The fauna of the Ross Sea. 7. Pycnogonida,
I. Colossendeidae, Pycnogonidae, Endeidae,
Ammotheidae. Bull. N.Z.D.S.I.R. 198 (Mem.
N.Z. Oceanogr. Inst. 49); 139 pp.

Ascorhynchus glaber
Fage, L., 1956.
Les pycnogonides (excl. le genre Nymphon).
Galathea Rept., 2:167-183.

Ascorhynchus glaberrimum
Utinomi, Huzio, 1962
Pycnogonida of Sagami Bay - Supplement.
Publ. Seto Mar. Biol. Lab., 10(1) (Article 5):
91-104.

Ascorhynchus glaberrimum
Utinomi, H., 1955.
Report on the Pycnogonida collected by the Soyo-
maru expeditions made on the continental shelf
bordering Japan during the years 1926-1930.
Publ. Seto Mar. Biol. Lab., 5(1):1-42.

Ascorhynchus glaberrimum
Utinomi, Huzio, 1952
Pycnogonida of Sagami Bay.
Publ. Seto Mar. Biol. Lab., 7(2):197-222.

Ascorhynchus glaboides
Utinomi, H., 1955.
Report on the Pycnogonida collected by the Soyo-
maru expeditions made on the continental shelf
bordering Japan during the years 1926-1930.
Publ. Seto Mar. Biol. Lab., 5(1):1-42.

Ascorhynchus glabroides
Utinomi, H., 1951.
On some pycnogonids from the sea around the Kii
Peninsula.
Publ. Seto Mar. Biol. Sta., Kyoto Univ., 1(4):
159-168.

Ascorhynchus inflatum n.sp.
Stock, J.H., 1963
South African deep-sea Pycnogonida, with
descriptions of five new species.
Annals. South African Mus., 46(12):321-340.

Ascorhynchus inflatum
Turpaeva, E.P., 1971.
The deep-water Pantopoda collected in the
Kurile-Kamchatka Trench. (In Russian;
English abstract). Trudy Inst. Okeanol. P.P.
Shirshov 92: 274-291.

Ascorhynchus insularum n.sp.
Clark, W.C. 1971.
Pycnogonida of the Snares Islands.
N.Z. Jl mar. Freshwat. Res. 5(2): 329-341.

Ascorhynchus japonicum
Stock, J. H. 1954.
Pycnogonida from Indo-West Pacific,
Australian and New Zealand waters. Papers
from Dr. Th. Mortensen's Pacific Expedition,
1914-1916. LXXVII.
Vidensk. Medd. Dansk Naturhist. Foren.,
København, 116: 1-168.

Ascorhynchus japonicum
Utinomi, H., 1955.
Report on the Pycnogonida collected by the
Soyo-maru expeditions made on the continental
shelf bordering Japan during the years 1926-1930
Publ. Seto Mar. Biol. Lab., 5(1):1-42.

Ascorhynchus japonicum
Utinomi, Huzio, 1952
Pycnogonida of Sagami Bay.
Publ. Seto Mar. Biol. Lab., 7(2):197-222.

Ascorhynchus japonicus
Utinomi, H., 1951.
On some pycnogonids from the sea around the Kii
Peninsula.
Publ. Seto Mar. Biol. Sta., Kyoto Univ., 1(4):
159-168.

Ascorhynchus laterospinum
Child, C. Allan, and Joel W. Hedgpeth, 1971.
Pycnogonida of the Galápagos Islands. J. nat.
Hist. 5(6): 609-634.

Ascorhynchus latipes
Fage, L., 1952.
Sur quelques Pycnogonides de Dakar. Bull. Mus.
Nat. Hist. Nat., 2 ser., 24(6):530-533, 1 text fig.

Ascorhynchus latipes
Hedgpeth, J.W., 1948
The Pycnogonida of the western North
Atlantic and the Caribbean. Proc. U.S. Nat.
Mus. 97:157-342, figs.4-53, charts 1-3.

Asorhynchus latum
Stock, J. H. 1954.
Pycnogonida from Indo-West Pacific,
Australian and New Zealand waters. Papers
from Dr. Th. Mortensen's Pacific Expedition,
1914-1916. LXXVII.
Vidensk. Medd. Dansk Naturhist. Foren.,
København, 116: 1-168.

Ascorhynchus losinalosinskii n. sp.
Turpaeva, E.P., 1971.
The deep-water Pantopoda collected in the
Kurile-Kamchatka Trench. (In Russian;
English abstract). Trudy Inst. Okeanol. P.P.
Shirshov 92: 274-291.

Ascorhynchus mariae n. sp.
Turpaeva, E.P., 1971.
The deep-water Pantopoda collected in the
Kurile-Kamchatka Trench. (In Russian;
English abstract). Trudy Inst. Okeanol. P.P.
Shirshov 92: 274-291.

Ascorhynchus melwardi
Stock, Jan H., 1968.
Pycnognida collected by the Galathea and Anton
Bruun in the Indian and Pacific oceans.
Vidensk. Meddr. dansk naturh. Foren., 131: 7-65.

Asorhynchus melwardi
Stock, J. H. 1954.
Pycnogonida from Indo-West Pacific, Australian and New Zealand waters. Papers from Dr. Th. Mortensen's Pacific Expedition, 1914-1916. LXXVII.
Vidensk. Medd. Dansk Naturhist. Foren., København, 116: 1-168.

Ascorhynchus minutus
Haswell, W. A., 1885.
On the Pycnogonida of the Australian Coast.
Proc. Linn. Soc., N.S. Wales, 9:1-21-1034, Pls. 54-57.

Asorhynchus minutum
Stock, J. H. 1954.
Pycnogonida from Indo-West Pacific, Australian and New Zealand waters. Papers from Dr. Th. Mortensen's Pacific Expedition, 1914-1916. LXXVII.
Vidensk. Medd. Dansk Naturhist. Foren., København, 116: 1-168.

Asorhynchus mucosa
Stock, J. H. 1954.
Pycnogonida from Indo-West Pacific, Australian and New Zealand waters. Papers from Dr. Th. Mortensen's Pacific Expedition, 1914-1916. LXXVII.
Vidensk. Medd. Dansk Naturhist. Foren., København, 116: 1-168.

Anoplodactylus pygmaeus
Stock, Jan H. 1970.
The Pycnogonida collected off northwestern Africa during the cruise of the Meteor.
Meteor - Forsch. Ergebn. (D) 5: 6-10.

Ascorhynchus pyrginospinum n.sp.
McClosky, L.R., 1967.
New and little-known benthic pycnogonids from North Carolina.
J. nat. Hist., 1:119-132.

Ascorhynchus ramipes
Utinomi, Huzio, 1962
Pycnogonida of Sagami Bay - Supplement.
Publ. Seto Mar. Biol. Lab., 10(1) (Article 5): 91-104.

Ascorhynchus ramipes
Utinomi, Huzio, 1952
Pycnogonida of Sagami Bay.
Publ. Seto Mar. Biol. Lab., 7(2):197-222.

Ascorhynchus serratum n.sp.
Hedgpeth, J.W., 1948
The Pycnogonida of the western North Atlantic and the Caribbean. Proc. U.S. Nat. Mus. 97:157-342, figs.4-53, charts 1-3.

Ascorhynchus simile
Stock, J.H., 1957.
Pantopoden aus dem Zoologischen Museum Hamburg.
Mitt. Hambrugischen Zool. Mus. u. Inst., 55:81-106.

Ascorhyohus simile
Stock, Jan H., et Jacques Soyer, 1965.
Sur quelques pycnogonides rares de Banyuls-Sur-mer.
Vie et Milieu, Bull. Lab. Arego. (B) 16 (1):415-

Ascorhynchus tridens n.sp.
Meinert, Fr., 1899
Pycnogonida. Danish Ingolf Exped. 3(1): 71 pp., 5 pls., 2 text figs., 1 charts, list of stations.

Athernopycnon meridionalis n.gen.
Fry, William G. and Joel W. Hedgpeth, 1969.
The fauna of the Ross Sea. 7. Pycnogonida, I. Colossendeidae, Pycnogonidae, Endeidae, Ammotheidae. Bull. N.Z.D.S.I.R. 198 (Mem. N.Z. Oceanogr. Inst. 49): 139 pp.

Austrodecus sp.
Arnaud, Françoise 1970.
Pycnogonides subantarctiques des îles Crozet.
Bull. Mus. natn. Hist. nat. (2) 41 (6): 1423-1428.

Austrodecus sp.
Stock, J.H., 1957.
The pycnogonid family Austrodecidae.
Beaufortia, 6(68):1-81.

Austrodecus breviceps
Hedgpeth, J. W., 1950
Pycnogonida of the United States Navy Antarctic Expedition, 1947-48. Proc. U.S. Nat. Mus. 100 (3260):147-160, 19 text figs.

Austrodecus brevipes
Stock, J.H., 1957.
The pycnogonid family Austrodecidae.
Beaufortia, 6(68):1-81.

Austrodecus calcaricauda n.sp.
Stock, J.H., 1957.
The pycnogonid family Austrodecidae.
Beaufortia 6(68):1-81.

Austrodecus breviceps
Arnaud, Françoise 1972.
Pycnogonides des îles Kerguelen (Sud Océan Indien): matériel nouveau et révision des spécimens du Museum national d'Histoire naturelle de Paris.
Bull. Mus. natn. Hist. nat. Paris, (3) 65 (Zool. 5):801-815.

Austrodecus confusum n.sp.
Stock, J.H., 1957.
The pycnogonid family Austrodecidae.
Beaufortia 6(68):1-81.

Austrodecus curtipes
Arnaud, Françoise 1972.
Pycnogonides des îles Kerguelen (Sud Océan Indien): matériel nouveau et révision des spécimens du Museum national d'Histoire naturelle de Paris.
Bull. Mus. natn. Hist. nat. Paris, (3) 65 (Zool. 5):801-815.

Austrodecus curtipes n.sp.
Stock, J.H., 1957.
The pycnogonid family Austrodecidae.
Beaufortia 6(68):1-81.

Austrodecus elegans n.sp.
Stock, J.H., 1957.
The pycnogonid family Austrodecidae.
Beaufortia 6(68):1-81.

Austrodecus ensoi n.sp.
Clark, W.C. 1971.
Pycnogonida of the Snares Islands.
N.Z. Jl Mar. Freshwat. Res. 5(2): 329-341.

Austrodecus fagei n.sp.
Stock, J.H., 1957.
The pycnogonid family Austrodecidae.
Beaufortia 6(68):1-81.

Austrodecus frigorifugum
Stock, Jan H., 1968.
Pycnogonida collected by the Galathea and Anton Brunn in the Indian and Pacific oceans.
Vidensk. Meddr dansk naturh. Foren. 131: 7-65.

Austrodecus frigorifugum
Stock, J.H., 1957.
The pycnogonid family Austrodecidae.
Beaufortia 6(68):1-81.

Austrodecus frigorifrigum n.sp.
Stock, J. H. 1954.
Pycnogonida from Indo-West Pacific, Australian and New Zealand waters. Papers from Dr. Th. Mortensen's Pacific Expedition, 1914-1916. LXXVII.
Vidensk. Medd. Dansk Naturhist. Foren., København, 116: 1-168.

Austrodecus glabrum n.sp.
Stock, J.H., 1957.
The pycnogonid family Austrodecidae.
Beaufortia 6(68):1-81.

Austrodecus glaciale
Fage, L., 1952.
II. Pycnogonides. Missions du Bâtiment Polaire "Commandant Charcot", récoltés faites en Terre Adélie (1950) par Paul Tchernia. Bull. Mus. Nat. Hist. Nat., 2 ser., 24:180-186, 2 textfigs.

Austrodecus glaciale
Fage, L., 1952.
Pycnogonides de la Terre Adélie. Echantillons rapportés par le Docteur Sapin-Jaloustre, Medecin de la Première Expédition en Terre Adélie (1949-1951). Expéditions Polaires Françaises, Missions Paul-Emile Victor).
Bull. Mus. Nat. Hist. Nat., 2 ser., 24:263-273, 2 textfigs. —Biologist

Austrodecus glaciale
Fry, William G., 1965.
The feeding mechanisms and preferred foods of three species of Pycnogonida.
Bull. Brit. Mus. (N.H.), Zool., 12(6):197-223.

Austrodecus glaciale
Gordon, I., 1932.
Pycnogonida. Discovery Repts., 6:1-138.

Austrodecus glaciale
Hedgpeth, J. W., 1950
Pycnogonida of the United States Navy Antarctic Expedition, 1947-48. Proc. U.S. Nat. Mus. 100 (3260):147-160, 19 text figs.

Austrodecus glaciale
Hodgson, T.V., 1907.
Pycnogonida. Nat. Antarctic Exped., 1901-1904, 3:72 pp., 10 pls.

Austrodecus glaciale
Stock, J.H., 1957.
The pycnogonid family Austrodecidae.
Beaufortia 6(68):1-81.

Austrodecus glacialis
Stock, J. H. 1954.
Pycnogonida from Indo-West Pacific, Australian and New Zealand waters. Papers from Dr. Th. Mortensen's Pacific Expedition, 1914-1916. LXXVII.
Vidensk. Medd. Dansk Naturhist. Foren., København, 116: 1-168.

Austrodecus gordonae n.sp.
Stock, J. H. 1954.
Pycnogonida from Indo-West Pacific, Australian and New Zealand waters. Papers from Dr. Th. Mortensen's Pacific Expedition, 1914-1916. LXXVII.
Vidensk. Medd. Dansk Naturhist. Foren., København, 116: 1-168.

Austrodecus goughense n.sp.
Stock, J.H., 1957.
The pycnogonid family Austrodecidae.
Beaufortia, 6(68):1-81.

Austrodecus longispinum n.sp.
Stock, J.H., 1957.
The pycnogonid family Austrodecidae.
Beaufortia, 6(68):1-81.

Austrodecus minutum n.sp.

Clark, W.C. 1971.
Pycnogonida of the Snares Islands.
N.Z. Jl Mar. Freshwat. Res. 5(2): 329-341.

Austrodecus pentamerum n. sp.

Stock, Jan H., 1968.
Pycnogonida collected by the Galathea and Anton Brunn in the Indian and Pacific oceans.
Vidensk. Meddr dansk. naturh. Foren. 131: 7-65.

Austrodecus profundum n.sp.
Stock, J.H., 1957.
The pycnogonid family Austrodecidae.
Beaufortia, 6(68):1-81.

Austrodecus simulans n.sp.
Stock, J.H., 1957.
The pycnogonid family Austrodecidae.
Beaufortia 6(68):1-81.

Austrodecus sinuatum n.sp.
Stock, J.H., 1957.
The pycnogonid family Austrodecidae.
Beaufortia 6(68):1-81.

Austrodecus tristanense
Stock, J.H., 1957.
The pycnogonid family Austrodecidae.
Beaufortia, 6(68):1-81.

Austrodecus tubiferum n.sp.
Stock, J.H., 1957.
The pycnogonid family Austrodecidae.
Beaufortia, 6(68):1-81.

Austropallene brachyura
Gordon, I., 1932.
Pycnogonida. Discovery Repts., 6:1-138.

Austropallene cornigera
Hedgpeth, J. W., 1950
Pycnogonida of the United States Navy Antarctic Expedition, 1947-48. Proc. U.S. Nat. Mus. 100 (3260):147-160, 19 text figs.

Austropallene cornigera
Gordon, I., 1932.
Pycnogonida. Discovery Repts., 6:1-138.

Austropallene cornigera
Stephensen, K., 1947
Tanaidacea, Isopoda, Amphipoda, and Pycnogonida. Sci. Res. Norweg. Antarctic Exp. 1927-1928 et seq. Date Norske Videnskaps-Akademi i Oslo, II, No.27:1-90, 26 text figs.

Austropallene cristata
Gordon, I., 1932.
Pycnogonida. Discovery Repts., 6:1-138.

Austropallene tcherniai n.sp.
Fage, L., 1952.
II. Pycnogonides. Missions du Bâtiment Polaire "Commandant Charcot", récoltés faites en Terre Adélie (1950) par Paul Tchernia. Bull. Mus. Nat. Hist. Nat., 2 ser., 24: 180-186, 2 textfigs.

Austropallene tibiana
Hedgpeth, J. W., 1950
Pycnogonida of the United States Navy Antarctic Expedition, 1947-48. Proc. U.S. Nat. Mus. 100 (3260):147-160, 19 text figs.

Austroraptus calcaratus
Fry, William G. and Joel W. Hedgpeth, 1969.
The fauna of the Ross Sea. 7. Pycnogonida, I. Colossendeidae, Pycnogonidae, Endeidae, Ammotheidae. Bull. N.Z.D.S.I.R. 198 (Mem. N.Z. Oceanogr. Inst. 49): 139 pp.

Austroraptus juvenilis
Fage, L., 1952.
Pycnogonides de la Terre Adélie. Échantillons rapportés par le Docteur Sapin-Jaloustre, Medecin-Biologiste de la Première Expédtion en Terre Adélie (1949-1951). (Expéditions Polaires Françaises, Missions Paul-Émile Victor). Bull. Mus. Nat. Hist. Nat., 2 ser., 24:263-273, 2 textfigs.

Austroraptus juvenilis
Fry, William G. and Joel W. Hedgpeth, 1969.
The fauna of the Ross Sea. 7. Pycnogonida, I. Colossendeidae, Pycnogonidae, Endeidae, Ammotheidae. Bull. N.Z.D.S.I.R. 198 (Mem. N.Z. Oceanogr. Inst. 49): 139 pp.

Austroraptus juvenilis
Gordon, I., 1932.
Pycnogonida. Discovery Repts., 6:1-138.

Austroraptus polaris
Fry, William G. and Joel W. Hedgpeth, 1969.
The fauna of the Ross Sea. 7. Pycnogonida, I. Colossendeidae, Pycnogonidae, Endeidae, Ammotheidae. Bull. N.Z.D.S.I.R. 198 (Mem. N.Z. Oceanogr. Inst. 49): 139 pp.

Austroraptus polaris
Gordon, I., 1932.
Pycnogonida. Discovery Repts., 6:1-138.

Austroraptus polaris
Hedgpeth, J. W., 1950
Pycnogonida of the United States Navy Antarctic Expedition, 1947-48. Proc. U.S. Nat. Mus. 100 (3260):147-160, 19 text figs.

Austroraptus polaris
Hodgson, T.V., 1907.
Pycnogonida. Nat. Antarctic Exped., 1901-1904, 3:72 pp., 10 pls.

Austroraptus praecox
Fry, William G. and Joel W. Hedgpeth, 1969.
The fauna of the Ross Sea. 7. Pycnogonida, I. Colossendeidae, Pycnogonidae, Endeidae, Ammotheidae. Bull. N.Z.D.S.I.R. 198 (Mem. N.Z. Oceanogr. Inst. 49): 139 pp.

Austroraptus praecox
Gordon, I., 1932.
Pycnogonida. Discovery Repts., 6:1-138.

Austroraptus sicarius n.sp.
Fry, William G. and Joel W. Hedgpeth, 1969.
The fauna of the Ross Sea. 7. Pycnogonida, I. Colossendeidae, Pycnogonidae, Endeidae, Ammotheidae. Bull. N.Z.D.S.I.R. 198 (Mem. N.Z. Oceanogr. Inst. 49): 139 pp.

Austroraptus thermophilus n.sp.
Barnard, K.H., 1946.
Diagnoses of new species and a new genus of Pycnogonida, in the South African Museum.
Ann. Mag. Nat. Hist. (11)13:60-63.

? Boehmia dubia n.sp.
Hedgpeth, J. W., 1950
Pycnogonida of the United States Navy Antarctic Expedition, 1947-48. Proc. U.S. Nat. Mus. 100 (3260):147-160, 19 text figs.

Boehmia longirostris n.sp.
Stock, J.H., 1957.
Pantopoden aus dem Zoologischen Museum Hamburg.
Mitt. Hamburgischen Zool. Mus. u. Inst., 55:81-106.

Boreonymphon abyssorum
Just, Jean 1972.
Revision of the genus Boreonymphon G.O. Sars (Pycnogonida) with a description of two new species, B. ossiansarsi Knaben and B. compactum Just.
Sarsia 49:1-27.

Boreonymphon compactum n.sp.
Just, Jean 1972.
Revision of the genus Boreonymphon G.O. Sars (Pycnogonida) with a description of two new species, B. ossiansarsi Knaben and B. compactum Just.
Sarsia 49:1-27.

Boreonymphon ossiansarsi
Just, Jean 1972.
Revision of the genus Boreonymphon G.O. Sars (Pycnogonida) with a description of two new species B. ossiansarsi Knaben and B. compactum Just.
Sarsia 49:1-27.

Boreonymphon robustum
Just, Jean 1972.
Revision of the genus Boreonymphon G.O. Sars (Pycnogonida) with a description of two new species B. ossiansarsi Knaben and B. compactum Just.
Sarsia 49:1-27.

Boreonymphon robustum
Nesis, K.N., 1960.
Littoral Pantopoda of the eastern Murmansk coast.
Trudy Murmansk. Morsk. Biol. Inst., 2(6):137-161.

Callipallene ap.
Barnard, K.H., 1954.
III(3):South African Pycnogonida.
Ann. S. African Mus., 41:81-158, 34 textfigs.

Callipallene spec.
Stock, Jan H., 1965.
Pycnogonida from the southwest Indian Ocean.
Beaufortia, 13(151):13-33.

Callipallene acus
Hedgpeth, J.W., 1948
The Pycnogonida of the western North Atlantic and the Caribbean. Proc. U.S. Nat. Mus. 97:157-342, figs.4-53, charts 1-3.

Callipallene acus
Stock, J.H., 1964.
Deep-sea Pycnogonida collected by the "Cirrus" in the northern Atlantic.
Beaufortia, Zool. Mus., Amsterdam, 11(135):45-52.

Callipallene brevirostris
Bourdillon, A., 1954.
Les pycnogonides de Marseille et ses environs.
Recueil Trav. Stat. Mar. Endoume, 12:145-158.

Callipallene brevirostris
Arnaud, Francoise 1972.
Pycnogonides des récifs coralliens de Madagascar. 3. Famille des Callipallenidae.
Tethys Suppl. 3:157-164.

Callipallene brevirostris
Bourdillon A., 1955.
Les Pycnogonides de la croisière 1951 du "President Theodore Tissier".
Rev. Trav. Inst. Pêches Marit., 19(4):581-609.

Callipallene brevirostris
Hedgpeth, J.W., 1948
The Pycnogonida of the western North Atlantic and the Caribbean. Proc. U.S. Nat. Mus. 97:157-342, figs. 4-53, charts 1-3.

Callipallene brevirostris
Stock, J. H. 1954.
Pycnogonida from Indo-West Pacific, Australian and New Zealand waters. Papers from Dr. Th. Mortensen's Pacific Expedition, 1914-1916. LXXVII.
Vidensk. Medd. Dansk Naturhist. Foren., København, 116: 1-168.

Callipallene brevirostris brevirostris
Stock, Jan N. 1970.
The Pycnogonida collected off northwestern Africa during the cruise of the Meteor.
Meteor-Forsch. Ergebn. (D) 5: 6-10.

Callipallene brevirostris novae-zealaniae
Stock, J. H. 1954.
Pycnogonida from Indo-West Pacific, Australian and New Zealand waters. Papers from Dr. Th. Mortensen's Pacific Expedition, 1914-1916. LXXVII.
Vidensk. Medd. Dansk Naturhist. Foren., København, 116: 1-168.

Callipallene conirostris n.sp.
Stock, J. H. 1954.
Pycnogonida from Indo-West Pacific, Australian and New Zealand waters. Papers from Dr. Th. Mortensen's Pacific Expedition, 1914-1916. LXXVII.
Vidensk. Medd. Dansk Naturhist. Foren., København, 116: 1-168.

Callipallene cuspidata n.sp.
Stock, J. H. 1954.
Pycnogonida from Indo-West Pacific, Australian and New Zealand waters. Papers from Dr. Th. Mortensen's Pacific Expedition, 1914-1916. LXXVII.
Vidensk. Medd. Dansk Naturhist. Foren., København, 116: 1-168.

Callipallene dubiosa
Stock, J.H., 1957.
Pantopoden aus dem Zoologischen Museum Hamburg.
Mitt. Hamburgischen Zool. Mus. u. Inst., 55:81-106.

Callipallene dubiosa
Stock, J. H. 1954.
Pycnogonida from Indo-West Pacific, Australian and New Zealand waters. Papers from Dr. Th. Mortensen's Pacific Expedition, 1914-1916. LXXVII.
Vidensk. Medd. Dansk Naturhist. Foren., København, 116: 1-168.

Callipallene (?) echinata n.sp.
Calman, W. T., 1938
Pycnogonida. John Murray Exped. 1933-34, Sci. Repts. 5(6):147-166, 10 text figs.

Callipallene emaciata
de Haro, Andrés, 1965.
Picnogonidos de la fauna español. Comunidad de picnogonidos en el alga parda Halopteris Scoparia (L.).
Bol. R. Soc. Española Hist. Nat. (Biol.), 63: 213-218.

Callipallene emaciata
Hedgpeth, J.W., 1948
The Pycnogonida of the western North Atlantic and the Caribbean. Proc. U.S. Nat. Mus. 97:157-342, figs. 4-53, charts 1-3.

Callipallene emaciata
Knapp, Franz 1973.
Pycnogonida from Pantellaria and Catania, Sicily.
Beaufortia 21 (277): 55-74.

Callipallene emaciata ssp micracantha n.subsp.
Stock, J. H. 1954.
Pycnogonida from Indo-West Pacific, Australian and New Zealand waters. Papers from Dr. Th. Mortensen's Pacific Expedition, 1914-1916. LXXVII.
Vidensk. Medd. Dansk Naturhist. Foren., København, 116: 1-168.

Calliphallene pectinata
Calman, W. T., 1938
Pycnogonida. John Murray Exped. 1933-34, Sci. Repts. 5(6):147-166, 10 text figs.

Callipallene pectinata
Stock, Jan. H., 1968.
Pycnogonida collected by the Galathea and Anton Bruun in the Indian and Pacific oceans.
Vidensk. Meddr. dansk naturh. Foren., 131: 7-65.

Callipallene pectinata
Stock, J.H., 1964.
Report on the Pycnogonida of the Israel South Red Sea Expedition.
Sea Fish Res. Inst., Israel, Bull., No. 35:27-34.

Callipallene phantoma
Hedgpeth, J.W., 1948
The Pycnogonida of the western North Atlantic and the Caribbean. Proc. U.S. Nat. Mus. 97:157-342, figs. 4-53, charts 1-3.

Callipallene phantoma
Knapp, Franz 1973.
Pycnogonida from Pantellaria and Catania, Sicily.
Beaufortia 21 (277): 55-74.

Callipallene phantoma
Utinoma, Huzio, 1962
Pycnogonida of Sagami Bay - Supplement.
Publ. Seto Mar. Biol. Lab., 10(1) (Article 5) 91-194.

Callipallene phantoma ssp amaxana
Stock, Jan H., 1968.
Pycnogonida collected by the Galathea and Anton Bruun in the Indian and Pacific oceans.
Vidensk. Meddr. dansk naturh. Foren., 131: 7-65.

Calypaspycnon georgiae n. gen. & n.sp.
Hedgpeth, J.W., 1948
The Pycnogonida of the western North Atlantic and the Caribbean. Proc. U.S. Nat. Mus. 97:157-342, figs. 4-53, charts 1-3.

Chaetonymphon australe
Hodgson, T.V., 1907.
Pycnogonida. Nat. Antarctic Exped., 1901-1904, 3:72 pp., 10 pls.

Chaetonymphon biarticulatum
Hodgson, T.V., 1907.
Pycnogonida. Nat. Antarctic Exped., 1901-1904, 3:72 pp., 10 pls.

Chaetonymphon hirtipes
Stephensen, K., 1917
Zoogeographical investigations of certain fjords in southern Greenland with special reference to Crustacea, Pycnogonida and Echinodermata including a list of Alcyonaria and Pisces, Medd. om Grønland, 53(3):229-378.

Chaetonymphon hirtipes
Stephensen, K., 1912
Report on the Malacostraca, Pycnogonida and some Entomostraca collected by the Danmark-Expedition to North East Greenland. Medd. om Grønland, 45(11):501-630, Pls. 39-43.

Chaetonymphon mendosum
Hodgson, T.V., 1907.
Pycnogonida. Nat. Antarctic Exped., 1901-1904, 3:72 pp., 10 pls.

Chaetonymphon villosum
Hodgson, T.V., 1907.
Pycnogonida. Nat. Antarctic Exped., 1901-1904, 3:72 pp., 10 pls.

Cheilopallene trappa n. sp.
Clark, W.C. 1971.
Pycnogonida of the Snares Islands.
N.Z. Jl mar. Freshwat. Res. 5(2): 329-341.

Cilunculus acanthus n.sp.
Fry, William G. and Joel W. Hedgpeth, 1969.
The fauna of the Ross Sea. 7. Pycnogonida, I. Colossendeidae, Pycnogonidae, Endeidae, Ammotheidae. Bull. N.Z.D.S.I.R. 198 (Mem. N.Z. Oceanogr. Inst. 49): 139 pp.

Cilunculus armatus
Utinomi, H., 1955.
Report on the Pycnogonida collected by the Soyomaru on the continental shelf bordering Japan during 1926-1930.
Publ. Seto Mar. Biol. Lab., 5(1):1-42.
(Expedition made)

Cilunculus armatus
Utinomi, Huzio, 1952
Pycnogonida of Sagami Bay.
Publ. Seto Mar. Biol. Lab., 7(2):197-222.

Cilunculus cactoides n.sp.
Fry, William G. and Joel W. Hedgpeth, 1969.
The fauna of the Ross Sea. 7. Pycnogonida, I. Colossendeidae, Pycnogonidae, Endeidae, Ammotheidae. Bull. N.Z.D.S.I.R. 198 (Mem. N.Z. Oceanogr. Inst. 49): 139 pp.

Cilunculus hirsutus
Stock, Jan H., 1968.
Pycnogonida collected by the Galathea and Anton Bruun in the Indian and Pacific oceans.
Vidensk. Meddr dansk naturh. Foren., 131:7-65.

Cilunculus sewelli n.sp.
Calman, W. T., 1938
Pycnogonida. John Murray Exped. 1933-34, Sci. Repts. 5(6):147-166, 10 text figs.

Cilunculus sewelli
Stock, Jan H., 1968.
Pycnogonida collected by the Galathea and Anton Bruun in the Indian and Pacific oceans.
Vidensk. Meddr. dansk naturh. Foren., 131:7-65.

Colossendeis sp.
Fage, L., 1956.
Les pycnogonides (excl. le genre Nymphon).
Galathea Rept., 2:167-183.

Colossendeis spp.
Fry, William G. and Joel W. Hedgpeth, 1969.
The fauna of the Ross Sea. 7. Pycnogonida, I. Colossendeidae, Pycnogonidae, Endeidae, Ammotheidae. Bull. N.Z.D.S.I.R. 198 (Mem. N.Z. Oceanogr. Inst. 49): 139 pp.

Colossendeis sp.
Turpaeva, E.P., 1971.
The deep-water Pantopoda collected in the Kurile-Kamchatka Trench. (In Russian; English abstract). Trudy Inst. Okeanol. P.P. Shirshov 92: 274-291.

Colossendeis angusta
Calman, W. T., 1938
Pycnogonida. John Murray Exped. 1933-34, Sci. Repts. 5(6):147-166, 10 text figs.

Colossendeis angusta
Fry, William G. and Joel W. Hedgpeth, 1969.
The fauna of the Ross Sea. 7. Pycnogonida, I. Colossendeidae, Pycnogonidae, Endeidae, Ammotheidae. Bull. N.Z.D.S.I.R. 198 (Mem. N.Z. Oceanogr. Inst. 49): 139 pp.

Colossendeis angusta
Hedgpeth, J.W., 1948
The Pycnogonida of the western North Atlantic and the Caribbean. Proc. U.S. Nat. Mus. 97:157-342, figs.4-53, charts 1-3.

Colossendeis angusta
Meinert, Fr., 1899
Pycnogonida. Danish Ingolf Exped. 3(1): 71 pp., 5 pls., 2 text figs., 1 charts, list of stations.

Colossendeis angusta
Nesis, K.N., 1960.
Littoral Pantopoda of the eastern Murmansk coast.
Trudy Murmansk. Morsk. Biol. Inst., 2(6):137-161.

Colossendeis australis
Fry, William G. and Joel W. Hedgpeth, 1969.
The fauna of the Ross Sea. 7. Pycnogonida, I. Colossendeidae, Pycnogonidae, Endeidae, Ammotheidae. Bull. N.Z.D.S.I.R. 198 (Mem. N.Z. Oceanogr. Inst. 49): 139 pp.

Colossendeis australis
Gordon, I., 1932.
Pycnogonida. Discovery Repts., 6:1-138.

Colossendeis australis
Hodgson, T.V., 1907.
Pycnogonida. Nat. Antarctic Exped., 1901-1904, 3:72 pp., 10 pls.

Colossendeis australis
Stephensen, K., 1947
Tanaidacea, Isopoda, Amphipoda, and Pycnogonida. Sci. Res. Norweg. Antarctic Exp. 1927-1928 et seq. Date Norske Videnskaps-Akademi i Oslo, II, No.27:1-90, 26 text figs.

Colossendeis avidus n.sp.
Pushkin, A.F. 1970
New species of the genus Colossendeis (Pantopoda). (In Russian; English abstract).
Zool. Zh., 59(10): 1488-1496

Colossendeis bicornis n.sp.
Losina-Losinsky, L.K., and E.P. Turpaeva, 1958.
The genus Colossendeis (Pantopoda) in the northern Pacific Ocean.
Biull. Moskovsk. Obshsh. Isp. Prirody, Biol., 1:23-34.

Colossendeis brevitarsus n.sp.
Losina-Losinsky, L.K., and E.P. Turpaeva, 1958.
The genus Colossendeis (Pantopoda) in the northern Pacific Ocean.
Biull. Moskovsk. Obshsh. Isp. Prirody, Biol., 1:23-34.

Colossendeis bruuni n.sp.
Fage, L., 1956.
Les pycnogonides (excl. le genre Nymphon).
Galathea Rept. 2:167-183.

Colossendeis chitinosa
Losina-Losinsky, L.K., and E.P. Turpaeva, 1958.
The genus Colossendeis (Pantopoda) in the northern Pacific Ocean.
Biull. Moskovsk. Obshsh. Isp. Prirody, Biol., 1:23-34.

Colossendeis chitinosa
Stock, J. H. 1954.
Pycnogonida from Indo-West Pacific, Australian and New Zealand waters. Papers from Dr. Th. Mortensen's Pacific Expedition, 1914-1916. LXXVII.
Vidensk. Medd. Dansk Naturhist. Foren., København, 116: 1-168.

Colossendeis chitinosa
Utinomi, Huzio, 1962
Pycnogonida of Sagami Bay - Supplement.
Publ. Seto Mar. Biol. Lab., 10(1) (Article 5) 91-104.

Colossendeis chitinosa
Utinomi, H., 1955.
Report on the Pycnogonida collected by the Soyo-maru expeditions made on the continental shelf bordering Japan during the years 1926-1930.
Publ. Seto Mar. Biol. Lab., 5(1):1-42.

Colossendeis clavata
Hedgpeth, J.W., 1948
The Pycnogonida of the western North Atlantic and the Caribbean. Proc. U.S. Nat. Mus. 97:157-342, figs.4-53, charts 1-3.

Colossendeis clavata n.sp.
Meinert, Fr., 1899
Pycnogonida. Danish Ingolf Exped. 3(1): 71 pp., 5 pls., 2 text figs., 1 charts, list of stations.

Colossendeis colossea
Barnard, K.H., 1954.
III(3): South African Pycnogonida.
Ann. S. African Mus., 41:81-158, 34 textfigs.

Colossendeis colossea
Fage, L., 1956.
Les pycnogonides (excl. le genre Nymphon).
Galathea Rept., 2:167-183.

Colossendeis colossea
Fry, William G. and Joel W. Hedgpeth, 1969.
The fauna of the Ross Sea. 7. Pycnogonida, I. Colossendeidae, Pycnogonidae, Endeidae, Ammotheidae. Bull. N.Z.D.S.I.R. 198 (Mem. N.Z. Oceanogr. Inst. 49): 139 pp.

Collossendeis collossea
Henry, L.M., 1953.
The nervous system of the Pycnogonida.
Microentomology 18(1):16-36, figs.

Colossendeis colossea
Meinert, Fr., 1899
Pycnogonida. Danish Ingolf Exped. 3(1): 71 pp., 5 pls., 2 text figs., 1 charts, list of stations.

Colossendeis colossea
Turpaeva, E.P., 1971.
The deep-water Pantopoda collected in the Kurile-Kamchatka Trench. (In Russian; English abstract). Trudy Inst. Okeanol. P.P. Shirshov 92: 274-291.

Colossendeis cucurbita
Fage, L., 1956.
Les pycnogonides (excl. le genre Nymphon).
Galathea Rept., 2:167-183.

Colessendeis curtirostris n.sp.
Stock, J.H., 1963
South African deep-sea Pycnogonida, with descriptions of five new species.
Annals, South African Mus., 46(12):321-340.

Colossendeis dofleini
Utinomi, H., 1951.
On some pycnogonids from the sea around the Kii Peninsula.
Publ. Seto Mar. Biol. Sta., Kyoto Univ., 1(4):159-168.

Colossendeis dofleini
Utinomi, H., 1955.
Report on the Pycnogonida collected by the Soyo-maru expeditions made on the continental shelf bordering Japan during the years 1926-1930.
Publ. Seto Mar. Biol. Lab., 5(1):1-42.

Colossendeis drakei
Fry, William G. and Joel W. Hedgpeth, 1969.
The fauna of the Ross Sea. 7. Pycnogonida, I. Colossendeidae, Pycnogonidae, Endeidae, Ammotheidae. Bull. N.Z.D.S.I.R. 198 (Mem. N.Z. Oceanogr. Inst. 49): 139 pp.

Colossendeis drakei
Gordon, I., 1932.
Pycnogonida. Discovery Repts., 6:1-138.

Colossendeis gigas-leptorhynchus
Haswell, W.A., 1885.
On the Pycnogonida of the Australian coast. Proc. Linn. Soc., N.S. Wales, 9:1021-1034, Pls. 54-57.

Colossendeis frigida
Gordon, I., 1932.
Pycnogonida. Discovery Repts., 6:1-138.

Colossendeis frigida
Hodgson, T.V., 1907.
Pycnogonida. Nat. Antarctic Exped., 1901-1904, 3:72 pp., 10 pls.

Colossendeis geoffroyi
Stock, J.H., 1966.
Pycnogonida. Campagne de la Calypso au large des côtes Atlantiques de l'Amerique du Sud (1961-1962). Ann. Inst. Oceanogr., Monaco, n.s. 44: 385-406.

Colossendeis glacialis
Gordon, I., 1932.
Pycnogonida. Discovery Repts., 6:1-138.

Colossendeis glacialis
Hodgson, T.V., 1907.
Pycnogonida. Nat. Antarctic Exped., 1901-1904, 3:72 pp., 10 pls.

Colossendeis gracilis
Stock, Jan H., 1968.
Pycnogonida collected by the Galathea and Anton Bruun in the Indian and Pacific oceans.
Vidensk. Meddr dansk naturh. Foren.

Colessendeis gracilis
Stock, J.H., 1963
South African deep-sea Pycnogonida, with descriptions of five new species.
Annals, South African Mus., 46(12):321-340.

Colossendeis hoeki
Fry, William G. and Joel W. Hedgpeth, 1969.
The fauna of the Ross Sea. 7. Pycnogonida, I. Colossendeidae, Pycnogonidae, Endeidae, Ammotheidae. Bull. N.Z.D.S.I.R. 198 (Mem. N.Z. Oceanogr. Inst. 49): 139 pp.

Colossendeis japonica
Fage, L., 1956.
Les pycnogonides (excl. le genre Nymphon).
Galathea Rept. 2:167-183.

Colossendeis lilliei

Fry, William G. and Joel W. Hedgpeth, 1969.
The fauna of the Ross Sea. 7. Pycnogonida,
I. Colossendeidae, Pycnogonidae, Endeidae,
Ammotheidae. Bull. N.Z.D.S.I.R. 198 (Mem.
N.Z. Oceanogr. Inst. 49): 139 pp.

Colossendeis longirostris

Fry, William G. and Joel W. Hedgpeth, 1969.
The fauna of the Ross Sea. 7. Pycnogonida,
I. Colossendeidae, Pycnogonidae, Endeidae,
Ammotheidae. Bull. N.Z.D.S.I.R. 198 (Mem.
N.Z. Oceanogr. Inst. 49): 139 pp.

Colossendeis longirostris

Utinomi, Huzio, 1959
Pycnogonida of the Japanese Antarctic
Research Expeditions 1956-1958. Biol. Res.
Japan Antarctic Res. Exped., 8(Spec. Publ.
Seto Mar. Biol. Lab.): 12 pp.

Colossendeis macerrima

Barnard, K.H., 1954.
III(3):South African Pycnogonida.
Ann. S. African Mus. 41:81-158, 34 textfigs.

Colossendeis macerrima

Fage, L., 1956.
Les pycnogonides (excl. le genre Nymphon).
Galathea Rept., 2:167-183.

Colossendeis macerrima

Fry, William G. and Joel W. Hedgpeth, 1969.
The fauna of the Ross Sea. 7. Pycnogonida,
I. Colossendeidae, Pycnogonidae, Endeidae,
Ammotheidae. Bull. N.Z.D.S.I.R. 198 (Mem.
N.Z. Oceanogr. Inst. 49): 139 pp.

Colossendeis macerrima

Hedgpeth, J.W., 1948
The Pycnogonida of the western North
Atlantic and the Caribbean. Proc. U.S. Nat.
Mus. 97:157-342, figs.4-53, charts 1-3.

Colossendeis macerrima

Meinert, Fr., 1899
Pycnogonida. Danish Ingolf Exped. 3(1):
71 pp., 5 pls., 2 text figs., 1 charts, list of
stations.

Colessendeis macerrima

Stock, J.H., 1963
South African deep-sea Pycnogonida, with
descriptions of five new species.
Annals, South African Mus., 46(12):321-340.

Colossendeis macerrima

Turpaeva, E.P., 1971.
The deep-water Pantopoda collected in the
Kurile-Kamchatka Trench. (In Russian;
English abstract). Trudy Inst. Okeanol. P.P.
Shirshov 92: 274-291.

Colossendeis megalonyx

Fage, L., 1956.
Les pycnogonides (excl. le genre Nymphon).
Galathea Rept., 2:167-183.

Colossendeis megalonyx, var.

Fry, William G. and Joel W. Hedgpeth, 1969.
The fauna of the Ross Sea. 7. Pycnogonida,
I. Colossendeidae, Pycnogonidae, Endeidae,
Ammotheidae. Bull. N.Z.D.S.I.R. 198 (Mem.
N.Z. Oceanogr. Inst. 49): 139 pp.

Colossendeis megalonyx megalonyx

Arnaud, Françoise 1972.
Pycnogonides des Iles Kerguelen
(Sud Océan Indien): matériel nouveau et
révision des spécimens du Muséum
national d'Histoire naturelle de Paris.
Bull. Mus. natn. Hist. nat. Paris, (3)
65 (Zool. 51):801-815.

Colossendeis mica n.sp.

Pushkin, A.F. 1970
New species of the genus Colossendeis
(Pantopoda). (In Russian; English abstract).
Zool. Zh., 59(10): 1488-1496

Colossendeis michaelsarsi

Fage, L., 1956.
Les pycnogonides (excl. le genre Nymphon).
Galathea Rept., 2:167-183.

Colossendeis michaelsarsi

Hedgpeth, J.W., 1948
The Pycnogonida of the western North
Atlantic and the Caribbean. Proc. U.S. Nat.
Mus. 97:157-342, figs.4-53, charts 1-3.

Colossendeis minuta

Hedgpeth, J.W., 1948
The Pycnogonida of the western North
Atlantic and the Caribbean. Proc. U.S. Nat.
Mus. 97:157-342, figs.4-53, charts 1-3.

Colessendeis minuta

Stock, J.H., 1963
South African deep-sea Pycnogonida, with
descriptions of five new species.
Annals, South African Mus., 46(12):321-340.

Colessendeis nasuta

Utinomi, H., 1955.
Report on the Pycnogonida collected by the Soyo-
maru expeditions made on the continental shelf
bordering Japan during the years 1926-1930.
Publ. Seto Mar. Biol. Lab., 5(1):1-42.

Colessendeis oculifera n. sp.

Stock, J.H., 1963
South African deep-sea Pycnogonida, with
descriptions of five new species.
Annals, South African Mus., 46(12):321-340.

Colessendeis orcadensis

Stock, J.H., 1963
South African deep-sea Pycnogonida, with
descriptions of five new species.
Annals, South African Mus., 46(12):321-340.

Colossendeis orientalis n.sp.

Losina-Losinsky, L.K., and E.P. Turpaeva, 1958.
The genus Colossendeis (Pantopoda) in the northern part of the Pacific Ocean.
Biull. Moskovsk. Obshsh. Isp. Prirodа, Biol.,
1:23-34.

Colossendeis pennatum n. sp.

Pushkin, A.F. 1970
New species of the genus Colossendeis
(Pantopoda). (In Russian; English abstract).
Zool. Zh., 59(10): 1488-1496

Colossendeis proboscidea

Meinert, Fr., 1899
Pycnogonida. Danish Ingolf Exped. 3(1):
71 pp., 5 pls., 2 text figs., 1 charts, list of
stations.

Colossendeis proboscidea

Nesis, K.N., 1960.
Littoral Pantopoda of the eastern Murmansk
coast.
Trudy Murmansk. Morsk. Biol. Inst., 2(6):137-161

Colossendeis robusta

Fry, William G. and Joel W. Hedgpeth, 1969.
The fauna of the Ross Sea. 7. Pycnogonida,
I. Colossendeidae, Pycnogonidae, Endeidae,
Ammotheidae. Bull. N.Z.D.S.I.R. 198 (Mem.
N.Z. Oceanogr. Inst. 49): 139 pp.

Colossendeis rugosa

Hodgson, T.V., 1907.
Pycnogonida. Nat. Antarctic Exped., 1901-1904,
III 3:72 pp., 10 pls.

Colossendeis scoresbii n.sp.

Gordon, I., 1932.
Pycnogonida. Discovery Repts., 6:1-138.

Colossendeis scotti

Fry, William G. and Joel W. Hedgpeth, 1969.
The fauna of the Ross Sea. 7. Pycnogonida,
I. Colossendeidae, Pycnogonidae, Endeidae,
Ammotheidae. Bull. N.Z.D.S.I.R. 198 (Mem.
N.Z. Oceanogr. Inst. 49): 139 pp.

Colossendeis scotti

Gordon, I., 1932.
Pycnogonida. Discovery Repts., 6:1-138.

Colossendeis scotti

Stephensen, K., 1947
Tanaidacea, Isopoda, Amphipoda, and
Pycnogonida. Sci. Res. Norweg. Antarctic
Exp. 1927-1928 et seq. Date Norske Videns-
kaps-Akademi i Oslo, II, No.27:1-90, 26
text figs.

Colossendeis spei n.sp.

Pushkin, A.F. 1970
New species of the genus Colossendeis
(Pantopoda). (In Russian; English abstract).
Zool. Zh., 59(10): 1488-1496

Colossendeis stramenti n.sp.

Fry, William G. and Joel W. Hedgpeth, 1969.
The fauna of the Ross Sea. 7. Pycnogonida,
I. Colossendeidae, Pycnogonidae, Endeidae,
Ammotheidae. Bull. N.Z.D.S.I.R. 198 (Mem.
N.Z. Oceanogr. Inst. 49): 139 pp.

Colossendeis tenuissima n.sp.

Haswell, W.A., 1885.
On the Pycnogonida of the Australian coast.
Proc. Linn. Soc., N.S. Wales, 9:1021-1034, Pls.
54-57.

Colossendeis tortipalpis

Fry, William G. and Joel W. Hedgpeth, 1969.
The fauna of the Ross Sea. 7. Pycnogonida,
I. Colossendeidae, Pycnogonidae, Endeidae,
Ammotheidae. Bull. N.Z.D.S.I.R. 198 (Mem.
N.Z. Oceanogr. Inst. 49): 139 pp.

Colossendeis tortipalpis n.sp.

Gordon, I., 1932.
Pycnogonida. Discovery Repts., 6:1-138.

Colossendeis wilsoni

Fry, William G. and Joel W. Hedgpeth, 1969.
The fauna of the Ross Sea. 7. Pycnogonida,
I. Colossendeidae, Pycnogonidae, Endeidae,
Ammotheidae. Bull. N.Z.D.S.I.R. 198 (Mem.
N.Z. Oceanogr. Inst. 49): 139 pp.

Colossendeis wilsoni

Gordon, I., 1932.
Pycnogonida. Discovery Repts., 6:1-138.

Cordylochele brevicollis

Nesis, K.N., 1960.
Littoral Pantopoda of the eastern Murmansk
coast.
Trudy Murmansk. Morsk. Biol. Inst., 2(6):
137-161.

Cordylochele longicollis

Meinert, Fr., 1899
Pycnogonida. Danish Ingolf Exped. 3(1):
71 pp., 5 pls., 2 text figs., 1 charts, list of
stations.

Cordylochele longirostris

Hedgpeth, J.W., 1948
The Pycnogonida of the western North
Atlantic and the Caribbean. Proc. U.S. Nat.
Mus. 97:157-342, figs.4-53, charts 1-3.

Cordylochele malleolata
Meinert, Fr., 1899
Pycnogonida. Danish Ingolf Exped. 3(1): 71 pp., 5 pls., 2 text figs., 1 charts, list of stations.

Cordylochele malleolata
Nesis, K.N., 1960.
Littoral Pantopoda of the eastern Murmansk coast. Trudy Murmansk. Morsk. Biol. Inst., 2(6):137-161.

Decolopoda antarctica
Gordon, I., 1932.
Pycnogonida. Discovery Repts., 6:1-138.

Decolopoda australis
Arnaud, Francoise 1972.
Pycnogonids des îles Kerguelen (Sud Océan Indien): matériel nouveau et révision des spécimens du Museum national d'Histoire naturelle de Paris. Bull. Mus. natn. Hist. nat. Paris (3) 65 (Zool. 51):801-815

Decolopoda australis
Fry, William G. and Joel W. Hedgpeth, 1969.
The fauna of the Ross Sea. 7. Pycnogonida, I. Colossendeidae, Pycnogonidae, Endeidae, Ammotheidae. Bull. N.Z.D.S.I.R. 198 (Mem. N.Z. Oceanogr. Inst. 49): 139 pp.

Decolopoda australis
Gordon, I., 1932.
Pycnogonida. Discovery Repts., 6:1-138.

Decolopoda australis
Stephensen, K., 1947
Tanaidacea, Isopoda, Amphipoda, and Pycnogonida. Sci. Res. Norweg. Antarctic Exp. 1927-1928 et seq. Date Norske Videnskaps-Akademi i Oslo, II, No.27:1-90, 26 text figs.

Dodecolopoda mawsoni
Fry, William G. and Joel W. Hedgpeth, 1969.
The fauna of the Ross Sea. 7. Pycnogonida, I. Colossendeidae, Pycnogonidae, Endeidae, Ammotheidae. Bull. N.Z.D.S.I.R. 198 (Mem. N.Z. Oceanogr. Inst. 49): 139 pp.

Ecleipsothremma spinosa n.gen.
Fry, William G. and Joel W. Hedgpeth, 1969.
The fauna of the Ross Sea. 7. Pycnogonida, I. Colossendeidae, Pycnogonidae, Endeidae, Ammotheidae. Bull. N.Z.D.S.I.R. 198 (Mem. N.Z. Oceanogr. Inst. 49): 139 pp.

Endeis australis
Fage, L., 1952.
II. Pycnogonides. Missions du Bâtiment Polaire "Commandant Charcot", récoltes faites en Terre Adélie (1950) par Paul Tchernia, Bull. Mus. Nat. Hist. Nat., 2 ser., 24:180-186, 2 textfigs.

Endeis australis
Fry, William G. and Joel W. Hedgpeth, 1969.
The fauna of the Ross Sea. 7. Pycnogonida, I. Colossendeidae, Pycnogonidae, Endeidae, Ammotheidae. Bull. N.Z.D.S.I.R. 198 (Mem. N.Z. Oceanogr. Inst. 49): 139 pp.

Endeis australis
Gordon, I., 1932.
Pycnogonida. Discovery Repts., 6:1-138.

Endeis australis
Stephensen, K., 1947
Tanaidacea, Isopoda, Amphipoda, and Pycnogonida. Sci. Res. Norweg. Antarctic Exp. 1927-1928 et seq. Date Norske Videnskaps-Akademi i Oslo, II, No.27:1-90, 26 text figs.

Endeis biseriata n.sp.
Stock, Jan H., 1968.
Pycnogonida collected by the Galathea and Anton Brunn in the Indian and Pacific oceans. Vidensk. Meddr dansk naturh. Foren. 131: 7-65.

Endeis charybdea
Stock, Jan H. 1970.
The Pycnogonida collected off northwestern Africa during the cruise of the Meteor. Meteor-Forsch. Ergebn. (D) 5: 6-10.

Endeis charybdaea
Stock, Jan H., et Jacques Soyer, 1965.
Sur quelques pycnogonides rares de Banyuls-Sur-mer. Vie et Milieu, Bull. Lab. Arego. (B) 16(1):415-

Endeis clipeata
Barnard, K.H., 1954.
III(3):South African Pycnogonida. Ann. S. African Mus., 41:81-158, 34 textfigs.

Endeis clipeata
Stock, Jan H., 1965.
Pycnogonida from the southwest Indian Ocean. Beaufortia, 13(151):13-33.

Endeis flaccida
Stock, Jan H., 1968.
Pycnogonida collected by the Galathea and Anton Brunn in the Indian and Pacific oceans. Vidensk. Meddr dansk naturh. Foren. 131: 7-65.

Endeis meridionali
Arnaud, Françoise 1972 (1973).
Pycnogonides des récifs coralliens de Madagascar. 4. Colossendeidae, Phoxichilidiidae et Endeidae. Tethys 4(4): 953-960.

Endeis meridionalis
Clark W.C. 1973.
New species of Pycnogonida from New Britain and Tonga. Pacific Sci. 27(1): 28-33.

Endeis meridionalis
Stock, Jan H., 1965.
Pycnogonida from the southwest Indian Ocean. Beaufortia, 13(151):13-33.

Endeis mollis
Barnard, K.H., 1954.
III(3):South African Pycnogonida. Ann. S. African Mus., 41:81-158, 34 textfigs.

Endeis mollis
Bourdillon, A., 1954.
Contribution à l'étude des Pycnogonides de Tunisie. Sta. Oceanogr., Salammbo, Notes, 35: 8 pp., 1 pl.

Endeis mollis
Calman, W. T., 1938
Pycnogonida. John Murray Exped. 1933-34, Sci. Repts. 5(6):147-166, 10 text figs.

Endeis mollis
Calman, W.T., 1927.
Zoological results of the Cambridge Expedition to the Suez Canal. XXVIII. Report on the Pycnogonida Trans. Zool. Soc., London, 22:403-410, textfigs. 102-104.

Endeis mollis
Stock, Jan H., 1966.
Pycnogonida from West Africa. Atlantide Rep., 9:45-57.

Endeis mollis
Stock, Jan H., 1965.
Pycnogonida from the southwest Indian Ocean. Beaufortia, 13(151):13-33.

Endeis mollis
Stock, J.H., 1957.
Pantopoden aud dem Zoologischen Museum Hamburg. Mitt. Hamburgischen Zool. Mus. u. Inst., 55:81-106.

Endeis pauciporosa n. sp.
Stock, Jan H., 1970.
A new species of Endeis and other pycnogonid records from the Gulf of Aqaba. Bull. Zool. Mus. Amsterdam, 2(1): 1-4

Endeis procera
Stock, J. H. 1954.
Pycnogonida from Indo-West Pacific, Australian and New Zealand waters. Papers from Dr. Th. Mortensen's Pacific Expedition, 1914-1916. LXXVII. Vidensk. Medd. Dansk Naturhist. Foren., København, 116: 1-168.

Endeis spinosa
de Haro, Andrés, 1965.
Picnogonidos de la fauna español. Comunidad de picnogonidos en el alga parda Halopteris Scoperia (L.). Bol. R. Soc. Espanola Hist. Nat. (Biol.) 63: 213-218.

Endeis spinosa
Hedgpeth, J.W., 1948
The Pycnogonida of the western North Atlantic and the Caribbean. Proc. U.S. Nat. Mus. 97:157-342, figs.4-53, charts 1-3.

Endeis spinosa
Krapp, Franz 1973.
Pycnogonida from Pantelleria and Catania, Sicily. Beaufortia 21 (277):55-74.

Endeis spinosus
Lebour, M.V., 1945.
Notes on the Pycnogonida of Plymouth. J.M.B.A. 26:139-165, 7 textfigs.

Endeis spinosa
Stock, Jan H., 1966.
Pycnogonida from West Africa. Atlantide Rep., 9:45-57.

Endeis spinosa
Stock, J.H., 1957.
Pantopoden aud dem Zoologischen Museum Hamburg. Mitt. Hamburgischen Zool. Mus. u. Inst., 55:81-106.

Ephyrogymna circularis
Hedgpeth, J.W., 1948
The Pycnogonida of the western North Atlantic and the Caribbean. Proc. U.S. Nat. Mus. 97:157-342, figs.4-53, charts 1-3.

Eurycyde clitellaria
McClosky, L.R., 1967.
New and little-known benthic pycnogonids from North Carolina. J. nat. Hist., 1:119-132.

Eurycyde encantada n. sp.
Child, C. Allan, and Joel W. Hedgpeth, 1971.
Pycnogonida of the Galápagos Islands. J. nat. Hist. 5(6): 609-634.

Eurycyde extenuata n.sp.
Calman, W. T., 1938
Pycnogonida. John Murray Exped. 1933-34, Sci. Repts. 5(6):147-166, 10 text figs.

Eurycyde hispida
Stephensen, K., 1912
Report on the Malacostraca, Pycnogonida and some Entomostraca collected by the Danmark-Expedition to North East Greenland. Medd. om Grønland, 45(11):501-630, Pls. 39-43.

Eurycyde raphiaster
Fage, L., 1952.
Sur quelques Pycnogonides de Dakar. Bull. Mus. Nat. Hist. Nat., 2 ser., 24(6):530-533, 1 textfig.

Eurydyce raphiaster
Fage, Louis (posthumous) et Jan H. Stock, 1966.
Pycnogonides, Campagne de la Calypso aux Îles du Cap Vert (1959). Ann. Inst. Océanogr., Monaco, n.s., 44: 315-328.

Eurycyde raphiaster
Hedgpeth, J.W., 1948
The Pycnogonida of the western North Atlantic and the Caribbean. Proc. U.S. Nat. Mus. 97:157-342, figs.4-53, charts 1-3.

Halosoma robustum
Hedgpeth, J.W., 1948
The Pycnogonida of the western North Atlantic and the Caribbean. Proc. U.S. Nat. Mus. 97:157-342, figs.4-53, charts 1-3.

Hannonia typica
Barnard, K.H., 1954.
III(3):South African Pycnogonida. Ann. S. African Mus., 41:81-158, 34 textfigs.

Hemichela micrasterias n.gen.n.sp.
Stock, J. H. 1954.
Pycnogonida from Indo-West Pacific, Australian and New Zealand waters. Papers from Dr. Th. Mortensen's Pacific Expedition, 1914-1916. LXXVII. Vidensk. Medd. Dansk Naturhist. Foren., København, 116: 1-168.

Heterofragilia amica
Stock, J. H. 1954.
Pycnogonida from Indo-West Pacific, Australian and New Zealand waters. Papers from Dr. Th. Mortensen's Pacific Expedition, 1914-1916. LXXVII. Vidensk. Medd. Dansk Naturhist. Foren., København, 116: 1-168.

Heterofragilia amica
Utinomi, H., 1955.
Report on the Pycnogonida collected by the Soyamaru expeditions made on the continental shelf bordering Japan during 1926-1930. (the years)
Publ. Seto Mar. Biol. Lab., 5(1):1-42.

Heterofragilia fimbriata
Hedgpeth, J.W., 1948
The Pycnogonida of the western North Atlantic and the Caribbean. Proc. U.S. Nat. Mus. 97:157-342, figs.4-53, charts 1-3.

Heteronymphon abyssale
Turpaeva, E.P., 1970.
Belonging of Nymphon abyssale to the genus Heteronymphon (Pantopoda). (In Russian)
Zool. Zh. 49(11), 1723-1725.

Heteronymphon bioculatum
Turpaeva, E.P., 1971.
The deep-water Pantopoda collected in the Kurile-Kamchatka Trench. (In Russian; English abstract). Trudy Inst. Okeanol. P.P. Shirshov 92: 274-291.

Heteronymphon kempi n.gen., n.sp.
Gordon, I., 1932.
Pycnogonida. Discovery Repts., 6:1-138.

Heteronymphon profundum
Turpaeva, E.P., 1971.
The deep-water Pantopoda collected in the Kurile-Kamchatka Trench. (In Russian; English abstract). Trudy Inst. Okeanol. P.P. Shirshov 92: 274-291.

Lecythorhynchus hedgpethi
Utinomi, Huzio, 1962
Pycnogonida of Sagami Bay - Supplement. Publ. Seto Mar. Biol. Lab., 10(1) (Article 5): 91-104.

Lecythorhynchus hedgpethi n.sp.
Utinomi, Huzio, 1952
Pycnogonida of Sagami Bay. Publ. Seto Mar. Biol. Lab., 7(2):197-222.

Lecythorhynchus hilgendorfi
Utinomi, H., 1954.
The fauna of Akkeshi Bay. XIX. Littoral Pycnogonida. Publ. Akkeshi Mar. Biol. Sta. No. 3:1-28, 1 pl., 11 textfigs.

Contr. 221 Seto Mar. Biol. Lab.

Lecythorhynchus hilgendorfi syn. n.
Utinomi, Huzio, 1952
Pycnogonida of Sagami Bay. Publ. Seto Mar. Biol. Lab., 7(2):197-222.

Lecythorhynchus hilgendorfi
Utinomi, H., 1951.
On some pycnogonids from the sea around the Kii Peninsula. Publ. Seto Mar. Biol. Sta., Kyoto Univ., 1(4): 159-168.

Lecythorhynchus marginatum
Stock, J. H. 1954.
Pycnogonida from Indo-West Pacific, Australian and New Zealand waters. Papers from Dr. Th. Mortensen's Pacific Expedition, 1914-1916. LXXVII. Vidensk. Medd. Dansk Naturhist. Foren., København, 116: 1-168.

Leionymphon australe
Hodgson, T.V., 1907.
Pycnogonida. Nat. Antarctic Exped., 1901-1904, 3:72 pp., 10 pls.

Leionymphon glaciale
Hodgson, T.V., 1907.
Pycnogonida. Nat. Antarctic Exped., 1901-1904, 3:72 pp., 10 pls.

Leionymphon grande
Hodgson, T.V., 1907.
Pycnogonida. Nat. Antarctic Exped., 1901-1904, 3:72 pp., 10 pls.

Leionymphon minus
Hodgson, T.V., 1907.
Pycnogonida. Nat. Antarctic Exped., 1901-1904, 3:72 pp., 10 pls.

Leionymphon spinosum
Hodgson, T.V., 1907.
Pycnogonida. Nat. Antarctic Exped., 1901-1904, 3:72 pp., 10 pls.

Magnammothea gigantia n.gen.
Fry, William G. and Joel W. Hedgpeth, 1969.
The fauna of the Ross Sea. 7. Pycnogonida, I. Colossendeidae, Pycnogonidae, Endeidae, Ammotheidae. Bull. N.Z.D.S.I.R. 198 (Mem. N.Z. Oceanogr. Inst. 49): 139 pp.

Melloleitanianus candidoi n.gen.n.sp.
DeMello-Leitas, A.C. de G., 1955.
Novo Pantopoda da Baia de Guanabara (Melloleitanianus candidoi, Phoxichilidiidae) Rev. Biol. Mar., 5(1952):119-129.

Metapallene dubitans
Barnard, K.H., 1954.
III(3):South African Pycnogonida. Ann. S. African Mus. 41:81-158, 34 textfigs.

Nanymphon grasslei n.gen.n.sp.
McClosky, L.R., 1967.
New and little-known benthic pycnogonids from North Carolina. J. nat. Hist., 1:119-132.

Neopallene antipoda n.sp.
Stock, J. H. 1954.
Pycnogonida from Indo-West Pacific, Australian and New Zealand waters. Papers from Dr. Th. Mortensen's Pacific Expedition, 1914-1916. LXXVII. Vidensk. Medd. Dansk Naturhist. Foren., København, 116: 1-168.

Nymphon sp.
Gordon, I., 1932.
Pycnogonida. Discovery Repts., 6:1-138.

Nymphon abyssale n.sp.
Stock, Jan H., 1968.
Pycnogonida collected by the Galathea and Anton Bruun in the Indian and Pacific oceans. Vidensk. Meddr dansk naturh. Foren. 131: 7-65.

Nymphon sp.
Stock, J.H., 1966.
Pycnogonida. Campagne de la Calypso au large des côtes Atlantiques de l'Amerique du Sud (1961-1962). Ann. Inst. Océanogr. Monaco. n.s. 44: 385-496.

Nymphon abyssale
Turpaeva, E.P., 1970.
Belonging of Nymphon abyssale to the genus Heteronymphon (Pantopoda). (In Russian)
Zool. Zh. 49(11), 1723-1725.

Nymphon adami
Fage, L., 1956.
Sur deux espèces de Pycnogonides du Sierra Leone. Bull. Mus. Nat. Hist. Nat., (2)28(3):290-295.

Nymphon adami
Stock, Jan H., 1966.
Pycnogonida from West Africa. Atlantide. Rep. 9:45-57.

Nymphon adareanum
Fage, L., 1952.
Pycnogonides de la Terre Adélie. Échantillons rapportés par le Docteur Sapin-Jaloustre, Medecin-Biologiste de la Première Expédition en Terre Adélie (1949-1951). (Expéditions Polaires Françaises, Missions Paul-Émile Victor).
Bull. Mus. Nat. Hist. Nat., 2 ser., 24:263-273, 2 textfigs.

Nymphon adareanum
Fage, L., 1952.
II. Pycnogonides. Missions du Bâtiment Polaire "Commandant Charcot", récoltes faites en Terre Adélie (1950) par Paul Tchernia. Bull. Mus. Nat. Hist. Nat., 2 ser., 24:180-186, 2 textfigs.

Nymphon adareanum
Hedgpeth, J. W., 1950
Pycnogonida of the United States Navy Antarctic Expedition, 1947-48. Proc. U. S. Nat. Mus. 100 (3260):147-160, 19 text figs.

Nymphon adareanum
Hodgson, T.V., 1907.
Pycnogonida. Nat. Antarctic Exped., 1901-1904, 3:72 pp., 10 pls.

Nymphon aequidigitatum n.sp.
Haswell, W.A., 1885.
On the Pycnogonida of the Australian coast.
Proc. Linn. Soc., N.S. Wales, 9:1021-1034, Pls. 54-57.

Nymphon aequidigitatum
Stock, Jan H. 1973.
Pycnogonids from south-eastern Australia.
Beaufortia 20 (266), 99-127

Nymphon affinis
Barnard, K.H., 1954.
III(3): South African Pycnogonida.
Ann. S. African Mus., 41:81-158, 34 textfigs.

Nymphon andamanense
Calman, W. T., 1938
Pycnogonida. John Murray Exped. 1933-34,
Sci. Repts. 5(6):147-166, 10 text figs.

Nymphon angolense n.sp.
Gordon, I., 1932.
Pycnogonida. Discovery Repts., 6:1-138.

Nymphon arabicum n.sp.
Calman, W. T., 1938
Pycnogonida. John Murray Exped. 1933-34,
Sci. Repts. 5(6):147-166, 10 text figs.

Nymphon arabicum
Stock, Jan H., 1965.
Pycnogonida from the southwest Indian Ocean.
Beaufortia, 13(151):13-33.

Nymphon articulare
Gordon, I., 1932.
Pycnogonida. Discovery Repts., 6:1-138.

Nymphon australe
Barnard, K.H., 1954.
III(3): South African Pycnogonida.
Ann. S. African Mus., 41:81-158, 34 textfigs.

Nymphon australe
Fage, L., 1952.
II. Pycnogonides. Missions du Bâtiment Polaire "Commandant Charcot", récoltés faites en Terre Adélie (1950) par Paul Tchernia. Bull. Mus. Nat. Hist. Nat., 2 ser., 24:180-186, 2 textfigs.

Nymphon australe
Fage, L., 1952.
Pycnogonides de la Terre Adélie. Echantillons rapportés par le Docteur Sapin-Jaloustre, Medecin-Biologiste de la Première Expédition en Terre Adélie (1949-1951). (Expéditions Polaires Françaises, Missions Paul-Emile Victor).
Bull. Mus. Nat. Hist. Nat., 2 ser., 24:263-273, 2 textfigs.

Nymphon australe
Gordon, I., 1932.
Pycnogonida. Discovery Repts., 6:1-138.

Nymphon australe
Hedgpeth, J. W., 1950
Pycnogonida of the United States Navy Antarctic Expedition, 1947-48. Proc. U. S. Nat. Mus. 100 (3260):147-160, 19 text figs.

Nymphon australe
Stephensen, K., 1947
Tanaidacea, Isopoda, Amphipoda, and Pycnogonida. Sci. Res. Norweg. Antarctic Exp. 1927-1928 et seq. Date Norske Videns-kaps-Akademi i Oslo, II, No.27:1-90, 26 text figs.

Nymphon biarticulatum
Arnaud, Françoise 1972.
Pycnogonides des Îles Kerguelen (Sud Océan Indien): matériel nouveau et révision des spécimens du Muséum national d'Histoire naturelle de Paris.
Bull. Mus. natn. Hist. nat. Paris, (3) 65 (Zool. 51): 801-815.

Nymphon biarticulatum
Gordon, I., 1932.
Pycnogonida. Discovery Repts., 6:1-138.

Nymphon birsteini n.sp.
Turpajeva, E.P., 1955.
New Pantopoda sp of the Kurile-Kamchatka Trench.
Trudy Inst. Oceanol., 12:322-327.

Nymphon Bouvieri n.sp.
Gordon, I., 1932.
Pycnogonida. Discovery Repts., 6:1-138.

Nymphon bouvieri ?
Stephensen, K., 1947
Tanaidacea, Isopoda, Amphipoda, and Pycnogonida. Sci. Res. Norweg. Antarctic Exp. 1927-1928 et seq. Date Norske Videns-kaps-Akademi i Oslo, II, No.27:1-90, 26 text figs.

Nymphon braschnikour
Utinomi, H., 1955.
Report on the Pycnogonida collected by the Soyo-maru expeditions on the continental shelf bordering Japan during the years 1926-1930.
Publ. Seto Mar. Biol. Lab., 5(1):1-42.
(mad?)

Nymphon brevicaudatum
Arnaud, Françoise 1972.
Pycnogonides des Îles Kerguelen (Sud Océan Indien): matériel nouveau et révision des spécimens du Muséum national d'Histoire naturelle de Paris.
Bull. Mus. natn. Hist. nat. Paris, (3) 65 (Zool. 51): 801-815.

Nymphon brevicaudatum
Gordon, I., 1932.
Pycnogonida. Discovery Repts., 6:1-138.

Nymphon brevirostrum
Lebour, M.V., 1945.
Notes on the Pycnogonida of Plymouth. J.M.B.A. 26:139-165, 7 textfigs.

Nymphon brevirostris glaciale
Nesis, K.N., 1960.
Littoral Pantopoda of the eastern Murmansk coast.
Trudy Murmansk. Morsk. Biol. Inst., 2(6):137-161.

Nymphon brevitarse
Meinert, Fr., 1899
Pycnogonida. Danish Ingolf Exped. 3(1): 71 pp., 5 pls., 2 text figs., 1 charts, list of stations.

Nymphon calypso
Stock, Jan H., 1966.
Pycnogonida from West Africa.
Atlantide Rep. 9:45-57.

Nymphon capense? = phasmatodes
Barnard, K.H., 1954.
III(3):South African Pycnogonida.
Ann. S. African Mus. 41:81-158, 34 textfigs.

Nymphon charcoti
Gordon, I., 1932.
Pycnogonida. Discovery Repts., 6:1-138.

Nymphon chaetochir n. sp.
Utinomi, Hugio 1971.
Nymphon chaetochir n.sp., a new deepsea pycnogonid from Chatham Rise, east of New Zealand.
J. R. Soc. N.Z. 1 (3/4): 197-202.

Nymphon clarencei n.sp.
Gordon, I., 1932.
Pycnogonida. Discovery Repts., 6:1-138.

Nymphon comes
Barnard, K.H., 1954.
III(3):South African Pycnogonida.
Ann. S. African Mus., 41:81-158, 34 textfigs.

Nyphon conirostrum n.sp.
Stock, Jan H. 1973.
Pycnogonids from south-eastern Australia.
Beaufortia 20 (266), 99-127

Nymphon coxi
Stock, Jan H., 1968.
Pycnogonida collected by the Galathea and Anton Bruun in the Indian and Pacific oceans.
Vidensk. Meddr. dansk naturh. Foren., 131: 7-65.

Nymphon crenatiunguis
Barnard, K.H., 1954.
III(3): South African Pycnogonida.
Ann. S. African Mus. 41:81-158, 34 textfigs.

Nymphon crenatiunguis n.sp.
Barnard, K.H., 1946.
Diagnoses of new species and a new genus of Pycnogonida in the South African Museum.
Ann. Mag. Nat. Hist., (11)13:60-63.

Nymphon crosnieri n. sp.
Stock, Jan H., 1965.
Pycnogonida from the southwest Indian Ocean.
Beaufortia, 13 (151):13-33.

Nymphopsis deomedaria n.sp.
McClosky, L.R., 1967.
New and little-known benthic pycnogonids from North Carolina.
J. nat. Hist., 1:119-132.

Nymphon distensum
Barnard, K.H., 1954.
III(3): South African Pycnogonica.
Ann. S. African Mus., 41:81-158, 34 textfigs.

Nymphon dubitabile
Stock, Jan H. 1973.
Pycnogonids from south-eastern Australia.
Beaufortia 20 (266), 99-127

Nymphon elegans
Meinert, Fr., 1899
Pycnogonida. Danish Ingolf Exped. 3(1): 71 pp., 5 pls., 2 text figs., 1 charts, list of stations.

Nymphon falcatum n.sp.
Utinomi, H., 1955.
Report on the Pycnogonida collected by the Soyo-maru expeditions made on the continental shelf bordering Japan during the years 1926-1930.
Publ. Seto Mar. Biol. Lab., 5(1):1-42.

Nymphon elongatum
Stock, J. H. 1954.
Pycnogonida from Indo-West Pacific, Australian and New Zealand waters. Papers from Dr. Th. Mortensen's Pacific Expedition, 1914-1916. LXXVII.
Vidensk. Medd. Dansk Naturhist. Foren., København, 116: 1-168.

Nymphon femorale n.sp.
Fage, L., 1956.
Les pycnogonides du genre Nymphon.
Galathea Repts., 2:159-165.

Nymphon femorale
Stock, Jan H., 1968.
Pycnogonida collected by the Galathea and Anton Bruun in the Indian and Pacific oceans.
Vidensk. Meddr. dansk naturh. Foren., 131:7-65.

Nymphon foresti n.sp.
Fage, L., 1953.
Duex pycnogonides nouveaux de la cote occidentale d'Afrique. Bull. Mus. Nat.YHist. Nat., 2e ser., 25:376-382, 4 textfigs.

Nymphon foxi
Calman, W. T., 1938
Pycnogonida. John Murray Exped. 1933-34, Sci. Repts. 5(6):147-166, 10 text figs.

Nymphon foxi n.sp.
Calman, W.T., 1927.
Zoological results of the Cambridge Expedition to the Suez Canal. XXVIII. Report on the Pycnogonida
Trans. Zool. Soc., London, 22: 403-410, textfigs. 102-104.

Nymphon foxi
Stock, Jan H., 1968.
Pycnogonida collected by the Galathea and Anton Bruun in the Indian and Pacific oceans.
Vidensk, Meddr dansk naturh. Foren. 131: 7-65.

Nymphon foxi
Stock, J.H., 1957.
Pantopoden aus dem Zoologischen Museum Hamburg.
Mitt. Hamburgischen Zool. Mus. u. Inst., 55:81-106.

Nymphon frigidum
Hodgson, T.V., 1907.
Pycnogonida. Nat. Antarctic Exped., 1901-1904, 3:72 pp., 10 pls.

Nymphon galatheae n.sp.
Fage, L., 1956.
Les pycnogonides du genre Nymphon.
Galathea Repts.,, 2:159-165.

Nymphon giltayi n.sp.
Hedgpeth, J.W., 1948
The Pycnogonida of the western North Atlantic and the Caribbean. Proc. U.S. Nat. Mus. 97:157-342, figs.4-53, charts 1-3.

Nymphon giraffa
Utinomi, Huzio, 1962
Pycnogonida of Sagami Bay - Supplement.
Publ. Seto Mar. Biol. Lab., 10(1) (Article 5): 91-104.

Nymphon glaciale
Giltray, L., 1942.
New records of Pycnogonida from the Canadian Atlantic coast. J. Fish. Res. Bd., Canada, 5(5): 459-460.

Nymphon gracile
Bourdillon, A., 1954.
Les pycnogonides de Marseille et ses environs.
Recueil Trav. Stat. Mar. Endoume, 12:145-158.

Nymphon gracile
King, P.E. and J.H. Jarvis, 1970.
Egg development in a littoral pycnogonid Nymphon gracile. Mar. Biol., 7(4): 294-304.

Nymphon gracile
Lebour, M.V., 1945.
Notes on the Pycnogonida of Plymouth. J.M.B.A. 26:139-165, 7 textfigs.

Nymphon gracile
Morgan, Elfed, A. Nelson-Smith and E. W. Knight-Jones, 1964.
Responses of Nymphon gracile (Pycnogonida) to pressure cycles of tidal frequency.
J. Exp. Biol., 41(4):825-836.

Nymphon gracilipes
Arnaud, Francoise 1972.
Pycnogonides des îles Kerguelen (Sud Océan Indien): matériel nouveau et révision des spécimens du Muséum national d'Histoire naturelle de Paris.
Bull. Mus. natn. Hist. nat. Paris, (3) 65 (Zool. 51):801-815.

Nymphon gracillimum
Gordon, I., 1932.
Pycnogonida. Discovery Repts., 6:1-138.

Nymphon groenlandica n.sp.
Meinert, Fr., 1899
Pycnogonida. Danish Ingolf Exped. 3(1): 71 pp., 5 pls., 2 text figs., 1 charts, list of stations.

Nymphon grossipes
Hedgpeth, J.W., 1948
The Pycnogonida of the western North Atlantic and the Caribbean. Proc. U.S. Nat. Mus. 97:157-342, figs.4-53, charts 1-3.

Nymphon grossipes
Meinert, Fr., 1899
Pycnogonida. Danish Ingolf Exped. 3(1): 71 pp., 5 pls., 2 text figs., 1 charts, list of stations.

Nymphon grossipes
Nesis, K.N., 1960.
Littoral Pantopoda of the eastern Murmansk coast.
Trudy Murmansk. Morsk. Biol. Inst., 2(6):137-161.

Nymphon grossipes
Stephensen, K., 1917
Zoogeographical investigations of certain fjords in southern Greenland with special reference to Crustacea, Pycnogonida and Echinodermata including a list of Alcyonaria and Pisces, Medd. om Grønland, 53(3):229-378.

Nymphon Grossipes
Turpaeva, E.P., 1971.
The deep-water Pantopoda collected in the Kurile-Kamchatka Trench. (In Russian; English abstract). Trudy Inst. Okeanol. P.P. Shirshov 92: 274-291.

Nymphon grossipes
Utinomi, H., 1955.
Report on the Pycnogonida collected by the Soyomaru expeditions made on the continental shelf bordering Japan during the years 1926-1930.
Publ. Seto Mar. Biol. Lab., 5(1):1-42.

Nymphon grossipes
Utinomi, Huzio, 1952
Pycnogonida of Sagami Bay.
Publ. Seto Mar. Biol. Lab., 7(2):197-222.

Nymphon gruveli
Stock, Jan H., 1966.
Pycnogonida from West Africa.
Atlentide Rep., 9:45-57.

Nymphon hiemale
Gordon, I., 1932.
Pycnogonida. Discovery Repts., 6:1-138.

Nymphon haemale
Hodgson, T.V., 1907.
Pycnogonida. Nat. Antarctic Exped., 1901-1904, 3:72 pp., 10 pls.

Nymphon hiemale
Stephensen, K., 1947
Tanaidacea, Isopoda, Amphipoda, and Pycnogonida. Sci. Res. Norweg. Antarctic Exp. 1927-1928 et seq. Date Norske Videnskaps-Akademi i Oslo, II, No.27:1-90, 26 text figs.

Nymphon hirtipes
Bourdillon, A., 1955.
Les Pycnogonides de la croisière 1951 du "President Theodore Tissier".
Rev. Trav. Inst. Pêches Marit., 19(4):581-609.

Nymphon hirtipes
Hedgpeth, J.W., 1948
The Pycnogonida of the western North Atlantic and the Caribbean. Proc. U.S. Nat. Mus. 97:157-342, figs.4-53, charts 1-3.

Nymphon hirtum
Nesis, K.N., 1960.
Littoral Pantopoda of the eastern Murmansk coast.
Trudy Murmansk. Morsk. Biol. Inst., 2(6):137-161.

Nymphon hodgsoni
Turpaeva, E.P., 1971.
The deep-water Pantopoda collected in the Kurile-Kamchatka Trench. (In Russian; English abstract). Trudy Inst. Okeanol. P.P. Shirshov 92: 274-291.

Nymphon hoeckii n.sp.
Meinert, Fr., 1899
Pycnogonida. Danish Ingolf Exped. 3(1): 71 pp., 5 pls., 2 text figs., 1 charts, list of stations.

Nymphon immane n.sp.
Stock, J. H. 1954.
Pycnogonida from Indo-West Pacific, Australian and New Zealand waters. Papers from Dr. Th. Mortensen's Pacific Expedition, 1914-1916. LXXVII.
Vidensk. Medd. Dansk Naturhist. Foren., København, 116: 1-168.

Nymphon inerme n.sp.
Fage, F., 1956.
Les pycnogonides du genre Nymphon.
Galathea Rept., 2:159-165.

Nymphon japonicum
Utinomi, Huzio, 1962
Pycnogonida of Sagami Bay - Supplement.
Publ. Seto Mar. Biol. Lab., 10(1) (Article 5): 91-104.

Nymphon japonicum
Utinomi, H., 1955.
Report on the Pycnogonida collected by the Soyomaru expeditions made on the continental shelf duringxx bordering Japan during the years 1926-1930. Publ. Seto Mar. Biol. Lab., 5(1):1-42.

Nymphon japonicum
Utinomi, Huzio, 1952
Pycnogonida of Sagami Bay.
Publ. Seto Mar. Biol. Lab., 7(2):197-222.

Nymphon japonicus
Utinomi, H., 1951.
On some pycnogonids from the sea around the Kii Peninsula.
Publ. Seto Mar. Biol. Sta., Kyoto Univ., 1(4):159-168

Nymphon kodanii
Stock, J. H. 1954.
Pycnogonida from Indo-West Pacific, Australian and New Zealand waters. Papers from Dr. Th. Mortensen's Pacific Expedition, 1914-1916. LXXVII. Vidensk. Medd. Dansk Naturhist. Foren., København, 116: 1-168.

Nymphon Kodanii
Turpaeva, E.P., 1971.
The deep-water Pantopoda collected in the Kurile-Kamchatka Trench. (In Russian; English abstract). Trudy Inst. Okeanol. P.P. Shirshov 92: 274-291.

Nymphon kodanii
Utinomi, Huzio, 1962
Pycnogonida of Sagami Bay - Supplement. Publ. Seto Mar. Biol. Lab., 10(1) (Article 5) 91-104.

Nymphon kodanii
Utinomi, H., 1955.
Report on the Pycnogonida collected by the Soyo-maru expeditions made on the continental shelf bordering Japan during the years 1926-1930. Publ. Seto Mar. Biol. Lab., 5(1):1-42.

Nymphon kurilo kamchaticum n. sp.
Turpaeva, E.P., 1971.
The deep-water Pantopoda collected in the Kurile-Kamchatka Trench. (In Russian; English abstract). Trudy Inst. Okeanol. P.P. Shirshov 92: 274-291.

Nymphon lanare
Hodgson, T.V., 1907.
Pycogonida. Nat. Antarctic Exped., 1901-1904, 3:72 pp., 10 pls.

Nymphon laterospinum n. sp.
Stock, J.H., 1963.
South African deep-sea Pycnogonida, with descriptions of five new species. Annals. South African Mus., 46(12):321-340.

Nymphon leptocheles
Meinert, Fr., 1899
Pycnogonida. Danish Ingolf Exped. 3(1): 71 pp., 5 pls., 2 text figs., 1 charts, list of stations.

Nymphon lobatum
Stock, Jan H., 1966.
Pycnogojida from West Africa. Atlantide Rep., 9:45-57.

Nymphon lobatum n.sp.
Stock, J.H., 1962.
Second list of Pycnogonida of the University of Cape Town ecological survey. Trans. R. Soc., S. Africa, 36(4):273-286.

Nymphon longicaudatum
Stock, Jan H., 1968.
Pycnogonida collected by the Galathea and Anton Bruun in the Indian and Pacific oceans. Vidensk. Meddr dansk naturh. Foren., 131: 7-65.

Nymphon longicoxa
Clark, W.C., 1958.
Some Pycnogonida from Cook Strait, New Zealand, with descriptions of two new species. Victoria Univ., Wellington, Zool. Publ. 23:7 pp.

Nymphon longitarse
Hedgpeth, J.W., 1948
The Pycnogonida of the western North Atlantic and the Caribbean. Proc. U.S. Nat. Mus. 97:157-342, figs.4-53, charts 1-3.

Nymphon longitarse
Meinert, Fr., 1899
Pycnogonida. Danish Ingolf Exped. 3(1): 71 pp., 5 pls., 2 text figs., 1 charts, list of stations.

Nymphon longitarse
Nesis, K. N., 1960.
[Littoral Pantopoda of the eastern Murmansk coast.] Trudy Murmansk. Morsk. Biol. Inst., 2(6):137-161.

Nymphon longitarse
Stephensen, K., 1917
Zoogeographical investigations of certain fjords in southern Greenland with special reference to Crustacea, Pycnogonida and Echinodermata including a list of Alcyonaria and Pisces, Medd. om Grønland, 53(3):229-378.

Nymphon longitarse caecum n. subsp.
Turpaeva, E.P., 1971.
An addition to the pantopod fauna of deep-sea trenches in the northwestern part of the Pacific Ocean. (In Russian; English abstract). Trudy Inst. Okeanol. P.P. Shirshov 92: 292-297.

Nymphon macronyx
Meinert, Fr., 1899
Pycnogonida. Danish Ingolf Exped. 3(1): 71 pp., 5 pls., 2 text figs., 1 charts, list of stations.

Nymphon macronyx
Nesis, K. N., 1960.
[Littoral Pantopoda of the eastern Murmansk coast.] Trudy Murmansk. Morsk. Biol. Inst., 2(6):137-161.

Nymphon macrum
Meinert, Fr., 1899
Pycnogonida. Danish Ingolf Exped. 3(1): 71 pp., 5 pls., 2 text figs., 1 charts, list of stations.

Nymphon macrum
Nesis, K.N., 1960.
[Littoral Pantopoda of the eastern Murmansk coast.] Trudy Murmansk. Morsk. Biol. Inst., 2(6):137-161.

Nymphon macrum
Stephensen, K., 1917
Zoogeographical investigations of certain fjords in southern Greenland with special reference to Crustacea, Pycnogonida and Echinodermata including a list of Alcyonaria and Pisces, Medd. om Grønland, 53(3):229-378.

Nymphon maculatum
Stock, J.H., 1964.
Report on the Pycnogonida of the Israel South Red Sea Expedition. Sea Fish Res. Inst., Israel, Bull., No. 35:27-34.

Nymphon maoriana n.sp.
Clark, W.C., 1958.
Some Pycnogonida from Cook Strait, New Zealand, with descriptions of two new species. Victoria Univ., Wellington, Zool. Publ., 23:7 pp.

Nymphon megalops
Meinert, Fr., 1899
Pycnogonida. Danish Ingolf Exped. 3(1): 71 pp., 5 pls., 2 text figs., 1 charts, list of stations.

Nymphon mendosum
Fage, L., 1952.
Pycnogonides de la Terre Adélie. Echantillons rapportés par le Docteur Sapin-Jaloustre, Medecin-Biologiste de la Première Expédition en Terre Adélie (1949-1951). (Expéditions Polaires Francaises, Missions Paul-Emile Victor). Bull. Mus. Nat. Hist. Nat., 2 ser., 24:263-273, 2 text figs.

Nymphon mendosum ?
Stephensen, K., 1947
Tanaidacea, Isopoda, Amphipoda, and Pycnogonida. Sci. Res. Norweg. Antarctic Exp. 1927-1928 et seq. Date Norske Videnskaps-Akademi i Oslo, II, No.27:1-90, 26 text figs.

Nymphon Microctenatus
Barnard, K.H., 1954.
III(3):South African Pycnogonida. Ann. S. African Mus. 41:81-158, 34 textfigs.

Nymphon microctenatus n.sp.
Barnard, K.H., 1946.
Diagnoses of new species and a new genus of Pycnogonida in the South African Museum. Ann. Mag. Nat. Hist. (11)13:60-63.

Nymphon micropedes
Utinomi, H., 1955.
Report on the Pycnogonida collected by the Soyo-maru expeditions made on the continental shelf bordering Japan during the years 1926-1930. Publ. Seto Mar. Biol. Lab., 5(1):1-42.

Nymphon mixtum
Nesis, K.N., 1960.
[Littoral Pantopoda of the eastern Murmansk coast.] Trudy Murmansk. Morsk. Biol. Inst., 2(6):137-161.

Nymphon mixtum
Stephensen, K., 1912
Report on the Malacostraca, Pycnogonida and some Entomostraca collected by the Danmark-Expedition to North East Greenland. Medd. om Grønland, 45(11):501-630, Pls. 39-43.

Nymphon molleri
Stock, Jan H. 1973.
Pycnogonida from south-eastern Australia. Beaufortia 20 (266):99-127

Nymphon cf. molleri
Stock, Jan H., 1968.
Pycnogonida collected by the Galathea and Anton Brunn in the Indian and Pacific oceans. Vidensk. Meddr dansk naturh. Foren. 131: 7-65.

Nymphon multidens n.sp.
Gordon, I., 1932.
Pycnogonida. Discovery Repts., 6:1-138.

Nymphon multidens
Hedgpeth, J. W., 1950
Pycnogonida of the United States Navy Antarctic Expedition, 1947-48. Proc. U.S. Nat. Mus. 100 (3260):147-160, 19 text figs.

Nymphon multidens
Stephensen, K., 1947
Tanaidacea, Isopoda, Amphipoda, and Pycnogonida. Sci. Res. Norweg. Antarctic Exp. 1927-1928 et seq. Date Norske Videnskaps-Akademi i Oslo, II, No.27:1-90, 26 text figs.

Nymphon natalense
Barnard, K.H., 1954.
III(3):South African Pycnogonida. Ann. S. African Mus., 41:81-158, 34 textfigs.

Nymphon neumayri n.sp.
Gordon, I., 1932.
Pycnogonida. Discovery Repts., 6:1-138.

Nymphon novaehollandiae
Stock, Jan H., 1968.
Pycnogonida collected by the Galathea and Anton Brunn in the Indian and Pacific oceans. Vidensk. Meddr dansk naturh. Foren. 131: 7-65.

Nymphon nugex n.sp.
Stock, Jan H., 1966.
Pycnogonida from West Africa.
Atlantide Rep., 9:45-57.

Nymphon orcadense
Gordon, I., 1932.
Pycnogonida. Discovery Repts., 6:1-138.

Nymphon ortmanni
Stock, J. H. 1954.
Pycnogonida from Indo-West Pacific, Australian and New Zealand waters. Papers from Dr. Th. Mortensen's Pacific Expedition, 1914-1916. LXXVII.
Vidensk. Medd. Dansk Naturhist. Foren., København, 116: 1-168.

Nymphon ortmanni
Utinomi, Huzio, 1962
Pycnogonida of Sagami Bay - Supplement.
Publ. Seto Mar. Biol. Lab., 10(1) (Article 5): 91-104.

Nymphon ortmanni
Utinomi, H., 1955.
Report on the Pycnogonida collected by the Soyo-maru expeditions made on the continental shelf bordering Japan during the years 1926-1930.
Publ. Seto Mar. Biol. Lab., 5(1):1-42.

Nymphon paucidens n.sp.
Gordon, I., 1932.
Pycnogonida. Discovery Repts., 6:1-138.

Nymphon pfefferi
Gordon, I., 1932.
Pycnogonida. Discovery Rept., 6:1-138.

Nymphon pilosum
Stock, Jan H., 1968.
Pycnogonida collected by the Galathea and Anton Bruun in the Indian and Pacific oceans.
Vidensk. Meddr dansk naturh. Foren., 131: 7-65.

Nymphon pixellae
Henry, L.M., 1953.
The nervous system of the Pycnogonidae.
Microentomology 18(1):16-36, figs.

Nymphon pleodon, n.sp.
Stock, J.H., 1962.
Second list of Pycnogonida of the University of Cape Town ecological survey.
Trans. R. Soc., S. Africa, 36(4):273-286.

Nymphon primacoxa n.sp.
Stock, Jan H., 1968.
Pycnogonida collected by the Galathea and Anton Bruun in the Indian and Pacific oceans.
Vidensk. Meddr. dansk naturh. Foren., 131: 7-65.

Nymphon proceroides
Gordon, I., 1932.
Pycnogonida. Discovery Repts., 6:1-138.

Nymphon procerum
Fage, L., 1951.
Sur un pycnogonide de l'expedition suedoise des grands fonds 1947-48. Repts. Swedish Deep-Sea Exped. 2, Zool. (7):95-97, 2 textfigs.

Nymphon procerum
Turpaeva, E.P., 1971.
The deep-water Pantopoda collected in the Kurile-Kamchatka Trench. (In Russian; English abstract). Trudy Inst. Okeanol. P.P. Shirshov 92: 274-291.

Nymphon prolatum
Stock, Jan H., 1966.
Pycnogonida from West Africa.
Atlantide Rep., 9:45-57.

Nymphon proximum
Utinomi, Huzio, 1959
Pycnogonida of the Japanese Antarctic Research Expeditions 1956-1958. Biol. Res. Japan Antarctic Res. Exped., 8(Spec. Publ. Seto Mar. Biol. Lab.): 12 pp.

Nymphon puellula n.sp.
Knapp, Franz 1973.
Pycnogonida from Pantellaria and Catania Sicily.
Beaufortia 21 (277): 55-74.

Nymphon residuum n.sp.
Stock, Jan H. 1972.
A new deep-sea Nymphon (Pycnogonida) collected by the Galathea off Kenya.
Zool. Mededel. Leiden 47: 257-260

Nymphon robustum
Meinert, Fr., 1899
Pycnogonida. Danish Ingolf Exped. 3(1): 71 pp., 5 pls., 2 text figs., 1 charts, list of stations.

Nymphon rubrum
Hedgpeth, J.W., 1948
The Pycnogonida of the western North Atlantic and the Caribbean. Proc. U.S. Nat. Mus. 97:157-342, figs.4-53, charts 1-3.

Nymphon rubrum
Lebour, M.V., 1945.
Notes on the Pycnogonida of Plymouth. J.M.B.A. 26:139-165, 7 textfigs.

Nymphon rubrum
Nesis, K.N., 1960.
[Littoral Pantopoda of the eastern Murmansk coast.]
Trudy Murmansk. Morsk. Biol. Inst., 2(6):137-161

Nymphon sarsii n.sp.
Meinert, Fr., 1899
Pycnogonida. Danish Ingolf Exped. 3(1): 71 pp., 5 pls., 2 text figs., 1 charts, list of stations.

Nymphon serratum
Giltray, L., 1942.
New records of Pycnogonida from the Canadian Atlantic coast. J. Fish. Res. Bd., Canada, 5(5): 459-460.

Nymphon serratum
Meinert, Fr., 1899
Pycnogonida. Danish Ingolf Exped. 3(1): 71 pp., 5 pls., 2 text figs., 1 charts, list of stations.

Nymphon serratum
Nesis, K.N., 1960.
[Littoral Pantopoda of the eastern Murmansk coast.]
Trudy Murmansk. Morsk. Biol. Inst., 2(6):137-161

Nymphon serratum
Stephensen, K., 1912
Report on the Malacostraca, Pycnogonida and some Entomostraca collected by the Danmark-Expedition to North East Greenland. Medd. om Grønland, 45(11):501-630, Pls. 39-43.

Nymphon setimanus
Barnard, K.H., 1954.
III(3):South African Pycnogonida.
Ann. S. African Mus., 41:81-158, 34 textfigs.

Nymphon setimanus n.sp.
Barnard, K.H., 1946.
Diagnoses of new species and a new genus of Pycnogonida in the South African Museum.
Ann. Mag. Nat. Hist. (11)13:60-63.

Nymphon setimanus
Stock, Jan H., 1965.
Pycnogonida from the southwest Indian Ocean.
Beaufortia, 13(151):13-33.

Nymphon sluiteri
Meinert, Fr., 1899
Pycnogonida. Danish Ingolf Exped. 3(1): 71 pp., 5 pls., 2 text figs., 1 charts, list of stations.

Nymphon sluiteri
Nesis, K.N., 1960.
[Littoral Pantopoda of the eastern Murmansk coast.]
Trudy Murmansk. Morsk. Biol. Inst., 2(6):137-161

Nymphon signatum
Barnard, K.H., 1954.
III(3):South African Pycnogonida.
Ann. S. African Mus., 41:81-158, 34 textfigs.

Nymphon singulare n.sp.
Stock, J. H. 1954.
Pycnogonida from Indo-West Pacific, Australian and New Zealand waters. Papers from Dr. Th. Mortensen's Pacific Expedition, 1914-1916. LXXVII.
Vidensk. Medd. Dansk Naturhist. Foren., København, 116: 1-168.

Nymphon soyae n.sp.
Utinomi, Huzio, 1959
Pycnogonida of the Japanese Antarctic Research Expeditions 1956-1958. Biol. Res. Japan Antarctic Res. Exped., 8(Spec. Publ. Seto Mar. Biol. Lab.): 12 pp/

Nymphon soyoi n.sp.
Utinomi, H., 1955.
Report on the Pycnogonida collected by the Soyo-maru expeditions made on the continental shelf bordering Japan during the years 1926-1930.
Publ. Seto Mar. Biol. Lab., 5(1):1-42.

Nymphon spinossimum
Hedgpeth, J.W., 1948
The Pycnogonida of the western North Atlantic and the Caribbean. Proc. U.S. Nat. Mus. 97:157-342, figs.4-53, charts 1-3.

Nymphon spinosum hirtipes
Nesis, K.N., 1960.
[Littoral Pantopoda of the eastern Murmansk coast.]
Trudy Murmansk. Morsk. Biol. Inst., 2(6):137-161.

Nymphon spinosum
Meinert, Fr., 1899
Pycnogonida. Danish Ingolf Exped. 3(1): 71 pp., 5 pls., 2 text figs., 1 charts, list of stations.

Nymphon stocki n.sp.
Utinomi, H., 1955.
Report on the Pycnogonida collected by the Soyo-maru expeditions made on the continental shelf bordering Japan during the years 1926-1930.
Publ. Seto Mar. Biol. Lab., 5(1):1-42.

Nymphon striatum
Utinomi, H., 1954.
The fauna of Akkeshi Bay. XIX. Littoral Pycnogonida. Publ. Akkeshi Mar. Biol. Sta. No. 3:1-28, 1 pl., 11 textfigs.

Contr. 221 Seto Mar. Biol. Lab.

Nymphon strömi
Hedgpeth, J.W., 1948
The Pycnogonida of the western North Atlantic and the Caribbean. Proc. U.S. Nat. Mus. 97:157-342, figs.4-53, charts 1-3.

Nymphon stroemii
Meinert, Fr., 1899
Pycnogonida. Danish Ingolf Exped. 3(1): 71 pp., 5 pls., 2 text figs., 1 charts, list of stations.

Nymphon Stroemii
Stephensen, K., 1917
Zoogeographical investigations of certain fjords in southern Greenland with special reference to Crustacea, Pycnogonida and Echinodermata including a list of Alcyonaria and Pisces, Medd. om Grønland, 53(3):229-378.

Nymphon Stroemi
Stephensen, K., 1912
Report on the Malacostraca, Pycnogonida and some Entomostraca collected by the Danmark-Expedition to North East Greenland. Medd. om Grønland, 45(11):501-630, Pls. 39-43.

Nymphon strömii strömii
Nesis, K.N., 1960.
Littoral Pantopoda of the eastern Murmansk coast. Trudy Murmansk. Morsk. Biol. Inst., 2(6):137-161

Nymphon subtile
Gordon, I., 1932.
Pycnogonida. Discovery Repts., 6:1-138.

Nymphon tenellum
Meinert, Fr., 1899
Pycnogonida. Danish Ingolf Exped. 3(1): 71 pp., 5 pls., 2 text figs., 1 charts, list of stations.

Nymphon tenuipes
Gordon, I., 1932.
Pycnogonida. Discovery Repts., 6:1-138.

Nymphon tripectinatum n.sp.
Turpaeva, E.P., 1971.
An addition to the pantopod fauna of deep-sea trenches in the northwestern part of the Pacific Ocean. (In Russian; English abstract). Trudy Inst. Okeanol. P.P. Shirshov 92: 292-297.

Nymphon validum n.sp.
Haswell, W.A., 1885.
On the Pycnogonida of the Australian coast. Proc. Linn. Soc., N.S. Wales, 9:1021-1034, Pls. 54-57.

Nymphon villosum
Fage, L., 1952.
Pycnogonides de la Terre Adélie. Échantillons rapportés par le Docteur Sapin-Jaloustre, Medecin-Biologiste de la Première Expédtion en Terre Adélie (1949-1951). (Expéditions Polaires Françaises, Missions Paul-Émile Victor). Bull. Mus. Nat. Hist. Nat., 2 ser., 24:263-273, 2 textfigs.

Nymphonella lecalvezi n.sp.
Guille A. et J. Soyer, 1967.
Nouvelle signalisation du genre Nymphonella Ohshima à Banyuls-sur-mer: Nymphonella lecalvezi n.sp.
Vie Milieu (A) 18(2): 345-350

Nymphonella tapetis
Le Calvez, J., 1950.
Un pycnogonide nouveau pour la Méditerranée: Nymphonella tapetis Ohshima. Arch. Zool. Exper. et Gen., Notes et Rev., 86(3):114-113.

Nymphopsis anarthra
Hedgpeth, J.W., 1948
The Pycnogonida of the western North Atlantic and the Caribbean. Proc. U.S. Nat. Mus. 97:157-342, figs.4-53, charts 1-3.

Nymphopsis armatus n.gen., n.sp.
Haswell, W.A., 1885.
On the Pycnogonida of the Australian coast. Proc. Linn. Soc., N.S. Wales, 9:1021-1034, Pls. 54-57.

Nymphopsis denticulata n.sp.
Gordon, I., 1932.
Pycnogonida. Discovery Repts., 6:1-138.

Nymphopsis dromedaria n.sp.
McCloskey, L.R. 1967.
New and little-known pycnogonids from North Carolina.
J. nat. Hist. 1(1):119-134.

Nymphopsis duodorsospinosa
Hedgpeth, J.W., 1948
The Pycnogonida of the western North Atlantic and the Caribbean. Proc. U.S. Nat. Mus. 97:157-342, figs.4-53, charts 1-3.

Nymphopsis mucosa
Stock, J. H. 1954.
Pycnogonida from Indo-West Pacific, Australian and New Zealand waters. Papers from Dr. Th. Mortensen's Pacific Expedition, 1914-1916. LXXVII. Vidensk. Medd. Dansk Naturhist. Foren., København, 116: 1-168.

Nymphopsis muscosa
Utinomi, Huzio, 1952
Pycnogonida of Sagami Bay. Publ. Seto Mar. Biol. Lab., 7(2):197-222.

Nymphopsis varipes n.sp.
Stock, J.H., 1962.
Second list of Pycnogonida of the University of Cape Town ecological survey. Trans. R. Soc., S. Africa, 36(4):273-286.

Pallene acus n.sp.
Meinert, Fr., 1899
Pycnogonida. Danish Ingolf Exped. 3(1): 71 pp., 5 pls., 2 text figs., 1 charts, list of stations.

Pallene australiensis
Haswell, W.A., 1885.
On the Pycnogonida of the Australian coast. Proc. Linn. Soc., N.S. Wales, 9:1021-1034, Pls. 54-57.

Pallene brevirostris
Lebour, M.V., 1945.
Notes on the Pycnogonida of Plymouth. J.M.B.A. 26:139-165, 7 textfigs.

Pallene chiragra
Haswell, W.A., 1885.
On the Pycnogonida of the Australian coast. Proc. Linn. Soc., N.S. Wales, 9:1021-1034, Pls. 54-57.

Pallene hastata n.sp.
Meinert, Fr., 1899
Pycnogonida. Danish Ingolf Exped. 3(1): 71 pp., 5 pls., 2 text figs., 1 charts, list of stations.

Pallene laevis
Haswell, W.A., 1885.
On the Pycnogonida of the Australian coast. Proc. Linn. Soc., N.S. Wales, 9:1021-1034, Pls. 54-57.

Pallene languida
Haswell, W.A., 1885.
On the Pycnogonida of the Australian coast. Proc. Linn. Soc., N.S. Wales, 9:1021-1034, Pls. 54-57.

Pallene lappa
Barnard, K.H., 1954.
III(3): South African Pycnogonida. Ann. S. African Mus., 41:81-158, 34 textfigs.

Pallene margarita n.sp.
Gordon, I., 1932.
Pycnogonida. Discovery Repts., 32:1-138.

Pallene pachycheira n.sp.
Haswell, W.A., 1885.
On the Pycnogonida of the Australian coast. Proc. Linn. Soc., N.S. Wales, 9:1021-1034, Pls. 54-57.

Pallene spectrum
Lebour, M.V., 1945.
Notes on the Pycnogonida of Plymouth. J.M.B.A. 26:139-165, 7 textfigs.

Pallenoides magnicollis
Barnard, K.H., 1954.
I(3):South African Pycnogonida. Ann. S. African Mus., 41:81-158, 34 textfigs.

Pallenoides opuntia n.sp.
Stock, Jan H., 1965.
Pycnogonida from the southwest Indian Ocean. Beaufortia, 13(151):13-33.

Pallenoides proboscideum n.sp.
Barnard, K. H., 1955.
Additions to the fauna-list of South African Crustacea and Pycnogonida. Ann. S. African Mus., 43(1): 1-107.

Pallenoides (?) spinulosa
Capriles, Víctor A. 1970.
Note on the occurrence of the pycnogonid Pallenoides (?) spinulosa Stock in Puerto Rican waters.
Carib. J. Sci. 10(1/2):105.

Pallenoides stylinostrum n.sp.
Stock, Jan H. 1973.
Pycnogonida from south-eastern Australia.
Beaufortia 20 (266):99-127

Pallenopsis sp.
Clark, W.C., 1958.
Some Pycnogonida from Cook Strait, New Zealand, with descriptions of two new species. Victoria Univ., Wellington, Zool. Publ., 23:7 pp.

? Pallenopsis sp.
Gordon, I., 1932.
Pycnogonida. Discovery Repts., 6:1-138.

Pallenopsis sp.
Stock, Jan H. 1973.
Pycnogonida from south-eastern Australia.
Beaufortia 20 (266):99-127

Pallenopsis sp.
Stock, Jan H., 1968.
Pycnogonida collected by the Galathea and Anton Bruun in the Indian and Pacific oceans. Vidensk. Meddr dansk naturh. Foren. 131: 7-65.

Pallenopsis (Pallenopsodon) antipoda n.sp.
Clark, W.C. 1971.
Pycnogonida of the Antipodes Islands.
N.Z. Jl. mar. Freshwat. Res. 5(3/4):427-452.

Pallenopsis bicuspidata n.sp.
Stock, Jan H., 1968.
Pycnogonida collected by the Galathea and Anton Bruun in the Indian and Pacific oceans.
Vidensk. Meddr dansk naturh. Foren. 131: 7-65.

Pallenopsis brevidigitata
Barnard, K.H., 1954.
III(3):South African Pycnogonida.
Ann. S. African Mus., 41:81-158, 34 textfigs.

Pallenopsis brevidigitata
Calman, W. T., 1938
Pycnogonida. John Murray Exped. 1933-34,
Sci. Repts. 5(6):147-166, 10 text figs.

Pallenopsis calcanea
Hedgpeth, Joel W., 1962
A bathypelagic pycnogonid.
Deep-Sea Res., 9(5):487-491.

Pallenopsis calcanea
Hedgpeth, J.W., 1948
The Pycnogonida of the western North Atlantic and the Caribbean. Proc. U.S. Nat. Mus. 97:157-342, figs.4-53, charts 1-3.

Pallenopsis macneilli
Stock, Jan H. 1973.
Pycnogonida from south-eastern Australia.
Beaufortia 20 (266), 99-127

Pallenopsis calcanea
Stock, Jan H., 1968.
Pycnogonida collected by the Galathea and Anton Bruun in the Indian and Pacific oceans.
Vidensk. Meddr. dansk naturh. Foren., 131: 7-65.

Pallenopsis calcanea
Turpaeva, E.P., 1971.
The deep-water Pantopoda collected in the Kurile-Kamchatka Trench. (In Russian; English abstract). Trudy Inst. Okeanol. P.P. Shirshov 92: 274-291.

Pallenopsis candidoi
Stock, J.H., 1966.
Pycnogonida. Campagne de la Calypso au large des côtes Atlantiques de l'Amerique du Sud (1961-1962). Ann. Inst. Océanogr., Monaco, n.s. 44: 385-406.

Pallenopsis candidoi
Stock, J.H., 1957.
Pantopoden aus dem Zoologischen Museum Hamburg.
Mitt. Hamburgischen Zool. Mus. u. Inst., 55:81-108.

Pallenopsis capensis
Barnard, K.H., 1954.
III(3):South African Pycnogonida.
Ann. S. African Mus., 41:81-158, 34 textfigs.

Pallenopsis capensis n.sp.
Barnard, K.H., 1946.
Diagnoses of new species and a new genus of Pycnogonida in the South African Museum.
Ann. Mag. Nat. Hist. (11)13:60-63.

Pallenopsis crosslandi
Arnaud, Françoise 1972
Pycnogonides des récifs coralliens de Madagascar. 3. Famille des Callipallenidae.
Tethys Suppl. 3: 157-164.

Pallenopsis crosslandi
Barnard, K.H., 1954.
III(3):South African Pycnogonida.
Ann. S. African Mus., 41:81-158, 34 textfigs.

Pallenopsis gippslandiae n.sp.
Stock, J. H. 1954.
Pycnogonida from Indo-West Pacific, Australian and New Zealand waters. Papers from Dr. Th. Mortensen's Pacific Expedition, 1914-1916. LXXVII.
Vidensk. Medd. Dansk Naturhist. Foren., København, 116: 1-168.

Pallenopsis fluminensis
Meinert, Fr., 1899
Pycnogonida. Danish Ingolf Exped. 3(1): 71 pp., 5 pls., 2 text figs., 1 charts, list of stations.

Pallenopsis forficifer
Hedgpeth, J.W., 1948
The Pycnogonida of the western North Atlantic and the Caribbean. Proc. U.S. Nat. Mus. 97:157-342, figs.4-53, charts 1-3.

Pallenopsis glabra
Hodgson, T.V., 1907.
Pycnogonida. Nat. Antarctic Exped., 1901-1904, 3:72 pp., 3 pls.

Pallenopsis hiemalis
Fage, L., 1952.
Pycnogonides de la Terre Adélie. Echantillons rapportés par le Docteur Sapin-Jaloustre, Medecin-Biologiste de la Première Expédition en Terre Adélie (1949-1951). (Expéditions Polaires Françaises, Missions Paul-Emile Victor).
Bull. Mus. Nat. Hist. Nat., 2 ser., 24:263-273, 2 textfigs.

Pallenopsis hiemalis
Hodgson, T.V., 1907.
Pycnogonida. Nat. Antarctic Exped., 1901-1904, 3:82 pp., 10 pls.

Pallenopsis hodgsoni
Hedgpeth, J. W., 1950
Pycnogonida of the United States Navy Antarctic Expedition, 1947-48. Proc. U. S. Nat. Mus. 100 (3260):147-160, 19 text figs.

Pallenopsis intermedia
Barnard, K.H., 1954.
III(3):South African Pycnogonida.
Ann. S. African Mus., 41:81-158, 34 textfigs.

Pallenopsis (Pallenopsodon) juttingae n.sp.
Stock, J.H., 1964.
Deep-sea Pycnogonida collected by the "Cirrus" in the northern Atlantic.
Beaufortia, Zool. Mus., Amsterdam, 11(135):45-52.

Pallenopsis kupei n.sp.
Clark, W.C. 1971.
Pycnogonida of the Antipodes Islands.
N.Z. Jl mar. Freshwat. Res. 5(3/4): 427-452.

Pallenopsis longicoxa n.sp.
Stock, J.H., 1966.
Pycnogonida. Campagne de la Calypso au large des côtes Atlantiques de l'Amerique du Sud (1961-1962). Ann. Inst. Océanogr., Monaco, n.s. 44: 385-406.

Pallenopsis longirostris
Giltray, L., 1942.
New records of Pycnogonida from the Canadian Atlantic coast. J. Fish. Res. Bd., Canada, 5(5): 459-460.

Pallenopsis longirostris
Hedgpeth, J.W., 1948
The Pycnogonida of the western North Atlantic and the Caribbean. Proc. U.S. Nat. Mus. 97:157-342, figs.4-53, charts 1-3.

Pallenopsis longiseta n.sp.
Turpaeva, E.P., 1958.
[Die neue Arten der Pantopoda des Genus Pallenopsis aus dem Nord-Westlichen Teil des Stillen Ozeans.] Trudy Inst. Okeanol., 27:356-361.

Pallenopsis macneilli
Stock, Jan H., 1968.
Pycnogonida collected by the Galathea and Anton Bruun in the Indian and Pacific oceans.
Vidensk. Meddr. dansk naturh. Foren., 131: 7-65.

Pallenopsis mauii
Clark, W.C. 1971.
Pycnogonida of the Antipodes Islands.
N.Z. Jl mar. Freshwat. Res. 5(3/4): 427-452.

Pallenopsis maui n.sp.
Clark, W.C., 1958.
Some Pycnogonida from Cook Strait, New Zealand, with descriptions of two new species.
Victoria Univ., Wellington, Zool. Publ., 23:7 pp.

Pallenopsis mollissima
Fage, L., 1956.
Les pycnogonides (excl. le genre Nymphon).
Galathea Rept., 2:167-183.

Pallenopsis obliqua
Clark, W.C. 1971.
Pycnogonida of the Antipodes Islands.
N.Z. Jl mar. Freshwat. Res. 5(3/4): 427-452.

Pallenopsis obliqua
Clark, W.C. 1971.
Pycnogonida of the Snares Islands.
N.Z. Jl mar. Freshwat. Res. 5(2): 329-341.

Pallenopsis obliqua
Clark, W.C., 1958.
Some Pycnogonida from Cook Strait, New Zealand, with descriptions of two new species.
Victoria Univ., Wellington, Zool. Publ., 23:7 pp.

Pallenopsis obliqua
Stock, J. H. 1954.
Pycnogonida from Indo-West Pacific, Australian and New Zealand waters. Papers from Dr. Th. Mortensen's Pacific Expedition, 1914-1916. LXXVII.
Vidensk. Medd. Dansk Naturhist. Foren., København, 116: 1-168.

Pallenopsis oscitans
Barnard, K.H., 1954.
III(3):South African Pycnogonida.
Ann. S. African Mus., 41:81-158, 34 textfigs.

Pallenopsis ovalis
Barnard, K.H., 1954.
III(3):South African Pycnogonida.
Ann. S. African Mus., 41:81-158, 34 textfigs.

Pallenopsis ovalis
Stock, Jan H., 1968.
Pycnogonida collected by the Galathea and Anton Bruun in the Indian and Pacific oceans.
Vidensk. Meddr dansk naturh. Foren. 131: 7-65.

Pallenopsis ovalis
Stock, J. H. 1954.
Pycnogonida from Indo-West Pacific, Australian and New Zealand waters. Papers from Dr. Th. Mortensen's Pacific Expedition, 1914-1916. LXXVII.
Vidensk. Medd. Dansk Naturhist. Foren., København, 116: 1-168.

Pallenopsis patagonica
Gordon, I., 1932.
Pycnogonida. Discovery Repts., 6:1-138.

Pallenopsis patagonica
Hedgpeth, J. W., 1950
Pycnogonida of the United States Navy
Antarctic Expedition, 1947-48. Proc. U. S.
Nat. Mus. 100 (3260):147-160, 19 text figs.

Pallenopsis patagonica
Stock, J.H., 1957.
Pantopoden aus dem Zoologischen Museum Hamburg.
Mitt. Hamburgischen Zool. Mus. u. Inst., 55:81-106.

Pallenopsis pilosa
Gordon, I., 1932.
Pycnogonida. Discovery Repts., 6:1-138.

Pallenopsis pilosa
Hodgson, T.V., 1907.
Pycnogonida. Nat. Antarctic Exped., 1901-1904,
3:72 pp., 10 pls.

Pallenopsis plumipes n.sp.
Meinert, Fr., 1899
Pycnogonida. Danish Ingolf Exped. 3(1):
71 pp., 5 pls., 2 text figs., 1 charts, list of stations.

Pallenopsis scoparia n.sp.
Fage, L., 1956.
Les pycnogonides (excl) le genre Nymphon).
Galathea Rept., 2:167-183.

Pallenopsis sibogae
Stock, J. H. 1954.
Pycnogonida from Indo-West Pacific,
Australian and New Zealand waters. Papers
from Dr. Th. Mortensen's Pacific Expedition,
1914-1916. LXXVII.
Vidensk. Medd. Dansk Naturhist. Foren.,
København, 116: 1-168.

Pallenopsis sibogae
Utinomi, Huzio, 1962
Pycnogonida of Sagami Bay - Supplement.
Publ. Seto Mar. Biol. Lab., 10(1) (Article 5)
91-104.

Pallenopsis sibogae
Utinomi, H., 1955.
Report on the Pycnogonida collected by the Soyo-
maru expeditions made on the continental
shelf bordering Japan during the years 1926-1930.
Publ. Seto Mar. Biol. Lab., 5(1):1-42.

Pallenopsis sibogae
Utinomi, Huzio, 1952
Pycnogonida of Sagami Bay.
Publ. Seto Mar. Biol. Lab., 7(2):197-222.

Pallenopsis spicata
Gordon, I., 1932.
Pycnogonida. Discovery Repts., 6:1-138.

Pallenopsis spinipes
Stock, J.H., 1957.
Pantopoden aus dem Zoologischen Museum Hamburg.
Mitt. Hamburgischen Zool. Mus. u. Inst., 55:81-106.

Pallenopsis stschapovae n.sp
Turpaeva, E.P., 1958.
Die neue Arten der Pantopoden des Genus
Pallenopsis aus dem Nord-Westlichen Teil des
Stillen Ozeans. Trudy Inst. Okeanol., 27:356-361.

Pallenopsis temperans
Stock, J. H. 1954.
Pycnogonida from Indo-West Pacific,
Australian and New Zealand waters. Papers
from Dr. Th. Mortensen's Pacific Expedition,
1914-1916. LXXVII.
Vidensk. Medd. Dansk Naturhist. Foren.,
København, 116: 1-168.

Pallenopsis temperans
Stock, Jan H., 1968.
Pycnogonida collected by the Galathea and Anton
Bruun in the Indian and Pacific oceans.
Vidensk. Meddr dansk naturh. Foren., 131: 7-65.

Pallenopsis tongaensis n.sp.
Clark W.C. 1973.
New species of Pycnogonida from
New Britain and Tonga.
Pacific Sci. 27(1): 28-33.

Pallenopsis triregia n.sp.
Clark, W.C., 1962.
Pallenopsis triregia, a new species of Pycnogon-
ida, from near Three Kings Islands, New Zealand.
New Zealand J. Sci., 5(4):517-520.

Pallenopsis tumidula
Stock, J.H., 1966.
Pycnogonida. Campagne de la Calypso au large
des côtes Atlantiques de l'Amerique du Sud
(1961-1962). Ann. Inst. Océanogr., Monaco,
n.s. 44: 385-406.

Pallenopsis tumidula
Stock, J.H., 1957.
Pantopoden aus dem Zoologischen Museum Hamburg.
Mitt. Hamburgischen Zool. Mus. u. Inst., 55:81-106.

Pallenopsis tydemani
Utinomi, H., 1951.
On some pycnogonids from the sea around the Kii
Peninsula.
Pub. Seto Mar. Biol. Lab., Kyoto Univ., 1(4):
159-168.

Pallenopsis vanhöffeni
Fage, L., 1952.
II. Pycnogonides. Missions du Bâtiment Polaire
"Commandant Charcot", récoltés faites en Terre Ad
Adélie (1950) par Paul Tchernia. Bull. Mus. Nat.
Hist. Nat., 2 ser., 24:180-186, 2 textfigs.

Pallenopsis vanhöffeni
Fage, L., 1952.
Pycnogonides de la Terre Adélie. Échantillons
rapportés par le Docteur Sapin-Jaloustre,
Medecin-Biologiste de la Première Expédition en
Terre Adélie (1949-1951). (Expéditions Polaires
Françaises, Missions Paul-Émile Victor).
Bull. Mus. Nat. Hist. Nat., 2 ser., 24:263-273,
2 testfigs.

Pallenopsis vanhöffeni
Hedgpeth, J. W., 1950
Pycnogonida of the United States Navy
Antarctic Expedition, 1947-48. Proc. U. S.
Nat. Mus. 100 (3260):147-160, 19 text figs.

Pallenopsis verrucosa
Stock, Jan H., 1968.
Pycnogonida collected by the Galathea and Anton
Bruun in the Indian and Pacific oceans.
Vidensk. Meddr dansk naturh. Foren. 131: 7-65.

Pallenopsis villosa
Hodgson, T.V., 1907.
Pycnogonida. Nat. Antarctic Exped., 1901-1904,
3:72 pp., 10 pls.

Pallenopsis virgata
Utinomi, Huzio, 1952
Pycnogonida of Sagami Bay.
Publ. Seto Mar. Biol. Lab., 7(2):197-222.

Pallenopsis virgatus
Utinomi, H., 1951.
On some pycnogonids from the sea around the Kii
Peninsula.
Publ. Seto Mar. Biol. Sta., Kyoto Univ., 1(4):159-168.

Pantopipetta nom nov.
Stock, J.H., 1963
South African deep-sea Pycnogonida, with
descriptions of five new species.
Annals, South African Mus., 46(12):321-340.

Pantopipetta sp.
Stock, J.H., 1963
South African deep-sea Pycnogonida, with
descriptions of five new species.
Annals, South African Mus., 46(12):321-340.

Pantopipetta auxiliata n.sp.
Stock, Jan H., 1968.
Pycnogonida collected by the Galathea and Anton
Bruun in the Indian and Pacific oceans.
Vidensk. Meddr dansk naturh. Foren. 131: 7-65.

Pantopipetta brevicauda n.sp.
Stock, J.H., 1963
South African deep-sea Pycnogonida, with
descriptions of five new species.
Annals, South African Mus., 46(12):321-340.

Pantopipetta capensis
Stock, J.H., 1963
South African deep-sea Pycnogonida, with
descriptions of five new species.
Annals, South African Mus., 46(12):321-340.

Pantopipetta longituberculata
Turpaeva, E.P., 1971.
The deep-water Pantopoda collected in the
Kurile-Kamchatka Trench. (In Russian;
English abstract). Trudy Inst. Okeanol. P.P.
Shirshov 92: 274-291.

Pantopipetta oculata n.sp.
Stock, Jan H. 1968.
Pycnogonida collected by the Galathea and Anton
Bruun in the Indian and Pacific oceans.
Vidensk. Meddr dansk naturh. Foren. 131: 7-65.

Paranymphon spinosum
Meinert, Fr., 1899
Pycnogonida. Danish Ingolf Exped. 3(1):
71 pp., 5 pls., 2 text figs., 1 charts, list of stations.

Paranymphon Spinosum
Stock, Jan H., 1966
Sur quelques pycnogonides de la région de
Banyuls. (3° note).
Vie Milieu (B)17 (1):407-417.

Paranymphon spinosum
Stock, Jan H., et Jacques Soyer, 1965.
Sur quelques pycnogonides rares de Banyuls-Sur-mer.
Vie et Milieu, Bull. Lab.Arego, (B)16(1): 415-

Parapallene? aculeata n.sp.
Stock, J. H. 1954.
Pycnogonida from Indo-West Pacific,
Australian and New Zealand waters. Papers
from Dr. Th. Mortensen's Pacific Expedition,
1914-1916. LXXVII.
Vidensk. Medd. Dansk Naturhist. Foren.,
København, 116: 1-168.

Parapallene algoae
Barnard, K.H., 1954.
III(3): South African Pycnogonida.
Ann. S. African Mus., 41:81-158, 34 textfigs.

Parapallene algoae n.sp.
Barnard, K.H., 1946.
Diagnoses of new species and a new genus of Pycnogonidae in the South African Museum.
Ann. Mag. Nat. Hist. (11)13:60-63.

Parapallene australiensis
Stock, Jan N. 1973.
Pycnogonids from south-eastern Australia.
Beaufortia 20 (266):99-127

Parapallene australiensis
Stock, J. H. 1954.
Pycnogonida from Indo-West Pacific, Australian and New Zealand waters. Papers from Dr. Th. Mortensen's Pacific Expedition, 1914-1916. LXXVII.
Vidensk. Medd. Dansk Naturhist. Foren., København, 116: 1-168.

Parapallene avida n.sp.
Stock, Jan N. 1973.
Pycnogonids from south-eastern Australia.
Beaufortia 20 (266):99-127

Parapallene bermudiensis, n. sp.
Lebour, M. V., 1949.
Two new Pycnogonids from Bermuda. Proc. Zool. Soc., London, 118(4):929-932, 3 textfigs.

Parapallene calmani
Barnard, K.H., 1954.
III(3): South African Pycnogonida.
Ann. S. African Mus., 41:81-158, 34 textfigs.

Parapallene capilliata n.sp.
Stock, J. H. 1954.
Pycnogonida from Indo-West Pacific, Australian and New Zealand waters. Papers from Dr. Th. Mortensen's Pacific Expedition, 1914-1916. LXXVII.
Vidensk. Medd. Dansk Naturhist. Foren., København, 116: 1-168.

Parapallene exigua n.sp.
Stock, J. H. 1954.
Pycnogonida from Indo-West Pacific, Australian and New Zealand waters. Papers from Dr. Th. Mortensen's Pacific Expedition, 1914-1916. LXXVII.
Vidensk. Medd. Dansk Naturhist. Foren., København, 116: 1-168.

Parapalleni hodgsoni
Barnard, K.H., 1954.
III(s):South African Pycnogonida.
Ann. S. African Mus., 41:81-158, 34 textfigs.

Parapallene hodgsoni n.sp.
Barnard, K.H., 1946.
Diagnoses of new species and a new genus of Pycnogonida in the South African Museum.
Ann. Mag. Nat. Hist. (11)13:60-63.

Parapallene hodgsoni
Stock, Jan H., 1968.
Pycnogonida collected by the Galathea and Anton Bruun in the Indian and Pacific oceans.
Vidensk. Meddr. dansk naturh. Foren., 131: 7-65.

Parapallene hodgsoni
Stock, Jan H., 1965.
Pycnogonida from the southwest Indian Ocean.
Beaufortia. 13(151):13-33.

Parapallene longipes n.sp.
Calman, W. T., 1938
Pycnogonida. John Murray Exped. 1933-34, Sci. Repts. 5(6):147-166, 10 text figs.

Parapallene nierstraszi
Barnard, K.H., 1954.
III(3):South African Pycnogonida.
Ann. S. African Mus., 41:81-158, 34 textfigs.

Parapallene nierstraszi
Stock, J. H. 1954.
Pycnogonida from Indo-West Pacific, Australian and New Zealand waters. Papers from Dr. Th. Mortensen's Pacific Expedition, 1914-1916. LXXVII.
Vidensk. Medd. Dansk Naturhist. Foren., København, 116: 1-168.

Parapallene nierstraszi
Utinomi, H., 1955.
Report on the Pycnogonida collected by the Soyo-maru expeditions made on the continental shelf bordering Japan during the years 1926-1930.
Publ. Seto Mar. Biol. Lab., 5(1):1-42.

Parapallene spinosus
Barnard, K.H., 1954.
III(3):South African Pycnogonida.
Ann. S. African Mus., 41:81-158, 34 textfigs.

Pentacolossendeis reticulata
Hedgpeth, J.W., 1948
The Pycnogonida of the western North Atlantic and the Caribbean. Proc. U.S. Nat. Mus. 97:157-342, figs.4-53, charts 1-3.

Pentanymphon antarcticum
Fage, L., 1952.
Pycnogonides de la Terre Adélie. Echantillons rapportés par le Docteur Sapin-Jaloustre, Medecin-Biologiste de la Première Expédition en Terre Adélie (1949-1951). (Expéditions Polaires Françaises, Missions Paule Emile Victor
Bull. Mus. Nat. Hist. Nat., 2 ser., 24:263-273, 2 textfigs.

Pentanymphon antarcticum
Gordon, I., 1932.
Pycnogonida. Discovery Repts., 6:1-138.

Pentanymphon antarcticum
Hedgpeth, J. W., 1950
Pycnogonida of the United States Navy Antarctic Expedition, 1947-48. Proc. U. S. Nat. Mus. 100 (3260):147-160, 19 text figs.

Pentanymphon antarcticum
Hodgson, T.V., 1907.
Pycnogonida. Nat. Antarctic Exped., 1901-1904, 3:72 pp., 10 pls.

Pentanymphon antarcticum
Stephensen, K., 1947
Tanaidacea, Isopoda, Amphipoda, and Pycnogonida. Sci. Res. Norweg. Antarctic Exp. 1927-1928 et seq. Date Norske Videns-kaps-Akademi i Oslo, II, No.27:1-90, 26 text figs.

Pentanymphon minutum
Fage, L., 1952.
Pycnogonides de la Terre Adélie. Echantillons rapportés par le Docteur Sapin-Jaloustre, Medecin-Biologiste de la Première Expédition en Terre Adélie (1949-1951). (Expéditions Polaires Françaises, Missions Paul-Emile Victor).
Bull. Mus. Nat. Hist. Nat., 2 ser., 24:263-273, 2 textfigs.

Pentapycnon charcoti
Fry, William G. and Joel W. Hedgpeth, 1969.
The fauna of the Ross Sea. 7. Pycnogonida, I. Colossendeidae, Pycnogonidae, Endeidae, Ammotheidae. Bull. N.Z.D.S.I.R. 198 (Mem. N.Z. Oceanogr. Inst. 49): 139 pp.

Pentapycnon geayi
Hedgpeth, J.W., 1948
The Pycnogonida of the western North Atlantic and the Caribbean. Proc. U.S. Nat. Mus. 97:157-342, figs.4-53, charts 1-3.

Phoxichilidium sp.
Child, C. Allan, and Joel W. Hedgpeth, 1971.
Pycnogonida of the Galápagos Islands. J. nat. Hist. 5(6): 609-634.

Phoxichilidium australe
Fage, L., 1952.
II. Pycnogonides. Missions du Bâtiment Polaire "Commandant Charcot", récoltés faites en Terre Adélie (1950) par Paul Tchernia. Bull. Mus. Nat. Hist. Nat., 2 ser., 24:180-186, 2 textfigs.

Phoxichilidium australe
Fage, L., 1952.
Pycnogonides de la Terre Adélie. Echantillons rapportés par le Docteur Sapin-Jaloustre, Medecin-Biologiste de la Première Expédition en Terre Adélie (1949-1951). (Expéditions Polaires Françaises, Missions Paul-Emile Victor).
Bull. Mus. Nat. Hist. Nat., 2 ser., 24:263-273, 2 textfigs.

Phoxichilidium australe
Gordon, I., 1932.
Pycnogonida. Discovery Repts., 6:1-138.

Phoxochilidium capense
Barnard, K.H., 1954.
III(3): South African Pycnogonida.
Ann. S. African Mus., 41:81-158, 34 textfigs.

Phoxichilidium femoratum
Hedgpeth, J.W., 1948
The Pycnogonida of the western North Atlantic and the Caribbean. Proc. U.S. Nat. Mus. 97:157-342, figs.4-53, charts 1-3.

Phoxichilidium femoratum
Lebour, M.V., 1945.
Notes on the Pycnogonida of Plymouth. J.M.B.A. 26:139-165, 7 textfigs.

Phoxichilidium femoratum
Stephensen, K., 1917
Zoogeographical investigations of certain fjords in southern Greenland with special reference to Crustacea, Pycnogonida and Echinodermata including a list of Alcyonaria and Pisces, Medd. om Grønland, 53(3):229-378.

Phoxichilidium Hoeckii
Haswell, W.A., 1885.
On the Pycnogonida of the Australian coast.
Proc. Linn. Soc., N.S. Wales, 9:1021-1034, Pls. 54-57.

Phoxichilidium hokkaidoense n.sp.
Utinomi, H., 1954.
The fauna of Akkeshi Bay. XIX. Littoral Pycnogonida. Pub. Akkeshi Mar. Biol. Sta. No. 3:1-28, 1 pl., 11 textfigs.

Contr. 221 Seto Mar. Biol. Lab.

Phoxichilidium robustum
Lebour, M.V., 1945.
Notes on the Pycnogonida of Plymouth. J.M.B.A. 26:139-165, 7 textfigs.

Phoxichilidium tubiferum n.sp.
Haswell, W.A., 1885.
On the Pycnogonida of the Australian coast.
Proc. Linn. Soc., N.S. Wales, 9:1-21-1034, Pls. 54-57.

Phoxichilidium tubulariae n.sp.
Lebour, M.V., 1945.
Notes on the Pycnogonida of Plymouth. J.M.B.A. 26:139-165, 7 textfigs.

Phoxichilidium ungellatum
Stock, J. H. 1954.
Pycnogonida from Indo-West Pacific, Australian and New Zealand waters. Papers from Dr. Th. Mortensen's Pacific Expedition, 1914-1916. LXXVII.
Vidensk. Medd. Dansk Naturhist. Foren., København, 116: 1-168.

Phoxichilidium ungellatum
Utinomi, H., 1955.
Report on the Pycnogonida collected by the Soyo-maru expeditions made on the continental shelf bordering Japan during the years 1926-1930.
Publ. Seto Mar. Biol. Lab., 5(1):1-42.

Phoxichilidium virescens
Bourdillon, A., 1954.
Les pycnogonides de Marseille et ses environs.
Recueil Trav. Sta. Mar., Endoume, 12:145-158.

Phoxichilidium virescens
Lebour, M.V., 1945.
Notes on the Pycnogonida of Plymouth. J.M.B.A. 26:139-165, 7 textfigs.

Phoxichilus australis
Hodgson, T.V., 1907.
Pycnogonida. Nat. Antarctic Exped., 1901-1904, 3:72 pp., 10 pls.

Phoxichilus charybdaeus?
Haswell, W.A., 1885.
On the Pycnogonida of the Australian Coast.
Proc. Linn. Soc., N.S. Wales, 9:1021-1034, Pls. 54-57.

Pigrogromitus n.gen.
Calman, W.T., 1927.
Zoological results of the Cambridge Expedition to the Suez Canal. XXVIII. Report on the Pycnogonida.
Trans. Zool. Soc., London, 22: 403-410, textfigs.

Pigrogromitus timsanus
Arnaud, Françoise 1972
Pycnogonides des récifs coralliens de Madagascar. 3. Famille des Callipallenidae.
Téthys Suppl. 3: 157-164.

Pigrogromitus timsanus n.sp.
Calman, W.T., 1927.
Zoological results of the Cambridge Expedition to the Suez Canal. XXVIII. Report on the Pycnogonida.
Trans. Zool. Soc., London, 22:403-410, textfigs. 102-104.

Pigromitus timsanus
Hedgpeth, J.W., 1948
The Pycnogonida of the western North Atlantic and the Caribbean. Proc. U.S. Nat. Mus. 97:157-342, figs.4-53, charts 1-3.

Pigrogromitus timsanus
Stock, Jan H., 1968.
Pycnogonida collected by the Galathea and Anton Bruun in the Indian and Pacific oceans.
Vidensk. Meddr dansk naturh. Foren. 131: 7-65.

Pipetta australis
Turpajeva, E.P., 1955.
[New Pantopoda sp of the Kurile-Kamchatka Trench]
Trudy Inst. Oceanol., 12:322-327.

Pipetta capensis
Barnard, K.H., 1954.
III(3):South African Pycnogonida.
Ann. S. African Mus. 41:81-158, 34 textfigs.

Pipetta capensis n.sp.
Barnard, K.H., 1946.
Diagnoses of new species and a new genus of Pycnogonida in the South African Museum.
Ann. Mag. Nat. Hist. (11)13:60-63.

Pipetta longitubarculata n.sp.
Turpajeva, E.P., 1955.
[New Pantopoda sp. of the Kurile-Kamchatka Trench]
Trudy Inst. Oceanol., 12:322-327.

Pipetta weberi
Turpajeva, E.P., 1955.
[New Pantopoda sp of the Kurile-Kamchatka Trench]
Trudy Inst. Oceanol., 12:322-327.

Propallene crassimanus
Stock, Jan H., 1968.
Pycnogonida collected by the Galathea and Anton Bruun in the Indian and Pacific oceans.
Vidensk. Meddr dansk naturh. Foren. 131: 7-65.

Propallene crinipes n. sp.
Stock, Jan H., 1968.
Pycnogonida collected by the Galathea and Anton Bruun in the Indian and Pacific oceans.
Vidensk. Meddr. dansk naturh. Foren., 131: 7-65.

Propallene kempi
Arnaud, Françoise 1972
Pycnogonides des récifs coralliens de Madagascar. 3. Famille des Callipallenidae.
Téthys Suppl. 3: 157-164.

Propallene kempi
Stock, J. H. 1954.
Pycnogonida from Indo-West Pacific, Australian and New Zealand waters. Papers from Dr. Th. Mortensen's Pacific Expedition, 1914-1916. LXXVII.
Vidensk. Medd. Dansk Naturhist. Foren., København, 116: 1-168.

Propallene longiceps
Stock, J. H. 1954.
Pycnogonida from Indo-West Pacific, Australian and New Zealand waters. Papers from Dr. Th. Mortensen's Pacific Expedition, 1914-1916. LXXVII.
Vidensk. Medd. Dansk Naturhist. Foren., København, 116: 1-168.

Propallene longiceps
Utinomi, Huzio, 1962
Pycnogonida of Sagami Bay - Supplement.
Publ. Seto Mar. Biol. Lab., 10(1) (Article 5) 91-104.

Propallene longiceps
Utinomi, Huzio, 1952
Pycnogonida of Sagami Bay.
Publ. Seto Mar. Biol. Lab., 7(2):197-222.

Propallene similis n.sp.
Barnard, K. H., 1955.
Additions to the fauna-list of South African Crustacea and Pycnogonida.
Ann. S. African Mus., 43(1): 1-107.

Propallene stocki n.sp.
Fage, L., 1956.
Sur deux espèces de Pycnogonides du Sierra Leone.
Bull. Mus. Nat. Hist. Nat., (2)28(3):290-295.

Pseudopallene ambigua
Stock, Jan H. 1973.
Pycnogonida from south-eastern Australia.
Beaufortia 20 (266). 99-127

Pseudopallene australis
Hodgson, T.V., 1907.
Pycnogonida. Nat. Antarctic Exped., 1901-1904, 3:72 pp., 3 pls.

Pseudopallene circularis
Hedgpeth, J.W., 1948
The Pycnogonida of the western North Atlantic and the Caribbean. Proc. U.S. Nat. Mus. 97:157-342, figs.4-53, charts 1-3.

Pseudopallene circularis
Meinert, Fr., 1899
Pycnogonida. Danish Ingolf Exped. 3(1): 71 pp., 5 pls., 2 text figs., 1 charts, list of stations.

Pseudopallene circularis
Nesis, K.N., 1960.
[Littoral Pantopoda of the eastern Murmansk coast.]
Trudy Murmansk. Morsk. Biol. Inst., 2(6):137-161

Pseudopalleni circularis
Stephensen, K., 1917
Zoogeographical investigations of certain fjords in southern Greenland with special reference to Crustacea, Pycnogonida and Echinodermata including a list of Alcyonaria and Pisces, Medd. om Grønland, 53(3):229-378.

Pseudopallene cornigera
Hodgson, T.V., 1907.
Pycnogonida. Nat. Antarctic Exped., 1901-1904, 3:72 pp., 10 pls.

Pseudopallene gilchristi
Barnard, K.H., 1954.
III(3):South African Pycnogonida.
Ann. S. African Mus. 41:81-158, 34 textfigs.

Pseudopallene gilchristi
Stock, Jan H., 1968.
Pycnogonida collected by the Galathea and Anton Bruun in the Indian and Pacific oceans.
Vidensk. Meddr dansk naturh. Foren. 131: 7-65.

Pseudopallene spinipes
Nesis, K.N., 1960.
[Littoral Pantopoda of the eastern Murmansk coast.]
Trudy Murmansk. Morsk. Biol. Inst., 2(6):137-161

Pseudopallene zamboagea
Stock, J. H. 1954.
Pycnogonida from Indo-West Pacific, Australian and New Zealand waters. Papers from Dr. Th. Mortensen's Pacific Expedition, 1914-1916. LXXVII.
Vidensk. Medd. Dansk Naturhist. Foren., København, 116: 1-168.

Pycnogonum africanum n.sp.
Calman, W. T., 1938
Pycnogonida. John Murray Exped. 1933-34, Sci. Repts. 5(6):147-166, 10 text figs.

Pycnogonum (Nulloviger) africanum
Stock, Jan H., 1968.
Pycnogonida collected by the Galathea and Anton Bruun in the Indian and Pacific oceans.
Vidensk. Meddr dansk naturh. Foren. 131: 7-65.

Pycnogonum arbustum nsp.
Stock, Jan H., 1966.
Pycnogonida from West Africa.
Atlantide Rep., 9:45-57.

Pycnogonum aurilineatum
Stock, Jan H. 1973.
Pycnogonida from south-eastern Australia.
Beaufortia 20 (266). 99-127

Pycnogonum australe
Haswell, W.A., 1885.
On the Pycnogonida of the Australian coast.
Proc. Linn. Soc., N.S. Wales, 9:1-21-1034, Pls. 54-57.

Pycnogonum cessaci (Fig.)
Fage, L., 1952.
Sur quelques Pycnogonides de Dakar. Bull. Mus. Nat. Hist. Nat., 2 ser, 24(6):530-533, 1textfig.

Pycnogonum cessaci
McClosky, L.R., 1967.
New and little-known benthic pycnogonids from North Carolina.
J. nat. Hist., 1:119-132.

Pycnogonum crassirostre
Meinert, Fr., 1899
Pycnogonida. Danish Ingolf Exped. 3(1): 71 pp., 5 pls., 2 text figs., 1 charts, list of stations.

Pycnogonum elephas
Stock, J.H., 1966.
Pycnogonida. Campagne de la Calypso au large des côtes Atlantiques de l'Amerique du Sud (1961-1962). Ann. Inst. Oceanogr., Monaco, n.s. 44: 385-406.

Pycnogonum eltanin n.sp.
Fry, William G. and Joel W. Hedgpeth, 1969.
The fauna of the Ross Sea. 7. Pycnogonida, I. Colossendeidae, Pycnogonidae, Endeidae, Ammotheidae. Bull. N.Z.D.S.I.R. 198 (Mem. N.Z. Oceanogr. Inst. 49): 139 pp.

Pycnogonum gaini
Fry, William G. and Joel W. Hedgpeth, 1969.
The fauna of the Ross Sea. 7. Pycnogonida, I. Colossendeidae, Pycnogonidae, Endeidae, Ammotheidae. Bull. N.Z.D.S.I.R. 198 (Mem. N.Z. Oceanogr. Inst. 49): 139 pp.

Pycnogonum hancocki
Child, C. Allan, and Joel W. Hedgpeth, 1971.
Pycnogonida of the Galápagos Islands. J. nat. Hist. 5(6): 609-634.

Pycnogonum littorale
Hedgpeth, J.W., 1948
The Pycnogonida of the western North Atlantic and the Caribbean. Proc. U.S. Nat. Mus. 97:157-342, figs.4-53, charts 1-3.

Pycnogonum littorale
Jarvis, J.H. and P.E. King, 1972.
Reproduction and development in the pycnogonid Pycnogonum littorale. Mar. Biol. 13(2): 146-154.

Pycnogonium littorale
Lebour, M.V., 1945.
Notes on the Pycnogonida of Plymouth. J.M.B.A. 26:139-165, 7 textfigs.

Pycnogonum littorale
Nesis, K.N., 1960.
[Littoral Pantopoda of the eastern Murmansk coast.]
Trudy Murmansk. Morsk. Biol. Inst., 2(6):137-161.

Pycnogonum littorale
Stephensen, K., 1917
Zoogeographical investigations of certain fjords in southern Greenland with special reference to Crustacea, Pycnogonida and Echinodermata including a list of Alcyonaria and Pisces, Medd. om Grønland, 53(3):229-378.

Pycnogonum madagascariensis
Arnaud, Françoise 1971.
Pycnogonides des récifs coralliens de Madagascar. II Redescription de Pycnogonum madagascariensis Bouvier 1911.
Tethys Suppl. 1: 161-164.

Pycnogonum magellanicum
Gordon, I., 1932.
Pycnogonida. Discovery Repts., 6:1-138.

Pycnogonum magellanicum
Stephensen, K., 1947
Tanaidacea, Isopoda, Amphipoda, and Pycnogonida. Sci. Res. Norweg. Antarctic Exp. 1927-1928 et seq. Date Norske Videnskaps-Akademi i Oslo, II, No.27:1-90, 26 text figs.

Pycnogonum magellanicum
Stock, J.H., 1966.
Pycnogonida. Campagne de la Calypso au large des côtes Atlantiques de l'Amerique du Sud (1961-1962). Ann. Inst. Oceanogr., Monaco, n.s. 44: 385-406.

Pycnogonum sp.cf.microps
Barnard, K.H., 1955.
Additions to the fauna-list of South African Crustacea and Pycnogonida.
Ann. S. African Mus., 43(1): 1-107.

Pycnogonum minutum n.sp.
Lozina-Lozinsky, L.K. and L.M. Kopaneva 1973.
New Pantopoda of the Guinean Republic coast. (In Russian, English abstract).
Zool. Zh. 52(7): 1083-1084.

Pycnogonum nodulosum
Barnard, K.H., 1955.
Additions to the fauna-list of South African Crustacea and Pycnogonida.
Ann. S. African Mus., 43(1): 1-107.

Pycnogonum nodulosum
Bourdillon, A., 1954.
Les pycnogonides de Marseille et ses environs.
Recueil Trav. Sta. Mar., Endoume, 12:145-158.

Pycnogonum occa
Stock, Jan H., 1968.
Pycnogonida collected by the Galathea and Anton Bruun in the Indian and Pacific oceans.
Vidensk. Meddr. dansk naturh. Foren., 131: 7-65.

Pycnogonum planum n.sp.
Stock, J.H. 1954.
Pycnogonida from Indo-West Pacific, Australian and New Zealand waters. Papers from Dr. Th. Mortensen's Pacific Expedition, 1914-1916. LXXVII.
Vidensk. Medd. Dansk Naturhist. Foren., København, 116: 1-168.

Pycnogonum platylophum
Arnaud, Françoise 1970.
Pycnogonides subantarctiques des îles Crozet.
Bull. Mus. natn. Hist. nat. (2) 41 (6): 1423-1428.

Pycnogonum platylophum
Fry, William G. and Joel W. Hedgpeth, 1969.
The fauna of the Ross Sea. 7. Pycnogonida, I. Colossendeidae, Pycnogonidae, Endeidae, Ammotheidae. Bull. N.Z.D.S.I.R. 198 (Mem. N.Z. Oceanogr. Inst. 49): 139 pp.

Pycnogonum plumipes, n.sp.
Stock, J.H., 1960.
Pycnogonum plumipes, n.sp., nouveau pycnogonide de la région de Banyuls.
Vie et Milieu, 11(1):124-126.

Pycnogonum portus n.sp.
Barnard, K.H., 1946.
Diagnoses of new species and a new genus of Pycnogonida in the South African Museum.
Ann. Mag. Nat. Hist. (11)13:60-63.

Pycnogonum pusillum
Bourdillon, A., 1954.
Les pycnogonides de Marseille et ses environs.
Recueil Trav. Sta. Mar., Endoume, 12:145-158.

Pycnogonum pusillum
Stock, J.H., 1964.
Report on the Pycnogonida of the Israel South Red Sea Expedition.
Sea Fish Res. Inst., Israel, Bull. No. 35:27-34.

Pycnogonum reticulatum n.sp.
Hedgpeth, J.W., 1948
The Pycnogonida of the western North Atlantic and the Caribbean. Proc. U.S. Nat. Mus. 97:157-342, figs.4-53, charts 1-3.

Pycnogonum rhinoceros
Fry, William G. and Joel W. Hedgpeth, 1969.
The fauna of the Ross Sea. 7. Pycnogonida, I. Colossendeidae, Pycnogonidae, Endeidae, Ammotheidae. Bull. N.Z.D.S.I.R. 198 (Mem. N.Z. Oceanogr. Inst. 49): 139 pp.

Pycnogonum rhinoceros
Gordon, I., 1932.
Pycnogonida. Discovery Repts., 6:1-138.

Pycnogonum stearnsi
Fry, William G., 1965.
The feeding mechanisms and preferred foods of three species of Pycnogonida.
Bull. Brit. Mus. (N.H.), Zool., 12(6):197-223.

Pycnogonum tenua
Stock, J.H. 1954.
Pycnogonida from Indo-West Pacific, Australian and New Zealand waters. Papers from Dr. Th. Mortensen's Pacific Expedition, 1914-1916. LXXVII.
Vidensk. Medd. Dansk Naturhist. Foren., København, 116: 1-168.

Pycnogonum tenue
Utinomi, H., 1955.
Report on the Pycnogonida collected by the Soyomaru expeditions made on the continental shelf bordering Japan during the years 1926-1930.
Publ. Seto Mar. Biol. Lab., 5(1): 1-42.

Pycnogonum tenue
Utinomi, Huzio, 1952
Pycnogonida of Sagami Bay.
Publ. Seto Mar. Biol. Lab., 7(2):197-222.

Pycnogonum tessellatum n.sp.
Stock, Jan H., 1968.
Pycnogonida collected by the Galathea and Anton Bruun in the Indian and Pacific oceans.
Vidensk. Meddr. dansk naturh. Foren., 131: 7-65.

Pycnothea flynni
Arnaud, Françoise 1972.
Pycnogonides des récifs coralliens de Madagascar. 3. Famille des Callipallenidae.
Tethys, Suppl. 3: 157-164.

Pycnothea flynni
Stock, Jan H. 1973.
Pycnogonida from south-eastern Australia.
Beaufortia 20 (266): 99-127

Queubus jamesanus n.sp
Barnard, K.H., 1946.
Diagnoses of new species and a new genus of Pycnogonida in the South African Museum.
Ann. Mag. Nat. Hist., (11)13:60-63.

Rhopalorhynchus articulatus
Stock, Jan H., 1968.
Pycnogonida collected by the Galathea and Anton Brunn in the Indian and Pacific oceans.
Vidensk. Meddr dansk naturh. Foren. 131: 7-65.

Rhopalorhynchus atlanticum n.sp.
Stock, Jan H. 1970.
The Pycnogonida collected off northwestern Africa during the cruise of the Meteor.
Meteor-Forsch. Ergebn. (D) 5: 6-10.

Rhopalorhynchus aff. clavipes
Monod, Théodore 1971.
A propos d'un Rhopalorhynchus australien (Pycnogonida).
Bull. Mus. natn. Hist. nat. (2) 42 (6): 1263-1267.

Rhopalorhynchus clavipes
Stock, Jan H., 1968.
Pycnogonida collected by the Galathea and Anton Brunn in the Indian and Pacific oceans.
Vidensk. Meddr dansk naturh. Foren. 131: 7-65.

Rhopalorhynchus kröyeri
Barnard, K.H., 1954.
III(3): South African Pycnogonida.
Ann. S. African Mus. 41:81-158, 34 textfigs.

Rhopalorhynchus kröyeri
Calman, W. T., 1938
Pycnogonida. John Murray Exped. 1933-34.
Sci. Repts. 5(6):147-166, 10 text figs.

Rhopalorhynchus kröyeri
Stock, J. H. 1954.
Pycnogonida from Indo-West Pacific, Australian and New Zealand waters. Papers from Dr. Th. Mortensen's Pacific Expedition, 1914-1916. LXXVII.
Vidensk. Medd. Dansk Naturhist. Foren., København, 116: 1-168.

Rhopalorhynchus kroeyeri
Stock, J.H., 1957.
Pantopoden aus dem Zoologischen Museum Hamburg.
Mitt. Hamburgischen Zoo. Mus. u. Inst., 55:81-106.

Rhopalorhynchus lomani [a]
Arnaud, Françoise 1972 (1973).
Pycnogonides des récifs coralliens de Madagascar. 4. Colossendeidae, Rhynchothoriidae et Endeidae.
Téthys 4(4): 953-960.

Rhopalorhynchus lomani
Stock, Jan H., 1965.
Pycnogonida from the southwest Indian Ocean.
Beaufortia, 13(151):13-33.

Rhopalorhynchus mortensi
Stock, Jan H., 1968.
Pycnogonida collected by the Galathea and Anton Brunn in the Indian and Pacific oceans.
Vidensk. Meddr dansk naturh. Foren., 131: 7-65.

Rhopalorhynchus pedunculatum n.sp.
Stock, J.H., 1957.
Pycnogonida from the Gulf of Aqaba. Contr. Knowl. Red Sea, 2. Bull. Sea Fish. Res. Sta., Israel, No. 13:13-14.

Rhynchothorax anophthalmus n.sp.
Arnaud, Françoise 1972
Un nouveau pycnogonide de Méditerranée nord-occidentale. Rhynchothorax anophthalmus n.sp. et redécouverte de Rhynchothorax mediterraneus Costa 1861.
Téthys 3(2): 405-409

Rhynchothorax articulatus n.sp.
Stock, Jan H., 1968.
Pycnogonida collected by the Galathea and Anton Brunn in the Indian and Pacific oceans.
Vidensk. Meddr dansk naturh. Foren., 131: 7-65.

Rhynchothorax australis
Fray, William G., 1965.
The feeding mechanisms and preferred foods of three species of Pycnogonida.
Bull. Brit. Mus.(N.H.), Zool., 12(6):197-223.

Rhyncothorax australis
Gordon, I., 1932.
Pycnogonida. Discovery Repts., 6:1-138.

Rhynchothorax australis
Hedgpeth, J. W., 1950
Pycnogonida of the United States Navy Antarctic Expedition, 1947-48. Proc. U. S. Nat. Mus. 100 (3260):147-160, 19 text figs.

Rhynchothorax australis
Hodgson, T.V., 1907.
Pycnogonida. Nat. Antarctic Exped., 1901-1904, 3:72 pp., 10 pls.

Rhynchothorax barnardi n. sp.
Child, C. Allan, and Joel W. Hedgpeth, 1971.
Pycnogonida of the Galápagos Islands. J. nat. Hist. 5(6): 609-634.

Rhynchothorax malaccensis n. sp.
Stock, Jan H., 1968.
Pycnogonida collected by the Galathea and Anton Brunn in the Indian and Pacific oceans.
Vidensk. Meddr dansk naturh. Foren., 131: 7-65.

Rhynchothorax mediterraneus [a]
Arnaud, Françoise 1972.
Un nouveau pycnogonide de Méditerranée nord-occidentale. Rhynchothorax anophthalmus n.sp. et redécouverte de Rhynchothorax mediterraneus Costa, 1861.
Téthys 3(2): 405-409

Rhynchothorax mediterraneus [a]
Arcifa Zago, Marlene Sofia, 1970.
Sôbre o pantópodo Rhynchothorax mediterraneus Costa, 1861.
Contrções Inst. oceanogr. São Paulo, sér. Oceanogr. Biol. 21:1-5.

Rhynchothorax mediterraneus
Zilberberger, F., 1963
Notes on Pantopoda.
Bol. Inst. Oceanogr., Sao Paulo, 13(2):21-32.

Rhyncothorax unicornis n.sp.
Fage, Louis (posthumous) et Jan H. Stock, 1966.
Pycnogonides. Campagne de la Calypso aux îles du Cap Vert (1959). Ann. Inst. Océanogr., Monaco, n.s., 44: 315-328.

Rhynchothorax Voxorinum n.sp.
Stock, Jan H., 1966.
Sur quelques pycnogonides de la région de Banyuls. (3º note).
Vie Milieu (B) 17(1):407-417.

Scipiolus bifidus n. sp.
Stock, Jan H., 1968.
Pycnogonida collected by the Galathea and Anton Brunn in the Indian and Pacific oceans.
Vidensk. Meddr dansk naturh. Foren. 131: 7-65.

Scipiolus spinosus n.sp.
Utinomi, H., 1955.
Report on the Pycnogonida collected by the Soyo-maru expeditions made on the continental shelf bordering Japan during the years 1926-1930.
Publ. Seto Mar. Biol. Lab., 5(1):1-42.

Scipiolus validus n.sp.
Stock, J.H., 1957.
Pantopoden aus dem Zoologischen Museum Hamburg.
Mitt. Hamburgischen Zool. Mus. u. Inst., 55:81-106.

Sericosura nitrata n. gen. [a]
Fry, William G. and Joel W. Hedgpeth, 1969.
The fauna of the Ross Sea. 7. Pycnogonida. I. Colossendeidae, Pycnogonidae, Endeidae, Ammotheidae. Bull. N.Z.D.S.I.R. 198 (Mem. N.Z. Oceanogr. Inst. 49): 139 pp.

Siphopallene tubirostris
Stock, Jan H., 1968.
Pycnogonida collected by the Galathea and Anton Brunn in the Indian and Pacific oceans.
Vidensk. Meddr dansk naturh. Foren., 131: 7-65.

Spasmopallene clarki n.sp.
Stock, Jan H., 1968.
Pycnogonida collected by the Galathea and Anton Brunn in the Indian and Pacific oceans.
Vidensk. Meddr dansk naturh. Foren., 131: 7-65.

Spasmopallene reflexa n.gen. n. sp.
Stock, Jan H., 1968.
Pycnogonida collected by the Galathea and Anton Brunn in the Indian and Pacific oceans.
Vidensk. Meddr dansk naturh. Foren., 131: 7-65.

Stylopallene cheilorhynchus
Stock, Jan H. 1973.
Pycnogonida from south-eastern Australia.
Beaufortia 20 (266). 99-127

Stylopallene longicauda n.sp.
Stock, Jan H. 1973.
Pycnogonida from south-eastern Australia.
Beaufortia 20 (266). 99-127

Tanystylum spec.
Stock, Jan H., 1965.
Pycnogonida from the southwest Indian Ocean.
Beaufortia, 13(151):13-33.

Tanystylum acuminatum n. sp. [a]
Stock, J. H. 1954
Four new Tanystylum species, and other Pycnogonida from the West Indies. Studies Fauna Curaçao and other Carib. Is. 5(24):115-129, 7 pls.
West Indies

Tanystylum anthomasthi
Utinomi, H., 1954.
The fauna of Akkeshi Bay. XIX. Littoral Pycnogonida. Publ. Akkeshi Mar. Biol. Sta. No. 3:1-28, 1 pl., 11 textfigs.

Contr. 221 Seto Mar. Biol. Lab.

Tanystylum bigibbosum n.sp.
Fage, Louis (posthumous) et Jan H. Stock, 1966.
Pycnogonides. Campagne de la Calypso aux îles du Cap Vert (1959). Ann. Inst. Océanogr., Monaco, n.s., 44: 315-328.

Tantystylum bredini n.sp.
Child, C. Allan, 1970.
Pycnogonida of the Smithsonian-Bredin Pacific Expedition, 1957.
Proc. Biol. Soc. Wash. 82(27): 287-308.

Tantystylum brevicaudatum n.sp.
Fage, Louis (posthumous) et Jan H. Stock, 1966.
Pycnogonides. Campagne de la Calypso aux îles du Cap Vert (1959). Ann. Inst. Océanogr., Monaco, n.s., 44: 315-328.

Tanystylum brevipes
Arnaud, Françoise 1971.
Pycnogonides littoraux des îles Saint-Paul et Amsterdam (Océan Indien).
Tethys 3(1): 159-166.

Tanystylum calicirostra
Hedgpeth, J.W., 1948
The Pycnogonida of the western North Atlantic and the Caribbean. Proc. U.S. Nat. Mus. 97:157-342, figs.4-53, charts 1-3.

Tanystylum cavidorsum
Arnaud, Françoise 1970.
Pycnogonides subantarctiques des îles Crozet.
Bull. Mus. natn. Hist. nat. (2) 41 (6): 1423-1428.

Tanystylum cavidorsum
Clark, W.C. 1971.
Pycnogonida of the Antipodes Islands.
N.Z. Jl mar. Freshwat. Res. 5(3/4): 427-452.

Tanystylum cavidorsum
Clark, W.C. 1971.
Pycnogonida of the Snares Islands.
N.Z. Jl mar. Freshwat. Res. 5(2): 329-341.

Tanystylum cavidorsum n.sp.
Stock, J.H., 1957.
Pantopoden aus dem Zoologischen Museum Hamburg.
Mitt. Hamburgischen Zool. Mus. u. Inst., 55:81-106.

Tanystylum conirostrum
Krapp, Franz 1973.
Pycnogonida from Pantelleria and Catania, Sicily.
Beaufortia 21 (277): 55-74.

Tanystylum distinctum n. sp.
Child, C. Allan, and Joel W. Hedgpeth, 1971.
Pycnogonida of the Galápagos Islands. J. nat. Hist. 5(6): 609-634.

Tanystylum excuratum n.sp.
Stock, J. H. 1954.
Pycnogonida from Indo-West Pacific, Australian and New Zealand waters. Papers from Dr. Th. Mortensen's Pacific Expedition, 1914-1916. LXXVII.
Vidensk. Medd. Dansk Naturhist. Foren., København, 116: 1-168.

Tanystylum geminum
Bourdillon, A., 1955.
Les Pycnogonides de la croisière 1951 du "President Theodore Tissier".
Rev. Trav. Inst. Pêches Marit., 19(4):581-609.

Tanystylum geminum n. sp.
Stock, J.H. 1954
Four new Tanystylum species, and other Pycnogonida from the West Indies.
Studies Fauna Curaçao and Other Carib. Is. 5(24):115-129, 7 pls.
West Indies

Tanystylum hummelincki n. sp.
Stock, J.H. 1954
Four new Tanystylum species, and other Pycnogonida from the West Indies.
Studies Fauna Curaçao and Other Carib. Is. 5(24):115-129, 7 pls.
West Indies

Tanystylum isthmiacum difficile
Fage, Louis (posthumous) et Jan H. Stock, 1966.
Pycnogonides. Campagne de la Calypso aux îles du Cap Vert (1959). Ann. Inst. Océanogr., Monaco, n.s., 44: 315-328.

Tanystylum isthmiacum difficile n.subsp.
Stock, J.H., 1966.
Pycnogonida. Campagne de la Calypso au large des côtes Atlantiques de l'Amérique du Sud (1961-1962). Ann. Inst. Océanogr., Monaco, n.s. 44: 385-406.

Tanystylum neorhetum
Arnaud, Françoise 1972.
Pycnogonides des îles Kerguelen (Sud Océan Indien): matériel nouveau et révision des spécimens du Museum national d'Histoire naturelle de Paris.
Bull. Mus. natn. Hist. nat. Paris, (3) 65 (Zool. 51): 801-815.

Tanystylum neorhetum
Stock, J. H. 1954.
Pycnogonida from Indo-West Pacific, Australian and New Zealand waters. Papers from Dr. Th. Mortensen's Pacific Expedition, 1914-1916. LXXVII.
Vidensk. Medd. Dansk Naturhist. Foren., København, 116: 1-168.

Tanystylum neorhetum
Stock, J.H., 1957.
Pantopoden aus dem Zoologischen Museum Hamburg.
Mitt. Hamburgischen Zool. Mus. u. Inst., 55:81-106.

Tantystylum neriotes n.sp.
Child, C. Allan, 1970.
Pycnogonida of the Smithsonian-Bredin Pacific Expedition, 1957.
Proc. Biol. Soc. Wash. 83(27): 287-308.

Tanystylum oculospinosum
Child, C. Allan, and Joel W. Hedgpeth, 1971.
Pycnogonida of the Galápagos Islands. J. nat. Hist. 5(6): 609-634.

Tanystylum oedinotum
Arnaud, Françoise 1972.
Pycnogonides des îles Kerguelen (Sud Océan Indien): matériel nouveau et révision des spécimens du Museum national d'Histoire naturelle de Paris.
Bull. Mus. natn. Hist. nat. Paris, (3) 65 (Zool. 51): 801-815.

Tanystylum orbiculare
Bourdillon, A., 1955.
Les Pycnogonides de la croisière 1951 du "President Theodore Tissier".
Rev. Trav. Inst. Pêches Marit., 19(4):581-609.

Tanystylum orbiculare
Bourdillon, A., 1954.
Les pycnogonides de Marseille et ses environs.
Recueil Trav. Sta. Mar., Endoume, 12:145-158.

Tanystylum orbiculare
Fage, L., 1952.
Sur quelques Pycnogonides de Dakar. Bull. Mus. Nat. Hist. Nat., 2 ser., 24(6):530-533, 1 textfig.

Tanystylum orbiculare
Fage, Louis (posthumous) et Jan H. Stock, 1966
Pycnogonides. Campyne de la Calypso aux îles du Cap Vert (1959)
Ann. Inst. Océanogr., Monaco, n.s., 44:315-328.

Tanystylum orbiculare
Hedgpeth, J.W., 1948
The Pycnogonida of the western North Atlantic and the Caribbean. Proc. U.S. Nat. Mus. 97:157-342, figs.4-53, charts 1-3.

Tanystylum orbiculare
Krapp, Franz 1973.
Pycnogonida from Pantelleria and Catania, Sicily.
Beaufortia 21 (277): 55-74.

Tanystylum orbiculare
Stock, Jan H., 1966.
Pycnogonida from West Africa.
Atlantide Rep., 9:45-57.

Tanystylum orbiculare
Stock, J. H. 1954.
Pycnogonida from Indo-West Pacific, Australian and New Zealand waters. Papers from Dr. Th. Mortensen's Pacific Expedition, 1914-1916. LXXVII.
Vidensk. Medd. Dansk Naturhist. Foren., København, 116: 1-168.

Tanystylum orbiculare
Stock, J.H., 1958
Pycnogonida from the Mediterranean coast of Israel.
Bull. Res. Counc., Israel, (B) 7(3/4):137-142.
Reprinted in:
Bull. Sea Fish. Res. Sta., Haifa, No. 19:

Tanystylum pfefferi
Gordon, I., 1932.
Pycnogonida. Discovery Repts., 6:1-138.

Tantystylum rehderi n.sp.
Child, C. Allan, 1970.
Pycnogonida of the Smithsonian-Bredin Pacific Expedition, 1957.
Proc. Biol. Soc. Wash. 83(27): 287-308.

Tanystylum scutator n.sp.
Stock, J. H. 1954.
Pycnogonida from Indo-West Pacific, Australian and New Zealand waters. Papers from Dr. Th. Mortensen's Pacific Expedition, 1914-1916. LXXVII.
Vidensk. Medd. Dansk Naturhist. Foren., København, 116: 1-168.

Tanystylum styligerum
Arnaud, Françoise 1972.
Pycnogonides des îles Kerguelen (Sud Océan Indien): matériel nouveau et révision des spécimens du Museum national d'Histoire naturelle de Paris.
Bull. Mus. natn. Hist. nat. Paris, (3) 65 (Zool. 51): 801-815.

Tanystylum styligerum
Gordon, I., 1932.
Pycnogonida. Discovery Repts., 6:1-138.

Tanystylum thermophilum
Stock, J.H., 1957.
Pantopoden aus dem Zoologischen Museum Hamburg. Mitt. Hamburgischen Zool. Mus. u. Inst., 55:81-106.

Tanystylum tubirostra
Bourdillon, A., 1955.
Les pycnogonides de la croisière 1951 du "President Theodore Tissier". Rev. Trav. Inst. Pêches Marit., 19(4):581-609.

Tanystylum tubirostre n. sp.
Stock, J.H. 1954
Four new Tanystylum species, and other Pycnogonida from the West Indies. Studies Fauna Curaçao and other Carib. Is. 5(24):115-129, 7 pls.
West Indies

Thaumatopycnon profunda
Fry, William G. and Joel W. Hedgpeth, 1969.
The fauna of the Ross Sea. 7. Pycnogonida, I. Colossendeidae, Pycnogonidae, Endeidae, Ammotheidae. Bull. N.Z.D.S.I.R. 198 (Mem. N.Z. Oceanogr. Inst. 49): 139 pp.

Thaumastopycnon striata n.gen.
Fry, William G. and Joel W. Hedgpeth, 1969.
The fauna of the Ross Sea. 7. Pycnogonida, I. Colossendeidae, Pycnogonidae, Endeidae, Ammotheidae. Bull. N.Z.D.S.I.R. 198 (Mem. N.Z. Oceanogr. Inst. 49): 139 pp.

Trygaeus communis
Bourdillon, A., 1954.
Les pycnogonides de Marseille et ses environs. Recueil Trav. Sta. Mar., Endoume, 12:145-158.

Trygaeus communis
Krapp, Franz 1973.
Pycnogonida from Pantellaria and Catania, Sicily.
Beaufortia 21 (277):55-74.

Trygaeus communis
Stock, Jan H., 1966.
Sur quelques pycnogonides de la région de Banyuls. (3°note).
Vie Milieu (B)17(1):407-417.

SCHIZOPODA
Chiefly under euphausids mysids

schizopods
Banner, A.H., 1954.
Some schizopod crustaceans from the deeper water off California.
Allan Hancock Found. Publ., Occ. Papers, No. 15:1-48.

schizopodes
Caullery, M., 1896
Crustacés Schizopodes et Décapodes. Res. camp. sci. "Caudan", Golfe de Gascogne, fasc. 2, pp. 365-419, pls. 13-17

schizopods
Esterly, C.O., 1914.
The Schizopoda of the San Diego region. Univ. Calif. Publ., Zool., 13(1):1-20, Pls. 1-2.

schizopods
Holt, E.W.L., and W.M. Tattersall, 1906.
Schizopodous Crustacea from the north-east Atlantic slope. Supplement.
Fish., Ireland, Sci. Invest., 1904, No. 5:1-50, 5 pls.

Schizopods
Holt, E.W.L., and W.M. Tattersall, 1905.
Schizopodous Crustacea from the northeast Atlantic slope.
Sci. Invest., Dept. Agric. & Tech., Fish. Ireland, 1902-1903, 4(1):99-152.

schizopoda
Kramp, P.L. 1913.
Schizopoda. Bull. Trim. Cons. Int. Explor. Mer, pp. 539-556

Schizopoda
Pillai, N.K., 1957.
Pelagic Crustacea of Travancore. II. Schizopoda. III. Amphipoda.
Bull. Central Res. Inst., Univ. Travancore. Trivandrum, (C) Nat. Sci., 5(1):1-68.

Schizopoda
Sars, G. O., 1885
Report on the Schizopoda collected by H.M.S. "Challenger" during the years 1873-76. Challenger Repts., Zool. 13:1-225, 38 pls.

Schizopoda
Savage, R.E., 1937
The food of North Sea herring 1930-1934. Ministry of Agriculture and Fisheries. Fish. Invest. Ser. II, 15(5):1-60; 16 text figs.

Schizopoda
Stephensen, K., 1933
The Godthaab expedition 1928. Schizopoda. Medd. om Grønland, 79(9):20 pp.

Schizopoda
Verrill, A. E., 1923.
Crustacea of Bermuda - Schizopoda, Cumacea, Stomatopoda, and Phyllocarida. Trans Conn. Acad. Arts Sci., 26:181-211.

Stomatopoda
Armstrong, J. C., 1941.
The Caridea and Stomatopoda of the Second Templeton Crocker-American Museum Expedition to the Pacific Ocean. Amer. Mus. Novitates, 1137:1-14, 4 figs.

Stomatopod
Barnard, K. H., 1950
Descriptive list of South African Stomatopod Crustacea (Mantis Shrimps). Ann. S. African Mus. 38:838-864, 4 text figs.

stomatopoda
Bigelow, R. P., 1931.
Stomatopoda of the Southern and Eastern Pacific Ocean and the Hawaiian Islands. Bull. M. C. Z., 72(4):105-191.

stomatopods
Bigelow, R. P., 1902.
The Stomatopoda of Porto Rico. Bull. U.S. Fish Comm., for 1900, 20:151-160.

stomatopoda
Bigelow, R. P., 1894.
Report on the Crustacea of the Order Stomatopoda collected by the Steamer "Albatross" between 1885 and 1891, and on other specimens in the U. S. National Museum. Proc. U.S. Nat. Mus., 17:489-550.

stomatopoda
Bigelow, R. P., 1893.
The Stomatopoda of Bimini. Johns Hopkins Univ. Circ., 106:102-103.

stomatopoda
Bigelow, R. P., 1893.
Preliminary notes on the Stomatopods of the Albatross Collections and on other Specimens in the National Museum. Johns Hopkins Univ. Circ., 106:100-102.

stomatopoda
Brooks, W. K., 1886.
Report of the Stomatopoda Collected by H. M. S. Challenger, 1873-76. Sci. Rept., etc., Zool., 16, II

Stomatopoda
Calman, W.T., 1927.
Zoological results of the Cambridge Expedition to the Suez Canal. XXVII. Report on the Phyllocaridea, Cumacea and Stomatopoda. Trans. Zool. Soc., London, 22:399-401, textfig. 101.

Stomatopoda
Calman, W. T., 1917.
Stomatopoda, Cumacea, Phyllocarida, and Cladocera. Brit. Antarct. (Terra Nova) Exped., III(5):137-162.

stomatopods
Chace, F.A., Jr., 1956.
Crustaceos decapodos y stomatopodos del Archipelago de Los Roques y Isla de la Orchila. In: El Archipelago de Los Roques y La Orchila, Editorial Sucre, Caracas:145-168.

stomatopods
Chace, F.A., jr., 1951.
The number of species of decapod and stomatopod Crustacea. J. Washington Acad. Sci. 41(11):370-372.

Stomatopoda
Chopra, B., 1939
Stomatopoda. John Murray Exped., 1933-34, Sci. Repts. 6(3):137-181, 13 text figs.

stomatopods
Dawydoff, C., 1952.
Contribution à l'etude des invertébrés de la faune marine bentheque de l'Indochine. Bull. Biol., France Belg., Suppl., 37:1-158.

stomatopods
de Man, J. G., 1897.
Bericht über die von Herrn Schiffscapitän STORM zu Atjeh, an den westlichen Küsten von Malakka, Borneo, und Celebes, sowie in der Java-Ses gesammelten Decapoden und Stomatopoden. Pt. 5. Zool. Jahrb., Abth. Syst., 9:725-790, Pls. 12-14.

stomatopods
de Man, J.G., 1896.
Berichte über die von Herrn Schiffscapitän STORM zu Atjeh, an den westlichen Küsten von Malakka, Borneo und Celebes sowie in der Java-See gesammelten Decapoden und Stomatopoden. Pt. 3. Zool. Jahrb., Abth. Syst., 9:339-386.

Stomatopoden
de Man, J.G., 1896.
Bericht über die von Herrn Schiffscapitän STORM zu Atjeh, an den westlichen Küsten von Malakka, Borneo, und Celebes, sowie in der Java-See gesammelten Decapoden und Stomatopoden. t. 4. Zool. Jahrb., Abth. Syst., 9:459-514.

Stomatopods
Edmondson, C. H., 1921.
Stomatopoda in the B. P. Bishop Museum. Occ. Papers, B. P. Bishop Mus., 7(13):279-302.

Stomatopod
Foxon, G. E. H., 1932.
Report on Stomatopod Larvae, Cumacea, and Cladocera. Gt. Barrier Reef Exped., 1928-1929. Sci. Rep., IV(11):375-398.

stomatopods
Hansen, H. J., 1926.
Stomatopoda of the Siboga Expedition. Siboga Exped. Mon. 35, (Livr. 104):

Stomatopoda
Hansen, H. J., 1895.
Isopoden, Cumaceen und Stomatopoden der Plankton Expedition, pp. 64-103, Taf. VIII.

Stomatopod

Hansen, H. J., 1895
Isopoden, Cumaceen und Stomatopoden der Plankton Expedition. Ergeb. Plankton Exped. II, Gc: 105 pp., 8 pls.

stomatopod

Holthuis, L.B., 1941
Note on some Stomatopoda from the Atlantic Coasts of Africa and America with the description of a new species. Zool. Mededeel. Leyden 23:31-43. 1 fig.

stomatopoda

Kemp, S. W., 1913.
An account of the Crustacea Stomatopoda of the Indo Pacific Region based on the Collection in the Indian Museum. Mem. Ind. Mus., 4:1-217

stomatopods

Kemp, S. W., and B. Chopra, 1921.
Notes on Stomatopoda. Records of the Indian Mus., 22:

stomatopods

Lanchester, W. F., 1903.
Marine Crustacea. Stomatopoda, etc. Gardiners Fauna and Geography of the Maldive and Laccadive 1:444-459.

stomatopods

Lunz, G. R. jr., 1935.
Stomatopods (Mantis shrimps) of the Carolinas. J. Elisha Mitschell Soc., 51(1):151-159.

Stomatopods

Manning, Raymond B., 1968.
Stomatopod Crustacea from Madagascar. Proc.U.S. nat.Mus.124(3641):1-61.

Stomatopods

Manning, Raymond B., 1966.
Stomatopod Crustacea. Campagne de la Calypso au large des côtes Atlantiques de l'Amerique du Sud (1961-1962)-1.
Ann. Inst. Océanogr., Monaco, n.s., 44:359-384

stomatopods

Manning, R.B., 1963.
Preliminary revision of the genera Pseudosquilla and Lysiosquilla with descriptions of six new genera (Crustacea: Stomatopoda).
Bull. Mar. Sci., Gulf and Caribbean, 13(2):308-328.

stomatopods

Miers, E. J., 1880.
On the Squillidae. Ann. Mag. Nat. Hist., (5), 5:1-30; 108-127.

stomatopodes

Monod, T., 1927,
Sur les Stomatopodes de la cote occidentale d'Afrique. Bull. Soc. Sci. Nat., 5(3):86-93

stomatopods

Parisi, B., 1940
Gli Stomatopodi raccolti del Prof. L. Sanzo nella Campagna idrografica nel Mar Rosso della R. N. Ammiraglio Magnaghi 1923-1924. (Mem. 16a della Campagne). Mem. R. Com. Talass. Ital. CCLXXV:7 pp.

stomatopods

Ramos, F. de P., 1951.
Nota preliminar sobre alguno Stomatopoda da Costa Brasileira. Bol. Inst. Paulista Ocean. 2 (1):139-150, 1 pl.

stomatopods

Rathbun, M. J., 1900.
Results of the Branner-Agassiz Expedition to Brazil. Decapod and Stomatopod Crustacea. Proc. Washington Acad. Sci., 2:135-156.

stomatopods

Schmitt, W. L., 1940
The stomatopods of the west coast of America, based on collections made by the Allan Hancock Expeditions, 1933-1938. Rept. Allan Hancock Pacific Exped. 5(4):129-226, 33 figs.

stomatopoda

Schmitt, W. L., 1929.
Chinese stomatopoda collected by S. F. Light. Lingnan Sci. Jour., 8:127-148.

stomatopoda

Schmitt, W. L., 1924.
Bijdragen tot de Kennis der Fauna van Curacao The Macruran, Anomuran, and Stomatopod Crustacea. Bijdragen tot de Dierkunde, Natura Artis Magistra te Amsterdam, 23:61-81.

stomatopoda

Schmitt, W. L., 1924.
Report on the Macrura, Anomura, and Stomatopoda collected by the Barbados-Antigua Expedition from the University of Iowa in 1918. Univ. Iowa Studies Nat. Hist., 10(4):65-99.

Stomatopoda

Stephenson, W., 1953.
Three new Stomatopoda (Crustacea from eastern Australia. Australian J. Mar. Freshwater Res. 4(1):201-218, 4 textfigs.

stomatopods

Stephenson, W., 1952.
Faunistic records from Queensland. Pt. I. General Introduction. Pt. II. Adult Stomatopoda (Crustacea). Univ. Queensland, Dept. Zool., Papers, 1(1):1-15.

Stomatopoda

Steuer, A., 1938
The fishery grounds near Alexandria. XVI Cumacea, Stomatopoda, Leptostraca. Hydrobiol. & Fish. Dir., Notes & Mem. No.26: 16 pp., 16 figs.

stomatopods

Townsley, S.J., 1953.
Adult and larval stomatopod crustaceans occurring in Hawaiian waters. Pacific Science 7(4):399-437, 399-437, 28 textfigs.

Stomatopods

van der Baan, S. M., and L. B. Holthuis, 1966.
on the occurrence of Stomatopoda in the North Sea, with special reference to larvae from the surface plankton near the lightship "Texel".
Netherlands J. Sea Res., 3(1):1-12.

Stomatopoda

Verrill, A. E., 1923.
Crustacea of Bermuda - Schizopoda, Cumacea, Stomatopoda, and Phyllocarida. Trans. Conn. Acad. Arts Sci., 26:181-211.

stomatopod

Vilela, H., 1936
Crustaceos Decapodes. Estomatopodes. Trav. St. Biol. Mar., Lisbonne, No. 40:215-242.

Stomatopoda

Wood-Mason, J., 1895.
Figures and Descriptions of nine species of Squillidae from the Collection in the Indian Museum.

stomatopods, anat.-physiol.

Alexandrowicz, J.S., 1957.
Notes on the nervous system in the Stomatopoda. Pubbl. Staz. Zool., Napoli, 29:213-225.

stomatopods, anat.-phys

Alexamdrowicz, J.S., 1954.
Notes on the nervous system in the Stomatopoda. IV. Muscle receptor organs.
Pubbl. Staz. Zool., Napoli, 25(1):94-111, Pls. 3-4. 3 textfigs.

Stomatopod, anat.-physiol.

Brown, Hilary F., 1964.
Electrophysiological investigations of the heart of Squilla mantis. 1. The ganglionic nerve trunk. 2. The heart muscle. 3. The mode of action of pericardial organ extract on the heart.
J. Exp. Biol., 41(4):689-700; 701-722; 723-734.

stomatopods, anat.

Burnett, Bryan R. 1972.
Notes on the lateral arteries of two stomatopods.
Crustaceana 23(3):303-305

stomatopods, anat.-physiol.

Irisawa, H., and A.F. Irisawa, 1957.
The electrocardiogram of a stomatopod. Biol. Bull., 112(3):358-362.

stomatopods, anat.-phys.

Jacques, Francoise 1970.
La glande de mue chez les larves de stomatopodes.
C. r. hebd. Séanc. Acad. Sci., Paris (D) 270 (24): 2965-2968.

stomatopods, anatomy

Lauterbach, Karl-Ernst 1972.
Zur Kenntnis von Carapax und Thorax der Stomatopoda (Crustacea).
Zool. Anz. 188 (1/2): 75-78

stomatopods, anat.-physiol.

Mayrat, A., 1958.
Le coeur et les artères latérales des Stomatopodes. Historique et interprétation (Recherches sur l'appareil circulatoire des Crustacés IV). Bull. Soc., Zool., 83(5/6):462-477.

stomatopods, anat.-physi

Oka, T.B., 1953.
On the anal-gland found in Squilla oratoria. Nat. Sci. Rept., Ochanomizu Univ., 4(1):119-127, 16 textfigs.

stomatopod, anat.

Pilgrim, R.L.C., 1964.
Stretch receptor organs in Squilla mantis Latr. (Crustacea: Stomatopoda).
J. Exp. Biol., 41(4):793-804.

stomatopoda, anat., physiol.

Serène, R., 1954.
Observations biologiques sur les stomatopodes. Mem. Inst. Ocean., Nhatrang, No. 8:1-93, 10 pls., 15 textfigs.

stomatopod physiology

Serene, R., 1951.
Sur la circulation d'eau à la surface du corps des Stomatopodes. Bull. Soc. Zool., France, 76(3) :137-143, 3 textfigs., 1 pl.

stomatopod, anat.-phys.

Snodgrass, R.E., 1956.
Crustacean metamorphosis. Smithsonian Misc. Coll., 131(10):1-78.

stomatopods, anatomy

Yanase, T., Y. Okuno and K. Fujimoto 1972.
Fine structure of the compound eye in mantis crab, Squilla oratoria.
Zool. Mag. (Dobuts. Zasshi) 81 (3): 211-216

Stomatopod distribution

Blumstein, R.G. 1972.
A contribution to the geographic distribution of Stomatopoda. (In Russian; English abstract).
Zool. Zh. 51 (8): 1165-1170.

stomatopods, lists of spp.

Alikunhi, K.H., 1967.
An account of the post-larval development, moulting and growth of the common stomatopods of the Madras coast.
Proc. Symp. Crustacea, Ernakulam, Jan. 12-15, 1965, 2: 824-939

Stomatopods, lists of spp. (figs.)

Chirichigno Fonseca, Norma, 1970
Lista de crustáceos del Perú (Decapoda y Stomatopoda) con datos de su distribución geográfica.
Informe Inst. Mar, Perú, 32: 95 pp.

stomatopods, check list

Forest, J., and D. Guinot, 1956.
Sur une collection de crustacés décapodes et stomatopodes des mers tunisiennes.
Bull. Sta. Océan., Salammbo, 53:24-43.

Stomatopods, lists of spp.

Holthuis, 1967.
The Stomatopod Crustacea collected by the 1962 and 1965 Israel South Red Sea expeditions.
Israel J. Zool., 16(1):1-45.

stomatopods, lists of spp.

Indonesia, Institute of Marine Science 1972.
Biological and hydrological observations in the Piru Bay, Ambon Bay and Buton Strait
Oceanogr. Cruise Rept. 7: 27 pp.

stomatopods, lists of spp.

Ingle, R.W., and N. Della Croce, 1967.
Stomatopod larvae from the Mozambique Channel.
Boll. Mus. Ist. biol. Univ. Genova, 35:55-70

Stomatopods, lists of spp.

Manning, Raymond B. 1970.
The R/V Pillsbury Deep-Sea Biological Expedition to the Gulf of Guinea, 1964-65. 13. The Stomatopod Crustacea.
Studies trop. Oceanogr. Miami 4(2): 256-275.

Stomatopods, lists of spp.

Manning, Raymond B. 1970.
Some stomatopod crustaceans from Tulear Madagascar.
Bull. Mus. netn. Hist. nat. (2) 41 (6): 1429-1441.

Acanthosquilla septemspinosa

Manning, Raymond B. 1970.
The R/V Pillsbury Deep-Sea Biological Expedition to the Gulf of Guinea, 1964-65. The Stomatopod Crustacea.
Studies trop. Oceanogr. Miami 4(2): 256-275.

stomatopods, lists of spp.

Manning, R.B., 1968.
Stomatopod Crustacea collected by the Yale Seychelles Expedition, 1957-1958.
Postilla, Yale Univ., 68:15 pp.

stomatopods

Serene, R., 1953.
Sur la collection des stomatopodes de l'Institut Oceanographique de l'Indochine.
Proc. Seventh Pacific Sci. Congr. 4:506-508.

list of species only.

stomatopods, lists of spp.

Springer, S., and H.R. Bullis, Jr., 1956.
Collections by the Oregon in the Gulf of Mexico. List of Crustaceans, mollusks and fishes identified from collections made by the Exploratory Fishing Vessel Oregon in the Gulf of Mexico and adjacent seas, 1950 through 1955.
U.S.F.W.S. Spec. Sci. Rept.: Fish. No. 197:134 pp.

stomatopods, lists of spp.

Stephenson, W., 1962.
5. Some interesting Stomatopoda - mostly from Western Australia.
J.R. Soc., Western Australia, 45(2):33-43.

Acanthosquilla n. gen.

Manning, R.B., 1963.
Preliminary revision of the genera Pseudosquilla and Lysiosquilla with descriptions of six new genera (Crustacea: Stomatopoda).
Bull. Mar. Sci. Gulf and Caribbean, 13(2):308-328.

Lysiosquilla acanthocarpus   L. floridensis
L. septemspinosa
L. biminiensis
L. multifasciata (type)
L. tigrina
L. vicina
L. digueti

Acanthosquilla acanthocarpus

Alikunhi, K.H., 1967.
An account of the post-larval development, moulting and growth of the common stomatopods of the Madras coast.
Proc. Symp. Crustacea, Ernakulam, Jan. 12-15, 1965, 2: 824-939

Acanthosquilla acanthocarpus

Tirmizi, Nasima M., and Raymond B. Manning, 1968.
Stomatopod Crustacea from West Pakistan.
Proc. U.S. nat. Mus., 125(3666):1-48.

(has good labelled diagrams of parts).

Acanthosquilla biminiensis

Manning, Raymond B., 1969
Stomatopod Crustacea of the Western Atlantic.
Stud. Trop. Oceanogr., Univ. Miami, 8:380 pp.

Acanthosquilla derijardi n. sp.

Manning, Raymond B., 1970
Some stomatopod crustaceans from Tulear Madagascar.
Bull. Mus. netn. Hist. Net. (2) 41 (6): 1429-1441

Acanthosquilla derijardi

Moosa M. Kasim 1973.
The Stomatopod Crustacea collected by the Mariel King Memorial Expedition in Maluku waters in 1970.
Penelit. Laut Indonesia (Mar. Res. Indonesia) 13:3-30.

Acanthosquilla floridensis

Manning, Raymond B., 1969
Stomatopod Crustacea of the Western Atlantic.
Stud. Trop. Oceanogr., Univ. Miami, 8:380 pp.

Acanthosquilla humesi n. sp.

Manning, Raymond B., 1968.
Stomatopod Crustacea from Madagascar.
Proc. U.S. nat. Mus. 124(3641):1-61.

Acanthosquilla multifasciata

Alikunhi, K.H., 1967.
An account of the post-larval development, moulting and growth of the common stomatopods of the Madras coast.
Proc. Symp. Crustacea, Ernakulam, Jan. 12-15, 1965, 2: 824-939

Acanthosquilla multifasciata

Holthuis, L.B., 1967.
The Stomatopod Crustacea collected by the 1962 and 1965 Israel South Red Sea expeditions.
Israel J. Zool., 16(1):1-45.

Acanthosquilla multifasciata

Ingle, R.W., and N. della Croce, 1967.
Stomatopod larvae from the Mozambique Channel.
Boll. Musei Ist. biol. Univ. Genova, 35(226):55-70.

Acanthosquilla multifasciata

Moosa M. Kasim 1973.
The Stomatopod Crustacea collected by the Mariel King Memorial Expedition in Maluku waters in 1970.
Penelit. Laut Indonesia (Mar. Res. Indonesia) 13:3-30.

Acanthosquilla tigrina

Alikunhi, K.H., 1967.
An account of the post-larval development, moulting and growth of the common stomatopods of the Madras coast.
Proc. Symp. Crustacea, Ernakulam, Jan. 12-15, 1965, 2: 824-939

Acanthosquilla vicina

Holthuis, L.B., 1967.
The Stomatopod Crustacea collected by the 1962 and 1965 Israel South Red Sea expeditions.
Israel J. Zool., 16(1):1-45.

Acanthosquilla wilsoni n. sp.

Moosa M. Kasim 1973.
The Stomatopod Crustacea collected by the Mariel King Memorial Expedition in Maluku waters in 1970.
Penelit. Laut Indonesia (Mar. Res. Indonesia) 13:3-30.

Alima sp.

Foxon, G. E. H., 1939
Stomatopod larvae. John Murray Exped., 1933-34, Sci. Repts. 6(6):251-266, 4 text figs.

Alima

Manning, Raymond B., 1968.
A revision of the Family Squillidae (Crustacea, Stomatopoda), with the description of eight new genera.
Bull. mar. Sci., Miami, 18(1):105-142.

Alima bigelowi

Hansen, H. J., 1895
Isopoden, Cumacean und Stomatopoden der Plankton Expedition. Ergeb. Plankton Exped. II, Gc:105 pp., 8 pls.

Alima ctenura

Philippi, R.A., 1857.
Kurze Beschreibung einiger neuen Crustaceen.
Arch. f. Naturgesch., Berlin, 23:319-329

Alima dilatata n. sp.

Hansen, H. J., 1895
Isopoden, Cumacean und Stomatopoden der Plankton Expedition. Ergeb. Plankton Exped. II, Gc:105 pp., 8 pls.

Alima dubia

Hansen, H. J., 1895
Isopoden, Cumacean und Stomatopoden der Plankton Expedition. Ergeb. Plankton Exped. II, Gc:105 pp., 8 pls.

Alima hieroglyphica

Manning, Raymond B., 1969
Stomatopod Crustacea of the Western Atlantic.
Stud. Trop. Oceanogr., Univ. Miami, 8:380 pp.

Alima hyalina

Manning, Raymond B. 1970.
Some stomatopod crustaceans from Tuléar, Madagascar.
Bull. Mus. natn. Hist. nat. (2) 41 (6): 1429-1441.

Alima hyalina

Manning, Raymond B., 1969
Stomatopod Crustacea of the Western Atlantic.
Stud. Trop. Oceanogr., Univ. Miami, 8:380 pp.

Alima hyalina

Manning, R.B., 1962.
Alima hyalina Leach, the pelagic larva of the stomatopod crustacean, Squilla alba Bigelow.
Bull. Mar. Sci., Gulf and Caribbean, 12(3):496-504.

Alima paradoxa

Barnard, K. H., 1950
Descriptive list of South African Stomatopod Crustacea (Mantis Shrimps). Ann. S. African Mus. 38:838-864, 4 text figs.

Alima trivialis n.sp

Hansen, H. J., 1895
Isopoden, Cumaceen und Stomatopoden der Plankton Expedition. Ergeb. Plankton Exped. II, Gc:105 pp., 8 pls.

Alima valdiviana

Philippi, R.A., 1857.
Kurze Beschreibung einiger neuen Crustaceen.
Arch. f. Naturgesch., Berlin, 23:319-329

Anchisquilla n-gen.

Manning, Raymond B., 1968.
A revision of the Family Squillidae (Crustacea, Stomatopoda), with the description of eight new genera.
Bull. mar. Sci., Miami, 18(1):105-142.

Anchisquilla fasciata

Ghosh, A.C. 1975.
A note on two species of stomatopods from the Arabian Sea collected by the "John Murray" Expedition 1933-1934.
Crustaceana 24(1):143-144.

Anchisquilla fasciata [a]

Moosa M. Kasim 1973.
The stomatopod Crustacea collected by the Mariel King Memorial Expedition in Maluku waters in 1970.
Penelit. Laut Indonesia (Mar. Res. Indonesia) 13:3-30.

Anchisquilla punctata n. sp.

Blumstein, Radda, 1970.
New stomatopod crustaceans from the Gulf of Tonkin, South China Sea.
Crustaceana 18(2): 218-224.

Austrosquilla new subgenus

Manning, Raymond B., 1966.
Notes on some Australian and New Zealand stomatopod Crustacea, with an account of the species collected by the Fisheries Investigation Ship ENDEAVOUR.
Rec. Aust. Mus., 27(4):79-137.

Austrosquilla litoralis n. sp.

Michel, Alain, and Raymond B. Manning 1971.
A new Austrosquilla (Stomatopoda) from the Marquesas Islands.
Crustaceana 20(3): 237-240

Bathysquilla

Manning, Raymond B., 1968.
A revision of the Family Squillidae (Crustacea, Stomatopoda), with the description of eight new genera.
Bull. mar. Sci., Miami, 18(1):105-142.

Bathysquilla n.gen.

Manning, R.B., 1963.
Preliminary revision of the genera Pseudosquilla and Lysiosquilla with descriptions of six new genera (Crustacea: Stomatopoda).
Bull. Mar. Sci., Gulf and Caribbean, 13(2):308-328.

Lysiosquilla crassispinosa
L. microps (type)

Bathysquilla crassispinosa

Ingle, Raymond W., and Nigel R. Merrett 1971.
A stomatopod crustacean from the Indian Ocean, Indosquilla manihinei gen. et sp. nov. (Family Bathysquillidae) with remarks on Bathysquilla crassispinosa (Fukuda, 1910)
Crustaceana 20(2):192-198.

Cancer mantis

Holthuis, L.B., 1969.
Indication of a neotype for Cancer mantis L., 1758 (Stomatopoda, Squillidae).
Crustaceana 16(2): 221-223.

Bathysquilla microps

Manning, Raymond B., 1969
Stomatopod Crustacea of the Western Atlantic.
Stud. Trop. Oceanogr., Univ. Miami, 8:380 pp.

Carinosquilla n-gen.

Manning, Raymond B., 1968.
A revision of the Family Squillidae (Crustacea, Stomatopoda), with the description of eight new genera.
Bull. mar. Sci., Miami, 18(1):105-142.

Carinosquilla multicarinata [a]

Moosa M. Kasim 1973.
The stomatopod Crustacea collected by the Mariel King Memorial Expedition in Maluku waters in 1970.
Penelit. Laut Indonesia (Mar. Res. Indonesia) 13:3-30.

Chorisquilla brooksi [a]

Moosa M. Kasim 1973.
The stomatopod Crustacea collected by the Mariel King Memorial Expedition in Maluku waters in 1970.
Penelit. Laut Indonesia (Mar. Res. Indonesia) 13:3-30.

Chloridella aculeata

Lunz, R. G., jr., 1937.
Stomatopoda of the Bingham Oceanographic Collection. Bull. Bingham Ocean. Coll., V(5): 1-19, 10 textfigs.

Chloridella edentata n.sp.

Lunz, G. R., jr. 1937.
Stomatopoda of the Bingham Oceanographic Collection. Bull. Bingham Ocean. Coll., V(5): 1-19, 10 textfigs.

Chloridella empusa

Lunz, G. R., jr. 1937.
Stomatopoda of the Bingham Oceanographic Collection. Bull. Bingham Ocean. Coll., V(5): 1-19, 10 textfigs.

Chloridella heptacantha n.sp.

Chace, F.A., jr., 1939
Reports on the scientific results of the First Atlantis Expedition to the West Indies, under the joint auspices of the University of Havana and Harvard University. Preliminary descriptions of one new genus and seventeen new species of decapod and stomatopod crustacea. Mem. Soc. Cub. His. Nat. 13(1):31-54.

Chloridella neglecta

Lunz, G. R., jr. 1937.
Stomatopoda of the Bingham Oceanographic Collection. Bull. Bingham Ocean. Coll., V(5): 1-19, 10 textfigs.

Chloridella panamensis var. A

Lunz, G. R., jr. 1937
Stomatopoda of the Bingham Oceanographic Collection. Bull. Bingham Ocean. Coll., V(5): 1-19, 10 textfigs.

Chloridella rugosa pinensis n.var.

Lunz, G. R., jr. 1937.
Stomatopoda of the Bingham Oceanographic Collection. Bull. Bingham Ocean. Coll., V(5): 1-19, 10 textfigs.

Clorida

Manning, Raymond B., 1968.
A revision of the Family Squillidae (Crustacea, Stomatopoda), with the description of eight new genera.
Bull. mar. Sci., Miami, 18(1):105-142.

Clorida chlorida

Manning, Raymond B., 1968.
Stomatopod Crustacea from Madagascar.
Proc. U.S. nat. Mus. 124(3641):1-61.

Clorida clorida [a]

Moosa M. Kasim 1973.
The stomatopod Crustacea collected by the Mariel King Memorial Expedition in Maluku waters in 1970.
Penelit. Laut Indonesia (Mar. Res. Indonesia) 13:3-30.

Clorida depressa

Manning, Raymond B., 1966.
Notes on some Australian and New Zealand stomatopod Crustacea, with an account of the species collected by the Fisheries Investigation Ship ENDEAVOUR.
Rec. Aust. Mus., 27(4):79-137.

Clorida fallax

Manning, Raymond B., 1968.
Stomatopod Crustacea from Madagascar.
Proc. U.S. nat. Mus. 124(3641):1-61.

Clorida latreillei

Manning, Raymond B., 1969
Notes on some stomatopod Crustacea from southern Africa.
Smithson. Contrib. Zool. 1:1-17.

Clorida malaccensis moluccensis n.subsp. [a]

Moosa M. Kasim 1973.
The stomatopod Crustacea collected by the Mariel King Memorial Expedition in Maluku waters in 1970.
Penelit. Laut Indonesia (Mar. Res. Indonesia) 13:3-30.

Clorida microphthalma
Manning, Raymond B. 1966.
Notes on some Australian and New Zealand stomatopod Crustacea, with an account of the species collected by the Fisheries Investigation Ship ENDEAVOUR.
Rec. Aust. Mus., 27(4):79-137.

clorida microphthalma
Tirmizi, Nasima M., and Raymond B. Manning, 1968.
Stomatopod Crustacea from West Pakistan.
Proc. U.S. nat. Mus., 125(3666):1-48.

(has good labelled diagrams of parts).

Clorida miersi n.sp.
Manning, Raymond B. 1968.
Stomatopod Crustacea from Madagascar.
Proc. U.S. nat. Mus. 124(3641):1-61.

Clorida miersi
Moosa M. Kasim 1973.
The Stomatopod Crustacea collected by the Mariel King Memorial Expedition in Maluku waters in 1970.
Penelit. Laut Indonesia (Mar. Res. Indonesia) 13:3-30.

Clorida pelamidae n. sp.
Blumstein, Radda, 1970.
New stomatopod crustaceans from the Gulf of Tonkin, South China Sea.
Crustaceana 18(2): 218-224.

Clorida seversi n.sp.
Moosa M. Kasim 1973.
The Stomatopod Crustacea collected by the Mariel King Memorial Expedition in Maluku waters in 1970.
Penelit. Laut Indonesia (Mar. Res. Indonesia) 13:3-30.

Cloridopsis, n-gen
Manning, Raymond B., 1968.
A revision of the Family Squillidae (Crustacea, Stomatopoda), with the description of eight new genera.
Bull. mar. Sci., Miami, 18(1):105-142.

Cloridopsis dubia
Manning, Raymond B., 1969
Stomatopod Crustacea of the Western Atlantic.
Stud. Trop. Oceanogr., Univ. Miami, 8:380 pp.

cloridopsis emmaculata
Tirmizi, Nasima M., and Raymond B. Manning, 1968.
Stomatopod Crustacea from West Pakistan.
Proc. U.S. nat. Mus., 125(3666):1-48.

(has good labelled diagrams of parts).

Cloridopsis scorpio
Tirmizi, Nasima M., and Raymond B. Manning, 1968.
Stomatopod Crustacea from West Pakistan.
Proc. U.S. nat. Mus., 125(3666):1-48.

(has good labelled diagrams of parts).

Coronida sp.
Ingle, R.W., and N. della Croce, 1967.
Stomatopod larvae from the Mozambique Channel.
Boll. Musei Ist. biol. Univ. Genova, 35(226):55-70.

Coronida
Manning, R.B., 1963.
Preliminary revision of the genera Pseudosquilla and Lysiosquilla with descriptions of six new genera (Crustacea: Stomatopoda).
Bull. Mar. Sci., Gulf and Caribbean, 13(2):308-328.

Squilla hedyi (type)
Gonodactylus trachurus
Squilla multituberculata
Coronida sinuosa

Coronida armata
Manning, Raymond B. 1970.
The R/V Pillsbury Deep-Sea Biological Expedition to the Gulf of Guinea, 1964-65. 13. The Stomatopod Crustacea.
Studies trop. Oceanogr. Miami 4(2): 256-275.

Coronida bradyi
Hansen, H. J., 1895
Isopoden, Cumaceen und Stomatopoden der Plankton Expedition. Ergeb. Plankton Exped. II, Gc: 105 pp., 8 pls.

Coronida cocosiana n.sp.
Manning, Raymond B. 1972.
Three new stomatopod crustaceans of the family Lysiosquillidae from the eastern Pacific region.
Proc. Biol. Soc. Wash. 85 (21): 271-278

Coronida sinuosa
Townsley, S.J., 1953.
Adult and larval stomatopod crustaceans occurring in Hawaiian waters. Pacific Science 7(4):399-437, 28 textfigs.

Coronida trachurus ?
Foxon, G. E. H., 1939
Stomatopod larvae. John Murray Exped., 1933-34, Sci. Repts. 6(6):251-266, 4 text figs.

Coronida trachura
Ingle, R.W., 1963.
Crustacee Stomatopoda from the Red Sea and Gulf of Aden.
Sea Fish. Res. Sta., Israel, Bull. No. 33:1-69.

Coronis
Manning, R.B., 1963.
Preliminary revision of the genera Pseudosquilla and Lysiosquilla with descriptions of six new genera (Crustacea: Stomatopoda).
Bull. Mar. Sci., Gulf and Caribbean, 13(2):308-328.

Coronis scolopendra (type)
Lysiosquilla excavatrix

Coronis excavatrix
Manning, Raymond B., 1969
Stomatopod Crustacea of the Western Atlantic.
Stud. Trop. Oceanogr., Univ. Miami, 8:380 pp.

Coronis scolopendra
Manning, Raymond B., 1969
Stomatopod Crustacea of the Western Atlantic.
Stud. Trop. Oceanogr., Univ. Miami, 8:380 pp.

Coronidopsis bicuspis
Manning, R.B., 1963.
Preliminary revision of the genera Pseudosquilla and Lysiosquilla with descriptions of six new genera (Crustacea: Stomatopoda).
Bull. Mar. Sci., Gulf and Caribbean, 13(2):308-328.

Coronidopsis serenei n.sp.
Moosa M. Kasim 1973.
The Stomatopod Crustacea collected by the Mariel King Memorial Expedition in Maluku waters in 1970.
Penelit. Laut Indonesia (Mar. Res. Indonesia) 13:3-30.

Coroniderichthus armatu
Hansen, H. J., 1895
Isopoden, Cumaceen und Stomatopoden der Plankton Expedition. Ergeb. Plankton Exped. II, Gc: 105 pp., 8 pls.

Dictyosquilla n. gen
Manning, Raymond B., 1968.
A revision of the Family Squillidae (Crustacea, Stomatopoda), with the description of eight new genera.
Bull. mar. Sci., Miami, 18(1):105-142.

Euacanthus longispinus n.gen.
Philippi, R.A., 1857.
Kurze Beschreibung einiger neuen Crustaceen.
Arch. f. Naturgesch., Berlin, 23:319-329, Pl. 14

Eurysquilla n.gen.
Manning, R.B., 1963.
Preliminary revision of the genera Pseudosquilla and Lysiosquilla with descriptions of six new genera (Crustacea: Stomatopoda).
Bull. Mar. Sci., Gulf and Caribbean, 13(2):308-328.

Lysiosquilla plumata (type)
Lysiosquilla maiaguesensis
Lysiosquilla veuleti
Pseudosquilla veleronis

Eurysquilla chacei n.sp.
Manning, Raymond B., 1969
Stomatopod Crustacea of the Western Atlantic.
Stud. Trop. Oceanogr., Univ. Miami, 8:380 pp.

Eurysquilla holthuisi n. sp.
Manning, Raymond B., 1969
Stomatopod Crustacea of the Western Atlantic.
Stud. Trop. Oceanogr., Univ. Miami, 8:380 pp.

Eurysquilla maiaguesensis
Manning, Raymond B., 1969
Stomatopod Crustacea of the Western Atlantic.
Stud. Trop. Oceanogr., Univ. Miami, 8:380 pp.

Eurysquilla plumata
Manning, Raymond B., 1969
Stomatopod Crustacea of the Western Atlantic.
Stud. Trop. Oceanogr., Univ. Miami, 8:380 pp.

Eurysquilla plumata
Manning, Raymond B., 1966.
Stomatopod Crustacea. Campagne de la Calypso au large des côtes Atlantiques de l'Amerique du Sud (1961-1962)-1.
Ann. Inst. Océanogr., Monaco, n.s., 44:359-384

Eurysquilla solari n.sp.
Manning, Raymond B. 1970.
Nine new American Stomatopod crustaceans.
Proc. Biol. Soc. Wash. 83(8): 99-114.

Eurysquilla veleronis
Manning, Raymond B. 1971.
Eastern Pacific expeditions of the New York Zoological Society. Stomatopod Crustacea.
Zoologica, N.Y. 56(3), 95-113

Eurysquilloides n.gen.
Manning, R.B., 1963.
Preliminary revision of the genera Pseudosquilla and Lysiosquilla with descriptions of six new genera (Crustacea: Stomatopoda).
Bull. Mar. Sci., Gulf and Caribbean, 13(2):308-328.

Squilla sibogae (type)

Gonodactylus
Holthuis, L.B., and Raymond B. Manning, 1964.
Proposed use of the plenary powers (A) to designate a type-species for the genera Pseudosquilla Dana, 1852, and Gonodactylus Berthold 1827, and (B) for the suppression of the generic name Smerdis Leach, 1817 (Crustacea, Stomatopoda)
Z.N. (S.) 1609.
Bull. Zool. Nomencl., 21(2):137-143.

Gonodactylus austrinus n. sp.
Manning, Raymond B., 1969
Stomatopod Crustacea of the Western Atlantic.
Stud. Trop. Oceanogr., Univ. Miami, 8:380 pp.

Gonodactylus bahiahondensis
Manning, Raymond B. 1971.
Eastern Pacific expeditions of the New York Zoological Society. Stomatopod Crustacea.
Zoologica, N.Y. 56(2): 95-113.

Gonodactylus bahiahondensis
Schmitt, W. L., 1940
The stomatopods of the west coast of America, based on collections made by the Allan Hancock Expeditions, 1933-1938. Rept. Allan Hancock Pacific Exped. 5(4):129-226, 33 figs.

Gonodactylus bicarinatus n. sp.
Manning, Raymond B., 1968.
Stomatopod Crustacea from Madagascar.
Proc. U.S. nat. Mus., 124(3641):1-61.

Gonodactylus bredini
Dingle, Hugh and Roy L. Caldwell, 1972.
Reproductive and maternal behavior of the mantis shrimp Gonodactylus bredini Manning (Crustacea: Stomatopoda). Biol. Bull. mar. biol. Lab. Woods Hole 142(3): 417-426.

Gonodactylus bredini
Dingle, Hugh, and Roy L. Caldwell, 1969.
The aggressive and territorial behaviour of the mantis shrimp Gonodactylus bredini Manning (Crustacea: Stomatopoda).
Behaviour 33(1/2): 115-136.

Gonodactylus bredini n. sp.
Manning, Raymond B., 1969
Stomatopod Crustacea of the Western Atlantic.
Stud. Trop. Oceanogr., Univ. Miami, 8:380 pp.

Gonodactylus brevisquematus
Ingle, R.W., 1963.
Crustacea Stomatopoda from the Red Sea and Gulf of Aden.
Sea Fish. Res. Sta., Israel, Bull. No. 33:1-69.

Gonodactylus brevisquamatus
Manning, R.B., 1962.
Stomatopod Crustacea collected by the Yale Seychelles Expedition, 1957-1958.
Postilla, Yale Univ., 68:15 pp.

Gonodactylus childi n. sp.
Manning, Raymond B. 1971.
Two new species of Gonodactylus (Crustacea, Stomatopoda) from Eniwetok Atoll, Pacific Ocean.
Proc. Biol. Soc. Wash., 84(10): 73-80.

Gonodactylus chiragra
Barnard, K. H., 1950
Descriptive list of South African Stomatopod Crustacea (Mantis Shrimps). Ann. S. African Mus. 38:838-864, 4 text figs.

Gonodactylus chiragra
Bedot, M., 1909
Sur la faune de l'Archipel Malais (resume).
Rev. Suisse Zool. 17:143-169.

Gonodactylus chiragra
Chopra, B., 1939
Stomatopoda. John Murray Exped., 1933-34, Sci. Repts. 6(3):137-181, 13 text figs.

Gonodactylus chiragra
Holthuis, L.B., 1967.
The Stomatopod Crustacea collected by the 1962 and 1965 Israel South Red Sea expeditions.
Israel J. Zool., 16(1):1-45.

Gonodactylus chiragra
Ingle, R.W., 1963.
Crustacea Stomatopoda from the Red Sea and Gulf of Aden.
Sea Fish. Res. Sta., Israel, Bull. No. 33:1-69.

Gonydactylus chiragra
Johnson, M.W., 1943
Underwater Sounds of Biological Origin.
UCDWR No.U28, Sec. No.6.1-sr30-412, File No. 01.33, 20 pp. (mimeographed), 1 chart (photostat), 2 figs. (photo). 15 Feb. 1943.

Gonodactylus chiragra
Manning, Raymond B., 1969
Notes on some stomatopod Crustacea from southern Africa.
Smithson. Contrib. Zool. 1:1-17.

Gonodactylus chiragra
Manning, Raymond B., 1968.
Stomatopod Crustacea from Madagascar.
Proc. U.S. nat. Mus., 124(3641):1-61.

Gonodactylus chiragra
Manning, Raymond B., 1967.
Stomatopoda in the Vanderbilt Marine Museum.
Crustaceana, 12(1):102-106.

Gonodactylus chiragra
Manning, Raymond B., 1966.
Notes on some Australian and New Zealand stomatopod Crustacea, with an account of the species collected by the Fisheries Investigation Ship ENDEAVOUR.
Rec. Aust. Mus. 27(4):79-137.

Gonodactylus chiragra
Manning, R.M., 1962.
Stomatopod Crustacea collected by the Yale Seychelles Expedition, 1957-1958.
Postilla, Yale Univ., 68:15 pp.

Gonodactylus chiraga
Moosa M. Kasim 1973.
The Stomatopod Crustacea collected by the Mariel King Memorial Expedition in Maluku waters in 1970. Penelit. Laut Indonesia (Mar. Res. Indonesia) 13:3-30.

Gonodactylus chiragra
Parisi, B., 1940
Gli Stomatopodi raccolti del Prof. L. Sanzo nella Campagna idrografica nel Mar Rosso della R. N. Ammiraglio Magnaghi 1923-1924. (Mem. 16ª della Campagne). Mem. R. Com. Talass. Ital. CCLXXV:7 pp.

Gonodactylus chiragra
Serène, R., 1954.
Observations biologiques sur les stomatopodes.
Mém. Inst. Ocean., Nhatrang, No. 8:1-93, 10 pls., 15 textfigs.

Gonodactylus chiragra
Serène, R., 1951.
Sur la circulation d'eau à la surface du corps des Stomatopodes. Bull. Soc. Zool., France, 76(3):137-143, 3 textfigs., 1 pl.

Gonodactylus chiragra
Serène, R., 1950.
Cas de malformations chez les stomatopodes. Bull. Mus., Indochine, 2nd ser., 22(3):341-343, 2 pls.

Gonodactylus chiragra
Stephenson, W., 1952.
Faunistic records from Queensland. Pt. I. General introduction. Pt. II. Adult Stomatopoda (Crustacea).
Univ. Queensland, Dept. Zool., Papers, 1(1):1-15

Gonodactylus chiragra
Tirmizi, Nasima M., and Raymond B. Manning, 1968.
Stomatopod Crustacea from West Pakistan.
Proc. U.S. nat. Mus., 125(3666):1-48.

(has good labelled diagrams of parts.

Gonodactylus chiragra platysoma
Lunz, G. R., Jr., 1937.
Stomatopoda of the Bingham Oceanographic Collection. Bull. Bingham Ocean. Coll., V(5):1-19, 10 textfigs.

Gonodactylus choprai n.sp.
Manning, Raymond B. 1967.
Notes on the demanii section of genus Gonodactylus Berthold with description of three new species (Crustacea: Stomatopoda).
Proc. U.S. nat. Mus. 123(3618):1-27.

Gonodactylus crinatus n.sp.
Manning, R.B., 1962.
Stomatopod Crustacea collected by the Yale Seychelles Expedition, 1957-1958.
Postilla, Yale Univ., 68:15 pp.

Gonodactylus crosnieri n.sp.
Manning, Raymond B., 1968.
Stomatopod Crustacea from Madagascar.
Proc. U.S. nat. Mus., 124(3641):1-61.

Gonodactylus curacaoensis
Manning, Raymond B., 1969
Stomatopod Crustacea of the Western Atlantic.
Stud. Trop. Oceanogr., Univ. Miami, 8:380 pp.

Gonodactylus demanii
Barnard, K. H., 1950
Descriptive list of South African Stomatopod Crustacea (Mantis Shrimps). Ann. S. African Mus. 38:838-864, 4 text figs.

Gonodactylus demanii
Holthuis, L.B., 1967.
The Stomatopod Crustacea collected by the 1962 and 1965 Israel South Red Sea expeditions.
Israel J. Zool., 16(1):1-45.

Gonodactylus demani - var.
Ingle, R.W., 1963.
Crustacea Stomatopoda from the Red Sea and Gulf of Aden.
Sea Fish. Res. Sta., Israel, Bull. No. 33:1-69.

Gonodactylus demani ? espinosus
Chopra, B., 1939
Stomatopoda. John Murray Exped., 1933-34, Sci. Repts. 6(3):137-181, 13 text figs.

Gonodactylus demani spinosus
Chopra, B., 1939
Stomatopoda. John Murray Exped., 1933-34, Sci. Repts. 6(3):137-181, 13 text figs.

Gonodactylus demanii
Manning, Raymond B., 1968.
Stomatopod Crustacea from Madagascar.
Proc. U.S. nat. Mus., 124(3641):1-61.

Gonodactylus demanii
Manning, Raymond B. 1967.
Notes on the demanii section of genus Gonodactylus Berthold with description of three new species (Crustacea: Stomatopoda).
Proc. U.S. nat. Mus. 123(3618):1-27.

Gonodactylus Demani
Parisi, B., 1940
Gli Stomatopodi raccolti del Prof. L. Sanzo nella Campagna idrografica nel Mar Rosso della R. N. Ammiraglio Magnaghi 1923-1924. (Mem. 16ª della Campagne). Mem. R. Com. Talass. Ital. CCLXXV:7 pp.

Gonodactylus demani
Serène, R., 1954.
Observations biologiques sur les stomatopodes.
Mém. Inst. Océan., Nhatrang, No. 8:1-93, 10 pls., 15 textfigs.

Gonodactylus demanii
Tirmizi, Nasima M., and Raymond B. Manning, 1968.
Stomatopod Crustacea from West Pakistan.
Proc.U.S.nat.Mus., 125(3666):1-48.

Gonodactylus falcatus
Holthuis, L.B., 1967.
The Stomatopod Crustacea collected by the 1962 and 1965 Israel South Red Sea expeditions.
Israel J. Zool., 16(1):1-45.

Gonodactylus falcatus
Ingle, R.W., 1963.
Crustacea Stomatopoda from the Red Sea and Gulf of Aden.
Sea Fish.Res.Sta.,Israel,Bull.No.33:1-69.

Gonodactylus falcatus
Kinzie, Robert A., III, 1968.
The ecology of the replacement of Pseudosquilla ciliata (Fabricius by Gonodactylus falcatus (Forskål)(Crustacea; Stomatopoda) recently introduced into the Hawaiian islands.
Pacif.Sci., 22(4):465-475.

Gonodactylus falcatus
Manning, Raymond B., 1969
Notes on some stomatopod Crustacea from southern Africa.
Smithson. Contrib. Zool. 1:1-17.

Gonodactylus falcatus
Manning, Raymond B., 1968.
Stomatopod Crustacea from Madagascar.
Proc. U.S. nat. Mus., 124(3641):1-61.

Gonodactylus falcatus
Manning, Raymond B., 1967.
Stomatopoda in the Vanderbilt Marine Museum
Crustaceana, 12(1):102-106.

Gonodactylus falcatus
Manning, Raymond B., 1966.
Notes on some Australian and New Zealand stomatopod Crustacea, with an account of the species collected by the Fisheries Investigation Ship ENDEAVOUR.
Rec. Aust. Mus., 27(4):79-137.

Gonodactylus falcatus
Manning, R.B., 1962.
Stomatopod Crustacea collected by the Yale Seychelles Expedition, 1957-1958.
Postilla, Yale Univ., 68:15 pp.

Gonodactylus falcatus
Moosa, M. Kasim 1973.
The Stomatopod Crustacea collected by the Mariel King Memorial Expedition in Maluku waters in 1970.
Penelit. Laut Indonesia (Mar. Res. Indonesia) 13:3-30.

Gonodactylus falcatus
Serène, R., 1954.
Observations biologiques sur les stomatopodes.
Mém. Inst. Océan., Nhatrang, No. 8:1-93, 10 pls., 15 textfigs.

Gonodactylus falcatus
Serène, R., 1951.
Sur la circulation d'eau à la surface du corps des Stomatopodes. Bull. Soc. Zool., France 76(3): 137-143, 3 textfigs., 1 pl.

Gonodactylus falcatus
Serène, R., 1950.
Cas de malformations chez les Stomatopodes. Bull. Mus., Indochine, 2nd ser., 22(3):341-343, 2 pls.

Gonodactylus falcatus = G. glabrous
Stephenson, W., 1952.
Faunistic records from Queensland. Pt. I. General introduction. Pt. II. Adult Stomatopoda (Crustacea).
Univ. Queensland, Dept. Zool., Papers 1(1):1-15.

Gonodactylus festae
Manning, Raymond B. 1971.
Eastern Pacific expeditions of the New York Zoological Society. Stomatopod Crustacea.
Zoologica, N.Y. 56(2):95-113

Gonodactylus festae subsp. libertadensis
Schmitt, W. L., 1940
The stomatopods of the west coast of America, based on collections made by the Allan Hancock Expeditions, 1933-1938. Rept. Allan Hancock Pacific Exped. 5(4):129-226, 33 figs.

Gonodactylus fimbriatus
Manning, R.B., 1962.
Stomatopod Crustacea collected by the Yale Seychelles Expedition, 1957-1958.
Postilla, Yale Univ., 68:15 pp.

Gonodactylus folinii
Hansen, H. J., 1895
Isopoden, Cumaceen und Stomatopoden der Plankton Expedition. Ergeb. Plankton Exped. II, GC:105 pp., 8 pls.

Gonodactylus glabrous
Barnard, K. H., 1950
Descriptive list of South African Stomatopod Crustacea (Mantis Shrimps). Ann. S. African Mus. 38:838-864, 4 text figs.

Gonodactylus glabrus
Calman, W.T., 1927.
Zoological results of the Cambridge Expedition to the Suez Canal. XXVII. Report on the Phyllocaridea, Cumacea, and Stomatopoda. Trans. Zool. Soc., London, 22:399-401, textfig. 101.

Gonodactylus glabrous
Parisi, B., 1940
Gli Stomatopodi raccolti del Prof. L. Sanzo nella Campagna idrografica nel Mar Rosso della R. N. Ammiraglio Magnaghi 1923-1924. (Mem. 16ª della Campagne). Mem. R. Com. Talass. Ital. CCLXXV:7 pp.

Gonodactylus glyptocerus
Stephenson, W., 1952.
Faunistic records from Queensland. Pt. I. General introduction. Pt. II. Adult Stomatopoda (Crustacea).
Univ. Queensland, Dept. Zool., Papers 1(1):1-15.

Gonodactylus glyptocercus
Tiwari, K.K., 1951.
On two new species of the genus Squilla Fabr., with notes on other stomatopods in the collections of the Zoological Survey of India.
Rec. Indian Mus. 49(3/4):349-363, 5 textfigs.

Gonodactylus graphurus
Bedot, M., 1909
Sur la faune de l'Archipel Malais (resume).
Rev. Suisse Zool. 17:143-169.

Gonodactylus graphurus
Ingle, R.W. 1971.
On the nomenclature of Gonodactylus graphurus White 1847 (nomen nudem), Miers, 1875 (Stomatopoda Gonodactylidae).
Crustaceana 21(2):220-221.

Gonodactylus graphurus
Manning, Raymond B., 1966.
Notes on some Australian and New Zealand stomatopod Crustacea, with an account of the species collected by the Fisheries Investigation Ship ENDEAVOUR.
Rec. Aust. Mus., 27(4):79-137.

Gonodactylus graphurus
Stephenson, W., 1952.
Faunistic records from Queensland. Pt. I. General introduction. Pt. II. Adult Stomatopoda (Crustacea).
Univ. Queensland, Dept. Zool., Papers 1(1):1-15.

Gonodactylus guerini
Manning, R.B., 1962.
Stomatopod Crustacea collected by the Yale Seychelles Expedition, 1957-1958.
Postilla, Yale Univ., 68:15 pp.

Gonodactylus guerini
Townsley, S.J., 1953.
Adult and larval stomatopod crustaceans occurring in Hawaiian waters. Pacific Science 7(4):399-437, 28 textfigs.

Gonodactylus gyrosus
Manning, R.B., 1962.
Stomatopod Crustacea collected by the Yale Seychelles Expedition, 1957-1958.
Postilla, Yale Univ., 68:15 pp.

Gonodactylus gyrosus
Tiwari, K.K., 1951.
On two new species of the genus Squilla Fabr., with notes on other stomatopods in the collections of the Zoological Survey of India.
Rec. Indian Mus. 49(3/4):349-363, 5 textfigs.

Gonodactylus hendersoni n.sp.
Manning, Raymond B., 1967.
Notes on the demanii section of genus Gonodactylus Berthold with description of three new species (Crustacea: Stomatopoda).
Proc. U.S nat. Mus. 123(3618):1-27.

Gonodactylus incipiens
Manning, Raymond B. 1967.
Notes on the demanii section of genus Gonodactylus Berthold with description of three new species (Crustacea: Stomatopoda).
Proc. U.S. nat. Mus. 123(3618):1-27.

Gonodactylus lacunatus
Manning, Raymond B., 1969
Stomatopod Crustacea of the Western Atlantic.
Stud. Trop. Oceanogr., Univ. Miami, 8:380 pp.

Gonodactylus lacunatus n. sp.
Manning, Raymond B., 1966.
Stomatopod Crustacea. Campagne de la Calypso au large des côtes Atlantiques de l'Amerique du Sud (1961-1962)-1.
Ann. Inst. Océanogr., Monaco, n.s., 44:359-384

Gonodactylus laliberta densis
Manning, Raymond B. 1971.
Eastern Pacific expeditions of the New York Zoological Society. Stomatopod Crustacea.
Zoologica, N.Y. 56(2):95-113

Gonodactylus lanchesteri
Manning, Raymond B., 1968.
Stomatopod Crustacea from Madagascar.
Proc. U.S. nat. Mus., 124(3641):1-61.

Gonodactylus lankesteri n.sp.
Manning, Raymond B. 1967.
Notes on the demanii section of the genus Gonodactylus Berthold with description of three new species (Crustacea: Stomatopoda).
Proc. U.S. nat. Mus. 123(3618):1-27.

Gonodactylus lanchesteri
Tirmizi, Nashima M., and Raymond B. Manning, 1968.
Stomatopod Crustacea from West Pakistan.
Proc.U.S.nat.Mus., 125(3666):1-48.

Oceanographic Index: Marine Organisms Cumulation, 1946-1973

Gonodactylus lenzi
Ingle, R.W., 1963.
Crustacea Stomatopoda from the Red Sea and Gulf of Aden.
Sea Fish.Res.Ste., Israel, Bull. No. 33:1-69.

Gonodactylus lenzi
Manning, R.B., 1962.
Stomatopod Crustacea collected by the Yale Seychelles Expedition, 1957-1958.
Postilla, Yale Univ., 68:15 pp.

Gonodactylus lenzi
Serène, R., 1954.
Observations biologiques sur les stomatopodes.
Mem. Inst. Ocean. Nhatrang, No. 8:1-93, 10 pls., 15 textfigs.

Gonodactylus lenzi
Tiwari, K.K., 1951.
On two new species of the genus Squilla Fabr., with notes on other stomatopods in the collections of the Zoological Survey of India.
Rec. Indian Mus. 49(3/4):349-363, 5 textfigs.

Gonodactylus micronesica n. sp.
Manning, Raymond B., 1971
Two new species of Gonodactylus (Crustacea, Stomatopoda) from Eniwetok Atoll, Pacific Ocean.
Proc. Biol. Soc. Wash., 84(10): 73-80

Gonodactylus minutus n. sp.
Manning, Raymond B., 1969
Stomatopod Crustacea of the Western Atlantic.
Stud. Trop. Oceanogr., Univ. Miami, 8:380 pp.

Gonodactylus oerstedii
Lemos de Castro, A., 1955.
Contribuicao ao conhecimento dos crustaceos da Ordem Stomatopoda do litoral brasileiro (Crustacea, Hoplocarida).
Bol. Mus. Nac., Rio de Janeiro, n.s., Zool., No. 128:1-68, 56 figs.

Gonodactylus oerstedii
Lunz, G. R., Jr., 1937.
Stomatopoda of the Bingham Oceanographic Collection. Bull. Bingham Ocean. Coll., V(5): 1-19, 10 textfigs.

Gonodactylus oerstedii
Manning, Raymond B., 1969
Stomatopod Crustacea of the Western Atlantic.
Stud. Trop. Oceanogr., Univ. Miami, 8:380 pp.

Gonodactylus oerstedii
Manning, Raymond B., 1967.
Stomatopoda in the Vanderbilt Marine Museum.
Crustaceana, 12(1):102-106.

Gonodactylus oerstedii
Manning, Raymond B., 1966.
Stomatopod Crustacea. Campagne de la Calypso au large des côtes Atlantiques de l'Amerique du Sud (1961-1962)-1.
Ann. Inst. Océanogr., Monaco, n.s., 44:359-384

Gonodactylus oerstedii
Manning, Raymond B., 1963.
Notes on the embryology of the stomatopod crustacean Gonodactylus oerstedii Hansen.
Bull. Mar. Sci., Gulf and Caribbean, 13(3):422-432.

Gonodactylus oerstedii
Manning, Raymond B., and Anthony J. Provenzano, Jr., 1963.
Studies on development of stomatopod Crustacea. 1. Early larval stages of Gonodactylus oerstedii Hansen.
Bull. Mar. Sci., Gulf and Caribbean, 13(3):467-487.

Gonodactylolus paulus n. gen. n. sp.
Manning, Raymond B. 1970.
A new genus and species of stomatopod crustacean from Madagascar.
Bull. Mus. nat. Hist. nat. (2) 42(1):206-209.

GONODACTYLUS PETILUS N.SP.
Manning, Raymond B., 1970.
Nine new American stomatopod crustaceans
Proc. Biol. Soc. Wash., 83(5): 99-114

Gonodactylus platysoma
Manning, Raymond B., 1968.
Stomatopod Crustacea from Madagascar.
Proc. U.S. nat. Mus., 124(3641):1-61.

Gonodactylus platysoma
Manning, Raymond B., 1967.
Stomatopoda in the Vanderbilt Marine Museum.
Crustaceana, 12(1):102-106.

Gonodactylus platysoma
Manning, Raymond B., 1966.
Notes on some Australian and New Zealand stomatopod Crustacea, with an account of the species collected by the Fisheries Investigation Ship ENDEAVOUR.
Rec. Aust. Mus., 27(4):79-137.

Gonodactylus platysoma
Manning, R.B., 1962.
Stomatopod Crustacea collected by the Yale Seychelles Expedition, 1957-1958.
Postilla, Yale Univ., 68:15 pp.

Gonodactylus pulchellus
Ingle, R.W., 1963.
Crustacea Stomatopoda from the Red Sea and Gulf of Aden.
Sea Fish Res.Ste., Israel, Bull.No. 33:1-69.

Gonodactylus pulchellus
Parisi, B., 1940
Gli Stomatopodi raccolti del Prof. L. Sanzo nella Campagna idrografica nel Mar Rosso della R. N. Ammiraglio Magnaghi 1923-1924. (Mem. 16ª della Campagne). Mem. R. Com. Talass. Ital. CCLXXV:7 pp.

Gonodactylus pulchellus
Serène, R., 1950.
Cas de malformations chez les Stomatopodes. Bull. Mus., Indochine, 2nd ser., 22(3):341-343, 2 pls.

GONODACTYLUS PUMILUS N.SP
Manning, Raymond B., 1970.
Nine new American stomatopod crustaceans
Proc. Biol. Soc. Wash., 83(5): 99-114

Gonodactylus segregatus
Manning, Raymond B., 1968.
Stomatopod Crustacea from Madagascar.
Proc. U.S. nat. Mus., 124(3641):1-61.

Gonodactylus segregatus
Moosa M. Kasim 1973.
The stomatopod Crustacea collected by the Mariel King Memorial Expedition in Maluku waters in 1970.
Penelit. Laut Indonesia (Mar. Res. Indonesia) 13:3-30.

Gonodactylus smithii
Holthuis, L.B., 1967.
The Stomatopod Crustacea collected by the 1962 and 1965 Israel South Red Sea expeditions.
Israel J. Zool., 16(1):1-45.

Gonodactylus smithii
Manning, Raymond B., 1968.
Stomatopod Crustacea from Madagascar.
Proc. U.S. nat. Mus., 124(3641):1-61.

Gonodactylus smithii
Manning, Raymond B., 1966.
Notes on some Australian and New Zealand stomatopod Crustacea, with an account of the species collected by the Fisheries Investigation Ship ENDEAVOUR.
Rec. Aust. Mus., 27(4):79-137.

Gonodactylus smithii
Moosa M. Kasim 1973.
The stomatopod Crustacea collected by the Mariel King Memorial Expedition in Maluku waters in 1970.
Penelit. Laut Indonesia (Mar. Res. Indonesia) 13:3-30.

Gonodactylus smithi
Serène, R., 1954.
Observations biologiques sur les stomatopodes.
Mem. Inst. Océan., Nhatrang, No. 8:1-93, 10 pls., 15 textfigs.

Gonodactylus smithii
Tirmizi, Nasima M., and Raymond B. Manning, 1968.
Stomatopod Crustacea from west Pakistan.
Proc. U.S. nat. Mus., 125(3666):1-48.

Gonodactylus spinosocarinatus
Serène, R., 1952.
Etude d'une collection de stomatopodes de l'Australian Museum de Sydney. Rec. Australian Mus. 23(1):1-28, 33 textfigs., Pls. 1-3.

Gonodactylus spinosissimus
Ingle, R.W., 1963.
Crustacea Stomatopoda from the Red Sea and Gulf of Aden.
Sea Fish.Res.Ste., Israel, Bull.No.33:1-69.

Gonodactylus spinosus
Holthuis, L.B., 1967.
The Stomatopod Crustacea collected by the 1962 and 1965 Israel South Red Sea expeditions.
Israel J. Zool., 16(1):1-45.

Gonodactylus spinosus
Manning, Raymond B. 1967.
Notes on the demanii section of the genus Gonodactylus Berthold with description of three new species (Crustacea: Stomatopoda)
Proc. U.S. nat. Mus. 123(3618): 1-27.

Gonodactylus spinosus
Manning, R.B., 1962.
Stomatopod Crustacea collected by the Yale Seychelles Expedition, 1957-1958.
Postilla, Yale Univ., 68:15 pp.

Gonodactylus spinosus
Parisi, B., 1940
Gli Stomatopodi raccolti del Prof. L. Sanzo nella Campagna idrografica nel Mar Rosso della R. N. Ammiraglio Magnaghi 1923-1924. (Mem. 16ª della Campagne). Mem. R. Com. Talass. Ital. CCLXXV:7 pp.

Gonodactylus spinulosus
Manning, Raymond B. 1972.
Gonodactylus spinulosus Schmitt, a West Indian stomatopod new to Bermuda.
Crustaceana 23(3): 315.

Gonodactylus spinulosus
Manning, Raymond B., 1969
Stomatopod Crustacea of the Western Atlantic.
Stud. Trop. Oceanogr., Univ. Miami, 8:380 pp.

Gonodactylus spinulosus
Manning, Raymond B., 1966.
Stomatopod Crustacea. Campagne de la Calypso au large des côtes Atlantiques de l'Amerique du Sud (1961-1962)-1.
Ann. Inst. Océanogr., Monaco, n.s., 44:359-384

Gonodactylus stanschi

Manning, Raymond B. 1971
Eastern Pacific expeditions of the New York Zoological Society. Stomatopod Crustacea.
Zoologica, N.Y. 56(2): 95-113

Gonodactylus stanschi n.sp.

Schmitt, W. L., 1940
The stomatopods of the west coast of America, based on collections made by the Allan Hancock Expeditions, 1933-1938. Rept. Allan Hancock Pacific Exped. 5(4):129-226, 33 figs.

Gonodactylus strigatus

Serène, R., 1949.
Observations sur le Gonodactylus strigatus Hansen (Crustacé Stomatopode). Bull. Soc. Zool., France, 74(4/5):225-231, 2 textfigs.

Gonodactylus torus n. sp.

Manning, Raymond B., 1969
Stomatopod Crustacea of the Western Atlantic.
Stud. Trop. Oceanogr., Univ. Miami, 8:380 pp.

Gonodactylus tweedei n.sp.

Serène, R., 1952.
Étude d'une collection de stomatopodes de l'Australian Museum de Sydney. Rec. Austalian Mus. 23(1):1-28, 33 textfigs., Pls. 1-3.

Gonodactylus tweedei n.sp.

Serène, R., 1950.
Deux nouvelles espèces Indopacifiques de Stomatopodes. Bull. Mus. Nat. Hist. Nat., 2e ser., 22(5):571-572.

Gonodactylus tweedei

Stephensen, W., 1952.
Faunistic records from Queensland. Pt. I. General introduction. Pt. II. Adult Stomatopoda (Crustacea).
Univ. Queensland, Dept. Zool., Papers 1(1):1-15.

Gonodactylus zacae nsp

Manning, Raymond B. 1971
Eastern Pacific expeditions of the New York Zoological Society. Stomatopod Crustacea.
Zoologica, N.Y. 56(2): 95-113

Hadrosquilla n. gen.

Manning, Raymond B., 1966.
Notes on some Australian and New Zealand stomatopod Crustacea, with an account of the species collected by the Fisheries Investigation Ship ENDEAVOUR.
Rec. Aust. Mus., 27(4):79-137.

Hadrosquilla perpasta n. comb.

Manning, Raymond B., 1966.
Notes on some Australian and New Zealand stomatopod Crustacea, with an account of the species collected by the Fisheries Investigation Ship ENDEAVOUR.
Rec. Aust. Mus., 27(4):79-137.

Haptosquilla nefanda

Moosa M. Kasim 1973.
The stomatopod Crustacea collected by the Mariel King Memorial Expedition in Maluku waters in 1970. Penelit. Laut Indonesia (Mar. Res. Indonesia) 13:3-30.

Haptosquilla pulchella

Moosa M. Kasim 1973.
The stomatopod Crustacea collected by the Mariel King Memorial Expedition in Maluku waters in 1970. Penelit. Laut Indonesia (Mar. Res. Indonesia) 13:3-30.

Haptosquilla stoliura

Moosa M. Kasim 1973.
The stomatopod Crustacea collected by the Mariel King Memorial Expedition in Maluku waters in 1970. Penelit. Laut Indonesia (Mar. Res. Indonesia) 13:3-30.

Haptosquilla tuberosa

Moosa M. Kasim 1973.
The stomatopod Crustacea collected by the Mariel King Memorial Expedition in Maluku waters in 1970. Penelit. Laut Indonesia (Mar. Res. Indonesia) 13:3-30.

Harpiosquilla n.gen.

Holthuis, L.B., 1964.
Preliminary note on two new genera of Stomatopoda.
Crustaceana, 7(2):140-141.

Harpiosquilla

Manning, Raymond B., 1968.
A revision of the Family Squillidae (Crustacea, Stomatopoda), with the description of eight new genera.
Bull. mar. Sci., Miami, 18(1):105-142.

Harpiosquilla annandalei

Manning, Raymond B. 1969.
A review of the genus Harpiosquilla (Crustacea, Stomatopoda) with descriptions of three new species.
Smithson. Contrib. Zool. 36:1-41.

Harpiosquilla annandalei

Manning, Raymond B. 1966.
Stomatopoda from the collection of His Majesty, The Emperor of Japan.
Crustaceana 9(3): 249-262

Harpiosquilla harpax

Holthuis, L.B., 1967.
The Stomatopod Crustacea collected by the 1962 and 1965 Israel South Red Sea expeditions.
Israel J. Zool., 16(1):1-45.

Harpiosquilla harpax

Manning, Raymond B., 1968.
Stomatopod Crustacea from Madagascar.
Proc. U.S. nat. Mus. 124(3641):1-61.

Harpiosquilla harpax

Manning, Raymond B. 1969.
A review of the genus Harpiosquilla (Crustacea, Stomatopoda) with descriptions of three new species.
Smithson. Contrib. Zool. 36:1-41.

Harpiosquilla harpax

Manning, Raymond B., 1969
Notes on some stomatopod Crustacea from southern Africa.
Smithson. Contrib. Zool. 1:1-17.

Harpiosquilla harpax

Manning, Raymond B., 1969.
Notes on some Australian and New Zealand stomatopod Crustacea, with an account of the species collected by the Fisheries Investigation Ship ENDEAVOUR.
Rec. Aust. Mus., 27(4):79-137.

Harpiosquilla harpax

Manning, Raymond B., 1967.
Stomatopods in the Vanderbilt Marine Museum.
Crustaceana, 12(1):102-106.

Harpiosquilla harpex

Tirmizi, Nasima M., and Raymond B. Manning, 1968.
Stomatopod Crustacea from West Pakistan.
Proc. U.S. nat. Mus., 125(3666):1-48.

Harpiosquilla indica n.sp.

Manning, Raymond B. 1969.
A review of the genus Harpiosquilla (Crustacea, Stomatopoda) with descriptions of three new species.
Smithson. Contrib. Zool. 36:1-41.

Harpiosquilla japonica n.sp.

Manning, Raymond B. 1969.
A review of the genus Harpiosquilla (Crustacea, Stomatopoda) with descriptions of three new species.
Smithson. Contrib. Zool. 36:1-41.

Harpiosquilla melanoura

Manning, Raymond B. 1969.
A review of the genus Harpiosquilla (Crustacea, Stomatopoda) with descriptions of three new species.
Smithson. Contrib. Zool. 36:1-41.

Harpiosquilla melanoura n.sp.

Manning, Raymond B., 1968.
Stomatopod Crustacea from Madagascar.
Proc. U.S. nat. Mus. 124(3641):1-61.

Harpiosquilla raphidea

Alikunhi, K.H. 1967.
An account of the post-larval development, moulting and growth of the common stomatopods of the Madras coast.
Proc. Symp. Crustacea, Ernakulam, Jan.12-15, 1965, 2: 824-939

Harpiosquilla raphidea

Manning, Raymond B. 1969.
A review of the genus Harpiosquilla (Crustacea, Stomatopoda) with descriptions of three new species.
Smithson. Contrib. Zool. 36:1-41.

Harpiosquilla raphidea

Tirmizi, Nasima M., and Raymond B. Manning, 1968.
Stomatopod Crustacea from West Pakistan.
Proc. U.S. nat. Mus., 125(3666):1-48.

Hemisquilla bigelowi

Manning, R.B., 1963.
Hemisquilla ensigera (Owen, 1832) an earlier name for H. bigelowi (Rathbun, 1910) (Stomatopoda).
Crustaceana, 5(4):315-317.

Hemisquilla braziliensis

Lemos de Castro, A., 1955.
Contribuicao ao conhecimento dos crustaceos da Ordem Stomatopoda do litoral brasileiro (Crustacea, Hoplocarida).
Bol. Mus. Nac., Rio de Janeiro, n.s., Zool., No. 128:1-68, 56 figs.

Hemisquilla braziliensis

Manning, Raymond B., 1969
Stomatopod Crustacea of the Western Atlantic.
Stud. Trop. Oceanogr., Univ. Miami, 8:380 pp.

Hemisquilla braziliensis

Manning, Raymond B., 1966.
Stomatopod Crustacea. Campagne de la Calypso au large des côtes Atlantiques de l'Amerique du Sud (1961-1962)-1.
Ann. Inst. Océanogr., Monaco, n.s., 44:359-384

Hemisquilla braziliensis
Ramos, F. de P., 1951.
Nota preliminar sôbre alguns Stomatopoda da Costa Brasileira. Bol. Inst. Paulista Ocean. 2(1):139-150, 1 pl.

Hemisquilla ensigera
Burnett, Bryan R. 1972.
Notes on the lateral arteries of two stomatopods.
Crustaceana 23(3): 303-305.

Hemisquilla ensigera
Manning, Raymond B., 1966.
Notes on some Australian and New Zealand stomatopod Crustacea, with an account of the species collected by the Fisheries Investigation Ship ENDEAVOUR.
Rec. Aust. Mus., 27(4):79-137.

Hemisquilla ensigera
Manning, R.B., 1963.
Hemisquilla ensigera (Owen, 1832) an earlier name for H. bigelowi (Rathbun, 1910)(Stomatopoda).
Crustaceana, 5(4):315-317.

Hemisquilla ensigera
Stephenson, William, 1967.
A comparison of Australasian and American specimens of Hemisquilla ensigera (Owen, 1832) (Crustacea: Stomatopoda).
Proc. U.S. nat. Mus., 120(3564):18 pp.

Hemisquilla ensigera californiensis
Manning, Raymond B. 1971.
Eastern Pacific expeditions of the New York Zoological Society. Stomatopod Crustacea.
Zoologica, N.Y. 56(3), 95-113

Hemisquilla stylifera
Dahl, E., 1954.
Reports of the Lund University Chile Expedition, 1948-49. 15. Stomatopoda. Lunds Univ. Arrskrift. N.F., Avd. 2, 49(17):12 pp., 1 textfig.
also:
K. Fysiografiska Sällskapets Handl., N.F., 64(17)

Hemisquilla stylifera
Stephensen, W., 1955.
Notes on stomatopod Crustacea from Victoria and Tasmania. Mem. Nat. Mus., Victoria, No. 19:7-10.

Heterosquilla n. gen.
Manning, R.B., 1963.
Preliminary revision of the genera Pseudosquilla and Lysiosquilla with descriptions of six new genera (Crustacea: Stomatopoda).
Bull. Mar. Sci., Gulf and Caribbean, 13(2):308-328.

Squilla eusebia
S. latifrons
Coronis spinosa
Lysiosquilla armata
L. polydactyla
L. platensis (type)
Lysiosquilla insignis
L. osculens
L. perpasta
L. vercoi
L. mccullochae
L. enodis
L. insolita

Heterosquilla (Heterosquilloides) africana n. sp.
Manning, Raymond B. 1970.
The R/V Pillsbury Deep-Sea Biological Expedition to the Gulf of Guinea, 1964-65. 13. The Stomatopod Crustacea.
Studies trop. Oceanogr. Miami 4(2): 256-275.

Heterosquilla (Heterosquilla) armata
Manning, Raymond B., 1969
Stomatopod Crustacea of the Western Atlantic. Stud. Trop. Oceanogr., Univ. Miami, 8:380 pp.

Heterosquilla (Heterosquilloides) brazieri
Manning, Raymond B., 1966.
Notes on some Australian and New Zealand stomatopod Crustacea, with an account of the species collected by the Fisheries Investigation Ship ENDEAVOUR.
Rec. Aust. Mus., 27(4):79-137.

Heterosquilla (Heterosquilloides) brazieri
Michel, A. 1969.
Dernier stade larvaire pélagique et post-larve de Heterosquilla (Heterosquilloides) brazieri (Miers 1880) (Crustacés, Stomatopodes).
Bull. Mus. natn. Hist. Nat. Paris (2) 40 (5): 992-997.

Heterosquilla eusebia
van der Baan, S. M., and L. B. Holthuis, 1966.
On the occurrence of Stomatopoda in the North Sea, with special reference to larvae from the surface plankton near the lightship "Texel".
Netherlands J. Sea Res., 3(1):1-12.

Heterosquilla eusebia
Verwey, J., 1966.
The origin of the stomatopod larvae of the southern North Sea.
Netherlands J. Sea Res., 3(1):13-20.

Haptosquilla glyptocerca
Moosa M. Kasim 1973.
The stomatopod Crustacea collected by the Mariel King Memorial Expedition in Maluku waters in 1970.
Penelit. Laut Indonesia (Mar. Res. Indonesia) 13:3-30.

Heterosquilla (Heterosquilloides) insueta n.sp.
Manning, Raymond B., 1970.
Two new stomatopod crustaceans from Australia.
Rec. Austral. Mus. 28(4): 77-85.

Heterosquilla (Heterosquilloides) insolita
Manning, Raymond B., 1969
Stomatopod Crustacea of the Western Atlantic. Stud. Trop. Oceanogr., Univ. Miami, 8:380 pp.

Heterosquilla jonesi n. sp.
Shanbhogue, S.L. 1970 (1971)
A new species of Heterosquilla (Crustacea-Stomatopoda) from Indian Seas.
J. mar. biol. Ass. India 12 (1+2): 100-104.

Heterosquilla (Heterosquilla) mccullochae
Manning, Raymond B., 1969
Stomatopod Crustacea of the Western Atlantic. Stud. Trop. Oceanogr., Univ. Miami, 8:380 pp.

Heterosquilla (Heterosquilla) platensis
Manning, Raymond B., 1969
Stomatopod Crustacea of the Western Atlantic. Stud. Trop. Oceanogr., Univ. Miami, 8:380 pp.

Heterosquilla polydactyla
Manning, Raymond B. 1971.
The postlarva of the stomatopod crustacean Heterosquilla polydactyla (Von Martens)
Proc. biol. Soc. Wash. 84 (33): 265-270.

Heterosquilla (Heterosquilla) polydactyla
Manning, Raymond B., 1969
Stomatopod Crustacea of the Western Atlantic. Stud. Trop. Oceanogr., Univ. Miami, 8:380 pp.

Heterosquilla (Austrosquilla) osculens
Manning, Raymond B., 1966.
Notes on some Australian and New Zealand stomatopod Crustacea, with an account of the species collected by the Fisheries Investigation Ship ENDEAVOUR.
Rec. Aust. Mus., 27(4):79-137.

Heterosquilla (Heterosquilla) tricarinata
Manning, Raymond B., 1966.
Notes on some Australian and New Zealand stomatopod Crustacea, with an account of the species collected by the Fisheries Investigation Ship ENDEAVOUR.
Rec. Aust. Mus., 27(4):79-137.

Heterosquilla (Austrosquilla) vercoi
Manning, Raymond B., 1966.
Notes on some Australian and New Zealand stomatopod Crustacea, with an account of the species collected by the Fisheries Investigation Ship ENDEAVOUR.
Rec. Aust. Mus., 27(4):79-137.

Heterosquilloides new subgenus
Manning, Raymond B., 1966.
Notes on some Australian and New Zealand stomatopod Crustacea, with an account of the species collected by the Fisheries Investigation Ship ENDEAVOUR.
Rec. Aust. Mus., 27(4):79-137.

stomatopod Hoplites longirostris n.gen.
Philippi, R.A., 1857.
Kurze Beschreibung einiger neuen Crustaceen.
Arch. f. Naturgesch. 23:319-329, Pl. 14.
Berlin.

Hoplosquilla n.gen.
Holthuis, L.B., 1964.
Preliminary note on two new genera of Stomatopoda Crustaceana, 7(2):140-141.

Indosquilla manihinei n.gen. n.sp.
Ingle, Raymond W., and Nigel R. Merrett 1971
A stomatopod crustacean from the Indian Ocean, Indosquilla manihinei gen. et sp. nov. (Family Bathysquillidae) with remarks on Bathysquilla crassispinosa (Fukuda, 1910).
Crustaceana 20 (2): 192-198.

Leptosquilla
Manning, Raymond B., 1968.
A revision of the Family Squillidae (Crustacea, Stomatopoda), with the description of eight new genera.
Bull. mar. Sci., Miami, 18(1):105-142.

Leptosquilla schmeltzii
Holthuis, L.B., 1967.
The Stomatopod Crustacea collected by the 1962 and 1965 Israel South Red Sea expeditions.
Israel J. Zool., 16(1):1-45.

Leptosquilla schmeltzii
Manning, Raymond B. 1970.
Some stomatopod crustaceans from Tulear Madagascar
Bull. Mus. natn. Hist. nat. (2) 41 (6): 1429-1441.

Lophosquilla, n.gen
Manning, Raymond B., 1968.
A revision of the Family Squillidae (Crustacea, Stomatopoda), with the description of eight new genera.
Bull. mar. Sci., Miami, 18(1):105-142.

Lophosquilla paulocarinata n. sp.
Blumstein, Radda 1972.
A new species of Lophosquilla (Stomatopoda) from the Indian Ocean.
Crustaceana 22 (1): 64-66.

Lysierichthus edwardsii

Hansen, H. J., 1895
Isopoden, Cumaceen und Stomatopoden der Plankton Expedition. Ergeb. Plankton Exped. II, Gc: 105 pp., 8 pls.

Lysierichthus minutus

Hansen, H. J., 1895
Isopoden, Cumaceen und Stomatopoden der Plankton Expedition. Ergeb. Plankton Exped. II, Gc: 105 pp., 8 pls.

Lysierichthus ophthalmicus n.sp.

Hansen, H. J., 1895
Isopoden, Cumaceen und Stomatopoden der Plankton Expedition. Ergeb. Plankton Exped. II, Gc: 105 pp., 8 pls.

Lysierichthus vitreus

Hansen, H. J., 1895
Isopoden, Cumaceen und Stomatopoden der Plankton Expedition. Ergeb. Plankton Exped. II, Gc: 105 pp., 8 pls.

Lysiosquilla sp.

Jacques, Françoise, et Alain Thiriot 1967.
Larves de stomatopodes du plancton de la région de Banyuls-sur-Mer.
Vie Milieu (B) 18(2B): 367-380.

Lysiosquilla sp.

Ingle, R.W., and N. della Croce, 1967.
Stomatopod larvae from the Mozambique Channel.
Boll. musei Ist. biol. Univ. Genova, 35(226):55-70.

Lysiosquilla

Manning, Raymond B., 1966
Stomatopod Crustacea. Campagne de la Calypso au large des côtes Atlantiques de l'Amerique du Sud (1961-1962)-1.
Ann. Inst. Océanogr., Monaco, n.s., 44:359-384

Manning, R.B., 1963. Lysiosquilla
Preliminary revision of the genera Pseudosquilla and Lysiosquilla with descriptions of six new genera (Crustacea: Stomatopoda).
Bull. Mar. Sci., Gulf and Caribbean, 13(2):308-328.

Squilla maculata   Lysiosquilla maculata tridecimdentata
Squilla glabriuscula
Squilla scabricauda   Lysiosquilla aulacorhynchus
Squilla desaussurei   Lysiosquilla campechiensis
Lysiosquilla capensis
Lysiosquilla cultrirostris   Tura (was L. monodontus)

Lysiosquilla sp.

Michel, A., 1968.
Dérive des larves de stomatopodes de l'est de l'Ocean Indien.
Cah. O.R.S.T.O.M., ser. Océanogr., VI(1):13-41.

Lysiosquilla acanthocarpus

Alikunhi, K.H., 1951.
An account of the stomatopod larvae of the Madras plankton. Rec. Indian Mus. 49(3/4):239-319, 25 textfigs.

Lysiosquilla acanthocarpus

Barnard, K.H., 1962.
New records of marine Crustacea from the East African region.
Crustaceana, 3(3):239-245.

Lysiosquilla acanthocarpus

Serène, R., 1952.
Étude d'une collection de stomatopodes de l'Australian Museum de Sydney. Rec. Australian Mus. 23(1):1-24, 33 textfigs., Pls. 1-3.

Lysiosquilla acanthocarpus

Tiwari, K.K., 1951.
On two new species of the genus Squilla Fabr., with notes on other stomatopods in the collections of the Zoological Survey of India.
Rec. Indian Mus. 49(3/4):349-363, 5 textfigs.

Lysiosquilla aulacorhynchus n.sp.

Cadenat, J., 1957.
Lysiosquilla aulacorhynchus, espèce nouvelle de stomatopode de la côte occidentale d'Afrique.
Bull. I.F.A.N., 19(1):126-133.

Diagnose préliminaire

Lysiosquilla biminiensis pacificus

Manning, R.B., 1962.
A redescription of Lysiosquilla biminiensis pacificus Borradaile (Stomatopoda).
Crustaceana, 4(4):301-306.

Lysiosquilla campechiensis

Manning, Raymond B., 1969
Stomatopod Crustacea of the Western Atlantic.
Stud. Trop. Oceanogr., Univ. Miami, 8:380 pp.

Lysiosquilla capensis

Barnard, K.H., 1950
Descriptive list of South African Stomatopod Crustacea (Mantis Shrimps). Ann. S. African Mus. 38:838-864, 4 text figs.

Lysiosquilla capensis

Manning, Raymond B., 1969
Notes on some stomatopod Crustacea from southern Africa.
Smithson. Contrib. Zool. 1:1-17.

Lysioaquilla chilensis n.sp.

Dahl, E., 1954.
Reports of the Lund University Chile Expedition, 1948-49. 15. Stomatopoda. Lunds Univ. Arrskrift. N.F., Avd. 2, 49(17):12 pp., 1 textfig.

also:
K. Fysiografiska Sällskapets Handl., N.F., 64(17)

Lysiosquilla crassispinosa

Barnard, K.H., 1950
Descriptive list of South African Stomatopod Crustacea (Mantis Shrimps). Ann. S. African Mus. 38:838-864, 4 text figs.

Lysiosquilla desaussurei

Manning, Raymond B. 1971.
Eastern Pacific expeditions of the New York Zoological Society. Stomatopod Crustacea.
Zoologica, N.Y. 56(2), 95-113

Lysiosquilla excavatrix

Lemos de Castro, A., 1955.
Contribuicao ao conhecimento dos crustaceos da Ordem Stomatopoda do litoral brasileiro (Crustacea, Hoplocarida).
Bol. Mus. Nac., Rio de Janeiro, n.s., Zool., No. 128:1-68, 56 figs.

Lysiosquilla glabriuscula

Hansen, H. J., 1895
Isopoden, Cumaceen und Stomatopoden der Plankton Expedition. Ergeb. Plankton Exped. II, Gc: 105 pp., 8 pls.

Lysiosquilla glabriuscula

Lemos de Castro, A., 1955.
Contribuicao ao conhecimento dos crustaceos da Ordem Stomatopoda do litoral brasileiro (Crustacea, Hoplocarida).
Bol. Mus. Nac., Rio de Janeiro, n.s., Zool. No. 128: 1-68, 56 figs.

Lysioquilla glabriuscula

Manning, Raymond B., 1969
Stomatopod Crustacea of the Western Atlantic.
Stud. Trop. Oceanogr., Univ. Miami, 8:380 pp.

Lysiosquilla grayi n.sp.

Chace, F.A., Jr., 1958.
A new stomatopod crustacean of the genus Lysiosquilla from Cape Cod, Massachusetts.
Biol. Bull., 114(2):141-145.

Lysiosquilla insignis

Barnard, K. H., 1950
Descriptive list of South African Stomatopod Crustacea (Mantis Shrimps). Ann. S. African Mus. 38:838-864, 4 text figs.

Lysiosquilla insolita, n.sp

Manning, R.B., 1963.
A new species of Lysiosquilla (Crustacea, Stomatopoda) from the northern Straits of Florida.
Bull. Mar. Sci., Gulf and Caribbean, 13(1):54-57.

Lysiosquilla latifrons brasieri

Stephenson, W., 1962.
5. Some interesting Stomatopoda - mostly from Western Australia.
J. R. Soc., Western Australia, 45(2):33-43.

Lysiosquilla maculata

Alikunhi, K.N., 1967.
An account of the post-larval development, moulting and growth of the common stomatopods of the Madras coast.
Proc. Symp. Crustacea, Ernakulam, Jan. 12-15, 1965, 2: 824-939

Lysioaquilla maculata

Alikunhi, K.H., 1951.
An account of the stomatopod larvae of the Madras plankton. Rec. Indian Mus. 49(3/4):239-319, 25 textfigs.

Lysiosquilla maculata

Barnard, K. H., 1950
Descriptive list of South African Stomatopod Crustacea (Mantis Shrimps). Ann. S. African Mus. 38:838-864, 4 text figs.

Lysiosquilla maculata

Boone, L., 1938
Scientific results of the world cruises of the yachts "Ara", 1928-1929, and "Alva" 1931-1932, "Alva" Mediterranean cruise, 1933, and "Alva" South American cruise, 1935, William K. Vanderbilt, commanding---- Bull. Vanderbilt Mus. 7:372 pp., 152 pls., 22 text figs.

Lysiosquilla maculata

Boss, Kenneth J., 1965.
A new mollusk (Bivalvia, Erycinidae) commensal on the stomatopod crustacean Lysiosquilla.
Amer. Mus. Novitates, No. 2215:11 pp.

Lysiosquilla maculata

Cadenat, J., 1950.
Sur quelques espèces des Squilles des côtes du Sénégal. C.R. Première Conf. Int. des Africanistes de l'Ouest, 1:192-193.

Lysiosquilla maculata

Chopra, B., 1939
Stomatopoda. John Murray Exped., 1933-34, Sci. Repts. 6(3):137-181, 13 text figs.

Lysiosquilla maculata

Foxon, G. E. H., 1939
Stomatopod larvae. John Murray Exped., 1933-34, Sci. Repts. 6(6):251-266, 4 text figs.

Lysiosquilla maculata

Ingle, R.W., 1963.
Crustacea Stomatopoda from the Red Sea and Gulf of Aden.
Sea Fish Res. Sta., Israel, Bull. No. 33:1-69.

Lysiosquilla maculata
Ingle, R.W., and N. della Croce, 1967.
Stomatopod larvae from the Mozambique Channel.
Boll. Musei Isr. biol.Univ.Genova, 35(226):55-70.

Lysiosquilla maculata
Manning, Raymond B., 1968.
Stomatopod Crustacea from Madagascar.
Proc. U.S. nat. Mus. 124(3641):1-61.

Lysiosquilla maculata
Manning, Raymond B., 1967.
Stomatopoda in the Vanderbilt Marine Museum.
Crustaceana, 12(1):102-106.

Lysiosquilla maculata
Manning, R.B., 1962.
Stomatopod Crustacea collected by the Yale Seychelles Expedition, 1957-1958.
Postilla, Yale Univ., 68:15 pp.

Lysiosquilla maculata
Parisi, B., 1940
Gli Stomatopodi raccolti del Prof. L. Sanzo nella Campagna idrografica nel Mar Rosso della R. N. Ammiraglio Magnaghi 1923-1924. (Mem. 16ª della Campagne). Mem. R. Com. Talass. Ital. CCLXXV:7 pp.

Lysiosquilla maculata
Serène, R., 1954.
Observations biologiques sur les stomatopodes.
Mém. Inst. Océan., Nhatrang, No. 8:1-93, 10 pls., 15 textfigs.

Lysiosquilla maculata
Stephenson, W., 1952.
Faunistic records from Queensland. Pt. I. General introduction. Pt. II. Adult Stomatopoda (Crustacea).
Univ. Queensland, Dept. Zool., Papers 1(1):1-15.

Lysiosquilla maculata (incl. larvae)
Townsley, S.J., 1953.
Adult and larval stomatopod crustaceans occurring in Hawaiian waters. Pacific Science 7(4):399-437, 28 textfigs.

Lysiosquilla mccullochae n.sp.
Schmitt, W. L., 1940
The stomatopods of the west coast of America, based on collections made by the Allan Hancock Expeditions, 1933-1938. Rept. Allan Hancock Pacific Exped. 5(4):129-226, 33 figs.

Lysiosquilla multifasciata
Alikunhi, K.H., 1951.
An account of the stomatopod larvae of the Madras plankton. Rec. Indian Mus. 49(3/4):239-319, 25 textfigs.

Lysiosquilla multifasciata
Chopra, B., 1939
Stomatopoda. John Murray Exped., 1933-34, Sci. Repts. 6(3):137-181, 13 text figs.

Lysiosquilla multifasciata ?
Foxon, G. E. H., 1939
Stomatopod larvae. John Murray Exped., 1933-34, Sci. Repts. 6(6):251-266, 4 text figs.

Lysiosquilla multifasciata
Ingle, R.W., 1963.
Crustacea Stomatopoda from the Red Sea and Gulf of Aden.
Sea Fish Res.Sta., Israel,Bull.No.33:1-69.

Lysiosquilla multifasciata
Serène, R., 1952.
Étude d'une collection de stomatopodes de l' Australian Museum de Sydney. Rec. Australian Mus. 23(1):1-24, 33 textfigs., Pls. 1-3.

Lysiosquilla multifasciata
Tiwari, K.K., 1951.
On two new species of the genus Squilla Fabr., with notes on other stomatopods in the collections of the Zoological Survey of India.
Rec. Indian Mus. 49(3/4):349-363, 5 textfigs.

Lysiosquilla occulta
Jacques, Françoise 1970.
La glande de mue chez les larves de stomatopodes.
C. r. hebd. Séanc. Acad. Sci. Paris (D) 270 (24): 2965-2968.

Lysiosquilla occulata
Jacques, Françoise 1969.
Ajtogénèse des pédoncules oculaires des larves des stomatopodes.
Vie Milieu (A) 20(3): 565-593.

Lysiosquilla osculans
Stephensen, W., 1955.
Notes on stomatopod Crustacea from Victoria and Tasmania. Mem. Nat. Mus., Victoria, No. 19:7-10.

Lysiosquilla panamica n.sp.
Manning, Raymond B., 1971.
Lysiosquilla panamica, a new stomatopod crustacean from the eastern Pacific region.
Proc. biol. Soc. Wash. 84 (24): 225-230.

Lysiosquilla perpasta
Stephensen, W., 1955.
Notes on stomatopod Crustacea from Victoria and Tasmania. Mem. Nat. Mus., Victoria, No. 19:7010.

Lysiosquilla perpasta
Stephenson, W., 1952.
Faunistic records from Queensland. Pt. I. General introduction. Pt. II. Adult Stomatopoda (Crustacea).
Univ. Queensland, Dept. Zool., Papers 1(1):1-15.

Lysiosquilla polydactyla
Bahamonde N., Nibaldo, 1957.
Sobre la distribucion geografica de Lysiosquilla polydactyla Von Martens 1881 (Crustacea, Stomatopoda). Invest. Zool. Chilenas 3(5/6/7):119-121.

Lysiosquilla scabricauda
Dennell, R., 1950.
The occurrence of Lysiosquilla scabricauda (Lamarck) in Bermuda. Proc. Linn. Soc., London, 162(1):63-64.

Lysiosquilla scabricauda
Hansen, H. J., 1895
Isopoden, Cumaceen und Stomatopoden der Plankton Expedition. Ergeb. Plankton Exped. II, Gc: 105 pp., 8 pls.

Lysiosquilla scabricauda
Lemos de Castro, A., 1955.
Contribucao ao conhecimento dos crustaceos da Ordem Stomatopoda do litoral brasileiro (Crustacea, Hoplocarida).
Bol. Mus. Nac. Rio de Janeiro, n.s., Zool., No. 128:1-68, 56 figs.

Lysiosquilla scabricauda
Lunz, G. R., jr., 1937.
Stomatopoda of the Bingham Oceanographic Collection. Bull. Bingham Ocean. Coll., V(5):1-19, 10 textfigs.

Lysiosquilla scabricauda
Manning, Raymond B., 1969
Stomatopod Crustacea of the Western Atlantic.
Stud. Trop. Oceanogr., Univ. Miami, 8:380 pp.

Lysiosquilla scabricauda
Manning, Raymond B., 1967.
Stomatopoda in the Vanderbilt Marine Museum.
Crustaceana, 12(1):102-106.

Lysiosquilla scabricauda
Manning, Raymond B., 1961
Sexual dimorphism in Lysiosquilla scabricauda (Lamarck) a stomatopod crustacean.
Q.J. Florida Acad. Sci., 24(2):101-107.

Lysiosquilla scabricauda
Ramos, F. de P., 1951.
Nota preliminar sôbre alguno Stomatopoda da Costa Brasileira. Bol. Inst. Paulista Ocean. 2(1):139-150, 1 pl.

Lysiosquilla sewelli n.sp.
Chopra, B., 1939
Stomatopoda. John Murray Exped., 1933-34, Sci. Repts. 6(3):137-181, 13 text figs.

Lysiosquilla sulcirostris
Alikunhi, K.H., 1967.
An account of the post-larval development, moulting and growth of the common stomatopods of the Madras coast.
Proc. Symp. Crustacea, Ernakulam, Jan. 12-15 1965, 2: 824-939

Lysiosquilla sulcirostris
Alikunhi, K.H., 1951.
An account of the stomatopod larvae of the Madras plankton. Rec. Indian Mus. 49(3/4):239-319, 25 textfigs.

Lysiosquilla sulcirostris
Manning, Raymond B. 1970.
Some stomatopod crustaceans from Tuléar Madagascar.
Bull. Mus. natn. Hist. nat. (2) 41 (6): 1429-1441.

Lysiosquilla sulcirostris
Michel, A., 1968.
Dérive des larves de stomatopodes de l'est de l'Océan Indien.
Cah.O.R.S.T.O.M.,ser.Océanogr., VI(1):13-41.

Lysiosquilla tigrina
Alikunhi, K.H., 1951.
An account of the stomatopod larvae of the Madras plankton. Rec. Indian Mus. 49(3/4):239-319, 25 textfigs.

Lysiosquilla tredecimdentata
Manning, Raymond B., 1969
Notes on some stomatopod Crustacea from southern Africa.
Smithson. Contrib. Zool. 1:1-17.

Lysiosquilla tredecimdentata
Manning, Raymond B.,1968.
Stomatopod Crustacea from Madagascar.
Proc. U.S. nat.Mus., 124(3641):1-61.

Lysiosquilla tredecimdentata
Tirmizi, Nasima M., und Raymond B.Manning,1968.
Stomatopod Crustacea from west Pakistan.
Proc.U.S.nat.Mus.,125(3666):1-48.

Lysiosquilla veroci
Stephensen, W., 1955.
Notes on stomatopod Crustacea from Victoria and Tasmania. Mem. Nat. Mus., Victoria, No. 19:7-10.

Lysiosquilla vicina
Ingle, R.W., 1963.
Crustacea Stomatopoda from the Red Sea and Gulf of Aden.
Sea Fish. Res.Sta., Israel,Bull.No.33:1-69.

Manningia nov. gen.
Serène, R., 1962.
Revision du genre Pseudosquilla (Stomatopoda) et définition de genres nouveaux.
Bull. Inst. Océanogr., Monaco, 59(1241):27 pp.

Manningia amabilis n.sp.
Holthuis, L.B., 1967.
The Stomatopod Crustacea collected by the 1962 and 1965 Israel South Red Sea expeditions.
Israel J. Zool., 16(1):1-45.

Manningia amabilis
Tirmizi, Nasima M., and Raymond B. Manning, 1968.
Stomatopod Crustacea from West Pakistan.
Proc. U.S. nat. Mus., 125(3666):1-48.

Manningia australiensis n.sp.
Manning, Raymond B., 1970.
Two new stomatopod crustaceans from Australia.
Rec. Austral. Mus. 28(4):77-85.

Manningia notialis n.sp.
Manning, Raymond B., 1966.
Notes on some Australian and New Zealand stomatopod Crustacea, with an account of the species collected by the Fisheries Investigation Ship ENDEAVOUR.
Rec. Aust. Mus., 27(4):79-137.

Manningia pilaensis
Manning, Raymond B., 1967.
Notes on the genus Manningia with description of a new species (Crustacea: Stomatopoda).
Proc. U.S. nat. Mus., 122(3589):1-13.

Manningia pilaensis
Manning, R.B., 1963.
Preliminary revision of the genera Pseudosquilla and Lysiosquilla with descriptions of six new genera (Crustacea: Stomatopoda).
Bull. Mar. Sci., Gulf and Caribbean 13(2):308-328.

Manningia pilaensis
Serène, R., 1962.
Revision du genre Pseudosquilla (Stomatopoda) et définition de genres nouveaux.
Bull. Inst. Océanogr., Monaco, 59(1241):27 pp.

Manningia serenei n.sp.
Manning, Raymond B., 1967.
Notes on the genus Manningia with description of a new species (Crustacea: Stomatopoda).
Proc. U.S. nat. Mus., 122(3589):1-13.

Meiosquilla, n-gen
Manning, Raymond B., 1968.
A revision of the Family Squillidae (Crustacea, Stomatopoda), with the description of eight new genera.
Bull. mar. Sci., Miami, 18(1):105-142.

MEIOSQUILLA DAWSONI N.SP.
Manning, Raymond B., 1970.
Nine new American stomatopod crustaceans
Proc. Biol. Soc. Wash., 83(5):99-114

Meiosquilla desmareoti
Manning, Raymond B., 1969
Notes on some stomatopod Crustacea from southern Africa.
Smithson. Contrib. Zool. 1:1-17.

Meiosquilla lebouri
Manning, Raymond B., 1969
Stomatopod Crustacea of the Western Atlantic.
Stud. Trop. Oceanogr., Univ. Miami, 8:380 pp.

Meiosquilla oculinova
Manning, Raymond B., 1971.
Eastern Pacific expeditions of the New York Zoological Society. Stomatopod Crustacea.
Zoologica, N.Y. 56(3), 95-113

?Meiosquilla pallida
Manning, Raymond B., 1970.
The R/V Pillsbury Deep-Sea Biological Expedition to the Gulf of Guinea, 1964-65. 13. The Stomatopod Crustacea.
Studies trop. Oceanogr. Miami 4(2): 256-275.

Meiosquilla polita
Manning, Raymond B., 1971.
Eastern Pacific expeditions of the New York Zoological Society. Stomatopod Crustacea.
Zoologica, N.Y. 56(3), 95-113

Meiosquilla quadridens
Manning, Raymond B., 1969
Stomatopod Crustacea of the Western Atlantic.
Stud. Trop. Oceanogr., Univ. Miami, 8:380 pp.

Meiosquilla randalli
Manning, Raymond B., 1969
Stomatopod Crustacea of the Western Atlantic.
Stud. Trop. Oceanogr., Univ. Miami, 8:380 pp.

Meiosquilla schmitti
Manning, Raymond B., 1969
Stomatopod Crustacea of the Western Atlantic.
Stud. Trop. Oceanogr., Univ. Miami, 8:380 pp.

Meiosquilla swetti
Manning, Raymond B., 1971.
Eastern Pacific expeditions of the New York Zoological Society. Stomatopod Crustacea.
Zoologica, N.Y. 56(3), 95-113

Meiosquilla tricarinata
Manning, Raymond B., 1969
Stomatopod Crustacea of the Western Atlantic.
Stud. Trop. Oceanogr., Univ. Miami, 8:380 pp.

Mesacturus brevisquamatus
Holthuis, L.B., 1967.
The Stomatopod Crustacea collected by the 1962 and 1965 Israel South Red Sea expeditions.
Israel J. Zool., 16(1):1-45.

Nannosquilla n.gen.
Manning, R.B., 1963.
Preliminary revision of the genera Pseudosquilla and Lysiosquilla with descriptions of six new genera (Crustacea: Stomatopoda).
Bull. Mar. Sci., Gulf and Caribbean, 13(2): 308-328
Lysiosquilla decemspinosa   L. antillensis
L. occulta   L. hancocki
L. varicosa   L. schmitti
L. chilensis
L. hystricotelson
L. grayi (type)
L. californiensis

Nannosquilla anomala
Manning, Raymond B., 1967.
Nannosquilla anomala, a new stomatopod crustacean from California.
Proc. biol. Soc. Wash. 80:147-150.

Nannosquilla antillensis
Manning, Raymond B., 1969
Stomatopod Crustacea of the Western Atlantic.
Stud. Trop. Oceanogr., Univ. Miami, 8:380 pp.

NANNOSQUILLA CAROLINENSIS N.SP.
Manning, Raymond B., 1970.
Nine new American stomatopod crustaceans
Proc. Biol. Soc. Wash., 83(5):99-114

NANNOSQUILLA DACOSTAI N.SP.
Manning, Raymond B., 1970.
Nine new American stomatopod crustaceans
Proc. Biol. Soc. Wash., 83(5):99-114

Nannosquilla galapagensis n.sp.
Manning, Raymond B., 1972.
Three new stomatopod crustaceans of the family Lysiosquillidae from the eastern Pacific region.
Proc. Biol. Soc. Wash. 85(21), 271-278

Nannosquilla grayi
Manning, Raymond B., 1969
Stomatopod Crustacea of the Western Atlantic.
Stud. Trop. Oceanogr., Univ. Miami, 8:380 pp.

Nannosquilla hancocki
Manning, Raymond B., 1969
Stomatopod Crustacea of the Western Atlantic.
Stud. Trop. Oceanogr., Univ. Miami, 8:380 pp.

Nannosquilla hystricotelson
Holthuis, L.B., 1967.
The Stomatopod Crustacea collected by the 1962 and 1965 Israel South Red Sea expeditions.
Israel J. Zool., 16(1):1-45.

Nannosquilla schmitti
Manning, Raymond B., 1969
Stomatopod Crustacea of the Western Atlantic.
Stud. Trop. Oceanogr., Univ. Miami, 8:380 pp.

Nannosquilla similis n.sp.
Manning, Raymond B., 1972.
Three new stomatopod crustaceans of the family Lysiosquillidae from the eastern Pacific region.
Proc. Biol. Soc. Wash. 85(21), 271-278

Nannosquilla taylori n. sp.
Manning, Raymond B., 1969
Stomatopod Crustacea of the Western Atlantic.
Stud. Trop. Oceanogr., Univ. Miami, 8:380 pp.

Odonterichthus tenuicornis
Ingle, R.W., and N. della Croce, 1967.
Stomatopod larvae from the Mozambique Channel.
Boll. Musei Ist. biol. Univ. Genova, 35(226):55-70.

Odondactylus sp.
Michel, A., 1968.
Dérive des larves de stomatopodes de l'est de l'Océan Indien.
Cah. O.R.S.T.O.M., ser. Océanogr., VI(1):13-41.

Odontodactylus brevirostris
Ingle, R.W., and N. della Croce, 1967.
Stomatopod larvae from the Mozambique Channel.
Boll. Musei Ist. biol. Univ. Genova, 35(226):55-70.

Odontodactylus brevirostris
Manning, Raymond B., 1969
Stomatopod Crustacea of the Western Atlantic.
Stud. Trop. Oceanogr., Univ. Miami, 8:380 pp.

Odontodactylus brevirostris
Michel, A., 1970.
Larves pélagiques et post-larves du genre Odontodactylus (Crustaces: Stomatopodes) dans le Pacifique tropical sud et équatorial. Cah. ORSTOM sér. Océanogr., 8(2): 111-126.

Odontodactylus brevirostris [a]
Moosa, M. Kasim 1973.
The Stomatopod Crustacea collected by the Mariel King Memorial Expedition in Maluku waters in 1970. Penelit. Laut Indonesia (Mar. Res. Indonesia) 13:3-30.

Ondontodactylus cavirostris n.sp.
Chace, F.A., jr., 1942.
Six new species of decapod and stomatopod crustacea from the Gulf of Mexico. Proc. N.E. Zool. Club 19:79-92, 6 pls.

Odontodactylus cultrifer
Manning, Raymond B., 1966.
Notes on some Australian and New Zealand stomatopod Crustacea, with an account of the species collected by the Fisheries Investigation Ship ENDEAVOUR. Rec. Aust. Mus., 27(4):79-137.

Odontodactylus cultrifer [a]
Moosa, M. Kasim 1973.
The Stomatopod Crustacea collected by the Mariel King Memorial Expedition in Maluku waters in 1970. Penelit. Laut Indonesia (Mar. Res. Indonesia) 13:3-30.

Odontodactylus cultrifer
Serène, R., 1954.
Observations biologiques sur les stomatopodes. Mém. Inst. Océan., Nhatrang, No. 8:1-93, 10 pls., 15 textfigs.

Odontodactylus cultrifer
Stephenson, W., 1952.
Faunistic records from Queensland. Pt. I. General introduction. Pt. II. Adult Stomatopoda (Crustacea). Univ. Queensland, Dept. Zool., Papers 1(1):1-15.

Oratosquilla gonypetes [a]
Moosa, M. Kasim 1973.
The Stomatopod Crustacea collected by the Mariel King Memorial Expedition in Maluku waters in 1970. Penelit. Laut Indonesia (Mar. Res. Indonesia) 13:3-30.

Odontodactylus hanseni
Townsley, S.J., 1953.
Adult and larval stomatopod crustaceans occurring in Hawaiian waters. Pacific Science 7(4):399-437, 28 textfigs.

Odontodactylus havanensis
Lunz, G.R., jr., 1937.
Stomatopoda of the Bingham Oceanographic Collection. Bull. Bingham Ocean. Coll., V(5): 1-19, 10 textfigs.

Odontodactylus hawaiiensis [a]
Michel, A., 1970.
Larves pélagiques et post-larves du genre Odontodactylus (Crustaces: Stomatopodes) dans le Pacifique tropical sud et équatorial. Cah. ORSTOM sér. Océanogr., 8(2): 111-126.

Odontodactylus japonicus
Manning, Raymond B., 1968.
Stomatopod Crustacea from Madagascar. Proc.U.S. nat. Mus., 124(3641):1-61.

Odontodactylus japonica
Manning, Raymond B. 1965.
Stomatopoda from the collection of His Majesty, The Emperor of Japan. Crustaceana 9(3): 249-262.

Odontodactylus paleatus
Manning, Raymond B. 1965.
Stomatopoda from the collection of His Majesty, the Emperor of Japan. Crustaceana 9(3): 249-262.

Oratosquilla quadraticauda [a]
Moosa, M. Kasim 1973.
The Stomatopod Crustacea collected by the Mariel King Memorial Expedition in Maluku waters in 1970. Penelit. Laut Indonesia (Mar. Res. Indonesia) 13:3-30.

Odontodactylus scyllarus
Fukuda, T., 1910.
Further report on Japanese Stomatopoda with descriptions of two new species. Annot. Zool. Japon. 7:285-290, Pl. 11.

Odontodactylus scyllarus
Manning, Raymond B., 1969
Notes on some stomatopod Crustacea from southern Africa. Smithson. Contrib. Zool. 1:1-17.

Odontodactylus scyllarus
Manning, Raymond B., 1968.
Stomatopod Crustacea from Madagascar. Proc.U.S. nat. Mus., 124(3641):1-61.

Odontodactylus scyllarus [a]
Michel, A., 1970.
Larves pélagiques et post-larves du genre Odontodactylus (Crustaces: Stomatopodes) dans le Pacifique tropical sud et équatorial. Cah. ORSTOM sér. Océanogr., 8(2): 111-126.

Odontodactylus scyllarus [a]
Moosa, M. Kasim 1973.
The Stomatopod Crustacea collected by the Mariel King Memorial Expedition in Maluku waters in 1970. Penelit. Laut Indonesia (Mar. Res. Indonesia) 13:3-30.

Oratosquilla n. gen
Manning, Raymond B., 1968.
A revision of the Family Squillidae (Crustacea, Stomatopoda), with the description of eight new genera. Bull. mar. Sci., Miami, 18(1):105-142.

Oratosquilla calumnia [a]
Manning, Raymond B., 1971.
Keys to the species of Oratosquilla (Crustacea: Stomatopoda), with descriptions of new species. Smithson. Contrib. Zool. 71:16pp. (two)

Oratosquilla gonypetes
Manning, Raymond B., 1969
Notes on some stomatopod Crustacea from southern Africa. Smithson. Contrib. Zool. 1:1-17.

Orarosquilla hesperia
Tirmizi, Nasima M., and Raymond B. Manning, 1968.
Stomatopod Crustacea from West Pakistan. Proc.U.S. nat.Mus., 125(3666):1-48.

Oratosquilla holoschista
Manning, Raymond B., 1969
Notes on some stomatopod Crustacea from southern Africa. Smithson. Contrib. Zool. 1:1-17.

Oratosquilla interrupta
Tirmizi, Nasima M., and Raymond B. Manning, 1968.
Stomatopod Crustacea from West Pakistan. Proc.U.S. nat.Mus., 125(3666):1-48.

Oratosquilla investigatoris
Losse, G.F. and N.R. Merrett, 1971.
The occurrence of Oratosquilla investigatoris (Crustacea: Stomatopoda) in the pelagic zone of the Gulf of Aden and the equatorial western Indian Ocean. Mar. Biol. 10(3): 244-253.

Oratosquilla mikado
Manning, Raymond B., 1969
Notes on some stomatopod Crustacea from southern Africa. Smithson. Contrib. Zool. 1:1-17.

Oratosquilla nepa
Tirmizi, Nasima M., and Raymond B. Manning, 1968.
Stomatopod Crustacea from West Pakistan. Proc.U.S. nat.Mus., 125(3666):1-48.

Oratosquilla oratoria [a]
Manning, Raymond B., 1971.
Keys to the species of Oratosquilla (Crustacea: Stomatopoda), with descriptions of new species. Smithson. Contrib. Zool. 71:16pp. (two)

Oratosquilla ornata n. sp. [a]
Manning, Raymond B., 1971.
Keys to the species of Oratosquilla (Crustacea: Stomatopoda), with descriptions of new species. Smithson. Contrib. Zool. 71:16pp. (two)

Oratosquilla stridulans
Ghosh, A.C. 1975.
A note on two species of stomatopods from the Arabian Sea collected by the John Murray Expedition 1933-1934. Crustaceana 24(1):143-144.

Oratosquilla tweedii n.sp.
Manning, Raymond B. 1971.
Keys to the species of Oratosquilla (Crustacea: Stomatopoda), with descriptions of two new species. Smithson. Contrib. Zool. 71: 16pp.

Parasquilla
Manning, R.B., 1963.
Preliminary revision of the genera Pseudosquilla and Lysiosquilla with descriptions of six new genera (Crustacea: Stomatopoda). Bull. Mar. Sci., Gulf and Caribbean, 13(2):308-328.

Squilla ferussaci
Pseudosquilla laevi
Parasquilla meridionalis (type)
Parasquilla coccinea

Parasquilla (Parasquilla) coccinea
Manning, Raymond B., 1969
Stomatopod Crustacea of the Western Atlantic.
Stud. Trop. Oceanogr., Univ. Miami, 8:380 pp.

Parasquilla coccinea n.sp.
Manning, R.B., 1962.
A new species of Parasquilla (Stomatopoda) from the Gulf of Mexico with a redescription of Squilla ferussaci Roux.
Crustaceana, 4(3):180-190.

Parasquilla (Parasquilla) meridionalis
Manning, Raymond B., 1969
Stomatopod Crustacea of the Western Atlantic.
Stud. Trop. Oceanogr., Univ. Miami, 8:380 pp.

Parvisquilla n.gen.
Manning, Raymond B. 1972.
Preliminary definition of a new genus of Stomatopoda.
Crustaceana 23(3):299-300

Platysquilla
Manning, Raymond B., 1968.
A revision of the Family Squillidae (Crustacea, Stomatopoda), with the description of eight new genera.
Bull. mar. Sci., Miami, 18(1):105-142.

Platysquilla
Manning, Raymond B., 1967.
Preliminary account of a new genus and a new family of Stomatopoda.
Crustaceana, 13(2):238-239.

Platysquilla enodis
Manning, Raymond B., 1969
Stomatopod Crustacea of the Western Atlantic.
Stud. Trop. Oceanogr., Univ. Miami, 8:380 pp.

Platysquilla eusebia
Ceidigh, Padraig O. 1970.
The occurrence of Platysquilla eusebia (Risso, 1816) on the west coast of Ireland (Stomatopoda).
Crustaceana 19(2):205-206

Platysquilla horologii
Camp, David K., 1971.
Platysquilla horologii (Stomatopoda, Lysiosquillidae) a new species from the Gulf of Mexico, with an emendation of the generic definition.
Proc. Biol. Soc. Wash., 84(15):119-128

Protosquilla stoliura
Bedot, M., 1909
Sur la faune de l'Archipel Malais (resume).
Rev. Suisse Zool. 17:143-169.

Protosquilla
Manning, Raymond B., 1966.
Notes on some Australian and New Zealand stomatopod Crustacea, with an account of the species collected by the Fisheries Investigation Ship ENDEAVOUR.
Rec. Aust. Mus., 27(4):79-137.

Protosquilla folini
Manning, Raymond B. 1970.
The R/V Pillsbury Deep-Sea Biological Expedition to the Gulf of Guinea, 1964-65. 13. The stomatopod Crustacea.
Studies trop. Oceanogr. Miami 4(2):256-275

Protosquilla lenzi
Holthuis, L.B., 1967.
The Stomatopod Crustacea collected by the 1962 and 1965 Israel South Red Sea expeditions.
Israel J. Zool., 16(1):1-45.

Protosquilla lenzi
Manning, Raymond B., 1968.
Stomatopod Crustacea from Madagascar.
Proc. U.S. nat. Mus. 124(3641):1-61.

Protosquilla lenzi
Tirmizi, Nasima M., and Raymond B. Manning, 1968.
Stomatopod Crustacea from West Pakistan.
Proc. U.S. nat. Mus., 125(3666):1-48.

(Has good labelled diagrams of parts).

Protosquilla pulchella
Manning, Raymond B., 1968.
Stomatopod Crustacea from Madagascar.
Proc. U.S. nat. Mus., 124(3641):1-61.

Protosquilla pulchella
Tirmizi, Nasima M., and Raymond B. Manning, 1968.
Stomatopod Crustacea from west Pakistan.
Proc. U.S. nat. Mus., 125(3666):1-48.

Protosquilla spinosissima
Manning, Raymond B., 1968.
Stomatopod Crustacea from Madagascar.
Proc. U.S. nat. Mus., 124(3641):1-61.

Protosquilla tanensis n.sp.
Fukuda, T., 1910.
Further report on Japanese Stomatopoda with descriptions of two new species.
Annot. Zool. Japon. 7:285-290, Pl. 11.

Pseuderichthus communis n.sp.
Hansen, H. J., 1895
Isopoden, Cumaceen und Stomatopoden der Plankton Expedition. Ergeb. Plankton Exped. II, Gc: 105 pp., 8 pls.

Pseuderichthus distinguendus n.sp.
Hansen, H. J., 1895
Isopoden, Cumaceen und Stomatopoden der Plankton Expedition. Ergeb. Plankton Exped. II, Gc: 105 pp., 8 pls.

Pseudosquilla
Holthuis, L.B., and Raymond B. Manning, 1964.
Proposed use of the plenary powers (A) to designate a type-species for the genera Pseudosquilla Dana, 1852, and Gonodactylus Berthold 1827, and (B) for the suppression of the generic name Smerdis Leach, 1817 (Crustacea, Stomatopoda) Z.N. (S.) 1609.
Bull. Zool. Nomencl., 21(2):137-143.

Pseudosquilla spp.
Ingle, R.W., and N. della Croce, 1967.
Stomatopod larvae from the Mozambique Channel.
Boll. Musei Ist. biol. Univ. Genova, 35(226):55-70.

Pseudosquilla
Manning, R.B., 1963.
Preliminary revision of the general Pseudosquilla and Lysiosquilla with descriptions of six new genera (Crustacea: Stomatopoda).
Bull. Mar. Sci., Gulf and Caribbean, 13(2):308-328.

Contains:
Squilla ciliata (type)
Squilla oculata
Pseudosquilla ornata
Pseudosquilla megalophthalma
Pseudosquilla oxyrhyncha
Pseudosquilla oculata

Pseudosquilla sp.
Michel, A., 1968.
Derive des larves de stomatopodes de l'est de l'Océan Indien.
Cah. O.R.S.T.O.M., ser. Océanogr., VI(1):13-41.

Pseudosquilla
Serène, R., 1962.
Révision du genre Pseudosquilla (Stomatopoda) et définition de genres nouveaux.
Bull. Inst. Océanogr., Monaco, 59(1241):27 pp.

Pseudosquilla adiastalta
Manning, Raymond B. 1971.
Eastern Pacific expeditions of the New York Zoological Society. Stomatopod Crustacea.
Zoologica, N.Y. 56(2):95-113

Pseudosquilla asiastalta n.sp.
Manning, Raymond B., 1964.
A new west American species of Pseudosquilla (Stomatopoda).
Crustaceana, 6(4):303-308.

Pseudosquilla ciliata
Barnard, K. H., 1950
Descriptive list of South African Stomatopod Crustacea (Mantis Shrimps). Ann. S. African Mus. 38:838-864, 4 text figs.

Pseudosquilla ciliata
Chopra, B., 1939
Stomatopoda. John Murray Exped., 1933-34, Sci. Repts. 6(3):137-181, 13 text figs.

Pseudosquilla ciliata
Foxon, G. E. H., 1939
Stomatopod larvae. John Murray Exped., 1933-34, Sci. Repts. 6(6):251-266, 4 text figs.

Pseudosquilla ciliata
Hansen, H. J., 1895
Isopoden, Cumaceen und Stomatopoden der Plankton Expedition. Ergeb. Plankton Exped. II, Gc: 105 pp., 8 pls.

Pseudosquilla ciliata
Holthuis, L.B., 1967.
The Stomatopod Crustacea collected by the 1962 and 1965 Israel South Red Sea expeditions.
Israel J. Zool., 16(1):1-45.

Pseudosquilla ciliata
Ingle, R.W., 1963.
Crustacea stomatopoda from the Red Sea and Gulf of Aden.
Sea Fish. Res. Sta., Israel, Bull. No. 33:1-69.

Pseudosquilla ciliata
Ingle, R.W., and N. della Croce, 1967.
Stomatopod larvae from the Mozambique Channel.
Boll. Musei Ist. biol. Univ. Genova, 35(226):55-70.

Pseudosquilla ciliata
Kinzie, Robert A., III, 1968.
The ecology of the replacement of Pseudosquilla ciliata (Fabricius) by Gonodactylus falcatus (Forskal)(Crustacea; Stomatopoda) recently introduced into the Hawaiian islands.
Pacif. Sci. 22(4):465-475.

Pseudosquilla ciliata
Lemos de Castro, A., 1955.
Contribuicao ao conhecimento dos crustaceos da Ordem Stomatopoda do litoral brasileiro (Crustacea, Hoplocarida).
Bol. Mus. Nac., Rio de Janeiro, n.s., Zool., No. 128:1-68, 56 figs.

Pseudosquilla ciliata
Lunz, G. R., jr., 1937.
Stomatopoda of the Bingham Oceanographic Collection. Bull. Bingham Ocean. Coll., V(5):

Pseudosquilla ciliata
Manning, Raymond B. 1970.
The R/V Pillsbury Deep-Sea Biological Expedition to the Gulf of Guinea, 1964-65. 13. The stomatopod Crustacea.
Studies trop. Oceanogr. Miami 4(2):256-275

Pseudosquilla ciliata
Manning, Raymond B., 1969
Stomatopod Crustacea of the Western Atlantic.
Stud. Trop. Oceanogr., Univ. Miami, 8:380 pp.

Pseudosquilla ciliata
Manning, Raymond B., 1968.
Stomatopod Crustacea from Madagascar.
Proc.U.S. nat. Mus., 124(3641):1-61.

Pseudosquilla ciliata
Manning, Raymond B., 1967.
Stomatopoda in the Vanderbilt Marine Museum.
Crustaceana, 12(1):102-106.

Pseudosquilla ciliata
Manning, Raymond B., 1966.
Stomatopod Crustacea. Campagne de la Calypso au large des côtes Atlantiques de l'Amerique du Sud (1961-1962)-1.
Ann. Inst. Océanogr., Monaco, n.s., 44:359-384

Pseudosquilla ciliata
Manning, R.B., 1962.
Stomatopod Crustacea collected by the Yale Seychelles Expedition, 1957-1958.
Postilla, Yale Univ., 68:15 pp.

Pseudosquilla ciliata
Michel, A., 1968.
Derive des larves de stomatopodes de l'est de l'Océan Indien.
Cah.O.R.S.T.O.M., ser.Océanogr., VI(1):13-41.

Pseudosquilla ciliata
Moosa, M. Kasim 1973.
The Stomatopod Crustacea collected by the Mariel King Memorial Expedition in Maluku waters in 1970.
Penelit. Laut Indonesia (Mar. Res. Indonesia) 13:3-30.

Pseudosquilla ciliata
Parisi, B., 1940
Gli Stomatopodi raccolti del Prof. L. Sanzo nella Campagna idrografica nel Mar Rosso della R. N. Ammiraglio Magnaghi 1923-1924. (Mem. 16ª della Campagne). Mem. R. Com. Talass. Ital. CCLXXV:7 pp.

Pseudosquilla ciliata
Serène, R., 1950.
Cas de malformations chez les stomatopodes. Bull Mus., Indochine, 2nd ser., 22(3):341-343, 2 pls.

Pseudosquilla ciliata
Stephenson, W., 1952.
Faunistic records from Queensland. Pt. I. General introduction. Pt. II. Adult Stomatopoda (Crustacea).
Univ. Australia, Dept. Zool., Papers 1(1):1-15.

Pseudosquilla ciliata
Townsley, S.J., 1953.
Adult and larval stomatopod crustaceans occurring in Hawaiian waters. Pacific Science 7(4):399-437,

Pseudosquilla ciliata
Van Weel, P.B., J.E.Randall and M. Takata, 1954.
Observations on the oxygen consumption of certain marine Crustacea. Pacific Science 8(2):209-218, 7 textfigs.

Pseudosquilla ferussaci
Figueiredo, M.J., 1962.
Un stomatopode nouveau pour la faune portugaise et pour l'Ocean atlantique (Pseudosquilla ferussaci (Roux)).
Notas e Estudos, Inst. Biol. Marit, Lisboa, 25: 9 pp.

Pseudosquilla guttata n.sp.
Manning, Raymond B. 1972.
Two new species of Pseudosquilla (Crustacea, Stomatopoda) from the Pacific Ocean.
Novitates, Am. Mus. 2484: 11 pp.

Pseudosquilla hieroglyphica n.sp.
Manning, Raymond B. 1972.
Two new species of Pseudosquilla (Crustacea, Stomatopoda) from the Pacific Ocean.
Novitates Am. Mus. 2484: 11 pp.

Pseudosquilla lessoni
Dahl, E., 1954.
Reports of the Lund University Chile Expedition, 1948-49.15 Stomatopoda. Lunds Univ. Arrskrift., N.F., Avd. 2, 49(17):12 pp., 1 textfig.

also:
K. Fysiografiska Sällskapets Handl., N.F., 64(17)

Pseudosquilla megalophthalma
Ingle, R.W. 1963.
Crustacea Stomatopoda from the Red Sea and Gulf of Aden.
Sea Fish. Res. Sta. Israel, Bull. No. 33:1-69.

Pseudosquilla oculata
Barnard, K. H., 1950
Descriptive list of South African Stomatopod Crustacea (Mantis Shrimps). Ann. S. African Mus. 38:838-864, 4 text figs.

Pseudosquilla oculata?
Foxon, G. E. H., 1939
Stomatopod larvae. John Murray Exped., 1933-34, Sci. Repts. 6(6):251-266, 4 text figs.

Pseudosquilla oculata
Hansen, H. J., 1895
Isopoden, Cumaceen und Stomatopoden der Plankton Expedition. Ergeb. Plankton Exped. II, Gc: 105 pp., 8 pls.

Pseudosquilla oculata
Lemos de Castro, A., 1955.
Contribucao ao conhecimento dos crustaceos da Ordem Stomatopoda do litoral brasileiro (Crustacea, Hoplocarida).
Bol. Mus. Nac., Rio de Janeiro, n.s,, Zool., No. 1-68, 56 figs.

Pseudosquilla oculata
Manning, Raymond B. 1970.
The R/V Pillsbury Deep-Sea Biological Expedition to the Gulf of Guinea, 1964-65. 13. The stomatopod Crustacea.
Studies trop. Oceanogr. Miami 4(2): 256-275.

Pseudosquilla oculata
Manning, Raymond B. 1970.
Some Stomatopod crustaceans from Tuléar, Madagascar.
Bull. Mus. natn. Hist. nat. (2) 41(6): 1429-1441.

Pseudosquilla oculata
Manning, Raymond B., 1969
Stomatopod Crustacea of the Western Atlantic.
Stud. Trop. Oceanogr., Univ. Miami, 8:380 pp.

Pseudosquilla oculata
Michel, A., 1968.
Dérive des larves de stomatopodes de l'est de l'Océan Indien.
Cah.O.R.S.T.O.M., ser.Océanogr., VI(1):13-41.

Pseudosquilla oculata
Townsley, S.J., 1953.
Adult and larval stomatopod crustaceans occurring in Hawaiian waters. Pacific Science 7(4):399-437, 28 textfigs.

Pseudosquilla pilaensis
Ingle, R.W., 1963.
Crustacea Stomatopoda from the Red Sea and Gulf of Aden.
Sea Fish.Res.Sta.,Israel,Bull.No.33:1-69.

Pseudosquilla ornata
Barnard, K. H., 1950
Descriptive list of South African Stomatopod Crustacea (Mantis Shrimps). Ann. S. African Mus. 38:838-864, 4 text figs.

Pseudosquilla ornata
Bedot, M., 1909
Sur la faune de l'Archipel Malais (resume). Rev. Suisse Zool. 17:143-169.

Pseudosquilla ornata
Lunz, G. R., jr. 1937.
Stomatopoda of the Bingham Oceanographic Collection. Bull. Bingham Ocean. Coll., V(5): 1-19, 10 textfigs.

Pseudosquilla ornata
Manning, Raymond B., 1967.
Stomatopoda in the Vanderbilt Marine Museum.
Crustaceana, 12(1):102-106.

Pseudosquilla sewelli
Ingle, R.W., 1963.
Crustacea Stomatopoda from the Red Sea and Gulf of Aden.
Sea Fish.Res.Sta.,Israel,Bull.No. 33:1-69.

Pseudosquilla veleronis n.sp.
Schmitt, W. L., 1940
The stomatopods of the west coast of America, based on collections made by the Allan Hancock Expeditions, 1933-1938. Rept. Allan Hancock Pacific Exped. 5(4):129-226, 33 figs.

Pseudosquillopsis
Manning, R.B., 1963.
Preliminary revision of the genera Pseudosquilla and Lysiosquilla with descriptions of six new genera (Crustacea: Stomatopoda).
Bull. Mar. Sci., Gulf and Caribbean, 13(2):308-328.

Squilla cerisii (type)
Squilla lessoni
Pseudosquilla dofleini

Pseudosquillopsis n.gen.
Serène, R., 1962.
Révision du genre Pseudosquilla (Stomatopoda) et définition de genres nouveaux.
Bull. Inst. Océanogr., Monaco, 59(1241):27 pp.

Pseudosquillopsis lessonii
Manning, Raymond B., 1969.
The postlarvae and juvenile stages of two species of Pseudosquillopsis (Crustacea, Stomatopoda) from eastern Pacific region.
Proc. biol. Soc., Wash., 82: 525-538.

Pseudosquillopsis marmorata
Manning, Raymond B. 1971.
Eastern Pacific expeditions of the New York Zoological Society. Stomatopod Crustacea.
Zoologica, N.Y. 56(3), 95-113

Pseudosquillopsis marmorata
Manning, Raymond B. 1969
The postlarva and juvenile stages of two species of Pseudosquillopsis (Crustacea, Stomatopoda) from the eastern Pacific region.
Proc. biol. Soc. Wash. 82: 525-538.

Pterygosquilla
Manning, Raymond B., 1968.
A revision of the Family Squillidae (Crustacea, Stomatopoda), with the description of eight new genera.
Bull. mar. Sci., Miami, 18(1):105-142.

Pterygosquilla armata

Manning, Raymond B., 1969
Stomatopod Crustacea of the Western Atlantic.
Stud. Trop. Oceanogr., Univ. Miami, 8:380 pp.

Pterygosquilla armata capensis n. subsp.

Manning, Raymond B., 1969
Notes on some stomatopod Crustacea from southern Africa.
Smithson. Contrib. Zool. 1:1-17.

Smerdis

Holthuis, L.B., and Raymond B. Manning, 1964.
Proposed use of the plenary powers (A) to designate a type-species for the genera Pseudosquilla Dana 1853, and Gonodactylus Berthold, 1827, and (B) for the suppression of the generic name Smerdis Leach, 1817 (Crustacea, Stomatopoda) Z. N. (S) 1609.
Bull. Zool. Nomencl., 21(2):137-143.

Squilla

Calman, W. T., 1916.
A new species of Crustacean, Genus Squilla.
Ann. Mag. Nat. Hist. (8), 18:373-376.

Squilla spp.

Foxon, G. E. H., 1939
Stomatopod larvae. John Murray Exped., 1933-34, Sci. Repts. 6(6):251-266, 4 text figs.

Squilla spp.

Ingle, R.W., and N. della Croce, 1967.
Stomatopod larvae from the Mozambique Channel.
Boll. Musei Ist.biol.Univ.Genova, 35(226):55-70.

Squilla

Manning, Raymond B., 1968.
A revision of the Family Squillidae (Crustacea, Stomatopoda), with the description of eight new genera.
Bull. mar. Sci., Miami, 18(1):105-142.

Squilla, spp.

Michel, A., 1968.
Dérive des larves de stomatopodes de l'est de l'Océan Indien.
Cah.O.R.STT.O.M., ser. Océanogr., VI(1):13-41.

Squilla aculeata aculeata

Manning, Raymond B. 1971.
Eastern Pacific expeditions of the New York Zoological Society. Stomatopod Crustacea
Zoologica, N.Y. 56(3):95-113

Squilla aculeata calmani

Manning, Raymond B. 1970.
The R/V Pillsbury Deep-Sea Biological Expedition to the Gulf of Guinea, 1964-65. 13. The stomatopod Crustacea.
Studies trop. Oceanogr. Miami 4(2):256-275.

Squilla africana

Cadenat, J., 1950.
Sur quelques espèces des Squilles des côtes du Sénégal. C.R. Première Conf. Int. des Africanistes de l'Ouest, 1:192-193.

Squilla africana

Steuer, A., 1938
The fishery grounds near Alexandria. XVI Cumacea, Stomatopoda, Leptostraca. Hydrobiol, & Fish. Dir., Notes & Mem. No.26: 16 pp., 16 figs.

Squilla alba

Manning, R.B., 1962.
Alima hyalina Leach, the pelagic larva of the stomatopod crustacean Squilla alba Bigelow.
Bull. Mar. Sci., Gulf and Caribbean, 12(3):496-504.

Squilla alba

Michel, A., 1968.
Dérive des larves de stomatopodes de l'est de l'Océan Indien.
Cah.O.R.S.T.O.M., ser. Océanogr., VI(1):13-41.

Squilla alba

Townsley, S.J., 1953.
Adult and larval stomatopod crustaceans occurring in Hawaiian waters. Pacific Science 7(4):399-437, 28 textfigs.

Squilla annandalei

Chopra, B., 1939
Stomatopoda. John Murray Exped., 1933-34, Sci. Repts. 6(3):137-181, 13 text figs.

Squilla anomala

Stephenson, W., 1952.
Faunistic records from Queensland. Pt. I. General Introduction. Pt. II. Adult Stomatopoda (Crustacea).
Univ. Queensland, Dept. Zool., Papers 1(1):1-15.

Squilla armata

Barnard, K. H., 1950
Descriptive list of South African Stomatopod Crustacea (Mantis Shrimps). Ann. S. African Mus. 38:838-864, 4 text figs.

Squilla armata

Dahl, E., 1954.
Reports of the Lund University Chile Expedition, 1948-49. 15. Stomatopoda. Lunds Univ. Årsskrift. N.F., Avd. 2, 49(17):12 pp., 1 textfig.
also:
K. Fysiografiska Sällskapets Handl., N.F., 64(17)

Squilla armata

Davies, D. H., 1949
Preliminary investigations on the foods of South African fishes (with notes on the general fauna of the area surveyed). Fish & Mar. Biol. Survey Div., Dept. Comm. & Industries, S. Africa, Invest. Rept. No.11, Commerce & Industry, Jan. 1949:1-36, 4 pls.

Squilla armata

Manning, Raymond B., 1966.
Notes on some Australian and New Zealand stomatopod Crustacea, with an account of the species collected by the Fisheries Investigation Ship ENDEAVOUR.
Rec. Aust. Mus. 27(4):79-137.

Squilla armata

Richardson, L.R., 1953.
Variation in Squilla armata M. Edw. (Stomatopoda) suggesting a distinct form in New Zealand waters.
Trans. R.Soc., New Zealand, 81(2):315-317, 1 textfig

Squilla bengalensis n.sp.

Tiwari, K.K., 1951.
On two new species of the genus Squilla Fabr., with notes on other stomatopods in the collections of the Zoological Survey of India.
Rec. Indian Mus. 49(3/4):349-363, 6 textfigs.

Squilla bigelowi

Manning, Raymond B., 1967.
Stomatopoda in the Vanderbilt Marine Museum.
Crustaceana, 12(1):102-106.

Squilla bigelowi

Manning, R.B., 1962.
A striking abnormality in Squilla bigelowi Schmitt (Stomatopoda).
Crustaceana, 4(3):243.

Squilla bigelowi n.sp.

Schmitt, W. L., 1940
The stomatopods of the west coast of America, based on collections made by the Allan Hancock Expeditions, 1933-1938. Rept. Allan Hancock Pacific Exped. 5(4):129-226, 33 figs.

Squilla bombeuensis n.sp.

Chhapgar, B.F., and S.R. Sane, 1967.
Two new species of Squilla (Stomatopoda) from Bombay.
Crustaceana, 12(1):1-8.

Squilla boops

Alikunhi, K.H., 1967.
An account of the post-larval development, molting and growth of the common stomatopods of the Madras coast.
Proc. Symp. Crustacea, Ernakulam, Jan. 12-15, 1965, 2:824-939

Squilla boops

Alikunhi, K.H., 1951.
An account of the stomatopod larvae of the Madras plankton. Rec. Indian Mus. 49(3/4): 239-319, 25 textfigs.

Squilla boops

Townsley, S.J., 1953.
Adult and larval stomatopod crustaceans occurring in Hawaiian waters. Pacific Science 7(4):399-437, 28 textfigs.

Squilla brasiliensis

Lemos de Castro, A., 1955.
Contribuicao ao conhecimento dos crustaceos da Ordem Stomatopoda do litoral brasileiro (Crustacea, Hoplocarida).
Bol. Mus. Nac., Rio de Janeiro, n.s., Zool., No.

Squilla brasiliensis

Manning, Raymond B., 1969
Stomatopod Crustacea of the Western Atlantic.
Stud. Trop. Oceanogr., Univ. Miami, 8:380 pp.

Squilla brasiliensis

Manning, Raymond B., 1966.
Stomatopod Crustacea. Campagne de la Calypso au large des côtes Atlantiques de l'Amerique du Sud (1961-1962)-1.
Ann. Inst. Océanogr., Monaco, n.s., 44:359-384

Squilla brasiliensis

Ramos, F. de P., 1951.
Nota preliminar sobre alguno Stomatopoda da Costa Brasileira.
Bol. Inst. Paulista Ocean. 2(1):139-150, 1 pl.

Squilla cadenati n.sp.

Manning, Raymond B. 1970.
The R/V Pillsbury Deep-Sea Biological Expedition to the Gulf of Guinea, 1964-65. 13. The stomatopod Crustacea.
Studies trop. Oceanogr. Miami 4(2):256-275.

Squilla calumnia n.sp.

Townsley, S.J., 1953.
Adult and larval stomatopod crustaceans occurring in Hawaiian waters. Pacific Science 7(4):399-437, 28 textfigs.

Squilla caribaea n. sp.

Manning, Raymond B., 1969
Stomatopod Crustacea of the Western Atlantic.
Stud. Trop. Oceanogr., Univ. Miami, 8:380 pp.

Squilla carinata

Ingle, R.W., 1963.
Crustacea Stomatopoda from the Red Sea and Gulf of Aden.
Sea Fish Res.Sta., Israel, Bull. No. 33:1-69.

Squilla carinata

Manning, Raymond B., 1968.
Stomatopod Crustacea from Madagascar.
Proc.U.S. nat.Mus.124(3641):1-61.

Squilla carinatus n.sp.
Serène, R., 1950.
Deux nouvelles espèces IndoPacifiques de Stomatopodes. Bull. Mus. Nat. Hist. Nat., 2e ser., 22(5): 571-572.

Squilla chydaea
Manning, Raymond B., 1969

Stomatopod Crustacea of the Western Atlantic. Stud. Trop. Oceanogr., Univ. Miami, 8:380 pp.

Squilla costata
Liu, J. Y., 1949
On some species of Squilla (Crustacea Stomatopoda) from China Coasts. Contrib. Inst. Zool., Nat. Acad., Peiping, 5(1):27-47, 6 pls.

Squilla costata
Tiwari, K.K., 1951.
On two new species of the genus Squilla Fabr. with notes on other stomatopods in the collections of the Zoological Survey of India. Rec. Indian Mus. 49(3/4):349-363, 5 textfigs.

Squilla deceptrix n. sp.
Manning, Raymond B., 1969

Stomatopod Crustacea of the Western Atlantic. Stud. Trop. Oceanogr., Univ. Miami, 8:380 pp.

SQUILLA DECIMDENTATA N.SP.
Manning, Raymond B., 1970.
New American stomatopod crustaceans
Proc. Biol. Soc. Wash., 83(5):99-114

Squilla denticauda n.sp.
Chhapgar, B.F., and S.R. Sane, 1967.
Two new species of Squilla (Stomatopoda) from Bombay.
Crustaceana, 12(1):1-8.

Squilla depressa
Serène, R., 1952.
Étude d'une collection de stomatopodes de l' Australian Museum de Sydney. Rec. Australian Mus. 23(1):1-24, 33 textfigs., Pls. 1-3.

Squilla depressa
Stephenson, W., 1952.
Faunistic records from Queensland. Pt. 1. General Introduction. Pt. II. Adult Stomatopoda (Crustacea).
Univ. Queensland, Dept. Zool., Papers 1(1):1-15.

Squilla desmarestii
Barnard, K. H., 1950
Descriptive list of South African Stomatopod Crustacea (Mantis Shrimps). Ann. S. African Mus. 38:838-864, 4 text figs.

Squilla desmaresti
Jacques, Françoise 1970.
La glande de mue chez les larves de Stomatopodes.
C. r. hebd. Séanc. Acad. Sci. Paris (D) 270 (24):2965-2968.

Squilla desmaresti
Jacques, Françoise, et Alain Thériot 1967.
Larves de stomatopodes du plancton de la région de Banyuls-sur-Mer.
Vie Milieu (B) 18 (2B):367-380.

Squilla desmaresti
van der Baan, S. M., and L. B. Holthuis, 1966.
On the occurrence of Stomatopoda in the North Sea, with special reference to larvae from the surface plankton near the lightship "Texel".
Netherlands J. Sea Res., 3(1):1-12.

Squilla desmaresti
Verwey, J., 1966.
The origin of the stomatopod larvae of the southern North Sea.
Netherlands J. Sea Res., 3(1):13-20.

Squilla discors
Manning, Raymond B., 1969

Stomatopod Crustacea of the Western Atlantic. Stud. Trop. Oceanogr., Univ. Miami, 8:380 pp.

Squilla dubia
Hansen, H. J., 1895
Isopoden, Cumaceen und Stomatopoden der Plankton Expedition. Ergeb. Plankton Exped. II, Gc:105 pp., 8 pls.

Squilla dubia
Lemos de Castro, A., 1955.
Contribuçao ao conhecimento dos crustaceos da Ordem Stomatopoda do litoral brasileiro (Crustacea, Hoplocarida).
Bol. Mus. Nac., Rio de Janeiro, n.s., Zool., No. 128:1-68, 56 figs.

Squilla dubia
Manning, Raymond B., 1962.
Stomatopoda in the Vanderbilt Marine Museum.
Crustaceana, 12(1):102-106.

Squilla dubia
Ramos, F. de P., 1951.
Nota preliminar sôbre alguno Stomatopoda da Costa Brasileira. Bol. Inst. Paulista Ocean. 2(1):139-150, 1 pl.

Squilla edentata aushelis n. sub sp.
Manning, Raymond B., 1969

Stomatopod Crustacea of the Western Atlantic. Stud. Trop. Oceanogr., Univ. Miami, 8:380 pp.

Squilla edentata edentata
Manning, Raymond B., 1969

Stomatopod Crustacea of the Western Atlantic. Stud. Trop. Oceanogr., Univ. Miami, 8:380 pp.

Squilla empusa
Brooks, W. K., 1879.
The larval stages of Squilla empusa Say. Chesapeake Zool. Lab., Sci. Res. for 1878: 143-170, Pls. 9-13.

Squilla empusa
Gunter, G., 1950
Seasonal population changes and distribution as related to salinity, of certain invertebrates of the Texas coast including the commercial shrimp. Publ. Inst. Mar. Sci. 1(2):7-51, 8 text figs.

Squilla empusa
Manning, Raymond B., 1969

Stomatopod Crustacea of the Western Atlantic. Stud. Trop. Oceanogr., Univ. Miami, 8:380 pp.

Squilla ampusa
Whitten, H. L., H. F. Rosene, and J. W. Hedgpeth, 1950
The invertebrate fauna of Texas coast jetties; a preliminary survey. (Systematic Appendix). Publ. Inst. Mar. Sci. 1(2):53-86, 1 pl., 4 text figs.

Squilla ? fallax
Holthuis, L.B., 1967.
The Stomatopod Crustacea collected by the 1962 and 1965 Israel South Red Sea expeditions.
Israel J. Zool., 16(1):1-45.

Squilla fasciata
Alikunhi, K.H., 1967.
An account of the post-larval development, molting and growth of the common stomatopods of the Madras coast.
Proc. Symp. Crustacea, Ernakulam, Jan. 12-15 1965, 2:824-939

Squilla fasciata
Alikunhi, K.H., 1951.
An account of the stomatopod larvae of the Madras plankton. Rec. Indian Mus. 49(3/4):239-319, 25 textfigs.

Squilla fasciata
Chopra, B., 1939
Stomatopoda. John Murray Exped., 1933-34, Sci. Repts. 6(3):137-181, 13 text figs.

Squilla fasciata
Holthuis, L.B., 1967.
The Stomatopod Crustacea collected by the 1962 and 1965 Israel South Red Sea Expeditions.
Israel J. Zool., 16(1):1-45.

Squilla fasciata
Serène, R., 1954.
Observations biologiques sur les stomatopodes. Mém. Inst. Océan. Nhatrang, No. 8:1-93, 10 pls., 15 textfigs.

Squilla fasciata
Stephenson, W., 1952.
Faunistic records from Queensland. Pt. I. General Introduction. Pt. II. Adult Stomatopoda (Crustacea).
Univ. Queensland, Dept. Zool., Papers 1(1):1-15.

Squilla ferussaci
Manning, R.B., 1962
A new species of Parasquilla (Stomatopoda) from the Gulf of Mexico, with a redescription of Squilla ferussaci Roux
Crustaceana, 4(3):180-190.

Squilla foveolata
Liu, J. Y., 1949
On some species of Squilla (Crustacea Stomatopoda) from China Coasts. Contrib. Inst. Zool., Nat. Acad., Peiping, 5(1):27-47, 6 pls.

Squilla foveolata
Manning, Raymond B., 1966.
Notes on some Australian and New Zealand stomatopod Crustacea, with an account of the species collected by the Fisheries Investigation Ship ENDEAVOUR.
Rec. Aust. Mus., 27(4):79-137.

Squilla gilesi
Boone, L., 1938
Scientific results of the world cruises of the yachts "Ara", 1928-1929, and "Alva" 1931-1932, "Alva" Mediterranean cruise, 1933, and "Alva" South American cruise, 1935, William K. Vanderbilt, commanding----Bull. Vanderbilt Mus. 7:372 pp., 152 pls., 22 text figs.

Squilla gilesi
Chopra, B., 1939
Stomatopoda. John Murray Exped., 1933-34, Sci. Repts. 6(3):137-181, 13 text figs.

Squilla gilesi
Holthuis, L.B., 1967.
The Stomatopod Crustacea collected by the 1962 and 1965 Israel South Red Sea expeditions.
Israel J. Zool., 16(1):1-45.

Squilla gonypetes
Alikunhi, K.H., 1967.
An account of the post-larval development, molting and growth of the common stomatopods of the Madras coast.
Proc. Symp. Crustacea, Ernakulam, Jan. 12-15 1965, 2:824-939

Squilla gonypetes
Alikunhi, K.H., 1951.
An account of the stomatopod larvae of the Madras plankton. Rec. Indian Mus. 49(3/4):239-319, 25 textfigs.

Squilla gonypetes
Chopra, B., 1939
Stomatopoda. John Murray Exped., 1933-34, Sci. Repts. 6(3):137-181, 13 text figs.

Squilla gonypetes
Ingle, R.W., 1963.
Crustacea Stomatopoda from the Red Sea and Gulf of Aden.
Sea Fish. Res. Sta., Israel, Bull. No. 33:1-69.

Squilla gonypetes
Manning, Raymond B., 1968.
Stomatopod Crustacea from Madagascar.
Proc. U.S. nat. Mus. 124(3641):1-61.

Squilla gonypetes
Manning, Raymond B. 1965.
Stomatopoda from the collection of His Majesty, the Emperor of Japan.
Crustaceana 9(3):249-262.

Squilla granti n. sp.
Stephenson, W., 1953.
Three new Stomatopoda (Crustacea) from eastern Australia. Australian J. Mar. Freshwater Res. 4(1):201-218, 4 textfigs.

Squilla grenadensis n. sp.
Manning, Raymond B., 1969
Stomatopod Crustacea of the Western Atlantic.
Stud. Trop. Oceanogr., Univ. Miami, 8:380 pp.

Squilla hancocki
Manning, Raymond B. 1971.
Eastern Pacific expeditions of the New York Zoological Society. Stomatopod Crustacea. Zoologica, N.Y. 56(3), 95-113

Squilla harpax
Barnard, K. H., 1955.
Additions to the fauna-list of South African Crustacea and Pycnogonida.
Ann. S. African Mus., 43(1):1-107.

Squilla harpax
Ingle, R.W., 1963.
Crustacea Stomatopoda from the Red Sea and Gulf of Aden.
Sea Fish Res. Sta., Israel, Bull. No. 33:1-69.

Squilla hesperia n. sp.
Manning, Raymond B., 1968.
Stomatopod Crustacea from Madagascar.
Proc. U.S. nat. Mus. 124(3641):1-61.

Squilla heptacantha
Manning, Raymond B., 1969
Stomatopod Crustacea of the Western Atlantic.
Stud. Trop. Oceanogr., Univ. Miami, 8:380 pp.

Squilla harpax
Tiwari, K.K., 1951.
On two new species of the genus Squilla Fabr., with notes on other stomatopoda in the collections of the Zoological Survey of India.
Rec. Indian Mus. 49(3/4):349-363, 5 textfigs.

Squilla hieroglyphica
Alikunhi, K.H., 1967.
An account of the post-larval development, moulting and growth of the common stomatopods of the Madras coast.
Proc. Symp. Crustacea, Ernakulam, Jan. 12-15, 1965, 2:824-939

Squilla hieroglyphica
Alikunhi, K.H., 1951.
An account of the stomatopod larvae of the Madras plankton. Rec. Indian Mus. 49(3/4):239-319, 25 textfigs.

Squilla hieroglyphica
Barnard, K. H., 1950
Descriptive list of South African Stomatopod Crustacea (Mantis Shrimps). Ann. S. African Mus. 38:838-864, 4 text figs.

Squilla hieroglyphica
Kurian, C.V. 1947
On the occurrences of Squilla hieroglyphica Kemp. (Crustacea, Stomatopoda) in the Coastal Waters of Travancore. Current Sci. 16(4):124.

Squilla hildebrandi
Schmitt, W. L., 1940
The stomatopods of the west coast of America, based on collections made by the Allan Hancock Expeditions, 1933-1938. Rept. Allan Hancock Pacific Exped. 5(4):129-226, 33 figs.

Squilla holoschista
Alikunhi, K.H., 1967.
An account of the post-larval development, moulting and growth of the common stomatopods of the Madras coast.
Proc. Symp. Crustacea, Ernakulam, Jan. 12-15, 1965, 2:824-939

Squilla holoschista
Alikunhi, K.H., 1951.
An account of the stomatopod larvae of the Madras plankton. Rec. Indian Mus. 49(3/4):239-319, 25 textfigs.

Squilla holoschista
Barnard, K. H., 1950
Descriptive list of South African Stomatopod Crustacea (Mantis Shrimps). Ann. S. African Mus. 38:838-864, 4 text figs.

Squilla holoschista
Michel, A., 1968.
Dérive des larves de stomatopodes de l'est de l'Océan Indien.
Cah. O.R.S.T.O.M., ser. Océanogr., VI(1):13-41.

Squilla holoschista
Daniel, Ruby, 1958.
Neurosecretion in Crustacea. 1. The neurosecretory organs of the eyestalk of Squilla holoschista.
J. Madras Univ., (b) 28(1):49-63.

Squilla hyalina
Manning, Raymond B., 1967.
Stomatopoda in the Vanderbilt Marine Museum.
Crustaceana, 12(1):102-106.

Squilla imperialis n. sp.
Manning, Raymond B. 1965.
Stomatopoda from the collection of His Majesty, the Emperor of Japan.
Crustaceana 9(3):249-262.

Squilla indica
Chopra, B., 1939
Stomatopoda. John Murray Exped., 1933-34, Sci. Repts. 6(3):137-181, 13 text figs.

Squilla inermis
Manning, Raymond B. 1965.
Stomatopoda from the collection of His Majesty, the Emperor of Japan.
Crustaceana 9(3):249-262.

Squilla inornata
Notes on Stomatopod Crustacea from Victoria and Tasmania. Mem. Nat. Mus., Victoria, No. 19:7-10.

Squilla inornata
Manning, Raymond B., 1966.
Notes on some Australian and New Zealand stomatopod Crustacea, with an account of the species collected by the Fisheries Investigation Ship ENDEAVOUR.
Rec. Aust. Mus., 27(4):79-137.

Squilla intermedia
Ingle, R.W., 1959
Squilla labadiensis n. sp and Squilla intermedia Bigelow; two stomatopod crustaceans new to the West African coast. Ann. & Mag. Nat. Hist., 13 Ser. 2(21):565-576.

Squilla intermedia
Manning, Raymond B., 1969
Stomatopod Crustacea of the Western Atlantic.
Stud. Trop. Oceanogr., Univ. Miami, 8:380 pp.

Squilla interrupta
Alikunhi, K.H., 1967.
An account of the post-larval development, moulting and growth of the common stomatopods of the Madras coast.
Proc. Symp. Crustacea, Ernakulam, Jan. 12-15, 1965, 2:824-939

Squilla interrupta
Alikuhni, K.H., 1951.
An account of the stomatopod larvae of the Madras plankton. Rec. Indian Mus. 49(3/4):239-319, 25 textfigs.

Squilla interrupta
Liu, J. Y., 1949
On some species of Squilla (Crustacea Stomatopoda) from China Coasts. Contrib. Inst. Zool., Nat. Acad., Peiping, 5(1):27-47, 6 pls.

Squilla interrupta
Manning, Raymond B., 1967.
Stomatopoda in the Vanderbilt Marine Museum.
Crustaceana, 12(1):102-106.

Squilla interrupta
Manning, Raymond B., 1966.
Notes on some Australian and New Zealand stomatopod Crustacea, with an account of the species collected by the Fisheries Investigation Ship ENDEAVOUR.
Rec. Aust. Mus., 27(4):79-137.

Squilla interrupta
Serène, R., 1950.
Cas de malformations chez les stomatopodes. Bull. Mus., Indochine, 2nd ser., 22(3):341-343, 2 pls.

Squilla interrupta
Stephenson, W., 1952.
Faunistic records from Queensland. Pt. 1. General introduction. Pt. II. Adult Stomatopoda (Crustacea).
Univ. Queensland, Dept. Zool., Papers 1(1):1-15.

Squilla investigatoris
Barnard, K. H., 1950
Descriptive list of South African Stomatopod Crustacea (Mantis Shrimps). Ann. S. African Mus. 38:838-864, 4 text figs.

Squilla investigatoris
Chopra, B., 1939
Stomatopoda. John Murray Exped., 1933-34, Sci. Repts. 6(3):137-181, 13 text figs.

Squilla investigatoris
Ingle, R.W., 1963.
Crustacea Stomatopoda from the Red Sea and Gulf of Aden.
Sea Fish. Res. Sta., Israel, Bull. No. 33:1-69.

Squilla kempi
Liu, J. Y., 1949
On some species of Squilla (Crustacea Stomatopoda) from China Coasts. Contrib. Inst. Zool., Nat. Acad., Peiping, 5(1):27-47, 6 pls.

Squilla labadiensis
Ingle, R.W., 1959
Squilla labadiensis n. sp and Squilla intermedia Bigelow; two stomatopod crustaceans new to the West African coast. Ann. & Mag. Nat. Hist., 13, Ser. 2(21): 565-576.

Squilla laevis
Manning, Raymond B., 1966.
Notes on some Australian and New Zealand stomatopod Crustacea, with an account of the species collected by the Fisheries Investigation Ship ENDEAVOUR. Rec. Aust. Mus., 27(4):79-137.

Squilla laevis
Stephenson, W., 1952.
Faunistic records from Queensland. Pt. I. General introduction. Pt. II. Adult Stomatopoda (Crustacea). Univ. Queensland, Dept. Zool., Papers 1(1):1-15.

Squilla lata
Alikunhi, K.H., 1967.
An account of the post-larval development, moulting and growth of the common stomatopods of the Madras coast. Proc. Symp. Crustacea, Ernakulam, Jan. 12-15, 1965, 2: 824-939.

Squilla lata
Alikunhi, K.H., 1951.
An account of the stomatopod larvae of the Madras plankton. Rec. Indian Mus. 49(3/4):239-319, 25 textfigs.

Squilloides lata
Manning, Raymond B., 1969
Notes on some stomatopod Crustacea from southern Africa.

Squilla lata
Manning, Raymond B., 1965.
Stomatopoda from the collection of His Majesty, the Emperor of Japan. Crustaceana 9(3):249-262.

Squilla latreillei
Alikunhi, K.H., 1967.
An account of the post-larval development, moulting and growth of the common stomatopods of the Madras coast. Proc. Symp. Crustacea, Ernakulam, Jan. 12-15, 1965, 2: 824-939.

Squilla latreillei
Alikunhi, K.H., 1951.
An account of the stomatopod larvae of the Madras plankton. Rec. Indian Mus. 49(3/4):239-319, 25 textfigs.

Squilla latreillei
Barnard, K.H., 1950
Descriptive list of South African Stomatopod Crustacea (Mantis Shrimps). Ann. S. African Mus. 38:838-864, 4 text figs.

Squilla latreilllii
Ingle, R.W., 1963.
Crustacea Stomatopoda from the Red Sea and Gulf of Aden. Sea Fish. Res. Sta., Israel, Bull. No. 33:1-69.

Squilla leptosquilla
Rao, P. Vedavyasa, M. J. Sebastian and P. Karunakaran Nair, 1965.
On the occurrence of Squilla leptosquilla Brooks (Crustacea, Stomatopoda) in the west coast of India. J. mar. biol. Ass. India 7(2): 468-469.

Squilla lijdingi
Fausto Filho, José, 1966.
Sôbre a ocorrência de Squilla lijdingi Holthuis, 1959 no litoral brasileiro (Crustacea Stomatopoda). Arquivos Estação Biol. Marinha, Univ. Fed. Ceará 6(2): 139-141.

Squilla lijdingi
Manning, Raymond B., 1969
Stomatopod Crustacea of the Western Atlantic. Stud. Trop. Oceanogr., Univ. Miami, 8:380 pp.

Squilla mantis
Alexandrowicz, J.S., 1957.
Notes on the nervous system in the Stomatopoda. Pubbl. Staz. Zool., Napoli, 29:213-225.

Squilla mantis
Alexandrowicz, J.S., 1954.
Notes on the nervous system in the Stomatopoda. IV. Muscle receptor organs. Pubbl. Staz. Zool., Napoli, 25(1):94-111, Pls. 3-4, 3 textfigs.

Squilla mantis
Brown, Hilary F., 1964.
Electrophysiological investigations of the heart of Squilla mantis. 1. The ganglionic nerve trunk. 2. The heart muscle. 3. The mode of action of pericardial organ extract on the heart. J. Exp. Biol., 41(4):689-700; 701-722; 723-734.

Squilla mantis
Cadenat, J., 1950.
Sur quelques espèces des Squilles des côtes du Sénégal. C.R. Première Conf. Int. des Africaniste de l'Ouest, 1:192-193.

Squilla mantis
Carlisle, D.B., and L.M. Passano, 1953.
The x-organ of Crustacea. Nature 171(4363):1070-1071, 3 figs.

Squilla mantis
Hanstroem, B., 1948
The brain, the sense organs and the incretory organs of the head in the Crustacea Malacostraca. Suppl. 33 to Bull. Biol. de France et de Belgique: 98-125, 11 text figs.

Squilla mantis
Holthuis, L.B., 1961
Report on a collection of Crustacea Decapoda and Stomatopoda from Turkey and the Balkans. Zool. Verhandel., Rijksmuseum Nat. Hist., Leiden. No. 47: 1-67.

Squilla mantis
Jacques, Françoise, 1970.
La glande de mue chez les larves de Stomatopodes. C.R. hebd. Séanc. Acad. Sci. Paris (D) 270(24): 2965-2968.

Squilla mantis
Jacques, Françoise, et Alain Thiriot, 1967.
Larves de stomatopodes du plancton de la région de Banyuls-sur-Mer. Vie Milieu (B) 18 (2B): 367-380.

Squilla mantis
Manfrin, G., e C. Piccinetti, 1970.
Osservazioni etologiche su Squilla mantis L. Note Lab. Biol. Mar. Pesca-Fano, annesso Ist. Zool. Univ. Bologna, 3(5): 93-104.

Squilla mantis
Manning, Raymond B., 1970.
The R/V Pillsbury Deep-Sea Biological Expedition to the Gulf of Guinea, 1964-65. 13. The Stomatopod Crustacea. Studies trop. oceanogr., Miami 4(2): 256-275.

Squilla mantis
Parisi, B., 1940
Gli Stomatopodi raccolti del Prof. L. Sanzo nella Campagna idrografica nel Mar Rosso della R. N. Ammiraglio Magnaghi 1923-1924. (Mem. 16ª della Campagne). Mem. R. Com. Talass. Ital. CCLXXV:7 pp.

Squilla mantis
Pilgrim, R.L.C., 1964.
Stretch receptor organs in Squilla mantis Latr. (Crustacea: Stomatopoda). J. Exp. Biol., 41(4):793-804.

Squilla mantis
Pilgrim, Robert, 1964.
Observations on the anatomy of Squilla mantis Latr. (Crustacea, Stomatopoda). Pubbl. Staz. Zool., Napoli, 34:9-42.

Squilla mantis
Steuer, A., 1938
The fishery grounds near Alexandria. XVI Cumacea, Stomatopoda, Leptostraca. Hydrobiol. & Fish. Dir., Notes & Mem. No.26: 16 pp., 16 figs.

Squilla mantis
van der Baan, S.M., and L.B. Holthuis, 1966.
on the occurrence of Stomatopoda in the North Sea, with special reference to larvae from the surface plankton near the lightship "Texel". Netherlands J. Sea Res., 3(1):1-12.

Squilla mantis
Vilela, H., 1936
Crustaceos Decapodes. Estomatopodes. Trav. St. Biol. Mar., Lisbonne, No. 40:215-242.

Squilla mantoides
Manning, Raymond B., 1968.
Correction of the type locality of Squilla mantoidea Bigelow (Stomatopoda). Crustaceana, 14(1):107.

Squilla massavensis
Holthuis, L.B., 1967.
The Stomatopod Crustacea collected by the 1962 and 1965 Israel South Red Sea expeditions. Israel J. Zool., 16(1):1-45.

Squilla massavensis
Holthuis, L.B., 1961
Report on a collection of Crustacea Decapoda and Stomatopoda from Turkey and the Balkans. Zool. Verhandel., Rijksmuseum Nat. Hist., Leiden. No. 47: 1-67.

Squilla massavensis
Ingle, R.W., 1963.
Crustacea Stomatopoda from the Red Sea and Gulf of Aden. Sea Fish. Res. Sta., Israel, Bull. No. 33:1-69.

Squilla massauensis
Parisi, B., 1940
Gli Stomatopodi raccolti del Prof. L. Sanzo nella Campagna idrografica nel Mar Rosso della R. N. Ammiraglio Magnaghi 1923-1924. (Mem. 16ª della Campagne). Mem. R. Com. Talass. Ital. CCLXXV:7 pp.

Squilla massavensis
Serène, R., 1954.
Observations biologiques sur les stomatopodes. Mem. Inst. Océan., Nhatrang, No. 8:1-93, 10 pls., 15 textfigs.

Squilla massavensis
Serene, R., 1951.
Sur la circulation d'eau à la surface du corps des Stomatopodes. Bull. Soc. Zool., France 76(3): 137-143, 1 pl., 3 textfigs.

Squilla massavensis
Steuer, A., 1938
The fishery grounds near Alexandria. XVI Cumacea, Stomatopoda, Leptostraca. Hydrobiol. & Fish. Dir., Notes & Mem. No.26: 16 pp., 16 figs.

Squilla mauritiana
Manning, Raymond B., 1968.
Stomatopod Crustacea from Madagascar. Proc.U.S. nat.Mus.124(3641):1-61.

Squilla mcneilli
Manning, Raymond B., 1966.
Notes on some Australian and New Zealand stomatopod Crustacea, with an account of the species collected by the Fisheries Investigation Ship ENDEAVOUR. Rec. Aust. Mus., 27(4):79-137.

Squilla mcneilli n.sp.
Stephenson, W., 1953.
Three new Stomatopoda (Crustacea) from eastern Australia. Australian J. Mar. Freshwater Res. 4(1):201-218, 4 textfigs.

Squilla merguiensis n.sp.
Tiwari, K.K., 1951.
On two new species of the genus Squilla Fabr., with notes on other stomatopods in the collections of the Zoological Survey of India. Rec. Indian Mus. 49(3/4):349-363, 5 textfigs.

Squilla microphthalma
Stephenson, W., 1962.
5. Some interesting Stomatopoda - mostly from Western Australia. J. R. Soc., Western Australia, 45(2):33-43.

Squilla microphthalma
Tiwari, K.K., 1951.
On two new species of the genus Squilla Fbr., with notes on other stomatopods in the collection of the Zoological Survey of India. Rec. Indian Mus. 49(3/4):349-363, 5 textfigs.

Squilla mikado
Barnard, K. H., 1950
Descriptive list of South African Stomatopod Crustacea (Mantis Shrimps). Ann. S. African Mus. 38:838-864, 4 text figs.

Squilla mikado
Manning, Raymond B. 1965.
Stomatopods from the Collection of His Majesty, The Emperor of Japan. Crustaceana 9(3):249-262.

Squilla miles
Stephensen, W., 1955.
Notes on Stomatopod Crustacea from Victoria and Tasmania. Mem. Nat. Mus., Victoria, No. 19:7-10

Squilla multicarinata
Liu, J. Y., 1949
On some species of Squilla (Crustacea Stomatopoda) from China Coasts. Contrib. Inst. Zool., Nat. Acad., Peiping, 5(1):27-47, 6 pls.

Squilla neglecta
Lemos de Castro, A., 1955.
Contribucao ao conhecimento des crustaceos da Ordem Stomatopoda do litoral brasileiro (Crustacea, Hoplpcarida). Bol. Mus. Nac., Rio de Janeiro, n.s., Zool., No. 128:1-68, 56 figs.

Squilla neglecta
Lunz, G. R., Jr. 1933.
Rediscovery of Squilla neglecta Gibbes. Charleston (S.C.) Museum Leaflet, 5:1-8.

Squilla neglecta
Manning, Raymond B., 1969
Stomatopod Crustacea of the Western Atlantic. Stud. Trop. Oceanogr., Univ. Miami, 8:380 pp.

Squilla neglecta
Manning, Raymond B., 1966.
Stomatopod Crustacea. Campagne de la Calypso au large des côtes Atlantiques de l'Amerique du Sud (1961-1962)-1. Ann. Inst. Océanogr., Monaco, n.s., 44:359-384

Squilla nepa
Alikunhi, K.H., 1967.
An account of the post-larval development, moulting and growth of the common stomatopods of the Madras coast. Proc. Symp. Crustacea, Ernakulam, Jan.12-15, 1965, 2:824-939

Squilla nepa
Alikunhi, K.H., 1951.
An account of the stomatopod larvae of the Madras plankton. Rec. Indian Mus. 49(3/4):239-319, 25 textfigs.

Squilla nepa
Barnard, K. H., 1950
Descriptive list of South African Stomatopod Crustacea (Mantis Shrimps). Ann. S. African Mus. 38:838-864, 4 text figs.

Squilla nepa
Holthuis, L.B., 1967.
The Stomatopod Crustacea collected by the 1962 and 1965 Israel South Red Sea expeditions. Israel J. Zool., 16(1):1-45.

Squilla nepa
Manning, Raymond B., 1968.
Stomatopod Crustacea from Madagascar. Proc. U.S. nat.Mus.124(3641):1-61.

Squilla nepa
Serene, R., 1951.
Sur la circulation d'eau à la surface du corps des Stomatopodes. Bull. Soc. Zool., France 76(3):137-143, 1 pl., 3 textfigs.

Squilla obtusa
Manning, Raymond B., 1969
Stomatopod Crustacea of the Western Atlantic. Stud. Trop. Oceanogr., Univ. Miami, 8:380 pp.

Squilla obtusa
Manning, Raymond B., 1966.
Stomatopod Crustacea. Campagne de la Calypso au large des côtes Atlantiques de l'Amerique du Sud (1961-1962)-1. Ann. Inst. Océanogr., Monaco, n.s., 44:359-384

Scylla oceanica
Serene, R., 1952.
Les espèces du Genre Scylla à Nhatrang (Vietnam) Indo-Pacific Fish. Counc., Proc. 3rd Meet., 1-16 Feb. 1951, Sects. 2/2:133-137, 2 pls., 1 textfig.

? Squilla oratoria
Koma, T., 1932.
An enormous swarm of stomatopod larvae. Annot. Zool., Jap., 13:351-354, Fig. A.

Squilla oratoria
Kubo, I., S. Hori, M. Kumemura, M. Naganawa and J. Soedjono, 1959.
A biological study on a Japanese edible mantis-shrimp Squilla oratoria De Haan. J. Tokyo Univ. Fish., 45(1):1-25.

Squilla oratoria
Liu, J. Y., 1949
On some species of Squilla (Crustacea Stomatopoda) from China Coasts. Contrib. Inst. Zool., Nat. Acad., Peiping, 5(1):27-47, 6 pls.

Squilla oratoria inornata
Liu, J. Y., 1949
On some species of Squilla (Crustacea Stomatopoda) from China Coasts. Contrib. Inst. Zool., Nat. Acad., Peiping, 5(1):27-47, 6 pls.

Squilla oratoria
Manning, Raymond B. 1965.
Stomatopods from the Collection of His Majesty, The Emperor of Japan. Crustaceana 9(3):249-262.

Squilla oratoria
Oka, T.B., 1953.
On the anal-gland found in Squilla oratoria. Nat. Sci. Rept., Ochanomizu Univ, 4(1):119-127, 16 textfigs.

Squilla oratoria
Senta, Tetsushi 1967
Seasonal abundance and diurnal migration of alima larva of Squilla oratoria in the Seto Inland Sea. (In Japanese English abstract) Bull. Jap. Soc. scient. Fish. 33(6):508-512

Squilla oratoria
Shiino, S.M., 1942.
Studies on the embryology of Squilla oratoria de Haan. Mem. Coll. Sci., Kyoto Imp. Univ., Ser. B, 17(1):77-174, figs.

Squilla oratoria (incl. larvae ?)
Townsley, S.J., 1953.
Adult and larval stomatopod crustaceans occurring in Hawaiian waters. Pacific Science 7(4):399-437, 28 textfigs.

Squilla oratoria
Yanase, T., Y. Okuno and K. Fujimoto 1972
Fine structure of the compound eye in mantis crab, Squilla oratoria. Zool. Mag. (Dobuts. Zasshi) 81(3):211-216

Squilla ornata
Irisawa, H., and Aya Funaishi Irisawa, 1957.
The electrocardiogram of a stomatopod. Biol. Bull 112(3):358-362.

Squilla panamensis
Manning, Raymond B. 1971.
Eastern Pacific expeditions of the New York Zoological Society. Stomatopod Crustacea. Zoologica, N.Y. 56(3):95-113

Squilla parva
Manning, Raymond B. 1971.
Eastern Pacific expeditions of the New York Zoological Society. Stomatopod Crustacea. Zoologica, N.Y. 56(3):95-113

Squilla perpensa
Manning, Raymond B., 1967.
Stomatopoda in the Vanderbilt Marine Museum. Crustaceana, 12(1):102-106.

Squilla perpensa
Manning, Raymond B., 1966.
Notes on some Australian and New Zealand stomatopod Crustacea, with an account of the species collected by the Fisheries Investigation Ship ENDEAVOUR. Rec. Aust. Mus., 27(4):79-137.

Squilla prasinolineata
Lemos de Castro, A., 1955.
Contribuicao ao conhecimento dos crustaceos da Ordem Stomatopoda do litoral brasileiro (Crustacea, Hoplocarida). Bol. Mus. Nac., Rio de Janeiro, n.s., Zool., No. 128:1-68, 56 figs.

Squilla prasinolineata
Manning, Raymond B., 1969
Stomatopod Crustacea of the Western Atlantic. Stud. Trop. Oceanogr., Univ. Miami, 8:380 pp.

Squilla prasinolineata

Manning, Raymond B., 1967.
Stomatopodes in the Vanderbilt Marine Museum.
Crustaceana, 12(1):102-106.

Squilla quadraticauda n.sp.

Fukuda, T., 1910.
Further report on Japanese Stomatopoda with descriptions of two new species.
Annot. Zool. Japon. 7:285-290, Pl. 11.

Squilla quadridens

Hansen, H. J., 1895
Isopoden, Cumaceen und Stomatopoden der Plankton Expedition. Ergeb. Plankton Exped. II, Gc:105 pp., 8 pls.

Squilla quinquedentata

Alikunhi, K.H., 1967.
An account of the post-larval development, moulting and growth of the common stomatopods of the Madras coast.
Proc. Symp. Crustacea, Ernakulam, Jan. 12-15, 1965, 2:824-939

Squilla quinquedentata

Alikunhi, K.H., 1951.
An account of the stomatopod larvae of the Madras plankton. Rec. Indian Mus. 49(3/4):239-319, 25 textfigs.

Squilla raphidea

Alikunhi, K.H., 1951.
An account of the stomatopod larvae of the Madras plankton. Rec. Indian Mus. 49(3/4):239-319, 25 textfigs.

Squilla raphidea

Barnard, K. H., 1950
Descriptive list of South African Stomatopod Crustacea (Mantis Shrimps). Ann. S. African Mus. 38:838-864, 4 text figs.

Squilla raphidea

Chopra, B., 1939
Stomatopoda. John Murray Exped., 1933-34, Sci. Repts. 6(3):137-181, 13 text figs.

Squilla raphidea

Liu, J. Y., 1949
On some species of Squilla (Crustacea Stomatopoda) from China Coasts. Contrib. Inst. Zool., Nat. Acad., Peiping, 5(1):27-47, 6 pls.

Squilla raphidea

Michel, A. 1968.
Dérive des larves de stomatopodes de l'est de l'Océan Indien.
Cah. O.R.S.T.O.M., ser. Océanogr., VI(1):13-41.

Squilla raphidea

Serène, R., 1954.
Observations biologiques sur les stomatopodes.
Mém. Inst. Océan., Nhatrang, No. 8:1-93, 10 pls.

Squilla raphidea

Stephenson, W., 1952.
Faunistic records from Queensland. Pt. I. General introduction. Pt. II. Adult Stomatopoda (Crustacea).
Univ. Queensland, Dept. Zool., Papers, 1(1):1-15.

Squilla raphidea

Tiwari, K.K., 1951.
On two new species of the genus Squilla Fabr., with notes on other stomatopods in the collections of the Zoological Survey of India.
Rec. Indian Mus. 49(3/4):349-363, 5 textfigs.

Squilla rugosa

Manning, Raymond B., 1969
Stomatopod Crustacea of the Western Atlantic.
Stud. Trop. Oceanogr., Univ. Miami, 8:380 pp.

Squilla schmitti n.sp.

Lemos de Castro, A., 1955.
Contribuicao ao conhecimento dos crustaceos da Ordem Stomatopoda do litoral brasileiro (Crustacea, Hoplocarida).
Bol. Mus. Nac. Rio de Janeiro, n.s., Zool., No. 128:1-68, 56 figs.

Squilla schmitti

Manning, Raymond B., 1966.
Stomatopod Crustacea. Campagne de la Calypso au large des côtes Atlantiques de l'Amerique du Sud (1961-1962)-1.
Ann. Inst. Océanogr., Monaco, n.s., 44:359-384

Squilla scorpio

Alikunhi, K.H., 1967.
An account of the post-larval development, moulting and growth of the common stomatopods of the Madras coast.
Proc. Symp. Crustacea, Ernakulam, Jan. 12-15, 1965, 2:824-939

Squilla scorpio

Alikunhi, K.H., 1951.
An account of the stomatopod larvae of the Madras plankton. Rec. Indian Mus. 49(3/4):239-319, 25 textfigs.

Squilla scorpio

Liu, J. Y., 1949
On some species of Squilla (Crustacea Stomatopoda) from China Coasts. Contrib. Inst. Zool., Nat. Acad., Peiping, 5(1):27-47, 6 pls.

Squilla scorpio immaculata

Liu, J. Y., 1949
On some species of Squilla (Crustacea Stomatopoda) from China Coasts. Contrib. Inst. Zool., Nat. Acad., Peiping, 5(1):27-47, 6 pls.

Squilla scorpio

Tiwari, K.K., 1951.
On two new species of the genus Squilla Fabr., with notes on other stomatopods in the collections of the Zoological Survey of India.
Rec. Indian Mus. 49(3/4):349-363, 5 textfigs.

Scylla serrata

Serène, R., 1952.
Les espèces du Genre Scylla à Nhatrang (Vietnam).
Indo-Pacific Fish. Counc., Proc., 3rd Meeting, 1-16 Feb. 1951, Sects. 2/3:133-137, 2 pls., 1 textfig.

Scylla serrata paramamosain

Serène, R., 1952.
Les espèces du Genre Scylla à Nhatrang (Vietnam).
Indo-Pacific Fish. Counc., Proc., 3rd Meeting, 1-16 Feb. 1951, Sects. 2/3:133-137, 2 pls., 1 textfig.

Squilla simulans n.sp.

Holthuis, L.B., 1967.
The Stomatopod Crustacea collected by the 1962 and 1965 Israel South Red Sea expeditions.
Israel J. Zool., 16(1):1-45.

Squilla stylifera

Clark, G., 1869.
On the Squill of Mauritius (Squilla stylifera).
Proc. Zool. Soc., London, :3-4.

Squilla surinamica

Manning, Raymond B., 1969

Stomatopod Crustacea of the Western Atlantic.
Stud. Trop. Oceanogr., Univ. Miami, 8:380 pp.

Squilla swetti n.sp.

Schmitt, W. L., 1940
The stomatopods of the west coast of America, based on collections made by the Allan Hancock Expeditions, 1933-1938. Rept. Allan Hancock Pacific Exped. 5(4):129-226, 33 figs.

Squilla terrareginensis n.sp.

Stephenson, W., 1953.
Three new Stomatopoda (Crustacea) from eastern Australia. Australian J. Mar. Freshwater Res. 4(1):201-218, 4 textfigs.

Squilla tiburonensis

Manning, Raymond B., 1971.
Eastern Pacific expeditions of the New York Zoological Society. Stomatopod Crustacea. Zoologica, N.Y. 56(2):95-113

Squilla tiburonensis n.sp.

Schmitt, W. L., 1940
The stomatopods of the west coast of America, based on collections made by the Allan Hancock Expeditions, 1933-1938. Rept. Allan Hancock Pacific Exped. 5(4):129-226, 33 figs.

Scylla tranquebarica

Serène, R., 1952.
Les espèces du Genre Scylla à Nhatrang (Vietnam).
Indo-Pacific Fish. Counc., Proc., 3rd Meet., 1-16 Feb. 1951, Sects. 2/3:133-137, Pls. 1-2, 1 textfig.

Squilla tricarinata

Holthuis, L.B., 1941
Note on some Stomatopoda from the Atlantic Coasts of Africa and America with the description of a new species. Zool. Mededeel. Leyden 23:31-43.1 fig.

Squilla tricarinata

Lemos de Castro, A., 1955.
Contribuicao ao conhecimento dos crustaceos da Ordem Stomatopoda do litoral brasileiro (Crustaceo, Hoplocarida).
Bol. Mus. Nac., Rio de Janeiro, n.s., Zool., No. 128:1-68, 56 figs.

Squilla tricarinata

Manning, Raymond B., 1966.
Stomatopod Crustacea. Campagne de la Calypso au large des côtes Atlantiques de l'Amerique du Sud (1961-1962)-1.
Ann. Inst. Océanogr., Monaco, n.s., 44:359-384

Squilla wood-masoni

Alikunhi, K.H., 1967.
An account of the post-larval development, moulting and growth of the common stomatopods of the Madras coast.
Proc. Symp. Crustacea, Ernakulam, Jan. 12-15, 1965, 2:824-939

Squilla wood-masoni

Alikunhi, K.H., 1954.
An account of the stomatopod larvae of the Madras plankton. Rec. Indian Mus. 49(3/4):239-319, 25 textfigs.

Squilla woodmasoni

Barnard, K.H., 1962.
New records of marine Crustacea from the East African region.
Crustaceana, 3(3):239-245.

Squilla woodmasoni

Barnard, K. H., 1950
Descriptive list of South African Stomatopod Crustacea (Mantis Shrimps). Ann. S. African Mus. 38:838-864, 4 text figs.

Squilla wood-masoni

Liu, J. Y., 1949
On some species of Squilla (Crustacea Stomatopoda) from China Coasts. Contrib. Inst. Zool., Nat. Acad., Peiping, 5(1):27-47, 6 pls.

Squilla woodmasoni

Michel, A., 1968.
Dérive des larves de stomatopodes de l'est de l'Océan Indien.
Cah. O.R.S.T.O.M., ser. Océanogr., VI(1):13-41.

Squilla wood-masoni

Stephenson, W., 1952.
Faunistic records from Queensland. Pt. I. General introduction. Pt. II. Adult Stomatopoda (Crustacea).
Univ. Queensland, Dept. Zool., Papers 1(1):1-15.

Squilla woodmasoni

Manning, Raymond B. 1966.
Notes on some Australian and New Zealand stomatopod Crustacea, with an account of the species collected by the Fisheries Investigation Ship ENDEAVOUR.
Rec. Aust. Mus., 27(4):79-137.

Squilla zanzibarica n.sp.

Chopra, B., 1939
Stomatopoda. John Murray Exped., 1933-34, Sci. Repts. 6(3):137-181, 13 text figs.

Squilla zanzibarica

Manning, Raymond B. 1965.
Stomatopoda from the collection of His Majesty, The Emperor of Japan.
Crustaceana 9(3): 249-262.

Squilloides n. gen

Manning, Raymond B., 1968.
A revision of the Family Squillidae (Crustacea, Stomatopoda), with the description of eight new genera.
Bull. mar. Sci., Miami, 18(1):105-142.

Squilloides latus spinosus n. subsp.

Blumstein, Radda, 1970.
New stomatopod crustaceans from the Gulf of Tonkin, South China Sea.
Crustaceana 18(2): 218-224.

Spelaeogriphacea (close to Tanaidacea)

Grindley, J.R., and R.R. Hessler, 1971.
The respiratory mechanism of Spelaeogriphus and its phylogenetic significance (Spelaeogriphacea).
20(2): 141-144

Tanaidaceans

Bacescu, Mihai, 1961
Contribution à la connaissance des tanaidacés de la Méditerranée Orientale. 1. Les Apseudidae et Kalliapseudidae des côtes d'Israel.
Bull. Res. Counc., Israel, 10B(4):137-170.

tanaidacea, anatomy

Haffer, Klaus, 1965.
Zur Morphologie der Malacostraca: der Kaumagen der Mysidacea im Vergleich zu dem verschiedener Peracarida und Eucarida.
Helgoländer wiss. Meeresuntersuch., 12(½):156-206.

Tanadaicea

Hansen, H.J., 1913
Crustacea Malacostraca II. IV Tanaidacea Ingolf Exped. III(3): 152 pp., 12 pls., list of stations.

Tanaidacea

Hansen, H. J., 1913
Crustacea Malacostraca (II). IV Tanaidacea.
Danish Ingolf Exped. 3(3):145 pp., 12 pls.

Tanaidaceans

Lang, Karl, 1967.
Taxonomische und phylogenetische Untersuchungen über die Tanaidaceen. 3. Der Umfang der Familien Tanaidae Sars, Lang und Paratanaidae Lang nebst Bemerkungen über den taxonomischen Wert der Mandibeln und Maxillulae. Dazu eine taxonomisch-monographische Darstellung der Gattung Tanaopsis Sars.
Arkiv för Zoologi. 19(18):243-268.

Tanaidacea

Lang, K., 1950.
Contribution to the systematics and synonymies of the Tanaidacea. Ark. Zool. 42(3)(18):14 pp.

Tanaidacea

Larwood, H.J., 1940
The fishery grounds near Alexandria. XXI Tanaidacea and Isopoda. Fouad I. Inst. Hydrobiol. & Fish., Notes & Mem. No.35:72 pp., 17 figs.

Tanaidacea

Omer-Cooper, J., 1927.
Zoological results of the Cambridge Expedition to the Suez Canal. XII. Report on the Crustacea Tanaidacea and Isopoda. Trans. Zool. Soc., London, 22: 201-209.

Tanaceans

Sars, G. O., 1886
Nye Bidrag til Kundskaben om Middelhavets Invertebrat fauna. III. Middelhavets Saxisopoder (Isopoda chelifera). Archiv for Mathem. og Natur. 263-368, 15 pls.

Tanaidacea

Shiino, Sueo M., 1963.
Tanaidacea collected by Naga Expedition in the Bay of Nha-Trang, South Viet-Nam.
Rept., Fac. Fish., Pref. Univ. Of Mie, 4(3):437-507.

Tanaids, anat.

Siewing, E., 1954.
Morphologische Untersuchungen an Tanaidaceen und Lophogastriden. Zeits. Wiss. Zool., A, 157(3/4):333-426, 44 textfigs.

Tanaidacea

Soares Moreira Plinio, 1965.
On the distribution and ecology of Isopoda and Tanadacea at Ubatuba, Sao Paulo state coast. (Abstract).
Anais Acad. bras. Cienc, 37(Supl.):260.

Tanaidacea

Stephensen, K., 1943.
The zoology of East Greenland. Leptostraca, Mysidacea, Cumacea, Tanaidacea, Isopoda, and Euphausiacea. Medd. om Grønland, 121 (10): 1-82, 11 maps.

Tanaidacea

Stephenson, K., 1937
Marine Isopoda and Tanaidacea. Zool. Iceland. 3(27):26 pp.

Tanaidacea

Stephensen, K., 1915
Isopoda, Tanaidacea, Cumacea, Amphipoda (excl. Hyperiidea). Rept. Danish Oceanogr. Exped. 1908-10, to the Mediterranean and adjacent seas, Vol.II, Biol. D. 1, 53 pp., 33 text figs.

Tanaidacea

Tattersall, W.M., 1921
Crustacea. Part VI. Tanaidacea and Isopoda. Brit. Antarctic "Terra Nova" Exped., 1910, Nat. Hist. Rept., Zool. 3(8):191-258, 2 text figs., 11 pls.

Tanaidacea

Wolff, T., 1956.
Crustacea Tanaidacea from depths exceeding 6000 meters. Galathea Repts., 2:187-241.

Tanaidacea, behaviour of

Greve, Lita 1969.
On the tube building of some Tanaidacea.
Sarsia 29: 295-298.

tanaidaceans, lists of spp.

Brattegard, Torleiv, and Wim Vader 1972.
A collection of Peracarida from Møre and Romsdal, northwestern Norway.
Sarsia 49: 33-40.

tanaidaces, lists of spp.

Chardy, Pierre 1970
Ecologie des Crustacés Peracarides des fonds rocheux de Banyuls-sur-Mer. Amphipodes, isopodes tanaidacés, cumacés, infra et circalittoraux.
Vie Milieu (B) 21(3): 657-727

tanaids, lists of spp.

Greve, Lita, 1972.
Some new records of Tanaidacea from Norway.
Sarsia 48: 33-38.

Tanaidacea, lists of spp.

Greve, Lita, 1968.
Tanaidacea from Hardangerfjorden, western Norway.
Sarsia, 36:77-84.

Tanaidacea, lists of spp

Greve, Lita, 1965.
New records of some Tanaidacea (Crustacea) from the vicinity of Tromsö.
Asterte. No. 27: 6 pp.

Tanaidacea, lists of spp.

Zavodnik, Dusan, 1967.
Über die Scherenasseln (Tanaidacea) der Umgebung von Rovinj (Nördl. Adria).
Thalassia jugosl., 3(1/6):115-119.

Agathotanaidae n. fam.

Lang, Karl, 1970.
Taxonomische und phylogenetische Untersuchungen über die Tanaidaceen. 6. Revision der Gattung Paranarthrura Hansen, 1913, und Aufstellung von zwei neuen Familien, vier neuen Gattungen und zwei neuen Arten. Ark. Zool. (2) 23(6): 361-401. 2 plates.

Agathotanais hanseni n.sp.

Lang, Karl 1971.
Die Gattungen Agathotanais Hansen und Paragathotanais n.gen. (Tanaidacea).
Crustaceana 21(1): 57-71.

Agathotanais n. gen.

Hansen, H. J., 1913
Crustacea Malacostraca (II). IV Tanaidacea.
Danish Ingolf Exped. 3(3):145 pp., 12 pls.

Agathotanais ingolfi n.gen., n.sp.

Hansen, H.J., 1913
Crustacea Malacostraca II. IV Tanaidacea. Ingolf Exped. III(3): 152 pp., 12 pls., list of stations.

Agathotanais ingolfi n.sp.

Hansen, H. J., 1913
Crustacea Malacostraca (II). IV Tanaidacea.
Danish Ingolf Exped. 3(3):145 pp., 12 pls.

Anarthrura simplex

Greve, Lita, 1965.
The biology of some Tanaidacea from Raunefjorden, western Norway.
Sarsia, 20:43-54.

Anarthrura simplex

Sars, G.O., 1896-1899
An account of the Crustacea of Norway with short descriptions and figures of all species. Vol. II Isopoda. Pts.I,II. Apseudidae, Tanaidae:1-40, 16 pls. Pts.III,IV. Anthuridae, Gnathiidae, Aegidae, Cirolanidae, Limnoriidae:41-80, pls.17-32, Pts.V,VI. Idotheidae, Arcturidae, Asellidae, Ianiridae, Munnidae, 81-116, Pls.33-48. Pts.VII,V,Desmosomidae, Munnopsidae (part):117-144, pls. 49-64. Pts.IX,X. Munnopsidae (concluded), Ligiidae, Trichoniscidae, Oniscidae (part): 145-184, Pls.65-89. Pts.X,XII. Oniscidae (concluded), Bopyridae, Dajidae:185-232, pls.
(over)

Anarthruridae n. fam.

Lang, Karl, 1970.
Taxonomische und phylogenetische Untersuchungen über die Tanaidaceen. 6. Revision der Gattung Paranarthrura Hansen, 1913, und Aufstellung von zwei neuen Familien, vier neuen Gattungen und zwei neuen Arten. Ark. Zool. (2) 23(6): 361-401. 2 plates.

Anarthruropsis galatheae n.sp.
Lang, Karl, 1968.
Deep-Sea Tanaidacea.
Galathea Rept., 9:23-209.

Anatanais n.sp.
(aff. A. normani)
Kussakin, O.G., and L.A. Tzareva 1972.
Tanaidacea from the coastal zones of the middle Kurile Islands
Crustaceana Suppl.3: 237-245.

Anatanais normani
Miller, Milton A., 1968
Isopoda and Tanaidacea from buoys in coastal waters of the continental United States, Hawaii and the Bahamas
Proc. U.S. natn. Mus. 125(3652):1-53

Apseudella typica n.gen., n.sp.
Lang, Karl, 1968.
Deep-Sea Tanaidacea.
Galathea Rept., 9:23-209.

Apseudes
Brown, A.C., 1956.
Additions to the genus Apseudes (Crustacea: Tanaidacea) from South Africa.
Ann. Mag. Nat. Hist. (12) 9:705-709.

Apseudes sp.
Stebbing, T.R.R., 1910
Isopoda from the Indian Ocean and British East Africa. Trans. Linn. Soc., London, Zool., 2nd ser. 14:83-122, Pls. 5-11.

Apseudes acutifrons
Sars, G.O., 1886
Nye Bidrag til Kundskaben om Middelhavets Invertebrat fauna. III. Middelhavets Saxisopoder (Isopoda chelifera). Archiv for Mathem. og Natur. 263-368, 15 pls.

Apseudes africanus orientalis n.ssp.
Băcescu, Mihai, 1961
Contribution à la connaissance des Tanaidacés de la Mediterranée Orientale - 1. Les Apseudidae et Kalliapseudidae des côtes d'Israel.
Bull. Res. Council, Israel, (B. Zool.), 10(4):137-170.

Apseudes anomalus
Sars, G.O., 1896-1899
An account of the Crustacea of Norway with short descriptions and figures of all species. Vol.II Isopoda. Pts.I,II. Apseudidae, Tanaidae:1-40, 16 pls. Pts.III,IV. Anthuridae, Gnathiidae, Aegidae, Cirolanidae, Limnoriidae:41-80, pls.17-32, Pts.V,VI. Idotheidae, Ancthuridae, Asellidae, Ioniridae, Munnidae, 81-116, Pls.33-48. Pts.VII,VIII.Desmosomidae, Munnopsidae (part):117-144, pls. 49-64. Pts.IX,X. Munnopsidae (concluded), Ligiidae, Trichoniscidae, Oniscidae (part): 145-184, Pls.65-80. Pts.X,XII. Oniscidae (concluded), Bopyridae, Dajidae:185-232,pls. (over)

Apseudes caeruleus n.sp. a
Boesch, Donald F. 1973.
Three new tanaids (Crustacea, Tanaidacea) from southern Queensland.
Pacific Sci. 27(2): 168-188

Apseudes cooperi
Brown, A.C., 1956.
Additions to the genus Apseudes (Crustacea: Tanaidacea) from South Africa).
Ann. Mag. Nat. Hist., (12) 9:705-709.

Apseudes digitalis n.sp.
Brown, A.C., 1956.
Additions to the genus Apseudes (Crustacea: Tanaidacea) from South Africa.
Ann. Mag. Nat. Hist., (12) 9:705-709.

Apseudes diversus n.sp.
Lang, Karl, 1968.
Deep-Sea Tanaidacea.
Galathea Rept., 9:23-209.

Apseudes echinatus
Sars, G.O., 1886
Nye Bidrag til Kundskaben om Middelhavets Invertebrat fauna. III. Middelhavets Saxisopoder (Isopoda chelifera). Archiv for Mathem. og Natur. 263-368, 15 pls.

Apseudes echinatus
Stephensen, K., 1915
Isopoda, Tanaidacea, Cumacea, Amphipoda (excl. Hyperiidea). Rept. Danish Oceanogr. Exped. 1908-10, to the Mediterranean and adjacent seas, Vol.II, Biol. D. 1, 53 pp., 33 text figs.

Apseudes elisae n.sp.
Băcescu, Mihai, 1961
Contribution à la connaissance des Tanaidacés de la Méditerranée Orientale - 1. Les Apseudidae et Kalliapseudidae des côtes d'Israel.
Bull. Res. Council, Israel, (B. Zool.), 10(4):137-170.

Apseudes espinosus n.sp.
Moore, H.F., 1901
Report on the Porto Rican Isopoda.
U.S. Fish Comm. Bull. for 1900, Vol.2: 161-176, pls.7-11.

Apseudes estuarius n.sp. a
Boesch, Donald F. 1973.
Three new tanaids (Crustacea, Tanaidacea) from southern Queensland.
Pacific Sci. 27(2): 168-188

Apseudes galathea n.sp.
Wolff, T., 1956.
Crustacea Tanaidacea from depths exceeding 6000 meters. Galathea Repts., 2:187-241.

Apseudes gallardoi n. sp.
Shiino, Sueo M., 1963.
Tanaidacea collected by Naga Expedition in the Bay of Nha-Trang, South Viet-Nam.
Rept., Fac. Fish., Pref. Univ. of Mie, 4(3):437-507.

Apseudes gracilis
Hansen, H.J., 1913
Crustacea Malacostraca (II). IV Tanaidacea.
Danish Ingolf Exped. 3(3):145 pp., 12 pls.

Apseudes gracilis
Wolff, T., 1956.
Crustacea Tanaidacea from depths exceeding 6000 meters. Galathea Repts., 2:187-241.

Apseudes gracillimus n.sp
Hansen, H.J., 1913
Crustacea Malacostraca (II). IV Tanaidacea.
Danish Ingolf Exped. 3(3):145 pp., 12 pls.

Apseudes graciloides n.sp.
Stephensen, K., 1915
Isopoda, Tanaidacea, Cumacea, Amphipoda (excl. Hyperiidea). Rept. Danish Oceanogr. Exped. 1908-10, to the Mediterranean and adjacent seas, Vol.II, Biol. D. 1, 53 pp., 33 text figs.

Apseudes grossimanus
Lang, Karl, 1968.
Deep-Sea Tanaidacea.
Galathea Rept., 9:23-209.

Apseudes grossimanus
Stephensen, K., 1915
Isopoda, Tanaidacea, Cumacea, Amphipoda (excl. Hyperiidea). Rept. Danish Oceanogr. Exped. 1908-10, to the Mediterranean and adjacent seas, Vol.II, Biol. D. 1, 53 pp., 33 text figs.

Apseudes grossimanus
Tattersall, W.M., 1905
The Marine Fauna of the Coast of Ireland. Part V Isopoda. Dept. Agric. and Tech. Instr. for Ireland, Fisheries Branch, Sci. Invest. 1904. No.II:1-90, pls.1-11, 1 chart.

Apseudes gymnophobia
Balasubrahmanyan, K., 1962
Apseudidae (Isopoda-Crustacea) from the Vellar Estuary and inshore waters off Porto Nov.
Proc. First All-India Congr., Zool., 1959, Sci. Pap., (2):279-285.
Also in:
Annamalai Univ. Mar. Biol. Sta., Porto Novo, S. India, Publ., 1961-1962.

Apseudes hibernicus
Tattersall, W.M., 1905
The Marine Fauna of the Coast of Ireland. Part V Isopoda. Dept. Agric. and Tech. Instr. for Ireland, Fisheries Branch, Sci. Invest. 1904. No.II:1-90, pls.1-11, 1 chart.

Apseudes holthuisi (for A. sarsi-nomen praeoccupatum) Syn: A talpa
Băcescu, Mihai, 1961
Contribution à la connaissance des Tanaidacés de la Méditerranée Orientale -1. Les Apseudidae et Kalliapseudidae des côtes d'Israel.
Bull. Res. Council, Israel, (B.Zool.), 10(4):137-170.

Apseudes intermedius
Băcescu, Mihai, 1961
Contribution à la connaissance des Tanaidacés de la Méditerranée Orientale - 1. Les Apseudidae et Kalliapseudidae des côtes d'Israel.
Bull. Res. Council, Israel, (B. Zool.), 10(4) 137-170.

Apseudes intermedius n.sp.
Hansen, H.J., 1895
Isopoden, Cumaceen und Stomatopoden der Plankton Expedition. Ergeb. Plankton Exped. II, Gc: 105 pp., 8 pls.

Apseudes intermedius
Larwood, H.J., 1940
The fishery grounds near Alexandria. XXI Tanaidacea and Isopoda. Fouad I. Inst. Hydrobiol. & Fish., Notes & Mem. No.35:72 pp., 17 figs.

Apseudes killaiyensis n.sp.
Balasubrahmanyan, K., 1962
Apseudidae (Isopoda-Crustacea) from the Vellar Estuary and inshore waters off Porto Novo.
Proc. First All-India Congr., Zool., 1959, Sci. Pap., (2):279-285.
Also in:
Annamalai Univ. Mar. Biol. Sta., Porto Novo, S. India, Publ. 1961-1962.

Apseudes lagenirostris n.sp.
Lang, Karl, 1968.
Deep-Sea Tanaidacea.
Galathea Rept., 9:23-209.

Apseudes latreillii
Larwood, H.J., 1940
The fishery grounds near Alexandria. XXI Tanaidacea and Isopoda. Fouad I. Inst. Hydrobiol. & Fish., Notes & Mem. No.35:72 pp., 17 figs.

Apseudes latreilli
Sars, G.O., 1886
Nye Bidrag til Kundskaben om Middelhavets Invertebrat fauna. III. Middelhavets Saxisopoder (Isopoda chelifera). Archiv for Mathem. og Natur. 263-368, 15 pls.

Apseudes latreilli mediterraneus n.ssp
Băcescu, Mihai, 1961
Contribution à la connaissance des Tanaidacés de la Méditerranée Orientale - 1. Les Apseudidae et Kalliapseudidae des côtes d'Israel.
Bull. Res. Council, Israel, (B. Zool.), 10 (4):137-170.

Apseudes littoralis
Shiino, S.M., 1952.
A new genus and two new species of the order Tanaidacea found at Seto.
Publ. Seto Mar. Biol. Lab., Kyoto Univ., 2(2): 53-68, 7 textfigs.

Apseudes minutus n.sp.
Brown, A.C., 1956.
Additions to the genus Apseudes (Crustacea: Tanaidacea) from South Africa.
Ann. Mag. Nat. Hist., (12) 9:705-709.

Apseudes nagae, n. sp.
Shiino, Sueo M., 1963.
Tanaidacea collected by Naga Expedition in the Bay of Nha-Trang, South Viet-Nam.
Rept., Fac. Fish., Pref. Univ. of Mie, 4(3):437-507.

Apseudes nhatrangensis n.sp.
Shiino, Sueo M., 1963.
Tanaidacea collected by Naga Expedition in the Bay of Nha-Trang, South Viet-Nam.
Rept., Fac. Fish., Pref. Univ. of Mie, 4(3):437-507.

Apseudes nigrifrons n. sp.
Shiino, Sueo M., 1963.
Tanaidacea collected by Naga Expedition in the Bay of Nha-Trang, South Viet-Nam.
Rept., Fac. Fish., Pref. Univ. of Mie, 4(3):437-507.

Apseudes nipponicus f. hermaphroditicus
Shiino, Sueo M. 1970.
Paratanaidae collected in Chile Bay, Greenwich Island by the XXII Chilean Antarctic Expedition, with an Apseudes from Porvenir Point, Tierra del Fuego Island.
Inst. Antart. Chileno, Ser. Cient. 1(2):79-122.

Apseudes portnovensis n.sp.
Balasubrahmanyan, K., 1962
Apseudidae (Isopoda-Crustacea) from the Vellar Estuary and inshore waters off Porto Novo.
Proc. First All-India Congr., Zool., 1959, Sci. Pap., (2):279-285.

Also in:
Annamalai Univ. Mar. Biol. Sta., Porto Novo., S. India, Publ. 1961-1962.

Apseudes retusifrons
Stephensen, K., 1915
Isopoda, Tanaidacea, Cumacea, Amphipoda (excl. Hyperiidea). Rept. Danish Oceanogr. Exped. 1908-10, to the Mediterranean and adjacent seas, Vol.II, Biol. D. 1, 53 pp., 33 text figs.

Apseudes robustus
Larwood, H.J., 1940
The fishery grounds near Alexandria. XXI Tanaidacea and Isopoda. Fouad I. Inst. Hydrobiol. & Fish., Notes & Mem. No.35:72 pp., 17 figs.

Apseudes robustus
Sars, G.O., 1886
Nye Bidrag til Kundskaben om Middelhavets Invertebrat fauna. III. Middelhavets Saxisopoder (Isopoda chelifera). Archiv for Mathem. og Natur. 263-368, 15 pls.

Apseudes robustus israeliticus n.ssp.
Bǎcescu, Mihai, 1961
Contribution à la connaissance des Tanaidacés de la Méditerranée Orientale -1. Les Apseudidae et Kalliapseudidae des côtes d'Israel.
Bull. Res. Council, Israel, (B. Zool.), 10(4) 137-170.

Apseudes sarsi
Bacescu, Mihai, 1961
Contribution à la connaissance des tanaidacés de la Méditerranée Orientale. 1. Les Apseudidae et Kalliapseudidae des côtes d'Israel.
Bull. Res. Counc., Israel, 10B(4):137-170.

Apseudes setosus n.sp.
Lang, Karl, 1968.
Deep-Sea Tanaidacea.
Galathea Rept., 9:23-209.

Apseudes spectabilis
Stephensen, K., 1947
Tanaidacea, Isopoda, Amphipoda, and Pycnogonida. Sci. Res. Norweg. Antarctic Exp. 1927-1928 et seq. Date Norske Videnskaps-Akademi i Oslo, II, No.27:1-90, 26 text figs.

Apseudes spinosus
Dahl, E., 1941
Tanaidacea from the Sound and Skälderviken. Kungl. Fysiografiska Sällskapets i Lund Förhandlingar, 11(7):1-7, 1 text fig.

Apseudes spinosus
Hansen, H.J., 1913
Crustacea Malacostraca (II). IV Tanaidacea.
Danish Ingolf Exped. 3(3):145 pp., 12 pls.

Apseudes spinosa
Hanstroem, B., 1948
The brain, the sense organs and the incretory organs of the head in the Crustacea Malacostraca. Suppl. 33 to Bull. Biol. de France et de Belgique: 98-125, 11 text figs.

Apseudes spinosus
Sars, G.O., 1896-1899
An account of the Crustacea of Norway with short descriptions and figures of all species. Vol.II Isopoda. Pts.I,II. Apseudidae, Tanaidae:1-40, 16 pls. Pts.III,IV. Anthuridae, Gnathiidae, Aegidae, Cirolanidae, Limnoriidae:41-80, pls.17-32, Pts.V,VI. Idotheidae, Ancthuridae, Asellidae, Ioniridae, Munnidae, 81-116, Pls.33-48. Pts.VII,VIII Desmosomidae, Munnopsidae (part):117-144, pls. 49-64. Pts.IX,X. Munnopsidae (concluded), Ligiidae, Trichoniscidae, Oniscidae (part): 145-194, Pls.65-83. Pts.X,XII. Oniscidae (concluded), Bopyridae, Dajidae:185-232,pls. (over)

Apseudes spinosus
Sars, G.O., 1886
Crustacea II. Zool. Norwegian North-Atlantic Exped., 1876-1878: 96 pp., 1 map.

Apseudes spinosus
Stephensen, K., 1937
Marine Isopoda and Tanaidacea. Zool. Iceland. 3(27):26 pp.

Apseudes spinosus
Tattersall, W.M., 1905
The Marine Fauna of the Coast of Ireland. Part V Isopoda. Dept. Agric. and Tech. Instr. for Ireland, Fisheries Branch, Sci. Invest. 1904. No.II:1-90, pls.1-11, 1 chart.

Apseudes spinosus
Wolff, T., 1956.
Crustacea Tanaidacea from depths exceeding 6000 meters. Galathea Repts., 2:187-241.

Apseudes talpa
Bacescu, Mihai, 1961
Contribution à la connaissance des tanaidacés de la Méditerranée Orientale. 1. Les Apseudidae et Kalliapseudidae des côtes d'Israel.
Bull. Res. Counc., Israel, 10B(4):137-170.

Apseudes talpa
Sars, G.O., 1896-1899
An account of the Crustacea of Norway with short descriptions and figures of all species. Vol.II Isopoda. Pts.I,II. Apseudidae, Tanaidae:1-40, 16 pls. Pts.III,IV. Anthuridae, Gnathiidae, Aegidae, Cirolanidae, Limnoriidae:41-80, pls.17-32, Pts.V,VI. Idotheidae, Ancthuridae, Asellidae, Ioniridae, Munnidae, 81-116, Pls.33-48. Pts.VII,VIII Desmosomidae, Munnopsidae (part):117-144, pls. 49-64. Pts.IX,X. Munnopsidae (concluded), Ligiidae, Trichoniscidae, Oniscidae (part): 145-194, Pls.65-83. Pts.X,XII. Oniscidae (concluded), Bopyridae, Dajidae:185-232,pls. (over)

Apseudes talpa
Sars, G.O., 1886
Nye Bidrag til Kundskaben om Middelhavets Invertebrat fauna. III. Middelhavets Saxisopoder (Isopoda chelifera). Archiv for Mathem. og Natur. 263-368, 15 pls.

Apseudes tenuicorporeus n. sp.
Shiino, Sueo M., 1963.
Tanaidacea collected by Naga Expedition in the Bay of Nha-Trang, South Viet-Nam.
Rept., Fac. Fish., Pref. Univ. of Mie, 4(3):437-507.

Apseudes tenuimanus
Sars, G.O., 1886
Nye Bidrag til Kundskaben om Middelhavets Invertebrat fauna. III. Middelhavets Saxisopoder (Isopoda chelifera). Archiv for Mathem. og Natur. 263-368, 15 pls.

Apseudes tenuimanus gottliebi nssp
Bacescu, Mihai, 1961
Contribution à la connaissance des Tanaidacés de la Méditerranée Orientale - 1. Les Apseudidae et Kalliapseudidae des côtes d'Israel.
Bull. Res. Council, Israel, (B. Zool.), 10(4):137-170.

Apseudes tenuis n.sp.
Hansen, H.J., 1913
Crustacea Malacostraca (II). IV Tanaidacea.
Danish Ingolf Exped. 3(3):145 pp., 12 pls.

Apseudes tuberculatus n.sp.
Lang, Karl, 1968.
Deep-Sea Tanaidacea.
Galathea Rept., 9:23-209.

Apseudes vicinus n.sp.
Hansen, H.J., 1913
Crustacea Malacostraca (II). IV Tanaidacea.
Danish Ingolf Exped. 3(3):145 pp., 12 pls.

Apseudes genkevitchi n. sp.
Kudinova-Pasternak, R.K., 1966.
Tanaidacea (Crustacea) of the Pacific ultra-abyssal zone (In Russian; English Abstract).
Zool. Zhurn., Akad. Nauk SSSR 44(4): 518-535.

Anthrura andriashevi n. gen., n. sp.
Kudinova-Pasternak 1967.
On a new abyssal tanaidacean from the Pacific, Anthrura andriashevi n. gen. n.sp.
Crustaceana 12(2):257-260.

bathytanais bathybrotes
Lang, Karl 1972.
Bathytanais bathybrotes (Beddard) und Leptognathia dissimilis n.sp. (Tanaidacea).
Crustaceana Suppl.3:221-236

Carpoapseudes longissimus, n.sp.
Lang, Karl, 1968.
Deep-Sea Tanaidacea.
Galathea Rept., 9:23-209.

Carpoapseudes oculicornutus n.sp.
Lang, Karl, 1968.
Deep-Sea Tanaidacea.
Galathea Rept., 9:23-209.

Carpoapseudes serratispinosus n.sp.
Lang, Karl, 1968.
Deep-Sea Tanaidacea.
Galathea Rept., 9:23-209.

Cirratodactylus floridensis n.gen.n.sp.
Gardiner, Lion F. 1973
A new species and genus of a new monokonophoran family (Crustacea: Tanaidacea), from southeastern Florida.
J. Zool. Lond. 169(2):237-253.

Cryptocope abbreviata
Greve, Lita, 1965.
The biology of some Tanadacea from Raunefjorden, western Norway.
Sarsia, 20:43-54.

Cryptocope abbreviata
Sars, G.O., 1896-1899
An account of the Crustacea of Norway with short descriptions and figures of all species. Vol.II Isopoda. Pts.I,II. Apseudidae, Tanaidae:1-40, 16 pls. Pts.III,IV. Anthuridae, Gnathiidae, Aegidae, Cirolanidae, Limnoriidae:41-80, pls.17-32, Pts.V,VI. Idotheidae, Ancthuridae, Asellidae, Ianiridae, Munnidae, 81-116, Pls.33-48. Pts.VII,VIII,Desmosomidae, Munnopsidae (part):117-144, pls. 49-64. Pts.IX,X. Munnopsidae (concluded), Ligiidae, Trichoniscidae, Oniscidae (part): 145-184, Pls.65-88. Pts.X,XII. Oniscidae (concluded), Bopyridae, Dajidae:185-232,pls.
(over)

Cryptocope agringii
Hansen, H.J., 1913
Crustacea Malacostraca II. IV Tanaidacea Ingolf Exped. III(3): 152 pp., 12 pls., list of stations.

Cryptocope arctica
Hansen, H. J., 1913
Crustacea Malacostraca (II). IV Tanaidacea.
Danish Ingolf Exped. 3(3):145 pp., 12 pls.

Cryptocope arctica
Sars, G.O., 1896-1899
An account of the Crustacea of Norway with short descriptions and figures of all species. Vol.II Isopoda. Pts.I,II. Apseudidae, Tanaidae:1-40, 16 pls. Pts.III,IV. Anthuridae, Gnathiidae, Aegidae, Cirolanidae, Limnoriidae:41-80, pls.17-32, Pts.V,VI. Idotheidae, Ancthuridae, Asellidae, Ianiridae, Munnidae, 81-116, Pls.33-48. Pts.VII,VIII,Desmosomidae, Munnopsidae (part):117-144, pls. 49-64. Pts.IX,X. Munnopsidae (concluded), Ligiidae, Trichoniscidae, Oniscidae (part): 145-184, Pls.65-88. Pts.X,XII. Oniscidae (concluded), Bopyridae, Dajidae:185-232,pls.
(over)

Cryptocope arctica
Stephensen, K., 1912
Report on the Malacostraca, Pycnogonida and some Entomostraca collected by the Danmark-Expedition to North East Greenland. Medd. om Grønland, 45(11):501-630, Pls. 39-43.

Cryptocope arctophylax
Hansen, H. J., 1913
Crustacea Malacostraca (II). IV Tanaidacea.
Danish Ingolf Exped. 3(3):145 pp., 12 pls.

Cryptocope Vøringii
Hansen, H. J., 1913
Crustacea Malacostraca (II). IV Tanaidacea.
Danish Ingolf Exped. 3(3):145 pp., 12 pls.

Cryptocope vøringii n.gen., n.sp.
Sars, G.O., 1886
Crustacea II. Zool. Norwegian North-Atlantic Exped., 1876-1878: 96 pp., 1 map.

Cryptocope vøringii n.sp.
Sars, G.O. 1885
Crustacea 1. Zool., Norwegian North-Atlantic Exped. 1876-1878: 280 pp., 21 pls. 1 map.

Cyclopoapseudes dicenson n.sp.
Gardiner, Lion F. 1973.
New species of the genera Synapseudes and Cyclopoapseudes with notes on morphological variation, postmarsupial development and phylogenetic relationships within the family Metapseudidae (Crustacea: Tanaidacea).
Zool. J. Linn. Soc. 53(1):25-58

Dalapseudes pedispinus n.gen., n.sp.
Boone, P. L., 1923.
New marine tanaid and isopod Crustacea. Proc. Biol. Soc., Washington, 36:147-156.

Exspina typica n.sp.
Lang, Karl, 1968.
Deep-Sea Tanaidacea.
Galathea Rept., 9:23-209.

Haplocope angusta
Greve, Lita, 1965.
The biology of some Tanaidacea from Raunefjorden, western Norway.
Sarsia, 20:43-54.

Haplocope angusta
Sars, G.O., 1896-1899
An account of the Crustacea of Norway with short descriptions and figures of all species. Vol.II Isopoda. Pts.I,II. Apseudidae, Tanaidae:1-40, 16 pls. Pts.III,IV. Anthuridae, Gnathiidae, Aegidae, Cirolanidae, Limnoriidae:41-80, pls.17-32, Pts.V,VI. Idotheidae, Ancthuridae, Asellidae, Ianiridae, Munnidae, 81-116, Pls.33-48. Pts.VII,VIII,Desmosomidae, Munnopsidae (part):117-144, pls. 49-64. Pts.IX,X. Munnopsidae (concluded), Ligiidae, Trichoniscidae, Oniscidae (part): 145-184, Pls.65-88. Pts.X,XII. Oniscidae (concluded), Bopyridae, Dajidae:185-232,pls.
(over)

Haplocope linearis n.sp.
Hansen, H. J., 1913
Crustacea Malacostraca (II). IV Tanaidacea.
Danish Ingolf Exped. 3(3):145 pp., 12 pls.

Herpotanais kirkegaardi n.gen., n.sp.
Wolff, T., 1956.
Crustacea Tanaidacea from depths exceeding 6000 meters. Galathea Repts., 2:187-241.

Heterotanais anomalus
Sars, G.O., 1896-1899
An account of the Crustacea of Norway with short descriptions and figures of all species. Vol.II Isopoda. Pts.I,II. Apseudidae, Tanaidae:1-40, 16 pls. Pts.III,IV. Anthuridae, Gnathiidae, Aegidae, Cirolanidae, Limnoriidae:41-80, pls.17-32, Pts.V,VI. Idotheidae, Ancthuridae, Asellidae, Ianiridae, Munnidae, 81-116, Pls.33-48. Pts.VII,VIII,Desmosomidae, Munnopsidae (part):117-144, pls. 49-64. Pts.IX,X. Munnopsidae (concluded), Ligiidae, Trichoniscidae, Oniscidae (part): 145-184, Pls.65-88. Pts.X,XII. Oniscidae (concluded), Bopyridae, Dajidae:185-232,pls.
(over)

Heterotanais anomalus
Sars, G.O., 1886
Nye Bidrag til Kundskaben om Middelhavets Invertebrat fauna. III. Middelhavets Saxisopoder (Isopoda chelifera). Archiv for Mathem. og Natur. 263-368, 15 pls.

?Heterotanais anomalus
Stebbing, T. R. R., 1910
Isopoda from the Indian Ocean and British East Africa. Trans. Linn. Soc., London, Zool., 2nd ser. 14:83-122, Pls. 5-11.

Heterotanais groenlandicus n.sp.
Hansen, H. J., 1913
Crustacea Malacostraca (II). IV Tanaidacea.
Danish Ingolf Exped. 3(3):145 pp., 12 pls.

Kussakin, O.G. and L.A. Tzareva 1972.
Heterotanais modestus n.sp.
Tanaidacea from the coastal zones of the middle Kurile Islands.
Crustaceana Suppl. 3:237-245.

Heterotanais oerstedi
Buckle Ramirez, I.F., 1965.
Untersuchungen über die Biologie von Heterotanais oerstedi Kroyer (Crustacea: Tanaidacea.)
Z. Morph. Ökol. Tiere, 55:641-655.

Heterotanais orstedii
Dahl, E., 1944
The Swedish Brackish Water Malacostraca.
Kungl. Fysiografiska Sällskapets i Lund Förhandlingar, 14(9):1-17.

Heterotanais örstedii
Dahl, E., 1941
Tanaidacea from the Sound and Skälderviken. Kungl. Fysiografiska Sällskapets i Lund Förhandlingar, 11(7):1-7, 1 text figs.

Heterotanais Ørstedi
Sars, G.O., 1896-1899
An account of the Crustacea of Norway with short descriptions and figures of all species. Vol.II Isopoda. Pts.I,II. Apseudidae, Tanaidae:1-40, 16 pls. Pts.III,IV. Anthuridae, Gnathiidae, Aegidae, Cirolanidae, Limnoriidae:41-80, pls.17-32, Pts.V,VI. Idotheidae, Ancthuridae, Asellidae, Ianiridae, Munnidae, 81-116, Pls.33-48. Pts.VII,VIII,Desmosomidae, Munnopsidae (part):117-144, pls. 49-64. Pts.IX,X. Munnopsidae (concluded), Ligiidae, Trichoniscidae, Oniscidae (part): 145-184, Pls.65-88. Pts.X,XII. Oniscidae (concluded), Bopyridae, Dajidae:185-232,pls.
(over)

Heterotanais oerstedi
Siewing, R., 1954.
Morphologische Untersuchungen an Tanaidaceen und Lophogastriden. Zeits. Wiss. Zool., A, 157(3/4): 333-426, 44 textfigs.

Kalliapseudes dentatus n.sp.
Brown, A.C., 1956.
A new species of Kalliapseudes (Tanadacea) from South Africa. Ann. Mag. Nat. Hist., (12)9(104): 582-586.

Kalliapseudes makrothrix
Balasubrahmanyan, K., 1962
Apseudidae (Isopoda-Crustacea) from the Vellar Estuary and inshore waters off Porto Novo. Proc. First All-India Congr. Zool., 1959, Sci. Pap., (2):279-285.
Also in:
Annamalai Univ. Mar. Biol. Sta., Porto Novo, S. India, Publ. 1961-1962.

Kalliapseudes makrothrix n. sp.
Stebbing, T. R. R., 1910
Isopoda from the Indian Ocean and British East Africa. Trans. Linn. Soc., London, Zool., 2nd ser. 14:83-122, Pls. 5-11.

Kalliapseudes omer-cooperi
Bacescu, Mihai, 1961
Contribution à la connaissance des tanaidacés de la Méditerranée Orientale. 1. Les Apseudidae et Kalliapseudidae des côtes d'Israel. Bull. Res. Counc., Israel, 10B(4):137-170.

Kalliapseudes schubarti
Lang, K., 1956.
Tanaidacea aus Brasilien gesammelt von Professor A. Remane und Dr. S. Gerlach. Kieler Meeresf., 12(2):249-260.

Kalliapseudes schubartii n.sp.
Mane-Garzon, F., 1949.
Un nuevo Tanaidacea ciego de Sud América, Kalliapseudes schubartii nov. sp.
Comun. Zool. Mus. Hist. Nat., Montevideo, 3(52): 1-6, 1 pl.

Kalliapseudes(Kalliapseudes) tomiokaensis n. sp.
Shiino, Sueo M.,1966.
On Kalliapseudes(Kalliapseudes) tomiokaensis sp. nov. (Crustacea:Tanaidacea) from Japanese waters. Rep.Fac.Fish.,pref.Univ.Mie,5(3):473-488.

Leiopus sp.
Sars, G.O., 1896-1899
An account of the Crustacea of Norway with short descriptions and figures of all species. Vol.II Isopoda. Pts.I,II. Apseudidae, Tanaidae:1-40, 16 pls. Pts.III,IV. Anthuridae, Gnathiidae, Aegidae, Cirolanidae, Limnoriidae:41-80, pls.17-32, Pts.V,VI. Idotheidae, Ancthuridae, Asellidae, Ianiridae, Munnidae, 81-116, Pls.33-48. Pts.VII,VIII,Desmosomidae, Munnopsidae (part):117-144, pls. 49-64. Pts.IX,X. Munnopsidae (concluded), Ligiidae, Trichoniscidae, Oniscidae (part): 145-184, Pls.65-88. Pts.X,XII. Oniscidae (concluded), Bopyridae, Dajidae:185-232,pls.
(over)

Leiopus aberrans n.sp.
Lang, Karl, 1968.
Deep-Sea Tanaidacea.
Galathea Rept. 9:23-209.

Leiopus conspicuus n.sp.
Lang, Karl, 1968.
Deep-Sea Tanaidacea.
Galathea Rept., 9:23-209.

Leiopus galatheae
Lang, Karl, 1968.
Deep-Sea Tanaidacea.
Galathea Rept., 9:23-209.

Leiopus gracilis
Lang, Karl, 1968.
Deep-Sea Tanaidacea.
Galathea Rept., 9:23-209.

Leiopus gracillimus
Lang, Karl, 1968.
Deep-Sea Tanaidacea.
Galathea Rept., 9:23-209.

Leiopus hanseni n. nom.
Lang, Karl, 1968.
Deep-Sea Tanaidacea.
Galathea Rept., 9:23-209.

Leiopus leptodactylus
Lang, Karl, 1968.
Deep-Sea Tanaidacea.
Galathea Rept., 9:23-209.

Leiopus shiinoi n. sp.
Lang, Karl, 1968.
Deep-Sea Tanaidacea.
Galathea Rept. 9:23-209.

Leiopus sibogae
Lang, Karl, 1968.
Deep-Sea Tanaidacea.
Galathea Rept., 9:23-209.

Leiopus weberi
Lang, Karl, 1968.
Deep-Sea Tanaidacea.
Galathea Rept., 9:23-209.

Leiopus wolffi n. sp.
Lang, Karl, 1968.
Deep-Sea Tanaidacea.
Galathea Rept., 9:23-209.

Leptochelia affinis n. sp.
Hansen, H. J., 1895
Isopoden, Cumaceen und Stomatopoden der Plankton Expedition. Ergeb. Plankton Exped. II, Gc: 105 pp., 8 pls.

Leptochelia australis
Sars, G.O., 1896-1899
An account of the Crustacea of Norway with short descriptions and figures of all species. Vol.II Isopoda. Pts.I,II. Apseudidae, Tanaidae:1-40, 16 pls. Pts.III,IV. Anthuridae, Gnathiidae, Aegidae, Cirolanidae, Limnoriidae:41-80, pls.17-32, Pts.V,VI. Idotheidae, Ancthuridae, Asellidae, Ianiridae, Munnidae, 81-116, Pls.33-48. Pts.VII,VIII, Desmosomidae, Munnopsidae (part):117-144, pls. 49-64. Pts.IX,X. Munnopsidae (concluded), Ligiidae, Trichoniscidae, Oniscidae (part): 145-184, Pls.65-88. Pts.X,XII. Oniscidae (concluded), Bopyridae, Dajidae:185-232, pls. (over)

Leptochelia coeca
Sars, G.O., 1896-1899
An account of the Crustacea of Norway with short descriptions and figures of all species. Vol.II Isopoda. Pts.I,II. Apseudidae, Tanaidae:1-40, 16 pls. Pts.III,IV. Anthuridae, Gnathiidae, Aegidae, Cirolanidae, Limnoriidae:41-80, pls.17-32, Pts.V,VI. Idotheidae, Ancthuridae, Asellidae, Ianiridae, Munnidae, 81-116, Pls.33-48. Pts.VII,VIII, Desmosomidae, Munnopsidae (part):117-144, pls. 49-64. Pts.IX,X. Munnopsidae (concluded), Ligiidae, Trichoniscidae, Oniscidae (part): 145-184, Pls.65-88. Pts.X,XII. Oniscidae (concluded), Bopyridae, Dajidae:185-232, pls. (over)

Leptochelia danica
Dahl, E., 1941
Tanaidacea from the Sound and Skälderviken. Kungl. Fysiografiska Sällskapets i Lund Förhandlingar, 11(7):1-7, 1 text figs.

Leptochelia dubia
Larwood, H.J., 1940
The fishery grounds near Alexandria. XXI Tanaidacea and Isopoda. Fouad L. Inst. Hydrobiol. & Fish., Notes & Mem. No.35:72 pp., 17 figs.

Leptochelia dubia
Miller, Milton A., 1968
Isopoda and Tanaidacea from buoys in coastal waters of the continental United States, Hawaii and the Bahamas
Proc. U.S. natn. Mus. 125(3652):1-53

Leptochelia dubia
Sars, G.O., 1886
Nye Bidrag til Kundskaben om Middelhavets Invertebrat fauna. III. Middelhavets Saxisopoder (Isopoda chelifera). Archiv for Mathem. og Natur. 263-368, 15 pls.

Leptochelia dubia
Tattersall, W.M., 1905
The Marine Fauna of the Coast of Ireland. Part V Isopoda. Dept. Agric. and Tech. Instr. for Ireland, Fisheries Branch, Sci. Invest. 1904. No.II:1-90, pls.1-11, 1 chart.

Leptochelia incerta n.sp.
Moore, H.F., 1901
Report on the Porto Rican Isopoda. U.S. Fish Comm. Bull. for 1900, Vol.2: 161-176, pls.7-11.

Leptochelia limicola
Sars, G.O., 1896-1899
An account of the Crustacea of Norway with short descriptions and figures of all species. Vol.II Isopoda. Pts.I,II. Apseudidae, Tanaidae:1-40, 16 pls. Pts.III,IV. Anthuridae, Gnathiidae, Aegidae, Cirolanidae, Limnoriidae:41-80, pls.17-32, Pts.V,VI. Idotheidae, Ancthuridae, Asellidae, Ianiridae, Munnidae, 81-116, Pls.33-48. Pts.VII,VIII, Desmosomidae, Munnopsidae (part):117-144, pls. 49-64. Pts.IX,X. Munnopsidae (concluded), Ligiidae, Trichoniscidae, Oniscidae (part): 145-184, Pls.65-88. Pts.X,XII. Oniscidae (concluded), Bopyridae, Dajidae:185-232, pls. (over)

Leptochelia longimana, n. sp.
Shiino, Sueo M., 1963.
Tanaidacea collected by Naga Expedition in the Bay of Nha-Trang, South Viet-Nam.
Rept., Fac. Fish., Pref. Univ. of Mie, 4(3):437-507.

Leptochelis mirabilis
Barnard, K. H., 1955.
Additions to the fauna-list of South African Crustacea and Pycnogonida.
Ann. S. African Mus., 43(1): 1-107.

Leptochelia neapolitana
Sars, G.O., 1886
Nye Bidrag til Kundskaben om Middelhavets Invertebrat fauna. III. Middelhavets Saxisopoder (Isopoda chelifera). Archiv for Mathem. og Natur. 263-368, 15 pls.

Leptochelia savignyi
Barnard, K. A., 1925
10. Contributions to the Crustacean fauna of South Africa. No. 9. Further additions to the list of Isopoda. Ann. S. African Mus. 20: 381-342, 6 text figs.

Leptochelia Savignyi
Sars, G.O., 1886
Nye Bidrag til Kundskaben om Middelhavets Invertebrat fauna. III. Middelhavets Saxisopoder (Isopoda chelifera). Archiv for Mathem. og Natur. 263-368, 15 pls.

Leptochelia tenuis
Hurley, D.E., 1957.
Some Amphipoda, Isopoda and Tanaidacea from Cook Strait. Publ. Zool., Victoria Univ. Coll., No. 21:1-20.

Leptognatha sp.
Wiborg, K.F., 1948
Some observation on the food of Cod (Gadus callarias L.) of the 0-II group from deep water and the littoral zone in northern Norway and from deep water at Spitzbergen. Repts. Norwegian Fishery and Marine Invest., 9(4):19 pp., 5 text figs.

Leptognatha acanthifera n.sp.
Hansen, H. J., 1913
Crustacea Malacostraca (II). IV Tanaidacea.
Danish Ingolf Exped. 3(3):145 pp., 12 pls.

Leptognatha alba n.sp.
Hansen, H. J., 1913
Crustacea Malacostraca (II). IV Tanaidacea.
Danish Ingolf Exped. 3(3):145 pp., 12 pls.

Leptognatha Amdrupii n.sp.
Hansen, H. J., 1913
Crustacea Malacostraca (II). IV Tanaidacea.
Danish Ingolf Exped. 3(3):145 pp., 12 pls.

Leptognatha armata n.sp.
Hansen, H. J., 1913
Crustacea Malacostraca (II). IV Tanaidacea.
Danish Ingolf Exped. 3(3):145 pp., 12 pls.

Leptognatha armata
Kudinov-Pasternak, R.K., 1965.
Deep-Sea Tanaidacea from the Bougainville Trench of the Pacific.
Crustaceana, 8(1):75-91.

Leptognathia armata
Menzies, R.J., and J.L. Mohr, 1962
Benthic Tanaidacea and Isopoda from the Alaskan Arctic and the Polar basin.
Crustaceana, 3(3):192-202.

Leptognatha brachiata n.sp.
Hansen, H. J., 1913
Crustacea Malacostraca (II). IV Tanaidacea.
Danish Ingolf Exped. 3(3):145 pp., 12 pls.

Leptognathia birsteini
Kudinova-Pasternak, R. K., 1966.
Tanaidacea (Crustacea) of the Pacific ultra-abyssal zone (In Russian; English Abstract).
Zool. Zhurn., Akad. Nauk SSSR 44(4): 518-535.

Leptognathia birsteini n.sp.
Kudinova-Pasternak, R.K., 1965.
Deep-sea Tanaidacea from the Bougainville Trench of the Pacific.
Crustaceana, 8(1):75-91.

Leptognathia brevimanu
Dahl, E., 1941
Tanaidacea from the Sound and Skälderviken. Kungl. Fysiografiska Sällskapets i Lund Förhandlingar, 11(7):1-7, 1 text figs.

Leptognatia brevimanu
Greve, Lita, 1965.
The biology of some Tanaidacea from Raunefjorden, western Norway.
Sarsia, 20:43-54.

Leptognathia brevimana
Sars, G.O., 1896-1899
An account of the Crustacea of Norway with short descriptions and figures of all species. Vol.II Isopoda. Pts.I,II. Apseudidae, Tanaidae:1-40, 16 pls. Pts.III,IV. Anthuridae, Gnathiidae, Aegidae, Cirolanidae, Limnoriidae:41-80, pls.17-32, Pts.V,VI. Idotheidae, Ancthuridae, Asellidae, Ianiridae, Munnidae, 81-116, Pls.33-48. Pts.VII,VIII, Desmosomidae, Munnopsidae (part):117-144, pls. 49-64. Pts.IX,X. Munnopsidae (concluded), Ligiidae, Trichoniscidae, Oniscidae (part): 145-184, Pls.65-88. Pts.X,XII. Oniscidae (concluded), Bopyridae, Dajidae:185-232, pls. (over)

Leptognatha brevimana
Sars, G.O., 1886
Nye Bidrag til Kundskaben om Middelhavets Invertebrat fauna. III. Middelhavets Saxisopoder (Isopoda chelifera). Archiv for Mathem. og Natur. 263-368, 15 pls.

Leptognathia breviremis
Greve, Lita, 1965.
The biology of some Tanaidacea from Raunefjorden, western Norway.
Sarsia, 20:43-54.

Leptognatha breviremis
Hansen, H. J., 1913
Crustacea Malacostraca (II). IV Tanaidacea.
Danish Ingolf Exped. 3(3):145 pp., 12 pls.

Leptognathia breviremis
Sars, G.O., 1896-1899
An account of the Crustacea of Norway with short descriptions and figures of all species. Vol.II Isopoda. Pts.I,II. Apseudidae, Tanaidae:1-40, 16 pls. Pts.III,IV. Anthuridae, Gnathiidae, Aegidae, Cirolanidae, Limnoriidae;41-80, pls.17-32, Pts.V,VI. Idotheidae, Ancthuridae, Asellidae, Ianiridae, Munnidae, 81-116, Pls.33-48. Pts.VII,V,Desmosomidae, Munnopsidae (part):117-144, pls. 49-64. Pts.IX,X. Munnopsidae (concluded), Ligiidae, Trichoniscidae, Oniscidae (part): 145-184, Pls.65-88. Pts.X,XII. Oniscidae (concluded), Bopyridae, Dajidae:185-232,pls.
(over)

Leptognathia breviremis
Tattersall, W.M., 1905
The Marine Fauna of the Coast of Ireland. Part V Isopoda. Dept. Agric. and Tech. Instr. for Ireland, Fisheries Branch, Sci. Invest. 1904. No.II:1-90, pls.1-11, 1 chart.

Leptognatha caudata n.sp.
Kudinova-Pasternak, R.K., 1965.
Deep-sea Tanaidacea from the Bougainville Trench of the Pacific.
Crustaceana, 8(1):75-91.

Leptognatha crassa n.sp.
Hansen, H. J., 1913
Crustacea Malacostraca (II). IV Tanaidacea.
Danish Ingolf Exped. 3(3):145 pp., 12 pls.

Leptognatha dentifera
Greve, Lita, 1965.
The biology of some Tanaidacea from Raunefjorden, western Norway.
Sarsia, 20:43-54.

Leptognatha dentifera
Greve, Lita, 1964.
The records of Leptognatha dentifera G.O. Sars (Tanaidacea).
Sarsia, 15:71.

Leptognathia dentifera n. sp.
Sars, G.O., 1896-1899
An account of the Crustacea of Norway with short descriptions and figures of all species. Vol.II Isopoda. Pts.I,II. Apseudidae, Tanaidae:1-40, 16 pls. Pts.III,IV. Anthuridae, Gnathiidae, Aegidae, Cirolanidae, Limnoriidae;41-80, pls.17-32, Pts.V,VI. Idotheidae, Ancthuridae, Asellidae, Ianiridae, Munnidae, 81-116, Pls.33-48. Pts.VII,V,Desmosomidae, Munnopsidae (part):117-144, pls. 49-64. Pts.IX,X. Munnopsidae (concluded), Ligiidae, Trichoniscidae, Oniscidae (part): 145-184, Pls.65-88. Pts.X,XII. Oniscidae (concluded), Bopyridae, Dajidae:185-232,pls.
(over)

Leptognathia dissimilis n.sp.
Lang, Karl 1972.
Bathytanais bathybrotes (Beddard) und Leptognathia dissimilis n.sp. (Tanaidacea).
Crustaceana Suppl. 3: 221-236

Leptognathia elegans n.sp.
Kudinova-Pasternak, R.K., 1965.
Deep-sea Tanaidacea from the Bougainville Trench of the Pacific.
Crustaceana, 8(1):75-91.

Leptognathia elongata n.sp.
Shiino, Sueo M. 1970.
Paratanaidae collected in Chile Bay, Greenwich Island by the XXII Chilean Antarctic Expedition, with an Apseudes from Porvenir Point, Tierra del Fuego Island.
Inst. Antart. Chileno, Ser. Cient. 1(2):79-122.

Leptognathia filiformis
Dahl, E., 1941
Tanaidacea from the Sound and Skälderviken. Kungl. Fysiografiska Sällskapets i Lund Förhandlingar, 11(7):1-7, 1 text figs.

Leptognathia filiformis
Sars, G.O., 1896-1899
An account of the Crustacea of Norway with short descriptions and figures of all species. Vol.II Isopoda. Pts.I,II. Apseudidae, Tanaidae:1-40, 16 pls. Pts.III,IV. Anthuridae, Gnathiidae, Aegidae, Cirolanidae, Limnoriidae;41-80, pls.17-32, Pts.V,VI. Idotheidae, Ancthuridae, Asellidae, Ianiridae, Munnidae, 81-116, Pls.33-48. Pts.VII,V,Desmosomidae, Munnopsidae (part):117-144, pls. 49-64. Pts.IX,X. Munnopsidae (concluded), Ligiidae, Trichoniscidae, Oniscidae (part): 145-184, Pls.65-88. Pts.X,XII. Oniscidae (concluded), Bopyridae, Dajidae:185-232,pls.
(over)

Leptognathia filiformis
Greve, Lita, 1965.
The biology of some Tanaidacea from Raunefjorden, western Norway.
Sarsia, 20:43-54.

Leptognathia forcifera n.sp.
Lang, Karl, 1968.
Deep-Sea Tanaidacea.
Galathea Rept. 9:23-209.

Leptognathia gallardoi n.sp.
Shiino, Sueo M. 1970.
Paratanaidae collected in Chile Bay, Greenwich Island by the XXII Chilean Antarctic Expedition, with an Apseudes from Porvenir Point, Tierra del Fuego Island.
Inst. Antart. Chileno, Ser. Cient. 1(2):79-122.

Leptognatha glacialis n.sp.
Hansen, H. J., 1913
Crustacea Malacostraca (II). IV Tanaidacea.
Danish Ingolf Exped. 3(3):145 pp., 12 pls.

Leptognathia gracilis
Greve, Lita, 1965.
The biology of some Tanaidacea from Raunefjorden, western Norway.
Sarsia, 20:43-54.

Leptognathia gracilis
Hansen, H. J., 1913
Crustacea Malacostraca (II). IV Tanaidacea.
Danish Ingolf Exped. 3(3):145 pp., 12 pls.

Leptognathia gracilis
Shiino, Sueo M. 1970.
Paratanaidae collected in Chile Bay, Greenwich Island by the XXII Chilean Antarctic Expedition, with an Apseudes from Porvenir Point, Tierra del Fuego Island.
Inst. Antart. Chileno, Ser. Cient. 1(2):79-122.

Leptognathia gracilis
Stephensen, K., 1937
Marine Isopoda and Tanaidacea. Zool. Iceland. 3(27):26 pp.

Leptognathia graciloides
Dahl, E., 1941
Tanaidacea from the Sound and Skälderviken. Kungl. Fysiografiska Sällskapets i Lund Förhandlingar, 11(7):1-7, 1 text figs.

Leptognatha hanseni
Hansen, H. J., 1913
Crustacea Malacostraca (II). IV Tanaidacea.
Danish Ingolf Exped. 3(3):145 pp., 12 pls.

Leptognatha hastata n.sp.
Hansen, H. J., 1913
Crustacea Malacostraca (II). IV Tanaidacea.
Danish Ingolf Exped. 3(3):145 pp., 12 pls.

Leptognatha hastata
Menzies, R.J., 1957.
The Tanaidacean Leptognatha hastata from abyssal depths in the Atlantic. Ann. Mag. Nat. Hist., (12) 10(109):68-69.

Leptognatha inermis
Hansen, H. J., 1913
Crustacea Malacostraca (II). IV Tanaidacea.
Danish Ingolf Exped. 3(3):145 pp., 12 pls.

Leptognathia laticaudata
Sars, G.O., 1896-1899
An account of the Crustacea of Norway with short descriptions and figures of all species. Vol.II Isopoda. Pts.I,II. Apseudidae, Tanaidae:1-40, 16 pls. Pts.III,IV. Anthuridae, Gnathiidae, Aegidae, Cirolanidae, Limnoriidae;41-80, pls.17-32, Pts.V,VI. Idotheidae, Ancthuridae, Asellidae, Ianiridae, Munnidae, 81-116, Pls.33-48. Pts.VII,V,Desmosomidae, Munnopsidae (part):117-144, pls. 49-64. Pts.IX,X. Munnopsidae (concluded), Ligiidae, Trichoniscidae, Oniscidae (part): 145-184, Pls.65-88. Pts.X,XII. Oniscidae (concluded), Bopyridae, Dajidae:185-232,pls.
(over)

Leptognatha laticaudata
Sars, G. O., 1886
Nye Bidrag til Kundskaben om Middelhavets Invertebrat fauna. III. Middelhavets Saxisopoder (Isopoda chelifera). Archiv for Mathem. og Natur. 263-368, 15 pls.

Leptognathia latiremis n.sp.
Hansen, H.J., 1913
Crustacea Malacostraca II. IV Tanaidacea Ingolf Exped. III(3): 152 pp., 12 pls., list of stations.

Leptognatha latiremis n.sp.
Hansen, H. J., 1913
Crustacea Malacostraca (II). IV Tanaidacea.
Danish Ingolf Exped. 3(3):145 pp., 12 pls.

Leptognathia longiremis
Greve, Lita, 1965.
The biology of some Tanaidacea from Raunefjorden, western Norway.
Sarsia, 20:34-45.

Leptognathia longiremis
Hansen, H. J., 1913
Crustacea Malacostraca (II). IV Tanaidacea.
Danish Ingolf Exped. 3(3):145 pp., 12 pls.

Leptognathia longiremis
Sars, G.O., 1896-1899
An account of the Crustacea of Norway with short descriptions and figures of all species. Vol.II Isopoda. Pts.I,II. Apseudidae, Tanaidae:1-40, 16 pls. Pts.III,IV. Anthuridae, Gnathiidae, Aegidae, Cirolanidae, Limnoriidae;41-80, pls.17-32, Pts.V,VI. Idotheidae, Ancthuridae, Asellidae, Ianiridae, Munnidae, 81-116, Pls.33-48. Pts.VII,V,Desmosomidae, Munnopsidae (part):117-144, pls. 49-64. Pts.IX,X. Munnopsidae (concluded), Ligiidae, Trichoniscidae, Oniscidae (part): 145-184, Pls.65-88. Pts.X,XII. Oniscidae (concluded), Bopyridae, Dajidae:185-232,pls.
(over)

Leptognathia longiremis
Sars, G.O., 1886
Crustacea II. Zool. Norwegian North-Atlantic Exped., 1876-1878: 96 pp., 1 map.

Leptognathia longiremis
Sars, G.O. 1885
Crustacea 1. Zool., Norwegian North-Atlantic Exped. 1876-1878: 280 pp., 21 pls. 1 map.

Leptognathia longiremis
Tattersall, W.M., 1905
The Marine Fauna of the Coast of Ireland. Part V Isopoda. Dept. Agric. and Tech. Instr. for Ireland, Fisheries Branch, Sci. Invest. 1904. No.II:1-90, pls.1-11, 1 chart.

Leptognatha manca n.sp.
Hansen, H. J., 1913
Crustacea Malacostraca (II). IV Tanaidacea.
Danish Ingolf Exped. 3(3):145 pp., 12 pls.

Leptognathia manca
Sars, G.O., 1896-1899
An account of the Crustacea of Norway with short descriptions and figures of all species. Vol.II Isopoda. Pts.I,II. Apseudidae, Tanaidae:1-40, 16 pls. Pts.III,IV. Anthuridae, Gnathiidae, Aegidae, Cirolanidae, Limnoriidae;41-80, pls.17-32, Pts.V,VI. Idotheidae, Ancthuridae, Asellidae, Ianiridae, Munnidae, 81-116, Pls.33-48. Pts.VII,V,Desmosomidae, Munnopsidae (part):117-144, pls. 49-64. Pts.IX,X. Munnopsidae (concluded), Ligiidae, Trichoniscidae, Oniscidae (part): 145-184, Pls.65-88. Pts.X,XII. Oniscidae (concluded), Bopyridae, Dajidae:185-232,pls.
(over)

Leptognathia manca n.sp.
Hansen, H.J., 1913
Crustacea Malacostraca II. IV Tanaidacea Ingolf Exped. III(3): 152 pp., 12 pls., list of stations.

Leptognathia manca
Stephensen, K., 1937
Marine Isopoda and Tanaidacea. Zool. Iceland. 3(27):26 pp.

Leptognatha multiserrata n.sp.
Hansen, H. J., 1913
Crastacea Malacostraca (II). IV Tanaidacea.
Danish Ingolf Exped. 3(3):145 pp., 12 pls.

Leptognathia parabrevimanu n.sp.
Lang, Karl, 1968.
Deep-Sea Tanaidacea.
Galathea Rept., 9:23-209.

Leptognathia paraforcifera n.sp.
Lang, Karl, 1968.
Deep-Sea Tanaidacea.
Galathea Rept., 9:23-209.

Leptognatha polita n.sp.
Hansen, H. J., 1913
Crastacea Malacostraca (II). IV Tanaidacea.
Danish Ingolf Exped. 3(3):145 pp., 12 pls.

Leptognatha profunda n.sp.
Hansen, H. J., 1913
Crastacea Malacostraca (II). IV Tanaidacea.
Danish Ingolf Exped. 3(3):145 pp., 12 pls.

Leptognatha sarsii
Hansen, H. J., 1913
Crastacea Malacostraca (II). IV Tanaidacea.
Danish Ingolf Exped. 3(3):145 pp., 12 pls.

Leptognathia sarsi
Stephenson, K., 1937
Marine Isopoda and Tanaidacea. Zool. Iceland. 3(27):26 pp.

Leptognatha subaequalis n.sp.
Hansen, H. J., 1913
Crastacea Malacostraca (II). IV Tanaidacea.
Danish Ingolf Exped. 3(3):145 pp., 12 pls.

Leptognatha tenella n.sp.
Hansen, H. J., 1913
Crastacea Malacostraca (II). IV Tanaidacea.
Danish Ingolf Exped. 3(3):145 pp., 12 pls.

Leptognatha tuberculata n.sp.
Hansen, H. J., 1913
Crastacea Malacostraca (II). IV Tanaidacea.
Danish Ingolf Exped. 3(3):145 pp., 12 pls.

Leptognatha uncinata n.sp.
Hansen, H. J., 1913
Crastacea Malacostraca (II). IV Tanaidacea.
Danish Ingolf Exped. 3(3):145 pp., 12 pls.

Leptognatha ventralis n.sp.
Hansen, H. J., 1913
Crastacea Malacostraca (II). IV Tanaidacea.
Danish Ingolf Exped. 3(3):145 pp., 12 pls.

Leptognatha vincina n.sp.
Hansen, H. J., 1913
Crastacea Malacostraca (II). IV Tanaidacea.
Danish Ingolf Exped. 3(3):145 pp., 12 pls.

Leptognathiella n. gen.
Hansen, H. J., 1913
Crastacea Malacostraca (II). IV Tanaidacea.
Danish Ingolf Exped. 3(3):145 pp., 12 pls.

Leptognathiella abyssi n.sp.
Hansen, H. J., 1913
Crastacea Malacostraca (II). IV Tanaidacea.
Danish Ingolf Exped. 3(3):145 pp., 12 pls.

Libanius nonacanthus
Lang, Karl, 1970.
Taxonomische und phylogenetische Untersuchungen über die Tanaidaceen. 6. Revision der Gattung Paranarthrura Hansen, 1913, und Aufstellung von zwei neuen Familien, vier neuen Gattungen und zwei neuen Arten. Ark. Zool. (2) 23(6): 361-401. 2 plates.

Libanius pulcher n.gen., n.sp.
Lang, Karl, 1970.
Taxonomische und phylogenetische Untersuchungen über die Tanaidaceen. 6. Revision der Gattung Paranarthrura Hansen, 1913, und Aufstellung von zwei neuen Familien, vier neuen Gattungen und zwei neuen Arten. Ark. Zool. (2) 23(6): 361-401. 2 plates.

Macrinella clavipes n. gen.
Lang, Karl, 1970.
Taxonomische und phylogenetische Untersuchungen über die Tanaidaceen. 6. Revision der Gattung Paranarthrura Hansen, 1913, und Aufstellung von zwei neuen Familien, vier neuen Gattungen und zwei neuen Arten. Ark. Zool. (2) 23(6): 361-401. 2 plates.

Metapseudea albidus n.sp.
Shiino, S.M., 1951.
On two new species of the Family Apseudidae found at Seto. Rept. Faculty Fish., Pref. Univ. Mie, 1(1):12-25, 6 textfigs.

Seto Mar. Biol. Lab. Contrib. 170.

Metapseudes auklandiae
Gardiner, Lion F. 1973.
New species of the genera Synapseudes and Cyclopoapseudes with notes on morphological variation, postmarsupial development, and phylogenetic relationships within the family Metapseudidae (Crustacea: Tanaidacea).
Zool. J. Linn. Soc. 53(1):25-58

Metatanais cylindricus n.gen., n.sp.
Shiino, S.M., 1952.
A new genus and two new species of the order Tanaidacea found at Seto. Publ. Seto Mar. Biol. Lab., Kyoto Univ., 2(2): 53-68, 7 textfigs.

Neotanais sp.
Kudinova-Pasternak, R. K., 1966.
Tanaidacea (Crustacea) of the Pacific ultra-abyssal zone (In Russian; English Abstract). Zool. Zhurn., Akad. Nauk SSSR 44(4): 518-535.

Neotanais deflexirostris n. sp.
Lang, Karl, 1968.
Deep-Sea Tanaidacea.
Galathea Rept., 9:23-209.

Neotanais giganteus n.sp.
Hansen, H. J., 1913
Crastacea Malacostraca (II). IV Tanaidacea.
Danish Ingolf Exped. 3(3):145 pp., 12 pls.

Neotanais giganteus
Wolff, T., 1956.
Crustacea Tanaidacea from depths exceeding 6000 meters. Galathea Repts., 2:187-241.

Neotanais hastiger
Hansen, H. J., 1913
Crastacea Malacostraca (II). IV Tanaidacea.
Danish Ingolf Exped. 3(3):145 pp., 12 pls.

Neotanais peculiaris n.sp.
Lang, Karl, 1968.
Deep-Sea Tanaidacea.
Galathea Rept., 9:23-209.

Neotanais pfaffioides n.sp.
Lang, Karl, 1968.
Deep-Sea Tanaidacea.
Galathea Rept., 9:23-209.

Neotanais serratospinosus
Hansen, H. J., 1913
Crastacea Malacostraca (II). IV Tanaidacea.
Danish Ingolf Exped. 3(3):145 pp., 12 pls.

Neotanais serratispinosus
Kudinova-Pasternak, R.K., 1965.
Deep-Sea Tanadacea from the Bougainville Trench of the Pacific.
Crustaceana, 8(1):75-91.

Neotanais serratispinosa hadalis n.subsp.
Wolff, T., 1956.
Crustacea Tanaidacea from depths exceeding 6000 meters. Galathea Rept., 2:187-241.

Neotanais wolffi n. sp.
Kudinova-Pasternak, R. K., 1966.
Tanaidacea (Crustacea) of the Pacific ultra-abyssal zone (In Russian; English Abstract).
Zool. Zhurn., Akad. Nauk SSSR 44(4): 518-535.

Nesotanais lacustris
Lang, Karl, 1970.
Taxonomische und phylogenetische Untersuchungen über die Tanaidaceen. 6. Revision der Gattung Paranarthrura Hansen, 1913, und Aufstellung von zwei neuen Familien, vier neuen Gattungen und zwei neuen Arten. Ark. Zool. (2) 23(6): 361-401. 2 plates.

Nototanais antarcticus
Shiino, Sueo M. 1970.
Paratanaidae collected in Chile Bay, Greenwich Island by the XXII Chilean Antarctic Expedition, with an Apseudes from Porvenir Point, Tierra del Fuego Island. Inst. Antart. Chileno, Ser. Cient. 1(2): 79-122.

Nototanais antarctica
Stephensen, K., 1947
Tanaidacea, Isopoda, Amphipoda, and Pycnogonida. Sci. Res. Norweg. Antarctic Exp. 1927-1928 et seq. Date Norske Videnskaps-Akademi i Oslo, II, No.27:1-90, 26 text figs.

Nototanais dimorphus
Shiino, Sueo M. 1970.
Paratanaidae collected in Chile Bay, Greenwich Island by the XXII Chilean Antarctic Expedition, with an Apseudes from Porvenir Point, Tierra del Fuego Island. Inst. Antart. Chileno, Ser. Cient. 1(2): 79-122.

Nototanais dimorphus
Tattersall, W.M., 1921
Crustacea. Part VI. Tanaidacea and Isopoda. Brit. Antarctic "Terra Nova") Exped., 1910, Nat. Hist. Rept., Zool. 3(8):191-258, 2 text figs., 11 pls.

Neotanais serratospinosus
Kudinova-Pasternak, R.K., 1965.
On the proterogyny in Tanaidacea (Crustacea).
Zool. Zhurn., Akad. Nauk, SSSR, 44(3):458-459.

Pagurapseudopsis gracilipes n. gen., n. sp.
Shiino, Sueo M., 1963.
Tanaidacea collected by Naga Expedition in the Bay of Nha-Trang, South Viet-Nam.
Rept., Fac. Fish., Pref. Univ. of Mie, 4(3):437-507.

Pancolus
Lang, K., 1951.
The genus Pancolus Richardson and some remarks on Paratanais euelpis Barnard (Tanaidacea). Ark. f. Zool., And. Ser., 1(4/5):357-360, 1 pl., text-figs.

Paragathotanais typicus n. gen. n. sp.
Lang, Karl 1971.
Die Gattungen Agathotanais Hansen und Paragathotanais n.gen. (Tanaidacea). Crustaceana 21(1):57-71.

Paranarthrura n. gen.
Hansen, H. J., 1913
Crastacea Malacostraca (II). IV Tanaidacea. Danish Ingolf Exped. 3(3):145 pp., 12 pls.

Paranarthrura clavipes n. sp.
Hansen, H. J., 1913
Crastacea Malacostraca (II). IV Tanaidacea. Danish Ingolf Exped. 3(3):145 pp., 12 pls.

Paranarthrura insignis n.sp.
Hansen, H. J., 1913
Crastacea Malacostraca (II). IV Tanaidacea. Danish Ingolf Exped. 3(3):145 pp., 12 pls.

Paranarthrura insignis
Lang, Karl, 1970.
Taxonomische und phylogenetische Untersuchungen über die Tanaidaceen. 6. Revision der Gattung Paranarthrura Hansen, 1913, und Aufstellung von zwei neuen Familien, vier neuen Gattungen und zwei neuen Arten. Ark. Zool. (2) 23(6): 361-401. 2 plates.

Paranarthrura subtilis n.sp.
Hansen, H. J., 1913
Crastacea Malacostraca (II). IV Tanaidacea. Danish Ingolf Exped. 3(3):145 pp., 12 pls.

Paranarthrura similis n.sp.
Lang, Karl, 1970.
Taxonomische und phylogenetische Untersuchungen über die Tanaidaceen. 6. Revision der Gattung Paranarthrura Hansen, 1913, und Aufstellung von zwei neuen Familien, vier neuen Gattungen und zwei neuen Arten. Ark. Zool. (2) 23(6): 361-401. 2 plates.

Paranarthrura subtilis
Lang, Karl, 1970.
Taxonomische und phylogenetische Untersuchungen über die Tanaidaceen. 6. Revision der Gattung Paranarthrura Hansen, 1913, und Aufstellung von zwei neuen Familien, vier neuen Gattungen und zwei neuen Arten. Ark. Zool. (2) 23(6): 361-401. 2 plates.

Paranarthrura undulata n.sp.
Lang, Karl, 1968.
Deep-Sea Tanaidacea. Galathea Rept., 9:23-209.

Parapseudes sp.
Sars, G.O., 1896-1899
An account of the Crustacea of Norway with short descriptions and figures of all species. Vol.II Isopoda. Pts.I,II. Apseudidae, Tanaidae:1-40, 16 pls. Pts.III,IV. Anthuridae, Gnathiidae, Aegidae, Cirolanidae, Limnoriidae:41-80, pls.17-32, Pts.V,VI. Idotheidae, Ancthuridae, Asellidae, Ianiridae, Munnidae, 81-116, Pls.33-48. Pts.VII,V³Desmosomidae, Munnopsidae (part):117-144, pls. 49-64. Pts.IX,X. Munnopsidae (concluded), Ligiidae, Trichoniscidae, Oniscidae (part): 145-184, Pls.65-88. Pts.X,XII. Oniscidae (concluded), Bopyridae, Dajidae:185-232,pls.
(over)

Parapseudes hirsutus n. sp.
Stebbing, T. R. R., 1910
Isopoda from the Indian Ocean and British East Africa. Trans. Linn. Soc., London, Zool., 2nd ser. 14:83-122, Pls. 5-11.

Parapseudes latifrons
Larwood, H.J., 1940
The fishery grounds near Alexandria. XXI Tanaidacea and Isopoda. Fouad I. Inst. Hydrobiol. & Fish., Notes & Mem. No.35:72 pp., 17 figs.

Parapseudes latifrons
Sars, G. O., 1886
Nye Bidrag til Kundskaben om Middelhavets Invertebrat fauna. III. Middelhavets Saxisopoder (Isopoda chelifera). Archiv for Mathem. og Natur. 263-368, 15 pls.

Paratanais atlanticus
Belloc, G., 1960.
Catalogue des types d'Isopodes du Musée Océanographique de Monaco. Bull. Inst. Océanogr., Monaco, 57(1188):1-7.

Paratanais batei
Hansen, H. J., 1913
Crastacea Malacostraca (II). IV Tanaidacea. Danish Ingolf Exped. 3(3):145 pp., 12 pls.

Paratanais batei
Hansen, H.J., 1913
Crustacea Malacostraca II. IV Tanaidacea Ingolf Exped. III(3): 152 pp., 12 pls., list of stations.

Paratanais Batei
Sars, G.O., 1896-1899
An account of the Crustacea of Norway with short descriptions and figures of all species. Vol.II Isopoda. Pts.I,II. Apseudidae, Tanaidae:1-40, 16 pls. Pts.III,IV. Anthuridae, Gnathiidae, Aegidae, Cirolanidae, Limnoriidae:41-80, pls.17-32, Pts.V,VI. Idotheidae, Ancthuridae, Asellidae, Ianiridae, Munnidae, 81-116, Pls.33-48. Pts.VII,V³Desmosomidae, Munnopsidae (part):117-144, pls. 49-64. Pts.IX,X. Munnopsidae (concluded), Ligiidae, Trichoniscidae, Oniscidae (part): 145-184, Pls.65-88. Pts.X,XII. Oniscidae (concluded), Bopyridae, Dajidae:185-232,pls.
(over)

Paratanais Batei
Sars, G. O., 1886
Nye Bidrag til Kundskaben om Middelhavets Invertebrat fauna. III. Middelhavets Saxisopoder (Isopoda chelifera). Archiv for Mathem. og Natur. 263-368, 15 pls.

Paratanais batei
Stephenson, K., 1937
Marine Isopoda and Tanaidacea. Zool. Iceland. 3(27):26 pp.

Paratanais cornutus
Sars, G.O., 1896-1899
An account of the Crustacea of Norway with short descriptions and figures of all species. Vol.II Isopoda. Pts.I,II. Apseudidae, Tanaidae:1-40, 16 pls. Pts.III,IV. Anthuridae, Gnathiidae, Aegidae, Cirolanidae, Limnoriidae:41-80, pls.17-32, Pts.V,VI. Idotheidae, Ancthuridae, Asellidae, Ianiridae, Munnidae, 81-116, Pls.33-48. Pts.VII,V³Desmosomidae, Munnopsidae (part):117-144, pls. 49-64. Pts.IX,X. Munnopsidae (concluded), Ligiidae, Trichoniscidae, Oniscidae (part): 145-184, Pls.65-88. Pts.X,XII. Oniscidae (concluded), Bopyridae, Dajidae:185-232,pls.
(over)

Paratanais elongatus
Sars, G.O., 1896-1899
An account of the Crustacea of Norway with short descriptions and figures of all species. Vol.II Isopoda. Pts.I,II. Apseudidae, Tanaidae:1-40, 16 pls. Pts.III,IV. Anthuridae, Gnathiidae, Aegidae, Cirolanidae, Limnoriidae:41-80, pls.17-32, Pts.V,VI. Idotheidae, Ancthuridae, Asellidae, Ianiridae, Munnidae, 81-116, Pls.33-48. Pts.VII,V³Desmosomidae, Munnopsidae (part):117-144, pls. 49-64. Pts.IX,X. Munnopsidae (concluded), Ligiidae, Trichoniscidae, Oniscidae (part): 145-184, Pls.65-88. Pts.X,XII. Oniscidae (concluded), Bopyridae, Dajidae:185-232,pls.
(over)

Paratanais euelpis
Lang, K., 1951.
The genus Pancolus Richardson and some remarks on Tanais euelpis Barnard (Tanaidacea). Ark. f. Zool. And. Ser., 1(4/5):357-360, 1 pl., text-fig.

Paratanais forcipatus
Sars, G.O., 1896-1899
An account of the Crustacea of Norway with short descriptions and figures of all species. Vol.II Isopoda. Pts.I,II. Apseudidae, Tanaidae:1-40, 16 pls. Pts.III,IV. Anthuridae, Gnathiidae, Aegidae, Cirolanidae, Limnoriidae:41-80, pls.17-32, Pts.V,VI. Idotheidae, Ancthuridae, Asellidae, Ianiridae, Munnidae, 81-116, Pls.33-48. Pts.VII,V³Desmosomidae, Munnopsidae (part):117-144, pls. 49-64. Pts.IX,X. Munnopsidae (concluded), Ligiidae, Trichoniscidae, Oniscidae (part): 145-184, Pls.65-88. Pts.X,XII. Oniscidae (concluded), Bopyridae, Dajidae:185-232,pls.
(over)

Paratanais impressus n.sp.
Kussakin, O.G. and L.A. Tzareva 1972
Tanaidacea from the coastal zones of the middle Kurile Islands. Crustaceana Suppl. 3: 237-245.

Paratanais linearis
Sars, G.O., 1896-1899
An account of the Crustacea of Norway with short descriptions and figures of all species. Vol.II Isopoda. Pts.I,II. Apseudidae, Tanaidae:1-40, 16 pls. Pts.III,IV. Anthuridae, Gnathiidae, Aegidae, Cirolanidae, Limnoriidae:41-80, pls.17-32, Pts.V,VI. Idotheidae, Ancthuridae, Asellidae, Ianiridae, Munnidae, 81-116, Pls.33-48. Pts.VII,V³Desmosomidae, Munnopsidae (part):117-144, pls. 49-64. Pts.IX,X. Munnopsidae (concluded), Ligiidae, Trichoniscidae, Oniscidae (part): 145-184, Pls.65-88. Pts.X,XII. Oniscidae (concluded), Bopyridae, Dajidae:185-232,pls.
(over)

Paratanais rigidus
Sars, G.O., 1896-1899
An account of the Crustacea of Norway with short descriptions and figures of all species. Vol.II Isopoda. Pts.I,II. Apseudidae, Tanaidae:1-40, 16 pls. Pts.III,IV. Anthuridae, Gnathiidae, Aegidae, Cirolanidae, Limnoriidae:41-80, pls.17-32, Pts.V,VI. Idotheidae, Ancthuridae, Asellidae, Ianiridae, Munnidae, 81-116, Pls.33-48. Pts.VII,V³Desmosomidae, Munnopsidae (part):117-144, pls. 49-64. Pts.IX,X. Munnopsidae (concluded), Ligiidae, Trichoniscidae, Oniscidae (part): 145-184, Pls.65-88. Pts.X,XII. Oniscidae (concluded), Bopyridae, Dajidae:185-232,pls.
(over)

Paratanais tenuis
Sars, G.O., 1896-1899
An account of the Crustacea of Norway with short descriptions and figures of all species. Vol.II Isopoda. Pts.I,II. Apseudidae, Tanaidae:1-40, 16 pls. Pts.III,IV. Anthuridae, Gnathiidae, Aegidae, Cirolanidae, Limnoriidae:41-80, pls.17-32, Pts.V,VI. Idotheidae, Ancthuridae, Asellidae, Ianiridae, Munnidae, 81-116, Pls.33-48. Pts.VII,V³Desmosomidae, Munnopsidae (part):117-144, pls. 49-64. Pts.IX,X. Munnopsidae (concluded), Ligiidae, Trichoniscidae, Oniscidae (part): 145-184, Pls.65-88. Pts.X,XII. Oniscidae (concluded), Bopyridae, Dajidae:185-232,pls.
(over)

Psammokalliapseudes mirabilis n.gen., n.sp.
Lang, K., 1956.
Tanaidacea aus Brasilen gesammelt von Professor A. Remane und Dr. S. Gerlach. Kieler Meeresf., 12(2):249-260.

Pseudotanais abyssi n.sp.
Hansen, H. J., 1913
Crastacea Malacostraca (II). IV Tanaidacea. Danish Ingolf Exped. 3(3):145 pp., 12 pls.

Pseudotanais affinis
Hansen, H. J., 1913
Crastacea Malacostraca (II). IV Tanaidacea. Danish Ingolf Exped. 3(3):145 pp., 12 pls.

Pseudotanais affinis
Sars, G.O., 1896-1899
An account of the Crustacea of Norway with short descriptions and figures of all species. Vol.II Isopoda. Pts.I,II. Apseudidae, Tanaidae:1-40, 16 pls. Pts.III,IV. Anthuridae, Gnathiidae, Aegidae, Cirolanidae, Limnoriidae:41-80, pls.17-32, Pts.V,VI. Idotheidae, Ancthuridae, Asellidae, Ianiridae, Munnidae, 81-116, Pls.33-48. Pts.VII,V³Desmosomidae, Munnopsidae (part):117-144, pls. 49-64. Pts.IX,X. Munnopsidae (concluded), Ligiidae, Trichoniscidae, Oniscidae (part): 145-184, Pls.65-88. Pts.X,XII. Oniscidae (concluded), Bopyridae, Dajidae:185-232,pls.
(over)

Pseudotanais crassicornis

Sars, G.O., 1896-1899
An account of the Crustacea of Norway with short descriptions and figures of all species. Vol.II Isopoda. Pts.I,II. Apseudidae, Tanaidae:1-40, 16 pls. Pts.III,IV. Anthuridae, Gnathiidae, Aegidae, Cirolanidae, Limnoriidae:41-80, pls.17-32, Pts.V,VI. Idotheidae, Ancthuridae, Asellidae, Ianiridae, Munnidae, 81-116, Pls.33-48. Pts.VII,V,Desmosomidae, Munnopsidae (part):117-144, pls. 49-64. Pts.IX,X. Munnopsidae (concluded), Ligiidae, Trichoniscidae, Oniscidae (part): 145-184, Pls.65-88. Pts.X,XII. Oniscidae (concluded), Bopyridae, Dajidae:185-232, pls.
(over)

Pseudotanais forcipatus

Dahl, E., 1941
Tanaidacea from the Sound and Skälderviken. Kungl. Fysiografiska Sällskapets i Lund Förhandlingar, 11(7):1-7, 1 text figs.

Pseudotanais forcipatus

Hansen, H. J., 1913
Crastacea Malacostraca (II). IV Tanaidacea. Danish Ingolf Exped. 3(3):145 pp., 12 pls.

Pseudotanais forcipatus

Sars, G.O., 1896-1899
An account of the Crustacea of Norway with short descriptions and figures of all species. Vol.II Isopoda. Pts.I,II. Apseudidae, Tanaidae:1-40, 16 pls. Pts.III,IV. Anthuridae, Gnathiidae, Aegidae, Cirolanidae, Limnoriidae:41-80, pls.17-32, Pts.V,VI. Idotheidae, Ancthuridae, Asellidae, Ianiridae, Munnidae, 81-116, Pls.33-48. Pts.VII,V,Desmosomidae, Munnopsidae (part):117-144, pls. 49-64. Pts.IX,X. Munnopsidae (concluded), Ligiidae, Trichoniscidae, Oniscidae (part): 145-184, Pls.65-88. Pts.X,XII. Oniscidae (concluded), Bopyridae, Dajidae:185-232, pls.
(over)

Pseudotanais forcipatus

Stephenson, K., 1937
Marine Isopoda and Tanaidacea. Zool. Iceland. 3(27):26 pp.

Pseudotanais lilljeborgii

Hansen, H. J., 1913
Crastacea Malacostraca (II). IV Tanaidacea. Danish Ingolf Exped. 3(3):145 pp., 12 pls.

Pseudotanais lilljeborgii

Sars, G.O., 1896-1899
An account of the Crustacea of Norway with short descriptions and figures of all species. Vol.II Isopoda. Pts.I,II. Apseudidae, Tanaidae:1-40, 16 pls. Pts.III,IV. Anthuridae, Gnathiidae, Aegidae, Cirolanidae, Limnoriidae:41-80, pls.17-32, Pts.V,VI. Idotheidae, Ancthuridae, Asellidae, Ianiridae, Munnidae, 81-116, Pls.33-48. Pts.VII,V,Desmosomidae, Munnopsidae (part):117-144, pls. 49-64. Pts.IX,X. Munnopsidae (concluded), Ligiidae, Trichoniscidae, Oniscidae (part): 145-184, Pls.65-88. Pts.X,XII. Oniscidae (concluded), Bopyridae, Dajidae:185-232, pls.
(over)

Pseudotanais lilljeborgi

Stephenson, K., 1937
Marine Isopoda and Tanaidacea. Zool. Iceland. 3(27):26 pp.

Pseudotanais longipes n.sp.

Hansen, H. J., 1913
Crastacea Malacostraca (II). IV Tanaidacea. Danish Ingolf Exped. 3(3):145 pp., 12 pls.

Pseudotanais macrocheles

Greve, Lita, 1965.
The biology of some Tanaidacea from Raunefjorder western Norway. Sarsia, 20:43-54.

Pseudotanais macrocheles

Sars, G.O., 1896-1899
An account of the Crustacea of Norway with short descriptions and figures of all species. Vol.II Isopoda. Pts.I,II. Apseudidae, Tanaidae:1-40, 16 pls. Pts.III,IV. Anthuridae, Gnathiidae, Aegidae, Cirolanidae, Limnoriidae:41-80, pls.17-32, Pts.V,VI. Idotheidae, Ancthuridae, Asellidae, Ianiridae, Munnidae, 81-116, Pls.33-48. Pts.VII,V,Desmosomidae, Munnopsidae (part):117-144, pls. 49-64. Pts.IX,X. Munnopsidae (concluded), Ligiidae, Trichoniscidae, Oniscidae (part): 145-184, Pls.65-88. Pts.X,XII. Oniscidae (concluded), Bopyridae, Dajidae:185-232, pls.
(over)

Pseudotanais mediterraneus

Sars, G.O., 1896-1899
An account of the Crustacea of Norway with short descriptions and figures of all species. Vol.II Isopoda. Pts.I,II. Apseudidae, Tanaidae:1-40, 16 pls. Pts.III,IV. Anthuridae, Gnathiidae, Aegidae, Cirolanidae, Limnoriidae:41-80, pls.17-32, Pts.V,VI. Idotheidae, Ancthuridae, Asellidae, Ianiridae, Munnidae, 81-116, Pls.33-48. Pts.VII,V,Desmosomidae, Munnopsidae (part):117-144, pls. 49-64. Pts.IX,X. Munnopsidae (concluded), Ligiidae, Trichoniscidae, Oniscidae (part): 145-184, Pls.65-88. Pts.X,XII. Oniscidae (concluded), Bopyridae, Dajidae:185-232, pls.
(over)

Pseudotanais mediterraneus

Sars, G. O., 1886
Nye Bidrag til Kundskaben om Middelhavets Invertebrat fauna. III. Middelhavets Saxisopoder (Isopoda chelifera). Archiv for Mathem. og Natur. 263-368, 15 pls.

Pseudotanais oculatus n.sp

Hansen, H. J., 1913
Crastacea Malacostraca (II). IV Tanaidacea. Danish Ingolf Exped. 3(3):145 pp., 12 pls.

Pseudotanais vitjazi n. sp.

Kudinova-Pasternak, R. K., 1966.
Tanaidacea (Crustacea) of the Pacific ultra-abyssal zone (In Russian; English Abstract). Zool. Zhurn., Akad. Nauk SSSR 44(4): 518-535.

Pseudotanais Willemoesi

Sars, G.O., 1896-1899
An account of the Crustacea of Norway with short descriptions and figures of all species. Vol.II Isopoda. Pts.I,II. Apseudidae, Tanaidae:1-40, 16 pls. Pts.III,IV. Anthuridae, Gnathiidae, Aegidae, Cirolanidae, Limnoriidae:41-80, pls.17-32, Pts.V,VI. Idotheidae, Ancthuridae, Asellidae, Ianiridae, Munnidae, 81-116, Pls.33-48. Pts.VII,V,Desmosomidae, Munnopsidae (part):117-144, pls. 49-64. Pts.IX,X. Munnopsidae (concluded), Ligiidae, Trichoniscidae, Oniscidae (part): 145-184, Pls.65-88. Pts.X,XII. Oniscidae (concluded), Bopyridae, Dajidae:185-232, pls.
(over)

Siphonolabrum mirabile n. gen., a n. sp.

Lang Karl, 1972.
Siphonolabrum mirabile n.gen., n.sp. (Tanaidacea).
Crustaceana Suppl. 3: 214-220.

Sphyrapus anomalus

Hansen, H. J., 1913
Crastacea Malacostraca (II). IV Tanaidacea. Danish Ingolf Exped. 3(3):145 pp., 12 pls.

Sphyrapus anomalus

Menzies, R.J., and J.L. Mohr, 1962
Benthic Tanaidacea and Isopoda from the Alaskan Arctic and the Polar basin. Crustaceana, 3(3):192-202.

Sphyrapus anomalus

Sars, G.O., 1896-1899
An account of the Crustacea of Norway with short descriptions and figures of all species. Vol.II Isopoda. Pts.I,II. Apseudidae, Tanaidae:1-40, 16 pls. Pts.III,IV. Anthuridae, Gnathiidae, Aegidae, Cirolanidae, Limnoriidae:41-80, pls.17-32, Pts.V,VI. Idotheidae, Ancthuridae, Asellidae, Ianiridae, Munnidae, 81-116, Pls.33-48. Pts.VII,V,Desmosomidae, Munnopsidae (part):117-144, pls. 49-64. Pts.IX,X. Munnopsidae (concluded), Ligiidae, Trichoniscidae, Oniscidae (part): 145-184, Pls.65-88. Pts.X,XII. Oniscidae (concluded), Bopyridae, Dajidae:185-232, pls.
(over)

Sphyrapus anomalus

Sars, G.O., 1886
Crustacea II. Zool. Norwegian North-Atlantic Exped., 1876-1878: 96 pp., 1 map.

Sphyrapus anomalus

Stephenson, K., 1937
Marine Isopoda and Tanaidacea. Zool. Iceland. 3(27):26 pp.

Sphyrapus dispar n.sp.

Lang, Karl, 1968.
Deep-Sea Tanaidacea. Galathea Rept., 9:23-209.

Sphyrapus malleolus

Sars, G.O., 1896-1899
An account of the Crustacea of Norway with short descriptions and figures of all species. Vol.II Isopoda. Pts.I,II. Apseudidae, Tanaidae:1-40, 16 pls. Pts.III,IV. Anthuridae, Gnathiidae, Aegidae, Cirolanidae, Limnoriidae:41-80, pls.17-32, Pts.V,VI. Idotheidae, Ancthuridae, Asellidae, Ianiridae, Munnidae, 81-116, Pls.33-48. Pts.VII,V,Desmosomidae, Munnopsidae (part):117-144, pls. 49-64. Pts.IX,X. Munnopsidae (concluded), Ligiidae, Trichoniscidae, Oniscidae (part): 145-184, Pls.65-88. Pts.X,XII. Oniscidae (concluded), Bopyridae, Dajidae:185-232, pls.
(over)

Sphyrapus serratus

Hansen, H. J., 1913
Crastacea Malacostraca (II). IV Tanaidacea. Danish Ingolf Exped. 3(3):145 pp., 12 pls.

Sphyrapus serratus

Sars, G.O., 1896-1899
An account of the Crustacea of Norway with short descriptions and figures of all species. Vol.II Isopoda. Pts.I,II. Apseudidae, Tanaidae:1-40, 16 pls. Pts.III,IV. Anthuridae, Gnathiidae, Aegidae, Cirolanidae, Limnoriidae:41-80, pls.17-32, Pts.V,VI. Idotheidae, Ancthuridae, Asellidae, Ianiridae, Munnidae, 81-116, Pls.33-48. Pts.VII,V,Desmosomidae, Munnopsidae (part):117-144, pls. 49-64. Pts.IX,X. Munnopsidae (concluded), Ligiidae, Trichoniscidae, Oniscidae (part): 145-184, Pls.65-88. Pts.X,XII. Oniscidae (concluded), Bopyridae, Dajidae:185-232, pls.
(over)

Sphyrapus serratus n.sp.

Sars, G.O., 1886
Crustacea II. Zool. Norwegian North-Atlantic Exped., 1876-1878: 96 pp., 1 map.

Sphyrapus tudes

Hansen, H. J., 1913
Crastacea Malacostraca (II). IV Tanaidacea. Danish Ingolf Exped. 3(3):145 pp., 12 pls.

Sphyrapus tudes

Sars, G.O., 1896-1899
An account of the Crustacea of Norway with short descriptions and figures of all species. Vol.II Isopoda. Pts.I,II. Apseudidae, Tanaidae:1-40, 16 pls. Pts.III,IV. Anthuridae, Gnathiidae, Aegidae, Cirolanidae, Limnoriidae:41-80, pls.17-32, Pts.V,VI. Idotheidae, Ancthuridae, Asellidae, Ianiridae, Munnidae, 81-116, Pls.33-48. Pts.VII,V,Desmosomidae, Munnopsidae (part):117-144, pls. 49-64. Pts.IX,X. Munnopsidae (concluded), Ligiidae, Trichoniscidae, Oniscidae (part): 145-184, Pls.65-88. Pts.X,XII. Oniscidae (concluded), Bopyridae, Dajidae:185-232, pls.
(over)

Sphyrapus serratus n.sp.

Sars, G.O. 1885
Crustacea 1. Zool., Norwegian North-Atlantic Exped. 1876-1878: 280 pp., 21 pls. 1 map.

Sphyrapus tudes

Hansen, H.J., 1913
Crustacea Malacostraca II. IV Tanaidacea Ingolf Exped. III(3): 152 pp., 12 pls., list of stations.

Strongylura antarctica

Lang, Karl, 1970
Taxonomische und phylogenetische Untersuchungen über die Tanaidaceen. 7. Revision der Gattung Strongylura G.O. Sars, 1882, nebst Beschreibung einer neuen Art dieser Gattung. Ark. Zool. (2)23(7): 403-415. 2 plates.

Strongylura arctophylax

Sars, G.O., 1896-1899
An account of the Crustacea of Norway with short descriptions and figures of all species. Vol.II Isopoda. Pts.I,II. Apseudidae, Tanaidae:1-40, 16 pls. Pts.III,IV. Anthuridae, Gnathiidae, Aegidae, Cirolanidae, Limnoriidae:41-80, pls.17-32, Pts.V,VI. Idotheidae, Ancthuridae, Asellidae, Ianiridae, Munnidae, 81-116, Pls.33-48. Pts.VII,V,Desmosomidae, Munnopsidae (part):117-144, pls. 49-64. Pts.IX,X. Munnopsidae (concluded), Ligiidae, Trichoniscidae, Oniscidae (part): 145-184, Pls.65-88. Pts.X,XII. Oniscidae (concluded), Bopyridae, Dajidae:185-232, pls.
(over)

Strongylura cylindrata

Hansen, H. J., 1913
Crustacea Malacostraca (II). IV Tanaidacea.
Danish Ingolf Exped. 3(3):145 pp., 12 pls.

Strongylura cylindrata

Lang, Karl, 1970
Taxonomische und phylogenetische Untersuchungen über die Tanaidaceen. 7. Revision der Gattung Strongylura G.O. Sars, 1882, nebst Beschreibung einer neuen Art dieser Gattung. Ark. Zool. (2)23(7): 403-415. 2 plates.

Strongylura cylindrata

Sars, G.O., 1896-1899
An account of the Crustacea of Norway with short descriptions and figures of all species. Vol.II Isopoda. Pts.I,II. Apseudidae, Tanaidae:1-40, 16 pls. Pts.III,IV. Anthuridae, Gnathiidae, Aegidae, Cirolanidae, Limnoriidae:41-80, pls.17-32, Pts.V,VI. Idotheidae, Ancthuridae, Asellidae, Ianiridae, Munnidae, 81-116, Pls.33-48. Pts.VII,VIII.Desmosomidae, Munnopsidae (part):117-144, pls. 49-64. Pts.IX,X. Munnopsidae (concluded), Ligiidae, Trichoniscidae, Oniscidae (part): 145-184, Pls.65-88. Pts.X,XII. Oniscidae (concluded), Bopyridae, Dajidae:185-232, pls.
(over)

Strongylura indivisa n.gen., n.sp.

Hansen, H.J., 1913
Crustacea Malacostraca II. IV Tanaidacea Ingolf Exped. III(3): 152 pp., 12 pls., list of stations.

Strongylura minima n.sp.

Hansen, H. J., 1913
Crustacea Malacostraca (II). IV Tanaidacea.
Danish Ingolf Exped. 3(3):145 pp., 12 pls.

Strongylura minima

Lang, Karl, 1970
Taxonomische und phylogenetische Untersuchungen über die Tanaidaceen. 7. Revision der Gattung Strongylura G.O. Sars, 1882, nebst Beschreibung einer neuen Art dieser Gattung. Ark. Zool. (2)23(7): 403-415. 2 plates.

Strongylura vermiformis n.sp.

Lang, Karl, 1970
Taxonomische und phylogenetische Untersuchungen über die Tanaidaceen. 7. Revision der Gattung Strongylura G.O. Sars, 1882, nebst Beschreibung einer neuen Art dieser Gattung. Ark. Zool. (2)23(7): 403-415. 2 plates.

Strongylurella n. gen.

Hansen, H. J., 1913
Crustacea Malacostraca (II). IV Tanaidacea.
Danish Ingolf Exped. 3(3):145 pp., 12 pls.

Strongylurella indivisa n.sp.

Hansen, H. J., 1913
Crustacea Malacostraca (II). IV Tanaidacea.
Danish Ingolf Exped. 3(3):145 pp., 12 pls.

Synapseudes idios n.sp.

Gardiner, Lion F. 1973.
New species of the genera Synapseudes and Cyclopoapseudes with notes on morphological variation postmarsupial development, and phylogenetic relationships within the family Metapseudidae (Crustacea: Tanaidacea).
Zool. J. Linn. Soc. 53(1): 25-58

Synapseudes setoensis n.sp.

Shiino, S.M., 1951.
On two new species of the Family Apseudidae found at Seto. Rept. Faculty Fish., Pref. Univ. Mie, 1(1):12-25, 6 textfigs.
Seto Mar. Biol. Lab. Contrib. 170

Tanaella ochracea n.sp.

Hansen, H. J., 1913
Crustacea Malacostraca (II). IV Tanaidacea.
Danish Ingolf Exped. 3(3):145 pp., 12 pls.

Tanaella unguicillata

Hansen, H. J., 1913
Crustacea Malacostraca (II). IV Tanaidacea.
Danish Ingolf Exped. 3(3):145 pp., 12 pls.

Tanais sp

Miller, Milton A., 1968
Isopoda and Tanaidacea from buoys in coastal waters of the continental United States, Hawaii and the Bahamas
Proc. U.S. natn. Mus. 125(3652):1-53

Tanais abbreviatus

Sars, G.O., 1896-1899
An account of the Crustacea of Norway with short descriptions and figures of all species. Vol.II Isopoda. Pts.I,II. Apseudidae, Tanaidae:1-40, 16 pls. Pts.III,IV. Anthuridae, Gnathiidae, Aegidae, Cirolanidae, Limnoriidae:41-80, pls.17-32, Pts.V,VI. Idotheidae, Ancthuridae, Asellidae, Ianiridae, Munnidae, 81-116, Pls.33-48. Pts.VII,VIII.Desmosomidae, Munnopsidae (part):117-144, pls. 49-64. Pts.IX,X. Munnopsidae (concluded), Ligiidae, Trichoniscidae, Oniscidae (part): 145-184, Pls.65-88. Pts.X,XII. Oniscidae (concluded), Bopyridae, Dajidae:185-232, pls.
(over)

Tanais aeqviremis

Sars, G.O., 1896-1899
An account of the Crustacea of Norway with short descriptions and figures of all species. Vol. II Isopoda. Pts. I,II. Apseudidae, Tanaidae:1-40, 16 pls. Pts.III,IV. Anthuridae, Gnathiidae, Aegidae, Cirolanidae, Limnoriidae:41-80, pls.17-32, Pts.V,VI. Idotheidae, Ancthuridae, Asellidae, Ianiridae, Munnidae, 81-116, pls.33-48. Pts.VII,VIII. Desmosomidae, Munnopsidae (part):117-144, pls. 49-64. Pts.IX,X. Munnopsidae (concluded), Ligiidae, Trichoniscidae, Oniscidae (part): 145-184, pls.65-88. Pts.X,XII. Oniscidae (concluded), Bopyridae, Dajidae:185-232, pls.
(over)

Tanais balthicus

Sars, G.O., 1896-1899
An account of the Crustacea of Norway with short descriptions and figures of all species. Vol.II Isopoda. Pts.I,II. Apseudidae, Tanaidae:1-40, 16 pls. Pts.III,IV. Anthuridae, Gnathiidae, Aegidae, Cirolanidae, Limnoriidae:41-80, pls.17-32, Pts.V,VI. Idotheidae, Ancthuridae, Asellidae, Ianiridae, Munnidae, 81-116, Pls.33-48. Pts.VII,VIII.Desmosomidae, Munnopsidae (part):117-144, pls. 49-64. Pts.IX,X. Munnopsidae (concluded), Ligiidae, Trichoniscidae, Oniscidae (part): 145-184, Pls.65-88. Pts.X,XII. Oniscidae (concluded), Bopyridae, Dajidae:185-232, pls.
(over)

Tanais brevicornis

Sars, G.O., 1896-1899
An account of the Crustacea of Norway with short descriptions and figures of all species. Vol.II Isopoda. Pts.I,II. Apseudidae, Tanaidae:1-40, 16 pls. Pts.III,IV. Anthuridae, Gnathiidae, Aegidae, Cirolanidae, Limnoriidae:41-80, pls.17-32, Pts.V,VI. Idotheidae, Ancthuridae, Asellidae, Ianiridae, Munnidae, 81-116, Pls.33-48. Pts.VII,VIII.Desmosomidae, Munnopsidae (part):117-144, pls. 49-64. Pts.IX,X. Munnopsidae (concluded), Ligiidae, Trichoniscidae, Oniscidae (part): 145-184, Pls.65-88. Pts.X,XII. Oniscidae (concluded), Bopyridae, Dajidae:185-232, pls.
(over)

Tanais brevimanus

Sars, G.O., 1896-1899
An account of the Crustacea of Norway with short descriptions and figures of all species. Vol.II Isopoda. Pts.I,II. Apseudidae, Tanaidae:1-40, 16 pls. Pts.III,IV. Anthuridae, Gnathiidae, Aegidae, Cirolanidae, Limnoriidae:41-80, pls.17-32, Pts.V,VI. Idotheidae, Ancthuridae, Asellidae, Ianiridae, Munnidae, 81-116, Pls.33-48. Pts.VII,VIII.Desmosomidae, Munnopsidae (part):117-144, pls. 49-64. Pts.IX,X. Munnopsidae (concluded), Ligiidae, Trichoniscidae, Oniscidae (part): 145-184, Pls.65-88. Pts.X,XII. Oniscidae (concluded), Bopyridae, Dajidae:185-232, pls.
(over)

Tanais breviremis

Sars, G.O., 1896-1899
An account of the Crustacea of Norway with short descriptions and figures of all species. Vol.II Isopoda. Pts. I,II. Apseudidae, Tanaidae:1-40, 16 pls. Pts.III,IV. Anthuridae, Gnathiidae, Aegidae, Cirolanidae, Limnoriidae:41-80, pls.17-32, Pts.V,VI. Idotheidae, Ancthuridae, Asellidae, Ianiridae, Munnidae, 81-116, pls.33-48. Pts.VII,VIII. Desmosomidae, Munnopsidae (part):117-144, pls. 49-64. Pts.IX,X. Munnopsidae (concluded), Ligiidae, Trichoniscidae, Oniscidae (part): 145-184, pls.65-88. Pts.X,XII. Oniscidae (concluded), Bopyridae, Dajidae:185-232, pls.
(over)

Tanais cavolini

Cléret, J.J., 1967.
Le pigmentation chez deux espèces de Tanais (Crustacea Tanaidacea) des côtes de la Manche.
Archs Zool.exp.gén., 107(4):677-691.

Tanais cavolinii

Dahl, E., 1941
Tanaidacea from the Sound and Skälderviken. Kungl. Fysiografiska Sällskapets i Lund Förhandlingar, 11(7):1-7, 1 text figs.

Tanais cavolinii

Hansen, H. J., 1913
Crustacea Malacostraca (II). IV Tanaidacea.
Danish Ingolf Exped. 3(3):145 pp., 12 pls.

Tanais cavolinii

Iacobescu, Valentin 1970.
Les particularités morphologiques de quelques tanides de la mer Noire. Trav. Mus. Hist. nat. "Grigore Antipa" 10:25-31.

Tanais cavolinii

Larwood, H.J., 1940
The fishery grounds near Alexandria. XXI Tanaidacea and Isopoda. Fouad I. Inst. Hydrobiol. & Fish., Notes & Mem. No.35:72 pp., 17 figs.

Tanais Cavolinii

Sars, G. O., 1886
Nye Bidrag til Kundskaben om Middelhavets Invertebrat fauna. III. Middelhavets Saxisopoder (Isopoda chelifera). Archiv for Mathem. og Natur. 263-368, 15 pls.

Tanais chevreuxi

Cléret, J.J., 1967.
Le pigmentation chez deux espèces de Tanais (Crustacea Tanaidacea) des côtes de la Manche.
Archs Zool. exp.gén., 107(4):677-691.

Tanais chevreuxi

Gamble, J.C., 1970.
Anaerobic survival of the crustaceans Corophium volutator, C. arenarium and Tanais chevreuxi. J. mar. biol. Ass., U.K., 50(3): 657-671.

Tanais depressus

Sars, G.O., 1896-1899
An account of the Crustacea of Norway with short descriptions and figures of all species. Vol.II Isopoda. Pts. I,II. Apseudidae, Tanaidae:1-40, 16 pls. Pts.III,IV. Anthuridae, Gnathiidae, Aegidae, Cirolanidae, Limnoriidae:41-80, pls.17-32, Pts.V,VI. Idotheidae, Ancthuridae, Asellidae, Ianiridae, Munnidae, 81-116, pls.33-48. Pts.VII,VIII. Desmosomidae, Munnopsidae (part):117-144, pls. 49-64. Pts.IX,X. Munnopsidae (concluded), Ligiidae, Trichoniscidae, Oniscidae (part): 145-184, pls.65-88. Pts.X,XII. Oniscidae (concluded), Bopyridae, Dajidae:185-232, pls.
(over)

Tanais Dulongii

Sars, G.O., 1896-1899
An account of the Crustacea of Norway with short descriptions and figures of all species. Vol.II Isopoda. Pts. I,II. Apseudidae, Tanaidae:1-40, 16 pls. Pts.III,IV. Anthuridae, Gnathiidae, Aegidae, Cirolanidae, Limnoriidae:41-80, pls.17-32, Pts.V,VI. Idotheidae, Ancthuridae, Asellidae, Ianiridae, Munnidae, 81-116, pls.33-48. Pts.VII,VIII. Desmosomidae, Munnopsidae (part):117-144, pls. 49-64. Pts.IX,X. Munnopsidae (concluded), Ligiidae, Trichoniscidae, Oniscidae (part): 145-184, pls.65-88. Pts.X,XII. Oniscidae (concluded), Bopyridae, Dajidae:185-232, pls.
(over)

Tanais filiformis

Sars, G.O., 1896-1899
An account of the Crustacea of Norway with short descriptions and figures of all species. Vol.II Isopoda. Pts. I,II. Apseudidae, Tanaidae:1-40, 16 pls. Pts.III,IV. Anthuridae, Gnathiidae, Aegidae, Cirolanidae, Limnoriidae:41-80, pls.17-32, Pts.V,VI. Idotheidae, Ancthuridae, Asellidae, Ianiridae, Munnidae, 81-116, pls.33-48. Pts.VII,VIII. Desmosomidae, Munnopsidae (part):117-144, pls. 49-64. Pts.IX,X. Munnopsidae (concluded), Ligiidae, Trichoniscidae, Oniscidae (part): 145-184, pls.65-88. Pts.X,XII. Oniscidae (concluded), Bopyridae, Dajidae:185-232, pls.
(over)

Tanais forcipatus
Sars, G.O., 1896-1899
An account of the Crustacea of Norway with short descriptions and figures of all species. Vol.II Isopoda. Pts. I,II. Apseudidae, Tanaidae:1-40, 16 pls. Pts.III,IV. Anthuridae, Gnathiidae, Aegidae, Cirolanidae, Limnoriidae:41-80, pls.17-32, Pts.V,VI. Idotheidae, Ancthuridae, Asellidae, Ianiridae, Munnidae, 81-116, pls.33-48. Pts.VII,VIII. Desmosomidae, Munnopsidae (part):117-144, pls. 49-64. Pts.IX,X. Munnopsidae (concluded), Ligiidae, Trichoniscidae, Oniscidae (part): 145-184, pls.65-88. Pts.X,XII. Oniscidae (concluded), Bopyridae, Dajidae:185-232, pls.
(over)

Tanais gracilis
Sars, G.O., 1896-1899
An account of the Crustacea of Norway with short descriptions and figures of all species. Vol.II Isopoda. Pts. I,II. Apseudidae, Tanaidae:1-40, 16 pls. Pts.III,IV. Anthuridae, Gnathiidae, Aegidae, Cirolanidae, Limnoriidae:41-80, pls.17-32, Pts.V,VI. Idotheidae, Ancthuridae, Asellidae, Ianiridae, Munnidae, 81-116, pls.33-48. Pts.VII,VIII. Desmosomidae, Munnopsidae (part):117-144, pls. 49-64. Pts.IX,X. Munnopsidae (concluded), Ligiidae, Trichoniscidae, Oniscidae (part): 145-184, pls.65-88. Pts.X,XII. Oniscidae (concluded), Bopyridae, Dajidae:185-232, pls.
(over)

Tanais gracilis
Stephensen, K., 1947
Tanaidacea, Isopoda, Amphipoda, and Pycnogonida. Sci. Res. Norweg. Antarctic Exp. 1927-1928 et seq. Date Norske Videnskaps-Akademi i Oslo, II, No.27:1-90, 26 text figs.

Tanais gracilis
Tattersall, W.M., 1921
Crustacea. Part VI. Tanaidacea and Isopoda. Brit. Antarctic "Terra Nova") Exped., 1910, Nat. Hist. Rept., Zool. 3(8):191-258, 2 text figs., 11 pls.

Tanais graciloides
Sars, G.O., 1896-1899
An account of the Crustacea of Norway with short descriptions and figures of all species. Vol.II Isopoda. Pts. I,II. Apseudidae, Tanaidae:1-40, 16 pls. Pts.III,IV. Anthuridae, Gnathiidae, Aegidae, Cirolanidae, Limnoriidae:41-80, pls.17-32, Pts.V,VI. Idotheidae, Ancthuridae, Asellidae, Ianiridae, Munnidae, 81-116, pls.33-48. Pts.VII,VIII. Desmosomidae, Munnopsidae (part):117-144, pls. 49-64. Pts.IX,X. Munnopsidae (concluded), Ligiidae, Trichoniscidae, Oniscidae (part): 145-184, pls.65-88. Pts.X,XII. Oniscidae (concluded), Bopyridae, Dajidae:185-232, pls.
(over)

Tanais Grimaldii
Belloc, G., 1960.
Catalogue des types d'Isopodes du Musée Océanographique de Monaco. Bull. Inst. Océanogr., Monaco, 57(1188):1-7.

Tanais hirticaudatus
Sars, G.O., 1896-1899
An account of the Crustacea of Norway with short descriptions and figures of all species. Vol.II Isopoda. Pts. I,II. Apseudidae, Tanaidae:1-40, 16 pls. Pts.III,IV. Anthuridae, Gnathiidae, Aegidae, Cirolanidae, Limnoriidae:41-80, pls.17-32, Pts.V,VI. Idotheidae, Ancthuridae, Asellidae, Ianiridae, Munnidae, 81-116, pls.33-48. Pts.VII,VIII. Desmosomidae, Munnopsidae (part):117-144, pls. 49-64. Pts.IX,X. Munnopsidae (concluded), Ligiidae, Trichoniscidae, Oniscidae (part): 145-184, pls.65-88. Pts.X,XII. Oniscidae (concluded), Bopyridae, Dajidae:185-232, pls.
(over)

Tanais islandicus
Sars, G.O., 1896-1899
An account of the Crustacea of Norway with short descriptions and figures of all species. Vol.II Isopoda. Pts. I,II. Apseudidae, Tanaidae:1-40, 16 pls. Pts.III,IV. Anthuridae, Gnathiidae, Aegidae, Cirolanidae, Limnoriidae:41-80, pls.17-32, Pts.V,VI. Idotheidae, Ancthuridae, Asellidae, Ianiridae, Munnidae, 81-116, pls.33-48. Pts.VII,VIII. Desmosomidae, Munnopsidae (part):117-144, pls. 49-64. Pts.IX,X. Munnopsidae (concluded), Ligiidae, Trichoniscidae, Oniscidae (part): 145-184, pls.65-88. Pts.X,XII. Oniscidae (concluded), Bopyridae, Dajidae:185-232, pls.
(over)

Tanais longiremis
Sars, G.O., 1896-1899
An account of the Crustacea of Norway with short descriptions and figures of all species. Vol.II Isopoda. Pts. I,II. Apseudidae, Tanaidae:1-40, 16 pls. Pts.III,IV. Anthuridae, Gnathiidae, Aegidae, Cirolanidae, Limnoriidae:41-80, pls.17-32, Pts.V,VI. Idotheidae, Ancthuridae, Asellidae, Ianiridae, Munnidae, 81-116, pls.33-48. Pts.VII,VIII. Desmosomidae, Munnopsidae (part):117-144, pls. 49-64. Pts.IX,X. Munnopsidae (concluded), Ligiidae, Trichoniscidae, Oniscidae (part): 145-184, pls.65-88. Pts.X,XII. Oniscidae (concluded), Bopyridae, Dajidae:185-232, pls.
(over)

Tanais novae-zealandiae
Sars, G.O., 1896-1899
An account of the Crustacea of Norway with short descriptions and figures of all species. Vol.II Isopoda. Pts. I,II. Apseudidae, Tanaidae:1-40, 16 pls. Pts.III,IV. Anthuridae, Gnathiidae, Aegidae, Cirolanidae, Limnoriidae:41-80, pls.17-32, Pts.V,VI. Idotheidae, Ancthuridae, Asellidae, Ianiridae, Munnidae, 81-116, pls.33-48. Pts.VII,VIII. Desmosomidae, Munnopsidae (part):117-144, pls. 49-64. Pts.IX,X. Munnopsidae (concluded), Ligiidae, Trichoniscidae, Oniscidae (part): 145-184, pls.65-88. Pts.X,XII. Oniscidae (concluded), Bopyridae, Dajidae:185-232, pls.
(over)

Tanais novae-zealandice
Tattersall, W.M., 1921
Crustacea. Part VI. Tanaidacea and Isopoda. Brit. Antarctic "Terra Nova") Exped., 1910, Nat. Hist. Rept., Zool. 3(8):191-258, 2 text figs., 11 pls.

Tanais robustus
Larwood, H.J., 1940
The fishery grounds near Alexandria. XXI Tanaidacea and Isopoda. Fouad I. Inst. Hydrobiol. & Fish., Notes & Mem. No.35:72 pp., 17 figs.

?Tanais robustus
Omer-Cooper, J., 1927.
Zoological results of the Cambridge Expedition to the Suez Canal. XII. Report on the Crustacea Tanaidacea and Isopoda. Trans. Zool. Soc., London, 22:201-209.

Tanais robustus
Stephensen, K., 1915
Isopoda, Tanaidacea, Cumacea, Amphipoda (excl. Hyperiidea). Rept. Danish Oceanogr. Exped. 1908-10, to the Mediterranean and adjacent seas, Vol.II, Biol. D. 1, 53 pp., 33 text figs.

Tanais rhynchites
Sars, G.O., 1896-1899
An account of the Crustacea of Norway with short descriptions and figures of all species. Vol.II Isopoda. Pts. I,II. Apseudidae, Tanaidae:1-40, 16 pls. Pts.III,IV. Anthuridae, Gnathiidae, Aegidae, Cirolanidae, Limnoriidae:41-80, pls.17-32, Pts.V,VI. Idotheidae, Ancthuridae, Asellidae, Ianiridae, Munnidae, 81-116, pls.33-48. Pts.VII,VIII. Desmosomidae, Munnopsidae (part):117-144, pls. 49-64. Pts.IX,X. Munnopsidae (concluded), Ligiidae, Trichoniscidae, Oniscidae (part): 145-184, pls.65-88. Pts.X,XII. Oniscidae (concluded), Bopyridae, Dajidae:185-232, pls.
(over)

Tanais stanfordi
Lang, K., 1956.
Tanaidacea aus Brasilien gesammelt von Professor A. Remane und Dr. S. Gerlach. Kieler Meeresf., 249-360.

Tanais tenuimanus
Sars, G.O., 1896-1899
An account of the Crustacea of Norway with short descriptions and figures of all species. Vol.II Isopoda. Pts. I,II. Apseudidae, Tanaidae:1-40, 16 pls. Pts.III,IV. Anthuridae, Gnathiidae, Aegidae, Cirolanidae, Limnoriidae:41-80, pls.17-32, Pts.V,VI. Idotheidae, Ancthuridae, Asellidae, Ianiridae, Munnidae, 81-116, pls.33-48. Pts.VII,VIII. Desmosomidae, Munnopsidae (part):117-144, pls. 49-64. Pts.IX,X. Munnopsidae (concluded), Ligiidae, Trichoniscidae, Oniscidae (part): 145-184, pls.65-88. Pts.X,XII. Oniscidae (concluded), Bopyridae, Dajidae:185-232, pls.
(over)

Tanais tomentosus
Sars, G.O., 1896-1899
An account of the Crustacea of Norway with short descriptions and figures of all species. Vol.II Isopoda. Pts. I,II. Apseudidae, Tanaidae:1-40, 16 pls. Pts.III,IV. Anthuridae, Gnathiidae, Aegidae, Cirolanidae, Limnoriidae:41-80, pls.17-32, Pts.V,VI. Idotheidae, Ancthuridae, Asellidae, Ianiridae, Munnidae, 81-116, pls.33-48. Pts.VII,VIII. Desmosomidae, Munnopsidae (part):117-144, pls. 49-64. Pts.IX,X. Munnopsidae (concluded), Ligiidae, Trichoniscidae, Oniscidae (part): 145-184, pls.65-88. Pts.X,XII. Oniscidae (concluded), Bopyridae, Dajidae:185-232, pls.
(over)

Tanais vittatus
Sars, G.O., 1896-1899
An account of the Crustacea of Norway with short descriptions and figures of all species. Vol.II Isopoda. Pts. I,II. Apseudidae, Tanaidae:1-40, 16 pls. Pts.III,IV. Anthuridae, Gnathiidae, Aegidae, Cirolanidae, Limnoriidae:41-80, pls.17-32, Pts.V,VI. Idotheidae, Ancthuridae, Asellidae, Ianiridae, Munnidae, 81-116, pls.33-48. Pts.VII,VIII. Desmosomidae, Munnopsidae (part):117-144, pls. 49-64. Pts.IX,X. Munnopsidae (concluded), Ligiidae, Trichoniscidae, Oniscidae (part): 145-184, pls.65-88. Pts.X,XII. Oniscidae (concluded), Bopyridae, Dajidae:185-232, pls.
(over)

Tanais Voeringi
Sars, G.O., 1896-1899
An account of the Crustacea of Norway with short descriptions and figures of all species. Vol.II Isopoda. Pts. I,II. Apseudidae, Tanaidae:1-40, 16 pls. Pts.III,IV. Anthuridae, Gnathiidae, Aegidae, Cirolanidae, Limnoriidae:41-80, pls.17-32, Pts.V,VI. Idotheidae, Ancthuridae, Asellidae, Ianiridae, Munnidae, 81-116, pls.33-48. Pts.VII,VIII. Desmosomidae, Munnopsidae (part):117-144, pls. 49-64. Pts.IX,X. Munnopsidae (concluded), Ligiidae, Trichoniscidae, Oniscidae (part): 145-184, pls.65-88. Pts.X,XII. Oniscidae (concluded), Bopyridae, Dajidae:185-232, pls.
(over)

Tanaopsis antarctica n.sp.
Lang, Karl, 1967.
Taxonomische und phylogenetische Untersuchungen über die Tanaidaceen. 3. Der Umfang der Familien Tanaidae Sars, Lang und Paratanaidae Lang nebst Bemerkungen über den taxonomischen Wert der Mandibeln und Maxillulae. Dazu eine taxonomischmonographische Darstellung der Gattung Tanaopsis Sars. Arkiv for Zoolog: 19(18):243-268.

Tanaopsis graciloides
Lang, Karl., 1967.
Taxonomische und phylogenetische Untersuchungen über die Tanaidaceen. 3. Der Umfang der Familien Tanaidae Sars, Lang und Paratanaidae Lang nebst Bemerkungen über den taxonomischen Wert der Mandibeln und Maxillulae. Dazu eine taxonomischmonographische Darstellung der Gattung Tanaopsis Sars. Arkiv for Zoolog: 19(18):243-268.

Tanaopsis laticaudata
Sars, G.O., 1896-1899
An account of the Crustacea of Norway with short descriptions and figures of all species. Vol.II Isopoda. Pts. I,II. Apseudidae, Tanaidae:1-40, 16 pls. Pts.III,IV. Anthuridae, Gnathiidae, Aegidae, Cirolanidae, Limnoriidae:41-80, pls.17-32, Pts.V,VI. Idotheidae, Ancthuridae, Asellidae, Ianiridae, Munnidae, 81-116, pls.33-48. Pts.VII,VIII. Desmosomidae, Munnopsidae (part):117-144, pls. 49-64. Pts.IX,X. Munnopsidae (concluded), Ligiidae, Trichoniscidae, Oniscidae (part): 145-184, pls.65-88. Pts.X,XII. Oniscidae (concluded), Bopyridae, Dajidae:185-232, pls.
(over)

Tanaopsis laticaudata
Tattersall, W.M., 1905
The Marine Fauna of the Coast of Ireland. Part V Isopoda. Dept. Agric. and Tech. Instr. for Ireland, Fisheries Branch, Sci. Invest. 1904. No.II:1-90, pls.1-11, 1 chart.

Tanaopsis profunda n. sp.
Lang, Karl, 1967.
Taxonomische und phylogenetische Untersuchungen über die Tanaidaceen. 3. Der Umfang der Familien Tanaidae Sars, Lang und Paratanaidae Lang nebst Bemerkungen über den taxonomischen Wert der Mandibeln und Maxillulae. Dazu eine taxonomischmonographische Darstellung der Gattung Tanaopsis Sars. Arkiv for Zoolog., 19(18):243-268.

Tanaissus lilljeborgi
Siewing, R., 1954.
Morphologische Untersuchungen an Tanaidaceen und Lophogastriden. Zeits. Wiss. Zool., A, 157(3/4): 333-426, 44 textfigs.

Teleotanais gerlachi n.gen. n.sp.
Lang, K., 1956.
Tanaidacea aus Brasilien gesammelt von Professor A. Remane und Dr. S. Gerlach. Kieler Meeresf., 12(2):249-260.

Typhlapseudes sp.
Sars, G.O., 1896-1899
An account of the Crustacea of Norway with short descriptions and figures of all species. Vol.II Isopoda. Pts. I,II. Apseudidae, Tanaidae:1-40, 16 pls. Pts.III,IV. Anthuridae, Gnathiidae, Aegidae, Cirolanidae, Limnoriidae:41-80, pls.17-32, Pts.V,VI. Idotheidae, Ancthuridae, Asellidae, Ianiridae, Munnidae, 81-116, pls.33-48. Pts.VII,VIII. Desmosomidae, Munnopsidae (part):117-144, pls. 49-64. Pts.IX,X. Munnopsidae (concluded), Ligiidae, Trichoniscidae, Oniscidae (part): 145-184, pls.65-88. Pts.X,XII. Oniscidae (concluded), Bopyridae, Dajidae:185-232, pls.
(over)

Typhlotanais, lists of spp.
Lang, Karl, 1970
Taxonomische und phylogenetische Untersuchungen über die Tanaidaceen. 5. Die Gattung Typhlotanais G.O. Sars, 1882, nebst Beschreibung einer neuen Art dieser Gattung. Dazu eine Berichtigung des Dornenzahl des Enditen der Maxillulae bei T. peculiaris Lang, 1968. Ark. Zool. (2)23(4):267-291.

Typhlotanais sp.

Morino, Hiroshi, 1974.
Record of Typhlotanais a tube-building paratanaid from Seto (Crustacea: Malacostraca).
Publ. Seto mar. biol. Lab. 18(5):349-354.

Typhlotanais aequiremis

Greve, Lita, 1965.
The biology of some Tanaidacea from Raunefjorden western Norway.
Sarsia, 20:43-54.

Typhlotanais aequiremis

Lang, Karl, 1970
Taxonomische und phylogenetische Untersuchungen über die Tanaidaceen. 5. Die Gattung Typhlotanais G.O. Sars, 1882, nebst Beschreibung einer neuen Art dieser Gattung. Dazu eine Berichtigung der Dornenzahl des Enditen der Maxillulae bei T. peculiaris Lang, 1968. Ark. Zool. (2)23(4):267-291.

Typhlotanais aeqviremis

Sars, G.O., 1896-1899
An account of the Crustacea of Norway with short descriptions and figures of all species. Vol.II Isopoda. Pts. I,II. Apseudidae, Tanaidae:1-40, 16 pls. Pts.III,IV. Anthuridae, Gnathiidae, Aegidae, Cirolanidae, Limnoriidae:41-80, pls.17-32, Pts.V,VI. Idotheidae, Ancthuridae, Asellidae, Ianiridae, Munnidae, 81-116, pls.33-48. Pts.VII,VIII. Desmosomidae, Munnopsidae (part):117-144, pls. 49-64. Pts.IX,X. Munnopsidae (concluded), Ligiidae, Trichoniscidae, Oniscidae (part): 145-184, pls.65-88. Bts.X,XII. Oniscidae (concluded), Bopyridae, Dajidae:185-232, pls. (over)

Typhlotanais aeqviremis

Sars, G.O., 1886
Crustacea II. Zool. Norwegian North-Atlantic Exped., 1876-1878: 96 pp., 1 map.

Typhlotanais angularis n. sp.

Kudinova-Pasternak, R. K., 1966.
Tanaidacea (Crustacea) of the Pacific ultra-abyssal zone. (In Russian; English Abstract).
Zool. Zhurn., Akad. Nauk SSSR 44(4): 518-535.

Typhlotanais assimilis

Sars, G.O., 1896-1899
An account of the Crustacea of Norway with short descriptions and figures of all species. Vol.II Isopoda. Pts. I,II. Apseudidae, Tanaidae:1-40, 16 pls. Pts.III,IV. Anthuridae, Gnathiidae, Aegidae, Cirolanidae, Limnoriidae:41-80, pls.17-32, Pts.V,VI. Idotheidae, Ancthuridae, Asellidae, Ianiridae, Munnidae, 81-116, pls.33-48. Pts.VII,VIII. Desmosomidae, Munnopsidae (part):117-144, pls. 49-64. Pts.IX,X. Munnopsidae (concluded), Ligiidae, Trichoniscidae, Oniscidae (part): 145-184, pls.65-88. Bts.X,XII. Oniscidae (concluded), Bopyridae, Dajidae:185-232, pls. (over)

Typhlotanais brachyurus

Sars, G.O., 1896-1899
An account of the Crustacea of Norway with short descriptions and figures of all species. Vol.II Isopoda. Pts. I,II. Apseudidae, Tanaidae:1-40, 16 pls. Pts.III,IV. Anthuridae, Gnathiidae, Aegidae, Cirolanidae, Limnoriidae:41-80, pls.17-32, Pts.V,VI. Idotheidae, Ancthuridae, Asellidae, Ianiridae, Munnidae, 81-116, pls.33-48. Pts.VII,VIII. Desmosomidae, Munnopsidae (part):117-144, pls. 49-64. Pts.IX,X. Munnopsidae (concluded), Ligiidae, Trichoniscidae, Oniscidae (part): 145-184, pls.65-88. Bts.X,XII. Oniscidae (concluded), Bopyridae, Dajidae:185-232, pls. (over)

Typhlotanais brevicornis

Greve, Lita, 1965.
The biology of some Tanaidacea from Raunefjorden, western Norway.
Sarsia, 20:34-54.

Typhlotanais brevicornis

Sars, G.O., 1896-1899
An account of the Crustacea of Norway with short descriptions and figures of all species. Vol.II Isopoda. Pts. I,II. Apseudidae, Tanaidae:1-40, 16 pls. Pts.III,IV. Anthuridae, Gnathiidae, Aegidae, Cirolanidae, Limnoriidae:41-80, pls.17-32, Pts.V,VI. Idotheidae, Ancthuridae, Asellidae, Ianiridae, Munnidae, 81-116, pls.33-48. Pts.VII,VIII. Desmosomidae, Munnopsidae (part):117-144, pls. 49-64. Pts.IX,X. Munnopsidae (concluded), Ligiidae, Trichoniscidae, Oniscidae (part): 145-184, pls.65-88. Bts.X,XII. Oniscidae (concluded), Bopyridae, Dajidae:185-232, pls. (over)

Typhlotanais compactus n. sp.

Kudinova-Pasternak, R. K., 1966.
Tanaidacea (Crustacea) of the Pacific ultra-abyssal zone (In Russian; English Abstract).
Zool. Zhurn., Akad. Nauk SSSR 44(4): 518-535.

Typhlotanais cornutus

Hansen, H. J., 1913
Crustacea Malacostraca (II). IV Tanaidacea.
Danish Ingolf Exped. 3(3):145 pp., 12 pls.

Typhlotanais cornutus

Sars, G.O., 1896-1899
An account of the Crustacea of Norway with short descriptions and figures of all species. Vol.II Isopoda. Pts. I,II. Apseudidae, Tanaidae:1-40, 16 pls. Pts.III,IV. Anthuridae, Gnathiidae, Aegidae, Cirolanidae, Limnoriidae:41-80, pls.17-32, Pts.V,VI. Idotheidae, Ancthuridae, Asellidae, Ianiridae, Munnidae, 81-116, pls.33-48. Pts.VII,VIII. Desmosomidae, Munnopsidae (part):117-144, pls. 49-64. Pts.IX,X. Munnopsidae (concluded), Ligiidae, Trichoniscidae, Oniscidae (part): 145-184, pls.65-88. Bts.X,XII. Oniscidae (concluded), Bopyridae, Dajidae:185-232, pls. (over)

Typhlotanais cornutus n. sp.

Sars, G.O., 1886
Crustacea II. Zool. Norwegian North-Atlantic Exped., 1876-1878: 96 pp., 1 map.

Typhlotanais cornutus n.sp.

Sars, G.O. 1885
Crustacea 1. Zool., Norwegian North-Atlantic Exped. 1876-1878: 280 pp., 21 pls. 1 map.

Typhlotanais eximius n.sp.

Hansen, H. J., 1913
Crustacea Malacostraca (II). IV Tanaidacea.
Danish Ingolf Exped. 3(3):145 pp., 12 pls.

Typhlotanais finmarchicus

Hansen, H. J., 1913
Crustacea Malacostraca (II). IV Tanaidacea.
Danish Ingolf Exped. 3(3):145 pp., 12 pls.

Typhlotanais finmarchicus

Sars, G.O., 1896-1899
An account of the Crustacea of Norway with short descriptions and figures of all species. Vol.II Isopoda. Pts. I,II. Apseudidae, Tanaidae:1-40, 16 pls. Pts.III,IV. Anthuridae, Gnathiidae, Aegidae, Cirolanidae, Limnoriidae:41-80, pls.17-32, Pts.V,VI. Idotheidae, Ancthuridae, Asellidae, Ianiridae, Munnidae, 81-116, pls.33-48. Pts.VII,VIII. Desmosomidae, Munnopsidae (part):117-144, pls. 49-64. Pts.IX,X. Munnopsidae (concluded), Ligiidae, Trichoniscidae, Oniscidae (part): 145-184, pls.65-88. Bts.X,XII. Oniscidae (concluded), Bopyridae, Dajidae:185-232, pls. (over)

Typhlotanais finmarchicus

Stephenson, K., 1937
Marine Isopoda and Tanaidacea. Zool. Iceland. 3(27):26 pp.

Typhlotanais gracilipes n.sp.

Hansen, H. J., 1913
Crustacea Malacostraca (II). IV Tanaidacea.
Danish Ingolf Exped. 3(3):145 pp., 12 pls.

Typhlotanais grandis n.sp.

Hansen, H. J., 1913
Crustacea Malacostraca (II). IV Tanaidacea.
Danish Ingolf Exped. 3(3):145 pp., 12 pls.

Typhlotanais grandis

Kudinova-Pasternak, R. K., 1966.
Tanaidacea (Crustacea) of the Pacific ultra-abyssal zone (In Russian; English Abstract).
Zool. Zhurn., Akad. Nauk SSSR 44(4): 518-535.

Typhlotanais greenwichensis n.sp.

Shiino, Sueo M. 1970.
Paratanaidae collected in Chile Bay, Greenwich Island by the XII Chilean Antarctic Expedition, with an Apseudes from Porvenir Point, Tierra del Fuego Island.
Inst. Antart. Chileno, Ser. Cient. 1(2):79-122.

Typhlotanais inaequipes n.sp.

Hansen, H. J., 1913
Crustacea Malacostraca (II). IV Tanaidacea.
Danish Ingolf Exped. 3(3):145 pp., 12 pls.

Typhlotanais inermis n.sp.

Hansen, H. J., 1913
Crustacea Malacostraca (II). IV Tanaidacea.
Danish Ingolf Exped. 3(3):145 pp., 12 pls.

Typhlotanais irregularis n.sp.

Hansen, H. J., 1913
Crustacea Malacostraca (II). IV Tanaidacea.
Danish Ingolf Exped. 3(3):145 pp., 12 pls.

Typhlotanais kerguelensis

Sars, G.O., 1896-1899
An account of the Crustacea of Norway with short descriptions and figures of all species. Vol.II Isopoda. Pts. I,II. Apseudidae, Tanaidae:1-40, 16 pls. Pts.III,IV. Anthuridae, Gnathiidae, Aegidae, Cirolanidae, Limnoriidae:41-80, pls.17-32, Pts.V,VI. Idotheidae, Ancthuridae, Asellidae, Ianiridae, Munnidae, 81-116, pls.33-48. Pts.VII,VIII. Desmosomidae, Munnopsidae (part):117-144, pls. 49-64. Pts.IX,X. Munnopsidae (concluded), Ligiidae, Trichoniscidae, Oniscidae (part): 145-184, pls.65-88. Bts.X,XII. Oniscidae (concluded), Bopyridae, Dajidae:185-232, pls. (over)

Typhlotanais longidactylus n.sp.

Shiino, Sueo M. 1970.
Paratanaidae collected in Chile Bay, Greenwich Island by the XII Chilean Antarctic Expedition, with an Apseudes from Porvenir Point, Tierra del Fuego Island.
Inst. Antart. Chileno, Ser. Cient. 1(2):79-122.

Typhlotanais macrocephala n.sp.

Hansen, H. J., 1913
Crustacea Malacostraca (II). IV Tanaidacea.
Danish Ingolf Exped. 3(3):145 pp., 12 pls.

Typhlotanais magnificus

Kudinova-Pasternak, R.K. 1969.
A case of extramarsupial development of eggs in Typhlotanais magnificus (Crustacea, Tanaidacea) living in a tube. (In Russian; English abstract).
Zool.Zh. 48(11):1737-1738.

Typhlotanais messinensis

Sars, G.O., 1896-1899
An account of the Crustacea of Norway with short descriptions and figures of all species. Vol.II Isopoda. Pts. I,II. Apseudidae, Tanaidae:1-40, 16 pls. Pts.III,IV. Anthuridae, Gnathiidae, Aegidae, Cirolanidae, Limnoriidae:41-80, pls.17-32, Pts.V,VI. Idotheidae, Ancthuridae, Asellidae, Ianiridae, Munnidae, 81-116, pls.33-48. Pts.VII,VIII. Desmosomidae, Munnopsidae (part):117-144, pls. 49-64. Pts.IX,X. Munnopsidae (concluded), Ligiidae, Trichoniscidae, Oniscidae (part): 145-184, pls.65-88. Bts.X,XII. Oniscidae (concluded), Bopyridae, Dajidae:185-232, pls. (over)

Typhlotanais messinensis

Sars, G.O., 1886
Nye Bidrag til Kundskaben om Middelhavets Invertebrat fauna. III. Middelhavets Saxisopoder (Isopoda chelifera). Archiv for Mathem. og Natur. 263-368, 15 pls.

Typhlotanais microcheles

Sars, G.O., 1896-1899
An account of the Crustacea of Norway with short descriptions and figures of all species. Vol.II Isopoda. Pts. I,II. Apseudidae, Tanaidae:1-40, 16 pls. Pts.III,IV. Anthuridae, Gnathiidae, Aegidae, Cirolanidae, Limnoriidae:41-80, pls.17-32, Pts.V,VI. Idotheidae, Ancthuridae, Asellidae, Ianiridae, Munnidae, 81-116, pls.33-48. Pts.VII,VIII. Desmosomidae, Munnopsidae (part):117-144, pls. 49-64. Pts.IX,X. Munnopsidae (concluded), Ligiidae, Trichoniscidae, Oniscidae (part): 145-184, pls.65-88. Bts.X,XII. Oniscidae (concluded), Bopyridae, Dajidae:185-232, pls. (over)

Oceanographic Index: Marine Organisms Cumulation, 1946-1973

Typhlotanais mixtus n.sp.
Hansen, H. J., 1913
Crastacea Malacostraca (II). IV Tanaidacea.
Danish Ingolf Exped. 3(3):145 pp., 12 pls.

Typhlotanais mucronatus n.sp.
Hansen, H. J., 1913
Crastacea Malacostraca (II). IV Tanaidacea.
Danish Ingolf Exped. 3(3):145 pp., 12 pls.

Typhlotanais peculiaris n.sp.
Lang, Karl, 1968.
Deep-Sea Tanaidacea.
Galathea Rept., 9:23-209.

Typhlotanais penicillatus n.sp.
Hansen, H. J., 1913
Crastacea Malacostraca (II). IV Tanaidacea.
Danish Ingolf Exped. 3(3):145 pp., 12 pls.

Typhlotanais penicillatus
Sars, G.O., 1896-1899
An account of the Crustacea of Norway with short descriptions and figures of all species. Vol.II Isopoda. Pts. I,II. Apseudidae, Tanaidae:1-40, 16 pls. Pts.III,IV. Anthuridae, Gnathiidae, Aegidae, Cirolanidae, Limnoriidae:41-80, pls.17-32, Pts.V,VI. Idotheidae, Ancthuridae, Asellidae, Ianiridae, Munnidae, 81-116, pls.33-48. Pts.VII,VIII. Desmosomidae, Munnopsidae (part):117-144, pls. 49-64. Pts.IX,X. Munnopsidae (concluded), Ligiidae, Trichoniscidae, Oniscidae (part): 145-184, pls.65-88. Pts.X,XII. Oniscidae (concluded), Bopyridae, Dajidae:185-232, pls. (over)

Typhlotanais plebejus n.sp.
Hansen, H. J., 1913
Crastacea Malacostraca (II). IV Tanaidacea.
Danish Ingolf Exped. 3(3):145 pp., 12 pls.

Typhlotanais proctagon
Tattersall, W.M., 1905
The Marine Fauna of the Coast of Ireland. Part V Isopoda. Dept. Agric. and Tech. Instr. for Ireland, Fisheries Branch, Sci. Invest. 1904. No.II:1-90, pls.1-11, 1 chart.

Typhlotanais profundus n.sp.
Hansen, H. J., 1913
Crastacea Malacostraca (II). IV Tanaidacea.
Danish Ingolf Exped. 3(3):145 pp., 12 pls.

Typhlotanais pulcher n.sp.
Hansen, H. J., 1913
Crastacea Malacostraca (II). IV Tanaidacea.
Danish Ingolf Exped. 3(3):145 pp., 12 pls.

Typhlotanais rectus n. sp.
Kudinova-Pasternak, R. K., 1966.
Tanaidacea (Crustacea) of the Pacific ultra-abyssal zone (In Russian; English Abstract).
Zool. Zhurn., Akad. Nauk SSSR 44(4): 518-535.

Typhlotanais Richardii
Belloc, G., 1960.
Catalogue des types d'Isopodes du Musée Océanographique de Monaco.
Bull. Inst. Océanogr., Monaco, 57(1188):1-7.

Typhlotanais Richardi
Tattersall, W.M., 1905
The Marine Fauna of the Coast of Ireland. Part V Isopoda. Dept. Agric. and Tech. Instr. for Ireland, Fisheries Branch, Sci. Invest. 1904. No.II:1-90, pls.1-11, 1 chart.

Typhlotanais rotundirostris n.sp.
*Lang, Karl, 1970
Taxonomische und phylogenetische Untersuchungen über die Tanaidaceen. 5. Die Gattung Typhlotanais G.O. Sars, 1882, nebst Beschreibung einer neuen Art dieser Gattung. Dazu eine Berichtigung der Dornenzahl des Enditen der Maxillulae bei T. peculiaris Lang, 1968. Ark. Zool. (2)23(4):267-291.

Typhlotanais setosus n. sp.
Kudinova-Pasternak, R. K., 1966.
Tanaidacea (Crustacea) of the Pacific ultra-abyssal zone (In Russian; English Abstract).
Zool. Zhurn., Akad. Nauk SSSR 44(4): 518-535.

Typhlotanais solidus n.sp.
Hansen, H. J., 1913
Crastacea Malacostraca (II). IV Tanaidacea.
Danish Ingolf Exped. 3(3):145 pp., 12 pls.

Typhlotanais tenuicornis
Sars, G.O., 1896-1899
An account of the Crustacea of Norway with short descriptions and figures of all species. Vol.II Isopoda. Pts. I,II. Apseudidae, Tanaidae:1-40, 16 pls. Pts.III,IV. Anthuridae, Gnathiidae, Aegidae, Cirolanidae, Limnoriidae:41-80, pls.17-32, Pts.V,VI. Idotheidae, Ancthuridae, Asellidae, Ianiridae, Munnidae, 81-116, pls.33-48. Pts.VII,VIII. Desmosomidae, Munnopsidae (part):117-144, pls. 49-64. Pts.IX,X. Munnopsidae (concluded), Ligiidae, Trichoniscidae, Oniscidae (part): 145-184, pls.65-88. Pts.X,XII. Oniscidae (concluded), Bopyridae, Dajidae:185-232, pls. (over)

Typhlotanais spinicauda n.sp.
Hansen, H. J., 1913
Crastacea Malacostraca (II). IV Tanaidacea.
Danish Ingolf Exped. 3(3):145 pp., 12 pls.

Typhlotanais spiniventris
Belloc, G., 1960.
Catalogue des types d'Isopodes du Musée Océanographique de Monaco.
Bull. Inst. Océanogr., Monaco, 57(1188):1-7.

Typhlotanais tenuicornis
Greve, Lita, 1965. some
The biology of Tanaidacea from Raunefjorden, western Norway.
Sarsia, 20: 34-54.

Typhlotanais tenuimanus
Sars, G.O., 1886
Crustacea II. Zool. Norwegian North-Atlantic Exped., 1876-1878: 96 pp., 1 map.

Typhlotanais tenuicornis
Tattersall, W.M., 1905
The Marine Fauna of the Coast of Ireland. Part V Isopoda. Dept. Agric. and Tech. Instr. for Ireland, Fisheries Branch, Sci. Invest. 1904. No.II:1-90, pls.1-11, 1 chart.

Typhlotanais tenuimanus
Sars, G.O., 1896-1899
An account of the Crustacea of Norway with short descriptions and figures of all species. Vol.II Isopoda. Pts. I,II. Apseudidae, Tanaidae:1-40, 16 pls. Pts.III,IV. Anthuridae, Gnathiidae, Aegidae, Cirolanidae, Limnoriidae:41-80, pls.17-32, Pts.V,VI. Idotheidae, Ancthuridae, Asellidae, Ianiridae, Munnidae, 81-116, pls.33-48. Pts.VII,VIII. Desmosomidae, Munnopsidae (part):117-144, pls. 49-64. Pts.IX,X. Munnopsidae (concluded), Ligiidae, Trichoniscidae, Oniscidae (part): 145-184, pls.65-88. Pts.X,XII. Oniscidae (concluded), Bopyridae, Dajidae:185-232, pls. (over)

Typhlotanais trispinosus n.sp.
Hansen, H. J., 1913
Crastacea Malacostraca (II). IV Tanaidacea.
Danish Ingolf Exped. 3(3):145 pp., 12 pls.

Typhlotanais variabilis n.sp.
Hansen, H. J., 1913
Crastacea Malacostraca (II). IV Tanaidacea.
Danish Ingolf Exped. 3(3):145 pp., 12 pls.

Whiteleggia stephensoni n.sp.
Boesch, Donald F. 1973.
Three new tanaids (Crustacea, Tanaidacea) from southern Queensland.
Pacific Sci. 27(2): 168-188

Tardigrads
Rudescu, L. 1969.
Die Tardigraden des Schwarzen Meeres.
Hydrobiologia, Romania 10:3-12.

Tardigrada, lists of spp.
Mehlen, Ronal H. 1972.
Eutardigrada: distribution at Eniwetok Atoll, Marshall Islands.
Pacific Sci. 26(2):223-225.

Tardigrada, lists of spp.
Schuster, Robert O., and Albert A. Grigarick, 1966.
Tardigrada from the Galápagos and Cocos Islands.
Proc. Calif. Acad. Sci. (4), 34(5):315-328.

Tardigrada, ecology of
deZio, Susanna, and Piero Grimaldi, 1966.
Ecological aspects of Tardigrada distribution in south Adriatic beaches.
Veröff. Inst. Meeresforsch., Bremerh., Sonderband II:87-94.

Actinarctus doryphorus ocellatus n. subsp.
Renaud-Mornant, Jeanne 1970 (1971).
Campagne d'essais du Jean Charcot (3-8 décembre 1968). 8. Méiobenthos. II. Tardigrades.
Bull. Mus. nat. Hist. nat. (2)42(5): 957-969

Batillipes acaudatus n. sp.
Pollock, Leland W., 1971.
On some British marine Tardigrada, including two new species of Batillipes. J. mar. biol. Ass. U.K., 51(1): 93-103.

Batillipes gilmartini, n. sp.
McGinty, Marine, 1969.
Batillipes gilmartini a new marine tardigrade from a California beach.
Pacific Sci. 23(3):394-396.

Batillipes mirus
Pollock, Leland W., 1971.
On some British marine Tardigrada, including two new species of Batillipes. J. mar. biol. Ass. U.K., 51(1): 93-103.

Batillipes phreaticus
Pollock, Leland W., 1971.
On some British marine Tardigrada, including two new species of Batillipes. J. mar. biol. Ass. U.K., 51(1): 93-103.

Batillipes tubernatus sp.nov.
Pollock, Leland W., 1971.
On some British marine Tardigrada, including two new species of Batillipes. J. mar. biol. Ass. U.K., 51(1): 93-103.

Floractis heimi n.gen., n.sp.
Deboutteville, Claude Delamare et Jeanne Renaud-Mornant, 1966.
Un nouveau genre de tardigrades des sables détritique coralliens de Nouvelle-Calédonie.
Cah. pacif., No.9:149-156.

Halechiniscus perfectus
Renaud-Mornant, Jeanne 1970 (1971).
Campagne d'essais du Jean Charcot (3-8 décembre 1968). 8. Méiobenthos. II. Tardigrades.
Bull. Mus. nat. Hist. nat. (2)42(5): 957-969

Halechiniscus remanei

Renaud-Mornant, Jeanne 1970 (1971).
Campagne d'essais du Jean Charcot
(3-8 décembre 1968). 8. Méiobenthos.
II. Tardigrades.
Bull. Mus. nat. Hist. nat. (2) 42 (5):
957-969

Hypsibius (Diphascon) pinguis

Mitchell, David 1973.
Hypsibius (Diphascon) pinguis Marcus,
a tardigrade new to the British Isles.
In. nat. Jl 17(11): 395.

Hypsibius (Isohypsibius) renaudi n.sp.

Ramazzotti Giuseppe 1972.
Tardigradi delle isole Kerguelen e
descrizione della nuova specie
Hypsibius (I.) renaudi.
Mem. Ist. Ital. Idrobiol. 29:141-144.

Orzeliscus belopus

Pollock, Leland W., 1971.
On some British marine Tardigrada, including
two new species of Batillipes. J. mar. biol.
Ass., U.K., 51(1): 93-103.

Pleocola limnoriae

Renaud-Mornant, Jeanne 1970 (1971).
Campagne d'essais du Jean Charcot
(3-8 décembre 1968). 8. Méiobenthos.
II. Tardigrades.
Bull. Mus. nat. Hist. nat. (2) 42 (5):
957-969

XIPHOSURA

Limulus polyphemus
Baptist, J. P., 1953.
Record of a hermaphroditic horseshoe crab,
Limulus polyphemus L. Breviora No. 14:4 pp., 2 pl

Limulus
Cooper, Carol Davis and George Gordon Brown, 1972.
Immunological studies of the sperm and seminal
fluid in the horseshoe crab Limulus polyphemus
L. (Merostomata). Biol. Bull. mar. biol. Lab.
Woods Hole 142(3): 397-406.

Limulus
Fein, Alan and Richard A. Cone, 1973
Limulus rhodopsin: rapid return of
transient intermediates to the thermally
stable state. Science 182(4111):495-497.

Limulus polyphemus
Lang, Fred, 1971.
Induced myogenic activity in the neurogenic
heart of Limulus polyphemus. Biol. Bull. mar.
biol. Lab. Woods Hole 141(2): 269-277.

Limulus
Murray, George C., 1966.
Intracellular absorption difference spectrum of
Limulus extra-ocular photolabile pigment.
Science, 154(3753):1182-1183.

Limulus polyphemus
Millecchia, Ronald, Jack Bradbury and Alexander Mauro, 1966.
Simple photoreceptors in Limulus polyphemus.
Science, 154(3753):1199-1201.

Limulus polyphemus
Page, Charles H., 1973
Localization of Limulus polyphemus
oxygen sensitivity. Biol. Bull. 144
(2): 383-390.

Limulus polyphemus
Knudsen, Eric I., 1973
Muscular activity underlying ventilation
and swimming in the horseshoe
crab, Limulus polyphemus (Linnaeus).
Biol. Bull. 144(2): 355-367.

Tachypleus hoeveni
Waterman Talbot H., 1958.
On the doubtful validity of Tachypleus hoeveni
Pocock, an Indonesian horshoe crab (Xiphosura).
Postilla, Yale Peabody Mus., No. 36:17 pp., 3 pls

Chaetognaths
Aida, T., 1897.
Chaetognaths of Misaki Harbour. Ann. Zool.,
Japon., 1:13-21, Pl. 3.

chaetognaths
Alvariño, Angeles, 1966
Zoogeografia de California: quetognatos Revista
Soc. Mexicana Hist. nat. 27: 199-243

chaetognaths
Alvariño, Angeles, 1964.
Zoogeografía de los quetognatos, e specialmente
de la región de California.
Ciencia, Mexico, 23:51-74.

chaetognaths
Alvarino, Angeles, 1963
Quetognatos epiplanctonicos del Mar de Cortes.
Revista, Soc. Mexicana, Hist. Nat., 24:97-202.

chaetognaths
Aurich, Horst J., 1971.
Die Verbreitung der Chaetognathen
im Gebiet des Nordatlantischen Strom-
Systems.
Ber. dt. wiss. Komm. Meeresforsch. 22(1):1-30

Sagittae
Bal, D. V., and L. B. Pradhan, 1945
A preliminary note on the plankton of Bombay
harbour. Current. Sci., 14(8):211-212

chaetognaths
Béranek, E., 1895.
Les chétonates de la Baie d'Amboina. Rev. Suisse
Zool. 3:137-159, Pl. IV.

chaetognatha
Bieri, Robert 1959
The distribution of planktonic chaetognatha
in the Pacific and their relationship to
water masses.
Limnol. and Ocean. Vol. 4 No. 1 pp. 1-28

Chaetognath
Bieri, Robert, 1957.
The chaetognath fauna of Peru in 1941.
Pacific Science 11(3):255-264.

chaetognaths
Bumpus, D.F., and E.L. Pierce, 1955.
Hydrography and the distribution of chaetognaths
over the continental shelf of North Carolina.
Pap. Mar. Biol. and Oceanogr., Deep-Sea Res.,
Suppl., to Vol. 3:92-109.

chaetognaths
Burfield, S.T. 1930.
Chaetognatha. Brit. Antarctic ("Terra Nova")
Exped., 1910, Zool., 7(4):203-228, 3 maps.

chaetognatha
Burfield, S.T. 1927.
Zoological results of the Cambridge Expedition to
the Suez Canal. XXI. Report on the Chaetognatha.
Trans. Zool. Soc., London, 22:335-356.

Sagitta
Burfield, S. T., 1927.
Sagitta. L.M.B.C. Mem:XXVIII
Findin Vol. XLI of L.B.S.

chaetognaths
Castillo, I. Moreno, 1963.
Sobre los terminos a emplear en el estudio
morfologico de los Quetognatos.
Bol. R. Soc., Espanola Hist. Nat., (B), 61:5-30.

sagittae
Clarke, G. L., E.L. Pierce and D.F. Bumpus, 1943
The distribution and reproduction of Sagitta
elegans on Georges Bank in relation to the
hydrographical conditions. Biol. Bull., 85(3):
201-226; 10 textfigs.

Chaetognatha
Coleman, J. S., 1959.
The "Rosaura" Expedition 1937-1938:
Chaetognatha.
Bull. Brit. Mus. (Natural History)
Zool. 5(8):219-253.

chaetognaths
Conant, F.S., 1896.
23. Notes on the chaetognaths. Ann. Mag. Nat.
Hist., 6th ser., 18:201-214.

chaetognaths
David, P.M., 1965.
The Chaetognatha of the Southern Ocean.
In: Biogeography and ecology in Antarctica,
P. Van Oye and J. Van Mieghen, Dr. W. Junk, The Hague,
296-323.

chaetognaths
David, Peter M., 1964
The distribution of Antarctic chaetognaths.
In: Biologie Antarctique, Proc. S.C.A.R. Symposium, Paris, 2-8 September 1962, Hermann, Paris,
253-258.

chaetognaths
David, P.M., 1962.
The distribution of Antarctic chaetognaths.
(Abstract for Symposium Antarctic Biol., Paris,
2-8 Sept. 1962).
Polar Record, 11(72):324.
Also: SCAR Bull. #12

chaetognaths
Fowler, G.H., 1907.
Chaetognatha, with a note on those collected by
H.M.S. Challenger in Subantarctic and Antarctic
waters. Nat. Antarctic Exped., "Discovery", 1901-1904, 3:6 pp., 1 chart.

chaetognaths
Fowler, G.H., 1906.
The Chaetognatha of the Siboga Expedition, with a
discussion of the synonymy and distribution of th
the group. Rept. Siboga Exped., 21:1-88, 3 pls.,
6 charts.

chaetognaths
Fowler, G.H., 1905.
Biscayan plankton collected during a cruise of
H.M.S. "Research", 1900. Pt. III. The Chaeto-
gnatha. Trans. Linn. Soc., London 10(3):55-87,
Pls. 4-7.

chaetognatha
Fraser, J. H., 1939
The distribution of Chaetognatha in
Scottish Waters in 1937. J. du Cons. 14(1):
25-34, 3 charts in text.

chaetognaths

Furuhashi, Kenzo, 1959.
[On the pelagic chaetognatha collected from the Kuroshio warm current region south of Honshu. 1. Notes on some chaetognaths as indicator of "Kuroshio" area and cold water region.]
Umi to Sora 35(4), 34(12): 81-84.

chaetognaths

Furnestin, M.L., 1953.
Sur quelques chaetognathes d'Israel.
Bull. Res. Counc., Israel, 2(4):411-414.

S. friderici
S. bipunctata
S. enflata
S. minima
S. serratodentata

chaetognaths

Furnestin, M.L., 1953.
Chaetognathes récoltés en Méditerranée par le "Président Théodor Tissier" aux mois de juin et juillet 1950. Bull. Trav. Stat. d'Aquicult. Pêche de Castiglione, n.s., 4:275-317.

chaetognaths

Furnestin, M.L., 1958.
Quelques échantillons de zooplankton du Golfe d'Eylath (Akaba).
Haifa, Sea Fish. Res. Sta., Bull., 16:6-14.

sagittae

Fraser, J. H., 1937
The distribution of Chaetognatha in Scottish waters during 1936, with notes on the Scottish indicator species. J. Cons., XII(3):311-320

Chaetognatha

Fraser, J. H., 1949
The occurrence of unusual species of Chaetognatha in Scottish plankton collections.
JMBA 28(2):489-491.

Chaetognaths

Fraser, J.H., 1952.
The Chaetognatha and other zooplankton of the Scottish area and their value as biological indicators of hydrographical conditions.
Scottish Home Dept., Mar. Res., 1952(2):1-52, 3 pls., 21 charts.

chaetognaths

Ganapathy, P.N., and T.S. Rao, 1954.
Studies on the Chaetognatha of the Visakapetnam coast. Andhra Univ. Ocean. Mem. 1:143-148.

chaetognaths

Germain, L. and L. Joubin, 1916
Chétognathes provenant des campagnes des yachts Hirondelle et Princesse-Alice (1885-1910) Rés. Camp. Sci., Monaco, 49:118 pp., 7 pls., 7 maps.

Chaetognaths

George, P.C., 1952.
A systematic account of the Chaetognatha of the Indian coastal waters, with observations on their seasonal fluctuations along the Malabar coast.
Proc. Nat. Inst. Sci., India, 18(6):657-689.

chaetognaths

Ghirardelli, E., 1952.
Osservazioni biologiche e sistematiche sui Chetognati del Golfo di Napoli.
Pubbl. Staz. Zool., Napoli, 23(2/3):296-312.

sagittae

Ghirardelli, E., 1950.
Osservazioni biologiche e sistematiche sui chetognati della Baia di Villafranche-sur-mer. (Trav. Sta. Zool. Villefranche-sur-mer 10(9)). Bol. Pesca, Piscic. Idrobiol., n.s., 5(1):5-27, 7 textfigs.

chaetognaths

Ghirardelli, E., 1950.
Osservazioni biologiche e sistematiche sui chetognati della baia di Villefranche sur mer. Bol. Pesca, Piscicol. e Idrobiol., n.s., 5(1):105-127, 7 textfigs.

chaetognaths

Ghirardelli, Elvezio, 1947
Chetognati raccolti nel Mar Rosso e nel l'Oceano Indiano dalla Nave "Cherso", Bollettino di Pesca, Piscicoltura, e Idrobiologia Anno 23, Vol. 2(n.s.) (2):253-270, 2 pls., 9 text figs.

Chaetognaths

Grant, George C., 1963
Chaetognatha from inshore coastal waters off Delaware, and a northward extension of the known range of Sagitta tenuis
Chesapeake Science. 4(1):38-42.

Sagitta

Henderson, G. T. D.; and N. B. Marshall, 1944
Continuous plankton records: The zooplankton (other than copepods and young fish) in the southern North Sea, 1932-1937. Hull. Bull. Mar. Ecol., 1(6):255-275, 24 pls.

chaetognaths

Hosoe, K., 1956.
Chaetognaths from the Isles of Fernando de Noronha.
Contr. Avulsas Inst. Oceanogr., Sao Paulo, No. 3: 8 pp.

sagittae

Huntsman, A.G. and M.E. Reid, 1921
The success of reproduction of Sagitta elegans in the Bay of Fundy and the Gulf of St. Lawrence. Trans. Roy. Canadian Inst., 13:99-112

chaetognaths

Japan, Hakodate Marine Observatory, 1967.
Report of the oceanographic observations in the Okhotsk Sea, east of the Kurile islands and Hokkaido and east of the Tohoku district from August to September, 1964. (In Japanese).
Bull. Hakodate mar. Obs., 13:7-19.

chaetognaths

Japan, Hakodate Marine Observatory, 1967.
Report of the oceanographic observations in the sea southeast of Hokkaido in February, 1964. (In Japanese).
Bull. Hakodate mar. Obs., 13:3-9.

chaetognaths

Japan, Nagasaki Marine Observatory, Oceanographic section, 1965.
Report of the oceanographic observations in the sea west of Japan from February to March 1963, -- from July to August 1963.
Res. Mar. Meteorol. Oceanogr., Japan, Meteorol Agency, 33:39-58; 34:53-80.
Also in :- Oceanogr. Meteorol. Nagasaki, 15 (227-228).

Sagitta

John, C. C., 1931.
On the anatomy of the head of Sagitta. Proc. Zool. Soc., London, :1607.

chaetognaths

Kado, Y., 1953.
The chaetognath fauna of the Inland Sea of Japan especially on the distribution of Sagitta enflata and S. crassa. Dobuts. Zasshi 62(10):

also:
S. bedoti
S. robusta
S. regularis
S. minima
Pterosagitta draco

Chaetognatha

Kramp, P.L., 1939
Chaetognatha. Godthaab Exped. 1928
Medd. om Grønland. 80(5):40 pp.

chaetognath

Kramp, P.L., 1938.
Chaetognatha. Zool. Iceland 4(71):4 pp.

Chaetognatha

Kramp, P.L., 1933
Coelenterata, Ctenophora and Chaetognatha. 2nd East Greenland Exped. 1932. Medd. om Grønland, 104(11):20 pp.

chaetognaths

Kramp, P.L., 1915.
Medusae, Ctenophora and Chaetognathi from the Great Belt and the Kattegat in 1909.
Medd. Komm. Havundersøgelser, Ser. Plankton, 1(12):20 pp.

sagittae

Kuhl, W., 1928.
Chaetognatha. In: Grimpe u. Wagler, Tierwelt Nord- u. Ostsee. Leipzig.

chaetognaths

Lea, H.E., 1955.
The chaetognaths of western Canadian coastal waters. J. Fish. Res. Bd., Canada, 12(4):593-617.

sagittae

Lucas, C.E., N.B. Marshall and C.B. Rees 1942
Continuous Plankton Records: The Faeroe-Shetland Channel, 1939. Hull Bull. Mar. Ecol., II(10):71-94

Chaetognatha

Michael, E. L., 1911
Classification and vertical distribution of the Chaetognatha of the San Diego Region including redescriptions of some doubtful species of the group. Univ. Calif. Publ. Zool. 8(3):21-186, Pls. 1-8.

chaetognaths

Moore, H.B., 1955.
Variations in temperature and light response within a plankton population. Biol. Bull. 108(2):175-181, 4 textfigs.

chaetognaths

Moore, H.B., H. Owre, F.C. Jones and T. Dow, 1953
Plankton of the Florida Current. III. The control of the vertical distribution of zooplankton in the daytime by light and temperature.
Bull. Mar. Sci., Gulf and Caribbean, 3(2):83-95, 1 textfig.

chaetognaths

Owre, H.B., 1960
Plankton of the Florida Current. VI. The Chaetognatha. Bull. Mar. Sci., Gulf and Caribbean, 10(3): 255-322.

chaetognaths

Pierce, E.L., 1958.
The Chaetognatha of the inshore waters of North Carolina. Limnol. & Oceanogr., 3(2):166-170.

chaetognaths

Pierce, E.L., 1953.
The Chaetognatha over the continental shelf of North Carolina with attention to their relation to the hydrography of the area. J. Mar. Res. 12(1):75-92, 4 textfigs.

chaetognaths

Pierce, E.L., 1951.
The Chaetognatha of the west coast of Florida.
Biol. Bull. 100(3):206-228, 5 textfigs.

chaetognaths

Ramult, M., and M. Rose, 1945
Recherches sur les Chetognathes de la Baie d'Alger. Bull. Soc. Hist. Nat. Afrique du Nord 36:45-71, 39 figs.

chaetognaths

Rao, T.S. Satyanarayana, and P.N. Ganapati, 1958

Studies on Chaetognatha in the Indian Seas. III. Systematics and distribution in the water off Visakhapaynam. Andhra Univ. Mem. Oceanogr. 2:147-163.

chaetognaths

Redfield, A. C. and A. Beale, 1940.
Factors determining the distribution of populations of chaetognaths in the Gulf of Maine. Biol Bull., 79(3):459-487, 11 text figs.

Chaetognaths
Ritter-Záhony, R. von, 1914
Chaetognaths. Danish Ingolf Exped.
IV (3):6 pp.

Chaetognath
Ritter-Záhony, R. von, 1911
Revision der Chätognathen. Deutsche
Südpolar Exped., 1901-1903, XIII (Zool.V).,
Pt.1:71 pp., 51 text figs.

chaetognaths
Ritter-Zahony, R. von, 1910.
Chaetognatha. Die Fauna Sudwest Australiens 3:
123-126.

chaetognaths
Rose, M., and M. Hamon, 1953.
Nouvelle note complémentaire sur les chétognathes
de la Baie d'Alger. Bull. Soc. Hist. Nat.,
Afrique du Nord, 44(5/6):167-171.

chaetognaths
Russell, F. S., 1939.
Chaetognatha. Fiches d'Ident. Zooplancton. Cons.
Perm. Int. Expl. Mer, 1:4 pp., 12 textfigs.

sagittae
Russell, F. S., 1933
On the biology of Sagittae. IV. Observations
on the natural history of Sagitta elegans
Verrill and Sagitta setosa J. Müller in the
Plymouth area. JMBA 18:559-574

sagittae
Russell, F. S. 1932
On the biology of Sagitta. The breeding and
growth of Sagitta elegans Verrill in the
Plymouth area. JMBA, 18:131-146
1930-1931

chaetognaths
Russell, F. S. and J. S. Colman, 1934
The Zooplankton. II. The Composition of
the Zooplankton of the Barrier Reef Lagoon.
Brit. Mus. (Nat. Hist.), Great Barrier Reef
Expedition, 1928-1929, Sci. Repts. II(6):159-176, 186-201, 11 text figs., tables 2-10.

Chaetognaths
Scaccini, A., e E. Ghirardelli, 1941.
Chaetognathes collected on the coast of Rio de
Oro. Not. Ist. Biol., Rovigno, 2(21):3-15.

chaetognaths
Scaccini, A., e E. Ghirardelli, 1941.
Chaetognaths of Adriatic Sea near Rovigno.
Not. Ist. Biol., Rovigno, 2(22):3-16.

chaetognaths
Suarez Caabro, J.A., 1955.
Quetognatos de los mares cubanos.
Mem. Soc. Cubana Hist. Nat., 22(2):125-180, 9 pls

chaetognaths
Suarez-Caabro, Jose A., and Juan E. Madruga, 1960.
The Chaetognatha of the northeastern coast of
Honduras Central America.
Bull. Mar. Sci., Gulf & Caribbean, 10(4):421-429

Chaetognatha
Sund, Paul N., 1959.
A key to the Chaetognatha of the tropical
Eastern Pacific Ocean.
Pacific Science, 13(3):269-285.

chaetognatha
Sund, P. N., and J. A. Renner, 1959.
The chaetognatha of the Eastropic Expedition, with
notes as to their possible value as indicators of
hydrographic conditions.
Inter-Amer. Trop. Tuna Comm., Bull., 3(9):395-436.

chaetognaths
Tchindonova, J.G., 1955.
[Chaetognatha of the Kurile-Kamchatka Trench]
Trudy Inst. Oceanol., 12:298-310.

chaetognatha
Thomson, J.M., 1948
Some chaetognatha from western Australia.
J. Roy. Soc. Western Australia, Inc., 31 (1944-1945), 17-18.

Chaetognatha
Thomson, J.M., 1947
The Chaetognatha of southeastern Australia.
Australia Counc. Sci. and Indust. Res. Bull.
222:1-43.

Chaetognatha
Thomson, J. M., 1947
The Chaetognatha of South-eastern Australia.
Counc. Sci. & Ind. Res., Australia, Bull. No.222,
(Div. Fish. Rept. 14), 43 pp., 8 text figs.

Chaetognatha
Tokioka, Takasi, 1965.
Supplementary notes on the systematics of
Chaetognatha.
Publs Seto mar.biol.Lab., 13(3):231-242.

Chaetognatha
Tokioka, Takasi, 1965.
The taxonomic outline of Chaetognatha.
Publs Seto mar.biol.Lab., 12(5):335-357.

chaetognaths
Tokioka, Takasi, 1962
The outline of the investigations made on
chaetognaths in the Indian Ocean. (In Japanese
and English).
Info. Bull. Planktology, Japan, No. 8:5-11.

chaetognaths
Tokioka, Takasi, 1959
Observations on the taxonomy and distribution of chaetognaths of the North
Pacific.
Publ. Seto Mar. Biol. Lab., 7(3):349-456.

chaetognaths
Tokioka, T., 1955.
Notes on some chaetognaths from the Gulf of Mexico. Bull. Mar. Sci., Gulf & Caribbean, 5(1):52-65.

chaetognaths
Tokioka, T., 1954.
Droplets from the plankton net. XVI. On a small
collection of chaetognaths from the central
Pacific. Publ. Seto Mar. Biol. Lab. 4(1):99-102,
1 textfig.

chaetognaths
Tokioka, T., 1952.
Chaetognaths of the Indo-Pacific.
Annot. Zool. Japon. 25(1/2):307-316.

chaetognaths
Tokioka, T., 1951.
Pelagic tunicates and chaetognaths collected
during the cruises to the New Yamato Bank in the
Sea of Japan. Publ. Seto Mar. Biol. Labl. 2(1):
1-26, 1 chart, 6 tables, 12 textfigs.

chaetognaths
Tokioka, T., 1950.
Notes on the development of the eye and the
vertical distribution of Chaetognatha. Nature
et Kultura, Kyoto, 1:

chaetognaths
Tregouboff, G., 1956.
Rapport sur les travaux concernant le
plancton Mediterranéen publiés entre
Novembre 1952 et Novembre 1954.
Rapp. Proc. Verb., Comm. Int. Expl.
Sci., Mer Mediterranee, 13:65-100

chaetognaths
Vannucci, M., and K. Hosoe, 1952.
Resultados cientificos do Cruzeiro do "Baependi"
e do "Vega" a Ilha da Trinidada. Bol. Inst.
Ocean., Univ. Sao Paulo, 3(1/2):5-30, 4 pls.

chaetognaths
Varadarajan, S. and P.I. Chacko, 1943.
On the arrow-worms of Krusadai. Proc.
Nat. Inst. Sci., India, 9(2):245-248, 2 figs.

chaetognaths
von Ritter-Zahony, R., 1914.
Chaetognaths. Danish Ingolf-Exped. 4(3):4 pp.

Chaetognatha
Wiborg, K.F., 1944
The production of plankton in a land
locked fjord. The Nordåsvatn near Bergen
in 1941-1947. With special reference to
copepods. Fiskeridirektoratets Skrifter
Sorie Havundersøkelser (Rept. Norwegian
Fish. and Marine Invest.) 7(7):1-83, Map.

chaetognath, anatomy
Aida, T., 1897.
On the growth of the ovum in chaetognaths.
Ann. Zool., Japon., 1:77-81, Pl. 4.

chaetognaths, physiol.
Beers, John R., 1964.
Ammonia and inorganic phosphorus excretion by
the planktonic chaetognath, Sagitta hispida
Conant. J. du Cons., 29(2):123-129.

chaetognaths, physiol.
Beers, John R., 1962
Ammonia and inorganic phosphorus excretion
of some marine zooplankton. (Abstract).
Assoc. Island Mar. Labs. 4th Meet. Curacao,
18-21 Nov., 1962: 1.

chaetognaths, anat.-physiol.
Bieri, Robert, 1966.
The function of the "wings" of Pterosagitta
draco and the so-called tangoreceptors in other
species of Chaetognatha.
Publs Seto mar.biol.Lab., 14(1):23-26.

chaetognaths, anat.-phys.
Bogorov, B.G., 1940.
[On the biology of Euphausiidae and Chaetognatha
in the Barents Sea]
Biull. Moskovskoi Obshchestvo Ispytatelei Prirody
Otdel Biol., n.s., 49(2):3-13, diagrams.

chaetognaths, anat.
Bordas, M., 1920.
Estudio de la ovogenesis en la Sagitta bipunctata Quoy
et Gaimard.
Trab. Mus. Nac. Ciencias Nat. Madrid, Ser. Zool., 42:
5-

chaetognaths, anat.
Buchner, P., 1910.
Die Schicksale des Keimplasmas der Sagitten in Reifung,
Befruchtung, Keimbahn, Ovogenese und Spermatogenese.
Festschr. f. R. Hertwig, 1: 233-

chaetognaths, anat.
Cosper, T.C. and M.R. Reeve, 1970.
Structural details of the mouthparts of a
chaetognath, as revealed by scanning electron
microscopy. Bull. mar. Sci., 20(2): 441-445.

chaetognathes, anat. physiol.
Dallot, Serge, 1970.
L'anatomie du tube digestif dans la
phylogénie et la systématique des
Chaetognathes.
Bull. Mus. natn. Hist. nat. (2) 42(3): 549-565.

Oceanographic Index: Marine Organisms Cumulation, 1946-1973

chaetognaths, characteristics of
Dallot, S. et F. Ibanez, 1972
Etude préliminaire de la morphologie et de l'évolution chez les Chaetognathes. Inv. Pesq. Barcelona 36(1): 31-41.

Chaetognaths, anat.-physiol.
David, P.M., 1958.
A new species of Eukrohnia from the Southern Ocean, with a note on fertilization. Proc. Zool. Soc., London, 131:597-606.

chaetognaths, anat.-physiol
Della Croce, N., 1963.
Osservazioni sull'alimentazione di Sagitta. Rapp. Proc. Verb., Réunions, Comm. Int. Expl. Sci. Mer Méditerranée, Monaco, 17(2):627-630.

chaetognaths, anat.
Faure, M. L., 1952.
Contribution a l'etude biologique et morphologique de deux chaetognaths des eaux atlantiques du Maroc: Sagitta friderici Ritter-Zahony et Sagitta bipunctata Quoy et Gaimard. Vie et Milieu 3:25.

chaetognathes, anat.
Furnestin, Marie-Louise, 1967.
Contribution a l'étude histologique des chaetognathes. Rev. Trav. Inst. Pêches marit., 31(4):383-392.

chaetognatha, anat.-physiol.
Furnestin, M.-L., 1959.
Sur la coloration du tube digestif de certaines chaetognathes. Bull. Soc. Zool. de France, 84(2-3):132-135.

chaetognaths, anat-physiol.
Ghirardelli, Elvezio, 1968.
Some aspects of the biology of the chaetognaths. Advances in Marine Biology, 6:271-375.

chaetognaths, anat.-physiol.
Ghirardelli, E., 1962.
Ambiente e biologia della riproduzione nei Chetognati. Metodi di valutazione degli stadi di maturità e loro importanza nelle ricerche ecologiche. Problemi ecologici delle zone litorale del Mediterraneo, 17-23 luglio 1961. Pubbl. Staz. Zool., Napoli, 32(Suppl.):380-399.

chaetognath, anatomy
Ghirardelli, E., 1961.
Histologie et cytologie des stades de maturité chez les chetognathes. Rapp. Proc. Verb., Réunions, Comm. Int. Expl. Sci. Mer Méditerranée, Monaco, 16(2):103-110.

chaetognaths anat. physiol.
Ghirardelli, E., 1960.
Habitat e biologia della riproduzione nei Chetognati. Arch. Oceanogr. e Limnol., 11(3):287-304.

Reprinted in: Trav. Sta. Zool. Villefranche-sur-Mer, 18(11):

Chaetognaths, anat.
Ghirardelli, Elvezio, 1960.
Istologia e citologia degli stadi di maturita nei Chetognati. Boll. Pesca, Piscicolt. e Idrobiol., n.s., 15(1)

chaetognaths, anat-physiol.
Ghirardelli, Elvezio, 1959.
Osservazioni sulla corona ciliata nei Chetognati. Boll. Zool., Unione Zool. Italiana, 26(2):413-421.
Also in: Trav. Sta. Zool., Villefranche-sur-Mer, 19(1960).

chaetognaths, anat. phys.
Ghiradelli, Elvezio, 1959.
L'apparato riproduttore femminile nei Chetognati. Rendiconti Accad. Naz. dei XL (4) 10:46 pp., 14 pls.

Reprinted in: Trav. Sta. Zool., Villefranche-sur-Mer, 18(18).

chaetognaths, anat-phys.
Ghirardelli, E., 1959.
Osservazioni sulla deficienza dei poteri rigenerativi nei Cetognati. Trav. Sta. Zool., Villefranche-sur-Mer, 18(8):

Reprinted from: Atti, Accad. Sci., Ist. Bolognia, Cl. Sci. Fis., (11) 6:1-15.

chaetognaths, anat.
Ghirardelli, E., 1954.
Osservazioni sul corredo cromosomico di Sagitta inflata Grassi. Scientia Genetica, 4(4):336-343.

In: Trav. Sta. Zool. Villefranche-sur-mer, 13.

chaetognaths, anat. physiol.
Ghirardelli, E., 1954.
Studi sul determinante germinale (d.g.) nei chetognati: R cerche sperimentali su Spadella cephaloptera Busch. Pubbl. Staz. Zool., Napoli, 25:444-453, 1 textfig.

chaetognaths, anat
Ghiradelli, E., 1953.
Appunti sulla morfologia dell'apparecchio riproduttore femminile e sulla biologia della riproduzione in Pterosagitta draco Krohn. Monitore Zoologico Italiano, 61(2/3):71-79.

chaetognaths - anat.-physiol.
Ghirardelli, E., 1953.
L'accoppiamento in Spadella cephaloptera Busch. Pubbl. Staz. Zool., Napoli, 24(3):345-354, 1 textfigs.

chaetognaths, anat.-phys.
Ghirardelli, E., 1953.
Osservazioni sul determinante germinale (d.g.) e su altre formazioni citoplasmatiche nelle uova di Spadella cephaloptera Busch (Chaetognata). Pubbl. Staz. Zool., Napoli, 24(3):332-344, Pl.

chaetognaths, anat.
Ghirardelli, E., 1951.
Cicli di maturita sessuale nelle gonadi di Sagitta inflata Grassi del Golfo di Napoli. Boll. Zool., Unione Zool. Ital. 18(4/5/6):149-161, 10 textfigs.

chaetognaths, biol.
Heinrich, A.K., 1956.
Dimensional composition of Chaetognatha and their terms of propagation in the western regions of the Bering Sea. Dokl. Akad. Nauk, SSSR, 110(6):1105-1107.

chaetognathes, anat-physiol
Horridge, G.A., and P.S. Boulton, 1967.
Prey detection by Chaetognatha via a vibration sense. Proc. R. Soc., (B)168(1013):413-419.

Chaetognath anat. physiol
Hyguet, D., 1969.
Contribution à l'étude de la structure du ganglion nerveaux ventral des Sagitta (chaetognathes). Bull. Mus. natn. Hist. Nat., (2) 40 (5): 1031-1042.

chaetognaths, physiol.
Jagerstan, G., 1940.
Zur Kenntnis der Physiologie der Zeugung bei Sagitta. Zool. Bidrag., Uppsala, 18:397-413.

chaetognaths, anat.
Nagasawa, Sachiko and Ryuzo Marumo, 1972
Structure of grasping spines of six chaetognath species observed by scanning electron microscopy. Bull. Plankt. Soc. Japan 19(2):5-16 (63-74).

chaetognaths, anat. physiol.
Reeve, M.R. and M.A. Walter, 1972.
Observations and experiments on methods of fertilization in the chaetognath Sagitta hispida. Biol. Bull. mar. biol. Lab. Woods Hole 143 (1): 207-214.

chaetognaths, physiol.
Sameoto, D.D. 1972
Yearly respiration rate and estimated energy budget for Sagitta elegans. J. Fish. Res. Bd Can. 29 (7): 987-996.

Chaetognatha, anat.-physiol.
Tokioka, T., 1959.
Chaetognaths collected chiefly from the bays of Sagami and Suruga with some notes on the shape and structure of the seminal vesicles. Rec. Oceanogr. Wks., Japan, 10(2):123-150.

Chaetognaths, biomass
Japan, Hakodate Marine Observatory 1970
Report of the oceanographic observations in the sea east of Honshu and Hokkaido from February to March 1966. (In Japanese) Bull. Hakodate mar. Obs. 15:3-10.

chaetognaths, biomass
Nair, Vijayalakshmi R., 1969.
A preliminary report on the biomass of chaetognaths in the Indian Ocean comparing the south-west and north-east monsoon periods. Bull. natn. Inst. Sci. India 38(2): 747-752. Also in: Coll. Repr. Nat. Inst. Oceanogr. Goa India 2 (1968-1970).

breeding
Clarke, G. L., and E. L. Pierce and D. F. Bumpus 1943
The distribution and reproduction of Sagitta elegans on Georges Bank in relation to the hydrographical conditions. Biol. Bull., 85(3): 201-226, 10 textfigs.

breeding
Huntsman, A.H., and M.E. Reid, 1921
The success of reproduction of Sagitta elegans in the Bay of Fundy and the Gulf of St. Lawrence Trans. Roy. Canadian Inst., 13:99-112

breeding
Pierce, E. L., 1941
The occurrence and breeding of Sagitta elegans Verrill and Sagitta setosa J. Müller in parts of the Irish Sea. JMBA, 25:113-124

breeding
Russell, F.S. 1932
On the biology of Sagitta. The breeding and growth of Sagitta elegans Verrill in the Plymouth Area, 1930-1931. JMBA, 18:131-146

chaetognaths, chemistry
Reeve, M.R., J.E.G. Raymont and J.K.B. Raymont, 1970.
Seasonal biochemical composition and energy sources of Sagitta hispida. Marine Biol., 6(4): 357-364.

chaetognaths, chemistry of
Moreno, Isabel 1972
La ausencia de quitina en los quetognatos. Bol. R. Soc. Española Hist. Nat. (Biol.) 70 (1/2): 127-130

chaetognaths, color of
Bieri, Robert, 1966.
A pale blue chaetognath from Tanabe Bay. Publs Seto mar. biol. Lab., 14(1):21-22.

chaetognaths, color of
Tokioka, Takasi, and Robert Bieri, 1966.
The colour pattern of Spadella anguleta Tokioka. Publs Seto mar. biol. Lab., 14(4):323-326.

chaetognaths, distribution
Fowler, G.H., 1906.
The Chaetognatha of the Siboga Expedition with a discussion of the synonymy and distribution of the group. Rept. Siboga Exped. 21:1-88, 3 pls., 6 charts.

chaetognaths (distr.)
Furnestin, M.-L., 1957.
Chaetognathes et zooplancton du secteur Atlantique marocain. Ann. Biol., (3)33(7/8):345-366.

sagittae (distr. only)
Jespersen, P. (posthumous).1954
On the quantities of macroplankton in the North Atlantic. Medd. Danmarks Fisk. og Havundersøgel., n.s., 1(2):1-12.

chaetognaths (seasonal distribution)
Sudarsan, D., 1961.
Observations on the Chaetognatha of the waters around Mandapam.
Indian J. Fish., 8(2):364-382.

chaetognatha, ecology
Rao, T.S. Satyanarayana, 1958
Studies on Chaetognatha in the Indian Seas. IV. Distribution in relation to currents.
Andhra Univ. Mem. Oceanogr., 2: 164-167.

chaetognaths, egg production [a]
Reeve, Michael R., 1970
The biology of Chaetognatha. I. Quantitative aspects of growth and egg production in Sagitta hispida.
In: Marine Food Chains, J.H. Steele, editor, Oliver and Boyd, 168-189.

feeding, chaetognaths
Reeve, M.R., 1964.
Feeding of zooplankton with special reference to some experiments with Sagitta.
Nature, 201(4915):211-213.

chaetognaths fluctuations (seasonal) [a]
Nair, Vijayalakshmi R. 1971.
Seasonal fluctuations of chaetognaths in the Cochin Backwater.
J. mar. biol. Ass. India 13(2): 226-233.

chaetognaths, growth of [a]
Reeve, Michael R., 1970
The biology of Chaetognatha. I. Quantitative aspects of growth and egg production in Sagitta hispida.
In: Marine Food Chains, J.H. Steele, editor, Oliver and Boyd, 168-189.

chaetognaths, lists of spp.
Alvariño, Angeles 1969.
Zoogeografía del Mar de Cortés: Quetognatos, sifonóforos y medusas.
An. Inst. Biol. Nal. Autón. México Ser. Cienc. Mar. Limnol. 40(1): 11-54.

chaetognaths, lists of spp.
Alvariño, Angeles 1968.
Los quetognatos, sifonóforos y medusas en la region del Atlantico ecuatorial bajo la influencia del Amazonas.
An. Inst. Biol. Univ. Nat. Autón. México Ser. Cienc. Mar. Limnol. 39(1): 41-76.

chaetognaths, lists of spp.
Alvarino, Angeles, 1964
Bathymetric distribution of chaetognaths.
Pacific Science, 18(1):64-82.

chaetognaths, lists of spp.
Alvariño, Angeles, 1964.
The Chaetognatha of the MONSOON Expedition in the Indian Ocean.
Pacific Science, 18(3):336-348.

chaetognaths, lists of spp. [a]
Beaudouin, Jacqueline 1971.
Données écologiques sur quelques groupes planctoniques indicateurs dans le Golfe de Gascogne.
Revue Trav. Inst. Pêches marit. 35(4): 375-414.

chaetognaths, lists of spp.
Belloc, G., 1961.
Catalogue des types de chétognathes du Musée Océanographique de Monaco.
Bull. Inst. Océanogr., Monaco, 58(1216):3 pp.

Chaetognathes, lists of spp. [a]
Casanova, Bernadette, Francoise Ducret et Jeannine Rampal 1973
Zooplancton de Méditerranée orientale et de mer Rouge (Chaetognathes, Euphausiacés, Pteropodes). Rapp. Proc.-v. Reun. Comm. int. Explor. scient. Mer Medit. Monaco 21(8):515-519.

chaetognaths, lists of spp.
David, P.M., 1963
Some aspects of speciation in the chaetognaths.
Systematics Assoc., Publ. (Speciation in the Sea), No. 5:129-143.
Also in:
Collected Reprints, N.I.O., Vol. 11(435).1963.

Chaetognatha, lists of spp.
David, Peter M., 1959
Chaetognatha.
Brit.-Australian-New Zealand Antarctic Res. Exp. (B) 8:73-79.

chaetognaths, lists of spp.
de Almeida Prado, M.S., 1968.
Distribution and annual occurrence of Chaetognatha of Cananeia and Santos Coast (São Paulo, Brazil).
Bolm Inst oceanogr. S. Paulo 17(1):33-55.

chaetognaths
de Almeida Prado, M.S., 1961
Chaetognatha encontrados em aguas brasileiras.
Bol. Inst. Oceanogr., Univ. Sao Paulo, 11(2): 31-55.

chaetognaths, lists of spp. [a]
Ducret, Francoise, 1973.
Contribution a l'etude des chaetognaths de la mer Rouge. Beaufortia 20 (268): 135-153.

chaetognaths, lists of spp.
Ducret, Francoise, 1968.
Chaetognathes des campagnes de l'Ombango dans les eaux equatoriales et tropicales africaines.
Cah. ORSTOM., Sér.Océanogr.6(1):95-141.

chaetognaths, lists of spp. [a]
Fives, Julie M., 1971.
Investigations of the plankton of the west coast of Ireland. V. Chaetognatha recorded from the inshore plankton off Co. Galway.
Proc. R. Ir. Acad. (B) 71 (9): 119-138.

Chaetognaths, lists of spp. [a]
Furnestin, Marie Louise 1971.
Chaetognathes des Campagnes danoises en Méditerranée et en mer Noir.
Rapp. P.-V. Comm. int. Explor. mer Medit. 20(3): 421-424.

Chaetognaths, lists of spp.
Furnestin, Marie-Louise 1970
Chaetognathes des campagnes danoises dans l'Atlantique nord: notes biologiques et biogéographiques.
Dana Rept. 80:1-7.

chaetognaths, lists of species
Furnestin, M.-L., 1956.
Chaetognathes recueillis par l'Élie Monnier au large des côtes du Sénégal. Bull. d'IFAN, 8(2): 406-409.

chaetognathes, lists of spp
Furnestin, M.-L., et J. Radiguet, 1964.
Chaetognathes de Madagascar (Secteur de Nosy-Be).
Cahiers, O.R.S.T.R.O.M., Océanogr., 11(4):55-98.

chaetognaths, lists of spp.
Furuhashi, Kenzo, 1961
On the distribution of chaetognaths in the waters off the south-eastern coast of Japan. (JEDS-3).
Publ. Seto Mar. Biol. Lab.,9(1)(2):17-30.
Also in: Repts. of JEDS, Vol. 2(1961)

Chaetognaths, lists of spp.
Furuhashi, Kenzo, 1961.
On the distributions of some plankton animals in the Kuroshio region of Honshu, Japan, with notes on the nature and origin of the cold water mass appearing in the region. 1. The distribution of copepods and chaetognaths.
Umi to Sora, 37(3):73-80.

Chaetognaths, lists of spp.
Furuhashi, Kenzo, 1961.
On the distribution of some plankton animals in the Kuroshio region south of Honshu, Japan, with notes on the nature and origin of the cold water mass appearing in the region. 1. The distribution of copepods and chaetognaths.
Umi to Sora, 37(4):100-111.

Chaetognaths, lists of spp
Ghirardelli Elvezio, 1968.
Chaetognathes récoltés par l'Argonaut en haute Adriatique.
Rapp. Proc.-verb. Réun. Comm. int. Explor. scient. Mer Méditerranée 19(3): 475-477.

chaetognaths, lists of spp.
Hamon, M., 1956.
Chétognathes recueillis dans la Baie de Nhatrang-Cauda (Viet-Nam).
Bull. Mus. Nat. Hist. Nat., 28(5):466-473.

chaetognaths, lists of spp.
Hida, T.S., 1957.
Chaetognaths and pteropods as biological indicators in the North Pacific.
USFWS Spec. Sci. Rept., Fish., No. 215:13 pp.

chaetognaths, lists of spp.
Hoenigman, J., I. Gasparoviv and J. Kavao, 1961.
Cladocères et chétognathes provenant d'une station au large de l'Île de Mljet (Adriatique).
Rapp. Proc. Verb., Réunions, Comm. Int. Expl. Sci. Mer Méditerranée, Monaco, 16(2):117-121.

chaetognaths, lists of spp
Ibanez, F. et S. Dallot, 1969.
Etude du cycle annuel des Chaetognathes planctoniques de la rade de Villefranche par la methode d'analyse des composantes principales. [Study of the annual cycle of planktonic chaetognaths in Villefranche Bay, by the method of principal component analysis]. Marine Biol., 3(1): 11-17.

chaetognaths, lists of spp

Japan, Hakodate Marine Observatory, 1969.
Report of the oceanographic observations in the sea east of Hokkaido and the Kuril Islands, and in the Okhotsk Sea from July to September, 1965.
Bull. Hakodate mar. Obs. 14: 3-15.

chaetognaths, lists of spp.

Japan, Maizuru Marine Observatory 1972.
Data of the oceanographic observations (1966-1970), 235pp.

chaetognaths, lists of spp.

Japan, Maizuru Marine Observatory, 1965.
Report of the oceanographical observations in the Japan Sea from August to September, 1964. (In Japanese).
Bull. Maizuru mar. Obs., 10:74-86.

Chaetognaths, lists of spp.

Japan Meteorological Agency 1969
The results of marine meteorological and oceanographical observations, Jan.-June 1966, 39: 349 pp. (multilithed)

chaetognaths, lists of spp.

Japan Meteorological Agency, 1968.
The results of marine meteorological and oceanographical observations, July-December, 1965, 38: 404 pp. (multilithed)

chaetognaths, lists of spp.

Japan, Japan Meteorological Agency. 1965.
The results of marine meteorological and oceanographical observations. July-December 1963, No. 34: 360 pp.

chaetognaths, lists of spp. (quant.)

Japan, Japan Meteorological Agency, 1962
The results of marine meteorological and oceanographical observations, January-June 1961, No. 29: 284 pp.

chaetognaths, lists of spp.

Japan, Kobe Marine Observatory, Oceanographical Section, 1962
Report of the oceanographical observations in the sea south of Honshu from July to August 1961. (In Japanese).
Res. Mar. Meteorol. & Oceanogr., 30:39-48.
Also in:
Bull. Kobe Mar. Obs., No. 173(3). 1964.

chaetognaths, lists of spp.

Japan, Maizuru Marine Observatory, 1965.
Report of the oceanographic observations in the central part of the Japan Sea from February to March, 1962.---in the Japan Sea from June to July, 1962.---in the western part of Wakasa Bay from January to April, 1962.---in the central part of the Japan Sea from September to October 1962.---in the western part of Wakasa Bay from May to November, 1962.---in the central part of the Japan Sea in March, 1963.---in the Japan Sea in June, 1963.---in Wakasa Bay in July, 1963.---in the central part of the Japan Sea in October, 1963. (In Japanese).
Bull. Maizuru Mar. Obs., No.9:67-73;74-88;89-95; 71-80;81-87;59-65;66-77;80-84;85-91.

chaetognaths, lists of spp.

Kado, Yoichi, 1957
The seasonal change of the chaetognath and pelagic copepod fauna of Hiroshima Bay in the Inland Sea of Japan, with special reference to the appearance of oceanic species.
J. Sci. Hiroshima Univ., (B) (1):17(9):122-129.
Also:
Contrib. Mukaishima Mar. Biol. Sta., No. 54.

chaetognaths, lists of spp.

Katori, Moriyuki, 1972
Vertical distribution of chaetognaths in the northern North Pacific Ocean and Bering Sea.
In: Biological oceanography of the northern North Pacific Ocean, A.Y. Takenouti, Chief Editor, Idemitsu Shoten, Tokyo, 291-308.

chaetognaths, lists of spp.

Kawarada, Y., M. Kitou, K. Furuhashi and A. Sano, 1969
Distribution of plankton in the waters neighboring Japan in 1966 (CSK)
Oceanogr. Mag. 20(2): 187-212.

chaetognaths, lists of spp.

Kawarada, Yutaka, Masataka Kitou, Kenzo Furuhashi, Akira Sano, Kohei Karohji, Kazunori Kuroda, Osamu Asaoka, Masao Matsuzaki, Mamoru Ohwada and Futomi Ogawa 1966.
Distribution of plankton collected on board the Research Vessels of J.M.A. in 1965 (CSK).
Oceanogr. Mag. Jap. Met. Soc. 18(1/2): 91-112.

chaetognaths, lists of spp.

Kinzer, Johannes, 1963.
Untersuchungen über das Makroplankton bei Ischia und Capri und im Golf von Neapel im Mai 1962. 1. Hydrographie und quantitative Verbreitung einiger Zooplankter.
Pubbl. Staz. Zool., Napoli, 33:141-162.

chaetognaths, lists of spp.

Kitou, Masataka, 1966.
Chaetognaths collected on the Fifth Cruise of the Japanese Expedition of Deep Seas
La Mer, Bull. Soc. franco-japon. Océanogr., 4(3):169-177.

Also in:
Reps JEDS, 5-7 (1966)

chaetognaths, lists of spp.

Kitou, Masataka, 1966.
Chaetognaths collected on the Sixth Cruise of the Japanese Expedition of Deep Seas.
La Mer, Bull. franco-japon. Océanogr., 4(4):261-265.

Also in:
Reps JEDS, 5-7 (1966)

chaetognaths, lists of spp.

Kitou, Masataka, 1963.
On chaetognaths collected in the Japan Trench. 1. The fourth cruise of the Japanese Expedition of Deep Seas.
Oceanogr. Mag., 15(1):63-66.

JEDS Contrib. No. 52.

Chaetognathes, lists of spp.

Kolesnikov, A.N. 1969.
Migraciones diarias del zooplancton en la región occidental del golfo de México.
Serie Oceanologia, Inst. Oceanol. Acad. Cienc., Cuba 4:4-14.

Chaetognaths, list of spp.

Kolesnikov, A.N. y A. Alfonso 1969
Datos preliminares del zooplancton de la región oriental del Golfo de México y el Estrecho de la Florida.
Serie Oceanologia, Inst. Oceanol. Acad. Cienc. Cuba 4:15-20.

chaetognaths, lists of spp.

Kuroda Kazunori 1973.
Macroplankton biomass and chaetognath fauna in the Kuroshio south of Japan in the summer of 1971 and the winter of 1972. (In Japanese; English abstract)
Bull. Kobe mar. Obs. 159:57-64.

chaetognaths, lists of spp.

Liaw, Wen Kuang, 1967.
On the occurrence of chaetognaths in the Tanshui River estuary of northern Taiwan (Formosa).
Publs. Seto mar. biol. Lab., 15(1):5-18.

chaetognaths, lists of spp.

Marukina, N.P., 1969.
The distribution of chaetognaths in the Kuroshio area. (In Russian).
Izv. Tichookean. nauchno issled. Inst. ribn. khoz. okeanogr. (TINRO) 68: 174-185

chaetognaths, lists of species

Massuti, M., 1961
Note préliminaire à l'étude des chétognaths de la Méditerranée occidentale. Campagne du "Xauen" X-6911.
Rapp. Proc. Verb. Réunions. Comm. Int. Expl. Sci. Mer Méditerranée, Monaco, 16(2):237-244.

Chaetognaths, lists of spp.

Owre, Harding B., 1960.
Plankton of the Florida Current. VI. The Chaetognatha.
Bull. Mar. Sci., Gulf & Caribbean, 10(3):255-322

chaetognaths, lists of spp.

Owre, H. B. (Michel) and Maria Foyo, 1972.
Studies on Caribbean zooplankton. Description of the program and results of the first cruise. Bull. mar. Sci. Miami 22(2): 483-521.

Chaetognaths, lists of spp.

Park, Joo-Suck, 1970.
The chaetognaths of Korean waters. (In Korean; English abstract)
Bull. Fish. Res. Dev. Agency Pusan 6:1-174

chaetognaths, lists of spp.

Park, Joo Suck 1968.
Chaetognaths and plankton in Korean waters. II. The distribution of chaetognaths in the southern waters and their relation to the character of the water masses in the summer of 1967. (In Korean; English abstract)
Bull. Fish. Res. Develop. Agency, Korea 3:83-102.

Chaetognaths, lists of spp.

Park, Joo Suck 1967.
Note sur les chaetognathes indicateurs planctoniques dans la mer coréenne en hiver 1967. (Korean abstract)
J. oceanogr. Soc. Korea 2 (1/2): 34-41.

chaetognaths, lists of spp.

Park, Joo Suck, 1967.
Chaetognaths and plankton in the Korean waters. 1. The distribution of chaetognaths in the Korean waters and their relation to the character of water masses in summer 1966 and winter 1967. (In Korean; English abstract).
Bull. Fish. Res. Develop. Agency, 1:35-63.

Chaetognaths, lists of spp.

Pathansali, D., 1968.
Some observations on the distribution of Chaetognatha west of Penang Island. Publs. Seto mar. Biol. Lab., 15(5): 391-397.

chaetognaths, lists of spp.

Pereiro, José A. 1972.
Ciclo anual de los quetognatos epiplanctónicos de las aguas de Castellón.
Bol. Inst. esp., Madrid 153: 23pp.

chaetognaths, lists of spp.
Rao, T.S. Satyanarayana, 1958
Studies on chaetognatha in the Indian seas. II
The Chaetognata of the Lawson's Bay, Waltair.
Andhra Univ., Mem. Oceanogr., 2:137-146.

chaetognaths, lists of spp.
Srinivasan M. 1971.
Biology of the chaetognaths of
the estuarine waters of India
J. mar. biol. Ass. India 13(2): 173-181.

Chaetognatha, lists of spp.
Stone, James H., 1969.
The Chaetognatha community of the
Agulhas Current: its structure and
related properties.
Ecol. Monogr. 39: 433-463

Chaetognatha, lists of spp.
Timonin, A.G., 1968.
Distribution of Chaetognatha in the Southern
Ocean. (In Russian; English abstract).
Okeanologiia, Akad. Nauk, SSSR. 8(5):878-887.

chaetognaths, phylogeny
Tokioka, Takasi, 1965.
Supplementary notes on the systematics of
Chaetognatha.
Publs Seto mar. biol. Lab., 13(3):231-242.

chaetognaths, lists of spp.
Tokioka, T., 1960
Droplets from the plankton net. XIX. A
glimpse upon chaetognaths and pelagic tunicates collected in the lagoon water near
Noumea, New Calendonia. Publ. Seto Mar. Biol.
Lab., 8(1): 51-54.

chaetognaths, lists of spp.
Tokioka, T., 1957.
Chaetognaths collected by the Soyo-maru in the
years 1934 and 1937-1939.
Publ. Seto Mar. Biol. Lab., 6(2):137-146.

chaetognaths, lists of spp.
Tokioka, T., 1956.
On chaetognaths and appendicularians collected by
Mr. Z. Sagara in the Arafura Sea in May-August
1955. Publ. Seto Mar. Biol. Lab., 5(2)(11):203-208.

chaetognaths, lists of spp.
Tokioka, T., 1956.
On chaetognaths and appendicularians collected in
the central part of the Indian Ocean. Publ. Seto
Mar. Biol. Lab., 5(2)(10):197-202.

chaetognaths (lists of species)
Tokioka, T., 1955.
Droplets from a plankton net. 17. A small collection of chaetognaths and pelagic tunicates from
the northeastern part of the Indian Ocean. 18.
Short notes on a few appendicularians in the
"Kurosio" off Sionomisaki.
Publ. Seto Mar. Biol. Lab., 5(1):75-80.

Good figure of krohnita subtilis

chaetognatha, lists of species
Tokioka, T., 1955.
On some plankton animals collected by the
Syunkotu-maru in May-June 1954. 1. Chaetognatha.
Publ. Seto Mar. Biol. Lab. 4(2/3):219-225.

chaetognaths, lists of spp.
Tokioka, T., 1954.
Droplets from the plankton nets. 13&14.
Publ. Seto Mar. Biol. Lab., 3(3):359-368.

Chaetognatha, lists of spp.
Vega Rodriguez, Filiberto 1965.
Distribución de Chaetognatha en Veracruz,
Ver.
Anales, Inst. Biol. Univ. Mex. 36 (1/2): 229-247

Chaetognaths, lists of spp
Venter, G.E., 1969.
The distribution of some chaetognaths
and their relation to hydrographical
conditions with special reference to the
South West African region of the Benguela
Current.
Invest'l Rept. SWAR. mar. res. Lab. 16: 73pp.

chaetognaths, parasites of
Rebecq, Jacques, 1965.
Considerations sur la place des trematodes dans le
zooplancton marin. Ann., Fac. Sci., Marseille, 38:61-84.

chaetognaths, parasites of
Vitiello, P., J. Beurois et D. Gouedard, 1970.
Stade larvaire de Thynnascaris sp.
(Nematode Anisakidae) chez Sagitta setosa.
Vie Milieu (A) 21 (1-A): 257-260

Chaetognatha, quantitative
Alvarino, Angeles, 1965.
Distributional atlas of Chaetognatha in the
California Current region.
California Coop. Oceanic Fish. Invest., Atlas,
No. 3:291 charts

Chaetognaths, quantitative
Japan, Kobe Marine Observatory 1967.
Report of the oceanographic observations
in the sea south of Honshu from July to
August 1963. (In Japanese).
Bull. Kobe Mar. Obs. No. 178: 31-40.

Chaetognaths, quantitative
Japan, Kobe Marine Observatory 1967.
Report of the oceanographic observations
in the sea south of Honshu from February
to March 1964. (In Japanese)
Bull. Kobe Mar. Obs. No. 178: 27-

chaetognaths, quantitative
Japan, Kobe Marine Observatory,
Oceanographical Section, 1964.
Report of the oceanographic observations
in the sea south of Honshu from July to
August, 1962. Res. Mar. Meteorol. and
Oceanogr., Japan, Meteorol. Agency, 32:
32-40. (In Japanese).

Also in: Bull. Kobe Mar. Obs. 175?

chaetognaths, quantitative
Japan, Kobe Marine Observatory,
Oceanographical Section, 1964.
Report of the oceanographic observations
in the sea south of Honshu from February
to March, 1963. Res. Mar. Meteorol. and
Oceanogr., Japan Meteorol. Agency, 33:
27-32.

Also in: Bull. Kobe Mar. Obs., 175. 1965.

chaetognaths, quantitative
Japan, Kobe Marine Observatory, Oceanographical Observatory 1962
Report of the oceanographic observations in
the cold water region off Enshu Nada in May,
1961. (In Japanese).
Res. Mar. Meteorol. and Oceanogr. Obs., Jan.-June, 1961, No. 29:28-35.

chaetognaths, quantitative
Japan, Kobe Marine Observatory, Oceanographical Section, 1962
Report of the oceanographic observations in
the sea south of Honshu from February to
March, 1961. (In Japanese).
Res. Mar. Meteorol. and Oceanogr. Obs., Jan.-June, 1961, No. 29:22-27.

chaetognaths (quantitative)
Japan, Maizuru Marine Observatory, 1963
Report of the oceanographic observations in
the Japan Sea in June 1961. (In Japanese).
Bull. Maizuru Mar. Obs., No. 8:59-79.

chaetognath, quantitative
Japan, Maizuru Marine Observatory and Hakodate Marine Observatory, Oceanographical
Sections, 1962.
Report of the oceanographic observations in
the Japan Sea in June, 1961. (In Japanese).
Res. Mar. Meteorol. and Oceanogr. Obs., Jan.-June, 1961, No. 29:59-79.

Chaetognathes, numbers
Ressac, Josette, 1969.
Chaetognathes récoltés par le Ludovic Pierre
pendant la campagne d'assistance à la
flotille Thonière (Année 1967).
Trav. Fac. Sci. Rennes, Sér. Océanogr. Biol.
2: 53-54

Chaetognaths, quantitative
Venter, G.E., 1969.
The distribution of some chaetognaths
and their relation to hydrographical
conditions with special reference to the
South West African region of the Benguela
Current.
Invest'l Rept. SWAR. mar. res. Lab. 16: 73pp.

chaetognaths, standing stock of
Polo, Francisco Pineda, 1971.
The relationship between chaetognaths, water
masses, and standing stock off the Colombia
Pacific Coast. (English and Portuguese abstracts)
In: Fertility of the Sea, John D. Costlow,
editor, Gordon Breach, 2: 309-335.

Chaetognaths, vertical distribution
Alvariño, Angeles, 1967.
Bathymetric distribution of Chaetognatha
Siphonophorae, Medusae and Ctenophorae off San
Diego, California.
Pacif. Sci., 21(4):474-485.

chaetognaths, vertical distribution
Alvarino, Angeles, 1964
Bathymetric distribution of chaetognaths.
Pacific Science, 18(1):64-82.

chaetognathes, vertical distributing
Ducret, Françoise, 1968.
Chaetognathes des campagnes de l'"Ombango" dans
les eaux équatoriales et tropicales africaines.
Cah. O.R.S.T.O.M., ser. Océanogr., VI(1):95-141.

chaetognaths, vertical distribution
Fagetti, E., 1972
Bathymetric distribution of chaetognaths in
the south eastern Pacific Ocean. Mar. Biol.
17(1): 7-29.

Chaetognaths, vertical distribution
Kotori, Moriyuki, 1969.
Vertical distribution of chaetognaths in the
northern North Pacific and Bering Sea. Bull.
Plankt. Soc. Japan, 16(1): 52-57.
(In Japanese; English abstract)

chaetognaths, vertical migrations
Kolosova, E.G. 1972.
Vertical distribution and diel migrations of
chaetognaths in the tropical Pacific. (In
Russian; English abstract). Okeanologiia 12
(1): 129-136.

chaetognaths, vertical migration

Pearre, Sifford, Jr., 1973
Vertical migration and feeding in
Sagitta elegans Verrill. Ecology
54(2): 300-314.

chaetognaths, vertical migration

Schmidt, H.-E., 1973
The vertical distribution and diurnal
migration of some zooplankton in the
Bay of Eilat (Red Sea). Helgoländer
wiss. Meeresunters 24(1/4):333-340.

chaetognaths, vertical distribution

Sund, Paul N., 1961.
Some features of the autoecology and distributions of Chaetognatha in the Eastern Tropical Pacific.
Inter-American Tropical Tuna Comm., Bull., 5(4): 307-340.

chaetognaths, vertical migration of

Sund, P.N., and K.C. Cummings, 1966.
Observations of vertical migrations of Chaetognatha in the Gulf of Guinea.
Bull. Inst. fondament. Afr. noire, 28(4):1322-1331.

Aidanosagitta

Tokioka, Takashi, and D. Pathansali, 1963.
Another new chaetognath from Malay waters, with a proposal of grouping some species of Sagitta into subgenera.
Publ. Seto Mar. Biol. Lab., 11(1)(8):119-123.

Aidanosagitta delicata

Fukume, Toshihide, and Koji Shimizu, 1966.
Preliminary notes on the distribution of Aidanosagitta delicata (Tokioka) in Tanabe Bay.
Publs Seto mar. biol. Lab., 14(3):171-175.

Aidanosagitta delicata

Nakayama, Yumiko, 1970.
Successive changes of the vertical distribution of Aidanosagitta delicata (Tokioka) at a fixed station in a cove of Tanabe Bay. Publ. Seto mar. biol. Lab., 18(3): 207-213.

Eukrohnia

Furnestin, M.-L., 1965.
Variations morphologiques des crochets au cours du développement dans le genre Eukrohnia.
Rev. Trav. Inst. Pêches marit., 29(3):275-284.

EUKROHNIA BATHYANTARCTICA

Alvariño, Angeles, 1969
Los quetognatos del Atlántico: distribución y notas esenciales de sistemática.
Trab. Inst. esp. Oceanogr., 37: 290 pp.

Eukrohnia bathyantarctica

Alvariño, Angeles, 1968.
Egg pouches and other reproductive structures in pelagic Chaetogratha.
Pacif. Sci., 22(4):488-492.

Eukrohnia bathyantarctica n.sp.

David, P.M., 1958.
A new species of Eukrohnia from the Southern Ocean with a note on fertilization.
Proc. Zool. Soc., London, 131(4):597-606.

Eukrohnia bathyantarctica n.sp.

David, P.M., 1958.
A new species of Eukrohnia from the Southern Ocean with a note on fertilization.
Proc. Zool. Soc., London, 131(4):597-605.

Eukrohnia bathyantarctica

Ducret, F., 1965.
Les espèces du genre Eukrohnia dans les eaux equatoriales et tropicales africaines. Cahiers, O.R.S.T.R.O.M., Océanogr., 3(1): 63-78.

Eukrohnia bathyantarctica

Fagetti G., Elda, 1968.
New record of Eukrohnia bathyantarctica David, 1958, from the Gulf of Mexico and Caribbean Sea. Bull. mar. Sci., Miami, 18(2): 383-387.

Eukrohnia bathyantarctica

Owre, Harding B., 1972.
Some temperatures, salinities, and depths of collection of Eukrohnia bathyantarctica (Chaetognatha) in the Caribbean Sea. Bull. mar. Sci., Miami 22(1): 94-99.

EUKROHNIA BATHYPELAGICA

Alvariño, Angeles, 1969
Los quetognatos del Atlántico: distribución y notas esenciales de sistemática.
Trab. Inst. esp. Oceanogr., 37: 290 pp.

Eukrohnia bathypelagica

Alvariño, Angeles, 1967.
The Chaetognatha of the NAGA Expedition (1959-1961) in the South China Sea and the Gulf of Thailand.
NAGA Rept., 4(2):1-87. Scripps Inst. Oceanogr.,

Eukrohnia bathypelagica n. sp.

Alvarino, Angeles, 1962
Two new Pacific chaetognaths: their distribution and relationship to allied species.
Bull. Scripps Inst. Oceanogr., 8(1):1-50.

Eukrohnia bathypelagica — growth stage of E. hamata

Aurich, Horst J., 1971.
Die Verbreitung der Chaetognathen im Gebiet des Nordatlantischen Strom-Systems.
Ber. dt. wiss. Kommn. Meeresforsch. 22(6):1-30

Eukrohnia bathypelagica

Ducret, Françoise, 1968.
Chaetognathes des campagnes de l'"Ombango" dans les eaux équatoriales et tropicales africaines.
Cah. ORSTOM., Sér. Océanogr., 6(1):95-141.

Eukrohnia bathypelagica

Ducret, Françoise, 1968.
Chaetognathes des eaux superficielles et profondes de la zone équatoriale et tropicale friderici africaine.
Thèse Fac. Sci., Marseille, 99 pp., 25 figs. In: Recueil Trav. publiées de 1965 à 1968, Océanogr. ORSTROM.

Eukrohnia bathypelagica

Ducret, Françoise, 1968.
Chaetognathes des campagnes de l'"Ombango" dans les eaux équatoriales et tropicales africaines.
Cah. O.R.S.T.O.M., ser. Océanogr., VI(1):95-141.

Eukrohnia bathypelagica

Figueira, Armando J.G. 1972.
Occurrence of Eukrohnia bathypelagica Alvariño 1962 (Chaetognatha) in the Atlantic waters of Canada.
J. Fish. Res. Bd Can. 29(2):213-214.

Eukrohnia bathypelagica

Ducret, F., 1965.
Les espèces du genre Eukrohnia dans les eaux equatoriales et tropicales africaines. Cahiers, O.R.S.T.R.O.M., Océanogr., 3(1): 63-78.

EUKROHNIA FOWLERI

Alvariño, Angeles, 1969
Los quetognatos del Atlántico: distribución y notas esenciales de sistemática.
Trab. Inst. esp. Oceanogr., 37: 290 pp.

Eukrohnia fowleri

Alvariño, Angeles, 1967.
The Chaetognatha of the NAGA Expedition (1959-1961) in the South China Sea and the Gulf of Thailand.
NAGA Rept., 4(2):1-87. Scripps Inst. Oceanogr.

Eukrohnia fowleri

Alvariño, Angeles, 1966
Zoogeografia de California: quetognatos. Revista Soc. Mexicana Hist. nat. 27: 199-243

Eukrohnia fowleri

Alvarino, Angeles, 1965.
Distributional atlas of Chaetognatha in the California Current region.
California Coop. Oceanic Fish. Invest., Atlas, No. 3:291 charts

Eukrohnia fowleri

Bainbridge, V., 1963.
Continuous plankton records: contribution toward a plankton atlas of the North Atlantic and the North Sea. VIII. Chaetognatha.
Bull. Mar. Ecol., 6(2):40-51.

Eukrohnia fowleri

Bieri, Robert, 1966.
A pale blue chaetognath from Tanabe Bay.
Publs Seto mar. biol. Lab., 14(1):21-22.

Eukrohnia fowleri

Coleman, J. S., 1959.
The "Rosaura" Expedition 1937-1938: Chaetognatha.
Bull. Brit. Mus. (Natural History) Zool. 5(8):219-253.

Eukrohnia fowleri

David, P.M., 1958.
A new species of Eukrohnia from the Southern Ocean with a note on fertilization.
Proc. Zool. Soc., London, 131(4):597-605.

Eukrohnia fowleri

de Saint-Bon, Marie-Catherine, 1963.
Complement à l'étude des chaetognathes de la Côte d'Ivoire (espèces profondes).
Rev. Trav. Inst. Pêches Marit., 27(4):403-415.

Eukrohnia fowleri

Ducret, Françoise, 1968.
Chaetognathes des campagnes de l'"Ombango" dans les eaux équatoriales et tropicales africaines.
Cah. O.R.S.T.O.M., ser. Océanogr., VI(1):95-141.

Eukrohnia fowleri

Ducret, Françoise, 1968.
Chaetognathes des eaux superficielles et profondes de la zone équatoriale et tropicale africaine.
Thèse Fac. Sci., Marseille, 99 pp., 25 figs. In: Recueil Trav. publiées de 1965 à 1968, Océanogr. ORSTROM.

Eukrohnia fowleri

Ducret, Françoise, 1968.
Chaetognathes des campagnes de l'Ombango dans les eaux équatoriales et tropicales africaines.
Cah. ORSTOM., Sér. Océanogr., 6(1):95-141.

Eukrohnia fowleri

Ducret, F., 1965.
Les espèces du genre Eukrohnia dans les eaux equatoriales et tropicales africaines. Cahiers, O.R.S.T.R.O.M., Océanogr., 3(1): 63-78.

Eukrohnia fowleri

Fraser, J.H., 1952.
The Chaetognatha and other zooplankton of the Scottish area and their value as biological indicators of hydrographical conditions.
Scottish Home Dept., Mar. Res., 1952(2):1-52, 3 pls., 21 charts.

good for distinguishing characters

Eukrohnia fowleri
Fraser, J. H., 1949
The occurrence of unusual species of Chaetognatha in Scottish plankton collections.
JMBA 28(2):489-491.

Eukrohnia fowleri
Furnestin, Marie-Louise, 1966.
Chaetognathes des eaux africaines.
Atlantide Rep. 9:105-135.

Eukrohnia fowleri
Germain, L. and L. Joubin, 1916
Chétognathes provenant des campagnes des yachts Hirondelle et Princesse-Alice (1885-1910)
Rés. Camp. Sci., Monaco, 49:118 pp., 7 pls., 7 maps.

Eukrohnia fowleri
Kramp, P. L., 1939
Chaetognatha. The Godthaab Expedition 1928. Medd. om Grønland, 80(5):1-40, 7 text figs.

Eukrohnia? fowleri
Moore, H. B., 1949
The zooplankton of the upper waters of the Bermuda area of the North Atlantic. Bull. Bingham Ocean. Coll. 12(2):97, 208 text figs.

Eukrohnia fowleri
Ritter-Záhony, R. von, 1911
Revision der Chätognathen. Deutsche Südpolar Exped., 1901-1903, XIII (Zool.V), Pt.1:71 pp., 51 text figs.

Eukrohnia fowleri
Russell, F.S., 1939.
Chaetognatha. Fiches d'Ident. Zooplancton, Cons. Perm. Int. Expl. Mer, 1:4 pp., 12 textfigs.

Eukrohnia fowleri
Schilp, H., 1964.
Chaetognatha of the genus Eukrohnia von Ritter-Zahony in the material of the 'Snellius' Expedition.
Zool. Mededel., 39:533-549.

Eukrohnia fowleri
Silas, E.G., and M. Srinivasan, 1968.
A new species of Eukrohnia from the Indian seas with notes on three other species of Chaetognatha.
J. mar. biol. Ass., India, 10(1):1-33

Eukrohnia fowleri
Sund, Paul N., 1964.
Los quetognatos en las aguas de la region del Peru.
Inter-Amer. Tropical Tuna Comm., Bull., 9(3):115-216.

Eukrohnia fowleri
Sund, Paul N., 1961.
Some features of the autecology and distributions of Chaetognatha in the Eastern Tropical Pacific.
Inter-American Tropical Tuna Comm., Bull., 5(4):307-340

Eukrohnia fowleri
Tchindonova, J.G., 1955.
Chaetognatha of the Kurile-Kamchatka Trench.
Trudy Inst. Oceanol., 12:298-310.

Eukrohnia fowleri
Truveller, K.A., 1966.
Chaetognatha of the Davis and Denmark Straits as indicators of water masses. (In Russian).
Mater. Ribokhoz. Issled. severn. Basseina, Poliarn. Nauchno-Issled. Proektn. Inst. Morsk. Ribn. Khoz. Okeanogr. (PINRO), 7:114-124.

Eukrohnia fowleri
von Ritter-Zahony, R., 1914.
Chaetognaths. Danish Ingolf-Exped. 4(3):4 pp.

Eukrohnia fowleri
von Ritter-Záhony, R., 1911
Vermes. Chaetognathi. Das Tierreich 29:34 pp., 16 text figs.

EUKROHNIA HAMATA
Alvariño, Angeles, 1969
Los quetognatos del Atlántico: distribución y notas esenciales de sistemática.
Trab. Inst. esp. Oceanogr., 37: 290 pp.

Eukrohnia hamata
Alvariño, Angeles, 1967.
The Chaetognatha of the NAGA Expedition (1959-1961) in the South China Sea and the Gulf of Thailand.
NAGA Rept., 4(2):1-87.
Scripps Inst. Oceanogr.

Eukrohnia hamata
Alvarino, Angeles, 1965.
Distributional atlas of Chaetognatha in the California Current region.
California Coop. Oceanic Fish. Invest., Atlas, No. 3:291 charts

Eukrohnia hamata
Alvariño, A., 1957.
Estudio del zooplancton del Mediterráneo occidental. Zooplancton del Atlántico Ibérico. Campanas del "Xauen" en el verano del 1954.
Bol. Inst. Español Ocean., Nos. 81-82:1-26; 1-51.

Eukrohnia hamata
Alvariño, A., 1956.
Estudio del zooplancton recogido en la campana "Vendaval" en Terranova, marzo, abril y mayo de 1953. Zooplancton de Terranova, febrero, marzo y junio de 1955. Bol. Inst. Espanol de Ocean., Nos., 76/77:28 pp. y 18 pp.

Eukrohnia hamata
Bainbridge, V., 1963.
Continuous plankton records: contribution toward a plankton atlas of the North Atlantic and the North Sea. VIII. Chaetognatha.
Bull. Mar. Ecol., 6(2):40-51.

Eukrohnia hamata
Beaudovin, Jacqueline, 1967.
Oeufs et larves de poissons récoltés par le "Thalassa" dans le Détroit de Danemark et le Nord de la Mer d'Irminger (NORWESTLANT I- 20 mars 8 Mai 1963: relations avec l'hydrologie et le zooplancton.
Revue Trav. Inst. (scient. tech.) Pêches marit. 31 (3):307-326.

Eukrohnia hamata
Bieri, Robert, 1966.
A pale blue chaetognath from Tanabe Bay.
Publs Seto mar. biol. Lab., 14(1):21-22.

Eukrohnia hamata
Bieri, Robert 1959
The distribution of planktonic chaetognatha in the Pacific and their relationship to water masses.
Limnol. and Ocean. Vol. 4 No. 1 pp. 1-28

Eukrohnia hamata
Bigelow, H. B., 1922
Exploration of the coastal water off the northeastern United States in 1916 by the U.S. Fisheries Schooner Grampus. Bull. M.C.Z. 65 (5):85-188, 53 text figs.

Eukrohnia hamata
Bigelow, H.B., and M. Sears, 1939
Studies of the waters of the continental shelf, Cape Cod to Chesapeake Bay. III. A volumetric study of the zooplankton. Mem. M.C.Z. 54(4):183-378, 42 text figs.

Eukrohnia hamata
Burfield, S.T., 1930.
Chaetognatha. Brit. Antarctic ("Terra Nova") Exped., 1910, Zool., 7(4):203-228, 3 maps.

Eukrohnia hamata
Burfield, S.T., and E.J.W. 1926.
The Chaetognatha of the "Sealark" Expedition.
Trans. Linn. Soc., London, Ser. 2, 19:93-119, Pls. 4-7.

Eukrohnia hamata
Coleman, J. S., 1959.
The "Rosaura" Expedition 1937-1938: Chaetognatha.
Bull. Brit. Mus. (Natural History) Zool. 5(8):219-253.

Eukrohnia hamata
David, Peter M., 1959.
Chaetognatha.
British-Australian-New Zealand Antarctic Res. Exp. (B) 8:73-79.

Eukrohnia hamata
David, P.M., 1958.
A new species of Eukrohnia from the Southern Ocean, with a note on fertilization.
Proc. Zool. Soc., London, 131:597-606.

Eukrohnia hamata
Dawson, John Kayl, 1968.
Chaetognaths from the Arctic Basin, including the description of a new species of Heterokrohnia.
Bull. Sth. Calif. Acad. Sci., 67(2):112-124

Eukrohnia hamata
Ducret, Françoise, 1968.
Chaetognathes des campagnes de l'"Ombango" dans les eaux équatoriales et tropicales africaines.
Cah. O.R.S.T.O.M., ser. Océanogr., VI(1):95-141.

Eukrohnia hamata
Ducret, Françoise, 1968.
Chaetognathes des eaux superficielles et profondes de la zone équatoriale et tropicale africaine.
Thèse Fac. Sci., Marseille, 99 pp., 25 figs. In: Receuil Trav. publiées de 1965 à 1968, Océanogr. ORSTOM.

Eukrohnia hamata
Ducret, Françoise, 1968.
Chaetognathes des campagnes de l'Ombango dans les eaux équatoriales et tropicales africaines.
Cah. ORSTOM. Sér. Océanogr., 6(1):95-141.

Eukrohnia hamata
Ducret, F., 1965.
Les espèces du genre Eukrohnia dans les eaux équatoriales et tropicales africaines. Cahiers, O.R.S.T.R.O.M., Océanogr., 3(1): 63-78.

Eukrohnia hamata
Dunbar, M.J., 1942.
Marine macroplankton from the Canadian Eastern Arctic. II. Medusae, Siphonophora, Ctenophora, Pteropoda, and Chaetognatha. Canad. J. Res., D, 20:71-77.

Eukrohnia hamata
Fagetti Guaita, Elda, 1958.
Investigaciones sobre quetognatos colectados, especialmente, frente a la costa central y norte de Chile. Revista Biol. Mar., Valparaiso, 8(1-3):25-82.

Eukrohnia hamata
Fraser, J.H., 1952.
The Chaetognatha and other zooplankton of the Scottish area and their value as biological indicators of hydrographical conditions.
Scottish Home Depts. Mar. Res., 1952(2):1-52, 3 pls., 21 charts.

good for distinguishing characters.

Eukrohnia hamata
Fraser, J. H., 1949
The occurrence of unusual species of Chaetognatha in Scottish plankton collections.
JMBA 28(2):489-491.

Eukrohnia hamata
Fraser, J. H., 1949
Plankton of the Faroe-Shetland Channel and the Faroes, June and August 1947. Ann. Biol., Int. Cons., 4:27-28, text fig. 10.

Eukrohnia hamata
Fraser, J. H., 1939
The distribution of Chaetognatha in Scottish Waters in 1937. J. du Cons. 14(1): 25-34, 3 charts in text.

Eukrohnia hamata
Fraser, J. H. and A. Saville, 1949
Macroplankton in the Faroe Channel, 1948. Ann. Biol. 5:29-30, text figs.

Eukrohnia hamata
Fraser, J. H., and A. Saville, 1949
Plankton distribution in Scottish and adjacent waters in 1948. Ann. Biol. 5:61-62.

Eukrohnia hamata
Furnestin, Marie-Louise, 1966.
Chaetognathes des eaux africaines. Atlantide Rep. 9:105-135.

Eukrohnia hamata
Germain, L. and L. Joubin, 1916
Chétognathes provenant des campagnes des yachts Hirondelle et Princesse-Alice (1885-1910) Rés. Camp. Sci., Monaco, 49:118 pp., 7 pls., 7 maps.

Eukronia (sic!) hamata
Hansen, Vagn Kr., 1960.
Investigations on the quantitative and qualitative distribution of zooplankton in the southern part of the Norwegian Sea. Medd. Danmarks Fiskeri- og Havundersøgelser, n.s., 2(23):1-53.

Eukrohnia hamata
Heinrich, A.K., 1956.
Dimensional composition of Chaetognatha, and the terms of their propagation in the western regions of the Bering Sea. Dokl. Akad. Nauk, SSSR, 110(6):1105-1107.

Eukrohnia hamata
Kielhorn, W.V., 1952
The biology of the surface zone zooplankton of a Boreo-Arctic Atlantic Ocean area. J. Fish Res. Bd., Canada 9 (5): 223-264, 13 text figs.

Eukrohnia hamata
Kitou, Masataka, 1967.
Distribution of Eukrohnia hamata (Chaetognatha) in the western North Pacific. Inf. Bull. Planktol. Japan, Co.m.No.Dr.Y. Matsue, 91-96.

Eukrohnia hamata
Kitou, Masataka, 1966.
Chaetognaths collected on the Sixth Cruise of the Japanese Expedition of Deep Seas. La Mer, Bull. franco-japon. Oceanogr., 4(8):261-265.
Also in:
Reps JEDS, 5-7 (1966)

Eukrohnia hamata
Kramp, P. L., 1939
Chaetognatha. The Godthaab Expedition 1928. Medd. om Grønland, 80(5):1-40, 7 text figs.

Eukrohnia hamata
Kramp, P.L., 1938
Chaetognatha. Zool. Iceland 4(71):4 pp.

Eukrohnia hamata
Lea, H.E., 1955.
The chaetognaths of western Canadian coastal waters. J. Fish. Res. Bd., Canada, 12(4):593-617.

Eukrohnia hamata
Mackintosh, N.A., 1934
Distribution of the Macroplankton in the Atlantic Sector of the Antarctic. Discovery reports, Vol.9:65-160, 48 text figs.

Eukrohnia hamata
Michael, E.L., 1919
Contributions to the Biology of the Philippine Archipelago and Adjacent regions. Report on the chaetognatha collected by the United States Fisheries Steamer "Albatross" during the Philippine Expedition, 1907-1910. Bull. U.S. Nat. Mus. No.100, Vol.1(4):235-277, pls.34-38.

Eukrohnia hamata
Michael, E. L., 1911
Classification and vertical distribution of the Chaetognatha of the San Diego Region including redescriptions of some doubtful species of the group. Univ. Calif. Publ. Zool. 8(3):21-186, Pls. 1-8.

Eukrohnia hamata
Mostajo, Elena 1973.
Quetognatos colectados en el Atlantico sudoccidental entre los 44°44' y 52°36' de latitud sur. Neotropica 19(59): 94-100.

Eukrohnia hamata
Rae, K. M., 1949
Plankton. Some broad changes in the plankton round the north of the British Isles in 1948. Ann. Biol. 5:56-60, 12 text figs.

Eukrohnia hamata
Redfield, A. C. and A. Beale, 1940.
Factors determining the distribution of populations of chaetognaths in the Gulf of Maine. Biol Bull., 79(3):459-487, 11 text figs.

Eukrohnia hamata
Reyssac, J., 1963.
Chaetognathes du plateau continental européen (de la baie ibéro-marocaine à la Mer Celtique). Rev. Trav. Inst. Pêches Marit., 27(3):245-299.

Eukrohnia hamata
Ritter-Záhony, R. von, 1911
Revision der Chätognathen. Deutsche Südpolar Exped., 1901-1903, XIII (Zool.V). Pt.1:71 pp., 51 text figs.

Eukrohnia hamata
von Ritter-Záhony, R., 1911
Vermes. Chaetognathi. Das Tierreich 29:34 pp., 16 text figs.

Eukrohnia hamata
Russell, F.S., 1939.
Chaetognatha. Fiches d'Ident. Zooplancton, Cons. Perm. Int. Expl. Mer, 1:4 pp., 12 textfigs.

Eukrohnia hamata
Schilp, H., 1964.
Chaetognatha of the genus Eukrohnia von Ritter-Zahony in the material of the 'Snellius' Expedition. Zool. Mededel. 39:533-549.

Eukrohnia hamata
Sund, Paul N., 1964.
Los quetognatos en las aguas de la region del Peru. Inter-Amer. Tropical Tuna Comm., Bull., 9(3):115-216.

Eukrohnia hamata
Sund, Paul N., 1961.
Some features of autecology and distributions of Chaetognatha in the eastern Tropical Pacific. Inter-American Tropical Tuna Comm., Bull., 5(4):307-340.

Eukrohnia hamata
Sund, Paul N., 1959.
A key to the Chaetognatha of the tropical Eastern Pacific Ocean. Pacific Science, 13(3):269-285.

Eukrohnia hamata
Sund, P.N., 1959
The distribution of Chaetognatha in the Gulf of Alaska in 1954 and 1956. J. Fish. Res. Bd., Canada, 16(3):351-361.

Eukrohnia hamata
Tchindonova, J.G., 1955.
Chaetognatha of the Kurile-Kamchatka Trench. Trudy Inst. Oceanol., 12:298-310.

Eukrohnia hamata
Thiel, M.E., 1938
Die Chaetognathen - Bevölkerung des Südatlantischen Ozeans. Biologische Sonderuntersuchungen. 1st Lief. Wissenschaftliche Ergebnisse der Deutschen Atlantischen Expedition auf dem Forschungs. und Vermessungsschiff "Meteor" 1925-1927, 13:1-110, 62 text figs.

Eukrohnia hamata
Thomson, J. M., 1947
The Chaetognatha of South-eastern Australia. Counc. Sci. & Ind. Res., Australia, Bull. No.222, (Div. Fish. Rept. 14), 43 pp., 8 text figs.

Eukrohnia hamata
Tokioka, Takasi, 1959
Observations on the taxonomy and distribution of chaetognaths of the North Pacific. Publ. Seto Mar. Biol. Lab., 7(3):349-456.

Eukrohnia hamata
Tokioka, T., 1939.
Chaetognaths collected chiefly in the bays of Sagami and Suruga with some notes on the shape and structure of the seminal vesicles. Rec. Oceanogr. Wks., Japan, 10(2):123-150.

Eukrohnia hamata
Truveller, K.A., 1966.
Chaetognatha of the Davis and Denmark Straits as indicators of water masses. (In Russian). Mater. Ribokhoz. Issled. severn. Basseina, Polliarn. Nauchno-Issled. Proektn. Inst. Morsk. Ribn. Khoz. Okeanogr. (PINRO), 7:114-124.

Eukrohnia hamata
von Ritter Zahony, R., 1914.
Chaetognaths. Danish Ingolf-Exped. 4(3):4 pp.

Eukrohnia hamata
Wiborg, K.F., 1955.
Zooplankton in relation to hydrography in the Norwegian Sea. Rept. Norwegian Fish. Mar. Invest. 11(4):66 pp.

Eukrohnia hamata f. antarctica
Mackintosh, N.A., 1934
Distribution of the Macroplankton in the Atlantic Sector of the Antarctic. Discovery reports, Vol.9:65-160, 48 text figs.

Eukrohnia minuta n.sp
Silas, E.G., and M. Srinivasan 1968.
A new species of Eukrohnia from the Indian seas with notes on three other species of Chaetognatha. J. mar. biol. Ass. India, 10(1):1-33

Eukrohnia pacifica
Michael, E. L., 1911
Classification and vertical distribution of the Chaetognatha of the San Diego Region including redescriptions of some doubtful species of the group. Univ. Calif. Publ. Zool. 8(3):21-186, Pls. 1-8.

Eukrohnia proboscidea
Ducret, Françoise, 1968.
Chaetognathes des Campagnes de l'"Ombango" dans les eaux équatoriales et tropicales africaines.
Cah. O.R.S.T.O.M., ser. Oceanogr., VI(1):95-141.

Eukrohnia proboscida
Ducret, Françoise, 1968.
Chaetognathes des eaux superficielles et profondes de la zone équatoriale et tropicale africaine.
Thèse Fac. Sci., Marseille, 99 pp., 25 figs. In: Receuil Trav. publiées de 1965 à 1968, Océanogr. ORSTOM.

Eukrohnia proboscida
Ducret, Françoise, 1968.
Chaetognathes des campagnes de l'Ombango dans les eaux équatoriales et tropicales africaines.
Cah. ORSTOM., Sér. Océanogr., 6(1):95-141.

Eukrohnia proboscidea n.sp.
Ducret, F., 1965.
Les espèces du genre Eukrohnia dans les eaux équatoriales et tropicales africaines. Cahiers, O.R.S.T.R.O.M., Océanogr., 3(1): 63-78.

Eukrohnia proboscidea n.sp.
Furnestin, M.-L., et F. Ducret, 1965.
Eukrohnia proboscidea, nouvelle espèce de chaetognathe.
Rev. Trav. Inst. Pêches marit., 29(3):271-273.

Eukrohnia Richardi
Belloc, G., 1961.
Catalogue des types de chétognathes du Musée Océanographique de Monaco.
Bull. Inst. Océanogr., Monaco, 58(1216):3 pp.

good diagram

Eukrohnia Richardi
Germain, L. and L. Joubin, 1916
Chétognathes provenant des campagnes des yachts Hirondelle et Princesse-Alice (1885-1910)
Rés. Camp. Sci., Monaco, 49:118 pp., 7 pls., 7 maps.

Eukrohnia richardi n.sp
Germain, L., and L. Joubin, 1912.
Note sur quelques chétognathes nouveaux des croisières de S.S.S. le Prince de Monaco. Bull. Inst. Océan., Monaco, No. 228:1-14, 15 textfigs.

Eukrohnia richardi
Michael, E.L., 1919
Contributions to the Biology of the Philippine Archipelago and Adjacent regions. Report on the chaetognatha collected by the United States Fisheries Steamer "Albatross" during the Philippine Expedition, 1907-1910.
Bull. U.S. Nat. Mus. No.100, Vol.1(4):235-277, pls.34-38.

Eukrohnia richardi
Schilp, H., 1964.
Chaetognatha of the genus Eukrohnia von Ritter-Zahony in the material of the 'Snellius' Expedition.
Zool. Mededel., 39:533-549.

Eukrohnia subtilis
Michael, E. L., 1911
Classification and vertical distribution of the Chaetognatha of the San Diego Region including redescriptions of some doubtful species of the group. Univ. Calif. Publ. Zool. 8(3):21-186, Pls. 1-8.

Heterokrohnia bathybia b.sp.
Marumo, Ryuzo, and Masataka Kitou, 1966.
A new species of Heterokrohnia (Chaetognatha) from the western North Pacific.
Le Mer, Bull. Soc. franco-japon. Océanogr., 4(3):178-183.

Heterokrohnia involucrum n.sp.
Dawson, John Kayl, 1968.
Chaetognaths from the Arctic Basin, including the description of a new species of Heterokrohnia.
Bull. Sth. Calif. Acad. Sci., 67(2):112-124

Heterokrohnia mirabilis
Bieri, Robert 1959
The distribution of planktonic chaetognatha in the Pacific and their relationship to water masses.
Limnol. and Ocean. Vol. 4 No. 1 pp. 1-28

Heterokrohnia mirabilis
Dawson, John Kayl, 1968.
Chaetognaths from the Arctic Basin, including the description of a new species of Heterokrohnia.
Bull. Sth. Calif. Acad. Sci., 67(2):112-124

Heterokrohnia mirabilis
Marumo, Ryuzo, and Masataka Kitou, 1966.
A new species of Heterokrohnia (Chaetognatha) from the western North Pacific.
La Mer, Bull. Soc. franco-japon. Océanogr., 4(3):178-183.

Also in:
Reps JEDS, 5-7(1966)

Heterokrohnia mirabilis
von Ritter-Záhony, R., 1911
Vermes. Chaetognathi. Das Tierreich 29:34 pp., 16 text figs.

Heterokrohnia mirabilis
Ritter-Záhony, R. von, 1911 n.gen., n.sp.
Revision der Chätognathen. Deutsche Südpolar Exped., 1901-1903, XIII (Zool.V)., Pt.1:71 pp., 51 text figs.

Heterokrohnia mirabilis
Tchindonova, J.G., 1955.
Chaetognatha of the Kurile-Kamchatka Trench.
Trudy Inst. Oceanol., 12:298-310.

Krohnia foliacea n.sp.
Aida, T., 1897.
Chaetognaths of Misaki Harbour.
Ann. Zool., Japon., 1:13-21, Pl. 3.

Krohnia foliacea
Fowler, G.H., 1906.
The Chaetognatha of the Siboga Expedition with a discussion of the synonymy and distribution of the group. Rept. Siboga Exped. 21:1-88, 3 pls., 6 charts.

Krohnia hamata
Conant, F.S., 1896.
23. Notes on the chaetognaths. Ann. Mag. Nat. Hist., 6th ser., 18:201-214.

Krohnia hamata
Fowler, G.H., 1907.
Chaetognatha, with a note on those collected by H.M.S. "Challenger" in Subantarctic and Antarctic waters. Nat. Antarctic Exped., 1901-1904, 3:6 pp. 1 chart.

Krohnia hamata
Fowler, G.H., 1906.
The Chaetognatha of the Siboga Expedition, with a discussion of the synonymy and distribution of the group. Rept. Siboga Exped. 21:1-88, 3 pls., 6 charts.

Krohnia hamata
Fowler, G.H., 1905.
Biscayan plankton collected during a cruise of H.M.S. "Research", 1900. Pt. III. The Chaetognatha. Trans. Linn. Soc., London, 10(3):55-87, Pls. 4-7.

Krohnia pacifica n.sp.
Aida, T., 1897.
Chaetognaths of Misaki Harbour.
Ann. Zool., Japon., 1:13-21, Pl. 3.

Krohnia pacifica
Fowler, G.H., 1906.
The Chaetognatha of the Siboga Expedition with a discussion of the synonymy and distribution of the group. Rept. Siboga Exped. 21:1-88, 3 pls., 6 charts.

Krohnia pacifica
Varadarajan, S. and P.I. Chacko, 1943.
On the arrow-worms of Krusadai. Proc. Nat. Inst. Sci., India, 9(2):245-248, 2 figs.

Krohnia subtilis
Fowler, G.H., 1906.
The Chaetognatha of the Siboga Expedition with a discussion of the synonymy and distribution of the group. Rept. Siboga Exped. 21:1-88, 3 pls., 6 charts.

Krohnia subtilis
Fowler, G.H., 1905.
Biscayan plankton collected during a cruise of H.M.S. "Research", 1900. Pt. III. The Chaetognatha. Trans. Linn. Soc., London, 10(3):55-87, Pls. 4-7.

Krohnia viridis
Ritter-Záhony, R. von, 1911
Revision der Chätognathen. Deutsche Südpolar Exped., 1901-1903, XIII (Zool.V)., Pt.1:71 pp., 51 text figs.

KROHNITTA mutabbii
Alvariño, Angeles, 1969
Los quetognatos del Atlántico: distribución y notas esenciales de sistemática.
Trab. Inst. esp. Oceanogr., 37: 290 pp.

Krohnitta pacifica
Alvariño, Angeles, 1967.
The Chaetognatha of the NAGA Expedition (1959-1961) in the South China Sea and the Gulf of Thailand.
NAGA Rept.,4(2):1-87.
Scripps Inst. Oceanogr.,

Krohnitta pacifica
Alvariño, Angeles, 1966
Zoogeografia de California: quetognatos. Revista Soc. Mexicana Hist. nat. 27: 199-243

Krohnitta pacifica
Alvarino, Angeles, 1965
Distributional atlas of Chaetognatha in the California Current region.
California Coop. Oceanic Fish. Invest., Atlas, No. 3:291 charts

Krohnitta pacifica
Alvariño, Angeles, 1964.
Zoogeografia de los quetognatos, especialmente de la region de California.
Ciencia, Mexico, 23:51-74.

Krohnitta pacifica
Alvarino, Angeles, 1963
Quetognatos epiplanctonicos del Mar de Cortes.
Revista. Soc. Mexicana. Hist. Nat., 24:97-202.

Krohnitta pacifica
Bieri, Robert, 1966.
A pale blue chaetognath from Tanabe Bay.
Publs Seto mar. biol. Lab., 14(1):21-22.

Krohnitta, pacifica
Bieri, Robert 1959
The distribution of planktonic chaetognatha in the Pacific and their relationship to water masses.
Limnol. and Ocean. Vol. 4 No. 1 pp. 1-28

Krohnitta pacifica
Bieri, Robert, 1957
The chaetognath fauna of Peru in 1941.
Pacific Science 11(3): 255-264.

Krohnita pacifica
Bumpus, D.F., and E.L. Pierce, 1955.
Hydrography and the distribution of chaetognaths over the continental shelf of North Carolina.
Pap. Mar. Biol. and Oceanogr., Deep-Sea Res., Suppl. to Vol. 3:92-109.

Krohnitta pacifica
de Saint-Bon, Marie Catherine, 1963.
Les chaetognathes de la Côte d'Ivoire (espèces de surface).
Rev. Trav. Inst. Pêches Marit., 27(3):301-346.

Krohnitta pacifica
Ducret, Françoise, 1968.
Chaetognathes des campagnes de l'Ombango dans les eaux équatoriales et tropicales africaines.
Cah. ORSTOM., Sér. Océanogr., 6(1):95-141.

Krohnitta pacifica
*Ducret, Françoise, 1968.
Chaetognathes des campagnes de l'"Ombango" dans les eaux équatoriales et tropicales africaines.
Cah.O.R.S.T.O.M., ser.Océanogr., VI(1):95-141.

Krohnitta pacifica
Ducret, Françoise, 1968.
Chaetognathes des eaux superficielles et profondes de la zone équatoriale et tropicale africaine.
Thèse Fac. Sci., Marseille, 99 pp., 25 figs. In: Recueil Trav. publiées de 1965 à 1968, Océanogr. ORSTROM.

Krohnitta pacifica
Furnestin, Marie-Louise, 1966.
Chaetognathes des eaux africaines.
Atlantide Rep. 9:105-135.

Krohnitta pacifica
Furnestin, M.L., et J. Balança, 1968
Chaetognathes de la Mer Rouge (Archipel Dallac). Israel South Red Sea Expédition, 1962, Reports, No.32.
Bull. Sea. Fish. Res. Sta. 52:3-20

Krohnitta pacifica
Furnestin, M.-L. et J.-C. Codaccioni, 1968.
Chaetognathes du nord-ouest de l'Océan Indien (golfe d'Aden, mer d'Arabie, golfe d'Oman, golfe Persique).
Cah.O.R.S.T.O.M., ser. Océanogr., VI(1):143-171.

Krohnitta pacifica
George, P.C., 1952.
A systematic account of the Chaetognatha of the Indian coastal waters, with observations on their seasonal fluctuations along the Malabar coast.
Proc. Nat. Inst. Sci., India, 18(6):657-689.

Krohnitta pacifica
Grant, George C., 1963
Investigations of inner continental shelf waters off lower Chesapeake Bay. IV. Descriptions of the Chaetognatha and a key to their identification.
Chesapeake Science, 4(3):109-119.

Krohnitta pacifica
Hosoe, K., 1956.
Chaetognaths from the Isles of Fernando de Noronha.
Contr. Avulsas, Inst. Oceanogr., Sao Paulo, No. 3: 8 pp.

Krohnitta pacifica
Legare, J. E. Henri, and E. Zoppi, 1961.
Notas sobre la abundancia y distribucion de Chaetognatha en las aguas del Oriente de Venezuela. (In English).
Bol. Inst. Oceanograf., Univ. Oriente, Cumana, Venezuela, 1(1):149-171.

Krohnita pacifica
Nair, Vijayalakshmi R. 1971.
Seasonal fluctuations of chaetognaths in the Cochin Backwater.
J. mar. biol. Ass. India 13(2): 226-233.

Krohnita pacifica
Pierce, E.L., 1953.
The Chaetognatha over the continental shelf of North Carolina with attention to their relation to the hydrography of the area. J. Mar. Res. 12(1):75-92, 4 textfigs.

Krohnitta pacifica
Pierce, E.L., 1951.
The occurrence and breeding of the Chaetognatha along the Gulf coast of Florida. Proc. Gulf and Caribbean Fish. Inst., 3rd session:128-127.

Krohnitta pacifica
Pierce, E.L., 1951.
The Chaetognatha of the west coast of Florida.
Biol. Bull. 100(3):206-228, 5 textfigs.

Krohnitta pacifica
Rao, T.S. Satyanarayana, and P.N. Ganapati, 1958
Studies on Chaetognatha in the Indian Seas. III Systematics and distribution in the waters off Visakhapaynam. Andhra Univ. Mem. Oceanogr., 2:147-163.

Krohnitta pacifica
Suarez Caabro, J.A., 1955.
Quetognatos de los mares cubanos.
Mem. Soc. Cubana Hist. Nat., 22(2):125-180, 9 pls

Krohnitta pacifica
Suarez-Caabro, Jose A., and Juan E. Madruga, 1960.
The Chaetognatha of the northeastern coast of Honduras Central America.
Bull. Mar. Sci., Gulf & Caribbean, 10(4):421-429

Krohnitta pacifica
Sudarsan, D., 1961.
Observations on the Chaetognatha of the waters around Mandapam.
Indian J. Fish., 8(2):364-382.

Krohnitta pacifica
Sund, Paul N., 1961.
Some features of the autecology and distributions of Chaetognatha in the Eastern Tropical Pacific.
Inter-American Tropical Tuna Comm., Bull., 5(4): 307-340.

Krohnitta pacifica
Sund, Paul N., 1959.
A key to the Chaetognatha of the tropical Eastern Pacific Ocean.
Pacific Science, 13(3):269-285.

Krohnita pacifica
Thomson, J. M., 1947
The Chaetognatha of South-eastern Australia.
Counc. Sci. & Ind. Res., Australia, Bull. No.222, (Div. Fish. Rept. 14), 43 pp., 8 text figs.

Krohnitta pacifica
Tokioka, Takasi, 1959
Observations on the taxonomy and distribution of chaetognaths of the North Pacific.
Publ. Seto Mar. Biol. Lab., 7(3):349-456.

Krohnitta pacifica
Tokioka, T., 1951.
Pelagic tunicates and chaetognaths collected during the cruises to the New Yamato Bank in the Sea of Japan. Publ. Seto Mar. Biol. Lab. 2(1): 1-26, 1 chart, 6 tables, 12 textfigs.

Krohnitta pacifica
Tokioka, T., 1942
Systematic studies of the plankton organisms occurring in Iwayama Bay, Palao. III Chaetognaths from the bay and adjacent waters. Palao Tropical Biol. Sta. Studies, II(3):527-548, pls.5-7, 11 text figs.

Krohnitta pacifica
Tokioka, T., 1940.
The chaetognath fauna of the waters of western Japan, Rec. Oceanogr. Wks., Japan, 12(1):1-22, 3 charts, 4 tables. (Contrib. 87, Seto M.B.L.)

Krohnitta pacifica
Tokioka, T., 1939.
Chaetognaths collected chiefly in the bays of Sagami and Suruga with some notes on the shape and structure of the seminal vesicles.
Rec. Oceanogr. Wks., Japan, 10(2):123-150.

Krohnitta pacifica
Tsuruta, Arao, 1963
Distribution of plankton and its characteristics in the oceanic fishing grounds, with special reference to their relation to fishery.
J. Shimonoseki Univ., Fish., 12(1):13-214.

Krohnitta subtilis
Alvariño, Angeles, 1969
Los quetognatos del Atlántico: distribución y notas esenciales de sistemática.
Trab. Inst. esp. Oceanogr., 37: 290 pp.

Krohnitta subtilis
Alvariño, Angeles, 1967.
The Chaetognatha of the NAGA Expedition (1959-1961) in the South China Sea and the Gulf of Thailand.
NAGA Rept., 4(2):1-87.
Scripps Inst. Oceanogr.

Krohnitta subtilis
Alvariño, Angeles, 1964.
Zoogeografía de los quetognatos, especialmente de la región de California.
Ciencia, México, 23:51-74.

Krohnitta subtilis
Alvarino, Angeles, 1963
Quetognatos epiplanctonicos del Mar de Cortes.
Revista. Soc. Mexicana, Hist. Nat., 24:97-202.

Krohnitta subtilis
Alvariño, A., 1957.
Estudio del zooplancton del Mediterráneo occidental. Zooplancton del Atlántico Ibérico. Campañas del "Xauen" en el verano del 1954.
Bol. Inst. Español Oceanogr., Nos., 81-82:1-26; 1-51.

Krohnitta subtilis
Bieri, Robert 1959
The distribution of planktonic chaetognatha in the Pacific and their relationship to water masses.
Limnol. and Ocean. Vol. 4 No. 1 pp. 1-28

Krohnita subtilis
Bigelow, H.B., and M. Sears, 1939
Studies of the waters of the continental shelf, Cape Cod to Chesapeake Bay. III. A volumetric study of the zooplankton. Mem. M.C.Z. 54(4):183-378, 42 text figs.

Krohnita subtilis
Bumpus, D.F., and E.L. Pierce, 1955.
Hydrography and the distribution of chaetognaths over the continental shelf of North Carolina.
Pap. Mar. Biol. and Oceanogr., Deep-Sea Res., Suppl. to Vol. 3:92-109.

Krohnita subtilis
Burfield, S.T., 1930.
Chaetognatha. Brit. Antarctic ("Terra Nova") Exped., 1910, Zool., 7(4):203-228, 3 maps.

Krohnitta subtilis
Burfield, S.T., 1927.
Zoological results of the Cambridge Expedition to the Suez Canal. XXI. Report on the Chaetognatha.
Trans. Zool. Soc., London, 22:355-356.

Krohnita subtilis
Burfield, S.T., and E.J.W. Harvey, 1926.
The Chaetognatha of the "Sealark" Expedition.
Trans. Linn. Soc., London, Ser. 2, 19:93-119, Pls. 4-7.

Krohnitta subtilis
Coleman, J. S., 1959.
The "Rosaura" Expedition 1937-1938: Chaetognatha.
Bull. Brit. Mus. (Natural History) Zool. 5(8):219-253.

Oceanographic Index: Marine Organisms Cumulation, 1946-1973

Krohnitta subtilis
Ducret, Françoise, 1968.
Chaetognathes des campagnes de l'"Ombango" dans les eaux equatoriales et tropicales africaines.
Cah. O.R.S.T.O.M., ser Oceanogr., VI(1):95-141.

Krohnitta subtilis
Ducret, Françoise, 1968.
Chaetognathes des eaux superficielles et profondes de la zone équatoriale et tropicale africaine.
Thèse Fac. Sci., Marseille, 99 pp., 25 figs. In: Receuil Trav. publiées de 1965 à 1968, Océanogr. ORSTROM.

Krohnita subtilis
Ducret, Françoise, 1968.
Chaetognathes des campagnes de l'Ombango dans les eaux équatoriales et tropicales africaines.
Cah. ORSTOM., Ser.Océanogr.,6(1):95-141.

Krohnitta subtilis
Fagetti Guaita, Elda, 1958.
Investigaciones sobre quetognatos colectados, especialmente, frente a la costa central y norte de Chile. Revista Biol. Mar., Valparaiso, 8(1-3):25-82.

Krohnitta subtilis
Fraser, J.H., 1952.
The Chaetognatha and other zooplankton of the Scottish area and their value as biological indicators of hydrographical conditions.
Scottish Home Dept., Mar. Res., 1952(2):1-52

Krohnitta (Eukrohnia) subtilis
Fraser, J. H., 1949
The occurrence of unusual species of Chaetognatha in Scottish plankton collections.
JMBA 28(2):489-491.

Krohnitta subtilis
Fraser, J. H. and A. Saville, 1949
Plankton distribution in Scottish and adjacent waters in 1948. Ann. Biol. 5:61-62.

Krohnitta subtilis
Fraser, J. H. and A. Saville, 1949
Plankton distribution in Scottish and adjacent waters in 1948. Ann. Biol. 5:61-62.

Krohnitta subtilis
Fraser, J. H. and A. Saville, 1949
List of rare exotic species found in the plankton by the Scottish Vessel "Explorer" in 1948. Ann. Biol. 5:62-64.

Krohnitta subtilis
Furnestin, Marie-Louise, 1970.
Chaetognathes des campagnes du Thor (1908-11) en Méditerranée et en Mer Noire.
Dana - Rept. 79:1-51

Krohnitta subtilis
Furnestin, Marie-Louise,1966.
Chaetognathes des eaux africaines.
Atlantide Rep., 9:105-135.

Krohnitta subtilis
Furnestin, M.L., 1958.
Quelques échantillons de zooplancton du Golfe d'Eylath (Akaba).
Haifa, Sea Fish. Res. Sta., Bull., 16:6-14.

Krohnitta subtilis
George, P.C., 1952.
A systematic account of the Chaetognatha of the Indian coastal waters, with observations of their seasonal fluctuations along the Malabar coast.
Proc. Nat. Inst. Sci., India, 18(6):657-689.

Krohnitta subtilis
Germain, L. and L. Joubin, 1916
Chétognathes provenant des campagnes des yachts Hirondelle et Princesse-Alice (1885-1910)
Res. Camp. Sci., Monaco, 49:118 pp., 7 pls., 7 maps.

Khronitta subtilis
Hure, J., 1961
Dneva migracija i sezonska vertikalna raspodjela zooplanktona dubljeg mora. [Migration journalière et distribution saisonnière verticale du zooplancton dans la région profunde de l'Adriatique]
Acta Adriatica, 9(6):1-59.

Krohnita subtilis
Michael, E.L., 1919
Contributions to the Biology of the Philippine Archipelago and Adjacent regions. Report on the chaetognatha collected by the United States Fisheries Steamer "Albatross" during the Philippine Expedition, 1907-1910.
Bull. U.S. Nat. Mus. No.100, Vol.1(4):235-277, pls.34-38.

Krohnitta subtilis
Moore, H. B., 1949
The zooplankton of the upper waters of the Bermuda area of the North Atlantic. Bull. Bingham Ocean. Coll. 12(2):97, 208 text figs.

Krohnita pacifica
Nogueira Paranaguá, Maryse (1963-1964) 1966.
Sobre o plancton da região comprendida entre 3° Lat. S e 13° Lat. S ao largo do Brasil.
Trabhs Inst. Oceanogr. Univ. Recife 5(5/6):125-139.

Krohnita subtilis
Pierce, E.L., 1953.
The Chaetognatha over the continental shelf of North Carolina with attention to the relation to the hydrography of the area. J. Mar. Res. 12(1):75-92, 4 textfigs.

Krohnitta subtilis
Pierce, E.L., and M.L. Wass, 1962.
Chaetognatha from the Florida Current and coastal waters of the southeastern Atlantic states.
Bull. Mar. Sci., Gulf and Caribbean, 12(3):403-431.

Khrohnitta subtilis
Ramult, M., and M. Rose, 1945
Recherches sur les Chetognathes de la Baie d'Alger. Bull. Soc. Hist. Nat. Afrique du Nord 36:45-71, 39 figs.

Krohnitta subtilis
Reyssac, J., 1963.
Chaetognathes du plateau continental européen (de la baie ibéro-marocaine à la Mer Celtique).
Rev. Trav. Inst. Pêches Marit., 27(3):245-299.

Krohnitta subtilis
Ritter-Záhony, R. von, 1911
Revision der Chätognathen. Deutsche Südpolar Exped., 1901-1903, XIII (Zool.V)., Pt.1:71 pp., 51 text figs.

Krohnitta subtilis
von Ritter-Záhony, R., 1911
Vermes. Chaetognathi. Das Tierreich 29:34 pp., 16 text figs.

Krohnitta subtilis
Rose, M., and M. Hamon, 1953.
Nouvelle note complémentaire sur les chétognathes de la Baie d'Alger.
Bull. Soc. Hist. Nat. Afrique du Nord, 44(5/6):167-171.

Krohnitta subtilis
Suarez Caabro, J.A., 1955.
Quetognatos de los mares cubanos.
Mem. Soc. Cubana Hist. Nat., 22(2):125-180, 9 pls

Krohnitta subtilis
Suarez-Caabro, Jose A., and Juan E. Madruga, 1960.
The Chaetognatha of the northeastern coast of Honduras Central America.
Bull. Mar. Sci., Gulf & Caribbean, 10(4):421-429.

Krohnitta subtilis
Sund, Paul N., 1964.
Los quetognatos en las aguas de la region del Peru.
Inter-Amer. Tropical Tuna Comm., Bull., 9(3):115-216.

Krohnitta subtilis
Sund, Paul N., 1961.
Some features of the autecology and distributions of Chaetognatha in the Eastern Tropical Pacific.
Inter-American Tropical Tuna Comm., Bull., 5(4):307-340.

Krohnitta subtilis
Thiel, M.E., 1938
Die Chaetognathen - Bevölkerung des Südatlantischen Ozeans. Biologische Sonderuntersuchungen. 1st Lief. Wissenschaftliche Ergebnisse der Deutschen Atlantischen Expedition auf dem Forschungs, und Vermessungsschiff "Meteor" 1925-1927, 13:1-110, 62 text figs.

Krohnita subtilis
Thomson, J. M., 1947
The Chaetognatha of South-eastern Australia.
Counc. Sci. & Ind. Res., Australia, Bull. No.222, (Div. Fish. Rept. 14), 43 pp., 8 text figs.

Krohnitta subtilis
Tokioka, Takasi, 1959
Observations on the taxonomy and distribution of chaetognaths of the North Pacific.
Publ. Seto Mar. Biol. Lab., 7(3):349-456.

Krohnitta subtilis
Tokioka, T., 1942
Systematic studies of the plankton organisms occurring in Iwayama Bay, Palao. III Chaetognaths from the bay and adjacent waters. Palao Tropical Biol. Sta. Studies, II(3):527-548, pls.5-7, 11 text figs.

Krohnitta subtilis
Tokioka, T., 1940.
The chaetognath fauna of the waters of western Japan. Rec. Oceanogr. Wks., Japan, 12(1):1-22, 3 charts, 4 tables. (Contrib. 87, Seto M.B.L.)

Krohnitta subtilis
Tokioka, T., 1939?
Chaetognaths collected chiefly in the bays of Sagami and Suruga with some notes on the shape and structure of the seminal vesicles.
Rec. Oceanogr. Wks., Japan, 10(2):123-150.

Krohnitta subtilis
Tsuruta, Arao, 1963
Distribution of plankton and its characteristics in the oceanic fishing grounds, with special reference to their relation to fishery.
J. Shimonoseki Univ., Fish., 12(1):13-214.

Krohnitta subtilis
Vannucci, M., and K. Hosoe, 1952.
Resultados cientificos do Cruzeiro do "Baependi" e do "Vega" a Ilha da Trinidada. Bol. Inst. Ocean., Univ. Sao Paulo, 3(1.2):5-30, 4 pls.

Krohnitta subtilis
Vučetic, T., 1961
Sur la répartition des chaetognathes en Adriatique et leur utilisation comme indicateurs biologiques des conditions hydrographiques.
Rapp. Proc. Verb., Réunions, Comm. Int. Expl. Sci. Mer Méditerranée, Monaco, 16(2):111-116.

Krohnittella n.gen.
Germain, L., and L. Joubin, 1912.
Note sur quelques chétognathes nouveaux des croisières de S.A.S. le Prince de Monaco.
Bull. Inst. Océan., Monaco, No. 228:1-14, 15 textfigs.

Khronittella Bourcei
Belloc, G., 1961.
Catalogue des types de chétognathes du Musée Océanographique de Monaco.
Bull. Inst. Océanogr., Monaco, 58(1216): 3 pp.

Krohnittella bourcei n.sp.
Germain, L., and L. Joubin, 1912.
Note sur quelques chétognathes nouveaux des croisières de S.A.S. le Prince de Monaco.
Bull. Inst. Océan., Monaco, No. 228:1-14, 15 textfigs.

Leptosagitta n.gen.
Kassatkina, A.P. 1973.
A new genus Leptosagitta and its status in the system Chaetognatha.
Zool. Zh. 52(8):1202-1207.
(In Russian; English abstract)

Leptosagitta collariata n.sp.
Kassatkina, A.P. 1973.
A new genus Leptosagitta and its status in the system Chaetognatha.
Zool. Zh. 52(8):1202-1207.
(In Russian; English abstract)

Leptosagitta rudata n.sp.
Kassatkina, A.P. 1973.
A new genus Leptosagitta and its status in the system Chaetognatha.
Zool. Zh. 52(8):1202-1207.
(In Russian; English abstract)

Leptosagitta uschakovi n.sp.
Kassatkina, A.P. 1973.
A new genus Leptosagitta and its status in the system Chaetognatha.
Zool. Zh. 52(8):1202-1207.
(In Russian; English abstract)

Parasagitta elegans
Rakusa-Suszczewski, S., 1968.
Predation of Chaetognatha by Tomopteris helgolandica Greff.
J.Cons.perm.int.Explor.Mer, 32(2):226-231.

Parasagitta litturata n.sp.
Kassatkina, A.P. 1973.
A new species of the genus Parasagitta (Chaetognatha) from the far eastern seas.
(In Russian; English abstract)
Zool. Zh. 52(7):1097-1101

Parasagitta litturata maculata n. subsp.
Kassatkina, A.P. 1973.
A new species of the genus Parasagitta (Chaetognatha) from the far eastern seas.
(In Russian; English abstract)
Zool. Zh. 52(7):1097-1101

Pseudosagitta n.gen.
Germain, L., and L. Joubin, 1912.
Note sur quelques chétognathes nouveaux des croisières de S.A.S. le Prince de Monaco.
Bull. Inst. Océan., Monaco, No. 228:1-14, 15 textfigs.

Pseudosagitta Grimaldii
Belloc, G., 1961.
Catalogue des types de chétognathes du Musée Océanographique de Monaco.
Bull. Inst. Océanogr., Monaco, 58(1216):3 pp.
good diagrams

Pseudosagitta grimaldii n.sp.
Germain, L., and L. Joubim 1912.
Note sur quelques chétognathes nouveaux des croisières de S.A.S. le Prince de Monaco.
Bull. Inst. Océan., Monaco, No. 228:1-14, 15 textfigs.

Sagitta draco
Varadarajan, S. and P.I. Chacko, 1943.
On the arrow-worms of Krusadai. Proc. Nat. Inst. Sci., India, 9(2):245-248, 2 figs.

Pterosagitta besnardi
Vannuci, M., and K. Hosoe, 1956.
Pterosaggita besnardi Van. & Hosoe 1952, synonym of P. draco (Krohn 1853).
Bol. Inst. Ocean. Sao Paulo, 7(1/2):195-197.

Pterosagitta besnardi n.sp.
Vannucci, M., and K. Hosoe, 1952.
Resultados cientificos do Cruzeiro do "Baependi" e do "Vega" a Ilha da Trinidada. Bol. Inst.Ocean. Univ. Sao Paulo, 3(1/2):5-30, 4 pls.

Pterosagitta draco
Alvariño, Angeles, 1969
Los quetognatos del Atlántico: distribución y notas esenciales de sistemática.
Trab. Inst. esp. Oceanogr., 37: 290 pp.

Pterosagitta draco
Alvariño, Angeles, 1967.
The Chaetognatha of the NAGA Expedition (1959-1961) in the South China Sea and the Gulf of Thailand.
NAGA Rept., 4(2):1-87.
Scripps Inst. Oceanogr.

Pterosagitta draco
Alvariño, Angeles, 1966
Zoogeografía de California: quetognatos. Revista Soc. Mexicana Hist. nat. 27: 199-243

Pterosagitta draco
Alvarino, Angeles, 1965
Distributional atlas of Chaetognatha in the California Current region.
California Coop. Oceanic Fish. Invest., Atlas, No. 3:291 charts

Pterosagitta draco
Alvariño, Angeles, 1964.
Zoogeografía de los quetognatos, especialmente de la región de California.
Ciencia, México, 23:51-74.

Pterosagitta draco
Alvarino, A., 1957.
Estudio del zooplancton del Mediterráneo occidental. Zooplancton del Atlántico Ibérico. Campañas del "Xauen" en el verano del 1954.
Bol. Inst. Español Oceanogr., Nos., 81-82:1-26; 1-51.

Pterosagitta draco
Bieri, Robert, 1966.
The function of the "wings" of Pterosagitta draco and the so-called tangoreceptors in other species of Chaetognatha.
Publs Seto mar.biol. Lab., 14(1):23-26.

Pterosagitta draco
Bieri, Robert 1959
The distribution of planktonic chaetognatha in the Pacific and their relationship to water masses.
Limnol. and Ocean. Vol. 4 No. 1 pp. 1-28

Pterosagitta draco
Bieri, Robert, 1957.
The chaetognath fauna of Peru in 1941.
Pacific Science 11(3):255-264.

Pterosagitta draco
Bigelow, H.B., and M. Sears, 1939
Studies of the waters of the continental shelf, Cape Cod to Chesapeake Bay. III. A volumetric study of the zooplankton. Mem. M.C.Z. 54(4):183-378, 42 text figs.

Pterosagitta draco
Bumpus, D.F., and E.L. Pierce, 1955.
Hydrography and the distribution of chaetognaths over the continental shelf of North Carolina.
Pap. Mar. Biol. and Oceanogr., Deep-Sea Res., Suppl. to Vol. 3:92-109.

Pterosagitta draco
Burfield, S.T., 1930.
Chaetognatha. Brit. Antarctic ("Terra Nova") Exped., 1910, Zool., 7(4):203-228, 3 maps.

Pterosagitta draco
Burfield, S.T., and E.J.W. Harvey, 1926.
The Chaetognatha of the "Sealark" Expedition.
Trans. Linn. Soc., London, Ser. 2, 19:93-119, Pls. 4-7.

Pterosagitta draco
Coleman, J. S., 1959.
The "Rosaura" Expedition 1937-1938: Chaetognatha.
Bull. Brit. Mus. (Natural History) Zool. 5(8):219-253.

Pterosagitta draco
de Saint-Bon, Marie Catherine, 1963.
Les chaetognathes de la Côte d'Ivoire (espèces de surface).
Rev. Trav. Inst. Pêches Marit., 27(3):301-346.

Pterosagitta draco
Ducret, Françoise, 1968.
Chaetognathes des campagnes de l'Ombango dans les eaux équatoriales et tropicales africaines.
Cah. ORSTOM. Sér.Océanogr.,6(1):95-141.

Pterosagitta draco
Ducret, Françoise, 1968.
Chaetognathes des campagnes de l'"Ombango" dans les eaux équatoriales et tropicales africaines.
Cah.O.R.S.T.O.M.,ser.Océanogr., VI(1):95-141.

Pterosagitta draco
Ducret, Françoise, 1968.
Chaetognathes des eaux superficielles et profondes de la zone équatoriale et tropicale africaine.
Thèse Fac. Sci., Marseille, 99 pp., 25 figs. In: Receuil Trav. publiées de 1965 à 1968, Océanogr. ORSTROM.

Pterosagitta draco
Fagetti Guaita, Elda, 1958.
Investigaciones sobre quetognatos colectados, especialmente, frente a la costa central y norte de Chile. Revista Biol. Mar., Valparaiso, 8(1-3):25-82.

Pterosagitta draco
Fraser, J.H., 1952.
The Chaetognatha and other zooplankton of the Scottish area and their value as biological indicators of hydrographical conditions.
Scottish Home Dept., Mar. Res., 1952(2):1-52, 3 pls., 21 charts.

Pterosagitta draco
Furnestin, Marie-Louise, 1970.
Chaetognathes des campagnes du Thor (1908-11) en Méditerranée et en Mer Noire.
Dana - Rept. 79:1-51

Pterosagitta draco
Furnestin, Marie-Louise, 1966.
Chaetognathes des eaux africaines.
Atlantide Rep., 9:105-135.

Pterosagitta draco
Furnestin, Marie-Louise, 1965.
Chaetognathes de quelques récoltes dans la mer des Antilles et l'Atlantique ouest tropical.
Inst. r. Sci. Nat. Belg., Bull., 41(9):15 pp.

Pterosagitta draco
Furnestin, Marie-Louise, 1963.
Les chaetognathes atlantiques en Méditerranée.
Rev. Trav. Inst. Pêches Marit., 27(2):157-160.

Pterosagitta draco
Furnestin, M. L., 1957.
Chaetognathes et zooplancton du secteur atlantique marocain. Rev. Trav. Inst. Pêches Marit. 21(1/2): 9-356.

Pterosagitta draco
Furnestin, M.-L., et J.-C.Codaccioni, 1968.
Chaetognathes du nord-ouest de l'Océan Indien (golfe d'Aden, mer d'Arabie, golfe d'Oman, golfe Persique).
Cah. ORSTOM., Sér.Océanogr., 6(1):143-171.

Pterosagitta draco
George, P.C., 1952.
A systematic account of the Chaetognatha of the Indian coastal waters, with observations of their seasonal fluctuations along the Malabar coast.
Proc. Nat. Inst. Sci., India, 18(6):657-689.

Pterosagitta draco
Germain, L. and L. Joubin, 1916
Chetognathes provenant des campagnes des yachts Hirondelle et Princesse-Alice (1885-1910)
Rés. Camp. Sci., Monaco, 49:118 pp., 7 pls., 7 maps.

Pterosagitta draco
Ghirardelli, E., 1960.
Habitat e biologia della riproduzione nei chetognati.
Arch. Oceanogr. e Limnol., 11(3):287-304.
Reprinted in:
Trav. Sta. Zool., Villefranche-sur-Mer, 18(11):

Pterosagitta draco
Ghirardelli, Elvezio, 1959.
Osservazioni sulla corona ciliata nei Chetognati.
Boll. Zoo., Unione Zool. Italiana, 26(2):413-421.
Also in:
Trav. Sta. Zool., Villefranche-sur-Mer, 19(1960).

Pterosagitta draco
Ghiradelli, E., 1953.
Appunti sulla morfologia dell'apparecchio riproduttore femminile e sulla biologia della riproduzione in Pterosagitta draco Krohn.
Monitore Zoologico Italiano, 61(2/3):71-79.

Pterosagitta draco
Ghirardelli, E., 1952.
Osservazioni biologiche e sistematiche sui Chetognati del Golfo di Napoli.
Pubbl. Staz. Zool., Napoli, 23(2/3):296-312.

Pterosagitta draco
Ghirardelli, E., 1950.
Osservazioni biologiche e sistematiche sui chetognati della Baia di Villefranche-sur-mer. (Trav. Sta. Zool. Villefranche-sur-mer 10(9). Bol. Pesca, Piscicult., Idrobiol. 5(1):5-27, 7 text-

Pterosagitta draco
Ghirardelli, E., 1950.
Osservazioni biologiche e sistematiche sui chetognati della baia di Villafranche sur mer. Bol. Pesca, Piscicol. e Idrobiol., n.s., 5(1):105-127, 7 textfigs.

Pterosagitta draco
Ghirardelli, Elvezio, 1947
Chetognati raccolti nel Mar Rosso e nel l'Oceano Indiano dalla Nave "Cherso", Bolletino di Pesca, Piscicoltura, e Idrobiologia Anno 23, Vol. 2(n.s.) (2):253-270, 2 pls., 9 text figs.

Pterosagitta draco
Hosoe, K., 1956.
Chaetognaths from the Iles of Fernando de Noronha.
Contr. Avulsas, Inst. Oceanogr., Sao Paulo, No. 3:8 pp.

Pterosagitta draco
Japan, Kobe Marine Observatory, 1963.
Report of the oceanographical observations in the sea south of Honshu from February to March 1961. Report of the oceanographical observations in the cold water region off Enshu Nada in May 1961.
(In Japanese)
Bull. Kobe Mar. Obs., 171(4):22-35.

Pterosagitta draco
Japan, Kobe Marine Observatory, Oceanographical Section, 1962
Report on the oceanographic observations in the sea south of Honshu from July to August, 1959.
Bull. Kobe Mar. Obs., No. 169(11):37-43.
(In Japanese)

Pterosagitta draco
Kado, Y., 1953.
The chaetognath fauna of the Inland Sea of Japan, especially on the distribution of Sagitta enflata and S. crassa. Dobuts. Zasshi 62(10):337-342.

Pterosagitta draco
Legare, J. E. Henri, and E. Zoppi, 1961.
Notas sobre la abundancia y distribucion de Chaetognatha en las aguas del Oriente de Venezuela. (In English).
Bol. Inst. Oceanograf., Univ. Oriente, Cumana, Venezuela, 1(1):149-171.

Pterosagitta draco
Michael, E.L., 1919
Contributions to the Biology of the Philippine Archipelago and Adjacent regions. Report on the chaetognatha collected by the United States Fisheries Steamer "Albatross" during the Philippine Expedition, 1907-1910.
Bull. U.S. Nat. Mus. No.100, Vol.1(4):235-277, pls.34-38.

Pterosagitta draco
Moore, H. B., 1949
The zooplankton of the upper waters of the Bermuda area of the North Atlantic. Bull. Bingham Ocean. Coll. 12(2):97, 208 text figs.

Pterosagitta draco
Nogueira Paranaguá, Maryse (1963-1964) 1966.
Sôbre o plancton da região compreendida entre 3° Lat.S e 18° Lat.S ao largo do Brasil.
Trabs Inst. Oceanogr. Univ. Recife 5(5/6): 125-139

Pterosagitta draco
Pierce, E.L., 1953.
The Chaetognatha over the continental shelf of North Carolina with attention to the relation to the hydrography of the area. J. Mar. Res. 12(1): 75-92, 4 textfigs.

Pterosagitta draco
Ramult, M., and M. Rose, 1945
Recherches sur les Chetognathes de la Baie d'Alger. Bull. Soc. Hist. Nat. Afrique du Nord 36:45-71, 39 figs.

Pterosagitta draco
Rao, T.S. Satyanarayana, and P.N. Ganapati, 1958.
Studies on Chaetognatha in the Indian Seas. III. Systematics and distribution in the waters off Visakhapaynam. Andhra Univ. Mem. Oceanogr. 2:147-163.

Pterosagitta draco
Reyssac, J., 1963.
Chaetognathes du plateau continental européen (de la baie ibero-marocaine à la Mer Celtique).
Rev. Trav. Inst. Pêches Marit., 27(3):245-299.

Pterosagitta draco
Ritter-Záhony, R. von, 1911
Revision der Chätognathen. Deutsche Südpolar Exped., 1901-1903, XIII (Zool.V)., Pt.1:71 pp., 51 text figs.

Pterosagitta draco
Rose, M., and M. Hamon, 1953.
Nouvelle note complémentaire sur les chétognathes de la Baie d'Alger.
Bull. Soc. Hist. Nat., Afrique du Nord, 44(5/6):167-171.

Pterosagitta draco
Suarez Caabro, J.A., 1955.
Quetognatos de los mares cubanos.
Mem. Soc. Cubana Hist. Nat., 22(2):125-180, 9 pls.

Pterosagitta draco
Suarez-Caabro, Jose A., and Juan E. Madruga, 1960.
The Chaetognatha of the northeastern coast of Honduras Central America.
Bull. Mar. Sci., Gulf & Caribbean, 10(4):421-429

Pterosagitta draco
Sund, Paul N., 1961.
Some features of the autecology and distributions of Chaetognatha in the Eastern Tropical Pacific.
Inter-American Tropical Tuna Comm., Bull. 5(4):302-340.

Pterosagitta draco
Sund, Paul N., 1959.
A key to the Chaetognatha of the tropical Eastern Pacific Ocean.
Pacific Science, 13(3):269-285.

Pterosagitta draco
Thiel, M.E., 1938
Die Chaetognathen - Bevölkerung des Südatlantischen Ozeans. Biologische Sonderuntersuchungen. 1st Lief. Wissenschaftliche Ergebnisse der Deutschen Atlantischen Expedition auf dem Forschungs. und Vermessungschiff "Meteor" 1925-1927, 13:1-110, 62 text figs.

Pterosagitta draco
Thomson, J. M., 1947
The Chaetognatha of South-eastern Australia.
Counc. Sci. & Ind. Res., Australia, Bull. No.222, (Div. Fish. Rept. 14), 43 pp., 8 text figs.

Pterosagitta draco
Tokioka, Takasi, 1959
Observations on the taxonomy and distribution of chaetognaths of the North Pacific.
Publ. Seto Mar. Biol. Lab., 7(3):349-456.

Pterosagitta draco
Tokioka, T., 1954.
Droplets from the plankton net. XVI. On a small collection of chaetognaths from the central Pacific.
Publ. Seto Mar. Biol. Lab. 4(1):99-102, 1 textfig.

Pterosagitta draco
Tokioka, T., 1951.
Pelagic tunicates and chaetognaths collected during the cruises to the New Yamato Bank in the Sea of Japan. Publ. Seto Mar. Biol. Lab. 2(1): 1-26, 1 chart, 6 tables, 12 textfigs.

Pterosagitta draco
Tokioka, T., 1942
Systematic studies of the plankton organisms occurring in Iwayama Bay, Palao. III Chaetognaths from the bay and adjacent waters. Palao Tropical Biol. Sta. Studies, II(3):527-548, pls.5-7, 11 text figs.

Pterosagitta draco
Tokioka, T., 1940.
The chaetognath fauna of the waters of western Japan. Rec. Oceanogr. Wks., Japan, 12(1):1-22, 3 charts, 4 tables. (Contrib. 87, Seto. M.B.)

Pterosagitta draco
Tokioka, T., 1939.
Chaetognaths collected chiefly in the bays of Sagami and Suruga with some notes on the shape and structure of the seminal vesicles.
Rec. Oceanogr. Wks., Japan, 10(2):123-150.

Pterosagitta draco
Tsuruta, Arao, 1963
Distribution of plankton and its characteristics in the oceanic fishing grounds, with special reference to their relation to fishery.
J. Shimonoseki Univ. Fish., 12(1):13-214.

Pterosagitta draco
Vannuci, M., and K. Hosoe, 1956.
Pterosagitta besnardi Van. & Hosoe 1952, synonym of P. draco (Krohn 1853). Bol. Inst. Ocean., Sao Paulo, 7(1/2):195-197.

Pterosagitta draco
von Ritter-Záhony, R., 1911
   Vermes. Chaetognathi. Das Tierreich
29:34 pp., 16 text figs.

Sagitta sp.
Deevey, G. B., 1948
   The zooplankton of Tisbury Great Pond.
Bull. Bingham Oceanogr. Coll., 12(1):44 pp.,
14 figs.

Sagitta
Furnestin, J., 1938
Influence de la salinité sur la répartition du
genre Sagitta dans l'Atlantique Nord-Est
(Juillet-Août-Septembre, 1936).
Rev. Trav. Off. Pêches Marit. 11(3):425-437.

Sagitta
Huguet D., 1969.
Contribution à l'étude de la structure
du ganglion nerveux ventral des
Sagitta (chaetognathes). Paris
Bull. Mus. natn. Hist. Nat. (2) 40 (5):
1031-1042.

Sagitta
Jagersten, G., 1940.
Zur Kenntnis der Physiologie der Zeugung bei
Sagitta. Zool. Bidrag., Uppsala, 18:397-413.

Sagitta sp.
Japan, Hokkaido University, Faculty of Fisheries, 1962
II. The "Oshoro Maru" cruise 48 to the Bering
Sea, and northwestern North Pacific in June-
July 1961.
Data Record Oceanogr. Obs., Expl. Fish., 6:
22-149.

Sagitta spp.
Johnson, M. W., 1949
   Zooplankton as an index of water exchange
between Bikini lagoon and the open sea. Trans.
Am. Geophys, Un. 30(2):238-214, 15 text figs.

Sagitta sp.
Ogilvie, H.S., 1934
   A preliminary account of the food of
the herring in the North-Western North Sea.
Rapp. Proc. Verb.89(3):85-92, 2 text figs.

Sagitta sp.
Tokioka, T., 1954
   Droplets from the plankton net. XVI. On a small
collection of chaetognaths from the central
Pacific. Publ. Seto Mar. Biol. Lab. 4(1):99-102,
1 textfig.

The figure is of this Sagitta and is perhaps a
new species.

sagitta
Van Oye, P., 1918.
Untersuchungen über die Chaetognathen des
Javameeres. Contributions à la Faunes des Indes
Neerlandaises, IV:

sagittae
von Ritter-Záhony, R., 1911
   Vermes. Chaetognathi. Das Tierreich
29:34 pp., 16 text figs.

Sagitta sp.
Woodmansee, R.A., 1958
The seasonal distribution of the zooplankton off
Chicken Key in Biscayne Bay, Florida.
Ecology, 39(2):247-261.

Sagitta ai
Sund, Paul N., 1961.
Some features of the autecology and distributions
of Chaetognatha in the Eastern Tropical Pacific.
Inter-American Tropical Tuna Comm., Bull., 5(4):
307-340.

Sagitta ai
Tokioka, T., 1942
   Systematic studies of the plankton
organisms occurring in Iwayama Bay, Palao.
III Chaetognaths from the bay and adjacent
waters. Palao Tropical Biol. Sta. Studies,
II(3):527-548, pls.5-7, 11 text figs.

Sagitta ai
Tokioka, T., 1940.
The chaetognath fauna of the waters of western
Japan. Rec. Oceanogr. Wks., Japan, 12(1):1-22,
3 charts, 4 tables. (Contrib. 87, Seto M.B.L.).

Sagitta ai n.sp.
Tokioka, T., 1939.
Chaetognaths collected chiefly from the bays of
Sagami and Suruga with some notes on the shape
and structure of the seminal vesicles.
Rec. Oceanogr. Wks., Japan, 10(2):123-150.

Sagitta arctica
Fowler, C.H., 1906.
The Chaetognatha of the Siboga Expedition with a
discussion of the synonymy and distribution of
the group. Rept. Siboga Exped. 21:1-88, 3 pls.,
6 charts.

Sagitta arctica
Germain, L. and L. Joubin, 1916
Chétognathes provenant des campagnes des
yachts Hirondelle et Princesse-Alice (1885-1910)
Rés. Camp. Sci., Monaco, 49:118 pp., 7 pls.,
7 maps.

Sagitta arctica
von Ritter-Záhony, R., 1911
   Vermes. Chaetognathi. Das Tierreich
29:34 pp., 16 text figs.

Sagitta bedfordii Doncaster 1903
Tokioka, T., 1942
   Systematic studies of the plankton
organisms occurring in Iwayama Bay, Palao.
III Chaetognaths from the bay and adjacent
waters. Palao Tropical Biol. Sta. Studies,
II(3):527-548, pls.5-7, 11 text figs.

Sagitta baltica
von Ritter-Záhony, R., 1911
   Vermes. Chaetognathi. Das Tierreich
29:34 pp., 16 text figs.

Sagitta batava n.sp.
Biersteker, Rinie H., and S. van der Spoel, 1966.
Sagitta batava s.sp. from the Scheldt Estuary,
The Netherlands (Chaetognatha).
Beaufortia, 14(167):61-69.

Sagitta bedfordii
Alvariño, Angeles, 1967.
The Chaetognatha of the NAGA Expedition (1959-
1961) in the South China Sea and the Gulf of
Thailand.
NAGA Rept., 4(2):1-87.
   Scripps Inst. Oceanogr.,

Sagitta bedfordi
Tokioka, Takasi, 1959.
Observations on the taxonomy and distribution of chaetognaths of the North Pacific.
Publ. Seto Mar. Biol. Lab., 7(3): 349-456.

Sagitta bedfordii
Tsuruta, Arao, 1963
Distribution of plankton and its
characteristics in the oceanic fishing
grounds, with special reference to their
relation to fishery.
J. Shimonoseki Univ., Fish., 12(1):13-214.

Sagitta bedoti
Alvariño, Angeles, 1967.
The Chaetognatha of the NAGA Expedition (1959-
1961) in the South China Sea and the Gulf of
Thailand.
NAGA Rept., 4(2):1-87.
   Scripps Inst. Oceanogr.,

Sagitta bedoti
Alvariño, Angeles, 1966
Zoogeografia de California: quetognatos, Revista
Soc. Mexicana Hist. nat. 27: 199-243

Sagitta bedoti
Alvariño, Angeles, 1965.
Distributional atlas of Chaetognatha in the
California Current region.
California Coop. Oceanic Fish. Invest., Atlas,
No. 3:291 charts

Sagitta bedoti
Alvariño, Angeles, 1964.
Zoogeografia de los quentgnatos, especialmente
de la región de California.
Ciencia, México, 23:51-74.

Sagitta bedoti
Alvariño, Angeles, 1963
Quetognatos epiplanctonicos del Mar de Cortes.
Revista, Soc. Mexicana, Hist. Nat., 24:97-202.

Sagitta bedoti n.sp.
Béranek, E., 1895.
Les chétonates de la Baie d'Amboina. Rev. Suisse
Zool. 3:137-159, Pl. IV.

Sagitta bedoti
Bieri, Robert 1959
The distribution of planktonic chaetognatha
in the Pacific and their relationship to
water masses.
Limnol. and Ocean. Vol. 4 No. 1 pp. 1-28

Sagitta bedoti
Bieri, Robert, 1957.
The chaetognath fauna of Peru in 1941.
Pacific Science 11(3):255-264.

Sagitta bedoti
Burfield, S.T. 1930.
Chaetognatha. Brit. Antarctic ("Terra Nova")
Exped., 1910, Zool., 7(4):203-228, 3 maps.

Sagitta bedoti
Burfield, S.T., and E.J.W. Harvey, 1926.
The Chaetognatha of the "Sealark" Expedition.
Trans. Linn. Soc., London, Ser. 2, 19:93-119,
Pls. 4-7.

Sagitta bedoti
Dallot, Serge, 1970.
L'anatomie du tube digestif dans la
phylogénie et la systématique des
Chaetognathes.
Bull. Mus. natn. Hist. nat. (2) 42(3): 549-565.

Sagitta bedoti
Delsman, H. C., 1939.
Preliminary plankton investigations in the Java
Sea. Treubia, 17:139-181, 8 maps, 41 figs.

Sagitta bedoti
Ducret, Françoise, 1968.
Chaetognathes des eaux superficielles et
profondes de la zone équatoriale et tropicale
africaine.
Thèse Fac. Sci., Marseille, 99 pp., 25
figs. In: Receuil Trav. publiées de
1965 à 1968, Océanogr. ORSTOM.

Sagitta bedotu
Ducret, Françoise, 1968.
Chaetognathes des campagnes de l'"Ombango" dans
les eaux équatoriales et tropicales africaines.
Cah. O.R.S.T.O.M., ser.Océanogr., VI(1):95-141.

Sagitta bedoti
Ducret, Françoise, 1968.
Chaetognathes des campagnes de l'Ombango dans
les eaux équatoriales et tropicales africaines.
Cah. ORSTOM., sér.Océanogr., 6(1):95-141.

Sagitta bedoti
Enomoto Y., 1962
Studies on the food base in the Yellow and
the East China seas. 1. Plankton survey in
summer of 1956.
Bull. Japan Soc. Sci. Fish., 28(8):759-765.

Sagitta bedoti
Fowler, G.H., 1906.
The Chaetognatha of the Siboga Expedition, with a discussion of the synonymy and distribution of the group. Rept. Siboga Exped. 21:1-88, 3 pls., 6 charts.

Sagitta bedoti
Furnestin, M-L. et J. Balancé 1965
Chaetognathes de la mer Rouge (Archipel Dahlac) Israel South Red Sea Expedition 1962, Reports No. 32.
Bull. Sea Fish. Res. Sta. 52:3-20.

Sagitta bedoti
Furnestin, M.-L., et J.-C. Coduccioni, 1968.
Chaetognathes du nord-ouest de l'Océan Indien (golfe d'Aden, mer d'Arabie, golfe d'Oman, golfe Persique).
Cah. ORSTOM., Sér. Oceanogr., 6(1):143-171.

Sagitta bedoti
Furnestin, M.-L., et J. Radiguet, 1964.
Chaetognathes de Madagascar (Secteur de Nosy-Bé).
Cahiers, O.R.S.T.R.O.M., Océanogr., 11(4):55-98.

Sagitta bedoti
George, M.J., 1958.
Observations on the plankton of the Cochin backwaters.
Indian J. Fish., 5(2):375-401.

Sagitta bedoti
George, P.C., 1952.
A systematic account of the Chaetognatha of the Indian coastal waters, with observations on their seasonal fluctuations, along the Malabar coast.
Proc. Nat. Inst. Sci., India, 18(6):657-689.

Sagitta bedoti
Ghirardelli, Elvezio, 1947
Chetognati raccolti nel Mar Rosso e nel l'Oceano Indiano dalla Nave "Cherso", Bolletino di Pesca, Piscicoltura, e Idrobiologia Anno 23, Vol. 2(n.s.) (2):253-270, 2 pls., 9 text figs.

Sagitta bedoti
Hamada, Takeo, 1967.
Studies on the distribution of Chaetognatha in the Harima-nada and Oseka Bay, with special reference to Sagitta enflata. (In Japanese; English abstract).
Bull. Jap. Soc. scient. Fish., 33(2):98-103.

Sagitta bedoti
Japan, Kobe Marine Observatory, 1963.
Report of the oceanographic observations in the sea south of Honshu from July to August, and from the cold water region south of Enshu Nada October to November 1960. (In Japanese).
Bull. Kobe Mar. Obs., 171(3):36-52.

Sagitta bedoti
Japan, Kobe Marine Observatory, 1962
Report of the oceanographic observations in the sea south of Honshu from July to August 1960. (In Japanese).
Res. Mar. Meteorol. and Oceanogr., July-Dec. 1960, Japan Meteorol. Agency, No. 28:36-42.

Sagitta bedoti
Japan, Kobe Marine Observatory, Oceanographical Section, 1962
Report of the oceanographic observations in the sea south of Honshu in May, 1960. (In Japanese).
Bull. Kobe Mar. Obs., No. 169(12):27-33.

Sagitta bedoti
Japan, Kobe Marine Observatory, Oceanographical Section, 1962
Report of the oceanographic observations in the sea south of Honshu from October to November, 1959. (In Japanese).
Bull. Kobe Mar. Obs., No. 169(11):44-50.

Sagitta bedoti
Japan, Kobe Marine Observatory, Oceanographical Section, 1962
Report on the oceanographic observations in the sea south of Honshu from July to August, 1959.
Bull. Kobe Mar. Obs., No. 169(11):37-43. (In Japanese)

Sagitta bedoti
Kado, Y., 1953.
The chaetognath fauna of the Inland Sea of Japan, especially on the distribution of Sagitta englata and S. crassa. Dobuts. Zasshi 62(10):337-342.

Sagitta bedoti
Michael, E.L., 1919
Contributions to the Biology of the Philippine Archipelago and Adjacent regions. Report on the chaetognatha collected by the United States Fisheries Steamer "Albatross" during the Philippine Expedition, 1907-1910.
Bull. U.S. Nat. Mus. No. 100, Vol.1(4):235-277, pls. 34-38.

Sagitta bedoti
Michael, E. L., 1911
Classification and vertical distribution of the Chaetognatha of the San Diego Region including redescriptions of some doubtful species of the group. Univ. Calif. Publ. Zool. 8(3):21-186, Pls. 1-8.

Sagitta bedoti
Nair, Vijayalakshmi R. 1971.
Seasonal fluctuations of chaetognaths in the Cochin Backwater.
J. mar. biol. Ass. India 13(2):226-233.

Sagitta bedoti
Rao, T.S. Satyanarayana, and P.N. Ganapati, 1958
Studies on Chaetognatha in the Indian Seas. III. Systematics and distribution in the waters off Visakhapaynam. Andhra Univ. Mem. Oceanogr., 2:147-163.

Sagitta bedoti
Ritter-Záhony, R. von, 1911
Revision der Chätognathen. Deutsche Südpolar Exped., 1901-1903, XIII (Zool.V)., Pt.1:71 pp., 51 text figs.

Sagitta bedoti
Subrahmanyam, M. K., 1940.
Sagitta bedoti Beraneck in Madras plankton. Curr Sci. 9:379-380.

Sagitta bedoti
Sudarsan, D., 1961.
Observations on the Chaetognatha of the waters around Mandapam.
Indian J. Fish., 8(2):364-382.

Sagitta bedoti
Sund, Paul N., 1964.
Los quetognatos en las aguas de la region del Peru.
Inter-Amer. Tropical Tuna Comm., Bull., 9(3):115-216.

Sagitta bedoti
Sund, Paul N., 1961.
Some features of the autecology and distributions of Chaetognatha in the Eastern Tropical Pacific.
Inter-American Tropical Tuna Comm., Bull., 5(4):307-340.

Sagitta bedoti
Sund, Paul N., 1959.
A key to the Chaetognatha of the tropical Eastern Pacific Ocean.
Pacific Science, 13(3):269-285.

Sagitta bedoti
Thomson, J. M., 1947
The Chaetognatha of South-eastern Australia. Counc. Sci. & Ind. Res., Australia, Bull. No. 222, (Div. Fish. Rept. 14), 43 pp., 8 text figs.

Sagitta bedoti
Tokioka, Takasi, 1959
Observations on the taxonomy and distribution of chaetognaths of the North Pacific.
Publ. Seto Mar. Biol. Lab., 7(3):349-456.

Sagitta bedoti
Tokioka, T., 1951.
Pelagic tunicates and chaetognaths collected during the cruises to the New Yamato Bank in the Sea of Japan. Publ. Seto Mar. Biol. Lab. 2(1):1-26, 1 chart, 6 tables, 12 textfigs.

Sagitta bedoti
Tokioka, T., 1942
Systematic studies of the plankton organisms occurring in Iwayama Bay, Palao. III Chaetognaths from the bay and adjacent waters. Palao Tropical Biol. Sta. Studies, II(3):527-548, pls.5-7, 11 text figs.

Sagitta bedoti
Tokioka, T., 1940.
The chaetognath fauna of the waters of western Japan. Rec. Oceanogr. Wks., Japan, 12(1):1-22, 3 charts, 4 tables. (Contrib. 87, Seto M.B.L.).

Sagitta bedoti
Tokioka, T., 1939.
Chaetognaths collected chiefly from the bays of Sagami and Suruga with notes on the shape and structure of the seminal vesicles.
Rec. Oceanogr. Wks., Japan, 10(2):123-150.

Sagitta bedoti
Tsuruta, Arao, 1963
Distribution of plankton and its characteristics in the oceanic fishing grounds, with special reference to their relation to fishery.
J. Shimonoseki Univ. Fish., 12(1):13-214

Sagitta bedoti
von Ritter-Záhony, R., 1911
Vermes. Chaetognathi. Das Tierreich 29:34 pp., 16 text figs.

Sagitta bedoti forma littoralis nov.
Tokioka, T., and D. Pathansali, 1965.
A new form of Sagitta bedoti Beraneck found in the littoral waters near Penang.
Bull. natn. Mus. Singapore (33-1):1-5.

Sagitta betesios
Alvariño, Angeles, 1966
Zoographia de California: quetognatos. Revista Soc. Mexicana Hist. nat. 27: 199-243
Revta Soc. mex. Hist. nat.

Sagitta bierii
Alvariño, Angeles, 1966
Zoogeografia de California: quetognatos. Revista Soc. Mexicana Hist. nat. 27: 199-243

Sagitta bierii
Alvarino, Angeles, 1965.
Distributional atlas of Chaetognatha in the California Current region.
California Coop. Oceanic Fish. Invest., Atlas, No. 3:291 charts

Sagitta bieri
Alvariño, Angeles, 1964.
Zoogeografia de los quetognatos, especialmente de la region de California.
Ciencia, Mexico, 23:15-74.

Sagitta bierii
Alvarino, Angeles, 1963
Quetognatos epiplanctonicos del Mar de Cortes.
Revista, Soc. Mexicana, Hist. Nat., 24:97-202.

Sagitta bieri n.sp.
Alvarino, Angeles, 1961.
Two new chaetognaths from the Pacific.
Pacific Science, 15(1):67-77.

sagitta bierii
Ducret, Françoise, 1968.
Chaetognathes des campagnes de l'"Ombango" dans les eaux équatoriales et tropicales africaines.
Cah. O.R.S.T.O.M., ser. Océanogr., VI(1):95-141.

Sagitta bieri
Ducret, Françoise, 1968.
Chaetognathes des eaux superficielles et profondes de la zone équatoriale et tropicale africaine.
Thèse Fac. Sci., Marseille, 99 pp., 25 figs. In: Receuil Trav. publiées de 1965 à 1968, Océanogr. ORSTROM.

Sagitta bieri
Furnestin, Marie-Louise, 1966.
Chaetognathes des eaux africaines.
Atlantide Rep., 9:105-135.

Sagitta bierii
Sund, Paul N., 1961.
Some features of the autoecology and distributions of Chaetognatha in the Eastern Tropical Pacific.
Inter-American Tropical Tuna Comm., Bull., 5(4):307-340.

Alvariño, Angeles, 1969 SAGITTA BIPUNCTATA
Los quetognatos del Atlántico: distribución y notas esenciales de sistemática.
Trab. Inst. esp. Oceanogr., 37: 290 pp.

Sagitta bipunctata
Alvariño, Angeles, 1967.
The Chaetognatha of the NAGA Expedition (1959-1961) in the South China Sea and the Gulf of Thailand.
NAGA Rept., 4(2):1-87.
Scripps Inst. Oceanogr.

Alvarino, Angeles, 1965. Sagitta bipunctata
Distributional atlas of Chaetognatha in the California Current region.
California Coop. Oceanic Fish. Invest., Atlas, No. 3:291 charts

Sagitta bipunctata
Aida, T., 1897.
Chaetognaths of Misaki Harbour.
Ann. Zool., Japon., 1:13-21, Pl. 3.

Sagitta bipunctata
Aida, T., 1897.
On the growth of the ovum in chaetognaths.
Ann. Zool., Japon., 1:77-81, Pl. 4.

Sagitta bipunctata
Alvariño, Angeles, 1964.
Zoogeografía de los quetognatos, especialmente de la región de California.
Ciencia, México, 23:51-74.

Sagitta bipunctata
Alvarino, A., 1957.
Estudio del zooplancton del Mediterráneo occidental. Zooplancton del Atlántico Ibérico. Campañas del "Xauen" en el verano del 1954.
Bol. Inst. Español Oceanogr., Nos. 81-82:1-26;

Sagitta bipunctata
Béranek, E., 1895.
Les chétonates de la Baie d'Amboina. Rev. Suisse Zool. 3:137-159, Pl. IV.

Sagitta bipunctata
Bigelow, H.B., and M. Leslie, 1930
Reconnaissance of the waters and plankton of Monterey Bay, July 1928.
Bull. M.C.Z., 70(5):429-481, 43 text figs.

Sagitta bipunctata
Bordas, M., 1920.
Estudio de la ovogénesis en la Sagitta bipunctata Quoy et Gaimard.
Trab. Mus. Nac. Ciencias Nat. Madrid, Ser. Zool., 42: 5-

Sagitta bipunctata
Browne, E.T., 1905.
Notes on the pelagic faunas of the Firth of Clyde (1901-1902). Proc. R. Soc., Edinburgh, 25(9):779-791.

Sagitta bipunctata
Bumpus, D.F., and E.L. Pierce, 1955.
Hydrography and the distribution of chaetognaths over the continental shelf of North Carolina.
Pap. Mar. Biol. and Oceanogr., Deep-Sea Res., Suppl. to Vol. 3:92-109.

Sagitta bipunctata
Burfield, S.T. 1930.
Chaetognatha. Brit. Antarctic ("Terra Nova") Exped., 1910, Zool., 7(4):203-228, 3 maps.

Sagitta bipunctata
Burfield, S.T., 1927.
Zoological results of the Cambridge Expedition to the Suez Canal. XXI. Report on the Chaetognatha.
Trans. Zool. Soc., London, 22:355-356.

Sagitta bipunctata
Burfield, S.T., and E.J.W. Harvey, 1926.
The Chaetognatha of the "Sealark" Expedition.
Trans. Linn. Soc., London, Ser. 2, 19:93-119, Pls. 4-7.

Sagitta bipunctata
Coleman, J. S., 1959.
The "Rosaura" Expedition 1937-1938: Chaetognatha.
Bull. Brit. Mus. (Natural History) Zool. 5(8):219-253.

Sagitta bipunctata
de Saint-Bon, Marie Catherine, 1963.
Les chaetognathes de la Côte d'Ivoire (espèces de surface).
Rev. Trav. Inst. Pêches Marit., 27(3):301-346.

Sagitta bipunctata
Ducret, Françoise, 1968.
Chaetognathes des eaux superficielles et profondes de la zone équatoriale et tropicale africaine.
Thèse Fac. Sci., Marseille, 99 pp., 25 figs. In: Receuil Trav. publiées de 1965 à 1968, Océanogr. ORSTROM.

Sagitta bipunctata
Ducret, Françoise, 1968.
Chaetognathes des campagnes de l'Ombango" dans les eaux équatoriales et tropicales africaines.
Cah. O.R.S.T.O.M., ser. Océanogr., VI(1):95-141.

Sagitta bipunctata
Ducret, Françoise, 1968.
Chaetognathes des campagnes de l'Ombango dans les eaux équatoriales et tropicales africaines.
Cah. ORSTOM., Sér. Océanogr., 6(1):95-141.

Sagitta bipunctata
Fagetti Guaita, Elda, 1958.
Investigaciones sobre quetognatos colectados, especialmente, frente a la costa central y norte de Chile. Revista Biol. Mar., Valparaiso, 8(1-3):25-82.

Sagitta bipunctata
Faure, M. L., 1952.
Contribution a l'étude biologique et morphologique de deux chaetognaths des eaux atlantiques du Maroc: Sagitta friderici Ritter-Zahony et Sagitta bipunctata Quoy et Gaimard. Vie et Milieu 3:25.

Sagitta bipunctata
Fowler, G.H., 1906.
The Chaetognatha of the Siboga Expedition with a discussion of the synonymy and distribution of the group. Rept. Siboga Exped. 21:1-88, 3 pls., 6 charts.

Sagitta bipunctata
Fowler, G.H., 1905.
Biscayan plankton collected during a cruise of H.M.S. "Research", 1900. Pt. III. The Chaetognatha. Trans. Linn. Soc., London, 10(3):55-87, Pls. 4-7.

Sagitta bipunctata
Fraser, J.H., 1952.
The Chaetognatha and other zooplankton of the Scottish area and their value as biological indicators of hydrographical conditions.
Scottish Home Dept., Mar. Res., 1952(2):1-52, 3 pls., 21 charts.

Sagitta bipunctata
Fraser, J. H., 1949
The occurrence of unusual species of Chaetognatha in Scottish plankton collections.
JMBA 28(2):489-491.

Sagitta bipunctata
Furnestin, Marie-Louise, 1970.
Chaetognathes des campagnes du Thor (1908-11) en Méditerranée et en Mer Noire.
Dana - Rept. 79:1-51

Sagitta bipunctata
Furnestin, Marie-Louise, 1966.
Chaetognathes des eaux africaines.
Atlantide Rep., 9:105-135.

Sagitta bipunctata
Furnestin, Marie-Louise, 1965.
Chaetognathes de quelques récoltes dans la mer des Antilles et l'Atlantique ouest tropical.
Inst. r. Sci. Nat. Belg., Bull., 41(9):15 pp.

Sagitta bipunctata
Furnestin, M. L., 1957.
Chaetognathes et zooplancton du secteur atlantique marocain. Rev. Trav. Inst. Pêches Marit. 21(1/2):9-356.

Sagitta bipunctata
Furnestin, M.-L., 1953.
Sur quelques chaetognathes d'Israel.
Bull. Res. Counc., Israel, 2(4):411-414.

Sagitta bipunctata
Furnestin, M.-L., et J. Radiguet, 1964.
Chaetognathes de Madagascar (Secteur de Nosy-Be).
Cahiers, O.R.S.T.R.O.M., Océanogr., 11(4):55-98.

Sagitta bipunctata
Gamulin, T., 1948
Prilog poznavanju Zooplanktona Srednjedalmatinskog Otočnog Područja. (Contrib. a la connaissance du zooplankton de la zone insulaire de la Dalmatie Moyenne). Acta Adriatica 3(7):38 pp., 6 tables, 1 map.

Sagitta bipunctata
Germain, L. and L. Joubin, 1916
Chétognathes provenant des campagnes des yachts Hirondelle et Princesse-Alice (1885-1910)
Rés. Camp. Sci., Monaco, 49:118 pp., 7 pls., 7 maps.

Sagitta bipunctata
Ghirardelli, E., 1960.
Habitat e biologia della riproduzione nei chetognati.
Arch. Oceanogr. é Limnol., 11(3):287-304.
Reprinted in:
Trav. Sta. Zool., Villefranche-sur-Mer, 18(11):

Sagitta bipunctata
Ghirardelli, Elvezio, 1959.
Osservazioni sulla corona ciliata nei Chetognati.
Boll. Zool., Unione Zool. Italiana, 26(2):413-421
Also in:
Trav. Sta. Zool., Villefranche-sur-Mer, 19(1960).

Sagitta bipunctata
Ghirardelli, E., 1952.
Osservazioni biologiche e sistematiche sui Chetognati del Golfo di Napoli.
Pubbl. Staz. Zool., Napoli, 23(2/3):296-312.

Sagitta bipunctata
Ghirardelli, E., 1950.
Osservazioni biologiche e sistematiche sui chetognati della baia di Villefranche sur mer. Bol. Pesca, Piscicol. e Idrobiol., n.s., 5(1):105-127, 7 textfigs.

Sagitta bipunctata
Ghiradelli, E., 1950.
Osservazioni biologiche e sistematiche sui chetognati della Baia di Villefranche-sur-mer. (Trav. Sta. Zool. Villefranche-sur-mer 10(9)). Bol. Pesca, Piscicult. Idrobiol., n.s., 5(1):5-27, 7 textfigs.

Sagitta bipunctata
Grant, George C., 1963
Investigations of inner continental shelf waters off lower Chesapeake Bay. IV. Descriptions of the Chaetognatha and a key to their identification.
Chesapeake Science, 4(3):109-119.

Sagitta bipunctata
Legare, J. E. Henri, and E. Zoppi, 1961.
Notas sobre la abundancia y distribucion de Chaetognatha en las aguas del Oriente de Venezuela. (In English).
Bol. Inst. Oceanograf., Univ. Oriente, Cumana, Venezuela, 1(1):149-171.

Sagitta bipunctata
Macdonald, R., 1933
An examination of plankton hauls made in the Suez Canal during the year 1928. Fish. Res. Dis., Notes & Mem. No.3, 11 pp., 1 chart.

Sagitta bipunctata
Marukawa, H., 1921
Plankton lists and some new species of copepods, from the northern waters of Japan. Bull. Inst. Ocean., No.384, 15 pp., 3 pls., 1 chart. Monaco

Sagitta bipunctata
Massuti, M., 1961
Note préliminaire a l'etude des chétognathos de la Méditerranee occidentale. Campagne du "Xauen" X-6911.
Rapp. Proc. Verb., Réunions, Comm. Int. Expl. Sci. Mer Méditerranée, Monaco, 16(2):237-244.

Sagitta bipunctata
Massuti Oliver, M., 1954.
Sobre la biologia de las Sagitta del plancton del Levante espanol. Publ. Inst. Biol. Aplic. 16:137-148, 4 textfigs.

Sagitta bipunctata
Massuti Oliver, M., 1951.
Sobre la biologia de las Sagitta del plancton del levante español. Publ. Inst. Biol. Aplic., Barcelona, 8:71-82, 3 textfigs.

Sagitta bipunctata
Michael, E. L., 1911
Classification and vertical distribution of the Chaetognatha of the San Diego Region including redescriptions of some doubtful species of the group. Univ. Calif. Publ. Zool. 8(3):21-186, Pls. 1-8.

Sagitta bipunctata
Moore, H. B., 1949
The zooplankton of the upper waters of the Bermuda area of the North Atlantic. Bull. Bingham Ocean. Coll. 12(2):97, 208 text figs.

Sagitta bipunctata
Oliver, M.M., 1954.
Sobre la Biologia de los Sagitta del plancton del Levante espanol. Publ. Inst. Biol. Aplic., 16:137-148.

Sagitta bipunctata
Pierce, E.L., 1953.
The Chaetognatha over the continental shelf of North Carolina with attention to their relation to the hydrography of the area. J. Mar. Res. 12(1):75-92, 4 textfigs.

Sagitta bipunctata
Ramult, M., and M. Rose, 1945
Recherches sur les Chetognathes de la Baie d'Alger. Bull. Soc. Hist. Nat. Afrique du Nord 36:45-71, 39 figs.

Sagitta bipunctata
Rao, T.S. Satyanarayana, and P.N. Ganapati, 1958
Studies on Chaetognatha in the Indian Seas. III Systematics and distribution in the waters off Visakhapaynam. Andhra Univ. Mem. Oceanogr., 2:147-163.

Sagitta bipunctata
Reyssae, J., 1963.
Chaetognathes du plateau continental européen (de la baie ibero-marocaine à la Mer Celtique). Rev. Trav. Inst. Peches Marit., 27(3):245-299.

Sagitta bipunctata
Ritter-Záhony, R. von, 1911
Revision der Chätognathen. Deutsche Südpolar Exped., 1901-1903, XIII (Zool.V)., Pt.1:71 pp., 51 text figs.

Sagitta bipunctata
Ritter-Záhony, R. von, 1911
Revision der Chätognathen. Deutsche Südpolar Exped., 1901-1903, XIII (Zool.V)., Pt.1:71 pp., 51 text figs.

Sagitta bipunctata
Ritter-Záhony, R. von, 1911
Revision der Chätognathen. Deutsche Südpolar Exped., 1901-1903, XIII (Zool.V)., Pt.1:71 pp., 51 text figs.

Sagitta bipunctata
Rose, M., 1925.
Contribution à l'etude de la biologie du plankton. Le problème des migrations verticales journalières. Arch. Zool. expér. et gén. 64: 387-542, 41 textfigs.

Sagitta bipunctata
Rose, M., and M. Hamon, 1953.
Nouvelle note complémentaire sur les chétognathes de la Baie d'Alger.
Bull. Soc. Hist. Nat., Afrique du Nord, 44(5/6): 167-171.

Sagitta bipunctata
Roubault, A., 1946
Observations sur la Répartition du Plancton. Bull. Mus. Inst. Ocean., No.902: 4 pp.

Sagitta bipunctata
Scaccini, A., e E. Ghirardelli, 1941.
Chaetognathes collected on the coast of Rio de Oro. Not. Ist. Biol., Rovigno, 2(21);3-15.

Sagitta bipunctata
Scaccini, A. e E. Ghirardelli, 1941.
Chaetognatesa of Adriatic Sea near Rovigno. Not. Ist. Biol., Rovigno, 2(22):3-16.

Sagitta bipunctata
Schodduyn, M., 1926
Observations faites dans la baie d'Ambleteuse (Pas de Calais). Bull. Inst. Ocean., Monaco, No. 482: 64 pp.

Sagitta bipunctata
Smidt, E.L.B., 1944
Biological Studies of the Invertebrate Fauna of the Harbor of Copehagen. Vidensk. Medd. fra Dansk naturh. Foren. 107:235-316, 23 text figs.

Sagitta bipunctata
Suarez Caabro, J.A., 1955.
Quetognatos de los mares cubanos.
Mem. Soc. Cubana Hist. Nat., 22(2):125-180, 9 pls.

Sagitta bipunctata
Sund, Paul N., 1961.
Some features of the autecology and distribution of Chaetognatha in the Eastern Tropical Pacific.
Inter-American Tropical Tuna Comm., Bull., 5(4): 307-340.

Sagitta bipunctata
Sund, Paul N., 1961.
Two new species of Chaetognatha from the waters off Peru.
Pacific Science, 15(1):105-111.

Sagitta bipunctata
Sund, Paul N., 1959.
A key to the Chaetognatha of the tropical Eastern Pacific Ocean.
Pacific Science, 13(3):269-285.

Sagitta bipunctata
Thiel, M.E., 1938
Die Chaetognathen - Bevölkerung des Südatlantischen Ozeans. Biologische Sonderuntersuchungen. 1st Lief. Wissenschaftliche Ergebnisse der Deutschen Atlantischen Expedition auf dem Forschungs, und Vermessungsschiff "Meteor" 1925-1927, 13:1-110, 62 text figs.

Sagitta bipunctata
Thomson, J.M., 1948
Some chaetognatha from western Australia.
J. Roy. Soc. Western Australia, Inc., 31 (1944-1945), 17-18.

Sagitta bipunctata
Thomson, J. M., 1947
The Chaetognatha of South-eastern Australia.
Counc. Sci. & Ind. Res., Australia, Bull. No.222, (Div. Fish. Rept. 14), 43 pp., 8 text figs.

Sagitta bipunctata
Tokioka, Takasi, 1959.
Observations on the taxonomy and distribution of chaetognaths of the North Pacific.
Publ. Seto Mar. Biol. Lab., 7(3): 349-456.

Sagitta bipunctata
Tokioka, T., 1954.
Droplets from the plankton, XVI. On a small collection of chaetognaths from the central Pacific. Publ. Seto Mar. Biol. Lab. 4(3):99-102, 1 textfig.

Sagitta bipunctata
Tokioka, T., 1951.
Pelagic tunicates and chaetognaths collected during the cruises to the New Yamato Bank in the Sea of Japan. Publ. Seto Mar. Biol. Lab. 2(1): 1-26, 1 chart, 6 tables, 12 textfigs.

Sagitta bipunctata
Tokioka, T., 1940.
The chaetognath fauna of the waters of western Japan. Rec. Oceanogr. Wks., Japan, 12(1):1-22, 3 charts, 4 tables. (Contrib. 87, Seto M.B.L.).

Sagitta bipunctata
Tokioka, T., 1939.
Chaetognaths collected chiefly from the bays of Sagami and Suruga with some notes on the shape and structure of the seminal vesicles.
Rec. Oceanogr. Wks., Japan, 10(2):123-150.

Sagitta bipunctata
Tsuruta, Arao, 1963
Distribution of plankton and its characteristics in the oceanic fishing grounds, with special reference to their relation to fishery.
J. Shimonoseki Univ., Fish., 12(1):13-214

Sagitta bipunctata
Vannucci, M., and K. Hosoe, 1952.
Resultados cientificos do Cruzeiro do "Baependi" e do "Vega" a Ilha da Trindada. Bol. Inst. Ocean., Univ. Sao Paulo, 3(1/2):5-30, 4 pls.

Sagitta bipunctata
von Ritter-Záhony, R., 1911
Vermes, Chaetognathi, Das Tierreich 29:34 pp., 16 text figs.

Sagitta bipunctata

Vuceti, T., 1961
Sur la repartition des chaetognathes en Adriatique et leur utilisation comme indicateurs biologiques des conditions hydrographiques.
Rapp. Proc. Verb., Réunions. Comm. Int. Expl. Sci. Mer Méditerranée, Monaco, 16(2):111-116.

Sagitta bombayensis

Silas, E.G., and M. Srinivasan, 1967(1968).
On the little known Chaetognatha Sagitta bombayensis Lele and Gae (1936) from Indian waters.
J.mar.biol.Ass.India, 9(1):84-95.

Sagitta bruuni n.sp.

Alvariño, Angeles, 1967.
The Chaetognatha of the NAGA Expedition (1959-1961) in the South China Sea and the Gulf of Thailand.
NAGA Rept., 4(2):1-87.
Scripps Inst. Oceanogr.

Sagitta britannica

Ritter-Záhony, R. von, 1911
Revision der Chätognathen. Deutsche Südpolar Exped., 1901-1903, XIII (Zool.V), Pt.1:71 pp., 51 text figs.

Sagitta californica

Bieri, Robert 1959
The distribution of planktonic chaetognatha in the Pacific and their relationship to water masses.
Limnol. and Ocean. Vol. 4 No. 1 pp. 1-28

Sagitta californica

Sund, Paul N., 1961.
Some features of the autecology and distributions of Chaetognatha in the Eastern Tropical Pacific.
Inter-American Tropical Tuna Comm., Bull., 5(4):307-340.

Sagitta coreana

von Ritter-Záhony, R., 1911
Vermes. Chaetognathi. Das Tierreich 29:34 pp., 16 text figs.

Sagitta coreana

Ritter-Záhony, R. von, 1911
Revision der Chätognathen. Deutsche Südpolar Exped., 1901-1903, XIII (Zool.V), Pt.1:71 pp., 51 text figs.

Sagitta crassa

Alvariño, Angeles, 1967.
The Chaetognatha of the NAGA Expedition (1959-1961) in the South China Sea and the Gulf of Thailand.
NAGA Rept., 4(2):1-87.
Scripps Inst. Oceanogr.

Sagitta crassa

Bieri, Robert 1959
The distribution of planktonic chaetognatha in the Pacific and their relationship to water masses.
Limnol. and Ocean. Vol. 4 No. 1 pp. 1-28

Sagitta crassa

Hamada, Takao, 1967.
Studies on the distribution of Chaetognatha in the Harima-nada and Osaka Bay, with special reference to Sagitta enflata. (In Japanese; English abstract).
Bull.Jap.Soc.scient.Fish., 33(2):98-103.

Sagitta crassa

Hirota, Reiichiro, 1961
Zooplankton investigations in the Bingo-Nada region of the Setonaikai (Inland Sea of Japan)
J. Sci. Hiroshima Univ., (B) Div. 1, 20: 83-145.
Also in:
Contrib., Mukaishima Mar. Biol. Sta., Hiroshima Univ., No. 67,

Sagitta crassa

Hirota, Reiichiro, 1959
[On the morphological variation of Sagitta crassa.] (4)
J. Oceanogr. Soc., Japan, 15: 191-202.
Also:
Contrib. Mukaishima Mar. Biol. Sta., No. 63.
English resume

Sagitta crassa

Kado, Y., 1954.
Notes on the seasonal variation of Sagitta crassa.
Ann. Zool. Japon., 27(1):52-55.

Sagitta crassa

Kado, Y., 1953.
The chaetognath fauna of the Inland Sea of Japan, especially on the distribution of Sagitta enflata and S. crassa. Dobuts. Zasshi 62(10):337-342

Sagitta crassa

Kado, Yoichi, and Reiichiro Hirota, 1957
Further studies on the seasonal variation of Sagitta crassa.
J. Sci., Hiroshima Univ., (B1), 17(10):131-136.
Also: Contrib. Mukaishima Mar. Biol. Sta., No. 55

Sagitta crassa

Murakami, Akio, 1959
Marine biological study on the planktonic Chaetognaths in the Seto Inland Sea.
Bull. Naikai Reg. Fish. Res. Lab., Fish. Agency No. 12: 1-186.

Sagitta crassa

Masui, T., T. Namibe, and D. Miyade, 1944.
[Gulf factors and the type of gulf fauna.]
Physiol. and Ecol. Contr. Otsu Hydrobiol. Exp. Sta., Kyoto Univ., 3:1-20.

Sagitta crassa

Murakami, Akio, 1966.
Rearing experiments of a chaetognath, Sagitta crassa. (In Japanese; English abstract).
Inf. Bull. Planktol.Japan, No. 13:62-65.

Sagitta crassa

Murakami, Akio, 1957
[Value of chaetognaths preferring low salinity as indicator forms of water masses.]
Info. Bull., Plankton., Japan, (5):8-10.

Sagitta crassa

Tokioka, T., 1940.
The chaetognath fauna of the waters of western Japan. Rec. Oceanogr. Wks., Japan, 12(1):1-22, 3 charts, 4 tables. (Contrib. 87, Seto M.B.L.)

Sagitta crassa naikaiensis

Tokioka, Takasi, 1959
Observations on the taxonomy and distribution of chaetognaths of the North Pacific.
Publ. Seto Mar. Biol. Lab., 7(3):349-456.

Sagitta crassa naikaiensis

Tokioka, T., 1951.
Pelagic tunicates and chaetognaths collected during the cruises to the New Yamato Bank in the Sea of Japan. Publ. Seto Mar. Biol. Lab. 2(1):1-26, 1 chart, 6 tables, 12 textfigs.

Sagitta darwini

Ritter-Záhony, R. von, 1911
Revision der Chätognathen. Deutsche Südpolar Exped., 1901-1903, XIII (Zool.V), Pt.1:71 pp., 51 text figs.

SAGITTA DECIPIENS

Alvariño, Angeles, 1969
Los quetognatos del Atlántico: distribución y notas esenciales de sistemática.
Trab. Inst. esp. Oceanogr., 37: 290 pp.

Sagitta decipiens

Alvariño, Angeles, 1967.
The Chaetognatha of the NAGA Expedition (1959-1961) in the South China Sea and the Gulf of Thailand.
NAGA Rept., 4(2):1-87.
Scripps Inst. Oceanogr.

Sagitta decipiens

Alvariño, Angeles, 1966
Zoogeografía de California: quetognatos. Revista Soc. Mexicana Hist. nat. 27: 199-243

Sagitta decipiens

Alvariño, Angeles, 1965.
Distributional atlas of Chaetognatha in the California Current region.
California Coop. Oceanic Fish. Invest., Atlas, No. 3:291 charts

Sagitta decipiens

Alvariño, Angeles, 1964.
Zoogeografía de los quetognatos, especialmente de la región de California.
Ciencia, Mexico, 23:51-74.

Sagitta decipiens

Alvariño, Angeles, 1963
Quetognatos epiplanctonicos del Mar de Cortes.
Revista, Soc. Mexicana, Hist. Nat., 24:97-202.

Sagitta decipiens

Bieri, Robert 1959
The distribution of planktonic chaetognatha in the Pacific and their relationship to water masses.
Limnol. and Ocean. Vol. 4 No. 1 pp. 1-28

Sagitta decipiens

Burfield, S.T., and E.J.W. Harvey, 1926.
The Chaetognatha of the "Sealark" Expedition.
Trans. Linn. Soc., London, Ser. 2, 19:93-119, Pls. 4-7.

Sagitta decipiens

Coleman, J. S., 1959.
The "Rosaura" Expedition 1937-1938: Chaetognatha.
Bull. Brit. Mus. (Natural History) Zool. 5(8):219-253.

Sagitta decipiens

Dallot, Serge 1970.
L'Anatomie du tube digestif dans la phylogénie et la systématique des Chaetognathes.
Bull. Mus. natn. Hist. nat. (2) 42(3):549-565.

Sagitta decipiens

Dallot, Serge et Françoise Ducret, 1968.
A propos de Sagitta decipiens Fowler et de Sagitta neodecipiens Tokioka
Rapp. Proc.-verb. Réun., Comm. int. Explor. scient. Mer Méditerranée 19(3): 433-435.

Sagitta decipiens

Ducret, Françoise, 1968.
Chaetognathes des campagnes de l'"Ombango" dans les eaux équatoriales et tropicales africaines.
Cah. O.R.S.T.O.M., ser.Oceanogr., VI(1):95-141.

Sagitta decipiens

Ducret, Françoise, 1968.
Chaetognathes des eaux superficielles et profondes de la zone équatoriale et tropicale africaine.
Thèse Fac. Sci., Marseille, 99 pp., 25 figs. In: Receuil Trav. publiées de 1965 à 1968, Oceanogr. ORSTROM.

Sagitta decipiens
Ducret, Françoise, 1968.
Chaetognathes des campagnes de l'Ombango dans les eaux équatoriales et tropicales africaines.
Cah. ORSTOM., Sér. Océanogr., 6(1):95-141.

Sagitta decipiens
Fagetti Guaita Elda, 1958.
Investigaciones sobre quetognatos colectados, especialmente, frente a la costa central y norte de Chile. Revista Biol. Mar., Valparaiso, 8(1-3):25-82.

Sagitta decipiens n.sp.
Fowler, G.H., 1905.
Biscayan plankton collected during a cruise of H.M.S. "Research", 1900. Pt. III. The Chaetognatha. Trans. Linn. Soc., London, 10(3):55-87, Pls. 4-7.

Sagitta decipiens
Fraser, J.H., 1952.
The Chaetognatha and other zooplankton of the Scottish area and their value as biological indicators of hydrographical conditions.
Scottish Home Dept., Mar. Res., 1952(2):1-52, 3 pls., 21 charts.

Sagitta decipiens
Fraser, J. H., 1949
The occurrence of unusual species of Chaetognatha in Scottish plankton collections.
JMBA 28(2):489-491.

Sagitta decipiens
Furnestin, Marie-Louise, 1970.
Chaetognathes des campagnes du Thor (1908-11) en Méditerranée et en Mer Noire.
Dana - Rept. 79:1-51

Sagitta decipiens
Furnestin, Marie-Louise, 1966.
Chaetognathes des eaux africaines.
Atlantide Rep. 9:105-135.

Sagitta decipiens
Furnestin, M.-L., et J.-C. Codaccioni, 1968.
Chaetognathes du nord-ouest de l'Océan Indien (Golfe d'Aden, mer d'Arabie, golfe d'Oman, golfe Persique).
Cah. ORSTOM., Sér. Océanogr., 6(1):143-171.

Sagitta decipiens
Hamon, M., 1950.
Deux nouveaux Chétognathes de la Baie d'Alger.
Bull. Soc. Hist. Nat., Afrique du Nord, 21(1/4): 10-14, 6 textfigs.

Sagitta decipiens
Hure, J., 1961
Dneva migracija i sezonaka vertikalna raspodjela zooplanktona dubljeg mora. [Migration journalière et distribution saisonnière verticale du zooplancton dans la région profunde de l'Adriatique]
Acta Adriatica, 9(6):1-59.

Sagitta decipiens
Lea, H.E., 1955.
The chaetognaths of western Canadian coastal waters. J. Fish. Res. Bd., Canada, 12(4):593-617.

Sagitta decipiens
Michael, E.L., 1919
Contributions to the Biology of the Philippine Archipelago and Adjacent regions. Report on the chaetognatha collected by the United States Fisheries Steamer "Albatross" during the Philippine Expedition, 1907-1910. Bull. U.S. Nat. Mus. No.100, Vol.1(4):235-277, pls.34-38.

Sagitta decipiens
Michael, E. L., 1911
Classification and vertical distribution of the Chaetognatha of the San Diego Region including redescriptions of some doubtful species of the group. Univ. Calif. Publ. Zool. 8(3):21-186, Pls. 1-8.

Sagitta decipiens
Rao, T.S. Satyanarayana, and P.N. Ganapati, 1958
Studies on Chaetognatha in the Indian Seas. III Systematics and distribution in the waters off Visakhapaynam. Andhra Univ. Mem. Oceanogr., 2:147-163.

Sagitta decipiens
Rose, M., and M. Hamon, 1953.
Nouvelle note complémentaire sur les chétognathes de la Baie d'Alger.
Bull. Soc. Hist. Nat., Afrique du Nord, 44(5/6): 167-171.

Sagitta decipiens
Russell, F.S., 1939.
Chaetognatha. Fiches d'Ident. Zooplancton, Cons. Perm. Int. Expl. Mer, 1:4 pp., 12 textfigs.

Sagitta decipiens
Sund, Paul N., 1964.
Los quetognatos en las aguas de la region del Peru.
Inter-Amer. Tropical Tuna Comm., Bull., 9(3): 115-216.

Sagitta decipiens
Sund, Paul N., 1961.
Some features of the autecology and distributions of Chaetognatha in the Eastern Tropical Pacific.
Inter-American Tropical Tuna Comm., Bull., 5(4): 307-340.

Sagitta decipiens
Sund, Paul N., 1959.
A key to the Chaetognatha of the tropical Eastern Pacific Ocean.
Pacific Science, 13(3):269-285.

Sagitta decipiens
Thomson, J. M., 1947
The Chaetognatha of South-eastern Australia.
Counc. Sci. & Ind. Res., Australia, Bull. No.222, (Div. Fish. Rept. 14), 43 pp., 8 text figs.

Sagitta decipiens
Tokioka, T., 1940.
The chaetognath fauna of the waters of western Japan. Rec. Oceanogr. Wks., Japan, 12(1):1-22, 3 charts, 4 tables. (Contrib. 87, Seto M.B.L.)

Sagitta decipiens
Tokioka, Takasi, 1959.
Observations on the taxonomy and distribution of chaetognaths of the North Pacific.
Publ. Seto Mar. Biol. Lab., 7(3): 349-456.

Sagitta decipiens
Tokioka, T., 1939.
Chaetognaths collected chiefly from the bays of Sagami and Suruga with some notes on the shape and structure of the seminal vesicles.
Rec. Oceanogr. Wks., Japan, 10(2):123-150.

Sagitta decipiens
Ritter-Záhony, R. von, 1911
Revision der Chätognathen. Deutsche Südpolar Exped., 1901-1903, XIII (Zool.V), Pt.1:71 pp., 51 text figs.

Sagitta decipiens
von Ritter-Záhony, R., 1911
Vermes. Chaetognathi. Das Tierreich 29:34 pp., 16 text figs.

Sagitta decipiens
Vučetić, Tamara, 1969.
Contribution to the knowledge of biologic indicators of water masses in the Mediterranean. (In Jugoslavian; English and Italian abstracts).
Thalassia Jugoslavica, 5:435-441.

Sagitta decipiens
Vučetić, Tamara 1969.
Distribution of Sagitta decipiens and identification of Mediterranean water masses circulation.
Bull. Inst. océanogr. Monaco 69 (1398): 12pp.

Sagitta decipiens
Vučetić, T., 1961
Sur la répartition des chaetognathes en Adriatique et leur utilisation comme indicateurs biologiques des conditions hydrographiques.
Rapp. Proc. Verb., Réunions, Comm. Int. Expl. Sci. Mer Méditerranée, Monaco, 16(2):111-116.

Sagitta delicata
Tokioka, T., 1951.
Pelagic tunicates and chaetognaths collected during the cruises to the New Yamato Bank in the Sea of Japan. Publ. Seto Mar. Biol. Lab. 2(1): 1-26, 1 chart, 6 tables, 12 textfigs.

Sagitta demipenna n.sp.
Tokioka, Takashi, and D. Pathansali, 1963.
Another new chaetognath from Malay waters, with a proposal of grouping some species of Sagitta into subgenera.
Publ. Seto Mar. Biol. Lab., 11(1)(8):119-123.

Sagitta diptera Orbigny
Ritter-Záhony, R. von, 1911
Revision der Chätognathen. Deutsche Südpolar Exped., 1901-1903, XIII (Zool.V), Pt.1:71 pp., 51 text figs.

Alvariño, Angeles, 1969 SAGITTA ELEGANS
Los quetognatos del Atlántico: distribución y notas esenciales de sistemática.
Trab. Inst. esp. Oceanogr., 37: 290 pp.

Sagitta elegans
Alvariño, Angeles, 1967.
The Chaetognatha of the NAGA Expedition (1959-1961) in the South China Sea and the Gulf of Thailand.
NAGA Rept. 4(2):1-87.
Scripps Inst. Oceanogr.

Sagitta elegans
Alvariño, Angeles, 1964.
Zoogeografía de los quetognatos, especialmente de la región de California.
Ciencia, México, 23:51-74.

Sagitta elegans
Alvariño, A., 1956.
Estudio del zooplancton recogido en la campaña "Vendaval" en Terranova, marzo, abril y mayo de 1953. Zooplancton de Terranova, febrero, marzo y junio de 1955. Bol. Inst. Español de Oceanogr., Nos. 76/77: 28 pp. y 18 pp.

Sagitta elegans
Bainbridge, V., 1963.
Continuous plankton records: contribution toward a plankton atlas of the North Atlantic and the North Sea. VIII. Chaetognatha.
Bull. Mar. Ecol., 6(2):40-51.

Sagitta elegans
Bary, B. McK., 1963.
Temperature, salinity and plankton in the eastern North Atlantic and coastal waters of Britain, 1957. II. The relationships between species and water bodies.
J. Fish. Res. Bd., Canada, 20(4):1031-1065.

Sagitta elegans
Beaudouin, Jacqueline, 1967.
Oeufs et larves de poissons récoltés par la "Thalassa" dans le Detroit de Danemark et le Nord de la Mer d'Irminger (NORWESTLANT I-20mars-8 Mai 1963: relations avec l'hydrologie et le zooplancton.
Revue Trav. Inst. (scient. tech.) Pêches marit. 31 (3): 307-326.

Sagitta elegans
Bieri, Robert 1959
The distribution of planktonic chaetognatha in the Pacific and their relationship to water masses.
Limnol. and Ocean. Vol. 4 No. 1 pp. 1-28

Sagitta elegans

Bigelow, H. B., 1922
Exploration of the coastal water off the northeastern United States in 1916 by the U.S. Fisheries Schooner Grampus. Bull. M.C.Z. 65(5):85-188, 53 text figs.

Sagitta elegans

Bigelow, H.B., and M. Sears, 1939
Studies of the waters of the continental shelf, Cape Cod to Chesapeake Bay. III. A volumetric study of the zooplankton. Mem. M.C.Z. 54(4):183-378, 42 text figs.

Sagitta elegans

Bogorov, B. G., 1938
Diurnal vertical distribution of plankton under polar conditions (in the southeastern part of the Barents Sea). Trudy Poliarnii Nauchno-Issledovatelskii Institut Morskogo Ribogo Khoziastva i Okeanografii im Pochetnogo Chlena Akademii Nauk SSSR. Prof. N.M. Knipovicha g. Murmansk, 2:93-106, 4 text figs. (In Russian with English Summary). [Trans. Polar Sci. Res. Inst. Mar. Fish & Oceanogr.]

Sagitta elegans

Bull, H.O. 1966.
Chaetognatha.
Rept. Dove Mar. Lab. (3) 15: 17-20.

Sagitta elegans

Coleman, J. S., 1959.
The "Rosaura" Expedition 1937-1938: Chaetognatha.
Bull. Brit. Mus. (Natural History) Zool. 5(8):219-253.

Sagitta elegans

Conant, F.S., 1896.
23. Notes on chaetognaths. Ann. Mag. Nat. Hist., 6th ser., 18:201-214.

Sagitta elegans

Corbin, P.G., 1948.
On the seasonal abundance of young fish. IX. The year 1947. J.M.B.A., n.s., 27:718-722, 3 textfigs

Sagitta elegans

Corbin, P.G., 1947.
The spawning of mackerel, Scomber scombrus L., and pilchard, Clupea pilchardus Walbaum, in the Celtic Sea in 1937-39 with observations on the zooplankton indicator species, Sagitta and Muggiaea. J.M.B.A., n.s., 27:65-132, 21 textfigs.

Sagitta elegans

Cronin, L. Eugene, Joanne C. Daiber and Edward M. Hulburt, 1962
Quantitative seasonal aspects of zooplankton in the Delaware River estuary. Chesapeake Science, 3(2):63-93.

Sagitta elegans

Dallot, Serge, 1970.
L'anatomie du tube digestif dans la phylogénie et la systématique des Chaetognathes.
Bull. Mus. natn. Hist. nat. (2) 42(3): 549-565.

Sagitta elegans

Farran, G.P., 1947
Vertical distribution of plankton (Sagitta Calanus and Metridia) off the south coast of Ireland. Proc. Roy. Irish Acad. Sect. B., 51(6):121-131

Sagitta elegans

Dunbar, M.J., 1962
The life cycle of Sagitta elegans in Arctic and subarctic seas and the modifying effects of hydrographic differences in the environment.
J. Mar. Res., 20(1):76-91.

Sagitta elegans

Fish, C.J., and M.W. Johnson, 1937
The biology of the zooplankton population in the Bay of Fundy and the Gulf of Maine with special reference to production and distribution. J. Biol. Bd., Canada 3(3):189-322, 29 tables, 45 text figs.

Sagitta elegans

Fowler, G.H., 1906.
The Chaetognatha of the Siboga Expedition with a discussion of the synonymy and distribution of the group. Rept. Siboga Exped. 21:1-88, 3 pls., 6 charts.

Sagitta elegans

Fraser, J. H., 1962.
Plankton.
Proc. R. Soc., London, (A), 265(1322):335-341.

Sagitta elegans

Fraser, J.H., 1961
The survival of larval fish in the northern North Sea according to the quality of the water.
J. Mar. Biol. Assoc., U.K., 41:305-312.

Sagitta elegans

Fraser, J.H., 1952.
The Chaetognatha and other zooplankton of the Scottish area and their value as biological indicators of hydrographical conditions. Scottish Home Dept., Mar. Res., 1952(2):1-52, 3 pls., 21 charts.

S. elegans elegans
S. elegans arctica
S. elegans baltica
good distinguishing characters

Sagitta elegans

Fraser, J. H., 1949
Plankton investigations from the Scottish Research Vessel. Ann. Biol., Int. Cons., 4: 66-67.

Sagitta elegans

Fraser, J. H., 1949
Plankton of the Faroe-Shetland Channel and the Faroes, June and August 1947. Ann. Biol., Int. Cons., 4:27-28, text fig. 10.

Sagitta elegans

Fraser, J. H., and A. Saville, 1949
Plankton distribution in Scottish and adjacent waters in 1948. Ann. Biol. 5:61-62.

Sagitta elegans

Furnestin, J., 1938.
Influence de la salinité sur la répartition du genre Sagitta dans l'Atlantique Nord-Est (Juillet-Août-Septembre, 1936).
Rev. Trav. Off. Pêches Marit. 11(3):425-437.

Sagitta elegans

Ghirardelli, E., 1960.
Habitat e biologia della riproduzione nei chetognati.
Arch. Oceanogr. e Limnol., 11(3):287-304.

Reprinted in:
Trav. Sta. Zool., Villefranche-sur-Mer, 18(11):

Sagitta elegans

Grant, George C., 1963
Investigations of inner continental shelf waters off lower Chesapeake Bay. IV. Descriptions of the Chaetognatha and a key to their identification.
Chesapeake Science, 4(3):109-119.

Sagitta elegans

Hansen, Vagn Kr., 1960.
Investigations on the quantitative and qualitative distribution of zooplankton in the southern part of the Norwegian Sea.
Medd. Danmarks Fiskeri- og Havundersøgelser, n.s., 2(23):1-53.

Sagitta elegans

Hansen, K.V. 1951.
On the diurnal migration of zooplankton in relation to the discontinuity layer. J. du Cons. 17(3):231-241, 9 textfigs.

Sagitta elegans

Heinrich, A.K., 1956.
Dimensional composition of Chaetognatha and the terms of their propagation in the western regions of the Bering Sea. Dokl. Akad. Nauk, SSSR, 110(6):1105-1107.

Sagitta elegans

Huntsman, A. G. and M.E. Reid, 1921
The success of reproduction of Sagitta elegans in the Bay of Fundy and the Gulf of St. Lawrence. Trans. Roy. Canadian Inst., 13: 99-112

Sagitta elegans

Jakobsen, Tore 1971.
On the biology of Sagitta elegans Verrill and Sagitta setosa J. Müller in inner Oslofjord.
Norw. J. Zool. 19(2): 201-225.

Sagitta elegans

Khan, M.A. and D.I. Williamson, 1970.
Seasonal changes in the distribution of chaetognatha and other plankton in the eastern Irish Sea. J. exp. mar. Biol. Ecol., 5(3): 285-303.

Sagitta elegans

Kielhorn, W.V., 1952
The biology of the surface zone zooplankton of a Boreo-Arctic Atlantic Ocean area. J. Fish Res. Bd., Canada 9(5): 223-264, 13 text figs.

Sagitta elegans

Kramp, P. L., 1939
Chaetognatha. The Godthaab Expedition 1928. Medd. om Grønland, 80(5):1-40, 7 text figs.

Sagitta elegans

Lacroix, Guy, et Pierre Morisset, 1962.
Observations sur les migrations verticales de Sagitta elegans Verrill. Sta. Biol. Mar. Grande-Rivière, Québec, Cahiers d'Inform., (14):33-39. (multilithed).

From Ann. Rept., 1961 (1963).

Sagitta elegans

Lea, H.E., 1955.
The chaetognaths of western Canadian coastal waters. J. Fish. Res. Bd., Canada, 12(4):593-517.

Sagitta elegans

Lebour, M.V., 1947
Notes on the inshore plankton of Plymouth JMBA 26(4):527-547, 1 textfig.

Sagitta elegans

Lucas, C. E., 1949
Notes on continuous plankton records at 10 m depth in the North Sea and Northeastern Atlantic during 1946-1947. Ann. Biol., Int. Cons., 4:63-66, text fig. 4.

Sagitta elegans

Marshall, N. B., 1948
Continuous plankton records: Zooplankton (other than Copepoda and young fish) in the North Sea 1938-1939. Hull Bull. Mar. Ecol. 2(13):173-213, Pls. 89-108.

Sagitta elegans

Marumo, Ryuzo, 1966.
Sagitta elegans in the Oyashio undercurrent.
J. Oceanogr. Soc., Japan, 22(4):129-137.

Sagitta elegans

Massuti, M., 1961
Note préliminaire à l'étude des chétognathos de la Méditerranée occidentale. Campagne du "Xauen" X-6911.
Rapp. Proc. Verb., Réunions, Comm. Int. Expl. Sci. Mer Méditerranée, Monaco, 16(2):237-244.

Sagitta elegans

McLaren, Ian A., 1966.
Adaptive significance of large size and long life of the chaetognath Sagitta elegans in the Arctic.
Ecology, 47(5):852-855.

Sagitta elegans

McLaren, Ian A., 1963.
Effects of temperature on growth of zooplankton and the adaptive value of vertical migration.
J. Fish. Res. Bd., Canada, 20(3):685-727.

Sagitta elegans

McLaren, I.A., 1961
The hydrography and zooplankton of Ogac Lake, a landlocked fjord on Baffin Island.
Fish. Res. Bd., Canada, MSS Rept. Ser. (Biol.) No. 709: 167 pp. (multilithed).

Sagitta elegans

Meek, A., 1928.
Sagitta elegans and Sagitta setosa from the Northumberland coast. Proc. Zool. Soc., London, :743.

Sagitta elegans

Michael, E. L., 1911
Classification and vertical distribution of the Chaetognatha of the San Diego Region including redescriptions of some doubtful species of the group. Univ. Calif. Publ. Zool. 8(3):21-186, Pls. 1-8.

Sagitta elegans

Murakami, Akio, 1957
[Value of chaetognaths preferring low salinity as indicator forms of water masses.]
Info. Bull., Plankton., Japan, (5):8-10.

Sagitta elegans

Pearre, Sifford, Jr., 1973
Vertical migration and feeding in Sagitta elegans Verrill. Ecology 54(2): 300-314.

Sagitta elegans

Pierce, E. L. 1941
The occurrence and breeding of Sagitta elegans Verrill and Sagitta setosa J. Müller in parts of the Irish Sea. JMBA 25:113-124

Sagitta elegans

Rae, K. M., 1949
Plankton. Some broad changes in the plankton round the north of the British Isles in 1948. Ann. Biol. 5:56-60, 12 text figs.

Sagitta elegans

Rakusa-Suszczewski, S.,1968.
Predation of Chaetognatha by Tomopteris helgolandica Greff.
J.Cons.perm.int.Explor.Mer,32(2):226-231.

Sagitta elegans

Rakusa-Suszczewski, S., 1967.
The use of chaetognath and copepod population age-structures as an indication of similarity between water masses. J. Cons. perm. int. Explor. Mer, 31(1): 46-55.

Sagitta elegans

Redfield, A. C. and A. Beale, 1940.
Factors determining the distribution of populations of chaetognaths in the Gulf of Maine. Biol Bull., 79(3):459-487, 11 text figs.

Sagitta elegans

Reyssac, J., 1963.
Chaetognathes du plateau continental européen (de la baie ibéro-marocaine à la Mer Celtique). Rev. Trav. Inst. Pêches Marit., 27(3):245-299.

Sagitta elegans

Riley, G.A., 1947
A theoretical analysis of the zooplankton population of Georges Bank. J. Mar. Res. 6(2):104-113, text figs. 27-31.

Sagitta elegans

Ritter-Záhony, R. von, 1911
Revision der Chätognathen. Deutsche Südpolar Exped., 1901-1903, XIII (Zool.V)., Pt.1:71 pp., 51 text figs.

Sagitta elegans

Russell, F.S., 1939.
Chaetognatha. Fiches d'Ident. Zooplancton, Cons. Perm. Int. Expl. Mer, 1:4 pp., 12 textfigs.

Sagitta elegans

Russell, F. S. 1933
On the biology of Sagitta. IV. Observations on the natural history of Sagitta elegans Verrill and Sagitta setosa J. Müller in the Plymouth area. JMBA, 18:559-574

Sagitta elegans

Russell, F. S. 1932
On the biology of Sagitta. The breeding and growth of Sagitta elegans Verrill in the Plymouth Area, 1930-1931. JMBA, 18:131-146

Sagitta elegans

Sameoto, D.D. 1973.
Annual life cycle and production of the chaetognath Sagitta elegans in Bedford Basin, Nova Scotia.
J. Fish. Res. Bd Can. 30(3):333-344.

Sagitta elegans

Sameoto, D.D. 1972.
Yearly respiration rate and estimated energy budget for Sagitta elegans.
J. Fish. Res. Bd Can. 29(7):987-996.

Sagitta elegans

Sameoto, D.D. 1971.
Life history, ecological production, and an empirical mathematical model of the population of Sagitta elegans in St. Margaret's Bay, Nova Scotia.
J. Fish. Res. Bd Can. 28(7):971-985

Sagitta elegans

Sameoto, D.D. 1971.
Macrozooplankton biomass measurements in Bedford Basin 1969-1971.
Techn. Rept. Fish. Res. Bd Can. 282: 238 pp.

Sagitta elegans

Sherman, Kenneth and Everett G. Schaner,1968.
Observations on the distribution and breeding of Sagitta elegans (Chaetognatha) in coastal waters of the Gulf of Maine.
Limnol.Oceanogr., 13(4):618-625.

Sagitta elegans

Sund, P. N., 1959
The distribution of Chaetognatha in the Gulf of Alaska in 1954 and 1956.
J. Fish. Res. Bd., Canada, 16(3):351-361.

Sagitta elegans

Tchindonova, J.G., 1955.
[Chaetognatha of the Kurile-Kamchatka Trench.]
Trudy Inst. Oceanol., 12:298-310.

Sagitta elegans

Thiel, M.E., 1938
Die Chaetognathen - Bevölkerung des Südatlantischen Ozeans. Biologische Sonderuntersuchungen. 1st Lief. Wissenschaftliche Ergebnisse der Deutschen Atlantischen Expedition auf dem Forschungs. und Vermessungsschiff "Meteor" 1925-1927, 13:1-110, 62 text figs.

Sagitta elegans

Tokioka, Takasi, 1959
Observations on the taxonomy and distribution of chaetognaths of the North Pacific.
Publ. Seto Mar. Biol. Lab., 7(3):349-456.

Sagitta elegans

Tokioka, T., 1951.
Pelagic tunicates and chaetognaths collected during the cruises to the New Yamato Bank in the Sea of Japan. Publ. Seto Mar. Biol. Lab. 2(1):

Sagitta elegans

Tokioka, T., 1940.
The chaetognath fauna of the waters of western Japan. Rec. Oceanogr. Wks., Japan, 12(1):1-22, 3 charts, 4 tables. (Contrib. 87, Seto M.B.L.)

Sagitta elegans

Tokioka, T., 1939.
Chaetognaths collected chiefly in the bays of Sagami and Suruga with some notes on the shape and structure of the seminal vesicles.
Rec. Oceanogr. Wks., Japan, 10(2):123-150.

Sagitta elegans

Truveller, K.A., 1966.
Chaetognatha of the Davis and Denmark Straits as indicators of water masses. (In Russian).
Mater. Ribokhoz. Issled. severn. Basseina, Poliarn. Nauchno-Issled. Proektn. Inst. Morsk. Ribn. Khoz. Okeanogr. (PINRO), 7: 114-124.

Sagitta elegans

von Ritter-Záhony, R., 1914.
Chaetognaths. Danish Ingolf-Exped. 4(3):4 pp.

Sagitta elegans

von Ritter-Záhony, R., 1911
Vermes. Chaetognathi. Das Tierreich 29:34 pp., 16 text figs.

Sagitta elegans

Vučetić, T., 1961.
Vertikalna raspodjela zooplanktona u Velikom Jezeru otoka Mljeta. [Vertical distribution of plankton in the Bay Veliko Jezero on the island of Mljet.]
Acta Adriatica, 6(9):1-14 (in Jugoslavian). 15-20 'in English)

Sagitta elegans

Weinstein, Martin, 1967.
Endoparasitism of the chaetognath Sagitta elegans in the Gulf of St. Lawrence.
Rapp. Stn Biol. Mar. Grande-Rivière, 1966:47-53.

Sagitta elegans

Weinstein, Martin, 1966.
Parasites of the chaetognath Sagitta elegans Verrill in the Gulf of St. Lawrence.
Rapp. Stn Biol. Mar. Grande-Rivière,1965:55-59.

Sagitta elegans

Wiborg, K.F., 1955.
Zooplankton in relation to hydrography in the Norwegian Sea.
Rept. Norwegian Fish. Mar. Invest. 11(4):66 pp.

Sagitta elegans

Wiborg, K.F., 1944
The production of zooplankton in a landlocked fjord, the Nordåsvatn near Bergen, in 1941-42, with special reference to the copepods. (Repts. Norwegian Fish. and Mar. Invest.) 7(7):83 pp., 40 text figs.

Sagitta elegans arctica

Alvariño, A., 1956.
Estudio del zooplancton recogido en la campana "Vendaval" en Terranova, marzo, abril y mayo de 1953. Zooplancton de Terranov, febrero, marzo y junio de 1955. Bol. Inst. Espanol de Oceanogr., Nos., 76/77:28 pp., y 18 pp.

Sagitta elegans var. arctica

Dunbar, M. J., 1942,
Marine macroplankton from the Canadian Eastern Arctic. II. Medusae, Siphonophora, Ctenophora, Pteropoda, and Chaetognatha. Canad. J. Res., D, 20:71-77.

Sagitta elegans arctica
Dunbar, M.J., 1940.
On the size distribution and breeding cycles of four marine planktonic animals from the Arctic. J. An. Ecol. 9(2):215-226, 5 textfigs.

Sagitta elegans arctica
Fraser, J. H., 1949
The occurrence of unusual species of Chaetognatha in Scottish plankton collections. JMBA 28(2):489-491.

Sagitta elegans arctica
Fraser, J. H., 1939
The distribution of Chaetognatha in Scottish Waters in 1937. J. du Cons. 14(1):25-34, 3 charts in text.

Sagitta elegans arctica
Fraser, J. H. and A. Saville, 1949
Macroplankton in the Faroe Channel, 1948. Ann. Biol. 5:29-30, text figs.

Sagitta elegans var. arctica
Kramp, P. L., 1939
Chaetognatha. The Godthaab Expedition 1928. Medd. om Grønland, 80(5):1-40, 7 text figs.

Sagitta elegans arctica
Kramp, P.L., 1938.
Chaetognatha. Zool. Iceland 4(71):4 pp.

Sagitta elegans arctica
Zaika, V.E., and A.N. Kolesnikov 1967.
On mass infection of Sagitta elegans arctica Aurivillus by sexually mature trematodes. (In Russian; English abstract). Zool. Zh. 47(7):1121-1124.

Sagitta elegans baltica
Mankowski, W., 1950.
Makroplankton of the Gulf of Gdansk in 1947. (In Polish). Bull. Inst. Pêches Maritimes, Gdynia, No. 5:45-63, 6 textfigs.

Sagitta elegans baltica
Mankowski, W., 1950.
Plankton investigations in the southern Baltic, 1948. Bull. Inst. Pêches Maritimes, Gdynia, No. 5:71-101, 6 textfigs. (In Polish; English summary).

Sagitta elegans baltica
Mankowski, W., 1948
Plankton investigations in the middle Baltic during the summer of 1938. Bull. Lab. mar., Gdynia, No. 4: 93-120, 2 text figs.

Sagitta elegans baltica
Mankowski, W., 1948.
Macroplankton investigations in the Gulf of Gdansk in June-July period 1946. Bull. Lab. Mar., Gdynia, No. 4:121-137, 6 textfigs.

Sagitta elegans balticus
Mankowski, W., 1937.
Notatka o Zooplanktonie Zatoki Gdanskiej. (Note sur le zooplankton du Golf de Dantzig). Biul. Stacji Morskiej Helu (Bull. Sta. Mar. Hel) I(1):23-25.

Sagitta elegans elegans
Fraser, J. H., 1949
The occurrence of unusual species of Chaetognatha in Scottish plankton collections. JMBA 28(2):489-491.

Sagitta elegans elegans
Fraser, J. H., 1939
The distribution of Chaetognatha in Scottish Waters in 1937. J. du Cons. 14(1):25-34, 3 charts in text.

Sagitta elegans elegans
Fraser, J. H. and A. Saville, 1949
Macroplankton in the Faroe Channel, 1948. Ann. Biol. 5:29-30, text figs.

Sagitta elegans elegans
Künne, Cl., 1937
Über die Verbreitung der Leitformen des Großplanktons in der südlichen Nordsee im Winter. Ber. Deutschen Wissenschaflichen Kommision für Meeresforschung, n.f. VIII (3):131-164, 9 text figs., 4 fold-ins.

Sagitta enflata
Aida, T., 1897.
Chaetognaths of Misaki Harbour. Ann. Zool., Japon., 1:13-21, Pl. 3.

Sagitta enflata
Alvariño, Angeles, 1969 SAGITTA ENFLATA
Los quetognatos del Atlántico: distribución y notas esenciales de sistemática. Trab. Inst. esp. Oceanogr., 37: 290 pp.

Sagitta enflata
Alvariño, Angeles, 1967.
The Chaetognatha of the NAGA Expedition (1959-1961) in the South China Sea and the Gulf of Thailand. NAGA Rept. 4(2):1-87. Scripps Inst. Oceanogr.,

Sagitta enflata
Alvariño, Angeles, 1966
Zoogeografía de California: quetognatos Revista Soc. Mexicana Hist. nat. 27: 199-243

Sagitta enflata
Alvarino, Angeles, 1965.
Distributional atlas of Chaetognatha in the California Current region. California Coop. Oceanic Fish. Invest., Atlas, No. 3: 291 charts

Sagitta enflata
Alvariño, Angeles, 1964.
Zoogeografía de los quetognatos, especialmente de la región de California. Ciencia, México, 23:51-74.

Sagitta enflata
Alvariño, Angeles, 1964.
The Chaetognatha of the MONSOON Expedition in the Indian Ocean. Pacific Science, 18(3):336-348.

Sagitta enflata
Alvarino, Angeles, 1963
Quetognatos epiplanctonicos del Mar de Cortes. Revista, Soc. Mexicana, Hist. Nat., 24:97-202.

Sagitta enflata
Alvarino, A., 1957.
Estudio del zooplancton del Mediterráneo occidental. Zooplancton del Atlántico Ibérico. Campañas del "Xauen" en el verano del 1954. Bol. Inst. Espanol Oceanogr., Nos. 81-82:1-26; 1-51.

Sagitta enflata
Araya, H., and T. Otsuki, 1955.
Predation pressure on the squid larva by the arrow worm. Bull. Hokkaido Reg. Fish. Res. Lab., No. 12:40-42, 1 pl., 1 textfig.

Sagitta enflata
Béranek, E., 1895.
Les chétonates de la Baie d'Amboina. Rev. Suisse Zool. 3:137-159, Pl. IV.

Sagitta enflata
Bieri, Robert 1959
The distribution of planktonic chaetognatha in the Pacific and their relationship to water masses. Limnol. and Ocean. Vol. 4 No. 1 pp. 1-28

Sagitta enflata
Bieri, Robert, 1957
The chaetognath fauna of Peru in 1941. Pacific Science 11(3):255-264.

Sagitta enflata
Bigelow, H. B., 1922
Exploration of the coastal water off the northeastern United States in 1916 by the U.S. Fisheries Schooner Grampus. Bull. M.C.Z. 65(5):85-188, 53 text figs.

Sagitta enflata
Bigelow, H.B., and M. Sears, 1939
Studies of the waters of the continental shelf, Cape Cod to Chesapeake Bay. III. A volumetric study of the zooplankton. Mem. M.C.Z. 54(4):183-378, 42 text figs.

Sagitta enflata
Bumpus, D.F., and E.L. Pierce, 1955.
Hydrography and the distribution of chaetognaths over the continental shelf of North Carolina. Pap. Mar. Biol. and Oceanogr., Deep-Sea Res., Suppl. to Vol. 3:92-109.

Sagitta enflata
Burfield, S.T., 1930.
Chaetognatha. Brit. Antarctic ("Terra Nova") Exped., 1910, Zool., 7(4):203-228, 3 maps.

Sagitta enflata
Burfield, S.T., 1927.
Zoological results of the Cambridge Expedition to the Suez Canal. XXI. Report on the Chaetognatha. Trans. Zool. Soc., London, 22:355-356.

Sagitta enflata
Burfield, S.T., and E.J.W. Harvey, 1926.
The Chaetognatha of the "Sealark" Expedition. Trans. Linn. Soc., London, Ser. 2; 19:93-119, Pls. 4-7.

Sagitta enflata
Coleman, J. S., 1959.
The "Rosaura" Expedition 1937-1938: Chaetognatha. Bull. Brit. Mus. (Natural History) Zool. 5(8):219-253.

Sagitta enflata
Cronin, L. Eugene, Joanne C. Daiber and Edward M. Hulburt, 1962
Quantitative seasonal aspects of zooplankton in the Delaware River estuary. Chesapeake Science, 3(2):63-93.

Sagitta enflata
Delsman, H.C., 1936.
Preliminary plankton investigations in the Java Sea. Treubia 17:139-181, 8 maps, 41 figs.

Sagitta enflata
de Saint-Bon, Marie Catherine, 1963.
Les chaetognathes de la Côte d'Ivoire (espèces de surface). Rev. Trav. Inst. Pêches Marit., 27(3):301-346.

Sagitta enflata
Enomoto, Y., 1962
Studies on the food base in the Yellow and the East China seas. 1. Plankton survey in summer of 1956. Bull. Japan Soc. Sci. Fish., 28(8):759-765.

Sagitta enflata
Fagetti Guaita, Elda, 1958.
Investigaciones sobre quetognatos colectados, especialmente, frente a la costa central y norte de Chile. Revista Biol. Mar., Valparaiso, 8(1-3):25-82.

Sagitta enflata
Ferreria da Costa, Pedro 1970.
Nota preliminar sobre a ocorrência de Sagitta friderici e S. enflata (Chaetognatha) na Baía de Guanabara. Publ. Inst. Pesquisas Marinha, Brasil, 47: 10 pp.

Sagitta inflata
Fish, C.J., and M.W. Johnson, 1937
The biology of the zooplankton population in the Bay of Fundy and the Gulf of Maine with special reference to production and distribution. J. Biol. Bd., Canada 3(3):189-322, 29 tables, 45 text figs.

Sagitta enflata
Fowler, G.H., 1906.
The Chaetognatha of the Siboga Expedition with a discussion of the synonymy and distribution of the group. Rept. Siboga Exped. 21:1-88, 3 pls., 6 charts.

Sagitta enflata
Fraser, J.H., 1952.
The Chaetognatha and other zooplankton of the Scottish area and their value as biological indicators of hydrographical conditions. Scottish Home Dept., Mar. Res., 1952(2):1-52, 3 pls., 21 charts.

Sagitta enflata
Furnestin, Marie-Louise, 1965.
Chaetognathes de quelques récoltes dans la mer des Antilles et l'Atlantique ouest tropical. Inst. r. Sci. Nat. Belg., Bull., 41(9): 15 pp.

Sagitta enflata
Furnestin, M.L., 1958.
Quelques échantillons de zooplancton du Golfe d'Eylath(akaba). Haifa, Sea Fish. Res. Sta., Bull., 16:6-14.

Sagitta enflata
Furnestin, M. L., 1957.
Chaetognathes et zooplancton du secteur atlantique marocain. Rev. Trav. Inst. Pêches Marit. 21(1/2): 9-356.

Sagitta enflata
Furnestin, M.-L., 1953.
Sur quelques chaetognathes d'Israel. Bull. Res. Counc., Israel, 2(4):411-414.

Sagitta inflata
Furnestin, M.-L. et J.-C. Codaccioni, 1968.
Chaetognathes du nord-ouest de l'Océan Indien (golfe d'Aden, mer d'Arabie, golfe d'Oman, golfe Persique). Cah. O.R.S.T.O.M., ser. Océanogr., VI(1):143-171.

Sagitta enflata
Furnestin, M.-L., et J. Radiguet, 1964.
Chaetognathes de Madagascar (Secteur de Nosy-Bé). Cahiers, O.R.S.T.R.O.M., Océanogr., 11(4):55-98.

Sagitta enflata
Gamulin, T., 1948
Prilog poznavanju Zooplanktona Srednjedalmatinskog Otocnog Podrucja. (Contrib. a la connaissance du zooplankton de la zone insulaire de la Dalmatie Moyenne). Acta Adriatica 3(7):38 pp., 6 tables, 1 map.

Sagitta enflata
George, M.J., 1958.
Observations on the plankton of the Cochin backwaters. Indian J. Fisheries, 5(2):375-401.

Sagitta enflata
George, P.C., 1952.
A systematic account of the Chaetognatha of the Indian coastal waters, with observations of their seasonal fluctuations along the Malabar coast. Proc. Nat. Inst. Sci., India, 18(6):657-689.

Sagitta enflata
Ghirardelli, Elvezio, 1959.
Osservazioni sulla corona ciliata nei Chetognati. Boll. Zool., Unione Zool. Italiana, 26(2):413-421
Also in:
Trav. Sta. Zool., Villefranche-sur-Mer, 19(1960).

Sagitta inflata
Ghirardelli, Elvezio, et Marion Kerbiriou, 1965.
Chaetoghethes et cladoceres du Golfe de Trieste. Rapp. Proc. Verb., Réunions, Comm. Int. Expl. Sci., Mer Méditerranée, Monaco, 18(2):403-407.

Sagitta enflata
Grant, George C., 1963
Investigations of inner continental shelf waters off lower Chesapeake Bay. IV. Descriptions of the Chaetognatha and a key to their identification. Chesapeake Science, 4(3):109-119.

Sagitta enflata
Hamada, Takao, 1967.
Studies on the distribution of Chaetognatha in the Harima-nada and Osaka Bay, with special reference to Sagitta enflata. (In Japanese; English abstract). Bull. Jap. Soc. scient. Fish., 33(2):98-103.

Sagitta enflata
Hamada, Takao, Shozo Iwai and Hanji Moriwaki 1971.
The hydrological conditions for the entry of Sagitta enflata into Osaka Bay. II. In the case of appearance of a cold water mass. (In Japanese; English abstract). Bull. Jap. Soc. scient. Fish. 37(5): 357-363.

Sagitta enflata
Hosoe, K., 1956.
Chaetognaths from the Isles of Fernando de Noronha. Contr. Avulsas, Inst. Oceanogr., Sao Paulo, No. 3 :8 pp.

Sagitta enflata
Hure, J., 1961
Dneva migracija i sezonaka vertikalna raspodjela zooplanktona dubljeg mora. [Migration journalière et distribution saisonnaière verticale du zooplancton dans la région profunde de l'Adriatique] Acta Adriatica, 9(6):1-59.

Sagitta enflata
Japan, Kobe Marine Observatory, 1963.
Report of the oceanographical observations in the sea south of Honshu from February to March 1961. Report of the oceanographical observations in the cold water region off Enshu Nada in May 1961. (In Japanese). Bull. Kobe Mar. Obs., 171(4):22-35.

Sagitta enflata
Japan, Kobe Marine Observatory, Oceanographical Section, 1962
Report of the oceanographic observations in the sea south of Honshu from October to November, 1959. (In Japanese). Bull. Kobe Mar. Obs., No. 169(11):44-50.

Sagitta enflata
Japan, Kobe Marine Observatory, Oceanographical Section, 1962
Report of the oceanographic observations in the sea south of Honshu in May, 1960. (In Japanese). Bull. Kobe Mar. Obs., No. 169(12):27-33.

Sagitta enflata
Kado, Y., 1953.
The chaetognath fauna of the Inland Sea of Japan especially on the distribution of Sagitta enflata and S. crassa. Dobuts. Zasshi 62(10): 337-342.

Sagitta enflata
Laguarda Figueras, Alfredo 1965.
Contribución al conocimiento de los quetognatos de Sinaloa. Anales. Inst. Biol. Univ. Mex. 36(1/2):215-228.

Sagitta enflata
Legare, J. E. Henri, and E. Zoppi, 1961.
Notas sobre la abundancia y distribución de Chaetognatha en las aguas del Oriente de Venezuela. (In English). Bol. Inst. Oceanograf., Univ. Oriente, Cumana, Venezuela, 1(1):149-171.

Sagitta enflata
Marukawa, H., 1921
Plankton lists and some new species of copepods, from the northern waters of Japan. Bull. Inst. Ocean., No.384, 15 pp., 3 pls., 1 chart. Monaco

Sagitta enflata
Massuti, M., 1961
Note préliminaire à l'étude des chétognathos de la Méditerranée occidentale. Campagne du "Xauen" X-6911. Rapp. Proc. Verb., Réunions, Comm. Int. Expl. Sci. Mer Méditerranée, Monaco, 16(2):237-244.

Sagitta enflata
Massuti, M., 1958.
Estudio del crecimiento reletivo de Sagitta enflata Grassi del plancton de Castellón. Investigacion Pesquera, Barcelona, 13:37-48.

Sagitta enflata
Massuti Oliver, M., 1951.
Sobre la biologia de las Sagitta del plancton del levante español. Publ. Inst. Biol. Aplic., Barcelona, 8:71-82, 3 textfigs.

Sagitta enflata
Michael, E.L., 1919
Contributions to the Biology of the Philippine Archipelago and Adjacent regions. Report on the chaetognatha collected by the United States Fisheries Steamer "Albatross" during the Philippine Expedition, 1907-1910. Bull. U.S. Nat. Mus. No.100, Vol.1(4):235-277, pls.34-38.

Sagitta enflata
Michael, E. L., 1911
Classification and vertical distribution of the Chaetognatha of the San Diego Region including redescriptions of some doubtful species of the group. Univ. Calif. Publ. Zool. 8(3):21-186, Pls. 1-8.

Sagitta enflata
Moore, H. B., 1949
The zooplankton of the upper waters of the Bermuda area of the North Atlantic. Bull. Bingham Ocean. Coll. 12(2):97, 208 text figs.

Sagitta enflata
Murakami, Akio, 1957
[Value of chaetognaths preferring low salinity as indicator forms of water masses.] Info. Bull., Plankton., Japan, (5):8-10.

Sagitta enflata
Nair, Vijayalakshmi R. 1971.
Seasonal fluctuations of chaetognaths in the Cochin Backwater. J. mar. biol. Ass. India 13(2): 226-233.

Sagitta enflata
Nogueira Paranaguá, Maryse (1963-1964) 1966.
Sobre o plancton da região comprendida entre 3° Lat. S e 13° Lat. S ao largo do Brasil. Trabs Inst. Oceanogr. Univ. Recife 5(5/6):123-139.

Sagitta enflata
Pereiro, José A., 1972.
Análisis de la correlación de caracteres en el quetognato Sagitta enflata Grassi. Inv. Pesq. Barcelona 36(1): 15-22.

Sagitta enflata
Pierce, E.L., 1953.
The Chaetognatha over the continental shelf of North Carolina with attention to the relation of to the hydrography of the area. J. Mar. Res. 12(1):75-92, 4 textfigs.

Sagitta enflata
Pierce, E.L., 1951.
The Chaetognatha of the west coast of Florida.
Biol. Bull. 100(3):206-228, 5 textfigs.

Sagitta enflata
Pierce, E.L., 1951.
The occurrence and breeding of the Chaetognatha along the Gulf coast of Florida. Proc. Gulf and Caribbean Fish. Inst., 3rd session:126-127.

Sagitta enflata
Rao, T.S. Satyanarayana, and P.N. Ganapati, 1958
Studies on Chaetognatha in the Indian Seas. III. Systematics and distribution in the waters off Visakhapaynam. Andhra Univ. Mem. Oceanogr., 2:147-163.

Sagitta enflata
Raymont, J.E.G., and Eileen Linford, 1966.
A note on the biochemical composition of some Mediterranean zooplankton.
Int. Rev. ges. Hydrobiol., 51(3):485-488.

Sagitta enflata
Ritter-Záhony, R. von, 1911
Revision der Chätognathen. Deutsche Südpolar Exped., 1901-1903, XIII (Zool.V)., Pt.1:71 pp., 51 text figs.

Sagitta enflata
Scaccini, A., e E. Ghirardelli, 1941.
Chaetognathes collected on the coast of Rio de Oro. Not. Ist. Biol., Rovigno, 2(21):3-15.

Sagitta enflata
Stone, J.H., 1966.
The distribution and fecundity of Sagitta enflata Grassi in the Agulhas current.
J. Anim. Ecol., 35(3):533-541.

Sagitta enflata
Suarez Caabro, J.A., 1955.
Quetognatos de los mares cubanos.
Mem. Soc. Cubana Hist. Nat., 22(2):125-180, 9 pls.

Sagitta enflata
Suarez-Caabro, Jose A., and Juan E. Madruga, 1960.
The Chaetognatha of the northeastern coast of Honduras Central America.
Bull. Mar. Sci., Gulf & Caribbean 10(4):421-429.

Sagitta enflata
Sudarsan, D., 1961.
Observations on the Chaetognatha of the waters around Mandapam.
Indian J. Fish., 8(2):364-382.

Sagitta enflata
Sund, Paul N., 1961.
Some features of the autecology and distributions of Chaetognatha in the Eastern Tropical Pacific.
Inter-American Tropical Tuna Comm., Bull., 5(4):307-340.

Sagitta enflata
Sund, Paul N., 1959.
A key to the Chaetognatha of the tropical Eastern Pacific Ocean.
Pacific Science, 13(3):269-285.

Sagitta enflata
Thiel, M.E., 1938
Die Chaetognathen - Bevölkerung des Südatlantischen Ozeans. Biologische Sonderuntersuchungen. 1st Lief. Wissenschaftliche Ergebnisse der Deutschen Atlantischen Expedition auf dem Forschungs, und Vermessungsschiff "Meteor" 1925-1927, 13:1-110, 62 text figs.

Sagitta enflata
Thomson, J.M., 1948
Some chaetognatha from western Australia.
J. Roy. Soc. Western Australia, Inc., 31 (1944-1945), 17-18.

Sagitta enflata
Thomson, J.M., 1947
The Chaetognatha of South-eastern Australia.
Counc. Sci. & Ind. Res., Australia, Bull. No.222, (Div. Fish. Rept. 14), 43 pp., 8 text figs.

Sagitta enflata
Tokioka, Takasi, 1959
Observations on the taxonomy and distribution of chaetognaths of the North Pacific.
Publ. Seto Mar. Biol. Lab., 7(3):349-456.

Sagitta enflata
Tokioka, T., 1954.
Droplets from a plankton net. XVI. On a small collection of chaetognaths from the central Pacific. Publ. Seto Mar. Biol. Lab. 4(1):99-102, 1 textfig.

Sagitta enflata
Tokioka, T., 1951.
Pelagic tunicates and chaetognaths collected during the cruises to the New Yamato Bank in the Sea of Japan. Publ. Seto Mar. Biol. Lab. 2(1):1-26, 1 chart, 6 tables, 12 textfigs.

Sagitta enflata
Tokioka, T., 1942
Systematic studies of the plankton organisms occurring in Iwayama Bay, Palao. III Chaetognatha from the bay and adjacent waters. Palao Tropical Biol. Sta. Studies, II(3):527-548, pls.5-7, 11 text figs.

Sagitta enflata
Tokioka, T., 1940.
The chaetognath fauna of the waters of western Japan. Rec. Oceanogr. Wks., Japan, 12(1):1-22, 3 charts, 4 tables. (Contrib. 87, Seto M.B.L.).

Sagitta enflata
Tokioka, T., 1939.
Chaetognaths collected chiefly from the bays of Sagami and Suruga with some notes on the shape and structure of the seminal vesicles.
Rec. Oceanogr. Wks., Japan, 10(2):123-150.

Sagitta enflata
Vannucci, M., and K. Hosoe, 1952.
Resultados cientificos do Cruzeiro do "Baependi" e do "Vega" a Ilha da Trinidada. Bol. Inst. Ocean., Univ. Sao Paulo, 3(1/2):5-30, 4 pls.

Sagitta enflata
von Ritter-Záhony, R., 1911
Vermes. Chaetognathi. Das Tierreich 29:34 pp., 16 text figs.

Sagitta enflata
Vucetic, T., 1961
Sur la repartition des chaetognathes en Adriatique et leur utilisation comme indicateurs biologiques des conditions hydrographiques.
Rapp. Proc. Verb., Réunions, Comm. Int. Expl. Sci. Mer Méditerranée, Monaco, 16(2):111-116.

Sagitta enflata
Wickstead, J., 1959
A predatory copepod.
J. Animal Ecology, 28(1):69-72.

Sagitta enflata gardineri
Tokioka, Takasi, 1959
Observations on the taxonomy and distribution of chaetognaths of the North Pacific.
Publ. Seto Mar. Biol. Lab., 7(3):349-456.

Sagitta enflata
Tsuruta, Arao, 1963
Distribution of plankton and its characteristics in the oceanic fishing grounds, with special reference to their relation to fishery.
J. Shimonoseki Univ. Fish., 12(1):13-214

Sagitta euneritica
Alvariño, Angeles, 1967.
The Chaetognatha of the NAGA Expedition (1959-1961) in the South China Sea and the Gulf of Thailand.
NAGA Rept., 4(2):1-87.
Scripps Inst. Oceanogr.,

Sagitta euneritica
Alvariño, Angeles, 1966
Zoogeografía de California: quetognatos Revista
Soc. Mexicana Hist. nat. 27: 199-243

Sagitta euneritica
Alvarino, Angeles, 1965.
Distributional atlas of Chaetognatha in the California Current region.
California Coop. Oceanic Fish. Invest., Atlas, No. 3:291 charts

Sagitta euneritica
Alvariño, Angeles, 1964.
Zoogeografía de los quetognatos, especialmente de la región de California.
Ciencia, Mexico, 23:51-74.

Sagitta euneritica
Alvarino, Angeles, 1963
Quetognatos epiplanctonicos del Mar de Cortes.
Revista, Soc. Mexicana, Hist. Nat., 24:97-202.

Sagitta euneritica n.sp.
Alvarino, Angeles, 1961.
Two new chaetognaths from the Pacific.
Pacific Science, 15(1):67-77.

Spadella cephaloptera
DiMarcotullio, Angelo, 1965.
Rapporti fra neurosecrezione e stadi di maturità sessuale in Spadella cephaloptera Busch (Chaetognatha).
Bolletino Zoologia, 32:671-683.

Sagitta euneritica
Laguarda Figueras, Alfredo, 1965.
Contribución al conocimiento de los quetognatos de Sinaloa.
Anales Inst. Biol. Univ. Mex. 36(1/2): 215-228.

Sagitta euxina
Alvariño, Angeles, 1967.
The Chaetognatha of the NAGA Expedition (1959-1961) in the South China Sea and the Gulf of Thailand.
NAGA Rept., 4(2):1-87.
Scripps Inst. Oceanogr.,

Sagitta euxina
Furnestin, M. L., 1961.
Complements à l'étude de Sagitta euxina, variété de Sagitta setosa.
Rapp. Proc. Verb., Réunions, Comm. Int. Expl. Mer Méditerranée, Monaco, 16(2):97-101.

Sagitta euxina
Ritter-Záhony, R. von, 1911
Revision der Chätognathen. Deutsche Südpolar Exped., 1901-1903, XIII (Zool.V)., Pt.1:71 pp., 51 text figs.

Sagitta euxina
von Ritter-Záhony, R., 1911
Vermes. Chaetognathi. Das Tierreich 29:34 pp., 16 text figs.

Sagitta exaptera
Ritter-Záhony, R. von, 1911
Revision der Chätognathen. Deutsche Südpolar Exped., 1901-1903, XIII (Zool.V)., Pt.1:71 pp., 51 text figs.

SAGITTA FEROX
Alvariño, Angeles, 1969
Los quetognatos del Atlántico: distribución y notas esenciales de sistemática.
Trab. Inst. esp. Oceanogr., 37: 290 pp.

Sagitta ferox
Alvariño, Angeles, 1967.
The Chaetognatha of the NAGA Expedition (1959-1961) in the South China Sea and the Gulf of Thailand.
NAGA Rept., 4(2):1-87.
Scripps Inst. Oceanogr.,

Sagitta ferox
Alvariño, Angeles, 1966
Zoogeografía de California: quetognatos, Revista
Soc. Mexicana Hist. nat. 27: 199-243

Sagitta ferox
Alvarino, Angeles, 1965.
Distributional atlas of Chaetognatha in the California Current region.
California Coop. Oceanic Fish. Invest., Atlas, No. 3:291 charts

Sagitta ferox
Alvariño, Angeles, 1964.
Zoogeografía de los quetognatos, especialmente de la región de California.
Ciencia, México, 23:51-74.

Sagitta ferox
Alvariño, Angeles, 1964.
The Chaetognatha of the Monsoon Expedition in the Indian Ocean.
Pacific Science, 18(3):336-348.

Sagitta ferox
Alvarino, Angeles, 1962
Taxonomic revision of Sagitta robusta and Sagitta ferox Doncaster, and notes on their distribution in the Pacific.
Pacific Science, 16(2):186-201.

Sagitta ferox
Bieri, Robert 1959
The distribution of planktonic chaetognatha in the Pacific and their relationship to water masses.
Limnol. and Ocean. Vol. 4 No. 1 pp. 1-28

Sagitta ferox
Bieri, Robert, 1957
The chaetognath fauna of Peru in 1941.
Pacific Science 11(3): 255-264.

Sagitta ferox
Fowler, G.H., 1906.
The Chaetognatha of the Siboga Expedition, with a discussion of the synonymy and distribution of the group. Rept. Siboga Exped. 21:1-88, 3 pls., 6 charts.

Sagitta ferox
Furnestin, M.-L., et J. Radiguet, 1964.
Chaetognathes de Madagascar (Secteur de Nosy-Be).
Cahiers, O.R.S.T.R.O.M., Océanogr., 11(4):55-98.

Sagitta ferox
Michael, E.L., 1919
Contributions to the Biology of the Philippine Archipelago and Adjacent regions. Report on the chaetognatha collected by the United States Fisheries Steamer "Albatross" during the Philippine Expedition, 1907-1910.
Bull. U.S. Nat. Mus. No.100, Vol.1(4):235-277, pls.34-38.

Sagitta ferox
Michael, E. L., 1911
Classification and vertical distribution of the Chaetognatha of the San Diego Region including redescriptions of some doubtful species of the group. Univ. Calif. Publ. Zool. 8(3):21-186, Pls. 1-8.

Sagitta ferox
Rao, T.S. Satyanarayana and Sarada Kelly, 1964.
Studies on the chaetognatha of the Indian Seas. Pt. VIII. On the occurrence of Sagitta ferox Doncaster and S. hexaptera d'Orbigny in the waters off Visakhapatnam. Some aspects of Plankton Research Proc. Seminar. Mar. Biol. Sta. Porto Novo. Mar. 23-25 1964: 10-13.

Sagitta ferox
Sund, Paul S., 1964.
Los quetognatos en las aguas de la region del Peru.
Inter-Amer. Tropical Tuna Comm., Bull., 9(3):115-216.

Sagitta ferox
Sund, Paul N., 1961.
Some features of the autecology and distributions of Chaetognatha in the Eastern Tropical Pacific.
Inter-American Tropical Tuna Comm., Bull., 5(4):307-340.

Sagitta ferox
Sund, Paul N., 1959.
A key to the Chaetognatha of the tropical Eastern Pacific Ocean. Pacific Science, 13(3): 269-285.

Sagitta ferox
Thomson, J. M., 1947
The Chaetognatha of South-eastern Australia.
Counc. Sci. & Ind. Res., Australia, Bull. No.222, (Div. Fish. Rept. 14), 43 pp., 8 text figs.

Sagitta ferox
Tokioka, Takasi, 1959
Observations on the taxonomy and distribution of chaetognaths of the North Pacific.
Publ. Seto Mar. Biol. Lab., 7(3):349-456.

Sagitta ferox
Tokioka, T., 1954.
Droplets from the plankton net. XVI. On a small collection of chaetognaths from the central Pacific. Publ. Seto Mar. Biol. Lab. 4(1 99-102, 1 textfig.

Sagitta ferox
Tokioka, T., 1951.
Pelagic tunicates and chaetognaths collected during the cruises to the New Yamato Bank in the Sea of Japan. Publ. Seto Mar. Biol. Lab. 2(1):1-26, 1 chart, 6 tables, 12 textfigs.

Sagitta ferox
Tsuruta, Arao, 1963
Distribution of plankton and its characteristics in the oceanic fishing grounds, with special reference to their relation to fishery.
J. Shimonoseki Univ., Fish., 12(1):13-214

Sagitta ferox americana n. subsp.
Tokioka, Takasi, 1959
Observations on the taxonomy and distribution of chaetognaths of the North Pacific.
Publ. Seto Mar. Biol. Lab., 7(3):349-456.

Sagitta flaccida n.sp.
Conant, F.S., 1896.
23. Notes on the chaetognaths. Ann. Mag. Nat. Hist., 6th ser., 18:201-214.

Sagitta flaccida
Ritter-Záhony, R. von, 1911
Revision der Chätognathen. Deutsche Südpolar Exped., 1901-1903, XIII (Zool.V), Pt.1:71 pp., 51 text figs.

Sagitta fowleri
Bieri, Robert 1959
The distribution of planktonic chaetognatha in the Pacific and their relationship to water masses.
Limnol. and Ocean. Vol. 4 No. 1 pp. 1-28

Sagitta friderici
Alvariño, Angeles, 1969
Los quetognatos del Atlántico: distribución y notas esenciales de sistemática.
Trab. Inst. esp. Oceanogr., 37: 290 pp.

Sagitta friderici
Alvariño, Angeles, 1967.
The Chaetognatha of the NAGA Expedition (1959-1961) in the South China Sea and the Gulf of Thailand.
NAGA Rept., 4(2):1-87.
Scripps Inst. Oceanogr.,

Sagitta friderici
Alvarino, Angeles, 1961.
Two new chaetognaths from the Pacific.
Pacific Science, 15(1):67-77.

Sagitta friderici
Alvarino, A., 1957.
Estudio del zooplancton del Mediterráneo occidental. Zooplancton del Atlántico Ibérico. Campañas del "Xauen" en el verano del 1954.
Bol. Inst. Espanol Oceanogr., Nos. 81-82:1-26, 1-51.

Sagitta friderici
Bieri, Robert 1959
The distribution of planktonic chaetognatha in the Pacific and their relationship to water masses.
Limnol. and Ocean. Vol. 4 No. 1 pp. 1-28

Sagitta friderici
Coleman, J. S., 1959.
The "Rosaura" Expedition, 1937-1938: Chaetognatha.
Bull. Brit. Mus. (Natural History) Zool. 5(8):219-253.

Sagitta friderici (fig.)
de Almeida Prado, M.S., 1961
Chaetognatha encontrados em aguas brasileiras
Bol. Inst. Oceanogr., Univ. Sao Paulo, 11(2):31-55.

Sagitta friderici
de Saint-Bon, Marie catherine, 1963.
Les chaetognathes de la Côte 'Ivoire (espèces de surface).
Rev. Trav. Inst. Pêches Marit., 27(3):301-346.

Sagitta friderici
Ducret, Françoise, 1968.
Chaetognathes des campagnes de l'"Ombango" dans les eaux équatoriales et tropicales africaines.
Cah. O.R.S.T.O.M., sér. Océanogr., VI(1):95-141.

Sagitta friderici
Ducret, Françoise, 1968.
Chaetognathes des eaux superficielles et profondes de la zone équatoriale et tropicale africaine.
Thèse Fac. Sci., Marseille, 99 pp., 25 figs. In: Receuil Trav. publiées de 1965 à 1968, Océanogr. ORSTROM.

Sagitta friderici
Ducret, Françoise, 1968.
Chaetognathes des campagnes de l'Ombango dans les eaux équatoriales et tropicales africaines.
Cah.ORSTOM, Sér.Océanogr.,6(L):95-141.

Sagitta friderici
Faure, M. L., 1952.
Contribution a l'etude biologique et morphologique de deux chaetognaths des eaux atlantiques du Maroc: Sagitta friderici Ritter-Zahony et Sagitta bipunctata Quoy et Gaimard. Vie et Milieu 3:25.

Sagitta friderici
Ferreira da Costa, Pedro, 1970.
Nota preliminar sobre a ocorrência de Sagitta friderici e S. enflata (Chaetognatha) na Baía de Guanabara
Publ. Inst. Pesquisas Marinha, Brasil, 47:10 pp.

Sagitta friderici
Fraser, J.H., 1952.
The Chaetognatha and other zooplankton of the Scottish area and their value as biological indicators of hydrographical conditions.
Scottish Home Dept., Mar. Res., 1952(2):1-52, 3 pls., 21 charts.

Sagitta friderici
Furnestin, Marie-Louise, 1970.
Chaetognathes des campagnes du Thor (1908-11) en Méditerranée et en Mer Noire.
Dana - Rept. 79:1-51

Sagitta friderici
Furnestin, Marie-Louise, 1965.
Chaetognathes de quelques récoltes dans la mer des Antilles et l'Atlantique ouest tropical.
Inst. r. Sci. Nat. Belg., Bull., 41(9):15 pp.

Sagitta friderici
Furnestin, Marie-Louise, 1963.
Les chaetognathes atlantiques en Méditerranée.
Rev. Trav. Inst. Pêches Marit., 27(2):157-160.

Sagitta friderici
Furnestin, M. L., 1957.
Chaetognathes et zooplancton du secteur atlantique marocain. Rev. Trav. Inst. Pêches Marit. 21(1/2): 9-356.

Sagitta friderici
Furnestin, M.-L., 1953.
Sur quelques chaetognathes d'Israel.
Bull. Res. Counc., Israel, 2(4):411-414.

Sagitta friderici
Ghirardelli, E., 1960.
Habitat e biologia della riproduzione nei chetognati.
Arch. Oceanogr., e Limnol., 11(3):287-304.

Reprinted in:
Trav. Sta. Zool., Villefranche-sur-Mer, 18(11):

Sagitta friderici
Halim, Youssef, et Shoukry K. Guerguess 1973
Chaetognathes du plancton d'Alexandrie 1. Generalities. S. friderici R.E. Rapp.
Proc.-v. Reun. Comm. int. Explor. scient. Mer Medit. Monaco 21(8):493-496.

Sagitta friderici
Mostajo, Elena 1973.
Quetognatus colectados en el Atlantico sudoccidental entre los 44°44' y 52°36' de latitud sur.
Neotropica 19(59):94-100.

Sagitta friderici
Pierce, E.L., 1951.
The Chaetognatha of the west coast of Florida.
Biol. Bull. 100(3):206-228, 5 textfigs.

See: S. Eenuis

Sagitta friderici
Reyssac, J., 1963.
Chaetognathes du plateau continental européen (de la baie ibero-marocaine à la Mer Celtique).
Rev. Trav. Inst. Pêches Marit., 27(3):245-299.

Sagitta friderici n.sp.
Ritter-Záhony, R. von, 1911
Revision der Chätognathen. Deutsche Südpolar Exped., 1901-1903, XIII (Zool.V)., Pt.1:71 pp., 51 text figs.

Sagitta friderici
Rose, M., and M. Hamon, 1953.
Nouvelle note complémentaire sur les chétognathes de la Baie d'Alger.
Bull. Soc. Hist. Nat., Afrique du Nord, 44(5/6): 167-171.

Sagitta friderici
Scaccini, A., e E. Ghirardelli, 1941.
Chaetognathes collected on the coast of Rio de Oro. Not. Ist. Biol., Rovigno, 2(21):3-15.

Sagitta friderici
Sund, Paul N., 1961.
Two new species of Chaetognatha from the waters off Peru.
Pacific Science, 15(1):105-111.

Sagitta tenuis-friderici
Sund, Paul N., 1959
A key to the Chaetognatha of the tropical Eastern Pacific Ocean.
Pacific Science, 13(3):269-285

Sagitta friderici
Thiel, M.E., 1938
Die Chaetognathen - Bevölkerung des Südatlantischen Ozeans. Biologische Sonderuntersuchungen. 1st Lief. Wissenschaftliche Ergebnisse der Deutschen Atlantischen Expedition auf dem Forschungs, und Vermessungsschiff "Meteor" 1925-1927, 13:1-110, 62 text figs.

Sagitta friderici
Tokioka, Takasi, 1961.
Notes on Sagitta friderici Ritter-Zahony collected off Peru Postilla, Yale Peabody Mus. Nat. Hist., 55:16 pp.

Sagitta friderici
Tokioka, Takasi, 1959
Observations on the taxonomy and distribution of chaetognaths of the North Pacific.
Publ. Seto Mar. Biol. Lab., 7(3):349-456.

Sagitta friderici
Tokioka, T., 1955.
Notes on some chaetognaths from the Gulf of Mexico. Bull. Mar. Sci., Gulf & Caribbean, 5(1): 52-65.

Sagitta friderici
Vannucci, M., and K. Hosoe, 1952.
Resultados cientificos do Cruzeiro do "Baependi" e do "Vega" a Ilha da Trinidada. Bol. Inst. Ocean., Univ. Sao Paulo, 3(1/2):5-30, 4 pls.

Sagitta friderici
von Ritter-Záhony, R., 1911
Vermes. Chaetognathi. Das Tierreich 29:34 pp., 16 text figs.

?Sagitta furcata
Fowler, G.H., 1906.
The Chaetognatha of the Siboga Expedition with a discussion of the synonymy and distribution of the group. Rept. Siboga Exped. 21:1-88, 3 pls., 6 charts.

Sagitta furcata
Fowler, G.H., 1905.
Biscayan plankton collected during a cruise of H.M.S. "Research", 1900. Pt III. The Chaetognatha. Trans. Linn. Soc., London, 10(3):55-87, Pls. 4-7.

Sagitta galerita
Ducret, Françoise 1973.
Contribution à l'Etude des chaetognathes de la mer Rouge.
Beaufortia 20(268): 135-153.

Sagitta galerita n.sp.
Dallot, Serge, 1971
Les Chaetognathes de Nosy Bé: description de Sagitta galerita sp.n. Bull. Zool. Mus. Universiteit van Amsterdam 2(3):13-18, 1 table, 6 figs.

SAGITTA GAZELLAE
Alvariño, Angeles, 1969
Los quetognatos del Atlántico: distribución y notas esenciales de sistemática.
Trab. Inst. esp. Oceanogr., 37: 290 pp.

Sagitta helenae
Alvariño, Angeles, 1967.
The Chaetognatha of the NAGA Expedition (1959-1961) in the South China Sea and the Gulf of Thailand.
NAGA Rept., 4(2):1-87.
Scripps Inst. Oceanogr.

Sagitta gazellae
Alvariño, Angeles, 1964.
Zoogeografía de los quetognatos, especialmente de la región de California.
Ciencia, Mexico, 23:51-74.

Sagitta gazellae
Alvariño, Angeles, 1964.
The Chaetognatha of the MONSOON Expedition in the Indian Ocean.
Pacific Science, 18(3):336-348.

Sagitta gazellae
Boden, B.P., 1950.
Plankton organisms in the deep scattering layers.
USNEL Rept. No. 186:29 pp., textfigs.

Sagitta gazellae
Burfield, S.T., 1930.
Chaetognatha. Brit. Antarctic ("Terra Nova") Exped., 1910, Zool., 7(4):203-228, 3 maps.

Sagitta gazellae
David, Peter M., 1959.
Chaetognatha.
Brit-Australian-New Zealand Antarctic Res. Exp. (B) 8:73-79.

Sagitta gazellae
David, P.M., 1955.
The distribution of Sagitta gazellae Ritter-Zahony. Discovery Repts., 27:235-278.

Sagitta gazellae
Foxton, P., 1964
Seasonal variations in the plankton of Antarctic waters.
In: Biologie Antarctique, Proc. S.C.A.R. Symposium, Paris, 2-8 September 1962, Hermann, Paris, 311-318.

zooplankton, standing crop
zooplankton, seasonal variation
Salpa thompsoni
Calanoides acutus
plant pigments

Also in: Collected Reprints, Nat. Inst. Oceanogr., Wormley, 12. 1964

Sagitta gazellae
Ghirardelli, E., 1960.
Habitat e biologia della riproduzione nei chetognati.
Arch. Oceanogr. e Limnol., 11(3):287-304.

Reprinted in:
Trav. Sta. Zool., Villefranche-sur-Mer, 18(11):

Sagitta gazellae
Hamon, M., 1952.
Note complémentaire sur les chétognathes de la baie d'Alger. Bull. Soc. Hist. Nat., Afrique du Nord, 43(4/6):50-52, 1 textfig.

Sagitta gazellae
Mackintosh, N.A., 1964.
Distribution of the plankton in relation to the Antarctic Convergence.
Proc. R. Soc., London, (B), 281(1384):21-38.

Sagitta gazellae
Mackintosh, N.A., 1934
Distribution of the Macroplankton in the Atlantic Sector of the Antarctic. Discovery reports, Vol.9:65-160, 48 text figs.

Sagitta gazelle
Michael, E. L., 1911
Classification and vertical distribution of the Chaetognatha of the San Diego Region including redescriptions of some doubtful species of the group. Univ. Calif. Publ. Zool. 8(3):21-186, Pls. 1-8.

Sagitta gazellae
Mostajo, Elena 1973.
Quetognatus colectados en el Atlantico sudoccidental entre los 44°44' y 52°36' de latitud sur.
Neotropica 19(59): 94-100.

Sagitta gazellae
Ritter-Záhony, R. von, 1911
Revision der Chätognathen. Deutsche Südpolar Exped., 1901-1903, XIII (Zool.V), Pt.1:71 pp., 51 text figs.

Sagitta gazellae
von Ritter-Záhony, R., 1911
Vermes. Chaetognathi. Das Tierreich 29:34 pp., 16 text figs.

Sagitta gegenbauri
Ritter-Záhony, R. von, 1911
Revision der Chätognathen. Deutsche Südpolar Exped., 1901-1903, XIII (Zool.V), Pt.1:71 pp., 51 text figs.

Sagitta helenae
Alvariño, Angeles, 1969 SAGITTA HELENZE
Los quetognatos del Atlántico: distribución y notas esenciales de sistemática.
Trab. Inst. esp. Oceanogr., 37: 290 pp.

Sagitta helenae
Bumpus, D.F., and E.L. Pierce, 1955.
Hydrography and the distribution of chaetognaths over the continental shelf of North Carolina.
Pap. Mar. Biol. and Oceanogr., Deep-Sea Res., Suppl. to Vol. 3:92-109.

Sagitta helenae
Grant, George C., 1963
Investigations of inner continental shelf waters off lower Chesapeake Bay. IV. Descriptions of the Chaetognatha and a key to their identification.
Chesapeake Science, 4(3):109-119.

Sagitta helenae
Legare, J. E. Henri, and E. Zoppi, 1961.
Notas sobre la abundancia y distribucion de Chaetognatha en las aguas del Oriente de Venezuela. (In English).
Bol. Inst. Oceanograf., Univ. Oriente, Cumana, Venezuela, 1(1):149-171.

Sagitta helenae
Moore, H. B., 1949
The zooplankton of the upper waters of the Bermuda area of the North Atlantic. Bull. Bingham Ocean. Coll. 12(2):97, 208 text figs.

Sagitta helenae
Pierce, E.L., 1953.
The Chaetognatha over the continental shelf of North Carolina with attention to the relation to the hydrography of the area. J. Mar. Res. 12(1):75-92, 4 textfigs.

Sagitta helenae
Pierce, E.L., 1951.
The Chaetognatha of the west coast of Florida.
Biol. Bull. 100(3):206-228, 5 textfigs.

Sagitta helenae
Pierce, E.L., 1951.
The occurrence and breeding of the Chaetognatha along the Gulf coast of Florida. Proc. Gulf and Caribbean Fish. Inst., 3rd session:126-127.

Sagitta helenae
Pierce, E.L., and M.L. Wass, 1962.
Chaetognatha from the Florida Current and coastal waters of the southeastern Atlantic states.
Bull. Mar. Sci., Gulf and Caribbean, 12(3):403-431.

Sagitta helenae
Ritter-Záhony, R. von, 1911
Revision der Chätognathen. Deutsche Südpolar Exped., 1901-1903, XIII (Zool.V), Pt.1:71 pp., 51 text figs.

Sagitta helenae
Tokioka, T., 1955.
Notes on some chaetognaths from the Gulf of Mexico. Bull. Mar. Sci., Gulf & Caribbean, 5(1):52-65.

Sagitta helenae
von Ritter-Záhony, R., 1911
Vermes. Chaetognathi. Das Tierreich 29:34 pp., 16 text figs.

Sagitta helgolandica
Ritter-Záhony, R. von, 1911
Revision der Chätognathen. Deutsche Südpolar Exped., 1901-1903, XIII (Zool.V), Pt.1:71 pp., 51 text figs.

Sagitta hexaptera
Aida, T., 1897.
Chaetognaths of Misaki Harbour.
Ann. Zool., Japon. 1:13-21, Pl. 3.

Sagitta hexaptera
Alvariño, Angeles, 1969 SAGITTA HEXEPTERA
Los quetognatos del Atlántico: distribución y notas esenciales de sistemática.
Trab. Inst. esp. Oceanogr., 37: 290 pp.

Sagitta hexaptera
Alvariño, Angeles, 1967.
The Chaetognatha of the NAGA Expedition (1959-1961) in the South China Sea and the Gulf of Thailand.
NAGA Rept., 4(2):1-87.
Scripps Inst. Oceanogr.,

Sagitta hexoptera
Alvariño, Angeles, 1966
Zoogeografia de California: quetognatos, Revista
Soc. Mexicana Hist. nat. 27: 199-243

Sagitta hexaptera
Alvarino, Angeles, 1965.
Distributional atlas of Chaetognatha in the California Current region.
California Coop. Oceanic Fish. Invest., Atlas, No. 3:291 charts

Sagitta hexaptera
Alvariño, Angeles, 1964.
Zoogeografía de los quetognatos, especialmente de la región de California.
Ciencia, Mexico, 23:51-74.

Sagitta hexaptera
Alvariño, Angeles, 1964.
The Chaetognatha of the MONSOON Expedition in the Indian Ocean.
Pacific Science, 18(3):336-348.

Sagitta hexaptera
Alvarino, Angeles, 1963
Quetognatos epiplanctonicos del Mar de Cortes.
Revista, Soc. Mexicana, Hist. Nat., 24:97-202.

Sagitta hexaptera
Bainbridge, V., 1963.
Continuous plankton records: contribution toward a plankton atlas of the North Atlantic and the North Sea. VIII. Chaetognatha.
Bull. Mar. Ecol., 6(2):40-51.

Sagitta hexaptera
Alvarino, A., 1957.
Estudio del zooplancton del Mediterráneo occidental. Zooplankton del Atlántico Ibérico. Campañas del "Xauen" en el verano del 1954.
Bol. Inst. Espanol Oceanogr. Nos. 81-82:1-26; 1-51.

Sagitta hexaptera
Bieri, Robert 1959
The distribution of planktonic chaetognatha in the Pacific and their relationship to water masses.
Limnol. and Ocean. Vol. 4 No. 1 pp. 1-28

Sagitta hexaptera
Bieri, Robert, 1957.
The chaetognath fauna of Peru in 1941.
Pacific Science 11(3): 255-264.

Sagitta hexaptera
Bigelow, H.B., and M. Leslie, 1930
Reconnaissance of the waters and plankton of Monterey Bay, July 1928.
Bull. M.C.Z., 70(5):429-481, 43 text figs.

Sagitta hexaptera
Bigelow, H.B., and M. Sears, 1939
Studies of the waters of the continental shelf, Cape Cod to Chesapeake Bay. III. A volumetric study of the zooplankton. Mem. M.C.Z. 54(4):183-378, 42 text figs.

Sagitta hexaptera
Boden, B.P., 1950.
Plankton organisms in the deep scattering layers.
USNEL Rept. 186:29 textfigs.

Sagitta hexaptera
Bumpus, D.F., and E.L. Pierce, 1955.
Hydrography and the distribution of chaetognaths over the continental shelf of North Carolina.
Pap. Mar. Biol. and Oceanogr., Deep-Sea Res., Suppl. to Vol. 3:92-109.

Sagitta hexaptera
Burfield, S.T., 1930.
Chaetognatha. Brit. Antarctic ("Terra Nova") Exped., 1910, Zool., 7(4):203-228, 3 maps.

Sagitta hexaptera
Burfield, S.T., and E.J.W. Harvey, 1926.
The Chaetognatha of the "Sealark" Expedition.
Trans. Linn. Soc., London, Ser. 2, 19:93-119, Pls. 4-7.

Sagitta hexaptera
Coleman, J. S., 1959.
The "Rosaura" Expedition 1937-1938: Chaetognatha.
Bull. Brit. Mus. (Natural History) Zool. 5(8):219-253.

Sagitta hexaptera
Conant, F.S., 1896.
23. Notes on the chaetognaths. Ann. Mag. Nat. Hist., 6th ser., 18:201-214.

Sagitta hexaptera
de Saint-Bon, Marie-Catherine, 1963.
Complement à l'étude des chaetognathes de la Côte d'Ivoire (espèces profondes).
Rev. Trav. Inst. Pêches Marit., 27(4):403-415.

Sagitta hexaptera
Ducret, Françoise, 1968.
Chaetognathes des eaux superficielles et profondes de la zone équatoriale et tropicale africaine.
Thèse Fac. Sci., Marseille, 99 pp., 25 figs. In: Recueil Trav. publiées de 1965 à 1968, Océanogr. ORSTROM.

Sagitta hexaptera
Ducret, Françoise, 1968.
Chaetognathes des campagnes de l'Ombango dans les eaux équatoriales et tropicales africaines.
Cah. ORSTOM., Sér. Océanogr., 6(1):95-141.

Sagitta hexaptera
Ducret, Françoise, 1968.
Chaetognathes des campagnes de l'"Ombango" dans les eaux équatoriales et tropicales africaines.
Cah. O.R.S.T.O.M., ser. Oceanogr., VI(1):95-141.

Sagitta hexaptera
Fagetti Guaita, Elda, 1958.
Investigaciones sobre quetognatos colectados, especialmente, frente a la costa central y norte de Chile. Revista Biol. Mar., Valparaiso, 8(1-3):25-82.

Sagitta hexaptera

Fowler, G.H., 1907.
Chaetognatha, with a note on those collected by H.M.S. "Challenger" in Subantarctic and Antarctic waters. Nat. Antarctic Exped., 1901-1904, 3:6 pp., 1 chart.

Sagitta hexaptera

Fowler, G.H., 1906.
The Chaetognatha of the Siboga Expedition, with a discussion of the synonymy and distribution of the group. Rept. Siboga Exped. 21:1-88, 3 pls., 6 charts.

Sagitta hexaptera

Fraser, J.H., 1952.
The Chaetognatha and other zooplankton of the Scottish area and their value as biological indicators of hydrographical conditions. Scottish Home Dept., Mar. Res., 1952(2):1-52, 3 pls., 21 charts.

good for distinguishing characters

Sagitta hexaptera

Fraser, J. H., 1949
The occurrence of unusual species of Chaetognatha in Scottish plankton collections. JMBA 28(2):489-491.

Sagitta hexaptera

Fraser, J. H., 1939
The distribution of Chaetognatha in Scottish Waters in 1937. J. du Cons. 14(1):25-34, 3 charts in text.

Sagitta hexaptera

Fraser, J. H. and A. Saville, 1949
List of rare exotic species found in the plankton by the Scottish Vessel "Explorer" in 1948. Ann. Biol. 5:62-64.

Sagitta hexaptera

Furnestin, Marie-Louise 1971.
Au sujet de la "variété" magna de Sagitta hexaptera (chaetognath)
Rapp. P.-v. Comm. int. Explor. scient. mer Medit. 20(3): 355-358.

Sagitta hexaptera

Furnestin, Marie-Louise, 1970.
Chaetognathes des campagnes du Thor (1908-11) en Méditerranée et en Mer Noire.
Dana - Rept. 79:1-51

Sagitta hexaptera

Furnestin, Marie-Louise, 1966.
Chaetognathes des eaux africaines.
Atlantide Rep., 9:105-135.

Sagitta hexaptera

Furnestin, Marie-Louise, 1965.
Chaetognathes de quelques récoltes dans la mer des Antilles et l'Atlantique ouest tropical.
Inst. r. Sci. Nat. Belg., Bull., 41(9):15 pp.

Sagitta hexaptera

Furnestin, M.L., 1958.
Quelques échantillons du zooplancton du Golfe d'Eylath (Akaba).
Haifa, Sea Fish. Res. Sta., Bull., 16:6-14.

Sagitta hexaptera

Furnestin, M. L., 1957.
Chaetognathes et zooplancton du secteur atlantique marocain. Rev. Trav. Inst. Peches Marit. 21(1/2): 9-356.

Sagitta hexaptera

Furnestin, M-L.et J.-C.Codaccioni,1968.
Chaetognathes du nord-ouest de l'Océan Indien (golfe d'Aden,mer d'Arabie,golfe d'Oman,golfe Persique).
Cah. O.R.S.T.O.M.,ser.Océanogr.,VI(1):143-171.

Sagitta hexaptera

Furnestin, M-L., et J. Radiguet, 1964.
Chaetognathes de Madagascar (Secteur de Nosy-Bé).
Cahiers, O.R.S.T.R.O.M., Océanogr., 11(4):55-98.

Sagitta hexaptera

Gamulin, T., 1948
Prilog poznavanju Zooplanktona Srednjedalmatinskog Otocnog Područja. (Contrib. a la connaissance du zooplankton de la zone insulaire de la Dalmatie Moyenne). Acta Adriatica 3(7):38 pp., 6 tables, 1 map.

Sagitta hexaptera

Germain, L. and L. Joubin, 1916
Chétognathes provenant des campagnes des yachts Hirondelle et Princesse-Alice (1885-1910) Rés. Camp. Sci., Monaco, 49:118 pp., 7 pls., 7 maps.

Sagitta hexaptera

Ghirardelli, E., 1960.
Habitat e biologia della riproduzione nei chetognati.
Arch. Oceanogr. e Limnol., 11(3):287-304.
Reprinted in:
Trav. Sta. Zool., Villefranche-sur-Mer, 18(11):

Sagitta hexaptera

Ghirardelli, E., 1952.
Osservazioni biologiche e sistematiche sui Chetognati del Golfo di Napoli.
Pubbl. Staz. Zool., Napoli, 23(2/3):296-312.

Sagitta hexaptera

Ghirardelli, E., 1950.
Osservazioni biologiche e sistematiche sui chetognati della baia di Villefranche sur mer. Bol. Pesca, Piscicol. e Idrobiol., n.s., 5(1):105-127, 7 textfigs.

Sagitta hexaptera

Ghirardelli, E., 1950.
Osservazioni biologiche e sistematiche sui chetognati della Baia di Villefranche-sur-mer. (Trav. Sta. Zool. Villefranche-sur-mer 10(9)). Bol. Pesca, Piscicul. Idrobiol., n.s., 5(1):5-27, 7 textfigs.

Sagitta hexaptera

Ghirardelli, Elvezio, 1947
Chetognati raccolti nel Mar Rosso e nell'Oceano Indiano dalla Nave "Cherso", Bollettino di Pesca, Piscicoltura, e Idrobiologia Anno 23, Vol. 2(n.s.) (2):253-270, 2 pls., 9 text figs.

Sagitta hexaptera

Hosoe, K., 1956.
Chaetognaths from the Isles of Fernando de Noronha.
Contr. Avulsas Inst. Oceanogr., Sao Paulo, No. 3: 8 pp.

Sagitta hexaptera

Japan, Kobe Marine Observatory, 1963.
Report of the oceanographic observations in the sea south of Honshu from July to August, and from the cold water region south of Enshu Nada October to November 1960. (In Japanese).
Bull. Kobe Mar. Obs., 171(3):36-52.

Sagitta hexaptera

Japan, Kobe Marine Observatory, 1962
Report of the oceanographic observations in the sea south of Honshu from July to August, 1960. (In Japanese).
Res. Mar. Meteorol. and Oceanogr., July-Dec., 1960. Japan Meteorol. Agency, No. 28:36-42.

Sagitta hexaptera

Japan, Kobe Marine Observatory, Oceanographical Section, 1962
Report of the oceanographic observations in the sea south of Honshu from October to November, 1959. (In Japanese).
Bull. Kobe Mar. Obs., No. 169(11):44-50.

Sagitta hexaptera

Legare, J. E. Henri, and E. Zoppi, 1961.
Notas sobre la abundancia y distribucion de Chaetognatha en las aguas del Oriente de Venezuela. (In English).
Bol. Inst. Oceanograf., Univ. Oriente, Cumana, Venezuela, 1(1):149-171.

Sagitta hexaptera

Marukawa, H., 1921
Plankton lists and some new species of copepods, from the northern waters of Japan.
Bull. Inst. Ocean., No.384, 15 pp., 3 pls., 1 chart. Monaco

Sagitta hexaptera

Massuti, M., 1961
Note préliminaire à l'étude des chétognathos de la Méditerranée occidentale. Campagne du "Xauen" X-6911.
Rapp. Proc. Verb., Réunions, Comm. Int. Expl. Sci. Mer Méditerranée, Monaco, 16(2):237-244.

Sagitta hexaptera

Michael, E.L., 1919
Contributions to the Biology of the Philippine Archipelago and Adjacent regions. Report on the chaetognatha collected by the United States Fisheries Steamer "Albatross" during the Philippine Expedition, 1907-1910.
Bull. U.S. Nat. Mus. No.100, Vol.1(4):235-277, pls.34-38.

Sagitta hexaptera

Michael, E. L., 1911
Classification and vertical distribution of the Chaetognatha of the San Diego Region including redescriptions of some doubtful species of the group. Univ. Calif. Publ. Zool. 8(3):21-186, Pls. 1-8.

Sagitta hexaptera

Moore, H. B., 1949
The zooplankton of the upper waters of the Bermuda area of the North Atlantic. Bull. Bingham Ocean. Coll. 12(2):97, 208 text figs.

Sagitta hexaptera

Nogueira Paranagua, Maryse (1963-1964)
1966
Sôbre o plancton de região comprendida entre 3° Lat. S e 13° Lat. S ao largo do Brasil.
Trabhs Inst. Oceanogr. Univ. Recife 5(5/6): 125-139.

Sagitta hexaptera

Pierce, E.L., 1953.
The Chaetognatha over the continental shelf of North Carolina with attention to the relation to the hydrography of the area. J. Mar. Res. 12(1): 75-92, 4 textfigs.

Sagitta hexaptera

Pierce, E.L., and M.L. Wass, 1962.
Chaetognatha from the Florida Current and coastal waters of the southeastern Atlantic states.
Bull. Mar. Sci., Gulf and Caribbean, 12(3):403-431.

Sagitta hexaptera

Rao, T. S. Satyanarayana and Sarada Kelly, 1964.
Studies on the chaetognatha of the Indian Seas. Pt. VIII. On the occurrence of Sagitta ferox Doncaster and S. hexaptera d'Orbigny in the waters off Visakhapatnam. Some aspects of Plankton Research. Proc. Seminar, Mar. Biol. Sta. Porto Novo. Mar. 23-25 1964: 10-13.

Sagitta hexaptera

Reyssac, J., 1963.
Chaetognathes du plateau continental européen (de la baie ibèro-marocaine à la Mer Celtique).
Rev. Trav. Inst. Peches Marit., 27(3):245-299.

Sagitta hexaptera

Ritter-Záhony, R. von, 1911
Revision der Chätognathen. Deutsche Südpolar Exped., 1901-1903, XIII (Zool.V), Pt.1:71 pp., 51 text figs.

Sagitta hexaptera
von Ritter-Záhony, R., 1911
Vermes. Chaetognathi. Das Tierreich
29:34 pp., 16 text figs.

Sagitta hexaptera
Rose, M., and M. Hamon, 1953.
Nouvelle note complémentaire sur les chétognathes de la Baie d'Alger.
Bull. Soc. Hist. Nat., Afrique du Nord, 44(5/6): 167-171.

Sagitta hexaptera
Russell, F.S., 1939.
Chaetognatha. Fiches d'Ident. Zooplancton, Cons. Perm. Int. Expl. Mer, 1:4 pp., 12 textfigs.

Sagitta hexaptera
Suarez Caabro, J.A., 1955.
Quetognatos de los mares cubanos.
Mem. Soc. Cubana Hist. Nat., 22(2):125-180, 9 pls

Sagitta hexaptera
Sund, Paul N., 1964.
Los quetognatos en las aguas de la region del Peru.
Inter-Amer. Tropical Tuna Comm., Bull., 9(3):115-216.

Sagitta hexaptera
Sund, Paul N., 1961.
Some features of the autoecology and distributions of Chaetognatha in the Eastern Tropical Pacific.
Inter-American Tropical Tuna Comm., Bull., 5(4): 307-340.

Sagitta hexaptera
Sund, Paul N., 1959.
A key to the Chaetognatha of the tropical Eastern Pacific Ocean.
Pacific Science, 13(3):269-285.

Sagitta hexaptera
Thiel, M.E., 1938
Die Chaetognathen - Bevölkerung des Südatlantischen Ozeans. Biologische Sonderuntersuchungen. 1st Lief. Wissenschaftliche Ergebnisse der Deutschen Atlantischen Expedition auf dem Forschungs, und Vermessungsschiff "Meteor" 1925-1927, 13:1-110, 62 text figs.

Sagitta hexaptera
Thomson, J. M., 1947
The Chaetognatha of South-eastern Australia.
Counc. Sci. & Ind. Res., Australia, Bull. No.222, (Div. Fish. Rept. 14), 43 pp., 8 text figs.

Sagitta hexaptera
Tokioka, Taskasi, 1959.
Observations on the taxonomy and distribution of chaetognaths of the North Pacific.
Publ. Seto Mar. Biol. Lab., 7(3): 349-456.

Sagitta hexaptera
Tokioka, T., 1954.
Droplets from the plankton net. XVI. On a small collection of chaetognaths from the central Pacific. Publ. Seto Mar. Biol. Lab. 4(1):99-102, 1 textfig.

Sagitta hexaptera
Tokioka, T., 1942
Systematic studies of the plankton organisms occurring in Iwayama Bay, Palao. III Chaetognaths from the bay and adjacent waters. Palao Tropical Biol. Sta. Studies, II(3):527-548, pls.5-7, 11 text figs.

Sagitta hexaptera
Tokioka, T., 1940.
The chaetognath fauna of the waters of western Japan, Rec. Oceanogr. Wks., Japan, 12(1):1-22, 3 charts, 4 tables. (Contrib. 87, Seto M.B.L.).

Sagitta hexaptera
Tokioka, T., 1939.
Chaetognaths collected chiefly from the bays of Sagami and Suruga with some notes on the shape and structure of the seminal vesicles.
Rec. Oceanogr. Wks., Japan, 10(2):123-150.

Sagitta hexaptera
Tsuruta, Arao, 1963
Distribution of plankton and its characteristics in the oceanic fishing grounds, with special reference to their relation to fishery.
J. Shimonoseki Univ., Fish., 12(1):13-214

Sagitta hexaptera
Vannucci, M., and K. Hosoe, 1952.
Resultados cientificos do Cruzeiro do "Baependi" e do "Vega" a Ilha da Trinidada. Bol. Inst. Ocean Univ. Sao Paulo, 3(1/2):5-30, 4 pls.

Sagitta hexaptera
Vučetic, T., 1961
Sur la répartition des chaetognathes en Adriatique et leur utilisation comme indicateurs biologiques des conditions hydrographiques.
Rapp. Proc. Verb., Réunions, Comm. Int. Expl. Sci. Mer Méditerranée, Monaco, 16(2):111-116.

Sagitta hexaptera
Welsh, J.H., F.A. Chace, jr. and R.F. Nunnemacher, 1937
The diurnal migration of deep-water animals. Biol. Bull. 73(2):185-196, 7 text figs.

Sagitta hispida
Aida, T., 1897.
Chaetognaths of Misaki Harbour.
Ann. Zool., Japon., 1:13-21, Pl. 3.

SAGITTA HISPIDA
Alvariño, Angeles, 1969
Los quetognatos del Atlántico: distribución y notas esenciales de sistemática.
Trab. Inst. esp. Oceanogr., 37: 290 pp.

Sagitta hispida
Alvariño, Angeles, 1967.
The Chaetognatha of the NAGA Expedition (1959-1961) in the South China Sea and the Gulf of Thailand.
NAGA Rept., 4(2):1-87.
Scripps Inst. Oceanogr.,

Sagitta hispida
Beers, John R., 1964.
Ammonia and inorganic phosphorus excretion by the planktonic chaetognath, Sagitta hispida Conant. J. du Cons., 29(2):123-129.

Sagitta hispida
Beers, John R., 1962
Ammonia and inorganic phosphorus excretion of some marine zooplankton. (Abstract).
Assoc. Island Mar. Labs., 4th Meet., Curacao, 18-21 Nov., 1962: 1.

Sagitta hispida
Bumpus, D.F., and E.L. Pierce, 1955.
Hydrography and the distribution of chaetognaths over the continental shelf of North Carolina.
Pap. Mar. Biol. and Oceanogr., Deep-Sea Res., Suppl. to Vol. 3:92-109.

Sagitta hispida
Burfield, S.T., and E.J.W. Harvey, 1926.
The Chaetognatha of the "Sealark" Expedition.
Trans. Linn. Soc., London, Ser. 2, 19:93-119, Pls. 4-7.

Sagitta hispida
Conant, F.S., 1896.
23. Notes on the chaetognaths. Ann. Mag. Nat. Hist., 6th ser., 18:201-214.

Sagitta hispida
Cosper, T.C. and M.R. Reeve, 1970.
Structural details of the mouthparts of a chaetognath, as revealed by scanning electron microscopy. Bull. mar. Sci., 20(2): 441-445.

Sagitta hispida
Dallot, Serge, 1971
Les Chaetognathes de Nosy Bé: description de Sagitta galerita sp.n. Bull. Zool. Mus. Universiteit van Amsterdam 2(3):13-18, 1 table, 6 figs.

Sagitta hispida (fig.)
de Almeida Prado, M.S., 1961
Chaetognatha encontrados em aguas brasileiras Bol. Inst. Oceanogr., Univ. Sao Paulo, 11(2): 31-55.

Sagitta hispida
de Saint-Bon, Marie Catherine, 1963.
Les chaetognathes de la côte d'Ivoire (espèces de surface).
Rev. Trav. Inst. Pêches Marit., 27(3):301-346.

Sagitta hispida
Ducret, Françoise, 1968.
Chaetognathes des eaux superficielles et profondes de la zone équatoriale et tropicale africaine.
Thèse Fac. Sci., Marseille, 99 pp., 25 figs. In: Receuil Trav. publiées de 1965 à 1968, Océanogr. ORSTOM.

Sagitta hispida
Ducret, Françoise, 1968.
Chaetognathes des campagnes de l'Ombango dans les eaux équatoriales et tropicales africaines.
Cah. ORSTOM, Sér. Océanogr., 6(1):95-141.

Sagitta hispoda
Ducret, Françoise, 1968.
Chaetognathes des campagnes de l'"Ombango" dans les eaux equatoriales et tropicales africaines.
Cah. O.R.S.T.O.M., ser. Océanogr., VI(1):95-141.

Sagitta hispida
Fowler, G.H., 1906.
The Chaetognatha of the Siboga Expedition with a discussion of the synonymy and distribution of the group. Rept. Siboga Exped. 21:1-88, 3 pls., 6 charts.

Sagitta hispida
Furnestin, Marie-Louise, 1966.
Chaetognathes des eaux africaines.
Atlantide Rep., 9:105-135.

Sagitta hispida
Furnestin, Marie-Louise, 1965.
Chaetognathes de quelques récoltes dans la mer des Antilles et l'Atlantique ouest tropical.
Inst. r. Sci. Nat. Belg., Bull., 41(9):15 pp.

Sagitta hispida
Furnestin, M. L., 1957.
Chaetognathes et zooplancton du secteur atlantique marocain. Rev. Trav. Inst. Pêches Marit, 21(1/2): 9-356.

Sagitta hispida
Furnestin, M.L. et J. Brémec, 1968
Chaetognathes de la Mer Rouge (Archipel Dallac). Israel South Red Sea Expedition, 1962, Reports, N.32
Bull. Sea. Fish. Res. Sta. 52:3-20

Sagitta hispida
Furnestin, M.-L. et J.-C. Codaccioni, 1968.
Chaetognathes du nord-ouest de l'ocean Indien (golfe d'Aden, mer d'Arabie, golfe d'Oman, golfe Persique).
Cah. O.R.S.T.O.M., ser. Océanogr., VI(1):143-171.

Sagitta hispida
Furnestin, M.-L., et J. Radiguet, 1964.
Chaetognathes de Madagascar (Secteur de Nosy-Be).
Cahiers, O.R.S.T.O.M., Oceanogr., 11(4):55-98.

Sagitta hispida

Ghirardelli, Elvezio, 1947
Chetognati raccolti nel Mar Rosso e nel l'Oceano Indiano dalla Nave "Cherso", Bollettino di Pesca, Piscicoltura, e Idrobiologia Anno 23, Vol. 2(n.s.) (2):253-270, 2 pls., 9 text figs.

Sagitta hispida

George, P.C., 1952.
A systematic account of the Chaetognatha of the Indian coastal waters, with observations on their seasonal fluctuations along the Malabar coast.
Proc. Nat. Inst. Sci., India, 18(6):657-689.

Sagitta hispida

Grant, George C., 1963
Investigations of inner continental shelf waters off lower Chesapeake Bay. IV. Descriptions of the Chaetognatha and a key to their identification.
Chesapeake Science, 4(3):109-119.

Sagitta hispida

Legare, J. E. Henri, and E. Zoppi, 1961.
Notas sobre la abundancia y distribucion de Chaetognatha en las aguas del Oriente de Venezuela. (In English).
Bol. Inst. Oceanograf., Univ. Oriente, Cumana, Venezuela, 1(1):149-171.

Sagitta hispida

Michael, E. L., 1911
Classification and vertical distribution of the Chaetognatha of the San Diego Region including redescriptions of some doubtful species of the group. Univ. Calif. Publ. Zool. 8(3):21-186, Pls. 1-8.

Sagitta hispida = (S. robusta)

Pierce, E.L., 1951.
The chaetognatha of the west coast of Florida.
Biol. Bull. 100(3):206-228, 5 textfigs.

Sagitta hispida

Pierce, E.L., 1951.
The occurrence and breeding of the Chaetognatha along the Gulf coast of Florida. Proc. Gulf and Caribbean Fish. Inst., 3rd session:126-127.

Sagitta hispida

Pierce, E.L., and M.L. Wass, 1962.
Chaetognatha from the Florida Current and coastal waters of the southeastern Atlantic states.
Bull. Mar. Sci., Gulf and Caribbean, 12(3):403-431.

Sagitta hispida

Rao, T.S. Satyanarayana, and P.N. Ganapati, 1958
Studies on Chaetognatha in the Indian Seas. III. Systematics and distribution in the waters off Visakhapaynam. Andhra Univ. Mem. Oceanogr., 2:147-163.

Sagitta hispida

Reeve, Michael R., 1970
The biology of Chaetognatha. I. Quantitative aspects of growth and egg production in Sagitta hispida.
In: Marine Food Chains, J.H. Steele, editor, Oliver and Boyd, 168-189.

Sagitta hispida

Reeve, M.R., 1970.
Complete cycle of development of a pelagic chaetognath in culture.
Nature, Lond., 227(5256): 381.

Sagitta hispida

Reeve, M.R., 1966.
Observation on the biology of a chaetognath.
In: Some contemporary studies in marine science, H. Barnes, editor, George Allen & Unwin, Ltd., 613-630.

Sagitta hispida

Reeve, M.R., 1964.
Feeding of zooplankton with special reference to some experiments with Sagitta.
Nature, 201(4915):211-213.

Sagitta hispida

Reeve, M.R., 1964.
Studies on the seasonal variation of the zooplankton in a marine sub-tropical in-shore environment.
Bull. Mar. Sci., Gulf and Caribbean, 14(1):103-122.

Sagitta hispida

Reeve, M.R., J.E.G. Raymont and J.K.B. Raymont, 1970.
Seasonal biochemical composition and energy sources of Sagitta hispida. Marine Biol., 6(4): 357-364.

Sagitta hispida

Reeve, M.R. and M.A. Walter, 1972.
Observations and experiments on methods of fertilization in the chaetognath Sagitta hispida. Biol. Bull. mar. biol. Lab. Woods Hole 143 (1): 207-214.

Sagitta hispida

Reeve, M.R. and M.A. Walter, 1972
Conditions of culture, food-size selection, and the effects of temperature and salinity on growth rate and generation time in Sagitta hispida Conant. J. exp. mar. Biol. Ecol. 9(2): 191-200.

Sagitta hispida

Ritter-Záhony, R. von, 1911
Revision der Chätognathen. Deutsche Südpolar Exped., 1901-1903, XIII (Zool.V), Pt.1:71 pp., 51 text figs.

Sagitta hispida

Scaccini, A. e E. Ghirardelli, 1941.
Chaetognathes collected on the coast of Rio de Oro. Not. Ist. Biol., Rovigno, 2(21):3-15.

Sagitta hispida

Silas, E.G., and M. Srinivasan, 1968.
A new species of Eukrohnia from the Indian seas with notes on three other species of Chaetognatha.
J. mar. biol. Ass., India, 10(1): 1-33

Sagitta hispida

Suarez Caabro, J.A., 1955.
Quetognatos de los mares cubanos.
Mem. Soc. Cubana Hist. Nat., 22(2):125-180, 9 pls.

Sagitta hispida

Suarez-Caabro, J.A., and J.E. Madruga, 1960
The Chaetognatha of the northeastern coast of Honduras, Central America. Bull. Mar. Sci., Gulf and Caribbean, 10(4): 421-429.

Sagitta hispida

Sund, Paul N., 1961.
Two new species of Chaetognatha from the waters off Peru.
Pacific Science, 15(1):105-111.

Sagitta hispida

Tokioka, T., 1955.
Notes on some chaetognaths from the Gulf of Mexico. Bull. Mar. Sci., Gulf & Caribbean, 5(1): 52-65.

S. inflata

Clarke, G. L., E. L. Pierce and D. F. Bumpus, 1943
The distribution and reproduction of Sagitta elegans on Georges Bank in relation to the hydrographical conditions. Biol. Bull., 85(3): 201-226; 10 textfigs.

Sagitta inflata

Ducret, Françoise, 1968.
Chaetognathes des eaux superficielles et profondes de la zone équatoriale et tropicale africaine.
Thèse Fac. Sci., Marseille, 99 pp., 25 figs. In: Recueil Trav. publiées de 1965 à 1968, Océanogr. ORSTROM.

Sagitta inflata

Ducret, Françoise, 1968.
Chaetognathes des campagnes de l'"Ombango" dans les eaux équatoriales et tropicales africaines.
Cah. O.R.S.T.O.M., ser. Oceanogr., VI(1):95-141.

Sagitta inflata

Ducret, Françoise, 1968.
Chaetognathes des campagnes de l'Ombango dans les eaux équatoriales et tropicales africaines.
Cah. ORSTOM., Sér. Océanogr., 6(1):95-141.

Sagitta inflata

Furnestin, Marie-Louise, 1970.
Chaetognathes des campagnes du Thor (1908-11) en Méditerranée et en Mer Noire.
Dana - Rept. 79:1-51

Sagitta inflata

Furnestin, Marie-Louise, 1966.
Chaetognathes des eaux africaines.
Atlantide Rep. 9:105-135.

Sagitta inflata

Furnestin, M.L., et J. Belang2, 1968
Chaetognathes de la Mer Rouge (Archipel Dahlac). Israel South Red Sea Expedition, 1962, Reports, No. 32
Bull. Sea. Fish. Res. Sta. 52:3-20

Sagitta inflata

Furnestin, M.-L., et J.-C. Codaccioni, 1968.
Chaetognathes du nord-ouest de l'Océan Indien (golfe d'Aden, mer d'Arabie, golfe d'Oman, golfe Persique).
Cah. ORSTOM., Sér. Océanogr., 6(1):143-171

Sagitta inflata

Germain, L. and L. Joubin, 1916
Chétognathes provenant des campagnes des yachts Hirondelle et Princesse-Alice (1885-1910)
Rés. Camp. Sci., Monaco, 49:118 pp., 7 pls., 7 maps.

Sagitta inflata

Ghirardelli, E., 1962.
Ambiente e biologia della riproduzione nei Chetognati. Metodi di valutazione degli stadi di maturità e loro importanza nelle ricerche ecologiche. Problemi ecologici delle zone litorale del Mediterraneo, 17-23 luglio, 1961.
Pubbl. Staz. Zool., Napoli, 32 (Suppl.):380-399.

Sagitta inflata

Ghirardelli, E., 1961.
Histologie et cytologie des stades de maturité chez les chétognathes.
Rapp. Proc. Verb., Reunions, Comm. Int. Expl. Sci. Mer Méditerranée, Monaco, 16(2):103-110.

Sagitta inflata

Ghirardelli, E., 1960.
Habitat e biologia della riproduzione nei chetognati.
Arch. Oceanogr. e Limnol., 11(3):287-304.

Reprinted in:
Trav. Sta. Zool., Villefranche-sur-Mer, 18(11):

Sagitta inflata

Ghirardelli, Elvezio, 1960.
Istologia e citologia degli stadi di maturita nei Chetognati
Boll. Pesca, Piscicolt. e Idrobiol., n.s., 15(1)

Sagitta inflata

Ghirardelli, E., 1954.
Osservazioni sul corredo cromosomico di Sagitta inflata Grassi.
Scientia Genetica, 4(4):336-343.

In: Trav. Sta. Zool. Villefranche-sur-mer, 13.

Sagitta inflata

Ghirardelli, E., 1952.
Osservazioni biologiche e sistematiche sui Chetognati del Golfo di Napoli.
Pubbl. Staz. Zool., Napoli, 23(2/3):296-312.

Sagitta inflata
Ghirardelli, E., 1951.
Cicli di maturita sessuale nelle gonadi di Sagitta inflata Grassi del Golfo di Napoli. Boll. Zool., Unione Zool. Ital. 18(4/5/6):149-161, 10 textfigs.

Sagitta inflata
Ghirardelli, E., 1950.
Osservazioni biologiche e sistematiche sui chetognati della baia di Villefranche sur mer. Bol. Pesca, Piscicol. e Idrobiol., n.s., 5(1):105-127, 7 textfigs.

Sagitta inflata
Ghirardelli, E., 1950.
Osservazioni biologiche e sistematiche sui chetognati della Baia di Villefranche-sur-mer. (Trav. Sta. Zool. Villefranche-sur-mer 10(9)). Bol. Pesca, Piscicul. Idrobiol., n.s., 5(1):5-27, 7 textfigs.

Sagitta inflata
Ghirardelli, Elvezio, 1947
Chetognati raccolti nel Mar Rosso e nel l'Oceano Indiano dalla Nave "Cherso". Bollettino di Pesca, Piscicoltura, e Idrobiologia Anno 23, Vol. 2(n.s.) (2):253-270, 2 pls., 9 text figs.

Sagitta inflata
Massuti Oliver, M., 1954.
Sobre la biologia de las Sagitta del plancton del Levante espanol. Publ. Inst. Biol. Aplic. 16:137-148, 4 textfigs.

Sagitta inflata
Oliver, M.M., 1954.
Sobre la biologie de los Sagitta del plancton del Levante espanol. Publ. Inst. Biol. Aplic., 16:137-148.

Sagitta inflata
Ramult, M., and M. Rose, 1945
Recherches sur les Chetognathes de la Baie d'Alger. Bull. Soc. Hist. Nat. Afrique du Nord 36:45-71, 39 figs.

Sagitta inflata
Rose, M., and M. Hamon, 1953.
Nouvelle note complémentaire sur les chetognathes de la Baie d'Alger. Bull. Soc. Hist. Nat., Afrique du Nord, 44(5/6): 167-171.

Sagitta inflata
Scaccini, A., e E. Ghirardelli, 1941.
Chaetognathes of Adriatic Sea near Rovigno. Not. Ist. Biol., Rovigno, 2(22):3-16.

Sagitta inflata
Varadarajan, S. and P.I. Chacko, 1943.
On the arrow-worms of Krusadai. Proc. Nat. Inst. Sci., India, 9(2):245-248, 2 figs.

Sagitta izuensis n.sp.
Kitou, Masataka, 1966.
A new species of Sagitta Chaetognatha) collected off the Izu Peninsula. La Mer, Bull. franco-japon. Oceanogr., 4(4):238-239.
Also in: Reps JEDS 5-6 (1966).

Sagitta johorensis
Alvariño, Angeles, 1967.
The Chaetognatha of the NAGA Expedition (1959-1961) in the South China Sea and the Gulf of Thailand. NAGA Rept., 4(2):1-87.
Scripps Inst. Oceanogr.,

Sagitta johorensis n.sp.
Pathansali, D., and T. Tokioka, 1963.
A new chaetognath, Sagitta johorensis n.sp., from Malay waters. Publ. Seto Mar. Biol. Lab., 11(1)(6):105-107.

Sagitta lacunae n.sp.
Tokioka, T., 1942
Systematic studies of the plankton organisms occurring in Iwayama Bay, Palao. III Chaetognaths from the bay and adjacent waters. Palao Tropical Biol. Sta. Studies, II(3):527-548, pls.5-7, 11 text figs.

Sagitta levis
Ritter-Záhony, R. von, 1911
Revision der Chätognathen. Deutsche Südpolar Exped., 1901-1903, XIII (Zool.V), Pt.1:71 pp., 51 text figs.

Sagitta laevis
von Ritter-Záhony, R., 1911
Vermes. Chaetognathi. Das Tierreich 29:34 pp., 16 text figs.

Sagitta longicauda
Ritter-Záhony, R. von, 1911
Revision der Chätognathen. Deutsche Südpolar Exped., 1901-1903, XIII (Zool.V)., Pt.1:71 pp., 51 text figs.

Sagitta lyra
Aida, T., 1897.
Chaetognaths of Misaki Harbour. Ann. Zool., Japon., 1:13-21, Pl. 3.

Sagitta lyra
Alvariño, Angeles, 1969
Los quetognatos del Atlántico: distribución y notas esenciales de sistemática. Trab. Inst. esp. Oceanogr., 37: 290 pp.

Sagitta lyra
Alvariño, Angeles, 1967.
The Chaetognatha of the NAGA Expedition (1959-1961) in the South China Sea and the Gulf of Thailand. NAGA Rept., 4(2):1-87.
Scripps Inst. Oceanogr.,

Sagitta lyra
Alvariño, Angeles, 1964.
The Chaetognatha of the MONSOON Expedition in the Indian Ocean. Pacific Science, 18(3):336-348.

Sagitta lyra
Alvarino, A., 1957.
Estudio del zooplancton del Mediterráneo occidental. Zooplancton del Atlántico Ibérico. Campañas del "Xauen" en el verano del 1954. Bol. Inst. Espanol. Oceanogr., Nos. 81-82: 1-26; 1-51.

Sagitta lyra
Bieri, Robert 1959
The distribution of planktonic chaetognatha in the Pacific and their relationship to water masses. Limnol. and Ocean. Vol. 4 No. 1 pp. 1-28

Sagitta lyra
Bigelow, H.B., and M. Leslie, 1930
Reconnaissance of the waters and plankton of Monterey Bay, July 1928. Bull. M.C.Z., 70(5):429-481, 43 text figs.

Sagitta lyra
Boden, B.P., 1950.
Plankton organisms in the deep scattering layers. USNEL REPT. 186:29 pp., textfigs.

Sagitta lyra
Bumpus, D.F., and E.L. Pierce, 1955.
Hydrography and the distribution of chaetognaths over the continental shelf of North Carolina.
Pap. Mar. Biol. and Oceanogr., Deep-Sea Res., Suppl. to Vol. 3:92-109.

Sagitta lyra
Burfield, S.T., 1930.
Chaetognatha. Brit. Antarctic ("Terra Nova") Exped., 1910, Zool., 7(4):203-228, 3 maps.

Sagitta lyra
Burfield, S.T., and E.J.W. Harvey, 1926.
The Chaetognatha of the "Sealark" Expedition. Trans. Linn. Soc., London, Ser. 2;19:93-119, Pls. 4-7.

Sagitta lyra
Casanova, Jean-Paul et Françoise Ducret 1971.
Contribution à l'étude morphologique du chaetognathe Sagitta lyra (Krohn 1853)
Rapp. P.-v. Comm. int. Explor. scient. mer Medit. 20(3):359-361

Sagitta lyra
Coleman, J. S., 1959.
The "Rosaura" Expedition 1937-1938: Chaetognatha. Bull. Brit. Mus. (Natural History) Zool. 5(8):219-253.

Sagitta lyra
Corbin, P.G., 1947.
The spawning of mackerel, Scomber scombrus L., and pilchard, Clupea pilchardus Walbaum, in the Celtic Sea in 1937-39 with observations on the zooplankton indicator species, Sagitta and Muggiaea. J.M.B.A., n.s., 27:65-132, 21 textfigs.

Sagitta lyra
de Saint-Bon, Marie-Catherine, 1963.
Complement à l'étude des chaetognathes de la Côte d'Ivoire (espèces profondes). Rev. Trav. Inst. Pêches Marit., 27(4):403-415.

Sagitta lyra
Ducret, Françoise, 1968.
Chaetognathes des eaux superficielles et profondes de la zone équatoriale et tropicale africaine.
Thèse Fac. Sci., Marseille, 99 pp., 25 figs. In: Receuil Trav. publiées de 1965 à 1968, Oceanogr. ORSTOM.

Sagitta lyra
Ducret, Françoise, 1968.
Chaetognathes des campagnes de l'"Ombango" dans les eaux équatoriales et tropicales africaines. Cah. O.R.S.T.O.M., ser. Océanogr., VI(1):95-141.

Sagitta lyra
Ducret, Françoise, 1968.
Chaetognathes des campagnes de l'Ombango dans les eaux équatoriales et tropicales africaines. Cah. ORSTOM., Sér. Océanogr., 6(1):95-141.

Sagitta lyra
Fagetti Guaita, Elda, 1958.
Investigaciones sobre quetognatos colectados, especialmente, frente a la costa central y norte de Chile. Revista Biol. Mar., Valparaiso, 8(1-3):25-82.

Sagitta lyra
Fowler, G.H., 1906.
The Chaetognatha of the Siboga Expedition with a discussion of the synonymy and distribution of the group. Rept. Siboga Exped. 21:1-88, 3 pls., 6 charts.

Sagitta lyra
Fraser, J.H., 1952.
The Chaetognatha and other zooplankton of the Scottish area and their value as biological indicators of hydrographical conditions. Scottish Home Dept., Mar. Res., 1952:1-52, 3 pls., 21 charts.
(2)

Sagitta lyra
Fraser, J. H., 1949
The occurrence of unusual species of Chaetognatha in Scottish plankton collections. JMBA 28(2):489-491.

Sagitta lyra
Fraser, J. H., 1949
Plankton of the Faroe-Shetland Channel and the Faroes, June and August 1947. Ann. Biol., Int. Cons., 4:27-28, text fig. 10.

Sagitta lyra
Fraser, J. H. and A. Saville, 1949
Macroplankton in the Faroe Channel, 1948. Ann. Biol. 5:29-30, text figs.

Sagitta lyra
Fraser, J. H., and A. Saville, 1949
Plankton distribution in Scottish and adjacent waters in 1948. Ann. Biol. 5:61-62.

Sagitta lyra
Furnestin, Marie-Louise, 1970.
Chaetognathes des campagnes du Thor (1908-11) en Méditerranée et en Mer Noire.
Dana - Rept. 79:1-51

Sagitta lyra
Furnestin, Marie-Louise, 1966.
Chaetognathes des eaux africaines.
Atlantide Rep., 9:105-135.

Sagitta lyra
Furnestin, M. L., 1957.
Chaetognathes et zooplancton du secteur atlantique marocain. Rev. Trav. Inst. Pêches Marit. 21(1/2): 9-356.

Sagitta lyra
Germain, L. and L. Joubin, 1916
Chétognathes provenant des campagnes des yachts Hirondelle et Princesse-Alice (1885-1910) Rés. Camp. Sci., Monaco, 49:118 pp., 7 pls., 7 maps.

Sagitta lyra
Ghirardelli, E., 1960.
Habitat e biologia della riproduzione nei chetognati.
Limnol. e Oceanogr., 11(3):287-304.
Arch.
Reprinted in:
Trav. Sta. Zool., Villefranche-sur-Mer, 18(11):

Sagitta lyra
Ghirardelli, E., 1952.
Osservazioni biologiche e sistematiche sui Chetognati del Golfo di Napoli,
Pubbl. Staz. Zool., Napoli, 23(2/3):296-312.

Sagitta lyra
Ghirardelli, E., 1950.
Osservazioni biologiche e sistematiche sui chetgnati della Baia di Villefranche-sur-mer. (Trav. Sta. Zool. Villefranche-sur-mer 10(9)). Bol. Pesca, Piscicul. Idrobiol., n.sp, 5(1):5-27, 7 textfigs.

Sagitta lyra
Ghirardelli, E., 1950.
Osservazioni biologiche e sistematiche sui chetognati della Baia di Villefranche sur mer. Bol. Pesca, Piscicol. e Idrobiol., n.s., 5(1):105-127, 7 textfigs.

Sagitta lyra
Hamon, M., 1952.
Note complémentaire sur les chétognathes de la baie d'Alger. Bull. Soc. Hist. Nat., Afrique du Nord, 43(4/6):50-52, 1 textfig.

Sagitta lyra
Hure, J., 1961.
Dneva migracija i sezonaka vertikalna raspodjela zooplantona dubljeg mora. (Migration journalière et distribution saisonnaire verticale du zooplancton dans la region profunde de l'Adriatique). Acta Adriatica, 9(6):1-59.

Sagitta lyra
Lea, H.E., 1955.
The chaetognaths of western Canadian coastal waters. J. Fish. Res. Bd., Canada, 12(4):593-617.

Sagitta lyra
LeBrasseur, R. J., 1959.
Sagitta lyra, a biological indicator species in the subarctic waters of the eastern Pacific Ocean.
J. Fish. Res. Bd., Canada, 16(6):795-805.

Sagitta lyra
Massuti, M., 1961
Note préliminaire à l'étude des chétognathos de la Méditerranée occidentale. Campagne du "Xauen" X-6911.
Rapp. Proc. Verb., Réunions. Comm. Int. Expl. Sci. Mer Méditerranée, Monaco, 16(2):237-244.

Sagitta lyra
Michael, E. L., 1911
Classification and vertical distribution of the Chaetognatha of the San Diego Region including redescriptions of some doubtful species of the group. Univ. Calif. Publ. Zool. 8(3):21-186, Pls. 1-8.

Sagitta lyra
Moore, H. B., 1949
The zooplankton of the upper waters of the Bermuda area of the North Atlantic. Bull. Bingham Ocean. Coll. 12(2):97, 208 text figs.

Sagitta lyra
Pierce, E.L., 1953.
The Chaetognatha over the continental shelf of North Carolina with attention to their relation to the hydrography of the area. J. Mar. Res. 12(1):75-92, 4 textfigs.

Sagitta lyra
Pierce, E.L., and Sagitta lyra M. Wass, 1962.
Chaetognatha from the Florida Current and coastal waters of the southeastern Atlantic states.
Bull. Mar. Sci., Gulf and Caribbean, 12(3):403-431.

Sagitta lyra
Ramult, M., and M. Rose, 1945
Recherches sur les Chetognathes de la Baie d'Alger. Bull. Soc. Hist. Nat. Afrique du Nord 36:45-71, 39 figs.

Sagitta lyra
Redfield, A. C. and A. Beale, 1940.
Factors determining the distribution of populations of chaetognaths in the Gulf of Maine. Biol Bull. 79(3):459-487, 11 text figs.

Sagitta lyra
Ritter-Záhony, R. von, 1911
Revision der Chätognathen. Deutsche Südpolar Exped., 1901-1903, XIII (Zool.V), Pt.1:71 pp., 51 text figs.

Sagitta lyra
von Ritter-Záhony, R., 1911
Vermes. Chaetognathi. Das Tierreich 29:34 pp., 16 text figs.

Sagitta lyra
Rose, M., and M. Hamon, 1953.
Nouvelle note complémentaire sur les chétognathes de la Baie d'Alger.
Bull. Soc. Hist. Nat., Afrique du Nord, 44(5/6):167-171.

Sagitta lyra
Russell, F.S., 1939.
Chaetognatha. Fiches d'Ident. Zooplancton, Cons. Perm. Int. Expl. Mer, 1:4 pp., 12 textfigs.

Silas, E.G., and M. Srinivasan, 1968 Sagitta lyra
A new species of Eukrohnia from the Indian seas with notes on three other species of Chaetognatha.
J. mar. biol. Ass. India, 10(1):1-33

Sagitta lyra
Sund, Paul N., 1959.
A key to the Chaetognatha of the tropical Eastern Pacific Ocean.
Pacific Science, 13(3):269-285.

Sagitta lyra
Sund, P.N., 1959
The distribution of Chaetognatha in the Gulf of Alaska in 1954 and 1956.
J. Fish. Res. Bd., Canada, 16(3):351-361.

Sagitta lyra
Tchindonova, J.G., 1955.
[Chaetognatha of the Kurile-Kamchatka Trench.]
Trudy Inst. Oceanol., 12:298-310.

Sagitta lyra
Thomson, J. M., 1947
The Chaetognatha of South-eastern Australia. Counc. Sci. & Ind. Res., Australia, Bull. No.222, (Div. Fish. Rept. 14), 43 pp., 8 text figs.

Sagitta lyra
Tokioka, Takasi, 1959
Observations on the taxonomy and distribution of chaetognaths of the North Pacific.
Publ. Seto Mar. Biol. Lab., 7(3):349-456.

Sagitta lyra
Tokioka, T., 1940.
The chaetognath fauna of the waters of western Japan. Rec. Oceanogr. Wks., Japan, 12(1):1-22, 3 charts, 4 tables. (Contrib. 87, Seto M.B.L.).

Sagitta lyra
Tokioka, T., 1939.
Chaetognaths collected chiefly in the bays of Sagami and Suruga with some notes on the shape and structure of the seminal vesicles.
Rec. Oceanogr. Wks., Japan, 10(2):123-150.

Sagitta lyra
Tsuruta, Arao, 1963
Distribution of plankton and its characteristics in the oceanic fishing grounds, with special reference to their relation to fishery.
J. Shimonoseki Univ., Fish., 12(1):13-214

Sagitta lyra
von Ritter-Zahony, R., 1914.
Chaetognaths. Danish Ingolf-Exped. 4(3):4 pp.

Sagitta lyra
Vucetic, T., 1961
Sur la repartition des chaetognathes en Adriatique et leur utilisation comme indicateurs biologiques des conditions hydrographiques.
Rapp. Proc. Verb., Réunions, Comm. Int. Expl. Sci. Mer Méditerranée, Monaco, 16(2):111-116.

Sagitta lyra-gazellae
Hamon, M., 1952.
Note complémentaire sur les chétognathes de la baie d'Alger. Bull. Soc. Hist. Nat., Afrique du Nord, 43(4/6):50-52, 1 textfig.

SAGITTA MACROCEPHALA
Alvariño, Angeles, 1969
Los quetognatos del Atlántico: distribución y notas esenciales de sistemática.
Trab. Inst. esp. Oceanogr., 37: 290 pp.

Sagitta macrocephala
Alvariño, Angeles, 1967.
The Chaetognatha of the NAGA Expedition (1959-1961) in the South China Sea and the Gulf of Thailand.
NAGA Rept., 4(2):1-87.
Scripps Inst. Oceanogr.,

Sagitta macrocephala

Alvariño, Angeles, 1966
Zoogeografía de California: quetognatos. Revista
Soc. Mexicana Hist. nat. 27: 199-243

Sagitta macrocephala

Alvarino, Angeles, 1965. Sagitta macrocephala
Distributional atlas of Chaetognatha in the
California Current region.
California Coop. Oceanic Fish. Invest., Atlas,
No. 3:291 charts

Sagitta macrocephala

Bieri, Robert 1959

The distribution of planktonic chaetognatha
in the Pacific and their relationship to
water masses.
Limnol. and Ocean. Vol. 4 No. 1 pp. 1-28

Sagitta macrocephala

Coleman, J. S., 1959.

The "Rosaura" Expedition 1937-1938:
Chaetognatha.
Bull. Brit. Mus. (Natural History)
Zool. 5(8):219-253.

Sagitta macrocephala

de Saint-Bon, Marie-Catherine, 1963.
Complement à l'étude des chaetognathes de la
Côte d'Ivoire (espèces profondes).
Rev. Trav. Inst. Pêches Marit., 27(4):403-415.

Sagitta macrocephale

Ducret, Françoise, 1968.
Chaetognathes des campagnes de l'Ombango dans
les eaux équatoriales et tropicales africaines.
Cah. ORSTOM., Sér. Océanogr., 6(1):95-141.

Sagitta macrocephala

Ducret, Françoise, 1968.
Chaetognathes des eaux superficielles et
profondes de la zone équatoriale et tropicale
africaine.
Thèse Fac. Sci., Marseille, 99 pp., 25
figs. In: Receuil Trav. publiées de
1965 à 1968, Océanogr. ORSTROM.

Sagitta macrocephala

Ducret, Françoise, 1968.
Chaetognathes des campagnes de l'"Ombango" dans
les eaux équatoriales et tropicales africaines.
Cah. O.R.S.T.O.M., ser. Océanogr., VI(1):95-141.

Sagitta macrocephala

Fowler, G.H., 1906.
The Chaetognatha of the Siboga Expedition, with a
discussion of the synonymy and distribution of
the group. Rept. Siboga Exped. 21:1-88, 3 pls.,
6 charts.

Sagitta macrocephala n.sp.

Fowler, G.H., 1905.
Biscayan plankton collected during a cruise of
H.M.S. "Research", 1900. Pt. III. The Chaetognatha. Trans. Linn. Soc., London 10(3):55-87, Pls. 4-7.

Sagitta macrocephala

Fraser, J.H., 1952.
The Chaetognatha and other zooplankton of the
Scottish area and their value as biological
indicators of hydrographical conditions.
Scottish Home Dept., Mar. Res., 1952(2):1-52,
3 pls., 21 charts.

Sagitta macrocephala

Fraser, J. H., 1949
The occurrence of unusual species of
Chaetognatha in Scottish plankton collections.
JMBA 28(2):489-491.

Sagitta macrocephala

Furnestin, Marie-Louise, 1966.
Chaetognathes des eaux africaines.
Atlantide Rep., 9:105-135.

Sagitta macrocephala

Germain, L. and L. Joubin, 1916
Chétognathes provenant des campagnes des
yachts Hirondelle et Princesse-Alice (1885-1910)
Res. Camp. Sci., Monaco, 49:118 pp., 7 pls.,
7 maps.

Sagitta macrocephala

Michael, E.L., 1919
Contributions to the Biology of the
Philippine Archipelago and Adjacent regions.
Report on the chaetognatha collected by the
United States Fisheries Steamer "Albatross"
during the Philippine Expedition, 1907-1910.
Bull. U.S. Nat. Mus. No.100, Vol.1(4):235-277,
pls.34-38.

Sagitta macrocephala

Michael, E. L., 1911
Classification and vertical distribution
of the Chaetognatha of the San Diego Region
including redescriptions of some doubtful
species of the group. Univ. Calif. Publ. Zool.
8(3):21-186, Pls. 1-8.

Sagitta macrocephala

Russell, F.S., 1939.
Chaetognatha. Fiches d'Ident. Zooplancton, Cons.
Perm. Int. Expl. Mer, 1:4 pp., 12 textfigs.

Sagitta macrocephala

Srinivasan, M. 1971.
Two new records of meso- and
bathy-planktonic chaetognaths
from the Indian seas.
J. mar. biol. Ass. India 13(1): 130-133

Sagitta macrocephala

Sund, Paul N., 1961.
Some features of the autecology and distributions of Chaetognatha in the Eastern Tropical
Pacific.
Inter-American Tropical Tuna Comm., Bull., 5(4):
307-340.

Sagitta macrocephala

Tchindonova, J.G., 1955.
[Chaetognaths of the Kurile-Kamchatka Trench.]
Trudy Inst. Oceanol., 12:298-310.

Sagitta macrocephala

Thiel, M.E., 1938
Die Chaetognathen - Bevölkerung des
Südatlantischen Ozeans. Biologische Sonderuntersuchungen. 1st Lief. Wissenschaftliche
Ergebnisse der Deutschen Atlantischen Expedition auf dem Forschungs- und Vermessungsschiff
"Meteor" 1925-1927, 13:1-110, 62 text figs.

Sagitta macrocephala

Tokioka, Takasi, 1959
Observations on the taxonomy and distribution of chaetognaths of the North
Pacific.
Publ. Seto Mar. Biol. Lab., 7(3):349-456.

Sagitta macrocephala

Tokioka, T., 1939.
Chaetognaths collected chiefly in the bays of
Sagami and Suruga with some notes on the shape
and structure of the seminal vesicles.
Rec. Oceanogr. Wks., Japan, 10(2):123-150.

Sagitta macrocephala

Truveller, K.A., 1966.
Chaetognatha of the Davis and Denmark Straits
as indicators of water masses. (In Russian).
Mater. Ribokhoz. Issled. severn. Basseina,
Poliarn. Nauchno-Issled. Proektn. Inst. Morsk.
Ribn. Khoz. Okeanogr. (PINRO), 7:114-124.

Sagitta macrocephala

Ritter-Záhony, R. (von), 1911
Revision der Chätognathen. Deutsche
Südpolar Exped., 1901-1903, XIII (Zool.V),
Pt.1:71 pp., 51 text figs.

Sagitta macrocephala

von Ritter-Zahony, R., 1914.
Chaetognaths. Danish Ingolf-Exped. 4(3):4 pp.

Sagitta macrocephala

von Ritter-Záhony, R., 1911
Vermes. Chaetognathi. Das Tierreich
29:34 pp., 16 text figs.

Sagitta marri

Alvariño, Angeles, 1969
Los quetognatos del Atlántico: distribución
y notas esenciales de sistemática.
Trab. Inst. esp. Oceanogr., 37: 290 pp.

Sagitta marri

Alvariño, Angeles, 1967.
The Chaetognatha of the NAGA Expedition (1959-1961) in the South China Sea and the Gulf of
Thailand.
NAGA Rept., 4(2):1-87.
Scripps Inst. Oceanogr.,

Sagitta marri

Dallot, Serge, 1970.
L'anatomie du tube digestif dans la
phylogénie et la systématique des
Chaetognathes.
Bull. Mus. natn. Hist. nat. (2) 42(3): 549-565.

Sagitta marri

David, Peter M., 1959.
Chaetognatha.
Brit.-Australia-New Zealand Antarctic Res. Exped.
(B) 8:73-79.

Sagitta marri n.sp.

David, P.M., 1956.
Sagitta planctonis and related forms.
Bull. Brit. Mus. (N.H.), Zool., 4(8):435-451,
Pl. 11, 7 textfigs.

SAGITTA MAXIMA

Alvariño, Angeles, 1969
Los quetognatos del Atlántico: distribución
y notas esenciales de sistemática.
Trab. Inst. esp. Oceanogr., 37: 290 pp.

Sagitta maxima

Alvariño, Angeles, 1966
Zoogeografía de California: quetognatos. Revista
Soc. Mexicana Hist. nat. 27: 199-243

Sagitta maxima

Alvarino, Angeles, 1965.
Distributional atlas of Chaetognatha in the
California Current region.
California Coop. Oceanic Fish. Invest., Atlas,
No. 3:291 charts

Sagitta maxima

Bainbridge, V., 1963.
Continuous plankton records: contribution toward
a plankton atlas of the North Atlantic and the
North Sea. VIII. Chaetognatha.
Bull. Mar. Ecol., 6(2):40-51.

Sagitta maxima

Beaudouin, Jacqueline, 1967.
Oeufs et larves de poissons récoltés par la
"Thalassa" dans le Détroit de Danemark et le
Nord de la Mer d'Irminger (NORWESTLANT I-20 mars-
8 Mai 1963: relations avec l'hydrologie et le
zooplancton.
Revue Trav. Inst. (scient. tech.) Pêches marit.
31 (3):307-326.

Sagitta maxima

Bieri, Robert 1959

The distribution of planktonic chaetognatha
in the Pacific and their relationship to
water masses.
Limnol. and Ocean. Vol. 4 No. 1 pp. 1-28

Sagitta maxima

Bigelow, H. B., 1922
Exploration of the coastal water off the
northeastern United States in 1916 by the U.S.
Fisheries Schooner Grampus. Bull. M.C.Z. 65
(5):85-188, 53 text figs.

Sagitta maxima
Bigelow, H.B., and M. Sears, 1939
Studies of the waters of the continental shelf, Cape Cod to Chesapeake Bay. III. A volumetric study of the zooplankton. Mem. M.C.Z. 54(4):183-378, 42 text figs.

Sagitta maxima
Burfield, S.T., 1930.
Chaetognatha. Brit. Antarctic ("Terra Nova") Exped., 1910, Zool. 7(4):203-228, 3 maps.

Sagitta maxima
Coleman, J. S., 1959.
The "Rosaura" Expedition 1937-1938: Chaetognatha.
Bull. Brit. Mus. (Natural History) Zool. 5(8):219-253.

Sagitta maxima
David, Peter M., 1959.
Chaetognatha.
Brit.-Australia-New Zealand Antarctic Res. Exped. (B) 8:73-79.

Sagitta maxima
Dawson, John Kayl, 1968.
Chaetognaths from the Arctic Basin, including the description of a new species of Heterokrohnia.
Bull. Sth. Calif. Acad. Sci., 67(2):112-124

Sagitta maxima
Ducret, Françoise, 1968.
Chaetognathes des eaux superficielles et profondes de la zone équatoriale et tropicale africaine.
Thèse Fac. Sci., Marseille, 99 pp., 25 figs. In: Recueil Trav. publiées de 1965 à 1968, Océanogr. ORSTROM.

Sagitta maxima
Fish, C.J., and M.W. Johnson, 1937
The biology of the zooplankton population in the Bay of Fundy and the Gulf of Maine with special reference to production and distribution. J. Biol. Bd., Canada 3(3):189-322, 29 tables, 45 text figs.

Sagitta maxima
Fraser, J. H., 1949
The occurrence of unusual species of Chaetognatha in Scottish plankton collections. JMBA 28(2):489-491.

Sagitta maxima
Fraser, J.H., 1952.
The Chaetognatha and other zooplankton of the Scottish area and their value as biological indicators of hydrographical conditions.
Scottish Home Dept., Mar. Res., 1952:1-52, 3 pls., 21 charts.

Sagitta maxima
Fraser, J. H., 1949
Plankton of the Faroe-Shetland Channel and the Faroes, June and August 1947. Ann. Biol., Int. Cons., 4:27-28, text fig. 10.

Sagitta maxima
Fraser, J. H., 1939
The distribution of Chaetognatha in Scottish Waters in 1937. J. du Cons. 14(1): 25-34, 3 charts in text.

Sagitta maxima
Fraser, J. H., and A. Saville, 1949
Plankton distribution in Scottish and adjacent waters in 1948. Ann. Biol. 5:61-62.

Sagitta maxima
Hansen, Vagn Kr., 1960.
Investigations on the quantitative and qualitative distribution of zooplankton in the southern part of the Norwegian Sea.
Medd. Danmarks Fiskeri- og Havundersøgelser, n.s., 2(23):1-53.

Sagitta maxima
Kielhorn, W.V., 1952
The biology of the surface zone zooplankton of a Boreo-Arctic Atlantic Ocean area. J.Fish Res. Bd., Canada 9 (5): 223-264, 13 text figs.

Sagitta maxima
Kramp, P. L., 1939
Chaetognatha. The Godthaab Expedition 1928. Medd. om Grønland, 80(5):1-40, 7 text figs.

Sagitta maxima
Kramp, P.L., 1938.
Chaetognatha. Zool. Iceland 4(71):4 pp.

Sagitta maxima
Mackintosh, N.A., 1934
Distribution of the Macroplankton in the Atlantic Sector of the Antarctic. Discovery reports, Vol.9:65-160, 48 text figs.

Sagitta maxima
Rae, K. M., 1949
Plankton. Some broad changes in the plankton round the north of the British Isles in 1948. Ann. Biol. 5:56-60, 12 text figs.

Sagitta maxima
Redfield, A. C. and A. Beale, 1940.
Factors determining the distribution of populations of chaetognaths in the Gulf of Maine. Biol Bull. 79(3):459-487, 11 text figs.

Sagitta maxima
Russell, F.S., 1939.
Chaetognatha. Fiches d'Ident. Zooplancton, Cons. Perm. Int. Expl. Mer, 1:4 pp., 12 textfigs.

Sagitta prox. maxima
Srinivasan, M. 1971.
Two new records of meso- and bathy-planktonic chaetognaths from the Indian seas.
J. mar. biol. Ass. India 13(1):130-133.

Sagitta maxima
Sund, Paul N., 1964.
Los quetognatos en las aguas de la region del Peru.
Inter-Amer. Tropical Tuna Comm., Bull., 9(3):115-216.

Sagitta maxima
Sund, Paul N., 1961.
Some features of the autecology and distributions of Chaetognatha in the Eastern Tropical Pacific.
Inter-American Tropical Tuna Comm., Bull., 5(4): 307-340.

Sagitta maxima
Thiel, M.E., 1938
Die Chaetognathen - Bevölkerung des Südatlantischen Ozeans. Biologische Sonderuntersuchungen. 1st Lief. Wissenschaftliche Ergebnisse der Deutschen Atlantischen Expedition auf dem Forschungs. und Vermessungsschiff "Meteor" 1925-1927, 13:1-110, 62 text figs.

Sagitta maxima
Truveller, K.A., 1966.
Chaetognatha of the Davis and Denmark Straits as indicators of water masses. (In Russian). Mater. Ribokhoz. Issled. severn. Basseina, Poliarn. Nauchno-Issled. Proektn. Inst. Morsk. Ribn. Khoz. Okeanogr. (PINRO), 7: 114-124.

Sagitta maxima
von Ritter-Zahony, R., 1914.
Chaetognaths. Danish Ingolf-Exped. 4(3):4 pp.

Sagitta maxima
Ritter-Záhony, R. von, 1911
Revision der Chätognathen. Deutsche Südpolar Exped., 1901-1903, XIII (Zool.V)., Pt.1:71 pp., 51 text figs.

Sagitta maxima
von Ritter-Záhony, R., 1911
Vermes. Chaetognathi. Das Tierreich 29:34 pp., 16 text figs.

Sagitta megalophthalma n.sp.
Dallot S. et F. Ducret 1969.
Un Chaetognathe mésoplanctonique: Sagitta megalophthalma sp.n.
Beaufortia 17(224):13-20

Sagitta megalophthalma
Dallot, Serge, 1970.
L'anatomie du tube digestif dans la phylogénie et la systématique des Chaetognathes.
Bull. Mus. natn. Hist. nat. (2) 42(3):549-565.

Sagitta megalophthalma
Ören Enver, 1970.
Sagitta megalophthalma S. Dallot and F. Ducret 1969 in Turkish waters and in the Bay of Naples.
Hidrobiol, Araştırma Enst. Yayınlarından, Istanbul Univ. Fen Fakült, 6(3/4):27-32.

Sagitta minima
Aida, T., 1897.
Chaetognaths of Misaki Harbour.
Ann. Zool., Japon., 1:13-21, Pl. 3.

SAGITTA MINIMA
Alvariño, Angeles, 1969
Los quetognatos del Atlántico: distribución y notas esenciales de sistemática.
Trab. Inst. esp. Oceanogr., 37: 290 pp.

Sagitta minima
Alvariño, Angeles, 1967.
The Chaetognatha of the NAGA Expedition (1959-1961) in the South China Sea and the Gulf of Thailand.
NAGA Rept.,4(2):1-87.
Scripps Inst. Oceanogr.,

Sagitta minima
Alvariño, Angeles, 1966
Zoogeografía de California: quetognatos. Revista Soc. Mexicana Hist. nat. 27: 199-243

Sagitta minima
Alvarino, Angeles, 1965.
Distributional atlas of Chaetognatha in the California Current region.
California Coop. Oceanic Fish. Invest., Atlas, No. 3:291 charts

Sagitta minima
Alvariño, Angeles, 1964.
Zoogeografía de los quetognatos, especialmente de la region de California.
Ciencia, Mexico, 23:51-74.

Sagitta minima
Alvarino, Angeles, 1963
Quetognatos epiplanctonicos del Mar de Cortes. Revista. Soc. Mexicana, Hist. Nat., 24:97-202.

Sagitta minima
Alvariño, A., 1957.
Estudio del zooplankton del Mediterráneo occidental. Zooplancton del Atlántico Ibérico. Campañas del "Xauen" en el verano, del 1954.
Bol. Inst. Español Oceanogr., Nos. 81-82:1-26; 1-51.

Sagitta minima
Bieri, Robert 1959
The distribution of planktonic chaetognatha in the Pacific and their relationship to water masses.
Limnol. and Ocean. Vol. 4 No. 1 pp. 1-28

Sagitta minima
Bumpus, D.F., and E.L. Pierce, 1955.
Hydrography and the distribution of chaetognaths over the continental shelf of North Carolina.
Pap. Mar. Biol. and Oceanogr., Deep-Sea Res., Suppl. to Vol. 3:92-109.

Sagitta minima (fig.)
de Almeida Prado, M.S., 1961
Chaetognatha encontrados em aguas brasileiras.
Bol. Inst. Oceanogr., Univ. Sao Paulo, 11(2): 31-55.

Sagitta minima
de Saint-bon, Marie Catherine, 1963.
Les chaetognathes de la Côte d'Ivoire (espèces de surface).
Rev. Trav. Inst. Pêches Marit., 27(3):301-346.

Sagitta minima
Ducret, Françoise, 1968.
Chaetognathes des campagnes de l'Ombango dans les eaux équatoriales et tropicales africaines.
Cah. ORSTOM., Sér. Oceanogr., 6(1):95-141.

Sagitta minima
Ducret, Françoise, 1968.
Chaetognathes des campagnes de l'"Ombango" dans les eaux équatoriales et tropicales africaines.
Cah. O.R.S.T.O.M., ser. Oceanogr., VI(1):95-141.

Sagitta minima
Ducret, Françoise, 1968.
Chaetognathes des eaux superficielles et profondes de la zone équatoriale et tropicale africaine.
Thèse Fac. Sci., Marseille, 99 pp., 25 figs. In: Receuil Trav. publiées de 1965 à 1968, Océanogr. ORSTROM.

Sagitta minima
Fagetti Guaita, Elda, 1958.
Investigaciones sobre quetognatos colectados, especialmente, frente a la costa central y norte de Chile. Revista Biol. Mar., Valparaiso, 8(1-3):25-82.

Sagitta minima
Furnestin, Marie-Louise, 1970.
Chaetognathes des campagnes du Thor (1908-11) en Méditerranée et en Mer Noire.
Dana - Rept. 79:1-51

Sagitta minima
Furnestin, Marie-Louise, 1966.
Chaetognaths des eaux africaines.
Atlantide Rep., 9:105-135.

Sagitta minima
Furnestin, M. L., 1957.
Chaetognathes et zooplancton du secteur atlantique marocain. Rev. Trav. Inst. Pêches Marit. 21(1/2): 9-356.

Sagitta minima
Furnestin, M.-L., 1953.
Sur quelques chaetognathes d'Israel.
Bull. Res. Counc., Israel, 2(4):411-414.

Sagitta minima
Furnestin, M.-L., et J. Radiguet, 1964.
Chaetognathes de Madagascar (Secteur de Nosy-Bé).
Cahiers, O.R.S.T.R.O.M., Oceanogr., 11(4):55-98.

Sagitta minima
Germain, L. and L. Joubin, 1916
Chétognathes provenant des campagnes des yachts Hirondelle et Princesse-Alice (1885-1910)
Rés. Camp. Sci., Monaco, 49:118 pp., 7 pls., 7 maps.

Sagitta minima
Ghirardelli, E., 1960.
Habitat e biologia della riproduzione nei chetognati.
Arch. Oceanogr. e Limnol., 11(3):287-304.

Reprinted in:
Trav. Sta. Zool., Villefranche-sur-Mer, 18(11):

Sagitta minima
Ghirardelli, E., 1952.
Osservazioni biologiche e sistematiche sui Chetognati del Golfo di Napoli.
Pubbl. Staz. Zool., Napoli, 23(2/3):296-312.

Sagitta minima
Ghirardelli, E., 1950.
Osservazioni biologiche e sistematiche sui chetognati della baia di Villafranche sur mer. Bol. Pesca, Piscicol. e Idrobiol., n.s., 5(1):105-127, 7 textfigs.

Sagitta minima
Ghirardelli, E., 1950.
Osservazioni biologiche e sistematiche sui chetognati della Baia di Villefranche-sur-mer. (Trav. Sta. Zool. Villefranche-sur-mer 10(9)). Bol. Pesca, Piscicult. Idrobiol., n.sp, 5(1):5-27, 7 textfigs.

Sagitta minima
Ghirardelli, Elvezio, et Mario Specchi, 1965.
Chaetoghethes et cladoceres du Golfe de Trieste.
Rapp. Proc. Verb. Réunions, Comm. Int. Expl. Sci. Mer Méditerranée, Monaco, 18(2):403-407.

Sagitta minima
Grant, George C., 1963
Investigations of inner continental shelf waters off lower Chesapeake Bay. IV. Descriptions of the Chaetognatha and a key to their identification.
Chesapeake Science, 4(3):109-119.

Sagitta minima
Hameda, Takao, 1967.
Studies on the distribution of Chaetognatha in the Harima-nada and Osaka Bay, with special reference to Sagitta enflata. (In Japanese; English abstract).
Bull. Jap. Soc. scient. Fish., 33(2):98-103.

Sagitta minima
Hamon, M., 1950.
Deux nouveaux Chétognathes de la Baie d'Alger.
Bull. Soc. Hist. Nat. Afrique du Nord, 21(1/4): 10-14, 6 textfigs.

Sagitta minima
Hure, J., 1961
Dneva migracija i sezonaka vertikalna raspodjela zooplanktona dubljeg mora. [Migration journalière et distribution saisonnaire verticale du zooplancton dans la région profunde de l'Adriatique]
Acta Adriatica, 9(6):1-59.

Sagitta minima
Kado, Y., 1953.
The chaetognath fauna of the Inland Sea of Japan, especially on the distribution of Sagitta enflata and S. crassa. Dobuts. Zasshi 62(10):337-342.

Sagitta minima
Michael, E.L., 1919
Contributions to the Biology of the Philippine Archipelago and Adjacent regions. Report on the chaetognatha collected by the United States Fisheries Steamer "Albatross" during the Philippine Expedition, 1907-1910.
Bull. U.S. Nat. Mus. No.100, Vol.1(4):235-277, pls.34-38.

Sagitta minima
Marukawa, H., 1921
Plankton lists and some new species of copepods, from the northern waters of Japan.
Bull. Inst. Ocean., No.384, 15 pp., 3 pls., 1 chart. Monaco

Sagitta minima
Pierce, E.L., 1953.
The Chaetognatha over the continental shelf of North Carolina with attention to the relation to the hydrography of the area. J. Mar. Res. 12(1): 75-92, 4 textfigs.

Sagitta minima
Reyssac, J., 1963.
Chaetognathes du plateau continental européen (de la baie ibéro-marocaine à la Mer Celtique).
Rev. Trav. Inst. Pêches Marit., 27(3):245-299.

Sagitta minima
Ritter-Záhony, R. von, 1911
Revision der Chätognathen. Deutsche Südpolar Exped., 1901-1903, XIII (Zool.V)., Pt.1:71 pp., 51 text figs.

Sagitta minima
von Ritter-Záhony, R., 1911
Vermes. Chaetognathi. Das Tierreich 29:34 pp., 16 text figs.

Sagitta minima
Rose, M., and M. Hamon, 1953.
Nouvelle note complémentaire sur les chétognathes de la Baie d'Alger.
Bull. Soc. Hist. Nat., Afrique du Nord, 44(5/6): 167-171.

Sagitta minima
Sund, Paul N., 1964.
Los quetognatos en las aguas de la region del Peru.
Inter-Amer. Tropical Tuna Comm., Bull., 9(3):115-216.

Sagitta minima
Sund, Paul N., 1961.
Some features of the autecology and distributions of the Chaetognatha in the Eastern Tropical Pacific.
Inter-American Tropical Tuna Comm., Bull., 5(4): 307-340.

Sagitta minima
Sund, Paul N., 1959.
A key to the Chaetognatha of the tropical Eastern Pacific Ocean.
Pacific Science, 13(3):269-285.

Sagitta minima
Thiel, M.E., 1938
Die Chaetognathen - Bevölkerung des Südatlantischen Ozeans. Biologische Sonderuntersuchungen. 1st Lief. Wissenschaftliche Ergebnisse der Deutschen Atlantischen Expedition auf dem Forschungs, und Vermessungsschiff "Meteor" 1925-1927, 13:1-110, 62 text figs.

Sagitta minima
Thomson, J.M., 1948
Some chaetognatha from western Australia.
J. Roy. Soc. Western Australia, Inc., 31 (1944-1945), 17-18.

Sagitta minima
Thomson, J. M., 1947
The Chaetognatha of South-eastern Australia.
Counc. Sci. & Ind. Res., Australia, Bull. No.222, (Div. Fish. Rept. 14), 43 pp., 8 text figs.

Sagitta minima
Tokioka, Takasi, 1959
Observations on the taxonomy and distribution of chaetognaths of the North Pacific.
Publ. Seto Mar. Biol. Lab., 7(3):349-456.

Sagitta minima
Tokioka, T., 1951.
Pelagic tunicates and chaetognaths collected during the cruises to the New Yamato Bank in the Sea of Japan. Publ. Seto Mar. Biol. Lab. 2(1): 1-26, 1 chart, 6 tables, 12 textfigs.

Sagitta minima
Tokioka, T., 1940.
The chaetognath fauna of the waters of western Japan. Rec. Oceanogr. Wks., Japan, 12(1):1-22, 3 charts, 4 tables. (Contrib. 87, Seto. M.B.L.)

Sagitta minima
Tokioka, T., 1939.
Chaetognaths collected chiefly from the bays of Sagami and Suruga with some notes on the shape and structure of the seminal vesicles.
Rec. Oceanogr. Wks., Japan, 10(2):123-150.

Sagitta minima

Tsuruta, Arao, 1963
Distribution of plankton and its characteristics in the oceanic fishing grounds, with special reference to their relation to fishery.
J. Shimonoseki Univ., Fish., 12(1):13-214.

Sagitta minima

Vucetic, T., 1961
Sur la répartition des chaetognathes en Adriatique et leur utilisation comme indicateurs biologiques des conditions hydrographiques.
Rapp. Proc. Verb., Réunions, Comm. Int. Expl. Sci. Mer Mediterranée, Monaco, 16(2):111-116.

Sagitta nagae n.sp.

Alvariño, Angeles, 1967.
The Chaetognatha of the NAGA Expedition (1959-1961) in the South China Sea and the Gulf of Thailand.
NAGA Rept., 4(2):1-87.
Scripps Inst. Oceanogr.

Sagitta nagae

Nagasawa, Sachiko and Ryuzo Marumo 1972.
Feeding of a pelagic chaetognath, Sagitta nagae Alvariño in Suruga Bay, central Japan.
J. oceanogr. Soc. Japan, 28(5): 181-186

Sagitta neglecta n.sp.

Aida, T., 1897.
Chaetognaths of Misaki Harbour.
Ann. Zool., Japon., 1:13-21, Pl. 3.

Sagitta neglecta

Alvariño, Angeles, 1966
Zoogeografía de California: quetognatos, Revista
Soc. Mexicana Hist. nat. 27: 199-243

Sagitta neglecta

Alvarino, Angeles, 1965.
Distributional atlas of Chaetognatha in the California Current region.
California Coop. Oceanic Fish. Invest., Atlas, No. 3:291 charts

Sagitta neglecta

Alvariño, Angeles, 1964.
Zoogeografía de los quetognatos, especialmente de la región de California.
Ciencia, México, 23:51-74.

Sagitta neglecta

Alvarino, Angeles, 1963
Quetognatos epiplanctonicos del Mar de Cortes.
Revista, Soc. Mexicana, Hist. Nat., 24:97-202.

Sagitta neglecta

Bieri, Robert 1959
The distribution of planktonic chaetognatha in the Pacific and their relationship to water masses.
Limnol. and Ocean. Vol. 4 No. 1 pp. 1-28

Sagitta neglecta

Burfield, S.T., 1927.
Zoological results of the Cambridge Expedition to the Suez Canal. XXI. Report on the Chaetognatha.
Trans. Zool. Soc., London, 22:355-356.

Sagitta neglecta

Burfield, S.T., and E.J.W., 1926.
The Chaetognatha of the "S.alark" Expedition.
Trans. Linn. Soc., London, Ser. 2, 19:93-119.

Sagitta neglecta

Coleman, J. S., 1959.
The "Rosaura" Expedition 1937-1938: Chaetognatha.
Bull. Brit. Mus. (Natural History) Zool. 5(8):219-253.

Sagitta neglecta

Delsman, H. C., 1939.
Preliminary plankton investigations in the Java Sea. Treubia, 17:139-181, 8 maps, 41 figs.

Sagitta neglecta

Fowler, G.H., 1906.
The Chaetognatha of the Siboga Expedition with a discussion of the synonymy and distribution of the group. Rept. Siboga Exped. 21:1-88, 3 pls., 6 charts.

Sagitta neglecta

Furnestin, M.-L. et J.-C. Cédaccioni, 1968.
Chaetognathes du nord-ouest de l'Océan Indien (golfe d'Aden, mer d'Arabie, golfe d'Oman, golfe Persique).
Cah. O.R.S.T.O.M., ser. Océanogr., VI(1):143-171.

Sagitta neglecta

George, P.C., 1952.
A systematic account of the Chaetognatha of the Indian coastal waters, with observations of their seasonal fluctuations along the Malabar coast.
Proc. Nat. Inst. Sci., India, 18(6):657-689.

Sagitta neglecta

Germain, L. and L. Joubin, 1916
Chétognathes provenant des campagnes des yachts Hirondelle et Princesse-Alice (1885-1910)
Res. Camp. Sci., Monaco, 49:118 pp., 7 pls., 7 maps.

Sagitta neglecta

Guerguess, Shoukry K. et Youssef Halim 1973.
Chaetognathes du plancton d'Alexandrie II. Un spécimen mûr de Sagitta neglecta Aida en Méditerranée. Rapp. Proc.-v. Réun. Commn int. Explor. scient. Mer Medit. Monaco 21(8):497-498.

Sagitta neglecta

Laguarda Figueras, Alfredo 1965.
Contribución al conocimiento de los quetognatos de Sinaloa.
Anales, Inst. Biol. Univ. Mex. 36(1/2):215-228.

Sagitta neglecta

Michael, E.L., 1919
Contributions to the Biology of the Philippine Archipelago and Adjacent regions. Report on the chaetognatha collected by the United States Fisheries Steamer "Albatross" during the Philippine Expedition, 1907-1910.
Bull. U.S. Nat. Mus. No.100, Vol.1(4):235-277, pls.34-38.

Sagitta neglecta

Michael, E. L., 1911
Classification and vertical distribution of the Chaetognatha of the San Diego Region including redescriptions of some doubtful species of the group. Univ. Calif. Publ. Zool. 8(3):21-186, Pls. 1-8.

Sagitta neglecta

Rajagopal, P.K., 1962.
Respiration of some marine planktonic organisms.
Proc. Indian Acad. Sci., (B) 55(2):76-81.

Sagitta neglecta

Ramult, M., and M. Rose, 1945
Recherches sur les Chétognathes de la Baie d'Alger. Bull. Soc. Hist. Nat. Afrique du Nord 36:45-71, 39 figs.

Sagitta neglecta

Rao, T.S. Satyanarayana, and P.N. Ganapati, 1958
Studies on Chaetognatha in the Indian Seas. III. Systematics and distribution in the waters off Visakhapaynam. Andhra Univ. Mem. Oceanogr 2:147-163.

Sagitta neglecta

Ritter-Záhony, R. von, 1911
Revision der Chätognathen. Deutsche Südpolar Exped., 1901-1903, XIII (Zool.V), Pt.1:71 pp., 51 text figs.

Sagitta neglecta

von Ritter-Záhony, R., 1911
Vermes. Chaetognathi. Das Tierreich 29:34 pp., 16 text figs.

Sagitta neglecta

Sudarsan, D., 1961.
Observations on the Chaetognatha of the waters around Mandapam.
Indian J. Fish., 8(2):364-382.

Sagitta neglecta

Sund, Paul N., 1964.
Los quetognatos en las aguas de la region del Peru.
Inter-Amer. Tropical Tuna Comm., Bull., 9(3):115-216.

Sagitta neglecta

Sund, Paul N., 1961.
Some features of the autecology and distributions of Chaetognatha in the Eastern Tropical Pacific.
Inter-American Tropical Tuna Comm., Bull., 5(4): 307-340.

Sagitta neglecta

Sund, Paul N., 1961.
Two new species of Chaetognatha from the waters off Peru.
Pacific Science, 15(1):105-111.

Sagitta neglecta

Sund, Paul N., 1959.
A key to the Chaetognatha of the tropical Eastern Pacific Ocean.
Pacific Science, 13(3):269-285.

Sagitta neglecta

Thomson, J. M., 1947
The Chaetognatha of South-eastern Australia.
Counc. Sci. & Ind. Res., Australia, Bull. No.222, (Div. Fish. Rept. 14), 43 pp., 8 text figs.

Sagitta neglecta

Tokioka, Takasi, 1959
Observations on the taxonomy and distribution of chaetognaths of the North Pacific.
Publ. Seto Mar. Biol. Lab., 7(3):349-456.

Sagitta neglecta

Tokioka, T., 1951.
Pelagic tunicates and chaetognaths collected during the cruises to the New Yamato Bank in the Sea of Japan. Publ. Seto Mar. Biol. Lab. 2(1): 1-26, 1 chart, 5 tables, 12 textfigs.

Sagitta neglecta

Tokioka, T., 1942
Systematic studies of the plankton organisms occurring in Iwayama Bay, Palao. III Chaetognaths from the bay and adjacent waters. Palao Tropical Biol. Sta. Studies, II(3):527-548, pls.5-7, 11 text figs.

Sagitta neglecta

Tokioka, T., 1940.
The chaetognath fauna of the waters of western Japan. Rec. Oceanogr. Wks., Japan, 12(1):1-22, 3 charts, 4 tables. (Contrib. 87, Seto M.B.L.)

Sagitta neglecta

Tokioka, T., 1939.
Chaetognaths collected chiefly in the bays of Sagami and Suruga with some notes on the shape and structure of the seminal vesicles.
Rec. Oceanogr. Wks., Japan, 10(2):123-150.

Sagitta neglecta

Tsuruta, Arao, 1963
Distribution of plankton and its characteristics in the oceanic fishing grounds, with special reference to their relation to fishery.
J. Shimonoseki Univ., Fish., 12(1):13-214.

Sagitta neglecta

Varadarajan, S. and P.I. Chacko, 1943.
On the arrow-worms of Krusadai. Proc.
Nat. Inst. Sci., India, 9(2):245-248, 2 figs.

Sagitta neodecipiens

Dallot, Serge, 1970.
L'anatomie du tube digestif dans la
phylogénie et la systématique des
Chaetognathes.
Bull. Mus. natn. Hist. nat. (2) 42(3): 549-565.

Sagitta neodecipiens

Dallot, Serge et Françoise Ducret, 1968.
A propos de Sagitta bipiens Fowler
et de Sagitta neodecipiens Tokioka
Rapp. Proc.-verb. Réun., Comm. int. Explor.
scient. Mer Méditerranée 19(3): 433-435.

Sagitta neodecipiens

Ducret, Françoise, 1968.
Chaetognathes des campagnes de l'Ombango dans
les eaux équatoriales et tropicales africaines.
Cah. ORSTOM., Sér. Océanogr., 6(1):95-141.

Sagitta neodecipiens

Ducret, Françoise, 1968.
Chaetognathes des eaux superficielles et
profondes de la zone équatoriale et tropicale
africaine.
Thèse Fac. Sci., Marseille, 99 pp., 25
figs. In: Receuil Trav. publiées de
1965 à 1968, Océanogr. ORSTROM.

Sagitta neodecipiens

Ducret, Françoise, 1968.
Chaetognathes des campagnes de l'"Ombango" dans
les eaux équatoriales et tropicales africaines.
Cah. O.R.S.T.O.M., ser. Océanogr., VI(1):95-141.

Sagitta neodecipiens

Furnestin, Marie-Louise, 1970.
Chaetognathes des campagnes du
Thor (1908-11) en Méditerranée
et en Mer Noire.
Dana - Rept. 79:1-51

Sagitta neodecipiens

Furnestin, Marie-Louise, 1966.
Chaetognathes des eaux africaines.
Atlantide Rep., 9:105-135.

Sagitta neodecipiens

Furnestin, M.-L. J.-C. Codaccioni, 1968.
Chaetognathes du nord-ouest de l'Océan Indien
(golfe d'Aden, mer d'Arabie, golfe d'Oman, golfe
Persique).
Cah. O.R.S.T.O.M., ser. Océanogr., VI(1):143-171.

Sagitta neodecipiens

Kitou, Masataka, 1966.
Chaetognaths collected on the Sixth Cruise
of the Japanese Expedition of Deep Seas.
La Mer, Bull. franco-japon. Oceanogr., 4(4):261-265.

Also in:
Reps JEDS, 5-7 (1966)

Sagitta neodecipiens n.sp.

Tokioka, Takasi, 1959
Observations on the taxonomy and dis-
tribution of chaetognaths of the North
Pacific.
Publ. Seto Mar. Biol. Lab., 7(3):349-456.

Sagitta oceania

Alvariño, Angeles, 1967.
The Chaetognatha of the NAGA Expedition (1959-
1961) in the South China Sea and the Gulf of
Thailand.
NAGA Rept., 4(2):1-87.
Scripps Inst. Oceanogr.,

Sagitta oceanica

Tokioka, Takasi, 1959
Observations on the taxonomy and dis-
tribution of chaetognaths of the North
Pacific.
Publ. Seto Mar. Biol. Lab., 7(3):349-456.

Sagitta orientalis

Ritter-Záhony, R. von, 1911
Revision der Chätognathen. Deutsche
Südpolar Exped., 1901-1903, XIII (Zool.V),
Pt.1:71 pp., 51 text figs.

Sagitta orientalis

von Ritter-Záhony, R., 1911
Vermes. Chaetognathi. Das Tierreich
29:34 pp., 16 text figs.

SAGITTA PACIFICA

Alvariño, Angeles, 1969
Los quetognatos del Atlántico: distribución
y notas esenciales de sistemática.
Trab. Inst. esp. Oceanogr., 37: 290 pp.

Sagitta pacifica

Alvariño, Angeles, 1967.
The Chaetognatha of the NAGA Expedition (1959-
1961) in the South China Sea and the Gulf of
Thailand.
NAGA Rept., 4(2):1-87.
Scripps Inst. Oceanogr.,

Sagitta pacifica

Alvarino, Angeles, 1965.
Distributional atlas of Chaetognatha in the
California Current region.
California Coop. Oceanic Fish. Invest., Atlas,
No. 3:291 charts

Sagitta pacifica

Alvariño, Angeles, 1964.
The Chaetognatha of the MONSOON Expedition in
the Indian Ocean.
Pacific Science, 18(3):336-348.

Sagitta pacifica

Alvariño, Angeles, 1964.
Zoogeografía de los quetognatos, especialmente
de la región de California.
Ciencia, México, 23:51-74.

Sagitta pacifica

Alvarino, Angeles, 1963
Quetognatos epiplanctonicos del Mar de Cortes.
Revista, Soc. Mexicana, Hist. Nat., 24:97-202.

Sagitta pacifica

Alvarino, Angeles, 1961.
Two new chaetognaths from the Pacific.
Pacific Science, 15(1):67-77.

Sagitta pacifica (Tokioka) Serrato dentata group

Bieri, Robert 1959
The distribution of planktonic chaetognatha
in the Pacific and their relationship to
water masses.
Limnol. and Ocean. Vol. 4 No. 1 pp. 1-28

Sagitta pacifica

Bieri, Robert, 1957.
The chaetognath fauna of Peru in 1941.
Pacific Science 11(3):255-264.

Sagitta pacifica

Fagetti Guaita, Elda, 1958.
Investigaciones sobre quetognatos colectados,
especialmente, frente a la costa central y norte de
Chile. Revista Biol. Mar., Valparaiso, 8(1-3):25-82.

Sagitta pacifica

Furnestin, M.L., 1958.
Quelques échantillons de zooplancton du Golfe
d'Eylath (Akaba).
Haifa, Sea Fish. Res. Sta., Bull., 16:6-14.

Sagitta pacifica

Furnestin, M.-L. et J.-C. Codaccioni, 1968.
Chaetognathes du nord-ouest de l'Océan Indien
(golfe d'Aden, mer d'Arabie, golfe d'oman, golfe
Persique).
Cah. O.R.S.T.O.M., ser. Océanogr., VI(1):143-171.

Sagitta pacifica

Kuroda Kazunori 1973.
Macroplankton biomass and chaeto-
gnath fauna in the Kuroshio south of
Japan in the summer of 1971 and the
winter of 1972. (In Japanese; English abstract)
Bull. Kobe mar. Obs. 159:57-64.

Sagitta pacifica

Sund, Paul N., 1964.
Los quetognatos en las aguas de la region del
Peru.
Inter-Amer. Tropical Tuna Comm., Bull., 9(3):115-216.

Sagitta pacifica

Sund, Paul N., 1961.
Some features of the autecology and distribu-
tions of Chaetognatha in the Eastern Tropical
Pacific.
Inter-American Tropical Tuna Comm., Bull., 5(4):307-340.

Sagitta peruviana

Alvariño, Angeles, 1967.
The Chaetognatha of the NAGA Expedition (1959-
1961) in the South China Sea and the Gulf of
Thailand.
NAGA Rept., 4(2):1-87.
Scripps Inst. Oceanogr.,

Sagitta peruviana

Sund, Paul N., 1964.
Los quetognatos en las aguas de la region del
Peru.
Inter-Amer. Tropical Tuna Comm., Bull., 9(3):115-216.

Sagitta peruviana n.sp.

Sund, Paul N., 1961.
Two new species of Chaetognatha from the waters
off Peru.
Pacific Science, 15(1):105-111.

Sagitta philippine n.sp.

Michael, E.L., 1919
Contributions to the Biology of the
Philippine Archipelago and Adjacent regions.
Report on the chaetognatha collected by the
United States Fisheries Steamer "Albatross"
during the Philippine Expedition, 1907-1910.
Bull. U.S. Nat. Mus. No.100, Vol.1(4):235-277,
pls.34-38.

SAGITTA PLANCTONIS

Alvariño, Angeles, 1969
Los quetognatos del Atlántico: distribución
y notas esenciales de sistemática.
Trab. Inst. esp. Oceanogr., 37: 290 pp.

Sagitta planctonis

Alvariño, Angeles, 1967.
The Chaetognatha of the NAGA Expedition (1959-
1961) in the South China Sea and the Gulf of
Thailand.
NAGA Rept., 4(2):1-87.
Scripps Inst. Oceanogr.,

Sagitta planctonis

Alvariño, Angeles, 1964.
Zoogeografía de los quetognatos, especialmente
de la región de California.
Ciencia, México, 23:15-74.

Sagitta planctonis

Bieri, Robert 1959
The distribution of planktonic chaetognatha
in the Pacific and their relationship to
water masses.
Limnol. and Ocean. Vol. 4 No. 1 pp. 1-28

Sagitta planctonis
Burfield, S.T., 1930.
Chaetognatha. Brit. Antarctic ("Terra Nova") Exped., 1910, Zool., 7(4):203-228, 3 maps.

Sagitta planctonis
Burfield, S.T., and E.J.W. Harvey, 1926.
The Chaetognatha of the "Sealark" Expedition. Trans. Linn. Soc., London, Ser. 2, 19:93-119, Pls. 4-7.

Sagitta planctonis
Coleman, J. S., 1959.
The "Rosaura" Expedition 1937-1938: Chaetognatha.
Bull. Brit. Mus. (Natural History) Zool. 5(8):219-253.

Sagitta planctonis
Dallot, Serge, 1970.
L'anatomie du tube digestif dans la phylogénie et la systématique des Chaetognathes.
Bull. Mus. natn. Hist. nat. (2) 42(3): 549-565.

Sagitta planctonis
David, P.M., 1956.
Sagitta planctonis and related forms.
Bull. Brit. Mus. (N.H.), Zool., 4(8):435-451, Pl. 11, 7 textfigs.

Sagitta planctonis
Delsman, H.C., 1939.
Preliminary plankton investigations in the Java Sea. Treubia 17:139-181, 8 maps, 41 figs.

Sagitta planctonis
de Saint-Bon, Marie-Catherine, 1963.
Complement à l'étude des chaetognathes de la Côte d'Ivoire (especes profondes).
Rev. Trav. Inst. Pêches Marit., 27(4):403-415.

Sagitta planctonis
Ducret, Françoise, 1968.
Chaetognathes des eaux superficielles et profondes de la zone équatoriale et tropicale africaine.
Thèse Fac. Sci. Marseille, 99 pp., 25 figs. In: Receuil Trav. publiées de 1965 à 1968, Océanogr. ORSTROM.

Sagitta planctonis
Ducret, Françoise, 1968.
Chaetognathes des campagnes de l'Ombango dans les eaux équatoriales et tropicales africaines.
Cah. ORSTOM, Sér. Océanogr., 6(1):95-141.

Sagitta planctonis
Ducret, Françoise, 1968.
Chaetognathes des campagnes de l'"Ombango" dans les eaux équatoriales et tropicales africaines. Cah. O.R.S.T.O.M., ser. Océanogr., VI(1):95-141.

Sagitta planctonis
Fagetti Guaita, Elda, 1958.
Investigaciones sobre quetognatos colectados, especialmente, frente a la costa central y norte de Chile. Revista Biol. Mar., Valparaiso, 8(1-3):25-82.

?Sagitta planctonis
Fowler, G.H., 1906.
The Chaetognatha of the Siboga Expedition with a discussion of the synonymy and distribution of the group. Rept. Siboga Exped. 21:1-88, 3 pls., 6 charts.

Sagitta ? planctonis
Fowler, G.H., 1905.
Biscayan plankton collected during a cruise of H.M.S. "Research", 1900. Pt. III. The Chaetognatha. Trans. Linn. Soc., London, 10(3):55-87, Pls. 4-7.

Sagitta planctonis
Fraser, J.H., 1952.
The Chaetognatha and other zooplankton of the Scottish area and their value as biological indicators of hydrographical conditions. Scottish Home Dept., Mar. Res., 1952(2):1-52, 3 pls., 21 charts.

good for distinguishing characters

Sagitta planctonis
Fraser, J. H., 1949
The occurrence of unusual species of Chaetognatha in Scottish plankton collections. JMBA 28(2):489-491.

Sagitta planctonis
Fraser, J. H. and A. Saville, 1949
List of rare exotic species found in the plankton by the Scottish Vessel "Explorer" in 1948. Ann. Biol. 5:62-64.

Sagitta planctonis
Fraser, J. H. and A. Saville, 1949
Plankton distribution in Scottish and adjacent waters in 1948. Ann. Biol. 5:61-62.

Sagitta planctonis
Furnestin, Marie-Louise, 1970.
Chaetognathes des campagnes du Thor (1908-11) en Méditerranée et en Mer Noire.
Dana - Rept. 79:1-51

Sagitta planctonis
Furnestin, Marie-Louise, 1966.
Chaetognathes des eaux africaines.
Atlantide Rep., 9:105-135.

Sagitta planctonis
Furnestin, Marie-Louise, 1963.
Les chaetognathes atlantiques en Méditerranée.
Rev. Trav. Inst. Pêches Marit., 27(2):157-160.

Sagitta planctonis
Furnestin, M.-L., et J. Radiguet, 1964.
Chaetognathes de Madagascar (Secteur de Nosy-Bé).
Cahiers, O.R.S.T.R.O.M., Océanogr., 11(4):55-98.

Sagitta planctonis
George, P.C., 1952.
A systematic account of the Chaetognatha of the Indian coastal waters, with observations on their seasonal fluctuations along the Malabar coast. Proc. Nat. Inst. Sci., India, 18(6):657-689.

Sagitta planctonis
Germain, L. and L. Joubin, 1916
Chétognathes provenant des campagnes des yachts Hirondelle et Princesse-Alice (1885-1910) Rés. Camp. Sci., Monaco, 49:118 pp., 7 pls., 7 maps.

Sagitta planctonis
Kramp, P. L., 1939
Chaetognatha. The Godthaab Expedition 1928. Medd. om Grønland, 80(5):1-40, 7 text figs.

Sagitta planctonis
Mackintosh, N.A., 1934
Distribution of the Macroplankton in the Atlantic Sector of the Antarctic. Discovery reports, Vol.9:65-160, 48 text figs.

Sagitta planctonis
Michael, E.L., 1919
Contributions to the Biology of the Philippine Archipelago and Adjacent regions. Report on the chaetognatha collected by the United States Fisheries Steamer "Albatross" during the Philippine Expedition, 1907-1910. Bull. U.S. Nat. Mus. No.100, Vol.1(4):235-277, pls.34-38.

Sagitta planctonis
Michael, E. L., 1911
Classification and vertical distribution of the Chaetognatha of the San Diego Region including redescriptions of some doubtful species of the group. Univ. Calif. Publ. Zool. 8(3):21-186, Pls. 1-8.

Sagitta planctonis
Moore, H. B., 1949
The zooplankton of the upper waters of the Bermuda area of the North Atlantic. Bull. Bingham Ocean. Coll. 12(2):97, 208 text figs.

Sagitta planctonis
Pierrot-Bults, A.C. 1970.
Variability in Sagitta planctonis Steinhaus 1896 (Chaetognatha) from West-African waters in comparison to North Atlantic specimens.
Atlantide Rept. 11: 141-149.

Sagitta planctonis
Pierrot-Bults, A.C., 1969.
The synonymy of Sagitta planctonis and Sagitta zetesios (Chaetognatha).
Bull. Zool. Mus. Univ. Amsterdam 1(10):125-129.

Sagitta planctonis
Russell, F.S., 1939.
Chaetognatha. Fiches d'Ident. Zooplancton, Cons. Perm. Int. Expl. Mer, 1:4 pp., 12 textfigs.

Sagitta planctonis
Sund, Paul N., 1961.
Some features of the autecology and distributions of Chaetognatha in the eastern Tropical Pacific.
Inter-American Tropical Tuna Comm., Bull., 5(4):307-340.

Sagitta planctonis
Tchindonova, J.G., 1955.
[Chaetognatha of the Kurile-Kamchatka Trench.]
Trudy Inst. Oceanol., 12:298-310.

Sagitta planctonis
Thiel, M.E., 1938
Die Chaetognathen - Bevölkerung des Südatlantischen Ozeans. Biologische Sonderuntersuchungen. 1st Lief. Wissenschaftliche Ergebnisse der Deutschen Atlantischen Expedition auf dem Forschungs. und Vermessungsschiff "Meteor" 1925-1927, 13:1-110, 62 text figs.

Sagitta planctonis
Thomson, J. M., 1947
The Chaetognatha of South-eastern Australia. Counc. Sci. & Ind. Res., Australia, Bull. No.222, (Div. Fish. Rept. 14), 43 pp., 8 text figs.

Sagitta planctonis
Tokioka, Takasi, 1959
Observations on the taxonomy and distribution of chaetognaths of the North Pacific.
Publ. Seto Mar. Biol. Lab., 7(3):349-456.

Sagitta planctonis
Tokioka, T., 1939.
Chaetognaths collected chiefly in the bays of Sagami and Suruga with some notes on the shape and structure of the seminal vesicles. Rec. Oceanogr. Wks., Japan, 10(2):123-150.

Sagitta planctonis
von Ritter-Záhony, R., 1914.
Chaetognaths. Danish Ingolf-Exped. 4(3):4 pp.

Sagitta planctonis
Ritter-Záhony, R. von, 1911
Revision der Chätognathen. Deutsche Südpolar Exped., 1901-1903, XIII (Zool.V), Pt.1:71 pp., 51 text figs.

Sagitta planctonis
von Ritter-Záhony, R., 1911
Vermes, Chaetognathi. Das Tierreich 29:34 pp., 16 text figs.

Sagitta popovicii
Alvariño, Angeles, 1967.
The Chaetognatha of the NAGA Expedition (1959-1961) in the South China Sea and the Gulf of Thailand.
NAGA Rept., 4(2):1-87.
Scripps Inst. Oceanogr.,

Sagitta popocicii n.sp.
Sund, Paul N., 1961.
Two new species of Chaetognatha from the waters off Peru.
Pacific Science, 15(1):105-111.

Sagitta profunda
Ritter-Záhony, R. von, 1911
Revision der Chätognathen. Deutsche Südpolar Exped., 1901-1903, XIII (Zool.V)., Pt.1:71 pp., 51 text figs.

Sagitta pseudoserratodentata
Alvarino, Angeles, 1965.
Distributional atlas of Chaetognatha in the California Current region.
California Coop. Oceanic Fish. Invest., Atlas, No. 3:291 charts

Sagitta pseudoserratodentata
Alvarino, Angeles, 1964.
Zoogeografia de los quetognatos, especialmente de la región de California.
Ciencia, México, 23:51-74.

Sagitta pseudoserratodentata
Alvarino, Angeles, 1963
Quetognatos epiplanctonicos del Mar de Cortes.
Revista, Soc. Mexicana, Hist. Nat., 24:97-202.

Sagitta pseudoserratodentata
Alvarino, Angeles,
Two new chaetognaths from the Pacific.
Pacific Science, 15(1):67-77.

Sagitta pseudo serrato dentata
Bieri, Robert 1959
The distribution of planktonic chaetognatha in the Pacific and their relationship to water masses.
Limnol. and Ocean. Vol. 4 No. 1 pp. 1-28

Sagitta pseudoserratodenta
Ducret, Françoise, 1968.
Chaetognathes des eaux superficielles et profondes de la zone équatoriale et tropicale africaine.
Thèse Fac. Sci., Marseille, 99 pp., 25 figs. In: Receuil Trav. publiées de 1965 à 1968, Océanogr. ORSTROM.

Sagitta pseudoserratodentata
Furnestin, M.-L., 1963.
Les Chaetognathes du groupe serratodentata en Mediterranee.
Rapp. Proc. Verb., Réunions, Comm. Int. Expl.Sci. Mer Méditerranée, Monaco, 17(2):631-634.

Sagitta pseudoserratodentata
Sund, Paul N., 1959.
A key to the Chaetognatha of the tropical Eastern Pacific Ocean.
Pacific Science, 13(3):269-285.

Sagitta pseudoserratodentata n.sp.
Tokioka, T., 1939.
Chaetognaths collected chiefly from the bays of Sagami and Suruga with some notes on the shape and structure of the seminal vesicles.
Rec. Oceanogr. Wks., Japan, 10(2):123-150.

SAGITTA PULCHRA
Alvariño, Angeles, 1969
Los quetognatos del Atlántico: distribución y notas esenciales de sistemática.
Trab. Inst. esp. Oceanogr., 37: 290 pp.

Sagitta pulchra
Alvariño, Angeles, 1967.
The Chaetognatha of the NAGA Expedition (1959-1961) in the South China Sea and the Gulf of Thailand.
NAGA Rept., 4(2):1-87.
Scripps Inst. Oceanogr.,

Sagitta pulchra
Alvariño, Angeles, 1966
Zoogeografía de California: quetognatos. Revista
Soc. Mexicana Hist. nat. 27: 199-243

Sagitta pulchra
Alvarino, Angeles, 1965.
Distributional atlas of Chaetognatha in the California Current region.
California Coop. Oceanic Fish. Invest., Atlas, No. 3:291 charts

Sagitta pulchra
Alvariño, Angeles, 1964.
Zoogeografía de los quetognatos, especialmente de la región de California.
Ciencia, México, 23:51-74.

Sagitta pulchra
Bieri, Robert 1959
The distribution of planktonic chaetognatha in the Pacific and their relationship to water masses.
Limnol. and Ocean. Vol. 4 No. 1 pp. 1-28

Sagitta pulchra
Burfield, S.T., and E.J.W. Harvey, 1926.
The Chaetognatha of the "Sealark" Expedition.
Trans. Linn. Soc., London, Ser. 2, 19:93-119, Pls. 4-7.

Sagitta pulchra
Coleman, J. S., 1959.
The "Rosaura" Expedition 1937-1938: Chaetegnatha.
Bull. Brit. Mus. (Natural History) Zool. 5(8):219-253.

Sagitta pulchra
Delsman, H.C., 1936.
Preliminary plankton investigations in the Java Sea. Treubia 17:139-181, 8 maps, 41 figs.

Sagitta pulchra
Fowler, G.H., 1906.
The Chaetognatha of the Siboga Expedition with a discussion of the synonymy and distribution of the group. Rept. Siboga Exped. 21:1-88, 3 pls., 6 charts.

Sagitta pulchra
Furnestin, M.-L., et J.-C.Codaccioni,1968.
Chaetognathes du nord-ouest de l'Océan Indien (golfe d'Aden, mer d'Arabie, golfe d'Oman, golfe Persique).
Cah.ORSTOM. Sér.Océanogr.,6(1):143-171.

Sagitta pulchra
Furnestin, M.-L., et J. Radiguet, 1964.
Chaetognathes de Madagascar (Secteur de Nosy-Bé).
Cahiers, O.R.S.T.R.O.M., Océanogr., 11(4):55-98.

Sagitta pulchra
Ghirardelli, Elvezio, 1947
Chetognati raccolti nel Mar Rosso e nel l'Oceano Indiano dalla Nave "Cherso", Bolletino di Pesca, Piscicoltura, e Idrobiologia Anno 23, Vol. 2(n.s.) (2):253-270, 2 pls., 9 text figs.

Sagitta pulchra
George, P.C., 1952.
A systematic account of the Chaetognatha of the Indian coastal waters, with observations of their seasonal fluctuations along the Malabar coast.
Proc. Nat. Inst. Sci., India, 18(6):657-689.

Sagitta pulchra
Michael, E.L., 1919
Contributions to the Biology of the Philippine Archipelago and Adjacent regions. Report on the chaetognatha collected by the United States Fisheries Steamer "Albatross" during the Philippine Expedition, 1907-1910.
Bull. U.S. Nat. Mus. No.100, Vol.1(4):235-277, pls.34-38.

Sagitta pulchra
Michael, E. L., 1911
Classification and vertical distribution of the Chaetognatha of the San Diego Region including redescriptions of some doubtful species of the group. Univ. Calif. Publ. Zool. 8(3):21-186, Pls. 1-8.

Sagitta pulchra
Rao, T.S. Satyanarayana, and P.N. Ganapati, 1958
Studies on Chaetognatha in the Indian Seas.III. Systematics and distribution in the waters off Visakhapaynam. Andhra Univ. Mem. Oceanogr.,2: 147-163.

Sagitta pulchra
Ritter-Záhony, R. von, 1911
Revision der Chätognathen. Deutsche Südpolar Exped., 1901-1903, XIII (Zool.V)., Pt.1:71 pp., 51 text figs.

Sagitta pulchra
von Ritter-Záhony, R., 1911
Vermes. Chaetognathi. Das Tierreich 29:34 pp., 16 text figs.

Sagitta pulchra
Sund, Paul N., 1964.
Los quetognatos en las aguas de la region del Peru.
Inter-Amer. Tropical Tuna Comm., Bull., 9(3):115-216.

Sagitta pulchra
Sund, Paul N., 1961.
Some features of the autecology and distribution of Chaetognatha in the Eastern Tropical Pacific.
Inter-American Tropical Tuna Comm., Bull., 5(4): 307-340.

Sagitta pulchra
Sund, Paul N., 1959.
A key to the Chaetognatha of the tropical Eastern Pacific Ocean.
Pacific Science, 13(3):269-285.

Sagitta pulchra
Thomson, J. M., 1947
The Chaetognatha of South-eastern Australia.
Counc. Sci. & Ind. Res., Australia, Bull. No.222, (Div. Fish. Rept. 14), 43 pp., 8 text figs.

Sagitta pulchra
Tokioka, Takasi, 1959
Observations on the taxonomy and distribution of chaetognaths of the North Pacific.
Publ. Seto Mar. Biol. Lab., 7(3):349-456.

Sagitta pulchra
Tokioka, T., 1942
Systematic studies of the plankton organisms occurring in Iwayama Bay, Palao. III Chaetognaths from the bay and adjacent waters. Palao Tropical Biol. Sta. Studies, II(3):527-548, pls.5-7, 11 text figs.

Sagitta pulchra
Tokioka, T., 1940.
The chaetognath fauna of the waters of western Japan. Rec. Oceanogr. Wks., Japan, 12(1):1-22, 3 charts, 4 tables. (Contrib. 87, Seto M.B.L.).

Sagitta pulchra
Tokioka, T., 1939.
Chaetognaths collected chiefly in the bays of Sagami and Suruga with some notes on the shape and structure of the seminal vesicles.
Rec. Oceanogr. Wks., Japan, 10(2):123-150.

Sagitta regularis n.sp.
Aida, T., 1897.
Chaetognaths of Misaki Harbour.
Ann. Zool., Japon., 1:13-21, Pl. 3.

Sagitta regularis
Alvariño, Angeles, 1966
Zoogeografía de California: quetognatos, Revista
Soc. Mexicana Hist. nat. 27: 199-243

Sagitta regularis
Alvarino, Angeles, 1965.
Distributional atlas of Chaetognatha in the California Current region.
California Coop. Oceanic Fish. Invest., Atlas, No. 3:291 charts

Sagitta regularis
Alvariño, Angeles, 1964.
Zoogeografía de los quetognatos, especialmente de la región de California.
Ciencia, México, 23:51-74.

Sagitta regularis
Alvarino, Angeles, 1963
Quetognatos epiplanctonicos del Mar de Cortes.
Revista, Soc. Mexicana, Hist. Nat., 24:97-202.

Sagitta regularis
Bieri, Robert 1959
The distribution of planktonic chaetognatha in the Pacific and their relationship to water masses.
Limnol. and Ocean. Vol. 4 No. 1 pp. 1-28

Sagitta regularis
Bieri, Robert, 1957
The chaetognath fauna of Peru in 1941.
Pacific Science 11(3): 255-264.

Sagitta regularis
Burfield, S.T., and E.J.W. Harvey, 1926.
The Chaetognatha of the "S alark" Expedition.
Trans. Linn. Soc., London, Ser. 2,19:93-119, Pls. 4-7.

Sagitta regularis
Delsman, H. C., 1939.
Preliminary plankton investigations in the Java Sea. Treubia, 17:139-181, 8 maps, 41 figs.

Sagitta regularis
Fowler, G.H., 1906.
The Chaetognatha of the Siboga Expedition with a discussion of the synonymy and distribution of the group. Rept. Siboga Exped. 21:1-88, 3 pls., 6 charts.

Sagitta regularis
Furnestin, M.L., 1958.
Quelques échantillons de zooplancton du Golfe d'Eylath (Akaba).
Haifa, Sea Fish. Res. Sta., Bull., 16:6-14.

Sagitta regularis
Furnestin, M.L., et J. Balança, 1968.
Chaetognathes de la Mer Rouge (Archipel Dahlac).
Israel South Red Sea Expedition, 1962. Reports, No. 32
Bull. Sea. Fish. Res. Sta. 52:3-20

Sagitta regularis
Furnestin, M.-L. et J.-C. Codaccioni, 1968.
Chaetognathes du nord-ouest de l'Océan Indien (golfe d'Aden, mer d'Arabie, golfe d'Oman, golfe Persique).
Cah. O.R.S.T.O.M., ser. Océanogr., VI(1):143-171.

Sagitta regularis
Ghirardelli, Elvezio, 1947
Chetognati raccolti nel Mar Rosso e nel l'Oceano Indiano dalla Nave "Cherso", Bollettino di Pesca, Piscicoltura, e Idrobiologia Anno 23, Vol. 2(n.s.) (2):253-270, 2 pls., 9 text figs.

Sagitta regularis
George, P.C., 1952.
A systematic account of the Chaetognatha of the Indian coastal waters, with observations on their seasonal fluctuations along the Malabar coast.
Proc. Nat. Inst. Sci., India, 18(6):657-689.

Sagitta regularis
Kado, Y., 1953.
The chaetognath fauna of the Inland Sea of Japan especially on the distribution of Sagitta enflata and S. crassa. Dobuts. Zasshi 62(10): 337-342.

Sagitta regularis
Michael, E. L., 1911
Classification and vertical distribution of the Chaetognatha of the San Diego Region including redescriptions of some doubtful species of the group. Univ. Calif. Publ. Zool. 8(3):21-186, Pls. 1-8.

Sagitta regularis
Rao, T.S. Satyanarayana, and P.N. Ganapati, 1958
Studies on Chaetognatha in the Indian Seas. III. Systematics and distribution in the waters off Visakhapaynam. Andhra Univ. Mem. Oceanogr., 2: 147-163.

Sagitta regularis
Ritter-Záhony, R. von, 1911
Revision der Chätognathen. Deutsche Südpolar Exped., 1901-1903, XIII (Zool.V)., Pt.1:71 pp., 51 text figs.

Sagitta regularis
von Ritter-Záhony, R., 1911
Vermes. Chaetognathi. Das Tierreich 29:34 pp., 16 text figs.

Sagitta regularis
Sund, Paul N., 1964.
Los quetognatos en las aguas de la region del Peru.
Inter-Amer. Tropical Tuna Comm., Bull., 9(3):115-216.

Sagitta regularis
Sund, Paul N., 1961.
Some features of the autecology and distributions of Chaetognatha in the Eastern Tropical Pacific.
Inter-American Tropical Tuna Comm., Bull., 5(4): 307-340.

Sagitta regularis
Sund, Paul N., 1959.
A key to the Chaetognatha of the tropical Eastern Pacific Ocean.
Pacific Science, 13(3):269-285.

Sagitta regularis
Thomson, J. M., 1947
The Chaetognatha of South-eastern Australia.
Counc. Sci. & Ind. Res., Australia, Bull. No. 222, (Div. Fish. Rept. 14), 43 pp., 8 text figs.

Sagitta regularis
Tokioka, Takasi, 1959
Observations on the taxonomy and distribution of chaetognaths of the North Pacific.
Publ. Seto Mar. Biol. Lab., 7(3):349-456.

Sagitta regularis
Tokioka, T., 1954.
Droplets from the plankton net. XVI. On a small collection of chaetognaths from the central Pacific. Publ. Seto Mar. Biol. Lab. 4(1):99-102, 1 textfig.

Sagitta regularis
Tokioka, T., 1951.
Pelagic tunicates and chaetognaths collected during the cruises to the New Yamato Bank in the Sea of Japan. Publ. Seto Mar. Biol. Lab. 2(1): 1-26, 1 chart, 6 tables, 12 textfigs.

Sagitta regularis
Tokioka, T., 1942
Systematic studies of the plankton organisms occurring in Iwayama Bay, Palao. III Chaetognaths from the bay and adjacent waters. Palao Tropical Biol. Sta. Studies, II(3):527-548, pls.5-7, 11 text figs.

Sagitta regularis
Tokioka, T., 1940.
The chaetognath fauna of the waters of western Japan. Rec. Oceanogr. Wks., Japan, 12(1):1-22, 3 charts, 4 tables. (Contrib.87, Seto M.B.L.)

Sagitta regularis
Tokioka, T., 1939.
Chaetognaths collected chiefly from the bays of Sagami and Suruga with some notes on the shape and structure of the seminal vesicles.
Rec. Oceanogr. Wks., Japan, 10(2):123-150.

Sagitta regularis
Tsuruta, Arao, 1963
Distribution of plankton and its characteristics in the oceanic fishing grounds, with special reference to their relation to fishery.
J. Shimonoseki Univ. Fish., 12(1):13-214.

Sagitta robusta
Alvariño, Angeles, 1969
Los quetognatos del Atlántico: distribución y notas esenciales de sistemática.
Trab. Inst. esp. Oceanogr., 37: 290 pp.

Sagitta robusta
Alvariño, Angeles, 1967.
The Chaetognatha of the NAGA Expedition (1959-1961) in the South China Sea and the Gulf of Thailand.
NAGA Rept., 4(2):1-87.
Scripps Inst. Oceanogr.,

Sagitta robusta
Alvariño, Angeles, 1966
Zoogeografía de California: quetognatos, Revista
Soc. Mexicana Hist. nat. 27: 199-243

Sagitta robusta
Alvarino, Angeles, 1965.
Distributional atlas of Chaetognatha in the California Current region.
California Coop. Oceanic Fish. Invest., Atlas, No. 3:291 charts

Sagitta robusta
Alvariño, Angeles, 1964.
Zoogeografía de los quetognatos, especialmente de la región de California.
Ciencia, México, 23:51-74.

Sagitta robusta
Alvariño, Angeles, 1964.
The Chaetognatha of the MONSOON Expedition in the Indian Ocean.
Pacific Science, 18(3):336-348.

Sagitta robusta
Alvarino, Angeles, 1962
Taxonomic revision of Sagitta robusta and Sagitta ferox Doncaster, and notes on their distribution in the Pacific.
Pacific Science, 16(2):186-201.

Sagitta robusta
Araya, H., and T. Otsuki, 1955.
Predation pressure on the squid larva by the arrow worm.
Bull. Hokkaido Reg. Fish. Res. Lab., No. 12: 40-42, 1 pl., 1 textfig.

Sagitta robustor
Bieri, Robert 1959
The distribution of planktonic chaetognatha in the Pacific and their relationship to water masses.
Limnol. and Ocean. Vol. 4 No. 1 pp. 1-28

Sagitta robusta
Bieri, Robert, 1957.
The chaetognath fauna of Peru in 1941.
Pacific Science 11(3): 255-264.

Sagitta robusta
Burfield, S.T., 1930.
Chaetognatha. Brit. Antarctic ("Terra Nova")
Exped., 1910, Zool., 7(4):203-228, 3 maps.

Sagitta robusta
Burfield, S.T., and E.J.W. Harvey, 1926.
The Chaetognatha of the "Sealark" Expedition.
Trans. Linn. Soc., London, Ser. 2, 19:93-119,
Pls. 4-7.

Sagitta robusta
Coleman, J. S., 1959.
The "Rosaura" Expedition 1937-1938:
Chaetognatha.
Bull. Brit. Mus. (Natural History)
Zool. 5(8):219-253.

Sagitta robusta
Dallot, Serge, 1971
Les Chaetognathes de Nosy Bé: description de Sagitta galerita sp.n. Bull.
Zool. Mus. Universiteit van Amsterdam
2(3):13-18, 1 table, 6 figs.

Sagitta robusta
Delsman, H. C., 1939.
Preliminary plankton investigations in the Java
Sea. Treubia, 17:139-181, 8 maps, 41 figs.

Sagitta robusta
Ducret, Françoise 1973.
Contribution à l'Étude des chaetognathes de la mer Rouge.
Beaufortia 20 (268): 135-153

Sagitta robusta
Fowler, G.H., 1906.
The Chaetognatha of the Siboga Expedition with a discussion of the synonymy and distribution of the goup. Rept. Siboga Exped. 21:1-88, 3 pls., 6 charts.

Sagitta robusta
Furnestin, M.-L.et J.-C.Codaccioni,1968.
Chaetognathes du nord-ouest de l'Océan Indien
(golfe d'Aden, mer d'Arabie, golfe d'Oman, golfe Persique).
Cah. O.R.S.T.O.M., ser. Océanogr., VI(1):143-171.

Sagitta robusta
Furnestin, M.-L., et J. Radiguet, 1964.
Chaetognathes de Madagascar (Secteur de Nosy-Bé).
Cahiers, O.R.S.T.O.M., Océanogr., 11(4):55-98.

Sagitta robusta
George, M.J., 1958.
Observations on the plankton of the Cochin backwaters.
Indian J. Sci., 5(2):375-401.

Sagitta robusta
George P.C., 1952.
A systematic account of the Chaetognatha of the Indian coastal waters, with observations on their seasonal fluctuations along the Malabar coast.
Proc. Nat. Inst. Sci., India, 18(6):657-689.

Sagitta robusta
Kado, Y., 1953.
The chaetognath fauna of the Inland Sea of Japan, especially on the distribution of Sagitta enflata and S. crassa. Dobuts. Zasshi 62(10):337-342.

Sagitta robusta
Marukawa, H., 1921
Plankton lists and some new species of copepods, from the northern waters of Japan.
Bull. Inst. Ocean. No.384, 15 pp., 3 pls., 1 chart. Monaco

Sagitta robusta
Moore, H. B., 1949
The zooplankton of the upper waters of the Bermuda area of the North Atlantic. Bull. Bingham Ocean. Coll. 12(2):97, 208 text figs.

Sagitta robusta
Nair, Vijayalakshmi R. 1971.
Seasonal fluctuations of chaetognaths in the Cochin Backwater.
J. mar. biol. Ass. India 13(2): 226-253

Sagitta robusta
Pierce, E.L., 1951.
The Chaetognatha of the west coast of Florida.
Biol. Bull 100(3):206-228, 5 textfigs.
See S. hispida.

Sagitta robusta
Rao, T.S. Satyanarayana, and P.N. Ganapati, 1958
Studies on Chaetognatha in the Indian Seas. III Systematics and distribution in the waters off Visakhapaynam. Andhra Univ. Mem. Oceanogr., 2: 147-163.

Sagitta robusta
Sudarsan, D., 1961.
Observations on the Chaetognatha of the waters around Mandapam.
Indian J. Fish., 8(2):364-382.

Sagitta robusta
Sund, Paul N., 1961.
Some features of the autecology of and distributions of the Eastern Tropical Pacific.
Inter-American Tropical Tuna Comm., Bull., 5(4): 307-340.

Sagitta robusta
Sund, Paul N., 1959.
A key to the Chaetognatha of the tropical Eastern Pacific Ocean.
Pacific Science, 13(3):269-285.

Sagitta robusta
Thiel, M.E., 1938
Die Chaetognathen - Bevölkerung des Südatlantischen Ozeans. Biologische Sonderuntersuchungen. 1st Lief. Wissenschaftliche Ergebnise der Deutschen Atlantischen Expedition auf dem Forschungs. und Vermessungsschiff "Meteor" 1925-1927, 13:1-110, 62 text figs.

Sagitta robusta
Thomson, J.M., 1948
Some chaetognatha from western Australia.
J. Roy. Soc. Western Australia, Inc., 31 (1944-1945), 17-18.

Sagitta robusta
Thomson, J. M., 1947
The Chaetognatha of South-eastern Australia.
Counc. Sci. & Ind. Res., Australia, Bull. No.222, (Div. Fish. Rept. 14), 43 pp., 8 text figs.

Sagitta robusta
Tokioka, Takasi, 1959
Observations on the taxonomy and distribution of chaetognaths of the North Pacific.
Publ. Seto Mar. Biol. Lab., 7(3):349-456.

Sagitta robusta
Tokioka, T., 1954.
Droplets from the plankton net. XVI. On a small collection of chaetognaths from the central Pacific. Publ. Seto Mar. Biol. Lab. 4(1):99-102, 1 textfig.

Sagitta robusta
Tokioka, T., 1942
Systematic studies of the plankton organisms occurring in Iwayama Bay, Palao. III Chaetognatha from the bay and adjacent waters. Palao Tropical Biol. Sta. Studies, II(3):527-548, pls.5-7, 11 text figs.

Sagitta robusta
Tokioka, T., 1940.
The chaetognath fauna of the waters of western Japan. Rec. Oceanogr. Wks., Japan, 12(1):1-22, 3 charts, 4 tables. (Contrib. 87, Seto M.B.L.).

Sagitta robusta
Tokioka, T., 1939.
Chaetognaths collected chiefly from the bays of Sagami and Suruga with some notes on the shape and structure of the seminal vesicles.
Rec. Oceanogr. Wks., Japan, 10(2):123-150.

Sagitta robusta
Tsuruta, Arao, 1963
Distribution of plankton and its characteristics in the oceanic fishing grounds, with special reference to their relation to fishery.
J. Shimonoseki Univ., Fish., 12(1):13-214

Sagitta robusta
Varadarajan, S. and P.I. Chacko, 1943.
On the arrow-worms of Krusadai. Proc.
Nat. Inst. Sci., India, 9(2):245-248, 2 figs.

Sagitta robusta
Ritter-Záhony, R. von, 1911
Revision der Chätognathen. Deutsche Südpolar Exped., 1901-1903, XIII (Zool.V)., Pt.1:71 pp., 51 text figs.

Sagitta robusta
von Ritter-Záhony, R., 1911
Vermes. Chaetognathi. Das Tierreich
29:34 pp., 16 text figs.

Sagitta rotrata
Ritter-Záhony, R. von, 1911
Revision der Chätognathen. Deutsche Südpolar Exped., 1901-1903, XIII (Zool.V)., Pt.1:71 pp., 51 text figs.

Sagitta scrippsae
Alvariño, Angeles, 1966
Zoogeografia de California: quetognatos. Revista Soc. Mexicana Hist. nat. 27: 199-243

Sagitta scrippsae
Alvarino, Angeles, 1965.
Distributional atlas of Chaetognatha in the California Current region.
California Coop. Oceanic Fish. Invest., Atlas, No. 3:291 charts

Sagitta scrippsae
Alvariño, Angeles, 1964.
Zoogeografia de los quetognatos, especialmente de la region de California.
Ciencia, México, 23:51-74.

Sagitta scrippsae n. sp.
Alvarino, Angeles, 1962
Two new Pacific chaetognaths: their distribution and relationship to allied species.
Bull. Scripps Inst. Oceanogr., 8(1):1-50.

Sagitta scrippsae
Kotori, Moriyuki and Akhiko Hara, 1972
On the Chaetognatha in the Bering Sea, with special reference to a new record of Sagitta scrippsae. Bull. Plankt. Soc. Japan 19(1): 5-12.

Sagitta selkirki n. sp.
Fagetti Guaita, Elda, 1958
Quetognato nuevo procedente del Archipielago de Jaun Fernandez. Revista de Biol. Mar., Valparaiso, 8 (1-3): 125-131.

Sagitta septata
Alvariño, Angeles, 1967.
The Chaetognatha of the NAGA Expedition (1959-1961) in the South China Sea and the Gulf of Thailand.
NAGA Rept., 4(2):1-87.
Scripps Inst. Oceanogr.,

Sagitta serratodentata
Aida, T., 1897.
Chaetognaths of Misaki Harbour.
Ann. Zool., Japon., 1:13-21, Pl. 3.

Sagitta serratodentata
Alvariño, Angeles, 1969 SAGITTA SERRATODENTATA
Los quetognatos del Atlántico: distribución y notas esenciales de sistemática.
Trab. Inst. esp. Oceanogr., 37: 290 pp.

Sagitta serratodentata
Alvariño, Angeles, 1966
Zoogeografía de California: quetognatos. Revista
Soc. Mexicana Hist. nat. 27: 199-243

Sagitta serratodentata
Alvarino, Angeles, 1961.
Two new species from the Pacific.
Pacific Science, 15(1):67-77.

Sagitta serratodentata
Alvariño, A., 1957.
Estudio del zooplancton del Mediterráneo occidental. Zooplancton del Atlántico Ibérico. Campañas del "Xauen" en el verano del 1954.
Bol. Inst. Español Ocean., Nos. 81-82:1-26;1-51.

Sagitta serratodentata
Bainbridge, V., 1963.
Continuous plankton records: contribution toward a plankton atlas of the North Atlantic and the North Sea. VIII. Chaetognatha.
Bull. Mar. Ecol., 6(2):40-51.

Sagitta serratodentata
Bary, B. McK., 1963.
Temperature, salinity and plankton in the eastern North Atlantic and coastal waters of Britain 1957 II. The relationship between species and water bodies.
J. Fish. Res. Bd., Canada, 20(4):1031-1065.

Sagitta serratodentata
Béranek, E., 1895.
Les chetonates de la Baie d'Amboina. Rev. Suisse Zool. 3:1370159, Pl. IV.

Sagitta serratodentata
Bigelow, H. B., 1922
Exploration of the coastal water off the northeastern United States in 1916 by the U.S. Fisheries Schooner Grampus. Bull. M.C.Z. 65 (5):85-188, 53 text figs.

Sagitta serratodentata
Bigelow, H.B., and M. Leslie, 1930
Reconnaissance of the waters and plankton of Monterey Bay, July 1928.
Bull. M.C.Z., 70(5):429-481, 43 text figs.

Sagitta serratodentata
Bigelow, H.B., and M. Sears, 1939
Studies of the waters of the continental shelf, Cape Cod to Chesapeake Bay. III. A volumetric study of the zooplankton. Mem. M.C.Z. 54(4):183-378, 42 text figs.

Sagitta serratodentata
Bumpus, D.F., and E.L. Pierce, 1955.
Hydrography and the distribution of chaetognaths over the continental shelf of North Carolina.
Pap. Mar. Biol. and Oceanogr., Deep-Sea Res., Suppl. to Vol. 3:92-109.

Sagitta serratodentata
Burfield, S.T., 1930.
Chaetognatha. Brit. Antarctic ("Terra Nova") Exped., 1910, Zool., 7(4):203-228, 3 maps.

Sagitta serratodentata
Burfield, S.T., and E.J.W. Harvey, 1926.
The Chaetognatha of the "Sealark" Expedition.
Trans. Linn. Soc., London, Ser. 2, 19:93-119, Pls. 4-7.

S. serratodentata
Clarke, G. L., E. L. Pierce and D. F. Bumpus, 1943
The distribution and reproduction of Sagitta elegans on Georges Bank in relation to the hydrographical conditions. Biol. Bull., 85(3): 201-226.

Sagitta serrato dentata
Coleman, J. S., 1959.
The "Rosaura" Expedition 1937-1938: Chaetognatha.
Bull. Brit. Mus. (Natural History) Zool. 5(8):219-253.

Sagitta serratodentata
Corbin, P.G., 1948.
On the seasonal abundance of young fish IX. The year 1947. J.M.B.A., n.s., 27:718-722, 3 textfig

Sagitta serratodentata
Corbin, P.G., 1947.
The spawning of mackerel, Scomber scombrus L., and pilchard, Clupea pilchardus Walbaum, in the Celtic Sea in 1937-39 with observations on the zooplankton indicator species, Sagitta and Muggiaea. J.M.B.A., n.s., 27:65-132, 21 textfigs.

Sagitta serratodentata
Ducret, Françoise, 1968.
Chaetognathes des eaux superficielles et profondes de la zone équatoriale et tropicale africaine.
Thèse Fac. Sci., Marseille, 99 pp., 25 figs. In: Receuil Trav. publiées de 1965 à 1968, Océanogr. ORSTROM.

Sagitta serratodentata
Ducret, Françoise, 1968.
Chaetognathes des campagnes de l'Ombango" dans les eaux équatoriales et tropicales africaines.
Cah. O.R.S.T.O.M., ser.Océanogr., VI(1):95-141.

Sagitta serratodentata
Fagetti Guaita, Elda, 1958.
Investigaciones sobre quetognatos colectados, especialmente, frente a la costa central y norte de Chile. Revista Biol. Mar., Valparaiso, 8(1-3):25-82.

Sagitta serratodentata
Fish, C.J., and M.W. Johnson, 1937
The biology of the zooplankton population in the Bay of Fundy and the Gulf of Maine with special reference to production and distribution. J. Biol. Bd., Canada 3(3):189-322, 29 tables, 45 text figs.

Sagitta serrato-dentata
Fowler, G.H., 1907.
Chaetognatha, with a note on those collected by H.M.S. "Challenger" in Subantarctic and Antarctic waters. Nat. Antarctic Exped., 1901-1904, 3:6 pp., 1 chart.

Sagitta serratodentata
Fowler, G.H., 1906.
The Chaetognatha of the Siboga Expedition with a discussion of the synonymy and distribution of the group. Rept. Siboga Exped. 21:1-88, 3 pls., 6 charts.

Sagitta serratodentata
Fowler, G.H., 1905.
Biscayan plankton collected during a cruise of H.M.S. "Research", 1900. Pt. III. The Chaetognatha. Trans. Linn. Soc., London, 10(3):55-87, Pls. 4-7.

Sagitta serratodentata
Fraser, J.H., 1952.
The Chaetognatha and other zooplankton of the Scottish area and their value as biological indicators of hydrographical conditions.
Scottish Home Dept., Mar. Res., 1952:1-52, 3 pls., 21 charts.

Sagitta serratodentata
Fraser, J. H., 1949
Plankton of the Faroe-Shetland Channel and the Faroes, June and August 1947. Ann. Biol., Int. Cons., 4:27-28, text fig. 10.

Sagitta serratodentata
Fraser, J. H., 1949
The occurrence of unusual species of Chaetognatha in Scottish plankton collections.
JMBA 28(2):489-491.

Sagitta serratodentata
Fraser, J. H., 1949
Plankton investigations from the Scottish Research Vessel. Ann. Biol., Int. Cons., 4: 66-67.

Sagitta serratodentata
Fraser, J. H., 1939
The distribution of Chaetognatha in Scottish Waters in 1937. J. du Cons. 14(1): 25-34, 3 charts in text.

Sagitta serratodentata
Fraser, J. H., and A. Saville, 1949
Plankton distribution in Scottish and adjacent waters in 1948. Ann. Biol. 5:61-62.

Sagitta serratodentata
Fraser, J. H. and A. Saville, 1949
Macroplankton in the Faroe Channel, 1948. Ann. Biol. 5:29-30, text figs.

Sagitta serratodentata
Frost, N., S.T. Lindsay, and H. Thompson, 1934.
III. Hydrographic and biological investigations. B. Plankton. Ann. Rept., Newfoundland Fish. Res. Comm. 2(2):47-59, Textfigs. 6-11.

Sagitta serratodentata
Frost, N., S.T. Lindsay, and H. Thompson, 1933.
III. Hydrographic and biological investigations. B. Plankton more abundant in 1932 than 1931. Ann. Rept., Newfoundland Fish. Res. Comm. 2(1): 58-74, Textfigs. 17-27.

Sagitta serratodentata
Furnestin, Marie-Louise, 1966.
Chaetognathes des eaux africaines.
Atlantide Rep., 9:105-135.

Sagitta serratodentata
Furnestin, Marie-Louise, 1965.
Chaetognathes de quelques récoltes dans la mer des Antilles et l'Atlantique ouest tropical.
Inst. r. Sci. Nat. Belg. Bull., 41(9):15 pp.

Sagitta serratodentata
Furnestin, M. L., 1957.
Chaetognathes et zooplancton du secteur atlantique marocain. Rev. Trav. Inst. Pêches Marit. 21(1/2): 9-356.

Sagitta serratodentata
Furnestin, M.-L., 1953.
Sur quelques Chaetognathes d'Israel.
Bull. Res. Counc., Israel, 2(4):411-414.

Sagitta serratodentata
Furnestin, M.-L., 1953.
Contribution à l'étude morphologique, biologique et systématique de Sagitta serrato-dentata Krohn des eaux atlantiques du Maroc. Bull. Inst. Océan Monaco, No. 1025:39 pp., 11 textfigs.

Sagitta serratodentata
Furnestin, Marie-Louise, 1970.
Chaetognathes des campagnes du Thor (1908-11) en Méditerranée et en Mer Noire.
Dana - Rept. 79:1-51

Sagitta serratodentata
Furnestin, J., 1938.
Influence de la salinité sur la répartition du genre Sagitta dans l'Atlantique Nord-Est (Juillet-Août-Septembre, 1936).
Rev. Trav. Off. Pêches Marit. 11(3):425-437.

Sagitta serratodentata
Gamulin, T., 1948
Prilog poznavanju Zooplanktona Srednjedalmatinskog Otocnog Područja. (Contrib. a la connaissance du zooplankton de la zone insulaire de la Dalmatie Moyenne). Acta Adriatica 3(7):38 pp., 6 tables, 1 map.

Sagitta serratodentata
Germain, L. and L. Joubin, 1916
Chétognathes provenant des campagnes des yachts Hirondelle et Princesse-Alice (1885-1910) Rés. Camp. Sci., Monaco, 49:118 pp., 7 pls., 7 maps.

Sagitta serratodentata
Ghirardelli, E., 1952.
Osservazioni biologiche e sistematiche sui Chetognati del Golfo di Napoli.
Pubbl. Staz. Zool., Napoli, 23(2/3):296-312.

Sagitta serratodentata
Ghirardelli, E., 1950.
Osservazioni biologiche e sistematiche sui chetgnati della baia di Villefranche sur mer. Bol. Pesca, Piscicol. e Idrobiol., n.s., 5(1):105-127, 7 textfigs.

Sagitta serratodentata
Ghirardelli, E., 1950.
Osservazioni biologiche e sistematiche sui chetognati della Baia di Villefranche-sur-mer. (Trav. Sta. Zool. Villefranche-sur-mer 10(9)). Bol. Pesca, Piscicult. Idrobiol. 5(1):5-27, 7 textfigs.

Sagitta serratodentata
Ghirardelli, Elvezio, 1947
Chetognati raccolti nel Mar Rosso e nel l'Oceano Indiano dalla Nave "Cherso", Bolletino di Pesca, Piscicoltura, e Idrobiologia Anno 23, Vol. 2(n.s.) (2):253-270, 2 pls., 9 text figs.

Sagitta serratodentata
Grant, George C., 1963
Investigations of inner continental shelf waters off lower Chesapeake Bay. IV. Descriptions of the Chaetognatha and a key to their identification.
Chesapeake Science, 4(3):109-119.

Sagitta serratodentata
Hansen, Vagn Kr., 1960.
Investigations on the quantitative and qualitative distribution of zooplankton in the southern part of the Norwegian Sea.
Medd. Danmarks Fiskeri- og Havundersøgelser, n.s., 2(23):1-53.

Sagitta serratodentata
Hure, J., 1961.
Dneva migracija i sezonaka vertikalna raspodjela zooplanktona dubljeg mora. (Migration journalière et distribution saisonnière verticale du zooplancton dans la région profunde de l'Adriatique). Acta Adriatica, 9(6): 1-59.

Sagitta serratodentata
Kielhorn, W.V., 1952
The biology of the surface zone zooplankton of a Boreo-Arctic Atlantic Ocean area. J. Fish Res. Bd., Canada 9 (5): 223-264, 13 text figs.

Sagitta serratodentata
Legare, J. E. Henri, and E. Zoppi, 1961.
Notas sobre la abundancia y distribucion de Chaetognatha en las aguas del Oriente de Venezuela. (In English).
Bol. Inst. Oceanograf.; Univ. Oriente, Cumana, Venezuela, 1(1):149-171.

Sagitta serratodentata
Marshall, N. B., 1948
Continuous plankton records: Zooplankton (other than Copepoda and young fish) in the North Sea 1938-1939. Hull Bull. Mar. Ecol. 2(13):173-213, Pls. 89-108.

Sagitta serratodentata
Michael, E.L., 1919
Contributions to the Biology of the Philippine Archipelago and Adjacent regions. Report on the chaetognatha collected by the United States Fisheries Steamer "Albatross" during the Philippine Expedition, 1907-1910. Bull. U.S. Nat. Mus. No.100, Vol.1(4):235-277, pls.34-38.

Sagitta serratodentata
Michael, E. L., 1911
Classification and vertical distribution of the Chaetognatha of the San Diego Region including redescriptions of some doubtful species of the group. Univ. Calif. Publ. Zool. 8(3):21-186, Pls. 1-8.

Sagitta serratodentata
Moore, H. B., 1949
The zooplankton of the upper waters of the Bermuda area of the North Atlantic. Bull. Bingham Ocean. Coll. 12(2):97, 208 text figs.

Sagitta serratodentata
Nogueiro Paranagua, Maryse (1963-1964) 1964
Sobre o plancton da região comprendida entre 3° Lat. S e 13° Lat. S ao largo do Brasil
Trabs Inst. Oceanogr. Univ. Recife 5(5/6): 125-139.

Sagitta serratodentata
Pierce, E.L., 1953.
The Chaetognatha over the continental shelf of North Carolina with attention to the relation to the hydrography of the area. J. Mar. Res. 12(1): 75-92, 4 textfigs.

Sagitta serratodentata
Rae, K. M., 1949
Plankton. Some broad changes in the plankton round the north of the British Isles in 1948. Ann. Biol. 5:56-60, 12 text figs.

Sagitta serratodentata
Ramult, M., and M. Rose, 1945
Recherches sur les Chetognathes de la Baie d'Alger. Bull. Soc. Hist. Nat. Afrique du Nord 36:45-71, 39 figs.

Sagitta serratodentata
Raymont, J.E.G., and Eileen Linford, 1966.
A note on the biochemical composition of some Mediterranean zooplankton.
Int. Rev. ges. Hydrobiol., 51(3):485-488.

Sagitta serratodentata
Redfield, A. C. and A. Beale, 1940.
Factors determining the distribution of populations of chaetognaths in the Gulf of Maine. Biol Bull., 79(3):459-487, 11 text figs.

Sagitta serratodentata
Ritter-Záhony, R. von, 1911
Revision der Chätognathen. Deutsche Südpolar Exped., 1901-1903, XIII (Zool.V), Pt.1:71 pp., 51 text figs.

Sagitta serratodentata
Rose, M., and M. Hamon, 1953.
Nouvelle note complémentaire sur les chétognathes de la Baie d'Alger.
Bull. Soc. Hist. Nat., Afrique du Nord, 44(5/6): 167-171.

Sagitta serratodentata
Russell, F.S., 1939.
Chaetognatha. Fiches d'Ident. Zooplancton, Cons. Perm. Int. Expl. Mer, 1:4 pp., 12 textfigs.

Sagitta serratodentata
Scaccini, A., e E. Ghirardelli, 1941.
Chaetognathes collected on the coast of Rio de Oro. Not. Ist. Biol., Rovigno, 2(21):3-15.

Sagitta serratodentata
Suarez Caabro, J.A., 1955.
Quetognatos de los mares cubanos.
Mem. Soc. Cubana Hist. Nat., 22(2):125-180, 9 pls.

Sagitta serratodentata
Suarez-Caabro, Jose A., and Juan E. Madruga, 1960.
The Chaetognatha of the northeastern coast of Honduras Central America.
Bull. Mar. Sci., Gulf & Caribbean, 10(4):421-429.

Sagitta serratodentata
Sund, Paul L., 1961.
Some features of the autecology and distributions of Chaetognatha in the Eastern Tropical Pacific.
Inter-American Tropical Tuna Comm., Bull., 5(4): 307-340.

Sagitta serratodentata
Thiel, M.E., 1938
Die Chaetognathen - Bevölkerung des Südatlantischen Ozeans. Biologische Sonderuntersuchungen. 1st Lief. Wissenschaftliche Ergobnisse der Deutschen Atlantischen Expedition auf dem Forschungs. und Vermessungsschiff "Meteor" 1925-1927, 13:1-110, 62 text figs.

Sagitta serratodentata
Thomson, J. M., 1947
The Chaetognatha of South-eastern Australia.
Counc. Sci. & Ind. Res., Australia, Bull. No.222, (Div. Fish. Rept. 14), 43 pp., 8 text figs.

Sagitta serratodentata
Tokioka, T., 1951.
Pelagic tunicates and chaetognaths collected during the cruises to the New Yamato Bank in the Sea of Japan. Publ. Seto Mar. Biol. Lab. 2(1): 1-26, 1 chart, 6 tables, 12 textfigs.

Sagitta serratodentata
Tokioka, T., 1942
Systematic studies of the plankton organisms occurring in Iwayama Bay, Palao. III Chaetognaths from the bay and adjacent waters. Palao Tropical Biol. Sta. Studies, II(3):527-548, pls.5-7, 11 text figs.

Sagitta serratodentata
Tokioka, T., 1940.
The chaetognath fauna of the waters of western Japan. Rec. Oceanogr. Wks., Japan, 12(1):1-22, 3 charts, 4 tables. (Contrib. 87, Seto M.B.L.)

Sagitta serratodentata
Tokioka, T., 1939.
Chaetognaths collected chiefly from the bays of Sagami and Suruga with some notes on the shape and structure of the seminal vesicles.
Rec. Oceanogr. Wks., Japan, 10(2):123-150.

Sagitta serratodentata
Tsuruta, Arao, 1963
Distribution of plankton and its characteristics in the oceanic fishing grounds, with special reference to their relation to fishery.
J. Shimonoseki Univ., Fish., 12(1):13-214.

Sagitta serratodentata
Vannucci, M., and K. Hosoe, 1952.
Resultados cientificos do Cruzeiro do "Baependi" e do "Vega" a Ilha da Trinidada. Bol. Inst. Ocean., Univ. Sao Paulo, 3(1/2):5-30, 4 pls.

Sagitta serratodentata
von Ritter-Záhony, R., 1911
Vermes. Chaetognathi. Das Tierreich 29:34 pp., 16 text figs.

Sagitta serratodentata
Vučetić, T., 1961
Sur la répartition des chaetognathes en Adriatique et leur utilisation comme indicateurs biologiques des conditions hydrographiques.
Rapp. Proc. Verb. Réunions. Comm. Int. Expl. Sci. Mer Méditerranée, Monaco, 16(2):111-116.

Sagitta sp. serratodentata group
Bieri, Robert 1959
The distribution of planktonic chaetognatha in the Pacific and their relationship to water masses.
Limnol. and Ocean. Vol. 4 No. 1 pp. 1-28

Sagitta serratodentata atlantica
de Saint-Bon, Marie Catherine, 1963.
Les chaetognathes de la Côte d'Ivoire (espèces de surface).
Rev. Trav. Inst. Pêches Marit., 27(3):301-346.

Sagitta serratodentata atlantica
Ghirardelli, E., 1960.
Habitat e biologia della riproduzione nei chetognati.
Arch. Oceanogr. e Limnol., 11(3):287-304.

Reprinted in:
Trav. Sta. Zool., Villefranche-sur-Mer, 18(11):

Sagitta serratodentata atlantica
Hosoe, K., 1956.
Chaetognaths from the Isles of Fernando de Noronha.
Contr. Avulsas Inst. Oceanogr., Sao Paulo, No. 3: 8 pp.

Sagitta serratodentata atlantica
Reyssae, J., 1963.
Chaetognathes du plateau continental européen (de la baie ibéro-marocaine à la Mer Celtique).
Rev. Trav. Inst. Pêches Marit. 27(3):245-299.

Sagitta serratodentata pacifica
Furnestin, M.-L., et J. Radiguet, 1964.
Chaetognathes de Madagascar (Secteur de Nosy-Bé).
Cahiers, O.R.S.T.R.O.M., Océanogr., 11(4):55-98.

Sagitta serratodentata pacifica
Rao, T.S. Satyanarayana, and P.N. Ganapati, 1958
Studies on Chaetognatha in the Indian Seas. III. Systematics and distribution in the waters off Visakhapaynam. Andhra Univ. Mem. Oceanogr., 2: 147-163.

Sagitta serratodentata pacifica
Sund, Paul N., 1959.
A key to the Chaetognatha of the tropical Eastern Pacific Ocean.
Pacific Science, 13(3):269-285.

Sagitta serratodentata pacifica
Tokioka, Takasi, 1959
Observations on the taxonomy and distribution of chaetognaths of the North Pacific.
Publ. Seto Mar. Biol. Lab., 7(3):349-456.

Sagitta serratodentata pacifica
Tokioka, T., 1954.
Droplets from the plankton net. XVI. On a small collection of chaetognaths from the central Pacific. Publ. Seto Mar. Biol. Lab. 4(1):99-102, 1 textfig.

Sagitta serratodentata pacifica
Tsuruta, Arao, 1963
Distribution of plankton and its characteristics in the oceanic fishing grounds, with special reference to their relation to fishery.
J. Shimonoseki Univ. Fish., 12(1):13-214.

Sagitta serratodentata pseudoserratodentata
Tokioka, Takasi, 1959
Observations on the taxonomy and distribution of chaetognaths of the North Pacific.
Publ. Seto Mar. Biol. Lab., 7(3):349-456.

Sagitta serratodentata tasmanica
Reyssae, J., 1963.
Chaetognathes du plateau continental européen (de la baie ibéro-marocaine à la Mer Celtique).
Rev. Trav. Inst. Pêches Marit., 27(3):245-299.
Inst.

Sagitta serratodentata tasmanica
Tokioka, Takasi, 1959
Observations on the taxonomy and distribution of chaetognaths of the North Pacific.
Publ. Seto Mar. Biol. Lab., 7(3):349-456.

Alvariño, Angeles, 1969 SAGITTA SETASA
Los quetognatos del Atlántico: distribución y notas esenciales de sistemática.
Trab. Inst. esp. Oceanogr., 37: 290 pp.

Sagitta setosa
Alvariño, Angeles, 1967
The Chaetognatha of the NAGA Expedition (1959-1961) in the South China Sea and the Gulf of Thailand.
NAGA Rept., 4(2):1-87.
Scripps Inst. Oceanogr.,

Sagitta setosa
Alvarino, Angeles, 1961.
Two new chaetognaths from the Pacific.
Pacific Science, 15(1):67-77.

Sagitta setosa
Bainbridge, V., 1963.
Continuous plankton records: contribution toward a plankton atlas of the North Atlantic and the North Sea. VIII. Chaetognaths.
Bull. Mar. Ecol., 6(2):40-51.

Sagitta setosa
Barnes, H., 1950.
Sagitta setosa J. Müller in the Clyde. Nature 166(4219):447.

Sagitta setosa
Bary, B. McK., 1963.
Temperature, salinity and plankton in the eastern North Atlantic and coastal waters of Britain 1957. II. The relationships between species and water bodies.
J. Fish. Res. Bd., Canada, 20(4):1031-1065.

Sagitta setosa
Bull, H.O. 1866.
Chaetognatha.
Rept. Dove mar. Lab., (3) 15:17-20.

Sagitta setosa
Corbin, P.G., 1948.
On the seasonal abundance of young fish. IX The year 1947. J.M.B.A., n.s., 27:718-722, 3 textfigs.

Sagitta setosa
Corbin, P.G., 1927.
The spawning of mackerel, Scomber scombrus L., and pilchard, Clupea pilchardus Walbaum, in the Celtic Sea in 1937-39 with observations on the zooplankton indicator species, Sagitta and Muggiaea. J.M.B.A., n.s., 27:65-132, 21 textfigs.

Sagitta setosa
Dallot, Serge, 1968.
Observations préliminaires sur la reproduction en élevage du chaetognathe planctonique Sagitta Setosa Müller
Rapp. Proc.-verb. Réun., Comm. int. Explor. scient. Mer Méditerranée 19(3): 521-523.

Sagitta setosa
Dallot, Serge 1967
La reproduction du chaetognathe planctonique Sagitta setosa Müller, en été, dans la rade de Villefranche.
C. r. hebd. Séanc. Acad. Sci. Paris (D) 264(7): 972-975.

Sagitta setosa
Fraser, J.H., 1961
The survival of larval fish in the northern North Sea according to the quality of the water.
J. Mar. Biol. Assoc., U.K., 41:305-312.

Sagitta setosa
Fraser, J.H., 1952.
The Chaetognatha and other zooplankton of the Scottish area and their value as biological indicators of hydrographical conditions.
Scottish Home Dept., Mar. Res., 1952:1-52, 3 pls., 21 charts.

good distinguishing characters

Sagitta setosa
Fraser, J. H., 1949
The occurrence of unusual species of Chaetognatha in Scottish plankton collections.
JMBA 28(2):489-491.

Sagitta setosa
Fraser, J. H., 1949
Plankton investigations from the Scottish Research Vessel. Ann. Biol., Int. Cons., 4: 66-67.

Sagitta setosa
Fraser, J. H., 1939
The distribution of Chaetognatha in Scottish Waters in 1937. J. du Cons. 14(1): 25-34, 3 charts in text.

Sagitta setosa
Fraser, J. H., and A. Saville, 1949
Plankton distribution in Scottish and adjacent waters in 1948. Ann. Biol. 5:61-62.

Sagitta setosa
Furnestin, Marie-Louise, 1970.
Chaetognathes des campagnes du Thor (1908-11) en Méditerranée et en Mer Noire.
Dana - Rept. 79:1-51

Sagitta setosa
Furnestin, Marie-Louise, 1963.
Les chaetognathes atlantiques en Méditerranée.
Rev. Trav. Inst. Pêches Marit., 27(2):157-160.

Sagitta setosa
Furnestin, M.L., 1961
Compléments à l'étude de Sagitta euxina, variété de Sagitta setosa.
Rapp. Proc. Verb., Réunions, Comm. Int. Expl. Sci. Mer Méditerranée, Monaco, 16(2):97-101.

Sagitta setosa
Furnestin, M. L., 1957.
Chaetognathes et zooplancton du secteur atlantique marocain. Rev. Trav. Inst. Pêches Marit. 21(1/2): 9-356.

Sagitta setosa
Furnestin, J., 1938.
Influence de la salinité sur la répartition du genre Sagitta dans l'Atlantique Nord-Est (Juillet-Août-Septembre, 1936).
Rev. Trav. Off. Pêches Marit. 11(3):425-437.

Sagitta setosa
Ghirardelli, E., 1960.
Habitat e biologia della riproduzione nei chetognati.
Arch. Oceanogr. e Limnol., 11(3):287-304.

Reprinted in:
Trav. Sta. Zool., Villefranche-sur-Mer, 18(11):

Sagitta setosa
Ghirardelli, E., 1952.
Osservazioni biologiche e sistematiche sui
Chetognati del Golfo di Napoli.
Pubbl. Staz. Zool., Napoli, 23(2/3):296-312.

Sagitta setosa
Ghirardelli, E., 1950.
Osservazioni biologiche e sistematiche sui chetognati della baia di Villafranche sur mer. Bol. Pesca, Piscicol. e Idrobiol., n.s., 5(1):105-127, 7 textfigs.

Sagitta setosa
Ghirardelli, E., 1950.
Osservazioni biologiche e sistematiche sui chetognati della Baia di Villefranche-sur-mer. (Trav. Sta. Zool., Villefranche-sur-mer) 10(9)). Bol. Pesca, Piscicult. Idrobiol. 5(1):5-27, 7 textfigs.

Sagitta setosa
Ghirardelli, Elvezio, et Mario Specchi, 1965.
Chaetognathes et cladoceres du Golfe de Trieste.
Rapp. Proc. Verb., Réunions, Comm. Int. Expl. Sci., Mer Méditerranée, Monaco, 18(2):403-407.

Sagitta setosa
Hamon, M., 1952.
Note complémentaire sur les chétognathes de la baie d'Alger. Bull. Soc. Hist. Nat., Afrique du Nord, 43(4/6):50-52, 1 textfig.

Sagitta setosa
Jakobsen, Tore 1971.
On the biology of Sagitta elegans Verrill and Sagitta setosa J. Müller in inner Oslofjord. Norw. J. Zool. 19 (2): 201-225

Sagitta setosa
Jespersen, P., 1949
Investigations on the occurrence and quantity of holoplankton animals in the Isefjord, 1940-1943. Medd. Komm. Danmarks Fiskeri - og Havundersøgelser, ser. Plankton, 5(3):18 pp., 10 text figs.

Sagitta setosa
Khan, M.A. and D.I. Williamson, 1970.
Seasonal changes in the distribution of chaetognatha and other plankton in the eastern Irish Sea. J. exp. mar. Biol. Ecol., 5(3): 285-303.

Sagitta setosa
Kielhorn, W.V., 1952
The biology of the surface zone zooplankton of a Boreo-Arctic Atlantic Ocean area. J.Fish Res. Bd., Canada 9 (5): 223-264, 13 text figs.

Sagitta setosa
Kuhl, Willi, und Gertrud Kuhl, 1965.
Die Dynamik der Frühentwicklung von Sagitta setosa. Lauf- und Teilbild-Analysen von Zeitrafferfilm aufnehmen.
Helgoländer wiss Meeresunters, 12(3):260-301.

Sagitta setosa
Lebour, M.V., 1947
Notes on the inshore plankton of Plymouth
JMBA 26(4):527-547, 1 textfig.

Sagitta setosa
Lucas, C. E., 1949
Notes on continuous plankton records at 10 m depth in the North Sea and Northeastern Atlantic during 1946-1947. Ann. Biol., Int. Cons., 4:63-66, text fig. 4.

Sagitta setosa
Marshall, N. B., 1948
Continuous plankton records: Zooplankton (other than Copepoda and young fish) in the North Sea 1938-1939. Hull Bull. Mar. Ecol. 2(13):173-213, Pls. 89-108.

Sagitta setosa
Mayzaud, P. et S. Dallot, 1973
Respiration et excrétion azotée du zooplancton. I. Evaluation des niveaux métaboliques de quelques espèces de Méditerranée occidentale. Mar. Biol. 19(4): 307-314.

Sagitta setosa
Meek, A., 1928.
Sagitta elegans and Sagitta setosa from the Northumberland coast. Proc. Zool. Soc., London, :743

Sagitta setosa
Murakami, Akio, 1957
[Value of chaetognaths preferring low salinity as indicator forms of water masses.]
Info. Bull., Plankton., Japan, (5):8-10.

Sagitta setosa
Parry, D.A., 1947
Structure and function of the gut in Spadella cephaloptera and Sagitta setosa. JMBA 26(1):16-36, 14 text figs.

Sagitta setosa
Parry, D. A., 1944.
Structure and function of the gut in Spadella cephaloptera and Sagitta setosa. JMBA XXVI(1):16-36.

Sagitta setosa
Pierce, E. L. 1941
The occurrence and breeding of Sagitta elegans Verrill and Sagitta setosa J. Müller in parts of the Irish Sea. JMBA, 25:113-124

Sagitta setosa
Rae, K. M., 1949
Plankton. Some broad changes in the plankton round the north of the British Isles in 1948. Ann. Biol. 5:56-60, 12 text figs.

Sagitta setosa
Rakusa-Suszczewski, S., 1968.
Predation of Chaetognatha by Tomopteris helgolandica Greff.
J.Cons.perm.Int.Explor.Mer, 32(2):226-231.

Sagitta setosa
Rakusa-Suszczewski, S. 1967.
The use of chaetognath and copepod population age-structures as an indication of similarity between water masses.
J. Cons. perm. int. Explor. Mer, 31(1):46-55.

Sagitta setosa
Reyssac, J., 1963.
Chaetognathes du plateau continental européen (de la baie ibero-marocaine à la Mer Celtique). Rev. Trav. Inst. Pêches Marit. 27(3):245-299.

Sagitta setosa
Ritter-Záhony, R. von, 1911
Revision der Chätognathen. Deutsche Südpolar Exped., 1901-1903, XIII (Zool.V.), Pt.1:71 pp., 51 text figs.

Sagitta setosa
Rose, M., and M. Hamon, 1953.
Nouvelle note complémentaire sur les chétognathes de la Baie d'Alger.
Bull. Soc. Hist. Nat., Afrique du Nord, 44(5/6): 167-171.

Sagitta setosa
Russell, F.S., 1939.
Chaetognatha. Fiches d'Ident. Zooplancton, Cons. Perm. Int. Expl. Mer, 1:4 pp., 12 textfigs.

Sagitta setosa
Russell, F. S. 1933
On the biology of Sagitta. IV. Observations on the natural history of Sagitta elegans Verrill and Sagitta setosa J. Müller in the Plymouth area. JMBA, 18:559-574

Sagitta setosa
Savage, R.E., 1937
The food of North Sea herring 1930-1934. Ministry of Agriculture and Fisheries. Fish. Invest. Ser. II, 15(5):1-60; 16 text figs.

Sagitta setosa
Scaccini, A., e E. Ghirardelli, 1941.
Chaetognathes of Adriatic Sea near Rovigno. Nat. Ist. Biol., Rovigno, 2(22):3-16.

Sagitta setosa
Vitiello, P., J.Beurois et D.Gouedard, 1970.
Stade larvaire de Thynnascaris sp. (Nematode Anisakidae) chez Sagitta setosa.
Vie Milieu (A) 21 (1-A): 257-260

Sagitta setosa
von Ritter-Záhony, R., 1914.
Chaetognaths. Danish Ingolf-Exped. 4(3):4 pp.

Sagitta setosa
von Ritter-Záhony, R., 1911
Vermes. Chaetognathi. Das Tierreich 29:34 pp., 16 text figs.

Sagitta setosa
Ritter-Záhony, R. von, 1911
Revision der Chätognathen. Deutsche Südpolar Exped., 1901-1903, XIII (Zool.V.), Pt.1:71 pp., 51 text figs.

Sagitta setosa
Vučetić, T., 1961
Quelques données préliminaires sur la répartition verticale du zooplancton dans la Baie Veliko Jezero de l'ile de Mljet pendant l'été.
Rapp. Proc. Verb., Réunions, Comm. Int. Expl. Sci. Mer Méditerranée, Monaco, 16(2):149-151.

Sagitta setosa
Vučetić, T., 1961
Sur la répartition des chaetognathes en Adriatique et leur utilisation comme indicateurs biologiques des conditions hydrographiques.
Rapp. Proc. Verb., Réunions, Comm. Int. Expl. Sci. Mer Méditerranée, Monaco, 16(2):111-116.

Sagitta setosa
Wiborg, K.F., 1944
The production of zooplankton in a landlocked fjord, the Nordåsvatn near Bergen, in 1941-42, with special reference to the copepods... (Repts. Norwegian Fish. and Mar. Invest.) 7(7):83 pp., 40 text figs.

Sagitta sibogae n.sp.
Fowler, G.H., 1906.
The Chaetognatha of the Siboga Expedition with a discussion of the synonymy and distribution of the group. Rept. Siboga Exped. 21:1-88, 3 pls., 6 charts.

Sagitta sibogae
Michael, E. L., 1911
Classification and vertical distribution of the Chaetognatha of the San Diego Region including redescriptions of some doubtful species of the group. Univ. Calif. Publ. Zool. 8(3):21-186, Pls. 1-8.

SAGITTA TASMANICA
Alvariño, Angeles, 1969
Los quetognatos del Atlantico: distribucion y notas esenciales de sistematica.
Trab. Inst. esp. Oceanogr., 37: 290 pp.

Sagitta tasmanica
Alvariño, Angeles, 1964.
Zoogeografía de los quetognatos, especialmente de la region de California.
Ciencia, México, 23:51-74.

Sagitta tasmanica
Alvariño, Angeles, 1964.
The Chaetognatha of the MONSOON Expedition in the Indian Ocean.
Pacific Science, 18(3):336-348.

Sagitta tasmanica
Alvarino, Angeles, 1961.
Two new chaetognaths from the Pacific.
Pacific Science, 15(1):67-77.

Sagitta tasmanica
Ducret, Françoise, 1968.
Chaetognathes des eaux superficielles et profondes de la zone équatoriale et tropicale africaine.
Thèse Fac. Sci., Marseille, 99 pp., 25 figs. In: Recueil Trav. publiées de 1965 à 1968, Océanogr. ORSTROM.

Sagitta tasmanica
Ducret, Françoise, 1968.
Chaetognathes des campagnes de l'"Ombango" dans les eaux équatoriales et tropicales africaines.
Cah.O.R.S.T.O.M., ser.Océanogr., VI(1):95-141.

Sagitta tasmanica
Fagetti Guaita, Elda, 1958.
Investigaciones sobre quetognatos colectados, especialmente, frente a la costa central y norte de Chile. Revista Biol. Mar., Valparaiso, 8(1-3):25-82.

Sagitta tasmanica
Furnestin, Marie-Louise, 1970.
Chaetognathes des campagnes du Thor (1908-11) en Méditerranée et en Mer Noire.
Dana - Rept. 79:1-51

Sagitta tasmanica
Furnestin, Marie-Louise, 1963.
Les chaetognathes atlantiques en Méditerranée.
Rev. Trav. Inst. Pêches Marit., 27(2):157-160.

Sagitta tasmanica
Furnestin, M.-L., 1963.
Les Chaetognathes du groupe serratodentata en Mediterranée.
Rapp. Proc. Verb., Réunions, Comm. Int. Expl. Sci., Mer Méditerranée, Monaco, 17(2):631-634.

Sagitta tasmanica
Mostajo, Elena 1973.
Quetognatos colectados en el Atlantico sudoccidental entre los 44°44' y 52°36' de latitud sur.
Neotropica 19(59): 94-100.

Sagitta tenuis
Alvariño, Angeles, 1969 SAGITTA TENUIS
Los quetognatos del Atlántico: distribución y notas esenciales de sistemática.
Trab. Inst. esp. Oceanogr., 37: 290 pp.

Sagitta tenuis
Alvariño, Angeles, 1967.
The Chaetognatha of the NAGA Expedition (1959-1961) in the South China Sea and the Gulf of Thailand.
NAGA Rept., 4(2):1-87.
Scripps Inst. Oceanogr.

Sagitta tenuis
Bieri, Robert, 1957.
The chaetognath fauna of Peru in 1941.
Pacific Science 11(3): 255-264.

Sagitta tenuis
Bumpus, D.F., and E.L. Pierce, 1955.
Hydrography and the distribution of chaetognaths over the continental shelf of North Carolina.
Pap. Mar. Biol. and Oceanogr., Deep-Sea Res., Suppl. to Vol. 3:92-109.

Sagitta tenuis
Coleman, J. S., 1959.
The "Rosaura" Expedition 1937-1938: Chaetognatha.
Bull. Brit. Mus. (Natural History) Zool. 5(8):219-253.

Sagitta tenuis n.sp.
Conant, F.S., 1896.
23. Notes on the chaetognaths. Ann. Mag. Nat. Hist., 6th ser., 18:201-214.

Sagitta tenuis (fig.)
de Almeida Prado, M.S., 1961
Chaetognatha encontrados em aguas brasileiras.
Bol. Inst. Oceanogr., Univ. Sao Paulo, 11(2):31-55.

Sagitta tenuis
Furnestin, Marie-Louise, 1965.
Chaetognathes de quelques recoltes dans la mer des Antilles et l'Atlantique ouest tropical.
Inst. r. Sci. Nat. Belg., Bull., 41(9):15 pp.

Sagitta tenuis
Furnestin, Marie-Louise, 1966.
Chaetognathes des eaux africaines.
Atlantide Rep. 9:105-135.

Sagitta tenuis
George, P.C., 1952.
A systematic account of the Chaetognatha of the Indian coastal waters, with observations on their seasonal fluctuations along the Malabar coast. Proc. Nat. Inst. Sci., India, 18(6):657-689.

Sagitta tenuis
Grant, George C., 1963
Chaetognatha from inshore coastal waters off Delaware, and a northward extension of the known range of Sagitta tenuis.
Chesapeake Science, 4(1):38-42.

Sagitta tenuis
Grant, George C., 1963
Investigations of inner continental shelf waters off lower Chesapeake Bay. IV. Descriptions of the Chaetognatha and a key to their identification.
Chesapeake Science, 4(3):109-119.

Sagitta tenius
Legare, J. E. Henri, and E. Zoppi, 1961.
Notas sobre la abundancia y distribucion de Chaetognatha en las aguas del Oriente de Venezuela. (In English).
Bol. Inst. Oceanograf., Univ. Oriente, Cumana, Venezuela, 1(1):149-171.

Sagitta tenuis
Pierce, E.L., 1953.
The Chaetognatha over the continental shelf of North Carolina with attention to the relation to the hydrography of the area. J. Mar. Res. 12(1):75-92, 4 textfigs.

Sagitta tenuis =(S. friderici)
Pierce, E.L, 1951.
The Chaetognatha of the west coast of Florida.
Biol. Bull. 100(3):206-228, 5 textfigs.

Sagitta tenuis
Pierce, E.L., 1951.
The occurrence and breeding of the Chaetognatha along the Gulf coast of Florida. Proc. Gulf and Caribbean Fish. Inst., 3rd session:126-127.

Sagitta tenuis
Pierce, E.L., and M.L. Wass, 1962.
Chaetognatha from the Florida Current and coastal waters of the southeastern Atlantic states.
Bull. Mar. Sci., Gulf and Caribbean, 12(3):403-431.

Sagitta tenuis
Ritter-Záhony, R. von, 1911
Revision der Chätognathen. Deutsche Südpolar Exped., 1901-1903, XIII (Zool.V), Pt.1:71 pp., 51 text figs.

Sagitta triptera
Ritter-Záhony, R. von, 1911
Revision der Chätognathen. Deutsche Südpolar Exped., 1901-1903, XIII (Zool.V), Pt.1:71 pp., 51 text figs.

Sagitta tenuis
Suarez-Caabro, Jose A., and Juan E. Madruga, 1960.
The Chaetognatha of the northeastern coast of Honduras Central America.
Bull. Mar. Sci., Gulf & Caribbean, 10(4):421-429

Sagitta tenuis
Sudarsan, D., 1961.
Observations on the Chaetognatha of the waters sound Mandapam.
Indian J. Fish., 8(2):364-382.

Sagitta tenuis
Tokioka, Takasi, 1961.
Notes on Sagitta friderici Ritter-Zahony collected off Peru
Postilla, Yale Peabody Mus. Nat. Hist., 55:16 pp.

Sagitta tenuis
Tokioka, Takasi, 1959
Observations on the taxonomy and distribution of chaetognaths of the North Pacific.
Publ. Seto Mar. Biol. Lab., 7(3):349-456.

Sagitta tenuis
Varadarajan, S. and P.I. Chacko, 1943.
On the arrow-worms of Krusadai. Proc. Nat. Inst. Sci., India, 9(2):245-248, 2 figs.

Sagitta tenuis
Germain, L. and L. Joubin, 1916
Chaetognathes provenant des campagnes des yachts Hirondelle et Princesse-Alice (1885-1910)
Rés. Camp. Sci., Monaco, 49:118 pp., 7 pls., 7 maps.

Sagitta tenuis
Michael, E. L., 1911
Classification and vertical distribution of the Chaetognatha of the San Diego Region including redescriptions of some doubtful species of the group. Univ. Calif. Publ. Zool. 8(3):21-186, Pls. 1-8.

Sagitta tenuis
Suarez Caabro, J.A., 1955.
Quetognatos de los mares cubanos.
Mem. Soc. Cubana Hist. Nat., 22(2):125-180, 9 pl

Sagitta tenuis = friderici
Sund, Paul N., 1959.
A key to the Chaetognatha of the tropical Eastern Pacific Ocean.
Pacific Science, 13(3):269-285.

Sagitta tenuis
Tokioka, T., 1955.
Notes on some chaetognaths from the Gulf of Mexico. Bull. Mar. Sci., Gulf and Caribbean, 5(1):52-65.

Sagitta tokiokai n.sp.
Alvariño, Angeles, 1967.
The Chaetognatha of the NAGA Expedition (1959-1961) in the South China Sea and the Gulf of Thailand.
NAGA Rept., 4(2):1-87.
Scripps Inst. Oceanogr.

Sagitta tropica

Ducret, Françoise 1973.
Contribution à l'Etude des chaetognathes de la mer Rouge.
Beaufortia 20 (268): 135-153.

Sagitta tropica n.sp.

Tokioka, T., 1942
Systematic studies of the plankton organisms occurring in Iwayama Bay, Palao. III Chaetognaths from the bay and adjacent waters. Palao Tropical Biol. Sta. Studies, II(3):527-548, pls.5-7, 11 text figs.

SAGITTA ZETESIOS

Alvariño, Angeles, 1969
Los quetognatos del Atlántico: distribución y notas esenciales de sistemática.
Trab. Inst. esp. Oceanogr., 37: 290 pp.

Sagitta zetesios

Alvariño, Angeles, 1967.
The Chaetognatha of the NAGA Expedition (1959-1961) in the South China Sea and the Gulf of Thailand.
NAGA Rept., 4(2):1-87.
Scripps Inst. Oceanogr.,

Sagitta zetesios

Alvarino, Angeles, 1965.
Distributional atlas of Chaetognatha in the California Current region.
California Coop. Oceanic Fish. Invest., Atlas, No. 3:291 charts

Sagitta zetesios

Alvariño, Angeles, 1964.
Zoogeografía de los quetognatos, especialmente de la region de California.
Ciencia, México, 23:51-74.

Sagitta zetesios = S. planctonis

Aurich, Horst J., 1971.
Die Verbreitung der Chaetognathen im Gebiet des Nordatlantischen Strom-Systems.
Ber. dt. wiss. Kommn. Meeresforsch. 22(1):1-30

Sagitta zetesios

Coleman, J. S., 1959.
The "Rosaura" Expedition 1937-1938: Chaetognatha.
Bull. Brit. Mus. (Natural History) Zool. 5(8):219-253.

Sagitta zetesios

Dallot, Serge, 1970.
L'anatomie du tube digestif dans la phylogénie et la systématique des Chaetognathes.
Bull. Mus. natn. Hist. nat. (2) 42(3): 549-565.

Sagitta zetesios

David, P.M., 1956.
Sagitta planctonis and related forms.
Bull. Brit. Mus. (N.H.), Zool., 4(8):435-451, Pl. 11, 7 textfigs.

Sagitta zetesios

de Saint-Bon, Maria-Catherine, 1963.
Complement à l'étude des chaetognathes de la Côte d'Ivoire (espèces profondes).
Rev. Trav. Inst. Pêches Marit., 27(4):403-415.

Sagitta zetesios

Ducret, Françoise, 1968.
Chaetognathes des eaux superficielles et profondes de la zone équatoriale et tropicale africaine.
Thèse Fac. Sci., Marseille, 99 pp., 25 figs. In: Recueil Trav. publiées de 1965 à 1968, Océanogr. ORSTROM.

Sagitta zetesios

Ducret, Françoise, 1968.
Chaetognathes des campagnes de l'"Ombango" dans les eaux équatoriales et tropicales africaines.
Cah. O.R.S.T.O.M., ser. Oceanogr., VI(1):95-141.

Sagitta zetesios

Ducret, Françoise, 1968.
Chaetognathes des campagnes de l'Ombango dans les eaux équatoriales et tropicales africaines.
Cah. ORSTOM., Sér. Oceanogr., 6(1):95-141.

Sagitta zetesios

Fowler, G.H., 1907.
Chaetognatha, with a note on those collected by H.M.S. "Challenger" in Subantarctic and Antarctic waters. Nat. Antarctic Exped., 1901-1904, 3:6 pp., 1 chart.

Sagitta zetesios

Fowler, G.H., 1906.
The Chaetognatha of the Siboga Expedition with a discussion of the synonymy and distribution of the group. Rept. Siboga Exped. 21:1-88, 3 pls., 6 charts.

Sagitta zetesios n.sp.

Fowler, G.H., 1905.
Biscayan plankton collected during a cruise of H.M.S. "Research". Pt. III. The Chaetognatha.
Trans. Linn. Soc., London, 10(3):55-87, Pls. 4-7
(1900)

Sagitta zetesios

Furnestin, Marie-Louise, 1966.
Chaetognathes des eaux africaines.
Atlantide Rep., 9:105-135.

Sagitta zetesios

Kitou, Masataka, 1966.
Chaetognaths collected on the Sixth Cruise of the Japanese Expedition of Deep Seas.
La Mer, Bull. franco-japon. Oceanogr., 4(4):261-265.

Also in:
Reps JEDS, 5-7 (1966)

Sagitta zetesios

Pierrot-Bults, A.C. 1970.
Variability in Sagitta planctonis Steinhaus 1896 (Chaetognatha) from West-African waters in Comparison to North Atlantic specimens.
Atlantide Rept. 11: 141-149.

Sagitta zetesios

Pierrot-Bults, A.C., 1969.
The synonymy of Sagitta planctonis and Sagitta zetesios (Chaetognatha).
Bull. Zool. Mus. Univ. Amsterdam 1(10):125-129.

Sagitta zetesios

Truveller, K.A., 1966.
Chaetognatha of the Davis and Denmark Straits as indicators of water masses. (In Russian).
Mater. Ribokhoz. Issled. severn. Basseina, Poliarn. Nauchno-Issled. Proektn. Inst. Morsk. Ribn. Khoz. Okeanogr. (PINRO) 7: 114-124.

Serratosagitta

Tokioka, Takasi, and D. Pathansali, 1963.
Another new chaetognath from Malay waters, with a proposal of grouping some species of Sagitta into subgenera.
Publ. Seto Mar. Biol. Lab., 11(1)(8):119-123.

Serratosagitta pacifica

Bieri, Robert, 1966.
A pale blue chaetognath from Tanabe Bay.
Publs Seto mar. biol. Lab., 14(1):21-22.

Spadella spp.

Bieri, Robert, 1966.
A pale blue chaetognath from Tanabe Bay.
Publs Seto mar. biol. Lab. 14(1):21-22.

Spadella

Mawson, B.M., 1944.
Some species of the chaetognath genus Spadella from New South Wales. Trans. Roy. Soc., S. Australia, 68(2):327-333, 16 textfigs.

Spadella angulata

Tokioka, Takasi, and Robert Bieri, 1966.
The colour pattern of Spadella angulata Tokioka.
Publs Seto mar. biol. Lab., 14(4):323-326.

Spadella angulata n. sp.

Tokioka, Takasi, and D. Pathansali, 1964.
Spadella cephaloptera forma angulata raised to the rank of species.
Publ. Seto Mar. Biol. Lab., 12(2)(9):145-148.

Spadella cephaloptera

Bull, H.D. 1966.
Chaetognatha
Rept. Dove Mar. Lab. (3) 15:17-20.

Spadella cephaloptera

Burfield, S.T., 1927.
Zoological results of the Cambridge Expedition to the Suez Canal. XXI. Report on the Chaetognatha.
Trans. Zool. Soc., London, 22:335-356.

Spadella cephaloptera

Fowler, G.H., 1906.
The Chaetognatha of the Siboga Expedition with a discussion of the synonymy and distribution of the group. Rept. Siboga Exped. 21:1-88, 3 pls., 6 charts.

Spadella cephaloptera

Fraser, J.H., 1952.
The Chaetognatha and other zooplankton of the Scottish area and their value as biological indicators of hydrographical conditions.
Scottish Home Dept., Mar. Res., 1952(2):1-52, 3 pls., 21 charts.

good for distinguishing characters

Spadella cephaloptera

Fraser, J. H., 1949
The occurrence of unusual species of Chaetognatha in Scottish plankton collections.
JMBA 28(2):489-491.

Spadella Cephaloptera

Furnestin, Marie-Louise, et Michel Brunet, 1965.
Sur une nouvelle mention de Spadella cephaloptera dans le golfe de Marseille
Rapp. Proc.-verb. Réun., Comm. int. Explor. scient. Mer Méditerranée 19(3): 471-473.

Spadella cephaloptera

Furnestin, M.-L., et Brunet, 1965.
Sur une station à Spadella cephaloptera dans le Golfe de Marseille.
Rapp. Proc. Verb., Reunions, Comm. Int. Expl. Sci., Mer Méditerranée, 18(2):445-450.

Spadella cephaloptera

Germain, L. and L. Joubin, 1916
Chétognathes provenant des campagnes des yachts Hirondelle et Princesse-Alice (1885-1910)
Res. Camp. Sci., Monaco, 49:118 pp., 7 pls., 7 maps.

Spadella cephaloptera

Ghirardelli, E., 1963.
Stades di maturité sexuelle chez les Chaetognathes. Observations préliminaires sur Spadella cephaloptera.
Rapp. Proc. Verb., Réunions, Comm. Int. Expl. Sci., Mer Méditerranée, Monaco, 17(2):621-626.

Spadella cephaloptera

Ghirardelli, E., 1962.
Ambiente e biologia della riproduzione nei Chetognati. Metodi di valutazione degli stadi di maturità e loro importanza nelle ricerche ecologiche. Problemi ecologici delle zone litorale del Mediterraneo, 17-23 luglio, 1961.
Pubbl. Staz. Zool., Napoli, 32(Suppl.):380-399.

Spadella cephaloptera
Ghirardelli, E., 1960.
Habitat e biologia della riproduzione nei chetognati.
Arch. Oceanogr. e Limnol., 11(3):287-304.

Reprinted in:
Trav. Sta. Zool., Villefranche-sur-Mer, 18(11)

Spadella cephaloptera
Ghirardelli, Elvezio 1959.
Osservazioni sulla corona ciliata nei Chetognati.
Boll. Zool., Unione Zool. Italiana 26(2):413-421

Also in:
Trav. Sta. Zool., Villefranche-sur-Mer, 19(1960).

Spadella cephaloptera
Ghirardelli, E., 1956.
L'apparato riprodottore feminile e la deposizione delle uova in Spadella cephaloptera Busch.
Atti. Accad. Sc. Isti. Bologna, Rendi Conti, (11) 3:1-17.

Spadella cephaloptera
Ghirardelli, E., 1954.
Studi sul determinante germinale (d.g.) nei chetognati: ricerche sperimentali su Spadella cephaloptera Busch. Pubbl. Staz. Zool., Napoli, 25:444-453, 1 textfig.

Spadella cephaloptera
Ghirardelli, E., 1953.
Osservazioni sul determinante germinale (d.g.) e su altre formazioni citoplasmatiche nelle uova di Spadella cephaloptera Busch (Chaetognata).
Pubbl. Staz. Zool., Napoli, 24(3):332-334, Pl. 16.

Spadella cephaloptera
Ghirardelli, E., 1953.
L'accoppiamento in Spadella cephaloptera Busch.
Pubbl. Staz. Zool., Napoli, 24(3):345-354, 1 textfigs.

Spadella cephaloptera
Ghirardelli, E., 1950.
Osservazioni biologiche e sistematiche sui chetognati della Baia di Villefranche-sur-mer. (Trav. Sta. Zool. Villefranche-sur-mer 10(9)). Bol. Pesca, Piscicult., Idrobiol. 5(1):5-27, 7 textfigs.

Spadella cephaloptera
Ghirardelli, E., 1950.
Osservazioni biologiche e sistematiche sui chetognati della baia di Villafranche sur mer. Bol. Pesca, Piscicol. e Idrobiol., n.s., 5(1):105-127, 7 textfigs.

Spadella cephaloptera
Ghirardelli, Elvezio, and Luisa Brandi, 1961.
Osservazioni sull'accrescimento degli ovociti di Spadella cephaloptera.
Atti Accad. Sci., Ist. Bologna, Cl. Sci. Fis., Anno 249, Rendiconti, (11) 8:14 pp.

Also in:
Trav. Sta. Zool., Villefranche-sur-Mer, 20.

Spadella cephaloptera
Horridge, G.A., and P.S. Boulton, 1967.
Prey detection by Chaetognatha via a vibration sense.
Proc. R. Soc. (B)168(1013):413-419.

Spadella cephaloptera
John, C. C., 1933.
Habits, structure, and development of Spadella cephaloptera. Quart. Jour. Micr. Sci., 75:625

Spadella cephaloptera
Michael, E. L., 1911
Classification and vertical distribution of the Chaetognatha of the San Diego Region including redescriptions of some doubtful species of the group. Univ. Calif. Publ. Zool. 8(3):21-186, Pls. 1-8.

Spadella cephaloptera
Owre, Harding B. 1972.
Marine biological investigations in The Bahamas. 18. The genus Spadella and other Chaetognatha.
Sarsia 49: 49-58.

Spadella cephaloptera
Parry, D.A., 1947
Structure and function of the gut in Spadella cephaloptera and Sagitta setosa. JMBA 26(1):16-36, 14 text figs.

Spadella cephaloptera
Parry, D. A., 1944.
Structure and function of the gut in Spadella cephaloptera and Sagitta setosa. JMBA XXVI (1): 16-36.

Spadella cephaloptera
Russell, F.S., 1939.
Chaetognatha. Fiches d'Ident. Zooplancton, Cons. Perm. Int. Expl. Mer, 1:4 pp., 12 textfigs.

Spadella cephaloptera
Rose, M., and M. Hamon, 1953.
Nouvelle note complémentaire sur les chétognathes de la Baie d'Alger.
Bull. Soc. Hist. Nat., Afrique du Nord, 44(5/6): 167-171.

Spadella cephaloptera
Scaccini, A., e E. Ghirardelli, 1941.
Chaetognathes of Adriatic Sea near Rovigno.
Not. Ist. Biol., Rovigno, 2(22):3-16.

Spadella cephaloptera
Thomson, J. M., 1947
The Chaetognatha of South-eastern Australia.
Counc. Sci. & Ind. Res., Australia, Bull. No.222, (Div. Fish. Rept. 14), 43 pp., 8 text figs.

Spadella cephaloptera
Tokioka, T., 1939.
Chaetognaths collected chiefly in the bays of Sagami and Suruga with some notes on the shape and structure of the seminal vesicles.
Rec. Oceanogr. Wks., Japan, 10(2):123-150.

Spadella cephaloptera
Varadarajan, S. and P.I. Chacko, 1943.
On the arrow-worms of Krusadai. Proc. Nat. Inst. Sci., India, 9(2):245-248, 2 figs.

Spadella cephaloptera
van Deurs Bo 1972.
On the ultrastructure of the mature spermatozoan of a chaetognath, Spadella cephaloptera.
Acta Zoologica 53 (1): 93-104

Spadella cephaloptera
von Ritter-Záhony, R., 1911
Vermes. Chaetognathi. Das Tierreich 29:34 pp., 16 text figs.

Spadella cephaloptera
Ritter-Záhony, R. von, 1911
Revision der Chätognathen. Deutsche Südpolar Exped., 1901-1903, XIII (Zool.V), Pt.1:71 pp., 51 text figs.

Spadella cephaloptera angulata (fig.)
Tokioka, T., 1951.
Pelagic tunicates and chaetognaths collected during the cruises to the New Yamato Bank in the Sea of Japan. Publ. Seto Mar. Biol. Lab. 2(1): 1-26, 1 chart, 6 tables, 12 textfigs.

Spadella draco
Aida, T., 1897.
Chaetognaths of Misaki Harbour.
Ann. Zool., Japon., 1:13-21, Pl. 3.

Spadella draco
Béranek, E., 1895.
Les chétonates de la Baie d'Amboina. Rev. Suisse Zool. 3:137-159, Pl. IV.

Spadella draco
Conant, F.S., 1896.
23. Notes on the chaetognaths. Ann. Mag. Nat. Hist., 6th ser., 18:201-214.

Spadella draco
Fowler, G.H., 1906.
The Chaetognatha of the Siboga Expedition with a discussion of the synonymy and distribution of the group. Rept. Siboga Exped. 21:1-88, 3 pls., 6 charts.

Spadella draco
Michael, E. L., 1911
Classification and vertical distribution of the Chaetognatha of the San Diego Region including redescriptions of some doubtful species of the group. Univ. Calif. Publ. Zool. 8(3):21-186, Pls. 1-8.

Spadella hummelincki n.sp.
Alvariño, Angeles 1970.
A new species of Spadella (benthic Chaetognatha).
Studies Fauna Curaçao Other Caribb. Is. 125: 73-89.

Spadella johnstoni n.sp.
Mawson, P.M., 1944.
Some species of the chaetognath genus Spadella from New South Wales. Trans. Roy. Soc. S. Australia, 68(2):327-333, 16 textfigs.

Spadella maxima n.sp.
Conant, F.S., 1896.
23. Notes on the chaetognaths. Ann. Mag. Nat. Hist., 6th ser., 18:201-214.

Spadella nana
Owre Harding B. 1972.
Marine biological investigations in The Bahamas. 18. The genus Spadella and other Chaetognatha.
Sarsia 49: 49-58.

Spadella nana n.sp.
Owre, Harding B., 1963.
The genus Spadella (Chaetognatha) in the western North Atlantic Ocean, with descriptions of two new species.
Bull. Mar. Sci., Gulf and Caribbean, 13(3):378-390.

Spadella profunda
von Ritter-Záhony, R., 1911
Vermes. Chaetognathi. Das Tierreich 29:34 pp., 16 text figs.

Spadella pulchella
Owre, Harding B. 1972.
Marine biological investigations in The Bahamas. 18. The genus Spadella and other Chaetognatha.
Sarsia 49: 49-58.

Spadella pulchella n.sp.
Owre, Harding B., 1963.
The genus Spadella (Chaetognatha) in the western North Atlantic Ocean, with descriptions of two new species.
Bull. Mar. Sci., Gulf and Caribbean, 13(3):378-390.

Spadella schizoptera
Conant, F.S., 1896.
23. Notes on the chaetognaths. Ann. Mag. Nat. Hist., 6th ser., 18:201-214.

Spadella schizoptera
Owre, Harding B. 1972.
Marine biological investigations in The Bahamas. 18. The genus Spadella and other Chaetognatha.
Sarsia 49: 49-58.

Spadella schizoptera
Owre, Harding B., 1963.
The genus Spadella (Chaetognatha) in the western North Atlantic Ocean, with descriptions of two new species.
Bull. Mar. Sci., Gulf and Caribbean, 13(3):378-390.

Spadella schizoptera
von Ritter-Záhony, R., 1911
Vermes. Chaetognathi. Das Tierreich
29:34 pp., 16 text figs.

Spadella schizoptera
Ritter-Záhony, R. von, 1911
Revision der Chätognathen. Deutsche
Südpolar Exped., 1901-1903, XIII (Zool.V.),
Pt.1:71 pp., 51 text figs.

Spadella sheardi n.sp.
Mawson, P.M., 1944.
Some species of the chaetognath genus Spadella
from New South Wales. Trans. Roy. Soc., S. Australia, 68(2):327-333, 16 textfigs.

Spadella schizoptera
Mawson, P.M., 1944.
Some species of the chaetognath genus Spadella from New South Wales. Trans. Roy. Soc., S. Australia, 68(2):327-333, 16 textfigs.

Spadella vougai n.sp.
Béranek, E., 1895.
Les chétonates de la Baie d'Amboina. Rev. Suisse Zool. 3:137-159, Pl. IV.

Rotatoria
Berzins, Bruno, 1960
Rotatoria I, Order Monogononta, Sub-order: Ploima, Family: Synchaetidae, Genus:Synchaeta.
Fiches d'Ident., Cons. Perm. Int. Expl. Mer, Zoopl., Sheet 84: 7 pp.

Rotatoria
Berzins, Bruno, 1960
Rotatoria II. Order: Monogononta, Sub-order: Ploima, Family: Trichocercidae, Genus: Trichocerca.
Fiches d'Ident., Cons. Perm. Int. Expl. Mer, Zoopl. Sheet, 85: 3 pp.

Rotatoria
Berzins, Bruno, 1960
Rotatoria III. Order: Monogonenta(sic), Sub-Order:Ploima, Family: Brachionidae; Genus: Keratella.
Fiches d'Ident., Cons. Perm. Int. Expl. Mer, Zoopl. Sheet, 86: 4 pp.

rotatoria
Berzins, Bruno, 1960
Rotatoria IV. Order: Monogononta, Sub-order: Ploima, Family: Brachionidae (Cont.), Genera: Brachionus, Kellicottia, Argonotholca, Notholca, Pseudonotholca, Euchlanis, Tripleuchlanis.
Fiches d'Ident., Cons. Perm. Int. Expl. Mer, Zool. Sheet, 87: 5 pp.

rotifers
De Ridder, Margaretha, 1963.
Recherches sur les Rotifères des eaux saumâtres. X. Les rotifères planctoniques de Nieuport et environs.
Inst. R. Sci. Nat. Belg. Bull., 39(4):39 pp.

brackish!

Rotatoria
Berzins, Bruno, 1960
Rotatoria, V. Order: Monogononta, Sub-Order: Ploima, (i) Family: Asphanchnidae, Genus: Asplanchna. (ii) Family: Synchaetidae,Genera: Ploesoma, Polyarthra.
Fiches d'Ident., Cons. Perm. Int. Expl. Mer, Zool. Sheet, 88: 4 pp.

Rotatoria
Berzins, Bruno, 1960
Rotatoria VI, Order: Monogononta, (1) Sub-Order: Flosculariaceae, (i) Family: Testudinellidae, Genera: Testudinella, Filinia, Hexarthra, (ii) Family: Conochildae, Genus: Conochilus. (2) Sub-Order: Collothecaceae, Family: Collothecidae, Genus Collotheca.
Fiches d'Ident., Cons. Perm. Int. Expl. Mer, Zoopl. Sheet, 89:4 pp.

Argonotholca foliacea
Berzins, Bruno, 1960
Rotatoria IV. Order: Monogononta, Sub-order: Ploima, Family: Brachionidae (Cont.), Genera: Brachionus, Kellicottia, Argonotholca, Notholca, Pseudonotholca, Euchlanis, Tripleuchlanis.
Fiches d'Ident., Cons. Perm. Int. Expl. Mer, Zool. Sheet, 87: 5 pp.

Asplanchna priodonta
Berzins, Bruno, 1960
Rotatoria, V. Order: Monogononta, Sub-Order: Ploima, (i) Family: Asplanchnidae, Genus: Asphanchna. (ii) Family: Synchaetidae, Genera Ploesoma, Polyarthra.
Fiches d'Ident., Cons. Perm. Int. Expl. Mer, Zool. Sheet, 88: 4 pp.

Brachionus calyciflorus
Berzins, Bruno, 1960
Rotatoria IV. Order: Monogononta, Sub-order: Ploima, Family: Brachionidae (Cont.), Genera: Brachionus, Kellicottia, Argonotholca, Notholca, Pseudonotholca, Euchlanis, Tripleuchlanis.
Fiches d'Ident., Cons. Perm. Int. Expl. Mer, Zool. Sheet, 87: 5 pp.

Brachionus plicatilis
Berzins, Bruno, 1960
Rotatoria IV. Order: Monogononta, Sub-order: Ploima, Family: Brachionidae (Cont.), Genera: Brachionus, Kellicottia, Argonotholca, Notholca, Pseudonotholca, Euchlanis, Tripleuchlanis.
Fiches d'Ident., Cons. Perm. Int. Expl. Mer, Zool. Sheet, 87: 5 pp.

Brachionus quadridentatus brevispinus
Berzins, Bruno, 1960
Rotatoria IV. Order: Monogononta, Sub-order: Ploima, Family: Brachionidae (Cont.), Genera: Brachionus, Kellicottia, Argonotholca, Notholca, Pseudonotholca, Euchlanis, Tripleuchlanis.
Fiches d'Ident., Cons. Perm. Int. Expl. Mer, Zool. Sheet, 87: 5 pp.

Brachionus urceolaris urceolaris
Berzins, Bruno, 1960
Rotatoria IV. Order: Monogononta, Sub-order: Ploima, Family: Brachionidae (Cont.), Genera: Brachionus, Kellicottia, Argonotholca, Notholca, Pseudonotholca, Euchlanis, Tripleuchlanis.
Fiches d'Ident., Cons. Perm. Int. Expl. Mer, Zool. Sheet, 87: 5 pp.

Collotheca mutabilis
Berzins, Bruno, 1960
Rotatoria VI, Order: Monogononta, (1) Sub-Order: Flosculariaceae, (i) Family: Testudinellidae, Genera: Testudinella, Filinia, Hexarthra, (ii) Family: Conochildae, Genus: Conochilus. (2) Sub-Order: Collothecaceae, Family: Collothecidae, Genus Collotheca.
Fiches d'Ident., Cons. Perm. Int. Expl. Mer, Zoopl. Sheet, 89: 4 pp.

Collotheca pelagica
Berzins, Bruno, 1960
Rotatoria VI, Order: Monogononta, (1) Sub-Order: Flosculariaceae, (i) Family: Testudinellidae, Genera: Testudinella, Filinia, Hexarthra, (ii) Family: Conochildae, Genus: Conochilus. (2) Sub-Order: Collothecaceae, Family: Collothecidae, Genus Collotheca.
Fiches d'Ident., Cons. Perm. Int. Expl. Mer, Zoopl. Sheet, 89: 4 pp.

Conochilus hippocrepis
Berzins, Bruno, 1960
Rotatoria VI, Order: Monogononta, (1) Sub-Order: Flosculariaceae, (i) Family: Testudinellidae, Genera: Testudinella, Filinia, Hexarthra, (ii) Family: Conochildae, Genus: Conochilus. (2) Sub-Order: Collothecaceae, Family: Collothecidae, Genus Collotheca.
Fiches d'Ident., Cons. Perm. Int. Expl. Mer, Zoopl. Sheet, 89: 4 pp.

Conochilus unicornis
Berzins, Bruno, 1960
Rotatoria VI, Order: Monogononta, (1) Sub-Order: Flosculariaceae, (i) Family: Testudinellidae, Genera: Testudinella, Filinia, Hexarthra, (ii) Family: Conochildae, Genus: Conochilus. (2) Sub-Order: Collothecaceae, Family: Collothecidae, Genus: Collotheca.
Fiches d'Ident., Cons. Perm. Int. Expl. Mer, Zoopl. Sheet, 89: 4 pp.

Euchlanis dilatata dilatata
Berzins, Bruno, 1960
Rotatoria IV. Order: Monogononta, Sub-order: Ploima, Family: Brachionidae (Cont.), Genera: Brachionus, Kellicottia, Argonotholca, Notholca, Pseudonotholca, Euchlanis, Tripleuchlanis.
Fiches d'Ident., Cons. Perm. Int. Expl. Mer, Zool. Sheet, 87: 5 pp.

Filinia longiseta
Berzins, Bruno, 1960
Rotatoria VI, Order: Monogononta, (1) Sub-Order: Flosculariaceae, (i) Family: Testudinellidae, Genera: Testudinella, Filinia, Hexarthra, (ii) Family: Conochildae, Genus: Conochilus. (2) Sub-Order: Collothecaceae, Family: Collothecidae, Genus Collotheca.
Fiches d'Ident., Cons. Perm. Int. Expl. Mer, Zoopl. Sheet, 89:4 pp.

Hexarthra fennica fennica
Berzins, Bruno, 1960
Rotatoria VI, Order: Monogononta, (1) Sub-Order: Flosculariaceae, (i) Family: Testudinellidae, Genera: Testudinella, Filinia, Hexarthra, (ii) Family: Conochildae, Genus: Conochilus. (2) Sub-Order: Collothecaceae, Family: Collothecidae, Genus Collotheca.
Fiches d'Ident., Cons. Perm. Int. Expl. Mer, Zoopl. Sheet, 89: 4 pp.

Hexarthra fennica oxyuris
Berzins, Bruno, 1960
Rotatoria VI, Order: Monogononta, (1) Sub-Order: Flosculariaceae, (i) Family: Testudinellidae, Genera: Testudinella, Filinia, Hexarthra, (ii) Family: Conochildae, Genus: Conochilus. (2) Sub-Order: Collothecaceae, Family: Collothecidae, Genus Collotheca.
Fiches d'Ident., Cons. Perm. Int. Expl. Mer, Zoopl. Sheet, 89: 4 pp.

Kellicottia longispina
Berzins, Bruno, 1960
Rotatoria IV. Order: Monogononta, Sub-order: Ploima, Family: Brachionidae (Cont.), Genera: Brachionus, Kellicottia, Argonotholca, Notholca, Psudonotholca, Euchlanis, Tripleuchlanis.
Fiches d'Ident., Cons. Perm. Int. Expl. Mer, Zool. Sheet, 87: 5 pp.

Keratella cochlearis cochlearis
Berzins, Bruno, 1960
Rotatoria III. Order: Monogonenta (sic), Sub-Order:Ploima, Family: Brachionidae; Genus: Keratella.
Fiches d'Ident., Cons. Perm. Int. Expl. Mer, Zoopl. Sheet, 86: 4 pp.

Keratella cochlearis recurvispina
Berzins, Bruno, 1960
Rotatoria III. Order: Monogonenta (sic), Sub-Order: Ploima, Family: Brachionidae; Genus: Keratella.
Fiches d'Ident., Cons. Perm. Int. Expl. Mer, Zoopl. Sheet, 86: 4 pp.

Keratella cruciformis cruciformis
Berzins, Bruno, 1960
Rotatoria III. Order: Monogonenta (sic), Sub-Order: Ploima, Family: Brachionidae; Genus: Keratella.
Fiches d'Ident., Cons. Perm. Int. Expl. Mer, Zoopl. Sheet, 86: 4 pp.

Keratella cruciformis eichwaldi
Berzins, Bruno, 1960
Rotatoria III. Order: Monogonenta (sic), Sub-Order: Ploima, Family: Brachionidae; Genus: Keratella.
Fiches d'Ident., Cons. Perm. Int. Expl. Mer, Zoopl. Sheet, 86: 4 pp.

Keratella quadrata platei
Berzins, Bruno, 1960
Rotatoria III. Order: Monogonenta (sic), Sub-Order: Ploima, Family: Brachionidae; Genus: Keratella.
Fiches d'Ident., Cons. Perm. Int. Expl. Mer, Zoopl. Sheet, 86: 4 pp.

Keratella quadrata quadrata
Berzins, Bruno, 1960
Rotatoria III. Order: Monogonenta (sic), Sub-Order: Ploima, Family: Brachionidae; Genus: Keratella.
Fiches d'Ident., Cons. Perm. Int. Expl. Mer, Zoopl. Sheet, 86: 4 pp.

Notolca striata
Berzins, Bruno, 1960
Rotatoria IV. Order: Monogononta, Sub-order: Ploima, Family: Brachionidae (Cont.), Genera: Brachionus, Kellicottia, Argonotholca, Notholca, Pseudonotholca, Euchlanis, Tripleuchlanis.
Fiches d'Ident., Cons. Perm. Int. Expl. Mer, Zool. Sheet, 87: 5 pp.

Ploesoma hudsoni
Berzins, Bruno, 1960
Rotatoria, V. Order: Monogononta, Sub-Order: Ploima, (i) Family: Asplanchnidae, Genus: Asplanchna. (ii) Family: Synchaetidae, Genera: Ploesoma, Polyarthra.
Fiches d'Ident., Cons. Perm. Int. Expl. Mer, Zool. Sheet, 88: 4 pp.

Ploesoma truncatum
Berzins, Bruno, 1960
Rotatoria, V. Order: Monogononta, Sub-Order: Ploima, (i) Family: Asplanchnidae, Genus: Asplanchna. (ii) Family: Synchaetidae, Genera: Ploesoma, Polyarthra.
Fiches d'Ident., Cons. Perm. Int. Expl. Mer, Zool. Sheet, 88: 4 pp.

Polyarthra dolichoptera
Berzins, Bruno, 1960
Rotatoria, V. Order: Monogononta, Sub-Order: Ploima, (i) Family: Asplanchnidae, Genus: Asplanchna. (ii) Family: Synchaetidae, Genera: Ploesoma, Polyarthra.
Fiches d'Ident., Cons. Perm. Int. Expl. Mer, Zool. Sheet, 88: 4 pp.

Polyarthra remata
Berzins, Bruno, 1960
Rotatoria, V. Order: Monogononta, Sub-Order: Ploima, (i) Family: Asplanchnidae, Genus: Asplanchna. (ii) Family: Synchaetidae, Genera: Ploesoma, Polyarthra.
Fiches d'Ident., Cons. Perm. Int. Expl. Mer, Zool. Sheet, 88: 4 pp.

Polyarthra vulgaris
Berzins, Bruno, 1960
Rotatoria, V. Order: Monogononta, Sub-Order: Ploima (i) Family: Asplanchnidae, Genus: Asplanchna. (ii) Family: Synchaetidae, Genera: Ploesoma, Polyarthra.
Fiches d'Ident., Cons. Perm. Int. Expl. Mer, Zool. Sheet, 88: 4 pp.

Pseudonotholca japonica
Berzins, Bruno, 1960
Rotatoria IV. Order: Monogononta, Sub-order: Ploima, Family: Brachionidae (Cont.), Genera: Brachionus, Kellicottia, Argonotholca, Notholca, Pseudonotholca, Euchlanis, Tripleuchlanis.
Fiches d'Ident., Cons. Perm. Int. Expl. Mer, Zool. Sheet, 87: 5 pp.

Synchaeta atlantica
Berzins, Bruno, 1960
Rotatoria I, Order Monogononta, Sub-order: Ploima, Family: Synchaetidae, Genus:Synchaeta.
Fiches d'Ident., Cons. Perm. Int. Expl. Mer, Zoopl., Sheet 84: 7 pp.

Synchaeta baccilifera
Berzins, Bruno, 1960
Rotatoria I, Order Monogononta, Sub-order: Ploima, Family: Synchaetidae, Genus:Synchaeta.
Fiches d'Ident., Cons. Perm. Int. Expl. Mer, Zoopl., Sheet 84: 7 pp.

Synchaeta baltica
Berzins, Bruno, 1960
Rotatoria I, Order Monogononta, Sub-order: Ploima, Family: Synchaetidae, Genus:Synchaeta.
Fiches d'Ident., Cons. Perm. Int. Expl. Mer, Zoopl., Sheet 84: 7 pp.

Synchaeta bicornis
Berzins, Bruno, 1960
Rotatoria I, Order Monogononta, Sub-order: Ploima, Family: Synchaetidae, Genus:Synchaeta.
Fiches d'Ident., Cons. Perm. Int. Expl. Mer, Zoopl., Sheet 84: 7 pp.

Synchaeta cecilia
Berzins, Bruno, 1960
Rotatoria I, Order Monogononta, Sub-order: Ploima, Family: Synchaetidae, Genus:Synchaeta.
Fiches d'Ident., Cons. Perm. Int. Expl. Mer, Zoopl., Sheet 84: 7 pp.

Synchaeta curvata
Berzins, Bruno, 1960
Rotatoria I, Order Monogononta, Sub-order: Ploima, Family: Synchaetidae, Genus:Synchaeta.
Fiches d'Ident., Cons. Perm. Int. Expl. Mer, Zoopl., Sheet 84: 7 pp.

Synchaeta fennica
Berzins, Bruno, 1960
Rotatoria I, Order Monogononta, Sub-order: Ploima, Family: Synchaetidae, Genus:Synchaeta.
Fiches d'Ident., Cons. Perm. Int. Expl. Mer, Zoopl., Sheet 84: 7 pp.

Synchaeta glacialis
Berzins, Bruno, 1960
Rotatoria I, Order Monogononta, Sub-order: Ploima, Family: Synchaetidae, Genus: Synchaeta.
Fiches d'Ident., Cons. Perm. Int. Expl. Mer, Zoopl., Sheet 84: 7 pp.

Synchaeta grimpei
Berzins, Bruno, 1960
Rotatoria I, Order Monogononta, Sub-order: Ploima, Family: Synchaetidae, Genus: Synchaeta.
Fiches d'Ident., Cons. Perm. Int. Expl. Mer, Zoopl., Sheet 84: 7 pp.

Synchaeta gyrina
Berzins, Bruno, 1960
Rotatoria I, Order Monogononta, Sub-order: Ploima, Family: Synchaetidae, Genus:Synchaeta.
Fiches d'Ident., Cons. Perm. Int. Expl. Mer, Zoopl., Sheet 84: 7 pp.

Synchaeta hyperborea
Berzins, Bruno, 1960
Rotatoria I, Order Monogononta, Sub-order: Ploima, Family: Synchaetidae, Genus:Synchaeta.
Fiches d'Ident., Cons. Perm. Int. Expl. Mer, Zoopl., Sheet 84: 7 pp.

Synchaeta johanseni
Berzins, Bruno, 1960
Rotatoria I, Order Monogononta, Sub-order: Ploima, Family: Synchaetidae, Genus:Synchaeta
Fiches d'Ident., Cons. Perm. Int. Expl. Mer, Zoopl., Sheet 84: 7 pp.

Synchaeta littoralis
Berzins, Bruno, 1960
Rotatoria I, Order Monogononta, Sub-order: Ploima, Family: Synchaetidae, Genus:Synchaeta.
Fiches d'Ident., Cons. Perm. Int. Expl. Mer, Zoopl., Sheet 84: 7 pp.

Synchaeta monopus
Berzins, Bruno, 1960
Rotatoria I, Order Monogononta, Sub-order: Ploima, Family: Synchaetidae, Genus:Synchaeta
Fiches d'Ident., Cons. Perm. Int. Expl. Mer, Zoopl., Sheet 84: 7 pp.

Synchaeta neapolitana
Berzins, Bruno, 1960
Rotatoria I, Order Monogononta, Sub-order: Ploima, Family: Synchaetidae, Genus:Synchaeta.
Fiches d'Ident., Cons. Perm. Int. Expl. Mer, Zoopl., Sheet 84: 7 pp.

Synchaeta pectinata
Berzins, Bruno, 1960
Rotatoria I, Order Monogononta, Sub-order: Ploima, Family: Synchaetidae, Genus:Synchaeta.
Fiches d'Ident., Cons. Perm. Int. Expl. Mer, Zoopl., Sheet 84: 7 pp.

Synchaeta stylata
Berzins, Bruno, 1960
Rotatoria I, Order Monogononta, Sub-order: Ploima, Family: Synchaetidae, Genus:Synchaeta
Fiches d'Ident., Cons. Perm. Int. Expl. Mer, Zoopl., Sheet 84: 7 pp.

Synchaeta tamara
Berzins, Bruno, 1960
Rotatoria I, Order Monogononta, Sub-order: Ploima, Family: Synchaetidae, Genus:Synchaeta.
Fiches d'Ident., Cons. Perm. Int. Expl. Mer, Zoopl., Sheet 84: 7 pp.

Synchaeta tavina
Berzins, Bruno, 1960
Rotatoria I, Order Monogononta, Sub-order: Ploima, Family: Synchaetidae, Genus:Synchaeta
Fiches d'Ident., Cons. Perm. Int. Expl. Mer, Zoopl., Sheet 84: 7 pp.

Synchaeta tremula
Berzins, Bruno, 1960
Rotatoria I, Order Monogononta, Sub-order: Ploima, Family: Synchaetidae, Genus:Synchaeta.
Fiches d'Ident., Cons. Perm. Int. Expl. Mer, Zoopl., Sheet 84: 7 pp.

Synchaeta triophthalma
Berzins, Bruno, 1960
Rotatoria I, Order Monogononta, Sub-order: Ploima, Family: Synchaetidae, Genus:Synchaeta.
Fiches d'Ident., Cons. Perm. Int. Expl. Mer, Zoopl., Sheet 84: 7 pp.

Synchaeta vorax
Berzins, Bruno, 1960
Rotatoria I, Order Monogononta, Sub-order: Ploima, Family: Synchaetidae, Genus:Synchaeta.
Fiches d'Ident., Cons. Perm. Int. Expl. Mer, Zoopl., Sheet 84: 7 pp.

Testudinella clypeata clypeata
Berzins, Bruno, 1960
Rotatoria VI, Order: Monogononta, (1) Sub-Order: Flosculariaceae, (i) Family: Testudinellidae, Genera: Testudinella, Filinia, Hexarthra, (ii) Family: Conochildae, Genus: Conochilus. (2) Sub-Order: Collothecaceae, Family: Collothecidae, Genus Collotheca.
Fiches d'Ident.,Cons. Perm. Int. Expl. Mer, Zoopl. Sheet, 89: 4 pp.

Testudinella clypeata crassa
Berzins, Bruno, 1960
Rotatoria VI, Order: Monogononta, (1) Sub-Order: Flosculariaceae, (i) Family: Testudinellidae, Genera: Testudinella, Filinia, Hexarthra, (ii) Family: Conochildae, Genus: Conochilus. (2) Sub-Order: Collothecaceae, Family Collothecidae, Genus Collotheca.
Fiches d'Ident., Cons. Perm. Int. Expl. Mer, Zoopl. Sheet, 89: 4 pp.

Trichocerca capuchina
Berzins, Bruno, 1960
Rotatoria II. Order: Monogononta, Sub-order: Ploima, Family: Trichocercidae, Genus: Trichocerca.
Fiches d'Ident., Cons. Perm. Int. Expl. Mer, Zoopl. Sheet, 85: 3 pp.

Trichocerca marina curvata
Berzins, Bruno, 1960
Rotatoria II. Order: Monogononta, Sub-order: Ploima, Family: Trichocercidae, Genus: Trichocerca.
Fiches d'Ident., Cons. Perm. Int. Expl. Mer, Zool. Sheet, 85: 3 pp.

Trichocerca marina dubius
Berzins, Bruno, 1960
Rotatoria II. Order: Monogononta, Sub-order: Ploima, Family: Trichocercidae, Genus: Trichocerca.
Fiches d'Ident., Cons. Perm. Int. Expl. Mer, Zool. Sheet, 85: 3 pp.

Trichocerca marina lie-petterseni
Berzins, Bruno, 1960
Rotatoria II. Order: Monogononta, Sub-order: Ploima, Family: Trichocercidae, Genus: Trichocerca.
Fiches d'Ident., Cons. Perm. Int. Expl. Mer, Zoopl. Sheet, 85: 3 pp.

Trichocerca marina marina
Berzins, Bruno, 1960
Rotatoria II. Order: Monogononta, Sub-order: Ploima, Family: Trichocercidae, Genus: Trichocerca.
Fiches d'Ident., Cons. Perm. Int. Expl. Mer, Zool. Sheet, 85: 3 pp.

Tripleuchlanis plicata
Berzins, Bruno, 1960
Rotatoria IV. Order: Monogononta, Sub-order: Ploima, Family: Brachionidae (Cont.), Genera: Brachionus, Kellicottia, Argonotholca, Nothol Pseudonotholca, Euchlanis, Tripleuchlanis.
Fiches d'Ident., Cons. Perm. Int. Expl. Mer, Zool. Sheet, 87: 5 pp.

rotifers
Edmondson, W.T., 1964.
The rate of egg production by rotifers and copepods in natural populations as controlled by food and temperature.
Verh. Internat. Verein. Limnol., 15:673-675.

rotifers
Focke, Eberhard, 1961.
Die Rotatoriengattung Notholca und ihr Verhalten im Salzwasser.
Kieler Meeresf., 17(2):190-205.

rotifers
Margaleff, Ramón, 1956
Rotifers marinos del plancton de la ría de Vigo. Invest. Pesquera 4: 133-135.

rotifers
Edmondson, W.T., 1965.
Reproductive rate of planktonic rotifers as related to food and temperature in nature.
Ecological Monographs, 35:61-111.

rotifers
Thane-Fenchel, Anne, 1966.
Proales paguri sp. nov., a rotifer living on the gills of the hermit crab Pagurus bernhardus (L.).
Ophelia, 3:93-97.

Kinorhynch
Gerlach, Sebastian A., 1969.
Cantrie submersa sp. n. ein Cryptorhagen Kinorhynch aus dem sublitoralen Mesopsammal der Nordsee.
Veröff. Inst. Meeresforsch. Bremerh. 12(2): 161-168.

chordates
Barrington, E.J.W., 1965.
The biology of Hemichordata & Protochordata.
W.H. Freeman and Company, San Francisco, 176 pp. (paperback). $2.50.

Hemichordata
Stebbing, A.R.D., 1970.
Aspects of the reproduction and life cycle of Rhabdopleura compacta (Hemichordata).
Mar. Biol., 5(3): 205-212.

cephalochordates
Wickstead, J.H., 1970.
On a small collection of Acrania (Phylum Chordata) from New Caledonia.
Cah. Pacifique 14: 237-243.

Appendicularia
Aida, T., 1907
Appendicularia of Japanese Waters. J. Coll. Sci., Tokyo, XXIII (5):1-25, pls.i-iv.

appendicularians
Björnberg, T.K.S., and L. Forneris, 1955.
Resultados cientificos do cruzeiro do "Baependi" e do "Vega" à ilha de Trindade.
Contrib. Avulsas, Inst. Oceanograf. (Ocean. Biol.), Sao Paulo, No. 1:1-68.

appendicularia
Buckman, A., 1945.
Appendicularia I-III. Fiches d'Ident. Zooplancton, Cons. Perm. Int. Expl. Mer, 7:8 pp., 16 textfigs.

appendicularia
Essenberg, C.E., 1926
Copelata from the San Diego Region, Univ. Calif. Pub. Zool. 28:399-421, 170 textfigs.

Appendicularia
Essenberg, C.E., 1922
The Seasonal Distribution of the Appendicularia in the Region of San Diego, California. Ecology 3:55-64, 3 textfigs.

appendicularins
Fenaux, R., 1969.
Sur l'état de conservation des appendiculaires dans les matériel des expéditions.
Bull. Mus. natn. Hist. Nat. Paris (2) 40(5): 934-937.

appendicularians, distribution of
Fenaux, F., 1961
Existence d'un ordre cyclique d'abondance relative maximale chez les Appendiculaires de surface (Tuniciers pelagiques).
Comptes Rendus. Acad. Sci. Paris. 253: 2271-2273.
Also in:
Trav. Sta. Zool., Villefranche-sur-Mer, 20.

appendicularians
Fenaux, R., 1960.
Sur quelques appendiculaires d'Israel. Contribution to the knowledge of the Red Sea, No. 17.
Sea Fish. Res. Sta., Haifa, Bull., No. 29:3-7.

appendicularians
Forneris, Liliana, 1965.
Appendicularian species groups and southern Brazil water masses.
Bolm Inst. Oceanogr., S Paulo, 14(1):54-94.

appendicularians
Forneris, Liliana, 1965.
Appendicularia species groups and southern Brazil water masses. (abstract).
Anais Acad. bras. Cienc., 37(Supl.):239.

Appendicularian
Ihle, J.E.W., 1908
Die Appendicularian der Siboga-Expedition Siboga Exped., Monog. LVI c, 1-123, pls. 1-iv, 10 textfigs.

appendicularians
Last, J.M., 1972
Egg development, fecundity and growth of Oikopleura dioica Fol in the North Sea.
J. Cons. int. Explor. Mer 34(2): 232-237.

appendicularians
Matsumura, M., 1965.
Studies on seasonal variation and relative abundance of Appendiculariae at 25° Lat. S. (Abstract).
Anais Acad. bras. Cienc., 37 (Supl.):257.

appendicularia
Russell, F.S., and J.S. Colman, 1935
The Zooplankton. IV. The occurrence and seasonal distribution of the Tunicata, Mollusca and Coelenterata (siphonophora). Brit. Mus. (N.H.) Great Barrier Reef Expedition 1928-1929, Sci. Repts., 2(7):203-276, 30 text figs.

appendicularians
Tokioka, T., 1955.
General considerations on Japanese appendicularian fauna. Publ. Seto Mar. Biol. Lab., 4(2/3): 251-261, 6 textfigs.

appendicularians
Tokioka, T., 1955.
On some plankton animals collected by the Syunkotu maru in May-June 1954. III. Appendicularians. Bull. Biogeogr. Soc., Japan, Vols. 16-19, 251-255.

appendicularians
Tokioka, T., and J.A. Suarez Caabro, 1956.
Apendicularias de los mares cubanos.
Mem. Soc. Cubana. Hist. Nat., 23(1):37-80, 15 pl.

appendicularians
Tregouboff, G., 1956.
Rapport sur les travaux concernant le plancton Méditerranéen publiés entre Novembre 1952 et Novembre 1954.
Rapp. Proc. Verb., Comm. Int. Expl. Sci., Mer Mediterranee, 13:65-100

appendicularians
Udvardy, M.D.F., 1954.
Distribution of appendicularians in relation to the Straits of Belle Isle.
J. Fish. Res. Bd., Canada, 11(4):431-453.

Copelata
Wiborg, K.F., 1944
The production of plankton in a land locked fjord. The Nordåsvatn near Bergen in 1941-1947. With special reference to copepods. Fiskeridirektoratets Skrifter Serie Havundersøkelser (Rept. Norwegian Fish. and Marine Invest.) 7(7):1-83, Map.

appendicularians, anat. physiol.
Alldredge, Alice L., 1972.
Abandoned larvacean houses: a unique food source in the pelagic environment. Science 177(4052): 885-887.

appendicularians, anatomy
Berrill, N.J., 1950.
The Tunicata with an account of the British species. Ray Soc., No. 133:354 pp., 120 textfigs.

Appendicularia, anat.-physiol.
Bogoraze, Dimitri, et Odette Tuzet 1969.
Ultrastructure du muscle de la queue de l'appendiculaire Oikopleura longicauda Vogt. Les limites cellulaires; les disques intercalaires.
Cah. Biol. mar. 10(4): 365-374.

appendicularia, lists of spp.
Buckmann, A., 1970.
Die Verbreitung der Kaltwasser - und der Warmwasserfauna der Appendicularien im nördlichen nordatlantischen Ozean im Spätwinter und Spätommer 1958. Mar. Biol., 5(1): 35-56.

appendicularians, anat.-physiol
Fenaux, Robert, 1963.
Écologie et biologie des appendiculaires méditerranéens.
Vie et Milieu, Suppl., (16):142 pp.

appendicularians, anat.
Fenaux, R., 1971
La couche oïkoplastique de l'Appendiculaire Oikopleura albicans (Leuckart) (Tunicata). Z. Morph. Tiere 69:184-200.

appendicularians, anat.-physiol.
Fenaux, R., 1961
Rôle du pylore chez Fritillaria pellucida Buch (Appendiculaire).
C.R. Acad. Sci., Paris, 252(19):2936-2939.

appendicularians, "house"
Fenaux, Robert, et Bertrand Hirel 1972.
Cinétique du déploiement de la logette chez l'appendiculaire Oikopleura dioica Fol, 1872.
C. r. hebd. Séanc. Acad. Sci. Paris (D) 275(3): 449-452

appendicularians, anat.-physiol.
Olssen, R., 1965.
Comparative morphology and physiology of the Oikopleura notochord.
Israel J. Zool., 14(1/4):213-220.

appendicularians, anat.

Van Gansen P., 1960
Adaptations structurelles des animaux filtrants.
Ann. Soc. R. Zool. Belgique, 90(2): 161-231.

appendicularian, biomass

Degterjeva, A.A., 1966.
Zooplankton in the southwest of the Barents and northeast of the Norwegian seas in 1959-1961. (In Russian).
Mater.Ribokhoz. Issled. severn. Bassoina, Poliarn. Nauchno-Issled. Proektn. Inst. Morsk. Ribn. Khoz. Okeanogr. (PINRO), 7:70-83.

appendicularians, ecology

Fenaux, R., 1959.
Observations écologiques sur les appendiculaires du plancton de surface dans la Baie de Ville-franche-sur-Mer.
Bull. l'Inst. Ocean., Monaco, 1141:26 pp.

appendicularians, lists of spp.

Bhavanarayana, P.V., and P.N. Ganapati 1972.
Distribution of pelagic tunicates in the western part of the Bay of Bengal.
Proc. Indian Acad. Sci. (B) 75 (1): 1-14.

appendicularians, lists of spp.

Björnberg T.K.S., and L. Forneris, 1956.
On the uneven distribution of the Copelata of the Alcatrazes area.
Bol. Inst. Ocean., Sao Paulo, 7(1/2):113-115.

appendicularians, lists of spp

Björnberg T.K.S., and L. Forneris, 1956.
On the uneven distribution of the Copelata of the Fernando de Noronha area.
Bol. Inst. Ocean., Sao Paulo, 7(1/2):105-111.

appendicularians, lists of spp. [a]

Bückmann, A., 1973
Sorted samples and quantitative counts in appendicularian catches. Mar. Biol. 21(4):349-353.

appendicularians, lists of spp.[a]

Bückmann, Adolf, 1972
Die Appendicularien von den Fahrten der Meteor, der Anton Bruun und der Discovery in das Arabische Meer im Rahmen der IIOE. Meteor Forsch. Ergebn. D10:1-45. Also in: Gesammelte Sonderdrucke, Inst. Hydrobiol. Fischereiwiss. Univ. Hamburg 5 (1973).

appendicularians, lists of spp

Bückmann, A. 1967.
Untersuchungen über das Makroplankton bei Ischia und Capri und im Golf von Neapel im Mai 1962. III. Die Appendicularien.
Pubbl. Stag. zool. Napoli 35(2): 215-238.

appendicularians, lists of spp.[a]

Fenaux, R., 1972
A historical survey of the appendicularians from the area covered by the IIOE. Mar. Biol. 16(3): 230-235.

appendicularians, lists of sp. [a]

Fenaux, R., 1972
Variations saisonnières des Appendiculaires de la Région Nord Adriatique. Mar. Biol. 16(4): 310-319.

appendicularians, lists of spp.[a]

Fenaux, R., 1971.
Sur les appendiculaires de la Méditerranée orientale.
Bull. Mus. natn. Hist. nat. (2) 42(6): 1208-1211.

appendicularians, lists of spp.

Fenaux R., 1970.
Deuxième note faunistique sur les appendiculaires de la mer Rouge.
Bull. Mus. natn. Hist. nat. (2) 41(5): 1150-1152.

appendicularians, lists of spp.

Fenaux, R. 1969.
Les Appendiculaires du Golfe du Bengale. Expedition Internationale de l'Ocean Indien (croisieres du Kistna, juin-aout 1964). Marine Biol. 2(3): 252-263.

Appendiculaires, lists of spp. [a]

Fenaux, R., 1969.
Les Appendiculaires de Madagascar (Région de Nosy-Bé) variations saisonnieres. Cah. O.R.S.T.O.M., sér. Océanogr., 7(4): 29-37.

appendicularians, lists of sp.

Fenaux R. 1968.
Algunas apendicularias de la costa peruana.
Bol. Inst. Mar Peru 1(9): 535-552

appendicularians, lists of spp.

Fenaux, R. 1966.
Les appendiculaires de la mer Rouge.
Bull. Mus. natn. Hist. nat. Paris (2) 38(6): 784-785.

appendicularians, lists of spp

Fenaux, Robert, 1966.
Synonymie et distribution géographique des appendiculaires.
Bull. Inst. Océanogr., Monaco, 66(1363):23pp.

appendicularians, lists of spp

Fenaux, Robert, 1964.
Les appendiculaires de la troisième campagne du Commandant Robert Giraud en mer d'Arabie.
Bull. Inst. Océanogr., Monaco, 62(1302):14 pp.

appendicularians, lists of spp.

Fenaux, R., 1959.
Observations écologiques sur les appendiculaires du plancton de surface de la Baie de Villefranche-sur-Mer (novembre 1957- octobre 1958).
Trav. Sta. Zool., Villefranche, 18(5):
Reprinted from:
Bull. Inst. Océanogr., Monaco, No. 1141:26 pp.

appendicularians, lists of spp.

Fenaux, R. et J. Godeaux, 1970.
Répartition verticale des tuniciers pélagiques au large d'Eilat (Golfe d'Aqaba). Bull. Soc. R. Sci. de Liège, 39(3-4): 200-209. Also in: Trav. Stn zool., Villefranche-sur-Mer, 32, 1970.

appendicularians, lists of spp

Tokioka, Takasi 1967.
Pacific Tunicata of the United States National Museum.
Bull. U.S. natn. Mus. 251: 1-247.

appendicularians, lists of spp

Tokioka, Takashi, 1962.
Appendicularians of the Japanese Antarctic Research Expedition.
Bull. Mar. Biol. Sta., Asamushi, Tohoku Univ., 10(4):241-245.

appendicularians, lists of spp.

Tokioka, Takasi, 1960
Studies on the distribution of appendicularians and some thaliaceans of the North Pacific with some morphological notes.
Publ. Seto Mar. Biol. Lab., 8(2) (27):351-443, tables.

appendicularians, lists of sp.

Tokioka, T., 1956.
On chaetognaths and appendicularians collected by Mr. Z. Sagara in the Arafura Sea in May-August 1955. Publ. Seto Mar. Biol. Lab., 5(2)(11):203-208.

appendicularians, lists of spp.

Tokioka, T., 1956.
On chaetognaths and appendicularians collected in the central part of the Indian Ocean.
Publ. Seto Mar. Biol. Lab., 5(2)(10):197-202.

appendicularians

Tokioka, T., 1940
Some additional notes on the Japanese appendicularian fauna.
Rec. Oceanogr. Wks., Japan, 11(1):1-26.

appendicularians, lists of spp. [a]

Tundisi, T.M., 1970.
On the seasonal occurrence of appendicularians in waters off the coast of São Paulo State.
Bolm Inst. oceanogr. São Paulo, 19: 131-144.

appendicularians, lists of spp. [a]

Zoppi de Roa, Evelyn 1971.
Apendicularias de la region oriental de Venezuela.
Studies Fauna Curaçao and other Carib. Is. 38(132):77-109

appendicularians, parasites of

Cachon, Jean, et Monique Cachon 1966.
Ultrastructure d'un péridinian parasite d'appendiculaires, Neresheimeria catenata (Neresheimer).
Protistologica 2(4): 17-25.

appendicularians synonomy

Fenaux, Robert, 1966.
Synonymie et distribution géographique des appendiculaires.
Bull. Inst. Oceanogr., Monaco, 66(1363):23pp.

appendicularians, Vertical distribution

Fenaux, R., 1968.
Quelques aspects de la distribution verticale chez les appendicularies en Méditerranée.
Cah. Biol.mar. 9(1):23-29.

Appendicularians, vertical distribution

Fenaux Robert, 1968.
Distribution verticale de la fréquence chez quelques appendiculaires.
Rapp. Proc.-verb. Réun., Comm. int. Explor. scient. Mer Méditerranée 19(3): 513-515.

appendicularians, vertical distribution

Fenaux, R. et J. Godeaux, 1970.
Répartition verticale des tuniciers pélagiques au large d'Eilat (Golfe d'Aqaba). Bull. Soc. R. Sci. de Liège, 39(3-4): 200-209. Also in: Trav. Stn zool., Villefranche-sur-Mer, 32, 1970.

appendicularia, vertical migration [a]

Schmidt, H.-E., 1973
The vertical distribution and diurnal migration of some zooplankton in the Bay of Eilat (Red Sea). Helgoländer wiss. Meeresunters 24(1/4):333-340.

Althoffia pacifica n.sp.

Essenberg, C. E., 1926
Copelata from the San Diego Region. Univ. Calif. Publ. Zool. 28(22):399-521, 170 text figs.

Althoffia tumida

Buckman, A., 1945.
Appendicularia I-III. Fiches d'Ident. Zooplankton, Cons. Perm. Int. Expl. Mer, 7:8 pp., 16 textfigs.

Oceanographic Index: Marine Organisms Cumulation, 1946-1973

Althoffia tumida
Tokioka, T., 1955.
On some plankton animals collected by the Syunkotu maru in May-June 1954. III. Appendicularians. Bull. Biogeogr. Soc., Japan, Vols. 16-19:251-255.

Appendicularia sicula
Bernard, Michelle, 1958
Systématique et distribution saisonnière des tuniciers pélagiques d'Alger.
Rapp. Proc. Verb., Comm. Int. Expl. Sci. Mer Medit., n.s., 14:211-231.

Appendicularia sicula
Berrill, N.J., 1950.
The Tunicata with an account of the British species. Ray Soc. No. 133:354 pp., 120 textfigs.

Appendicularia sicula
Buckman, A., 1945.
Appendicularia I-III. Fiches d'Ident. Zooplancton, Cons. Perm. Int. Expl. Mer, 7:8 pp., 16 textfigs.

Appendicularia sicula
Essenberg, C. E., 1926
Copelata from the San Diego Region. Univ. Calif. Publ. Zool. 28(22):399-521, 170 text figs.

Appendicularia sicula
Fenaux, R., 1967
5. Appendiculaires; Campagne de la Calypso au large des Cotes at antiques de l'Amerique du Sud (1961-1962)
Annls Inst. Oceanogr. n.s. 45 (2): 34-46

Appendicularia sicula
Forneris, Liliana, 1965.
Appendicularian species groups and southern Brazil water masses.
Bolm Inst. Oceanogr., S Paulo, 14(1):54-94.

Appendicularia sicula
Tokioka, T., 1942
Systematic studies of the plankton organisms occurring in Iwayama Bay, Palao VII A preliminary report on the Appendicularian fauna of the bay and adjacent waters. Palao Tropical Biological Sta. Studies, 2(3): 613-616 (Contrib. 106, Seto Mar. Biol. Lab.)

Appendicularia sicula
Tokioka, T., 1940
Some additional notes on the Japanese appendicularian fauna.
Rec. Oceanogr. Wks., Japan, 11(1):1-26.

Appendicularia tregouboffi n.sp.
Fenaux, R., 1960.
Un appendiculaire nouveau, Appendicularia tregouboffi n.sp. récolté dans le plancton de Villefranche-sur-Mer.
Bull. Soc. Zool., France, 85(1):L20-122.

Also in:
Trav. Sta. Zool., Villefranche-sur-Mer, 19(1960)

Chunopleura microgaster
Tokioka, T., 1940
Some additional notes on the Japanese appendicularian fauna.
Rec. Oceanogr. Wks., Japan, 11(1):1-26.

Folia gracilis
Tokioka, Takasi, 1959.
Further notes on some appendicularians from the eastern Pacific.
Publ. Seto Mar. Biol. Lab., 7(1):1-17.

Fritillaria sp.
Bigelow, H.B., and M. Sears, 1939
Studies of the waters of the continental shelf, Cape Cod to Chesapeake Bay. III. A volumetric study of the zooplankton. Mem. M.C.Z. 54(4):183-378, 42 text figs.

Fritillaria sp.
Fenaux, R., 1965.
A propos des expansions cuticulaires du tronc de quelques fritillaires.
Rapp. Proc. Verb., Reunions, Comm. Int. Expl. Sci., Mer Mediterranee, Monaco. 18(2):455-456.

Fritillaria sp.
Johnson, M.W. (undated)
The production and distribution of Zooplankton in the surface waters of Bering Sea and Bering Strait, with special reference to copepods, echinoderms, mollusks and annelids. Report of oceanographic cruise of U.S. Coast Guard Cutter Chelan 1934 Part II (B):45-85, fig.1, table 1.

Fritillaria sp.
Lebour, M.V., 1947
Notes on the inshore plankton of Plymouth. JMBA 26(4):527-547.

Fritillaria aberrans
Buckman, A., 1945.
Appendicularia I-III. Fiches d'Ident. Zooplancton, Cons. Perm. Int. Expl. Mer, 7:8 pp., 16 textfigs.

Fritillaria aberrans
Tokioka, Takasi, 1959.
Further notes on some appendicularians from the eastern Pacific.
Publ. Seto Mar. Biol. Lab., 7(1):1-17.

Fritillaria abjornseni
Tokioka, T., 1955.
General considerations on Japanese appendicularian fauna. Publ. Seto Mar. Biol. Lab., 4(2/3): 251-261, 6 textfigs.

Fritillaria abjorsensi
Tokioka, T., 1942
Systematic studies of the plankton organisms occurring in Iwayama Bay, Palao VII A preliminary report on the Appendicularian fauna of the bay and adjacent waters. Palao Tropical Biological Sta. Studies, 2(3): 613-616 (Contrib. 106, Seto Mar. Biol. Lab.)

Fritillaria aequatorialis
Bernard, Michelle, 1958
Systématique et distribution saisonnière des tuniciers pélagiques d'Alger.
Rapp. Proc. Verb., Comm. Int. Expl. Sci., Mer Medit., n.s., 14:211-231.

Fritillaria aequatorialis
Björnberg, T.K.S., and L. Forneris, 1955.
Resultados cientificos do cruzeiro do "Baependi" e do "Vega" à ilha Trindade.
Contrib. Avulsas, Inst. Oceanograf. (Ocean. Biol.), Sao Paulo, No. 1:1-68.

Fritillaria aequatorialis
Tokioka, T., 1957.
Two new appendicularians deom the astern Pacific with notes on the morphology of Fritillaria aequatorialis and Tectillaria fertilis. Amer. Microsc. Soc., 76(4):359-365.

Fritillaria amygdala n.sp.
Essenberg, C. E., 1926
Copelata from the San Diego Region. Univ. Calif. Publ. Zool. 28(22):399-521, 170 text figs.

Fritillaria angularis n.sp.
Essenberg, C. E., 1926
Copelata from the San Diego Region. Univ. Calif. Publ. Zool. 28(22):399-521, 170 text figs.

Fritillaria antarctica
Thompson, Harold, 1954.
Pelagic tunicates. Identifications and relevant notes. B.A.N.Z. Antarctic Res. Exped., 1929-1931 Repts., B(Zool. Bot.), 1(4):183-185.

Fritillaria (Acroecerous) antarctica
Tokioka, Takasi, 1964.
Taxonomic studies of appendicularians collected by the Japanese Antarctic Research Expedition, 1957.
Japan. Antarct. Res. Exped., 1956-1962, Sci. Repts. (E)(JARE Sci. Repts., Biol.), No. 2:16pp.

Fritillaria aplostoma
Björnberg, T.K.S., and L. Forneris, 1955.
Resultados cientificos do cruzeiro "Baependi" e do "Vega" à ilha Trindade.
Contrib. Avulsas, Inst. Oceanograf. (Ocean. Biol.), Sao Paulo, No. 1:1-68.

Fritillaria arafoera n.sp.
Tokioka, T., 1956.
Fritillaria arafoera n.sp., a form of the sibling species Fritillaria haplostoma-complex. (Appendicularia:Chordata). Pacific Science 10(4):403-406.

Fritillaria artus n.sp.
Essenberg, C. E., 1926
Copelata from the San Diego Region. Univ. Calif. Publ. Zool. 28(22):399-521, 170 text figs.

Fritillaria bicornis
Tokioka, T., 1942
Systematic studies of the plankton organisms occurring in Iwayama Bay, Palao VII A preliminary report on the Appendicularian fauna of the bay and adjacent waters. Palao Tropical Biological Sta. Studies, 2(3): 613-616 (Contrib. 106, Seto Mar. Biol. Lab.)

Fritillaria borealis
Berrill, N.J., 1950.
The Tunicata with an account of the British species. Ray Soc. No. 133:354 pp., 120 textfigs.

Fritillaria borealis
Björnberg, T.K.S., and L. Forneris, 1955.
Resultados cientificos do cruzeiro do "Baependi" e do "Vega" à ilha Trindade.
Contrib. Avulsas, Inst. Oceanograf. (Ocean. Biol.), Sao Paulo, No. 1:1-68.

Fritillaria borealis
Ciszewski, P., 1962
Southern Baltic zooplankton.
Prace Morsk. Inst. Ryback., Gdyni, Oceanogr -Ichtiol., 11(A):37-58.

Fritillaria borealis
Essenberg, C. E., 1926
Copelata from the San Diego Region. Univ. Calif. Publ. Zool. 28(22):399-521, 170 text figs.

Fritillaria borealis
Fenaux, Robert, 1963.
Ecologie et biologie des appendiculaires méditerranéens.
Vie et Milieu, Suppl., (16):142 pp.

Fritillaria borealis
Fenaux, R., 1959.
Observations écologiques sur les appendiculaires du plancton de surface de la Baie de Villefranche-sur-Mer (novembre 1957-octobre 1958).
Trav. Sta. Zool., Villefranche, 18(5):

Reprinted from:
Bull. Inst. Océanogr., Monaco, No. 1141:26 pp.

Fritillaria borealis
Fish, C.J., and M.W. Johnson, 1937
The biology of the zooplankton population in the Bay of Fundy and the Gulf of Maine with special reference to production and distribution. J. Biol. Bd., Canada 3(3):189-322, 29 tables, 45 text figs.

Fritillaria borealis
Forneris, Liliana, 1965.
Appendicularian species groups and southern Brazil water masses.
Bolm Inst. Oceanogr., S Paulo, 14(1):54-94.

Fritillaria borealis
Fraser, J.H., 1961
The oceanic bathypelagic plankton of the North-East Atlantic and its possible significance to fisheries.
Dept. Agric. & Fish., Scotland, Marine Research (4):1-48.

Fritillaria borealis
Fraser, J. H., 1949
Plankton investigations from the Scottish Research Vessel. Ann. Biol., Int. Cons., 4: 66-67.

Fritillaria borealis
Frost, N., S.T. Lindsay, and H. Thompson, 1934.
III. Hydrographic and biological investigations.
B. Plankton. Ann. Rept., Newfoundland Fish. Res. Comm. 2(2):47-49, Textfigs. 6-11.

Fritillaria borealis
Frost, N., S.T. Lindsay, and H. Thompson, 1933.
III. Hydrography and biological investigations.
B. Plankton more abundant in 1932 than 1931.
Ann. Rept. Newfoundland Fish. Res. Comm. 2(1):58-74, Textfigs. 17-27.

Fritillaria borealis
Furnestin, M. L., 1957.
Chaetognathes et zooplancton du secteur atlantique marocain. Rev. Trav. Inst. Pêches Marit. 21(1/2):9-356.

Fritillaria borealis
Hansen, Vagn Kr., 1960.
Investigations on the quantitative and qualitative distribution of zooplankton in the southern part of the Norwegian Sea.
Medd. Danmarks Fiskeri- og Havundersøgelser, n.s., 2(23):1-53.

Fritillaria borealis
Levander, K.M., 1947
Plankton gesammelt in den Jahren 1899-1910 an den Küsten Finnlands.
Finnländische Hydrographisch-Biologishhe Untersuchunger (aus dem Wasserbiologischen Laboratorin der Societas Scientiarum Fennica) No.11:40 pp., 6 diagrams, 13 pls., tables.

Fritillaria borealis
Lindquist, Armin, 1959
Studien über das Zooplankton der Bottensee II. Zur Verbreitung und Zusammensetzung des Zooplanktons.
Inst. Mar. Res., Lysekil, Ser. Biol., Rept., No. 11:136 pp.

Fritillaria borealis
Tokioka, T., 1940
Some additional notes on the Japanese appendicularian fauna.
Rec. Oceanogr. Wks., Japan, 11(1):1-26.

Fritillaria borealis
Wiborg, K.F., 1944
The production of zooplankton in a landlocked fjord, the Nordåsvatn near Bergen, in 1941-42, with special reference to the copepods... (Repts. Norwegian Fish. and Mar. Invest.) 7(7):83 pp., 40 text figs.

Fritillaria borealis acuta
Buckman, A., 1945.
Appendicularia I-III. Fiches d'Ident. Zooplancton, Cons. Perm. Int. Expl. Mer, 7:8 pp., 16 textfigs.

Fritillaria borealis f. intermedia
Gamulin, T., 1948
Prilog poznavanju Zooplanktona Srednjedalmatinskog Otocnog Podrucja. (Contrib. a la connaissance du zooplankton de la zone insulaire de la Dalmatie Moyenne). Acta Adriatica 3(7):38 pp., 6 tables, 1 map.

Fritillaria borealis f. intermedia
Tokioka, T., 1942
Systematic studies of the plankton organisms occurring in Iwayama Bay, Palao VII A preliminary report on the Appendicularian fauna of the bay and adjacent waters. Palao Tropical Biological Sta. Studies, 2(3):613-616 (Contrib. 106, Seto Mar. Biol. Lab.)

Fritillaria (Eurycercus) borealis intermedia
Tokioka, T., and J.A. Suarez Caabro, 1956.
Apendicularias de los mares cubanos.
Mem. Soc. Cubana Hist. Nat., 23(1):37-80, 15 pls.

Fritillaria borealis ritteri
Delsman, H.C., 1939.
Preliminary plankton investigations in the Java Sea. Treubia 17:139-181, 8 maps, 41 figs.

Fritillaria borealis sargassi
Fenaux, R., 1967
5. Appendiculaires; Campagne de la Calypso au large des Cotes at antiques de l'Amerique du Sud (1961-1962)
Annls Inst. Oceanogr. n.s. 45 (2): 34-46

Fritillaria (Eurycercus) borealis f. sargassi
Tokioka, T., 1955.
On some plankton animals collected by the Syunkotu maru in May-June 1954. III. Appendicularians. Bull. Biogeogr. Soc., Japan, Vols., 16-19:251-255.

Fritillaria borealis sargassi
Tokioka, T., 1950.
Droplets from the plankton net. Publ. Seto Mar. Biol. Lab. 1(3):152-157, 9 textfigs.

Fritillaria borealis f. sargassi
Tokioka, T., 1942
Systematic studies of the plankton organisms occurring in Iwayama Bay, Palao VII A preliminary report on the Appendicularian fauna of the bay and adjacent waters. Palao Tropical Biological Sta. Studies, 2(3):613-616 (Contrib. 106, Seto Mar. Biol. Lab.)

Fritillaria (Eurycercus) borealis sargassi
Tokioka, T., and J.A. Suarez Caabro, 1956.
Apendicularias de los mares cubanos.
Mem. Soc. Cubana Hist. Nat., 23(1):37-80, 15 pls.

Fritillaria (Eurycercus) borealis f. typica
Tokioka, Takasi, 1964
Taxonomic studies of appendicularians collected by the Japanese Antarctic Research Expedition, 1957.
Japan. Antarct. Res. Exped., 1956-1962, Sci. Repts. (E)(JARE Sci. Repts., Biol.), No. 2:16 pp.

Fritillaria borealis
Wiktor, K., 1963.
Zooplankton of the Pomeranian Bay. (In Polish; Russian and English summary).
Prace Morsk. Inst. Ryback. w Gdyni, 12A:51-78.

Fritillaria borealis acuta
Strømgren, Tor 1973.
Zooplankton investigations in Borgenfjorden, 1967-1969.
Miscellanea, Norske Vidensk. Selsk. Mus. Trondheim 9:37pp.

Fritillaria borealis truncata
Buckman, A., 1945.
Appendicularia I-III. Fiches d'Ident. Zooplancton, Cons. Perm. Int. Expl. Mer, 7:8 pp., 16 textfigs.

Fritillaria borealis truncata f. intermedia
Russell, F.S., and J.S. Colman, 1935
The Zooplankton. IV. The occurrence and seasonal distribution of the Tunicata, Mollusca and Coelenterata (siphonophora). Brit. Mus. (N.H.) Great Barrier Reef Expedition 1928-1929, Sci. Repts., 2(7):203-276, 30 text figs.

Fritillaria brevicollis n.sp.
Essenberg, C. E., 1926
Copelata from the San Diego Region. Univ. Calif. Publ. Zool. 28(22):399-521, 170 text figs.

Fritillaria campila n.sp.
Essenberg, C. E., 1926
Copelata from the San Diego Region. Univ. Calif. Publ. Zool. 28(22):399-521, 170 text figs.

Fritillaria charybdea (fig
Tokioka, T., 1951.
Pelagic tunicates and chaetognaths collected during the cruises to the New Yamato Bank in the Sea of Japan. Publ. Seto Mar. Biol. Lab. 2(1):1-26, 1 chart, 6 tables, 12 textfigs.

Fritillaria claudaria n.sp.
Essenberg, C. E., 1926
Copelata from the San Diego Region. Univ. Calif. Publ. Zool. 28(22):399-521, 170 text figs.

Fritillaria clava n.sp.
Essenberg, C. E., 1926
Copelata from the San Diego Region. Univ. Calif. Publ. Zool. 28(22):399-521, 170 text figs.

Fritillaria delicata n.sp.
Essenberg, C. E., 1926
Copelata from the San Diego Region. Univ. Calif. Publ. Zool. 28(22):399-521, 170 text figs.

Fritillaria diafana n.sp.
Essenberg, C. E., 1926
Copelata from the San Diego Region. Univ. Calif. Publ. Zool. 28(22):399-521, 170 text figs.

Fritillaria dispara n.sp.
Essenberg, C. E., 1926
Copelata from the San Diego Region. Univ. Calif. Publ. Zool. 28(22):399-521, 170 text figs.

Fritillaria drygalski
Bückmann, A. 1967.
Untersuchungen über das Makroplankton bei Ischia und Capri und im Golf von Neapel im Mai 1962. III Die Appendicularien.
Pubbl. Staz. zool. Napoli 35(2):215-298.

Fritillaria exilis n.sp.
Essenberg, C. E., 1926
Copelata from the San Diego Region. Univ. Calif. Publ. Zool. 28(22):399-521, 170 text figs.

Fritillaria fagei n.sp.
Fenaux, R., 1961
Fritillaria fagei n.sp. un appendiculaire nouveau découvert dans le plancton de Villefranche-sur-Mer.
Rapp. Proc. Verb., Réunions, Comm. Int. Expl. Sci. Mer Méditerranée, Monaco, 16(2):147-148.

Fritillaria formica
Björnberg, T.K.S., and L. Forneris, 1955.
Resultados cientificos do cruzeiro do "Baépendi" e do "Vega" à ilha Trindade.
Contrib. Avulsas, Inst. Oceanograf. (Ocean. Biol.), Sao Paulo, No. 1:1-68.

Fritillaria formica
Essenberg, C. E., 1926
Copelata from the San Diego Region. Univ. Calif. Publ. Zool. 28(22):399-521, 170 text figs.

Fritillaria formica
Fenaux, R., 1959.
Observations écologiques sur les appendiculaires du plancton de surface de la Baie de Villefranche-sur-Mer (novembre 1957-octobre 1958).
Trav. Sta. Zool., Villefranche, 18(5):
Reprinted from:
Bull. Inst. Océanogr., Monaco, No. 1141:26 pp.

Fritillaria formica
Forneris, Liliana, 1965.
Appendicularien species groups and southern Brazil water masses.
Bolm Inst. Oceanogr., S Paulo, 14(1):54-94.

Fritillaria formica
Gamulin, T., 1948
Prilog poznavanju Zooplanktona Srednjedalmatinskog Otocnog Podrucja. (Contrib. a la connaissance du zooplankton de la zone insulaire de la Dalmatie Moyenne). Acta Adriatica 3(7):38 pp., 6 tables, 1 map.

Fritillaria formica

Hure, J., 1961
Dneva migracija i sezonska vertikalna raspodjela zooplanktona dubljeg mora. [Migration journalière et distribution saisonnaière verticale du zooplancton dans la région profunde de l'Adriatique]
Acta Adriatica, 9(6):1-59.

Fritillaria formica

Sewell, R.B.S., 1953.
The pelagic Tunicata. Brit. Mus. (N.H.), John Murray Exped., 1933-34, Sci. Repts. 10(1):1-90, 1 pl., 32 textfigs.

Fritillaria (Acrocercus) formica

Tokioka Takasi, 1964.
Taxonomic studies of appendicularians collected by the Japanese Antarctic Research Expedition, 1957.
Japan. Antarct. Res. Exped., 1956-1962, Sci. Repts. (E)(JARE Sci. Repts., Biol.), No. 2:16pp.

Fritillaria formica

Tokioka, T., 1942
Systematic studies of the plankton organisms occurring in Iwayama Bay, Palao VII A preliminary report on the Appendicularian fauna of the bay and adjacent waters.
Palao Tropical Biological Sta. Studies, 2(3): 613-616 (Contrib. 106, Seto Mar. Biol. Lab.)

Fritillaria formica

Tokioka, T., 1940
Some additional notes on the Japanese appendicularian fauna.
Rec. Oceanogr. Wks., Japan, 11(1):1-26.

Fritillaria (Acrocercus) formica

Tokioka, T., and J.A. Suarez Caabro, 1956.
Apendicularias de los mares cubanos.
Mem. Soc. Cubana Hist. Nat., 23(1):37-80, 15 pls.

Fritillaria formica f. digitata

Tokioka, T., 1942
Systematic studies of the plankton organisms occurring in Iwayama Bay, Palao VII A preliminary report on the Appendicularian fauna of the bay and adjacent waters.
Palao Tropical Biological Sta. Studies, 2(3): 613-616 (Contrib. 106, Seto Mar. Biol. Lab.)

Fritillaria formica tuberculata

Fenaux, R., 1967
5. Appendiculaires; Campagne de la Calypso au large des Cotes at antiques de l'Amerique du Sud (1961-1962)
Annls Inst. Oceanogr. n.s. 45 (2): 34-46

Fritillaria fraudax

Fenaux, R., 1959.
Observations écologiques sur les appendiculaires du plancton de surface de la Baie de Villefranche-sur-Mer (novembre 1957-octobre 1958).
Trav. Sta. Zool., Villefranche, 18(5):
Reprinted from:
Bull. Inst. Oceanogr., Monaco, No. 1141:26 pp.

Fritillaria fraudax

Tokioka, T., 1940
Some additional notes on the Japanese appendicularian fauna.
Rec. Oceanogr. Wks., Japan, 11(1):1-26.

Fritillaria furcata

Browne, E.T., 1905.
Notes on the pelagic fauna of the Firth of Clyde (1901-1902). Proc. R. Soc., Edinburgh, 25(9): 779-791.

Fritillaria gigas n.sp.

Essenberg, C. E., 1926
Copelata from the San Diego Region. Univ. Calif. Publ. Zool. 28(22):399-521, 170 text figs.

Fritillaria gracilis

Forneris, Liliana, 1965.
Appendicularian species groups and southern Brazil water masses.
Bolm Inst. Oceanogr., S Paulo, 14(1):54-94.

Fritillaria gracilis

Tokioka, T., 1955.
General considerations on Japanese appendicularian fauna. Publ. Seto Mar. Biol. Lab., 4(2/3): 251-261, 6 textfigs.

Fritillaria gracilis

Tokioka, T., 1942
Systematic studies of the plankton organisms occurring in Iwayama Bay, Palao VII A preliminary report on the Appendicularian fauna of the bay and adjacent waters.
Palao Tropical Biological Sta. Studies, 2(3): 613-616 (Contrib. 106, Seto Mar. Biol. Lab.)

Fritillaria haplostoma

Essenberg, C. E., 1926
Copelata from the San Diego Region. Univ. Calif. Publ. Zool. 28(22):399-521, 170 text figs.

Fritillaria haplostoma

Fenaux, R., 1967
5. Appendiculaires; Campagne de la Calypso au large des Cotes at antiques de l'Amerique du Sud (1961-1962)
Annls Inst. Oceanogr. n.s. 45 (2): 34-46

Fritillaria haplostoma

Fenaux, R., 1959.
Observations écologiques sur les appendiculaires du plancton de surface de la Baie de Villefranche-sur-Mer (novembre 1957-octobre 1958).
Trav. Sta. Zool., Villefranche, 18(5):
Reprinted from:
Bull. Inst. Oceanogr., Monaco, No. 1141:26 pp.

Fritillaria haplostoma

Forneris, Liliana, 1965.
Appendicularian species groups and southern Brazil water masses.
Bolm Inst. Oceanogr., S Paulo, 14(1):54-94.

Fritillaria haplostoma

Gamulin, T., 1948
Prilog poznavanju Zooplanktona Srednjedalmatinskog Otocnog Podrucja. (Contrib. a la connaissance du zooplankton de la zone insulaire de la Dalmatie Moyenne). Acta Adriatica 3(7):38 pp., 6 tables, 1 map.

Fritillaria haplostoma

Hastings, A. B., 1931
Tunicata. Great Barrier Reef Exped., 1928-29. Sci. Repts. 4(3):69-100, 17 figs-top, 5plo.

Fritillaria haplostoma

Russell, F.S., and J.S. Colman, 1935 Fol.
The Zooplankton. IV. The occurrence and seasonal distribution of the Tunicata, Mollusca and Coelenterata (siphonophora). Brit. Mus. (N.H.) Great Barrier Reef Expedition 1928-1929, Sci. Repts., 2(7):203-276, 30 text figs.

Fritilleria (Acrocercus) haplostoma f. glandularis, nov.

Tokioka, Takasi, 1964.
Taxonomic studies of appendicularians collected by the Japanese Antarctic Research Expedition 1957.
Japan. Antarct. Res. Exped., 1956-1962, Sci. Repts., (E)(JARE Sci. Repts., Biol.), No. 2:16pp.

Fritillaria haplostoma

Tokioka, T., 1956.
Fritillaria arafoera n.sp., a form of the sibling species Fritillaria haplostoma-complex (Appendicularia:Chordata). Pacific Science 10(4):403-406.

Fritillaria (Acrocercus) haplostoma

Tokioka, T., 1955.
On some plankton animals collected by the Syunkotu maru in May-June 1954. III. Appendicularians. Bull. Biogeogr. Soc., Japan, Vols. 16-19:251-255.

Fritillaria haplostoma

Tokioka, T., 1942
Systematic studies of the plankton organisms occurring in Iwayama Bay, Palao VII A preliminary report on the Appendicularian fauna of the bay and adjacent waters.
Palao Tropical Biological Sta. Studies, 2(3): 613-616 (Contrib. 106, Seto Mar. Biol. Lab.)

Fritillaria haplostoma

Tokioka, T., 1940
Some additional notes on the Japanese appendicularian fauna.
Rec. Oceanogr. Wks., Japan, 11(1):1-26.

Fritillaria (Acrocercus) haplostoma

Tokioka, T., and J.A. Suarez Caabro, 1956.
Apendicularias de los mares cubanos.
Mem. Soc. Cubana Hist. Nat., 23(1):37-80, 15 pls.

Fritillaria inverta n.sp.

Essenberg, C. E., 1926
Copelata from the San Diego Region. Univ. Calif. Publ. Zool. 28(22):399-521, 170 text figs.

Fritillaria juncea n.sp.

Essenberg, C. E., 1926
Copelata from the San Diego Region. Univ. Calif. Publ. Zool. 28(22):399-521, 170 text figs.

Fritillaria limpida n.sp.

Essenberg, C. E., 1926
Copelata from the San Diego Region. Univ. Calif. Publ. Zool. 28(22):399-521, 170 text figs.

Fritillaria lohmanni n.sp.

Essenberg, C. E., 1926
Copelata from the San Diego Region. Univ. Calif. Publ. Zool. 28(22):399-521, 170 text figs.

Fritillaria lubicila n.sp.

Essenberg, C. E., 1926
Copelata from the San Diego Region. Univ. Calif. Publ. Zool. 28(22):399-521, 170 text figs.

Fritillaria macrotrachela n.sp.

Essenberg, C. E., 1926
Copelata from the San Diego Region. Univ. Calif. Publ. Zool. 28(22):399-521, 170 text figs.

Fritillaria magna

Tokioka, Takasi, 1958.
Further notes on some appendicularians from the eastern Pacific.
Publ. Seto Mar. Biol. Lab., 7(1):1-17.

Fritillaria magna

Tokioka, T., 1951.
Pelagic tunicates and chaetognaths collected during the cruises to the New Yamato Bank in the Sea of Japan. Publ. Seto Mar. Biol. Lab. 2(1): 1-26, 1 chart, 6 tables, 12 textfigs.

? Fritillaria magna

Tokioka, T., 1942
Systematic studies of the plankton organisms occurring in Iwayama Bay, Palao VII A preliminary report on the Appendicularian fauna of the bay and adjacent waters.
Palao Tropical Biological Sta. Studies, 2(3): 613-616 (Contrib. 106, Seto Mar. Biol. Lab.)

Fritillaria megachile

Björnberg, T.K.S., and L. Forneris, 1955
Resultados científicos do cruzeiro do "Baependi" e do "Vega" à ilha Trindade.
Contrib. Avulsas, Inst. Oceanograf. (Ocean. Biol.), Sao Paulo, No. 1:1-68.

Fritillaria megachile

Fenaux, R., 1967
5. Appendiculaires; Campagne de la Calypso au large des Cotes at antiques de l'Amerique du Sud (1961-1962)
Annls Inst. Oceanogr. n.s. 45 (2): 34-46

Fritillaria megachile
Fenaux, R., 1956.
Observations écologiques sur les appendiculaires du plancton de surface de la Baie de Villefranche-sur-Mer (novembre 1957-octobre 1958).
Trav. Sta. Zool., Villefranche, 18(5):

Reprinted from:
Bull. Inst. Océanogr., Monaco, No. 1141:26 pp.

Fritillaria megachile
Forneris, Liliana, 1965.
Appendicularian species groups and southern Brazil water masses.
Bolm Inst. Oceanogr., S Paulo. 14(1):54-94.

Fritillaria megachile
Furnestin, M. L., 1957.
Chaetognathes et zooplancton du secteur atlantique marocain. Rev. Trav. Inst. Pêches Marit. 21(1/2): 9-356.

Fritillaria megachile
Tokioka, T., 1942
Systematic studies of the plankton organisms occurring in Iwayama Bay, Palao VII A preliminary report on the Appendicularian fauna of the bay and adjacent waters.
Palao Tropical Biological Sta. Studies, 2(3): 613-616 (Contrib. 106, Seto Mar. Biol. Lab.)

Fritillaria megachile
Tokioka, T., 1940
Some additional notes on the Japanese appendicularian fauna.
Rec. Oceanogr. Wks., Japan, 11(1):1-26.

Fritillaria (Eurycercus) megachile
Tokioka, T., and J.A. Suarez Caabro, 1956.
Apendicularias de los mares cubanos.
Mem. Soc. Cubana Hist. Nat., 23(1):37-80, 15 pls.

Fritillaria messanensis
Björnberg, T.K.S., and L. Forneris, 1955.
Resultados científicos do cruzeiro do "Baependi" e do "Vega" à ilha Trindade.
Contrib. Avulsas, Inst. Oceanograf. (Ocean. Biol.), Sao Paulo, No. 1:1-68.

Fritillaria nitida n.sp.
Essenberg, C. E., 1926
Copelata from the San Diego Region. Univ. Calif. Publ. Zool. 28(22):399-521, 170 text figs.

Fritillaria pacifica n.sp.
Tokioka, Takasi, 1958.
Further notes on some appendicularians from the eastern Pacific.
Publ. Seto Mar. Biol. Lab., 7(1):1-17.

Fritillaria pellucida
Björnberg, T.K.S., and L. Forneris, 1955.
Resultados científicos do cruzeiro do "Baependi" e do "Vega" à ilha Trindade.
Contrib. Avulsas, Inst. Oceanograf. (Ocean. Biol.), Sao Paulo, No. 1:1-68.

Fritillaria pellucida
Bückmann, A., 1967.
Untersuchungen über das Macroplankton bei Ischia und Capri und im Golf von Neapel im Mai 1962. II Die Appendicularien.
Publ. Staz. Zool. Napoli, 35 (2), 215-235.

Fritillaria pellucida
Cachon, Jean, et Monique Cachon-Enjumet, 1964.
Cycle évolutif et cytologie de Neresheimeria catenata Neresheimer, péridinien parasite d'appendiculaires.
Ann. Sci. Nat., Zool. et Biol. Animale, (12), 4: 779-800.

Fritillaria pellucida
Essenberg, C. E., 1926
Copelata from the San Diego Region. Univ. Calif. Publ. Zool. 28(22):399-521, 170 text figs.

Fritillaria pellucida
Fenaux, R., 1967
5. Appendiculaires; Campagne de la Calypso au large des Cotes at antiques de l'Amerique du Sud (1961-1962)
Annls Inst. Oceanogr. n.s. 45 (2): 34-46

Fritillaria pellucida
Fenaux, R., 1966.
Une variété de Fritillaria pellucida (Busch), 1851, récoltée dans la Mer d'Oman.
Cah. ORSTOM, Sér. Oceanogr., 4(2):147-151.

Fritillaria pellucida
Fenaux, Robert, 1963.
Ecologie et biologie des appendiculaires méditerranéens.
Vie et Milieu, Suppl., (16):142 pp.

Fritillaria pellucida
Fenaux, R., 1961.
Rôle du pylore chez Fritillaria pellucida Buch (Appendiculaire).
C.R. Acad. Sci., Paris, 252(19):2936-2939.

Fritillaria pellucida
Fenaux, R., 1959.
Observations écologiques sur les appendiculaires du plancton de surface de la Baie de Villefranche-sur-Mer (novembre 1957-octobre 1958).
Trav. Sta. Zool., Villefranche, 18(5):

Reprinted from:
Bull. Inst. Océanogr., Monaco, No. 1141:26 pp.

Fritillaria pellucida
Forneris, Liliana, 1965.
Appendicularian species groups and southern Brazil water masses.
Bolm Inst. Oceanogr., S Paulo, 14(1):54-94.

Fritillaria pellucida
Fraser, J.H., 1961
The oceanic bathypelagic plankton of the North-East Atlantic and its possible significance to fisheries.
Dept. Agric. & Fish., Scotland, Marine Research (4):1-48.

Fritillaria pellucida
Furnestin, M. L., 1957.
Chaetognathes et zooplancton du secteur atlantique marocain. Rev. Trav. Inst. Pêches Marit. 21(1/2): 9-356.

Fritillaria pellucida
Gamulin, T., 1948
Prilog poznavanju Zooplanktona Srednjedalmatinskog Otocnog Podrucja. (Contrib. a la connaissance du zooplankton de la zone insulaire de la Dalmatie Moyenne). Acta Adriatica 3(7):38 pp., 6 tables, 1 map.

Fritillaria pellucida
Hure, J., 1961
Dneva migracija i sezonaka vertikalna raspodjela zooplanktona dubljeg mora. [Migration journalière et distribution saisonnière verticale du zooplancton dans la région profunde de l'Adriatique]
Acta Adriatica, 9(6):1-59.

Fritillaria pellucida
Russell, F.S., and J.S. Colman, 1935   Busch
The Zooplankton. IV. The occurrence and seasonal distribution of the Tunicata, Mollusca and Coelenterata (siphonophora). Brit. Mus. (N.H.) Great Barrier Reef Expedition 1928-1929, Sci. Repts., 2(7):203-276, 30 text figs.

Fritillaria pellucida
Sewell, R.B.S., 1953.
The pelagic Tunicata. Brit. Mus. (N.H.), John Murray Exped., 1933-34, Sci. Repts. 10(1):1-90, 1 pl., 32 textfigs.

Fritillaria pellucida
Tokioka, T., 1942
Systematic studies of the plankton organisms occurring in Iwayama Bay, Palao VII A preliminary report on the Appendicularian fauna of the bay and adjacent waters.
Palao Tropical Biological Sta. Studies, 2(3): 613-616 (Contrib. 106, Seto Mar. Biol. Lab.)

Fritillaria pellucida
Tokioka, T., 1940
Some additional notes on the Japanese appendicularian fauna.
Rec. Oceanogr. Wks., Japan, 11(1):1-26.

Fritillaria (Eurycercus) pellucida
Tokioka, T., and J.A. Suarez Caabro, 1956.
Apendicularias de los mares cubanos.
Mem. Soc. Cubana Hist. Nat., 23(1):37-80, 15 pls.

Fritillaria plana n.sp.
Essenberg, C. E., 1926
Copelata from the San Diego Region. Univ. Calif. Publ. Zool. 28(22):399-521, 170 text figs.

Fritillaria pulchrituda n.sp.
Essenberg, C. E., 1926
Copelata from the San Diego Region. Univ. Calif. Publ. Zool. 28(22):399-521, 170 text figs.

Fritillaria ritteri
Essenberg, C. E., 1926
Copelata from the San Diego Region. Univ. Calif. Publ. Zool. 28(22):399-521, 170 text figs.

Fritillaria sargassi
Essenberg, C. E., 1926
Copelata from the San Diego Region. Univ. Calif. Publ. Zool. 28(22):399-521, 170 text figs.

Fritillaria tacita n.sp.
Essenberg, C. E., 1926
Copelata from the San Diego Region. Univ. Calif. Publ. Zool. 28(22):399-521, 170 text figs.

Fritillaria taeniogona n. sp.
Tokioka, T., 1957.
Two new appendicularians from the Eastern Pacific with notes on the morphology of Fritillaria aequatorialis and Tectillaria fertilis. Trans. Amer. Microsc. Soc., 76(4):359-365.

Fritillaria tenebra n.sp.
Essenberg, C. E., 1926
Copelata from the San Diego Region. Univ. Calif. Publ. Zool. 28(22):399-521, 170 text figs.

Fritillaria tenella
Buckman, A., 1945.
Appendicularia I-III. Fiches d'Ident. Zooplancton Cons. Perm. Int. Expl. Mer, 7:8 pp., 16 textfigs.

Fritillaria tenella
Fenaux, R., 1967
5. Appendiculaires; Campagne de la Calypso au large des Cotes at antiques de l'Amerique du Sud (1961-1962)
Annls Inst. Oceanogr. n.s. 45 (2): 34-46

Fritillaria tenella
Fenaux, R., 1959.
Observations écologiques sur les appendiculaires du plancton de surface de la Baie de Villefranche-sur-Mer (novembre 1957-octobre 1958).
Trav. Sta. Zool., Villefranche, 18(5):

Reprinted from:
Bull. Inst. Océanogr., Monaco, No. 1141:26 pp.

Fritillaria (Eurycercus) tenella
Tokioka, Takasi, 1964.
Taxonomic studies of appendicularians collected by the Japanese Antarctic Research Expedition, 1957.
Japan. Antarct. Res. Exped., 1956-1962, Sci. Repts. (E)JARE Sci. Repts., Biol.), No. 2:16 pp.

Fritillaria tenella (fig.)
Tokioka, T., 1951.
Pelagic tunicates and chaetognaths collected during the cruises to the New Yamato Bank in the Sea of Japan. Publ. Seto Mar. Biol. Lab. 2(1):1-26, 1 chart, 6 tables, 12 textfigs.

Fritillaria tenella
Tokioka, T., 1942
Systematic studies of the plankton organisms occurring in Iwayama Bay, Palao VII A preliminary report on the Appendicularian fauna of the bay and adjacent waters. Palao Tropical Biological Sta. Studies, 2(3): 613-616 (Contrib. 106, Seto Mar. Biol. Lab.)

Fritillaria tenella
Tokioka, T., 1940
Some additional notes on the Japanese appendicularian fauna.
Rec. Oceanogr. Wks., Japan, 11(1):1-26.

Fritillaria tereta nsp.
Essenberg, C. E., 1926
Copelata from the San Diego Region. Univ. Calif. Publ. Zool. 28(22):399-521, 170 text figs.

Fritillaria trigonis n.sp.
Essenberg, C. E., 1926
Copelata from the San Diego Region. Univ. Calif. Publ. Zool. 28(22):399-521, 170 text figs.

Fritillaria truncata n.sp.
Essenberg, C. E., 1926
Copelata from the San Diego Region. Univ. Calif. Publ. Zool. 28(22):399-521, 170 text figs.

Fritillaria velocita n.sp.
Essenberg, C. E., 1926
Copelata from the San Diego Region. Univ. Calif. Publ. Zool. 28(22):399-521, 170 text figs.

Fritillaria venusta
Berrill, N.J., 1950.
The Tunicata with an account of the British species. Ray Soc., No. 133:354 pp., 120 textfigs.

Fritillaria venusta
Buckman, A., 1945.
Appendicularia I-III. Fiches d'Ident. Zooplancton, Cons. Perm. Int. Expl. Mer, 7:8 pp., 16 textfigs.

Fritillaria venusta
Essenberg, C. E., 1926
Copelata from the San Diego Region. Univ. Calif. Publ. Zool. 28(22):399-521, 170 text figs.

Fritillaria venusta
Fenaux, R., 1967
5. Appendiculaires; Campagne de la Calypso au large des Cotes et antiques de l'Amerique du Sud (1961-1962)
Annls Inst. Oceanogr. n.s. 45 (2): 34-46

Fritillaria venusta (fig)
Tokioka, T., 1951.
Pelagic tunicates and chaetognaths collected during the cruises to the New Yamato Bank in the Sea of Japan. Publ. Seto Mar. Biol. Lab. 2(1):1-26, 1 chart, 6 tables, 12 textfigs.

Fritillaria venusta
Tokioka, T., 1942
Systematic studies of the plankton organisms occurring in Iwayama Bay, Palao VII A preliminary report on the Appendicularian fauna of the bay and adjacent waters. Palao Tropical Biological Sta. Studies, 2(3): 613-616 (Contrib. 106, Seto Mar. Biol. Lab.)

Fritillaria venusta
Tokioka, T., 1940
Some additional notes on the Japanese appendicularian fauna.
Rec. Oceanogr. Wks., Japan, 11(1):1-26.

Kowalevskia tenuis
Essenberg, C. E., 1926
Copelata from the San Diego Region. Univ. Calif. Publ. Zool. 28(22):399-521, 170 text figs.

Kowalevskia tenuis
Fenaux, R., 1959.
Observations écologiques sur les appendiculaires du plancton de surface de la Baie de Villefranche-sur-Mer (novembre 1957-octobre 1958).
Trav. Sta. Zool., Villefranche, 18(5):
Reprinted from: Bull. Inst. Océanogr., Monaco, No. 1141:26 pp.

Kowaleskaia tenuis
Forneris, Liliana, 1965.
Appendicularian species groups and southern Brazil water masses.
Bolm Inst. Oceanogr., S Paulo, 14(1):54-94.

Kowalevskaia tenuis
Gamulin, T., 1948
Prilog poznavanju Zooplanktona Srednjedalmatinskog Otocnog Podrucja. (Contrib. a la connaissance du zooplankton de la zone insulaire de la Dalmatie Moyenne). Acta Adriatica 3(7):38 pp., 6 tables, 1 map.

Kowalevskaia tenuis
Tokioka, Takasi, 1964.
Taxonomic studies of appendicularians collected by the Japanese Antarctic Research Expedition, 1957.
Japan Antarct. Res. Exped., 1956-1962, Sci. Repts. (E)(JARE Sci. Repts., Biol.), No. 2:16 pp.

Kowalevskia tenuis
Tokioka, T., 1942
Systematic studies of the plankton organisms occurring in Iwayama Bay, Palao VII A preliminary report on the Appendicularian fauna of the bay and adjacent waters. Palao Tropical Biological Sta. Studies, 2(3): 613-616 (Contrib. 106, Seto Mar. Biol. Lab.)

Kowalevskaia tenuis
Tokioka, T., 1940
Some additional notes on the Japanese appendicularian fauna.
Rec. Oceanogr. Wks., Japan, 11(1):1-26.

Megalocercus abyssorum
Bernard, M.F., 1955.
Capture de Megalocercus abyssorum Chun (Oikopleuridae) dans la Baie d'Alger.
Bull. Soc. Hist. Nat., Afrique du Nord, 45(7/8): 344-347, 2 textfigs.

Megalocercus abyssorum
Fenaux, R., 1964.
Contribution à la connaissance d'un appendiculaire peu commun Megalocercus abyssorum Chun, 1888.
Vie et Milieu, 15(4):979-991.
Also in: Trav. Sta. Zool., Villefranche-sur-Mer, 25.

Megalocercus abyssorum
Tokioka, Takasi, 1958.
Further notes on some appendicularians from the eastern Pacific.
Publ. Seto Mar. Biol. Lab., 7(1):1-17.

Megalocercus diegensis n.sp.
Essenberg, C. E., 1926
Copelata from the San Diego Region. Univ. Calif. Publ. Zool. 28(22):399-521, 170 text figs.

Megalocercus huxleyi
Delsman, H. C., 1939.
Preliminary plankton investigations in the Java Sea. Treubia, 17:139-181, 8 maps, 41 figs.

Megalocercus huxleyi
Fenaux, R., 1960.
Sur quelques appendiculaires d'Israel. Contribution to the knowledge of the Red Sea, No. 17.
Sea Fish. Res. Sta., Haifa, Bull., No. 29:3-7.

Megalocercus huxleyi
Hastings, A. B., 1931
Tunicata. Great Barrier Reef Exped., 1928-29. Sci. Repts. 4(3):69-109, 17 text figs., 3 pls.

Megalocercus huxleyi
Russell, F.S., and J.S. Colman, 1935
The Zooplankton. IV. The occurrence and seasonal distribution of the Tunicata, Mollusca and Coelenterata (siphonophora). Brit. Mus. (N.H.) Great Barrier Reef Expedition 1928-1929, Sci. Repts., 2(7):203-276, 30 text figs.

Megalocercus huxleyi
Tokioka, T., 1942
Systematic studies of the plankton organisms occurring in Iwayama Bay, Palao VII A preliminary report on the Appendicularian fauna of the bay and adjacent waters. Palao Tropical Biological Sta. Studies, 2(3): 613-616 (Contrib. 106, Seto Mar. Biol. Lab.)

Megalocercus huxleyi
Tokioka, T., 1940
Some additional notes on the Japanese appendicularian fauna.
Rec. Oceanogr. Wks., Japan, 11(1):1-26.

Oikopleura sp.
Bernard, Michelle, 1958
Systématique et distribution saisonnière des tuniciers pélagiques d'Alger.
Rapp. Proc. Verb., Comm. Int. Expl. Sci., Mer Medit., n.s., 14:211-231.

Oikopleura sp.
Delsman, H. C., 1939.
Preliminary plankton investigations in the Java Sea. Treubia, 17:139-181, 8 maps, 41 figs.

Oikopleura
Henderson, G. T. D., and N. B. Marshall, 1944
Continuous plankton records: The zooplankton (other than copepoda and young fish) in the southern North Sea 1932-1937. Hull Bull. Mar. Ecol., 1(6):255-275, 24 pls.

Oikoplura sp.
Johnson, M.W. (undated)
The production and distribution of Zooplankton in the surface waters of Bering Sea and Bering Strait, with special reference to copepods, echinoderms, mollusks and annelids. Report of oceanographic cruise of U.S. Coast Guard Cutter Chelan 1934 Part II (B):45-85, fig.1, table 1.

Oikopleura sp.
Kokubo, S., and S. Sato., 1947
Plankters in Ju-San Gata. Physiol. and Ecol. (Japan) 1(4):1-16, 3 text figs., tables.

Oikopleura sp.
Marshall, N. B., 1948
Continuous plankton records: Zooplankton (other than Copepoda and young fish) in the North Sea 1938-1939. Hull Bull. Mar. Ecol. 2(13):173-213, Pls. 89-108.

Oikopleura sp.
Ogilvie, H.S., 1934
A preliminary account of the food of the herring in the North-Western North Sea. Rapp. Proc. Verb. 89(3):85-92, 2 text figs.

Oikopleura
Roubault, A., 1946
Observations sur la Répartition du Plancton. Bull. Mus. Inst. Ocean., No.902: 4 pp.

Oikopleura sp.
Savage, R.E., 1937
The food of North Sea herring 1930-1934. Ministry of Agriculture and Fisheries. Fish. Invest. Ser. II, 15(5):1-60; 16 text figs.

Oikopleura albicans
Buckman, A., 1945.
Appendicularis I-III. Fiches d'Ident. Zooplancton.
Cons. Perm. Int. Expl. Mer, 7:8 pp., 16 textfigs.

Oikopleura albicans
Essenberg, C. E., 1926
Copelata from the San Diego Region. Univ. Calif. Publ. Zool. 28(22):399-521, 170 text figs.

Oikopleura albicans
Fenaux, R., 1971
La couche oikoplastique de l'Appendiculaire Oikopleura albicans (Leuckart) (Tunicata). Z. Morph. Tiere 69:184-200.

Oikopleura (Vexillaria) albicans
Flores Coto, Cesar, 1965.
Notas preliminares sobre la identificacion de las aguas Veracruzanas.
Anales, Inst. Biol., Univ. Mex., 36 (1/2):293-296.

Oikopleura albicans
Forneris, Liliana, 1965.
Appendicularian species groups and southern Brazil water masses.
Bolm Inst. Oceanogr., S Paulo, 14(1):54-94.

Oikopleura albicans
Furnestin, M. L., 1957.
Chaetognathes et zooplancton du secteur atlantique marocain. Rev. Trav. Inst. Pêches Marit. 21(1/2): 9-356.

Oikopleura albicans
Tokioka, T., 1940
Some additional notes on the Japanese appendicularian fauna.
Rec. Oceanogr. Wks., Japan, 11(1):1-26.

Oikopleura californica n.sp.
Essenberg, C. E., 1926
Copelata from the San Diego Region. Univ. Calif. Publ. Zool. 28(22):399-521, 170 text figs.

Oikopleura cophocerca
Björnberg, T.K.S, and L. Forneris, 1955.
Resultados cientificos do cruzeiro do "Baependi" e do "Vega" à ilha Trindade.
Contrib. Avulsas, Inst. Oceanograf. (Ocean. Biol.) Sao Paulo, No. 1:1-68.

Oikopleura cophocerca
Bückmann, A. 1967.
Untersuchungen über das Makroplankton bei Ischia und Capri und im Golf von Neapel im Mai 1962. III. Die Appendicularien. Pubbl. Staz. zool. Napoli 35(2): 215-238.

Oikopleura cophocerca
Buckman, A., 1945.
Appendicularia I-III. Fiches d'Ident. Zooplancton Cons. Perm. Int. Expl. Mer 7:8 pp., 16 textfigs.

Oikopleura cophocerca
Essenberg, C. E., 1926
Copelata from the San Diego Region. Univ. Calif. Publ. Zool. 28(22):399-521, 170 text figs.

Oikopleura cophocerca
Fenaux, R., 1967
5. Appendiculaires; Campagne de la Calypso au large des Cotes at antiques de l'Amerique du Sud (1961-1962)
Annls Inst. Oceanogr. n.s. 45 (2): 34-46

Oikopleura cophocera
Fenaux, Robert, 1963.
Ecologie et biologie des appendiculaires méditerranéens.
Vie et Milieu, Suppl., (16):142 pp.

Oikopleura cophocerca
Fenaux, R., 1959.
Observations écologiques sur les appendiculaires du plancton de surface de la Baie de Villefranche-sur-Mer (novembre 1957- octobre 1958).
Trav. Sta. Zool., Villefranche, 18(5):
Reprinted from:
Bull. Inst. Oceanogr., Monaco, No. 1141:26 pp.

Oikopleura dioica
Fenaux, Robert, et Bertrand Hirel 1972.
Cinétique du déploiement de la logette chez l'appendiculaire Oikopleura dioica Fol, 1872.
C. r. hebd. Seanc. Acad. Sci. Paris (D) 275 (3):449-452

Oikopleura (Vexillaria) Cophocerca
Flores Coto, Cesar 1965.
Notas preliminares sobre la identificacion de las apendicularios de las aguas Veracruzanas.
Anales, Inst. Biol. Univ. Mex. 36 (1/2):293-296.

Oikopleura cophocerca
Forneris, Liliana, 1965.
Appendicularian species groups and southern Brazil water masses.
Bolm Inst. Oceanogr., S Paulo, 14(1):54-94.

Oikopleura cophocerca
Furnestin, M. L., 1957.
Chaetognathes et zooplancton du secteur atlantique marocain. Rev. Trav. Inst. Pêches Marit. 21(1/2): 9-356.

Oikopleura cophocerca
Gamulin, T., 1948
Prilog poznavanju Zooplanktona Srednjedalmatinskog Otocnog Podrucja. (Contrib. a la connaissance du zooplankton de la zone insulaire de la Dalmatie Moyenne). Acta Adriatica 3(7):38 pp., 6 tables, 1 map.

Oikopleura cophocerca
Tokioka, Takasi, 1960
Studies on the distribution of appendicularians and some thaliaceans of the North Pacific, with some morphological notes.
Publ. Seto Mar. Biol. Lab., 8(2) (27):351-443. tables.

Oikopleura cophocerca
Tokioka, T., 1951.
Pelagic tunicates and chaetognaths collected during the cruises to the New Yamato Bank in the Sea of Japan. Publ. Seto Mar. Biol. Lab. 2(1): 1-26, 1 chart, 6 tables, 12 textfigs.

Oikopleura cophocerca
Tokioka, T., 1942
Systematic studies of the plankton organisms occurring in Iwayama Bay, Palao VII A preliminary report on the Appendicularian fauna of the bay and adjacent waters.
Palao Tropical Biological Sta. Studies, 2(3): 513-516 (Contrib. 106, Seto Mar. Biol. Lab.)

Oikopleura cophocera
Tokioka, T., 1940
Some additional notes on the Japanese appendicularian fauna.
Rec. Oceanogr. Wks., Japan, 11(1):1-26.

Oikopleura (Vexillaria) cophocerca
Tokioka, T., and J.A. Suarez Caabro, 1956.
Apendicularias de los mares cubanos.
Mem. Soc. Cubana Hist. Nat., 23(1):37-80, 15 pls.

Oikopleura cornutogastra
Tokioka, T., 1942
Systematic studies of the plankton organisms occurring in Iwayama Bay, Palao VII A preliminary report on the Appendicularian fauna of the bay and adjacent waters.
Palao Tropical Biological Sta. Studies, 2(3): 513-516 (Contrib. 106, Seto Mar. Biol. Lab.)

Oikopleura dioica
Bary, B.M., 1960.
Notes on ecology, distribution and systematics of pelagic Tunicata from New Zealand.
Pacific Science 14:101-121.

Oikopleura dioica
Bernard, Michelle, 1958
Systematique et distribution saisonnière des tuniciers pélagiques d'Alger.
Rapp. Proc. Verb., Comm. Int. Expl. Sci., Mer Medit., n.s., 14:211-231.

Oikopleura dioica
Berrill, N.J., 1950.
The Tunicata with an account of the British species. Ray Soc. No. 133:354 pp., 120 textfigs.

Oikopleura dioica
Bigelow, H.B., and M. Leslie, 1930
Reconnaissance of the waters and plankton of Monterey Bay, July 1928.
Bull. M.C.Z., 70(5):429-481, 43 text figs.

Oikopleura dioica
Bigelow, H.B., and M. Sears, 1939
Studies of the waters of the continental shelf, Cape Cod to Chesapeake Bay. III. A volumetric study of the zooplankton. Mem. M.C.Z. 54(4):183-378, 42 text figs.

Oikopleura dioica
Björnberg, T.K.S, and L. Forneris, 1955.
Resultados cientificos do cruzeiro do "Baependi" e do "Vega" à ilha Trindade.
Contrib. Avulsas, Inst. Oceanograf. (Ocean. Biol.), Sao Paulo, No. 1:1-68.

Oikopleura dioica
Browne, E.T., 1905.
Notes on the pelagic fauna of the Firth of Clyde (1901-1902). Proc. R. Soc., Edinburgh, 25(9): 779-791.

Oikopleura dioica
Bückmann, A. 1967.
Untersuchungen über das Makroplankton bei Ischia und Capri und im Golf von Neapel im Mai 1962 III. Die Appendicularien. Pubbl. Staz. zool. Napoli 35(2): 215-238.

Oikopleura dioica
Buckman, A., 1945.
Appendicularia I-III. Fiches d'Ident. Zooplancton Cons. Perm. Int. Expl. Mer, 7:8 pp., 16 textfigs.

Oikopleura dioica
Chiba, T., 1949
On the distribution of the plankton in the eastern China Sea and Yellow Sea. 1. Plankton composition in the spring. J. Shimonoseki Coll. Fisheries, 1(1):57-63, 1 fig.

Oikopleura dioica
Cronin, L. Eugene, Joanne C. Daiber and Edward M. Hulburt, 1962
Quantitative seasonal aspects of zooplankton in the Delaware River estuary.
Chesapeake Science, 3(2):63-93.

Oikopleura dioica
Essenberg, C. E., 1926
Copelata from the San Diego Region. Univ. Calif. Publ. Zool. 28(22):399-521, 170 text figs.

Oikopleura dioica
Fenaux, R., 1967
5. Appendiculaires; Campagne de la Calypso au large des Cotes at antiques de l'Amerique du Sud (1961-1962)
Annls Inst. Oceanogr. n.s. 45 (2): 34-46

Oceanographic Index: Marine Organisms Cumulation, 1946-1973

Oikopleura dioica
Fenaux, Robert, 1963.
Ecologie et biologie des appendiculaires méditerranéens.
Vie et Milieu, Suppl., (16):142 pp.

Oikopleura dioica
Fenaux, R., 1960.
Sur quelques appendiculaires d'Israel.
Contributions to the knowledge of the Red Sea, No. 17.
Sea Fish. Res. Sta., Haifa, Bull., No. 29:3-7.

Oikopleura dioica
Fenaux, R., 1959.
Observations écologiques sur les appendiculaires du plancton de surface de la Baie de Villefranche-sur-Mer (novembre 1957-octobre 1958).
Trav. Sta. Zool., Villefranche, 18(5):

Reprinted from:
Bull. Inst. Océanogr., Monaco, No. 1141:26 pp.

Oikopleura (Vexillaria) dioica
Flores Coto, Cesar 1965.
Notas preliminares sobre la identificacion de los apendicularias de las aguas Veracruzanas.
Anales, Inst. Biol. Univ. Mex. 36(1/2):293-296.

Oikopleura dioica
Forneris, Liliana, 1965.
Appendicularian species groups and southern Brazil water masses.
Bolm Inst. Oceanogr., S Paulo, 14(1):54-94.

Oikopleura dioica
Fraser, J.H., 1961
The oceanic bathypelagic plankton of the North-East Atlantic and its possible significance to fisheries.
Dept. Agric. & Fish., Scotland, Marine Research (4):1-48.

Oikopleura dioica
Frost, N., S.T. Lindsay, and H. Thompson, 1934.
III. Hydrographic and biological investigations.
B. Plankton. Ann. Rept., Newfoundland Fish. Res. Comm. 2(2):47-59, Textfigs. 6-11.

Oikopleura dioica
Frost, N., S.T. Lindsay, and H. Thompson, 1933.
III. Hydrographic and biological investigations.
B. Plankton more abundant in 1932 than in 1931.
Ann. Rept., Fish. Res. Comm. 2(1):58-74, Textfigs. 17-27.
Newfoundland

Oikopleura dioica
Furnestin, M. L., 1957.
Chaetognathes et zooplancton du secteur atlantique marocain. Rev. Trav. Inst. Pêches Marit. 21(1/2):9-356.

Oikopleura dioica
Gamulin, T., 1948
Prilog poznavanju Zooplanktona Srednjedalmatinskog Otocnog Podrucja. (Contrib. a la connaissance du zooplankton de la zone insulaire de la Dalmatie Moyenne). Acta Adriatica 3(7):38 pp., 6 tables, 1 map.

Oikopleura dioica
Jespersen, P., 1949
Investigations on the occurrence and quantity of holoplankton animals in the Isefjord, 1940-1943. Medd. Komm. Danmarks Fiskeri - og Havundersøgelser, ser. Plankton, 5(3):18 pp., 10 text figs.

Oikopleura doica
Last, J.M., 1972
Egg development, fecundity and growth of Oikopleura dioica Fol in the North Sea.
J. Cons. int. Explor. Mer 34(2):232-237.

Oikopleura dioica
Lebour, M.V., 1947
Notes on the inshore plankton of Plymouth
JMBA 26(4):527-547, 1 textfig.

Oikopleura dioica
Mankowski, W., 1950.
Plankton investigations in the southern Baltic, 1948. Bull. Inst. Pêches Maritimes, Gdynia, No. 5:71-101, 6 textfigs. (In Polish; English summary)

Oikopleura dioica
Masui, T., T. Namibe, and D. Miyadi, 1944.
[Gulf factors and the type of gulf fauna.]
Physiol. and Ecol. Contr. Otsu Hydrobiol. Exp. Sta., Kyoto Univ., 3:1-20.

Oikopleura dioica
Olsson, R., 1965.
Comparative morphology and physiology of the Oikopleura notochord.
Israel J. Zool., 14(1/4):213-220.

Oikopleura dioica
Schodduyn, M., 1926
Observations faites dans la baie d'Ambleteuse (Pas de Calais). Bull. Inst. Ocean., Monaco, No. 482: 64 pp.

Oikopleura dioica
Sewell, R.B.S., 1953.
The pelagic Tunicata. Brit. Mus. (N.H.), John Murray Exped., 1933-34, Sci. Rept. 10(1):1-90, 1 pl., 32 textfigs.

Oikopleura dioica
Shelbourne, J.E., 1962.
A predator-prey relationship for plaice larvae feeding on Oikopleura.
J. Mar. Biol. Assoc., U.K., 42(2):243-252.

Oikopleura dioica Fab.
Smidt, E.L.B., 1944
Biological Studies of the Invertebrate Fauna of the Harbor of Copenhagen. Vidensk. Medd. fra Dansk naturh. Foren. 107:235-316, 23 text figs.

Oikopleura dioica
Tokioka, T., 1942
Systematic studies of the plankton organisms occurring in Iwayama Bay, Palao
VII A preliminary report on the Appendicularian fauna of the bay and adjacent waters.
Palao Tropical Biological Sta. Studies, 2(3):613-616 (Contrib. 106, Seto Mar. Biol. Lab.)

Oikopleura dioica
Tokioka, T., 1940
Some additional notes on the Japanese appendicularian fauna.
Rec. Oceanogr. Wks., Japan, 11(1):1-26.

Oikopleura (Vexillaria) dioica
Tokioka, T., and J.A. Suarez Caabro, 1956.
Apendicularias de los mares cubanos.
Mem. Soc. Cubana Hist. Nat., 23(1):37-80, 15 pls.

Oikopleura dioica
Vučetić, T., 1961.
Vertikalna raspodjela zooplanktona u Velikom Jezeru otoka Mljeta, [Vertical distribution of zooplankton in the Bay Veliko Jezero on the island of Mljet.]
Acta Adriatica, 6(9):1-14 (in Jugoslavian) 15-20 (in English).

Oikopleura dioica
White, Ray J., 1968.
Importance of appendicularians as food of larval plaice (Pleuronectes platessa L.) off Helgoland.
Ber. dt. will. Komm. Meeresforsch. 19(4):288-291.

Oikopleura dioica
Wiborg, K.F., 1944
The production of zooplankton in a landlocked fjord, the Nordåsvatn near Bergen, in 1941-42, with special reference to the copepods. (Repts. Norwegian Fish. and Mar. Invest.) 7(7):83 pp., 40 text figs.

Oikopleura fusiformis
Bary, B.M., 1960.
Notes on ecology, distribution and systematics of pelagic Tunicata from New Zealand.
Pacific Science, 14:101-121.
(2)

Oikopleura fusiformis
Bernard, Michelle, 1958
Systématique et distribution saisonnière des tuniciers pélagiques d'Alger.
Rapp. Proc. Verb., Comm. Int. Expl. Sci. Mer Medit. n.s., 14:211-231.

Oikopleura fusiformis
Berrill, N.J., 1950.
The Tunicata with an account of the British species. Ray Soc. No. 133:354 pp., 120 textfigs.

Oikopleura fusiformis
Björnberg, T.K.S., and L. Forneris, 1955.
Resultados científicos do cruzeiro do "Baependi" e do "Vega" à ilha Trindade.
Contrib. Avulsas, Inst. Oceanograf. Ocean. Biol.), Sao Paulo, No. 1:1-68.

Oikopleura fusiformis
Buckman, A., 1945.
Appendicularia I-III. Fiches d'Ident. Zooplancton Cons. Perm. Int. Expl. Mer, 7:8 pp., 16 textfigs.

Oikopleura fusiformis
Essenberg, C. E., 1926
Copelata from the San Diego Region. Univ. Calif. Publ. Zool. 28(22):399-521, 170 text figs.

Oikopleura fusiformis
Fenaux, R., 1967
5. Appendiculaires; Campagne de la Calypso au large des Cotes at antiques de l'Amerique du Sud (1961-1962)
Annls Inst. Oceanogr. n.s. 45 (2): 34-46

Oikopleura fusiformis
Fenaux, Robert, 1963.
Ecologie et biologie des appendiculaires méditerranéens.
Vie et Milieu, Suppl., (16):142 pp.

Oikopleura fusiformis
Fenaux, R., 1959.
Observations écologiques sur les appendiculaires du plancton de surface de la Baie de Villefranche-sur-Mer (novembre 1957-octobre 1958).
Trav. Sta. Zool., Villefranche, 18(5):

Reprinted from:
Bull. Inst. Océanogr., Monaco, No. 1141:26 pp.

Oikopleura fusiformis
Forneris, Liliana, 1965.
Appendicularian species groups and southern Brazil water masses.
Bolm Inst. Oceanogr., S Paulo, 14(1):54-94.

Oikopleura fusiformis
Fraser, J.H., 1961
The oceanic bathypelagic plankton of the North-East Atlantic and its possible significance to fisheries.
Dept. Agric. & Fish., Scotland, Marine Research (4):1-48.

Oikopleura fusiformis
Gamulin, T., 1948
Prilog poznavanju Zooplanktona Srednjedalmatinskog Otocnog Podrucja. (Contrib. a la connaissance du zooplankton de la zone insulaire de la Dalmatie Moyenne). Acta Adriatica 3(7):38 pp., 6 tables, 1 map.

Oikopleura fusiformis
Hastings, A. B., 1931
Tunicata. Great Barrier Reef Exped., 1928-29. Sci. Repts. 4(3):69-109, 17 Rat figs; 3 pls.

Oikopleura fusiformis
Russell, F.S., and J.S. Colman, 1935 Fol.
The Zooplankton. IV. The occurrence and seasonal distribution of the Tunicata, Mollusca and Coelenterata (siphonophora). Brit. Mus. (N.H.) Great Barrier Reef Expedition 1928-1929, Sci. Repts., 2(7):203-276, 30 text figs.

Oikopleura fusiformis
Tokioka, T., 1942
Systematic studies of the plankton organisms occurring in Iwayama Bay, Palao VII A preliminary report on the Appendicularian fauna of the bay and adjacent waters. Palao Tropical Biological Sta. Studies, 2(3):613-616 (Contrib. 106, Seto Mar. Biol. Lab.)

Oikopleura fusiformis
Tokioka, T., 1940
Some additional notes on the Japanese appendicularian fauna.
Rec. Oceanogr. Wks., Japan, 11(1):1-26.

Oikopleura (Coecaria) fusiformis
Tokioka, T., and J.A. Suarez Caabro, 1956.
Apendicularias de los mares cubanos.
Mem. Soc. Cubana Hist. Nat., 23(1):37-80, 15 pls.

Oikopleura fusiformis cornugaster
Tokioka, T., 1951. (figs.)
Pelagic tunicates and chaetognaths collected during the cruises to the New Yamato Bank in the Sea of Japan. Publ. Seto Mar. Biol. Lab. 2(1):1-26, 1 chart, 6 tables, 12 textfigs.

Oikopleura (Coecaria) fusiformis cornutogastra
Tokioka, T., and J.A. Suarez Caabro, 1956.
Apendicularias de los mares cubanos.
Mem. Soc. Cubana Hist. Nat., 23(1):37-80, 15 pls.

Oikopleura gaussica
Thompson, Harold, 1954.
Pelagic tunicates. Identifications and relevant notes.
B.A.N.Z. Antarctic Res. Exped., 1929-1931 Repts., B(Zool. Bot.), 1(4):183-185.

Oikopleura (Vexillaria) gaussica
Tokioka, Takasi, 1964.
Taxonomic studies of appendicularians collected by the Japanese Antarctic Research Expedition, 1957.
Japan. Antarct. Res. Exped., 1956-1962, Sci. Repts. (E)(JARE Sci Repts., Biol.), No. 2:16 pp., 9 pls.

Oikopleura gracilis
Bückmann A. 1967.
Untersuchungen über das Makroplankton bei Ischia und Capri und im Golf von Neapel im Mai 1962. III Die Appendicularien.
Pubbl. Staz. zool. Napoli 35(2): 215-238.

Oikopleura gracilis
Forneris, Liliana, 1965.
Appendicularian species groups and southern Brazil water masses.
Bolm Inst. Oceanogr., S Paulo, 14(1):54-94.

Oikopleura gracilis (fig.)
Tokioka, T., 1951.
Pelagic tunicates and chaetognaths collected during the cruises to the New Yamato Bank in the Sea of Japan. Publ. Seto Mar. Biol. Lab. 2(1):1-26, 1 chart, 6 tables, 12 textfigs.

Oikopleura graciloides
Fenaux, R., 1967
5. Appendiculaires; Campagne de la Calypso au large des Cotes at antiques de l'Amerique du Sud (1961-1962)
Annls Inst. Oceanogr. n.s. 45 (2): 34-46

Oikopleura graciloides
Forneris, Liliana, 1965.
Appendicularian species groups and southern Brazil water masses.
Bolm Inst. Oceanogr., S Paulo, 14(1):54-94.

Oikopleura graciloides
Tokioka, T., 1955.
General considerations on Japanese appendicularian fauna. Publ. Seto Mar. Biol. Lab. 4(2/3):251-261, 6 textfigs.

Oikopleura graciloides
Tokioka, T., 1942
Systematic studies of the plankton organisms occurring in Iwayama Bay, Palao VII A preliminary report on the Appendicularian fauna of the bay and adjacent waters. Palao Tropical Biological Sta. Studies, 2(3):613-616 (Contrib. 106, Seto Mar. Biol. Lab.)

Oikopleura (Coecaria) graciloides
Tokioka, T., and J.A. Suarez Caabro, 1956.
Apendicularias de los mares cubanos.
Mem. Soc. Cubana Hist. Nat., 23(1):37-80, 15 pls.

Oikopleura huxleyi
Ritter, R. E. and E. S. Byxbee, 1905
VIII. The pelagic Tunicata. Reports on the Scientific results of the expedition to the Tropical Pacific, in charge of Alexander Agassiz, in the U. S. Fish Commission Steamer "Albatross", from August, 1899 to March, 1900, Commander Jefferson F. Moser, U. S. N., commanding. Mem. M.C.Z. 26(5):195-211, 2 pls.

Oikopleura intermedia
Bernard, Michelle, 1958
Systématique et distribution saisonnière des tuniciers pélagiques d'Alger.
Rapp. Proc. Verb., Comm. Int. Expl. Sci., Mer Medit., n.s., 14:211-231.

Oikopleura intermedia ?
Bigelow, H.B., and M. Leslie, 1930
Reconnaissance of the waters and plankton of Monterey Bay, July 1928.
Bull. M.C.Z., 70(5):429-481, 43 text figs.

Oikopleura intermedia
Fenaux, R., 1967
5. Appendiculaires; Campagne de la Calypso au large des Cotes at antiques de l'Amerique du Sud (1961-1962)
Annls Inst. Oceanogr. n.s. 45 (2): 34-46

Oikopleura intermedia
Fenaux, R., 1959.
Observations écologiques sur les appendiculaires du plancton de surface de la Baie de Villefranche-sur-Mer (novembre 1957- octobre 1958).
Trav. Sta. Zool., Villefranche, 18(5):

Reprinted from:
Bull. Inst. Océanogr., Monaco, No. 1141:26 pp.

Oikopleura intermedia
Forneris, Liliana, 1965.
Appendicularian species groups and southern Brazil water masses.
Bolm Inst. Oceanogr., S Paulo, 14(1):54-94.

Oikopleura intermedia
Gamulin, T., 1948
Prilog poznavanju Zooplanktona Srednjedalmatinskog Otocnog Područja. (Contrib. a la connaissance du zooplankton de la zone insulaire de la Dalmatie Moyenne). Acta Adriatica 3(7):38 pp., 6 tables, 1 map.

Oikopleura (Coecaria) intermedia
Tokioka, T., 1955.
On some plankton animals collected by the Syunkotu maru in May-June 1954. III. Appendicularians. Bull. Biogeogr. Soc., Japan, Vols. 16-18:351-355.

Oikopleura intermedia
Tokioka, T., 1942
Systematic studies of the plankton organisms occurring in Iwayama Bay, Palao VII A preliminary report on the Appendicularian fauna of the bay and adjacent waters. Palao Tropical Biological Sta. Studies, 2(3):613-616 (Contrib. 106, Seto Mar. Biol. Lab.)

Oikopleura intermedia
Tokioka, T., 1940
Some additional notes on the Japanese appendicularian fauna.
Rec. Oceanogr. Wks., Japan, 11(1):1-26.

Oikopleura labradoriensis
Berrill, N.J., 1950.
The Tunicata with an account of the British species. Ray Soc. No. 133:354 pp., 120 textfigs.

Oikopleura labradoriensis
Bigelow, H.B., and M. Leslie, 1930
Reconnaissance of the waters and plankton of Monterey Bay, July 1928.
Bull. M.C.Z., 70(5):429-481, 43 text figs.

Oikopleura labradoriensis
Bigelow, H.B., and M. Sears, 1939
Studies of the waters of the continental shelf, Cape Cod to Chesapeake Bay. III. A volumetric study of the zooplankton. Mem. M.C.Z. 54(4):183-378, 42 text figs.

Oikopleura labradoriensis
Buckman, A., 1945.
Appendicularia I-III. Fiches d'Ident. Zooplancton, Cons. Perm. Int. Expl. Mer, 7:8 pp., 16 textfigs.

Oikopleura labradoriensis
Essenberg, C. E., 1926
Copelata from the San Diego Region. Univ. Calif. Publ. Zool. 28(22):399-521, 170 text figs.

Oikopleura labradoriensis
Fish, C.J., and M.W. Johnson, 1937
The biology of the zooplankton population in the Bay of Fundy and the Gulf of Maine with special reference to production and distribution. J. Biol. Bd., Canada 3(3):189-322, 29 tables, 45 text figs.

Oikopleura labradoriensis
Fraser, J.H., 1961
The oceanic bathypelagic plankton of the North-East Atlantic and its possible significance to fisheries.
Dept. Agric. & Fish., Scotland, Marine Research (4):1-48.

Oikopleura labradoriensis
Frost, N., S.T. Lindsay, and H. Thompson, 1934.
III. Hydrographic and biological investigations. B. Plankton. Ann. Rept., Newfoundland Fish. Res. Comm. 2(2):47-59, Textfigs. 6-11.

Oikopleura labradoriensis
Frost, N., S.T. Lindsay, and H. Thompson, 1933.
III. Hydrographic and biological investigations. B. Plankton more abundant in 1932 than 1931. Ann. Rept., Newfoundland Fish. Res. Comm. 2(1):58-74, Textfigs. 17-27.

Oikopleura labradoriensis
Hansen, Vagn Kr., 1960.
Investigations on the quantitative and qualitative distribution of zooplankton in the southern part of the Norwegian Sea.
Medd. Danmarks Fiskeri- og Havundersøgelser, n.s., 2(23):1-53.

Oikopleura labradoriensis
Kielhorn, W.V., 1952
The biology of the surface zone zooplankton of a Boreo-Arctic Atlantic Ocean area. J.Fish Res. Bd., Canada 9 (5): 223-264, 13 text figs.

Oikopleura labradoriensis
Künne, Cl., 1937
Über die Verbreitung der Leitformen des Großplanktons in der südlichen Nordsee im Winter. Ber. Deutschen Wissenschaflichen Kommission für Meeresforschung, n.f. VIII (3):131-164, 9 text figs., 4 fold-ins.

Oikopleura labradoriensis
Tokioka, T., 1940
Some additional notes on the Japanese appendicularian fauna.
Rec. Oceanogr. Wks., Japan, 11(1):1-26.

Oikopleura labradoriensis
Udvardy, M.D.F., 1954.
Distribution of appendicularians in relation to the Straits of Belle Isle.
J. Fish. Res. Bd., Canada, 11(4):431-453.

Oikopleura longicauda
Bernard, Michelle, 1958
Systématique et distribution saisonnière des tuniciers pélagiques d'Alger.
Rapp. Proc. Verb., Comm. Int. Expl. Sci., Mer Medit., n.s., 14:211-231.

Oikopleura longicauda
Bogoraze, Dimitri et Odette Tuzet, 1949
Ultrastructure du muscle de la queue de l'appendiculaire Oikopleura longicauda Vogt. Les limites cellulaires; les disques intercalaires.
Cah. Biol. mar. 10(4):365-374

Oikopleura longicauda
Bückmann, A. 1967.
Untersuchungen über das Makroplankton bei Ischia und Capri und im Golf von Neapel im Mai 1962. III Die Appendicularien
Pubbl. Staz. zool. Napoli 35(2): 215-238.

Oikopleura longicauda
Buckman, A., 1945.
Appendicularia I-III. Fiches d'Ident. Zooplancton, Cons. Perm. Int. Expl. Mer, 7:8 pp., 16 textfigs.

Oikopleura longicauda
Essenberg, C. E., 1926
Copelata from the San Diego Region. Univ. Calif. Publ. Zool. 28(22):399-521, 170 text figs.

Oikopleura longicauda
Fenaux, R., 1967
5. Appendiculaires; Campagne de la Calypso au large des Cotes at antiques de l'Amerique du Sud (1961-1962)
Annls Inst. Oceanogr. n.s. 45 (2): 34-46

Oikopleura longicauda
Fenaux, M.R., 1963.
Composition annuelle de la population de Oikopleura longicauda (Appendiculaire).
Rapp. Proc. Verb., Réunions, Comm. Int. Expl. Sci. Mer Méditerranée, Monaco, 17(2):635-636.

Oikopleura longicauda
Fenaux, Robert, 1963.
Ecologie et biologie des appendiculaires méditerranéens.
Vie et Milieu, Suppl., (16):142 pp.

Oikopleura longicauda
Fenaux, R., 1960.
Sur quelques appendiculaires d'Israel.
Contributions to the knowledge of the Red Sea, No. 17.
Sea Fish. Res. Sta., Haifa, Bull., No. 29:3-7.

Oikopleura longicauda
Fenaux, R., 1959.
Observations écologiques sur les appendiculaires du plancton de surface de la Baie de Villefranche-sur-Mer (novembre 1957-octobre 1958).
Trav. Sta. Zool., Villefranche, 18(5):
Reprinted from:
Bull. Inst. Océanogr. Monaco, No. 1141:26 pp.

Oikopleura (Coecaria) longicauda
Flores Coto, Cesar 1965.
Notas preliminares sobre la identificacion de las apendicularias de las aguas Veracruzanas.
Anales, Inst. Biol. Univ. Mex. 36(1/2):293-296.

Oikopleura longicauda
Forneris, Liliana, 1965.
Appendicularian species groups and southern Brazil water masses.
Bolm Inst. Oceanogr., S Paulo 14(1):54-94.

Oikopleura longicauda
Furnestin, M.L., 1957
Chaetognathes et zooplancton du secteur atlantique marocain. Rev. Trav. Inst. Pêches Marit. 21(1/2): 9-356.

Oikopleura longicauda
Gamulin, T., 1948
Prilog poznavanju Zooplanktona Srednjedalmatinskog Otocnog Podrucja. (Contrib. a la connaissance du zooplankton de la zone insulaire de la Dalmatie Moyenne). Acta Adriatica 3(7):38 pp., 6 tables, 1 map.

Oikopleura longicauda
Hastings, A. B., 1931
Tunicata. Great Barrier Reef Exped., 1928-29. Sci. Repts. 4(3):69-109, 17 text figs, 3 pl.

Oikopleura longicauda
Hure, J., 1961
Dneva migracija i sezonaka vertikalna raspodjela zooplanktona dubljeg mora. (Migration journalière et distribution saisonnaire verticale du zooplancton dans la région profunde de l'Adriatique)
Acta Adriatica, 9(6):1-59.

Oikopleura longicauda
Owen, Robert W., Jr.,
Small-scale, horizontal vortices in the surface layer of the sea.
J. Mar. Res., 24(1):56-66.

Oikopleura longicauda
Russell, F.S., and J.S. Colman, 1935 (Vogt)
The Zooplankton. IV. The occurrence and seasonal distribution of the Tunicata, Mollusca and Coelenterata (siphonophora). Brit. Mus. (N.H.) Great Barrier Reef Expedition 1928-1929, Sci. Repts., 2(7):203-276, 30 text figs.

Oikopleura longicauda
Tokioka, T., 1942
Systematic studies of the plankton organisms occurring in Iwayama Bay, Palao VII A preliminary report on the Appendicularian fauna of the bay and adjacent waters.
Palao Tropical Biological Sta. Studies, 2(3): 613-616 (Contrib. 106, Seto Mar. Biol. Lab.)

Oikopleura longicauda
Tokioka, T., 1940
Some additional notes on the Japanese appendicularian fauna.
Rec. Oceanogr. Wks., Japan, 11(1):1-26.

Oikopleura (Coecaria) longicauda
Tokioka, T., and J.A. Suarez Caabro, 1956.
Apendicularias de los mares cubanos.
Mem. Soc. Cubana Hist. Nat., 23(1):37-80, 15 pls

Oikopleura mediterranea
Bückmann, A. 1967.
Untersuchungen über das Makroplankton bei Ischia und Capri und im Golf von Neapel im Mai 1962. III. Die Appendicularien
Pubbl. Staz. zool. Napoli 35(2):215-238.

Oikopleura parva
Björnberg, T.K.S., and L. Forneris, 1955.
Resultados cientificos do cruzeiro do "Baependi" e do "Vega" à ilha Trindade.
Contrib. Avulsas, Inst. Oceanograf. (Ocean. Biol.), Sao Paulo, No. 1:1-68.

Oikopleura parva
Bückmann, A. 1967.
Untersuchungen über das Makroplankton bei Ischia und Capri und im Golf von Neapel im Mai 1962. III. Die Appendicularien
Pubbl. Staz. zool. Napoli 35(2):215-238.

Oikopleura parva
Buckman, A., 1945.
Appendicularia I-III. Fiches d'Ident. Zooplancton Cons. Perm. Int. Expl. Mer, 7:8 pp., 16 textfigs.

Oikopleura parva
Fenaux, R., 1967
5. Appendiculaires; Campagne de la Calypso au large des Cotes at antiques de l'Amerique du Sud (1961-1962)
Annls Inst. Oceanogr. n.s. 45 (2): 34-46

Oikopleura parva
Sebastian, V.O., 1968.
The occurrence of Oikopleura parva Lohmann at Porto Novo Coast. Bull. Dept. mar. Biol. Oceanogr. Univ. Kerala, 4: 155-157.

Oikopleura (Vexillaria) parva
Tokioka, T., 1955.
On some plankton animals collected by the Syunkotu maru in May-June 1954. III. Appendicularians. Bull. Biogeogr. Soc., Japan, Vols. 16-19:251-255.

Oikopleura parva (figs.)
Tokioka, T., 1951.
Pelagic tunicates and chaetognaths collected during the cruises to the New Yamato Bank in the Sea of Japan. Publ. Seto Mar. Biol. Lab. 2(1): 1-26, 1 chart, 6 tables, 12 textfigs.

Oikopleura (Vexillaria) parva
Tokioka, T., and J.A. Suarez Caabro, 1956.
Apendicularias de los mares cubanos.
Mem. Soc. Cubana Hist. Nat., 23(1):37-80, 15 pls.

Oikopleura rufescens
Björnberg, T.K.S., and L. Forneris, 1955.
Resultados cientificos do cruzeiro do "Baependi" e do "Vega" à ilha Trindade.
Contrib. Avulsas, Inst. Oceanograf. (Ocean. Biol.), Sao Paulo, No. 1:1-68.

Oikopleura rufescens
Essenberg, C. E., 1926
Copelata from the San Diego Region. Univ. Calif. Publ. Zool. 28(22):399-521, 170 text figs.

Oikopleura rufescens
Fenaux, R., 1967
5. Appendiculaires; Campagne de la Calypso au large des Cotes at antiques de l'Amerique du Sud (1961-1962)
Annls Inst. Oceanogr. n.s. 45 (2): 34-46

Oikopleura rufescens
Fenaux, R., 1960.
Sur quelques appendiculaires d'Israel. Contribution to the knowledge of the Red Sea, No. 17.
Sea Fish. Res. Sta., Haifa, Bull., No. 29:3-7.

Oikopleura (Vexillaria) rufescens
Flores Coto, Cesar 1965.
Notas preliminares sobre la identificacion de las apendicularias de las aguas Veracruzanas.
Anales, Inst. Biol. Univ. Mex. 36(1/2):293-296.

Oikopleura rufescens
Forneris, Liliana, 1965.
Appendicularian species groups and southern Brazil water masses.
Bolm Inst. Oceanogr., S Paulo 14(1):54-94.

Oikopleura rufescens

Furnestin, M. L., 1957.
Chaetognathes et zooplancton du secteur atlantique marocain. Rev. Trav. Inst. Pêches Marit. 21(1/2): 9-356.

Oikopleura rufescens

Hastings, A. B., 1931
Tunicata. Great Barrier Reef Exped., 1928-29. Sci. Repts. 4(3):69-109, 17 text-figs, 3 pls.

Oikopleura rufescens

Russell, F.S., and J.S. Colman, 1935 Fol.
The Zooplankton. IV. The occurrence and seasonal distribution of the Tunicata, Mollusca and Coelenterata (siphonophora). Brit. Mus. (N.H.) Great Barrier Reef Expedition 1928-1929, Sci. Repts., 2(7):203-276, 30 text figs.

Oikopleura rufescens

Sewell, R.B.S. 1953.
The pelagic Tunicata. Brit. Mus. (N.H.), John Murray Exped., 1933-34, Sci. Repts. 10(1):1-90, 1 pl., 32 textfigs.

Oikopleura rufescens (fig.)

Tokioka, T., 1951.
Pelagic tunicates and chaetognaths collected during the cruises to the New Yamato Bank in the Sea of Japan. Publ. Seto Mar. Biol. Lab. 2(1): 1-26, 1 chart, 6 tables, 12 textfigs.

Oikopleura rufescens

Tokioka, T., 1942
Systematic studies of the plankton organisms occurring in Iwayama Bay, Palao VII A preliminary report on the Appendicularian fauna of the bay and adjacent waters. Palao Tropical Biological Sta. Studies, 2(3): 613-616 (Contrib. 106, Seto Mar. Biol. Lab.)

Oikopleura rufescens

Tokioka, T., 1940
Some additional notes on the Japanese appendicularian fauna.
Rec. Oceanogr. Wks., Japan, 11(1):1-26.

Oikopleura (Vexillaria) rufescens

Tokioka, T., and J.A. Suarez Caabro, 1956.
Apendicularias de los mares cubanos.
Mem. Soc. Cubana Hist. Nat., 23(1):37-80, 15 pls.

Oikopleura valdiviae

Thompson, Harold, 1954.
Pelagic tunicates. Identifications and relevant notes. B.A.N.Z. Antarctic Res. Exped., 1929-1931 Repts., B(Zool. Bot.), 1(4):183-185.

Oikopleura vanhöffeni

Berrill, N.J., 1950.
The Tunicata with an account of the British species. Ray Soc. No. 133:354 pp., 120 textfigs.

Oikopleura vanhoeffeni

Bogorov, B. G., 1938
Diurnal vertical distribution of plankton under polar conditions (in the southeastern part of the Barents Sea). Trudy Poliarnii Nauchno-Issledovatelskii Institut Morskogo Ribogo Khoziastva i Okeanografii im Pochetnogo Chlena Akademii Nauk SSSR. Prof. N.M. Knipovicha g. Murmansk, 2:93-106, 4 text figs. (In Russian with English Summary). Trans. Polar Sci. Res. Inst. Mar. Fish & Oceanogr.

Oikopleura vanhoeffeni

Buckman, A., 1945.
Appendicularia I-III. Fiches d'Ident. Zooplancton Cons. Perm. Int. Expl. Mer, 7:8 pp., 16 textfigs.

Oikopleura vanhöffeni

Essenberg, C. E., 1926
Copelata from the San Diego Region. Univ. Calif. Publ. Zool. 28(22):399-521, 170 text figs.

Oikopleura vanhoffeni

Fraser, J. H., 1949
Plankton of the Faroe-Shetland Channel and the Faroes, June and August 1947. Ann. Biol., Int. Cons., 4:27-28, text fig. 10.

Oikopleura vanhoffeni

Fraser, J. H. and A. Saville, 1949
List of rare exotic species found in the plankton by the Scottish Vessel "Explorer" in 1948. Ann. Biol. 5:62-64.

Oikopleura vanhoffeni

Fraser, J. H, and A. Saville, 1949
Macroplankton in the Faroe Channel, 1948. Ann. Biol. 5:29-30, text figs.

Oikopleura vanhöffeni

Frost, N., S.T. Lindsay, and H. Thompson, 1934.
III. Hydrographic and biological investigations. B. Plankton. Ann. Rept. Newfoundland Fish. Res. Comm. 2(2):47-59, Textfigs. 6-11.

Oikopleura vanhöffeni

Frost, N., S.T. Lindsay, and H. Thompson, 1933.
III. Hydrographic and biological investigations. B. Plankton more abundant in 1932 than 1931. Ann. Rept., Newfoundland Fish. Res. Comm. 2(1): 58-74, Textfigs. 17-27.

Oikopleura vanhoeffeni

Hansen, Vagn Kr., 1960.
Investigations on the auantitative and qualitative distribution of zooplankton in the southern part of the Norwegian Sea. Medd. Danmarks Fiskeri- og Havundersøgelser, n.s., 2(23):1-53.

Oikopleura vanhöffeni

Udvardy, M.D.F., 1954.
Distribution of appendicularians in relation to the Straitis of Belle Isle.
J. Fish. Res. Bd., Canada, 11(4):431-453.

Pelagopleura haranti

Bernard, Michelle, 1958
Systématique et distribution saisonnière des tuniciers pélagiques d'Alger.
Rapp. Proc. Verb., Comm. Int. Expl. Sci., Mer Medit., n.s., 14:211-231.

Pelagopleura haranti

Bückmann, A. 1967.
Untersuchungen über das Makroplankton bei Ischia und Capri und im Golf von Neapel im Mai 1962. III. Die Appendicularien. Pubbl. Staz. zool. Napoli 35(2): 215-238.

Pelagopleura magna

Tokioka, Takesi, 1964.
Taxonomic studies of appendicularians collected by the Japanese Antarctic Research Expedition, 1957.
Japan. Antarct. Res. Exped., 1956-1962, Sci. Repts. (E), (JARE Sci Repts., Biol.), No. 2:16pp.

Pelagopleura verticalis

Tokioka, Takasi, 1960
Studies on the distribution of appendicularians and some thaliaceans of the North Pacific, with some morphological notes.
Publ. Seto Mar. Biol. Lab., 8(2) (27):351-443. tables.

Pelagopleura verticalis

Tokioka, T., 1955.
General considerations on Japanese appendicularian fauna. Publ. Seto Mar. Biol. Lab., 4(2/3): 251-261, 6 textfigs.

Pelagopleura verticalis

Tokioka, T., 1942
Systematic studies of the plankton organisms occurring in Iwayama Bay, Palao VII A preliminary report on the Appendicularian fauna of the bay and adjacent waters. Palao Tropical Biological Sta. Studies, 2(3): 613-616 (Contrib. 106, Seto Mar. Biol. Lab.)

Sinisteroffia scrippsi

Tokioka, Takasi, 1964.
Taxonomic studies of appendicularians collected by the Japanese Antarctic Research Expedition, 1957.
Japan. Antarct. Res. Exped. 1956-1962, Sci. Repts., (E)(JARE Sci. Repts., Biol.), No. 2:16pp.

Sinisteroffia scrippsi n.gen., n.sp.

Tokioka, T., 1957.
Two new appendicularians from the Eastern Pacific with notes on the morphology of Fritillaria aequatorialis and Tectillaria fertilis, Trans. Amer. Microsc. Soc., 76(4):359-365.

Stegosoma magnum

Björnberg, T.K.S., and L. Forneris, 1955.
Resultados cientificos do cruzeiro do "Baependi" e do "Vega" à ilha Trindade.
Contrib. Avulsas, Inst. Oceanograf. (Ocean. Biol.), Sao Paulo, No. 1:1-68.

Stegosoma magnum

Delsman, H. C., 1939.
Preliminary plankton investigations in the Java Sea. Treubia, 17:139-181, 8 maps, 41 figs.

Stegosoma magnum

Essenberg, C. E., 1926
Copelata from the San Diego Region. Univ. Calif. Publ. Zool. 28(22):399-521, 170 text figs.

Stegosoma magnum

Fenaux, R., 1967
5. Appendiculaires; Campagne de la Calypso au large des Cotes at antiques de l'Amerique du Sud (1961-1962)
Annls Inst. Oceanogr. n.s. 45 (2): 34-46

Stegosoma magnum

Forneris, Liliane, 1965.
Appendicularian species groups and southern Brazil water masses.
Bolm Inst. Oceanogr., S Paulo, 14(1):54-94.

Stegosoma magnum

Gamulin, T., 1948
Prilog poznavanju Zooplanktona Srednjedalmatinskog Otocnog Podrucja. (Contrib. a la connaissance du zooplankton de la zone insulaire de la Dalmatie Moyenne). Acta Adriatica 3(7):38 pp., 6 tables, 1 map.

Stegosoma magnum

Hastings, A. B., 1931
Tunicata. Great Barrier Reef Exped., 1928-29. Sci. Repts. 4(3):69-109, 17 text-figs, 3 pls.

Stegosoma magnum

Russell, F.S., and J.S. Colman, 1935 (Langerhans)
The Zooplankton. IV. The occurrence and seasonal distribution of the Tunicata, Mollusca and Coelenterata (siphonophora). Brit. Mus. (N.H.) Great Barrier Reef Expedition 1928-1929, Sci. Repts., 2(7):203-276, 30 text figs.

Stegosoma magnum

Tokioka, T., 1942
Systematic studies of the plankton organisms occurring in Iwayama Bay, Palao VII A preliminary report on the Appendicularian fauna of the bay and adjacent waters. Palao Tropical Biological Sta. Studies, 2(3): 613-616 (Contrib. 106, Seto Mar. Biol. Lab.)

Stegosoma magnum

Tokioka, T., 1940
Some additional notes on the Japanese appendicularian fauna.
Rec. Oceanogr. Wks., Japan, 11(1):1-26.

Stegosoma magnum

Tokioka, T., and J.A. Suarez Caabro, 1956.
Apendicularias de los mares cubanos.
Mem. Soc. Cubana Hist. Nat., 23(1):37-80, 15 pls.

Tectillaria fertilis

Fenaux, R., 1967
5. Appendiculaires; Campagne de la Calypso au large des Cotes at antiques de l'Amerique du Sud (1961-1962)
Annls Inst. Oceanogr. n.s. 45 (2): 34-46

Oceanographic Index: Marine Organisms Cumulation, 1946-1973

Tectillaria fertilis
Forneris, Liliana, 1965.
Appendicularian species groups and southern Brezil water masses.
Bolm Inst. Oceanogr., S Paulo. 14(1):54-94.

Tectillaria fertilis
Tokioka, T., 1957.
Two new appendicularians from the Eastern Pacific with notes on the morphology of Fritillaria aequatorialis and Tectillaria fertilis. Trans. Amer. Microsc. Soc., 76(4):359-365.

Tectillaria fertilis
Tokioka, T., 1942
Systematic studies of the plankton organisms occurring in Iwayama Bay, Palao VII A preliminary report on the Appendicularian fauna of the bay and adjacent waters.
Palao Tropical Biological Sta. Studies, 2(3): 613-616 (Contrib. 106, Seto Mar. Biol. Lab.)

Tectillaria fertilis
Tokioka, T., 1940
Some additional notes on the Japanese appendicularian fauna.
Rec. Oceanogr. Wks., Japan, 11(1):1-26.

Tectillaria taeniogona
Tokioka, Takasi, 1959.
Further notes on some appendicularians from the eastern Pacific.
Publ. Seto Mar. Biol. Lab., 7(1):1-17.

ascidians
Diehl, Manfred 1970.
Die neue ökologisch extreme Sand-Ascidie von der Josephine-Bank: Seriocarpa rhizoides Diehl 1969 (Ascidiacea, Styelidae).
Meteor Forsch.-Ergebnisse (D) 7:43-58.

ascidians, taxonomy
Eldredge, L.G., 1966. (1967)
A taxonomic review of Indo-Pacific didemnid ascidians. Micronesica, J. Coll. Guam, 2(2): 161-262.

ascidians
Millar, R.H., 1970.
Ascidians including specimens from the deep sea, collected by the R.V. Vema and now in the American Museum of Natural History.
Zool. J. Linn. Soc. 49(2): 99-159.

tunicates (ascidians)
Monniot, Claude 1970.
Campagnes d'essais du Jean Charcot (3-8 décembre 1968) 3. Ascidies.
Bull. Mus. natn. Hist. nat. (2) 41(5): 1146-1149.

tunicates
Monniot, Claude, 1969.
Ascidies récoltées par la Thalassa sur la pente du plateau continental du Golfe de Gascogne (3-12 août 1967).
Bull. Mus. natn. Hist. Nat. (2) 41(1): 155-186.

tunicates
Monniot, Claude et Françoise Monniot, 1970.
Les Ascidies des grandes profondeurs récoltées par les navires Atlantis, Atlantis II et Chain (2 ème note). Deep-Sea Res., 17(2): 317-336.

ascidians
Monniot, Claude et Françoise Monniot 1968.
Les ascidies de grandes profondeurs récoltées par le navire océanographique américain Atlantis II (première note).
Bull. Inst. océanogr. Monaco 67(1379): 45pp.

ascidians
Monniot, C., and F. Monniot, 1963.
Présence à Bergen et Roscoff de Pyuridae psammicoles du genre Heterostigma.
Sarsia, 13:51-57.

tunicates
Monniot, Françoise 1969.
Sur une collection d'ascidies composées de Dakar.
Bull. Mus. natn. Hist. Nat. (2) 41(2): 426-457.

ascidians
Monniot, Francoise, 1966.
Ascidies interstitielles.
Veröff. Inst. Meeresforsch., Bremerh., Sonderband II:161-164.

ascidians
Tokioka, Takasi, 1967.
On a small collection of ascidians from the vicinity of Nhatrang, Viet Nam.
Publs. Seto mar. biol. Lab., 14(5):391-402.

ascidians
Vasseur, Pierre, 1970.
Contribution à l'Etude des ascidies de Madagascar (Région de Tuléar). III. La faune ascidiologique des herbiers de phanérogames marines
Rec. Trav. Sta. mar. Endoume, hors sér. Suppl. 10:209-221

ascidians
Yamaguchi, Masashi, 1970.
Spawning periodicity and settling time in ascidians, Ciona intestinalis and Styela plicata
Rec. oceanogr. Wks. Japan, n.s., 10(2): 147-155.

Tunicates, anatomy
Pérès, J.M. 1943
Recherches sur le sang et les organes neuvraux des Tuniciers. Ann. de l'Inst. Océano, Monaco. 21(5):229-359.

tunicates, anatomy
Schiller, Joseph, 1968.
Controverses autour de certaines structures chez les tuniciers au XIX Siecle.
Bull. Inst. Océanogr., Monaco, No. special 2: 387-396.

tunicates, chemistry of
Culkin, F. and R.J. Morris, 1970.
The fatty acid composition of two marine filter-feeders in relation to a phytoplankton diet.
Deep-Sea Res., 17(5): 861-865.

tunicates, chemistry
Strohal, Petar, Josip Tuta and Zvonimir Kolar, 1969.
Investigations of certain microconstituents in two tunicates. Limnol. Oceanogr., 14(2): 265-268.

tunicates
Millar, R.H., 1966.
Evolution in ascidians.
In: Some contemporary studies in marine science. H. Barnes, editor, George Allen & Unwin, Ltd., 519-534.

ascidians, lists of spp.
Monniot, Claude et Françoise Monniot, 1970.
Les Ascidies des grandes profondeurs récoltées par les navires Atlantis, Atlantis II et Chain (2 ème note). Deep-Sea Res., 17(2): 317-336.

ascidians, list of spp.
Monniot, Claude, 1970.
Ascidies récoltées par la Thalassa sur la pente du plateau continental du Golfe de Gascogne (15-25 octobre 1968).
Bull. Mus. Natn Hist. nat. (2) 41(5):1131-1145.

Amaroucium pro? Baron
Pérès, J.M. 1943
Recherches sur le sang et les organes neuvraux des Tuniciers. Ann. de l'Inst. Océano, Monaco. 21(5):229-359.

Ascidia aspersa
Pérès, J.M. 1943
Recherches sur le sang et les organes neuvraux des Tuniciers. Ann. de l'Inst. Océano, Monaco. 21(5):229-359.

Botryllus schlosseri
Pérès, J.M. 1943
Recherches sur le sang et les organes neuvraux des Tuniciers. Ann. de l'Inst. Océano, Monaco. 21(5):229-359.

Ciona intestinalis
Pérès, J.M. 1943
Recherches sur le sang et les organes neuvraux des Tuniciers. Ann. de l'Inst. Océano, Monaco. 21(5):229-359.

Halocynthia papillose
Pérès, J.M. 1943
Recherches sur le sang et les organes neuvraux des Tuniciers. Ann. de l'Inst. Océano, Monaco. 21(5):229-359.

Microcosmus sulcatus
Pérès, J.M. 1943
Recherches sur le sang et les organes neuvraux des Tuniciers. Ann. de l'Inst. Océano, Monaco. 21(5):229-359.

Molgula
Pérès, J.M. 1943
Recherches sur le sang et les organes neuvraux des Tuniciers. Ann. de l'Inst. Océano, Monaco. 21(5):229-359.

Polycarpa pomaria
Pérès, J.M. 1943
Recherches sur le sang et les organes neuvraux des Tuniciers. Ann. de l'Inst. Océano, Monaco. 21(5):229-359.

Polycitor lepadiformis
Pérès, J.M. 1943
Recherches sur le sang et les organes neuvraux des Tuniciers. Ann. de l'Inst. Océano, Monaco. 21(5):229-359.

Styela clava
Saito, T., 1931
Researches in fouling organisms of the Ships' bottom. Zosen Kiokai (T. Soc. Naval Arch.) 47pp:13-64, 51 figs., 9 graphs.

Styela plicata
Pérès, J.M. 1943
Recherches sur le sang et les organes neuvraux des Tuniciers. Ann. de l'Inst. Océano, Monaco. 21 (5):229-359.

Styela plicata
Saito, T., 1931
Researches in fouling organisms of the Ships' bottom. Zosen Kiokai (T. Soc. Naval Arch.) 47pp:13-64, 51 figs., 9 graphs.

salps
Apstein, C., 1906
Salpen der deutschen Tiefsee-Expedition.
Wiss. Ergebn. "Valdivia" 1898-1899, STII:245-290, pls. xxvi-xxii, 15 textfigs.

salps
Apstein, C., 1904
Salpes d'Amboine. Rev. Suisse Zool. 12:649-656, Pl. 12.

See Bedot (1909), ibid. 17:143-169 for list of indexed species.

Salps
Apstein, C., 1894
Die Thaliacea der Plankton-Expedition B. Vertheilung der Salpen. Ergebn. Plankton-Exped. Humboldt-Stiftung II, E.a.B., pp.1-68, pls.ii-iv, 14 textfigs.

Salps
Bal, D.V., and L.B. Pradhan, 1945.
A preliminary note on the plankton of Bombay harbour. Current Sci., 14(8): 211-212.

Thaliacea
Barnes, Beatrice I., 1961
Contribution towards a plankton atlas of the north-eastern Atlantic and the North Sea. 4. Thaliacea.
Bulls. Mar. Ecology, 5(42): 102-104, Pl. 29.

salps
Belloc, G., 1938.
Liste des Tuniciers pélagiques capturés au cours de la cinquième croisière. Rev. Trav. Off. Pêches Marit. 11(3):315-334, 20 textfigs.

salps
Berrill, N.J., 1950.
Budding and development in Salpa. J. Morph. 83(3):553-606, 16 textfigs.

salps
Borgert, A., 1894
Die Thaliacea der Plankton-Expedition. C. Vertheilung der Doliolen. Ergebn. Plankton-Exped. Humboldt-Stiftung, II, E.a.C., pp.1-68, pls.V-VIII, 2 textfigs.

salps
Caldwell, Melba C., 1966.
The distribution of pelagic tunicates Family Salpidae in Antarctic and Subantarctic waters.
Bull. S. Calif. Acad. Sci., 65(1):1-16.

salps, effect of
Fraser, J.H., 1962.
19. The role of ctenophores and salps in zooplankton production and standing crop. Contributions to symposium on zooplankton production, 1961.
Rapp. Proc. Verb., Cons. Perm. Int. Expl. Mer, 153:121-123.

Thaliacea
Fraser, J. H., 1949.
The distribution of Thaliacea (Salps and Doliolids) in Scottish Waters 1920 to 1939. Scottish Home Dept., Fish. Div., Sci. Invest. 1949(1): 44 pp., 16 textfigs.

salps
Fraser, J.H., 1948.
Thaliacea II. Family: Doliolidae. Fiches d'Ident. Zooplancton, Cons. Perm. Int. Expl. Mer, 10:4 pp., 13 textfigs.

salps
Fraser, J.H., 1947.
Thaliacea I. Family: Salpidae. Fiches d'Ident. Zooplancton, Cons. Perm. Int. Expl. Mer, 9:4 pp., 17 textfigs.

tunicates
Godeaux, G., 1964.
Tuniciers pélagiques.
Res. Sci. Exped. Océanogr. Belge, Eaux Cotières Afr. Atlant Sud., 3(7):1-32. (not seen).

tunicates
Godeaux, J., 1960.
Tuniciers pelagiques du Golfe d'Eylath. Contribution to the knowledge of the Red Sea, No. 18.
Sea Fish. Res. Sta., Haifa, Bull., No. 29:9-15.

tunicates (salps)
Harant, H., and P. Vernières, 1934.
Tuniciers pélagiques provenant des croisières du Prince Albert 1er de Monaco. Rés. camp. sci., Monaco, 88:47 pp., 5 textfigs.

Tunicata
Hastings, A. B., 1931
Tunicata. Great Barrier Reef Exped., 1928-29. Sci. Repts. 4(3):69-109, 17 text-figs, 3 pls.

Tunicata
Herdman, W.A., 1888
Report upon the Tunicata collected during the voyage of H.M.S. "Challenger" during the years 1873-1876. Rep. Sci. Res. "Challenger" Exped., Zool., XXVII (76) 11-166, pls. 1-xi, 28 textfigs.

salps
Hunt, H.G., 1968.
Continuous plankton records: contribution towards a plankton atlas of the North Atlantic and the North Sea. II. The seasonal and annual distribution of Thaliacea.
Bull. mar. Ecol. 6(7):225-249.

salps
Ihle, J.E.W., 1912
Tunicata. Salpae I. Desmomyaria. Das Tierreich. 32:66 pp., 68 text figs.

salps
Ihle, J.E.W., 1911.
Über die Nomenklatur der Salpen. Zool. Anz. 38: 585-589.

salps
Ihle, J. E. W., and M. E. Ihle-Landenberg, 1935.
Über eine kleine Salpen Sammlung aus der Javasee. Zool. Anz., 110:

salps
Lucas, C.E., N.B. Marshall and C.B. Rees 1942
Continuous plankton records; The Faeroe-Shetland Channel 1939. Hull Bull. Mar. Ecol., II(10):71-94

salps
McKenzie, R.A., and R.E.S. Homans, 1938.
Rare and interesting fishes and salps in the Bay of Fundy and off Nova Scotia.
Proc. Nova Scotian Inst. Sci., 19:277-281.

Salpidae
Metcalf, M.M., 1919
Contributions to the biology of the Philippine Archipelago and adjacent regions. The Salpidae collected by the United States Fisheries Steamer "Albatross" in Philippine waters during the years 1908 and 1909.
Bull. U. S. Nat. Mus. No.100, Vol. 2(1):1-4.

salps
Metcalf, M.M., and M.M. Bell, 1918
The Salpidae: A Taxonomic Study. Bull. U.S. Nat. Mus. 100, II:5-193, pls. 1-xiv, 15 textfigs.

salps
Nair, R.V., 1949.
The Thaliacea of the Madras plankton. Bull. Madras Govt. Mus., n.s., Nat. Hist. Sect., 6(1): 41 pp., 6 pls.

salps
Neumann, G., 1913
Tunicata Salpae II: Cyclomyaria et Pyrosomida. 40:36 pp., 19 text figs.
(Das Tierreich)

salps
Queiroz, Debora, 1965.
On the distribution of the Thaliacea in the region of Cananeia. (Abstract).
Anais Acad. bras Cienc., 37 (Supl.):256.

salps
Ritter, W. E. 1905
The pelagic Tunicata of the San Diego Region, excepting the Larvacea. Univ. Calif. Pub. Zool. II, 51-112, pls. 11-111, 31 textfigs.

Tunicata
Ritter, R. E. and E. S. Byxbee, 1905
VIII. The pelagic Tunicata. Reports on the Scientific results of the expedition to the Tropical Pacific, in charge of Alexander Agassiz, in the U. S. Fish Commission Steamer "Albatross", from August, 1899 to March, 1900, Commander Jefferson F. Moser, U. S. N., commanding. Mem. M.C.Z. 26(5):195-211, 2 pls.

salps
Russell, F.S., and J.S. Colman, 1935
The Zooplankton. IV. The occurrence and seasonal distribution of the Tunicata, Mollusca and Coelenterata (siphonophora). Brit. Mus. (N.H.) Great Barrier Reef Expedition 1928-1929, Sci. Repts., 2(7):203-276, 30 text figs.

salps
Russell, F. S. and A. B. Hastings, 1933
On the occurrence of pelagic tunicates (Thaliacea) in the waters of the English Channel off Plymouth. JMBA 18(2):635-640.

tunicates
Sewell, R.B.S., 1953.
The pelagic Tunicata. Brit. Mus. (N.H.), John Murray Exped., 1933-34, Sci. Repts. 10(1):1-90, 1 pl., 32 textfigs.

salps
Seymour Sewell, R.B. 1926
The Salps of Indian Seas. Rec. Ind. Museum XXVIII:65-126, 43 textfigs.

tunicates
Thompson, Harold, 1954.
Pelagic tunicates. Identifications and relevant notes.
B.A.N.Z. Antarctic Res. Exped., 1929-1931 Repts., B(Zool. Bot.), 1(4):183-185.

tunicates, pelagic
Tokioka, Takasi, 1960
Studies on the distribution of appendicularians and some thaliaceans of the North Pacific, with some morphological notes.
Publ. Seto Mar. Biol. Lab., 8(2) (27):351-443. tables.

tunicates
Tokioka, T., 1951.
Pelagic tunicates and chaetognaths collected during the cruises to the New Yamato Bank in the Sea of Japan. Publ. Seto Mar. Biol. Lab. 2(1):1-26, 1 chart, 6 tables, 12 textfigs.

salps
Tokioka, T., 1937.
Notes on salps and doliolums. Annot. Zool. Jap., 16(3):219-232.

salps
Van Zyl, R. P., 1960.
A preliminary study of the salps and doliolids off the west and south coasts of South Africa. Union S. Africa, Dept. Comm. & Ind., Div. Fish., Invest. Rept. 40:31 pp.

tunicates
Yount, J.L., 1954.
The taxonymy of the Salpidae (Tunicata) of the central Pacific Ocean. Pacific Science 8(3): 276-330, 30 textfigs.

salps, anatomy
Berrill, N.J., 1950.
The Tunicata with an account of the British species. Ray Soc. 133:354 pp., 120 textfigs.

tunicates
Bone, Q., 1959.
Observations upon the nervous systems of pelagic tunicates.
Quart. J. Microscop. Sci., 100(2):167-182.

salps, anat.-physiol.
Brooks, W.K., 1893.
The genus Salpa. Mem. Biol. Lab., Johns Hopkins Univ., 2:1-306, 46 pls.

salps, physiology
Carlisle, D.B., 1950.
Alcune osservazioni sulla meccanica dell' alimentazione della Salpa. Pubbl. Staz. Zool. Napoli, 22(2):146-154, 3 textfigs.

Oceanographic Index: Marine Organisms Cumulation, 1946-1973

---

salps, anat.-physiol.
Ebara, A., 1954.
The periodic reversal of the heart-beat in Salpa fusiformis. Sci. Rept., Tokyo, Bunrika Daigaku, (B), 110:

Salps, anat. physiol
Godeaux, J., 1965.
Observations sur le tunique des tuniciers pélagiques.
Rapp. Proc. Verb., Réunions, Comm. Int. Expl. Sci. Mer Méditerranée, Monaco, 18(2):457-460.

Thalacea, physiol.-anat.
Godeaux, J., 1961.
L'éléoblaste des Thalacés.
Rapp. Proc. Verb., Reunions, Comm. Int. Expl. Sci. Mer Mediterranee, Monaco, 16(2):143-145.

Thalacea, anat.-physiol.
Godeaux, J., 1957-1958.
Contribution à la connaissance des Thalacés. Embryogénèse et blastogénèse du complexe neural. Constitution et développement du stolon prolifère.
Ann. Soc. Roy. Zool. Belgique, 88:5-285.

---

salps, anat.
Johnson, M.E., 1910.
A quantitative study of the development of the Salpa chain in Salpa fusiformis-nunciata. Univ. Calif. Publ., Zool. 6(7):145-176, 15 textfigs.

---

salps, anat.-physiol.
Metcalf, M.M., 1893.
Pt. 4. The eyes and subneural gland of Salpa.
Mem. Biol. Lab., Johns Hopkins Univ. 2:307-371, Pls. 47-57.

---

salps, anat., physiol.
Sawicki, R.M., 1966
Development of the stolen in Salpa fusiformis Cuvier and Salpa aspera Chamisso.
Discovery Repts., 33:335-383.

---

salps, ecol., lists of spp.
Yount, J.L., 1958.
Distribution and ecological aspects of central Pacific Salpidae (Tunicata). Pacific Science 12(2):111-130.

---

salps, lists of spp.
Beaudouin, Jacqueline 1971.
Données écologiques sur quelques groupes planctoniques indicateurs dans le Golfe de Gascogne.
Revue Trav. Inst. Pêches marit. 35(4): 375-414.

---

tunicates, lists of spp.
Bernard, Michelle, 1958.
Systématique et distribution saisonnière des tuniciers pélagiques d'Alger.
Rapp. Proc. Verb., Comm. Int. Expl. Sci., Mer Medit., n.s., 14:211-231.

---

salps, lists of spp
Berner, Leo D., 1967.
Distributional atlas of Thaliacea in the California Current region.
Atlas, Calif. Coop. Ocean. Fish. Invest., 8:1-322

---

salps, lists of spp.
Esnal, Graciela B. 1970.
Contribución al conocimiento de los salpas del Atlantico Sur, en especial Thalia democratica.
Neotropica 16(51):121-134

Salps, lists of spp
Esnal, Graciela B., 1968.
Salpas colectadas por El Austral y el Walther Herwig en el Océano Atlántico.
Revta Mus. argent. Cienc. nat. Bernardino Ricadavia, Hidrobiol., 2(8):257-277.

---

salps, lists of spp.
Fenaux, R., et J. Godeaux 1970.
Répartition verticale des tuniciers pélagiques au large d'Eilat (Golfe d'Aqaba).
Bull. Soc. roy. Sci. Liège 39(3/4):200-209
Also in: Trav. Sta. Zool. Villefranche-sur-Mer 32

Salps, lists of spp.
Godeaux, Jean 1973.
Distribution des Thaliacés dans les mers bordant le nord de l'Afrique. Rapp. Proc.-v. Reun. Commn int. Explor. scient. Mer Medit. Monaco 21(8):489-491.

---

salps, lists of spp.
Godeaux, J., 1963.
Tuniciers pélagiques recoltés sur la côte occidentale d'Israel.
Sea Fish.Res.Sta., Israel, Bull., No.34:3-4.

Salpa, lists of spp.
Godeaux, J., et G. Goffinet 1968
Données sur la faune pélagique vivant au large des côtes du Gabon du Congo et de l'Angola (0-18° lat. S. et 5-12° long. E.): Tuniciers pélagiques: I. Salpidae. Ann. Soc. r. zool. Belg. 98(1): 49-86. In: Receuil Trav. publiées de 1965 a 1966, Oceanogr. ORSTROM.

---

salps, lists of spp.
Hubbard, Lyle T. Jr. and William G. Pearcy 1972.
Geographic distribution and relative abundance of Salpidae off the Oregon coast.
J. Fish. Res. Bd Can. 28(12):1831-1836.

Thalacea, lists of spp.
Tokioka, Takasi 1967.
Pacific Tunicata of the United States National Museum.
Bull. U.S. natn. Mus. 251:1-247.

---

tunicates, lists of spp.
Tokioka, T., 1960
Droplets from the plankton net. XIX. A glimpse upon chaetognaths and pelagic tunicates collected in the lagoon water near Noumea, New Calendonia. Publ. Seto Mar. Biol. Lab., 8(1): 51-54.

---

tunicates (pelagic) (lists of sp
Tokioka, T., 1955.
Droplets from a plankton net. 17. A small collection of chaetognaths and pelagic tunicates from the north eastern part of the Indian Ocean. 18. Short notes on a few appendicularians collected in the "Kurosio" off Siono-misaki. Publ. Seto Mar. Biol. Lab., 5(1):75-80.

Figure of palp of Thalia democratica
Figure of Fritillaria venusta
Figures of tails of appendicularians helpful in identification.

---

salps, methods
Foxton, P., 1965.
An aid to the detailed examination of salps (Tunidata:Salpidae).
J. mar. biol. Ass., U.K., 45(3):679-681.

---

salps, parasites of
Ormières, René, 1966.
Recherches sur les sporozoaires parasites des tuniciers.
Vie et Milieu, Bull. Lab. Arago, Univ. Paris 15(4):823-946.

---

salps, vertical migration
Hunt, H.G., 1968.
Continuous plankton records: contribution towards a plankton atlas of the North Atlantic and the North Sea.II.The seasonal and annual distribution of Thaliacea.
Bull.mar.Ecol.6(7):225-249.

Apsteinia asymmetrica
Belloc, G., 1938.
Liste des Tuniciers pélagiques capturés au cours de la cinquième croisière. Rev. Trav. Off. Pêches Marit. 11(3):315-334, 20 textfigs.

Apsteinia punctata
Belloc, G., 1938.
Liste des Tuniciers pélagiques capturés au cours de la cinquième croisière. Rev. Trav. Off. Pêches Marit. 11(3):315-334, 20 textfigs.

Brooksia rostrata
Amor, Analía, 1966.
Tunicados pelágicos de la "Operación Convergencia" en el Atlántico Sur (1961).
Physis, Buenos Aires, 26(71):163-179.

Brooksia rostrata
Delsman, H. C., 1939.
Preliminary plankton investigations in the Java Sea. Treubia, 17:139-181, 8 maps, 41 figs.

Brooksia rostrata
Esnal, Graciela B., 1968.
Salpas colectadas por El Austral y el Walther Herwig en el Océano Atlántico.
Revta Mus. argent. Cienc. nat. Bernardino Ricadavia, Hidrobiol., 2(8):257-277.

Brooksia rostrata
Godeaux, J., 1960.
Tuniciers pélagiques du Golfe d'Eylath. Contribution to the knowledge of the Red Sea, No. 18.
Sea Fish. Res. Sta., Haifa, Bull., No. 29:9-15.

Brooksia rostrata
Godeaux, J., et G. Goffinet 1968
Données sur la faune pélagique vivant au large des côtes du Gabon du Congo et de l'Angola (0-18° lat. S. et 5-12° long. E.): Tuniciers pélagiques: I. Salpidae. Ann. Soc. r. zool. Belg. 98(1): 49-86. In: Receuil Trav. publiées de 1965 a 1966, Oceanogr. ORSTROM.

Brooksia rostrata
Nair, R. Velappan, 1949.
The Thaliacea of the Madras plankton. Bull. Madras Govt. Mus., n.s., Nat. Hist. Sect., 6(1): 1-41, 6 pls.

Brooksia rostrata
Tokioka, T., 1954.
Descriptions on the aggregated form of Brooksia rostrata (Traustedt), an insufficiently known Salpa. Publ. Seto Mar. Biol. Lab. 4(1):147-153, Pls. 9-10, 4 textfigs.

Brooksia rostrata
Yount, J.L., 1954.
The taxonomy of the Salpidae (Tunicata) of the central Pacific Ocean. Pacific Science 8(3): 276-330, 30 textfigs.

Cyclosalpa affinis
Amor, Analía, 1966.
Tunicados pelágicos de la "Operación Convergencia" en el Atlántico Sur (1961).
Physis, Buenos Aires, 26(71):163-179.

Cyclosalpa affinis
Godeaux, J., et G. Goffinet 1968
Données sur la faune pélagique vivant au large des côtes du Gabon du Congo et de l'Angola (0-18° lat. S. et 5-12° long. E.): Tuniciers pélagiques: I. Salpidae. Ann. Soc. r. zool. Belg. 98(1): 49-86. In: Receuil Trav. publiées de 1965 a 1966, Oceanogr. ORSTROM.

Cyclosalpa affinis
Ihle, J.E.W., 1912
Tunicata. Salpae I. Desmomyaria. Das Tierreich. 32:66 pp., 68 text figs.

Cyclosalpa affinis
Ritter, W.E., 1905.
III. The pelagic Tunicata of the San Diego region, excepting the Larvacea. Univ. Calif. Publ. Zool. 2(3):51-112, Pls. 2-3.

Cyclosalpa affinis
Yount, J.L., 1954.
The taxonomy of the Salpidae (Tunicata) in the central Pacific Ocean. Pacific Science 8(3): 276-330, 30 textfigs.

Cyclosalpa affinis
Tokioka, T., 1937.
Notes on salpas and doliolums. Annot. Zool. Jap., 16(3):219-232.

Cyclosalpa bakeri
Fraser, J. H., 1949.
The distribution of Thaliacea (Salps and Doliolids) in Scottish Waters 1920 to 1939. Scottish Home Dept., Fish. Div., Sci. Invest. 1949(1):44 pp., 16 textfigs.

Cyclosalpa bakeri
Fraser, J.H., 1947.
Thaliacea I. Family: Salpidae. Fiches d'Ident. Zooplancton, Cons. Perm. Int. Expl. Mer, 9:4 pp., 17 textfigs.

Cyclosalpa bakeri
Godeaux, J., et G. Goffinet 1968
Données sur la faune pélagique vivant au large des côtes du Gabon du Congo et de l'Angola (0-18° lat. S. et 5-12° long. E.): Tuniciers pélagiques: I. Salpidae. Ann. Soc. r. zool. Belg. 98(1): 49-86. In: Receuil Trav. publiées de 1965 a 1966, Oceanogr. ORSTROM.

Cyclosalpa bakeri
Ihle, J.E.W., 1912
Tunicata. Salpae I. Desmomyaria. Das Tierreich. 32:66 pp., 68 text figs.

Cyclosalpa bakeri n.sp.
Ritter, W.E., 1905.
III. The pelagic Tunicata of the San Diego region, excepting the Larvacea. Univ. Calif. Publ. Zool. 2(3):51-112, Pls. 2-3.

Cyclosalpa bakeri
Ritter, R. E. and E. S. Byxbee, 1905
VIII. The pelagic Tunicata. Reports on the Scientific results of the expedition to the Tropical Pacific, in charge of Alexander Agassiz, in the U. S. Fish Commission Steamer "Albatross", from August, 1899 to March, 1900, Commander Jefferson F. Moser, U. S. N., commanding. Mem. M.C.Z. 26(5):195-211, 2 pls.

Cyclosalpa bakeri
Tokioka, T.,,1951.
Droplets from the plankton net. IX. Records of Cyclosalpa bakeri from Japanese waters. X. Rexex record of Creseis chierchiae (BOAS) from the Palao Islands. Publ. Seto Mar. Biol. Lab. 1(4):183-184.

Cyclosalpa bakeri
Yount, J.L., 1954.
The taxonomy of the Salpidae (Tunicata) in the central Pacific Ocean. Pacific Science 8(3):276-330, 30 textfigs.

Cyclosalpa floridana
Delsman, H. C., 1939.
Preliminary plankton investigations in the Java Sea. Treubia, 17:139-181, 8 maps, 41 figs.

Cyclosalpa floridana
Godeaux, J., et G. Goffinet 1968
Données sur la faune pélagique vivant au large des côtes du Gabon du Congo et de l'Angola (0-18° lat. S. et 5-12° long. E.): Tuniciers pélagiques: I. Salpidae. Ann. Soc. r. zool. Belg. 98(1): 49-86. In: Receuil Trav. publiées de 1965 a 1966, Oceanogr. ORSTROM.

Cyclosalpa floridana
Ihle, J.E.W., 1912
Tunicata. Salpae I. Desmomyaria. Das Tierreich. 32:66 pp., 68 text figs.

Cyclosalpa floridana
Yount, J.L., 1954.
The taxonomy of the Salpidae (Tunicata) in the central Pacific Ocean. Pacific Science 8(3):276-330, 30 textfigs.

Cyclosalpa irregula
Van Zyl, R. P., 1960.
A preliminary study of the salps and doliolids off the west and south coasts of South Africa. Union S. Africa, Dept. Comm. & Ind., Div. Fish., Invest. Rept. 40:31 pp.

Cyclosalpa kamaii
Tokioka, T., 1937.
Notes on salpas and doliolums. Annot. Zool. Jap. 16(3):219-232.

Cyclosalpa pinnata
Amor, Analía, 1966.
Tunicados pelágicos de la "Operación Convergencia" en el Atlántico Sur (1961). Physis, Buenos Aires, 26(71):163-179.

Cyclosalpa pitnata (sic)
Belloc, G., 1938.
Liste des Tuniciers pélagiques capturés au cours de la cinquième croisière. Rev. Trav. Off. Pêches Marit. 11(3):315-334, 20 textfigs.

Cyclosalpa pinnata
Berrill, N.J., 1950.
Budding and development in Salpa. J. Morph. 83(3) 553-606, 16 textfigs.

Cyclosalpa pinnata
Esnal, Graciela B., 1968.
Salpas colectadas por El Austral y el Walther Herwig en el Océano Atlántico. Revta Mus. argent. Cienc. nat. Bernardino Ricadavia, Hidrobiol., 2(8):257-277.

Cyclosalpa pinnata
Godeaux, J., et G. Goffinet 1968
Données sur la faune pélagique vivant au large des côtes du Gabon du Congo et de l'Angola (0-18° lat. S. et 5-12° long. E.): Tuniciers pélagiques: I. Salpidae. Ann. Soc. r. zool. Belg. 98(1): 49-86. In: Receuil Trav. publiées de 1965 a 1966, Oceanogr. ORSTROM.

Cyclosalpa pinnata
Harant, H., and P. Vernières, 1934.
Tuniciers pélagiques provenant des croisères du Prince Albert 1er de Monaco. Rés. camp. sci., Monaco, 88:47 pp., 5 textfigs.

Cyclosalpa pinnata
Hastings, A. B., 1931
Tunicata. Great Barrier Reef Exped., 1928-29. Sci. Repts. 4(3):69-109, 17 text-figs., 3 pls.

Cyclosalpa pinnata
Ihle, J.E.W., 1912
Tunicata. Salpae I. Desmomyaria. Das Tierreich. 32:66 pp., 68 text figs.

Cyclosalpa pinnata
Metcalf, M.M., 1893.
Pt. 4. The eyes and subneural gland of Salpa. Mem. Biol. Lab., Johns Hopkins Univ., 2:307-371, Pls. 47-57.

Cyclosalpa pinnata
Russell, F.S., and J.S. Colman, 1935
The Zooplankton. IV. The occurrence and seasonal distribution of the Tunicata, Mollusca and Coelenterata (siphonophora). Brit. Mus. (N.H.) Great Barrier Reef Expedition 1928-1929, Sci. Repts., 2(7):203-276, 30 text figs.

Cyclosalpa pinnata
Tokioka, T., 1937.
Notes on salpas and doliolums. Annot. Zool. Jap., 16(3):219-232.

Cyclosalpa pinnata
Yount, J.L., 1954.
The taxonomy of the Salpidae (Tunicata) in the central Pacific Ocean. Pacific Science 8(3):276-330, 30 textfigs.

Cyclosalpa pinnata parallela n.subsp.
Kashkina A.A. 1973.
A contribution to the fauna of Salpidae (Tunicata) in the Indian Ocean. (In Russian. English abstract) Zool. Zh. 52(2): 215-219.

Cyclosalpa pinnata polae
Nair, R. Velappan, 1949.
The Thaliacea of the Madras plankton. Bull. Madras Govt. Mus., n.s., Nat. Hist. Sect., 6(1):1-41, 6 pls.

Cyclosalpa pinnata quadriluminis n.spbsp.
Berner, L.D., 1955.
Two new pelagic tunicates from the eastern Pacific Ocean. Pacific Sci. 9(2):247-253, 8 textfigs.

Cyclosalpa pinnata sewelli
Nair, R. Velappan, 1949.
The Thaliacea of the Madras plankton. Bull. Madras Govt. Mus., n.s., Nat. Hist. Sect., 6(1):1-41, 6 pls.

Cyclosalpa polae
Ihle, J.E.W., 1912
Tunicata. Salpae I. Desmomyaria. Das Tierreich. 32:66 pp., 68 text figs.

Cyclosalpa pinnata var sewelli
Sewell, R.B.S., 1953.
The pelagic Tunicata. Brit. Mus. (N.H.), John Murray Exped., 1933-34, Sci. Repts. 10(1):1-90, 1 pl., 32 textfigs.

Cyclosalpa polae n.sp.
Sigl, A., 1912.
Cyclosalpa polae n.sp. aus dem östlichen Mittelmeere. Zool. Anz. 39(2):66-74, 9 pls.

Cyclosalpa strongylanteron n.sp.
Berner, L.D., 1955.
Two new pelagic tunicates from the eastern Pacific Ocean. Pacific Sci. 9(2):247-253, 8 textfigs.

Cyclosalpa virgula
Belloc, G., 1938.
Liste des Tuniciers pélagiques capturés au cours de la cinquième croisière. Rev. Trav. Off. Pêches Marit. 11(3):315-334, 20 textfigs.

Cyclosalpa virgula
Godeaux, J., et G. Goffinet 1968
Données sur la faune pélagique vivant au large des côtes du Gabon du Congo et de l'Angola (0-18° lat. S. et 5-12° long. E.): Tuniciers pélagiques: I. Salpidae. Ann. Soc. r. zool. Belg. 98(1): 49-86. In: Receuil Trav. publiées de 1965 a 1966, Oceanogr. ORSTROM.

Cyclosalpa virgula
Ihle, J.E.W., 1912
Tunicata. Salpae I. Desmomyaria. Das Tierreich. 32:66 pp., 68 text figs.

Dipleurosoma elliptica
Neumann, G., 1913
Tunicata Salpae II: Cyclomyaria et Pyrosomida. 40:36 pp., 19 text figs. (Das Tierreich)

Helicosalpa komai
Yount, J.L., 1954.
The taxonomy of the Salpidae (Tunicata) in the Central Pacific Ocean. Pacific Science 8(3):276-330, 30 textfigs.

Helicosalpa virgula
Yount, J.L., 1954.
The taxonomy of the Salpidae (Tunicata) of the Central Pacific Ocean. Pacific Science 8(3):276-330, 30 textfigs.

Helicosalpa virgula younti n.subsp.
Kashkina A.A. 1973.
A contribution to the fauna of Salpidae (Tunicata) in the Indian Ocean. (In Russian. English abstract) Zool. Zh. 52(2): 215-219.

Iasis zonaria
Barnes, Beatrice I., 1961
Contribution towards a plankton atlas of the north-eastern Atlantic and the North Sea. 4. Thalacea. Bulls. Mar. Ecology, 5(42): 102-104, Pl. 29.

Iasis zonaria
Bary, B.M., 1960.
Notes on ecology, distribution and systematics of pelagic Tunicata from New Zealand. Pacific Science, 14:101-221.
(2)

Iasis zonaria
Belloc, G., 1938.
Liste des Tuniciers pélagiques capturés au cours de la cinquième croisière. Rev. Trav. Off. Pêches Marit. 11(3):315-334, 20 textfigs.

Iasis zonaria
Bigelow, H.B., and M. Sears, 1939
Studies of the waters of the continental shelf, Cape Cod to Chesapeake Bay. III. A volumetric study of the zooplankton. Mem. M.C.Z. 54(4):183-378, 42 text figs.

Iasis zonaria
Caldwell, Melba C., 1966.
The distribution of pelagic tunicates Family Salpidae in Antarctic and Subantarctic waters. Bull. S. Calif. Acad. Sci. 65(1):1-16.

Iasis zonaria
Fraser, J. H., 1949.
The distribution of Thaliacea (Salps and Doliolids) in Scottish Waters 1920 to 1939. Scottish Home Dept., Fish. Div., Sci. Invest. 1949(1):44 pp., 16 textfigs.

Iasis zonaria
Fraser, J.H., 1947.
Thaliacea I. Family: Salpidae. Fiches d'Ident. Zooplancton, Cons. Perm. Int. Expl. Mer, 9:4 pp., 17 textfigs.

Iasis zonaria
Godeaux, J., et G. Goffinet 1968
Données sur la faune pélagique vivant au large des côtes du Gabon du Congo et de l'Angola (0-18° lat. S. et 5-12° long. E.): Tuniciers pélagiques: I. Salpidae. Ann. Soc. r. zool. Belg. 98(1): 49-86. In: Receuil Trav. publiées de 1965 a 1966, Oceanogr. ORSTROM.

Iasis zonaria
Harant, H., and P. Vernières, 1934.
Tuniciers pélagiques provenant des croisières du Prince Albert 1er de Monaco. Rés. camp. sci., Monaco, 88:47 pp., 5 textfigs.

Iasis zonaria
Hunt, H.G., 1968.
Continuous plankton records: contribution towards a plankton atlas of the North Atlantic and the North Sea.II. The seasonal and annual distributio of Thaliacea. Bull.mar.Ecol.6(7):225-249.

Iasis zonaria
Moore, H. B., 1949
The zooplankton of the upper waters of the Bermuda area of the North Atlantic. Bull. Bingham Ocean. Coll. 12(2):97, 208 text figs.

Jasis zonaria
Nair, R. Velappan, 1949.
The Thaliacea of the Madras plankton. Bull. Madras Govt. Mus., n.s., Nat. Hist. Sect., 6(1): 1-41, 6 pls.

Iasis zonaria
Sewell, R.B.S., 1953.
The pelagic Tunicata. Brit. Mus. (N.H.), John Murray Exped., 1933-34, Sci. Repts., 10(1):1-90, 1 pl., 32 textfigs.

Iasis zonaria
Thompson, Harold, 1954.
Pelagic tunicates. Identifications and relevant notes. B.A.N.Z. Antarctic Res. Exped., 1929-1931 Repts. B(Zool. Bot.), 1(4):183-185.

Iasis zonaria
Tokioka, T., 1937.
Notes on salpas and doliolums. Annot. Zool. Jap. 16(3):219-232.

Iasis zonaria
Yount, J.L., 1954.
The taxonomy of the Salpidae (Tunicata) of the central Pacific Ocean. Pacific Science 8(3): 276-330, 30 textfigs.

Ihlea asymmetrica
Barnes, Beatrice I., 1961
Contribution towards a plankton atlas of the north-eastern Atlantic and the North Sea. 4. Thaliacea. Bulls. Mar. Ecology, 5(42): 102-104, Pl. 29.

Ihlea asymmetrica
Berrill, N.J., 1950.
The Tunicata with an account of the British species. Ray Soc. No. 133:354 pp., 120 textfigs.

Ihlea asymmetrica
Fraser, J.H., 1949.
The distribution of Thaliacea (Salps and Doliolids) in Scottish Waters 1920 to 1939. Scottish Home Dept., Fish. Div., Sci. Invest. 1949(1):44 pp., 16 textfigs.

Ihlea (Salpa) asymmetrica
Fraser, J. H., 1949
Plankton of the Faroes-Shetland Channel and the Faroes, June and August 1947. Ann. Biol., Int. Cons., 4:27-28, text fig. 10.

Ihlea asymmetrica
Fraser, J.H., 1947.
Thaliacea I. Family: Salpidae. Fiches d'Ident. Zooplancton, Cons. Perm. Int. Expl. Mer, 9:4 pp., 17 textfigs.

Ihlea asymmetrica
Fraser, J. H. and A. Saville, 1949
Plankton distribution in Scottish and adjacent waters in 1948. Ann. Biol. 5:61-62.

Ihlea assymmetrica
Fraser, J. H. and A. Saville, 1949
List of rare exotic species found in the plankton by the Scottish Vessel "Explorer" in 1948. Ann. Biol. 5:62-64.

Ihlea asymmetrica
Hunt, H.C., 1968.
Continuous plankton records: contribution towards a plankton atlas of the North Atlantic and the North Sea.II.The seasonal and annual distribution of Thaliacea. Bull.mar.Ecol.6(7):225-249.

Ihlea asymmetrica
Tokioka, T., 1937.
Notes on salpas and doliolums. Annot. Zool. Jap. 16(3):219-232.

Ihlea magalhanica
Bary, B.M., 1960.
Notes on ecology, distribution and systematics of pelagic Tunicata from New Zealand. Pacific Science, 14:101-121.
(2)

Ihlea magalhanica
Foxton, P. 1971.
On Ihlea magalhanica (Apstein) (Tunicata: Salpidae) and Ihlea racovitzai (Van Beneden). Discovery Rept. 35: 179-198.

Ihlea punctata
Amor, Analía, 1966.
Tunicados pelágicos de la "Operación Convergencia" en el Atlántico Sur (1961). Physis, Buenos Aires, 26(71):163-179.

Ihlea punctata
Esnal, Graciela B., 1968.
Salpas colectadas por El Austral y el Walther Herwig en el Océano Atlántico. Revta Mus. argent. Cienc. nat. Bernardino Ricadavia, Hidrobiol., 2(8):257-277.

Ihlea punctata
Godeaux, J., et G. Goffinet 1968
Données sur la faune pélagique vivant au large des côtes du Gabon du Congo et de l'Angola (0-18° lat. S. et 5-12° long. E.): Tuniciers pélagiques: I. Salpidae. Ann. Soc. r. zool. Belg. 98(1): 49-86. In: Receuil Trav. publiées de 1965 a 1966, Oceanogr. ORSTROM.

Ihlea punctata
Kashkina A.A. 1973.
A contribution to the fauna of Salpidae (Tunicata) in the Indian Ocean. (In Russian, English abstract) Zool. Zh. 52(2). 215-219.

Ihlea punctata
Yount, J.L., 1954.
The taxonomy of the Salpidae (Tunicata) of the central Pacific Ocean. Pacific Science 8(3): 276-330, 30 textfigs.

Ihlea racovitzai
Foxton, P. 1971.
On Ihlea magalhanica (Apstein) (Tunicata: Salpidae) and Ihlea racovitzai (Van Beneden). Discovery Rept. 35: 179-198.

Ihlea vagina -- Salpa tilesii- Thetys vagina
Belloc, G., 1938.
Listes des Tuniciers pélagiques capturés au cours de la cinquième croisière. Rev. Trav. Off. Pêches Marit. 11(3):315-334, 20 textfigs.

Metcalfina hexagona
Nair, R. Velappan, 1949.
The Thaliacea of the Madras plankton. Bull. Madras Govt. Mus., n.s., Nat. Hist. Sect., 6(1): 1-41, 6 pls.

Metcalfina Hexagona
Sewell, R.B.S., 1953.
The pelagic Tunicata. Brit. Mus., (N.H.), John Murray Exped., 1933-34, Sci. Repts. 10(1):1-90, 1 pl., 32 textfigs.

Metcalfina hexagona
Yount, J.L., 1954.
The taxomomy of the Salpidae (Tunicata) of the central Pacific Ocean. Pacific Science 8(3): 276-330, 30 textfigs.

Monophora noctiluca
Neumann, G., 1913
Tunicata Salpae II: Cyclomyaria et Pyrosomida. 40:36 pp., 19 text figs. (Das Tierreich)

Octacnemus bythius
Ihle, J.E.W., 1912
Tunicata. Salpae I. Desmomyaria. Das Tierreich. 32:66 pp., 68 text figs.

Octacnemus herdmani
Ihle, J.E.W., 1912
Tunicata. Salpae I. Desmomyaria. Das Tierreich. 32:66 pp., 68 text figs.

Octacnemus patagoniensis
Ihle, J.E.W., 1912
Tunicata. Salpae I. Desmomyaria. Das Tierreich. 32:66 pp., 68 text figs.

Pegea confoederata
Bary, B.M., 1960.
Notes on ecology, distribution and systematics of pelagic Tunicata from New Zealand. Pacific Science, 14:101-121.

Pegea confederata
Belloc, G., 1938.
Liste des Tuniciers pélagiques capturés au cours de la cinquième croisière. Rev. Trav. Off. Pêches Marit. 11(3):315-334, 20 textfigs.

Pegea confoederata
Casanova, Jean-Paul, 1966.
Pêches planctoniques superficielles et profondes en Méditerranée occidentale (Campagne de la Thalassa - janvier 1961 - entre les îles Baléares, la Sardaigne et l'Algérois) VII. Thaliacés.
Revue Trav. Inst. (Scient. tech) marit., 30(4):385-390.

Pegea confoederata
Fraser, J.H., 1947.
Thaliacea I. Family: Salpidae. Fiches d'Ident. Zooplancton, Cons. Perm. Int. Expl. Mer, 9:4 pp., 17 textfigs.

Pegea confoederata
Godeaux, J., et G. Goffinet 1968
Données sur la faune pélagique vivant au large des côtes du Gabon du Congo et de l'Angola (0-18° lat. S. et 5-12° long. E.): Tuniciers pélagiques: I. Salpidae. Ann. Soc. r. zool. Belg. 98(1): 49-86. In: Receuil Trav. publiées de 1965 a 1966, Oceanogr. ORSTROM.

Pegea confoederata
Haneda, Y., and T. Tokioka, 1954.
Droplets from the plankton net. 15. Records of a caudate form of Pegea confoederata from the Japanese waters with some notes on its luminescence. Publ. Seto Mar. Biol. Lab., 3(3):369-371.

Pegea confoederata
Harant, H., and P. Vernières, 1934.
Tuniciers pélagiques provenant des croisières du Prince Albert 1er de Monaco. Res. camp. sci., Monaco, 88:47 pp., 5 textfigs.

Pegea confoederata
Lal Mohan, R.L., 1965.
On a swarm of salps, Pegea confoederata (Forskål), from the Gujarat coast.
J. mar. biol. Ass., India, 7(1):201-202.

Pegea confoederata
Moore, H. B., 1949
The zooplankton of the upper waters of the Bermuda area of the North Atlantic. Bull. Bingham Ocean. Coll. 12(2):97, 208 text figs.

Pegea confoederata
Nair, R. Velappan, 1949.
The Thaliacea of the Madras plankton. Bull. Madras Govt. Mus., n.s., Nat. Hist. Sect., 6(1): 1-41, 6 pls.

Pegea confoederata
Sewell, R.B.S., 1953.
The pelagic Tunicata. Brit. Mus. (N.H.), John Murray Exped., 1933-34, Sci. Repts. 10(1):1-90, 1 pl., 32 textfigs.

Pegea confoederata
Tokioka, T., 1937.
Notes on salps and doliolums. Annot. Zool. Jap., 16(3):219-232.

Pegea confoederata
Van Zyl, R. P., 1960.
A preliminary study of the salps and doliolids off the west and south coasts of South Africa.
Union S. Africa, Dept. Comm. & Ind., Div. Fish., Invest. Rept: 40-31 pp.

Pegea confoederata
Yount, J.L., 1954.
The taxonomy of the Salpidae (Tunicata) of the central Pacific Ocean. Pacific Science 8(3): 276-330, 30 textfigs.

Pegea confoederata bicaudata
Furuhashi, Kenzo, and Takasi Tokioka, 1966.
Droplets from the plankton net. XXII. Observation on a nine-individual chain of Pegea confoederata bicaudata (Q & G,).
Publs Seto mar.biol.Lab., 14(2):117-122.

Pyrosomisarum sp.
Neumann, G., 1913
Tunicata Salpae II: Cyclomyaria et Pyrosomida. 40:36 pp., 19 text figs.
Das Tierreich.

Ritteria retracta
Belloc, G., 1938.
Liste des Tuniciers pélagiques capturés au cours de la cinquième croisière. Rev. Trav. Off. Pêches Marit. 11(3):315-334, 20 textfigs.

Ritteriella (Ritteria) amboinsis
Godeaux, J., 1960.
Tuniciers pélagiques du Golfe d'Eylath. Contribution to the knowledge of the Red Sea, No. 18.
Sea Fish. Res. Sta., Haifa, Bull., No. 29:9-15.

Ritteriella amboinensis
Nair, R. Velappan, 1949.
The Thaliacea of the Madras plankton. Bull. Madras Govt. Mus., n.s., Nat. Hist. Sect., 6(1): 1-41, 6 pls.

Ritteriella amboinensis
Fraser, J.H., 1954.
Warm-water species in the plankton off the English Channel entrance. J.M.B.A., U.K., 33: 345-346.

Ritteriella amboinensis
Sewell, R.B.S., 1953.
The pelagic Tunicata. Brit. Mus. (N.H.), John Murray Exped., 1933-34, Sci. Repts., 10(1):1-90, 1 pl., 32 textfigs.

Ritteriella amboinensis
Tokioka, T., 1937.
Notes on salpas and doliolums. Annot. Zool. Jap., 16(3):219-232.

Ritteriella amboinensis
Yount, J.L., 1954.
The taxonomy of the Salpidae (Tunicata) of the central Pacific Ocean. Pacific Science 8(3): 276-330, 30 textfigs.

Ritteriella picteti
Berner, L., 1954.
On the previously undescribed aggregate form of the pelagic tunicate Ritteriella picteti (Apstein) (1904). Pacific Science 8:121-124.

Ritteriella picteti
Fraser, J.H., 1955.
The salp, Ritteriella off the English coast - a correction. J.M.B.A. 34(2):247-248.

Ritteriella picteti
Fraser, J.H., 1954.
Warm-water species in the plankton off the English Channel entrans. J.M.B.A., U.K., 33: 345-346.

Ritteriella picteti
Tokioka, T., 1937.
Notes on salpas and doliolums. Annot. Zool. Jap. 16(3):219-232.

Ritteriella picteti
Yount, J.L., 1954.
The taxonomy of the Salpidae (Tunicata) of the central Pacific Ocean. Pacific Science 8(3): 276-330, 30 textfigs.

Salpa sp.
Davies, D. H., 1949
Preliminary investigations on the foods of South African fishes (with notes on the general fauna of the area surveyed). Fish & Mar. Biol. Survey Div., Dept. Comm. & Industries, S. Africa, Invest. Rept. No.11, Commerce & Industry, Jan. 1949:1-36, 4 pls.

Salpa spp.
Osterberg, Charles, 1962
Zn 65 content of salps and euphausiids. Limnol. and Oceanogr., 7(4):478-479.

Salpa
Waal, J.P., 1966.
Salpa Edwards, 1771 (Pisces): proposed suppression under the plenary powers in favor of Salpa Forskal, 1775 together with the designation of a type-species for Thalia Blumenbach, 1798 (Tunicata, Thaliacea): Z.N.(S.) 1651.
Bull. Zool. Nomencl., 23(5):232-234.

Salpa affinis
Berrill, N. J., 1961
Salpa.
Scientific American, 204(1): 150 - 160

Salpa africana
Brooks, W.K., 1893.
The genus Salpa. Mem. Biol. Lab., Johns Hopkins Univ. 2:1-306, 48 pls.

Salpa africana-maxima
Carlisle, D. B., 1950.
Alcune osservazioni sulla meccanica dell' alimentazione della Salpa. Pubbl. Staz. Zool., Napoli, 22(2):146-154, 3 textfigs.

Salpa amboinensis n.sp.
Bedot, M., 1909
Sur la faune de l'Archipel Malais (resume). Rev. Suisse Zool. 17:143-169.

Salpa amboinensis
Ihle, J.E.W., 1912
Tunicata. Salpae I. Desmomyaria. Das Tierreich. 32:66 pp., 68 text figs.

Salpa (Ritteria) amboinensis
Metcalf, M.M., 1919
Contributions to the biology of the Philippine Archipelago and adjacent regions. The Salpidae collected by the United States Fisheries Steamer "Albatross" in Philippine waters during the years 1908 and 1909.
Bull. U. S. Nat. Mus. No.100, Vol. 2(1):1-4.

Salpa antheliophora
Ihle, J.E.W., 1912
Tunicata. Salpae I. Desmomyaria. Das Tierreich. 32:66 pp., 68 text figs.

Salpa aspera
Amor, Analía, 1966.
Tunicados pelágicos de la "Operación Convergencia" en el Atlántico Sur (1961). Physis, Buenos Aires, 26(71): 163-179.

Salpa aspera
Esnal, Graciela B., 1968.
Salpas colectadas por El Austral y el Walther Herwig en el Océano Atlántico.
Revta Mus. argent. Cienc. nat. Bernardino Ricadavia, Hidrobiol., 2(8):257-277.

Salpa aspera
Foxton, P., 1961.
Salpa fusiformis Cuvier and related species. Discovery Repts., 32:1-32, Pls. 1-2.

Salpa aspera
Sawicki, R.M., 1966.
Development of the stolon in Salpa fusiformis Cuvier and Salpa aspera Chamisso.
Discovery Repts., 33:335-383.

Salpa asymmetrica
Ihle, J.E.W., 1912
Tunicata. Salpae I. Desmomyaria. Das Tierreich. 32:66 pp., 68 text figs.

Salpa atlantica
Ihle, J.E.W., 1912
Tunicata. Salpae I. Desmomyaria. Das Tierreich. 32:66 pp., 68 text figs.

Salpa (Cyclosalpa) bakeri
Metcalf, M.M., 1919
Contributions to the biology of the Philippine Archipelago and adjacent regions. The Salpidae collected by the United States Fisheries Steamer "Albatross" in Philippine waters during the years 1908 and 1909. Bull. U. S. Nat. Mus. No.100, Vol. 2(1):1-4.

Salpa bicornis
Ihle, J.E.W., 1912
Tunicata. Salpae I. Desmomyaria. Das Tierreich. 32:66 pp., 68 text figs.

Salpa chamissonis
Brooks, W.K., 1893.
The genus Salpa. Mem. Biol. Lab., Johns Hopkins Univ. 2:1-306, 48 pls.

Salpa (Pegea) confederata
Berrill, N.J., 1950.
The Tunicata with an account of the British species. Ray Soc. No. 133:354 pp., 120 textfigs.

Salpa confoederata
Hastings, A. B., 1931
Tunicata. Great Barrier Reef Exped., 1928-29. Sci. Repts. 4(3):69-109, 17 text-figs, 3 pls.

Salpa confoederata
Ihle, J.E.W., 1912
Tunicata. Salpae I. Desmomyaria. Das Tierreich. 32:66 pp., 68 text figs.

Salpa (Pegea) confederata
Metcalf, M.M., 1919
Contributions to the biology of the Philippine Archipelago and adjacent regions. The Salpidae collected by the United States Fisheries Steamer "Albatross" in Philippine waters during the years 1908 and 1909. Bull. U. S. Nat. Mus. No.100, Vol. 2(1):1-4.

Salpa confoederata
Russell, F.S., and J.S. Colman, 1935
The Zooplankton. IV. The occurrence and seasonal distribution of the Tunicata, Mollusca and Coelenterata (siphonophora). Brit. Mus. (N.H.) Great Barrier Reef Expedition 1928-1929, Sci. Repts., 2(7):203-276, 30 text figs.

Salpa confoederata-scutigera
Ritter, W.E., 1905.
III. The pelagic Tunicata of the San Diego region excepting the Larvacea. Univ. Calif. Publ. Zool. 2(3):51-112, Pls. 2-3.

Salpa confoederata-scutigera
Ritter, R. E. and E. S. Byxbee, 1905
VIII. The pelagic Tunicata. Reports on the Scientific results of the expedition to the Tropical Pacific, in charge of Alexander Agassiz, in the U. S. Fish Commission Steamer "Albatross", from August, 1899 to March, 1900, Commander Jefferson F. Moser, U. S. N., commanding. Mem. M.C.Z. 26(5):195-211, 2 pls.

Salpa cordiformis
Brooks, W.K., 1893.
The genus Salpa. Mem. Biol. Lab., Johns Hopkins Univ. 2:1-306, 48 pls.

Salpa costata
Brooks, W.K., 1893.
The genus Salpa. Mem. Biol. Lab., Johns Hopkins Univ. 2:1-306, 48 pls.

Salpa cylindrica
Belloc, G., 1938.
Liste des Tuniciers pélagiques capturés au cours de la cinquième croisière. Rev. Trav. Off. Pêches Marit. 11(3):315-334, 20 textfigs.

Salpa cylindrica
Brooks, W.K., 1893.
The genus Salpa. Mem. Biol. Lab. Johns Hopkins Univ. 2:1-306, 48 pls.

Salpa cylindrica
Culkin, F. and R.J. Morris, 1970.
The fatty acid composition of two marine filter-feeders in relation to a phytoplankton diet. Deep-Sea Res., 17(5): 861-865.

Salpa cylindrica
Delsman, H. C., 1939.
Preliminary plankton investigations in the Java Sea. Treubia, 17:139-181, 8 maps, 41 figs.

Salpa cylindrica
Godeaux, J., 1960.
Tuniciers pelagiques du Golfe d'Eylath. Contribution to the knowledge of the Red Sea, No. 18.
Sea Fish. Res. Sta., Haifa, Bull., No. 29:9-15.

Salpa cylindrica
Godeaux, J., et G. Goffinet 1968
Données sur la faune pélagique vivant au large des côtes du Gabon du Congo et de l'Angola (0-18° lat. S. et 5-12° long. E.): Tuniciers pélagiques: I. Salpidae. Ann. Soc. r. zool. Belg. 98(1): 49-86. In Receuil Trav. publiées de 1965 a 1966, Oceanogr. ORSTROM.

Salpa cylindrica
Harant, H., and P. Vernières, 1934.
Tuniciers pélagiques provenant des croisières du Prince Albert 1er de Monaco. Rés. camp. sci., Monaco, 88:47 pp., 5 textfigs.

Salpa cylindrica
Hastings, A. B., 1931
Tunicata. Great Barrier Reef Exped., 1928-29. Sci. Repts. 4(3):69-109, 17 text-figs, 3 pls.

Salpa cylindrica
Ihle, J.E.W., 1912
Tunicata. Salpae I. Desmomyaria. Das Tierreich. 32:66 pp., 68 text figs.

Salpa cylindrica
Johnson, M.E., 1910.
A quantitative study of the development of the Salpa chain in Salpa fusiformis-nunciata. Univ. Calif. Publ., Zool. 6(7):145-176, 15 textfigs.

Salpa cylindrica
Metcalf, M.M., 1919
Contributions to the biology of the Philippine Archipelago and adjacent regions. The Salpidae collected by the United States Fisheries Steamer "Albatross" in Philippine waters during the years 1908 and 1909. Bull. U. S. Nat. Mus. No.100, Vol. 2(1):1-4.

Salpa cylindrica
Metcalf, M.M., 1893.
Pt. 4. The eyes and subneural gland of Salpa. Mem. Biol. Lab., Johns Hopkins Univ. 2:307-371, Pls. 47-57.

Salpa cylindrica
Moore, H. B., 1949
The zooplankton of the upper waters of the Bermuda area of the North Atlantic. Bull. Bingham Ocean. Coll. 12(2):97, 208 text figs.

Salpa cylindrica
Nair, R.V., 1949.
The Thaliacea of the Madras plankton. Bull. Madras Govt. Mus., n.s., Nat. Hist. Sect., 6(1): 41 pp., 6 pls.

Salpa cylindrica
Nair, R. Velappan, 1949.
The Thaliacea of the Madras plankton. Bull. Madras Govt. Mus., n.s., Nat. Hist. Sect., 6(1): 1-41, 6 pls.

Salpa cylindrica
Ritter, W.E., 1905.
III. The pelagic Tunicata of the San Diego region, excepting the Larvacea. Univ. Calif. Publ. Zool. 2(3):51-112, Pls. 2-3.

Salpa cylindrica
Ritter, R. E. and E. S. Byxbee, 1905
VIII. The pelagic Tunicata. Reports on the Scientific results of the expedition to the Tropical Pacific, in charge of Alexander Agassiz, in the U. S. Fish Commission Steamer "Albatross", from August, 1899 to March, 1900, Commander Jefferson F. Moser, U. S. N., commanding. Mem. M.C.Z. 26(5):195-211, 2 pls.

Salpa cylindrica
Russell, F.S., and J.S. Colman, 1935
The Zooplankton. IV. The occurrence and seasonal distribution of the Tunicata, Mollusca and Coelenterata (siphonophora). Brit. Mus. (N.H.) Great Barrier Reef Expedition 1928-1929, Sci. Repts., 2(7):203-276, 30 text figs.

Salpa cylindrica
Sewell, R.B.S., 1953.
The pelagic Tunicata. Brit. Mus. (N.H.), John Murray Exped., 1933-34, Sci. Repts. 10(1):1-90, 1 pl., 32 textfigs.

Salpa cylindrica
Tokioka, T., 1937.
Notes on salpas and doliolums. Annot. Zool. Jap. 16(3):219-232.

Salpa cymbiola
Ihle, J.E.W., 1912
Tunicata. Salpae I. Desmomyaria. Das Tierreich. 32:66 pp., 68 text figs.

Salpa (Thalia) democratica
Berrill, N.J., 1950.
The Tunicata with an account of the British species. Ray Soc. No. 133:345 pp., 120 textfigs.

Salpa democratica
Berrill, N.J., 1950.
Budding and development in Salpa. J. Morph. 83(3): 553-606, 16 textfigs.

Salpa democratica
Brooks, W.K., 1893.
The genus Salpa. Mem. Biol. Lab., Johns Hopkins Univ. 2:1-306, 48 pls.

Salpa democratica
Hastings, A. B., 1931
Tunicata. Great Barrier Reef Exped., 1928-29. Sci. Repts. 4(3):69-109, 17 text-figs, 3 pls.

Salpa democratica
Ihle, J.E.W., 1912
Tunicata. Salpae I. Desmomyaria. Das Tierreich. 32:66 pp., 68 text figs.

Salpa (Thetys) democratica
Metcalf, M.M., 1919
Contributions to the biology of the Philippine Archipelago and adjacent regions. The Salpidae collected by the United States Fisheries Steamer "Albatross" in Philippine waters during the years 1908 and 1909. Bull. U. S. Nat. Mus. No.100, Vol. 2(1):1-4.

Salpa democratica
Michael, E. L., 1918.
Differentials in behavior of Salpa democratica relative to the temperature of the sea. Univ. California Publ. Zool. 18(12):239-298, Pls. 9-11, 1 textfig.

Salpa democratica
Roubault, A., 1946
Observations sur la Répartition du Plancton. Bull. Mus. Inst. Ocean., No.902: 4 pp.

Salpa democratica
Russell, F.S., and J.S. Colman, 1935
The Zooplankton. IV. The occurrence and seasonal distribution of the Tunicata, Mollusca and Coelenterata (siphonophora). Brit. Mus. (N.H.) Great Barrier Reef Expedition 1928-1929, Sci. Repts., 2(7):203-276, 30 text figs.

Salpa democratica
Russell, F. S. and A. B. Hastings, 1933
On the occurrence of pelagic tunicates (Thaliacea) in the waters of the English Channel off Plymouth. JMBA 18(2):635-640.

Salpa democratica - mucronata
Bedot, M., 1909
Sur la faune de l'Archipel Malais (resume). Rev. Suisse Zool. 17:143-169.

Salpa democratica-mucronata
Fish, C.J., and M.W. Johnson, 1937
The biology of the zooplankton population in the Bay of Fundy and the Gulf of Maine with special reference to production and distribution. J. Biol. Bd., Canada 3(3):189-322, 29 tables, 45 text figs.

Salpa democratica-mucronata
Ritter, W.E. 1905.
III. The pelagic Tunicata of the San Diego region, excepting the Larvacea. Univ. Calif. Publ. Zool. 2(3):51-112, Pls. 2-3.

Salpa democratica-mucronata
Ritter, R. E. and E. S. Byxbee, 1905
VIII. The pelagic Tunicata. Reports on the Scientific results of the expedition to the Tropical Pacific, in charge of Alexander Agassiz, in the U. S. Fish Commission Steamer "Albatross", from August, 1899 to March, 1900, Commander Jefferson F. Moser, U. S. N., commanding. Mem. M.C.Z. 26(5):195-211, 2 pls.

Salpa dubia
Ihle, J.E.W., 1912
Tunicata. Salpae I. Desmomyaria. Das Tierreich. 32:66 pp., 68 text figs.

Salpa emarginata
Ihle, J.E.W., 1912
Tunicata. Salpae I. Desmomyaria. Das Tierreich. 32:66 pp., 68 text figs.

Salpa fusiformis
Alvariño, A., 1957.
Estudio del zooplancton del Mediterráneo occidental. Zooplancton del Atlántico Ibérico. Campañas del "Xauen" en el verano del 1954. Bol. Inst. Espanol Oceanogr., 81-82:1-26; 1-51

Salpa fusiformis (fig)
Amor, Anulía, 1966.
Tunicados pelágicos de la "Operación Convergencia" en el Atlántico Sur (1961). Physis, Buenos Aires, 26(71):163-179.

Salpa fusiformis
Barnes, Beatrice I., 1961
Contribution towards a plankton atlas of the north-eastern Atlantic and the North Sea. 4. Thalacea. Bulls. Mar. Ecology, 5(42): 102-104, Pl. 29.

Salpa fusiformis
Belloc, G., 1938.
Liste des Tuniciers pélagiques capturés au cours de la cinquième croisière. Rev. Trav. Off. Peches Marit. 11(3):315-334, 20 textfigs.

Salpa fusiformis
Berkeley, E., and C. Berkeley, 1960
Some further records of pelagic Polychaeta from the northeast Pacific, north of Latitude 40°N. and east of Longitude 175°W., together with records of Siphonophora, Mollusca and Tunicata from the same region. Canadian J. Zoology. 38:787-799.

Salpa fusiformis
Berrill, N. J., 1961
Salpa. Scientific American, 204(1): 150 - 160

Salpa fusiformis
Berrill, N.J., 1950.
The Tunicata with an account of the British species. Ray Soc. No. 133:354 pp., 120 textfigs.

Salpa fusiformis
Berrill, N.J., 1950.
Budding and development in Salpa. J. Morph. 83(3) 553-606, 16 textfigs.

Salpa fusiformis
Bigelow, H. B., 1922
Exploration of the coastal water off the northeastern United States in 1916 by the U.S. Fisheries Schooner Grampus. Bull. M.C.Z. 65 (5):85-188, 53 text figs.

Salpa fusiformis
Braconnot, Jean-Claude 1971.
Contribution à l'étude biologique et écologique des tuniciers pélagiques salpides et doliolides. I. Hydrologie et écologie des salpides. Vie Milieu (B) 22 (2): 257-286.

Salpa fusiformis
Braconnot, J.C., 1963.
Étude du cycle annuel des Salpes et Doliales en rade de Villefranche-sur-mer. Journal du Conseil, 28(1):21-36.

Salpa fusiformis
Brattström Hans 1972.
On Salpa fusiformis Cuvier (Thaliacea) in Norwegian coastal and offshore waters. Sarsia 48: 71-90.

Salpa fusiformis
Delsman, H. C., 1939.
Preliminary plankton investigations in the Java Sea. Treubia, 17:139-181, 8 maps, 41 figs.

Salpa fusiformis
Ebara, E., 1954.
The periodic reversal of the heart-beat in Salpa fusiformis. Sci. Rept., Tokyo, Bunriku Daigaku, (B) 110:

Salpa fusiformis
Esnal, Graciela B., 1968.
Salpas colectadas por El Austral y el Walther Herwig en el Océano Atlántico. Revta Mus. argent. Cienc. nat. Bernardino Ricadavia, Hidrobiol., 2(8):257-277.

Salpa fusiformis
Fish, C.J., and M.W. Johnson, 1937
The biology of the zooplankton population in the Bay of Fundy and the Gulf of Maine with special reference to production and distribution. J. Biol. Bd., Canada 3(3):189-322, 29 tables, 45 text figs.

Salpa fusiformis
Foxton, P., 1961.
Salpa fusiformis Cuvier and related species. Discovery Repts., 32:1-32, Pls. 1-2.

Salpa fusiformis
Fraser, J. H., 1949.
The distribution of Thaliacea (Salps and Doliolids) in Scottish Waters 1920 to 1939. Scottish Home Dept., Fish. Div., Sci. Invest. 1949(1):44 pp., 16 textfigs.

Salpa fusiformis
Fraser, J.H., 1947.
Thaliacea I. Family: Salpidae. Fiches d'Ident. Zooplancton, Cons. Perm. Int. Expl. Mer, 9:4 pp., 17 textfigs.

Salpa fusiformis
Furnestin, M.L., 1957.
Chaetognathes et zooplancton du secteur atlantique marocain. Rev. Trav. Inst. Pêches Marit. 21(1/2): 9-356.

Salpa fusiformis
Glover, R.S., J.M. Colebrook and G.A. Robinson, 1964.
The continuous plankton recorder survey: plankton around the British Isles during 1962. Ann. Biol., Cons. Perm. Int. Expl. Mer, 1962, 19: 65-69.

Salpa fusiformis
Glover, R., and G.A. Robinson, 1965.
The continuous plankton recorder survey: plankton around the British Isles during 1963. Ann. Biol., Cons. Perm. Int. Expl. Mer, 1963, 20:93-97.

Salpa fusiformis
Glover, R.S., 1957.
An ecological survey of the drift-net herring fishery off the north-east coast of Scotland. 2. The planktonic environment of the herring. Bull. Mar. Ecol., 5(39):43 pp.

Salpa fusiformis
Glover, R.S., and G.A. Robinson, 1966.
The continuous plankton recorder survey: Plankton around the British Isles during 1964. Annls biol., Copenh., 21:56-60.

Salpa fusiformis
Godeaux, J., et G. Goffinet 1968
Données sur la faune pélagique vivant au large des côtes du Gabon du Congo et de l'Angola (0-18° lat. S. et 5-12° long. E.): Tuniciers pélagiques: I. Salpidae. Ann. Soc. r. zool. Belg. 98(1): 49-86. In: Receuil Trav. publiées de 1965 a 1966, Oceanogr. ORSTROM.

Salpa fusiformis
Harant, H., and P. Vernières, 1934.
Tuniciers pélagiques provenant des croisières du Prince Albert 1er de Monaco. Rés. camp. sci., Monaco, 88:47 pp., 5 textfigs.

Salpa fusiformis
Hunt, H.G., 1968.
Continuous plankton records: contribution towards a plankton atlas of the North Atlantic and the North Sea.II. The seasonal and annual distribution of Thaliacea. Bull.mar.Ecol. 6(7):225-249.

Salpa fusiformis
Hunt, H.G., 1966.
Salpa fusiformis in continuous plankton records during 1964. Annls biol., Copenh., 21:60-61.

Salpa fusiformis
Ihle, J.E.W., 1912
Tunicata. Salpae I. Desmomyaria. Das Tierreich. 32:66 pp., 68 text figs.

Salpa fusiformis
Lebour, M.V., 1947
Notes on the inshore plankton of Plymouth. JMBA 26(4):527-547.

Salpa fusiformis
Marukawa, H., 1921
Plankton lists and some new species of copepods, from the northern waters of Japan. Bull. Inst. Ocean., No.384, 15 pp., 3 pls. 1 chart. Monaco

Salpa fusiformis
Massuti, Miguel, 1959.
Estudio de los taliáceos del plancton de Castellón durante el año 1954.
Inv. Pesq., Barcelona, 14:53-64.

Salpa fusiformis
Mayzaud, P. et S. Dallot, 1973
Respiration et excrétion azotée du zooplancton. I. Evaluation des niveaux metaboliques de quelques espèces de Méditerranée occidentale. Mar. Biol. 19(4): 307-314.

Salpa fusiformis
Metcalf, M.M., 1919
Contributions to the biology of the Philippine Archipelago and adjacent regions. The Salpidae collected by the United States Fisheries Steamer "Albatross" in Philippine waters during the years 1908 and 1909.
Bull. U. S. Nat. Mus. No.100, Vol. 2(1):1-4.

Salpa fusiformis
Moore, H. B., 1949
The zooplankton of the upper waters of the Bermuda area of the North Atlantic. Bull. Bingham Ocean. Coll. 12(2):97, 208 text figs.

Salpa fusiformis
Pavshtics, E.A., 1965.
Distribution of plankton and summer feeding of herring in the Norwegian Sea and on Georges Ban.
ICNAF Spec. Publ., 6:583-589.

Salpa fusiformis
Russell, F. S. and A. B. Hastings, 1933
On the occurrence of pelagic tunicates (Thaliacea) in the waters of the English Channel off Plymouth. JMBA 18(2):635-640.

Salpa fusiformis
Sawicki, R.M., 1966.
Development of the Stolon in Salpa fusiformis Cuvier and Salpa aspera Chamisso.
Discovery Repts., 33:335-383.

Salpa fusiformis
Sewell, R.B.S., 1953.
The pelagic Tunicata. Brit. Mus.(N.H.), John Murray Exped., 1933-34, Sci. Repts., 10(1):1-90, 1 pl., 32 textfigs.

Salpa fusiformis
Thompson, Harold, 1954.
Pelagic tunicates. Identifications and relevant notes.
B.A.N.Z. Antarctic Res. Exped., 1929-1931 Repts., B(Zool. Bot.), 1(4):183-185.

Salpa fusiformis
Tokioka, T., 1937.
Notes on salpas and doliolums. Annot. Zool. Jap. 16(3):219-232.

Salpa fusiformis
*Van Soest, R.W.M. 1972.
Latitudinal variation in Atlantic Salpa fusiformis Cuvier, 1804 (Tunicata, Thaliacea). Beaufortia 20(262):59-68.

Salpa fusiformis
Yount, J.L., 1954.
The taxonomy of the Salpidae (Tunicata) of the central Pacific Ocean. Pacific Science 8(3): 276-330, 30 textfigs.

Salpa fusiformis f. aspera
Bary, B.M., 1960.
Notes on ecology, distribution and systematics of pelagic Tunicata from New Zealand.
Pacific Science, 14:101-121.
(2)

Salpa fusiformis aspera
Fraser, J.H., 1947.
Thaliacea I. Family: Salpidae. Fiches d'Ident. Zooplancton, Cons. Perm. Int. Expl. Mer, 9:4 pp., 17 textfigs.

Salpa fusiformis aspera
Ihle, J.E.W., 1912
Tunicata. Salpae I. Desmomyaria. Das Tierreich. 32:66 pp., 68 text figs.

Salpa fusiformis-runcinata
Johnson, M.E., 1910.
A quantitative study of the development of the Salpa chain in Salpa fusiformis-runcinata. Univ. Calif. Publ., Zool. 6(7):145-176, 15 textfigs.

Salpa fusiformis-runcinata
Ritter, W.E., 1905.
III. The pelagic Tunicata of the San Diego region, excepting the Larvacea. Univ. Calif. Publ. Zool. 2(3):51-112, Pls. 2-3.

Salpa fusiformis-runcinata
Ritter, R. E. and E. S. Byxbee, 1905
VIII. The pelagic Tunicata. Reports on the Scientific results of the expedition to the Tropical Pacific, in charge of Alexander Agassiz, in the U. S. Fish Commission Steamer "Albatross", from August, 1899 to March, 1900, Commander Jefferson F. Moser, U. S. N., commanding. Mem. M.C.Z. 26(5):195-211, 2 pls.

Salpa fusiformis-runcinata forma echinata
Ritter, R. E. and E. S. Byxbee, 1905
VIII. The pelagic Tunicata. Reports on the Scientific results of the expedition to the Tropical Pacific, in charge of Alexander Agassiz, in the U. S. Fish Commission Steamer "Albatross", from August, 1899 to March, 1900, Commander Jefferson F. Moser, U. S. N., commanding. Mem. M.C.Z. 26(5):195-211, 2 pls.

Salpa gerlachei
Caldwell, Melba C., 1966.
The distribution of pelagic tunicates Family Salpidae in Antarctic and Subantarctic waters. Bull. S. Calif. Acad. Sci., 65(1):1-16.

Salpa gerlachei
Foxton, P., 1966.
The distribution and life-history of Salpa thompsoni Foxton with observations on a related species Salpa gerlachei Foxton.
Discovery Repts., 34:1-116.

Salpa gerlachei n.sp.
Foxton, P., 1961.
Salpa fusiformis Cuvier and related species.
Discovery Repts., 32:1-32, Pls. 1-2.

Salpa henseni
Bedot, M., 1909
Sur la faune de l'Archipel Malais (resume).
Rev. Suisse Zool. 17:143-169.

Salpa herculea
Ihle, J.E.W., 1912
Tunicata. Salpae I. Desmomyaria. Das Tierreich. 32:66 pp., 68 text figs.

Salpa hexagona
Bedot, M., 1909
Sur la faune de l'Archipel Malais (resume).
Rev. Suisse Zool. 17:143-169.

Salpa hexagona
Brooks, W.K., 1893.
The genus Salpa. Mem. Biol. Lab., Johns Hopkins Univ. 2:1-306, 48 pls.

Salpa hexagona
Ihle, J.E.W., 1912
Tunicata. Salpae I. Desmomyaria. Das Tierreich. 32:66 pp., 68 text figs.

Salpa (Ritteria) hexagona
Metcalf, M.M., 1919
Contributions to the biology of the Philippine Archipelago and adjacent regions. The Salpidae collected by the United States Fisheries Steamer "Albatross" in Philippine waters during the years 1908 and 1909.
Bull. U. S. Nat. Mus. No.100, Vol. 2(1):1-4.

Salpa hexagona
Ritter, R. E. and E. S. Byxbee, 1905
VIII. The pelagic Tunicata. Reports on the Scientific results of the expedition to the Tropical Pacific, in charge of Alexander Agassiz, in the U. S. Fish Commission Steamer "Albatross", from August, 1899 to March, 1900, Commander Jefferson F. Moser, U. S. N., commanding. Mem. M.C.Z. 26(5):195-211, 2 pls.

Salpa informis
Ihle, J.E.W., 1912
Tunicata. Salpae I. Desmomyaria. Das Tierreich. 32:66 pp., 68 text figs.

Salpa longicauda
Ihle, J.E.W., 1912
Tunicata. Salpae I. Desmomyaria. Das Tierreich. 32:66 pp., 68 text figs.

Salpa magalhanica
Ihle, J.E.W., 1912
Tunicata. Salpae I. Desmomyaria. Das Tierreich. 32:66 pp., 68 text figs.

Salpa maxima
Belloc, G., 1938.
Liste des Tuniciers pélagiques capturés au cours de la cinquième croisière. Rev. Trav. Off. Pêches Marit. 11(3): 315-334, 20 textfigs.

Salpa maxima
Berrill, N. J., 1961
Salpa.
Scientific American, 204(1): 150 - 160

Salpa maxima
Berrill, N.J., 1950.
Budding and development in Salpa. J. Morph. 83(3):553-606, 16 textfigs.

Salpa maxima
Denton, E.J., and T.I. Shaw, 1961.
The buoyancy of gelatinous marine animals.
J. Physiol., 161:14-15P.
Also:
Trav. Sta. Zool., Villefranche-sur-Mer, 21(3). (1962).

Salpa maxima
Fraser, J.H., 1947.
Thaliacea I. Family: Salpidae. Fiches d'Ident. Zooplancton, Cons. Perm. Int. Expl. Mer, 9:4 pp., 17 textfigs.

Salpa maxima
Furnestin, M. L., 1957
Chaetognathes et zooplancton du secteur atlantique marocain. Rev. Trav. Inst. Pêches Marit. 21(1/2): 9-356.

Salpa maxima
Godeaux, J., 1960.
Tuniciers pelagiques du Golfe d'Eylath. Contribution to the knowledge of the Red Sea, No. 18.
Sea Fish. Res. Sta., Haifa, Bull., No. 29:9-15.

Salpa maxima
Godeaux, J., et G. Goffinet 1968
Données sur la faune pélagique vivant au large des côtes du Gabon du Congo et de l'Angola (0-18° lat. S. et 5-12° long. E.): Tuniciers pélagiques: I. Salpidae. Ann. Soc. r. zool. Belg. 98(1): 49-86. In: Recueil Trav. publiées de 1965 a 1966, Oceanogr. ORSTROM.

Salpa maxima
Harant, H., and P. Vernières, 1934.
Tuniciers pélagiques provenant des croisières du Prince Albert 1er de Monaco. Rés. camp. sci., Monaco, 88:47 pp., 5 textfigs.

Salpa maxima
Ihle, J.E.W., 1912
Tunicata. Salpae I. Desmomyaria. Das Tierreich. 32:66 pp., 68 text figs.

Salpa maxima
Metcalf, M.M., 1919
Contributions to the biology of the Philippine Archipelago and adjacent regions. The Salpidae collected by the United States Fisheries Steamer "Albatross" in Philippine waters during the years 1908 and 1909. Bull. U. S. Nat. Mus. No.100, Vol. 2(1):1-4.

Salpa maxima
Nair, R. Velappan, 1949.
The Thaliacea of the Madras plankton. Bull. Madras Govt. Mus., n.s., Nat. Hist. Sect., 6(1): 1-41, 6 pls.

Salpa maxima tuberculata
Nair, R. Velappan, 1949.
The Thaliacea of the Madras plankton. Bull. Madras Govt. Mus., n.s., Nat. Hist. Sect., 6(1): 1-41, 6 pls.

Salpa maxima
Sewell, R.B.S., 1953.
The pelagic Tunicata. Brit. Mus. (N.H.), John Murray Exped., 1933-34, Sci. Repts. 10(1):1-90.

Salpa maxima
Tokioka, T., 1937.
Notes on salpas and doliolums. Annot. Zool. Jap. 16(3):219-232.

Salpa maxima
Van Zyl, R. P., 1960.
A preliminary study of the salps and doliolids off the west and south coasts of South Africa. Union S. Africa, Dept. Comm. & Ind., Div. Fish., Invest. Rept. 40:31 pp.

Salpa maxima
Yount, J.L., 1954.
The taxonomy of the Salpidae (Tunicata) of the Central Pacific Ocean. Pacific Science 8(3): 276-330, 30 textfigs.

Salpa mollis
Ihle, J.E.W., 1912
Tunicata. Salpae I. Desmomyaria. Das Tierreich. 32:66 pp., 68 text figs.

Salpa moniliformis
Ihle, J.E.W., 1912
Tunicata. Salpae I. Desmomyaria. Das Tierreich. 32:66 pp., 68 text figs.

Salpa mucronata
Frost, N., S.T. Lindsay, and H. Thompson, 1933.
III. Hydrographic and biological investigations. B. Plankton more abundant in 1932 than 1931. Ann. Rept. Newfoundland Fish. Res. Comm. 2(1): 58-74, Textfigs. 17-27.

Salpa mucronata
Russell, F. S. and A. B. Hastings, 1933
On the occurrence of pelagic tunicates (Thaliacea) in the waters of the English Channel off Plymouth. JMBA 18(2):635-640.

Salpa multitentaculata
Ihle, J.E.W., 1912
Tunicata. Salpae I. Desmomyaria. Das Tierreich. 32:66 pp., 68 text figs.

Salpa nucleata
Ihle, J.E.W., 1912
Tunicata. Salpae I. Desmomyaria. Das Tierreich. 32:66 pp., 68 text figs.

Salpa picteti n.sp.
Bedot, M., 1909
Sur la faune de l'Archipel Malais (resume). Rev. Suisse Zool. 17:143-169.

Salpa picteti
Ihle, J.E.W., 1912
Tunicata. Salpae I. Desmomyaria. Das Tierreich. 32:66 pp., 68 text figs.

Salpa (Cyclosalpa) pinnata
Bedot, M., 1909
Sur la faune de l'Archipel Malais (resume). Rev. Suisse Zool. 17:143-169.

Salpa pinnata
Berrill, N. J., 1961
Salpa.
Scientific American, 204(1): 150 - 160

Salpa pinnata
Brooks, W.K., 1893.
The genus Salpa. Mem. Biol. Lab., Johns Hopkins Univ. 2:1-306, 46 pls.

Salpa (Cyclosalpa) pinnata
Metcalf, M.M., 1919
Contributions to the biology of the Philippine Archipelago and adjacent regions. The Salpidae collected by the United States Fisheries Steamer "Albatross" in Philippine waters during the years 1908 and 1909. Bull. U. S. Nat. Mus. No.100, Vol. 2(1):1-4.

Salpa punctata
Carlisle, D. B., 1950.
Alcune osservazioni sulla meccanica dell' alimentazione della Salpa. Pubbl. Staz. Zool., Napoli, 22(2):146-154, 3 textfigs.

Salpa punctata
Ihle, J.E.W., 1912
Tunicata. Salpae I. Desmomyaria. Das Tierreich. 32:66 pp., 68 text figs.

Salpa ? punctata
Moore, H. B., 1949
The zooplankton of the upper waters of the Bermuda area of the North Atlantic. Bull. Bingham Ocean. Coll. 12(2):97, 208 text figs.

Salpa pyramidalis
Ihle, J.E.W., 1912
Tunicata. Salpae I. Desmomyaria. Das Tierreich. 32:66 pp., 68 text figs.

Salpa retracta
Ihle, J.E.W., 1912
Tunicata. Salpae I. Desmomyaria. Das Tierreich. 32:66 pp., 68 text figs.

Salpa rostrata
Hastings, A. B., 1931
Tunicata. Great Barrier Reef Exped., 1928-29. Sci. Repts. 4(3):69-109, 17 text-figs., 3 pls.

Salpa rostrata
Ihle, J.E.W., 1912
Tunicata. Salpae I. Desmomyaria. Das Tierreich. 32:66 pp., 68 text figs.

Salpa rostrata
Russell, F.S., and J.S. Colman, 1935
The Zooplankton. IV. The occurrence and seasonal distribution of the Tunicata, Mollusca and Coelenterata (siphonophora). Brit. Mus. (N.H.) Great Barrier Reef Expedition 1928-1929, Sci. Repts., 2(7):203-276, 30 text figs.

Salpa rubiolineata
Ihle, J.E.W., 1912
Tunicata. Salpae I. Desmomyaria. Das Tierreich. 32:66 pp., 68 text figs.

Salpa runcinata
Brooks, W.K., 1893.
The genus Salpa. Mem. Biol. Lab., Johns Hopkins Univ. 2:1-306, 46 pls.

Salpa scutigera-confoederata
Bedot, M., 1909
Sur la faune de l'Archipel Malais (resume). Rev. Suisse Zool. 17:143-169.

Salpa scutigera
Brooks, W.K., 1893.
The genus Salpa. Mem. Biol. Lab., Johns Hopkins Univ. 2:1-306, 46 pls.

Salpa thompsoni
Amor, Analía, 1968.
Tunicados pelágicos de la "Operación Convergencia" en el Atlántico Sur (1961). Physis, Buenos Aires, 26(71):163-179.

Salpa Thompsoni
Caldwell, Melba C., 1966.
The distribution of pelagic tunicates Family Salpidae in Anarctic and Subanterctic waters. Bull. S. Calif. Acad. Sci., 65(1):1-16.

Salpa thompsoni
Esnal, Graciela B., 1968.
Salpas colectadas por El Austral y el Walther Herwig en el Océano Atlántico. Revta Mus. argent. Cienc. nat. Bernardino Ricadavia, Hidrobiol., 2(8):257-277.

Salpa thompsoni
Foxton, P., 1966.
The distribution and life-history of Salpa thompsoni Foxton with observations on a related species Salpa gerlachei Foxton. Discovery Repts., 34:1-116.

Salpa thompsoni
Foxton, P., 1964
Seasonal variations in the plankton of Antarctic waters.
In: Biologie Antarctique, Proc. S.C.A.R. Symposium, Paris, 2-8 September 1962, Hermann, Paris, 311-318.

Also in:
Collected Reprints, Nat. Inst. Oceanogr., Wormley, 12. 1964

Salpa thompsoni n.sp.
Foxton, P., 1961.
Salpa fusiformis Cuvier and related species. Discovery Repts., 32:1-32, Pls. 1-2.

Salpa tilesii
Moore, H. B., 1949
The zooplankton of the upper waters of the Bermuda area of the North Atlantic. Bull. Bingham Ocean. Coll. 12(2):97, 208 text figs.

Salpa tilesii

Ritter, W.E., 1905.
III. The pelagic Tunicata of the San Diego region excepting the Larvacea. Univ. Calif. Publ. Zool. 2(3):51-112, Pls. 2-3.

Salpa tilesii-costatata

Ritter, R. E. and E. S. Byxbee, 1905
VIII. The pelagic Tunicata. Reports on the Scientific results of the expedition to the Tropical Pacific, in charge of Alexander Agassiz, in the U. S. Fish Commission Steamer "Albatross", from August, 1899 to March, 1900, Commander Jefferson F. Moser, U. S. N., commanding. Mem. M.C.Z. 26(5):195-211, 2 pls.

Salpa tricuspidata

Ihle, J.E.W., 1912
Tunicata. Salpae I. Desmomyaria. Das Tierreich. 32:66 pp., 68 text figs.

Salpa (Thetys) vagina

Berrill, N.J., 1950.
The Tunicata with an account of the British species. Ray Soc. No. 133:354 pp., 120 textfigs.

Salpa vagina

Ihle, J.E.W., 1912
Tunicata. Salpae I. Desmomyaria. Das Tierreich. 32:66 pp., 68 text figs.

Salpa (Thetys) vagina

Metcalf, M.M., 1919
Contributions to the biology of the Philippine Archipelago and adjacent regions. The Salpidae collected by the United States Fisheries Steamer "Albatross" in Philippine waters during the years 1908 and 1909. Bull. U. S. Nat. Mus. No.100, Vol. 2(1):1-4.

Salpa vaginata

Ihle, J.E.W., 1912
Tunicata. Salpae I. Desmomyaria. Das Tierreich. 32:66 pp., 68 text figs.

Salpa younti n.sp.

Van Soest, R.W.M. 1973.
A new species in the genus Salpa Forskål, 1775 (Tunicata, Thaliacea). Beaufortia 21 (273): 9-15.
(Soest, R.W.M. van)

Salpa (Iasis) zonaria

Berrill, N.J., 1950.
The Tunicata with an account of the British species. Ray Soc., No. 133:354 pp., 120 textfigs.

Salpa zonaria

Bigelow, H. B., 1922
Exploration of the coastal water off the northeastern United States in 1916 by the U.S. Fisheries Schooner Grampus. Bull. M.C.Z. 65 (5):85-188, 53 text figs.

Salpa zonaria

Fish, C.J., and M.W. Johnson, 1937
The biology of the zooplankton population in the Bay of Fundy and the Gulf of Maine with special reference to production and distribution. J. Biol. Bd., Canada 3(3):189-322, 29 tables, 45 text figs.

Salpa zonaria

Frost, N., S.T. Lindsay, and H. Thompson, 1933.
III. Hydrographic and biological investigations. B. Plankton more abundant in 1932 than 1931. Ann. Rept., Newfoundland Fish. Res. Comm. 2(1): 58-74, Textfigs. 17-27.

Salpa zonaria

Hastings, A. B., 1931
Tunicata. Great Barrier Reef Exped., 1928-29. Sci. Repts. 4(3):69-109, 17 text figs., 3 pls.

Salpa zonaria

Ihle, J.E.W., 1912
Tunicata. Salpae I. Desmomyaria. Das Tierreich. 32:66 pp., 68 text figs.

Salpa zonaria

Russell, F.S., and J.S. Colman, 1935
The Zooplankton. IV. The occurrence and seasonal distribution of the Tunicata, Mollusca and Coelenterata (siphonophora). Brit. Mus. (N.H.) Great Barrier Reef Expedition 1928-1929, Sci. Repts., 2(7):203-276, 30 text figs.

Salpa zonaria

Russell, F. S. and A. B. Hastings, 1933
On the occurrence of pelagic tunicates (Thaliacea) in the waters of the English Channel off Plymouth. JMBA 18(2):635-640.

Salpa zonaria-cordiformis

Johnson, M.E., 1910.
A quantitative study of the development of the Salpa chain in Salpa fusiformis-nunciata. Univ. Calif. Publ., Zool. 6(7):145-176, 15 textfigs.

Salpa zonaria-cordiformis

Ritter, W.E., 1905.
III. The pelagic Tunicata of the San Diego region, excepting the Larvacea. Univ. Calif. Publ. Zool. 2(3):51-112, Pls. 2-3.

Salpa zonaria-cordiformis

Ritter, R. E. and E. S. Byxbee, 1905
VIII. The pelagic Tunicata. Reports on the Scientific results of the expedition to the Tropical Pacific, in charge of Alexander Agassiz, in the U. S. Fish Commission Steamer "Albatross", from August, 1899 to March, 1900, Commander Jefferson F. Moser, U. S. N., commanding. Mem. M.C.Z. 26(5):195-211, 2 pls.

Stephanosalpa polyzona

Ihle, J.E.W., 1912
Tunicata. Salpae I. Desmomyaria. Das Tierreich. 32:66 pp., 68 text figs.

Thalia cicar n.sp.

Van Soest, R.W.M. 1973.
The genus Thalia Blumenbach, 1798 (Tunicata, Thaliacea), with descriptions of two new species. Beaufortia 20(271):193-212.

Thalia democratica

Alvariño, A., 1957.
Estudio del Zooplancton del Mediterraneo occidental. Zooplancton del Atlántico Ibérico. Campañas del "Xauen" en el verano del 1954. Bol. Inst. Español Oceanogr., 81-82:1-26; 1-51.

Thalia democratica (fig)

Amor, Analía, 1966.
Tunicados pelágicos de la "Operación Convergencia" en el Atlántico Sur (1961). Physis, Buenos Aires, 26(71):163-179.

Thalia democratica

Barnes, Beatrice I., 1961
Contribution towards a plankton atlas of the north-eastern Atlantic and the North Sea. 4. Thaliacea. Bulls. Mar. Ecology, 5(42): 102-104, Pl. 29.

Thalia democratica

Bary, B.M., 1960.
Notes on ecology, distribution and systematics of pelagic Tunicata from New Zealand. Pacific Science, 14:101-121.

Thalia democratica

Belloc, G., 1938.
Liste des Tuniciers pélagiques capturés au cours de la cinquième croisière. Rev. Trav. Off. Pêches Marit. 11(3):315-334, 20 textfigs.

Thalia democratica

Bernard, Michelle, 1958
Systématique et distribution saisonnière des tuniciers pélagiques d'Alger. Rapp. Proc. Verb., Comm. Int. Expl. Sci. Mer Medit., n.s., 14:211-231.

Thalia democratica

Berrill, N. J., 1961
Salpa. Scientific American, 204(1): 150-160.

Thalia democratica

Braconnot, Jean-Claude 1971.
Contribution à l'étude biologique et écologique des tuniciers pélagiques salpides et doliolides. I. Hydrologie et écologie des salpides. Vie Milieu (B) 22 (2): 257-286.

Thalia democratica

Braconnot, J.C., 1963.
Étude du cycle annuel des Salpes et Dolioles en rade de Villefranche-sur-Mer. Journal du Conseil, 28(1):21-36.

Thalia democratica

Caldwell, Melba C., 1966.
The distribution of pelagic tunicates Family Salpidae in Antarctic and Subantarctic waters. Bull. S. Calif. Acad. Sci., 65(1):1-16.

Thalia democratica

Casanova, Jean-Paul, 1966.
Pêches planctoniques superficielles et profondes en Méditerranée occidentale (Campagne de la Thalassa - janvier 1961 - entre les Îles Baléares, la Sardaigne et l'Algérois) VII. Thaliacés. Revue Trav. Inst. (Scient. tech.) marit., 30(4):385-390.

Thalia democratica

Delsman, H. C., 1939.
Preliminary plankton investigations in the Java Sea. Treubia. 17:139-181, 8 maps, 41 figs.

Thalia democratica

Denton, E.J., and T.I. Shaw, 1961.
The buoyancy of gelatinous marine animals. J. Physiol., 161:14-15P.
Also:
Trav. Sta. Zool., Villefranche-sur-Mer, 21(3), (1962).

Thalia democratica

Esnal, Graciela B. 1970.
Contribución al conocimiento de las salpas del Atlántico Sur, en especial Thalia democratica. Neotropica 16 (51): 121-134.

Thalia democratica

Esnal, Graciela B., 1968.
Salpas colectadas por El Austral y el Walther Herwig en el Océano Atlántico. Revta Mus. argent. Cienc. nat. Bernardino Rivadavia, Hidrobiol., 2(8):257-277.

Thalia democratica

Fraser, J.H., 1947.
Thaliacea I. Family: Salpidae. Fiches d'Ident. Zooplancton, Cons. Perm. Int. Expl. Mer, 9:4 pp. 17 textfigs.

Thalia democratica

Furnestin, M. L., 1957.
Chaetognathes et zooplancton du secteur atlantique marocain. Rev. Trav. Inst. Pêches Marit. 21(1/2): 9-356.

Thalia democratica
Gamulin, T., 1948
Prilog poznavanju Zooplanktona Srednjedalmatinskog Otoonog Podrucja. (Contrib. a la connaissance du zooplankton de la zone insulaire de la Dalmatie Moyenne). Acta Adriatica 3(7):38 pp., 6 tables, 1 map.

Thalia democratica
Glover, R.S. and Beatrice I. Barnes, 1958(1960)
The continuous plankton recorder survey: Plankton around the British Isles during 1958. Cons. Perm. Int. l'Expl. Mer., Ann. Biol. 15: 58-61.

Thalia democratica
Godeaux, J., 1960.
Tuniciers pélagiques du Golfe d'Eylath. Contribution to the knowledge of the Red Sea, No. 18.
Sea Fish. Res. Sta., Haifa, Bull., No. 29:9-15.

Thalia democratica
Godeaux, J., et G. Goffinet 1968
Données sur la faune pélagique vivant au large des côtes du Gabon du Congo et de l'Angola (0-18° lat. S. et 5-12° long. E.): Tuniciers pélagiques: I. Salpidae. Ann. Soc. r. zool. Belg. 98(1): 49-86. In: Receuil Trav. publiées de 1965 a 1966, Oceanogr. ORSTROM.

Thalia democratica
Harant, H., and P. Vernières, 1934.
Tuniciers pélagiques provenant des croisières du Prince Albert 1er de Monaco. Rés. camp. sci., Monaco, 88:47 pp., 5 textfigs.

Thalia democratica
Heron, A.C., 1973
A specialized predator-prey relationship between the copepod Sapphirina angusta and the pelagic Tunicate Thalia democratica. J. mar. biol. Ass. U.K. 53(2):429-435.

Thalia democratica
Hunt, H.G., 1968.
Continuous plankton records: contribution towards a plankton atlas of the North Atlantic and the North Sea. II. The seasonal and annual distribution of Thaliacea.
Bull. mar. Ecol. 6(7):225-249.

Thalia democratica
Massuti, Miguel, 1959.
Estudio de los taliaceos del plancton de Castellón durante el año 1954.
Inv. Pesq., Barcelona, 14:53-64.

Thalia democratica
Mayzaud, P. et S. Dallot, 1973
Respiration et excrétion azotée du zooplancton. I. Evaluation des niveaux métaboliques de quelques espèces de Méditerranée occidentale. Mar. Biol. 19(4): 307-314.

Thalia democratica
Moore, H. B., 1949
The zooplankton of the upper waters of the Bermuda area of the North Atlantic. Bull. Bingham Ocean. Coll. 12(2):97, 208 text figs.

Thalia democratica
Nair, R. Velappan, 1949.
The Thaliacea of the Madras plankton. Bull. Madras Govt. Mus., n.s., Nat. Hist. Sect., 6(1): 1-41, 6 pls.

Thalia democratica
Rajagopal, P.K., 1962
Respiration of some marine phanktonic organisms.
Proc. Indian Acad. Sci., (B) 55(2):76-81.

Thalia democratica
Sewell, R.B.S., 1953.
The pelagic Tunicata. Brit. Mus. (N.H.), John Murray Exped., 1933-34, Sci. Repts. 10(1):1-90, 1 pl., 32 textfigs.

Thalia democratica
Tavares, D.Q., 1967.
Occurrence of doliolids and salps during 1958, 1959 and 1960 off the São Paulo Coast.
Bolm Inst. oceanogr. S. Paulo, 16(1):87-97.

Thalia democratica
Tokioka, Takasi, 1960
Studies on the distribution of appendicularians and some thaliaceans of the North Pacific, with some morphological notes.
Publ. Seto Mar. Biol. Lab., 8(2) (27):351-443. tables.

Thalia democratica
Yount, J.L., 1954.
The taxonomy of the Salpidae (Tunicata) of the central Pacific Ocean. Pacific Science 8(3): 276-330, 30 textfigs.

Thalia democratica [a]
Van Soest, R.W.M. 1973.
The genus Thalia Blumenbach, 1798 (Tunicata, Thaliacea), with descriptions of two new species.
Beaufortia 20(271):193-212.

Thalia democratica
Van Zyl, R. P., 1960.
A preliminary study of the salps and doliolids off the west and south coasts of South Africa.
Union S. Africa, Dept. Comm. & Ind., Div. Fish., Invest. Rept. 40:31 pp.

Thalia democratica, forma orientalis
Seymour Sewell, R.B., 1953
The pelagic Tunicata.
Brit. Mus. (N.H.), John Murray Exped., 1933-1934, Rept., 10(1): 1-90.

Thalia democratica (figs.) (var. orientalis n.var.)
Tokioka, T., 1937.
Notes on salpas and doliolums. Annot. Zool. Jap. 16(3):219-232.

Thalia longicauda (fig)
Amar, Analía, 1966.
Salpas de la Operución Drake IV y secciones (abril-mayo de 1965).
Physis Buenos Aires, 26(72):331-339.

Thalia longicauda
Beklemishev, C.W., 1958.
Plankton stops a ship. (In Russian).
Priroda, (11):105-106.

Thalia longicauda
Esnal, Graciela B., 1968.
Salpas colectadas por El Austral y el Walther Herwig en el Océano Atlántico.
Revta Mus. argent. Cienc. nat. Bernardino Ricadavia, Hidrobiol., 2(8):257-277.

Thalia longicauda
Sewell, R.B.S., 1953.
The pelagic Tunicata. Brit. Mus. (N.H.), John Murray Exped., 1933-34, Sci. Repts., 10(1):1-90, 1 pl., 32 textfigs.

Thalia longicauda [a]
Van Soest, R.W.M. 1973.
The genus Thalia Blumenbach, 1798 (Tunicata, Thaliacea), with descriptions of two new species.
Beaufortia 20(271):193-212.

Thalia longicauda
Van Zyl, R. P., 1960.
A preliminary study of the salps and doliolids off the west and south coasts of South Africa.
Union S. Africa, Dept. Comm. & Ind., Div. Fish., Invest. Rept. 40:31 pp.

Thalia longicaudata
Beklemishev, C.W., 1958
[Plankton stops a ship.]
Priroda (11): 105-106.

Thalia orientalis [a]
Van Soest, R.W.M. 1973.
The genus Thalia Blumenbach, 1798 (Tunicata, Thaliacea), with descriptions of two new species.
Beaufortia 20(271):193-212.

Thalia rhomboides [a]
Van Soest, R.W.M. 1973.
The genus Thalia Blumenbach, 1798 (Tunicata, Thaliacea), with descriptions of two new species.
Beaufortia 20(271):193-212.

Thalia sibogae n.sp. [a]
Van Soest, R.W.M. 1973.
The genus Thalia Blumenbach, 1798 (Tunicata, Thaliacea), with descriptions of two new species.
Beaufortia 20(271):193-212.

Thetys vagina
Fraser, J.H., 1947.
Thaliacea, I. Family: Salpidae. Fiches d'Ident. Zooplancton, Cons. Perm. Int. Expl. Mer, 9:4 pp., 17 textfigs.

Thetys vagina
Godeaux, J., et G. Goffinet 1968
Données sur la faune pélagique vivant au large des côtes du Gabon du Congo et de l'Angola (0-18° lat. S. et 5-12° long. E.): Tuniciers pélagiques: I. Salpidae. Ann. Soc. r. zool. Belg. 98(1): 49-86. In: Receuil Trav. publiées de 1965 a 1966, Oceanogr. ORSTROM.

Thetys vagina
Harant, H., and P. Vernières, 1934.
Tuniciers pélagiques provenant des croisières du Prince Albert 1er de Monaco. Rés. camp. sci., Monaco, 88:47 pp., 5 textfigs.

Thetys vagina
Moore, H. B., 1949
The zooplankton of the upper waters of the Bermuda area of the North Atlantic. Bull. Bingham Ocean. Coll. 12(2):97, 208 text figs.

Thetys vagina
Sewell, R.B.S., 1953.
The pelagic Tunicata. Brit. Mus. (N.H.), John Murray Exped., 1933-34, Sci. Repts. 10(1):1-90, 1 pl., 32 textfig.

Thetys vagina
Thompson, Harold, 1954.
Pelagic tunicates. Identifications and relevant notes.
B.A.N.Z. Antarctic Res. Exped., 1929-1931 Repts., B(Zool. Bot.), 1(4):183-185.

Thetys vagina
Yount, J.L., 1954.
The taxonomy of the Salpidae (Tunicata) of the central Pacific Ocean. Pacific Science 8(3): 276-330, 30 textfigs.

Traustedtia multitentaculata
Belloc, G., 1938.
Liste des Tuniciers pélagiques capturés au cours de la cinquième croisière. Rev. Trav. Off. Pêches Marit. 11(3):315-334, 20 textfig.

Traustedtia multitentaculata
Godeaux, J., et G. Goffinet 1968
Données sur la faune pélagique vivant au large des côtes du Gabon du Congo et de l'Angola (0-18° lat. S. et 5-12° long. E.): Tuniciers pélagiques: I. Salpidae. Ann. Soc. r. zool. Belg. 98(1): 49-86. In: Recueil Trav. publiées de 1965 a 1966, Oceanogr. ORSTROM.

Traustadtia multitenaculata
Harant, H. and P. Vernieres, 1934
Tuniciers pelagiques provenant des croisieres du Prince Albert ler de Monaco. Res. Camp. Sci. 88:48 pp.

Traustedtia multitentaculata
Moore, H. B., 1949
The zooplankton of the upper waters of the Bermuda area of the North Atlantic. Bull. Bingham Ocean. Coll. 12(2):97, 208 text figs.

Traustedtia multidentata
Nair, R. Velappan, 1949.
The Thaliacea of the Madras plankton. Bull. Madras Govt. Mus., n.s., Nat. Hist. Sect., 6(1): 1-41, 6 pls.

Traustedtia multitentaculata
Harant, H., and P. Vernières, 1934.
Tuniciers pélagiques provenant des croisières du Prince Albert Ier de Monaco. Rés. camp. sci., Monaco, 88:48 pp., 5 textfigs.

Traustedtia multitentaculata
Oka, O., 1921.
Über Traustedtia multitentaculata (Quoy and Gaimard), eine seltene Salpe. Annotat. Zool. Japon. 10(1):1-14, 5 textfigs.

Traustedtia multitesticulata
Tokioka, T., 1937.
Notes on salpas and doliolums. Annot. Zool. Jap., 16(3):219-232.

Traustedtia multidentata
Yount, J.L., 1954.
The taxonomy of the Salpidae (Tunicata) of the central Pacific Ocean. Pacific Science 8(3): 276-330, 30 textfigs.

Weelia n.gen.
Yount, J.L., 1954.
The taxonomy of the Salpidae (Tunicata) of the central Pacific Ocean. Pacific Science 8(3): 276-330, 30 textfigs.

Weeli cylindrica
Amor, Analía, 1966.
Tunicados pelágicos de la "Operación Convergencia" en el Atlántico Sur (1961). Physis, Buenos Aires, 26(71):163-179.

Weelia cylindrica
Esnal, Graciela B., 1968.
Salpas colectadas por El Austral y el Walther Herwig en el Océano Atlántico. Revta Mus. argent. Cienc. nat. Bernardino Ricadavia, Hidrobiol., 2(8):257-277.

Weelia cylindrica
Yount, J.L., 1954.
The taxonomy of the Salpidae (Tunicata) of the central Pacific Ocean. Pacific Science 8(3): 276-330, 30 textfigs.

doliolids, lists of spp.
Berner, Leo D., 1967.
Distributional atlas of Thaliacea in the California Current region. Atlas, Calif. Coop. Ocean. Fish. Invest., 8:1-322.

Doliolids
Lucas, C. E. 1933
Occurrence of Dolioletta gegenbauri (Ulyanin) in the North Sea. Nature CXXXII:858

Doliolids
Neumann, G., 1906
Doliolum. Wiss. Ergebn. "Valdivia" 1898-1899, XII:93-243, pls. xi-xxv, 20 textfigs.

tunicates, anat.
Godeaux, Jean, 1971
L'ultrastructure de l'endostyle des Doliolides (Tuniciers Cyclomyaires). C.r. Acad. Sci. Paris 272:592-595.

doliolids, vertical distribution
Tregouboff, G., 1965.
La distribution verticale des doliolides au large de Villefranche-sur-Mer. Bull. Inst. Oceanogr., Monaco, 64(1333):47 pp.

Also in:
Trav. Sta. Zool., Villefranche-sur-Mer, 25.

Dolioletta chuni
Neumann, G., 1913
Tunicata Salpae II: Cyclomyaria et Pyrosomida. 40:36 pp., 19 text figs.
(Das Tierreich,)

Dolioletta denticulatum
Neumann, G., 1913
Tunicata Salpae II: Cyclomyaria et Pyrosomida. 40:36 pp., 19 text figs.
(Das Tierreich,)

Dolioletta gegenbauri
Barnes, Beatrice I., 1961
Contribution towards a plankton atlas of the north-eastern Atlantic and the North Sea. 4. Thalacea. Bulls. Mar. Ecology, 5(42): 102-104, Pl. 29.

Dolioletta gegenbauri
Braconnot, Jean-Claude, 1970.
Contribution à l'étude des stades successifs dans le cycle des tuniciers pélagiques doliolides I, Les stades larvaire, oozoide, nourrice et gastrozoide. Arch. Zool. exp. gén, 111: 629-668. Also in: Trav. Stn zool., Villefranche-sur-Mer, 32, 1970.

Dolioletta gegenbauri
Fraser, J. H., 1949
Plankton investigations from the Scottish Research Vessel. Ann. Biol., Int. Cons., 4: 66-67.

Dolioletta gegenbauri
Fraser, J. H., 1949
Plankton of the Faroe-Shetland Channel and the Faroes, June and August 1947. Ann. Biol., Int. Cons., 4:27-28, text fig. 10.

Dolioletta gegenbauri
Fraser, J. H., 1949.
The distribution of Thaliacea (Salps and Doliolids) in Scottish Waters 1920 to 1939. Scottish Home Dept., Fish. Div., Sci. Invest. 1949(1):44 pp., 16 textfigs.

Dolioletta gegenbauri
Fraser, J. H. and A. Saville, 1949
Plankton distribution in Scottish and adjacent waters in 1948. Ann. Biol. 5:61-62.

Dolioletta gegenbauri
Glover, R.S., J.M. Colebrook and G.A. Robinson, 1964.
The continuous plankton recorder survey: plankton around the British Isles during 1962. Ann. Biol., Cons. Perm. Int. Expl. Mer, 1962, 19: 65-69.

Dolioletta gegenbauri
Glover, R.S., and G.A. Robinson, 1966.
The continuous plankton recorder survey: plankton around the British Isles during 1964. Annls biol., Copenh., 21:56-60.

Dolioletta gegenbauri
Glover, R. and G.A. Robinson, 1965.
The continuous plankton recorder survey: plankton around the British Isles during 1963. Ann. Biol., Cons. Perm. Int. Expl. Mer, 1963, 20:93-97.

Dolioletta gegenbauri
Hunt, H.G., 1968.
Continuous plankton records: contribution towards a plankton atlas of the North Atlantic and the North Sea. II. The seasonal and annual distribution of Thaliacea. Bull. mar. Ecol. 6(7):225-249.

Dolioletta gegenbauri
Lucas, C. E., 1949
Notes on continuous plankton records at 10 m depth in the North Sea and Northeastern Atlantic during 1946-1947. Ann. Biol., Int. Cons., 4:63-66, text fig. 4.

Dolioletta gegenbauri
Nair, R.V., 1949.
The Thaliacea of the Madras plankton. Bull. Madras Govt. Mus., n.s., Nat. Hist. Sect., 6(1): 41 pp., 6 pls.

Dolioletta gegenbauri
Neumann, G., 1913
Tunicata Salpae II: Cyclomyaria et Pyrosomida. 40:36 pp., 19 text figs.
(Das Tierreich,)

Dolioletta gegenbauri
Tavares, D.Q., 1967.
Occurrence of doliolids and salps during 1958, 1959 and 1960 off the São Paulo Coast. Bolm Inst. oceanogr. S.Paulo, 16(1):87-97.

Dolioletta gegenbauri
Tregouboff, G., 1965.
La distribution verticale des doliolides au large de Villefranche-sur-Mer. Bull. Inst. Oceanogr., Monaco, 64(1333):47 pp.

Also in:
Trav. Sta. Zool., Villefranche-sur-Mer, 25.

Dolioletta mirabilis
Alvariño, A., 1957.
Estudio del zooplancton del Mediterráneo occidental. Zooplancton del Atlántico Ibérico. Campañas del "Xauen" en el verano del 1954. Bol. Inst. Español. Oceanogr., 81-82:1-26; 1-51.

Dolioletta mirabile
Neumann, G., 1913
Tunicata Salpae II: Cyclomyaria et Pyrosomida. 40:36 pp., 19 text figs.
(Das Tierreich,)

Dolioletta nationalis
Neumann, G., 1913
Tunicata Salpae II: Cyclomyaria et Pyrosomida. 40:36 pp., 19 text figs.
(Das Tierreich,)

Dolioletta tritonis
Fraser, J. H., 1949.
The distribution of Thaliacea (Salps and Doliolids) in Scottish Waters 1920 to 1939. Scottish Home Dept., Fish. Div., Sci. Invest. 1949(1):44 pp., 16 textfigs.

Dolioletta tritonis
Neumann, G., 1913
Tunicata Salpae II: Cyclomyaria et Pyrosomida. 40:36 pp., 19 text figs.
(Das Tierreich,)

Dolioletta valdiviae
Bary, B.M., 1960.
Notes on ecology, distribution and systematics of pelagic Tunicata from New Zealand. Pacific Science, 14:101-121.
(2)

Dolioletta valdiviae
Neumann, G., 1913
Tunicata Salpae II: Cyclomyaria et Pyrosomida. 40:36 pp., 19 text figs.
(Das Tierreich,)

Doliolidarum affine

Neumann, G., 1913
Tunicata Salpae II: Cyclomyaria et Pyrosomida. 40:36 pp., 19 text figs.
Das Tierreich

Doliolidarum ehrenbergii

Neumann, G., 1913
Tunicata Salpae II: Cyclomyaria et Pyrosomida. 40:36 pp., 19 text figs.
Das Tierreich

Doliolina intermedia

Tokioka, Takasi, and Leo Berner, 1958.
On certain Thaliacea (Tunicata) from the Pacific Ocean, with descriptions of two new species of doliolids.
Pacific Science 12(4):317-326.

Doliolina mülleri

Barnes, Beatrice I., 1961
Contribution towards a plankton atlas of the north-eastern Atlantic and the North Sea. 4. Thalacea.
Bulls. Mar. Ecology, 5(42): 102-104, Pl. 29.

Doliolina mülleri

Braconnot, Jean-Claude, 1970.
Contribution à l'étude des stades successifs dans le cycle des tuniciers pélagiques doliolides I, Les stades larvaire, oozoïde, nourrice et gastrozoïde. Arch. Zool. exp. gén., 111: 629-668. Also in: Trav. Stn zool., Villefranche-sur-Mer, 32, 1970.

Doliolina mülleri

Hunt, H.G., 1968.
Continuous plankton records: contribution towards a plankton atlas of the North Atlantic and the North Sea. II. The seasonal and annual distribution of Thaliacea.
Bull.mar.Ecol.6(7):225-249.

Doliolina muelleri

Tregouboff, G., 1965.
La distribution verticale des doliolides au large de Villefranche-sur-Mer.
Bull. Inst. Oceanogr., Monaco, 64(1333):47 pp.
Also in:
Trav. Sta. Zool., Villefranche-sur-Mer, 25.

Doliolina obscura n.sp.

Tokioka, Takasi, and Leo Berner, 1958.
On certain Thaliacea (Tunicata) from the Pacific Ocean, with descriptions of two new species of doliolids. Pacific Science 12(4): 317-326.

Doliolina separata n.sp.

Tokioka, Takasi, and Leo Berner, 1958.
On certain Thaliacea (Tunicata) from the Pacific Ocean, with descriptions of two new species of doliolids. Pacific Science 12(4): 317-326.

Doliolina undulata

Tokioka, Takasi, and Leo Berner, 1958.
On certain Thaliacea (Tunicata) from the Pacific Ocean, with descriptions of two new species of doliolids. Pacific Science 12(4): 317-326.

Doliolina undulatum n.sp.

Tokioka, T., and L. Berner, 1958.
Two new doliolids from the eastern Pacific Ocean.
Pacific Science, 12(2):135-138.

Dolioloides rarum

Tregouboff, G., 1965.
La distribution verticale des doliolides au large de Villefranche-sur-Mer.
Bull. Inst. Oceanogr., Monaco, 64(1333):47 pp.
Also in:
Trav. Sta. Zool., Villefranche-sur-Mer, 25.

Doliolum sp.

Bigelow, H.B., and M. Leslie, 1930
Reconnaissance of the waters and plankton of Monterey Bay, July 1928.
Bull. M.C.Z., 70(5):429-481, 43 text figs.

Doliolum sp.

Bigelow, H.B., and M. Sears, 1939
Studies of the waters of the continental shelf, Cape Cod to Chesapeake Bay. III. A volumetric study of the zooplankton. Mem. M.C.Z. 54(4):183-378, 42 text figs.

Doliolum

Fraser, J.H., 1947.
Thaliacea. II. Family: Doliolidae. Fiches d'Ident. Zooplancton, Cons. Perm. Int. Expl. Mer, 10:4 pp., 13 textfigs.

Doliolum sp.

Godeaux, J., 1955.
Stades larvaires du Doliolum.
Bull. Cl. Sci., Acad. R. du Belgique, (5) 41:769-787.

Doliolum sp.

Harant, H. and P. Vernieres, 1934
Tuniciers pelagiques provenant des croisieres du Prince Albert 1er de Monaco. Res. Camp. Sci. 88:48 pp.

Doliolum

Lebour, M.V., 1947
Notes on the inshore plankton of Plymouth
JMBA 26(4):527-547, 1 textfig.

Doliolum sp.

Massuti, Miguel, 1959.
Estudio de los taliáceos del plancton de Castellón durante el año 1954.
Inv. Pesq., Barcelona, 14:53-64.

Doliolum sp.

Pavshtics, E.A., 1965.
Distribution of plankton and summer feeding of herring in the Norwegian Sea and on Georges Ban.
ICNAF Spec. Publ., 6:583-589.

Doliolum sp.

Raymont, J.E.G., and Eileen Linford, 1966.
A note on the biochemical composition of some Mediterranean zooplankton.
Int. Rev. ges. Hydrobiol., 51(3):485-488.

Doliolum affine

Belloc, G., 1938.
Liste des Tuniniers pélagiques capturés au cours de la cinquième croisière. Rev. Trav. Off. Pêches Marit. 11(3):315-334, 20 textfigs.

Doliolum denticulatum

Alvariño, A., 1957.
Estudio del zooplancton del Mediterráneo occidental. Zooplancton del Atlántico Ibérico. Campañas del "Xauen" en el verano del 1954.
Bol. Inst. Español. Oceanogr., 81-82:1-26; 1-58

Doliolum denticulatum

Bedot, M., 1909
Sur la faune de l'Archipel Malais (resume).
Rev. Suisse Zool. 17:143-169.

Doliolum denticulatum

Berner, L. D., 1960
Unusual features in the distribution of pelagic tunicates in 1957 and 1958. Cal. Coop. Ocean. Fish. Invest. Rept. Vol. 7: 133-135.

Doliolum denticulatum

Berner, Leo D., and Joseph L. Reid, Jr., 1961
On the response to changing temperature of the temperature-limited plankter Doliolum denticulatum Quoy and Gaimard 1835.
Limnol. & Oceanogr., 6(2): 205-215.

Doliolum denticulatum

Borgert, A., 1896.
Die Doliolum-Ausbeute des "Vettor Pisani". Zool. Jahrb., Abth. Syst., 9:414-419.

Doliolum denticulatum

Braconnot, Jean-Claude, 1970.
Contribution à l'étude des stades successifs dans le cycle des tuniciers pélagiques doliolides I, Les stades larvaire, oozoïde, nourrice et gastrozoïde. Arch. Zool. exp. gén., 111: 629-668. Also in: Trav. Stn zool., Villefranche-sur-Mer, 32, 1970.

Doliolum gegenbauri

Braconnot, Jean-Claud, 1968.
Sur le développement de la larve du tunicier pélagique doliolide:Doliolum (Dolioletta) gegenbaum Ulj.1884.
C.r.hebd.Séanc., Acad.Sci.,Paris,(D)267 (6):629-630.

Doliolum denticulatum

Braconnot, Jean-Claude, 1964.
Sur le developpement de la larve de Doliolum denticulatum Q & G.
C.R., Acad. Sci., Paris, 259(23):4361-4363.

Doliolum denticulatum

Braconnot, J.C., 1963.
Étude du cycle annuel des Salpes et Dolioles en rade de Villefranche-sur-mer.
Journal du Conseil, 28(1):21-36.

Doliolum denticulatum

Casanova, Jean-Paul, 1966.
Pêches planctoniques superficielles et profondes en Méditerranée occidentale (Campagne de la Thalassa - janvier 1961- entre les îles Baléares, la Sardaigne et l'Algérois). VII. Thaliacés.
Revue.Trav.Inst.(Scient.Tech.) marit.,30(4):385-390.

Doliolum denticulatum

Fraser, J.H., 1947.
Thaliacea II. Family: Doliolidae. Fiches d'Ident. Zooplancton, Cons. Perm. Int. Expl. Mer, 10:4 pp., 13 textfigs.

Doliolum denticulatum

Godeaux, J., 1960.
Tuniciers pelagiques du Golfe d'Eylath. Contribution to the knowledge of the Red Sea, No. 18.
Sea Fish. Res. Sta., Haifa, Bull., No. 29:9-15.

Doliolum denticulatum

Harant, H., and P. Vernières, 1934.
Tuniciers pelagiques provenant des croisières du Prince Albert 1er de Monaco. Res. camp. sci., Monaco, 88:47 pp., 5 textfigs.

Doliolum denticulatum

Hastings, A. B., 1931
Tunicata. Great Barrier Reef Exped., 1928-29. Sci. Repts. 4(3):69-109, 17 text-figs, 3 pls.

Doliolum denticulatum

Keferstein, Wilhelm und Ernst Ehlers, 1861
Zoologische Beitrage gesammelt im Winter 1859/60 in Neapel und Messina. Wilhelm Engelmann, Leipzig, 112 pp., 15 pls.

Doliolum denticulatum

Nair, R. Velappan, 1949.
The Thaliacea of the Madras Plankton. Bull. Madras Govt. Mus., n.s., Nat. Hist. Sect., 6(1): 1-41, 6 pls.

Doliolum denticulatum

Neumann, Günther, 1913.
Die Pyrosomen und Doliolioden der Deutschen Sudpolar-Expedition, 1901-1903.
Deutsche Südpolar-Exped., 14(1)(Zool. 6):34 pp.

Doliolum denticulatum
Russell, F.S., and J.S. Colman, 1935
The Zooplankton. IV. The occurrence and seasonal distribution of the Tunicata, Mollusca and Coelenterata (siphonophora). Brit. Mus. (N.H.) Great Barrier Reef Expedition 1928-1929, Sci. Repts., 2(7):203-276, 30 text figs.

Doliolum (Dolioletta) denticulatum
Sewell, R.B.S., 1953.
The pelagic Tunicata. Brit. Mus. (N.H.), John Murray Exped., 1933-34, Sci. Repts. 10(1):1-90, 1 pl., 32 textfigs.

Doliolum (Dolioletta) denticulatum
Tokioka, T., 1937.
Notes on salpas and doliolums. Annot. Zool. Jap. 16(3):219-232.

Doliolum denticulatum
Trogouboff, G., 1965.
La distribution verticale des doliolides au large de Villefranche-sur-Mer.
Bull. Inst. Oceanogr., Monaco, 64(1333):47 pp.
Also in:
Trav. Sta. Zool., Villefranche-sur-Mer, 25.

Doliolum denticulatum
Van Zyl, R. P., 1960.
A preliminary study of the salps and doliolids off the west and south coasts of South Africa.
Union S. Africa, Dept. Comm. & Ind., Div. Fish., Invest. Rept. 40:31 pp.

Doliolum ehrenbergi
Belloc, G., 1938.
Liste des Tuniciers pélagiques capturés au cours de la cinquième croisière. Rev. Trav. Off. Pêches Marit. 11(3):315-334, 20 textfigs.

Doliolum ehrenbergii
Brooks, W.K., 1893.
The genus Salpa. Mem. Biol. Lab., Johns Hopkins Univ. 2:1-306, 48 pls.

Doliolum ehrenbergii
Ritter, W.E., 1905.
III. The pelagic Tunicata of the San Diego region, excepting the Larvacea. Univ. Calif. Publ. Zool. 2(3):51-112, Pls. 2-3.

Doliolum gegenbauri
Alvariño, A., 1957.
Estudio del zooplancton del Mediterráneo occidental. Zooplancton del Atlántico Ibérico. Campañas del "Xauen" en el verano del 1954.
Bol. Inst. Español Oceanogr., 81-82:1-26; 1-51.

Doliolum (Dolioletta) gegenbauri
Berkeley, E., and C. Berkeley, 1960
Some further records of pelagic Polychaeta from the northeast Pacific, north of Latitude 40°N. and east of Longitude 175°W., together with records of Siphonophora, Mollusca and Tunicata from the same region.
Canadian J. Zoology. 38:787-799.

Doliolum gegenbauri
Berner, L. D., 1960
Unusual features in the distribution of pelagic tunicates in 1957 and 1958. Cal. Coop. Ocean. Fish. Invest. Rept. Vol. 7: 133-135.

Doliolum (Dolioletta) gegenbauri
Berrill, N.J., 1950.
The Tunicata with an account of the British species. Ray Soc., No. 133:354 pp., 120 textfigs.

Doliolum (Dolioletta) gegenbauri
Fraser, J.H., 1947.
Thaliacea II. Family: Doliolidae. Fiches d'Ident. Zooplancton, Cons. Perm. Int. Expl. Mer, 10:4 pp. 13 figs.

Dolioletta gegenbauri
Hansen, Vagn Kr., 1960.
Investigations on the quantitative and qualitative distribution in the southern part of the Norwegian Sea.
Medd. Danmarks Fiskeri- og Havundersøgelser, n.s., 2(23):1-53.
(of zooplankton)

Doliolum gegenbauri
Neumann, Günther, 1913.
Die Pyrosomen und Dolioliden der Deutschen Südpolar-Expedition, 1901-1903.
Deutsche Südpolar-Exped., 14(1)(Zool.6):34 pp.

Doliolum gegenbauri
Russell, F.S., and J.S. Colman, 1935
The Zooplankton. IV. The occurrence and seasonal distribution of the Tunicata, Mollusca and Coelenterata (siphonophora). Brit. Mus. (N.H.) Great Barrier Reef Expedition 1928-1929, Sci. Repts., 2(7):203-276, 30 text figs.

Doliolum gegenbauri
Russell, F. S. and A. B. Hastings, 1933
On the occurrence of pelagic tunicates (Thaliacea) in the waters of the English Channel off Plymouth. JMBA 18(2):635-640.

Doliolum (Dolioletta) gegenbauri
Sewell, R.B.S., 1953.
The pelagic Tunicata. Brit. Mus. (N.H.), John Murray Exped., 1933-34, Sci. Repts. 10(1):1-90, 1 pl., 32 textfigs.

Doliolum gegenbauri tritonis
Godeaux, J., 1960.
Tuniciers pelagiques du Golfe d'Eylath. Contribution to the knowledge of the Red Sea, No. 18.
Sea Fish. Res. Sta, Haifa, Bull., No. 29:9-15.

Doliolum gegenbauri tritonis
Fraser, J. H., 1947.
Thaliacea II. Family: Doliolidae. Fiches d'Ident. Zooplancton, Cons. Perm. Int. Expl. Mer, 10:4 pp. 13 textfigs.

Doliolim (Doliolina) indicum
Sewell, R.B.S., 1953.
The pelagic Tunicata. Brit. Mus. (N.H.), John Murray Exped., 1933-34, Sci. Repts., 10(1):1-90,

Doliolum indicum
Neumann, G., 1913
Tunicata Salpae II: Cyclomyaria et Pyrosomida. 40:36 pp., 19 text figs.
Das Tierreich

Doliolim (Doliolina) intermedium
Fraser, J. H., 1947.
Thaliacea II. Family: Doliolidae. Fiches d'Ident. Zooplancton, Cons. Perm. Int. Expl. Mer, 10:4 pp. 13 textfigs.

Doliolum intermedium
Neumann, G., 1913
Tunicata Salpae II: Cyclomyaria et Pyrosomida. 40:36 pp., 19 text figs.
Das Tierreich

Doliolum krohni
Borgert, A., 1896.
Die Doliolum-Ausbeute des "Vetter-Pisani". Zool. Jahrb., Abth. Syst., 9:414-419.

Doliolum (Doliolina) krohni
Fraser, J.H., 1947.
Thaliacea II. Family: Doliolidae. Fiches d'Ident. Zooplancton, Cons. Perm. Int. Expl. Mer, 10:4 pp. 13 textfigs.

Doliolum Krohni ?
Marukawa, H., 1921
Plankton lists and some new species of copepods, from the northern waters of Japan. Bull. Inst. Ocean., No.384, 15 pp., 3 pls., 1 chart.
Monaco

Doliolum krohni
Neumann, G., 1913
Tunicata Salpae II: Cyclomyaria et Pyrosomida. 40:36 pp., 19 text figs.
Das Tierreich

Doliolum krohni
Neumann, Günther, 1913.
Die Pyrosomen und Dolioliden der Deutschen Südpolar-Expedition, 1901-1903.
Deutsche Südpolar-Exped., 14(1)(Zool. 6):34 pp.

Doliolum (Dolioletta) mirabilis
Fraser, J.H., 1947.
Thaliacea II. Family: Doliolidae. Fiches d'Ident. Zooplancton, Cons. Perm. Int. Expl. Mer, 10:4 pp. 13 textfigs.

Doliolum (Dolioletta) mirabilis
Sewell, R.B.S., 1953.
The pelagic Tunicata. Brit. Mus. (N.H.), John Murray Exped., 1933-34, Sci. Repts., 10(1):1-90, 1 pl., 32 textfigs.

Doliolina mülleri
Casanova, Jean-Paul, 1966.
Pêches planctoniques superficielles et profondes en Méditerranée occidentale (Campagne de la Thalassa - janvier 1961 -entre les îles Baléares, la Sardaigne et l'Algérois). VII. Thaliacés.
Revue Trav. Inst.(Scient.tech.) marit.,30(4):385-390.

Doliolum (Doliolina) mülleri
Fraser, J.H., 1947.
Thaliacea, II. Family: Doliolidae. Fiches d'Ident. Zooplancton, Cons. Perm. Int. Expl. Mer, 10:4 pp., 13 textfigs.

Doliolum mülleri
Godeaux, J., 1960.
Tuniciers pelagiques du Golfe d'Eylath. Contribution to the knowledge of the Red Sea, No. 18.
Sea Fish. Res. Sta., Haifa, Bull., No. 29:9-15.

Doliolum müllerii
Keferstein, Wilhelm und Ernst Ehlers, 1861
Zoologische Beitrage gesammelt im Winter 1859/60 in Neapel und Messina. Wilhelm Engelmann, Leipzig, 112 pp., 15 pls.

Doliolum mulleri
Neumann, Günther, 1913.
Die Pyrosomen und Dolioliden der Deutschen Südpolar-Expedition, 1901-1903.
Deutsche Südpolar-Exped., 14(1)(Zool. 6):34 pp.

Doliolum mülleri
Neumann, G., 1913
Tunicata Salpae II: Cyclomyaria et Pyrosomida. 40:36 pp., 19 text figs.
Das Tierreich

Doliolum mülleri
Ritter, W. E., 1905.
III. The pelagic Tunicata of the San Diego region, excepting the Larvacea. Univ. Calif. Publ. Zool. 2(3):51-112, Pls. 2-3.

Doliolum mülleri krohni
Fraser, J.H., 1947.
Thaliacea II. Family: Doliolidae. Fiches d'Ident. Zooplancton, Cons. Perm. Int. Expl. Mer, 10:4 pp. 13 textfigs.

Doliolum nationalis
Alvariño, A., 1957.
Estudio del zooplancton del Mediterráneo occidental. Zooplancton del Atlántico Ibérico. Campañas del "Xauen" en el verano del 1954.
Bol. Inst. Español Oceanogr., 81-82:1-26; 1-51.

Doliolum nationalis
Anon., 1951.
Bull. Mar. Biol. Sta., Asamushi 4(3/4):15 pp.

Doliolum nationalis
Barnes, Beatrice I., 1961
Contribution towards a plankton atlas of the north-eastern Atlantic and the North Sea. 4. Thalacea.
Bulls. Mar. Ecology, 5(42): 102-104, Pl. 29.

Doliolum nationalis
Berrill, N.J., 1950.
The Tunicata with an account of the British species. Ray Soc. No. 133:354 pp., 120 textfigs.

Doliolum nationalis
Borgert, A., 1896.
Die Doliolum-Ausbeute des "Vettor Pisani". Zool. Jahrb., Abth. Syst., 9:414-419.

Doliolum nationalis
Braconnet, Jean-Claude 1967
Sur la possibilité d'un cycle court de développement chez le tunicier pélagique
C.r. hebd. Séanc. Acad. Sci. Paris (D) 264(11): 1434-1437.

Doliolum nationalis
Braconnot, J.C., 1963.
Étude du cycle annuel des Salpes et Dolioles en rade de Villefranche-sur-mer.
Journal du Conseil, 28(1):21-36.

Doliolum nationalis
Casanova, Jean-Paul, 1966.
Pêches planctoniques superficielles et profondes en Méditerranée occidentale (Campagne de la Thalassa - janvier 1961 - entre les îles Baléares, la Sardaigne et l'Algérois) VII. Thaliacés.
Revue Trav. Inst. (Scient. Tech.) marit., 30(4): 385-390.

Doliolum nationalis
Braconnet, Jean-Claude, et Jean-Paul Casanova, 1967.
Sur le tunicier pélagique Doliolum nationalis Borgert 1893 in Méditerrannée occidentale (campagne du Président-Théodore-Tissier, septembre-Octobre 1958).
Rev. Trav. Inst. Pêches marit., 31(4):393-402.

Doliolum nationalis
Chiba, T., 1949
On the distribution of the plankton in the eastern China Sea and Yellow Sea. 1. Plankton composition in the spring. J. Shimonoseki Coll. Fisheries, 1(1):57-63, 1 fig.

Doliolum nationalis
Enomoto, Y., 1962
Studies on the food base in the Yellow and the East China seas. 1. Plankton survey in summer of 1956.
Bull. Japan Soc. Sci. Fish., 28(8):759-765.

Doliolum nationalis
Fraser, J.H., 1947.
Thaliacea II. Family: Doliolidae. Fiches d'Ident. Zooplancton, Cons. Perm. Int. Expl. Mer, 10:4 pp., 13 textfigs.

Doliolum nationalis
Frost, N., S.T. Lindsay, and H. Thompson, 1933. III. Hydrographic and biological investigations. B. Plankton more abundant in 1932 than 1931.
Ann. Rept., Newfoundland Fish. Res. Comm. 2(1): 58-74, Textfigs. 17-27.

Doliolum nationalis a
Godeaux, Jean, 1971
L'ultrastructure de l'endostyle des Doliolides (Tuniciers Cyclomyaires).
C.r. Acad. Sci. Paris 272:592-595.

Doliolum nationalis
Harant, H., and P. Vernières, 1934.
Tuniciers pélagiques provenant des croisières du Prince Albert 1er de Monaco. Rés. camp. sci., Monaco, 88:47 pp., 5 textfigs.

Doliolum nationalis
Hunt, H.G., 1968.
Continuous plankton records: contribution towards a plankton atlas of the North Atlantic and the North Sea. II. The seasonal and annual distribution of Thaliacea.
Bull. mar. Ecol. 6(7):225-249.

Doliolum nationalis
Neumann, Günther, 1913.
Die Pyrosomen und Dolioliden der Deutschen Südpolar-Expedition, 1901-1903.
Deutsche Südpolar Exped., 14(1)(Zool. 6):34 pp.

Doliolum nationalis
Russell, F.S. and A.B. Hastings, 1933
On the occurrence of pelagic tunicates (Thaliacea) in the waters of the English Channel off Plymouth. JMBA 18(2):635-640.

Doliolum nationalis
Tavares, D.Q., 1967.
Occurrence of doliolids and salps during 1958, 1959 and 1960 off the São Paulo Coast.
Bolm Inst. oceanogr. S. Paulo, 16(1):87-97.

Doliolum nationalis
Terry, Robert M., 1961
Investigations of inner continental shelf waters off lower Chesapeake Bay. III. The Phorozooid stage of the tunicate, Doliolum nationalis.
Chesapeake Science, 2(1/2):60-64.

Doliolum nationalis
Tregouboff, G., 1965.
La distribution verticale des doliolides au large de Villefranche-sur-Mer.
Bull. Inst. Oceanogr., Monaco, 64(1333):47 pp.
Also in:
Trav. Sta. Zool., Villefranche-sur-Mer, 25.

Doliolum nationalis
Van Zyl, R.P., 1960.
A preliminary study of the salps and doliolids off the west and south coast of South Africa.
Union S. Africa, Dept. Comm. & Ind., Div. Fish., Invest. Rept. 40:31 pp.

Doliolum (Dolioloides) rarum
Fraser, J.H., 1947.
Thaliacea - II. Family: Doliolidae. Fiches d'Ident. Zooplancton, Cons. Perm. Int. Expl. Mer, 10:4 pp., 13 textfigs.

Doliolum rarum
Neumann, G., 1913
Tunicata Salpae II: Cyclomyaria et Pyrosomida. 40:36 pp., 19 text figs.
(Das Tierreich)

Doliolum resistible n.sp.
Neumann, Günther, 1913.
Die Pyrosomen und Dolioliden der Deutschen Südpolar-Expedition, 1901-1903.
Deutsche Südpolar-Exped., 14(1)(Zool. 6):34 pp.

Doliolum resistibile
Neumann, G., 1913
Tunicata Salpae II: Cyclomyaria et Pyrosomida. 40:36 pp., 19 text figs.
(Das Tierreich)

Doliopsis savigniana
Neumann, G., 1913
Tunicata Salpae II: Cyclomyaria et Pyrosomida. 40:36 pp., 19 text figs.
(Das Tierreich)

Doliolum tritonis
Bedot, M., 1909
Sur la faune de l'Archipel Malais (resume).
Rev. Suisse Zool. 17:143-169.

Doliolum tritonis
Belloc, G., 1938.
Liste des Tuniciers pélagiques capturés au cours de la cinquième croisière. Rev. Trav. Off. Pêches Marit. 11(3):315-334, 20 textfigs.

Doliolum tritonis
Borgert, A., 1896.
Die Doliolum-Ausbeute des "Vettor Pisani". Zool. Jahrb., Abth. Syst., 9:414-419.

Doliolum tritonis
Hastings, A.B., 1931
Tunicata. Great Barrier Reef Exped., 1928-29. Sci. Repts. 4(3):69-109, 17 textfigs, 6pls.

Doliolum tritonis
Ritter, W.E., 1905.
III. The pelagic Tunicata of the San Diego region, excepting the Larvacea. Univ. Calif. Publ. Zool. 2(3):51-112, Pls. 2-3.

Doliolum tritonis
Russell, F.S., and J.S. Colman, 1935
The Zooplankton. IV. The occurrence and seasonal distribution of the Tunicata, Mollusca and Coelenterata (siphonophora). Brit. Mus. (N.H.) Great Barrier Reef Expedition 1928-1929, Sci. Repts., 2(7):203-276, 30 text figs.

Doliolum (Dolioletta) tritonis
Tokioka T., 1937.
Notes on salpas and doliolums. Annot. Zool. Jap. 16(3):219-232.

Doliolum tritonis
Van Zyl, R.P., 1960.
A preliminary study of the salps and doliolids off the west and south coast of South Africa.
Union S. Africa, Dept. Comm. & Ind., Div. Fish., Invest. Rept. 40:31 pp.

Doliopsoides horizoni
Tokioka, Takasi, and Leo Berner, 1954.
On certain Thaliacea (Tunicata) from the Pacific Ocean, with descriptions of two new species of doliolids.
Pacific Science 12(4):317-326.

Doliopsoides horizoni n.sp.
Tokioka, T., and L. Berner, 1958.
Two new doliolids from the eastern Pacific Ocean.
Pacific Science 12(2):135-138.

Pyrosomids
Hopkins, H.S., 1943
Biological results of the last cruise of the Carnegie. VIII. Miscellaneous determinations. The Pyrosomida. Sci. Res. of Cruise VII of the Carnegie during 1928-1929 under command of Captain J.P. Ault., Carnegie Inst., Washington, Publ.555:92.

pyrosomids
Metcalf, M.M., and H.S. Hopkins, 1919.
Pyrosoma: a taxonomic study based on the collections of the United States Bureau of Fisheries and the United States National Museum.
Bull. U.S. Nat. Mus. 100, Vol. 2:195-275, pls. 15-36.

Pyrosoma
Berrill, N.J., 1950.
Budding in Pyrosomma. J. Morph. 83(3):537-553, 3 textfigs.

Pyrosoma
Cowper, T.R., 1960
Occurrence of Pyrosoma on the continental slope. Nature, 187(4740): 878-879.

Pyrosoma congregates at depths of 160-170 m.

Pyrosoma
Culkin, F. and R.J. Morris, 1970.
The fatty acid composition of two marine filter-feeders in relation to a phytoplankton diet.
Deep-Sea Res., 17(5): 861-865.

Pyrosoma sp.
Davies, D. H., 1949
Preliminary investigations on the foods of South African fishes (with notes on the general fauna of the area surveyed). Fish & Mar. Biol. Survey Div., Dept. Comm. & Industries, S. Africa, Invest. Rept. No.11, Commerce & Industry, Jan. 1949:1-36, 4 pls.

Pyrosoma
Fournier, Robert O., 1973
Studies on pigmented microorganisms from aphotic marine environments. III. Evidence of apparent utilization by benthic and pelagic tunicata. Limnol. Oceanogr. 18(1): 38-43.

Pyrosoma sp.
Harant, H. and P. Vernieres, 1934
Tuniciers pelagiques provenant des croisieres du Prince Albert ler de Monaco. Res. Camp. Sci. 88:48 pp.

Pyrosoma sp.
Keferstein, Wilhelm und Ernst Ehlers, 1861
Zoologische Beitrage gesammelt im Winter 1859/60 in Neapel und Messina. Wilhelm Engelmann, Leipzig, 112 pp., 15 pls.

Pyrosoma - anatomy
Neumann, G., 1912.
Über Bau und Entwicklung des Stolo prolifer der Pyrosomen. Zool. Anz. 39:13-21, 10 textfigs.

Pyrosoma
Puisségur, C., 1946.
Les Êtres vivants lumineux. Sci.et Vie, Vol. LXX (348), translated by R. Widmer, David Taylor Model Basin Translation 221, dated May 1927 (as Luminous Living Organisms), 17 pp., 9 text figs.

Pyrosoma sp.
Sebastian, V.O., 1968.
Flat and encrusting colony of Pyrosoma from the Kerala Coast of India. Bull. Dept. mar. Biol. Oceanogr. Univ. Kerala, 4: 158-160.

Pyrosoma agassizi
Harant, H., and P. Vernières, 1934.
Tuniciers pelagiques provenant des croisières du Prince Albert 1er de Monaco. Rés. camp. sci., Monaco, 88:47 pp., 5 textfigs.

Pyrosoma agassizi
Metcalf, M.M., and H.S. Hopkins, 1919.
Pyrosoma: a taxonomic study based on the collections of the United States Bureau of Fisheries and the United States National Museum. Bull. U.S. Nat. Mus. 100, Vol. 2:195-275, pls. 15-36.

Pyrosoma agassizi
Neumann, G., 1913
Tunicata Salpae II: Cyclomyaria et Pyrosomida. 40:36 pp., 19 text figs.
(Das Tierreich)

Pyrosoma agassizi
Neumann, Günther, 1913.
Die Pyrosomen und Dolioliden der Deutschen Südpolar-Expedition, 1901-1903.
Deutsche Südpolar-Exped., 14(1)(Zool. 6):34 pp.

Pyrosoma agassizi n.sp.
Ritter, R. E. and E. S. Byxbee, 1905
VIII. The pelagic Tunicata. Reports on the Scientific results of the expedition to the Tropical Pacific, in charge of Alexander Agassiz, in the U. S. Fish Commission Steamer "Albatross", from August, 1899 to March, 1900, Commander Jefferson F. Moser, U. S. N., commanding. Mem. M.C.Z. 26(5):195-211, 2 pls.

Pyrosoma aherniosum
Metcalf, M.M., and H.S. Hopkins, 1919.
Pyrosoma: a taxonomic study based on the collections of the United States Bureau of Fisheries and the United States National Museum. Bull. U.S. Nat. Mus., 100, Vol. 2:195-275, Pls. 15-36.

Pyrosoma aherniosum
Neumann, G., 1913
Tunicata Salpae II: Cyclomyaria et Pyrosomida. 40:36 pp., 19 text figs.
(Das Tierreich)

Pyrosoma aherniosum
Neumann, Günther, 1913
Die Pyrosomen und Dolioliden der Deutschen Südpolar-Expedition, 1901-1903.
Deutsche Südpolar-Exped., 14(1)(Zool. 6):34 pp.

Pyrosoma aherniosum
Sewell, R.B.S., 1953.
The pelagic Tunicata. Brit. Mus. (N.H.), John Murray Exped., 1933-34, Sci. Repts. 10(1):1-90, 1 pl., 32 textfigs.

Pyrosoma ambulata
Hopkins, H.S., 1943
Biological results of the last cruise of the Carnegie. VIII. Miscellaneous determinations. The Pyrosomida. Sci. Res. of Cruise VII of the Carnegie during 1928-1929 under command of Captain J.P. Ault, Carnegie Inst., Washington, Publ.555:92.

Pyrosoma atlanticum
Bary, B.M., 1960.
Notes on ecology, distribution and systematics of pelagic Tunicata from New Zealand. Pacific Science, 14:101-121.

Pyrosoma atlanticum
Casanova, Jean-Paul, 1966.
Pêches planctoniques superficielles et profondes en Mediterranée occidentale (Campagne de la Thalassa - Janvier 1961 - entre les Îles Baléares, la Sardaigne et l'Algérois) VII. Thaliacés.
Revue Trav.Inst.(scient.tech.)marit.,30(4):385-390.

Pyrosoma atlanticum
Godeaux, J., 1956.
Blastogenese du systeme nerveux chez le Pyrosome. Ann. Soc. R. Zool., Belgique, 86(2):281-301.

Pyrosoma atlantica
Godeaux, J., 1953.
Note sur le complexe neural du pyrosome. Ann. Soc. R. Zool., Belgique, 84(1):61-70.

Pyrosoma atlantica
Godeaux, J., 1953.
Sur l'ebauche nerveuse du cyathozoide. Ann. Soc. R. Zool., Belgique, 84(1):71-85.

Pyrosoma atlanticum
Harant, H., and P. Vernières, 1934.
Tuniciers pelagiques provenant des croisières du Prince Albert 1er de Monaco. Rés. camp. sci., Monaco, 88:47 pp., 5 textfigs.

Pyrosoma atlanticum
Hardy, A.C., and R.H. Kay, 1964
Experimental studies of plankton luminescence Jour. Mar. Biol. Assoc. U.K., 44(2):435-484.

Pyrosoma atlanticum
Neumann, Günther, 1913.
Die Pyrosomen und Dolioliden der Deutschen Südpolar-Expedition, 1901-1903.
Deutsche Südpolar-Exped., 14(1)(Zool.6):34 pp.

Pyrosoma atlanticum
Neumann, G., 1913
Tunicata Salpae II: Cyclomyaria et Pyrosomida. 40:36 pp., 19 text figs.
(Das Tierreich)

Pyrosoma atlanticum
Ritter, R. E. and E. S. Byxbee, 1905
VIII. The pelagic Tunicata. Reports on the Scientific results of the expedition to the Tropical Pacific, in charge of Alexander Agassiz, in the U. S. Fish Commission Steamer "Albatross", from August, 1899 to March, 1900, Commander Jefferson F. Moser, U. S. N., commanding. Mem. M.C.Z. 26(5):195-211, 2 pls.

Pyrosoma atlanticum
Thompson, Harold, 1954.
Pelagic tunicates. Identifications and relevant notes. B.A.N.Z. Antarctic Res. Exped. 1929-1931 Repts., B(Zool. Bot.), 1(4):183-185.

Pyrosoma atlanticum atlanticum
Harant, H. and P. Vernieres, 1934
Tuniciers pelagiques provenant des croisieres du Prince Albert ler de Monaco. Res. Camp. Sci. 88:48 pp.

Pyrosoma atlanticum & subsp.
Metcalf, M.M., and H.S. Hopkins, 1919.
Pyrosoma: a taxonomic study based on the collections of the United States Bureau of Fisheries and the United States National Museum. Bull. U.S. Nat. Mus., 100, Vol. 2:195-275, Pls. 15-36.

Pyrosoma atlanticum
Sewell, R.B.S., 1953.
The pelagic Tunicata. Brit. Mus. (N.H.), John Murray Exped., 1933-34, Sci. Repts., 10(1):1-90, 1 pl., 32 textfigs.

Pyrosoma benthica
Monniot, Claude, et Francoise Monniot,1966.
Un pyrosome benthique: Pyrosoma benthica n.sp. C.R. hebd. Séanc. Acad. Sci., Paris, (D)263(4): 368-270.

Pyrosoma atlanticum giganteum
Berrill, N.J., 1950.
The Tunicata with an account of the British species. Ray Soc. No. 133:354 pp., 120 textfigs.

Pyrosoma atlanticum levatum
Neumann, G., 1913
Tunicata Salpae II: Cyclomyaria et Pyrosomida. 40:36 pp., 19 text figs.
(Das Tierreich)

Pyrosoma elegans
Neumann, G., 1913
Tunicata Salpae II: Cyclomyaria et Pyrosomida. 40:36 pp., 19 text figs.
(Das Tierreich)

Pyrosoma ellipticum n.sp.
Metcalf, M.M., and H.S. Hopkins 1919
Pyrosoma: a taxonomic study based on the collections of the United States Bureau of Fisheries and the United States National Museum. Bull. U.S. Nat. Mus., 100, Vol. 2:195-275, Pls. 15-36.

Pyrosoma fixata
Sewell, R.B.S., 1953.
The pelagic Tunicata. Brit. Mus. (N.H.), John Murray Exped., 1933-34, Sci. Repts., 10(1):1-90, 1 pl., 32 textfigs.

Pyrosoma giganteum
Belloc, G., 1938.
Liste des Tuniciers pélagiques capturés au cours de la cinquième croisière. Rev. Trav. Off. Pêches Marit. 11(3):315-334, 20 textfigs.

Pyrosoma giganteum
Harant, H. and P. Vernieres, 1934
Tuniciers pelagiques provenant des croisieres du Prince Albert ler de Monaco. Res. Camp. Sci. 88:48 pp.

Pyrosoma giganteum

Ritter, R. E. and E. S. Byxbee, 1905
VIII. The pelagic Tunicata. Reports on the Scientific results of the expedition to the Tropical Pacific, in charge of Alexander Agassiz, in the U. S. Fish Commission Steamer "Albatross", from August, 1899 to March, 1900, Commander Jefferson F. Moser, U. S. N., commanding. Mem. M.C.Z. 26(5):195-211, 2 pls.

Pyrosoma hybridum n. sp.

Metcalf, M.M., and H.S. Hopkins, 1919.
Pyrosoma: a taxonomic study based on the collections of the United States Bureau of Fisheries and the United States National Museum.
Bull. U.S. Nat. Mus., 100, Vol. 2:195-275, Pls. 15-36.

Pyrosoma minimum

Neumann, G., 1913
Tunicata Salpae II: Cyclomyaria et Pyrosomida. 40:36 pp., 19 text figs.
Das Tierreich,

Pyrosoma operculatum

Metcalf, M.M., and H.S. Hopkins, 1919.
Pyrosoma: a taxonomic study based on the collections of the United States Bureau of Fisheries and the United States National Museum.
Bull. U.S. Nat. Mus., 100, Vol. 2:195-275, Pls. 15-36.

Pyrosoma operculatum

Neumann, G., 1913
Tunicata Salpae II: Cyclomyaria et Pyrosomida. 40:36 pp., 19 text figs.
Das Tierreich,

Pyrosoma pygmaea

Thompson, J.V., 1830.
On the luminosity of the ocean with descriptions of some remarkable species of luminous animals (Pyrosoma pygmaea and Sapphirina indicator), and particularly of the four genera, Noctiluca (sic), Cynthis, Lucifer, and Podopsis of the Schizopoda (sic). (Addenda to Memoir 1. Addendum to Memoir 2 Zoological researches and illustrations on natural history of nondescript or imperfectly known animals in a series of memoirs. Vol. 1, Mem. 3: 110 pp., 14 pls.

Pyrosoma pygmaea

Neumann, G., 1913
Tunicata Salpae II: Cyclomyaria et Pyrosomida. 40:36 pp., 19 text figs.
Das Tierreich,

Pyrosoma ovatum

Metcalf, M.M., and H.S. Hopkins, 1919.
Pyrosoma: a taxonomic study based on the collections of the United States Bureau of Fisheries and the United States National Museum.
Bull. U.S. Nat. Mus., 100, Vol. 2:195-275, Pls. 15-36.

Pyrosoma ovatum

Neumann, G., 1913
Tunicata Salpae II: Cyclomyaria et Pyrosomida. 40:36 pp., 19 text figs.
Das Tierreich,

Pyrosoma rufum

Neumann, G., 1913
Tunicata Salpae II: Cyclomyaria et Pyrosomida. 40:36 pp., 19 text figs.
Das Tierreich,

Pyrosoma sedentarium n.sp.

Sebastian, V.O., 1971.
Pyrosoma sedentarium n. sp. Bull. Dept. mar. Biol. Oceanogr. Univ. Cochin 5: 77-79.

Pyrosoma spinosum

Baker, Alan N., 1971
Pyrosoma spinosum Herdman, a giant pelagic tunicate new to New Zealand waters.
Rec. Dominion Mus. N.Z. 7(12):107-117.

Pyrosoma spinosum

Bary, B.M., 1960.
Notes on ecology, distribution and systematics of pelagic Tunicata from New Zealand.
Pacific Science, 14:101-121.
(a)

Pyrosoma (Pyrostremma) spinosum

Berrill, N.J., 1950.
The Tunicata with an account of the British species. Ray. Soc. No. 133:354 pp., 120 textfigs.

Pyrosoma spinosum

Grace, Roger V. 1971.
Giant Pyrosoma seen in New Zealand seas.
Australian Nat. Hist., Dec.1971: 118-119.

Pyrosoma spinosum

Griffin, D.J.G., and J.C. Yaldwyn 1970.
Giant colonies of pelagic tunicates (Pyrosoma spinosum) from SE Australia and New Zealand.
Nature, Lond., 226 (5244): 464-465.

Pyrosoma spinosum

Harant, H., and P. Vernières, 1934.
Tuniciers pélagiques provenant des croisières du Prince Albert 1er de Monaco. Rés. camp. sci., Monaco, 88:47 pp., 5 textfigs.

Pyrosoma spinosum

Hopkins, H.S., 1943
Biological results of the last cruise of the Carnegie. VIII. Miscellaneous determinations The Pyrosomida. Sci. Res. of Cruise VII of the Carnegie during 1928-1929 under command of Captain J.P. Ault., Carnegie Inst., Washington, Publ.555:92.

Pyrosoma spinosum

Metcalf, M.M., and H.S. Hopkins, 1919.
Pyrosoma: a taxonomic study based on the collections of the United States Bureau of Fisheries and the United States National Museum.
Bull. U.S. Nat. Mus. 100, Vol. 2:195-275, Pls. 15-36.

Pyrosoma spinosum

Neumann, G., 1913
Tunicata Salpae II: Cyclomyaria et Pyrosomida. 40:36 pp., 19 text figs.
Das Tierreich,

Pyrosoma spinosum

Neumann, Günther, 1913.
Die Pyrosomen und Dolioliden der Deutschen Südpolar-Expedition, 1901-1903.
Deutsche Südpolar-Exped., 14(1)(Zool. 6):34 pp.

Pyrosoma spinosum

Sewell, R.B.S., 1953.
The pelagic Tunicata. Brit. Mus. (N.H.), John Murray Exped., 1933-34, Sci. Repts. 10(1):1-90, 1 pl., 32 textfigs.

Pyrosoma triangulum

Neumann, G., 1913
Tunicata Salpae II: Cyclomyaria et Pyrosomida. 40:36 pp., 19 text figs.
Das Tierreich,

Pyrosoma verticillatum

Metcalf, M.M., and H.S. Hopkins, 1919.
Pyrosoma: a taxonomic study based on the collections of the United States Bureau of Fisheries and the United States National Museum.
Bull. U.S. Nat. Mus. 100, Vol. 2:195-275, Pls. 15-36.

Also P. verticillatum cylindricum n.subsp.

Pyrosoma verticillatum

Neumann, G., 1913
Tunicata Salpae II: Cyclomyaria et Pyrosomida. 40:36 pp., 19 text figs.
Das Tierreich,

Pyrosoma verticillatum

Neuman, Günther, 1913.
Die Pyrosomen und Dolioliden der Deutschen Südpolar-Expedition, 1901-1903.
Deutsche Südpolar-Exped., 14(1)(Zool. 6):34 pp.

Pyrosoma verticillatum

Sewell, R.B., 1953.
The pelagic Tunicata. Brit. Mus. (N.H.), John Murray Exped., 1933-34, Sci. Repts. 10(1):1-90, 1 pl., 32 textfigs.

the two subspecies:
cylindricum
hybridum

MOLLUSCS

molluscs

Allen, J.A., 1965.
Records of Mollusca from the northwest Atlantic obtained by Canadian fishery research vessels, 1946-61.
J. Fish. Res. Bd., Canada, 22(4):977-997.

molluscs

Allen, J.A., 1963
Ecology and functional morphology of molluscs. In: Oceanography and Marine Biology, H. Barnes Edit., George Allen & Unwin, 1:253-288.

molluscs

Barnard, K.H., 1963.
Deep sea Mollusca from the region south of Madagascar.
Rept. South Africa, Dept. Comm. Industr., Div. Sea Fish., Invest. Rept., (44):19 pp.

molluscs

Barnard, K.H., 1963
Deep sea Mollusca from west of Cape Point, South Africa.
Annals, S. African Mus., 46(17):407-452.

molluscs

Bayer, Frederick M., 1971.
Biological results of the University of Miami deep-sea expeditions. 79. New and unusual mollusks collected by R/V John Elliott Pillsbury and R/V Gerda in the tropical western Atlantic. Bull. mar. Sci. 21(1): 111-236.

Molluscs

Beltrán, Vicente, 1965.
Sobre tres raros micromoluscos del Mediterráneo español.
Bd. R. Soc. Española Hist. Nat. (Biol), 63:205-212.

molluscs, lists of spp.

Bernard, F., 1967.
Prodrome for a distributional check-list and bibliography of the recent marine Mollusca of the west coast of Canada.
Tech.Rep.Fish.Res.Bd.,Can.,2:261 pp. (mimeographed).

molluscs

Binder, E. 1968.
Répartition des mollusques dans la lagune Ébrié (Côte d'Ivoire).
Cah. ORSTOM, Sér. Hydrobiol. 2(3/4): 3-34.

molluscs

Cate, J.M., 1961.
A discussion of Vexillum regina (Bowerby, 1825) and related species with description of a new subspecies.
Veliger, 4(2):76-85.

mollusks, deep

Clarke, Arthur H., Jr. 1962
On the composition, zoogeography, origin and age of the deep-sea mollusk fauna.
Deep-Sea Res., 9(4):291-306.

molluscs

Clarke, A.H., Jr., 1961
Abyssal mollusks from the South Atlantic Ocean.
Bull. Mus. Comp. Zool., 125(12): 345-387.

molluscs

Clench, William J., 1964.
The genera *Pedipes* and *Laemodonta* in the western Atlantic.
Johnsonia, Mus. Comp. Zool., Harvard Coll., 4(42):117-128.

molluscs

Clench, William J., and Ruth Turner, 1964.
The subfamilies Volutinae, Zidoninae, Odontocymbiolinae and Calliotectinae in the western Atlantic.
Johnsonia, Mus. Comp. Zool., Harvard Coll., 4(43):129-180.

molluscs

D'Asaro, Charles N., 1970.
Egg capsules of prosobranch mollusks from South Florida and the Bahamas and notes on spawning in the laboratory. Bull. mar. Sci., 20(2): 414-440.

molluscs

Dell, R.K., 1964.
Marine Mollusca from MacQuarie and Heard islands.
Rec. Dominion Mus., Wellington, 4(20):267-301.

molluscs

Dell, R.K., 1963.
The littoral marine Mollusca of the Snares Islands.
Rec. Dom. Mus., New Zealand, 4(15):221-229.

molluscs

Edmunds, Malcom, 1968.
Eolid Mollusca from Ghana, with further details of two west Atlantic species.
Bull. mar. Sci., Miami, 18(1): 203-219.

Molluscs

Eisma D., 1966.
The distribution of benthic marine molluscs off the main Dutch coast.
Netherlands J. Sea Res., 3(1):107-163.

molluscs

Feininger, Andreas, and William K. Emerson 1972
Shells. Viking Press 295 pp. $27.50

molluscs

Fretta, Vera, 1968.
Studies in the structure, physiology and ecology of molluscs.
Symp. Zool. Soc., Malacol. Soc., London, 22:377.

(mostly fresh waterforms)

molluscs

Gabriel, Charles J., 1962.
Additions to the marine molluscan fauna of south eastern Australia including descriptions of new genus *Pillarginella*, six new species and two subspecies.
Mem. Nat. Mus., Victoria, (25):177-210.

molluscs

Gonor, J.J., 1961.
Observations on the biology of *Hermaeina smithi*: a sacoglossan opisthobranch from the west coast of North America.
Veliger, 4(2):85-98.

Mollusca

Greenhill, Julia F., 1965.
New records of Marine Mollusca from Tasmania.
Papers and Proc., R. Soc., Tasmania, 99:67-69.

molluscs

Kaneko, Sueo, 1947
Marine molluscan Fauna from the Tashufany District, Kashsiunghsien. Bull. Ocean. Inst., Taiwan 1(2):33-51.

molluscs

Kornicker, Louis S., Charles D. Wise and Juanita M. Wise, 1963.
Factors affecting the distribution of opposing mollusk valves.
J. Sed. Petr., 33(3):703-712.

molluscs

MacPherson, J. Hope, and C.J. Gabriel, 1962
Marine molluscs of Victoria.
Melbourne Univ. Press, 475 pp.

molluscs

Moore, H.B., 1931
The systematic Value of a Study of Molluscan faeces. Proc. Malacological Soc., 19(6):281-290, 22 figs.

molluscs

Muraika, James S., 1965.
Deep-ocean boring mollusk.
BioScience, A.I.B.S., 15(3):191.

molluscs

Olsson, Axel A., 1971.
Biological results of the University of Miami deep-sea expeditions. 77. Mollusks from the Gulf of Panama collected by R/V *John Elliott Pillsbury*, 1967. Bull. mar. Sci. 21(1): 35-92.

molluscs

Powell, A.W.B., 1961.
Mollusca from the Kermadec Islands, a marginal Indo-Pacific area. (Abstract).
Proc. Ninth. Pacific Sci. Congr., Pacific Sci. Assoc., 1957, Fish., 10:8.

molluscs

Ranson, M. Gilbert, 1943
Titres et travaux scientifiques. Paris: Masson et Cie, Editeurs, Libraries de l'Academie de Medecine, 120 Boulevard Saint Germain. 88 pp.

molluscs

Rice, Winnie H., and Louis S. Kornicker, 1965.
Mollusks from the deeper waters of the northwestern Campeche Bank, Mexico.
Publ. Inst. Mar. Sci., Port Aransas, 10:108-172.

molluscs

Rosewater, Joseph, 1961.
The familty Pinnidae in the Indo-Pacific.
Indo-Pacific Mollusca, 1(4):175-226.

mollusca

Russell, F.S., and J.S. Colman, 1935
The Zooplankton. IV. The occurrence and seasonal distribution of the Tunicata, Mollusca and Coelenterata (siphonophora). Brit. Mus. (N.H.) Great Barrier Reef Expedition 1928-1929, Sci. Repts., 2(7):203-276, 30 text figs.

molluscs

Stuardo B., Jose, 1964.
Distribucion en los moluscos marinos litorales en Latinoamerica.
Bol. Inst. Biol. Mar., Mar del Plata, Argentina, No. 7:79-91.

molluscs

Towe, K.M., H.A. Lowenstam and M.H. Nesson, 1963
Invertebrate ferritin: occurrence in Mollusca.
Science, 142(3588):63-64.

molluscs

Turner, R.D., 1959.
The genera Hemitona and Diodora in the western Atlantic.
Johnsonia, Mus. Comp. Zool., 3(39):334-344.

molluscs

Vallentine, James W., 1966.
Numerical analysis of marine molluscan ranges on the estratropical northeastern Pacific shelf.
Limnol. Oceanogr., 11(2):198-211.

molluscs

Villamar C., Alejandro, 1965.
Fauna malacoloqica de la Bahia de la Paz, B.C., con no as ecologicas.
Anales Inst. nac. Invest. biol.-pesqueras, México, 1: 115-152.

molluscs

Woodring, W.P., 1965.
Endemism in middle Miocene Caribbean molluscan faunas.
Science, 148(3672):961-963.

mollusks

Work, Robert C., 1969.
Systematics, ecology, and distribution of the mollusks of Los Roques, Venezuela. Bull. mar. Sci. 19(3): 614-711.

MOLLUSCS
Anatomy and/or physiology

molluscs, anat. physiol.

Andrews, J.T. 1972.
Recent and fossil growth rates of marine bivalves, Canadian Arctic, and Late-Quaternary Arctic marine environments.
Palaeogr., Palaeoclimatol. Palaeoecol. 11(3):157-176.

mollusca, anat.-physiol

Clark, George R., II, 1968.
Mollusk shell: daily growth lines.
Science, 161(3843):800-802.

molluscs, physiol.

Coughlan, John, and Alan D. Ansell, 1964.
A direct method for determining the pumping rate of siphonate bivalves.
J. du Cons., 29(2):205-213.

molluscs, anat. physiol

Hancock, D.A. 1965.
Adductor muscle size in Danish and British mussels and its relation to starfish predation.
Ophelia 2(2): 253-267.

mollusca, anat., physiol.

Mutvei, H., 1964.
On the shells of *Nautilus* and *Spirula* with notes on the secretion in non-cephalopod molluscs.
Arkiv för Zoologi, andra ser., 16(3):221-278.

molluscs, anat.

Narayan, Ramesh 1972.
X-ray patterns of mollusc shells from Indian waters.
Current Sci. 39(3): 51-54.

molluscs, anat-physiol

Potts, W.T.W., 1967.
Excretion in the molluscs.
Biol. Rev., Cambridge Phil. Soc., 42(1):1-41.

molluscs, physiol.

Turner, H.J., 1953.
The drilling mechanism of the Naticidae.
Ecology 34(1):222-223

molluscs

Wilbur, Karl M., and C.M. Yonge, editors, 1966.
Physiology of Mollusca.
Academic Press, Vol. 2: 645 pp. $22.00

mollusca, anat.physiol.

Wilbur, Karl M., and C. M. Yonge, Editors, 1964.
Physiology of Mollusca.
Academic Press, xiii 473 pp. 11d s 6 d

molluscs, anatomy

Yonge, C.M., 1947
The pallial organs in the Aspidobranch Gastropoda and their evolution throughout the Mollusca. Phil. Trans. Roy. Soc., London, Ser.B. Biol. Sci. 232(591):443-518, 40 textfigs., 1 pl.

molluscs, benthic

Dell, R.K., 1962.
Zoogeography of Antarctic benthic Mollusca.
(Abstract for Symposium Antarctic Biol., Paris, 2-8 Sept., 1962).
Polar Record, 11(2):327.
Also: SCAR Bull. #12.

Mollusca, biology of

Purchon, R.D. 1968.
The Biology of the Mollusca.
Pergamon Press, 560 pp.

MOLLUSCS chemistry of

See also: Subject Cumulation under chemistry

molluscs, mercury content

Establier, Rafael, 1973
Contenido en mercurio de los mejillones (Mytilus edulis) silvestres y cultivados de la zona noroeste española. Invest. pesq. Barcelona 37(1):101-106.

molluscs, chemistry

Gardner, Doris and J.P. Riley, 1972
The component fatty acids of the lipids of some species of marine and freshwater molluscs.
J. mar. biol. Ass. U.K. 52(4): 827-838.

molluscs, chemistry

Ghiselin, Michael T., Egon T. Degens and Derek W. Spencer, 1967.
A phylogenetic survey of molluscan shell matrix proteins.
Breviora, No. 262:35 pp.

molluscs, chemistry

Giese, Arthur C., 1969.
A new approach to the biochemical composition of the mollusc body. Oceanogr. Mar. Biol. Ann. Rev. H. Barnes, editor, George Allen and Unwin, Ltd., 7: 175-229.

molluscs, chemistry

Price, N.B., and A. Hallam, 1967.
Variation of strontium content within shells of Recent Nautilus and Sepia.
Nature, 215(5107):1272-1274.

molluscs, chemistry

Rao, S.R., S.M. Shah and R. Viswanathan, 1968.
Calcium, strontium and radium contents of molluscan shells.
J. mar. biol. Ass., India, 10(1): 159-165.

molluscs, chemistry

Shah, S.M., V.N. Sastry and the late Y.M. Bhatt 1973
Trace element distribution in some Mollusca from Bombay coast.
Current Sci. 42(17): 589-592.

molluscs, chemistry

Stegeman, J.J. and J.M. Teal 1973
Accumulation, release and retention of petroleum hydrocarbons by the oyster Crassostrea virginica. Mar. Biol. 22(1):37-44.

molluscs, biochemistry of

Stenzel, H.B., 1963.
Aragonite and calcite as constituents of adult oyster shells. Science, 142(3589):232-233.

molluscs, chemistry of

Strusi, Angelo, 1964.
Su alcuni caratteri chimici dei mitili (Mytilus galloprovincialis Lamarck) coltivati nel Mar Piccolo e nel Mar Grande (Golfo di Taranto).
Boll. Pesca, Piscicolt. Idrobiol., n.s., 19(2): 199-218.
English summary

molluscan shells, chemistry of

Turekian, Karl K., and Richard L. Armstrong, 1960
Magnesium, strontium and barium concentrations and calcite-aragonite ratios of some recent molluscan shells. J. Mar. Res., 18(3): 133-151.

molluscs

Cheng, Thomas C., 1967.
Marine molluscs as hosts for symbioses with a review of known parasites of commercially important species.
Adv. mar. Biol., Sir Frederick S. Russell, editor, Academic Press, 424 pp.

molluscs, deep-sea

Allen, J.A., and H.L. Sanders 1973.
Studies on deep-sea Protobranchia (Bivalvia); The families Siliculidae and Lametilidae.
Bull. Mus. Comp. Zool. 145(6): 263-310.

molluscs, ecology

Anderson, A., 1971.
Intertidal activity, breeding and the floating habit of Hydrobia ulvae in the Ythan Estuary.
J. mar. biol. Ass. U.K. 51(2): 423-437.

MOLLUSCS learning

molluscs, learning of

Wells, M. J., 1965.
Learning by marine invertebrates.
In: Advances in Marine Biology, Sir Frederick S. Russell, editor, Academic Press, 3:1-62.

Molluscs lists of species

molluscs, lists of spp.

Aguayo, C.G., 1962.
Notas sobre moluscos. Antillanos III.
Caribbean J. Science, 2(3):108-112.

molluscs, lists of spp.

Dell, R.K., 1964.
A list of Mollusca and Brachiopoda collected by N.Z.O.I. from Milford Sound.
New Zealand Dept. Sci. Ind. Res., Bull., No. 157:
New Zealand Oceanogr. Inst., Memoir, No. 17: 91-92.

molluscs, lists of spp.

Demetropoulos Andreas 1969.
Marine molluscs of Cyprus. A. Placophora, Gastropoda, Scaphopoda, Cephalopoda.
Fish. Bull. Cyprus 2: 15pp.

molluscs, lists of spp.

Indonesia Institute of Marine Science 1972.
Biological and hydrological observations in the Pitu Bay, Ambon Bay and Buton Strait
Oceanogr. Cruise Rept. 7: 27pp.

molluscs, lists of spp.

Keith, M.L., and R.H. Parker, 1965.
Local variation of ^{13}C and ^{18}O content of mollusk shells and relatively minor temperature effect in marginal marine environments.
Marine geology Elsevier Publ. Co., 3(1/2): 115-129.

molluscs, lists of spp.

Kundu, H.L., 1965.
On the marine fauna of the Gulf of Kutch. III. Pelecypods (Cont.)
J. Bombay nat. hist. Soc., 62(2):211-236.

molluscs, lists of spp.

Okutani, Takashi, 1963
Preliminary notes on Molluscan assemblages of the submarine banks around the Izu Islands.
Pacific Science, 17(1):73-89.

molluscs, lists of spp.

O'Riordon, C.E. 1972.
Some noteworthy Crustacea and Mollusca from the Dingle Bay area
Ir. Nat. J. 17(8): 252-255

molluscs, lists of spp.

Robertson, Robert, 1969.
On some molluscs collected from southwest Ceylon during the international Indian Ocean Expedition 1964. Spolia zeylanica, 31(2): 413-420.

molluscs, lists of spp.

Vilela, H. 1970.
Aperçu général sur les crustacés et mollusques.
Rapp. P.-v. Réun. Cons. int. Explor. Mer 159: 119-125.

molluscs, nomenclature

Clench, W.J., and R.D. Turner, 1962.
New names introduced by H.A. Pilsbry in the Mollusca and Crustacea.
Acad. Nat. Sci., Philadelphia, Spec. Publ., 4: 218 pp.
Reviewed: Smith, R., 1963. Veliger, 5(3):124.

molluscs, phylogeny

Ghiselin, Michael T., Egon T. Degens and Derek W. Spencer, 1967.
A phylogenetic survey of molluscan shell matrix proteins.
Breviora, No. 262:35 pp.

mollusc productivity

Zaika, V.E. 1970.
Productivity of marine molluscs as dependent on their lifetime. (In Russian; English abstract).
Okeanologiia 10(4): 702-708.

molluscs, ultra-abyssal

Filatova, Z.A., 1971.
On some mass species of bivalve molluscs from the ultra-abyssal zone of the Kurile-Kamchatka Trench. (In Russian; English abstract). Trudy Inst. Okeanol. P.P. Shirshov 92: 46-60.

Molluscs, ultra-abyssal

Lus, V.J., 1971.
A new genus and species of gastropod molluscs (family Buccinidae) from the ultra-abyssal zone of the Kurile-Kamchatka Trench. (In Russian; English abstract). Trudy Inst. Okeanol. P.P. Shirshov 92: 61-72.

Montacuta tenella

Ockelmann, Kurt W., 1965.
Redescription, distribution, biology and dimorphous sperm of Montacuta tenella Lovén (Mollusca, Leptonacea).
Ophelia, 2(1):211-221.

Neopilina

Tebble, Norman, 1967.
A Neopilina from the Gulf of Aden.
Nature, Lond. 215 (5101): 663-664

Neopilina (Neopilina) veleronis

Menzies, Robert James, and William Leyton, Jr., 1962.
A new species of monoplacophoran mollusc Neopilina (Neopilina) veleronis from the slope of the Cedros Trench, Mexico.
Ann. Mag. Nat. Hist., (13), 5(55):401-406.

MOLLUSCS
Amphineurans
chitons

molluscs, chitons

Leloup, Eugène 1968.
44. Acanthochitons de la côte atlantique Africaine.
Mém. Junta Invest. Ultram. (2ª) 54: 55-84.

molluscs

Leloup, Eugène, 1968.
Chitons le la côte africaine occidentale.
Atlantide Rept., 10:7-31.

molluscs

Leloup, E., 1960.
Amphineures du Golfe d'Aqaba et de la Peninsule Sinaï. Contribution to the knowledge of the Red Sea.
Sea Fish. Res. Sta., Haifa, Bull., No. 29:29-55.

Placellozona laqueata
Acanthochiton curvisetosus n.sp.
Anthochiton penicillatus
Cryptoplax sykesi
Ischnochiton yerburyi   Onithochiton lyelli
Chiton corallinus
Chiton olicaceus
Chiton platei
Acanthopleura haddoni
Tonicia perligera
Tonicia suezensis

Mopalia mucosa

Boolootian, Richard A., 1964.
On growth, feeding and reproduction in the chiton Mopalia mucosa of Santa Monica Bay.
Helgoländer Wiss. Meeresuntersuchungen, 11(3/4): 186-199.

MOLLUSCS
Bivalves = lamellibranchs, pelecypods etc.
i.e., clams, oysters, etc.

oysters

Azouz, Abderrazak, 1966.
Étude des peuplements et des possibilités d'ostreiculture du Lac de Bizerte.
Inst.nat.Sci.tech.Océanogr.Pêche,Salammbo,Ann. 15: 69 pp.

molluscs

Barnard, K.H., 1964
Contributions to the knowledge of South African marine Mollusca. V. Lamellibranchiata
Ann. South African Mus., 47(3):361-593.

bivalve mollusks

Coe, W. R., 1948
Nutrition, environmental conditions, and growth of marine bivalve mollusks. J. Mar. Res. 7(3):586-601, 2 text figs.

molluscs, bivalve

Comps, Michel 1970.
Observations sur les causes d'une mortalité anormale des huîtres plates dans le bassin de Marennes.
Rev. Trav. Inst. Pêches marit. 34(3): 317-326.

molluscs

Filatova, Z.A., 1957.
[Some new molluscs of family Astartidae (Bivalvia) from Far Eastern Seas.] Trudy Inst. Okeanol., 23: 297-302.

molluscs

Klappenbach, Miguel A., 1965.
Lista preliminar e los Mytilidae brasileiros con claves para su determinacion y notas sobre su distribucion.
Anais Acad. bras., Cienc., 37(Supl.):327-352/

oysters

Korringa, P., 1952.
Recent advances in oyster biology. Q. Rev. Biol. 27:266-308, 339-365.

molluscs

Lassig, Julius, 1965.
The distribution of marine and brackishwater lamellibranchs in the northern Baltic area.
Commentationes Biol., Soc. Sci. Fennica, 28(5): 1-41.

molluscs

Lever, J., A. Kessler, A.P. van Overbeeke and R. Thijssen, 1961.
Quantitative beach research. II. The "hole effect": a second mode of sorting of lamellibranch valves on sandy beaches.
Netherlands J. Sea Res., 1(3):339-358.

molluscs, bivalves

Longwell, A. Crosby, and S. S. Stiles 1970.
The genetic system and breeding potential of the commercial American oyster.
Endeavour 29 (107): 94-99.

molluscs

Menzel, R.W., 1961 (1963)
Seasonal growth of the northern quahog Mercenaria mercenaria and the southern quahog, M. capechiensis in Alligator Harbor, Florida.
Proc. Nat. Shellfish Assoc., 52:37-46.

molluscs, bivalves

Millar, R. H. 1968.
Changes in the population of oysters in Loch Ryan between 1957 and 1967.
Mar. Res. Dept. Agric. Fish. Scotland (1): 8 pp

molluscs

Nevesskaia, L.A., 1962.
[Some peculiar features in the development of bivalve molluscs of the Black Sea during Late Quaternary.]
Doklady Akad. Nauk, SSSR, 143(5):1170-1172.

molluscs

Nicol, David, 1965.
A new Thyasira (Pelecypoda) from the Ross Sea, Antarctica.
The Nautilus, 78(3):79-80.

molluscs, pelecypods

Nicol, David, 1963.
Lack of shell-attached pelecypods in Arctic and Antarctic waters.
Nautilus, 77(3):92-93.

oysters

Numachi, Ken-Ichi, 1962.
Serological studies of species and races in oysters.
American Naturalist, 96(889):211-217.
(Paper presented at 10th Pacific Science Congress)

oysters

Numachi, Ken-ichi, Juichi Oizumi, Shigeru Sato, and Takeo Imai, 1965.
Studies on the mass mortality of the oyster in Matsushima Bay. III. The pathological changes of the oyster caused by gram-positive bacteria and the frequency of their injection. (In Japanese:English abstract).
Bull. Tohoku Reg. Fish Res. Lab.,No. 25:39-44.

molluscs (deep)

Ōkutani, T., 1962
Report on the archibenthal and abyssal lamellibranchiate Mollusca mainly collected from Sagami Bay and adjacent waters by the R.V. Soyo-Maru during the years 1955-1960. (Japanese abstract).
Bull. Tokai Reg. Fish. Res. Lab., Tokyo, No. 32: 1-40.

oysters

Padilla, Miguel, y Julio Orrego, 1967.
La fijación larval de ostras sobre colectores experimentales en Quetalmahue, 1966-67.
Bol. cient., Inst. Fomento Pesquero,Chile, 5: 15 pp.

molluscs

Pojeta, John, Jr., Bruce Runnegar, Noel J. Morris and Norman D. Newell, 1972.
Rostroconchia: a new class of bivalved mollusks.
Science 177(4045): 264-267.

clams

Russell, H.D., 1955.
A new clam industry in New England. Nautilus 69(2):53-56.

oysters

Shaw, W.N., 1961(1963).
Index of condition and per cent solids of raft-grown oysters in Massachusetts.
Proc. Nat. Shellfish Assoc., 52:47-52.

molluscs (oysters)

Shaw, William N., and James A. McCann, 1963.
Comparison of growth of four strains of oysters raised in Taylors pond, Chatham, Massachusetts.
U.S.Fish and Wildlife Service, Fish. Bull., 63(1):11-18.

molluscs

Squires, H.J., 1962.
Giant scallops in Newfoundland coastal waters.
Fish. Res. Bd., Canada, Bull., No. 135:29 pp.

molluscs

Stenzel, H.B., 1963.
Aragonite and calcite as constituents of adult oyster shells.
Science, 142(3589):232-233.

pearl oysters

Wada, S., 1960
Occurrences of maxima pearl oyster in the Oshima Strait, Amami-Oshima, a northern limit of the distribution.
Mem. Fac. Fish., Kagoshima Univ., 9: 79-85.

oysters

Wada, S.K., 1953.
Larviparous oysters from the tropical west Pacific. Rec. Ocean. Wks., Japan, n.s., 1(2): 66-72.

MOLLUSCS
Bivalves, anatomy and/or physiology

molluscs, physiology
Coe, W. R., 1948
Nutrition, environmental conditions, and growth of marine bivalve mollusks. J. Mar. Res. 7(3):586-601, 2 text figs.

oysters
Collier, A., S. Ray, and W. Magnitzky, 1950.
A preliminary note on naturally occurring organic substances in sea water affecting the feeding of oysters. Science 111:151-152, 1 textfig.

oysters
Collignont, J., 1967.
La croissance des huitres dans les lagunes marocains.
Bull. Inst. Pêches marit., Maroc, 15:49-57.

molluscs, anat. physiol.
Crisp, D.J.
Chemical factors inducing settlement in Crassostrea virginica (Gmelin).
J. anim. Ecol. 36(2):329-335.

molluscs, anatomy
Bunachie, J.F., 1963.
The periostracum of "Mytilus edulis".
Trans. R. Soc., Edinburgh, 65(15):382-411.

molluscs, anat.-physiol.
Kuenzler, E.J., 1961.
Phosphorus budget of a mussel population.
Limnol. & Oceanogr., 6(4):400-415.

oysters
Mackin, J.G., & S. H. Hopkins. 1961
Studies on oysters in relation to the oil industry.
Publ. Inst. Mar. Sci., 7: 1-314.

mollusca, anat.-physiol.
Merrill, Arthus S., and Ruth D. Turner, 1963.
Nest building in the bivalve mollusk genera Musculus and Lima.
Veliger, 6(2):55-59.

oysters, phys.
Pedersen, E., 1947.
Østersens respirasjon. Undersøkelser utfort ved Statens Utklekninsanstalt Flødevigen. Rept. Norwegian Fish. & Mar. Invest. 8(10):51 pp., 4 figs.

Molluscs
Read, Kenneth R.H., 1964.
Ecology and environment physiology of some Puerto Rican bivalve molluscs and a comparison with boreal forms.
Caribbean J. Sci., 4(4):459-465.

molluscs, physiology
Read, Kenneth R.H., 1963.
Thermal inactivation of preparations of aspartic/glutamic transaminase from species of bivalved molluscs from the sublittoral and intertidal zones.
Comp. Biochem. Physiol., 9(3):161-180.

molluscs, anat.-physiol.
Reid, R.G.B., 1965.
The structure and function of the stomach in bivalve molluscs.
J. Zool., 147(2):156-184.

molluscs, anat. physiol.
Reshöft, K., 1961.
Untersuchungen zur zellulären osmotischen und thermischen Resistenz verschiedener Lamellibranchier der deutschen Kustengewässer.
Kiel. Meeresf., 17(1):65-84.

oysters, anat.-physiol.
Shinkawa, H., 1961.
The relation between the ciliary activity of some species of Japanese oysters and the concentration of sea water.
Sci. Repts., Tohoku Univ. (4-Biol.) 27(1):47-56.

molluscs, physiol.
van Dam, L., 1954.
On the respiration in scallops (Lamellibranchiata). Biol. Bull. 107(2):192-202, 7 textfigs.

oysters, anat-physiol.
Wells, H.W., 1961
The fauna of oyster beds with special reference to the salinity factor.
Ecolog. Mon., 31(3):239-266.

mussels
Wilson, B.R., and E.P. Hodgkin, 1967.
A comparative account of the reproductive cycles of five species of marine mussels (Bivalvia: Mytilidae) in the vicinity of Fremantle, western Australia.
Aust. J. mar. Freshwat. Res., 18(2):175-203.

mollusc hybrids
Woodburn, K.D., 1961(1963).
Survival and growth of laboratory-reared northern clams (Mercenaria mercenaria) and hybrids (M. mercenaria x M. campechiensis) in Florida waters.
Proc. Nat. Shellfish Assoc., 52:31-36.

oysters
Yonge, C.M., 1960
Oysters. Collins, St. Jame's Place, London, 209 pp.

molluscs - physiology
Yonge, C.M., 1946
Digestion of animals by lamellibranchs
Nature 157:729

MOLLUSCS
Bivalves, chemistry of
See also: Subject Cumulation, under chemistry

Mollusca, chemistry of
Antunes, S.A. and Yasugo Ita 1968.
Chemical composition of oysters from São Paulo and Paraná Brazil.
Bolm Inst. oceanogr. S. Paulo 17(1):71-88.

oysters, chemical composition
Bordovskiy, O.K., 1965.
Accumulation and transformation of organic substance in marine sediments. 1. Summary and introduction. 2. Sources of organic matter in marine basins. 3. Accumulation of organic matter in bottom sediments. 4. Transformation of organic matter in marine sediments.
Marine Geology, Elsevier Publ. Co., 3(½):3-4; 5-31; 33-82;83-114.

molluscs, chemistry of
Hiltz, Doris Fraser 1970.
Occurrence of trigonelline (N-methyl nicotinic acid) in the adductor muscle of a lamellibranch, the sea scallop (Placopecten magellanicus).
J. Fish. Res. Bd Can. 27(3):604-606.

molluscs, chemistry of
Marinković-Roje, Marija 1968.
Les variations saisonières de la composition chimique des moules (Mytilus galloprovincialis Lmk) de Canal de Lim.
Thalassia Jugoslavica 4:69-85

MOLLUSCS
Bivalves, culture and cultivation
See also: Subject Cumulation under cultures, molluscs farming, shellfish

molluscs, bivalves
Crossland, C., 1957. (posthumous)
The cultivation of the mother-of-pearl oyster in the Red Sea. Australian J. Mar. Freshwater Res. 8(2):111-130.

oyster culture
Shaw, W.N., 1960.
A fiberglass raft for growing oysters off the bottom.
Progressive Fish-Cult., 22(4):154.

molluscs, deep-sea
Sanders, H.L., and J.A. Allen 1973.
Studies on deep-sea Protobranchia (Bivalvia); prologue and the Pristiglomidae.
Bull. Mus. Comp. Zool. 145(5):237-262.

molluscs, ecology
Coe, W. R., 1948
Nutrition, environmental conditions, and growth of marine bivalve mollusks. J. Mar. Res. 7(3):586-601, 2 text figs.

molluscs, ecology of
Oertzen, J.-A. von, 1972.
Cycles and rates of reproduction of six Baltic Sea bivalves of different zoogeographical origin. Mar. Biol. 14(2):143-149.

molluscs, lists of spp.
Keen, A.M., 1962.
A new west Mexican subgenus and a new species of Montacutidae (Mollusca: Pelecypoda) with a list of Mollusca from Bahia de San Quintin.
Pacific Naturalist, 3(9):321-328.

molluscs, lists of spp.
Pasteur-Humbert, C., 1962.
Les mollusques marins testacés du Maroc. Catalogue non critique. II. Les lamellibranches et les scaphopodes.
Trav. Inst. Sci., Chérifien, Zool., 28:180 pp.

MOLLUSCS
Bivalves, mass mortality of
See also: Subject Cumulation under mass mortality

bivalves, mass mortality
Kon-no, Hisashi, Minoru Saseki, Yasuo Sakurai, Tsuyoshi Watenabe and Kenzo Suzuki, 1965.
Studies on the mass mortality of the oyster in Matsushima Bay. I. General aspects of the mass mortality of the oyster in Matsushima Bay and its environmental conditions. (In Japanese; English abstract).
Bull. Tohoku Reg. Fish Res. Lab. No. 25:1-26.

bivalves, mass mortality
Imai, Takeo, Ken-ichi Numachi, Juichi Oizumi and Shigeru Sato, 1965.
Studies on the mass mortality of the oyster in Matsushima Bay. II. Search for the cause of mass mortality and the possibility to prevent it by transplantation experiment. (In Japanese; English abstract).
Bull. Tohoku Reg. Fish. Res. Lab., No. 25:27-38.

oysters
Shinkawa, H., 1961.
Studies on the vertical distribution of Japanese oysters.
Sci. Rept. Tohoku Univ., (4, Biol.) 27(1):19-30.

Aequipecten irradians

Cooper, Richard A., and Nelson Marshall, 1963.
Condition of the bay scallop, *Aequipecten irradians*, in relation to age and the environment.
Chesapeake Science, 4(3):126-134.

Aequipecten irradians

Davis, R.L., and N. Marshall, 1963.
The feeding of the bay scallop *Aequipecten irradians*.
Proc. Nat. Shellfish. Assoc., 1961, 52:25-29.

Aequipecten irradians

Sastry, A.N., 1966.
Temperature effects in reproduction of the bay scallop, *Aequipecten irradians* Lamarck.
Biol. Bull., 130(1):118-134.

Aequipecten irradians

Sastry, Akella N. and Norman J. Black, 1971.
Regulation of gonad development in the bay scallop, *Aequipecten irradians* Lamarck.
Biol. Bull. mar. biol. Lab. Woods Hole. 140(2): 274-283.

Aloidis (Corbula) Gibba

Yonge, C.M., 1946
On the habits and adaptations of *Aloidis (Corbula) Gibba*. JMBA XXVI: 358-376, 14 textfigs.

Crassostrea gigas

Bae, Gyung Man and Pyung Arm Bae, 1972
Study on spat collection of oyster. (In Korean; English abstract). Bull. Fish. Res. Dev. Agency, Korea 9: 47-54.
Crassostrea gigas (Thunberg)

Crassostrea gigas

Imai, T. & S. Sakai, 1961
Study of breeding of Japanese oyster, *Crassostrea gigas*.
Tohoku J. Agric. Res., 12(2): 124-172.

Crassostrea gigas

Sparks, Albert K., Gilbert B. Pauley, Richard R. Bates and Clyde S. Sayce, 1964.
A tumor like fecal impaction in a Pacific oyster, *Crassostrea gigas* (Thunberg).
J. Insect Pathol., 6(4):453-456.

Crassostrea gigas

Sparks, Albert K., Gilbert B. Pauley, Richard R. Bates and Clyde S. Sayce, 1964.
A mesenchymal tumor in a Pacific oyster, *Crassostrea gigas* (Thunberg).
J. Insect Pathol., 6:448-452.

Crassostrea rhizophorae

Pora, E.A., C. Wittenberger, G. Suárez and N. Portilla, 1969.
The resistance of *Crassostrea rhizophorae* to starvation and asphyxia. Marine Biol., 3(1): 18-23.

Crassostrea rhizophorae

Saenz, Braulio A., 1965.
El ostion antillano: *Crassostrea rhizopnorae* Guilding y su cultivo experimental en Cuba.
Centro Invest. Pesq., Cuba, Nota sobre Invest. No. 6:34 pp.

Crassostrea virginica

Dame, R.F., 1972
The ecological energies of growth, respiration and assimilation in the intertidal American oyster *Crassostrea virginica*. Mar. Biol. 17(3) 243-250.

Crassostrea virginica

Galtsoff, Paul S., 1964.
The American oyster, *Crassostrea virginica* Gmelin.
Fishery Bull., 64:480 pp.

Crossostrea virginica

Loosanoff, Victor L., 1966.
Time and intensity of setting of the oyster, *Crassostrea virginica*, in Long Island Sound.
Biol. Bull., 130(2):211-227.

Donax

Wobber, Frank J., 1967.
The orientation of *Donax* on an Atlantic coast beach.
J. sedim. Petrol. 37(4):1233-1251.

Donax denticulatus

Wade, Barry A., 1968.
Studies on the biology of the West Indian beach clam, *Donax denticulatus* Linne. 2. Life history. Bull. mar. Res. 18(4): 876-901.

Echinocardium cordatum

Moore, H.B., 1936
The biology of *Echinocardium crodatum*.
JMBA 20(3):655-572, 5 textfigs., Pls.I,II.

molluscs, Macoma baltica
Segerstrale, Sven G., 1965.
Biotic factors affecting the vertical distribution and abundance of the bivalve, *Macoma baltica* (L.), in the Baltic Sea.
Botanica Gothoburgensia, Proc. Fifth Mar. Biol. Symp., 3:195-204.

Mercenaria mercenaria

Ansell, A.D., 1968.
The rate of growth of the hard clam *Mercenaria mercenaria* (L) throughout the geographic range.
J. Cons. perm. int. Explor. Mer, 31(3):364-409.

Mercenaria mercenaria

Hamwi, Adel, and Harold H. Haskin, 1969.
Oxygen comsumption and pumping rates in the hard clam *Mercenaria mercenaria*: a direct method.
Science, 163(3869):823-824.

Mya arenaria

Gomoiu, M.T. and I.I. Porumb, 1969.
Mya arenaria L., a bivalve recently penetrated into the Black Sea.
Revue Roumaine Biol. (Zool.) 14(3): 199-202.

Mytilus californianus

Moon, Thomas W. and A.W. Pritchard, 1970.
Metabolic adaptations in vertically-separated populations of *Mytilus californianus* Conrad.
J. exp. mar. Biol. Ecol., 5(1): 35-46.

Mytilus edulis

Bayne, B.L., 1965.
Growth and the delay of metamorphosis of the larvae of *Mytilus edulis* (L.).
Ophelia, 2(1):1-47.

Mytilus edulis

Boëtius, I., 1962.
Temperature and growth in a population of *Mytilus edulis* (L.) from the northern harbour of Copenhagen (The Sound).
Medd. Dansk. Fisk. Havundersøgelser, n.s., 3 (7-12):339-346.

Mytilus edulis

Drzycimski, Idzi, 1963.
Nutrition, chemical composition and possibilities for utilization of the sea mussel, *Mytilus edulis* L., from the Southern Baltic.
Ann. Biol., Cons. Perm. Int. Expl. Mer, 1961, 18: 209-210.

Mytilus edulis

Fish, C.J., and M.W. Johnson, 1937
The biology of the zooplankton population in the Bay of Fundy and the Gulf of Maine with special reference to production and distribution. J. Biol. Bd., Canada 3(3):189-322, 29 tables, 45 text figs.

Mytilus edulis L.

Newcombe, C.L., 1935
A study of the community relationships of the sea mussel, *Mytilus edulis* L.
Ecology 16:234-243.

Mytilus edulis

Scattergood, L.W., and C.C. Taylor, 1950.
The mussel resources of the North Atlantic region. Pt. 1. The survey to discover the locations and areas of the North Atlantic mussel-producing beds. Pt. 2. Observations on the biology and the methods of collecting and processing the mussel. Pt. 3. Development of the fishery and the possible need for conservation measures. Fishery Leaflet 364:33 pp., 7 figs. (multilith).

Mytilus edulis

Theede, Hans, 1963
Experimentelle Untersuchungen uber die Filtrationsleistung der Miesmuschel *Mytilus edulis* L.
Kieler Meeresf., 19(1):20-41.

German and English abstract

Mytilus edulis

Thompson, R.J. and B.L. Bayne, 1972.
Active metabolism associated with feeding in the mussel *Mytilus edulis* L. J. exp. mar. Biol. Ecol. 9(1): 111-124.

Mytilus edulis (larvae)

Wiborg, K.F., 1944
The production of zooplankton in a landlocked fjord, the Nordåsvatn near Bergen, in 1941-42, with special reference to the copepods... (Repts. Norwegian Fish. and Mar. Invest.) 7(7):83 pp., 40 text figs.

Mytilus galloprovincialis

Genovese, S., 1961.
Analisi biometrica di una popolazione di *Mytilus galloprovincialis* Lamarck (Moll. Lam.) vivente nella Lagune Veneta.
Rapp. Proc. Verb., Réunions, Comm. Int. Expl. Sci., Mer Méditerranée, Monaco, 16(3):799-809.

Mytilus galloprovincialis

Hrs-Brenko, Mirjana 1968.
Biometrical analyses of the mussel (*Mytilus galloprovincialis* LMK) along the eastern coast of the Adriatic.
Thalassia Jugoslavica 4: 19-30.

Mytilus galloprovincialis

Renzoni, A., and C. Sacchi, 1961.
Notes sur l'écologie de la moule (*Mytilus galloprovincialis* Lam.) dans le lac Fusaro (Naples).
Rapp. Proc. Verb., Réunions, Comm. Int. Expl. Sci., Mer Méditerranée, Monaco, 16(3):811-814.

Lima (Plicacesta) sphoni n.sp.

Hertlein, L.G., 1963.
A new species of giant *Lima* from off Southern California (Mollusca: Pelecypoda).
Occ. Papers California Assoc. Sci., 40:6 pp.

Ostrea

Ranson, Gilbert, 1967.
les espèces d'huitres vivant actuellement dans le monde définies par leurs coquilles larvaires ou prodissoconques: études des collections de quelques-uns des grands musées d'histoire naturelle.
Revue Trav. Inst. (Scient. Tech.) Pêches marit., 31 (3):205-274.

Ostrea edulis

Bayne, B.L., 1969.
The gregarious behaviour of the larvae of *Ostrea edulis* L. at settlement. J. mar. biol. Ass. U.K., 49(2): 327-356.

Ostrea edulis

Davis, Harry C., and Alan D. Ansell, 1962.
Survival and growth of larvae of the European oyster, *O. edulis*, at lowered salinities.
Biol. Bull., 122(1):33-39.

Ostrea edulis

Davis, Harry C. and Anthony Calabrese, 1969.
Survival and growth of larvae of the European oyster (*Ostrea edulis* L.) at different temperatures. Biol. Bull., mar. biol. Lab., Woods Hole, 136(2): 193-199.

Ostrea edulis

Loosanoff, V. L., 1962.
Gametogenesis and spawning of the European oyster, O. edulis, in waters of Maine.
Biol. Bull., 122(1):86-94.

Ostrea edulis

Sheldon, R.W., 1968.
The effect of high population density on the growth and mortality of oysters (Ostrea edulis).
J. Cons. perm. int. Explor. Mer, 31(3):352-363.

Ostrea gigas

Saito, T., 1931
Researches in fouling organisms of the Ships' bottom. Zosen Kiokai (T. Soc. Naval Arch.) 47pp:13-64, 51 figs., 9 graphs.

Ostrea plicatura

Saito, T., 1931
Researches in fouling organisms of the Ships' bottom. Zosen Kiokai (T. Soc. Naval Arch.) 47pp:13-64, 51 figs., 9 graphs.

Ostrea taurica

Goromosova, S.A., 1968.
Seasonal changes of chemical composition of the Black Sea Ostrea taurica. (In Russian).
Gidrobiol. Zh., 4(3): 72-76.

Ostrea taurica

Ivanova, A.I., 1966.
Studies of oyster (Ostrea taurica Kryn.) growth in the Black Sea. (In Russian; English abstract).
Okeanologiia, Akad. Nauk, SSSR, (5):869-876.

Placopecten magellanicus

Dickie, L.M., and J.C. Medcof, 1963.
Causes of mass mortalities of scallops (Placopecten magellanicus) in the southwestern Gulf of St. Lawrence.
J. Fish. Res. Bd., Canada, 20(2):451-482.

Spinula (Bathyspinula) viyyazi n.sp.

Filatova, Z.A., 1964.
A new species of bivalve mollusc from the ultra-abyssal of the Pacific. (In Russian English abstract).
Zool. Zhurn., Akad. Nauk, SSSR, 43(12):1866-1868.

Spisula polynyma

Chamberlin, J.L., and F. Stearns, 1963.
A geographic study of the clam, Spisula polynyma (Stimpson). Serial Atlas, Mar. Environ., Amer. Geogr. Soc., 3:12 pp., 6 pls. (quarto).

Tellina martinicensis

Penzias, Leslie P., 1969.
Tellina martinicensis (Mollusca: Bivalvia): biology and productivity. Bull. mar. Sci., 19(3): 568-579.

molluscs, Tridacna [a]

Valentine, James W., Dennis Hedgecock, Gary S. Zumwalt and Francisco J. Ayala 1973.
Mass extinctions and genetic polymorphism in the "killer clam" Tridacna.
Bull. Geol. Soc. Am. 84(10): 3411-3414.

Tridacna [a]

Wells, J.M., A.H. Wells and J.G. VanDerwalker 1973.
In situ studies of metabolism in benthic reef communities. Helgoländer wiss. Meeresunters 24(1/4): 78-81.

Tridacna crocea

Jeffrey, S.W., and Kazuo Shibata, 1969.
Some spectral characteristics of chlorophyll from Tridacna crocea zoozanthellae.
Biol. Bull., mar. biol. Lab., Woods Hole, 136(1):54-62.

Tridacna gigas

Bonham, Kilshaw, 1965.
Growth rate of giant clam Tridacna gigas at Bikini Atoll as revealed by radioautography.
Science, 149(3681):300-302.

Turtonia minuta

Ockelmann, Kurt W., 1964
Turtonia minuta (Fabricius), a neotenous veneracean bivalve.
Ophelia, Mar. Biol. Lab., Helsingør, 1(1): 121-146.

Venus mercenaria

Ansell, Alan D., 1964.
Some parameters of growth of mature Venus mercenaria L.
J. du Cons., 29(2):214-220.

Yoldia amygdalea [a]

Lande, Eirik 1971.
A new southern record of Yoldia amygdalea Valenciennes (Mollusca Pelecypoda) in Pölen, North Tröndelag.
Skr. K. norske Vidensk. Selsk. (16):1-4.

MOLLUSCS Cephalopods

molluscs, cephalopod [a]

Abbes, René 1970.
Remarques sur quelques céphalopodes mésopélagiques du Golfe de Gascogne.
Rev. Trav. Inst. Pêches marit. 34(2):195-20

cephalopods

Adam, William, 1967.
Cephalopda from the Mediterranean Sea.
Bull. Sea Fish. Res. Stn. Israel, 45:65-78.

squid, giant

Aldrich, Frederick A., 1968.
The distribution of giant squids (Cephalopoda, Architeuthidae) in the North Atlantic and particularly about the shores of Newfoundland.
Sarsia, 34:393-398.

squid

Baker, A. de C., 1960.
Observations of squid at the surface in the NE Atlantic.
Deep-Sea Res., 6:206-210.

squid

Boycott, B.B., 1965.
A comparison of living Sepia Sepioidea and Doryteuthis plei with other squids, and with Sepia officinalis.
J. Zool. 147(3):344-351.

cephalopoda (octopus)

Boycott, Brian B., 1965.
Learning in the octopus.
Scientific American, 212(3):42-50.

cephalopods

Clarke, Malcolm R., 1969.
Cephalopoda collected on the SOND cruise.
J. mar. biol. Ass. U.K., 49(4): 961-976.

cephalopods

de Castellanos, Zulma, J.A. y Roberto C. Menni 1969
Sobre dos pulpos costeros de la Argentina.
Neotropica 15(47):89-94.

molluscs, cephalopods

Dhamniyom, Dinex, and Surapol Vadhanakul 1970.
Results of the trawling survey in the inner Gulf of Thailand carried out by R.V. Asa from June 1968-May 1969. (In Thai; English abstract)
Contrib. Mar. Fish. Lab. Bangkok, 17:66 pp.

molluscs, cephalopods [a]

Fields, W. Gordon, and Veronica A. Gauley 1971.
Preliminary description of an unusual gonatid squid (Cephalopoda: Oegopsida) from the North Pacific.
J. Fish. Res. Bd Can. 28(11):1796-1801

squid

Filippova, J.A., 1969.
Squids of the South Atlantic. (In Russian; English abstract).
Zool. Zh., 48(1): 51-63.

cephalopods

Gemulin-Bride, Helena, et Vesna Ilijanic, 1965.
Note sur quelques espèces de céphalopodes rares en Adriatique, déposées dans le Musée de Zoologie de Zagreb.
Rapp. Proc. Verb. Réunions, Comm. Int. Expl. Sci., Mer Méditerranée, Monaco, 18(2):207-210.

squid

Haefner, Paul A., Jr., 1964
Morphometry of the common Atlantic squid, Loligo pealei, and the brief squid, Lolliguncula brevis, in Chesapeake Bay.

squid

Hamabe, M., and T. Shimizu, 1959
Littoral aggregation of the squid at the Oki Islands. II. Ann. Rept., Japan Sea Res. Fish. Res. Lab., (5): 19-28.

molluscs, cephalopods

Inoue, Hiroo, Yoshiaki Tanaka and Kiyoshi Fukuda 1970.
On water exchange in a shallow marine fishfarm. II. Hamachi fishfarm at Tanoura. (In Japanese; English abstract)
Bull. Jap. Soc. scient. Fish. 36(8):776-782.

squid

Katoh, Gendi, 1964
A few comments on the biological grouping of the common squid derived from its ecological aspect. (In Japanese; English abstract).
Bull. Jap. Sea Reg. Fish. Res. Lab., (13): 31-42.

squid, giant

Kjennerud, J., 1959.
Description of a giant squid, Architeuthis stranded on the west coast of Norway.
Univ. Bergen Årbok, 1958, Natur Rekke, (9):14 pp.

cephalopods

Legac, Mirjana, 1964.
Contribution a la connaissance des cephalopodes de la region des canaux de l'Adriatique septentrionale. (In Jugoslavian; French resume).
Acta Adriatica, 11(1):181-188.

molluscs, cephalopods [a]

Lumare Febo 1970:
Nota sulla distribuzione di alcuni cefalopodi del mar Tirreno.
Boll. Pesca Piscic. Idrobiol. 25(2):313-344.

cephalopods
Lumare, Febo, 1968.
Ricerche sui cefalopodi dell'alto Tirreno.
Studio Oceanograf. Limnol. Programma
Ricerca Risorse mar. Fondo mar., (B)(21):29 pp.
Comm.

squid
Machinaka, Shigeru, 1959
Migration of the squid in the Japan Sea as
studies by tagging experiments.
Bull. Jap. Sea Reg. Fish. Res. Sta., (7): 57-66.

cephalopods
Mangold-Wirz, K., 1963.
Dimensions et croissance relatives de quelques
Ommastrephidés méditerranéens.
Rapp. Proc. Verb., Réunions, Comm. Int. Expl.
Sci., Mer Méditerranée, Monaco, 17(2):401-405.

squid
Muus, B.J., 1956.
Development and distribution of a North Atlantic
pelagic squid, Family Cranchidae.
Medd. Danmarks Fisk. Havundersøgelser, n.s.,
1(15):15 pp.

molluscs, cephalopods
Nesis, K.N. 1972.
Two new species of gonatid squids
from the North Pacific. (In Russian;
English abstract).
Zool. Zh. 51(9):1300-1307

molluscs, cephalopods
Nesis, K.N. 1972.
A review of the squid genera Taonius
and Belonella (Oegopsida, Cranchiidae).
Zool. Zh. 51(3): 341-350

cephalopods
Park, Joo Suck, and Joo Youl Lim, 1967.
On the results of the tagging experiment on
squids in the Korean waters. (In Korean;English
abstract).
Repts.Fish.Resources,7:29-40.

cephalopods
Pearcy, William G., 1963.
Distribution of oceanic cephalopods off Oregon,
U.S.A.
Proc. XVI Int. Congr. Zool., Washington, D.C.,
Aug. 20-27, 1963, 1:69.

cephalopods
Pearcy, William G. and Alan Beal, 1973
Deep-sea cirromorphs (Cephalopoda) photographed in the Arctic Ocean. Deep-Sea Res.
20(1): 107-108.

molluscs, cephalopods (orientation)
Raup, David M. 1973.
Depth inferences from vertically
imbedded cephalopods.
Lethaia 6(3): 217-226

molluscs, cephalopod
Roper, Clyde F.E., and Walter L. Brundage jr.
1972
Cirrate octopods with associated deep-sea
organisms: new biological data based on
deep benthic photographs (Cephalopoda).
Smithson. Contrib. Zool. 121: 46 pp.

molluscs, cephalopods
Schevtsov, G.A. 1969.
Preliminary data on the specific
composition of larvae, Cephalopoda, their
distribution and the biology of
Onychoteuthis banksi Leach. (In Russian)
Izv. Tichookean. nauchno-issled. Inst.
rybn. Khoz. Okeanogr. (TINRO) 68:186-192.

cephalopods
Shimizu, Torao and Mototsugu Hamabe, 1966.
On tagging experiment on the common squid in the
waters around the Tsushima islands, Nagasaki
Prefecture. (In Japanese; English abstract)
Bull. Jap. Sea Reg. Fish. Res. Lab., (16):7-12.

octopuses
Taki, Iwao, 1961.
On two new Eledonid octopods from the Antarctic
Sea.
J. Fac. Fish. and Animal Husbandry, Hiroshima
Univ., 3(2):297-316.

cephalopods
Testa, Gilbert, 1964.
Les céphalopodes des collections du Musée
Océanographique de Monaco.
Bull. Inst. Océanogr., Monaco, 62(1298):8 pp.

molluscs, cephalopod
Tortonèse, E., et M. Demir 1965.
Rapport sur les travaux récents
(1962-64) concernant les vertébrés marins
et les céphalopodes de la Méditerranée et
de ses dépendences.
Rapp. Proc.-Verb. Réunions Comm int.
Expl. scient. Mer Médit. Monaco 18(3):
575-590.

cephalopods
Tortonèse, E., and M. Demir, 1963.
Rapport sur les travaux récents (1960-62) sur
les vertébrés marins et les céphalopodes de la
Méditerranée et de ses dépendances.
Rapp. Proc. Verb., Réunions, Comm. Int. Expl.
Sci. Mer Méditerranée, Monaco, 17(2):287-302.

molluscs, cephalopods
Voss, Gilbert L. 1971.
Biological results of the University of Miami
deep-sea expeditions. 76. Cephalopods collected
by the R/V John Elliott Pillsbury in the Gulf
of Panama in 1967. Bull. mar. Sci. 21(1): 1-34.

molluscs (octopuses)
Voss, Gilbert L., 1968.
Octopoda from the R/V Pillsbury southwestern
Caribbean cruise, 1966, with a description of a
new species, Octopus zonatus.
Bull.mar.Sci., Miami, 18(3):645-659.

cephalopods
Voss, Gilbert L., 1967.
The biology and bathymetric distribution
of deep-sea cephalopods.
Stud. trop. Oceanogr., Miami, 5:511-535.

cephalopods
Voss, Gilbert L., 1967.
Some bathypelagic cephalopods from South
African waters.
Ann. S. Afr. Mus., 50(5):61-88.

cephalopods
Voss, Gilbert L., 1967.
Systematics and distribution of Antarctic
cephalopods.
Antarct. Jl. U.S.A., 2(5):202-203.

cephalopods
Voss, G.L., 1962.
South African cephalopods.
Trans. R. Soc., South Africa, 36(4):245-272.

cephalopods
Voss, G.L., 1958.
The cephalopods collected by the R.V. Atlantis
during the West Indian cruise of 1954.
Bull. Mar. Sci., Gulf & Caribbean, 8(4):369-389.

cephalopods
Voss, Nancy A., 1969.
Biological investigations of the deep sea. 47.
A monograph of the Cephalopoda of the North
Atlantic. The family Histioteuthidae. Bull.
mar. Sci., 19(4): 713-867.

molluscs, cephalopods
Young, Richard Edward 1973.
Information feedback from
photophores and ventral counter-
shading in mid-water squid.
Pacific Sci. 27(1):1-7.

molluscs, cephalopods
Young, Richard Edward 1972.
The systematics and areal distribution
of pelagic cephalopods from the seas off
Southern California.
Smithson. Contrib. Zool. 97: 159 pp.

cephalopods
Young, Richard E., and Clyde F.E. Roper
1969.
A monograph of the Cephalopoda of the
North Atlantic: the family Joubiniteuthidae.
Smithson. Contrib. Zool. 15: 1-10.

cephalopods
Young, Richard E., and Clyde F. Roper, 1969.
A monograph of the Cephalopoda of the
North Atlantic: the family Cycloteuthidae.
Smithson. Contrib. Zool. 15, 24 pp.

cephalopods
Zuev, G.V., 1965.
Main features and adaptive significance of the
shell evolution in cephalopods. (In Russian;
English abstract).
Zool. Zhurn., Akad. Nauk, SSSR, 44(2):284-286.

MOLLUSCS
Cephalopods, anatomy and/or physiology

cephalopods, anat.physiol.
Alexandrowicz, J.S., 1965.
The neurosecretory system of the vena cava in
Cephalopoda. II. Sepia officinalis and Octopus
vulgaris.
Jour. Mar. Biol. Assoc., U.K., 45(1):209-228.

cephalopods, anat.phys.
Alexandrowicz, J.S., 1964.
The neurosecretory system of the vena cava in
Cephalopoda. 1. Eledone cirrosa.
J. Mar. Biol. Assoc., U.K., 44(1):111-132.

cephalopods, anat.-physiol.
Arnold, John M., 1965.
Observations on the mating behavior of the
squid, Sepioteuthis sepioidea.
Bull. Mar. Sci., 15(1):216-222.

cephalopods
Bidder, A.M., 1962.
Use of the tentacles, swimming and buoyancy
control in the pearly nautilus.
Nature, 196(4853):451-454.

cephalopods, anat.-physical.
Clarke, M.R., 1965.
Large light organs on the dorsal surfaces of the squids
Ommastrephes pteropus, 'Symplectoteuthis oualaniensis'
and 'Dosidicus gigas'.
Proc. Malac. Soc., Lond., 36:319-321.
Also in:
Collected Reprints, Nat. Inst. Oceanogr., Vol. 13

cephalopods, anat. physiol.
Clarke, M.R., 1962.
Respiratory and swimming movements in the
cephalopod, Cranchia scabra.
Nature, 196:351-352.
Also in:
Collected Reprints, NIO, Vol. 11(418). 1963.

cephalopods, anat.
Clarke, Malcolm R., 1962.
The identification of cephalopod "beaks" and the
relationship between beak size and total body
weight.
Bull. Brit. Mus.(Nat. Hist.) (Zool.), 8(10):
421-480, Pls. 13-22.

cephalopods
Clarke, Malcolm R., 1962
Significance of cephalopod beaks.
Nature, 193:560-561.

Also in:
Collected Reprints. Nat. Inst. Oceanogr.,

cephalopods
Cole, K.S. and D.L. Gilbert, 1970.
Jet propulsion of squid. Biol. Bull., mar. biol. Lab., Woods Hole, 138(3): 245-246.

cephalopod, anatomy
Denton, E.J., 1962
Some recently discovered buoyancy mechanisms in marine animals.
Proc. R. Soc., London, (A), 265(1322):366-370.

cephalopods, anat.-physiol.
Denton, E.J., and D.W. Taylor, 1964.
The composition of gas in the chambers of the cuttlebone of Sepia officinalis.
J. Mar. Biol. Assoc., U.K., 44(1):203-207.

cephalopods, anat.-physiol.
Fields, W. Gordon, 1965.
The structure, development, food relations, reproduction, and life history of the squid, Loligo opalescens Berry.
Calif. Resources Agency, Dept. Fish Game, Fish. Bull., 131:108 pp.

molluscs, cephalopod mandibles
Gaskin, D.E., and M.W. Cawthorn 1967.
Squid mandibles from the stomachs of sperm whales (Physeter catodon L.) captured in the Cook Strait region of New Zealand.
N.Z. Jl. mar. Freshwat. Res. 1(1):59-70.

molluscs, cephalopods, anat. physiol.
Hara, Tomiyuki, and Reiko Hara 1967.
Rhodopsin and retinochrome in the squid retina.
Nature, Lond. 214 (5088): 573-575.

Molluscs, octopus
Hara, Tomiyuki, Reiko Hara and Jitsuzo Takeuchi 1967.
Rhodopsin and retinochrome in the octopus retina.
Nature, Lond. 214 (5088): 572-573.

molluscs, cephalopods
Hayashi, Yasuyuki 1971.
Studies on the maturity of the common squid. III. Ponderal index and weight indices of internal organs during maturation and exhaustion.
Bull. Jap. Soc. scient. Fish. 37(10): 960-963. (In Japanese; English abstract)

cephalopods, anat.-physiol.
Jander, Rudolf, Karl Daumer and Talbot H. Waterman, 1963
Polarized light orientation by two Hawaiian decapod cephalopods.
Zeits. Vergl. Physiol., 46:383-394.

molluscs, cephalopod mandibles
Mangold, Katharina, et Pio Fioroni 1967.
Morphologie et biométrie des mandibules de quelques céphalopodes méditerranéens.
Vie Milieu (A) 17 (3-A): 1139-1196.

cephalopods, anat.-physiol
Martin, Anton W., and Frederick A. Aldrich, 1970.
Comparison of hearts and branchial heart appendages in some cephalopods.
Can. J. Zool. 47(4): 751-756.

molluscs, cephalopod (anat. physiol)
Mauro, Alexander, and Ove Sten-Knudsen 1972.
Light-evoked impulses from extra-ocular photoreceptors in the squid Todarodes.
Nature, Lond. 237 (5354): 342-343.

cephalopoda
Slepzov, M.M., 1955.
On the biology of cephalopod molluscs of the far eastern seas and the north-western Pacific
Trudy Inst. Oceanol., 18:69-77.

cephalopods
Tittel, K., 1964.
Saugnapf-, epi- und hypofasciale Armmuskulatur der Kephalopoden - ein Beitrag zur funktionellen Annatomie greibeweglicher Skelettmuskelkörper.
Gegenbaurs Morphol. Jahrb., 106(1):89-115.

molluscs (cephalopods) (anat. Physiol.)
Ward, Diana Valiela 1972.
Locomotory function of the squid mantle.
J. Zool. Lond. 167(4): 487-499.

molluscs cephalopods) (anat. physiol.)
Ward, Diana Valiela, and Stephen A. Wainwright 1972.
Locomotry aspects of squid mantle structure.
J. Zool. Lond. 167(4): 437-449.

octopus, anat.physiol.
Young, J.Z., M.F. Moody, J.R. Parriss and N.S. Sutherland, 1960.
The visual system of Octopus.
Nature, 186(4728):836-843.

cephalopods, anat-physiol.
Young, Richard Edward, 1972.
Function of extra-ocular photoreceptors in bathypelagic cephalopods. Deep-Sea Res. 19 (9): 651-660.

molluscs, cephalopods
Young, Richard Edward 1972.
Brooding in a bathypelagic octopus.
Pacific Sci. 26(4): 400-404.

cephalopods, chemistry
Culkin, F. and R.J. Morris, 1970.
The fatty acids of some cephalopods. Deep-Sea Res., 17(1): 171-174.

chemistry, squid
Ishikawa, Senji, Kunitsugu Kitabayashi, 1964.
Biochemical studies on squid. XXIII. On the total nitrogen and arginine nitrogen in hot water muscle extract. (In Japanese; English abstract).
Bull. Hokkaido Reg. Fish. Res. Lab., (28):65-70.

chemistry, squid
Kitabayashi, Kunitsugu, and Senji Ishikawa, 1964.
Biochemical studies on squid. XXIV On dilute hydrogen chloride solution soluble protein. (In Japanese; English abstract).
Bull. Hokkaido Reg. Fish. Res. Lab., (28):71-73.

MOLLUSCS
Cephalopod fishery

See also: Subject Cumulation under fisheries, mollusc.

cephalopod fishery
Araya, Hisao, 1969.
Relationship between migration and fishing ground formation of squid, Todarodes pacificus Steenstrup in the Japan Sea Coast of northern Tohoku and southern Hokkaido districts in summer. (In Japanese; English abstract). Bull. Jap. Soc. fish. Oceanogr. Spec. No. (Prof. Uda Commem. Pap.): 269-273.

MOLLUSCS
Cephalopods, learning

cephalopods, learning of
Wells, M.J., 1965.
Learning by marine invertebrates.
In: Advances in Marine Biology, Sir Frederick S. Russell, editor, Academic Press, 3:1-62.

MOLLUSCS
Cephalopods, lists of species

cephalopods
Adam, W., 1964.
Considerations sur la systematique des Sepiidae (Cephalopoda).
Zoolog. Mededel., 39:263-278.

cephalopods
Adam, William, 1960.
Cephalopoda from the gulf of Aqaba. Contributions to the Knowledge of the Red Sea, No. 16.
Sea Fish. Res. Sta., Haifa, Bull., No. 26:1-26.

Sepia pharonis
Sepioteuthis lessoniana
Symplectoteuthis oualaniensis
?Enoploteuthis dubia n.sp.
Octopus macropus
Octopus cyaneus
Octopus aegina

Cephlelopods, lists of spp.
Bonnet, M., 1965,
Remarques sur l'écologie des céphalopodes des côtes de Sardaigne et de Corse captures par la "Thalassa" en novembre et décembre 1963.
Rapp. Proc. Verb. Réunions, Comm. Int. Expl. Sci., Mer Méditerranee, Monaco, 18(2)235-240.

Cephalopoda
Bruun, A. Fr., 1945.
Cephalopoda. Zool. Iceland 4(64):15 pp., 4 text-figs.

cephalopods
Clarke, M.R., 1963.
Economic importance of North Atlantic squids.
New Scientist, 17:567-570.

Also in:
Collected Reprints, N.I.O., Vol. 11(419):1963.

cephalopods
Dees, L.T., 1961.
Cephalopods, cuttlefish, octopuses, squids.
U.S.F.W.S., Fishery Leaflet, 524:10 pp.

cephalopods
Donovan, D.T., 1964.
Cephalopod phylogeny and classification.
Biol. Rev., 39(3):259-287.

cephalopods
Gamulin-Brida, H., 1963.
Quelques renseignements statistiques sur les Céphalopodes adriatiques.
Rapp. Proc. Verb. Réunions, Comm. Int. Expl. Sci., Mer Méditerranee, Monaco, 17(2):387-400.

molluscs, cephalopods
Gamulin-Brida, Helena et Vesna Ilijanić 1972.
Contribution à la connaissance des céphalopodes de l'Adriatique.
Acta adriatica 14(6): 12pp.
(In Jugoslavian; French abstract)

cephalopods, lists of spp.
McGowan, John A., 1967.
Distributional atlas of pelagic molluscs in the Califronia Current region.
Calif. Coop. Ocean. Fish. Invest., 6:1-218.

cephalopods
Morales, E., 1962
Cefalopodos de Cataluna II.
Inv. Pesq., Barcelona, 21:97-111.

molluscs (cephalopods) (lists of spp.)
Nesis, K.N., 1972.
Oceanic cephalopods of the Peru current: horizontal and vertical distribution.
(In Russian; English abstract). Okeanologiia 12(3): 506-519.

cephalopods, lists of spp.
Nishimura, Saburo, 1968.
A preliminary list of the pelagic Cephalopda from the Japan Sea.
Publ. Seto mar. biol. Lab., 16(1):71-83.

molluscs (cephalopods)
Oommen, Varghese P., 1967
New records of Octopods from the Arabian Sea.
Bull. Dept. Mar. Biol. Oceanogr. Univ. Kerala 3: 29-32.

cephalopods
Park, Ju Seok, 1962.
Some results of the tagging experiments on squids in the east coast waters of Korea.
(In Japanese).
Repts. Res. Fish. Res., 5:101-112.

cephalopods, 1 lists of spp.
Pearcy, William G., 1965.
Species composition and distribution of pelagic cephalopods from the Pacific Ocean off Oregon.
Pacific Science, 19(2):261-266.

molluscs, cephalopods
Roeleveld, Martina A. 1972.
A review of the Sepiidae (Cephalopoda) of southern Africa.
Ann. S. Afr. Mus. 59 (10): 193-313.

molluscs, cephalopod
Ruby, G. and J. Knudsen, 1972
Cephalopoda from the eastern Mediterranean.
Israel J. Zool. 21(2): 83-97.

cephalopods
Voss, Gilbert L., 1963.
Cephalopods of the Philippine Islands.
U.S. Nat. Mus., Smithsonian Inst., Bull., 234:1-180.

cephalopods
Voss, G.L., 1956.
A checklist of the cephalopods of Florida.
Q.J. Florida Academy of Sciences, 19(4):274-282.

cephalopods
Voss, G.L., 1956.
A review of the cephalopods of the Gulf of Mexico
Bull. Mar. Sci., Gulf and Caribbean, 6(2):85-178.

cephalopods, lists of spp.
Adam, W., 1962.
Cephalopodes de l'Archipel du Cap Vert, de l'Angola et de Mozambique.
Trab. Cent. Biol. Pisc., No. 32:9-64.

Also, Mem. Junta Invest. Ultram., (2)No. 33

cephalopods, lists of spp.
Mangold-Wirz, K., 1963.
Biologie des Céphalopodes benthiques et nectonoques de la Mer Catalane.
Vie et Milieu, Suppl., 13:1-295.

cephalopods, lists of sp
Mangold-Wirz, K., 1961.
La migration des céphalopodes méditerranéens.
Rapp. Proc. Verb., Réunions, Comm. Int. Expl. Sci. Mer Méditerranée, Monaco, 16(2):299-304.

cephalopods
Hartmann, Jürgen, 1970.
Tagesgang und vertikale Mikroverteilung von Cephalopoden des Neuston westlich von Madeira.
Ber. dt. wiss. Komm. Meeresforsch. 21(1/4): 494-499.

molluscs, cephalopods, migrations of
Kasahara, Syogo, and Sukekata Ito, 1968.
Studies on the migration of common squids in the Japan Sea. II. Migrations and some biological aspects of squids having occurred in the offshore regions of the Japan Sea during the autumn season of 1966 and 1967. (In Japanese; English abstract).
Bull. Jap. Sea. Reg. Fish. Res. Lab., 20:49-69.

cephalopods
Kawana, Takeshi, Yoshio Takemura and Takamiki Yamane, 1970.
Tagging experiments of the common squid along the coast of the Izu Peninsula. (In Japanese; English abstract). Bull. Tokai reg. Fish. Res. Lab, 63: 11-15.

cephalopods, migrations
Mangold-Wirz, K., 1961.
La migration des céphalopodes méditerranéens.
Rapp. Proc. Verb., Réunions, Comm. Int. Expl. Sci. Mer Méditerranée, Monaco, 16(2):299-304.

cephalopods, tagging of
Park, Joo Suck, and Joo Youl Lim, 1967.
On the results of the tagging experiment on squids in the Korean waters. (In Korean; English abstract).
Repts Fish. Resources, Fish. Res. Develop. Agency, Korea, 7:29-39.

octopus
Lane, F.W., 1960.
Kingdom of the octopus.
Sheridan House, New York, 300 pp.

cephalopods, parasites of
Brown, Elizabeth L., and William Threlfall, 1968.
A quantitative study of the helminth parasites of the Newfoundland short-finned squid, Illex illecebrosus illecebrosus (Le Sueur)(Cephalopoda Decapoda).
Can. J. Zool., 46(6):1087-1093.

Architeuthis
Dell, R.K., 1970.
A specimen of the giant squid Architeuthis from New Zealand.
Rec. Dominion Mus., N.Z. 7(4): 25-36

Architeuthis sp
Frost, N., 1934.
Notes on a giant squid (Architeuthis sp) captured at Dildo, Newfoundland, in December 1933. Ann. Rept., Newfoundland Fish. Res. Comm. 2(2):100-113 5 textfigs., 3 pls.

Architeuthis
Roper, Clyde F.E., and Richard E. Young 1972
First records of juvenile giant squid Architeuthis (Cephalopoda: Oegopsida).
Proc. Biol. Soc. Wash. 85 (16): 205-222.

Argonauta argo
Nishimura, Saburo, 1968.
Glimpse of the biology of Argonauta argo Linnaeus (Cephalopoda:Octopodida) in Japanese waters.
Publ. Seto mar. biol. Lab., 16(1):61-70.

Ascocranchia joubini
Voss, G.L., 1962.
Ascocranchia joubini, a new genus and species of cranchiid squid from the North Atlantic.
Bull. Inst. Océanogr., Monaco, 59(1242):6 pp.

Bathothauma
Young, J.Z. 1970.
The stalked eyes of Bathothauma (Mollusca Cephalopoda)
J. Zool. Lond. 162 (4): 437-447.

Bathyteuthis
Roper, Clyde F.E. 1969.
Systematics and zoogeography of the worldwide bathypelagic squid Bathyteuthis (Cephalopoda: Oegopsida).
Bull. U.S. natn. Mus. 291: 210 pp.

molluscs- cephalopods Bathyteuthis
Roper, Clyde F.E. 1968.
Preliminary descriptions of two new species of the bathypelagic squid Bathyteuthis (Cephalopoda: Oegopsidae).
Proc. Biol. Soc. Wash. 81: 161-172.

molluscs, cephalopods Berrya
Oommen, Varghese P. 1966.
The Octopoda of the Kerala coast. I. A new species of the genus Berrya Adam 1943.
Bull. Dept. Mar. Biol. Oceanogr. Univ. Kerala 2: 51-59.

Cranchia scabra
Clarke, M.R., 1962.
Respiratory and swimming movements in the cephalopod Cranchia scabra.
Nature, 196(4852):351.

Doryteuthis plei
Waller, Richard A., and Robert I. Wicklund, 1968.
Observations from a research submersible-mating and spawning of the squid. Doryteuthis plei.
Bio Science, 18(2):110-111.

Dosidicus gigas
Nesis, K.N., 1970.
Biology of the Peruvian-Chilean gigantic squid Dosidicus gigas (Orbigny). (In Russian; English abstract). Okeanologiia, 10(1): 140-152.

Enoploteuthis anapsis
Roper, Clyde E.E., 1964.
Enoploteuthis anapsis, a new species of enoploteuthid squid (Cephalopoda:Oegopsida) from the Atlantic Ocean.
Bull. Mar. Sci., Gulf and Caribbean, 14(1):140-148.

Enoploteuthis leptura cephalopods
Roper, Clyde F.E. 1966.
A study of the genus Enoploteuthis (Cephalopoda: Oegopsida) in the Atlantic Ocean with a redescription of the type species, E. leptura (Leach, 1817).
Dana Rept., No. 66: 46 pp.

Euprymna morsei
Rao, P.J. Sanjeeva, and N. Kalyani 1971.
Euprymna morsei (Verrill, 1881) (Sepiolidae: Cephalopoda) from the Indian coast.
J. mar. biol. Ass. India 13(1):135-137.

Gonatopsis japonicus n.sp.
Okiyama, M., 1969.
A new species of Gonatopsis from the Japan Sea with the record of a specimen referable to Okutani's Gonatopsis sp., 1967 (Cephalopoda: Oegopsida, Gonatidae). Publ. Seto mar. biol. Lab., 17(1): 19-32.

Gonatopsis octopedatus

Okiyama Muneo 1970.
A record of the eight-armed squid Gonatopsis octopedatus Sasaki from the Japan Sea (Cephalopoda, Oegopsida, Gonatidae).
Bull. Jap. Sea Reg. Fish. Res. Lab. (22): 71-80.

Gonatus anonychus n.sp.

Pearcy, William G., and Gilbert L. Voss, 1963.
A new species of gonatid squid from the northeastern Pacific.
Proc. Biol. Soc., Washington, 76:105-112.

Gonatus fabricii

Nesis, K.N. 1971.
Gonatus fabricii (Licht.) in the centre of the Arctic Basin. (In Russian).
Gidrobiol. Zh. 7(1): 93-96.

Hapalochlaena maculosa

Tranter, D.J. and O. Augustine, 1973
Observations on the life history of the blue-ringed octopus Hapalochlaena maculosa.
Mar. Biol. 18(2): 115-128.

Illex spp.

Aldrich, Frederick A., and C.C. Lu, 1968.
A reconsideration of forms of squids of the genus Illex (Illicinae, Ommastrephidae) in Newfoundland waters.
Can. J. Zool., 46(5):815-818.

Illex spp

Roper, Clyde F.E., C.C. Lu, and Katharina Mangold, 1969.
A new species of Illex from the western Atlantic and distributional aspects of other Illex species (Cephalopoda: Oegopsida).
Proc. Biol. Soc. Wash. 82: 295-322.

Illex illecebrosus

de Castellanos, Zulma J. A., 1964.
Contribucion al conocimiento biologico del calamar Argentino Illex illecebrosus argentinus.
Bol. Inst. Biol. Mar., Mar del Plata, Argentina, No 8: 35 pp., 2 pls.

Illex illecebrosus

Jangaard, Peter M., and R.G. Ackman, 1965.
Lipids and component fatty acids of the Newfoundland squid, Illex illecebrosus (LeSueur).
J. Fish. Res. Bd., Canada, 22(1):131-137.

Illex illecebrosus

Squires H.J. 1966
Feeding habits of the squid Illex illecebrosus.
Nature, 211 (5055): 1321

Illex illecebrosus illecebrosus

Squires H.J. 1967
Growth and hypothetical age of the Newfoundland bait squid Illex illecebrosus illecebrosus.
J. Fish. Res. Bd Can. 24 (6): 1209-1217.

Lepidoteuthis grimaldii

Clarke, M.R., 1960
Lepidoteuthis grimaldii - a squid with scales.
Nature, 188(4754): 955-956.

Lepidoteuthis grimaldi

Clarke, Malcolm R., and G.E. Maul, 1962
A description of the 'scaled' squid Lepidoteuthis grimaldi Joubin 1895.
Proc. Zool. Soc., London, 139:97-118.
Also in:
Collected Reprints, Nat. Inst. Oceanogr., Wormley, 10.

Loligo brasiliensis

de Castellanos, Zulma J.A., 1967.
Contribucion al estudio biologico de Loligo brasiliensis Bl.
Bol. Inst. Biol. mar. Mar del Plata, 14: 33 pp.

Loligo corolliflora

Voss, Gilbert L., 1965.
A note on Loligo corolliflora Tilesius, 1829, a long forgotten squid from eastern seas.
Proc. Biol. Soc., Washington, 78:155-158.

Loligo pealii

Austin, C.R., Cecelia Lutwak-Mann, and T. Mann, 1964.
Spermatophores and Spermatozoa of the squid, Loligo pealii.
Proc. R. Soc., London, (B), 161(983):143-152.

Loligo pealei

Mercer M.C. 1970.
Sur la limite septentrionale du calmar Loligo pealei Lesueur.
Naturaliste can. 97 (6): 823-824

Loligo pealei

Summers, William C., 1971.
Age and growth of Loligo pealei, a population study of the common Atlantic Coast squid.
Biol. Bull. mar. biol. Lab. Woods Hole 141(1): 189-201.

Loligo pealei

Summers, William C., 1969.
Winter population of Loligo pealei in the mid-Atlantic bight.
Biol. Bull., mar. biol. Lab., Woods Hole, 137(1):202-216.

Loligo pealei

Summers, William C., 1968.
The growth and size distribution of current year class Loligo pealei.
Biol. Bull., mar. biol. Lab., Woods Hole, 135 (2): 366-377.

Loligo pealei

Summers, William C. and John J. McMahon, 1970.
Survival of unfed squid, Loligo pealei, in an aquarium. Biol. Bull. mar. biol. Lab., Woods Hole, 138(3): 389-396.

Loligo vulgaris

Porebski, J. 1970.
Observations on the occurrence of Cephalopoda in the waters of the NW African shelf, with particular regard to Loligo vulgaris (Lamarck).
Rapp. P.-v. Réun. Cons. int. Explor. Mer 159: 142-145.

Loligo vulgaris

Weber, W., 1973
Peculiarities of innervation in chromatophore muscle cells of Loligo vulgaris.
Mar. Biol. 19(3): 224-226.

Loliolus rhomboidalis

Burgess, Lourdes Alvina, 1967.
Loliolus rhomboidalis, a new species of loliginid squid from the Indian Ocean.
J. mar. Sci., Miami, 17(2):319-329.

Lolliguncula brevis

Dragovich, Alexander, and John A. Kelly, Jr., 1967.
Occurrence of the squid, Lolliguncula brevis, in some coastal waters of western Florida.
Bull. mar. Sci., Miami, 17(4): 840-844.

Lolliguncula brevis

Dragovich, A., and J.A. Kelly, Jr., 1963.
A biological study and some economic aspects of squid in Tampa Bay, Florida.
Proc., 15th Ann. Sess., Gulf and Caribbean Fish. Inst., Univ. Miami, 87-102.

Mastigoteuthis hjorti

Rancurel, P., 1973
Mastigoteuthis hjorti Chun 1913 description de trois échantillons provenant du Golfe de Guinée. (Cephalopoda-Oegopsida). Cah. O.R.S.T.O.M. sér, Océanogr, 11(1): 27-32.

Nautilus

Mutvei, H., 1964.
On the shells of Nautilus and Spirula with notes on the secretion in non-cephalopod mollusca.
Arkiv för Zoologi, andra Ser., 16(3):221-278.

Nautilus, anatomy of

Young, J.Z., 1965.
The central nervous system of Nautilus.
Phil. Trans., R. Soc., London, (b), 249(754):1-25.

Nautilus pompilius

Price, N.B., and A. Hallam, 1967.
Variation of strontium content within shells of Recent Nautilus and Sepia.
Nature, 215(5107):1272-1274.

Nautilus pompilius

Toriyama, Ryuzo, Tadashi Sato, Takashi Hamada, and Pumwarn Komalarjun, 1965.
Nautilus pompilius drifts on the west coast of Thailand.
Japan. J. Geol. Geogr., Trans., 36(2/4):149-161.

Nototodarus sloani sloani

Kawakami, Takehiko, Yasuo Sasagawa and Motosugu Hamabe, 1972
A preliminary note on the ecology of the ommastrephid squid Nototodarus sloani sloani (Gray) in New Zealand waters. (In Japanese; English abstract). Bull. Tokai reg. Fish. Res. Lab. 70:1-23.

Octopus

Young, J.Z., 1965.
The buccal nervous system of Octopus. The centres for touch discrimination in Octopus.
Phil. Trans., R. Soc., London, (B), 249(755):27-44; 45-67.

Octopus cyanea

Boucher-Rodoni, R., 1973
Vitesse de digestion d'Octopus cyanea (Cephalopoda: Octopoda). Mar. Biol. 18(3): 237-242.

Octopus cyanea

Van Heukelem, William F. 1973.
Growth and life-span of Octopus cyanea (Mollusca: Cephalopoda).
J. Zool. Lond. 169(3): 299-315.

Octopus cyanea

Wells, M.J., and J. Wells 1970.
Observations on the feeding, growth rate and habits of newly settled Octopus cyanea.
J. Zool. Lond. 161(1): 65-74

Octopus defilippi

Voss, Gilbert L., 1964.
Octopus defilippi Verany, 1851, an addition to the cephalopod fauna of the western Atlantic.
Bull. Mar. Sci., Gulf and Caribbean, 14(4):554-560.

Octopus dofleini
Harrison, F.M., and A.W. Martin, 1965.
Excretion in the cephalopod, Octopus dofleini.
J. Exp. Biol., 42(1):71-98.

Octopus dofleini
Pickford, Grace E., 1964.
Octopus dofleini (Wülker).
Bull. Bingham Oceanogr. Coll., 19(1):70 pp.

Octopus hummelincki
Burgess, Lourdes Alvina, 1966.
A study of the morphology and biology of Octopus hummelincki Adam, 1936 (Mollusca: Cephalopoda).
Bull. mar. Sci., Miami, 16(4):762-813.

Octopus maya
Voss, Gilbert L., and Manuel Solis Ramirez, 1966
Octopus maya, a new species from the Bay of Campeche, Mexico.
Bull. Mar. Sci., 16(3):615-625.

Octopus Varunae
Oommen, Varghese P., 1971.
Octopus Varunae, a new species from the west coast of India. Bull. Dept. Mar. Biol. Oceanogr. Univ. Cochin 5: 69-76.

molluscs, cephalopods
Altman J.S. and M. Nixon 1970.
Use of the beaks and radula by Octopus vulgaris in feeding.
J. Zool. Lond. 161 (1): 25-38.

Octopus vulgaris
Boycott, B.B., 1960
The functioning of the statocysts of Octopus vulgaris. Proc. Roy. Soc., London, Ser. B., 152(946): 78-87.

Octopus vulgaris
Dilly, Noel, and Marion Dixon, 1965.
Further observations on forces exerted by Octopus vulgaris.
Pubbl. Staz. Zool., Napoli, 34:340-345.

Octopus vulgaris
Dilly, Noel, and Marion Dixon, 1965.
Further observations on forces by Octopus vulgaris.
Pubbl. Staz. Zool., Napoli, 34:240-245.

Octopus vulgaris
Froesch, D., 1973
Projection of chromatophore nerves on the body surface of Octopus vulgaris.
Mar. Biol. 19(2): 153-155.

Octopus vulgaris
Katsutani, Kunio, 1968.
Tagging and recapture experiments on octopi, Octopus vulgaris Cuvier. (In Japanese).
Bull. Fish. Exp. Sta., Okayama Pref., 42:115-117.

Octopus vulgaris
Lund, R.D., 1965.
The staining of degeneration in the nervous system of the octopus by modified silver methods.
Q.J. Microsc. Soc., 106(1):115-117.

Octopus vulgaris
Mangold, K. and S. von Boletzky, 1973.
New data on reproductive biology and growth of Octopus vulgaris. Mar. Biol. 19(1): 7-12.

molluscs, cephalopods
Nixon, Marion 1973.
Beak and radula growth in Octopus vulgaris.
J. Zool. Lond. 170 (4): 451-462.

Octopus vulgaris
Nixon, Marion, 1971.
Some parameters of growth in Octopus vulgaris.
J. Zool. Lond., 163(3): 277-284

Octopus vulgaris
Nixon, Marion, 1969.
Growth of the beak and radule of Octopus vulgaris.
J. Zool. Lond., 159 (3): 363-379.

Octopus Vulgaris
Nixon, Marion, 1966.
Changes in body weight and intake of food by Octopus Vulgaris.
J. Zool., London, 150(1):1-9.

Octopus vulgaris
Nixon, Marion, 1965.
Some observations on the food intake and learning in Octopus vulgaris.
Pubbl. Staz. Zool., Napoli, 34:329-339.

Octopus vulgaris
Wodinsky, J., 1973
Ventilation rate and copulation in Octopus vulgaris. Mar. Biol. 20(2): 154-164.

Octopus vulgaris
Wodinsky, J., 1972.
Breeding season of Octopus vulgaris. Mar. Biol. 16(1): 59-63.

Octopus vulgaris
Young, J.Z., 1964.
A model of the brain.
Oxford University Press, vii 348 pp. 35 shillings.

Octopus vulgaris
Young, J.Z., 1960.
The statocysts of Octopus vulgaris.
Proc. R. Soc., London, (B), 152(946):3-29.

Ommastrephes bartrami
Shevtsov, G.A., 1972.
Nutrition of Ommastrephes bartrami Lesueur in the Kuril-Khokkaido region. (In Russian).
Gidrobiol. Zh. 8(3): 97-101.

Ommastrephes pteropus
Roper, C.F.E., 1963.
Observations on bioluminescence in Ommastrephes pteropus (Steenstrup, 1855) with notes on its occurrence in the family Ommastrephidae (Mollusca: Cephalopoda).
Bull. Mar. Sci., Gulf and Caribbean, 13(2):343-353.

Ommatostrephes pteropus
Zuev, G.V. 1973.
Some features of biology and distribution of the squid Ommatostrephes pteropus in the Caribbean Sea. (In Russian; English abstract)
Zool. Zh. 52(2):180-184

Ommastrephes sloani
Okutani, Takashi, 1962
Diet of the common squid, Ommastrephes sloani pacificus landed around Ito port, Shizuoka Prefecture. (In English; Japanese abstract).
Bull. Tokai Reg. Fish. Res. Lab., Tokyo, No. 32: 41-47.

Ommastrephes sloani pacificus
Araya, Hisao, and Seiwa Kawasaki, 1962
On the locomotion of the squids, Ommastrephes sloani pacificus (Steenstrup), in the Okhotsk Sea along the coast of Hokkaido during the autumn of 1961. (In Japanese; English summary
Bull. Hokkaido Reg. Fish. Res. Lab., (25): 11-19.

Ommastrephes sloani pacificus
Hamabe, Mototsugu, 1964.
Study on the migration of squid (Ommastrephes sloani pacificus Steenstrup) with reference to age of the moon. (In Japanese; English abstract).
Bull. Jap. Soc. Sci. Fish., Tokyo, 30(3):209-215.

Ommastrephes sloani pacificus
Hamabe Mototsugu, 1963.
Spawning experiments of the common squid, Ommastrephes sloani pacificus Steenstrup in an indoor aquarium. (In Japanese; English abstract).
Bull. Jap. Soc. Sci. Fish., 29(10):930-934.

Ommastrephes sloani pacificus
Hamabe, Mototsugu, 1962.
[Experimental studies on breeding habit and larval development of the squid, Ommastrephes sloani pacificus Steenstrum. V. Formation of the fourth arm and the tentacle in the Rhynchoteuthis larvae.]
Dobuts. Zashi, 71(3):65-70.

(Ommastrephes sloani pacificus)
Katoh, Gendi, 1964
A few comments on the biological grouping of the common squid derived from its ecological aspect. (In Japanese; English abstract).
Bull. Jap. Sea Reg. Fish. Res. Lab., (13): 31-42.

Ommastrephes sloani pacificus
Lim, Joo Youl, 1967.
Ecological studies on common squids, Ommastrephes sloani pacificus Steenstrup in the eastern water of Korea.
Repts. Fish. Resources, 7:41-49.

Ommastrephes sloani pacificus
Lim, Joo Youl, 1967.
Ecological studies on common squids, Ommastrephes sloani pacificus Steenstrup in the eastern water of Korea. (In Korean; English abstract).
Repts. Fish. Resources, Fish Res. Develop. Agency, Korea, 7:41-49.

Ommastrephes sloani pacificus
Shuntov, V.P., 1964.
Distribution and migration of the Pacific squid. (Ommatostrephes sloani pacificus Steenstrup) in the Japan Sea. (In Russian).
Izv. Tokhookean. Nauchno-Issled. Inst. Ribn. Khoz. i Okeanogr., 50:147-157.

Ommastrephes sloani pacificus
Suzuki, Tsuneyoshi, 1963
Studies on the relationship between current boundary zones in waters to the southeast of Hokkaido and migration of the squid. Ommastrephes sloani pacificus (Steenstrup).
Mem. Fac. Fish., Hokkaido Univ., 11(2):153 pp.

Phasmatopsis cymoctypus
Clarke, Malcolm R., 1962
A large member of the squid family Cranchiidae, Phasmatopsis cymoctypus de Rochebrune 1884.
Proc. Malac. Soc., London, 35:27-42.

Also in:
Collected Reprints, Nat. Inst. Oceanogr., Wormley, 10.

Phasmatopsis lucifer
Voss, G.L., 1963.
A new species of cranchid squid, Phasmatopsis lucifer from the Gulf of Mexico.
Bull. Mar. Sci., Gulf and Caribbean, 13(1):77-88.

Rossia macrosoma

von Boletzky, S. and M.V. von Boletzky 1973
Observations on the embryonic and early post-embryonic development of Rossia macrosoma (Mollusca, Cephalopoda).
Helgoländer wiss. Meeresunters 25(1): 135-161.

Semirossia

Boletzky, S. v., 1970.
Biological results of the University of Miami deep sea expeditions. 54. On the presence of light organs in Semirossia Steenstrup 1887 (Mollusca: Cephalopoda). Bull. mar. Sci., 20 (2): 374-388.

Sepia spp.

Garcia Cabrera, R.C. 1970.
Espèces du genre Sepia du Sahara espagnol.
Rapp. P.-v. Réun. Cons. int. Explor. Mer. 159: 132-139.

Sepia officinalis

Boucaud-Camou, E., 1971.
Constituants lipidiques du foie de Sepia officinalis. Marine Biol., 8(1): 66-69.

Sepia officinalis

Decleir, W., J. Lemaire and A. Richard, 1970.
Determination of copper in embryos and very young specimens of Sepia officinalis. Mar. Biol., 5(3): 256-258.

Sepia officinalis

Mangold, Katharina 1966
Sepia officinalis de la mer catalane.
Vie Milieu (A) 17 (2-A): 961-1012.

Sepia officinalis

Price, N.B., and A. Hallem, 1967.
Variation of strontium content within shells of Recent Nautilus and Sepia.
Nature, 215(5107): 1272-1274.

Sepioteuthis bilineata

Larcombe, M.F. and B.C. Russell 1971.
Egg laying behaviour of the broad squid, Sepioteuthis bilineata.
N.Z. Jl mar. Freshwat. Res. 5(1): 3-11

Sepioteuthis sepioidea

La Roe, E.T., 1971.
The culture and maintenance of the loliginid squids Sepioteuthis sepioidea and Doryteuthis plei. Mar. Biol. 9(1): 9-25.

Sepioteuthis sepioidea

Mercer, M.C. 1970.
The tropical loliginid squid Sepioteuthis sepioidea from the northwest Atlantic.
J. Fish. Res. Bd Can. 27 (10): 1892-1893.

Spirula

Colman, J.S., 1954.
The "Rosaura" Expedition, 1937-1938. I. Gear, narrative and station list.
Bull. Brit. Mus. (N.H.), Zool., 6(2): 119-130.

molluscs, cephalopods - Spirula

Denton, E.J. and J.B. Gilpin-Brown, 1971.
Further observations on the buoyancy of Spirula.
J. mar. biol. Ass. U.K. 51(2): 363-373.

Spirula

Mutvei, H., 1964.
On the shells of Nautilus and Spirula with notes on the secretion in non-cephalopod molluscs.
Arkiv för Zoologi, andra ser., 16(3): 221-278.

Spirula

Voss, Gilbert L., 1967.
Some bathypelagic cephalopods from South African waters.
Ann. S. Afr. Mus., 50(5): 61-88.

Spirula Spirula

Bruun, A. Fr., 1943
The Biology of Spirula Spirula (L.)
Dana Rept. No. 24, 46 pp.

Spirula spirula

Clarke, M.R., 1970.
Growth and development of Spirula spirula.
J. mar. biol. Ass. U.K., 50(1): 53-64.

Spirula Spirula

Denton, E.J., J.B. Gilpin-Brown and J.V. Howarth, 1967.
On the buoyancy of Spirula spirula.
J. mar. biol. Ass. U.K., 47(1): 181-191.

Taningia danae

Clarke, Malcolm R., 1967.
A deep-sea squid, Taningia danae Joubin, 1931. In: Aspects of Marine Zoology, N.B. Marshall, editor, Symp. Zool. Soc., Lond., 19: 127-143.

molluscs, cephalopods - Taonius megalops

Dilly, P.N. 1972.
Taonius megalops, a squid that rolls up into a ball.
Nature, Lond. 237 (5355): 403-404.

Tetronychoteuthis dussmieri

Rees, E.I.S., and M.R. Clarke, 1963.
First records of Tetronychoteuthis dussmieri (d'Orbigny)(Cephalopoda: Onychoteuthidae) from the northwest Atlantic.
J. Fish. Res. Bd., Canada, 20(3): 853-854.

Thydanoteuthis rhombus

Nishimura, Saburo, 1966.
Notes on the occurence and biology of the oceanic squid, Thydanoteuthis rhombus Troschel, in Japan.
Publs Seto mar. biol. Lab., 14(4): 327-349.

Todarodes pacificus

Akabane, Mitsuaki and Seigo Kubota, 1972.
Some considerations for the migration and the formation of fishery grounds of the common squid, Todarodes pacificus Steenstrup in the northeastern sea area of Japan. Bull Tohoku reg. Fish. Res. Lab. 32: 47-58. (In Japanese; English abstract).

Todarodes pacificus

Hamabe, Mototsugu and Torao Shimizu, 1966.
Ecological studies on the common squid, Todarodes pacificus Steenstrup, mainly in the southwestern waters of the Japan Sea. (In Japanese; English abstract).
Bull. Jap. Sea Reg. Fish. Res. Lab., (16): 13-55.

Todarodes pacificus

Kasahara, Syogo, 1968.
A primary record of the mating behavior of the common squid, Todarodes pacificus Steenstrup, observed in the offshore region of the Japan Sea.
Bull. Japan Sea reg. Fish. Res. Lab., 19: 65-67.
(In Japanese).

Todarodes pacificus

Kasahara, Shogo, Akira Ogino and Tadashi Hamaya, 1969.
Considerations on the biology of the common squid, Todarodes pacificus Steenstrup, collected from Toyama Bay in the winter season of 1967. (In Japanese; English abstract). Bull. Jap. Sea Reg. Fish. Res. Lab., 21: 55-65.

Todarodes pacificus

Kawasaki, Tsuyoshi, 1971.
Population structure of the common squid, Todarodes pacificus Steenstrup, distributed in waters east of the Izu Peninsula. Bull. Tokai reg. Fish. Res. Lab. 67: 81-88.
(In Japanese; English abstract)

Todarodes pacificus

Murata, Mamoru and Hisao Araya, 1970.
Ecological studies on squid, Todarodes pacificus Steenstrup, in the waters off the north-east coast of Hokkaido in 1968.
Bull. Hokkaido reg. Fish. Res. Lab., 36: 1-17.
(In Japanese; English abstract)

molluscs, cephalopods

Murata, Mamoru, Yutaka Onoda, Masaaki Tashiro and Yoshihiro Yamagishi 1971
Ecological studies on the squid Todarodes pacificus Steenstrup in the northern waters of the Japan Sea in 1970 (In Japanese; English abstract).
Bull. Hokkaido reg. Fish. Res. Lab. 37: 10-31.

Todarodes pacificus

Shojima, Yoichi 1972.
The common squid Todarodes pacificus, in the East China Sea. (In Japanese; English abstract).
Bull Seikai reg. Fish. Res. Lab. 42: 25-58
II. Eggs, larvae + spawning ground.

Todarodes pacificus

Shojima, Yoichi 1971.
The common squid Todarodes pacificus Steenstrup, in the East China Sea. I. Distribution and annual abundance.
Bull. Seikai reg. Fish. Res. Lab. 41: 21-44.

Todarodes pacificus

Tashiro, Masatoki, Yoshihiro Yamagishi and Takayuki Suzuuchi 1972.
Results of the tagging experiments of a common squidfish Todarodes pacificus (Steenstrup) made in off-shore area of the northern Japan Sea during a summer season of 1970 — an attempt to grouping of the tagged squids by size. (In Japanese; English abstract).
Scient. Repts Hokkaido Fish. exp. Stn. 14: 1-16.

Tremoctopus violaceus

Jones, Everett C., 1963
Tremoctopus violaceus uses Physalia tentacles as weapons.
Science, 139(3556): 764-766.

MOLLUSCS
Gastropods

Mollusca

Barnard, K.H., 1963
Contributions to the knowledge of South African marine Mollusca. Part III. Gastropoda: Prosobranchiata: Taenioglossa.
Annals. S. African Mus., 47(1): 1-199.

Oceanographic Index: Marine Organisms Cumulation, 1946-1973

nudibranchs
Bonnevie, K., (1931) 1946.
Pelagic nudibranchs from the "Michael Sars" North Atlantic Deep-sea Expedition, 1910. Rept. Sci. Res "Michael Sars" N. Atlantic Deep-sea Exped., 1910, 5(3):9 pp., 4 pls.

molluscs
Burn, Robert, 1962.
Descriptions of Victorian nudibranchiate Mollusca with a comprehensive review of the Eolidacea.
Mem. Nat. Mus., Victoria, (25):95-128.

molluscs
Burn, Robert, 1962.
Notes on a collection of nudibranchs (Gastropoda Dorididae and Dendrodorididae) from South Australia with remarks on the species of Basedow and Hedley, 1905.
Mem. Nat. Mus., Victoria, (25):149-170.

molluscs
Burn, Robert, 1962.
On the new pleurobranch subfamily Berthellinae (Mollusca: Gastropoda); a revision and new classification of the species of New South Wales and Victoria.
Mem. Nat. Mus., Victoria, (25):129-146.

molluscs
Burn, R., 1961.
A new doridid nudibranch from Torquay, Victoria.
Veliger, 4(2):55-56.

molluscs
Engel, H., and C.J. Van Eeken, 1962
Red Sea Opisthobranchia from the coast of Israel and Sinai. Contributions to the knowledge of the Red Sea, No. 22.
Sea Fish. Res. Sta., Haifa, Israel, Bull. No. 30: 15-34.

molluscs
Kawaguti, S., and T. Yamasu, 1961.
The shell structure of the bivalved gastropod with a note on the mantle.
Biol. J. Okayama Univ., 7(1/2):1-16.

molluscs
Keen, A. Myra, 1961.
A proposed reclassification of the gastropod family Vermetidae.
Bull. Brit. Mus. (Nat. Hist.) Zool., 7(3):183-212.

molluscs
Lance, J.R., 1961.
A distributional list of southern California opisthobranchs.
Veliger, 4(2):64-69.

molluscs
Moskalev, L.I., 1964.
The life form of gastropod molluscs Docoglossa in the litoral waters of the northwestern part of the Pacific Ocean. (In Russian).
Okeanologiia, Akad. Nauk, SSSR, 4(6):1073-1078.

molluscs
Nybakken, James, 1971.
Biological results of the University of Miami deep-sea expeditions. 78. The Conidae of the Pillsbury Expedition to the Gulf of Panama.
Bull. mar. Sci. 21(1): 93-110.

molluscs
Powell, A.W.B., 1964.
The family Turridae in the Indo-Pacific.
Indo-Pacific Mollusca, 1(5):227-346.

molluscs
Steinberg, J.E., 1961.
Notes on the opisthobranchs of the west coast of North America. 1. Nomenclatural changes in the order Nudibranchia (southern California).
Veliger, 4(2):57-63.

molluscs
Swennen, C., 1961.
Data on distribution, reproduction and ecology of the nudibranchiate molluscs in the Netherlands.
Netherlands J. Sea Res., 1(1/2):191-240.

gastropods, anat. physiol.
Anderson, D.T., 1962.
The reproduction and early life histories of the gastropods, Bembicium auratum (Quoy and Gaimard) (Fam. Littorinidae), Cellana tranoserica (Sower.) (Fam. Patellidae) and Melanerita melanotragus (Smith) (Fam. Neritidae).
Proc. Linn. Soc., New South Wales, 87(1)(398): 62-68.

Mollusks
D'Asaro, Charles N., 1970.
Egg capsules of prosobranch mollusks from South Florida and the Bahamas and notes on spawning in the laboratory. Bull. mar. Sci., 20(2): 414-440.

MOLLUSCS
Abalone
Cox, K.W., 1960.
Review of the abalone of California.
California Fish and Game, 46(4):381-406.

abalone
Newman, P.G., 1969.
Distribution of the abalone (Haliotis midae) and the effect of temperature on productivity.
Invest. Rept. Div. Sea Fish. S.Af. 74: 7pp

GASTROPODS, anat.-physiol.
Radwin, George E., and Harry W. Wells, 1968.
Comparative radular morphology and feeding habits of muricid gastropods from the Gulf of Mexico.
Bull. mar. Sci., Miami, 18(1):72-85.

gastropods, classification of
Taylor, D.W., and N.F. Sohl, 1962.
An outline of gastropod classification.
Malacologia, 1(1):7-32.

gastropods, effect of
McLean, Roger F., 1967.
Measurements of beachrock erosion by some tropical marine gastropods.
Bull. mar. Sci., Miami, 17(3):551-561.

molluscs, zoogeography
Franz, D.R., 1970.
Zoogeography of northwest Atlantic opisthobranch molluscs.
Marine Biol., 7(2):171-180.

Berghia
Tardy, J., 1962.
A propos des espèces de Berghia (Gastéropodes Nudibranches) des côtes de France et de leur biologie.
Bull. Inst. Oceanogr., Monaco, 59(1255):20 pp.

Berthelina carribea n.sp.
Edmunds, M., 1963.
Berthelina caribbea n.sp., a bivalved gastropod from the West Atlantic.
J. Linnean Soc., London, 44(302):731-739.

molluscs
Buccinum
Mateeva, T.A., 1966.
Some aspects of the biology of the genus Buccinum along the eastern Murmansk. (In Russian).
Trudy murmansk biol. Inst., 11(15): 122-139.

Busycon canaliculatum
Mirolli, Maurizio, 1965.
Tritium: distribution in Busycon canaliculatum (L.) injected with labeled reserpine.
Science, 149(3691):1503-1504.

Cassis madagascariensis
D'Asaro, Charles N., 1969.
The spawn of the emperor helmet shell, Cassis madagascariensis Lamarck, from South Florida.
Bull. mar. Sci., 19(4): 905-910.

Cellana radiata
Rao, M.B. and P.N. Ganapati, 1971.
Ecological studies on a tropical limpet Cellana radiata. Mar. Biol. 9(2): 109-114.

Cellana radiata
Sukumaran, Sukunda, and S. Krishnaswamy, 1961.
Reactions of Cellana radiata (Bom)(Gastropoda) to salinity changes.
Proc. Indian Acad. Sci., 54:122-129.

Crepidula fornicata
Cole, H.A., and H.H. Baird, 1953.
The American slipper limpet (Crepidula fornicata) in Milford Haven. Nature 172(4380):687.

Conus flavidus
Alexander, C.G., 1970.
The osphradium of Conus flavidus. Mar. Biol., 6(3): 236-240.

Conus gloria maris
Bruun, A. Fr., 1945.
On the type specimen of Conus gloria maris.
Vidensk. Medd. Dansk. Naturh. Foren. 108:95-101, 1 pl.

Dactylopus michaelsarsi n.gen. n.sp.
Bonnevie, K., (1931) 1946.
Pelagic nudibranchs from the "Michael Sars" North Atlantic Deep-sea Expedition, 1910. Rept. Sci. Res. "Michael Sars" North Atlantic Deep-sea Exped., 1910, 5(3):9 pp., 4 pls.

Dactylopus preoccupied. Nectophyllirhoē therefore suggested.

Glaucus atlanticus
Bonnevie, K., (1931) 1946.
Pelagic nudibranchs from the "Michael Sars" North Atlantic Deep-sea Expedition, 1910. Rept. Sci. Res. "Michael Sars" N. Atlantic Deep-sea Exped., 1910, 5(3):9 pp., 4 pls.

Littorina littorea
Moore, H.B., 1940
The biology of Littorina littorea. Part II. Zonation in relation to other gastropods. on stony and muddy shores. JMBA 24: 227-237, 8 textfigs.

Littorina littorea
Moore, H.B., 1937
The Biology of Littorina littorea. Part 1. Growth of the Shell and Tissues, Spawning, Length of Life and Mortality. JMBA XXI (2), 721-742, 10 textfigs., Pl.III.

Littorina littorea
Newell, G.E., 1965.
The eye of Littorina littorea.
Proc. Zool. Soc., London, 144(1):75-86.

Littorina littorea
Newell, R.C., V.I. Pye and M. Ahsanullah, 1971.
Factors affecting the feeding rate of the winkle Littorina littorea. Mar. Biol. 9(2): 138-144.

Littorina littorea
Spjeldnaes, Nils, and Kari E. Henningsmoen, 1963
Littorina littorea: an indicator of Norse settlement in North America?
Science, 141(3577):275-276.

Littorina littorea
Wells, Harry W., 1965.
Maryland records of the gastropod, Littorina littorea, with a discussion of factors controlling its southern distribution.
Chesapeake Science, 6(1):38-42.

Littorina obtusata

Colman, J., 1932
A statistical test of the species concept of Littorina. Biol. Bull. 62(3):223-243, 11 text figs.

Littorina obtusata

Sacchi, C.F., 1963.
Contribution à l'étude des rapports écologie - polychromatisme chez un prosobranche intercotidal Littorina obtusata (L.). III. Données expérimentales et diverses.
Cahiers Biol. Mar., Roscoff, 4(3):299-313.

Littorina punctata

Evans, F., 1961.
Responses to disturbance of the periwinkle Littorina punctata (Gmelin) on a shore in Ghana.
Proc. Zool. Soc., London, 137(3):393-402.

Nassarius nitida

Collyer, D.M., 1961
Differences revealed by paper partition chromatography between the gastropod Nassarius reticulatus (L.) and specimens believed to be N. nitida (Jeffreys).
J. Mar. Biol. Ass., U.K., 41(3):683-694.

Nassarius obsoletus

Crisp, Mary 1969.
Studies on the behavior of Nassarius obsoletus (Say) (Mollusca, Gastropoda). Biol. Bull. mar. biol. Lab., Woods Hole, 136(3): 355-373.

Nassarius obsoletus

Nagabhushanam, R., and R. Sarojni, 1963.
Resistance of the mud snail, Nassarius obsoletus, to high temperatures.
Indian J. Exp. Biol., 1(3):160-161.

Nassarius obsoletus

Paulson, Theodora C., and Rudolf S. Scheltema, 1968.
Selective feeding on algal cells by the veliger larvae of Nassarius obsoletus (Gastropoda, Prosobranchia). Biol. Bull. mar. biol. Lab., Woods Hole, 134,(3): 481-489.

Nassarius obsoletus

Scheltema, Rudolf S., 1964.
Feeding habits and growth in the mud-snail Nassarius obsoletus.
Chesapeake Science, 5(4):161-166.

Nassarius reticulatus

Collyer, D.M., 1961
Differences revealed by paper partition chromatography between the gastropod Nassarius reticulatus (L.) and specimens believed to be N. nitida (Jeffreys).
J. Mar. Biol. Ass., U.K., 41(3):683-694.

Nectophyllirhoë n.gen.

Bonnevie, K., (1931) 1946.
Pelagic nudibranchs from the "Michael Sars" North Atlantic Deep-sea Expedition, 1910. Rept. Sci. Res. "Michael Sars" N. Atlantic Deep-sea Exped., 1910, 5(3):9 pp., 4 pls.

to replace Dactylopus which was preoccupied.

Neptunea antiqua

Pearce, Jack B., and Gunnar Thorson, 1967.
The feeding and reproductive biology of the red whelk, Neptunea antiqua (L.) (Gastropoda, Prosobranchia).
Ophelia, 4:277-314.

molluscs, Nerita

Hughes, R.N., 1971.
Notes on the Nerita (Archaeogastropoda) populations of Aldabra Atoll, Indian Ocean. Mar. Biol. 9(4): 290-299.

Ocinebra japonica

Chew, Kenneth K., 1960.
Study of food preference and rate of feeding of Japanese oyster drill, Ocinebra japonica (Dunker).
U.S.F.W.S., Spec. Sci. Rept., Fish., No. 365: 1-27.

Patella lamanonii

Moskalev, L.I., 1957.
[Systematical position of the Patella lamanonii Schrenk (Gastropoda, Prosobranchia).]
Trudy Inst. Okeanol., 23:303-305.

Patella vulgata

Moore, H.B., 1934.
On "Ledging" in Shells at Port Erin. The Relation of Shell Growth to Environment in Patella vulgata. Proc. Malacol. Soc. 21(3):213-222, pls. 22-25.

Phyllirhoë atlantica

Bonnevie, K., (1931) 1946.
Pelagic nudibranchs from the "Michael Sars" North Atlantic Deep-sea Expedition 1910. Rept. Sci. Res. "Michael Sars" N. Atlantic Deep-sea Exped., 1910, 5(3):9 pp., 4 pls.

Phylliroe cf. atlantica

van der Spoel, S., 1970.
The pelagic Mollusca from the Atlantide and Galathea expeditions collected in the East Atlantic.
Atlantide Rept. 11:99-139.

Phylliroe bucephala

van der Spoel, S., 1970.
The pelagic Mollusca from the Atlantide and Galathea expeditions collected in the East Atlantic.
Atlantide Rept. 11:99-139.

Pleuroploca gigantea

Paine, R.T., 1963.
Feeding rate of a predaceous gastropod, Pleuroploca gigantea.
Ecology, 44(2):402-403.

Pneumoderma peronii

Engel, H., and C.J. Van Eeken, 1962
Red Sea Opisthobranchia from the coast of Israel and Sinai. Contributions to the knowledge of the Red Sea, No. 22.
Sea Fish. Res. Sta., Haifa, Israel, Bull., No. 30:15-34.

Purpura lapillus

Moore, H.B., 1938.
The biology of Purpura lapillus. Part II. Growth. Part III. Life history and relation to environmental factors.
J. Mar. Biol. Assoc., U.K., 23:57-74.

Purpura lapillus

Moore, H.B., 1936
The biology of Purpura lapillus. I. Shell variation in Relation to Environment. JMBA 21(1):61-89, 11 textfigs.

Strombus gigas

Little, Colin, 1965.
Notes on the anatomy of the queen conch, Strombus gigas.
Bull. Mar. Sci. 15(2):338-358.

Strombus gigas

D'Asaro, Charles N., 1965.
Organogenesis, development and metamorphosis in the queen conch, Strombus gigas, with notes on breeding habits.
Bull. Mar. Sci.,15(2):359-416/

Tritonia hombergi

Thompson, T.E., 1962.
Studies on the ontogeny of Tritonia hombergi Cuvier (Gastropoda Opisthobranchia).
Phil. Trans. R. Soc., London, (B), 245 (722):171-218.

Turritella communis

Yonge, C.M., 1946
On the habits of Turritella communis Risso. JMBA XXVI:377-380, 1 textfig.

Urosalpinx

Carriker, Melbourne R., Dirk Van Zandt and Garry Charlton, 1967.
Gastropod Urosalpinx: pH of accessory boring organ while boring.
Science, 158(3803):920-922.

Urosalpinx

Medcof, J.C. and M.L.H. Thomas, 1964.
Canadian Atlantic oyster drills (Urosalpinx) - distribution and industrial importance.
J. Fish. Res. Bd. Can., 26(5):1121-1131

Urosalpinx cinerea

Adams, J.R., 1947
The oyster drill in Canada. Fish. Res. Bd., Canada, Prog. Repts. 37:14-18.

heteropods

Chen, Chin, and Davis B. Ericson, 1965.
Holoplanktonic Gastrpoda in the southern oceans.
Antarctic Jl., U.S.A., 2(5):200.

heteropods

Dales, R.P., 1952.
The distribution of some heteropod molluscs off the Pacific coast of North America.
Proc. Zool. Soc., London, 122:1007-1015, 7 textfigs.

heteropods

Franc, A., 1949
Hétéropodes, et autres Gastropodes planctoniques de Méditerranée occidentale.
J. de Conchyliologia 89: 209-230, 19 text figs. Trav. Sta. Zool. Villefranche-sur-Mer 10(3)

heteropods

Frontier, S., 1963.
Hétéropodes et ptéropodes récoltés dans le plancton de Nosy-Bé.
Trav. Centre Océanogr., Nosy-Bé, Cahiers, O.R.S.T.R.O.M., Océanogr., Paris, No. 6:213-227.

heteropods, anat.

Gabe, M. and M. Prenant, 1950.
Recherches sur la gaine radulaires des mollusques. 2. Données histologiques sur l'appareil radulaire des hétéropodes. (Trav. Sta. Zool., Villefranche-sur-mer 10(7)). Bull. Soc. Zool., France, 75(4):176-184, 2 textfigs., 1 pl.

heteropods

Ralph, Patricia M., 1957.
A guide to the New Zealand heteropod molluscs.
Tuatara 6(3):116-120.

Heteropods

Richter, Gotthard, 1968.
Heteropoden und Heteropodenlarven im Oberflachenplankton des Golfs von Neapel.
Pubbl. Staz. zool. Napoli, 36(3):346-400.

heteropods

Russell, F.S., and J.S. Colman, 1935
The Zooplankton. IV. The occurrence and seasonal distribution of the Tunicata, Mollusca and Coelenterata (siphonophora). Brit. Mus. (N.H.) Great Barrier Reef Expedition 1928-1929, Sci. Repts., 2(7):203-276, 30 text figs.

heteropods

Okutani, T., 1957.
On ptertrachean fauna in Japanese waters.
Bull. Tokai Reg. Fish. Res. Lab., 16:15-22.

molluscs, heteropod

Saxthivel, M. 1972.
Studies on Desmopterus Chun 1889 species in the Indian Ocean.
Meteor Forsch.-Ergebn. (D) 10: 46-57.

Heteropoda

Tesch, J. J., 1949
Heteropoda. Dana Report No. 34: 53 pp., 44 text figs., 5 pls.

Heteropoda

Tesch, J.J., 1910
X Pteropoda and Heteropoda. Percy Sladen Trust Expedition to the Indian Ocean in 1915. Trans. Linn. Soc., London, Zool., 2nd ser.14: 165-189, pls. 12-14.

heteropods

Tesch, J.J., 1906
Die Heteropoden der Siboga-Expedition.
Res. Expl. Siboga, 29:112 pp., 14 pls.

Monograph 51

heteropods

Tokioka, Takasi, 1961
The structure of the operculum of the species of Atlantidae (Gastropoda: Heteropoda) as a taxonomic criterion, with records of some pelagic mollusks of the North Pacific.
Publ. Seto Mar. Biol. Lab., Kyoto Univ., 9(2):267-332.

heteropods

Tokioka, T., 1955.
On some plankton animals collected by the Syunkotu-maru in May-June 1954. II. Shells of the Atlantidae (Heteropoda).
Publ. Seto Mar. Biol. Lab. 4(2/3):227-238, Pls. 15-16, 5 textfigs.

The drawings are beautiful.

heteropods

Tokioka, T., 1955.
Shells of Atlantidae (Heteropoda) collected by the Soyo-maru in the southern waters of Japan.
Publ. Seto Mar. Biol. Lab. 4(2/3):237-250, Pls. 17-18, 10 textfigs.

beautiful figures.

heteropods

Vayssière, A., 1904.
Mollusques Hétéropodes provenant des campagnes des Yachts Hirondelle et Princesse Alice.
Rés. Camp. Sci., Monaco, 26:1-65, Pls. 1-6.

Heteropods

Vayssière, A., 1927.
Note sur les Hétéropodes et sur les Euptéropodes recueillis de 1902 à 1924 près de la Principauté de Monaco. Bull. Inst. Océan., Monaco, No. 493:4 pp.

molluscs, heteropods, anat.-physiol.

Gabe, M. 1966.
Contribution à l'histologie de Firoloida desmaresti Lesueur.
Vie Milieu (A) 17 (2-A): 845-959.

heteropods, anat. physiol. [a]

Ipsel, S. van der 1972.
Notes on the identification and speciation of Heteropoda (Gastropoda).
Zool. Mededel. Leiden, 47: 545-560

molluscs, heteropods, lists of spp.

Aravindakshan, P.N., 1969
Preliminary report on the geographical distribution of the species of Carinariidae and Pterotracheidae (Heteropoda, Mollusca) from the International Indian Ocean Expedition. Bull. natn. Inst. Sci. India 38(2): 575-584. Also in: Coll. Repr. Nat. Inst. Oceanogr. Goa India 2(1968-1970).

pteropods, lists of spp.

Barth, Rudolf, e Tristão Alencar Pereira Oleivo 1968
Contribuição ao estudo dos moluscos planctônicos da região de Cabo Frio-RJ
Publ. Inst. Pesquis. Marinha, Brasil 029: 1-17.

Heteropods, lists of spp.

Frontier, S., 1965.
Données sur la faune pélagique vivant au large des côtes du Gabon du Congo et de l'Angola (0 à 15°S, 6°E à la côte): Hétéropodes et pteropodes.
ORSTOM Centre de Pointe Noire, Doc. 417: 11pp (mimeographed)
In: Recueil Trav. publiés de 1965 à 1966, Oceanogr. ORSTOM

heteropods, lists of spp.

Frontier, S., 1966.
Zooplancton de la région de Nosy Bé. 1. Programme des récoltes et techniques d'études. 2. Plancton de surface aux stations 5 et 10.
Cah. ORSTOM, Océanogr., 4(3):1-36.

pteropods, lists of spp.

Leal Rodriguez, Dora G. 1965.
Distribución de pteropodos en Veracruz, Ver.
Anales, Inst. Biol. Univ. Mex. 36 (1/2): 249-251.

heteropods, lists of spp.

McGowan, John A., 1967.
Distributional atlas of pelagic molluscs in the California Current region.
Calif. Coop. Ocean. Fish. Invest., 6:1-218.

heteropods, lists of spp.

Okutani, Takashi, 1965.
Heteropoda and Thecosomata in the Kuroshio. (In Japanese; English abstract).
Inf. Bull. Plankton., Japan, No. 12:37-39.

heteropods, lists of spp. [a]

Ipsel, S. van der 1972.
Notes on the identification and speciation of Heteropoda (Gastropoda).
Zool. Mededel. Leiden, 47: 545-560

heteropods, lists of spp.

Taki, Iwao, and Takashi Okutani, 1962
Reports on the biology of the "Umitaka-Maru" Expedition, Part 2. Planktonic gastropods collected by the training vessel "Umitaka-Maru" from the Pacific and Indian Oceans in the course of her Antarctic Expedition, 1956. J. Fac. Fish. and Animal Husband., Hiroshima Univ., 4(1/2):81-97.

heteropods, lists of spp. [a]

Tanaka, Tsuneo 1971.
Pteropoda and Heteropoda (Gastropoda, Mollusca) collected in the western Pacific Ocean in the northern summer 1968.
Kaiyo Rept. No.3: 27-36.

molluscs, heteropods, lists of spp. (annotated) [a]

Taylor, Danny D. and Leo Berner, Jr., 1970.
The Heteropoda (Mollusca: Gastropoda). Oceanogr Stud. Texas A & M Univ. 1: 231-244.

heteropods, lists of spp.

Thiriot-Quiévreux, Catherine, 1968.
Variations saisonnières des mollusques dans le plancton de la région de Banyuls-sur-Mer (zone sud du Golfe du Lion) novembre 1965-décembre 1967.
Vie Milieu, 19(1-B):35-83.

Atlanta sp.

Berkeley, E., and C. Berkeley, 1960
Some further records of pelagic Polychaeta from the northeast Pacific, north of Latitude 40°N. and east of Longitude 175°W., together with records of Siphonophora, Mollusca and Tunicata from the same region.
Canadian J. Zoology. 38:787-799.

Atlanta sp.

Tesch, J.J., 1906
Die Heteropoden der Siboga-Expedition.
Res. Expl. Siboga, 29:112 pp., 14 pls.

Monograph 51

Atlanta sp

Vane, F.R., and J.M. Colebrook, 1962
Continuous plankton records: Contributions towards a plankton atlas of the north-eastern Atlantic and North Sea. VI. The seasonal and annual distributions of the Gastropoda.
Bull. Mar. Ecology, 5(50):247-253.

Atlanta canicula

Tesch, J.J., 1906
Die Heteropoden der Siboga-Expedition.
Res. Expl. Siboga, 29:112 pp., 14 pls.

Monograph 51

Atlanta depressa

Tesch, J.J., 1906
Die Heteropoden der Siboga-Expedition.
Res. Expl. Siboga, 29:112 pp., 14 pls.

Monograph 51

Atlanta fusca

Franc, A., 1949
Hétéropodes, et autres Gastropodes planctoniques de Méditerranée occidentale.
J. de Conchyliologia 89: 209-230, 19 text figs. Trav. Sta. Zool. Villefranche-sur-Mer 10(3)

Atlanta fusca

Frontier, Serge, 1966.
Notes morphologiques sur les Atlanta récoltés dans le plancton de Nosy Bé (Madagascar).
Cah. ORSTOM, Sér. Océanogr., 4(2):131-139.

Atlanta fusca

Frontier, S., 1963.
Hétéropodes et ptéropodes récoltés dans le plancton de Nosy-Bé.
Trav. Centre Océanogr., Nosy-Bé, Cahiers, O.R.S.T.R.O.M., Océanogr., Paris, No. 6:213-227.

Atlanta fusca

Issel, R., 1915.
Atlantidae e Carinaria.
R. Com. Talassogr. Ital., Mem. 52:1-26, 3 pls.

Atlanta fusca

Richter, Gotthard. 1968.

Heteropoden und Heteropodenlarven im Oberflachenplankton des Golfs von Neapel.

Pubbl. Staz. zool. Napoli, 36(3):346-400.

Atlanta fusca

Tesch, J.J., 1949
Heteropoda. Dana Report No. 34: 53 pp., 44 text figs., 5 pls.

Atlanta fusca

Tesch, J.J., 1906
Die Heteropoden der Siboga-Expedition.
Res. Expl. Siboga, 29:112 pp., 14 pls.

Monograph 51

Atlanta fusca

Tokioka, Takasi, 1961
The structure of the operculum of the species of Atlantidae (Gastropoda: Heteropoda) as a taxonomic criterion, with records of some pelagic mollusks of the North Pacific.
Publ. Seto Mar. Biol. Lab., Kyoto Univ., 9(2):267-332.

Atlanta fusca

Tokioka, T., 1955.
Shells of Atlantidae (Heteropoda) collected by the Soyo-maru in the southern waters of Japan.
Publ. Seto Mar. Biol. Lab. 4(2/3):237-250, Pls. 17-18, 10 textfigs.

Atlanta fusca

van der Spoel, S., 1970.
The pelagic Mollusca from the Atlantide and Galathea expeditions collected in the East Atlantic.
Atlantide Rept. 11:99-139.

Atlanta gaudichaudi

Bedot, M., 1909
Sur la faune de l'Archipel Malais (resume).
Rev. Suisse Zool. 17:143-169.

Atlanta gaudichaudi

Frontier, Serge, 1966.
Notes morphologiques sur les Atlanta récoltées dans le plancton de Nosy Bé (Madagascar).
Cah. ORSTOM, Sér. Océanogr., 4(2):131-139.

Atlanta gaudichaudi

van der Spoel, S., 1970.
The pelagic Mollusca from the Atlantide and Galathea expeditions collected in the East Atlantic.
Atlantide Rept. 11:99-139.

Atlanta gaudichaudi

Tesch, J. J., 1949
Heteropoda. Dana Report No. 34: 53 pp., 44 text figs., 5 pls.

Atlanta gaudichaudi

Tesch, J.J., 1906
Die Heteropoden der Siboga-Expedition.
Res. Expl. Siboga, 29:112 pp., 14 pls.
Monograph 51

Atlanta gaudichaudi

Tokioka, Takasi, 1961
The structure of the operculum of the species of Atlantidae (Gastropoda: Heteropoda) as a taxonomic criterion, with records of some pelagic mollusks of the North Pacific.
Publ. Seto Mar. Biol. Lab., Kyoto Univ., 9(2):267-332.

Atlanta gaudichaudi

Tokioka, T., 1955.
Shells of Atlantidae (Heteropoda collected by the Soyo-maru in the southern waters of Japan.
Publ. Seto Mar. Biol. Lab., 4(2/3):237-250, Pls. 17-18, 10 textfigs.

Atlanta gaudichaudi

Tokioka, T., 1955.
On some plankton animals collected by the Syunkotu-maru in May-June 1954. II Shells of the Atlantidae (Heteropoda).
Publ. Seto Mar. Biol. Lab. 4(2/3):227-238, Pls. 15-16, 5 textfigs.

Atlanta gibbosa

Tesch, J. J., 1910
X Pteropoda and Heteropoda. Percy Sladen Trust Expedition to the Indian Ocean in 1915.
Trans. Linn. Soc., London, Zool., 2nd ser.14: 165-189, pls. 12-14.

Atlanta gibbosa

Tesch, J.J., 1906
Die Heteropoden der Siboga-Expedition.
Res. Expl. Siboga, 29:112 pp., 14 pls.
Monograph 51

Atlanta helicialis

Tesch, J.J., 1906
Die Heteropoden der Siboga-Expedition.
Res. Expl. Siboga, 29:112 pp., 14 pls.
Monograph 51

Atlanta helicinoides

Franc, A., 1949
Hétéropodes, et autres Gastropodes planctoniques de Méditerranée occidentale.
J. de Conchyliologia 89: 209-230, 19 text figs. Trav. Sta. Zool. Villefranche-sur-Mer 10(3)

Atlanta helicinoides

Frontier, Serge, 1966.
Notes morphologiques sur les Atlanta récoltées dans le plancton de Nosy Bé (Madagascar).
Cah. ORSTOM, Sér. Océanogr., 4(2): 131-139.

Atlanta helicinoides

Tesch, J. J., 1949
Heteropoda. Dana Report No. 34: 53 pp., 44 text figs., 5 pls.

Atlanta helicinoides

Tesch, J. J., 1910
X Pteropoda and Heteropoda. Percy Sladen Trust Expedition to the Indian Ocean in 1915.
Trans. Linn. Soc., London, Zool., 2nd ser.14: 165-189, pls. 12-14.

Atlanta helicinoides

Tesch, J.J., 1906
Die Heteropoden der Siboga-Expedition.
Res. Expl. Siboga, 29:112 pp., 14 pls.
Monograph 51

Atlanta helicinoides

Tokioka, Takasi, 1961
The structure of the operculum of the species of Atlantidae (Gastropoda: Heteropoda) as a taxonomic criterion, with records of some pelagic mollusks of the North Pacific.
Publ. Seto Mar. Biol. Lab., Kyoto Univ., 9(2):267-332.

Atlanta helicinoides

van der Spoel, S., 1970.
The pelagic Mollusca from the Atlantide and Galathea expeditions collected in the East Atlantic.
Atlantide Rept. 11:99-139.

Atlanta inclinata

Franc, A., 1949
Hétéropodes, et autres Gastropodes planctoniques de Méditerranée occidentale.
J. de Conchyliologia 89: 209-230, 19 text figs. Trav. Sta. Zool. Villefranche-sur-Mer 10(3)

Atlanta inclinata

Frontier, Serge, 1966.
Notes morphologiques sur les Atlanta récoltées dans le plancton de Nosy Bé (Madagascar).
Cah. ORSTOM, Sér. Océanogr., 4(2): 131-139.

Atlanta inclinata

Moore, H. B., 1949
The zooplankton of the upper waters of the Bermuda area of the North Atlantic. Bull. Bingham Ocean. Coll. 12(2):97, 208 text figs.

Atlanta inclinata

Tesch, J. J., 1949
Heteropoda. Dana Report No. 34: 53 pp., 44 text figs., 5 pls.

Atlanta inclinata

Tesch, J. J., 1910
X Pteropoda and Heteropoda. Percy Sladen Trust Expedition to the Indian Ocean in 1915.
Trans. Linn. Soc., London, Zool., 2nd ser.14: 165-189, pls. 12-14.

Atlanta inclinata

Tesch, J.J., 1906
Die Heteropoden der Siboga-Expedition.
Res. Expl. Siboga, 29:112 pp., 14 pls.
Monograph 51

Atlanta inclinata

Tokioka, Takasi, 1961
The structure of the operculum of the species of Atlantidae (Gastropoda: Heteropoda) as a taxonomic criterion, with records of some pelagic mollusks of the North Pacific.
Publ. Seto Mar. Biol. Lab., Kyoto Univ., 9(2):267-332.

Atlanta inclinata

Tokioka, T., 1955.
On some planktonic animals collected by the Syunkotu-maru in May-June 1954. II. Shells of the Atlantidae (Heteropoda).
Publ. Seto Mar. Biol. Lab. 4(2/3):227-238, Pls. 15-16. 5 textfigs.

Atlanta inclinata

Tokioka, T., 1955.
Shells of Atlantidae (Heteropoda) collected by the Soyo-maru in the southern waters of Japan.
Publ. Seto Mar. Biol. Lab. 4(2/3):237-250, Pls. 17-18, 10 textfigs.

Atlanta inclinata

van der Spoel, S., 1970.
The pelagic Mollusca from the Atlantide and Galathea expeditions collected in the East Atlantic.
Atlantide Rept. 11:99-139.

Atlanta inclinata

Vayssière, A., 1904.
Mollusques hétéropodes provenant des campagnes des yachts Hirondelle et Princesse Alice.
Rés. Camp. Sci., Monaco, 26:1-65, Pls. 1-6.

Atlanta inflata

Frontier, Serge, 1966.
Notes morphologiques sur les Atlanta récoltées dans le plancton de Nosy Bé (Madagascar).
Cah. ORSTOM, Sér. Océanogr., 4(2): 131-139.

Atlanta inflata

Issel, R., 1915.
Atlantidae e Carinaria.
R. Com. Talassogr. Ital., Mem. 52:1-26, 3 pls.

Atlanta inflata

Richter, Gotthard, 1968.
Heteropoden und Heteropodenlarven im Oberflachenplankton des Golfs von Neapel.
Pubbl. Staz. zool. Napoli, 36(3):346-400.

Atlanta inflata

Tesch, J. J., 1949
Heteropoda. Dana Report No. 34: 53 pp., 44 text figs., 5 pls.

Atlanta inflata

Tesch, J. J., 1910
X Pteropoda and Heteropoda. Percy Sladen Trust Expedition to the Indian Ocean in 1915.
Trans. Linn. Soc., London, Zool., 2nd ser.14: 165-189, pls. 12-14.

Atlanta inflata

Tesch, J.J., 1906
Die Heteropoden der Siboga-Expedition.
Res. Expl. Siboga, 29:112 pp., 14 pls.
Monograph 51

Atlanta inflata

Tokioka, Takasi, 1961
The structure of the operculum of the species of Atlantidae (Gastropoda: Heteropoda) as a taxonomic criterion, with records of some pelagic mollusks of the North Pacific.
Publ. Seto Mar. Biol. Lab., Kyoto Univ., 9(2):267-332.

Atlanta inflata

Tokioka, T., 1955.
On some plankton animals collected by the Syunkotu-maru in May-June 1954. II. Shells of the Atlantidae (Heteropoda).
Publ. Seto Mar. Biol. Lab. 4(2/3):227-238, Pls. 15-16, 5 textfigs.

Atlanta inflata
Tokioka, T., 1955.
Shells of Atlantidae (Heteropoda) collected by the Soyo-maru in the southern waters of Japan.
Publ. Seto Mar. Biol. Lab., 4(2/3):237-250, Pls. 17-18, 10 textfigs.

Atlanta involuta
Tesch, J.J., 1906
Die Heteropoden der Siboga-Expedition.
Res. Expl. Siboga, 29:112 pp., 14 pls.
Monograph 51

Atlanta Lamanoni
Tesch, J.J., 1906
Die Heteropoden der Siboga-Expedition.
Res. Expl. Siboga, 29:112 pp., 14 pls.
Monograph 51

Atlanta lesueuri
Frontier, Serge, 1966.
Notes morphologiques sur les Atlanta récoltées dans le plancton de Nosy Bé (Madagascar).
Cah. ORSTOM, Sér. Océanogr., 4(2):131-139.

Atlanta lesueuri
Issel, R., 1915.
Atlantidae e Carineria. R. Com. Talassogr. Ital. Mem. 52:1-26, 3 pls.

Atlanta lesuerii
Richter, Gotthard, 1968.
Heteropoden und Heteropodenlarven im Oberflachenplankton des Golfs von Neapel.
Pubbl. Staz. zool. Napoli, 36(3):346-400.

Atlanta lesueuri
Tesch, J. J., 1949
Heteropoda. Dana Report No. 34: 53 pp., 44 text figs., 5 pls.

Atlanta lesueuri
Tesch, J. J., 1910
X Pteropoda and Heteropoda. Percy Sladen Trust Expedition to the Indian Ocean in 1915.
Trans. Linn. Soc., London, Zool., 2nd ser.14: 165-189, pls. 12-14.

Atlanta lesueuri
Tesch, J.J., 1906
Die Heteropoden der Siboga-Expedition.
Res. Expl. Siboga, 29:112 pp., 14 pls.
Monograph 51

Atlanta lesueuri
Thiriot-Quiévreux, Catherine, 1969.
Organogenèse larvaire du genre Atlanta (mollusque hétéropode).
Vie Milieu (A) 20(2): 347-395.

Atlanta lesueuri
Tokioka, Takasi, 1961
The structure of the operculum of the species of Atlantidae (Gastropoda: Heteropoda) as a taxonomic criterion, with records of some pelagic mollusks of the North Pacific.
Publ. Seto Mar. Biol. Lab., Kyoto Univ., 9(2):267-332.

Atlanta leseuri
Tokioka, T., 1955.
On some plankton animals collected by the Syunkotu-maru in May-June 1954. II. Shells of the Atlantidae (Heteropoda).
Publ. Seto Mar. Biol. Lab. 4(2/3):227-238, Pls. 15-16, 5 textfigs.

Atlanta lesueuri
Tokioka, T., 1955.
Shells of Atlantidae (Heteropoda) collected by the Soyo-maru in the southern waters of Japan.
Publ. Seto Mar. Biol. Lab., 4(2/3):237-250, Pls. 17-18, 10 textfigs.

Atlanta leseuri
van der Spoel, S., 1970.
The pelagic Mollusca from the Atlantide and Galathea expeditions collected in the East Atlantic.
Atlantide Rept. 11:99-139.

Atlanta lesueuri
Vayssière, A., 1904.
Mollusques hétéropodes provenant des campagnes des yachts Hirondelle et Princesse Alice.
Rés. Camp. Sci., Monaco, 26:1-65, Pls. 1-6.

Atlanta mediterranea
Tesch, J.J., 1906
Die Heteropoden der Siboga-Expedition.
Res. Expl. Siboga, 29:112 pp., 14 pls.
Monograph 51

Atlanta pacifica n.sp.
Tokioka, T., 1955.
Shells of Atlantidae (Heteropoda) collected by the Soyo-maru in the southern waters of Japan.
Publ. Seto Mar. Biol. Lab., 4(2/3):237-250, Pls. 17-18, 10 textfigs.

Atlanta peresi n.sp.
Frontier, Serge, 1966.
Notes morphologiques sur les Atlanta récoltées dans le plancton de Nosy Bé (Madagascar).
Cah. ORSTOM, Sér. Océanogr., 4(2):131-139.

Atlanta peroni
Franc, A., 1949
Hétéropodes, et autres Gastropodes planctoniques de Méditerranée occidentale.
J. de Conchyliologia 89: 209-230, 19 text figs. Trav. Sta. Zool. Villefranche-sur-Mer 10(3)

Atlanta peroni
Frontier, Serge, 1966.
Notes morphologiques sur les Atlanta récoltées dans le plancton de Nosy Bé (Madagascar).
Cah. ORSTOM, Sér. Océanogr., 4(2):131-139.

Atlanta peroni
Issel, R., 1915.
Atlantidae e Carineria.
Mem. R. Com. Talassogr. Ital. 52:1-26, 3 pls.

Atlanta peronii
Moore, H. B., 1949
The zooplankton of the upper waters of the Bermuda area of the North Atlantic. Bull. Bingham Ocean. Coll. 12(2):97, 208 text figs.

Atlanta peronii
Richter, Gotthard, 1968.
Heteropoden und Heteropodenlarven im Oberflachenplankton des Golfs von Neapel.
Pubbl. Staz. zool. Napoli, 36(3):346-400.

Atlanta peroni Leseur
Russell, F.S., and J.S. Colman, 1935
The Zooplankton. IV. The occurrence and seasonal distribution of the Tunicata, Mollusca and Coelenterata (siphonophora). Brit. Mus. (N.H.) Great Barrier Reef Expedition 1928-1929, Sci. Repts., 2(7):203-276, 30 text figs.

Atlanta peroni
Morton, J.E., 1954.
The pelagic Mollusca of the Benguela Current. Pt. 1. First survey, R.R.S. William Scoresby, March 1951, with an account of the reproductive system and sexual succession of Limacina bulmodes.
Discovery Report 24:163-200.

Atlanta peroni
Tesch, J. J., 1949
Heteropoda. Dana Report No. 34: 53 pp., 44 text figs., 5 pls.

Atlanta peroni
Tesch, J. J., 1910
X Pteropoda and Heteropoda. Percy Sladen Trust Expedition to the Indian Ocean in 1915.
Trans. Linn. Soc., London, Zool., 2nd ser.14: 165-189, pls. 12-14.

Atlanta peroni
Tesch, J.J., 1906
Die Heteropoden der Siboga-Expedition.
Res. Expl. Siboga, 29:112 pp., 14 pls.
Monograph 51

Atlanta peroni
Tokioka, Takasi, 1961
The structure of the operculum of the species of Atlantidae (Gastropoda: Heteropoda) as a taxonomic criterion, with records of some pelagic mollusks of the North Pacific.
Publ. Seto Mar. Biol. Lab., Kyoto Univ., 9(2):267-332.

Atlanta peroni
van der Spoel, S., 1970.
The pelagic Mollusca from the Atlantide and Galathea expeditions collected in the East Atlantic.
Atlantide Rept. 11:99-139.

Atlanta peroni
Vayssière, A., 1904.
Mollusques hétéropodes provenant des campagnes des yachts Hirondelle et Princesse Alice.
Rés. Camp. Sci., Monaco, 26:1-65, Pls. 1-6.

Atlanta peroni
Tokioka, T., 1955.
On some plankton animals collected by the Syunkotu-maru in May-June 1954. II. Shells of the Atlantidae (Heteropoda).
Publ. Seto Mar. Biol. Lab. 4(2/3):227-238, Pls. 15-16, 5 textfigs.

Atlanta planorboides
Tesch, J.J., 1906
Die Heteropoden der Siboga-Expedition.
Res. Expl. Siboga, 29:112 pp., 14 pls.
Monograph 51

Atlanta primitia
Tesch, J.J., 1906
Die Heteropoden der Siboga-Expedition.
Res. Expl. Siboga, 29:112 pp., 14 pls.
Monograph 51

Atlanta quoyana
Franc, A., 1949
Hétéropodes, et autres Gastropodes planctoniques de Méditerranée occidentale.
J. de Conchyliologia 89: 209-230, 19 text figs. Trav. Sta. Zool. Villefranche-sur-Mer 10(3)

Atlanta quoyana
Tesch, J.J., 1906
Die Heteropoden der Siboga-Expedition.
Res. Expl. Siboga, 29:112 pp., 14 pls.
Monograph 51

Atlanta quoyana
Vayssière, A., 1904.
Mollusques hétéropodes provenant des campagnes des yachts Hirondelle et Princesse Alice.
Rés. Camp. Sci., Monaco, 26:1-65, Pls. 1-6.

Atlanta rosea
Franc, A., 1949
Hétéropodes, et autres Gastropodes planctoniques de Méditerranée occidentale.
J. de Conchyliologia 89: 209-230, 19 text figs. Trav. Sta. Zool. Villefranche-sur-Mer 10(3)

Atlanta rosea
Tesch, J.J., 1906
Die Heteropoden der Siboga-Expedition.
Res. Expl. Siboga, 29:112 pp., 14 pls.
Monograph 51

Atlanta souleyeti

Tesch, J.J., 1906
Die Heteropoden der Siboga-Expedition.
Res. Expl. Siboga, 29:112 pp., 14 pls.

Monograph 51

Atlanta steindachneri

Tesch, J.J., 1906
Die Heteropoden der Siboga-Expedition.
Res. Expl. Siboga, 29:112 pp., 14 pls.

Monograph 51

Atlanta tessellata

Tesch, J.J., 1906
Die Heteropoden der Siboga-Expedition.
Res. Expl. Siboga, 29:112 pp., 14 pls.

Monograph 51

Atlanta turriculata

Frontier, Serge, 1966.
Notes morphologiques sur les Atlanta récoltées dans le plancton de Nosy Bé (Madagascar).
Cah. ORSTOM. Sér. Océanogr., 4(2):131-139.

Atlanta turriculata

Tesch, J.J., 1949
Heteropoda. Dana Report No. 34: 53 pp., 44 text figs., 5 pls.

Atlanta turriculata

Tesch, J.J., 1910
X Pteropoda and Heteropoda. Percy Sladen Trust Expedition to the Indian Ocean in 1915.
Trans. Linn. Soc., London, Zool., 2nd ser.14: 165-189, pls. 12-14.

Atlanta turriculata

Tesch, J.J., 1906
Die Heteropoden der Siboga-Expedition.
Res. Expl. Siboga, 29:112 pp., 14 pls.

Monograph 51

Atlanta turriculata

Tokioka, Takasi, 1961
The structure of the operculum of the species of Atlantidae (Gastropoda: Heteropoda) as a taxonomic criterion, with records of some pelagic mollusks of the North Pacific.
Publ. Seto Mar. Biol. Lab., Kyoto Univ., 9(2):267-332.

Atlanta turriculata

Tokioka, T., 1955.
On some plankton animals collected by the Syunkotu-maru in May-June 1954. II. Shells of the Atlantidae (Heteropoda).
Publ. Seto Mar. Biol. Lab. 4(2/3):227-238, Pls. 15-16, 5 textfigs.

Atlanta turruculata

Tokioka, T., 1955.
Shells of Atlantidae (Heteropoda) collected by the Soyo-maru in the southern waters of Japan.
Publ. Seto Mar. Biol. Lab., 4(2/3):237-250, Pls. 17-18, 10 textfigs.

Atlanta violacea

Tesch, J.J., 1906
Die Heteropoden der Siboga-Expedition.
Res. Expl. Siboga, 29:112 pp., 14 pls.

Monograph 51

Cardiapoda sp.

Tesch, J.J., 1906
Die Heteropoden der Siboga-Expedition.
Res. Expl. Siboga, 29:112 pp., 14 pls.

Monograph 51

Cardiapoda sp.

Tokioka, Takasi, 1961
The structure of the operculum of the species of Atlantidae (Gastropoda: Heteropoda) as a taxonomic criterion, with records of some pelagic mollusks of the North Pacific.
Publ. Seto Mar. Biol. Lab., Kyoto Univ., 9(2):267-332.

Cardiapoda acuta

Tesch, J.J., 1910
X Pteropoda and Heteropoda. Percy Sladen Trust Expedition to the Indian Ocean in 1915.
Trans. Linn. Soc., London, Zool., 2nd ser.14: 165-189, pls. 12-14.

Cardiapoda carinata

Tesch, J.J., 1906
Die Heteropoden der Siboga-Expedition.
Res. Expl. Siboga, 29:112 pp., 14 pls.

Monograph 51

Cardiapoda caudina

Tesch, J.J., 1906
Die Heteropoden der Siboga-Expedition.
Res. Expl. Siboga, 29:112 pp., 14 pls.

Monograph 51

Cardiapoda pedunculata

Tesch, J.J., 1906
Die Heteropoden der Siboga-Expedition.
Res. Expl. Siboga, 29:112 pp., 14 pls.

Monograph 51

Cardiapoda placenta

Bedot, M., 1909
Sur la faune de l'Archipel Malais (resume).
Rev. Suisse Zool. 17:143-169.

Cardiapoda placenta

Tesch, J.J., 1949
Heteropoda. Dana Report No. 34: 53 pp., 44 text figs., 5 pls.

Cardiapoda placenta

Tesch, J.J., 1906
Die Heteropoden der Siboga-Expedition.
Res. Expl. Siboga, 29:112 pp., 14 pls.

Monograph 51

Cardiapoda placenta

Tokioka, Takasi, 1961
The structure of the operculum of the species of Atlantidae (Gastropoda: Heteropoda) as a taxonomic criterion, with records of some pelagic mollusks of the North Pacific.
Publ. Seto Mar. Biol. Lab., Kyoto Univ., 9(2):267-332.

Cardiopoda placenta

van der Spoel, S., 1970.
The pelagic Mollusca from the Atlantide and Galathea expeditions collected in the East Atlantic.
Atlantide Rept. 11:99-139.

Cardiapoda richardi

Moore, H. B., 1949
The zooplankton of the upper waters of the Bermuda area of the North Atlantic. Bull. Bingham Ocean. Coll. 12(2):97, 208 text figs.

Cardiapoda richardi

Tesch, J. J., 1949
Heteropoda. Dana Report No. 34: 53 pp., 44 text figs., 5 pls.

Cardiapoda richardi

Tesch, J.J., 1906
Die Heteropoden der Siboga-Expedition.
Res. Expl. Siboga, 29:112 pp., 14 pls.

Monograph 51

Cardiopoda richardi n.sp.

Vayssière, A., 1904.
Mollusques hétéropodes provenant des campagnes des Yachts Hirondelle et Princesse Alice.
Rés. Camp. Sci., Monaco, 26:1-65, Pls. 1-6.

Cardiapoda sublaevis

Tesch, J. J., 1910
X Pteropoda and Heteropoda. Percy Sladen Trust Expedition to the Indian Ocean in 1915.
Trans. Linn. Soc., London, Zool., 2nd ser.14: 165-189, pls. 12-14.

Cardiapoda trachydermon

Tesch, J. J., 1910
X Pteropoda and Heteropoda. Percy Sladen Trust Expedition to the Indian Ocean in 1915.
Trans. Linn. Soc., London, Zool., 2nd ser.14: 165-189, pls. 12-14.

Carinaria sp.

Moore, H. B., 1949
The zooplankton of the upper waters of the Bermuda area of the North Atlantic. Bull. Bingham Ocean. Coll. 12(2):97, 208 text figs.

Carineria sp.

Tesch, J. J., 1910
X Pteropoda and Heteropoda. Percy Sladen Trust Expedition to the Indian Ocean in 1915.
Trans. Linn. Soc., London, Zool., 2nd ser.14: 165-189, pls. 12-14.

Carinaria sp.

Tesch, J.J., 1906
Die Heteropoden der Siboga-Expedition.
Res. Expl. Siboga, 29:112 pp., 14 pls.

Monograph 51

Carinaria atlantica

Tesch, J.J., 1906
Die Heteropoden der Siboga-Expedition.
Res. Expl. Siboga, 29:112 pp., 14 pls.

Monograph 51

Carinaria australia

Tesch, J.J., 1906
Die Heteropoden der Siboga-Expedition.
Res. Expl. Siboga, 29:112 pp., 14 pls.

Monograph 51

Carinaria australis

Vayssière, A., 1904.
Mollusques hétéropodes provenant des campagnes des Yachts Hirondelle et Princesse Alice.
Rés. Camp. Sci., Monaco, 26:1-65, Pls. 1-6.

Carinaria cithara

Tesch, J. J., 1949
Heteropoda. Dana Report No. 34: 53 pp., 44 text figs., 5 pls.

Carinaria cithara

Tesch, J.J., 1906
Die Heteropoden der Siboga-Expedition.
Res. Expl. Siboga, 29:112 pp., 14 pls.

Monograph 51

Carinaria cithara procumbens n.var.

Tesch, J. J., 1949
Heteropoda. Dana Report No. 34: 53 pp., 44 text figs., 5 pls.

Carinaria cornucopia

Tesch, J.J., 1906
Die Heteropoden der Siboga-Expedition.
Res. Expl. Siboga, 29:112 pp., 14 pls.

Monograph 51

Carinaria cristata

Okutani, Takashi, 1961.
Notes on the genus Carinaria (Heteropoda) from Japanese and adjacent waters.
Publ. Seto Mar. Biol. Lab., Kyoto Univ., 9(2):333-352, Pls. 12-13.

Carinaria cristata

Tesch, J. J., 1949
Heteropoda. Dana Report No. 34: 53 pp., 44 text figs., 5 pls.

Carinaria cristata

Tesch, J.J., 1906
Die Heteropoden der Siboga-Expedition.
Res. Expl. Siboga, 29:112 pp., 14 pls.

Monograph 51

Carinaria (cristata var.?) japonica

Okutani, Takashi, 1961.
Notes on the genus Carinaria (Heteropoda) from Japanese and adjacent waters.
Publ. Seto Mar. Biol. Lab., Kyoto Univ., 9(2):333-352, Pls. 12-13.

Carinaria depressa

Tesch, J.J., 1906
Die Heteropoden der Siboga-Expedition.
Res. Expl. Siboga, 29:112 pp., 14 pls.

Monograph 51

Carinaria fragilis

Tesch, J.J., 1906
Die Heteropoden der Siboga-Expedition.
Res. Expl. Siboga, 29:112 pp., 14 pls.

Monograph 51

Carinaria galea

Okutani, Takashi, 1961.
Notes on the genus Carinaria (Heteropoda) from Japanese and adjacent waters.
Publ. Seto Mar. Biol. Lab., Kyoto Univ., 9(2):333-352, Pls. 12-13.

Carinaria galea

Tesch, J. J., 1949
Heteropoda. Dana Report No. 34: 53 pp., 44 text figs., 5 pls.

Carinaria galea

Tesch, J.J., 1906
Die Heteropoden der Siboga-Expedition.
Res. Expl. Siboga, 29:112 pp., 14 pls.

Monograph 51

Carinaria galea

Tokioka, Takasi, 1961
The structure of the operculum of the species of Atlantidae (Gastropoda: Heteropoda) as a taxonomic criterion, with records of some pelagic mollusks of the North Pacific.
Publ. Seto Mar. Biol. Lab., Kyoto Univ., 9(2):267-332.

Carinaria gaudichaudi

Tesch, J.J., 1906
Die Heteropoden der Siboga-Expedition.
Res. Expl. Siboga, 29:112 pp., 14 pls.

Monograph 51

Carinaria grimaldi

Tesch, J.J., 1906
Die Heteropoden der Siboga-Expedition.
Res. Expl. Siboga, 29:112 pp., 14 pls.

Monograph 51

Carinaria grimaldii n.sp.

Vayssière, A., 1904.
Mollusques hétéropodes provenant des campagnes des yachts Hirondelle et Princesse Alice.
Rés. Camp. Sci., Monaco, 26:1-65, Pls. 1-6.

Carinaria lamarcki

Alvariño, A., 1957.
Estudio del zooplancton del Mediterráneo occidental. Zooplancton del Atlántico Ibérico. Campañas del "Xauen" en el verano del 1954.
Bol. Inst. Oceanol., 81-82:1-26; 1-51.

(Español)

Carinaria lamarki

Dales, R.P., 1953.
The distribution of some heteropod molluscs of the Pacific coast of North America.
Proc. Zool. Soc., London, 122(4):1007-1015, 7 textfigs.

Carinaria lamarki

Dales, R.P., 1952.
The distribution of some heteropod molluscs off the Pacific coast of North America.
Proc. Zool. Soc., London, 122:1007-1015, 7 textfigs.

Carinaria lamarcki

Issel, R., 1915.
Atlantidae e Carinaria.
Mem. R. Com. Talassogr., 52:1-26, 3 pls.

(Ital.)

Carinaria Lamarcki

Okutani, Takashi, 1961.
Notes on the genus Carinaria (Heteropoda) from Japanese and adjacent waters.
Publ. Seto Mar. Biol. Lab., Kyoto Univ., 9(2):333-352, Pls. 12-13.

Carinaria lamarckii

Richter, Gotthard, 1968.
Heteropoden und Heteropodenlarven im Oberflachenplankton des Golfs von Neapel.
Pubbl. Staz. zool. Napoli, 36(3):346-400.

Carinaria lamarcki

Tesch, J. J., 1949
Heteropoda. Dana Report No. 34: 53 pp., 44 text figs., 5 pls.

Carinaria lamarcki

Tesch, J.J., 1906
Die Heteropoden der Siboga-Expedition.
Res. Expl. Siboga, 29:112 pp., 14 pls.

Monograph 51

Carinaria Lamarcki challengeri

Okutani, Takashi, 1961.
Notes on the genus Carinaria (Heteropoda) from Japanese and adjacent waters.
Publ. Seto Mar. Biol. Lab., Kyoto Univ., 9(2):333-352, Pls. 12-13.

Carinaria mediterranea

Franc, A., 1949
Hétéropodes, et autres Gastropodes planctoniques de Méditerranée occidentale.
J. de Conchyliologia 89: 209-230, 19 text figs. Trav. Sta. Zool. Villefranche-sur-Mer 10(3)

Carinaria mediterranea

Gabe, M., and M. Prenant, 1950.
Recherches sur la gaine radulaires des mollusques. 2. Données histologiques sur l'appareil radulaire des hétéropodes. (Trav. Sta. Zool., Villefranche-sur-mer 10(7); Bull. Soc. Zool., France, 75(4):176-184, 2 textfigs., 1 pl.

Carinaria mediterranea

Vayssière, A., 1904.
Mollusques hétéropodes provenant des campagnes des yachts Hirondelle et Princesse Alice.
Rés. Camp. Sci., Monaco, 26:1-65, Pls. 1-6.

Carinaria pseudo-rugosa

Tesch, J.J., 1906
Die Heteropoden der Siboga-Expedition.
Res. Expl. Siboga, 29:112 pp., 14 pls.

Monograph 51

Carinaria pseudo-rugosa n.sp.

Vayssière, A., 1904.
Mollusques hétéropodes provenant des campagnes des yachts Hirondelle et Princesse Alice.
Rés. Camp. Sci., Monaco, 26:1-65, Pls. 1-6.

Carinaria punctata

Tesch, J.J., 1906
Die Heteropoden der Siboga-Expedition.
Res. Expl. Siboga, 29:112 pp., 14 pls.

Monograph 51

Firola coronata

Gabe, M., and M. Prenant, 1950.
Recherches sur la gaine radulaires des mollusques 2. Données histologiques sur l'appareil radulaire des Hétéropodes. (Trav. Sta. Zool., Villefranche-sur-mer 10(7)). Bull. Soc. Zool., France, 75(4):

Firola coronata

Vayssière, A., 1927.
Notes sur les Hétéropodes et sur les Euptéropodes recueillis de 1902 à 1924 près de la Principauté de Monaco. Bull. Inst. Océan., Monaco, No. 493:4 pp.

Firola coronata

Vayssière, A., 1904.
Mollusques hétéropodes provenant des campagnes des yachts Hirondelle et Princesse Alice.
Rés. Camp. Sci., Monaco, 26:1-65, Pls. 1-6.

Firola desmaresti

Moore, H. B., 1949
The zooplankton of the upper waters of the Bermuda area of the North Atlantic. Bull. Bingham Ocean. Coll. 12(2):97, 208 text figs.

Firoloïda desmaresti

Vayssière, A., 1904.
Mollusques hétéropodes provenant des campagnes des yachts Hirondelle et Princesse Alice.
Rés. Camp. Sci., Monaco, 26:1-65, Pls. 1-6.

Firola gegenbauri

Moore, H. B., 1949
The zooplankton of the upper waters of the Bermuda area of the North Atlantic. Bull. Bingham Ocean. Coll. 12(2):97, 208 text figs.

Firola gegenbauri n.sp.

Vayssière, A., 1904.
Mollusques hétéropodes provenant des campagnes des yachts Hirondelle et Princesse Alice.
Rés. Camp. Sci., Monaco, 26:1-65, Pls. 1-6.

Firola hippocampus

Moore, H. B., 1949
The zooplankton of the upper waters of the Bermuda area of the North Atlantic. Bull. Bingham Ocean. Coll. 12(2):97, 208 text figs.

Firola hippocampus

Vayssière, A., 1927.
Notes sur les Hétéropodes et sur les Euptéropodes recueillis de 1902 à 1924 près de la Principauté de Monaco. Bull. Inst. Océan., Monaco, No 493:4 pp.

Firola hippocampus

Vayssière, A., 1904.
Mollusques hétéropodes provenant des campagnes des yachts Hirondelle et Princesse Alice.
Rés. Camp. Sci., Monaco, 26:1-65, Pls. 1-6.

Firola mutica

Gabe, M., and M. Prenant, 1950.
Recherches sur la gaine radulaires des mollusques 2. Données histologiques sur l'appareil radulaire des hétéropodes. (Trav. Sta. Zool., Villefranche-sur-mer 10(7)). Bull. Soc. Zool., France, 75(4): 176-184, 2 textfigs., 1 pl.

Firola mutica

Vayssière, A., 1927.
Note sur les Hétéropodes et sur les Euptéropodes recueillis de 1902 à 1924 près de la Principauté de Monaco. Bull. Inst. Océan., Monaco, No. 493:4 pp.

Firola mutica

Vayssière, A., 1904.
Mollusques hétéropodes provenant des campagnes des yachts Hirondelle et Princesse Alice.
Rés. Camp. Sci., Monaco, 26:1-65, Pls. 1-6.

Firola Souleyeti
Vayssière, A., 1927.
Note sur les Hétéropodes et sur les Euptéropodes recueillis de 1902 à 1924 près de la Principauté de Monaco. Bull. Inst. Océan., Monaco, No. 493:4 pp.

Firola souleyeti n.sp.
Vayssière, A., 1904.
Mollusques hétéropodes provenant des campagnes des yachts Hirondelle et Princesse Alice. Rés. Camp. Sci., Monaco, 26:1-65, Pls. 1-6.

Firola talismani
Vayssière, A., 1904.
Mollusques hétéropodes provenant des campagnes des yachts Hirondelle et Princesse Alice. Rés. Camp. Sci., Monaco, 26:1-65, Pls. 1-6.

Firoloida sp.
Tesch, J.J., 1906
Die Heteropoden der Siboga-Expedition. Res. Expl. Siboga, 29:112 pp., 14 pls.
Monograph 51

Firoloida aculeata
Tesch, J.J., 1906
Die Heteropoden der Siboga-Expedition. Res. Expl. Siboga, 29:112 pp., 14 pls.
Monograph 51

Firoloida blainvilleana
Tesch, J.J., 1906
Die Heteropoden der Siboga-Expedition. Res. Expl. Siboga, 29:112 pp., 14 pls.
Monograph 51

Firoloida desmaresti
Alvariño A., 1957.
Estudio del zooplancton del Mediterráneo occidental. Zooplancton del Atlántico Ibérico. Campañas del "Xauen" en el verano del 1954. Bol. Inst. Español Oceanogr., 81-82:1-26;1-51.

Firoloidea desmaresti
Bedot, M., 1909
Sur la faune de l'Archipel Malais (resume). Rev. Suisse Zool. 17:143-169.

Firoloidea desmarestia
Bigelow, H.B., and M. Sears, 1939
Studies of the waters of the continental shelf; Cape Cod to Chesapeake Bay. III. A volumetric study of the zooplankton. Mem. M.C.Z. 54(4):183-378, 42 text figs.

Firoloidea desmaresti
Dales, R.P., 1953.
The distribution of some heteropod molluscs of the Pacific coast of North America. Proc. Zool. Soc., London, 122(4):1007-1015, 7 textfigs.

Firoloidea desmaresti
Dales, R.P., 1952.
The distribution of some heteropod molluscs off the Pacific coast of North America. Proc. Zool. Soc., London, 122:1007-1015, 7 textfigs.

Firoloida desmaresti
Franc, A., 1949
Hétéropodes, et autres Gastropodes planctoniques de Méditerranée occidentale. J. de Conchyliologia 89: 209-230, 19 text figs. Trav. Sta. Zool. Villefranche-sur-Mer 10(3)

molluscs, heteropods-
Firoloida desmaresti

Gabe, M. 1966.
Contribution à l'histologie de Firoloida desmaresti Lesueur. Vie Milieu (A) 17 (2-A): 845-959.

Firoloida desmaresti
Owre, Harding B., 1964.
Observations on development of the heteropod molluscs Pterotrachea hippocampus and Firoloidea desmaresti. Bull. Mar. Sci., Gulf and Caribbean, 14(4):529-538.

Firoloida desmarestii
Richter, Gotthard, 1968.
Heteropoden und Heteropodenlarven im Oberflachenplankton des Golfs von Neapel.
Pubbl. Staz. zool. Napoli, 36(3):346-400.

Firoloida desmareste
Tesch, J.J., 1949
Heteropoda. Dana Report No. 34: 53 pp., 44 text figs., 5 pls.

Firoloida demarestia
Tesch, J.J., 1906
Die Heteropoden der Siboga-Expedition. Res. Expl. Siboga, 29:112 pp., 14 pls.
Monograph 51

Firoloida desmaresti
Thomopoulos, T., 1952.
Notes sur le plancton de la Baie de Banyuls. Vie et Milieu, Bull. Lab. Arago, Univ. Paris, 3(3):327-335, 8 textfigs.

Firoloida desmaresti
Tokioka, Takasi, 1961
The structure of the operculum of the species of Atlantidae (Gastropoda: Heteropoda) as a taxonomic criterion, with records of some pelagic mollusks of the North Pacific. Publ. Seto Mar. Biol. Lab. Kyoto Univ., 9(2):267-332.

Firoloidea desmaresti
van der Spoel, S., 1970.
The pelagic Mollusca from the Atlantide and Galathea expeditions collected in the East Atlantic. Atlantide Rept. 11:99-139.

Firoloida gaimardi
Tesch, J.J., 1906
Die Heteropoden der Siboga-Expedition. Res. Expl. Siboga, 29:112 pp., 14 pls.
Monograph 51

Firoloida gracilis
Tesch, J.J., 1906
Die Heteropoden der Siboga-Expedition. Res. Expl. Siboga, 29:112 pp., 14 pls.
Monograph 51

Firoloida kowalewskyi
Tesch, J.J., 1910
X Pteropoda and Heteropoda. Percy Sladen Trust Expedition to the Indian Ocean in 1915. Trans. Linn. Soc., London, Zool., 2nd ser.14:165-189, pls. 12-14.

Firoloida kowalewskyi
Tesch, J.J., 1906
Die Heteropoden der Siboga-Expedition. Res. Expl. Siboga, 29:112 pp., 14 pls.
Monograph 51

Firoloida kowalewskyi n.sp.
Vayssière, A., 1904.
Mollusques hétéropodes provenant des campagnes des yachts Hirondelle et Princesse Alice. Rés. Camp. Sci., Monaco, 26:1-65, Pls. 1-6.

Firoloida lesueuri
Tesch, J.J., 1906
Die Heteropoden der Siboga-Expedition. Res. Expl. Siboga, 29:112 pp., 14 pls.
Monograph 51

Firoloidea lesueuri
Vannucci, M., 1951.
Resultados cientificos do cruzeiro do "Baependi" e do "Vega" à Ilha de Trinidade. O genero Firoidoida, Prosobranchia Heteropoda. Bol. Inst. Paulista Ocean. 2(2):73-89, 2 pls.

Firoloida vigilans
Tesch, J.J., 1906
Die Heteropoden der Siboga-Expedition. Res. Expl. Siboga, 29:112 pp., 14 pls.
Monograph 51

Oxygyrus n.sp.
Tesch, J.J., 1906
Die Heteropoden der Siboga-Expedition. Res. Expl. Siboga, 29:112 pp., 14 pls.
Monograph 51

Oxygyrus inflatus
Tesch, J.J., 1906
Die Heteropoden der Siboga-Expedition. Res. Expl. Siboga, 29:112 pp., 14 pls.
Monograph 51

Oxygyrus keraudreni
Franc, A., 1949
Hétéropodes, et autres Gastropodes planctoniques de Méditerranée occidentale. J. de Conchyliologia 89: 209-230, 19 text figs. Trav. Sta. Zool. Villefranche-sur-Mer 10(3)

Oxygyrus keraudreni
Issel, R., 1915.
Atlantidae e Carinaria. R. Com. Talassogr. Ital., Mem:52:1-26, 3 pls.

Oxygyrus keraudreni
Richter, Gotthard, 1968.
Heteropoden und Heteropodenlarven im Oberflachenplankton des Golfs von Neapel.
Pubbl. Staz. zool. Napoli, 36(3):346-400.

Oxygyrus keraudreni
Tesch, J.J., 1949
Heteropoda. Dana Report No. 34: 53 pp., 44 text figs., 5 pls.

Oxygyrus keraudreni
Tesch, J.J., 1906
Die Heteropoden der Siboga-Expedition. Res. Expl. Siboga, 29:112 pp., 14 pls.
Monograph 51

Oxygyrus keraudreni
Tokioka, Takasi, 1961
The structure of the operculum of the species of Atlantidae (Gastropoda: Heteropoda) as a taxonomic criterion, with records of some pelagic mollusks of the North Pacific. Publ. Seto Mar. Biol. Lab. Kyoto Univ., 9(2):267-332.

Oxygyrus keraudreni
Tokioka, T., 1955.
Shells of Atlantidae (Heteropoda) collected by the Soyo-maru in the southern waters of Japan. Publ. Seto Mar. Biol. Lab. 4(2/3):237-250, Pls. 17-18, 10 textfigs.

Oxygyrus keraudreni
van der Spoel, S., 1970.
The pelagic Mollusca from the Atlantide and Galathea expeditions collected in the East Atlantic. Atlantide Rept. 11:99-139.

Oxygyrus keraudreni
Vayssière, A., 1927.
Note sur les Hétéropodes et sur les Euptéropodes recueillis de 1902 à 1927 près de la Principauté de Monaco. Bull. Inst. Océan, Monaco, No. 493:4 pp.

Oxygyrus Keraudreni
Vayssière, A., 1908.
Mollusques hétéropodes provenant des campagnes des yachts Hirondelle et Princesse Alice.
Rés. Camp. Sci., Monaco, 26:1-65, Pls. 1-6.

Oxygyrus rangi
Tesch, J. J., 1910
X Pteropoda and Heteropoda. Percy Sladen Trust Expedition to the Indian Ocean in 1915.
Trans. Linn. Soc., London, Zool., 2nd ser.14: 165-189, pls. 12-14.

Oxygyrus rangi
Tesch, J.J., 1906
Die Heteropoden der Siboga-Expedition.
Res. Expl. Siboga, 29:112 pp., 14 pls.

Monograph 51

Protatlanta sculpta mediterranea n.var.
Issel, R., 1915.
Atlantidae e Carinaria.
Mem. R. Com. Talassogr. 52:1-26, 3 pls.

Protatlanta souleyetii
Richter, Gotthard, 1968.
Heteropoden und Heteropodenlarven im Oberflachenplankton des Golfs von Neapel.
Pubbl. Staz. zool. Napoli, 36(3):346-400.

Protatlanta souleyeti
Tesch, J. J., 1949
Heteropoda. Dana Report No. 34: 53 pp., 44 text figs., 5 pls.

Protatlanta souleyeti
Tesch, J. J., 1910
X Pteropoda and Heteropoda. Percy Sladen Trust Expedition to the Indian Ocean in 1915.
Trans. Linn. Soc., London, Zool., 2nd ser.14: 165-189, pls. 12-14.

?Protatlanta souleyeti
Tokioka, Takasi, 1961
The structure of the operculum of the species of Atlantidae (Gastropoda: Heteropoda) as a taxonomic criterion, with records of some pelagic mollusks of the North Pacific.
Publ. Seto Mar. Biol. Lab., Kyoto Univ., 9(2):267-332.

Pterosoma
Tesch, J.J., 1906
Die Heteropoden der Siboga-Expedition.
Res. Expl. Siboga, 29:112 pp., 14 pls.

Monograph 51

Pterosoma planum
Tesch, J. J., 1949
Heteropoda. Dana Report No. 34: 53 pp., 44 text figs., 5 pls.

Pterosoma planum
Tesch, J. J., 1910
X Pteropoda and Heteropoda. Percy Sladen Trust Expedition to the Indian Ocean in 1915.
Trans. Linn. Soc., London, Zool., 2nd ser.14: 165-189, pls. 12-14.

Pterosoma planum
Tokioka, Takasi, 1961
The structure of the operculum of the species of Atlantidae (Gastropoda: Heteropoda) as a taxonomic criterion, with records of some pelagic mollusks of the North Pacific.
Publ. Seto Mar. Biol. Lab., Kyoto Univ., 9(2):267-332.

Pterotrachea sp.
Tesch, J. J., 1910
X Pteropoda and Heteropoda. Percy Sladen Trust Expedition to the Indian Ocean in 1915.
Trans. Linn. Soc., London, Zool., 2nd ser.14: 165-189, pls. 12-14.

Pterotrachea sp.
Tesch, J.J., 1906
Die Heteropoden der Siboga-Expedition.
Res. Expl. Siboga, 29:112 pp., 14 pls.

Monograph 51

Pterotrachea sp.
Tokioka, Takasi, 1961
The structure of the operculum of the species of Atlantidae (Gastropoda: Heteropoda) as a taxonomic criterion, with records of some pelagic mollusks of the North Pacific.
Publ. Seto Mar. Biol. Lab., Kyoto Univ., 9(2):267-332.

Pterotrachea adamastor
Tesch, J.J., 1906
Die Heteropoden der Siboga-Expedition.
Res. Expl. Siboga, 29:112 pp., 14 pls.

Monograph 51

Pterotrachea coronata
Dales, R.P., 1953.
The distribution of some heteropod molluscs of the Pacific coast of North America.
Proc. Zool. Soc., London, 122(4):1007-1015, 7 textfigs.

Pterotrachea coronata
Dales, R.P., 1952.
The distribution of some heteropod molluscs off the Pacific coast of North America.
Proc. Zool. Soc., London, 122:1007-1015, 7 textfigs.

Pterotrachea (Firola) coronata
Franc, A., 1949
Hétéropodes, et autres Gastropodes planctoniques de Méditerranée occidentale.
J. de Conchyliologia 89: 209-230, 19 text figs. Trav. Sta. Zool. Villefranche-sur-Mer 10(3)

Pterotrachea coronata
Okutani, T., 1957.
On pterotrachean fauna in Japanese waters.
Bull. Tokai Reg. Fish. Res. Lab., 16:15-22.

Pterotrachea coronata
Okutani, Takashi, and Tadashige Habe, 1960.
Pterotrachea coronata (Forskål) a heteropod mollusc from South African waters.
Ann. Natal Mus., 14(3):513-515.

Petrotrachea coronata
Richter, Gotthard, 1968.
Heteropoden und Heteropodenlarven im Oberflachenplankton des Golfs von Neapel.
Pubbl. Staz. zool. Napoli, 36(3):346-400.

Pterotrachea coronata
Tesch, J. J., 1949
Heteropoda. Dana Report No. 34: 53 pp., 44 text figs., 5 pls.

Pterotrachea coronata
Tesch, J.J., 1906
Die Heteropoden der Siboga-Expedition.
Res. Expl. Siboga, 29:112 pp., 14 pls.

Monograph 51

Pterotrachea coronata
Tokioka, Takasi, 1961
The structure of the operculum of the species of Atlantidae (Gastropoda: Heteropoda) as a taxonomic criterion, with records of some pelagic mollusks of the North Pacific.
Publ. Seto Mar. Biol. Lab., Kyoto Univ., 9(2):267-332.

Pterotrachea cuviera
Tesch, J.J., 1906
Die Heteropoden der Siboga-Expedition.
Res. Expl. Siboga, 29:112 pp., 14 pls.

Monograph 51

Pterotrachea edwardsi
Tesch, J.J., 1906
Die Heteropoden der Siboga-Expedition.
Res. Expl. Siboga, 29:112 pp., 14 pls.

Monograph 51

Pterotrachea forskalia
Tesch, J.J., 1906
Die Heteropoden der Siboga-Expedition.
Res. Expl. Siboga, 29:112 pp., 14 pls.

Monograph 51

Pterotrachea frederica
Tesch, J.J., 1906
Die Heteropoden der Siboga-Expedition.
Res. Expl. Siboga, 29:112 pp., 14 pls.

Monograph 51

Pterotrachea frederici
Tesch, J.J., 1906
Die Heteropoden der Siboga-Expedition.
Res. Expl. Siboga, 29:112 pp., 14 pls.

Monograph 51

Pterotrachea frederici(ana)
Tesch, J.J., 1906
Die Heteropoden der Siboga-Expedition.
Res. Expl. Siboga, 29:112 pp., 14 pls.

Monograph 51

Pterotrachea fredericia
Tesch, J.J., 1906
Die Heteropoden der Siboga-Expedition.
Res. Expl. Siboga, 29:112 pp., 14 pls.

Monograph 51

Pterotrachea gegenbauri
Tesch, J.J., 1906
Die Heteropoden der Siboga-Expedition.
Res. Expl. Siboga, 29:112 pp., 14 pls.

Monograph 51

Pterotrachea gibbosa
Tesch, J.J., 1906
Die Heteropoden der Siboga-Expedition.
Res. Expl. Siboga, 29:112 pp., 14 pls.

Monograph 51

Pterotrachea hippocampus
Dales, R.P., 1953.
The distribution of some heteropod molluscs of the Pacific coast of North America.
Proc. Zool. Soc., London, 122(4):1007-1015, 7 textfigs.

Pterotrachea hippocampus
Dales, R.P., 1952.
The distribution of some heteropod molluscs off the Pacific coast of North America.
Proc. Zool. Soc., London, 122:1007-1015, 7 textfigs.

Pterotrachea hippocampus
Franc, A., 1949
Hétéropodes, et autres Gastropodes planctoniques de Méditerranée occidentale.
J. de Conchyliologia 89: 209-230, 19 text figs. Trav. Sta. Zool. Villefranche-sur-Mer 10(3)

Pterotrachea hippocampus
Okutani, T., 1957.
On pterotrachean fauna in Japanese waters.
Bull. Tokai Reg. Fish. Res. Lab., 16:15-22.

Pterotrachea hippocampus
Owre, Harding B., 1964.
Observations on development of the heteropod molluscs Pterotrachea hippocampus and Firoloidea desmaresti.
Bull. Mar. Sci., Gulf and Caribbean, 14(4):529-538.

Petrotrachea hippocampus
Richter, Gotthard, 1968.

Heteropoden und Heteropodenlarven im Oberflachenplankton des Golfs von Neapel.

Pubbl. Staz. zool. Napoli, 36(3):346-400.

Pterotrachea hippocampus
Tesch, J. J., 1949
Heteropoda. Dana Report No. 34: 53 pp., 44 text figs., 5 pls.

Pterotrachea hippocampus
Tesch, J.J., 1906
Die Heteropoden der Siboga-Expedition.
Res. Expl. Siboga, 29:112 pp., 14 pls.

Monograph 51

Pterotrachea keraudreni
Tesch, J.J., 1906
Die Heteropoden der Siboga-Expedition.
Res. Expl. Siboga, 29:112 pp., 14 pls.

Monograph 51

Pterotrachea lesueuri
Tesch, J.J., 1906
Die Heteropoden der Siboga-Expedition.
Res. Expl. Siboga, 29:112 pp., 14 pls.

Monograph 51

Pterotrachea microptera
Franc, A., 1949
Hétéropodes, et autres Gastropodes planctoniques de Méditerranée occidentale.
J. de Conchyliologia 89: 209-230, 19 text figs. Trav. Sta. Zool. Villefranche-sur-Mer 10(3)

Pterotrachea microptera ?
Tesch, J. J., 1910
X Pteropoda and Heteropoda. Percy Sladen Trust Expedition to the Indian Ocean in 1915.
Trans. Linn. Soc., London, Zool., 2nd ser.14: 165-189, pls. 12-14.

Pterotrachea minuta
Dales, R.P., 1953.
The distribution of some heteropod molluscs of the Pacific coast of North America.
Proc. Zool. Soc., London, 122(4):1007-1015, 7 textfigs.

Pterotrachea minuta
Dales, R.P., 1952.
The distribution of some heteropod molluscs off the Pacific coast of North America.
Proc. Zool. Soc., London, 122:1007-1015, 7 textfigs.

Pterotrachea minuta
Okutani, T., 1957.
On pterotrachean fauna in Japanese waters.
Bull. Tokai Reg. Fish. Res. Lab., 16:15-22.

Petrotrachea minuta
Richter, Gotthard, 1968.

Heteropoden und Heteropodenlarven im Oberflachenplankton des Golfs von Neapel.

Pubbl. Staz. zool. Napoli, 36(3):346-400.

Pterotrachea minuta
Tesch, J. J., 1949
Heteropoda. Dana Report No. 34: 53 pp., 44 text figs., 5 pls.

Pterotrachea minuta
van der Spoel, S., 1970.
The pelagic Mollusca from the Atlantide and Galathea expeditions collected in the East Atlantic.
Atlantide Rept. 11:99-139.

Pterotrachea (Euryops) mutabilis
Russell, F.S., and J.S. Colman, 1935 Tesch.
The Zooplankton. IV. The occurrence and seasonal distribution of the Tunicata, Mollusca and Coelenterata (siphonophora). Brit. Mus. (N.H.) Great Barrier Reef Expedition 1928-1929, Sci. Repts., 2(7):203-276, 30 text figs.

Pterotrachea mutabilis
Tesch, J. J., 1910
X Pteropoda and Heteropoda. Percy Sladen Trust Expedition to the Indian Ocean in 1915.
Trans. Linn. Soc., London, Zool., 2nd ser.14: 165-189, pls. 12-14.

Pterotrachea mutica
Franc, A., 1949
Hétéropodes, et autres Gastropodes planctoniques de Méditerranée occidentale.
J. de Conchyliologia 89: 209-230, 19 text figs. Trav. Sta. Zool. Villefranche-sur-Mer 10(3)

Pterotrachea mutica
Tesch, J.J., 1906
Die Heteropoden der Siboga-Expedition.
Res. Expl. Siboga, 29:112 pp., 14 pls.

Monograph 51

Pterotrachea (Firola) mutica
Thomopoulos, T., 1952.
Notes sur le plancton de la Baie de Banyuls.
Vie et Milieu, Bull. Lab. Arago, Univ. Paris, 327-335, 8 textfigs.

Pterotrachea peronia
Tesch, J.J., 1906
Die Heteropoden der Siboga-Expedition.
Res. Expl. Siboga, 29:112 pp., 14 pls.

Monograph 51

Pterotrachea quoyana
Tesch, J. J., 1906
Die Heteropoden der Siboga Expedition.
Res. Expl. Siboga, 29: 112pp. 14 pls.

(Monograph 51)

Pterotrachea rufa
Tesch, J.J., 1906
Die Heteropoden der Siboga-Expedition.
Res. Expl. Siboga, 29:112 pp., 14 pls.

Monograph 51

Pterotrachea scutata
Dales, R.P., 1952.
The distribution of some heteropod molluscs off the Pacific coast of North America.
Proc. Zool. Soc., Amer., 122:1007-1015, 7 textfigs.

Pterotrachea scutata
Okutani, T., 1957.
On pterotrachean fauna in Japanese waters.
Bull. Tokai Reg. Fish. Res. Lab., 16:15-22.

Petrotrachea scutata
Richter, Gotthard, 1968.

Heteropoden und Heteropodenlarven im Oberflachenplankton des Golfs von Neapel.

Pubbl. Staz. zool. Napoli, 36(3):346-400.

Pterotrachea scutata
Tesch, J. J., 1949
Heteropoda. Dana Report No. 34: 53 pp., 44 text figs., 5 pls.

Pterotrachea scutata
Tesch, J. J., 1910
X Pteropoda and Heteropoda. Percy Sladen Trust Expedition to the Indian Ocean in 1915.
Trans. Linn. Soc., London, Zool., 2nd ser.14: 165-189, pls. 12-14.

Pterotrachea scutata
Tesch, J.J., 1906
Die Heteropoden der Siboga-Expedition.
Res. Expl. Siboga, 29:112 pp., 14 pls.

Monograph 51

Pterotrachea souleyeti
Franc, A., 1949
Hétéropodes, et autres Gastropodes planctoniques de Méditerranée occidentale.
J. de Conchyliologia 89: 209-230, 19 text figs. Trav. Sta. Zool. Villefranche-sur-Mer 10(3)

Pterotrachea souleyeti
Tesch, J.J., 1906
Die Heteropoden der Siboga-Expedition.
Res. Expl. Siboga, 29:112 pp., 14 pls.

Monograph 51

Pterotrachea talismani
Tesch, J.J., 1906
Die Heteropoden der Siboga-Expedition.
Res. Expl. Siboga, 29:112 pp., 14 pls.

Pterotrachea umbilicata
Tesch, J.J., 1906
Die Heteropoden der Siboga-Expedition.
Res. Expl. Siboga, 29:112 pp., 14 pls.

Monograph 51

Pterotrachea zenoptera
Franc, A., 1949
Hétéropodes, et autres Gastropodes planctoniques de Méditerranée occidentale.
J. de Conchyliologia 89: 209-230, 19 text figs. Trav. Sta. Zool. Villefranche-sur-Mer 10(3)

pteropods
Chen, Chin, and David B. Ericson, 1965.
Holoplanktonic Gastropoda in the southern oceans.
Antarctic Jl., U.S.A., 2(5):200.

Pteropoda
Dall, W. H., 1925.
The Pteropoda collected by the Canadian Arctic Expedition 1913-18, with description of a new species from the North Pacific. Rept. Canadian Arctic Exped., 1913-1918:8(B):9-12.

pteropods
Eliot, C., 1907.
Mollusca. VI. Pteropoda. Nat. Antarctic Exped., "Discovery", 1901-1904, 3:15 pp., 2 pls.

pteropods
Franc, A., 1949
Hétéropodes, et autres Gastropodes planctoniques de Méditerranée occidentale.
J. de Conchyliologia 89: 209-230, 19 text figs. Trav. Sta. Zool. Villefranche-sur-Mer 10(3)

pteropods
Frontier, S., 1963.
Zooplancton récolté en Mer d'Arabie, Golfe Persique et Golfe d'Aden. (8 campagne océanographique du "Commandant Robert Giraud" avril à juin 1961). II. Ptéropodes: systématique et répartition.
Trav. Centre Océanogr., Nosy-Bé, Cahiers, O.R.S.T.R.O.M., Océanogr., Paris, No. 6:233-254.

pteropods
Frontier, 1963.
Hétéropodes et ptéropodes récoltés dans le plancton de Nosy-Bé.
Trav. Centre Océanogr., Nosy-Bé, Cahiers, O.R.S.T.R.O.M., Océanogr., Paris, No. 6:213-227.

pteropods
Kramp, P.L., 1961.
Pteropoda. The Godthaab Expedition, 1928.
Medd. om Grønland, 81(4):3-13.

pteropods
Lucas, C. E., N.B. Marshall and C.B. Rees 1942
Continuous plankton records: The Faeroe-Shetland Channel 1939. Hull Bull. Mar. Ecol., II(10):71-94

pteropods
Massy, A.L., 1932.
Mollusca: Gastropoda Thecostomata and Gymnostomata. Discovery Repts., 3:267-296.

pteropods
Meisenheimer, Johannes, 1906
Die Pteropoden der Deutschen Südpolar-Expedition 1901-1903. Deutsche Südpolar Exped., 1901-1903, 9(Zool. 1): 96-153, Tables 5-7.

pteropods
Meisenheimer, Johannes, 1905
Pteropoda. Wiss. Ergeb., Deutschen Tiefsee-Exped. "Valdivia", 1898-1899: 314 pp., with atlas with 27 pls.

pteropods
Menzies, R.J., 1958.
Shell-bearing pteropod gastropods from Mediterranean plankton (Cavoliniidae). Publ. Sta. Zool., Napoli, 30(3):381-402.

pteropods
Moore, H.B., H. Owre, E.C. Jones and T. Dow, 1953.
Plankton of the Florida Current. III. The control of the vertical distribution of zooplankton in the daytime by light and temperature. Bull. Mar. Sci., Gulf and Caribbean, 3(2):83-95, 1 textfig.

pteropods, etc.
Morton, J.E., 1954.
The pelagic Mollusca of the Benguela Current. Pt. 1. First survey, R.R.S. William Scoresby, March 1951, with an account of the reproductive system and sexual succession of Limacina bulmodes. Discovery Rept. 24:163-200.

pteropods
Rampal, J., 1963.
Pteropodes thécosomes de pêches par paliers entre les Baléares, la Sardaigne et la côte nord-africaine. Rapp. Proc. Verb., Réunions, Comm. Int. Expl. Sci., Mer Méditerranée, Monaco, 17(2):637-639.

pteropods
Rucker, James B., 1962.
Key to principal pteropod species. H.O. Informal Mss. Repts., No. 0-50-62:17 pp. (duplicated) (Unpublished manuscript).

pteropods
Russell, F.S., and J.S. Colman, 1935
The Zooplankton. IV. The occurrence and seasonal distribution of the Tunicata, Mollusca and Coelenterata (siphonophora). Brit. Mus. (N.H.) Great Barrier Reef Expedition 1928-1929, Sci. Repts., 2(7):203-276, 30 text figs.

Pteropoda
Stubbings, H. G., 1938
Pteropoda. John Murray Exped., 1933-34, Sci. Repts. 5(2):15-33, 2 text figs.

pteropods
Tesch, J.J., 1950.
The Gymnosomata. II. Dana Rept. No. 36: 55 pp., 37 textfigs.

pteropods
Tesch, J. J., 1948
The Thecosomatous pteropods. II. The Indo-Pacific. Dana Rept. No. 30: 45 pp., 34 text figs., 3 pls.

Pteropoda
Tesch, J. J., 1946
The Thecosomatous Pteropods. Dana Rept. No. 28:82 pp., 34 text figs., 8 pls.

pterppoda
Tesch, J.J., 1913
Mollusca. Pteropoda. Das Tierreich. 36:154 pp., 108 text figs.

Pteropoda
Tesch, J. J., 1910
X Pteropoda and Heteropoda. Percy Sladen Trust Expedition to the Indian Ocean in 1915. Trans. Linn. Soc., London, Zool., 2nd ser.14:165-189, pls. 12-14.

pteropods
Tesch, J.J., 1904
The Thecosomata and Gymnosomata of the Siboga Expedition. Siboga Exped. Monogr. LII:1-92, pls.i-vi.

molluscs, pteropods
Thiriot-Quiévreux, Catharine 1967
Variations saisonnières qualitatives des gastéropodes dans le plancton de la région de Banyuls-sur-Mer (novembre 1965-novembre 1966).
Vie Milieu (B) 18 (2-B):331-342.

pteropods
Tokioka, T., 1955.
On some plankton animals collected by the Syunkotu-maru in May-June 1954. 4. Thecosomatous pteropods. Publ. Seto Mar. Biol. Lab. 5(1):59-74
good illustrations which would be helpful in identification

pterppods
Vayssière, A., 1927.
Note sur les Hétéropodes et sur les Euptéropodes recueillis de 1902 à 1924 près de la Principauté de Monaco. Bull. Inst. Océan., Monaco, No. 493:4 pp.

pteropods
Vayssière, A., 1915
Mollusques Eutéropodes (Ptéropodes Thécosomes) provenant des campagnes des yachts HIRONDELLE et PRINCESSE-ALICE (1885-1913). Rés. Camp. Sci., Monaco, 47:226 pp., 14 pls.

pteropods
Wormelle, Ruth L., 1962
A survey of the standing crop of plankton of the Florida Current. VI. A study of the distribution of the pteropods of the Florida Current.
Bull. Mar. Sci., Gulf & Caribbean, 12(1):95-136.

pteropods, anat. physiol.
Van Gransen, P., 1960
Adaptations structurelles des animaux filtrants Ann. Soc. R. Zool. Belgique, 90(2): 161-231.

pteropods, anat.-physiol.
Smith, K.L., Jr. and J.M. Teal, 1973
Temperature and pressure effects on respiration of thecosomatous pteropods. Deep-Sea Res. 20(9):853-858.

pteropods, anat.
Van der Spoel, S., 1968.
The shell and its shape in Cavoliniidae (Pteropoda, Gastropoda). Beaufortia, 15(206):185-189.

pteropods, anat.-physiol
Van der Spoel, S.,1967.
Euthecosomata, a group with remarkable developmental stages (Gastropoda, Pteropoda). J. Noorduijn en Zoom,N.V., Gorinchem, 375 pp.

pteropod, anatomy
Vitagliano, G., 1950.
Osservazioni sul comportamento delle cellule follicolari nella spermatogenesi di Cavolinia tridentata FORSKAL (Mollusca, Pteropoda). Pubbl. Staz. Zool. Napoli, 22:367-377, Pls. 11-12, 1 textfig.

pteropods
Herman, Yvonne and P.E. Rosenberg, 1969.
Pteropods as bathymetric indicators. Marine Geol. 7(2):169-173.

pteropods, chemistry
Krinsley, D., and R. Bieri, 1959.
Changes in the chemical composition of pteropod shells after deposition on the sea floor.
J. Paleont., 33(4):682-684.

pteropod shells, chemistry of
Pyle, Thomas E., and Thomas T. Tieh 1970.
Strontium, vanadium, zinc in the shells of pteropods.
Limnol. Oceanogr. 15(1):153-154.

pteropods, chemistry
Turekian, Karl K., Amitai Katz and Lui Chan, 1973
Trace element trapping in pteropod tests. Limnol. Oceanogr. 18(2): 240-249.

pteropods, distribution of
Chen, Chin, 1971.
Distribution of shell-bearing pteropods in the oceans. (abstract only). In: Micropalaeontology of oceans, B.M. Funnell and W.R. Riedel, editors Cambridge Univ. Press, 161.

Pteropoda, ecology of
Hillman, Norman S., 1966.
Ecology of skeletal plankton.
Antarctic J., United States, 1(5):214-215.

pteropods, lists of spp.
Casanova, Bernadette, Françoise Ducret et Jeannine Rampal 1973
Zooplancton de Méditerranée orientale et de mer Rouge (Chaetognathes, Euphausiacés, Ptéropodes). Rapp. Proc.-v. Reun. Commn int. Explor. scient. Mer Medit. Monaco 21(8):515-519.

pteropods, lists of spp.
Chen, Chin,1966.
Calcareous zooplankton in the Scotia Sea and Drake Passage.
Nature, 212(5063):678-681.

pteropods, lists of spp.
Chen, Chin, and Allan W.H. Bé, 1964.
Seasonal distributions of euthecosomatous pteropods in the surface waters of five stations in the western North Atlantic.
Bull. Mar. Sci., Gulf and Caribbean, 14(2):185-220.

Pteropods, lists of spp.
Chen, Chin and Norman S. Hillman, 1970.
Shell-bearing pteropods as indicators of water masses off Cape Hatteras, North Carolina. Bull. mar. Sci., 20(2): 350-367.

pteropods, lists of spp.
Della Croce, N., and S. Frontier,1966.
Thecosomatous pteropods from the Mozambique Channel.
Boll. Musei Ist.biol.Univ.Genova, 34:107-113.

Pteropods, lists of spp
Frontier, S., 1968.
Données sur la faune pélagique vivant au large des côtes du Gabon du Congo et de l'Angola (6 à 15°S, 6°E à la côte): Hétéropodes et Ptéropodes.
ORSTOM Centre de Pointe Noire, Doc. 417:1pp (mimeographed)
In: Recueil Trav. publiés de 1965 à 1966, Océanogr. ORSTOM

pteropods, lists of spp.

Frontier, S., 1966.
Zooplancton de la region de Nosy Bé. 1. Programme des récoltes et techniques d'études. 2. Plancton de surface aux stations 5 et 10.
Cah. ORSTOM, Océanogr., 4(3):1-36.

pteropods, lists of species

Frontier, S., 1966.
Liste complémentaire des ptéropodes du plancton de Nosy Bé (Madagascar).
Cah. ORSTOM, Sér. Océanogr., 4(2):141-146.

pteropods, lists of spp.

Fu-sui, Zhang, 1966.
The pelagic molluscs of the China coast II. On the ecology of the pelagic molluscs of the Yellow Sea and the East China Sea. (In Chinese; English abstract).
Oceanologia et Limnologia Sinica, 8(1):13-28.

pteropods, lists of spp.

Hida, T.S., 1957.
Chaetognaths and pteropods as biological indicators in the North Pacific.
U.S.F.W.S. Spec. Sci. Rept., Fish., No. 215:13 pp.

pteropods, lists of spp.

McGowan, John A., 1967.
Distributional atlas of pelagic molluscs in the California Current region.
Calif. Coop. Ocean. Fish. Invest., 6:1-218.

pteropods, lists of spp.

Okutani, Takashi, 1965.
Heteropoda and Thecosomata in the Kuroshio. (In Japanese; English abstract).
Inf. Bull. Planktol., Japan, No. 12:37-39.

pteropods, lists of spp.

Rampal, Jeannine, 1965.
Ptéropodes thécosomes indicateurs hydrologiques.
Rev. Trav. Inst. Pêches marit., 29(4):393-400.

pteropods, lists of spp.

Sakthivel, M., 1969.
A preliminary report on the distribution and relative abundance of Euthecosomata with a note on the seasonal variation of Limacina species in the Indian Ocean. Bull. natn. Inst. Sci. India 38(2): 700-717. Also in: Coll. Repr. Nat. Inst. Oceanogr. Goa India 2(1968-1970).

pteropods, lists of spp.

Taki, Iwao, and Takashi Okutani, 1962
Reports on the biology of the "Umitaka-Maru" Expedition, Part 2. Planktonic gastropods collected by the training vessel "Umitaka-Maru" from the Pacific and Indian Oceans in the course of her Antarctic Expedition, 1956.
J. Fac. Fish. and Animal Husband., Hiroshima Univ., 4(1/2):81-97.

Pteropods, lists of spp.

Tanaka, Tsuneo 1971.
Pteropoda and Heteropoda (Gastropoda Mollusca) collected in the western Pacific Ocean in the northern summer 1968.
Kaiyo Rept. No. 3: 27-36.

pteropods, lists of spp.

Thiriot-Quiévreux, Catherine, 1970
Cycles annuels des populations planctoniques de mollusques en 1968 dans la région de Banyuls-sur-Mer. Comparaison avec les années précédentes 1965-1967.
Vie Milieu 20 (2-B): 311-335.

pteropods, lists of spp.

Thiriot-Quiévreux, Catherine, 1968.
Étude du plancton de la région de Banyuls-sur-Mer. Variations saisonnières des mollusques (juillet 1965 - juillet 1966)
Rapp. Proc.-verb. Réun., Comm. int. Explor. scient. Mer Méditerranée 19(3): 469-470.

pteropods, lists of spp.

Thiriot-Quiévreux, Catherine, 1968.
Variations saisonnières des mollusques dans le plancton de la région de Banyuls-sur-Mer (zone sud du Golfe du Lion) novembre 1965-décembre 1967).
Vie Milieu, 19(1-B):35-83.

pteropods, lists of spp.

Vicente, N., et J.P. Ehrhardt, 1964.
Gastéropodes pélagiques du zooplancton de la Mer tyrrhenienne.
Rec. Trav. Sta. Mar., Endoume, 51(35):259-267.

pteropods, lists of spp

Vinogradov, M.E., and N.M. Voronina, 1963.
Quantitative distribution of plankton in the upper layers of the Pacific equatorial currents. 1. The distribution of standing crop and the horizontal distribution of some species. Biological investigations of the ocean (Plankton). (In Russian; English abstract).
Trudy Inst. Okeanol., Akad. Nauk, SSSR, 71:22-59.

See author card for those species with figures of distribution, horizontal and vertical.

pteropods (distribution)

Van der Spoel, S., 1964.
Notes on some pteropods from the North Atlantic.
Beaufortia, 10(121):167-176.

pteropods, vertical distribution

Wormelle, Ruth L., 1962
A survey of the standing crop of plankton of the Florida Current. VI. A study of the distribution of the pteropods of the Florida Current.
Bull. Mar. Sci., Gulf & Caribbean, 12(1): 95-136.

Amacina (Munthea) cochlostyloides

Van der Spoel, S., 1967.
Euthecosomata, a group with remarkable developmental stages (Gastropoda, pteropoda).
J. Noorduijn en Zoom, N.V., Gorinchem, 275 pp.

Argivora parva

Tesch, J.J., 1913
Mollusca. Pteropoda. Das Tierreich. 36:154 pp., 108 text figs.

Cavolinia sp.

Boltovskoy, Demetrio 1971.
Pterópodos thecosomados del Atlántico sudoccidental.
Malacologia 11(1): 121-140.

Cavolinia affinis

Tesch, J.J., 1913
Mollusca. Pteropoda. Das Tierreich. 36:154 pp., 108 text figs.

Cavolinia angulata

Tesch, J.J., 1913
Mollusca. Pteropoda. Das Tierreich. 36:154 pp., 108 text figs.

Cavolinia gibbosa

Bedot, M., 1909
Sur la faune de l'Archipel Malais (resume).
Rev. Suisse Zool. 17:143-169.

Cavolinia gibbosa

Massy, A.L., 1932.
Mollusca: Gastropoda Thecostomata and Gymnostomata. Discovery Repts., 3:267-296.

Cavolinia gibbosa

Meisenheimer, Johannes, 1906
Die Pteropoden der Deutschen Südpolar-Expedition 1901-1903. Deutsche Südpolar Exped., 1901-1903, 9(Zool. 1): 96-153, Tables 5-7.

Cavolinia gibbosa

Meisenheimer, Johannes, 1905
Pteropoda. Wiss. Ergeb., Deutschen Tiefsee-Exped. "Valdivia", 1898-1899: 314 pp., with atlas with 27 pls.

Cavolinia gibbosa

Menzies, R.J., 1958.
Shell-bearing pteropod gastropods from Mediterranean plankton (Cavoliniidae).
Pub. Staz. Zool., Napoli, 30(3):381-402.

Cavolinia gibbosa

Moore, H. B., 1949
The zooplankton of the upper waters of the Bermuda area of the North Atlantic. Bull. Bingham Ocean. Coll. 12(2):97, 208 text figs.

Cavolinia gibbosa

Tesch, J. J., 1948
The Thecosomatous pteropods. II. The Indo-Pacific. Dana Rept. No. 30: 45 pp., 34 text figs., 3 pls.

Cavolinia gibbosa

Tesch, J. J., 1946
The Thecosomatous Pteropods. Dana Rept. No. 28:82 pp., 34 text figs., 8 pls.

Cavolinia gibbosa

Tesch, J.J., 1913
Mollusca. Pteropoda. Das Tierreich. 36:154 pp., 108 text figs.

Cavolinia gibbosa

Tesch, J.J., 1910
X Pteropoda and Heteropoda. Percy Sladen Trust Expedition to the Indian Ocean in 1915.
Trans. Linn. Soc., London, Zool., 2nd ser.14: 165-189, pls. 12-14.

Cavolinia gibbosa

Tokioka, Takasi, 1961
The structure of the operculum of the species of Atlantidae (Gastropoda: Heteropoda) as a taxonomic criterion, with records of some pelagic mollusks of the North Pacific.
Publ. Seto Mar. Biol. Lab., Kyoto Univ., 9(2):267-332.

Cavolinia gibbosa

van der Spoel, S., 1970.
The pelagic Mollusca from the Atlantide and Galathea expeditions collected in the East Atlantic.
Atlantide Rept. 11:99-139.

Cavolinia gibbosa

Vayssière, A., 1915
Mollusques Eutéropodes (Ptéropodes Thécosomes) provenant des campagnes des yachts HIRONDELLE et PRINCESSE-ALICE (1885-1913).
Rés. Camp. Sci. Monaco, 47:226 pp., 14 pls.

Cavolinia gibbosa

Wormelle, Ruth L., 1962
A survey of the standing crop of plankton of the Florida Current. VI. A study of the distribution of the pteropods of the Florida Current.
Bull. Mar. Sci., Gulf & Caribbean, 12(1): 95-136.

Cavolina gibbosa + forms

Van der Spoel, S., 1968.
Euthecosomata, a group with remarkable developmental stages (Gastropoda, Pteropoda).
J. Noorduijn en Zoom, N.V., Gorinchem, 375 pp.

Cavolinia gibbosa plana
Tesch, J.J., 1913
  Mollusca. Pteropoda. Das Tierreich.
36:154 pp., 108 text figs.

Cavolinia globulosa
Aron, William 1962
The distribution of animals in the eastern North Pacific and its relationship to physical and chemical conditions.
J. Fish. Res. Bd., Canada, 19(2):271-314.

Cavolinia globulosa
Bedot, M., 1909
Sur la faune de l'Archipel Malais (resume).
Rev. Suisse Zool. 17:143-169.

Cavolinia globulosa
Meisenheimer, Johannes, 1905
Pteropoda. Wiss. Ergeb., Deutschen Tiefsee-Exped. "Valdivia", 1898-1899: 314 pp., with atlas with 27 pls.

Cavolinia globulosa
Stubbings, H. G., 1938
  Pteropoda. John Murray Exped., 1933-34.
Sci. Repts. 5(2):15-33, 2 text figs.

Cavolinia globulosa
Tesch, J. J., 1948
  The Thecosomatous pteropods. II. The Indo-Pacific. Dana Rept. No. 30: 45 pp., 34 text figs., 3 pls.

Cavolinia globulosa
Tesch, J. J., 1946
  The Thecosomatous Pteropods. Dana Rept. No. 28:82 pp., 34 text figs., 8 pls.

Cavolinia globulosa
Tesch, J.J., 1913
  Mollusca. Pteropoda. Das Tierreich.
36:154 pp., 108 text figs.

Cavolinia globulosa
Tesch, J. J., 1910
X Pteropoda and Heteropoda. Percy Sladen Trust Expedition to the Indian Ocean in 1915.
Trans. Linn. Soc., London, Zool., 2nd ser.14: 165-189, pls. 12-14.

Cavolinia globulosa
Tokioka, Takasi, 1961
The structure of the operculum of the species of Atlantidae (Gastropoda: Heteropoda) as a taxonomic criterion, with records of some pelagic mollusks of the North Pacific.
Publ. Seto Mar. Biol. Lab., Kyoto Univ., 9(2):267-332.

Cavolinia globulosa
Tokioka, T., 1955.
On some plankton animals collected by the Syunkotu-maru in May-June 1954. 4. Thecosomatous pteropods. Publ. Seto Mar. Biol. Lab., 5(1):59-74, Pls. 7-13.

Cavolinia globulosa
van der Spoel, S., 1970.
  The pelagic Mollusca from the Atlantide and Galathea expeditions collected in the East Atlantic.
  Atlantide Rept. 11:99-139.

Cavolinia globulosa
Van der Spoel, S., 1967.
Euthecosomata, a group with remarkable developmental stages (Gastropoda, Pteropoda).
J. Noorduijn en Zoom, N.V., Gorinchem, 375 pp.

Cavolinia inflexa
Franc, A., 1949
  Hétéropodes, et autres Gastropodes planctoniques de Méditerranée occidentale.
J. de Conchyliologia 89: 209-230, 19 text figs. Trav. Sta. Zool. Villefranche-sur-Mer 10(3)

Cavolinia inflexa (Fig.)
Frontier, S., 1963.
Hétéropodes et ptéropodes récoltés dans le plancton de Nosy-Bé.
Trav. Centre Océanogr., Nosy-Bé, Cahiers, O.R.S.T.R.O.M., Océanogr., Paris, No. 6:213-227.

Cavolinia inflexa
Massy, A.L., 1932.
Mollusca: Gastropoda Thecostomata and Gymnostomata. Discovery Repts., 3:267-296.

Cavolinia inflexa
Meisenheimer, Johannes, 1906
Die Pteropoden der Deutschen Südpolar-Expedition 1901-1903. Deutsche Südpolar Exped., 1901-1903, 9(Zool. 1): 96-153, Tables 5-7.

Cavolinia inflexa
Meisenheimer, Johannes, 1905
Pteropoda. Wiss. Ergeb., Deutschen Tiefsee-Exped. "Valdivia", 1898-1899: 314 pp., with atlas with 27 pls.

Cavolinia inflexa longa
Menzies, R.J., 1958.
Shell-bearing pteropod gastropods from Mediterranean plankton (Cavoliniidae).
Pub. Staz. Zool., Napoli, 30(3):381-402.

Cavolinia inflexa
Morton, J.E., 1954.
The pelagic Mollusca of the Benguela Current. Pt. 1. First survey, R.R.S. William Scoresby, March 1951, with an account of the reproductive system and sexual succession of Limacina bulmodes.
Discovery Rept. 24:163-200.

Cavolinia inflexa
Rampal, Jeannine, 1967.
Répartition quantitative et bathymétrique des ptéropodes Thécosomes récoltés en Méditerranée occidentale au nord du 40° parallèle: remarques morphologiques sur certaines espèces.
Rev. Trav. Inst. Pêches marit., 31(4):304-416.

Cavolinia inflexa
Rampal, Jeannine, 1966.
Pêches planctoniques, superficielles et profondes, en Méditerranée occidentale (Campagne de la Thalessa- janvier 1961 -entre les îles Baléares, la Sardaigne et l'Algérois) VI. Ptéropodes.
Revue Trav. Inst.(Scient.tech.)marit.30(4):375-383.

Cavolinia inflexa
Rampal, J., 1963.
Ptéropodes thécosomes de pêches par paliers entre les Baléares, la Sardaigne et la côte nord-africaine.
Rapp. Proc. Verb., Réunions, Comm. Int. Expl. Sci., Mer Méditerranée, Monaco, 17(2):637-639.

Cavolinia inflexa (Les.)
Russell, F.S., and J.S. Colman, 1935
  The Zooplankton. IV. The occurrence and seasonal distribution of the Tunicata, Mollusca and Coelenterata (siphonophora). Brit. Mus. (N.H.) Great Barrier Reef Expedition 1928-1929, Sci. Repts., 2(7):203-276, 30 text figs.

Cavolinia inflexa
Stubbings, H. G., 1938
  Pteropoda. John Murray Exped., 1933-34.
Sci. Repts. 5(2):15-33, 2 text figs.

Cavolinia inflexa
Tesch, J. J., 1948
  The Thecosomatous pteropods. II. The Indo-Pacific. Dana Rept. No. 30: 45 pp., 34 text figs., 3 pls.

Cavolinia inflexa
Tesch, J. J., 1946
  The Thecosomatous Pteropods. Dana Rept. No. 28:82 pp., 34 text figs., 8 pls.

Cavolinia inflexa
Tesch, J.J., 1913
  Mollusca. Pteropoda. Das Tierreich.
36:154 pp., 108 text figs.

Cavolinia inflexa
Tesch, J. J., 1910
X Pteropoda and Heteropoda. Percy Sladen Trust Expedition to the Indian Ocean in 1915.
Trans. Linn. Soc., London, Zool., 2nd ser.14: 165-189, pls. 12-14.

Cavolinia inflexa
Tokioka, Takasi, 1961
The structure of the operculum of the species of Atlantidae (Gastropoda: Heteropoda) as a taxonomic criterion, with records of some pelagic mollusks of the North Pacific.
Publ. Seto Mar. Biol. Lab., Kyoto Univ., 9(2):267-332.

Cavolinia inflexa
van der Spoel, S., 1970.
  The pelagic Mollusca from the Atlantide and Galathea expeditions collected in the East Atlantic.
  Atlantide Rept. 11:99-139.

Cavolinia inflexa
Vayssière, A., 1927.
Note sur les Hétéropodes et sur les Eupteropodes recueillis de 1902 à 1924 près de la Principauté de Monaco. Bull. Inst. Océan., Monaco, No. 493:4 pp.

Cavolinia inflexa
Vayssière, A., 1915
Mollusques Eutéropodes (Ptéropodes Thécosomes) provenant des campagnes des yachts HIRONDELLE et PRINCESSE-ALICE (1885-1913).
Rés. Camp. Sci. Monaco, 47:226 pp., 14 pls.

Cavolinia inflexa
Wormelle, Ruth L., 1962
A survey of the standing crop of plankton of the Florida Current. VI. A study of the distribution of the pteropods of the Florida Current.
Bull. Mar. Sci. Gulf & Caribbean, 12(1): 95-136.

Cavolinia inflexa + forms
Van der Spoel, S., 1967.
Euthecosomata, a group with remarkable developmental stages (Gastropoda, pteropoda).
J. Noorduijn en Zoom, N.V., Gorinchem, 375 pp.

Cavolinia inflexa labiata
Tokioka, T., 1955.
On some plankton animals collected by the Syunkotu-maru in May-June 1954. 4. Thecosomatous pteropods. Publ. Seto Mar. Biol. Lab., 5(1):59-74, Pls. 7-13.

Cavolinia kransi
Tesch, J.J., 1913
  Mollusca. Pteropoda. Das Tierreich.
36:154 pp., 108 text figs.

Cavolinia labiata
Tesch, J.J., 1913
  Mollusca. Pteropoda. Das Tierreich.
36:154 pp., 108 text figs.

Cavolinia laevigata
Vayssière, A., 1915
Mollusques Eutéropodes (Ptéropodes Thécosomes) provenant des campagnes des yachts HIRONDELLE et PRINCESSE-ALICE (1885-1913).
Rés. Camp. Sci. Monaco, 47:226 pp., 14 pls.

Cavolinia longirostris
Delsman, H. C., 1939.
Preliminary plankton investigations in the Java Sea. Treubia, 17:139-181, 8 maps, 41 figs.

Cavolinia longirostris
Frontier, S., 1963.   (Fig.)
Hétéropodes et ptéropodes récoltés dans le plancton de Nosy-Bé.
Trav. Centre Océanogr., Nosy-Bé, Cahiers, O.R.S.T.R.O.M., Océanogr., Paris, No. 6:213-227.

Cavolinia longirostris
Massy, A.L., 1932.
Mollusca: Gastropoda Thecostomata and Gymnostomata. Discovery Repts., 3:267-296.

Cavolinia longirostris
Meisenheimer, Johannes, 1906
Die Pteropoden der Deutschen Südpolar-Expedition 1901-1903. Deutsche Südpolar Exped., 1901-1903, 9(Zool. 1): 96-153, Tables 5-7.

Cavolinia longirostris
Meisenheimer, Johannes, 1905
Pteropoda. Wiss. Ergeb., Deutschen Tiefsee-Exped. "Valdivia", 1898-1899: 314 pp., with atlas with 27 pls.

Cavolinia longirostris
Moore, H. B., 1949
The zooplankton of the upper waters of the Bermuda area of the North Atlantic. Bull. Bingham Ocean. Coll. 12(2):97, 208 text figs.

Cavolinia longirostris
Russell, F.S., and J.S. Colman, 1935
The Zooplankton. IV. The occurrence and seasonal distribution of the Tunicata, Mollusca and Coelenterata (siphonophora). Brit. Mus. (N.H.) Great Barrier Reef Expedition 1928-1929, Sci. Repts., 2(7):203-276, 30 text figs.

Cavolinia longirostris
Stubbings, H. G., 1938
Pteropoda. John Murray Exped., 1933-34, Sci. Repts. 5(2):15-33, 2 text figs.

Cavolinia longirostris
Tesch, J. J., 1948
The Thecosomatous pteropods. II. The Indo-Pacific. Dana Rept. No. 30: 45 pp., 34 text figs., 3 pls.

Cavolinia longirostris
Tesch, J. J., 1946
The Thecosomatous Pteropods. Dana Rept. No. 28:82 pp., 34 text figs., 8 pls.

Cavolinia longirostris
Tesch, J.J., 1913
Mollusca. Pteropoda. Das Tierreich. 36:154 pp., 108 text figs.

Cavolinia longirostris
Tesch, J. J., 1910
X Pteropoda and Heteropoda. Percy Sladen Trust Expedition to the Indian Ocean in 1915. Trans. Linn. Soc., London, Zool., 2nd ser.14: 165-189, pls. 12-14.

Cavolinia longirostris
Tokioka, Takasi, 1961
The structure of the operculum of the species of Atlantidae (Gastropoda: Heteropoda) as a taxonomic criterion, with records of some pelagic mollusks of the North Pacific.
Publ. Seto Mar. Biol. Lab., Kyoto Univ., 9(2):267-332.

Cavolinia longirostris
van der Spoel, S., 1970.
The pelagic Mollusca from the Atlantide and Galathea expeditions collected in the East Atlantic.
Atlantide Rept. 11:99-139.

Cavolinia longirostris
Vayssière, A., 1915
Mollusques Eutéropodes (Ptéropodes Thécosomes) provenant des campagnes des yachts HIRONDELLE et PRINCESSE-ALICE (1885-1913). Rés. Camp. Sci., Monaco, 47:226 pp., 14 pls.

Cavolinia longirostris
Wormelle, Ruth L., 1962
A survey of the standing crop of plankton of the Florida Current. VI. A study of the distribution of the pteropods of the Florida Current.
Bull. Mar. Sci., Gulf & Caribbean, 12(1): 95-136.

Cavolinia longirostris + forms
Van der Spoel, S., 1967.
Euthecosomata, a group with remarkable developmental stages (Gastropoda, pteropoda).
J. Noorduijn en Zoom. N.V., Gorichem, 375 pp.

Cavolinia longirostris longirostris
Frontier, S., 1966.
Liste complémentaire des ptéropodes du plancton de Nosy Bé (Madagascar).
Cah. ORSTOM, Sér. Océanogr., 4(2):141-146.

Cavolinia longirostris longirostris
Cavolinia longirostris angulata
Tokioka, T., 1955.
On some plankton animals collected by the Syunkotu-maru in May-June 1954. 4. Thecosomatous pteropods. Publ. Seto Mar. Biol. Lab., 5(1):59-74, pls. 7-13, 1 textfig.

Cavolinia quadridentata
Russell, F.S., and J.S. Colman, 1935
The Zooplankton. IV. The occurrence and seasonal distribution of the Tunicata, Mollusca and Coelenterata (siphonophora). Brit. Mus. (N.H.) Great Barrier Reef Expedition 1928-1929, Sci. Repts., 2(7):203-276, 30 text figs.

Cavolinia quadridentata
Vayssière, A., 1915
Mollusques Eutéropodes (Ptéropodes Thécosomes) provenant des campagnes des yachts HIRONDELLE et PRINCESSE-ALICE (1885-1913). Rés. Camp. Sci., Monaco, 47:226 pp., 14 pls.

Cavolinia strangulata
Tesch, J.J., 1913
Mollusca. Pteropoda. Das Tierreich. 36:154 pp., 108 text figs.

Cavolinia tridentata
Capuis, G., and G. Ciaccio, 1946.
Osservazioni sulla gametogenesis di Cavolinia tridentata (Forskal). Arch. Zool. Ital., 31: 57-63, Fig. 1.

Cavolinia tridentata (Fig.)
Frontier, S., 1963.
Hétéropodes et ptéropodes récoltés dans le plancton de Nosy-Bé.
Trav. Centre Océanogr., Nosy-Bé, Cahiers, O.R.S.T.R.O.M., Océanogr., Paris, No. 6:213-227.

Cavolinia tridentata
Massy, A.L., 1932.
Mollusca: Gastropoda Thecostomata and Gymnostomata. Discovery Repts., 3:267-296.

Cavolinia tridentata
Meisenheimer, Johannes, 1906
Die Pteropoden der Deutschen Südpolar-Expedition 1901-1903. Deutsche Südpolar Exped., 1901-1903, 9(Zool. 1): 96-153, Tables 5-7.

Cavolinia tridentata
Moore, H. B., 1949
The zooplankton of the upper waters of the Bermuda area of the North Atlantic. Bull. Bingham Ocean. Coll. 12(2):97, 208 text figs.

Cavolinia tridentata
Rampal, Jeannine, 1966.
Pêches planctoniques, superficielles et profondes, en Méditerranée occidentale (Campagne de la Thalassa- janvier 1961-entre les îles Baléares, la Sardaigne et l'Algérois) VI. Ptéropodes.
Revue Trav. Inst. (Scient. tech) marit., 30(4):375-38

Cavolinia tridentata (Forsk)
Russell, F.S., and J.S. Colman, 1935
The Zooplankton. IV. The occurrence and seasonal distribution of the Tunicata, Mollusca and Coelenterata (siphonophora). Brit. Mus. (N.H.) Great Barrier Reef Expedition 1928-1929, Sci. Repts., 2(7):203-276, 30 text figs.

Cavolinia tridentata
Stubbings, H. G., 1938
Pteropoda. John Murray Exped., 1933-34, Sci. Repts. 5(2):15-33, 2 text figs.

Cavolinia tridentata
Tesch, J. J., 1948
The Thecosomatous pteropods. II. The Indo-Pacific. Dana Rept. No. 30: 45 pp., 34 text figs., 3 pls.

Cavolinia tridentata
Tesch, J. J., 1946
The Thecosomatous Pteropods. Dana Rept. No. 28:82 pp., 34 text figs., 8 pls.

Cavolinia tridentata
Tesch, J.J., 1913
Mollusca. Pteropoda. Das Tierreich. 36:154 pp., 108 text figs.

Cavolinia tridentata
Tesch, J. J., 1910
X Pteropoda and Heteropoda. Percy Sladen Trust Expedition to the Indian Ocean in 1915. Trans. Linn. Soc., London, Zool., 2nd ser.14: 165-189, pls. 12-14.

Cavolinia tridentata
Tokioka, Takasi, 1961
The structure of the operculum of the species of Atlantidae (Gastropoda: Heteropoda) as a taxonomic criterion, with records of some pelagic mollusks of the North Pacific.
Publ. Seto Mar. Biol. Lab., Kyoto Univ., 9(2):267-332.

Cavolinia tridentata
van der Spoel, S., 1970.
The pelagic Mollusca from the Atlantide and Galathea expeditions collected in the East Atlantic.
Atlantide Rept. 11:99-139.

Cavolinia tridentata
Vayssière, A., 1927.
Notes sur les Hétéropodes et sur les Euptéropodes recueillis de 1902 à 1924 près de la Principauté de Monaco. Bull. Inst. Océan, Monaco, No. 493: 4 pp.

Cavolinia tridentata
Vayssière, A., 1915
Mollusques Eutéropodes (Ptéropodes Thécosomes) provenant des campagnes des yachts HIRONDELLE et PRINCESSE-ALICE (1885-1913). Rés. Camp. Sci., Monaco, 47:226 pp., 14 pls.

Cavolinia tridentata
Vitagliano, G., 1950.
Osservazioni sul comportamento delle cellule follicolari nella spermatogenesi di Cavolinia tridentata FORSKAL (Mollusca, Pteropoda). Pubbl. Staz. Zool., Napoli, 22:367-377, Pls. 11-12, 1 textfig.

Cavolinia tridentata
Wormelle, Ruth L., 1962
A survey of the standing crop of plankton of the Florida Current. VI. A study of the distribution of the pteropods of the Florida Current.
Bull. Mar. Sci., Gulf & Caribbean, 12(1): 95-136.

Cavolinia tridentata tridentata
Tokioka, T., 1955.
On some plankton animals collected by the Syunkotu-maru in May-June 1954. 4. Thecosomatous pteropods. Publ. Seto Mar. Biol. Lab., 5(1):59-74, Pls. 7-13, 1 textfig.

Cavolinia tridentata + forms
Van der Spoel, S., 1967.
Euthecosomata, a group with remarkable developmental stages (Gastropoda, pteropoda).
J. Noorduijn en Zoom, N.V., Gorinchem. 375 pp.

Cavolinia uncinata
Bedot, M., 1909
Sur la faune de l'Archipel Malais (resume).
Rev. Suisse Zool. 17:143-169.

Cavolinia uncinata
Engel, H., and C.J. Van Eeken, 1962
Red Sea Opisthobranchia from the coast of Israel and Sinai. Contributions to the knowledge of the Red Sea, No. 22.
Sea Fish. Res. Sta., Haifa, Israel, Bull., No. 30:15-34.

Cavolinia uncinata
Massy, A.L., 1932.
Mollusca: Gastropoda Thecostomata and Gymnostomata. Discovery Repts., 3:267-296.

Cavolinia uncinata
Meisenheimer, Johannes, 1906
Die Pteropoden der Deutschen Südpolar-Expedition 1901-1903. Deutsche Südpolar Exped., 1901-1903, 9(Zool. 1): 96-153, Tables 5-7.

Cavolinia uncinata
Meisenheimer, Johannes, 1905
Pteropoda. Wiss. Ergeb., Deutschen Tiefsee-Exped. "Valdivia", 1898-1899: 314 pp., with atlas with 27 pls.

Cavolinia uncinata
Moore, H. B., 1949
The zooplankton of the upper waters of the Bermuda area of the North Atlantic. Bull. Bingham Ocean. Coll. 12(2):97, 208 text figs.

Cavolinia uncinata
Russell, F.S., and J.S. Colman, 1935
The Zooplankton. IV. The occurrence and seasonal distribution of the Tunicata, Mollusca and Coelenterata (siphonophora). Brit. Mus. (N.H.) Great Barrier Reef Expedition 1928-1929, Sci. Repts., 2(7):203-276, 30 text figs.

Cavolinia uncinata
Stubbings, H. G., 1938
Pteropoda. John Murray Exped., 1933-34, Sci. Repts. 5(2):15-33, 2 text figs.

Cavolina uncinata
Tesch, J. J., 1948
The Thecosomatous pteropods. II. The Indo-Pacific. Dana Rept. No. 30: 45 pp., 34 text figs., 3 pls.

Cavolinia uncinata
Tesch, J. J., 1946
The Thecosomatous Pteropods. Dana Rept. No. 28:82 pp., 34 text figs., 8 pls.

Cavolinia uncinata
Tesch, J.J., 1913
Mollusca. Pteropoda. Das Tierreich. 36:154 pp., 108 text figs.

Cavolinia uncinata
Tesch, J. J., 1910
X Pteropoda and Heteropoda. Percy Sladen Trust Expedition to the Indian Ocean in 1915. Trans. Linn. Soc., London, Zool., 2nd ser.14: 165-189, pls. 12-14.

Cavolinia uncinata
Tokioka, Takasi, 1961
The structure of the operculum of the species of Atlantidae (Gastropoda: Heteropoda) as a taxonomic criterion, with records of some pelagic mollusks of the North Pacific.
Publ. Seto Mar. Biol. Lab., Kyoto Univ., 9(2):267-332.

Cavolinia uncinata
Tokioka, T., 1955.
On some plankton animals collected by the Syunkotu-maru in May-June 1954. 4. Thecosomatous pteropods. Publ. Seto Mar. Biol. Lab., 5(1):59-74, Pls. 7-13, 1 textfig.

Cavolinia uncinata
van der Spoel, S., 1970.
The pelagic Mollusca from the Atlantide and Galathea expeditions collected in the East Atlantic.
Atlantide Rept. 11:99-139.

Cavolinia uncinata
Van der Spoel, S., 1967.
Euthecosomata, a group with remarkable developmental stages (Gastropoda, pteropoda).
J. Noorduijn en Zoom, N.V., Gorinchem. 375 pp.

Cavolinia uncinata
Wormelle, Ruth L., 1962
A survey of the standing crop of plankton of the Florida Current. VI. A study of the distribution of the pteropods of the Florida Current.
Bull. Mar. Sci., Gulf & Caribbean, 12(1): 95-136.

Cephalobrachia macrochaeta
Tesch, J.J., 1950.
The Gymnosomata II. Dana Rept. No. 36:55 pp., 37 textfigs.

Cephalobranchaea macrochaeta
van der Spoel, S., 1970.
The pelagic Mollusca from the Atlantide and Galathea expeditions collected in the East Atlantic.
Atlantide Rept. 11:99-139.

Cleodora compressa
Vayssière, A., 1915
Mollusques Eutéropodes (Ptéropodes Thécosomes) provenant des campagnes des yachts HIRONDELLE et PRINCESSE-ALICE (1885-1913).
Rés. Camp. Sci., Monaco, 47:226 pp., 14 pls.

Cleodora curvata
Bedot, M., 1909
Sur la faune de l'Archipel Malais (resume).
Rev. Suisse Zool. 17:143-169.

Cleodora curvata
Vayssière, A., 1915
Mollusques Eutéropodes (Ptéropodes Thécosomes) provenant des campagnes des yachts HIRONDELLE et PRINCESSE-ALICE (1885-1913).
Rés. Camp. Sci., Monaco, 47:226 pp., 14 pls.

Cleodora cuspidata
Vayssière, A., 1915
Mollusques Eutéropodes (Ptéropodes Thécosomes) provenant des campagnes des yachts HIRONDELLE et PRINCESSE-ALICE (1885-1913).
Rés. Camp. Sci., Monaco, 47:226 pp., 14 pls.

Cleodora pyramidata
Russell, F.S., and J.S. Colman, 1935
The Zooplankton. IV. The occurrence and seasonal distribution of the Tunicata, Mollusca and Coelenterata (siphonophora). Brit. Mus. (N.H.) Great Barrier Reef Expedition 1928-1929, Sci. Repts., 2(7):203-276, 30 text figs.

Cleodora pyramidata
Vayssière, A., 1915
Mollusques Eutéropodes (Ptéropodes Thécosomes) provenant des campagnes des yachts HIRONDELLE et PRINCESSE-ALICE (1885-1913).
Rés. Camp. Sci., Monaco, 47:226 pp., 14 pls.

?Cléodora sulcata
Mackintosh, N.A., 1934
Distribution of the Macroplankton in the Atlantic Sector of the Antarctic. Discovery reports, Vol.9:65-160, 48 text figs.

Cleodora trifilis
Tesch, J.J., 1913
Mollusca. Pteropoda. Das Tierreich. 36:154 pp., 108 text figs.

Clio sp.
Lucas, C. E., 1949
Notes on continuous plankton records at 10 m depth in the North Sea and Northeastern Atlantic during 1946-1947. Ann. Biol., Int. Cons., 4:63-66, text fig. 4.

Clio andreae
Meisenheimer, Johannes, 1906
Die Pteropoden der Deutschen Südpolar-Expedition 1901-1903. Deutsche Südpolar Exped., 1901-1903, 9(Zool. 1): 96-153, Tables 5-7.

Clio andreae
Meisenheimer, Johannes, 1905
Pteropoda. Wiss. Ergeb., Deutschen Tiefsee-Exped. "Valdivia", 1898-1899: 314 pp., with atlas with 27 pls.

Clio andreae
Tesch, J.J., 1913
Mollusca. Pteropoda. Das Tierreich. 36:154 pp., 108 text figs.

Clio andreae
Van der Spoel, S., 1967.
Euthecosomata, a group with remarkable developmental stages (Gastropoda, pteropoda).
J. Noorduijn en Zoom, N.V., Gorinchem. 375 pp.

Clio antaratica
Tesch, J.J., 1913
Mollusca. Pteropoda. Das Tierreich. 36:154 pp., 108 text figs.

Clio australis
Meisenheimer, Johannes, 1906
Die Pteropoden der Deutschen Südpolar-Expedition 1901-1903. Deutsche Südpolar Exped., 1901-1903, 9(Zool. 1): 96-153, Tables 5-7.

Clio australis
Meisenheimer, Johannes, 1905
Pteropoda. Wiss. Ergeb., Deutschen Tiefsee-Exped. "Valdivia", 1898-1899: 314 pp., with atlas with 27 pls.

Clio balantium

Meisenheimer, Johannes, 1906
Die Pteropoden der Deutschen Südpolar-
Expedition 1901-1903. Deutsche Südpolar
Exped., 1901-1903, 9(Zool. 1): 96-153,
Tables 5-7. (2)

Clio balantium

Meisenheimer, Johannes, 1905
Pteropoda. Wiss. Ergeb., Deutschen Tiefsee-
Exped. "Valdivia", 1898-1899: 314 pp., with
atlas with 27 pls.

Clio borealis

Marukawa, H., 1921
Plankton lists and some new species of
copepods, from the northern waters of Japan.
Bull. Inst. Ocean., No.384, 15 pp., 3 pls.,
1 chart. Monaco

Clio campylura

Van der Spoel, S., 1967.
Euthecosomata, a group with remarkable develop-
mental stages (Gastropoda, pteropoda).
J. Noorduijn en Zoom, N.V., Gorinchem. 375 pp.

Clio capensis

Tesch, J.J., 1913
Mollusca. Pteropoda. Das Tierreich.
36:154 pp., 108 text figs.

Clio chaptali

Meisenheimer, Johannes, 1906
Die Pteropoden der Deutschen Südpolar-
Expedition 1901-1903. Deutsche Südpolar
Exped., 1901-1903, 9(Zool. 1): 96-153,
Tables 5-7. (2)

Clio chaptali n.sp.

Meisenheimer, Johannes, 1905
Pteropoda. Wiss. Ergeb., Deutschen Tiefsee-
Exped. "Valdivia", 1898-1899: 314 pp., with
atlas with 27 pls.

Clio chaptali

Stubbings, H. G., 1938
Pteropoda. John Murray Exped., 1933-34,
Sci. Repts. 5(2):15-33, 2 text figs.

Clio chaptalii

Tesch, J.J., 1913
Mollusca. Pteropoda. Das Tierreich.
36:154 pp., 108 text figs.

Clio chaptalii

Van der Spoel, S., 1967.
Euthecosomata, a group with remarkable develop-
mental stages (Gastropoda, pteropoda).
J. Noorduijn en Zoom, N.V., Gorinchem. 375 pp.

Clio convexa

Tesch, J.J., 1913
Mollusca. Pteropoda. Das Tierreich.
36:154 pp., 108 text figs.

Clio cuspidata

Boltovskoy, Demetrio 1971.
Pteropodos thecosomados del Atlantico
sudoccidental.
Malacologia 11(6): 121-140.

Clio cuspidata

Fraser, J. H., 1949
List of rare exotic species found in the
plankton by the Scottish Vessel "Explorer" in
1947. Ann. Biol., Int. Cons., 4:68-69.

Clio cuspidata

Fraser, J. H. and A. Saville, 1949
List of rare exotic species found in
the plankton by the Scottish Vessel "Explorer"
in 1948. Ann. Biol. 5:62-64.

Clio cuspidata

Meisenheimer, Johannes, 1906
Die Pteropoden der Deutschen Südpolar-
Expedition 1901-1903. Deutsche Südpolar
Exped., 1901-1903, 9(Zool. 1): 96-153,
Tables 5-7.

Clio cuspidata

Meisenheimer, Johannes, 1905
Pteropoda. Wiss. Ergeb., Deutschen Tiefsee-
Exped. "Valdivia", 1898-1899: 314 pp., with
atlas with 27 pls.

Clio cuspidata

Stubbings, H. G., 1938
Pteropoda. John Murray Exped., 1933-34,
Sci. Repts. 5(2):15-33, 2 text figs.

Clio cuspidata

Tesch, J.J., 1913
Mollusca. Pteropoda. Das Tierreich.
36:154 pp., 108 text figs.

Clio cuspidata

Tesch, J. J., 1910
X Pteropoda and Heteropoda. Percy Sladen
Trust Expedition to the Indian Ocean in 1915.
Trans. Linn. Soc., London, Zool., 2nd ser.14:
165-189, pls. 12-14.

Clio cuspidata

van der Spoel, S., 1970.
The pelagic Mollusca from the Atlantide
and Galathea expeditions collected in the
East Atlantic.
Atlantide Rept. 11:99-139.

Clio cuspidata

Van der Spoel, S., 1967.
Euthecosomata, a group with remarkable develop-
mental stages (Gastropoda, pteropoda).
J. Noorduijn en Zoom, N.V., Gorinchem. 375 pp.

Clio cuspidata

Wormelle, Ruth L., 1962
A survey of the standing crop of plankton
of the Florida Current. VI. A study of the
distribution of the pteropods of the Florida
Current.
Bull. Mar. Sci., Gulf & Caribbean, 12(1):
95-136.

Clio falcata

Meisenheimer, Johannes, 1906
Die Pteropoden der Deutschen Südpolar-
Expedition 1901-1903. Deutsche Südpolar
Exped., 1901-1903, 9(Zool. 1): 96-153,
Tables 5-7.

Clio lanceolata

Tesch, J.J., 1913
Mollusca. Pteropoda. Das Tierreich.
36:154 pp., 108 text figs.

Clio limacina

Marukawa, H., 1921
Plankton lists and some new species of
copepods, from the northern waters of Japan.
Bull. Inst. Ocean., No.384, 15 pp., 3 pls.,
1 chart. Monaco

Clione limacina

Meisenheimer, Johannes, 1906
Die Pteropoden der Deutschen Südpolar-
Expedition 1901-1903. Deutsche Südpolar
Exped., 1901-1903, 9(Zool. 1): 96-153,
Tables 5-7.

Clio orthotheca

Van der Spoel, S., 1967.
Euthecosomata, a group with remarkable develop-
mental stages (Gastropoda, pteropoda).
J. Noorduijn en Zoom, N.V., Gorinchem. 375 pp.

Clio polita

Meisenheimer, Johannes, 1905
Pteropoda. Wiss. Ergeb., Deutschen Tiefsee-
Exped. "Valdivia", 1898-1899: 314 pp., with
atlas with 27 pls.

Clio polita

Tesch, J.J., 1913
Mollusca. Pteropoda. Das Tierreich.
36:154 pp., 108 text figs.

Clio polita

Van der Spoel, S., 1967.
Euthecosomata, a group with remarkable develop-
mental stages (Gastropoda, pteropoda).
J. Noorduijn en Zoom, N.V., Gorinchem. 375 pp.

Clio polita

Wormelle, Ruth L., 1962
A survey of the standing crop of plankton
of the Florida Current. VI. A study of the
distribution of the pteropods of the Florida
Current.
Bull. Mar. Sci., Gulf & Caribbean, 12(1):
95-136.

Clio pyramidalis

Tesch, J.J., 1913
Mollusca. Pteropoda. Das Tierreich.
36:154 pp., 108 text figs.

Clio pyramidata

Boltovskoy, Demetrio 1971.
Pteropodos thecosomados del Atlantico
sudoccidental.
Malacologia 11(6): 121-140.

Clio pyramidata

Bedot, M., 1909
Sur la faune de l'Archipel Malais (resume).
Rev. Suisse Zool. 17:143-169.

Clio pyramidata

Fraser, J. H., 1949
Plankton of the Faroe-Shetland Channel and
the Faroes, June and August 1947. Ann. Biol.,
Int. Cons., 4:27-28, text fig. 10.

Clio pyramidata

Fraser, J. H. and A. Saville, 1949
Macroplankton in the Faroe Channel,
1948. Ann. Biol. 5:29-30, text figs.

Clio pyramidata

Fraser, J. H., and A. Saville, 1949
Plankton distribution in Scottish and
adjacent waters in 1948. Ann. Biol. 5:61-
62.

Clio (Euclio) pyramidata

Kramp, P.L., 1961.
Pteropoda. The Godthaab Expedition, 1928.
Medd. om Grønland, 81(4):3-13.

Clio pyramidata

Meisenheimer, Johannes, 1906
Die Pteropoden der Deutschen Südpolar-
Expedition 1901-1903. Deutsche Südpolar
Exped., 1901-1903, 9(Zool. 1): 96-153,
Tables 5-7.

Clio pyramidata

Meisenheimer, Johannes, 1905
Pteropoda. Wiss. Ergeb., Deutschen Tiefsee-Exped. "Valdivia", 1898-1899: 314 pp., with atlas with 27 pls.

Clio pyramidata

Moore, H. B., 1949
The zooplankton of the upper waters of the Bermuda area of the North Atlantic. Bull. Bingham Ocean. Coll. 12(2):97, 208 text figs.

Clio pyramidata

Österberg, Charles, W.G Pearcy and Herbert Curl, Jr., 1964
Radioactivity and its relationship to oceanic food chains.
J. Mar. Res., 22(1):2-12.

Clio pyramidata

Stubbings, H. G., 1938
Pteropoda. John Murray Exped., 1933-34, Sci. Repts. 5(2):15-33, 2 text figs.

Clio pyramidata

Tesch, J.J., 1913
Mollusca. Pteropoda. Das Tierreich. 36:154 pp., 108 text figs.

Clio pyramidata

Tesch, J. J., 1910
X Pteropoda and Heteropoda. Percy Sladen Trust Expedition to the Indian Ocean in 1915. Trans. Linn. Soc., London, Zool., 2nd ser.14: 165-189, pls. 12-14.

Cliopyramidata

van der Spoel, S., 1970.
The pelagic Mollusca from the Atlantide and Galathea expeditions collected in the East Atlantic.
Atlantide Rept. 11:99-139.

Clio pyramidata

Van der Spoel, S. 1969.
The shell of Clio pyramidata L., 1767 forma lanceolata (Lesueur, 1813) and forma convexa (Boas, 1886) (Gastropoda, Pteropoda). Vidensk Meddr dansk naturh. Foren. 132: 95-114.

Clio pyramidata + forms

Van der Spoel, S., 1967.
Euthecosomata, a group with remarkable developmental stages (Gastropoda, pteropoda).
J. Noorduijn en Zoom.N.V., Gorinchem, 375 pp.

Clio pyramidata antarctica

Spoel, S.V.D. 1962.
Aberrant forms of the genus Clio Linnaeus 1767 with a review of the genus Proclio Hubendick 1951 (Gastropoda, Pteropoda).
Beaufortia, Zool. Mus., Amsterdam, 9(107):173-200.

Clio pyramidata convexa

van der Spoel, S. 1973.
Clio pyramidata Linnaeus, 1767 forma convexa (Boas, 1886) (Mollusca, Pteropoda). Bull. Zool. Mus. Univ. Amsterdam, 3(3): 15-20

Clio pyramidata sulcata

Spoel, S. v.d., 1962.
Aberrant forms of the genus Clio Linnaeus 1767, with a review of the genus Proclio Hubendick 1951 (Gastropoda, Pteropoda).
Beaufortia, Zool. Mus., Amsterdam, 9(107):173-200.

Clio recurva

Tesch, J.J., 1913
Mollusca. Pteropoda. Das Tierreich. 36:154 pp., 108 text figs.

Clio recurva

van der Spoel, S., 1970.
The pelagic Mollusca from the Atlantide and Galathea expeditions collected in the East Atlantic.
Atlantide Rept. 11:99-139.

Clio recurva

Van der Spoel, S., 1967.
Euthecosomata, a group with remarkable developmental stages (Gastropoda, pteropoda).
J. Noorduijn en Zoom.N.V., Gorinchem, 375 pp.

Clio recurva

Wormelle, Ruth L., 1962
A survey of the standing crop of plankton of the Florida Current. VI. A study of the distribution of the pteropods of the Florida Current.
Bull. Mar. Sci., Gulf & Caribbean, 12(1): 95-136.

Clio scheelei

Meisenheimer, Johannes, 1906
Die Pteropoden der Deutschen Südpolar-Expedition 1901-1903. Deutsche Südpolar Exped., 1901-1903, 9(Zool. 1): 96-153, Tables 5-7.

Clio scheelei

Meisenheimer, Johannes, 1905
Pteropoda. Wiss. Ergeb., Deutschen Tiefsee-Exped. "Valdivia", 1898-1899: 314 pp., with atlas with 27 pls.

Clio scheelei

Tesch, J.J., 1913
Mollusca. Pteropoda. Das Tierreich. 36:154 pp., 108 text figs.

Clio scheelei

Van der Spoel, S., 1967.
Euthecosomata, a group with remarkable developmental stages (Gastropoda, pteropoda).
J. Noorduijn en Zoom.N.V., Gorinchem, 375 pp.

Clio sulcata

Eliot, C., 1907.
Mollusca. VI. Pteropoda. Nat. Antarctic Exped., "Discovery", 1901-1904, 3:15 pp., 2 pls.

Clio sulcata

Meisenheimer, Johannes, 1906
Die Pteropoden der Deutschen Südpolar-Expedition 1901-1903. Deutsche Südpolar Exped., 1901-1903, 9(Zool. 1): 96-153, Tables 5-7.

Clio sulcata

Meisenheimer, Johannes, 1905
Pteropoda. Wiss. Ergeb., Deutschen Tiefsee-Exped. "Valdivia", 1898-1899: 314 pp., with atlas with 27 pls.

Clio sulcata

Tesch, J.J., 1913
Mollusca. Pteropoda. Das Tierreich. 36:154 pp., 108 text figs.

Clione

Henderson, G. T. D., and N. B. Marshall, 1944
Continuous plankton records: The zooplankton (other than copepoda and young fish) in the southern North Sea 1932-1937. Hull Bull. Mar. Ecol., 1(6):255-275, 24 pls.

Clione sp.

Lucas, C. E., 1949
Notes on continuous plankton records at 10 m depth in the North Sea and Northeastern Atlantic during 1946-1947. Ann. Biol., Int. Cons., 4:63-66, text fig. 4.

Clione

Meisenheimer, Johannes, 1906
Die Pteropoden der Deutschen Südpolar-Expedition 1901-1903. Deutsche Südpolar Exped., 1901-1903, 9(Zool. 1): 96-153, Tables 5-7.

Clione antarctica

Eliot, C., 1907.
Mollusca. VI. Pteropoda. Nat. Antarctic Exped., "Discovery", 1901-1904, 3:15 pp., 2 pls.

Clione antarctica

Mackintosh, N.A., 1934
Distribution of the Macroplankton in the Atlantic Sector of the Antarctic. Discovery reports, Vol.9:65-160, 48 text figs.

Clione antarctica

Massy, A.L., 1932.
Mollusca: Gastropoda Thecostomata and Gymnostomata. Discovery Repts., 3:267-296.

Clione antarctica

Tesch, J.J., 1950.
The Gymnosomata II. Dana Rept. No. 36:55 pp., 37 textfigs.

Clione borealis

Dall, W.H., 1925.
The Pteropoda collected by the Canadian Arctic Expedition 1913-18, with description of a new species from the North Pacific. Rept. Canadian Arctic Expedition, 1913-18, 8(B):9-12.

Clione flavescens

Tesch, J.J., 1913
Mollusca. Pteropoda. Das Tierreich. 36:154 pp., 108 text figs.

Clione gracilis

Tesch, J.J., 1913
Mollusca. Pteropoda. Das Tierreich. 36:154 pp., 108 text figs.

Clione gracilis

Van der Spoel, S., 1964.
Notes on some pteropods from the North Atlantic. Beaufortia, 10(121):167-176.

Clione limacina

Bigelow, H. B., 1922
Exploration of the coastal water off the northeastern United States in 1916 by the U.S. Fisheries Schooner Grampus. Bull. M.C.Z. 65 (5):85-188, 53 text figs.

Clione limacina

Bigelow, H.B., and M. Sears, 1939
Studies of the waters of the continental shelf, Cape Cod to Chesapeake Bay. III. A volumetric study of the zooplankton. Mem. M.C.Z. 54(4):183-378, 42 text figs.

Clione limacina

Conovor, R.J., and C.M.Lalli, 1972
Feeding and growth in Clione limacina (Phipps) a pteropod mollusc. J. exp. mar. Biol. Ecol. 9(3): 279-302.

Clione limacina

Dunbar, M. J., 1942.
Marine macroplankton from the Canadian Eastern Arctic. II. Medusae, Siphonophora, Ctenophora, Pteropoda, and Chaetognatha. Canad. J. Res., D, 20:71-77.

Clione limacina

Fish, C.J., and M.W. Johnson, 1937
The biology of the zooplankton population in the Bay of Fundy and the Gulf of Maine with special reference to production and distribution. J. Biol. Bd., Canada 3(3):189-322, 29 tables, 45 text figs.

Clione limacina
Glover, R.S., 1957.
An ecological survey of the drift-net herring fishery off the north-east coast of Scotland. 2. The planktonic environment of the herring.
Bull. Mar. Ecol., 5(39):43 pp.

Clione limacina
Glover, R.S., and G.A. Robinson, 1966.
The continuous plankton recorder survey: Plankton around the British Isles during 1964.
Annls biol., Copenh., 21:56-60.

Clione limacina
Glover, R., and G.A. Robinson, 1965.
The continuous plankton recorder survey: plankton around the British Isles during 1963.
Ann. Biol., Cons. Perm. Int. Expl. Mer, 1963, 20:93-97.

Clione limacina
Hansen, Vagn Kr., 1960.
Investigations on the quantitative and qualitative distribution of zooplankton in the southern part of the Norwegian Sea.
Medd. Danmarks Fiskeri- og Havundersøgelser, n.s., 2(23):1-53.

Clione limacina Phipps
Holm, Th., 1889
Om de paa Fyllas Togt i 1884 foretagne zoologiske Undersøgelser i Grønland.
Medd. Grønland. 8(5):151-171.

Clione limacina
Kielhorn, W.V., 1952
The biology of the surface zone zooplankton of a Boreo-Arctic Atlantic Ocean area. J.Fish Res. Bd., Canada 9 (5): 223-264, 13 text figs.

Clione limacina
Kramp, P.L., 1961.
Pteropoda. The Godthaab Expedition, 1928.
Medd. om Grønland, 81(4):3-13.

Clione limacina
Künne, Cl., 1937
Über die Verbreitung der Leitformen des Grossplanktons in der südlichen Nordsee im Winter. Ber. Deutschen Wissenschaflichen Kommission für Meeresforschung, n.f. VIII (3):131-164, 9 text figs., 4 fold-ins.

Clione limacina
Marshall, N. B., 1948
Continuous plankton records: Zooplankton (other than Copepoda and young fish) in the North Sea 1938-1939. Hull Bull. Mar. Ecol. 2(13):173-213, Pls. 89-108.

Clione limacina
Meisenheimer, Johannes, 1905
Pteropoda. Wiss. Ergeb., Deutschen Tiefsee-Exped. "Valdivia", 1898-1899: 314 pp., with atlas with 27 pls.

Clione limacina
Mileikovsky, S.A., 1970.
Breeding and larval distribution of the pteropod Clione limacina in the North Atlantic, Subarctic and North Pacific Oceans. Marine Biol., 6(4): 317-334.

Clione limacina
Ogilvie, H.S., 1934
A preliminary account of the food of the herring in the North-Western North Sea.
Rapp. Proc. Verb. 89(3):85-92, 2 text figs.

Clione limacina
Rae, K. M., 1949
Plankton. Some broad changes in the plankton round the north of the British Isles in 1948. Ann. Biol. 5:56-60, 12 text figs.

Clione limacina
Tesch, J.J., 1950.
The Gymnosomata II. Dana Rept. No. 36:55 pp., 37 textfigs.

Clione limacina
Tesch, J.J., 1913
Mollusca. Pteropoda. Das Tierreich. 36:154 pp., 108 text figs.

Clione limacina
Van der Spoel, S., 1964.
Notes on some pteropods from the North Atlantic.
Beaufortia, 10(121):167-176.

Clione limacina
Vane, F.R., 1961
Contribution towards a plankton atlas of the north-eastern Atlantic and the North Sea. 3. Gastropoda.
Bulls. Mar. Ecol., 5(42): 98-101, Pl. 28.

Clione limacina
Vane, F.R., and J.M. Colebrook, 1962
Continuous plankton records: Contributions towards a plankton atlas of the north-eastern Atlantic and North Sea. VI. The seasonal and annual distributions of the Gastropoda.
Bull. Mar. Ecology, 5(50):247-253.

Clione limacina antarctig
Tesch, J.J., 1913
Mollusca. Pteropoda. Das Tierreich. 36:154 pp., 108 text figs.

Clione longicaudata
Meisenheimer, Johannes, 1905
Pteropoda. Wiss. Ergeb., Deutschen Tiefsee-Exped. "Valdivia", 1898-1899: 314 pp., with atlas with 27 pls.

Clione longicaudata
Tesch, J.J., 1913
Mollusca. Pteropoda. Das Tierreich. 36:154 pp., 108 text figs.

Clione minuta
Tesch, J.J., 1950.
The Gymnosomata II. Dana Rept. No. 36:55 pp., 37 textfigs.

Clione minuta
Van der Spoel, S., 1964.
Notes on some pteropods from the North Atlantic.
Beaufortia, 10(121):167-176.

Clione schmidti
Christomanos, Anastasios A., 1963.
Eigenschaften des Farbstoffes des Schwarmes Clione schmidtii.
Res. Proc. Mar. Lab., Thessaloniki, 1(5):2 pp.

Clionina flavescens
Tesch, J.J., 1950.
The Gymnosomata. II. Dana Rept. No. 36:55 pp., 37 textfigs.

Clionina longicaudata
Stubbings, H. G., 1938
Pteropoda. John Murray Exped., 1933-34, Sci. Repts. 5(2):15-33, 2 text figs.

Clionina longicaudata
Tesch, J.J., 1950.
The Gymnosomata II. Dana Rept. No. 36:55 pp., 37 textfigs.

Clionopsis grandis
Meisenheimer, Johannes, 1905
Pteropoda. Wiss. Ergeb., Deutschen Tiefsee-Exped. "Valdivia", 1898-1899: 314 pp., with atlas with 27 pls.

Clionopsis krohni
Meisenheimer, Johannes, 1905
Pteropoda. Wiss. Ergeb., Deutschen Tiefsee-Exped. "Valdivia", 1898-1899: 314 pp., with atlas with 27 pls.

Cliopsis grandis
Tesch, J.J., 1913
Mollusca. Pteropoda. Das Tierreich. 36:154 pp., 108 text figs.

Cliopsis krohni
Berkeley, E., and C. Berkeley, 1960
Some further records of pelagic Polychaeta from the northeast Pacific, north of Latitude 40°N. and east of Longitude 175°W., together with records of Siphonophora, Mollusca and Tunicata from the same region.
Canadian J. Zoology. 38:787-799.

Cliopsis krohni
Franc, A., 1949
Hétéropodes, et autres Gastropodes planctoniques de Méditerranée occidentale. J. de Conchyliologia 89: 209-230, 19 text figs. Trav. Sta. Zool. Villefranche-sur-Mer 10(3)

Cliopsis krohni
Massy, A.L., 1932.
Mollusca: Gastropoda Thecostomata and Gymnostomata. Discovery Repts., 3:267-296.

Cliopsis krohni
Tesch, J.J., 1950.
The Gymnosomata II. Dana Rept. No. 36:55 pp., 37 textfigs.

Cliopsis krohnii
Tesch, J.J., 1913
Mollusca. Pteropoda. Das Tierreich. 36:154 pp., 108 text figs.

Cliopsis krohnii
van der Spoel, S., 1970.
The pelagic Mollusca from the Atlantide and Galathea expeditions collected in the East Atlantic.
Atlantide Rept. 11:99-139.

Cliopsis microcephala
Tesch, J.J., 1913
Mollusca. Pteropoda. Das Tierreich. 36:154 pp., 108 text figs.

Cliopsis modesta
Tesch, J.J., 1913
Mollusca. Pteropoda. Das Tierreich. 36:154 pp., 108 text figs.

Corolla
Meisenheimer, Johannes, 1906
Die Pteropoden der Deutschen Südpolar-Expedition 1901-1903 Deutsche Südpolar Exped., 1901-1903, 9(Zool. 1): 96-153, Tables 5-7.

Corolla sp.
Tokioka, Takasi, 1961
The structure of the operculum of the species of Atlantidae (Gastropoda: Heteropoda) as a taxonomic criterion, with records of some pelagic molluscs of the North Pacific.
Publ. Seto Mar. Biol. Lab., Kyoto Univ., 9(2):267-332.

Corolla calceola
Bigelow, H.B., and M. Sears, 1939
Studies of the waters of the continental shelf, Cape Cod to Chesapeake Bay. III. A volumetric study of the zooplankton. Mem. M.C.Z. 54(4):183-378, 42 text figs.

Corolla calceola

Meisenheimer, Johannes, 1905
Pteropoda. Wiss. Ergeb., Deutschen Tiefsee-Exped. "Valdivia", 1898-1899: 314 pp., with atlas with 27 pls.

Corolla calceola

Tesch, J. J., 1946
The Thecosomatous Pteropods. Dana Rept. No. 28:82 pp., 34 text figs., 8 pls.

Corolla calceola

Tesch, J.J., 1913
Mollusca. Pteropoda. Das Tierreich. 36:154 pp., 108 text figs.

Corolla calceolus

Wormelle, Ruth L., 1962
A survey of the standing crop of plankton of the Florida Current. VI. A study of the distribution of the pteropods of the Florida Current.
Bull. Mar. Sci., Gulf & Caribbean, 12(1): 95-136.

Corolla intermedia

Meisenheimer, Johannes, 1905
Pteropoda. Wiss. Ergeb., Deutschen Tiefsee-Exped. "Valdivia", 1898-1899: 314 pp., with atlas with 27 pls.

Corolla intermedia

Tesch, J. J., 1946
The Thecosomatous Pteropods. Dana Rept. No. 28:82 pp., 34 text figs., 8 pls.

Corolla intermedia

Tesch, J.J., 1913
Mollusca. Pteropoda. Das Tierreich. 36:154 pp., 108 text figs.

Corolla intermedia

Tokioka, T., 1955.
On some plankton animals collected by the Syunkotu-maru in May-June 1954. 4. Thecosomatous pteropods. Publ. Seto Mar. Biol. Lab. 5(1):59-

Corolla ovata

Meisenheimer, Johannes, 1905
Pteropoda. Wiss. Ergeb., Deutschen Tiefsee-Exped. "Valdivia", 1898-1899: 314 pp., with atlas with 27 pls.

Corolla ovata

Tesch, J.J., 1913
Mollusca. Pteropoda. Das Tierreich. 36:154 pp., 108 text figs.

Corolla ovata

Tokioka, Takasi, 1961
The structure of the operculum of the species of Atlantidae (Gastropoda: Heteropoda) as a taxonomic criterion, with records of some pelagic mollusks of the North Pacific.
Publ. Seto Mar. Biol. Lab., Kyoto Univ., 9(2):267-332.

Corolla ovata

Tokioka, T., 1955.
On some plankton animals collected by the Syunkotu-maru in May-June 1954.4. Thecosomatous pteropods. Publ. Seto Mar. Biol. Lab., 5(1):59-74, Pls. 7-13.

Corolla spectabilis

Meisenheimer, Johannes, 1905
Pteropoda. Wiss. Ergeb., Deutschen Tiefsee-Exped. "Valdivia", 1898-1899: 314 pp., with atlas with 27 pls.

Corolla spectabilis

Tesch, J.J., 1913
Mollusca. Pteropoda. Das Tierreich. 36:154 pp., 108 text figs.

Corolla cf. spectabilis

van der Spoel, S., 1970.
The pelagic Mollusca from the Atlantide and Galathea expeditions collected in the East Atlantic.
Atlantide Rept. 11:99-139.

Corolla spectabilis

Wormelle, Ruth L., 1962
A survey of the standing crop of plankton of the Florida Current. VI. A study of the distribution of the pteropods of the Florida Current.
Bull. Mar. Sci., Gulf & Caribbean, 12(1): 95-136.

Creseis sp.

Boltovskoy, Demetrio 1971.
Pteropodos thecosomados del Atlantico sudoccidental.
Malacologia 11(1): 121-140.

Creseis sp.

Johnson, M. W., 1949
Zooplankton as an index of water exchange between Bikini lagoon and the open sea. Trans. Am. Geophys. Un. 30(2):238-214, 15 text figs.

Creseis acicula

Bedot, M., 1909
Sur la faune de l'Archipel Malais (resume). Rev. Suisse Zool. 17:143-169.

Criseis acicula

Bigelow, H.B., and M. Sears, 1939
Studies of the waters of the continental shelf, Cape Cod to Chesapeake Bay. III. A volumetric study of the zooplankton. Mem. M.C.Z. 54(4):183-378, 42 text figs.

Creseis acicula

Burgi, A., and C. Devos, 1962.
Accumulation exceptionnelle de Creseis acicula, au long des cotes dans la region de Banyuls-sur-Mer.
Vie et Milieu, 13(2):391-392.

Creseis acicula

Delsman, H. C., 1939
Preliminary plankton investigations in the Java Sea. Treubia, 17:139-181, 8 maps, 41 figs.

Creseis acicula

Franc, A., 1949
Hétéropodes, et autres Gastropodes planctoniques de Méditerranée occidentale.
J. de Conchyliologia 89: 209-230, 19 text figs. Trav. Sta. Zool. Villefranche-sur-Mer 10(3)

Creseis acicula

Frontier, S., 1965.
Le problème des Creseis.
Cahiers, O.R.S.T.R.O.M., Océanogr., 3(2):11-17.

Creseis acicula (fig)

Frontier, S., 1963.
Hétéropodes et ptéropodes récoltés dans le plancton de Nosy-Bé.
Trav. Centre Océanogr., Nosy-Bé, Cahiers, O.R.S.T.R.O.M., Océanogr., Paris, No. 6:213-227.

Creseis acicula

Hutton, Robert F., 1960.
Marine dermatosis. Notes on "Seabather's Eruption" with Creseis acicula Rang (Mollusca: Pteropoda) as the cause of a particular type of sea sting along the west coast of Florida.
Archives of Dermatology, 82:951-956.

Creseis acicula

Meisenheimer, Johannes, 1906
Die Pteropoden der Deutschen Südpolar-Expedition 1901-1903. Deutsche Südpolar Exped., 1901-1903, 9(Zool. 1): 96-153, Tables 5-7.

Creseis acicula

Meisenheimer, Johannes, 1905
Pteropoda. Wiss. Ergeb., Deutschen Tiefsee-Exped. "Valdivia", 1898-1899: 314 pp., with atlas with 27 pls.

Creseis acicula

Menzies, R.J., 1958.
Shell-bearing pteropod gastropods from Mediterranean plankton (Cavoliniidae). Pub. Staz. Zool., Napoli, 30(3):381-402.

Creseis acicula

Nishimura, Saburo, 1965.
Droplets from the plankton net. XX. "Sea stings" caused by Creseis acicula Rang (Mollusca: Pteropoda) in Japan.
Publ. Seto Mar. Biol. Lab., 13(4):287-290.

Creseis acicula

Rampal, Jeannine, 1967.
Répartition quantitative et bathymétrique des ptéropodes Thécosomes récoltés en Méditerranée occidentale au nord du 40° parallèle: remarques morphologiques sur certaines espèces.
Rev. Trav. Inst. Pêches marit., 31(4):403-416.

Creseis acicula

Rampal, Jeannine, 1966.
Pêches planctoniques, superficielles et profondes, en Méditerranée occidentale (Campagne de la Thalassa- janvier 1961 - entre les îles Baléares, la Sardaigne et l'Algérois) VI. Ptéropodes.
Revue Trav. Inst. (Scient.Tech.)marit.,30(4):375-

Creseis acicula

Rampal, J., 1963.
Ptéropodes thécosomes de pêches par palier entre les Baléares, la Sardaigne et la côte nord-africaine.
Rapp. Proc. Verb., Réunions, Comm. Int. Expl. Sci., Mer Méditerranée, Monaco, 17(2):637-639.

Creseis acicula

Moore, H. B., 1949
The zooplankton of the upper waters of the Bermuda area of the North Atlantic. Bull. Bingham Ocean. Coll. 12(2):97, 208 text figs.

Creseis acicula

Russell, F.S., and J.S. Colmann, 1935
The Zooplankton. IV. The occurrence and seasonal distribution of the Tunicata, Mollusca and Coelenterata (siphonophora). Brit. Mus. (N.H.) Great Barrier Reef Expedition 1928-1929, Sci. Repts., 2(7):203-276, 30 text figs.

Creseis acicula

Stubbings, H. G., 1938
Pteropoda. John Murray Exped., 1933-34, Sci. Repts. 5(2):15-33, 2 text figs.

Creseis acicula

Tesch, J. J., 1948
The Thecosomatous pteropods. II. The Indo-Pacific. Dana Rept. No. 30: 45 pp., 34 text figs., 3 pls.

Creseis acicula

Tesch, J. J., 1946
The Thecosomatous Pteropods. Dana Rept. No. 28:82 pp., 34 text figs., 8 pls.

Creseis acicula

Tesch, J.J., 1913
Mollusca. Pteropoda. Das Tierreich. 36:154 pp., 108 text figs.

Creseis acicula
Tesch, J. J., 1910
X Pteropoda and Heteropoda. Percy Sladen Trust Expedition to the Indian Ocean in 1915. Trans. Linn. Soc., London, Zool., 2nd ser.14: 165-189, pls. 12-14.

Creseis acicula
Tokioka, Takasi, 1961
The structure of the operculum of the species of Atlantidae (Gastropoda: Heteropoda) as a taxonomic criterion, with records of some pelagic mollusks of the North Pacific.
Publ. Seto Mar. Biol. Lab., Kyoto Univ., 9(2):267-332.

Creseis acicula and vars.
Tokioka, T., 1955.
On some plankton animals collected by the "Syunkotu-maru in May-June 1954. 4. Thecosomatous pteropods.
Publ. Seto Mar. Biol. Lab., 5(1):59-74, Pls. 7-13.

Creseis acicula
van der Spoel, S., 1970.
The pelagic Mollusca from the Atlantide and Galathea expeditions collected in the East Atlantic.
Atlantide Rept. 11:99-139.

Creseis acicula
Van der Spoel, S., 1967.
Euthecosomata, a group with remarkable developmental stages (Gastropoda, pteropoda).
J. Noorduijn en Zoom, N.V., Gorinchem. 375 pp.

Creseis acicula
Vayssière, A., 1927.
Note sur les Hétéropodes et sur les Euptéropodes recueillis de 1902 à 1924 près de la Principauté de Monaco. Bull. Inst. Océan., Monaco, No. 493: 4 pp.

Creseis acicula
Vayssière, A., 1915
Mollusques Eutéropodes (Ptéropodes Thécosomes) provenant des campagnes des yachts HIRONDELLE et PRINCESSE-ALICE (1885-1913).
Rés. Camp. Sci. Monaco, 47:226 pp., 14 pls.

Creseis acicula
Vinogradov, M.E., and N.M. Voronina, 1964.
Quantitative distribution of the plankton in the upper layers of the Pacific equatorial currents. II. Vertical distribution of several species. Regularity of the distribution of oceanic plankton. (In Russian; English abstract).
Trudy Inst. Okeanol., Akad. Nauk, SSSR, 65:58-76.

Creseis caliciformis
Tesch, J.J., 1913
Mollusca. Pteropoda. Das Tierreich.
36:154 pp., 108 text figs.

Creseis caliciformis
Van der Spoel, S., 1967.
Euthecosomata, a group with remarkable developmental stages (Gastropoda, pteropoda).
J. Noorduijn en Zoom, N.V., Gorinchem. 375 pp.

Creseis chierchia (= C. virgula constricta = C. caliciformis)
Frontier, S. 1965.
Le problème des Creseis.
Cah. O.R.S.T.O.M. Océanogr. 3(2): 11-17.

Creseis chierchiae
Frontier, S., 1963.
Présence de Creseis chierchiae (Boas) dans l'Océan Indien.
Trav. Centre Océanogr., Nosy-Bé, Cahiers, O.R.S.T.R.O.M., Océanogr., Paris, No. 6:229-232.

Creseis chierchiae
Meisenheimer, Johannes, 1905
Pteropoda. Wiss. Ergeb., Deutschen Tiefsee-Exped. "Valdivia", 1898-1899: 314 pp., with atlas with 27 pls.

Creseis chierchiae
Menzies, R.J., 1958.
Shell-bearing pteropod gastropods from Mediterranean plankton (Cavoliniidae).
Pub. Staz. Zool., Napoli, 30(3):381-402.

Creseis chierchiae
Tesch, J.J., 1913
Mollusca. Pteropoda. Das Tierreich.
36:154 pp., 108 text figs.

Creseis chierchiae
Tokioka, T., 1951.
Droplets from the plankton net. IX. Records of Cyclosalpa bakeri from Japanese waters. X. Record of Creseis chierchiae (Boas) from the Palao Islands. Publ. Seto Mar. Biol. Lab. 1(4):183-184.

Creseis chierchiae
Van der Spoel, S., 1967.
Euthecosomata, a group with remarkable developmental stages (Gastropoda, pteropoda).
J. Noorduijn en Zoom, N.V., Gorinchem. 375 pp.

Creseis clava
Frontier, S., 1965.
Le problème des Creseis.
Cahiers, O.R.S.T.R.O.M., Océanogr., 3(2):11-17.

Creseis clava
Tesch, J.J., 1913
Mollusca. Pteropoda. Das Tierreich.
36:154 pp., 108 text figs.

Creseis conica
Bigelow, H.B., and M. Sears, 1939
Studies of the waters of the continental shelf, Cape Cod to Chesapeake Bay. III. A volumetric study of the zooplankton. Mem. M.C.Z. 54(4):183-378, 42 text figs.

Creseis conica
Meisenheimer, Johannes, 1905
Pteropoda. Wiss. Ergeb., Deutschen Tiefsee-Exped. "Valdivia", 1898-1899: 314 pp., with atlas with 27 pls.

Creseis conica
Tesch, J.J., 1913
Mollusca. Pteropoda. Das Tierreich.
36:154 pp., 108 text figs.

Creseis (Hyalocylix) striata
Russell, F.S., and J.S. Colman, 1935
The Zooplankton. IV. The occurrence and seasonal distribution of the Tunicata, Mollusca and Coelenterata (siphonophora). Brit. Mus. (N.H.) Great Barrier Reef Expedition 1928-1929, Sci. Repts., 2(7):203-276, 30 text figs.

Creseis (Styliola) subula
Russell, F.S., and J.S. Colman, 1935
The Zooplankton. IV. The occurrence and seasonal distribution of the Tunicata, Mollusca and Coelenterata (siphonophora). Brit. Mus. (N.H.) Great Barrier Reef Expedition 1928-1929, Sci. Repts., 2(7):203-276, 30 text figs.

Creseis virgula
Bedot, M., 1909
Sur la faune de l'Archipel Malais (resume).
Rev. Suisse Zool. 17:143-169.

Creseis virgula
Bigelow, H.B., and M. Sears, 1939
Studies of the waters of the continental shelf, Cape Cod to Chesapeake Bay. III. A volumetric study of the zooplankton. Mem. M.C.Z. 54(4):183-378, 42 text figs.

Contribution No.194 of the Woods Hole Oceanographic Institution.

Creseis virgula
Frontier, S. 1965.
Le problème des Creseis.
Cah. O.R.S.T.O.M. Océanogr. 3(2):11-17.

Creseis virgula (fig.)
Frontier, S., 1963.
Hétéropodes et ptéropodes récoltés dans le plancton de Nosy-Bé.
Trav. Centre Océanogr., Nosy-Bé, Cahiers, O.R.S.T.R.O.M., Océanogr., Paris, No. 6:213-227.

Creseis virgula
Meisenheimer, Johannes, 1906
Die Pteropoden der Deutschen Südpolar-Expedition 1901-1903. Deutsche Südpolar Exped., 1901-1903, 9(Zool. 1): 96-153, Tables 5-7.

Creseis virgula
Meisenheimer, Johannes, 1905
Pteropoda. Wiss. Ergeb., Deutschen Tiefsee-Exped. "Valdivia", 1898-1899: 314 pp., with atlas with 27 pls.

Creseis virgula
Menzies, R.J., 1958.
Shell-bearing pteropod gastropods from Mediterranean plankton (Cavoliniidae).
Pub. Staz. Zool., Napoli, 30(3):381-402.

Creseis virgula
Moore, H. B., 1949
The zooplankton of the upper waters of the Bermuda area of the North Atlantic. Bull. Bingham Ocean. Coll. 12(2):97, 208 text figs.

Creseis virgula
Rampal, Jeannine, 1967.
Répartition quantitative et bathymétrique des ptéropodes Thécosmoes récoltés en Méditerranée occidentale au nord du 40° parallèle: remarques morphologiques sur certaines espèces.
Rev.Trav.Inst.Pêches marit., 31(4):403-416.

Creseis virgula
Rampal, Jeannine, 1966.
Pêches planctoniques, superficielles et profondes, en Méditerranée occidentale (Campagne de la Thalassa - janvier 1961 - entre les îles Baléares, la Sardaigne et l'Algérois).VI. Ptérophodes.
Revue Trav.Inst.(Scient.tech.)marit. 30(4):375-383.

Creseis virgula
Rampal, J., 1963.
Ptéropodes thécosomes de pêches par paliers entre les Baléares, la Sardaigne et la côte nord-africaine.
Rapp. Proc. Verb., Réunions, Comm. Int. Expl. Sci. Mer Méditerranée, Monaco, 17(2):637-639.

Creseis virgula
Russell, F.S., and J.S. Colman, 1935
The Zooplankton. IV. The occurrence and seasonal distribution of the Tunicata, Mollusca and Coelenterata (siphonophora). Brit. Mus. (N.H.) Great Barrier Reef Expedition 1928-1929, Sci. Repts., 2(7):203-276, 30 text figs.

Creseis virgula
Stubbings, H. G., 1938
Pteropoda. John Murray Exped., 1933-34, Sci. Repts. 5(2):15-33, 2 text figs.

Creseis virgula
Tesch, J. J., 1948
The Thecosomatous pteropods. II. The Indo-Pacific. Dana Rept. No. 30: 45 pp., 34 text figs., 3 pls.

Creseis virgula
Tesch, J. J., 1946
The Thecosomatous Pteropods. Dana Rept. No. 28:82 pp., 34 text figs., 8 pls.

Creseis virgula

Tesch, J.J., 1913
Mollusca. Pteropoda. Das Tierreich.
36:154 pp., 108 text figs.

Creseis virgula

Tesch, J. J., 1910
X Pteropoda and Heteropoda. Percy Sladen
Trust Expedition to the Indian Ocean in 1915.
Trans. Linn. Soc., London, Zool., 2nd ser.14:
165-189, pls. 12-14.

Creseis virgula

Tokioka, Takasi, 1961
The structure of the operculum of the
species of Atlantidae (Gastropoda:
Heteropoda) as a taxonomic criterion,
with records of some pelagic mollusks
of the North Pacific.
Publ. Seto Mar. Biol. Lab., Kyoto Univ.,
9(2):267-332.

Creseis virgula

van der Spoel, S., 1970.
The pelagic Mollusca from the Atlantide
and Galathea expeditions collected in the
East Atlantic.
Atlantide Rept. 11:99-139.

Creseis virgula

Van der Spoel, S., 1967.
Euthecosomata, a group with remarkable developmental stages (Gastropoda, pteropoda).
J. Noorduijn en Zoom, N.V., Gorinchem, 375 pp.

Creseis virgula

Vayssière, A., 1915
Mollusques Eutéropodes (Ptéropodes Thécosomes) provenant des campagnes des yachts
HIRONDELLE et PRINCESSE-ALICE (1885-1913).
Rés. Camp. Sci., Monaco, 47:226 pp., 14 pls.

Creseis virgula

Vinogradov, M.E., and N.M. Voronina, 1964.
Quantitative distribution of the plankton in the
upper layers of Pacific equatorial currents.
II. Vertical distribution of several species.
Regularity of the distribution of oceanic
plankton. (In Russian; English abstract).
Trudy Inst. Okeanol., Akad. Nauk, SSSR, 65:58-76.

Creseis virgula

Wormelle, Ruth L., 1962
A survey of the standing crop of plankton
of the Florida Current. VI. A study of the
distribution of the pteropods of the Florida
Current.
Bull. Mar. Sci. Gulf & Caribbean, 12(1):
95-136.

Creseis virgula conica

Tokioka, T., 1955.
On some plankton animals collected by the
Syunkotu-maru in May-June 1954. 4. Thecosomatous
pteropods. Publ. Seto Mar. Biol. Lab., 5(1):59-74, Pls., 7-13, 1 textfig.

Creseis virgula constricta n.subsp.

Chen, Chin, and Allan W.H. Bé, 1964.
Seasonal distributions of euthecosomatous pteropods in the surface waters of five stations in
the western North Atlantic.
Bull. Mar. Sci., Gulf and Caribbean, 14(2):185-220.

Creseis virgula virgula

Tokioka, T., 1955.
On some plankton animals collected by the
Syunkotu-maru in May-June 1954. 4. Thecosomatous
pteropods. Publ. Seto Mar. Biol. Lab., 5(1):59-74, Pls. 7-13, 1 textfig.

Creseis virgula virgula

Vinogradov, M.E. and N.M. Voronina, 1965.
Some peculiarites of plankton distribution
in the Pacific and Indian equatorial current
areas. (In Russian; English abstract).
Okeanolog. Issled. Rez. Issled. po Programme
Mezhd. Geofoz. Goda, Mezhd. Geofiz. Komitet,
Prezidiume Akad. Nauk, SSSR, No.13:128-136.

Crucibranchaea macrochira

Tesch, J.J., 1950.
The Gymnosomata II. Dana Rept. No. 36:55 pp.,
37 textfigs.

Cuvierina columnella

Frontier, S., 1966.
Liste complémentaire des ptéropodes du plancton
de Nosy Bé (Madagascar).
Cah. ORSTOM, Sér. Océanogr., 4(2):141-146.

Cuvierina columnella (fig.)

Frontier, S., 1963.
Hétéropodes et ptéropodes récoltés dans le
plancton de Nosy-Bé.
Trav. Centre Océanogr., Nosy-Bé, Cahiers,
O.R.S.T.R.O.M., Océanogr., Paris, No. 6:213-227.

Cuvierina columnella

Meisenheimer, Johannes, 1906
Die Pteropoden der Deutschen Südpolar-Expedition 1901-1903. Deutsche Südpolar
Exped., 1901-1903, 9(Zool. 1): 96-153,
Tables 5-7.

Cuvierina columnella

Meisenheimer, Johannes, 1905
Pteropoda. Wiss. Ergeb., Deutschen Tiefsee-Exped. "Valdivia", 1898-1899: 314 pp., with
atlas with 27 pls.

Cuvierina columnella

Moore, H. B., 1949
The zooplankton of the upper waters of
the Bermuda area of the North Atlantic. Bull.
Bingham Ocean. Coll. 12(2):97, 208 text figs.

Cuvierina columnella

Tesch, J. J., 1910
X Pteropoda and Heteropoda. Percy Sladen
Trust Expedition to the Indian Ocean in 1915.
Trans. Linn. Soc., London, Zool., 2nd ser.14:
165-189, pls. 12-14.

Cuvierina columnella

Tesch, J. J., 1948
The Thecosomatous pteropods. II. The
Indo-Pacific. Dana Rept. No. 30: 45 pp.,
34 text figs., 3 pls.

Cuvierina columnella

Tesch, J. J., 1946
The Thecosomatous Pteropods. Dana Rept.
No. 28:82 pp., 34 text figs., 8 pls.

Cuvierina columnella

Tesch, J.J., 1913
Mollusca. Pteropoda. Das Tierreich.
36:154 pp., 108 text figs.

Cuvierina columnella

Tokioka, T., 1955.
On some plankton animals collected by the
Syunkotu-maru in May-June 1954. 4. Thecosomatous
pteropods. Publ. Seto Mar. Biol. Lab., 5(1):59-74, Pls. 7-13, 1 textfig.

Cuvierina columnella

van der Spoel, S., 1970.
The pelagic Mollusca from the Atlantide
and Galathea expeditions collected in the
East Atlantic.
Atlantide Rept. 11:99-139.

Cuvierina columnella

Vayssière, A., 1915
Mollusques Eutéropodes (Ptéropodes Thécosomes) provenant des campagnes des yachts
HIRONDELLE et PRINCESSE-ALICE (1885-1913).
Rés. Camp. Sci., Monaco, 47:226 pp., 14 pls.

Cuvierina columnella

Wormelle, Ruth L., 1962
A survey of the standing crop of plankton
of the Florida Current. VI. A study of the
distribution of the pteropods of the Florida
Current.
Bull. Mar. Sci. Gulf & Caribbean, 12(1):
95-136.

Cuvierina columnella + forms

Van der Spoel, S., 1967.
Euthecosomata, a group with remarkable developmental stages (Gastropoda, pteropoda).
J. Noorduijn en Zoom, N.V., Gorinchem. 375 pp.

Cuvierina urceolaris

Tesch, J.J., 1913
Mollusca. Pteropoda. Das Tierreich.
36:154 pp., 108 text figs.

Cymbulia

Meisenheimer, Johannes, 1906
Die Pteropoden der Deutschen Südpolar-Expedition 1901-1903. Deutsche Südpolar
Exped., 1901-1903, 9(Zool. 1): 96-153,
Tables 5-7.

Cymbulia sp.

Tesch, J. J., 1948
The Thecosomatous pteropods. II. The
Indo-Pacific. Dana Rept. No. 30: 45 pp.,
34 text figs., 3 pls.

Cymbulia sp.

Tesch, J. J., 1946
The Thecosomatous Pteropods. Dana Rept.
No. 28:82 pp., 34 text figs., 8 pls.

Cymbulia sp.

Tesch, J. J., 1910
X Pteropoda and Heteropoda. Percy Sladen
Trust Expedition to the Indian Ocean in 1915.
Trans. Linn. Soc., London, Zool., 2nd ser.14:
165-189, pls. 12-14.

Cymbulia parvidentata

Meisenheimer, Johannes, 1905
Pteropoda. Wiss. Ergeb., Deutschen Tiefsee-Exped. "Valdivia", 1898-1899: 314 pp., with
atlas with 27 pls.

Cymbulia parvidentata

Tesch, J.J., 1913
Mollusca. Pteropoda. Das Tierreich.
36:154 pp., 108 text figs.

Cymbulia peroni

Daumas, Raoul, et Hubert J. Ceccaldi, 1965.
Contribution à l'étude biochimique d'organismes
marins. 1 Acides amines libres et proteiques
chez Beroe ovata (Eschscholtz), Ciona intestinalis (L.), Cymbulia peroni (De Blainville)
et Rhizostoma pulmo (Agassiz).
Rec. Trav. Sta. Mar. Endoume, 38(54):3-14.

Also in:
Trav. Sta. Zool., Villefranche-sur-Mer, 25.

Cymbulia peroni

Franc, A., 1949
Hétéropodes, et autres Gastropodes
planctoniques de Méditerranée occidentale.
J. de Conchyliologia 89: 209-230, 19 text
figs. Trav. Sta. Zool. Villefranche-sur-Mer
10(3)

Cymbulia peroni

Magaldi, Norman N. 1971.
Sobre la presencia en el Atlantico
sudoccidental de Cymbulia de Blainville,
1818 (Pteropoda Pseudothecosomata).
Neotropica 17 (53): 92-94

Cymbulia peroni

Massy, A.L., 1932.
Mollusca: Gastropoda Thecostomata and Gymnostomata. Discovery Repts., 3:267-296.

Cymbulia peroni

Meisenheimer, Johannes, 1905
Pteropoda. Wiss. Ergeb., Deutschen Tiefsee-Exped. "Valdivia", 1898-1899: 314 pp., with
atlas with 27 pls.

Cymbulia peroni
Morton, J.E., 1954.
The pelagic Mollusca of the Benguela Current. Pt. 1. First survey, R.R.S. William Scoresby, March 1951, with an account of the reproductive system and sexual succession of Limacina bulmodes. Discovery Rept. 24:163-200.

Cymbulia peroni
Rampal, Jeannine, 1967.
Répartition quantitative et bathymétrique des ptéropodes Thécosmoes récoltés en Méditerranée occidentale au nord du 40° parallèle: remarques morphologiques sur certaines espèces. Rev. Trav. Inst. Pêches marit., 31(4):403-416.

? Cymbulia peroni
Stubbings, H. G., 1938
Pteropoda. John Murray Exped., 1933-34, Sci. Repts. 5(2):15-33, 2 text figs.

Cymbulia peronii
Tesch, J.J., 1913
Mollusca. Pteropoda. Das Tierreich. 36:154 pp., 108 text figs.

Cymbulia peroni
van der Spoel, S., 1970.
The pelagic Mollusca from the Atlantide and Galathea expeditions collected in the East Atlantic. Atlantide Rept. 11:99-139.

Cymbulia peroni
Vayssière, A., 1915
Mollusques Eutéropodes (Ptéropodes Thécosomes) provenant des campagnes des yachts HIRONDELLE et PRINCESSE-ALICE (1885-1913). Rés. Camp. Sci., Monaco, 47:226 pp., 14 pls.

Cymbulia peronii
Wormelle, Ruth L., 1962
A survey of the standing crop of plankton of the Florida Current. VI. A study of the distribution of the pteropods of the Florida Current. Bull. Mar. Sci., Gulf & Caribbean, 12(1):95-136.

Cymbulia radiata
Tesch, J.J., 1913
Mollusca. Pteropoda. Das Tierreich. 36:154 pp., 108 text figs.

Cymbulia sibogae
Meisenheimer, Johannes, 1905
Pteropoda. Wiss. Ergeb., Deutschen Tiefsee-Exped. "Valdivia", 1898-1899: 314 pp., with atlas with 27 pls.

Cymbulia sibogae
Stubbings, H. G., 1938
Pteropoda. John Murray Exped., 1933-34, Sci. Repts. 5(2):15-33, 2 text figs.

Cymbulia sibogae
Tesch, J.J., 1913
Mollusca. Pteropoda. Das Tierreich. 36:154 pp., 108 text figs.

Cymbulia sibogae
Tokioka, Takasi, 1961
The structure of the operculum of the species of Atlantidae (Gastropoda: Heteropoda) as a taxonomic criterion, with records of some pelagic mollusks of the North Pacific. Publ. Seto Mar. Biol. Lab., Kyoto Univ., 9(2):267-332.

Cymbulia sibogae
Tokioka, T., 1955.
On some plankton animals collected by the Syunkotu-maru in May-June 1954. 4. Thecosomatous pteropods. Publ. Seto Mar. Biol. Lab. 5(1):59-74, Pls. 7-13.

Cymbuliopsis calceola
Bedot, M., 1909
Sur la faune de l'Archipel Malais (resume). Rev. Suisse Zool. 17:143-169.

Cymbuliopsis intermedia
Massy, A.L., 1932.
Mollusca: Gastropoda Thecostomata and Gymnostomata. Discovery Repts., 3:267-296.

Desmopterus gardineri
Frontier, S., 1963.
Zooplancton récolté en Mer d'Arabie, Golfe Persique et Golfe d'Aden. (3e campagne océanographique du "Commandant Robert Giraud" avril à juin 1961) II. Ptéropodes: systématique et répartition. Trav. Centre Océanogr., Nosy-Bé, Cahiers, O.R.S.T.R.O.M., Océanogr., Paris, No. 6:233-254.

Desmopterus papilio
Sakthivel, M. 1972. Studies on Desmopterus Chun 1889 species in the Indian Ocean. Meteor Forsch.-Ergebn. (D) 10: 46-57.

Desmopterus gardineri
Tesch, J.J., 1913
Mollusca. Pteropoda. Das Tierreich. 36:154 pp., 108 text figs.

Desmopterus gardineri n.sp.
Tesch, J. J., 1910
X Pteropoda and Heteropoda. Percy Sladen Trust Expedition to the Indian Ocean in 1915. Trans. Linn. Soc., London, Zool., 2nd ser.14:165-189, pls. 12-14.

Desmopterus gardenii
Wormelle, Ruth L., 1962
A survey of the standing crop of plankton of the Florida Current. VI. A study of the distribution of the pteropods of the Florida Current. Bull. Mar. Sci., Gulf & Caribbean, 12(1):95-136.

Desmopterus intermedia
Wormelle, Ruth L., 1962
A survey of the standing crop of plankton of the Florida Current. VI. A study of the distribution of the pteropods of the Florida Current. Bull. Mar. Sci., Gulf & Caribbean, 12(1):95-136.

Desmopterus papilio
Bedot, M., 1909
Sur la faune de l'Archipel Malais (resume). Rev. Suisse Zool. 17:143-169.

Desmopterus papilio
Franc, A., 1949
Hétéropodes, et autres Gastropodes planctoniques de Méditerranée occidentale. J. de Conchyliologia 89: 209-230, 19 text figs. Trav. Sta. Zool. Villefranche-sur-Mer 10(3)

Desmopterus papilio
Frontier, S., 1963.
Zooplancton récolté en Mer d'Arabie, Golfe Persique et Golfe d'Aden. (3e campagne océanographique du "Commandant Robert Giraud", avril à juin 1961) II. Ptéropodes: systématique et répartition. Trav. Centre Océanogr., Nosy-Bé, Cahiers, O.R.S.T.R.O.M., Océanogr., Paris, No. 6:233-254.

Desmopterus papilio
Meisenheimer, Johannes, 1906
Die Pteropoden der Deutschen Südpolar-Expedition 1901-1903. Deutsche Südpolar Exped., 1901-1903, 9(Zool. 1): 96-153, Tables 5-7.

Desmopterus papilio
Meisenheimer, Johannes, 1905
Pteropoda. Wiss. Ergeb., Deutschen Tiefsee-Exped. "Valdivia", 1898-1899: 314 pp., with atlas with 27 pls.

Desmopterus papilio
Tesch, J.J., 1946
The Thecosomatous Pteropods. Dana Rept. No. 28:82 pp., 34 text figs., 8 pls.

Desmopterus papilio
Tesch, J.J., 1913
Mollusca. Pteropoda. Das Tierreich. 36:154 pp., 108 text figs.

Desmopterus papilio
Tesch, J.J., 1910
X Pteropoda and Heteropoda. Percy Sladen Trust Expedition to the Indian Ocean in 1915. Trans. Linn. Soc., London, Zool., 2nd ser.14:165-189, pls. 12-14.

Desmopterus papilio
Tokioka, Takasi, 1961
The structure of the operculum of the species of Atlantidae (Gastropoda: Heteropoda) as a taxonomic criterion, with records of some pelagic mollusks of the North Pacific. Publ. Seto Mar. Biol. Lab., Kyoto Univ., 9(2):267-332.

Desmopteris papilio
Tokioka, T., 1955.
On some plankton animals collected by the Syunkotu-maru in May-June 1954. 4. Thecosomatous pteropods. Publ. Seto Mar. Biol. Lab., 5(1):59-74, Pls. 7-13.

Desmopterus papilio
van der Spoel, S., 1970.
The pelagic Mollusca from the Atlantide and Galathea expeditions collected in the East Atlantic. Atlantide Rept. 11:99-139.

Desmopterus papilio
Wormelle, Ruth L., 1962
A survey of the standing crop of plankton of the Florida Current. VI. A study of the distribution of the pteropods of the Florida Current. Bull. Mar. Sci., Gulf & Caribbean, 12(1):95-136.

Diacria sp.
Boltovskoy, Demetrio 1971. Pteropodos thecosomados del Atlantico Sudoccidental. Malacologia 11(1): 121-140.

Diacria costata
Tesch, J.J., 1913
Mollusca. Pteropoda. Das Tierreich. 36:154 pp., 108 text figs.

Diacria major
Tesch, J.J., 1913
Mollusca. Pteropoda. Das Tierreich. 36:154 pp., 108 text figs.

Diacria quadridentata
Delsman, H. C., 1939.
Preliminary plankton investigations in the Java Sea. Treubia, 17:139-181

Diacria quadridenta
Frontier, S., 1966.
Liste complémentaire des ptéropodes du plancton de Nosy Bé (Madagascar). Cah. ORSTOM, Sér. Océanogr., 4(2):141-146.

Diacria quadridentata

Frontier, S., 1963.
Hétéropodes et ptéropodes récoltés dans le plancton de Nosy-Bé. Cahiers Trav. Centre Oceanogr., Nosy-Bé, O.R.S.T.O.M. Oceanogr., Paris, No. 6:213-227.

Diacria quadridentata

Meisenheimer, Johannes, 1906
Die Pteropoden der Deutschen Südpolar-Expedition 1901-1903. Deutsche Südpolar Exped., 1901-1903, 9(Zool. 1): 96-153, Tables 5-7.

Diacria quadridentata

Meisenheimer, Johannes, 1905
Pteropoda. Wiss. Ergeb., Deutschen Tiefsee-Exped. "Valdivia", 1898-1899: 314 pp., with atlas with 27 pls.

Diacria quadridentata

Moore, H. B., 1949
The zooplankton of the upper waters of the Bermuda area of the North Atlantic. Bull. Bingham Ocean. Coll. 12(2):97, 208 text figs.

Diacria quadridentata

Stubbings, H. G., 1938
Pteropoda. John Murray Exped., 1933-34, Sci. Repts. 5(2):15-33, 2 text figs.

Diacria quadridentata

Tesch, J. J., 1948
The Thecosomatous pteropods. II. The Indo-Pacific. Dana Rept. No. 30: 45 pp., 34 text figs., 3 pls.

Diacria quadridentata

Tesch, J. J., 1946
The Thecosomatous Pteropods. Dana Rept. No. 28:82 pp., 34 text figs., 8 pls.

Diacria quadridentata

Tesch, J.J., 1913
Mollusca. Pteropoda. Das Tierreich. 36:154 pp., 108 text figs.

Diacria quadridentata

Tesch, J. J., 1910
X Pteropoda and Heteropoda. Percy Sladen Trust Expedition to the Indian Ocean in 1915. Trans. Linn. Soc., London, Zool., 2nd ser.14: 165-189, pls. 12-14.

Diacria quadridentata

Tokioka, Takasi, 1961
The structure of the operculum of the species of Atlantidae (Gastropoda: Heteropoda) as a taxonomic criterion, with records of some pelagic mollusks of the North Pacific.
Publ. Seto Mar. Biol. Lab., Kyoto Univ., 9(2):267-332.

Diacria quadridentata

van der Spoel, S., 1970.
The pelagic Mollusca from the Atlantide and Galathea expeditions collected in the East Atlantic.
Atlantide Rept. 11:99-139.

Diacria quadridentata

Van der Spoel, S., 1968.
A new form of Diacria quadridentata (Blainville, 1821) and shell growth of this species (Gastropoda, Pteropoda).
Vidensk, Meddr dansk naturh. Foren. 131: 217-224.

Diacria quadridentata

Wormelle, Ruth L., 1962
A survey of the standing crop of plankton of the Florida Current. VI. A study of the distribution of the pteropods of the Florida Current.
Bull. Mar. Sci. Gulf & Caribbean, 12(1): 95-136.

Diacria quadridentata + forms

Van der Spoel, S., 1967.
Euthecosomata, a group with remarkable developmental stages (Gastropoda, pteropoda).
J. Noorduijn en Zoom, N.V., Gorinchem, 375 pp.

Diacria quadridentata quadridenta

Tokioka, T., 1955.
On some animals collected by the Syunkotu-maru in May-June 1954. 4. Thecosomatous pteropods.
Publ. Seto Mar. Biol. Lab., 5(1):59-74, Pls. 7-13.

Diacria trispinosa

Fraser, J. H., 1949
List of rare exotic species found in the plankton by the Scottish Vessel "Explorer" in 1947. Ann. Biol., Int. Cons., 4:68-69.

Diacria trispinosa

Fraser, J. H., 1949
Plankton of the Faroe-Shetland Channel and the Faroes, June and August 1947. Ann. Biol., Int. Cons., 4:27-28, text fig. 10.

Diacria trispinosa

Fraser, J. H., and A. Saville, 1949
Plankton distribution in Scottish and adjacent waters in 1948. Ann. Biol. 5:61-62.

Diacria trispenosa

Fraser, J. H. and A. Saville, 1949
Macroplankton in the Faroe Channel, 1948. Ann. Biol. 5:29-30, text figs.

Diacria trispinosa

Fraser, J. H. and A. Saville, 1949
List of rare exotic species found in the plankton by the Scottish Vessel "Explorer" in 1948. Ann. Biol. 5:62-64.

Diacria trispinosa

Meisenheimer, Johannes, 1906
Die Pteropoden der Deutschen Südpolar-Expedition 1901-1903. Deutsche Südpolar Exped., 1901-1903, 9(Zool. 1): 96-153, Tables 5-7.

Diacria trispinosa

Meisenheimer, Johannes, 1905
Pteropoda. Wiss. Ergeb., Deutschen Tiefsee-Exped. "Valdivia", 1898-1899: 314 pp., with atlas with 27 pls.

Diacria trispinosa

Menzies, R.J., 1958.
Shell-bearing pteropod gastropods from Mediterranean plankton (Cavoliniidae).
Pub. Staz. Zool., Napoli, 30(3):381-402.

Diacria trispinosa

Moore, H. B., 1949
The zooplankton of the upper waters of the Bermuda area of the North Atlantic. Bull. Bingham Ocean. Coll. 12(2):97, 208 text figs.

Diacria trispinosa

Morton, J.E., 1954.
The pelagic Mollusca of the Benguela Current. Pt. 1. First survey, R.R.S. William Scoresby, March 1951, with an account of the reproductive system and sexual succession of Limacina bulmodes. Discovery Rept. 24:163-200.

Diacria trispinosa

Russell, F.S., and J.S. Colman, 1935
The Zooplankton. IV. The occurrence and seasonal distribution of the Tunicata, Mollusca and Coelenterata (siphonophora). Brit. Mus. (N.H.) Great Barrier Reef Expedition 1928-1929, Sci. Repts., 2(7):203-276, 30 text figs.

Diacria trispinosa

Stubbings, H. G., 1938
Pteropoda. John Murray Exped., 1933-34, Sci. Repts. 5(2):15-33, 2 text figs.

Diacria trispinosa

Tesch, J. J., 1948
The Thecosomatous pteropods. II. The Indo-Pacific. Dana Rept. No. 30: 45 pp., 34 text figs., 3 pls.

Diacria trispinosa

Tesch, J. J., 1946
The Thecosomatous Pteropods. Dana Rept. No. 28:82 pp., 34 text figs., 8 pls.

Diacria trispinosa

Tesch, J.J., 1913
Mollusca. Pteropoda. Das Tierreich. 36:154 pp., 108 text figs.

Diacria trispinosa

Tesch, J.J., 1910
X Pteropoda and Heteropoda. Percy Sladen Trust Expedition to the Indian Ocean in 1915. Trans. Linn. Soc., London, Zool., 2nd ser.14: 165-189, pls. 12-14.

Diacria trispinosa

Tokioka, Takasi, 1961
The structure of the operculum of the species of Atlantidae (Gastropoda: Heteropoda) as a taxonomic criterion, with records of some pelagic mollusks of the North Pacific.
Publ. Seto Mar. Biol. Lab., Kyoto Univ., 9(2):267-332.

Diacria trispinosa

van der Spoel, S., 1970.
The pelagic Mollusca from the Atlantide and Galathea expeditions collected in the East Atlantic.
Atlantide Rept. 11:99-139.

Diacria trispinosa

Vane, F.R., 1961
Contribution towards a plankton atlas of the north-eastern Atlantic and the North Sea. 3. Gastropoda.
Bulls. Mar. Ecol., 5(42): 98-101, Pl. 28.

Diacria trispinosa

Vane, F.R., and J.M. Colebrook, 1962
Continuous plankton records: Contributions towards a plankton atlas of the north-eastern Atlantic and North Sea. VI. The seasonal and annual distributions of the Gastropoda. Bull. Mar. Ecology, 5(50):247-253.

Diacria trispinosa

Vayssière, A., 1915
Mollusques Eutéropodes (Ptéropodes Thécosomes) provenant des campagnes des yachts HIRONDELLE et PRINCESSE-ALICE (1885-1913). Rés. Camp. Sci., Monaco, 47:226 pp., 14 pls.

Diacria trispinosa

Wormelle, Ruth L., 1962
A survey of the standing crop of plankton of the Florida Current. VI. A study of the distribution of the pteropods of the Florida Current.
Bull. Mar. Sci. Gulf & Caribbean, 12(1): 95-136.

Diacria trispinosa +forma
Van der Spoel, S., 1967.
Euthecosomata, a group with remarkable developmental stages (Gastropoda, pteropoda).
J. Noorduijn en Zoom, N.V., Gorinchem, 375 pp.

Diacria trispinosa trispinosa
Tokioka, T., 1955.
On some plankton animals collected by the Syunkotu-maru in May-June 1954. 4. Thecosomatous pteropods. Publ. Seto Mar. Biol. Lab., 5(1):59-74, Pls. 7-13, 1 textfig.

Diacria trispinosa trispinosa
Van der Spoel, S., 1964.
Notes on some pteropods from the North Atlantic.
Beaufortia, 10(121):167-176.

Elysia viridis
Franc, A., 1949
Hétéropodes, et autres Gastropodes planctoniques de Méditerranée occidentale.
J. de Conchyliologia 89: 209-230, 19 text figs. Trav. Sta. Zool. Villefranche-sur-Mer 10(3)

Euclio spp.
Vane, F.R., 1961
Contribution towards a plankton atlas of the north-eastern Atlantic and the North Sea. 3. Gastropoda.
Bulls. Mar. Ecol., 5(42): 98-101, Pl. 28.

Euclio spp
Vane, F.R., and J.M. Colebrook, 1962
Continuous plankton records: Contributions towards a plankton atlas of the north-eastern Atlantic and North Sea. VI. The seasonal and annual distributions of the Gastropoda.
Bull. Mar. Ecology, 5(50):247-253.

Euclio antarctica
Tesch, J. J., 1948
The Thecosomatous pteropods. II. The Indo-Pacific. Dana Rept. No. 30: 45 pp., 34 text figs., 3 pls.

Euclio balantium
Tesch, J. J., 1948
The Thecosomatous pteropods. II. The Indo-Pacific. Dana Rept. No. 30: 45 pp., 34 text figs., 3 pls.

Euclio balantium
Tesch, J. J., 1946
The Thecosomatous Pteropods. Dana Rept. No. 28:82 pp., 34 text figs., 8 pls.

Euclio balantium
Tokioka, Takasi, 1961
The structure of the operculum of the species of Atlantidae (Gastropoda: Heteropoda) as a taxonomic criterion, with records of some pelagic mollusks of the North Pacific.
Publ. Seto Mar. Biol. Lab., Kyoto Univ., 9(2):267-332.

Euclio campybura n.sp.
Tesch, J. J., 1948
The Thecosomatous pteropods. II. The Indo-Pacific. Dana Rept. No. 30: 45 pp., 34 text figs., 3 pls.

Euclio chaptali
Okutani, Takashi, 1964.
The cosmatous pteropods collected during the Second Cruise of the Japanese Expedition of Deep Sea (JEDS-2).
Venus, 22(4):336-341.

Euclio chaptali
Tesch, J. J., 1948
The Thecosomatous pteropods. II. The Indo-Pacific. Dana Rept. No. 30: 45 pp., 34 text figs., 3 pls.

Euclio chaptali
Tesch, J. J., 1946
The Thecosomatous Pteropods. Dana Rept. No. 28:82 pp., 34 text figs., 8 pls.

Euclio cuspidata
Menzies, R.J., 1958.
Shell-bearing pteropod gastropods from Mediterranean plankton (Cavoliniidae).
Pub. Staz. Zool., Napoli, 30(3):381-402.

Euclio cuspidata
Rampal, Jeannine, 1967.
Répartition quantitative et bathymétrique des ptéropodes Thécosomes récoltes en Méditerranée occidentale au nord du 40° parallèle: remarques morphologiques sur certaines espèces.
Rev. Trav. Inst. Pêches marit., 31(4):403-416.

Euclio cuspidata
Rampal, Jeannine, 1965.
Varitations morphologiques au cours de la croissance d'Euclio Cuspidata (Bosc) (Ptéropode thécosome).
Bull. Inst. Océanogr., Monaco, 65(1360):12pp.

Euclio cuspidata
Rampal, Jeannine, 1965.
Variations morphologiques au cours de la Croissance d'Euclio cuspidata (Bosc) (Ptéropode thécosome).
Repp. Proc. Verb. Réunions, Comm. Int. Expl. Sci Mer Méditerranée, Monaco, 18(2):465.

Euclio cuspidata
Tesch, J. J., 1948
The Thecosomatous pteropods. II. The Indo-Pacific. Dana Rept. No. 30: 45 pp., 34 text figs., 3 pls.

Euclio cuspidata
Tesch, J. J., 1946
The Thecosomatous Pteropods. Dana Rept. No. 28:82 pp., 34 text figs., 8 pls.

Euclio cuspidata
Tokioka, Takasi, 1961
The structure of the operculum of the species of Atlantidae (Gastropoda: Heteropoda) as a taxonomic criterion, with records of some pelagic mollusks of the North Pacific.
Publ. Seto Mar. Biol. Lab., Kyoto Univ., 9(2):267-332.

Euclio arthotheca n.sp.
Tesch, J. J., 1948
The Thecosomatous pteropods. II. The Indo-Pacific. Dana Rept. No. 30: 45 pp., 34 text figs., 3 pls.

Euclio polita
Menzies, R.J., 1958.
Shell-bearing pteropod gastropods from Mediterranean plankton (Cavoliniidae).
Pub. Sta. Zool., Napoli, 30(3):381-402.

Euclio polita
Tesch, J. J., 1946
The Thecosomatous Pteropods. Dana Rept. No. 28:82 pp., 34 text figs., 8 pls.

Euclio polita
Rampal, Jeannine, 1966.
Pêches planctoniques, superficielles et profondes, en Méditerranée occidentale (Campagne de la Thalassa - janvier 1961 - entre les Îles Baléares, le Sardaigne et l'Algérois) VI. Ptéropodes.
Revue Trav. Inst. (Scient. tech.) marit., 30(4):375-383.

Euclio pyramidata
Aron, William, 1962
The distribution of animals in the eastern North Pacific and its relationship to physical and chemical conditions.
J. Fish. Res. Bd., Canada, 19(2):271-314.

Euclio pyramidata
Berkeley, E., and C. Berkeley, 1960
Some further records of pelagic Polychaeta from the northeast Pacific, north of Latitude 40°N. and east of Longitude 175°W., together with records of Siphonophora, Mollusca and Tunicata from the same region.
Canadian J. Zoology, 38:787-799.

Euclio pyramidata
Franc, A., 1949
Hétéropodes, et autres Gastropodes planctoniques de Méditerranée occidentale.
J. de Conchyliologia 89: 209-230, 19 text figs. Trav. Sta. Zool. Villefranche-sur-Mer 10(3)

Euclio pyramidata
Jespersen, P., 1954 (posthumous).
On the quantities of macroplankton in the North Atlantic. Medd. Danmarks Fisk. og Havundersøgel., n.s., 1(2):1-12.

Euclio pyramidata
Menzies, R.J., 1958.
Shell-bearing pteropod gastropods from Mediterranean plankton (Cavoliniidae).
Pub. Staz. Zool., Napoli, 30(3):381-402.

Euclio pyramidata
Morton, J.E., 1954.
The pelagic Mollusca of the Benguela Current. Pt. 1. First survey, R.R.S. William Scoresby, March 1951, with an account of the reproductive system and sexual succession of Limacina bulmodes.
Discovery Rept., 24:163-200.

Euclio pyramidata
Okutani, Takashi, 1964.
The cosomatous pteropods collected during the Second Cruise of the Japanese Expedition of Deep Sea (JEDS-2).
Venus, 22(4):336-341.

Euclio pyramidata
Rampal, Jeannine, 1967.
Répartition quantitative et bathymétrique des ptéropodes Thécosomes récoltes en Méditerranée occidentale au nord du 40° parallèle: remarques morphologiques sur certaines espèces.
Rev. Trav. Inst. Pêches marit., 31(4):403-416.

Eucleo pyramidata
Rampal, Jeannine, 1966.
Pêches planctoniques, superficielles et profondes, en Méditerranée occidentale (Campagne de la Thalassa - janvier 1961 entre les Îles Baléares, le Sardaigne et l'Algérois) VI. Ptéropodes.
Revue Trav. Inst. (Scient. Tech.) marit., 30(4):375-383.

Euclio pyramidata
Tesch, J. J., 1948
The Thecosomatous pteropods. II. The Indo-Pacific. Dana Rept. No. 30: 45 pp., 34 text figs., 3 pls.

Euclio pyramidata
Tesch, J. J., 1946
The Thecosomatous Pteropods. Dana Rept. No. 28:82 pp., 34 text figs., 8 pls.

Euclio pyramidata
Tokioka, Takasi, 1961
The structure of the operculum of the species of Atlantidae (Gastropoda: Heteropoda) as a taxonomic criterion, with records of some pelagic mollusks of the North Pacific.
Publ. Seto Mar. Biol. Lab., Kyoto Univ., 9(2):267-332.

Euclio pyramidata lanceolata
Tokioka, T., 1955.
On some plankton animals collected by the Syunkotu-maru in May-June 1954. 4. Thecosomatous pteropods. Publ. Seto Mar. Biol. Lab., 5(1):59-74, Pls. 7-13, 1 textfig.

Fowleri zetesios
Tesch, J.J., 1913
Mollusca. Pteropoda. Das Tierreich. 36:154 pp., 108 text figs.

Fowlerina
Tesch, J.J., 1950.
The Gymnosomata II. Dana Rept. No. 36:55 pp., 37 textfigs.

Gastropteron (pacificum?) cinereum n.sp.
Dall, W. H., 1925.
The Pteropoda collected by the Canadian Arctic Expedition 1913-18, with description of a new species from the North Pacific. Rept. Canadian Arctic Exped., 1913-1918, 8(B):9-12.

Gleba sp.
Russell, F.S., and J.S. Colman, 1935
The Zooplankton. IV. The occurrence and seasonal distribution of the Tunicata, Mollusca and Coelenterata (siphonophora). Brit. Mus. (N.H.) Great Barrier Reef Expedition 1928-1929, Sci. Repts., 2(7):203-276, 30 text figs.

Gleba chrysistica
Meisenheimer, Johannes, 1905
Pteropoda. Wiss. Ergeb., Deutschen Tiefsee-Exped. "Valdivia", 1898-1899: 314 pp., with atlas with 27 pls.

Gleba chrysosticta
Tesch, J.J., 1913
Mollusca. Pteropoda. Das Tierreich. 36:154 pp., 108 text figs.

Gleba cordata
Franc,A.,1949
Hétéropodes, et autres Gastropodes planctoniques de Méditerranée occidentale. J. de Conchyliologia 89: 209-230, 19 text figs. Trav. Sta. Zool. Villefranche-sur-Mer 10(3)

Gleba cordata
Meisenheimer, Johannes, 1905
Pteropoda. Wiss. Ergeb., Deutschen Tiefsee-Exped. "Valdivia", 1898-1899: 314 pp., with atlas with 27 pls.

Gleba cordata
Moore, H. B., 1949
The zooplankton of the upper waters of the Bermuda area of the North Atlantic. Bull. Bingham Ocean. Coll. 12(2):97, 208 text figs.

Gleba cordata
Tesch, J. J., 1946
The Thecosomatous Pteropods. Dana Rept. No. 28:82 pp., 34 text figs., 8 pls.

Gleba cordata
Tesch, J.J., 1913
Mollusca. Pteropoda. Das Tierreich. 36:154 pp., 108 text figs.

Gleba cordata
Vayssière, A., 1915
Mollusques Eutéropodes (Ptéropodes Thécosomes) provenant des campagnes des yachts HIRONDELLE et PRINCESSE-ALICE (1885-1913). Rés. Camp. Sci., Monaco, 47:226 pp., 14 pls.

Halopsyche gaudichaudi
Meisenheimer, Johannes, 1905
Pteropoda. Wiss. Ergeb., Deutschen Tiefsee-Exped. "Valdivia", 1898-1899: 314 pp., with atlas with 27 pls.

Halopsyche gaudichaudii
Tesch, J.J., 1913
Mollusca. Pteropoda. Das Tierreich. 36:154 pp., 108 text figs.

Hyalaea rugosa
Tesch, J.J., 1913
Mollusca. Pteropoda. Das Tierreich. 36:154 pp., 108 text figs.

Hyalaea tridentata
Davies, D. H., 1949
Preliminary investigations on the foods of South African fishes (with notes on the general fauna of the area surveyed). Fish & Mar. Biol. Survey Div., Dept. Comm. & Industries, S. Africa, Invest. Rept. No.11, Commerce & Industry, Jan. 1949:1-36, 4 pls.

Hyalaea truncata
Tesch, J.J., 1913
Mollusca. Pteropoda. Das Tierreich. 36:154 pp., 108 text figs.

Hyalocyclis striata
Delsman, H. C., 1939.
Preliminary plankton investigations in the Java Sea. Treubia, 17:139-181, 8 maps, 41 figs.

Hyalocylix striata
Franc,A.,1949
Hétéropodes, et autres Gastropodes planctoniques de Méditerranée occidentale. J. de Conchyliologia 89: 209-230, 19 text figs. Trav. Sta. Zool. Villefranche-sur-Mer 10(3)

Hyalocylix striata
Meisenheimer, Johannes, 1906
Die Pteropoden der Deutschen Südpolar-Expedition 1901-1903. Deutsche Südpolar Exped., 1901-1903, 9(Zool. 1): 96-153, Tables 5-7.

Hyalocyclix striata
Meisenheimer, Johannes, 1905
Pteropoda. Wiss. Ergeb., Deutschen Tiefsee-Exped. "Valdivia", 1898-1899: 314 pp., with atlas with 27 pls.

Hyalocyclix striata
Menzies, R.J., 1958.
Shell-bearing pteropod gastropods from Mediterranean plankton (Cavoliniidae). Pub. Staz. Zool., Napoli, 30(3):381-402.

Hyalocylis striata
Moore, H. B., 1949
The zooplankton of the upper waters of the Bermuda area of the North Atlantic. Bull. Bingham Ocean. Coll. 12(2):97, 208 text figs.

Hyalocylix striata
Rampal,Jeannine,1967.
Répartition quantitative et bathymétrique des ptéropodes Thécosomes récoltés en Méditerranée occidentale au nord du 40° parallèle: remarques morphologiques sur certaines espèces. Rev.Trav.Inst.Pêches marit., 31(4):403-416.

Hyalocylix striata
Rampal,Jeannine,1966.
Pêches planctoniques, superficielles et profondes,en Méditerranée occidental (Campagne de la Thalassa- janvier 1961 -entre les Iles Baléares, la Sardaigne et l'Algerois)VI. Pteropodes. Revue Trav.Inst.(Scient.Tech.)marit., 30(4):375-383.

Hyalocylis striata
Stubbings, H. G., 1938
Pteropoda, John Murray Exped., 1933-34, Sci. Repts. 5(2):15-33, 2 text figs.

Hyalocylis striata
Tesch, J. J., 1948
The Thecosomatous pteropods. II. The Indo-Pacific. Dana Rept. No. 30: 45 pp., 34 text figs., 3 pls.

Hyalocylis striata
Tesch, J. J., 1946
The Thecosomatous Pteropods. Dana Rept. No. 28:82 pp., 34 text figs., 8 pls.

Hyalocyclis striata
Tesch, J.J., 1913
Mollusca. Pteropoda. Das Tierreich. 36:154 pp., 108 text figs.

Hyalocylis striata
Tesch, J. J., 1910
X Pteropoda and Heteropoda. Percy Sladen Trust Expedition to the Indian Ocean in 1915. Trans. Linn. Soc., London, Zool., 2nd ser.14: 165-189, pls. 12-14.

Hyalocylis striata
Tokioka, Takasi, 1961
The structure of the operculum of the species of Atlantidae (Gastropoda: Heteropoda) as a taxonomic criterion, with records of some pelagic mollusks of the North Pacific. Publ. Seto Mar. Biol. Lab., Kyoto Univ., 9(2):267-332.

Hyalocylis striata
Tokioka, T., 1955.
On some plankton animals collected by the Syunkotu-maru in May-June 1954. Publ. Seto Mar. Biol. Lab., 5(1):59-74, Pls. 7-13.

Hyalocyclis striata
van der Spoel, S., 1970.
The pelagic Mollusca from the Atlantide and Galathea expeditions collected in the East Atlantic. Atlantide Rept. 11:99-139.

Hyalocylis striata
Van der Spoel,S., 1967.
Euthecosomata, a group with remarkbale developmental stages (Gastropoda,Pteropoda). J.Noorduijn en Zoom. N.V., Gorinchem, 375 pp.

Hyalocylix striata
Vayssière, A., 1915
Mollusques Eutéropodes (Ptéropodes Thécosomes) provenant des campagnes des yachts HIRONDELLE et PRINCESSE-ALICE (1885-1913). Rés. Camp. Sci., Monaco, 47:226 pp., 14 pls.

Hyalocylix striata
Wormelle, Ruth L., 1962
A survey of the standing crop of plankton of the Florida Current. VI. A study of the distribution of the pteropods of the Florida Current. Bull. Mar. Sci. Gulf & Caribbean, 12(1):95-136.

Hydromyles globulosa
Tesch, J.J., 1950.
The Gymnosomata II. Dana Rept. No. 36:55 pp., 37 textfigs.

Hydromyles globulosa
Tokioka, Takasi, 1961
The structure of the operculum of the species of Atlantidae (Gastropoda: Heteropoda) as a taxonomic criterion, with records of some pelagic mollusks of the North Pacific. Publ. Seto Mar. Biol. Lab., Kyoto Univ., 9(2):267-332.

Ianthina globosa
Morton, J.E., 1954.
The pelagic Mollusca of the Benguela Current. Pt. 1. First survey, R.R.S. William Scoresby, March 1951, with an account of the reproductive system and sexual succession of Limacina bulmodes. Discovery Rept. 24:163-200.

Ianthina ianthina
Morton, J.E., 1954.
The pelagic Mollusca of the Benguela Current. Pt. 1. First survey, R.R.S. William Scoresby, March 1951, with an account of the reproductive system and sexual succession of Limacina bulmodes. Discovery Rept. 24:163-200.

Laginiopsis triloba
Tesch, J.J., 1950.
The Gymnosomata II. Dana Rept. No. 36:55 pp., 37 textfigs.

Limacina sp.
Delsman, H.C., 1939.
Preliminary plankton investigations in the Java Sea. Treubia, 17:139-181, 8 maps, 41 figs.

Limacina
Henderson, G.T.D., and N.B. Marshall, 1944
Continuous plankton records: the zooplankton (other than copepoda and young fish) in the southern North Sea 1932-1937. Hull Bull. Mar. Ecol., 1(6):255-275, 24 pls.

Limacina sp.
Ogilvie, H.S., 1934
A preliminary account of the food of the herring in the North-Western North Sea. Rapp. Proc. Verb.89(3):85-92, 2 text figs.

Limacina antarctica
Eliot, C., 1907.
Mollusca. VI. Pteropoda. Nat. Antarctic Exped., "Discovery", 1901-1904, 3:15 pp., 2 pls.

Limacina antarctica
Meisenheimer, Johannes, 1905
Pteropoda. Wiss. Ergeb., Deutschen Tiefsee-Exped. "Valdivia", 1898-1899: 314 pp., with atlas with 27 pls.

Limacina antarctica
Tesch, J.J., 1913
Mollusca. Pteropoda. Das Tierreich. 36:154 pp., 108 text figs.

Limacina arctica
Marukawa, H., 1921
Plankton lists and some new species of copepods, from the northern waters of Japan. Bull. Inst. Ocean., No.384, 15 pp., 3 pls., 1 chart. Monaco

Limacina australis
Meisenheimer, Johannes, 1905
Pteropoda. Wiss. Ergeb., Deutschen Tiefsee-Exped. "Valdivia", 1898-1899: 314 pp., with atlas with 27 pls.

Limacina australis
Tesch, J.J., 1913
Mollusca. Pteropoda. Das Tierreich. 36:154 pp., 108 text figs.

Limacina balea
Bigelow, H.B., 1922
Exploration of the coastal water off the northeastern United States in 1916 by the U.S. Fisheries Schooner Grampus. Bull. M.C.Z. 65(5):85-188, 53 text figs.

Limacina balea
Korotkevich, V.S., and K.V. Beklemishev, 1960
[Zooplankton research.] Arktich. i Antarkt. Nauchno-Issled. Inst., Sovetsk. Antarkt. Exped., Mezhd. Geofiz. God, Vtoraia Morsk. Exped., "Ob", 1956-1957, 7: 111-125.

Limacina balea
Mackintosh, N.A., 1934
Distribution of the Macroplankton in the Atlantic Sector of the Antarctic. Discovery reports, Vol.9:65-160, 48 text figs.

Limacina balea
Massy, A.L., 1932.
Mollusca: Gastropoda Thecostomata and Gymnostomata. Discovery Repts., 3:267-296.

Limacina balea
Smidt, E.L.B., 1944
Biological Studies of the Invertebrate Fauna of the Harbor of Copenhagen. Vidensk. Medd. fra Dansk naturh. Foren. 107:235-316, 23 text figs.

Limacina balea
Tesch, J.J., 1913
Mollusca. Pteropoda. Das Tierreich. 36:154 pp., 108 text figs.

Limacina bulimoides
Boltovskoy, Demetrio 1971.
Pteropodos thecosomados del Atlantico sudoccidental.
Malacologia 11(6): 121-140.

Limacina bulimoides
Franc, A., 1949
Hétéropodes, et autres Gastropodes planctoniques de Méditerranée occidentale. J. de Conchyliologia 89: 209-230, 19 text figs. Trav. Sta. Zool. Villefranche-sur-Mer 10(3)

Limacina bulimoides
Massy, A.L., 1932.
Mollusca: Gastropoda Thecostomata and Gymnostomata. Discovery Repts., 3:267-296.

Limacina bulimoides
Meisenheimer, Johannes, 1906
Die Pteropoden der Deutschen Südpolar-Expedition 1901-1903. Deutsche Südpolar Exped., 1901-1903, 9(Zool. 1): 96-153, Tables 5-7.

Limacina bulimoides
Meisenheimer, Johannes, 1905
Pteropoda. Wiss. Ergeb., Deutschen Tiefsee-Exped. "Valdivia", 1898-1899: 314 pp., with atlas with 27 pls.

Limacina bulimoides
Moore, H.B., 1949
The zooplankton of the upper waters of the Bermuda area of the North Atlantic. Bull. Bingham Ocean. Coll. 12(2):97, 208 text figs.

Limacina bulmodes
Morton, J.E., 1954.
The pelagic Mollusca of the Benguela Current. Pt. 1. First survey, R.R.S. William Scoresby, March 1951, with an account of the reproductive system and sexual succession of Limacina Bulmodes. Discovery Rept. 24:163-200.

Limacina bulimoides
Tesch, J.J., 1946
The Thecosomatous Pteropods. Dana Rept. No. 28:82 pp., 34 text figs., 8 pls.

Limacina bulimoides
Tesch, J.J., 1913
Mollusca. Pteropoda. Das Tierreich. 36:154 pp., 108 text figs.

Limacina bulimoides
Tesch, J.J., 1910
X Pteropoda and Heteropoda. Percy Sladen Trust Expedition to the Indian Ocean in 1915. Trans. Linn. Soc., London, Zool., 2nd ser.14: 165-189, pls. 12-14.

Limacina bulimoides
Tokioka, Takasi, 1961
The structure of the operculum of the species of Atlantidae (Gastropoda: Heteropoda) as a taxonomic criterion, with records of some pelagic mollusks of the North Pacific. Publ. Seto Mar. Biol. Lab., Kyoto Univ., 9(2):267-332.

Limacina bulimoides
Tokioka, T., 1955.
On some plankton animals collected by the Syunkotu-maru in May-June 1954. 4. Thecosomatous pteropods. Publ. Seto Mar. Biol. Lab., 5(1):59-74, Pls. 7-13, 1 textfig.

Limacina bulimoides
van der Spoel, S., 1970.
The pelagic Mollusca from the Atlantide and Galathea expeditions collected in the East Atlantic.
Atlantide Rept. 11:99-139.

Limacina (Munthea) bulimodes
Van der Spoel, S., 1967.
Euthecosomata, a group with remarkable developmental stages (Gastropoda, pteropoda).
J. Noorduijn en Zoon, N.V., Gorinchem, 375 pp.

Limacina bulimoides
Vayssière, A., 1915
Mollusques Eutéropodes (Ptéropodes Thécosomes) provenant des campagnes des yachts HIRONDELLE et PRINCESSE-ALICE (1885-1913). Rés. Camp. Sci. Monaco, 47:226 pp., 14 pls.

Limacina bulimoides
Wormelle, Ruth L., 1962
A survey of the standing crop of plankton of the Florida Current. VI. A study of the distribution of the pteropods of the Florida Current.
Bull. Mar. Sci., Gulf & Caribbean, 12(1): 95-136.

Limacina cochlostyloides
Tesch, J.J., 1913
Mollusca. Pteropoda. Das Tierreich. 36:154 pp., 108 text figs.

Limacina helicina
Aron, William, 1962
The distribution of animals in the eastern North Pacific and its relationship to physical and chemical conditions.
J. Fish. Res. Bd., Canada, 19(2):271-314.

Limacina helicina
Boltovskoy, Demetrio 1971.
Pteropodos thecosomados del Atlantico sudoccidental.
Malacologia 11(6): 121-140.

Limacina helicina
Boltovskoy, Demetrio 1971.
Contribución al conocimiento de los pteropodos thecasomados sobre la plataforma continental Bonaerense.
Revista Mus. La Plata, Univ. Nac. La Plata, 11 (Zool. 100): 121-136.

Limacina helicina
Dunbar, M.J., 1942.
Marine macroplankton from the Canadian Eastern Arctic. II. Medusae, Siphonophora, Ctenophora, Pteropoda, and Chaetognatha. Canad. J. Res., D. 20:71-77.

Limacina helicina
Fish, C.J., and M.W. Johnson, 1937
The biology of the zooplankton population in the Bay of Fundy and the Gulf of Maine with special reference to production and distribution. J. Biol. Bd., Canada 3(3):189-322, 29 tables, 45 text figs.

Limacina helicina
Holm, Th., 1889
Om de paa Fylla's Togt i 1884 foretagne zoologiske Undersøgelser i Grønland. Medd. Grønland. 8(5):151-171.

Limacina helicina
Kramp, P.L., 1961.
Pteropoda. The Godthaab Expedition, 1928. Medd. om Grønland, 81(4):3-13.

Limacina helicina
Mackintosh, N.A., 1934
Distribution of the Macroplankton in the Atlantic Sector of the Antarctic. Discovery reports, Vol.9:65-160, 48 text figs.

Limacina helicina
Massy, A.L., 1932.
Mollusca: Gastropoda Thecostomata and Gymnostomata. Discovery Repts., 3:267-296.

Limacina helicina
Meisenheimer, Johannes, 1906
Die Pteropoden der Deutschen Südpolar-Expedition 1901-1903. Deutsche Südpolar Exped., 1901-1903, 9(Zool. 1): 96-153, Tables 5-7.

Limacina helicina
McGowan, John A., 1963.
Geographical variation in Limacina helicina in the North Pacific.
In: Speciation in the sea, Systematics Association Publication, No. 5:109-128.

Limacina helicina
Nishimura, S., 1957.
On some plankton animals occurring in spring off Wajima, Noto Peninsula, Japan Sea with special reference to their vertical distribution. Ann. Rept., Japan Sea Reg. Fish. Res. Lab., No. 3:61-71.

Limacina helicina
Meisenheimer, Johannes, 1905
Pteropoda. Wiss. Ergeb., Deutschen Tiefsee-Exped. "Valdivia", 1898-1899: 314 pp., with atlas with 27 pls.

Limacina helicina helicina
Okutani, Takeshi, 1964.
The cosomatous pteropods collected during the Second Cruise of the Japanese Expedition of Deep Sea (JEDS-2). Venus, 22(4):336-341.

Limacina helicina
Tesch, J.J., 1948
The Thecosomatous pteropods. II. The Indo-Pacific. Dana Rept. No. 30: 45 pp., 34 text figs., 3 pls.

Limacina helicina
Tesch, J.J., 1946
The Thecosomatous Pteropods. Dana Rept. No. 28:82 pp., 34 text figs., 8 pls.

Limacina helicina
Tesch, J.J., 1913
Mollusca. Pteropoda. Das Tierreich. 36:154 pp., 108 text figs.

Limacina helicina
Tokioka, Takasi, 1961
The structure of the operculum of the species of Atlantidae (Gastropoda: Heteropoda) as a taxonomic criterion, with records of some pelagic mollusks of the North Pacific. Publ. Seto Mar. Biol. Lab., Kyoto Univ., 9(2):267-332.

Limacina helicina
Vayssière, A., 1915
Mollusques Eutéropodes (Ptéropodes Thécosomes) provenant des campagnes des yachts HIRONDELLE et PRINCESSE-ALICE (1885-1913). Rés. Camp. Sci., Monaco, 47:226 pp., 14 pls.

Limacina helicina and subspecies
Van der Spoel, S., 1967.
Euthecosomata, a group with remarkable developmental stages (Gastropoda, Pteropoda). J. Noorduijn en Zoom, N.V., Gorinchem, 375 pp.

Limacina helicoides
Massy, A.L., 1932.
Mollusca: Gastropoda Thecostomata and Gymnostomata. Discovery Repts., 3:267-296.

Limacina helicoides
Tesch, J.J., 1948
The Thecosomatous pteropods. II. The Indo-Pacific. Dana Rept. No. 30: 45 pp., 34 text figs., 3 pls.

Limacina helicoides
Tesch, J.J., 1946
The Thecosomatous Pteropods. Dana Rept. No. 28:82 pp., 34 text figs., 8 pls.

Limacina helicoides
Tesch, J.J., 1913
Mollusca. Pteropoda. Das Tierreich. 36:154 pp., 108 text figs.

Limacina helicoides
Van der Spoel, S., 1964.
Notes on some pteropods from the North Atlantic. Beaufortia, 10(121):167-176.

Limacina helicoides
Vayssière, A., 1915
Mollusques Eutéropodes (Ptéropodes Thécosomes) provenant des campagnes des yachts HIRONDELLE et PRINCESSE-ALICE (1885-1913). Rés. Camp. Sci., Monaco, 47:226 pp., 14 pls.

Limacina helicoides
van der Spoel, S., 1970.
The pelagic Mollusca from the Atlantide and Galathea expeditions collected in the East Atlantic. Atlantide Rept. 11:99-139.

Limacina (Thilea) helicoides
Van der Spoel, S., 1967.
Euthecosomata, a group with remarkable developmental stages (Gastropoda, Pteropoda). J. Noorduijn en Zoom, N.V., Gorinchem, 375 pp.

Limacina inflata
Boltovskoy, Demetrio 1971.
Pteropodos thecosomados del Atlantico sudoccidental. Malacologia 11(1):121-140.

Limacina inflata
Franc, A., 1949
Hétéropodes, et autres Gastropodes planctoniques de Méditerranée occidentale. J. de Conchyliologia 89: 209-230, 19 text figs. Trav. Sta. Zool. Villefranche-sur-Mer 10(3)

Limacina inflata
Meisenheimer, Johannes, 1906
Die Pteropoden der Deutschen Südpolar-Expedition 1901-1903. Deutsche Südpolar Exped., 1901-1903, 9(Zool. 1): 96-153, Tables 5-7.

Limacina inflata
Meisenheimer, Johannes, 1905
Pteropoda. Wiss. Ergeb., Deutschen Tiefsee-Exped. "Valdivia", 1898-1899: 314 pp., with atlas with 27 pls.

Limacina inflata
Moore, H. B., 1949
The zooplankton of the upper waters of the Bermuda area of the North Atlantic. Bull. Bingham Ocean. Coll. 12(2):97, 208 text figs.

Limacina inflata
Morton, J.E., 1954.
The pelagic Mollusca of the Benguela Current. Pt. 1. First survey, R.R.S. William Scoresby, March 1951, with an account of the reproductive system and sexual succession of Limacina bulmodes. Discovery Rept. 24:163-200.

Limacina inflata
Stubbings, H. G., 1938
Pteropoda. John Murray Exped., 1933-34, Sci. Repts. 5(2):15-33, 2 text figs.

Limacina inflata
Tesch, J.J., 1946
The Thecosomatous Pteropods. Dana Rept. No. 28:82 pp., 34 text figs., 8 pls.

Limacina inflata
Tesch, J.J., 1913
Mollusca. Pteropoda. Das Tierreich. 36:154 pp., 108 text figs.

Limacina inflata
Tesch, J.J., 1910
X Pteropoda and Heteropoda. Percy Sladen Trust Expedition to the Indian Ocean in 1915. Trans. Linn. Soc., London, Zool., 2nd ser. 14:165-189, pls. 12-14.

Limacina inflata
Tokioka, Takasi, 1961
The structure of the operculum of the species of Atlantidae (Gastropoda: Heteropoda) as a taxonomic criterion, with records of some pelagic mollusks of the North Pacific. Publ. Seto Mar. Biol. Lab., Kyoto Univ., 9(2):267-332.

Limacina inflata
Tokioka, T., 1955.
On some plankton animals collected by the Syunkotu-maru in May-June 1954. 4. Thecosomatous pteropods. Publ. Seto Mar. Biol. Lab., 5(1):59-74, Pls. 7-13, 1 textfig.

Limacina inflata
van der Spoel, S., 1970.
The pelagic Mollusca from the Atlantide and Galathea expeditions collected in the East Atlantic. Atlantide Rept. 11:99-139.

Limacina (Thilea) inflata
Van der Spoel, S., 1967.
Euthecosomata, a group with remarkable developmental stages (Gastropoda, Pteropoda). J. Noorduijn en Zoom, N.V., Gorinchem, 375 pp.

Limacina inflata
Vayssière, A., 1915
Mollusques Eutéropodes (Ptéropodes Thécosomes) provenant des campagnes des yachts HIRONDELLE et PRINCESSE-ALICE (1885-1913). Rés. Camp. Sci., Monaco, 47:226 pp., 14 pls.

Limacina inflata
Vinogradov, M.E., and N.M. Voronina, 1964.
Quantitative distribution of the plankton in the upper layers of the Pacific equatorial currents. II. Vertical distribution of several species. Regularity of the distribution of oceanic plankton. (In Russian; English abstract). Trudy Inst. Okeanol., Akad. Nauk, SSSR, 65:58-76.

Limacina inflata

Wormelle, Ruth L., 1962.
A survey of the standing crop of plankton of the Florida Current. VI. A study of the distribution of the pteropods of the Florida Current.
Bull. Mar. Sci. Gulf & Caribbean, 12(1):95-136.

Limacina Lesueuri

Meisenheimer, Johannes, 1906
Die Pteropoden der Deutschen Südpolar-Expedition 1901-1903. Deutsche Südpolar Exped., 1901-1903, 9(Zool. 1): 96-153, Tables 5-7.

Limacina Lesueuri

Meisenheimer, Johannes, 1905
Pteropoda. Wiss. Ergeb., Deutschen Tiefsee-Exped. "Valdivia", 1898-1899: 314 pp., with atlas with 27 pls.

Limacina lesueurii

Moore, H. B., 1949
The zooplankton of the upper waters of the Bermuda area of the North Atlantic. Bull. Bingham Ocean. Coll. 12(2):97, 208 text figs.

Limacina lesueuri

Tesch, J. J., 1948
The Thecosomatous pteropods. II. The Indo-Pacific. Dana Rept. No. 30: 45 pp., 34 text figs., 3 pls.

Limacina lesueuri

Tesch, J. J., 1946
The Thecosomatous Pteropods. Dana Rept. No. 28:82 pp., 34 text figs., 8 pls.

Limacina lesueurii

Tesch, J.J., 1913
Mollusca. Pteropoda. Das Tierreich. 36:154 pp., 108 text figs.

Limacina lesueuri

Tokioka, Takasi, 1961
The structure of the operculum of the species of Atlantidae (Gastropoda: Heteropoda) as a taxonomic criterion, with records of some pelagic mollusks of the North Pacific.
Publ. Seto Mar. Biol. Lab., Kyoto Univ., 9(2):267-332.

Limacina lesueuri

Tokioka, T., 1955.
On some plankton animals collected by the Syunkotu-maru in May-June 1954. 4. Thecosomatous pteropods. Publ. Seto Mar. Biol. Lab., 5(1):59-74, Pls. 7-13, 1 textfig.

Limacina lesueurii

van der Spoel, S., 1970.
The pelagic Mollusca from the Atlantide and Galathea expeditions collected in the East Atlantic.
Atlantide Rept. 11:99-139.

Limacina (Thilia) lesueurii

Van der Spoel, S., 1967.
Euthecostomata, a group with remarkable developmental stages (Gastropoda, Pteropoda).
J. Noorduijn en Zoom, N.V., Gorinchem, 375 pp.

Limacina lesueuri

Vayssière, A., 1915
Mollusques Eutéropodes (Ptéropodes Thécosomes) provenant des campagnes des yachts HIRONDELLE et PRINCESSE-ALICE (1885-1913).
Rés. Camp. Sci. Monaco, 47:226 pp., 14 pls.

Limacina lesueurii

Wormelle, Ruth L., 1962
A survey of the standing crop of plankton of the Florida Current. VI. A study of the distribution of the pteropods of the Florida Current.
Bull. Mar. Sci. Gulf & Caribbean, 12(1):95-136.

Limacina parium

Boltovskoy, Demetrio 1971.
Contribución al conocimiento de los pteropodos thecosomados sobre la plataforma continental Bonaerense.
Revista Mus. La Plata, Univ. Nac. La Plata, 11 (Zool. 100): 121-136

Limacina rangii

Meisenheimer, Johannes, 1906
Die Pteropoden der Deutschen Südpolar-Expedition 1901-1903. Deutsche Südpolar Exped., 1901-1903, 9(Zool. 1): 96-153, Tables 5-7.

Limacina rangii

Tesch, J.J., 1913
Mollusca. Pteropoda. Das Tierreich. 36:154 pp., 108 text figs.

Limacina retroversa

Bigelow, H.B., and M. Sears, 1939
Studies of the waters of the continental shelf, Cape Cod to Chesapeake Bay. III. A volumetric study of the zooplankton. Mem. M.C.Z. 54(4):183-378, 42 text figs.

Limacina retroversa

Boltovskoy, Demetrio 1971.
Pteropodos thecosomados del Atlantico sudoccidental.
Malacologia 11 (1): 121-140.

Limacina retroversa

Boltovskoy, Demetrio 1971.
Contribucion al conocimiento de los pteropodos thecosomados sobre la plataforma continental bonaerense.
Revista Mus. La Plata n.s. (Zool.) 11:121-136.

Limacina retroversa

Eliot, C., 1907.
Mollusca. VI. Pteropoda. Nat. Antarctic Exped., "Discovery", 1901-1904, 3:15 pp., 2 pls.

Limacina retroversa

Fish, C.J., and M.W. Johnson, 1937
The biology of the zooplankton population in the Bay of Fundy and the Gulf of Maine with special reference to production and distribution. J. Biol. Bd., Canada 3(3):189-322, 29 tables, 45 text figs.

Limacina retroversa

Fraser, J. H., 1949
Plankton investigations from the Scottish Research Vessel. Ann. Biol., Int. Cons., 4: 66-67.

Limacina retroversa

Hansen, Vagn Kr., 1960.
Investigations on the quantitative and qualitative distribution of zooplankton in the southern part of the Norwegian Sea.
Medd. Danmarks Fiskeri- og Havundersøgelser, n.s., 2(23):1-53.

Limacina retroversa

Jespersen, P., 1954 (posthumous).
On the quantities of macroplankton in the North Atlantic. Medd Danmarks Fisk. og Havundersøgel., n.s., 1(2):1-12.

Limacina retroversa

Kielhorn, W.V., 1952
The biology of the surface zone zooplankton of a Boreo-Arctic Atlantic Ocean area. J.Fish Res. Bd., Canada 9 (5): 223-264, 13 text figs.

Limacina retroversa

Kramp, P.L., 1961.
Pteropoda. The Godthaab Expedition, 1928. Medd. om Grønland, 81(4):3-13.

Limacina retroversa

Künne, Cl., 1937
Über die Verbreitung der Leitformen des Großplanktons in der südlichen Nordsee im Winter. Ber. Deutschen Wissenschaflichen Kommission für Meeresforschung, n.f. VIII (3):131-164, 9 text figs., 4 fold-ins.

Limacina retroversa

Marshall, N. B., 1948
Continuous plankton records: Zooplankton (other than Copepoda and young fish) in the North Sea 1938-1939. Hull Bull. Mar. Ecol. 2(13):173-213, Pls. 89-108.

Limacina retroversa

Massy, A.L., 1932.
Mollusca; Gastropoda Thecostomata and Gymnostomata. Discovery Repts., 3:267-296.

Limacina retroversa

Meisenheimer, Johannes, 1906
Die Pteropoden der Deutschen Südpolar-Expedition 1901-1903. Deutsche Südpolar Exped., 1901-1903, 9(Zool. 1): 96-153, Tables 5-7.

Limacina retroversa

Meisenheimer, Johannes, 1905
Pteropoda. Wiss. Ergeb., Deutschen Tiefsee-Exped. "Valdivia", 1898-1899: 314 pp., with atlas with 27 pls.

Limacina retroversa

Morton, J.E., 1954.
The biology of Limacina retroversa. J.M.B.A. 33(2):297-312.

Limacina retroversa

Rae, K. M., 1949
Plankton. Some broad changes in the plankton round the north of the British Isles in 1948. Ann. Biol. 5:56-60, 12 text figs.

Limacina

Redfield, A. C., 1941
The effect of the circulation of water on the distribution of the calanoid community in the Gulf of Maine. Biol. Bull., LXXX(1):86-110.
See Fig. 7

Limacina retroversa

Redfield, A.C., 1939.
The history of a population of Limacina retroversa during its drift across the Gulf of Maine. Biol. Bull. 76(1):26-47, 10 textfigs.

Spiratella retroversa (see footnote for priority of names).

Limacina retroversa

Savage, R.E., 1937
The food of North Sea herring 1930-1934. Ministry of Agriculture and Fisheries. Fish. Invest. Ser. II, 15(5):1-60; 16 text figs.

Limacina retroversa

Strømgren, Tor 1973.
Zooplankton investigations in Borgenfjorden, 1967-1969.
Miscellanea, Norske Vidensk. Selsk. Mus. Trondheim 9:37pp.

Limacina retroversa

Tesch, J. J., 1946
The Thecosomatous Pteropods. Dana Rept. No. 28:82 pp., 34 text figs., 8 pls.

Limacina retroversa

Tesch, J.J., 1913
Mollusca. Pteropoda. Das Tierreich. 36:154 pp., 108 text figs.

Limacina retroversa

Wormelle, Ruth L., 1962
A survey of the standing crop of plankton of the Florida Current. VI. A study of the distribution of the pteropods of the Florida Current.
Bull. Mar. Sci., Gulf & Caribbean, 12(1): 95-136.

Limacina retroversa and subsp.

Van der Spoel, S., 1967.
Euthecosomata, a group with remarkable developmental stages (Gastropoda, Pteropoda).
J. Noorduijn en Zoom.N.V.,Gorinchem, 375 pp.

Limacina triacantha

Meisenheimer, Johannes, 1905
Pteropoda. Wiss. Ergeb., Deutschen Tiefsee-Exped. "Valdivia", 1898-1899: 314 pp., with atlas with 27 pls.

Limacina trochiformis

Bedot, M., 1909
Sur la faune de l'Archipel Malais (resume).
Rev. Suisse Zool. 17:143-169.

Limacina trochiformis

Franc,A.,1949
Hétéropodes, et autres Gastropodes planctoniques de Méditerranée occidentale.
J. de Conchyliologia 89: 209-230, 19 text figs. Trav. Sta. Zool. Villefranche-sur-Mer 10(3)

Limacina trochiformis

Meisenheimer, Johannes, 1906
Die Pteropoden der Deutschen Südpolar-Expedition 1901-1903. Deutsche Südpolar Exped., 1901-1903, 9(Zool. 1): 96-153, Tables 5-7.

Limacina trochiformis

Meisenheimer, Johannes, 1905
Pteropoda. Wiss. Ergeb., Deutschen Tiefsee-Exped. "Valdivia", 1898-1899: 314 pp., with atlas with 27 pls.

Limacina trochiformis

Moore, H. B., 1949
The zooplankton of the upper waters of the Bermuda area of the North Atlantic. Bull. Bingham Ocean. Coll. 12(2):97, 208 text figs.

Limacina trochiformis

Stubbings, H. G., 1938
Pteropoda. John Murray Exped., 1933-34, Sci. Repts. 5(2):15-33, 2 text figs.

Limacina trochiformis

Tesch, J. J., 1946
The Thecosomatous Pteropods. Dana Rept. No. 28:82 pp., 34 text figs., 8 pls.

Limacina trochiformis

Tesch, J. J., 1910
X Pteropoda and Heteropoda. Percy Sladen Trust Expedition to the Indian Ocean in 1915.
Trans. Linn. Soc., London, Zool., 2nd ser.14: 165-189, pls. 12-14.

Limacina trochiformis

Tokioka, Takasi, 1961
The structure of the operculum of the species of Atlantidae (Gastropoda: Heteropoda) as a taxonomic criterion, with records of some pelagic mollusks of the North Pacific.
Publ. Seto Mar. Biol. Lab., Kyoto Univ. 9(2):267-332.

Limacina trochiformis

Tesch, J.J., 1913
Mollusca. Pteropoda. Das Tierreich. 36:154 pp., 108 text figs.

Limacina trochiformis

Tokioka, T., 1955.
On some plankton animals collected by the Syunkotu-maru in May-June 1954. 4. Thecosomatous pteropods. Publ. Seto Mar. Biol. Lab., 5(1):59-74, Pls. 7-13, 1 textfig.

Limacina trochiformis

van der Spoel, S., 1970.
The pelagic Mollusca from the Atlantide and Galathea expeditions collected in the East Atlantic.
Atlantide Rept. 11:99-139.

Limacina (Munthea) trochiformis

Van der Spoel,S., 1967.
Euthecosomata, a group with remarkable developmental stages (Gastropoda,pteropoda).
J. Noorduijn en Zoom,N.V., Gorinchem, 375 pp.

Limacina trochiformis

Wormelle, Ruth L., 1962
A survey of the standing crop of plankton of the Florida Current. VI. A study of the distribution of the pteropods of the Florida Current.
Bull. Mar. Sci., Gulf & Caribbean, 12(1): 95-136.

Limapontia cupitata

Franc,A.,1949
Hétéropodes, et autres Gastropodes planctoniques de Méditerranée occidentale.
J. de Conchyliologia 89: 209-230, 19 text figs. Trav. Sta. Zool. Villefranche-sur-Mer 10(3)

Massya longecirrata

Tesch, J.J., 1950.
The Gymnosomata II. Dana Rept. No. 36:55 pp., 37 textfigs.

Massya longicirrata

van der Spoel, S., 1970.
The pelagic Mollusca from the Atlantide and Galathea expeditions collected in the East Atlantic.
Atlantide Rept. 11:99-139.

Notobranchaea grandis

Tesch, J.J., 1950.
The Gymnosomata II. Dana Rept. No. 36:55 pp., 37 textfigs.

Notobranchaea inopinata

Meisenheimer, Johannes, 1905
Pteropoda. Wiss. Ergeb., Deutschen Tiefsee-Exped. "Valdivia", 1898-1899: 314 pp., with atlas with 27 pls.

Notobranchaea inopinata

Tesch, J.J., 1913
Mollusca. Pteropoda. Das Tierreich. 36:154 pp., 108 text figs.

Notobranchaea macdonald-i

Berkeley, E., and C. Berkeley, 1960
Some further records of pelagic Polychaeta from the northeast Pacific, north of Latitude 40°N. and east of Longitude 175°W., together with records of Siphonophora, Mollusca and Tunicata from the same region.
Canadian J. Zoology. 38:787-799.

Notobranchaea macdonaldi

Tesch, J.J., 1950.
The Gymnosomata. II. Dana Rept. No. 36:55 pp., 37 textfigs.

Notobranchaea macdonaldi

Meisenheimer, Johannes, 1905
Pteropoda. Wiss. Ergeb., Deutschen Tiefsee-Exped. "Valdivia", 1898-1899: 314 pp., with atlas with 27 pls.

Notobranchaea macdonaldi

Tesch, J.J., 1913
Mollusca. Pteropoda. Das Tierreich. 36:154 pp., 108 text figs.

Notobranchaea macdonaldi

van der Spoel, S., 1970.
The pelagic Mollusca from the Atlantide and Galathea expeditions collected in the East Atlantic.
Atlantide Rept. 11:99-139.

Notobranchaea valdiviae n.sp.

Meisenheimer, Johannes, 1905
Pteropoda. Wiss. Ergeb., Deutschen Tiefsee-Exped. "Valdivia", 1898-1899: 314 pp., with atlas with 27 pls.

Notobranchaea valdiviae

Tesch, J.J., 1913
Mollusca. Pteropoda. Das Tierreich. 36:154 pp., 108 text figs.

Paedoclione doliiformis

Lalli, Carol M., 1972.
Food and feeding of Paedoclione doliiformis Danforth, a neotenous gymnosomatous pteropod.
Biol. Bull. mar. biol. Lab. Woods Hole 143: 392-402.

Paedoclione doliiformis

Tesch, J.J., 1950.
The Gymnosomata II. Dana Rept. No. 36:55 pp., 37 textfigs.

Paedoclione doliiformis

Tesch, J.J., 1913
Mollusca. Pteropoda. Das Tierreich. 36:154 pp., 108 text figs.

Paraclimne pelseneeri

Meisenheimer, Johannes, 1905
Pteropoda. Wiss. Ergeb., Deutschen Tiefsee-Exped. "Valdivia", 1898-1899: 314 pp., with atlas with 27 pls.

Paraclione pelseneeri

Russell, F.S., and J.S. Colman, 1935
The Zooplankton. IV. The occurrence and seasonal distribution of the Tunicata, Mollusca and Coelenterata (siphonophora). Brit. Mus. (N.H.) Great Barrier Reef Expedition 1928-1929, Sci. Repts., 2(7):203-276, 30 text figs.

Paraclione pelseneeri

Tesch, J.J., 1913
Mollusca. Pteropoda. Das Tierreich. 36:154 pp., 108 text figs.

Peraclis

Meisenheimer, Johannes, 1906
Die Pteropoden der Deutschen Südpolar-Expedition 1901-1903. Deutsche Südpolar Exped., 1901-1903, 9(Zool. 1): 96-153, Tables 5-7.

Peraclis sp.

Rampal,Jeannine,1966.
Pêches planctoniques, superficielles et profondes, en Méditerranée occidentale (Campagne de la Thalassa- janvier 1961 -entre les îles Baléares,la Sardaigne et l'Algérois)VI. Pteropodes.
Revue Trav.Inst.(Scient.tech.) marit.30(4):375-383.

Peraclis sp.
Russell, F.S., and J.S. Colman, 1935
The Zooplankton. IV. The occurrence and seasonal distribution of the Tunicata, Mollusca and Coelenterata (siphonophora). Brit. Mus. (N.H.) Great Barrier Reef Expedition 1928-1929, Sci. Repts., 2(7):203-276, 30 text figs.

Peraclis apicifulva
Magaldi, Norman N. 1972
Moluscos holoplanctonicos del Atlantico II. Presencia de Peraclis apicifulva Meisenheimer, 1906, frente a la costa Argentina (Pteropoda, Pseudothecosomata). Neotropica 18(57): 118-120.

Peraclis apicifulva n.sp.
Meisenheimer, Johannes, 1906
Die Pteropoden der Deutschen Südpolar-Expedition 1901-1903. Deutsche Südpolar Exped., 1901-1903, 9(Zool. 1): 96-153, Tables 5-7.

Peraclis apicifulva
Tesch, J.J., 1948
The Thecosomatous pteropods. II. The Indo-Pacific. Dana Rept. No. 30: 45 pp., 34 text figs., 3 pls.

Peraclis apicifulva
Tesch, J.J., 1946
The Thecosomatous Pteropods. Dana Rept. No. 28:82 pp., 34 text figs., 8 pls.

Peracle apicifulva
Tesch, J.J., 1913
Mollusca. Pteropoda. Das Tierreich. 36:154 pp., 108 text figs.

Peraclis apicifulva
Tokioka, Takasi, 1961
The structure of the operculum of the species of Atlantidae (Gastropoda: Heteropoda) as a taxonomic criterion, with records of some pelagic mollusks of the North Pacific. Publ. Seto Mar. Biol. Lab., Kyoto Univ., 9(2):267-332.

Peraclis apicifulva
Tokioka, T., 1955.
On some plankton animals collected by the Syunkotu-maru in May-June 1954. 4. Thecosomatous pteropods. Publ. Seto Mar. Biol. Lab., 5(1):59-74, Pls. 7-13.

Peraclis apicifulva
van der Spoel, S., 1970.
The pelagic Mollusca from the Atlantide and Galathea expeditions collected in the East Atlantic. Atlantide Rept. 11:99-139.

Peraclis bispinosa
Meisenheimer, Johannes, 1906
Die Pteropoden der Deutschen Südpolar-Expedition 1901-1903. Deutsche Südpolar Exped., 1901-1903, 9(Zool. 1): 96-153, Tables 5-7.

Peraclis bispinosa
Meisenheimer, Johannes, 1905
Pteropoda. Wiss. Ergeb., Deutschen Tiefsee-Exped. "Valdivia", 1898-1899: 314 pp., with atlas with 27 pls.

Peraclis bispinosa
Tesch, J.J., 1946
The Thecosomatous Pteropods. Dana Rept. No. 28:82 pp., 34 text figs., 8 pls.

Peracle bispinosa
Tesch, J.J., 1913
Mollusca. Pteropoda. Das Tierreich. 36:154 pp., 108 text figs.

Peraclis bispinosa
van der Spoel, S., 1970.
The pelagic Mollusca from the Atlantide and Galathea expeditions collected in the East Atlantic. Atlantide Rept. 11:99-139.

Peraclis bispinosa
Vayssière, A., 1915
Mollusques Eutéropodes (Ptéropodes Thécosomes) provenant des campagnes des yachts HIRONDELLE et PRINCESSE-ALICE (1885-1913). Rés. Camp. Sci. Monaco, 47:226 pp., 14 pls.

Peracle brevispira
Tesch, J.J., 1913
Mollusca. Pteropoda. Das Tierreich. 36:154 pp., 108 text figs.

Peraclis depressa n.sp.
Meisenheimer, Johannes, 1906
Die Pteropoden der Deutschen Südpolar-Expedition 1901-1903. Deutsche Südpolar Exped., 1901-1903, 9(Zool. 1): 96-153, Tables 5-7.

Peracle depressa
Tesch, J.J., 1913
Mollusca. Pteropoda. Das Tierreich. 36:154 pp., 108 text figs.

Peraclis depressa
Vayssière, A., 1915
Mollusques Eutéropodes (Ptéropodes Thécosomes) provenant des campagnes des yachts HIRONDELLE et PRINCESSE-ALICE (1885-1913). Rés. Camp. Sci. Monaco, 47:226 pp., 14 pls.

Peraclis moluccensis
Meisenheimer, Johannes, 1906
Die Pteropoden der Deutschen Südpolar-Expedition 1901-1903. Deutsche Südpolar Exped., 1901-1903, 9(Zool. 1): 96-153, Tables 5-7.

Peraclis moluccensis
Meisenheimer, Johannes, 1905
Pteropoda. Wiss. Ergeb., Deutschen Tiefsee-Exped. "Valdivia", 1898-1899: 314 pp., with atlas with 27 pls.

Peraclis moluccensis
Tesch, J.J., 1948
The Thecosomatous pteropods. II. The Indo-Pacific. Dana Rept. No. 30: 45 pp., 34 text figs., 3 pls.

Peraclis moluccensis
Tesch, J.J., 1946
The Thecosomatous Pteropods. Dana Rept. No. 28:82 pp., 34 text figs., 8 pls.

Peracle moluccensis
Tesch, J.J., 1913
Mollusca. Pteropoda. Das Tierreich. 36:154 pp., 108 text figs.

Peraclis moluccensis
van der Spoel, S., 1970.
The pelagic Mollusca from the Atlantide and Galathea expeditions collected in the East Atlantic. Atlantide Rept. 11:99-139.

Peraclis reticulata
Franc, A., 1949
Hétéropodes, et autres Gastropodes planctoniques de Méditerranée occidentale. J. de Conchyliologia 89: 209-230, 19 text figs. Trav. Sta. Zool. Villefranche-sur-Mer 10(3)

Peraclis reticulata
Meisenheimer, Johannes, 1906
Die Pteropoden der Deutschen Südpolar-Expedition 1901-1903. Deutsche Südpolar Exped., 1901-1903, 9(Zool. 1): 96-153, Tables 5-7.

Peraclis reticulata
Meisenheimer, Johannes, 1905
Pteropoda. Wiss. Ergeb., Deutschen Tiefsee-Exped. "Valdivia", 1898-1899: 314 pp., with atlas with 27 pls.

Peracle reticulata
Moore, H. B., 1949
The zooplankton of the upper waters of the Bermuda area of the North Atlantic. Bull. Bingham Ocean. Coll. 12(2):97, 208 text figs.

Peraclis reticulata
Tesch, J.J., 1948
The Thecosomatous pteropods. II. The Indo-Pacific. Dana Rept. No. 30: 45 pp., 34 text figs., 3 pls.

Peraclis reticulata
Tesch, J.J., 1946
The Thecosomatous Pteropods. Dana Rept. No. 28:82 pp., 34 text figs., 8 pls.

Peracle reticulata
Tesch, J.J., 1913
Mollusca. Pteropoda. Das Tierreich. 36:154 pp., 108 text figs.

Peraclis reticulata
Tesch, J. J., 1910
X Pteropoda and Heteropoda. Percy Sladen Trust Expedition to the Indian Ocean in 1915. Trans. Linn. Soc., London, Zool., 2nd ser.14: 165-189, pls. 12-14.

Peraclis reticulata
Tokioka, Takasi, 1961
The structure of the operculum of the species of Atlantidae (Gastropoda: Heteropoda) as a taxonomic criterion, with records of some pelagic mollusks of the North Pacific. Publ. Seto Mar. Biol. Lab., Kyoto Univ., 9(2):267-332.

Peraclis reticulata
Tokioka, T., 1955.
On some plankton animals collected by the Syunkotu-maru in May-June 1954. 4. Thecosomatous pteropods. Publ. Seto Mar. Biol. Lab., 5(1):59-74, Pls. 7-13.

Peraclis reticulata
van der Spoel, S., 1970.
The pelagic Mollusca from the Atlantide and Galathea expeditions collected in the East Atlantic. Atlantide Rept. 11:99-139.

Peraclis reticulata
Vayssière, A., 1915
Mollusques Eutéropodes (Ptéropodes Thécosomes) provenant des campagnes des yachts HIRONDELLE et PRINCESSE-ALICE (1885-1913). Rés. Camp. Sci. Monaco, 47:226 pp., 14 pls.

Peraclis reticulata
Wormelle, Ruth L., 1962
A survey of the standing crop of plankton of the Florida Current. VI. A study of the distribution of the pteropods of the Florida Current. Bull. Mar. Sci. Gulf & Caribbean, 12(1): 95-136.

Peraclis rissoides

Meisenheimer, Johannes, 1905
Pteropoda. Wiss. Ergeb., Deutschen Tiefsee-Exped. "Valdivia", 1898-1899: 314 pp., with atlas with 27 pls.

Peracle rissoides

Tesch, J.J., 1913
Mollusca. Pteropoda. Das Tierreich. 36:154 pp., 108 text figs.

Peraclis triacantha

Fraser, J. H. and A. Saville, 1949
List of rare exotic species found in the plankton by the Scottish Vessel "Explorer" in 1948. Ann. Biol. 5:62-64.

Paraclis triacantha

Fraser, J. H. and A. Saville, 1949
Plankton distribution in Scottish and adjacent waters in 1948. Ann. Biol. 5:61-62.

Peraclis triacantha

Meisenheimer, Johannes, 1906
Die Pteropoden der Deutschen Südpolar-Expedition 1901-1903. Deutsche Südpolar Exped., 1901-1903, 9(Zool. 1): 96-153, Tables 5-7.

Moore, H. B., 1949 Peracle triacantha
The zooplankton of the upper waters of the Bermuda area of the North Atlantic. Bull. Bingham Ocean. Coll. 12(2):97, 208 text figs.

Peraclis triacantha

Tesch, J. J., 1946
The Thecosomatous Pteropods. Dana Rept. No. 28:82 pp., 34 text figs., 8 pls.

Peracle triacantha

Tesch, J.J., 1913
Mollusca. Pteropoda. Das Tierreich. 36:154 pp., 108 text figs.

Peraclis triacantha

van der Spoel, S., 1970.
The pelagic Mollusca from the Atlantide and Galathea expeditions collected in the East Atlantic.
Atlantide Rept. 11:99-139.

Peraclis triacantha

Vayssière, A., 1915
Mollusques Eutéropodes (Ptéropodes Thécosomes) provenant des campagnes des yachts HIRONDELLE et PRINCESSE-ALICE (1885-1913). Rés. Camp. Sci. Monaco, 47:226 pp., 14 pls.

Peraclis valdiviae

Tesch, J. J., 1948
The Thecosomatous pteropods. II. The Indo-Pacific. Dana Rept. No. 30: 45 pp., 34 text figs., 3 pls.

Pleuropus hargeri

Tesch, J.J., 1913
Mollusca. Pteropoda. Das Tierreich. 36:154 pp., 108 text figs.

Pneumoderma

Meisenheimer, Johannes, 1906
Die Pteropoden der Deutschen Südpolar-Expedition 1901-1903. Deutsche Südpolar Exped., 1901-1903, 9(Zool. 1): 96-153, Tables 5-7.

Pneumoderma atlanticum

Franc, A., 1949
Hétéropodes, et autres Gastropodes planctoniques de Méditerranée occidentale. J. de Conchyliologia 89: 209-230, 19 text figs. Trav. Sta. Zool. Villefranche-sur-Mer 10(3)

Pneumoderma atlanticum

Massy, A.L., 1932.
Mollusca: Gastropoda Thecostomata and Gymnostomata. Discovery Repts., 3:267-296.

Pneumoderma atlanticum

van der Spoel, S., 1970.
The pelagic Mollusca from the Atlantide and Galathea expeditions collected in the East Atlantic.
Atlantide Rept. 11:99-139.

Pneumoderma boasi

Meisenheimer, Johannes, 1905
Pteropoda. Wiss. Ergeb., Deutschen Tiefsee-Exped. "Valdivia", 1898-1899: 314 pp., with atlas with 27 pls.

Pneumoderma boasi

Tesch, J.J., 1913
Mollusca. Pteropoda. Das Tierreich. 36:154 pp., 108 text figs.

Pneumoderma eurycotylum n.sp.

Meisenheimer, Johannes, 1905
Pteropoda. Wiss. Ergeb., Deutschen Tiefsee-Exped. "Valdivia", 1898-1899: 314 pp., with atlas with 27 pls.

Pneumoderma mediterraneum

Meisenheimer, Johannes, 1905
Pteropoda. Wiss. Ergeb., Deutschen Tiefsee-Exped. "Valdivia", 1898-1899: 314 pp., with atlas with 27 pls.

Pneumoderma eurycotylum

Tesch, J.J., 1913
Mollusca. Pteropoda. Das Tierreich. 36:154 pp., 108 text figs.

Pneumoderma heterocotylum

Tesch, J.J., 1913
Mollusca. Pteropoda. Das Tierreich. 36:154 pp., 108 text figs.

Pneumoderma mediterraneum

Russell, F.S., and J.S. Colman, 1935 Van Beneden
The Zooplankton. IV. The occurrence and seasonal distribution of the Tunicata, Mollusca and Coelenterata (siphonophora). Brit. Mus. (N.H.) Great Barrier Reef Expedition 1928-1929, Sci. Repts., 2(7):203-276, 30 text figs.

Pneumoderma mediterraneum

Tesch, J.J., 1913
Mollusca. Pteropoda. Das Tierreich. 36:154 pp., 108 text figs.

Pneumoderma mediterraneum

van der Spoel, S., 1970.
The pelagic Mollusca from the Atlantide and Galathea expeditions collected in the East Atlantic.
Atlantide Rept. 11:99-139.

Pneumoderma pacificum

Meisenheimer, Johannes, 1905
Pteropoda. Wiss. Ergeb., Deutschen Tiefsee-Exped. "Valdivia", 1898-1899: 314 pp., with atlas with 27 pls.

? Pneumoderma pacificum

Stubbings, H. G., 1938
Pteropoda. John Murray Exped., 1933-34, Sci. Repts. 5(2):15-33, 2 text figs.

Pneumoderma pacificum

Tesch, J.J., 1913
Mollusca. Pteropoda. Das Tierreich. 36:154 pp., 108 text figs.

Pneumoderma peroni

Meisenheimer, Johannes, 1905
Pteropoda. Wiss. Ergeb., Deutschen Tiefsee-Exped. "Valdivia", 1898-1899: 314 pp., with atlas with 27 pls.

Pneumoderma peronii

Tesch, J.J., 1913
Mollusca. Pteropoda. Das Tierreich. 36:154 pp., 108 text figs.

Pneumoderma pygmaeum

Tesch, J.J., 1913
Mollusca. Pteropoda. Das Tierreich. 36:154 pp., 108 text figs.

Pneumoderma souleyeti

Tesch, J.J., 1913
Mollusca. Pteropoda. Das Tierreich. 36:154 pp., 108 text figs.

Pneumoderma violaceum

Meisenheimer, Johannes, 1905
Pteropoda. Wiss. Ergeb., Deutschen Tiefsee-Exped. "Valdivia", 1898-1899: 314 pp., with atlas with 27 pls.

Pneumoderma violaceum

Tesch, J.J., 1913
Mollusca. Pteropoda. Das Tierreich. 36:154 pp., 108 text figs.

Pneumodermon pellucidum

Tesch, J.J., 1913
Mollusca. Pteropoda. Das Tierreich. 36:154 pp., 108 text figs.

Pneumodermon ruber

Tesch, J.J., 1913
Mollusca. Pteropoda. Das Tierreich. 36:154 pp., 108 text figs.

Pneumodermopsis

Meisenheimer, Johannes, 1906
Die Pteropoden der Deutschen Südpolar-Expedition 1901-1903. Deutsche Südpolar Exped., 1901-1903, 9(Zool. 1): 96-153, Tables 5-7.

Pneumodermopsis atlanticum

Tesch, J.J., 1950.
The Gymnosomata II. Dana Rept. No. 36:55 pp., 37 textfigs.

Pneumodopsis canephora

Franc, A., 1949
Hétéropodes, et autres Gastropodes planctoniques de Méditerranée occidentale. J. de Conchyliologia 89: 209-230, 19 text figs. Trav. Sta. Zool. Villefranche-sur-Mer 10(3)

Pneumodermopsis (Pneumodermopsis) Canephora

van der Spoel, S., 1970.
The pelagic Mollusca from the Atlantide and Galathea expeditions collected in the East Atlantic.
Atlantide Rept. 11:99-139.

Pneumodermopsis ciliata
Cooper, G.A., and D.C.T. Forsyth, 1963
Continuous plankton records: contribution towards a plankton atlas of the North Atlantic and the North Sea. VII. The seasonal and annual distributions of the pteropod Pneumodopsis Keferstein.
Bull. Mar. Ecol., 6(1):31-38, Pls. VI-IX.

Pneumodopsis ciliata
Franc, A., 1949
Hétéropodes, et autres Gastropodes planctoniques de Méditerranée occidentale.
J. de Conchyliologia 89: 209-230, 19 text figs. Trav. Sta. Zool. Villefranche-sur-Mer 10(3)

Pneumodermopsis ciliata
Fraser, J. H. and A. Saville, 1949
List of rare exotic species found in the plankton by the Scottish Vessel "Explorer" in 1948. Ann. Biol. 5:62-64.

Pneumodermopsis ciliata
Fraser, J. H., and A. Saville, 1949
Plankton distribution in Scottish and adjacent waters in 1948. Ann. Biol. 5:61-62.

Pneumodermopsis ciliata
Meisenheimer, Johannes, 1905
Pteropoda. Wiss. Ergeb., Deutschen Tiefsee-Exped. "Valdivia", 1898-1899: 314 pp., with atlas with 27 pls.

Pneumodermopsis ciliata
Tesch, J.J., 1950.
The Gymnosomata II. Dana Rept. No. 36:55 pp., 37 textfigs.

Pneumodermopsis ciliata
Tesch, J.J., 1913
Mollusca. Pteropoda. Das Tierreich. 36:154 pp., 108 text figs.

Pneumodermopsis (Pneumodermopsis) ciliata
van der Spoel, S., 1970.
The pelagic Mollusca from the Atlantide and Galathea expeditions collected in the East Atlantic.
Atlantide Rept. 11:99-139.

Pneumodermopsis ciliata
Van der Spoel, S., 1964.
Notes on some pteropods from the North Atlantic.
Beaufortia, 10(121):167-176.

Pneumodermopsis macrochira
Massy, A.L., 1932.
Mollusca: Gastropoda Thecostomata and Gymnostomata. Discovery Repts., 3:267-296.

Pneumodermopsis macrochira n.sp.
Meisenheimer, Johannes, 1905
Pteropoda. Wiss. Ergeb., Deutschen Tiefsee-Exped. "Valdivia", 1898-1899: 314 pp., with atlas with 27 pls.

Pneumodermopsis macrochira
Tesch, J.J., 1913
Mollusca. Pteropoda. Das Tierreich. 36:154 pp., 108 text figs.

Pneumodermopsis (Crucibranchaea) macrochira
van der Spoel, S., 1970.
The pelagic Mollusca from the Atlantide and Galathea expeditions collected in the East Atlantic.
Atlantide Rept. 11:99-139.

Pneumodermopsis mediterraneum
Tesch, J.J., 1950.
The Gymnosomata II. Dana Rept. No. 36:55 pp., 37 textfigs.

Pneumodermopsis michaelsarsi
Van der Spoel, S., 1964.
Notes on some pteropods from the North Atlantic.
Beaufortia, 10(121):167-176.

Pneumodermopsis minuta
Meisenheimer, Johannes, 1905
Pteropoda. Wiss. Ergeb., Deutschen Tiefsee-Exped. "Valdivia", 1898-1899: 314 pp., with atlas with 27 pls.

Pneumodermopsis minuta
Tesch, J.J., 1913
Mollusca. Pteropoda. Das Tierreich. 36:154 pp., 108 text figs.

Pneumodermopsis paucidens
Cooper, G.A., and D.C.T. Forsyth, 1963
Continuous plankton records: contribution towards a plankton atlas of the North Atlantic and the North Sea. VII. The seasonal and annual distributions of the pteropod Pneumodopsis Keferstein.
Bull. Mar. Ecol., 6(1):31-38, Pls. VI-IX.

Pneumodermopsis paucidens
Franc, A., 1949
Hétéropodes, et autres Gastropodes planctoniques de Méditerranée occidentale.
J. de Conchyliologia 89: 209-230, 19 text figs. Trav. Sta. Zool. Villefranche-sur-Mer 10(3)

Pneumodermopsis paucidens
Meisenheimer, Johannes, 1905
Pteropoda. Wiss. Ergeb., Deutschen Tiefsee-Exped. "Valdivia", 1898-1899: 314 pp., with atlas with 27 pls.

Pneumodermopsis paucidens
Morton, J.E., 1954.
The pelagic Mollusca of the Benguela Current. Pt. 1. First survey, R.R.S. William Scoresby, March 1951, with an account of the reproductive system and sexual succession of Limacina bulmodes.
Discovery Rept. 24:163-200.

Pneumodermopsis paucidens
Sentz-Braconnot, E., 1965.
Sur la capture des proies par le pteropode gymnosome Pneumodermopsis paucidens (Boas).
Cahiers de Biol. Mar., Roscoff, 6(2):191-194.

Pneumodermopsis (Pneumodermopsis) paucidens [a]
Stock, Jan H. 1973
Nannallecto fusii n.gen., n. sp. a copepod parasitic on the pteropod Pneumodermopsis.
Bull. Zool. Mus. Univ. Amsterdam 3(4): 21-24

Pneumodermopsis paucidens
Tesch, J.J., 1913
Mollusca. Pteropoda. Das Tierreich. 36:154 pp., 108 text figs.

Pneumodermopsis (Pneumodermopsis) paucidens
van der Spoel, S., 1970.
The pelagic Mollusca from the Atlantide and Galathea expeditions collected in the East Atlantic.
Atlantide Rept. 11:99-139.

Pneumodermopsis polycotyla
Meisenheimer, Johannes, 1905
Pteropoda. Wiss. Ergeb., Deutschen Tiefsee-Exped. "Valdivia", 1898-1899: 314 pp., with atlas with 27 pls.

Pneumodermopsis polycotyla
Tesch, J.J., 1913
Mollusca. Pteropoda. Das Tierreich. 36:154 pp., 108 text figs.

Pneumodermopsis simplex
Meisenheimer, Johannes, 1905
Pteropoda. Wiss. Ergeb., Deutschen Tiefsee-Exped. "Valdivia", 1898-1899: 314 pp., with atlas with 27 pls.

Pneumodermopsis simplex
Tesch, J.J., 1913
Mollusca. Pteropoda. Das Tierreich. 36:154 pp., 108 text figs.

Pneumodermopsis teschi n.sp. [a]
Spoel, S. van der 1973
Pneumodermopsis teschi n. sp. and notes on some other Pteropoda of the Thor expeditions 1905-1910 (Gastropoda).
Bull. Zool. Mus. Univ. Amsterdam, 3 (9): 53-64

Prionoglossa hjortii
van der Spoel, S., 1970.
The pelagic Mollusca from the Atlantide and Galathea expeditions collected in the East Atlantic.
Atlantide Rept. 11:99-139.

Prionoglossa tetrabranchiata
Tesch, J.J., 1950.
The Gymnosomata II. Dana Rept. No. 36:55 pp., 37 textfigs.

Prionoglassa tetrabranchiata
van der Spoel, S., 1970.
The pelagic Mollusca from the Atlantide and Galathea expeditions collected in the East Atlantic.
Atlantide Rept. 11:99-139.

Proclio
Spoel, S.V.D., 1962
Aberrant forms of the genus Clio Linnaeus 1767, with a review of the genus Proclio Hubendick 1951 (Gastropoda; Pteropoda).
Beaufortia, Zool. Mus., Amsterdam, 9(107):173-200.

Procymbulia valdiviae
Massy, A.L., 1932.
Mollusca: Gastropoda Thecostomata and Gymnostomata. Discovery Repts., 3:267-296.

Procymbulis Valdiviae n.gen., n.sp.
Meisenheimer, Johannes, 1905
Pteropoda. Wiss. Ergeb., Deutschen Tiefsee-Exped. "Valdivia", 1898-1899: 314 pp., with atlas with 27 pls.

Procymbulia valdiviae
Tesch, J.J., 1913
Mollusca. Pteropoda. Das Tierreich. 36:154 pp., 108 text figs.

Pruvotella danae
Tesch, J.J., 1950.
The Gymnosomata II. Dana Rept. No. 36:55 pp., 37 textfigs.

Pruvotella pellucida
Tesch, J.J., 1950.
The Gymnosomata II. Dana Rept. No. 36:55 pp., 37 textfigs.

Schizobrachium polycotylum
Massy, A.L., 1932.
Mollusca: Gastropoda Thecostomata and Gymnostomata. Discovery Repts., 3:267-296.

Schizobrachium polycotylum n.gen., n.sp.

Meisenheimer, Johannes, 1905
Pteropoda. Wiss. Ergeb., Deutschen Tiefsee-Exped. "Valdivia", 1898-1899: 314 pp., with atlas with 27 pls.

Schizobranchium polycotylum

Tesch, J.J., 1950.
The Gymnosomata II. Dana Rept. No. 36:55 pp., 37 textfigs.

Schizobrachium polycotylum

Tesch, J.J., 1913
Mollusca. Pteropoda. Das Tierreich. 36:154 pp., 108 text figs.

Schizobrachium polycotylum

van der Spoel, S., 1970.
The pelagic Mollusca from the Atlantide and Galathea expeditions collected in the East Atlantic.
Atlantide Rept. 11:99-139.

Sonogiobranchaea australis

Meisenheimer, Johannes, 1905
Pteropoda. Wiss. Ergeb., Deutschen Tiefsee-Exped. "Valdivia", 1898-1899: 314 pp., with atlas with 27 pls.

Spiratella helicoides

Rampal, Jeannine, 1967.
Répartition quantitative et bathymétrique des ptéropodes Thécosomes récoltés en Méditerranée occidentale au nord du 40° parallèle: remarques morphologiques sur certaines espèces.
Rev.Trav.Inst.Pêches marit., 31(4):403-416.

Spiratella inflata

Rampal, Jeannine, 1967.
Répartition quantitative et bathymétrique des ptéropodes Thécosmoes récoltés en Méditerranée occidentale au nord du 40° parallèle:remarques morphologiques sur certaines espèces.
Rev.Trav.Inst.Pêches marit., 31(4):403-416.

Spiratella (= Limacina) inflata

Rampal, Jeannine, 1964.
Etude de l'opercule de Spiratella (= Limacina) inflata (D'Orbigny) 1836.
Bull. Inst. Océanogr., Monaco, 61(1285):12 pp.

Spiratella inflata

Rampal, J., 1963.
Ptéropodes thécosomes de pêches par paliers entre les Baléares, la Sardaigne et la côte nord-africaine.
Rapp. Proc. Verb., Réunions, Comm. Int. Expl. Sci., Mer Méditerranée, Monaco, 17(2):637-639.

Spiratella inflata

Sentz-Braconnot, 1965.
Répartition des ptéropodes Spiratella inflata (d'Orbigny) et S. trochiformis (d'Orbigny) dans la rade de Villefranche-sur-Mer.
Rapp. Proc.-verb. Réun., Comm. int. Explor. scient. Mer Méditerranée 19(3): 463-467.

Spiratella lesueuri

Rampal, Jeannine, 1966.
Pêches planctoniques, superficielles et profondes, en Méditerranée occidentale (Campagne de la Thalassa- janvier 1961 - entre les îles Baléares, la Sardaigne et l'Algérois) VI. Ptéropodes.
Revue Trav.Inst.(Scient.tech.) marit. 30(4):375-383.

Spiratella lesueuri

Rampal, Jeannine, 1965.
Ptéropodes thécosomes indicateurs hydrologiques.
Rev. Trav. Inst. Pêches Marit., 29(4):393-400.

?Spiratella pacifica

Dall, W.H., 1925.
The Pteropoda collected by the Canadian Arctic Expedition 1913-18, with description of a new species deom the North Pacific. Rept. Canadian Arctic Exped., 1913-1918, 8(B):9-12.

Spiratella retroversa

Bary, B. McK., 1963.
Temperature, salinity and plankton in the eastern North Atlantic and coastal waters of Britain 1957 II. The relationships between species and water bodies.
J. Fish. Res. Bd., Canada, 20(4):1031-1065.

formerly Limacina retroversa

Spiratella retroversa

Beaudovin, Jacqueline, 1967.
Oeufs et larves de poissons récoltés par la "Thalassa" dans le Détroit de Danemark et le Nord de la Mer d'Irminger (NORWESTLANT I-20 mars-8 Mai 1963: relations avec l'hydrologie et le zooplancton.
Revue Trav. Inst. (scient. tech.) Pêches marit. 31 (3):307-326.

Spiratella retroversa

Glover,R.S., and G.A. Robinson,1966.
The continuous plankton recorder survey: plankton around the British Isles during 1964.
Annls biol., Copenh., 21-56-60.

Spiratella retroversa

Glover, R., and G.A. Robinson, 1965.
The continuous plankton recorder survey: plankton around the British Isles during 1963.
Ann. Biol., Cons. Perm. Int. Expl. Mer, 1963, 20:93-97.

Spiratella retroversa

Redfield, A.C., 1939.
The history of a population of Limacina retroversa during its drift across the Gulf of Maine.
Biol. Bull. 76(1):26-47, 10 textfigs.

see footnote for priority of this name.

Spiratella retroversa (=Limacina retroversa)

Vane, F.R., 1961
Contribution towards a plankton atlas of the north-eastern Atlantic and the North Sea. 3. Gastropoda.
Bulls. Mar. Ecol., 5(42): 98-101, Pl. 28.

Spiratella retroversa (=Limacina)

Vane, F.R., and J.M. Colebrook, 1962
Continuous plankton records: Contributions towards a plankton atlas of the north-eastern Atlantic and North Sea. VI. The seasonal and annual distributions of the Gastropoda.
Bull. Mar. Ecology, 5(50):247-253.

Spiratella inflata

Rampal,Jeannine,1966.
Pêches planctoniques, superficielles et profondes,en Méditerranée occidentale(Campagne de la Thalassa - janvier 1961-entre les îles Baléares la Sardaigne et l'Algérois).VI. Ptéropodes.
Revue Trav.Inst.(Scient.tech.)Pech.marit.,30(4):375-383.

Spiratella trochiformis

Rampal,Jeannine,1967.
Répartition quantitative et bathymétrique des ptéropodes Thécosmoes récoltés en Méditerranée occidentale au nord du 40° parallèle:remarques morphologiques sur certaines espèces.
Rev.Trav.Inst.Pêches marit., 31(4):403-416.

Spiratella trochiformis

Rampal,Jeannine,1966.
Pêches planctoniques, superficielles et profondes, en Méditerranée occidentale (Campagne de la Thalassa - janvier 1961- entre les îles Baléares la Sardaigne et l'Algérois). VI. Ptéropodes.
Revue Trav.Inst.(Scient. Tech.)marit.,30(4):375-383.

Spiratella trochiformis

Rampal, J., 1963.
Ptéropodes thécosomes de pêches par paliers entre les Baléares, la Sardaigne et la côte nord-africaine.
Rapp. Proc. Verb., Réunions, Comm. Int. Expl. Sci., Mer Méditerranée, Monaco, 17(2):637-639.

Spiratella trochiformis

Sentz-Braconnot, 1965.
Répartition des ptéropodes Spiratella inflata (d'Orbigny) et S. trochiformis (d'Orbigny) dans la rade de Villefranche-sur-Mer.
Rapp. Proc.-verb. Réun.. Comm. int. Explor. scient. Mer Méditerranée 19(3): 463-467.

Spongiobranchaea australis

Eliot, C., 1907.
Mollusca. VI. Pteropoda. Nat. Antarctic Exped., "Discovery", 1901-1904, 3:15 pp., 2 pls.

Spongiobranghaea australis

Meisenheimer, Johannes, 1906
Die Pteropoden der Deutschen Südpolar-Expedition 1901-1903. Deutsche Südpolar Exped., 1901-1903, 9(Zool. 1): 96-153, Tables 5-7.

Spongiobranchaea australis

Tesch, J.J., 1950.
The Gymnosomata II. Dana Rept. No. 36:55 pp., 37 textfigs.

Spongiobranchia australis

Tesch, J.J., 1913
Mollusca. Pteropoda. Das Tierreich. 36:154 pp., 108 text figs.

Spongiobranchaea australis

van der Spoel, S., 1970.
The pelagic Mollusca from the Atlantide and Galathea expeditions collected in the East Atlantic.
Atlantide Rept. 11:99-139.

Spongioderma intermedia

Massy, A.L., 1932.
Mollusca: Gastropoda Thecostomata and Gymnostomata. Discovery Repts., 3:267-296.

Spongiobranchaea intermedia

Tesch, J.J., 1950.
The Gymnosomata II. Dana Rept. No. 36:55 pp., 37 textfigs.

Styliola subula

Franc, A., 1949
Hétéropodes, et autres Gastropodes planctoniques de Méditerranée occidentale.
J. de Conchyliologia 89: 209-230, 19 text figs. Trav. Sta. Zool. Villefranche-sur-Mer 10(3)

Styliola subula

Meisenheimer, Johannes, 1906
Die Pteropoden der Deutschen Südpolar-Expedition 1901-1903. Deutsche Südpolar Exped., 1901-1903, 9(Zool. 1): 96-153, Tables 5-7.

Styliola subula

Meisenheimer, Johannes, 1905
Pteropoda. Wiss. Ergeb., Deutschen Tiefsee-Exped. "Valdivia", 1898-1899: 314 pp., with atlas with 27 pls.

Stylida subula

Menzies, R.J., 1958.
Shell-bearing pteropod gastropods from Mediterranean plankton (Cavoliniidae).
Pub. Staz. Zool., Napoli, 30(3):381-402.

Styliola subula
Moore, H. B., 1949
The zooplankton of the upper waters of the Bermuda area of the North Atlantic. Bull. Bingham Ocean. Coll. 12(2):97, 208 text figs.

Styliola subula
Rampal, Jeannine, 1967.
Répartition quantitative et bathymétrique des ptéropodes Thécosomes récoltés en Méditerranée occidentale au nord du 40° parallèle: remarques morphologiques sur certaines espèces.
Rev. Trav. Inst. Pêches marit., 31(4):403-416.

Styliola subula
Rampal, Jeannine, 1966.
Pêches planctoniques, superficielles et profondes en Méditerranée occidentale (Campagne de la Thalassa - janvier 1961 - entre les îles Baléares la Sardaigne et l'Algérois) VI. Ptéropodes.
Revue Trav. Inst. (Scient. tech.) marit., 30(4):375-383.

Styliola subula
Rampal, J., 1963.
Ptéropodes thécosomes de pêches par paliers entre les Baléares, la Sardaigne et la côte nord-africaine.
Rapp. Proc. Verb., Reunions, Comm. Int. Expl. Sci., Mer Méditerranée, Monaco, 17(2):637-639.

Styliola subula
Stubbings, H. G., 1938
Pteropoda. John Murray Exped., 1933-34, Sci. Repts. 5(2):15-33, 2 text figs.

Styliola subula
Tesch, J. J., 1948
The Thecosomatous pteropods. II. The Indo-Pacific. Dana Rept. No. 30: 45 pp., 34 text figs., 3 pls.

Styliola subula
Tesch, J. J., 1946
The Thecosomatous Pteropods. Dana Rept. No. 28:82 pp., 34 text figs., 8 pls.

Styliola subula
Tesch, J. J., 1913
Mollusca. Pteropoda. Das Tierreich. 36:154 pp., 108 text figs.

Styliola subula
Tesch, J. J., 1910
X Pteropoda and Heteropoda. Percy Sladen Trust Expedition to the Indian Ocean in 1915. Trans. Linn. Soc., London, Zool., 2nd ser. 14: 165-189, pls. 12-14.

Styliola subula
Tokioka, Takasi, 1961
The structure of the operculum of the species of Atlantidae (Gastropoda: Heteropoda) as a taxonomic criterion, with records of some pelagic mollusks of the North Pacific.
Publ. Seto Mar. Biol. Lab., Kyoto Univ., 9(2):267-332.

Styliola subula
Tokioka, T., 1955.
On some plankton animals collected by the Syunkotu-maru in May-June 1954. 4. Thecosomatous pteropods.
Publ. Seto Mar. Biol. Lab., 5(1):59-74, Pls. 7-13

Styliola subula
van der Spoel, S., 1970.
The pelagic Mollusca from the Atlantide and Galathea expeditions collected in the East Atlantic.
Atlantide Rept. 11:99-139.

Styliola subula
Van der Spoel, S., 1967.
Euthecosomata, a group with remarkable developmental stages (Gastropoda, pteropoda).
J. Noorduijn en Zoom, N.V., Gorinchem, 375 pp.

Styliola subula
Vayssière, A., 1915
Mollusques Eutéropodes (Ptéropodes Thécosomes) provenant des campagnes des yachts HIRONDELLE et PRINCESSE-ALICE (1885-1913). Rés. Camp. Sci., Monaco, 47:226 pp., 14 pls.

Styliola subula
Wormelle, Ruth L., 1962
A survey of the standing crop of plankton of the Florida Current. VI. A study of the distribution of the pteropods of the Florida Current.
Bull. Mar. Sci., Gulf & Caribbean, 12(1): 95-136.

Thalassopterus zancleus
Franc, A., 1949
Hétéropodes, et autres Gastropodes planctoniques de Méditerranée occidentale.
J. de Conchyliologia 89: 209-230, 19 text figs. Trav. Sta. Zool. Villefranche-sur-Mer 10(3)

Thalassopterus zancleus
Tesch, J.J., 1950.
The Gymnosomata II. Dana Rept. No. 36:55 pp., 37 textfigs.

Thalassopterus zancleus
Tesch, J.J., 1913
Mollusca. Pteropoda. Das Tierreich. 36:154 pp., 108 text figs.

Thilea procera
Tesch, J.J., 1913
Mollusca. Pteropoda. Das Tierreich. 36:154 pp., 108 text figs.

Thliptodon sp.
Berkeley, E., and C. Berkeley, 1960
Some further records of pelagic Polychaeta from the northeast Pacific, north of Latitude 40°N. and east of Longitude 175°W., together with records of Siphonophora, Mollusca and Tunicata from the same region.
Canadian J. Zoology. 38:787-799.

Thliptodon akatsukai
Tokioka, Takasi, 1961
The structure of the operculum of the species of Atlantidae (Gastropoda: Heteropoda) as a taxonomic criterion, with records of some pelagic mollusks of the North Pacific.
Publ. Seto Mar. Biol. Lab., Kyoto Univ., 9(2):267-332.

Thliptodon diaphanus
Bedot, M., 1909
Sur la faune de l'Archipel Malais (resume). Rev. Suisse Zool. 17:143-169.

Thliptodon diaphanus
Meisenheimer, Johannes, 1906
Die Pteropoden der Deutschen Südpolar-Expedition 1901-1903. Deutsche Südpolar Exped., 1901-1903, 9(Zool. 1): 96-153, Tables 5-7.

Thliptodon diaphanus n.sp.
Meisenheimer, Johannes, 1905
Pteropoda. Wiss. Ergeb., Deutschen Tiefsee-Exped. "Valdivia", 1898-1899: 314 pp., with atlas with 27 pls.

Thliptodon diaphanus
Morton, J.E., 1954.
The pelagic Mollusca of the Benguela Current. Pt 1. First survey, R.R.S. William Scoresby, March 1951, with an account of the reproductive system and sexual succession of Limacina bulmodes. Discovery Rept. 24:163-200.

Thliptodon diaphanes
Tesch, J.J., 1950.
The Gymnosomata II. Dana Rept. No. 36:55 pp., 37 textfigs.

Thliptodon diapanus
Tesch, J.J., 1913
Mollusca. Pteropoda. Das Tierreich. 36:154 pp., 108 text figs.

Thliptodon diaphanus
van der Spoel, S., 1970.
The pelagic Mollusca from the Atlantide and Galathea expeditions collected in the East Atlantic.
Atlantide Rept. 11:99-139.

Thliptodon diaphenus
Wormelle, Ruth L., 1962
A survey of the standing crop of plankton of the Florida Current. VI. A study of the distribution of the pteropods of the Florida Current.
Bull. Mar. Sci., Gulf & Caribbean, 12(1): 95-136.

Thliptodon gegenbauri
Meisenheimer, Johannes, 1905
Pteropoda. Wiss. Ergeb., Deutschen Tiefsee-Exped. "Valdivia", 1898-1899: 314 pp., with atlas with 27 pls.

Thliptodon gegenbauri
Tesch, J.J., 1913
Mollusca. Pteropoda. Das Tierreich. 36:154 pp., 108 text figs.

Thliptodon rotundatus
Tesch, J.J., 1950.
The Gymnosomata II. Dana Rept. No. 36:55 pp., 37 textfigs.

Tiedemmannia scyllae
Tesch, J.J., 1913
Mollusca. Pteropoda. Das Tierreich. 36:154 pp., 108 text figs.

Tiedemmania charybdis
Tesch, J.J., 1913
Mollusca. Pteropoda. Das Tierreich. 36:154 pp., 108 text figs.

larvae
Banse, K., 1955.
Über das Verhalten von meroplanktischen Larven in geschichtetem Wasser. Kieler Meeresf. 11(2): 188-200.

chiefly larvae of bottom forms.

larvae, benthos
Carriker, Melbourne R., 1967.
Ecology of estuarine benthic invertebrates: a perspective.
In: Estuaries, G.H. Lauff, editor, Publs Am. Ass. Advmt Sci., 83:442-487.

larvae
Costlow, John D., Jr., 1963.
Larval development.
(Special Issue on Marine Biology), AIBS Bull., 13(5):63-65.

larvae (benthonic invertebrates)
Fish, C.J., and M.W. Johnson, 1937
The biology of the zooplankton population in the Bay of Fundy and the Gulf of Maine with special reference to production and distribution. J. Biol. Bd., Canada 3(3):189-322, 29 tables, 45 text figs.

larvae, invertebrate
Holland, D.L. and P.A. Gabbott, 1971.
A micro-analytical scheme for the determination of protein, carbohydrate, lipid and RNA levels in marine invertebrate larvae. J. mar. biol. Ass. U.K., 51(3): 659-668.

larvae
Knight-Jones, E.W., 1954.
Notes on invertebrate larvae observed at Naples, during May and June. Publ. Staz. Zool., Napoli, 25(1):135-144.

larval forms
Lebour, M.V., 1947
Notes on the Inshore Plankton of Plymouth. JMBA 26(4):527-547, 1 textfig.

larvae, littoral animals
Marumo, R., and M. Kitou, 1956.
Distribution of pelagic larvae of littoral animal-a in the open sea.
Bull. Jap. Soc. Sci. Fish., 22(4):225-234.

larvae, pelagic
Mileikovsky, S.A., 1973
"Pelagic larvation" and its biological role in the life of the sea. (In Russian; English abstract). Okeanologiia 13(2):346-347.

larvae, pelagic
Mileikovsky, S.A., 1972.
The "pelagic larvation" and its role in the biology of the World Ocean, with special reference to pelagic larvae of marine bottom invertebrates. Mar. Biol. 16(1): 13-21.

larvae, benthic
Mileikovsky S.A., 1971.
Types of larval development in marine bottom invertebrates, their distribution and ecological significance: a re-evaluation. Mar. Biol. 10(3): 193-213.

Pelagic larvae
Mileikovsky, S.A., 1970.
Seasonal and daily dynamics in pelagic larvae of marine shelf bottom invertebrates in nearshore waters of Kandalaksha Bay (White Sea). Mar. Biol., 5(3): 180-194.

larvae, pelagic
Mileikovsky, Simon A., 1968.
Some common features in the drift of pelagic larvae and juvenile stages of bottom invertebrates with marine currents in temperate regions. Sarsia, 34:209-216.

larvae, planktonic
Prasad, R.R., 1954.
Observations on the distribution and fluctuations of planktonic larvae of Mandapam.
Symp. Mar. & Fresh-water Plankton, Indo-Pacific, Bangkok, Jan., 25-26, 1954:21-34.
FAO+UNESCO

larvae of benthos
Schram, Thomas A., 1968.
Studies on the meroplankton in the inner Oslofjord. 1. Composition of the plankton at Nakkholmen during a whole year.
Ophelia, 5(2):221-243.

larvae, planktonic
Sentz-Braconnot, Eveline, 1964.
Données écologiques sur quelques groupes de larves planctoniques de la rade de Villefranche-sur-Mer.
Vie et Milieu, 15(3): 503-545.
Also in:
Trav. Sta. Zool. Villefranche-sur-Mer, 24(1964).

larvae
Smidt, E.L.B., 1944
Biological Studies of the Invertebrate Fauna of the Harbor of Copenhagen. Vidensk. Medd. fra Dansk naturh. Foren. 107:235-316, 23 text figs.

larvae, settlement of
Thorson, Gunnar, 1964
Light as an ecological factor in the dispersal and settlement of larvae of marine bottom invertebrates.
Ophelia, Mar. Biol. Lab., Helsingør, 1(1): 167-208.

Abstract 167-168

larvae
Thorson, G., 1950.
Reproductive and larval ecology of marine bottom invertebrates. Biol. Rev. 25(1):"6"textfigs

larvae
Thorson, G., 1944.
Reproduction and larval development of Danish marine bottom invertebrates with special reference to the planktonic larvae in the sound (Øresund). Medd. Komm. Danmarks Fiskeri- og Havundersøgelser, Serie: Plankton, 4(1):523 pp., 199 textfigs.

larval forms
Thorson, G., 1936.
The larval development, growth and metabolism of Arctic Marine bottom Invertebrates, compared with those of other seas. Medd. om Grønland 100 (6): 156 pp.

larval forms
Thorson, G., 1934.
On the reproduction and larval stages of the Brittle stars Ophiocten sericeum (Forbes) and Ophiurarobusta Ayres in East Greenland. Medd. om Grønland 100 (4):28 pp.

larvae
Vannuci, M., 1959
Catalogue of marine larvae.
Inst. Oceanogr., Univ. de Sao Paulo, 44 pp.

larvae
Wiborg, K.F., 1944
The production of plankton in a land locked fjord. The Nordåsvatn near Bergen in 1941-1947. With special reference to copepods. Fiskeridirektoratets Skrifter Serie Havundersøkelser (Rept. Norwegian Fish. and Marine Invest.) 7(7):1-83, Map.

larvae, invertebrate
Konstantinove, M.I., 1966.
A characteristic of the movement of pelagic larvae of marine invertebrates. (In Russian). Doklady, Akad. Nauk, SSSR, 170(3):726-729.

larvae, pelagic
Mileikovsky, S.A., 1966.
The range of dispersal of the pelagic larvae of bottom invertebrates by ocean currents and its distributional role on the example of Gastropoda and Lamellibranchia. (In Russian; English abstract).
Okeanologiia, Akad. Nauk, SSSR, 6(3):482-492.

larvae, dispersal of
Mileikovskii, S.A., 1960.
About the range of dispersal of pelagic larvae of bottom invertebrates with marine currents. (On the example of Limapontia capitata Müll. (Gastropoda Opisthobranchia) of Norwegian and Barents seas)
Doklady Akad. Nauk, SSSR, 135(4):965-967.

larval drift
Cushing, D.H., 1969.
Migration and abundance. Bull. Jap. Soc. fish. Oceanogr. Spec. No. (Prof. Uda Commem. Pap.): 207-212.

larval drift
Serafy, Donald Keith 1971.
A new species of Clypeaster (Echinodermata, Echinoidea) from San Felix Island, with a key to the Recent species of the eastern Pacific Ocean.
Pacific Sci. 25(2): 165-170.

larval, dispersal
Scheltema, Rudolf S., 1971.
Larval dispersal as a means of genetic exchange between geographically separated populations of shallow-water benthic marine gastropods.
Biol. Bull. mar. biol. Lab., Woods Hole, 140(2): 284-322.

larvae, transport of
Scheltema, Rudolf S., 1966.
Evidence for trans-Atlantic transport of gastropod larvae belonging to the genus Cymatium. Deep-Sea Research, 13(1):83-95.

larvae, transport of
Scheltema, Rudolf S., 1972.
Eastward and westward dispersal across the tropical Atlantic Ocean of larvae belonging to the genus Bursa (Prosobranchia, Mesogastropoda, Bursidae).
Int. Revue ges. Hydrobiol. 57(6): 863-873.

larvae, distributing of
Scheltema, Rudolf S., 1968.
Dispersal of larvae by equatorial ocean currents and its importance to the zoogeography of shoal-water tropical species.
Nature, 217(5134):1159-1162.

larvae, pelagic
Mileikovsky, S.A. 1968.
Distribution of pelagic larvae of bottom invertebrates of the Norwegian and Barents seas.
Mar. Biol. 1(3): 161-167.

larvae, planktonic
Vives, F. 1967.
Sobre la ecología de las larvas planctónicas de animales litorales.
Bol. R. Soc. esp. Hist. nat. (Biol.) 65(3/4): 291-300.

larvae, amphioxus
Della Croce, Norberto, 1960.
Sulla presenze di larve di Anfiosso nel plancton delle acque sud-orientali Sarde.
Boll. Musei e Ist. Biol., Univ. Genova, 30:15-17.

larvae, Amphioxus
Wickstead, J.H., 1964.
Acraniate larvae from the Zanzibar area of the Indian Ocean.
J. Linn. Soc. (Zool.), 45(205):191-199.

larvae, amphioxus
Wickstead, J.H., 1964.
On the status of the 'Amphiozides' larva.
J. Linn Soc. (Zool.), 45(305):201-207.

larvae, annelids
Abe, N., 1943.
The ecological observation on Spirorbis, especially on the post-larval development of Spirorbis argutus Bush. Sci. Repts., Tohoku Imp. Univ., 4th ser., Biol., Sendai, 18(4):327-351, 14 textfigs.

larvae, annelids
Åkesson, Bertil 1967.
On the biology and larval morphology of Ophryotrocha puerilis Claparède & Metschnikov (Polychaeta).
Ophelia, 4(1):111-119

larvae, annelids
Åkesson, Bertil, 1962.
The embryology of Tomopteris helgolandica (Polychaeta).
Acta Zoologica, 43(2/3):135-199.

larvae, annelid
Åkesson, B., and Y. Melander, 1967.
A preliminary report on the early development of the polychaete *Tomopteris helgolandica*.
Arkiv Zoologi, 20(5):141-146.

larvae, annelids
Anderson, D.T., 1966.
The comparative embryology of the Polychaeta
Acta Zool. 4F(½):1-42.

larvae, annelids
Bhaud, M., 1972.
Identification de larves d'Amphinomidae (Annélides Polychètes) recueillies près de Nosy-Bé (Madagascar) et problèmes biologiques connexes. Cah. ORSTOM, sér. Océanogr. 10(2): 203-216.

larvae, annelids
Bhaud, M., 1972.
Quelques données sur la biologie des invertébrés benthiques en climat tropical étude parallèle des larves pélagiques. Cah. ORSTOM sér. Océanogr. 10(2): 161-188.

larvae, annelids
Bhaud, M., 1969.
Etude de la migration verticale quotidienne des larves de *Mesochaetopterus sagittarius* à Nosy-Bé (Madagascar). Mar. Biol., 4(1): 28-35.

larvae, annelids
Bhaud, Michel, 1968.
Etude du plancton de la région de Banyuls-sur-Mer: les larves d'annélides polychètes: Repartition saisonnière et Comparaison avec les régions septentrionales.
Rapp. Proc.-verb. Réun., Comm. int. Explor. scient. Mer Méditerranée 19(3): 483-485.

larvae, annelids
Bhaud, Michel 1967.
Contribution à l'écology des larves pélagiques d'annélides polychètes à Banyuls-sur-Mer: comparaison avec les régions septentrionales.
Vie Milieu (B) 18(2B): 273-315.

larvae, annelids
Cazaux, Claude, 1969.
Etude morphologique du développement larvaire d'annélides polychètes (Bassin d'Arcachon). II Phyllodocidae, Syllidae, Nereidae.
Archives Zool. exp. gén. 110(2): 145-202.

larvae, annelid
Cazaux, Claude, 1964.
Développement larvaire de *Sabellaria alveolata* (Linné).
Bull. Inst. Océanogr., Monaco, 62(1296):15 pp.
Also in:
Bull. Sta. Biol. Arcachon, No. 16 (1964).

larvae, annelids (polychaets)
Dean, David, 1965.
On the reproduction and larval development of *Streblospio benedicti* Webster.
Biol. Bull., 128(1):67-76.

larvae, annelids
Gee, J.M., and E.W. Knight-Jones, 1962
The morphology and larval behaviour of a new species of *Spirorbis* (Serpulidae).
J. Mar. Biol. Assoc., U.K., 42(3):641-654.

larvae, annelids
Hopkins, Sewell H., 1958.
The planktonic larvae of *Polydora websteri* Hartman (Annelida: Polychaeta) and their settling on oysters.
Bull. Mar. Sci., Gulf and Caribbean, 8(3):268-277

larvae, annelid
Knight-Jones, E.W., P. Knight-Jones and P.J. Vine, 1972.
Anchorage of embryos in Spirorbinae (Polychaeta)
Mar. Biol. 12(4): 289-294.

larvae, annelids
Konstantinova M.I., 1970.
On the motion of polychaete larvae. (In Russian).
Dokl. Akad. Nauk SSSR, 188(4): 942-945

larvae, annelids
Lebour, M.V., 1947
Notes on the inshore plankton of Plymouth. JMBA 26(4):527-547.

larvae, polychaeta
Lindquist, Armin, 1959.
Studien über das zooplankton der Bottensee II. Zur Verbreitung und Zusammensetzung des Zooplanktons. Inst. Mar. Res., Lysekil, Ser. Biol. Rept. No. 11: 136 pp.

larvae, annelid
Lyster, I.H.J., 1965.
The salinity tolerance of polychaete larvae.
J. An. Ecol., 34(3):517-527.

larvae, annelid
Marodin, J.R., 1960.
Polychaetous annelids from the shallow waters around Barbados and other islands of the West Indies with notes on larval forms.
Canadian J. Zool., 38(5):989-1020.

larvae, annelids
Mileikovsky, S.A., 1968.
Morphology of larvae and systematics of Polychaeta. (In Russian; English abstract).
Zool. Zh., 47(1):49-59.

larvae, annelids
Mileikouskii, S.A., 1967.
On the larval development of the polychaet family Sphaerodoridae and some thoughts about the systematics. (In Russian).
Dokl. Akad. Nauk, SSSR, 177(2):471-474.

larvae, annelids
Orth, Robert J., 1971.
Observations on the planktonic larvae of *Polydora ligni* Webster (Polychaeta: Spionidae) in the York River, Virginia.
Chesapeake Sci. 12(3): 121-124

larvae, annelids
Sentz-Braconnot, Eveline, 1964.
Sur le developpement des Serpulidae *Hydroides norvegica* (Gunnerus) et *Serpula concharum* Langerhans.
Cahiers de Biol. Mar., Roscoff, 5:385-389.

larvae, annelids
Simon, Joseph L., 1967.
Reproduction and larval development of *Spio setosa* (Spionidae: Polychaeta).
Bull. mar. Sci., 17(2):398-341.

larvae, annelids
Srinivasagam, Theodore, 1966.
Effect of biological conditioning of sea water on development of larvae of a sedentary polychaete.
Nature, 212(5063):742-743.

larvae, annelids
Srinivasagam, R. Theodore, 1964.
The larval development of the serpulid, *Hydroides norvegica* (Gunn) in different latitudes. Some Aspects of Plankton Research, Proc. Seminar, Mar. Biol. Sta., Porto Novo, Mar. 23-25, 1964: 35-41.

larvae, annelids
Wilson, Douglas P., 1970.
The larvae of *Sabellaria spinulosa* and their settlement behaviour. J. mar. biol. Ass. U.K., 50(1): 33-52.

larvae, annelids
Wilson, Douglas P., 1970.
Additional observations on larval growth and settlement of *Sabellaria alveolata*. J. mar. biol. Ass. U.K., 50(1): 1-31.

larvae, annelid
Wilson, Douglas P., 1968.
The settlement behaviour of the larvae of *Sabellaria alveolata* (L.).
J. mar. biol. Ass., U.K., 48(2):387-435.

larvae, annelids
Wilson, Douglas P., 1968.
Some aspects of the development of eggs and larvae of *Sabellaria alveolata* (L.).
J. mar. biol. Ass., U.K., 48(2):367-386.

larvae, annelids
Wisely, B., 1958.
The development and settling of a serpulid worm, *Hydroides norvegica* Gunnerus (Polychaeta). Australian J. Mar. Freshwater Res., 9(3):351-361

larvae, polychaete
Ganapati, P.N., and Y. Radhakrishna, 1958
Studies on the polychaete larvae in the plankton off Waltair coast. Andhra Univ. Mem. Ocean 2: 210-237.

larvae, polychaetes
Kessler, Margit, 1963.
Die Entwicklung von *Lanice conchilega* (Pallas) mit besonderer Berücksichtung der Lebensweise.
Helgoländer Wiss. Meeresunters., 8(4):425-476.

larvae, polychaetes
Sentz, E., 1963.
Etude comparative de la richesse en larves planctoniques de différents points de la rade de Villefranche.
Rapp. Proc. Verb., Réunions, Comm. Int. Expl. Sci., Mer Méditerranée, Monaco, 18(2):581-584.

larvae, arachnids
Seriguchi, Koichi, 1970.
On the embryonic moultings of the Japanese horse-shoe crab, *Tachypleus tridentatus*.
Sci. Repts. Tokyo Kyoiku Daigaku 14(212): 121-128.

larvae, brachiopod
Bagetti G., Elda, 1964.
Nota sobre larvas de Brachiopoda Discinidae la costa chilena. Montemar, Chile, 4: 195-200.

larvae, bryozoa
Nielsen, Claus, 1967.
The larvae of *Loxostoma pectinaricola* and *Loxostoma elegans* (Entoprocta).
Ophelia, 4:203-206.

larvae, bryozoa
* Nielsen, Claus, 1967.
Metamorphosis of the larva of *Loxosomella murmanica* (Nilus) (Entoprocta).
Ophelia, 4(1): 85-89.

larvae, bryozoa
Ryland, J.S., 1965.
Polyzoa (Bryozoa), order Cheilostomata, Cyphonautes larvae.
Fiches d'Ident., Zooplancton, Cons. Perm. Int. Expl. Mer, 107:5 pp.

larvae, bryozoan
Sentz, E., 1963.
Étude comparative de la richesse en larves planctoniques de différents points de la rade de Villefranche.
Rapp. Proc. Verb., Réunions, Comm. Int. Expl. Sci., Mer Méditerranée, Monaco, 17(2):581-584.

larvae, chaetognaths
Kuhl, Willi, und Gertrud Kuhl, 1965.
Die Dynamik der Frühentwicklung von Sagitta setosa. Lauf- und Teilbild-Analysen von Zeitrafferfilm aufnahmen.
Helgoländer wiss Meeresunters. 12(3):260-301.

larvae, siphonophore
Barrois, J., 1927.
Stade médusoïde des Velelles. C.R. Acad. Sci., Paris, 184:1280-1281.

larvae, medusae
Beadle, L.C., I.F. Thomas and D.F.G. Poole, 1960,
Early development of Limnocnida victoriae Gunther (Limnomedusae).
Proc. Zool. Soc. London, 134(2):217-219.

larvae, coelenterate
Berrill, N.J., 1950.
Development and medusa-bud formation in the hydromedusae. Quart. Rev. Biol. 25(3):292-316, 10 textfigs.

COELENTERATE LARVAE
Berrill, N. J., 1949.
The polymorphic transformations of Obelia.
Q.J. Microsc. Sci., 3rd Ser., No. 11, 90(3):235-264, 12 textfigs.

larvae, coelenterate
Berrill, N. J., 1949.
Developmental analysis of Scyphomedusae.
Biol. Rev. 24(4):393-410, 7 textfigs.

larvae, coelenterate
Brinckmann, Anita, 1964.
Observations on the biology and development of Staurocladia portmanni, sp. n. (Anthomedusae, Eleutheridae).
Canadian J. Zool., 42(4):693-705.

larvae, siphonophore
Carré, Claude, et Danièle Carré 1969.
Le developpement larvaire de Lilyopsis rosea (Chun, 1885), siphonophore calycophore, Prayidae.
Cah. Biol. mar. 10(4):359-364.

larvae, siphonophore
Carré, Danièle 1971.
Étude du développement d'Halistemma rubrum (Vogt 1852), siphonophore physonecte Agalmidae.
Cah. Biol. mar. 12:77-93

larvae, siphonophore
Carré, Danièle, 1969.
Étude du dévelopement larvaire de Sphaeronectes gracilis (Claus, 1873) et de Sphaeronectes irregularis (Claus, 1873), siphonophores calycophores.
Cah. Biol. mar., 10(1):31-34.

larvae, siphonophores
Carré, Danièle, 1968.
Sur le développement post-larvaire d'Hippopodius hippopus (Forskål).
Cah. Biol. mar. 9(4):417-420.

larvae, coelenterates
Carré, Daniele, 1967.
Étude du developpement larvaire de deux Siphonophores: Lensia conoidea (calycophore) et Forskalia edwardsi (physonecte).
Cah. Biol. mar. 8(3):233-251.

larvae, medusae
Coyne, Jerry A. 1973.
An investigation of the dynamics of population growth and control in scyphistomae of the scyphozoan, Aurelia aurita.
Chesapeake Sci. 14(1). 55-58.

larvae, coelenterates
Gohar, H.A.F., and A.M. Eisawy, 1961.
The development of Cassiopea andromeda (Scyphomedusae).
Publ. Mar. Biol. Sta., Al-Ghardaqa (Red Sea), 11:147-189.

larvae, coelenterate
Hargitt, C.W., 1910.
The organization and early development of the eggs of hydromedusae. Proc. 7th Int. Zool. Congr. Aug. 19-24, 1907:7 pp.

larvae, medusae
Hargitt, C.W., and G.T., 1910.
Studies in the development of Scyphomedusae.
J. Morph., 21(2):217-262.

larvae (medusae)
Kandler, R., 1949
Jahreszeitliches Vorkommen und unperiodisches Auftreten von Fischbrut, Medusen und Decapodenlarven in Fehmarnbelt in den Jahren 1934-1943. Ber. Deutschen Wiss. Komm. f. Meeresf. nf. 12(1):49-85, 6 figs.

larvae, medusae
Kaufman, Z.S., 1956.
The difference in the rate of metamorphosis of the anterior or posterior halves in Scyphomedusa planulae. Dokl. Akad. Nauk, SSSR, 110(3):473-475.

larvae, coelenterate
Kuhl, Willi, und Gertrud Kuhl, 1967.
Regenerations- und Heilungsversuche an der Proactinula von Ectopleura dumortieri (Athecatae-Anthomedusae) unter Anwendung der Zeittransformation.
Helgoländer wiss Meeresunters., 16(½):75-91.

larvae, coelenterate
Maas, O., 1917.
Über den Bau des Meduseneis.
Verhandl. Deutschen Zool. Gesellschaft, 1908:114-

Aequoride juv.
Maas, O., 1905
Die craspedoten Medusen der Siboga-Expedition. Res. Siboga Expeditie. Mon. X, Livr. XXVI:84 pp., 14 pls.

larvae, coelenterate
Picard, J., 1956.
Le premier stade de l'hydroméduse Pandea conica issu de l'hydropolype Campaniclava cleodorae.
Bull. Inst. Océan., Monaco, No. 1086:11 pp.

larvae, medusae
Rajan, C.T., 1963.
On the larval stages of Solmundella biterteculata Browne.
J Mar. biol. Ass., India, 5(2):314-316.

larvae - medusae
Rees, W. J., 1941.
On the life history and developmental stages of the medusa Podocoryne borealis. JMBA, 25(2):307-316, 1 fig.

Russell, F.S., 1938.
On the development of Muggiaea atlantica Cunningham. J.M.B.A. 22(2):441-446, textfig.

larvae, coelenterate
Sars, M., 1857.
Einige Worte über die Entwickelung der Medusen. Vortrag gehalten in der Versammlung der Skandinavischen Naturforscher in Christiania im Juli 1856. Arch. f. Naturgesch., Berlin, 23:117-123.

larvae, coelenterate
Thiel, Hjalmar, 1962
Untersuchungen über die Strobilisation von Aurelia aurita Lam. an einer Population der Kieler Förde.
Kieler Meeresf. 18(2): 198-230.

larvae, siphonophores
Totton, A.K., 1955.
Development and metamorphosis of the larva of Agalma elegans (Sars)(Siphonophora Physonectae). Pap. Mar. Biol. and Oceanogr. Deep-Sea Res., Suppl. to Vol. 3:239-241.

larvae, medusae
Vanhöffen, E., 1906
Acraspede Medusen. Nordisches Plankton XI:40-64, 37 text figs.

larvae, siphonophores
Vanhöffen, E., 1906
Siphonophoren. Nordisches Plankton XI:9-39, 65 text figs.

LARVAE, COELENTERATE
Werner, B., 1955.
On the development and reproduction of the anthomedusan Margelopsis haeckeli Hartlaub.
Ann. N.Y. Acad. Sci., 62(1):1-30, 9 textfigs.

Arachnactis bournei = larva
Wiborg, K.F., 1944 of Cerianthus lloydi
The production of zooplankton in a landlocked fjord, the Nordåsvatn near Bergen, in 1941-42, with special reference to the copepods. (Repts. Norwegian Fish. and Mar. Invest.) 7(7):83 pp., 40 text figs.

larvae, coelenterate
Widersten, Bernt 1968.
On the morphology and development in some cnidarian larvae.
Zool. Bidrag, Uppsala 37 (2): 140-182.

larvae, ctenophore
Dawydoff, C., 1946.
Contribution à la connaissance des cténophores pélagiques des eaux de l'Indochine.
Bull. Biol. France Belg. 80:113-170, 35 textfigs.

larvae, ctenophore
Dawydoff, C., 1940.
Incubation des oeufs et viviparité chez les Coeloplanides indochinois. C.R. Acad. Sci., Paris 211(8):146-148.

larvae, ctenophores
Pylilo I.V. 1973.
Phénomènes de régénération chez les cténaires.
Cah. Biol. mar. 14 (3): 391-406

larvae ctenophores
Remane, A., 1956.
Zur Biologie des Jugendstadiums der Ctenophore Pleurobrachia pileus O. Müller. Kieler Meeresf. 12(1):72-75.

larvae, ctenophores
Reverberi, G., 1966.
Quelques nouvelles recherches expérimentales sur le développement des cténophores.
Ann. Biol. 5 (7/8):375-390.

larvae, ctenophores
Reverberi, Giuseppe, e Giuseppina Ortoleni, 1965.
Nuove ricerche sullo sviluppo dell uovo di Ctenofori.
Rivista Biol., Univ. Perugia, 58(2/3):113-137.

larvae - crustacean
Behre, E. H., 1941.
The recognition of crustacean larvae by their pigment patterns. Anat. Rec., 81(4):suppl. 116.

larvae, Crustacea
Elofsson, Rolf, 1971.
Some observations on the internal morphology of Hansen's nauplius y (Crustacea).
Sarsia 46: 23-40.

larvae, crustacean
Williamson, D.I. 1967.
Some recent advances and outstanding problems in the study of larval Crustacea.
Proc. Symp. Crustacea Ernakulam, Jan. 12-15, 1965, 2: 815-823.

larvae, amphipods
Bregazzi, P.K. 1973.
Embryological development in Tryphosella kergueleni (Miers) and Cheirimedon femoratus (Pfeffer) (Crustacea: Amphipoda). Bull.
Brit. Antarct. Surv. 32: 63-74.

larvae, amphipod
Clemens, H.P., 1950.
Life cycle and ecology of Gammarus fasciatus Say. Franz Theodore Stone Inst. Hydrobiol. Contrib. No. 12:63 pp., 22 textfigs.

larvae, amphipod
Gottlieb, E., 1962.
The benthonic Amphipoda of the Mediterranean coast of Israel. 1. Notes on the geographical distribution. 2. Ecology and life history
Sea Fish. Res. Sta. Haifa. Israel. Bull., No. 31:157-166; 71-90.

Originally published in: Bull Res. Council, Israel: 9B: 157-166
11B(1/2):71-90

larvae, amphipods
Kane, Jasmine E., 1963.
Stages in the early development of Parathemisto gaudichaudi (Guer.)(Crustacea Amphipoda: Hyperiidea), the development of secondary sexual characters and of the ovary.
Trans. R. Soc., New Zealand, 3:35-45.
Also in:
Collected Reprints, N.I.O., Vol. 11(447):1963.

larvae, amphipods
Kinne, O., 1959.
Ecological data on the amphipod Gammarus duebeni. A monograph.
Veröff. Inst. Meeresf., Bremerhavn, 6(1):177-203.

larvae, amphipods
Laval, Philippe 1968.
Développement en Élevage et systématique d'Hyperia schizogeneios Stbb. (amphipode hypéride).
Arch. Zool. exp. gén. 109(1): 25-67.

larvae, amphipod
Laval, Philippe, 1966.
Bougisia ornata, genre et espèce nouveaux de la famille des Hyperiidae (Amphipoda, Hyperiides).
Crustaceana, 10(2):210-

larvae, amphipods
Laval, Philippe, 1963.
Sur la biologie et les larves de Vibilia armata Bov. et de V. propinqua Stebb. Amphipodes Hypérides.
Comptes Rendus, Acad. Sci., Paris, 257(6):1389-1392.

larvae, amphipods
Rappaport, R., Jr., 1960.
The origin and formation of blastoderm cells of gammarid Crustacea.
J. Exper. Zool., 144(1):43-59.

larvae, amphipod
Sexton, E.W., and D. M. Reid, 1951.
The life history of the multiform species Jassa falcata (Montagu) (Crustacea Amphipoda) with a review of the bibliography of the species.
J. Linn. Soc., London, 42(283):29-91, Pls. 3-40.

larvae, amphipods
Shih, Chang-tai, 1969.
The systematics and biology of the family Phronimidae (Crustacea: Amphipoda).
Dana Report 74: 99 pp.

larvae, amphipod
Vlasblom, A.G., and Gerda Bolier 1970.
Tolerance of embryos of Marinogammarus marinus and Orchestia gammarellus (Amphipoda) to lowered salinities.
Netherl. J. Sea Res. 5(3): 334-341.

larvae, amphipod
Weygoldt, P., 1958.
Die Embryonalentwicklung des Amphipoden Gammarus pulex pulex (L).
Zool. Jahrb. Anat., 77(1):51-110, 70 textfigs.

larvae, Cephalocarida
Sanders, Howard L., and Robert R. Hessler, 1964.
The larval development of Lightiella incisa Gooding (Cephalocarida).
Crustaceana, 7(2):81-97.

larvae, cirripede
Bainbridge, V., and J. Roskell, 1966.
A redescription of the larvae of Lepas fascicularis Ellis and Solander with observations on the distribution of Lepas nauplii in the north-eastern Atlantic.
In: Some contemporary studies in marine biology, 67-81.

larvae, cirripeda
Barnard, K. H., 1924
Contributions to the crustacean fauna of South Africa. No. 7. Cirripedia. So. African Mus., Ann. 20:1-103,1pl.

Larvae, cirripede
Barnes, H., 1965.
Studies on the biochemistry of cirripede eggs. 1. Changes in the general biochemical composition during development of Balanus balanoides and B. balanus.
Jour. Mar. Biol. Assoc., U.K., 45(2):321-339.

larvae, cirriped
Barnes, H., 1962.
Note on variations in the release of nauplii of Balanus balanoides with special reference to the spring diatom outburst.
Crustaceana, 4(2):118-122.

larvae, cirripeds.
Barnes, H., 1953.
Size variations in the cyprids of some common barnacles. J.M.B.A. 32(2):297-304, figs.

larvae, cirriped
Barnes, H., 1953.
The effect of lowerd salinity on some barnacle nauplii. J. Animal Ecology 22(2):328-330, 1 textfig.

larvae, cirriped
Barnes, H., 1950.
A note on the barnacle larvae of the Clyde Sea Area as sampled by the Hardy Continuous Plankton Recorder. J. M. B. A. 29(1):73-80

larvae, cirripede
Barnes, H., & M. Barnes, 1959.
Note on stimulation of cirripede nauplii.
Oikos, 10(1):19-23.

larvae, cirripeds
Barnes, H., and Margaret Barnes, 1965.
Egg size, nauplius size and their variation with local geographical and specific factors in some common cirripedes.
J. Animal Ecol., 34(2):391-402.

larvae, cirriped
Barnes, H., and M. Barnes, 1959.
The naupliar stages of Balanus hesperius Pilsbry. Canadian J. Zool., 37:237-244.

larvae, cirriped
Barnes, H., and J.D. Costlow, Jr., 1961.
The larval stages of Balanus balanus (L.) Da Costa.
J.M.B.A., U.K., 41(1):59-68.

larvae, cirripede
Barnes, H. and Waltraud Klepal, 1972
Phototaxis in stage I nauplius larvae of two cirripedes. J. exp. mar. Biol. Ecol. 10(3): 267-273.

larvae, cirriped
Bassindale, R., 1936.
The developmental stages of three English barnacles, Balanus balanoides (Linn.), Chthamalus stellatus (Poli), and Verruca stroemia (O.F. Müller). Proc. Zool. Soc., London, 1936:57-74, 9 textfigs.

larvae, cirriped
Buchholz, H., 1951.
Die Larvenformen von Balanus improvisus. Beiträge zur Kenntnis des Larvenplanktonts. Kieler Meeresf. 8(1):49-57, Pls. 16-17.

larvae, cirripeda
Costlow, John D., Jr. and C. G. Bookhout, 1959.
Larval development of Balanus amphitrite var. denticulata Broch reared in the laboratory.
Biol. Bull., 114(3):284-305.

larvae, cirriped
Costlow, J.D., Jr., and C.G. Bookhout, 1957.
Larval development of Balanus eburneus in the laboratory. Biol. Bull., 112(3):313-324.

larvae, cirriped
Crisp, D.J., 1962.
The larval stages of Balanus hameri (Ascanius, 1767).
Crustaceana, 4(2):123-130.

larvae, cirripeds
Crisp, D.J., 1962.
The planktonic stages of the Cirripedia Balanus balanoides (L.) and Balanus balanus (L.) from north temperate waters.
Crustaceana, 3(3):207-221.

larvae, cirriped
Crisp, D.J., 1959.
The rate of development of Balanus balanoides (L.) embryos in vitro.
Ecology 28:119-132.

LARvae, cirripede
Crisp, D.J., and P.S. Meadows, 1963.
Adsorbed layers: the stimulus to settlement in barnacles.
Proc. R. Soc., London, (B), 158(972):364-387.

larvae, cirripede
Crisp, D.J., and P.S. Meadows, 1962.
The chemical basis of gregariousness in cirripedes.
Proc. R. Soc., London, (B), 156(965):500-520.

larvae, cirripedes
Crisp, D.J., and B. S. Patel, 1960
The moulting cycle in Balanus balanoides L.
Biol. Bull. 118(1): 31-47.

larvae, cirriped
Crisp, D. J. and C. P. Spencer, 1958.
The control of the hatching process in Barnacles
Proc. Roy. Soc. Ser. B, 149(935):278-299.

larvae, cirriped
Daniel, A., 1958.
The development and metamorphosis of three species of sessile barnacles.
J. Madras Univ., (B), 28(1):23-47.

larvae, cirriped
Daniel, A., 1953.
The attachment of barnacle cyprids to different types of South Indian wood. J. Madras Univ., B, 23(3):227-231.

larvae cirripede
Dawson, R.M.C., and H. Barnes, 1966.
Studies in the biochemistry of cirripede eggs. II. Changes in lipid composition during development of Balanus balanoides and B. balanus. J. Mar. Biol. Assoc., U.K., 46(2):249-261.

larvae, cirriped
De Wolf, P., 1973.
Ecological observations on the mechanisms of dispersal of barnacle larvae during planktonic life and settling.
Netherl. J. Sea Res. 6(1/2): 1-129.

larvae, cirriped
Doochin, H.G., 1951.
The morphology of Balanus improvisus Darwin and Balanus amphitrite niveus Darwin during initial attachment and morphosis. Bull. Mar. Sci., Gulf and Caribbean 1(1):15-39, 7 textfigs.

larvae, cirriped
Drummond, Ainslie H. Jr. 1968.
The "foul" barnacle.
Sea Frontiers 14(3): 158-165 (popular)

larvae, cirriped
Erokhin, V.E., 1970.
On possibility of sorbtion accumulation of macromolecules solved in sea water by copepods Tigriopus brevicornis O.F. Müller and larvae of Balanus improvisus Darw. (In Russian).
Gidrobiol. Zh., 6(6): 94-98.

larvae, cirriped
Foster, B.A. 1967.
The early stages of some New Zealand shore barnacles.
Tane, J. Auckland Univ. Fld Club 13: 33-42.

larvae, cirriped
Fudinami, M., and H. Kasahara, 1942.
[Rearing and metamorphosis of Balanus amphitrite hawaiiensis Bloch] Zool. Mag., Tokyo, 54(3):108-118.

larvae, cirriped
Gruvel, A., 1909
Die Cirripedien der Deutschen Südpolar-Expedition 1901-1903. Deutsche Südpolar-Exped. 11, Zool. 3(2):193-230. Pls. 23-26.

larvae, barnacles
Hirano, R., 1953.
On the rearing and metamorphosis of four important barnacles in Japan.
J. Ocean. Soc., Japan, 8(3/4):139-143, 3 textfigs.

larvae, cirriped
Hirano, R., and J. Okushi, 1952.
[Studies on sedentary marine organisms. 1. Seasonal variations in the attachment and growth rates of barnacle cyprids in Aburatsubo Bay, near Misaki.] Bull. Jap. Soc. Sci. Fish. 18(11):639-654, 4 textfigs. (English abstract).

larvae, cirripedes
Hoenigman, J., 1961.
Sur quelques espèces nouvelles pour l'Adriatique.
Rapp. Proc. Verb., Réunions, Comm. Int. Expl. Sci. Mer, Méditerranée, Monaco, 16(2):217-218.

larvae, cirriped
Jones, L.W.G., and D.J. Crisp, 1954.
The larval stages of the barnacle, Balanus improvisus Darwin. Proc. Zool. Soc. London, 123(4):765-780, 6 textfigs.

larvae, cirripeds
Knight-Jones, E.W., and G.D. Waugh, 1949.
On the larval development of Elminius modestus Darwin. J.M.B.A. 28(2):413-428, 6 textfigs.

larvae, cirripede
Le Reste, L., 1965
Contribution à l'étude des lerves de cirripèdes dans le Golfe de Marseille.
Rec. Trav. Sta. Mar. Endoume, Bull., 38(54): 33-121.

larvae, cirriped
Molenock, Joane, and Edgardo D. Gomez 1972.
Larval stages and settlement of the barnacle Balanus (Conopea) galeatus (L.) (Cirripedia Thoracica).
Crustaceana 23 (1): 100-108.

larval forms
Moore, H. B. and J. A. Kitching, 1939
The biology of Chthamalus stellatus (Poli).
JMBA 23:521-541.

larvae cirripede
Moyse, John, 1961
The larval stages of Acasta spongites and Pyrgoma anglicum (Cirripedia).
Proc. Zool. Soc. London, 137(3):371-392.

larvae, cirripede
Moyse, J., 1960.
Mass rearing of barnacle cyprids in the laboratory.
Nature, 185(4706):120.

larvae, cirriped
Moyse, J., and E.W. Knight-Jones 1967.
Biology of cirripede larvae.
Proc. Symp. Crustacea, Ernakulam, Jan. 12-15, 1965, 2: 595-611.

larvae, cirriped
Newman, William A., 1965.
Prospectus on larval cirriped setation formulae.
Crustaceana, 9(1):51-56.

larvae, cirripedes
Norris, E., and D.J. Crisp, 1953.
The distribution and planktonic stages of the cirripede Balanus perforatus Brugiere.
Proc. Zool. Soc., London, 123(2):393-409, 7 textfigs.

larvae, cirripeds
Norris, E., L.W.G. Jones, T. Lovegrove, and D.J. Crisp, 1951.
Variability in larval stages of cirripeds. Nature, 167(4246):444-445.

larvae, cirripedes
Patel, Bhupendra, and D.J. Crisp, 1960.
Rates of development of the embryos of several species of barnacles.
Physiol. Zool., 33(2):104-119.

larvae, cirripede
Polischuk, L.N., 1966.
On the distribution and numerical estimation of Balanus sp. larvae in the near-surface layer of the Caspian Sea. (In Russian).
Okeanologiia. Akad. Nauk, SSSR, 6 (1):148-150.

larvae, cirripeds
Pyefinch, K. A., 1949.
The larval stages of Balanus crenatus Brugiere.
Proc. Zool. Soc., London, 118(4):916-923.

larvae, cirriped
Pyefinch, K. A., 1949.
Short period fluctuations in the numbers of barnacle larvae, with notes on comparisons between pump and net plankton hauls. J.M.B.A. 28(2):353-369, 7 textfigs.

larvae, cirriped
Pyefinch, K. A., 1948.
Methods of identification of the larvae of Balanus balanoides (L.), B. crenatus Brug. and Verruca stroemia O. F. Müller. J.M.B.A., 27(2):451-463, 6 figs.

larvae, cirriped
Rosenberg, Rutger 1972.
Salinity tolerance of larvae of Balanus balanoides (L.).
Ophelia 10(1): 11-15.

larvae, cirripeds
Runnström, S., 1925.
Zur Biologie und Entwicklung von Balanus balanoides (Linné). Bergens Mus. Aarbok, 1924-25, Naturvidensk. raekke No. 5:46 pp., 22 text figs.

larvae, cirriped
Rzepishevsky, I.K., 1962.
Conditions of the mass liberation of the nauplii of the common barnacle Balanus balanoides (L.) in the eastern Murman.
Int. Revue Ges. Hydrobiol., 47(3):471-479.

larvae, cirripede
Rzhepishevskii, I.K., 1958.
Certain diagnostic features of the nauplii of three Balanus species in the Barents Sea. (In Russian).
Doklady, Akad. Nauk, SSSR, 120(5):1159-

larvae, cirripede
Sandison, Eyvor E., 1967.
The nauplier stages of Balanus pallidus Stutsburi Darwin and chthamalus aestuarii stubbings (Cirripedia Thoracica).
Crustaceana, 13(2):161-174.

larvae, cirripeds
Sandison, E.E., 1950.
Nauplius longispinus, a new larval form of barnacle. Trans. Roy. Soc., S. Africa, 32(3):301-313, figs.

larvae, cirripede
Schäfer, 1952.
Biologische Bedeutung der Ortswahl bei Balanidenlarven. Senckenbergiana 33(4/6):235-246, 4 textfigs.

larvae, cirripedes
Sentz, E., 1963.
Étude comparative de la richesse en larves planctoniques de différents points de la rade de Villefranche.
Rapp. Proc. Verb., Réunions, Com. Int. Expl. Sci., Mer Méditerranée, Monaco, 17(2):581-584.

larvae, cirripedes
Takesita, Isao, 1962
Electrical shocking effect on larvae of Balanus.
J. Fac. Fish and Animal Husband., Hiroshima Univ., 4(1/2):1-6.

larvae, cirripede

Thiriot-Quiévreux, Catherine, 1965.
Description de Spirorbis (L'aeospira) pseudomilitaris n. sp., polychète Spirorbinae, et de sa larve.
Bull. Mus. Nat. Hist. Nat., (2)37(3): 495-502.

larvae, cirripeds

Turquier, Yves, 1972.
Contribution à la connaissance des cirripèdes acrothoraciques.
Arch. Zool. exp. gén. 113 (4): 499-551.

larvae cirripede

Utinomi, Huzio, 1961.
Studies on the Cirripedia Acrothoracica. III. Development of the female and male of Berndtia purpurea Utinomi.
Publ. Seto Mar. Biol. Lab., Kyoto Univ., 9(2): 413-446.

larvae, cirriped

Utinomi, H., 1943.
The larval stages of Creusia, the barnacle inhabiting coral reefs. Annotationes Zool. Japon. 22(1):15-22, 4 textfigs.

larvae, cirripede

Walker, G., 1971.
A study of the cement apparatus of the cypris larva of the barnacle Balanus balanoides.
Mar. Biol. 9(3): 205-212.

larvae, cirripede

Wisely, B., 1960
Experiments on rearing the barnacle Elminius modestus Darwin to the settling stage in the laboratory.
Australian J. Marine and Freshwater Research 11(1): 42-54.

larvae, cirriped.

Wisely, B., and R.A.P. Blick, 1964.
Seasonal abundance of first stage nauplii in 10 species of barnacles at Sydney.
Australian J. Mar. Freshw. Res., 15(2):162-171.

larvae, cirripeds

Yanagimachi, R., 1961.
The life cycle of Peltogasterella (Cirripedia, Rhizocephala).
Crustaceana, 2(3):183-186.

larvae, cirripeds

Yanagimachi, R., 1960.
The life cycle of Peltogasterella gracilis (Rhizocephala, Cirripedia).
Bull. Mar. Biol. Sta. Asamushi, Tohoku Univ., 10(2):104-110.

larvae, cladocerans

Della Croce, N., e E. Gaino, 1970.
Osservazioni sulla biologia del maschio di Penilia avirostris Dana (1).
Cah. Biol. mar. 11(4): 361-365.

larvae, cladocerans

Gaino, Elda, 1971.
Dimorfismo sessuale a livello embrionale in Penilia avirostris Dana.
Cah. Biol. mar. 12(3):283-289.

larvae, copepod.

Alvarez V., and N. G. Kewalramani, 1970.
Naupliar development of Pseudodiaptomus ardjuna Brehm (Copepoda).
Crustaceana 18(5): 269-276.

larvae, copepods

Andrews, Keith, J.H., 1966.
The distribution and life-history of Calanoides acutus (Giesbrecht).
Discovery Repts., 34:117-162.

larvae, copepods

Barnett, Peter R.O. 1970.
The life cycles of two species of Platychelipus Brady (Harpacticoida) on an intertidal mudflat.
Int. Revue ges. Hydrobiol. 55(2):169-195.

larvae, copepod

Barnett, P.R.O., 1966.
The comparative development of two species of Platychelipus Brady (Harpacticoida).
In: Some contemporary studies in marine biology. H. Barnes, editor, George Allen & Unwin, Ltd., 113-127.

larvae, copepods

Battaglia, B., 1957.
Richerche sul ciclo biologico di Tisbe gracilis (T. Scott), (Copepoda, Harpacticoida), studiato in condizioni di Laboratorio. Arch. Ocean. e Limnol., 11(1):29-46.

larvae, copepods

Battaglia, B., and P. Talamini, 1957.
Osservazioni sullo sviluppo larvale di Tisbe reticulata Bocquet (Copepoda, Harpacticoida).
Arch. Ocean. e Limnol., 11(1):63-68.

larvae, copepod

Bernard, Michelle 1970.
Quelques aspects de la biologie du copépode pélagique Temora stylifera en Méditerranée: essai d'écologie expérimentale.
Pelagos, Alger 11: 196 pp.

larvae, copepod

Bernard, Michelle, 1965.
Description du mâle et du premier stade nauplien de Corina granulosa (Giesbrecht), copépode pélagique (Sapphirinidae).
Pelagos, Bull. Inst. Océanogr., Alger, 2(4):45-50.

larvae, copepods

Bernard, Michelle, 1964.
Le développement nauplien de deux copépodes carnivores: Euchaeta marina (Prestandr.) et Candacia armata (Boeck) et observations sur le cycle de l'astaxanthine au cours de l'ontogénèse
Pelagos, Bull. Inst. Océanogr., Alger, 2(1): 51-71.

larvae, copepods

Bernard, Michelle, 1963.
Le cycle vital en laboratoire d'un copépode pélagique de Méditerranée, Euterpina acutifrons Claus.
Pelagos, Bull. Inst. Océanogr., Alger, 1(2):35-48.

larvae, copepods

Björnberg, Tagea K.S. 1972.
Developmental stages of some tropical and subtropical planktonic marine copepods.
Studies Fauna Curaçao and other Carib. Is. 40(136):1-185.

larvae, copepods

Björnberg, Tagea K.S. 1966.
The developmental stages of Undinula vulgaris (Dana) (Copepoda).
Crustaceana 11(1): 65-76.

larvae, copepods

Bjornberg, T.K.S., 1965.
The study of plantonic copepods in the south west Atlantic.
Anais Acad. bras. Cienc., 37(Supl.):219-230.

larvae, copepod

Björnberg, Tagea K.S., 1967.
The larvae and young forms of Eucalanus Dana (Copepoda) from tropical Atlantic waters.
Crustaceana, 12(1):59-73.

Larvae, copepod

Bjornberg, Tagea K.S., 1965.
Observations on the development and the biology of the Miracidae Dana (Copepoda: Crustacea).
Bull. Mar. Sci., 15(2):512-520.

larvae, copepods

Bresciani, J., 1961.
Some features of the larval development of Stenhalia (Delavalia) palustris Brady 1868 (Copepoda Harpacticoida).
Vidensk. Meddel. Dansk Naturhist. Foret., København, 123:237-247.

larval stages

Campbell, M.H., 1934.
The life history and post-embryonic development of the copepods, Calanus tomsus Brady and Euchaeta japonica. J. Biol. Board, Canada, 1(1) 65 pp., 18 textfigs.

larvae, copepods

Candeias, A., e Inácia de Paiva, 1967.
Nota sobre algumas formas de copepoditos de Longipedia Claus, 1863.
Notas mimeogr. Centro Biol. aquat. trop., Lisboa, 8: 19 pp.

larvae, copepods

Chiba, T., 1956.
Studies on the development and the systematics of Copepoda. J. Shimonoseki Coll. Fish., 6(1):1-90.
young stages of Eurytemora pacifica included

larvae, copepod

Chiba, T., A. Tsuruta and H. Maeda, 1955.
Report on zooplankton samples hauled by larva-net during the cruise of Bikini-Expedition, with special reference to copepods.
J. Shimonoseki Coll. Fish. 5(3):189-213.

larvae, copepod

Conover, R.J., 1956.
Oceanography of Long Island Sound, 1952-1954. 6. Biology of Acartia clausi and A. tonsa.
Bull. Bingham Oceanogr. Coll., 15:156-233.

larvae, copepod

Corkett, C.J., 1968.
Observations sur les stades larvaires de Pseudocalanus elongatus Boeck et Temora longicornis O.F. Muller. Pelagos, 8: 51-57.

larvae, copepod

Corkett, C.J. 1967.
The copepodid stages of Temora longicornis (O.F. Müller, 1792) (Copepoda).
Crustaceana 12(2): 261-273.

larvae, copepod

Crisafi, Pietro,
Les copépodes du détroit de Messine: oeufs, stades naupliens et segmentation du corps du copépode pélagique Pontella mediterranea Claus.
Rapp. Proc. Verb. Reunions, Comm. Int. Expl. Sci., Mer Mediterranee, Monaco, 18(2):411-416.

larvae, copepod

Davis, C. C. 1959.
Osmotic hatching in the eggs of some fresh-water copepods
Biol. Bull. 116(1):15-29.

larvae, copepods

Davis, C.C. 1943.
The larval stages of the calanoid copepod, Eurytemora hirundoides (Nordquist). Bd. Nat. Res., Maryland, Publ. 58:52 pp., 9 pls.

larvae-copepoda

Davis, C. C., 1943.
The larval stages of the calanoid copepod Eurytemora hirundoides (Nordquist). State of Maryland, Dept. Res. Educ. Publ., 58:1-52, 71 figs.

Oceanographic Index: Marine Organisms Cumulation, 1946-1973

---

larvae, copepod

Dinet, Alain 1972.
Reproduction, développement et croissance de Bulbamphiascus imus (Brady) et Halectinosoma herdmani (T. + A. Scott) Copepoda, Harpacticoida.
Téthys 4(2): 437-444.

---

larvae, copepod

Elgmork, K., and A.L. Langeland 1970.
The number of naupliar instars in Cyclopoida (Copepoda).
Crustaceana 18(3): 277-282.

---

larvae, copepods

El-Maghraby, A.M., 1964.
The developmental stages and occurrence of the copepod *Euterpina acutifrons* Dana in the marine environment.
Ann. Mag. Nat. Hist., (3) 7(76):223-233.

---

larvae, copepod, lists of spp.

Faber, Daniel J., 1966.
Seasonal occurence and abundance of free-swimming copepod nauplii in Narragansett Bay.
J. Fish. Res. Bd., Canada, 23(3):415-422.

---

larvae, copepod

Faber, Daniel J., 1966.
Free-swimming copepod nauplii of Narragansett Bay with a key to their identification.
J. Fish. Res. Bd., Canada, 23(2):189-205.

---

larvae, copepod

Fanta, Edith Susana 1972
Anatomy of the nauplii of *Euterpina acutifrons* (Dana) (Copepoda, Harpacticoida).
Crustaceana 23(2):165-181.

---

larvae, copepods

Furuhashi, Kenzo, 1966.
Droplets from the plankton net. XXIII. Record of *Sapphirina salpae* Giesbrecht from the North Pacific, with notes on its copepodite stages.
Publs Seto mar. biol. Lab. 14(2):123-127.

---

larvae, copepods

Garber, B.I., 1951.
[The observation for development and propagation of *Calanipeda aquae dulcio* Kritsch (Copepoda Calanoida).] Trudy Karadagsk Biol. Sta., 11:3-55.

---

larvae, copepods

Gaudy, R., 1962.
Biologie des copépodes pélagiques du Golfe de Marseille.
Rec. Trav., Sta. Mar. Endoume, 42(27):93-184.

Calanus helgolandicus
Calanus gracilis
Calanus minor
Euchaeta marina
Temora longicornis
Temora stylifera
Pleuromamma abdominalis
Centropages spp

---

larvae, copepod

Grice, George D. 1971.
The developmental stages of *Eurytemora americana* Williams, 1906, and *Eurytemora herdmani* Thompson and Scott, 1897 (Copepoda, Calanoida).
Crustaceana 20(2):145-158.

---

larvae, copepod

Grice, George D. 1969.
The developmental stages of *Pseudodiaptomus coronatus* Williams (Copepoda, Calanoida).
Crustaceana 16(3):291-301.

---

larvae, copepods

Gurney, R., 1934.
The development of Rhincalanus. Discovery Repts. 9:209-214, 7 textfigs.

---

larvae, copepod

Hanaoka, T., 1952.
[On nauplius of *Oncaea mediterranea* and *Corycaeus* sp.] Bull. Nakai Regional Fish. Res. Lab. No. 1: 37-41, 7 textfigs.

---

larvae, copepod

Hanaoka, T., 1952.
[Study on free-living Copepoda-nauplius.]
Bull. Nakai Regional Fish. Res. Lab. No. 1:1-36, 16 textfigs.

compares generic differences of nauplii. Appears to be excellent.

---

larvae, copepods

Haq, S.M., 1965.
Development of the copepod *Euterpina acutifrons* with special reference to dimorphism in the male.
Proc. Zool. Soc., London, 144(2):175-201.

---

larvae, copepod

Haq, S.M., 1965.
The larval development of *Oithonina nana*.
J. Zool., Proc. Zool. Soc., London, 146(4):555-566.

---

larvae, copepod

Harding, J.P., and S.M. Marshall, 1955.
Triploid nauplii of *Calanus finmarchicus*.
Nature 175(4447):175, 5 textfigs.

---

larvae, copepods (parasitic)

Izawa, Kunihiko, 1969.
Life history of *Caligus spinosus* Yamaguti, 1939 obtained from cultured yellowtail, *Seriola quinqueradiata* T. + S. (Crustacea: Caligoida)
Rept. Fac. Fish. Pref. Univ. Mie, 6(3):127-157.

---

larvae, copepod

Johnson, Martin W., 1966.
The nauplius larvae of *Eurytemora herdmani* Thompson & Scott, 1897 (Copepoda, Calanoida).
Crustaceana, 11(3):307-313.

---

larvae, copepods

Johnson, M.W., 1938.
The postembryonic development of the copepod *Pseudodiaptomus euryhalinus* Johnson, and its phylogenetic significance.
Trans. Am. Microsc. Soc. 67(4):319-330.

Larvae, copepod

Katona, S.K. 1971.
The developmental stages of *Eurytemora affinis* (Poppe, 1880) (Copepoda, Calanoida) raised in laboratory cultures, including a comparison with the larvae of *Eurytemora americana* Williams, 1906, and *Eurytemora herdmani* Thompson & Scott, 1897.
Crustaceana 21(1):5-20.

---

larvae, copepod

Koga, Fumihiro 1970.
On the life history of *Tigriopus japonicus* Mori (Copepoda). (In Japanese, English abstract).
J. oceanogr. Soc. Japan 26(1):11-21.

---

larvae, copepods

Koga, Fumihiro, 1970.
On the nauplius of *Centropages yamadai* Mori (Copepoda).
J. oceanogr. Soc. Japan, 26(4):195-202.

---

larvae, copepod

Kogo, Fumihiro, 1968.
On the nauplius of *Undinula vulgaris* (Dana) (Copepoda, Calanoida).
J. oceanogr. Soc., Japan, 24(4):173-177.

Larvae, copepods

Koga, Fumihiro, 1960.
The developmental stages of nauplius larvae of *Pareuchaeta russelli* (Farran).
Bull. Jap. Soc. Sci. Fish. 26(8):792-795.
Also in:
Contrib. Dept. Fish. and Fish. Res. Lab., Kyushu Univ., No. 6

Larvae, copepod

Kogo, Fumihiro, 1960.
[The nauplius larvae of *Centropages abdominalis* Sato.]
Bull. Japan. Soc. Sci. Fish., 26(9):877-881.
Also in:
Contrib. Dept. Fish. and Fish. Res. Lab., Kyushu Univ., No. 6.

---

larvae, copepod

Krishnaswamy, S., 1951.
The development of harpacticioid copepod *Macrosetella gracilis* (Dana).
J. Madras Univ. B, 21:256-271.

---

larvae, copepods

Krishnaswamy, S., 1950.
Larval stages of some copepods in the Madras plankton and their seasonal fluctuation.
J. Madras Univ., Sect. B, 19:33-58, 1 table, 65 textfigs.

---

larvae, copepod

Lawson, Thomas J., and George D. Grice 1973.
The developmental stages of *Paracalanus crassirostris* Dahl, 1894 (Copepoda, Calanoida).
Crustaceana 24(1):43-56.

---

larvae, copepod

Lawson, Thomas J., and George D. Grice 1970.
The developmental stages of *Centropages typicus* Kröyer (Copepoda, Calanoida).
Crustaceana 18(2):167-208.

---

larvae, copepods

Lewis, Alan G., 1972
Hydrographic conditions and the seasonal effect of an enrichment medium on the early developmental stages of *Euchaeta japonica* (Crustacea, Copepoda, Calanoida). In: Biological oceanography of the northern North Pacific Ocean, A.Y. Takenouti, Chief Editor, Idemitsu Shoten, Tokyo, 395-401.

---

larvae, copepod

Lewis, Alan G., 1967.
An enrichment solution for culturing the early developmental stages of the planktonic marine copepod *Euchaeta japonica* Marukawa.
Limnol. Oceanogr., 12(1):147-148.

---

larvae, copepods

Lewis, A.G., and A. Ramnarine, 1969.
Some chemical factors affecting the early developmental stages of *Euchaeta japonica* (Crustacea: Copepoda: Calanoida) in the laboratory.
J. Fish. Res. Bd. Can., 26(5):1347-1362.

---

larvae, copepods

Lindquist, A., 1959.
Studien uber das zooplankton der Bottensee 1. Nauplien und Copepoditen von *Limnocalanus grimaldii* (de Guerne) (Copepoda, Calanoida).
Rept. Inst. Mar. Res., Lysekil, Ser. Biol., Rept. 10:

---

larvae, copepods

Matthews, J.B.L., 1964.
On the biology of some bottom-living copepods (Aetideidae and Phaennidae) from western Norway.
Sarsia, Univ. i Bergen, (16):1-46.

---

larvae, copepods

Mazza, Jacques, 1965.
Le développement de quelques copépodes en Méditerranée. II. Les stades jeunes de *Gaetanus kruppi* Giesb., *Euchirella messinensis* Cl., *Chiridius poppei* Giesb., *Pseudaetideus armatus* (Boeck) et *Heterorhabdus spinifrons*
Rev. Trav. Inst. Pêches marit., 29(3):285-320.

---

larvae, copepods

Mazza, J., 1964.
Le développement de quelques copépodes en Méditerranée. 1. Les stades jeunes d'*Euchaeta acuta* Giesbrecht et d'*E. spinosa* Giesbrecht.
Rev. Trav. Inst. Pêches Marit., 28(3):271-292.

larvae, copepod

Mullin, Michael M. and Elain R. Brooks, 1972
The vertical distribution of juvenile *Calanus* (Copepoda) and phytoplankton within the upper 50 m of water off La Jolla California. In: Biological oceanography of the northern North Pacific Ocean, A.Y. Takenouti, Chief Editor, Idemitsu Shoten, Tokyo, 347-354.

larvae, copepods

Murphy, H.E., 1923.
The life cycle of *Oithona nana* reared experimentally.
Univ. California Publ. Zool., 22:449-454.

Larvae, copepod

Nakai, Zinziro, 1969.
Note on mature female, floating egg and nauplius of *Calanus cristatus* Kröyer (Crustacea, Copepoda) - A suggestion on biological indicator for tracing of movement of the Oyashio water.
Bull. Soc. Jap. fish. Oceanogr. Spec. No. (Prof. Uda Commem. Pap.): 183-191.

larvae, copepods

Nicholls, A.G., 1941.
The developmental stages of *Metis jousseaumei* (Richard) (Copepoda, Harpacticoida). Ann. Mag. Nat. Hist., ser. 11, 7:317-328, 5 textfigs.

larvae, copepods

Nicholls, A.G., 1935.
The larval stages of *Longipedia coronata* Claus, *L. scotti*, G.O. Sars, and *L. minor* T. and A. Scott, with a description of the male of *L. Scotti*. J.M.B.A. 20(1):29-46, 8 textfigs.

larvae, copepods

Nicholls, A.G., 1934.
The developmental stages of *Euchaeta norvegica*, Boeck. Proc. Roy. Soc., Edinburgh 54, Pt. 1, No. 4:31-50, 8 textfigs.

larvae, copepod (effect of)

Nozu, Junji 1966.
Seasonal variation of copepod nauplii and the occurrence of eggs and larvae of *Engraulis japonica* shoals in Bungo Channel. (In Japanese; English abstract). Bull. Jap. Soc. scient. Fish. 32(3):

larvae, copepods

Oberg, Max, 1905.
Die metamorphose der Plankton-Copepoden der Kieler Bucht.
Wiss. Meer. Abt. Kiel, N.F., 9:37-103.

larvae, copepods

Ogilvie, H.S. 1953.
Copepod nauplii (1). Fiches d'Ident., Cons. Perm. Int. Expl. Mer, Sheet 50:4 pp.

larvae, copepods

Paffenhöfer, G.-A., 1971.
Grazing and ingestion rates of nauplii, copepodids and adults of the marine planktonic copepod *Calanus helgolandicus*. Mar. Biol. 11 (3): 286-298.

larvae, copepod (quantitative)

Pavshtics, E.A., 1965.
Distribution of plankton and summer feeding of herring in the Norwegian Sea and on Georges Ban.
ICNAF Spec. Publ., 6:583-589.

larvae, copepod

Pillai, P. Parameswaran 1971.
On the post-naupliar development of the calanoid copepod *Labidocera pectinata* Thompson and Scott (1903).
J. mar. biol. Ass. India 13(1): 66-77.

larvae, copepods

Prygunkova, R.V., 1968.
On the development of *Calanus glacialis* Iaschnov in the White Sea. (In Russian).
Dokl. Akad. Nauk. SSSR, 182 (6):1447-1450.

larvae, copepod

Rao, V.R., 1958.
Development of a cyclopoid copepod, *Oithona rigida* (Giesbrecht).
Andhra Univ. Mem., Oceanogr., 2:128-131.

larvae, copepod

Rao, V.R., 1958.
The development of a marine copepod, *Euterpina acutifrons* (Dana).
Andhra Univ. Mem., Oceanogr., 2:132-136.

larvae, copepod

Ravera, O., 1953.
Gli stadi di sviluppo dei copepodi pelagici del Lago Maggiore. Mem. Ist. Ital. Idrobiol., "Dott. Marco De Marchi", Pallanza, 7:129-151, 13 pls.

larvae, copepods

Sagshina, L.E., 1961.
Development of Black Sea copepods. II. Naupliar stages of *Calanus helgolandicus* (Claus).
Trudy Sevastopol Biol. Sta., (14):102-108.

larvae, copepoda (Rhincalanus)

Schmaus, P. Heinrich, und Karl Lehnhofer, 1927.
Copepoda 4: *Rhincalanus* Dana 1852 der Deutschen Tiefsee-Expedition, Systematik und Verbreitung der Gattung.
Wiss. Ergebn., Deutschen Tiefsee-Exped., "Valdivia", 1898-1899, 23(8):355-399.

larvae, copepods

Tokioka, Takasi, and Robert Bieri, 1966.
Juveniles of *Macrosetella gracilis* (Dana) from clumps of *Trichodesmium* in the vicinity of Seto.
Publs Seto mar. biol. Lab., 14(3):177-184.

larvae, copepods

Ummerkutty, A.N.P., 1964.
Studies on Indian copepods. 6. The post-embryonic developments of two calanoid copepods, *Pseudodiaptomus aurivilli* Cleve and *Labidocera bengalensis* Krishnaswamy.
J. Mar. biol. Ass., India, 6(1):48-60.

larvae, copepods

Ummerkutty, A.N.P., 1960.
Studies on Indian copepods. 2. An account of the morphology and life history of a harpacticoid copepod *Tisbintra jonesi* sp. nov. from the Gulf of Mannar.
J. Mar. Biol. Assoc., India, 2(2):149-164.

larvae, copepod

Vilela, Maria Helena 1972.
The developmental stages of the marine calanoid copepod *Acartia grani* Sars bred in the laboratory.
Notas Estudos Inst. Biol. marit. Lisboa 40: 20 pp., 18 pls

larvae Cumacea

Harada, Isokiti, 1967.
Post larval development and growth-stages in Cumacean Crustacea.
Jap. J. Zool., 15(3):343-347.

larvae Cypris

Schram, Thomas A. 1970
Marine biological investigations in the Bahamas. 14. *Cypris* y, a later developmental stage of nauplius y Hansen.
Sarsia 44: 9-24.

larvae, decapod

Aikawa, H., 1942.
Systematic studies of the plankton organisms occurring in Iwayama Bay, Palao. VI. On Brachyuran larvae from the Palao Islands (South Sea Islands) Palao Trop. Biol. Sta. Studies 2(3):585-611

larvae, decapod

Aikawa, H., and K. Isobe, 1955.
On the first larva of *Parribacus ursus-major* (Herbst). Rec. Ocean. Wks., Japan, 2(2):113-114.

larvae, decapods

Al-Kholy, A.A., 1961.
Larvae of some macruran Crustacea (from the Red Sea).
Publ. Mar. Biol. Sta., Al-Ghardaqa (Red Sea), 11:73-93.

larvae decapod

Al-Kholy, A.A., 1959
Larval stages of four brachyuran crustacea. (From the Red Sea). Publ. Mar. Biol. Sta. Al-Ghardaqa (Red Sea) 10: 239-246.

larvae, decapod

Al-Kholy, A.A., 1959
Larval stages of three anomuran crustacea. (From the Red Sea). Publ. Mar. Biol. Sta. Al-Ghardaqa (Red Sea), 83-90.

larvae decapod

Al-Kholy, A.A. and M.M. El-Hawary, 1970.
Some penaeids of the Red Sea. Bull. Inst. Oceanogr. Fish., Cairo, 1: 341-377.

larvae, decapod

Al-Kholy, A.A. and M.M. El-Hawary, 1970.
Some penaeids of the Red Sea. Bull. Inst. Oceanogr. Fish., U.A.R. 1: 339-378.

larvae, decapod

Al-Kholy, A.A., and M. Fikry Mahmoud, 1967.
Some larval stages of *Sergestes* sp. and *Synalpheus biunguiculatus* (Stimpson).
Publ. Mar. Biol. Sta. Ghardaqa, 14:167-175.

larvae, decapod

Al-Kholy, A.A., and M. Fikry Mahmoud, 1967.
Some larval stages of *Callianassa* sp. and *Hippolyte* sp.
Publ. Mar. Biol. Sta. Ghardaqa, 14:55-76.

larvae, decapod

Atkins, D., 1955.
The post-embryonic development of British *Pinnotheres* (Crustacea). Proc. Zool. Soc., London 124:687-715.

larvae, decapod

Atkins, D., 1954.
Leg disposition in the brachyuran megalopa when swimming. J.M.B.A., u.k., 33:627-636, 2 textfigs.

larvae, decapod

Austin, Herbert M. 1972.
Notes on the distribution of phyllosoma of the spiny lobster, *Panulirus* ssp., in the Gulf of Mexico.
Proc. Nat. Shellfish. Ass. 62:26-30.

larvae, decapod

Avila, Quinto, y Harold Loesch, 1965.
Identificacion de los camarones (Peneidae). juveniles de los esteros del Ecuador.
Bol. Cient. y Tecnico, Inst. Nacional. Pesca, Ecuador, 1(3):24 pp.

larvae, decapod

Baisre, Julio A., 1969.
A note on the phyllosoma of *Justitia longimanus* (H. Milne Edwards) (Decapoda, Palinuridea). Crustaceana 16(2): 182-184.

larvae, decapod

Baisre, Julio A. 1966.
Desarrollo larval en *Scyllarus* sp. (Crustacea, Decapoda) con notas sobre la abundancia y distribución de sus estadios.
Estudios, Inst. Oceanol. Acad. Cienc. Cuba 1: 5-34 (not seen)

Oceanographic Index: Marine Organisms Cumulation, 1946-1973

---

larvae, decapod
Baisre, Julio A., 1964.
Sobre los estados larvales de la langosta común *Panulirus argus*.
Inst. Nac. Pesca, Centro de Invest. Pesqueras, No. 19: 37 pp.

---

larvae, decapod
Batham, E.J., 1967.
The first three larval stages and feeding behaviour of phyllosoma of the New Zealand palinurid crayfish *Jasus edwardsii* (Hutton 1875)
Trans. R. Soc. N.Z. Zool., 9(6):53-64.

---

larvae, decapod
Baxter, K.N., 1963.
Abundance of postlarval shrimp - one index of future shrimping success.
Proc. 15th Ann. Sess., Gulf and Caribbean Fish. Inst., Galveston, Texas, Nov. 1962, Univ. Miami, Inst. Mar. Sci., 79-87.

---

larvae, decapod
Bensam, P., and K.N. Rasachandra Kartha 1967.
Notes on the eggs and early larval stages of *Hippolysmata ensirostris* Kemp.
Proc. Symp. Crustacea, Ernakulam, Jan. 12-15, 1965, 2:736-743.

---

larvae, decapod
Berkeley, A., 1930.
The post-embryonic development of the common pandalids of British Columbia.
Contr. Canadian Biol. Fish., n.s. 6(6):79-163, 43 textfigs.

---

larvae, decapod
Bocquet, Charles, 1965.
Stades larvaires et juvéniles de *Tritodynamia Atlantica* (Th. Monod) (= *Asthenognathus atlanticus* Th. Monod) et position systematique de ce crabe.
Cahiers Biol. Mar. 6(4):407-418.

---

larvae, decapod
Bookhout, C.G. 1972.
Larval development of the hermit crab *Pagurus alatus* Fabricius, reared in the laboratory (Decapoda, Paguridae).
Crustaceana 22(3):215-238.

---

larvae, decapod
Bookhout, C.G. and J.D. Costlow, Jr., 1970.
Nutritional effects of *Artemia* from different locations on larval development of crabs.
Helgoländer wiss. Meeresunters, 20(1/4): 435-442.

---

larvae, decapod
Boschi, E.E., y M.A. Scelzo, 1969.
El desarrollo larval de los crustáceos decápodos.
Cienc. Invest. 25(6):146-154.

---

larvae, decapod
Boschi, E.E., B. Goldstein y M.A. Scelzo, 1968.
Metamorfosis del crustáceo *Blepharipoda doelloi* Schmitt de las aguas de la Provincia de Buenos Aires (Decapoda, Anomura, Albuneidae).
Physis 27(75):291-311.

---

larvae, decapod
Boschi, E.E., M.A. Scelzo y B. Goldstein, 1969.
Desarrollo larval del cangrejo, *Halicarcinus planatus* (Fabricius) (Crustacea, Decapoda, Hymenosomidae), en el laboratorio, con observaciones sobre la distribución de la especie.
Bull. Mar. Sci., 19(1): 225-242.

---

larvae, decapod
Boschii, Enrique E., Marcelo A. Scelzo y Beatriz Goldstein, 1967.
Desarrollo larval de dos especies de crustaceos decapodos en el laboratorio. *Pachycheles haigae* Rodrigues Da Costa (Porcellanidae) y *Chasmagnathus granulata* Dana (Grapsidae).
Bol. Inst. Biol. mar., Mar del Plata, 12:5-46.

---

larvae, decapod
Bourdillon-Casanova, L., 1960.
Le meroplancton du Golfe de Marseille: les larves de crustacés décapodes.
Rec. Trav. Sta. Mar. Endoume, 30(18):286 pp.

---

larvae, decapod
Bouvier, M.E.L. 1914
Recherches sur le developpement postembryonnaire de la langouste commune (*Palinurus vulgaris*). J.M.B.A., n.s., 10:179-193, 6 figs.

---

larvae, decapod
Boyd, Carl M., 1960
The larval stages of *Pleuroncodes planipes* Stimpson (Crustacea, Decapoda, Galatheidae).
Biol. Bull., 118(1): 17-30.

---

larvae, decapod
Boyd, Carl M., and Martin W. Johnson, 1963
Variations in the larval stages of a decapod crustacean, *Pleuroncodes planipes* Stimpson (Galatheidae).
Biological Bulletin, 124(2):141-152.

---

larvae, decapod
Brattegard, T., and C. Sankarankutty 1967.
On prezoea and zoea of *Geryon tridens*.
Sarsia 26:7-12.

---

larvae, decapod
Broad, A.C., 1957.
Larval development of *Palaemonetes pugio*. Biol. Bull., 112(2):144-161.
Holthuis

---

larvae, decapod
Broad, A.C., and Jerry H. Hubschman, 1963.
The larval development of *Palaemonetes kadiakensis* M.J. Rathbun in the laboratory.
Trans. Amer. Microsc. Soc., 82(2):185-197.

---

larvae - decapoda
Broekhuysen, G.J., 1941.
The life of *Cyclograpsus punctatus*, breeding and growth. Trans. Roy. Soc. S. Africa, 28(4):331-365, 11 figs.

---

larvae, decapod
Buchanan, David V. and Raymond E. Millemann, 1969.
The prezoeal stage of the dungeness crab, *Cancer magister* Dana. Biol. Bull., mar. biol. Lab., Woods Hole, 137(2): 250-255.

---

larvae, decapod
Cabrera J, Jorge A, 1965
Contribuciones carcinológicas. I. El primer estadio zoea en *Gecarcinus lateralis* (Freminville) (Brachyura Gecarcinidae) procedente de Veracruz, México.
Anales Inst. Biol., Univ. Mex., 36(1/2):173-187.

---

larvae, decapod
Caroli, E., 1946.
Di un "puerulus" di *Palinurus vulgaris* pescato nel Golfo di Napoli. Pubbl. Staz. Zool., Napoli, 20(2):152-157, Fig. 2, Pl. 1.

---

larvae decapod
Ceccaldi, Hubert J., 1968.
Évolution des oeufs et cycle de reproduction chez *Plesionika edwardsi* (Brandt).
Rec. Trav. Sta. mar. Endoume, 44(60):403-412.

---

larvae, decapod
Chamberlain, N.A., 1961.
Studies on the larval development of *Neopanope texana sayi* (Smith) and other crabs of the Family Xanthidae (Brachyura).
Chesapeake Bay Inst., Ref. 61-1:37 pp., 16 pls. (multilithed)

---

larvae, decapod
Chhapgar, B.F., 1956.
On the breeding habits and larval stages of some crabs of Bombay.
Rec. Indian Mus., 54(1-2):33-52.

---

larvae, decapod
Choudhury, P.C., 1971.
Complete larval development of the palaemonid shrimp *Macrobrachium carcinus* (L.), reared in the laboratory (Decapoda, Palaemonidae).
Crustaceana 20(1):51-69.

---

larvae, decapod
Choudhury, P.C., 1970.
Complete larval development of the palaemonid shrimp *Macrobrachium acanthurus* (Wiegmann 1836), reared in the laboratory.
Crustaceana 18(2):113-132.

---

larvae, decapod
Christensen, A.M., and J.J. McDermott, 1958.
Life history and biology of the oyster crab, *Pinnotheres ostreum* Say. Biol. Bull., 114(2):145-179.

---

larvae, decapod
Christiansen, Marit E., 1973
The complete larval development of *Hyas araneus* (Linnaeus) and *Hyas coarctatus* Leach (Decapoda, Brachyura, Majidae) reared in the laboratory. Norw. J. Zool. 21(2):63-89.

---

larvae, decapod
Christiansen, Marit E. 1971.
Larval development of *Hyas araneus* (Linnaeus) with and without antibiotics (Decapoda, Brachyura, Majidae).
Crustaceana 21(3):307-315.

---

larvae, decapod
Christmas, J.Y., Gordon Gunter and Patricia Musgrave, 1966.
Studies of annual abundance of postlarval penaeid shrimp in the estuarine waters of Mississippi as related to subsequent commercial catches.
Gulf. Res. Repts. Ocean Springs, Miss., 2(2):177-212.

---

larvae, decapod
Churchill, E.P., 1942.
The zoeal stages of the blue crab, *Callinectes sapidus* Rathbun. Bd. Nat. Res., Chesapeake Biol. Lab., Publ. No. 49:26 pp., 4 pls.

---

larvae - decapoda
Churchill, E.P., 1941.
The zoeal stages of the blue crab, *Callinectes sapidus*. Anat. Rec., 81(4):suppl. 37-38.

---

larvae decapod
Cobb, J. Stanley, 1968.
Delay of moult by the larvae of *Homarus americanus*.
J. Fish. Res. Bd. Can., 25(10):2251-2253.

---

larvae, decapod
Coffin, Harold B., 1960.
The ovulation, embryology and developmental stages of the hermit crab (*Pagurus samuelis* Stimpson).
Walla Walla Coll. Publ., Dept. Biol. Sci. and Biol. Sta., No. 25:30 pp.

larvae, decapod
Cook, H.L. 1969.
A method of rearing penaeid shrimp larvae for experimental studies.
FAO Fish. Repts. 3(57) (FRm/R 57.3 (Trm)): 709-715. (mimeographed).

larvae, decapod
Cook, Harry L., 1966.
A generic key to the protozoean mysis and postlarval stages of the littoral Penaeidae of the northwestern Gulf of Mexico.
Fish. Bull., U.S. Fish and Wildlife Service, 65(2):437-447.

larvae, decapod
Cook, Harry L., and M. Alice Murphy 1971.
Early developmental stages of the brown shrimp, Penaeus aztecus Ives, reared in the laboratory.
Fish. Bull. U.S. Dept. Comm. 69(1): 223-239.

larvae, decapods
Cook, Harry L., and M. Alice Murphy, 1969.
The culture of larval penaeid shrimp.
Trans. Am. Fish. Soc. 98(4): 751-754

larvae, decapod
Cook, Harry L., and M. Alice Murphy, 1965.
Early developmental stages of the rock shrimp, Sicyonia brevirostris Stimpson, reared in the laboratory. 1.
Tulane Studies Zool., 12(4):109-127.

larvae, decapod
Costlow, John D., Jr., 1966.
The effect of eyestalk extirpation on larval development of the crab, Sesarma reticulatum Say.
In: Some contemporary studies in marine biology, H. Barnes, editor, George Allen & Unwin, Ltd., 209-224.

larvae, decapod
Costlow, John D., Jr., 1965.
Variability in larval stages of the blue crab, Callinectes sapidus.
Biol. Bull., 128(1):58-66.

larvae, decapod
Costlow, J.D., Jr., 1963.
Regeneration and metamorphosis in larvae of the blue crab Callinectes sapidus Rathbun.
J. Exp. Zool., 152(3):219-227.

larvae, decapod
Costlow, John D., Jr., and C.G. Bookhout, 1967.
Larval stages of the crab, Neopanope packardii (Kingsley), in the laboratory.
Bull. mar. Sci., Miami, 17(1):52-63.

larvae, decapod
Costlow, John D., and C.G. Bookhout, 1966.
The larval development of Ovalipes ocellatus (Herbst) under laboratory conditions.
J. Elisha Mitchell Sci. Soc., 82(2):160-171.

larvae, decapod
Costlow, John D. Jr., and C.G. Bookhout, 1966.
Larval stages of the crab, Pinnotheres maculatus, under laboratory conditions.
Chesapeake Sci., 7(3):157-163.

larvae, decapod
Costlow, John D., Jr., and C.G. Bookhout, 1966.
Larval development of the crab, Hexapanopeus angustifrons.
Chesapeake Sci., 7(3):148-156.

larvae, decapod
Costlow, John D., and C.G. Bookhout, 1965.
The effect of environmental factors on larval development of crabs.
Biological Problems in Water Pollution, Third Seminar, August 13-17, 1962, Environmental Health Series, Water Supply and Pollution Control, U.S. Publ. Health Serv., 77-86.

larvae, decapod (ecology)
Costlow, John D., Jr., and C.G. Bookhout, 1964.
An approach to the ecology of marine invertebrate larvae.
Narragansett Mar. Lab., Univ. Rhode Island, Occ. Publ., 2:69-75.

larvae, decapod
Costlow, J.D., Jr., and C.G. Bookhout, 1962
The larval development of Sesarma reticulatum Say reared in the laboratory.
Crustaceana, 4(4):281-294.

larvae, decapod
Costlow, J.D., Jr., and C.G. Bookhout, 1961.
The larval development of Eurypanopeus depressus (Smith) under laboratory conditions.
Crustaceana, 2(1):6-15.

larvae, decapod
Costlow, J.D., Jr., and C.G. Bookhout, 1960
The complete larval development of Sesarma cinereum (Bosc) reared in the laboratory.
Biol. Bull., 118(2): 203-214.

larvae, decapod
Costlow, John D., Jr. and C.G. Bookhout, 1960.
A method for developing brachyuran eggs in vitro. Limnol. & Oceanogr., 5(2): 212-215.

larvae, decapod
Costlow, John D., Jr. and C.G. Bookhout, 1959
The larval development of Callinectes sapidus Rathbun reared in the laboratory. Biol. Bull. 116(3): 373-396.

larvae, decapods
Costlow, John D., Jr., C.G. Bookhout and R.J. Monroe, 1966.
Studies on the larval development of the crab Rhithropanopeus harrisii (Gould). I The effect of salinity and temperature on larval development.
Physiol. Zool., 39(2):81-100.

larvae, decapod
Costlow, J.D., Jr., C.G. Bookhout and R. Monroe, 1962.
Salinity-temperature effects on the larval development of the crab Panopeus herbstii Milne-Edwards, reared in the laboratory.
Physiol. Zool., 35(1):79-93.

larvae, decapod
Costlow, J.D., Jr., C.G. Bookhout and R. Monroe 1960.
The effect of salinity and temperature on larval development of Sesarma cinereum (Bosc) reared in the laboratory. Biol. Bull., 118(2): 183-202.

larvae, decapod
Costlow, John D., Jr., and Elda Fagetti, 1967.
The larval development of the crab, Cyclograpsus cinereus Dana, under laboratory conditions.
Pacif. Sci., 21(2):166-177.

larvae, decapod
Costlow, J.D., Jr., G.H. Rees and C.G. Bookhout, 1959.
Preliminary notes on the complete larval development of Callinectes sapidus Rathbun under laboratory conditions.
Limnol. & Oceanogr., 4(2):222-223.

Phyllosoma commune
Couch, Jonathan, 1858.
Note on the occurrence of Phyllosoma commune on the coast of Cornwall.
J. Linn. Soc., London, Zool., 2:146-149.

larvae - decapod
Crane, J., 1941.
Eastern Pacific Expeditions of the New York Zoological Society. XXIX. On the growth and ecology of brachyuran crabs of the genus Ocypoda. Zoologica (N.Y.), 26(4):297-310, 2 pls.

larvae, decapod
Crosnier, A., 1972
Naupliosoma, phyllosomes et pseudibacus de Scyllarides herklotsi (Herklots) (Crustacea, Decapoda, Scyllaridae) récoltés par lomgango dans le sud du Golfe de Guinée. Cah. ORSTOM, sér. Océanogr. 10(2): 139-149.

larvae, decapod
Crosnier, A., 1971.
Ponte et développement de la langouste verte Panulirus regius de Brito Capello dans le sud du Golfe de Guinée. Cah. ORSTOM, sér. Océanogr. 9(3): 339-361.

larvae, decapod
Cunningham, J.T., 1892.
On the development of Palinurus vulgaris, the rock lobster or sea crayfish. J.M.B.A., n.s., 2:141-150, Pls. 8 and 9.

larvae, decapod
Cviic, V., 1960
Contribution à la connaissance du role des bactéries dans l'alimentation des larves de langoûtes (Palinurus vulgaris Latr.).
Comm. Int. Expl. Sci. Mer Medit., Monaco, Rapp. Proc. Verb., 15(3): 45-47.

larvae, decapod
Davis, Charles C., 1965.
A study of the hatching process in aquatic invertebrates. XX. The blue crab, Callinectes sapidus, Rathbun. XXI. The nemertean, Carcinonemertes carcinophila (Kölliker).
Chesapeake Science, 6(4):201-208.

larvae, decapod
DeBacker, Jean, 1963.
Evolution de l'organe hematopoietique chez l'embryon et la larve de Palaemonetes varians Leach var. microgenitor Boas.
La Cellule 64(1):7-26.

larvae, decapod
Dechancé, M., 1961.
Nombre et caractères des stades larvaires dans le genre Dardanus (Crustacé Décapode Paguride).
C.R. Acad. Sci., Paris, 253(3):529-531.

larvae, decapod
Dechancé, M., and J. Forest, 1962.
Sur Anapagurus bicorniger A. Milne Edwards et E.L. Bouvier et A. petiti sp. nov. (Crustacea Decapoda Paguridae).
Bull. Mus. Nat. Hist. Nat., (2)34(3):293-307.

larvae, decapod
Decharné, M., 1962.
Remarques sur les premiers stades larvaires de plusieurs espèces indopacifiques du genre Dardanus (Crustacés, décapodes Pagurides).
Bull. Must Nat. Hist. Nat., Paris, (2), 34(1): 82-94.

larvae, decapod
Demirhindi, U., 1961.
Preliminary notes on the Jaxea nocturna Nardo (1847) larvae from Turkish waters.
Rapp. Proc. Verb., Réunions, Comm. Int. Expl. Sci. Mer Méditerranée, Monaco, 16(2):219-222.

larvae, decapod
Deshmukh, Shriniwas, 1966.
The puerulus of the spiny lobster Panulirus polyphagus (Herbst) and its metamorphosis into the post-puerulus.
Crustaceana, 10(2):137-150.

larvae, decapod
Dexter, Deborah M. 1972.
Molting and growth in laboratory reared phyllosomes of the California spiny lobster, Panulirus interruptus.
Calif. Fish Game 58(2): 107-115

larvae, decapod
Diaz, H. and J.D. Costlow, 1972.
Larval development of Ocypode quadrata (Brachyura: Crustacea) under laboratory conditions. Mar. Biol. 15(2): 120-131.

larvae, decapod
Dobkin, Sheldon 1971.
A contribution to knowledge of the larval development of *Macrobrachium acanthurus* (Wiegmann, 1836) (Decapoda, Palaemonidae).
Crustaceana 21(3): 294-297

larvae, decapod
Dobkin, Sheldon, 1971.
The larval development of *Palaemonetes cummingi* Chace 1954 (Decapoda, Palaemonidae) reared in the laboratory.
Crustaceana 20(3): 285-297.

larvae, decapod
Dobkin, Sheldon 1970.
Manual de métodos para el estudio de larvas y primeras postlarvas de camarones y gambas.
Instruct. Inst. Nac. Invest. biol. pesq. México, ser. Divulgación 4:82pp.

larvae, decapod
Dobkin, Sheldon, 1965.
The early larval stages of *Glyphocrangon spinicauda* A. Milne Edwards.
Bull. Mar. Sci., 15(4): 872-884.

Larvae, decapod
Dobkin, Sheldon, 1965.
The first post-embryonic stage of *Synalpheus brooksi* Coutiere.
Bull. Mar. Sci., 15(2):450-462.

larvae, decapod
Dobkin, Sheldon, 1963.
The larval development of *Palaemonetes paludosus* (Gibbes, 1850) (Decapoda, Palaemonidae) reared in the laboratory.
Crustaceana, 6(1):41-61.

larvae, decapod
Dobkin, Sheldon, 1961.
Early developmental stages of pink shrimp *Penaeus duorarum* from Florida waters.
U.S. Fish and Wildlife Service, Fish. Bull., 61(190):321-349.

larvae, decapod
Dotsu, Yoshie, Kuniyoshi Seno and Shunji Inoue, 1966.
Rearing experiments on early phyllosomas of *Ibacus ciliatus* (von Siebold) and *I. novemdentatus* Gibbes (Crustacea: Reptantia). (In Japanese; English abstract).
Bull. Fac. Fish. Nagasaki Univ., No. 21:181-194.

larvae-decapod
Edmondson, C. H., 1949
Some Brachyuran megalopa. Occ. Pep. Bernice P. Bishop Mus. 19(12):233-246, 5 text figs.

larvae, decapod [a]
Eldred, Bonnie, Jean Williams George T. Martin, Edward A. Joyce Jr. 1965.
Seasonal distribution of penaeid larvae and postlarvae of the Tampa Bay area, Florida.
Techn. Ser. Fla Bd Conserv. 44:47pp.

larvae, decapod
Elofsson, Rolf, 1961.
The larvae of *Pasiphaea multidentata* (Esmark) and *Pasiphaea tarda* (Krøyer).
Sarsia, 4:43-53.

Larvae, decapod
Ewald, Joseph Jay, 1965.
The laboratory rearing of pink shrimp, *Penaeus duorarum* Burkenroad.
Bull. Mar. Sci, 15(2):437-449.

larvae, decapod [a]
Fagetti G., Elda 1970.
Desarrollo larval en el laboratorio de *Homalaspis plana* (Milne-Edwards) (Crustacea Brachyura: Xanthidae).
Rev. Biol. mar. Valparaiso 14(2):29-49.

larvae, decapod
Fagetti Elda 1969.
The larval development of the spider crab *Libidoclaea granaria* H. Milne Edwards and Lucas under laboratory conditions (Decapoda, Brachyura, Majidae, Pisinae).
Crustaceana 17(2):131-140.

larvae, decapod
Fagetti G., E. 1969.
Larval development of the spider crab *Pisoides edwardsi* (Decapoda, Brachyura) under laboratory conditions. Marine Biol., 4(2): 160-165.

larvae, decapod
Fagetti Guaitas, Elda, 1960.
Primer estudio larval de cuatro crustaceos braquiuros de la Bahia de Valparaiso.
Rev. Biol. Mar., 10(1/3):143-154.

larvae, decapod [a]
Fagetti, E. and I. Campodonico, 1973
Larval development of *Pilumnoides perlatus* (Brachyura: Xanthidae) under laboratory conditions. Mar. Biol. 18(2): 129-139.

larvae, decapod
Fagetti Elda and Italo Campodonico 1971.
The larval development of the crab *Cyclograpsus punctata* H. Milne Edwards, under laboratory conditions (Decapoda, Brachyura, Grapsidae, Sesarminae).
Crustaceana 21(2): 183-195.

larvae, decapod
Fagetti, E. and I. Campodonico, 1971.
Larval development of the red crab *Pleuroncodes monodon* (Decapoda Anomura: Galatheidae) under laboratory conditions. Marine Biol., 8(1): 70-81.

larvae, decapod
Fagetti G., Elda, e I. Campodonico G., 1971.
Desarrollo larval en el laboratorio de *Taliepus dentatus* (Milne-Edwards) (Crustacea Brachyura: Majidae, Acanthonychinae).
Rev. Biol. mar. Valparaiso 14(3):1-14.

larva, decapod [a]
Fagetti G., Elda, e Italo Campodonico G., 1970.
Desarrollo larval en el laboratorio de *Acanthocyclus gayi* Milne-Edwards et Lucas. (Crustacea Brachyura: Atelecyclidae, Acanthocyclinae).
Rev. Biol. mar. Valparaiso, 14(2):63-78

larvae, decapod
Feliciano, C., 1956.
A prenaupliosoma stage in the larval development of the spiny lobster, *Panulirus argus* (Latreille) from Puerto Rico. Bull. Mar. Sci., Gulf and Caribbean, 6(4):341-345.

larvae, decapod
Fielder D.R. 1970.
The larval development of *Macrobrachium australiense* Holthuis, 1950 (Decapoda, Palaemonidae), reared in the laboratory.
Crustaceana 18(1):60-74.

decapod, life history
Figueroa, M. Cardenas, 1951.
Ciclo evolutivo de tres peneidos del noroeste de Mexico. Rev. Sci. Mexicana Hist. Nat. 12(1/4):229-258, 1 graph.

larvae, decapod
Foxton, P., and P.J. Herring 1970.
Recent records of *Physetocaris microphthalma* Chace with notes on the male and description of the early larvae (Decapoda, Carides).
Crustaceana 18(1):93-104.

larvae, decapod
Fukataki, Hiroshi, 1969.
Occurrence and distribution of planktonic larvae of edible crabs belonging to the genus *Chionoecetes* (Majidae, Brachyura) in the Japan sea. (In Japanese; English abstract)
Bull. Jap. Sea Reg. Fish. res. Lab. (21):35-54

larvae, decapod
Garcia Pinto, Lope 1971.
Identificación de las postlarvas del camarón (genero *Penaeus*) en el occidente de Venezuela y observaciones sobre su crecimiento en el laboratorio.
Informe tecn. Proyecto Invest. Desarrollo prog. MAC-PNUD-FAO, Caracas, 39: 23pp.

larvae, decapod
George, M.J., 1969.
Genus *Metapenaeus* Wood-Mason & Alcock 1891.
Bull. Cent. mar. Fish. Res. Inst., India, 14:75-125 (mimeographed)

larvae, decapod
George, M.J., 1962.
On the breeding of penaeids and the recruitment of their postlarvae into the backwaters of Cochin.
Indian Jour. Fish., (A), 9(1):110-116.

larvae, decapod
Gilet, R., 1952.
Métazoe de *Dorippe lanata* (Linné) et se megalope.
Vie et Milieu 3(4): 415-420.

larvae, decapod [a]
Goldstein, Beatriz, 1970 (1971).
Développement larvaire de *Macropipus marmoreus* (Leach) en laboratoire (Crustacea, Decapoda, Portunidae).
Bull. Mus. nat. Hist. nat. (2) 42(5): 919-943.

larva, decapoda [a]
Goldstein, Beatriz, and C.G. Bookhout 1972.
The larval development of *Pagurus prideauxi* Leach 1814, under laboratory conditions (Decapoda, Paguridea).
Crustaceana 23(3):263-281.

larvae, decapod [a]
Gonor, S.L. and J.J. Gonor 1973.
Feeding, cleaning and swimming behavior in larval stages of porcellanid crabs (Crustacea: Anomura).
Fish. Bull. U.S. Dept. Comm. 71(1):225-234.

larvae, decapod (lists of spp.) [a]
Gonor, S.L. and J.J. Gonor 1973.
Descriptions of the larvae of four North Pacific Porcellanidae (Crustacea: Anomura).
Fish. Bull. U.S. Dept. Comm. 71(1):189-223

larvae, decapod

Gopalakrishnan, K., and R. Michael Laurs, 1971
Enetmocaris corniger Bate larvae from the eastern tropical Pacific Ocean (Caridea, Hippolytidae).
Crustaceana, 20(1):9-18.

larvae, decapod

Gordon I., 1960
On *Problemacaris* (Stebbing), a rare caridean larva of uncertain position (Crustacea Decapoda) Crustaceana 1(1): 39-46.
Collected by A.C. Hardy from Discovery II, 11 Sept. 1954

larvae, decapod

Gordon, I., 1953.
On the puerulus stage of some spiny lobsters (Palinuridae). Bull. Brit. Mus., (N.H.), Zool., 2(2):17-42, figs.

larvae, decapod

Gore, Robert H., 1973
Pachycheles monilifer (Dana, 1852): the development in the laboratory of larvae from an Atlantic specimen with a discussion of some larval characters in the genus (Crustacea: Decapoda: Anomura). Biol. Bull. mar. biol. Lab. Woods Hole 144(1): 132-150.

larvae, decapod

Gore, Robert H., 1972.
Petrolisthes platymerus: the development of larvae in laboratory culture (Crustacea: Decapoda: Porcellanidae). Bull. mar. Sci. Miami 22(2): 336-354

larvae, decapod

Gore, Robert H., 1971.
Petrolisthes tridentatus: the development of larvae from a Pacific specimen in laboratory culture with a discussion of larval characters in the genus (Crustacea: Decapoda: Porcellanidae). Biol. Bull. mar. biol. Lab. Woods Hole 141(3): 485-501.

larvae, decapod

Gore, Robert H. 1971.
Megalobrachium poeyi (Crustacea, Decapoda, Porcellanidae): Comparison between larval development in Atlantic and Pacific specimens reared in the laboratory.
Pacific Sci. 25(3):404-425

larvae, decapods

Gore, R.H., 1971.
The complete larval development of *Porcellana sigsbeiana* (Crustacea: Decapoda) under laboratory conditions. Mar. Biol. 11(4): 344-355.

larvae, decapod

Gore, Robert H., 1970.
Petrolisthes armatus: a redescription of larval development under laboratory conditions (Decapoda, Porcellanidae).
Crustaceana, 18(1):75-89.

larvae, decapod

Gore, Robert H., 1968.
The larval development of the commensal crab *Polyonyx gibbesi* Haig, 1956 (Crustacea: Decapoda). Biol. Bull. mar. biol. Lab., Woods Hole, 135(1): 111-129.

larvae, decapod

Green, P.A., and D.T. Anderson 1973.
The first zoea larvae of the estuarine crabs *Sesarma erythrodactyla* Hess, *Helograpsus haswellianus* (Whitelegge) and *Chasmagnathus laevis* Dana (Brachyura, Grapsidae, Sesarminae).
Proc. Linn. Soc. N.S.W. 98(1): 13-28.

larvae, decapod

Greenwood, J.G., 1966.
Some larval stages of *Pagurus novae-zealandiae* (Dana), 1852 (Decapoda Anomura).
New Zealand J. Sci., 9(3):545-558.

larvae, decapod

Greenwood, J.G., 1965.
The larval development of *Petrolisthes elongatus* (H. Milne Edwards) and *Petrolisthes novaezelandiae* Filhol (Anomura, Porcellanidae) with notes on breeding.
Crustaceana, 8(3):285-307.

larvae, decapod

Gurney, R., 1950.
A remarkable penaeid larva. Proc. Zool. Soc., London, 119(4):803-806, figs.

decapods, larvae

Gurney, R., 1949.
The larval stages of the snapping shrimp, *Synalpheus goodei* Coutiere. Proc. Zool. Soc., London, 119(2):293-295, 9 figs.

larvae, decapods

Gurney, R., 1939.
Bibliography of the larvae of decapod Crustacea. Ray Soc., London, 123 pp.

larvae, decapoda

Gurney, R., 1938.
Larvae of decapod Crustacea. V. Nephropsidea and Thalassinidea. Discovery Rept., 17:293-343.

larvae, decapod

Gurney, R., 1938.
The larvae of the decapod Crustacea Palaemonidae and Alpheidae. Sci. Repts. Gt. Barrier Reef Exped., 1928-29, B.M.(N.H.), 6(1):60 pp., 265 textfigs.

larvae, decapod

Gurney, R., 1936.
Larvae of decapod Crustacea. 1. Stenopidea. 2. Amphionidae. 3. Phyllosoma. Discovery Repts., 12:377-440.

decapod larvae

Gurney, R., 1927.
Zoological results of the Cambridge Expedition to the Suez Canal. XV. Report on the larvae of the Crustacea Decapoda. Trans. Zool. Soc., London, 22:231-286, textfigs. 49-76.

larvae - decapods

Gurney, R., 1924.
The Larval Development of some British Prawns (Palaemonidae). II. *Leander longirostris* and *Leander squilla*. Proc. Zool. Soc., London, 1924, 961-982.

larvae - decapods

Gurney, R., 1924.
The larval development of some British Prawns (Palaemonidae). I. *Palaemonetes varians*. Proc. Zool. Soc., London, 1924:297-328.

larvae - decapod

Gurney, R., 1924.
British Antarctic ("Terra Nova") Expedition, 1910. Nat. hist. rep., Zoology, Vol. VIII(2): 37-302. Crustacea, Part IX, Decapod larvae.

larvae, decapod

Gurney, R., and M.V. Lebour, 1940
Larvae of decapod crustacea. Pt. VI. The genus *Sergestes*. Discovery Repts. 20:68 pp., 56 text figs.

larvae, decapod

Hanson, Arthur J., 1969.
The larval development of the sand crab *Hippa cubensis* (De Saussure) in the laboratory (Decapoda Anomura).
Crustaceana, 16(2):143-157.

larvae, decapod

Hansen, H.J., 1922.
Crustacés décapodes (Sergestides) provenant des yachts HIRONDELLE et PRINCESSE-ALICE (1885-1915). Rés. camp. sci. Monaco, 64:229 pp., 11 pls. (Campagnes des)

larvae, decapod

Harada, Eiji, 1959
A study of the productivity of Tanabe Bay (II). V. Occurrence of the first stage phyllosoma larva of *Panulirus japonicus* (von Siebold) in Tanabe Bay. Rec. Oceanogr. Wks., Japan, Spec. No. 3: 57-60.

larvae, decapod

Harada, Eiji, 1958.
Notes on the naupliosoma and newly hatched phyllosoma of *Ibacus ciliatus* (Von Siebold).
Publ. Seto Mar. Biol. Lab., 7(1):173-180.

larvae, decapod

Hardy, A.C., and R. Bainbridge, 1951.
Effect of pressure on the behavior of decapod larvae (Crustacea). Nature 167(4244):354-355, 2 textfigs.

larvae, decapod

Hart, Josephine F.L., 1960.
The larval development of British Columbia brachyura. II. Majidae, subfamily Oregoniinae. Canadian J. Zool., 38(3):539

larvae, decapod

Hartnoll, R.G., 1964.
The zoeal stages of the spider crab *Microphrys bicornutus* (Latr.).
Ann. Mag. Nat. Hist., (13) 7(76):241-246.

larvae, decapod

Hashmi, Syed Salahuddin 1970.
The brachyuran larvae of W. Pakistan hatched in the laboratory. II Portunidae: *Charybdis* (Decapoda: Crustacea).
Pakistan J. Sci. Ind. Res. 12(3):272-278.

larvae, decapod

Hashmi, Syed Salahuddin, 1970.
The larval development of *Philyra corallicola* (Alcock) under laboratory conditions (Brachyura, Decapoda).
Pakistan J. Zool. 2(2):219-233

larvae, decapod

Heegaard, Poul 1971.
Larval stages and growth in the decapods.
Vidensk. Meddr dansk naturh. Foren. 134:119-126.

larvae, decapod

Heegaard, Poul 1969.
Larvae of decapod Crustacea: The Amphionidae.
Dana Rept 77:82 pp.

larvae, decapod

Heegaard, Poul, 1969.
The first larval stage of *Chlorotocus crassicornis* (Decapoda, Pandalidae).
Crustaceana, 17(2):151-158

larvae, decapod

Heegard, Poul, 1966.
Larvae of decapod Crustacea. The oceanic penaeids, *Solenocera*, *Cerataspis*, *Cerataspides*.
Dana Rept., No. 67:147 pp.

Solenocera membranacea
Solenocera muelleri
Solenocera sp. (aequatorialis)
Solenocera sp. (danae)
Solenocera sp. (sumatransis)
Solenocera sp. (nodulosa)
Solenocera sp. (elongata)
Solenocera sp. (karista)

larvae, decapod

Heegaard, P.E., 1953.
Observations on spawning and larval history of the shrimp, *Penaeus setiferus* (L.).
Publ. Inst. Mar. Sci. 3(1):73-105, 12 pls.

larvae, decapod
Hart, Josephine F.L., 1965.
Life history and larval development of Cryptolithodes typicus Brandt (Decapoda, Anomura) from British Columbia.
Crustaceana, 8(3):255-276.

larvae, decapod
Heldt, J.H., 1955.
Contributions à l'étude de la biologie des crevettes pénéides Aristiomorpha foliacea (Risso) et Aristeus antennatus (Risso) (formes larvaires).
Bull. Soc. Sci. Nat., Tunisie, 8(1/2):9-30, 17 pls.

larvae, decapod
-Heldt, J.H., 1955.
Contribution à l'étude de la biologie des crevettes pénéides. Formes larvaires de Solenocera membranacea (H.M.Edw.).
Bull. Sta. Océan., Salammbô, No. 51:29-55, textfigs.

larvae, decapod
Heldt, J., 1954.
Stades larvaires d'Aristeomorphe foliacea (Risso) et Aristeus antennatus (Risso), décapodes pénéides. C.R. Acad. Sci., Paris, 239(17):1080-1082.

larvae, decapod
Herring, P.J., 1967.
Observations on the early larvae of three species of Acanthephyra (Crustacea, Decapoda, Caridea).
Deep-Sea Res., 14(325-329.

larvae, decapod (lists)
Hoese, H.D., 1960
Juvenile penaeid shrimp in the shallow Gulf of Mexico. Ecology 41(3): 592-593.

larvae, decapod
Hoffman, Ethelwyn G., 1968.
Description of laboratory-reared larvae of Paralithodes platypus (Decapoda, Anomura, Lithodidae).
J. Fish. Res. Bd Can. 25(3): 439-455.

larvae - decapoda
Höglund, H., 1943.
On the biology and larval development of Leander squilla (L.) forma typica de Man.
Svenska Hydrografisk-Biologiska Kommissionens Skrifter (n.s.,) Biologi, II(6):3-44, 4 pls., 26 textfigs.

larvae, decapod
Hood, M.R., 1962.
Studies on the larval development of Rithropanopeus harrissii (Gould of the family Xanthidae (Brachyura).
Gulf Res. Rept., 1(3):122-130.

larvae-decapoda
Hopkins, S. H., 1944.
The external morphology of the third and fourth zoeal stages of the blue crab Callinectes sapidus Rathbun. Biol. Bull., 87(2):145-152

larvae, decapod
Hopkins, S.H., 1943.
The external morphology of the first and second zoeal stages of the blue crab, Callinectes sapidus Rathbun. Trans. Amer. Micro. Soc. 62:85-90.

larvae, decapod
Hudinaga, Motosaku, and Ziro Kittaka, 1966.
Studies on food and growth of larval stage of a prawn, Penaeus japonicus, with reference to the application to practical mass culture.
(In Japanese; English abstract).
Inf. Bull.Planktol.Japan, No.13:83-94.

larvae, decapod [a]
Hue, Jong Soo, Keuk Soon Bang, and Yong Kil Rho 1972.
Studies on the growth and artificial rearing of the larval blue-crab, Portunus trituberculatus (Miers).
Bull. Fish. Res. Dev. Agency 9:55-70.
(In Korean; English abstract)

larvae, decapod
Hughes, D.A., 1966.
Investigations of the "nursery areas" and habitat preferences of juvenile penaeid prawns in Mozambique.
J. Appl. Ecol., 3(2):349-354.

larvae, decapod
Hughes, John T., and George C. Mattiessen, 1967.
Observations on the biology of the American lobster (Homerus americanus).
Techn. Ser. Mass.Div.Mar.Fish.,(Publ.#595)No.2: 21 pp. (multilithed).

larval stages -decapods
Hyman, O. W., 1925
Studies on the larvae of crabs of the family Xanthidae. Proc. U. S. Nat. Mus. 67(3):1-22, 14 pls.

larval stages - decapods
Hyman, O. W., 1924
Studies on larvae of crabs of the family Pinnotheridae. Proc. U. S. Nat. Mus. 64(7): 1-9, 6 pls.

larvae
Illig, G., 1927
Die Sergestiden der Deutschen Tiefsee-Expedition. Wiss. Ergeb. Deutschen Tiefsee-Exped., "Valdivia" 1898-1899, 23(7):277-354, 29 text figs.

larvae decapod, quantitative
India, Indian Ocean Biological Center, Cochin, and National Institute of Oceanography, 1970.
Distribution of Copepoda and decapod larvae in the Indian Ocean.
Int. Indian Ocean Exped., Plankton Atlas, 2 (1): 11 charts.

larvae, decapod
Ingle, Robert M., Bonnie Eldred, Harold W. Sims, and Eric A. Eldred, 1963
On the possible Caribbean origin of Florida's spiny lobster populations.
State of Florida, Bd., Conserv., Div. Salt Water Fish., Techn. Ser., No. 40:12 pp.

larvae, decapod [a]
Ingle, R.W., and A.L. Rice, 1971.
The larval development of the masked crab, Corystes cassivelaunus (Pennant) (Brachyura, Corystides) reared in the laboratory.
Crustaceana 20(3): 271-284

larvae, decapod
Inoue, Masaaki, 1965.
On the relation of amount of food taken to the density and size of food and water temperature in rearing the phyllosoma of the Japanese spiny lobster, Panulirus japonicus (V. Siebold).
(In Japanese; English abstract).
Bull. Jap. Soc., Sci. Fish., 31(11):902-906.

Larvae, decapod
Irvine, John, and Harold G. Coffin, 1960.
Laboratory culture and early stages of Fabia subquadrata (Dana), (Crustacea, Decapoda). Walla Walla Coll., Publ., Dept. Biol. Sci. and Biol. Sta., No. 28:24 pp.

larvae, decapod
Ito, Katsuchiyo, 1968.
Observation on the primary features of the newly hatched zoeal larvae of the zuwai-crab, chionoecetes opilio O. Fabricius. (In Japanese)
Bull.Jap.Sea Reg.Fish.Res.Lab., 20:91-93.

larvae, decapod [a]
Ito, Katsuchiyo, and Koji Ikehara 1971.
Observations on the occurrence and distribution of the planktonic larvae of the queen crabs, Chionoecetes spp., in the neighboring waters of Sado Island.
(In Japanese; English abstract)
Bull. Jap. Sea Reg. Fish. Res. Lab. (23): 83-100.

larvae, decapod
Ivanov, B.G. 1971.
Larvae of some Far East shrimp in relation to their taxonomic status. (In Russian; English abstract)
Zool. Zh. 50(5): 657-665.

larvae, decapod
Ivanov, B.G. 1968.
Larvae of some Far Eastern common shrimps of the family Crangonidae (Crustacea, Decapoda). (In Russian; English abstract)
Zool. Zh. 47(4): 534-540.

larvae, decapod
Ivanov, B.G., 1965.
A description of the first larva of the Far Eastern shrimp (Pandalus goniurus). (In Russian; English summary)
Zool. Zhurn., Akad. Nauk, SSSR, 44(8):1255-1257.

larvae, decapod [a]
Iwata, Fumio, 1970.
Studies on the development of the crab Pugettia quadridens (De Haan): 1. Hatching and zoeae.
Publ. Seto mar. biol. Lab., 18(3): 189-197.

larvae, decapod.
Jensen, J.P., 1958.
Studies in the life history of the prawn Leander adspersus (Rathke) and the Danish fishery on this species.
Medd. Dansk. Fisk. Havundersøgelser, n.s., 2(18):1-28.

larvae, decapod [a]
Johnson, Martin W. 1971.
The palinurid and scyllarid lobster larval of the tropical eastern Pacific and their distribution as related to the prevailing hydrography
Bull. Scripps. Inst. Oceanogr. 19: 36 pp.

larvae, decapod
Johnson, Martin W. 1971.
The phyllosoma larva of Scyllarus delfini (Bouvier) (Decapoda, Palinuridea).
Crustaceana 21(2): 161-164.

larvae decapod
Johnson, Martin W., 1971.
On palinurid and scyllarid lobster larvae and their distribution in the South China Sea (Decapoda, Palinuridea).
Crustaceana 21(3): 247-282.

larvae, decapod [a]
Johnson, Martin W., 1971.
The phyllosoma larvae of slipper lobsters from the Hawaiian Islands and adjacent areas (Decapoda, Scyllaridae).
Crustaceana 20(1): 77-103.

larvae, decapod
Johnson, Martin W., 1970.
On the phyllosoma larvae of the genus Scyllarides Gill (Decapoda, Scyllaridae).
Crustaceana, 18(1): 13-20.

larvae, decapod
Johnson, Martin W., 1969.
Two chelate palinurid larvae from Hawaiian and Philippine waters (Decapoda, Palinuridea).
Crustaceana, 16(2): 113-118.

larvae, decapod
Johnson, Martin W., and Margaret Knight, 1966.
The phyllosoma larvae of the spring lobster Panulirus inflatus (Bouvier).
Crustaceana, 10(1):31-47.

Oceanographic Index: Marine Organisms Cumulation, 1946-1973

---

larvae, decapod
Johnson, M.W. 1960.
Production and distribution of larvae of the spiny lobster Panulirus interruptus (Randall) with records on P. gracilis Streets.
Bull. S.I.O., 7(6):413-462.

larvae, decapod
Johnson, M.W., 1956.
The larval development of the California spiny lobster, Panulirus interruptus (Randall), with notes on Panulirus gracilis Streets.
Proc. Calif. Acad. Sci., 29(1):1-19.

larvae, decapod
Johnson, M.W., and W.M. Lewis, 1942.
Pelagic larval stages of the sand crabs, Emerita analoga (Stimpson), Blepharipoda occidentalis Randall, and Lepidope myops Stimpson. Biol. Bull. 83(1):67-87.

larvae, decapod
Kalber, F.A., 1970.
Osmoregulation in decapod larvae as a consideration in culture techniques. Helgoländer wiss. Meeresunters. 20(1/4): 697-706.

decapod larvae
Kändler, R., 1961
Über das Vorkommen von Fischbrut, Decapodenlarven und medusen in der Kieler Förde (5. Beitrag über langfristige Beobachtungen in der Kieler Förde).
Kieler Meeresf., 17(1):48-64.

Decapodenlarven
Kandler, R., 1949
Jahreszeitliches Vorkommen und unperiodisches Auftreten von Fischbrut, Medusen und Decapodenlarven in Fehmarnbelt in den Jahren 1934-1943. Ber. Deutschen Wiss. Komm. f. Meeresf. nf. 12(1):49-85, 6 figs.

larvae, decapod.
Karlovac, O., 1953.
An ecological study of Nephrops norvegicus (L.) of the high Adriatic. "Hvar" Repts. 5(2C):3-50, 14 textfigs., 1 chart.
has figures of larvae.

larvae, decapod
Kensler, Craig B. 1967.
Notes on laboratory rearing of juvenile spiny lobsters Jasus edwardsii (Hutton) (Crustacea: Decapoda: Palinuridae).
N.Z. Jl mar. Freshwat. Res. 1(1):71-75.

larvae, decapod
Kircher, Ann B., 1970.
The zoeal stages and glaucothoe of Hypoconcha arcuata Stimpson (Decapoda: Dromidae) reared in the laboratory. Bull. mar. Sci., 20(3): 769-792.

larvae decapod
Knight Margaret D. 1970.
The larval development of Lepidopa myops Stimpson (Decapoda, Albuneidae) reared in the laboratory and the zoeal stages of another species of the genus from California and the Pacific coast of Baja California, Mexico.
Crustaceana 19(2): 125-156.

larvae, decapod
Knight, Margaret D., 1967.
The larval development of the sand crab, Emerita rathbunae Schmitt (Decapoda, Hippidae).
Pacif. Sci., 21(1):58-76.

larvae, decapod
Knight, Margaret D., 1966.
The larval development of Polyonyx quadriungulatus Glassell and Pachycheles rudis Stimpson (Decapoda, Porcellanidae) Cultured in the laboratory.
Crustaceana, 10(1):75-97.

---

larvae, decapod
Knowlton, Robert E. 1973
Larval development of the snapping shrimp Alpheus heterochaelis Say, reared in the laboratory.
J. nat. Hist. 7(3):273-306.

larvae, decapods
Knudsen, Jens W., 1960.
Reproduction, life history and larval ecology of the California Xanthidae, the pebble crabs.
Pacific Science, 14(1):3-18.

larvae, decapod
Kon, Tohshi, 1970.
Fisheries biology of the Tanner crab. IV. The duration of planktonic stages estimated by rearing experiments of larvae. (In Japanese; English abstract)
Bull. Jap. Soc. scient. Fish. 36(3):219-224.

larvae decapod
Kon, Tohshi, 1967.
Fisheries biology of the Tanner Crab, Chionoecetes Opilio. I. On the prezoeal larva.
Bull. Jap. Soc. scient. Fish., 33(8):726-730.

larvae, decapods
Krishna Menon, M., Menon P. Gopala and V.T. Paulinose, 1969.
Preliminary notes on the decapod larvae of the Arabian Sea. Bull. natn. Inst. Sci. India 38(2): 753-757. Also in: Coll. Repr. Nat. Inst. Oceanogr. Goa India 2(1968-1970)

larvae, decapod
Krishna Pillai, N., 1955.
Pelagic Crustacea of Travancore. 1. Decapod larvae.
Bull. Central Res. Inst., Univ. Travancore, Trivandrum, 4(1):47-102.

larvae, decapod
Kubo, I., 1951.
Bionomics of the prawn, Pandalus kessleri Czerniavski. J. Tokyo Univ. Fish. 38(1):1-26, 9 textfigs

Larvae, decapod
Kurata, Hiroshi, 1969.
Larvae of Decapoda Brachyura of Arasaki, Sagami Bay - IV. Majidae. Bull. Tokai reg. Fish. Res. Lab., 57: 81-125. (In Japanese; English abstract)

larvae, decapod
Kurata, Hiroshi 1968.
Larvae of Decapoda Natantia of Arasaki, Sagami Bay - IV Palaemoninae. (In Japanese; English abstract).
Bull. Tokai reg. Fish. Res. Lab. 56: 143-159.

larvae, decapod
Kurata, Hiroshi, 1968.
Larvae of Decapoda Anomura of Arasaki, Sagami Bay - III. Paguristes digitalis (Stimpson) (Diogenidae). (In Japanese; English abstract).
Bull. Tokai reg. Fish Res. Lab., 56: 181-186.

larvae, decapod
Kurata, Hiroshi, 1968.
Larvae of Decapoda Anomura of Arasaki, Sagami Bay - II. Dardanus arrosor (Herbst) (Diogenidae) (In Japanese; English abstract).
Bull. Tokai reg. Fish. Res. Lab., 56: 173-180.

larvae, decapod
Kurata, Hiroshi, 1968.
Larvae of Decapoda Brachyura of Arasaki, Sagami Bay - III. Carcinoplax longimanus (De Haan) (Goneplacidae). (In Japanese; English abstract). Bull Tokai reg. Fish. Res. Lab., 56: 167-172.

---

larvae, decapod
Kurata, Hiroshi, 1968.
Larvae of Decapoda Brachyura of Arasaki, Sagami Bay - II. Hemigrapsus sanguineus (De Haan) (Grapsidae). (In Japanese; English abstract). Bull. Tokai reg. Fish Res. Lab., 56, 161-165.

larvae, decapod
Kurata, Hiroshi, 1968.
Larvae of Decapoda Natantia of Arasaki, Sagami Bay - III. Heptacarpus geniculatus (Stimpson) (Hippolytidae). (In Japanese; English abstract). Bull. Tokai reg. Fish Res. Lab. 56: 137-142.

larvae, decapod
Kurata, Hiroshi, 1968.
Larvae of Decapoda Anomura of Arasaki, Sagami Bay. I. Pagurus samuelis (Stimpson)(Paguridae).
Bull. Tokai reg. Fish. Res. Lab. 55:265-269.
(In Japanese; English abstract)

Larvae, decapod
Kurata, Hiroshi, 1968.
Larvae of Decapoda Brachyura of Arasaki, Sagami Bay. I. Acmaeopleura parvula (Stimpson (Grapsidae). (In Japanese; English abstract). Bull. Tokai reg. Fish. Res. Lab., 55:259-263.

Larvae, decapod
Kurata, Hiroshi, 1968.
Larvae of Decapoda Macrura of Arasaki, Sagami Bay. II. Heptacarpus futilirostris (Bate) (Hippolytidae). (In Japanese; English abstract). Bull. Tokai reg. Fish. Res. Lab., 55:253-258.

larvae, decapod
Kurata, Hiroshi, 1968.
Larvae of Decapoda Macrura of Arasaki, Sagami Bay. I. Eualus gracilirostris (Stimpson) (Hippolytidae). (In Japanese; English abstract). Bull. Tokai reg. Fish. Res. Lab., 55:245-251.

larvae, decapod
Kurata, Hiroshi, 1964.
Larvae of decapod Crustacea of Hokkaido. 5. Paguridae (Anomura). 6. Lithodidae (Anomura). 7. Porcellanidae (Anomura). 8. Dorippidae (Brachyura).
Bull. Hokkaido Reg. Fish. Res. Lab., Fish. Agency No. 69:24-48; 49-65; 66-70; 71.

Pagurus middendorfii
Dermaturus mandti
Paralithodes camtschatica
P. brevipes
P. platypus
Pachycheles stevensii
Dorippe granulata

larvae, decapod
Kurata, Hiroshi, 1964.
Larvae of decapod Crustacea of Hokkaido. 4. Crangonidae and Glyphocrangonidae. (In Japanese; English abstract).
Bull. Hokkaido Reg. Fish. Res. Lab., (28):35-50.

Crangon affinin (species A-F)
Glyphocrangon sp.

larvae, decapods
Kurata, Hiroshi, 1964.
Larvae of decapod Crustacea of Hokkaido. 3. Pandalidae. (In Japanese; English abstract).
Bull. Hokkaido Reg. Fish. Res. Lab., (28):23-34.

Pandalus borealis
Pandalus hypsinotus
Pandalopsis coccinata

larvae, decapod
Kurata, H., 1963.
Larvae of Decapoda Crustacea of Hokkaido. 2. Majidae (Pisinae). (In Japanese; English abstract
Bull. Hokkaido Reg. Fish. Res. Lab., No. 27:25-31.

larvae, decapods
Kurata, H., 1963.
Larvae of Decapoda Crustacea of Hokkaido. 1. Atelecyclidae (Atelecyclinae). (In Japanese; English summary).
Bull. Hokkaido Reg. Fish. Res. Lab., No. 27:13-24

larvae, decapod
Kurata, Hiroshi, 1960.
Last stage zoea of Paralithodes with intermediate form between normal last stage zoea and glaucothoe.
Bull. Hokkaido Reg. Fish. Res. Lab., No. 22: 49-56.

larvae, decapod

Kurata, Hiroshi, 1960.
[Studies on the larva and post-larva of *Paralithodes camtschatica*. II. Feeding habits of the zoea. III. The influence of temperature and salinity on the survival and growth of the larva.]
Bull. Hokkaido Reg. Fish. Res. Lab., No. 21:1-8; 9-14. (English summaries)

larvae, decapod

Kurata, H., 1956.
The larval stages of *Paralithodes brevipes* (Decapoda, Anomura). Bull. Hokkaido Reg. Fish. Res. Lab., Fish. Agency, No. 14:1-24.

larvae, decapod

Kurata, H., 1955.
The post-embryonic development of the prawn *Pandalus kessleri*. Bull. Hokkaido Reg. Fish. Res. Lab. No. 12:1-15, 12 textfigs.

larvae, decapod

Kurata, Hiroshi and Harumi Omi, 1968.
The larval stages of a swimming crab, *Charybdis acuta*. (In Japanese; English abstract). Bull. Tokai reg. Fish Res. Lab., 57: 129-136.

larvae, decapod

Kurian, C.V., 1956.
Larvae of decapod Crustacea from the Adriatic Sea. Acta Adriatica, 6(3):108 pp.

larvae, decapod

Kutkuhn, J.H., H.L. Cook and K.N. Baxter 1969.
Distribution and density of prejuvenile *Penaeus* shrimp in Galveston entrance and the nearby Gulf of Mexico (Texas). FAO Fish. Rept. 3(57) (FRm/57.3 (Thm)): 1075-1099. (mimeographed)

larvae, decapod

Kuwatani, Yukimasa, Takuya Wakui and Takashi Nakanishi 1971.
Studies on the larvae and the post-larvae of a Tanner crab, *Chionoecetes opilio elongatus* Rathbun. I. On the protozoeal larvae. (In Japanese; English abstract). Bull. Hokkaido reg. Fish. Res. Lab. 37: 32-40.

larvae, decapod

Lazarus, B.I., 1967.
The occurrence of Phyllosomata of the Cape with special reference to *Jasus lalandii*. Investl. Rep. Div.Sea.Fish.Un.S. Afr. 63: 38 pp.

larvae, decapods

Lebour, Marie V., 1959.
The larval decapod Crustacea of Tropical West Africa. Atlantide Rept., Sci. Res., Danish Exped., Coasts of Tropical West Africa, 1945-1946, 5: 119-143.

larvae, decapod

LeBour, M.V., 1955.
First-stage larvae hatched from New Zealand decapod Crustacea. Ann. Mag. Nat. Hist. (12)8: 43-48.

larvae, decapods

Lebour, M.V., 1950.
Notes on some larval decapods from Bermuda. Proc. Zool. Soc., London, 120(2):369-379.

larvae

Lebour, M.V., 1947
The larval stages of *Portumnus* (Crustacea Brachyura) with notes on some other genera. JMBA 26(1):7-15, 5 text figs.

larvae, decapod

Lebour, M.V., 1954.
The planktonic decapod Crustacea and Stomatopoda of the Benguela Current. 1. First Survey, R.R.S. 'William Scoresby', March, 1950. Discovery Rept., 27:219-234.

larvae, decapod

Lebour, M.V., 1947
Notes on the inshore plankton of Plymouth. JMBA 26(4):527-547.

larvae - decapoda

Lebour, M.V., 1944
Larval crabs from Bermuda. Zoologica, 29(3):113-128, 19 figs.

larvae - decapoda

Lebour, M. V., 1944
The larval stages of *Portumnus* (Crustacea Brachyura) with notes on some other genera. JMBA XXVI(1):7-15.

decapod, larvae

Lebour, M. V., 1943
The larvae of the genus *Porcellana* (Crustacea Decapoda) and related forms. J.M.B.A., n.s., 25:721-737, 12 figs.

larvae, decapod

Lebour, M.V., 1940.
The larvae of the British species of *Spirontocaris* and their relation to *Thor* (Crustacea Decapoda). J.M.B.A. 24(2): 505-514.

larvae, decapod

Lebour, M. V., 1931
The larvae of the Plymouth Caridea. I. The larvae of the Crangonidae. II. The larvae of the Hippolytidae. Proc. Zool. Soc., London, 1931:1-9, pls. 1-3.

larvae

Lebour, M.V., 1928
The larval stages of the Plymouth Brachyura. Proc. Zool. Soc., London, 1928: 473-560, 5 text figs.

larvae decapod

Lee, Byung Don, and Sung Yun Hong 1970.
The larval development and growth of decapod crustaceans of Korean waters. II. *Pagurus similis* Ortmann (Paguridae, Anomura). Publ. mar. Lab. Pusan Fish. Coll. 3:13-26

larvae decapod

Lee, Byung Don, and Sung Yun Hong 1970.
The larval development and growth of decapod crustaceans of Korean waters. I. *Carcinoplax vestitus* (De Haan) (Goneplacidae, Brachyura). (Korean abstract) Publ. mar. Lab. Pusan Fish. Coll. 3: 1-11.

larvae decapod

Lee, Byung Don, and Taek Yuil Lee 1970.
Studies on the rearing of larvae and juveniles of *Metapenaeus joyneri* (Miers) under various feeding regimes. (In Korean; English abstract)
Publ. mar. Lab. Pusan Fish. Coll. 3: 27-35.

larvae, decapod

Lee, Byung D., and Taek Yuil Lee, 1969.
Studies on the larval development of *Metapenaeus joyneri* (Miers). Metamorphosis and growth. (In Korean; English abstract). Publ.mar.Lab.Pusan Fish.Coll. 2:19-25.

larvae, decapod

Lee, Byung Don, and Taek Yuil Lee, 1968.
Larval development of the penaeidean shrimp *Metapenaeus joyneri* (Miers). Publ. Haewundae mar Lab., 1:1-18.

larvae, decapod

Lee, Byung Don, and Taek Yuil Lee, 1968.
Larval development of the penaeidean shrimp *Metapenaeus joyneri* (Miers). Publ. mar.Lab., Pusan Fish Coll. 1:1-18.

larvae, decapod

Le Reste, Louis, 1971.
Rythme saisonnier de la reproduction, migration et croissance des postlarves et des jeunes chez la crevette *Penaeus indicus* H. Milne Edwards de la baie d'Ambaro, Côte N.O. de Madagascar. Cah. ORSTOM, sér. Océanogr. 9(3): 279-292.

larvae, decapod

Le Roux, Auguste 1966.
Contribution à l'étude du développement larvaire de *Clibanarius erythropus* (Latreille) (Crustacé Décapode Anomure Diogénidé). Cah. Biol. mar. Roscoff, 7(2): 225-230.

larvae, decapod

Le Roux, Auguste, 1966.
Le développement larvaire de *Porcellana longicornis* Pennant (Crustacé Décapode Anomoure Galathéidae). Cahiers. Biol. Mar. 7(1):69-78.

larvae, decapod

Le Roux, A., 1963.
Contribution à l'étude du développement larvaire d'*Hippolyte inermis* Leach (Crustacé décapode macroure). Comptes Rendus, Acad. Sci., Paris, 256(16):3499-3511.

larvae, decapod

Lewis, J.B., 1951.
The phyllosoma larvae of the spiny lobster, *Panulirus argus*. Bull. Mar. Sci., Gulf and Caribbean 1(2):89-103, 5 textfigs.

larvae, decapod

Lewis, John B., and Janet Ward, 1965.
Developmental stages of the palaemonid shrimp *Macrobrachium carcinus* (Linnaeus, 1758). Crustaceana, 9(2):137-148.

larvae, decapod

Ling, S.W., and A.B.O. Merican, 1962.
Notes on the life and habits of the adults and larval stages of *Macrobrachium rosenbergi* (DeMan) Indo-Pacific Fish. Council, FAO, Proc., 9th Sess., (2/3):55-61.

larvae, decapod

Little, Georgianna, 1969.
The larval development of the shrimp *Palaemon macrodactylus* Rathbun reared in the laboratory and the effect of eyestalk extirpation on development. Crustaceana, 17(1): 69-82

larvae, decapod

Loesch, Harold, y Quinto Avila 1966
Observaciones sobre la presencia de camarones juveniles en dos esteros de la costa del Ecuador. Bol. cient. tecn. Inst. nac. Pesca Ecuador 1(5): 30pp.

larvae, decapod

Longhurst, Alan, 1967.
The pelagic phase of *Pleuroncodes planipes* Stimpson (Crustacea, Galatheidae) in the California Current. Rep. Calif. Coop. Oceanic Fish.Invest., 11:142-154.

larvae, decapod

Lucas, J.S., 1972
The larval stages of some Australian species of *Halicarcinus* (Crustacea, Brachyura, Hymenosomatidae). II. Physiology. Bull. mar. Sci. 22(4): 824-840.

larvae, decapod

Lucas, J.S. 1971.
The larval stages of some Australian species of Halicarcinus (Crustacea, Brachyura, Hymenosomatidae). 1. Morphology. Bull. mar. Sci. Miami 21(2): 471-490.

larvae, decapod

Lucas, J.S., and E.P. Hodgkin, 1970.
Growth and reproduction of Halicarcinus australis (Haswell) (Crustacea, Brachyura) in the Swan Estuary, western Australia.
Aust. J. mar. Freshwat. Res. 21(2):163-173)
II. Larval stages

larvae, decapod

Lund, William A. Jr. and Lance L. Stewart 1970.
Abundance and distribution of larval lobsters, Homarus americanus, off the coast of southern New England.
1969 Proc. Nat. Shellfish. Ass. 60: 40-49.

larvae, decapod

Lyons, William G. 1970.
Scyllarid lobsters (Crustacea Decapoda).
Mem. Hourglass Cruises, Mar. Res. Lab., Fla. Dept. Nat. Res. 1(4): 1-74.

larvae, decapod

MacDonald, J.D., R.B. Pike and D.I. Williamson, 1957.
Larvae of the British species of Diogenes, Pagurus, Anapagurus and Lithodes (Crustacea: Decapoda). Proc. Zool. Soc., London, 128:209-257.

larvae, decapod

MacMillan, Floy E., 1972.
The larval development of northern California Porcellanidae (Decapoda, Anomura). I. Pachycheles pubescens Holmes in comparison to Pachycheles rudis Stimpson. Biol. Bull. mar. biol. Lab. Woods Hole 142(1): 57-70.

larvae (decapod)

Makarov, R.R. 1973.
Larval development of Notocrangon antarcticus (Decapoda, Crangonidae)
Zool. Zh. 52(8): 1149 -1155
(In Russian; English abstract)

larvae, decapod

Makarov, R.R., 1969.
Transport and distribution of decapod larvae in the plankton of the western Kamchatka Shelf. (In Russian; English abstract). Okeanologiia 9(2): 306-317.

decapod larvae

Marshall, N. B., 1948
Continuous plankton records: Zooplankton (other than Copepoda and young fish) in the North Sea 1938-1939. Hull Bull. Mar. Ecol. 2(13):173-213, Pls. 89-108.

larvae, decapods

Mayrat, A., 1959.
Remarques sur le développement de la crevette caramote, Penaeus kerathurus (Forskål).
Bull. Inst. Francais Afrique Noire (A), 21(2): 554-564.

larvae, decapod

Menon, M. K., 1949.
The larval stages of Periclimenes (Periclimenes) indicus Kemp. Proc. Indian Acad. Sci., Sect. B, 30(3):121-133.

larvae, decapod

Menon, P. Gopala 1972.
Decapod Crustacea from the International Indian Ocean Expedition: The larval development of Heterocarpus (Caridea).
J. Zool. Lond. 167(3): 371-397.

larvae, decapod

Menon, P. Gopala, and D.I. Williamson 1971.
Decapod Crustacea from The International Indian Ocean Expedition. The species of Thalassocaris (Caridea) and their larvae.
J. Zool. Lond. 165(1): 27-51.

larvae, decapod

Michel, A., 1971.
Note sur les puerulus de Palinuridae et les larves phyllosomes de Panulirus homarus (L). Clef de détermination des larves phyllosomes recoltées dans le Pacifique équatorial et sud-tropical (Décapodes). Cah. ORSTOM, sér. Océanogr. 9(4): 459-473.

Larvae, =decapod

Michel A. 1970.
Les larves phyllosomes du genre Palinurellus Von Martens (Crustacés Décapodes: Palinuridae).
Bull. Mus. natn. Hist. nat. (2)41(5):1228-1237.

larvae, decapod

Michel, A., 1968.
Les larves phyllosomes et la post-larvae de Scyllarides squamosus (H. Milne Edwards) - Scyllaridae (Crustacés Decapodes). Cah. ORSTOM. Océanogr., 6(3/4): 47-53.

Larvae, decapod

Miller, Paul Emanuel, and Harold G. Coffin, 1961.
A laboratory study of the developmental stages of Hapalogaster mertensii (Brandt), (Crustacea, Decapoda). Walla Walla College, Dept. Biol. Sci., and Biol. Sta., Publ., No. 30:18 pp.

larvae, decapod

Mito, S. 1972.
Investigations on the pursuit of artificially produced prawn larvae liberated into the sea.
Proc. Indo-Pacific Fish Counc. FAO 13(3): 215-223.

larvae, decapod

Modin, John C., and Keith W. Cox,1967.
Post-embryonic development of laboratory-reared ocean shrimp, Pandalus jordani Rathbun.
Crustaceana,13(2):197-218.

-larvae decapod-

Mohamed, K.H., P. Vedavyasa Rao and C. Suseelan 1971.
The first Phyllosoma stage of the Indian deep-sea spiny lobster, Puerulus sewelli Ramadan.
Proc. Indian Acad. Sci. (B) 74(4):208-215.

decapod larvae

Moisan, G., and J.-L. Tremblay, 1949.
Elevage des larves de homard. Année 1948. Sta. Biol., Saint-Laurent, 8th Rapport, App. 1:20-32, 5 textfigs.

larvae, decapod

Monod, Th., 1965.
Sur une mégalope de Raninidé (Crust. Brachyura)
Bull. Inst. Francais Afrique Noire. 27(4):1237-1244.

larvae, decapod

Morris, M.S., and I. Bennett, 1952.
The life history of a penaeid prawn (Metapenaeus) breeding in a coastal lake (Tuggerah, New South Wales). Proc. Linnean Soc., N.S.W. 76(5/6): 164-182, Pl. 12, 96 textfigs.

larvae, decapod

Mukai, Hiroshi, 1969.
Life histories of the shrimps in the Sargassum region. Bull. Biol. Soc. Hiroshima Univ., 35: 7-13. Also in: Contrib. Mukaishima Mar. Biol. Sta. 98

larvae, decapod

Munro, J.L., A.C. Jones and D. Dimitriou 1968.
Abundance and distribution of the larvae of the pink shrimp (Penaeus duorarum) on the Tortugas shelf of Florida, August 1962-October 1964.
Fish. Bull., Bur. Comm. Fish., U.S.F.W.S. 67(6): 165-181

larvae, decapod

Murano, Masaaki, 1967.
Preliminary notes on the ecological study of the phyllosoma larvae of the Japanese spiny lobster.
Inf. Bull.Planktol. Japan, Comm.No. Dr.Y.Matsue, 129-137.

larvae decapod

Nagabhushanam, R., and R. Sarojini 1968.
Chromatophore physiology of the zoea of the mud shrimp Upogebia affinis.
Broteria 37(3/4): 119-124.

larvae, decapod

Naylor, E. and M.J. Isaac 1972
Behavioural significance of pressure responses in megalopa larvae of Callinectes sapidus and Macropipus sp.
Mar. Behav. Physiol. 1(4): 341-350.

larvae, decapod

Needler, A.B., 1941.
Larval stages of Crago septemspinosus Say.
Trans. R. Canadian Inst., 23(50):193-199, 2 figs.

larvae, decapod

Nichols, Paul R., and Peggy M. Keney, 1963.
Crab larvae (Callinectes), in plankton collections from cruise of M/V Theodore N. Gill, South Atlantic coast of the United States, 1953-54.
U.S. Fish and Wildlife Service, Spec. Sci. Repts. Fish., No. 448:1-14.

Callinectes, probably several species.

marginatus
sapidus   } probably
ornatus
danae

larvae, decapod

Nonaka, Tadashi, Yasuo Ohshima and Reijiro Hirano, 1959.
On the culture and ecdysis of the Phyllosoma in the spiny lobster, Panulirus japonicus.
The Aquiculture 5(3):13-15.
Resume in: Rec. Res., Fac. Agric., Univ. Tokyo, No. VIII (1957-1958), p. 54.

larvae, decapod

Nyblade, Carl F., 1970.
Larval development of Pagurus annulipes (Stimpson, 1862) and Pagurus pollicaris Say, 1817 reared in the laboratory. Biol. Bull., 139:557-573.

decapod larvae

Ogilvie, H.S., 1934
A preliminary account of the food of the herring in the North-Western North Sea.
Rapp. Proc. Verb.89(3):85-92, 2 text figs.

larvae, decapod

Oka, Masao, 1968.
Studies on Penaeus orientalis Kishinouye. 9. Development of the post-larva. (In Japanese; English abstract).
Bull. Fac. Fish., Nagasaki Univ., 26: 1-23.

Oceanographic Index: Marine Organisms Cumulation, 1946-1973

larvae, decapod
Oke, Masao, 1967.
Studies on Penaeus orientalis Kishinouye. III. Structure of ovary and mechanism of ovulation. IV. Physiological mechanism of ovulation. V. Fertilization and development. VI. Some influences on metemorphosis, growth and feeding inclination. (In Japanese; English abstract).
Bull. Fac. Fish., Nagasaki Univ., 23: 43-56; 57-69; 70-87; 89-100.

larva, decapod
Omori, M., 1971.
Preliminary rearing experiments on the larvae of Sergestes lucens (Penaedia, Natantia, Decapoda). Mar. Biol. 9(3): 228-234.

larvae, decapod
Pearson, J.C., 1939.
The early life histories of some American Penaeidae, chiefly the commercial shrimp, Penaeus setiferus (Linn.). Bull. Bur. Fish. No. 30:73 pp., 67 textfigs.

larvae, decapod
Perkins, Herbert C. 1973.
The larval stages of the deep sea red crab, Geryon quinquedens Smith, reared under laboratory conditions (Decapoda: Brachyrhyncha).
Fish. Bull. U.S. Dept. Comm. 71(1): 69-82.

larvae, decapod
Pike, R.B., 1954.
Notes on the growth and biology of the prawn Spirontocaris lilljeborgii (Danielssen).
J.M.B.A. 33:739-747, 3 textfigs.

larvae, decapod
Pike, R.B., and D.I. Williamson, 1966.
The first zoeal stage of Campylonotus rathbunae Schmitt and its bearing on the systematic position of the Campylonotidae (Decapoda, Cerides).
Trans. R. Soc. New Zealand, Zool., 7(16):209-213.

larvae, decapod
Pandian, T.J. and S. Katre, 1972.
Effect of hatching time on larval mortality and survival of the prawn Macrobrachium idae.
Mar. Biol. 13(4): 330-337.

larvae, decapod
Rautsch, Fryderyk 1967.
Pigmentation and colour change in decapod larvae.
Proc. Symp. Crustacea, Ernakulam, Jan. 12-15, 1965, 3: 1108-1123.

larvae, decapods
Perkins, Herbert C. 1972.
Developmental rates at various temperatures of embryos of the northern lobster (Homarus americanus Milne-Edwards).
Fish. Bull. U.S. Nat. Fish. Serv. 70(1): 95-99.

larvae, decapods
Pike, R.B., 1961.
Larval variation in Philocheras bispinosus (Hailstone) (Decapoda, Crangonidae).
Crustaciana, 2(1):21-25.

larvae, decapod
Pike, Richard B., and Robert G. Wear 1969
Newly hatched larvae of the genera Gastroptychus and Uroptychus (Crustacea, Decapoda, Galatheidea) from New Zealand waters.
Trans. R. Soc. N.Z., Biol. Sci. 11(13): 189-195.

larvae, decapods
Pike, R.B., and D.I. Williamson, 1964.
The larvae of some species of Pandalidae (Decapoda).
Crustaceana, 6(4):265-284.

Pandalina brevirostris
Dichelopandalus bonnieri
Pandalus propinquus
Pandalus montagui

larvae, decapod
Pike, R.B., and D.I. Williamson, 1961.
The larvae of Spirontocaris and related genera (Decapoda, Hippolytidae).
Crustaceana, 2(3):187-208.

LARVAE, decapod
Pike, R.B., and D.I. Williamson, 1960
Larvae of decapod Crustacea of the Families Diogenidae and Paguridae from the Bay of Naples.
Pubbl. Staz. Zool., Napoli, 30(1/3):493-552.

larvae, decapod
Pike, R.B., and D.I. Williamson, 1960
Larvae of decapod Crustacea of the Families Dromiidae and Homolidae from the Bay of Naples.
Pubbl. Staz. Zool., Napoli, 30(1/3):553-563.

larvae, decapod
Pike, R.B., and D.I. Williamson, 1958(1959)
Crustacea, Decapoda: Larvae. XI. Paguridea, Coenobitidea, Dromiidea and Homolidea.
Fiches d'Ident., Cons. Perm. Int. Expl. Mer, Sheet 81: 9 pp.

larvae, decapod
Poole, Richard L., 1966.
A description of laboratory-reared zoeae of Cancer magister Dana, and megalopae taken under natural conditions (Decapoda Brachyura).
Crustaceana, 11(6): 83-97.

larvae, decapod
Porter, Hugh J., 1960.
Zoeal stages of the stone crab, Menippe mercenaria Sat.
Chesapeake Science, 1(3/4):168-177.

larvae, decapods
Powell, B.L., 1962.
Types, distribution and rhythmical behavior of the chromatophores of juvenile Carcinus maenas (L.)
J. Animal Ecology, 31(2):251-261.

larvae, decapod
Price, Vincent A., and Kenneth K. Chew 1972.
Laboratory rearing of spot shrimp larvae (Pandalus platyceros) and description of stages.
J. Fish. Res. Bd. Can. 29(4): 413-422.

larvae, decapod
Provenzano, Anthony J., Jr., 1971.
Biological results of the University of Miami deep-sea expeditions. 74. Rediscovery of Munidopagurus macrocheles (A. Milne-Edwards, 1880) (Crustacea, Decapoda, Paguridae), with a description of the first zoeal stage. Bull. mar. Sci. 21(1): 256-266.

larvae, decapod
Provenzano, Anthony J., Jr., 1971.
Biological results of the University of Miami deep-sea expeditions. 73. Zoeal development of Pylopaguropsis atlantica Wass, 1963, and evidence from larval characters of some generic relationships within the Paguridae. Bull. mar. Sci. 21(1): 237-255.

larvae, decapod
Provenzano, Anthony J., Jr., 1968.
Lithopagurus yucatanicus, a new genus and species of hermit crab with a distinctive larva.
Bull. mar. Sci., Miami, 18(3):627-644.

larvae, decapod
Provenzano, Anthony J., 1968.
The complete larval development of the West Indian hermit crab Petrochirus diogenes (L.) (Decapoda, Diogenidae) reared in the laboratory. Bull. mar. Sci., Miami, 18(1): 143-181.

larvae, decapod
Provenzano, Anthony J., Jr. 1967.
Recent advances in the laboratory culture of decapod larvae.
Proc. Symp. Crustacea, Ernakulam, Jan. 12-15, 1965, 2: 940-945.

larvae, decapod
Provenzano, Anthony J., Jr., 1967.
The zoeal stages and glaucothoe of the tropical eastern Pacific hermit crab, Trizopagurus magnificus (Bouvier, 1898) (Decapoda; Diogenidae) reared in the laboratory.
Pacif. Sci., 21(4):457-473.

larvae, decapod
Provenzano, A.J., Jr., 1963.
The glaucothoe stage of Dardanus venosus (H. Milne Edwards) (Decapoda: Anomura).
Bull. Mar. Sci., Gulf and Caribbean, 13(1):11-22.

larvae, decapods
Provenzano, A.J., Jr., 1963.
The glaucothoes of Petrochirus diogenes (L.) and two species of Dardanus (Decapoda: Diogenidae)
Bull. Mar. Sci. Gulf and Caribbean, 13(2):242-261.

larvae, decapods
Provenzano, Anthony J., 1962
The larval development of Calcinus tibicen Herbst (Crustacea, Anomura) in the laboratory.
Biol. Bull., 123(1):179-202.

larvae, decapod
Provenzano, Anthony J., Jr., and Anthony L. Rice, 1966.
Juvenile morphology and the development of taxonomic characters in Paguristes sericeus A. Milne Edwards (Decapoda, Diogenidae).
Crustaceana, 10(1):53-69.

larvae, decapod
Provenzano, Anthony J., Jr., and Anthony L. Rice, 1964.
The larval stages of Pagurus marshi Benedict (Decapoda, Anomura) reared in the laboratory.
Crustaceana, 7(3):217-235.

larvae, decapod
Raghu Prasad, R., and P.R.S. Tampi 1965.
A preliminary report on the phyllosomas of the Indian Ocean collected by the Dana Expedition 1928-30.
J. mar. biol. Ass. India 7(2): 277-283.

phyllosoma larva
Raghu Prasad, R., and P.R.S. Tampi, 1960
On the newly hatched phyllosoma of Scyllarus sordidus (Stimpson).
J. Mar. Biol. Assoc., India, 2(2):250-252.

larvae, decapod
Raghu Prasad, R., and P.R.S. Tampi, 1960.
Phyllosomas of scyllarid lobsters from the Arabian Sea.
J. Mar. Biol. Assoc., India, 2(2):241-249.

larvae, decapod
Raja Bai, K.G., 1960.
Studies on the larval development of Brachyura. II. Development of Philyra scabriuscula (Fabricius) and Ixa cylindrus (Fabricius) of the family Leucosiidae.
Crustaceana, 1(1):1-8.

larvae, decapods,
Raja Bai Naidu, K.G., 1955.
The early development of Scylla serrata (Forsk.) de Haan and Neptunus sanguinolentus (Herbst.).
Indian J. Fish., 2(1):67-76.

larvae, decapod
Raja Bai Naidu, K.G., 1954.
The post-larval development of the shore crab, Ocypoda platytarsis M. Edwards and Ocypoda cordimana Desmarest.
Proc. Indian Acad. Sci., Sect. B, 40(4):89-101, 30 textfigs.

larvae, decapods
Rajyalakshmi, K., 1960.
Observations on the embryonic and larval development of some estuarine palaemonid prawns.
Proc. N. I. Sci., India, B, 26(6):395-408.

larvae, decapod
Ramirez Granados, Rodolfo, 1963
Langostas (Crustacea Decapoda), Identificacion, distribucion, comercio.
Trabajos de Divulgacion, Sec. Indust. y Comercio, Dir. Gen. de Pesca, 4(46):39 pp. (mimeographed).

larvae, decapods
Rao, P. Vedavyasa, 1969.
On the identification of juveniles of three species of Metapenaeus (Decapoda, Penaeidae).
Ind. J. Fish. 16 (1/2): 51-55.

larvae, decapod
Rao, P. Vedavyasa 1968.
A new species of shrimp, Acetes cochinensis (Crustacea: Decapoda, Sergestidae) from southwest coast of India with an account of its larval development.
J. mar. biol. Ass. India 10(2): 298-320.

larvae, decapod
Rao, R. Mallikarjuna, and V. Gopalakrishnan 1970.
Identification of juveniles of the prawns Penaeus monodon Fabricius and P. indicus H.M. Edwards.
Proc. Indo-Pacific Fish. Counc. F.A.O. 13(2): 128-131.

larvae, decapod
Reed, Paul H. 1969.
Culture methods and effects of temperature and salinity on survival and growth of Dungeness crab (Cancer magister) larvae in the laboratory.
J. Fish. Res. Bd Can. 26(2): 389-397.

larvae, decapod
Rees, C.B., 1955.
Continuous plankton records: the decapod larvae in the North Sea, 1950-51.
Bull. Mar. Ecol. 4(29):69-80, Pls. 13-14.

larvae, decapod
Rees, C.B., 1952.
Continuous plankton records: the decapod larvae in the North Sea, 1947-1949. Hull Bull. Mar. Ecol. 3(22):157-184, Pls. 17-19.

larvae, decapod
Rees, G.H., 1963.
Progress on blue crab research in the South Atlantic.
Proc. 15th Ann. Sess., Gulf and Caribbean Fish. Inst., Univ. Miami, 110-115.

larvae, decapod
Rees, George H., 1959.
Larval development of the sand crab, Emerita talpoida (Say) in the laboratory.
Biol. Bull., 117(2):356-370.

larvae, decapod
Reeve, M.R. 1969.
The laboratory culture of the prawn Palaemon serratus.
Fish. Invest. Min. Agric. Fish. Food (2) 26 (1): 1-38.

larvae decapod
Reeve, M.R., 1969.
Growth, metamorphosis and energy conversion in the larvae of the prawn, Palaemon serratus.
J. mar. biol. Ass., U.K. 49(1):77-96.

larvae, decapod
Regnault, Michèle 1972.
Développement de l'estomac chez les larves de Crangon septemspinosa Say (Crustacea, Decapoda, Crangonidae); son influence sur le mode de nutrition.
Bull. Mus. natn. Hist. nat. Paris (3) 67 (Zool. 53):841-856.

larvae, decapods
Regnault, M., 1971.
Acides aminés libres chez les larves de Crangon septemspinosa (Caridea). Variation de leur taux de l'éclosion à la métamorphose. Leur rôle au cours du développement et leur importance dans la nutrition. Mar. Biol. 11(1): 35-44.

larvae, decapod
Regnault, Michèle 1970 (1971).
Croissance au laboratoire de Crangon septemspinosa Say (Crustacea Decapoda, Natantia), et la metamorphose à la maturité sexuelle.
Bull. Mus. nat. Hist. nat. (2) 42 (5): 1108-1126.

larvae, decapod
Regnault, Michèle 1969.
Recherche du mode de nutrition d'Hippolyte inermis Leach (Decapoda, Carides) au début de sa vie larvaire: structure et rôle des pièces buccales.
Crustaceana 17(3): 253-264.

larvae, decapod
Regnault, Michèle, 1969.
Influence de la température et de l'origine de l'eau de mer sur le développement larvaire au laboratoire d'Hippolyte inermis Leach (Decapode-Natantia).
Vie Milieu 20(1A): 137-152.

larvae, decapod
Regnault, Michele, 1969.
Etude expérimentale de la nutrition d'Hippolyte inermis Leach (Decapode-Natantia) au cours de son développement larvaire, au laboratoire.
Int. Revue ges. Hydrobiol., 54(5): 749-764.
Also in: Trav. Sta. zool. Villefranche, 30 (1969).

larvae, decapod
Renfro, William C., and Harry L. Cook, 1963.
Early larval stages of the seabob, Xiphopeneus kroyeri (Heller).
U.S. Fish and Wildlife Service, Fish. Bull., 63(1):165-177.

larvae, decapod
Rice, Anthony Leonard 1970.
Decapod Crustacean larvae collected during the International Indian Ocean Expedition. Families Raninidae and Homolidae.
Bull. Brit. Mus. (N.H.) Zool. 21(1): 1-24.

larvae, decapod
Rice, A.L., R.W. Ingle and Elizabeth Allen 1970.
The larval development of the sponge crab, Dromia personata (L.) (Crustacea, Decapoda, Dromiidea) reared in the laboratory.
Vie Milieu (A) 21(1-A): 223-246.

larvae, decapod
Rice, A.L. and Anthony J. Provenzano, Jr., 1970.
Biological results of the University of Miami deep-sea expeditions. 55. The larval stages of Homola barbata (Fabricius) (Crustacea, Decapoda, Homolidae) reared in the laboratory.
Bull. mar. Sci., 20(2): 446-471.

larvae, decapod
Rice, Anthony L., and Anthony J. Provenzano, Jr., 1965.
The zoeal stages and glaucothoe of Paguristes sericeus A. Milne-Edwards (Anomura, Diogenidae).
Crustaceana, 8(3):239-254.

larvae, decapod
Ritz, D.A., 1972.
Factors affecting the distribution of rock-lobster larvae (Panulirus longipes cygnus) with reference to variability of plankton-net catches. Mar. Biol. 13(4): 309-317.

larvae, decapod
Ritz, D.A., 1972
Behavioural response to light of the newly hatched phyllosoma larvae of Panulirus longipes cygnus George (Crustacea: Decapoda: Palinuridae). J. exp. mar. Biol. Ecol. 10(2): 105-114.

larvae, decapod
Ritz, D.A., and L.R. Thomas 1973.
The larval and postlarval stages of Ibacus peronii Leach (Decapoda, Reptantia, Scyllaridae).
Crustaceana 24 (1): 5-16.

larvae, decapod
Roberts, Morris H., Jr. 1973.
Larval development of Pagurus acadianus Benedict, 1901, reared in the laboratory (Decapoda, Anomura).
Crustaceana 24 (3): 303-317.

larvae decapod
Roberts, Morris H., Jr., 1971.
Larval development of Pagurus longicarpus Say reared in the laboratory. III. Behavioral responses to salinity discontinuities. Biol. Bull. mar. biol. Lab. Woods Hole, 140(3): 489-501.

larvae, decapod
Roberts, Morris H., Jr., 1971.
Larval development of Pagurus longicarpus Say reared in the laboratory. IV. Aspects of the ecology of the megalopa. Biol. Bull. mar. biol. Lab. Woods Hole 141(1): 162-166.

larvae, decapod
Roberts, Morris H., Jr., 1971.
Larval development of Pagurus longicarpus Say reared in the laboratory. II. Effects of reduced salinity on larval development. Biol. Bull. mar. biol. Lab. Woods Hole, 140(1): 104-116.

larvae, decapod
Roberts, Morris H., Jr., 1970.
Larval development of Pagurus longicarpus Say reared in the laboratory. I. Description of larval instars. Biol. Bull. mar. biol. Lab., Woods Hole, 139(1): 188-202.

larvae, decapod
Roberts, Morris H., Jr., 1969.
Larval development of Bathynectes superba (Costa) reared in the laboratory. Biol. Bull., mar. biol. Lab., Woods Hole, 137(2): 338-351.

larvae, decapod
Roberts, P.E. 1972.
Larvae of *Porcellanopagurus edwardsi* Filhol, 1885 (Crustacea: Decapoda, Paguridae) from Perseverance Harbour, Campbell Island.
Jl R. Soc. N.Z. 2(3): 383-391.

larvae, decapod
Roberts, P.E., 1971.
Zoea larvae of *Pagurus campbelli* Filhol 1885, from Perseverance Harbour, Campbell Island (Decapoda: Paguridae).
J. R. Soc. N.Z. 1(3/4): 187-196.

larvae, decapod
Robertson, Philip B. 1972.
A unique scyllarid phyllosoma larva from the Straits of Florida (Decapoda, Palinurdea).
Crustaceana 22(3): 309-312.

larvae, decapod
Robertson, P., 1969.
Biological investigations of the deep sea. No. 48. Phyllosoma larvae of a scyllarid lobster, *Arctides guineensis*, from the western Atlantic. Marine Biol., 4(2): 143-151.

larvae, decapod
Robertson, Philip B., 1969.
The early larval development of the scyllarid lobster *Scyllarides aequinoctialis* (Lund) in the laboratory, with a revision of the larval characters of the genus. Deep-Sea Res 16(6): 557-586.

larvae, decapod (Jasus)
Robertson, Philip B. 1969.
Rock lobster *Jasus*: similarity of first phyllosoma larva to that of certain scyllarid lobsters (Decapoda, Palinuridea).
Crustaceana 17(3): 311-314.

larvae, decapod
Robertson, Philip B., 1969.
Biological investigations of the deep sea. 49. Phyllosoma larvae of a palinurid lobster, *Justitia longimana* (H. Milne Edwards), from the western Atlantic. Bull. mar. Sci., 19(4): 922-944.

larvae, decapod
Robertson, Philip B., 1968.
The complete larval development of the sand lobster, *Scyllarus americanus* (Smith), (Decapoda, Scyllaridae) in the laboratory, with notes on larvae from the plankton.
Bull. mar. Sci., Miami, 18(2): 294-342.

larvae, decapod
Roessler, M.A., A.C. Jones and J.L. Munro, 1969.
Larval and postlarval pink shrimp *Penaeus duorarum* in South Florida.
FAO Fish. Repts. 3(57) (FRm/57.3 (Trm)): 859-866. (mimeographed)

larvae, decapod
Rogers, Bruce A., J. Stanley Cobb and Nelson Marshall 1967.
Size comparisons of inshore and offshore larvae of the lobster, *Homarus americanus*, off southern New England.
Proc. nat. Shellfish. Ass. 58: 78-81.

decapod larvae
Russell, F.S. and J.S. Colman, 1934
The Zooplankton. II. The Composition of the Zooplankton of the Barrier Reef Lagoon.
Brit. Mus. (Nat. Hist.), Great Barrier Reef Expedition, 1928-1929, Sci. Repts. II(6):159-176, 186-201, 11 text figs., tables 2-10.

larvae, decapod
Saint Laurent-Dechance, Michele, 1964.
Développement et position systématique du genre Parapagurus Smith (Crustacea Decapoda Paguridae). 1. Description des stades larvaires. Bull. Inst. Oceanogr., Monaco, 64(1321):26 pp.

larvae, decapod
Saisho, Toshio, 1966.
Studies on the phyllosoma larvae with reference to oceanographical conditions. (In Japanese; English abstract).
Mem. Fac. Fish., Kagoshima Univ., 15:177-239.

larvae, decapod
Saisho, Toshio, 1966.
A note on the phyllosoma stages of spiny lobster. (In Japanese; English abstract).
Inf. Bull. Planktol. Japan. No.13:69-71.

larvae, decapod
Saisho, Toshio, 1964.
Notes on the first stage phyllosoma of Scyllarid lobster, *Scyllarus bicuspidatus*.
Mem. Fac. Fish., Kagoshima Univ., 13:1-4.

larvae, decapod
Saisho, Toshio, 1964.
The first phyllosoma of the spiny lobster, *Panulirus*. (In Japanese; English abstract).
Mem. Fac. Fish., Kagoshima Univ., 12(2):127-134.

larvae, decapod
Saisho, Tosio, 1962
Notes on the early development of a scyllarid lobster, *Parribacus antarcticus* (Lund).
Memoirs. Fac. Fish., Kagoshima Univ., 11(2): 174-178.

larvae, decapod
Saisho, Toshio, 1962
Notes on the early development of phyllosoma of *Panulirus japonicus*. (In Japanese; English abstract).
Mem., Fac. Fish., Kagoshima Univ., 11(1):18-23.

larvae, decapods
Saisho, T., and K. Nakahara, 1960.
On the early development of phyllosomas of *Ibacus ciliatus* (von Siebold) and *Panulirus longipes* (A, Milne Edwards).
Mem. Fac. Fish., Kogoshima Univ., 9:84-90.

larvae, decapod
Saisho, Toshio and Motonori Sone, 1971.
Notes on the early development of a scyllarid lobster, *Scyllarides squamosus* (H. Milne-Edwards). (In Japanese; English abstract).
Mem. Kagoshima Univ. Fac. Fish. 20(1): 191-196.

larvae, decapod
Sakai, Katsushi, 1971.
The larval stages of *Ranina ranina* (Linnaeus) (Crustacea, Decapoda, Raninidae) reared in the laboratory, with a review of uncertain zoeal larvae attributed to *Ranina*. Publ. Seto mar. biol. Lab. 19(2/3): 123-156.

larvae, decapod
Samuelsen, Tor J. 1972.
Larvae of *Munidopsis tridentata* (Esmark) (Decapoda, Anomura) reared in the laboratory.
Sarsia 48: 91-98.

larvae, decapod
Samuelsen, Tor J., 1972.
Larvae of *Pagurus variabilis* Milne-Edwards & Bouvier (Decapoda, Anomura) reared in the laboratory.
Sarsia 48: 1-11.

larvae, decapod
Sandifer, Paul A. 1973.
Effects of temperature and salinity on larval development of grass shrimp, *Palaemonetes vulgaris* (Decapoda, Carides).
Fish. Bull. U.S. Dept. Comm. 71(1): 115-123.

larvae, decapod
Sandifer, Paul A. 1973.
Mud shrimp (*Callianassa*) larvae (Crustacea, Decapoda, Callianassidae) from Virginia plankton.
Chesapeake Sci. 14(3): 149-159.

larvae, decapod
Sandifer, Paul A. 1973.
Larvae of the burrowing shrimp, *Upogebia affinis*, (Crustacea, Decapoda, Upogebiidae) from Virginia plankton.
Chesapeake Sci. 14(2): 98-104.

larvae, decapod
Sandifer, Paul A. 1971.
The first two phyllosomas of the sand lobster, *Scyllarus depressus* (Smith) (Decapoda, Scyllaridae).
J. Elisha Mitchell scient. Soc. 87(4): 183-187.

larvae, decapod
Sandifer, Paul A. and Willard A. van Engel 1972
Larval stages of the spider crab, *Anasimus latus* Rathbun 1894 (Brachyura, Majidae, Inachinae) obtained in the laboratory.
Crustaceana 23(2): 141-151.

larvae, decapod
Sandoz, M., and S.H. Hopkins, 1944.
Zoeal larvae of the blue crab *Callinectes sapidus*. J. Wash. Acad. Sci., 34:132-133.

larvae, decapod
Sankarankutty, C. 1963.
On three species of porcellanids (Crustacea, Anomura) from the Gulf of Manaar.
J. mar. biol. Ass. India 5(2): 273-279.

larvae, decapod
Sankarankutty, C. 1958.
Larvae of an unrecorded pagurid (Crustacea Paguridea) from western Norway.
Sarsia 31: 57-62.

larvae, decapod
Sankolli, K.N., 1967.
Studies on larval development in Anomura (Crustacea, Decapoda)-1.
Proc. Symp. Crustacea, Ernakulam, Jan.12-5, 1965, 2: 744-776.

larvae, decapod
Sankolli, K.N. 1961.
On the early larval stages of two leucosiid crabs, *Philyra corallicola* Alcock and *Arcania septemspinosa* (Fabricius).
J. Mar. Biol. Assoc., India, 3(1/2):87-91.

larvae, decapod
Sankolli, K.N., and H.G. Kewalramani, 1962.
Larval development of Saron marmoratus (Olivier) in the laboratory.
J. Mar. Biol. Assoc., India, 4(1):106-120.

larvae, decapod
Sankolli, K.N., and Shakuntala S. Shenoy 1972.
On the occurrence of the hippolytid prawn, Angasia armata (Paulson) (Decapoda, Crustacea) in Bombay waters, its cannibalistic behaviour and its larvae.
J. Bombay Nat. Hist. Soc. 69(2):369-377.

larvae, decapod
Sankolli, K.N., and Shakuntala Shenoy, 1967(1968).
Larval development of a dromiid crab, Conchoecetes artificiosus (Fabr.) (Decapoda, Crustacea) in the laboratory.
J. mar. biol. Ass. India, 9(1):96-110.

Larvae, decapod
Sars, G.O., 1900.
Account of the postembryonal development of Pandalus borealis Krøyer with remarks on the development of other Pandali and description of the adult of Pandalus borealis.
Rept. Norwegian Fish. Mar. Invest. 1(3):45 pp., 10 pls.

larvae, decapod
Sarojini, R., and R. Nagabushanam 1968.
Larval development of Diogenes bicristimanus in the laboratory.
J. mar. biol. Ass. India 10(1):71-77.

larvae, decapod
Sars, G. O., 1874.
Om Hummerens postembryonale Udvikling.
Forhandl. Vidensk. Selsk. i Christiana for 1874: 27 pp., 2 pls. (fold-in).

larvae, decapod
Sastry, A.N., and J.F. McCarthy 1973.
Diversity in metabolic adaptation of pelagic larval stages of two sympatric species of brachyuran crabs.
Netherl. J. Sea Res. 7: 434-446.

larvae, decapod
Sato, S., 1958.
Studies on the larval development and fishery biology of the king crab, Paralithodes camtschatica (Tilesius). Bull. Hokkaido Reg. Fish. Res. Lab., Fish. Agency, No. 17:1-102, 10 pls.

larvae decapod
Scarratt, D.J., 1968.
Distribution of lobster larvae (Homarus americanus) of Pictou, Nova Scotia.
J. Fish. Res. Bd., Can. 25(2):427-430.

larvae, decapod
Scarratt, D.J., 1964.
Abundance and distribution of lobster larvae (Homarus americana) in Northumberland Strait.
J. Fish. Res. Bd., Canada, 21(4):661-680.

larvae decapod
Scarratt, D.J., and G.E. Reine 1966.
Avoidance of low salinity by newly hatched lobster larvae.
J. Fish. Res. Bd Can. 24(6):1403-1406.

larvae, decapod
Scattergood, L. W., 1949.
A bibliography of lobster culture. U. S. Fish and Wildlife Service, Spec. Sci. Rept. 64:24pp.

larvae, decapod
Scelzo, Marcelo, y Enrique E. Boschi, 1969.
Desarrollo larval del cangrejo ermitaño Pagurus exilis (Benedict) en laboratorio (Crustacea Anomura Paguridae).
Physis 29(78):165-184

larvae, decapod, lists of spp.
Seguin, Gerard, 1966.
Note sur la répartition annuelle des larves de Crustacés Décapodes des eaux neritiques de Dakar (Sénégal).
Bull. I.F.A.N., (A), 28(2):576-582.

larvae, decapod
Serène, R., 1961.
A megalopa commensal in a squid.
Proc. Ninth Pacific Sci. Congr., Pacific Sci. Assoc., 1957, Fish., 10:35-36.

larvae, decapod
Seridji R., 1971.
Contribution à l'étude des larves Crustacés décapodes en baie d'Alger.
Pelagos, Bull. Inst. océanogr. Alger 3(2):1-105

larvae, decapod, lists of spp.
Seridji, Ratiba 1968.
Note préliminaire sur la répartition saisonnière des larves de crustacés décapodes en Baie d'Alger.
Pelagos 10:91-108.

larvae, decapod
Shenoy, Shakuntala, 1967.
Studies on larval development in Anomura (Crustacea Decapoda) - II.
Proc. Symp. Crustacea, Ernakulam, Jan. 12-15, 1965, 2: 777-804

larvae, decapod
Shenoy, Shakuntala and K.N. Sankolli, 1967.
Studies on larval development in Anomura (Crustacea, Decapoda) - III.
Proc. Symp. Crustacea, Ernakulam, Jan. 12-15, 1965, 2:805-814.

larvae, decapod
Sherman, Kenneth, and Robert D. Lewis, 1967.
Seasonal occurrence of larval lobsters in coastal waters of central Maine.
Proc. natn. Shellfish. Ass., 57:27-30.

larvae, decapod
Shield, Pamela 1973.
The chromatophores of Emerita talpoida (Say) zoeae considered as a diagnostic character.
Chesapeake Sci. 14(1): 41-47

larvae, decapod
Sick, Lowell V., 1970.
Larval distribution of commercially important Penaeidae in North Carolina.
J. Elisha Mitchell scient. Soc. 86(3): 118-127.

larvae, decapod
Silberbauer, B.I., 1970.
The biology of the South African rock lobster Jasus lalandii (H. Milne Edwards). 1. Development.
Invest. Rept. Div. Sea Fish. SAfr. 92: 70pp.

larvae, decapod
Sims, Harold W., Jr. 1966.
Notes on the newly hatched phyllosoma of the sand lobster Scyllarus americanus (Smith).
Crustaceana, 11(3):288-290.

larvae, decapod
Sims, Harold W., Jr. 1966.
The phyllosoma larvae of the spiny lobster Palinurellus gundlachii Von Martens (Decapoda Palinuridae).
Crustaceana, 11(2):205-215.

larvae, decapod
Sims, Harold W., Jr. 1965.
The phyllosoma larvae of Parribacus.
Q.J. Florida Acad. Sci. 28(2):142-172.

larvae, decapod
Sims, Harold W., Jr. 1965.
Notes on the occurrence of prenaupliosoma larvae of spiny lobsters in the plankton.
Bull. Mar. Sci. 15(1):223-227.

larvae, decapod
Sims, Harold W., Jr. 1964.
Four giant scyllarid phyllosoma larvae from the Florida Straits with notes on smaller specimens.
Crustaceana, 7(4):259-266.

larvae, decapods
Sin, Ong Keh, 1966.
The early developmental stages of Scylla serrata Forskol (Crustacea Portunidae, reared in the laboratory.
Indo-Pacific Fish. Counc. Proc. 11th Sess., 135-146.

LARVAE-decapod
Sollaud, E., 1923.
Le développement larvaire des "Palaemoninae".
Bull. Biol. de la France et de la Belgique, Paris, LVII:509-603.

larvae, decapod
Squires, H.J., 1965.
Larvae and megalopa of Argis dentata (Crustacea: Decapoda) from Ungava Bay.
J. Fish. Res. Bd., Canada, 22(1):69-82.

larvae (decapod)
Stephensen, K., 1923
Decapoda-Macrura excl. Sergestidae. Rep. Danish Oceanogr. Exped. Medit. 2 Biol., D.3: 1-85, 27 text figs., 8 charts.

larvae, spirontocari
Stephensen, K., 1917
Zoogeographical investigations of certain fjords in southern Greenland with special reference to Crustacea, Pycnogonida and Echinodermata including a list of Alcyonaria and Pisces, Medd. om Grønland, 53(3):229-378.

larvae, decapod
Subrahmanyam, Chebium B. 1971.
The relative abundance and distribution of penaeid shrimp larvae of the coast of Mississippi.
Gulf Res. Repts 3(2): 291-345

larvae, decapod
Subrahmanyam, C.B. 1971.
Description of shrimp larvae (family Penaeidae) off the Mississippi coast.
Gulf Res. Repts 3(2):241-258

larvae, decapod
Subrahmanyam, C.B., and Gordon Gunter 1970.
New penaeid shrimp larvae from the Gulf of Mexico (Decapoda Penaeidae).
Crustaceana 19(1): 94-98.

larvae, decapod
Subrahmanyam, M., and K. Janardhana Rao 1970.
Observations on the postlarval prawns (Penaeidae) in the Pulicat Lake with notes on their utilization in capture and culture fisheries.
Proc. Indo-Pacific Fish. Counc. FAO 13(2): 113-119.

larvae, decapod
Sukô, T., 1958.
Studies on the development of the crayfish. VI. The reproductive cycle.
Sci. Repts., Saitama Univ., (B), 3(1):77-92.

larvae, decapod
Sukô, T., 1958.
Studies on the development of the crayfish. V. The histological changes of the developmental ovaries influenced by the condition of darkness.
Sci. Repts., Saitama Univ. (B), 3(1):67-78.

larvae, decapod
Sutcliffe, W.H., Jr., 1957.
Observations on the growth rate of the immature Bermuda spiny lobster, Panulirus argus. Ecology 38(3):526-529.

larvae, decapod
Tabeta, O. and S. Kanamaru, 1970.
On the post larva of Munida gregaria (Crustacea Galatheida) in Penas Bay, Chile, with reference to mass occurrence in 1969. Sci. Bull. Fac. Agr. Kyushu Univ. 24(4): 227-230. Also in: Contrib. Dept. Fish Fish. Res. Lab., Kyushu Univ. 16(1970). (In Japanese; English summary).

larvae, decapod
Tagatz, Marlin E., 1968.
Growth of juvenile blue crabs, Callinectes sapidus Rathbun, in the St. Johns River, Florida.
Fish. Bull. U.S. Dept. Comm. 67(2): 281-288

larvae decapod
Takeuchi, Isamu, 1967.
On the distribution of decapod Anomura larvae off the west coast of the Kamtchatka Peninsula in 1962. (In Japanese; English abstract).
Bull. Hokkaido reg. Fish. Res. Lab. 33:64-71.

larvae, decapod
Tanase, Hidetomo, 1967.
Preliminary notes on zoea and megalopa of the giant spider crab, Macrocheira Kaempferi de Haan. Publ. Seto mar. biol. Lab. 15(4): 303-309.

larvae, decapod
Temple, Robert F., and Clarence C. Fischer, 1968.
Seasonal distribution and relative abundance of planktonic-stage shrimp (Penaeus spp.) in the northwestern Gulf of Mexico, 1961.
Fishery Bull. Fish Wildl. Serv. U.S. 66(2):323-334.

larvae, decapod
Temple, Robert F., and Clarence C. Fischer, 1965.
Vertical distribution of the planktonic stages of penaeid shrimp.
Publ. Inst. Mar. Sci., Port Aransas, 10:59-67.

larvae, decapod
Templeman, W., 1948.
Body form and stage identification in the early stages of the American lobster. Bull. Newfoundland Gov't Lab. No. 18:12-25, 5 textfigs.

larvae, decapod
Templeman, W., and S. N. Tibbo, 1945.
Lobster investigations in Newfoundland 1938 to 1941. Res. Bull. (Fisheries), No. 16: 98 pp., 20 textfigs.

larvae, decapod
Terao, A., 1919.
On the development of Panulirus japonicus (v. Siebold). J. Tokyo Univ., Fish., 14(5):1-7, 4 pls.

larvae, decapod
Tesmer, Charlsie Ann, and A.C. Broad, 1964.
The larval development of Crangon septemspinosa (Say).
Ohio J. Sci., 64(4):239-250.

larvae, decapods
Thiriot, Alain 1973.
Stades larvaires de Parthenopidae méditerranéens: Heterocrypta maltzani Miers et Parthenope massena (H. Milne-Edwards).
Cah. Biol. mar. 14(2): 111-134

larvae, decapod
Tokioka, T., 1954.
Droplets from the plankton net, 13 &14.
Publ. Seto Mar. Biol. Lab., 3(3):359-368.

larvae, decapod
Trask Thomas 1970.
A description of laboratory-reared larvae of Cancer productus Randal (Decapoda, Brachyura) and a comparison to larvae of Cancer magister Dana.
Crustaceana 18(2): 133-146.

larvae, decapod
Tsurnamal, M., 1963.
Larval development of the prawn Palaemon elegans Rathke (Crustacea, Decapoda) from the coast of Israel.
Israel J. Zool., 12(1-4):117-141.

larvae, decapod
Uno, Yutaka, and Kwon chin Soo, 1969.
Larval development of Macrobrachium rosenbergii (DeMan) reared in the laboratory.
J. Tokyo Univ. Fish. 55(2):179-190

larvae, decapod
Vameuchi, Koji, 1965.
Hatching and rearing of Crago affinis (de haen), and on the utilization of the larvae esfond. (In Japanese; English abstract).
Bull. Jap. Soc., Sci. Fish., 31(11):907-915.

larvae, decapod
van der Baan, S. M., L.B. Holthuis and B. Schrieken 1972.
Decapoda and decapod larvae in the surface plankton from the southern North Sea near Texel lightship.
Bijdr. Faun. Nederland 2(13): 75-97 (Zool. Bijdr.).

larvae, decapod
Van Wormhoudt, A., 1973
Variation des protéases, des amylases et des protéines solubles au cours du développement larvaire chez Palaemon serratus. Mar. Biol. 19(3): 245-248.

larvae decapod
Villaluz, D.K., Antonio Villaluz, Bienvenido Ladrera, Madid Sheik and Alejandro Gonzaga 1969 (1972).
Reproduction, larval development, and cultivation of sugpo (Penaeus monodon Fabricius).
Philippine J. Sci. 98(3/4): 205-234.

larvae, decapod
Wear, Robert G. 1970.
Some larval stages of Petalomera wilsoni (Fulton + Grant, 1902) (Decapoda, Dromiidae).
Crustaceana 18(1):1-12.

larvae, decapod
*Wear, Robert G., 1968.
Life-history studies on New Zealand Brachyura. 3. Family Ocypodidae. First stage zoea larva of Hemiplax hirtipes (Jacquinot, 1853).
N.Z. Jl. mar. Freshwat. Res. 2(4):698-702.

larvae, decapod
Wear, Robert G., 1967.
Life history studies of New Zealand Brachyra. 1. Embryonic and post-embryonic development of Pilumnus novaezealandiae Filhol, 1886, and of P. lumpinus Bennett, 1964 (Xanthidae, pilumninae).
N.Z. Jl mar. Freshwat. Res. 1(4):482-535.

larvae, decapod
Wear, Robert G., 1966.
Pre-zoea larva of Petrocheles spinosus Miers, 1876 (Crustacea, Decapoda, Anomura).
Trans. R. Soc. N.Z. Zool., 8(10):119-124.

larvae, decapod
Wear, R.G., 1965.
Zooplankton of Wellington Harbour, New Zealand.
Zoology Publ., Victoria Univ., Wellington, No. 38: 31 pp.

larvae, decapod
Wear, Robert G., and J.C. Yaldwyn, 1966.
Studies on Thalassinid Crustacea (Decapoda, Macrura Reptantia) with a description of a new Jaxea from New Zealand and an account of its larval development.
Zoology Publs. Vict. Univ. Coll., No. 41:27 pp.

larvae, decapod
Wellershaus, Stefan 1972.
Larval development of an unknown crab (Brachyura Decapoda) in the Cochin Backwater (South India).
Veröff. Inst. Meeresforsch. Bremerh. 13(2): 275-284.

larvae, decapoda
Welsh, J.H., 1932.
Temperature and light as factors influencing the rate of swimming of larvae of the mussel crab, Pinnotheres maculatus Say. Biol. Bull. 63:310-325, 6 textfigs.

larvae, decapod
Weymouth, F.W., M.J. Lindner, and W.W. Anderson, 1933.
Preliminary report on the life history of the common shrimp Penaeus setiferus (Linn.). Bull. U.S.B.F., No. 14:1-26, 11 figs.

larvae, decapod
Whitney, J. O'C. 1969.
Absence of sterol synthesis in larvae of the mud crab Rhithropanopeus harrisii and of the spider crab Libinia emarginata. Marine Biol., 3(2): 134-195.

larvae, decapod
Wickens, J.F., 1972
The food value of brine shrimp, Artemia salina L. to larvae of the prawn, Palaemon serratus Pennant. J. exp. mar. Biol. Ecol. 10(2): 151-170.

larvae, decapod
Williams, Austin B. 1972.
A ten-year study of meroplankton in North Carolina estuaries: juvenile and adult Ogyrides (Caridea: Ogyrididae).
Chesapeake Sci. 13(2): 145-159.

larvae, decapod
Williams, Austin B., 1971
A ten-year study of meroplankton in North Carolina estuaries: annual occurrence of some brachyuran developmental stages.
Chesapeake Sci., 12(2): 53-61

larvae, decapod
Williams, Austin B., 1959
Spotted and brown shrimp post larvae (Penaeus) in North Carolina. Bull. Mar. Sci., Gulf and Caribbean, 9(3): 281-290.

larvae decapod
Williams, Barbara G., 1968.
Laboratory rearing of the larval stages of Carcinus maenas (L.)(Crustacea:Decapoda).
J. nat. Hist., 2(1):121-126.

larvae, decapod
Williamson, D.I., 1969.
Names of larvae in the Decapoda and Euphausiacea.
Crustaceana, 16(2): 210-213.

larvae, decapod
Williamson, D.I. 1967.
Crustacea Decapoda: larvae. IV Carides, families Pandalidae and Alpheidae.
Fiches d'Identification, Cons. perm. int. Expl. Mer, Zooplankton 109:

larvae, decapod
Williamson, D.I., 1967.
The megalopa stage of the homolid crab Latreillia australiensis Henderson and comments on other homolid megalopas.
Aust. Zool., 14(2): 206-211.

larvae, decapod
Williamson, D.I., 1967.
On a collection of planktonic Decapoda and Stomatopoda(Crustacea) from the Mediterranean coast of Israel.
Bull. Sea Fish.Res. Stn Israel,45:32-64.

larvae, decapod
Williamson, D.I., 1965.
Some larval stages of three Australian crabs belonging to the families Homolidae and Raninidae, and observations on the affinities of these families(Crustacea, Decapoda).
Australian J. Mar. Freshwater Res.,16(3):369-398.

larvae, decapod
Williamson, D.I., 1962.
Crustacea, Decapoda: larvae. III. Caridea, Families Oplophoridae, Nematocarcinidae and Pasiphaeidae.
Fiches d'Ident., Cons. Perm. Int. Expl. Mer, Zooplankton, Sheet, 92:5 pp.

larvae, decapod
Williamson, D.I., 1960.
A remarkable zoea, attributed to the Majidae (Decapoda, Brachyura).
Ann. Mag. Nat. Hist., (13), 3:141-144.

Not sure just what!
Very spiny!

larvae, decapod
Williamson, D.I., 1960.
Crustacea, Decapoda: Larvae. VII. Caridea, Family Crangonidae. Stenopodidea.
Fiches d'Identification, Cons. Perm. Int. Expl. Mer, Sheet 90: 5 pp.

larvae, decapod
Williamson, D.I., 1960.
Larval stages of Pasiphaea sivado and some other Pasiphaeidae (Decapoda).
Crustaceana, 1(4):331-341.

larvae, decapod
Williamson, D.I., and K.G.von Levetzow, 1967.
Larvae of Paragalus diogenes (Whitelegge) and some related species (Decapoda, Anomura).
Crustaceana, 12(2):179-192.

larvae, decapod
Winstanley, R.H., 1970.
Rock lobster larvae in the Tasman Sea
Tasman. Fish. Res. 4(1): 11-12

larvae, decapod
Woodburn, Kenneth D., Bonnie Eldred, Eugenie Clark, Robert F. Hutton and Robert M. Ingle, 1957.
The live bait industry of the west coast of Florida (Cedar Key to Naples).
Florida State Bd., Conserv. Mar. Lab., Techn. Ser. No. 21:33 pp.

Also in:
Collected Papers, Cape Haze Mar. Lab., Sarasota, 1957-1963, Vol. 1.

Penaeus setiferus

larvae, decapod
Woodmansee, Robert A., 1966.
Daily vertical migration of Lucifer (Decapoda, Sergestidae). Egg development, oviposition and hatching.
Int. Revue ges. Hydrobiol., 51(5):689-698.

larvae, decapod
Yang, Won Tack, 1971.
The larval and postlarval development of Parthenope serrata reared in the laboratory and the systematic position of the Parthenopinae (Crustacea, Brachyura). Biol. Bull. mar. biol. Lab., Woods Hole, 140(1): 166-189.

larvae, decapods
Zein-Eldin, Zoula P., and George W. Griffith, 1969.
An appraisal of the effects of salinity and temperature on growth and survival of postlarval penaeids.
FAO Fish. Repts. 3 (57)(FRm/57.3 (Trn)): 1015-1026 (mimeographed).

larvae, decapod
Zein-Eldin, Zoula P., and George W. Griffith, 1966.
The effect of temperature upon the growth of laboratory-held postlarval Penaeus aztecus.
Biol. Bull., 131(1):186-196.

larvae, euphausids
Bary, B.M., 1956.
Notes on ecology, systematics and development of some Mysidacea and Euphausiacea (Crustacea) from New Zealand. Pacific Science 10(4):431-467.

larvae, euphausids
Boden, Brian P., 1961.
Euphausiacea (Crustacea) from Tropical West Africa.
Atlantide Rept., Sci. Res., Danish Exped., Coasts of Tropical West Africa, 1945-1946, 6: 251-262

larvae, euphausids
Boden, B.P., 1955.
Euphausiacea of the Benguela Current. First survey, R.R.S. "William Scoresby", March 1950.
Discovery Repts., 27:337-376.

Nyctiphanes capensis
Euphausia lucens
E. tenera
Nematoscelis megalops

larvae, euphausids
Boden, B.P., 1951.
The egg and larval stages of Nyctiphanes simplex, a euphausid crustacean from California
Proc. Zool. Soc., London, 121(3):515-527, 5 figs

larvae, euphausid
Boden, B.P., 1950.
The post-nauplius stages of the crustacean Euphausia pacifica. Trans. Am. Microsc. Soc. 69(4):373-386, 3 pls.

larvae, euphausids
Casanova-Soulier, Bernadette, 1968.
Clé de détermination des larves furcilie des euphausiacés de la Méditerranée.
Rapp. Proc.-verb. Réun. Comm. int. Explor. scient. Mer Méditerranée 19(3): 527-529

larvae, euphausiid
Casanova-Soulier, Bernadette, 1968.
Une série larvaire dans le genre Nematoscelis (Euphausiacés).
Cah.Biol.mar., 9(1):1-12.

larvae, euphausids
Einarsson, H., 1945.
Euphausiacea. 1. Northern Atlantic species.
"Dana" Rept. No. 27:191 pp., 84 textfigs.

larvae, euphausid
Fraser, F.C., 1936.
On the development and distribution of the young stages of krill (Euphausia superba). Discovery Rept. 14:1-192, 76 textfigs.

larvae, euphausiids
Gopalakrishnan, K. 1973.
Developmental and growth studies of the euphausiid Nematoscelis difficilis (Crustacea) based on rearing.
Bull. Scripps. Inst. Oceanogr. 20:87pp.

larvae, euphausiids
Gopalakrishnan, Kakkala 1972.
A note on developmental and growth studies of the euphausiid Nematoscelis difficilis (Crustacea) based on rearing.
Mahasagar, Bull. Nat. Inst. Oceanogr. India 5(1): 31-35.

larvae, euphausids
Gurney, R., 1947.
Some notes on the development of the Euphausiacea. Proc. Zool. Soc., London, 117(1):49-64, 8 textfigs.

larval stages
Heegaard, P., 1948
Larval stages of Meganyctiphanes (Euphausiacea) and some general phylogenetic remarks. Medd. Komm. Dan. Fisk og Havundersøgelser, Ser. Plankton, 5:27 pp., 4 pls.

larvae - Thysanopoda
Illig, G., 1930
Die Schizopoden der Deutschen Tiefsee-Expedition. Wiss. Ergeb. Deutschen Tiefsee Exped., "Valdivia" 1898-1899: 22(6):399-625, 215 text figs.

larvae, euphausid
John, D.D., 1936.
The southern species of the genus Euphausia.
Discovery Rept. 14:193-324, 40 textfigs.

larvae euphausiid
Jones, Lester T. 1968.
Occurrence of the larvae of Meganyctiphanes norvegica (Crustacea, Euphausiacea) off West Greenland.
J. Fish. Res. Bd. Can. 25(5): 1071-1073.

larvae, euphausiids
Knight, Margaret D. 1971.
The nauplius II, metanauplius and calyptopsis stages of Thysanopoda tricuspidata Milne-Edwards (Euphausiacea).
Fish. Bull. U.S. Dept. Comm. 71 (1): 53-67.

larvae, euphausiid
Komaki, Yuzo, 1967.
On the early metamorphosis of Nematoscelis difficilis Hansen (Euphausiacea, Crustacea).
Inf.Bull.Planktol.Japan,Comm.No.Dr.Y.Matsue. 101-108.

Larvae, euphausids
Lacroix, Guy, 1961.
Les migrations verticales journalieres des euphausides a l'entree de la Baie de Chaleurs.
Contrib. Depart. Pecheries, Quebec, No. 83:257-316.

larvae, euphausid
Lebour, M.V., 1950.
Some euphausids from Bermuda. Proc. Zool. Soc., London, 119(4):823-837, 7 textfigs.

larvae, euphausids
Lewis, J.B., 1955.
Some larval euphausids of the genus Stylocheiron from the Florida Current.
Bull. Mar. Sci., Gulf and Caribbean 5(3):190-202

larvae, euphausid
MacDonald, R., 1927.
Irregular development in the larval history of Meganyctiphanes norvegica. J.M.B.A., n.s., 14(3): 785-795.

larvae, euphausid

Mathew, K.J. 1971.
Studies on the larval stages of Euphausiacea from the Indian Seas.
1. Diagnostic characters of post-naupliar stages of Euphausia diomedeae Ortmann and E. distinguenda Hansen.
J. mar. biol. Ass. India 13(1): 52-60

larvae, euphausids

Mauchline, J., 1965.
The larval development of the euphausiid, Thysanoessa raschii (M. Sars).
Crustaceana, 9(1):31-40.

larvae, euphausid

Mauchline, J., 1959.
The development of the Euphausiacea (Crustacea) especially that of Meganyctiphanes norvegica (M. Sars)
Proc. Zool. Soc. Lon., 132(4):627-639.

larvae, euphausiids

Mauchline, John, and the late Leonard R. Fisher, 1969
The biology of euphausiids
Adv. mar. Biol., J.S. Russell and Maurice Yonge, editors, Academic Press, 7:454pp. $17.50

larval stages

Ruud, J.T., 1932.
On the biology of southern Euphausiidae.
Hvalrådets Skrifter No. 2:1-105, 37 text figs.

larvae, euphausids

Sheard, K., 1953.
Taxonomy, distribution and development of the Euphausiacea (Crustacea).
B.A.N.Z. Antarctic Res. Exped., 1929-31, Repts., B, 8(1):1-72, 17 textfigs.

larvae, euphausiid

Soulier, Bernadette, 1965.
Essai d'harmonisation de la nomenclature des larves d'euphausiacés.
Rev. Trav. Inst. Pêches Marit., 29(2):191-195.

larvae, euphausid

Wear, R.G., 1965
Zooplankton of Wellington Harbour, New Zealand.
Zoology Publ., Victoria Univ., Wellington, No. 38 :31 pp.

larvae, euphausiid

Williamson, D.I., 1969.
Names of larvae in the Decapoda and Euphausiacea.
Crustaceana, 16(2): 210-213.

larvae, isopod

Davis, Charles C., 1964.
A study of the hatching process in aquatic invertebrates. IX. Hatching within the brood sac of the ovoviviparous isopod, Cirolana sp. (Isopoda.Cirolanidae).
Pacific Science, 18(4):378-381.

larvae, isopod

Hansen, H.J., 1895
Isopoden, Cumaceen und Stomatopoden der Plankton Expedition. Ergeb. Plankton Exped. II, Gc: 105 pp., 8 pls.

larvae, isopod

Kjennerud, J., 1950.
Ecological observations on Idothea neglecta G.O. Sars. Univ. Bergen Årbok, 1950, Naturvitenskapelig rekke No. 7:47 pp., 15 textfigs.

larvae, isopod

Lemercier, Annie, 1957.
Sur le developpement in vitro des embryons d'un Crustacé Isopode Asellote: Jaera marina (Fabr.) C. R. Acad. Sci., Paris, 244:1280-1283.

larvae, isopod

Menzies, R.J., and R.J. Waidzunas, 1948.
Postembryonic growth changes in the isopod Pentidotea resicata (Simpson) with remarks on their taxonomic significance. Biol. Bull. 95(1):107-113, 20 figs.

larvae, isopod

Naylor, E., 1955.
The life cycle of the isopod Idotea emarginata (Fabricius). J. An. Ecol., 24(2):270-281.

larvae, isopod

Nishimura, Saburo, 1968.
A larval gnathiid from Seto, Japan (Crustacea, Isopoda).
Publ. Seto mar. biol. Lab., 16(1):7-9.

larvae, isopods

Stoll, C., 1962.
Cycle evolutif de Paragnathia formica (Hesse) (Isopode-Gnathiidae).
Cahiers Biol. Mar., Roscoff, 3(4):401-415.

larvae, isopods

Strömberg, Jarl-ove, 1972.
Cyathura polita (Crustacea, Isopoda), some embryological notes. Bull. mar. Sci. Miami 22(2): 463-482.

larvae, isopod

Strömberg, Jarl-Ove, 1971.
Contribution to the embryology of bopyrid isopods with special reference to Bopyroides, Hemiarthrus, and Pseudione (Isopoda, Epicaridea).
Sarsia 47:1-46.

larvae, isopod

Strömberg, J.-O., 1965.
On the embryology of the isopod Idotea.
Arkiv för Zoologi, n.s., 17(5):421-473.

larvae, isopods

Tattersall, W.M., 1905
The Marine Fauna of the Coast of Ireland. Part V Isopoda. Dept. Agric. and Tech. Instr. for Ireland, Fisheries Branch, Sci. Invest. 1904. No.II:1-90, pls.1-11, 1 chart.

larvae, isopod

Tsikhon-Lukanina, E.A. and T.A. Lukasheva, 1970.
On food energy transformation by some marine isopod juveniles. (In Russian; English abstract)
Okeanologiia, 10(4): 709-713.

larvae, mysids

Davis, Charles C., 1966.
A study of the hatching process in aquatic invertebrates. XXII. Multiple membrane shedding in Mysidium columbiae (Zimmer) (Crustacea. Mysidacea). Bull. Mar. Sci. 16(1): 124-131.

larvae, mysid

Matsudaira, C., T. Kariya and T. Tsuda, 1952.
On the biology of a mysid, Gastrosaccus vulgaris Nakazawa. Tohoku J. Agricult. Res. 3(1):155-174, 11 textfigs.

larvae, ostracods

Angel M.V. 1970.
Bathyconchoecia subrufa n. sp. and B. septemspinosa n.sp., two new halocyprids (Ostracoda, Myodocopida) from the tropical North Atlantic and the description of the larval development of B. subrufa.
Crustaceana, 19(2): 181-199.

larvae, ostracod

Elofson, O. 1941
Zur Kenntnis der marinen Ostracoden Schwedens mit besonderer Berücksichtigung des Skageraks. Zool. Bidrag. Uppsala 19: 215-534; 42 maps, 52 figs.

larvae, ostracods

Leveau, Michel, 1965.
Contribution à l'étude des ostracodes et cladocères du Golfe de Marseille.
Rec. Trav. Stn. mar. Endoume 37 (53): 161-246.

larvae, ostracod

Poulsen, Erik M., 1962
Ostracoda-Myodocopa. Part 1. Cypridiniformes-Cypridinidae.
Dana-Report No. 57:414 pp.

larvae, ostracod

Theisen, Bent Friis, 1967.
The life history of seven species of ostracods from a Danish brackish-water locality.
Meddelelser Danmarks Fiskeri- og Havundersøgelser, n.s. 4(8):215-

larvae, pycnogonid

Gnanamuthu, C.P., 1950.
Notes on the morphology and development of a pycnogonid, Propallene kempi (Calman), from Madras plankton. Proc. Zool. Soc. Bengal, 3(1):39-47, 4 textfigs.

larvae, pycnogonids

King, P.E. and J.H. Jarvis, 1970.
Egg development in a littoral pycnogonid Nymphon gracile. Mar. Biol., 7(4): 294-304.

LARVAE, PYCNOGONID

Stock, J.H., 1957.
The pycnogonid family Austrodecidae.
Beaufortia, 6(68):1-81.

larvae, stomatopod

Alikunhi, K.H., 1967.
An account of the post-larval development, moulting and growth of the common stomatopods of the Madras coast.
Proc. Symp. Crustacea, Ernakulam, Jan. 12-15, 1965. 2: 824-939

larvae, stomatopod

Alikunhi, K.H., 1951.
An account of the stomatopod larvae of the Madras plankton. Rec. Indian Mus. 49(3/4):239-319, 25 textfigs.

larvae, stomatopod

Brooks, W.K., 1879.
The larval stages of Squilla empusa Say.
Chesapeake Zool. Lab. Sci. Repts. for 1878: 143-170, Pls. 9-13.

larvae, stomatopod

Couch, Jonathan, 1858.
Note on the occurrence of Phyllosoma commune on the coast of Cornwall.
J. Lin. Soc., London, Zool. 2:146-149.

Stomatopod larvae

Foxon, G.E.H., 1939
Stomatopod larvae. John Murray Exped., 1933-34, Sci. Repts. 6(6):251-266, 4 text figs.

larvae - stomatopod

Foxon, G.E.H., 1932.
Report on Stomatopod Larvae, Cumacea, and Cladocera. Gt. Barrier Reef Exped., 1928-1929. Sci. Rept. IV(11):375-398.

larvae, stomatopod

Hansen, H.J., 1895
Isopoden, Cumaceen und Stomatopoden der Plankton Expedition. Ergeb. Plankton Exped. II, Gc: 105 pp., 8 pls.

larvae, stomatopod

Ingle, R.W., and N. della Croce, 1967.
Stomatopod larvae from the Mozambique Channel.
Boll. Musei Ist. biol. Univ. Genova, 35(226):55-70.

Oceanographic Index: Marine Organisms Cumulation, 1946-1973

larvae, stomatopod

Jacques, Françoise, 1969.
Histogenèse des pédoncules oculaires des larves des Stomatopodes.
Vie Milieu (A) 20(3): 565-593.

larvae, stomatopod

Jacques, Françoise, 1969.
Existence d'un organe sensoriel dans le pédoncule oculaire des larves de stomatopodes (Crustacés).
C.r.hebd.Séanc,Acad.Sci.Paris,268(1):89-90.

larvae, stomatopod

Jacques,Françoise,1968.
Note complémentaire sur les larves de stomatopodes présentes a Banyuls-sur-Mer.
Vie Milieu 19(1-A):209-210.

larvae, stomatopod

Jacques, Françoise, et Alain Thiriot 1967.
Larves de Stomatopodes du plancton de la région de Banyuls-sur-Mer.
Vie Milieu (B) 18(2B): 367-380.

larvae, stomatopod

Johnson, M.W., 1951.
A giant phyllosoma larva of a loricate crustacean from the tropical Pacific. Trans. Amer. Microsc. Soc. 70(3):274-277, 2 pls.

larvae, stomatopod

Koma, T., 1932.
An enormous swarm of stomatopod larvae. Annot. Zool., Jap., 13:351-354, Fig. A.

larvae, stomatopod

Lebour, M.V., 1954.
The planktonic decapod Crustacea and Stomatopoda of the Benguela Current. 1. First survey, R.R.S. 'William Scoresby', March, 1950. Discovery Repts 27:219-234.

larvae, stomatopod

Manning, Raymond B. 1969.
The postlarvae and juvenile stages of two species of Pseudosquilla (Crustacea, Stomatopoda) from the eastern Pacific region.
Proc. biol. Soc. Wash. 82: 525-538.

larvae, stomatopod

Manning, Raymond B., 1963.
Notes on the embryology of the stomatopod crustacean Gonodactylus oerstedii Hansen.
Bull. Mar. Sci., Gulf and Caribbean, 13(3):422-432.

larvae, stomatopod

Manning, R.B., 1962.
Alima hyalina Leach, the pelagic larva of the stomatopod crustacean, Squilla alba Bigelow.
Bull. Mar. Sci., Gulf and Caribbean, 12(3):496-504.

larvae, stomatopod

Manning, Raymond B., and Anthony J. Provenzano, Jr., 1963.
Studies on development of stomatopod Crustacea. 1. Early larval stages of Gonodactylus oerstedii Hansen.
Bull. Mar. Sci., Gulf and Caribbean, 13(3):467-487.

stomatopod larvae

Marshall, N. B., 1948
Continuous plankton records: Zooplankton (other than Copepoda and young fish) in the North Sea 1938-1939. Hull Bull. Mar. Ecol. 2(13):173-213, Pls. 89-108.

larvae, stomatopod

Michel, A., 1970.
Larves pélagiques et post-larves du genre Lysiosquilla (Crustacés Stomatopodes) dans le Pacifique tropical sud et équatorial. Cah. O.R. S.T.O.M. sér. Océanogr. 8(3):53-75.

larvae, stomatopod

Michel, A., 1970.
Larves pélagiques et post-larves du genre Odontodactylus (Crustaces: Stomatopodes) dans le Pacifique tropical sud et équatorial.
Cah. ORSTOM sér. Océanogr., 8(2): 111-126.

larvae, stomatopod

Michel, A. 1969.
Dernier stade larvaire pélagique et post-larve de Heterosquilla (Heterosquilloides) brazieri (Miers, 1880) (Crustacés, Stomatopodes).
Bull. Mus. natn. Hist. Nat. Paris (2) 40(5): 992-997.

larvae, stomatopod

Michel,A.,1968.
Dérive des larves de stomatopodes de l'est de l'Océan Indien.
Cah.O.R.S.T.O.M., ser. Océanogr., VI(1):13-41.

Larvae, stomatopod

Serène, R., 1954.
Observations biologiques sur les stomatopodes. Mém. Inst. Océan., Nhatrang, No. 8:1-93, 10 pls., 15 textfigs.

larvae, stomatopod

Shiino, S.M., 1942.
Studies on the embryology of Squilla oratoria de Haan. Mem. Coll. Sci., Kyoto Imp. Univ., Ser. B, 17(1):77-174, figs.

larvae, stomatopod

Tokioka, Takasi, and Eiji Harada, 1963
Further notes on Phyllosoma utivaebi Tokioka. Publ. Seto Mar. Biol. Lab., Kyoto Univ., 11(2): 425-434. (255-264).

larvae, stomatopod

Townsley, S.J., 1953.
Adult and larval stomatopod crustaceans occurring in Hawaiian waters. Pacific Science 7(4):399-437, 28 textfigs.

larvae, stomatopod

Van der Baan, S.M., and L.B. Holthuis 1969
Second note on the occurrence of stomatopod larvae in the North Sea near the lightship Texel.
Neth. J. Sea Res. 4(3): 350-353.

Larvae, stomatopod

van der Baan, S. M., and L. B. Holthuis, 1966.
On the occurrence of Stomatopoda in the North Sea, with special reference to larvae from the surface plankton near the lightship "Texel".
Netherlands J. Sea Res., 3(1):1-12.

Larvae, stomatopod

Verwey, J., 1966.
The origin of the stomatopod larvae of the southern North Sea.
Netherlands J. Sea Res., 3(1):13-20.

larvae, stomatopod

Wear, R.G., 1965.
Zooplankton of Wellington Harbour, New Zealand. Zoology Publ., Victoria Univ., Wellington, No. 38 :31 pp.

larvae, tanaids

Sieg, Jürgen 1972.
Untersuchungen über Tanaidaceen
1. Bemerkungen über die post-marsupiale Entwicklung der Tanaidaceen.
Kieler Meeresf. 28(2): 232-236

larvae, echinoderm

Agassiz, A., 1864.
1. On the embryology of echinoderms. Mem. Am. Acad. Arts and Sci., n.s., 9:30 pp., 31 figs. (white on black).

larvae, echinoderm

Atwood, David G., 1973
Larval development in the asteroid Echinaster echinophorus. Biol. Bull. mar. biol. Lab, Woods Hole 144(1): 1-11.

larvae, echinoderm

Bougis, Paul, 1964.
Sur le développement des plutéus in vitro et l'interprétation du test de Wilson.
C.R. Acad. Sci., Paris, 259(5):1250-1253.

larvae, echinoderm

Chia, Fu-Shiang, 1968.
The embryology of a brooding starfish, Leptasteria hexactis (Stimpson).
Acta Zoologica, 49(3):321-364.

larvae, echinoderms

Chia, Fu-Shiang and J.B. Buchanan,1969.
Larval development of Cucumaria elongata (Echinodermata:Holothuroidea).
J.mar.biol.Ass., U.K., 49(1):151-159.

larvae echinoderms

Dix, Trevor G., 1969.
Larval life span of the echinoid Evechinus chloroticus (Val.).
N.Z. Jl mar. Freshwat. Res. 3(1): 13-16

larvae, echinoderm

Devaney, Dennis M., 1973
Zoogeography and faunal composition of south-eastern Polynesian asterozoan echinoderms. In: Oceanography of the South Pacific 1972, Ronald Fraser, Compiler, N.Z. Nat. Commn. UNESCO, 357-366.

larvae, echinoderm

Fenaux, Lucienne, 1970.
Maturation of the gonads and seasonal cycle of the planktonic larvae of the Ophiuroid Amphiura chiajei Forbes. Biol. Bull. mar. biol. Lab., Woods Hole, 138(3): 262-271.

larvae, echinoderm

Fenaux, Lucienne, 1969.
Le développement larvaire chez Ophioderma longicauda (Retzius).
Cah.Biol.mar., 10(1):59-62.

larvae, echinoderm

Fenaux, Lucienne, 1969.
Les échinoplutéus de la Méditerranée.
Bull. Inst. Océanogr. Monaco, 68 (1394): 28 pp.

larvae, echinoderm

Geiger, S.R., 1964.
Echinodermata: larvae, classes: Ophiuroidea and Echinoidea (plutei).
Fiches d'Ident., Zooplancton, Cons. Perm. Int. Expl. Mer, 105:5 pp.

larvae, sea urchins

Greenhouse, Gerald A., Richard O Hynes and Paul R. Gross, 1971.
Sea urchin embryos are permeable to actinomycin. Science, 171 (3972): 686-689.

larvae echinoderm

Hsaio, Sidney C., 1965.
Kinetic studies on alkaline phosphatase from echinoplutei.
Limnol. and Oceanogr., Redfield Vol., Suppl. to 10:R129-R136.

larvae, echinoderms

Pearse, J.S., 1969.
Slow developing demersal embryos and larvae of the antarctic sea star Odontaster validus.
Marine Biol. 3(2): 110-116.

larvae, echinoderm

Rees, C.b., 1954.
Continuous plankton records: the distribution of echinoderm and other larvae in the North Sea, 1947-51. Bull. Mar. Ecol. 4(28):47-67, Pls. 11-12, 12 textfigs.

includes identification of larvae.

larvae, echinoderm

Schoener, Amy, 1969.
Atlantic ophiuroids: some post-larval forms.
Deep-Sea Res., 16(2): 127-140.

larvae, echinoderms

Schoener, Amy, 1967.
Post-larval development of five deep-sea ophiuroids.
Deep-Sea Res., 14(6):645-660.

larvae, brittle star

Semenova, T.N., S.A. Mileikovsky, and K.N. Nesis, 1964.
The morphology, distribution and seasonal occurrence of the brittle star larvae Ophiooten sericeum (Forbes) s.l. in the plankton of the northwest Atlantic, Norwegian and Barents seas. (In Russian).
Okeanologiia, Akad. Nauk, SSSR, 4(4):669-683.

larvae, fish

Aboussouan, A. 1969.
Sur une petite collection de larves de téléostéens récoltée au large de Brésil (campagne "Calypso" 1962).
Vie Milieu (A) 20 (3): 595-610.

larvae, fish

Aboussouan A. 1965.
Oeufs et larves de téléostéens de l'ouest africain. II. Distribution verticale.
Bull. Inst. français Afrique Noire 27(4): 1504-1521.

larvae, fish

Anraku, Masateru, and Masanori Azeta 1967.
The feeding activities of yellowtail larvae Seriola quinqueradiata Temminck et Schlegel, associated with floating seaweeds. (In Japanese; English abstract).
Bull. Seikai reg. Fish.Res. Lab. 35:41-50.

Anraku,Masateru, and Azeta Masanori, 1966.
The feeding habits of larvae and juveniles of the yellowtail associated with floating seaweeds. (Abstract only.)
Second, Int. Oceanogr. Congr., 30 May-9 June 1966. Abstract,Moscow:8-9.

larvae, fish

Beaudouin, Jacqueline, 1967.
Oeufs et larves de poissons récoltés par la "Thalassa" dans le Détroit de Danemark et le Nord de la Mer d'Irminger (NORWESTLANT I-20 mars-8 Mai 1963: relations avec l'hydrologie et le zooplancton.
Revue Trav. Inst. (scient. tech.) Pêches marit. 31 (3): 307-326.

larvae, fish

Bergeron, Julien, et Guy Lacroix, 1963.
Prélèvement de larves de poissons dans le sud-ouest du Golfe Saint-Laurent en 1962.
Rapp. Ann., 1962, Sta. Biol. Mar. Grande-Rivière, Canada, 69-79. (multilithed).

fish larvae distribution

Bishai, H. M., 1960
The effect of water currents on the survival and distribution of fish larvae. J. du Cons., 25(2):134-146.

larvae, fish

Blaxter, J.H.S., 1963
The behaviour and physiology of herring and other clupeids.
In: Advances in Marine Biology, F.S. Russell, Editor, Academic Press, London and New York, 1:261-393.

young fish

Burd, A.C., and A.J. Lee, 1951.
The sonic scattering layer in the sea. Nature 167(4251):624-626, 1 fig.

larvae, fish (leptocephali)

Castle, P.H.J. 1970.
Ergebnisse der Forschungsreisen des FFS Walther Herwig nach Südamerika. II. The Leptocephali.
Arch. FischWiss. 21 (1):1-21.

larvae, leptocephali

Castle, P.H.J., 1966.
Les leptocéphales dans le Pacifique sud-ouest.
Cah. ORSTOM, Sér. Océanogr., 4(4): 51-71.

larvae, eel

Castle, P.H.J., 1964.
Eels and eel-larvae of the Tiri Oceanographic Cruise, 1962, to the South Fiji Basin.
Trans. R. Soc., New Zealand, Zool., 5(7):71-84.

larvae, fish

Conand, F. et E. Fagetti, 1971.
Description et distribution saisonnière des larves de sardinelles des côtes du Sénégal et de la Gambie en 1968 et 1969. Cah. ORSTOM sér. Océanogr. 9(3): 293-318.

fish larvae

Dannevig, A., and G. Dannevig, 1950.
Factors affecting the survival of fish larvae.
J. du Cons. 16(2):211-215.

larvae fish

Dechnik, T.V., L.A. Duka and V.I. Sinyukova 1970.
Food supply and the causes of mortality among the larvae of some common Black Sea fishes.
J. Ichthyol.10(3):304-310 (Translation from Voprosy Ichth. but original reference not given) Scripta Publ. Corp. for Am. Fish. Soc.)

larvae fish (anchovies)

Dechnik, T.V., and V.I. Siniukova, 1964.
The distribution of pelagic eggs and larvae of fish in the Mediterranean sea. (In Russian).
Trudy Sevastopol Biol. Sta., 7:77-115.

E. encrasicholus

larvae, fish

Demir, Necla and F.S. Russell, 1971.
On the postlarva of the Goby lebetus. J. mar. biol. Ass. U.K. 51(3): 669-678.

larvae fish

Detwyler, R. and E.D. Houde, 1970.
Food selection by laboratory-reared larvae of the scaled sardine Harengula pensacolae (Pisces, Clupeidae) and the bay anchovy Anchoa mitchilli (Pisces, Engraulidae). Marine Biol., 7(3): 214-222.

larvae,fish

deSylva,Donald P., 1963.
Systematics and life history of the Great Barracuda Sphyraena barracuda (Welbaum).
Stud. trop. Oceanogr., Miami, 1:viii 179 pp., 32 tables,36 figs. $2.50

Larvae, fish

Duka, L.A., 1969.
Feeding of pelagic larvae of some common species of fish in the Mediterranean Sea. (In Russian)
Voprosy. Ichtiol. 9(6): (not seen)
Translation: 9(6) 547-554.

larvae, fish (anchovies)

Duka, L.A., 1964.
Intensity of the feeding and the increase in weight of the larvae of Engraulis encrasicholus ponticus Alex in the Black Sea during the spawning season. (In Russian).
Trudy Sevastopol Biol. Sta., 7:276-292.

fish, larvae

Duka, L.A. and A.D. Gordina, 1973
On the ichthyoplankton number and larval fish nutrition in the western part of the Mediterranean and adjoining Atlantic area. Gidrobiol. Zh. 9(2):87-93. (In Russian; English abstract).

larvae, fish

Ehrenbaum, E. 1905, 1909
Eier und Larven von Fischen.
Nordisches Plankton 4: 10:

larvae, fish

Eldred Bonnie 1971.
First records of Anguilla rostrata larvae in the Gulf of Mexico and Yucatan Straits Leaflet Ser., Immature Vertebrates, Mar. Res. Lab., Fla. Dep. Nat. Res. 4 (1) (19): 3pp.

larvae, fish

Eldred, Bonnie 1968.
Larvae of the marbled moray eel, Uropterygius juliae (Tommasi, 1960).
Leaflet Ser. Immature Vertebrates, Mar. Res.Lab. Fla. Bd Conserv. 4 (1 Pisces) (8): 1-4.

larvae, fish

Eldred, Bonnie 1968.
The larval development and taxonomy of the pygmy moray eel Anarchias yoshiae Kanazawa 1952.
Leaflet Ser. Immature Vertebrates, Mar. Res.Lab. Fla. Bd Conserv. 4 (1 Pisces)(10): 8pp.

larvae, fish

Fives, Julie M., 1970.
Investigations of the plankton of the west coast of Ireland - IV. Larval and post-larval stages of fishes taken from the plankton of the west coast in surveys during the years 1958-1966.
Proc. R. Ir. Acad., (B) 70(3): 15-92.

larvae, fish

Hattori, Shigemasa, 1965.
Fish eggs and larvae in the Kuroshio. (In Japanese; English abstract).
Inf. Bull. Planktol., Japan, No. 12:40-48.

larvae, fish
Kuroshio (diagram of branches)

larvae, fish

Herman, Sidney S., 1963
Planktonic fish eggs and larvae of Narraganset Bay.
Limnology and Oceanography, 8(1):103-109.

larvae, fish

Hiemstra, W.H., 1962.
A correlation table as an aid for identifying fish eggs in plankton samples.
J. Conseil, 27(1):100-108.

larvae, fish

Higgins, Bruce E. 1970.
Juvenile tunas collected by midwater trawling in Hawaiian waters, July-September 1967.
Trans. Am. Fish. Soc. 99(1): 60-69.

larvae, fish

India, Indian Ocean Biological Center, Cochin, 1970.
Distribution of fish eggs and larvae in the Indian Ocean.
Int. Indian Ocean Exped., Plankton Atlas, 2(2): 10 charts.

larvae, fish

Ishiyama, R., and K. Okada, 1957.
[Postlarval form of the skipjack (Katsuwonus pelamis) from the Phoenix Islands.]
J. Shimonoseki College of Fish., 7(1): 141-147.

larvae, fish

Karlovac, Jozica, 1967.
Étude de l'écologie de la Sardine, Sardina pilchardus Walb., dans la phase planctonique de sa vie en Adriatique moyenne.
Acta adriat., 13(2): 1-109.

larvae, fish

Klawe, W.L., J.J. Pella and W.S. Leet 1970.
The distribution, abundance and ecology of larval tunas from the entrance to the Gulf of California. (In English and Spanish).
Bull. Int.-Am. Trop. Tuna Commn 14(4): 507-544.

larvae, fish

Kramer, David, 1970.
Distributional atlas of fish eggs and larvae in the California Current region: Pacific sardine Sardinops caerulea (Girard), 1951 through 1966.
Calcofi Atlas 12: v + 277 charts.

fish larvae (Engraulis mordax)

Kramer, David and Elbert H. Ahlstrom, 1968.
Distributional atlas of fish larvae in the California Current region: northern anchovy, Engraulis mordax Girard, 1951-through 1965.
Atlas, Calif. Coop. Ocean. Fish. Invest., 9: 1-269.

larvae, fish

Kuntz, A. 1915.
The embryology and larval development of Bairdiella chrysura and Anchovia mitchelli.
Bull. U.S. Bur. Fish. 33: 3-19.

larvae, fish

Langham, N.P.E., 1971.
The distribution and abundance of larval sand-eels (Ammodytidae) in Scottish waters. J. mar. biol. Ass. U.K. 51(3): 697-707.

larvae, fish

Lelchikova L.I. 1969.
The distribution of certain bathypelagic fish, fry, and larvae in the area of the Kuroshio Current in the summer of 1965. (In Russian).
Izv. Tichookean. nauchno-issled. Inst. ribn. Khoz. Okeanogr. (TINRO) 68: 193-202.

fish larvae

Longhurst, Alan R., 1968.
Distribution of the larvae of Pleuroncodes planipes in the California Current. Limnol. Oceanogr., 13(1): 143-155.

larvae, fish (good illustrations)

Marinaro, J.-Y., 1971.
Contribution à l'étude des oeufs et larves pélagiques de poissons méditerranéens V. Oeufs pélagiques de la baie d'Alger.
Pelagos 3(1): 118 pp; 27 pls.

larvae, fish

Matsui, I., 1957.
On the records of a leptocephalous and catadromous eels of Anguilla japonica in the waters around Japan with a presumption of their spawning places. J. Shimonoseki Coll. Fish., 7(1): 151-167.

larvae, fish

Matsumoto, Walter M., 1962.
Identification of larvae of four species of tuna from the Indo-Pacific region. 1.
Dana Rept., No. 55: 16 pp.

larvae, fish

Mercado Silgado, Jorge E. 1971.
Notas sobre los estados larvales del sábalo Megalops atlanticus Valenciennes, con comentarios sobre su importancia comercial.
Bol. Mus. Mar, Bogota, 2: 5-28.

larvae, fish

Mercado S., Jorge E. y Alejandro Ciardelli, 1972.
Contribución a la morfología y organogénesis de los leptocéfalos del sábalo Megalops atlanticus (Pisces: Megalopidae). Bull. mar. Sci., Miami 22(1): 153-184.

fish larvae

Miller, D., J.B. Colton, Jr., and R.R. Marak, 1963.
A study of the vertical distribution of larval haddock.
J. du Conseil, 28(1): 37-49.

larvae, fish

Mito, Satoshi, 1967.
Some ecological notes on the planktonic of fish larvae. (In Japanese; English abstract).
Inf. Bull. Planktol. Japan, 14: 33-49.

larvae, fish

Nakamura, Eugene L., and Walter M. Matsumoto 1966.
Distribution of larval tunas in Marquesan waters.
Fish. Bull. Fish Wildlife Service 66(1): 1-12.

fish larvae

NORPAC Committee, 1960.
The NORPAC Atlas. Oceanic observations of the Pacific, 1955.

larva, fish

O'Connell, C.P. and L.P. Raymond, 1970.
The effect of food density on survival and growth of early post yolk-sac larvae of the northern anchovy (Engraulis mordax Girard) in the laboratory. J. exp. mar. Biol. Ecol., 5(2): 187-197.

larvae, fish

Richards, William J. 1969.
Distribution and relative apparent abundance of larval tunas collected in the tropical Atlantic during Equalant surveys I and II.
Actes Symp. Oceanogr. Ressources halieut. Atlant. trop., Abidjan, 20-28 Oct. 1966, UNESCO, 289-315.

larva, fish

Richards, William J., and David C. Simmons 1971.
Distribution of tuna larvae (Pisces Scombridae) in the northwestern Gulf of Guinea and off Sierra Leone.
Fish. Bull. nat. mar. fish. Serv. NOAA, 69(3): 555-580.

larvae, fish

Russell, F.S., 1973
A summary of the observations on the occurrence of planktonic stages of fish off Plymouth 1924-1972. J. mar. biol. Ass. U.K. 53(2): 347-355.

fish larvae

Russell, F. S., 1926.
The vertical distribution of marine macroplankton. 3. Diurnal observations of the pelagic young of teleostean fishes in the Plymouth area.
J.M.B.A., n.s., 14(2): 387-414, Figs. 1-8.

larvae, fish

Russell, F. S., 1926.
The vertical distribution of marine macroplankton. 2. The pelagic young of teleostean fishes in the daytime in the Plymouth area, with a note on the eggs of certain species. J.M.B.A., n.s., 14(1): 101-159, figs. 1-5.

larvae, fish

Saville, Alan, 1964.
Clupeoides.
Fiches d'Ident., Oeux et Larves, Poissons, Cons. Perm. Int. Expl. Mer, 1: 5 pp.

larvae, fish

Serebryakov, V.P., 1962
Some notes on ichthyoplankton of Newfoundland and Labrador areas. (In Russian; English summary).
Sovetskie Ribochoz. Issledov. Severo-Zapad. Atlant. Okeana, VNIRO-PINRO, 227-233.

larvae, Anguilla

Smith, David G., 1968.
The occurrence of larvae of the American eel, Anguilla rostrata, in the Straits of Florida and nearby areas. J. mar. Sci. Miami, 18(2): 280-293.

fish larvae, oceanic

SunZsi Gen., 1960. Larvae and fry of tuna, sailfish and sword fish collected in the western and central Pacific. Trudy Inst. Okeanol., 41: 175-191.

larvae, fish

Tabeta, O., 1970.
A giant leptocephalus from the sea off northern Peru. Japan. J. Ichthyol. 17(2): 80-81. Also in: Contrib. Dept. Fish. Fish. Res. Lab. Kyushu, Univ. 16 (1970).

larvae, fish

Takeuchi, Isamu, 1972
Some observations of eggs and larvae of the Alaska pollack, Theragra chalcogramma (Pallas) off the west coast of Kamchatka. In: Biological oceanography of the northern North Pacific Ocean, A.Y. Takenouti, Chief Editor, Idemitsu Shoten, Tokyo, 613-620.

larvae, fish

Taylor, F.H.C., 1964.
Life history and present status of British Columbia herring stocks.
Fish. Res. Bd., Canada, Bull., No. 143: 1-81.

larvae, fish

Tibbo, S.N., and L.M. Lauzier 1969
On the origin and distribution of larval swordfish Xiphius gladius L. in the western Atlantic.
Techn. Rept. Fish. Res. Bd Can. 136: 20 pp.

larvae, fish
Ueyanagi, Shoji, 1966.
On the red pigmentation of larval tuna and its usefulness in species identification.
Rep. Nankai reg. Fish.Res.Lab., No.24:41-48.

larvae, fish
Wheeler, J.F.G. 1924
The growth of the egg in the dab (Pleuronectes limanda).
Quart. J. microsc. Soc. 68:

larvae, fish
Yasuie, Shigeki and Isamu Mitani, 1969.
Anchovy eggs and larvae in Bisan-Seto, the Seto Inland Sea, 1969. Bull. Fish. exp. Sta. Okayama, 1969: 14-16. (In Japanese).

larvae, fish
Zijlstra, J.J., 1970.
Herring larvae in the central North Sea.
Ber. dt. wiss. Komm. Meeresforsch. 21 (1/4): 92-115

fish larvae and eggs
Zuta, Salvador, y Jorge Mejía, 1968.
Informe preliminar del Crucero Unanue 6708, 24 agosto- 25 de setiembre, inverno 1967.
Informe, Inst. Mar Peru 25: 23 pp.

larvae, infusoriform
Bresciani José, and Tom Fenchel, 1967.
Studies on dicyemid Mesozoa. II
The fine structure of the infusoriform larva.
Ophelia, 4 (1): 1-17.

larvae, invertebrate
Scheltema, Rudolf S., 1964.
Origin and dispersal of invertebrate larvae in the North Atlantic. (Abstract).
American Zoologist, 4(3):299.

larvae, leptocephali
Blache, J., 1972
Larves leptocéphales des poissons anguilliformes dans le golfe de Guinée (zone sud). 2e note: les espèces adultes de xenocongridae et leurs larves. Cah. ORSTOM Sér. Océanogr. 10(3): 219-241.

larvae, eel
Blache, J., 1964.
Note préliminaire sur les larves leptocéphales d'apodes du Golfe de Guinée.
Trav. Centre Océanogr., Pointe-Noire, Cahiers O.R.S.T.R.O.M., Océanogr., No. 5 (Paris), 5-55.

larvae, eels
Castle, P.H.J., 1968.
Larval development of the congrid eel Gnathophis capensis (Kaup), off Southern Africa, with notes on the identity of Congermuraena australis Barnard.
Zoologica africana 3(2):139-154.

larvae, eel
Castle, P.H.J., 1967.
Two remarkable eel-larvae from off southern Africa.
Spec. Pub. Dept. Ichthyol., Rhodes Univ. 1:1-12.

larvae, fish, leptocephali
Castle, P.H.J., 1965.
Moringuid leptocephali in Australasian waters.
Trans. R. Soc., New Zealand, Zool., 7(7):125-133.

larvae fish, leptocephali
Castle, P.H.J., 1965.
Muraenid Leptocephali in Australian waters.
Trans. R. Soc., New Zealand, Zool., 7(3):57-84.

larvae, fish, leptocephali
Castle, P.H.J., 1965.
Ophichthid leptocephali in Australian waters.
Trans. R. Soc., New Zealand, Zool., 7(6):97-123.

larvae, eel
Castle, P.H.J., 1964
Congrid leptocephali in Australasian waters with the descriptions of Conger wilsoni (Bl. and Schn.) and C. verreauxi Kaup.
Zool. Publ., Univ. Wellington, No. 37:45 pp.

larvae, eel
Castle, P.H.J., 1963.
Anguillid leptocephali in the southwest Pacific.
Zool. Publ., Victoria Univ., Wellington, No. 33: 1-14.

leptocephalus
Cohen, D. M., 1959.
A remarkable leptocephalus from off the coast of Washington.
Deep-Sea Research 5(3):238-240.

larvae, leptocephali
Della Croce, N., and P.H.J.Castle,1966.
Leptocephali from the Mozambique Channel.
Boll.Musei Ist. biol. Univ. Genova, 34:149-164.

larvae, fish
Eldred, Bonnie, 1968.
Larvae and glass eels of the American freshwater eel Anguilla rostrata (Lesueur, 1817) in Florida waters.
Leaflet Ser. Immature Vertebrates, Mar. Res. Lab. Fla. Bd Conserv. 4 (1=Pisces) (9): 4 pp.

leptocephali
Hulet, William H., Joseph Fischer and Barbara J. Rietberg, 1972.
Electrolyte composition of anguilliform leptocephali from the Straits of Florida.
Bull. mar. Sci. Miami 22(2): 432-448.

larvae, eel
Macer, C.T., 1965.
The distribution of larval sand eels (Ammodytidae) in the southern North Sea.
Jour. Mar. Biol. Assoc., U.K., 45(1):187-207.

leptocephali
Matsui, Isao, and Toru Takai 1971
Leptocephalus of the eel, Anguilla japonica, found in the waters of Ryukyu Deep.
J. Shimonoseki Univ. Fish. 20 (1): 13-18.

larvae, fish
Matsui, Isao, Toru Takai and Akiyoshi Kataoka, 1968.
Anguillid leptocephalus found in the Japan Current and its adjacent waters.
J. Shimonoseki Univ. Fish., 17(1): 17-23.

Leptocephali
Nair, R. Velappan, 1948
Leptocephali of the Gulf of Manaar. Proc. Indian Acad. Sci. Sect.B. 27(4), 87-91, 2 text fig.

leptocephali
Nair, R. Velappan, 1947
On the metamorphosis of two leptocephali from the Madras Plankton.
Proc. Indian Acad. Sci., Section B XXV (1):1-14; 2 pls.

larvae, fish (leptocephalus)
Nair, R.V. 1946
On the leptocephalus of Uroconger lepturus (Richardson) from the Madras plankton.
Current Sci. 15 (11): 318-319.

larvae, eel
Nielsen, Jorgen G., and Verner Larsen, 1970
Remarks on the identity of the giant Dana eel larva with notes on some related notacanthi form larvae
Vidensk Meddr dansk naturh. Foren. 133:149-157

leptocephalus larva
Tabeta, O., 1970.
A giant leptocephalus from the sea off northern Peru. Japan. J. Ichthyol. 17(2): 80-81. Also in: Contrib. Dept. Fish. Res. Lab., Kyushu Univ. 16(1970).

larvae, fish
Uchida, Kazuyoshi, Akiyoshi Kataoka and Toru Takai,1968.
On the congrid leptocaphali in Ise Bay. (In Japanese; English abstract).
J. Shimonoseki Univ.Fish., 17(1):25-34.

larvae, eel
Yada, Shigeaki, 1964.
On some specimens of Leptocephalus in the Philippine Sea. (In Japanese; English abstract).
Bull. Fac. Fish., Nagasaki Univ., 15:85-91.

Larvae, Mollusc
D'Asaro, Charles N.,1965.
Organogenesis, development and metamorphosis in the queen conch, Strombus Gigas, with notes on breeding habits.
Bull. Mar Sci.,15(2):359-416.

larvae, mollusc
Horikoshi, Masuoki 1966.
Reproduction, larval features and life history of Philine denticulata (J. Adams) (Mollusca - Tectibranchiata).
Ophelia 4 (1): 43-84

larvae, mollusc
Lebour, M.V., 1947
Notes on the inshore plankton of Plymouth. JMBA 26(4):527-547.

larvae, mollusc
Mileikovsky, S.A., 1966.
The range of dispersal of the pelagic larvae of bottom invertebrates by ocean currents and its distributional role on the example of Gastropoda and Lamellibranchia. (In Russian; English abstract).
Okeanologiia, Akad. Nauk, SSSR, 6(3):482-492.

larvae, molluscs
Poggiani, Luciano, 1968.
Note sulle larve planctoniche di alcuni Molluschi dell'Adriatico medio-occidentale e sviluppo post-larvale di alcuni di essi. Note Lab. Biol. mar. Pesca, Fano, Univ. Bologna, 2(8): 137-180.

larvae, mollusc (vertical migration)
Schmidt, H.-E., 1973.
The vertical distribution and diurnal migration of some zooplankton in the Bay of Eilat (Red Sea). Helgoländer wiss. Meeresunters 24(1/4): 333-340.

larvae, mollusc
Thiriot-Quiévreux, Catherine,1968.
Variations saisonnières des mollusques dans le plancton de la région de Banyuls-sur-Mer(zone sud du Golfe du Lion) novembre 1965-décembre 1967
Vie Milieu, 19(1-B)L35-83.

larvae, mollusc
Verwey, J., 1966.
The role of some external factors in the vertical migrations of marine animals.
Neth. J. Sea Res., 3(2):245-266.

LARVAE
Mollusc - cephalopod

larvae, cephalopod
Allan, J., 1945.
Planktonic cephalopod larvae from the eastern Australian coast. Rec. Australian Mus., 21(6): 317-350, Pls. 24-27, map.

larvae, molluscs

Arnold, John M., 1965.
Normal embryonic stages of the squid, Loligo pealii (Lesueur).
Biol. Bull., 128(1):24-32.

larvae, mollusc

Boletzky, S. von, M.V. von Boletzky, D. Frösch and V. Gätzi, 1971.
Laboratory rearing of Sepiolinae (Mollusca: Cephalopoda). Marine Biol., 8(1): 82-87.

larvae, cephalopods

Clarke, M.R., 1964.
Young stages of Lepidoteuthis grimaldi (Cephalopoda, Decapoda).
Proc. Malac. Soc., London, 36:69-78.

Also in:
Collected Reprints, Nat. Inst. Oceanogr., Wormley, 12. 1964.

larvae, squid

Hall, John R., 1970.
Description of egg capsules and embryos of the squid, Lolliguncula brevis from Tampa Bay, Florida. Bull. mar. Sci., 20(3): 762-768.

larvae, mollusc

Mangold, K., S. von Boletzky and D. Frösch, 1971.
Reproductive biology and embryonic development of Eledone cirrosa (Cephalopoda: Octopoda). Marine Biol., 8(2): 109-117.

larvae, mollusc

McMahon, John J. and William C. Summers, 1971.
Temperature effects on the developmental rate of squid (Loligo pealei) embryos. Biol. Bull. mar. biol. Lab. Woods Hole 141(3): 561-567.

larvae, molluscs

Okutani, Takashi, 1969.
Studies on early life history of decapodan mollusca - IV. Bull. Tokai Reg. Fish. Res. Lab., 58: 83-96. (In Japanese; English abstract).
Squid larvae collected by oblique hauls of a larva net from the Pacific Coast of eastern Honshu, during the winter seasons, 1965-1968.

larvae, mollusca (cephalopods)

Okutani, Takashi, 1968.
Studies on early life history of decapodan Mollusca. III. Systematics and distribution of larvae of decapod cephalopods collected from the sea surface on the Pacific coast of Japan, 1960-1965.
Bull. Tokai reg. Fish. Res. Lab., 55:9-57.

larvae, cephalopods

Okutani, Takashi, 1966.
Studies on early life history of decapodan Mollusca. II. Planktonic larvae of decapodan cephalopods from the northern North Pacific in summer seasons during 1952-1959.
Bull. Tokai Reg. Fish. Res. Lab., No. 45:61-79.

larvae, cephalopod

Voss, G.L., 1962.
South African cephalopods.
Trans. R. Soc., S. Africa, 36(4):245-272.

larvae, mollusc

Watanabe, Taisuke, 1965.
Ecological distribution of Rhynchoteuthion larvae of common squid Todarodes pacificus Steenstrup in the southwestern waters off Japan during the winters, 1959-1962. (In Japanese; English abstract).
Bull. Tokai Reg. Fish. Res. Lab., No. 43: 1-12.

LARVAE
Mollusc - gastropod

larvae, gastropod

Amio, Masaru, 1963
A comparative embryology of marine gastropods, with ecological considerations. (In Japanese; English summary).
J. Shimonoseki Univ. Fish., 12(2/3):15-358.

larvae, gastropoda

Christiansen, Marit E., 1964.
Some observations on the larval stages of the gastropod Nassarius pygmaeus (Lamarck)
Pubbl. Staz. Zool., Napoli, 34:1-8.

larval shells

Fretter, Vera and Margaret C. Pilkington, 1971.
The larval shell of some prosobranch gastropods.
J. mar. biol. Ass., U.K., 51(1): 49-62.

larvae, molluscs

Hadfield, Michael G., 1964.
Opisthobranchia. The veliger larvae of the Nudibranchia.
Fiches d'Ident., Zooplancton, Cons. Perm. Int. Expl., Mer, 106:3 pp.

larvae - molluscs

Lebour, M.V., 1937
The eggs and larvae of the British Prosobranchs with special reference to those living in plankton. JMBA. 22:105-166, 4 text figs.

larvae, molluscs

Scheltema, Amélie H., 1969.
Pelagic larvae of New England gastropods. II Anachis translirata and Anachis avara (Columbellidae, Prosobranchia).
Vie Milieu 20 (1A): 95-103

larvae, gastropod

Scheltema, R.S., 1973
Dispersal of the protozoan Folliculina simplex Dons (Chiliophora, Heterotricha) throughout the North Atlantic Ocean on the shells of gastropod veliger larvae. J. mar. Res, 31(1): 11-20.

larvae, mollusc

Scheltema, Rudolf S., 1972.
Eastward and westward dispersal across the tropical Atlantic Ocean of larvae belonging to the genus Bursa (Prosobranchia, Mesogastropoda, Bursidae).
Int. Revue ges. Hydrobiol. 57(6): 863-873

larvae, molluscs

Scheltema, Rudolf S., 1967.
The relationship of temperature to the larval development of Nassarius obsoletus (Gastropoda).
Biol. Bull., mar. biol. Lab., Woods Hole, 132 (2):253-265.

larvae, molluscs

Scheltema, R.S., and A.H. Scheltema, 1965.
Pelagic larvae of New England intertidal gastropods.
Hydrobiologia, 25(3/4):321-329.

larvae, gastropods

Scheltema, Rudolf S., and Amelie H. Scheltema, 1963.
Pelagic larvae of New England intertidal gastropods. II. Anachis avara.
Hydrobiologia, 22(1/2):85-91.

larvae, gastropod

Thiriot-Quiévreux, Catherine, 1969.
Caractéristiques morphologiques des véligères planctoniques de gastéropodes de la région de Banyuls-sur-Mer.
Vie Milieu (B) 20(2): 333-366

larvae, mollusc

Thiriot-Quiévreux, Catherine, 1967.
Descriptions de quelques véligères planctoniques de gastéropodes.
Vie Milieu (A) 18(2): 303-315.

LARVAE
Mollusc - heteropod

Larvae, heteropoda

Okutani, Takashi, 1961.
Notes on the genus Carinaria (Heteropoda) from Japanese and adjacent waters.
Publ. Seto Mar. Biol. Lab., Kyoto Univ., 9(2):333-352, Pls. 12-13.

larvae, heteropod

Owre, Harding B., 1964.
Observations on development of the heteropod molluscs Pterotrachea hippocampus and Firoloida desmaresti.
Bull. Mar. Sci., Gulf and Caribbean, 14(4):529-538.

larval heteropods

Richter, Gotthard, 1968.
Heteropoden und Heteropodenlarven im Oberflachenplankton des Golfs von Neapel.
Pubbl. Staz. zool. Napoli, 36(3):346-400.

Larvae, heteropods

Thiriot-Quiévreux, Catherine, 1969.
Organogénèse larvaire du genre Atlanta (mollusque hétéropode).
Vie Milieu (A) 20(2): 347-395.

larvae, heteropod

Thomopoulos, T., 1952.
Notes sur le plancton de la Baie de Banyuls.
Vie et Milieu, Bull. Lab. Arago, Univ. Paris, 3(3):327-335, 8 textfigs.

LARVAE
Mollusc - lamellibranch

larvae, molluscs

Allen, John A. and Rudolf S. Scheltema, 1972.
The functional morphology and geographical distribution of Planktomya henseni, a supposed neotenous pelagic bivalve. J. mar. biol. Ass. U.K. 52(1): 19-31.

larvae, molluscs

Bayne, B.L., 1969
The gregarious behaviour of the larvae of Ostrea edulis L. at settlement. J. mar. biol. Ass., U.K., 49(2): 327-356.

larvae, molluscs

Brenko, M.Hrs. and A. Calabrese, 1969.
The combined effects of salinity and temperature on larvae of the mussel Mytilus edulis.
Marine Biol., 4(3): 224-226.

larvae mollusc

Calabrese, Anthony, 1969.
Individual and combines effects of salinity and temperature on embryos and larvae of the coot clam, Mulinia lateralis (Say). Biol. Bull. mar. biol. Lab., Woods Hole, 137(3): 417-428.

larvae, mollusc.

Carriker, M.R., 1961.
Interrelation of functional morphology, behavior and autecology in early stages of the bivalve Mercenaria mercenaria.
J. Elisha Mitchell Sci. Soc., 77(2):168-241.

larvae, mollusc.
Chanley, Paul E., 1965.
Larval development of a boring clam, Barnea truncata.
Chesapeake Sci., 6(3):162-164.

larvae, molluscs
Culliney, John L., 1971.
Laboratory rearing of the larvae of the mahogany date mussel Lithophaga bisulcata.
Bull. mar. Sci. Miami 21(2): 591-602.

larvae, mollusc
Davis, Harry C. and Anthony Calabrese, 1969.
Survival and growth of larvae of the European oyster (Ostrea edulis L.) at different temperatures. Biol. Bull., mar. biol. Lab., Woods Hole, 136(2): 193-199.

larvae, mollusc
Forbes, Milton L., 1971.
Habitats and substrates of Ostrea frons and distinguishing features of early spat.
Bull. mar. Sci. Miami 21(2): 613-625.

larvae, molluscs
Hayashi, Tadaniko, and Katsuji Terai, 1964.
Study on the larvae and young of Japanese surf clam, Spisula (S.) sachaliensis (Schrenck) at Shikuzu, Muroran City. 1. Taxonomy of the Pelecypoda veliger larvae in plankton. (In Japanese; English abstract).
Sci. Repts., Hokkaido Fish. Exper. Sta., (2):7-38.

larvae, molluscs
Hrs-Brenko, Mirjana, 1969.
Contribution to the larval stages of Bivalvia in Limoxi Canal. (In Jugoslavian; English abstract).
Thalassia Jugoslavica 5:113-120.

larvae, molluscs
Ivanov, A.I., 1965.
The effect of water of different salinity on the survival of larvae of Black Sea oysters (Ostrea taurica Kryn.).
Doklady, Akad. Nauk, SSSR, 163(5):1256-1258.

larvae, mollusk
Loosanoff, Victor L., and Harry C. Davis, 1963
Rearing of bivalve mollusks. In: Advances in Marine Biology, F.S. Russell, Editor, Academic Press, London and New York, 1:1-136.

larvae, mollusc
Lough, R.G. and J.J. Gonor, 1971.
Early embryonic stages of Adula californiensis (Pelecypoda: Mytilidae) and the effect of temperature and salinity on developmental rate.
Marine Biol., 8(2): 118-125.

larvae, molluscs
Millar, R.H., and P.J. Hollis, 1963.
Abbreviated pelagic life of Chilean and New Zealand oysters.
Nature, 197(4866):512-513.

larvae, mollusc (oyster)
Millar, R.H., and J.M. Scott, 1968.
An effect of water quality on the growth of cultured larvae of the oyster Ostrea edulis L.
J. Cons. perm. int. Explor. Mer, 32(1):123-130.

larvae, molluscs
Millar, R.H., and J.M. Scott, 1967.
The larva of the oyster Ostrea edulis during starvation.
J. mar. biol. Ass., U.K., 47(3):475-484.

larvae, mollusc
Miyazaki, I., 1962.
On the identification of lamellibranch larvae. (In Japanese; English abstract).
Bull. Jap. Soc., Sci. Fish., 28(10):955-966.

larvae, molluscs
Paranaguá, M.N., 1970.
Primeiros resultados sôbre o desenvolvimento larvar de Mytella falcata (d'Orbigny, 1846) (Mollusca Pelecipoda). Trabhs oceanogr. Univ. Fed. Pernambuco 9/11: 275-284.

larvae, lamellibranch
Rees, C.B., 1954.
Continuous plankton records: the distribution of lamellibranch larvae in the North Sea, 1950-51.
Bull. Mar. Ecol. 4(27):21-46, Pls. 7-10.

includes identification of.

larvae, mollusc
Rees, C.B., 1950.
The identification and classification of Lamellibranch larvae. Hull Bull. Mar. Ecol. 3(19):73-104, 5 pls., 4 textfigs.

Larvae, Mollusc
Sastry, A.M., 1965.
The development and external morphology of pelagic larval and post-larval stages of the bay scallop, Aequipecten irradians concentricus Say, reared in the laboratory.
Bull. Mar. Sci., 15(2):417-435.

larvae, mollusc
Scheltema, R.S., 1971.
Dispersal of phytoplanktotrophic shipworm larvae (Bivalvia: Teredinidae) over long distances by ocean currents. Mar. Biol. 11(1): 5-11.

larvae, lamellibranch
Sentz-Braconnot, 1968.
Relation entre les larves planctoniques et les jeunes stades fixés chez les lamellibranches, dans la rade de Villifranche-sur-Mer (Alpes-Maritimes).
Vie Milieu, 19(1-B):85-108.

larvae, oysters
Stenzel, H.B., 1964
Oysters: composition of the larval shell.
Science, 145(3628):155-156.

larvae, mollusc
Walne, P.R., 1966.
Experiments in the large-scale culture of the larvae of Ostrea edulis L.
Minist. Agric., Fish., Food, Fish Invest., Great Britain (2)25(4):1-53.

LARVAE
Mollusc - pteropod

larvae, pteropods
Frontier, S., 1963.
Hétéropodes et ptéropodes récoltés dans le plancton de Nosy-Bé.
Trav. Centre Océanogr., Nosy-Bé, Cahiers, O.R.S.T.R.O.M., Océanogr., Paris, No. 6:213-227.

LARVAE
Molluscs - arranged by genus (incomplete)

Alvania cimicoides
Lebour, M.V., 1937
The eggs and larvae of the British Prosobranchs with special reference to those living in plankton. JMBA. 22:105-166, 4 text figs.

Alvania crassa
Lebour, M.V., 1937
The eggs and larvae of the British Prosobranchs with special reference to those living in plankton. JMBA. 22:105-166, 4 text figs.

Alvania jeffreysii
Lebour, M.V., 1937
The eggs and larvae of the British Prosobranchs with special reference to those living in plankton. JMBA. 22:105-166, 4 text figs.

Alvania punctura
Lebour, M.V., 1937
The eggs and larvae of the British Prosobranchs with special reference to those living in plankton. JMBA. 22:105-166, 4 text figs.

Aporrhais pespelicani
Lebour, M.V., 1937
The eggs and larvae of the British Prosobranchs with special reference to those living in plankton. JMBA. 22:105-166, 4 text figs.

Baleis alba
Lebour, M.V., 1937
The eggs and larvae of the British Prosobranchs with special reference to those living in plankton. JMBA. 22:105-166, 4 text figs.

Baleis devians
Lebour, M.V., 1937
The eggs and larvae of the British Prosobranchs with special reference to those living in plankton. JMBA. 22:105-166, 4 text figs.

Barlecia unifasciata
Lebour, M.V., 1937
The eggs and larvae of the British Prosobranchs with special reference to those living in plankton. JMBA. 22:105-166, 4 text figs.

Beringius turtoni
Lebour, M.V., 1937
The eggs and larvae of the British Prosobranchs with special reference to those living in plankton. JMBA. 22:105-166, 4 text figs.

Bittium reticulatum
Lebour, M.V., 1937
The eggs and larvae of the British Prosobranchs with special reference to those living in plankton. JMBA. 22:105-166, 4 text figs.

Buccinum phresianum
Lebour, M.V., 1937
The eggs and larvae of the British Prosobranchs with special reference to those living in plankton. JMBA. 22:105-166, 4 text figs.

Buccinum undatum
Lebour, M.V., 1937
The eggs and larvae of the British Prosobranchs with special reference to those living in plankton. JMBA. 22:105-166, 4 text figs.

Caecum imperforatum
Lebour, M.V., 1937
The eggs and larvae of the British Prosobranchs with special reference to those living in plankton. JMBA. 22:105-166, 4 text figs.

Calliostoma papillosum
Lebour, M.V., 1937
The eggs and larvae of the British Prosobranchs with special reference to those living in plankton. JMBA. 22:105-166, 4 text figs.

Calliostoma zizyphinum
Lebour, M.V., 1937
The eggs and larvae of the British Prosobranchs with special reference to those living in plankton. JMBA. 22:105-166, 4 text figs.

Cantharidus clelondi
Lebour, M.V., 1937
The eggs and larvae of the British Prosobranchs with special reference to those living in plankton. JMBA. 22:105-166, 4 text figs.

Cantharidus exasperatus
Lebour, M.V., 1937
The eggs and larvae of the British Prosobranchs with special reference to those living in plankton. JMBA. 22:105-166, 4 text figs.

Cantharidus montagui
Lebour, M.V., 1937
The eggs and larvae of the British Prosobranchs with special reference to those living in plankton. JMBA. 22:105-166, 4 text figs.

Cantharidus striatus
Lebour, M.V., 1937
The eggs and larvae of the British Prosobranchs with special reference to those living in plankton. JMBA. 22:105-166, 4 text figs.

Capulus ungaricus
Lebour, M.V., 1937
The eggs and larvae of the British Prosobranchs with special reference to those living in plankton. JMBA. 22:105-166, 4 text figs.

Cerithiopsis barleei
Lebour, M.V., 1937
The eggs and larvae of the British Prosobranchs with special reference to those living in plankton. JMBA. 22:105-166, 4 text figs.

Cerithiopsis jeffreysi
Lebour, M.V., 1937
The eggs and larvae of the British Prosobranchs with special reference to those living in plankton. JMBA. 22:105-166, 4 text figs.

Cerithiopsis tubercularis
Lebour, M.V., 1937
The eggs and larvae of the British Prosobranchs with special reference to those living in plankton. JMBA. 22:105-166, 4 text figs.

Chauvetia brunnea
Lebour, M.V., 1937
The eggs and larvae of the British Prosobranchs with special reference to those living in plankton. JMBA. 22:105-166, 4 text figs.

Chrysallida decussata
Lebour, M.V., 1937
The eggs and larvae of the British Prosobranchs with special reference to those living in plankton. JMBA. 22:105-166, 4 text figs.

Cingula cingullus
Lebour, M.V., 1937
The eggs and larvae of the British Prosobranchs with special reference to those living in plankton. JMBA. 22:105-166, 4 text figs.

Cingula fulgida
Lebour, M.V., 1937
The eggs and larvae of the British Prosobranchs with special reference to those living in plankton. JMBA. 22:105-166, 4 text figs.

Cingula semistriata
Lebour, M.V., 1937
The eggs and larvae of the British Prosobranchs with special reference to those living in plankton. JMBA. 22:105-166, 4 text figs.

Clathrus clathrus
Lebour, M.V., 1937
The eggs and larvae of the British Prosobranchs with special reference to those living in plankton. JMBA. 22:105-166, 4 text figs.

Colus gracilis
Lebour, M.V., 1937
The eggs and larvae of the British Prosobranchs with special reference to those living in plankton. JMBA. 22:105-166, 4 text figs.

Colus howsei
Lebour, M.V., 1937
The eggs and larvae of the British Prosobranchs with special reference to those living in plankton. JMBA. 22:105-166, 4 text figs.

Colus islandicus
Lebour, M.V., 1937
The eggs and larvae of the British Prosobranchs with special reference to those living in plankton. JMBA. 22:105-166, 4 text figs.

Colus jeffresianus
Lebour, M.V., 1937
The eggs and larvae of the British Prosobranchs with special reference to those living in plankton. JMBA. 22:105-166, 4 text figs.

Crepidula fornicata
Lebour, M.V., 1937
The eggs and larvae of the British Prosobranchs with special reference to those living in plankton. JMBA. 22:105-166, 4 text figs.

Diodora apertura
Lebour, M.V., 1937
The eggs and larvae of the British Prosobranchs with special reference to those living in plankton. JMBA. 22:105-166, 4 text figs.

Erato voluta
Lebour, M.V., 1937
The eggs and larvae of the British Prosobranchs with special reference to those living in plankton. JMBA. 22:105-166, 4 text figs.

Gibbula cineraria
Lebour, M.V., 1937
The eggs and larvae of the British Prosobranchs with special reference to those living in plankton. JMBA. 22:105-166, 4 text figs.

Gibbula magus
Lebour, M.V., 1937
The eggs and larvae of the British Prosobranchs with special reference to those living in plankton. JMBA. 22:105-166, 4 text figs.

Gibbula tumida
Lebour, M.V., 1937
The eggs and larvae of the British Prosobranchs with special reference to those living in plankton. JMBA. 22:105-166, 4 text figs.

Gibbula umbilicalis
Lebour, M.V., 1937
The eggs and larvae of the British Prosobranchs with special reference to those living in plankton. JMBA. 22:105-166, 4 text figs.

Haliotis tuberculata
Lebour, M.V., 1937
The eggs and larvae of the British Prosobranchs with special reference to those living in plankton. JMBA. 22:105-166, 4 text figs.

Hydrobia ulvae
Lebour, M.V., 1937
The eggs and larvae of the British Prosobranchs with special reference to those living in plankton. JMBA. 22:105-166, 4 text figs.

Hydrobia ventrosa
Lebour, M.V., 1937
The eggs and larvae of the British Prosobranchs with special reference to those living in plankton. JMBA. 22:105-166, 4 text figs.

Ianthina britannica
Lebour, M.V., 1937
The eggs and larvae of the British Prosobranchs with special reference to those living in plankton. JMBA. 22:105-166, 4 text figs.

Lacuna pallidula
Lebour, M.V., 1937
The eggs and larvae of the British Prosobranchs with special reference to those living in plankton. JMBA. 22:105-166, 4 text figs.

Lacuna vincta
Lebour, M.V., 1937
The eggs and larvae of the British Prosobranchs with special reference to those living in plankton. JMBA. 22:105-166, 4 text figs.

? Lamellaria latens
Lebour, M.V., 1937
The eggs and larvae of the British Prosobranchs with special reference to those living in plankton. JMBA. 22:105-166, 4 text figs.

Liomesus ovum
Lebour, M.V., 1937
The eggs and larvae of the British Prosobranchs with special reference to those living in plankton. JMBA. 22:105-166, 4 text figs.

Littorina littoralis
Lebour, M.V., 1937
The eggs and larvae of the British Prosobranchs with special reference to those living in plankton. JMBA. 22:105-166, 4 text figs.

Littorina littorea (larvae)
Wiborg, K.F., 1944
The production of zooplankton in a landlocked fjord, the Nordåsvatn near Bergen, in 1941-42, with special reference to the copepods. (Repts. Norwegian Fish. and Mar. Invest.) 7(7):83 pp., 40 text figs.

Littorina neritoides
Lebour, M.V., 1937
The eggs and larvae of the British Prosobranchs with special reference to those living in plankton. JMBA. 22:105-166, 4 text figs.

Littorina saxatilis
Lebour, M.V., 1937
The eggs and larvae of the British Prosobranchs with special reference to those living in plankton. JMBA. 22:105-166, 4 text figs.

Lora turricula
Lebour, M.V., 1937
The eggs and larvae of the British Prosobranchs with special reference to those living in plankton. JMBA. 22:105-166, 4 text figs.

Mangelia nebula
Lebour, M.V., 1937
The eggs and larvae of the British Prosobranchs with special reference to those living in plankton. JMBA. 22:105-166, 4 text figs.

Margarites helicinus
Lebour, M.V., 1937
The eggs and larvae of the British Prosobranchs with special reference to those living in plankton. JMBA. 22:105-166, 4 text figs.

Monodonta lineata
Lebour, M.V., 1937
The eggs and larvae of the British Prosobranchs with special reference to those living in plankton. JMBA. 22:105-166, 4 text figs.

Mya arenaria

Corbeil, H.-E., 1949.
Travail sur les mollusques, Mya arenaria L. et Ostrea virginica (Gmelin). Sta. Biol. Saint-Laurent, 8th Rapport, App. 2:45-54.

Mytilus californianus

Chadwick, W.L., F.S. Clark, and D.L. Fox, 1950.
Thermal control of marine fouling at Redondo Steam Station of the Southern California Edison Company. Trans. ASME, Feb. 1950:1270131, 3 text-figs.

Mytilus edulis

Chadwick, W.L., F.S. Clark, and D.L. Fox, 1950.
Thermal control of marine fouling at Redondo Steam Station of the Southern California Edison Company. Trans. ASME, Feb, 1950:127-131, 3 text-figs.

Nassarius incrassatus

Lebour, M.V., 1937
The eggs and larvae of the British Prosobranchs with special reference to those living in plankton. JMBA. 22:105-166, 4 text figs.

Nassarius pygmaeus

Lebour, M.V., 1937
The eggs and larvae of the British Prosobranchs with special reference to those living in plankton. JMBA. 22:105-166, 4 text figs.

Nassarius reticulatus

Lebour, M.V., 1937
The eggs and larvae of the British Prosobranchs with special reference to those living in plankton. JMBA. 22:105-166, 4 text figs.

larvae, gastropod

Scheltema, Rudolf S., 1961.
Metamorphosis of the veliger larvae of Nassarius obsoletus (Gastropoda) in response to bottom sediment.
Biol. Bull., 120(1):92-109.

Natica catena

Lebour, M.V., 1937
The eggs and larvae of the British Prosobranchs with special reference to those living in plankton. JMBA. 22:105-166, 4 text figs.

Natica pallida

Lebour, M.V., 1937
The eggs and larvae of the British Prosobranchs with special reference to those living in plankton. JMBA. 22:105-166, 4 text figs.

Natica poliana

Lebour, M.V., 1937
The eggs and larvae of the British Prosobranchs with special reference to those living in plankton. JMBA. 22:105-166, 4 text figs.

Neptunea antiqua

Lebour, M.V., 1937
The eggs and larvae of the British Prosobranchs with special reference to those living in plankton. JMBA. 22:105-166, 4 text figs.

Nucella lapillus

Lebour, M.V., 1937
The eggs and larvae of the British Prosobranchs with special reference to those living in plankton. JMBA. 22:105-166, 4 text figs.

Ocenebra erinacea

Lebour, M.V., 1937
The eggs and larvae of the British Prosobranchs with special reference to those living in plankton. JMBA. 22:105-166, 4 text figs.

Odostomia eulimoides

Lebour, M.V., 1937
The eggs and larvae of the British Prosobranchs with special reference to those living in plankton. JMBA. 22:105-166, 4 text figs.

Omalogyra atomus

Lebour, M.V., 1937
The eggs and larvae of the British Prosobranchs with special reference to those living in plankton. JMBA. 22:105-166, 4 text figs.

Ostrea edulis

Cole, H.A., and E.W. Knight-Jones, 1939.
Some observations and experiments on the setting behaviour of larvae of Ostrea edulis. J. du Cons. 14:86-105.

Ostrea virginica

Burkenroad, M.D., 1937.
The sex-ratio in alternational hermaphrodites, with especial reference to the determination of rate reversal of sexual phase in oviparous oysters. J. Mar. Res. 1(1):75-84.

Ostrea virginica

Corbeil, H.-E., 1949.
Travail sur les mollusques, Mya arenaria L. et Ostrea virginica (Gmelin). Sta. Biol. Saint-Laurent, 8th Rapport, App. 2:45-54.

Patella vulgata

Lebour, M.V., 1937
The eggs and larvae of the British Prosobranchs with special reference to those living in plankton. JMBA. 22:105-166, 4 text figs.

Patella vulgata (larvae)

Wiborg, K.F., 1944
The production of zooplankton in a landlocked fjord, the Nordåsvatn near Bergen, in 1941-42, with special reference to the copepods... (Repts. Norwegian Fish. and Mar. Invest.) 7(7):83 pp., 40 text figs.

Patelloida tessullata

Lebour, M.V., 1937
The eggs and larvae of the British Prosobranchs with special reference to those living in plankton. JMBA. 22:105-166, 4 text figs.

Patelloida virginea

Lebour, M.V., 1937
The eggs and larvae of the British Prosobranchs with special reference to those living in plankton. JMBA. 22:105-166, 4 text figs.

Patina pellucida

Lebour, M.V., 1937
The eggs and larvae of the British Prosobranchs with special reference to those living in plankton. JMBA. 22:105-166, 4 text figs.

Pelseneeria stylifera

Lebour, M.V., 1937
The eggs and larvae of the British Prosobranchs with special reference to those living in plankton. JMBA. 22:105-166, 4 text figs.

Philbertia asperrima

Lebour, M.V., 1937
The eggs and larvae of the British Prosobranchs with special reference to those living in plankton. JMBA. 22:105-166, 4 text figs.

Philbertia gracilis

Lebour, M.V., 1937
The eggs and larvae of the British Prosobranchs with special reference to those living in plankton. JMBA. 22:105-166, 4 text figs.

Philbertia leufroyi

Lebour, M.V., 1937
The eggs and larvae of the British Prosobranchs with special reference to those living in plankton. JMBA. 22:105-166, 4 text figs.

Philbertia linearis

Lebour, M.V., 1937
The eggs and larvae of the British Prosobranchs with special reference to those living in plankton. JMBA. 22:105-166, 4 text figs.

Philbertia purpurea

Lebour, M.V., 1937
The eggs and larvae of the British Prosobranchs with special reference to those living in plankton. JMBA. 22:105-166, 4 text figs.

Philbertia teres

Lebour, M.V., 1937
The eggs and larvae of the British Prosobranchs with special reference to those living in plankton. JMBA. 22:105-166, 4 text figs.

Rissoa albella

Lebour, M.V., 1937
The eggs and larvae of the British Prosobranchs with special reference to those living in plankton. JMBA. 22:105-166, 4 text figs.

Rissoa guerini

Lebour, M.V., 1937
The eggs and larvae of the British Prosobranchs with special reference to those living in plankton. JMBA. 22:105-166, 4 text figs.

Rissoa inconspicua

Lebour, M.V., 1937
The eggs and larvae of the British Prosobranchs with special reference to those living in plankton. JMBA. 22:105-166, 4 text figs.

Rissoa membranacea

Lebour, M.V., 1937
The eggs and larvae of the British Prosobranchs with special reference to those living in plankton. JMBA. 22:105-166, 4 text figs.

Rissoa parva

Lebour, M.V., 1937
The eggs and larvae of the British Prosobranchs with special reference to those living in plankton. JMBA. 22:105-166, 4 text figs.

Rissoa sarsii

Lebour, M.V., 1937
The eggs and larvae of the British Prosobranchs with special reference to those living in plankton. JMBA. 22:105-166, 4 text figs.

Rissoella diaphana

Lebour, M.V., 1937
The eggs and larvae of the British Prosobranchs with special reference to those living in plankton. JMBA. 22:105-166, 4 text figs.

Rissoella opalina

Lebour, M.V., 1937
The eggs and larvae of the British Prosobranchs with special reference to those living in plankton. JMBA. 22:105-166, 4 text figs.

Skenea serpuloides

Lebour, M.V., 1937
The eggs and larvae of the British Prosobranchs with special reference to those living in plankton. JMBA. 22:105-166, 4 text figs.

Skeneopsis planorbis

Lebour, M.V., 1937
The eggs and larvae of the British Prosobranchs with special reference to those living in plankton. JMBA. 22:105-166, 4 text figs.

Simnia patula

Lebour, M.V., 1937
The eggs and larvae of the British Prosobranchs with special reference to those living in plankton. JMBA. 22:105-166, 4 text figs.

Tornus subcarinatus
Lebour, M.V., 1937
The eggs and larvae of the British Prosobranchs with special reference to those living in plankton. JMBA. 22:105-166, 4 text figs.

Tricolia pullus
Lebour, M.V., 1937
The eggs and larvae of the British Prosobranchs with special reference to those living in plankton. JMBA. 22:105-166, 4 text figs.

Trivia arctica
Lebour, M.V., 1937
The eggs and larvae of the British Prosobranchs with special reference to those living in plankton. JMBA. 22:105-166, 4 text figs.

Trivia arctica (larvae)
Wiborg, K.F., 1944
The production of zooplankton in a landlocked fjord, the Nordåsvatn near Bergen, in 1941-42, with special reference to the copepods.... (Repts. Norwegian Fish. and Mar. Invest.) 7(7):83 pp., 40 text figs.

Trivia monarcha
Lebour, M.V., 1937
The eggs and larvae of the British Prosobranchs with special reference to those living in plankton. JMBA. 22:105-166, 4 text figs.

Trophon muricatus
Lebour, M.V., 1937
The eggs and larvae of the British Prosobranchs with special reference to those living in plankton. JMBA. 22:105-166, 4 text figs.

Triphora perversa
Lebour, M.V., 1937
The eggs and larvae of the British Prosobranchs with special reference to those living in plankton. JMBA. 22:105-166, 4 text figs.

Turbonilla elegantissima
Lebour, M.V., 1937
The eggs and larvae of the British Prosobranchs with special reference to those living in plankton. JMBA. 22:105-166, 4 text figs.

Turbonilla fenestrata
Lebour, M.V., 1937
The eggs and larvae of the British Prosobranchs with special reference to those living in plankton. JMBA. 22:105-166, 4 text figs.

Turritella communis
Lebour, M.V., 1937
The eggs and larvae of the British Prosobranchs with special reference to those living in plankton. JMBA. 22:105-166, 4 text figs.

Urosalpinx cinerea
Lebour, M.V., 1937
The eggs and larvae of the British Prosobranchs with special reference to those living in plankton. JMBA. 22:105-166, 4 text figs.

Velutina velutina
Lebour, M.V., 1937
The eggs and larvae of the British Prosobranchs with special reference to those living in plankton. JMBA. 22:105-166, 4 text figs.

Volutopsius norvegicus
Lebour, M.V., 1937
The eggs and larvae of the British Prosobranchs with special reference to those living in plankton. JMBA. 22:105-166, 4 text figs.

---

larvae, nemerteans
Davis, Charles C., 1965.
A study of the hatching process in aquatic invertebrates. XX. The blue crab, Callinectes sapidus, Rathbun. XXI. The nemertean, Carcinonemertes carcinophila (Kölliker). Chesapeake Science, 6(4):201-208.

larvae, nematode
Shimazu, Takeshi and Tomoo Oshima, 1972
Some larval nematodes from euphausiid crustaceans. In: Biological oceanography of the northern North Pacific Ocean, A.Y. Takenouti, Chief Editor, Idemitsu Shoten, Tokyo, 403-409.

larvae, pogonophora
Ivanov, A.V., 1957.
[Materials on the embryonic development of Pogonophora.] Zool. Zhurn., 36(8):1127-1144.

larvae, Pogonophora
Webb, Michael, 1964.
The larvae of Siboglinum fiordicum and a reconsideration of the adult body regions (Pogonophora). Sarsia, 15:57-68.

larvae, sponge
Bergquist, P.R. and M.E. Sinclair, 1973
Seasonal variation in settlement and spiculation of sponge larvae. Mar. Biol. 20(1):35-44.

larvae, sipunculid
Murine, V.V., 1965.
Some data on the structure of pelafospheres-Sipunculid larvae. (In Russian; English Abstract).
Zool. Zhurn., Akad. Nauk. SSSR. 44(11):1610-1619

sipunculids
Stephen A.C., 1966.
A collection of Sipuncula taken on the summit of Vema sea-mount, South Atlantic Ocean.
Ann. Mag. Nat. Hist., (13), 9 (97-98-99):145-145

larvae, Planctosphaera
Scheltema R.S. 1970
Two new records of Planctosphaera larvae (Hemichordata: Planctosphaeroidea).
Mar. Biol. 7(1): 47-48.

larvae, tunicates
Braconnot, Jean-Claude, 1970.
Contribution à l'étude des stades successifs dans le cycle des tuniciers pélagiques doliolide I, Les stades larvaire, oozoide, nourrice et gastrozoide. Arch. Zool. exp. gén, 111: 629-668. Also in: Trav. Stn zool., Villefranche-sur-Mer, 32, 1970.

larvae, tunicates
Braconnot, Jean-Claude, 1968.
Sur le développement de la larve du tunicier pélagique doliolide: Doliolum Dolioletta gegenbauri Ulj. 1884.
C.r.hebd.Séanc., Acad.Sci., Paris, (D)267 (6):629-630.

larvae, salps
Braconnet, Jean-Claude, 1964.
Sur le developpement de la larve de Doliolum denticulatum Q & G.
C.R., Acad. Sci., Paris, 259(23):4361-4363.

larvae, salps
Godeaux, J. 1968.
Observations sur le développement embryonaire des Doliolidae.
Rapp. Proc.-verb. Réun. Commn int. Explor. scient. Mer Médit. 19(3):535-536.

---

larvae, tunicates
Godeaux, J., 1961.
L'éléoblaste des Thaliacés.
Rapp. Proc. Verb., Reunions, Comm. Int. Expl. Sci. Mer Mediterranee, Monaco, 16(2):143-145.

larvae, tunicates
Godeaux, J., 1957-1958.
Contribution à la connaissance des Thaliacés. Embryogénèse et blastogénèse du complexe neural. Constitution et développement du stolon prolifère
Ann. Soc. Roy. Zool., Belgique, 88:5-285.

Larvae, tunicate
Godeaux, J., 1955
Stades larvaires du Doliolum.
Bull. Cl. Sci., Acad. R. du Belgique, (5):41:769-787.

ALGAE
Casual references only, but enough to start a search.
For uses, etc. etc. see: Subject Cumulation

algae
Aleem, A.A., 1956.
Quantitative underwater study of benthic communities inhabiting kelp beds off California.
Science 123(3188):183.

algae
Almodovar, L.R., 1964.
The marine algae of Bahia de Jobos, Puerto Rico.
Nova Hedwigia, 7(1/2):33-52.

all attached except for
Sargassum natans
Sargassum polyceratium

algae
Blinks, L.R., and Curtis V. Givan, 1961.
The absence of daily photosynthesis rhythm in some littoral marine algae.
Biol. Bull., 121(2):230-233.

algae
Blinova, E.I., 1966.
The basic types of littoral algae on the Murmansk coast. (In Russian).
Okeanologiia. Akad. Nauk. SSSR, 6(1):151-158.

algae
Blum, J.L., and J.T. Conover, 1953.
New or noteworthy Vaucheriae from New England salt marshes. Biol. Bull. 105(3):395-401, 29 textfigs.

algae
Braarud, Trygve, and N.A. Sörensen, 1956.
Second International Seaweed Symposium, Trondheim, July, 1955, Pergamon Press, 220 pp.

Papers not separately indexed. Sections on:
Chemical composition
Practical uses
Analytical methods
Microbiology

algae
Brandão Joly, Aylthon, 1965.
Marine flora of the tropical and subtropical western South Atlantic.
Anais Acad. bras. Cienc., 37 (Supl.):279-282.

algae
Brandao Joly, Aylthon, 1964.
Flora marinha do litoral norte do estado de São Paulo e regiões circunvizinhas.
Fac. Filos., Ciências e Lett., Univ. Sao Paulo, Boletim, No. 294 (Botânica, No. 21):393 pp., 59 pl.

algae, red
Burns, Richard L. and Arthus C. Mathieson, 1972.
Ecological studies of economic red algae. II. Culture studies of Chondrus crispus Stackhouse and Gigartina stellata (Stackhouse) Batters. J. exp. mar. Biol. Ecol, 8(1):1-6.

algae

Butcher, R.W., 1952.
Contributions to our knowledge of the smaller marine algae. J.M.B.A. 31(1):175-191, 2 color pls.

algae

Câmara Neto, C., 1971
Primeira contribuição ao inventário das algas marinhas bentônicas do litoral do Rio Grande do Norte. Bolm Inst. Biol. marinha, Univ. Fed. Rio Grande do Norte, Brasil 5:137-154.

Algae

Chapman, D. J., and V. J., 1961.
Life histories in the algae.
Annals of Botany, n.s., 25(100):547-561.

algae

Chapman, V.J., 1964
The Chlorophyta.
In: Oceanography and Marine Biology, Harold Barnes, Editor, George Allen & Unwin, Ltd., 2: 193-228.

algae

Chapman, V.J., 1961
The marine algae of Jamaica. 1. Myxophyceae and Chlorophyceae.
Published by the Institute of Jamaica, 159 pp.

algae

Chapman, V.J., 1957.
Marine algal ecology.
Botan. Rev., 23(5):320-350.

algae

Colinveaux, Llewellyn Hillis 1970
Marine algae of eastern Canada: a seasonal study in the Bay of Fundy.
Nova Hedwigia 19 (1/2): 139-157

algae

Cribb, A.B., 1960.
Records of marine algae from south-eastern Queensland V.
Univ. Queensland, Dept. Botany, Pap., 4(1):31 pp

algae

Collins, F.S., 1915.
Some algae from the Chincha Islands. Rhodora 17:89-96.

algae

Craigie, J.S., and J. McLachlan, 1964.
Excretion of colored ultraviolet-absorbing substances by marine algae.
Canadian J. Botany, 42(1):23-33.

algae

Cummings, William C. 1969
The biggest plants in the sea.
Oceans 2(2): 26-27 (popular)

algae

Dawson, E. Y, 1959.
Some marine algae from Canton Atoll
Atoll Research Bull., No. 65: 1-6.
Issued-Pac. Sci. Bd. Nat. Akad. Sci. - N R C.

algae

Dawson, E. Yale, 1957.
Marine algae from the Pacific Costa Rica gulfs.
Los Angeles City Mus., Contr. in Sci., 15:28 pp.

algae

Dawson, E. Yale, 1957.
Notes on eastern Pacific insular marine algae.
Los Angeles City Mus. Contrib. Sci., 8:8 pp.

algae

Dawson, E. Yale, 1956.
An annotated list of marine algae from Eniwetok Atoll.
Pacific Science, 11(1):92-132.

algae

Díaz-Piferrer, Manuel, 1967.
Las algas superiores y fanerógamas marinas.
In: Ecología marina, Monogr. Fundación La Salle de Ciencias Naturales, Caracas. 14:273-307.

algae

Diaz-Piferrer, M., 1964.
Adiciones a la flora marina de las Antillas Holandesas Curazao y Bonaire.
Caribbean J. Sci., 4(4):513-543.

algae

Diaz-Piferrer, M., 1964.
Biogeografia de las algas marinas tropicales de la costa Atlantica de America. Resumen.
Bol. Inst. Biol. Mar., Mar del Plata, Argentina, No. 7:25.

algae

Dickinson, C.I., 1963.
British seaweeds.
London: Eyre and Spottiswoode, 232 pp. (25s.)
Reviewed: Newton, L., 1963. Nature, 199(4892):417

algae

Dixon, P.S., 1963
The Rhodophyta: some aspects of their biology
In: Oceanography and Marine Biology, H. Barnes Edit., George Allen & Unwin, 1:177-196.

algae

Earle Sylvia A. 1969.
Phaeophyta of the eastern Gulf of Mexico.
Phycologia 7(2): 71-254.

algae

Edelstein, Tikvah, 1964.
On the sublittoral algae of the Haifa bay area.
Vie et milieu, Bull. Lab. Arago, 15(1):177-212.

algae

Edwards Peter, 1970.
Illustrated guide to the seaweeds and sea grasses in the vicinity of Port Aransas, Texas.
Contrib. mar. Sci., Port Aransas, 15 (Suppl.): 128pp

Kelp

Emery, K.O., and R.H. Tschudy, 1941
Transportation of rock by Kelp. Bull. G.S.A. 52:855-862.

algae

Ercegovic, Ante, 1964.
Division verticale et horizontale de la vegetation des algues Adriatiques et ses facteurs. (In Jugoslavian; French resume).
Acta Adriatica, 11(1):75-84.

algae

Ercegovic, Ante, 1964.
La flore Adriatique des algues vertes, brunes et rouges en fonction de la profondeur. (In Jugoslavian; French resume).
Acta Adriatica, 11(1):71-74.

algae

Etcheverry Daza, Hector, 1964.
Distribucion geografica de las algas del Pacifico.
Bol. Inst. Biol. Mar., Mar del Plata, Argentina, No. 7:17-23.

algae, red

Ferreira-Correia, M.M., 1969.
Epifitas de Digenia simplex (Wulfen) C. Agardh no estado do Ceará (Rhodophyta, Rhodomelaceae).
Arq. Ciên. Mar, Fortaleza, Ceará, Brasil, 9(1): 63-69

algae

Fritch, F.E., 1935
The structure and reproduction of the Algae. I. XVIII 791. Cambridge, England.

algae

Funk, G., 1955.
Beiträge zur Kenntnis der Meeresalgen von Neapel zugleich mikrophotographischer Atlas.
Pubbl. Staz. Zool., Napoli, 25(Suppl.):1-178, 30 pls., 36 textfigs.

algae

Generalova, V.H., 1950.
Seaweed of the Black Sea in the vicinity of the Karadagsk Biological Station.
Trudy Karadagsk Biol. Sta., 10:106-148.

algae

Giaccone, G., 1972.
Struttura, ecologia e corologia dei popolamenti a laminare dello Stretto di Messina e del Mare de Alboran. Mem. Biol. mar. Oceanogr. Messina (n.s.) 2(2): 37-59.

algae

Ginés, Hno., y R. Margalef, editores, 1967.
Ecología marina.
Fundación La Salle de Ciencias Naturales, Caracas Monografía 14: 711 pp.

algae

Hammer, Lieselotte, y Fritz Gessner, 1967.
La taxonomia de la vegetacion marina en la costa oriental de Venezuela.
Bol. Inst. Oceanogr., Univ. Oriente, 6(12):186-265.

algae

Howe, M.A., 1914.
The marine algae of Peru. Mem. Torrey Bot. Club, 15:185

algae

Humm, H.J., 1962.
Marine algae of Virginia as a source of agar and agaroids.
Virginia Inst. Mar. Sci., Spec. Sci. Rept., 37: 13 pp.

algae

Humm, H.J., & S.E. Taylor, 1961
Marine chlorophyta of the upper West Coast of Florida.
Bull. Mar. Sci. Gulf & Caribbean, 11(3): 321-380.

algae

John D.M., 1971.
The distribution and net productivity of sublittoral populations of attached macrophytic algae in an estuary on the Atlantic coast of Spain. Mar. Biol. 11(1): 90-97.

algae

Joly, Aylthon B., 1964.
Extensao da flora marinha tropical no sul do Brasil.
Bol. Inst. Biol. Mar., Mar del Plata, Argentina, No. 7:11-15.

algae

Jonsson, S., 1962.
Recherches sur des cladophoracées marines (structure, reproduction, cycles comparés, conséquences systématiques)(Parties 1-3).
Ann. Sci. Nat. Bot., 12(3)(1):25-191.

algae

Jorde, Ingerid, and N. Klavestad, 1963.
The natural history of the Hardangerfjord. 4. The benthoic algal vegetation.
Sarsia, 9:99 pp.

Oceanographic Index: Marine Organisms Cumulation, 1946-1973

algae
Juhl-Noodt, H., 1959
Las algas marinas de la costa peruana y las posibilidades de su utilizacion. Bol. Comp. Adm. del Guano, 35(5,6): 10-16; 16-23.

algae
Kanazawa, Akio, 1961
Studies on the vitamin B-complex in marine algae. 1. On vitamin contents. (In Japanese). Memoirs, Fac. Fish., Kagoshima Univ. 10: 38-69.

algae
Kingsbury, John M. 1969
Seaweeds of Cape Cod and the islands. Chatham Press, Inc. 212 pp.

algae
Kireeva, M.S., 1960.
Distribution and supply of macro-algae along the southern coasts. Biology of the seas. (In Russian).
Trudy Okeanogr. Komissii, Akad. Nauk, SSSR, 10(4):71-74.

algae
Kireeva, M.S., and T.F. Stcharova, 1957.
Materials on systematical composition and biomass of the seaweeds and the highest vegetation of the Caspian Sea. Trudy Inst. Okeanol., 23:125-137.

algae
Kuckuck, Paul, 1963.
Ectocarpaceen Studien. VIII. Einige Arten aus warmen Meeren.
Helgoländer Wiss. Meeresuntersuch., 8(4):361-382.
Posthumous.
Prefaced by Peter Kornmann

algae
Kuhnemann, Oscar, 1964.
Importancia de la vegetacion en biogeografia marina.
Bol. Inst. Biol. Mar., Mar del Plata, Argentina, No. 7:27-35.

algae
Kulebakina, L.G. and G.G. Polikarpov, 1967.
On algal radioecology of the Black Sea shelf. (In Russian; English abstract).
Okeanologiia, Akad. Nauk, SSSR, 7(2):279-286.

algae
Langen, P., 1961.
Über Unterschiede im Gehalt an labilem Phosphat zwischen Rot- und Braunalgen.
Pubbl. Staz. Zool. Napoli, 32(1):130-133.

algae
Lee, R.K.S. 1973.
General ecology of the Canadian Arctic benthic marine algae.
Arctic, 26(1):32-43

algae
Lemoine, Marie, 1965.
Algues calcaires (Melobesiées) recueillie par le Professeur Drach (croisière de la Calypso en mer Rouge, 1952).
Bull. Inst. Oceanogr., Monaco, 64(1331):20 pp.

algae
Levring, T., 1947.
Submarine daylight and the photosynthesis of marine algae. Medd. Oceanogr. Inst., Göteborg 14 (Göteborgs Kungl. Vetenskaps- och Vitterhets-samhälles Handl., Sjätte Följden, Ser. B, 5(6)): 1-89, 32 textfigs.

algae
Lindauer, V.W., V.J. Chapman, & M. Aiken, 1961
The marine algae of New Zealand. Part II. Phaeophyceae.
Nova Hedwigia, 3(2 + 3): 129 - 350.

algae
Margalef, Ramón, 1967.
Las algas inferiores.
In: Ecología marina. Monogr. Fundación La Salle de Ciencias Naturales, Caracas, 14:230-272.

Kelp
McLean, James H., 1962
Sublittoral ecology of kelp beds of the open coast area near Carmel, California.
Biol. Bull., 122(1):95-114.

algae
Meyer, K.I., 1961.
On the phylogeny of green algae (Chlorophycophyta).
Botan. Zhurn., Akad. Nauk, SSSR, 46(8):1073-1086.

algae
Migita, Seiji, 1964.
Freeze-preservation of Porphyra thalli in viable state. 1. Viability of Porphyra tenera preserved at low temperature after freezing in the sea water and freezing under half-dried conditions. (In Japanese; English abstract).
Bull., Fac., Fish., Nagasaki Univ., No. 17:44-54.

algae
Munda, I., 1963.
Kulturversuche mit Ascophyllum nodosum (L.) Le Jol. und Fucus vesiculosus L. in Median von verschneidenem Salzgehalt.
Botanica Marina, 5(2/3):84-96.

Algae
Neushul, M., 1965.
Diving observations of sub-tidal Antarctic marine vegetation.
Botanica Marina, 8(2/4):234-243.

algae
Nizamuddin, Mohammed, and Fritz Gessner, 1970.
The marine algae of the northern part of the Arabian Sea and of the Persian Gulf.
Meteor Forsch. Ergebn. (D) (6):1-42

algae
Ogino, C., 1955.
Biochemical studies on the nitrogen compounds of algae. J. Tokyo Univ., Fish., 41(2):107-152.

algae
Okumura, Ayako, Keiichi Oishi and Kiichi Murata, 1963.
Quality of Kombu, one of the edible seaweeds belonging to the Laminariaceae. VII. Water-extracting conditions of total and amino nitrogens.
Bull. Jap. Soc. Sci. Fish., 29(12):1089-1091.
(In Japanese; English abstract).

algae
Orvokki, Ravanko, 1965.
Algae collected in deep water on a cruise with m/s Aranda in the northern part of the Gulf of Bothnia.
Annales Botan. Fennica, 2(2):171-173.

Algae
Overbeck, Jürgen, 1965.
Die Meeresalgen und ihre Gesellschaften an den Küsten der Insel Hiddensee (Ostsee).
Botanica Marina, 8(2/4):218-233.

algae
Papenfuss, G.F., 1962.
Problems in the taxonomy and geographic distribution of Antarctic marine algae. (Abstract for Symposium Antarctic Biol., Paris, 2-8 Sept. 1962)
Polar Record, 11(72):320.
Also: SCAR Bull. #12.

algae
Parkes, H.M., 1958.
A general survey of the marine algae of Mulroy Bay, Co. Donegal. II.
The Irish Naturalist, 12(12):324-329.

algae
Pérès, J.M., 1967.
Les biocoenoses benthiques dans le système phytal
Recl. Trav. Stn. mar. Endoume 42(58):3-113.

algae
Petrov, K.M., 1967.
The vertical distribution of phytobenthos in the Black and Caspian seas. (In Russian; English abstract).
Okeanologiia, Akad. Nauk, SSSR, 7(2):314-320.

algae
Phillips, Ronald, C., 1960
Ecology and distribution of marine algae found in Tampa Bay, Boca Ciega Bay and at Tarpon Springs, Florida. Quart. J. Fla. Acad. Sc., 23(3): 222-260.

algae
Prescott G.W., 1968.
The algae: a review.
Houghton Mifflin Co., Boston, 436 pp.

algae
Printz, H., 1957
Norwegian occurrences of driftweed.
Norw. Inst. Seaweed Res. 15: 49 pp.

algae
Priou, M.-L., 1962.
Étude expérimentale et écologique des fluctuations de la teneur en eau chez quelques Fucacées.
Bull. Lab. Marit. Dinard, (48):3-112.

Algae
Salim, K.M., 1965.
The distribution of marine algae along Karachi coast.
Botanica Marina, 8(2/4):183-198.

algae
Saoane-Cambra, Juan, 1965.
Estudios sobre las algas bentónicas en la costa sur de la Peninsula Iberica (litoral de Cadiz)
Inv. pesq., Barcelona, 29:3-216.

algae
Scagel, Robert F. 1966.
Marine algae of British Columbia and northern Washington. 1. Chlorophyceae (Green Algae).
Nat. Mus. Canada, Bull. 207: 257 pp

algae
Scagel, Robert F., 1966.
The Phaeophyceae in perspective.
In: Oceanography and marine biology, H. Barnes, editor, George Allen & Unwin, Ltd., 4:123-194.

algae
Scagel, Robert F., 1963
Distribution of attached marine algae in relation to oceanographic conditions in the northeast Pacific. In: Marine Distributions.
Roy. Soc., Canada, Spec. Publ. No. 5:37-50.

algae
Schwenke, Heinz 1964.
Untersuchungen zur marinen Vegetationskunde. 1. Über den Aufbau der marinen Benthosvegetation im Westteil der Kieler Bucht (westliche Ostsee).
Kieler Meeresforsch. 22(2): 163-170.

algae
Schwenke, Heinz, 1964.
Vegetation and vegetationsbedingungen in der westlichen Ostsee (Kieler Bucht).
Kieler Meeresf., 20(2):157-168.

algae

Schwenke, Heinz, 1960
Neuere Erkenntnisse über die Beziehungen zwischen den Lebensfunktionen mariner Pflanzen und dem Salzgehalt des Meer-und Brackwassers. Kieler Meersf., 16(1): 28-37.

algae

Segawa, S., T. Sawada, M. Higaki and T. Yoshida, 1961
Studies on the floating seaweeds. VIII. The drifting movement of the floating seaweeds off northern coast of Kyushu.
Sci. Bull. Fac. Agric., Kyushu Univ., 19(1): 135-
In Japanese; English summary
Also in: Contrib. Dept. Fish., Fish. Res. Lab. Kyushu Univ., No. 7.

algae

Segawa, S., T. Sawada, M. Higaki and T. Yoshida, 1961
Studies on the floating seaweeds. VII. Seasonal changes in the amount of the floating seaweeds found on Iki and Tusima Passages.
Sci. Bull. Fac. Agric., Kyushi Univ., 19(1): 125-133.
Also in: Contrib. Dept. Fish., Fish Res. Lab., Kyushu Univ., No. 7.

algae

Segawa, S., Sawada, T., M. Higaki, T. Yoshida, and S. Kamura, 1961.
Studies on the floating seaweeds. VI. The floating seaweeds of the west Kyushu region.
Sci. Bull. Fac. Agric., Kyushu Univ., 18(4): 411-417.
In Japanese; English summary

Contrib. Dept. Fish., Fish. Res. Lab., Kyushu, Univ. No. 7.

algae

Segawa, Sokichi, Takeo Sawada, Masahiro Higaki, Tadao Yoshida and Shintoku Kamura, 1961
The floating seaweeds of the sea to the west of Kyushu.
Rec. Oceanogr. Wks., Japan, Spec. No. 5:179-186.

Alga

Segawa, S., T. Sawada, M. Higaki and T. Yoshida, 1960.
Studies on the floating seaweeds. IV. Growth of some sargasseous algae based on the material secured from floating seaweeds.
Sci. Bull. Fac. Agric., Kyushu Univ., 17(4):429-435.

Also in:
Contrib. Dept. Fish., and Fish. Res. Lab., Kyushu Univ., No. 6.

Algae

Segawa, S., T. Sawada, and T. Yoshida, 1960.
Studies on floating seaweeds. V. Seasonal change in amount of floating seaweeds off the coast of Tsuyazaki.
Sci. Bull. Fac. Agric., Kyushu Univ., 17(4):437-441.
Also in:
Contrib. Dept. Fish. and Fish. Res. Lab., Kyushu Univ., No. 6.

algae

Shepherd, S.A., and N.B.S. Womersley, 1970.
The sublittoral ecology of West Island, South Australia. 1. Environmental features and the algal ecology.
Trans. R. Soc. S. Austr. 94: 105-137

algae

Sieburth, John McN., 1968.
A bacteriologist looks at the production of Norwegian seaweed meal.
Maritimes, Univ. R.I., 12(1):7-10 (popular)

algae

Span, Ante, 1964
Preliminary quantitative investigations of Cystoseirae in the Split area. (In Jugoslavian; English resume).
Acta Adriatica, 11(1):255-260.

algae

Steentoft, Margaret 1967.
A revision of the marine algae of São Tomé and Principe (Gulf of Guinea).
J. Linn. Soc. (Bot) 60 (382): 99-146

algae, blue-green

Stewart, W.D.P., 1962.
Fixation of elemental nitrogen by marine blue-green algae.
Annals of Botany, 26(103):439-446.

algae

Stschapova, T.F., and V.B. Voszhinskaya, 1960
[The algal flora of the littoral of the western Sakhalin coast.] Trudy Inst. Okeanol, 34: 123-146.

algae

Stscharova, T.F., and N.M. Selitskaya, 1957.
[Distribution of the seaweeds on the littoral of Moneron Island (Japan Sea)] Trudy Inst. Okeanol., 23:112-124.

algae

Tanaka, Takesi, 1965.
Studies on some marine algae from southern Japan VI.
Mem. Fac. Fish., Kagoshima Univ., 14:52-71.

algae

Taylor, Wm. Randolph, 1972.
Marine algae of the Smithsonian-Bredin Expedition to Yucatan - 1960. Bull. mar. Sci., Miami 22(1): 34-44.

algae

Taylor, Wm. Randolph, 1967.
Caulerpas of the Israel South Red Sea Expedition 1962.
Rep. Israel South Red Sea Exped., Sea Fish. Res. Stn, 24:13-17.
(Bull. Sea Fish. Res. Stn. Israel 43).

algae

Taylor, W.R., 1961.
Notes on three Bermudian marine algae.
Hydrobiologia, 18(4):277-283.

algae

Taylor, W.R., 1960
Marine algae of the Eastern Tropical and Subtropical coasts of the Americas.
Univ. Michigan Press, Ambassador Books, Ltd., Toronto, x + 870 pp., 14 text figs., 80 pls. $19.50.

Reviewed by:
A.J. Bernatowicz, Limnol. & Oceanogr., 6(1): 99-100.

ALGAE

Taylor, W. R., 1937.
Marine algae of the northeastern coast of North America. Univ. Michigan Press, 427 pp., 60 pls.

algae

Terekhova, T.K., 1972.
The effect of surf strength degree and flow rate on growth and development of the White Sea Fucales at summer. Gidrobiol. Zh. 8(2): 22-27. (In Russian; English abstract).

algae

del Val, M. Jesus, and M. Dolores Garcia Pineda, 1949.
Ensayos de algas marinas industriales.
Bol. Inst. Espanol Ocean., No. 13: 15 pp.

algae

Umamaheswararao, M., and T. Sreeramulu, 1964.
An ecological study of some intertidal algae of the Visakhapatnam coast.
J. Ecology, 52(3):595-616.

Algae

University of Southern California, Allan Hancock Foundation, 1965.
An oceanographic and biological survey of the southern California mainland shelf.
State of California, Resources Agency, State Water Quality Control Board, Publ. No. 27:232 pp. Appendix, 445 pp.

alga

Van Overbeck, J. and R.E. Crist, 1947
The role of a tropical green alga in beach sand formation. Am. J. Bot. 34:299-300.

algae

Villot, J.P., 1963.
Contribution à l'étude des Ulvacées de la region de Rabat (Station des Oudaias).
Trav. Inst. Sci. Chérifien, Ser. Bot., (26):49 pp.

algae

Von Brandt, A., 1956.
Algenwertung an der bretonischen Küste.
Die Fischwirtschaft, 8(7):168-169.

algae

von Stosch, H.A., 1962
Kulturexperiment und Oekologie bei Algen.
Kieler Meeresf., 18(3) (Sonderheft):13-27.

algae

Vozzhinskaya, V.B., 1967.
Studies of the ecology and vertical distribution of benthic algae in the Kandalaksha Bay of the White Sea. (In Russian; English abstract).
Okeanologiia, Akad. Nauk, SSSR, 7(6):1108-1118.

algae

Vozzhinskaya, V.B., 1965.
The distribution of algae near the shores of western Kamchatka. (In Russian).
Okeanologiia, Akad. Nauk, SSSR, 5(2):348-353.

algae

Vozzhinskaya, V.B., 1964
Floating algae in the western Pacific. (In Russian).
Okeanologiia, Akad. Nauk, SSSR, 4(5):876-883.

algae

Werner, M., 1948.
Répartition des algues sur le côte Atlantique du Maroc. Com. Océan. et d'Études de Côtes du Maroc, Ann. 1948, 3 fasc.:13-15.

algae

Widdowson, Thomas B. 1971.
Changes in the intertidal algal flora of the Los Angeles area since the survey by E. Yale Dawson in 1956-1959.
Bull. S. Calif. Acad. Sci. 70(1): 2-16.

algae

Widdowson, Thomas B., 1965.
A survey of the distribution of intertidal algae along a coast transitional in respect to salinity and tidal factors.
J. Fish Res. Bd., Canada, 22(6):1425-1454.

red alga

Womersley, H. B. S., and R. E. Norris, 1959.
A free floating marine red alga.
Nature, 184(4689):828-829.

algae

Zinova, A.D., 1957.
[Sea weeds in the eastern part of the Soviet sector of the Arctic] Trudy Inst. Okeanol., 23: 146-167.

algae, antibacterial effect

Glombitza, K.-W., 1969.
Antibakerielle Inhaltsstoffe in Algen. 1. Mitteilung. Helgolander wiss. Meersunters. 19(3): 376-384.

algae biomass

Vozzhinskaya, V.B. and N.M. Selitskaya 1970.
The composition, distribution and crops of the fucoid algae in the Okhotsk Sea. (In Russian; English abstract).
Trudy Inst. Okeanol. Akad. Nauk SSSR 88: 281-288.

algae caloric values

Paine, R.T. and R.L. Vadas, 1969.
Calorific values of benthic marine algae and their postulated relation to invertebrate food preference. Marine Biol., 4(2): 79-86.

algae, checklists for

Parke, Mary, and Peter S. Dixon, 1968.
Checklist of British marine algae - second revision.
J.mar.biol.Ass., U.K., 48(3):783-832.

algae, chemistry

Gryzhankova, L.N., N.V. Laktionova, E.A. Boichenko and A.V. Karyakin, 1973
Distribution of polyvalent metals in different types of algae. (In Russian; English abstract). Okeanologiia 13(4): 611-614.

ALGAE, Chemistry of

Krishnamurthy, V., editor, 1967.
Proceedings of the Seminar on Sea, Salt and Plants held at CSMCRI-Bhavnagar on December 20-23, 1965, Central Salt + Marine Chemicals Research Institute, 372 pp.

algae, chemistry of

Rao, M. Umamaheswara 1970.
The economic seaweeds of India.
Bull. Centr. mar. Fish. Res. Inst. 20: 68 pp. (mimeographed).

algae, brown

Ryndina, D.D., 1973
A role of some high molecular compounds of the brown algae in the strontium-90 extraction from the sea water. Gidrobiol. Zh. 9(2):34-39. (In Russian; English abstract).

algae, chemistry

Strogonov, A.A. and I.K. Ivanova, 1970.
Calcium in the Black Sea Cystoseira. (In Russian). Gidrobiol. Zh., 6(6): 99-101.

algae, chemistry of

Takagi, Mitsuzo, 1970.
Low molecular nitrogen compounds of marine algae.
Bull. Fac. Fish. Hokkaido Univ., 21(3): 227-233.

algae, chemistry of

Yamamoto, Toshio, Tetsuo Fujita and Masayoshi Ishibashi 1970.
Chemical studies on the seaweeds (25).
Vanadium and titanium content in seaweeds.
Rec. oceanogr. Wks, Japan 10(2):125-135.

algae, chemistry (boron)

Yamamoto, Toshio, Teruko Yamaoka, Tetsuo Fujita and Chikako Isoda 1971.
Chemical studies on the seaweeds (26).
Boron content in seaweeds.
Rec. oceanogr. Wks. Japan no. 11(1):7-13.

algae, chemistry

Youngblood, W.W. and M. Blumer 1973
Alkanes and alkenes in marine benthic algae. Mar. Biol. 21(3):163-172.

algae, chromosomes

Yabu, Hiroshi, 1973
Alternation of chromosomes in the life history of Laminaria japonica Aresch.
Bull. Fac. Fish. Hokkaido Univ. 23(4): 171-176.

algae, classification of

Lewin, Ralph A., Editor, 1962.
Physiology and biochemistry of algae.
Academic Press, New York and London, 929 pp.

algal communities

Nienhuis, P.H., 1970.
The benthic algal communities of flats and salt marshes in the Grevelingen, a sea-arm in the southwestern Netherlands.
Netherl. J. Sea Res. 5(1):20-49.

algae, coralline

Adey, Walter H. 1971.
The sublittoral distribution of crustose corallines on the Norwegian coast.
Sarsia 46: 41-58.

algae, coralline

Milliman, John D., Manfred Gastner and Jens Müller 1971.
Utilization of magnesium in coralline algae.
Bull. geol. Soc. Am. 82(3):573-580

algae, coralline

Schlanger, S.O., 1957.
Dolomite growth in coralline algae.
J. Sediment. Petrol., 27(2):181-186.

algal crusts

Conover, John T., 1962
Algal crusts and the formation of lagoon sediments.
The Environmental Chemistry of Marine Sediments, Proc. Symp., Univ. R.I., Jan. 13, 1962, Occ. Papers, Narragansett Mar. Lab., No. 1:69-76.

algae, cultures

Strand, John A., Joseph T. Cummins and Burton E. Vaughn, 1966.
Artificial culture of marine sea weeds in recirculation aquarium systems.
Biol.Bull., mar.biol.Lab., Woods Hole, 131(3): 487-500.

algae, ecology of

Fuller, Stephen W. and Arthur C. Mathieson, 1972.
Ecological studies of economic red algae. IV.
Variations of carrageenan concentration and properties in Chondrus crispus Stackhouse.
J. exp. mar. Biol. Ecol. 10(1): 49-58.

algae, ecology

Yanagida, Katsuhiko, Masahiro Kakiuchi and Yasuaki Tsuji 1971
Ecological study on Laminaria japonica var. ochotensis (Miyabe) Okamura in the vicinity of Monbetsu on the Okhotsk Sea coast of Hokkaido, Japan. (In Japanese; English abstract).
Scient. Repts Hokkaido Fish. exp. Sta. 13:1-18

algae, effect of

Ben-Avraham, Zvi, 1971.
Accumulation of stones on beaches by Codium fragile. Limnol. Oceanogr. 16(3): 553-554.

algae (endozoic)

Lewin, Ralph A., Editor, 1962.
Physiology and biochemistry of algae.
Academic Press, New York and London, 929 pp.

algae, exudations

Sieburth, John McN., 1969.
Studies on algal substances in the sea. III. The production of extracellular organic matter by littoral marine algae.
J. exp.mar.Biol.Ecol., 3(3):290-309.

algae (floating)

Yoshida, Tadao, 1963.
Studies on the distribution and drift of the floating seaweeds. (In Japanese; English abstract)
Bull. Tohoku Reg. Fish. Res. Lab., (23):141-186.

algae, life cycle

*Lorch, Jacob, 1968.
The history of the sexuality of marine algae.
Bull. Inst. oceanogr., Monaco, No. special 2: 397-406.

algae, lists of genera

Díaz-Piferrer, M., 1971
Distribución de la flora marina bentónica del mar Caribe. Symp. Investigations and resources of the Caribbean Sea and adjacent regions, UNESCO, 18-26 Nov. 1968, Curaçao; 385-397.

algae, lists of spp.

Abbott, I.A., 1961.
A check list of marine algae from Ifaluk Atoll, Caroline Islands.
Atoll Res. Bull., No. 77:5 pp.

algae

*Cardinal, André, 1967.
Inventaire des algues marines benthiques de la Baie des Chaleurs et de la Baie de Gaspé (Québec).II. Chlorophycées.
Naturaliste Canadien, 94(4):447-469.

algae, list of species

Cardinal, André, 1966.
Additions à la liste des algues benthiques de la Baie des Chaleurs.
Rapp. Stn. Biol. Mar., Grande-Rivière, 1965:35-43.

algae, lists of spp.

Cardinal, André, 1965.
Liste préliminaire des algues benthiques de la Baie-des-Chaleurs.
Rapp. Ann., Sta. Biol. Mar., Grande Rivière, 1964:41-51.

algae, lists of spp.

Celan, M., et A. Bavaru 1973.
Aperçu général sur les groupements algaux des côtes roumaines de la mer Noire. Rapp. Proc.-v. Réun Commn int. Explor. scient. Mer Médit. Monaco 21(9):655-656.

algae, lists of spp.

Dawson, E. Yale, 1962.
Additions to the marine flora of Costa Rica and Nicaragua.
Pacific Naturalist, 2(13):375-395.

algae, lists of spp.

Dawson, E.Y., 1962.
Benthic marine exploration of Bahia de San Quintin, Baja California, 1960-1961. No. 7.
Marine and marsh vegetation.
Pacific Naturalist, 3(6/7):275-280.

algae, lists of spp.

Dawson, E. Yale, 1959.
Marine algae from the 1958 cruise of the Stella Polaris in the Gulf of California.
L.A. County Mus., Contrib. in Sci., No. 27:3-39.

algae, lists of spp.

de la Campa de Guzmán, Sara, 1965.
Notas preliminares sobre un reconocimiento de la flora marina del estado de Veracruz.
Anales Inst. nac. Invest.biol.-pesqueras, México, 1: 9-49.

algae, lists of spp.
de Virville. Ad. Davy
La flore marine de la presqu'île de Quiberon.
Rev. Gén. Botan., Paris, 69(814):89-143.

algae, lists of spp.
Díaz-Piferrer, M., 1971
Distribución de la flora marina bentónica del mar Caribe. Symp. Investigations and resources of the Caribbean Sea and adjacent regions, UNESCO, 18-26 Nov. 1968, Curaçao: 385-397.

algae, lists of spp.
Díaz-Piferrer, M., 1970.
Adiciones a la flora marina de Venezuela.
Carib. J. Sci., 10(3-4):159-198.

algae, lists of spp.
El-Sayed, S.Z., W.M. Sackett, L.M. Jeffrey, A.D. Fredericks, R.P. Saunders, P.S. Conger, G.A. Fryxell, K.A. Steidinger and S.A. Earle 1972.
Chemistry, primary productivity and benthic algae of the Gulf of Mexico.
Ser. Atlas Mar. Environm. Am. Geogr. Soc. 22: 29 pp., 6 pls. (quarto)

algae, lists of spp.
Ferreira-Correia, M.M., e F. Pinheiro-Vieira, 1969
Terceira contribuição ao inventário das algas marinhas bentônicas do nordeste Brasileiro.
Arq. Ciên. Mar, Fortaleza, Ceará, Brasil, 9(1): 21-26.

algae, lists of spp.
Humm, Harold J., 1963.
Some new records and range extensions of Florida marine algae.
Bull. Mar. Sci. Gulf and Caribbean, 13(4):516-526

algae, lists of spp.
Karlström, O., D. Callieri and K. Bäck, 1965.
Studies on vitamin B12 in algae.
Arkiv för Kemi, 16(3/4):299-307.

algae, lists of spp.
Labanca, Leda, 1970.
Contribuição ao conhecimento da flora algológica marinha do nordeste Brasileiro. Trabhs oceanogr. Univ. Fed. Pernambuco 9/11: 325-436.

algae, lists of spp.
Lackey, James B., 1967.
The microbiota of estuaries and their roles.
In: Estuaries, G.H. Lauff, editor, Publs Am. Ass. Advmt Sci., 83:291-302.

algae, lists of spp.
Mountain, Joe A. 1972.
Further thermal addition studies at Crystal River, Florida, with an annotated check list of marine fishes collected 1969-1971.
Prof. Pap. Ser., Fla. Dept. Nat. Resources, St. Petersburg. 20: 103 pp.

algae, lists of spp.
Neushal, M., 1967.
Studies on sibtidal marine vegetation in western Washington.
Ecology, 48(1):83-94.

algae, lists of spp.
Russell, G., 1962.
Observations on the marine algae of the Isle of May.
Trans. Proc. Bot. Soc., Edinburgh, 39(3):271-289.

algae, lists of spp.
Scannell, M.J.P., 1969.
Unpublished records of marine algae made mainly in County Waterford by Thomas Johnson and Matilda Knowles.
Ir. Nat. J., 16(7): 192-198

algae, lists of spp.
Taylor, W.R., 1962.
Marine algae from the tropical Atlantic Ocean: V. Algae from the Lesser Antilles.
U.S. Nat. Mus., Contr. U.S. Nat. Herbarium, 36(2):43-62.

algae, lists of spp.
Tsuda, Roy T., 1966.
Marine benthic algae from The Leeward Hawaiian Group.
Atoll Res. Bull., 115:13 pp. (mimeographed).

algae, lists of spp.
Vozzhinskaya, V.B., and E.I. Blinova, 1970.
On the composition and distribution of algae off the Kamchatka (Okhotsk Sea). (In Russian; English abstract).
Trudy Inst. Okeanol. Akad. Nauk SSSR 88: 299-307.

algae, lists of spp.
Vozzhinskaya, V.B., and N.M. Selitskaya 1970.
The bottom flora of the Big Shantar Isle (Okhotsk Sea). (In Russian; English abstract).
Trudy Inst. Okeanol. Akad. Nauk SSSR 88: 289-298.

algae, lists of spp.
Zaneveld, Jacques S., and William D. Barnes, 1965.
Reproductive periodicities of some benthic algae in Lower Chesapeake Bay.
Chesapeake Science, 6(1):17-32.

algae, lists of spp.
Zinova, A.D., 1959.
List of marine algae of southern Sakhalin and the Southern Kurile islands. (In Russian).
Issled. Dal'nevostochnykh Morei, SSSR, 6:146-161.

Translation: USN Oceanogr. Off. TRANS. 310. (M. Slessers). 1967.

algal meal
Bastos, José Raimundo, Francisca Pinheiro-Vieira, Gustavo Hitzschky Fernandes Vieira, 1971
Informação preliminar sobre a farinha de algas marinhas. (In Portuguese; English summary). Arquivos Ciências Mar. Univ. Fed. Ceará, Brasil 11(2): 159-160.

algae
Augier, Henry, 1965.
Contribution à l'étude des facteurs de croissance des algues rouges.
Bull. Inst. Océanogr., Monaco, 65(1341):18 pp.

algae, physiol
Baslavskaya, S.S., 1961.
Inorganic nutrients as a factor in increasing the intensity and productivity of photosynthesis of algae. (In Russian).
Pervichnaya Produkt. Morey i Vnutrennikh Vod, 319-327

USN-HO-TRANS 203
M. Slessers 1963
P.O. 39010

algae, physiol.
Biebl, R., 1962.
Temperaturrestenz tropischer Meeresalgen.
Botanica Marina, 4(3/4):241-254.

algae, physiol.
Droop, M.R., and Susanne McGill, 1966.
The carbon nutrition of some algae: the inability to utilize glycollic acid for growth.
Jour. mar. biol. Assoc., U.K., 46(3):679-684.

algae, physiology of
Feldmann, N.L., and M.I. Lutova, 1963.
Variations de la thermostabilité cellulaire des algues en fonction des changements de la température du milieu.
Cahiers Biol. Mar., Roscoff, 4(4):435-458.

Fucus filiformis
Fucus vesiculosus
Fucus distichus
Fucus serratus

algae, physiol.
Halldal, Per, 1964
Ultraviolet action spectra of photosynthesis and photosynthetic inhibition in a green and a red alga.
Physiol. Plantarum, 17:414-421.

algae, physiology of
Kanwisher, John W., 1966.
Photosynthesis and respiration in some sea weeds.
In: Some contemporary studies in marine biology, H. Barnes, editor, George Allen & Unwin, Ltd., 407-420.

algal substances
Lefèvre, M., H. Jacob, and M. Nisbet, 1951.
Compatibilités et antagonismes entre algues d'eau douce dans les collections d'eau naturelles.
Trav. Assoc. Int. Limnol. Théorique et Appliquée, 11:224-229.

algal substances
Lefèvre, M., H. Jakob, and M. Nisbet, 1950.
Sur la sécrétion par certaines Cyanophytes de substances algostatiques dans les collections d'eau naturelles. C.R. Acad. Sci., Paris, 230: 2226-2227.

Algae, Physiol.
Linskens, H.F., 1963.
Oberflächenspannung an marinen Algen.
Proc. K. Nederl. Akad. Wetens., (C), 66(2):205-217.

algae, physiol.
Ogata, E. and W. Schramm, 1971.
Some observations on the influence of salinity on growth and photosynthesis in Porphyra umbilicalis. Mar. Biol. 10(1): 70-76.

algae, physiology
Parker, Bruce C., 1963
Translocation in the giant kelp Macrocystis.
Science, 140(3569):891-892.

algae (physiol-ecol)
Provosoli, Luigi, 1958
Nutrition and ecology of protozoa and algae.
Ann. Rev. Microbiol., 12: 279-308.

algae, physiol.
Rhee, G.-Yull, 1972
Competition between an alga and an aquatic bacterium for phosphate. Limnol. Oceanogr. 17(4): 505-514.

algae, physiol.
Wood, E.J. Ferguson, 1969.
Algae live without light.
Sea Frontiers 15(5): 278-283 (popular).

algae, physiology

Ukeles, R., 1961.
The effect of temperature on the growth and survival of several marine algal species. Biol. Bull. 120(2):255-264.

algae, productivity of

Mann, K.H., 1972.
Ecological energetics of the sea-weed zone in a marine bay on the Atlantic coast of Canada. II. Productivity of the seaweeds. Mar. Biol. 14(3): 199-209.

algae, primary productivity

Wassman, E.R. and J. Ramus, 1973
Primary-production measurements for the green seaweed Codium fragile in Long Island Sound. Mar. Biol. 21(4):289-297.

algae, radioactivity of

Unni, C.K. 1967.
Natural radioactivity of marine algae. In: Proc. Seminar, Sea, Salt and Plants, V. Krishnamurthy, editor, Bhavnagar, India, 264-272.

algae, species succession

Goldman, Joel C., 1973
Carbon dioxide and pH: effect on species succession of algae. Science 182(4109): 306-307.

algae, standing crops

Doty, Maxwell S., 1971.
Antecedent event influence on benthic marine algal standing crops in Hawaii. J. exp. mar. Biol. Ecol. 6(3): 161-166.

algal substances

Lefèvre, M., and H. Jakob, 1949.
Sur quelques propriétés des substances actives tirées des cultures d'Algues d'eau douce. C.R. Acad. Sci. Paris, 229:234-236.

algal substances

Lefèvre, M., M. Nisbet, E. Jakob, 1949.
Action des substances excrétées en culture par certaines espèces d'algues, sur le métabolisme d'autres espèces d'algues. Verhandl. Int. Verein. f. Theoret. u. Angewandte Limnol. 10:259-264.

algae, unicellular (effect of)

Kadota, Hajima, and Yuzaburo Ishida 1968.
Evolution of volatile sulfur compounds from unicellular marine algae. Bull. Misaki mar. biol. Inst. Kyoto Univ. 12: 35-48.

algae, vertical

Lüning, K., 1970.
Tauchuntersuchungen zur Vertikalverteilung der sublitoralen Helgoländer Algenvegetation. Helgoländer wiss. Meeresunters. 21(3): 271-291.

algae, zonation of

Delépine, R., et J.C. Hureau, 1963.
La végétation marine dans l'Archipel de Pointe Géologie (Terre Adélie). Bull. Mus. Nat. Hist. Nat., Paris, (2)35(1):108-115.

algae, zonation

Wulff, Barry L., and Kenneth L. Webb, 1969.
Intertidal zonation of marine algae at Gloucester Point, Virginia; Chesapeake Sci., 10(1): 29-35.

Anadyomene menziesii

Humm, H.J. 1956.
Rediscovery of Anadyomene menziesii, a deep-water green alga from the Gulf of Mexico. Bull. Mar. Sci., Gulf and Caribbean, 6(4):346-348

algae, Ascophyllum nodosum

Kohlmeyer, Jan, and Erika Kohlmeyer 1972.
Is Ascophyllum nodosum lichenized? Botanica Marina 15(2):109-112.

Ascophyllum nodosum

Sundene, Ove, 1973
Growth and reproduction in Ascophyllum nodosum (Phaeophyceae). Norwegian J. Bot. 20(2/3):249-255.

Ascophyllum nodosum

Yentsch, Charles S., and Carol A. Reichert, 1962.
The interrelationship between water-soluble yellow substances and chloroplastic pigments in marine algae. Botanica Marina, 3(3/4):65-74.

Botryocladia

Feldman, Genevieve, et Marcel Bodard, 1965.
Une nouvelle espèce de Botryocladia des côtes du Sénégal. Bull. Inst. Océanogr., Monaco. 65(1342):14 pp.

Caulerpa charoides

Thivy, Francesca, and V. Visalakshmi, 1963.
Caulerpa charoides (Harv. ex W.v.Bosse) nov. comb., a new record for the Indian Ocean region. Botanica Marina, 5(4):101-104.

Caulerpa olliverieri

Hine, A.E. and H.J. Humm, 1971.
Caulerpa olliverieri in the Gulf of Mexico. Bull. mar. Sci. Miami, 21(2): 552-555.

Chondrus crispus

Cardinal, André, 1966.
La mousse d'Irlande (Chondrus crispus) dans la Baie des Chaleurs et la Baie de Gaspé. Rapp. Stn.Biol. Mar., Grande-Rivière,1965:113-116.

Chondrus crispus

Lilly, George R. 1968.
Some aspects of the ecology of Irish moss Chondrus crispus (L.) Stack, in Newfoundland waters. Techn. Rept. Fish. Res. Bd., Can. 43: 44 pp. (unpublished manuscript).

Cladophora

Söderström, Jahan, 1965.
Remarks on some species of Cladophora in the sense of Van den Hoek and of Söderström. Botanica Marine, 8(2/4):169-182.

Clathromorphum compactum

Chave, Keith E., and Bradner D. Wheeler, Jr., 1965.
Mineralogic changes during growth in the red alga, Clathromorphum compactum. Science, 147(3658):621.

algae-Codium

Coffin, Gareth W., and Alden P. Stickney,1967.
Codium enters Maine waters. Fish.Bull.,Fish Wildl.Serv.,U.S., 66(1):159-161.

algae, Codium fragile

Wassman, E.R. and J. Ramus, 1973
Primary-production measurements for the green seaweed Codium fragile in Long Island Sound. Mar. Biol. 21(4):289-297.

Coelosphaerium sp.

Braarud, T., 1945
A phytoplankton survey of the polluted waters of inner Oslo Fjord. Hvalrådets Skrifter, No.28, 142 pp., 19 text figs., 17 tables.

Cosmarium sp.

Braarud, T., 1945
A phytoplankton survey of the polluted waters of inner Oslo Fjord. Hvalrådets Skrifter, No.28, 142 pp., 19 text figs., 17 tables.

Crucigenia tetrapeda

Braarud, T., 1945
A phytoplankton survey of the polluted waters of inner Oslo Fjord. Hvalrådets Skrifter, No.28, 142 pp., 19 text figs., 17 tables.

algae (Cystoseira)

Celan, M., et A. Bavaru. 1968.
Quelques observations sur l'embryologie des espèces de Cystoseira de la Mer Noire. In: Lucrările Sesiunii Stiintifice a Statiunii de Cercetari Marine "Prof Ioan Borcea", Agigea, (1-2 Noiembrie 1966), Volum Festiv, Iași, 1968: 95-100.

Cystoseira barbata

Müller, G.I., H.V. Skolka și N. Bodeanu 1969.
Date preliminare asupra populației lor algae și animale asociate vegetației de Cystoseira barbata de la litoralul românesc al Mării Negre. Hydrobiologia, Romania, 10:279-289.

Dasyclonium

Scagel, Robert F., 1962
The Genus Dasyclonium J. Agardh. Canadian J. Botany, 40:1017-1040.

Egregia laevigata

Chapman, V.J., 1961.
A contribution to the ecology of Egregia laevigata Setchell. Botanica Marina, 3(2):33-55.

Enteromorpha sp.

Saito, T., 1931
Researches in fouling organisms of the Ships' bottom. Zosen Kiokai (T. Soc. Naval Arch.) 47pp:13-64, 51 figs., 9 graphs.

Fucus

Faure, L., 1954.
Croissance des Fucus vesicularis et Fucus serratus. Sci. et Peches, Off. Sci. Tech. Peches Marit., 1(18):1-3.

used as fertilizer
harvested twice a year.

Fucus vesiculosus

Bidwell, R.G.S., and N.R. Ghosh, 1963
Photosynthesis and metabolism of marine algae. V. Respiration and metabolism of C14-labelled glucose and organic acids supplied to Fucus vesiculosus. Canadian J. Botany, 41(1):155-164.

Goniolithon sp.

Schmalz, R.F., 1965.
Brucite in carbonate secreted by the red alga Goniolithon sp. Science, 149(3687):993-996.

Goniolithon

Weber, Jon N., and John W. Kaufman, 1965.
Brucite in the calcareous alga Goniolithon. Science, 149(3687):996-997.

Goniotrichum elegans

Fries, L., 1959
Goniotrichum elegans: a marine red alga requiring Vitamin B 12. Nature 183(4660)558-559.

Halimeda batanensis n.sp.

Taylor, Wm. Randolph 1973.
A new Halimeda (Chlorophyceae, Codiaceae) from the Philippines. Pacific Sci. 27(1): 34-36.

Halimeda tuna

Beth, K., 1962.
Reproductive phases in populations of Halimeda tuna in the Bay of Naples. Problemi ecologici delle zone litorali del Mediterraneo, 17-23 luglio 1961.
Pubbl. Staz. Zool., Napoli, 32 (Suppl.):515-534.

Laminaria cloustoni

Walker, F.T., and W.D. Richardson, 1957.
Perennial changes of Laminaria cloustoni Edm. on the coasts of Scotland. J. du Cons., 22(3): 298-308.

Algae, Laminaria cucullata [a]

Svendsen, Per, and Joanna M. Kain (Mrs. N.S. Jones) 1971.
The taxonomic status, distribution, and morphology of Laminaria cucullata sensu Jorde and Klavestad.
Sarsia 46:1-21.

Laminaria hyperborea

Kain, Joanna M. (Mrs. N.S. Jones), 1963
Aspects of the biology of Laminaria hyperborea II. Age, weight and length.
J. Mar. Biol. Assoc., U.K. 43(1):129-151.

Laminaria hyperborea

Kain, Joanna M. 1962
Aspects of the biology of Laminaria hyperborea 1. Vertical distribution.
J. Mar. Biol. Assoc. U.K. 42(2):377-385.

algae, Laminaria hyperborea [a]

Larkum, A.W.D., 1972.
Frond structure and growth in Laminaria hyperborea. J. mar. biol. Ass. U.K. 52(2): 405-418.

Lithothamnion [a]

Gessner, Fritz, 1970.
Lithothamnion-Terrassen im Karibischen Meer.
Int. Revue ges. Hydrobiol., 55(5): 757-762.

Macrocystis pyrifera

Wohnus, J.F., 1942
The development of the sporophyte of Macrocystis pyrifera. Turtox News. 20:133-136.

Phormidium persicinum

Pintner, I.J., and L. Provosoli, 1958.
Artificial cultivation of a red-pigmented marine blue-green alga, Phormidium persicinum.
J. Gen. Microbiol., 18(1):190-197.

Pediastrum boryanum

Kokubo, S., and S. Sato, 1947
Plankters in Jû-San Gata. Physiol. and Ecol. (Japan) 1(4):1-16, 3 text figs., tables.

Pediastrum duplex

Kokubo, S., and S. Sato, 1947
Plankters in Jû-San Gata. Physiol. and Ecol. (Japan) 1(4):1-16, 3 text figs., tables.

Phyllophora nervosa

Chernov, G.L., 1972
Some aspects of external metabolism of the Black Sea alga Phyllophora nervosa. (In Russian).
Gidrobiol. Zh. 8(5): 100-103.

Porphyra yezoensis

Terumoto, Isao, 1965.
Freezing and drying in a red marine alga Porphyra yezoensis Veda. (In Japanese; English Summary).
Low Temp. Sci., Hokkaido Univ., (B) 23:11-20.

Pseudobryopsis

Cassie Vivienne 1969.
A free-floating Pseudobryopsis (Chlorophyceae) from New Zealand.
Phycologia 8(2):71-77.

Pseudobryopsis

Diaz-Piferrer, M., 1965.
A new species of Pseudobryopsis from Puerto Rico.
Bull. Mar. Sci., 15(2):463-474.

Rhodella maculata

Paasche, E. and J. Throndsen, 1970.
Rhodella maculata Evans (Rhodophyceae, Porphyridiales) isolated from the plankton of the Oslo Fjord.
Nytt Mag. Bot. 17(3/4): 209-212.

Sargassum

Segawa, Sokichi, Takeo Sawada, Masahiro Higaki, Tadao Yoshida and Shintoku Kamura, 1961
The floating seaweeds of the sea to the west of Kyushu.
Rec. Oceanogr. Wks., Japan, Spec. No. 5: 179-186.

Sargassum

Segawa, S., T. Sawada, M. Higaki and T. Yoshida, 1960.
Studies on the floating seaweeds. IV. Growth of some sargasseous algae based on the material secured from floating seaweeds.
Sci. Bull. Fac. Agric., Kyushu Univ., 17(4):429-435.

Also in:
Contrib. Dept. Fish., and Fish. Res. Lab., Kyushu Univ., No. 6.

Sargassum

Segawa, S., T. Sawada and T. Yoshida, 1960.
Studies on floating seaweeds. V. Seasonal change in amount of floating seaweeds off the coast of Tsuyazaki.
Sci. Bull. Fac. Agric., Kyushu Univ., 17(4):437-441.

Also in:
Contrib. Dept. Fish., and Fish. Res. Lab., Kyushu Univ., No. 6.

sargassum

Yoshida, T., 1961
A brief study on the Sargassum vegetation around Ushibuka, west Kyushu, Japan.
Japan J. Ecology, 11(5):191-194.
In Japanese; English summary.
Also in Contrib. Dept., Fish., Fish. Res. Lab. Kyushu Univ., No. 7.

Sargassum fluitans

Parr, A.E., 1939
Quantitative observations on the pelagic sargassum vegetation of the North Atlantic.
Bull. Bingham Ocean. Coll. 6(7):1-94.

Sargassum hawaiiensis [a]

DeWreede, Robert E. and Everet C. Jones 1973
New records of Sargassum hawaiiensis Doty and Newhouse (Sargassaceae, Phaeophyta), a deep water species.
Phycologia 12 (1/2): 59-62.

Sargassum merrifieldii

Chauhen, V.D. and Francesca Thivy, 1965.
Sargassum merrifieldii J. Ag., new to the shores of India.
Phykos 4(2):69-70.

sargassum muticum

Scagel, R., 1956.
Introduction of a Japanese Sargassum muticum alga, into the northeast Pacific.
Washington Dept. Fish., Fish. Res. Papers, 1(4):49-58.

Sargassum natans

Parr, A.E., 1939
Quantitative observations on the pelagic sargassum vegetation of the North Atlantic.
Bull. Bingham Ocean. Coll. 6(7):1-94.

Scenedesmus sp.

Braarud, T., 1945
A phytoplankton survey of the polluted waters of inner Oslo Fjord. Hvalrådets Skrifter, No.28, 142 pp., 19 text figs., 17 tables.

Ulva pertusa

Saito, T., 1931
Researches in fouling organisms of the Ships' bottom. Zosen Kiokai (T. Soc. Naval Arch.) 47pp:13-64, 51 figs., 9 graphs.

Ulva mutabilis

Føyn, B., 1961.
Globose, a recessive mutant in Ulva mutabilis.
Botanica Marina, 3(2):60-64.

Whidbeyella cartilaginea

Scagel, Robert F., 1962
A morphological study of the red alga. Whidbeyella cartilaginea Setchell et Gardner.
Canadian J. Botany, 40: 1217-1222.

ALGAE, Blue-Green

marine forms - or those recorded from sea areas

Blue-green Algae

Forest, H.S., H.L. Change, M.M. Davis, 1959.
The possible application of bacterial cytological technique to the taxonomy of Blue-green algae.
Rev. Algologique, n.s. 4(3):170-180.

algae, blue green

Golubic, S., 1961
Die "Seebälle" - ein seltsamer Standort der Blaualgen.
Hydrobiologie, 18(1-2): 109-120.

blue green algae

Bruia, Lucian, 1968.
Contributions to the knowledge of the Cyanophyceae of the Romanian Black Sea littoral. I.
Trav. Mus. hist. nat. "Grigore Antipa" 8(1): 217-224.

algae, blue green, physiol

Halldal, Per, 1958.
Pigment formation and growth in blue-green algae in crossed gradients of light intensity and temperature.
Physiol. Plant., 11:401-420.

algae, blue-green [a]

Jayaraman, R. 1972.
On the occurrence of blooms of blue-green Algae and the associated oceanographic conditions in the northern Indian Ocean.
In: Taxonomy and Biology of Blue-Green Algae, T.V. Desikachary, editor, Univ. Madras, 428-432. (not seen)

algae (blue green)

Kalbe, Lothar, 1963.
Oscillatoria lanceaeformis nov. spec., eine Alge des baltischen Seenplanktons.
Arch. Protist., 106(4):591-592.

cyanophyceae, lists of spp.

Komarovsky, B., and Tikva Edelstein, 1960
Diatomeae and Cyanophyceae occurring on deep-water algae in the Haifa Bay Area.
Bull. Res. Counc., Israel, 9D(2): 73-92.

blue-green algae

Margalef, Ramón, 1967.
Las algas inferiores.
In: Ecologia marina. Monogr. Fundación La Salle de Ciencias Naturales, Caracas, 14:230-272.

algae, bluegreen

Van Baalen, C., 1962
Studies on marine blue-green algae.
Botanica Marina, 4(1/2): 129-139.

blue green algae

Van Baalen, C., 1961
Vitamin B_{12} requirement of a marine blue-green alga.
Science, 133(3468): 1922-1923.

Agmenellum quadruplicatum

Parsons, T.R., 1961
On the pigment composition of eleven species of marine phytoplankton.
J. Fish. Res. Bd., Canada, 18(6):1017-1025.

Agmenellum quadruplicatum

Parsons, T.R., K. Stephens and J.D.H. Strickland, 1961
On the chemical composition of eleven species of marine phytoplankters.
J. Fish. Res. Bd., Canada, 18(6):1001-1016.

Gloeocapsa sp.

Lillick, L.C., 1938
Preliminary report of the phytoplankton of the Gulf of Maine. Am. Mid. Nat. 20(3):624-640, 1 text fig, 37 tables.

Glaeocapsa

Lillick, L.C., 1937
Seasonal studies of the phytoplankton off Woods Hole, Massachusetts. Biol. Bull. LXXIII (3):488-503, 3 text figs.

Merismopedia sp.

Braarud, T., 1945
A phytoplankton survey of the polluted waters of inner Oslo Fjord. Hvalrådets Skrifter, No.28, 142 pp., 19 text figs., 17 tables.

Merismopedia sp.

Lillick, L.C., 1938
Preliminary report of the phytoplankton of the Gulf of Maine. Am. Mid. Nat. 20(3):624-640, 1 text fig, 37 tables.

Oscillatoria

Lillick, L.C., 1937
Seasonal studies of the phytoplankton off Woods Hole, Massachusetts. Biol. Bull. LXXIII (3):488-503, 3 text figs.

Oscillatoria

Sournia, A. 1968.
La cyanophycée Oscillatoria (= Trichodesmium) dans le plancton marin: taxonomie et observations dans le Canal de Mozambique.
Nova Hedwigia 15(1):1-12.

Oscillatoria (Trichodesmium) erythraea

Yentsch, Clarice M., Charles S. Yentsch and James P. Perras, 1972.
Alkaline phosphatase activity in the tropical marine blue-green alga, Oscillatoria erythraea ("Trichodesmium"). Limnol. Oceanogr. 17(5): 772-774.

Oscillatoria (Trichodesmium) thiebautii

Carpenter, Edward J., 1973
Nitrogen fixation by Oscillatoria (Trichodesmium) thiebautii in the southwestern Sargasso Sea. Deep-Sea Res. 20(3): 285-288.

Trichodesmium

Barth, Rudolf, 1967.
Observações sôbre ocorrência em massa de Cyanophyceae.
Publ. Inst. Pesquis. Marinha, Brasil 006: 5 pp.

Trichodesmium

Delsman, H.C., 1939.
Preliminary plankton investigations in the Java Sea. Treubia 17:139-181, 8 maps, 41 figs.

Trichodesmium

Dugdale, Richard C., John J. Goering and John H. Ryther, 1964
High nitrogen fixation rates in the Sargasso Sea and the Arabian Sea.
Limnology and Oceanography, 9(4):507-510.

Trichodesmium sp.

Galtsoff, P.S., 1948
Red Tide. Progress Report on the investigations of the cause of the mortality of fish along the west coast of Florida conducted by the U.S. Fish and Wildlife Service and Cooperating Organizations, Fish and Wildlife Service, Special Scientific Rept. No.46, 44 pp. (mimeographed), 9 figs.

Trichodesmium

Goering, John J., Richard C. Dugdale and David W. Menzel, 1966.
Estimates of in situ rates of nitrogen uptake by Trichodesmium sp. in the tropical Atlantic Ocean. Limnol. Oceanogr.,11(4):614-620.

Trichodesmium

Graham, H.W., J.M. Amison and K.T. Marvin, 1954.
Phosphorus content of waters along the west coast of Florida. U.S.F.W.S. Spec. Sci. Rept. 122:1-43.

Trichodesmium sp.

Gunter, G., R.H. Williams, C.C. Davis, and F.G. Walton Smith, 1948.
Catastrophic mass mortality of marine animals and coincident phytoplankton bloom on the West Coast of Florida, November 1946 to August 1947. Ecol. Mon., 18:309-324, 2 text figs.

Trichodesmium en masa!

Margalef, R., F. Cervignon and G. Yepez T., 1960
Exploracion preliminar de las caracteristicas hidrograficas y de la distribucion del fitoplancton en el area de la Isla Margarita (Venezuela).
Mem. Soc., Ciencias Nat. de la Salle, 22(57): 210-221.
Contribucion No. 2, Estacion de Investigaciones Marinas de Margarita, Fundacion La Salle de Ciencias Naturales.

Trichodesmium

Menon, M.A.S., 1945.
Observations on the seasonal distribution of the plankton, Trivandrum Coast. Proc. Indian Acad. Sci., Sect. B, 22(2):31-62, 1 textfig.

Trichodesmium

Mohler, W.A., 1941.
Eenblauwuicren-phenomeen aan het strand van Balikpapan. Naturw. Tijdschr. Ned. Indie 101(3):

mass mortality

Trichodesmium

*Nagasawa,Sachiko,and Ryuzo Marumo,1967.
Taxonomy and distribution of Trichodesmium (Cyanophyceae)in the Kuroshio water. (In Japanese: English abstract).
Inf.Bull.Planktol.Japan,Comm.No.Dr.Y.Matsue, 139-144.

Trichodesmium

Qasim, S.Z. 1972.
Some observations on Trichodesmium blooms.
In: Taxonomy and Biology of Blue-Green Algae, T.V. Desikachary, editor, Univ. Madras, 433-438 (not seen)

Trichodesmium

Riley, Gordon A., Peter J. Wangersky and Denise Van Hemert, 1964
Organic aggregates in tropical and subtropical surface waters of the North Atlantic Ocean.
Limnology and Oceanography, 9(4):546-550.

Trichodesmium

Satô, Shigekatsu, Maryse Nogueira Paranaguá and Enide Eskinazi (1963-1964) 1966.
On the mechanism of red tide of Trichodesmium in Recife, northeastern Brazil, with some considerations of the relation to the human disease "Tamandaré fever."
Trabs Inst. Oceanogr. Univ. Recife 5(5/6): 7-49.

Trichodesmium

Sournia, A. 1968.
La cyanophycée Oscillatoria (= Trichodesmium) dans le plancton marin: taxonomie et observations dans le Canal de Mozambique.
Nova Hedwigia 15(1):1-12.

Trichodesmium

Steven, D.M., and R. Glombitza 1972.
Oscillatory variation of a phytoplankton population in a tropical ocean.
Nature, Lond. 237 (5350):105-107.

Trichodesmium

Tokioka,Takasi, and Robert Bieri, 1966.
Juveniles of Macrosetella gracilis (Dana) from clumps of Trichodesmium in the vicinity of Seto. Publs Seto mar. biol. Lab., 14(3):177-184.

Trichodesmium sp.

Wheeler, J.E.G., 1939
Plankton investigations. Bermuda Biological Station. Second Report. October 1939. 7 pp. (typed), 5 figs. Plymouth, Oct. 23, 1939.

Trichodesmium erythraeum

Bowman, Thomas E., and L.J. Lancaster, 1965.
A bloom of the planktonic blue-green alga, Trichodesmium erythraeum in the Tonga Islands. Limnology and Oceanography, 10(2):291-293.

Trichodesmium erythraeum

Chidambaram, K., and M.M. Unny, 1944.
Note on the swarming of the planktonic algae Trichodesmium erythraeum in the Pamban area and its effect on the fauna. Current Science 13(10): 263.

Trichodesmium erythreum

Dangeard, P., 1927
Phytoplankton de la croisière du "Sylvana". Ann. Inst. Ocean., Monaco, n.s., 4(8):286-401, 54 text figs. (Feirer-Juin 1913).

Trichodesmium erythraeum

Galtsoff, P.S., 1949
The mystery of the red tide. Sci. Mon. LXVIII (2):108-117.

Trichodesmium erythraeum

* Iizuka, Shoji, and Haruhiko Irie, 1968.
Discoloration phenomena by microalgae in Nagasaki Pref. in 1966 and ecology of causative organisms, Olisthodiscus. (In Japanese; English abstract).
Bull. Fac. Fish., Nagasaki Univ., 26: 25-35.

Trichodesmium erythraeum

Nagabhushanam, A.K., 1967.
On an unusually dense phytoplankton 'bloom' around Minicoy Island (Arabian Sea) and its effect on the local tuna fisheries.
Current. Sci., India, 36(22):611-612.

Trichodesmium erythreum

Prabhu, M.S., S. Ramamurthy, M.H. Dhulkhed and N.S. Radhakrishnan, 1971.
Trichodesmium bloom and the failure of oil sardine fishery. Mahasagar, CSIR, India, June 1971: 62-64.

Trichodesmium erythraeum

Prabhu, M.S., S. Ramamurthy, M.D.K. Kuthalingam and M.H. Dhulkhed, 1965.
On an unusual swarming of the planktonic blue-green lagae Trichodesmium spp., off Mangalore. Current Science, 34(3):95.

Trichodesmium erythraeum

Qasim, S.Z., 1970.
Some characteristics of a Trichodesmium bloom in the Laccadives. Deep-Sea Res., 17(3): 655-660.

Trichodesmium erythraeum

Ramamurthy, V.D., 1972.
Procedures adopted for the laboratory cultivation of Trichodesmium erythraeum. Mar. Biol. 14(3): 232-234.

Trichodesmium erythraeum

Ramamurthy, V.D., 1970.
Antibacterial activity of the marine blue-green alga Trichodesmium erythraeum in the gastro-intestinal contents of the sea gull Larus brunicephalus. Marine Biol., 6(1): 74-76.

Trichodesmium erythaeum

Ramamurthy, V.D., 1970.
Antibacterial activity traceable to the marine blue green alga Trichodesmium erythraeum in the gastrointestinal contents of two pelagic fishes.
Hydrobiologia, 36(1): 159-163

Trichodesmium erythraeum

Ramamurthy, V.D., 1970.
Experimental study relating to red tide. Mar. Biol., 5(3): 203-204.

Trichodesmium erythraeum

Ramamurthy, V.D., and S. Krishnamurthy, 1967.
The antibacterial properties of marine blue-green alga Trichodesmuim erytheaeum (Ehr.). Current Science, 36(19): 524-525.

Trichodesmium erythraeum

Ramamurthy, V.D., and S. Krishnamurthy 1967
Effects of N:P ratios on the uptake of nitrate and phosphate by laboratory cultures of Trichodesmium erythraeum (Ehr.)
Proc. Indian Acad. Sci. (B) 45(2):43-48.

Trichodesmium erythraeum

Ramamurthy, V.D., R. Alfred Selvakumar and R.M.S. Bhargava 1972.
Studies on the blooms of Trichodesmium erythraeum (Ehr.) in the waters of the central west coast of India.
Current Sci. 41(22): 803-804.

Trichodesmium erythraeum

Ramamurthy, V.D., and R. Seshadri, 1966.
Effects of gibberellic acid (GA) on laboratory cultures of Trichodesmium erythraeum (Ehr.) and Melosira sulcata (Ehr.).
Proc. Indian Acad. Sci., (B), 64(3):146-151.

Trichodesmium erythaeum

Ramamurthy, V.D., and R. Seshadri, 1966.
Phosphorus concentration during red-water phenomenon in the near shore waters of Port Novo (S. India).
Current Science, 35(4):100-101.

Trichodesmium erythraeum

Veenhuyzen, J.C., 1879.
Communication on: Trichodesmium erythraeum. Naturw. Tijdschr. Ned. Indie, 38:150-151.

causes mass mortality along shore and in tide pools.

Trichodesmium hildebrantii

Prabhu, M.S., S. Ramamurthy, M.D.K. Kuthalingam and M.H. Dhulkhed, 1965.
On an unusual swarming of the planktonic blue-green algae Trichodesmium spp., off Mangalore. Current Science, 34(3):95.

Trichodesmium thiebautii

Calef, George W., and George D. Grice, 1966.
Relationship between the blue-green alga Trichodesmium thiebautii and copepod Macrosetella gracilis in the plankton off northeastern South America.
Ecology, 47(5):855-856.

Trichodesmium Thiebauitii

Dangeard, P., 1927
Phytoplankton de la croisière du "Sylvana". Ann. Inst. Ocean. Monaco, n.s., 4(8):286-401, 54 text figs. (Feirer-Juin 1913).

Trichodesmium thiebauti

Farran, G.P., 1932.
The occurrence of Trichodesmium thiebauti of the south coast of Ireland. Cons. Perm. Int. Expl. Mer, Rapp. Proc. Verb. 87:60-64, 1 textfig.

Trichodesmium thibauti

Japan, Kobe Marine Observatory 1967.
Report of the oceanographic observations in the sea south of Honshu from July to August 1963. (In Japanese)
Bull. Kobe Mar. Obs. 178:31-40.

Trichodesmium Thiebautii

Marukawa, H., 1921
Plankton lists and some new species of copepods, from the northern waters of Japan. Bull. Inst. Ocean. No.384, 15 pp., 3 pls., 1 chart. Monaco

Trichodesmium Thiebaultii

Schröder, B., 1900
Phytoplankton des Golfes von Neapel nebst vergleichenden Ausblicken auf das atlantischen Ozean. Mitt. Zool. Stat. Neapel, 14:1-38.

phytoplankton

Allen, E.J., 1919.
A contribution to the quantitative study of plankton. J.M.B.A. 12(1):1-8.

phytoplankton

Allen, W. E., 1945
Occurrences and abundances of marine plankton diatoms offshore in Southern California. Trans. Amer. Microsc. Soc., 64(1):21-24.

phytoplankton

Allen, W. E., 1939(1940).
Summary of results of twenty years of researches on marine phytoplankton. Proc. Sixth Pacific Sci. Congr., 3:577-583.

phytoplankton

Allen, W.E., 1939
Micropopepoda in marine phytoplankton catches Science 89:532-533.

phytoplankton

Allen, W.E., 1938
The Templeton Crocker Expedition to the Gulf of California in 1935 - the Phytoplankton. Trans. Amer. Micro. Soc., 67: 328-335.

phytoplankton

Allen, W.E., 1928.
Review of five years of studies of phytoplankton at Southern California piers, 1920-1924 inclusive. Bull. S.I.O., tech. ser., 1:357-401, 5 text figs.

phytoplankton

Anderson, George C., 1965.
Fractionation of phytoplankton communities off the Washington and Oregon coasts. Limnol. Oceanogr., 10(3):477-480.

phytoplankton

Anon., 1951
Bulletin of the Marine Biological Station of Asamushi 4(3/4): 15 pp.

phytoplankton (lists)

Anon. 1914
Ice observation, meteorology and oceanography in the North Atlantic Ocean. Rept. on the work carried out by the S.S. "Scotia", 1913, 139 pp.

phytoplankton

Atkins, W.R.G., 1945.
Conditions for the vernal increase in the phytoplankton and a supposed lag in the process. Nature 156:599.

phytoplankton

Atkins, W.R.G., 1928
Seasonal variation in the phosphate and silicate content of sea water during 1926 and 1927 in relation to the phytoplankton crop. JMBA, 15: 191-205.

phytoplankton

Atkins, W.R.G., and P.G. Jenkins, 1956.
Factors affecting the vernal phytoplankton outburst in the English Channel. Nature 177(4522): 1218-1219.

phytoplankton

Atkins, W.R.G., and P.G. Jenkins, 1953.
Seasonal changes in the phytoplankton during the year 1951-52 as indicated by spectrophotometric chlorophyll estimations. J.M.B.A. 31(3):495-508, 8 textfigs.

phytoplankton

Atkins, W.R.G., and M. Parke, 1951.
Seasonal changes in phytoplankton as indicated by chlorophyll estimation. J.M.B.A. 29(3):609-618.

nannoplankton

Aurich, H. J., 1949.
Die Verbreitung des Nannoplanktons im Oberflächenwasser vor der Nordfriesischen Küste. Ber. Deustchen Wiss. Komm. f. Meeresf., n.f., II(4): 403-405, 2 figs.

Phytoplankton

Averina, I.A., 1967.
Characteristics and distribution of the phytoplankton in the West African coastal waters in spring and summer 1960-61. (In Russian)
Atlant. nauchno- issled. Inst. rybn. khoz. okeanogr. (AtlantNIRO). Materialy Konferentsii po Rezul'tatam Okeanograficheskikh Issledovanii v Atlanticheskom Okeane, 150-157.

phytoplankton

Bainbridge, R., 1953.
Studies on the inter-relationships of zooplankton and phytoplankton. J.M.B.A. 32(2):385-447.

phytoplankton

Bal., D.V. and Pradhan, L.B., 1945.
A preliminary note on the plankton of Bombay harbour. Current Sci., 14(8): 211-212.

phytoplankton
Ballester Nolla, Antonio, 1962.
Métodos experimentales en el estudio del desarrollo y dinámica de las poblaciones fitoplanctónicas. (Abstract).
Assoc. Island Mar. Labs., 4th Meet., Curacao, 18-21 Nov., 1962:2.

phytoplankton
Banse, Karl, 1960
Bemerkungen zu meereskundlichen Beobachtungen vor der Ostküste von Indien.
Kieler Meeresf., 16(1): 214-220.

phytoplankton
Barg, Traute, 1943
Über den Fettgehalt der Diatomeen.
Ber. deutsch. bot. Ges., 61(1):13-27

phytoplankton
Barlow, J. P., 1958
Spring changes in the phytoplankton abundance in a deep estuary, Hood Canal, Washington. J. Mar. Res., 17: 53-67.

phytoplankton
Baron, G., 1938.
Étude du plancton dans le bassin de Marennes.
Rev. Trav. Off. Pêches Marit. 11(2):167-188, 2 textfigs.

phytoplankton
Barreda, M., 1957.
El plancton de la Bahía de Pisco.
Bol. C.A.G. 33(9):7-24
33(10):7-19
33(11):10-18.

phytoplankton
Beklemishev, C.W., 1960
[Phytoplankton research.]
Arktich. i Antarktich. Nauchno-Issled. Inst., Mezhd. Geofiz. God, Sovetsk. Antarkt. Eksped., Vtoraia Morsk. Ekspied., "Ob", 1956-1957, 7: 143-152.

Phytoplankton
Beklemishev, K. V., 1958.
[The dependence of the phytoplankton on the hydrological conditions in the Indian Sector of the Antarctic.] DAN, SSSR, 119(4):694-697.
Translation NIOT

phytoplankton
Beklevmishev, C.W., 1958
The interrelations between phyto- and zooplankton in the Bering and Okhotsk seas. (Abstract).
Proc. Ninth Pacific Sci. Congr., Pacific Sci. Assoc., 1957, 16(Oceanogr.):218.

phytoplankton
Beklemishev, K. V., 1958.
On the latitudinal zonation in the distribution of Antarctic phytoplankton.
Inform. Biull. Sovetsk. Antarkt. Exped., (3):35-36.

phytoplankton
Beklemishev, Constantin W. and Aida P. Nakonechnaya, 1972
Plankton of the North Pacific current.
In: Biological oceanography of the northern North Pacific Ocean, A.Y. Takenouti, Chief Editor, Idemitsu Shoten, Tokyo, 367-371.

phytoplankton
Bernard, Francis, 1967.
Research on phytoplankton and pelagic Protozoa in the Mediterranean Sea from 1953-1966.
Oceanogr. Mar.Biol., Ann.Rev., H.Barnes, editor, George Allen and Unwin, Ltd., 5:205-229.

phytoplankton
Bernard, F., 1961
Données sur la quantités moyennes de flagellés en sept regions de la Méditerranée comparées avec l'Atlantique tropical et l'Océan Indien.
Rapp. Proc. Verb., Réunions, Comm. Int. Expl. Sci. Mer Mediterranee, Monaco, 16(2):123-128.

phytoplankton
Bernard, Francis, 1960.
Rapports caractéristiques entre les principaux Unicellulaires du plancton, dénombrés en sept regions des mers chaudes.
C.R. Acad. Sci., Paris, 251:1585-1587.

phytoplankton
Bernard, F., 1939.
Étude sur les variations de fertilité des eaux Méditerranéenes. Climat et Nanoplancton à Monaco en 1937-38. J. du Cons. 14:228-241.

phytoplankton
Bernard, F., 1939.
Variations du nanoplancton et des sals nutritifs en mer profonde: les eaux cotières de Monaco en 1938. Cons. Perm. Int. Expl. de la Mer, Rapp. et Proc. Verb. 109(3):51-59.

phytoplankton
Bernard, M. F., 1938
Recherches récentes sur la densité du plancton méditerranéen. Rap. Proc. Verb des Réunion, Comm. Int. l'Expl. Sci. de la Méditerranée, n.s., XI:289-300.

phytoplankton
Bernard, F., 1937.
Resultats d'une annee de recherches quantitatives sur le phytoplancton de Monaco. Cons. Perm. Int. Expl. Mer, Rapp. Proc. Verb. 105:28-31, 5 textfig.

Phytoplankton
Bernard, F., and J. Lecal, 1960.
Plancton unicellulaire récolté dans l'océan Indien par le Charcot (1950) et le Norsel (1955-56)
Bull. Inst. Oceanogr., Monaco, No. 1166:59 pp.

phytoplankton
Bernhard, M. and L. Rampi, 1965.
Horizontal microdistribution of marine phytoplankton in the Ligurian Sea.
Botanica Gothoburgensia 3: 13-24
Also: Com. Naz. Energia Nucleare RT/Bio (66).7.

phytoplankton
Besnard, W., 1950.
Considerações gerais em torno de região lagunar de Cananéia-Iguape. Bol. Inst. Paulista Oceanogr. 1(2):3-28.

phytoplankton
Bigelow, H. B., 1926.
Plankton of the offshore waters of the Gulf of Maine. Bull. USBF, 40(II):1-509.

phytoplankton
Bigelow, H. B., 1922
Exploration of the coastal water off the northeastern United States in 1916 by the U.S. Fisheries Schooner Grampus. Bull. M.C.Z. 65(5):85-188, 53 text figs.

phytoplankton
Bigelow, H.B., and M. Leslie, 1930
Reconnaissance of the waters and plankton of Monterey Bay, July 1928.
Bull. M.C.Z. 70(5):429-481, 43 text figs.

phytoplankton
Bigelow, H.B., L.C. Lillick, and M. Sears, 1940.
Phytoplankton and planktonic protozoa of the offshore waters of the Gulf of Maine. Pt. 1. Numerical distribution. Trans. Am. Phil. Soc., n.s., 31(3):149-191, 10 textfigs.

phytoplankton
Blackburn, M., et al., 1962.
Tuna oceanography in the eastern Tropical Pacific.
U.S.F.W.S. Spec. Sci. Rept. Fish., No. 400:48 pp.

Phytoplankton
Blanc, F. et M. Leveau, 1970.
Effets de l'eutrophie et de la dessalure sur les populations phytoplanctoniques.
Marine Biol., 5(4): 283-293.

phytoplankton
Bodeanu N., 1968.
Recherches sur la répartition du phytoplancton dans la zone de petite profondeur de la Côte roumaine de la Mer Noire.
Trav. Mus. Hist. nat. "Grigore Antipa", 8 (6):199-205.

Phytoplankton
Bodeanu, Nicolae, 1968.
Quelques caractéristiques du phytoplancton de la zone de faible profondeur du littoral roumain de la mer Noire.
Rapp. Proc-verb. Réun. Comm. int. Explor. scient. Mer Méditerranée, 19(3):561-565.

phytoplankton
Bogorov, V.G. 1967.
The Pacific Ocean: Biology of the Pacific Ocean. 1. Plankton. (In Russian)
Akad. Nauk SSSR, Inst. Okeanol. Isdatel. "Nauka", Moskva 266pp.

phytoplankton
Bogorov, B.G., 1938.
Biological seasons of the Arctic Sea. C.R. (Doklady) de l'Acad. des Sci. de l'URSS, 19(8): 641-643, 1 textfig.

phytoplankton
Bogorov, V.G., and K.V. Beklemishev, 1955.
[The production of phytoplankton in the N.W. part of the Pacific Ocean.] Dokl. Akad. Nauk SSSR, 104(1):141-

phytoplankton
Borde, J., 1938.
Étude du plancton au bassin d'Arcachon, des rivières et du golfe de Morbihan.
Rev. Trav. Off. Pêches Marit. 11(4):523-542.

phytoplankton
Bougis, P., P. Nival et S. Nival, 1968.
Distribution quantitative comparée du phytoplancton et des copépodes dans les eaux superficielles de la rade de Villefranche.
J. exp.mar.Biol.Ecol.,2(3):239-251.

phytoplankton
Braarud, T., 1945
A phytoplankton survey of the polluted waters of inner Oslo Fjord. Hvalrådets Skrifter, No.28, 142 pp., 19 text figs., 17 tables.

phytoplankton
Braarud, T., 1944
Planteplanktonets høstmaksimum ved Norskekysten. Blyttia 2:57-64, 1 textfig.

phytoplankton
Braarud, T., 1939
Observations on the phytoplankton of the Oslo Fjord, March-April, 1937. Nytt Magasin for Naturvidenskapene, 80:211-218, 1 text fig.

phytoplankton
Braarud, T., 1934
A note on the phytoplankton of the Gulf of Maine in the summer of 1933. Biol. Bull., 67:76-82

phytoplankton
Braarud, T., and Bjørg Føyn, 1958.
Phytoplankton observations in a brackish water locality of south-east Norway.
Nytt Mag. Botanik, 6:47-73.

phytoplankton

Braarud, T. Bjørg Føyn and Grethe Rytter Hasle 1958
The marine and fresh-water phytoplankton of the Dramsfjord and the adjacent part of the Oslofjord March-December 1951. Hvalradets Skrifter No. 43: 102 pp.

phytoplankton

Braarud, T., K.R. Gaarder, and J. Grøntved, 1953.
The phytoplankton of the North Sea and adjacent waters in May 1948. Rapp. Proc. Verb, Cons. Perm. Int. Expl. Mer. 133:1-87, 29 tables, Pls. A-B, 18 textfigs.

phytoplankton

Braarud, T., K. Ringdal Gaarder & O. Nordli, 1958
Seasonal changes in the phytoplankton at various points off the Norwegian west coast. (Observations at the permanent oceanographic stations 1945-46). Fisheridirek Skrifter 12(3): 5-77.

phytoplankton

Braarud, T., and B. Hope, 1952.
The annual phytoplankton cycle of a landlocked fjord near Bergen (Nordåsvatn). Rep. Norwegian Fish. Mar. Invest. 9(16):26 pp., 4 textfigs.

phytoplankton

Braarud, T. and J. T. Ruud, 1937
The Hydrographic conditions and aeration of the Oslo Fjord, 1933-1934. Hvalrådets Skr. No. 15:56 pp., 24 figs.

Reviewed: J. du Cons. XIV(3):406-408. J. N. Carruthers.

phytoplankton

Brandes, C. -H., 1939(1951).
Über die räumlichen und zeitlichen Unterschiede in der Zusammensetzunk des Ostseeplanktons. Mitt. Hamburg Zool. Mus. u. Inst. 48:1-47, 23 textfigs.

phytoplankton

Brewin, B.I., 1952.
Seasonal changes in the microplankton of the Otago Harbour during the years 1944 and 1945. Trans. and Proc. R.Soc., N.Z., 79(4):614-627, 5 textfigs.

phytoplankton

Brunel, J., 1962
Le phytoplancton de la Baie de Chaleurs. Inst. Botan., Univ. Montréal, Contrib. No. 77: 365 pp., 66 pls.

phytoplankton

Buch, K., 1952.
The cycle of nutrient salts and marine production. Rapp. Proc. Verb., Cons. Perm. Int. Expl. Mer, 36-46, 4 textfigs.

phytoplankton

Burkholder, Paul, R., 1960.
A survey of the microplankton of Lake Erie. Limnological survey of eastern and central Lake Erie, 1928-1929. USFWS Spec. Sci. Rept., Fish., No. 334 :123-144.

phytoplankton

Burkholder, Paul R., and John M. Sieburth, 1961.
Phytoplankton and chlorophyll in the Gerlache and Bransfield Straits of Antarctica, Limnol. & Oceanogr., 6(1):45-52.

phytoplankton

Burlini, G., e D. Voltolina 1967.
Nota preliminare sulla distribuzione quantitiva e qualitiva del fitoplancton in alto Adriatico.
Arch. Oceanogr. Limnol. 15(1):85-92.

phytoplankton

California, Humboldt State College, 1964.
An oceanographic study between the points of Trinidad Head and Eel River.
State Water Quality Control Bd., Resources Agency, California, Sacramento, Publ., No. 25: 136 pp.

phytoplankton

Carpenter, Edward J., 1971.
Effects of phosphorus mining wastes on the growth of phytoplankton in the Pamlico River Estuary.
Chesapeake Sci., 12(2): 85-94

phytoplankton

Caspers, H., 1951.
Quantitative Untersuchungen über die Bodentierwelt des Schwarzen Meeres im bulgarischen Küstenbereich. Arch. f. Hydrobiol. 45(1/2):192 pp., 66 textfigs.

phytoplankton

Cattley, J.G., 1954.
Zoo- and phytoplankton of the Flamborough Line, 1950-53. Ann. Biol., Cons. Perm. Int. Expl. Mer, 10:101-103, Fig. 38.

phytoplankton

Cheng, C., 1941
Ecological relations between the herring and the plankton off the north-east coast of England Hull Bull. Mar. Ecol., 1(5):239-254, 8 textfigs.

phytoplankton

Chu, S. P. 1946
Note on the technique of making bacteria-free cultures of marine diatoms. JMBA XXVI, No. 3, pp. 296-302

phytoplankton

Chu, S. P., 1946.
The utilization of organic phosphorus by phytoplankton. JMBA XXVI(3):285-295

phytoplankton

Clarke, G.L. and R.H. Oster, 1934
The penetration of the blue and red components of daylight into Atlantic coastal waters and its relation to phytoplankton metabolism. Biol. Bull., 47:59-75.

phytoplankton

Cleve, P.T., 1900
Plankton from the Red Sea. Öfversigt af Kongl. Vetenskaps-Akademiens Forhendlinges, No.9

phytoplankton

Cleve, P.T., 1897
A treatise on the phytoplankton of the Northern Atlantic and its tributaries. Upsala

phytoplankton

Coblentz-Michke, O.I., 1957.
On the production of phytoplankton in the northwestern part of the Pacific in the spring of 1955.
Doklady Akad. Nauk, SSSR, 116(6):1029-1032.

phytoplankton

Colebrook, J.M., and G.A. Robinson, 1965.
Continuous plankton records: Seasonal cycles of phytoplankton and copepods in the northeastern North Atlantic and the North Sea.
Bull. Mar. Ecol., 6(5):123-139, Pls. 42-43.

phytoplankton

Copenhagen, W. J. and L. D., 1949
Variation in the phytoplankton of Table Bay, October 1934 to October 1935. With a note on the calorific value of Chaetoceros spp. Trans. Roy. Soc. S. Africa, 32(2):113-123, 2 text figs.

phytoplankton

Corlett, J., 1953.
Net phytoplankton at Ocean Weather Stations I and J. J. du Cons. 19(2):178-190, 4 textfigs.

phytoplankton

Crawford, D.A., 1949.
A phytoplankton season in Cook Strait. Trans. R. Soc., N.Z., 77(5):173-177.

phytoplankton

Cushing, D.H., 1963
Studies on a Calanus patch. II. The estimation of algal productive rates.
J. Mar. Biol. Assoc. U.K., 43(2):339-347.

phytoplankton

Cushing, D.H., 1956.
Phytoplankton and the herring. 5.
Min. Afric., Fish., Food, Fish. Invest. (2) 20(4):1-19.

phytoplankton

Cushing, D.H., 1955.
Production and a pelagic fishery.
Fish. Invest., Min. Agric., Fish., & Food, (2), 18(7):1-104.

phytoplankton

Dajoz, R., 1959.
La mer Noire, la Caspienne et leurs annexes. Année Biol. (3), 35:1-41.

phytoplankton

Dandonneau, Y., 1971.
Étude du phytoplancton sur le plateau continental de côte d'Ivoire. 1. Groupes d'espèces associées. Cah. ORSTOM, sér. Océanogr. 9(2): 247-265.

phytoplankton

Das, S. M., 1954.
Submarine illumination in relation to phytoplankton. Science and Culture 19:528-534.

phytoplankton

Davidson, V.M., 1931.
Biological and oceanographic conditions in Hudson Bay. Contr. Canadian Biol. Fish., n.s., 6(26): 497-509, 7 textfigs.

phytoplankton

De Angelis, Costanzo M., 1961
Osservazioni sulle condizioni fisico-chimiche e sul fitoplancton del Canale Calambrone e del Fosso dei Navicelli (Livorno).
Boll. Pesca, Piscicolt. e Idrobiol, n.s., 16(2):307-332.

phytoplankton

DeDecker, A., 1964.
Observations on the ecology and distribution of Copepoda in the marine plankton of South Africa. S. Africa, Dept. Comm. & Industr., Div. Sea Fish. Invest. Rept., No. 49:33 pp.

Original:
Zur Ökologie und Verbreitung der Copepoden aus dem Meeresplankton Südafrikas.

Biol. Jaarboek, Dodonaea, Vol. 30, (1962), Antwerp & The Hague

phytoplankton

Defant, A., G. Böhnecke, H. Wattenberg, 1936.
I. Plan und Reiseberichte die Tiefenkarte das Beobachtungsmaterial. Die Ozeanographischen Arbeiten des Vermessungsschiffes "Meteor" in der Dänemarkstrasse und Irmingersee während der Fischereischutzfahrten 1929, 1930, 1933 und 1935. Veroffentlichungen des Instituts für Meereskunde, n.f., A. Geogr.-naturwiss. Reihe, 32:1-152 pp., 7 text figs., 1 plate.

phytoplankton

Degtyaryova, A.A., and E.A. Pavstiks, 1963.
Results of Soviet investigations on plankton in the Norwegian Sea and the Barents Sea, 1961.
Ann. Biol., Cons. Perm. Int. Expl. Mer, 1961, 18: 60-62.

phytoplankton

de Jager, B. v. D., 1957.
Variations in the phytoplankton of the St. Helena Bay area during 1954.
Union of S. Africa, Div. Fish., Invest. Rept., No. 25:78 pp.

phytoplankton
Delsman, H.C. 1939.
Preliminary plankton investigations in the Java Sea, Treubia, 17:139-181, 8 maps, 41 figs.

phytoplankton
De Mort, Carole L., Robert Lowry, Ian Tinsley and H.K. Phinney 1972
The biochemical analysis of some estuarine phytoplankton species. I. Fatty acid composition.
J. Phycol. 8(3):211-216.

phytoplankton
Denisenko, V.V., 1963.
Some data on the phytoplankton of the Adriatic Sea in July 1960. (In Russian).
Trudy Sevastopol Biol. Sta., 16:107-112.

phytoplankton
Denisenko, V.V., 1963.
Certain data on the seasonal and diurnal changes of phytoplankton in the Adriatic Sea. (In Russian).
Doklady, Akad. Nauk, SSSR, 151(5):1193-1194.
Abstr. in:
Soviet Bloc Res., Geophys., Astron., and Space, 1963(68):14.

phytoplankton
de Sousa e Silva, E., 1956.
Contribution à l'étude du microplancton de Dakar et des regions maritimes voisines. Bull. I.F.A.N 8(2):335-371, 7 pls.

phytoplankton
DeSousa e Silva, E., 1956.
Contribução para o estudo do microplancton marinho de Moçambique. Colect. Junta Invest. Ultramar Miss. Biol. Mar., 1(8):97 pp.

phytoplankton
Desrosieres, R., 1971.
Quelques stations de phytoplancton entre les îles Tuamotu et les îles Marquises (océan Pacifique central). Cah. ORSTOM, ser. Océanogr. 9(2):119-124.

phytoplankton
Desrosieres, Roger, 1969.
Surface macrophytoplankton of the Pacific Ocean along the Equator.
Limnol. Oceanogr. 14(4):626-632.

phytoplankton
Desrosieres, R., 1965.
Observations sur le phytoplancton superficiel de l'Océan Indien oriental. (no abstract).
Océanographie, Cahiers, O.R.S.T.R.O.M., 3(4):31-37.

phytoplankton
Digby, Peter S.B., 1960
Midnight-sun illumination above and below the sea surface in the Sörgat, N.W. Spitzbergen, and its significance to plankton.
J. Animal Ecology, 29:273-297.

phytoplankton (incl. lists of species)
Digby, P.S.B., 1953.
Plankton production in Scoresby Sound, East Greenland. J. Animal Ecology 22(2):289-322, 5 textfigs.

phytoplankton
Elizarov, A.A. and O.A. Movchan, 1973
Specific features of the vertical circulation and phytoplankton distribution in the north-western part of the Atlantic Ocean (Grand Newfoundland Bank). (In Russian; English abstract). Okeanologiia 13(4):662-668.

phytoplankton
El-Sayed, Sayed Z. 1973.
Biological oceanography.
Antarctic Jl U.S.A. 8(3):93-100.

phytoplankton
El-Sayed, Sayed Z., 1969.
Ecological studies of Antarctic marine phytoplankton.
Antarctic J., U.S. 4(5):193-194

phytoplankton
Enomoto, Y., 1957.
Studies on plankton in the west of Kyushu. 1. On the general seasonal succession of phytoplankton and zooplankton chiefly in 1954.
Bull. Seikai Res. Fish. Res. Lab. No. 11:1-10.

phytoplankton
Ercegovic, A., 1940
Weitere Untersuchungen über einige hydrographische Verhältnisse und über die Phytoplanktonproduktion in den Gewässern der östlichen Mitteladria. Acta Adriatica 2(3):95-134, 8 text figs.

phytoplankton
Ercegovic, A., 1936
Etudes qualitative et quantitatives du phytoplancton dans les eaux cotières de l'Adriatique oriental moyen au cours de l'année 1934. Acta Adriatica 1(9):1-126

phytoplankton
Ferguson Wood, E.J., 1951.
Phytoplankton studies in eastern Australia. Proc. Indo-Pacific Fish. Counc., 17-28 Apr. 1950, Cronulla N.S.W., Australia, Sects. II-III:60-63.

phytoplankton
Ferguson Wood, E.J., and P.S. Davis, 1956.
Importance of small phytoplankton elements. Nature 177(4505):438.

phytoplankton
Ferrando, H.J., 1958.
Red para fitoplancton con copo intercambiable.
Contra. Avulsas Inst. Ocean., Sao Paulo, 1:5 pp.

phytoplankton
Fish, C.J., 1925
Seasonal distribution of the plankton of the Woods Hole region. Bull. U.S.B.F., 41:91-179.

phytoplankton
Forti, A., 1922
Ricerche sulla flora pelagica (fitoplancton) di Quarto dei Mille. Mem. R. Com. Talass. Ital. 97:248 pp., 13 pls.

phytoplankton
Forti, A., 1922
Ricerhe su la flora pelagica (fitoplancton) di Quarto dei Mille (Mare Ligure). R. Com. Talass. It. Lab. Mar. di quarto dei Mille Presso Genova, Mem. XCVII

phytoplankton
Fedorov, V.D., 1973
Connection of the phytoplankton species diversity with changes in the mineral nutrition conditions. (In Russian; English abstract). Gidrobiol. Zh. 9(3):21-24.

phytoplankton
Føyn, B.R., 1929.
Investigations of the phytoplankton of Lofoten, March-April, 1922-1927. Skr. Norsk Vid.-Akad., Oslo, I. Math.-Naturvid. Klasse, 1928, No. 10.

phytoplankton
Føyn, Bjorg, 1953.
Svartediket, et vest-norsk oligotroft ferskvann. Fytoplankton-undersøgelser 1942-1943.
Univ. Bergen Aarbok, Naturv. rekke, 1952(18):1-19, figs.

phytoplankton
Fraser, J.H., 1959.
Scottish plankton investigations.
Ann. Biol., Cons. Perm. Int. Expl. Mer, 12:82-84.

phytoplankton
Fraser, J.H., 1956.
Plankton collected by the Scottish Research vessels.
Ann. Biol., Cons. Perm. Int. Expl. Mer, 11:52-54.

phytoplankton
Friedrich, H., 1950.
Versuch einer Darstellung den relativen Besiedlungsdichte in den Oberflächenschichten des Atlantischen Ozeans. Kieler Meeresf. 7(2):108-121, 4 textfigs.

phytoplankton
Gaarder, Karen Ringdal, 1938.
Phytoplankton studies from the Tromsø district, 1930-31. Tromsø Mus. Årshefter, Naturhist. Avd., 11, 55(1):159 pp., 4 fold-in pls., 12 textfigs.

phytoplankton, physiol.
Gabrielsen, E.K., and E. Steemann Nielsen, 1938.
Kohlensäureassimilation und Lichtqualität bei den marinen Planktondiatomeen (Vorlaufige Mitteilung) Cons. Perm. Int. Expl. Mer, Rapp. Proc. Verb. 108(2):20-21, 2 textfigs.

phytoplankton
Galtsoff, P.S., ed., 1954.
Gulf of Mexico, its origin, waters and marine life. Fish. Bull., Fish and Wildlife Service, 55:1-604, 74 textfigs.

phytoplankton
Gardiner, A.C., 1943.
Measurement of phytoplankton population by the pigment extraction method. J.M.B.A., n.s. 25:739-744.

phytoplankton
George, M.J., 1958.
Observations on the plankton of the Cochin backwaters.
Indian J. Fish., 5(2):375-401.

phytoplankton
Ghazzawi, F.M., 1939
Plankton of the Egyptian waters. A study of the Suez Canal Plankton. (A) Phytoplankton. Preliminary Report 83 pp. Notes and Mémoires, Min. Commerce-Industry, Egypt, Hydrobiol. & Fish. 65 figs.

phytoplankton
Giacomelli, Anna Maria, 1965.
Ricerche planctonologiche ital i ane dell' Anno geofisico internazionale 1957-58.
II Varaizioni stagionali del plancton presso Palermo.
Arch. Oceanogr. e Limnol. 14(2):265-307.

phytoplankton
Gibor, A., 1957.
Conversion of phytoplankton to zooplankton.
Nature 179(4573):1304.

phytoplankton
Gilbert, J.Y., 1942
The errors of the Sedgwick-Rafter counting chamber in the enumeration of phytoplankton
Trans. Am. Micros. Soc. 61:217-226.

phytoplankton
Gilbert, J.Y., and W.E. Allen, 1943
The phytoplankton of the Gulf of California obtained by the "E.W. Scripps" in 1939 and 1940. J. Mar. Res. V(2):89-110, figs. 30-31.

phytoplankton
Gillbricht, M., 1959
Das Phytoplankton im nördlichen Nordatlantischen Ozean im Spätwinter und Spätsommer 1958.
Deutsche Hydrogr. Zeits., Ergänzungsheft. Reihe B, 4(3):90-92.

Oceanographic Index: Marine Organisms Cumulation, 1946-1973

phytoplankton
Gillbricht, M., 1959.
Die Planktonverteilung in der Irminger See im Juni 1955.
Ber. Deutsch. Wiss. Komm. Meeresf., n.f., 15(3):260-275.

phytoplankton
Gillbricht, M., 1955.
Wucherungen von Phytoplankton in einem abgeschlossenen Hafenbecken.
Helgolander Wissenschaftliche Meeresuntersuchungen 5(2):141-168, 32 textfigs.

phytoplankton
Ginés, Hno., y R. Margalef, editores, 1967.
Ecologia marina.
Fundacion La Salle de Ciencias Naturales, Caracas Monografía 14: 711 pp.

phytoplankton
Glover, R.S., and G.A. Robinson, 1966.
The continuous plankton recorder survey: plankton around the British Isles during 1964.
Annls biol., Copenh., 21:56-60.

phytoplankton
Goering, J.J., D.M. Nelson and J.A. Carter, 1973
Silicic acid uptake by natural populations of marine phytoplankton. Deep-Sea Res. 20(9):777-789.

phytoplankton
Gonzalves, E.A., 1947.
Variations in the seasonal composition of the phytoplankton in Bombay Harbour. Current. Sci. 16(10):304-305.

Phytoplankton
Gogoleva, N.A., 1967.
The distribution of phytoplankton in the Norwegian Sea according to data obtained during June surveys in 1962-1965. (In Russian).
Mater. Rybokhoz. Issled. Severn. Basseina, 10:43-50. (PINRO)

phytoplankton
Gostan, Jacques, and Paul Nival, 1963.
Distribution hivernale des caractéristiques hydrologiques en mer Ligure et estimation de l'abondance du phytoplancton par la méthode des pigments.
Comptes Rendus, Acad. Sci. Paris, 257(19):2872-2875.

phytoplankton
Graham, H. W., 1943.
Last cruise of the Carnegie: Biological results of the Carnegie: The phytoplankton. Scientific Results of the cruise VII of the Carnegie 1928-1929, Biology IV:1-14, 4 tables.

phytoplankton
Graham, H. W., 1943.
Chlorophyll content of marine plankton.
J. Mar. Res., 5:153-160.

phytoplankton
Gran, H. H., 1933
Studies on the biology and the chemistry of the Gulf of Maine. II. Distribution of the phytoplankton in August 1932. Biol. Bull., 64(2):159-182.

phytoplankton
Gran, H. H., 1932
Phytoplankton. Methods and Problems. J. Cons., 7:343-358

phytoplankton
Gran H.H., 1927.
The production of plankton in the coastal waters off Bergen, March-April, 1922. Rept. Norwegian Fish. Mar. Invest. 3(8):74 pp., 8 textfigs.

phytoplankton
Gran H.H., 1915
The plankton production in the North European Waters in the spring of 1912. Bull. Plankt. Cons Int. Explor. Mer 1912

phytoplankton
Gran, H. H., 1912.
In: Murray and Hjort, "Depths of the Ocean"

phytoplankton
Gran, H. H., 1902.
Das Plankton des Norwegischen Nordmeeres von Biologischen und Hydrographischen Gesichtspunkten behandelt. Rep. Norw. Fish. Mar. Invest., II(5): 36-39, 173-175, pl. 1, figs. 1-9

phytoplankton
Gran, H.H., 1900.
Hydrographic-biological studies of the North Atlantic and the coast of Nordland.
Rept. Norwegian Fish. and Mar. Invest. 1(5):1-89, 2 textfigs., 39 hydro. tables, 13 plankton tables (Ocean)

phytoplankton
Gran, H.H., and T. Braarud, 1935
A quantitative study of the phytoplankton in the Bay of Fundy and the Gulf of Maine (including observations on hydrography, chemistry, and turbidity). J. Biol. Bd., Canada, 1(5):279-467, 69 text figs.

phytoplankton
Grøntved, J., 1962.
Preliminary report on the productivity of microbenthos and phytoplankton in the Danish Wadden Sea.
Medd. Dansk Fisk. Havundersøgelser, n.s., 3(7-12):347-371.

phytoplankton
Grøntved, J., 1952.
Investigations on the phytoplankton in the southern North Sea in May 1947. Medd. Komm. Danmarks Fisk.- og Havundersøgelser, Plankton Ser 5(5):1-49, 1 pl., 21 tables, 24 textfigs.

phytoplankton
Grøntved, J., 1949
Investigations on the phytoplankton in the Danish Waddensea in July 1941. Medd. Komm. Danmarks Fiskeri og Havundersøgelser, ser. Plankton, 5(2):55 pp., 2 pls., 38 text figs.

phytoplankton
Grøntved, J., 1949(1950).
Investigations on the phytoplankton in the Danish Waddensee in July 1941. Medd. Komm. Danmarks Fiskeri- og Havundersøgelser, Ser. Plankton, 5(2)(Medd. Skalling Lab. 10):55 pp., 2pls., 38 textfigs.

phytoplankton
Grøntved, J., 1940(1950).
Das Wattenmeer bei Skallinen. Physiographisch-Biologisch Untersuchung eines Dänischen Tidengebietes. No. 2. Quantitativen und qualitativen Untersuchungen des Mikroplanktons während der Gezeiten. Folia Geogr. Danica II(2)(Medd. Skalling Lab. 10):1-67, 33 textfigs.

phytoplankton
Gueorguieva, L.V., et I.S. Gueorguiev, 1968.
Répartition du phytoplancton dans la région près du Bosphore dans la mer Noire en rapport avec les conditions hydrologiques et hydrochimiques.
Rapp. Proc.-verb. Reun., Comm. int. Explor. scient. Mer Méditerranée 19 (3): 407-408.

phytoplankton
Guillén, Oscar, Blanca Rojas de Mendiola and Raquel Izaguirre de Rondán, 1973
Primary productivity and phytoplankton in the coastal Peruvian waters. In: Oceanography of the South Pacific 1972, Ronald Fraser, Compiler, N.Z. Nat. Commn. UNESCO, 405-418.

Phytoplankton
Gunnerson, C. G., and K. O. Emery, 1962.
Suspended sediment and plankton over San Pedro Basin, Calif. Limnol. & Oceanogr., 7(1):14-20.

phytoplankton
Gunter, G., R.H. Williams, C.C. Davis, and F.G. Walton Smith, 1948
Catastrophic mass mortality of marine animals and coincident phytoplankton bloom on the West Coast of Florida, November 1946 to August 1947. Ecol. Mon., 18:309-324, 2 text figs.

phytoplankton
Gusev, K.A., 1961.
Factors determining the development of phytoplankton in water basins. (In Russian)
Pervichnaya Produkt. Morey i Vnutrennikh Vod, 301-307.
USN-HO-TRANS 200
M. Slessers 1963.
P.O. 39010

phytoplankton
Hagmeier, E., 1961
Plankton-Aquivalente (Auswertung von chemischen und mikrokopischen Analysen0. Kieler Meeresf., 17(1):32-47.

phytoplankton
Halldal, P., 1953.
Phytoplankton investigations from Weather Ship M in the Norwegian Sea, 1948-49 (including observations during the "Armauer Hansen" cruise, July 1949). Hvalrådets Skrifter No. 38:91 pp., 20 tables, 21 textfigs.

phytoplankton
Halldal, Per and Kari Halldal, 1973
Phytoplankton, chlorophyll, and submarine light conditions in Kings Bay, Spitsbergen, July 1971. Norwegian J. Bot. 20(2/3):99-108.

phytoplankton
Hardy, A. C., G.T.D. Henderson, C.E. Lucas, and J.H. Fraser, 1936.
The ecological relations between the herring and the plankton investigated with the plankton indicator. JMBA, ns, XXI:147-291

Phytoplankton
Hargraves, Paul E., Robert W. Brody and Paul R. Burkholder, 1970.
A study of phytoplankton in the Lesser Antilles region. Bull. mar. Sci., 20(2): 331-349.

phytoplankton
Hart, T.J., 1963
Speciation in marine phytoplankton.
Systematics Association Publ. (Speciation in the Sea), No. 5:145-155.
Also in:
Collected Reprints, N.I.O., Vol. 11(442).1963.

phytoplankton
Hart, T.J., 1953.
Plankton of the Benguela Current. Nature 171 (4354):631-632, 2 textfigs.

phytoplankton
Hart, T. J., 1942.
Phytoplankton periodicity in Antarctic Surface waters. Discovery Reports, XXI: 261-356.

phytoplankton
Hart, T.J., 1934
On the phytoplankton of the southwest Atlantic and the Bellingshausen Sea, 1929-1931. Discovery Repts., 8:1-268.

phytoplankton
Harvey, H.W., 1947
Manganese and the growth of phytoplankton JMBA 26 (4):562-579, 2 text figs.

phytoplankton
Harvey, H. W., 1934
Measurement of phytoplankton population. JMBA, 19:761-773.

phytoplankton
Hasle, G.R., 1960
Phytoplankton and ciliate species from the tropical Pacific.
Skrif. Norske Vidensk.-Akad., Oslo, 1(2): 50 pp.

phytoplankton
Hasle, G.R., 1956.
Phytoplankton and hydrography of the Pacific part of the Antarctic Ocean. Nature 177(4509): 616-617.

phytoplankton
Hasle, G.R., 1954.
The reliability of single observations in phytoplankton surveys. Nytt Mag. Bot. 2:121-137, 3 textfigs.

phytoplankton
Hentschel, E., 1951.
Untersuchungen über das Plankton des Bornholm-beckens. Ber. Deutschen Wiss. Komm. f. Meeresf. n.s., 12(3):215-315, 49 textfigs., 3 fold-ins, 24 tables.

phytoplankton
Hentschel, E., 1942.
Die Planktonbevölkerung der Meere um Island. Ber. Deutsch. Wiss. Komm. Meeresf. n.f., 10(2): 117-194, 65 textfigs.

phytoplankton
Hjort, J., and H.H. Gran, 1900.
Hydrographic-biological investigations of the Skagerrak and the Christiana Fiord.
Rept. Norwegian Fish. and Mar. Invest. 1(2):41 pp.

phytoplankton
Holmes, R. W., 1958.
Size fractionation of photosynthesizing phytoplankton
U. S. F. W. S., Sp. Sci. Rept. Fisheries No. 279, pt. 2: p. 69-71.

phytoplankton
Holmes, R. W., and F. T. Haxo, 1958.
Diurnal variation in the photosynthesis of natural phytoplankton populations in artificial light.
U. S. F. W. Ser, Sp. Sci. Rept. Fisheris No. 279, Pt. 2, 73-76.

phytoplankton
Holmes, R.W., and T.M. Widrig, 1956.
The enumeration and collection of marine phyto-plankton. J. du Cons. 22(1):21-32.

phytoplankton
Hulburt, Edward M. 1963
The diversity of phytoplanktonic populations in oceanic, coastal and estuarine regions.
J. Mar. Res., 21(2):81-93.

phytoplankton
Hulburt, E.M., 1956.
The phytoplankton of Great Pond, Massachusetts.
Biol. Bull., 110(2):157-168.

phytoplankton
Hulburt, Edward M., and Nathaniel Corwin, 1970.
Relation of the phytoplankton to turbulence and nutrient renewal in Casco Bay, Maine.
J. Fish. Res. Bd. Can., 27(11): 2081-2090.

phytoplankton a
Ibanez, F., 1972
Interprétation de données écologiques par l'analyse des composantes principales: ecologie planctonique de la Mer du Nord.
J. Cons. int. Explor. Mer 34(3): 323-340.

phytoplankton, maxima
Iizuka, Shoiji, 1963
Records of maximum occurrence of phytoplankton in various localities. (In Japanese; English abstract).
Info. Bull. Plankt. Japan, 9:5-9.

phytoplankton
Indelli, E., 1944.
Il microplancton de superficie del Golfo di Napoli. Acta Pontificia Acad. Sci. 8(11):91-100.

not seen.

plankton (remarks only)
India, Naval Headquarters, New Delhi, 1958.
Indian oceanographic station list, Ser. No. 2: unnumbered pp. (mimeographed).

phytoplankton
C. Iselin, 1930
A report on the coastal waters of Labrador based on explorations of the "Chance" during the summer of 1926. Proc. Am. Acad. Arts and Sci., 66(1):27. pp., 14 textfigs.

phytoplankton
Ivanov, A. I., 1959.
Some peculiarities of Antarctic water phytoplankton in the area of whaling of the "Slava" fleet in 1957/58.
Info. Biull., Sovetsk. Antarkt. Exped., No. 10:29-32.

Phytoplankton (data only)
Japan, Central Meteorological Observatory, 1955.
The results of marine meteorological and oceanograph-ical observations. Pt. 1. Oceanography January-June 1955, No. 16:120 pp.

phytoplankton
Japan, Central Meteorological Observatory, 1953
The results of Marine meteorological and oceanographical observations. Jan.-June 1952, No. 11:362, 1 fig.

phytoplankton
Japan, Central Meteorological Observatory, 1952.
The Results of Marine Meteorological and oceanographical observations. Jan. - June 1950. No. 7: 220 pp.

phytoplankton
Japan, Central Meteorological Observatory, 1951.
The results of marine meteorological and oceanographic observations. July - Dec. 1949. No. 6: 423 pp.

phytoplankton
Japan, Hakodate Marine Observatory, 1961
[Report of the oceanographic observations in the sea east of Tohoku District from February to March 1959.]
Bull. Hakodate Mar. Obs., (8):3-7.

phytoplankton
Japan, Hakodate Marine Observatory, 1957.
[Report of the oceanographic observations in the sea east of Tohoku District from February to March, 1956.]
Bull. Hakodate Mar. Obs., No. 4:49-57.
1-9.

phytoplankton
Japan, Hakodate Marine Observatory, 1957
[Report of the oceanographic observations in the sea east of the Tohoku district in August 1956.]
Bull. Hakodate Mar. Obs., No. 4: 1-12.

phytoplankton
Japan, Hakodate Marine Observatory, 1957
[Report of the oceanographic observations in the sea east of Tohoku district from October to November 1956.]
Bull. Hakodate Mar. Obs., No. 4: 1-8.

phytoplankton
Japan, Hakodate Marine Observatory, 1957
[Report of the oceanographic observations in the Tsugaru Straits in July 1956.]
Bull. Hakodate Mar. Obs., No. 4: 13-21.

phytoplankton
Japan, Hakodate Marine Observatory, 1957.
[Report of the oceanographic observations of the Okhotsk Sea from April to May 1956.]
Bull. Hakodate Marine Observatory, No. 4:105-112.
1-8.

phytoplankton
Japan, Hakodate Marine Observatory, 1957.
[Report of the oceanographic observations in the sea east of Tohoku District from May to June 1956]
Bull. Hakodate Mar. Obs., No. 4:113-119.
9-15.

phytoplankton
Japan, Hakodate Marine Observatory, 1956.
[Report of the oceanographic observations in the sea off the Sanriku District in Feb.-Mar., 1955.]
Bull. Hakodate Mar. Obs., (3):11-16.

phytoplankton
Japan, Hakodate Marine Observatory, 1956.
[Report of the oceanographical observations east off Sanriku District in 1954.]
Bull. Hakodate Mar. Obs., (3):25-45.

phytoplankton
Japan, Hakodate Marine Observatory, 1956.
[Report of the oceanographic observations east off Tohoku District from August to September 1955.]
Bull. Hakodate Mar. Obs., (3):13-21.

phytoplankton
Japan, Hakodate Marine Observatory, 1956.
[Report of the oceanographic observations east of Tohoku District in November 1955.]
Bull. Hakodate Mar. Obs., (3):11-12.

phytoplankton
Japan, Hakodate Marine Observatory, 1956.
[Report of the oceanographical observations in the sea east off Tohoku District in May-June 1955.]
Bull. Hakodate Mar. Obs., (3):11-16.

phytoplankton
Japan, Kobe Marine Observatory, 1961
[Report of the oceanographic observations in the sea south of Honshu in March 1958.]
Bull. Kobe Mar. Obs., No. 167(21-22):30-36.

--from May to June, 1958(21-22):37-42
--from July to September, 1958(23-24):34-40
--from October to December, 1958(23-24):41-47
--from February to March, 1959(25-26):33-47.

phytoplankton
Japan, Kobe Marine Observatory, 1955.
[The outline of the oceanographical observations on the Southern Sea of Japan on board R.M.S. "Syunpu-maru" (Jan. 1955).]
J. Ocean., Kobe Mar. Obs., (2)6(1):1-19, 10 figs.

phytoplankton
Japan, Kobe Marine Observatory, 1955.
The results of the regular monthly surface observations in Osaka Bay.
J. Ocean., Kobe Mar. Obs. (2)6(1):20-27, 4 figs.

phytoplankton (surface)
Japan, Kobe Marine Observatory, 1954.
[The result of the regular monthly surface obser-vations on board the M.S. "Hayanami" in Osaka Bay]

Jan. 14-15, 1954 - J. Ocean. (2)5(1):20-21.
Feb.-Apr. 1954 - J. Ocean. (2) 5(4):1-5, 3 figs.
May 11, 1954 - J. Ocean. (2)5(5):12-13, 12 figs.

phytoplankton
Japan, Kobe Marine Observatory, 1954.
[The outline of the oceanographical observations off Shionomisaki on board the R.M.S. "Syunpu-maru" (May 1954).] J. Ocean. (2)5(5):1-11, 14 figs.

phytoplankton
Japan, Kobe Marine Observatroy, 1954.
The outline of the oceanographical observations in the southern area of Honshu on board the R.M.S. "Syunpu-maru".
Aug.-Sept.-1954 - J. Ocean. (2)5(9):1-44.
Oct. 1954 - J. Ocean. (2)5(10):1-44.
Oct. 1954 (Pt. 2) - (2)11(1-6).
J. Ocean.

phytoplankton
Japan, Kobe Marine Observatory, 1954.
The results of the regular monthly oceanographical observations on board the R.M.S. "Syunpu-maru" in the Kii Suido and Osaka Wan. J. Ocean. (2)5(1): 1-19, 16 figs.

phytoplankton
Japan, Kobe Marine Observatory, 1953.
The results of the regular monthly oceanographical observations on board the R.M.S. "Syunpu-maru" in the Kii Suido and Osaka Wan.
Jan. 1953 - J. Ocean. (2)4(1):1-10, 9 figs.
Feb. 1953 - J. Ocean. (2)4(2):1-9, 8 figs.
Mar. 1953 - J. Ocean. (2)4(3):1-10, 9 figs.
Apr. 1953 - J. Ocean. (2)4(5):1-9, 9 figs.
May 12-19, 1953 - J. Ocean. (2)4(6):1-12, 9 figs.
June 15-22, 1953 - J. Ocean. (2)4(6):13-22, 8 figs.
July 1953 - J. Ocean. (2)4(8):1-25, 29 figs.
Aug. 1953 - J. Ocean. (2)4(9):1-30, 27 figs.
Sept. 1953 - J. Ocean. (2)4(10):1-13, 9 figs.
Oct. 1953 - J. Ocean. (2)4(11):1-21, 15 figs.
Nov. 11-18, 1953 - J. Ocean. (2)4(12):1-15, 12 figs.

phytoplankton
Japan, Kobe Marine Observatory, 1953.
Report of the plankton collected by the R.M.S. "Syunpu maru" in Osaka Bay (Dec. 16-18, 1952).
J. Ocean. (2)4(1):11-12.

phytoplankton
Johnson, M.W., 1954.
Plankton of northern Marshall Islands. Bikini and nearby atolls, Marshall Islands.
Geol. Survey, Prof. Pap. 260-F:301-314, Fig. 101.

phytoplankton
Jones, Margaret and C.P. Spencer, 1970.
The phytoplankton of the Menai Straits. J. Cons. int. Explor. Mer, 33(2): 169-180.

phytoplankton
Karsten, G. 1905.
Das Phytoplankton des Antarktischen Meeres nach dem Material der deutschen Tiefsee-Expedition, 1898-1899. Wiss. Ergebn. Deutsch. Tiefsee Exped. 'Valdivia', Bd. II.

phytoplankton
Kawamura, Akito, 1967.
Observations of phytoplankton in the Arctic Ocean in 1964.
Inf.Bull.Planktol.Japan,Comm.No.Dr.Y.Matsue, 71-89.

phytoplankton
Kawareda, Y., 1956.
On the plankton association in Japan Sea. (1).
Bull. Hakodate Mar. Obs., No. 2:1-8.
95-102.

phytoplankton
Kawareda, Y., 1956.
Plankton associations in Japan Sea, and Tsugaru Straits in May, 1950.
Bull. Hakodate Mar. Obs., No. 2:1-6.
103-105.

phytoplankton
Kawarada, Y., 1953.
Plankton associations in Japan Sea and Tsugaru Straits in May 1950. J. Ocean. Soc., Japan, 9(2):103-108, 5 textfigs.

phytoplankton
Kawarada, Y., 1953.
On the plankton association in Japan Sea.
J. Ocean. Soc., Japan, 9(2):95-102, 7 textfigs.

phytoplankton
Ketchum, B.H., 1951.
Ch. 17. Plankton algae and their biological significance. Pp.335-346 in : Manual of Phycology - An Introduction to the Algae and their Biology. G.M. Smith, ed., Vol. 27 of "A New Series of Plant Science Books", Chronica Botanica Co.

phytoplankton
Kiselev, I.A., 1959.
The qualitative and quantitative composition of phytoplankton and its distribution in the waters of South Sakhalin and the southern Kurile Islands. (In Russian).
Issledovaniya dal' Nevostoachnykh Morey, 6:58-72.
Translation 305 Noo (M. Slessers).

phytoplankton
Kisselew, I. A., 1959.
The quantitative and qualitative composition of the phytoplankton and its distribution in sea waters of the South Sakhalin and South Kuril Islands.
Issledovaniya Dalievostochibikh Morei, 6:58-77.

phytoplankton
Kitou, M., and O. Asaoka, 1957.
Seasonal variation of phytoplankton at the Ocean Weather Station "X" (1950-1952 - II.
J. Ocean. Soc., Japan, 13(1):23-28.

phytoplankton
Kitou, M., and O. Asaoka, 1956.
Seasonal variation of phytoplankton at the ocean weather station "X" (1950-1952).
J. Ocean. Soc., Japan, 12(4):125-128.

phytoplankton
Koblentz-Mishke, O.J., 1961
Phytoplankton (taxonomy and primary production) in the northeastern Pacific in the winter of 1958-1959.
Trudy Inst. Okeanol., Akad. Nauk, SSSR, 45: 172-189.

phytoplankton
Koblentz-Mishke, O.I., 1958.
The distribution of certain varieties of phytoplankton in relation to the principal currents in the western part of the Pacific Ocean.
Doklady Akad. Nauk, SSSR, 121(6):1012-1014.

Translation NIOT/23

Phytoplankton
Koblenz-Mishke, O.I. and M.V. Kozlieninov, 1966.
Vertical distribution of phytoplankton and of transparency in the northern part of the Pacific. (In Russian).
Doklady, Akad. Nauk, SSSR, 166(2):459-461.

phytoplankton
Kokubo, S., 1952
Results of the observations on the plankton and oceanography of Mutsu Bay during 1950, reference being made also to the period 1946-1950. Bull Mar.Biol.Sta., Asamushi 5(1/4): 1-54, 3 tables,(fold-in), 1 fold-in.

phytoplankton
Kokubo, S., 1939(1940).
Quantitative studies of the neritic littoral microplankton of Japan collected at sixteen stations ranging from Saghalin to Formosa, 1931-1933. Proc. Sixth Pacific Sci. Congr., 3:541-564, 2 maps.

phytoplankton
Kon, H., 1956.
On the distribution of phytoplankton in the north-western part of the Pacific (Tohoku region) in autumn (Oct.-Nov., 1951).
Bull. Hakodate Mar. Obs., No. 2:1-6.
109-114.

phytoplankton
Kon, H., 1953.
On the distribution of the phytoplankton in the northeastern part of the Pacific (Tohoku district) in autumn. (Oct.-Nov. 1951).
J. Ocean. Soc., Japan, 9(2):109-114, 4 textfigs.

phytoplankton
Kondrat'eva, T.M., 1963.
Drainage fluctuations of phytoplankton in the Black Sea. (In Russian).
Trudy Sevastopol Biol. Sta., 16:53-70.

phytoplankton
Koshevoy, V.V., 1959
Phytoplankton observations near the Karadag shore, Black Sea.
Biull. Okeanograf. Komissii, Akad. Nauk, SSSR, (3): 40-45.

phytoplankton
Krey, J., 1954.
Beziehungen zwischen Phytoplankton, Temperatursprungschicht und Trübungsschirm in der Nordsee im August 1952. Kieler Meeresf. 10(1):3-18, 11 textfigs.

phytoplankton
Krishnamurthy, K., and A. Purushothaman 1971.
Diurnal variations in phytoplankton pigments in the Vellar Estuary.
J. mar. biol. Ass. India 13(2):271-274

phytoplankton
Kriss, A.E., E.A. Rukina and V.I. Biryusova, 1949.
Composition according to species of the microorganisms in the Black Sea. Tr. Sevastop. Biol. St. 7:

phytoplankton
Krylov, V.V., 1971.
Distribution of net phytoplankton in the East China Sea (February-March 1962). (In Russian; English abstract). Okeanologiia 11(6): 1105-1109.

phytoplankton
Kuzmenko, L.V., 1971
Phytoplankton of the Arabian Sea in summer period. (In Russian; English abstract).
Gidrobiol. Zh. 7(5): 25-31.

phytoplankton
Kuzmina, A.I., 1962
Certain data on the vernal phytoplankton of the North Atlantic (from the data of the II voyage of the e.s. "Lomonosov" in 1958)
Doklady Akad. Nauk, SSSR, 144(5): 1156-1159.

phytoplankton
Kuzmina, A.I., 1962.
Phytoplankton of the Kuril Straits as an indicator of different water bodies. (In Russian).
Issled. Dalinevostochnich Moreia, SSSR, Zool. Inst., Akad. Nauk, SSSR, 8:6990.

phytoplankton
Kuzh'mina, A.I., 1962.
Quantitative development and distribution of the phytoplankton in the northern part of the Greenland Sea. (In Russian)
Trudy Vses. Nauchno-Issledov. Inst. Morsk. Ribn. Chos. i Okean., VNIRO, 46:287-296.

phytoplankton
Lafon, M. M. Durchon and Y. Saudray, 1955.
Recherches sur les cycles saisonnieres du plancton. Ann. Inst. Ocean., 31(3):125-230.

phytoplankton
Landa, A., 1953.
Análisis de muestras diarias de fitoplancton superficial en Chimbote, julio 1951 a junio 1952.
Bol. Cient., C.A.G., 1(1):63-75, 4 tables, 5 graphs, 7 figs.

phytoplankton
Lecal, J., 1953.
Rapport entre phytoplancton et eaux de débordement de l'Oued Chéliff. Bull. Soc. Hist. Nat., Afrique du Nord, 44(3/4):120-133, 2 textfigs.

phytoplankton
Léger, Guy, 1964.
Les populations phytoplanctoniques en mer Ligure (radiale Monaco-Calvi) en juin 1963.
Bull. Inst. Océanogr., Monaco, 64(1326):32 pp.

phytoplankton
Leloup, E., editor, 1966.
Recherches sur l'ostreiculture dans le bassin d'Ostende en 1964. Minist. Agricult. Comm. T.W.O.Z., Groupe de Travail "Ostreiculture": 58 pp.

phytoplankton
Lenz, Jürgen, Heinz Schöne und Bernt Zeitschel, 1967.
Planktonologische Beobachtungen auf einem Schnitt durch die Nordsee von Cuxhaven nach Edinburgh.
Kieler Meeresforsch., 23(2):92-98.

phytoplankton
Lovegrove, T., 1961.
Plankton investigations from Aberdeen. Phytoplankton.
Ann. Biol., Cons. Perm. Int. Expl. Mer, 16:74, 21 75.

phytoplankton
Lovegrove, T., 1958(1960).
Plankton investigations from Aberdeen in 1958. Phytoplankton.
Ann. Biol., Cons. Perm. Int. Expl. Mer, 15:57.

phytoplankton
Lovegrove, T., 1958(1960).
Plankton investigations from Aberdeen in 1958.
Cons. Perm. Int. Expl. Mer, Ann. Biol., 15:55.

phytoplankton
Lillick, L.C., 1940
Phytoplankton and planktonic protozoa of the offshore waters of the Gulf of Maine. Pt.II. Qualitative Composition of the Planktonic Flora. Trans. Am. Phil. Soc., n.s., 31(3):193-237, 13 text figs.

phytoplankton
Lillick, L.C., 1938
Preliminary report of the phytoplankton of the Gulf of Maine. Amer. Midland Naturalist, Vol. 20(3):624-640, 37 tables, 1text fig.

phytoplankton
Lillick, L.C., 1937
Seasonal studies of the phytoplankton of Woods Hole, Massachusetts. Biol. Bull. 73(3): 488-503, 3 text figs., 2 tables.

phytoplankton
Lovegrove, T., 1963.
Plankton investigations from Aberdeen during 1960-1961. Phytoplankton. (A Distant Northern Sess)
Ann. Biol., Cons. Perm. Int. Expl. Mer, 1961, 18:65.

phytoplankton
Lovegrove, T., 1963.
Plankton investigations from Aberdeen during 1960-1961. Phytoplankton. (B. Near Northern Sess)
Ann. Biol. Cons. Perm. Int. Expl. Mer, 1961, 18:66-67.

phytoplankton
Lucas, C.E., 1942
Continuous plankton records: Phytoplankton in the North Sea 1938-1939 II, Dinoflagellates, Phaeocystis, etc. Hull. Bull. Mar. Ecol., II(9):47-70, 6 textfigs., 24 pls.

phytoplankton
Lucas, C.E. 1941
Continuous plankton records: Phytoplankton in the North Sea 1938-39. Pt. I. Diatoms. Hull Bull. Mar. Ecol., II(8):19-46, 5 textfigs., 33 pls.

phytoplankton
Lucas, C.E., 1940
Ecological investigations with the continuous plankton recorder: the phytoplankton in the southern North Sea, 1932-37. Hull Bull. Mar. Ecol., 1(3):73-170, 17 textfigs., 21 pls.

phytoplankton
Mandelli E.F. 1969.
The inhibitory effects of copper on marine phytoplankton.
Contrib. mar. Biol. Port Aransas 14:47-57.

phytoplankton
Mangin, L., 1915
Phytoplankton de L'Antartique. Deuxieme Exped. Ant. Francaise (1908-1910), 95 pp., 3 pls., 58 text figs.

Phytoplankton
Mangin, M. L., 1912
Phytoplancton de la croisière du "René" dans l'Atlantique (Septembre 1908). Ann. Inst. Ocean., n.s., 4(1):1-66, 2 pls., 41 text figs., 2 tables.

Phytoplankton
Mangin, L., 1910
Sur quelques algues nouvelles ou peu connues du Phytoplankton de l'Atlantique, Bull. Soc. Bot., France, 57:

phytoplankton
Margalef, Ramon 1971.
Composition et analyse par groupes du phytoplancton au large des côtes méditerranéennes espagnoles, en 1965-1967.
Rapp. p.-v. Comm. int. Explor. scient. mer Medit. 20(3): 307-310.

phytoplankton
Margalef, Ramon, 1965.
Distribution des especes du phytoplancton Méditerraneen par rapport aux differentes combinaisons des facteurs du milieu.
Rapp. Proc. Verb., Reunions, Comm. Int. Expl. Sci. Mer Mediterranée, Monaco, 18(2):349-352.

phytoplankton
Margalef, Ramon, 1963
Algunas regularidades en la distribución a escala pequeña y media de las poblaciones marinas de fitoplancton y en sus características funcionales.
Invest. Pesquera, Barcelona, 23:169-230.

English summary, pp. 224-225.

phytoplankton
Margalef, R., 1962.
Organisation spatiale et temporelle des populations de phytoplancton dans un secteur du littoral méditerranéen espagnol.
Pubbl. Staz. Zool., Napoli, 32 (Suppl.):336-348.

phytoplankton
Margalef, R., 1961
Correlations entre certains caractères synthétiques des populations de phytoplancton.
Hydrobiologie, 18(1-2): 155-164.

phytoplankton
Margalef, R., 1957.
Variacion local e internual en la secuencia de las poblaciones de fitoplancton de red en las aguas superficiales de la costa mediterranea espanole. Inv. Pesq., 9:65-96.

phytoplankton
Margalef, R., 1955.
Variaciones interanuales en el fitoplancton de Castellon. Reunion sobre Prod. Pesquerias (2): 48-51.

phytoplankton
Margalef, R., 1955.
Dinamica de las poblaciones de fitoplancton.
Reunion sobre Prod. Pesquerias (2):24-27.

phytoplankton
Margalef, R., 1951.
Ciclo anual del fitoplancton marino en la costa NE de la Peninsula Ibérica. Publ. Inst. Biol. Apl. 9:83-118, 11 textfigs.

phytoplankton
Margalef, R., 1951.
Plancton recogido por los Laboratorios Costeros. III. Fitoplancton de las costas de Castellón durante el año 1950. Publ. Inst. Biol. Aplic. 9: 49-62, 2 textfigs.

phytoplankton
Margalef, R., 1949.
Fitoplancton nerítico de la Costa Brava en 1947-48. Publ. Inst. Biol. Aplicada, 5: 41-51, 3 text figs.

phytoplankton
Margalef, R., 1948.
Fitoplancton neritico de la Costa Brava en 1947-48. Publ. Inst. Biol. Aplic. 5:41-52, 3 textfigs.

phytoplankton
Margalef, R., 1948.
Le phytoplancton estival de la "Costa Brava" catalane en 1946. Hydrobiol. 1(1):15-21.

phytoplankton
Margalef, R., 1948.
Fitoplancton neritico estival de Cadaqués (Mediterráneo catalán). Publ. Inst. Biol. Aplic. 2:89-95, 3 textfigs.

phytoplankton
Margalef, R., F. Cervignon and G. Yepez T., 1960
Exploracion preliminar de las caracteristicas hidrograficas y de la distribucion del fitoplancton en el area de la Isla Margarita (Venezuela).
Mem. Soc. Ciencias Nat. de la Salle, 22(57): 210-221.
Contribucion No. 2, Estacion de Investigaciones Marinas de Margarita, Fundacion La Salle de Ciencias Naturales.

phytoplankton
Margalef, R., F. Saiz, J. Rodriguez-Rode, R. Toll y J.M. Valles, 1952.
Plancton recogido por los laboratorios costeros. V. Fitoplancton de las costas de Castellon durante el año 1951. Publ. Inst. Biol. Aplic. 10: 133-143, 2 textfigs.

phytoplankton
Margalef, Ramón, y Francisco Vives, 1967.
La vida suspendida en las aguas.
In: Ecologie marina, Monogr. Fundación La Salle de Ciencias Naturales, Caracas, 14:493-562.

phytoplankton
Marshall Harold G., 1969.
Observations on the spatial concentrations of phytoplankton.
Castanea 34: 217-222.

phytoplankton
Marshall, N., 1956.
Chlorophylla a in the phytoplankton in coastal waters of the eastern Gulf of Mexico.
J. Mar. Res., 15(1):14-32.

microplankton
Marshall, S. M., 1933
The production of microplankton in the Great Barrier Reef Region. Brit. Mus. (N.H.) Great Barrier Reef Exped. 1928-29, Sci. Repts. II(5):111-157, 14 text figs.

phytoplankton
Marshall, S.M., and A.P. Orr, 1962
Carbohydrate as a measure of phytoplankton.
J. Mar. Biol. Assoc., U.K., 42(3):511-519.

phytoplankton
Marshall, S.M., and A.P. Orr, 1927
The relation of the Plankton to some Chemical and Physical Factors in the Clyde Sea Area. J.M.B.A., XIV:837-868.

phytoplankton
Marumo, R., 1955.
Analysis of water masses by distribution of the microplankton (1). Distribution of the microplankton and their relation to water masses in the North Pacific Ocean in the summer of 1954.
J. Ocean. Soc., Japan, 99(3):133-137.

Oceanographic Index: Marine Organisms Cumulation, 1946-1973

phytoplankton
Marumo, R., 1954.
Relation between planktonological and oceanographical conditions of a sea area west of Kinka zan in winter. J. Ocean. Soc., Japan, 10(2):77-84, 5 textfigs.

phytoplankton
Marumo, R., 1950.
Phytoplankton and sea conditions in summer at the fixed point 29° N., 135° E., south off Honshu. (In Japanese.)
Oceanogr. Rept., Tokyo, 1(3):153-159.

phytoplankton
Marumo, R., M. Kiton, and O. Asaoka, 1954.
The productivity and its seasonal variations of main plankton groups in the open sea.
J. Ocean. Soc., Japan, 10(4):209-215, 8 textfigs.

Phytoplankton
Mashtakova, G.P., 1964.
The influence of the land drainage on the development of the phytoplankton in the northwestern part of the Black Sea. (In Russian).
Trudy azov. chernomorsk nauchno-issled. Inst. morsk. ryb. khoz. okeanogr. 23:55-67.

phytoplankton
Massuti Algamora, M., 1949
Estudio de diez y seis muestras de plancton del Golfo de Nápoles. Publ. Inst. Biol. Appl. 5:85-94, 1 fold-in table.

phytoplankton
Matsudaira, C., 1940.
Seasonal change of chemical composition of sea water and phytoplankton in the Bay of Ise. (In Japanese.)
Suisangakkaiho, 8(2):148-162.

phytoplankton
Matsue, Y., 1950.
Phytoplankton and its oxidisability by permanganate. Bull. Jap. Soc. Sci. Fish. 15(12): 813-817.

phytoplankton
Maucha, R., 1948.
Die Photosynthese des Phytoplanktons von Gesichtspunkte der Quantenlehre. Hydrobiol. 1(1) 45-62.

phytoplankton
McAlice, B.J., 1970.
Observations on the small-scale distribution of estuarine phytoplankton.
Marine Biol., 7(2):101-111.

phytoplankton
McAllister C.D., 1969.
Aspects of estimating zooplankton production from phytoplankton production.
J. Fish. Res. Bd. Can., 26(2): 199-220

phytoplankton
McLaren, I.A., 1961
The hydrography and zooplankton of Ogac Lake, a landlocked fjord on Baffin Island.
Fish. Res. Bd., Canada, MSS Rept. Ser. (Biol.) No. 709: 167 pp. (multilithed).

phytoplankton
Meunier, A., 1910
Microplancton des Mers de Barents et de Kara. Duc d'Orléans, Campagne Arctique de 1907, Brussels.

phytoplankton
Michel, A., C. Colin, R. Desrosières et C. Oudot, 1971.
Observations sur l'hydrologie et le plancton des abords et de la zone des passes de l'atoll de Rangiroa (Archipel des Tuamotu, Océan Pacifique central). Cah. ORSTOM, ser. Océanogr. 9(3): 375-405.

phytoplankton
MINAS, H.J., A. TRAVERS, M. Travers et S. Maestrini, 1968.
Première utilisation à Villefranche-sur-mer de la Bouée Laboratoire du COMEXO pour l'étude de la distribution du microplancton et de certains facteurs écologiques.
Rec. Trav. Sta. mar. Endoume 44 (60):13-45.

phytoplankton
Moberg, E.G., and W.E. Allen, 1927.
Effect of tidal changes on physical, chemical, and biological conditions in the sea water of the San Diego region. 1. Observations on the effect of tidal changes on physical and chemical conditions of sea water in the San Diego region. 2. Half-hourly collections of marine microplankton taken at the Scripps Institution pier in 1923.
Bull. S.I.O., tech. ser., 1:1-17, 4 textfigs.

phytoplankton
Morales, E., 1952.
Plancton recogido por los laboratorios costeros. IV. Fitoplancton de Blanes durante los meses de julio de 1950 a julio de 1951. Publ. Inst. Biol. Aplic. 10:67-80, 4 textfigs.

phytoplankton
Morales, E., 1951.
Plancton recogidos por los Laboratorios costeras. II. Plancton de Blanes desde octubre de 1949 hasta juinio de 1950. Publ. Inst. Biol. Aplic., Barcelona, 8:121-125.

phytoplankton
Morozova-Vodyanitzkaya, N.V., 1950.
Numbers and biomass of phytoplankton of the Black Sea. Dokl. Akad. Nauk, SSSR, 73:821-824.

phytoplankton
Morozova-Vodyanitakaya, N.V., 1948.
Phytoplankton of the Black Sea. Pt. 1. Phytoplankton in the Sevastopol region and general outline of the phytoplankton of the Black Sea.
Trans. Sevastopol Biol. Sta., USSR Akad. Sci. 6: 39-172.

phytoplankton
Motoda, S., 1940.
Comparison of the conditions of water in the bay, lagoon and open sea in Palao.
Palao Trop. Biol. Sta. Studies 2(1):41-48, 2 textfigs.

phytoplankton
Movchan, O.A., 1970.
Qualitative composition of phytoplankton in the vicinity of Newfoundland. Okeanologiia, 10(3): 496-504.
(In Russian; English abstract)

phytoplankton
Movchan, O.A. 1969.
Phytoplankton from the Newfoundland area sampled in early spring. (In Russian)
Trudy vses. Nauchno-issled. Inst. morsk. ryb. khoz. okean. (VNIRO) 65: 164-177.

phytoplankton
Movchan, O.A., 1967.
Phytoplankton distribution and development in the vicinity of Newfoundland as dependent on seasonal changes of some abiotic factors. (In Russian; English abstract).
Okeanologiia, Akad. Nauk, SSSR, 7(6):1053-1067.

phytoplankton, lists of spp
Mulford, Richard A., 1963.
The net phytoplankton taken in Virgina tidal waters, January-December 1962.
Virginia Inst. Mar. Sci., Spec. Sci. Rept., (43): 22 pp., (unpublished manuscript).

phytoplankton, lists of spp.
Mulford, Richard A., and John J. Norcross 1971
Species composition and abundance of net phytoplankton in Virginian coastal waters, 1963-1964.
Chesapeake Sci. 12(3): 142-155.

phytoplankton
Munoz Sardon, Filipe, 1962
Estudio del fitoplancton del Golfo de Cadiz en relacion con el regimen regional de vientos.
Inv. Pesq. Barcelona, 21:165-188.

phytoplankton
Muñoz Sardon, F., 1960.
Estudios previos sobre el fitoplancton del Golfo de Cadiz.
Bol. Real Soc. Esp. Hist. Nat., Sec. Biol., 58(2):335-346.

phytoplankton
Muñoz, F., J. Herrera, and R. Margalef, 1956.
Fitoplancton de las costas de Castellón durante el ano 1954. Inv. Pesq., 3:75-90.

Phytoplankton
Muñoz, Felipe, y Jose M. San Feliu, 1965.
Hidrografia y fitoplancton de las costas de Castellón de agosto de 1962 a julio de 1963.
Inv. Pesq., Barcelona, 28:173-209.

phytoplankton
Murakami, A., 1954.
Oceanography of Kasaoka Bay in Seto Inland Sea.
Bull. Nakai Regional Fish. Res. Bay, 6:15-57, 42 textfigs.

phytoplankton
Nathansohn, A., 1906
Über die Bedeutung Vertikaler Wasserbewegungen für die Produktion des Planktons im Meere. Abh. Säch. Ges. Wiss., XXIX: 358-441.

phytoplankton
Newell, G.E., and R.C., 1963.
Marine plankton; a practical guide.
Hutchinseon Biological Monographys, 207 pp.

phytoplankton
Nikolaev, I.I., 1950.
Basic ecological and geographical complexes of the phytoplankton of the Baltic Sea and their distribution. Botan. Zhurn., 35(6):602-

phytoplankton
Norris, R.E., 1961.
Observations on phytoplankton organisms collected on the N.Z.O.I. Pacific Cruise, September 1958
N.Z. J. Sci., 4(1):162-188.

phytoplankton
Ogilvie, H.S., 1923.
Microplankton from the south coast of Iceland.
Rapp. Proc. Verb., Cons. Perm. Int. Expl. Mer, 29:30-71.

phytoplankton
Ohwada, M., and O. Asaoka, 1963.
A microplankton survey as a contribution to the hydrography of the North Pacific and adjacent seas. (1).
Oceanogr. Mag., Meteorol. Agency, Tokyo, 14(2):73-85.

Also printed in:
Bull. Hakodate Mar. Obs., (10)(8).

phytoplankton
Okul, A. V., 1941
Materialy po produktivnosti planktona Azovskogo morea (The productivity of plankton in the Sea of Azov.) (In Russian with English summary). Zoologicheskii Zhurnal 20(2):198-212, 8 figs.

phytoplankton
Ostenfeld, C. H. 1903
Phytoplankton from the sea around the Faeroes.
Botany of the Faeroes. II. Copenhagen

phytoplankton
Ostenfeld, C. H. 1931
Concluding remarks on the plankton collected on the quarterly cruises in the years 1902-1908.
Bull. Trim. Cons. Int. Explor. Mer., pp. 601-672

microplankton
Ostenfeld, C.H., and O. Paulsen, 1911
General Remarks on the Microplankton.
Danmark Exped. Medd. om Grønland, 43(11): 319-336.

phytoplankton
Owada, M., 1953.
The planktonological and oceanographical conditions in the vicinity of Funka-Bay (Volcanic Bay)
J. Ocean. Soc., Japan, 9(3/4):181-184, 9 textfigs.

phytoplankton
Paasche, E., 1961.
Notes on the phytoplankton from the Norwegian Sea.
Botanica Marina, 2(3/4):197-210.

phytoplankton
Pasternak, F.A., 1970.
Marine biology of the East Atlantic continental region. In: The geology of the East Atlantic continental margin, 1. General and economic papers, ICSU/SCOR Working Party 31 Symposium, Cambridge 1970, Rept. No. 70/13:67-77

phytoplankton
Patten, Bernard C., 1963
Plankton: Optimum diversity structure of a summer community.
Science, 140(3569):894-898.

phytoplankton
Patten, B.C., 1962
Species diversity in net phytoplankton of Raritan Bay.
J. Mar. Res., 20(1):57-75.

phytoplankton
Paulsen, O., 1930.
Études sur le microplancton de la mer d'Alboran.
Trab. Inst. Esp. Ocean. No. 4:1-108, 61 textfigs.

phytoplankton
Paulsen, O. 1918
Plankton and biological investigations in the sea around the Faeroes in 1913. Medd. Komm. Havunders. Plankton I, No. 13

phytoplankton
Paulsen, O. 1909
Plankton investigations round Iceland and in the North Atlantic in 1904. Medd. Komm. Havunders. Plankton I, No. 8

phytoplankton
Paulsen, O. 1904.
Plankton investigations in the waters round Iceland in 1903. Medd. Komm. Havunders., Plankton I, No. 1.

phytoplankton
Petrova, Vera J., 1965.
Conditions régissant le développement du phytoplancton sur le littoral bulgare de la mer Noire.
Rapp. Proc.-Verb. Réun. Comm. int. Explor. scient. Mer Méditerranée, 19(3): 583-585.

phytoplankton
Petrova, V.J., 1965.
Sur le phytoplancton de la mer Noire devant le littoral Bulgare.
Rapp. Proc. Verb. Réunions, Comm. Int. Expl. Sci. Mer Méditerranée, Monaco, 18(2):357-361.

Phytoplankton
Petrova, Vera, 1964.
Phytoplankton, in the Black Sea off the Bulgarian shores from 1958 1960. (In Russian).
Izv. Inst. Ribov'dst. i Ribol., Varna, 5:30-32.

phytoplankton
Petrova, V., 1964.
Day-and-night changes of the phytoplankton in the Black Sea along the Bulgarian coast. (In Bulgarian; Russian and English summaries).
Izv. Inst. Ribiov'dstvo Ribolov., Varna, Izdatel. Bilgarskata Akad. Naukite, Sofia, 4:5-23.

phytoplankton
Petsik, G.K., 1968.
Développement quantitatif du phytoplancton dans les mers du bassin méditerranéen.
Rapp. Proc.-Verb. Réun. Comm. int. Explor. scient. Mer Méditerranée, 19(3): 573.

phytoplankton
Posner, G.S., 1957.
The Peru Current.
Bull. Bingham Oceanogr. Coll., 16(2):106-153.

phytoplankton
Pratt, D. M., 1959.
The phytoplankton of Narragansett Bay
Limnol. & Ocean., 4(4):425-440.

phytoplankton
Pratt, D.M., 1950.
Experimental study of the phosphorus cycle in fertilized salt water. J. Mar. Res. 9(1):29-54, 10 textfigs.

phytoplankton
Pratt, D.M., 1949.
Experiments in the fertilization of a salt water pond. J. Mar. Res. 8(1):36-59, 8 textfigs.

Identification only to genera!

phytoplankton
Pshenin, L.N., 1961.
Concerning the connection between azobacter and phytoplankton.
Trudy Sevastopol Biol. Sta., (14):33-43.

phytoplankton
Pucher-Petkovic, Tereza, 1964.
Fluctuations, en pour-cent, de la composition des groupes phytoplanctoniques du large de l'Adriatique moyenne. (In Jugoslavian; French resume).
Acta Adriatica, 11(1):243-253.
resume, p. 252-253.

phytoplankton
Pucher-Petkovic, T., 1963.
Rapports quantitatifs entre les divers groupes de phytoplancton en Adriatique moyenne.
Rapp. Proc. Verb., Reunions, Comm. Int. Expl. Sci., Mer Mediterranee, Monaco, 17(2):479-485.

Phytoplancton
Pucher-Petkovic, T., 1960
Effet de la ferilisation artificielle sur le phytoplancton de la reion de Mljet.
Acta Adriatica 6 (8):24 pp

phytoplankton
Ramamurthy, S. and Rajindero M. Dhawan, 1963.
On the characteristics of the plankton at Kandla in the Gulf of Kutch during August 1958 - July 1960. Ind. J. Fish. (A)10(1): 94-101.

phytoplankton
Rampi, L., 1952.
Ricerche sul microplancton di superficie del Pacifico tropicale. Bull. Inst. Océan., Monaco, No. 1014:16 pp., 5 textfigs.

phytoplankton
Rampi, L., 1949.
Ricerche sul microplancton delle acque di Portofino (Mare Ligure). Centro Talassografico Tirreno, Publ. No. 2:7-14.

phytoplankton
Rampi, L., 1947.
Osservazioni sullo sviluppo quantitativo del fitoplancton nel Mare Mediterraneo. La Nuova Notarisia, Rivista di Algologia, n.s., 1(2): 5 pp.

phytoplankton
Rampi, L., 1945
Osservazioni sulla distribuzione qualitativa del fitoplancton nel mare Mediterraneo. Atti della Soc. Ital. di Sci. Nat. 84:105-113.

phytoplankton
Rampi, L., 1942
II Fitoplancton mediterraneo: Problemi ed affinita interoceaniche. Boll. di Pesca di Piscicoltura e di Idrobiologia, Anno 18, Fasc. 4:7-19.

phytoplankton
Reyssac, J., 1967.
Note sur les variations nycthémérales des diatomées et dinoflagellés en deux stations du littoral ivoirien. Doc. Centre Rech. Océanogr. Abidjan, 012: 4 pp.

phytoplankton
Reyssac, J., 1966.
Diatomées et dinoflagellés récoltés par le navire "Ombango dans les parages de l'ile Annobon. Doc. Centre Rech. Océanogr. Abidjan, 13: 14 pp.

phytoplankton
Reyssac, J., 1966.
Diatomées et dinoflagellés des eaux ivoiriennes pendant l'année 1965. Variations quantitatives Doc. Centre Rech. Océanogr. Abidjan, 010: 22 pp

phytoplankton
Reyssac, J., 1966.
Quelques données sur la composition et l'evolution annuelle du phytoplankton au large d'Abidjan (mai 1964 - mai 1965). Doc. Centre Rech. Océanogr. Abidjan, 003: 31 pp.

phytoplankton
Reyssac, J., 1966.
Le phytoplancton entre Abidjan et l'Équateur pendant la saison chaude. Doc. Centre Rech. Océanogr. Abidjan, 002: 11 pp.

phytoplankton
Reyssac, Josette, et Michel Prive, 1965.
Conditions hydrologiques et phytoplancton au large d'Abidjan, variations d'avril a juillet 1964. Cahiers. O.R.S.T.R.O.M. - Oceanogr., 3(1):67-69.

phytoplankton
Rice, T.R., 1953.
Phosphorus exchange in marine phytoplankton.
Fish. Bull. 80, Vol. 54:77-89, 3 textfigs.

phytoplankton (chlorophyll)
Riley, Gordon A., 1959
Oceanography of Long Island Sound, 1954-1955
Bull. Bingham Oceanogr. Coll., 17(1):9-30.

phytoplankton
Riley, G.A., 1957.
Phytoplankton of the North Central Sargasso Sea.
Limnol. & Oceanogr., 2(3):252-270.

phytoplankton
Riley, G.A., 1947
Seasonal fluctuations in the phytoplankton population in New England coastal waters.
J. Mar. Res. 6(2):114-125. text figs. 32-34

phytoplankton
Riley, G.A., 1946
Factors controlling phytoplankton populations on Georges Bank. J. Mar. Res. 6(1):54-73, figs. 14-21.

phytoplankton
Riley, G. A., 1938
The measurement of phytoplankton. Int. Rev. Ges. Hydrobiol. Hydrogr., 36:371-373.

phytoplankton
Riley, G. A., 1941.
Plankton studies. V. Regional summary. J. Mar. Res., 4(2):162-171, 1 fig.

phytoplankton

Riley, G. A., 1941.
Plankton studies. IV. Georges Bank. Bull. Bingham Ocean. Coll. 7(4):73 pp.

Riley, G. A. 1941
Plankton studies. III. Long Island Sound.
Bull. Bingham Ocean. Coll., 7(3):1-89

Riley, G. A., 1939.
Plankton studies. II. The western North Atlantic May-June, 1939. J. Mar. Res. 2(2): 145-162, Textfigs. 49-51, 4 tables.

Riley, G. A., H. Stommel, and D. F. Bumpus, 1949
Quantitative ecology of the plankton of the western North Atlantic. Bull. Bingham Ocean. Coll. 12(3):169 pp., 39 text figs.

Riley, G.A. and D.F. Bumpus, 1946
Phytoplankton-Zooplankton relationships on Georges Bank. J. Mar. Res. 6(1): 33-47, figs. 11-12

Riley, G. A., and R. von Arx., 1949.
Theoretical analysis of seasonal changes in the phytoplankton of Husan Harbor, Korea. J. Mar. Res. 8(1):60-72, 5 textfigs.

Riley, J.P. and D.A. Segar, 1969.
The pigments of some further marine phytoplankton species. J. mar. biol. Ass. U.K., 49(4): 1047-1056.

Rodhe, W., 1948
Environmental requirements of fresh-water plankton algae. Experimental studies in the ecology of phytoplankton. Symbolae Botanicae Upsalienses X(1):149 pp., 30 figs.

Rotschi, Henri, Michel Legand and Roger Desrosieres, 1961
Orsom III, Croisières diverses de 1960, physique chimie et biologie. ORSTOM, Inst. Français d'Océanie, Centre d'Océanogr., Noumea, Rapp. Sci., No. 20: 59 pp. (mimeographed).

Rouchiiajnen, M.I., 1960.
[Character of the phytoplankton development in May-June 1958 in the southern part of the Barents Sea.] Trudy Murmansk. Morsk. Biol. Inst., 2(6):59-67.

Roukhiyainen, M.I., 1966.
Vertical distribution of the phytoplankton in the southern Barents Sea. (In Russian).
Trudy murmansk. biol. Inst., 11(15): 24-33.

Roukhiiaynen, M.I., 1956.
[Some regularities in the vernal development of phytoplankton from the eastern part of the Murman Sea.] Dokl. Akad. Nauk, SSSR, 109(1):209-212.

Round, F.E., 1967.
The phytoplankton of the Gulf of California. I. Its composition, distribution and contribution to the sediments.
J. exp. mar. Biol. Ecol., 1(1):76-97.

Rumek, A., 1950.
Seasonal occurrence of phytoplankton in the Gulf of Gdansk. Bull. Inst. Pêches Maritimes, Gdynia, No. 5:145-149. (In Polish; English summary).

Rumkówna, A., 1948
[List of the phytoplankton species occurring in the superficial water layers in the Gulf of Gdansk] Bull. Lab. mar., Gdynia, No. 4: 139-141 with tables in back.

Ruud, B., 1926.
Quantitative investigations of plankton at Lofoten, March-April, 1922-1924. Preliminary report.
Rept. Norwegian Fish. Mar. Invest. 3(7):30 pp., 3 charts, 5 diagrams.

Ryther, J. H. and D. W. Menzel, 1959.
Light adaption by marine phytoplankton.
Limnol. & Ocean., 4(4):492-497.

Saijo, Yatsuka, 1964.
Size distribution of photosynthesizing phytoplankton in the Indian Ocean.
J. Oceanogr. Soc. Japan, 19(4):187-189.
Also in: Collected Papers, Sci. Atmosph., Hydrosph. Water Res. Lab. Nagoya Univ., 2(2).1964.

Sakshaug, Egil 1970.
Quantitative phytoplankton investigations in near-shore water masses.
Skr. K. norske Vidensk. Selsk. (3): 1-8.

Saunders, G.W., 1957.
Interrelations of dissolved organic matter and phytoplankton. Bot. Rev. 23(6):389-409.

Savage, R. E. and A. C. Hardy, 1935.
Phytoplankton and the herring: Pt. 1.
Min. Agric. Fish., Fish. Invest., ser. ii, XIV (2).

Savage, R. E. and Wimpenny, R. S., 1936
Phytoplankton and the herring. Pt. II, 1933 and 1934. Min. Agric. and Fish. Fish. Invest., Ser. 2, XV, No. 1,

Savich, M.S., 1971.
Some characteristic features of phytoplankton distribution in the Gulf of Aden depending on oceanographic conditions. (In Russian; English abstract). Okeanologiia 11(3): 471-474.

Savich, M.S., 1969.
Seasonal dynamics of phytoplankton of the Gulf of Aden in 1963. (In Russian; English abstract).
Okeanologiia, 9(6): 1056-1062.

Schröder, B., 1906
Beiträge zur Kenntnis des Phytoplanktons warmer Meere. Vierteljahrschr. d. naturforsch. Ges. in Zurich, Jahrg. 51:

Schröder, B., 1900
Phytoplankton des Golfes von Neapel nebst vergleichenden Ausblicken auf das atlantischen Ozean. Mitt. Zool. Stat. Neapel, 14:1-38.

Scripps Institution of Oceanography, 1949.
Marine life research program. Progress report, 1 May to 31 July 1949. 24 pp. (mimeographed), 16 figs. (ozalid).

Sears, M., 1944.
Qué es el Plankton y por qué debemos estudiarlo. Mar, Revista de la Liga Maritima de Chile, Ano XV, No. 101:5-16, textfigs.

Sears, M., 1941
Notes on the phytoplankton on Georges Bank in 1941. J. Mar. Res., IV(3):247-257; textfigs. 54-58.

Sears, M., 1941.
Que es el Plankton y por que debemos estudiarlo Bol. C.A.G., XVII(12):451-465, 3 textfigs.

Seaton, D.D. 1968.
Investigations from Aberdeen in 1966: phytoplankton.
Annls biol. Copenh. 1966, 23:90-91.

Seaton, D.D., 1967.
Scottish plankton investigations in the near northern seas 1965. Phytoplankton.
Annls. biol. Copenh. (1965)22:62-63.

Seaton, D.D., 1966.
Plankton investigations from Aberdeen 1964 in the distant northern seas. Phytoplankton.
Annls biol., Copenh., 21:56.

Seaton, D.D., 1966.
Investigations from Aberdeen 1964 in the near northern seas area. Phytoplankton.
Annls biol., Copenh., 21:61-62.

Seguin, Gerard, 1966.
Contribution à l'étude de la biologie du plancton de surface de la baie de Dakar (Sénégal). Etude quantitative et observations écologiques au cours d'un cycle annuel.
Inst. Francais d'Afrique Noire, (A), 28(1):1-90. Bull.

Semina, H.I., 1967.
Phytoplankton. (in Russian)
In: Biologiia Tichogo Okeana Plankton, Isdatel. Nauka, Moskva, 27-85.

Semina, H.J., 1961
[The phytoplankton of the mixing zone between Oyashio and Kuroshio in spring 1955.]
Trudy Inst. Okean., 51: 3-15.

Semina, H. J., 1960.
[Distribution of phytoplankton in the Central Pacific.] Trudy Inst. Okeanol., 41:17-30

Semina, H.J., 1960.
The influence of vertical circulation on the phytoplankton in the Bering Sea.
Int. Revue Ges. Hydrobiol., 45(1):1-10.

Semina, H.J., 1958
[Relation between phytogeographic zones in the pelagial of the northwestern Pacific Ocean with the distribution of water masses in the region.]
Trudy Inst. Okeanol., 27: 66-76.

Semina, H.J., 1958
The dependence of the amount of phytoplankton upon the vertical stability of the surface layers in the boreal region of the Pacific. (Abstract).
Proc. Ninth Pacific Sci. Congr., Pacific Sci. Assoc., 1957, 16(Oceanogr.):217.

phytoplankton

Semina, H.I. 1956.
Composition and distribution of phytoplankton in the north-western part of the Pacific during spring and autumn 1955. Dokl. Akad. Nauk, SSSR, 110(3):465-468.

phytoplankton

Semina, H.I. 1957.
Factors influencing the vertical distribution of phytoplankton in the sea.
Trudy Vses. Gidrobiol. Obsh., 8:119-129.

phytoplankton

Semina, H.I. 1955.
On the two zonal groupings of the phytoplankton (according to the pattern of the Barents Sea.) Doklady Akad. Nauk, SSSR, 101(2): 363-366.

phytoplankton

Semina, H.I. 1955.
The vertical distribution of phytoplankton in the Bering Sea. Doklady Akad. Nauk, SSSR, 100(5): 947-949.

phytoplankton

Shirshov, P. 1937.
Seasonal changes of the phytoplankton of the Polar seas in connection with the ice regime. (In Russian). Trudy Arktich. Antarktich. Nauchno-Issledov., 82: 42-112.

phytoplankton

Skolka, Vidor Hilarius, 1965.
Les pigments assimilateurs du phytoplancton du littoral roumain de la mer Noire.
Rapp. Proc.-Verb. Réun. Comm. int. Explor. scient. Mer. Méditerranée, 19(3): 567-570

phytoplankton

Skolka, V.H., 1968.
L'influence du débit du Danube sur la répartition du phytoplancton de la partie ouest de la Mer Noire.
Revue Roumaine Biol. (Zool.) 13(6): 453-459.

phytoplankton

Skolka, V. Hilarius, 1965.
Contributions à l'etude du phytoplancton de la partie nord-ouest de la mer Noire.
Rapp. Proc. Verb., Réunions, Comm. Int. Expl. Sci., Mer Mediterranee, Monaco, 18(2):363-366.

phytoplankton

Skolka, Vidor Hilarius, et Octavian Selariu 1971.
La répartition du phytoplancton de la mer Noire dans les conditions hydrologiques d'hiver.
Rapp. P.-v. Comm. int. Explor. scient. mer Medit. 20(3): 327-329.

phytoplankton

Sieburth, J.M., 1959.
Antibacterial activity of Antarctic marine phytoplankton.
Limnol. & Ocean., 4(4):419-424.

phytoplankton

Skabichevskii, A.P., 1950.
Influence of the duration of daily illumination on the development of the planktonic algae.
Dok. Akad. Nauk, BSSR, 72(1):141.

phytoplankton

Skuja, H., 1948.
Taxonomie des Phytoplanktons einiger Seen in Uppland Schweden. Symb. Bot. Upsaliensis 9(3): 398 pp., 39 pls.

phytoplankton

Sleggs, G.F., 1927.
Marine phytoplankton in the region of La Jolla, California during the summer of 1924.
Bull. S.I.O., tech. ser., 1:93-117, 8 textfigs.

phytoplankton

Smayda, T.J., 1957.
Phytoplankton studies in lower Narragansett Bay. Limnol. & Oceanogr., 2(4):342-359.

phytoplankton

Smirnova, L.I., 1956.
The phytoplankton of the northwestern part of the Pacific. Dokl. Akad. Nauk, SSSR, 109(4):649-652.

phytoplankton

Smirnova, L.I., 1949.
On the phytoplankton of the central Caspian Sea. Trudy Inst. Okeanol., 3:260-275.

phytoplankton

Sorokin, Yu. I., V.G. Snopkov and V.M. Grinberg, 1959
The determination of the relation between phytoplankton photosynthesis and submarine illumination in the waters of the Central part of the Atlantic.
Doklady Akad. Nauk, SSSR, 124(2):432-435.

Translation NIOT/39

phytoplankton

Stadel, Otto, 1968.
Das Phytoplankton der Nordsee auf zwei hydrographischen Schnitten im August 1953.
Ber. dt. wiss Komm. Meeresforsch. 19(4): 237-258.

phytoplankton

Steemann Nielsen, E., 1958.
The balance between phytoplankton and zooplankton in the sea. J. du Cons., 23(2):178-188.

phytoplankton, quantitat

Steemann-Nielsen, Einar, 1951
The marine vegetation of the Isefjord. A study on ecology and production. Medd. Komm. Danmarks Fiskeri-og Havundersøgelser. Ser. Plankton. 5(4); 114pp., 46 text figs.

phytoplankton

Steemann Nielsen, E., 1943.
Über das Frühlingsplankton bei Island und den Faröer-Inseln. Medd. Komm. Havundersøgelser, Ser. Plankton, 3(6):14 pp., 8n textfigs., 4 tables.

phytoplankton

Steemann Nielsen, E., 1935.
The production of phytoplankton at the Faroe Isles, Iceland, East Greenland and in the waters around.
Medd. Komm. Danmarks Fisk- of Havundersøgelser, Ser. Plankton, 3(1):1-93, 5 textfigs.

phytoplankton

Steeman-Nielsen, E. 1933.
Über quantitative Untersuchungen von marinem Plankton mit Untermöhls umgekehrtem Mikroscop.
J. Cons. VIII, p. 201.

phytoplankton

Steeman-Nielsen, E. 1935.
The production of phytoplankton at the Faeroe Isles, Iceland, East Greenland and in the waters around. Medd. Komm Havunders., Plankton III:1-93.

phytoplankton

Steemann-Nielsen, E. and Th. von Brand, 1934.
Quantitative zentrifugen Methoden zur Planktonbestimmung. Rapp. Proc. Verb. 89(3):99-100.

phytoplankton

Subrahmanyan, R., 1958.
Ecological studies on the marine phytoplankton on the west coast of India.
Mem. Indian Botan. Soc., 11:145-151.

phytoplankton

Strickland, J.D.H., R.W. Eppley y Blanca Rojas de Mendiola, 1969.
Poblaciones de fitoplancton, nutrientes y fotosintesis en aguas costeras peruanas.
Bol. Inst. Mar, Peru 2(1): 1-45

(In Spanish and English)

phytoplankton

Subrahmanyan, R., 1959.
Studies on the phytoplankton of the west coast of India. 1. Quantitative and qualitative fluctuation of the total phytoplankton crop, the zooplankton crop and their interrelationship, with remarks on the magnitude of the standing crop and production of matter and their relationship to fish landings. 2. Physical and chemical factors influencing the production of phytoplankton, with remarks on the cycle of nutrients and on the relationship of the phosphate content to fish landings.
Proc. Indian Acad. Sci., 1:113-252.

phytoplankton

Subrahmanyan, R., 1959.
Studies on the phytoplankton of the west coast of India. Pt. I. Quantitative and qualitative fluctuation of the total phytoplankton crop, the zooplankton crop and their interrelationship. Proc. Indian Acad. Sci., Sec. B., 50(3):113-187.

phytoplankton

Sukhanova, I.N., 1969.
Some data on the phytoplankton of the Red Sea and the western Gulf of Aden. (In Russian; English abstract). Okeanologiia, 9(2): 295-300.

phytoplankton

Sukhanova, I.N., 1964.
The phytoplankton of the northeastern part of the Indian Ocean in the season of the southwest monsoon. Regularity of the distribution of oceanic plankton. (In Russian; English abstract) Trudy Inst. Okeanol., Akad. Nauk, SSSR, 65:24-31

phytoplankton

Sukhanova, I.N., 1962.
The tropical phytoplankton of the Indian Ocean. (In Russian).
Doklady, Akad. Nauk, SSSR, 142(5):1162-1164.

Translation:
Soviet Oceanography, Issue 3, 4-6. (1964).

(Scripta Tecnica, Inc. for AGU)

phytoplankton

Sverdrup, H.U., 1953.
On conditions for the vernal blooming of phytoplankton. J. du Cons. 18(3):287-295, 2 textfigs.

phytoplankton

Talling, J.F., 1955.
The light-relations of phytoplankton populations. Verh. Int. Ver. Limnol. 12:141-142.

phytoplankton

Tanaka, Otohiko, Fumihiro Koga, Haruhiko Irie, Shozi Iizuka, Yosie Dotu, Keitaro Uchida, Satoshi Mito, Seiro Kimura, Osame Tabeta and Sadahiko Imai, 1962.
The fundamental investigations of the biological productivity in the north-western sea area of Kyushu II. Study on plankton productivity in the neighboring waters of Genkai-Nada region. Rec. Oceanogr. Wks., Japan, Spec. No. 6:1-20.

phytoplankton (net)

Teixeira, C., 1963
Relative rates of photosynthesis and standing stock of the net phytoplankton and nannoplankton.
Bol. Inst. Oceanogr., Sao Paulo, 13(2):53-60.

phytoplankton

Teixeira, Clovis, and Miryam B. Kutner, 1962
Plankton studies in a mangrove environment. 1. First assessment of standing stock and principal ecological factors.
Bol. Inst. Oceanogr., Sao Paulo, Brasil, 12(3):101-124.

phytoplankton

Teixeira, C., J. Tundisi, and J. Santoro, 1967.
Plankton studies in a mangrove environment. IV.
Size fractionation of the phytoplankton.
Bolm Inst. oceanogr. S. Paulo, 16(1):39-42.

phytoplankton

Thomas, William H., 1966.
Surface nitrogenous nutrients and phytoplankton in the northeastern tropical Pacific Ocean.
Limnol. Oceanogr. 11(3):393-400.

phytoplankton

Thorrington-Smith, Margaret, 1970.
Some new and little known phytoplankton forms from the West Indian Ocean. Brit. Phycol. Jl 5(1): 51-56.

phytoplankton

Tranter, D.J., and B.S. Newell, 1963
Enrichment experiments in the Indian Ocean.
Deep-Sea Res., 10(1/2):1-9.

phytoplankton

Travers, A., 1962.
Recherches sur le phytoplancton du Golfe de Marseille. II. Étude quantitative des populations phytoplanctoniques du Golfe de Marseille.
Rec. Trav. Sta. Mar. Endoume, Bull., 28(41):70-140.

phytoplankton

Travers, Marc. 1972.
Le microplancton du Golfe de Marseille: matériel et méthodes générales d'étude.
Téthys 4(2):313-358.

phytoplankton

Tregouboff, G., 1956.
Rapport sur les travaux concernant le plancton Méditerranéen publiés entre Novembre 1952 et Novembre 1954.
Rapp. Proc. Verb., Comm. Int. Expl. Sci., Mer Méditerranée, 13:65-100

phytoplankton

Tucker, A., 1949.
Pigment extraction as a method of quantitative analysis of phytoplankton. Trans. Am. Microsp. Soc. 68(1):21-33, 6 textfigs.

phytoplankton

University of Southern California, Allen Hancock Foundation, 1965.
An oceanographic and biological survey of the southern California mainland shelf.
State of California, Resources Agency, State Water Quality Control Board, Publ. No. 27:232 pp.
Appendix, 445 pp.

phytoplankton

Usachev, P.I., 1958.
General characteristics of phytoplankton distribution in far eastern seas (Abridged text of a lecture). Oceanographic investigations in the northwest part of the Pacific Ocean. (In Russian)
Trudy Okeanogr. Komissii, Akad. Nauk, SSSR, 3:75-78.

phytoplankton

Ushakov, P.V., 1958.
Phytoplankton investigations of the Zoological Institute, AN, SSSR in far eastern seas. Oceanographic investigations in the northwest part of the Pacific Ocean.
Trudy Okeanogr. Komissii, Akad. Nauk, SSSR, 3:102-108.

phytoplankton

Ussachev, P.I., 1948.
Quantitative fluctuations of phytoplankton in the North Caspian. Trudy Inst. Okeanol., 2:60-88.

phytoplankton

Vinberg, G.G., and T.N. Sivko, 1956.
Phytoplankton as an agent for self-purification of turbulent waters.
Trudy Vses. Gidrobiol. Obshshest. 7:5-24.

phytoplankton

Vinogradov, K.A., editor, 1967.
Biology of the northwestern Black Sea. (In Russian).
Naukova Dumka, Kiev, 268 pp.

phytoplankton

Vives, F., and A. Planas, 1952.
Plancton recogido por los laboratorios costeros. VI. Fitoplancton de las costas de Vinaroz, islas Columbretes y alrededores de la desembocadura del Ebro. Publ. Inst. Biol. Aplic. 11:141-156, 19 textfigs.

phytoplankton

Wallen, D.G. and G.H. Geen, 1971.
The nature of the photosynthate in natural phytoplankton populations in relation to light quality. Mar. Biol. 10(2): 157-168.

phytoplankton

Wauthy, B., R. Desrosières et J. Le Bourhis, 1967.
Importance présumée de l'ultraplancton dans les eaux tropicales oligotrophes du Pacifique central sud.
Cah. ORSTOM, Sér. Océanogr., 5(2):109-116.

phytoplankton

Whedon, W.F., 1939
A three year survey of the phytoplankton in the region of San Francisco, California
Internat. Rev. d. ges. Hydrobiol. 38:459-476.

phytoplankton

Williamson, D.I., 1950-52.
Distribution of plankton in the Irish Sea 1949 and 1950. Proc. Trans. Liverpool Biol. Soc., 58:1-46.

phytoplankton

Wilson-Barker, D., 1931.
Plankton changes on the coast of Ecuador.
Nature 127:975.

phytoplankton

Wood, E.J. Ferguson, 1971.
Phytoplankton study - an appraisal. J. Cons. int. Explor. Mer 34(1): 123-126.

phytoplankton

Wood, E.J. Ferguson, 1967.
Antarctic phytoplankton distribution.
Antarctic Jl., U.S.A., 2(5):190.

phytoplankton

Wyrtki, K., 1950.
Über die Beziehungen zwischen Trübung und ozeanographischen Aufbau. Kieler Meeresf. 7(2):87-107, 10 textfigs.

phytoplankton in inlet waters

Yamazi, I., 1954.
Plankton investigations along the coast of Japan XIV. The plankton of Turuga Bay on the Japan Sea coast. Publ. Seto Mar. Biol. Lab. 4(1):115-126, 11 textfigs.

phytoplankton

Yamazi, I., 1953.
Plankton investigations of inlet waters along the coast of Japan. X. The plankton of Kamaisi Bay on the eastern coast of Tohoku District.
Publ. Seto Mar. Biol. Lab. 3(2)(Article 18(:189-204, 16 textfigs.

phytoplankton

Yamazi, I., 1953.
Plankton investigations in inlet waters along the coast of Japan. IX. The plankton of Onagawa Bay on the eastern coast of Tohoku District.
Publ. Seto Mar. Biol. Lab. 3(2)(Article 17):173-187, 10 textfigs.

phytoplankton

Yamazi, I., 1953.
Plankton investigation in inlet waters along the coast of Japan. VII. The plankton collected during the cruises to the new Yamamoto Bank in the Sea of Japan. Publ. Seto Mar. Biol. Lab. 3(1):75-108, 19 textfigs.

phytoplankton

Yanagisawa, T., 1943.
The biological season and the microplankton. IV. Umi to Sora 23:283-292.

phytoplankton

Yanagisawa, T., 1943.
The biological season and the microplankton. III. Umi to Sora 23:255-272.

phytoplankton

Yanagisawa, T., 1943.
Biological season and microplankton. I. Umi to Sora 23:89-95.

phytoplankton (net)

Yentsch, C.S., and J.H. Ryther, 1959.
Relative significance of net phytoplankton and nanoplankton in the waters of Vineyard Sound.
J. du Conseil, 24(2):231-238.

phytoplankton

Yentsch, C.S., and R.F. Scagel, 1958
Diurnal study of phytoplankton pigments. An insitu study in East Sound, Washington. J. Mar. Res., 17:567-583.

phytoplankton (nitrogen)

Yentsch, Charles S., and Ralph F. Vaccaro, 1958.
Phytoplankton nitrogen in the oceans.
Limnol. & Oceanogr., 3(4):443-448.

phytoplankton

Zernova, V.V., 1964.
Distribution of the net phytoplankton in the tropical western Pacific. Regularity of the distribution of oceanic plankton. (In Russian; English abstract).
Trudy Inst. Okeanol., Akad. Nauk, SSSR, 65:32-48.

phytoplankton

Zobell, C.E., and J.H. Long, 1938
Studies on the isolation of bacteria-free cultures of marine phytoplankton. J. Mar. Res., 1:328-334.

phytoplankton, abundance

Dehadrai, P.V. and R.M.S. Bhargava, 1972
Distribution of chlorophyll, carotenoids and phytoplankton in relation to certain environmental factors along the central west coast of India. Mar. Biol. 17(1): 30-37.

phytoplankton abundance

Platt, Trevor, 1972.
Local phytoplankton abundance and turbulence.
Deep-Sea Res. 19(3): 183-187.

Phytoplankton abundance

Qasim, S.Z., P.M.A. Bhattathiri and V.P. Devassy, 1972.
The influence of salinity on the rate of photosynthesis and abundance of some tropical phytoplankton. Mar. Biol. 12(3): 200-206.

phytoplankton, airborne(!)

Stevenson, Robert E., and Albert Collier, 1962
Preliminary observations on the occurrence of airborne marine phytoplankton.
Lloydia, 25(2):89-93.

phytoplankton, annual cycle

Journia, Alain 1973
La production primaire planctonique en Méditerranée; essai de mise à jour.
Bull. Etude Commun Méditerranée, 5 (numéro spécial): 128pp.

phytoplankton, assimilation

Elster, H.J. 1965
Absolute and relative assimilation rates in relation to phytoplankton populations
Mem. Ist. Ital. Idrobiol. 18 (Suppl.): 77-103.

phytoplankton, assimilation

Niemi, Åke 1972
Effects of toxicants on brackish-water phytoplankton assimilation.
Comment. Biol. Soc. scient. Fennica 55: 19 pp.

phytoplankton, anat.-physiol

Eppley, Richard W., and P. R. Sloan, 1966.
Growth rates of marine phytoplankton: correlation with light absorption by cell chlorophyll
Physiologia Pl. 19:47-59.

phytoplankton, anat.-physiol.

Hellebust, Johan A., 1967.
Excretion of organic compounds by cultured and natural populations of marine phytoplankton.
In: Estuaries, G.H. Lauff, editor, Publs Am. Ass. Advmt Sci., 83:361-355.

phytoplankton, anat.-physiol.

Lewin, Joyce, and William F. Busby, 1967.
The sulfate requirements of some unicellular marine algae.
Phycologia, 6(4):211-217.

phytoplankton, growth of

McCombie, A.M., 1953.
Factors influencing the growth of phytoplankton.
J. Fish. Res. Bd., Canada, 10(5):253-282.

phytoplankton, anat., physiol

Sloan, P.R., and J.D.H. Strickland, 1966.
Heterotrophy of four marine phytoplankters at low substrate concentrations.
J. Phycol., 2:29-32.

phytoplankton, anat. physiol.

Steemann Nielsen, E., and E.G. Jørgensen 1968.
The adaptation of plankton algae III.
With special consideration of the importance in nature.
Physiol. Plant. 21(3): 647-654.

phytoplankton, anat.-physiol

Thomas, William H. 1967.
The nitrogen nutrition of phytoplankton in the northeastern tropical Pacific Ocean.
Studies Trop. Oceanogr. Miami 5: 280-289.

phytoplankton, anat.-physiol.

Thomas, William H., 1966.
Effects of temperature and illuminance on cell division rates of three species of tropical oceanic phytoplankton. J. Phycol. 2: 17-22.

phytoplankton, anat-physiol.

Watt, W. D., 1966.
Release of dissolved organic material from the cells of phytoplankton populations.
Proc. R. Soc., (B) 164:521-551.

phytoplankton, annual variations

Colebrook, J.M., and G.A. Robinson, 1964.
Continuous plankton records: annual variations of abundance of plankton 1948-1960.
Bull. Mar. Ecol., 6(3):52-69.

phytoplankton, annual cycle

Patten, B.C., R.A. Mulford and J.E. Warinner, 1963
An annual phytoplankton cycle in the lower Chesapeake Science, 4(1):1-20.

phytoplankton, biomass

Abbott, Donald P., and Richard Albrs, 1967.
Summary of thermal conditions and phytoplankton volumes measured in Monterey Bay, California, 1961-1965.
Rep. Calif. Coop. Oceanic Fish. Invest, 11:155-156.

phytoplankton, biomass

Băcescu, M., M.T. Gomoiu, N. Bodeanu, A. Petran, G.I. Muller și V. Chirila, 1965.
Dinamica populatiilor animale si vegetale din zona nisipurilor fine de la nord de Constanta în conditiile anilor 1962-1965.
In: Ecologie marinӑ, M. Bӑcescu, redactor, Edit. Acad. Republ. Pop. Române, București, 2: 7-167.

phytoplankton, biomass

Bӑcescu, M.T. Gomoiu, N. Bodeanu, A. Petran, G. Müller și V. Manea, 1965.
Studii asupra varietiei vietii marine în zona litoralӑ nisipoasӑ de la nord de Constanta (Cercetӑri efectate în anii 1960-61 la puncte fixe situate in dreptul statiunii Mamaie).
In: Ecologie marinӑ, M. Bӑcescu, redactor, Edit. Acad. Republ. Pop. Române, București, 1: 7-138.

phytoplankton biomass

Bakaev, V.G., editor, 1966.
Atlas Antarktiki, Sovetskaia Antarktioheskaia Ekspeditsiia. 1.
Glabnoe Upravlenie Geodezii i Kartografii. MG SSSR, Moskva-Leningrad, 225 charts.

phytoplankton, biomass

Berland, B.R., D.J. Bonin, P.L. Laborde et S.Y. Maestrini, 1972.
Variations de quelques facteurs estimatifs de la biomasse, et en particulier de l'ATP. chez plusieurs algues marines planctoniques.
Mar. Biol, 13(4): 338-345.

phytoplankton, biomass (average)

Herrara, Juan, and Ramon Margalef, 1961
Hidrografia y fitoplankton de las costas de Castellón de fulio de 1958 a junio de 1959.
Inv. Pesq., Bacelona, 20:17-63.

phytoplankton biomass

Hobbie, John E., Osmund Holm-Hansen, Theodore T. Packard, Lawrence R. Pomeroy, Raymond W. Sheldon, James P. Thomas and William J. Wiebe, 1972.
A study of the distribution and activity of microorganisms in ocean water. Limnol. Oceanogr 17(4): 544-555.

phytoplankton biomass

Jacques, Guy, 1970.
Aspects quantitatifs du phytoplancton de Banyuls-sur-Mer (Golfe du Lion) IV.
Biomasse et production, 1965-1969
Vie Milieu 21 (1B): 37-102

phytoplankton biomass

Koblentz-Mishke, O.I. and V.I. Vedernikov, 1973
A tentative comparison of primary production and phytoplankton quantities at the ocean surface. (In Russian; English abstract). Okeanologiia 13 (1): 75-84.

phytoplankton biomass

Kondratieva, T.M., and E.V. Belogorskaia, 1961.
Distribution of phytoplankton in the Black Sea and its relationship to hydrological conditions.
Trudy Sevastopol Biol. Sta., (14):44-63.

phytoplankton, biomass

Konovalova, G.V., 1972.
Seasonal characteristics of phytoplankton in the Amursky Bay of the Sea of Japan. (In Russian; English abstract). Okeanologiia 12(1): 123-128.

phytoplankton, biomass

Nakai, Zinziro, Tadashi Kubota, Takeshi Mizushimia and Yukio Yamada, 1969.
Abundance and distribution of plankton in the waters around the Shimizu Harbour. 1. Diurnal and vertical change of biomass in summer season.
J. Coll. mar. Sci. Technol., Tokai Univ. (3):35-66.

Phytoplankton biomass

Niemi, Åke 1972
Observations on phytoplankton in eutrophied and non-eutrophied archipelago waters of the southern coast of Finland.
Memoranda Soc. Fauna Flora Fennica 48: 63-74

phytoplankton biomass

Niemi, Åke 1971.
Late summer phytoplankton of the Kimito Archipelago (SW coast of Finland).
Merentutkimuslait. Julk. 233: 3-17

phytoplankton biomass

Orlando, Aldo Maria, Enrique F. Mandelli y Paul R. Burkholder, 1965.
El fitoplankton antártico y las variables fisico-químicas del medio.
Bol Servicio Hidrograf naval, 5(3): 201-212

phytoplankton, biomass

San Feliu, J.M., y F. Muñoz, 1965.
Hidrografía y plancton del puerto de Castellón de junio de 1961 a enero de 1963.
Inv. Pesq., Barcelona, 28:3-48.

phytoplankton, biomass

Semina, H.J., 1961
[The phytoplankton of the mixing zone between Oyashio and Kuroshio in spring 1955.]
Trudy Inst. Okeanol., 51:3-15.

phytoplankton, biomass

Skolka, H., 1965.
Consideratii asupra variatiilor calitative si cantitative ale fitoplanctonului litoralului Românesc al Mării Negre.
In: Ecologie marinӑ, M. Bӑcescu, redactor, Edit. Acad. Republ. Pop. Române, București, 2:193-293.

phytoplankton biomass

Vinogradov, M.E., and N.M. Voronina, 1965.
Some peculiarities of plankton distribution in the Pacific and Indian equatorial current areas. (In Russian; English abstract).
Okeanolog. Issled., Rez. Issled. po Programme Mezhd. Geofiz. Goda, Mezhd. Geofiz. Komitet, Prezidiume Akad. Nauk, SSSR, No.13:128-136.

phytoplankton, biomass

Zernova, V.V., 1970.
Phytoplankton in the waters of the Gulf of Mexico and the Caribbean Sea. (In Russian; English and Spanish abstracts). Okeanol. Issled. Rezult. Issled. Mezhd. Geofiz. Proekt. 20: 69-104.

phytoplankton biomass

Zernova, V.V., and J.A. Ivanov, 1964.
The dependence of distribution of phytoplankton upon the hydrological conditions in the northern part of the Indian Ocean. Investigations in the Indian Ocean (33rd Voyage of E/S "Vityaz"). (In Russian; English summary).
Trudy Inst. Okeanol., Akad. Nauk, SSSR, 64:257-264.

phytoplankton blooms

Burkholder, Paul R., Robert W. Brody and Arthur E. Dammann 1972.
Some phytoplankton blooms in the Virgin Islands.
Carib. J. Sci. 12 (1/2): 23-28.

phytoplankton blooms

Halim, Y., 1960
Observations on the Nile bloom of phytoplankton in the Mediterranean. J. du Cons. 26(1): 57-67.

phytoplankton blooms

Hickel, Wolfgang, 1967.
Untersuchungen über die Phytoplanktonblüte in der westlichen Ostsee.
Helgoländer wiss. Meeresunters., 16(½):1-66.

phytoplankton bloom

Jacques, Guy 1971.
Floraison printanière du phytoplankton à Banyuls (golfe du Lion) en 1968.
Rapp. p.-v. Comm. int. Explor. scient. mer Medit. 20(3), 311-313

phytoplankton, blooms

Jayaraman, R. 1972.
On the occurrence of blooms of blue-green Algae and the associated oceanographic conditions in the northern Indian Ocean.
In: Taxonomy and Biology of Blue-Green Algae, T.V. Desikachary, editor, Univ. Madras, 428-432. (not seen)

phytoplankton blooms

Jensen, Arne and Egil Sakshaug, 1973
Studies on the phytoplankton ecology of the Trondheimsfjord. II. Chloroplast pigments in relation to abundance and physiological state of the phytoplankton. J. exp. mar. Biol. Ecol. 11 (2):137-155.

phytoplankton bloom

Ketchum, Bostwick H., and Nathaniel Corwin, 1965
The cycle of phosphorus in a plankton bloom in the Gulf of Maine.
Limnol. and Oceanogr., Redfield Vol., Suppl. to 10:R148-R161.

phytoplankton, bloom

Krey, Johannes, Peter H. Koske and Karl-Heinz Szekielda, 1965.
Produktionsbiologische und hydrographische Untersuchungen in der Eckernförder Bucht.
Kieler Meeresforsch., 21(2):135-143.

blooms (phytoplankton)

Parker, Robert H., 1964.
Zoogeography and ecology of some macroinvertebrates, particularly mollusks, in the Gulf of California and the continental slope off Mexico.
Vidensk. Medd., Dansk Naturh. Foren., 126:1-178.

phytoplankton blooms

Parsons, T.R., R.J. LeBrasseur and J.D. Fulton 1967.
Some observations on the dependence of zooplankton grazing on the cell size and concentration of phytoplankton blooms.
J. oceanogr. Soc. Japan 23(1), 10-17.

phytoplankton bloom

Platt, Trevor, and D.V. Subba Rao 1970.
Energy flow and species diversity in a marine phytoplankton bloom.
Nature, Lond. 227 (5262): 1059-1060.

phytoplankton blooms

Qasim, S.Z. 1972.
Some observations on Trichodesmium blooms.
In: Taxonomy and Biology of Blue-Green Algae, T.V. Desikachary, editor, Univ. Madras, 433-438 (not seen)

phytoplankton blooms

Walsh, Gerald E., 1966.
Studies of dissolved carbohydrate in Cape Cod waters. III. Seasonal variation in Oyster Pond and Wequaquet Lake, Massachusetts.
Limnol. Oceanogr., 11(2):249-256.

abstract

Cape Cod area
carbohydrate
USA, east
chlorophyll a
phytoplankton blooms

phytoplankton, caloric content

Platt, Trevor and Brian Irwin, 1973
Caloric content of phytoplankton.
Limnol. Oceanogr. 18(2):306-310

phytoplankton cells

Semina, H.J. 1972.
The size of phytoplankton cells in the Pacific Ocean.
Int. Revue ges. Hydrobiol. 57(2): 177-215

phytoplankton, cell size

Semina, Halina J. and Irina A. Tarkhova, 1972
Ecology of phytoplankton in the North Pacific Ocean. In: Biological oceanography of the northern North Pacific Ocean, A.Y. Takenouti, Chief Editor, Idemitsu Shoten, Tokyo, 117-124.

phytoplankton, cell size

Semina, H.J., 1968.
Water movement and the size of phytoplankton cells.
Sarsia, 34:267-272.

phytoplankton, size of

Tundisi, J.G., 1971.
Size distribution of the phytoplankton and its ecological significance in tropical waters. (Portuguese and English abstracts). In: Fertility of the Sea, John D. Costlow, editor, Gordon Breach, 2: 603-612.

phytoplankton, cell area

Mullin, M.M., P.R. Sloan and R.W. Eppley, 1966.
Relationship between carbon content, cell volume and area in phytoplankton.
Limnol. Oceanogr., 11(2):307-311.

phytoplankton, cell volume

Mullin, M.M., P.R. Sloan and R.W. Eppley, 1966.
Relationship between carbon content, cell volume and area in phytoplankton.
Limnol. Oceanogr., 11(2):307-311.

phytoplankton, chemistry of

Ackman, R.G., C.S. Tocher and J. McLachlan 1968.
Marine phytoplankter fatty acids.
J. Fish. Res. Bd Can. 25 (8): 1603-1620.

phytoplankton, chemistry of

Ackman, R.G., C.S. Tocher and J. McLachlan, 1966.
Occurence of dimethyl-B- propiothetin in marine phytoplankton.
J. Fish. Res. Bd., Canada, 23(3):357-364.

phytoplankton, chemistry

Blumer, M., R.R.L. Guillard and T. Chase, 1971.
Hydrocarbons of marine phytoplankton. Marine Biol. 8(3): 183-189.

phytoplankton, chemistry of

Carlucci, A.F., and Peggy M. Bowes 1972.
Vitamin B₁₂ thiamine, and biotin contents of marine phytoplankton.
J. Phycol. 8(2): 133-137.

phytoplankton, chemistry of (amino acids)

Chau, Y.K., L. Chuecas and J.P. Riley, 1967.
The component combines amino acids of some marine phytoplankton species.
J. mar. biol. Ass., U.K., 47(3):543-554.

phytoplankton, chemistry

Droop, M.R. 1966.
Organic acids and bases and the lag phase in Nannochloris oculata.
Jour. Mar. biol. Assoc. U.K., 46(3):673-678.

phytoplankton, chemical analyses

Iwamura, Tatsuichi, Tamotsu Kanazawa, Kazuo Shibata, Yuji Morimura, Shun-ei Ichimura, Osamu Maeda, and Hiroshi Tamiya 1967.
Preliminary studies on the feasibility of microanalytic measurement of planktonic populations
J. oceanogr. Soc. Japan 23(5): 247-251.

phytoplankton, chemistry

Lee, R.F., J.C. Nevenzel and G.-A. Paffenhofer, 1971.
Importance of wax esters and other lipids in the marine food chain: phytoplankton and copepods. Mar. Biol. 9(2): 99-108.

phytoplankton, chemistry of

Martin, John H., and George A. Knauer 1973
The elemental composition of plankton.
Geochim. Cosmochim. Acta. 37(7): 1639-1653.

phytoplankton, chemistry

McCarthy, James J. 1972.
The uptake of urea by marine phytoplankton.
J. Phycol. 8(3): 216-222.

phytoplankton, carbon content

Mullin, M.M., P.R. Sloan and R.W. Eppley, 1966.
Relationship between carbon content, cell volume and area in phytoplankton.
Limnol. Oceanogr., 11(2):307-311.

phytoplankton, chemistry

Parsons, T.R., K. Stephens and J.D.H. Strickland, 1961
On the chemical composition of eleven species of marine phytoplankters.
J. Fish. Res. Bd., Canada, 18(6):1001-1016.

phytoplankton, chemistry of

Platt, Trevor and Brian Irwin, 1973
Caloric content of phytoplankton.
Limnol. Oceanogr. 18(2):306-310

phytoplankton, chemistry

Strathmann, Richard R., 1967.
Estimating the organic carbon content of phytoplankton from cell volume or plasma volume.
Limnol. Oceanogr., 12(3):411-412.

phytoplankton community structure

Fedorov, V.D. and T.I. Koltsova, 1973
Experimental study of the relative abundance of some phytoplankton species. (In Russian; English abstract). Okeanologiia 13(1): 85-94.

phytoplankton populations

Hart, T.J., 1966.
Some observations on the relative abundance of marine phytoplankton populations in nature.
In: Some contemporary studies in marine science H. Barnes, editor, George Allen & Unwin, Ltd., 375-393.

phytoplankton, communities

Margalef, Ramón, 1966.
Análisis y valor indicador de las comunidades de fitoplancton mediterráneo.
Inv. Pesq., Barcelona, 30:429-482.

phytoplankton cycle
Establier, R., and R. Margalef, 1964
Fitoplancton e hidrografía de las costas de
Cádiz (Barbate), de junio de 1961 a agosto de
1962.
Invest. Pesquera, Barcelona, 25:5-31.

phytoplankton development
Gillbricht, M., 1964
Einwirkungen des kalten Winters 1962/63 auf die
Phytoplanktonentwicklung bei Helgoland.
Helgoländer Wiss. Meeresuntersuch., 10(1/4):263-278.

phytoplankton, diurnal variation
Wood, E.J. Ferguson, and Eugene F. Corcoran, 1966.
Diurnal variation in phytoplankton.
Bull. Mar. Sci., 16(3):383-403.

phytoplankton, decomposition of
Grill, Edwin V., and Francis A. Richards, 1964
Nutrient regeneration from phytoplankton decomposing in seawater.
J. Mar. Res., 22(1):51-69.

phytoplankton, qualitative composition
Balech, E., 1960
The changes in the phytoplankton population off the California coast. Cal. Coop. Ocean. Fish. Invest. Rept. Vol. 7: 127-132.

phytoplankton, vertical distribution
Bucalossi, G., 1960
Etude quantitative des variations du phytoplankton dans la baie d'Alger en fonction du milieu (novembre 1959 à mai 1960).
Bull. Inst. Océanogr., Monaco, 57(1189):1-40.

vertical distribution (Phytoplankton)
Hasle, Grethe Rytter, 1959.
A quantitative study of phytoplankton from the equatorial Pacific.
Deep-Sea Res., 6(1):38-59.

phytoplankton, species distribution [a]
Hulburt, Edward M. and R.S. Mackenzie, 1971.
Distribution of phytoplankton species at the western margin of the North Atlantic Ocean.
Bull. mar. Sci. Miami 21(2): 603-612.

phytoplankton distribution [a]
Karohji, Kohei, 1972
Regional distribution of phytoplankton in the Bering Sea and western and northern Subarctic regions of the North Pacific Ocean in summer.
In: Biological oceanography of the northern North Pacific Ocean, A.Y. Takenouti, Chief Editor, Idemitsu Shoten, Tokyo, 99-115.

phytoplankton distribution [a]
Lebedeva, L.P., 1972
A model of the latitudinal distribution of the number of species of phytoplankton in the sea. J. Cons. int. Explor. Mer 34(3): 341-350.

phytoplankton, distribution (theoretical)
Margalef, R., 1961.
Distribution du phytoplancton dans une échelle moyenne de dimensions et signification de ses pigments assimilateurs dans l'interpretation de la dynamique des configurations.
Rapp. Proc. Verb., Réunions, Comm. Int. Expl. Sci. Mer Méditerranée, Monaco, 16(2):139-140.

phytoplankton distribution
Margalef, Ramón, 1961.
Distribución ecológica y geográfica de las especies del fitoplancton marino.
Inv. Pesq., Barcelona, 19:81-101.

phytoplankton, vertical distribution
Marumo, R., 1952.
On one case of the vertical distribution of phytoplankton in the ocean. (In Japanese).
Ocean. Rept., Central Meteorol. Obs., 2(3):279-281.

phytoplankton, vertical distribution
Pucher-Petkovic, Tereza, 1965.
Distribution verticale seisonnière du phytoplancton en adriatique moyenne orientale.
Rapp. Proc. Verb., Réunions, Comm. Int. Expl. Sci. Mer Mediterranee, Monaco, 18(2):353-356.

phytoplankton distribution
Ringer, Z., 1963.
The vertical and horizontal distribution of phytoplankton in the southern Baltic in 1956.
Ann. Biol., Cons. Perm. Int. Expl. Mer, 1961, 18:76-77.

phytoplankton, vertical distribution
Ryther, J.H., and E.M. Hulburt, 1960
On winter mixing and the vertical distribution of plankton. Limnol. & Oceanogr. 5(3):337-338.

phytoplankton distribution (by species) [a]
Scotland, Edinburgh, Oceanographic Laboratory 1973.
Continuous plankton records: a plankton atlas of the North Atlantic and the North Sea.
Bull. mar. Ecol. 7:1-174.

phytoplankton, distribution of
Semina, H.J., 1963.
The phytoplankton along 174°W in the central Pacific. Part. LI. Horizontal distribution of abundance. Biological investigations of the ocean (plankton). (In Russian: English abstract)
Trudy Inst. Okeanol., Akad. Nauk, SSSR, 71:5-21.

phytoplankton distribution
Semina, G.I., 1958.
Verbindung der phytogeographischen Zonen im Pelagial des Nord-Westlichen Teiles des Stillen Ozeans mit der Verteilung der Wassermasser in diesem Gebiet. Trudy Inst. Okeanol., 27:67-76.

phytoplankton diversity [a]
Margalef, Ramón 1972.
Regularidades en la distribución de la diversidad del fitoplancton en un área del mar Caribe.
Inv. pesq. Barcelona 36(2): 241-266.

phytoplankton, diversity of
Margalef, R., 1969.
Diversidad de fitoplancton de red en dos áreas del Atlántico. Investigación pesq. 33(1): 275-286.

phytoplankton diversity [a]
Travers, M., 1971.
Diversité du microplancton du Golfe de Marseille en 1964. Mar. Biol., 8(4): 308-343.

phytoplankton, environmental limits
Braarud, T., 1962
Species distribution in marine phytoplankton.
J. Oceanogr. Soc., Japan, 20th Ann. Vol., 628-649.

phytoplankton, ecology
Braarud, T., 1951.
Salinity as an ecological factor in marine phytoplankton. Physiol. Plantarum 4(1):28-34.

phytoplankton, ecology [a]
El-Sayed, Sayed, and William T. Dill 1972
Ecology of phytoplankton in the southwestern Pacific Ocean studied during Southwind cruise.
Antarctic Jl U.S.A. 7(4): 72-73.

phytoplankton ecology
Ryther, J.H., 1954.
The ecology of phytoplankton blooms in Moriches Bay and Great South Bay, Long Island, New York.
Biol. Bull. 106(2):198-209.

phytoplankton ecology [a]
Sakshaug, Egil and Sverre Myklestad, 1973.
Studies on the phytoplankton ecology of the Trondheimsfjord. III. Dynamics of phytoplankton blooms in relation to environmental factors, bioassay experiments and parameters for the physiological state of the populations. J. exp. mar. Biol. Ecol. 11(2):157-188.

phytoplankton ecology [a]
Jensen, Arne and Egil Sakshaug, 1973
Studies on the phytoplankton ecology of the Trondheimsfjord. II. Chloroplast pigments in relation to abundance and physiological state of the phytoplankton. J. exp. mar. Biol. Ecol. 11(2):137-155.

phytoplankton, ecology
Subrahmanyan, R., 1958.
Ecological studies on the marine phytoplankton on the west coast of India.
Mem. Indian Botan. Soc., 1:145-151.

phytoplankton, ecology of
Yentsch, Charles S., and Robert W. Lee, 1966.
A study of photosynthetic light reactions and a new interpretation of sun and shade phytoplankton.
J. mar. Res., 24(3):319-337.

phytoplankton, effect of [a]
Aubert, J., and Cl. Jorus 1971.
Action antibiotique de quelques espèces phytoplanctoniques marines vis-à-vis de différentes salmonelles.
Rev. int. Océanogr. Méd. 22-23:143-149

phytoplankton, effect of [a]
Edmondson, W.T., Gabriel W. Comita and George C. Anderson, 1962
Reproductive rate of copepods in nature and its relation to phytoplankton population.
Ecology, 43(4):625-634.

phytoplankton, effect of [a]
Jensen, Arne and Egil Sakshaug, 1970.
Producer-consumer relationships in the sea.
II. Correlation between Mytilus pigmentation and the density and composition of phytoplanktonic populations in inshore waters.
J. exp. mar. Biol. Ecol., 5(3): 246-253.

phytoplankton, effect of [a]
Kremser, Ulrich 1972.
Die Wirkung "nichtturbulenter" Prozesse auf die Konzentration eines Rhodamin-S-Fleckes im Meer unter besonderer Berücksichtigung des Einflusses von Phytoplankton suspendierten Sedimenten und Sonnenstrahlung.
Beitr. Meeresk. (30/31): 101-125.

phytoplankton, effect of

Lorenzen, Carl J., 1972
Extinction of light in the ocean by phytoplankton. J. Cons. int. Explor. Mer 34(2): 262-267.

phytoplankton, effect of

Moebus, K., 1972.
Studies on the influence of plankton on anti-bacterial activity of sea water.
Helgoländer wiss. Meersunters 23(2): 127-140.

phytoplankton, effect of

Nakamura, Nakaroku, 1958
On the seasonal variation of chlor. content, transparency and water-colour of the eel-culture ponds in Tokai district in relation to the particle size of phytoplankton. Bull. Jap. Soc. Sci. Fish., 24:495-600.

phytoplankton, effect of the

Ryther, J.H., 1954.
Inhibitory effects of phytoplankton upon feeding of Daphnia magna with reference to growth, reproduction and survival. Ecology 35(4):522-533.

phytoplankton, effect of

Thomas, J.P., 1971.
Release of dissolved organic matter from natural populations of marine phytoplankton. Mar. Biol. 11(4): 311-323.

phytoplankton, effect of

Wilson, W.B. and Albert Collier, 1972.
The production of surface-active material by marine phytoplankton cultures. J. mar. Res. 30(1): 15-26.

phytoplankton, effect of

Yentsch, C.S., 1960
The influence of phytoplankton on the colour of sea-water.
Deep-Sea Res., 7(1): 1-9.

phytoplankton, extinction of

Pitrat, Charles W., 1970.
Phytoplankton and the late Paleozoic wave of extinction
Palaeogeogr., Palaeoclimatol., Palaeoecol., 8(1): 49-65.

phytoplankton extinction

Tappan, Helen 1970.
Phytoplankton abundance and late Paleozoic extinctions: a reply.
Palaeogeogr. Palaeoclimatol. Palaeoecol. 8(1): 56-66.

Phytoplankton, floatation

Braarud, T., 1962
Species distribution in marine phytoplankton.
J. Oceanogr. Soc., Japan, 20th Ann. Vol., 628-649.

phytoplankton flowering

Gillbricht, Max, 1962
Die Frühjahrswucherung des Phytoplanktons in einen flachen Gezeitenmeer.
Kieler Meeresf., 18(3) (Sonderheft):151-156.

phytoplankton (as food)

Roushdy, H.M., and V.K. Hansen, 1960.
Ophiuroids feeding on phytoplankton.
Nature 188(4749):517-518.

phytoplankton, fluctuations of

Pucher-Petkovic, Tereza, 1965.
Fluctuations pluriannuelles du phytoplancton en relation avec certains facteurs météorologiques et hydrographiques.
Rapp. Proc.-verb. Reun., Comm. int. Explor. scient. Mer Méditerranée 19(3): 399-401.

phytoplankton growth

Dugdale, Richard C. and John J. Goering, 1971.
A model of nutrient-limited phytoplankton growth.
In: Impingement of man on the oceans, D.W. Hood, editor, Wiley Interscience: 589-600.

phytoplankton growth

Eppley, R.W., A.F. Carlucci, O. Holm-Hansen, D. Kiefer, J.J. McCarthy, Elizabeth Venrick and P.M. Williams, 1971.
Phytoplankton growth and composition in shipboard cultures supplied with nitrate, ammonium, or urea as the nitrogen source. Limnol. Oceanogr. 16(5): 741-751.

phytoplankton growth

Eppley, Richard W., Edward H. Renger, Elizabeth L. Venrick and Michael M. Mullin, 1973
A study of plankton dynamics and nutrient cycling in the central gyre of the North Pacific Ocean. Limnol. Oceanogr. 18(4):534-551.

phytoplankton growth

Eppley, R.W., and J.D.H. Strickland 1968.
Kinetics of marine phytoplankton growth.
In: Advances in Microbiology of the Sea, M.R. Droop and E.J.F. Wood, editors, Academic Press, 23-62.

phytoplankton growth

Eppley, Richard W. 1972
Temperature and phytoplankton growth in the sea.
Fish. Bull. U.S. nat. mar. Fish. Serv. NOAA 70(4):1063-1085.

phytoplankton growth

Golovkin, A.N., 1967.
The influence of sea colonial birds on phytoplankton growth. (In Russian; English abstract).
Okeanologiia, Akad. Nauk, SSSR, 7(4):672-682.

phytoplankton growth

Hood, Donald W., editor 1971
Impingement of man on the oceans.
Wiley-Interscience, 738 pp.

phytoplankton, growth kinetics

Qasim, S.Z., P.M.A. Bhattathiri and V.P. Devassy, 1973
Growth kinetics and nutrient requirements of two tropical marine phytoplankters. Mar. Biol. 21(4):299-304.

phytoplankton growth

Strickland J.D.H. 1968.
Synchrony in marine phytoplankton growth, some physiological and ecological observations.
Bull. Misaki mar. biol. Inst. Kyoto Univ. 12:1-2.

phytoplankton growth

Takahashi, M., K. Fujii and T.R. Parsons, 1973
Simulation study of phytoplankton photosynthesis and growth in the Fraser River Estuary. Mar. Biol. 19(2): 102-116.

phytoplankton interrelations

Beklemishev, K.V., 1957.
On the spatial relationships between marine zoo- and phytoplankton. Trudy Inst. Okeanol., 20:253-278.

phytoplankton, succession of spp.

Vives, F., and F. Fraga, 1961
Floristica y sucesión del fitoplancton en la Ria de Vigo.
Inv. Pesq., Barcelona, 19:17-36.

phytoplankton, lists of genera

Ferrando, Hugo J., 1959
Estudio del plancton en la zona de pesca de la merluza. Anales, Facultad de Veterinaria, Montevideo 8(6): 89-99.

phytoplankton, list of genera

Ward, Ronald W., Valerie Vreeland, Charles H. Southwick and Anthony J. Reading, 1965.
Ecological studies related to plankton productivity in two Chesapeake Bay estuaries.
Chesapeake Science, 6(4):214-225.

phytoplankton, lists of species

Acara, A., and U. Nalbandoglu, 1960
Preliminary report on the red tide outbreak in the Gulf of Izmir.
Comm. Inst. Expl. Sci. Mer Medit., Monaco, Rapp. Proc. Verb., 15(3): 33-38.

phytoplankton, lists of spp.

Akatsuka, K. (deceased), F. Uyeno, K. Mitani and M. Miyamura, 1960.
On the relation between the distribution of plankton and the annual changes of sea conditions in Ise Bay.
J. Oceanogr. Soc., Japan, 16(2):83-91.

phytoplankton, lists of spp.

Angot, Michel 1970.
Le phytoplancton des environs de Nosy-Bé (Madagascar) et ses variations au cours de 1965.
Annls Univ. Madagascar, Ser. Sci. Nat. Mat. 7:165-177.

phytoplankton, lists of spp.

Angot, Michel, 1965.
Le phytoplancton de surface pendant l'année 1964 dans la Baie d'Ambaro près de Nosy Bé.
conclusion p. 13.
Océanographie, Cahiers, O.R.S.T.R.O.M., 3(4):5-18.

phytoplankton, lists of spp.

Angot, M., 1964.
Phytoplancton et production primaire de la région de Nosy-Bé, decembre 1963 à mars 1964.
Cahiers O.R.S.T.R.O.M., Oceanogr., 11(4):99-125.

phytoplankton, lists of spp.

Angot, M., 1964.
Production primaire de la région de Nosy-Bé.
Août à novembre 1963.
Cahiers, O.R.S.T.R.O.M., Oceanogr., 11(4):27-53.

Phytoplankton, lists of spp.

Argentina, Secretaria de Marina, Servicio de Hidrografia Naval, 1962.
Plancton de las campanas oceanograficas DRAKE I y II.
Publico, H. 627:57.

phytoplankton, lists of spp.

Aubert, J., et J.P. Gambarotta 1972.
Etude de l'action antibactérienne d'espèces phytoplanctoniques marines vis-à-vis de germes anaérobies.
Rev. int. Océanogr. Méd. 25:39-47.

phytoplankton, lists of spp.

Australia, Commonwealth Scientific and Industrial Research Organization, Division of Fisheries and Oceanography, 1970.
Coastal investigations off Port Hacking, New South Wales in 1965.
Oceanogr. Sta. List 85: 124pp.

Oceanographic Index: Marine Organisms Cumulation, 1946-1973

phytoplankton, lists of spp
Australia, Commonwealth Scientific and Industrial Research Organization, 1963
Oceanographical observations in the Indian Ocean in 1961, H.M.A.S. Diamantina Cruise Dm 2/61.
Oceanogr. Cruise Rept., Div. Fish. and Oceanogr. No. (:155 pp., 14 figs.

phytoplankton, lists of spp.
Australia, Commonwealth Scientific and Industrial Research Organization, 1963
Oceanographical observations in the Pacific Ocean in 1961, H.M.A.S. Gascoyne, Cruise G 1/61.
Oceanogr. Cruise Rept., Div. Fish. and Oceanogr., No. 8:130 pp., 12 figs.

phytoplankton, lists of spp.
Australia, Commonwealth Scientific and Industrial Organization, 1962.
Oceanographic observations in the Indian Ocean in 1961.
Div. Fish. Oceanogr., Cruise DM 1/61:88 pp.

phytoplankton, lists of spp.
Australia, Commonwealth Scientific and Industrial Research Organization, Division of Fisheries and Oceanography, 1960.
F.R.V., "Derwent Hunter" Scientific report on Cruises 1-8, 1958: 57 pp. (Mimeographed)

phytoplankton, lists of spp.
Australia, C.S.I.R.O., Division of Fisheries and Oceanography, Marine Biological Laboratory, 1959. Cronulla
F.R.V. "Derwent Hunter", Scientific report of Cruise DH9/57, Aug. 19-25, 1957; Cruise DH10/57, Sept. 4-11, 1957; Cruise DH11/57, Sept. 18-21, 1957; Cruise DH12/57, Sept. 26-Oct. 1, 1957.
CSIRO, Div. Fish. & Ocean., Rept., No. 20:20 pp. (mimeographed)

phytoplankton, lists of spp.
Australia, C.S.I.R.O., Division of Fisheries and Oceanography, 1959.
Scientific reports of a cruise on H.M.A. Ships "Queenborough" and "Quickmatch", March 24-April 26, 1958. Rept. No. 24: 24 pp. (mimeographed).

phytoplankton, lists of spp.
Australia, C.S.I.R.O., Division of Fisheries and Oceanography, 1958/1959.
FVR "Derwent Hunter". Rept. No. 19:16 pp. No. 21:16 pp.
(mimeographed)

phytoplankton, lists of spp.
Azouz, Abderrazak, 1966.
Étude des peuplements et des possibilités d'ostreiculture du lac de Bizerte.
Inst.net.Sci.tech.Oceanogr.Pêche,Salammbo,Ann. 15: 69 pp.

phytoplankton, lists of species
Băcescu, M., 1961
Cercetari fizico-chimice și biologice romînesti la Marea Neagra, efectuate in perioda 1954-1959.
Hidrobiologia, Acad. Repub. Pop. Rom., (3): 17-46.

phytoplankton, lists of spp.
Băcescu, M., G. Müller, H. Skolka, A. Petran, V. Elian, M.T. Gomoiu, N. Bodeanu și S. Stănescu, 1965.
Cercetări de ecologie marină în sectorul predeltaic în condițiile anilor 1960-1961.
In: Ecologie marină, M. Băcescu, redactor, Edit. Acad. Republ. Pop. Romăne, Bucuresti, 1: 185-344.

phytoplankton, lists of spp.
Băcescu, M.T. Gomoiu, N. Bodeanu, A. Petran, G. Müller și V. Manea, 1965.
Studii asupra variatiei vietii marine în zona litoral nisipoasă de la nord de Constanța (Cercetări efectate în anii 1960-61 la puncte fixe situate în dreptul stațiunii Mamaia).
In: Ecologie marină, M. Băcescu, redactor, Edit. Acad. Republ. Pop. Romăne, Bucuresti, 1: 7-138.

phytoplankton, lists of spp.
Băcescu, M., M.T. Gomoiu, N. Bodeanu, A. Petran, G.I. Müller si V. Chirila, 1965.
Dinamica populațiilor animale și vegetale din zona nisipurilor fine de la nord de Constanța in condițiile anilor 1962-1965.
In: Ecologie marină, M. Băcescu, redactor, Edit. Acad. Republ. Pop. Romăne, Bucuresti, 2:7-167.

phytoplankton, lists of spp.
Bainbridge, V., 1960
The plankton of inshore waters off Freetown, Sierra Leone. Colonial Off., Fish. Publ., London, No. 13: 48 pp.

phytoplankton, lists of spp.
Balech, E., 1970
The distribution and endemism of some Antarctic microplankters
Antarctic Ecol. 1: 143-146

phytoplankton, lists of spp.
Balech, Enrique, 1964.
El plancton de Mar del Plata durante el periodo 1961-1962.
Bol. Inst. Biol. Mar., Buenos Aires, (4):56 pp.

phytoplankton, lists of species
Balech, E., 1954.
VII. Breves datos sobre la distribucion geografia y estacional del plancton marina de la Argentina. Rev. Biol. Mar., Valparaiso, 4(1/2/3):211-224, 2 figs.

phytoplankton, lists of spp
Balech, Enrique and Sayed Z. El-Sayed, 1965.
Microplankton of the Weddell Sea.
In: Biology of Antarctic seas. II.
Antarctic Res. Ser. Amer. Geophys. Union, 5:107-124

phytoplankton, lists of spp.
Balle, P., 1961.
Phytoplancton d'Ibiza et de la côte est et sud de la péninsule ibérique.
Rapp. Proc. Verb., Reunions, Comm. Int. Expl. Sci. Mer Mediterranée, Monaco, 16(2):231-236.

phytoplankton, lists of species
Balle, P., 1954.
Análisis cualitativo del fitoplancton de la bahía de Palma de Mallorca en 1953. Bol. Inst. Español Ocean., No. 68:1-3.

phytoplankton, lists of species
Balle, P., 1953.
Fitoplancton de la Bahía de Palma de Mallorca (Año 1942). Bol. Inst. Español. Oceanogr., No. 61:1-21, 1 textfig.

phytoplankton, lists of spp.
Bank, Keuk Soon, 1967.
Studies on the quantity and composition of the microplankton in a bay of Han San. (In Korean; English abstract).
Bull.Fish.Res.Develop.Agency,1:119-130.

phytoplankton, lists of spp
Becacos-Kontos, Theano and Lydia Ignatiades, 1970.
Preliminary biological, chemical and physical observations in the Corinth Canal area.
Cah. océanogr. 22(3):259-267.

phytoplankton (lists of species)
Bernard, F., and L. Fage, 1936.
Recherches quantitatives sur le plancton méditerranéen. Bull. Inst. Océan. Monaco, No. 701:1-20, 1 textfig.

phytoplankton, lists of spp.
Bernhard, M., and L. Rampi, 1967.
The annual cycle of the "Utermöhl-phytoplankton" in the Ligurian Sea in 1959 and 1962.
Pubbl. Staz. zool., Napoli, 35(2): 137-169.

phytoplankton, lists of spp
Bernhard, M., e L. Rampi, 1963
4. Botanica oceanografica.
Rapp. Attività Sci. e Tecn., Lab. Studio della Contaminazione Radioattiva del Mare. Fiasscherino, La Spezia (maggio 1959-maggio 1962), Comit. Naz. Energia Nucleare, Roma, RT/BIO(63) 8:57-123. (multilithed) plus Appendice 4.

phytoplankton, lists of spp
Bernhard, M., L. Rampi and A. Zattera, 1967.
A phytoplankton component not considered by the Utermöhl method.
Pubbl. Staz. zool., Napoli, 35(2):170-214.

phytoplankton, list of species
Birkenes, E., and T. Braarud, 1954.
Phytoplankton in the Oslo Fjord during a "Coccolithus huxleyi-summer".
Avhandl. Norske Vidnskaps-Akad., Oslo. 1. Mat.-Naturvid. Kl., 1952(2):1-23, 1 textfig.

phytoplankton, lists of spp. [a]
Björnberg, T.K.S., 1971
Distribution of plankton relative to the general circulation system in the area of the Caribbean Sea and adjacent regions. Symp. Investigations and resources of the Caribbean Sea and adjacent regions, UNESCO, 18-26 Nov. 1968, Curaçao: 343-356.

phytoplankton, lists of spp. [a]
Blasco, Dolores, 1971.
Composición y distribución del fitoplancton en la región del afloramiento de las costas peruanas. (In Spanish; English abstract).
Investigacion pesq. 35(1): 61-112.

phytoplankton, lists of spp.
Bodo, F., C. Razouls et A. Thiriot, 1965.
Etude dynamique et variations saisonnières du plankton de la région de Roscoff. II.
Cahiers Biol. Mar., Roscoff, 6(2):219-254.

phytoplankton, lists of spp.
Braarud, T., and Bjørg Føyn, 1958.
Phytoplankton observations in a brackish water locality of south-east Norway.
Nytt Mag. Botan., 6:47-73.

phytoplankton, lists of spp.
Braarud, Trygve, Bjørg Føyn and Grethe Rytter Hasle, 1958.
The marine and fresh-water phytoplankton of the Dramsfjord and the adjacent part of the Oslo-fjord, March - December 1951.
Hvalradets Skr., 43:102 pp.

phytoplankton, lists of spp
Buchanan, R.J., 1971.
Studies at Oyster Bay in Jamaica, West Indies. V. Qualitative observations on the planktonic algae and protozoa. Bull. mar. Sci., Coral Gables, 21(4): 914-937.

phytoplankton, lists of spp
Bursa, Adam, 1963.
Phytoplankton in coastal waters of the Arctic Ocean at Point Barrow, Alaska.
Arctic, 16(4):239-262.

phytoplankton, lists of spp. [a]
Carpenter, Edward J., 1973
Brackish-water phytoplankton response to temperature elevation. Estuarine coast. mar. Sci. 1(1): 37-44.

phytoplankton, lists of spp.
Carpenter, Edward J., 1971.
Annual phytoplankton cycle of the Cape Fear River Estuary, North Carolina.
Chesapeake Sci., 12(2): 95-100.

phytoplankton, lists of spp.
Cassie, Vivienne.
Distribution of surface phytoplankton between New Zealand and Antarctic, December 1957.
Trans. Antarctic Exped., 1955-1958, Sci. Repts., No. 7:11 pp.

phytoplankton, lists of spp.
Cassie, Vivienne, 1961.
Marine phytoplankton in New Zealand waters.
Botanica Marina, 2(Suppl.):54 pp., 8 pls.

phytoplankton, lists of spp.
Choe, Sang 1969.
Phytoplankton studies in Korean waters. IV. Phytoplankton in the adjacent seas of Korea. (In Korean; English abstract).
J. oceanol. Soc. Korea 4(2):49-67.

phytoplankton, lists of spp.
Choe, Sang 1969.
Phytoplankton studies in Korean waters. III. Surface phytoplankton survey of the north-eastern Korea Strait in May of 1967. (In Korean; English abstract).
J. oceanogr. Soc. Korea 4(1):1-8.

phytoplankton, lists of spp.
Choe, Sang 1967.
Phytoplankton studies in Korean waters. II. Phytoplankton in the coastal waters of Korea. (In Korean; English abstract).
J. oceanogr. Soc. Korea 2(1/2):1-12.

phytoplankton, lists of spp.
Choe, Sang, 1966.
Phytoplankton studies in Korean waters. 1. Phytoplankton survey of the surface in the Korea Strait in summer of 1965. (In Korean; English abstract).
J. oceanogr. Soc., Korea, 1(1/2):14-21.

phytoplankton, lists of species
Clark, R.M., 1905.
Plankton investigations. Fish. Bd., Scotland, North Sea Fish. Invest. Comm., 1902-1903 (Rept. No. 1):166-213.

PHYTOPLANKTON, LISTS OF SPP
Colebrook, J.M., and G.A. Robinson, 1964.
Continuous plankton records: annual variations of abundance of plankton 1948-1960.
Bull. Mar. Ecol., 6(3): 52-69.

phytoplankton, lists of spp.
Conover, S.A.M., 1956.
Oceanography of Long Island Sound, 1952-1954. 4. Phytoplankton. Bull. Bingham Oceanogr. Coll., 15:62-112.

phytoplankton, lists of spp.
De Angelis, C.M., 1961.
Report on the common species and characteristics of the phytoplankton of the Tyrrhenian brackish ponds.
Rapp. Proc. Verb., Reunions, Comm. Int. Expl. Sci. Mer Mediterranee, Monaco, 16(2):133-137.

phytoplankton, lists of spp.
de Angelis, C.M., 1956.
Ciclo annuale del fitoplancton del Golfo di Napoli. Boll. Pesca, Piscicolt. e Idrobiol., n.s. 11(1):37-55.

phytoplankton, lists of spp.
Denisenko, V.V., 1964.
On the phytoplankton of the Adriatic, Ionian, Aegaean and Black seas in August 1958. (In Russian).
Trudy Sevastopol Biol. Sta., 7:13-20.

phytoplankton, lists of spp.
Denisenko, V.V., 1962
Some data on the phytoplankton in the Adriatic Sea. (In Russian).
Okeanologiia. Akad. Nauk. SSSR, 2(4):699-704.

phytoplankton, lists of spp.
DeSousa e Silva, Estela, 1968.
Plancton da Lagoa de Óbidos (III). Abundância, variacões Sazonais e grandes "blooms".
Notas Estudos,Inst.Biol.marit.Lisboa 34:79pp.

phytoplankton, lists of spp.
Dragovich, A., 1963.
Hydrology and plankton of coastal waters at Naples, Florida.
Q.J. Florida Acad. Sci., 26(1):22-47.

phytoplankton (lists of spp.)
Dragovich, Alexander, 1961
Relative abundance of plankton off Naples, Florida, and associated hydrographic data, 1956-57.
USFWS Spec. Sci. Rept., Fish., No. 372:41 pp.

phytoplankton, lists of spp.
Durairatnam, M., 1964
Vertical distribution of phytoplankton in an area near Cocos-Keeling islands, Indian Ocean. (In Japanese; English abstract)
Inform. Bull. Planktol., Japan, No. 11:1-6.

phytoplankton, lists of spp.
Durairatnam, M.,1963.
Studies on the seasonal cycle of the sea surface temperatures, salinities and phytoplankton in Puttalem Lagoon, Dutch Bay and Portugal Bay along the west coast of Ceylon.
Bull. fish.res.Stn. Ceylon, 16(1):9-24.
Also in: Coll. Repr., Int. Indien Ocean Exped. 4(1967).

phytoplankton, lists of spp.
Durán, Miguel, Fernando Saiz, Manuel López-Benito y Ramón Margalef, 1956
El fitoplancton de la ría de Vigo, de abril de 1954 a junio de 1955.
Inv. Pesq., Barcelona, 4:67-96.

phytoplankton, lists of spp.
Ehrhardt, Jean-Paul,1967.
Contribution à l'étude du plancton superficiel et sub-superficiel du Canal de Sardaigne et de la mer Sud-Tyrrhenienne: campagne de l'Orbigny du 15 septembre au 14 octobre 1963, travaux du Laboratoire d'Océanographie biologique du Bureau d'Études Océanographiques - Toulon.
Cah. océanogr., 19(9):729-781.

phytoplankton, lists of spp
Ehrhardt,Jean-Paul, et Daniel Bonin,1968.
Contribution à l'etude du plancton dans le canal de Corse-Provence, campagne de l'Origny, 12 juin-4 juillet 1963.
Cah. océanogr., 20(2):133-156.

phytoplankton, lists of spp.
Ehrhardt, Jean-Paul, Félix Baudin-Laurencin, et Gérard Seguin, 1964
Contribution à l'étude du plancton dans le Cana C orse-Provence.
Cahiers Océanogr., C.C.O.E.C., 16(8):623-636.

phytoplankton, lists of spp. [a]
Elbrächter, Malte 1970.
4.2. Phytoplankton und Ciliaten
Chemische, mikrobiologische und planktologische Untersuchungen in der Schlei im Hinblick auf deren Abwasserbelastung. Kieler Meeresforsch 26(2): 193-203.

phytoplankton, lists of spp. [a]
Enomoto, Yoshimasa, 1971.
Oceanographic survey and biological study of shrimps in the waters adjacent to the eastern coasts of the State of Kuwait.
Bull. Tokai reg. Fish. Res. Lab. 66: 1-74.

phytoplankton, lists of spp.
Establier, R., and R Margalef, 1964
Fitoplancton e hidrografía de las costas de Cádiz (Barbate), de junio de 1961 a agosto de 1962.
Invest. Pesquera, Barcelona, 25:5-31.

phytoplankton, lists of spp. [a]
Estrada, Marta, 1972.
Analyse en composantes principales de données de phytoplancton de la zone côtière du Sud de l'Ebre. Inv. Pesq. Barcelona 36(1): 109-118.

Phytoplankton, lists of spp.
Fedorov, V. D., 1970.
The dominant forms of the White Sea phytoplankton. (In Russian).
Dokl Aked. Nauk SSSR 188(4): 909-912.

phytoplankton, lists of spp.
Gail, M.M., 1963.
The spring phytoplankton in the southeast part of the Tatar Strait. (In Russian).
Izv. Tikhookean, Nauchno-Issled. Inst. Ribn. Khoz. i Okeanogr., 49:137-158.

phytoplankton, lists of spp
Gladkikh, G.N. 1969.
Phytoplankton of one of the regions of the Kuroshio in summer of 1965. (In Russian)
Izv. Tichookean. nauchno-issled. Inst. ribn. Khz. okeanogr. 68: 86-92.
(TINRO)

phytoplankton, lists of spp. [a]
Gökalp, Nurettin 1972
A study of plankton conditions off Edremit, Bodrum and Iskenderun gulfs.
Hidrobiol. Arastirma Enst. Yayinlari, Istanbul Univ. Fen Fakült. 3: 71 pp.
(In Turkish; English abstract)

phytoplankton, lists of spp
Gomez-Aguirre, Samuel 1965.
Algunas consideraciones acerca del fitoplancton primaveral en la Boca de Paso Real, Campeche.
Anales Inst. Biol. Univ. Mex. 36(1/2): 65-69.

phytoplankton, lists of spp
Grant, B.R. and J.D. Kerr 1970.
Phytoplankton numbers and species at Port Hacking Station and their relationship to the physical environment.
Austr. J. mar. Freshwat. Res. 21(1): 35-45.

phytoplankton, lists of sp.
Grøntved, J., 1960-61
Planktological contributions. IV. Taxonomical and productional investigations in shallow coastal waters.
Medd. Dansk Fisk. Havundersøgelser,N.s.,3(1): 1-17.

phytoplankton, lists of spp.
Grøntved, J., 1958.
Planktonological contributions. III. Investigations on the phytoplankton and the primary production in an oyster culture in the Limfjord.
Medd. Dansk Fisk. Havundersøgelser, n.s., 2(17): 1-15.

phytoplankton, lists of speci
Grøntved, J., 1950.
Investigations of the geography and natural history of the Præstø Fiord, Zealand, 6. The phytoplankton of the Præstø Fiord.
Folia Geogr. Danica, 3(5):143-186, 6 textfigs., 8 tables.

phytoplankton, lists of spp.
Grøntved, J., and E. Steemann Nielsen, 1957.
Investigations on the phytoplankton in sheltered Danish marine localities.
Medd. Komm. Danmarks Fiskeri- og Havundersøgelser,
5(6):52 pp.
Ser. Plankton

phytoplankton, lists of spp.
Halim, Y., 1960
Observations on the Nile bloom of phytoplankton in the Mediterranean. J. du Cons. 26(1): 57-67.

phytoplankton (lists of spp.)
Hart, T. John, and Ronald I. Currie, 1960
The Benguela Current.
Discovery Repts., 31: 123-298.

phytoplankton, lists of sp.
Hasle, Grethe Rytter, 1969.
An analysis of the phytoplankton of the Pacific Southern Ocean: abundance, composition, and distribution during the Brategg Expedition, 1947-1948.
Hvalråd. Skr. 52: 168 pp.

phytoplankton, lists of spp.
Hasle, Grethe Rytter, 1960
Phytoplankton and ciliate species from the Tropical Pacific.
Skr. Norske Videnskaps-Akad., Oslo, 1.
Mat.-Nat. Kl., 1960(2): 1-50.

phytoplankton, lists of spp.
Hasle, Grethe Rytter, 1959.
A quantitative study of phytoplankton from the equatorial Pacific.
Deep-Sea Res., 6(1):38-59.

phytoplankton, lists of spp.
Hasle, Grethe Rytter, and Theodore J. Smayda, 1960
The annual phytoplankton cycle at Drøbak, Oslofjord.
Nytt Mag. Botanikk, 8:53-75.

phytoplankton, lists of spp.
Hellebust, Johan A., 1967.
Excretion of organic compounds by cultured and natural populations of marine phytoplankton.
In: Estuaries, G.H. Lauff, editor, Publs Am. Ass Advmt Sci., 83:361-355.

phytoplankton, lists of spp
Hellebust, J.A., 1965.
Excretion of some organic compounds by marine phytoplankton.
Limnology and Oceanography, 10(2):192-206.

phytoplankton, list of spp
Herrara, Juan, and Ramon Margalef, 1961
Hidrografía y fitoplancton de las costas de Castellón de julio de 1958 a junio de 1959.
Inv. Pesq., Bacelona, 20:17-63.

phytoplankton, lists of spp.
Herrera, J., and R. Margalef, 1957.
Hidrografia y fitoplancton de las costas de Castellón de julio de 1956 a juinio de 1957.
Inv. Pesq., 10:17-44.

phytoplankton, list of species
Herrera, J., F. Muñoz, and R. Margalef, 1955.
Fitoplancton de las costas de Castellón durante el año 1953. Invest. Pesq. 1:17-29.

phytoplankton, lists of spp.
Hirota, Reiichiro, and Takuo Ando, 1964.
On primary production in Bingo-Nada of Seto Inland Sea. II. Primary production and plankton. (In Japanese; English abstract).
J. Fac. Fish. and Animal Husbandry, Hiroshima Univ., 5(2):519-535.
Also in:
Contrib. Mukaishima Mar. Biol. Sta., Hiroshima Univ. 1963-1965. (77)

phytoplankton, lists of spp
Hoang, Quoc Truong, 1961
Preliminary plankton research in the bay of Nhatrang, Vietnam.
Ann. Fac. Sci. Saigon, 1961:91-100.
Also:
Contrib., Inst. Oceanogr., Nhatrang, No. 51.

phytoplankton, lists of species
Holmes, R.W., 1956.
The annual cycle of phytoplankton in the Labrador Sea, 1950-51. Bull. Bingham Oceanogr. Coll., 16(1):1-74.

phytoplankton (lists of species)
Hope, B., 1954.
Floristic and taxonomic observations on marine phytoplankton from Nordavatn, near Bergen.
Nytt Mag. f. Botanikk 2:149-183, 1 fig.

phytoplankton, lists of spp.
Hopkins, Thomas L., 1966.
The plankton of the St. Andrew Bay system, Florida.
Publs., Inst.mar.Sci., Univ Texas, Port Aransas, 11:12-64.

phytoplankton, lists of spp.
Hulburt, Edward M., 1968.
Phytoplankton observations in the western Caribbean Sea. Bull. mar. Sci., Miami, 18(2): 388-399.

phytoplankton, lists of spp.
Hulburt, Edward M., 1967.
Some notes on the phytoplankton off the southeastern coast of the United States.
Bull. mar. Sci., Miami, 17(2):330-337.

phytoplankton, lists of spp.
Hulburt, Edward M., 1966.
The distribution of phytoplankton, and its relationship to hydrography, between southern New England and Venezuela.
J. Mar. Res., 24(1):67-81.

phytoplankton, lists of spp.
Hulburt, Edward M., 1963
Distribution of phytoplankton in coastal waters of Venezuela.
Ecology, 44(1):169-171.

phytoplankton, lists of spp.
Hulburt, Edward M., 1962
Phytoplankton in the southwestern Sargasso Sea and North Equatorial Current, February 1961.
Limnol. and Oceanogr., 7(3):307-315.

phytoplankton, lists of spp.
Hulburt, E.M., 1962
A note on the horizontal distribution of phytoplankton in the open ocean.
Deep-Sea Res., 9(1): 72-74.

phytoplankton, lists of spp.
Hulburt, Edward M., and Nathaniel Corwin, 1972.
A note on the phytoplankton distribution central Gulf of Mexico. Carib. J. Sci. 12 (1/2): 29-38.

phytoplankton, lists of spp.
Hulburt, E.M., Nathaniel Corwin, 1969.
Influence of the Amazon River outflow on the ecology of the western tropical Atlantic. III. The planktonic flora between the Amazon River and the Windward Islands.
J. mar. Res., 27(1):55-72.

phytoplankton, lists of spp.
Hulburt, Edward M., and Janet Rodman, 1963
Distribution of phytoplankton species with respect to salinity between the coast of southern New England and Bermuda.
Limnol. and Oceanogr., 8(2):263-269.

phytoplankton, lists of spp.
Hulburt, E.M., J.H. Ryther and R.R.L. Guillard, 1960
The phytoplankton of the Sargasso Sea off Bermuda.
J. du Cons., 25(2):115-128.

phytoplankton, lists of spp.
Ivanov, A. I., 1960.
Characteristics of qualitative composition and quantitative distribution of phytoplankton in the northwestern Black Sea.
Trudy Vses. Gydrobiol. Obshch., Akad. Nauk. SSSR 10:182-196.

phytoplankton, lists of spp.
Jacques, G. 1967.
Aspects quantitatifs du phytoplancton de Banyuls-sur-Mer (Golfe de Lion)
1. Pigments et populations phytoplanctoniques dans le Golfe du Lion en mars 1966.
Vie Milieu (B) 18 (2B): 239-271.

phytoplankton, quantitative, lists of species (data only)
Japan, Central Meteorological Observatory, 1955.
The results of marine meteorological and oceanographical observations.
1. Oceanography, January-June 1954, No. 14:91 pp.
1. Oceanography, July-December, 1954, No. 15:134 pp.

phytoplankton, lists of species
Japan, Central Meteorological Observatory, 1954.
The results of marine meteorological and oceanographical observations 13(1):1-210.

phytoplankton (lists of species)
Japan, Central Meteorological Observatory, 1954.
The results of marine meteorological and oceanographical observations. Part 1, Oceanography, July-December, 1952, No. 12:138 pp.

phytoplankton, lists of species
Japan, Central Meteorological Observatory, 1952
The results of Marine Meteorological and oceanographical observations, July - Dec. 1951, No. 10:310 pp., 1 fig.

phytoplankton, lists of spp.
Japan, Hokkaido University, Faculty of Fisheries 1970.
Data record of oceanographic observations and exploratory fishing 13:406pp.

phytoplankton, lists of spp.
Japan, Hokkaido University, Faculty of Fisheries, 1968.
The Oshoro Maru cruise 21 to the Southern Sea of Japan, January 1967.
Data Record Oceanogr. Obs. Explor. Fish. 12: 1-97; 113-119.

phytoplankton, lists of spp.
Japan, Hokkaido University, Faculty of Fisheries, 1967.
The Oshoro Maru cruise 16 to the Great Australian Bight November 1965-February 1966.
Data Record Oceanogr. Obs. Explor. Fish., Fac. Fish., Hokkaido Univ. 11: 1-97; 113-119.

phytoplankton, lists of spp.
Japan, Faculty of Fisheries, Hokkaido University, 1961
Data record of oceanographic observations and exploratory fishing, No. 5:391 pp.

Oceanographic Index: Marine Organisms Cumulation, 1946-1973

Phytoplankton, lists of SPP

Japan, Hokkaido University, Faculty of Fisheries, 1966.

Data record of oceanographic observations and exploratory fishery, No 10: 388 pp.

---

phytoplankton, lists of spp.

Japan, Japan Meteorological Agency, 1970
The results of marine meterological and oceanographical observations, January-June 1968; July-December 1967, 43:289; 42:273pp. (multilithed)

Phytoplanton lists of spp.

Japan, Japan Meteorological Agency, 1970

The results of marine meteorological and oceanographical observations. (The results of the Japaneses Expedition of Deep Sea (JEDS-11); January-June 1967 41: 332 pp.

---

phytoplankton, lists of spp.
Japan, Japan Meteorological Agency 1970?
The results of marine meteorological and oceanographical observations, July-December 1966, 40:336pp.

---

phytoplankton, lists of spp.
Japan, Japan Meteorological Agency 1969.
The results of marine meteorological and oceanographical observations, Jan.-June 1966, 39: 349 pp. (multilithed).

---

phytoplankton, lists of spp. (quantitative)
Japan, Japan Meteorological Agency, 1968.
The results of the Japanese Expedition of Deep Sea (JEDS-10).
Res.mar.met.oceanogr.Observ., Jan-June 1965, 37: 385.

---

phytoplankton, lists of spp.
Japan, Japanese Meteorological Agency, 1967.
The results of marine meteorological and oceanographical observations, July-December, 1964, 36: 367 pp.

---

phytoplankton, lists of spp.
Japan, Japan Meteorological Agency, 1966.
The results of the Japanese Expedition of Deep Sea (JEDS-8).
Results mar.met.oceanogr.Obsns.Tokyo, 35:328 pp.

---

phytoplankton, lists of spp.
Japan, Japan Meteorological Agency, 1964.
Results of the Japanese Expedition of Deep Sea (JEDS-5).
Res. Mar. Meteorol. and Oceanogr. Obs., July-Dec. 1962, No. 32:328 pp.

---

phytoplankton, lists of spp.
Japan, Japan Meteorological Agency, 1964.
Oceanographic observations.
Res. Mar. Meteorol. & Oceanogr. Obs., (31):220 pp

---

phytoplankton, lists of sp.
Japan, Japan Meteorological Agency, 1962
The results of marine meteorological and oceanographical observations, July-December 1961, No. 30:326 pp.

---

phytoplankton, lists of spp.
Japan, Japan Meteorological Agency, 1960.
The results of marine meteorological and oceanographical observations, July-December, 1959, No. 26:256 pp.

---

phytoplankton, lists of spp.

Japan, Maizuru Marine Observatory, 1963
Data of the oceanographic observations (1960-1961) (35-36:115-272.

---

phytoplankton, lists of spp.
Japan, Meteorological Agency, 1962
The results of marine meteorological and oceanographical observations, July-December, 1960, No. 28: 304 pp.

---

phytoplankton, lists of spp.

Japan, Japan Meteorological Agency, 1961
The results of marine meteorological and oceanographical observations, January-June 1960. The results of the Japanese Expedition of Deep-Sea (JEDS-2, JEDS-3), No. 27: 257 pp.

---

phytoplankton, lists of spp

Japan, Japan Meteorological Agency, 1960
The results of marine meteorological and oceanographical observations, Jan.-June 1959, No. 25: 258 pp.

---

phytoplankton, lists of spp.

Japan, Japan Meteorological Agency, 1958
The results of marine meteorological and oceanographical observations, July-December 1957, No. 22: 183 pp.

---

phytoplankton, lists of spp.
Japan, Japan Meteorological Agency, 1958.
The results of marine meteorological and oceanographical observations, January-June, 1957, No. 21:168 pp.

---

phytoplankton, lists of spp.

Japan, Japanese Meteorological Agency, 1957

The results of marine meteorological and oceanographical observations, Jan.-June, 1956: 184 pp.
July-December, No. 20: 191 pp.

Phytoplankton lists of spp and percent

Japan, Japan Meteorological Agency, 1956.

The results of marine meteorological and oceanographical observations. Part 1. Oceanography, July-December, 1955. No. 18:90 pp.

---

phytoplankton (lists of spp)

Japan, Kobe Marine Observatory, 1961
Data of the oceanographic observations in the sea south of Honshu from February to March and in May, 1959.
Bull. Kobe Mar. Obs., No. 167(27):99-108; 127-130;149-152;161-164;205-218.

---

phytoplankton, lists of spp.
Japan, Maizuru Marine Observatory 1972.
Data of the oceanographic observations (1966-1970), 235pp.

---

phytoplankton (lists of species)
Jenkins, P.G., 1955.
Seasonal changes in the phytoplankton as indicated by spectrophotometric chlorophyll estimations 1952-53.
Pap. Mar. Biol. and Oceanogr., Deep-Sea Res., Suppl. to Vol. 3:58-67.

---

phytoplankton, lists of species
Kado, Y., 1955.
The seasonal change of plankton and hydrography of the neighboring Sea of Mukaishima.
J. Sci. Hiroshima Univ., B, 15:193-204, 5 textfigs.

---

phytoplankton, lists of spp.
Kawamura, Teruyoshi, 1966.
Distribution of phytoplankton populations in Sandy Hook Bay and adjacent areas in relation to hydrographic conditions in June 1962.
Tech.Pap.Bur.Sport Fish.Wildl.,U.S., (1):1-37. (multilithed).

---

phytoplankton, lists of spp
Kawarada, Y., M. Kitou, K. Furuhashi and A. Sano, 1969.
Distribution of plankton in the waters neighboring Japan in 1966 (CSK)
Oceanogr Mag. 20(2): 187-212.

---

phytoplankton lists of spp.
Kawarada, Yitaka, Masatake Kitou, Kenzo Furuhashi, and Akiro Sano, 1968.
Plankton in the western North Pacific in the winter of 1968 (CSK)
Oceanogrl Mag., 20(1):9-20.

---

phytoplankton, lists of spp.
Kawarada Yutaka, Masataka Kitou, Kenzo Furuhashi, Akira Sano, Kohei Karohji, Kazunori Kuroda, Osamu Asaoka, Masao Matsuzuka, Mamoru Ohwada and Futomi Ogawa, 1966.
Distribution of plankton collected on board the research vessels of J.M.A. in 1965 (CSK)
Oceanogrl Mag. 18(1|2): 91-112.

---

phytoplankton, lists of spp
Kawarada, Y., and M. Ohwada, 1957.
A contribution of microplankton observations to the hydrography of the northern North Pacific and adjacent seas. 1. Observations in the western North Pacific and Aleutian waters during the period from April to July 1954. Ocean. Mag., Toky 9(1):149-158.

---

phytoplankton, lists of spp.
Kilburn, Paul D., 1961.
Summer phytoplankton at Coos Bay, Oregon.
Ecology 42(1):165-166.

---

phytoplankton, lists of spp.
Kimor, B., and V. Berdugo, 1967.
Cruise to the eastern Mediterranean Cyprus-03, Plankton reports, 30.7.1964-15.8.1964.
Bull.Sea Fish Res.Stn Israel, 45:6-31.

---

phytoplankton, lists of spp.
Kiselev, I.A., 1959.
Phytoplankton composition in the sea water of southern Sakhalin and the Southern Kurile islands. (In Russian).
Issledovaniya Dal'Nevostochnykh Morei, SSSR, 6: 162-172.

Translation: USN Oceanogr. Off TRANS 311. (M. Slessers). 1967.

---

phytoplankton, lists of spp.
Konovalova, G.V., 1972.
Seasonal characteristics of phytoplankton in the Amursky Bay of the Sea of Japan. (In Russian; English abstract). Okeanologiia 12(1): 123-128.

Phytoplankton, lists of spp.

Korea, Republic of, Fisheries Research and Development Agency, 1964.

Annual Report of Oceanographic Observations, 1960, 9:184 pp.

Phytoplankton, lists of spp.

Korea, Republic of, Fisheries Research and Development Agency, 1964.

Annual Report of Oceanographic Observations, 1962, 11:203 pp.

---

diatoms, lists of spp.
Kozlova, O.G., and V.V. Mukhina, 1966.
Diatoms and silicoflagellates in suspension and in the bottom sediments of the Pacific Ocean. (In Russian).
In: Geochemistry of silica, N.M. Strakhov, editor, Isdatel. "Nauka", Moskva, 192-218.

---

phytoplankton (lists of spp)

Kuzmina, A.I., 1959

[Some data on the spring and summer phytoplankton of the North Kuril waters.]
Trudy Inst. Okeanol., 36: 215-229.

---

phytoplankton, lists of spp
Lalami-Taleb, R. 1971.
Facteurs de répartition verticale du phytoplancton au large d'Alger.
Bull. Inst. océanogr. Alger 3(3):1-186

phytoplankton, lists of spp.
Lacal, J., 1957.
Microplancton des stations algériennes occidentales de la croisière du "Professeur Lecaze-Duthiers" en 1952. Vie et Milieu, Suppl., No. 6:

phytoplankton, lists of spp.
Lee, Min Jai, Jae Hyung Shim and Chong Kyun Kim, 1967.
Studies on the plankton of the neighboring seas of Korea. 1. On the marine conditions and phytoplankton of the Yellow Sea in summer. Repts. Inst. mar. Biol., Seoul Nat. Univ. 1(6): 1-14.

phytoplankton (lists of spp.)
Légaré, J.E.H., 1957.
The qualitative and quantitative distribution of plankton in the Strait of Georgia in relation to certain oceanographical factors.
J. Fish. Res. Bd., Canada, 14(4):521-552.

phytoplankton, lists of spp.
Léger, Guy, 1971.
Les populations phytoplanctoniques au point φ= 42°27'N, G= 4°29'E Greenwich (Bouée laboratoire du COMEXO(CNEXO) A.- Généralités et premier séjour (21-27 février 1964).
Bull. Inst. océanogr. Monaco 69(1412A + 1412B): 42 pp. + tables

phytoplankton, lists of spp.
Louis, A. and R. Clarysse, 1971
Contribution à la connaissance du phytoplancton de l'Atlantique Nord-Est et de la mer du Nord. Biolog. Jaarb. Dodonaea 39: 261-337. Also in: Coll. Repr. Inst. Zeewetenschap. Onderzoek 2(1972).

phytoplankton, lists of spp.
Loya Rebolledo, Maria Eugenia 1965.
Notas acerca de la flora de diatomeas de la Laguna de Terminos, Campeche, Mexico.
Anales Inst. Biol. Univ. Mex. 36 (1/2): 61-64.

phytoplankton, lists of species
Lubet, P.E., 1955.
Notes sur le phytoplancton du Bassin d'Arcachon. Vie et Milieu, Bull. Lab. Arago 6(1):53-59.

phytoplankton, lists of spp
Lursinep, A., and S. Suvepun, 1966.
An analysis of plankton samples collected from the west coast of peninsular Thailand, 1963-1964.
Indo-Pacific Fish. Counc., Proc. 11th Sess., 1-11.

phytoplankton, lists of species
Maeda, H., 1955.
Studies on Yosa-Naikai. 4. Classification of phytoplankton communities and relation between communities and water masses.
J. Shimonoseki Coll. Fish. 4(2):301-310, 5 text-figs.

phytoplankton, lists of spp
Manguin, E., 1956.
Plancton de la Baie de Banyuls-sur-mer.
Vie et Milieu, Arago, 7(3):417-418.

phytoplankton, lists of species
Marchesoni, V., 1954.
Il trofismo della Laguna Veneta e la vivificazione marina. III. Ricerche sulle variazioni quantitative del fitoplancton.
Arch. Oceanogr. e Limnol. 9(3):153-184.

phytoplankton, lists of spp
Margalef, R., 1969.
Composición específica del fitoplancton de la costa catalano-levantina (Mediterráneo occidental) en 1962-1967. Investigación pesq. 33(1): 345-380.

phytoplankton, lists of spp.
Margaleff, Ramón, 1966.
Análisis y valor indicador de las comunidades de fitoplancton mediterráneo.
Inv. Pesq., Barcelona, 30:429-482.

phytoplankton, lists of spp.
Margalef, Ramón, 1965(1967).
Composición y distribución del fitoplancton.
Memoria Soc. Cienc.nat.La Salle, 25(70/71/72):141-205, numerous tables.

phytoplankton, lists of spp.
Margalef, R., 1965.
Distribution of phytoplankton above the Cariaco Trench. Informe de Progresso del Estudio Hidrografio de la Fosa de Cariaco. Fundación La Salle de Ciencias Naturales Estación de Investigación Marinas de Margarita, Caracas, Sept. 1965, (mimeographed): 13-23.

phytoplankton, lists of sp.
Margalef, Ramon, 1965.
Distribución ecológica de las especies del fitoplancton marino en un área del Mediterráneo occidental.
Inv. Pesq., Barcelona, 28:117-131.

phytoplankton, lists of spp.
Margalef, D. Ramon, 1963.
El ecosistema pelágico de un área costera del Mediterráneo occidental.
Memorias, Real Acad. Ciencias y Artes de Barcelona, 35(1):3-48.

phytoplankton, lists of spp.
Margalef, Ramón, 1961.
Distribución ecológica y geográfica de las especies del fitoplancton marino.
Inv. Pesq., Barcelona, 19:81-101.

phytoplankton, lists of spp.
Margalef, Ramón, 1961.
Fitoplancton atlántico de las costas de Mauritania y Senegal.
Inv. Pesq., Barcelona, 20:131-143.

phytoplankton, lists of spp.
Margalef, Ramón, 1961
Hidrografía y fitoplancton de un área marina de la costa meridional de Puerto Rico.
Inv. Pesq., Barcelona, 18:38-96.

phytoplankton, lists of spp.
Margalef, R., 1957.
Fitoplancton de las costas de Blanco (Gerona) de agosto de 1952 a junio de 1956. Inv. Perq., Barcelona, 8:89-95.

Phytoplankton, lists of spp.
Margalef, Ramon, 1957.
Fitoplancton de las costas de Puerto Rico.
Inv. Pesq., Barcelona, 6:39-52.

phytoplankton, lists of spp.
Margalef, Ramón, 1956
Estructura y dinámica de la "purga de mar" en la Ría de Vigo.
Inv. Pesq., Barcelona, 5:113-134.

phytoplankton, lists of spp.
Margalef, R., F. Cervigon, & G. Yépez T., 1960
Exploración preliminar de las características hidrográficas y de la distribución del fitoplancton en al área de la isla Margarita (Venezuela).
Mem. Soc. Cien. Nat. LaSalle, 20(57): 211-221.

phytoplankton, lists of species
Margalef, R., M. Duran, and F. Saiz, 1955.
El fitoplancton de la ría de Vigo de enero de 1953 a marzo de 1954. Invest. Pesq. 2:85-129.

phytoplankton, lists of spp
Margalef, R. y F. González Bernáldez, 1969.
Grupos de especies asociadas en el fitoplancton del mar Caribe (NE de Venezuela). Investigación pesq. 33(1): 287-312.

phytoplankton, lists of spp.
Margalef, Ramón, and Juan Herrera, 1963
Hidrografía y fitoplancton de las costas de Castellón, de julio de 1959 a junio de 1960.
Inv. Pesq., Barcelona, 22:49-109.

phytoplankton, lists of spp
Margalef, R., J. Herrera y E. Arias, 1959.
Hidrografia y fitoplancton de las costas de Castellon, de Julio de 1957 a junio de 1958.
Invest. Pesqueria 15:3-38.

phytoplankton, lists of species
Margalef, R. and E. Morales, 1960
Fitoplancton de las costas de Blanes (Gerona), de julio de 1956 a junio de 1959. Inves. Pesq. 16: 3-31. Barcelona

phytoplankton, lists of spp.
Margalef, R., F. Muñoz y J. Herrera, 1957.
Fitoplancton de las costas de Castellón de enero de 1955 a junio de 1956. Inv. Perquera, 8:3-31.

phytoplankton, lists of spp.
Marshall, Harold G. 1973.
Phytoplankton observations in the eastern Caribbean Sea.
Hydrobiologia 41 (1): 45-55

phytoplankton, lists of spp.
Marshall, Harold G., 1971.
Composition of phytoplankton off the southeastern coast of the United States. Bull. mar. Sci., Coral Gables 21(4): 806-825.

phytoplankton, lists of spp.
Marshall, Harold G. 1969.
Phytoplankton distribution off the North Carolina coast.
Am. Midland Nat. 82 (6): 241-257.

phytoplankton, lists of spp.
Marshall, Harold G., 1967.
Plankton in James River estuary, Virginia. 1. Phytoplankton in Willoughby Bay and Hampton Road. Chesapeake Sci., 8(2):90-101.

phytoplankton, lists of spp.
Marukawa, H., 1928.
On the plankton of the Japan Sea.
Annot. Oceanogr. Res., 2(1):9-13.

phytoplankton, lists of species
Marumo, R., 1955.
Distribution of plankton diatoms in relation to hydrographic conditions in the sea area east of Honshu in the summer of 1954.
Rec. Ocean. Wks., Japan, 2(2):115-119.

phytoplankton, lists of spp.
Marumo, R., M. Kitou and O. Asaoka, 1960
Plankton in the northwestern Pacific Ocean in summer of 1958.
Oceanogr. Mag., Tokyo, 12(1): 17-44.

phytoplankton, lists of spp.
Marumo, Ryuzo, Nobuo Taga and Toshisuke Nakai, 1971.
Neustonic bacteria and phytoplankton in surface microlayers of the equatorial waters. Bull. Plankt. Soc. Japan 18(2): 36-41.

phytoplankton, lists of spp
Matsudaira, Yasuo, 1964
Cooperative studies on primary productivity in the coastal waters of Japan, 1962-63. (In Japanese; English abstract).
Inform. Bull., Planktol., Japan, No. 11:24-73.

phytoplankton lists of spp
Menzel, D.W., E.M. Hulburt and J.H. Ryther, 1963
The effects of enriching Sargasso Sea water on the production and species composition of phytoplankton.
Deep-Sea Res., 10(3):209-219.

phytoplankton, lists of spp.
Michailov, A.A., and V.V. Denisenko, 1963.
On phytoplankton of the Aegaean Sea. (In Russian).
Trudy Sevastopol Biol. Sta., 16:90-106.

phytoplankton, lists of spp.
Morales, Enrique, 1956.
Fitoplancton de Blanes desde agosto de 1951 hasta julio de 1952.
Inv. Pesq., Barcelona, 4:47-48.

phytoplankton, lists of spp
Morales, E., y E. Arias, 1965.
Ecología del puerto de Barcelona y desarrollo de adherencias orgánicas sobre placas sumergidas.
Inv. Pesq., Barcelona, 28:49-79.

phytoplankton, lists of spp.
Morozova-Vodianitskaia, N.V., 1957.
[Phytoplankton in the Black Sea and its quantitative growth.]
Trudy Sevastopol. Biol. Stan., 9:3-13.

phytoplankton, lists of spp.
Motoda, Shigeru and Ryuzo Marumo, 1963.
Plankton of the Kuroshio water. Proc. Symposium on the Kuroshio, Oct. 29, 1963: 40-61.

phytoplankton, lists of spp.
Movchan, O.A., 1965.
On the seasonal changes in the composition and distribution of phytoplankton in the Newfoundland area. (In Russian).
Trudy vses. nauchno-issled. Inst. morsk.ryb. Khoz. Okeanogr. (VNIRO), 57:345-360.

phytoplankton, lists of spp.
Movchan, O.A., 1962.
Quantitative development of phytoplankton in Newfoundland and Flemish Cap areas and in adjacent waters in April-May, 1958.
Sovetskie Ribokh. Issledov. v Severo-Zapadnoi Atlant. Okeana, VNIRO-PINRO, Moskva, 211-218.

phytoplankton, lists of spp.
Movchan, O.A., 1962.
Spring phytoplankton in the western part of the North Atlantic. (In Russian).
Trudy Vses. Nauchno-Issledov. Inst. Morsk. Ribn. Chos. i Okean., VNIRO, 46:315-323.

phytoplankton, lists of spp.
Mulford, Richard A. 1972.
Phytoplankton of the Chesapeake Bay.
Chesapeake Sci. 13 (Suppl.): S74-S81

phytoplankton, lists of spp.
Muthu, M.S., 1964.
Phytoplankton in the Madras coastal waters.
Current Science, 33(18):559.

phytoplankton, lists of spp.
Nel, E.A., 1968.
The microplankton of the South-west Indian Ocean.
Investl Rep.Fish. Sea Fish.Div., Un.S.Afr.62: 106 pp.; 19 tables.

phytoplankton, lists of spp.
Niaussal, Pierre-Marie, and Roland Bourcart, 1963.
Contribution à l'étude du plancton dans les eaux de l'embouchure de la Gironde. Prédominance du dino-flagellé "Noctula miliaris".
Cahiers Océanogr., C.C.O.E.C., 15(10):722-725.

phytoplankton, lists of spp.
Niemi, Åke, Heinrichs Skuja and Torbjörn Willén 1970
Phytoplankton from the Pojoviken-Tvärminne area, S. coast of Finland.
Mem. Soc. Fauna Flora Fennica 46: 14-28.

phytoplankton, lists of spp
Norris, R.E., 1961.
Observations on phytoplankton organisms collected on the N.Z.O.I. Pacific Cruise, September 1958.
N.Z. J. Sci., 4(1):162-188.

phytoplankton, lists of spp.
Ohno, M., S. Imoto and K. Yatsuzuka 1971.
Oceanographic observation in the Uranouchi inlet, Tosa Bay (1968-1969).
Repts Usa mar. Biol. Sta. 18(1): 41 pp.
(In Japanese; English abstract)

phytoplankton, lists of spp.
Paasche, E., 1960
Phytoplankton distribution in the Norwegian Sea in June, 1954, related to hydrography and compared with primary production data.
Fiskeridirektoratets Skr., Ser. Havundersøgelser 12(11): 77 pp.

phytoplankton, list of spp.
Paasche, E., and A.M. Rom, 1961(1962)
On the phytoplankton vegetation of the Norwegian Sea in May 1958.
Nytt Mag. Botanikk, 9:33-60.

phytoplankton lists of spp.
Park, Joo Suck, and Jong Doo Kim, 1967.
A study on the "red water" caused at Chinhae Bay. (In Korean; English abstract).
Bull. Fish.Res.Develop.Agency, 1:65-79.

phytoplankton, lists of spp.
Park, Tai Soo, 1956.
A study on the quantity and composition of microplankton at Southern Sea of Korea in Summer 1956. Bull. Pusan Fish. Coll., 1(1):13-32

phytoplankton, lists of spp.
Park, Tai Soo, 1956.
On the seasonal change of the plankton at Korean Channel. Bull. Pusan Fish. Coll., 1(1): 1-12.

phytoplankton, lists of spp.
Parsons, T.R., 1960.
A data record and discussion of some observations made in 1958-1960 of significance to primary productivity.
Fish. Res. Bd., Canada, Manuscript Rept. Ser., (Oceanogr. & Limnol.), No. 81:19 pp. (multilithed).

phytoplankton, lists of spp.
Patten, B.C., 1966.
The biocoenetic process in an estuarine phytoplankton community.
Oak Ridge Nat. Lab., ORNL-3946 (UC-48-Biol.Med.): 97 pp. (Unpublished manuscript). $4.00.

phytoplankton, lists of spp.
Patten, Bernard C., 1963
Plankton: optimum diversity structure of a summer community.
Science, 140(3569):894-898.

phytoplankton, lists of spp
Patten, B.C., R.A. Mulford and J.E. Warinner, 1963
An annual phytoplankton cycle in the lower Chesapeake Bay.
Chesapeake Science, 4(1):1-20.

phytoplankton, lists of spp.
Paulmier, Gérard 1972.
Seston-phytoplancton et microphytobenthos en rivière d'Auray, leur rôle dans le cycle biologique des huîtres (Ostrea edulis L.).
Revue Trav. Inst. Pêches marit. 36(4): 375-506.

phytoplankton, lists of spp.
Paulmier, Gérard, 1971.
Cycle des matières organiques dissoutes, du plancton et du microphytoplancton dans l'estuaire du Belon: leur importance dans l'alimentation des huîtres.
Rev. Trav. Inst. Pêches marit. 35(2): 157-200.

phytoplankton, lists of spp.
Paulmier, Gérard, 1969.
Le microplancton des rivières de Morlaix et de la Penzé.
Revue Trav. Inst. Pêches marit. 33(3): 311-332.

phytoplankton, lists of spp.
Paulmier, Gérard, 1965.
Le microplancton de la Rivière d'Auray.
Rev. Trav. Inst. Pêches Marit., 29(2):211-224.

phytoplankton, lists of spp
Pieterse, F., and D.C. van der Post, 1967.
The pilchard of South West Africa (Sardinops ocellata): oceanographic conditions associated with red-tides and fish mortalities in the Walvis Bay region.
Investl Rept., Mar. Res. Lab., SWest Africa, 14: 1s5 pp.

phytoplankton, lists of spp.
Poukhilainen, M.I., 1962.
Seasonal distribution of the phytoplankton in coastal waters of the eastern Murman. (In Russian).
Trudy Murmansk. Morsk. Biol. Inst., 4(8):11-18.

phytoplankton
Priymachenko, A.D., 1961.
The current as a factor determining the development of phytoplankton in water basins. (In Russian).
Pervichnaya Produkt. Morey i Vnutrennikh Vod, 314-318.

USN-HO-TRANS 202
M. Slessers 1963.
P.O. 39010

phytoplankton, lists of spp.
Qasim, S.Z., and C.V.G. Reddy, 1967.
The estimation of plant pigments of Cochin backwater during the monsoon months.
Bull.mar.Sci., Miami, 17(1):95-110.

phytoplankton, lists of spp.
Republic of Korea, Central Fisheries Experimental Station, 1962.
Annual report of oceanographic observations, 1958 7:214 pp.

phytoplankton, lists of spp.
Reyes Vasquez, Gregorio, 1966.
Ch. 6. Fitoplancton.
Estudios hidrobiológicos en el Estuario de Maracaibo, Inst. Venezolano de Invest. Cient. 122-145.

phytoplankton, lists of spp.
Reyssac, J. et M. Roux, 1972.
Communautés phytoplanctoniques dans les eaux de Côte d'Ivoire. Groupes d'espèces associées.
Mar. Biol. 13(1): 14-33.

phytoplankton, lists of spp
Riera Tecla, y Dolores Blasco, 1967.
Plancton superficial del mar de Baleares en julio de 1966.
Investigación pesq. 31(3): 463-474

phytoplankton, lists of spp.
Riley, Gordon A., 1967.
The plankton of estuaries.
In: Estuaries, G.H. Lauff, editor, Publs Am. Ass. Advmt Sci., 83:316-326.

phytoplankton, lists of spp.
Robinson, G.A., 1961
Contribution towards an atlas of the northeastern Atlantic and the North Sea. 1. Phytoplankton.
Bulls. Mar. Ecology, 5(42): 81-89, Pls. 15-20.

phytoplankton, lists of spp.

Rotschi, Henri, Michel Angot, Michel Legand and Roget Desrosieres, 1961
Orsom III, Resultats de la Croisiere "Dillon", 2eme Partie. Chimie et Biologie.
ORSTOM, Inst. Francais d'Oceanie Centre d'Oceanogr., Rapp. Sci., No. 19: 105 pp. (mimeographed).

phytoplankton, lists of spp.

Rotschi, Henri, Michel Angot and Roger Desrosieres, 1960.
Orsom III, Resultats de la croisière "Choiseul" 2ème partie. Chimie, productivité, phytoplancton qualitatif.
Rapp. Sci., Noumeau, No. 16:91 pp. (mimeographed).

phytoplankton, lists of spp.

Roukhiyainen, M.I., 1966.
The qualitative composition of the phytoplankton of the Barents Sea. (In Russian)
Trudy murmansk. biol. Inst., 11(15): 3-23.

phytoplankton, lists of spp

Rouchiyainen, M.I., L.G. Senichkina and L.V. Georgieva 1971.
Review of the taxonomic composition of phytoplankton in the Central American Seas. (In Russian; Spanish and English abstracts).
Issled. Tsentral. Amerikansk. Morei, Akad. Nauk SSSR, Inst. Biol. Yuzhn'ch Morei, Akad. A.O. Kovalersk. 3: 16-49.

plankton, lists of spp

Saisho, T., 1957.
On the distribution of plankton collected along the meridian 130°E. Mem. Fac. Fish., Kagoshima Univ., 5:109-114.

phytoplankton, lists of spp.

Sakshaug, Egil 1972.
Phytoplankton investigations in Trondheimsfjord 1963-1966.
Skr. K. Norske Videnok. Selskab. (1): 56pp.

phytoplankton, lists of spp.

Salah, M. and G. Tamas, 1970.
General preliminary contribution to the plankton of Egypt. Bull. Inst. Oceanogr. Fish., U.A.R. 1: 305-338.

phytoplankton, lists of spp.

Sanina, L.V., 1969.
Phytoplankton sampled along 30°W in the Atlantic. (In Russian)
Trudy vses. nauchno-issled. Inst. morsk. rybn. Khoz. Okean (VNIRO) 65: 148-163.

phytoplankton, lists of spp.

Seguin, Gerard, 1968.
Le plancton de la côte nord de la Tunisie (note préliminaire).
Pelagos, 9:73-84.

phytoplankton, lists of spp.

Seguin, G., 1966.
Contribution à l'etude de la biologie du plancton de surface de la baie de Dakar (Sénégal). Etude quantitative, qualitative et observations écologiques au cours d'un cycle annuel.
Bull. Inst. Francais Afrique Noire, 28(1):1-90.

phytoplankton, lists of spp.

Seguin, Gérard, 1965.
Contribution a la connaissance du plancton des eaux Cotieres du Brésil (Copépodes et amphipodes exceptés) et comparaison avec celui du Sénégal (Campagne de la "Calypso". Janvier-Février 1962).
Pelagos, Bull. Inst. Océanogr., Alger. 2(3): 5-14.

phytoplankton, lists of spp.

Semina, H.J., 1962
Phytoplankton from the Central Pacific collected along the Meridian 174 W. 1. Method and taxonomy. (In Russian; English summary)
Trudy Inst. Okeanol.. Akad. Nauk. SSSR, 58: 3-26.

phytoplankton, lists of spp.

Semina, H.J., 1960
Distribution of phytoplankton in the central Pacific Ocean.
Trudy Inst. Okeanol., Akad. Nauk, SSSR, 41: 17-30.

phytoplankton(lists of spp)

Semina, G.I., 1959
Distribution of phytoplankton in Kronotsk Bay.
Trudy Inst. Okeanol., 36: 73-91.

phytoplankton, lists of spp.

Shimomura, T., 1957.
Planktonological study on the warm Tsushima Current regions. III. Plankton properties and their relation to oceanographic conditions of Noto Peninsula-Nyudo Saki region during the seasons from spring to winter, 1952.
Bull. Japan Sea Reg. Fish. Res. Lab., No. 6 (General survey of the warm Tsushima Current 1): 1-22.

also quantitative

plankton, lists of spp.

Shimomura, T., 1957.
Planktonological study on the warm Tsushima Current regions. IV. Plankton properties and their relation to oceanographic conditions of the offshore regions of the Japan Sea in the summer of 1955. Bull. Japan Sea Reg. Fish. Res. Lab., No. 6 (General survey of the warm Tsushima current): 129-138.

phytoplankton, lists of spp.

Skolka, H., 1965.
Considerații asupra variațiilor calitative si cantitative ale fitoplanctonului litoralului Romanesc al Marii Negre.
In: Ecologie marine, M. Băcescu, redactor, Edit. Acad. Republ. Pop. Romane, Bucaresti, 2:193-293.

phytoplankton, lists of spp.

Skolka, V.H., 1963.
La dynamique du phytoplancton près du littoral roumain de la Mer Noire pendant l'année 1961.
Rapp. Proc. Verb., Reunions, Comm. Int. Expl. Sci. Mer Mediterranée, Monaco, 17(2):467-477.

phytoplankton, lists of spp.

Smayda, Theodore J., 1963
A quantitative analysis of the phytoplankton of the Gulf of Panama. 1. Results of the regional phytoplankton surveys during July and November, 1957 and March 1958. (In English, Spanish resumé)
Inter-American Tropical Tuna Commission, Bull 7(3):193-253.

phytoplankton, lists of spp.

Smayda, Theodore J., 1958.
Phytoplankton studies around Jan Mayen Island, March-April, 1955.
Nytt. Magasin for Botanikk, 6:75-96.

phytoplankton, lists of spp.

Smayda, T.J., 1957. lower
Phytoplankton studies in Narragansett Bay.
Limnol. & Oceanogr., 2(4):342-357.

Phytoplankton, lists of spp.

Smirnova, L.I., 1959.
(Phytoplankton of the Okhotsk Sea and the neighboring Kurile region.) Trudy Inst. Okeanol. 30:3-51.

phytoplankton, lists of spp.

Sournia, A., 1966.
Premier inventaire du phytoplancton littoral de l'Ile Maurice.
Bull. Mus. Nat. Hist. Nat., (2)37(6):1046-1050.

phytoplankton, lists of spp

Spencer, C.P., 1964.
The estimation of phytoplankton pigments.
J. du Conseil, 28(3):327-334.

phytoplankton, lists of spp.

Suarez-Caabro, Jose A., 1959.
Salinidad, temperatura y planctón de las aguas costeras de Isla de Pinos.
Univ. Catolica de Santo Tomas de Villanueva, Lab. Biol. Mar., Monogr., 7:30 pp.

phytoplankton, lists genera

Suarez-Caabro, Jose A., y Samuel Gomez-Aguirre, 1965.
Observaciones sobre el plancton de la Laguna de Terminos, Campeche, Mexico.
Bull. Mar. Sci., 15(4):1072-1120.

phytoplankton, lists of spp.

Subrahmanyan, R., 1958.
Phytoplankton organisms of the Arabian Sea off the west coast of India.
J. Indian Botan. Soc., 37(4):435-441.

phytoplankton, lists of spp.

Szarejko-Lukaszewicz, D., 1957.
Qualitative investigations of phytoplankton of Firth of Fistula in 1953.
Prace Morsk. Inst. Ryback., Gdyni, No. 9:439-451.

phytoplankton, lists of spp.

Taguchi, Satoru and Kohki Nakajima, 1971.
Plankton and seston in the sea surface of three inlets of Japan. Bull. Plankt. Soc. Japan 18(2): 20-36.

phytoplankton, lists of spp

Tanaka, Otohiko, Haruhiko Irie, Shozi Iizuka and Fumihiro Koga, 1961
The fundamental investigation on the biological productivity in the north-west of Kyushu. 1. Rec. Oceanogr. Wks., Japan, Spec. No. 5: 1-58.

phytoplankton, lists of spp.

Tett, Paul, 1973
The use of log-normal statistics to describe phytoplankton populations from the Firth of Lorne area. J. exp. mar. Biol. Ecol. 11(2):121-136.

phytoplankton, lists

Thórdardóttir, Thórunn, 1973
Successive measurements of primary production and composition of phytoplankton at two stations west of Iceland. Norwegian J. Bot. 20(2/3): 256-270.

phytoplankton, lists of spp.

Thorrington-Smith, Margaret, 1969.
Phytoplankton studies in the Agulhas Current region of the Natal Coast.
Invest. Rept, Oceanogr. Res. Inst. S.Africa, 23. 24pp.

Phytoplankton, lists of spp.

Tellai, Salah, 1964.
Répartition géographique et saisonnière du microplancton dans la Baie d'Alger.
Pelagos, Bull. Inst. Oceanogr., Alger. 2(1):5-50

phytoplankton, lists of spp.

Travers, Anne 1966.
Microplancton récolté en un point fixe de la Mer Ligure (Bouée-Laboratoire du COMEXO) pendant l'année 1964.
Rec. Trav. Sta. mar. Endoume, Bull. 39(55): 11-50.

phytoplankton, lists of spp.

Travers, Anne, et Marc Travers 1969.
Le microplancton du Golfe de Gascogne au mois de juillet 1965.
Rec. Trav. Sta. Mar. Endoume 45(61):1-69.

phytoplankton, lists of spp.

Travers, A. et M., 1965.
Introduction à l'étude du phytoplancton et des tintinnides de la region de Tuléar (Madagascar).
Rec. Trav. Sta. Mar. Endoume, hors sér., Suppl. 125-162.

phytoplankton, lists of spp.

Tsuruta, Arao, 1963
Distribution of plankton and its characteristics in the oceanic fishing grounds, with special reference to their relation to fishery.
J. Shimonoseki Univ., Fish., 12(1):13-214.

phytoplankton, list of spp.

Tsuruta, Arao, 1962
The plankton distribution in the northeast waters of the Bering Sea during the early summer in 1961. In Japanese; English abstract.
J. Shimonoseki Coll., Fish., 11(3):577-586.

phytoplankton, lists of spp

Union of South Africa, Division of Fisheries, Department of Commerce and Industries, 1961.
Fisheries research in Natal waters.
Marine Studies off the Natal Coast, C.S.I.R. Symposium, No. S2:89-117 (multilithed).

phytoplankton lists of spp.

United States, National Oceanogrphic Data Center, 1965
Data report Equalant III.
Nat. Oceanogr. Data Cent., Gen. Ser., G-7: 339 pp. $5.00

phytoplankton, lists of spp.

Usachev, P.I., 1961
[Phytoplankton of the North Pole (Based on collections of P.P. Shirshov, first drifting station "The North Pole", 1937-1938, under the command of I.D. Papanin.]
Trudy Vses. Gidrobiol. Obshch., Akad. Nauk, SSSR, 11:189-208.

phytoplankton, lists of spp.

Uyeno, Fukuzo, 1961
Oceanographical and ecological studies on primary production of the sea, with special references to relationship between diatom production and temperature and chlorinity of water.
Rept., Fac. Fish., Pref. Univ., Mie, 4(1):1-64.

phytoplankton, quantitative

Vega Rodriguez, Filiberto, y Virgilio Arenas Fuentes 1965.
Resultados preliminares sobre la distribución del plancton y datos hidrográficos del Arrecife "La Blanquilla," Veracruz, Ver.
Anales Inst. Biol. Univ. Mexico 36(1/2): 53-59

phytoplankton, lists of spp.

Vinogradova, L.A., 1967.
Distribution of the phytoplankton in the different water masses of the Norwegian Sea in spring and autumn 1958-1959. (In Russ.)
Atlant. nauchno-issled. Inst. ryb. khoz. okeanogr. (AtlantNIRO). Materialy Konferentsii po Rezul'tatam Okeanologicheskikh Issledovanii v Atlanticheskom Okeane 130-149.

phytoplankton, lists of spp.

Vinogradova, L.A., 1963.
The development of the phytoplankton of the Norwegian Sea at 65°00'N in the spring-summer period. (In Russian).
Atlantich. Nauchno-Issled. Inst. Ribn. Khoz. i Okeanogr. (ATLANTNIRO), Trudy, 10:35-45.

phytoplankton, lists of spp.

Vinogradova, L.A., 1962
[Qualitative and quantitative distribution of phytoplankton in different water masses of the Norwegian Sea in October 1958.]
Mezhd. Geofiz. Komitet, Prezidiume Akad. Nauk, SSSR, Rezult. Issled. Programme Mezhd. Geofiz. Goda, Okeanol. Issled., No. 5:140-154.

phytoplankton, lists of spp.

Vives, Francisco, y Manuel López-Benito, 1958
El fitoplancton de la Ría de Vigo y su relación con los factores térmicos y energéticos.
Inv. Pesq., Barcelona, 13:87-124.

phytoplankton, list of spp.

Vives, Francisco, y Manuel López-Benito, 1957
El fitoplancton de la Ría de Vigo desde julio de 1955 a junio de 1956.
Inv. Pesq., Barcelona, 10:45-146.

phytoplankton, lists of spp.

Wood, E.J. Ferguson, 1966.
A phytoplankton study of the Amazon region.
Bull. Mar. Sci., 16(1):102-123.

phytoplankton, lists of spp (126)

Wood, E.J. Ferguson, 1960
Antarctic phytoplankton studies.
Proc. Linnean Soc., New South Wales, 85(2): 215-229.

phytoplankton, lists of spp

Whaley, Richard C. and W. Roland Taylor 1968.
A plankton survey of the Chesapeake Bay using a continuous underway sampling system.
Techn. Rept. Chesapeake Bay Inst. Ref. 68-4: 89 pp. (multilithed)

phytoplankton, lists of spp.

Williams, George C., Jeffry B. Mitton, Thomas H. Suchanek, Jr., Nancy Gebelein, Christine Grossman, Jack Pearce, James Young, Charles E. Taylor, Richard Mulstay and Charles D. Hardy 1971.
Studies on the effects of a steam-electric generating plant on the marine environment at Northport New York. Techn. Rept. Mar. Sci. Res. p.1-2 Cent. N.Y. State Univ. 9: 19 pp. (Unpublished)

phytoplankton, lists of spp.

Wood, E.J. Ferguson, 1971
Phytoplankton distribution in the Caribbean region Symp. Investigations and resources of the Caribbean Sea and adjacent regions, UNESCO, 18-26 Nov 1968, Curaçao:399-410.

phytoplankton, lists of spp.

Wood, E.J. Ferguson, 1968.
Studies of phytoplankton ecology in tropical and subtropical environments of the Atlantic Ocean. 3. Phytoplankton communities in the Providence Channel and the Tongue of the Ocean. Bull. mar. Sci., Miami, 18(2): 481-543.

phytoplankton, lists of spp.

Yamazi, Isamu, 1958.
Preliminary check list of plankton organisms found in Tanabe Bay and its environs.
Publ. Seto Mar. Biol. Lab., 7(1):111-163.

phytoplankton, lists of species

Yamazi, I., 1954.
Plankton investigation in inlet waters along the coast of Japan. 13. The plankton of Olama Bay on the Japan Sea coast.
Publ. Seto Mar. Biol. Lab. 4(1):103-114, 9 text-figs.

phytoplankton, lists of spp.

Zembrzuska, Donatylla, 1962
Szczecin Firth phytoplankton.
Prace Morsk. Inst. Ryback. Gdyni, Oceanolog. Ichtiol., 11(A):137-158.

phytoplankton, lists of spp.

Zernova, V.V., 1970.
Phytoplankton in the waters of the Gulf of Mexico and the Caribbean Sea. (In Russian; English and Spanish abstracts). Okeanol. Issled Rez. Issled. Mezhd. Geofiz. Proekt. Mezhd. Geofiz. Kom. Prezid. Akad. Nauk SSSR 20: 69-104.

phytoplankton, lists of spp.

Zernova, V.V., 1969.
The horizontal distribution of phyto plankton in the Gulf of Mexico. Okeanologiia, 9(4): 695-706.
(In Russian; English abstract)

phytoplankton, brackish water, lists of spp.

Iwai, Toshio, 1962.
Ecological studies on the phytoplankton of the brackish water ponds. (In Japanese; English abstract).
J. Fac. Fish., Prefect. Univ. Mie, 5(3):412-506.

phytoplankton, littoral

Skolka, H., 1965.
Consideratii asupra variatiilor calitative si cantitative ale fitoplanctonului litoralului Românesc al Mării Negre.
In: Ecologie marină, M. Băcescu, redactor, Edit. Acad. Republ. Pop. Române, Bucuresti, 2:193-293.

Phytoplankton, littoral

Vacelet, Eveline, 1969
Rôle des populations phytoplanctoniques et bactériennes dans le cycle du phosphore et de l'azote en mer et dans les flaques supralittorales du Golfe de Marseille.
Tethys 1(1): 5-118

phytoplankton, nutrient requirements

Qasim, S.Z., P.M.A. Bhattathiri and V.P. Devassy, 1973
Growth kinetics and nutrient requirements of two tropical marine phytoplankters. Mar. Biol. 21(4):299-304.

phytoplankton, photosynthetic rate

Stross, R.G., S.W. Chisholm and T.A. Downing, 1973
Causes of daily rhythms in photosynthetic rates of phytoplankton.
Biol. Bull. mar. biol. Lab. Woods Hole 145(1):200-209.

phytoplankton, physiol.

Clarke, G.L., and R.H. Oster, 1934.
The penetration of the blu and red components of daylight into Atlantic coastal waters and its relation to phytoplankton metabolism. Biol. Bull 67:59-75.

phytoplankton, physiol.

Droop, M.R., 1957.
Auxotrophy and organic compounds in the nutrition of marine phytoplankton. J. Gen. Microbiol., 16(1):286-293.

phytoplankton, physiol.

Fogg, G.E., 1957.
Relationships between metabolism and growth in plankton algae.
J. Gen. Microbiol., 16(1):294-297.

phytoplankton, photosynthesis by

Ichimura, Shun-ei, Yatsuka Saijo and Yusho Aruga, 1962.
Photosynthetic characteristics of marine phytoplankton and their ecological meaning in the chlorophyll method.
Botan. Mag., Tokyo, 75(888):212-220.
Also in:
Collected Papers on Science of Atmosphere and Hydrosphere, 1958-1963, Water Res. Lab., Nagoya Univ., 1(15).

phytoplankton, physiol.

Iizuka, Shoji, 1964
Depression of activity of the phytoplankton preserved in dark condition and the abnormal types of spectral absorption curve as its result.
(In Japanese; English abstract).
Bull. Fac. Fish., Nagasaki Univ., 15:100-115.

Oceanographic Index: Marine Organisms Cumulation, 1946-1973

phytoplankton, physiol
Ketchum, B.H., 1954.
Mineral nutrition of phytoplankton.
Ann. Rev. Plant Physiol. 5:55-74.

phytoplankton, physiol
Kuenzler, Edward J. 1970.
Dissolved organic phosphorus excretion by marine phytoplankton.
J. Phycol. 6(1):7-13.

phytoplankton, physiol
MacIsaac, J.J., and R.C. Dugdale, 1969.
The kinetics of nitrate and ammonia uptake by natural populations of marine phytoplankton.
Deep-Sea Res., 16(1):45-57.

phytoplankton, physiology
McAllister, C.D., N. Shah and J.D.H. Strickland, 1964.
Marine phytoplankton photosynthesis as a function of light intensity: a comparison of methods.
J. Fish. Res. Bd., Canada, 21(1):159-181.

phytoplankton physiology
Morris, Ian, Clarice M. Yentsch and Charles S. Yentsch, 1972.
The physiological state with respect to nitrogen of phytoplankton from low-nutrient subtropical water as measured by the effect of ammonium ion on dark carbon dioxide fixation. Limnol. Oceanogr. 16(6): 859-868.

phytoplankton, physiol.
Packard, Theodore T., 1973
The light dependence of nitrate reductase in marine phytoplankton.
Limnol. Oceanogr. 18(3):466-469.

phytoplankton, physiol.
Ryther, John H., and Dana D. Kramer, 1961
Relative iron requirement of some coastal and offshore algae.
Ecology, 42(2): 444-446.

phytoplankton, physiology
Steemann Nielsen, E., and Vagn Kr. Hansen, 1959.
Light adaptation in marine phytoplankton populations and its interrelation with temperature.
Physiologia Plantarum 12:353-370.

phytoplankton, physiology
Steemann Nielsen, E., and V. Kr. Hansen, 1959.
Measurements with the carbon-14 technique of the respiration rates in natural populations of phytoplankton.
Deep-Sea Res., 5:222-233.

phytoplankton, physiol.
Steemann Nielsen, E., and Tai Soo Park, 1964.
On the time course in adapting to low light intensities in marine phytoplankton.
Journal du Conseil, 29(1):19-24.

phytoplankton, physiol.
Taylor, W. Rowland, 1964.
Inorganic nutrient requirements of marine phytoplankton organisms.
Narragansett Mar. Lab., Univ. Rhode Island, Occ. Publ., No. 2:17-24.

phytoplankton, physiologie
Verdiun, J., 1950.
Quantum theory and phytoplankton photosynthesis.
Science 112(2905):260.

phytoplankton, physiol.
Zlobin, V.S. and Yu.G. Zhilin 1970
On the kinetic approach to the modelling of exchange processes in marine phytoplankton. (In Russian)
Poliarn. nauchno-issled. Proektn. Inst. morsk. Ribn. Khoz. Okeanogr. N.M. Knipovicha (PINRO) 4: 122-138.

phytoplankton, photosynthesis
Popescu, Dumitru, et Maria Stadniciuc, 1968.
Mensurations quantitatives de la photosynthèse à différentes profondeurs dans les eaux de la Mer Noire, en face du littoral d'Agigea. (In Roumanian; French abstract).
In: Lucrările Sesiunii Stiințifice a Stațiunii de Cercetări Marine "Prof. Ioan Borcea", Agigea, (1-2 Noiembrie 1966), Volum Festiv, Iaşi, 1968.

phytoplankton photosynthesis
Takahashi, M., K. Fujii and T.R. Parsons, 1973
Simulation study of phytoplankton photosynthesis and growth in the Fraser River Estuary. Mar. Biol. 19(2): 102-116.

phytoplankton - physiol.
Yentsch, C.S., and J.H. Ryther, 1957.
Short-term variations in phytoplankton chlorophyll and their significance. Limnol. & Oceanogr. 2(2):140-142.

phytoplankton, pigments of
Ehrhardt, Jean-Paul, et Daniel Bonin, 1968.
Contribution à l'étude du plancton dans le canal de Corse-Provence, campagne de L'Origny, 12 juin-
Cah. océanogr., 20(2):133-156.

plant pigment
Hart, T.J., 1962
Notes on the relation between transparency and plankton content of the surface waters of the Southern Ocean.
Deep-Sea Res., 9(2):109-114.

phytoplankton, pigments
Haxo, Francis T., 1961.
Some implications of recent studies on the role of accessory pigments to photosynthesis in submarine daylight. (Abstract).
Tenth Pacific Sci. Congr., Honolulu, 21 Aug.-6 Sept., 1961, Abstracts of Symposium Papers, 159.

pigments, plant
Humphrey, G.F., 1963.
Seasonal variations in plankton pigments in waters off Sydney.
Australian J. Mar. Freshwater Res., 14(1):24-36.

phytoplankton, pigments
Humphrey, G.F., 1961.
Phytoplankton pigments.
Tenth Pacific Sci. Congr., Honolulu, 21 Aug.-6 Sept., 1961, Abstracts of Papers:126.

phytoplankton, pigments
Jeffry, S.W., 1968.
Photosynthetic pigments of the phytoplankton of some coral reef waters. Limnol. Oceanogr., 13(2): 350-355.

phytoplankton pigments
Kutyurin, V.M., 1959
Determination of chlorophyll content in sea water and the spectral analysis of phytoplankton pigments. Arctic and Antarctic Sci. Res. Inst., Soviet Antarctic Exped., I.G.Y., 2nd Marine Exped. on the "Ob", 1956-1957, 5:173-175 (Publisher: Morskoi Transport, Leningrad).

phytoplankton, pigments
Margalef, R., 1961.
Nouveaux developpements dans la technique de l'extraction des pigments du phytoplancton.
Rapp. Proc. Verb. Réunions, Comm. Int. Expl. Sci. Mer Méditerranée, Monaco, 16(2):223-224.

phytoplankton pigments
Margaleff, R., 1958 or 1959?
Valeur indicatrice de la composition des pigments du phytoplancton sur la productivité, composition taxonomique et propriétés dynamiques des populations. Rapp. Proc. Verb. C.I.E.S.M.M. 15(2): 277-281.

phytoplankton, pigments
Riley, J.P., and T.R.S. Wilson, 1967.
The pigments of some marine phytoplankton species.
J. mar. biol. Ass., U.K., 47(2):351-362.

phytoplankton, pigments
McLeod, G.C., 1961.
The role of accessory pigments in photosynthesis. (Abstract).
Tenth Pacific Sci. Congr., Honolulu, 21 Aug.-6 Sept., 1961, Abstracts of Symposium Papers, 160.

plant pigments
Riley, G.A., 1941.
Plankton studies. IV. Georges Bank. Bull. Bingham Ocean. Coll. 7(4):73 pp.

plank pigments
Riley, G.A., 1939.
Plankton studies. II. The western North Atlantic, May-June, 1939. J. Mar. Res. 2(2):145-162, Text-figs. 49-51, 4 tables.

phytoplankton, pigments of
Riley, J.P., and T.R.S. Wilson, 1965.
The use of thin-layer chromatography for the separation and identification of phytoplankton pigments.
J. mar. biol. Ass., U.K., 45(3):583-591.

phytoplankton pigments
Rotschi, Henri, Michel Angot, Michel LeGand and H.R. Jitt, 1959
Chimie, productivité et zooplancton. "Orsom III". Résultats de la Croisière "Boussole". Résultats "production primaire" de la croisière 56-5.
Rapp. Sci. Inst. Francais d'Océanie, Centre d'Océanogr., No. 13.

phytoplankton pigments
Shah, N.M. 1968.
Certain features of diel variation of phytoplankton pigments and associated hydrographic conditions in the Laccadive Sea off Cochin: a comparison of two seasons.
Bull. Dept. Mar. Biol. Oceanogr. Univ. Kerala 4: 167-174

phytoplankton populations
Tett, Paul, 1973
The use of log-normal statistics to describe phytoplankton populations from the Firth of Lorne area. J. exp. mar. Biol. Ecol. 11(2):121-136.

phytoplankton productivity (data only)
Platt, Trevor, and Brian Irwin 1971.
Phytoplankton production and nutrients in Bedford Basin 1969-70.
Tech. Rept. Fish. Res. Bd Can. 247: 172 pp. (multilithed)

phytoplankton, quantitative (data only)
Australia, Commonwealth Scientific and Industrial Research Organization, 1963.
Coastal investigations at Port Hacking, New South Wales, 1960.
Div. Fish. and Oceanogr., Oceanogr. Sta. List, No. 52:135 pp.

A.D. Crooks, compiler

phytoplankton, quantitative
Australia, Commonwealth Scientific and Industrial Research Organization, 1960.
Coastal investigations at Port Hacking, New South Wales, 1958.
Oceanogr. Sta. List, Div. Fish. & Oceanogr., 42:99 pp.

phytoplankton, (chlorophyll), mean monthly, maximum and minimum
Berrit, G.R., 1964.
Observations océanographiques côtières à Pointe-Noire de 1953 à 1963.
Cahiers, O.R.S.T.R.O.M., Océanographie, Paris, 11(3):31-55.

Phytoplankton quantitative
Calvert, S.E., 1966.
Accumulation of diatomaceous silica in the sediments of the Gulf of California.
Geol. Soc., Am., Bull., 77(6):569-596.

Oceanographic Index: Marine Organisms Cumulation, 1946-1973

---

phytoplankton, quantitative
Cassie, Vivienne, 1966.
Diatoms, dinoflagellates and hydrology in the Hauraki Gulf, 1964-1965.
New Zealand J. Sci., 9(3):569-585.

---

phytoplankton, quantitative [a]
Chennubhotla, V.S. Krishnamurty, 1969.
Distribution of phytoplankton in the Arabian Sea between Cape Comorin and Cochin. Ind. J. Fish. 16(1/2): 129-136.

---

phytoplankton, quantitative
Choe Sang, 1969.
Phytoplankton studies in Korean waters. III. Surface phytoplankton survey of the North-eastern Korea Strait in May of 1967. (In Korean; English abstract).
J. oceanogr. Soc. Korea, 4(1):1-8.

---

phytoplankton, quantitative
Choe Sang 1967.
Phytoplankton studies in Korean waters. II. Phytoplankton in the coastal waters of Korea. (In Korean; English abstract).
J. oceanogr. Soc. Korea 2(1/2): 1-12.

---

phytoplankton, quantitative
Colton, John B., Jr., and Robert R. Marak, 1962.
Use of the Hardy plankton recorder in a fishery research program.
Bull. Mar. Ecology, 5(49):231-246.
(continuous)

---

phytoplankton, quantitative [a]
Dandonneau, Y., 1972
Aspects principaux des variations du phytoplancton sur le plateau continental ivoirien.
Doc. scient. Centre Rech. océanogr. ORSTOM Abidjan 3(2): 32-59.

---

phytoplankton, quantitative [a]
Dandonneau, Y., 1972
Etude du phytoplancton sur le plateau continental de côte d'Ivoire. 11. Représentativité de l'eau de surface pour la description et pour l'interprétation des phénomènes dynamiques. Cah. ORSTOM Sér. Océanogr. 10(3): 267-274.

---

phytoplankton, quantitative
Denisenko, V.V., 1962
Some data on the phytoplankton in the Adriatic Sea. (In Russian).
Okeanologiia, Akad. Nauk, SSSR, 2(4):699-704.

---

phytoplankton, quantitative [a]
Desrosières, R. et B. Wauthy, 1972
Distribution du phytoplancton et structure hydrologique dans la région des Tuamotu (Océan Pacifique Central). Cah. ORSTOM Sér. Océanogr. 10(3): 275-287.

---

phytoplankton, quantitative
Ehrhardt, Jean-Paul, Félix Baudin-Laurencin, et Gérard Seguin, 1964.
Contribution à l'étude du plancton dans le Canal Corse-Provence.
Cahiers Océanogr., C.C.O.E.C., 16(8):623-636.

---

phytoplankton, quantitative
Establier, R., y R. Margalef, 1964
Fitoplancton e hidrografía de las costas de Cádiz (Barbate) de junio de 1961 a agosto de 1962.
Inv. Pesq., Barcelona, 25:5-31.

---

phytoplankton, quantitative
Ganapati, P.N., and D.V. Subba Rao, 1958.
Quantitative study of plankton off Lawson's Bay, Waltair.
Proc. Indian Acad. Sci., (B), 48(4):189-209.

---

phytoplankton, quantitative
Gorgy, Samy, 1966.
Les pêcheries et le milieu marin dans le Secteur Méditerranéen de la Republique Arabe Unie.
Rev. Trav. Inst. Pêches Marit. 30(1):25-

---

phytoplankton, quantitative
Grall, Jean-René, et Guy Jacques, 1964.
Etude dynamique et variations saisonnières du plancton de la région de Roscoff. B. Phytoplancton.
Cahiers, Biol. Mar. Roscoff, 5:432-455.

---

phytoplankton, quantitative
Herrera, Juan, y Ramón Margalef, 1963
Hidrografía y fitoplancton de la costa comprendida entre Castellón y la desembocadura del Ebro, de julio de 1960 a junio de 1961.
Inv. Pesq., Barcelona, 24:33-112.

PHYTOPLANKTON QUANTITATIVE
Hickel, W., 1969.
Planktologische und hydrographisch-chemische Untersuchungen in der Eckernförder Bucht (westliche Ostsee) während und nach der Vereisung im extrem kalten Winter 1962/1963. Helgoländer wiss. Meeresunters, 19(2): 318-334.

---

phytoplankton, quantitative
Hulburt, Edward M., 1966.
The distribution of phytoplankton, and its relationship to hydrography, between southern New England and Venezuela.
J. Mar. Res., 24(1):67-81.

---

phytoplankton (quantitative)
Indonesia, National Institute of Oceanology, Institute of Marine Research 1971.
Hydrological and plankton observations in the Seribu islands, Banda Sea and Maluku Sea by R.V. Samudera, July 28-September 28, 1971.
Oceanogr. Cruise Rept. 5: 64 pp.

---

phytoplankton, quantitative
Irie, Haruhiko, and Syozi Iizuka, 1966.
Studies of the oceanographic characteristics of Haiki Channel and the adjacent waters and of effects of closing of the channel on pearl farms. 1. Present status of plankton-biota and its preseumptive changes. (In Japanese; English abstract).
Bull. Fac. Fish., Nagasaki Univ., No. 20:14-21.

---

phytoplankton, quantitative
Ivanov, A. I., 1960.
Characteristics of qualitative composition and quantitative distribution of phytoplankton in the northwestern Black Sea.
Trudy Vses. Gydrobiol. Obshch., Akad. Nauk, SSSR, 10: 182-196.

Phytoplankton, quantitative
Japan Hokkaido University, Faculty of Fisheries, 1966.
Data record of oceanographic observations and exploratory fishery, No 10: 388 pp.

Phytoplankton, quantitative
Japan, Japan Meteorological Agency, 1970
The results of marine meteorological and oceanographical observations. (The results of the Japaneses Expedition of Deep Sea (JEDS-11); January-June 1967 41: 332 pp.

---

phytoplankton, quantitative
Japan, Japan Meteorological Agency 1970?
The results of marine meteorological and oceanographical observations, July-December 1966, 40: 336 pp.

---

phytoplankton, quantitative
Japan, Japan Meteorological Agency, 1961
The results of marine meteorological and oceanographical observations, January-June 1960. The results of the Japanese Expedition of Deep-Sea (JEDS-2, JEDS-3), No. 27: 257 pp.

---

phytoplankton, quantitative
Japan, Japan Meteorological Agency, 1960.
The results of Marine meteorological and oceanographical observations, July-December, 1959, No. 26:256 pp.

---

phytoplankton quantitative
Japan, Japan Meteorological Agency, 1960.
The results of marine meteorological and oceanographical observations, Jan-June 1959, No. 25 258 pp.

---

phytoplankton, quantitative
Japan, Japan Meteorological Agency, 1958
The results of marine meteorological and oceanographical observations, July-December 1957, No. 22: 183 pp.

---

phytoplankton, quantitative
Japan, Maizuru Marine Observatory, 1963
Data of the oceanographic observations (1960-1961) (35-36):115-272.

---

phytoplankton, quantitative
Japan, Maizuru Marine Observatory, 1958
[Report of the oceanographic observations north of Sanin and Hokuriku districts in summer 1956.]
Bull. Maizuru Mar. Obs., No. 6: 157-unnumbered [53-85.]

---

phytoplankton, quantitative
Kawamura, Teruyoshi, 1966.
Distribution of phytoplankton populations in Sandy Hook Bay and adjacent areas in relation to hydrographic conditions in June 1962.
Tech. Pap. Bur. Sport Fish. Wildl., U.S., (1):1-37. (multilithed).

---

phytoplankton, quantitative
Körte, Friedrich (posthumous) 1966.
Plankton- und Detritusuntersuchungen zwischen Island und den Ferøer im Juni 1960.
Kieler Meeresforsch., 22(1):1-27.

---

phytoplankton, quantitative
Koshevoj, V. V., 1960.
Quantitative distribution of phytoplankton in the Black Sea. Trudy Vses. Gydrobiol. Obshch., Akad. Nauk, USSR, 10:197-200.

---

phytoplankton, quantitative [a]
Léger, Guy, 1972.
Les populations phytoplanctoniques au point φ=42°47'N, G= 7°29' E Greenwich (Bouée laboratoire du COMEXO/CNEXO). C. Troisième séjour (13-22 octobre 1964). Bull. Inst. océanogr. Monaco 70 (1415 A+B), 42 pp.

---

phytoplankton, quantitative
Léger, Guy 1971.
Les populations phytoplanctoniques au point φ=42°47'N, G= 7°29'E Greenwich (Bouée Laboratoire du COMEXO/CNEXO).
Bull. Inst. océanogr. Monaco 70 (1413 A+B), 41 pp., pls.

---

phytoplankton, quantitative
Leloup, E., Editor, with collaboration of L. Van Meel, Ph. Polk, R. Halewyck and A. Gryson, undated.
Recherches sur l'ostreiculture dans le Bassin de Chasse d'Ostende en 1962.
Ministere de l'Agriculture, Commission T.W.O.Z., Groupe de Travail - "Ostreiculture", 58 pp.

---

phytoplankton, quantitative
Lursinap, A., 1963.
Preliminary survey of marine plankton found in the Gulf of Thailand.
Indo-Pacific Fish. Council, Proc. 10th Sess., Seoul, Korea, 10-25 Oct., 1962, FAO, Bangkok, Sect. II:1-7.

---

phytoplankton, quantitative
Margalef, Ramón, 1965(1967).
Composición y distribución del fitoplancton.
Memoria Soc. Cienc. nat. La Salle, 25(70/71/72):141-205, numerous tables.

phytoplankton, quantitative
Margalef, Ramón, 1965.
Distribución ecológica de las especies del fitoplancton marino en un área del Mediterráneo occidental.
Inv. Pesq., Barcelona, 28:117-131.

phytoplankton, quantative
Margalef, Ramón, y Juan Herrera, 1964.
Hidrografía y fitoplancton de la costa comprendida entre Castellón y la desembocadura del Ebro, de julio de 1961 a julio de 1962.
Inv. Pesq., Barcelona, 26:49-90.

phytoplankton, quantitative
Matsudaira, Yasuo, 1964
Cooperative studies on primary productivity in the coastal waters of Japan, 1962-63. (In Japanese; English abstract).
Inform. Bull., Planktol., Japan, No. 11:24-73.

phytoplankton, quantitative
Movchan, O.A., 1962
Quantitative development of phytoplankton in Newfoundland Banks and Flemish Cap areas and in adjacent waters in April-May, 1958. (In Russian; English summary)
Sovetskie Ribochoz. Issledov. Severo-Zapad. Atlant. Okeans, VNIRO-PINRO, 211-218.

phytoplankton, quantitative
Ohwada, Mamoru, and Osamu Asaoka, 1963
A microplankton survey as a contribution to the hydrography of the North Pacific and adjacent seas. (I.) Distribution of the microplankton and their relation to the character of water masses and currents in the North Pacific Ocean in the summary of 1957.
Oceanogr. Magazine, Japan Meteorol. Agency, 14(2):73-85.

phytoplankton, quantitative
Ohwada, Mamoru, and Hisanori Kon, 1963
A microplankton survey as a contribution to the hydrography of the North Pacific and adjacent seas. (II). Distribution of the microplankton and their relation to the character of water masses in the Bering Sea and northern North Pacific Ocean in the summer of 1960.
Oceanogr. Magazine, Japan Meteorol. Agency, 14(2):87-99.

phytoplankton, quantitative
Paasche, E., 1960
Phytoplankton distribution in the Norwegian Sea in June, 1954, related to hydrography and compared with primary production data.
Fiskeridirektoratets Skr., Ser. Havundersøgelser 12(11): 77 pp.

phytoplankton, numerical
Reyes Vasquez, Gregorio, 1966.
Ch. 6. Fitoplancton.
Estudios hidrobiologicos en el Estuario de Maracaibo, Inst. Venezolano de Invest. Cient., 122-145.

phytoplankton, quantitative
Ryther, John H., D.W. Menzel and Nathaniel Corwin, 1967.
Influence of the Amazon River outflow on the ecology of the western tropical Atlantic. I. Hydrography and nutrient chemistry.
J. mar. Res., 25(1):69-83.

phytoplankton, quantitative
San Feliv, J.M. y F. Muñoz 1967.
Hidrografia y fitoplancton de las costas de Castellón, de mayo de 1965 a julio de 1966.
Invest. pesq. Barcelona, 31(3): 419-461.

phytoplankton, quantitative
Smayda, Theodore J., 1966.
A quantitative analysis of the phytoplankton of the Gulf of Panama. III General ecological conditions and the phytoplankton dynamics at 8o 45'N, 79o 23'W from November 1954 to May 1957.
Inter-Amer. Trop. Tuna Comm., Bull., 11(5): 355-612.

phytoplankton, quantitative
Smayda, Theodore J., 1963
A quantitative analysis of the phytoplankton of the Gulf of Panama. 1. Results of the regional phytoplankton surveys during July and November, 1957 and March 1958. (In English Spanish resume)
Inter-American Tropical Tuna Commission. Bull. 7(3):193-253.

phytoplankton, quantitative
Subrahmanyan, R., 1959.
Studies on the phytoplankton of the west coast of India. 1. Quantitative and qualitative fluctuation of the total phytoplankton crop, the zooplankton crop and their interrelationship, with remarks on the magnitude of the standing crop and production of matter and their relationship to fish landings. 2. Physical and chemical factors influencing the production of phytoplankton, with remarks on the cycle of nutrients and on the relationship of the phosphate content to fish landings.
Proc. Indian Acad. Sci., 1:113-252.

phytoplankton, quantitative
Takano, Hideaki, 1964.
Seasonal changes of microplankton quantity at Manazuru, Katsuura, and Hachijo Island.
Bull. Jap. Soc. Sci. Fish., Tokyo, 30(2):89-94.
(In Japanese; English summary)

phytoplankton, quantitative
Teixeira, C., J. Tundisi and M.B. Kutner, 1965.
Plankton studies in a mangrove environment. II. The standing stock and some ecological factors.
Bolm Inst. Oceanogr., S Paulo, 14(1):13-41.

phytoplankton, quantitative
Thorrington-Smith, M., 1971.
West Indian Ocean phytoplankton: a numerical investigation of phytohydrographic regions and their characteristic phytoplankton associations.
Mar. Biol. 9(2): 115-137.

phytoplankton quantitative
Uyeno, Fukuzo, 1961
Oceanographical and ecological studies on primary production of the sea, with special references to relationship between diatom production and temperature and chlorinity of water.
Rept., Fac. Fish., Pref. Univ., Mie, 4(1): 1-64.

phytoplankton, quantitative
Voltolina, Domenico 1971.
Distribuzione quantitativa del fitoplancton nell' Adriatico settentrionale. IV Primavera 1966.
Archo Oceanogr. Limnol. 17(2):169-177.

phytoplankton, quantitative
Voltolina Domenico, 1970.
Distribuzione quantitativa e qualitativa del fitoplancton nell' Adriatico Settentrionale. II. Autumno 1965.
Archo Oceanogr. Limnol. 16(3): 227-246.

phytoplankton, quantitative
Whaley, Richard C. and W. Rowland Taylor 1968.
A plankton survey of the Chesapeake Bay using a continuous underway sampling system.
Techn. Rept. Chesapeake Bay Inst. Ref. 68-4: 89 pp. (multilithed)

phytoplankton, quantitative
Yoshida, Yoichi and Masao Kimata, 1972
Effects of zooplankton on the changes in concentration of inorganic nitrogen compounds and phytoplankton number in natural sea water.
In: Biological oceanography of the northern North Pacific Ocean, A.Y. Takenouti, Chief Editor, Idemitsu Shoten, Tokyo, 535-540.

phytoplankton, species abundance
Estrada, Marta, et Dolores Blasco 1973.
Une méthode d'étude de la distribution temporelle de l'abondance des espèces phytoplanctoniques. Rapp. Proc.-v. Réun. Commn int. Explor. scient. Mer Medit. Monaco 21(8):433-436.

phytoplankton, quantitative (data only)
Indonesia Institute of Marine Research. 1972.
Hydrological, plankton and pigment observations in the South China Sea (C.S.K. Program, Cruise II) and around Seribu Islands.
Oceanogr. Cruise Rept. 6: 63 pp.

phytoplankton, refraction index of
Carder, Kendall L., Richard D. Tomlinson and George F. Beardsley, Jr., 1972
A technique for the estimation of indices of refraction of marine phytoplankters. Limnol. Oceanogr. 17(8): 833-839.

phytoplankton, respiration
Packard, T.T., 1971.
The measurement of respiratory electron-transport activity in marine phytoplankton. J. mar. Res. 29(3): 235-244.

phytoplankton, seasonal succession
Hasle, Grethe Rytter, 1969.
An analysis of the phytoplankton of the Pacific Southern Ocean: abundance, composition, and distribution during the Brategg Expedition, 1947-1948.
Hvalråd. Skr. 52: 168 pp.

phytoplankton, seasonal distribution
Lalami-Taleb, R. 1971.
Facteurs de répartition verticale du phytoplancton au large d'Alger.
Pelagos 3(3):7-186.

phytoplankton, seasonal cycle
Robinson, G.A., 1970.
Continuous plankton records: variation in the seasonal cycle of phytoplankton in the North Atlantic.
Bull. mar. Ecol. 6(9): 333-345

phytoplankton, siliceous
Orr, W.N. 1972.
Pacific northwest siliceous phytoplankton.
Palaeogeogr. Palaeoclimatol. Palaeoecol. 12(1,2): 95-114

phytoplankton sinking rates
Eppley, Richard W., Robert W. Holmes and John D.H. Strickland, 1967.
Sinking rates of marine phytoplankton measured with a fluorometer.
J. exp. mar. Biol. Ecol., 1(2):191-208.

phytoplankton, standing crop
Halim, Youssef, Shoukry K. Guergues and Hamed H. Saleh, 1967.
Hydrographic conditions and plankton in the south east Mediterranean during the last normal Nile flood (1964).
Int. Rev. ges. Hydrobiol., 52(3):401-425.

phytoplankton, standing crop
Hobson, Louis A., 1966.
Some influences of the Columbia River effluent on marine phytoplankton during January 1961.
Limnol. Oceanogr., 11(2):223-234.

Phytoplankton, standing crop.
Motoda, Shigeru, and Ryuzo Marumo, 1965.
Plankton of the Kuroshio water.
Proc. Symp., Kuroshio, Tokyo, Oct. 29, 1963, Oceanogr. Soc., Japan, and UNESCO, 40-61.

phytoplankton, standing crop
Ramamurthy, S., 1965.
Studies on the plankton of the North Canara coast in relation to the pelagic fishery.
J.mar.biol. Ass.India, 7(1):127-149.

phytoplankton, standing crop
Subrahmanyan, R., and A. H. Viswanatha Sarma 1965.
Studies on the phytoplankton of the west coast of India. IV Magnitude of the standing crop for 1955-1962, with observations on nanoplankton and its significance to fisheries.
J. mar. biol. Ass. India 7(2):406-419.

phytoplankton, standing stock
Teixeira, C., 1963.
Relative rates of photosynthesis and standing stock of the net phytoplankton and nannoplankton.
Bol. Inst. Oceanogr., Sao Paulo, 13(2):53-60.

phytoplankton, standing crop
Walsh, John J., 1969.
Vertical distribution of Antarctic phytoplankton. II. A comparison of phytoplankton standing crops in the Southern Ocean with that of the Florida Strait. Limnol. Oceanogr. 14(1): 86-94.

phytoplankton, seasonal variation
Comita, G.W., and G.C. Anderson, 1959.
The seasonal development of a population of Diaptomus ashlandi Marsh and related phytoplankton cycles in Lake Washington.
Limnol. & Oceanogr., 4(1):29-36.

phytoplankton, seasonal variation
Establier, R., y R. Margalef, 1964
Fitoplancton e hidrografía de las costas de Cádiz (Barbate) de junio de 1961 a agosto de 1962.
Inv. Pesq., Barcelona, 25:5-31.

phytoplankton succession
Johnston, R., 1963
Antimetabolites as an aid to the study of phytoplankyon nutrition.
J. Mar. Biol. Assoc. U.K., 43(2):409-425.

phytoplankton succession
Margalef, Ramón 1967.
The food web in the pelagic environment. Helgoländer wiss. Meeresunters. 15(1/4): 548-558.

phytoplankton succession [a]
Mulford, R.A. 1972.
An annual plankton cycle on the Chesapeake Bay in the vicinity of Calvert Cliffs, Maryland, June 1969- May 1970.
Proc. Acad. nat. Sci. Phila. 124(3):17-40.

phytoplankton, annual fluctuations
Reyes Vasquez, Gregorio, 1966.
Ch. 6. Fitoplancton.
Estudios hidrobiologicos en el Estuario de Maracaibo, Inst. Venezolano de Invest. Cient., 122-145.

Phytoplankton, Seasonal variation
Robinson, G.A., 1965.
Continuous plankton records: contribution towards a plankton atlas of the North Atlantic and the North Sea. IX. Seasonal cycles of phytoplankton.
Bulls. Mar. Ecol., Scottish Mar. Biol. Assoc.

Summary p. 120-121.

phytoplankton, succession
Smayda, Theodore J., 1963.
Succession of phytoplankton and the ocean as an holocoenotic environment.
Ch. 27 in: Symposium on Marine Microbiology, C.H. Oppenheimer, Editor, C.C. Thomas, Springfield, Illinois, 260-274.

phytoplankton, vertical distribution [a]
Konoplya, L.A., 1973
Vertical distribution of phytoplankton in the Karelian coastal waters of the White Sea. (In Russian; English abstract). Okeanologiia 13(2):314-320.

phytoplankton, vertical distribution [a]
Menshutkin, V.V. and T.I. Prikhodko, 1971.
Analogue investigation of phytoplankton vertical distribution and production. Gidrobiol. Zh., 7(2): 5-10. (In Russian; English abstract).

phytoplankton, vertical migration
Wheeler, Bernice, 1966.
Phototactic vertical migration in Exuviaella baltica.
Botanica Marina, 9 (1/2):15-17.

phytoplankton, vertical distribution of
Wood, E. J. Ferguson, 1965.
The vertical distribution of phytoplankton in tropical waters. (abstract).
Ocean Sci. and Ocean Eng., Mar. Techn. Soc., - Amer. Soc. Limnol. Oceanogr., 1:111-115.

phytoplankton, volume (settling) [a]
Indonesia, Institute of Marine Research 1973.
Hydrological, plankton and pigment observations in the South China Sea (C.S.K. Program, Cruise III) and around Seribu Islands by R.V. Samudera, July 22 - August 2, 1972.
Oceanogr. Cruise Rept. 9: 38pp.

phytoplankton, volume (settling) [a]
Indonesia, Institute of Marine Research 1973
Hydrological, plankton and pigment observations in the water around Seribu Islands, in the Seram Sea and in the Maluku Sea by R.V. Jalanidhi, April 27 - June 9, 1972.
Oceanogr. Cruise Rept. 8: 83pp.

phytoplankton-zooplankton relationship
Edmondson, W.T., 1962.
27. Food supply and reproduction of zooplankton in relation to phytoplankton population. Contributions to symposium on zooplankton production, 1961.
Rapp. Proc. Verb., Cons. Perm. Int. Expl. Mer, 153:137-141.

phytoplankton/zooplankton
Martin, John H., 1965.
Phytoplankton-zooplankton relationships in Narragansett Bay.
Limnology and Oceanography, 10(2):185-191.

phytoplankton-zooplankton relationship
Mashtova, G.P., 1962
The interrelationship of the phyto-and zooplankton during the spring-and-summer period in the North-Western portion of the Black Sea. (In Russian).
Okeanologiia, Akad. Nauk, SSSR, 2(6):1083-1084.

phytoplankton-zooplankton relationship
Sládeček, V., 1958.
A note on the phytoplankton-zooplankton relationship.
Ecology, 39(3):547-549.

phytoplankton - zooplankton
Smayda, Theodore J., 1966.
A quantitative analysis of the phytoplankton of the Gulf of Panama. III General ecological conditions and the phytoplankton dynamics at 8o 45'N, 79o 23'W from November 1954 to May 1957.
Inter-Amer. Trop. Tuna Comm. Bull., 11(5): 355-612.

phytoplankton-zooplankton relationship
Steemann Nielsen, Einar, 1962.
29. The relationship between phytoplankton and zooplankton in the sea. Contributions to symposium on zooplankton production, 1961.
Rapp. Proc. Verb., Cons. Perm. Int. Expl. Mer, 153:178-182.

phytoplankton-zooplankton relationship
Vives, F., and F. Fraga, 1961.
Florística y sucesión del fitoplanton en la Ria de Vigo.
Inv. Pesq., Barcelona, 19:17-36.

diatoms
Aleem, A.A., 1950.
Sur la répartition en zones des diatomées marines fixées. C.R. Acad. Sci., Paris, 231:924-926.

DIATOMS
Allen, W. E., 1945.
Vernal distribution of marine plankton diatoms offshore in Southern California in 1940. Bull. Scripps Inst. Oceanogr., Tech. ser., 5(4):335-370, 8 figs.

diatoms
Allen, W. E., 1945
Seasonal occurrence of marine plankton diatoms off southern California in 1938. Bull. Scripps Inst. Oceanogr., Tech. Ser., 5(3):293-334, 5 figs.

diatoms
Allen, W. E., 1942.
Diatoms. Rec. Observ., S.I.O., 1(1):25

Diatoms
Allen, W. E., 1941.
Depth relationships of plankton diatoms in sea water. J. Mar. Res., 4(2):107-111.

DIATOMS
Allen, W. E., 1941.
Offshore and depth distribution of marine plankton diatoms. Amer. J. Bot., 28(8):728

diatoms
Allen, W.E., 1941
Frustules of marine plankton diatoms. Chronica Botanica, 6:365-367.

diatoms
Allen, W.E., 1939
Surface distribution of Marine plankton diatoms in the Panama region in 1933. SIO Bull. tech. ser.4:181-196.

diatoms
Allen, W.E., 1938
Quantity collecting of planktonic diatoms. Science 87:171-172.

diatoms
Allen, W.E., 1937
Plankton diatoms of the Gulf of California obtained by the G. Allan Hancock Expedition of 1936. The Hancock Pacific Expeditions, Univ. So. Calif. Publ. 3:47-59, 1 fig.

diatoms
Allen, W.E., 1936.
Surface plankton diatoms in the North Pacific Ocean in 1934. Madroño 3(6):3 pp.

diatoms
Allen, W. E., 1934
Marine plankton diatoms of Lower California in 1931. Bot. Gaz. 95(3):485-492, 1 fig.

diatoms
Allen, W.E., 1934
The primary food supply of the sea.
Quart. Rev. Biol., 9:161-180

diatoms
Allen, W.E. 1932.
Problems of flotation and deposition of marine plankton diatoms. Trans. Am. Micros. Soc., 51(1): 7 pp.

diatoms
Allen, W.E., 1928.
Quantitative studies on inshore marine diatoms and dinoflagellates collected in Southern California in 1924. Bull. S.I.O., tech. ser., 1:347-356, 1 textfig.

diatoms
Allen, W.E., 1928.
Catches of marine diatoms and dinoflagellates taken by boat in Southern California waters in 1926. Bull. S.I.O., tech. ser., 1:201-246, 6 textfigs.

diatoms
Allen, W.E., 1927.
Surface catches of marine diatoms and dinoflagellates made by the U.S.S. "Pioneer" in Alaskan waters in 1923. Bull. S.I.O., tech. ser., 1:39-48, 2 textfigs.

diatoms
Allen, W.E., 1927.
Quantitative studies on inshore marine diatoms and dinoflagellates of Southern California in 1922. Bull. S.I.O., tech. ser., 1:31-38, 2 textfigs.

diatoms
Allen, W.E., 1927.
Quantitative studies on inshore marine diatoms and dinoflagellates of Southern California in 1921. Bull. S.I.O., tech. ser., 1:19-29, 2 textfigs.

diatoms
Allen, W. E., 1925
Statistical Studies of surface catches of marine diatoms and dinoflagellates made by the Yacht "Ohio" in tropical waters in 1924. Jan. Trans. Amer. Microscop. Soc.:24-30, 1 fig.

diatoms
Allen, W.E., 1925
Statistical studies of surface catches of marine diatoms and dinoflagellates made by the yacht 'Ohio' in tropical waters in 1924. Trans. Amer. Micros. Soc.:24-30, 1 fig.

diatoms
Allen, W.E., 1923
Observations on surface distribution of marine diatoms of Lower California in 1921. Proc. Cal Acad. Sci., 12:437-442.

diatoms
Allen, W.E., and E.E. Cupp, 1935
Plankton diatoms of the Java Sea. Annales du Jardin Botanique de Buitenzorg XLIV (2):101-174, figs.1-127.
(drawings of all species mentioned)

diatoms
Allen, W.E., and R. Lewis, 1927.
Surface catches of marine diatoms and dinoflagellates from Pacific high seas in 1925 and 1926. Bull. S.I.O., tech. ser., 1:197-200.

diatoms
Allen, E.J., and E.W.Nelson, 1910
On the artificial culture of marine plankton organisms. JMBA, 8:421-474

diatoms
Amano, M., 1957
[The distribution of plankton diatoms in relation to oceanographic conditions in the sea east of Kinkazan in 1952] J. Oceanogr. Soc., Japan, 13(4): 145-150.

diatoms
Andrade, e Clovis Teixeira, M.H. de, 1957.
Contribuição para o conhecimento das diatomáceas do Brasil.
Bol. Inst. Ocean., Univ. Sao Paulo, 8(1-2):171-225, 10 pls.

diatoms
Asaoka, O., 1955.
On the variation of the conditions of plankton diatoms and the sea at a pier on Jogashima Island in the period from March 1952 to May 1953.
J. Ocean. Soc., Japan, 11(2):69-74.

diatoms
Bailey, J.W., 1856
On microscopic forms in the Sea of Kamschatka. Am. J. Sci. and Arts, ser.2 Vol.22

diatoms
Bailey, J.W., 1851
Microscopical observations made in South Carolina, Georgia and Florida Smithson. Contr. Knowl. 2(8)

diatoms
Bailey, J.W., 1842
American Bacillariae, Part. II. Naviculaceae Am. J. Sci., 42.

diatom
Bainbridge, R., 1949
Movement of zooplankton in diatom gradients. Nature 163(4154):910-911, 2 text figs.

diatoms
Bal, D. V., and L. B. Pradhan, 1945.
A preliminary note on the plankton of Bombay harbour. Current. Sci., 14(8):211-212.

diatoms
Bandel, W., 1940.
Phytoplankton- und Nahrstoffgehalt der tsee im Gebiet der Darsser Schwelle. Internat. Rev. ges. Hydrobiol. u. Hydrogr., 40:249-304

diatoms
Barker, H. A., 1935.
Photosynthesis in diatoms. Archiv. Microbiol Zeitsch. Erforsch. Pflanzlich Microorgan. VI:141-156.

diatoms
Barnes, H., and E.A. Pyefinch, 1947.
Copper in diatoms. Nature, 160 (4055):97

diatoms
Bessemianova, H.P., 1957.
[Photosynthesis of some aspects of the forms of diatoms in the Black Sea.]
Trudy Sevastopol. Biol. Stan., 9:30-38.

diatoms
Boden, B.P., 1950
Some marine plankton diatoms from the west coast of South Africa. Trans. R.Soc. S. Africa. 32:321-434, 100 text figs.

diatoms
Behrend, H., 1950.
Notiz über die Wirkung intermittierenden Lichtes auf das Wachstum der Diatomeen. Arch. Mikrobiol. 14:531-533.

diatoms
Beklemishev, C. W., 1959.
Sur la colonialité des diatomées planctoniques. Internat. Rev. Gesamt. Hydrobiol. 44(1):11-26.

diatoms, misc.
Beklemishev, C.W., 1961
[On the role played by the colonies of plankton diatoms.]
Trudy Inst. Okean., 51: 16-30.

diatoms, misc.
Beklemishev, C.W., M.N. Petrikova, & H.J. Semina, 1961
[On the cause of the buoyancy of plankton diatoms.]
Trudy Inst. Okean., 51: 33-35.

diatoms
Belyaeva, T.V., 1970.
Taxonomy and distribution patterns of plankton diatoms in the equatorial Pacific. (In Russian; English abstract). Okeanologiia, 10(1): 132-139.

diatoms
Berland, Brigitte, 1966.
Contribution à l'étude des cultures de diatomees marines.
Recl. Trav.Stn.mar.,Endoume,40(56):3-82.

diatoms
Boyer, C.S., 1900
The biddulphoid forms of North American Diatomaceae. Proc. Acad. Phila., pp.685-748.

diatoms
Brandes, C. -H., 1939(1951).
Über die räumlichen und zeitlichen Unterschiede in der Zusammensetzunk des Ostseeplanktons. Mitt. Hamburg Zool. Mus. u. Inst. 48:1-47, 23 textfigs.

diatoms
Braarud, T., 1939.
Microspores in diatoms. Nature, 143:399

diatoms
Brebisson, A. de, 1854
Note sur quelques Diatomees marines rares ou peu connues, du littoral de Cherbourg. Mém. Soc. Sci. Nat., Cherb.2:241-258.

diatoms
Brightwell, T., 1856
On the filamentous long-horned Diatomaceae. Quart. J. Micr. Sci.4

diatoms
Brunel, J., 1962
Le phytoplancton de la Baie de Chaleurs. Inst. Botan., Univ. Montréal, Contrib. No. 77: 365 pp., 66 pls.

diatoms
Castracane degli Antelminelli, F., 1886.
1. Report on the Diatomaceae collected by H.M.S. Challenger during the years 1873-1876. Rept. Sci. Results, H.M.S. Challenger, Botany, Vol. II, 178 pp., 30 pls.

diatoms
Chin, T.G., 1951.
A list of Chinese diatoms from 1847 to 1946. Amoy Fish. Bull. 1(5):41-143.

diatoms
Chin, T.G., 1947.
A preliminary list of chinese diatoms. Biol. Bull. Fukien Christian Univ. 6:60-82.
Recorded from 1847-1946 - 100 genera 994 species Formosa not included.

diatoms
Chu, S.P., 1947
Note on the technique of making bacteria-free cultures of marine diatoms. JMBA 26(3): 296-302.

diatoms
Clarke, G.L., 1939.
The relation between diatoms and copepods as a factor in the productivity of the sea. Quart. Rev. Biol. 14(1):60-64.

diatoms
Clarke, G.L., 1937.
On securing large quantities of diatoms from the sea for chemical analysis. Science 86(2243):593-594.

diatoms
Cleve, P.T., 1894-1895
Synopsis of the naviculoid diatoms. K. Svenske Vet.-Ak Handl 26 (2):
27 (3):

diatoms
Cleve, P.T., 1883
Diatoms collected during the expedition of the "Vega". "Vega" Exp. Vetensk. Iaktt. 3:

diatoms
Cleve, P.T., 1881
On some new and little-known diatoms. K. Svenska Vet.-Ak Handl., 18(5):

diatoms
Cleve, P.T., 1878
Diatoms from the West Indian Archipelago. Bih. t. K. Svenska Vet.-Ak. Handl., 5(8):

diatoms
Cleve, P.T., 1873
Examination of diatoms found on the surface of the Sea of Java. Bih. t.K. Svenska Vet.-Ak Handl., Bd. 1(11)

diatoms
Cleve, P.T., 1873
An diatoms from the Arctic Sea. Bih. t.K. Svenska Vet.-Ak. Handl.,1(13):

diatoms
Cleve, P.T., and A. Grunow, 1880
Beiträge zur Kenntnis der Arktischen Diatomeen. K. Svenske Vet.- Ak. Handl. 17(2):

diatoms
Cleve-Euler, A., 1955.
Die Diatomeen von Schweden und Finnland. IV. Biraphideae, 2. K. Svenska Vetenskaps. Handl. Fjärde Ser., 5(4):1-232, 50 pls.

diatoms
Cleve-Euler, A., 1953.
Die Diatomeen von Schweden und Finnland. Teil 3. Monoraphideae, Biraphideae. 1. K. Svenska Vetenskapsakad. Handl., Fjärde Ser., 4(5):1-255, 41 pls.

diatoms
Cleve-Euler, A., 1953.
Die Diatomeen von Schweden und Finnland. Teil II Arraphideae, Brachyraphideae. K. Svenska Vetenskapsakad. Handl., Fjärde Ser., 4(1):1-158, pls. with 483 figs.

includes fresh water and fossil species
keys

diatoms (fresh and salt water)
Cleve-Euler, A., 1952.
Die Diatomeen von Schweden und Finnland. K. Svenska Vetenskapsakademiens Handl., Fjärde ser., 3(3):153 pp., 46 pls.

Diatoms
Cleve-Euler, A., 1951
Die Diatomeen von Schweden und Finnland. Kungl. Svenska Vetenskaps Akad. Handl., Fjärde Ser. 2(1): 161 pp., 6 pls.

diatoms
Conger, P.S., 1951.
Diatoms: their most important role. Sci. Mon. 73:315-323, figs.

diatoms
Cupp, Easter E., 1943
Marine plankton diatoms of the west coast of North America. Bull. S.I.O. 5(1):1-238, 5 pls., 168 text figs.

diatoms
Cupp, E.E., 1937
Seasonal distribution and occurrence of marine diatoms and dinoflagellates at Scotch Cap. Alaska. Bull. S.I.O. Tech. ser.4(3): 71-100, 7 textfigs.

diatom
Cupp, E.E., 1934
Analysis of marine diatom collections taken from the Canal Zone to California during March, 1933. Trans. Am. Micros. Soc. LIII (1):22-29, 1 map.

diatoms
Cupp, E., 1930
Quantitative Studies of miscellaneous series of surface catches of marine diatoms and dinoflagellates taken between Seattle and the Canal Zone from 1924 to 1928. Trans. Am. Micro. Soc., XLIX (3):238-245.

diatoms
Cupp, E.E. and Allen, W.E., 1938
Plankton diatoms of the Gulf of California obtained by Allan Hancock Pacific Expedition of 1937. The Hancock Pacific Expeditions, The Univ. So. Calif. Publ. 3: 61-74, 1 map, pls.4-15.

diatoms
Davidson, V. M., 1934
Fluctuations in the abundance of the planktonic diatoms in the Passamoquoddy Bay region, New Brunswick, from 1924-1931. Contrib. Canadian Biol. and Fish., n.s. 8(28):357-407; 33 textfigs.

diatoms
Dell, R.K., 1965.
Marine biology.
In: Antarctica, Trevor Hatherton, editor, Methuen & Co., Ltd., 129-152.

diatoms
Denffer, D. von, 1948.
Die planktische Massenkultur pinnater Grunddiatomeen. Arch. Mikrobiol. 14:159-202.

diatoms
Denffer, D.von, 1948.
Über einen Wachstumshemmstoff in alternden Diatomkulturen. Biol. Zentralb. 67:7-13.

diatoms
Donkin, A.S., 1858
On the marine Diatomaceae of Northumberland, with a description of eighteen new species. Trans. Micr. Soc. London, ns. 6:12-34.

diatoms
Dorman, H.P., 1927.
Quantitative studies on marine diatoms and dinoflagellates at four inshore stations on the coast of California in 1923. Bull. S.I.O., tech. ser., 1:73-89, 4 textfigs.

diatoms
Dorman, H.P., 1927.
Studies on marine diatoms and dinoflagellates caught in the Kofoid bucket in 1923. Bull. S.I.O. tech. ser., 1:49-61, 4 textfigs.

diatoms
Droop, M.R., and K.G.R. Elson, 1966.
Are pelagic diatoms free from bacteria? Nature, 211 (5053):1096-1097.

diatoms
Drum,Ryan,W., and Edgar Webber,1966.
Diatoms from a Massachusetts salt marsh. Botanica mar., 9(1/2):70-77.

diatoms
Ehrenberg, C.G., 1873.
Mikrogeologische Studien über das kleinste Leben der Meeres-Tiefgründe aller Zonen und dessen geologischen Einfluss. Abkandl. K. Akad. Wiss., Berlin, 1872:131-397, 12 pls., 1 map.

diatoms
Fleming, R. H. 1939
The control of diatom population by grazing. J. du Cons., 14:1-20

Diatomeae
Florin, M-B., 1948
9. Diatomeae in submarine cores from the Tyrrhenian Sea. Medd. Ocean. Inst., Göteborg, 15 (Göteborgs Kungl. Vetenskaps-och Viterrhets Samhälles Handlingar, Sjätte Foljden, Ser. B 5(13):80-88.

diatoms
Foged, N., 1955.
Diatoms from Peary Land, North Greenland collected by Kjeld Halmen. Sansk Pearyland Ekspedition, 1947-50. Medd. om Grønland 128(9):1-90, 14 pls.

fresh water only.

diatoms
Fox, M., 1957.
A first list of marine algae from Nigeria. J. Linn. Soc. London, 55(362):615-631.

diatoms
Freese, L.R., 1952.
Marine diatoms of the Rockport area. Texas J. Sci. 4(3):331-386, 49 textfigs.

diatoms
Frenguelli, J. and H. A. Orlando, 1958.
Diatomeas y silicoflagelados del Sector Antartico Sudamericano. Publ. Inst. Ant. Argentino, 5: 191 pp.

diatoms
Fujiya, M., 1952.
[On the seasonal variation of diatoms in Tokyo Bay.] Bull. Nakai Regional Fish. Res. Lab. 2: 27-33, 13 textfigs.

diatoms
Garstang, W., 1937
On the size of diatoms and their oceanographic significance. JMBA XXII:83-96

diatoms
Geitler, L., 1932.
Der Formwechsel der pennaten Diatomeen (Kieselalgen). Archw. Protistenkunde, 78: 1-226.

diatoms
Goryunova, S.V., and M.N. Ovsyannikova, 1961.
[Technique of cultivation of some marine forms of diatoms under laboratory conditions.] Mikrobiologiya, 30(6):995-997.

English Edit., (1962):810-811.

diatoms
Gran, H.H., 1944
Die Diatomeen der Arktischen Meere. 1. Teil: Die Diatomeen des Planktons. Fauna Arctica, Bd.3, Leif,3 Jena.

diatoms
Gran, H.H., 1912
The plankton production of the North European waters in the spring of 1912. Bull. Planktonique 1912:7-135.

diatoms
Gran, H. H., 1905.
Nordisches Plankton. XIX. Diatomeen.

diatoms
Gran, H.H., 1908
Diatomeen. Nordisches Plankton, Botanischer Teil pp. XIX.1-XIX 146; 178 text figs.

diatoms
Gran, H.H., 1897
Protophyta: Diatomaceae, Silico-flagellata and Cilioflagellata. Den Norske Nordhavs Expedition 1876-1878, h. 24, 36 pp., 4 pls.

diatoms
Gran, H. H. and E. C. Angst, 1931
Plankton diatoms of Puget Sound. Publ. Puget Sound Biol. Sta. 7:417-519, 95 text figs.

diatoms
Gran, H.H., and K. Yendo, 1914
Japanese diatoms: 1. On Chaetoceros; II. On Stephanopyxis. Vid. Skrift. I Mat.-Naturv. Klasse 1913., No.8

diatoms
Gregory, W., 1857
On new forms of marine Diatomaceae, found in the Firth of Clyde and in Loch Fine. Trans. Roy. Soc. Edinb. 21:

diatoms
Greville, R.K., 1866
Descriptions of new and rare diatoms. Ser. XVIII-XX. Trans. Micr. Soc., London, 14:

diatoms
Greville, R.K., 1865
Description of new genera and species of diatoms from Hongkong. Ann. Mag. Nat. Hist. Ser.3, (6) (91):

diatoms
Greville, R.K., 1865
Description of new and rare diatoms. Series XIV-XVII. Trans. Micr. Soc. London, n.s. 13:

diatoms
Greville, R.K., 1864
Description of new species of Diatoms from the South Pacific, Part III Trans. Bot. Soc., Edinburgh, 8:

diatoms
Greville, R.K., 1862
Description of new and rare diatoms. Series VII. Quart. J. Micr. Sci. ns. vol.2

diatoms
Greville, R.K., 1857
Description of some new diatomaceous forms from the West Indies. Quart. J. Micr. Soc. 5:

diatoms
Gross, F., 1937.
The life history of some marine plankton diatoms. Phil. Trans. Roy. Soc., B, 228:1-47

diatoms
Grunow, A., 1884
Die Diatomeen von Franz-Josefs-Land. Denkschr. d. Kais. Ak. d. Wissensch., Math.-Naturw. Klasse, 48:

diatoms
Grunow, A., 1879
New species and varieties of Diatomaceae from the Caspian Sea. J. Roy. Micr. Soc., 2:

diatoms
Grunow, A., 1863
Ueber einige neue und ungenügend bekannte Arten und Gattungen von Diatomaceen. Verhandl. d. K.K. Zool. Bot. Gesellsch., Vienna, 13: 137-162, pl.4-5 (Pl. 13-14).

diatoms
Harder, R., and H. von Witsh, 1942.
Über Massenkultur von Diatomeen. Ber. Deutsch. Bot. Ges. 60:146-152.

diatoms
Harper, M.A., 1969.
Marine diatoms past and present distributions. Tuatara 17(1): 30-34. Also in: Coll. Repr. N.Z. Oceanogr. Inst.

diatom
Harvey, H.W., 1939
Substances controlling the growth of a diatom. JMBA 23:499-520.

diatoms
Harvey, H. W., 1933
On the rate of diatom growth. JMBA, 19: 253-275

diatoms
Hendey, N.I., 1971.
Electron microscope studies and the classification of diatoms. In: Micropalaeontology of oceans, B.M. Funnell and W.R. Riedel, editors, Cambridge Univ. Press, 625-631.

diatoms
Hendy, N. Ingram, 1964
An introductory account of the smaller algae of British coastal waters. V. Bacillariophyceae (Diatoms). Her Majesty's Stationary Office, 317 pp., 45 pls.

diatoms (with lists of spp.)
Hendey, N-Ingram, 1958 [1957(Publ. 1958)]
Marine diatoms from some West African Ports. J. R. Microsc. Soc. (3) 77(1/2): 28-85.

diatoms
Hendey, N.I., 1958.
Diatoms from Indian Ocean cores. Nature 181(4614):953-954.
(equatorial)

diatoms
Hendey, N.I., 1951.
Littoral diatoms of Chichester Harbour with special reference to fouling. J.R. Microsc. Soc. 71(1):1-86, figs.

diatoms
Hendey, N.I., 1939
New species of diatoms. J. Roy. Microscop. Soc., Ser.III, 59:11-18.

diatoms
Hendey, N.I., 1937
The plankton diatoms of the southern seas. Discovery Repts. 16:151-364, pls.6-13.

diatoms
Hendy, N.I., D.H. Cushing and G.W. Ripley, 1954.
Electron microscope study of diatoms. J. Microsc. Soc. 74(3):22-34.
Abstr. in: Kodak Abstr. Bull. 41(3):163.

diatoms
Heurck, H. van, 1896
A treatise on Diatomaceae. London

diatoms
Hopkins, J. Trevor, 1967.
The diatom trail. Microscopy, Quekett Microsc. Club, 30(9):209-217.

diatoms
Hoshiai, Takao 1972.
Diatom distribution in sea ice near McMurdo and Syowa stations. Antarctic Jl U.S.A. 7(4): 84-85.

diatoms
Hoshiai, T., M. Kato, 1961
Ecological notes on the diatom community of the sea ice in Antarctica. Bull. Mar. Biol. Sta., Asamuchi, Tohoku Univ. 10(4): 221-230.

diatoms
Hustedt, F., 1950.
Die Diatomeedflora norddeutscher Seen mit besonderer Berücksichtigung des holsteinischen Seengebiets. V-VII. Seen im Meckelburg, Lauenburg, und Nordostdeutschland. Arch. Hydrobiol. 43(3/4):329-458, Pls. 35-41, 13 textfigs.

species not listed in index file
fresh water species.

diatoms - systematic
Hustedt, F. 1930.
Die Kieselalgen Deutschlands, Osterreichs und der Schweiz mit Berücksichtigung der übrigen Länder Europas sowie der angrenzenden Meeresgebiete. Dr. L. Rabenhorst's Kryptogamen-Flora von Deutschland, Österreich und der Schweiz, Vol. VII, 920 pp; 542 textfigs.

diatoms
Hustedt, F. and A.A. Aleem, 1951
Littoral diatoms from the Salstone near Plymouth. JMBA 30(1): 177-196.

diatoms
Ichiye, T., 1957.
On the relationship between the plankton and hydrographic conditions in the adjacent seas of Japan. Rec. Ocean. Wks., Japan, (Spec. No.):34-41.
distribution

diatoms
Ivashkin, M.V., 1958
[The skin film of diatoms on Antarctic fin whales] V.N.I.R.O., 33: 186-198.
Translation NIOT/33

diatoms
Janisch, C., 1891
The Diatoms of the Gazelle Expedition. 22 pl. photographs with mss. index.

diatoms (attached forms)
Jansson, Ann-Mari, 1966.
Diatoms and Meiofauna-producers and consumers in the Cladophora belt. Veroff. Inst. Meeresforsch., Bremerh., Sonderband II:281-288.

diatoms
Japan, Hakodate Marine Observatory, 1963.
Report of the oceanographic observations in the sea east of Tohoku District from February to March 1961:3-8.
Report of the oceanographic observations in the sea east of Tohoku District in May 1961, 9-12.
Report of the oceanographic observations in the Japan Sea in June 1961:59-79.
Report of the oceanographic observations in the western part of Wakasa Bay from January to May 1961:80.
Report of the oceanographic observations in the sea south of Hokkaido in July 1961:3-4.
In Japanese   OVER

diatoms
Japan, Hakodate Marine Observatory, 1962
Report of the oceanographic observations in the sea east of Tohoku District and in the Okhotsk Sea from August to September 1960. (In Japanese).
Results, Mar. Meteorol. and Oceanogr., July-Dec., 1960, Japan Meteorol. Agency, No. 28: 7-16.

diatoms
Japan, Hakodate Marine Observatory, 1962
Report of the oceanographic observations in the sea south of Hokkaido in July, 1960. (In Japanese).
Res. Mar. Meteorol. and Oceanogr., July-Dec., 1960, Japan Meteorol. Agency, No. 28: 3-6.

diatoms
Japan, Hakodate Marine Observatory, 1962
Report of the oceanographic observations in the sea west of Hiyama (Hokkaido) and in the sea east of Tohoku District in November 1960. (In Japanese).
Res. Mar. Meteorol. and Oceanogr., July-Dec., 1960, Japan Meteorol. Agency, No. 28:17-20.

diatoms
Japan, Hakodate Marine Observatory, 1961
Report of the oceanographic observations in the Okhotsk Sea and in the sea east of Tohoku District from May to June, 1959.
Bull. Hakodate Mar. Obs., (8):8-16.

diatoms
Japan, Hakodate Marine Observatory, 1961
[Report of the oceanographic observations in the sea east of Tohoku District from November to December 1959.]
Bull. Hakodate Mar. Obs., (8):15-19.

diatoms,
Japan, Kobe Marine Observatory, 1963.
Report of the oceanogrghical observations in the sea south of Honshu from February to March 1961. Report of the oceanographical observations in the cold water region off Enshu Nada in May 1961. (In Japanese).
Bull. Kobe Mar. Obs., 171(5):22-35.

diatoms
Japan, Kobe Marine Observatory, 1962
Report of the oceanographic observations in the cold water region south of Enshu Nada from October to November, 1960. (In Japanese).
Res. Mar. Meteorol. and Oceanogr., July-Dec., 1960, Japan Meteorol. Agency, No. 28: 43-51.

diatoms
Japan, Kobe Marine Observatory, 1962
Report of the oceanographic observations in the sea south of Honshu from July to August, 1960. (In Japanese).
Res. Mar. Meteorol. and Oceanogr., July-Dec., 1960, Japan Meteorol. Agency, No. 28:36-42.

diatoms-spp
Japan, Kobe Marine Observatory, Oceanographical Section, 1962
Report of the oceanographic observations in the sea south of Honshu from October to November, 1959. (In Japanese)
Bull. Kobe Mar. Obs., No. 169(11):44-50.

diatoms
Japan, Kobe Marine Observatory, Oceanographical Section, 1962
Report of the oceanographic observations in the sea south of Honshu in May, 1960. (In Japanese).
Bull. Kobe Mar. Obs., No. 169(12):27-33.

diatoms
Japan, Kobe Marine Observatory, Oceanographical Section, 1962
Report of the oceanographic observations in the sea south of Honshu in March, 1960. (In Japanese). Bull. Kobe Mar. Obs., No. 169(12):22-33.

diatoms
Japan, Kobe Marine Observatory, Oceanographical Section, 1962
Report on the oceanographic observations in the sea south of Honshu from July to August, 1959.
Bull. Kobe Mar. Obs., No. 169(11):37-43.
(In Japanese)

diatoms
Japan, Kobe Marine Observatory, 1958.
[Report of the oceanographic observations in the sea south of Honshu from November to December 1957.]
J. Oceanogr., Kobe Mar. Obs., (2) 10(1):21-28.

in May 1957.]
    Ibid., 9(2):69-78.
in August 1957.]
    Ibid., 9(2):79-86.

diatoms
Japan, Kobe Marine Observatory, 1956.
[Report of the oceanographic observations in the sea south of Honshu in August 1955.]
J. Ocean., Kobe, (2)7(3):23-32.

diatoms
Japan, Kobe Marine Observatory, 1956.
[Report of the oceanographical observations in the sea south off Honshu in March 1955.]
J. Ocean., Kobe (2)7(2):17-24.

diatoms
Japan, Maizuru Marine Observatory, 1962
Report of the oceanographic observations in the central part of the Japan Sea in September, 1960. (In Japanese).
Res. Mar. Meteorol. and Oceanogr., July-Dec., 1960, Japan Meteorol. Agency, No. 28:60-68.

diatoms
Japan, Maizuru Marine Observatory, 1962
Report of the oceanographic observations in the Wasaka Bay from July to August, 1960. (In Japanese).
Res. Mar. Meteorol. and Oceanogr., July-Dec., 1960, Japan Meteorol. Agency, No. 28:69-75.

diatoms
Japan, Maizuru Marine Observatory, 1962
Report of the oceanographic observations in the western part of Wakasa Bay from June to December 1960. (In Japanese).
Res. Mar. Meteorol. and Oceanogr., July-Dec., 1960, Japan Meteorol. Agency, No. 28:76-83.

diatoms
Japan, Maizuru Marine Observatory, Oceanographical Section, 1961
[Report of the oceanographic observations in the Japan Sea from June to July, 1959.]
Bull. Maizuru Mar. Obs., No. 7:57-64.

diatoms
Japan, Maizuru Marine Observatory, Oceanographical Section, 1961
Report of the oceanographic observations in the Japan Sea from June to July, 1958.
Bull. Maizuru Mar. Obs., No. 7:60-67.

diatoms
Japan, Maizuru Marine Observatory, Oceanographical Section, 1961.
[Report of the oceanographic observations in the sea north of San'in District in August, 1958.]
Bull. Maizuru Mar. Obs., No. 7: 64-68.

diatoms
Japan, Maizuru Marine Observatory, Oceanographical Section, 1961
[Report of the oceanographic observations in the Wakasa Bay from July to August, 1959.]
Bull. Maizuru Mar. Obs., No. 7:62-67.

diatoms
Japan, Maizuru Marine Observatory, Oceanographical Observatory, 1961
[Report of the Oceanographic observations in the western part of Wakasa Bay from August to December, 1959.]
Bull. Maizuru Mar. Obs., No. 7:68-74.

diatoms
Japan, Maizuru Marine Observatory, Oceanographical Section, 1961
[Report of the oceanographic observations in the western part of Wakasa Bay from January to March, 1958.]
Bull. Maizuru Mar. Obs., No. 7:50-54.

diatoms
Japan, Maizuru Marine Observatory, Oceanographical Section, 1961
[Report of the oceanographic observations off Kyoga-misaki in the Japan Sea in April and May, 1958.]
Bull. Maizuru Mar. Obs., No. 7:55-60.

diatoms
Japan, Maizuru Marine Observatory, Oceanographical Section, 1961
[Report of the oceanographic observations off Kyoga-misaki in Japan Sea from October to December, 1957.]
Bull. Maizuru Mar. Obs., No. 7:29-36.

diatoms
Japan Meteorological Agency, 1962
Report of the Oceanographic observations in the sea east of Honshu from August to September, 1960. (In Japanese).
Res. Mar. Meteorol. and Oceanogr., July-Dec., 1960, Japan Meteorol. Agency, No. 28:21-29.

diatoms
Japan, Nagasaki Marine Observatory, 1962
Report of the oceanographic observations in the sea west of Japan from October to November, 1960. (In Japanese).
Res. Mar. Meteorol. and Oceanogr., July-Dec., 1960, Japan Meteorol. Agency, No. 28:52-59.

diatoms
Jorgensen, E., 1900
Protophyten und Protozoën im Plankton aus der Norwegischen Westkerste. Bergens Mus. Aarb. 1899(6): 95 pp., 5 pls., 83 tables.

diatoms
Jousé, A.P., 1960
Les diatomées des dépôts de fond de la partie nord-ouest de l'Océan Pacifique. Deep-Sea Research 6(3): 187-192.

diatoms in bottom deposits
Jousé, A.P., and T.V. Seczina, 1955.
[Diatoms algae in the deposits of the Kurile-Kamchatka Trench.] Trudy Inst. Oceanol. 12:130-144.

diatoms
Karohji, Kohei, 1958.
Report from the "Oshoro Maru" on oceanographic and biological investigations in the Bering Sea and northern North Pacific in the summer of 1955 IV. Diatom standing crops and the major constituents of the populations as observed by net sampling.
Bull. Fac. Fish., Hokkaido Univ., 8(4):243-252.

diatoms
Karsten, G., 1899.
Die Diatomeen der Kieler Bucht. Wiss. Meeresuntersuch. (N. F.) Abt. Kiel, Bd. IV:17-205

diatoms
Ketchum, B. H., and A.C. Redfield, 1938
A method for maintaining a continuous supply of marine diatoms by culture. Biol. Bull., 75:165-169

diatoms
Kolbe, R.W., 1957.
Diatoms from Equatorial Indian Ocean cores.
Rept. Swedish Deep-Sea Exped., 1947-48, 9(1):3-50.

species not listed in my file.

diatoms
Kon, H., 1957.
[Seasonal distribution of microplankton in the Tsugaru Straits, northern Japan, with hydrographic bearing - observations on diatom communities in the years of 1952-1956. (1).]
Bull. Hakodate Mar. Obs., No. 4:247-253.
19-25.

diatoms
Krasske, G., 1941
Die Kieselalgen des chilenischen Küstenplankton (Aus dem südchilenischen Kusten gebiet IX). Arch. Hydrobiol. 38 (2):260-287.

diatoms
Krasske, G., and G.H. y E. Schwabe, 1950.
Diatomeas del Archipiélago de Formosa (Taiwan).
Bol. Soc. Biol., Concepción (Chile) 25:75-107.

diatoms
Kruger, D., 1950.
Variations quantitatives des protistes marins au voisinage du Port d'Alger durant l'hiver 1949-1950. Bull. Inst. Océan. Monaco, No. 978:20 pp., 5 text figs.

diatoms

Lauder, H.S., 1864
(a) On new diatoms. (b) Remarks on the marine Diatomaceae found at Hong-Kong, with descriptions of new species, with notes by J. Ralfs. Trans. Micro. Soc., ns. 12:

diatoms

Lebour, M. V., 1930.
The planktonic diatoms of northern seas. Ray Society, London.

diatoms

Leuduger-Fortmorel, G.M.D., 1892
Diatomées de la Malaisie. Annales du Jardin Bot. de Buitenzorg, 11:

diatoms

Lewis, R., 1927.
Surface catches of marine diatoms and dinoflagellates off the coast of Oregon by U.S.S. "Guide" in 1924. Bull. S.I.O., tech. ser., 1:189-196, 3 textfigs.

diatoms

Lillick, L.C., 1941
Habits and life history of the diatoms. J. of N.Y. Botanical Garden, 42(493):1-10 text figs.

Lillick, L.C., 1938
Preliminary report of the phytoplankton of the Gulf of Maine. Am. Mid. Nat. 20(3):624-640, 1 text figs 37 tables.

diatoms

Lohmann, K.E., 1960.
The ubiquitous diatom - a brief survey of the present state of knowledge. Amer. J. Sci. (Bradley vol.) 258A:180-191.

diatoms

Lucas, C. E. 1941
Continuous plankton records: Phytoplankton in the North Sea 1938-39. Pt. I. Diatoms Hull Bull. Mar. Ecol., II(2):19-46, 5 textfigs. 33 pls.

diatoms

Lucas, C.E., N.B. Marshall and C.B. Rees 1942
Continuous plankton records: The Faeroe-Shetland Channel, 1939. Hull Bull. Mar. Ecol. II(10):71-94.

diatoms

Lucas, C.E. and H.G. Stubbings, 1948
Size variations in diatoms and their ecological significance. Hull Bull. Mar. Ecol., 2(12):133-171, 14 text figs.

diatoms

Manguin, E., 1957.
Premier inventaire des diatomées de la Terre Adélie Antarctique. Espèces nouvelles. Rev. Algologique, n.s., 3(3):111-134.

diatoms

Mann, A., 1921.
The dependence of fishes on diatoms. Ecology, 2(2):79-83.

diatoms

Mann, A., 1907
Report on the diatoms of the Albatross voyages in the Pacific Ocean, 1888-1904. Contrib. U. S. Nat. Herb. 10(5):221-419, Pls. XLIV-LIV.

diatomaceae

Mann, A., 1893
List of Diatomaceae from a deep-sea dredging in the Atlantic Ocean off Delaware Bay by the U. S. Fish Commission Steamer Albatross. Proc. U. S. Nat. Mus. 16:303-312.

diatoms

Margalef, Ramón, 1967.
Las algas inferiores.
In: Ecologia marina, Monogr. Fundación la Salle de Ciencias Naturales, Caracas, 14:230-272.

diatoms

Marshall, S. M., and A.P. Orr, 1928
The photosynthesis of diatom cultures in the sea JMBA, 15:321-364.

diatoms population

Marumo, R., 1956.
Diatom populations in the western North Pacific Ocean in the summer of 1955. Ocean. Mag., Tokyo, 8(1):75-78.

diatoms

Marumo, R., 1953.
On the diatom plankton in the Antarctic in summer 1952. J. Ocean. Soc., Japan, 9(1):33-38, 4 textfigs.

diatoms

Matsue, Y., 1950.
[The variation of titratable base in sea water by the growth of diatoms.] J. Ocean. Soc., Tokyo, 6(1):32-38, 2 textfigs. (In Japanese with English abstract).

diatoms

Maxwell, B.E., 1956.
Note on diatoms in Wellington Harbour in 1953-54. Trans. R. Soc., N.Z., 84(1):197-200.

diatoms

Menon, M.A.S., 1945.
Observations on the seasonal distribution of the plankton, Trivandrum Coast. Proc. Indian Acad. Sci., Sect. B, 22(2):31-62, 1 textfig.

diatoms

Mereschowsky, C., 1901.
A list of California diatoms. Am. Mag. Nat. Hist., ser. 7, VII:474-480

diatoms

Meunier, A., 1915
Microplancton de la Mer Flamande. 2. Diatomées (excepté le genre Chaetoceros). Mem. Mus. Roy. Hist. Nat., Belgique, 7(3):1-118, Pls. VIII-XIV.

diatoms

Meunier, A., 1913
Microplancton de la Mer Flamande. 1. Chaetoceros. Mem. Mus. Roy. Hist. Nat., Belgique 7(2):1-55, 7 pls.

diatoms

Migita, Seiji, 1967.
Morphological and ecological studies on sexual reproduction of marine centric diatoms. (In Japanese; English abstract). Inf. Bull. Planktol. Japan, 14:13-22.

diatoms

Mispa, J.N., 1956.
A systematic account of some littoral marine diatoms from the west coast of India. J. Bombay Nat. Hist. Soc., 53(4):537-568.

diatoms

Moberg, E.G., 1928
The interrelation between diatoms their chemical environment, and and upwelling water in the sea, off the coast of southern California. Proc. Nat. Acad. Sci., 14(7): 511-518.

diatoms (very elementary-summary of chapters in a book "An Introduction to the Study of Diatoms")

Moreira Filho, Hermes, and Clovis Teixeira, 1963.
Noções gerais sôbre as diatomáceas (Chrysophyta-Bacillariophyceae).
Bol. Univ. Parana, Cons. Pesquisas, Botan., No. 11:1-26.

diatoms

Motoda, S., and K. Yutake, 1955.
Diatom communities in western Aleutian waters on the basis of net samples collected in May-June 1953. Bull. Fac. Fish., Hokkaido Univ., 6(3):191-200.

diatoms

Muhina, V.V., 1971.
Problems of diatom and silicoflagellate Quaternary stratigraphy in the equatorial Pacific Ocean. In: Micropaleontology of oceans, B.M. Funnell and W.R. Riedel, editors, Cambridge Univ. Press. 423-431.

diatoms

Müller Melchers, F.C., 1953.
New and little known diatoms from Uruguay and the South Atlantic coast.
Com. Bot. Mus. Hist. Nat., Montevideo, 3(30): 1-11, 8 pls.

diatoms

Nemoto, Takahisa, 1962
Some curious diatoms in the sea. (In Japanese English abstracts).
Info. Bull. Planktology. Japan, No. 8:3-5.

diatoms

Okamura, K., 1911
Some littoral diatoms of Japan. Rept. Imp. Fish. Inst., Tokyo, 7:

diatoms

Okuno, H., 1954.
Electron-microscope fine structure of some marine diatoms. Rev. Cytol. et Biol. Veget. 15(3):237-246.

diatoms

Oshide, T., 1954.
[On the diatoms from the bottom of Mutsu-Bay.] Studies from Geol. & Minerol. Inst., Tokyo Univ. Education (Mem. Vol. Prof. K. Kawada):159-165.

diatoms

Ostenfeld, C. H., 1913.
Bacillariales (Diatoms). Bull. Trim. Cons. Int. Explor. Mer:403-508.

diatoms

Ostenfeld, C.H., 1902
Marine Plankton Diatoms, in: Johs. Schmidt Flora of Koh Change. Contributions to the knowledge of the vegetation in the Gulf of Siam. Part VII. Botanisk Tidsskrift, Bd 25:1-27, 23 text figs.

diatoms

Paiva Carvalho, J., 1950
O plancton do Rio Maria Rodriques (Cananeis). 1. Diatomaceas e Dinoflagelados. Bol. Inst. Paulista Oceanogr. 1(1); 27-43, 2 fold-in tables, 2 figs.

diatoms

Patrick, Ruth, 1971.
The effects of increasing light and temperature on the structure of diatom communities. Limnol Oceanogr. 16(2): 405-421.

diatoms

Patrick, Ruth, 1967.
Diatom communities in estuaries.
In: Estuaries, G.H. Lauff, editor, Publs Am. As Advmt Sc., 83:311-315.

diatoms

Patrick, R., 1948.
Factors affecting the distribution of diatoms. Bot. Rev. 14:473-524.

diatoms

Patrick, Ruth and Charles W. Reimer, 1966.
The diatoms of the United States exclusive of Alaska and Hawaii. Fragilariaceae, Eunoticeae, Achnanthaceae, Naviculaceae.
Acad. Nat. Sci., Philadelphia, Monogr. No. 13: 688 pp.

diatoms

Pavillard, J., 1916
Recherches sur les Diatomées pélagiques du Golfe de Lion. Trav. de l'Inst. Bot. de Univ. de Montpellier, Mem. 5 Cette.

diatoms

Pavillard, J., 1913
Observations sur les Diatomées, 2° série. Bull. Soc. Bot. France, 60:

diatoms

Pavillard, J., 1905
Recherches sur la flore pelagique (Phytoplankton) de l'Etang de Thau. Theses presentees a la Fac. Sci., Paris, 116 pp., 3 pls.

diatoms

Peragallo, H., and Peragallo, M., 1897-1908. Diatomées marines de France et des districts maritimes voisins.

diatoms

Phifer, L.D. (undated)
The occurrence and distribution of plankton diatoms in Bering Sea and Bering Strait, July 26-August 24, 1934. Report of Oceanographic cruise of U.S. Coast Guard Cutter Chelan 1934, Part II(A):1-44 (mimeographed) plus fig.1 (after Pt.B)

diatoms

Politis, J., 1952.
[Diatomées marines de l'Ile de Rhodes.] Akad. Athen., Praktika Hellenik. Hidrobiol. Inst 6(2):5-17.

Diatomées

Politis, J., 1949
Diatomées marines de Bosphores et des îbes de la mer de Marmara. II Practica tou Hellenikou Hidrobiologikou Institutoutou 1929, Etoz 1929, 3(1):11-31.

diatoms

Pringsheim, E.G., 1951.
Über farblose Diatomeen. Arch. f. Mikrobiol. 16:18-27.

diatoms

Proschkina-Lavrenko, A.I., 1960.
[On the evolution of diatoms.]
Biull. M. O-VA ISP. Prirroda, Otd. Biol., 65(5):52-62.
In Russian with an English summary

diatoms

Ramamurthy, S., 1953.
Measurement of diatom population by pigment extraction method. J. Madras Univ., B, 23(2):164-173.

diatoms

Rampi, L., 1942
Ricerche sul fitoplancton del Mare Ligure 6. Le diatomee dolle acque di Sanremo. Nuovo Giornale Botanico Italiano, N.S., 49:252-268.

diatoms

Rampi, L., 1940
Diatomee del Mare Adriatico. Nuovo Giornale Botanico Italiano, n.s., 47:559-608.

diatoms

Ranson, M.G., 1937
Sur la soi-disant existence de plusieurs Diatomées bleues dans la nature. Revue Algologique.8:11 pp.

diatoms

Riley, G. A., 1942
The relationship of vertical turbulence and spring diatom flowerings. J. Mar. Res., 5(1):67-87

diatoms

Ross, R., and G. Abdin, 1949.
Notes on some diatoms from Norfolk. J. Roy. Micr. Soc., ser. 3, 69(4):225-230, 4 figs. on 1 pl.

diatoms

Salah, M.M., 1952(1953).
XII. Diatoms from Blakeney Point, Norfolk. New species and new records for Great Britain. J. R. Microsc. Soc., Ser. 3, 72(3):155-169, 3 pls

diatom

Sargent, M. S. and T. J. Walker, 1948
Diatom populations associated with eddies off southern California in 1941. J. Mar. Res. 7(3):490-505, 15 text figs.

diatoms, lists of spp.

Schrader, Hans-Joachim, 1973
Cenozoic diatoms from the northeast Pacific, leg 18. Initial Repts Deep Sea Drilling Proj, 18:673-797.

diatoms

Seki, Humitake, 1967.
Effect of organic nutrients on dark assimilation of carbon dioxide in the sea. II. Dark assimilation of marine diatoms. Inf.Bull.Planktol.Japan, 14:22-25.

diatoms

Silva, E. de S.E., and J.D. Santos Pinto, 1948. O plancton de Baia de S. Martinha de Porto. 1. Diatomaceas e Dinoflagelados. Bol. Soc. Portuguesa Ciencias Natur. 16(2):134-187.

diatoms

Simonsen, Reimer, 1960.
Neue Diatomeen aus der Ostsee. II. Kieler Meeresf., 16(1):126-130.

diatoms

Smith, W., 1853-1856
A synopsis of the British Diatomaceae Vol.I-II. London, Smith and Back

diatoms

Sournia, A., 1968.
Diatomees planctoniques du Canal de Mozambique et de l'Ile Maurice. Mem. ORSTOM. 31: 120 pp. (not seen)

diatoms

Sousa e Silva, E., 1949
Diatomaceas e Dinoflagelados de Baia de Cascais. Portugaliae Acta Biol. Volume: Julio Henriques, Ser. B: 300-383, 9 pls, 2 fold-in tables.

diatoms

Spencer, C.P., 1954.
Studies on the culture of a marine diatom. J.M.B.A. 33(1):265-290, 16 textfigs.

diatoms plankton

Subrahmanyan, R., 1946.
A systematic account of the marine diatoms of the Madras coast. Proc. Indian Acad. Sci. 24(4):85-197 Pl. 2, 440 textfigs.

diatom reproduction

von Stosch, 1950.
Oogamy in a centric diatom. Nature 165(4196):531-532.

diatoms

Takano, Hideaki, 1967.
Reproduction of diatoms (review). (In Japanese; English abstract). Inf.Bull.Planktol.Japan, 14:1-12.

diatoms

Takano, H., 1957.
Synonomy of pelagic diatoms. Info. Bull. Plankton, Japan, No. 4:10-13.

diatoms

Takano, H., 1954.
Preliminary report on the marine diatoms from Hachijo Island, Japan. Bull. Jap. Soc. Sci. Fish. 19(12):1189-1196, 4 textfigs.

diatoms

Tempère, J., and M. Peragallo, 1915
Diatomées du monde entier. 480 pp. (texte) 2nd edition.

diatoms

Tsubata, B., 1950.
[On the distribution of diatoms off southeast coast of Hokkaido with the consideration of the hydrographic conditions.] J. Ocean. Soc., Japan, (Nippon Kaiyo Gakkaisi) 6(2):33-38. (In Japanese English summary).

diatoms

Tsumura, K., 1956.
Diatomoj el la cirkaufoso de la restajo de la kastele de Odawara. J. Yokohama Municipal Univ., (C-14) No. 47:23 pp.

diatoms

Uyeno, Fukuzo, 1959
[Quantitative studies on the abundance and the seasons of bloom of some important diatom species in the neighboring Sea of Japan.] Rept. Fac. Fish., Prefect. Univ. of Mie, 3(2): 407-435.

diatoms

Van Heurck, H., 1880-1885
Synopsis des Diatomées de Belgique. Anvers.

diatoms

Waksman, S.A., J.L. Stokes, and M.R. Butler, 1937. Relation of bacteria to diatoms in sea water. J.M.B.A. 22:359-373.

diatoms

Wauthy, B., R. Desrosières et J. Le Bourhis, 1967. Importance presumée de l'ultraplancton dans les eaux tropicales oligotrophes du Pacifique central sud. Cah. ORSTOM, Sér.Océanogr., 5(2):109-116.

diatoms

Wimpenny, R. S., 1936
The size of diatoms. I. The diameter of Rhizosolenia styliformis Brightw. and R. alata Brightw. in particular and of pelagic marine diatoms in general. JMBA, XXI:29-60

diatoms

Wolle, F., 1894
Diatomaceae of North America. Bethlehem, Pa. The Comenius Press.

diatoms

Wood, E.J. Ferguson, 1964.
A note on diatoms occurring in Milford Sound. New Zealand Dept. Sci. Ind. Res., Bull., No. 157 New Zealand Oceanogr. Inst., Memoir, No. 17: 97.

diatoms (of mixed genera)

Wood, E. J. Ferguson, 1959
An unusual diatom from the Antarctic. Nature, 184: 1962-1963.

diatoms

Wood, E.J.F., 1956.
Diatoms in the ocean deeps. Pacific Science 10(4):377-381.

diatoms

Zanon, V., 1948
Diatomee marini di Sardegna e Pugillo di Alghe Marine della stressa. Boll. Pesca, Piscitutura e Idrobiologia, Anno 24, ns. 3(2):202-244, 27 figs. on 1 pl.

diatoms

Zanon, V., 1947.
Diatomee delle Isole Penziane - materiali per una florula del Mare Tirreno. Bol. Pesca, Piscicol. Idrobiol., n.s., 2(1):36-53, 1 pl. of 10 figs.

diatoms

Zanon, V., 1941.
Diatomee dell'Africa Occidentale Francese. Commentationes Pontificia Acad. Sci. 5:1-60, 3 pls

diatoms

Zanon, V., 1924.
Diatomee dello Stagno Palu (Ravigno). Correzioni ed aggiunte alle Florula Diatomologica dell' Adriatica. Thalassia 5(4):1-29.

diatoms

Zhuze, A.P., and G.I. Semina, 1955.
[General regularities in distribution of diatoms in plankton of Bering Sea and in the upper surface deposits.] Dokl. Akad. Nauk, SSSR, 100(3): 579-

diatoms, adaptation of

Jørgensen, Erik G., 1968.
The adaptation of plankton algae. II. Aspects of the temperature adaptation of Skeletonema costatum.
Physiol. Plant. 21(2): 423-427.

diatoms, adaptation of

Steemann Nielsen, E., and E.G. Jørgensen, 1968.
The adaptation of plankton algae. I. General part.
Physiol. Plant. 21(2): 401-413.

diatoms, antibiotic activity

Aubert, M., et D. Pesando, 1969.
Variations de l'action antibiotique de souches phytoplanctoniques en fonction de rythmes biologiques marins.
Rev. int. Océanogr. méd. 25-26: 29-37.

diatoms, antibiotic action

Gauthier, M. 1969.
Activité antibactérienne d'une diatomée marine: Asterionella notata (Grun).
Rev. int. Océanogr. méd. 25-26: 103-171.

diatoms, anat.-physiol.

Beklemishev, C.W., M.N. Petrikova and H.J. Semina, 1961
[On the cause of the buoyancy of plankton diatoms.]
Trudy Inst. Okeanol., 51:33-95.

diatoms, anat.-physiol

Berland, Brigitte R., et Serge Y. Maestrini, 1969.
Action de quelques antibiotiques sur le developpement de cinq diatomées en cultures.
J. mar. Biol. Ecol., 3(1):62-75.

diatoms, anat., physiol. a

Bunt, J.S., 1971.
Levels of dissolved oxygen and carbon fixation by marine microalgae. Limnol. Oceanogr. 16(3): 564-566.

diatoms, anat. physiol

Busby, William F., and Joyce Lewin, 1967.
Silicate uptake and silica shell formation by Synchronously dividing cells of the diatom Navicula pelliculosa (Breb.) Hilse.
J. Phycology, 3(3):127-131.

diatoms, anat. phys.

Callame, B., and Debyser, J., 1954.
Observations sur les mouvements des diatomées à la surface des sédiments marins de la zone inter-cotidade. Vie et Milieu 2:129-157.
Contr. C.R.E.O. Vol. 5.

diatoms, anat physiol

Chapman, G. and A.C. Rae, 1969.
Excretion of photosynthate by a benthic diatom.
Marine Biol., 3(4): 341-351.

diatoms, anatomy

Cleve-Euler, A., 1951
Die Diatomeen von Schweden und Finnland. Kungl. Svenska Vetenskaps Akad. Handl. Fjärde Ser. 2(1): 161 pp., 6 pls.

diatom, physiol.

Denffer, D., 1948.
Über einem Wachstumshemmstoff in alternden Diatomeenkulturen. Biol. Zentralb. 67(1/2):7-13.

diatoms, size variation

Egusa, S., 1949.
[Size variations in planktonic diatoms and some considerations of their ecological significance. 1. Skeletonema costatum and Biddulphia sinensis.] Bull. Japan. Soc. Sci. Fish. 15(7):332-336, 3 textfigs. (In Japanese).

diatoms, size reduction

Enomoto, Yoshimasa, 1959
[Size reduction of diatoms during the process of cell division.]
Info. Bull. Planktonology, Japan, No. 6:4-5.

diatoms, physiol.

Eppley, Richard W., and James L. Coatsworth 1968.
Uptake of nitrate and nitrite by Ditylum brightwelli - kinetics and mechanisms.
J. Phycol. 4(2): 151-156.

diatom - physiol.

Fogg, G.E., 1956.
Photosynthesis and formation of fats in a diatom. Ann. Botany 20(78):265-286.

diatoms, anat.-phys.

Follmann, G., 1958.
Plasmolyse-Verhalten und Vitalfaerbungs-Eigenschaften von Coscinodiscus granii Gough.
Bull. Inst. Ocean., Monaco, No. 1116:22 pp.

diatoms, anat. physiol.

Geissler, U., 1958.
Das Membran potential einiger Diatomeen u. seine Bedeutung fur die lebende Kieselalgenzelle.
Mikroskopie Band 13 No. 5/6 pp. 145-172.

diatoms, anatomy, etc.

Geitler, L., 1949-1951.
Die Auxosporenbildung von Nitzschia sigmoidea und die Geschlechtsbestimmung bei den Diatomeen. Portugales Acta Biol., Vol. R. Goldschmidt, Ser. A:79-87.

diatoms, physiology

Goldberg, E.D., 1952.
Iron assimilation by marine diatoms. Biol. Bull. 102(3):243-248.

diatom, physiol., etc.

Goldberg, E.D., T.J. Walker, and A. Whisenand, 1951.
Phosphate utilization by diatoms. Biol. Bull. 101(3):274-284, 5 textfigs.

diatoms, anat.-physiol.

Gross, F., and E. Zeuthen, 1952.
The buoyancy of plankton diatoms: a problem of cell physiology. Proc. R. Soc., London, B, 135:382-389.

diatoms, anat.-physiol.

Guillard, Robert R.L. 1968.
B₁₂ Specificity of marine centric diatoms.
J. Phycol. 4(1): 59-64.

diatoms, physiol.

Guillard, R.R.L., and Vivienne Cassie, 1963
Minimum cyanocobalamin requirements of some marine centric diatoms.
Limnol. and Oceanogr., 8(2):161-165.

diatoms, metabolism

Handa, N., 1969.
Carbohydrate metabolism in the marine diatom Skeletonema costatum. Mar Biol., 4(3): 208-214.

diatoms anat. physiol

Hellebust, Johan A., and Robert R.L. Guillard, 1967.
Uptake specificity for organic substrates by the marine diatom Melosira nummuloides.
J. Phycology, 3(3):132-136.

diatoms, anat. physiol.

Hendey, N. Ingram, 1959.
The structure of the diatom cell wall as revealed by the electron microscope.
J. Quekett Micros. Club. (4)5(6):147-175.

diatoms, anat.-physiol.

Höfler, Karl, und Luise Höfler, 1964.
Diatomeen des marinen Planktons als Isotonobionten Pubbl. Staz. Zool., Napoli, 33(3):315-330.

diatoms, anat.

Hustedt, Fr., 1952.
Die Struktur der Diatomeen und die Bedeutung des Elektronenmikroskops für ihre Analyse. II.
Arch. f. Hydrobiol. 47(2):295-301.

diatoms, anat.

Hustedt, Fr., 1945.
Die Struktur der Diatomeen und die Bedeutung des Elektronenmikroskops für ihre Analyse.
Arch. f. Hydrobiol. 41:315-332, Pl. 19, 1 textfig.

diatom, physiol.

Hutner, S.H., and L. Provosoli, 1953.
A pigmented marine diatom requiring Vitamin B12 and uracil. News Bull., Phycol., Soc., Amer., 6(18):7.

diatom, physiol-ecol.

Jørgensen, E.G., 1957.
Diatom periodicity and silicon assimilation, experimental and ecological investigations.
Dansk. Bot. Arkiv., 18(1):11-54.
Abstr. in
Biol. Abstr., 32(2):3947.

diatom physiol.

Jørgensen, E.G., 1953.
Silicate assimilation by diatoms. Physiol. Plant., 6:301-315.

diatoms, anat.-Physiol.

Kolbe, R.W., 1951.
Elektronenmikroskopische Untersuchen von Diatomeenmembranen II. Svenska Bot. Tidskr. 45(4): 636-647, Pls. 1-4.

diatoms, anat.

Kolbe, R.W., 1948.
Elektronenmikroskopische Untersuchungen von Diatomeenmembranen. Ark. f. Bot. 33A(17):1-21, 10 pls., 5 textfigs.

diatoms, physiology

Lewin, Joyce, 1966.
Boron as a growth requirement for diatoms.
J. Phycology, 2(4):160-163.

diatoms, anat. physiol

Lewin, Joyce, 1966.
Silicon metabolism in diatoms. V. Germanium dioxide, a specific inhibitor of diatom growth.
Phycologie, Int. Phycol. Soc., 6(1):1-12.

diatoms, anat.

Lewin, J.C., 1955.
The capsule of the diatom Navicula pelliculosa.
J. Gen. Microbiol. 13(1):162-169.

diatoms, anat.-physiol.

Lewin, Joyce C., and Robert R.L. Guillard, 1963.
Diatoms.
Annual Review of Microbiology, 17:373-414.

diatoms, anat.-physiol

Mandelli, Enrique F., 1967.
Enhanced photosynthetic assimilation ratios in Antarctic polar front (convergence) diatoms.
Limnol. Oceanogr., 12(3):484-491.

diatoms, anat.-physiol.

Matsue, Yoshiyuki, 1957.
On the absorption of nitrogen compounds in different forms in sea water by a marine plankton diatom, Skeletonema costatum (Grev.) Cleve.
Coll. Wks., Fish. Sci., Jubilee Publ., Prof. I. Amemiya, Tokyo Univ. Press:249-257.
Abstr. in:
Rec. Res., Fac. Agric., Tokyo Univ., Mar. 1958, 7(79):57.

diatoms, physiology
Matsue, Y., 1949.
The physiological analysis of brine and the preservation of phosphoric acid in Skeletonema costatum. J. Fish. Res. Inst. 2:34-49.

diatoms, physiology of
Matudavia, T., 1942.
On inorganic sulphides as a growth promoting ingredient for diatom.
Proc. Imp. Acad., Tokyo, 18(2):107-116.

diatoms, silicate uptake a
Paasche, E., 1973
Silicon and the ecology of marine plankton diatoms. II. Silicate-uptake kinetics in five diatom species. Mar. Biol. 19(3): 262-269.

diatoms, anatomy
Patrick, Ruth and Charles W. Reimer, 1966.
The diatoms of the United States exclusive of Alaska and Hawaii. Fragilariaceae, Eunotiaceae, Achnanthaceae Naviculaceae.
Acad. Nat. Sci., Philadelphia. Monogr. No. 13: 688 pp.

diatoms, anat. physiol.
Prasad, R.R., and K.N. Krishna Kartha, 1959
A note on the breeding of copepods and its relation to diatom cycle. J. Mar. Biol. Ass. India, 1(1): 77-84.

diatoms, physiology, etc.
Richards, F.A., 1952.
The estimation and characterization of plankton populations by pigment analyses. I. The absorption spectra of some pigments occurring in diatoms, dinoflagellates, and brown algae.
J. Mar. Res. 11(2):147-155.

diatoms, anat.
Schultz, Mary E., and Francis R. Trainor 1968.
Production of male gametes and auxospores in the centric diatoms Cyclotella meneghiniana and C. cryptica.
J. Phycol. 4(2): 85-88.

diatoms, anat.
Smayda, Theodore J., 1962.
Occurrence of unusual bodies in a marine pennate diatom.
Nature, 196(4850):191.

diatoms, anat.
Subrahmanyan, R., 1951.
Note on handling diatoms for cytological and life-history studies. Microscope (Nov.-Dec.):

diatoms, life history
Subrahmanyan, R., 1949.
On the occurrence of microspores in some centric diatoms of the Madras coast. J. Indian Bot. Soc. 25:61-66.

diatoms, anatomy, etc.
Subrahmanyan, R., 1945.
On the cell division and mitosis in some South Indian diatoms. Proc. Indian Acad. Sci., Sect. B, 22(6): 331-354, Pl. 39, 88 textfigs.

diatom, life history
Subrahmanyan, R., 1945.
On the formation of auxospores in Bacteriastrum. Curr. Sci. 14:154-155.

diatom, life history
Subrahmanyan, R., 1945.
On somatic division, reduction division, auxospore-formation and sex differentiation in Navicula halophila (Grunow)Cleve. Curr. Sci. 14: 75-77.

diatom, physiol.
Thomas, William H. and Anne N. Dodson, 1972
On nitrogen deficiency in tropical Pacific oceanic phytoplankton. II. Photosynthetic and cellular characteristics of a chemostat-grown diatom. Limnol. Oceanogr. 17(4): 515-523.

diatom growth
von Stosch, H.-A., 1951.
Entwicklungsgeschichtliche Untersuchungen an zentrischen Diatomeen. 1. Die Auxosporenbildung von Melosira varians. Arch. f. Mikrobiol. 16(2): 102-135, 2 pls. (29 figs).

diatoms, anat. a
von Sydow, Burkard, und Robert Christenhuss 1972.
Rasterelektronmikroskopische Untersuchungen der Hohlräume in der Schalenwand einiger centrischer Kieselalgen.
Arch. Protistenk 114 (3): 256-271.

diatoms, anat.-physiol
Zgurovskaya, L.N., and N.G. Kustenko, 1968.
The influence of different nitrite nitrogen concentrations on photosynthesis, pigment accumulation and cell division of Skeletonema costatum (Grev.)Cl. (In Russian; English abstract).
Okeanologiia, Akad. Nauk, SSSR, 8(6):;053-1058.

diatoms, benthic
Bisch W. K., M. Ricard, et R. Cantin, 1970.
Utilisation des diatomées benthiques comme indicateur de pollutions minières dans le bassin de la Miramichi N.W.
Tech. Rept. Fish. Res. Bd. Can. 202: 72pp. (multilithed)

diatoms, benthic, lists of spp. a
Luchini, Laura 1972.
Étude qualitative et quantitative d'une population de diatomées du microphytobenthe épilithe (Anse des Cuivres, Marseille).
Téthys 3(3): 459-505.

diatom (biomass)
Smayda, Theodore J., 1965.
A quantitative analysis of the phytoplankton of the Gulf of Panama. II. On the relationship between C14 assimilation and the diatom standing crop.
Inter-American Tropical Tuna Commission, Bull. 9(7):467-531.

diatom, blooms
Balle Cruellas, Pedro, 1965.
Note sur des floraisons anormales des diatomées bu large des iles Baleares.
Rapp. Proc. Verb. Reunions, Comm. Int. Expl. Sci., Mer Mediterranee, Monaco, 18(2):371-372.

diatom bloom a
El-Sayed, Sayed Z. 1971.
Observations on phytoplankton bloom in the Weddell Sea.
Biology of the Antarctic Seas IV, George A. Llano and I. Eugene Wallen editors, Antarct. Res. Ser. Am. Geophys. Un. 17: 301-312.

diatom blooms
Enomoto, Y., 1958.
Studies on plankton off the west coast of Kyushu-II. On conditions for the vernal blooming of phytoplankton.
Bull. Jap. Soc. Sci. Fish., 25(3):172-182.

diatom flowerings
Lund, J.W.G., 1949.
The dynamics of diatom outbursts, with special reference to Asterionella. Proc. Int. Assoc., Limnol. 10:275.

diatom flowerings
Ryther, John H., Charles S. Yentsch and E.M. Hulburt, 1958.
The dynamics of a diatom bloom.
Biol. Bull., 115(2):257-268.

diatom blooms
Vinogradova, L.A., 1962.
On the method for compiling charts for the "flowering" of the waters in the Norwegian Sea. (In Russian).
Trudy, Baltiisk, Nauchno-Issled. Inst. Morsk. Ribn. Khoz. i Okeanogr. (BALTNIRO), 8:253-256.

diatoms, cell size a
Parsons, T.R. and M. Takahashi, 1973
Environmental control of phytoplankton cell size. Limnol. Oceanogr. 18(4): 511-515.

diatoms, cell size a
Round, F.E. 1972.
The problem of reduction of cell size during diatom cell division.
Nova Hedwigia 23 (2/3): 291-303

diatoms, cell size
Semina, H.J., 1969.
The size of phytoplankton cells along 174°W in the Pacific Ocean. Okeanologiia 9(3): 479-487. (In Russian; English abstract)

diatoms, cell size
Semina, H.J. and V.V. Aratskaia 1970.
Main pycnocline, cell size and the distribution patterns of phytoplankton species. (In Russian).
Dokl. Akad. Nauk, SSSR, 191(2): 449-452

diatoms, characteristics of
Geissler Ursula, 1970
Die Schalenmerkmale der Diatomeen. Ursachen ihrer Variabilität und Bedeutung für die Taxonomie.
Beiheft Nova Hedwigia 31: 511-535

diatoms, chemistry of
Ackman, R.G., P.M. Jangaardm R.J. Hoyle and H. Brockerhoff, 1964.
Origin of marine fatty acids. 1. Analyses of the fatty acids produced by the diatom Skeletonema costatum.
J. Fish. Res. Bd., Canada, 21(4):747-756.

diatoms, chemistry of
Alfimov, N.N., 1966
On the biology and bio.chemistry of two mass marine diatoms, Coscinodiscus jonesianus (Grev?) Ostf. and Rnizosolenia calcar-avis M. Schultze from the Azov and Caspain seas. (In Russian)
Bot. Zh., 51(9): 1276-1283.

diatoms, chemistry of
Barashkov, G.K., 1962.
Chemistry of some marine plankton diatoms. (In Russian).
Trudy Murmansk. Morsk. Biol. Inst., 4(8):27-46.

diatoms, chemistry of
Baraskov, G.K., 1956.
On the carbohydrates of certain diatom species.
Dokl. Akad. Nauk, SSSR, 111(1):148-151.

chemistry, diatoms
Becking, L.B., C.F. Tolman, H.C. McMillan, H.C. Field, and T. Hashimoto, 1927.
Preliminary statement regarding the diatom "epidemic" at Copalis Beach, Washington, and an analysis of diatom oil. Econ. Geol. 22:356-368.

diatoms, chemical composition
Bordovskiy, O.K., 1965.
Accumulation and transformation of organic substance in marine sediments. 1. Summary and introduction. 2. Sources of organic matter in marine basins. 3. Accumulation of organic matter in bottom sediments. 4. Transformation of organic matter in marine sediments.
Marine Geology, Elsevier Publ. Co., 3(½):3-4; 5-31; 33-82; 83-114.

diatoms, chemistry of
Brongersma-Sanders, Margaretha, 1967.
Barium in pelagic sediments and in diatoms.
Proc. K. ned. Adak. Wet., (B), 70 (1): 93-99.

diatoms, chemistry
Chuecas, L. and J.P. Riley, 1969.
The component combined amino acids of some marine diatoms.
J. mar. biol. Ass., U.K., 49(1):117-120.

diatoms, chemistry of
Donchenko, N.S., 1970.
Effect of growing conditions of Melosira varians Ag. on its biochemical composition and fragrance. (In Russian). Gidrobiol. Zh., 6(6):90-93.

diatoms, chemistry
Hecky, R.E., K. Mopper, P. Kilham and E.T. Degens, 1973
The amino acid and sugar composition of diatom cell-walls. Mar. Biol. 19(4):323-331.

diatoms, chemistry of
Kates, M., 1966.
Lipid components of diatoms.
Biochimica et Biophysica Acta, 116(2):264-278.

diatoms, chemistry
Kesseler, Hanswerner, 1967.
Untersuchungen über die chemisch Zusammensetzung des Zellsaftes der Diatomee Coscinodiscus Wailesii (Bacillariophyceae,, Centrales).
Helgoländer wiss Meeresunters, 16(3):262-270.

diatoms, chemistry of
Lanskaia, L.A., and T.I. Pshenina, 1963.
Comparison of the chemical composition of some species of diatoms in culture and in the sea. (In Russian).
Trudy Sevastopol Biol. Sta., 16:457-462.

diatoms, chemistry of
Lewin, J.C., 1961.
The dissolution of silica from diatom walls.
Geochimica et Cosmochimica Acta, 21(2/3):182-198

diatoms, chemistry of
Lewin, Joyce C., and Robert R.L. Guillard, 1963.
Diatoms.
Annual Review of Microbiology, 17:373-414.

diatoms, chemistry of
Low, E.M., 1955.
Studies on some chemical constituents of diatoms.
J. Mar. Res. 14(2):199-204.

diatom colonies
Beklemishev, C.W., 1961
On the role played by the colonies of plankton diatoms.
Trudy Inst. Okeanol., 51:16-30.

diatoms, dead
Ohwada, M., 1960.
Vertical distribution of living and dead diatoms down to one thousand meters off Sanriku, northern Japan.
Bull. Hakodate Mar. Obs., 7(2):1-
Mem. Kobe Mar. Obs., 7(2):69-73.

diatoms, decomposition of
Kamatani, Akiyoshi, 1969.
Regeneration of inorganic nutrients from diatom decomposition.
J. oceanogr. Soc. Japan, 25(2):63-74.

diatom distributions
Belyaeva, T.V., 1971.
Quantitative distribution of planktonic diatoms in the western tropical Pacific. (In Russian; English abstract). Okeanologiia 11(4):687-694.

diatoms, vertical distribution of
Castenholz, Richard W., 1963.
An experimental study of the vertical distribution of littoral marine diatoms.
Limnology and Oceanography, 8(4):450-462.

diatoms
Edsbagge, Hans., 1966.
Grundlinien in der Verteilung von angehefteten Diatomeen an der schwedischen Westküste.
Veröff. Inst. Meeresforsch., Bremerh., Sonderband II:271-273.

diatom distribution
Kozlova, O.G., 1971.
The main features of diatom and silicoflagellate distribution in the Indian Ocean. In: Micropalaeontology of oceans, B.M. Funnell and W.R. Riedel, editors, Cambridge Univ. Press, 271-275.

diatoms, vertical distribution
Kucherova, Z.S., 1961.
Vertical distribution of diatoms from Sevastopol Bay.
Trudy Sevastopol Biol. Sta., (14):64-78.

diatoms, distribution
Uyeno, Fukuzo, 1957.
Relation between diatom quantity, temperature and chlorinity in the neighboring sea south of Honshu.
Umi to Sora, 33(4/5):70-74.

diatoms, distribution of
Uyeno, F., 1957.
The variation of diatom communities and the schematic explanation of their increase in Osaka Bay in summer. III. Schematic explanation of increase of diatoms in relation to the distribution of chlorinity.
J. Oceanogr. Soc., Japan, 13(3):107-110.

diatom distribution
Uyeno, Fukuzo, 1957.
Water movements and distribution of diatoms in a vertical section of the cold water region off Enshunada.
Umi to Sora, 33(4/5):75-81.

diatoms, division rates
Sakshaug, Egil 1970.
Quantitative phytoplankton investigations in near-shore water masses.
Skr. K. norske Vidensk. Selsk. (3):1-8.

diatoms, ecology of
Castenholz, Richard W. 1967.
Seasonal ecology of non-planktonic marine diatoms on the western coast of Norway.
Sarsia 29: 237-256.

diatoms, ecology of
Edsbagge, Hans, 1968.
Zur Ökologie der marinen angehefteten Diatomeen.
Botanica Gothoburgensia 6: 153 pp.

diatoms, ecol.
Faure-Fremiet, E., 1951.
The tidal rhythm of the diatom Hantzschia amphioxys. Biol. Bull. 100(3):173-177, 1 textfigs.

diatoms (bottom) ecology of
Hopkins, J. Trevor, 1963
A study of the diatoms of the Ouse Estuary, Sussex. 1. The movement of the mud-flat diatoms in response to some chemical and physical changes.
J. Mar. Biol. Assoc., U.K., 43(3):653-663.

diatoms
Pincemin, Jean-Marc, 1966.
Note préliminaire à l'étude écologique des dinoflagellés de la baie d'Alger et comparaison avec les diatomées.
Pelagos, (6):9-47.

diatom, ecol.
Uyeno, F., 1958.
Relation between diatom quantity, temperature and chlorinity in the neighboring sea near the coast of Honshu, Japan. Mem. Kobe Mar. Obs., 12(1):1-6.

diatoms, ecology
Uyeno, F., 1957.
The variation of diatom communities and the schematic explanation of their increase in Osaka Bay in summer. 1. Classification of communities and the fluctuation of their quantities. 2. Relation between the distribution of diatoms and sea conditions.
J. Oceanogr. Soc., Japan, 13(2):73-78; 79-84.

diatoms, ecol.
Uyeno, F., 1958.
Water movement and distribution of diatoms in a vertical section of the cold water region off Enshu-nada, Japan. Mem. Kobe Mar. Obs., 12(1):7-14.

diatoms, effect of
Buinitsky, V. Kh., 1968.
The influence of microscopic organisms on the structure and strength of Antarctic sea ice. (In Russian; English abstract).
Okeanologiia, Akad. Nauk, SSSR, 8(6):971-979.

diatoms, effect of
Gorbenko, Yu.A. and Z.S. Kucherova, 1960.
Correlation of diatoms and rod-shaped bacteria in primary films. (In Russian).
Trudy Sevastopol. Biol. Stants., Akad. Nauk, 15: 485-492.

diatoms, effect of
bacteria, effect of
fouling, films

diatoms, estuarine
Wulff, Barry L. and C. David McIntire, 1972.
Laboratory studies of assemblages of attached estuarine diatoms. Limnol. Oceanogr., 17(2):200-214.

diatoms, families
Hasle, Grethe R. 1973.
Thalassiosiraceae, a new diatom family.
Norw. J. Bot. 20(1):67-69

diatoms, flotation of
Smayda, Theodore J., and Brenda J. Boleyn, 1965.
Experimental observations on the flotation of marine diatoms. 1. Thalassiosira nana, Thalassiosira rotula and Nitzschia seriata.
Limnol. Oceanogr., 10(4):499-509.

diatoms, freshwater
Kolbe, R.W., 1957.
Fresh-water diatoms from Atlantic deep-sea sediments. Science, 126(3282):1052-1056.

diatoms, growth (stimulation of)
Prakash, A., M.A. Rashid, Arne Jensen and D.V. Subba Rao, 1973
Influence of humic substances on the growth of marine phytoplankton: diatoms.
Limnol. Oceanogr. 18(4):516-524.

diatoms, lists of spp.

Amossé, A., 1970.
Diatomées marines et Saumâtres du Sénégal et de la Côte d'Ivoire.
Bull. Inst. fond. Afr. Noire (A) 32(2): 289-311.

diatoms, lists of spp.
Angot, Michel, et Robert Gerard, 1966.
Hydrologie et phytoplancton de l'eau de surface en avril 1965 a Nosy Be.
Cah. ORSTOM, Ser. Oceanogr., 4(1):95-136.

diatoms, lists of spp.
Australia, Commonwealth Scientific and Industrial Research Organization, 1964.
Oceanographical observations in the Indian Ocean in 1961, H.M.A.S. Diamantina, Cruise Dm 3/61.
Div. Fish. and Oceanogr., Oceanogr. Cruise Rept., No. 11:215 pp.

diatoms, lists of spp.
Australia, Commonwealth Scientific and Industrial Research Organization, Division of Fisheries and Oceanography, 1963.
Oceanographical observations in the Pacific Ocean in 1960, H.M.A.S. Gascoyne, Cruise G 3/60.
Oceanographical Cruise Report, No. 6: 115 pp.

diatoms, lists of spp.
Australia, Commonwealth Scientific and Industrial Research Organization, 1962.
Oceanographic observations in the Pacific Ocean in 1960, H.M.A.S. Gascoyne, Cruises G 1/60 and G 2/60.
Oceanographical Cruise Report No. 5:255 pp.

diatoms (lists of spp.)
Australia, Commonwealth Scientific and Industrial Research Organization, Division of Fisheries and Oceanography, 1963.
Oceanographical observations in the Indian Ocean in 1960, H.M.A.S. Diamantina, Cruise Dm 2/60.
Oceanographical Cruise Report No. 3:347 pp.

diatoms, lists of spp.
Australia, Commonwealth Scientific and Industrial Organization, 1962.
Oceanographical observations in the Indian Ocean in 1960, H.M.A.S. Diamantina, Cruise Dm 1/60.
Oceanogr. Cruise Rept., Div. Fish. and Oceanogr., No. 2:128 pp.

diatoms, lists of spp.
Australia, Commonwealth Scientific and Industrial Organization, 1962.
Oceanographical observations in the Indian Ocean in 1959, H.M.A.S. Diamantina, Cruises Dm 1/59 and Dm 2/59.
Oceanographical Cruise Rept., Div. Fish and Oceanogr., No. 1:134 pp.

diatoms, lists of spp.
Australia, Commonwealth Scientific and Industrial Research Organization, 1961.
F.R.V. "Derwent Hunter".
C.S.I.R.O. Div. Fish. and Oceanogr. Rept. No. 32:56 pp.

diatoms, lists of spp.
Australia, Marine Biological Laboratory, Cronulla, 1960.
F.R.V. "Derwent Hunter", scientific report of cruises 19-20/58
C.S.I.R.O. Div. Fish. & Oceanogr., Rept., 30: 53 pp., numerous figs., (mimeographed).

See author card for complete "title".

diatoms, lists of spp.
Avaria P., Sergio 1970.
Fitoplancton de la expedición del Doña Berta en la zona Puerto Montt-Aysen.
Rev. Biol. mar. Valparaiso 14(2): 1-17.

diatoms, lists of spp.
Balech, Enrique, 1971.
Microplancton de la campaña oceanographica Productividad III.
Revta Mus. argent. Cienc. Nat. Bernadino Rivadavia, Hydrobiol. 3(1):1-202, 39 pls.

diatoms, lists of spp.
Belyaeva, T.V., 1972.
Distribution of large diatom algae in the southeastern Pacific. (In Russian; English abstract). Okeanologiia 12(3): 475-484.

diatoms, lists of spp.
Belyaeva, (Sechkina), T.V., 1961
[Diatoms in the upper layer of sediments of the north-western Pacific.]
Trudy Inst. Okeanol., 46:231-246.

diatoms, lists of spp.
Belyaeva (Sechkina), T.V., 1961
[Diatoms in the upper layer of sediments of the Sea of Japan.]
Trudy Inst. Okeanol., 46:247-262.

diatoms lists of spp
Bernard, Francis, 1967.
Contribution à l'étude du nannoplancton, de 0 à 3000 m, dans les zones atlantiques lusitanienne et mauritanienne (Campagnes de la Calypso, 1960, et du Coriolus, 1964).
Pelagos, Alger, 7:1-81.

diatoms, lists of spp.
Bernhard, M., L. Rampi e A. Zattera 1969.
La distribuzione del fitoplancton nel mar Ligure.
Pubbl. Staz. Zool. Napoli 37 (2 Suppl.): 73-114.

diatoms, lists of spp
Bucalossi, G., 1960.
Étude quantitative des variations du phytoplancton dans la baie d'Alger en fonction du milieu (novembre 1959 à mai 1960).
Bull. Inst. Océanogr., Monaco, 57(1189):1-40.

diatoms, lists of spp
Bursa, Adam, 1961.
Phytoplankton of the "Calanus" Expeditions in Hudson Bay, 1953 and 1954.
J. Fish. Res. Bd., Canada, 18(1):51-83.

diatoms, lists of spp.
Bursa, Adam S., 1961
The annual oceanographic cycle at Igloolik in the Canadian Arctic. II. The phytoplankton.
J. Fish. Res. Bd., Canada, 18(4):563-615.

diatoms, lists of spp
Cassie, Vivienne, 1966.
Distoms dinoflagellates and hydrology in the Hauraki Gulf, 1964-1965.
New Zealand J. Sci., 9(3):569-585.

diatoms, lists of spp
Cassie, Vivienne, 1960
Seasonal changes in diatoms and dinoflagellates off the east coast of New Zealand during 1957 and 1958. N.Z.J. Sci. 3(1): 137-172.

diatoms, lists of spp.
Cholnoky, B.J., 1968.
Die Diatomeenassoziationen der Santa-Lucia-Lagune in Natal (Südafrica).
Botanica mar, 11(Suppl.):121 pp.

diatoms, lists of spp.
Cholnoky, B.J., 1963.
Beiträge zur Kenntnis des marinen Litorals von Südafrika.
Botanica Marina, 5(2/3):38-83.

diatoms, lists of spp
Crosby, L. H., and E. J. Ferguson Wood, 1959.
Studies on Australian and New Zealand diatoms.
II. Normally epontic and benthic genera.
Trans. R. Soc., New Zealand, 86(1/2):1-58. Pls. 1-9.

diatoms, lists of spp.
Crosby, L. H., and E. J. Ferguson Wood, 1958.
Studies on Australian and New Zealand diatoms.
1. Planktonic and allied species. Trans. R. Soc. New Zealand, 85(4):483-530, Pls. 31-39.

diatoms, lists of spp.
Dureiretnam, M., 1964.
Some planktonic diatoms from the Indien Ocean.
Bull. fish Res. Stn., Ceylon, 17:159-168.
Also in: Coll. Rept. Int. Indien Ocean Exped. 4(1967).

diatoms, lists of spp.
Edsbagge, Hans, 1968.
The composition of the epiphytic diatom flora on the Swedish west coast.
Botanica mar., 11(1/4):68-71.

diatoms, lists of spp.
Edsbagge, Hans, 1968.
Distribution notes on some diatoms not earlier recorded from the Swedish west coast.
Botanica mar., 11(1/4):54-67.

diatoms, lists of spp.
El-Maghraby, A.M., and E.J. Perkins, 1956.
The marine diatoms of Whitstable.
Ann. Mag. Nat. Hist. (12)9(104):561-568.

diatoms, lists of spp.
El-Sayed, S.Z., W.M. Sackett, L.M. Jeffrey, A.D. Fredericks, R.P. Saunders, P.S. Congen, G.A. Fryxell, K.A. Steidinger and S.A. Earle 1972.
Chemistry primary productivity and benthic algae of the Gulf of Mexico.
Ser. Atlas Mar. Environm. Am. Geogr. Soc. 22: 29 pp., 6 pls. (quarts)

diatoms, lists of spp.
Eskinazi-Leça Enide, 1970.
Estudo da plataforma continental na área do Recife (Brasil). IIIa. Diatomáceas do Fitoplâncton. Trabhs oceanogr. Univ. Fed. Pernambuco 9/11: 159-172.

diatoms, lists of spp.
Eskinazi-Leca, Enide, 1970.
Shelf off Alagoas and Sergipe (northeastern Brazil). 3. Diatoms from the Sao Francisco River mouth. Trabhs oceanogr. Univ. Fed. Pernambuco 9/11: 181-192.

diatoms, lists of spp.
Eskinazi, Enide, 1965/1966.
Estudio da Barra das Jangadas. VI. Distribuição das diatomaceas.
Trabhs Inst. Oceanogr., Univ. Fed. Pernambuco, Recife, (7/8):17-32.

diatoms, lists of spp.
Establier, R., y R. Margalef, 1964
Fitoplancton e hidrografía de las costas de Cádiz (Barbate) de junio de 1961 a agosto de 1962.
Inv. Pesq., Barcelona, 25:5-31.

diatoms, lists of spp.
Falcão Paredes, Jorge 1969/70.
Subsidios para o conhecimento do plâncton marinho de Cabo Verde. 1. Diatomáceas Silicoflagelados e Dinoflagelados.
Memórias Inst. Invest. cient. Moçambique (A) 10: 3-107.

diatoms, lists of species
Ferguson Wood, E.J., 1959
Some aspects of marine microbiology. J. Mar. Biol. Ass. India, 1(1): 26-32.

diatoms, lists of spp.
Ferrando, H.J., 1957.
Hipotesis sobre productividad en el area bioceanografica correspondiente a los litorales maritimos de Argentina, Uruguay y sur del Brasil.
IV Reunion del Grupo de Trabajo de Ciencias del Mar, Montevideo, 22-24 de Mayo de 1957, Actas de Sesiones y Trabajos Presentados:71-94.

diatoms, lists of spp.
Fischer, H., 1964.
Verhalten und Resistenz mariner Diatomeen gegenüber Veränderungen der Salzkonzentration.
Helgoländer Wiss. Meeresuntersuch., 10(1/4):64-72.

diatoms, lists of spp.
Frenguelli, J., 1960.
Diatomeas y silicoflagelados recogidas en Tierra Adélia durante las Expediciones Polares Francesas de Paul-Emile VICTOR (1950-1952).
Revue Algologique, (1):1-47.

diatoms, lists of spp.
Frenguelli, Joaquin and Hector A. Orlando, 1959
Analisis de algunas muestras del Pleistoceno del fondo del Mar. Mediterraneo envidas por el Lamont Geological Observatory, Secretaria de Marina, Servicio de Hidrograficia Naval, Argentina, Publico, H. 1014: 1-7.

diatoms, lists of spp.
Frenguelli, Joaquin, and Hector Antonio Orlando, 1959.
Operacion MERLUZA. Diatomeas y silicoflagelados del plancton del "VI Crucero".
Servicio Hidrogr. Naval, Argentina, Publ. No. H. 619: 5-62.

diatoms, lists of spp.
Frenguelli, Joaquin, y Hector A. Orlando, 1958.
Diatomeas y silicoflagelados del sector Antartico Sudamericano.
Inst. Antartico Argentino, Publ., No. 5:191 pp.

diatoms, lists of spp.
Fukase, S., 1965.
Studies on diatoms of the Argentine coast. The Drake Passage, and the Bransfield Strait.
Oceanogr. Mag. Tokyo, 17(1/2): 1-10.
Also in: Oceanogr. Meteorol.Nagasaki Mer.Obs., 15 (226).

diatoms, lists of spp.
Fukase, Shigeru, and Sayed Z. El-Sayed, 1965.
Studies on diatoms of the Argentine coast, the Drake Passage and the Bransfield Strait.
Oceanogr. Mag., Tokyo, 17(1/2):1-10.

diatoms, lists of spp. a
Fukushima, Hiroshi, Tomohiko Watanuki and Tsuyako Ko-Bayashi 1973.
A preliminary report on the diatom from East Ongul Island. (In Japanese; English abstract).
Antarct. Rec., Tokyo 46:125-192.

diatoms, lists of spp. a
Furnestin, Marie-Louise 1972.
Phytoplancton et production primaire dans le secteur sud-occidental de la Méditerranée.
Rev. Trav. Inst. Pêches marit. 37(1):19-68.

diatoms, lists of spp. a
Genovese, S., G. Fangemi e F. De Domenio 1972.
Campagna estiva 1970 della n/o Bannock nel Mar Tirreno - misure di produzione primaria lungo la trasversale Palermo-Cagliari.
Boll. Pesca Piscic. Idrobiol. 27(1):139-157.

diatoms, lists of spp.
Gessner, Fritz, and Reimer Simonsen, 1967.
Marine diatoms in the Amazon?
Limnol.Oceanogr., 12(4):709-711.

diatoms, lists of spp. a
Giffen, Malcolm H. 1973.
Diatoms of the marine littoral of Steenberg's Cove in St. Helena Bay, Cape Province, South Africa.
Botanica marina 16(1): 32-48.

diatoms, lists of spp.
Giffen, Malcolm H. 1970.
Contributions to the diatom flora of South Africa. IV. The marine littoral diatoms of the estuary of the Kowie River, Port Alfred, Cape Province.
Beiheft Nova Hedwigia 31:259-312.

diatoms, lists of spp.
Giffen, Malcolm H. 1971.
Marine littoral diatoms from the Gordon's Bay region of False Bay, Cape Province, South Africa.
Botanica marina 14 (Suppl.):1-16pp.

diatoms, lists of spp.
Grall, Jean-René, et Guy Jacques, 1964.
Etude dynamique et variations saisonnières du plancton de la region de Roscoff. B. Phytoplancton.
Cahiers, Biol. Sta. Roscoff, 5:432-455.

diatoms, lists of spp. a
Heimdal, Berit R., Grethe R. Hasle and Jahn Throndsen 1973.
An annotated check-list of plankton algae from the Oslofjord, Norway (1951-1972).
Norw. J. Bot. 20(1): 13-19.

diatoms, lists of spp.
Hendey, N. Ingram 1971.
Some marine diatoms from the Galapagos Islands.
Nova Hedwigia 22(1/2):371-422.

diatoms, lists of spp.
Hendey, N. Ingram, 1970.
Some littoral diatoms of Kuwait.
Beiheft Nova Hedwigia 31: 107-167.

diatoms, lists of spp.
Herrera, Juan, y Ramón Margalef, 1963
Hidrografía y fitoplancton de la costa comprendida entre Castellón y la desembocadura del Ebro, de julio de 1960 a junio de 1961.
Inv. Pesq., Barcelona, 24:33-112.

diatoms, lists of spp.
Hopkins, J.T., 1964
A study of the diatoms of the Ouse Estuary, Sussex. III. The seasonal variation in the littoral epiphyte flora and the shore plankton.
Jour. Mar. Biol. Assoc., U.K., 44(3):613-644.

diatoms, lists of spp.
Hopkins, J.T., 1964.
A study of the diatoms of the Ouse Estuary, Sussex. II. The ecology of the mud-flat diatom flora.
J. Mar. Biol. Assoc., U.K., 44(2):333-341.

diatoms, lists of spp.
Hulburt, Edward M., 1964.
Succession and diversity in the plankton flora of the western North Atlantic.
Bull. Mar. Sci., Gulf and Caribbean, 14(1):33-44.

diatoms, lists of spp.
Jacques, G. 1969.
Aspects quantitatifs du phytoplancton de Banyuls-sur-Mer (Golfe du Lion) III. Diatomées et dinoflagellés de juin 1965 à juin 1968.
Vie Milieu (B) 20(1): 91-126.

diatoms, lists of spp.
Japan, Hakodate Marine Observatory 1970
Report of the oceanographic observations in the sea east of Honshu and Hokkaido from February to March 1966. (In Japanese)
Bull. Hakodate mar. Obs. 15:3-10.

diatoms, lists of spp.
Japan, Hakodate Marine Observatory, 1970.
Report of the oceanographic observations in the sea east of Honshu and Hokkaido and in the Tsugaru Straits from April to May, 1966. (In Japanese)
Bull. Hakodate mar. Obs. 15:11-16

diatoms, lists of spp.
Japan, Hakodate Marine Observatory, 1969.
Report of the oceanographic observations in the sea east of Hokkaido and the Kurile Islands from May to June 1965. (In Japanese).
Bull. Hakodate mar. Obs. 14: 12-17.

diatoms, lists of spp.
Japan, Hakodate Marine Observatory, 1969.
Report of the oceanographic observations in the sea east of Hokkaido and the Kurile Islands, and in the Okhotsk Sea from July to September, 1965.
Bull. Hakodate mar. Obs. 14: 3-15. (In Japanese)

diatoms, lists of spp.
Japan. Hakodate Marine Observatory, 1969.
Report of the oceanographic observations in the sea south of Hokkaido and in the Sea of Okhotsk from October to November 1965 (In Japanese)
Bull. Hakodate mar. Obs. 14:16-21

diatoms, lists of spp.
Japan, Hakodate Marine Observatory, 1969.
Report of the oceanographic observations in the sea east of Hokkaido and the Tohoku District from February to March, 1965. (In Japanese)
Bull. Hakodate mar. Obs. 14:3-9

Diatoms, lists of spp
Japan, Hakodate Marine Observatory, 1966
Sea ice observation along the Okhotsk coast of Hokkaido from January to April 1963.
Sea ice conditions in the sea off Hokkaido from December 1962 to 1963.
Local sea ice conditions based on coastal observations of the Okhotsk Sea based on the broadcasting of U.S.S.R., 1962-1963. (In Japanese; English abstracts)
Bull. Hakodate Mar. Obs. No.12;1, 1-35,38-52; 53-67.

diatoms, lists of spp.
Japan, Hakodate Marine Observatory, 1961.
Report of the oceanographic observations in the sea east of Tohoku District and in the Okhotsk Sea from August to September 1959.
Bull. Hakodate Mar. Obs., (8):6-14.

diatoms, lists of spp.
Japan, Hokkaido University, Faculty of Fisheries, 1967.
Data record of oceanographic observations and exploratory fishing,11:383 pp.

diatoms, lists of spp.
Japan Meteorological Agency 1969.
The results of marine meteorological and oceanographical observations, Jan.-June 1966, 39:349 pp. (multilithed)

diatoms, lists of spp.
Japan Meteorological Agency, 1968.
The results of marine meteorological and oceanographical observations, July-December, 1965, 38: 404 pp. (multilithed)

diatoms, lists of spp.
Japan, Japan Meteorological Agency. 1965. The results of marine meteorological and oceanographical observations, July-December 1963, No. 34: 360 pp.

diatoms, lists of spp.
Japan, Japan Meteorological Agency, 1964. The results of marine meteorological and oceanographical observations, January-June 1963, No. 33:289 pp.

diatoms, lists of spp.
Japan, Japan Meteorological Agency, Oceanographical Section, 1962
Report of the oceanographic observations in the sea east of Honshu from February to March, 1961. (In Japanese).
Res. Mar. Meteorol. and Oceanogr. Obs., Jan.-June, 1961, No. 29:13-21.

diatoms, lists of spp. (quant.)
Japan, Japan Meteorological Agency, 1962
The results of marine meteorological and oceanographical observations, January-June 1961, No. 29: 284 pp.

diatoms, lists of spp
Japan, Kobe Marine Observatory, 1967.
Report of the oceanographic observations in the sea south of Honshu from July to August 1965. (In Japanese).
Bull. Kobe Mar. Obs. No. 178: 31-40

diatoms, lists of spp.
Japan, Kobe Marine Observatory, 1967.
Report of the oceanographic observations in the sea south of Honshu from February to March 1964. (In Japanese)
Bull. Kobe Mar. Obs. No. 175: 23-

diatoms, lists of spp.
Japan, Kobe Marine Observatory, Oceanographical Section, 1964.
Report of the oceanographic observations in the Kuroshio and region east of Kyushu from October to November 1962. (In Japanese).
Res. Mar. Meteorol. and Oceanogr., Japan. Meteorol. Agency, 32: 41-50.

Also in: Bull. Kobe Mar. Obs., 175. 1965.

diatoms, list of spp.
Japan, Kobe Marine Observatory, Oceanographical Section, 1964.
Report of the oceanographic observations in the sea south of Honshu from February to March, 1963. Res. Mar. Meteorol. and Oceanogr., Japan Meteorol. Agency, 33: 27-32.

Also in: Bull. Kobe Mar. Obs., 175. 1965.

diatoms, lists of spp.
Japan, Kobe Marine Observatory, Oceanographical Section, 1964.
Report of the oceanographic observations in the sea south of Honshu from July to August, 1962. Res. Mar. Meteorol. and Oceanogr., Japan. Meteorol. Agency, 32: 32-40. (In Japanese).

Also in: Bull. Kobe Mar. Obs. 175½

diatoms, lists of spp
Japan, Kobe Marine Observatory, Oceanographical Section, 1962
Report of the oceanographical observations in the sea south of Honshu from July to August 1961. (In Japanese).
Res. Mar. Meteorol. & Oceanogr., 30:39-48.

Also in: Bull. Kobe Mar. Obs., No. 173(3) 1964.

diatoms, lists of spp.
Japan, Maizuru Marine Observatory, 1967.
Report of the oceanographic observations in the Japan Sea from October to November, 1964. (In Japanese). Bull. Maizuru mar. Obs., 10: 87-94.

diatoms, lists of spp.
Japan, Maizuru Marine Observatory, 1967.
Report of the oceanographical observations in the Japan Sea from August to September, 1964. (In Japanese).
Bull. Maizuru mar. Obs., 10:74-86.

diatoms, lists of spp.
Japan, Maizuru Marine Observatory, 1967.
Report of the oceanographic observations in the Japan Sea from May to June, 1964. Bull. Maizuru mar. Obs., 10:65-76. (In Japanese)

diatoms, lists of spp.
Japan, Maizuru Marine Observatory, 1965.
Report of the oceanographic observations in the central part of the Japan Sea from February to March, 1962.---in the Japan Sea from June to July 1962.---in the western part of Wakasa Bay from January to April, 1962.---in the central part of the Japan Sea from September to October, 1962.---in the western part of Wakasa Bay from May to November, 1962.---in the central part of the Japan Sea in June, 1963.---in Wakasa Bay in July, 1963.---in the central part of the Japan Sea in October, 1963. (In Japanese).
Bull. Maizuru Mar. Obs., No.9:67-73;74-88;89-95; 71-80;81-87;59-65;66-77;80-84;85-91.

diatoms, lists of spp.
Japan, Maizuru Marine Observatory and Hakodate Marine Observatory, Oceanographical Sections, 1962.
Report of the oceanographic observations in the Japan Sea in June, 1961. (In Japanese).
Res. Mar. Meteorol. and Oceanogr. Obs., Jan.-June, 1961, No. 29:59-79.

diatoms, lists of spp.
Japan, Nagasaki Marine Observatory 1971.
Report of the oceanographic observations in the sea west of Japan from June to August, 1965. (In Japanese)
Oceanogr. Met. Nagasaki Mar. Obs. 18: 59-74

diatoms, lists of spp.
Japan, Nagasaki Marine Observatory 1971.
Report of the oceanographic observations in the sea west of Japan from January to February 1966. (In Japanese)
Oceanogr. Met. Nagasaki Mar. Obs. 18: 34-49.

diatoms, lists of spp.
Japan, Nagasaki Marine Observatory, Oceanographic Section, 1965.
Report of the oceanographic observations in the sea west of Japan from February to March 1963,--from July to August 1963.
Res. Mar. Meteorol. Oceanogr., Japan, Meteorol Agency, 33:39-58; 34:53-80.

Also in: Oceanogr. Meteorol., Nagasaki, 15 (227-228).

diatoms, lists of spp.
Japan, Nagasaki Marine Observatory, Oceanographical Section, 1962
Report of the oceanographic observations in the sea west of Japan from April to May, 1961 (In Japanese).
Res. Mar. Meteorol. and Oceanogr. Obs., Jan.-June, 1961, No. 29: 45-53.

diatoms, lists of spp.
Japan, Nagasaki Marine Observatory, Oceanographical Section, 1962
Report of the oceanographic observations in the sea west of Japan from February to March, 1961. (In Japanese).
Res. Mar. Meteorol. and Oceanogr. Obs., Jan.-June, 1961, No. 29:36-44.

Diatoms, lists of spp.
Japan, Nagasaki Marine Observatory, Oceanographical Section 1960.
Report of the oceanographic observation in the sea west of Japan from January to February, 1960. Report of the Oceanographic observation in the sea north-west of Kyushu from April to May, 1960.
Results Mar. Meteorol. & Oceanogr., JMA, 27:42-50; 51-67.

Also in:
Oceanogr. & Meteorol., Nagasaki Mar. Obs., (1961) 11 (202).

diatoms, lists of spp.
Japan, Nagasaki Marine Observatory, Oceanographical Section, 1960
Report of the oceanographic observations in the sea west of Japan from June to July, 1959.
Res. Mar. Meteorol. & Oceanogr., J.M.A., 26: 51-57.

Also in:
Oceanogr. & Meteorol., Nagasaki Mar. Obs., (1961), 11(200).

diatoms, lists of spp.
Jouset, A.P., 1957.
[Diatoms in the surface layer of the Okhotsk Sea sediments.] Trudy Inst. Okeanol., 22:164-200.

diatoms, lists of spp.
Kadota, Sadami, and Hitomi Hirose, 1967.
Studies on the planktonological and oceanographical conditions in the entrance of Tokyo Bay. 1. Plankton diatoms in early summer. (In Japanese; English abstract).
Inf. Bull. Planktol. Japan, Comm. No. Dr. Y. Matsue, 47-52.

diatoms, lists of spp.
Karayeva, N.I. and I.V. Makarova, 1973
Specific features and origin of the Caspian Sea diatom flora. Mar. Biol., 21(4):269-275.

diatoms, lists of spp.
Karohji, Kohei, 1959.
Report from the "Oshoro Maru" on oceanographic and biological investigations in the Bering Sea and northern North Pacific in the summer of 1956. IV. Diatom associations as observed by underway samplings.
Bull. Fac. Fish., Hokkaido Univ., 9(4):259-282.

diatoms, lists of spp.
Karohji, K., 1957.
Associations of plankton diatoms around Japan as investigated by underway samplings abour the "Oshoro-maru" in October and December 1952.
Bull. Fac. Fish., Hokkaido Univ., 7(4):271-283.

diatoms, lists of spp.
Kawarada, Yutaka, 1965.
Diatoms in the Kuroshio waters neighboring Japan. (In Japanese; English abstract).
Inf. Bull. Planktol., Japan, No. 12:8-16.

diatoms, lists of spp.
Kawarada, Yutaka, 1960.
A contribution of microplankton observations to the hydrography of the northern North Pacific and adjacent seas. III. Plankton diatoms of the western Okhotsk Sea in the period from June to August 1957.
Mem. Kobe Mar. Obs., 14:74-80.

diatoms, lists of spp.
Kawarada, Y., 1957.
A contribution of microplankton observations to the hydrography of the northern North Pacific and adjacent seas. II. Plankton diatoms in the Bering Sea in the summer of 1955.
J. Oceanogr. Soc., Japan, 13(4):151-156.

diatoms, lists of spp.
Kimball, J.F., Jr., Eugene F. Corcoran and E.J. Ferguson Wood, 1963.
Chlorophyll-containing microorganisms in the aphotic zone of the oceans.
Bull. Mar. Sci., Gulf and Caribbean, 13(4): 574-577.

diatoms, lists of spp.
Kolbe, R.W., 1955.
Diatoms from Equatorial Atlantic cores.
Repts. Swedish Deep-Sea Exped., Sediment Cores from the N. Atlantic Ocean, 7(3):151-184, 2 pls.

diatoms, lists of species
Kolbe, R.W., 1954.
Diatoms from Equatorial Pacific cores.
Repts. Swedish Deep-Sea Exped., 1947-48, Sediments Cores for the W. Pacific, 6(1):3-49, 4 pls.

diatoms, lists of spp.
Komarovsky, B., and Tikva Edelstein, 1960
Diatomeae and Cyanophyceae occurring on deep-water algae in the Haifa Bay Area.
Bull. Res. Counc., Israel, 9D(2): 73-92.

diatoms, lists of spp.
Kon, Hisanori, 1963.
On the distribution of diatoms south of Shionomisaki. (In Japanese; English abstract)
Bull. Kobe Mar. Observ., No. 171(12):1-6.
Also in Umi to Sora 39(1): 1-6.

diatoms, lists of spp.
Kozlova, O.G. 1970.
Biogenic components of suspended matter (diatoms, Silicoflagellates, Coccolithine, Peridinées). (In Russian)
Ch. 4 in: Sedimentation in the Pacific Ocean P.L. Bezrukov, editor. Akad. Nauk SSSR, Inst. Okeanol. P.P. Shirshov, Izdatel "Nauka" Moskva 1: 127-144.

diatoms, lists of spp.
Kozlova, P.G., and V.V. Mukhina, 1967.
Diatoms and Silicoflagellates in suspension and floor sediments of the Pacific Ocean.
Int. Geol. Rev., 9(10):1322-1342.
(Translated from: Geokhimiya Kremnezema, NK 65-12 (51):192-218 (1966)).

diatoms, lists of spp.
Kucherova, Z.S., 1957.
Aspects of the composition and seasonal cycle of the blooming of marine diatoms.
Trudy Sevastopol. Biol. Stan., 9:28-29.

diatoms, lists of genera
Kurasige, Hidejiro, 1943.
Quantitative and qualitative characteristics of the marine diatoms in the coastal waters of Taiyato at the Yellow Sea side of Tyosen, in comparison with that of Tataho Bay at the southern coast of the peninsula.
Bull. Fish. Exper. Sta., Gov. Gen. Tyosen, (8):1-114.

diatoms, lists of spp.
Kuroda, Kazunori 1972
Surface distribution of pelagic diatoms before and after Typhoon 6734.
Bull. Kobe mar. Obs. 185-: 39-49.
(In Japanese; English abstract)

Diatoms, lists of species
Kuroda, K., 1969.
Short-term variations of the surface diatoms in the Kuroshio. Oceanogr. Mag., 21(2): 97-111.

diatoms, lists of spp.
Kusunoki, K., T. Minoda, K. Fujino and A. Kawamura 1967.
Data from oceanographic observations at Drift Station ARLIS I in 1964-1965.
Arctic Inst. N.A. unnumbered pp. (duplicated)

diatoms, lists of spp.
Lackey, James B., and Elsie W. Lackey, 1963
Microscopic algae and protozoa in the waters near Plymouth in August, 1962.
J. Mar. Biol. Assoc., U.K., 43(3):797-805.

diatoms, lists of spp.
Lapshina, V.I. 1971.
Characteristic plankton of the north tropical and equatorial zones in the eastern Pacific Ocean. (In Russian).
Izv. Tichookean. nauchno.-issled. Inst. ribn. Choz. Okean. 79: 100-126.
(TINRO)

diatoms, lists of spp.
Lecal, J. 1967.
Le nannoplancton des côtes d'Israel.
Hydrobiologia, 29(3/4):305-387

diatoms, lists of spp.
Magazzù, Giuseppe, e Carlo Andreoli 1971.
Trasferimenti fitoplanctonici attraverso lo Stretto di Messina in relazione alle condizioni idrologiche.
Boll. Pesca Piscic. Idrobiol. 26(1/2): 125-193

diatoms, lists of species
Manguin, E., 1956.
Les diatomées de l'estuaire de la Rance.
Bull. Lab. Marit. Dinard, (42):62-76.

diatoms, lists of spp.
Margalef, Ramón 1973.
Fitoplancton marino de la región de afloramiento del NW de África. II. Composición y distribución del fitoplancton (Campaña Sahara I del Cornide de Saavedra).
Res. Exped. cient. Cornide de Saavedra Madrid 2: 65-94.

diatoms, lists of spp.
Margalef, Ramón, 1964.
Fitoplancton de las costas de Blanes (provincia de Gerona, Mediterráneo Occidental), de julio de 1959 a junio de 1963.
Inv. Pesq., Barcelona, 26:49-90. 131-164.

diatoms, lists of spp.
Margalef, Ramón, y Juan Herrera, 1964.
Hidrografía y fitoplancton de la costa comprendida entre Castellón y la desembocadura del Ebro, de julio de 1961 a julio de 1962.
Inv. Pesq., Barcelona, 26:49-90.

diatoms, lists of spp.
Markina, N.P. 1971.
(COCTAB) Composition and distribution of plankton along the west and southern coasts of Australia in October-January 1962-63. (In Russian).
Izv. Tichookean. nauchno.-issled. Inst. ribn. Choz. Okean. (TINRO) 79:127-140.

diatoms, lists of spp.
Marumo, Ryuzo, 1967.
General features of diatom communities in the North Pacific Ocean in summer.
Inf. Bull. Planktol. Japan, Comm. No. Dr. Y. Matsue, 115-122.

diatoms, lists of spp.
Marumo, R., 1957.
The surface distribution of plankton diatoms in the western part of the Pacific Ocean and Antarctic Ocean in 1954 and 1955.
Ocean. Mag., Tokyo, 9(1):143-147.

diatoms, lists of spp.
Marumo, R., 1956.
Diatom communities in Bering Sea and its neighboring waters in the summer of 1954.
Ocean. Mag., Tokyo, 8(1):69-73.

diatoms, lists of species
Marumo, R., 1955.
Marine diatoms of the east coast of northern Japan in the winter from 1953 to 1954.
Ocean. Mag., Tokyo, 6(4):165-178, 8 textfigs.

diatoms, lists of species
Marumo, R., 1954.
Diatom plankton in the south of Cape Shionomisaki in 1953.
Ocean. Mag., Tokyo, 6(3):145-152, 7 textfigs.

diatoms, lists of species
Marumo, R., 1953.
On the diatom plankton of the Antarctic Ocean in the summers of 1946 to 1952. Rec. Ocean. Wks., Japan, n.s., 1(2):55-59.

diatoms, lists of spp.
McIntire, D. David and W. Scott Overton, 1971.
Distributional patterns in assemblages of attached diatoms from Yaquina Estuary, Oregon.
Ecology 52(5): 758-777.

diatoms, lists of spp.
Meguro, Hiroshi, Kuniyuki Ito and Hiroshi Fukushima, 1966.
Diatoms and the ecological conditions of their growth in sea ice in the Arctic Ocean.
Science, 152(3725):1089-1090.

diatoms, lists of spp.
Menzel, David W., and Jane P. Spaeth, 1962
Occurrence of Vitamin B12 in the Sargasso Sea.
Limnology and Oceanography, 7(2):151-154.

diatoms, lists of spp.
Mizushima, Takeshi, 1970.
Abundance and distribution of plankton in the waters around the Shimizu Harbour. II. Distribution of plankton diatoms in the winter 1964. (In Japanese; English abstract).
J. Coll. mar. Sci. Technol., Tokai Univ., 4:81-97

diatoms, lists of spp.
Mizushima, Takeshi 1972.
Abundance and distribution of plankton in the waters around the Shimizu Harbour. IV. Distribution of plankton diatoms in the summer 1966.
J. Coll. mar. Sci. Technol. Tokai Univ. 6: 105-122.

diatoms, lists of spp.
Mölder, Karl, 1962
Über die Diatomeenflora des Bottnischen Meerbusens und der Ostsee.
Havsforskningsinst. Skrift (Merent. Julk.), No. 203: 58 pp.

diatoms, lists of spp.
Mölder, Karl, 1943.
Studien über die Ökologie und Geologie der Bodendiatomeen in der Pojo-Bucht.
Annales Botanici Soc. Zool.-Bot. Fennicae Vanamo, 18(2):204 pp.

diatoms, lists of species (bottom forms, new species)
Møller, M., 1955.
Investigations of the geography and natural history of the Praestø Fiord, Zealand. 7. The diatoms of the Praestø Fiord. Folia Geogr. Danica 3(6): 187-237.

diatoms, lists of spp.
Moreira Filho, Hermes, 1965.
Contribuição ao estudo das diatomáceas da região de Cabo Frio (Estado do Rio de Janeiro - Brasil).
Anais Acad. bras. Cienc., 37(Supl.):231-238.

diatoms, lists of spp.
Moreira Filho, Hermes, 1960.
Diatomáceas no trato digestivo da Tegula viridula Gmelin (Chrysophyta - Bacillariophyceae).
Bol. Univ. Parana, Brasil, Botanica, No. 1:27 pp.

diatoms, lists of spp.
Moreira Filho, Hermes, Yassuko Maruo e Ita Moema V. Moreira, 1967.
Diatomáceas da enseada de Pôrto Belo (Estada de Santa Catarina, Brasil) (1).
Bol. Univ. fed., Paraná, Bot. 19:1-13.

diatoms, lists of spp.
Moreira Filho, Hermes, Ita Moema V. Moreira, Augusto Aldave Pajares y Irene I. M. Trippi, 1971.
Diatomáceas do Pôrto Salaverry (Provincia de Trujillo - Perú) (Chrysophyta - Bacillariophyceae).
Bolm Univ. Fed Parana, Brasil (Botânica) 26: 28 pp.

diatoms, lists of species
Motoda, S., and Y. Kawarada, 1955.
Diatom communities in western Aleutian waters on the basis of net samples collected in May-June 1953. Bull. Fac. Fish., Hokkaido Univ., 6(3): 191-200.

diatoms, lists of spp.
Moyse, John, 1963.
A comparison of the value of various flagellates and diatoms as food for barnacle larvae.
Journal du Conseil, 28(2):175-187.

diatoms, lists of spp.
Mulford, R.A. 1972
An annual plankton cycle on the Chesapeake Bay in the vicinity of Calvert Cliffs, Maryland, June 1969 - May 1970
Proc. Acad. nat. Sci Phila. 124(3):17-40.

diatoms, lists of spp.
Mulford, Richard A., 1962.
Diatoms from Virginia tidal waters, 1960 and 1961.
Virginia Inst. Mar. Sci., Gloucester Pt., Spec. Sci. Rept., No. 30:33 pp. (mimeographed).

diatoms, lists of spp.
Müller-Melchers, F.-C., 1959.
Plankton diatoms of the southern Atlantic Argentina and Uruguay coast.
Comunicaciones Botanicas del Museo Hist. Nat. Montevideo, III(38):1-45.

diatoms, lists of spp.
Müller-Melchers, F.C., 1957
Plankton diatoms of the "Toko-maru" voyage (Brazil Coast).
Bol. Inst. Ocean., Univ. Sao Paulo, 8(1/2): 111-126, 6 pls.

diatoms, lists of sp
Müller-Melchers, F.C., 1955.
Las diatomeas del plancton marino de las costas del Brasil. Bol. Inst. Oceanograf., Sao Paulo, 6(1/2):93-138.

diatoms, lists of spp.
Nel, E.A., 1968.
Diatoms in the aphotic zone of the southwest Indian Ocean. Fish. Bull. Misc. Contrib. Oceanogr. Fish. Biol. S. Afr., 5: 11-31.

diatoms, lists of spp
Ohwada, M., 1960.
Vertical distribution of living and dead diatoms down to one thousand meters off Sanreku, northern Japan.
Bull. Hakodate Mar. Obs., 7(2):1-
Mem. Kobe Mar. Obs., 14:69-73.

diatoms, lists of spp.
Ohwada, M., 1957.
Diatom communities in the Okhotsk Sea, principally on the west coast of Kamchatka, spring to summer, 1955. J. Ocean. Soc., Japan, 13(1):29-34.

diatoms, lists of spp.
Patten, B.C., R.A. Mulford and J.E. Warinner, 1963
An annual phytoplankton cycle in the lower Chesapeake Bay.
Chesapeake Science. 4(1):1-20.

diatoms, lists of spp
Plante-Cuny, Marie-Reine 1969.
Recherches sur la distribution qualitative et quantitative des diatomées benthiques de certains fonds meubles du Golfe de Marseille.
Recl Trav. Stn. mar. Endoume 45(61): 87-197.

diatoms, lists of spp.
Plante, M.R., 1966.
A perçu sur les peuplements de diatomées benthiques de quelques substrats meubles du Golfe de Marseille.
Recl. Trav. Stn. mar. Endoume, 40(56):83-101.

diatoms, lists of spp.
Politis, John, 1960
Sea diatoms of Greece. Praktika Hellenic Hydrobiol. Inst., 7(1):20 pp.

diatoms, lists of spp. [a]
Portugal Instituto Hidrográfico 1973
CAPEC-II Janeiro/Fevereiro -1971: resultados preliminares 9: 149 pp.

diatoms, lists of spp. [a]
Portugal Instituto Hidrográfico 1973
Companha oceanográfica para Apoio às Pescas do continente. CAPECI (12 Outubro a 15 de Novembro de 1970. Resultados Preliminares, 6:123pp.

diatoms, lists of spp. [a]
Pucher-Petković, Tereza, 1969.
Oceanographic conditions in the middle Adriatic area. IV. Seasonal changes and spatial distribution of diatomaceous populations. (Jugoslavian and Italian abstracts).
Thalassia Jugoslavica, 5: 267-275.

diatoms, lists of spp.
Pucher-Petković, Tereza 1966
Vegetation des diatomées pelagiques de l'Adriatique moyenne.
Acta adriat. 13 (1): 1-97.

diatoms, lists of spp.
Raghu Prasad, R., and P.V. Ramachandran Nair, 1960
Observations on the distribution and occurrence of diatoms in the inshore waters of the Gulf of Mannar and Palk Bay.
Indian J. Fish., 7(1):49-68.

diatoms, lists of species
Ramachandran Nair, P.V., 1959
New records of marine planktonic diatoms from the west coast of India. J. Mar. Biol. Asso., India 1(1): 96.

diatoms, lists of spp. [a]
Reyssac, Josette, 1972.
Phytoplankton récolté par le navire Ombango au large d'Angola (10-27 novembre 1968)
Bull. Inst. fond. Afr. Noire (A) 34(4): 796-808.

diatoms, lists of spp. [a]
Reyssac, Josette, 1972.
Premières observations sur le cycle annuel des diatomées et dinoflagellés dans la baie du Lévrier (Mauritanie).
Bull. Inst. fond. Afrique Noire 34(2):278-291.

diatoms, lists of spp
Reyssac, Josette, 1968.
Contribution a la connaissance des Diatomées des Baies de Nhatrang et de Cauda (Annam).
Revue algolog. n.s. 9(2):135-151.

diatoms, lists of spp.
Ricard, M., 1970.
Premier inventaire des diatomées et des dinoflagellés du plancton côtier de Tahiti.
Cah. Pacifique 14: 245-254.

diatoms, lists of spp.
Riley, Gordon A., and Shirley M. Conover 1967.
Phytoplankton of Long Island Sound, 1954-1955.
Bull. Bingham oceanogr. Coll. 19(2): 5-34.

diatoms, lists of spp.
Rivera R., Patricio, 1968.
Sinopsis de las diatomeas de la Bahia de Concepcion, Chile.
Gayana, Univ. Concepcion, Chile, 18: 111 pp.

diatoms, lists of spp. [a]
Round, F.E., 1968.
The phytoplankton of the Gulf of California. II. The distribution of phytoplanktonic diatoms in cores. J. exp. mar. Biol. Ecol. 2(1): 64-86.

diatoms, lists of species
Saisyo, T., 1955.
[Seasonal change of plankton diatoms in Kagoshima Bay.] Mem. Fac. Fish., Kagoshima Univ., 4:113-118.

diatoms, lists of spp. [a]
Saunders, Richard P., Bruce I. Birnhak, Joanne T. Davis and Carol L. Wahlquist, 1967.
Seasonal distribution of diatoms in Florida inshore waters from Tampa Bay to Caxambas Pass, 1963-1964. Prof. Pap. Ser. Fla. Bd. Conserv. 8: 48-78.

diatoms, lists of spp. [a]
Saunders, Richard P., and Donald A. Glenn 1969.
Diatoms.
Mem. Hourglass Cruises, Mar. Res. Lab. Fla. Dept. Nat. Res. 1(3):1-119

diatoms, lists of spp
Saunders, Richard P., and Carol L. Wahlquist, 1966.
Diatoms and their relationship to Gymnodinium breve Davis.
Florida Bd., Conserv., St. Petersburg. Mar. Lab. Prof. Papers Ser., No. 8:16-33.

diatoms, lists of occurrences
Simonsen, Reimer, 1962
Untersuchungen zur Systematik und Ökologie der Bodendiatomeen der westlichen Ostsee.
Inter. Rev. Gesamt. Hydrobiol., System. Beihefte 1:145 pp.

Bacillariophyta, list of species
Skolka, V.H. 1961
Données sur le phytoplancton des parages prébosphoriques de la Mer Noire.
Rapp. Proc. Verb. Réunions. Comm. Int. Expl. Sci. Mer. Méditerranée, Monaco, 16(2):129-132.

diatoms, lists of spp.

Semina, H. I., and A. P. Jouse, 1959.
[Diatom algae in the biocoenosis and thanacoenosis of the western part of the Bering Sea.] Trudy Inst. Okeanol. 30:52-67.

diatoms, lists of spp

Smayda, Theodore J., 1966.
A quantitative analysis of the phytoplankton of the Gulf of Panama. III General ecological conditions and the phytoplankton dynamics at 8o 45'N, 79o 23'W from November 1954 to May 1957.
Inter-Amer. Trop. Tuna Comm., Bull., 11(5): 355-612.

diatoms, lists of spp.

Smayda, Theodore J., 1965.
A quantitative analysis of the phytoplankton of the Gulf of Panama. II. On the relationship between C14 assimilation and the diatom standing crop.
Inter-American Tropical Tuna Commission, Bull., 9(7):467-531.

diatoms (list of)

Smyth, J.C., 1955.
A study of the benthic diatoms of Loch Sween (Argyll). J. Ecol. 43(1):149-171.

diatoms, lists of spp.

Sournia, A., 1970.
A checklist of planktonic diatoms and dinoflagellates from the Mozambique Channel.
Bull. mar. Sci., 20(3): 678-696.

diatoms, lists of spp.

Sournia, A. 1968.
Quelques nouvelles données sur le phytoplancton marin et la production primaire à Tuléar (Madagascar).
Hydrobiologia 31 (3/4): 545-560.

diatoms, lists of spp.

Stroukuna, V.G., 1950.
[Phytoplankton of the Black Sea in the vicinity of Karadaga and its seasonal dynamics.]
Trudy Karadagsk Biol. Sta., 10:38-82.

diatoms, lists of spp.

Takano, Hideaki, 1967.
Notes on marine littoral diatoms from Japan. III. Diatoms from Atashiri, Hokkaido.
Bull. Tokai Reg. Fish. Res. Lab., 49:1-9.

diatoms, lists of spp

Takano, Hideaki, 1964.
Notes on marine littoral diatoms from Japan II.
Bull. Tokai Reg. Fish. Res. Lab., No. 39:13-20.

Diatoms, (epiphytic), lists of spp.

Takano, Hideaki, 1961.
Epiphytic diatoms upon Japanese agar sea-weeds.
Bull. Tokai Reg. Fish. Res. Lab., No. 31:269-274.

diatoms, lists of spp.

Takano, H., 1960
Plankton diatoms in the eastern Caribbean Sea.
J. Oceanogr. Soc., Japan, 16(4): 180-184.

diatoms, lists of spp

Takano, H., 1959
Plankton diatoms western Aleutian waters in the summer, 1953.
Bull. Tokai Reg. Fish. Res. Lab., No. 23: 1-11. (In English).

diatoms, lists of species

Takano, H., 1955.
Plankton diatoms collected off Bozo District in August 1951. Bull. Jap. Soc. Sci. Fish., 21(2): 55-61.

diatoms (lists of species)

Takano, H., 1954.
Preliminary report on the marine diatoms from Hachijo Island, Japan. Bull. Jap. Soc. Sci. Fish. 19(12):1189-1196, 4 textfigs.

diatoms, lists of spp.

Taylor, F.J., 1970.
A preliminary annotated check list of diatoms from New Zealand coastal waters.
Trans. R. Soc. N.Z., Biol. Sci, 12(4): 153-174.

diatoms, lists of spp.

Tratet, Gérard, 1964.
Variations du phytoplancton à Tanger.
Trav. Inst. Sci., Cherifien, Rabat, Ser. Botan., (29):204 pp.

diatoms, lists of spp.

Travers, A., 1962.
Recherches sur le phytoplancton du Golfe de Marseille. 1. Étude qualitative des Diatomées et des Dinoflagellés du Golfe de Marseille.
Rec. Trav. Sta. Mar., Endoumes, Bull., 26(41):7-69.

diatoms, lists of spp.

Uhm, Kyu Baek, and Kwang Il Yoo, 1967.
Diatoms in the Korea Strait.
Repts Inst. mar. Biol., Seoul Nat. Univ., 1(5): 1-6.

diatoms, lists of spp.

Ünsal, İsmail, 1970.
Quelques espèces de diatomées du Golfe de Mersin.
Hidrobiol., Arastirma Enst. Yayinlarindan, Istanbul Univ. Fen Fakült., 6(3/4): 10-17.

diatoms, lists of spp.

Ünsal, İsmail, 1970.
Quelques espèces de diatomées du Golfe d'Iskenderun.
Hidrobiol., Arastirma Enst. Yayinlarindan, Istanbul Univ. Fen Fakült., 6(3/4): 18-26.

diatoms, lists of spp.

Uyeno, F., 1957.
[On the relations between the distribution of diatoms and the Kuroshio in the sea south off Shionomisaki.] Umi to Sora 33(1/2):1-10.

diatoms, lists of species

Van der Werff, A., 1954.
Diatoms in plankton samples of the William Barendsz-Expedition 1947. Hydrobiol., 6(3/4): 331-332.

diatoms, lists of spp.

Venkataraman, G.S., 1958.
A contribution to the knowledge of the Diatomaceae of Kanya Kumari (Cape Comorin), India, II. Proc. Nat. Inst. Sci., India, (B), 24(6):307-313.

diatoms, lists of spp

Venkataraman, G.S., 1957.
A contribution to the knowledge of the Diatomaceae of Kenya Kumari (Cape Comorin) India, 1.
Proc. Nat. Inst. Sci., India, 23(3/4):80-88.

diatoms, lists of spp. [a]

Venrick, E.L. 1972.
Small-scale distributions of oceanic diatoms.
Fish. Bull. U.S. nat. mar. Fish. Serv. NOAA 70(2): 363-372

diatoms, lists of spp.

Wawrik, F., 1961.
Die horizontale Verteilung der Planktondiatomeen im Golf von Neapel.
Int. Revue Ges. Hydrobiol., 46(3):460-479.

diatoms - lists of spp

Wawrik, F., 1957.
Interessante Florenelemente in der pelagischen Diatomeenvegetation des Golfes von Neapel.
Pubbl. Staz. Zool., Napoli, 30:269-278.

Diatoms, lists of spp.

Wood, E.J. Ferguson, 1965.
Protoplankton of the Benguela-Guinea current region.
Bull. Mar. Sci., 15(2):475-479.

diatoms, lists of spp.

Wood, E.J. Ferguson, 1963.
Studies on Australian and New Zealand diatoms. VI. Tropical and subtropical species.
Trans. R. Soc., New Zealand, 2(15):189-218.

diatoms, lists of spp.

Woods, E.J., L.H. Crosby and Vivienne Cassie, 1959.
Studies on Australian and New Zealand diatoms. III Descriptions of further discoid species. Trans. R. Soc., N Z., 87(3/4):211-219, pls. 15-17.

diatoms, lists of spp

Zhuze, A.P., 1960
[Diatoms in the surface layer of the Bering Sea sediments.] Trudy Inst. Okeanol., 32:171-205.

diatoms, littoral

Castenholz, Richard W., 1961
The effect of grazing on marine littoral diatom populations.
Ecology, 42(4):783-794.

diatoms littoral (lists of spp

Aleem Anwar A., 1949.
Zonal distribution of littoral diatoms at Cullercoats Northumberland (England) and their relation to the plankton.
Pubbl. Staz. zool. Napoli, 37(2): 332-365.

diatoms, "micro" distribution [a]

Venrick, E.L. 1972.
Small-scale distributions of oceanic diatoms.
Fish. Bull. U.S. nat. mar. Fish. Serv. NOAA 70(2): 363-372

diatoms, motility of

Lewin, Joyce C., and Robert R.L. Guillard, 1963.
Diatoms.
Annual Review of Microbiology, 17:373-414.

diatoms, movement of

Zauer, L.M., 1950.
[The movement of the diatom Eunotia lunaris (Ehr.) Grun. in connection with the question of the movement of diatoms generally.]
Doklady Akad. Nauk, SSSR, 72(6):1131-

diatom periodicity

Jørgensen, E.G., 1957.
Diatom periodicity and silicon assimilation, experimental and ecological investigations.
Dansk. Bot. Arkiv., 18(1):11-54.
Abstr. in Biol. Abstr., 32(2):3947.

diatoms, photosynthesis of

Taylor, W.R., 1964.
Light and photosynthesis in intertidal benthic diatoms.
Helgoländer Wiss. Meeresuntersuch., 10(1/4):29-37.

diatoms, phylogeny

Jurilj, A., 1956.
La phylogenese specifique d'un groupe de diatomees - Campylodiscoideae - et sa cause.
Hydrobiologia, Acta Hydrobiol., Hydrograph., et Protistol., 8(1/2):1-15.

diatom productivity

Uyeno, F., 1958.
[On the several problems concerning to the relation between the diatom quantity and the environmental factors - mainly about the relation between the sea conditions and the production] Umi to Sora, 34(1):1-10.

diatoms, quantitative
Hirote, Reiichiro, and Takuo Endo, 1965.
On primary production in the Seto Inland Sea.
II. Primary production and plankton, (In Japanese; English Abstract).
J. Fac. Fish., Animal Husbandry, Hiroshima Univ. 6(1):101-132.

diatoms (cells/l.)
Japan, Hakodate Marine Observatory, 1967.
Report of the oceanographic observations in the Okhotsk Sea and east of the Kurile islands and Hokkaido from October to November 1964. (In Japanese).
Bull. Hakodate mar. Obs., 13: 20-28.

diatoms (cells/l.)
Japan, Hakodate Marine Observatory, 1967.
Report of the oceanographic observations in the Okhotsk Sea, east of the Kurile islands and Hokkaido and east of the Tohoku district from August to September, 1964. (In Japanese).
Bull. Hakodate mar. Obs., 13:7-19.

diatoms (cells/l.)
Japan, Hakodate Marine Observatory, 1967.
Report of the oceanographic observations in the sea east of Hokkaido and in the southern part of the Okhotsk Sea from May to June, 1964. (In Japanese).
Bull. Hakodate mar. Obs., 13:10-17.

diatoms (cells/l.)
Japan, Hakodate Marine Observatory, 1967.
Report of the oceanographic observations in the sea southeast of Hokkaido in February, 1964. (In Japanese).
Bull. Hakodate mar. Obs., 13:3-9.

diatoms, quantitative
Japan, Hakodate Marine Observatory, 1964.
Report of the oceanographic observations in the sea east of the Tohoku District from February to March 1962. Report of the oceanographic observations in the Tsugaru Straits in April 1962.----in May 1962.----in June, 1962. from August to September 1962. Report of the oceanographic observations in the sea south of Hokkaido in June 1962. Report of the oceanographic observations in the sea west of Tsugaru Straits, the Tsugaru Straits and South of Hokkaido in May, 1962. (In Japanese).
Bull. Hakodate Mar. Obs.,
No. 11: misnumbered pp.

diatoms, quantitative
Japan, Hakodate Marine Observatory, Oceanographical Section, 1962
Report of the oceanographic observations in the sea east of Tohoku District from February to March, 1961. (In Japanese)
Res. Mar. Meteorol., and Oceanogr. Obs., Jan.-June, 1961, No. 29: 3-8.

diatoms, quantitative
Japan, Hakodate Marine Observatory, 1961.
Report of the oceanographic observations in the sea east of Tohoku District and in the Okhotsk Sea from August to September 1959.
Bull. Hakodate Mar. Obs., (8):6-17.

diatoms, numbers
Japan, Japan Meteorological Agency. 1965.
The results of marine meteorological and oceanographical observations, July-December 1963, No. 34: 360 pp.

diatoms, quantitative
Japan, Japan Meteorological Agency, 1964.
The results of marine meteorological and oceanographical observations, January-June 1963, No. 33:289 pp.

diatoms, quantitative
Japan, Japan Meteorological Agency, Oceanographical Section, 1962
Report of the oceanographic observations in the sea east of Honshu from February to March, 1961. (In Japanese).
Res. Mar. Meteorol. and Oceanogr. Obs., Jan.-June, 1961, No. 29: 13-21.

diatoms, quantitative
Japan, Kobe Marine Observatory, Oceanographical Section, 1963
Report of the oceanographic observations in the sea south of Honshu from February to March, 1962. (In Japanese).
Res. Mar. Met. & Ocean., J.M.A., 31:37-44.
Also in:
Bull. Kobe Mar. Obs., No. 173(4):1964.

diatoms quantitative
Japan, Kobe Marine Observatory, Oceanographical Section, 1962
Report of the oceanographic observations in the cold water region off Kii Peninsula from October to November 1961. (In Japanese).
Res. Mar. Meteorol. & Oceanogr., No. 30: 49-55.
Also in:
Bull. Kobe Mar. Obs., No. 173(3). 1964.

diatoms, quantitative
Japan, Kobe Marine Observatory, Oceanographical Observatory, 1962
Report of the oceanographic observations in the cold water region off Enshu Nada in May, 1961. (In Japanese).
Res. Mar. Meteorol. and Oceanogr. Obs., Jan.-June, 1961, No. 29: 28-35.

diatoms, quantitative
Japan, Kobe Marine Observatory, Oceanographical Section, 1962
Report of the oceanographic observations in the sea south of Honshu from February to March, 1961. (In Japanese).
Res. Mar. Meteorol. and Oceanogr. Obs., Jan.-June, 1961, No. 29:22-27.

diatoms (quantitative)
Japan, Maizuru Marine Observatory, 1963
Report of the oceanographic observations in the central part of the Japan Sea in October, 1961. (In Japanese).
Bull. Maizuru Mar. Obs., No. 8:78-88.

diatoms (quantitative)
Japan, Maizuru Marine Observatory, 1963
Report of the oceanographic observations in the central part of the Japan Sea in February 1961. (In Japanese).
Bull. Maizuru Mar. Obs., No. 8:54-58.

diatoms (quantitative)
Japan, Maizuru Marine Observatory, 1963
Report of the oceanographic observations in the central part of the Japan Sea in September 1960. (In Japanese).
Bull. Maizuru Mar. Obs., No. 8:60-68.

diatoms (quantitative)
Japan, Maizuru Marine Observatory, 1963
Report of the oceanographic observations in the Japan Sea in June 1961. (In Japanese).
Bull. Maizuru Mar. Obs., No. 8:59-79.

diatoms (quantitative)
Japan, Maizuru Marine Observatory, 1963
Report of the oceanographic observations in the Japan Sea from May to June, 1960. (In Japanese).
Bull. Maizuru Mar. Obs., No. 8:56-67.

diatoms (quantitative)
Japan, Maizuru Marine Observatory, 1963
Report of the oceanographic observations in the Wakasa Bay from July to August, 1960. (In Japanese).
Bull. Maizuru Mar. Obs., No. 8: 69-75.

diatoms (quantitative)
Japan, Maizuru Marine Observatory, 1963
Report of the oceanographic observations in the western part of Wakasa Bay from January to May, 1961. (In Japanese).
Bull. Maizuru Mar. Obs., No. 8:80-90.

diatoms (quantitative)
Japan, Maizuru Marine Observatory, 1963
Report of the oceanographic observations in the western part of Wakasa Bay from August to December 1961. (In Japanese).
Bull. Maizuru Mar. Obs., No. 8:96-102.

diatoms, (quantitative)
Japan, Maizuru Marine Observatory, 1963
Report of the oceanographic observations in the western part of Wakasa-Bay from January to June 1960. (In Japanese).
Bull. Maizuru Mar. Obs., No. 8:68-79, 59.

diatoms (quantitative)
Japan, Maizuru Marine Observatory, 1963
Report of the oceanographic observations in the western part of Wakasa Bay from June to December 1960. (In Japanese).
Bull. Maizuru Mar. Obs., No. 8:76-83.

diatoms (quantitative)
Japan, Maizuru Marine Observatory, 1963
Report of the oceanographic observations in the Wakasa Bay in August, 1961. (In Japanese).
Bull. Maizuru Mar. Obs., No. 8:89-95.

diatoms, quantitative
Japan, Maizuru Marine Observatory, Oceanographical Section, 1962
Report of the oceanographic observations in the central part of the Japan Sea in February 1961. (In Japanese).
Res. Mar. Meteorol. and Oceanogr. Obs., Jan.-June, 1961, No. 29:54-58.

diatoms, quantitative
Japan, Maizuru Marine Observatory and Hakodate Marine Observatory, Oceanographical Sections 1962.
Report of the oceanographic observations in the Japan Sea in June, 1961. (In Japanese).
Res. Mar. Meteorol. and Oceanogr. Obs., Jan.-June, 1961, No. 29:59-79.

diatoms, quantitative
Japan, Maizuru Marine Observatory, Oceanographical Section, 1962
Report of the oceanographic observations in the western part of Wakasa Bay from June to May, 1961. (In Japanese).
Res. Mar. Meteorol. and Oceanogr. Obs., Jan.-June, 1961, No. 29:80-90.

diatoms-numbers quantitative
Japan, Maizuru Marine Observatory, 1956
Report of the oceanographic observations off Kyoga-misaki during the latter half of 1955.
Bull. Maizuru Mar. Obs., (5): 31-37.

diatoms, numbers
Japan, Maizuru Marine Observatory, 1956
Report of the serial observations off Kyoga-misaki during the first half of 1955.
Bull. Maizuru Mar. Obs., (5): 27-32.

diatoms, quantitative
Kozlova, O.G., 1961.
The quantitative content of diatoms in the waters of the Indian Ocean sector of Antarctica.
Doklady Akad. Nauk, SSSR, 138(1):207-210.
OTS-61-11147-17   JPRS:8710:8-9

diatom numbers
Riley, Gordon A., and Shirley M. Conover 1967.
Phytoplankton of Long Island Sound, 1954-1955.
Bull. Bingham oceanogr. Coll. 19(2): 5-34.

diatoms, quantitative (data only)
Krauel, David P., 1969
Bedford Basin data report, 1967.
Techn. Rept. Fish. Res. Bd., Can., 120:84 pp (multilithed).

diatoms, seasonal distribution
Kucherova, Z.S., 1961.
Vertical distribution of diatoms from Sevastopol Bay.
Trudy Sevastopol Biol. Sta., (14):64-78.

diatoms, seasonal variation
Kwan-Kwoh, Li, and Hwang Shih-mei, 1956.
Seasonal variation of planktonic diatoms at Tsingtao. Acta Sci. Naturalum, 2(4):119-143.

diatoms, variation of
Proshkina-Lavrenko, A.I., 1961.
Variability of some Black Sea diatoms.
Botan. Zhurn., Akad. Nauk, SSSR, 46(12):1794-1897.

diatoms, silica content

Paasche, E., 1973
The influence of cell size on growth rate, silica content and some other properties of four marine diatom species. Norwegian J. Bot. 20(2/3): 199-204.

diatoms, sinking of

Sakamoto, Ichitaro, 1964.
Falling model on the planktonic diatoms.
Rept., Fac. Fish., Pref. Univ. Mie, 5(1):33-49.

diatoms, size of

Wimpenny, R.S., 1966.
The size of diatoms. IV. The cell diameter in Rhizosolenia styliformis var. oceanica.
J. mar. biol. Assoc., U.K., 46(3):541-546.

diatoms, standing crop

Kawarada, Y., M. Kitou, K. Furuhashi and A. Sano, 1969.
Distribution of plankton in the waters neighboring Japan in 1966 (CSK).
Oceanogr. Mag. 20(2): 187-212.

diatoms, standing crop

Kawarada, Yutaka, Masataka Kitou, Kenzo Furuhashi, Akira Sano, Kohei Karohji, Kazunori Kuroda, Osamu Asaoka, Masao Matsuzaki, Mamoru Ohwada and Futomi Ogawa 1966.
Distribution of plankton collected on board the research vessels of J.M.A. in 1965 (C.S.K).
Oceanogr. Mag., Jap. Met. Soc. 18(1/2): 91-112.

Diatom, standing crop

Kuroda, K., 1969.
Short-term variations of the surface diatoms in the Kuroshio. Oceanogr. Mag., 21(2): 97-111.

diatoms, standing crop

Smayda, Theodore J., 1965.
A quantitative analysis of the phytoplankton of the Gulf of Panama. II. On the relationship between C14 assimilation and the diatom standing crop.
Inter-American Tropical Tuna Commission, Bull., 9(7):467-531.

diatoms, surf zone

Lewin, Joyce and Thomas Hruby, 1973
Blooms of surf-zone diatoms along the coast of the Olympic Peninsula, Washington. II.[a] A diel periodicity in buoyancy shown by the surf-zone diatom species, Chaetoceros armatum T. West. Estuarine coast. mar. Sci. 1(1): 101-105.

diatoms, surf zone

Lewin, J. and D. Mackas, 1972.
Blooms of surf-zone diatoms along the coast of the Olympic Peninsula, Washington. 1. Physiological investigations of Chaetoceros armatum and Asterionella socialis in laboratory cultures.
Mar. Biol. 16(2): 171-181.

diatoms, synonomies

Takana, H., 1957.
Synonymy of pelagic diatoms. Info. Bull., Plankton, Japan, No. 4:10-13.

diatoms, taxonomy

Lewin, Joyce C., and Robert R.L. Guillard, 1963.
Diatoms.
Annual Review of Microbiology, 17:373-414.

diatoms, malformations

Ko-bayashi, Tsuyako, 1963
Variations on some pennate diatoms from Antarctica. 1.
Japan. Antarctic Res. Exped., 1956-1962. Sci. Repts., (E), No. 18:1-20, 16 pls.

diatoms, variations

Ko-bayashi, Tsuyako, 1963
Variations on some pennate diatoms from Antarctica. 1.
Japan. Antarctic Res. Exped., 1956-1962. Sci. Repts., (E), No. 18:1-20, 16 pls.

diatoms, vertical distribution

Ohwada, Mamoru, 1972
Vertical distribution of diatoms in the Sea of Japan. In: Biological oceanography of the northern North Pacific Ocean, A.Y. Takenouti, Chief Editor, Idemitsu Shoten, Tokyo, 145-163.

diatoms, vertical migration

Palmer, John D., and Frank E. Round, 1967.
Persistent, vertical-migration rhythms in benthic microflora. VI. The tidal and diurnal nature of the rhythm in the diatom Hantzschia virgata.
Biol. Bull. mar. biol. Lab. Woods Hole, 132(1):44-55.

Achnanthes sp.

Allen, W.E., 1937
Plankton diatoms of the Gulf of California obtained by the G. Allan Hancock Expedition of 1936. The Hancock Pacific Expeditions, Univ. So. Calif. Publ. 3:47-59, 1 fig.

Achnanthes sp.

Braarud, T., and Adam Bursa, 1939
On the phytoplankton of the Oslo Fjord, 1933-1934. Hvalrådets Skr. No.19:1-63; 9 text figs. Reviewed. J. du. Cons. 14(3): 418-420. A.C. Gardiner.

Achnanthes sp.

Cupp, E.E. and Allen, W.E., 1938
Plankton diatoms of the Gulf of California obtained by Allan Hancock Pacific Expedition of 1937. The Hancock Pacific Expeditions, The Univ. So. Calif. Publ. 3: 61-74, 1 map, pls. 4-15.

Achnanthes

Desikachary, T.V., 1956(1957).
Electron microscope studies on diatoms.
J.R. Microsc. Soc. (3)76(1/2):9-36.

Achnanthes sp.

Jørgensen, E., 1905
B. Protistplankton and the diatoms in bottom samples. Hydrographical and biological investigations in Norwegian fjords. Bergens Mus. Skr. 7: 49-225.

Achnanthes sp.

Meunier, A., 1915
Microplancton de la Mer Flamande. 2. Diatomées (excepté le genre Chaetoceros). Mem. Mus. Roy. Hist. Nat., Belgique, 7(3):1-118, Pls. VIII-XIV.

Achnanthes

Politis, J., 1949
Diatomees marines de Bosphores et des ibes de la mer de Marmara. II Practica tou Hellenikou Hidrobiologikou Institutoutou 1929, Etoz 1929, 3(1):11-31.

Achnanthes acus n.sp.

Simonsen, Reimer, 1960.
Neue Diatomeen aus der Ostsee. II.
Kieler Meeresf., 16(1):126-130.

Achnanthes angustata

Hendy, N. Ingram, 1964
An introductory account of the smaller algae of British coastal waters. V. Bacillariophyceae (Diatoms).
Her Majesty's Stationary Office, 317 pp., 45 pls.

Achnanthes angustata

Hendey, N.I., 1951
Littoral diatoms of Chicester Harbour with special reference to fouling. J. Roy. Microsc. Soc. 71(1): 1-86, 18 pls.

Achnanthes Bonae Aurae n.sp.

Zanon, D.V., 1949
Diatomee di Buenos Aires (Argentina)
Atti Accad. Naz. Lincei, Memorie, Cl. Sci. Fis., mat. e. nat., ser. 7, 11(3):59-151, 2 pls.

Achnanthes brevipes

Brunel, J., 1962
Le phytoplancton de la Baie de Chaleurs.
Inst. Botan., Univ. Montréal, Contrib. No. 77: 365 pp., 66 pls.

Achnanthes brevipes

Castenholz, Richard W., 1963.
An experimental study of the vertical distribution of littoral marine diatoms.
Limnology and Oceanography, 8(4):450-462.

Achnanthes brevipes

Hendy, N. Ingram, 1964
An introductory account of the smaller algae of British coastal waters. V. Bacillariophyceae (Diatoms).
Her Majesty's Stationary Office, 317 pp., 45 pls.

Achnanthes brevipes

Hendey, N.I., 1951
Littoral diatoms of Chicester Harbour with special reference to fouling. J. Roy. Microsc. Soc. 71(1): 1-86, 18 pls.

Achnanthes brevipes

Politis, J., 1949
Diatomees marines de Bosphores et des ibes de la mer de Marmara. II Practica tou Hellenikou Hidrobiologikou Institutoutou 1929, Etoz 1929, 3(1):11-31.

Achnanthes brevipes

Rampi, L., 1940
Diatomee del Mare Adriatico. Nuovo Giornale Botanico Italiano, n.s., 47:559-608.

Achnanthes brevipes

Rumkówna, A., 1948
List of the phytoplankton species occurring in the superficial water layers in the Gulf of Gdańsk. Bull. Lab. mar., Gdynia, No. 4: 139-141 with tables in back.

Achnanthes brevipes

Zanon, D.V., 1949
Diatomee di Buenos Aires (Argentina)
Atti Accad. Naz. Lincei, Memorie, Cl. Sci. Fis., mat. e. nat., ser. 7, 11(3):59-151, 2 pls.

Achnanthes brevipes

Zanon, V., 1948
Diatomee marini di Sardegna e Pugillo di Alghe Marine della stressa. Boll. Pesca, Piscitutura e Idrobiologia, Anno 24, ns. 3(2): 202-244, 27 figs. on 1 pl.

Achnanthes brevipes intermedia

Brunel, J., 1962
Le phytoplancton de la Baie de Chaleurs.
Inst. Botan., Univ. Montréal, Contrib. No. 77: 365 pp., 66 pls.

Achnanthes brevipes intermedia

Brunel, Jules, 1962
Le phytoplancton de la Baie des Chaleurs.
Contrib. Ministère de la Chasse et des Pêcheries, Province de Québec, No. 91: 365 pp.

Achnanthes brevipes intermedia

Iyengar, M.O.P. and G. Venkataraman, 1951.
The ecology and seasonal succession of the algae flora of the River Cooum at Madras with special reference to the Diatomaceae. J. Madras Univ. 21, Sect. B(1): 140-192, 1 pl of 4 figs., 11 text figs.

Achnanthes brevipes intermedia
Ko-bayashi, Tsuyako, 1963
Variations on some pennate diatoms from Antarctica. 1.
Japan. Antarctic. Res. Exped., 1956-1962, Sci. Repts., (E), No. 18:1-20, 16 pls.

Achnanthes brevipes
(var. brevipes, intermedia, seriata)
Takano, Hideaki, 1962
Notes on epiphytic diatoms upon sea-weeds from Japan.
J. Oceanogr. Soc., Japan, 18(1):29-33.

Achnanthes coarctata
Zanon, V., 1948
Diatomee marini di Sardegna e Pugillo di Alghe Marine della stressa. Boll. Pesca, Piscitutura e Idrobiologia, Anno 24, ns. 3(2): 202-244, 27 figs. on 1 pl.

Achnanthes crenulata
Tsumura, K., 1956.
Diatomoj el la cirkaufoso de la restajo de la kastele de Odawara. J. Yokohama Municipal Univ., (C-14) No. 47:23 pp.

+var arcuate var. nov.

Achnanthes curvirostrum
Hendey, N-Ingram, 1958 [1957(Publ. 1958)]
Marine diatoms from some West African Ports-J. R Microsc. Soc. (3) 77(1/2): 28-85.

Achnanthes danica
Politis, J., 1949
Diatomees marines de Bosphores et des Ibes de la mer de Marmara. II Practica tou Hellenikou Hidrobiologikou Institutoutou 1929, Etoz 1929, 3(1):11-31.

Achnanthes danica
Rampi, L., 1940
Diatomee del Mare Adriatico. Nuovo Giornale Botanico Italiano, n.s., 47:559-608.

Achnanthes dispar n.sp.
Mann, A., 1907
Report on the diatoms of the Albatross voyages in the Pacific Ocean, 1888-1904. Contrib. U. S. Nat. Herb. 10(5):221-419, Pls. XLIV-LIV.

Achnanthes exigua
Zanon, D. V., 1949
Diatomee di Buenos Aires (Argentina)
Atti Accad. Naz. Lincei, Memorie, Cl. Sci. fis., mat. e. nat., ser. 7, 11(3):59-151, 2 pls.

Achnanthes exigua
Zanon, V., 1948
Diatomee marini di Sardegna e Pugillo di Alghe Marine della stressa. Boll. Pesca, Piscitutura e Idrobiologia, Anno 24, ns. 3(2): 202-244, 27 figs. on 1 pl.

Achnanthes exiloides n.sp.
Cholnoky, B.J., 1963.
Beiträge zur Kenntnis des marinen Litorals von Südafrika.
Botanica Marina, 5(2/3):38-83.

Achnanthes fugeii
Carter, John R., 1963.
Some new diatoms from British waters.
Quekett Microscop. Club, 29(8):199-203.

Achnanthes glabrata n.sp.
Grunow, A., 1863
Ueber einige neue und ungenügend bekannte Arten und Gattungen von Diatomaceen. Verhandl. d. K.K. Zool. Bot. Gesellsch., Vienna, 13: 137-162, pl.4-5 (Pl. 13-14).

Achnanthes groenlandica
Zanon, V., 1948
Diatomee marini di Sardegna e Pugillo di Alghe Marine della stressa. Boll. Pesca, Piscitutura e Idrobiologia, Anno 24, ns. 3(2): 202-244, 27 figs. on 1 pl.

Achnanthes hauckiana
Hendy, N. Ingram, 1964
An introductory account of the smaller algae of British coastal waters. V. Bacillariophyceae (Diatoms).
Her Majesty's Stationary Office, 317 pp., 45 pls.

Achnanthes Hauckiana
Iyengar, M.O.P. and G.Venkataraman,1951.
The ecology and seasonal succession of the algae flora of the River Cooum at Madras with special reference to the Diatomaceae. J. Madras Univ. 21, Sect. B(1): 140-192, 1 pl of 4 figs., 11 text figs.

Achnanthes Hauckiana
Zanon, D. V., 1949
Diatomee di Buenos Aires (Argentina)
Atti Accad. Naz. Lincei, Memorie, Cl. Sci. fis., mat. e. nat., ser. 7, 11(3):59-151, 2 pls.

Achnanthes heteromorpha
Zanon, D. V., 1949
Diatomee di Buenos Aires (Argentina)
Atti Accad. Naz. Lincei, Memorie, Cl. Sci. fis., mat. e. nat., ser. 7, 11(3):59-151, 2 pls.

Achnanthes hungarica
Zanon, D. V., 1949
Diatomee di Buenos Aires (Argentina)
Atti Accad. Naz. Lincei, Memorie, Cl. Sci. fis., mat. e. nat., ser. 7, 11(3):59-151, 2 pls.

Achnanthes hungarica
Zanon, V., 1948
Diatomee marini di Sardegna e Pugillo di Alghe Marine della stressa. Boll. Pesca, Piscitutura e Idrobiologia, Anno 24, ns. 3(2): 202-244, 27 figs. on 1 pl.

Achnanthes inflata
Subrahmanyan, R., 1945.
On the cell division and mitosis in some South Indian diatoms. Proc. Indian Acad. Sci., Sect. B, 22(6):331-354, Pl. 39, 88 textfigs.

Achnanthes inflata
Zanon, D. V., 1949
Diatomee di Buenos Aires (Argentina)
Atti Accad. Naz. Lincei, Memorie, Cl. Sci. fis., mat. e. nat., ser. 7, 11(3):59-151, 2 pls.

Achnanthes inflata inflatissima n. var.
Zanon, D. V., 1949
Diatomee di Buenos Aires (Argentina)
Atti Accad. Naz. Lincei, Memorie, Cl. Sci. fis., mat. e. nat., ser. 7, 11(3):59-151, 2 pls.

Achnanthes kerguelensis n.sp.
Castracane degli Antelminelli, F., 1886
1. Report on the Diatomaceae collected by H.M.S. Challenger during the years 1873-1876. Rept. Sci. Results, H.M.S. Challenger, Botany Vol. II, 178 pp., 30 pls.

Achnanthes kerguelensis
Hendey, N.I., 1937
The plankton diatoms of the southern seas.
Discovery Repts. 16:151-364, pls.6-13.

Achnanthes kuwaitensis n.sp.
Hendey, N-Ingram, 1958 [1957(Publ. 1958)]
Marine diatoms from some West African Ports-J. R Microsc. Soc. (3) 77(1/2): 28-85.

Achnanthes lanceolata
Tsumura, K., 1956.
Diatomoj el la cirkaufoso de la restajo de la kastele de Odawara. J. Yokohama Municipal Univ., (C-14) No. 47:23 pp.

Achnanthes lanceolata
Zanon, D. V., 1949
Diatomee di Buenos Aires (Argentina)
Atti Accad. Naz. Lincei, Memorie, Cl. Sci. fis., mat. e. nat., ser. 7, 11(3):59-151, 2 pls.

Achnanthes lanceolata
Zanon, V., 1948
Diatomee marini di Sardegna e Pugillo di Alghe Marine della stressa. Boll. Pesca, Piscitutura e Idrobiologia, Anno 24, ns. 3(2): 202-244, 27 figs. on 1 pl.

Achnanthes lemmermanni var lineata n.var.
Salah, M.M., 1952(1953).
XII. Diatoms from Blakeney Point, Norfolk. New species and new records for Great Britain. J.R. Microsc. Soc., Ser. 3, 72(3):155-169, 3 pls.

Achnanthes Lilljeborgei
Hustedt, F. and A.A. Aleem,1951
Littoral diatoms from the Salstone near Plymouth. JMBA 30(1): 177.196.

Achnanthes longipes
Cupp, Easter E., 1943
Marine plankton diatoms of the west coast of North America. Bull. S.I.O. 5(1):1-238, 5 pls., 168 text figs.

Achnantes longipes
Ercegovic, A., 1936
Etudes qualitative et quantitatives du phytoplancton dans les eaux cotières de l'Adriatique oriental moyen au cours de l'année 1934. Acta Adriatica 1(9):1-126

Achnanthes longipes
Ghazzawi, F.M., 1939
Plankton of the Egyptian waters. A study of the Suez Canal Plankton. (A) Phytoplankton. Preliminary Report 83 pp. Notes and Memoires, Min. Commerce-Industry, Egypt, Hydrobiol. & Fish. 65 figs.

Achnanthes longipes
Hendy, N. Ingram, 1964
An introductory account of the smaller algae of British coastal waters. V. Bacillariophyceae (Diatoms).
Her Majesty's Stationary Office, 317 pp., 45 pls.

Achnanthes longipes
Hendey, N.I., 1951
Littoral diatoms of Chicester Harbour with special reference to fouling. J.Roy. Microsc. Soc. 71(1): 1-86, 18 pls.

Achnanthes longipes
Hustedt, F. and A.A. Aleem,1951
Littoral diatoms from the Salstone near Plymouth. JMBA 30(1): 177.196.

Achnanthes longipes
Morse, D.C., 1947
Some observations on seasonal variations in plankton population Patuxant River, Maryland 1943-1945. Bd. Nat. Res., Publ. No.65, Chesapeake Biol. Lab., 31, 3 figs.

Achnanthes longipes
Politis, J., 1949
Diatomees marines de Bosphores et des Ibes de la mer de Marmara. II Practica tou Hellenikou Hidrobiologikou Institutoutou 1929, Etoz 1929, 3(1):11-31.

Achnanthes longipes
Rampi, L., 1942
Ricerche sul fitoplancton del Mare Ligure 6. Le diatomee delle acque di Sanremo. Nuovo Giornale Botanico Italiano, N.S., 49:252-268.

Achnanthes longipes
Rampi, L., 1940
Diatomee del Mare Adriatico. Nuovo Giornale Botanico Italiano, n.s., 47:559-608.

Achnanthes longipes

Schodduyn, M., 1926
Observations faites dans la baie d'Ambleteuse (Pas de Calais). Bull. Inst. Ocean., Monaco, No. 482: 64 pp.

Achnanthes longipes (figs.)

Sousa e Silva, E., 1949
Diatomaceas e Dinoflagelados de Baia de Cascais. Portugaliae Acta Biol., Volume: Julio Henriques, Ser. B: 300-383, 9 pls, 2 fold-in tables.

Achnanthes longipes

Takano, Hideaki, 1962
Notes on epiphytic diatoms upon sea-weeds from Japan.
J. Oceanogr. Soc., Japan, 18(1):29-33.

Achnanthes longipes

Zanon, V., 1948
Diatomee marini di Sardegna e Pugillo di Alghe Marine della stressa. Boll. Pesca, Piscitutura e Idrobiologia, Anno 24, ns. 3(2): 202-244, 27 figs. on 1 pl.

Achnanthes Lorenziana

Rampi, L., 1940
Diatomee del Mare Adriatico. Nuovo Giornale Botanico Italiano, n.s., 47:559-608.

Achnanthes marginalis n.sp.

Hendey, N-Ingram, 1958 [1957(Publ. 1958)]
Marine diatoms from some West African Ports-J. R.Microsc. Soc. (3) 77(1/2): 28-85.

Achnanthes microcephala

Hendy, N. Ingram, 1964
An introductory account of the smaller algae of British coastal waters. V. Bacillariophyceae (Diatoms).
Her Majesty's Stationary Office, 317 pp., 45 pls.

Achnanthes microcephala

Hendey, N.I., 1951
Littoral diatoms of Chicester Harbour with special reference to fouling. J.Roy. Microscop. Soc. 71(1): 1-86, 18 pls.

Achnanthes minutissima

Zanon, D. V., 1949
Diatomee di Buenos Aires (Argentina) Atti Accad. Naz. Lincei, Memorie, Cl. Sci. fis., mat. e. nat., ser. 7, 11(3):59-151, 2 pls.

Achnanthes minutissima

Zanon, V., 1948
Diatomee marini di Sardegna e Pugillo di Alghe Marine della stressa. Boll. Pesca, Piscitutura e Idrobiologia, Anno 24, ns. 3(2): 202-244, 27 figs. on 1 pl.

Achnanthes parallela n.sp.

Castracane degli Antelminelli, F., 1886
1. Report on the Diatomaceae collected by H.M.S. Challenger during the years 1873-1876. Rept. Sci. Results, H.M.S. Challenger, Botany Vol. II, 178 pp., 30 pls.

Achnanthes parvula

Hendy, N. Ingram, 1964
An introductory account of the smaller algae of British coastal waters. V. Bacillariophyceae (Diatoms).
Her Majesty's Stationary Office, 317 pp., 45 pls.

Achnanthes pictii n.sp.

Carter, John R., 1963.
Some new diatoms from British waters. Quekett Microscop. Club, 29(8):199-203.

Achnanthes pseudogroenlandica

Hendy, N. Ingram, 1964
An introductory account of the smaller algae of British coastal waters. V. Bacillariophyceae (Diatoms).
Her Majesty's Stationary Office, 317 pp., 45 pls.

Achnanthes pseudobliqua n.sp.

Simonsen, Reimer, 1960.
Neue Diatomeen aus der Ostsee. II. Kieler Meeresf. 16(1):126-130.

Achnanthes Sapinii-Jaloustrei n.sp.

Manguin, E., 1957.
Premier inventaire des diatomées de la Terre Adélie Antarctique. Espèces nouvelles. Rev. Algologique, n.s., 3(3):111-134.

Achnanthes subhyalina n.sp.

Conger, Paul S., 1964.
A new species of marine pennate diatom from Honolulu Harbor.
Smithsonian Misc. Coll., (Publ. 4593), 146(7):1-5.

Achnanthes subsessilis

Hendy, N. Ingram, 1964
An introductory account of the smaller algae of British coastal waters. V. Bacillariophyceae (Diatoms).
Her Majesty's Stationary Office, 317 pp., 45 pls.

Achnanthes subsessiles

Hendey, N.I., 1951
Littoral diatoms of Chicester Harbour with special reference to fouling. J.Roy. Microscop. Soc. 71(1): 1-86, 18 pls.

Achnanthes subsessilis

Politis, J., 1949
Diatomees marines de Bosphores et des ibes de la mer de Marmara. II Practica tou Hellenikou Hidrobiologikou Institutoutou 1929, Etoz 1929, 3(1):11-31.

Achnanthes subsessilis

Rampi, L., 1940
Diatomee del Mare Adriatico. Nuovo Giornale Botanico Italiano, n.s., 47:559-608.

Achnanthes subsessilis

Schodduyn, M., 1926
Observations faites dans la baie d'Ambleteuse (Pas de Calais). Bull. Inst. Ocean., Monaco, No. 482: 64 pp.

Achnanthes taeniata

Gran, H.N., 1908
Diatomeen. Nordisches Plankton, Botanischer Teil pp. XIX. 1-XIX 146; 178 text figs.

Achnanthes taeniata

Gran, H.H., and T. Braerud, 1935
A quantitative study of the phytoplankton in the Bay of Fundy and the Gulf of Maine (including observations on hydrography, chemistry, and turbidity). J. Biol. Bd., Canada, 1(5):279-467, 69 text figs.

Achnanthes taeniata

Hendy, N. Ingram, 1964
An introductory account of the smaller algae of British coastal waters. V. Bacillariophyceae (Diatoms).
Her Majesty's Stationary Office, 317 pp., 45 pls.

Achnanthes taeniata

Iselin, C., 1930
A report on the coastal waters of Labrador based on explorations of the "Chance" during the summer of 1926. Proc. Am. Acad. Arts Sci., 66(1):1-37, 14 text figs.

Achnanthes taeniata

Levander, K.M., 1947
Plankton gesammelt in den Jahren 1899-1910 an den Küsten Finnlands. Finnländische Hydrographisch-Biologische Untersuchunger (aus dem Wasserbiologischen Laboratorin der Societas Scientiarum Fennica) No.11:40 pp., 6 diagrams, 13 pls., tables.

Achnanthes taeniata

Lillick, L.C., 1940
Phytoplankton and planktonic protozoa of the offshore waters of the Gulf of Maine. Pt.II. Qualitative Composition of the Planktonic Flora. Trans. Am. Phil. Soc., n.s., 31(3):193-237, 13 text figs.

Achnanthes taeniata

Lillick, L.C., 1938
Preliminary report of the phytoplankton of the Gulf of Maine. Am. Mid. Nat. 20(3):624-640, 1 text fig. 37 tables.

Achnanthes taeniata

Lillick, L.C., 1937
Seasonal studies of the phytoplankton off Woods Hole, Massachusetts. Biol. Bull. LXXIII (3):488-503, 3 text figs.

Achnanthes taeniata

Rothe, F., 1942.
Quantitativen Untersuchungen über die Planktonverteilung in der Östlichen Ostsee. Ber. Deutsch. Wiss. Komm. Meeresf., N.F., 10:291-368, 33 textfigs.

Achnanthes taeniata

Rothe, F., 1941
Quantitative Untersuchunger über die Plankton verteilung in der Östlichen Ostsee. Ber. Deut. Wiss. Komm. fur Meeresforschung. n.f. X(3):291-368, 33 text figs.

Achnanthes taeniata

Rumkówna, A., 1948
List of the phytoplankton species occurring in the superficial water layers in the Gulf of Gdańsk] Bull. Lab. mar., Gdynia, No. 4: 139-141 with tables in back.

Achnanthes taeniata

Stæmann-Nielsen, Einar, 1951
The marine vegetation of the Isefjord. A study on ecology and production. Medd. Komm. Danmarks Fiskeri-og Havundersøgelser. Ser. Plankton. 5(4); 114pp., 46 text figs.

Achnanthes vanhoeffeni

Boden, B.P., 1950
Some marine plankton diatoms from the west coast of South Africa. Trans. R.Soc. S. Africa. 32:321-434, 100 text figs.

Achnanthes Vicentii n.sp.

Manguin, E., 1957.
Premier inventaire des diatomées de la Terre Adélie Antarctique. Espèces nouvelles. Rev. Algologique, n.s., 3(3):111-134.

Achnanthidium hungaricum n.sp.

Grunow, A., 1863
Ueber einige neue und ungenügend bekannte Arten und Gattungen von Diatomaceen. Verhandl. d. K.K. Zool. Bot. Gesellsch., Vienna, 13: 137-162, pl.4-5 (Pl. 13-14).

Actinella

Desikachary, T.V., 1956(1957).
Electron microscope studies on diatoms. J.R.Microsc. Soc. (3)76(1/2):9-36.

Actinocyclus

Desikachary, T. V., 1956(1957).
Electron microscope studies on diatoms. J.R. Microsc. Soc. (3)76(1/2):9-36.

Actinocyclus Adeliae, n.sp.

Manguin, E., 1957.
Premier inventaire des diatomées de la Terre Adélie Antarctique. Espèces nouvelles. Rev. Algologique, n.s., 3(3):111-134.

Oceanographic Index: Marine Organisms Cumulation, 1946-1973

Actinocyclus alienus
Mann, A., 1907
Report on the diatoms of the Albatross voyages in the Pacific Ocean, 1888-1904. Contrib. U. S. Nat. Herb. 10(5):221-419, Pls. XLIV-LIV.

Actinocyclus (?) anceps
Castracane degli Antelminelli, F., 1886
1. Report on the Diatomaceae collected by H.M.S. Challenger during the years 1873-1876. Rept. Sci. Results, H.M.S. Challenger, Botany Vol. II, 178 pp., 30 pls.

Actinocyclus atlanticus n.sp.
Müller-Melchers, F-C., 1959.
Plankton diatoms of the southern Atlantic Argentina and Uruguay coast. Comunicaciones Botanicas del Museo Hist. Nat., Montevideo, III(38):1-45.

Actinocyclus bifrons
Boden, B. P., 1949.
The diatoms collected by the U.S.S. CACOPAN in the Antarctic in 1947. J. Mar. Res. 8(1):6-13, 3 textfigs.

Actinocyclus bifrons
Boden, Brian, 1948
Marine plankton diatoms on operation HIGHJUMP in: Some oceanographic observations on operation HIGHJUMP. By R.S. Dietz, USNEL Rept. No. 55, 97 pp., 41 figs. 7 July 1948.

Actinocyclus bifrons
Hendey, N.I., 1937
The plankton diatoms of the southern seas. Discovery Repts. 16:151-364, pls. 6-13.

Actinocyclus clevii n.sp.
Castracane degli Antelminelli, F., 1886
1. Report on the Diatomaceae collected by H.M.S. Challenger during the years 1873-1876. Rept. Sci. Results, H.M.S. Challenger, Botany Vol. II, 178 pp., 30 pls.

Actinocyclus complanatus n.sp.
Castracane degli Antelminelli, F., 1886
1. Report on the Diatomaceae collected by H.M.S. Challenger during the years 1873-1876. Rept. Sci. Results, H.M.S. Challenger, Botany Vol. II, 178 pp., 30 pls.

Actinocyclus complanatus
Hendey, N.I., 1937
The plankton diatoms of the southern seas. Discovery Repts. 16:151-364, pls. 6-13.

Actinocyclus corona
Hendey, N.I., 1937
The plankton diatoms of the southern seas. Discovery Repts. 16:151-364, pls. 6-13.

Actinocyclus crassus
Gran, H.H., 1908
Diatomeen. Nordisches Plankton, Botanischer Teil pp. XIX.1-XIX 146; 178 text figs.

Actinocyclus crassus
Jorgensen, E., 1900
Protophyten und Protozoën im Plankton aus der Norwegischen Westkerste. Bergens Mus. Aarb. 1899(6): 95 pp., 5 pls., 83 tables.

Actinocyclus crassus
Mann, A., 1907
Report on the diatoms of the Albatross voyages in the Pacific Ocean, 1888-1904. Contrib. U. S. Nat. Herb. 10(5):221-419, Pls. XLIV-LIV.

Actinocyclus crassus
Mann, A., 1893
List of Diatomaceae from a deep-sea dredging in the Atlantic Ocean off Delaware Bay by the U. S. Fish Commission Steamer Albatross. Proc. U. S. Nat. Mus. 16:303-312.

Actinocyclus curvulatus
Bigelow, H.B., and M. Leslie, 1930
Reconnaissance of the waters and plankton of Monterey Bay, July 1928. Bull. M.C.Z., 70(5):429-481, 43 text figs.

Actinocyclus curvatulus
Hasle, Grethe Rytter, 1960
Phytoplankton and ciliate species from the Tropical Pacific. Skr. Norske Videnskaps-Akad., Oslo, 1. Mat.-Nat. Kl., 1960(2): 1-50.

Actinocyclus curvatulus
Mann, A., 1907
Report on the diatoms of the Albatross voyages in the Pacific Ocean, 1888-1904. Contrib. U. S. Nat. Herb. 10(5):221-419, Pls. XLIV-LIV.

Actinocyclus (?) denticulatus n.sp.
Castracane degli Antelminelli, F., 1886
1. Report on the Diatomaceae collected by H.M.S. Challenger during the years 1873-1876. Rept. Sci. Results, H.M.S. Challenger, Botany Vol. II, 178 pp., 30 pls.

Actinocyclus Ehrenbergii
Bigelow, H.B., and M. Leslie, 1930
Reconnaissance of the waters and plankton of Monterey Bay, July 1928. Bull. M.C.Z., 70(5):429-481, 43 text figs.

Actinocyclus ehrenbergii
Cleve-Euler, A., 1951
Die Diatomeen von Schweden und Finnland. Kungl. Svenska Vetenskaps Akad. Handl., Fjärde Ser. 2(1): 161 pp., 6 pls.

Actinocyclus Ehrenbergii
Ercegovic, A., 1936
Etudes qualitative et quantitatives du phytoplancton dans les eaux cotières de l'Adriatique oriental moyen au cours de l'année 1934. Acta Adriatica 1(9):1-126

Actinocyclus Ehrenbergii
Florin, M-B., 1948
9. Diatomeae in submarine cores from the Tyrrhenian Sea. Medd. Ocean. Inst., Göteborg, 15 (Göteborgs Kungl. Vetenskaps-och Viterrhets Samhälles Handlingar, Sjätte Foljden, Ser. B 5(13):80-88.

Actinocyclus Ehrenbergii
Gran, H.H., 1908
Diatomeen. Nordisches Plankton, Botanischer Teil pp. XIX.1-XIX 146; 178 text figs.

Actinocyclus Ehrenbergi
Gran, H.H., and T. Braarud, 1935
A quantitative study of the phytoplankton in the Bay of Fundy and the Gulf of Maine (including observations on hydrography, chemistry, and turbidity). J. Biol. Bd., Canada, 1(5):279-467, 69 text figs.

Actinocyclus ehrenbergii
Gran, H. H. and E. C. Angst, 1931
Plankton diatoms of Puget Sound. Publ. Puget Sound Biol. Sta. 7:417-519, 95 text figs.

Actinocyclus ehrenbergii
Hasle, Grethe Rytter, 1960
Phytoplankton and ciliate species from the Tropical Pacific. Skr. Norske Videnskaps-Akad., Oslo, 1. Mat.-Nat. Kl., 1960(2): 1-50.

Actinocyclus ehrenbergi
Jørgensen, E., 1905
B. Protistplankton and the diatoms in bottom samples. Hydrographical and biological investigations in Norwegian fiords. Bergens Mus. Skr. 7: 49-225.

Actinocyclus ehrenbergii
Jorgensen, E., 1900
Protophyten und Protozoën im Plankton aus der Norwegischen Westkerste. Bergens Mus. Aarb. 1899(6): 95 pp., 5 pls., 83 tables.

Actinocyclus Ehrenberg
Levander, K.M., 1947
Plankton gesammelt in den Jahren 1899-1910 an den Küsten Finnlands. Finnländische Hydrographisch-Biologische Untersuchungen (aus dem Wasserbiologischen Laboratorium der Societas Scientiarum Fennica) No.11:40 pp., 6 diagrams, 13 pls., tables.

Actinocyclus Ehrenbergi
Lillick, L.C., 1940
Phytoplankton and planktonic protozoa of the offshore waters of the Gulf of Maine. Pt.II. Qualitative Composition of the Planktonic Flora. Trans. Am. Phil. Soc., n.s., 31(3):193-237, 13 text figs.

Actinocyclus Ehrenbergii
Mangin, M. L., 1912
Phytoplancton de la croisière du "René" dans l'Atlantique (Septembre 1908). Ann. Inst. Ocean., n.s., 4(1):1-66, 2 pls., 41 text figs., 2 tables.

Actinocyclus Ehrenbergii
Meunier, A., 1915
Microplancton de la Mer Flamande. 2. Diatomées (excepté le genre Chaetoceros). Mem. Mus. Roy. Hist. Nat., Belgique, 7(3):1-118, Pls. VIII-XIV.

Actinocyclus ehrenbergii
Pavillard, J., 1925
Bacillariales. Rept. on the Danish Oceangr. Exped., 1908-10 to the Mediterranean and adj. seas. Vol.II., Biol. J4:72 pp., 116 text figs.

Actinocyclus Ehrenbergi
Phifer, L.D. (undated)
The occurrence and distribution of plankton diatoms in Bering Sea and Bering Strait, July 26-August 24, 1934. Report of Oceanographic cruise of U.S. Coast Guard Cutter Chelan 1934, Part II(A):1-44 (mimeographed) plus fig.1 (after Pt.B)

Actinocyclus Ehrenberghii
Politis, J., 1949
Diatomées marines de Bosphores et des ibes de la mer de Marmara. II Practica tou Hellenikou Hidrobiologikou Institutoutou 1929, Etoz 1929, 3(1):11-31.

Actinocyclus Ehrenbergi
Rampi, L., 1942
Ricerche sul fitoplancton del Mare Ligure 6. Le diatomee delle acque di Sanremo. Nuovo Giornale Botanico Italiano, N.S., 49:252-268.

Actinocyclus Ehrenbergi
Rampi, L., 1940
Diatomee del Mare Adriatico. Nuovo Giornale Botanico Italiano, n.s., 47:559-608.

Actinocyclus Ehrenbergii
Rumkówna, A., 1948
[List of the phytoplankton species occurring in the superficial water layers in the Gulf of Gdańsk] Bull. Lab. mar. Gdynia, No. 4: 139-141 with tables in back.

Actinocyclus ehrenbergi
von Sydow, Burkard, und Robert Christenhuss 1972. Rasterelektronmikroskopische Untersuchungen der Hohlräume in der Schalenwand einiger centrischer Kieselalgen. Arch. Protistenk 114 (3): 256-271.

Actinocyclus Ehrenbergii
Zanon, D. V., 1949
Diatomee di Buenos Aires (Argentina) Atti Accad. Naz. Lincei, Memorie, Cl. Sci. fis., mat. e. nat., ser. 7, 11(3):59-151, 2 pls.

Actinocyclus Ehrenbergii

Zanon, V., 1948
Diatomee marini di Sardegna e Pugillo di Alghe Marine della stressa. Boll. Pesca, Piscitutura e Idrobiologia, Anno 24, ns. 3(2): 202-244, 27 figs. on 1 pl.

Actinocyclus elegans

Hendey, N.I., 1937
The plankton diatoms of the southern seas. Discovery Repts. 16:151-364, pls.6-13.

Actinocyclus (?) elongatus

Mann, A., 1907
Report on the diatoms of the Albatross voyages in the Pacific Ocean, 1888-1904. Contrib. U. S. Nat. Herb. 10(5):221-419, Pls. XLIV-LIV.

Actinocyclus excentricus

Frenguelli, Joaquin, and Hector Antonio Orlando, 1959.
Operacion MERLUZA. Diatomeas y silicoflagelados del plancton del "VI Crucero". Servicio Hidrogr. Naval., Argentina, Publ. No. H. 619: 5-62.

Actinocyclus fasciculatus

Castracane degli Antelminelli, F., 1886 n.sp.
1. Report on the Diatomaceae collected by H.M.S. Challenger during the years 1873-1876. Rept. Sci. Results, H.M.S. Challenger, Botany Vol. II, 178 pp., 30 pls.

Actinocyclus heptactis

Boden, B. P., 1949.
The diatoms collected by the U.S.S. CACOPAN in the Antarctic in 1947. J. Mar. Res. 8(1):6-13, 3 textfigs.

Actinocyclus heveticus

Cleve-Euler, A., 1951
Die Diatomeen von Schweden und Finnland. Kungl. Svenska Vetenskaps Akad. Handl., Fjärde Ser. 2(1): 161 pp., 6 pls.

Actinocyclus intermittens

Hendey, N.I., 1937
The plankton diatoms of the southern seas. Discovery Repts. 16:151-364, pls.6-13.

Actinocyclus interpunctatus

Mann, A., 1907
Report on the diatoms of the Albatross voyages in the Pacific Ocean, 1888-1904. Contrib. U. S. Nat. Herb. 10(5):221-419, Pls. XLIV-LIV.

Actinocyclus janus

Boden, B. P., 1949.
The diatoms collected by the U.S.S. CACOPAN in the Antarctic in 1947. J. Mar. Res. 8(1):6-13, 3 textfigs.

Actinocyclus janus

Boden, Brian, 1948
Marine plankton diatoms on operation HIGHJUMP in: Some oceanographic observations on operation HIGHJUMP. By R.S. Dietz. USNEL Rept. No.55, 97 pp., 41 figs. 7 July 1948.

Actinocyclus Janus

Hendey, N.I., 1937
The plankton diatoms of the southern seas. Discovery Repts. 16:151-364, pls.6-13.

Actinocyclus japonicus n.sp.

Castracane degli Antelminelli, F., 1886
1. Report on the Diatomaceae collected by H.M.S. Challenger during the years 1873-1876. Rept. Sci. Results, H.M.S. Challenger, Botany Vol. II, 178 pp., 30 pls.

Actinocyclus magnificus

Cleve-Euler, A., 1951
Die Diatomeen von Schweden und Finnland. Kungl. Svenska Vetenskaps Akad. Handl., Fjärde Ser. 2(1): 161 pp., 6 pls.

Actinocyclus minutus

Mann, A., 1907
Report on the diatoms of the Albatross voyages in the Pacific Ocean, 1888-1904. Contrib. U. S. Nat. Herb. 10(5):221-419, Pls. XLIV-LIV.

Actinocyclus moniliformis

Jorgensen, E., 1900
Protophyten und Protozoën im Plankton aus der Norwegischen Westkerste. Bergens Mus. Aarb. 1899(6): 95 pp., 5 pls., 83 tables.

Actinocyclus octonarius

Boden, B.P., 1950
Some marine plankton diatoms from the west coast of South Africa. Trans. R.Soc. S. Africa. 32:321-434, 100 text figs.

Actinocyclus octonarius & varieties

Hendy, N. Ingram, 1964
An introductory account of the smaller algae of British coastal waters. V. Bacillariophyceae (Diatoms).
Her Majesty's Stationary Office, 317 pp., 45 pls.

Actinocyclus octonarius

Hendey, N.I., 1937
The plankton diatoms of the southern seas. Discovery Repts. 16:151-364, pls.6-13.

Actinocyclus oliveranus

Castracane degli Antelminelli, F., 1886
1. Report on the Diatomaceae collected by H.M.S. Challenger during the years 1873-1876. Rept. Sci. Results, H.M.S. Challenger, Botany Vol. II, 178 pp., 30 pls.

Actinocyclus Oliveranus

Frenguelli, Joaquin, and Hector Antonio Orlando, 1959.
Operacion MERLUZA. Diatomeas y silicoflagelados del plancton del "VI Crucero". Servicio Hidrogr. Naval., Argentina, Publ. No. H. 619: 5-62.

Actinocylis oliverianus

Manguin, E., 1954
Diatomees marines provenant de l'ile Heard (Australian National Antarctic Expedition). Rev. Algol., n.s., 1: 14-24.

Actinocyclus oliverianus

Mann, A., 1907
Report on the diatoms of the Albatross voyages in the Pacific Ocean, 1888-1904. Contrib. U. S. Nat. Herb. 10(5):221-419, Pls. XLIV-LIV.

Actinocyclus ovatus n.sp.

Wood, E.J. Ferguson, 1963.
Studies on Australian and New Zealand diatoms. VI. Tropical and subtropical species. Trans. R. Soc., New Zealand, 2(15):189-218.

Actinocyclus parvus n.sp.

Hasle, Grethe Rytter, 1960
Phytoplankton and ciliate species from the Tropical Pacific.
Skr. Norske Videnskaps-Akad., Oslo, 1. Mat.-Nat. Kl., 1960(2): 1-50.

Actinocyclus pellucidus n.sp.

Castracane degli Antelminelli, F., 1886
1. Report on the Diatomaceae collected by H.M.S. Challenger during the years 1873-1876. Rept. Sci. Results, H.M.S. Challenger, Botany Vol. II, 178 pp., 30 pls.

Actinocyclus platensis

Hendey, N-Ingram, 1958 [1957(Publ. 1958)]
Marine diatoms from some West African Ports. J. R.Microsc. Soc. (3) 77(1/2): 28-85.

Actinocyclus pruinosus

Castracane degli Antelminelli, F., 1886
1. Report on the Diatomaceae collected by H.M.S. Challenger during the years 1873-1876. Rept. Sci. Results, H.M.S. Challenger, Botany Vol. II, 178 pp., 30 pls.

Actinocyclus pumilus n.sp.

Castracane degli Antelminelli, F., 1886
1. Report on the Diatomaceae collected by H.M.S. Challenger during the years 1873-1876. Rept. Sci. Results, H.M.S. Challenger, Botany Vol. II, 178 pp., 30 pls.

Actinocyclus punctulatus n.sp.

Castracane degli Antelminelli, F., 1886
1. Report on the Diatomaceae collected by H.M.S. Challenger during the years 1873-1876. Rept. Sci. Results, H.M.S. Challenger, Botany Vol. II, 178 pp., 30 pls.

Actinocyclus Ralfsii

Bigelow, H.B., and M. Leslie, 1930
Reconnaissance of the waters and plankton of Monterey Bay, July 1928. Bull. M.C.Z. 70(5):429-481, 43 text figs.

Actinocyclus Ralfsii

Gran, H.H., 1908
Diatomeen. Nordisches Plankton, Botanischer Teil pp. XIX.1-XIX 146; 178 text figs.

Actinocyclus ralfsii challengerensis n.var.

Castracane degli Antelminelli, F., 1886
1. Report on the Diatomaceae collected by H.M.S. Challenger during the years 1873-1876. Rept. Sci. Results, H.M.S. Challenger, Botany Vol. II, 178 pp., 30 pls.

Actinocyclus ralfsi

Jørgensen, E., 1905
B. Protistplankton and the diatoms in bottom samples. Hydrographical and biological investigations in Norwegian fjords. Bergens Mus. Skr. 7: 49-225.

Actinocyclus ralfsii

Jorgensen, E., 1900
Protophyten und Protozoën im Plankton aus der Norwegischen Westkerste. Bergens Mus. Aarb. 1899(6): 95 pp., 5 pls., 83 tables.

Actinocyclus ralfsii

Mann, A., 1907
Report on the diatoms of the Albatross voyages in the Pacific Ocean, 1888-1904. Contrib. U. S. Nat. Herb. 10(5):221-419, Pls. XLIV-LIV.

Actinocyclus Ralfsii

Mann, A., 1893
List of Diatomaceae from a deep-sea dredging in the Atlantic Ocean off Delaware Bay by the U. S. Fish Commission Steamer Albatross. Proc. U. S. Nat. Mus. 16:303-312.

Actinocyclus Ralfsii

Meunier, A., 1915
Microplancton de la Mer Flamande. 2. Diatomées (excepté le genre Chaetoceros). Mem. Mus. Roy. Hist. Nat., Belgique, 7(3):1-118, Pls. VIII-XIV.

Actinocyclus Ralfsii

Politis, J., 1949
Diatomees marines de Bosphores et des ibes de la mer de Marmara. II Practica tou Hellenikou Hidrobiologikou Institutoutou 1929, Etoz 1929, 3(1):11-31.

Actinocyclus roperii

Cleve-Euler, A., 1951
Die Diatomeen von Schweden und Finnland. Kungl. Svenska Vetenskaps Akad. Handl., Fjärde Ser. 2(1): 161 pp., 6 pls.

Actinocyclus roperi

Hendy, N. Ingram, 1964
An introductory account of the smaller algae of British coastal waters. V. Bacillariophyceae (Diatoms).
Her Majesty's Stationary Office, 317 pp., 45 pls.

Actinocyclus Roeperi

Schodduyn, M., 1926
Observations faites dans la baie d'Ambleteuse (Pas de Calais). Bull. Inst. Ocean., Monaco, No. 482: 64 pp.

Actinocyclus rotula

Boden, B.P., 1950
Some marine plankton diatoms from the west coast of South Africa. Trans. R.Soc. S. Africa. 32:321-434, 100 text figs.

Actinocyclus rotula

Hendey, N.I., 1937
The plankton diatoms of the southern seas. Discovery Repts. 16:151-364, pls.6-13.

Actinocyclus sparsus

Mann, A., 1907
Report on the diatoms of the Albatross voyages in the Pacific Ocean, 1888-1904. Contrib. U. S. Nat. Herb. 10(5):221-419, Pls. XLIV-LIV.

Actinocyclus sparsus

Mann, A., 1893
List of Diatomaceae from a deep-sea dredging in the Atlantic Ocean off Delaware Bay by the U. S. Fish Commission Steamer Albatross. Proc. U. S. Nat. Mus. 16:303-312.

Actinocyclus subocollatus

Jørgensen, E., 1905
B. Protistplankton and the diatoms in bottom samples. Hydrographical and biological investigations in Norwegian fjords. Bergens Mus. Skr. 7: 49-225.

Actinocyclus subtilis

Cleve-Euler, A., 1951
Die Diatomeen von Schweden und Finnland. Kungl. Svenska Vetenskaps Akad. Handl., Fjärde Ser. 2(1): 161 pp., 6 pls.

Actinocyclus subtilis

Gran, H.H., 1908
Diatomeen. Nordisches Plankton, Botanischer Teil pp. XIX 1-XIX 146; 178 text figs.

Actinocyclus subtilis

Hendy, N. Ingram, 1964
An introductory account of the smaller algae of British coastal waters. V. Bacillariophyceae (Diatoms). Her Majesty's Stationary Office, 317 pp., 45 pls.

Actinocyclus subtilis

Jørgensen, E., 1905
B. Protistplankton and the diatoms in bottom samples. Hydrographical and biological investigations in Norwegian fjords. Bergens Mus. Skr. 7: 49-225.

Actinocyclus subtilis

Mann, A., 1907
Report on the diatoms of the Albatross voyages in the Pacific Ocean, 1888-1904. Contrib. U. S. Nat. Herb. 10(5):221-419, Pls. XLIV-LIV.

Actinocyclus subtilis

Mann, A., 1893
List of Diatomaceae from a deep-sea dredging in the Atlantic Ocean off Delaware Bay by the U. S. Fish Commission Steamer Albatross. Proc. U. S. Nat. Mus. 16:303-312.

Actinocyclus subtilis

Pavillard, J., 1925
Bacillariales. Rept. on the Danish Oceangr. Exped., 1908-10 to the Mediterranean and adj. seas. Vol.II., Biol. J4:72 pp., 116 text figs.

Actinocyclus subtilis

Pavillard, J., 1905
Recherches sur la flore pelagique (Phytoplankton) de l'Etang de Thau. Theses presentees a la Fac. Sci., Paris, 116 pp., 3 pls.

Actinocyclus subtilis

Politis, J., 1949
Diatomees marines de Bosphores et des ibes de la mer de Marmara. II Practica tou Hellenikou Hidrobiologikou Institutoutou 1929, Etoz 1929, 3(1):11-31.

Actinocyclus subtilis

Rampi, L., 1942
Ricerche sul fitoplancton del Mare Ligure 6. Le diatomee delle acque di Sanremo. Nuovo Giornale Botanico Italiano, N.S., 49:252-268.

Actinocyclus subtilis

Rampi, L., 1940
Diatomee del Mare Adriatico. Nuovo Giornale Botanico Italiano, n.s., 47:559-608.

Actinocyclus subtilis

Schröder, B., 1900
Phytoplankton des Golfes von Neapel nebst vergleichenden Ausblicken auf das atlantischen Ozean. Mitt. Zool. Stat. Neapel, 14:1-38.

Actinocyclus subtilis

Takano, Hideaki, 1962
Notes on epiphytic diatoms upon sea-weeds from Japan. J. Oceanogr. Soc., Japan, 18(1):29-33.

Actinocyclus subtilis

Zanon, V., 1948
Diatomee marini di Sardegna e Pugillo di Alghe Marine della stressa. Boll. Pesca, Piscitutura e Idrobiologia, Anno 24, ns. 3(2): 202-244, 27 figs. on 1 pl.

Actinocyclus tenuissimus

Zanon, V., 1948
Diatomee marini di Sardegna e Pugillo di Alghe Marine della stressa. Boll. Pesca, Piscitutura e Idrobiologia, Anno 24, ns. 3(2): 202-244, 27 figs. on 1 pl.

Actinocyclus tesselatus

Mann, A., 1907
Report on the diatoms of the Albatross voyages in the Pacific Ocean, 1888-1904. Contrib. U. S. Nat. Herb. 10(5):221-419, Pls. XLIV-LIV.

Actinocyclus umbonatus n.sp.

Castracane degli Antelminelli, F., 1886
1. Report on the Diatomaceae collected by H.M.S. Challenger during the years 1873-1876. Rept. Sci. Results, H.M.S. Challenger, Botany Vol. II, 178 pp., 30 pls.

Actinocyclus umbonatus

Hendey, N.I., 1937
The plankton diatoms of the southern seas. Discovery Repts. 16:151-364, pls.6-13.

Actinoptychus

Neaverson, E., 1934
The sea-floor deposits. 1. General characteristics and distribution. Discovery Repts. 9: 297-349, Plates 17-22.

Acmoptychus adriaticus

Ercegovie, A., 1940
Weitere Untersuchungen über einige hydrographische Verhältnisse und über die Phytoplanktonproduktion in den Gewässern der östlichen Mitteladria. Acta Adriatica 2(3):95-134, 8 text figs.

Actinoptychus adriaticus

Grunow, A., 1863
Ueber einige neue und ungenügend bekannte Arten und Gattungen von Diatomaceen. Verhandl. d. K.K. Zool. Bot. Gesellsch., Vienna, 13: 137-162, pl.4-5 (Pl. 13-14).

Actinoptychus adriaticus

Rampi, L., 1940
Diatomee del Mare Adriatico. Nuovo Giornale Botanico Italiano, n.s., 47:559-608.

Actinoptychus adriaticus

Zanon, D. V., 1949
Diatomee di Buenos Aires (Argentina) Atti Accad. Naz. Lincei, Memorie, Cl. Sci. fis., mat. e. nat., ser. 7, 11(3):59-151, 2 pls.

Actinoptychus adriaticus

Zanon, V., 1948
Diatomee marini di Sardegna e Pugillo di Alghe Marine della stressa. Boll. Pesca, Piscitutura e Idrobiologia, Anno 24, ns. 3(2): 202-244, 27 figs. on 1 pl.

Actinoptychus alternans

Bigelow, H.B., and M. Leslie, 1930
Reconnaissance of the waters and plankton of Monterey Bay, July 1928. Bull. M.C.Z., 70(5):429-481, 43 text figs.

Actinoptychus alternans n.sp.

Mann, A., 1907
Report on the diatoms of the Albatross voyages in the Pacific Ocean, 1888-1904. Contrib. U. S. Nat. Herb. 10(5):221-419, Pls. XLIV-LIV.

Actinoptychus areolatus

Bigelow, H.B., and M. Leslie, 1930
Reconnaissance of the waters and plankton of Monterey Bay, July 1928. Bull. M.C.Z., 70(5):429-481, 43 text figs.

Actinoptychus bifrons

Hendey, N-Ingram, 1958 (1957 (Publ. 1958).
Marine diatoms from some West African Ports. J.R. Microsc. Soc. (3) 77 (1/2): 28-85.

Actinoptychus campanulifer

Hendey, N-Ingram, 1958 [1957(Publ. 1958)]
Marine diatoms from some West African Ports-J. R Microsc. Soc. (3) 77(1/2): 28-85.

Actinoptychus campanulifer

Müller Melchers, F.C., 1953.
New and little known diatoms from Uruguay and the South Atlantic coast. Com. Bot., Mus. Hist. Nat., Montevideo, 3(30): 1-11; 8 pls.

Actinoptychus erosus n.sp.

Castracane degli Antelminelli, F., 1886
1. Report on the Diatomaceae collected by H.M.S. Challenger during the years 1873-1876. Rept. Sci. Results, H.M.S. Challenger, Botany Vol. II, 178 pp., 30 pls.

Actinoptychus Frenguelli

Müller Melchers, F.C., 1953.
New and little known diatoms from Uruguay and the South Atlantic coast. Com. Bot., Mus. Hist. Nat., Montevideo, 3(30): 1-11, 8 pls.

Actinoptychus frenguellii n.sp.

Müller Melchers, F.C., 1951.
Actinoptychus frenguelli n.sp. (Diatomeas). Physis 20(58):320-323, 1 pl., with 5 figs.

Actinoptychus grundleri

Mann, A., 1907
Report on the diatoms of the Albatross voyages in the Pacific Ocean, 1888-1904. Contrib. U. S. Nat. Herb. 10(5):221-419, Pls. XLIV-LIV.

Actinoptychus hexagonus

Mann, A., 1893
List of Diatomaceae from a deep-sea dredging in the Atlantic Ocean off Delaware Bay by the U. S. Fish Commission Steamer Albatross. Proc. U. S. Nat. Mus. 16:303-312.

Actinoptychus janischii

Mann, A., 1907
Report on the diatoms of the Albatross voyages in the Pacific Ocean, 1888-1904. Contrib. U. S. Nat. Herb. 10(5):221-419, Pls. XLIV-LIV.

Actinoptychus lucidus

Cleve-Euler, A., 1951
Die Diatomeen von Schweden und Finnland. Kungl. Svenska Vetenskaps Akad. Handl., Fjärde Ser. 2(1): 161 pp., 6 pls.

Actinoptychus lucidus

Florin, M-B., 1948
9. Diatomeae in submarine cores from the Tyrrhenian Sea. Medd. Ocean. Inst., Göteborg, 15 (Göteborgs Kungl. Vetenskaps-och Vitterhets SamhÄlles Handlingar, Sjätte Foljden, Ser. B 5(13):80-88.

Actinoptychus mirans

Hendey, N-Ingram, 1958 [1957(Publ. 1958)]
Marine diatoms from some West African Ports- J. R Microsc. Soc. (3) 77(1/2): 28-85.

Actinoptychus mölleri

Mann, A., 1907
Report on the diatoms of the Albatross voyages in the Pacific Ocean, 1888-1904. Contrib. U. S. Nat. Herb. 10(5):221-419, Pls. XLIV-LIV.

Actinoptychus planus n.sp.

Mann, A., 1907
Report on the diatoms of the Albatross voyages in the Pacific Ocean, 1888-1904. Contrib. U. S. Nat. Herb. 10(5):221-419, Pls. XLIV-LIV.

Actinoptychus platensis n.sp.

Müller, Melchers, F.C., 1953.
New and little known diatoms from Uruguay and the South Atlantic coast. Com. Bot., Mus. Hist. Nat., Montevideo, 3(30): 1-11, 8 pls.

Actinoptychus punctulatus

Mann, A., 1907
Report on the diatoms of the Albatross voyages in the Pacific Ocean, 1888-1904. Contrib. U. S. Nat. Herb. 10(5):221-419, Pls. XLIV-LIV.

Actinoptychus radulus n.sp.

Mann, A., 1907
Report on the diatoms of the Albatross voyages in the Pacific Ocean, 1888-1904. Contrib. U. S. Nat. Herb. 10(5):221-419, Pls. XLIV-LIV.

Actinoptychus raganus n.sp.

Castracane degli Antelminelli, F., 1886
1. Report on the Diatomaceae collected by H.M.S. Challenger during the years 1873-1876. Rept. Sci. Results, H.M.S. Challenger, Botany Vol. II, 178 pp., 30 pls.

Actinoptychus senarius

Boden, B.P., 1950
Some marine plankton diatoms from the west coast of South Africa. Trans. R.Soc. S. Africa. 32:321-434, 100 text figs.

Actinoptychus senarius

Hendy, N. Ingram, 1964
An introductory account of the smaller algae of British coastal waters. V. Bacillariophyceae (Diatoms). Her Majesty's Stationary Office, 317 pp., 45 pls.

Actinoptychus senarius

Hendey, N.I., 1951
Littoral diatoms of Chicester Harbour with special reference to fouling. J. Roy. Microscop. Soc. 71(1): 1-86, 18 pls.

Actinoptychus senarius

Hendey, N.I., 1937
The plankton diatoms of the southern seas. Discovery Repts. 16:151-364, pls.6-13.

Actinoptychus splendens

Boden, B.P., 1950
Some marine plankton diatoms from the west coast of South Africa. Trans. R.Soc. S. Africa. 32:321-434, 100 text figs.

Actinoptychus splendens

Cleve-Euler, A., 1951
Die Diatomeen von Schweden und Finnland. Kungl. Svenska Vetenskaps Akad. Handl., Fjärde Ser. 2(1): 161 pp., 6 pls.

Actinoptychus splendens

Cupp, Easter E., 1943
Marine plankton diatoms of the west coast of North America. Bull. S.I.O. 5(1):1-238, 5 pls., 168 text figs.

Actinoptychus splendens

Dangeard, P., 1927
Phytoplankton de la croisière du "Sylvana". Ann. Inst. Ocean., Monaco, n.s., 4(8):286-401, 54 text figs, (Feirer-Juin 1913).

Actinoptychus splendens

Frenguelli, Joaquin, and Hector Antonio Orlando, 1959.
Operacion MERLUZA. Diatomeas y silicoflagelades del plancton del "VI Crucero". Servicio Hidrogr. Naval., Argentina, Publ. No. H. 619: 5-62.

Actinoptychus splendens

Gran, H.H., 1908
Diatomeen. Nordisches Plankton, Botanischer Teil pp. XIX.1-XIX 146; 178 text figs.

Actinoptychus splendens

Gran, H. H. and E. C. Angst, 1931
Plankton diatoms of Puget Sound. Publ. Puget Sound Biol. Sta. 7:417-519, 95 text figs.

Actinoptychus splendens

Hendy, N. Ingram, 1964
An introductory account of the smaller algae of British coastal waters. V. Bacillariophyceae (Diatoms). Her Majesty's Stationary Office, 317 pp., 45 pls.

Actinoptychus splendens

Hendey, N.I., 1937
The plankton diatoms of the southern seas. Discovery Repts. 16:151-364, pls.6-13.

Actinoptychus splendens

Mann, A., 1907
Report on the diatoms of the Albatross voyages in the Pacific Ocean, 1888-1904. Contrib. U. S. Nat. Herb. 10(5):221-419, Pls. XLIV-LIV.

Actinoptychus splendens

Mann, A., 1893
List of Diatomaceae from a deep-sea dredging in the Atlantic Ocean off Delaware Bay by the U. S. Fish Commission Steamer Albatross. Proc. U. S. Nat. Mus. 16:303-312.

Actinoptychus splendens

Meunier, A., 1915
Microplancton de la Mer Flamande. 2. Diatomées (excepté le genre Chaetoceros). Mem. Mus. Roy. Hist. Nat., Belgique, 7(3):1-118, Pls. VIII-XIV.

Actinoptychus splendens

Rampi, L., 1940
Diatomee del Mare Adriatico. Nuovo Giornale Botanico Italiano, n.s., 47:559-608.

Actinoptychus splendens

Zanon, D. V., 1949
Diatomee di Buenos Aires (Argentina) Atti Accad. Naz. Lincei, Memorie, Cl. Sci. fis., mat. e. nat., ser. 7, 11(3):59-151, 2 pls.

Actinoptychus splendens

Zanon, V., 1948
Diatomee marini di Sardegna e Pugillo di Alghe Marine della stressa. Boll. Pesca, Piscitutura e Idrobiologia, Anno 24, ns. 3(2): 202-244, 27 figs. on 1 pl.

Actinoptychus undulatus

Allen, W.E., 1937
Plankton diatoms of the Gulf of California obtained by the G. Allan Hancock Expedition of 1936. The Hancock Pacific Expeditions, Univ. So. Calif. Publ. 3:47-59, 1 fig.

Actinoptychus undulatus

Allen, W.E., and E.E. Cupp, 1935
Plankton diatoms of the Java Sea. Annales du Jardin Botanique de Buitenzorg XLIV (2):101-174, figs.1-127.

(drawings of all species mentioned)

Actinoptychus undulatus

Bigelow, H.B., and M. Leslie, 1930
Reconnaissance of the waters and plankton of Monterey Bay, July 1928. Bull. M.C.Z. 70(5):429-481, 43 text figs.

Actinoptychus undulatus

Cleve-Euler, A., 1951
Die Diatomeen von Schweden und Finnland. Kungl. Svenska Vetenskaps Akad. Handl., Fjärde Ser. 2(1): 161 pp., 6 pls.

Actinoptychus undulatus

Cupp, Easter E., 1943
Marine plankton diatoms of the west coast of North America. Bull. S.I.O. 5(1):1-238, 5 pls., 168 text figs.

Actinoptychus undulatus

Cupp, E.E. and Allen, W.E., 1938
Plankton diatoms of the Gulf of California obtained by Allan Hancock Pacific Expedition of 1937. The Hancock Pacific Expeditions, The Univ. So. Calif. Publ. 3: 61-74, 1 map, pls.4-15.

Actinoptychus undulatus

Dangeard, P., 1927
Phytoplankton de la croisière du "Sylvana". Ann. Inst. Ocean., Monaco, n.s., 4(8):286-401, 54 text figs, (Feirer-Juin 1913).

Actinoptychus undulatus

Frenguelli, Joaquin, and Hector Antonio Orlando, 1959.
Operacion MERLUZA. Diatomeas y silicoflagelades del plancton del "VI Crucero". Servicio Hidrogr. Naval., Argentina, Publ. No. H. 619: 5-62.

Actinoptychus undulatus

Gilbert, J.Y., and W.E. Allen, 1943
The phytoplankton of the Gulf of California obtained by the "E.W. Scripps" in 1939 and 1940. J. Mar. Res. V(2):89-110, figs.30-31.

Actinoptychus undulatus

Gran, H.H., 1908
Diatomeen. Nordisches Plankton, Botanischer Teil pp. XIX.1-XIX 146; 178 text figs.

Actinoptychus undulatus

Gran, H.H., and T. Braarud, 1935
A quantitative study of the phytoplankton in the Bay of Fundy and the Gulf of Maine (including observations on hydrography, chemistry, and turbidity). J. Biol. Bd., Canada, 1(5):279-467, 69 text figs.

Oceanographic Index: Marine Organisms Cumulation, 1946-1973

Actinoptychus undulatus
Gran, H. H. and E. C. Angst, 1931
Plankton diatoms of Puget Sound. Publ. Puget Sound Biol. Sta. 7:417-519, 95 text figs.

Actinoptychus undulatus
Grøntved, J., 1949
Investigations on the phytoplankton in the Danish Waddensea in July 1941. Medd. Komm. Danmarks Fiskeri og Havundersøgelser, ser. Plankton, 5(2):55 pp., 2 pls., 38 text figs.

Actinoptychus undulatus
Hustedt, F. and A.A. Aleem, 1951
Littoral diatoms from the Salstone near Plymouth. JMBA 30(1): 177.196.

Actinoptychus undulatus
Jørgensen, E., 1905
B.Protistplankton and the diatoms in bottom samples. Hydrographical and biological investigations in Norwegian fjords. Bergens Mus. Skr. 7: 49-225.

Actinoptychus undulata
Jorgensen, E., 1900
Protophyten und Protozoën im Plankton aus der Norwegischen Westkerste. Bergens Mus. Aarb. 1899(6): 95 pp., 5 pls., 83 tables.

Actinoptychus undulatus
Lafon, M., M. Durchon and Y. Saudray, 1955.
Recherches sur les cycles saisonnières du plankton. Ann. Inst. Océan., 31(3):125-230.

Actinoptychus undulatus
Lillick, L.C., 1940
Phytoplankton and planktonic protozoa of the offshore waters of the Gulf of Maine. Pt.II. Qualitative Composition of the Planktonic Flora. Trans. Am. Phil. Soc., n.s., 31(3):193-237, 13 text figs.

Actinoptychus undulatus
Lillick, L.C., 1938
Preliminary report of the phytoplankton of the Gulf of Maine. Am. Mid. Nat. 20(3):624-640, 1 text figs 37 tables.

Actinoptychus undulatus
Mangin, L., 1915
Phytoplancton de L'Antartique. Deuxieme Exped. Ant. Francaise (1908-1910), 95 pp., 3 pls., 58 text figs.

Actinoptychus undulatus
Mangin, M. L., 1912
Phytoplancton de la croisière du "René" dans l'Atlantique (Septembre 1908). Ann. Inst. Ocean., n.s., 4(1):1-66, 2 pls., 41 text figs., 2 tables.

Actinoptychus undulatus
Mann, A., 1907
Report on the diatoms of the Albatross voyages in the Pacific Ocean, 1888-1904. Contrib. U. S. Nat. Herb. 10(5):221-419, Pls. XLIV-LIV.

Actinoptychus undulatus
Mann, A., 1893
List of Diatomaceae from a deep-sea dredging in the Atlantic Ocean off Delaware Bay by the U. S. Fish Commission Steamer Albatross. Proc. U. S. Nat. Mus. 16:303-312.

Actinoptychus undulatus
Meunier, A., 1915
Microplancton de la Mer Flamande. 2. Diatomées (excepté le genre Chaetoceros). Mem. Mus. Roy. Hist. Nat., Belgique, 7(3):1-118, Pls. VIII-XIV.

Actinoptychus undulatus
Morse, D.C., 1947
Some observations on seasonal variations in plankton population Patuxant River, Maryland 1943-1945. Bd. Nat. Res., Publ. No.65, Chesapeake Biol. Lab., 31, 3 figs.

Actinoptychus undulatus
Paiva Carvalho, J., 1950
O plancton do Rio Maria Rodriques (Cananeis). 1. Diatomaceas e Dinoflagelados. Bol. Inst. Paulista Oceanogr. 1(1); 27-43, 2 fold-in tables, 2 figs.

Actinoptychus undulatus
Pavillard, J., 1925
Bacillariales. Rept. on the Danish Oceangr. Exped., 1908-10 to the Mediterranean and adj. seas. Vol.II, Biol. J4:72 pp., 116 text figs.

Actinoptychus undulatus
Rampi, L., 1942
Ricerche sul fitoplancton del Mare Ligure 6. Le diatomee delle acque di Sanremo. Nuovo Giornale Botanico Italiano, N.S., 49:252-268.

Actinoptychus undulatus
Rampi, L., 1940
Diatomee del Mare Adriatico. Nuovo Giornale Botanico Italiano, n.s., 47:559-608.

Actinoptychus undulatus
Sousa e Silva, E., 1949
Diatomaceas e Dinoflagelados de Baía de Cascais. Portugaliae Acta Biol., Volume: Julio Henriques, Ser. B: 300-383, 9 pls, 2 fold-in tables.

Actinoptychus undulatus
Sousa a Silva, E., and J. Dos Santos-Pinto, 1948
O Plancton da Baía de S. Martinho do Porto. 1. Diatomaceas e Dinoflagelados. Bol. Soc. Portuguese de Ciencias Naturais, 16(2):134-187, 6 pls. (Trav. Sta. Biol. Mar. de Lisbonne No. 52).

Actinoptychus undulatus
von Sydow, Burkard, und Robert Christenhuss 1972.
Rasterelektronmikroskopische Untersuchungen der Hohlräume in der Schalenwand einiger centrischer Kieselalgen. Arch. Protistenk 114 (3): 256-271.

Actinoptychus undulatus
Zanon, D. V., 1949
Diatomee di Buenos Aires (Argentina) Atti Accad. Naz. Lincei, Memorie, Cl. Sci. fis., mat. e. nat., ser. 7, 11(3):59-151, 2 pls.

Actinoptychus undulatus
Zanon, V., 1948
Diatomee marini di Sardegna e Pugillo di Alghe Marine della stressa. Boll. Pesca, Piscitutura e Idrobiologia, Anno 24, ns. 3(2): 202-244, 27 figs. on 1 pl.

Actinoptychus vulgaris
Boden, B.P., 1950
Some marine plankton diatoms from the west coast of South Africa. Trans. R.Soc. S. Africa. 32:321-434, 100 text figs.

Alloioneis antillarum
Castracane degli Antelminelli, F., 1886
1. Report on the Diatomaceae collected by H.M.S. Challenger during the years 1873-1876. Rept. Sci. Results, H.M.S. Challenger, Botany Vol. II, 178 pp., 30 pls.

Alloioneis japonica n.sp.
Castracane degli Antelminelli, F., 1886
1. Report on the Diatomaceae collected by H.M.S. Challenger during the years 1873-1876. Rept. Sci. Results, H.M.S. Challenger, Botany Vol. II, 178 pp., 30 pls.

Amphipleura
Desikachary, T.V., 1956(1957).
Electron microscope studies on diatoms. J.R. Microsc. Soc. (3)76(1/2):9-36.

Amphipleura frauenfeldii n.sp
Grunow, A., 1863
Ueber einige neue und ungenügend bekannte Arten und Gattungen von Diatomaceen. Verhandl. d. K.K. Zool. Bot. Gesellsch., Vienna, 13: 137-162, pl.4-5 (Pl. 13-14).

Amphipleura intermedia
Grunow, A., 1877
1. New Diatoms from Honduras. Monthly Micros. Jour., 18:165-186, pls. CXCIII-CXCVI.

Amphipleura Lindleimeri
Grunow, A., 1877
1. New Diatoms from Honduras. Monthly Micros. Jour., 18:165-186, pls. CXCIII-CXCVI.

Amphipleura Lindheimeri
Zanon, D. V., 1949
Diatomee di Buenos Aires (Argentina) Atti Accad. Naz. Lincei, Memorie, Cl. Sci.. fis., mat. e. nat., ser. 7, 11(3):59-151, 2 pls.

Amphipleura micans & var.
Hendy, N. Ingram, 1964
An introductory account of the smaller algae of British coastal waters. V. Bacillariophyceae (Diatoms). Her Majesty's Stationary Office, 317 pp., 45 pls.

Amphipleura Oregonuca
Grunow, A., 1877
1. New Diatoms from Honduras. Monthly Micros. Jour., 18:165-186, pls. CXCIII-CXCVI.

Amphipleura pellucida
Kolbe, R.W., 1951.
Elektronmikroskopische Untersuchungen von Diatomeenmembranen II. Svenska Bot. Tidskr. 45(4):636-647, Pls. 1-4.

Amphipleura rutilans
Castenholz, Richard W., 1963.
An experimental study of the vertical distribution of littoral marine diatoms. Limnology and Oceanography, 8(4):450-462.

Amphipleura rutilans
Hendy, N. Ingram, 1964
An introductory account of the smaller algae of British coastal waters. V. Bacillariophyceae (Diatoms). Her Majesty's Stationary Office, 317 pp., 45 pls.

Amphipleura rutilans
Hendey, N.I., 1951
Littoral diatoms of Chicester Harbour with special reference to fouling. J.Roy. Microscop. Soc. 71(1): 1-86, 18 pls.

Amphipleura rutilans
Hustedt, F. and A.A. Aleem, 1951
Littoral diatoms from the Salstone near Plymouth. JMBA 30(1): 177.196.

Amphipleura rutilans
Lewin, R. A., 1958.
The mucilage tubes of Amphipleura rutilans. Limnol. & Oceanogr., 3(1):111-113.

Amphipleura rutilans
Zanon, V., 1948
Diatomee marini di Sardegna e Pugillo di Alghe Marine della stressa. Boll. Pesca, Piscitutura e Idrobiologia, Anno 24, ns. 3(2): 202-244, 27 figs. on 1 pl.

Amphipleura Silvestrina n.sp.
Zanon, D. V., 1949
Diatomee di Buenos Aires (Argentina)
Atti Accad. Naz. Lincei, Memorie, Cl. Sci.
Fis., mat. e. nat., ser. 7, 11(3):59-151,
2 pls.

Amphiprora sp.
Braarud, T., and Adam Bursa, 1939
On the phytoplankton of the Oslo Fjord,
1933-1934. Hvalrådets Skr. No.19:1-63,
9 text figs. Reviewed. J. du. Cons. 14(3):
418-420. A.C. Gardiner.

Amphiprora
Desikachary, T.V., 1956(1957).
Electron microscope studies on diatoms.
J.R. Microsc. Soc. (3)76(1/2):9-36.

Amphiprora sp. (diatom)
Ogiino, C., 1963.
Studies on the chemical composition of some
natural foods of aquatic animals. (In Japanese;
English abstract).
Bull. Jap. Soc. Sci. Fish., 29(5):459-462.

Amphiprora aculeatas
Krasske, G., 1941
Die Kieselalgen des chilenischen
Küstenplankton (Aus dem südchilenischen
Kusten gebiet IX). Arch. Hydrobiol. 38
(2):260-287.

Amphiprora alata
Gran, H.H., and T. Braarud, 1935
A quantitative study of the phyto-
plankton in the Bay of Fundy and the
Gulf of Maine (including observations
on hydrography, chemistry, and turbidity).
J. Biol. Bd. Canada, 1(5):279-467, 69
text figs.

Amphiprora alata
Hendy, N. Ingram, 1964
An introductory account of the smaller algae
of British coastal waters. V. Bacillariophyceae (Diatoms).
Her Majesty's Stationary Office, 317 pp.,
45 pls.

Amphiprora alata
Hendey, N.I., 1951
Littoral diatoms of Chicester Harbour
with special reference to fouling. J.Roy.
Microscop. Soc. 71(1): 1-86, 18 pls.

Amphiprora alata
Jorgensen, E., 1900
Protophyten und Protozoën im Plank-
ton aus der Norwegischen Westkerste. Bergens
Mus. Aarb. 1899(6): 95 pp., 5 pls., 83 tables.

Amphiprora alata
Kokubo, S., and S. Sato, 1947
Plankters in Jū-San Gata. Physiol.
and Ecol. (Japan) 1(4):1-16, 3 text figs.,
tables.

Amphiprora alata
Lillick, L.C., 1940
Phytoplankton and planktonic
protozoa of the offshore waters
of the Gulf of Maine. Pt.II.
Qualitative Composition of the
Planktonic Flora. Trans. Am.
Phil. Soc. n.s., 31(3):193-237,
13 text figs.

Amphiprora alata
Rampi, L., 1942
Ricerche sul fitoplancton del Mare Ligure
6. Le diatomee delle acque di Sanremo. Nuovo
Giornale Botanico Italiano, N.S., 49:252-288.

Amphiprora alata
Rampi, L., 1940
Diatomee del Mare Adriatico. Nuovo
Giornale Botanico Italiano, n.s., 47:559-608.

Amphiprora alata
Schodduyn, M., 1926
Observations faites dans la baie
d'Ambleteuse (Pas de Calais). Bull. Inst.
Océan., Monaco, No. 482: 64 pp.

Amphiprora alata
Zanon, V., 1948
Diatomee marini di Sardegna e Pugillo
di Alghe Marine della stressa. Boll. Pesca,
Piscicoltura e Idrobiologia, Anno 24, ns. 3(2):
202-244, 27 figs. on 1 pl.

Amphiprora alata (var. japonica)
Takano, Hideaki, 1962
Notes on epiphytic diatoms upon sea-weeds
from Japan.
J. Oceanogr. Soc., Japan, 18(1):29-33.

Amphiprora angustata n.sp.
Hendy, N. Ingram, 1964
An introductory account of the smaller algae
of British coastal waters. V. Bacillariophyceae (Diatoms).
Her Majesty's Stationary Office, 317 pp.,
45 pls.

Amphiprora conspicua
Mann, A., 1907
Report on the diatoms of the Albatross
voyages in the Pacific Ocean, 1888-1904.
Contrib. U. S. Nat. Herb. 10(5):221-419, Pls.
XLIV-LIV.

Amphiprora costata
Kokubo, S., and S. Sato, 1947
Plankters in Jū-San Gata. Physiol.
and Ecol. (Japan) 1(4):1-16, 3 text figs.,
tables.

Amphiprora decussata
Morse, D.C., 1947
Some observations on seasonal variations in
plankton population Patuxant River, Maryland 1943-
1945. Bd. Nat. Res., Publ. No.65, Chesapeake
Biol. Lab., 31, 3 figs.

Amphiprora fimbriata n.sp.
Castracane degli Antelminelli, F., 1886
1. Report on the Diatomaceae collected
by H.M.S. Challenger during the years 1873-
1876. Rept. Sci. Results, H.M.S. Challenger,
Botany Vol. II, 178 pp., 30 pls.

Amphiprora gigantea
Rampi, L., 1940
Diatomee del Mare Adriatico. Nuovo
Giornale Botanico Italiano, n.s., 47:559-608.

Amphiprora gigantea var. sulcata
Allen, W.E., and E.E. Cupp, 1935
Plankton diatoms of the Java Sea.
Annales du Jardin Botanique de Buitenzorg
XLIV (2):101-174, figs.1-127.
(drawings of all species mentioned)

Amphiprora gigantea sulcata
Cupp, Easter E., 1943
Marine plankton diatoms of the west coast
of North America. Bull. S.I.O. 5(1):1-238,
5 pls., 168 text figs.

Amphiprora hyalina
Hendy, N. Ingram, 1964
An introductory account of the smaller algae
of British coastal waters. V. Bacillariophyceae (Diatoms).
Her Majesty's Stationary Office, 317 pp.,
45 pls.

Amphiprora hyalina
Hendey, N.I., 1951
Littoral diatoms of Chicester Harbour
with special reference to fouling. J.Roy.
Microscop. Soc. 71(1): 1-86, 18 pls.

Amphiprora hyperborea
Gran, H.H., 1908
Diatomeen. Nordisches Plankton, Botanis-
cher Teil pp. XIX.1-XIX 146; 178 text figs.

Amphiprora hyperborea
Hendy, N. Ingram, 1964
An introductory account of the smaller algae
of British coastal waters. V. Bacillariophyceae (Diatoms).
Her Majesty's Stationary Office, 317 pp.,
45 pls.

Amphiprora hyperborea
Ramsfjell, Einar, 1959
The occurrence of bi-flagellate swarmers within
cells of the pennate diatom Amphiprora hyper-
borea (Grunow) Gran.
Nytt Magasin for Botanikk, 7: 179-180.

Amphiprora kjellmanii
Hendy, N. Ingram, 1964
An introductory account of the smaller algae
of British coastal waters. V. Bacillariophyceae (Diatoms).
Her Majesty's Stationary Office, 317 pp.,
45 pls.

Amphiprora Kjellmanii
Hendey, N.I., 1937
The plankton diatoms of the southern seas.
Discovery Repts. 16:151-364, pls.6-13.

Amphiprora kjellmani
Jouse, A.P., G.S. Koroleva, G.A. Nagaeva,
1962
Diatoms in the surface layer of sediment in
the Indian sector of the Antarctic. Investigations of marine bottom sediments.
(In Russian; English summary).
Trudy Inst. Okeanol., Akad. Nauk, SSSR, 61:
19-92.

Amphiprora lauriensis, n.sp.
Prenguelli, Joaquin, y Hector A. Orlando, 1958.
Diatomeas y silicoflagelados del sector
Antartico Sudamericano.
Inst. Antartico Argentino, Publ., No. 5:191 pp.

Amphiprora lepidoptera
Grunow, A., 1863
Ueber einige neue und ungenügend bekannte
Arten und Gattungen von Diatomaceen. Verhandl.
d. K.K. Zool. Bot. Gesellsch., Vienna, 13:
137-162, pl.4-5 (Pl. 13-14).

Amphiprora lepidoptera
Jorgensen, E., 1900
Protophyten und Protozoën im Plank-
ton aus der Norwegischen Westkerste. Bergens
Mus. Aarb. 1899(6): 95 pp., 5 pls., 83 tables.

Amphiprora maxima
Jorgensen, E., 1900
Protophyten und Protozoën im Plank-
ton aus der Norwegischen Westkerste. Bergens
Mus. Aarb. 1899(6): 95 pp., 5 pls., 83 tables.

Amphiprora Oestrupi
Mangin, L., 1915
Phytoplancton de L'Antartique. Deuxieme
Exped. Ant. Francaise (1908-1910), 95 pp., 3 pls.,
58 text figs.

Amphiprora ornata
Mann, A., 1893
List of Diatomaceae from a deep-sea dredg-
ing in the Atlantic Ocean off Delaware Bay by
the U. S. Fish Commission Steamer Albatross.
Proc. U. S. Nat. Mus. 16:303-312.

Amphiprora ornata
Morse, D.C., 1947
Some observations on seasonal variations in
plankton population Patuxant River, Maryland 1943-
1945. Bd. Nat. Res., Publ. No.65, Chesapeake
Biol. Lab., 31, 3 figs.

Amphiprora ornata

Zanon, D. V., 1949
Diatomee di Buenos Aires (Argentina)
Atti Accad. Naz. Lincei, Memorie, Cl. Sci. fis., mat. e. nat., ser. 7, 11(3):59-151, 2 pls.

Amphiprora paludosa

Brunel, J., 1962
Le phytoplancton de la Baie de Chaleurs.
Inst. Botan., Univ. Montréal, Contrib. No. 77: 365 pp., 66 pls.

Amphiprora paludosa

Brunel, Jules, 1962
Le phytoplancton de la Baie des Chaleurs.
Contrib. Ministère de la Chasse et des Pêcheries, Province de Québec, No. 91: 365 pp.

Amphiprora paludosa

Hendey, N.I., 1951
Littoral diatoms of Chicester Harbour with special reference to fouling. J.Roy. Microscop. Soc. 71(1): 1-86, 18 pls.

Amphiprora paludosa

Hustedt, F. and A.A. Aleem, 1951
Littoral diatoms from the Salstone near Plymouth. JMBA 30(1): 177.196.

Amphiprora paludosa

Iyengar, M.O.P. and G.Venkataraman, 1951.
The ecology and seasonal succession of the algae flora of the River Cooum at Madras with special reference to the Diatomaceae. J. Madras Univ. 21, Sect. B(1): 140-192, 1 pl of 4 figs., 11 text figs.

Amphiprora paludosa

Rumkówna, A., 1948
[List of the phytoplankton species occurring in the superficial water layers in the Gulf of Gdańsk] Bull. Lab. mar., Gdynia, No. 4: 139-141 with tables in back.

Amphiprora paludosa

Zanon, D. V., 1949
Diatomee di Buenos Aires (Argentina)
Atti Accad. Naz. Lincei, Memorie, Cl. Sci. fis., mat. e. nat., ser. 7, 11(3):59-151, 2 pls.

Amphiprora paludosa

Zanon, V., 1948
Diatomee marini di Sardegna e Pugillo di Alghe Marine della stressa. Boll. Pesca, Piscitutura e Idrobiologia, Anno 24, ns. 3(2): 202-244, 27 figs. on 1 pl.

Amphiprora plicata var. japonica n.var.

Castracane degli Antelminelli, F., 1886
1. Report on the Diatomaceae collected by H.M.S. Challenger during the years 1873-1876. Rept. Sci. Results, H.M.S. Challenger, Botany Vol. II, 178 pp., 30 pls.

Amphiprora pulchra

Hendey, N. Ingram, 1964
An introductory account of the smaller algae of British coastal waters. V. Bacillariophyceae (Diatoms).
Her Majesty's Stationary Office, 317 pp., 45 pls.

Amphiprora pulchra

Hustedt, F. and A.A. Aleem, 1951
Littoral diatoms from the Salstone near Plymouth. JMBA 30(1): 177.196.

Amphiprora pulchra var. pulchella

Rampi, L., 1942
Ricerche sul fitoplancton del Mare Ligure 6. Le diatomee delle acque di Sanremo. Nuovo Giornale Botanico Italiano, N.S., 49:252-268.

Amphiprora sulcata

Hustedt, F. and A.A. Aleem, 1951
Littoral diatoms from the Salstone near Plymouth. JMBA 30(1): 177.196.

Amphiprora sulcata

Rampi, L., 1940
Diatomee del Mare Adriatico. Nuovo Giornale Botanico Italiano, n.s., 47:559-608.

Amphiprora surirelloides

Hendy, N. Ingram, 1964
An introductory account of the smaller algae of British coastal waters. V. Bacillariophyceae (Diatoms).
Her Majesty's Stationary Office, 317 pp., 45 pls.

Amphiprora surirelloides n.sp.

Hendey, N.I., 1951
Littoral diatoms of Chicester Harbour with special reference to fouling. J.Roy. Microscop. Soc. 71(1): 1-86, 18 pls.

Amphiprora venusta

de Sousa e Silva, E., 1956.
Contribution à l'étude du microplancton de Dakar et des régions maritimes voisines.
Bull. I.F.A.N., 8(2):335-371, 7 pls.

Amphiprora venusta

Rampi, L., 1942
Ricerche sul fitoplancton del Mare Ligure 6. Le diatomee delle acque di Sanremo. Nuovo Giornale Botanico Italiano, N.S., 49:252-268.

Amphiprora Vestrupii

Hendey, N.I., 1937
The plankton diatoms of the southern seas. Discovery Repts. 16:151-364, pls.6-13.

Amphora sp.

Brunel, J., 1962
Le phytoplancton de la Baie de Chaleurs.
Inst. Botan., Univ. Montréal, Contrib. No. 77: 365 pp., 66 pls.

Amphora

Desikachary, T.V., 1956(1957).
Electron microscope studies on diatoms.
J.R. Microsc. Soc. (3)76(1/2):9-36.

Amphora sp.

Kucherova, Z.S., 1961.
[Vertical distribution of diatoms from Sevastopol Bay.]
Trudy Sevastopol Biol. Sta., (14):64-78.

Amphora

O'Meara, E., 1871
On some new species of the genus Amphora.
Quart. Jour. Micro. Sci., 11:

Amphora sp.

Ukeles, R., 1961.
The effect of temperature on the growth and survival of several marine algal species.
Biol. Bull., 120(2):255-264.

Amphora abludens n.sp.

Simonsen, Reimer, 1960.
Neue Diatomeen aus der Ostsee. II.
Kieler Meeresf., 16(1):126-130.

Amphora acuta

Hendy, N. Ingram, 1964
An introductory account of the smaller algae of British coastal waters. V. Bacillariophyceae (Diatoms).
Her Majesty's Stationary Office, 317 pp., 45 pls.

Amphora acuta

Takano, Hideaki, 1963. Notes on marine littoral diatoms of Japan.
Bull. Tokai Reg. Fish. Res. Lab., No. 36:1-8.

Amphora acuta arcuata

Andrade e Clovis Teixeira, M.H. de, 1957.
Contribuição para o conhecimento das diatomáceas do Brasil.
Bol. Inst. Ocean., Univ. Sao Paulo, 8(1-2):171-225, 10 pls.

Amphora acuta var. acuata

Rampi, L., 1940
Diatomee del Mare Adriatico. Nuovo Giornale Botanico Italiano, n.s., 47:559-608.

Amphora acuta arcuata

Zanon, V., 1948
Diatomee marini di Sardegna e Pugillo di Alghe Marine della stressa. Boll. Pesca, Piscitutura e Idrobiologia, Anno 24, ns. 3(2): 202-244, 27 figs. on 1 pl.

Amphora acutiuscula

Zanon, V., 1948
Diatomee marini di Sardegna e Pugillo di Alghe Marine della stressa. Boll. Pesca, Piscitutura e Idrobiologia, Anno 24, ns. 3(2): 202-244, 27 figs. on 1 pl.

Amphora angularis

Zanon, V., 1948
Diatomee marini di Sardegna e Pugillo di Alghe Marine della stressa. Boll. Pesca, Piscitutura e Idrobiologia, Anno 24, ns. 3(2): 202-244, 27 figs. on 1 pl.

Amphora angusta

Hustedt, F. and A.A. Aleem, 1951
Littoral diatoms from the Salstone near Plymouth. JMBA 30(1): 177.196.

Amphora angusta

Zanon, V., 1948
Diatomee marini di Sardegna e Pugillo di Alghe Marine della stressa. Boll. Pesca, Piscitutura e Idrobiologia, Anno 24, ns. 3(2): 202-244, 27 figs. on 1 pl.

Amphora arcus

Hustedt, F. and A.A. Aleem, 1951
Littoral diatoms from the Salstone near Plymouth. JMBA 30(1): 177.196.

Amphora arcus

Zanon, V., 1948
Diatomee marini di Sardegna e Pugillo di Alghe Marine della stressa. Boll. Pesca, Piscitutura e Idrobiologia, Anno 24, ns. 3(2): 202-244, 27 figs. on 1 pl.

Amphora arenaria

de Sousa e Silva, E., 1956.
Contribution à l'étude du microplancton de Dakar et des régions maritimes voisines.
Bull. I.F.A.N., 8(2):335-371, 7 pls.

Amphora arenaria

Eskinazi Enide e Shigekatsv Satô (1963-1964) 1966.
Contribuição ao estudo das diatomaceas da Praia de Piedade.

Trabhs Inst. Oceanogr., Univ. Recife, 5 (5/6): 73-114.

Amphora arenaria

Hendy, N. Ingram, 1964
An introductory account of the smaller algae of British coastal waters. V. Bacillariophyceae (Diatoms).
Her Majesty's Stationary Office, 317 pp., 45 pls.

Amphora arenaria

Hendey, N.J., 1951
Littoral diatoms of Chicester Harbour with special reference to fouling. J.Roy. Microscop. Soc. 71(1): 1-86, 18 pls.

Amphora arenaria
Politis, J., 1949
Diatomees marines de Bosphores et des ibes de la mer de Marmara. II Practica tou Hellenikou Hidrobiologikou Institutoutou 1929, Etoz 1929, 3(1):11-31.

Amphora arenaria
Rampi, L., 1940
Diatomee del Mare Adriatico. Nuovo Giornale Botanico Italiano, n.s., 47:559-608.

Amphora arenaria
Zanon, V., 1948
Diatomee marini di Sardegna e Pugillo di Alghe Marine della stressa. Boll. Pesca, Piscitutura e Idrobiologia, Anno 24, ns. 3(2): 202-244, 27 figs. on 1 pl.

Amphora arenicola
Hendy, N. Ingram, 1964
An introductory account of the smaller algae of British coastal waters. V. Bacillariophyceae (Diatoms). Her Majesty's Stationary Office, 317 pp., 45 pls.

Amphora arenicola
Politis, J., 1949
Diatomees marines de Bosphores et des ibes de la mer de Marmara. II Practica tou Hellenikou Hidrobiologikou Institutoutou 1929, Etoz 1929, 3(1):11-31.

Amphora arenicola
Zanon, V., 1948
Diatomee marini di Sardegna e Pugillo di Alghe Marine della stressa. Boll. Pesca, Piscitutura e Idrobiologia, Anno 24, ns. 3(2): 202-244, 27 figs. on 1 pl.

Amphora baccata n.sp.
Mann, A., 1907
Report on the diatoms of the Albatross voyages in the Pacific Ocean, 1888-1904. Contrib. U. S. Nat. Herb. 10(5):221-419, Pls. XLIV-LIV.

Amphora Barrei n.sp.
Manguin, E., 1957.
Premier inventaire des diatomées de la Terre Adélie Antarctique. Espèces nouvelles. Rev. Algologique, n.s., 3(3):111-134.

Amphora bigibba
Andrade e Clovis Teixeira, M.H., 1957.
Contribuição para o conhecimento das diatomáceas do Brasil. Bol. Inst. Ocean., Univ. Sao Paulo, 8(1-2):171-225, 10 pls.

Amphora bigibba
Mann, A., 1893
List of Diatomaceae from a deep-sea dredging in the Atlantic Ocean off Delaware Bay by the U. S. Fish Commission Steamer Albatross. Proc. U. S. Nat. Mus. 16:303-312.

Amphora bigibba
Zanon, V., 1948
Diatomee marini di Sardegna e Pugillo di Alghe Marine della stressa. Boll. Pesca, Piscitutura e Idrobiologia, Anno 24, ns. 3(2): 202-244, 27 figs. on 1 pl.

Amphora binodis
Hendy, N. Ingram, 1964
An introductory account of the smaller algae of British coastal waters. V. Bacillariophyceae (Diatoms). Her Majesty's Stationary Office, 317 pp., 45 pls.

Amphora binodes
Politis, J., 1949
Diatomees marines de Bosphores et des ibes de la mer de Marmara. II Practica tou Hellenikou Hidrobiologikou Institutoutou 1929, Etoz 1929, 3(1):11-31.

Amphora binodis
Rampi, L., 1942
Ricerche sul fitoplancton del Mare Ligure 6. Le diatomee delle acque di Sanremo. Nuovo Giornale Botanico Italiano, N.S., 49:252-268.

Amphora binodis
Rampi, L., 1940
Diatomee del Mare Adriatico. Nuovo Giornale Botanico Italiano, n.s., 47:559-608.

Amphora cingulata
Mann, A., 1893
List of Diatomaceae from a deep-sea dredging in the Atlantic Ocean off Delaware Bay by the U. S. Fish Commission Steamer Albatross. Proc. U. S. Nat. Mus. 16:303-312.

Amphora clevei
Zanon, V., 1948
Diatomee marini di Sardegna e Pugillo di Alghe Marine della stressa. Boll. Pesca, Piscitutura e Idrobiologia, Anno 24, ns. 3(2): 202-244, 27 figs. on 1 pl.

Amphora clevia
Morse, D.C., 1947
Some observations on seasonal variations in plankton population Patuxant River, Maryland 1943-1945. Bd. Nat. Res., Publ. No.65, Chesapeake Biol. Lab., 31, 3 figs.

Amphora coffaeformis & varieties
Hendy, N. Ingram, 1964
An introductory account of the smaller algae of British coastal waters. V. Bacillariophyceae (Diatoms). Her Majesty's Stationary Office, 317 pp., 45 pls.

Amphora coffeaeformis
Hendey, N.I., 1951
Littoral diatoms of Chicester Harbour with special reference to fouling. J.Roy. Microscop. Soc. 71(1): 1-86, 18 pls.

Amphora coffaeiformis
Hustedt, F. and A.A. Aleem, 1951
Littoral diatoms from the Salstone near Plymouth. JMBA 30(1): 177.196.

Amphora coffeaeformis
Iyengar, M.O.P. and G.Venkataraman, 1951.
The ecology and seasonal succession of the algae flora of the River Cooum at Madras with special reference to the Diatomaceae. J. Madras Univ. 21, Sect. B(1): 140-192, 1 pl of 4 figs., 11 text figs.

Amphora coffaeiformis
Zanon, V., 1948
Diatomee marini di Sardegna e Pugillo di Alghe Marine della stressa. Boll. Pesca, Piscitutura e Idrobiologia, Anno 24, ns. 3(2): 202-244, 27 figs. on 1 pl.

Amphora coffeaeformis africana
Iyengar, M.O.P. and G.Venkataraman, 1951.
The ecology and seasonal succession of the algae flora of the River Cooum at Madras with special reference to the Diatomaceae. J. Madras Univ. 21, Sect. B(1): 140-192, 1 pl of 4 figs., 11 text figs.

Amphora commutata
Kokubo, S., and S. Sato., 1947
Plankters in JG-San Gata. Physiol. and Ecol. (Japan) 1(4):1-16, 3 text figs., tables.

Amphora commutata
Rumkówna, A., 1948
List of the phytoplankton species occurring in the superficial water layers in the Gulf of Gdańsk. Bull. Lab. mar., Gdynia, No. 4: 139-141 with tables in back.

Amphora commutata
Zanon, V., 1948
Diatomee marini di Sardegna e Pugillo di Alghe Marine della stressa. Boll. Pesca, Piscitutura e Idrobiologia, Anno 24, ns. 3(2): 202-244, 27 figs. on 1 pl.

Amphora contracta
Rampi, L., 1940
Diatomee del Mare Adriatico. Nuovo Giornale Botanico Italiano, n.s., 47:559-608.

Amphora costata
Andrade e Clovis Teixeira, M.H. de, 1957.
Contribuição para o conhecimento das diatomáceas do Brasil. Bol. Inst. Ocean., Univ. Sao Paulo, 8(1/2):171-225, 10 pls.

Amphora costata
Hendy, N. Ingram, 1964
An introductory account of the smaller algae of British coastal waters. V. Bacillariophyceae (Diatoms). Her Majesty's Stationary Office, 317 pp., 45 pls.

Amphora costata
Hustedt, F. and A.A. Aleem, 1951
Littoral diatoms from the Salstone near Plymouth. JMBA 30(1): 177.196.

Amphora costata
Politis, J., 1949
Diatomees marines de Bosphores et des ibes de la mer de Marmara. II Practica tou Hellenikou Hidrobiologikou Institutoutou 1929, Etoz 1929, 3(1):11-31.

Amphora costata
Rampi, L., 1940
Diatomee del Mare Adriatico. Nuovo Giornale Botanico Italiano, n.s., 47:559-608.

Amphora costata
Zanon, V., 1948
Diatomee marini di Sardegna e Pugillo di Alghe Marine della stressa. Boll. Pesca, Piscitutura e Idrobiologia, Anno 24, ns. 3(2): 202-244, 27 figs. on 1 pl.

Amphora crassa
de Sousa e Silva, E., 1956.
Contribution à l'étude du microplancton de Dakar et des region maritimes voisines. Bull. I.F.A.N., 8(2):335-371, 7 pls.

Amphora crassa
Hendy, N. Ingram, 1964
An introductory account of the smaller algae of British coastal waters. V. Bacillariophyceae (Diatoms). Her Majesty's Stationary Office, 317 pp., 45 pls.

Amphora crassa
Politis, J., 1949
Diatomees marines de Bosphores et des ibes de la mer de Marmara. II Practica tou Hellenikou Hidrobiologikou Institutoutou 1929, Etoz 1929, 3(1):11-31.

Amphora crassa
Rampi, L., 1940
Diatomee del Mare Adriatico. Nuovo Giornale Botanico Italiano, n.s., 47:559-608.

Amphora crassa
Zanon, V., 1948
Diatomee marini di Sardegna e Pugillo di Alghe Marine della stressa. Boll. Pesca, Piscitutura e Idrobiologia, Anno 24, ns. 3(2): 202-244, 27 figs. on 1 pl.

Amphora crescens n.sp.

Mann, A., 1907
Report on the diatoms of the Albatross voyages in the Pacific Ocean, 1888-1904. Contrib. U. S. Nat. Herb. 10(5):221-419, Pls. XLIV-LIV.

Amphora cuneata

Zanon, V., 1948
Diatomee marini di Sardegna e Pugillo di Alghe Marine della stressa. Boll. Pesca, Piscitutura e Idrobiologia, Anno 24, ns. 3(2): 202-244, 27 figs. on 1 pl.

Amphora cymbifera

Hendy, N. Ingram, 1964
An introductory account of the smaller algae of British coastal waters. V. Bacillariophyceae (Diatoms). Her Majesty's Stationary Office, 317 pp., 45 pls.

Amphora cymbiffera

Mann, A., 1893
List of Diatomaceae from a deep-sea dredging in the Atlantic Ocean off Delaware Bay by the U. S. Fish Commission Steamer Albatross. Proc. U. S. Nat. Mus. 16:303-312.

Amphora cymbilifera

Rampi, L., 1940
Diatomee del Mare Adriatico. Nuovo Giornale Botanico Italiano, n.s., 47:559-608.

Amphora decloitrei n.sp.

Amossé, A. 1970.
Diatomées marines et saumâtres du Sénégal et de la Côte d'Ivoire.
Bull. Inst. fond. Afr. Noire (A) 32 (2): 289-314.

Amphora decora n.sp.

Castracane degli Antelminelli, F., 1886
1. Report on the Diatomaceae collected by H.M.S. Challenger during the years 1873-1876. Rept. Sci. Results, H.M.S. Challenger, Botany Vol. II, 178 pp., 30 pls.

Amphora decussata

Allen, W.E., and E.E. Cupp, 1935
Plankton diatoms of the Java Sea. Annales du Jardin Botanique de Buitenzorg XLIV (2):101-174, figs.1-127.
(drawings of all species mentioned)

Amphora decussata

Andrade e Clovis Teixeira, M. H. de, 1957.
Contribuição para o conhecimento das diatomáceas do Brasil.
Bol. Inst. Ocean., Univ. Sao Paulo, 8(1/2):171-225, 10 pls.

Amphora decussata

Grunow, A., 1877
1. New Diatoms from Honduras. Monthly Micros. Jour., 18:165-186, pls. CXCIII-CXCVI.

Amphora decussata

Hendy, N. Ingram, 1964
An introductory account of the smaller algae of British coastal waters. V. Bacillariophyceae (Diatoms). Her Majesty's Stationary Office, 317 pp., 45 pls.

Amphora decussata

Rampi, L., 1940
Diatomee del Mare Adriatico. Nuovo Giornale Botanico Italiano, n.s., 47:559-608.

Amphora dubia

Zanon, V., 1948
Diatomee marini di Sardegna e Pugillo di Alghe Marine della stressa. Boll. Pesca, Piscitutura e Idrobiologia, Anno 24, ns. 3(2): 202-244, 27 figs. on 1 pl.

Amphora egregia

Rampi, L., 1940
Diatomee del Mare Adriatico. Nuovo Giornale Botanico Italiano, n.s., 47:559-608.

Amphora egregia

Zanon, V., 1948
Diatomee marini di Sardegna e Pugillo di Alghe Marine della stressa. Boll. Pesca, Piscitutura e Idrobiologia, Anno 24, ns. 3(2): 202-244, 27 figs. on 1 pl.

Amphora egregaria interrupta

Andrade e Clovis Teixeira, M.H., de, 1957.
Contribuição para o conhecimento das diatomáceas do Brasil.
Bol. Inst. Ocean., Univ. Sao Paulo, 8(1/2):171-225, 10 pls.

Amphora egregia

de Sousa e Silva, E., 1956.
Contribution à l'étude du microplancton de Dakar et des regions maritimes voisines. Bull. I.F.A.N., 8(2):335-371, 7 pls.

Amphora eunotia

Hendy, N. Ingram, 1964
An introductory account of the smaller algae of British coastal waters. V. Bacillariophyceae (Diatoms). Her Majesty's Stationary Office, 317 pp., 45 pls.

Amphora eunotia

Zanon, V., 1948
Diatomee marini di Sardegna e Pugillo di Alghe Marine della stressa. Boll. Pesca, Piscitutura e Idrobiologia, Anno 24, ns. 3(2): 202-244, 27 figs. on 1 pl.

Amphora exigua

Hendy, N. Ingram, 1964
An introductory account of the smaller algae of British coastal waters. V. Bacillariophyceae (Diatoms). Her Majesty's Stationary Office, 317 pp., 45 pls.

Amphora exigua

Hendey, N.T., 1951
Littoral diatoms of Chicester Harbour with special reference to fouling. J.Roy. Microscop. Soc. 71(1): 1-86, 18 pls.

Amphora exigua

Politis, J., 1949
Diatomees marines de Bosphores et des ibes de la mer de Marmara. II Practica tou Hellenikou Hidrobiologikou Institutoutou 1929, Etoz 1929, 3(1):11-31.

Amphora exigua

Zanon, V., 1948
Diatomee marini di Sardegna e Pugillo di Alghe Marine della stressa. Boll. Pesca, Piscitutura e Idrobiologia, Anno 24, ns. 3(2): 202-244, 27 figs. on 1 pl.

Amphora extensa

Hendy, N. Ingram, 1964
An introductory account of the smaller algae of British coastal waters. V. Bacillariophyceae (Diatoms). Her Majesty's Stationary Office, 317 pp., 45 pls.

Amphora flebilis n.sp.

Simonsen, Reimer, 1960.
Neue Diatomeen aus der Ostsee. II.
Kieler Meeresf., 16(1):126-130.

Amphora fluminensis

Grunow, A., 1863
Ueber einige neue und ungenügend bekannte Arten und Gattungen von Diatomaceen. Verhandl. d. K.K. Zool. Bot. Gesellsch., Vienna, 13: 137-162, pl.4-5 (Pl. 13-14).

Amphora formosa

Zanon, V., 1948
Diatomee marini di Sardegna e Pugillo di Alghe Marine della stressa. Boll. Pesca, Piscitutura e Idrobiologia, Anno 24, ns. 3(2): 202-244, 27 figs. on 1 pl.

Amphora Gallurae n.sp.

Zanon, V., 1948
Diatomee marini di Sardegna e Pugillo di Alghe Marine della stressa. Boll. Pesca, Piscitutura e Idrobiologia, Anno 24, ns. 3(2): 202-244, 27 figs. on 1 pl.

Amphora gigantea

Zanon, V., 1948
Diatomee marini di Sardegna e Pugillo di Alghe Marine della stressa. Boll. Pesca, Piscitutura e Idrobiologia, Anno 24, ns. 3(2): 202-244, 27 figs. on 1 pl.

Amphora graeffi minor

Hendy, N. Ingram, 1964
An introductory account of the smaller algae of British coastal waters. V. Bacillariophyceae (Diatoms). Her Majesty's Stationary Office, 317 pp., 45 pls.

Amphora Graeffii

Zanon, V., 1948
Diatomee marini di Sardegna e Pugillo di Alghe Marine della stressa. Boll. Pesca, Piscitutura e Idrobiologia, Anno 24, ns. 3(2): 202-244, 27 figs. on 1 pl.

Amphora granulata

Zanon, V., 1948
Diatomee marini di Sardegna e Pugillo di Alghe Marine della stressa. Boll. Pesca, Piscitutura e Idrobiologia, Anno 24, ns. 3(2): 202-244, 27 figs. on 1 pl.

Amphora Grevilleana

Andrade e Clovis Teixeira, M. H. de, 1957.
Contribuição para o conhecimento das diatomáceas do Brasil.
Bol. Inst. Ocean., Univ., Sao Paulo, 8(1/2):171-225, 10 pls.

Amphora grevilleana

Hendy, N. Ingram, 1964
An introductory account of the smaller algae of British coastal waters. V. Bacillariophyceae (Diatoms). Her Majesty's Stationary Office, 317 pp., 45 pls.

Amphora grevilleana

Zanon, V., 1948
Diatomee marini di Sardegna e Pugillo di Alghe Marine della stressa. Boll. Pesca, Piscitutura e Idrobiologia, Anno 24, ns. 3(2): 202-244, 27 figs. on 1 pl.

Amphora Gründleri

Andrade e Clovis Teixeira, M.H. de, 1957.
Contribuição para o conhecimento das diatomáceas do Brasil.
Bol. Inst. Ocean., Univ. Sao Paulo, 8(1/2):171-225, 10 pls.

Amphora gulfusiis n.sp.

Salah, M., and G. Tamás 1968.
Notes on new planktonic diatoms from Egypt.
Hydrobiologia 31 (2): 231-240.

Amphora helenensis n.sp.

Giffen, Malcolm H. 1973.
Diatoms of the marine littoral of Steenberg's Cove in S. Helena Bay, Cape Province, South Africa.
Botanica Marina 16 (1): 32-48.

Amphora holsatica

Iyengar, M.O.P. and G.Venkataraman,1951.
The ecology and seasonal succession of the algae flora of the River Cooum at Madras with special reference to the Diatomaceae. J. Madras Univ. 21, Sect. B(1): 140-192, 1 pl of 4 figs., 11 text figs.

Amphora honshuensis n.sp.
Mann, A., 1907
Report on the diatoms of the Albatross voyages in the Pacific Ocean, 1888-1904. Contrib. U. S. Nat. Herb. 10(5):221-419, Pls. XLIV-LIV.

Amphora hyalina
Hendy, N. Ingram, 1964
An introductory account of the smaller algae of British coastal waters. V. Bacillariophyceae (Diatoms). Her Majesty's Stationary Office, 317 pp., 45 pls.

Amphora hyalina
Hendey, N.I., 1951
Littoral diatoms of Chicester Harbour with special reference to fouling. J.Roy. Microscop. Soc. 71(1): 1-86, 18 pls.

Amphora hyalina
Hustedt, F. and A.A. Aleem, 1951
Littoral diatoms from the Salstone near Plymouth. JMBA 30(1): 177.196.

Amphora inflesca
Zanon, V., 1948
Diatomee marini di Sardegna e Pugillo di Alghe Marine della stressa, Boll. Pesca, Piscitutura e Idrobiologia, Anno 24, ns. 3(2): 202-244, 27 figs. on 1 pl.

Amphora inflexa
Rampi, L., 1940
Diatomee del Mare Adriatico. Nuovo Giornale Botanico Italiano, n.s., 47:559-608.

Amphora Janischii
Zanon, V., 1948
Diatomee marini di Sardegna e Pugillo di Alghe Marine della stressa, Boll. Pesca, Piscitutura e Idrobiologia, Anno 24, ns. 3(2): 202-244, 27 figs. on 1 pl.

Amphora Kamorthensis
Rampi, L., 1940
Diatomee del Mare Adriatico. Nuovo Giornale Botanico Italiano, n.s., 47:559-608.

Amphora laevis
Hendy, N. Ingram, 1964
An introductory account of the smaller algae of British coastal waters. V. Bacillariophyceae (Diatoms). Her Majesty's Stationary Office, 317 pp., 45 pls.

Amphora laevis
Hustedt, F. and A.A. Aleem, 1951
Littoral diatoms from the Salstone near Plymouth. JMBA 30(1): 177.196.

Amphora laevis
Rampi, L., 1942
Ricerche sul fitoplancton del Mare Ligure 6. Le diatomee delle acque di Sanremo. Nuovo Giornale Botanico Italiano, N.S., 49:252-268.

Amphora laevis
Zanon, V., 1948
Diatomee marini di Sardegna e Pugillo di Alghe Marine della stressa, Boll. Pesca, Piscitutura e Idrobiologia, Anno 24, ns. 3(2): 202-244, 27 figs. on 1 pl.

Amphora laevissima
Rampi, L., 1942
Ricerche sul fitoplancton del Mare Ligure 6. Le diatomee delle acque di Sanremo. Nuovo Giornale Botanico Italiano, N.S., 49:252-268.

Amphora lineolata var. chinensis
Allen, W.E., and E.E. Cupp, 1935
Plankton diatoms of the Java Sea. Annales du Jardin Botanique de Buitenzorg XLIV (2):101-174, figs.1-127.
(drawings of all species mentioned)

Amphora lineolata
Hustedt, F. and A.A. Aleem, 1951
Littoral diatoms from the Salstone near Plymouth. JMBA 30(1): 177.196.

Amphora lineata (figs.)
Sousa e Silva, E., 1949
Diatomaceas e Dinoflagelados de Baia de Cascais. Portugaliae Acta Biol., Volume: Julio Henriques, Ser. B: 300-383, 9 pls, 2 fold-in tables.

Amphora longiceps n.sp.
Simonsen, Reimer, 1960.
Neue Diatomeen aus der Ostsee. II. Kieler Meeresf., 16(1):126-130.

Amphora lunula
Rampi, L., 1940
Diatomee del Mare Adriatico. Nuovo Giornale Botanico Italiano, n.s., 47:559-608.

Amphora macilenta
Hendy, N. Ingram, 1964
An introductory account of the smaller algae of British coastal waters. V. Bacillariophyceae (Diatoms). Her Majesty's Stationary Office, 317 pp., 45 pls.

Amphora macilenta
Rampi, L., 1940
Diatomee del Mare Adriatico. Nuovo Giornale Botanico Italiano, n.s., 47:559-608.

Amphora macilenta
Zanon, V., 1948
Diatomee marini di Sardegna e Pugillo di Alghe Marine della stressa, Boll. Pesca, Piscitutura e Idrobiologia, Anno 24, ns. 3(2): 202-244, 27 figs. on 1 pl.

Amphora marina
Hustedt, F. and A.A. Aleem, 1951
Littoral diatoms from the Salstone near Plymouth. JMBA 30(1): 177.196.

Amphora marina
Zanon, V., 1948
Diatomee marini di Sardegna e Pugillo di Alghe Marine della stressa, Boll. Pesca, Piscitutura e Idrobiologia, Anno 24, ns. 3(2): 202-244, 27 figs. on 1 pl.

Amphora meneghiniana n.sp
Castracane degli Antelminelli, F., 1886
1. Report on the Diatomaceae collected by H.M.S. Challenger during the years 1873-1876. Rept. Sci. Results, H.M.S. Challenger, Botany Vol. II, 178 pp., 30 pls.

Amphora mexicana
Mann, A., 1907
Report on the diatoms of the Albatross voyages in the Pacific Ocean, 1888-1904. Contrib. U. S. Nat. Herb. 10(5):221-419, Pls. XLIV-LIV.

Amphora mexicana
Zanon, V., 1948
Diatomee marini di Sardegna e Pugillo di Alghe Marine della stressa, Boll. Pesca, Piscitutura e Idrobiologia, Anno 24, ns. 3(2): 202-244, 27 figs. on 1 pl.

Amphora novae-calidoniae Grun
Mann, A., 1893
List of Diatomaceae from a deep-sea dredging in the Atlantic Ocean off Delaware Bay by the U. S. Fish Commission Steamer Albatross. Proc. U. S. Nat. Mus. 16:303-312.

Amphora oblonga
Bigelow, H.B., and M. Leslie, 1930
Reconnaissance of the waters and plankton of Monterey Bay, July 1928. Bull. M.C.Z. 70(5):429-481, 43 text figs.

Amphora obtusa
Andrade e Clovis Teixeira, M.H. de, 1957.
Contribuição para o conhecimento das diatomáceas do Brasil. Bol. Inst. Ocean., Univ. Sao Paulo, 8(1/2):171-225, 10 pls.

Amphora obtusa
Hendy, N. Ingram, 1964
An introductory account of the smaller algae of British coastal waters. V. Bacillariophyceae (Diatoms). Her Majesty's Stationary Office, 317 pp., 45 pls.

Amphora obtusa
Mann, A., 1893
List of Diatomaceae from a deep-sea dredging in the Atlantic Ocean off Delaware Bay by the U. S. Fish Commission Steamer Albatross. Proc. U. S. Nat. Mus. 16:303-312.

Amphora obtusa
Zanon, V., 1948
Diatomee marini di Sardegna e Pugillo di Alghe Marine della stressa, Boll. Pesca, Piscitutura e Idrobiologia, Anno 24, ns. 3(2): 202-244, 27 figs. on 1 pl.

Amphora obtusa var. oceanica
Rampi, L., 1940
Diatomee del Mare Adriatico. Nuovo Giornale Botanico Italiano, n.s., 47:559-608.

Amphora oceanica n.sp.
Castracane degli Antelminelli, F., 1886
1. Report on the Diatomaceae collected by H.M.S. Challenger during the years 1873-1876. Rept. Sci. Results, H.M.S. Challenger, Botany Vol. II, 178 pp., 30 pls.

Amphora ocellata
de Sousa e Silva, E., 1956.
Contribution à l'étude du microplancton de Dakar et des regions maritimes voisines. Bull. I.F.A.N., 8(2):335-371, 7 pls.

Amphora ocellata
Hendy, N. Ingram, 1964
An introductory account of the smaller algae of British coastal waters. V. Bacillariophyceae (Diatoms). Her Majesty's Stationary Office, 317 pp., 45 pls.

Amphora ocellata
Hustedt, F. and A.A. Aleem, 1951
Littoral diatoms from the Salstone near Plymouth. JMBA 30(1): 177.196.

Amphora ocellata
Morse, D.C., 1947
Some observations on seasonal variations in plankton population Patuxant River, Maryland 1943-1945. Bd. Nat. Res., Publ. No.65, Chesapeake Biol. Lab., 31, 3 figs.

Amphora ocellata
Zanon, V., 1948
Diatomee marini di Sardegna e Pugillo di Alghe Marine della stressa, Boll. Pesca, Piscitutura e Idrobiologia, Anno 24, ns. 3(2): 202-244, 27 figs. on 1 pl.

Amphora ostrearia

Hendey, N.I., 1951
Littoral diatoms of Chicester Harbour with special reference to fouling. J. Roy. Microscop. Soc. 71(1): 1-86, 18 pls.

Amphora ostrearia

Hustedt, F. and A.A. Aleem, 1951
Littoral diatoms from the Salstone near Plymouth. JMBA 30(1): 177-196.

Amphora ostrearia & varieties

Hendy, N. Ingram, 1964
An introductory account of the smaller algae of British coastal waters. V. Bacillariophyceae (Diatoms).
Her Majesty's Stationary Office, 317 pp., 45 pls.

Amphora ostrearis var. vitrea

Rampi, L., 1942
Ricerche sul fitoplancton del Mare Ligure 6. Le diatomee delle acque di Sanremo. Nuovo Giornale Botanico Italiano, N.S., 49:252-268.

Amphora ostrearia var. vitrea

Rampi, L., 1940
Diatomee del Mare Adriatico. Nuovo Giornale Botanico Italiano, n.s., 47:559-608.

Amphora ostrearia vitrea

Zanon, V., 1948
Diatomee marini di Sardegna e Pugillo di Alghe Marine della stressa. Boll. Pesca, Piscitutura e Idrobiologia, Anno 24, ns. 3(2): 202-244, 27 figs. on 1 pl.

Amphora ovalis

Rumkówna, A., 1948
[List of the phytoplankton species occurring in the superficial water layers in the Gulf of Gdańsk] Bull. Lab. mar., Gdynia, No. 4: 139-141 with tables in back.

Amphora ovalis

Zanon, D. V., 1949
Diatomee di Buenos Aires (Argentina) Atti Accad. Naz. Lincei, Memorie, Cl. Sci. fis., mat. e. nat., ser. 7, 11(3):59-151, 2 pls.

Amphora ovalis

Zanon, V., 1948
Diatomee marini di Sardegna e Pugillo di Alghe Marine della stressa. Boll. Pesca, Piscitutura e Idrobiologia, Anno 24, ns. 3(2): 202-244, 27 figs. on 1 pl.

Amphora ovalis var. affinis

Rampi, L., 1942
Ricerche sul fitoplancton del Mare Ligure 6. Le diatomee delle acque di Sanremo. Nuovo Giornale Botanico Italiano, N.S., 49:252-268.

Amphora ovalis pediculus

Tsumura, K., 1956.
Diatomoj el la cirkaufoso de la restajo de la kastele de Odawara. J. Yokohama Municipal Univ., (C-14), No. 47:23 pp.

Amphora Pampaninii n.sp.

Zanon, V., 1948
Diatomee marini di Sardegna e Pugillo di Alghe Marine della stressa. Boll. Pesca, Piscitutura e Idrobiologia, Anno 24, ns. 3(2): 202-244, 27 figs. on 1 pl.

Amphora pellucida

Mann, A., 1907
Report on the diatoms of the Albatross voyages in the Pacific Ocean, 1888-1904. Contrib. U. S. Nat. Herb. 10(5):221-419, Pls. XLIV-LIV.

Amphora peragallorum

Boden, B. P., 1949.
The diatoms collected by the U.S.S. CACOPAN in the Antarctic in 1947. J. Mar. Res. 8(1):6-13, 3 textfigs.

Amphora peragallorum

Boden, Brian, 1948
Marine plankton diatoms on operation HIGHJUMP in: Some oceanographic observations on operation HIGHJUMP. By R.S. Dietz. USNEL Rept. No.55, 97 pp., 41 figs. 7 July 1948.

Amphora Peragallorum

Hendey, N.I., 1937
The plankton diatoms of the southern seas. Discovery Repts. 16:151-364, pls.6-13.

Amphora perpusilla

Hutner, S.H., and L. Provosoli, 1953.
A pigmented marine diatom requiring Vitamin B12 and uracil. News Bull., Phycol. Soc., Amer., 6(18):7.

Amphora perpusilla

Rumkówna, A., 1948
[List of the phytoplankton species occurring in the superficial water layers in the Gulf of Gdańsk] Bull. Lab. mar., Gdynia, No. 4: 139-141 with tables in back.

Amphora perpusilla

Zanon, D. V., 1949
Diatomee di Buenos Aires (Argentina) Atti Accad. Naz. Lincei, Memorie, Cl. Sci. fis., mat. e. nat., ser. 7, 11(3):59-151, 2 pls.

Amphora perpusilla

Zanon, V., 1948
Diatomee marini di Sardegna e Pugillo di Alghe Marine della stressa. Boll. Pesca, Piscitutura e Idrobiologia, Anno 24, ns. 3(2): 202-244, 27 figs. on 1 pl.

Amphora philippinica n.sp.

Castracane degli Antelminelli, F., 1886
1. Report on the Diatomaceae collected by H.M.S. Challenger during the years 1873-1876. Rept. Sci. Results, H.M.S. Challenger, Botany Vol. II, 178 pp., 30 pls.

Amphora polyzonata n.sp.

Castracane degli Antelminelli, F., 1886
1. Report on the Diatomaceae collected by H.M.S. Challenger during the years 1873-1876. Rept. Sci. Results, H.M.S. Challenger, Botany Vol. II, 178 pp., 30 pls.

Amphora porcellus

Mann, A., 1893
List of Diatomaceae from a deep-sea dredging in the Atlantic Ocean off Delaware Bay by the U. S. Fish Commission Steamer Albatross. Proc. U. S. Nat. Mus. 16:303-312.

Amphora proboscidea

Hendy, N. Ingram, 1964
An introductory account of the smaller algae of British coastal waters. V. Bacillariophyceae (Diatoms).
Her Majesty's Stationary Office, 317 pp., 45 pls.

Amphora proteus

Hendy, N. Ingram, 1964
An introductory account of the smaller algae of British coastal waters. V. Bacillariophyceae (Diatoms).
Her Majesty's Stationary Office, 317 pp., 45 pls.

Amphora proteus

Hustedt, F. and A.A. Aleem, 1951
Littoral diatoms from the Salstone near Plymouth. JMBA 30(1): 177-196.

Amphora proteus

Mann, A., 1893
List of Diatomaceae from a deep-sea dredging in the Atlantic Ocean off Delaware Bay by the U. S. Fish Commission Steamer Albatross. Proc. U. S. Nat. Mus. 16:303-312.

Amphora proteus

Morse, D.C., 1947
Some observations on seasonal variations in plankton population Patuxant River, Maryland 1943-1945. Bd. Nat. Res., Publ. No.65, Chesapeake Biol. Lab., 31, 3 figs.

Amphora proteus

Zanon, D. V., 1949
Diatomee di Buenos Aires (Argentina) Atti Accad. Naz. Lincei, Memorie, Cl. Sci. fis., mat. e. nat., ser. 7, 11(3):59-151, 2 pls.

Amphora proteus

Zanon, V., 1948
Diatomee marini di Sardegna e Pugillo di Alghe Marine della stressa. Boll. Pesca, Piscitutura e Idrobiologia, Anno 24, ns. 3(2): 202-244, 27 figs. on 1 pl.

Amphora pseudohyalina n.sp.

Simonsen, Reimer, 1960.
Neue Diatomeen aus der Ostsee. II. Kieler Meeresf., 16(1):126-130.

Amphora Pusio

Zanon, V., 1948
Diatomee marini di Sardegna e Pugillo di Alghe Marine della stressa. Boll. Pesca, Piscitutura e Idrobiologia, Anno 24, ns. 3(2): 202-244, 27 figs. on 1 pl.

Amphora rhombica

Rampi, L., 1940
Diatomee del Mare Adriatico. Nuovo Giornale Botanico Italiano, n.s., 47:559-608.

Amphora robusta

Hendy, N. Ingram, 1964
An introductory account of the smaller algae of British coastal waters. V. Bacillariophyceae (Diatoms).
Her Majesty's Stationary Office, 317 pp., 45 pls.

Amphora robusta

Zanon, V., 1948
Diatomee marini di Sardegna e Pugillo di Alghe Marine della stressa. Boll. Pesca, Piscitutura e Idrobiologia, Anno 24, ns. 3(2): 202-244, 27 figs. on 1 pl.

Amphora rostrata n.sp.

Zanon, V., 1948
Diatomee marini di Sardegna e Pugillo di Alghe Marine della stressa. Boll. Pesca, Piscitutura e Idrobiologia, Anno 24, ns. 3(2): 202-244, 27 figs. on 1 pl.

Amphora sabyii

Hendy, N. Ingram, 1964
An introductory account of the smaller algae of British coastal waters. V. Bacillariophyceae (Diatoms).
Her Majesty's Stationary Office, 317 pp., 45 pls.

Amphora scalaris, n.sp.

Castracane degli Antelminelli, F., 1886
1. Report on the Diatomaceae collected by H.M.S. Challenger during the years 1873-1876. Rept. Sci. Results, H.M.S. Challenger, Botany Vol. II, 178 pp., 30 pls.

Amphora securicula

Rampi, L., 1940
Diatomee del Mare Adriatico. Nuovo Giornale Botanico Italiano, n.s., 47:559-608.

Amphora speciosa n.sp.
Castracane degli Antelminelli, F., 1886
1. Report on the Diatomaceae collected by H.M.S. Challenger during the years 1873-1876. Rept. Sci. Results, H.M.S. Challenger, Botany Vol. II, 178 pp., 30 pls.

Amphora spectabilis
Mann, A., 1907
Report on the diatoms of the Albatross voyages in the Pacific Ocean, 1888-1904. Contrib. U. S. Nat. Herb. 10(5):221-419, Pls. XLIV-LIV.

Amphora spectabilis
Rampi, L., 1940
Diatomee del Mare Adriatico. Nuovo Giornale Botanico Italiano, n.s., 47:559-608.

Amphora spectabilis
Zanon, V., 1948
Diatomee marini di Sardegna e Pugillo di Alghe Marine della stressa. Boll. Pesca, Piscitutura e Idrobiologia, Anno 24, ns. 3(2): 202-244, 27 figs. on 1 pl.

Amphora staurophora n.sp.
Castracane degli Antelminelli, F., 1886
1. Report on the Diatomaceae collected by H.M.S. Challenger during the years 1873-1876. Rept. Sci. Results, H.M.S. Challenger, Botany Vol. II, 178 pp., 30 pls.

Amphora sulcata
Mann, A., 1893
List of Diatomaceae from a deep-sea dredging in the Atlantic Ocean off Delaware Bay by the U. S. Fish Commission Steamer Albatross. Proc. U. S. Nat. Mus. 16:303-312.

Amphora tenerrima n.sp.
Aleem, A.A., and Fr. Hustedt, 1951.
Einige neue Diatomeen von der Südküste Englands. Botaniska Notiser 1951(1):13-20, 6 textfigs.

Amphora tenuis
Zanon, V., 1948
Diatomee marini di Sardegna e Pugillo di Alghe Marine della stressa. Boll. Pesca, Piscitutura e Idrobiologia, Anno 24, ns. 3(2): 202-244, 27 figs. on 1 pl.

Amphora thaitiana n.sp.
Castracane degli Antelminelli, F., 1886
1. Report on the Diatomaceae collected by H.M.S. Challenger during the years 1873-1876. Rept. Sci. Results, H.M.S. Challenger, Botany Vol. II, 178 pp., 30 pls.

Amphora turgida
Hendy, N. Ingram, 1964
An introductory account of the smaller algae of British coastal waters. V. Bacillariophyceae (Diatoms). Her Majesty's Stationary Office, 317 pp., 45 pls.

Amphora turgida
Hendey, N.I., 1951
Littoral diatoms of Chicester Harbour with special reference to fouling. J.Roy. Microscop. Soc. 71(1): 1-86, 18 pls.

Amphora turgida
Hustedt, F. and A.A. Aleem, 1951
Littoral diatoms from the Salstone near Plymouth. JMBA 30(1): 177.196.

Amphora turgida
Rampi, L., 1942
Ricerche sul fitoplancton del Mare Ligure 6. Le diatomee delle acque di Sanremo. Nuovo Giornale Botanico Italiano, N.S., 49:252-268.

Amphora turgida
Rampi, L., 1940
Diatomee del Mare Adriatico. Nuovo Giornale Botanico Italiano, n.s., 47:559-608.

Amphora turgida
Zanon, D. V., 1949
Diatomee di Buenos Aires (Argentina) Atti Accad. Naz. Lincei, Memorie, Cl. Sci. fis., mat. e. nat., ser. 7, 11(3):59-151, 2 pls.

Amphora turgida
Zanon, V., 1948
Diatomee marini di Sardegna e Pugillo di Alghe Marine della stressa. Boll. Pesca, Piscitutura e Idrobiologia, Anno 24, ns. 3(2): 202-244, 27 figs. on 1 pl.

Amphora valida
Zanon, V., 1948
Diatomee marini di Sardegna e Pugillo di Alghe Marine della stressa. Boll. Pesca, Piscitutura e Idrobiologia, Anno 24, ns. 3(2): 202-244, 27 figs. on 1 pl.

Amphora veneta
Hendy, N. Ingram, 1964
An introductory account of the smaller algae of British coastal waters. V. Bacillariophyceae (Diatoms). Her Majesty's Stationary Office, 317 pp., 45 pls.

Amphora veneta
Hendey, N.I., 1951
Littoral diatoms of Chicester Harbour with special reference to fouling. J.Roy. Microscop. Soc. 71(1): 1-86, 18 pls.

Amphora veneta
Zanon, D. V., 1949
Diatomee di Buenos Aires (Argentina) Atti Accad. Naz. Lincei, Memorie, Cl. Sci. fis., mat. e. nat., ser. 7, 11(3):59-151, 2 pls.

Amphora veneta
Zanon, V., 1948
Diatomee marini di Sardegna e Pugillo di Alghe Marine della stressa. Boll. Pesca, Piscitutura e Idrobiologia, Anno 24, ns. 3(2): 202-244, 27 figs. on 1 pl.

Amphora ventricosa
Hendy, N. Ingram, 1964
An introductory account of the smaller algae of British coastal waters. V. Bacillariophyceae (Diatoms). Her Majesty's Stationary Office, 317 pp., 45 pls.

Amphora ventricosa
Hendey, N.I., 1951
Littoral diatoms of Chicester Harbour with special reference to fouling. J.Roy. Microscop. Soc. 71(1): 1-86, 18 pls.

Anaulus birostratus
Zanon, V., 1948
Diatomee marini di Sardegna e Pugillo di Alghe Marine della stressa. Boll. Pesca, Piscitutura e Idrobiologia, Anno 24, ns. 3(2): 202-244, 27 figs. on 1 pl.

Anaulus ellipticus n.sp.
Hendey, N.I., 1937
The plankton diatoms of the southern seas. Discovery Repts. 16:151-364, pls.6-13.

Anaulus mediterraneus intermedia
Hendy, N. Ingram, 1964
An introductory account of the smaller algae of British coastal waters. V. Bacillariophyceae (Diatoms). Her Majesty's Stationary Office, 317 pp., 45 pls.

Anaulus mediterraneus
Rampi, L., 1940
Diatomee del Mare Adriatico. Nuovo Giornale Botanico Italiano, n.s., 47:559-608.

Anaulus scalaris
Hendey, N.I., 1937
The plankton diatoms of the southern seas. Discovery Repts. 16:151-364, pls.6-13.

Anaulus uniseptus
Hendy, N. Ingram, 1964
An introductory account of the smaller algae of British coastal waters. V. Bacillariophyceae (Diatoms). Her Majesty's Stationary Office, 317 pp., 45 pls.

Anaulus weyprechtii
Cleve-Euler, A., 1951
Die Diatomeen von Schweden und Finnland. Kungl. Svenska Vetenskaps Akad. Handl. Fjärde Ser. 2(1): 161 pp., 6 pls.

Anomoeoneis
Desikachary, T.V., 1956(1957).
Electron microscope studies on diatoms. J.R. Microsc. Soc. (3)76(1/2):9-36.

Anomoeoneis exellii n.sp.
Salah, M.M., 1952(1953).
XII. Diatoms from Blakeney Point, Norfolk. New species and new records for Great Britain. J.R. Microsc. Soc., Ser. 3, 72(3):155-169, 3 pls.

Anomoeoneis okadae
Tsumura, K., 1956.
Diatomoj el la cirkaufoso de la restajo de la kastele de Odawara. J. Yokohama Municipal Univ., (C-14), No. 47:23 pp.

Anomoeoneis sculpta
Hendy, N. Ingram, 1964
An introductory account of the smaller algae of British coastal waters. V. Bacillariophyceae (Diatoms). Her Majesty's Stationary Office, 317 pp., 45 pls.

Anomoeoneis sculpta
Hendey, N.I., 1951
Littoral diatoms of Chicester Harbour with special reference to fouling. J.Roy. Microscop. Soc. 71(1): 1-86, 18 pls.

Anomoeoneis sphaeromorpha
Iyengar, M.O.P. and G.Venkataraman, 1951.
The ecology and seasonal succession of the algae flora of the River Cooum at Madras with special reference to the Diatomaceae. J. Madras Univ. 21, Sect. B(1): 140-192, 1 pl of 4 figs., 11 text figs.

Anomoeoneis sphaerophora
Zanon, D. V., 1949
Diatomee di Buenos Aires (Argentina) Atti Accad. Naz. Lincei, Memorie, Cl. Sci. fis., mat. e. nat., ser. 7, 11(3):59-151, 2 pls.

Anomoeoneis sphaerophora
Zanon, V., 1948
Diatomee marini di Sardegna e Pugillo di Alghe Marine della stressa. Boll. Pesca, Piscitutura e Idrobiologia, Anno 24, ns. 3(2): 202-244, 27 figs. on 1 pl.

Anomoeoneis (Navicula) subtilissima
Kolbe, R.W., 1951.
Elektronenmikroskopische Untersuchungen von Diatomeenmembranen II. Svenska Bot. Tidskr. 6360647, Pls. 1-4.

Anorthoneis excentrica
Hendy, N. Ingram, 1964
An introductory account of the smaller algae of British coastal waters. V. Bacillariophyceae (Diatoms). Her Majesty's Stationary Office, 317 pp., 45 pls.

Oceanographic Index: Marine Organisms Cumulation, 1946-1973

Antelminellia gigas
Pavillard, J., 1925
Bacillariales. Rept. on the Danish Oceangr. Exped., 1908-10 to the Mediterranean and adj. seas. Vol.II. Biol. J4:72 pp., 116 text figs.

Arachnoidiscus
Desikachary, T.V., 1956(1957).
Electron microscope studies on diatoms. J.R. Microsc. Soc. (3)76(1/2):9-36.

Arachnoidiscus sp.
Hendey, N.I., 1937
The plankton diatoms of the southern seas. Discovery Repts. 16:151-364, pls.6-13.

Arachnodiscus barbadensis
Cleve-Euler, A., 1951
Die Diatomeen von Schweden und Finnland. Kungl. Svenska Vetenskaps Akad. Handl., Fjärde Ser. 2(1): 161 pp., 6 pls.

Arachnoidiscus ehrenbergii
Cupp, Easter E., 1943
Marine plankton diatoms of the west coast of North America. Bull. S.I.O. 5(1):1-238, 5 pls., 168 text figs.

Arachnodiscus indicus
Cleve-Euler, A., 1951
Die Diatomeen von Schweden und Finnland. Kungl. Svenska Vetenskaps Akad. Handl., Fjärde Ser. 2(1): 161 pp., 6 pls.

Arachnoidiscus ornatus
Takano, Hideaki, 1961.
Epiphytic diatoms upon Japanese agar sea-weeds. Bull. Tokai Reg. Fish. Res. Lab., No. 31:269-274.

Ardissonia bacillaris
Rampi, L., 1940
Diatomee del Mare Adriatico. Nuovo Giornale Botanico Italiano, n.s., 47:559-608.

Ardissonia baculus
Rampi, L., 1940
Diatomee del Mare Adriatico. Nuovo Giornale Botanico Italiano, n.s., 47:559-608.

Ardissonia Brochmanni
Rampi, L., 1940
Diatomee del Mare Adriatico. Nuovo Giornale Botanico Italiano, n.s., 47:559-608.

Ardissonia crystallina
Rampi, L., 1940
Diatomee del Mare Adriatico. Nuovo Giornale Botanico Italiano, n.s., 47:559-608.

Ardissonia crystallina var. dalmatica
Rampi, L., 1940
Diatomee del Mare Adriatico. Nuovo Giornale Botanico Italiano, n.s., 47:559-608.

Ardissonia formosa
Rampi, L., 1940
Diatomee del Mare Adriatico. Nuovo Giornale Botanico Italiano, n.s., 47:559-608.

Ardissonia fulgens
Jorgensen, E., 1900
Protophyten und Protozoen im Plankton aus der Norwegischen Westkerste. Bergens Mus. Aarb. 1899(6): 95 pp., 5 pls., 83 tables.

Ardissonia fulgens
Rampi, L., 1940
Diatomee del Mare Adriatico. Nuovo Giornale Botanico Italiano, n.s., 47:559-608.

Ardissonia robusta
Rampi, L., 1940
Diatomee del Mare Adriatico. Nuovo Giornale Botanico Italiano, n.s., 47:559-608.

Asterocaelum algophilum n.gen., n.sp.
Canter, Hilda M. (Mrs. J.W.G. Lund) 1973.
A new primitive protozoan devouring centric diatoms in the plankton. Zool. J. Linn. Soc. 52(1): 63-83

Asterionella
Desikachary, T.V., 1956(1957).
Electron microscope studies on diatoms. J. R. Microsc. Soc. (3)76(1/2):9-36.

Asterionella Bleakeleyi
Gran, H.N., 1908
Diatomeen. Nordisches Plankton, Botanischer Teil pp. XIX.1-XIX 146; 178 text figs.

Asterionella bleakeleyi
Jorgensen, E., 1900
Protophyten und Protozoen im Plankton aus der Norwegischen Westkerste. Bergens Mus. Aarb. 1899(6): 95 pp., 5 pls., 83 tables.

Asterionella Bleakleyi
Lillick, L.C., 1937
Seasonal studies of the phytoplankton off Woods Hole, Massachusetts. Biol. Bull. LXXIII (3):488-503, 3 text figs.

Asterionella Bleakeleyi Wn Sm var. notata Grun
Grunow, A., 1877
1. New Diatoms from Honduras. Monthly Micros. Jour., 18:165-186, pls. CXCIII-CXCVI.

Asterionella formosa
Braarud, T., 1945
A phytoplankton survey of the polluted waters of inner Oslo Fjord. Hvalrådets Skrifter, No.28, 142 pp., 19 text figs., 17 tables.

Asterionella formosa
Jaworski G., and J.W.G. Lund 1970.
Drought resistance and dispersal of Asterionella formosa Hass. Beihefte Nova Hedwigia 31:37-48

Asterionella formosa
Kokubo, S., and S. Sato, 1947
Plankters in JU-San Gata. Physiol. and Ecol. (Japan) 1(4):1-16, 3 text figs., tables.

Asterionella formosa
Lund, J.W.G., 1950.
Studies on Asterionella formosa Hass. II. Nutrient depletion and the spring maximum. Pt. I. Observations on Windermere, Esthwaite water and Blelham Tarn. J. Ecol. 38(1):1-14, 7 text-figs.

Asterionella formosa
Lund, J.W., F.J.H. Mackereth and C. H. Mortimer, 1963.
Changes in depth and time of certain chemical and physical conditions and of the standing crop of Asterionella formosa Hass. in the North basin of Windermere in 1947. Phil. Trans. R. Soc., London, (B), 246(731):255-290.

Asterionella formosa
Mann, A., 1893
List of Diatomaceae from a deep-sea dredging in the Atlantic Ocean off Delaware Bay by the U. S. Fish Commission Steamer Albatross. Proc. U. S. Nat. Mus. 16:303-312.

Asterionella formosa
Rodhe, W., 1948
Environmental requirements of fresh-water plankton algae. Experimental studies in the ecology of phytoplankton. Symbolae Botanicae Upsalienses X(1):149 pp., 30 figs.

Asterionella formosa
Rumkówna, A., 1948
List of the phytoplankton species occurring in the superficial water layers in the Gulf of Gdansk. Bull. Lab. mar., Gdynia, No. 4: 139-141 with tables in back.

Asterionella formosa
Schodduyn, M., 1926
Observations faites dans la baie d'Ambleteuse (Pas de Calais). Bull. Inst. Ocean., Monaco, No. 482: 64 pp.

Asterionella formosa gracillima
Meunier, A., 1915
Microplancton de la Mer Flamande. 2. Diatomées (excepté le genre Chaetoceros). Mem. Mus. Roy. Hist. Nat., Belgique, 7(3):1-118, Pls. VIII-XIV.

Asterionella Frauenfeldii n.sp.
Grunow, A., 1863
Ueber einige neue und ungenügend bekannte Arten und Gattungen von Diatomaceen. Verhandl. d. K.K. Zool. Bot. Gesellsch., Vienna, 13: 137-162, pl.4-5 (Pl. 13-14).

Asterionella glacialis
Castracane degli Antelminelli, F., 1886 n.sp.
1. Report on the Diatomaceae collected by H.M.S. Challenger during the years 1873-1876. Rept. Sci. Results, H.M.S. Challenger, Botany Vol. II, 178 pp., 30 pls.

Asterionella glacialis (formerly japonica)
Lewin, Joyce, and Richard E Norris 1970.
Surf-zone diatoms of the coasts of Washington and New Zealand (Chaetoceros armatum T. West and Asterionella spp.) Phycologia 9(2):143-149

Asterionella gracillima
Castracane degli Antelminelli, F., 1886
1. Report on the Diatomaceae collected by H.M.S. Challenger during the years 1873-1876. Rept. Sci. Results, H.M.S. Challenger, Botany Vol. II, 178 pp., 30 pls.

Asterionella gracillima
Gran, H.H., 1908
Diatomeen. Nordisches Plankton, Botanischer Teil pp. XIX.1-XIX 146; 178 text figs.

Asterionella gracillima
Gran, H.H., and T. Braarud, 1935
A quantitative study of the phytoplankton in the Bay of Fundy and the Gulf of Maine (including observations on hydrography, chemistry, and turbidity). J. Biol. Bd., Canada, 1(5):279-467, 69 text figs.

Asterionella gracillima
Lillick, L.C., 1940
Phytoplankton and planktonic protozoa of the offshore waters of the Gulf of Maine. Pt.II. Qualitative Composition of the Planktonic Flora. Trans. Am. Phil. Soc., n.s., 31(3):193-237, 13 text figs.

Asterionella gracillima
Rumkówna, A., 1948
List of the phytoplankton species occurring in the superficial water layers in the Gulf of Gdansk. Bull. Lab. mar., Gdynia, No. 4: 139-141 with tables in back.

Asterionella japonica
Allen, W. E., 1938
The Templeton Crocker Expedition to the Gulf of California in 1935 - The Phytoplankton. Amer. Microsc. Soc., Trans. 57:328-335.

Asterionella japonica

Allen, W.E., 1937
Plankton diatoms of the Gulf of California obtained by the G. Allan Hancock Expedition of 1936. The Hancock Pacific Expeditions, Univ. So. Calif. Publ. 3:47-59, 1 fig.

Allen, W. E., 1936
Occurrence of marine plankton diatoms in a ten-year series of daily catches in southern California. Am. Jour. Bot. 23(1):60-63.

Allen, W. E., 1934
Marine plankton diatoms of Lower California in 1931. Bot. Gaz. 95(3):485-492, 1 fig.

Allen, W.E., 1928
Review of five years of studies of phytoplankton at Southern California piers, 1920-1924, inclusive. Bull. S.I.O., tech. ser., 1:357-401, 5 textfigs.

Allen, W.E., and E.E. Cupp, 1935
Plankton diatoms of the Java Sea. Annales du Jardin Botanique de Buitenzorg XLIV (2):101-174, figs.1-127.
(drawings of all species mentioned)

Aubert, M., et M. Gauthier, 1966.
Origine et nature des substances antibiotiques présentes dans le milieu marin. IV. Étude comparative de l'action antibiotique due à un organisme phytoplanctonique produit en culture axénique et en culture non axénique. Rev. Intern. Océanogr. Méd., 2:53-61.

Aubert, M., 1968.
Etude des effets des pollutions chimiques sur le phytoplancton. Rev. intern. Océanogr. Méd. 10: 81-91

Aubert, M., J. Aubert, M. Gauthier, D. Pesando et S. Daniel 1966.
Origine et nature des substances antibiotiques présentes dans le milieu marin. 6. Etude biochimique des substances antibactériennes extraites d'Asterionella japonica (Cleve). Rev. intern. Océanogr. Méd. 4: 23-32.

Aubert, M. et S. Daniel, 1968.
Eaux résiduaires et plancton. Rev. intern. Océanogr. Méd. 10: 93-110

Aubert, M., M. Gautier et S. Daniel, 1966
Origine et nature des substances antibiotiques présentes dans le milieu marin. 3. Activité antibactérienne d'une diatomée marine Asterionella japonica (Cleve). Revue Int. Océanogr. Médicale, C.E.R.B.O.M. 1: 35-43.

Aubert, M., D. Pesando et M. Gauthier, 1970.
Phénomènes d'antibiose d'origine phytoplanctonique en milieu marin. Substances antibactériennes produites par une diatomée Asterionella japonica (Cleve). Rev. int. Océanogr. méd. 18-19: 69-76

Aurich, H. J., 1949.
Die Verbreitung des Nannoplanktons im Oberflächenwasser vor der Nordfriesischen Küste. Ber. Deutschen Wiss. Komm. f. Meeresf., n.f., 11(4): 403-405, 2 figs.

Bigelow, H. B., 1922
Exploration of the coastal water off the northeastern United States in 1916 by the U.S. Fisheries Schooner Grampus. Bull. M.C.Z. 65 (5):85-188, 53 text figs.

Bigelow, H.B., and M. Leslie, 1930
Reconnaissance of the waters and plankton of Monterey Bay, July 1928. Bull. M.C.Z., 70(5):429-481, 43 text figs.

Boden, B.P., 1950
Some marine plankton diatoms from the west coast of South Africa. Trans. R.Soc. S. Africa. 32:321-434, 100 text figs.

Braarud, T., K.R. Gaarder and J. Grøntved, 1953.
The phytoplankton of the North Sea and adjacent waters in May 1948. Rapp. Proc. Verb., Cons. Perm Int. Expl. Mer. 133:1-87, 29 tables, Pls. A-B, 18 textfigs.

Brunel, J., 1962
Le phytoplancton de la Baie de Chaleurs. Inst. Botan., Univ. Montréal, Contrib. No. 77: 365 pp., 66 pls.

Brunel, Jules, 1962
Le phytoplancton de la Baie des Chaleurs. Contrib. Ministère de la Chasse et des Pêcheries, Province de Québec, No. 91: 365 pp.

Cupp, Easter E., 1943
Marine plankton diatoms of the west coast of North America. Bull. S.I.O. 5(1):1-238, 5 pls., 168 text figs.

Cupp, E.E., 1934
Analysis of marine diatom collections taken from the Canal Zone to California during March, 1933. Trans. Am. Micros. Soc. LIII (1):22-29, 1 map.

Cupp, E.E. and Allen, W.E., 1938
Plankton diatoms of the Gulf of California obtained by Allan Hancock Pacific Expedition of 1937. The Hancock Pacific Expeditions, The Univ. So. Calif. Publ. 3: 61-74, 1 map, pls.4-15.

Delegazione Italiana della Commissione Internazionale per l'Esplorazione Scientifica del Mediterraneo, 1941
Note sul plancton della Laguna veneta. Memoria CCLXXIX, Arch. di Ocean. e Limn. Anno I, Fasc. I, 1941 XIX: 31-57 pp.

Ercegovic, A., 1940
Weitere Untersuchungen über einige hydrographische Verhältnisse und über die Phytoplanktonproduktion in den Gewässern der Östlichen Mitteladria. Acta Adriatica 2(3):95-134, 8 text figs.

Eskinazi Enide e Shigekatsv Satô (1963-1964) 1966.
Contribuição ao estudo das diatomaceas da Praia de Piedade.
Trabhs Inst. Oceanogr., Univ. Recife, 5 (5/6): 73-114.

Forti, A., 1922
Ricerche sulla flora pelagica (fitoplancton) di Quarto dei Mille. Mem. R. Com. Talass. Ital. 97:248 pp., 13 pls.

Ghazzawi, F.M., 1939
Plankton of the Egyptian waters. A study of the Suez Canal Plankton. (A) Phytoplankton. Preliminary Report 83 pp. Notes and Memoires, Min. Commerce-Industry, Egypt, Hydrobiol. & Fish. 65 figs.

Gilson, H. C., 1937
Chemical and Physical Investigations. The nitrogen cycle. John Murray Exped., 1933-34, Sci. Repts., 2(2):21-81, 16 text figs.

Goldberg, E.D., 1952.
Iron assimilation by marine diatoms. Biol. Bull. 102(3):243-248.

Goldberg, E.D., T.J. Walker, and A. Whisenand, 1951.
Phosphate utilization by diatoms. Biol. Bull. 101(3):274-284, 5 textfigs.

Gran, H.H., 1908
Diatomeen. Nordisches Plankton, Botanischer Teil pp. XIX 1-XIX 146; 178 text figs.

Gran, H.H., and T. Braarud, 1935
A quantitative study of the phytoplankton in the Bay of Fundy and the Gulf of Maine (including observations on hydrography, chemistry, and turbidity). J. Biol. Bd., Canada, 1(5):279-467, 69 text figs.

Gran, H. H. and E. C. Angst, 1931
Plankton diatoms of Puget Sound. Publ. Puget Sound Biol. Sta. 7:417-519, 95 text figs.

Grøntved, J., 1949(1950).
Investigations on the phytoplankton in the Danish Waddensee in July 1941. Medd. Komm. Danmarks Fiskeri- og Havundersøgelser, Serie Plankton, 5(2) (Medd Skalling Lab. 10):55 pp., 2 pls., 38 textfigs.

Grøntved, J., 1949
Investigations on the phytoplankton in the Danish Waddensea in July 1941. Medd. Komm. Danmarks Fiskeri og Havundersøgelser, ser. Plankton, 5(2):55 pp., 2 pls., 38 text figs.

Hendy, N. Ingram, 1964
An introductory account of the smaller algae of British coastal waters. V. Bacillariophyceae (Diatoms). Her Majesty's Stationary Office, 317 pp., 45 pls.

Hendey, N.I., 1937
The plankton diatoms of the southern seas. Discovery Repts. 16:151-364, pls.6-13.

Kokubo, S., 1952
Results of the observations on the plankton and oceanography of Mutsu Bay during 1950, reference being made also to the period 1946-1950. Bull Mar.Biol.Sta., Asamushi 5(1/4): 1-54, 3 tables,(fold-in), 1 fold-in.

Lillick, L.C., 1940
Phytoplankton and planktonic protozoa of the offshore waters of the Gulf of Maine. Pt.II. Qualitative Composition of the Planktonic Flora. Trans. Am. Phil. Soc., n.s., 31(3):193-237, 13 text figs.

Asterionella japonica
Lillick, L.C., 1937
Seasonal studies of the phytoplankton off Woods Hole, Massachusetts. Biol. Bull. LXXIII (3):488-503, 3 text figs.

Asterionella formosa
Lund, J.W.G., 1950.
Studies on Asterionella formosa Hass. II. Nutrient depletion and the spring maximum. Pt. II Discussion. J. Ecol. 38(1):15-35.

Asterionella japonica
Massutí Algamora, M., 1949
Estudio de diez y seis muestras de plancton del Golfo de Nápoles. Publ. Inst. Biol. Appl. 5:85-94, 1 fold-in table.

Asterionella japonica
Meunier, A., 1915
Microplancton de la Mer Flamande. 2. Diatomées (excepté le genre Chaetoceros). Mem. Mus. Roy. Hist. Nat., Belgique, 7(3):1-118, Pls. VIII-XIV.

Asterionella japonica
Morse, D.C., 1947
Some observations on seasonal variations in plankton population Patuxant River, Maryland 1943-1945. Bd. Nat. Res., Publ. No.65, Chesapeake Biol. Lab., 31, 3 figs.

Asterionella japonica?
Murthy, V.S.R., and A. Venkataramaiah, 1958.
The diatom Asterionella in the Krishna estuary region. Nature, 181(4605):360-361.

Asterionella japonica
Paiva Carvalho, J., 1950
O plancton do Rio Maria Rodrigues (Cananeis). 1. Diatomaceas e Dinoflagelados. Bol. Inst. Paulista Oceanogr. 1(1); 27-43, 2 fold-in tables, 2 figs.

Asterionella japonica
Pavillard, J., 1925
Bacillariales. Rept. on the Danish Oceangr. Exped., 1908-10 to the Mediterranean and adj. seas. Vol.II., Biol. J4:72 pp., 116 text figs.

Asterionella japonica
Pavillard, J., 1905
Recherches sur la flore pelagique (Phytoplankton) de l'Etang de Thau. Theses presentees a la Fac. Sci., Paris, 116 pp., 3 pls.

Asterionella japonica
Desandes, D. 1972.
Etude chimique et structurale d'une substance lipidique antibiotique produite par une diatomée marine. Asterionella japonica.
Rev. int. Océanogr. Méd. 25:49-69

Asterionella japonica
Pincemin, J.-M. 1972.
Besoins en vitamines de trois organismes phytoplanctoniques, Asterionella japonica, Prorocentrum micans, Glenodinium monotis. Recherche du taux optimal de B12 pour Glenodinium monotis.
Rev. int. Oceanogr. Médic. 26:85-97.

Asterionella japonica
Pincemin, J.-M. 1971.
Télémediateurs chimiques et équilibre biologique océanique. 3. Etude in vitro de relations entre populations phytoplanctoniques.
Rev. int. Oceanogr. Méd. 22-23:165-196

Asterionella japonica
Rampi, L., 1942
Ricerche sul fitoplancton del Mare Ligure 6. Le diatomee delle acque di Sanremo. Nuovo Giornale Botanico Italiano, N.S., 49:252-268.

Asterionella japonica
Rao, D.V. Subba, 1969.
Asterionella japonica bloom and discoloration off Waltair, Bay of Bengal. Limnol. Oceanogr. 14(4):632-634.

Asterionella japonica
Robinson, G.A., 1965.
Continuous plankton records: contribution towards a plankton atlas of the North Atlantic and the North Sea. IX. Seasonal cycles of phytoplankton. Bulls. Mar. Ecol., Scottish Mar. Biol. Assoc., 6(4):104-122, pls.26-61.

Asterionella japonica
Robinson, G.A., 1961
Contribution towards an atlas of the north-eastern Atlantic and the North Sea. 1. Phytoplankton.
Bulls. Mar. Ecology, 5(42): 81-89, Pls. 15-20.

Asterionella japonica
Schodduyn, M., 1926
Observations faites dans la baie d'Ambleteuse (Pas de Calais). Bull. Inst. Ocean., Monaco, No. 482: 64 pp.

Asterionella japonica
Sleggs, G.F., 1927.
Marine phytoplankton in the region of La Jolla, California during the summer of 1924. Bull. S.I.O., tech. ser., 1:93-117, 8 textfigs.

Asterionella japonica
Sousa e Silva, E., 1949
Diatomaceas e Dinoflagelados de Baia de Cascais. Portugaliae Acta Biol., Volume: Julio Henriques, Ser. B: 300-383, 9 pls, 2 fold-in tables.

Asterionella japonica
Sousa e Silva, E., and J. Dos Santos-Pinto, 1948
O Plancton da Baia de S. Martinho do Porto. 1. Diatomaceas e Dinoflagelados. Bol. Soc. Portuguese de Ciencias Naturais, 16(2):134-187, 6 pls. (Trav. Sta. Biol. Mar. de Lisbonne No. 52).

Asterionella japonica
Steemann-Nielsen, Einar, 1951
The marine vegetation of the Isefjord. A study on ecology and production. Medd. Komm. Danmarks Fiskeri-og Havundersøgelser. Ser. Plankton. 5(4); 114 pp., 46 text figs.

Asterionella japonica
Uyeno, Fukuzo, 1961
Oceanographical and ecological studies on primary production of the sea, with special references to relationship between diatom production and temperature and chlority of water.
Rept., Fac. Fish., Pref. Univ., Mie, 4(1): 1-64.

Asterionella japonica
Yamazi, I., 1951.
Plankton investigations in inlet waters along the coast of Japan. II. The plankton of Hakodate Harbour and Yoichi Inlet in Hokkaido. Publ. Seto Mar. Biol. Sta., Kyoto Univ., 1(4): 185-194, 3 textfigs.

Asterionella kariana
Allen, W.E., 1938.
"Red water" along the west coast of the United States in 1938. Science 88:55-56.

Asterionella kariana
Cupp, Easter E., 1943
Marine plankton diatoms of the west coast of North America. Bull. S.I.O. 5(1):1-238, 5 pls., 168 text figs.

Asterionella kariana
Gran, H.H., 1908
Diatomeen. Nordisches Plankton, Botanischer Teil pp. XIX 1-XIX 146; 178 text figs.

Asterionella kariana
Gran, H.H., and T. Braarud, 1935
A quantitative study of the phytoplankton in the Bay of Fundy and the Gulf of Maine (including observations on hydrography, chemistry, and turbidity). J. Biol. Bd. Canada, 1(5):279-467, 69 text figs.

Asterionella kariana
Gran, H. H. and E. C. Angst, 1931
Plankton diatoms of Puget Sound. Publ. Puget Sound Biol. Sta. 7:417-519, 95 text figs.

Asterionella kariana
Hendy, N. Ingram, 1964
An introductory account of the smaller algae of British coastal waters. V. Bacillariophyceae (Diatoms).
Her Majesty's Stationary Office, 317 pp., 45 pls.

Asterionella kariana
Lillick, L.C., 1940
Phytoplankton and planktonic protozoa of the offshore waters of the Gulf of Maine. Pt.II. Qualitative Composition of the Planktonic Flora. Trans. Am. Phil. Soc., n.s., 31(3):193-237, 13 text figs.

Asterionella kariana
Meunier, A., 1915
Microplancton de la Mer Flamande. 2. Diatomées (excepté le genre Chaetoceros). Mem. Mus. Roy. Hist. Nat., Belgique, 7(3):1-118, Pls. VIII-XIV.

Asterionella kariana
Schodduyn, M., 1926
Observations faites dans la baie d'Ambleteuse (Pas de Calais). Bull. Inst. Ocean., Monaco, No. 482: 64 pp.

Asterionella notata
Ercegovic, A., 1940
Weitere Untersuchungen über einige hydrographische Verhältnisse und über die Phytoplanktonproduktion in den Gewässern der Östlichen Mitteladria. Acta Adriatica 2(3):95-134, 8 text figs.

Asterionella notata
Gauthier, M.J. 1972.
Note sur l'activité antibactérienne d'une diatomée marine Asterionella notata (Grun) vis-à-vis de staphylocoques pathogènes et de nombreux lysotypes de Salmonella typhi murium.
Rev. int. Océanogr. Méd. 25:35-38.

Asterionella notata
Gauthier, M. 1969.
Activité antibactérienne d'une diatomée marine: Asterionella notata (Grun).
Rev. int. Océanogr. méd. 25-26:103-171

Asterionella notata
Gran, H.H., 1908
Diatomeen. Nordisches Plankton, Botanischer Teil pp. XIX 1-XIX 146; 178 text figs.

Asterionella notata
Hendy, N. Ingram, 1964
An introductory account of the smaller algae of British coastal waters. V. Bacillariophyceae (Diatoms).
Her Majesty's Stationary Office, 317 pp., 45 pls.

Asterionella notata
Hendey, N.I., 1937
The plankton diatoms of the southern seas.
Discovery Repts. 16:151-364, pls.6-13.

Asterionella notata
Margalef, R., 1949
Fitoplancton nerítico de la Costa Brava en 1947-48. Publ. Inst. Biol. Aplicada, 5: 41-51, 3 text figs.

Asterionella notata
Pavillard, J., 1925
Bacillariales. Rept. on the Danish Oceangr. Exped., 1908-10 to the Mediterranean and adj. seas. Vol.II., Biol. J4:72 pp., 116 text figs.

Asterionella notata
Schröder, B., 1900
Phytoplankton des Golfes von Neapel nebst vergleichenden Ausblicken auf das atlantischen Ozean. Mitt. Zool. Stat. Neapel, 14:1-38.

Asterionella notata
Takano, Hideaki, 1963.
Notes on marine littoral diatoms of Japan. Bull. Tokai Reg. Fish. Res. Lab., No. 36:1-8.

Asterionella ralfsii
Ross, R., 1956.
Notulae Diatomologicae. Ann. Mag. Nat. Hist. (12)9:76-80.

Asterionella socialis
Lewin, Joyce, and Richard E Norris 1970.
Surf-zone diatoms of the coast of Washington and New Zealand (Chaetoceros armatum T. West and Asterionella spp.) Phycologia 9(2):143-149.

Asterionella spathulifera
Jorgensen, E., 1900
Protophyten und Protozoën im Plankton aus der Norwegischen Westkerste. Bergens Mus. Aarb. 1899(6): 95 pp., 5 pls., 83 tables

Asterolampra decora
Castracane degli Antelminelli, F., 1886
1. Rèport on the Diatomaceae collected by H.M.S. Challenger during the years 1873-1876. Rept. Sci. Results, H.M.S. Challenger, Botany Vol. II, 178 pp., 30 pls.

Asterolampa Grevillei
Ercegovic, A., 1936
Etudes qualitative et quantitatives du phytoplancton dans les eaux cotières de l'Adriatique oriental moyen au cours de l'année 1934. Acta Adriatica 1(9):1-126

Asterolampra Grevillei
Forti, A., 1922
Ricerche sulla flora pelagica (fitoplancton) di Quarto dei Mille. Mem. R. Com. Talass. Ital. 97:248 pp., 13 pls.

Asterolampra Grevillei
Hendey, N.I., 1937
The plankton diatoms of the southern seas. Discovery Repts. 16:151-364, pls.6-13.

Asterolampra grevillei
Pavillard, J., 1925
Bacillariales. Rept. on the Danish Oceangr. Exped., 1908-10 to the Mediterranean and adj. seas. Vol.II., Biol. J4:72 pp., 116 text figs.

Asterolampra Grevillei
Pavillard, J., 1905
Recherches sur la flore pelagique (Phytoplankton) de l'Etang de Thau. Theses presentees a la Fac. Sci., Paris, 116 pp., 3 pls.

Asterolampra Grevillei
Rampi, L., 1942
Ricerche sul fitoplancton del Mare Ligure 6. Le diatomee delle acque di Sanremo. Nuovo Giornale Botanico Italiano, N.S., 49:252-268.

Asterolampra Grevillei
Rampi, L., 1940
Diatomee del Mare Adriatico. Nuovo Giornale Botanico Italiano, n.s., 47:559-608.

Asterolampra Grevillei adriatica
Schröder, B., 1900
Phytoplankton des Golfes von Neapel nebst vergleichenden Ausblicken auf das atlantischen Ozean. Mitt. Zool. Stat. Neapel, 14:1-38.

Asterolampra grevillii exi-mia
Castracane degli Antelminelli, F., 1886
1. Rèport on the Diatomaceae collected by H.M.S. Challenger during the years 1873-1876. Rept. Sci. Results, H.M.S. Challenger, Botany Vol. II, 178 pp., 30 pls.

Asterolampra heptactis
Cupp, Easter E., 1943
Marine plankton diatoms of the west coast of North America. Bull. S.I.O. 5(1):1-238, 5 pls., 168 text figs.

Asterolampra marylandica
Allen, W.E., and E.E. Cupp, 1935
Plankton diatoms of the Java Sea. Annales du Jardin Botanique de Buitenzorg XLIV (2):101-174, figs.1-127.
(drawings of all species mentioned)

Asterolampra marylandica
Cupp, Easter E., 1943
Marine plankton diatoms of the west coast of North America. Bull. S.I.O. 5(1):1-238, 5 pls., 168 text figs.

Asterolampra marylandica
Delegazione Italiana della Commissione Internazionale per l'Esplorazione Scientifica del Mediterraneo, 1941
Note sul plancton della Laguna veneta. [Memoria CCLXXIX], Arch. di Ocean. e Limn. Anno I, Fasc. I, 1941 XIX: 31-57 pp.

Asterolampa marylandica
Ercegovic, A., 1936
Etudes qualitative et quantitatives du phytoplancton dans les eaux cotières de l'Adriatique oriental moyen au cours de l'année 1934. Acta Adriatica 1(9):1-126

Asterolampra Marylandica
Forti, A., 1922
Ricerche sulla flora pelagica (fitoplancton) di Quarto dei Mille. Mem. R. Com. Talass. Ital. 97:248 pp., 13 pls.

Asterolampra marylandica
Gilbert, J.Y., and W.E. Allen, 1943
The phytoplankton of the Gulf of California obtained by the "E.W. Scripps" in 1939 and 1940. J. Mar. Res. V(2):89-110, figs.30-31.

Asterolampra maylandica
Hendey, N.I., 1937
The plankton diatoms of the southern seas. Discovery Repts. 16:151-364, pls.6-13.

Asterolampra marylandica
Mann, A., 1907
Report on the diatoms of the Albatross voyages in the Pacific Ocean, 1888-1904. Contrib. U. S. Nat. Herb. 10(5):221-419, Pls. XLIV-LIV.

Asterolampra Marylandica
Mann, A., 1893
List of Diatomaceae from a deep-sea dredging in the Atlantic Ocean off Delaware Bay by the U. S. Fish Commission Steamer Albatross. Proc. U. S. Nat. Mus. 16:303-312.

Asterolampra marylandica
Pavillard, J., 1925
Bacillariales. Rept. on the Danish Oceangr. Exped., 1908-10 to the Mediterranean and adj. seas. Vol.II., Biol. J4:72 pp., 116 text figs.

Asterolampra marylandica
Pavillard, J., 1905
Recherches sur la flore pelagique (Phytoplankton) de l'Etang de Thau. Theses presentees a la Fac. Sci., Paris, 116 pp., 3 pls.

Asterolampra marylandica
Rampi, L., 1942
Ricerche sul fitoplancton del Mare Ligure 6. Le diatomee delle acque di Sanremo. Nuovo Giornale Botanico Italiano, N.S., 49:252-268.

Asterolampra marylandica
Rampi, L., 1940
Diatomee del Mare Adriatico. Nuovo Giornale Botanico Italiano, n.s., 47:559-608.

Asterolampra Marylandica
Zanon, V., 1948
Diatomee marini di Sardegna e Pugillo di Alghe Marine della stressa. Boll. Pesca, Piscitutura e Idrobiologia, Anno 24, ns. 3(2): 202-244, 27 figs. on 1 pl.

Asterolampra marylandica ausonia
Schröder, B., 1900
Phytoplankton des Golfes von Neapel nebst vergleichenden Ausblicken auf das atlantischen Ozean. Mitt. Zool. Stat. Neapel, 14:1-38.

Asterolampra marylandica var. major
Pavillard, J., 1925
Bacillariales. Rept. on the Danish Oceangr. Exped., 1908-10 to the Mediterranean and adj. seas. Vol.II., Biol. J4:72 pp., 116 text figs.

Asterolampra rotula
Schröder, B., 1900
Phytoplankton des Golfes von Neapel nebst vergleichenden Ausblicken auf das atlantischen Ozean. Mitt. Zool. Stat. Neapel, 14:1-38.

Asterolampra Vanheurcki
Hendey, N.I., 1937
The plankton diatoms of the southern seas. Discovery Repts. 16:151-364, pls.6-13.

Asterolampra Van Heurckii
Rampi, L., 1942
Ricerche sul fitoplancton del Mare Ligure 6. Le diatomee delle acque di Sanremo. Nuovo Giornale Botanico Italiano, N.S., 49:252-268.

Asteromphelus
Desikachary, T.V., 1956(1957).
Electron microscope studies on diatoms. J. R. Microsc. Soc. (3)76(1/2):9-36.

Asteromphalus
Neaverson, E., 1934
The sea-floor deposits. 1. General characteristics and distribution. Discovery Repts. 9: 297-349, Plates 17-22.

Asteromphalus antarcticus n. sp.
Castracane degli Antelminelli, F., 1886
1. Rèport on the Diatomaceae collected by H.M.S. Challenger during the years 1873-1876. Rept. Sci. Results, H.M.S. Challenger, Botany Vol. II, 178 pp., 30 pls.

Asteromphalus arachne

Mann, A., 1907
Report on the diatoms of the Albatross voyages in the Pacific Ocean, 1888-1904. Contrib. U. S. Nat. Herb. 10(5):221-419, Pls. XLIV-LIV.

Asteromphalus atlanticus

Gran, H.H., 1897
Protophyta: Diatomaceae, Silico-flagellata and Cilioflagellata. Den Norske Nordhavs Expedition 1876-1878, h. 24, 36 pp., 4 pls.

Asteromphalus beaumontii

Mann, A., 1907
Report on the diatoms of the Albatross voyages in the Pacific Ocean, 1888-1904. Contrib. U. S. Nat. Herb. 10(5):221-419, Pls. XLIV-LIV.

Asteromphalus brookei

Boden, B. P., 1949.
The diatoms collected by the U.S.S. CACOPAN in the Antarctic in 1947. J. Mar. Res. 8(1):6-13, 3 textfigs.

Asteromphalus brookei

Boden, Brian, 1948
Marine plankton diatoms on operation HIGHJUMP in: Some oceanographic observations on operation HIGHJUMP. By R.S. Dietz. USNEL Rept. No.55, 97 pp., 41 figs. 7 July 1948.

Asteromphalus brookei

Cleve-Euler, A., 1951
Die Diatomeen von Schweden und Finnland. Kungl. Svenska Vetenskaps Akad. Handl., Fjärde Ser. 2(1): 161 pp., 6 pls.

Asteromphalus Brookii

Mangin, L., 1915
Phytoplancton de L'Antartique. Deuxieme Exped. Ant. Francaise (1908-1910), 95 pp., 3 pls. 58 text figs.

Asteromphalus brookei

Mann, A., 1907
Report on the diatoms of the Albatross voyages in the Pacific Ocean, 1888-1904. Contrib. U. S. Nat. Herb. 10(5):221-419, Pls. XLIV-LIV.

Asteromphalus Brookei

Mann, A., 1893
List of Diatomaceae from a deep-sea dredging in the Atlantic Ocean off Delaware Bay by the U. S. Fish Commission Steamer Albatross. Proc. U. S. Nat. Mus. 16:303-312.

Asteromphalus challengerensis

Castracane degli Antelminelli, F., 1886 n.sp.
1. Report on the Diatomaceae collected by H.M.S. Challenger during the years 1873-1876. Rept. Sci. Results, H.M.S. Challenger, Botany Vol. II, 178 pp., 30 pls.

Asteromphalus Cleveanus

Allen, W.E., and E.E. Cupp, 1935
Plankton diatoms of the Java Sea. Annales du Jardin Botanique de Buitenzorg XLIV (2):101-174, figs.1-127.

(drawings of all species mentioned)

Asteromphalus elegans

Boden, B. P., 1949.
The diatoms collected by the U.S.S. CACOPAN in the Antarctic in 1947. J. Mar. Res. 8(1):6-13, 3 textfigs.

Asteromphalus elegans

Boden, Brian, 1948
Marine plankton diatoms on operation HIGHJUMP in: Some oceanographic observations on operation HIGHJUMP. By R.S. Dietz. USNEL Rept. No.55, 97 pp., 41 figs. 7 July 1948.

Asteromphalus elegans

Hendey, N.I., 1937
The plankton diatoms of the southern seas. Discovery Repts. 16:151-364, pls.6-13.

Asteromphalus elegans

Mann, A., 1907
Report on the diatoms of the Albatross voyages in the Pacific Ocean, 1888-1904. Contrib. U. S. Nat. Herb. 10(5):221-419, Pls. XLIV-LIV.

Asteromphalus flabellatus

Allen, W.E., and E.E. Cupp, 1935
Plankton diatoms of the Java Sea. Annales du Jardin Botanique de Buitenzorg XLIV (2):101-174, figs.1-127.

(drawings of all species mentioned)

Asteromphalus flabellatus

Dangeard, P., 1927.
Phytoplankton de la croisière du "Sylvana". Ann. Inst. Ocean., Monaco, n.s., 4(8):286-401, 54 text figs. (Feirer-Juin 1913).

Asteromphalus flabellatus

Ercegovic, A., 1936
Etudes qualitative et quantitatives du phytoplancton dans les eaux cotières de l'Adriatique oriental moyen au cours de l'année 1934. Acta Adriatica 1(9):1-126

Asteromphalus flabellatus

Mann, A., 1907
Report on the diatoms of the Albatross voyages in the Pacific Ocean, 1888-1904. Contrib. U. S. Nat. Herb. 10(5):221-419, Pls. XLIV-LIV.

Asteromphalus flabellatus

Mann, A., 1893
List of Diatomaceae from a deep-sea dredging in the Atlantic Ocean off Delaware Bay by the U. S. Fish Commission Steamer Albatross. Proc. U. S. Nat. Mus. 16:303-312.

Asteromphalus flabellatus

Pavillard, J., 1925
Bacillariales. Rept. on the Danish Oceangr. Exped., 1908-10 to the Mediterranean and adj. seas. Vol.II, Biol. J4:72 pp., 116 text figs.

Asteromphalus flabellatus

Pavillard, J., 1905
Recherches sur la flore pelagique (Phytoplankton) de l'Etang de Thau. Theses presentees a la Fac. Sci., Paris, 116 pp., 3 pls.

Asteromphalus flabellatus

Rampi, L., 1942
Ricerche sul fitoplancton del Mare Ligure 6. Le diatomee delle acque di Sanremo. Nuovo Giornale Botanico Italiano, N.S., 49:252-268.

Asteromphalus flabellatus

Rampi, L., 1940
Diatomee del Mare Adriatico. Nuovo Giornale Botanico Italiano, n.s., 47:559-608.

Asteromphalus heptactis

Allen, W.E., 1937
Plankton diatoms of the Gulf of California obtained by the G. Allan Hancock Expedition of 1936. The Hancock Pacific Expeditions, Univ. So. Calif. Publ. 3:47-59, 1 fig.

Asteromphalus heptactis

Bigelow, H.B., and M. Leslie, 1930
Reconnaissance of the waters and plankton of Monterey Bay, July 1928. Bull. M.C.Z., 70(5):429-481, 43 text figs.

Asteromphalus heptactis

Boden, B.P., 1950
Some marine plankton diatoms from the west coast of South Africa. Trans. R.Soc. S. Africa. 32:321-434, 100 text figs.

Asteromphalus heptactis

Boden, Brian, 1948
Marine plankton diatoms on operation HIGHJUMP in: Some oceanographic observations on operation HIGHJUMP. By R.S. Dietz. USNEL Rept. No.55, 97 pp., 41 figs. 7 July 1948.

Asteromphalus heptactis

Cleve-Euler, A., 1951
Die Diatomeen von Schweden und Finnland. Kungl. Svenska Vetenskaps Akad. Handl., Fjärde Ser. 2(1): 161 pp., 6 pls.

Asteromphalus heptactis

Cupp, E.E., 1937
Seasonal distribution and occurrence of marine diatoms and dinoflagellates at Scotch Cap, Alaska. Bull. S.I.O. Tech. ser.4(3):71-100, 7 textfigs.

Asteromphalus heptactis

Cupp, E., 1930
Quantitative Studies of miscellaneous series of surface catches of marine diatoms and dinoflagellates taken between Seattle and the Canal Zone from 1924 to 1928. Trans. Am. Micro. Soc., XLIX (3):238-245.

Asteromphalus heptactis

Cupp, E.E. and Allen, W.E., 1938
Plankton diatoms of the Gulf of California obtained by Allan Hancock Pacific Expedition of 1937. The Hancock Pacific Expeditions, The Univ. So. Calif. Publ. 3: 61-74, 1 map, pls.4-15.

Asteromphalus heptactis

Ercegovic, A., 1936
Etudes qualitative et quantitatives du phytoplancton dans les eaux cotières de l'Adriatique oriental moyen au cours de l'année 1934. Acta Adriatica 1(9):1-126

Asteromphalus heptactis

Frenguelli, Joaquin, and Hector Antonio Orlando, 1959.

Operacion MERLUZA. Diatomeas y silico-flagelados del plancton del "VI Crucero". Servicio Hidrogr. Naval., Argentina, Publ. No. H. 619: 5-62.

Asteromphalus heptactis

Gilbert, J.Y., and W.E. Allen, 1943
The phytoplankton of the Gulf of California obtained by the "E.W. Scripps" in 1939 and 1940. J. Mar. Res. V(2):89-110, figs.30-31.

Asteromphalus heptactis

Gran, H.H., 1908
Diatomeen. Nordisches Plankton, Botanischer Teil pp. XIX.1-XIX 146; 178 text figs.

Asteromphalus heptactis

Gran, H. H. and E. C. Angst, 1931
Plankton diatoms of Puget Sound. Publ. Puget Sound Biol. Sta. 7:417-519, 95 text figs.

Asteromphalus heptactis

Hendy, N. Ingram, 1964
An introductory account of the smaller algae of British coastal waters. V. Bacillariophyceae (Diatoms).
Her Majesty's Stationary Office, 317 pp., 45 pls.

Asteromphalus heptactis

Hendey, N.I., 1937
The plankton diatoms of the southern seas. Discovery Repts. 16:151-364, pls.6-13.

Asteromphalus heptactis

Jørgensen, E., 1905
B.Protistplankton and the diatoms in bottom samples. Hydrographical and biological investigations in Norwegian fjords. Bergens Mus. Skr. 7: 49-225.

Asteromphalus heptactis

Jorgensen, E., 1900
Protophyten und Protozoën im Plankton aus der Norwegischen Westkerste. Bergens Mus. Aarb. 1899(6): 95 pp., 5 pls., 83 tables.

Asteromphalus heptactis

Mann, A., 1907
Report on the diatoms of the Albatross voyages in the Pacific Ocean, 1888-1904. Contrib. U. S. Nat. Herb. 10(5):221-419, Pls. XLIV-LIV.

Asteromphalus hiltonianus

Mann, A., 1907
Report on the diatoms of the Albatross voyages in the Pacific Ocean, 1888-1904. Contrib. U. S. Nat. Herb. 10(5):221-419, Pls. XLIV-LIV.

Asteromphalus hookerii

Boden, B.P., 1950
Some marine plankton diatoms from the west coast of South Africa. Trans. R.Soc. S. Africa. 32:321-434, 100 text figs.

Asteromphalus Hookeri

Central Meteorological Observatory, 1949
Report on sea and weather observation on Antarctic Whaling Ground (1947-48). Ocean. Mag., Japan, 1(1):49-88, 17 text figs.

Asteromphalus Hookeri

Frenguelli, Joaquin, and Hector Antonio Orlando, 1959.
Operacion MERLUZA. Diatomeas y silicoflagelados del plancton del "VI Crucero". Servicio Hidrogr. Naval., Argentina, Publ. No. H. 619: 5-62.

Gran, H.H., 1908   Asteromphalus Hookeri
Diatomeen. Nordisches Plankton, Botanischer Teil pp. XIX.1-XIX 146; 178 text figs.

Hendy, N. Ingram, 1964   Asteromphalus hookeri
An introductory account of the smaller algae of British coastal waters. V. Bacillariophyceae (Diatoms). Her Majesty's Stationary Office, 317 pp., 45 pls.

Hendey, N.I., 1937   Asteromphalus Hookeri
The plankton diatoms of the southern seas. Discovery Repts. 16:151-364, pls.6-13.

Asteromphalus hookeri

Jouse, A.P., G.S. Koroleva, G.A. Nagaeva, 1962
Diatoms in the surface layer of sediment in the Indian sector of the Antarctic. Investigations of marine bottom sediments. (In Russian; English summary).
Trudy Inst. Okeanol., Akad. Nauk, SSSR, 61: 19-92.

Asteromphalus Hookeri

Mangin, L., 1915
Phytoplancton de L'Antartique. Deuxieme Exped. Ant. Francaise (1908-1910), 95 pp., 3 pls., 58 text figs.

Asteromphalus hookerii

Mann, A., 1907
Report on the diatoms of the Albatross voyages in the Pacific Ocean, 1888-1904. Contrib. U. S. Nat. Herb. 10(5):221-419, Pls. XLIV-LIV.

Asteromphalus hungaricus

Cleve-Euler, A., 1951
Die Diatomeen von Schweden und Finnland. Kungl. Svenska Vetenskaps Akad. Handl., Fjärde Ser. 2(1): 161 pp., 6 pls.

Asteromphalus hyalinus

Jouse, A.P., G.S. Koroleva, G.A. Nagaeva, 1962
Diatoms in the surface layer of sediment in the Indian sector of the Antarctic. Investigations of marine bottom sediments. (In Russian; English summary).
Trudy Inst. Okeanol., Akad. Nauk, SSSR, 61: 19-92.

Asteromphalus Leboimei n.sp.

Manguin, E., 1957.
Premier inventaire des diatomées de la Terre Adélie Antarctique. Espèces nouvelles. Rev. Algologique, n.s., 3(3):111-134.

Asteromphalus nanus n.sp.

Mann, A., 1907
Report on the diatoms of the Albatross voyages in the Pacific Ocean, 1888-1904. Contrib. U. S. Nat. Herb. 10(5):221-419, Pls. XLIV-LIV.

Asteromphalus ovatus n.sp.

Castracane degli Antelminelli, F., 1886
1. Report on the Diatomaceae collected by H.M.S. Challenger during the years 1873-1876. Rept. Sci. Results, H.M.S. Challenger, Botany Vol. II, 178 pp., 30 pls.

Asteromphalus parvulus

Boden, B. P., 1949.
The diatoms collected by the U.S.S. CACO-PAN in the Antarctic in 1947. J. Mar. Res. 8(1):6-13, 3 textfigs.

Boden, Brian, 1948   Asteromphalus parvulus
Marine plankton diatoms on operation HIGHJUMP in: Some oceanographic observations on operation HIGHJUMP. By R.S. Dietz. USNEL Rept. No.55, 97 pp., 41 figs. 7 July 1948.

Asteromphalus parvulus

Frenguelli, Joaquin, and Hector Antonio Orlando, 1959.
Operacion MERLUZA. Diatomeas y silicoflagelados del plancton del "VI Crucero". Servicio Hidrogr. Naval., Argentina, Publ. No. H. 619: 5-62.

Hendey, N.I., 1937   Asteromphalus parvulus
The plankton diatoms of the southern seas. Discovery Repts. 16:151-364, pls.6-13.

Asteromphalus parvulus

Jouse, A.P., G.S. Koroleva, G.A. Nagaeva, 1962
Diatoms in the surface layer of sediment in the Indian sector of the Antarctic. Investigations of marine bottom sediments. (In Russian; English summary).
Trudy Inst. Okeanol., Akad. Nauk, SSSR, 61: 19-92.

Actinocyclus radiatus

Jouse, A.P., G.S. Koroleva, G.A. Nagaeva, 1962
Diatoms in the surface layer of sediment in the Indian sector of the Antarctic. Investigations of marine bottom sediments. (In Russian; English summary).
Trudy Inst. Okeanol., Akad. Nauk, SSSR, 61: 19-92.

Asteromphalus Ralfsianus

Schröder, B., 1900
Phytoplankton des Golfes von Neapel nebst vergleichenden Ausblicken auf das atlantischen Ozean. Mitt. Zool. Stat. Neapel, 14:1-38.

Asteromphalus robustus

Ercegovic, A., 1936
Etudes qualitative et quantitatives du phytoplancton dans les eaux cotieres de l'Adriatique oriental moyen au cours de l'année 1934. Acta Adriatica 1(9):1-126

Asteromphalus robustus

Manguin, E., 1954
Diatomees marines provenant de l'ile Heard (Australian National Antarctic Expedition). Rev. Algol., n.s., 1: 14-24.

Asteromphalus robustus

Pavillard, J., 1925
Bacillariales. Rept. on the Danish Oceangr. Exped., 1908-10 to the Mediterranean and adj. seas. Vol.II., Biol. J4:72 pp., 116 text figs.

Asteromphalus robustus

Rampi, L., 1940
Diatomee del Mare Adriatico. Nuovo Giornale Botanico Italiano, n.s., 47:559-608.

Asteromphalus robustus

Schröder, B., 1900
Phytoplankton des Golfes von Neapel nebst vergleichenden Ausblicken auf das atlantischen Ozean. Mitt. Zool. Stat. Neapel, 14:1-38.

Asteromphalus roperianus

Boden, B. P., 1949.
The diatoms collected by the U.S.S. CACO-PAN in the Antarctic in 1947. J. Mar. Res. 8(1):6-13, 3 textfigs.

Boden, Brian, 1948   Asteromphalus roperianus
Marine plankton diatoms on operation HIGHJUMP in: Some oceanographic observations on operation HIGHJUMP. By R.S. Dietz. USNEL Rept. No.55, 97 pp., 41 figs. 7 July 1948.

Asteromphalus Roperianus

Hendey, N.I., 1937
The plankton diatoms of the southern seas. Discovery Repts. 16:151-364, pls.6-13.

Asteromphalus roperianus

Castracane degli Antelminelli, F., 1886
1. Report on the Diatomaceae collected by H.M.S. Challenger during the years 1873-1876. Rept. Sci. Results, H.M.S. Challenger, Botany Vol. II, 178 pp., 30 pls.
atlantica n.var.

Asteromphalus Roperianus

Mangin, L., 1915
Phytoplancton de L'Antartique. Deuxieme Exped. Ant. Francaise (1908-1910), 95 pp., 3 pls. 58 text figs.

Asteromphalus roperianus

Mann, A., 1907
Report on the diatoms of the Albatross voyages in the Pacific Ocean, 1888-1904. Contrib. U. S. Nat. Herb. 10(5):221-419, Pls. XLIV-LIV.

Asteromphalus shadboltianus

Mann, A., 1907
Report on the diatoms of the Albatross voyages in the Pacific Ocean, 1888-1904. Contrib. U. S. Nat. Herb. 10(5):221-419, Pls. XLIV-LIV.

Asteromphalus Shadboldtianus
Mann, A., 1893
List of Diatomaceae from a deep-sea dredging in the Atlantic Ocean off Delaware Bay by the U. S. Fish Commission Steamer Albatross. Proc. U. S. Nat. Mus. 16:303-312.

Asteromphalus van heurckii n.sp.
Mann, A., 1907
Report on the diatoms of the Albatross voyages in the Pacific Ocean, 1888-1904. Contrib. U. S. Nat. Herb. 10(5):221-419, Pls. XLIV-LIV.

Asteromphalus variabilis
Mann, A., 1907
Report on the diatoms of the Albatross voyages in the Pacific Ocean, 1888-1904. Contrib. U. S. Nat. Herb. 10(5):221-419, Pls. XLIV-LIV.

Asteromphalus wyvillii n.sp.
Castracane degli Antelminelli, F., 1886
1. Report on the Diatomaceae collected by H.M.S. Challenger during the years 1873-1876. Rept. Sci. Results, H.M.S. Challenger, Botany Vol. II, 178 pp., 30 pls.

Attheya
Desikachary, T.V., 1956(1957).
Electron microscope studies on diatoms. J. R. Microsc. Soc. (3)76(1/2):9-36.

Attheia decora
Rumkówna, A., 1948
[List of the phytoplankton species occurring in the superficial water layers in the Gulf of Gdańsk] Bull. Lab. mar., Gdynia, No. 4: 139-141 with tables in back.

Attheya Zacchariasi
Meunier, A., 1915
Microplancton de la Mer Flamande. 2. Diatomées (excepté le genre Chaetoceros). Mem. Mus. Roy. Hist. Nat., Belgique, 7(3):1-118, Pls. VIII-XIV.

Aulacodiscus sp.
Bigelow, H.B., and M. Leslie, 1930
Reconnaissance of the waters and plankton of Monterey Bay, July 1928. Bull. M.C.Z., 70(5):429-481, 43 text figs.

Aulacodiscus spp.
Ross, R., and Patricia A. Sims, 1970.
Studies of Aulacodiscus with the scanning electron microscope. Beihefts Nova Hedwigia 31:49-88.

Aulacodiscus argus
Cleve-Euler, A., 1951
Die Diatomeen von Schweden und Finnland. Kungl. Svenska Vetenskaps Akad. Handl., Fjärde Ser. 2(1): 161 pp., 6 pls.

Aulacodiscus argus
Grøntved, J., 1949
Investigations on the phytoplankton in the Danish Waddensea in July 1941. Medd. Komm. Danmarks Fiskeri og Havundersøgelser, ser. Plankton, 5(2):55 pp., 2 pls., 38 text figs.

Aulacodiscus argus
Hendy, N. Ingram, 1964
An introductory account of the smaller algae of British coastal waters. V. Bacillariophyceae (Diatoms). Her Majesty's Stationary Office, 317 pp., 45 pls.

Aulacodiscus argus
von Sydow, Burkard, und Robert Christenhuss 1972. Rasterelektronmikroskopische Untersuchungen der Hohlräume in der Schalenwand einiger centrischer Kieselalgen. Arch. Protistenk 114 (3): 256-271.

Aulacodiscus amoenus
Barker, J.W. and S.H. Meakin, 1948
New and rare diatoms. J. Quakett Micros. Club, ser.4, 2(5):233-235, pl.25.

Aulocodiscus bonei n.sp.
Woodard, John B., 1967.
A new Aulacodiscus from Nancowry, Nicobar islands. Microscopy. J. Quekett Microsc. Club, 30(12):307-308.

Aulacodiscus crux
Cleve-Euler, A., 1951
Die Diatomeen von Schweden und Finnland. Kungl. Svenska Vetenskaps Akad. Handl., Fjärde Ser. 2(1): 161 pp., 6 pls.

Aulacodiscus kittoni
Allen, W.E., 1938.
"Red water" along the west coast of the United States in 1938. Science 88:55-56.

Aulacodiscus kettoni
Becking, L.B., C.F. Tolman, H.C. McMillan, H.C. Field, and T. Hashimoto, 1927.
Preliminary statement regarding the diatom "epidemic" at Copalis Beach, Washington, and an analysis of diatom oil. Econ. Geol. 22:356-368.

Aulacodiscus kittoni
Cupp, Easter E., 1943
Marine plankton diatoms of the west coast of North America. Bull. S.I.O. 5(1):1-238, 5 pls., 168 text figs.

Aulacodiscus kittoni
Galtsoff, P.S., 1949
The mystery of the red tide. Sci. Mon. LXVIII(2):108-117.

Aulacodiscus kittoni
Hendy, N. Ingram, 1964
An introductory account of the smaller algae of British coastal waters. V. Bacillariophyceae (Diatoms). Her Majesty's Stationary Office, 317 pp., 45 pls.

Aulacodiscus Kittoni
Rampi, L., 1940
Diatomee del Mare Adriatico. Nuovo Giornale Botanico Italiano, n.s., 47:559-608.

Aulacodiscus oamaruensis
Barker, J.W. and S.H. Meakin, 1948
New and rare diatoms. J. Quakett Micros. Club, ser.4, 2(5):233-235, pl.25.

Aulacodiscus orbiculatus n.sp.
Subrahmanyan, R., 1946.
A systematic account of the marine plankton diatoms of the Madras coast. Proc. Indian Acad. Sci. 24(4):85-197, Pls. 2, 440 textfigs.

Aulacodiscus oregonus
Gran, H. H. and E. C. Angst, 1931
Plankton diatoms of Puget Sound. Publ. Puget Sound Biol. Sta. 7:417-519, 95 text figs.

Aulacodiscus pettersii
de Sousa e Silva, E., 1956.
Contribution à l'étude du microplancton de Dakar et des regions maritimes voisines. Bull. I.F.A.N., 8(2):335-371, 7 pls.

Aulacodiscus Petersii
Schröder, B., 1900
Phytoplankton des Golfes von Neapel nebst vergleichenden Ausblicken auf das atlantischen Ozean. Mitt. Zool. Stat. Neapel, 14:1-38.

Aulacodiscus punctata
de Sousa e Silva, E., 1956.
Contribution à l'étude du microplancton de Dakar et des regions maritimes voisines. Bull. I.F.A.N., 8(2):335-371, 7 pls.

Auliscus
Desikachary, T.V., 1956(1957).
Electron Microscope studies on diatoms. J. R. Microsc. Soc. (3)76(1/2):9-36.

Auliscus caelatus
Zanon, D. V., 1949
Diatomee di Buenos Aires (Argentina) Atti Accad. Naz. Lincei, Memorie, Cl. Sci. fis., mat. e. nat., ser. 7, 11(3):59-151, 2 pls.

Auliscus caelatus
Cleve-Euler, A., 1951
Die Diatomeen von Schweden und Finnland. Kungl. Svenska Vetenskaps Akad. Handl., Fjärde Ser. 2(1): 161 pp., 6 pls.

Auliscus caelatus
Mann, A., 1907
Report on the diatoms of the Albatross voyages in the Pacific Ocean, 1888-1904. Contrib. U. S. Nat. Herb. 10(5):221-419, Pls. XLIV-LIV.

Auliscus caelatus
Mann, A., 1893
List of Diatomaceae from a deep-sea dredging in the Atlantic Ocean off Delaware Bay by the U. S. Fish Commission Steamer Albatross. Proc. U. S. Nat. Mus. 16:303-312.

Auliscus caelatus
Rampi, L., 1942
Ricerche sul fitoplancton del Mare Ligure 6. Le diatomee delle acque di Sanremo. Nuovo Giornale Botanico Italiano, N.S., 49:252-268.

Auliscus caelatus
Rampi, L., 1940
Diatomee del Mare Adriatico. Nuovo Giornale Botanico Italiano, n.s., 47:559-608.

Auliscus confluens
Zanon, D. V., 1949
Diatomee di Buenos Aires (Argentina) Atti Accad. Naz. Lincei, Memorie, Cl. Sci. fis., mat. e. nat., ser. 7, 11(3):59-151, 2 pls.

Auliscus herdmanianus
Mann, A., 1907
Report on the diatoms of the Albatross voyages in the Pacific Ocean, 1888-1904. Contrib. U. S. Nat. Herb. 10(5):221-419, Pls. XLIV-LIV.

Auliscus insignis
Mann, A., 1907
Report on the diatoms of the Albatross voyages in the Pacific Ocean, 1888-1904. Contrib. U. S. Nat. Herb. 10(5):221-419, Pls. XLIV-LIV.

Auliscus pruinosus
Mann, A., 1907
Report on the diatoms of the Albatross voyages in the Pacific Ocean, 1888-1904. Contrib. U. S. Nat. Herb. 10(5):221-419, Pls. XLIV-LIV.

Auliscus reticulatus

Cleve-Euler, A., 1951
Die Diatomeen von Schweden und Finnland. Kungl. Svenska Vetenskaps Akad. Handl., Fjärde Ser. 2(1): 161 pp., 6 pls.

Auliscus sculptus

Cleve-Euler, A., 1951
Die Diatomeen von Schweden und Finnland. Kungl. Svenska Vetenskaps Akad. Handl., Fjärde Ser. 2(1): 161 pp., 6 pls.

Auliscus sculptus

Grøntved, J., 1949
Investigations on the phytoplankton in the Danish Waddensea in July 1941. Medd. Komm. Danmarks Fiskeri og Havundersøgelser, ser. Plankton, 5(2):55 pp., 2 pls., 38 text figs.

Auliscus sculptus

Hendy, N. Ingram, 1964
An introductory account of the smaller algae of British coastal waters. V. Bacillariophyceae (Diatoms).
Her Majesty's Stationary Office, 317 pp., 45 pls.

Auliscus sculptus

Hendey, N.I., 1951
Littoral diatoms of Chicester Harbour with special reference to fouling. J.Roy. Microscop. Soc. 71(1): 1-86, 18 pls.

Auliscus sculptus

Jørgensen, E., 1905
B. Protistplankton and the diatoms in bottom samples. Hydrographical and biological investigations in Norwegian fjords. Bergens Mus. Skr. 7: 49-225.

Auliscus sculptus

Meunier, A., 1915
Microplancton de la Mer Flamande. 2. Diatomées (excepté le genre Chaetoceros). Mem. Mus. Roy. Hist. Nat., Belgique, 7(3):1-118, Pls. VIII-XIV.

Auliscus sculptus

Zanon, V., 1948
Diatomee marini di Sardegna e Pugillo di Alghe Marine della stressa. Boll. Pesca, Piscitutura e Idrobiologia, Anno 24, ns. 3(2): 202-244, 27 figs. on 1 pl.

Auliscus stockhardtii

Mann, A., 1907
Report on the diatoms of the Albatross voyages in the Pacific Ocean, 1888-1904. Contrib. U. S. Nat. Herb. 10(5):221-419, Pls. XLIV-LIV.

Auricula adriatica

Zanon, V., 1948
Diatomee marini di Sardegna e Pugillo di Alghe Marine della stressa. Boll. Pesca, Piscitutura e Idrobiologia, Anno 24, ns. 3(2): 202-244, 27 figs. on 1 pl.

Auricula amphitritis

Rampi, L., 1940
Diatomee del Mare Adriatico. Nuovo Giornale Botanico Italiano, n.s., 47:559-608.

Auliscus caelatus

Zanon, V., 1948
Diatomee marini di Sardegna e Pugillo di Alghe Marine della stressa. Boll. Pesca, Piscitutura e Idrobiologia, Anno 24, ns. 3(2): 202-244, 27 figs. on 1 pl.

Auricula complexa

Hendy, N. Ingram, 1964
An introductory account of the smaller algae of British coastal waters. V. Bacillariophyceae (Diatoms).
Her Majesty's Stationary Office, 317 pp., 45 pls.

Auricula complexa

Hendey, N.I., 1951
Littoral diatoms of Chicester Harbour with special reference to fouling. J.Roy. Microscop. Soc. 71(1): 1-86, 18 pls.

Auricula complexa

Jørgensen, E., 1905
B. Protistplankton and the diatoms in bottom samples. Hydrographical and biological investigations in Norwegian fjords. Bergens Mus. Skr. 7: 49-225.

Auricula complexa

Jorgensen, E., 1900
Protophyten und Protozoën im Plankton aus der Norwegischen Westkerste. Bergens Mus. Aarb. 1899(6): 95 pp., 5 pls., 83 tables.

Auricula complexa

Pavillard, J., 1925
Bacillariales. Rept. on the Danish Oceangr. Exped., 1908-10 to the Mediterranean and adj. seas. Vol.II., Biol. J4:72 pp., 116 text figs.

Auricula complexa

Pavillard, J., 1905
Recherches sur la flore pelagique (Phytoplankton) de l'Etang de Thau. Theses presentees a la Fac. Sci., Paris, 116 pp., 3 pls.

Auricula complexa

Rampi, L., 1942
Ricerche sul fitoplancton del Mare Ligure 6. Le diatomee delle acque di Sanremo. Nuovo Giornale Botanico Italiano, N.S., 49:252-268.

Auricula dubia

Hendy, N. Ingram, 1964
An introductory account of the smaller algae of British coastal waters. V. Bacillariophyceae (Diatoms).
Her Majesty's Stationary Office, 317 pp., 45 pls.

Auricula dubia

Hustedt, F. and A.A. Aleem, 1951
Littoral diatoms from the Salstone near Plymouth. JMBA 30(1): 177-196.

Auliscus incertus

Zanon, V., 1948
Diatomee marini di Sardegna e Pugillo di Alghe Marine della stressa. Boll. Pesca, Piscitutura e Idrobiologia, Anno 24, ns. 3(2): 202-244, 27 figs. on 1 pl.

Auricula insecta

Jorgensen, E., 1900
Protophyten und Protozoën im Plankton aus der Norwegischen Westkerste. Bergens Mus. Aarb. 1899(6): 95 pp., 5 pls., 83 tables.

Auricula insecta

Pavillard, J., 1905
Recherches sur la flore pelagique (Phytoplankton) de l'Etang de Thau. Theses presentees a la Fac. Sci., Paris, 116 pp., 3 pls.

Auricula insecta

Rampi, L., 1942
Ricerche sul fitoplancton del Mare Ligure 6. Le diatomee delle acque di Sanremo. Nuovo Giornale Botanico Italiano, N.S., 49:252-268.

Auricula insecta

Rampi, L., 1940
Diatomee del Mare Adriatico. Nuovo Giornale Botanico Italiano, n.s., 47:559-608.

Auricula intermedia

Hendy, N. Ingram, 1964
An introductory account of the smaller algae of British coastal waters. V. Bacillariophyceae (Diatoms).
Her Majesty's Stationary Office, 317 pp., 45 pls.

Auricula minuta

Hendy, N. Ingram, 1964
An introductory account of the smaller algae of British coastal waters. V. Bacillariophyceae (Diatoms).
Her Majesty's Stationary Office, 317 pp., 45 pls.

Auricula ostrea

Hendey, N.I., 1951
Littoral diatoms of Chicester Harbour with special reference to fouling. J.Roy. Microscop. Soc. 71(1): 1-86, 18 pls.

Auriculaceae, n.fam.

Hendy, N. Ingram, 1964
An introductory account of the smaller algae of British coastal waters. V. Bacillariophyceae (Diatoms).
Her Majesty's Stationary Office, 317 pp., 45 pls.

Auriculopsis sparsipunctata n.gen., n.sp.

Hendy, N. Ingram, 1964
An introductory account of the smaller algae of British coastal waters. V. Bacillariophyceae (Diatoms).
Her Majesty's Stationary Office, 317 pp., 45 pls.

Bacillaria paradoxa

Allen, W.E., and E.E. Cupp, 1935
Plankton diatoms of the Java Sea. Annales du Jardin Botanique de Buitenzorg XLIV (2):101-174, figs.1-127.
(drawings of all species mentioned)

Bacillaria paradoxa

Forti, A., 1922
Ricerche sulla flora pelagica (fitoplancton) di Quarto dei Mille. Mem. R. Com. Talass. Ital. 97:248 pp., 13 pls.

Bacillaria paradoxa

Grunow, A., 1877
1. New Diatoms from Honduras. Monthly Micros. Jour., 18:165-186, pls. CXCIII-CXCVI.

Bacillaria paradoxa

Kokubo, S., and S. Sato, 1947
Plankters in Jū-San Gata. Physiol. and Ecol. (Japan) 1(4):1-16, 3 text figs., tables.

Bacillaria paradoxa

Levander, K.M., 1947
Plankton gesammelt in den Jahren 1899-1910 an den Küsten Finnlands. Finnländische Hydrographisch-Biologische Untersuchungen (aus dem Wasserbiologischen Laboratorin der Societas Scientiarum Fennica) No.11:40 pp., 6 diagrams, 13 pls., tables.

Bacillaria paradoxa

Margalef, R., 1949
Fitoplancton nerítico de la Costa Brava en 1947-48. Publ. Inst. Biol. Aplicada, 5: 41-51, 3 text figs.

Bacillaria paradoxa

Meunier, A., 1915
Microplancton de la Mer Flamande. 2. Diatomées (excepté le genre Chaetoceros). Mem. Mus. Roy. Hist. Nat., Belgique, 7(3):1-118, Pls. VIII-XIV.

Bacillaria paradoxa

Morse, D.C., 1947
Some observations on seasonal variations in plankton population Patuxent River, Maryland 1943-1945. Bd. Nat. Res., Publ. No.65, Chesapeake Biol. Lab., 31, 3 figs.

Bacillaria paradoxa

Pavillard, J., 1905
Recherches sur la flore pelagique (Phytoplankton) de l'Etang de Thau. Theses presentees a la Fac. Sci., Paris, 116 pp., 3 pls.

Bacillaria paradoxa
Rampi, L., 1942
  Ricerche sul fitoplancton del Mare Ligure
6. Le diatomee delle acque di Sanremo. Nuovo
Giornale Botanico Italiano, N.S., 49:252-268.

Bacillaria paradoxa
Zanon, D. V., 1949
  Diatomee di Buenos Aires (Argentina)
Atti Accad. Naz. Lincei, Memorie, Cl. Sci.
fis., mat. e. nat., ser. 7, 11(3):59-151,
2 pls.

Bacillaria paradoxa
Ercegovic, A., 1936
  Etudes qualitative et quantitatives du
phytoplancton dans les eaux cotières de
l'Adriatique oriental moyen au cours de l'année
1934. Acta Adriatica 1(9):1-126

Bacillaria paradoxa
Gran, H.H., 1908
  Diatomeen. Nordisches Plankton, Botanischer Teil pp. XIX.1-XIX 146; 178 text figs.

Bacillaria paradoxa
Jorgensen, E., 1900
  Protophyten und Protozoën im Plankton aus der Norwegischen Westkerste. Bergens Mus. Aarb. 1899(6): 95 pp., 5 pls., 83 tables.

Bacillaria paradoxa
Mangin, M. L., 1912
  Phytoplancton de la croisière du "René" dans l'Atlantique (Septembre 1908). Ann. Inst. Ocean., n.s., 4(1):1-66, 2 pls., 41 text figs., 2 tables.

Bacillaria paradoxa
Massutí Algamora, M., 1949
  Estudio de diez y seis muestras de plancton del Golfo de Nápoles. Publ. Inst. Biol. Appl. 5:85-94, 1 fold-in table.

Bacillaria paradoxica
Pavillard, J., 1925
  Bacillariales. Rept. on the Danish Oceangr. Exped., 1908-10 to the Mediterranean and adj. seas. Vol.II., Biol. J4:72 pp., 116 text figs.

Bacillaria paradoxa
Rumkówna, A., 1948
  [List of the phytoplankton species occurring in the superficial water layers in the Gulf of Gdańsk] Bull. Lab. mar., Gdynia, No. 4: 139-141 with tables in back.

Bacillaria paradoxa
Schodduyn, M., 1926
  Observations faites dans la baie d'Ambleteuse (Pas de Calais). Bull. Inst. Ocean., Monaco, No. 482: 64 pp.

Bacillaria paradoxa
Steemann-Nielsen, Einar, 1951
  The marine vegetation of the Isefjord. A study on ecology and production. Medd. Komm. Danmarks Fiskeri-og Havundersøgelser. Ser. Plankton. 5(4); 114pp., 46 text figs.

Bacillaria paradoxa
Zanon, V., 1948
  Diatomee marini di Sardegna e Pugillo di Alghe Marine della stressa. Boll. Pesca, Piscitutura e Idrobiologia, Anno 24, ns. 3(2): 202-244, 27 figs. on 1 pl.

Bacillaria paxillifer
Hendy, N. Ingram, 1964
  An introductory account of the smaller algae of British coastal waters. V. Bacillariophyceae (Diatoms). Her Majesty's Stationary Office, 317 pp., 45 pls.

Bacillaria paxillifer comb. nov.
Hendey, N.I., 1951
  Littoral diatoms of Chicester Harbour with special reference to fouling. J.Roy. Microscop. Soc. 71(1): 1-86, 18 pls.

Bacillaria socialis
Jørgensen, E., 1905
  B.Protistplankton and the diatoms in bottom samples. Hydrographical and biological investigations in Norwegian fiords. Bergens Mus. Skr. 7: 49-225.

Bacillaria socialis
Zanon, V., 1948
  Diatomee marini di Sardegna e Pugillo di Alghe Marine della stressa. Boll. Pesca, Piscitutura e Idrobiologia, Anno 24, ns. 3(2): 202-244, 27 figs. on 1 pl.

Bacillaria socialis var. indica var.
Castracane degli Antelminelli, F., 1886
  1. Rëport on the Diatomaceae collected by H.M.S. Challenger during the years 1873-1876. Rept. Sci. Results, H.M.S. Challenger, Botany Vol. II, 178 pp., 30 pls.

Bacteriastrum sp.
Allen, W. E., 1938
  The Templeton Crocker Expedition to the Gulf of California in 1935 - The Phytoplankton. Amer. Microsc. Soc., Trans. 57:328-335.

Bacteriastrum sp.
Allen, W.E., 1937
  Plankton diatoms of the Gulf of California obtained by the G. Allan Hancock Expedition of 1936. The Hancock Pacific Expeditions, Univ. So. Calif. Publ. 3:47-59, 1 fig.

Bacteriastrum sp.
Bigelow, H.B., 1922
  Exploration of the coastal water off the northeastern United States in 1916 by the U.S. Fisheries Schooner Grampus. Bull. M.C.Z. 65 (5):85-188, 53 text figs.

Bacteriastrum
Brunel, J., 1962
  Le phytoplancton de la Baie de Chaleurs. Inst. Botan., Univ. Montréal, Contrib. No. 77: 365 pp., 66 pls.

Bacteriastrum sp.
Cupp, E.E. and Allen, W.E., 1938
  Plankton diatoms of the Gulf of California obtained by Allan Hancock Pacific Expedition of 1937. The Hancock Pacific Expeditions, the Univ. So. Calif. Publ. 3: 61-74, 1 map, pls.4-15.

Bacteriastrum
Desikachary, T.V., 1956(1957).
  Electron microscope studies on diatoms. J. R. Microsc. Soc. (3)76(1/2):9-30.

Bacteriastrum sp.
Gilbert, J.Y., and W.E. Allen, 1943
  The phytoplankton of the Gulf of California obtained by the "E.W. Scripps" in 1939 and 1940. J. Mar. Res. V(2):89-110, figs.30-31.

Bacteriastrum
Ikari, J., 1927
  On Bacteriastrum of Japan. Bot. Mag. (Tokyo) 41 (486):

Bacteriastrum
Neaverson, E., 1934
  The sea-floor deposits. 1. General characteristics and distribution. Discovery Repts. 9: 297-349, Plates 17-22.

Bacteriastrum
Pavillard, J., 1924
  Observations sur les Diatomées (4 ser) Le genre Bacteriastrum. Bull. Soc. Bot. de France, 71:1084-1090.

Bacteriastrum
Subrahmanyan, R., 1945.
  On the formation of auxospores in Bacteriastrum. Curr. Sci. 14:154-155.

Bacteriastrum sp.
Wawrik, F., 1961
  Die horizontale Verteilung der Planktondiatomeen im Golf von Neapel. Int. Revue Ges. Hydrobiol., 46(3):460-479.

Bacteriastrum biconicum
Dangeard, P., 1927
  Phytoplankton de la croisière du "Sylvana". Ann. Inst. Ocean. Monaco, n.s., 4(8):286-401, 54 text figs. (Feirer-Juin 1913).

Bacteriastrum biconicum
Forti, A., 1922
  Ricerche sulla flora pelagica (fitoplancton) di Quarto dei Mille. Mem. R. Com. Talass. Ital. 97:248 pp., 13 pls.

Bacteriastrum biconicum
Pavillard, J., 1925
  Bacillariales. Rept. on the Danish Oceangr. Exped., 1908-10 to the Mediterranean and adj. seas. Vol.II., Biol. J4:72 pp., 116 text figs.

Bacteriastrum biconicum
Pavillard, J., 1924
  Observations sur les Diatomées (4 ser) Le genre Bacteriastrum. Bull. Soc. Bot. de France, 71:1084-1090.

Bacteriastrum biconicum
Rampi, L., 1942
  Ricerche sul fitoplancton del Mare Ligure 6. Le diatomee delle acque di Sanremo. Nuovo Giornale Botanico Italiano, N.S., 49:252-268.

Bacteriastrum brevispinum
Castracane degli Antelminelli, F., 1886 n.sp.
  1. Rëport on the Diatomaceae collected by H.M.S. Challenger during the years 1873-1876. Rept. Sci. Results, H.M.S. Challenger, Botany Vol. II, 178 pp., 30 pls.

Bacteriastrum comosum
Allen, W.E., and E.E. Cupp, 1935
  Plankton diatoms of the Java Sea. Annales du Jardin Botanique de Buitenzorg XLIV (2):101-174, figs.1-127.
  (drawings of all species mentioned)

Bacteriastrum comosum
Cupp, Easter E., 1943
  Marine plankton diatoms of the west coast of North America. Bull. S.I.O. 5(1):1-238, 5 pls., 168 text figs.

Bacteriastrum comosum
Cupp, E.E., 1934
  Analysis of marine diatom collections taken from the Canal Zone to California during March, 1933. Trans. Am. Micros. Soc. LIII (1):22-29, 1 map.

Bacteriastrum comosum
Dangeard, P., 1927
  Phytoplankton de la croisière du "Sylvana". Ann. Inst. Ocean. Monaco, n.s., 4(8):286-401, 54 text figs. (Feirer-Juin 1913).

Bacteriastrum comosum
Hendey, N.I., 1937
  The plankton diatoms of the southern seas. Discovery Repts. 16:151-364, pls.6-13.

Bacteriastrum comosum
Iyengar, M.O.P. and G.Venkataraman, 1951.
The ecology and seasonal succession of the algae flora of the River Cooum at Madras with special reference to the Diatomaceae. J. Madras Univ. 21, Sect. B(1): 140-192, 1 pl of 4 figs., 11 text figs.

Bacteriastrum comosum
Pavillard, J., 1925
Bacillariales. Rept. on the Danish Oceangr. Exped., 1908-10 to the Mediterranean and adj. seas. Vol.II., Biol. J4:72 pp., 116 text figs.

Bacteriastrum comosum
Pavillard, J., 1924
Observations sur les Diatomées (4 ser) Le genre Bacteriastrum. Bull. Soc. Bot. de France, 71:1084-1090.

Bacteriastrum comosum
Rampi, L., 1942
Ricerche sul fitoplancton del Mare Ligure 6. Le diatomee delle acque di Sanremo. Nuovo Giornale Botanico Italiano, N.S., 49:252-268.

Bacteriastrum comosum hispida
Kokubo, S., 1952
Results of the observations on the plankton and oceanography of Mutsu Bay during 1950, reference being made also to the period 1946-1950. Bull Mar.Biol.Sta., Asamushi 5(1/4): 1-54, 3 tables,(fold-in), 1 fold-in.

Bacteriastrum curvatum
Pavillard, J., 1924
Observations sur les Diatomées (4 ser) Le genre Bacteriastrum. Bull. Soc. Bot. de France, 71:1084-1090.

Bacteriastrum criophilum
Hendey, N.I., 1937
The plankton diatoms of the southern seas. Discovery Repts. 16:151-364, pls.6-13.

Bacteriastrum delicatulum
Allen, W.E., and E.E. Cupp, 1935
Plankton diatoms of the Java Sea. Annales du Jardin Botanique de Buitenzorg XLIV (2):101-174, figs.1-127.
(drawings of all species mentioned)

Bacteriastrum delicatulum
Cupp, Easter E., 1943
Marine plankton diatoms of the west coast of North America. Bull. S.I.O. 5(1):1-238, 5 pls., 168 text figs.

Bacteriastrum delicatulum
Cupp, E.E., 1937
Seasonal distribution and occurrence of marine diatoms and dinoflagellates at Scotch Cap, Alaska. Bull. S.I.O. Tech. ser.4(3):71-100, 7 textfigs.

Bacteriastrum delicatulum
Cupp, E.E., 1934
Analysis of marine diatom collections taken from the Canal Zone to California during March, 1933. Trans. Am. Micros. Soc. LIII (1):22-29, 1 map.

Bacteriastrum delicatulum
Dangeard, P., 1927
Phytoplankton de la croisière du "Sylvana". Ann. Inst. Ocean., Monaco, n.s., 4(8):286-401, 54 text figs. (Feirer-Juin 1913).

Bacteriastrum delicatulum
Delegazione Italiana della Commissione Internazionale per l'Esplorazione Scientifica del Mediterraneo, 1941
Note sul plancton della Laguna veneta. [Memoria CCLXXIX], Arch. di Ocean. e Limn. Anno I, Fasc. I, 1941 XIX: 31-57 pp.

Bacteriastrum delicatulum
de Sousa e Silver, E., 1956.
Contribution à l'étude du microplancton de Dakar et des regions maritimes voisines. Bull. I.F.A.N., 8(2):335-371, 7 pls.

Bacteriastrum delicatulum
Ercegovic, A., 1936
Etudes qualitative et quantitatives du phytoplancton dans les eaux cotières de l'Adriatique oriental moyen au cours de l'année 1934. Acta Adriatica 1(9):1-126

Bacteriastrum delicatulum
Eskinazi Enide e Shigekatsv Satô (1963-1964) 1966.
Contribuição ao estudo das diatomaceas da Praia de Piedade.
Trabhs Inst. Oceanogr., Univ. Recife, 5 (5/6): 73-114.

Bacteriastrum delicatulum
Frenguelli, Joaquin, and Hector Antonio Orlando, 1959.
Operacion MERLUZA. Diatomeas y silicoflagelados del plancton del "VI Crucero". Servicio Hidrogr. Naval., Argentina, Publ. No. H. 619: 5-62.

Bacteriastrum delicatulum
Ghazzawi, F.M., 1939
Plankton of the Egyptian waters. A study of the Suez Canal Plankton. (A) Phytoplankton. Preliminary Report 83 pp. Notes and Memoires, Min. Commerce-Industry, Egypt, Hydrobiol. & Fish. 65 figs.

Bacteriastrum delicatulum
Gran, H.H., 1908
Diatomeen. Nordisches Plankton, Botanischer Teil pp. XIX.1-XIX 146; 178 text figs.

Bacteriastrum delicatulum
Gran, H. H. and E. C. Angst, 1931
Plankton diatoms of Puget Sound. Publ. Puget Sound Biol. Sta. 7:417-519, 95 text figs.

Bacteriastrum delicatulum
Hendy, N. Ingram, 1964
An introductory account of the smaller algae of British coastal waters. V. Bacillariophyceae (Diatoms). Her Majesty's Stationary Office, 317 pp., 45 pls.

Bacteriastrum delicatulum
Hendey, N.I., 1937
The plankton diatoms of the southern seas. Discovery Repts. 16:151-364, pls.6-13.

Bacteriastrum delicatulum
Kokubo, S., 1952
Results of the observations on the plankton and oceanography of Mutsu Bay during 1950, reference being made also to the period 1946-1950. Bull Mar.Biol.Sta., Asamushi 5(1/4): 1-54, 3 tables,(fold-in), 1 fold-in.

Bacteriastrum delicatulum
Margalef, R., 1949
Fitoplancton nerítico de la Costa Brava en 1947-48. Publ. Inst. Biol. Aplicada, 5: 41-51, 3 text figs.

Bacteriastrum delicatulum
Massuti Algamora, M., 1949
Estudio de diez y seis muestras de plancton del Golfo de Nápoles. Publ. Inst. Biol. Appl. 5:85-94, 1 fold-in table.

Bacteriastrum delicatulum
Pavillard, J., 1925
Bacillariales. Rept. on the Danish Oceangr. Exped., 1908-10 to the Mediterranean and adj. seas. Vol.II., Biol. J4:72 pp., 116 text figs.

Bacteriastrum delicatulum
Pavillard, J., 1924
Observations sur les Diatomées (4 ser) Le genre Bacteriastrum. Bull. Soc. Bot. de France, 71:1084-1090.

Bacteriastrum delicatulum
Rampi, L., 1942
Ricerche sul fitoplancton del Mare Ligure 6. Le diatomee delle acque di Sanremo. Nuovo Giornale Botanico Italiano, N.S., 49:252-268.

Bacteriastrum delicatulum
Sousa e Silva, E., 1949
Diatomaceas e Dinoflagelados de Baia de Cascais. Portugaliae Acta Biol., Volume: Julio Henriques, Ser. B: 300-383, 9 pls, 2 fold-in tables.

Bacteriastrum elegans
Forti, A., 1922
Ricerche sulla flora pelagica (fitoplancton) di Quarto dei Mille. Mem. R. Com. Talass. Ital. 97:248 pp., 13 pls.

Bacteriastrum elegans
Pavillard, J., 1925
Bacillariales. Rept. on the Danish Oceangr. Exped., 1908-10 to the Mediterranean and adj. seas. Vol.II., Biol. J4:72 pp., 116 text figs.

Bacteriastrum elegans
Rampi, L., 1942
Ricerche sul fitoplancton del Mare Ligure 6. Le diatomee delle acque di Sanremo. Nuovo Giornale Botanico Italiano, N.S., 49:252-268.

Bacteriastrum elongatum
Allen, W.E., 1928.
Catches of marine diatoms and dinoflagellates taken by boat in Southern California waters in 1926. Bull. S.I.O., tech. ser., 1:201-246, 6 textfigs.

Bacteriastrum elongatum
Cupp, Easter E., 1943
Marine plankton diatoms of the west coast of North America. Bull. S.I.O. 5(1):1-238, 5 pls., 168 text figs.

Bacteriastrum elongatum
Cupp, E.E. and Allen, W.E., 1938
Plankton diatoms of the Gulf of California obtained by Allan Hancock Pacific Expedition of 1937. The Hancock Pacific Expeditions, The Univ. So. Calif. Publ. 3: 61-74, 1 map, pls.4-15.

Bacteriastrum elongatum
Ercegovic, A., 1936
Etudes qualitative et quantitatives du phytoplancton dans les eaux cotières de l'Adriatique oriental moyen au cours de l'année 1934. Acta Adriatica 1(9):1-126

Bacteriastrum elongatum
Forti, A., 1922
Ricerche sulla flora pelagica (fitoplancton) di Quarto dei Mille. Mem. R. Com. Talass. Ital. 97:248 pp., 13 pls.

Bacteriastrum elongatum
Gran, H.H., 1908
Diatomeen. Nordisches Plankton, Botanischer Teil pp. XIX.1-XIX 146; 178 text figs.

Bacteriastrum elongatum
Hendy, N. Ingram, 1964
An introductory account of the smaller algae of British coastal waters. V. Bacillariophyceae (Diatoms). Her Majesty's Stationary Office, 317 pp., 45 pls.

Bacteriastrum elongatum
Hendey, N.I., 1937
The plankton diatoms of the southern seas. Discovery Repts. 16:151-364, pls.6-13.

Bacteriastrum elongatum
Pavillard, J., 1925
Bacillariales. Rept. on the Danish Oceanogr. Exped., 1908-10 to the Mediterranean and adj. seas. Vol.II., Biol. J4:72 pp., 116 text figs.

Bacteriastrum elongatum
Pavillard, J., 1924
Observations sur les Diatomées (4 ser) Le genre Bacteriastrum. Bull. Soc. Bot. de France, 71:1084-1090.

Bacteriastrum elongatum
Pavillard, J., 1905
Recherches sur la flore pelagique (Phytoplankton) de l'Etang de Thau. Theses presentees a la Fac. Sci., Paris, 116 pp., 3 pls.

Bacteriastrum elongatum
Rampi, L., 1942
Ricerche sul fitoplancton del Mare Ligure 6. Le diatomee delle acque di Sanremo. Nuovo Giornale Botanico Italiano, N.S., 49:252-268.

Bacteriastrum elongatum
Schröder, B., 1900
Phytoplankton des Golfes von Neapel nebst vergleichenden Ausblicken auf das atlantischen Ozean. Mitt. Zool. Stat. Neapel, 14:1-38.

Bacteriastrum fragilis
Lillick, L.C., 1940
Phytoplankton and planktonic protozoa of the offshore waters of the Gulf of Maine. Pt.II. Qualitative Composition of the Planktonic Flora. Trans. Am. Phil. Soc., n.s., 31(3):193-237, 13 text figs.

Bacteriastrum furcatum
Pavillard, J., 1924
Observations sur les Diatomées (4 ser) Le genre Bacteriastrum. Bull. Soc. Bot. de France, 71:1084-1090.

Bacteriastrum hyalinum
Allen, W. E., 1936
Occurrence of marine plankton diatoms in a ten-year series of daily catches in southern California. Am. Jour. Bot. 23(1):60-63.

Bacteriastrum hyalinum
Allen, W.E., and E.E. Cupp, 1935
Plankton diatoms of the Java Sea. Annales du Jardin Botanique de Buitenzorg XLIV (2):101-174, figs.1-127.
(drawings of all species mentioned)

Bacteriastrum hyalinum
Boden, B.P., 1950
Some marine plankton diatoms from the west coast of South Africa. Trans. R.Soc. S. Africa. 32:321-434, 100 text figs.

Bacteriastrum hyalinum
Cupp, Easter E., 1943
Marine plankton diatoms of the west coast of North America. Bull. S.I.O. 5(1):1-238, 5 pls., 168 text figs.

Bacteriastrum hyalinum
Delsman, H. C., 1939.
Preliminary plankton investigations in the Java Sea. Treubia, 17:139-181, 8 maps, 41 figs.

Bacteriastrum hyalinum
Drebes, G., 1972.
The life history of the centric diatom Bacteriastrum hyalinum Lauder. Nova Hedwigia (Beih.)39:95-110.

Bacteriastrum hyalinum
Drebes, G. 1967
Bacteriastrum solitarium Mangin, a stage in the life history of the centric diatom Bacteriastrum hyalinum.
Mar. Biol. 1(1):40-42.

Bacteriastrum hyalinum
Ercegovic, A., 1936
Etudes qualitative et quantitatives du phytoplancton dans les eaux cotières de l'Adriatique oriental moyen au cours de l'année 1934. Acta Adriatica 1(9):1-126

Bacteriastrum hyalinum
Eskinazi Enide e Shigekatsv Sato (1963-1964) 1966.
Contribuição ao estudo das diatomaceas da Praia de Piedade.
Trabhs Inst. Oceanogr., Univ. Recife, 5 (5/6): 73-114.

Bacteriastrum hyalinum
Forti, A., 1922
Ricerche sulla flora pelagica (fitoplancton) di Quarto dei Mille. Mem. R. Com. Talass. Ital. 97:248 pp., 13 pls.

Bacteriastrum hyalinum
Frenguelli, Joaquin, and Hector Antonio Orlando, 1959,
Operacion MERLUZA. Diatomeas y silico-flagelados del plancton del "VI Crucero". Servicio Hidrogr. Naval, Argentina, Publ. No. H. 619: 5-62.

Bacteriastrum hyalinum
Ghazzawi, F.M., 1939
Plankton of the Egyptian waters. A study of the Suez Canal Plankton. (A) Phytoplankton. Preliminary Report 83 pp. Notes and Memoires, Min. Commerce-Industry, Egypt, Hydrobiol. & Fish. 65 figs.

Bacteriastrum hyalinum
Hendy, N. Ingram, 1964
An introductory account of the smaller algae of British coastal waters. V. Bacillariophyceae (Diatoms). Her Majesty's Stationary Office, 317 pp., 45 pls.

Bacteriastrum hyalinum
Kokubo, S., 1952
Results of the observations on the plankton and oceanography of Mutsu Bay during 1950, reference being made also to the period 1946-1950. Bull Mar.Biol.Sta., Asamushi 5(1/4): 1-54, 3 tables,(fold-in), 1 fold-in.

Bacteriastrum hyalinum
Lillick, L.C., 1937
Seasonal studies of the phytoplankton off Woods Hole, Massachusetts. Biol. Bull. LXXIII (3):488-503, 3 text figs.

Bacteriastrum hyalinum
Osorio Tafall, B.F., 1935.
La auxosporulacion en Bacteriastrum hyalinum Lauder. Bol. real Soc. Esp. Hist. Nat. 35:111-124.

Bacteriastrum hyalinum
Paiva Carvalho, J., 1950
O plancton do Rio Maria Rodriques (Cananeis). 1. Diatomaceas e Dinoflagelados. Bol. Inst. Paulista Oceanogr. 1(1); 27-43, 2 fold-in tables, 2 figs.

Bacteriastrum hyalinum
Pavillard, J., 1925
Bacillariales. Rept. on the Danish Oceanogr. Exped., 1908-10 to the Mediterranean and adj. seas. Vol.II., Biol. J4:72 pp., 116 text figs.

Bacteriastrum hyalinum
Pavillard, J., 1924
Observations sur les Diatomées (4 ser) Le genre Bacteriastrum. Bull. Soc. Bot. de France, 71:1084-1090.

Bacteriastrum hyalinum
Sousa e Silva, E., 1949
Diatomaceas e Dinoflagelados de Baia de Cascais. Portugalise Acta Biol., Volume: Julio Henriques, Ser. B: 300-383, 9 pls, 2 fold-in tables.

Bacteriastrum hyalinum princeps
Cupp, Easter E., 1943
Marine plankton diatoms of the west coast of North America. Bull. S.I.O. 5(1):1-238, 5 pls., 168 text figs.

Bacteriastrum hyalinum var. princeps
Hendey, N.I., 1937
The plankton diatoms of the southern seas. Discovery Repts. 16:151-364, pls.6-13.

Bacteriastrum hyalinum var. princeps
Rampi, L., 1942
Ricerche sul fitoplancton del Mare Ligure 6. Le diatomee delle acque di Sanremo. Nuovo Giornale Botanico Italiano, N.S., 49:252-268.

Bacteriastrum hyalinum var. princeps
Rampi, L., 1940
Diatomee del Mare Adriatico. Nuovo Giornale Botanico Italiano, n.s., 47:559-608.

Bacteriastrum hyalinum
Smayda, Theodore J., and Brenda J. Boleyn, 1966
Experimental observations on the flotation of marine diatoms. III. Bacteriastrum hyalinum and Chaetoceros lauderi. Limnol. and Oceanogr., 11(1):35-43.

Bacteriastrum hyalinum
Sousa a Silva, E., and J. Dos Santos-Pinto, 1948
O Plancton da Baia de S. Martinho do Porto. 1. Diatomaceas e Dinoflagelados. Bol. Soc. Portuguese de Ciencias Naturais, 16(2):134-187, 6 pls. (Trav. Sta. Biol. Mar. de Lisbonne No. 52).

Bacteriastrum mediterraneum
Ercegovic, A., 1936
Etudes qualitative et quantitatives du phytoplancton dans les eaux cotières de l'Adriatique oriental moyen au cours de l'année 1934. Acta Adriatica 1(9):1-126

Bacteriastrum mediterraneum
Forti, A., 1922
Ricerche sulla flora pelagica (fitoplancton) di Quarto dei Mille. Mem. R. Com. Talass. Ital. 97:248 pp., 13 pls.

Bacteriastrum mediterraneum
Pavillard, J., 1924
Observations sur les Diatomées (4 ser) Le genre Bacteriastrum. Bull. Soc. Bot. de France, 71:1084-1090.

Bacteriastrum mediterraneum
Rampi, L., 1942
Ricerche sul fitoplancton del Mare Ligure 6. Le diatomee delle acque di Sanremo. Nuovo Giornale Botanico Italiano, N.S., 49:252-268.

Bacteriastrum minus
Kokubo, S., 1952
Results of the observations on the plankton and oceanography of Mutsu Bay during 1950, reference being made also to the period 1946-1950. Bull Mar.Biol.Sta., Asamushi 5(1/4): 1-54, 3 tables,(fold-in), 1 fold-in.

Bacteriastrum solitarium
Drebes, G. 1967
Bacteriastrum solitarium Mangin, a stage in the life history of the centric diatom Bacteriastrum hyalinum.
Mar. Biol. 1(1):40-42.

Bacteriastrum solitarium

Frenguelli, Joaquin, and Hector Antonio Orlando, 1959.
Operacion MERLUZA. Diatomeas y silico-flagelados del plancton del "VI Crucero". Servicio Hidrogr. Naval., Argentina, Publ. No. H. 619: 5-62.

Bacteriastrum solitarium

Hendy, N. Ingram, 1964
An introductory account of the smaller algae of British coastal waters. V. Bacillariophyceae (Diatoms). Her Majesty's Stationary Office, 317 pp., 45 pls.

Bacteriastrum solitarium n.sp.

Mangin, M. L., 1912
Phytoplancton de la croisière du "René" dans l'Atlantique (Septembre 1908). Ann. Inst. Ocean., n.s., 4(1):1-66, 2 pls., 41 text figs., 2 tables.

Bacteriastrum spirillum n.sp.

Castracane degli Antelminelli, F., 1886
1. Report on the Diatomaceae collected by H.M.S. Challenger during the years 1873-1876. Rept. Sci. Results, H.M.S. Challenger, Botany Vol. II, 178 pp., 30 pls.

Bacteriastrum spirillum

Pavillard, J., 1924
Observations sur les Diatomées (4 ser.) Le genre Bacteriastrum. Bull. Soc. Bot. de France, 71:1084-1090.

Bacterioastrum tenue

Reinecke, Pandora, 1869
Anat on the diatom Bacteriastrum tenue Stuemann Nielsen.
JI S Afr. Bot. 35(3): 207-210.
(also in: Coll. Repr. Inst. Oceanogr. Univ. Cape Town 8.)

Bacteriastrum varians

Allen, W.E., and E.E. Cupp, 1935
Plankton diatoms of the Java Sea. Annales du Jardin Botanique de Buitenzorg XLIV (2):101-174, figs.1-127.
(drawings of all species mentioned)

Bateriastrum varians

Delsman, H. C., 1939.
Preliminary plankton investigations in the Java Sea. Treubia, 17, 139-181, 8 maps, 41 figs.

Bacteriastrum varians

Forti, A., 1922
Ricerche sulla flora pelagica (fitoplancton) di Quarto dei Mille. Mem. R. Com. Talass. Ital. 97:248 pp., 13 pls.

Bacteriastrum varians

Gran, H.H., 1908
Diatomeen. Nordisches Plankton, Botanischer Teil pp. XIX.1-XIX 146; 178 text figs.

Bacteriastrum varians

Hendey, N.I., 1937
The plankton diatoms of the southern seas. Discovery Repts. 16:151-364, pls.6-13.

Bacteriastrum varians

Jorgensen, E., 1900
Protophyten und Protozoën im Plankton aus der Norwegischen Westkerste. Bergens Mus. Aarb. 1899(6): 95 pp., 5 pls., 83 tables

Bacteriastrum varians

Kokubo, S., 1952
Results of the observations on the plankton and oceanography of Mutsu Bay during 1950, reference being made also to the period 1946-1950. Bull Mar.Biol.Sta., Asamushi 5(1/4): 1-54, 3 tables,(fold-in), 1 fold-in.

Bacteriastrum varians

Macdonald, R., 1933
An examination of plankton hauls made in the Suez Canal during the year 1928. Fish. Res. Dir., Notes & Mem. No.3, 11 pp., 1 chart.

Bacteriastrum varians

Mangin, M. L., 1912
Phytoplancton de la croisière du "René" dans l'Atlantique (Septembre 1908). Ann. Inst. Ocean., n.s., 4(1):1-66, 2 pls., 41 text figs., 2 tables.

Bacteriastrum varians

Marukawa, H., 1921
Plankton lists and some new species of copepods, from the northern waters of Japan. Bull. Inst. Ocean., No.384, 15 pp., 3 pls., 1 chart. Monaco

Bacteriastrum varians

Meunier, A., 1915
Microplancton de la Mer Flamande. 2. Diatomées (excepte le genre Chaetoceros). Mem. Mus. Roy. Hist. Nat., Belgique, 7(3):1-118, Pls. VIII-XIV.

Bacteriastrum varians

Pavillard, J., 1924
Observations sur les Diatomées (4 ser.) Le genre Bacteriastrum. Bull. Soc. Bot. de France, 71:1084-1090.

Bacteriastrum varians

Pavillard, J., 1905
Recherches sur la flore pelagique (Phytoplankton) de l'Etang de Thau. Theses presentees a la Fac. Sci., Paris, 116 pp., 3 pls.

Bacteriastrum varians

Schodduyn, M., 1926
Observations faites dans la baie d'Ambleteuse (Pas de Calais). Bull. Inst. Ocean., Monaco, No. 482: 64 pp.

Bacteriastrum varians

Schröder, B., 1900
Phytoplankton des Golfes von Neapel nebst vergleichenden Ausblicken auf das atlantischen Ozean. Mitt. Zool. Stat. Neapel, 14:1-38.

Bacteriastrum varians

Sleggs., G.F., 1927.
Marine phytoplankton in the region of La Jolla, California during the summer of 1924. Bull. S.I.O., tech. ser., 1:93-117, 8 textfigs.

Bacteriastrum varians

Uyeno, Fukuzo, 1961
Oceanographical and ecological studies on primary production of the sea, with special references to relationship between diatom production and temperature and chlorinity of water.
Rept., Fac. Fish., Pref. Univ., Mie, 4(1): 1-64.

Bacteriastrum varians var. hispida

Allen, W.E., and E.E. Cupp, 1935
Plankton diatoms of the Java Sea. Annales du Jardin Botanique de Buitenzorg XLIV (2):101-174, figs.1-127.
(drawings of all species mentioned)

Bacteriastrum varians princeps

Castracane degli Antelminelli, F., 1886 n.var.
1. Report on the Diatomaceae collected by H.M.S. Challenger during the years 1873-1876. Rept. Sci. Results, H.M.S. Challenger, Botany Vol. II, 178 pp., 30 pls.

Bacteriastrum wallichii var. hispida

Castracane degli Antelminelli, F., 1886
1. Report on the Diatomaceae collected by H.M.S. Challenger during the years 1873-1876. Rept. Sci. Results, H.M.S. Challenger, Botany Vol. II, 178 pp., 30 pls.

Bacteriastrum Wallichii var. hispida

Pavillard, J., 1924
Observations sur les Diatomées (4 ser.) Le genre Bacteriastrum. Bull. Soc. Bot. de France, 71:1084-1090.

Bacteriosira fragilis

Braarud, T., 1945
A phytoplankton survey of the polluted waters of inner Oslo Fjord. Hvalrådets Skrifter, No.28, 142 pp., 19 text figs., 17 tables.

Bacteriosira fragilis

Brunel, J., 1962
Le phytoplancton de la Baie de Chaleurs. Inst. Botan., Univ. Montréal, Contrib. No. 77: 365 pp., 66 pls.

Bacteriosira fragilis

Brunel, Jules, 1962
Le phytoplancton de la Baie des Chaleurs. Contrib. Ministère de la Chasse et des Pêcheries, Province de Québec, No. 91: 365 pp.

Bacteriosira fragilis

Cleve-Euler, A., 1951
Die Diatomeen von Schweden und Finnland. Kungl. Svenska Vetenskaps Akad. Handl., Fjärde Ser. 2(1): 161 pp., 6 pls.

Bacteriosira fragilis

Frenguelli, Joaquin, and Hector Antonio Orlando, 1959.
Operacion MERLUZA. Diatomeas y silico-flagelados del plancton del "VI Crucero". Servicio Hidrogr. Naval., Argentina, Publ. No. H. 619: 5-62.

Bacterosira fragilis

Gran, H.H., 1908
Diatomeen. Nordisches Plankton, Botanischer Teil pp. XIX.1-XIX 146; 178 text figs.

Bacteriosira fragilis

Gran, H.H., and T. Braarud, 1935
A quantitative study of the phytoplankton in the Bay of Fundy and the Gulf of Maine (including observations on hydrography, chemistry, and turbidity). J. Biol. Bd., Canada, 1(5):279-467, 69 text figs.

Bacteriosira fragilis

Hendy, N. Ingram, 1964
An introductory account of the smaller algae of British coastal waters. V. Bacillariophyceae (Diatoms). Her Majesty's Stationary Office, 317 pp., 45 pls.

Bacteriosera fragilis

Iselin, C., 1930
A report on the coastal waters of Labrador based on explorations of the "Chance" during the summer of 1926. Proc. Am. Acad. Arts Sci., 66(1):1-37, 14 text figs.

Bacterioaira fragilis

Jørgensen, E., 1905
B. Protistplankton and the diatoms in bottom samples. Hydrographical and biological investigations in Norwegian fjords. Bergens Mus. Skr. 7: 49-225.

Bellerochea indica

Hendey, N.I., 1937
The plankton diatoms of the southern seas. Discovery Repts. 16:151-364, pls.6-13.

Bellerochea malleus

Dangeard, P., 1927
Phytoplankton de la croisière du "Sylvana". Ann. Inst. Ocean., Monaco, n.s., 4(8):286-401, 54 text figs. (Feirer-Juin 1913).

Bellerochea malleus
Gran, H.H., 1908
Diatomeen. Nordisches Plankton, Botanischer Teil pp. XIX.1-XIX 146; 178 text figs.

Bellerochea malleus
Hendy, N. Ingram, 1964
An introductory account of the smaller algae of British coastal waters. V. Bacillariophyceae (Diatoms). Her Majesty's Stationary Office, 317 pp., 45 pls.

Bellerochea malleus
Robinson, G.A., 1961
Contribution towards an atlas of the northeastern Atlantic and the North Sea. 1. Phytoplankton. Bulls. Mar. Ecology. 5(42): 81-89, Pls. 15-20.

Bellarochia malleus
Sousa e Silva, E., 1949
Diatomaceas e Dinoflagelados de Baia de Cascais. Portugaliae Acta Biol., Volume: Julio Henriques, Ser. B: 300-383, 9 pls, 2 fold-in tables.

Bellerochea malleus form. biangulata
Allen, W.E., and E.E. Cupp, 1935
Plankton diatoms of the Java Sea. Annales du Jardin Botanique de Buitenzorg XLIV (2):101-174, figs.1-127.
(drawings of all species mentioned)

Bellerochea malleus
Ioannou, M.M., 1949.
Note sur Bellerochea malleus (Brightwell) Van Heurck, diatomée nouvelle pour la Mediterranée. Prak. Hellen. Hidrobiol. Inst., 1949, 3(1):33-38, 2 textfigs.

Bellerochea malleus
Meunier, A., 1915
Microplancton de la Mer Flamande. 2. Diatomées (excepté le genre Chaetoceros). Mem. Mus. Roy. Hist. Nat., Belgique, 7(3):1-118, Pls. VIII-XIV.

Bellerochea malleus
Schodduyn, M., 1926
Observations faites dans la baie d'Ambleteuse (Pas de Calais). Bull. Inst. Ocean., Monaco, No. 482: 64 pp.

Bellarochia malleus
Sousa e Silva, E., and J. Dos Santos-Pinto, 1948
O Plancton da Baia de S. Martinho do Porto. 1. Diatomaceas e Dinoflagelados. Bol. Soc. Portuguese de Ciencias Naturais, 16(2):134-187, 6 pls. (Trav. Sta. Biol. Mar. de Lisbonne No. 52).

Berkeleya sp.
Grunow, A., 1877
1. New Diatoms from Honduras. Monthly Micros. Jour., 18:165-186, pls. CXCIII-CXCVI.

Biddulphia sp.
Allen, W.E., and E.E. Cupp, 1935
Plankton diatoms of the Java Sea. Annales du Jardin Botanique de Buitenzorg XLIV (2):101-174, figs.1-127.
(drawings of all species mentioned)

Biddulphia sp.
Chin, T.G., C.F. Chen, S.C. Liu and S.S. Wu, 1965. Influence of temperature and salinity on the growth of three plankton diatom species. (In Chinese; English abstract). (Not seen). Oceanol. et Limnol. Sinica, 7(4):373-384.

Biddulphia sp.
Cupp, E.E. and Allen, W.E., 1938
Plankton diatoms of the Gulf of California obtained by Allan Hancock Pacific Expedition of 1937. The Hancock Pacific Expeditions, The Univ. So. Calif. Publ. 3: 61-74, 1 map, pls.4-15.

Biddulphia
Desikachary, T.V., 1956(1957).
Electron microscope studies o diatoms. J. R. Microsc. Soc. (3)76(1/2):9-36.

Biddulphia sp.
Gilbert, J.Y., and W.E. Allen, 1943
The phytoplankton of the Gulf of California obtained by the "E.W. Scripps" in 1939 and 1940. J. Mar. Res. V(2):89-110, figs.30-31.

Biddulphia
Lucas, C. E., 1940.
Ecological investigations with the continuous plankton recorder: the phytoplankton in the southern North Sea, 1932-37. Hull Bull. Mar. Ecol., 1(3):73-170, 17 textfigs, 21 pls.

Biddulphia agulhas
Dangeard, P., 1927
Phytoplankton de la croisière du "Sylvana". Ann. Inst. Ocean., Monaco, n.s., 4(8):286-401, 54 text figs. (Feirer-Juin 1913).

Biddulphia alaskiensis n.sp.
Mann, A., 1907
Report on the diatoms of the Albatross voyages in the Pacific Ocean, 1888-1904. Contrib. U. S. Nat. Herb. 10(5):221-419, Pls. XLIV-LIV.

Biddulphia alternans
Cupp, Easter E., 1943
Marine plankton diatoms of the west coast of North America. Bull. S.I.O. 5(1):1-238, 5 pls., 168 text figs.

Biddulphia alternans
Eskinazi Enide e Shigekatsv Satô (1963-1964) 1966.
Contribuição ao estudo das diatomaceas da Praia de Piedade.
Trabhs Inst. Oceanogr., Univ. Recife, 5 (5/6): 73-114.

Biddulphia alternans
Ghazzawi, F.M., 1939
Plankton of the Egyptian waters. A study of the Suez Canal Plankton. (A) Phytoplankton. Preliminery Report 83 pp. Notes and Memoires, Min. Commerce-Industry, Egypt, Hydrobiol. & Fish. 65 figs.

Biddulphia alternans
Gran, H.H., 1908
Diatomeen. Nordisches Plankton, Botanischer Teil pp. XIX.1-XIX 146; 178 text figs.

Biddulphia alternans
Gran, H.H., and T. Braarud, 1935
A quantitative study of the phytoplankton in the Bay of Fundy and the Gulf of Maine (including observations on hydrography, chemistry, and turbidity). J. Biol. Bd. Canada, 1(5):279-467, 69 text figs.

Biddulphia alternans
Hendy, N. Ingram, 1964
An introductory account of the smaller algae of British coastal waters. V. Bacillariophyceae (Diatoms). Her Majesty's Stationary Office, 317 pp., 45 pls.

Biddulphia alternans
Hendey, N.I., 1951
Littoral diatoms of Chicester Harbour with special reference to fouling. J. Roy. Microscop. Soc. 71(1): 1-86, 18 pls.

Biddulphia alternans
Meunier, A., 1915
Microplancton de la Mer Flamande. 2. Diatomées (excepté le genre Chaetoceros). Mem. Mus. Roy. Hist. Nat., Belgique, 7(3):1-118, Pls. VIII-XIV.

Biddulphia alternans
Schodduyn, M., 1926
Observations faites dans la baie d'Ambleteuse (Pas de Calais). Bull. Inst. Ocean., Monaco, No. 482: 64 pp.

Biddulphia antideluviana
Bigelow, H.B., and M. Leslie, 1930
Reconnaissance of the waters and plankton of Monterey Bay, July 1928. Bull. M.C.Z., 70(5):429-481, 43 text figs.

Biddulphia antediluvianus & varieties
Hendy, N. Ingram, 1964
An introductory account of the smaller algae of British coastal waters. V. Bacillariophyceae (Diatoms). Her Majesty's Stationary Office, 317 pp., 45 pls.

Biddulphia antediluviana
Hendey, N.I., 1951
Littoral diatoms of Chicester Harbour with special reference to fouling. J. Roy. Microscop. Soc. 71(1): 1-86, 18 pls.

Biddulphia antidiluvianum
Hendey, N.I., 1937
The plankton diatoms of the southern seas. Discovery Repts. 16:151-364, pls.6-13.

Biddulphia antediluviana
Mann, A., 1907
Report on the diatoms of the Albatross voyages in the Pacific Ocean, 1888-1904. Contrib. U. S. Nat. Herb. 10(5):221-419, Pls. XLIV-LIV.

Biddulphia arctica
Gran, H.H., 1908
Diatomeen. Nordisches Plankton, Botanischer Teil pp. XIX.1-XIX 146; 178 text figs.

Biddulphia arctica
Gran, H. H. and E. C. Angst, 1931
Plankton diatoms of Puget Sound. Publ. Puget Sound Biol. Sta. 7:417-519, 95 text figs.

Biddulphia arctica balaena
Gran, H.H., 1908
Diatomeen. Nordisches Plankton, Botanischer Teil pp. XIX.1-XIX 146; 178 text figs.

Biddulphia astrolabensis n.sp.
Hendey, N.I., 1937
The plankton diatoms of the southern seas. Discovery Repts. 16:151-364, pls.6-13.

Biddulphia aurita
Aurich, H. J., 1949.
Die Verbreitung des Nannoplanktons im Oberflächenwasser vor der Nordfriesischen Kuste. Ber. Deutschen Wiss. Komm. f. Meeresf., n.f., II(4): 403-405, 2 figs.

Biddulphia aurita
Bigelow, H.B., and M. Leslie, 1930
Reconnaissance of the waters and plankton of Monterey Bay, July 1928. Bull. M.C.Z., 70(5):429-481, 43 text figs.

Biddulphia aurita
Braarud, T., 1945
A phytoplankton survey of the polluted waters of inner Oslo Fjord. Hvalrådets Skrifter, No.28, 142 pp., 19 text figs., 17 tables.

Biddulphia aurita
Braarud, T., 1945
Experimental studies on marine plankton diatoms. Avhandlingar utgitt av Det Norske Videnskaps-Akademi i Oslo. 1. Mat.-Naturv. Klasse 1944. No.10, 1-16, 1 text fig.

Biddulphia aurita
Brunel, J., 1962
Le phytoplancton de la Baie de Chaleurs. Inst. Botan., Univ. Montréal, Contrib. No. 77: 365 pp., 66 pls.

Biddulphia aurita
Brunel, Jules, 1962
Le phytoplancton de la Baie des Chaleurs. Contrib. Ministère de la Chasse et des Pêcheries; Province de Québec, No. 91: 365 pp.

Biddulphia aurita
Cleve-Euler, A., 1951
Die Diatomeen von Schweden und Finnland. Kungl. Svenska Vetenskaps Akad. Handl., Fjarde Ser. 2(1): 161 pp., 6 pls.

Biddulphia aurita
Cupp, Easter E., 1943
Marine plankton diatoms of the west coast of North America. Bull. S.I.O. 5(1):1-238, 5 pls., 168 text figs.

Biddulphia aurita
Cupp, E.E. and Allen, W.E., 1938
Plankton diatoms of the Gulf of California obtained by Allan Hancock Pacific Expedition of 1937. The Hancock Pacific Expeditions, The Univ. So. Calif. Publ. 3: 61-74, 1 map, pls.4-15.

Biddulphia aurita
Dangeard, P., 1927
Phytoplankton de la croisière du "Sylvana". Ann. Inst. Ocean., Monaco, n.s., 4(8):286-401, 54 text figs. (Feirer-Juin 1913).

Biddulphia aurita
Eskinazi Enide e Shigekatsv Satô (1963-1964) 1966.
Contribuição ao estudo das diatomaceas da Praia de Piedade.
Trabhs Inst. Oceanogr., Univ. Recife, 5 (5/6): 73-114.

Biddulphia aurita
Gran, H.H., 1908
Diatomeen. Nordisches Plankton, Botanischer Teil pp. XIX.1-XIX 146; 178 text figs.

Biddulphia aurita
Gran, H.H., and T. Braarud, 1935
A quantitative study of the phytoplankton in the Bay of Fundy and the Gulf of Maine (including observations on hydrography, chemistry, and turbidity). J. Biol. Bd., Canada, 1(5):279-467, 69 text figs.

Biddulphia aurita
Gran, H. H. and E. C. Angst, 1931
Plankton diatoms of Puget Sound. Publ. Puget Sound Biol. Sta. 7:417-519, 95 text figs.

Biddulphia aurita
Grøntved, J., 1949(1950).
Investigations on the phytoplankton in the Danish Waddensee in July 1941. Medd. Komm. Danmarks Fiskeri- og Havundersøgelser, Serie Plankton, 5(2)(Medd. Skalling Lab. 10):55 pp., 2 pls., 38 textfigs.

Biddulphia aurita
Grøntved, J., 1949
Investigations on the phytoplankton in the Danish Waddensea in July 1941. Medd. Komm. Danmarks Fiskeri og Havundersøgelser, sor. Plankton, 5(2):55 pp., 2 pls., 38 text figs.

Biddulphia aurita
Hendy, N. Ingram, 1964
An introductory account of the smaller algae of British coastal waters. V. Bacillariophyceae (Diatoms).
Her Majesty's Stationary Office, 317 pp., 45 pls.

Biddulphia aurita
Hendey, N.I., 1951
Littoral diatoms of Chicester Harbour with special reference to fouling. J.Roy. Microscop. Soc. 71(1): 1-86, 18 pls.

Biddulphia aurita
Hustedt, F. and A.A. Aleem, 1951
Littoral diatoms from the Salstone near Plymouth. JMBA 30(1): 177-196.

Biddulphia aurita
Iselin, C., 1930
A report on the coastal waters of Labrador based on explorations of the "Chance" during the summer of 1926. Proc. Am. Acad. Arts Sci., 66(1):1-37, 14 text figs.

Biddulphia aurita
Jørgensen, E., 1905
B. Protistplankton and the diatoms in bottom samples. Hydrographical and biological investigations in Norwegian fjords. Bergens Mus. Skr. 7: 49-225.

Biddulphia aurita
Jorgensen, E., 1900
Protophyten und Protozoën im Plankton aus der Norwegischen Westkerste. Bergens Mus. Aarb. 1899(6): 95 pp., 5 pls., 83 tables.

Biddulphia aurita
Lillick, L.C., 1940
Phytoplankton and planktonic protozoa of the offshore waters of the Gulf of Maine. Pt.II. Qualitative Composition of the Planktonic Flora. Trans. Am. Phil. Soc., n.s., 31(3):193-237, 13 text figs.

Biddulphia aurita
Lillick, L.C., 1938
Preliminary report of the phytoplankton of the Gulf of Maine. Am. Mid. Nat. 20(3):624-640, 1 text figs 37 tables.

Biddulphia aurita
Lui, Nan S.T., and Oswald A. Roels 1972.
Nitrogen metabolism of aquatic organisms. II The assimilation of nitrate nitrite and ammonia by Biddulphia aurita.
J. Phycol. 8 (3): 259-264.

Biddulphia aurita
Mann, A., 1907
Report on the diatoms of the Albatross voyages in the Pacific Ocean, 1888-1904. Contrib. U. S. Nat. Herb. 10(5):221-419, Pls. XLIV-LIV.

Biddulphia aurita
Mann, A., 1893
List of Diatomaceae from a deep-sea dredging in the Atlantic Ocean off Delaware Bay by the U. S. Fish Commission Steamer Albatross. Proc. U. S. Nat. Mus. 16:303-312.

Biddulphia aurita
Meunier, A., 1915
Microplancton de la Mer Flamande. 2. Diatomées (excepté le genre Chaetoceros). Mem. Mus. Roy. Hist. Nat., Belgique, 7(3):1-118, Pls. VIII-XIV.

Biddulphia aurita
Misra, J.N., 1956.
A systematic account of some littoral marine diatoms from the west coast of India. J. Bombay Nat. Hist. Soc., 53(4):537-568.

Biddulphia aurita
Rampi, L., 1942
Ricerche sul fitoplancton del Mare Ligure 6. Le diatomee delle acque di Sanremo. Nuovo Giornale Botanico Italiano, N.S., 49:252-268.

Biddulphia aurita
Rampi, L., 1940
Diatomee del Mare Adriatico. Nuovo Giornale Botanico Italiano, n.s., 47:559-608.

Biddulphia aurita
Robinson, G.A., 1961
Contribution towards an atlas of the northeastern Atlantic and the North Sea. 1. Phytoplankton.
Bulls. Mar. Ecology, 5(42): 81-89. Pls. 15-20.

Biddulphia aurita
Schodduyn, M., 1926
Observations faites dans la baie d'Ambleteuse (Pas de Calais). Bull. Inst. Ocean., Monaco, No. 482: 64 pp.

Biddulphia aurita
Stæmann-Nielsen, Einar, 1951
The marine vegetation of the Isefjord. A study on ecology and production. Medd. Komm. Danmarks Fiskeri-og Havundersøgelser. Ser. Plankton. 5(4); 114pp., 46 text figs.

Biddulphia aurita
Zanon, V., 1948
Diatomee marini di Sardegna e Pugillo di Alghe Marine della stressa. Boll. Pesca, Piscitutura e Idrobiologia, Anno 24, ns. 3(2): 202-244, 27 figs. on 1 pl.

Biddulphia aurita obtusa
Brunel, J., 1962
Le phytoplancton de la Baie de Chaleurs. Inst. Botan., Univ. Montréal, Contrib. No. 77: 365 pp., 66 pls.

Biddulphia aurita obtusa
Brunel, Jules, 1962
Le phytoplancton de la Baie des Chaleurs. Contrib. Ministère de la Chasse et des Pêcheries; Province de Québec, No. 91: 365 pp.

Biddulphia aurita obtusa
Cupp, Easter E., 1943
Marine plankton diatoms of the west coast of North America. Bull. S.I.O. 5(1):1-238, 5 pls., 168 text figs.

Biddulphia aurita var. obtusa
Hendey, N.I., 1937
The plankton diatoms of the southern seas. Discovery Repts. 16:151-364, pls.6-13.

Biddulphia bicorne
Takano, Hideaki, 1962
Notes on epiphytic diatoms upon sea-weeds from Japan.
J. Oceanogr.Soc., Japan, 18(1):29-33.

Biddulphia biddulphiana
Cleve-Euler, A., 1951
Die Diatomeen von Schweden und Finnland. Kungl. Svenska Vetenskaps Akad. Handl., Fjarde Ser. 2(1): 161 pp., 6 pls.

Biddulphia biddulphiana
Dangeard, P., 1927
Phytoplankton de la croisière du "Sylvana". Ann. Inst. Ocean., Monaco, n.s., 4(8):286-401, 54 text figs. (Feirer-Juin 1913).

Biddulphia Bidduphiana
Gran, H.H., 1908
Diatomeen. Nordisches Plankton, Botanischer Teil pp. XIX.1-XIX 146; 178 text figs.

Biddulphia Biddulphiana
Meunier, A., 1915
Microplancton de la Mer Flamande. 2. Diatomées (excepté le genre Chaetoceros). Mem. Mus. Roy. Hist. Nat., Belgique, 7(3):1-118, Pls. VIII-XIV.

Biddulphia biddulphiana
Pavillard, J., 1925
Bacillariales. Rept. on the Danish Oceangr. Exped., 1908-10 to the Mediterranean and adj. seas. Vol.II., Biol. J4:72 pp., 116 text figs.

Biddulphia biddulphiana
Sousa e Silva, E., 1949
Diatomaceas e Dinoflagelados de Baia de Cascais. Portugaliae Acta Biol., Volume: Julio Henriques, Ser. B: 300-383, 9 pls, 2 fold-in tables.

Biddulphia biddulphiana
Sousa a Silva, E., and J. Dos Santos-Pinto, 1948
O Plancton da Baia de S. Martinho do Porto. 1. Diatomaceas e Dinoflagelados. Bol. Soc. Portuguese de Ciencias Naturais, 16(2):134-187, 6 pls. (Trav. Sta. Biol. Mar. de Lisbonne No. 52).

Biddulphia biquadrata
Mann, A., 1907
Report on the diatoms of the Albatross voyages in the Pacific Ocean, 1888-1904. Contrib. U. S. Nat. Herb. 10(5):221-419, Pls. XLIV-LIV.

Biddulphia birostris n.sp.
Grunow, A., 1863
Ueber einige neue und ungenügend bekannte Arten und Gattungen von Diatomaceen. Verhandl. d. K.K. Zool. Bot. Gesellsch., Vienna, 13: 137-162, pl.4-5 (Pl. 13-14).

Bidulphia catenata
Ercegovic, A., 1936
Etudes qualitative et quantitatives du phytoplancton dans les eaux cotières de l'Adriatique oriental moyen au cours de l'année 1934. Acta Adriatica 1(9):1-126

Biddulphia chinensis
Fox, M., 1957.
A first list of marine algae from Nigeria. J. Linn. Soc., London, 55(362):615-631.

Biddulphia chinensis
Frenguelli, Joaquin, and Hector Antonio Orlando, 1959.
Operacion MERLUZA. Diatomeas y silicoflagelados del plancton del "VI Crucero". Servicio Hidrogr. Naval., Argentina, Publ. No. H. 619: 5-62.

Biddulphia chinensis
Melchers, F.C.M., 1954.
Observaciones sobre Biddulphia sinensis. Rev. Biol. Mar, Valparaiso, 4(1/3):203-210.

Biddulphia chinensis
Müller Melchers, F.C., 1952.
Biddulphia chinensis Grev., an an indicator of ocean currents. Comm. Bot. Mus. Hist. Nat., Montevideo, 2(26):1-14, pls.

Biddulphia chinensis
Müller Melchers, F.C., 1954.
Observaciones sobre Biddulphia sinensis Grev. Rev. Biol. Mar., Valparaiso, 4(1/2/3):203-210, 16 figs.
Also printed: Bol. Soc. Biol. Concepcion, 34: 91-99, 1949.

Biddulphia chinensis
Müller Melchers, F.C., 1953.
New and little known diatoms from Uruguay and the South Atlantic coast. Com. Bot., Mus. Hist. Nat., Montevideo, 3(30):1-11, 8 pls.

Biddulphia connecta n.sp.
Wood, E.J. Ferguson, 1963.
Studies on Australian and New Zealand diatoms. VI. Tropical and subtropical species. Trans. R. Soc., New Zealand, 2(15):189-218.

Biddulphia consimilis
Mann, A., 1907
Report on the diatoms of the Albatross voyages in the Pacific Ocean, 1888-1904. Contrib. U. S. Nat. Herb. 10(5):221-419, Pls. XLIV-LIV.

Biddulphia culcitella n.sp.
Mann, A., 1907
Report on the diatoms of the Albatross voyages in the Pacific Ocean, 1888-1904. Contrib. U. S. Nat. Herb. 10(5):221-419, Pls. XLIV-LIV.

Biddulphia cylindrata n.sp.
Wood, E.J. Ferguson, L.H. Crosby and Vivienne Cassie, 1959
Studies on Australian and New Zealand diatoms. III. Descriptions of further discoid species. Trans. R. Soc., N.Z., 87(3/4):211-219, pls. 15-17.

Biddulphia delicatulum
Cleve-Euler, A., 1951
Die Diatomeen von Schweden und Finnland. Kungl. Svenska Vetenskaps Akad. Handl., Fjärde Ser. 2(1): 161 pp., 6 pls.

Biddulphia dubia
Allen, W.E., and E.E. Cupp, 1935
Plankton diatoms of the Java Sea. Annales du Jardin Botanique de Buitenzorg XLIV (2):101-174, figs.1-127.
(drawings of all species mentioned)

Biddulphia dubia
Cupp, Easter E., 1943
Marine plankton diatoms of the west coast of North America. Bull. S.I.O. 5(1):1-238, 5 pls., 168 text figs.

Biddulphia dubia
Mann, A., 1907
Report on the diatoms of the Albatross voyages in the Pacific Ocean, 1888-1904. Contrib. U. S. Nat. Herb. 10(5):221-419, Pls. XLIV-LIV.

Biddulphia edwardsii
Mann, A., 1907
Report on the diatoms of the Albatross voyages in the Pacific Ocean, 1888-1904. Contrib. U. S. Nat. Herb. 10(5):221-419, Pls. XLIV-LIV.

Biddulphia elongatum
Cleve-Euler, A., 1951
Die Diatomeen von Schweden und Finnland. Kungl. Svenska Vetenskaps Akad. Handl., Fjärde Ser. 2(1): 161 pp., 6 pls.

Biddulphia expedita - Odontella expedita
Zanon, V., 1947.
Diatomée delle Isole Ponziane - materiali per una florula diatomologica del Mare Tirreno. Bol. Pesca, Piscolt. Idrobiol., n.s., 2(1):36-53, 1 pl., of 10 figs.

Biddulphia extensa
Bigelow, H.B., and M. Leslie, 1930
Reconnaissance of the waters and plankton of Monterey Bay, July 1928. Bull. M.C.Z., 70(5):429-481, 43 text figs.

Biddulphia extansa n.sp.
Mann, A., 1907
Report on the diatoms of the Albatross voyages in the Pacific Ocean, 1888-1904. Contrib. U. S. Nat. Herb. 10(5):221-419, Pls. XLIV-LIV.

Biddulphia favus
Ghazzawi, F.M., 1939
Plankton of the Egyptian waters. A study of the Suez Canal Plankton. (A) Phytoplankton. Preliminery Report 83 pp. Notes and Memoires, Min. Commerce-Industry, Egypt, Hydrobiol. & Fish. 65 figs.

Biddulphia favus
Gran, H.H., 1908
Diatomeen. Nordisches Plankton, Botanischer Teil pp. XIX.1-XIX 146; 178 text figs.

Biddulphia favus
Mann, A., 1907
Report on the diatoms of the Albatross voyages in the Pacific Ocean, 1888-1904. Contrib. U. S. Nat. Herb. 10(5):221-419, Pls. XLIV-LIV.

Biddulphia favus
Meunier, A., 1915
Microplancton de la Mer Flamande. 2. Diatomées (excepté le genre Chaetoceros). Mem. Mus. Roy. Hist. Nat., Belgique, 7(3):1-118, Pls. VIII-XIV.

Biddulphia favus
Schodduyn, M., 1926
Observations faites dans la baie d'Ambleteuse (Pas de Calais). Bull. Inst. Ocean., Monaco, No. 482: 64 pp.

Biddulphia gladiorum n.sp.
Mann, A., 1907
Report on the diatoms of the Albatross voyages in the Pacific Ocean, 1888-1904. Contrib. U. S. Nat. Herb. 10(5):221-419, Pls. XLIV-LIV.

Biddulphia granulata
Cleve-Euler, A., 1951
Die Diatomeen von Schweden und Finnland. Kungl. Svenska Vetenskaps Akad. Handl., Fjärde Ser. 2(1): 161 pp., 6 pls.

Biddulphia granulata
Gran, H.H., 1908
Diatomeen. Nordisches Plankton, Botanischer Teil pp. XIX.1-XIX 146; 178 text figs.

Biddulphia granulata
Hendy, N. Ingram, 1964
An introductory account of the smaller algae of British coastal waters. V. Bacillariophyceae (Diatoms). Her Majesty's Stationary Office, 317 pp., 45 pls.

Biddulphia granulata
Mann, A., 1907
Report on the diatoms of the Albatross voyages in the Pacific Ocean, 1888-1904. Contrib. U. S. Nat. Herb. 10(5):221-419, Pls. XLIV-LIV.

Biddulphia granulata
Meunier, A., 1915
Microplancton de la Mer Flamande. 2. Diatomées (excepté le genre Chaetoceros). Mem. Mus. Roy. Hist. Nat., Belgique, 7(3):1-118, Pls. VIII-XIV.

Biddulphia granulata
Morse, D.C., 1947
Some observations on seasonal variations in plankton population Patuxent River, Maryland 1943-1945. Bd. Nat. Res., Publ. No. 65, Chesapeake Biol. Lab., 31, 3 figs.

Biddulphia grundleri
Mann, A., 1907
Report on the diatoms of the Albatross voyages in the Pacific Ocean, 1888-1904. Contrib. U. S. Nat. Herb. 10(5):221-419, Pls. XLIV-LIV.

Biddulphia heteroceros
Allen, W.E., and E.E. Cupp, 1935
Plankton diatoms of the Java Sea. Annales du Jardin Botanique de Buitenzorg XLIV (2):101-174, figs.1-127.
(drawings of all species mentioned)

Biddulphia hyalinum
Cleve-Euler, A., 1951
Die Diatomeen von Schweden und Finnland. Kungl. Svenska Vetenskaps Akad. Handl., Fjärde Ser. 2(1): 161 pp., 6 pls.

Biddulphia japonica n.sp.
Castracane degli Antelminelli, F., 1886
1. Report on the Diatomaceae collected by H.M.S. Challenger during the years 1873-1876. Rept. Sci. Results, H.M.S. Challenger, Botany Vol. II, 178 pp., 30 pls.

Biddulphia keeleyi
Mann, A., 1907
Report on the diatoms of the Albatross voyages in the Pacific Ocean, 1888-1904. Contrib. U. S. Nat. Herb. 10(5):221-419, Pls. XLIV-LIV.

Biddulphia laevis
Gran, H.H., 1908
Diatomeen. Nordisches Plankton, Botanischer Teil pp. XIX.1-XIX 146; 178 text figs.

Biddulphia laevis
Gran, H. H. and E. C. Angst, 1931
Plankton diatoms of Puget Sound. Publ. Puget Sound Biol. Sta. 7:417-519, 95 text figs.

Biddulphia levis
Heath, J. Brent, and W. Marshall Darley 1972.
Observations on the ultrastructure of the male gametes of Biddulphia levis Ehr.
J. Phycol. 8(1): 51-59

Biddulphia laevis
Hendy, N. Ingram, 1964
An introductory account of the smaller algae of British coastal waters. V. Bacillariophyceae (Diatoms). Her Majesty's Stationary Office, 317 pp., 45 pls.

Biddulphia levis
Kokubo, S., and S. Sato, 1947
Plankters in Jū-San Gata. Physiol. and Ecol. (Japan) 1(4):1-16, 3 text figs., tables.

Biddulphia laevis
Mann, A., 1907
Report on the diatoms of the Albatross voyages in the Pacific Ocean, 1888-1904. Contrib. U. S. Nat. Herb. 10(5):221-419, Pls. XLIV-LIV.

Biddulphia levis
Misra, J.N., 1956.
A systematic account of some littoral marine diatoms from the west coast of India. J. Bombay Nat. Hist. Soc., 53(4):537-568.

Biddulphia laevis
Zanon, D. V., 1949
Diatomee di Buenos Aires (Argentina) Atti Accad. Naz. Lincei, Memorie, Cl. Sci. fis., mat. e. nat., ser. 7, 11(3):59-151, 2 pls.

Biddulphia laevis minor
Meunier, A., 1915
Microplancton de la Mer Flamande. 2. Diatomées (excepté le genre Chaetoceros). Mem. Mus. Roy. Hist. Nat., Belgique, 7(3):1-118, Pls. VIII-XIV.

Biddulphia litigiosa
Balech, E., 1947
Contribution al conocimiento del plancton antarctico. Plankton del Mar de Bellinghausen. Physis 20:75-91, 76 figs. on 8 pls.

Biddulphia longicruris
Boden, B.P., 1950
Some marine plankton diatoms from the west coast of South Africa. Trans. R.Soc. S. Africa. 32:321-434, 100 text figs.

Biddulphia longicruris
Cupp, Easter E., 1943
Marine plankton diatoms of the west coast of North America. Bull. S.I.O. 5(1):1-238, 5 pls., 168 text figs.

Biddulphia longicruris
Eskinazi Enide e Shigekatsv Satô (1963-1964) 1966.
Contribuição ao estudo das diatomaceas da Praia de Piedade.
Trabhs Inst. Oceanogr., Univ. Recife, 5 (5/6): 73-114.

Biddulphia longicruris
Gran, H. H. and E. C. Angst, 1931
Plankton diatoms of Puget Sound. Publ. Puget Sound Biol. Sta. 7:417-519, 95 text figs.

Biddulphia longicruris
Hendey, N.I., 1937
The plankton diatoms of the southern seas. Discovery Repts. 16:151-364, pls.6-13.

Biddulphia longicruris var. hyalina n. var.
Cupp, Easter E., 1943
Marine plankton diatoms of the west coast of North America. Bull. S.I.O. 5(1):1-238, 5 pls., 168 text figs.

Biddulphia longicruris hyalina
Cupp, Easter E., 1943
Marine plankton diatoms of the west coast of North America. Bull. S.I.O. 5(1):1-238, 5 pls., 168 text figs.

Biddulphia longicruris var. hyalina n. var.
Cupp, E. E., 1943.
Marine plankton diatoms of the west coast of North America. Bull., S.I.O., 5(1):1-238, 5 pls., 168 textfigs.

Biddulphia luminosa
Mann, A., 1907
Report on the diatoms of the Albatross voyages in the Pacific Ocean, 1888-1904. Contrib. U. S. Nat. Herb. 10(5):221-419, Pls. XLIV-LIV.

Biddulphia mobiliensis
Allen, W. E., 1938
The Templeton Crocker Expedition to the Gulf of California in 1935 - The Phytoplankton. Amer. Microsc. Soc., Trans. 57:328-335.

Biddulphia mobiliensis
Allen, W.E., 1937
Plankton diatoms of the Gulf of California obtained by the G. Allan Hancock Expedition of 1936. The Hancock Pacific Expeditions, Univ. So. Calif. Publ. 3:47-59, 1 fig.

Biddulphia mobiliensis
Allen, W.E., and E.E. Cupp, 1935
Plankton diatoms of the Java Sea. Annales du Jardin Botanique de Buitenzorg XLIV (2):101-174, figs.1-127.
(drawings of all species mentioned)

Biddulphia mobiliensis
Bigelow, H.B., and M. Leslie, 1930
Reconnaissance of the waters and plankton of Monterey Bay, July 1928. Bull. M.C.Z. 70(5):429-481, 43 text figs.

Biddulphia mobiliensis
Boden, B.P., 1950
Some marine plankton diatoms from the west coast of South Africa. Trans. R.Soc. S. Africa. 32:321-434, 100 text figs.

Biddulphia mobiliensis
Cleve-Euler, A., 1951
Die Diatomeen von Schweden und Finnland. Kungl. Svenska Vetenskaps Akad. Handl., Fjärde Ser. 2(1): 161 pp., 6 pls.

Biddulphia mobiliensis
Cupp, Easter E., 1943
Marine plankton diatoms of the west coast of North America. Bull. S.I.O. 5(1):1-238, 5 pls., 168 text figs.

Biddulphia mobiliensis
Dangeard, P., 1927
Phytoplankton de la croisière du "Sylvana". Ann. Inst. Ocean. Monaco, n.s., 4(8):286-401, 54 text figs. (Feirer-Juin 1913).

Bidulphia mobilensis
Ercegovic, A., 1936
Etudes qualitative et quantitatives du phytoplancton dans les eaux cotières de l'Adriatique oriental moyen au cours de l'année 1934. Acta Adriatica 1(9):1-126

Biddulphia mobiliensis
Eskinazi Enide e Shigekatsv Satô (1963-1964) 1966.
Contribuição ao estudo das diatomaceas da Praia de Piedade.
Trabhs Inst. Oceanogr., Univ. Recife, 5 (5/6): 73-114.

Biddulphia mobiliensis
Forti, A., 1922
Ricerche sulla flora pelagica (fitoplancton) di Quarto dei Mille. Mem. R. Com. Talass. Ital. 97:248 pp., 13 pls.

Biddulphia mobiliensis
Frenguelli, Joaquin, and Hector Antonio Orlando, 1959.
Operacion MERLUZA. Diatomeas y silicoflagelados del plancton del "VI Crucero". Servicio Hidrogr. Naval., Argentina, Publ. No. H. 619: 5-62.

Biddulphia mobiliensis
Gran, H.H., 1908
Diatomeen. Nordisches Plankton, Botanischer Teil pp. XIX.1-XIX 146; 178 text figs.

Biddulphia mobiliensis
Gran, H. H. and E. C. Angst, 1931
Plankton diatoms of Puget Sound. Publ. Puget Sound Biol. Sta. 7:417-519, 95 text figs.

Oceanographic Index: Marine Organisms Cumulation, 1946-1973

Biddulphia mobiliensis
Ghazzawi, F.M., 1939
Plankton of the Egyptian waters. A study of the Suez Canal Plankton. (A) Phytoplankton. Preliminary Report 83 pp. Notes and Memoires, Min. Commerce-Industry, Egypt, Hydrobiol. & Fish. 65 figs.

Biddulphia mobiliensis
Hendy, N. Ingram, 1964
An introductory account of the smaller algae of British coastal waters. V. Bacillariophyceae (Diatoms).
Her Majesty's Stationary Office, 317 pp., 45 pls.

Biddulphia mobiliensis
Hendey, N.I., 1937
The plankton diatoms of the southern seas. Discovery Repts. 16:151-364, pls.6-13.

Biddulphia mobiliensis
Jørgensen, E., 1905
B. Protistplankton and the diatoms in bottom samples. Hydrographical and biological investigations in Norwegian fjords. Bergens Mus. Skr. 7: 49-225.

Biddulphia mobiliensis
Jorgensen, E., 1900
Protophyten und Protozoen im Plankton aus der Norwegischen Westkerste. Bergens Mus. Aarb. 1899(6): 95 pp., 5 pls., 83 tables.

Biddulphia mobiliensis
Macdonald, R., 1933
An examination of plankton hauls made in the Suez Canal during the year 1928. Fish. Res. Dis., Notes & Mem. No.3, 11 pp., 1 chart.

Biddulphia mobiliensis
Mangin, M. L., 1912
Phytoplancton de la croisière du "René" dans l'Atlantique (Septembre 1908). Ann. Inst. Ocean., n.s., 4(1):1-66, 2 pls., 41 text figs., 2 tables.

Biddulphia mobiliensis
Mann, A., 1907
Report on the diatoms of the Albatross voyages in the Pacific Ocean, 1888-1904. Contrib. U. S. Nat. Herb. 10(5):221-419, Pls. XLIV-LIV.

Biddulphia mobiliensis
Meunier, A., 1915
Microplancton de la Mer Flamande. 2. Diatomées (excepté le genre Chaetoceros). Mem. Mus. Roy. Hist. Nat., Belgique, 7(3):1-118, Pls. VIII-XIV.

Biddulphia mobiliensis
Morse, D.C., 1947
Some observations on seasonal variations in plankton population Patuxant River, Maryland 1943-1945. Bd. Nat. Res., Publ. No.65, Chesapeake Biol. Lab., 31, 3 figs.

Biddulphia mobiliensis
Paiva Carvalho, J., 1950
O plancton do Rio Maria Rodriques (Cananeis). 1. Diatomaceas e Dinoflagelados. Bol. Inst. Paulista Oceanogr. 1(1); 27-43, 2 fold-in tables, 2 figs.

Biddulphia mobiliensis
Pavillard, J., 1925
Bacillariales. Rept. on the Danish Oceangr. Exped., 1908-10 to the Mediterranean and adj. seas. Vol.II., Biol. J4:72 pp., 116 text figs.

Biddulphia mobiliensis
Pavillard, J., 1905
Recherches sur la flore pelagique (Phytoplankton) de l'Etang de Thau. Theses presentees a la Fac. Sci., Paris, 116 pp., 3 pls.

Biddulphia mobiliensis
Rampi, L., 1942
Ricerche sul fitoplancton del Mare Ligure 6. Le diatomee delle acque di Sanremo. Nuovo Giornale Botanico Italiano, N.S., 49:252-268.

Biddulphia mobiliensis
Schodduyn, M., 1926
Observations faites dans la baie d'Ambleteuse (Pas de Calais). Bull. Inst. Ocean., Monaco, No. 482: 64 pp.

Biddulphia mobiliensis
Schröder, B., 1900
Phytoplankton des Golfes von Neapel nebst vergleichenden Ausblicken auf das atlantischen Ozean. Mitt. Zool. Stat. Neapel, 14:1-38.

Biddulphia mobiliensis
Sousa e Silva, E., 1949
Diatomaceas e Dinoflagelados de Baia de Cascais. Portugaliae Acta Biol., Volume: Julio Henriques, Ser. B: 300-383, 9 pls, 2 fold-in tables.

Biddulphia mobiliensis
Sousa a Silva, E., and J. Dos Santos-Pinto, 1948
O Plancton da Baia de S. Martinho do Porto. 1. Diatomaceas e Dinoflagelados. Bol. Soc. Portuguese de Ciencias Naturais, 16(2):134-187, 6 pls. (Trav. Sta. Biol. Mar. de Lisbonne No. 52).

Biddulphia mobiliensis
Subrahmanyan, R., 1945.
On the cell division and mitosis in some South Indian diatoms. Proc. Indian Acad. Sci., Sect. B, 22(6):331-354, Pl. 39, 88 textfigs.

Biddulphia obtusa
Cleve-Euler, A., 1951
Die Diatomeen von Schweden und Finnland. Kungl. Svenska Vetenskaps Akad. Handl., Fjarde Ser. 2(1): 161 pp., 6 pls.

Biddulphia obtusa
Ghazzawi, F.M., 1939
Plankton of the Egyptian waters. A study of the Suez Canal Plankton. (A) Phytoplankton. Preliminary Report 83 pp. Notes and Memoires, Min. Commerce-Industry, Egypt, Hydrobiol. & Fish. 65 figs.

Biddulphia obtusa
Hendy, N. Ingram, 1964
An introductory account of the smaller algae of British coastal waters. V. Bacillariophyceae (Diatoms).
Her Majesty's Stationary Office, 317 pp., 45 pls.

Biddulphia obtusa
Sousa e Silva, E., 1949
Diatomaceas e Dinoflagelados de Baia de Cascais. Portugaliae Acta Biol., Volume: Julio Henriques, Ser. B: 300-383, 9 pls, 2 fold-in tables.

Biddulphia (Amphitetras) ornata hirsuta n. var.
Castracane degli Antelminelli, F., 1886
1. Report on the Diatomaceae collected by H.M.S. Challenger during the years 1873-1876. Rept. Sci. Results, H.M.S. Challenger, Botany Vol. II, 178 pp., 30 pls.

Biddulphia ovalis
Mann, A., 1907
Report on the diatoms of the Albatross voyages in the Pacific Ocean, 1888-1904. Contrib. U. S. Nat. Herb. 10(5):221-419, Pls. XLIV-LIV.

Biddulphia pacifica
Mann, A., 1907
Report on the diatoms of the Albatross voyages in the Pacific Ocean, 1888-1904. Contrib. U. S. Nat. Herb. 10(5):221-419, Pls. XLIV-LIV.

Biddulphia papillata
Mann, A., 1907
Report on the diatoms of the Albatross voyages in the Pacific Ocean, 1888-1904. Contrib. U. S. Nat. Herb. 10(5):221-419, Pls. XLIV-LIV.

Biddulphia parallela n.sp. (?)
Castracane degli Antelminelli, F., 1886
1. Report on the Diatomaceae collected by H.M.S. Challenger during the years 1873-1876. Rept. Sci. Results, H.M.S. Challenger, Botany Vol. II, 178 pp., 30 pls.

Biddulphia pelagica
Rampi, L., 1942
Ricerche sul fitoplancton del Mare Ligure 6. Le diatomee delle acque di Sanremo. Nuovo Giornale Botanico Italiano, N.S., 49:252-268.

Biddulphia pellucida n.sp.
Castracane degli Antelminelli, F., 1886
1. Report on the Diatomaceae collected by H.M.S. Challenger during the years 1873-1876. Rept. Sci. Results, H.M.S. Challenger, Botany Vol. II, 178 pp., 30 pls.

Biddulphia peruviana
Bigelow, H.B., and M. Leslie, 1930
Reconnaissance of the waters and plankton of Monterey Bay, July 1928. Bull. M.C.Z., 70(5):429-481, 43 text figs.

Biddulphia polymorpha
Cleve-Euler, A., 1951
Die Diatomeen von Schweden und Finnland. Kungl. Svenska Vetenskaps Akad. Handl., Fjarde Ser. 2(1): 161 pp., 6 pls.

Biddulphia polymorpha
Hendey, N.I., 1937
The plankton diatoms of the southern seas. Discovery Repts. 16:151-364, pls.6-13.

Biddulphia polymorpha n.sp.
Mangin, L., 1915
Phytoplancton de L'Antartique. Deuxieme Exped. Ant. Francaise (1908-1910), 95 pp., 3 pls., 58 text figs.

Biddulphia primordialis
Mann, A., 1907
Report on the diatoms of the Albatross voyages in the Pacific Ocean, 1888-1904. Contrib. U. S. Nat. Herb. 10(5):221-419, Pls. XLIV-LIV.

Biddulphia pulchella
Cupp, Easter E., 1943
Marine plankton diatoms of the west coast of North America. Bull. S.I.O. 5(1):1-238, 5 pls., 168 text figs.

Bidulphia pulchella
Ercegovic, A., 1936
Etudes qualitative et quantitatives du phytoplancton dans les eaux cotières de l'Adriatique oriental moyen au cours de l'année 1934. Acta Adriatica 1(9):1-126

Biddulphia pulchella
Eskinazi Enide e Shigekatsv Sato (1963-1964) 1966.
Contribuição ao estudo das diatomaceas da Praia de Piedade.
Trabhs Inst. Oceanogr., Univ. Recife, 5 (5/6): 73-114.

Biddulphia pulchella
Forti, A., 1922
Ricerche sulla flora pelagica (fitoplankton) di Quarto dei Mille. Mem. R. Com. Talass. Ital. 97:248 pp., 13 pls.

Biddulphia pulchella
Fox, M., 1957.
A first list of marine algae from Nigeria.
J. Linn. Soc., London, 55(362):615-631.

Biddulphia pulchella
Hendy, N. Ingram, 1964
An introductory account of the smaller algae of British coastal waters. V. Bacillariophyceae (Diatoms).
Her Majesty's Stationary Office, 317 pp., 45 pls.

Biddulphia pulchella
Hendey, N.I., 1951
Littoral diatoms of Chicester Harbour with special reference to fouling. J. Roy. Microscop. Soc. 71(1): 1-86, 18 pls.

Biddulphia pulchella
Mann, A., 1907
Report on the diatoms of the Albatross voyages in the Pacific Ocean, 1888-1904. Contrib. U. S. Nat. Herb. 10(5):221-419, Pls. XLIV-LIV.

Biddulphia pulchella
Misra, J.N., 1956.
A systematic account of some littoral marine diatoms from the west coast of India.
J. Bombay Nat. Hist. Soc., 53(4):537-568.

Biddulphia pulchella
Rampi, L., 1942
Ricerche sul fitoplancton del Mare Ligure 6. Le diatomee delle acque di Sanremo. Nuovo Giornale Botanico Italiano, N.S., 49:252-268.

Biddulphia pulchella
Rampi, L., 1940
Diatomee del Mare Adriatico. Nuovo Giornale Botanico Italiano, n.s., 47:559-608.

Biddulphia pulchella
Schröder, B., 1900
Phytoplankton des Golfes von Neapel nebst vergleichenden Ausblicken auf das atlantischen Ozean. Mitt. Zool. Stat. Neapel, 14:1-38.

Biddulphia pulchella
Takano, Hideaki, 1961.
Epiphytic diatoms upon Japanese agar sea-weeds.
Bull. Tokai Reg. Fish. Res. Lab., No. 31:269-274.

Biddulphia pulchella
Zanon, V., 1948
Diatomee marini di Sardegna e Pugillo di Alghe Marine della stressa. Boll. Pesca, Piscitutura e Idrobiologia, Anno 24, ns. 3(2): 202-244, 27 figs. on 1 pl.

Biddulphia pulchella major
Castracane degli Antelminelli, F., 1886 n.var.
1. Report on the Diatomaceae collected by H.M.S. Challenger during the years 1873-1876. Rept. Sci. Results, H.M.S. Challenger, Botany Vol. II, 178 pp., 30 pls.

Biddulphia pumila n.sp.
Castracane degli Antelminelli, F., 1886
1. Report on the Diatomaceae collected by H.M.S. Challenger during the years 1873-1876. Rept. Sci. Results, H.M.S. Challenger, Botany Vol. II, 178 pp., 30 pls.

Biddulphia regia
Boden, B.P., 1950
Some marine plankton diatoms from the west coast of South Africa. Trans. R. Soc. S. Africa. 32:321-434, 100 text figs.

Biddulphia regia
Dangeard, P., 1927
Phytoplankton de la croisière du "Sylvana". Ann. Inst. Ocean., Monaco, n.s., 4(8):286-401, 54 text figs. (Feirer-Juin 1913).

Biddulphia regia
Gilson, H. C., 1937
Chemical and Physical Investigations. The nitrogen cycle. John Murray Exped., 1933-34, Sci. Repts., 2(2):21-81, 16 text figs.

Biddulphia regia
Gran, H.H., and T. Braarud, 1935
A quantitative study of the phytoplankton in the Bay of Fundy and the Gulf of Maine (including observations on hydrography, chemistry, and turbidity). J. Biol. Bd., Canada, 1(5):279-467, 69 text figs.

Biddulphea regia
Grøntved, J., 1949
Investigations on the phytoplankton in the Danish Waddensea in July 1941. Medd. Komm. Danmarks Fiskeri og Havundersøgelser, ser. Plankton, 5(2):55 pp., 2 pls., 38 text figs.

Biddulphia regia
Hendy, N. Ingram, 1964
An introductory account of the smaller algae of British coastal waters. V. Bacillariophyceae (Diatoms).
Her Majesty's Stationary Office, 317 pp., 45 pls.

Biddulphia regia
Hendey, N.I., 1937
The plankton diatoms of the southern seas. Discovery Repts. 16:151-364, pls.6-13.

Biddulphia regia
Lillick, L.C., 1940
Phytoplankton and planktonic protozoa of the offshore waters of the Gulf of Maine. Pt.II. Qualitative Composition of the Planktonic Flora. Trans. Am. Phil. Soc., n.s., 31(3):193-237, 13 text figs.

Biddulphia regia
Morse, D.C., 1947
Some observations on seasonal variations in plankton population Patuxent River, Maryland 1943-1945. Bd. Nat. Res., Publ. No.65, Chesapeake Biol. Lab., 31, 3 figs.

Biddulphia regia
Pavillard, J., 1925
Bacillariales. Rept. on the Danish Oceangr. Exped., 1908-10 to the Mediterranean and adj. seas. Vol.II., Biol. J4:72 pp., 116 text figs.

Biddulphia regina
Gran, H.H., 1908
Diatomeen. Nordisches Plankton, Botanischer Teil pp. XIX.1-XIX 146; 178 text figs.

Biddulphia regina
Zanon, V., 1948
Diatomee marini di Sardegna e Pugillo di Alghe Marine della stressa. Boll. Pesca, Piscitutura e Idrobiologia, Anno 24, ns. 3(2): 202-244, 27 figs. on 1 pl.

Biddulphia reticulum
Allen, W.E., and E.E. Cupp, 1935
Plankton diatoms of the Java Sea. Annales du Jardin Botanique de Buitenzorg XLIV (2):101-174, figs.1-127.

(drawings of all species mentioned)

Biddulphia Reticulata
Cleve-Euler, A., 1951
Die Diatomeen von Schweden und Finnland. Kungl. Svenska Vetenskaps Akad. Handl., Fjärde Ser. 2(1): 161 pp., 6 pls.

Biddulphia reticulatum
Gran, H.H., 1908
Diatomeen. Nordisches Plankton, Botanischer Teil pp. XIX.1-XIX 146; 178 text figs.

Biddulphia reticulatum
Hendy, N. Ingram, 1964
An introductory account of the smaller algae of British coastal waters. V. Bacillariophyceae (Diatoms).
Her Majesty's Stationary Office, 317 pp., 45 pls.

Biddulphia reticulata
Mann, A., 1907
Report on the diatoms of the Albatross voyages in the Pacific Ocean, 1888-1904. Contrib. U. S. Nat. Herb. 10(5):221-419, Pls. XLIV-LIV.

Biddulphia reticulum
Meunier, A., 1915
Microplancton de la Mer Flamande. 2. Diatomées (excepté le genre Chaetoceros). Mem. Mus. Roy. Hist. Nat., Belgique, 7(3):1-118, Pls. VIII-XIV.

Biddulphia reticulata inermis
Castracane degli Antelminelli, F., 1886 n.var.
1. Report on the Diatomaceae collected by H.M.S. Challenger during the years 1873-1876. Rept. Sci. Results, H.M.S. Challenger, Botany Vol. II, 178 pp., 30 pls.

Biddulphia reversa
Gosden, R., 1966.
Note on Biddulphia reversa Grove and Sturt. J. Quekett Microsc. Club. 30 (5): 117.

Biddulphia rhombus
Aurich, H. J., 1949.
Die Verbreitung des Nannoplanktons im Oberflächenwasser vor der Nordfriesischen Kuste. Ber. Deutschen Wiss. Komm. f. Meeresf., n.f., 11(4): 403-405, 2 figs.

Biddulphia rhombus
Cleve-Euler, A., 1951
Die Diatomeen von Schweden und Finnland. Kungl. Svenska Vetenskaps Akad. Handl., Fjärde Ser. 2(1): 161 pp., 6 pls.

Biddulphia rhombus
Cupp, Easter E., 1943
Marine plankton diatoms of the west coast of North America. Bull. S.I.O. 5(1):1-238, 5 pls., 168 text figs.

Biddulphia rhombus
Ghazzawi, F.M., 1939
Plankton of the Egyptian waters. A study of the Suez Canal Plankton. (A) Phytoplankton. Preliminary Report 83 pp. Notes and Memoires, Min. Commerce-Industry, Egypt, Hydrobiol. & Fish. 65 figs.

Biddulphia rhombus
Gran, H.H., 1908
Diatomeen. Nordisches Plankton, Botanischer Teil pp. XIX.1-XIX 146; 178 text figs.

Biddulphia rhombus
Grøntved, J., 1949(1950).
Investigations on the phytoplankton in the Danish Waddensee in July 1941. Medd. Komm. Danmarks Fiskeri- og Havundersøgelser, Serie Plankton, 5(2)(Medd. Skalling Lab. 10):55 pp., 2 pls., 38 textfigs.

Biddulphia rhombus
Grøntved, J., 1949
Investigations on the phytoplankton in the Danish Waddensea in July 1941. Medd. Komm. Danmarks Fiskeri og Havundersøgelser, scr. Plankton, 5(2):55 pp., 2 pls., 38 text figs.

Biddulphia rhombus & varieties
Hendy, N. Ingram, 1964
An introductory account of the smaller algae of British coastal waters. V. Bacillariophyceae (Diatoms). Her Majesty's Stationary Office, 317 pp., 45 pls.

Biddulphia rhombus
Kokubo, S., and S. Sato., 1947
Plankters in Ju-San Gata. Physiol. and Ecol. (Japan) 1(4):1-16, 3 text figs., tables.

Biddulphia rhombus
Macdonald, R., 1933
An examination of plankton hauls made in the Suez Canal during the year 1928. Fish. Res. Dis., Notes & Mem. No.3, 11 pp., 1 chart.

Biddulphia rhombus
Meunier, A., 1915
Microplancton de la Mer Flamande. 2. Diatomées (excepté le genre Chaetoceros). Mem. Mus. Roy. Hist. Nat., Belgique, 7(3):1-118, Pls. VIII-XIV.

Biddulphia rhombus
Misra, J.N., 1956.
A systematic account of some littoral marine diatoms from the west coast of India. J. Bombay Nat. Hist. Soc., 53(4):537-568.

Biddulphia rhombus
Rampi, L., 1940
Diatomee del Mare Adriatico. Nuovo Giornale Botanico Italiano, n.s., 47:559-608.

Biddulphia rhombus
Schodduyn, M., 1926
Observations faites dans la baie d'Ambleteuse (Pas de Calais). Bull. Inst. Ocean., Monaco, No. 482: 64 pp.

Biddulphia rhombus
Zanon, V., 1948
Diatomee marini di Sardegna e Pugillo di Alghe Marine della stressa. Boll. Pesca, Piscitutura e Idrobiologia, Anno 24, ns. 3(2): 202-244, 27 figs. on 1 pl.

Biddulphia rhombus trigona
Gran, H.H., 1908
Diatomeen. Nordisches Plankton, Botanischer Teil pp. XIX.1-XIX 146; 178 text figs.

Biddulphia rigida
Cleve-Euler, A., 1951
Die Diatomeen von Schweden und Finnland. Kungl. Svenska Vetenskaps Akad. Handl., Fjarde Ser. 2(1): 161 pp., 6 pls.

Biddulphia robertsiana
Mann, A., 1907
Report on the diatoms of the Albatross voyages in the Pacific Ocean, 1888-1904. Contrib. U. S. Nat. Herb. 10(5):221-419, Pls. XLIV-LIV.

Biddulphia roperiana
Gran, H. H. and E. C. Angst, 1931
Plankton diatoms of Puget Sound. Publ. Puget Sound Biol. Sta. 7:417-519, 95 text figs.

Biddulphia roperiana
Mann, A., 1907
Report on the diatoms of the Albatross voyages in the Pacific Ocean, 1888-1904. Contrib. U. S. Nat. Herb. 10(5):221-419, Pls. XLIV-LIV.

Biddulphia Schroederiana
Margalef, R., 1949
Fitoplancton nerítico de la Costa Brava en 1947-48. Publ. Inst. Biol. Aplicada, 5: 41-51, 3 text figs.

Biddulphis scutellum n.sp.
Mann, A., 1907
Report on the diatoms of the Albatross voyages in the Pacific Ocean, 1888-1904. Contrib. U. S. Nat. Herb. 10(5):221-419, Pls. XLIV-LIV.

Biddulphia setigera
Mann, A., 1907
Report on the diatoms of the Albatross voyages in the Pacific Ocean, 1888-1904. Contrib. U. S. Nat. Herb. 10(5):221-419, Pls. XLIV-LIV.

Biddulphia shadboltiana
Mann, A., 1907
Report on the diatoms of the Albatross voyages in the Pacific Ocean, 1888-1904. Contrib. U. S. Nat. Herb. 10(5):221-419, Pls. XLIV-LIV.

Biddulphia sinensis
Allen, W.E., and E.E. Cupp, 1935
Plankton diatoms of the Java Sea. Annales du Jardin Botanique de Buitenzorg XLIV (2):101-174, figs.1-127.

(drawings of all species mentioned)

Biddulphia sinensis
Braarud, T., and Adam Bursa, 1939
On the phytoplankton of the Oslo Fjord, 1933-1934. Hvalrådets Skr. No.19:1-63; 9 text figs. Reviewed. J. du. Cons. 14(3): 418-420. A.C. Gardiner.

Biddulphia sinensis
Chiba, T., 1949
On the distribution of the plankton in the eastern China Sea and Yellow Sea. 1. Plankton composition in the spring. J. Shimonoseki Coll. Fisheries, 1(1):57-63, 1 fig.

Biddulphia sinensis
Cleve-Euler, A., 1951
Die Diatomeen von Schweden und Finnland. Kungl. Svenska Vetenskaps Akad. Handl., Fjarde Ser. 2(1): 161 pp., 6 pls.

Biddulphia sinensis
Corner, E.S.D., R.N. Head and C.C. Kilvington, 1972.
On the nutrition and metabolism of zooplankton. VIII. The grazing of Biddulphia cells by Calanus helgolandicus. J. mar. biol. Ass. U.K. 52(4): 847-861.

Biddulphia sinensis
Delsman, H. C., 1939.
Preliminary plankton investigations in the Java Sea. Treubia, 17:139-181, 8 maps, 41 figs.

Biddulphia sinensis
Egusa, S., 1949.
[Size variations in planktonic diatoms and some considerations of their ecological significance. 1. Skeletonema costatum and Biddulphia sinensis.] Bull. Japan. Soc. Sci. Fish. 15(7):332-336, 3 textfigs. (In Japanese).

Biddulphia sinensis
Gran, H.H., 1908
Diatomeen. Nordisches Plankton, Botanischer Teil pp. XIX.1-XIX 146; 178 text figs.

Biddulphia sinensis
Grøntved, J., 1949
Investigations on the phytoplankton in the Danish Waddensea in July 1941. Medd. Komm. Danmarks Fiskeri og Havundersøgelser, ser. Plankton, 5(2):55 pp., 2 pls., 38 text figs.

Biddulphia sinensis
Hendy, N. Ingram, 1964
An introductory account of the smaller algae of British coastal waters. V. Bacillariophyceae (Diatoms). Her Majesty's Stationary Office, 317 pp., 45 pls.

Biddulphia sinensis
Le Borgne-David, Annick 1972.
Polymorphisme de Biddulphia sinensis Greville dans le bassin de Marennes-Oléron.
Rev. Trav. Inst. Pêches marit. 36(3): 275-284

Biddulphia sinensis
Lucas, C. E., 1949
Notes on continuous plankton records at 10 m depth in the North Sea and Northeastern Atlantic during 1946-1947. Ann. Biol., Int. Cons., 4:63-66, text fig. 4.

Biddulphia sinensis
Lucas, C.E. and H.G. Stubbings, 1948
Size variations in diatoms and their ecological significance. Hull Bull. Mar. Ecol. 2(12):133-171, 14 text figs.

Biddulphia Sinensis
Meunier, A., 1915
Microplancton de la Mer Flamande. 2. Diatomées (excepté le genre Chaetoceros). Mem. Mus. Roy. Hist. Nat., Belgique, 7(3):1-118, Pls. VIII-XIV.

Biddulphia sinensis
Ostenfeld, C. H., 1908
On the immigration of Biddulphia sinensis Grev. and its occurrence in the North Sea during 1903-07. Medd. Komm. Havunders. Plankton, I:6

Biddulphia sinensis
Qasim, S.Z., P.M.A. Bhattathiri and V.P. Devassy, 1973
Growth kinetics and nutrient requirements of two tropical marine phytoplankters. Mar. Biol. 21(4):299-304.

Biddulphia sinensis
Robinson, G.A., 1961
Contribution towards an atlas of the northeastern Atlantic and the North Sea. 1. Phytoplankton.
Bulls. Mar. Ecology, 5(42): 81-89, Pls. 15-20.

Biddulphia sinensis
Schodduyn, M., 1926
Observations faites dans la baie d'Ambleteuse (Pas de Calais). Bull. Inst. Ocean., Monaco, No. 482: 64 pp.

Biddulphia sinensis
Wimpenny, R.S., 1956.
The size of diatoms. III. The cell width of Biddulphia sinensis Greville from the southern North Sea. J.M.B.A., 35(2):375-386.

Biddulphia smithii
Ghazzawi, F.M., 1939
Plankton of the Egyptian waters. A study of the Suez Canal Plankton. (A) Phytoplankton. Preliminary Report 83 pp. Notes and Memoires, Min. Commerce-Industry, Egypt, Hydrobiol. & Fish. 65 figs.

Biddulphia Smithii
Meunier, A., 1915
Microplancton de la Mer Flamande. 2. Diatomées (excepté le genre Chaetoceros). Mem. Mus. Roy. Hist. Nat., Belgique, 7(3):1-118, Pls. VIII-XIV.

Biddulphia striata
Hendey, N.I., 1937
The plankton diatoms of the southern seas. Discovery Repts. 16:151-364, pls.6-13.

Biddulphia striata

Mangin, L., 1915
Phytoplancton de L'Antartique. Deuxieme Exped. Ant. Francaise (1908-1910), 95 pp., 3 pls., 58 text figs.

Biddulphia subaequa

Cleve-Euler, A., 1951
Die Diatomeen von Schweden und Finnland. Kungl. Svenska Vetenskaps Akad. Handl., Fjarde Ser. 2(1): 161 pp., 6 pls.

Biddulphia subjuncta n.sp.

Mann, A., 1907
Report on the diatoms of the Albatross voyages in the Pacific Ocean, 1888-1904. Contrib. U. S. Nat. Herb. 10(5):221-419, Pls. XLIV-LIV.

Biddulphia titiana

Hofler, Karl, 1966.
On the physiology of Biddulphia titiana and other marine diatoms. Botanica mar., 9(1/2):27-32.

Biddulphia titiania

Höfler, Karl und Luise Höfler, 1964.
Diatomeen des marinen Planktons als Isotonebionten. Pubbl. Staz. Zool., Napoli, 33(3):315-330.

Biddulphia tuomey

Eskinazi Enide e Shigekatsv Satô (1963-1964) 1966.
Contribuição ao estudo das diatomaceas da Praia de Piedade.
Trabhs Inst. Oceanogr., Univ. Recife, 5 (5/6): 73-114.

Biddulphia Tuomayii

Mann, A., 1893
List of Diatomaceae from a deep-sea dredging in the Atlantic Ocean off Delaware Bay by the U. S. Fish Commission Steamer Albatross. Proc. U. S. Nat. Mus. 16:303-312.

Biddulphia Tuomeyi

Rampi, L., 1942
Ricerche sul fitoplancton del Mare Ligure 6. Le diatomee delle acque di Sanremo. Nuovo Giornale Botanico Italiano, N.S., 49:252-268.

Biddulphia Tuomey

Rampi, L., 1940
Diatomee del Mare Adriatico. Nuovo Giornale Botanico Italiano, n.s., 47:559-608.

Biddulphia tuomeyi

Srinivasan, K.S., 1954.
Report on Biddulphia tuomeyi (Bail) Roper — a diatom new to India. Current Science 23(7): 228-229.

Biddulphia Tuomeyi

Zanon, V., 1948
Diatomee marini di Sardegna e Pugillo di Alghe Marine della stressa. Boll. Pesca, Piscitutura e Idrobiologia, Anno 24, ns. 3(2): 202-244, 27 figs. on 1 pl.

Biddulphia turgida

Mann, A., 1907
Report on the diatoms of the Albatross voyages in the Pacific Ocean, 1888-1904. Contrib. U. S. Nat. Herb. 10(5):221-419, Pls. XLIV-LIV.

Biddulphia vesiculosa

Ghazzawi, F.M., 1939
Plankton of the Egyptian waters. A study of the Suez Canal Plankton. (A) Phytoplankton. Preliminary Report 83 pp. Notes and Memoires, Min. Commerce-Industry, Egypt, Hydrobiol. & Fish. 65 figs.

Biddulphia vesiculosa

Gran, H.H., 1908
Diatomeen. Nordisches Plankton, Botanischer Teil pp. XIX.1-XIX 146; 178 text figs.

Biddulphia vesiculosa

Meunier, A., 1915
Microplancton de la Mer Flamande. 2. Diatomees (excepté le genre Chaetoceros). Mem. Mus. Roy. Hist. Nat., Belgique, 7(3):1-118, Pls. VIII-XIV.

Biddulphia vesiculosa

Schodduyn, M., 1926
Observations faites dans la baie d'Ambleteuse (Pas de Calais). Bull. Inst. Ocean., Monaco, No. 482: 64 pp.

Biddulphia vesiculosa

Sousa e Silva, E., 1949
Diatomaceas e Dinoflagelados de Baia de Cascais. Portugaliae Acta Biol. Volume: Julio Henriques, Ser. B: 300-383, 9 pls, 2 fold-in tables.

Biddulphia vesiculosa

Sousa e Silva, E., and J. Dos Santos-Pinto, 1948
O Plancton da Baia de S. Martinho do Porto. 1. Diatomaceas e Dinoflagelados. Bol. Soc. Portuguese de Ciencias Naturais, 16(2):134-187, 6 pls. (Trav. Sta. Biol. Mar. de Lisbonne No. 52).

Biddulphia weissflogi

Balech, E., 1947
Contribution al conocimiento del plancton antartico. Plankton del Mar de Bellinghausen. Physis 20:75-91, 76 figs. on 8 pls.

Biddulphia weissflogii

Castracane degli Antelminelli, F., 1886
1. Report on the Diatomaceae collected by H.M.S. Challenger during the years 1873-1876. Rept. Sci. Results, H.M.S. Challenger, Botany Vol. II, 178 pp., 30 pls.

Biddulphia weissflogii

Jouse, A.P., G.S. Koroleva, G.A. Nagaeva, 1962
Diatoms in the surface layer of sediment in the Indian sector of the Antarctic. Investigations of marine bottom sediments. (In Russian; English summary).
Trudy Inst. Okeanol., Akad. Nauk, SSSR, 61: 19-92.

Biddulphia weissflogia

Manguin, E., 1954
Diatomees marines provenant de l'ile Heard (Australian National Antarctic Expedition). Rev. Algol., n.s., 1: 14-24.

Brebissonia boeckii

Hendy, N. Ingram, 1964
An introductory account of the smaller algae of British coastal waters. V. Bacillariophyceae (Diatoms).
Her Majesty's Stationary Office, 317 pp., 45 pls.

Brightwellia

Brunel, J., 1962
Le phytoplancton de la Baie de Chaleurs. Inst. Botan., Univ. Montréal, Contrib. No. 77: 365 pp., 66 pls.

Brightwellia murrayi n.sp.

Castracane degli Antelminelli, F., 1886
1. Report on the Diatomaceae collected by H.M.S. Challenger during the years 1873-1876. Rept. Sci. Results, H.M.S. Challenger, Botany Vol. II, 178 pp., 30 pls.

Caloneis

Desikachary, T.V., 1956(1957).
Electron microscope studies on diatoms. J.R. Microsc. Soc. (3)76(1/2):9-36.

Caloneis amphisbaena

Rumkówna, A., 1948
[List of the phytoplankton species occurring in the superficial water layers in the Gulf of Gdańsk] Bull. Lab. mar., Gdynia, No. 4: 139-141 with tables in back.

Caloneis bacillum

Tsumura, K., 1956.
Diatomoj el la cirkaufoso de la restajo de la kastele de Odawara. J. Yokohama Municipal Univ., (C-14)No. 47:23 pp.

Caloneis bivittata

Zanon, D. V., 1949
Diatomee di Buenos Aires (Argentina) Atti Accad. Naz. Lincei, Memorie, Cl. Sci. fis., mat. e. nat., ser. 7, 11(3):59-151, 2 pls.

Caloneis Bonae Aurae n.sp.

Zanon, D. V., 1949
Diatomee di Buenos Aires (Argentina) Atti Accad. Naz. Lincei, Memorie, Cl. Sci. fis., mat. e. nat., ser. 7, 11(3):59-151, 2 pls.

Caloneis brevis

Hendy, N. Ingram, 1964
An introductory account of the smaller algae of British coastal waters. V. Bacillariophyceae (Diatoms).
Her Majesty's Stationary Office, 317 pp., 45 pls.

Caloneis brevis

Hendey, N.I., 1951
Littoral diatoms of Chicester Harbour with special reference to fouling. J.Roy. Microscop. Soc. 71(1): 1-86, 18 pls.

Caloneis formosa

Hendey, N.I., 1951
Littoral diatoms of Chicester Harbour with special reference to fouling. J.Roy. Microscop. Soc. 71(1): 1-86, 18 pls.

Caloneis formosa

Zanon, D. V., 1949
Diatomee di Buenos Aires (Argentina) Atti Accad. Naz. Lincei, Memorie, Cl. Sci. fis., mat. e. nat., ser. 7, 11(3):59-151, 2 pls.

Caloneis formosa

Zanon, V., 1948
Diatomee marini di Sardegna e Pugillo di Alghe Marine della stressa. Boll. Pesca, Piscitutura e Idrobiologia, Anno 24, ns. 3(2): 202-244, 27 figs. on 1 pl.

Caloneis fusoides

Hendy, N. Ingram, 1964
An introductory account of the smaller algae of British coastal waters. V. Bacillariophyceae (Diatoms).
Her Majesty's Stationary Office, 317 pp., 45 pls.

Caloneis fusoides

Hendey, N.I., 1951
Littoral diatoms of Chicester Harbour with special reference to fouling. J.Roy. Microscop. Soc. 71(1): 1-86, 18 pls.

Caloneis fusoides

Rampi, L., 1940
Diatomee del Mare Adriatico. Nuovo Giornale Botanico Italiano, n.s., 47:559-608.

Caloneis hustedti n.sp.
Aleem, A.A., and F. Hustedt, 1951.
Einige neue Diatomeen von der Südküste Englands.
Botaniska Notiser, 1951(1):13-30, 6 textfigs.

Caloneis latuiscula subholstei
Rumkówna, A., 1948
[List of the phytoplankton species occurring in the superficial water layers in the Gulf of Gdańsk] Bull. Lab. mar., Gdynia, No. 4: 139-141 with tables in back.

Caloneis liber
Hendy, N. Ingram, 1964
An introductory account of the smaller algae of British coastal waters. V. Bacillariophyceae (Diatoms).
Her Majesty's Stationary Office, 317 pp., 45 pls.

Caloneis liber
Hendey, N.I., 1951
Littoral diatoms of Chicester Harbour with special reference to fouling. J.Roy. Microscop. Soc. 71(1): 1-86, 18 pls.

Caloneis liber
Hustedt, F. and A.A. Aleem, 1951
Littoral diatoms from the Salstone near Plymouth. JMBA 30(1): 177.196.

Caloneis liber
Rampi, L., 1940
Diatomee del Mare Adriatico. Nuovo Giornale Botanico Italiano, n.s., 47:559-608.

Caloneis liber
Zanon, V., 1948
Diatomee marini di Sardegna e Pugillo di Alghe Marine della stressa. Boll. Pesca, Piscitutura e Idrobiologia, Anno 24, ns. 3(2): 202-244, 27 figs. on 1 pl.

Caloneis linearis
Hendy, N. Ingram, 1964
An introductory account of the smaller algae of British coastal waters. V. Bacillariophyceae (Diatoms).
Her Majesty's Stationary Office, 317 pp., 45 pls.

Caloneis madraspatensis n.sp.
Subrahmanyan, R., 1946.
A systematic account of the marine plankton diatoms of the Madras coast. Proc. Indian Acad. Sci. 24(4):85-197, Pl. 2, 440 textfigs.

Caloneis maxima
Rampi, L., 1940
Diatomee del Mare Adriatico. Nuovo Giornale Botanico Italiano, n.s., 47:559-608.

Caloneis musca
Rampi, L., 1940
Diatomee del Mare Adriatico. Nuovo Giornale Botanico Italiano, n.s., 47:559-608.

Caloneis permagna
Hendy, N. Ingram, 1964
An introductory account of the smaller algae of British coastal waters. V. Bacillariophyceae (Diatoms).
Her Majesty's Stationary Office, 317 pp., 45 pls.

Caloneis pinnularia
Zanon, D. V., 1949
Diatomee di Buenos Aires (Argentina) Atti Accad. Naz. Lincei, Memorie, Cl. Sci. fis., mat. e. nat., ser. 7, 11(3):59-151, 2 pls.

Caloneis Powellii var. egyptiaca
Rampi, L., 1940
Diatomee del Mare Adriatico. Nuovo Giornale Botanico Italiano, n.s., 47:559-608.

Caloneis robusta
Rampi, L., 1940
Diatomee del Mare Adriatico. Nuovo Giornale Botanico Italiano, n.s., 47:559-608.

Caloneis schroderii
Tsumura, K., 1956.
Diatomoj el la cirkaufoso de la restajo de la kastele de Odawara. J. Yokohama Municipal Univ. (C-14) No. 47:23 pp.

Caloneis silicula & signata
Tsumura, K., 1956.
Diatomoj el la cirkaufoso de la restajo de la kastele de Odawara. J. Yokohama Municipal Univ., (C-14), No. 47:23 pp.

Caloneis silicula
Zanon, D. V., 1949
Diatomee di Buenos Aires (Argentina) Atti Accad. Naz. Lincei, Memorie, Cl. Sci. fis., mat. e. nat., ser. 7, 11(3):59-151, 2 pls.

Caloneis subsalina
Hendy, N. Ingram, 1964
An introductory account of the smaller algae of British coastal waters. V. Bacillariophyceae (Diatoms).
Her Majesty's Stationary Office, 317 pp., 45 pls.

Caloneis subsalina comb. nov.
Hendey, N.I., 1951
Littoral diatoms of Chicester Harbour with special reference to fouling. J.Roy. Microscop. Soc. 71(1): 1-86, 18 pls.

Caloneis Tigris n.sp.
Zanon, D. V., 1949
Diatomee di Buenos Aires (Argentina) Atti Accad. Naz. Lincei, Memorie, Cl. Sci. fis., mat. e. nat., ser. 7, 11(3):59-151, 2 pls.

Caloneis ventricosa
Tsumura, K., 1956.
Diatomoj el la cirkaufoso de la restajo de la kastele de Odawara. J. Yokohama Municipal Univ., (C-14), No. 47:23 pp.

Caloneis westii comb. nov.
Hendy, N. Ingram, 1964
An introductory account of the smaller algae of British coastal waters. V. Bacillariophyceae (Diatoms).
Her Majesty's Stationary Office, 317 pp., 45 pls.

Caloneis zachariasi
Rumkówna, A., 1948
[List of the phytoplankton species occurring in the superficial water layers in the Gulf of Gdańsk] Bull. Lab. mar., Gdynia, No. 4: 139-141 with tables in back.

Campylodiscus thuretii
Brunel, J., 1962
Le phytoplancton de la Baie de Chaleurs. Inst. Botan., Univ. Montréal, Contrib. No. 77: 365 pp., 66 pls.

Campylodiscus
Desikachary, T.V., 1956(1957).
Electron microscope studies on diatoms. J.R. Microsc. Soc. (3)76(1/2):9-3o.

Campylodiscus acheneis
Kokubo, S., and S. Sato., 1947
Plankterz in Ju-San Gata. Physiol. and Ecol. (Japan) 1(4):1-16, 3 text figs., tables.

Campylodiscus acheneis
Levander, K.M., 1947
Plankton gesammelt in den Jahren 1899-1910 an den Küsten Finnlands. Finnländische Hydrographisch-Biologische Untersuchunger (aus dem Wasserbiologischen Laboratorin der Societas Scientiarum Fennica) No.11:40 pp., 6 diagrams, 13 pls., tables.

Campylodiscus adriaticus
Schröder, B., 1900
Phytoplankton des Golfes von Neapel nebst vergleichenden Ausblicken auf das atlantischen Ozean. Mitt. Zool. Stat. Neapel, 14:1-38.

Campylodiscus Adriaticus
Zanon, V., 1948
Diatomee marini di Sardegna e Pugillo di Alghe Marine della stressa. Boll. Pesca, Piscitutura e Idrobiologia, Anno 24, ns. 3(2): 202-244, 27 figs. on 1 pl.

Campylodiscus adriaticus var. massiliensis
Rampi, L., 1940
Diatomee del Mare Adriatico. Nuovo Giornale Botanico Italiano, n.s., 47:559-608.

Campylodiscus anceps n.sp.
Castracane degli Antelminelli, F., 1886
1. Report on the Diatomaceae collected by H.M.S. Challenger during the years 1873-1876. Rept. Sci. Results, H.M.S. Challenger, Botany Vol. II, 178 pp., 30 pls.

Campylodiscus angularis
Hendy, N. Ingram, 1964
An introductory account of the smaller algae of British coastal waters. V. Bacillariophyceae (Diatoms).
Her Majesty's Stationary Office, 317 pp., 45 pls.

Campylodiscus angularis
Jørgensen, E., 1905
B. Protistplankton and the diatoms in bottom samples. Hydrographical and biological investigations in Norwegian fjords. Bergens Mus. Skr. 7: 49-225.

Campylodiscus biangulatus
Zanon, V., 1948
Diatomee marini di Sardegna e Pugillo di Alghe Marine della stressa. Boll. Pesca, Piscitutura e Idrobiologia, Anno 24, ns. 3(2): 202-244, 27 figs. on 1 pl.

Campylodiscus bicinctus n.sp.
Castracane degli Antelminelli, F., 1886
1. Report on the Diatomaceae collected by H.M.S. Challenger during the years 1873-1876. Rept. Sci. Results, H.M.S. Challenger, Botany Vol. II, 178 pp., 30 pls.

Campylodiscus Clevei
Rampi, L., 1942
Ricerche sul fitoplancton del Mare Ligure 6. Le diatomee delle acque di Sanremo. Nuovo Giornale Botanico Italiano, N.S., 49:252-258.

Campylodiscus clypeus
Rampi, L., 1940
Diatomee del Mare Adriatico. Nuovo Giornale Botanico Italiano, n.s., 47:559-608.

Campylodiscus clypeus
Rumkówna, A., 1948
[List of the phytoplankton species occurring in the superficial water layers in the Gulf of Gdańsk] Bull. Lab. mar., Gdynia, No. 4: 139-141 with tables in back.

Campylodiscus clypeus
Zanon, D. V., 1949
Diatomee di Buenos Aires (Argentina) Atti Accad. Naz. Lincei, Memorie, Cl. Sci. fis., mat. e. nat., ser. 7, 11(3):59-151, 2 pls.

Campylodiscus (?) cocconeiformis
Allen, W.E., and E.E. Cupp, 1935
Plankton diatoms of the Java Sea.
Annales du Jardin Botanique de Buitenzorg
XLIV (2):101-174, figs.1-127.
(drawings of all species mentioned)

Campylodiscus concinnus
Mann, A., 1907
Report on the diatoms of the Albatross voyages in the Pacific Ocean, 1888-1904.
Contrib. U. S. Nat. Herb. 10(5):221-419, Pls. XLIV-LIV.

Campylodiscus costatus
Smith, W., 1851
Notes on the Diatomaceae; with descriptions of British species included in the genera Campylodiscus, Surirella, and Cymatopleura.
Ann. Mag. Nat. Hist., 2nd ser., 7:1-14, 3 pls.

Campylodiscus cribrosus
Smith, W., 1851
Notes on the Diatomaceae; with descriptions of British species included in the genera Campylodiscus, Surirella, and Cymatopleura.
Ann. Mag. Nat. Hist., 2nd ser., 7:1-14, 3 pls.

Campylodiscus Daemelianus
Rampi, L., 1940
Diatomee del Mare Adriatico. Nuovo Giornale Botanico Italiano, n.s., 47:559-608.

Campylodiscus decorus
Pavillard, J., 1925
Bacillariales. Rept. on the Danish Oceanogr. Exped., 1908-10 to the Mediterranean and adj. seas. Vol.II., Biol. J4:72 pp., 116 text figs.

Campylodiscus decorus
Politis, J., 1949
Diatomees marines de Bosphores et des ibes de la mer de Marmara. II Practica tou Hellenikou Hidrobiologikou Institutoutou 1929, Etoz 1929, 3(1):11-31.

Campylodiscus decorus
Rampi, L., 1942
Ricerche sul fitoplancton del Mare Ligure 6. Le diatomee delle acque di Sanremo. Nuovo Giornale Botanico Italiano, N.S., 49:252-268.

Campylodiscus decorum
Rampi, L., 1940
Diatomee del Mare Adriatico. Nuovo Giornale Botanico Italiano, n.s., 47:559-608.

Campylodiscus decorus
Zanon, V., 1948
Diatomee marini di Sardegna e Pugillo di Alghe Marine della stressa. Boll. Pesca, Piscitutura e Idrobiologia, Anno 24, ns. 3(2): 202-244, 27 figs. on 1 pl.

Campylodiscus ecclesianus
Mann, A., 1907
Report on the diatoms of the Albatross voyages in the Pacific Ocean, 1888-1904.
Contrib. U. S. Nat. Herb. 10(5):221-419, Pls. XLIV-LIV.

Campylodiscus echinus
Hendy, N. Ingram, 1964
An introductory account of the smaller algae of British coastal waters. V. Bacillariophyceae (Diatoms).
Her Majesty's Stationary Office, 317 pp., 45 pls.

Campylodiscus (?) echeneis
Hendey, N.J., 1951
Littoral diatoms of Chicester Harbour with special reference to fouling. J.Roy. Microscop. Soc. 71(1): 1-86, 18 pls.

Campylodiscus echeneis
Rumkówna, A., 1948
List of the phytoplankton species occurring in the superficial water layers in the Gulf of Gdańsk. Bull. Lab. mar., Gdynia, No. 4: 139-141 with tables in back.

Campylodiscus echeneis (figs.)
Sousa e Silva, E., 1949
Diatomaceas e Dinoflagelados de Baia de Cascais. Portugaliae Acta Biol., Volume: Julio Henriques, Ser. B: 300-383, 9 pls, 2 fold-in tables.

Campylodiscus Echeneis
Zanon, V., 1948
Diatomee marini di Sardegna e Pugillo di Alghe Marine della stressa. Boll. Pesca, Piscitutura e Idrobiologia, Anno 24, ns. 3(2): 202-244, 27 figs. on 1 pl.

Campylodiscus echeneis var. dentatus comb. nov.
Hendey, N-Ingram, 1958 [1957(Publ. 1958)]
Marine diatoms from some West African Ports. J. R Microsc. Soc. (3) 77(1/2): 28-85.

Campylodiscus erosus n.sp.
Castracane degli Antelminelli, F., 1886
1. Report on the Diatomaceae collected by H.M.S. Challenger during the years 1873-1876. Rept. Sci. Results, H.M.S. Challenger, Botany Vol. II, 178 pp., 30 pls.

Campylodiscus eximius
Rampi, L., 1942
Ricerche sul fitoplancton del Mare Ligure 6. Le diatomee delle acque di Sanremo. Nuovo Giornale Botanico Italiano, N.S., 49:252-268.

Campylodiscus eximius
Rampi, L., 1940
Diatomee del Mare Adriatico. Nuovo Giornale Botanico Italiano, n.s., 47:559-608.

Campylodiscus eximius
Sousa e Silva, E., 1949
Diatomaceas e Dinoflagelados de Baia de Cascais. Portugaliae Acta Biol., Volume: Julio Henriques, Ser. B: 300-383, 9 pls, 2 fold-in tables.

Campylodiscus eximius
Zanon, V., 1948
Diatomee marini di Sardegna e Pugillo di Alghe Marine della stressa. Boll. Pesca, Piscitutura e Idrobiologia, Anno 24, ns. 3(2): 202-244, 27 figs. on 1 pl.

Campylodiscus fastuosus
Hendy, N. Ingram, 1964
An introductory account of the smaller algae of British coastal waters. V. Bacillariophyceae (Diatoms).
Her Majesty's Stationary Office, 317 pp., 45 pls.

Campylodiscus fastuosus
Hendey, N.J., 1951
Littoral diatoms of Chicester Harbour with special reference to fouling. J.Roy. Microscop. Soc. 71(1): 1-86, 18 pls.

Campylodiscus fastuosus
Ross, R., and G. Abdin, 1949.
Notes on some diatoms from Norfolk. J. Roy. Micr. Soc., (3 ser.), 69(4):225-230, 4 figs. on 1 pl.

Campylodiscus fluminensis
Rampi, L., 1940
Diatomee del Mare Adriatico. Nuovo Giornale Botanico Italiano, n.s., 47:559-608.

Campylodiscus galapagensis n.sp.
Mann, A., 1907
Report on the diatoms of the Albatross voyages in the Pacific Ocean, 1888-1904.
Contrib. U. S. Nat. Herb. 10(5):221-419, Pls. XLIV-LIV.

Campyloneis grevillei (figs.)
Sousa e Silva, E., 1949
Diatomaceas e Dinoflagelados de Baia de Cascais. Portugaliae Acta Biol., Volume: Julio Henriques, Ser. B: 300-383, 9 pls, 2 fold-in tables.

Campylodiscus hodgsoni
de Sousa e Silva, E., 1956.
Contribution à l'étude du microplancton de Dakar et des regions maritimes voisines.
Bull. I.F.A.N., 8(2):335-371, 7 pls.

Campylodiscus hodgsonii
Hendy, N. Ingram, 1964
An introductory account of the smaller algae of British coastal waters. V. Bacillariophyceae (Diatoms).
Her Majesty's Stationary Office, 317 pp., 45 pls.

Campylodiscus Hodgsonii
Hendey, N.J., 1951
Littoral diatoms of Chicester Harbour with special reference to fouling. J.Roy. Microscop. Soc. 71(1): 1-86, 18 pls.

Campylodiscus Hodgsoni
Zanon, D. V., 1949
Diatomee di Buenos Aires (Argentina)
Atti Accad. Naz. Lincei, Memorie, Cl. Sci. fis., mat. e. nat., ser. 7, 11(3):59-151, 2 pls.

Campylodiscus Hodgsonii
Zanon, V., 1948
Diatomee marini di Sardegna e Pugillo di Alghe Marine della stressa. Boll. Pesca, Piscitutura e Idrobiologia, Anno 24, ns. 3(2): 202-244, 27 figs. on 1 pl.

Campylodiscus horologium
Mann, A., 1907
Report on the diatoms of the Albatross voyages in the Pacific Ocean, 1888-1904.
Contrib. U. S. Nat. Herb. 10(5):221-419, Pls. XLIV-LIV.

Campylodiscus horologium
Rampi, L., 1942
Ricerche sul fitoplancton del Mare Ligure 6. Le diatomee delle acque di Sanremo. Nuovo Giornale Botanico Italiano, N.S., 49:252-268.

Campylodiscus horologium
Rampi, L., 1940
Diatomee del Mare Adriatico. Nuovo Giornale Botanico Italiano, n.s., 47:559-608.

Campylodiscus humilis n.sp.
Castracane degli Antelminelli, F., 1886
1. Report on the Diatomaceae collected by H.M.S. Challenger during the years 1873-1876. Rept. Sci. Results, H.M.S. Challenger, Botany Vol. II, 178 pp., 30 pls.

Campylodiscus hypodromus
Hendy, N. Ingram, 1964
An introductory account of the smaller algae of British coastal waters. V. Bacillariophyceae (Diatoms).
Her Majesty's Stationary Office, 317 pp., 45 pls.

Campylodiscus hypodromus
Hendey, N.J., 1951
Littoral diatoms of Chicester Harbour with special reference to fouling. J.Roy. Microscop. Soc. 71(1): 1-86, 18 pls.

Campylodiscus impressus
Rampi, L., 1940
Diatomee del Mare Adriatico. Nuovo Giornale Botanico Italiano, n.s., 47:559-608.

Campylodiscus impressus
Zanon, V., 1948
Diatomee marini di Sardegna e Pugillo di Alghe Marine della stressa. Boll. Pesca, Piscitutura e Idrobiologia, Anno 24, ns. 3(2): 202-244, 27 figs. on 1 pl.

Campylodiscus innominatus
Hendy, N. Ingram, 1964
An introductory account of the smaller algae of British coastal waters. V. Bacillariophyceae (Diatoms). Her Majesty's Stationary Office, 317 pp., 45 pls.

Campylodiscus innominatus, n. sp.
Ross, R., and G. Abdin, 1949.
Notes on some diatoms from Norfolk. J. Roy. Micr. Soc., ser. 3, 69(4):225-230, 4 figs. on 1 pl.

Campylodiscus japonicus
Castracane degli Antelminelli, F., 1886 n.sp.
1. Rēport on the Diatomaceae collected by H.M.S. Challenger during the years 1873-1876. Rept. Sci. Results, H.M.S. Challenger, Botany Vol. II, 178 pp., 30 pls.

Campylodiscus kinkeri
Mann, A., 1907
Report on the diatoms of the Albatross voyages in the Pacific Ocean, 1888-1904. Contrib. U. S. Nat. Herb. 10(5):221-419, Pls. XLIV-LIV.

Campylodiscus latus
Mann, A., 1907
Report on the diatoms of the Albatross voyages in the Pacific Ocean, 1888-1904. Contrib. U. S. Nat. Herb. 10(5):221-419, Pls. XLIV-LIV.

Campylodiscus lepidus n.sp.
Castracane degli Antelminelli, F., 1886
1. Rēport on the Diatomaceae collected by H.M.S. Challenger during the years 1873-1876. Rept. Sci. Results, H.M.S. Challenger, Botany Vol. II, 178 pp., 30 pls.

Campylodiscus limbatus
Rampi, L., 1942
Ricerche sul fitoplancton del Mare Ligure 6. Le diatomee delle acque di Sanremo. Nuovo Giornale Botanico Italiano, N.S., 49:252-268.

Campylodiscus limbatus
Rampi, L., 1940
Diatomee del Mare Adriatico. Nuovo Giornale Botanico Italiano, n.s., 47:559-608.

Campylodiscus limbatus (figs.)
Sousa e Silva, E., 1949
Diatomaceas e Dinoflagelados de Baia de Cascais. Portugaliae Acta Biol., Volume: Julio Henriques, Ser. B: 300-383, 9 pls, 2 fold-in tables.

Campylodiscus limbatus
Zanon, V., 1948
Diatomee marini di Sardegna e Pugillo di Alghe Marine della stressa. Boll. Pesca, Piscitutura e Idrobiologia, Anno 24, ns. 3(2): 202-244, 27 figs. on 1 pl.

Campylodiscus nitens n.sp.
Castracane degli Antelminelli, F., 1886
1. Rēport on the Diatomaceae collected by H.M.S. Challenger during the years 1873-1876. Rept. Sci. Results, H.M.S. Challenger, Botany Vol. II, 178 pp., 30 pls.

Campylodiscus noricus
Mann, A., 1907
Report on the diatoms of the Albatross voyages in the Pacific Ocean, 1888-1904. Contrib. U. S. Nat. Herb. 10(5):221-419, Pls. XLIV-LIV.

Camphylodiscus oceanicus
Castracane degli Antelminelli, F., 1886 n.sp.
1. Rēport on the Diatomaceae collected by H.M.S. Challenger during the years 1873-1876. Rept. Sci. Results, H.M.S. Challenger, Botany Vol. II, 178 pp., 30 pls.

Campylodiscus orbicularis
Castracane degli Antelminelli, F., 1886 n.sp.
1. Rēport on the Diatomaceae collected by H.M.S. Challenger during the years 1873-1876. Rept. Sci. Results, H.M.S. Challenger, Botany Vol. II, 178 pp., 30 pls.

Campylodiscus parvulus
Rampi, L., 1940
Diatomee del Mare Adriatico. Nuovo Giornale Botanico Italiano, n.s., 47:559-608.

Campylodiscus parvulus
Smith, W., 1851
Notes on the Diatomaceae; with descriptions of British species included in the genera Campylodiscus, Surirella, and Cymatopleura. Ann. Mag. Nat. Hist., 2nd ser., 7:1-14, 3 pls.

Campylodiscus phippinarum
Castracane degli Antelminelli, F., 1886 n.sp.
1. Rēport on the Diatomaceae collected by H.M.S. Challenger during the years 1873-1876. Rept. Sci. Results, H.M.S. Challenger, Botany Vol. II, 178 pp., 30 pls.

Campylodiscus ralfsii
Boden, B.P., 1950
Some marine plankton diatoms from the west coast of South Africa. Trans. R.Soc. S. Africa. 32:321-434, 100 text figs.

Campylodiscus ralfsii
Hendy, N. Ingram, 1964
An introductory account of the smaller algae of British coastal waters. V. Bacillariophyceae (Diatoms). Her Majesty's Stationary Office, 317 pp., 45 pls.

Camplyodiscus Ralfsii
Politis, J., 1949
Diatomees marines de Bosphores et des ibes de la mer de Marmara. II Practica tou Hellenikou Hidrobiologikou Institutoutou 1929, Etoz 1929, 3(1):11-31.

Campylodiscus samoensis
Rampi, L., 1942
Ricerche sul fitoplancton del Mare Ligure 6. Le diatomee delle acque di Sanremo. Nuovo Giornale Botanico Italiano, N.S., 49:252-268.

Campylodiscus samoensis
Rampi, L., 1940
Diatomee del Mare Adriatico. Nuovo Giornale Botanico Italiano, n.s., 47:559-608.

Campylodiscus Samoensis
Zanon, V., 1948
Diatomee marini di Sardegna e Pugillo di Alghe Marine della stressa. Boll. Pesca, Piscitutura e Idrobiologia, Anno 24, ns. 3(2): 202-244, 27 figs. on 1 pl.

Campylodiscus spiralis
Smith, W., 1851
Notes on the Diatomaceae; with descriptions of British species included in the genera Campylodiscus, Surirella, and Cymatopleura. Ann. Mag. Nat. Hist., 2nd ser., 7:1-14, 3 pls.

Campylodiscus subangulatus
Rampi, L., 1940
Diatomee del Mare Adriatico. Nuovo Giornale Botanico Italiano, n.s., 47:559-608.

Campylodiscus taeniatus
Mann, A., 1907
Report on the diatoms of the Albatross voyages in the Pacific Ocean, 1888-1904. Contrib. U. S. Nat. Herb. 10(5):221-419, Pls. XLIV-LIV.

Campylodiscus thuretii
Brunel, Jules, 1962
Le phytoplancton de la Baie des Chaleurs. Contrib. Ministère de la Chasse et des Pêcheries, Province de Québec, No. 91: 365 pp.

Campilodiscus thuretii
de Sousa e Silva, E., 1956.
Contribution à l'étude du microplancton de Dakar et des regions maritimes voisines. Bull. I.F.A.N., 8(2):335-371, 7 pls.

Campylodiscus thureti
Hustedt, F. and A.A. Aleem, 1951
Littoral diatoms from the Salstone near Plymouth. JMBA 30(1): 177.196.

Campylodiscus thuretii
Jørgensen, E., 1905
B.Protistplankton and the diatoms in bottom samples. Hydrographical and biological investigations in Norwegian fiords. Bergens Mus. Skr. 7: 49-225.

Campylodiscus thuretii
Jorgensen, E., 1900
Protophyten und Protozoën im Plankton aus der Norwegischen Westkerste. Bergens Mus. Aarb. 1899(6): 95 pp., 5 pls., 83 tables.

Campylodiscus Thuretii
Rampi, L., 1942
Ricerche sul fitoplancton del Mare Ligure 6. Le diatomee delle acque di Sanremo. Nuovo Giornale Botanico Italiano, N.S., 49:252-268.

Campylodiscus Thurethii
Rampi, L., 1940
Diatomee del Mare Adriatico. Nuovo Giornale Botanico Italiano, n.s., 47:559-608.

Camplyodiscus Thuretii
Politis, J., 1949
Diatomees marines de Bosphores et des ibes de la mer de Marmara. II Practica tou Hellenikou Hidrobiologikou Institutoutou 1929, Etoz 1929, 3(1):11-31.

Campylodiscus Thureti
Zanon, V., 1948
Diatomee marini di Sardegna e Pugillo di Alghe Marine della stressa. Boll. Pesca, Piscitutura e Idrobiologia, Anno 24, ns. 3(2): 202-244, 27 figs. on 1 pl.

Campylodiscus wallicheanus var. thaitiensis
Castracane degli Antelminelli, F., 1886 n.var.
1. Rēport on the Diatomaceae collected by H.M.S. Challenger during the years 1873-1876. Rept. Sci. Results, H.M.S. Challenger, Botany Vol. II, 178 pp., 30 pls.

Campylodiscus zebuanus
Castracane degli Antelminelli, F., 1886 n.sp.
1. Rēport on the Diatomaceae collected by H.M.S. Challenger during the years 1873-1876. Rept. Sci. Results, H.M.S. Challenger, Botany Vol. II, 178 pp., 30 pls.

Campyloneis Grevillei
Allen, W.E., and E.E. Cupp, 1935
Plankton diatoms of the Java Sea.
Annales du Jardin Botanique de Buitenzorg
XLIV (2):101-174, figs.1-127.

(drawings of all species mentioned)

Campyloneis grevillei
Eskinazi Enide e Shigekatsv Satô (1963-1964)
1966.
Contribuição ao estudo das diatomaceas da Praia de Piedade.

Trabhs Inst. Oceanogr., Univ. Recife, 5 (5/6):
73-114.

Campyloneis grevillei
Hendy, N. Ingram, 1964

An introductory account of the smaller algae of British coastal waters. V. Bacillariophyceae (Diatoms).
Her Majesty's Stationary Office, 317 pp., 45 pls.

Campyloneis Grevillei
Hustedt, F. and A.A. Aleem, 1951
Littoral diatoms from the Salstone near Plymouth. JMBA 30(1): 177.196.

Campyloneis Grevillei
Rampi, L., 1942
Ricerche sul fitoplancton del Mare Ligure 6. Le diatomee delle acque di Sanremo. Nuovo Giornale Botanico Italiano, N.S., 49:252-268.

Campyloneis Grevillei
Zanon, V., 1948
Diatomee marini di Sardegna e Pugillo di Alghe Marine della stressa. Boll. Pesca, Piscicutura e Idrobiologia, Anno 24, ns. 3(2):
202-244, 27 figs. on 1 pl.

Campyloneis scutellum
Jorgensen, E., 1900
Protophyten und Protozoën im Plankton aus der Norwegischen Westkerste. Bergens Mus. Aarb. 1899(6): 95 pp., 5 pls., 83 tables.

Campylosira cymbelliformis
Cupp, Easter E., 1943
Marine plankton diatoms of the west coast of North America. Bull. S.I.O. 5(1):1-238, 5 pls., 168 text figs.

Campylosira cymbelliformis
Grøntved, J., 1949(1950).
Investigations on the phytoplankton in the Danish Waddensee in July 1941. Medd. Komm. Danmarks Fiskeri- og Havundersøgelser, Serie Plankton, 5(2)(Medd. Skalling Lab. 10):55 pp., 2 pls., 38 textfigs.

Campylosira cymbelliformis
Hendy, N. Ingram, 1964

An introductory account of the smaller algae of British coastal waters. V. Bacillariophyceae (Diatoms).
Her Majesty's Stationary Office, 317 pp., 45 pls.

Campylosira cymbelliformis
Hendey, N.I. 1951
Littoral diatoms of Chicester Harbour with special reference to fouling. J. Roy. Microscop. Soc. 71(1): 1-86, 18 pls.

Campylosira cymbelliformis
Grøntved, J., 1949
Investigations on the phytoplankton in the Danish Waddensea in July 1941. Medd. Komm. Danmarks Fiskeri og Havundersøgelser, ser. Plankton, 5(2):55 pp., 2 pls., 38 text figs.

Campylosira cymbelliformis
Meunier, A., 1915
Microplancton de la Mer Flamande. 2. Diatomées (excepté le genre Chaetoceros). Mem. Mus. Roy. Hist. Nat., Belgique, 7(3):1-118, Pls. VIII-XIV.

Catenula adhaerens
Hendy, N. Ingram, 1964

An introductory account of the smaller algae of British coastal waters. V. Bacillariophyceae (Diatoms).
Her Majesty's Stationary Office, 317 pp., 45 pls.

Cerataulina sp.
Kitou, M., 1957.
Notes on a certain marine diatom.
J. Oceanogr. Soc., Japan, 13(3):111-114.

Cerataulina bergonii
Allen, W.E., 1937
Plankton diatoms of the Gulf of California obtained by the G. Allan Hancock Expedition of 1936. The Hancock Pacific Expeditions, Univ. So. Calif. Publ. 3:47-59, 1 fig.

Cerataulina Bergonii
Allen, W.E., and E.E. Cupp, 1935
Plankton diatoms of the Java Sea.
Annales du Jardin Botanique de Buitenzorg
XLIV (2):101-174, figs.1-127.

(drawings of all species mentioned)

Cerataulina Bergonii
Bigelow, H.B., and M. Leslie, 1930
Reconnaissance of the waters and plankton of Monterey Bay, July 1928.
Bull. M.C.Z. 70(5):429-481, 43 text figs.

Cerataulina Bergonii
Braarud, T., 1945
Experimental studies on marine plankton diatoms. Avhandlingar utgitt av Det Norske Videnskaps-Akademi i Oslo. 1. Mat.-Naturv. Klasse 1944. No.10, 1-16, 1 text fig.

Cerataulina Bergonii
Braarud, T., 1945
A phytoplankton survey of the polluted waters of inner Oslo Fjord. Hvalrådets Skrifter, No.28, 142 pp., 19 text figs., 17 tables.

Cerataulina Bergoni
Braarud, T., and Adam Bursa, 1939
On the phytoplankton of the Oslo Fjord, 1933-1934. Hvalrådets Skr. No.19:1-63; 9 text figs. Reviewed. J. du. Cons. 14(3): 418-420. A.C. Gardiner.

Cerataulina Bergonii
Chiba, T., 1949
On the distribution of the plankton in the eastern China Sea and Yellow Sea. 1. Plankton composition in the spring. J. Shimonoseki Coll. Fisheries, 1(1):57-63, 1 fig.

Cerataulina bergonii
Cleve-Euler, A., 1951
Die Diatomeen von Schweden und Finnland. Kungl. Svenska Vetenskaps Akad. Handl., Fjärde Ser. 2(1): 161 pp., 6 pls.

Cerataulina bergonii
Cupp, Easter E., 1943
Marine plankton diatoms of the west coast of North America. Bull. S.I.O. 5(1):1-238, 5 pls., 168 text figs.

Cerataulina bergonii
Cupp, E.E., 1934
Analysis of marine diatom collections taken from the Canal Zone to California during March, 1933. Trans. Am. Micros. Soc. LIII (1):22-29, 1 map.

Cerataulina bergonii
Cupp, E.E. and Allen, W.E., 1938
Plankton diatoms of the Gulf of California obtained by Allan Hancock Pacific Expedition of 1937. The Hancock Pacific Expeditions, The Univ. So. Calif. Publ. 3: 61-74, 1 map, pls.4-15.

Cerataulina Bergonii
Dangeard, P., 1927
Phytoplankton de la croisière du "Sylvana". Ann. Inst. Ocean., Monaco, n.s. 4(8):286-401, 54 text figs. (Feirer-Juin 1913).

Cerataulina bergoni
Delegazione Italiana della Commissione Internazionale per l'Esplorazione Scientifica del Mediterraneo, 1941
Note sul plancton della Laguna veneta.
[Memoria CCLXXXIX], Arch. di Ocean. e Limn. Anno I, Fasc. I, 1941 XIX: 31-57 pp.

Cerataulina bergonii
Delsman, H.C., 1939.
Preliminary plankton investigations in the Java Sea. Treubia, 17:139-181, 8 maps, 41 figs.

Cerataulina Bergoni
Ercegovic, A., 1936
Etudes qualitative et quantitatives du phytoplancton dans les eaux cotières de l'Adriatique oriental moyen au cours de l'année 1934. Acta Adriatica 1(9):1-126

Cerataulina Bergonii
Forti, A., 1922
Ricerche sulla flora pelagica (fitoplancton) di Quarto dei Mille. Mem. R. Com. Talass. Ital. 97:248 pp., 13 pls.

Cerataulina Bergoni
Gran, H.H., 1908
Diatomeen. Nordisches Plankton, Botanischer Teil pp. XIX.1-XIX 146; 178 text figs.

Cerataulina bergonii
Gran, H. H. and E. C. Angst, 1931
Plankton diatoms of Puget Sound. Publ. Puget Sound Biol. Sta. 7:417-519, 95 text figs.

Cerataulina Bergonii
Gran, H.H., and T. Braarud, 1935
A quantitative study of the phytoplankton in the Bay of Fundy and the Gulf of Maine (including observations on hydrography, chemistry, and turbidity). J. Biol. Bd. Canada, 1(5):279-467, 69 text figs.

Cerataulina bergonii
Jorgensen, E., 1900
Protophyten und Protozoën im Plankton aus der Norwegischen Westkerste. Bergens Mus. Aarb. 1899(6): 95 pp., 5 pls., 83 tables.

Cerataulina bergonii
Kokubo, S., 1952
Results of the observations on the plankton and oceanography of Mutsu Bay during 1950, reference being made also to the period 1946-1950. Bull Mar.Biol.Sta., Asamushi 5(1/4): 1-54, 3 tables,(fold-in), 1 fold-in.

Cerataulina bergonii
Lafon, M., M. Durchon and Y. Saudray, 1955.
Recherches sur les cycles saisonnières du plankton. Ann. Inst. Océan., 31(3):125-230.

Cerataulina Bergoni
Lillick, L.C., 1940
Phytoplankton and planktonic protozoa of the offshore waters of the Gulf of Maine. Pt.II. Qualitative Composition of the Planktonic Flora. Trans. Am. Phil. Soc., n.s., 31(3):193-237, 13 text figs.

Cerataulina Bergoni
Lillick, L.C. 1938
Preliminary report of the phytoplankton of the Gulf of Maine. Am. Mid. Nat. 20(3):624-640, 1 text figs 37 tables.

Cerataulina Bergonii
Lillick, L.C., 1937
 Seasonal studies of the phytoplankton off Woods Hole, Massachusetts. Biol. Bull. LXXIII (3):488-503, 3 text figs.

Cerataulina Bergonii
Mangin, M. L., 1912
 Phytoplancton de la croisière du "René" dans l'Atlantique (Septembre 1908). Ann. Inst. Ocean., n.s., 4(1):1-66, 2 pls., 41 text figs., 2 tables.

Cerataulina Bergonii
Meunier, A., 1915
 Microplancton de la Mer Flamande. 2. Diatomées (excepté le genre Chaetoceros). Mem. Mus. Roy. Hist. Nat., Belgique, 7(3):1-118, Pls. VIII-XIV.

Cerataulina bergonii
Morse, D.C., 1947
 Some observations on seasonal variations in plankton population Patuxant River, Maryland 1943-1945. Bd. Nat. Res., Publ. No.65, Chesapeake Biol. Lab., 31, 3 figs.

Cerataulina bergonii
Pavillard, J., 1925
 Bacillariales. Rept. on the Danish Oceangr. Exped., 1908-10 to the Mediterranean and adj. seas. Vol.II., Biol. J4:72 pp., 116 text figs.

Cerataulina Bergonii
Pavillard, J., 1905
 Recherches sur la flore pelagique (Phytoplankton) de l'Etang de Thau. Theses presentees a la Fac. Sci., Paris, 116 pp., 3 pls.

Cerataulina bergoni
Pettersson, H., 1934
 Scattering and extinction of light in sea-water. Meddelanden från Göteborgs Högskolas Oceanografiska Institut. 9 (Göteborgs Kungl. vetenskaps-och vitterhats-samhälles Handlingar. Femte Foljden, Ser.B, 4(4)):1-16.

Cerataulina Bergoni
Rampi, L., 1942
 Ricerche sul fitoplancton del Mare Ligure 6. Le diatomee delle acque di Sanremo. Nuovo Giornale Botanico Italiano, N.S., 49:252-268.

Cerataulina Bergonii
Schodduyn, M., 1926
 Observations faites dans la baie d'Ambleteuse (Pas de Calais). Bull. Inst. Ocean., Monaco, No. 482: 64 pp.

Cerataulina Bergonii
Schröder, B., 1900
 Phytoplankton des Golfes von Neapel nebst vergleichenden Ausblicken auf das atlantischen Ozean. Mitt. Zool. Stat. Neapel, 14:1-38.

Cerataulina bergonii (figs.)
Sousa e Silva, E., 1949
 Diatomaceas e Dinoflagelados de Baía de Cascais. Portugaliae Acta Biol., Volume: Julio Henriques, Ser. B: 300-383, 9 pls, 2 fold-in tables.

Cerataulina compacta
Allen, W.E., and E.E. Cupp, 1935
 Plankton diatoms of the Java Sea. Annales du Jardin Botanique de Buitenzorg XLIV (2):101-174, figs.1-127.
 (drawings of all species mentioned)

Cerataulina compacta
Dangeard, P., 1927
 Phytoplancton de la croisière du "Sylvana". Ann. Inst. Ocean., Monaco, n.s., 4(8):286-401, 54 text figs. (Feirer-Juin 1913).

Cerataulina compacta
Delsman, H. C., 1939
 Preliminary plankton investigations in the Java Sea. Treubia, 17:139-181, 8 maps, 41 figs.

Cerataulina curvata n.sp.
Wood, E.J. Ferguson, 1963.
 Studies on Australian and New Zealand diatoms. VI. Tropical and subtropical species. Trans. R. Soc., New Zealand, 2(15):189-218.

Cerataulina pelagica
Hendy, N. Ingram, 1964
 An introductory account of the smaller algae of British coastal waters. V. Bacillariophyceae (Diatoms). Her Majesty's Stationary Office, 317 pp., 45 pls.

Cerataulina pelagica
Hendey, N.I., 1937
 The plankton diatoms of the southern seas. Discovery Repts. 16:151-364, pls.6-13.

Cerataulina pelagica
Saunders, Richard P., 1968.
 Cerataulina pelagica (Cleve) Henday Leaflet Ser., I(Phytoplankton) (2-diatoms-5): 11 pp.

Cerataulus Chinensis
Grunow, A., 1877
 1. New Diatoms from Honduras. Monthly Micros. Jour., 18:165-186, pls. CXCIII-CXCVI.

Cerataulus laevis
de Sousa e Silva, E., 1956.
 Contribution à l'étude du microplancton de Dakar et des regions maritimes voisines. Bull. I.F.A.N., 8(2):335-371, 7 pls.

Cerataulus ? Reichardti n.sp.
Grunow, A., 1863
 Ueber einige neue und ungenügend bekannte Arten und Gattungen von Diatomaceen. Verhandl. d. K.K. Zool. Bot. Gesellsch., Vienna, 13: 137-162, pl.4-5 (Pl. 13-14).

Cerataulus smithii
Cleve-Euler, A., 1951
 Die Diatomeen von Schweden und Finnland. Kungl. Svenska Vetenskaps Akad. Handl., Fjarde Ser. 2(1): 161 pp., 6 pls.

Cerataulus Smithii
Gran, H.H., 1908
 Diatomeen. Nordisches Plankton, Botanischer Teil pp. XIX.1-XIX 146; 178 text figs.

Cerataulus smithii
Hendy, N. Ingram, 1964
 An introductory account of the smaller algae of British coastal waters. V. Bacillariophyceae (Diatoms). Her Majesty's Stationary Office, 317 pp., 45 pls.

Cerataulus Smithii
Rampi, L., 1942
 Ricerche sul fitoplancton del Mare Ligure 6. Le diatomee delle acque di Sanremo. Nuovo Giornale Botanico Italiano, N.S., 49:252-268.

Cerataulus Smithi
Zanon, V., 1948
 Diatomee marini di Sardegna e Pugillo di Alghe Marine della stressa. Boll. Pesca, Piscitutura e Idrobiologia, Anno 24, ns. 3(2): 202-244, 27 figs. on 1 pl.

Cerataulus Smithii
Sousa a Silva, E., and J. Dos Santos-Pinto, 1948
 O Plancton da Baía de S. Martinho do Porto. 1. Diatomaceas e Dinoflagelados. Bol. Soc. Portuguese de Ciencias Naturais, 16(2):134-187, 6 pls. (Trav. Sta. Biol. Mar. de Lisbonne No. 52).

Cerataulus titianus n.sp.
Grunow, A., 1863
 Ueber einige neue und ungenügend bekannte Arten und Gattungen von Diatomaceen. Verhandl. d. K.K. Zool. Bot. Gesellsch., Vienna, 13: 137-162, pl.4-5 (Pl. 13-14).

Cerataulus turgidus
Cleve-Euler, A., 1951
 Die Diatomeen von Schweden und Finnland. Kungl. Svenska Vetenskaps Akad. Handl., Fjarde Ser. 2(1): 161 pp., 6 pls.

Cerataulus turgidus
Gran, H.H., 1908
 Diatomeen. Nordisches Plankton, Botanischer Teil pp. XIX.1-XIX 146; 178 text figs.

Cerataulus turgidus
Grøntved, J., 1949
 Investigations on the phytoplankton in the Danish Waddensea in July 1941. Medd. Komm. Danmarks Fiskeri og Havundersøgelser, ser. Plankton, 5(2):55 pp., 2 pls., 38 text figs.

Cerataulus turgidus
Hendy, N. Ingram, 1964
 An introductory account of the smaller algae of British coastal waters. V. Bacillariophyceae (Diatoms). Her Majesty's Stationary Office, 317 pp., 45 pls.

Cerataulus turgidus
Sousa e Silva, E., 1949
 Diatomaceas e Dinoflagelados de Baía de Cascais. Portugaliae Acta Biol., Volume: Julio Henriques, Ser. B: 300-383, 9 pls, 2 fold-in tables.

Cerataulus turgidus
Sousa a Silva, E., and J. Dos Santos-Pinto, 1948
 O Plancton da Baía de S. Martinho do Porto. 1. Diatomaceas e Dinoflagelados. Bol. Soc. Portuguese de Ciencias Naturais, 16(2):134-187, 6 pls. (Trav. Sta. Biol. Mar. de Lisbonne No. 52).

Ceratulus turgidus polyceros n.var.
Castracane degli Antelminelli, F., 1886
 1. Report on the Diatomaceae collected by H.M.S. Challenger during the years 1873-1876. Rept. Sci. Results, H.M.S. Challenger, Botany Vol. II, 178 pp., 30 pls.

Cestodiscus convexus n.sp.
Castracane degli Antelminelli, F., 1886
 1. Report on the Diatomaceae collected by H.M.S. Challenger during the years 1873-1876. Rept. Sci. Results, H.M.S. Challenger, Botany Vol. II, 178 pp., 30 pls.

Cestodiscus coronatus n.sp.
Castracane degli Antelminelli, F., 1886
 1. Report on the Diatomaceae collected by H.M.S. Challenger during the years 1873-1876. Rept. Sci. Results, H.M.S. Challenger, Botany Vol. II, 178 pp., 30 pls.

Cestodiscus gemmifer n.sp.
Castracane degli Antelminelli, F., 1886
 1. Report on the Diatomaceae collected by H.M.S. Challenger during the years 1873-1876. Rept. Sci. Results, H.M.S. Challenger, Botany Vol. II, 178 pp., 30 pls.

Cestodiscus gemmifer decrescens
Castracane degli Antelminelli, F., 1886
 1. Report on the Diatomaceae collected by H.M.S. Challenger during the years 1873-1876. Rept. Sci. Results, H.M.S. Challenger, Botany Vol. II, 178 pp., 30 pls.

Cestodiscus parmula n.sp.
Castracane degli Antelminelli, F., 1886
1. Report on the Diatomaceae collected by H.M.S. Challenger during the years 1873-1876. Rept. Sci. Results, H.M.S. Challenger, Botany Vol. II, 178 pp., 30 pls.

Cestodiscus (?) rapax n.sp.
Castracane degli Antelminelli, F., 1886
1. Report on the Diatomaceae collected by H.M.S. Challenger during the years 1873-1876. Rept. Sci. Results, H.M.S. Challenger, Botany Vol. II, 178 pp., 30 pls.

Cestodiscus trochus n.sp.
Castracane degli Antelminelli, F., 1886
1. Report on the Diatomaceae collected by H.M.S. Challenger during the years 1873-1876. Rept. Sci. Results, H.M.S. Challenger, Botany Vol. II, 178 pp., 30 pls.

Charcotia
Desikachary, T.V., 1956(1957).
Electron microscope studies on diatoms. J.R.Microsc. Soc. (3)76(1/2):9-36.

Charcotia actinochilus
Jouse, A.P., G.S. Koroleva, G.A. Nagaeva, 1962
Diatoms in the surface layer of sediment in the Indian sector of the Antarctic. Investigations of marine bottom sediments. (In Russian; English summary).
Trudy Inst. Okeanol., Akad. Nauk, SSSR, 61: 19-92.

Charcotia bifrons
Boden, B.P., 1949.
The diatoms collected by the U.S.S. CACOPAN in the Antarctic in 1947. J. Mar. Res. 8(1):6-13, 3 textfigs.

Charcotia bifrons
Boden, Brian, 1948
Marine plankton diatoms on operation HIGHJUMP in: Some oceanographic observations on operation HIGHJUMP. By R.S. Dietz. USNEL Rept. No.55, 97 pp., 41 figs. 7 July 1948.

Charcotia bifrons
Hendey, N.I., 1937
The plankton diatoms of the southern seas. Discovery Repts. 16:151-364, pls.6-13.

Chaetocerus
Bajkov, A.D. and T.W. Robinson, 1947
The distribution and occurrence of marine plankton. Aero Medical Laboratory Army Air Forces. Air Material Command. Engineering Division, Memo. Rept., Ser. No. TSEAA-691-3F-1, Expenditure Order No.691-12, ADB/ml, 7 pp. (ozalid) 30 Apr. 1947.

Chaetoceros
Bandel, W. 1940
Phytoplankton- und Nahrstoffgehalt der Ostsee im Gebiet der Darsser Schwelle. Internat. Rev. ges. Hydrobiol. u. Hydrogr., 40:249-304.

Chaetoceros sp.
Braarud, T., 1945
A phytoplankton survey of the polluted waters of inner Oslo Fjord. Hvalrådets Skrifter, No.28, 142 pp., 19 text figs., 17 tables.

Chaetoceros spp.
Braarud, T., 1939
Observations on the phytoplankton of the Oslo Fjord, March-April, 1937. Nytt Magasin for Naturvidenskapene, 80:211-218, 1 text fig.

Chaetoceros sp.
Braarud, T., and Adam Bursa, 1939
On the phytoplankton of the Oslo Fjord, 1933-1934. Hvalrådets Skr. no.19:1-63; 9 text figs. Reviewed. J. du Cons. 14(3): 418-420. A.C. Gardiner.

Chaetoceros spp.
Brunel, J., 1962
Le phytoplancton de la Baie de Chaleurs. Inst. Botan., Univ. Montréal, Contrib. No. 77: 365 pp., 66 pls.

Chaetoceros sp.
Brunel, Jules, 1962
Le phytoplancton de la Baie des Chaleurs. Contrib. Ministère de la Chasse et des Pêcheries, Province de Québec, No. 91: 365 pp.

Chaetoceros sp.
Cassie, Vivienne, 1961
Marine phytoplankton in New Zealand waters. Botanica Marina, 2(Suppl.): 54 pp., 8 pls.

Chaetoceros spp
Copenhagen, W.J., and L.D. Copenhagen, 1949.
Variation in the phytoplankton of Table Bay, October 1934 to October 1935, With a note on the calorific value of Chaetoceros spp. Trans. Roy. Soc., S. Africa, 32(2):113-123.

Chaetoceros sp.
Cupp, E.E., 1934
Analysis of marine diatom collections taken from the Canal Zone to California during March, 1933. Trans. Am. Micros. Soc. LIII (1):22-29, 1 map.

Chaetoceros sp.
Cupp, E.E. and Allen, W.E., 1938
Plankton diatoms of the Gulf of California obtained by Allan Hancock Pacific Expedition of 1937. The Hancock Pacific Expeditions, The Univ. So. Calif. Publ. 3: 61-74, 1 map, pls.4-15.

Chaetoceros
Desikachary, T.V., 1956(1957).
Electron microscope studies on diatoms. J.R. Microsc. Soc. (3)76(1/2):9-36.

Chaetoceros sp.
Galtsoff, P.S., 1948
Red Tide. Progress Report on the investigations of the cause of the mortality of fish along the west coast of Florida conducted by the U.S. Fish and Wildlife Service and Cooperating Organizations. Fish and Wildlife Service, Special Scientific Rept. No.46, 44 pp. (mimeographed), 9 figs.

Chaetoceros sp.
Gilbert, J.Y., and W.E. Allen, 1943
The phytoplankton of the Gulf of California obtained by the "E.W. Scripps" in 1939 and 1940. J. Mar. Res. V(2):89-110, figs.30-31.

Chaetoceros sp.
Gunter, G., R.H. Williams, C.C. Davis, and F.G. Walton Smith, 1948
Catastrophic mass mortality of marine animals and coincident phytoplankton bloom on the West Coast of Florida, November 1946 to August 1947. Ecol. Mon., 18:309-324, 2 text figs.

Chaetoceros
Höfler, Karl, und Luise Höfler, 1964.
Diatomeen des marinen Planktons als Isotonobionten. Pubbl. Staz. Zool., Napoli, 33(3):315-330.

Chaetoceros spp.
Iizuka, Shoji, and Haruhiko Irie, 1964
Electron micrographic study on the marine diatoms especially Skeletonema costatum (Grev.) Cleve. Bull. Fac. Fish., Nagasaki Univ., 15:92-99.

Chaetoceros spp.
Lillick, L.C., 1937
Seasonal studies of the phytoplankton off Woods Hole, Massachusetts. Biol. Bull. LXXIII (3):488-503, 3 text figs.

Chaetoceros sp.
Mann, A., 1907
Report on the diatoms of the Albatross voyages in the Pacific Ocean, 1888-1904. Contrib. U.S. Nat. Herb. 10(5):221-419, Pls. XLIV-LIV.

Chaetoceros sp.
Mann, A., 1907
Report on the diatoms of the Albatross voyages in the Pacific Ocean, 1888-1904. Contrib. U.S. Nat. Herb. 10(5):221-419, Pls. XLIV-LIV.

Chaetoceros, lists of spp.
Сравнительный эколого-биогеографический анализ видов рода Chaetoceros некоторых южноевропейских морей.

Comparative ecological-biogeographical analysis of species in the genus Chaetoceros of some South European seas. (In Russian).
Trudy Sevastopol. Biol. Stants., Akad. Nauk, 15: 39-49.

Chaetoceros, lists of spp
Mulford, Richard A., 1964
Investigations of inner continental shelf waters off lower Chesapeake Bay. V. Seasonality of the diatom genus Chaetoceros. Limnology and Oceanography, 9(3):385-390.

Chaetoceros
Neaverson, E., 1934
The sea-floor deposits. 1. General characteristics and distribution. Discovery Repts. 9: 297-349, Plates 17-22.

Chaetoceros
Mangin, L., 1919
Sur les Chaetoceros du groupe Peruvianus Bstiv. Bull. Mus. Nat. d'Hist. 25:305 & 411.

Chaetoceros sp.
Marshall, S.M., 1947
An experiment in marine fish cultivation: III. The plankton of a fertilized loch. Proc. Roy. Soc., Edinburgh, Sect.B., 63, Pt.I(3):21-33, 7 text figs.

Chaetoceros
Meunier, A., 1913
Microplancton de la Mer Flamande. 1. Chaetoceros. Mem. Mus. Roy. Hist. Nat., Belgique 7(2):1-55, 7 pls.

Chaetoceros
Okamura, K., 1907
Some Chaetoceros and Peragallia of Japan. Bot. Mag., Tokyo, 21:

Chaetoceros sp.
Parsons, T.R., 1961
On the pigment composition of eleven species of marine phytoplankters. J. Fish. Res. Bd., Canada, 18(6):1017-1025.

Chaetoceros sp.
Parsons, T.R., K. Stephens and J.D.H. Strickland, 1961
On the chemical composition of eleven species of marine phytoplankters. J. Fish. Res. Bd., Canada, 18(6):1001-1016.

Chaetoceros, spp.
Proschkina-Lavrenko, A.I.,
Species Chaetoceros novae et curiosae mario Nigroi I. Not. System. Sect. Crypt. Inst. Bot. Nom. V.L.Komarovij Akad. Sci, SSSR (9):46-56.

cited in J. du Cons. 22(2) without date

Chaetoceros
Sears, M., 1941
Notes on the phytoplankton on Georges Bank in 1941. J. Mar. Res., IV(3):247-257; textfigs. 54-58.

Oceanographic Index: Marine Organisms Cumulation, 1946-1973

Chaetoceros sp.
Wheeler, J.E.G., 1939
Plankton investigations. Bermuda Biological Station. Second Report. October 1939. 7 pp. (typed), 5 figs. Plymouth, Oct. 23, 1939.

Chaetoceros sp.
Zgurovskaya, L.N., and N.G. Kustenko, 1968.
The action of ammonia nitrogen on cell division, photosynthesis and the accumulation of pigments in Skeletonema costatum (Grev.), Chaetoceros sp. and Prorocentrum micans. (In Russian; English abstract).
Okeanologiia, Akad. Nauk, SSSR, 8(1): 116-125.

Chaetoceros abnormis
Brunel, J., 1962
Le phytoplancton de la Baie de Chaleurs. Inst. Botan., Univ. Montréal, Contrib. No. 77: 365 pp., 66 pls.

Chaetoceros abnormis
Grøntved, J., 1960-61
Planktological contributions. IV. Taxonomical and productional investigations in shallow coastal waters.
Medd. Dansk Fisk. Havundersøgelser, n.s., 3(1): 1-17.

Chaetoceros adelianum n.sp.
Manguin, E., 1957.
Premier inventaire des diatomées de la Terre Adélie Antarctique. Espèces nouvelles.
Rev. Algologique, n.s., 3(3):111-134.

Chaetoceros adhaerens n.sp.
Mangin, M. L., 1912
Phytoplancton de la croisière du "René" dans l'Atlantique (Septembre 1908). Ann. Inst. Ocean., n.s., 4(1):1-66, 2 pls., 41 text figs., 2 tables.

Chaetoceros aequatorialis
Hasle, Grethe Rytter, 1960
Phytoplankton and ciliate species from the Tropical Pacific.
Skr. Norske Videnskaps-Akad., Oslo, 1. Mat.-Nat. Kl., 1960(2): 1-50.

Chaetoceros aequatoriale
Hendey, N.I., 1937
The plankton diatoms of the southern seas. Discovery Repts. 16:151-364, pls.6-13.

Chaetoceros affinis
Allen, W. E., 1938
The Templeton Crocker Expedition to the Gulf of California in 1935 - The Phytoplankton. Amer. Microsc. Soc., Trans. 57:328-335.

Chaetoceros affinis
Allen, W.E., 1937
Plankton diatoms of the Gulf of California obtained by the G. Allan Hancock Expedition of 1936. The Hancock Pacific Expeditions, Univ. So. Calif. Publ. 3:47-59, 1 fig.

Chaetoceros affine
Allen, W.E., and E.E. Cupp, 1935
Plankton diatoms of the Java Sea. Annales du Jardin Botanique de Buitenzorg XLIV (2):101-174, figs.1-127.
(drawings of all species mentioned)

Chaetoceros affinis
Braarud, T., and Adam Bursa, 1939
On the phytoplankton of the Oslo Fjord, 1933-1934. Hvalrådets Skr. No.19:1-63; 9 text figs. Reviewed. J. du. Cons. 14(3): 418-420. A.C. Gardiner.

Chaetoceras affinis
Bonin, Daniel J., 1969.
Influence de différents facteurs écologiques sur la croissance de la diatomée marine Chaetoceros affinis Lauder en culture.
Tethys 1(1): 173-255

Chaetoceros affinis
Braarud, T., 1945
Experimental studies on marine plankton diatoms. Avhandlingar utgitt av Det Norske Videnskaps-Akademi i Oslo. 1. Mat.-Naturv. Klasse 1944. No.10, 1-16, 1 text fig.

Chaetoceros affinis
Braarud, T., 1945
A phytoplankton survey of the polluted waters of inner Oslo Fjord. Hvalrådets Skrifter, No.28, 142 pp., 19 text figs., 17 tables.

Chaetoceros affinis
Brunel, J., 1962
Le phytoplancton de la Baie de Chaleurs. Inst. Botan., Univ. Montréal, Contrib. No. 77: 365 pp., 66 pls.

Chaetoceros affinis willei

Chaetoceros affinis
Brunel, Jules, 1962
Le phytoplancton de la Baie des Chaleurs. Contrib. Ministère de la Chasse et des Pêcheries, Province de Québec, No. 91: 365 pp.

Chaetoceros affinis
Cleve-Euler, A., 1951
Die Diatomeen von Schweden und Finnland. Kungl. Svenska Vetenskaps Akad. Handl., Fjärde Ser. 2(1): 161 pp., 6 pls.

Chaetoceros affinis
Cupp, Easter E., 1943
Marine plankton diatoms of the west coast of North America. Bull. S.I.O. 5(1):1-238, 5 pls., 168 text figs.

Chaetoceros affinis
Cupp, E.E., 1934
Analysis of marine diatom collections taken from the Canal Zone to California during March, 1933. Trans. Am. Micros. Soc. LIII (1):22-29, 1 map.

Chaetoceros affinis
Delegazione Italiana della Commissione Internazionale per l'Esplorazione Scientifica del Mediterraneo, 1941
Note sul plancton della Laguna veneta. [Memoria CCLXXXIX], Arch. di Ocean. e Limn. Anno I, Fasc. I, 1941 XIX: 31-57 pp.

Chaetoceros affinis
Ercegovic, A., 1936
Etudes qualitative et quantitatives du phytoplancton dans les eaux cotières de l'Adriatique oriental moyen au cours de l'année 1934. Acta Adriatica 1(9):1-126

Chaetoceros affinis
Forti, A., 1922
Ricerche sulla flora pelagica (fitoplancton) di Quarto dei Mille. Mem. R. Com. Talass. Ital. 97:248 pp., 13 pls.

Chaetoceros affinis
Frenguelli, Joaquin, and Hector Antonio Orlando, 1959.
Operacion MERLUZA. Diatomeas y silicoflagelados del plancton del "VI Crucero". Servicio Hidrogr. Naval., Argentina, Publ. No. H. 619: 5-62.

Chaetoceros affinis
Gilbert, J.Y., and W.E. Allen, 1943
The phytoplankton of the Gulf of California obtained by the "E.W. Scripps" in 1939 and 1940. J. Mar. Res. V(2):89-110, figs.30-31.

Chaetoceros affinis
Gilson, H. C., 1937
Chemical and Physical Investigations. The nitrogen cycle. John Murray Exped., 1933-34, Sci. Repts., 2(2):21-81, 16 text figs.

Chaetoceros affinis
Gran, H.H. and T. Braarud, 1935
A quantitative study of the phytoplankton in the Bay of Fundy and the Gulf of Maine (including observations on hydrography, chemistry, and turbidity). J. Biol. Bd. Canada, 1(5):279-467, 69 text figs.

Chaetoceros affinis
Gran, H. H. and E. C. Angst, 1931
Plankton diatoms of Puget Sound. Publ. Puget Sound Biol. Sta. 7:417-519, 95 text figs.

Chaetoceros affine
Hendy, N. Ingram, 1964
An introductory account of the smaller algae of British coastal waters. V. Bacillariophyceae (Diatoms).
Her Majesty's Stationary Office, 317 pp., 45 pls.

Chaetoceros affinis
Kokubo, S., 1952
Results of the observations on the plankton and oceanography of Mutsu Bay during 1950, reference being made also to the period 1946-1950. Bull Mar.Biol.Sta., Asamushi 5(1/4): 1-54, 3 tables,(fold-in), 1 fold-in.

Chaetoceros affinis
Kokubo, S., and S. Sato., 1947
Plankters in JQ-San Gata. Physiol. and Ecol. (Japan) 1(4):1-16, 3 text figs., tables.

Chaetoceros affinis
Lillick, L.C., 1940
Phytoplankton and planktonic protozoa of the offshore waters of the Gulf of Maine. Pt.II. Qualitative Composition of the Planktonic Flora. Trans. Am. Phil. Soc., n.s., 31(3):193-237, 13 text figs.

Chaetoceros affinis
Lillick, L.C., 1938
Preliminary report of the phytoplankton of the Gulf of Maine. Am. Mid. Nat. 20(3):624-640, 1 text fig, 37 tables.

Chaetoceros affinis
Lillick, L.C., 1937
Seasonal studies of the phytoplankton off Woods Hole, Massachusetts, Biol. Bull. LXXIII (3):488-503, 3 text figs.

Chaetoceros affine
Marukawa, H., 1921
Plankton lists and some new species of copepods, from the northern waters of Japan. Bull. Inst. Ocean., No.384, 15 pp., 3 pls., 1 chart. Monaco

Chaetoceros affinis
Massuti Algamora, M., 1949
Estudio de diez y seis muestras de plancton del Golfo de Nápoles. Publ. Inst. Biol. Appl. 5:85-94, 1 fold-in table.

Chaetoceros affinis
Michailova, N. F., 1964.
On the spreading of the habitat into the Black Sea of species of the genus Chaetoceros of northern seas and their biogeography. (In Russian).
Trudy Sevastopol Biol. Sta., 7:231-248.

Chaetoceros affinis
Morse, D.C., 1947
Some observations on seasonal variations in plankton population Patuxent River, Maryland 1943-1945. Bd. Nat. Res., Publ. No.65, Chesapeake Biol. Lab., 31, 3 figs.

Chaetoceros affine
Pavillard, J., 1925
Bacillariales. Rept. on the Danish Oceanogr. Exped., 1908-10 to the Mediterranean and adj. seas. Vol.II., Biol. J4:72 pp., 116 text figs.

Chaetoceros affinis
Rampi, L., 1942
Ricerche sul fitoplancton del Mare Ligure 6. Le diatomee delle acque di Sanremo. Nuovo Giornale Botanico Italiano, N.S., 49:252-268.

Chaetoceros affine
Sousa e Silva, E., 1949
Diatomaceas e Dinoflagelados de Baia de Cascais. Portugaliae Acta Biol., Volume: Julio Henriques, Ser. B: 300-383, 9 pls, 2 fold-in tables.

Chaetoceros affinis
Steemann-Nielsen, Einar, 1951
The marine vegetation of the Isefjord. A study on ecology and production. Medd. Komm. Danmarks Fiskeri-og Havundersøgelser. Ser. Plankton. 5(4); 114pp., 46 text figs.

Chaetoceros affinis
Talling, J. F., 1960.
Comparative laboratory and field studies of photosynthesis by a marine planktonic diatom. Limnol. & Oceanogr., 5(1):62-72.

Chaetoceros affinis
Tratet, Gérard, 1964.
Variations du phytoplancton à Tanger. Trav. Inst. Sci., Cherifien, Rabat, Ser. Botan., (29):204 pp.

Chaetoceros affinis
Uyeno, Fukuzo, 1961
Oceanographical and ecological studies on primary production of the sea, with special references to relationship between diatom production and temperature and chlorinity of water.
Rept., Fac. Fish., Pref. Univ., Mie, 4(1): 1-64.

Chaetoceros affinis circinalis
Cupp, Easter E., 1943
Marine plankton diatoms of the west coast of North America. Bull. S.I.O. 5(1):1-238, 5 pls., 168 text figs.

Chaetoceros affinis circinalis
Ercegovic, A., 1936
Etudes qualitative et quantitatives du phytoplancton dans les eaux cotières de l'Adriatique oriental moyen au cours de l'année 1934. Acta Adriatica 1(9):1-126

Chaetoceros affinis var. Schuttii
Ghazzawi, F.M., 1939
Plankton of the Egyptian waters. A study of the Suez Canal Plankton. (A) Phytoplankton. Preliminary Report 83 pp. Notes and Memoires, Min. Commerce-Industry, Egypt, Hydrobiol. & Fish. 65 figs.

Chaetoceros affinis singularis
Takano, H., 1960
Plankton diatoms in the eastern Caribbean Sea. J. Oceanogr. Soc., Japan, 16(4): 180-184.

Chaetoceros affinis singularis ferox subforma nov.
Takano, H., 1960
Plankton diatoms in the eastern Caribbean Sea. J. Oceanogr. Soc., Japan, 16(4): 180-184.

Chaetoceros affinis singularis extremis subforma nov.
Takano, H., 1960
Plankton diatoms in the eastern Caribbean Sea. J. Oceanogr. Soc., Japan, 16(4): 180-184.

Chaetoceros affinis willei
Brunel, Jules, 1962
Le phytoplancton de la Baie des Chaleurs. Contrib. Ministère de la Chasse et des Pêcheries; Province de Québec, No. 91: 365 pp.

Chaetoceros affinis willei
Cupp, Easter E., 1943
Marine plankton diatoms of the west coast of North America. Bull. S.I.O. 5(1):1-238, 5 pls., 168 text figs.

Chaetoceros affinis var. Willei
Lillick, L.C., 1940
Phytoplankton and planktonic protozoa of the offshore waters of the Gulf of Maine. Pt.II. Qualitative Composition of the Planktonic Flora. Trans. Am. Phil. Soc., n.s., 31(3):193-237, 13 text figs.

Chaetoceros affinis willei
Myklestad, Sverre and Arne Haug, 1972.
Production of carbohydrates by the marine diatom Chaetoceros affinis var. willei (Gran) Hustedt. 1. Effect of the concentration of nutrients in the culture medium. J. exp. mar. Biol. Ecol. 9(2): 125-136.

Chaetoceros affinis willei
Myklestad, Sverre, Arne Haug and Bjørn Larsen, 1972.
Production of carbohydrates by the marine diatom Chaetoceros affinis var. willei (Gran) Hustedt. II. Preliminary investigation of the extracellular polysaccharide. J. exp. mar. Biol. Ecol. 9(2): 137-144.

Chaetoceros amenita
Cleve-Euler, A., 1951.
Die Diatomeen von Schweden und Finnland. Kungl. Svenska Vetenskaps Akad. Handl., Fjärde Ser. 2(1): 161 pp., 6 pls.

Chaetoceros anastomosans
Anon., 1951.
Bulletin of the Marine Biological Station of Asamushi 4(3/4): 15 pp.

Chaetoceros anastomosans
Brunel, J., 1962
Le phytoplancton de la Baie de Chaleurs. Inst. Botan., Univ. Montréal, Contrib. No. 77: 365 pp., 66 pls.

Chaetoceros anastomosans
Cleve-Euler, A., 1951.
Die Diatomeen von Schweden und Finnland. Kungl. Svenska Vetenskaps Akad. Handl., Fjärde Ser. 2(1): 161 pp., 6 pls.

Chaetoceros anastomosans
Cupp, Easter E., 1943
Marine plankton diatoms of the west coast of North America. Bull. S.I.O. 5(1):1-238, 5 pls., 168 text figs.

Chaetoceros anastomosans
Ercegovic, A., 1936
Etudes qualitative et quantitatives du phytoplancton dans les eaux cotières de l'Adriatique oriental moyen au cours de l'année 1934. Acta Adriatica 1(9):1-126

Chaetoceros anastomosans
Gran, H.H., 1908
Diatomeen. Nordisches Plankton, Botanischer Teil pp. XIX.1-XIX 146; 178 text figs.

Chaetoceros anastomosans
Kokubo, S., 1952
Results of the observations on the plankton and oceanography of Mutsu Bay during 1950, reference being made also to the period 1946-1950. Bull Mar.Biol.Sta., Asamushi 5(1/4): 1-54, 3 tables,(fold-in), 1 fold-in.

Chaetoceros anastomosans
Meunier, A., 1913
Microplancton de la Mer Flamande. 1. Chaetoceros. Mem. Mus. Roy. Hist. Nat., Belgique 7(2):1-55, 7 pls.

Chaetoceros anastomosans
Michailova, N.F., 1964.
On the spreading of the habitat into the Black Sea of species of the genus Chaetoceros of of northern seas and their biogeography. (In Russian).
Trudy Sevastopol Biol. Sta., 7:231-248.

Chaetoceros anastomosans
Pavillard, J., 1925
Bacillariales. Rept. on the Danish Oceanogr. Exped., 1908-10 to the Mediterranean and adj. seas. Vol.II., Biol. J4:72 pp., 116 text figs.

Chaetoceros anastomosans
Rampi, L., 1942
Ricerche sul fitoplancton del Mare Ligure 6. Le diatomee delle acque di Sanremo. Nuovo Giornale Botanico Italiano, N.S., 49:252-268.

Chaetoceros anastomosans var. externa
Gran, H. H. and E. C. Angst, 1931
Plankton diatoms of Puget Sound. Publ. Puget Sound Biol. Sta. 7:417-519, 95 text figs.

Chaetoceros anastomosans
Mangin, M. L., 1912
Phytoplancton de la croisière du "René" dans l'Atlantique (Septembre 1908). Ann. Inst. Ocean., n.s., 4(1):1-66, 2 pls., 41 text figs., 2 tables.

Chaetoceros anastomosans
Schröder, B., 1900
Phytoplankton des Golfes von Neapel nebst vergleichenden Ausblicken auf das atlantischen Ozean. Mitt. Zool. Stat. Neapel, 14:1-38.

Chaetoceros anglicus
Forti, A., 1922
Ricerche sulla flora pelagica (fitoplancton) di Quarto dei Mille. Mem. R. Com. Talass. Ital. 97:248 pp., 13 pls.

Chaetoceros angulatum
Schröder, B., 1900
Phytoplankton des Golfes von Neapel nebst vergleichenden Ausblicken auf das atlantischen Ozean. Mitt. Zool. Stat. Neapel, 14:1-38.

Chaetoceros apendiculatus n.sp
Müller Melchers, F.C., 1953.
New and little known diatoms from Uruguay and the South Atlantic coast.
Com. Bot., Mus. Hist. Nat., Montevideo, 3(30): 1-11, 8 pls.

Chaetoceros approximatus

Cupp, E.E. and Allen, W.E., 1938
Plankton diatoms of the Gulf of California obtained by Allan Hancock Pacific Expedition of 1937. The Hancock Pacific Expeditions, The Univ. So. Calif. Publ. 3: 61-74, 1 map, pls.4-15.

Chaetoceros approximatus n.sp.

Gran, H. H. and E. C. Angst, 1931
Plankton diatoms of Puget Sound. Publ. Puget Sound Biol. Sta. 7:417-519, 95 text figs.

Chaetoceros armatum

Hendy, N. Ingram, 1964
An introductory account of the smaller algae of British coastal waters. V. Bacillariophyceae (Diatoms). Her Majesty's Stationary Office, 317 pp., 45 pls.

Chaetoceros armatum

Lewin, Joyce and Thomas Hruby, 1973
Blooms of surf-zone diatoms along the coast of the Olympic Peninsula, Washington. II.ª A diel periodicity in buoyancy shown by the surf-zone diatom species, Chaetoceros armatum T. West. Estuarine coast, mar. Sci. 1(1): 101-105.

Chaetoceros armatum

Lewin, Joyce, and Richard E. Norris 1970.
Surf-zone diatoms of the coasts of Washington and New Zealand (Chaetoceros armatum T. West and Asterionella spp.). Phycologia 9(2): 143-149.

Chaetoceros armatum

Meunier, A., 1913
Microplancton de la Mer Flamande. 1. Chaetoceros. Mem. Mus. Roy. Hist. Nat., Belgique 7(2):1-55, 7 pls.

Chaetoceros atlanticus

Allen, W. E., 1938
The Templeton Crocker Expedition to the Gulf of California in 1935 - The Phytoplankton. Amer. Microsc. Soc., Trans. 57:328-335.

Chaetoceros atlanticus

Allen, W.E., 1937
Plankton diatoms of the Gulf of California obtained by the G. Allan Hancock Expedition of 1936. The Hancock Pacific Expeditions, Univ. So. Calif. Publ. 3:47-59, 1 fig.

Chaetoceros atlanticus

Balech, E., 1947
Contribution al conocimiento del plancton antarctico. Plankton del Mar de Bellinghausen. Physis 20:75-91, 76 figs. on 8 pls.

Chaetoceros atlanticum

Bigelow, H.B., and M. Leslie, 1930
Reconnaissance of the waters and plankton of Monterey Bay, July 1928. Bull. M.C.Z., 70(5):429-481, 43 text figs.

Chaetoceros atlanticum

Boden, B.P., 1950
Some marine plankton diatoms from the west coast of South Africa. Trans. R.Soc. S. Africa. 32:321-434, 100 text figs.

Chaetoceros atlanticus

Brunel, J., 1962
Le phytoplancton de la Baie de Chaleurs. Inst. Botan., Univ. Montréal, Contrib. No. 77: 365 pp., 66 pls.

Chaetoceros atlanticus

Brunel, Jules, 1962
Le phytoplancton de la Baie des Chaleurs. Contrib. Ministère de la Chasse et des Pêcheries, Province de Québec, No. 91: 365 pp.

Chaetoceros atlanticum

Central Meteorological Observatory, 1949
Report on sea and weather observation on Antarctic Whaling Ground (1947-48). Ocean. Mag., Japan, 1(1):49-88, 17 text figs.

Chaetoceros atlanticum

Cupp, Easter E., 1943
Marine plankton diatoms of the west coast of North America. Bull. S.I.O. 5(1):1-238, 5 pls., 168 text figs.

Chaetoceros atlanticus

Cupp, E.E. and Allen, W.E., 1938
Plankton diatoms of the Gulf of California obtained by Allan Hancock Pacific Expedition of 1937. The Hancock Pacific Expeditions, The Univ. So. Calif. Publ. 3: 61-74, 1 map, pls.4-15.

Chaetoceros atlanticus

Davidson, V.M., 1931.
Biological and oceanographic conditions in Hudson Bay. Contr. Canadian Biol. Fish., n.s., 6(26): 497-509, 7 textfigs.

Chaetoceros atlanticus

Gilbert, J.Y., and W.E. Allen, 1943
The phytoplankton of the Gulf of California obtained by the "E.W. Scripps" in 1939 and 1940. J. Mar. Res. V(2):89-110, figs.30-31.

Chaetoceros atlanticum

Gran, H.H., 1908
Diatomeen. Nordisches Plankton, Botanischer Teil pp. XIX.1-XIX 146; 178 text figs.

Chaetoceros atlanticum

Gran, H.H., 1897
Protophyta: Diatomaceae, Silico-flagellata and Cilioflagellata. Den Norske Nordhavs Expedition 1876-1878, h, 24, 36 pp., 4 pls.

Chaetoceros atlanticus

Gran, H.H., and T. Braarud, 1935
A quantitative study of the phytoplankton in the Bay of Fundy and the Gulf of Maine (including observations on hydrography, chemistry, and turbidity). J. Biol. Bd., Canada, 1(5):279-467, 69 text figs.

Chaetoceros atlanticus

Gran, H. H. and E. C. Angst, 1931
Plankton diatoms of Puget Sound. Publ. Puget Sound Biol. Sta. 7:417-519, 95 text figs.

Chaetoceros atlanticum

Hendey, N.I., 1937
The plankton diatoms of the southern seas. Discovery Repts. 16:151-364, pls.6-13.

Chaetoceros atlanticum

Iselin, C., 1930
A report on the coastal waters of Labrador based on explorations of the "Chance" during the summer of 1926. Proc. Am. Acad. Arts Sci., 66(1):1-37, 14 text figs.

Chaetoceros atlanticus

Jorgensen, E., 1900
Protophyten und Protozoën im Plankton aus der Norwegischen Westkerste. Bergens Mus. Aarb. 1899(6): 95 pp., 5 pls., 83 table

Chaetoceros atlanticus

Kokubo, S., 1952
Results of the observations on the plankton and oceanography of Mutsu Bay during 1950, reference being made also to the period 1946-1950. Bull Mar.Biol.Sta., Asamushi 5(1/4): 1-54, 3 tables,(fold-in), 1 fold-in.

Chaetoceros atlanticus

Lillick, L.C., 1940
Phytoplankton and planktonic protozoa of the offshore waters of the Gulf of Maine. Pt.II. Qualitative Composition of the Planktonic Flora. Trans. Am. Phil. Soc., n.s., 31(3):193-237, 13 text figs.

Chaetoceros atlanticus

Lillick, L.C., 1938
Preliminary report of the phytoplankton of the Gulf of Maine. Am. Mid. Nat. 20(3):624-640, 1 text figs 37 tables.

Chaetoceros atlanticus

Mangin, L., 1915
Phytoplancton de L'Antartique. Deuxieme Exped. Ant. Francaise (1908-1910), 95 pp., 3 pls., 58 text figs.

Chaetoceros atlanticus

Manguin, E., 1954
Diatomées marines provenant de l'ile Heard (Australian National Antarctic Expedition). Rev. Algol., n.s., 1: 14-24.

Chaetoceros atlanticum

Marukawa, H., 1921
Plankton lists and some new species of copepods, from the northern waters of Japan. Bull. Inst. Ocean., No.384, 15 pp., 3 pls., 1 chart. Monaco

Chaetoceros atlanticum

Meunier, A., 1913
Microplancton de la Mer Flamande. 1. Chaetoceros. Mem. Mus. Roy. Hist. Nat., Belgique 7(2):1-55, 7 pls.

Chaetoceros atlanticus

Phifer, L.D. (undated)
The occurrence and distribution of plankton diatoms in Boring Sea and Bering Strait, July 26-August 24, 1934. Report of Oceanographic cruise of U.S. Coast Guard Cutter Chelan 1934, Part II(A):1-44 (mimeographed) plus fig.1 (after Pt.B)

Chaetoceros atlantica neapolitana

Ercegovic, A., 1936
Etudes qualitative et quantitatives du phytoplancton dans les eaux cotières de l'Adriatique oriental moyen au cours de l'année 1934. Acta Adriatica 1(9):1-126

Chaetoceros atlanticus var. neapolitana

Cupp, E.E., 1934
Analysis of marine diatom collections taken from the Canal Zone to California during March, 1933. Trans. Am. Micros. Soc. LIII (1):22-29, 1 map.

Chaetoceros atlanticum

Hendy, N. Ingram, 1964
An introductory account of the smaller algae of British coastal waters. V. Bacillariophyceae (Diatoms). Her Majesty's Stationary Office, 317 pp., 45 pls.

Chaetoceros atlanticum var. neopolitana

Hendey, N.I., 1937
The plankton diatoms of the southern seas. Discovery Repts. 16:151-364, pls.6-13.

Chaetoceros atlantidae

Brunel, J., 1962
Le phytoplancton de la Baie de Chaleurs. Inst. Botan., Univ. Montréal, Contrib. No. 77: 365 pp., 66 pls.

Chaetoceros atlantide nsp.

Müller Melchers, F.C., 1953.
New and little known diatoms from Uruguay and the South Atlantic coast. Com. Bot. Mus. Hist. Nat., Montevideo, 3(30): 1-11, 8 pls.

Chaetoceros audax

Cupp, Easter E., 1943
Marine plankton diatoms of the west coast of North America. Bull. S.I.O. 5(1):1-238, 5 pls., 168 text figs.

Chaetoceros bergonii

Steemann-Nielsen, Einar, 1951
The marine vegetation of the Isefjord. A study on ecology and production. Medd. Komm. Danmarks Fiskeri-og Havundersøgelser. Ser. Plankton. 5(4); 114 pp., 46 text figs.

Chaetoceros biconcavum n.sp.

Gran, H.H., 1897
Protophyta: Diatomaceae, Silico-flagellata and Cilioflagellata. Den Norske Nordhavs Expedition 1876-1878, h. 24, 36 pp., 4 pls.

Chaetoceros boreale

Bigelow, H.B., and M. Leslie, 1930
Reconnaissance of the waters and plankton of Monterey Bay, July 1928. Bull. M.C.Z., 70(5):429-481, 43 text figs.

Chaetoceros borealis

Braarud, T., 1945
Experimental studies on marine plankton diatoms. Avhandlingar utgitt av Det Norske Videnskaps-Akademi i Oslo. 1. Mat.-Naturv. Klasse 1944. No.10, 1-16, 1 text fig.

Chaetoceros borealis

Braarud, T., 1945
A phytoplankton survey of the polluted waters of inner Oslo Fjord. Hvalrådets Skrifter, No.28, 142 pp., 19 text figs., 17 tables.

Chaetoceros borealis

Braarud, T., and Adam Bursa, 1939
On the phytoplankton of the Oslo Fjord, 1933-1934. Hvalrådets Skr. No.19:1-63; 9 text figs. Reviewed. J. du. Cons. 14(3): 418-420. A.C. Gardiner.

Chaetoceros borealis

Brunel, J., 1962
Le phytoplancton de la Baie de Chaleurs. Inst. Botan., Univ. Montréal, Contrib. No. 77: 365 pp., 66 pls.

Chaetoceros borealis

Brunel, Jules, 1962
Le phytoplancton de la Baie des Chaleurs. Contrib. Ministère de la Chasse et des Pêcheries; Province de Québec, No. 91: 365 pp.

Chaetoceros borealis

Cleve-Euler, A., 1951
Die Diatomeen von Schweden und Finnland. Kungl. Svenska Vetenskaps Akad. Handl., Fjärde Ser. 2(1): 161 pp., 6 pls.

Chaetoceros borealis

Fraser, J.H., 1949
Plankton of the Faroe-Shetland Channel and the Faroes, June and August 1947. Ann. Biol., Int. Cons., 4:27-28, text fig. 10.

Chaetoceros borealis

Fraser, J.H., 1949
Plankton investigations from the Scottish Research Vessel. Ann. Biol., Int. Cons., 4: 66-67.

Chaetoceros boreale

Gran, H.H., 1908
Diatomeen. Nordisches Plankton, Botanischer Teil pp. XIX.1-XIX 146; 178 text figs.

Chaetoceros boreale

Gran, H.H., 1897
Protophyta: Diatomaceae, Silico-flagellata and Cilioflagellata. Den Norske Nordhavs Expedition 1876-1878, h. 24, 36 pp., 4 pls.

Chaetoceros borealis

Gran, H.H., and T. Braarud, 1935
A quantitative study of the phytoplankton in the Bay of Fundy and the Gulf of Maine (including observations on hydrography, chemistry, and turbidity). J. Biol. Bd., Canada, 1(5):279-467, 69 text figs.

Chaetoceros boreale

Hendy, N. Ingram, 1964
An introductory account of the smaller algae of British coastal waters. V. Bacillariophyceae (Diatoms). Her Majesty's Stationary Office, 317 pp., 45 pls.

Chaetoceros boreale

Hendey, N.I., 1937
The plankton diatoms of the southern seas. Discovery Repts. 16:151-364, pls.6-13.

Chaetoceros boreale

Iselin, C., 1930
A report on the coastal waters of Labrador based on explorations of the "Chance" during the summer of 1926. Proc. Am. Acad. Arts Sci., 66(1):1-37, 14 text figs.

Chaetoceros borealis

Jorgensen, E., 1900
Protophyten und Protozoën im Plankton aus der Norwegischen Westkerste. Bergens Mus. Aarb. 1899(6): 95 pp., 5 pls., 83 tables.

Chaetoceros borealis

Lillick, L.C., 1940
Phytoplankton and planktonic protozoa of the offshore waters of the Gulf of Maine. Pt.II. Qualitative Composition of the Planktonic Flora. Trans. Am. Phil. Soc., n.s., 31(3):193-237, 13 text figs.

Chaetoceros borealis

Lillick, L.C., 1938
Preliminary report of the phytoplankton of the Gulf of Maine. Am. Mid. Nat. 20(3):624-640, 1 text figs 37 tables.

Chaetoceros borealis

Lillick, L.C., 1937
Seasonal studies of the phytoplankton off Woods Hole, Massachusetts. Biol. Bull. LXXIII (3):488-503, 3 text figs.

Chaetoceros boreale

Meunier, A., 1913
Microplancton de la Mer Flamande. 1. Chaetoceros. Mem. Mus. Roy. Hist. Nat., Belgique 7(2):1-55, 7 pls.

Chaetoceros borealis

Pettersson, H., 1934
Scattering and extinction of light in sea-water. Meddelanden från Göteborgs Högskolas Oceanografiska Institut. 9 (Göteborgs Kungl. vetenskaps-och vitterhetssamhälles Handlingar. Femte Foljden, Ser.B, 4(4)):1-16.

Chaetoceros boreale

Schröder, B., 1900
Phytoplankton des Golfes von Neapel nebst vergleichenden Ausblicken auf das atlantischen Ozean. Mitt. Zool. Stat. Neapel, 14:1-38.

Chaetoceros boreale

Schulz, B., and A. Wulff, 1929
Hydrographie und Oberflächen plankton des westlichen Barentsmeeres im Sommer 1927. Ber. deutschen wissensch. Komm. F. Meeresforsch. n.s. 4(5):232-372, 13 tables, 25 text figs.

Chaetoceros borealis var. concavicornis

Lillick, L.C., 1938
Preliminary report of the phytoplankton of the Gulf of Maine. Am. Mid. Nat. 20(3):624-640, 1 text figs 37 tables.

Chaetoceros borealis var. concavicornis

Lillick, L.C., 1937
Seasonal studies of the phytoplankton off Woods Hole, Massachusetts. Biol. Bull. LXXIII (3):488-503, 3 text figs.

Chaetoceros borgei

Cleve-Euler, A., 1951
Die Diatomeen von Schweden und Finnland. Kungl. Svenska Vetenskaps Akad. Handl., Fjärde Ser. 2(1): 161 pp., 6 pls.

Chaetoceros borgei

Michailova, N.F., 1964.
On the spreading of the habitat into the Black Sea of species of the genus Chaetoceros of northern seas and their biogeography. (In Russian). Trudy Sevastopol Biol. Sta., 7:231-248.

Chaetoceros bottnicus

Jorgensen, E., 1900
Protophyten und Protozoën im Plankton aus der Norwegischen Westkerste. Bergens Mus. Aarb. 1899(6): 95 pp., 5 pls., 83 tables

Chaetoceros bottnicus

Levander, K.M., 1947
Plankton gesammelt in den Jahren 1899-1910 an den Küsten Finnlands. Finnländische Hydrographisch-Biologische Untersuchungen (aus dem Wasserbiologischen Laboratorin der Societas Scientiarum Fennica) No.11:40 pp., 6 diagrams, 13 pls., tables.

Chaetoceros breve

Allen, W.E., and E.E. Cupp, 1935
Plankton diatoms of the Java Sea. Annales du Jardin Botanique de Buitenzorg XLIV (2):101-174, figs.1-127.

(drawings of all species mentioned)

Chaetoceros brevis

Brunel, J., 1962
Le phytoplancton de la Baie de Chaleurs. Inst. Botan., Univ. Montréal, Contrib. No. 77: 365 pp., 66 pls.

Chaetoceros brevis

Brunel, Jules, 1962
Le phytoplancton de la Baie des Chaleurs. Contrib. Ministère de la Chasse et des Pêcheries; Province de Québec, No. 91: 365 pp.

Chaetoceros brevis

Cleve-Euler, A., 1951
Die Diatomeen von Schweden und Finnland. Kungl. Svenska Vetenskaps Akad. Handl., Fjärde Ser. 2(1): 161 pp., 6 pls.

Chaetoceros brevis

Cupp, Easter E., 1943
Marine plankton diatoms of the west coast of North America. Bull. S.I.O. 5(1):1-238, 5 pls., 168 text figs.

Chaetoceros brevis

Delegazione Italiana della Commissione Internazionale per l'Esplorazione Scientifica del Mediterraneo, 1941
Note sul plancton della Laguna veneta. [Memoria CCLXXXIX], Arch. di Ocean. e Limn. Anno I, Fasc. I, 1941 XIX: 31-57 pp.

Chaetoceros brevis
de Sousa e Silva, E., 1956.
Contribution à l'étude du microplancton de Dakar et des regions maritimes voisines.
Bull. I.F.A.N., 8(2):335-371, 7 pls.

Chaetoceros brevis
Ercegovic, A., 1940
Weitere Untersuchungen über einige hydrographische Verhältnisse und über die Phytoplanktonproduktion in den Gewässern der Östlichen Mitteladria. Acta Adriatica 2(3):95-134, 8 text figs.

Chaetoceros brevis
Eskinazi Enide e Shigekatsv Satō (1963-1964) 1966.
Contribuição ao estudo das diatomaceas da Praia de Piedade.
Trabhs Inst. Oceanogr., Univ. Recife, 5 (5/6): 73-114.

Chaetoceros brevis
Ghazzawi, F.M., 1939
Plankton of the Egyptian waters. A study of the Suez Canal Plankton. (A) Phytoplankton. Preliminary Report 83 pp. Notes and Memoires, Min. Commerce-Industry, Egypt, Hydrobiol. & Fish. 65 figs.

Chaetoceros breve
Gran, H.H., 1908
Diatomeen. Nordisches Plankton, Botanischer Teil pp. XIX.1-XIX 146; 178 text figs.

Chaetoceros brevis
Gran, H.H., and T. Braarud, 1935
A quantitative study of the phytoplankton in the Bay of Fundy and the Gulf of Maine (including observations on hydrography, chemistry, and turbidity).
J. Biol. Bd., Canada, 1(5):279-467, 69 text figs.

Chaetoceros breve
Hendy, N. Ingram, 1964
An introductory account of the smaller algae of British coastal waters. V. Bacillariophyceae (Diatoms).
Her Majesty's Stationary Office, 317 pp., 45 pls.

Chaetoceros breve
Hendey, N.I., 1937
The plankton diatoms of the southern seas.
Discovery Repts. 16:151-364, pls.6-13.

Chaetoceros brevis
Jorgensen, E., 1900
Protophyten und Protozoen im Plankton aus der Norwegischen Westkerste. Bergens Mus. Aarb. 1899(6): 95 pp., 5 pls., 83 tables.

Chaetoceros brevis
Lillick, L.C., 1940
Phytoplankton and planktonic protozoa of the offshore waters of the Gulf of Maine. Pt.II. Qualitative Composition of the Planktonic Flora. Trans. Am. Phil. Soc., n.s., 31(3):193-237, 13 text figs.

Chaetoceros brevis
Morse, D.C., 1947
Some observations on seasonal variations in plankton population Patuxant River, Maryland 1943-1945. Bd. Nat. Res., Publ. No.65, Chesapeake Biol. Lab., 31, 3 figs.

Chaetoceros brevis
Pettersson, H., 1934
Scattering and extinction of light in sea-water. Meddelanden från Göteborgs Högskolas Oceanografiska Institut. 9 (Göteborgs Kungl. vetenskaps-och vitterhetssamhälles Handlingar. Femte Foljden, Ser.B, 4(4):1-16.

Chaetoceros brevis
Phifer, L.D. (undated)
The occurrence and distribution of plankton diatoms in Bering Sea and Bering Strait, July 26-August 24, 1934. Report of Oceanographic cruise of U.S. Coast Guard Cutter Chelan 1934, Part II(A):1-44 (mimeographed) plus fig.1 (after Pt.B)

Chaetoceros brevis
Steemann-Nielsen, Einar, 1951
The marine vegetation of the Isefjord. A study on ecology and production. Medd. Komm. Danmarks Fiskeri-og Havundersøgelser. Ser. Plankton. 5(4); 114pp., 46 text figs.

Chaetoceros brevis
Vives, F., and A. Planas, 1952.
Plancton recogido por los laboratorios costeros. VI. Fitoplancton de las costas de Vinaroz, islas Columbretes y alrededores de la desembocaadura del Elroa. Publ. Inst. Biol. Aplic. 11:141-156, 19 textfigs.

Chaetoceros Brightwellii
Gran, H.H., 1897
Protophyta: Diatomaceae, Silico-flagellata and Cilioflagellata. Den Norske Nordhavs Expedition 1876-1878, h. 24, 36 pp., 4 pls.

Chaetoceros calcitrans
Takano, Hideaki, 1971.
Citation of the author's name on Chaetoceros calcitrans. Bull. Plankt. Soc. Japan 18(1): 103. (In Japanese and English).

chaetoceros calcitrans
Takano, Hideaki, 1968.
On the diatom Chaetoceros calcitrans (Paulsen) emend. and its dwarf form pumilus forma nov.
Bull. Tokai reg. Fish. Res. Lab., 55:1-7.

Chaetoceros calcitrans
Takeda, Keiji, 1970.
Culture experiments on the relative growth of a marine centric diatom, Chaetoceros calcitrans Takano, in various concentrations of nitrate nitrogen. (In Japanese; English abstract).
Bull. Plankt. Soc., Japan, 17(1): 11-19.

Chaetoceros calcitrans
Umebayashi, Osamu, 1972.
A technique for identifying a diatom, Chaetoceros calcitrans. Bull. Tokai reg. Fish. Res. Lab. 69: 63-64.

Chaetoceros calcitrans
Yoo, Sung Kyoo, 1968.
Studies on the growth of algal food, Cyclotella nana, Chaetoceros calcitrans and Monochrysis lutheri. (In Korean; English abstract).
Bull. Pusan Fish. Coll. (Nat. Sci.), 8(2):123-126.

Chaetoceros calcitrans pumilus
Takeda, Kieji, 1972
The growth of a marine centric diatom, Chaetoceros calcitrans f. pumilus Takano, in various concentration of iron. Bull. Plankt Soc. Japan 19(1): 34-41. (In Japanese; English abstract)

Chaetoceros calcitrans pumilis
Takeda, Keiji. 1972.
Relative cell growth of a marine diatom, Chaetoceros calcitrans f. pumilis Takano in media containing various concentration of Cobaltous Chloride. (In Japanese; English abstract).
Bull. Jap. Soc. scient. Fish. 38(5): 451-455

Chaetoceros calcitrans pumilus
Takeda, Keiji, 1970.
Relative growth of a marine centric diatom Chaetoceros calcitrans f. pumilus (Paulsen) Takano in media containing various concentration of manganese. Bull. plankt. Soc., Japan, 17(2): 77-83.

Chaetoceros capense
Boden, B.P., 1950
Some marine plankton diatoms from the west coast of South Africa. Trans. R.Soc. S. Africa. 32:321-434, 100 text figs.

Chaetoceros capense
Hendey, N.I., 1937
The plankton diatoms of the southern seas.
Discovery Repts. 16:151-364, pls.6-13.

Chaetoceros ceratospermum
Meunier, A., 1913
Microplancton de la Mer Flamande. 1. Chaetoceros. Mem. Mus. Roy. Hist. Nat., Belgique 7(2):1-55, 7 pls.

Chaetoceros ceratospermum major
Meunier, A., 1913
Microplancton de la Mer Flamande. 1. Chaetoceros. Mem. Mus. Roy. Hist. Nat., Belgique 7(2):1-55, 7 pls.

Chaetoceros ceratospermum minor
Meunier, A., 1913
Microplancton de la Mer Flamande. 1. Chaetoceros. Mem. Mus. Roy. Hist. Nat., Belgique 7(2):1-55, 7 pls.

Chaetoceros ceratosporus
Brunel, J., 1962
Le phytoplancton de la Baie de Chaleurs.
Inst. Botan., Univ. Montréal, Contrib. No. 77: 365 pp., 66 pls.

Chaetoceros ceratosporus
Cleve-Euler, A., 1951
Die Diatomeen von Schweden und Finnland. Kungl. Svenska Vetenskaps Akad. Handl., Fjärde Ser. 2(1): 161 pp., 6 pls.

Chaetoceros ceratosporus
Gran, H.H., and T. Braarud, 1935
A quantitative study of the phytoplankton in the Bay of Fundy and the Gulf of Maine (including observations on hydrography, chemistry, and turbidity).
J. Biol. Bd., Canada, 1(5):279-467, 69 text figs.

Chaetoceros ceratosporum
Hendy, N. Ingram, 1964
An introductory account of the smaller algae of British coastal waters. V. Bacillariophyceae (Diatoms).
Her Majesty's Stationary Office, 317 pp., 45 pls.

Chaetoceros ceratosporum
Lillick, L.C., 1940
Phytoplankton and planktonic protozoa of the offshore waters of the Gulf of Maine. Pt.II. Qualitative Composition of the Planktonic Flora. Trans. Am. Phil. Soc., n.s., 31(3):193-237, 13 text figs.

Chaetoceros ceratosporus
Morse, D.C., 1947
Some observations on seasonal variations in plankton population Patuxant River, Maryland 1943-1945. Bd. Nat. Res., Publ. No.65, Chesapeake Biol. Lab., 31, 3 figs.

Chaetoceros ceratosporus
Steemann-Nielsen, Einar, 1951
The marine vegetation of the Isefjord. A study on ecology and production. Medd. Komm. Danmarks Fiskeri-og Havundersøgelser. Ser. Plankton. 5(4); 114pp., 46 text figs.

Chaetoceros chilensis
Krasske, G., 1941
Die Kieselalgen des chilenischen Küstenplankton (Aus dem südchilenischen Kusten gebiet IX). Arch. Hydrobiol. 38 (2):260-287.

Chaetoceros chuni
Boden, B. P., 1949.
The diatoms collected by the U.S.S. CACOPAN in the Antarctic in 1947. J. Mar. Res. 8(1):6-13, 3 textfigs.

Chaetoceros chuni
Boden, Brian, 1948
Marine plankton diatoms on operation HIGHJUMP in: Some oceanographic observations on operation HIGHJUMP. By R.S. Dietz. USNEL Rept. No.55, 97 pp., 41 figs. 7 July 1948.

Chaetoceros Chunii
Hendey, N.I., 1937
The plankton diatoms of the southern seas. Discovery Repts. 16:151-364, pls.6-13.

Chaetoceros cinctus
Braarud, T., 1945
A phytoplankton survey of the polluted waters of inner Oslo Fjord. Hvalrådets Skrifter, No.28, 142 pp., 19 text figs., 17 tables.

Chaetoceros cinctus
Brunel, J., 1962
Le phytoplancton de la Baie de Chaleurs. Inst. Botan., Univ. Montréal, Contrib. No. 77: 365 pp., 66 pls.

Chaetoceros cinctus
Cleve-Euler, A., 1951
Die Diatomeen von Schweden und Finnland. Kungl. Svenska Vetenskaps Akad. Handl., Fjärde Ser. 2(1): 161 pp., 6 pls.

Chaetoceros cinctus
Cupp, Easter E., 1943
Marine plankton diatoms of the west coast of North America. Bull. S.I.O. 5(1):1-238, 5 pls., 168 text figs.

Chaetoceros cinctum
Gran, H.H., 1908
Diatomeen. Nordisches Plankton, Botanischer Teil pp. XIX.1-XIX 146; 178 text figs.

Chaetoceros cinctum n.sp.
Gran, H.H., 1897
Protophyta: Diatomaceae, Silico-flagellata and Cilioflagellata. Den Norske Nordhavs Expedition 1876-1878, h. 24, 36 pp., 4 pls.

Chaetoceros cinctus
Gran, H.H., and T. Braarud, 1935
A quantitative study of the phytoplankton in the Bay of Fundy and the Gulf of Maine (including observations on hydrography, chemistry, and turbidity). J. Biol. Bd., Canada, 1(5):279-467, 69 text figs.

Chaetoceros cinctum
Hendy, N. Ingram, 1964
An introductory account of the smaller algae of British coastal waters. V. Bacillariophyceae (Diatoms). Her Majesty's Stationary Office, 317 pp., 45 pls.

Chaetoceros cinctus
Lillick, L.C., 1940
Phytoplankton and planktonic protozoa of the offshore waters of the Gulf of Maine. Pt.II. Qualitative Composition of the Planktonic Flora. Trans. Am. Phil. Soc., n.s., 31(3):193-237, 13 text figs.

Chaetoceros cinctus var. hirtus
Krasske, G., 1941
Die Kieselalgen des chilenischen Küstenplankton (Aus dem südchilenischen Kusten gebiet IX). Arch. Hydrobiol. 38 (2):260-287.

Chaetoceros cinctum
Meunier, A., 1913
Microplancton de la Mer Flamande. 1. Chaetoceros. Mem. Mus. Roy. Hist. Nat., Belgique 7(2):1-55, 7 pls.

Chaetoceros coarctatus
Allen, W.E., 1937
Plankton diatoms of the Gulf of California obtained by the G. Allan Hancock Expedition of 1936. The Hancock Pacific Expeditions, Univ. So. Calif. Publ. 3:47-59, 1 fig.

Chaetoceros coarctatum
Allen, W.E., and E.E. Cupp, 1935
Plankton diatoms of the Java Sea. Annales du Jardin Botanique de Buitenzorg XLIV (2):101-174, figs.1-127.
(drawings of all species mentioned)

Chaetoceros coarctatus
Brunel, J., 1962
Le phytoplancton de la Baie de Chaleurs. Inst. Botan., Univ. Montréal, Contrib. No. 77: 365 pp., 66 pls.

Chaetoceros coarctatus
Cupp, Easter E., 1943
Marine plankton diatoms of the west coast of North America. Bull. S.I.O. 5(1):1-238, 5 pls., 168 text figs.

Chaetoceros coarctatus
Cupp, E.E. and Allen, W.E., 1938
Plankton diatoms of the Gulf of California obtained by Allan Hancock Pacific Expedition of 1937. The Hancock Pacific Expeditions, The Univ. So. Calif. Publ. 3: 61-74, 1 map, pls.4-15.

Chaetoceros coarctatum
Delsman, H. C., 1939.
Preliminary plankton investigations in the Java Sea. Treubia, 17:139-181, 8 maps, 41 figs.

Chaetoceros coarctatus
Eskinazi Enide e Shigekatsv Satô (1963-1964) 1966.
Contribuição ao estudo das diatomaceas da Praia de Piedade.
Trabhs Inst. Oceanogr., Univ. Recife, 5 (5/6): 73-114.

Chaetoceros coarctatum
Gran, H.H., 1908
Diatomeen. Nordisches Plankton, Botanischer Teil pp. XIX.1-XIX 146; 178 text figs.

Chaetoceros coarctatum
Hendy, N. Ingram, 1964
An introductory account of the smaller algae of British coastal waters. V. Bacillariophyceae (Diatoms). Her Majesty's Stationary Office, 317 pp., 45 pls.

Chaetoceros coarctatum
Hendey, N.I., 1937
The plankton diatoms of the southern seas. Discovery Repts. 16:151-364, pls.6-13.

Chaetoceros coarctatus
Kokubo, S., 1952
Results of the observations on the plankton and oceanography of Mutsu Bay during 1950, reference being made also to the period 1946-1950. Bull Mar.Biol.Sta., Asamushi 5(1/4): 1-54, 3 tables,(fold-in), 1 fold-in.

Chaetoceros coarctatus
Mann, A., 1907
Report on the diatoms of the Albatross voyages in the Pacific Ocean, 1888-1904. Contrib. U. S. Nat. Herb. 10(5):221-419, Pls. XLIV-LIV.

Chaetoceros coarctata
Mann, A., 1893
List of Diatomaceae from a deep-sea dredging in the Atlantic Ocean off Delaware Bay by the U. S. Fish Commission Steamer Albatross. Proc. U. S. Nat. Mus. 16:303-312.

Chaetoceros coarctatum
Marukawa, H., 1921
Plankton lists and some new species of copepods, from the northern waters of Japan. Bull. Inst. Ocean., No.384, 15 pp., 3 pls., 1 chart. Monaco

Chaetoceros compactum
Kokubo, S., and S. Sato., 1947
Plankters in Jū-San Gata. Physiol. and Ecol. (Japan) 1(4):1-16, 3 text figs., tables.

Chaetoceros compressus
Allen, W. E., 1938
The Templeton Crocker Expedition to the Gulf of California in 1935 - The Phytoplankton. Amer. Microsc. Soc., Trans. 57:328-335.

Chaetoceros compressus
Allen, W.E., 1937
Plankton diatoms of the Gulf of California obtained by the G. Allan Hancock Expedition of 1936. The Hancock Pacific Expeditions, Univ. So. Calif. Publ. 3:47-59, 1 fig.

Chaetoceros compressus
Allen, W. E., 1936
Occurrence of marine plankton diatoms in a ten-year series of daily catches in southern California. Am. Jour. Bot. 23(1):60-63.

Chaetoceros compressum
Allen, W.E., 1928.
Catches of marine diatoms and dinoflagellates taken by boat in Southern California waters in 1926. Bull. S.I.O., tech. ser., 1:201-246, 6 textfigs.

Chaetoceros compressum
Allen, W.E., 1927.
Quantitative studies on inshore marine diatoms and dinoflagellates of Southern California in 1922. Bull. S.I.O., tech. ser., 1:31-38, 2 text-figs.

Chaetoceros compressum
Allen, W. E., 1925
Statistical Studies of surface catches of marine diatoms and dinoflagellates made by the Yacht "Ohio" in tropical waters in 1924. Jan. Trans. Amer. Microscop. Soc.:24-30, 1 fig.

Chaetoceros compressum
Allen, W.E., and E.E. Cupp, 1935
Plankton diatoms of the Java Sea. Annales du Jardin Botanique de Buitenzorg XLIV (2):101-174, figs.1-127.
(drawings of all species mentioned)

Chaetoceros compressus
Braarud, T., 1945
Experimental studies on marine plankton diatoms. Avhandlingar utgitt av Det Norske Videnskaps-Akademi i Oslo. 1. Mat.-Naturv. Klasse 1944. No.10, 1-16, 1 text fig.

Chaetoceros compressus
Braarud, T., 1945
A phytoplankton survey of the polluted waters of inner Oslo Fjord. Hvalrådets Skrifter, No.28, 142 pp., 19 text figs., 17 tables.

Chaetoceros compressus
Braarud, T., and Adam Bursa, 1939
On the phytoplankton of the Oslo Fjord, 1933-1934. Hvalrådets Skr. No.19:1-63; 9 text figs. Reviewed. J. du. Cons. 14(3): 418-420. A.C. Gardiner.

Chaetoceros compressus
Brunel, J., 1962
Le phytoplancton de la Baie de Chaleurs. Inst. Botan., Univ. Montréal, Contrib. No. 77: 365 pp., 66 pls.

Chaetoceros compressus
Cleve-Euler, A., 1951
Die Diatomeen von Schweden und Finnland. Kungl. Svenska Vetenskaps Akad. Handl., Fjärde Ser. 2(1): 161 pp., 6 pls.

Chaetoceros compressus
Brunel, Jules, 1962
Le phytoplancton de la Baie des Chaleurs. Contrib. Ministère de la Chasse et des Pêcheries, Province de Québec, No. 91: 365 pp.

Chaetoceros compressus
Cupp, Easter E., 1943
Marine plankton diatoms of the west coast of North America. Bull. S.I.O. 5(1):1-238, 5 pls., 168 text figs.

Chaetoceros compressus
Cupp, E.E., 1934
Analysis of marine diatom collections taken from the Canal Zone to California during March, 1933. Trans. Am. Micros. Soc. LIII (1):22-29, 1 map.

Chaetoceros compressus
Cupp, E.E. and Allen, W.E., 1938
Plankton diatoms of the Gulf of California obtained by Allan Hancock Pacific Expedition of 1937. The Hancock Pacific Expeditions, The Univ. So. Calif. Publ. 3: 61-74, 1 map, pls.4-15.

Chaetoceros compressus
Cushing, D.H., 1963
Studies on a Calanus patch. II. The estimation of algal productive rates. J. Mar. Biol. Assoc. U.K., 43(2):339-347.

Chaetoceros compressum
Dangeard, P., 1927
Phytoplankton de la croisière du "Sylvana". Ann. Inst. Ocean., Monaco, n.s., 4(8):286-401, 54 text figs. (Feirer-Juin 1913).

Chaetoceros compressum
Davidson, V.M., 1931.
Biological and oceanographic conditions in Hudson Bay. Contr. Canadian Biol. Fish., n.s., 6(26): 497-509, 7 textfigs.

Chaetoceros compressus
Ercegovic, A., 1936
Etudes qualitative et quantitatives du phytoplankton dans les eaux côtières de l'Adriatique oriental moyen au cours de l'année 1934. Acta Adriatica 1(9):1-126

Chaetoceros compressus
Eskinazi Enide e Shigekatsv Satô (1963-1964) 1966.
Contribuição ao estudo das diatomaceas da Praia de Piedade.
Trabhs Inst. Oceanogr., Univ. Recife, 5 (5/6): 73-114.

Chaetoceros compressus
Frenguelli, Joaquin, and Hector Antonio Orlando, 1959.
Operacion MERLUZA. Diatomeas y silicoflagelados del plancton del "VI Crucero". Servicio Hidrogr. Naval., Argentina, Publ. No. H. 619: 5-62.

Chaetoceros compressus
Gilbert, J.Y., and W.E. Allen, 1943
The phytoplankton of the Gulf of California obtained by the "E.W. Scripps" in 1939 and 1940. J. Mar. Res. V(2):89-110, figs.30-31.

Chaetoceros compressus
Gran, H.H., and T. Braarud, 1935
A quantitative study of the phytoplankton in the Bay of Fundy and the Gulf of Maine (including observations on hydrography, chemistry, and turbidity). J. Biol. Bd., Canada, 1(5):279-467, 69 text figs.

Chaetoceros compressus
Gran, H. H. and E. C. Angst, 1931
Plankton diatoms of Puget Sound. Publ. Puget Sound Biol. Sta. 7:417-519, 95 text figs.

Chaetoceros compressum
Hendy, N. Ingram, 1964
An introductory account of the smaller algae of British coastal waters. V. Bacillariophyceae (Diatoms). Her Majesty's Stationary Office, 317 pp., 45 pls.

Chaetoceros compressus
Hendey, N.I., 1937
The plankton diatoms of the southern seas. Discovery Repts. 16:151-364, pls.6-13.

Chaetoceros compressus
Kokubo, S., 1952
Results of the observations on the plankton and oceanography of Mutsu Bay during 1950, reference being made also to the period 1946-1950. Bull Mar.Biol.Sta., Asamushi 5(1/4): 1-54, 3 tables.(fold-in), 1 fold-in.

Chaetoceros compressus
Lillick, L.C., 1940
Phytoplankton and planktonic protozoa of the offshore waters of the Gulf of Maine. Pt.II. Qualitative Composition of the Planktonic Flora. Trans. Am. Phil. Soc., n.s., 31(3):193-237, 13 text figs.

Chaetoceros compressus
Lillick, L.C., 1938
Preliminary report of the phytoplankton of the Gulf of Maine. Am. Mid. Nat. 20(3):624-640, 1 text fig. 37 tables.

Chaetoceros compressus
Lillick, L.C., 1937
Seasonal studies of the phytoplankton off Woods Hole, Massachusetts. Biol. Bull. LXXIII (3):488-503, 3 text figs.

Chaetoceros compressus
Michailova, N.F., 1964.
On the spreading of the habitat into the Black Sea of species of the genus Chaetoceros of northern seas and their biogeography. (In Russian).
Trudy Sevastopol Biol. Sta., 7:231-248.

Chaetoceros compressum
Moberg, E.G., and W. E. Allen, 1927.
Effect of tidal changes on physical, chemical, and biological conditions of sea water in the San Diego region, 1. Observations on the effect of tidal changes on physical and chemical conditions of sea water in the San Diego region. 2. Half-hourly collections of marine microplankton taken at the Scripps Institution pier in 1923. Bull. S.I.O., tech. ser., 1:1-17, 4 textfigs.

Chaetoceros compressum
Pavillard, J., 1925
Bacillariales. Rept. on the Danish Oceangr. Exped., 1908-10 to the Mediterranean and adj. seas. Vol.II., Biol. J4:72 pp., 116 text figs.

Chaetoceros compressus
Phifer, L.D. (undated)
The occurrence and distribution of plankton diatoms in Bering Sea and Bering Strait, July 26-August 24, 1934. Report of Oceanographic cruise of U.S. Coast Guard Cutter Chelan 1934, Part II(A):1-44 (mimeographed) plus fig.1 (after Pt.B)

Chaetoceros compressus
Sargent, M. S. and T. J. Walker, 1948
Diatom populations associated with eddies off southern California in 1941. J. Mar. Res. 7(3):490-505, 15 text figs.

Chaetoceros compressus (figs.)
Sousa e Silva, E., 1949
Diatomaceas e Dinoflagelados de Baia de Cascais. Portugaliae Acta Biol., Volume: Julio Henriques, Ser. B: 300-383, 9 pls, 2 fold-in tables.

Chaetoceros compressus
Steemann-Nielsen, Einar, 1951
The marine vegetation of the Isefjord. A study on ecology and production. Medd. Komm. Danmarks Fiskeri-og Havundersøgelser. Ser. Plankton. 5(4); 114pp., 46 text figs.

Chaetoceros compressus
Uyeno, Fukuzo, 1961
Oceanographical and ecological studies on primary production of the sea, with special references to relationship between diatom production and temperature and chlorinity of water.
Rept., Fac. Fish., Pref. Univ., Mie, 4(1): 1-64.

Chaetoceros compressus
Yamazi, I., 1951.
Plankton investigations in inlet waters along the coast of Japan. II. The plankton in Hakodate Harbour and Yoichi Inlet in Hokkaido. Publ. Seto Mar. Biol. Sta., Kyoto Univ., 1(4): 185-194, 3 textfigs.

Chaetoceros concavicornis
Allen, W. E., 1938
The Templeton Crocker Expedition to the Gulf of California in 1935 - The Phytoplankton. Amer. Microsc. Soc., Trans. 57:328-335.

Chaetoceros concavicornis
Allen, W.E., 1937
Plankton diatoms of the Gulf of California obtained by the G. Allan Hancock Expedition of 1936. The Hancock Pacific Expeditions, Univ. So. Calif. Publ. 3:47-59, 1 fig.

Chaetoceros concavicornis
Brunel, J., 1962
Le phytoplancton de la Baie de Chaleurs. Inst. Botan., Univ. Montréal, Contrib. No. 77: 365 pp., 66 pls.

Chaetoceros concavicornis
Brunel, Jules, 1962
Le phytoplancton de la Baie des Chaleurs. Contrib. Ministère de la Chasse et des Pêcheries, Province de Québec, No. 91: 365 pp.

Chaetoceros concavicornis
Cleve-Euler, A., 1951
Die Diatomeen von Schweden und Finnland. Kungl. Svenska Vetenskaps Akad. Handl., Fjärde Ser. 2(1): 161 pp., 6 pls.

Chaetoceros concavicornis
Cupp, Easter E., 1943
Marine plankton diatoms of the west coast of North America. Bull. S.I.O. 5(1):1-238, 5 pls., 168 text figs.

Chaetoceros concavicornis
Cupp, E.E. and Allen, W.E., 1938
Plankton diatoms of the Gulf of California obtained by Allan Hancock Pacific Expedition of 1937. The Hancock Pacific Expeditions, The Univ. So. Calif. Publ. 3: 61-74, 1 map, pls.4-15.

Chaetoceros concavicornis
Frenguelli, Joaquin, and Hector Antonio Orlando, 1959.
Operacion MERLUZA. Diatomeas y silico-flagelados del plancton del "VI Crucero". Servicio Hidrogr. Naval., Argentina, Publ. No. H. 619: 5-62.

Chaetoceros concavicornis
Gran, H. H. and E. C. Angst, 1931
Plankton diatoms of Puget Sound. Publ. Puget Sound Biol. Sta. 7:417-519, 95 text figs.

Chaetoceros concavicornis
Gran, H.H., and T. Braarud, 1935
A quantitative study of the phytoplankton in the Bay of Fundy and the Gulf of Maine (including observations on hydrography, chemistry, and turbidity). J. Biol. Bd., Canada, 1(5):279-467, 69 text figs.

Chaetoceros concavicorne
Hendy, N. Ingram, 1964
An introductory account of the smaller algae of British coastal waters. V. Bacillariophyceae (Diatoms).
Her Majesty's Stationary Office, 317 pp., 45 pls.

Chaetoceros concavicorni-s
Kokubo, S., 1952
Results of the observations on the plankton and oceanography of Mutsu Bay during 1950, reference being made also to the period 1946-1950. Bull Mar.Biol.Sta., Asamushi 5(1/4): 1-54, 3 tables,(fold-in), 1 fold-in.

Chaetoceros concavicornis
Lillick, L.C., 1940
Phytoplankton and planktonic protozoa of the offshore waters of the Gulf of Maine. Pt.II. Qualitative Composition of the Planktonic Flora. Trans. Am. Phil. Soc., n.s., 31(3):193-237, 13 text figs.

Chaetoceros concavicornis
Manguin, E., 1954
Diatomees marines provenant de l'ile Heard (Australian National Antarctic Expedition). Rev. Algol., n.s., 1: 14-24.

Chaetoceros concavicornis volans
Cupp, Easter E., 1943
Marine plankton diatoms of the west coast of North America. Bull. S.I.O. 5(1):1-238, 5 pls., 168 text figs.

Chaetoceros confertus n.sp.
Müller Melchers, F.C., 1953.
New and little known diatoms from Uruguay and the South Atlantic coast. Com. Bot., Mus. Hist. Nat., Montevideo, 3(30): 1-11, 8 pls.

Chaetoceros constrictum
Allen, W. E., 1938
The Templeton Crocker Expedition to the Gulf of California in 1935 - The Phytoplankton. Amer. Microsc. Soc., Trans. 57:328-335.

Chaetoceros constrictus
Allen, W.E., 1937
Plankton diatoms of the Gulf of California obtained by the G. Allan Hancock Expedition of 1936. The Hancock Pacific Expeditions, Univ. So. Calif. Publ. 3:47-59, 1 fig.

Chaetoceros constrictum
Allen, W.E., and E.E. Cupp, 1935
Plankton diatoms of the Java Sea. Annales du Jardin Botanique de Buitenzorg XLIV (2):101-174, figs.1-127.
(drawings of all species mentioned)

Chaetoceros constrictus
Anon., 1951
Bulletin of the Marine Biological Station of Asamushi 4(3/4): 15 pp.

Chaetoceros constrictum
Bigelow, H.B., and M. Leslie, 1930
Reconnaissance of the waters and plankton of Monterey Bay, July 1928. Bull. M.C.Z. 70(5):429-481, 43 text figs.

Chaetoceros constrictus
Boden, B.P., 1950
Some marine plankton diatoms from the west coast of South Africa. Trans. R.Soc. S. Africa. 32:321-434, 100 text figs.

Chaetoceros constrictus
Braarud, T., 1945
Experimental studies on marine plankton diatoms. Avhandlingar utgitt av Det Norske Videnskaps-Akademi i Oslo. 1. Mat.-Naturv. Klasse 1944. No.10, 1-16, 1 text fig.

Chaetoceros constrictus
Braarud, T., and Adam Bursa, 1939
On the phytoplankton of the Oslo Fjord, 1933-1934. Hvalrådets Skr. No.19:1-63; 9 text figs. Reviewed. J. du. Cons, 14(3): 418-420. A.C. Gardiner.

Chaetoceros constrictus
Brunel, J., 1962
Le phytoplancton de la Baie de Chaleurs. Inst. Botan., Univ. Montréal, Contrib. No. 77: 365 pp., 66 pls.

Chaetoceros constrictus
Brunel, Jules, 1962
Le phytoplancton de la Baie des Chaleurs. Contrib. Ministère de la Chasse et des Pêcheries, Province de Québec, No. 91: 365 pp.

Chaetoceros constrictus
Cleve-Euler, A., 1951
Die Diatomeen von Schweden und Finnland. Kungl. Svenska Vetenskaps Akad. Handl., Fjärde Ser. 2(1): 161 pp., 6 pls.

Chaetoceros constrictus
Cupp, Easter E., 1943
Marine plankton diatoms of the west coast of North America. Bull. S.I.O. 5(1):1-238, 5 pls., 168 text figs.

Chaetoceros constrictus
Cupp, E.E., 1934
Analysis of marine diatom collections taken from the Canal Zone to California during March, 1933. Trans. Am. Micros. Soc. LIII (1):22-29, 1 map.

Chaetoceros constrictus
Cupp, E.E. and Allen, W.E., 1938
Plankton diatoms of the Gulf of California obtained by Allan Hancock Pacific Expedition of 1937. The Hancock Pacific Expeditions, The Univ. So. Calif. Publ. 3: 61-74, 1 map, pls.4-15.

Chaetoceros constrictus
Delegazione Italiana della Commissione Internazionale per l'Esplorazione Scientifica del Mediterraneo, 1941
Note sul plancton della Laguna veneta. [Memoria CCLXXXIX], Arch. di Ocean. e Limn. Anno I, Fasc. I, 1941 XIX: 31-57 pp.

Chaetoceros constrictus
Frenguelli, Joaquin, and Hector Antonio Orlando, 1959.
Operacion MERLUZA. Diatomeas y silico-flagelados del plancton del "VI Crucero". Servicio Hidrogr. Naval., Argentina, Publ. No. H. 619: 5-62.

Chaetoceros constrictum
Gran, H.H., 1908
Diatomeen. Nordisches Plankton, Botanischer Teil pp. XIX.1-XIX 146; 178 text figs.

Chaetoceros constrictum n.sp.
Gran, H.H., 1897
Protophyta: Diatomaceae, Silico-flagellata and Cilioflagellata. Den Norske Nordhavs Expedition 1876-1878, h. 24, 36 pp., 4 pls.

Chaetoceros constrictus
Gran, H.H., and T. Braarud, 1935
A quantitative study of the phytoplankton in the Bay of Fundy and the Gulf of Maine (including observations on hydrography, chemistry, and turbidity). J. Biol. Bd., Canada, 1(5):279-467, 69 text figs.

Chaetoceros constrictum
Hendy, N. Ingram, 1964
An introductory account of the smaller algae of British coastal waters. V. Bacillariophyceae (Diatoms).
Her Majesty's Stationary Office, 317 pp., 45 pls.

Chaetoceros constrictum
Iselin, C., 1930
A report on the coastal waters of Labrador based on explorations of the "Chance" during the summer of 1926. Proc. Am. Acad. Arts Sci., 66(1):1-37, 14 text figs.

Chaetoceros constrictus
Jorgensen, E., 1900
Protophyten und Protozoën im Plankton aus der Norwegischen Westkerste. Bergens Mus. Aarb. 1899(6): 95 pp., 5 pls., 83 tables.

Chaetoceros constrictus
Kokubo, S., 1952
Results of the observations on the plankton and oceanography of Mutsu Bay during 1950, reference being made also to the period 1946-1950. Bull Mar.Biol.Sta., Asamushi 5(1/4): 1-54, 3 tables,(fold-in), 1 fold-in.

Chaetoceros constrictus
Lillick, L.C., 1940
Phytoplankton and planktonic protozoa of the offshore waters of the Gulf of Maine. Pt.II. Qualitative Composition of the Planktonic Flora. Trans. Am. Phil. Soc., n.s., 31(3):193-237, 13 text figs.

Chaetoceros constrictus
Lillick, L.C., 1938
Preliminary report of the phytoplankton of the Gulf of Maine. Am. Mid. Nat. 20(3):624-640, 1 text figs 37 tables.

Chaetoceros constrictus
Lillick, L.C., 1937
Seasonal studies of the phytoplankton off Woods Hole, Massachusetts. Biol. Bull. LXXIII (3):488-503, 3 text figs.

Chaetoceros constrictus
Mangin, M. L., 1912
Phytoplancton de la croisière du "René" dans l'Atlantique (Septembre 1908). Ann. Inst. Ocean., n.s., 4(1):1-66, 2 pls., 41 text figs., 2 tables.

Chaetoceros constrictus
Meunier, A., 1913
Microplancton de la Mer Flamande. 1. Chaetoceros. Mem. Mus. Roy. Hist. Nat., Belgique 7(2):1-55, 7 pls.

Chaetoceros constrictus
Michailova, N.P., 1964.
On the spreading of the habitat into the Black Sea of species of the genus Chaetoceros of northern seas and their biogeography. (In Russian).
Trudy Sevastopol Biol. Sta., 7:231-248.

Chaetoceros constrictus
Morse, D.C., 1947
Some observations on seasonal variations in plankton population Patuxant River, Maryland 1943-1945. Bd. Nat. Res., Publ. No.65, Chesapeake Biol. Lab., 31, 3 figs.

Chaetoceros constructus
Pettersson, H., 1934
Scattering and extinction of light in sea-water. Meddelanden från Göteborgs Högskolas Oceanografiska Institut: - 9 (Göteborgs Kungl. vetenskaps-och vitterhets-samhälles Handlingar. Femte Foljden, Ser.B, 4(4)):1-16.

Chaetoceros constrictum
Phifer, L.D. (undated)
The occurrence and distribution of plankton diatoms in Bering Sea and Bering Strait, July 26-August 24, 1934. Report of Oceanographic cruise of U.S. Coast Guard Cutter Chelan 1934, Part II(A):1-44 (mimeographed) plus fig.1 (after Pt.B)

Chaetoceros constrictus
Rampi, L., 1942
Ricerche sul fitoplancton del Mare Ligure 6. Le diatomee delle acque di Sanremo. Nuovo Giornale Botanico Italiano, N.S., 49:252-268.

Chaetoceros contortum
Bigelow, H.B., and M. Leslie, 1930
Reconnaissance of the waters and plankton of Monterey Bay, July 1928. Bull. M.C.Z., 70(5):429-481, 43 text figs.

Chaetoceros contortum
Gran, H.H., 1908
Diatomeen. Nordisches Plankton, Botanischer Teil pp. XIX.1-XIX 146; 178 text figs.

Chaetoceros contortum
Gran, H.H., 1897
Protophyta: Diatomaceae, Silico-flagellata and Cilioflagellata. Den Norske Nordhavs Expedition 1876-1878, h. 24, 36 pp., 4 pls.

Chaetoceros contortum
Iselin, C., 1930
A report on the coastal waters of Labrador based on explorations of the "Chance" during the summer of 1926. Proc. Am. Acad. Arts Sci., 66(1):1-37, 14 text figs.

Chaetoceros contortus
Jorgensen, E., 1900
Protophyten und Protozoën im Plankton aus der Norwegischen Westkerste. Bergens Mus. Aarb. 1899(6): 95 pp., 5 pls., 83 tables.

Chaetoceros contortua
Mangin, M. L., 1912
Phytoplancton de la croisière du "René" dans l'Atlantique (Septembre 1908). Ann. Inst. Ocean., n.s., 4(1):1-66, 2 pls., 41 text figs., 2 tables.

Chaetoceros contortum
Meunier, A., 1913
Microplancton de la Mer Flamande. 1. Chaetoceros. Mem. Mus. Roy. Hist. Nat., Belgique 7(2):1-55, 7 pls.

Chaetoceros contortum
Pavillard, J., 1905
Recherches sur la flore pelagique (Phytoplankton) de l'Etang de Thau. Theses presentees a la Fac. Sci., Paris, 116 pp., 3 pls.

Chaetoceros contortum
Schodduyn, M., 1926
Observations faites dans la baie d'Ambleteuse (Pas de Calais). Bull. Inst. Ocean., Monaco, No. 482: 64 pp.

Chaetoceros convexicornis
Brunel, J., 1962
Le phytoplancton de la Baie de Chaleurs. Inst. Botan., Univ. Montréal, Contrib. No. 77: 365 pp., 66 pls.

Chaetoceros convolutum
Boden, B.P., 1950
Some marine plankton diatoms from the west coast of South Africa. Trans. R.Soc. S. Africa. 32:321-434, 100 text figs.

Chaetoceros convolutus
Brunel, J., 1962
Le phytoplancton de la Baie de Chaleurs. Inst. Botan., Univ. Montréal, Contrib. No. 77: 365 pp., 66 pls.

Chaetoceros convolutus clavipes
Chaetoceros convolutus trisetosa

Chaetoceros convolutus
Brunel, Jules, 1962
Le phytoplancton de la Baie des Chaleurs. Contrib. Ministère de la Chasse et des Pêcheries, Province de Québec, No. 91: 365 pp.

Chaetoceros convolutum
Cassie, Vivienne, 1961
Marine phytoplankton in New Zealand waters. Botanica Marina, 2(Suppl.):54 pp., 8 pls.

Chaetoceros convolutum n.sp.
Castracane degli Antelminelli, F., 1886
1. Report on the Diatomaceae collected by H.M.S. Challenger during the years 1873-1876. Rept. Sci. Results, H.M.S. Challenger, Botany Vol. II, 178 pp., 30 pls.

Chaetoceros convolutus
Chiba, T., 1949
On the distribution of the plankton in the eastern China Sea and Yellow Sea. 1. Plankton composition in the spring. J. Shimonoseki Coll. Fisheries, 1(1):57-63, 1 fig.

Chaetoceros convolutus
Cleve-Euler, A., 1951
Die Diatomeen von Schweden und Finnland. Kungl. Svenska Vetenskaps Akad. Handl., Fjärde Ser. 2(1): 161 pp., 6 pls.

Chaetoceros convolutus
Cupp, E.E., 1937
Seasonal distribution and occurrence of marine diatoms and dinoflagellates at Scotch Cap, Alaska. Bull. S.I.O. Tech. ser.4(3):71-100, 7 textfigs.

Chaetoceros convolutum
Dangeard, P., 1927
Phytoplankton de la croisière du "Sylvana". Ann. Inst. Ocean., Monaco, n.s., 4(8):286-401, 54 text figs. (Feirer-Juin 1913).

Chaetoceros convolutus
Davidson, V.M., 1931.
Biological and oceanographic conditions in Hudson Bay. Contr. Canadian Biol. Fish., n.s., 6(26): 497-509, 7 textfigs.

Chaetoceros convolutus
Forti, A., 1922
Ricerche sulla flora pelagica (fitoplancton) di Quarto dei Mille. Mem. R. Com. Talass. Ital. 97:248 pp., 13 pls.

Chaetoceros convolutus
Fraser, J. H., and A. Saville, 1949
Plankton distribution in Scottish and adjacent waters in 1948. Ann. Biol. 5:61-62.

Chaetoceros convolutum
Gran, H.H., 1908
Diatomeen. Nordisches Plankton, Botanischer Teil pp. XIX.1-XIX 146; 178 text figs.

Chaetoceros convolutus
Gran, H.H., and T. Braerud, 1935
A quantitative study of the phytoplankton in the Bay of Fundy and the Gulf of Maine (including observations on hydrography, chemistry, and turbidity). J. B iol. Bd., Canada, 1(5):279-467, 69 text figs.

Chaetoceros convolutus
Gran, H. H. and E. C. Angst, 1931
Plankton diatoms of Puget Sound. Publ. Puget Sound Biol. Sta. 7:417-519, 95 text figs.

Chaetoceros convolutum
Hendy, N. Ingram, 1964
An introductory account of the smaller algae of British coastal waters. V. Bacillariophyceae (Diatoms).
Her Majesty's Stationary Office, 317 pp., 45 pls.

Chaetoceros convolutum
Hendey, N.I., 1937
The plankton diatoms of the southern seas. Discovery Repts. 16:151-364, pls.6-13.

Chaetoceros convolutus
Kokubo, S., 1952
Results of the observations on the plankton and oceanography of Mutsu Bay during 1950, reference being made also to the period 1946-1950. Bull Mar.Biol.Sta., Asamushi 5(1/4): 1-54, 3 tables,(fold-in), 1 fold-in.

Chaetoceros convolutus
Lillick, L.C., 1938
Preliminary report of the phytoplankton of the Gulf of Maine. Am. Mid. Nat. 20(3):624-640, 1 text figs 37 tables.

Chaetoceros convolutum
Meunier, A., 1913
Microplancton de la Mer Flamande. 1. Chaetoceros. Mem. Mus. Roy. Hist. Nat., Belgique 7(2):1-55, 7 pls.

Chaetoceros convolutus
Morse, D.C., 1947
Some observations on seasonal variations in plankton population Patuxant River, Maryland 1943-1945. Bd. Nat. Res., Publ. No.65, Chesapeake Biol. Lab., 31, 3 figs.

Chaetoceros convolutum
Pavillard, J., 1925
Bacillariales. Rept. on the Danish Oceangr. Exped., 1908-10 to the Mediterranean and adj. seas. Vol.II., Biol. J4:72 pp., 116 text figs.

Chaetoceros convolutus
Pettersson, H., 1934
Scattering and extinction of light in sea-water. Meddelanden från Göteborgs Högskolas Oceanografiska Institut: - 9 (Göteborgs Kungl. vetenskaps-och vitterhets-samhälles Handlingar. Femte Foljden, Ser.B, 4(4)):1-16.

Chaetoceros convolutus
Phifer, L.D. (undated)
The occurrence and distribution of plankton diatoms in Bering Sea and Bering Strait, July 26-August 24, 1934. Report of Oceanographic cruise of U.S. Coast Guard Cutter Chelan 1934, Part II(A):1-44 (mimeographed) plus fig.1 (after Pt.B)

Oceanographic Index: Marine Organisms Cumulation, 1946-1973

Chaetoceros convolutus
Rampi, L., 1942
Ricerche sul fitoplancton del Mare Ligure 6. Le diatomee delle acque di Sanremo. Nuovo Giornale Botanico Italiano, N.S., 49:252-268.

Chaetoceros convolutum
Schulz, B., and A. Wulff, 1929
Hydrographie und Oberflächen plankton des westlichen Barentsmeeres im Sommer 1927. Ber. deutschen wissensch. Komm. F. Meeres-forsch. n.s. 4(5):232-372, 13 tables, 25 text figs.

Chaetoceros convolutum
Sousa e Silva, E., 1949
Diatomaceas e Dinoflagelados de Baia de Cascais. Portugaliae Acta Biol., Volume: Julio Henriques, Ser. B: 300-383, 9 pls, 2 fold-in tables.

Chaetoceros convolutus clavipes m. f.
Paasche, E., 1961
Notes on phytoplankton from the Norwegian Sea. Botanica Marina, 2(3/4):197-210.

Chaetoceros convolutus trisetosa, forma nov.
Brunel, Jules, 1962
Le phytoplancton de la Baie des Chaleurs. Contrib. Ministère de la Chasse et des Pêcheries, Province de Québec, No. 91: 365 pp.

Chaetoceros coronatum
Bigelow, H.B., and M. Leslie, 1930
Reconnaissance of the waters and plankton of Monterey Bay, July 1928. Bull. M.C.Z., 70(5):429-481, 43 text figs.

Chaetoceros coronatus
Brunel, J., 1962
Le phytoplancton de la Baie de Chaleurs. Inst. Botan., Univ. Montréal, Contrib. No. 77: 365 pp., 66 pls.

Chaetoceros coronatus
Cleve-Euler, A., 1951
Die Diatomeen von Schweden und Finnland. Kungl. Svenska Vetenskaps Akad. Handl., Fjärde Ser. 2(1): 161 pp., 6 pls.

Chaetoceros coronatum
Gran, H.H., 1908
Diatomeen. Nordisches Plankton, Botanischer Teil pp. XIX.1-XIX 146; 178 text figs.

Chaetoceros coronatum n.sp.
Gran, H.H., 1897
Protophyta: Diatomaceae, Silico-flagellata and Cilioflagellata. Den Norske Nordhavs Expedition 1876-1878, h. 24, 36 pp., 4 pls.

Chaetoceros coronatum
Hendy, N. Ingram, 1964
An introductory account of the smaller algae of British coastal waters. V. Bacillariophyceae (Diatoms). Her Majesty's Stationary Office, 317 pp., 45 pls.

Chaetoceros coronatum
Meunier, A., 1913
Microplancton de la Mer Flamande. 1. Chaetoceros. Mem. Mus. Roy. Hist. Nat., Belgique 7(2):1-55, 7 pls.

Chaetoceros costatus
Allen, W. E., 1936
Occurrence of marine plankton diatoms in a ten-year series of daily catches in southern California. Am. Jour. Bot. 23(1):60-63.

Chaetoceros costatus
Allen, W. E., 1934
Marine plankton diatoms of Lower California in 1931. Bot. Gaz. 95(3):485-492, 1 fig.

Chaetoceros costatus
Brunel, J., 1962
Le phytoplancton de la Baie de Chaleurs. Inst. Botan., Univ. Montréal, Contrib. No. 77: 365 pp., 66 pls.

Chaetoceros costatus
Cupp, Easter E., 1943
Marine plankton diatoms of the west coast of North America. Bull. S.I.O. 5(1):1-238, 5 pls., 168 text figs.

Chaetoceros costatus
Cupp, E.E., 1934
Analysis of marine diatom collections taken from the Canal Zone to California during March, 1933. Trans. Am. Micros. Soc. LIII (1):22-29, 1 map.

Chaetoceros costatus
Cupp, E.E. and Allen, W.E., 1938
Plankton diatoms of the Gulf of California obtained by Allan Hancock Pacific Expedition of 1937. The Hancock Pacific Expeditions, The Univ. So. Calif. Publ. 3: 61-74, 1 map, pls.4-15.

Chaetoceros costatum
Glooschenko, Walter A., 1969.
Accumulation of 203Hg by the marine diatom Chaetoceros costatum. J. Phycol. 5(3): 224-226.

Chaetoceros costatus
Kokubo, S., 1952
Results of the observations on the plankton and oceanography of Mutsu Bay during 1950, reference being made also to the period 1946-1950. Bull Mar.Biol.Sta., Asamushi 5(1/4): 1-54, 3 tables, (fold-in), 1 fold-in.

Chaetoceros costatum
Pavillard, J., 1925
Bacillariales. Rept. on the Danish Oceangr. Exped., 1908-10 to the Mediterranean and adj. seas. Vol.II., Biol. J4:72 pp., 116 text figs.

Chaetoceros costatus
Rampi, L., 1942
Ricerche sul fitoplancton del Mare Ligure 6. Le diatomee delle acque di Sanremo. Nuovo Giornale Botanico Italiano, N.S., 49:252-268.

Chaetoceros costatum
Saunders, Richard P., 1967
Chaetoceros costatum Pavillard. Florida State Bd. Conserv., Mar. Lab, Leaflet Ser. 1 (Phytoplankton)(2)(3): 4 pp.

Chaetoceros costatus (fig.)
Vives, F., and A. Planas, 1952.
Plancton recogido por los laboratorios costeros. VI. Fitoplancton de las costas de Vinaroz, islas Columbretes y alrededores de la desembocadura del Elroa. Publ. Inst. Biol. Aplic. 11:141-156, 18 textfigs.

Chaetoceros crinitus
Bigelow, H.B., and M. Leslie, 1930
Reconnaissance of the waters and plankton of Monterey Bay, July 1928. Bull. M.C.Z., 70(5):429-481, 43 text figs.

Chaetoceros crinitus
Brunel, J., 1962
Le phytoplancton de la Baie de Chaleurs. Inst. Botan., Univ. Montréal, Contrib. No. 77: 365 pp., 66 pls.

Schütt emend Grøntved

Chaetoceros crinitus
Cleve-Euler, A., 1951
Die Diatomeen von Schweden und Finnland. Kungl. Svenska Vetenskaps Akad. Handl., Fjärde Ser. 2(1): 161 pp., 6 pls.

Chaetoceros crinitum
Gran, H.H., 1908
Diatomeen. Nordisches Plankton, Botanischer Teil pp. XIX.1-XIX 146; 178 text figs.

Chaetoceros crinitus
Gran, H.H., 1897
Protophyta: Diatomaceae, Silico-flagellata and Cilioflagellata. Den Norske Nordhavs Expedition 1876-1878, h. 24, 36 pp., 4 pls.

Chaetoceros crinitus
Grøntved, J., 1956.
Planktonological contributions. II. Taxonomical studies in some Danish coastal localities. Medd. Danmarks Fiskeri og Havundersøgelser n.s., 1(12):13 pp.

Chaetoceros crinitum
Hendy, N. Ingram, 1964
An introductory account of the smaller algae of British coastal waters. V. Bacillariophyceae (Diatoms). Her Majesty's Stationary Office, 317 pp., 45 pls.

Chaetoceros crinitus
Jorgensen, E., 1900
Protophyten und Protozoën im Plankton aus der Norwegischen Westkuste. Bergens Mus. Aarb. 1899(6): 95 pp., 5 pls., 83 tables.

Chaetoceros crinitus
Lillick, L.C., 1940
Phytoplankton and planktonic protozoa of the offshore waters of the Gulf of Maine. Pt.II. Qualitative Composition of the Planktonic Flora. Trans. Am. Phil. Soc., n.s., 31(3):193-237, 13 text figs.

Chaetoceros crinitus
Meunier, A., 1913
Microplancton de la Mer Flamande. 1. Chaetoceros. Mem. Mus. Roy. Hist. Nat., Belgique 7(2):1-55, 7 pls.

Chaetoceros crinitus
Pavillard, J., 1925
Bacillariales. Rept. on the Danish Oceangr. Exped., 1908-10 to the Mediterranean and adj. seas. Vol.II., Biol. J4:72 pp., 116 text figs.

Chaetoceros crinitus
Rampi, L., 1942
Ricerche sul fitoplancton del Mare Ligure 6. Le diatomee delle acque di Sanremo. Nuovo Giornale Botanico Italiano, N.S., 49:252-268.

Chaetoceros criophilum
Allen, W. E., 1925
Statistical Studies of surface catches of marine diatoms and dinoflagellates made by the Yacht "Ohio" in tropical waters in 1924. Jan. Trans. Amer. Microscop. Soc.:24-30, 1 fig.

Chaetoceros criophilus
Balech, E., 1947
Contribution al conocimiento del plancton antarctico. Plankton del Mar de Bellinghausen. Physis 20:75-91, 76 figs. on 8 pls.

Chaetoceros criophilum
Bigelow, H.B., and M. Leslie, 1930
Reconnaissance of the waters and plankton of Monterey Bay, July 1928. Bull. M.C.Z., 70(5):429-481, 43 text figs.

Chaetoceros criophilum
Boden, B. P., 1949.
The diatoms collected by the U.S.S. CACOPAN in the Antarctic in 1947. J. Mar. Res. 8(1):6-13, 3 textfigs.

Chaetoceros criophilum
Boden, Brian, 1948
Marine plankton diatoms on operation HIGHJUMP in: Some oceanographic observations on operation HIGHJUMP. By R.S. Dietz. USNEL Rept. No.55, 97 pp., 41 figs. 7 July 1948.

Chaetoceros criophilum n.sp.
Castracane degli Antelminelli, F., 1886
1. Report on the Diatomaceae collected by H.M.S. Challenger during the years 1873-1876. Rept. Sci. Results, H.M.S. Challenger, Botany Vol. II, 178 pp., 30 pls.

Chaetoceros criophilum
Central Meteorological Observatory, 1949
Report on sea and weather observation on Antarctic Whaling Ground (1947-48). Ocean. Mag., Japan, 1(1):49-88, 17 text figs.

Chaetoceros criophilus
Frenguelli, Joaquin, and Hector Antonio Orlando, 1959.
Operacion MERLUZA. Diatomeas y silicoflagelados del plancton del "VI Crucero". Servicio Hidrogr. Naval., Argentina, Publ. No. H. 619: 5-62.

Chaetoceros criophilum
Gran, H.H., 1908
Diatomeen. Nordisches Plankton, Botanischer Teil pp. XIX.1-XIX 146; 178 text figs.

Chaetoceros criophilum
Hendey, N.I., 1937
The plankton diatoms of the southern seas. Discovery Repts. 16:151-364, pls.6-13.

Chaetoceros criophilum
Iselin, C., 1930
A report on the coastal waters of Labrador based on explorations of the "Chance" during the summer of 1926. Proc. Am. Acad. Arts Sci., 66(1):1-37, 14 text figs.

Chaetoceros criophilus
Jorgensen, E., 1900
Protophyten und Protozoën im Plankton aus der Norwegischen Westkerste. Bergens Mus. Aarb. 1899(6): 95 pp., 5 pls., 83 tables.

Chaetoceros criophilum
Marukawa, H., 1921
Plankton lists and some new species of copepods, from the northern waters of Japan. Bull. Inst. Ocean., No.384, 15 pp., 3 pls., 1 chart. Monaco

Chaetoceros criophilum
Meunier, A., 1913
Microplancton de la Mer Flamande. 1. Chaetoceros. Mem. Mus. Roy. Hist. Nat., Belgique 7(2):1-55, 7 pls.

Chaetoceros criophilum
Mangin, L., 1917
Sur le Chaetoceros criophilus, espèce characteristique des mers antarctiques C.R. Acad. Sci., 164:7044770.

Chaetoceros criophilum
Mangin, L., 1915
Phytoplancton de L'Antartique. Deuxieme Exped. Ant. Francaise (1908-1910), 95 pp., 3 pls., 58 text figs.

Chaetoceros criophilum
Phifer, L.D. (undated)
The occurrence and distribution of plankton diatoms in Bering Sea and Bering Strait, July 26-August 24, 1934. Report of Oceanographic cruise of U.S. Coast Guard Cutter Chelan 1934, Part II(A):1-44 (mimeographed) plus fig.1 (after Pt.B)

Chaetoceros criophilum
Schulz, B., and A. Wulff, 1929
Hydrographie und Oberflächen plankton des westlichen Barentsmeeres im Sommer 1927. Ber. deutschen wissensch. Komm. F. Meeresforsch. n.s. 4(5):232-372, 13 tables, 25 text figs.

Chaetoceros criophilum
Sleggs, G.F., 1927.
Marine phytoplankton in the region of La Jolla, California during the summer of 1924. Bull. S.I.O., tech. ser., 1:93-117, 8 textfigs.

Chaetoceros cruciatum
Hendey, N.I., 1937
The plankton diatoms of the southern seas. Discovery Repts. 16:151-364, pls.6-13.

Chaetoceros crucifer n.sp.
Gran, H. H. and E. C. Angst, 1931
Plankton diatoms of Puget Sound. Publ. Puget Sound Biol. Sta. 7:417-519, 95 text figs.

Chaetoceros curvatum n.sp.
Castracane degli Antelminelli, F., 1886
1. Report on the Diatomaceae collected by H.M.S. Challenger during the years 1873-1876. Rept. Sci. Results, H.M.S. Challenger, Botany Vol. II, 178 pp., 30 pls.

Chaetoceros curvatum
Hendey, N.I., 1937
The plankton diatoms of the southern seas. Discovery Repts. 16:151-364, pls.6-13.

Chaetoceros curvatus
Mangin, L., 1915
Phytoplancton de L'Antartique. Deuxieme Exped. Ant. Francaise (1908-1910), 95 pp., 3 pls., 58 text figs.

Chaetoceros curvisetus
Allen, W. E., 1938
The Templeton Crocker Expedition to the Gulf of California in 1935 - The Phytoplankton. Amer. Microsc. Soc., Trans. 57:328-335.

Chaetoceros curvisetus
Allen, W.E., 1937
Plankton diatoms of the Gulf of California obtained by the G. Allan Hancock Expedition of 1936. The Hancock Pacific Expeditions, Univ. So. Calif. Publ. 3:47-59, 1 fig.

Chaetoceros curvisetus
Allen, W. E., 1936
Occurrence of marine plankton diatoms in a ten-year series of daily catches in southern California. Am. Jour. Bot. 23(1):60-63.

Chaetoceros curvisetum
Allen, W. E., 1925
Statistical Studies of surface catches of marine diatoms and dinoflagellates made by the Yacht "Ohio" in tropical waters in 1924. Jan. Trans. Amer. Microscop. Soc.:24-30, 1 fig.

Chaetoceros curvisetum
Bernhard, M., preparator, 1965.
Studies on the radioactive contamination of the sea, annual report 1964. Com. Naz. Energ. Nucleare, La Spezia, Rept., No. RT/BIO (65) 18:35 pp.

Chaetoceros curvisetus
Boden, B.P., 1950
Some marine plankton diatoms from the west coast of South Africa. Trans. R.Soc. S. Africa. 32:321-434, 100 text figs.

Chaetoceros curvisetus
Braarud, T., 1945
Experimental studies on marine plankton diatoms. Avhandlingar utgitt av Det Norske Videnskaps-Akademi i Oslo. 1. Mat.-Naturv. Klasse 1944. No.10, 1-16, 1 text fig.

Chaetoceros curvisetus
Braarud, T., 1945
A phytoplankton survey of the polluted waters of inner Oslo Fjord. Hvalrådets Skrifter, No.28, 142 pp., 19 text figs., 17 tables.

Chaetoceros curvisetus
Braarud, T., and Adam Bursa, 1939
On the phytoplankton of the Oslo Fjord, 1933-1934. Hvalrådets Skr. No.19:1-63; 9 text figs. Reviewed. J. du. Cons. 14(3): 418-420. A.C. Gardiner.

Chaetoceros curvisetus
Brunel, J., 1962
Le phytoplancton de la Baie de Chaleurs. Inst. Botan. Univ. Montréal, Contrib. No. 77: 365 pp., 66 pls.

Chaetoceros curvisetus
Cleve-Euler, A., 1951
Die Diatomeen von Schweden und Finnland. Kungl. Svenska Vetenskaps Akad. Handl. Fjärde Ser. 2(1): 161 pp., 6 pls.

Chaetoceros curvisetus
Cupp, Easter E., 1943
Marine plankton diatoms of the west coast of North America. Bull. S.I.O. 5(1):1-238, 5 pls., 168 text figs.

Chaetoceros curvisetus
Cupp, E.E., 1934
Analysis of marine diatom collections taken from the Canal Zone to California during March, 1933. Trans. Am. Micros. Soc. LIII (1):22-29, 1 map.

Chaetoceros curvisetus
Cupp, E.E. and Allen, W.E., 1938
Plankton diatoms of the Gulf of California obtained by Allan Hancock Pacific Expedition of 1937. The Hancock Pacific Expeditions, The Univ. So. Calif. Publ. 3: 61-74, 1 map, pls.4-15.

Chaetoceros curvisetus
Delegazione Italiana della Commissione Internazionale per l'Esplorazione Scientifica del Mediterraneo, 1941
Note sul plancton della Laguna veneta. [Memoria CCLXXXIX], Arch. di Ocean. e Limn. Anno I, Fasc. I, 1941 XIX: 31-57 pp.

Chaetoceros curvisetus
Ercegovic, A., 1936
Etudes qualitative et quantitatives du phytoplancton dans les eaux cotières de l'Adriatique oriental moyen au cours de l'année 1934. Acta Adriatica 1(9):1-126

Chaetoceros curvisetus
Eskinazi Enide e Shigekatsv Sato (1963-1964) 1966.
Contribuição ao estudo das diatomaceas da Praia de Piedade.

Trabhs Inst. Oceanogr., Univ. Recife, 5 (5/6): 73-114.

Chaetoceros curvisetum

Ghazzawi, F.M., 1939
Plankton of the Egyptian waters. A study of the Suez Canal Plankton. (A) Phytoplankton. Preliminary Report 83 pp. Notes and Memoires, Min. Commerce-Industry, Egypt, Hydrobiol. & Fish. 65 figs.

Chaetoceros curvisetus

Gilbert, J.Y., and W.E. Allen, 1943
The phytoplankton of the Gulf of California obtained by the "E.W. Scripps" in 1939 and 1940. J. Mar. Res. V(2):89-110, figs. 30-31.

Chaetoceros curvisetos

Gilson, H. C., 1937
Chemical and Physical Investigations. The nitrogen cycle. John Murray Exped., 1933-34, Sci. Repts., 2(2):21-81, 16 text figs.

Chaetoceros curvisetum

Gran, H.H., 1908
Diatomeen. Nordisches Plankton, Botanischer Teil pp. XIX.1-XIX 146; 178 text figs.

Chaetoceros curvisetum

Gran, H.H., 1897
Protophyta: Diatomaceae, Silico-flagellata and Cilioflagellata. Den Norske Nordhavs Expedition 1876-1878, h. 24, 36 pp., 4 pls.

Chaetoceros curvisetum

Hendy, N. Ingram, 1964
An introductory account of the smaller algae of British coastal waters. V. Bacillariophyceae (Diatoms). Her Majesty's Stationary Office, 317 pp., 45 pls.

Chaetoceros curvisitum

Iselin, C., 1930
A report on the coastal waters of Labrador based on explorations of the "Chance" during the summer of 1926. Proc. Am. Acad. Arts Sci., 66(1):1-37, 14 text figs.

Chaetoceros curvisetus

Jorgensen, E., 1900
Protophyten und Protozoën im Plankton aus der Norwegischen Westkerste. Bergens Mus. Aarb. 1899(6): 95 pp., 5 pls., 83 tables.

Chaetoceros curvisetus

Lillick, L.C., 1940
Phytoplankton and planktonic protozoa of the offshore waters of the Gulf of Maine. Pt.II. Qualitative Composition of the Planktonic Flora. Trans. Am. Phil. Soc., n.s., 31(3):193-237, 13 text figs.

Chaetoceros curvisetus

Lillick, L.C., 1937
Seasonal studies of the phytoplankton off Woods Hole, Massachusetts. Biol. Bull. LXXIII (3):488-503, 3 text figs.

Chaetoceros curvisetus

Mangin, M. L., 1912
Phytoplancton de la croisière du "René" dans l'Atlantique (Septembre 1908). Ann. Inst. Ocean., n.s., 4(1):1-66, 2 pls., 41 text figs., 2 tables.

Chaetoceros curvisetus

Massuti Algamora, M., 1949
Estudio de diez y seis muestras de plancton del Golfo de Nápoles. Publ. Inst. Biol. Appl. 5:85-94, 1 fold-in table.

Chaetoceros curvisetum

Meunier, A., 1913
Microplancton de la Mer Flamande. 1. Chaetoceros. Mem. Mus. Roy. Hist. Nat., Belgique 7(2):1-55, 7 pls.

Chaetoceros curvisetus

Michailova, N.F., 1964
On the spreading of the habitat into the Black Sea of species of the genus Chaetoceros of northern seas and their biogeography. (In Russian).
Trudy Sevastopol Biol. Sta., 7:231-248.

Chaetoceros curvisetum

Paiva Carvalho, J., 1950
O plancton do Rio Maria Rodriques (Cananeis). 1. Diatomaceas e Dinoflagelados. Bol. Inst. Paulista Oceanogr. 1(1); 27-43, 2 fold-in tables, 2 figs.

Chaetoceros curvisetum

Pavillard, J., 1925
Bacillariales. Rept. on the Danish Oceangr. Exped., 1908-10 to the Mediterranean and adj. seas. Vol.II., Biol. J4:72 pp., 116 text figs.

Chaetoceros curvisetum

Pavillard, J., 1905
Recherches sur la flore pelagique (Phytoplankton) de l'Etang de Thau. Theses presentees a la Fac. Sci., Paris, 116 pp., 3 pls.

Chaetoceros curvisetus

Pettersson, H., 1934
Scattering and extinction of light in sea-water. Meddelanden från Göteborgs Högskolas Oceanografiska Institut. 9 (Göteborgs Kungl. vetenskaps-och vitterhets-samhälles Handlingar. Femte Foljden, Ser.B, 4(4)):1-16.

Chaetoceros dadayi pacifica n.var.

Rampi, L., 1952
Ricerche sul microplancton di superficie del Pacifico tropicale. Bull. Ocean. Inst., Monaco, No. 1014:16 pp., 5 textfigs.

Chaetoceros curvisetum

Rampi, L., 1942
Ricerche sul fitoplancton del Mare Ligure 6. Le diatomee delle acque di Sanremo. Nuovo Giornale Botanico Italiano, N.S., 49:252-268.

Chaetoceros curvisetum

Rumkówna, A., 1948
[List of the phytoplankton species occurring in the superficial water layers in the Gulf of Gdańsk] Bull. Lab. mar., Gdynia, No. 4: 139-141 with tables in back.

Chaetoceros curvisetum (figs.)

Sousa e Silva, E., 1949
Diatomaceas e Dinoflagelados de Baia de Cascais. Portugaliae Acta Biol., Volume: Julio Henriques, Ser. B: 300-383, 9 pls, 2 fold-in tables.

Chaetoceros curvisetum

Sousa a Silva, E., and J. Dos Santos-Pinto, 1948
O Plancton da Baia de S. Martinho do Porto. 1. Diatomaceas e Dinoflagelados. Bol. Soc. Portuguese de Ciencias Naturais, 16(2):134-187, 6 pls. (Trav. Sta. Biol. Mar. de Lisbonne No. 52).

Chaetoceros curvisetus

Steemann-Nielsen, Einar, 1951
The marine vegetation of the Isefjord. A study on ecology and production. Medd. Komm. Danmarks Fiskeri-og Havundersøgelser. Ser. Plankton. 5(4); 114pp., 46 text figs.

Chaetoceros curvisetus

Uyeno, Fukuzo, 1961
Oceanographical and ecological studies on primary production of the sea, with special references to relationship between diatom production and temperature and chlorinity of water.
Rept., Fac. Fish., Pref. Univ., Mie, 4(1): 1-64.

Chaetoceros curvisetus

Zgurovskaya, L.N., 1969.
Effect of petroleum growth agent on the photosynthesis intensity and rate of Chaetoceros curvisetus Cl. cell division. (In Russian) Gidrobiol. Zh., 5(1): 55-59.

Chaetoceros dadayi

Brunel, J., 1962
Le phytoplancton de la Baie de Chaleurs. Inst. Botan., Univ. Montréal, Contrib. No. 77: 365 pp., 66 pls.

Chaetoceros dadayi

Cupp, Easter E., 1943
Marine plankton diatoms of the west coast of North America. Bull. S.I.O. 5(1):1-238, 5 pls., 168 text figs.

Chaetoceros dadayi

Cupp, E.E., 1934
Analysis of marine diatom collections taken from the Canal Zone to California during March, 1933. Trans. Am. Micros. Soc. LIII (1):22-29, 1 map.

Chaetoceros Dadayi

Delegazione Italiana della Commissione Internazionale per l'Esplorazione Scientifica del Mediterraneo, 1941
Note sul plancton della Laguna veneta. [Memoria CCLXXIX], Arch. di Ocean. e Limn. Anno I, Fasc. I, 1941 XIX: 31-57 pp.

Chaetoceros Daday

Ercegovic, A., 1936
Etudes qualitative et quantitatives du phytoplancton dans les eaux cotières de l'Adriatique oriental moyen au cours de l'année 1934. Acta Adriatica 1(9):1-126

Chaetoceros dadayi

Porti, A., 1922
Ricerche sulla flora pelagica (fitoplancton) di Quarto dei Mille. Mem. R. Com. Talass. Ital. 97:248 pp., 13 pls.

Chaetoceros dadayi

Pavillard, J., 1925
Bacillariales. Rept. on the Danish Oceangr. Exped., 1908-10 to the Mediterranean and adj. seas. Vol.II., Biol. J4:72 pp., 116 text figs.

Chaetoceros Dadayi

Rampi, L., 1942
Ricerche sul fitoplancton del Mare Ligure 6. Le diatomee delle acque di Sanremo. Nuovo Giornale Botanico Italiano, N.S., 49:252-268.

Chaetoceros danicus

Allen, W.E., 1937
Plankton diatoms of the Gulf of California obtained by the G. Allan Hancock Expedition of 1936. The Hancock Pacific Expeditions, Univ. So. Calif. Publ. 3:47-59, 1 fig.

Chaetoceros danicus

Braarud, T., 1945
A phytoplankton survey of the polluted waters of inner Oslo Fjord. Hvalrådets Skrifter, No.28, 142 pp., 19 text figs., 17 tables.

Chaetoceros danicus

Braarud, T., and Adam Bursa, 1939
On the phytoplankton of the Oslo Fjord, 1933-1934. Hvalrådets Skr. No.19:1-63; 9 text figs. Reviewed. J. du. Cons. 14(3): 418-420. A.C. Gardiner.

Chaetoceros danicum

Gran, H.H., 1908
Diatomeen. Nordisches Plankton, Botanischer Teil pp. XIX.1-XIX 146; 178 text figs.

Chaetoceros danicus

Brunel, J., 1962
Le phytoplancton de la Baie de Chaleurs. Inst. Botan., Univ. Montréal, Contrib. No. 77: 365 pp., 66 pls.

Chaetoceros danicus
Brunel, Jules, 1962
Le phytoplancton de la Baie des Chaleurs. Contrib. Ministère de la Chasse et des Pêcheries, Province de Québec, No. 91: 365 pp.

Chaetoceros danicum
Cassie, Vivienne, 1961
Marine phytoplankton in New Zealand waters. Botanica Marina, 2(Suppl.):54 pp., 8 pls.

Chaetoceros danicus
Chiba, T., 1949
On the distribution of the plankton in the eastern China Sea and Yellow Sea. 1. Plankton composition in the spring. J. Shimonoseki Coll. Fisheries, 1(1):57-63, 1 fig.

Chaetoceros danicus
Cleve-Euler, A., 1951
Die Diatomeen von Schweden und Finnland. Kungl. Svenska Vetenskaps Akad. Handl., Fjärde Ser. 2(1): 161 pp., 6 pls.

Chaetoceros danicus
Cupp, Easter E., 1943
Marine plankton diatoms of the west coast of North America. Bull. S.I.O. 5(1):1-238, 5 pls., 168 text figs.

Chaetoceros danicus
Cupp, E.E., 1934
Analysis of marine diatom collections taken from the Canal Zone to California during March, 1933. Trans. Am. Micros. Soc. LIII (1):22-29, 1 map.

Chaetoceros danicus
Cupp, E.E. and Allen, W.E., 1938
Plankton diatoms of the Gulf of California obtained by Allan Hancock Pacific Expedition of 1937. The Hancock Pacific Expeditions, The Univ. So. Calif. Publ. 3: 61-74, 1 map, pls.4-15.

Chaetoceros danicus
Forti, A., 1922
Ricerche sulla flora pelagica (fitoplancton) di Quarto dei Mille. Mem. R. Com. Talass. Ital. 97:248 pp., 13 pls.

Chaetoceros danicum
Gran, H.H., 1897
Protophyta: Diatomaceae, Silico-flagellata and Cilioflagellata. Den Norske Nordhavs Expedition 1876-1878, h. 24, 36 pp., 4 pls.

Chaetoceros danicus
Gran, H.H., and T. Braarud, 1935
A quantitative study of the phytoplankton in the Bay of Fundy and the Gulf of Maine (including observations on hydrography, chemistry, and turbidity). J. Biol. Bd., Canada, 1(5):279-467, 69 text figs.

Chaetoceros danicus
Gran, H. H. and E. C. Angst, 1931
Plankton diatoms of Puget Sound. Publ. Puget Sound Biol. Sta. 7:417-519, 95 text figs.

Chaetoceros danicum
Hendy, N. Ingram, 1964
An introductory account of the smaller algae of British coastal waters. V. Bacillariophyceae (Diatoms).
Her Majesty's Stationary Office, 317 pp., 45 pls.

Chaetoceros danicum
Hendey, N.I., 1937
The plankton diatoms of the southern seas. Discovery Repts. 16:151-364, pls.6-13.

Chaetoceroa danicus
Jorgensen, E., 1900
Protophyten und Protozoën im Plankton aus der Norwegischen Westkerste. Bergens Mus. Aarb. 1899(6): 95 pp., 5 pls., 83 tables.

Chaetoceros danicus
Levander, K.M., 1947
Plankton gesammelt in den Jahren 1899-1910 an den Küsten Finnlands. Finnländische Hydrographisch-Biologische Untersuchunger (aus dem Wasserbiologischen Laboratorin der Societas Scientiarum Fennica) No.11:40 pp., 6 diagrams, 13 pls., tables.

Chaetoceros danicus
Lillick, L.C., 1940
Phytoplankton and planktonic protozoa of the offshore waters of the Gulf of Maine. Pt.II. Qualitative Composition of the Planktonic Flora. Trans. Am. Phil. Soc., n.s., 31(3):193-237, 13 text figs.

Chaetoceros danicus
Mangin, M. L., 1912
Phytoplancton de la croisière du "René" dans l'Atlantique (Septembre 1908). Ann. Inst. Ocean., n.s., 4(1):1-66, 2 pls., 41 text figs., 2 tables.

Chaetoceros danicum
Meunier, A., 1913
Microplancton de la Mer Flamande. 1. Chaetoceros. Mem. Mus. Roy. Hist. Nat., Belgique 7(2):1-55, 7 pls.

Chaetoceros danicus
Michailova, N.F., 1964.
On the spreading of the habitat into the Black Sea of species of the genus Chaetoceros of northern seas and their biogeography. (In Russian).
Trudy Sevastopol Biol. Sta., 7:231-248.

Chaetoceros danicus
Morse, D.C., 1947
Some observations on seasonal variations in plankton population Patuxant River, Maryland 1943-1945. Bd. Nat. Res., Publ. No.65, Chesapeake Biol. Lab., 31, 3 figs.

Chaetoceros danicum
Pavillard, J., 1925
Bacillariales. Rept. on the Danish Oceangr. Exped., 1908-10 to the Mediterranean and adj. seas. Vol.II., Biol. J4:72 pp., 116 text figs.

Chaetoceros danicus
Rampi, L., 1942
Ricerche sul fitoplancton del Mare Ligure 6. Le diatomee delle acque di Sanremo. Nuovo Giornale Botanico Italiano, N.S., 49:252-268.

Chaetoceros danicus
Rothe, F., 1942.
Quantitativen Untersuchungen über die Planktonverteilung in der östlichen Ostsee. Ber. Deutsch. Wiss. Komm. Meeresf., N.F., 10:291-368, 33 text-figs.

Most identifications not to species

Chaetoceros danicus
Rothe, F., 1941
Quantitative Untersuchunger über die Plankton verteilung in der östlichen Ost see. Ber. Deut. Wiss. Komm. fur Meeresforschung. n.f. X(3):291-368, 33 text figs.

Chaetoceros danicum
Rumkówna, A., 1948
[List of the phytoplankton species occurring in the superficial water layers in the Gulf of Gdańsk] Bull. Lab. mar., Gdynia, No. 4: 139-141 with tables in back.

Chaetoceros dannicum
Sousa e Silva, E., 1949
Diatomaceas e Dinoflagelados de Baia de Cascais. Portugaliae Acta Biol., Volume: Julio Henriques, Ser. B: 300-383, 9 pls, 2 fold-in tables.

Chaetoceros dannicus
Sousa a Silva, E., and J. Dos Santos-Pinto, 1948
O Plancton da Baia de S. Martinho do Porto. 1. Diatomaceas e Dinoflagelados. Bol. Soc. Portuguese de Ciencias Naturais, 16(2):134-187, 6 pls. (Trav. Sta. Biol. Mar. de Lisbonne No. 52).

Chaetoceros danicus
Steemann-Nielsen, Einar, 1951
The marine vegetation of the Isefjord. A study on ecology and production. Medd. Komm. Danmarks Fiskeri-og Havundersøgelser. Ser. Plankton. 5(4); 114pp., 46 text figs.

Chaetoceros debilis
Allen, W. E., 1938
The Templeton Crocker Expedition to the Gulf of California in 1935 - The Phytoplankton. Amer. Microsc. Soc., Trans. 57:328-335.

Chaetoceros debilis
Allen, W.E., 1937
Plankton diatoms of the Gulf of California obtained by the G. Allan Hancock Expedition of 1936. The Hancock Pacific Expeditions, Univ. So. Calif. Publ. 3:47-59, 1 fig.

Chaetoceros debilis
Allen, W. E., 1936
Occurrence of marine plankton diatoms in a ten-year series of daily catches in southern California. Am. Jour. Bot. 23(1):60-63.

Chaetoceros debile
Allen, W.E., 1928.
Review of five years of studies of phytoplankton at Southern California piers, 1920-1924, inclusive. Bull. S.I.O., tech. ser., 1:357-401, 5 text figs.

Chaetoceros debile
Allen, W.E., 1927.
Quantitative studies on inshore marine diatoms and dinoflagellates of Southern California in 1922. Bull. S.I.O., tech. ser., 1:31-38, 2 text-figs.

Chaetoceros debile
Allen, W. E., 1925
Statistical Studies of surface catches of marine diatoms and dinoflagellates made by the Yacht "Ohio" in tropical waters in 1924. Jan. Trans. Amer. Microscop. Soc.:24-30, 1 fig.

Chaetoceros danicus
Bernhard, M., preparator, 1965.
Studies on the radioactive contamination of the sea, annual report 1964.
Com. Naz. Energ. Nucleare, La Spezia, Rept., No. RT/BIO (65) 18:35 pp.

Chaetoceros debile
Bigelow, H.B., and M. Leslie, 1930
Reconnaissance of the waters and plankton of Monterey Bay, July 1928. Bull. M.C.Z., 70(5):429-481, 43 text figs.

Chaetoceros debilis
Braarud, T., 1962
Species distribution in marine phytoplankton.
J. Oceanogr. Soc., Japan, 20th Ann. Vol. 628-649.

Chaetoceros debilis
Braarud, T., 1945
Experimental studies on marine plankton diatoms. Avhandlingar utgitt av Det Norske Videnskaps-Akademi i Oslo. 1. Mat.-Naturv. Klasse 1944. No.10, 1-16, 1 text fig.

Chaetoceros debilis
Braarud, T., 1945
A phytoplancton survey of the polluted waters of inner Oslo Fjord. Hvalrådets Skrifter, No.28, 142 pp., 19 text figs., 17 tables.

Chaetoceros debilis
Braarud, T., and Adam Bursa, 1939
On the phytoplankton of the Oslo Fjord, 1933-1934. Hvalrådets Skr. No.19:1-63; 9 text figs. Reviewed. J. du. Cons. 14(3): 418-420. A.C. Gardiner.

Chaetoceros debilis
Brunel, J., 1962
Le phytoplancton de la Baie de Chaleurs. Inst. Botan., Univ. Montréal, Contrib. No. 77: 365 pp., 66 pls.

Chaetoceros debilis
Brunel, Jules, 1962
Le phytoplancton de la Baie des Chaleurs. Contrib. Ministère de la Chasse et des Pêcheries, Province de Québec, No. 91: 365 pp.

Chaetoceros debilis
Cleve-Euler, A., 1951
Die Diatomeen von Schweden und Finnland. Kungl. Svenska Vetenskaps Akad. Handl., Fjärde Ser. 2(1): 161 pp., 6 pls.

Chaetoceros debilis
Cupp, Easter E., 1943
Marine plankton diatoms of the west coast of North America. Bull. S.I.O. 5(1):1-238, 5 pls., 168 text figs.

Chaetoceros debilis
Cupp, E.E., 1934
Analysis of marine diatom collections taken from the Canal Zone to California during March, 1933. Trans. Am. Micros. Soc. LIII (1):22-29, 1 map.

Chaetoceros debilis
Cupp, E.E. and Allen, W.E., 1938
Plankton diatoms of the Gulf of California obtained by Allan Hancock Pacific Expedition of 1937. The Hancock Pacific Expeditions, The Univ. So. Calif. Publ. 3: 61-74, 1 map, pls.4-15.

Chaetoceros debilis
Gilbert, J.Y., and W.E. Allen, 1943
The phytoplankton of the Gulf of California obtained by the "E.W. Scripps" in 1939 and 1940. J. Mar. Res. V(2):89-110, figs.30-31.

Chaetoceros debile
Gran, H.H., 1908
Diatomeen. Nordisches Plankton, Botanischer Teil pp. XIX.1-XIX 146; 178 text figs.

Chaetoceros debile
Gran, H.H., 1897
Protophyta: Diatomaceae, Silico-flagellata and Cilioflagellata. Den Norske Nordhavs Expedition 1876-1878, h. 24, 36 pp., 4 pls.

Chaetoceros debilis
Gran, H.H., and T. Braarud, 1935
A quantitative study of the phytoplankton in the Bay of Fundy and the Gulf of Maine (including observations on hydrography, chemistry, and turbidity). J. Biol. Bd., Canada, 1(5):279-467, 69 text figs.

Chaetoceros debilis
Gran, H. H. and E. C. Angst, 1931
Plankton diatoms of Puget Sound. Publ. Puget Sound Biol. Sta. 7:417-519, 95 text figs.

Chaetoceros debile
Hendy, N. Ingram, 1964
An introductory account of the smaller algae of British coastal waters. V. Bacillariophyceae (Diatoms). Her Majesty's Stationary Office, 317 pp., 45 pls.

Chaetoceros debile
Hendey, N.I., 1937
The plankton diatoms of the southern seas. Discovery Repts. 16:151-364, pls.6-13.

Chaetoceros debile
Iselin, C., 1930
A report on the coastal waters of Labrador based on explorations of the "Chance" during the summer of 1926. Proc. Am. Acad. Arts Sci., 66(1):1-37, 14 text figs.

Chaetoceros debilis
Jorgensen, E., 1900
Protophyten und Protozoën im Plankton aus der Norwegischen Westkerste. Bergens Mus. Aarb. 1899(6): 95 pp., 5 pls., 83 tables.

Chaetoceros debilis
Kokubo, S., 1952
Results of the observations on the plankton and oceanography of Mutsu Bay during 1950, reference being made also to the period 1946-1950. Bull Mar.Biol.Sta., Asamushi 5(1/4): 1-54, 3 tables,(fold-in), 1 fold-in.

Chaetoceros debilis
Lillick, L.C., 1940
Phytoplankton and planktonic protozoa of the offshore waters of the Gulf of Maine. Pt.II. Qualitative Composition of the Planktonic Flora. Trans. Am. Phil. Soc., n.s., 31(3):193-237, 13 text figs.

Chaetoceros debilis
Lillick, L.C., 1938
Preliminary report of the phytoplankton of the Gulf of Maine. Am. Mid. Nat. 20(3):624-640, 1 text figs 37 tables.

Chaetoceros debilis
Lillick, L.C., 1937
Seasonal studies of the phytoplankton off Woods Hole, Massachusetts. Biol. Bull. LXXIII (3):488-503, 3 text figs.

Chaetoceros debilis
Lovegrove, T., 1958(1960).
Plankton investigations from Aberdeen in 1958. Ann. Biol., Cons. Perm. Int. Expl. Mer, 15:55.

Chaetoceros debile
Marukawa, H., 1921
Plankton lists and some new species of copepods, from the northern waters of Japan. Bull. Inst. Ocean., No.384, 15 pp., 3 pls., 1 chart. Monaco

Chaetoceros debile
Meunier, A., 1913
Microplancton de la Mer Flamande. 1. Chaetoceros. Mem. Mus. Roy. Hist. Nat., Belgique 7(2):1-55, 7 pls.

Chaetoceros debilis
Morse, D.C., 1947
Some observations on seasonal variations in plankton population Patuxant River, Maryland 1943-1945. Bd. Nat. Res., Publ. No.65, Chesapeake Biol. Lab., 31, 3 figs.

Chaetoceros debilis
Paasche, E., 1961.
Notes on phytoplankton from the Norwegian Sea. Botanica Marine, 2(3/4):197-210.

Chaetoceros debilis
Paasche, E., and A.M. Rom, 1961(1962)
On the phytoplankton vegetation of the Norwegian Sea in May 1958. Nytt Mag. Botanikk, 9:33-60.

Chaetoceros debilis
Pettersson, H., F. Gross, and F. Koczy, 1939
Large scale plankton cultures. Medd. från Oceanografiska Institutet i Göteborg No.3 (Göteborgs Kungl. vetenskaps-och Vitterhets-Samhälles Handlingar Femte Följden. Ser. B) Vol.6(13):1-24.

1. The plankton shaft. H. Pettersson
2. Experiment with phytoplankton. F. Gross and F. Koczy
3. Experiments with zooplankton.

Chaetoceros debilis
Phifer, L.D. (undated)
The occurrence and distribution of plankton diatoms in Boring Sea and Bering Strait, July 26-August 24, 1934. Report of Oceanographic cruise of U.S. Coast Guard Cutter Chelan 1934, Part II(A):1-44 (mimeographed) plus fig.1 (after Pt.B)

Chaetoceros debilis
Sargent, M. S. and T. J. Walker, 1948
Diatom populations associated with eddies off southern California in 1941. J. Mar. Res. 7(3):490-505, 15 text figs.

Chaetoceros debile
Schodduyn, M., 1926
Observations faites dans la baie d'Ambleteuse (Pas de Calais). Bull. Inst. Ocean., Monaco, No. 482: 64 pp.

Chaetoceros debile
Schröder, B., 1900
Phytoplankton des Golfes von Neapel nebst vergleichenden Ausblicken auf das atlantischen Ozean. Mitt. Zool. Stat. Neapel, 14:1-38.

Chaetoceros debilis
Sousa e Silva, E., 1949
Diatomaceas e Dinoflagelados de Baia de Cascais. Portugaliae Acta Biol., Volume: Julio Henriques, Ser. B: 300-383, 9 pls, 2 fold-in tables.

Chaetoceros debilis
Sousa a Silva, E., and J. Dos Santos-Pinto, 1948
O Plancton da Baia de S. Martinho do Porto. 1. Diatomaceas e Dinoflagelados. Bol. Soc. Portuguese de Ciencias Naturais, 16(2):134-187, 6 pls. (Trav. Sta. Biol. Mar. de Lisbonne No. 52).

Chaetoceros debilis
Steemann-Nielsen, Einar, 1951
The marine vegetation of the Isefjord. A study on ecology and production. Medd. Komm. Danmarks Fiskeri-og Havundersøgelser. Ser. Plankton. 5(4); 114pp., 46 text figs.

Chaetoceros debilis
Uyeno, Fukuzo, 1961
Oceanographical and ecological studies on primary production of the sea, with special references to relationship between diatom production and temperature and chlorinity of water.
Rept., Fac. Fish., Pref. Univ., Mie, 4(1): 1-64.

Chaetoceros decipiens
Allen, W. E., 1938
The Templeton Crocker Expedition to the Gulf of California in 1935 - The Phytoplankton. Amer. Microsc. Soc., Trans. 57:328-335.

Chaetoceros decipiens

Allen, W.E., 1937
Plankton diatoms of the Gulf of California obtained by the G. Allan Hancock Expedition of 1936. The Hancock Pacific Expeditions, Univ. So. Calif. Publ. 3:47-59, 1 fig.

Chaetoceros decipiens

Allen, W. E., 1925
Statistical Studies of surface catches of marine diatoms and dinoflagellates made by the Yacht "Ohio" in tropical waters in 1924. Jan. Trans. Amer. Microscop. Soc.:24-30, 1 fig.

Chaetoceros decipiens

Bigelow, H. B., 1922
Exploration of the coastal water off the northeastern United States in 1916 by the U.S. Fisheries Schooner Grampus. Bull. M.C.Z. 65 (5):85-188, 53 text figs.

Chaetoceros decipiens

Bigelow, H.B., and M. Leslie, 1930
Reconnaissance of the waters and plankton of Monterey Bay, July 1928. Bull. M.C.Z., 70(5):429-481, 43 text figs.

Chaetoceros decipiens

Boden, B.P., 1950
Some marine plankton diatoms from the west coast of South Africa. Trans. R.Soc. S. Africa. 32:321-434, 100 text figs.

Chaetoceros decipiens

Braarud, T., 1945
Experimental studies on marine plankton diatoms. Avhandlingar utgitt av Det Norske Videnskaps-Akademi i Oslo. 1. Mat.-Naturv. Klasse 1944. No.10, 1-16, 1 text fig.

Chaetoceros decipiens

Braarud, T., 1945
A phytoplankton survey of the polluted waters of inner Oslo Fjord. Hvalrådets Skrifter, No.28, 142 pp., 19 text figs., 17 tables.

Chaetoceros decipiens

Braarud, T., 1934
A note on the phytoplankton of the Gulf of Maine in the summer of 1933. Biol. Bull. 67(1):76-82. (Contribution No.46 of the Woods Hole Oceanographic Institution)

Chaetoceros decipiens

Braarud, T., and Adam Bursa, 1939
On the phytoplankton of the Oslo Fjord, 1933-1934. Hvalrådets Skr. No.19:1-63; 9 text figs. Reviewed. J. du Cons. 14(3): 418-420. A.C. Gardiner.

Chaetoceros decipiens

Brunel, J., 1962
Le phytoplancton de la Baie de Chaleurs. Inst. Botan., Univ. Montréal, Contrib. No. 77: 365 pp., 66 pls.

Chaetoceros decipiens

Brunel, Jules, 1962
Le phytoplancton de la Baie des Chaleurs. Contrib. Ministère de la Chasse et des Pêcheries, Province de Québec, No. 91: 365 pp.

Chaetoceros decipiens

Bursa, Adam, 1963.
Phytoplankton in coastal waters of the Arctic Ocean at Point Barrow, Alaska. Arctic, 16(4):239-262.

Chaetoceros decipiens

Chiba, T., 1949
On the distribution of the plankton in the eastern China Sea and Yellow Sea. 1. Plankton composition in the spring. J. Shimonoseki Coll. Fisheries, 1(1):57-63, 1 fig.

Chaetoceros decipiens

Chu, S. P., 1949
Experimental studies on the environmental factors influencing the growth of phytoplankton. Sci. & Tech. in China 2(3):37-52.

Chaetoceros decipiens

Cleve-Euler, A., 1951
Die Diatomeen von Schweden und Finnland. Kungl. Svenska Vetenskaps Akad. Handl. Fjärde Ser. 2(1): 161 pp., 6 pls.

Chaetoceros decipiens

Copenhagen, W. J. and L. D., 1949
Variation in the phytoplankton of Table Bay, October 1934 to October 1935. With a note on the calorific value of Chaetoceros spp. Trans. Roy. Soc. S. Africa, 32(2):113-123, 2 text figs.

Chaetoceros decipiens

Cupp, Easter E., 1943
Marine plankton diatoms of the west coast of North America. Bull. S.I.O. 5(1):1-238, 5 pls., 168 text figs.

Chaetoceros decipiens

Cupp, E.E., 1937
Seasonal distribution and occurrence of marine diatoms and dinoflagellates at Scotch Cap, Alaska. Bull. S.I.O. Tech. ser4(3):71-100, 7 textfigs.

Chaetoceros decipiens

Cupp, E.E., 1934
Analysis of marine diatom collections taken from the Canal Zone to California during March, 1933. Trans. Am. Micros. Soc. LIII (1):22-29, 1 map.

Chaetoceros decipiens

Cupp, E.E. and Allen, W.E., 1938
Plankton diatoms of the Gulf of California obtained by Allan Hancock Pacific Expedition of 1937. The Hancock Pacific Expeditions, The Univ. So. Calif. Publ. 3: 61-74, 1 map, pls.4-15.

Chaetoceros decipiens

Dangeard, P., 1927
Phytoplankton de la croisière du "Sylvana". Ann. Inst. Ocean., Monaco, n.s., 4(8):286-401, 54 text figs. (Feirer-Juin 1913).

Chaetoceros decipiens

Davidson, V.M., 1931.
Biological and oceanographic conditions in Hudson Bay. Contr. Canadian Biol. Fish., n.s,, 6(26):

Chaetoceros decipiens

Delegazione Italiana della Commissione Internazionale per l'Esplorazione Scientifica del Mediterraneo, 1941
Note sul plancton della Laguna veneta. [Memoria CCLXXIX], Arch. di Ocean. e Limn. Anno I, Fasc. I, 1941 XIX: 31-57 pp.

Chaetoceros decipiens

Ercegovic, A., 1936
Etudes qualitative et quantitatives du phytoplancton dans les eaux cotières de l'Adriatique oriental moyen au cours de l'année 1934. Acta Adriatica 1(9):1-126

Chaetoceros decipiens

Fraser, J. H., 1949
Plankton of the Faroe-Shetland Channel and the Faroes, June and August 1947. Ann. Biol., Int. Cons., 4:27-28, text fig. 10.

Chaetoceros decipiens

Fraser, J. H., and A. Saville, 1949
Plankton distribution in Scottish and adjacent waters in 1948. Ann. Biol. 5:61-62.

Chaetoceros decipiens

Frenguelli, Joaquin, and Hector Antonio Orlando, 1959.
Operacion MERLUZA. Diatomeas y silico-flageladas del plancton del "VI Crucero". Servicio Hidrogr. Naval., Argentina, Publ. No. H. 619: 5-62.

Chaetoceros decipiens

Ghazzawi, F.M., 1939
Plankton of the Egyptian waters. A study of the Suez Canal Plankton. (A) Phytoplankton. Preliminary Report 83 pp. Notes and Memoires, Min. Commerce-Industry, Egypt, Hydrobiol. & Fish. 65 figs.

Chaetoceros decipiens

Gilson, H. C., 1937
Chemical and Physical Investigations. The nitrogen cycle. John Murray Exped., 1933-34, Sci. Repts., 2(2):21-81, 16 text figs.

Chaetoceros decipiens

Gran, H.H., 1908
Diatomeen. Nordisches Plankton, Botanischer Teil pp. XIX.1-XIX.146; 178 text figs.

Chaetoceros decipiens

Gran, H.H., 1897
Protophyta: Diatomaceae, Silico-flagellata and Cilioflagellata. Den Norske Nordhavs Expedition 1876-1878, h. 24, 36 pp., 4 pls.

Chaetoceros decipiens

Gran, H.H., and T. Braarud, 1935
A quantitative study of the phytoplankton in the Bay of Fundy and the Gulf of Maine (including observations on hydrography, chemistry, and turbidity). J. Biol. Bd., Canada, 1(5):279-467, 69 text figs.

Chaetoceros decipiens

Gran, H. H. and E. C. Angst, 1931
Plankton diatoms of Puget Sound. Publ. Puget Sound Biol. Sta. 7:417-519, 95 text figs.

Chaetoceros decipiens

Hendy, N. Ingram, 1964
An introductory account of the smaller algae of British coastal waters. V. Bacillariophyceae (Diatoms). Her Majesty's Stationary Office, 317 pp., 45 pls.

Chaetoceros decipiens

Hendey, N.I., 1937
The plankton diatoms of the southern seas. Discovery Repts. 16:151-364, pls.6-13.

Chaetoceros decipiens

Iselin, C., 1930
A report on the coastal waters of Labrador based on explorations of the "Chance" during the summer of 1926. Proc. Am. Acad. Arts Sci., 66(1):1-37, 14 text figs.

Chaetoceros decipiens

Iyengar, M.O.P. and G.Venkataraman,1951.
The ecology and seasonal succession of the algae flora of the River Cooum at Madras with special reference to the Diatomaceae. J. Madras Univ. 21, Sect. B(1): 140-192, 1 pl of 4 figs., 11 text figs.

Chaetoceros decipiens

Jorgensen, E., 1900
Protophyten und Protozoën im Plankton aus der Norwegischen Westkerste. Bergens Mus. Aarb. 1899(6): 95 pp., 5 pls., 83 table

Chaetoceros decipiens

Kokubo, S., 1952
Results of the observations on the plankton and oceanography of Mutsu Bay during 1950, reference being made also to the period 1946-1950. Bull Mar.Biol.Sta., Asamushi 5(1/4): 1-54, 3 tables,(fold-in), 1 fold-in.

Chaetoceros decipiens
Levring, T., 1945
Some culture experiments with marine plankton diatoms. Medd. Oceanografiska Institutet i Göteborg, No.9 (Göteborgs Kungl. Vetenskaps-och Vitterhets-Samhälles Handlingar, Sjätte Följden. Ser.B. Vol.3 (12), 1-17.

Chaetoceros decipiens
Lillick, L.C., 1940
Phytoplankton and planktonic protozoa of the offshore waters of the Gulf of Maine. Pt.II. Qualitative Composition of the Planktonic Flora. Trans. Am. Phil. Soc., n.s., 31(3):193-237, 13 text figs.

Chaetoceros decipiens
Lillick, L.C., 1938
Preliminary report of the phytoplankton of the Gulf of Maine. Am. Mid. Nat. 20(3):624-640, 1 text fig, 37 tables.

Chaetoceros decipiens
Lillick, L.C., 1937
Seasonal studies of the phytoplankton off Woods Hole, Massachusetts. Biol. Bull. LXXIII (3):488-503, 3 text figs.

Chaetoceros decipiens
Macdonald, R., 1933
An examination of plankton hauls made in the Suez Canal during the year 1928. Fish. Res. Dis., Notes & Mem. No.3, 11 pp., 1 chart.

Chaetoceros decipiens
Marukawa, H., 1921
Plankton lists and some new species of copepods, from the northern waters of Japan. Bull. Inst. Ocean., No.384, 15 pp., 3 pls., 1 chart. Monaco

Chaetoceros decipiens
Mangin, M. L., 1912
Phytoplancton de la croisière du "René" dans l'Atlantique (Septembre 1908). Ann. Inst. Ocean., n.s., 4(1):1-66, 2 pls., 41 text figs., 2 tables.

Chaetoceros decipiens
Massuti Algamora, M., 1949
Estudio de diez y seis muestras de plancton del Golfo de Nápoles. Publ. Inst. Biol. Appl. 5:85-94, 1 fold-in table.

Chaetoceros decipiens
Meunier, A., 1913
Microplancton de la Mer Flamande. 1. Chaetoceros. Mem. Mus. Roy. Hist. Nat., Belgique 7(2):1-55, 7 pls.

Chaetoceros decipiens
Morse, D.C., 1947
Some observations on seasonal variations in plankton population Patuxent River, Maryland 1943-1945. Bd. Nat. Res., Publ. No.65, Chesapeake Biol. Lab., 31, 3 figs.

Chaetoceros decipiens
Pavillard, J., 1925
Bacillariales. Rept. on the Danish Oceangr. Exped., 1908-10 to the Mediterranean and adj. seas. Vol.II., Biol. J4:72 pp., 116 text figs.

Chaetoceros decipiens
Pavillard, J., 1905
Recherches sur la flore pelagique (Phytoplankton) de l'Etang de Thau. Theses presentees a la Fac. Sci., Paris, 116 pp., 3 pls.

Chaetoceros decipiens
Pettersson, H., 1934
Scattering and extinction of light in sea-water. Meddelanden från Göteborgs Högskolas Oceanografiska Institut. 9 (Göteborgs Kungl. vetenskaps-och vitterhets-samhälles Handlingar. Femte Följden, Ser.B, 4(4)):1-16.

Chaetoceros decipiens
Pettersson, H., F. Gross, and F. Koczy, 1939
Large scale plankton cultures. Medd. från Oceanografiska Institutet i Göteborg No.3 (Göteborgs Kungl. vetenskaps-och Vitterhets-Samhälles Handlinger Femte Följden. Ser. B) Vol.6(13):1-24.
1. The plankton shaft. H. Pettersson
2. Experiment with phytoplankton. F. Gross and F. Koczy
3. Experiments with zooplankton.

Chaetoceros decipiens
Phifer, L.D. (undated)
The occurrence and distribution of plankton diatoms in Bering Sea and Bering Strait, July 26-August 24, 1934. Report of Oceanographic cruise of U.S. Coast Guard Cutter Chelan 1934, Part II(A):1-44 (mimeographed) plus fig.1 (after Pt.B)

Chaetoceros decipiens
Rampi, L., 1942
Ricerche sul fitoplancton del Mare Ligure 6. Le diatomee delle acque di Sanremo. Nuovo Giornale Botanico Italiano, N.S., 49:252-268.

Chaetoceros decipiens
Schodduyn, M., 1926
Observations faites dans la baie d'Ambleteuse (Pas de Calais). Bull. Inst. Ocean., Monaco, No. 482: 64 pp.

Chaetoceros decipiens
Schulz, B., and A. Wulff, 1929
Hydrographie und Oberflächen plankton des westlichen Barentsmeeres im Sommer 1927. Ber. deutschen wissensch. Komm. F. Meeresforsch. n.s., 4(5):232-372, 13 tables, 25 text figs.

Chaetoceros decipiens
Sousa e Silva, E., 1949
Diatomaceas e Dinoflagelados de Baia de Cascais. Portugalise Acta Biol., Volume: Julio Henriques, Ser. B: 300-383, 9 pls, 2 fold-in tables.

Chaetoceros decipiens
Sousa e Silva, E., and J. Dos Santos-Pinto, 1948
O Plancton da Baia de S. Martinho do Porto. 1. Diatomaceas e Dinoflagelados. Bol. Soc. Portuguese de Ciencias Naturais, 16(2):134-187, 6 pls. (Trav. Sta. Biol. Mar. de Lisbonne No. 52).

Chaetoceros decipiens
Stæmann-Nielsen, Einar, 1951
The marine vegetation of the Isefjord. A study on ecology and production. Medd. Komm. Danmarks Fiskeri-og Havundersøgelser. Ser. Plankton. 5(4); 114pp., 46 text figs.

Chaetoceros decipiens
Tratet, Gérard, 1964.
Variations du phytoplancton à Tanger. Trav. Inst. Sci., Cherifien, Rabat, Ser. Botan., (29):204 pp.

Chaetoceros decipiens
Uyeno, Fukuzo, 1961
Oceanographical and ecological studies on primary production of the sea, with special references to relationship between diatom production and temperature and chlorinity of water.
Rept., Fac. Fish., Pref. Univ., Mie, 4(1): 1-64.

Chaetoceros decipiens singularis
Cupp, Easter E., 1943
Marine plankton diatoms of the west coast of North America. Bull. S.I.O. 5(1):1-238, 5 pls., 168 text figs.

Chaetoceros Deflandrei n.sp.
Manguin, E., 1957.
Premier inventaire des diatomées de la Terre Adélie Antarctique. Espèces nouvelles. Rev. Algologique, n.s., 3(3):111-134.

Chaetoceros delicatulus
Brunel, J., 1962
Le phytoplancton de la Baie de Chaleurs. Inst. Botan., Univ. Montréal, Contrib. No. 77: 365 pp., 66 pls.

Chaetoceros delicatulum
Pavillard, J., 1905
Recherches sur la flore pelagique (Phytoplankton) de l'Etang de Thau. Theses presentees a la Fac. Sci., Paris, 116 pp., 3 pls.

Chaetoceros densum
Bigelow, H.B., and M. Leslie, 1930
Reconnaissance of the waters and plankton of Monterey Bay, July 1928. Bull. M.C.Z. 70(5):429-481, 43 text figs.

Chaetoceros densus
Braarud, T., 1945
A phytoplankton survey of the polluted waters of inner Oslo Fjord. Hvalrådets Skrifter, No.28, 142 pp., 19 text figs., 17 tables.

Chaetoceros densus
Brunel, J., 1962
Le phytoplancton de la Baie de Chaleurs. Inst. Botan., Univ. Montréal, Contrib. No. 77: 365 pp., 66 pls.

Chaetoceros densus
Cleve-Euler, A., 1951
Die Diatomeen von Schweden und Finnland. Kungl. Svenska Vetenskaps Akad. Handl., Fjärde Ser. 2(1): 161 pp., 6 pls.

Chaetoceros densa
Dangeard, P., 1927
Phytoplankton de la croisière du "Sylvana". Ann. Inst. Ocean., Monaco, n.s., 4(8):286-401, 54 text figs. (Feirer-Juin 1913).

Chaetoceros densus
Ercegovic, A., 1936
Etudes qualitative et quantitatives du phytoplancton dans les eaux cotières de l'Adriatique oriental moyen au cours de l'année 1934. Acta Adriatica 1(9):1-126

Chaetoceros densus
Forti, A., 1922
Ricerche sulla flora pelagica (fitoplancton) di Quarto dei Mille. Mem. R. Com. Talass. Ital. 97:248 pp., 13 pls.

Chaetoceros densus
Fraser, J. H., and A. Saville, 1949
Plankton distribution in Scottish and adjacent waters in 1948. Ann. Biol. 5:61-62.

Chaetoceros densum
Ghazzawi, F.M., 1939
Plankton of the Egyptian waters. A study of the Suez Canal Plankton. (A) Phytoplankton. Preliminary Report 83 pp. Notes and Memoires, Min. Commerce-Industry, Egypt, Hydrobiol. & Fish. 65 figs.

Chaetoceros densum
Gran, H.H., 1908
Diatomeen. Nordisches Plankton, Botanischer Teil pp. XIX.1-XIX 146; 178 text figs.

Chaetoceros densus
Gran, H.H., and T. Braarud, 1935
A quantitative study of the phytoplankton in the Bay of Fundy and the Gulf of Maine (including observations on hydrography, chemistry, and turbidity). J. Biol. Bd., Canada, 1(5):279-467, 69 text figs.

Chaetoceros densus

Grøntved, J., 1949
Investigations on the phytoplankton in the Danish Waddensea in July 1941. Medd. Komm. Danmarks Fiskeri og Havundersøgelser, ser. Plankton, 5(2):55 pp., 2 pls., 38 text figs.

Chaetoceros densum

Hendy, N. Ingram, 1964
An introductory account of the smaller algae of British coastal waters. V. Bacillariophyceae (Diatoms). Her Majesty's Stationary Office, 317 pp., 45 pls.

Chaetoceros densum

Iselin, C., 1930
A report on the coastal waters of Labrador based on explorations of the "Chance" during the summer of 1926. Proc. Am. Acad. Arts Sci., 66(1):1-37, 14 text figs.

Chaetoceros densus

Lillick, L.C., 1940
Phytoplankton and planktonic protozoa of the offshore waters of the Gulf of Maine. Pt.II. Qualitative Composition of the Planktonic Flora. Trans. Am. Phil. Soc., n.s., 31(3):193-237, 13 text figs.

Chaetoceros densus

Mangin, M. L., 1912
Phytoplancton de la croisière du "René" dans l'Atlantique (Septembre 1908). Ann. Inst. Ocean., n.s., 4(1):1-66, 2 pls., 41 text figs., 2 tables.

Chaetoceros densus

Margalef, R., 1949
Fitoplancton nerítico de la Costa Brava en 1947-48. Publ. Inst. Biol. Aplicada, 5:41-51, 3 text figs.

Chaetoceros densum

Meunier, A., 1913
Microplancton de la Mer Flamande. 1. Chaetoceros. Mem. Mus. Roy. Hist. Nat., Belgique 7(2):1-55, 7 pls.

Chaetoceros densus

Michailova, N.F., 1964
On the spreading of the habitat into the Black Sea of species of the genus Chaetoceros of northern seas and their biogeography. (In Russian).
Trudy Sevastopol Biol. Sta., 7:231-248.

Chaetoceros densa

Pavillard, J., 1925
Bacillariales. Rept. on the Danish Oceanogr. Exped., 1908-10 to the Mediterranean and adj. seas. Vol.II., Biol. J4:72 pp., 116 text figs.

Chaetoceros densum

Pavillard, J., 1905
Recherches sur la flore pelagique (Phytoplankton) de l'Etang de Thau. Theses presentees a la Fac. Sci., Paris, 116 pp., 3 pls.

Chaetoceros densus

Pettersson, H., 1934
Scattering and extinction of light in sea-water. Meddelanden från Göteborgs Högskolas Oceanografiska Institut 9 (Göteborgs Kungl. vetenskaps-och vitterhets-samhälles Handlingar. Femte Foljden, Ser.B, 4(4)):1-16.

Chaetoceros densus

Phifer, L.D. (undated)
The occurrence and distribution of plankton diatoms in Bering Sea and Bering Strait, July 26-August 24, 1934. Report of Oceanographic cruise of U.S. Coast Guard Cutter Chelan 1934, Part II(A):1-44 (mimeographed) plus fig.1 (after Pt.B)

Chaetoceros densus

Rampi, L., 1942
Ricerche sul fitoplancton del Mare Ligure 6. Le diatomee delle acque di Sanremo. Nuovo Giornale Botanico Italiano, N.S., 49:252-268.

Chaetoceros densum

Schodduyn, M., 1926
Observations faites dans la baie d'Ambleteuse (Pas de Calais). Bull. Inst. Ocean., Monaco, No. 482: 64 pp.

Chaetoceros densum

Sousa e Silva, E., 1949
Diatomaceas e Dinoflagelados de Baia de Cascais. Portugaliae Acta Biol., Volume: Julio Henriques, Ser. B: 300-383, 9 pls, 2 fold-in tables.

Chaetoceros densum

Sousa e Silva, E., and J. Dos Santos-Pinto, 1948
O Plancton da Baia de S. Martinho do Porto. 1. Diatomaceas e Dinoflagelados. Bol. Soc. Portuguese de Ciencias Naturais, 16(2):134-187, 6 pls. (Trav. Sta. Biol. Mar. de Lisbonne No. 52).

Chaetoceros densum solitaria

Pavillard, J., 1905
Recherches sur la flore pelagique (Phytoplankton) de l'Etang de Thau. Theses presentees a la Fac. Sci., Paris, 116 pp., 3 pls.

Chaetoceros denticulatum

Allen, W.E., and E.E. Cupp, 1935
Plankton diatoms of the Java Sea. Annales du Jardin Botanique de Buitenzorg XLIV (2):101-174, figs.1-127.

(drawings of all species mentioned)

Chaetoceros diadema

Bigelow, H.B., and M. Leslie, 1930
Reconnaissance of the waters and plankton of Monterey Bay, July 1928. Bull. M.C.Z., 70(5):429-481, 43 text figs.

Chaetoceros diadema

Braarud, T., 1945
Experimental studies on marine plankton diatoms. Avhandlingar utgitt av Det Norske Videnskaps-Akademi i Oslo. 1. Mat.-Naturv. Klasse 1944. No.10, 1-16, 1 text fig.

Chaetoceros diadema

Braarud, T., and Adam Bursa, 1939
On the phytoplankton of the Oslo Fjord, 1933-1934. Hvalrådets Skr. No.19:1-63; 9 text figs. Reviewed. J. du. Cons. 14(3):418-420. A.C. Gardiner.

Chaetoceros diadema

Brunel, J., 1962
Le phytoplancton de la Baie de Chaleurs. Inst. Botan., Univ. Montréal, Contrib. No. 77: 365 pp., 66 pls.

Chaetoceros diadema

Brunel, Jules, 1962
Le phytoplancton de la Baie des Chaleurs. Contrib. Ministère de la Chasse et des Pêcheries, Province de Québec, No. 91: 365 pp.

Chaetoceros diadema

Cleve-Euler, A., 1951
Die Diatomeen von Schweden und Finnland. Kungl. Svenska Vetenskaps Akad. Handl., Fjärde Ser. 2(1): 161 pp., 6 pls.

Chaetoceros diadema

Davidson, V.M., 1931
Biological and oceanographic conditions in Hudson Bay. Contr. Canadian Biol. Fish., n.s., 6(26):497-509, 7 textfigs.

Chaetoceros diadema

Forti, A., 1922
Ricerche sulla flora pelagica (fitoplancton) di Quarto dei Mille. Mem. R. Com. Talass. Ital. 97:248 pp., 13 pls.

Chaetoceros diadema

Gilbert, J.Y., and W.E. Allen, 1943
The phytoplankton of the Gulf of California obtained by the "E.W. Scripps" in 1939 and 1940. J. Mar. Res. V(2):89-110, figs.30-31.

Chaetoceros diadema

Gran, H.H., 1908
Diatomeen. Nordisches Plankton, Botanischer Teil pp. XIX.1-XIX 146; 178 text figs.

Chaetoceros diadema

Gran, H.H., 1897
Protophyta: Diatomaceae, Silico-flagellata and Cilioflagellata. Den Norske Nordhavs Expedition 1876-1878, h, 24, 36 pp., 4 pls.

Chaetoceros diadema

Gran, H.H., and T. Braarud, 1935
A quantitative study of the phytoplankton in the Bay of Fundy and the Gulf of Maine (including observations on hydrography, chemistry, and turbidity). J. Biol. Bd., Canada, 1(5):279-467, 69 text figs.

Chaetoceros diadema

Gran, H. H. and E. C., Angst, 1931
Plankton diatoms of Puget Sound. Publ. Puget Sound Biol. Sta. 7:417-519, 95 text figs.

Chaetoceros diadema

Hargraves, Paul E., 1972.
Studies on marine plankton diatoms. I. Chaetoceros diadema (Ehr.) Gran: life cycle, structural morphology, and regional distribution.
Phycologia 11 (3/4): 247-257.

Chaetoceros diadema

Hendy, N. Ingram, 1964
An introductory account of the smaller algae of British coastal waters. V. Bacillariophyceae (Diatoms). Her Majesty's Stationary Office, 317 pp., 45 pls.

Chaetoceros diadema

Iselin, C., 1930
A report on the coastal waters of Labrador based on explorations of the "Chance" during the summer of 1926. Proc. Am. Acad. Arts Sci., 66(1):1-37, 14 text figs.

Chaetoceros diadema

Jorgensen, E., 1900
Protophyten und Protozoën im Plankton aus der Norwegischen Westkerste. Bergens Mus. Aarb. 1899(6): 95 pp., 5 pls., 83 tables.

Chaetoceros diadema

Lillick, L.C., 1938
Preliminary report of the phytoplankton of the Gulf of Maine. Am. Mid. Nat. 20(3):624-640, 1 text figs 37 tables.

Chaetoceros diadema

Marukawa, H., 1921
Plankton lists and some new species of copepods, from the northern waters of Japan. Bull. Inst. Ocean., No.384, 15 pp., 3 pls., 1 chart. Monaco

Chaetoceros diadema

Meunier, A., 1913
Microplancton de la Mer Flamande. 1. Chaetoceros. Mem. Mus. Roy. Hist. Nat., Belgique 7(2):1-55, 7 pls.

Chaetoceros diadema
Pavillard, J., 1925
Bacillariales. Rept. on the Danish Oceanogr. Exped., 1908-10 to the Mediterranean and adj. seas. Vol.II., Biol. J4:72 pp., 116 text figs.

Chaetoceros diadema
Pettersson, H., 1934
Scattering and extinction of light in sea-water. Meddelanden från Göteborgs Högskolas Oceanografiska Institut. 9 (Göteborgs Kungl. vetenskaps-och vitterhets-samhälles Handlingar. Femte Foljden, Ser.B, 4(4)):1-16.

Chaetoceros diadema
Pettersson, H., H. Höglund, S. Landberg, 1934.
Submarine daylight and the photosynthesis of phytoplankton. Medd. fran Göteborgs Högskolas Oceanografiska Inst. 10 (Göteborgs Kungl. Vetenskaps-och Vitterhets-samhälles handlingar. Femte Följden, Ser. B 4(5):1-17.

Chaetoceros diadema
Phifer, L.D. (undated)
The occurrence and distribution of plankton diatoms in Bering Sea and Bering Strait, July 26-August 24, 1934. Report of Oceanographic cruise of U.S. Coast Guard Cutter Chelan 1934, Part II(A):1-44 (mimeographed) plus fig.1 (after Pt.B)

Chaetoceros diadema
Schodduyn, M., 1926
Observations faites dans la baie d'Ambleteuse (Pas de Calais). Bull. Inst. Ocean., Monaco, No. 482: 64 pp.

Chaetoceros diadema tenuis
Forti, A., 1922
Ricerche sulla flora pelagica (fitoplancton) di Quarto dei Mille. Mem. R. Com. Talass. Ital. 97:248 pp., 13 pls.

Chaetoceros dichaeta
Balech, E., 1947
Contribution al conocimiento del plancton antartico. Plankton del Mar de Bellinghausen. Physis 20:75-91, 76 figs. on 8 pls.

Chaetoceros dichaeta
Boden, B. P., 1949.
The diatoms collected by the U.S.S. CACOPAN in the Antarctic in 1947. J. Mar. Res. 8(1):6-13, 3 textfigs.

Chaetoceros dichaeta
Boden, Brian, 1948 Chaetoceros dichaeta
Marine plankton diatoms on operation HIGHJUMP in: Some oceanographic observations on operation HIGHJUMP. By R.S. Dietz. USNEL Rept. No.55, 97 pp., 41 figs. 7 July 1948.

Chaetoceros dichaeta
Brunel, J., 1962
Le phytoplancton de la Baie de Chaleurs. Inst. Botan., Univ. Montréal, Contrib. No. 77: 365 pp., 66 pls.

Chaetoceros dichaeta
Gran, H., 1908
Diatomeen. Nordisches Plankton, Botanischer Teil pp. XIX 1-XIX 146; 178 text figs.

Chaetoceros dichaeta
Central Meteorological Observatory, 1949
Report on sea and weather observation on Antarctic Whaling Ground (1947-48). Ocean. Mag., Japan, 1(1):49-88, 17 text figs.

Chaetoceros dichaeta
Cupp, Easter E., 1943
Marine plankton diatoms of the west coast of North America. Bull. S.I.O. 5(1):1-238, 5 pls., 168 text figs.

Chaetoceros dichaeta
Frenguelli, Joaquin, and Hector Antonio Orlando, 1959.
Operacion MERLUZA. Diatomeas y silico-flagelados del plancton del "VI Crucero". Servicio Hidrogr. Naval., Argentina, Publ. No. H. 619: 5-62.

Chaetoceros dichaeta
Hendy, N. Ingram, 1964
An introductory account of the smaller algae of British coastal waters. V. Bacillariophyceae (Diatoms). Her Majesty's Stationary Office, 317 pp., 45 pls.

Chaetoceros dichaeta
Hendey, N.I., 1937
The plankton diatoms of the southern seas. Discovery Repts. 16:151-364, pls.6-13.

Chaetoceros dichaeta
Mangin, L., 1915
Phytoplancton de L'Antartique. Deuxieme Exped. Ant. Francaise (1908-1910), 95 pp., 3 pls., 58 text figs.

Chaetoceros dichaeta
Manguin, E., 1954
Diatomees marines provenant de l'ile Heard (Australian National Antarctic Expedition). Rev. Algol., n.s., 1: 14-24.

Chaetoceros dicladia
Krasske, G., 1941
Die Kieselalgen des chilenischen Küstenplankton (Aus dem südchilenischen Kusten gebiet IX). Arch. Hydrobiol. 38 (2):260-287.

Chaetoceros dicladia
Manguin, E., 1954
Diatomees marines provenant de l'ile Heard (Australian National Antarctic Expedition). Rev. Algol., n.s., 1: 14-24.

Chaetoceros didymus
Allen, W. E., 1938
The Templeton Crocker Expedition to the Gulf of California in 1935 - The Phytoplankton. Amer. Microsc. Soc., Trans. 57:328-335.

Chaetoceros didymus
Allen, W.E., 1937
Plankton diatoms of the Gulf of California obtained by the G. Allan Hancock Expedition of 1936. The Hancock Pacific Expeditions, Univ. So. Calif. Publ. 3:47-59, 1 fig.

Chaetoceros didymum
Allen, W.E., and E.E. Cupp, 1935
Plankton diatoms of the Java Sea. Annales du Jardin Botanique de Buitenzorg XLIV (2):101-174, figs.1-127.
(drawings of all species mentioned)

Chaetoceros didymum
Bigelow, H.B., and M. Leslie, 1930
Reconnaissance of the waters and plankton of Monterey Bay, July 1928. Bull. M.C.Z., 70(5):429-481, 43 text figs.

Chaetoceros didymus
Blasco, Dolores, 1970.
Estudio de la morfología de Chaetoceros didymus al microscopio electrónico. Inv. Pesq. Barcelona 34(2):149-155

Chaetoceros didymus
Boden, B.P., 1950
Some marine plankton diatoms from the west coast of South Africa. Trans. R.Soc. S. Africa. 32:321-434, 100 text figs.

Chaetoceros didymum
Boden, B. P., 1949
The diatoms collected by the U.S.S. CACOPAN in the Antarctic in 1947. J. Mar. Res. 8(1):6-13, 3 textfigs.

Chaetoceros didymum
Boden, Brian, 1948 Chaetoceros didymum
Marine plankton diatoms on operation HIGHJUMP in: Some oceanographic observations on operation HIGHJUMP. By R.S. Dietz. USNEL Rept. No.55, 97 pp., 41 figs. 7 July 1948.

Chaetoceros didymus
Brunel, J., 1962
Le phytoplancton de la Baie de Chaleurs. Inst. Botan., Univ. Montréal, Contrib. No. 77: 365 pp., 66 pls.

Chaetoceros didymus
Cleve-Euler, A., 1951
Die Diatomeen von Schweden und Finnland. Kungl. Svenska Vetenskaps Akad. Handl., Fjärde Ser. 2(1): 161 pp., 6 pls.

Chaetoceros didymus
Cupp, Easter E., 1943
Marine plankton diatoms of the west coast of North America. Bull. S.I.O. 5(1):1-238, 5 pls., 168 text figs.

Chaetoceros didymus
Cupp, E.E., 1934
Analysis of marine diatom collections taken from the Canal Zone to California during March, 1933. Trans. Am. Micros. Soc. LIII (1):22-29, 1 map.

Chaetoceros didymus
Cupp, E.E. and Allen, W.E., 1938
Plankton diatoms of the Gulf of California obtained by Allan Hancock Pacific Expedition of 1937. The Hancock Pacific Expeditions, The Univ. So. Calif. Publ. 3: 61-74, 1 map, pls.4-15.

Chaetoceros didymum
Dangeard, P., 1927
Phytoplankton de la croisière du "Sylvana". Ann. Inst. Ocean., Monaco, n.s., 4(8):286-401, 54 text figs. (Feirer-Juin 1913).

Chaetoceros didymum
Eskinazi Enide e Shigekatsv Satô (1963-1964) 1966.
Contribuição ao estudo das diatomaceas da Praia de Piedade.
Trabhs Inst. Oceanogr., Univ. Recife, 5 (5/6): 73-114.

Chaetoceros didymus
Frenguelli, Joaquin, and Hector Antonio Orlando, 1959.
Operacion MERLUZA. Diatomeas y silico-flagelados del plancton del "VI Crucero". Servicio Hidrogr. Naval., Argentina, Publ. No. H. 619: 5-62.

Chaetoceros didymus
Ghazzawi, F.M., 1939
Plankton of the Egyptian waters. A study of the Suez Canal Plankton. (A) Phytoplankton. Preliminary Report 83 pp. Notes and Memoires, Min. Commerce-Industry, Egypt, Hydrobiol. & Fish. 65 figs.

Chaetoceros didymus
Gilbert, J.Y., and W.E. Allen, 1943
The phytoplankton of the Gulf of California obtained by the "E.W. Scripps" in 1939 and 1940. J. Mar. Res. V(2):89-110, figs.30-31.

Chaetoceros didymum
Gran, H.H., 1908
Diatomeen. Nordisches Plankton, Botanischer Teil pp. XIX.1-XIX.146; 178 text figs.

Chaetoceros didymum
Gran, H.H., 1897
Protophyta: Diatomaceae, Silico-flagellata and Cilioflagellata. Den Norske Nordhavs Expedition 1876-1878, h. 24, 36 pp., 4 pls.

Chaetoceros didymus
Gran, H.H., and T. Braarud, 1935
A quantitative study of the phytoplankton in the Bay of Fundy and the Gulf of Maine (including observations on hydrography, chemistry, and turbidity). J. Biol. Bd., Canada, 1(5):279-467, 69 text figs.

Chaetoceros didymus
Gran, H.H. and E.C. Angst, 1931
Plankton diatoms of Puget Sound. Publ. Puget Sound Biol. Sta. 7:417-519, 95 text figs.

Chaetoceros didymum
Hendy, N. Ingram, 1964
An introductory account of the smaller algae of British coastal waters. V. Bacillariophyceae (Diatoms). Her Majesty's Stationary Office, 317 pp., 45 pls.

Chaetoceros didymum
Hendey, N.I., 1937
The plankton diatoms of the southern seas. Discovery Repts. 16:151-364, pls.6-13.

Chaetoceros didymus
Jorgensen, E., 1900
Protophyten und Protozoën im Plankton aus der Norwegischen Westkerste. Bergens Mus. Aarb. 1899(6): 95 pp., 5 pls., 83 tables.

Chaetoceros didymus
Kokubo, S., 1952
Results of the observations on the plankton and oceanography of Mutsu Bay during 1950, reference being made also to the period 1946-1950. Bull. Mar. Biol. Sta., Asamushi 5(1/4): 1-54, 3 tables, (fold-in), 1 fold-in.

Chaetoceros didymus
Lillick, L.C., 1940
Phytoplankton and planktonic protozoa of the offshore waters of the Gulf of Maine. Pt. II. Qualitative Composition of the Planktonic Flora. Trans. Am. Phil. Soc., n.s. 31(3):193-237, 13 text figs.

Chaetoceros didymus
Lillick, L.C., 1938
Preliminary report of the phytoplankton of the Gulf of Maine. Am. Mid. Nat. 20(3):624-640, 1 text fig, 37 tables.

Chaetoceros didymus
Lillick, L.C., 1937
Seasonal studies of the phytoplankton off Woods Hole, Massachusetts. Biol. Bull. LXXIII (3):488-503, 3 text figs.

Chaetoceros didymus
Mangin, M.L., 1912
Phytoplankton de la croisière du "René" dans l'Atlantique (Septembre 1908). Ann. Inst. Ocean., n.s., 4(1):1-66, 2 pls., 41 text figs., 2 tables.

Chaetoceros didymus
Mann, A., 1907
Report on the diatoms of the Albatross voyages in the Pacific Ocean, 1888-1904. Contrib. U.S. Nat. Herb. 10(5):221-419, Pls. XLIV-LIV.

Chaetoceros didymum
Meunier, A., 1913
Microplancton de la Mer Flamande. 1. Chaetoceros. Mem. Mus. Roy. Hist. Nat., Belgique 7(2):1-55, 7 pls.

Chaetoceros didymus
Morse, D.C., 1947
Some observations on seasonal variations in plankton population Patuxant River, Maryland 1943-1945. Bd. Nat. Res., Publ. No. 65, Chesapeake Biol. Lab., 31, 3 figs.

Chaetoceros didynum
Pavillard, J., 1925
Bacillariales. Rept. on the Danish Oceangr. Exped., 1908-10 to the Mediterranean and adj. seas. Vol.II., Biol. J4:72 pp., 116 text figs.

Chaetoceros didymus
Phifer, L.D. (undated)
The occurrence and distribution of plankton diatoms in Bering Sea and Bering Strait, July 26-August 24, 1934. Report of Oceanographic cruise of U.S. Coast Guard Cutter Chelan 1934, Part II(A):1-44 (mimeographed) plus fig.1 (after Pt.B)

Chaetoceros didymus
Rampi, L., 1942
Ricerche sul fitoplancton del Mare Ligure 6. Le diatomee delle acque di Sanremo. Nuovo Giornale Botanico Italiano, N.S., 49:252-268.

Chaetoceros didymum
Schodduyn, M., 1926
Observations faites dans la baie d'Ambleteuse (Pas de Calais). Bull. Inst. Ocean., Monaco, No. 482: 64 pp.

Chaetoceros didymus
Schröder, B., 1900
Phytoplankton des Golfes von Neapel nebst vergleichenden Ausblicken auf das atlantischen Ozean. Mitt. Zool. Stat. Neapel, 14:1-38.

Chaetoceros didymus
Sousa e Silva, E., 1949
Diatomaceas e Dinoflagelados de Baía de Cascais. Portugaliae Acta Biol., Volume: Julio Henriques, Ser. B: 300-383, 9 pls, 2 fold-in tables.

Chaetoceros didymum
Sousa a Silva, E., and J. Dos Santos-Pinto, 1948
O Plancton da Baía de S. Martinho do Porto. 1. Diatomaceas e Dinoflagelados. Bol. Soc. Portuguese de Ciencias Naturais, 16(2):134-187, 6 pls. (Trav. Sta. Biol. Mar. de Lisbonne No. 52).

Chaetoceros didymus
Uyeno, Fukuzo, 1961
Oceanographical and ecological studies on primary production of the sea, with special references to relationship between diatom production and temperature and chlorinity of water. Rept. Fac. Fish., Pref. Univ. Mie, 4(1): 1-64.

Chaetoceros didymus anglica
Cupp, Easter E., 1943
Marine plankton diatoms of the west coast of North America. Bull. S.I.O. 5(1):1-238, 5 pls., 168 text figs.

Chaetoceros didymum anglica
Marukawa, H., 1921
Plankton lists and some new species of copepods from the northern waters of Japan. Bull. Inst. Ocean., No. 384, 15 pp., 3 pls., 1 chart. Monaco

Chaetoceros didymus protuberans
Cupp, Easter E., 1943
Marine plankton diatoms of the west coast of North America. Bull. S.I.O. 5(1):1-238, 5 pls., 168 text figs.

Chaetoceros didymus singularis forma nov.
Takano, H., 1960
Plankton diatoms in the eastern Caribbean Sea. J. Oceanogr. Soc., Japan, 16(4): 180-184.

Chaetoceros var. anglica didymus
Allen, W.E., and E.E. Cupp, 1935
Plankton diatoms of the Java Sea. Annales du Jardin Botanique de Buitenzorg XLIV (2):101-174, figs.1-127.
(drawings of all species mentioned)

Chaetoceros didymus anglica
Ercegovic, A., 1940
Weitere Untersuchungen über einige hydrographische Verhältnisse und über die Phytoplanktonproduktion in den Gewässern der östlichen Mitteladria. Acta Adriatica 2(3):95-134, 8 text figs.

Chaetoceros didymum var. protuberans
Allen, W.E., and E.E. Cupp, 1935
Plankton diatoms of the Java Sea. Annales du Jardin Botanique de Buitenzorg XLIV (2):101-174, figs.1-127.
(drawings of all species mentioned)

Chaetoceros difficile
Bigelow, H.B., and M. Leslie, 1930
Reconnaissance of the waters and plankton of Monterey Bay, July 1928. Bull. M.C.Z., 70(5):429-481, 43 text figs.

Chaetoceros difficile
Boden, B.P., 1950
Some marine plankton diatoms from the west coast of South Africa. Trans. R.Soc. S. Africa. 32:321-434, 100 text figs.

Chaetoceros difficilis
Brunel, J., 1962
Le phytoplancton de la Baie de Chaleurs. Inst. Botan., Univ. Montréal, Contrib. No. 77: 365 pp., 66 pls.

Chaetoceros difficilis
Cleve-Euler, A., 1951
Die Diatomeen von Schweden und Finnland. Kungl. Svenska Vetenskaps Akad. Handl., Fjärde Ser. 2(1): 161 pp., 6 pls.

Chaetoceros difficilis
Cupp, Easter E., 1943
Marine plankton diatoms of the west coast of North America. Bull. S.I.O. 5(1):1-238, 5 pls., 168 text figs.

Chaetoceros difficile
Gran, H.H., 1908
Diatomeen. Nordisches Plankton, Botanischer Teil pp. XIX.1-XIX.146; 178 text figs.

Chaetoceros difficile
Hendy, N. Ingram, 1964
An introductory account of the smaller algae of British coastal waters. V. Bacillariophyceae (Diatoms). Her Majesty's Stationary Office, 317 pp., 45 pls.

Chaetoceros dipyrenops
Brunel, J., 1962
Le phytoplancton de la Baie de Chaleurs. Inst. Botan., Univ. Montréal, Contrib. No. 77: 365 pp., 66 pls.

Chaetoceros dipyrenops sp. nov.
Meunier, A., 1913
Microplancton de la Mer Flamande. 1. Chaetoceros. Mem. Mus. Roy. Hist. Nat., Belgique 7(2):1-55, 7 pls.

Oceanographic Index: Marine Organisms Cumulation, 1946-1973

Chaetoceros dispar n.sp.

Castracane degli Antelminelli, F., 1886
 1. Report on the Diatomaceae collected by H.M.S. Challenger during the years 1873-1876. Rept. Sci. Results, H.M.S. Challenger, Botany Vol. II, 178 pp., 30 pls.

Chaetoceros distans

Allen, W.E., and E.E. Cupp, 1935
 Plankton diatoms of the Java Sea. Annales du Jardin Botanique de Buitenzorg XLIV (2):101-174, figs.1-127.

(drawings of all species mentioned)

Chaetoceros distans

Anon., 1951
Bulletin of the Marine Biological Station of Asamushi 4(3/4): 15 pp.

Chaetoceros distans

Kokubo, S., 1952
 Results of the observations on the plankton and oceanography of Mutsu Bay during 1950, reference being made also to the period 1946-1950. Bull Mar.Biol.Sta., Asamushi 5(1/4): 1-54, 3 tables,(fold-in), 1 fold-in.

Chaetoceros distans

Uyeno, Fukuzo, 1961
 Oceanographical and ecological studies on primary production of the sea, with special references to relationship between diatom production and temperature and chlorinity of water.
 Rept., Fac. Fish., Pref. Univ. Mie, 4(1): 1-64.

Chaetoceros diversicurvatus

Cleve-Euler, A., 1951
 Die Diatomeen von Schweden und Finnland. Kungl. Svenska Vetenskaps Akad. Handl., Fjärde Ser. 2(1): 161 pp., 6 pls.

Chaetoceros diversum

Allen, W.E., and E.E. Cupp, 1935
 Plankton diatoms of the Java Sea. Annales du Jardin Botanique de Buitenzorg XLIV (2):101-174, figs.1-127.

(drawings of all species mentioned)

Chaetoceros diversus

Brunel, J., 1962
 Le phytoplancton de la Baie de Chaleurs. Inst. Botan., Univ. Montréal, Contrib. No. 77: 365 pp., 66 pls.

Chaetoceros diversus

Cleve-Euler, A., 1951
 Die Diatomeen von Schweden und Finnland. Kungl. Svenska Vetenskaps Akad. Handl., Fjärde Ser. 2(1): 161 pp., 6 pls.

Chaetoceros diversus

Cupp, Easter E., 1943
 Marine plankton diatoms of the west coast of North America. Bull. S.I.O. 5(1):1-238, 5 pls., 168 text figs.

Chaetoceros diversus

Cupp, E.E., 1934
 Analysis of marine diatom collections taken from the Canal Zone to California during March, 1933. Trans. Am. Micros. Soc. LIII (1):22-29, 1 map.

Chaetoceros diversus

Delegazione Italiana della Commissione Internazionale per l'Esplorazione Scientifica del Mediterraneo, 1941
 Note sul plancton della Laguna veneta. [Memoria CCLXXXIX], Arch. di Ocean. e Limn. Anno I, Fasc. I, 1941 XIX: 31-57 pp.

Chaetoceros diversus

Ercegovic, A., 1936
 Etudes qualitative et quantitatives du phytoplancton dans les eaux cotières de l'Adriatique oriental moyen au cours de l'année 1934. Acta Adriatica 1(9):1-126

Chaetoceros diversus

Eskinazi Enide e Shigekatsv Sato (1963-1964) 1966.
 Contribuição ao estudo das diatomaceas da Praia de Piedade.
 Trabhs Inst. Oceanogr., Univ. Recife, 5 (5/6): 73-114.

Chaetoceros diversus

Frenguelli, Joaquin, and Hector Antonio Orlando, 1959.
 Operacion MERLUZA. Diatomeas y silicoflagelados del plancton del "VI Crucero". Servicio Hidrogr. Naval., Argentina, Publ. No. H. 619: 5-62.

Chaetoceros diversum

Gran, H.H., 1908
 Diatomeen. Nordisches Plankton, Botanischer Teil pp. XIX.1-XIX 146; 178 text figs.

Chaetoceros diversum

Hendy, N. Ingram, 1964
 An introductory account of the smaller algae of British coastal waters. V. Bacillariophyceae (Diatoms).
 Her Majesty's Stationary Office, 317 pp., 45 pls.

Chaetoceros diversus

Kokubo, S., 1952
 Results of the observations on the plankton and oceanography of Mutsu Bay during 1950, reference being made also to the period 1946-1950. Bull Mar.Biol.Sta., Asamushi 5(1/4): 1-54, 3 tables,(fold-in), 1 fold-in.

Chaetoceros diversum

Pavillard, J., 1925
 Bacillariales. Rept. on the Danish Oceangr. Exped., 1908-10 to the Mediterranean and adj. seas. Vol.II., Biol. J4:72 pp., 116 text figs.

Chaetoceros diversum

Pavillard, J., 1905
 Recherches sur la flore pelagique (Phytoplankton) de l'Etang de Thau. Theses presentees a la Fac. Sci., Paris, 116 pp., 3 pls.

Chaetoceros diversus

Rampi, L., 1942
 Ricerche sul fitoplancton del Mare Ligure 6. Le diatomee delle acque di Sanremo. Nuovo Giornale Botanico Italiano, N.S., 49:252-268.

Chaetoceros diversus

Rampi, L., 1940
 Diatomee del Mare Adriatico. Nuovo Giornale Botanico Italiano, n.s., 47:559-608.

Chaetoceros diversum mediterranea n. var.

Schröder, B., 1900
 Phytoplankton des Golfes von Neapel nebst vergleichenden Ausblicken auf das atlantischen Ozean. Mitt. Zool. Stat. Neapel, 14:1-38.

Chaetoceros eibenii

Allen, W. E., 1938
 The Templeton Crocker Expedition to the Gulf of California in 1935 - The Phytoplankton. Amer. Microsc. Soc., Trans. 57:328-335.

Chaetoceros eibenii

Allen, W.E., 1937
 Plankton diatoms of the Gulf of California obtained by the G. Allan Hancock Expedition of 1936. The Hancock Pacific Expeditions, Univ. So. Calif. Publ. 3:47-59, 1 fig.

Chaetoceros Eibenii

Allen, W.E., and E.E. Cupp, 1935
 Plankton diatoms of the Java Sea. Annales du Jardin Botanique de Buitenzorg XLIV (2):101-174, figs.1-127.

(drawings of all species mentioned)

Chaetoceros eibenii

Brunel, J., 1962
 Le phytoplancton de la Baie de Chaleurs. Inst. Botan., Univ. Montréal, Contrib. No. 77: 365 pp., 66 pls.

Chaetoceros eibenii

Cupp, Easter E., 1943
 Marine plankton diatoms of the west coast of North America. Bull. S.I.O. 5(1):1-238, 5 pls., 168 text figs.

Chaetoceros Eibenii

Dangeard, P., 1927
 Phytoplankton de la croisière du "Sylvana". Ann. Inst. Ocean. Monaco, n.s., 4(8):286-401, 54 text figs. (Feirer-Juin 1913).

Chaetoceros eibenii

Gran, H. H. and E. C. Angst, 1931
 Plankton diatoms of Puget Sound. Publ. Puget Sound Biol. Sta. 7:417-519, 95 text figs.

Chaetoceros eibenii

Hendy, N. Ingram, 1964
 An introductory account of the smaller algae of British coastal waters. V. Bacillariophyceae (Diatoms).
 Her Majesty's Stationary Office, 317 pp., 45 pls.

Chaetoceros Eibenii

Meunier, A., 1913
 Microplancton de la Mer Flamande. 1. Chaetoceros. Mem. Mus. Roy. Hist. Nat., Belgique 7(2):1-55, 7 pls.

Chaetoceros eibenii

Morse, D.C., 1947
 Some observations on seasonal variations in plankton population Patuxant River, Maryland 1943-1945. Bd. Nat. Res., Publ. No.65, Chesapeake Biol. Lab., 31, 3 figs.

Chaetoceros eibenii

Pavillard, J., 1925
 Bacillariales. Rept. on the Danish Oceangr. Exped., 1908-10 to the Mediterranean and adj. seas. Vol.II., Biol. J4:72 pp., 116 text figs.

Chaetoceros Eibenii

Rumkówna, A., 1948
 List of the phytoplankton species occurring in the superficial water layers in the Gulf of Gdańsk. Bull. Lab. mar., Gdynia, No. 4: 139-141 with tables in back.

Chaetoceros eibenii

Sousa e Silva, E., 1949
 Diatomaceas e Dinoflagelados de Baia de Cascais. Portugaliae Acta Biol., Volume: Julio Henriques, Ser. B: 300-383, 9 pls, 2 fold-in tables.

Chaetoceros Eibenii

Sousa a Silva, E., and J. Dos Santos-Pinto, 1948
 O Plancton da Baia de S. Martinho do Porto. 1. Diatomaceas e Dinoflagelados. Bol. Soc. Portuguese de Ciencias Naturais, 16(2):134-187, 6 pls. (Trav. Sta. Biol. Mar. de Lisbonne No. 52).

Chaetoceros elmorei

Brunel, J., 1962
 Le phytoplancton de la Baie de Chaleurs. Inst. Botan., Univ. Montréal, Contrib. No. 77: 365 pp., 66 pls.

Chaetoceros exospermum
Hendy, N. Ingram, 1964
An introductory account of the smaller algae of British coastal waters. V. Bacillariophyceae (Diatoms).
Her Majesty's Stationary Office, 317 pp., 45 pls.

Chaetoceros exospermum n.sp.
Meunier, A., 1913
Microplancton de la Mer Flamande. 1. Chaetoceros. Mem. Mus. Roy. Hist. Nat., Belgique 7(2):1-55, 7 pls.

Chaetoceros externum n.sp.
Gran, H.H., 1897
Protophyta: Diatomaceae, Silico-flagellata and Cilioflagellata. Den Norske Nordhavs Expedition 1876-1878, h. 24, 36 pp., 4 pls.

Chaetoceros externum
Hendy, N. Ingram, 1964
An introductory account of the smaller algae of British coastal waters. V. Bacillariophyceae (Diatoms).
Her Majesty's Stationary Office, 317 pp., 45 pls.

Chaetoceros filiferum
Hendey, N.I., 1937
The plankton diatoms of the southern seas. Discovery Repts. 16:151-364, pls.6-13.

Chaetoceros filiformis
Brunel, J., 1962
Le phytoplancton de la Baie de Chaleurs. Inst. Botan., Univ. Montréal, Contrib. No. 77: 365 pp., 66 pls.

Chaetoceros filiforme
Hendy, N. Ingram, 1964
An introductory account of the smaller algae of British coastal waters. V. Bacillariophyceae (Diatoms).
Her Majesty's Stationary Office, 317 pp., 45 pls.

Chaetoceros filiforme
Meunier, A., 1913
Microplancton de la Mer Flamande. 1. Chaetoceros. Mem. Mus. Roy. Hist. Nat., Belgique 7(2):1-55, 7 pls.

Chaetoceros flexuosus n.sp.
Mangin, L., 1915
Phytoplancton de L'Antartique. Deuxieme Exped. Ant. Francaise (1908-1910), 95 pp., 3 pls., 58 text figs.

Chaetoceros forcipatus
Mangin, L., 1915
Phytoplancton de L'Antartique. Deuxieme Exped. Ant. Francaise (1908-1910), 95 pp., 3 pls., 58 text figs.

Chaetoceros fragilis
Brunel, J., 1962
Le phytoplancton de la Baie de Chaleurs. Inst. Botan., Univ. Montréal, Contrib. No. 77: 365 pp., 66 pls.

Chaetoceros fragilis
Cleve-Euler, A., 1951
Die Diatomeen von Schweden und Finnland. Kungl. Svenska Vetenskaps Akad. Handl., Fjärde Ser. 2(1): 161 pp., 6 pls.

Chaetoceros fragile
Hendy, N. Ingram, 1964
An introductory account of the smaller algae of British coastal waters. V. Bacillariophyceae (Diatoms).
Her Majesty's Stationary Office, 317 pp., 45 pls.

Chaetoceros fragile
Hendey, N.I., 1937
The plankton diatoms of the southern seas. Discovery Repts. 16:151-364, pls.6-13.

Chaetoceros furca
Gran, H.H., 1908
Diatomeen. Nordisches Plankton, Botanischer Teil pp. XIX 1-XIX 146; 178 text figs.

Chaetoceros furca
Pavillard, J., 1905
Recherches sur la flore pelagique (Phytoplankton) de l'Etang de Thau. Theses presentees a la Fac. Sci., Paris, 116 pp., 3 pls.

Chaetoceros furca
Schröder, B., 1900
Phytoplankton des Golfes von Neapel nebst vergleichenden Ausblicken auf das atlantischen Ozean. Mitt. Zool. Stat. Neapel, 14:1-38.

Chaetoceros furca (figs.)
Sousa e Silva, E., 1949
Diatomacees e Dinoflagelados de Baia de Cascais. Portugaliae Acta Biol., Volume: Julio Henriques, Ser. B: 300-383, 9 pls, 2 fold-in tables.

Chaetoceros furcatus
Mann, A., 1907
Report on the diatoms of the Albatross voyages in the Pacific Ocean, 1888-1904. Contrib. U. S. Nat. Herb. 10(5):221-419, Pls. XLIV-LIV.

Chaetoceros furcellatus
Brunel, J., 1962
Le phytoplancton de la Baie de Chaleurs. Inst. Botan., Univ. Montréal, Contrib. No. 77: 365 pp., 66 pls.

Chaetoceros furcellatus
Brunel, Jules, 1962
Le phytoplancton de la Baie des Chaleurs. Contrib. Ministère de la Chasse et des Pêcheries, Province de Québec, No. 91: 365 pp.

Chaetoceros furcellatus
Cleve-Euler, A., 1951
Die Diatomeen von Schweden und Finnland. Kungl. Svenska Vetenskaps Akad. Handl., Fjärde Ser. 2(1): 161 pp., 6 pls.

Chaetoceros furcellatum
Gran, H.H., 1908
Diatomeen. Nordisches Plankton, Botanischer Teil pp. XIX 1-XIX 146; 178 text figs.

Chaetoceros furcellatum
Gran, H.H., 1897
Protophyta: Diatomaceae, Silico-flagellata and Cilioflagellata. Den Norske Nordhavs Expedition 1876-1878, h. 24, 36 pp., 4 pls.

Chaetoceros furcellatus
Gran, H.H., and T. Braarud, 1935
A quantitative study of the phytoplankton in the Bay of Fundy and the Gulf of Maine (including observations on hydrography, chemistry, and turbidity). J. Biol. Bd., Canada, 1(5):279-467, 69 text figs.

Chaetoceros furcellatum
Hendy, N. Ingram, 1964
An introductory account of the smaller algae of British coastal waters. V. Bacillariophyceae (Diatoms).
Her Majesty's Stationary Office, 317 pp., 45 pls.

Chaetoceros furcellatus
Lillick, L.C., 1940
Phytoplankton and planktonic protozoa of the offshore waters of the Gulf of Maine. Pt.II. Qualitative Composition of the Planktonic Flora. Trans. Am. Phil. Soc., n.s., 31(3):193-237, 13 text figs.

Chaetoceros furcellatus
Lillick, L.C., 1938
Preliminary report of the phytoplankton of the Gulf of Maine. Am. Mid. Nat. 20(3):624-640, 1 text figs 37 tables.

Chaetoceros furcellatum
Marukawa, H., 1921
Plankton lists and some new species of copepods, from the northern waters of Japan. Bull. Inst. Ocean., No.384, 15 pp., 3 pls., 1 chart. Monaco

Chaetoceros furcellatum
Meunier, A., 1913
Microplancton de la Mer Flamande. 1. Chaetoceros. Mem. Mus. Roy. Hist. Nat., Belgique 7(2):1-55, 7 pls.

Chaetoceros furcellatus
Phifer, L.D. (undated)
The occurrence and distribution of plankton diatoms in Bering Sea and Bering Strait, July 26-August 24, 1934. Report of Oceanographic cruise of U.S. Coast Guard Cutter Chelan 1934, Part II(A):1-44 (mimeographed) plus fig.1 (after Pt.B)

Chaetoceros galvestonensis n. sp.
Collier, A., and A. Murphy, 1962
Very small diatoms: preliminary notes and description of Chaetoceros galvestonensis. Science, 136(3518):780-781.

Chaetoceros glandazii
Hendy, N. Ingram, 1964
An introductory account of the smaller algae of British coastal waters. V. Bacillariophyceae (Diatoms).
Her Majesty's Stationary Office, 317 pp., 45 pls.

Chaetoceros Glandazi
Hendey, N.I., 1937
The plankton diatoms of the southern seas. Discovery Repts. 16:151-364, pls.6-13.

Chaetoceros Glandazi
Mangin, M. L., 1912
Phytoplancton de la croisière du "René" dans l'Atlantique (Septembre 1908). Ann. Inst. Ocean., n.s., 4(1):1-66, 2 pls., 41 text figs., 2 tables.

Chaetoceros gracile
Bigelow, H.B., and M. Leslie, 1930
Reconnaissance of the waters and plankton of Monterey Bay, July 1928. Bull. M.C.Z., 70(5):429-481, 43 text figs.

Chaetoceros gracilis
Boden, B.P., 1950
Some marine plankton diatoms from the west coast of South Africa. Trans. R.Soc. S. Africa. 32:321-434, 100 text figs.

Chatoceros gracilis
Brunel, J., 1962
Le phytoplancton de la Baie de Chaleurs. Inst. Botan., Univ. Montréal, Contrib. No. 77: 365 pp., 66 pls.

Chaetoceros gracilis
Cleve-Euler, A., 1951
Die Diatomeen von Schweden und Finnland. Kungl. Svenska Vetenskaps Akad. Handl., Fjärde Ser. 2(1): 161 pp., 6 pls.

Chaetoceros gracilis
Cupp, Easter E., 1943
Marine plankton diatoms of the west coast of North America. Bull. S.I.O. 5(1):1-238, 5 pls., 168 text figs.

Chaetoceros gracile
Gran, H.H., 1908
Diatomeen. Nordisches Plankton, Botanischer Teil pp. XIX.1-XIX 146; 178 text figs.

Chaetoceros gracile
Gran, H.H., 1897
Protophyta: Diatomaceae, Silico-flagellata and Cilioflagellata. Den Norske Nordhavs Expedition 1876-1878, h. 24, 36 pp., 4 pls.

Chaetoceros gracilis
Gran, H. H. and E. C. Angst, 1931
Plankton diatoms of Puget Sound. Publ. Puget Sound Biol. Sta. 7:417-519, 95 text figs.

Chaetoceros gracile
Hendy, N. Ingram, 1964
An introductory account of the smaller algae of British coastal waters. V. Bacillariophyceae (Diatoms). Her Majesty's Stationary Office, 317 pp., 45 pls.

Chaetoceros gracilis
Lillick, L.C., 1940
Phytoplankton and planktonic protozoa of the offshore waters of the Gulf of Maine. Pt.II. Qualitative Composition of the Planktonic Flora. Trans. Am. Phil. Soc., n.s., 31(3):193-237, 13 text figs.

Chaetoceros gracilis
Lillick, L.C., 1937
Seasonal studies of the phytoplankton off Woods Hole, Massachusetts. Biol. Bull. LXXIII (3):488-503, 3 text figs.

Chaetoceros gracilis
Marshall, S.M., 1947
An experiment in marine fish cultivation: III. The plankton of a fertilized loch. Proc. Roy. Soc. Edinburgh, Sect.B., 63, Pt.I(3):21-33, 7 text figs.

Chaetoceros gracilis
Morse, D.C., 1947
Some observations on seasonal variations in plankton population Patuxant River, Maryland 1943-1945. Bd. Nat. Res., Publ. No.65, Chesapeake Biol. Lab., 31, 3 figs.

Chaetoceros gracilis
Paulsen, O., 1905.
On some Peridineae and plankton diatoms. Medd. Komm. Havundersøgelser, Ser. Plankton, I(3):7 pp.

Chaetoceros gracilis
Sousa a Silva, E., and J. Dos Santos-Pinto, 1948
O Plancton da Baia de S. Martinho do Porto. 1. Diatomaceas e Dinoflagelados. Bol. Soc. Portuguese de Ciencias Naturais, 16(2):134-187, 6 pls. (Trav. Sta. Biol. Mar. de Lisbonne No. 52).

Chaetoceros gracilis
Thomas, William H. and Anne N. Dodson, 1972
On nitrogen deficiency in tropical Pacific oceanic phytoplankton. II. Photosynthetic and cellular characteristics of a chemostat-grown diatom. Limnol. Oceanogr. 17(4): 515-523.

Chaetoceros gracilis
Thomas, William H., and Anne N. Dodson, 1968.
Effects of phosphate concentration on cell division rates and yield of a tropical oceanic diatom. Biol.Bull.mar.biol.Lab.,Woods Hole,134(1):199-208.

Chaetoceros Hendeyi n.sp.
Manguin, E., 1957.
Premier inventaire des diatomées de la Terre Adélie Antarctique. Espèces nouvelles. Rev. Algologique, n.s., 3(3):111-134.

Chaetoceros holsaticus
Braarud, T., 1945
A phytoplankton survey of the polluted waters of inner Oslo Fjord. Hvalrådets Skrifter, No.28, 142 pp., 19 text figs., 17 tables.

Chaetoceros holsaticus
Brunel, J., 1962
Le phytoplancton de la Baie de Chaleurs. Inst. Botan., Univ. Montréal, Contrib. No. 77: 365 pp., 66 pls.

Chaetoceros holsaticus
Cleve-Euler, A., 1951
Die Diatomeen von Schweden und Finnland. Kungl. Svenska Vetenskaps Akad. Handl., Fjärde Ser. 2(1): 161 pp., 6 pls.

Chaetoceros holsaticus
Cupp, Easter E., 1943
Marine plankton diatoms of the west coast of North America. Bull. S.I.O. 5(1):1-238, 5 pls., 168 text figs.

Chaetoceros holasticus
Ghazzawi, F.M., 1939
Plankton of the Egyptian waters. A study of the Suez Canal Plankton. (A) Phytoplankton. Preliminery Report 83 pp. Notes and Memoires, Min. Commerce-Industry, Egypt, Hydrobiol. & Fish. 65 figs.

Chaetoceros holsaticum
Gran, H.H., 1908
Diatomeen. Nordisches Plankton, Botanischer Teil pp. XIX.1-XIX 146; 178 text figs.

Chaetoceros holsaticus
Hendy, N. Ingram, 1964
An introductory account of the smaller algae of British coastal waters. V. Bacillariophyceae (Diatoms). Her Majesty's Stationary Office, 317 pp., 45 pls.

Chaetoceros holsaticus
Levander, K.M., 1947
Plankton gesammelt in den Jahren 1899-1910 an den Küsten Finnlands. Finnländische Hydrographisch-Biologische Untersuchunger (aus dem Wasserbiologischen Laboratorin der Societas Scientiarum Fennica) No.11:40 pp., 6 diagrams, 13 pls., tables.

Chaetoceros holsaticum
Meunier, A., 1913
Microplancton de la Mer Flamande. 1. Chaetoceros. Mem. Mus. Roy. Hist. Nat., Belgique 7(2):1-55, 7 pls.

Chaetoceros holsaticum
Rumkówna, A., 1948
[List of the phytoplankton species occurring in the superficial water layers in the Gulf of Gdansk]. Bull. Lab. mar., Gdynia, No. 4: 139-141 with tables in back.

Chaetoceros horridus
Krasske, G., 1941
Die Kieselalgen des chilenischen Küstenplankton (Aus dem südchilenischen Kusten gebiet IX). Arch. Hydrobiol. 38 (2):260-287.

Chaetoceros imbricatum
Hendy, N. Ingram, 1964
An introductory account of the smaller algae of British coastal waters. V. Bacillariophyceae (Diatoms). Her Majesty's Stationary Office, 317 pp., 45 pls.

Chaetoceros imbricatus n.sp.
Mangin, M. L., 1912
Phytoplankton de la croisière du "René" dans l'Atlantique (Septembre 1908). Ann. Inst. Ocean., n.s., 4(1):1-66, 2 pls., 41 text figs., 2 tables.

Chaetoceros incurvum
Bigelow, H.B., and M. Leslie, 1930
Reconnaissance of the waters and plankton of Monterey Bay, July 1928. Bull. M.C.Z. 70(5):429-481, 43 text figs.

Chaetoceros incurvus
Jorgensen, E., 1900
Protophyten und Protozoen im Plankton aus der Norwegischen Westkerste. Bergens Mus. Aarb. 1899(6): 95 pp., 5 pls., 83 tables.

Chaetoceros indicus n.sp.
Subrahmanyan, R., 1946.
A systematic account of the marine plankton diatoms of the Madras coast. Proc. Indian Acad. Sci. 24(4):85-197, Pls2, &40 textfigs.

Chaetoceros ingolfianum
Bigelow, H.B., and M. Leslie, 1930
Reconnaissance of the waters and plankton of Monterey Bay, July 1928. Bull. M.C.Z. 70(5):429-481, 43 text figs.

Chaetoceros ingolfianus
Brunel, J., 1962
Le phytoplancton de la Baie de Chaleurs. Inst. Botan., Univ. Montréal, Contrib. No. 77: 365 pp., 66 pls.

Chaetoceros ingolfianus
Brunel, Jules, 1962
Le phytoplancton de la Baie des Chaleurs. Contrib. Ministère de la Chasse et des Pêcheries, Province de Québec, No. 91: 365 pp.

Chaetoceros ingolfianus
Cleve-Euler, A., 1951
Die Diatomeen von Schweden und Finnland. Kungl. Svenska Vetenskaps Akad. Handl., Fjärde Ser. 2(1): 161 pp., 6 pls.

Chaetoceros Ingolfianum
Gran, H.H., 1908
Diatomeen. Nordisches Plankton, Botanischer Teil pp. XIX.1-XIX 146; 178 text figs.

Chaetoceros ingolfianum
Hendy, N. Ingram, 1964
An introductory account of the smaller algae of British coastal waters. V. Bacillariophyceae (Diatoms). Her Majesty's Stationary Office, 317 pp., 45 pls.

Chaetoceros insignis n.sp.
Müller Melchers, F.C., 1953.
New and little known diatoms from Uruguay and the South Atlantic coast. Com. Bot., Mus. Hist. Nat., Montevideo, 3(30): 1-11, 8 pls.

Chaetoceros janischianum
Castracane degli Antelminelli, F., 1886 n.sp.
1. Report on the Diatomaceae collected by H.M.S. Challenger during the years 1873-1876. Rept. Sci. Results, H.M.S. Challenger, Botany Vol. II, 178 pp., 30 pls.

Chaetoceros Janischianus
Jorgensen, E., 1900
Protophyten und Protozoen im Plankton aus der Norwegischen Westkerste. Bergens Mus. Aarb. 1899(6): 95 pp., 5 pls., 83 tables

Chaetoceros karianus
Brunel, J., 1962
Le phytoplancton de la Baie de Chaleurs. Inst. Botan., Univ. Montréal, Contrib. No. 77: 365 pp., 66 pls.

Chaetoceros karianum
Hendy, N. Ingram, 1964
An introductory account of the smaller algae of British coastal waters. V. Bacillariophyceae (Diatoms). Her Majesty's Stationary Office, 317 pp., 45 pls.

Chaetoceros karianus
Phifer, L.D. (undated)
The occurrence and distribution of plankton diatoms in Bering Sea and Bering Strait, July 26-August 24, 1934. Report of Oceanographic cruise of U.S. Coast Guard Cutter Chelan 1934, Part II(A):1-44 (mimeographed) plus fig.1 (after Pt.B)

Chaetoceros laciniosus
Allen, W. E., 1938
The Templeton Crocker Expedition to the Gulf of California in 1935 - The Phytoplankton. Amer. Microsc. Soc., Trans. 57:328-335.

Chaetoceros laciniosus
Allen, W.E., 1937
Plankton diatoms of the Gulf of California obtained by the G. Allan Hancock Expedition of 1936. The Hancock Pacific Expeditions, Univ. So. Calif. Publ. 3:47-59, 1 fig.

Chaetoceros laciniosum
Allen, W.E., and E.E. Cupp, 1935
Plankton diatoms of the Java Sea. Annales du Jardin Botanique de Buitenzorg XLIV (2):101-174, figs.1-127.
(drawings of all species mentioned)

Chaetoceros laciniosum
Bigelow, H.B., and M. Leslie, 1930
Reconnaissance of the waters and plankton of Monterey Bay, July 1928. Bull. M.C.Z. 70(5):429-481, 43 text figs.

Chaetoceros laciniosus
Braarud, T., 1945
A phytoplankton survey of the polluted waters of inner Oslo Fjord. Hvalrådets Skrifter, No.28, 142 pp., 19 text figs., 17 tables.

Chaetoceros laciniosus
Braarud, T., and Adam Bursa, 1939
On the phytoplankton of the Oslo Fjord, 1933-1934. Hvalrådets Skr. No.19:1-63; 9 text figs. Reviewed. J. du. Cons. 14(3): 418-420. A.C. Gardiner.

Chaetoceros laciniosus
Brunel, J., 1962
Le phytoplancton de la Baie de Chaleurs. Inst. Botan., Univ. Montréal, Contrib. No. 77: 365 pp., 66 pls.

Chaetoceros laciniosus
Brunel, Jules, 1962
Le phytoplancton de la Baie des Chaleurs. Contrib. Ministère de la Chasse et des Pêcheries, Province de Québec, No. 91: 365 pp.

Chaetoceros laciniosus
Cleve-Euler, A., 1951
Die Diatomeen von Schweden und Finnland. Kungl. Svenska Vetenskaps Akad. Handl., Fjärde Ser. 2(1): 161 pp., 6 pls.

Chaetoceros laciniosus
Cupp, Easter E., 1943
Marine plankton diatoms of the west coast of North America. Bull. S.I.O. 5(1):1-238, 5 pls., 168 text figs.

Chaetoceros laciniosus
Cupp, E.E., 1934
Analysis of marine diatom collections taken from the Canal Zone to California during March, 1933. Trans. Am. Micros. Soc. LIII (1):22-29, 1 map.

Chaetoceros laciniosus
Cupp, E.E. and Allen, W.E., 1938
Plankton diatoms of the Gulf of California obtained by Allan Hancock Pacific Expedition of 1937. The Hancock Pacific Expeditions, The Univ. So. Calif. Publ. 3: 61-74, 1 map, pls.4-15.

Chaetoceros laciniosus
Davidson, V.M., 1931.
Biological and oceanographic conditions in Hudson Bay. Contr. Canadian Biol. Fish., n.s., 6(26): 497-509, 7 textfigs.

Chaetoceros laciniosus
Ercegovic, A., 1940
Weitere Untersuchungen über einige hydrographische Verhältnisse und über die Phytoplanktonproduktion in den Gewässern der östlichen Mitteladria. Acta Adriatica 2(3):95-134, 8 text figs.

Chaetoceros laciniosus
Frenguelli, Joaquin, and Hector Antonio Orlando, 1959.
Operacion MERLUZA. Diatomeas y silicoflagelados del plancton del "VI Crucero". Servicio Hidrogr. Naval., Argentina, Publ. No. H. 619: 5-62.

Chaetoceros laciniosus
Gilbert, J.Y., and W.E. Allen, 1943
The phytoplankton of the Gulf of California obtained by the "E.W. Scripps" in 1939 and 1940. J. Mar. Res. V(2):89-110, figs.30-31.

Chaetoceros laciniosum
Gran, H.H., 1908
Diatomeen. Nordisches Plankton, Botanischer Teil pp. XIX.1-XIX 146; 178 text figs.

Chaetoceros laciniosum
Gran, H.H., 1897
Protophyta: Diatomaceae, Silico-flagellata and Ciliaflagellata. Den Norske Nordhavs Expedition 1876-1878, h. 24, 36 pp., 4 pls.

Chaetoceros laciniosus
Gran, H.H., and T. Braarud, 1935
A quantitative study of the phytoplankton in the Bay of Fundy and the Gulf of Maine (including observations on hydrography, chemistry, and turbidity). J. Biol. Bd., Canada, 1(5):279-467, 69 text figs.

Chaetoceros laciniosus
Gran, H. H. and E. C. Angst, 1931
Plankton diatoms of Puget Sound. Publ. Puget Sound Biol. Sta. 7:417-519, 95 text figs.

Chaetoceros laciniosum
Hendy, N. Ingram, 1964
An introductory account of the smaller algae of British coastal waters. V. Bacillariophyceae (Diatoms). Her Majesty's Stationary Office, 317 pp., 45 pls.

Chaetoceros laciniosum
Hendey, N.I., 1937
The plankton diatoms of the southern seas. Discovery Repts. 16:151-364, pls.6-13.

Chaetoceros laciniosum
Iselin, C., 1930
A report on the coastal waters of Labrador based on explorations of the "Chance" during the summer of 1926. Proc. Am. Acad. Arts Sci., 66(1):1-37, 14 text figs.

Chaetoceros laciniosum
Jorgensen, E., 1900
Protophyten und Protozoen im Plankton aus der Norwegischen Westkerste. Bergens Mus. Aarb. 1899(6): 95 pp., 5 pls., 83 tables.

Chaetoceros laciniosus
Lillick, L.C., 1940
Phytoplankton and planktonic protozoa of the offshore waters of the Gulf of Maine. Pt.II. Qualitative Composition of the Planktonic Flora. Trans. Am. Phil. Soc., n.s., 31(3):193-237, 13 text figs.

Chaetoceros laciniosus
Lillick, L.C., 1938
Preliminary report of the phytoplankton of the Gulf of Maine. Am. Mid. Nat. 20(3):624-640, 1 text fig, 37 tables.

Chaetoceros laciniosus
Lillick, L.C., 1937
Seasonal studies of the phytoplankton off Woods Hole, Massachusetts. Biol. Bull. LXXIII (3):488-503, 3 text figs.

Chaetoceros laciniosum
Meunier, A., 1913
Microplancton de la Mer Flamande. 1. Chaetoceros. Mem. Mus. Roy. Hist. Nat., Belgique 7(2):1-55, 7 pls.

Chaetoceros laciniosus
Michailova, N.F., 1964.
On the spreading of the habitat into the Black Sea of species of the genus Chaetoceros of northern seas and their biogeography. (In Russian).
Trudy Sevastopol Biol. Sta., 7:231-248.

Chaetoceros laciniosum
Pavillard, J., 1925
Bacillariales. Rept. on the Danish Oceangr. Exped., 1908-10 to the Mediterranean and adj. seas. Vol.II., Biol. J4:72 pp., 116 text figs.

Chaetoceros laciniosum
Pavillard, J., 1905
Recherches sur la flore pelagique (Phytoplankton) de l'Etang de Thau. Theses presentees a la Fac. Sci., Paris, 116 pp., 3 pls.

Chaetoceros laciniosus
Pettersson, H., 1934
Scattering and extinction of light in sea-water. Meddelanden från Göteborgs Högskolas Oceanografiska Institut. 9 (Göteborgs Kungl. vetenskaps-och vitterhetssamhälles Handlingar. Femte Följden, Ser.B, 4(4)):1-16.

Chaetoceros laciniosus
Phifer, L.D. (undated)
The occurrence and distribution of plankton diatoms in Bering Sea and Bering Strait, July 26-August 24, 1934. Report of Oceanographic cruise of U.S. Coast Guard Cutter Chelan 1934, Part II(A):1-44 (mimeographed) plus fig.1 (after Pt.B)

Chaetoceros laciniosus
Rampi, L., 1942
Ricerche sul fitoplancton del Mare Ligure 6. Le diatomee delle acque di Sanremo. Nuovo Giornale Botanico Italiano, N.S., 49:252-268.

Chaetoceros laciniosus
Sousa e Silva, E., 1949
Diatomaceas e Dinoflagelados de Baia de Cascais. Portugalise Acta Biol. Volume: Julio Henriques, Ser. B: 300-383, 9 pls, 2 fold-in tables.

Chaetoceros laciniosus
Yamazi, I., 1951.
Plankton investigations in inlet waters along the coast of Japan. II. The plankton of Hakodate Harbour and Yoichi Inlet in Hokkaido.
Publ. Seto Mar. Biol. Sta., Kyoto Univ., 1(4): 185-194, 3 textfigs.

Chaetoceros lacimosum
Schulz, B., and A. Wulff, 1929
Hydrographie und Oberflächen plankton des westlichen Barentsmeeres im Sommer 1927. Ber. deutschen wissensch. Komm. F. Meeresforsch. n.s. 4(5):232-372, 13 tables, 25 text figs.

Chaetoceros laeve
Allen, W.E., 1937
Plankton diatoms of the Gulf of California obtained by the G. Allan Hancock Expedition of 1936. The Hancock Pacific Expeditions, Univ. So. Calif. Publ. 3:47-59, 1 fig.

Chaetoceros laeve
Allen, W.E., and E.E. Cupp, 1935
Plankton diatoms of the Java Sea. Annales du Jardin Botanique de Buitenzorg XLIV (2):101-174, figs.1-127.
(drawings of all species mentioned)

Chaetoceros laevis
Cupp, Easter E., 1943
Marine plankton diatoms of the west coast of North America. Bull. S.I.O. 5(1):1-238, 5 pls., 168 text figs.

Chaetoceros laevis
Cupp, E.E., 1934
Analysis of marine diatom collections taken from the Canal Zone to California during March, 1933. Trans. Am. Micros. Soc. LIII (1):22-29, 1 map.

Chaetoceros Laivii n.sp.
Manguin, E., 1954
Diatomées marines provenant de l'île Heard (Australian National Antarctic Expedition). Rev. Algol., n.s., 1: 14-24.

Chaetoceros Lauderi
Allen, W.E., and E.E. Cupp, 1935
Plankton diatoms of the Java Sea. Annales du Jardin Botanique de Buitenzorg XLIV (2):101-174, figs.1-127.
(drawings of all species mentioned)

Chaetoceros lauderi
Anon., 1951.
Bulletin of the Marine Biological Station of Asamushi 4(3/4): 15 pp.

Chaetoceros lauderi
Brunel, J., 1962
Le phytoplancton de la Baie de Chaleurs. Inst. Botan., Univ. Montréal, Contrib. No. 77: 365 pp., 66 pls.

Chaetoceros lauderi
Cleve-Euler, A., 1951
Die Diatomeen von Schweden und Finnland. Kungl. Svenska Vetenskaps Akad. Handl., Fjärde Ser. 2(1): 161 pp., 6 pls.

Chaetoceros lauderi
Cupp, Easter E., 1943
Marine plankton diatoms of the west coast of North America. Bull. S.I.O. 5(1):1-238, 5 pls., 168 text figs.

Chaetoceros Lauderi
Dangeard, P., 1927
Phytoplankton de la croisière du "Sylvana". Ann. Inst. Ocean., Monaco, n.s., 4(8):286-401, 54 text figs. (Feirer-Juin 1913).

Chaetoceros Lauderi
Delegazione Italiana della Commissione Internazionale per l'Esplorazione Scientifica del Mediterraneo, 1941
Note sul plancton della Laguna veneta. [Memoria CCLXXXI], Arch. di Ocean. e Limn. Anno I, Fasc. I, 1941 XIX: 31-57 pp.

Chaetoceros lauderi
Forti, A., 1922
Ricerche sulla flora pelagica (fitoplancton) di Quarto dei Mille. Mem. R. Com. Talass. Ital. 97:248 pp., 13 pls.

Chaetoceros lauderi
Ghazzawi, F.M., 1939
Plankton of the Egyptian waters. A study of the Suez Canal Plankton. (A) Phytoplankton. Preliminary Report 83 pp. Notes and Memoires, Min. Commerce-Industry, Egypt, Hydrobiol. & Fish. 65 figs.

Chaetoceros lauderi
Hendy, N. Ingram, 1964
An introductory account of the smaller algae of British coastal waters. V. Bacillariophyceae (Diatoms).
Her Majesty's Stationary Office, 317 pp., 45 pls.

Chaetoceros lauderi
Michailova, N.F., 1964.
On the spreading of the habitat into the Black Sea of species of the genus Chaetoceros of northern seas and their biogeography. (In Russian).
Trudy Sevastopol Biol. Sta., 7:231-248.

Chaetoceros lauderi
Mikhailova, N.F., 1962.
On the germination of resting spores of Chaetoceros lauderi Ralfs.
Doklady, Akad. Nauk, SSSR, 143(3):741-742.

Chaetoceros lauderi
Pavillard, J., 1925
Bacillariales. Rept. on the Danish Oceangr. Exped., 1908-10 to the Mediterranean and adj. seas. Vol.II., Biol. J4:72 pp., 116 text figs.

Chaetoceros lauderi
Rampi, L., 1942
Ricerche sul fitoplancton del Mare Ligure 6. Le diatomee delle acque di Sanremo. Nuovo Giornale Botanico Italiano, N.S., 49:252-268.

Chaetoceros lauderi
Smayda, Theodore J., and Brenda J. Boleyn, 1966
Experimental observations on the flotation of marine diatoms. III. Bacteriastrum hyalinum and Chaetoceros lauderi.
Limnol. and Oceanogr., 11(1):35-43.

Chaetoceros lauderi (figs.)
Sousa e Silva, E., 1949
Diatomaceas e Dinoflagelados de Baia de Cascais. Portugaliae Acta Biol., Volume: Julio Henriques, Ser. B: 300-383, 9 pls, 2 fold-in tables.

Chaetoceros longicrure
Pavillard, J., 1905
Recherches sur la flore pelagique (Phytoplankton) de l'Etang de Thau. Theses presentees a la Fac. Sci., Paris, 116 pp., 3 pls.

Chaetoceros lorenzianus
Allen, W. E., 1938
The Templeton Crocker Expedition to the Gulf of California in 1935 - The Phytoplankton. Amer. Microsc. Soc., Trans. 57:328-335.

Chaetoceros Lorenzianum
Allen, W.E., and E.E. Cupp, 1935
Plankton diatoms of the Java Sea. Annales du Jardin Botanique de Buitenzorg XLIV (2):101-174, figs.1-127.
(drawings of all species mentioned)

Chaetoceros lorenzianus
Boden, B.P., 1950
Some marine plankton diatoms from the west coast of South Africa. Trans. R.Soc. S. Africa. 32:321-434, 100 text figs.

Chaetoceros lorenzianus
Brunel, J., 1962
Le phytoplancton de la Baie de Chaleurs. Inst. Botan., Univ. Montréal, Contrib. No. 77: 365 pp., 66 pls.

Chaetoceros lorenzianus forceps

Chaetoceros lorenzianus
Brunel, Jules, 1962
Le phytoplancton de la Baie des Chaleurs. Contrib. Ministère de la Chasse et des Pêcheries; Province de Quebec, No. 91: 365 pp.

Chaetoceros lorenziana
Cleve-Euler, A., 1951
Die Diatomeen von Schweden und Finnland. Kungl. Svenska Vetenskaps Akad. Handl., Fjärde Ser. 2(1): 161 pp., 6 pls.

Chaetoceros lorenzianus
Cupp, Easter E., 1943
Marine plankton diatoms of the west coast of North America. Bull. S.I.O. 5(1):1-238, 5 pls., 168 text figs.

Chaetoceros lorenzianus
Cupp, E.E., 1934
Analysis of marine diatom collections taken from the Canal Zone to California during March, 1933. Trans. Am. Micros. Soc. LIII (1):22-29, 1 map.

Chaetoceros lorenzianus
Cupp, E.E. and Allen, W.E., 1938
Plankton diatoms of the Gulf of California obtained by Allan Hancock Pacific Expedition of 1937. The Hancock Pacific Expeditions, The Univ. So. Calif. Publ. 3: 61-74, 1 map, pls.4-15.

Chaetoceros lorenzianus
Delegazione Italiana della Commissione Internazionale per l'Esplorazione Scientifica del Mediterraneo, 1941
Note sul plancton della Laguna veneta. [Memoria CCLXXXI], Arch. di Ocean. e Limn. Anno I, Fasc. I, 1941 XIX: 31-57 pp.

Chaetoceros Lorenzianum
Dangeard, P., 1927
Phytoplankton de la croisière du "Sylvana". Ann. Inst. Ocean., Monaco, n.s., 4(8):286-401, 54 text figs. (Feirer-Juin 1913).

Chaetoceros lorenzianum
Delsman, H. C., 1939.
Preliminary plankton investigations in the Java Sea. Treubia, 17:139-181, 8 maps, 41 figs.

Chaetoceros lorenzianus
Ercegovic, A., 1936
Etudes qualitative et quantitatives du phytoplancton dans les eaux cotières de l'Adriatique oriental moyen au cours de l'année 1934. Acta Adriatica 1(9):1-126

Chaetoceros lorenzianus
Eskinazi Enide e Shigekatsv Sato (1963-1964) 1966.
Contribuição ao estudo das diatomaceas da Praia de Piedade.

Trabhs Inst. Oceanogr., Univ. Recife, 5 (5/6): 73-114.

Chaetoceros lorenzianus
Forti, A., 1922
Ricerche sulla flora pelagica (fitoplancton) di Quarto dei Mille. Mem. R. Com. Talass. Ital. 97:248 pp., 13 pls.

Chaetoceros Lorenzianus
Frenguelli, Joaquin, and Hector Antonio Orlando, 1959.
Operacion MERLUZA. Diatomeas y silicoflagelados del plancton del "VI Crucero". Servicio Hidrogr. Naval., Argentina, Publ. No. H. 619: 5-62.

Chaetoceros lorenzianum
Gilbert, J.Y., and W.E. Allen, 1943
The phytoplankton of the Gulf of California obtained by the "E.W. Scripps" in 1939 and 1940. J. Mar. Res. V(2):89-110, figs.30-31.

Chaetoceros lorenzianus
Gran, H. H. and E. C. Angst, 1931
Plankton diatoms of Puget Sound. Publ. Puget Sound Biol. Sta. 7:417-519, 95 text figs.

Chaetoceros Lorenzianus
Grunow, A., 1863
Ueber einige neue und ungenügend bekannte Arten und Gattungen von Diatomaceen. Verhandl. d. K.K. Zool. Bot. Gesellsch., Vienna, 13: 137-162, pl.4-5 (Pl. 13-14).

Chaetoceros lorenzianum
Hendy, N. Ingram, 1964
An introductory account of the smaller algae of British coastal waters. V. Bacillariophyceae (Diatoms).
Her Majesty's Stationary Office, 317 pp., 45 pls.

Chaetoceros Lorenzianum
Hendey, N.I., 1937
The plankton diatoms of the southern seas. Discovery Repts. 16:151-364, pls.6-13.

Chaetoceros Lorenzianus
Kokubo, S., and S. Sato, 1947
Plankters in Jū-San Gata. Physiol. and Ecol. (Japan) 1(4):1-16, 3 text figs., tables.

Chaetoceros lorenzianus
Lillick, L.C., 1940
Phytoplankton and planktonic protozoa of the offshore waters of the Gulf of Maine. Pt.II. Qualitative Composition of the Planktonic Flora. Trans. Am. Phil. Soc., n.s., 31(3):193-237, 13 text figs.

Chaetoceros lorenzianus
Lillick, L.C., 1938
Preliminary report of the phytoplankton of the Gulf of Maine. Am. Mid. Nat. 20(3):624-640, 1 text figs 37 tables.

Chaetoceros lorenzianum
Macdonald, R., 1933
An examination of plankton hauls made in the Suez Canal during the year 1928. Fish. Res. Dis., Notes & Mem. No.3, 11 pp., 1 chart.

Chaetoceros Lorenzianum
Marukawa, H., 1921
Plankton lists and some new species of copepods, from the northern waters of Japan. Bull. Inst. Ocean., No.384, 15 pp., 3 pls., 1 chart. Monaco

Chaetoceros Lorenzianum
Meunier, A., 1913
Microplancton de la Mer Flamande. 1. Chaetocerés. Mem. Mus. Roy. Hist. Nat., Belgique 7(2):1-55, 7 pls.

Chaetoceros lorenzianum
Pavillard, J., 1925
Bacillariales. Rept. on the Danish Oceangr. Exped., 1908-10 to the Mediterranean and adj. seas. Vol.II., Biol. J4:72 pp., 116 text figs.

Chaetoceros Lorenzianum
Pavillard, J., 1905
Recherches sur la flore pelagique (Phytoplankton) de l'Etang de Thau. Theses presentees a la Fac. Sci., Paris, 116 pp., 3 pls.

Chaetoceros Lorenzianus
Rampi, L., 1942
Ricerche sul fitoplancton del Mare Ligure 6. Le diatomee delle acque di Sanremo. Nuovo Giornale Botanico Italiano, N.S., 49:252-268.

Chaetoceros Lorenzianum
Schodduyn, M., 1926
Observations faites dans la baie d'Ambleteuse (Pas de Calais). Bull. Inst. Ocean., Monaco, No. 482: 64 pp.

Chaetoceros lorenzianus
Ryther, John H., and Dana D. Kramer, 1961
Relative iron requirement of some coastal and offshore algae. Ecology, 42(2): 444-446.

Chaetoceros lorenzianum
Schröder, B., 1900
Phytoplankton des Golfes von Neapel nebst vergleichenden Ausblicken auf das atlantischen Ozean. Mitt. Zool. Stat. Neapel, 14:1-38.

Chaetoceros lorenzianus forceps
Brunel, Jules, 1962
Le phytoplancton de la Baie des Chaleurs. Contrib. Ministère de la Chasse et des Pêcheries; Province de Québec, No. 91: 365 pp.

Chaetoceros lorenzianus singularis forma nov.
Takano, H., 1960
Plankton diatoms in the eastern Caribbean Sea. J. Oceanogr. Soc., Japan, 16(4): 180-184.

Chaetoceros mamillanus
Mangin, M. L., 1912
Phytoplancton de la croisière du "René" dans l'Atlantique (Septembre 1908). Ann. Inst. Ocean., n.s., 4(1):1-66, 2 pls., 41 text figs., 2 tables.

Chaetoceros messanensis
Allen, W. E., 1938
The Templeton Crocker Expedition to the Gulf of California in 1935 - The Phytoplankton. Amer. Microsc. Soc., Trans. 57:328-335.

Chaetoceros messanensis
Allen, W.E., 1937
Plankton diatoms of the Gulf of California obtained by the G. Allan Hancock Expedition of 1936. The Hancock Pacific Expeditions, Univ. So. Calif. Publ. 3:47-59, 1 fig.

Chaetoceros messanensis
Brunel, J., 1962
Le phytoplancton de la Baie de Chaleurs. Inst. Botan., Univ. Montréal, Contrib. No. 77: 365 pp., 66 pls.

Chaetoceros messanensis
Boden, B.P., 1950
Some marine plankton diatoms from the west coast of South Africa. Trans. R.Soc. S. Africa. 32:321-434, 100 text figs.

Chaetoceros messanensis
Cleve-Euler, A., 1951
Die Diatomeen von Schweden und Finnland. Kungl. Svenska Vetenskaps Akad. Handl., Fjärde Ser. 2(1): 161 pp., 6 pls.

Chaetoceros messanensis
Cupp, Easter E., 1943
Marine plankton diatoms of the west coast of North America. Bull. S.I.O. 5(1):1-238, 5 pls., 168 text figs.

Chaetoceros messanensis
Ercegovic, A., 1936
Etudes qualitative et quantitatives du phytoplancton dans les eaux cotières de l'Adriatique oriental moyen au cours de l'année 1934. Acta Adriatica 1(9):1-126

Chaetoceros messanensis
Forti, A., 1922
Ricerche sulla flora pelagica (fitoplancton) di Quarto dei Mille. Mem. R. Com. Talass. Ital. 97:248 pp., 13 pls.

Chaetoceros messanensis
Frenguelli, Joaquin, and Hector Antonio Orlando, 1959.
Operacion MERLUZA. Diatomeas y silicoflagelados del plancton del "VI Crucero". Servicio Hidrogr. Naval., Argentina, Publ. No. H. 619: 5-62.

Chaetoceros Messanensis
Gilbert, J.Y., and W.E. Allen, 1943
The phytoplankton of the Gulf of California obtained by the "E.W. Scripps" in 1939 and 1940. J. Mar. Res. V(2):89-110, figs.30-31.

Chaetoceros messanense
Hendy, N. Ingram, 1964
An introductory account of the smaller algae of British coastal waters. V. Bacillariophyceae (Diatoms).
Her Majesty's Stationary Office, 317 pp., 45 pls.

Chaetoceros messanense
Hendey, N.I., 1937
The plankton diatoms of the southern seas. Discovery Repts. 16:151-364, pls.6-13.

Chaetoceros messanensis
Kokubo, S., 1952
Results of the observations on the plankton and oceanography of Mutsu Bay during 1950, reference being made also to the period 1946-1950. Bull Mar.Biol.Sta., Asamushi 5(1/4): 1-54, 3 tables,(fold-in), 1 fold-in.

Chaetoceros messanense
Pavillard, J., 1925
Bacillariales. Rept. on the Danish Oceangr. Exped., 1908-10 to the Mediterranean and adj. seas. Vol.II., Biol. J4:72 pp., 116 text figs.

Chaetoceros messanensis
Rampi, L., 1942
Ricerche sul fitoplancton del Mare Ligure 6. Le diatomee delle acque di Sanremo. Nuovo Giornale Botanico Italiano, N.S., 49:252-268.

Chaetoceros messanensis
Zanon, V., 1948
Diatomee marini di Sardegna e Pugillo di Alghe Marine della stressa. Boll. Pesca, Piscitutura e Idrobiologia, Anno 24, ns. 3(2): 202-244, 27 figs. on 1 pl.

Chaetoceros meunieri
Brunel, J., 1962
Le phytoplancton de la Baie de Chaleurs. Inst. Botan., Univ. Montréal, Contrib. No. 77: 365 pp., 66 pls.

Chaetoceros mitra
Bigelow, H.B., and M. Leslie, 1930
Reconnaissance of the waters and plankton of Monterey Bay, July 1928. Bull. M.C.Z., 70(5):429-481, 43 text figs.

Chaetoceros mitra
Brunel, J., 1962
Le phytoplancton de la Baie de Chaleurs. Inst. Botan., Univ. Montréal, Contrib. No. 77: 365 pp., 66 pls.

Chaetoceros mitra
Brunel, Jules, 1962
Le phytoplancton de la Baie des Chaleurs. Contrib. Ministère de la Chasse et des Pêcheries, Province de Québec, No. 91: 365 pp.

Chaetoceros mitra
Cleve-Euler, A., 1951
Die Diatomeen von Schweden und Finnland. Kungl. Svenska Vetenskaps Akad. Handl., Fjärde Ser. 2(1): 161 pp., 6 pls.

Chaetoceros mitra
Gran, H.H., 1908
Diatomeen. Nordisches Plankton, Botanischer Teil pp. XIX.1-XIX.146; 178 text figs.

Chaetoceros mitra
Hendy, N. Ingram, 1964
An introductory account of the smaller algae of British coastal waters. V. Bacillariophyceae (Diatoms). Her Majesty's Stationary Office, 317 pp., 45 pls.

Chaetoceros mitra
Meunier, A., 1913
Microplancton de la Mer Flamande. 1. Chaetoceros. Mem. Mus. Roy. Hist. Nat., Belgique 7(2):1-55, 7 pls.

Chaetoceros mitra
Phifer, L.D. (undated)
The occurrence and distribution of plankton diatoms in Bering Sea and Bering Strait, July 26-August 24, 1934. Report of Oceanographic cruise of U.S. Coast Guard Cutter Chelan 1934, Part II(A):1-44 (mimeographed) plus fig.1 (after Pt.B)

Chaetoceros muelleri
Brunel, J., 1962
Le phytoplancton de la Baie de Chaleurs. Inst. Botan., Univ. Montréal, Contrib. No. 77: 365 pp., 66 pls.

Chaetoceros muelleri
Cleve-Euler, A., 1951
Die Diatomeen von Schweden und Finnland. Kungl. Svenska Vetenskaps Akad. Handl., Fjärde Ser. 2(1): 161 pp., 6 pls.

Chaetoceros nanodenticulatum
Allen, W.E., and E.E. Cupp, 1935
Plankton diatoms of the Java Sea. Annales du Jardin Botanique de Buitenzorg XLIV (2):101-174, figs.1-127.
(drawings of all species mentioned)

Chaetoceros natatum n.sp.
Manguin, E., 1957.
Premier inventaire des diatomées de la Terre Adélie Antarctique. Espèces nouvelles. Rev. Algologique, n.s., 3(3):111-134.

Chaetoceros neapolitanus
Allen, W.E., 1936.
Surface plankton diatoms in the North Pacific Ocean in 1934. Madroño 3(6):3 pp.

Chaetoceros neapolitanus
Allen, W.E., 1936
Occurrence of marine plankton diatoms in a ten-year series of daily catches in southern California. Am. Jour. Bot. 23(1):60-63.

Chaetoceros neapolitanum
Allen, W.E., 1928.
Catches of marine diatoms and dinoflagellates taken in Southern California waters in 1926. Bull. S.I.O., tech. ser., 1:201-246, 6 textfigs.

Chaetoceros neapolitana
Cupp, Easter E., 1943
Marine plankton diatoms of the west coast of North America. Bull. S.I.O. 5(1):1-238, 5 pls., 168 text figs.

Chaetoceros neappolitanum
Dangeard, P., 1927
Phytoplankton de la croisière du "Sylvana". Ann. Inst. Ocean., Monaco, n.s., 4(8):286-401, 54 text figs. (Feirer-Juin 1913).

Chaetoceros neapolitanum
Gran, H.H., 1908
Diatomeen. Nordisches Plankton, Botanischer Teil pp. XIX.1-XIX.146; 178 text figs.

Chaetoceros neapolitanum
Hendy, N. Ingram, 1964
An introductory account of the smaller algae of British coastal waters. V. Bacillariophyceae (Diatoms). Her Majesty's Stationary Office, 317 pp., 45 pls.

Chaetoceros neapolitanum
Pavillard, J., 1925
Bacillariales. Rept. on the Danish Oceanogr. Exped., 1908-10 to the Mediterranean and adj. seas. Vol.II., Biol. J4:72 pp., 116 text figs.

Chaetoceros neapolitanum n.sp.
Schröder, B., 1900
Phytoplankton des Golfes von Neapel nebst vergleichenden Ausblicken auf das atlantischen Ozean. Mitt. Zool. Stat. Neapel, 14:1-38.

Chaetoceros neglectum
Boden, B.P., 1949.
The diatoms collected by the U.S.S. CACOPAN in the Antarctic in 1947. J. Mar. Res. 8(1):6-13, 3 textfigs.

Chaetoceros neglectum
Boden, Brian, 1948
Marine plankton diatoms on operation HIGHJUMP in: Some oceanographic observations on operation HIGHJUMP. By R.S. Dietz. USNEL Rept. No.55, 97 pp., 41 figs. 7 July 1948.

Chaetoceros neglectum
Hendey, N.I., 1937
The plankton diatoms of the southern seas. Discovery Repts. 16:151-364, pls.6-13.

Chaetoceros neglectus
Mangin, L., 1915
Phytoplancton de L'Antartique. Deuxieme Exped. Ant. Francaise (1908-1910), 95 pp., 3 pls., 58 text figs.

Chaetoceros orientalis
Brunel, J., 1962
Le phytoplancton de la Baie de Chaleurs. Inst. Botan., Univ. Montréal, Contrib. No. 77: 365 pp., 66 pls.

Chaetoceros orientalis
Iyengar, M.O.P. and G.Venkataraman,1951.
The ecology and seasonal succession of the algae flora of the River Cooum at Madras with special reference to the Diatomaceae. J. Madras Univ. 21, Sect. B(1): 140-192, 1 pl of 4 figs., 11 text figs.

Chaetoceros pachyceros n.sp.
Margalef, R., and M. Duran, 1953.
Microplancton de Vigo de octubre 1951 a septiembre 1952. Publ. Inst. Biol. Aplic. 13:5-78, 22 figs.

Chaetoceros pacificus, n.sp.
Semina, H.J., 1961
[A new species of plankton diatom of the genus Chaetoceros Ehr. from the Central Pacific.]
Trudy Inst. Okeanol., 51: 31-32.

Chaetoceros paradoxum
Allen, W.E., and E.E. Cupp, 1935
Plankton diatoms of the Java Sea. Annales du Jardin Botanique de Buitenzorg XLIV (2):101-174, figs.1-127.
(drawings of all species mentioned)

Chaetoceros parallelis
Boden, B.P., 1950
Some marine plankton diatoms from the west coast of South Africa. Trans. R.Soc. S. Africa. 32:321-434, 100 text figs.

Chaetoceros paralellis
Vives, F., and A. Planas, 1952. (fig.)
Plancton recogido por los laboratorios costeros. VI. Fitoplancton de las costas de Vinaroz, islas Columbretes, y alrededores de la desembocadura del Elroa. Publ. Inst. Biol. Aplic. 11:141-156, 19 textfigs.

Chaetoceros parallelis
Boden, B.P., 1949.
A new diatom from South Africa - "Chaetoceros parallelis". Trans. R. Soc., S. Africa, 32(3): 1-2, 1 textfig.

Chaetoceros paulseni
Brunel, J., 1962
Le phytoplancton de la Baie de Chaleurs. Inst. Botan., Univ. Montréal, Contrib. No. 77: 365 pp., 66 pls.

Chaetoceros pelagica
Bigelow, H.B., and M. Leslie, 1930
Reconnaissance of the waters and plankton of Monterey Bay, July 1928. Bull. M.C.Z., 70(5):429-481, 43 text figs.

Chaetoceros pelagicum
Boden, B. P., 1949.
The diatoms collected by the U.S.S. CACOPAN in the Antarctic in 1947. J. Mar. Res. 8(1):6-13, 3 textfigs.

Chaetoceros pelagicum
Boden, Brian, 1948
Marine plankton diatoms on operation HIGHJUMP in: Some oceanographic observations on operation HIGHJUMP. By R.S. Dietz. USNEL Rept. No.55, 97 pp., 41 figs. 7 July 1948.

Chaetoceros pelagicus
Brunel, J., 1962
Le phytoplancton de la Baie de Chaleurs. Inst. Botan., Univ. Montréal, Contrib. No. 77: 365 pp., 66 pls.

Chaetoceros pelagicus
Cupp, Easter E., 1943
Marine plankton diatoms of the west coast of North America. Bull. S.I.O. 5(1):1-238, 5 pls., 168 text figs.

Chaetoceros pelagicum
Gran, H.H., 1908
Diatomeen. Nordisches Plankton, Botanischer Teil pp. XIX.1-XIX 146; 178 text figs.

Chaetoceros pelagicum
Hendey, N.I., 1937
The plankton diatoms of the southern seas. Discovery Repts. 16:151-364, pls.6-13.

Chaetoceros pelagicus
Lillick, L.C., 1940
Phytoplankton and planktonic protozoa of the offshore waters of the Gulf of Maine. Pt.II. Qualitative Composition of the Planktonic Flora. Trans. Am. Phil. Soc., n.s. 31(3):193-237, 13 text figs.

Chaetoceros pelagicus
Mangin, M. L., 1912
Phytoplancton de la croisière du "René" dans l'Atlantique (Septembre 1908). Ann. Inst. Ocean., n.s., 4(1):1-66, 2 pls., 41 text figs., 2 tables.

Chaetoceros pelagicus
Michailova, N.F., 1964.
On the spreading of the habitat into the Black Sea of species of the genus Chaetoceros of northern seas and their biogeography. (In Russian).
Trudy Sevastopol Biol. Sta., 7:231-248.

Chaetoceros pelagicum
Pettersson, H., F. Gross, and F. Koczy, 1939
Large scale plankton cultures. Medd. från Oceanografiska Institutet i Göteborg No.3 (Göteborgs Kungl. vetenskaps-och Vitterhets-Samhälles Handlinger Femte Följden. Ser. B) Vol.6(13):1-24.
1. The plankton shaft. H. Pettersson
2. Experiment with phytoplankton. F. Gross and F. Koczy
3. Experiments with zooplankton.

Chaetoceros pelagicus
Yentsch, Charles S., and Carol A. Reichert, 1962
The interrelationship between water-soluble yellow substances and chloroplastic pigments in marine algae.
Botanica Marina, 3(3/4):65-74.

Chaetoceros pendulum
Allen, W. E., 1938
The Templeton Crocker Expedition to the Gulf of California in 1935 - The Phytoplankton. Amer. Microsc. Soc., Trans. 57:328-335.

Chaetoderos pendulus
Cupp, Easter E., 1943
Marine plankton diatoms of the west coast of North America. Bull. S.I.O. 5(1):1-238, 5 pls., 168 text figs.

Chaetoceros pendulus
Cupp, E.E., 1934
Analysis of marine diatom collections taken from the Canal Zone to California during March, 1933. Trans. Am. Micros. Soc. LIII (1):22-29, 1 map.

Chaetoceros pendulum
Gilbert, J.Y., and W.E. Allen, 1943
The phytoplankton of the Gulf of California obtained by the "E.W. Scripps" in 1939 and 1940. J. Mar. Res. V(2):89-110, figs.30-31.

Chaetoceros pendulum
Hendey, N.I., 1937
The plankton diatoms of the southern seas. Discovery Repts. 16:151-364, pls.6-13.

Chaetoceros pendulus
Kokubo, S., 1952
Results of the observations on the plankton and oceanography of Mutsu Bay during 1950, reference being made also to the period 1946-1950. Bull Mar.Biol.Sta., Asamushi 5(1/4): 1-54, 3 tables,(fold-in), 1 fold-in.

Chaetoceros perpusillum
Hendy, N. Ingram, 1964
An introductory account of the smaller algae of British coastal waters. V. Bacillariophyceae (Diatoms).
Her Majesty's Stationary Office, 317 pp., 45 pls.

Chaetoceros perpusillus
Brunel, J., 1962
Le phytoplancton de la Baie de Chaleurs. Inst. Botan., Univ. Montréal, Contrib. No. 77: 365 pp., 66 pls.

Chaetoceros perpusillus
Cupp, Easter E., 1943
Marine plankton diatoms of the west coast of North America. Bull. S.I.O. 5(1):1-238, 5 pls., 168 text figs.

Chaetoceros peruvianus
Allen, W. E., 1938
The Templeton Crocker Expedition to the Gulf of California in 1935 - The Phytoplankton. Amer. Microsc. Soc., Trans. 57:328-335.

Chaetoceros peruvianus
Allen, W.E., 1937
Plankton diatoms of the Gulf of California obtained by the G. Allan Hancock Expedition of 1936. The Hancock Pacific Expeditions, Univ. So. Calif. Publ. 3:47-59, 1 fig.

Chaetoceros peruvianus
Allen, W.E., 1936.
Surface plankton diatoms in the North Pacific Ocean in 1934. Madroño 3(6):3 pp.

Chaetoceros peruvianum
Allen, W.E., and E.E. Cupp, 1935
Plankton diatoms of the Java Sea. Annales du Jardin Botanique de Buitenzorg XLIV (2):101-174, figs.1-127.
(drawings of all species mentioned)

Chaetoceros peruvianum
Bigelow, H.B., and M. Leslie, 1930
Reconnaissance of the waters and plankton of Monterey Bay, July 1928. Bull. M.C.Z. 70(5):429-481, 43 text figs.

Chaetoceros peruvianus
Boden, B.P., 1950
Some marine plankton diatoms from the west coast of South Africa. Trans. R.Soc. S. Africa. 32:321-434, 100 text figs.

Chaetoceros peruvianum
Boden, B. P., 1949.
The diatoms collected by the U.S.S. CACOPAN in the Antarctic in 1947. J. Mar. Res. 8(1):6-13, 3 textfigs.

Chaetoceros peruvianum
Boden, Brian, 1948
Marine plankton diatoms on operation HIGHJUMP in: Some oceanographic observations on operation HIGHJUMP. By R.S. Dietz. USNEL Rept. No.55, 97 pp., 41 figs. 7 July 1948.

Chaetoceros perpusillus
Cleve-Euler, A., 1951
Die Diatomeen von Schweden und Finnland. Kungl. Svenska Vetenskaps Akad. Handl., Fjärde Ser. 2(1): 161 pp., 6 pls.

Chaetoceros peruvianus
Cupp, Easter E., 1943
Marine plankton diatoms of the west coast of North America. Bull. S.I.O. 5(1):1-238, 5 pls., 168 text figs.

Chaetoceros peruvianus
Cupp, E.E., 1937
Seasonal distribution and occurrence of marine diatoms and dinoflagellates at Scotch Cap, Alaska. Bull. S.I.O. Tech. ser.4(3):71-100, 7 textfigs.

Chaetoceros peruvianus
Cupp, E.E., 1934
Analysis of marine diatom collections taken from the Canal Zone to California during March, 1933. Trans. Am. Micros. Soc. LIII (1):22-29, 1 map.

Chaetoceros peruvianus
Cupp, E.E. and Allen, W.E., 1938
Plankton diatoms of the Gulf of California obtained by Allan Hancock Pacific Expedition of 1937. The Hancock Pacific Expeditions, The Univ. So. Calif. Publ. 3: 61-74, 1 map, pls.4-15.

Chaetoceros peruvianum
Dangeard, P., 1927
Phytoplankton de la croisière du "Sylvana". Ann. Inst. Ocean., Monaco, n.s., 4(8):286-401, 54 text figs. (Feirer-Juin 1913).

Chaetoceros peruvianus
Delegazione Italiana della Commissione Internazionale per l'Esplorazione Scientifica del Mediterraneo, 1941
Note sul plancton della Laguna veneta. [Memoria CCLXXXIX], Arch. di Ocean. e Limn. Anno I, Fasc. I, 1941 XIX: 31-57 pp.

Chaetoceros peruvianum
Delsman, H. C., 1939.
Preliminary plankton investigations in the Java Sea. Treubia, 17:139-181; 8 maps, 41 figs.

Chaetoceros peruvianus
Ercegovic, A., 1936
Etudes qualitative et quantitatives du phytoplancton dans les eaux cotières de l'Adriatique oriental moyen au cours de l'année 1934. Acta Adriatica 1(9):1-126

Chaetoceros peruvianus
Eskinazi Enide e Shigekatsv Satô (1963-1964) 1966.
Contribuição ao estudo das diatomaceas da Praia de Piedade.
Trabhs Inst. Oceanogr., Univ. Recife, 5 (5/6): 73-114.

Chaetoceros peruvianus
Forti, A., 1922
Ricerche sulla flora pelagica (fitoplancton) di Quarto dei Mille. Mem. R. Com. Talass. Ital. 97:248 pp., 13 pls.

Chaetoceros peruvianus
Frenguelli, Joaquin, and Hector Antonio Orlando, 1959.
Operacion MERLUZA. Diatomeas y silicoflagelados del plancton del "VI Crucero". Servicio Hidrogr. Naval, Argentina, Publ. No. H. 619: 5-62.

Chaetoceros peruvianus
Gilbert, J.Y., and W.E. Allen, 1943
The phytoplankton of the Gulf of California obtained by the "E.W. Scripps" in 1939 and 1940. J. Mar. Res. V(2):89-110, figs.30-31.

Chaetoceros peruvianum
Gran, H.H., 1908
Diatomeen. Nordisches Plankton, Botanischer Teil pp. XIX.1-XIX 146; 178 text figs.

Chaetoceros peruvianus
(syn. = C. Chilensis Krasske 1941)
Hasle, Grethe Rytter, 1960
Phytoplankton and ciliate species from the Tropical Pacific.
Skr. Norske Videnskaps-Akad., Oslo, 1. Mat.-Nat. Kl., 1960(2): 1-50.

Chaetoceros peruvianum
Hendy, N. Ingram, 1964
An introductory account of the smaller algae of British coastal waters. V. Bacillariophyceae (Diatoms).
Her Majesty's Stationary Office, 317 pp., 45 pls.

Chaetoceros peruvianum
Hendey, N.I., 1937
The plankton diatoms of the southern seas. Discovery Repts. 16:151-364, pls.6-13.

Chaetoceros peruvianus
Kokubo, S., 1952
Results of the observations on the plankton and oceanography of Mutsu Bay during 1950, reference being made also to the period 1946-1950. Bull Mar.Biol.Sta., Asamushi 5(1/4):1-54, 3 tables,(fold-in), 1 fold-in.

Chaetoceros peruvianus
Mangin, M. L., 1912
Phytoplancton de la croisière du "René" dans l'Atlantique (Septembre 1908). Ann. Inst. Ocean., n.s., 4(1):1-66, 2 pls., 41 text figs., 2 tables.

Chaetoceros peruvianus
Manguin, E., 1954
Diatomées marines provenant de l'ile Heard (Australian National Antarctic Expedition). Rev. Algol., n.s., 1:14-24.

Chaetoceros peruvianus
Massutí Algamora, M., 1949
Estudio de diez y seis muestras de plancton del Golfo de Nápoles. Publ. Inst. Biol. Appl. 5:85-94, 1 fold-in table.

Chaetoceros peruvianus
Michailova, N.F., 1964.
On the spreading of the habitat into the Black Sea of species of the genus of Chaetoceros of northern seas and their biogeography. (In Russian).
Trudy Sevastopol Biol. Sta., 7:231-248.

Chaetoceros peruvianum
Pavillard, J., 1925
Bacillariales. Rept. on the Danish Oceangr. Exped., 1908-10 to the Mediterranean and adj. seas. Vol.II., Biol. J4:72 pp., 116 text figs.

Chaetoceros peruvianum
Pavillard, J., 1905
Recherches sur la flore pelagique (Phytoplankton) de l'Etang de Thau. Theses presentees a la Fac. Sci., Paris, 116 pp., 3 pls.

Chaetoceros peruvianus
Rampi, L., 1942
Ricerche sul fitoplancton del Mare Ligure 6. Le diatomee delle acque di Sanremo. Nuovo Giornale Botanico Italiano, N.S., 49:252-268.

Chaetoceros peruvianus
Rampi, L., 1940
Diatomee del Mare Adriatico. Nuovo Giornale Botanico Italiano, n.s., 47:559-608.

Chaetoceros peruvianum
Schröder, B., 1900
Phytoplankton des Golfes von Neapel nebst vergleichenden Ausblicken auf das atlantischen Ozean. Mitt. Zool. Stat. Neapel, 14:1-38.

Chaetoceros peruvianum
Sousa e Silva, E., 1949
Diatomaceas e Dinoflagelados de Baia de Cascais. Portugaliae Acta Biol. Volume: Julio Henriques, Ser. B: 300-383, 9 pls, 2 fold-in tables.

Chaetoceros Peruvianum form. robusta
Allen, W.E., and E.E. Cupp, 1935
Plankton diatoms of the Java Sea. Annales du Jardin Botanique de Buitenzorg XLIV (2):101-174, figs.1-127.
(drawings of all species mentioned)

Chaetoceros peruvianus Brightwell
Brunel, J., 1962
Le phytoplancton de la Baie de Chaleurs. Inst. Botan., Univ. Montréal, Contrib. No. 77: 365 pp., 66 pls.

Chaetoceros peruvianus currens
Forti, A., 1922
Ricerche sulla flora pelagica (fitoplancton) di Quarto dei Mille. Mem. R. Com. Talass. Ital. 97:248 pp., 13 pls.

Chaetoceros peruvianus gracilis
Cupp, Easter E., 1943
Marine plankton diatoms of the west coast of North America. Bull. S.I.O. 5(1):1-238, 5 pls., 168 text figs.

Chaetoceros peruvianus saltans
Forti, A., 1922
Ricerche sulla flora pelagica (fitoplancton) di Quarto dei Mille. Mem. R. Com. Talass. Ital. 97:248 pp., 13 pls.

Chaetoceros peruvianus volans
Forti, A., 1922
Ricerche sulla flora pelagica (fitoplancton) di Quarto dei Mille. Mem. R. Com. Talass. Ital. 97:248 pp., 13 pls.

Chaetoceros polygonum
Gran, H.H., 1908
Diatomeen. Nordisches Plankton, Botanischer Teil pp. XIX.1-XIX 146; 178 text figs.

Chaetoceros polygonum
Hendy, N. Ingram, 1964
An introductory account of the smaller algae of British coastal waters. V. Bacillariophyceae (Diatoms).
Her Majesty's Stationary Office, 317 pp., 45 pls.

Chaetoceros protuberans
Castracane degli Antelminelli, F., 1886
1. Report on the Diatomaceae collected by H.M.S. Challenger during the years 1873-1876. Rept. Sci. Results, H.M.S. Challenger, Botany Vol. II, 178 pp., 30 pls.

Chaetoceros protuberans
Mangin, M. L., 1912
Phytoplancton de la croisière du "René" dans l'Atlantique (Septembre 1908). Ann. Inst. Ocean., n.s., 4(1):1-66, 2 pls., 41 text figs., 2 tables.

Chaetoceros protuberans
Schröder, B., 1900
Phytoplankton des Golfes von Neapel nebst vergleichenden Ausblicken auf das atlantischen Ozean. Mitt. Zool. Stat. Neapel, 14:1-38.

Chaetoceros pseudo-breve
Dangeard, P., 1927
Phytoplankton de la croisière du "Sylvana". Ann. Inst. Ocean., Monaco, n.s., 4(8):286-401, 54 text figs. (Feirer-Juin 1913).

Chaetoceros pseudo-brevis
Delegazione Italiana della Commissione Internazionale per l'Esplorazione Scientifica del Mediterraneo, 1941
Note sul plancton della Laguna veneta. [Memoria CCLXXXIX], Arch. di Ocean. e Limn. Anno I, Fasc. I, 1941 XIX: 31-57 pp.

Chaetoceros pseudobreve
Pavillard, J., 1925
Bacillariales. Rept. on the Danish Oceangr. Exped., 1908-10 to the Mediterranean and adj. seas. Vol.II., Biol. J4:72 pp., 116 text figs.

Chaetoceros pseudocrinitum
Bigelow, H.B., and M. Leslie, 1930
Reconnaissance of the waters and plankton of Monterey Bay, July 1928. Bull. M.C.Z., 70(5):429-481, 43 text figs.

Chaetoceros pseudocrinitus
Brunel, J., 1962
Le phytoplancton de la Baie de Chaleurs. Inst. Botan., Univ. Montréal, Contrib. No. 77: 365 pp., 66 pls.

Chaetoceros pseudocrinitus
Cleve-Euler, A., 1951
Die Diatomeen von Schweden und Finnland. Kungl. Svenska Vetenskaps Akad. Handl., Fjärde Ser. 2(1): 161 pp., 6 pls.

Chaetoceros pseudocrinitum
Gran, H.H., 1908
Diatomeen. Nordisches Plankton, Botanischer Teil pp. XIX.1-XIX 146; 178 text figs.

Chaetoceros pseudocrinitus
Gran, H.H., and T. Braarud, 1935
A quantitative study of the phytoplankton in the Bay of Fundy and the Gulf of Maine (including observations on hydrography, chemistry, and turbidity). J. Biol. Bd., Canada, 1(5):279-467, 69 text figs.

Chaetoceros pseudocrinitus
Gran, H. H. and E. C. Angst, 1931
Plankton diatoms of Puget Sound. Publ. Puget Sound Biol. Sta. 7:417-519, 95 text figs.

Chaetoceros pseudocrinitum
Hendy, N. Ingram, 1964
An introductory account of the smaller algae of British coastal waters. V. Bacillariophyceae (Diatoms).
Her Majesty's Stationary Office, 317 pp., 45 pls.

Chaetoceros pseudocrinitum
Hendey, N.I., 1937
The plankton diatoms of the southern seas. Discovery Repts. 16:151-364, pls.6-13.

Chaetoceros pseudocrinitus
Kokubo, S., 1952
Results of the observations on the plankton and oceanography of Mutsu Bay during 1950, reference being made also to the period 1946-1950. Bull Mar.Biol.Sta., Asamushi 5(1/4):1-54, 3 tables,(fold-in), 1 fold-in.

Chaetoceros pseudocrinitus
Lillick, L.C., 1938
Preliminary report of the phytoplankton of the Gulf of Maine. Am. Mid. Nat. 20(3):624-640, 1 text figs 37 tables.

Chaetoceros pseudocrinitus
Morse, D.C., 1947
Some observations on seasonal variations in plankton population Patuxent River, Maryland 1943 1945. Bd. Nat. Res., Publ. No.65, Chesapeake Biol. Lab., 31, 3 figs.

Chaetoceros pseudocrinitum
Rumkówna, A., 1948
List of the phytoplankton species occurring in the superficial water layers in the Gulf of Gdańsk. Bull. Lab. mar., Gdynia, No. 4: 139-141 with tables in back.

Chaetoceros pseudocurvisetum
Allen, W.E., and E.E. Cupp, 1935
Plankton diatoms of the Java Sea. Annales du Jardin Botanique de Buitenzorg XLIV (2):101-174, figs.1-127.
(drawings of all species mentioned)

Chaetoceros pseudocurvisetus
Brunel, J., 1962
Le phytoplancton de la Baie de Chaleurs. Inst. Botan., Univ. Montréal, Contrib. No. 77: 365 pp., 66 pls.

Chaetoceros pseudocurvisetus
Cupp, Easter E., 1943
Marine plankton diatoms of the west coast of North America. Bull. S.I.O. 5(1):1-238, 5 pls., 168 text figs.

Chaetoceros pseudocurvisetus
Delegazione Italiana della Commissione Internazionale per l'Esplorazione Scientifica del Mediterraneo, 1941
Note sul plancton della Laguna veneta. [Memoria CCLXXXIX], Arch. di Ocean. e Limn. Anno I, Fasc. I, 1941 XIX: 31-57 pp.

Chaetoceros pseudocurvisatum
Delsman, H. C., 1939.
Preliminary plankton investigations in the Java Sea. Treubia, 17:139-181, 8 maps, 41 figs.

Chaetoceros pseudocurvisetus
Forti, A., 1922
Ricerche sulla flora pelagica (fitoplancton) di Quarto dei Mille. Mem. R. Com. Talass. Ital. 97:248 pp., 13 pls.

Chaetoceros pseudocurvisetum
Hendy, N. Ingram, 1964
An introductory account of the smaller algae of British coastal waters. V. Bacillariophyceae (Diatoms). Her Majesty's Stationary Office, 317 pp., 45 pls.

Chaetoceros pseudocurvisetum
Mangin, M. L., 1912
Phytoplancton de la croisière du "René" dans l'Atlantique (Septembre 1908). Ann. Inst. Ocean., n.s., 4(1):1-66, 2 pls., 41 text figs., 2 tables.

Chaetoceros pseudocurvisetum
Pavillard, J., 1925
Bacillariales. Rept. on the Danish Oceanogr. Exped., 1908-10 to the Mediterranean and adj. seas. Vol.II., Biol. J4:72 pp., 116 text figs.

Chaetoceros pseudocurvisetum
Rampi, L., 1942
Ricerche sul fitoplancton del Mare Ligure 6. Le diatomee delle acque di Sanremo. Nuovo Giornale Botanico Italiano, N.S., 49:252-268.

Chaetoceros pseudocurvisetum
Sousa e Silva, E., 1949
Diatomaceas e Dinoflagelados de Baia de Cascais. Portugaliae Acta Biol., Volume: Julio Henriques, Ser. B: 300-383, 9 pls, 2 fold-in tables.

Chaetoceros pseudosimilis
Cleve-Euler, A., 1951
Die Diatomeen von Schweden und Finnland. Kungl. Svenska Vetenskaps Akad. Handl., Fjärde Ser. 2(1): 161 pp., 6 pls.

Chaetoceros radians
Brunel, J., 1962
Le phytoplancton de la Baie de Chaleurs. Inst. Botan., Univ. Montréal, Contrib. No. 77: 365 pp., 66 pls.

Chaetoceros radians
Cleve-Euler, A., 1951
Die Diatomeen von Schweden und Finnland. Kungl. Svenska Vetenskaps Akad. Handl., Fjärde Ser. 2(1): 161 pp., 6 pls.

Chaetoceros radians
Hendy, N. Ingram, 1964
An introductory account of the smaller algae of British coastal waters. V. Bacillariophyceae (Diatoms). Her Majesty's Stationary Office, 317 pp., 45 pls.

Chaetoceros radians
Gran, H.H., 1908
Diatomeen. Nordisches Plankton, Botanischer Teil pp. XIX 1-XIX 146; 178 text figs.

Chaetoceros radians
Steemann-Nielsen, Einar, 1951
The marine vegetation of the Isefjord. A study on ecology and production. Medd. Komm. Danmarks Fiskeri-og Havundersøgelser. Ser. Plankton. 5(4); 114pp., 46 text figs.

Chaetoceros radicans
Allen, W. E., 1938
The Templeton Crocker Expedition to the Gulf of California in 1935 - The Phytoplankton. Amer. Microsc. Soc., Trans. 57:328-335.

Chaetoceros radicans
Allen, W.E., 1937
Plankton diatoms of the Gulf of California obtained by the G. Allan Hancock Expedition of 1936. The Hancock Pacific Expeditions, Univ. So. Calif. Publ. 3:47-59, 1 fig.

Chaetoceros radicans
Allen, W. E., 1936
Occurrence of marine plankton diatoms in a ten-year series of daily catches in southern California. Am. Jour. Bot. 23(1):60-63.

Chaetoceros radicans
Brunel, J., 1962
Le phytoplancton de la Baie de Chaleurs. Inst. Botan., Univ. Montréal, Contrib. No. 77: 365 pp., 66 pls.

Chaetoceros radicans
Brunel, Jules, 1962
Le phytoplancton de la Baie des Chaleurs. Contrib. Ministère de la Chasse et des Pêcheries, Province de Québec, No. 91: 365 pp.

Chaetoceros radicans
Cupp, Easter E., 1943
Marine plankton diatoms of the west coast of North America. Bull. S.I.O. 5(1):1-238, 5 pls., 168 text figs.

Chaetoceros radicans
Cupp, E.E., 1934
Analysis of marine diatom collections taken from the Canal Zone to California during March, 1933. Trans. Am. Micros. Soc. LIII (1):22-29, 1 map.

Chaetoceros radicans
Cupp, E.E. and Allen, W.E., 1938
Plankton diatoms of the Gulf of California obtained by Allan Hancock Pacific Expedition of 1937. The Hancock Pacific Expeditions, The Univ. So. Calif. Publ. 3: 61-74, 1 map, pls.4-15.

Chaetoceros radiaans
Frenguelli, Joaquin, and Hector Antonio Orlando, 1959.
Operacion MERLUZA. Diatomeas y silicoflagelados del plancton del "VI Crucero". Servicio Hidrogr. Naval., Argentina, Publ. No. H. 619: 5-62.

Chaetoceros radicans
Gilbert, J.Y., and W.E. Allen, 1943
The phytoplankton of the Gulf of California obtained by the "E.W. Scripps" in 1939 and 1940. J. Mar. Res. V(2):89-110, figs.30-31.

Chaetoceros radians
Gran, H.H., 1897
Protophyta: Diatomaceae, Silico-flagellata and Cilioflagellata. Den Norske Nordhavs Expedition 1876-1878, h. 24, 36 pp., 4 pls.

Chaetoceros radicans
Gran, H.H. and T. Braarud, 1935
A quantitative study of the phytoplankton in the Bay of Fundy and the Gulf of Maine (including observations on hydrography, chemistry, and turbidity). J. Biol. Bd., Canada, 1(5):279-467, 69 text figs.

Chaetoceros radians
Gran, H. H. and E. C. Angst, 1931
Plankton diatoms of Puget Sound. Publ. Puget Sound Biol. Sta. 7:417-519, 95 text figs.

Chaetoceros radicans
Hendy, N. Ingram, 1964
An introductory account of the smaller algae of British coastal waters. V. Bacillariophyceae (Diatoms). Her Majesty's Stationary Office, 317 pp., 45 pls.

Chaetoceros radicans
Kokubo, S., 1952
Results of the observations on the plankton and oceanography of Mutsu Bay during 1950, reference being made also to the period 1946-1950. Bull Mar.Biol.Sta., Asamushi 5(1/4): 1-54, 3 tables,(fold-in), 1 fold-in.

Chaetoceros radicans
Lillick, L.C., 1940
Phytoplankton and planktonic protozoa of the offshore waters of the Gulf of Maine. Pt.II. Qualitative Composition of the Planktonic Flora. Trans. Am. Phil. Soc., n.s., 31(3):193-237, 13 text figs.

Chaetoceros radians
Lillick, L.C., 1938
Preliminary report of the phytoplankton of the Gulf of Maine. Am. Mid. Nat. 20(3):624-640, 1 text figs 37 tables.

Chaetoceros radians
Meunier, A., 1913
Microplancton de la Mer Flamande. 1. Chaetoceros. Mem. Mus. Roy. Hist. Nat., Belgique 7(2):1-55, 7 pls.

Chaetoceros radicans
Phifer, L.D. (undated)
The occurrence and distribution of plankton diatoms in Bering Sea and Bering Strait, July 26-August 24, 1934. Report of Oceanographic cruise of U.S. Coast Guard Cutter Chelan 1934, Part II(A):1-44 (mimeographed) plus fig.1 (after Pt.B)

Chaetoceros radicans
Sargent, M. S. and T. J. Walker, 1948
Diatom populations associated with eddies off southern California in 1941. J. Mar. Res. 7(3):490-505, 15 text figs.

Chaetoceros radians
Sousa e Silva, E., and J. Dos Santos-Pinto, 1948
O Plancton da Baía de S. Martinho do Porto 1. Diatomaceas e Dinoflagelados. Bol. Soc. Portuguese de Ciencias Naturais, 16(2):134-187, 6 pls. (Trav. Sta. Biol. Mar. de Lisbonne No. 52).

Chaetoceros radiculum n.sp.
Castracane degli Antelminelli, F., 1886
1. Report on the Diatomaceae collected by H.M.S. Challenger during the years 1873-1876. Rept. Sci. Results, H.M.S. Challenger, Botany Vol. II, 178 pp., 30 pls.

Chaetoceros radiculum
Hendey, N.I., 1937
The plankton diatoms of the southern seas. Discovery Repts. 16:151-364, pls.6-13.

Chaetoceros radiculum
Mangin, L., 1915
Phytoplancton de L'Antartique. Deuxieme Exped. Ant. Francaise (1908-1910), 95 pp., 3 pls., 58 text figs.

Chaetoceros ralfsii var.
Brunel, J., 1962
Le phytoplancton de la Baie de Chaleurs. Inst. Botan., Univ. Montréal, Contrib. No. 77: 365 pp., 66 pls.

Chaetoceros Ralfsii
Forti, A., 1922
Ricerche sulla flora pelagica (fitoplancton) di Quarto dei Mille. Mem. R. Com. Talass. Ital. 97:248 pp., 13 pls.

Chaetoceros Ralfsii
Ghazzawi, F.M., 1939
Plankton of the Egyptian waters. A study of the Suez Canal Plankton. (A) Phytoplankton. Preliminary Report 83 pp. Notes and Memoires, Min. Commerce-Industry, Egypt, Hydrobiol. & Fish. 65 figs.

Chaetoceros Ralfsi
Hendey, N.I., 1937
The plankton diatoms of the southern seas. Discovery Repts. 16:151-364, pls. 6-13.

Chaetoceros rigidus
Brunel, J., 1962
Le phytoplancton de la Baie de Chaleurs. Inst. Botan., Univ. Montréal, Contrib. No. 77: 365 pp., 66 pls.

Chaetoceros rigidus
Michailova, N.F., 1964.
On the spreading of the habitat into the Black Sea of species of the genus Chaetoceros of northern seas and their biogeography. (In Russian).
Trudy Sevastopol Biol. Sta., 7:231-248.

Chaetoceros rostratum
Allen, W.E., and E.E. Cupp, 1935
Plankton diatoms of the Java Sea. Annales du Jardin Botanique de Buitenzorg XLIV (2):101-174, figs. 1-127.
(drawings of all species mentioned)

Chaetoceros rostratus
Anon., 1951
Bulletin of the Marine Biological Station of Asamushi 4(3/4): 15 pp.

Chaetoceros rostratus
Brunel, J., 1962
Le phytoplancton de la Baie de Chaleurs. Inst. Botan., Univ. Montréal, Contrib. No. 77: 365 pp., 66 pls.

Chaetoceros rostratum
Dangeard, P., 1927
Phytoplankton de la croisière du "Sylvana". Ann. Inst. Ocean., Monaco, n.s., 4(8):286-401, 54 text figs. (Feirer-Juin 1913).

Chaetoceros rostratus
Delegazione Italiana della Commissione Internazionale per l'Esplorazione Scientifica del Mediterraneo, 1941
Note sul plancton della Laguna veneta. [Memoria CCLXXXI], Arch. di Ocean. e Limn. Anno I, Fasc. I, 1941 XIX: 31-57 pp.

Chaetoceros rostratus
Ercegovic, A., 1936
Etudes qualitative et quantitatives du phytoplancton dans les eaux cotières de l'Adriatique oriental moyen au cours de l'année 1934. Acta Adriatica 1(9):1-126

Chaetoceros rostratus
Eskinazi Enide e Shigekatsv Satô (1963-1964) 1966.
Contribuição ao estudo das diatomaceas da Praia de Piedade.
Trabhs Inst. Oceanogr., Univ. Recife, 5 (5/6): 73-114.

Chaetoceros rostratum
Pavillard, J., 1925
Bacillariales. Rept. on the Danish Oceangr. Exped., 1908-10 to the Mediterranean and adj. seas. Vol.II., Biol. J4:72 pp., 116 text figs.

Chaetoceros rostratus
Rampi, L., 1942
Ricerche sul fitoplancton del Mare Ligure 6. Le diatomee delle acque di Sanremo. Nuovo Giornale Botanico Italiano, N.S., 49:252-268.

Chaetoceros rostratum
Sousa e Silva, E., 1949
Diatomaceas e Dinoflagelados de Baia de Cascais. Portugalise Acta Biol. Volume: Julio Henriques, Ser. B: 300-383, 9 pls, 2 fold-in tables.

Chaetoceros saltans
Brunel, J., 1962
Le phytoplancton de la Baie de Chaleurs. Inst. Botan., Univ. Montréal, Contrib. No. 77: 365 pp., 66 pls.

Chaetoceros saltans
Pavillard, J., 1925
Bacillariales. Rept. on the Danish Oceangr. Exped., 1908-10 to the Mediterranean and adj. seas. Vol.II., Biol. J4:72 pp., 116 text figs.

Chaetoceros Schimperianum
Hendey, N.I., 1937
The plankton diatoms of the southern seas. Discovery Repts. 16:151-364, pls. 6-13.

Chaetoceros Schimparianum
Mangin, L., 1915
Phytoplancton de L'Antartique. Deuxieme Exped. Ant. Francaise (1908-1910), 95 pp., 3 pls., 58 text figs.

Chaetoceros schimperianus
Manguin, E., 1954
Diatomees marines provenant de l'île Heard (Australian National Antarctic Expedition). Rev. Algol., n.s., 1: 14-24.

Chaetoceros Schuttei
Bigelow, H.B., and M. Leslie, 1930
Reconnaissance of the waters and plankton of Monterey Bay, July 1928. Bull. M.C.Z., 70(5):429-481, 43 text figs.

Chaetoceros shüttii
Cleve-Euler, A., 1951
Die Diatomeen von Schweden und Finnland. Kungl. Svenska Vetenskaps Akad. Handl., Fjärde Ser. 2(1): 161 pp., 6 pls.

Chaetoceros Schuettii
Forti, A., 1922
Ricerche sulla flora pelagica (fitoplancton) di Quarto dei Mille. Mem. R. Com. Talass. Ital. 97:248 pp., 13 pls.

Chaetoceros Schüttii
Gran, H.H., 1908
Diatomeen. Nordisches Plankton, Botanischer Teil pp. XIX.1-XIX 146; 178 text figs.

Chaetoceros Schüttii
Gran, H.H., 1897
Protophyta: Diatomaceae, Silico-flagellata and Cilioflagellata. Den Norske Nordhavs Expedition 1876-1878, h, 24, 36 pp., 4 pls.

Chaetoceros Schüttii
Jorgensen, E., 1900
Protophyten und Protozoen im Plankton aus der Norwegischen Westkerste. Bergens Mus. Aarb. 1899(6): 95 pp., 5 pls., 83 tables.

Chaetoceros Schüttei
Mangin, M.L., 1912
Phytoplankton de la croisière du "René" dans l'Atlantique (Septembre 1908). Ann. Inst. Ocean., n.s., 4(1):1-66, 2 pls., 41 text figs., 2 tables.

Chaetoceros Schüttii
Meunier, A., 1913
Microplancton de la Mer Flamande. 1. Chaetoceros. Mem. Mus. Roy. Hist. Nat., Belgique 7(2):1-55, 7 pls.

Chaetoceros Schuetti
Pavillard, J., 1905
Recherches sur la flore pelagique (Phytoplankton) de l'Etang de Thau. Theses presentees a la Fac. Sci., Paris, 116 pp., 3 pls.

Chaetoceros Schüttii circenalis var. nov
Meunier, A., 1913
Microplancton de la Mer Flamande. 1. Chaetoceros. Mem. Mus. Roy. Hist. Nat., Belgique 7(2):1-55, 7 pls.

Chaetoceros Schüttii genuina
Meunier, A., 1913
Microplancton de la Mer Flamande. 1. Chaetoceros. Mem. Mus. Roy. Hist. Nat., Belgique 7(2):1-55, 7 pls.

Chaetoceros Schüttii Willei
Meunier, A., 1913
Microplancton de la Mer Flamande. 1. Chaetoceros. Mem. Mus. Roy. Hist. Nat., Belgique 7(2):1-55, 7 pls.

Chaetoceros scolopendra
Allen, W.E., 1928.
Catches of marine diatoms and dinoflagellates taken by boat in Southern California waters in 1926. Bull. S.I.O., tech. ser., 1:201-246, 6 text figs.

Chaetoceros scolpendra
Allen, W.E., 1927.
Quantitative studies on inshore marine diatoms and dinoflagellates of Southern California in 1922. Bull. S.I.O., tech. ser., 1:31-38, 2 text-figs.

Chaetoceros scolopendra
Bigelow, H.B., and M. Leslie, 1930
Reconnaissance of the waters and plankton of Monterey Bay, July 1928. Bull. M.C.Z., 70(5):429-481, 43 text figs.

Chaetoceros scolopendra
Cleve-Euler, A., 1951
Die Diatomeen von Schweden und Finnland. Kungl. Svenska Vetenskaps Akad. Handl., Fjärde Ser. 2(1): 161 pp., 6 pls.

Chaetoceros scolopendra
Cupp, E., 1930
Quantitative Studies of miscellaneous series of surface catches of marine diatoms and dinoflagellates taken between Seattle and the Canal Zone from 1924 to 1928. Trans. Am. Micro. Soc., XLIX (3):238-245.

Chaetoceros scolopendra
Gran, H.H., 1908
Diatomeen. Nordisches Plankton, Botanischer Teil pp. XIX.1-XIX 146; 178 text figs.

Chaetoceros scolopendra
Gran, H.H., 1897
Protophyta: Diatomaceae, Silico-flagellata and Cilioflagellata. Den Norske Nordhavs Expedition 1876-1878, h. 24, 36 pp., 4 pls.

Chaetoceros scolopendra
Jorgensen, E., 1900
Protophyten und Protozoën im Plankton aus der Norwegischen Westkerste. Bergens Mus. Aarb. 1899(6): 95 pp., 5 pls., 83 tables

Chaetoceros Scolopendra
Meunier, A., 1913
Microplancton de la Mer Flamande. 1. Chaetoceros. Mem. Mus. Roy. Hist. Nat., Belgique 7(2):1-55, 7 pls.

Chaetoceros secundus
Forti, A., 1922
Ricerche sulla flora pelagica (fitoplancton) di Quarto dei Mille. Mem. R. Com. Talass. Ital. 97:248 pp., 13 pls.

Chaetoceros secundus
Gran, H. H. and E. C. Angst, 1931
Plankton diatoms of Puget Sound. Publ. Puget Sound Biol. Sta. 7:417-519, 95 text figs.

Chaetoceros secundum
Marukawa, H., 1921
Plankton lists and some new species of copepods, from the northern waters of Japan. Bull. Inst. Ocean., No.384, 15 pp., 3 pls., 1 chart. Monaco

Chaetoceros seiracanthus
Boden, B.P., 1950
Some marine plankton diatoms from the west coast of South Africa. Trans. R.Soc. S. Africa. 32:321-434, 100 text figs.

Chaetoceros seiracanthus
Brunel, J., 1962
Le phytoplancton de la Baie de Chaleurs. Inst. Botan., Univ. Montréal, Contrib. No. 77: 365 pp., 66 pls.

Chaetoceros seiracanthus
Cleve-Euler, A., 1951
Die Diatomeen von Schweden und Finnland. Kungl. Svenska Vetenskaps Akad. Handl., Fjärde Ser. 2(1): 161 pp., 6 pls.

Chaetoceros seiracanthus
Cupp, Easter E., 1943
Marine plankton diatoms of the west coast of North America. Bull. S.I.O. 5(1):1-238, 5 pls., 168 text figs.

Chaetoceros seiracanthum
Gran, H.H., 1908
Diatomeen. Nordisches Plankton, Botanischer Teil pp. XIX 1-XIX 146; 178 text figs.

Chaetoceros seiracanthum n sp.
Gran, H.H., 1897
Protophyta: Diatomaceae, Silico-flagellata and Cilioflagellata. Den Norske Nordhavs Expedition 1876-1878, h. 24, 36 pp., 4 pls.

Chaetoceros seiracanthum
Gran, H. H. and E. C. Angst, 1931
Plankton diatoms of Puget Sound. Publ. Puget Sound Biol. Sta. 7:417-519, 95 text figs.

Chaetoceros seiracanthus
Hendy, N. Ingram, 1964
An introductory account of the smaller algae of British coastal waters. V. Bacillariophyceae (Diatoms). Her Majesty's Stationary Office, 317 pp., 45 pls.

Chaetoceros seiracanthus
Jorgensen, E., 1900
Protophyten und Protozoën im Plankton aus der Norwegischen Westkerste. Bergens Mus. Aarb. 1899(6): 95 pp., 5 pls., 83 tables

Chaetoceros seiracanthus
Lillick, L.C., 1940
Phytoplankton and planktonic protozoa of the offshore waters of the Gulf of Maine. Pt.II. Qualitative Composition of the Planktonic Flora. Trans. Am. Phil. Soc., n.s., 31(3):193-237, 13 text figs.

Chaetoceros seiracanthum
Marukawa, H., 1921
Plankton lists and some new species of copepods, from the northern waters of Japan. Bull. Inst. Ocean., No.384, 15 pp., 3 pls., 1 chart. Monaco

Chaetoceros seiracanthum
Meunier, A., 1913
Microplancton de la Mer Flamande. 1. Chaetoceros. Mem. Mus. Roy. Hist. Nat., Belgique 7(2):1-55, 7 pls.

Chaetoceros septentrionalis
Brunel, J., 1962
Le phytoplancton de la Baie de Chaleurs. Inst. Botan., Univ. Montréal, Contrib. No. 77: 365 pp., 66 pls.

Chaetoceros septentrionalis
Brunel, Jules, 1962
Le phytoplancton de la Baie des Chaleurs. Contrib. Ministère de la Chasse et des Pêcheries, Province de Québec, No. 91: 365 pp.

Chaetoceros septentrionalis
Cleve-Euler, A., 1951
Die Diatomeen von Schweden und Finnland. Kungl. Svenska Vetenskaps Akad. Handl., Fjärde Ser. 2(1): 161 pp., 6 pls.

Chaetoceros septentrionale
Duke, Eleanor L., Joyce Lewin and Bernhard E.F. Reimann 1973
Light and electron microscope studies of diatom species belonging to the genus Chaetoceros Ehrenberg. 1. Chaetoceros septentrionale Oestrup.
Phycologia 12 (1/2): 1-9.

Chaetoceros septemtrionale
Hendy, N. Ingram, 1964
An introductory account of the smaller algae of British coastal waters. V. Bacillariophyceae (Diatoms). Her Majesty's Stationary Office, 317 pp., 45 pls.

Chaetoceros septentrionalis
Michailova, N.F., 1964.
On the spreading of the habitat into the Black Sea of species of the genus Chaetoceros of northern seas and thei biogeography. (In Russian).
Trudy Sevastopol Biol. Sta., 7:231-248.

Chaetoceros septentrionalis
Morse, D.C., 1947
Some observations on seasonal variations in plankton population Patuxent River, Maryland 1943-1945. Bd. Nat. Res., Publ. No.65, Chesapeake Biol. Lab. 31, 3 figs.

Chaetoceros sessile n.sp.
Grøntved, J., 1951.
Phytoplankton studies. 2. A new biological type within the genus Chaetoceros, Chaetoceros sessile sp. nov. K. Danske Widenskab. Selsk., Biol. Medd., 18(17):3-9, 7 textfigs.

Chaetoceros sessile
Hendy, N. Ingram, 1964
An introductory account of the smaller algae of British coastal waters. V. Bacillariophyceae (Diatoms). Her Majesty's Stationary Office, 317 pp., 45 pls.

Chaetoceros setoensis
Kokubo, S., 1952
Results of the observations on the plankton and oceanography of Mutsu Bay during 1950, reference being made also to the period 1946-1950. Bull Mar.Biol.Sta., Asamushi 5(1/4): 1-54, 3 tables,(fold-in), 1 fold-in.

Chaetoceros seychellarum
Hendey, N.I., 1937
The plankton diatoms of the southern seas. Discovery Repts. 16:151-364, pls.6-13.

Chaetoceros similis
Braarud, T., 1945
A phytoplankton survey of the polluted waters of inner Oslo Fjord. Hvalrådets Skrifter, No.28, 142 pp., 19 text figs., 17 tables.

Chaetoceros similis
Brunel, J., 1962
Le phytoplancton de la Baie de Chaleurs. Inst. Botan., Univ. Montréal, Contrib. No. 77: 365 pp., 66 pls.

Chaetoceros similis
Brunel, Jules, 1962
Le phytoplancton de la Baie des Chaleurs. Contrib. Ministère de la Chasse et des Pêcheries, Province de Québec, No. 91: 365 pp.

Chaetoceros similis
Cleve-Euler, A., 1951
Die Diatomeen von Schweden und Finnland. Kungl. Svenska Vetenskaps Akad. Handl., Fjärde Ser. 2(1): 161 pp., 6 pls.

Chaetoceros similis
Cupp, Easter E., 1943
Marine plankton diatoms of the west coast of North America. Bull. S.I.O. 5(1):1-238, 5 pls., 168 text figs.

Chaetoceros simile
Gran, H.H., 1908
Diatomeen. Nordisches Plankton, Botanischer Teil pp. XIX.1-XIX 146; 178 text figs.

Chaetoceros simile
Gran, H.H., 1897
Protophyta: Diatomaceae, Silico-flagellata and Cilioflagellata. Den Norske Nordhavs Expedition 1876-1878, h. 24, 36 pp., 4 pls.

Chaetoceros similis
Gran, H.H., and T. Braarud, 1935
A quantitative study of the phytoplankton in the Bay of Fundy and the Gulf of Maine (including observations on hydrography, chemistry, and turbidity). J. Biol. Bd., Canada, 1(5):279-467, 69 text figs.

Chaetoceros similis
Gran, H. H. and E. C. Angst, 1931
Plankton diatoms of Puget Sound. Publ. Puget Sound Biol. Sta. 7:417-519, 95 text figs.

Chaetoceros simile
Hendy, N. Ingram, 1964
An introductory account of the smaller algae of British coastal waters. V. Bacillariophyceae (Diatoms).
Her Majesty's Stationary Office, 317 pp., 45 pls.

Chaetoceros similis
Jorgensen, E., 1900
Protophyten und Protozoën im Plankton aus der Norwegischen Westkerste. Bergens Mus. Aarb. 1899(6): 95 pp., 5 pls., 83 tables.

Chaetoceros similis
Lillick, L.C., 1940
Phytoplankton and planktonic protozoa of the offshore waters of the Gulf of Maine. Pt.II. Qualitative Composition of the Planktonic Flora. Trans. Am. Phil. Soc., n.s., 31(3):193-237, 13 text figs.

Chaetoceros simile
Meunier, A., 1913
Microplancton de la Mer Flamande. 1. Chaetoceros. Mem. Mus. Roy. Hist. Nat., Belgique 7(2):1-55, 7 pls.

Chaetoceros similis & var.
Michailova, N.F., 1964.
On the spreading of the habitat into the Black Sea of species of the genus Chaetoceros of northern seas and their biogeography. (In Russian).
Trudy Sevastopol Biol. Sta., 7:231-248.

Chaetoceros similis
Steemann-Nielsen, Einar, 1951
The marine vegetation of the Isefjord. A study on ecology and production. Medd. Komm. Danmarks Fiskeri-og Havundersøgelser. Ser. Plankton. 5(4); 114pp., 46 text figs.

Chaetoceros simplex
Allen, W.E., 1937
Plankton diatoms of the Gulf of California obtained by the G. Allan Hancock Expedition of 1936. The Hancock Pacific Expeditions, Univ. So. Calif. Publ. 3:47-59, 1 fig.

Chaetoceros simplex
Brunel, J., 1962
Le phytoplancton de la Baie de Chaleurs. Inst. Botan., Univ. Montréal, Contrib. No. 77: 365 pp., 66 pls.

Chaetoceros simplex
Cleve-Euler, A., 1951
Die Diatomeen von Schweden und Finnland. Kungl. Svenska Vetenskaps Akad. Handl., Fjärde Ser. 2(1): 161 pp., 6 pls.

Chaetoceros simplex
Cupp, E.E. and Allen, W.E., 1938
Plankton diatoms of the Gulf of California obtained by Allan Hancock Pacific Expedition of 1937. The Hancock Pacific Expeditions, The Univ. So. Calif. Publ. 3: 61-74, 1 map, pls.4-15.

Chaetoceros simplex
Gold, Kenneth, 1965.
A note on the distribution of luminescent dinoflagellates and water constituents in Phosphorescent Bay, Puerto Rico.
Ocean Sci. and Ocean Eng., Mar. Techn. Soc.,- Amer. Soc. Limnol. Oceanogr., 1:77-80.

Chaetoceros simplex
Gran, H.H., and T. Braarud, 1935
A quantitative study of the phytoplankton in the Bay of Fundy and the Gulf of Maine (including observations on hydrography, chemistry, and turbidity). J. Biol. Bd., Canada, 1(5):279-467, 69 text figs.

Chaetoceros simplex
Hendy, N. Ingram, 1964
An introductory account of the smaller algae of British coastal waters. V. Bacillariophyceae (Diatoms).
Her Majesty's Stationary Office, 317 pp., 45 pls.

Chaetoceros simplex
Kanagawa, Akio, 1969.
On the vitamin B of a diatom Chaetoceros simplex, as the diet for the larva of marine animals. (In Japanese; English abstract).
Mem. Fac. Fish. Kagoshima Univ. 18: 93-97.

Chaetoceros simplex
Lillick, L.C., 1940
Phytoplankton and planktonic protozoa of the offshore waters of the Gulf of Maine. Pt.II. Qualitative Composition of the Planktonic Flora. Trans. Am. Phil. Soc., n.s., 31(3):193-237, 13 text figs.

Chaetoceros simplex
Marshall, S.M. and A.P. Orr, 1948
Further experiments on the fertilization of a sea loch (Loch Craiglin). The effect of different plant nutrients on the phytoplankton. J.M.B.A. 27(2):360-379, 10 text figs.

Chaetoceros simplex.
Michailova, N.F., 1964.
On the spreading of the habitat into the Black Sea of species of the genus Chaetoceros of northern seas and their biogeography. (In Russian).
Trudy Sevastopol Biol. Sta., 7:231-248.

Chaetoceros simplex
Ogiino, C., 1963.
Studies on the chemical composition of some natural foods of aquatic animals. (In Japanese; English summary).
Bull. Jap. Soc. Sci. Fish., 29(5):459-462.

Chaetoceros simplex
Pavillard, J., 1925
Bacillariales. Rept. on the Danish Oceangr. Exped., 1908-10 to the Mediterranean and adj. seas. Vol.II., Biol. J4:72 pp., 116 text figs.

Chaetoceros simplex
Pavillard, J., 1905
Recherches sur la flore pelagique (Phytoplankton) de l'Etang de Thau. Theses presentees a la Fac. Sci., Paris, 116 pp., 3 pls.

Chaetoceros simplex
Paulsen, O., 1905.
On some Peridineae and plankton diatoms.
Medd. Komm. Havundersøgelser, Ser. Plankton, 1(3):7 pp.

Chaetoceros simplex
Phifer, L.D. (undated)
The occurrence and distribution of plankton diatoms in Boring Sea and Bering Strait, July 26-August 24, 1934. Report of Oceanographic cruise of U.S. Coast Guard Cutter Chelan 1934, Part II(A):1-44 (mimeographed) plus fig.1 (after Pt.B)

Chaetoceros simplex
Rampi, L., 1942
Ricerche sul fitoplancton del Mare Ligure 6. Le diatomee delle acque di Sanremo. Nuovo Giornale Botanico Italiano, N.S., 49:252-268.

Chaetoceros simplex
Sato, Shigekatsu and Hisashi Kan-no 1967
Synchrony to the autospore phase in the cultured population of a centric diatom, Chaetoceros simplex var. calcitrans Paulsen. Bull. Tohoku reg. Fish. Res. Lab. 27: 101-109. (In Japanese; English abstract)

Chaetoceros simplex
Steemann-Nielsen, Einar, 1951
The marine vegetation of the Isefjord. A study on ecology and production. Medd. Komm. Danmarks Fiskeri-og Havundersøgelser. Ser. Plankton. 5(4); 114pp., 46 text figs.

Chaetoceros simplex calcitrans
Takano, Hideaki, 1963.
Notes on marine littoral diatoms of Japan. Bull. Tokai Reg. Fish. Res. Lab., No. 36:1-8.

Chaetoceros skeleton
Cupp, Easter E., 1943
Marine plankton diatoms of the west coast of North America. Bull. S.I.O. 5(1):1-238, 5 pls., 168 text figs.

Chaetoceros socialis
Allen, W.E., 1938
The Templeton Crocker Expedition to the Gulf of California in 1935 - The Phytoplankton. Amer. Microsc. Soc., Trans. 57:328-335.

Chaetoceros socialis
Allen, W.E., 1937
Plankton diatoms of the Gulf of California obtained by the G. Allan Hancock Expedition of 1936. The Hancock Pacific Expeditions, Univ. So. Calif. Publ. 3:47-59, 1 fig.

Chaetoceros socialis
Allen, W.E., 1936
Occurrence of marine plankton diatoms in a ten-year series of daily catches in southern California. Am. Jour. Bot. 23(1):60-63.

Chaetoceros socialis
Allen, W.E., 1934
Marine plankton diatoms of Lower California in 1931. Bot. Gaz. 95(3):485-492, 1 fig.

Chaetoceros socialis
Boden, B.P., 1950
Some marine plankton diatoms from the west coast of South Africa. Trans. R.Soc. S. Africa. 32:321-434, 100 text figs.

Chaetoceros socialis
Braarud, T., 1945
Experimental studies on marine plankton diatoms. Avhandlingar utgitt av Det Norske Videnskaps-Akademi i Oslo. 1. Mat.-Naturv. Klasse 1944. No.10, 1-16, 1 text fig.

Chaetoceros socialis
Braarud, T., 1945
A phytoplankton survey of the polluted waters of inner Oslo Fjord. Hvalrådets Skrifter, No.28, 142 pp., 19 text figs., 17 tables.

Chaetoceros socialis
Braarud, T., and Adam Bursa, 1939
On the phytoplankton of the Oslo Fjord, 1933-1934. Hvalrådets Skr. No.19:1-63; 9 text figs. Reviewed. J. du. Cons. 14(3): 418-420. A.C. Gardiner.

Chaetoceros socialis
Brunel, J., 1962
Le phytoplancton de la Baie de Chaleurs. Inst. Botan., Univ. Montréal, Contrib. No. 77: 365 pp., 66 pls.

Chaetoceros socialis
Brunel, Jules, 1962
Le phytoplancton de la Baie des Chaleurs. Contrib. Ministère de la Chasse et des Pêcheries; Province de Québec, No. 91: 365 pp.

Chaetoceros socialis
Cleve-Euler, A., 1951
Die Diatomeen von Schweden und Finnland. Kungl. Svenska Vetenskaps Akad. Handl., Fjärde Ser. 2(1): 161 pp., 6 pls.

Chaetoceros socialis
Copenhagen, W.J. and L.D., 1949
Variation in the phytoplankton of Table Bay, October 1934 to October 1935. With a note on the calorific value of Chaetoceros spp. Trans. Roy. Soc. S. Africa, 32(2):113-123, 2 text figs.

Chaetoceros socialis
Cupp, Easter E., 1943
Marine plankton diatoms of the west coast of North America. Bull. S.I.O. 5(1):1-238, 5 pls., 168 text figs.

Chaetoceros socialis
Cupp, E.E., 1934
Analysis of marine diatom collections taken from the Canal Zone to California during March, 1933. Trans. Am. Micros. Soc. LIII (1):22-29, 1 map.

Chaetoceros socialis
Cupp, E.E. and Allen, W.E., 1938
Plankton diatoms of the Gulf of California obtained by Allan Hancock Pacific Expedition of 1937. The Hancock Pacific Expeditions, The Univ. So. Calif. Publ. 3: 61-74, 1 map, pls.4-15.

Chaetoceros socialis
Ercegovic, A., 1936
Etudes qualitative et quantitatives du phytoplancton dans les eaux cotières de l'Adriatique oriental moyen au cours de l'année 1934. Acta Adriatica 1(9):1-126

Chaetoceros socialis
Forti, A., 1922
Ricerche sulla flora pelagica (fitoplancton) di Quarto dei Mille. Mem. R. Com. Talass. Ital. 97:248 pp., 13 pls.

Chaetoceros socialis
Gilbert, J.Y., and W.E. Allen, 1943
The phytoplankton of the Gulf of California obtained by the "E.W. Scripps" in 1939 and 1940. J. Mar. Res. V(2):89-110, figs.30-31.

Chaetoceros sociale
Gran, H.H., 1897
Protophyta: Diatomaceae, Silico-flagellata and Cilioflagellata. Den Norske Nordhavs Expedition 1876-1878, h. 24, 36 pp., 4 pls.

Chaetoceros socialis
Gran, H.H. and T. Braarud, 1935
A quantitative study of the phytoplankton in the Bay of Fundy and the Gulf of Maine (including observations on hydrography, chemistry, and turbidity). J. Biol. Bd., Canada, 1(5):279-467, 69 text figs.

Chaetoceros socialis
Gran, H.H. and E.C. Angst, 1931
Plankton diatoms of Puget Sound. Publ. Puget Sound Biol. Sta. 7:417-519, 95 text figs.

Chaetoceros sociale
Hendy, N. Ingram, 1964
An introductory account of the smaller algae of British coastal waters. V. Bacillariophyceae (Diatoms). Her Majesty's Stationary Office, 317 pp., 45 pls.

Chaetoceros sociale
Hendey, N.I., 1937
The plankton diatoms of the southern seas. Discovery Repts. 16:151-364, pls.6-13.

Chaetoceros sociale
Iselin, C., 1930
A report on the coastal waters of Labrador based on explorations of the "Chance" during the summer of 1926. Proc. Am. Acad. Arts Sci., 66(1):1-37, 14 text figs.

Chaetoceros socialis
Jorgensen, E., 1900
Protophyten und Protozoën im Plankton aus der Norwegischen Westkerste. Bergens Mus. Aarb. 1899(6): 95 pp., 5 pls., 83 tables.

Chaetoceros socialis
Kokubo, S., 1952
Results of the observations on the plankton and oceanography of Mutsu Bay during 1950, reference being made also to the period 1946-1950. Bull Mar.Biol.Sta., Asamushi 5(1/4): 1-54, 3 tables,(fold-in), 1 fold-in.

Chaetoceros socialis
Lillick, L.C., 1940
Phytoplankton and planktonic protozoa of the offshore waters of the Gulf of Maine. Pt.II. Qualitative Composition of the Planktonic Flora. Trans. Am. Phil. Soc., n.s., 31(3):193-237, 13 text figs.

Chaetoceros socialis
Lillick, L.C., 1938
Preliminary report of the phytoplankton of the Gulf of Maine. Am. Mid. Nat. 20(3):624-640, 1 text figs 37 tables.

Chaetoceros sociale
Macdonald, R., 1933
An examination of plankton hauls made in the Suez Canal during the year 1928. Fish. Res. Dis., Notes & Mem. No.3, 11 pp., 1 chart.

Chaetoceros socialis
Mangin, L., 1915
Phytoplancton de L'Antartique. Deuxieme Exped. Ant. Francaise (1908-1910), 95 pp., 3 pls. 58 text figs.

Chaetoceros socialis
Mangin, M.L., 1912
Phytoplancton de la croisière du "René" dans l'Atlantique (Septembre 1908). Ann. Inst. Ocean., n.s., 4(1):1-66, 2 pls., 41 text figs., 2 tables.

Chaetoceros sociale
Marukawa, H., 1921
Plankton lists and some new species of copepods, from the northern waters of Japan. Bull. Inst. Ocean., No.384, 15 pp., 3 pls., 1 chart. Monaco

Chaetoceros sociale
Meunier, A., 1913
Microplancton de la Mer Flamande. 1. Chaetoceros. Mem. Mus. Roy. Hist. Nat., Belgique 7(2):1-55, 7 pls.

Chaetoceros socialis & var.
Michailova, N.F., 1964
On the spreading of the habitat into the Black Sea of species of the genus Chaetoceros of northern seas and their biogeography. (In Russian). Trudy Sevastopol Biol. Sta., 7:231-248.

Chaetoceros socialis
Morse, D.C., 1947
Some observations on seasonal variations in plankton population Patuxent River, Maryland 1943-1945. Bd. Nat. Res., Publ. No.65, Chesapeake Biol. Lab., 31, 3 figs.

Chaetoceros socialis
Pettersson, H., 1934
Scattering and extinction of light in sea-water. Meddelanden från Göteborgs Högskolas Oceanografiska Institut. 9 (Göteborgs Kungl. vetenskaps-och vitterhets-samhälles Handlingar. Femte Foljden, Ser.B, 4(4)):1-16.

Chaetoceros sociale
Pettersson, H., F. Gross, and F. Koczy, 1939
Large scale plankton cultures. Medd. från Oceanografiska Institutet i Göteborg No.3 (Göteborgs Kungl. vetenskaps-och Vitterhets-Samhälles Handlingar Femte Följden. Ser. B) Vol.6(13):1-24.
1. The plankton shaft. H. Pettersson
2. Experiment with phytoplankton. F. Gross and F. Koczy
3. Experiments with zooplankton.

Chaetoceros socialis
Phifer, L.D. (undated)
The occurrence and distribution of plankton diatoms in Bering Sea and Bering Strait, July 26-August 24, 1934. Report of Oceanographic cruise of U.S. Coast Guard Cutter Chelan 1934, Part II(A):1-44 (mimeographed) plus fig.1 (after Pt.B)

Chaetoceros socialis
Rampi, L., 1942
Ricerche sul fitoplancton del Mare Ligure 6. Le diatomee delle acque di Sanremo. Nuovo Giornale Botanico Italiano, N.S., 49:252-258.

Chaetoceros socialis
Sargent, M.S. and T.J. Walker, 1948
Diatom populations associated with eddies off southern California in 1941. J. Mar. Res. 7(3):490-505, 15 text figs.

Chaetoceros sociale
Schodduyn, M., 1926
Observations faites dans la baie d'Ambleteuse (Pas de Calais). Bull. Inst. Ocean., Monaco, No. 482: 64 pp.

Chaetoceros sociale
Sousa e Silva, E., 1949
Diatomaceas e Dinoflagelados de Baia de Cascais. Portugaliae Acta Biol., Volume: Julio Henriques, Ser. B: 300-383, 9 pls, 2 fold-in tables.

Chaetoceros sociale
Sousa e Silva, E., and J. Dos Santos-Pinto, 1948
O Plancton da Baia de S. Martinho do Porto. 1. Diatomaceas e Dinoflagelados. Bol. Soc. Portuguese de Ciencias Naturais, 16(2):134-187, 6 pls. (Trav. Sta. Biol. Mar. de Lisbonne No. 52).

Chaetoceros strictum
Boden, B.P., 1950
Some marine plankton diatoms from the west coast of South Africa. Trans. R.Soc. S. Africa. 32:321-434, 100 text figs.

Chaetoceros subcompressum n.sp.
Schröder, B., 1900
Phytoplankton des Golfes von Neapel nebst vergleichenden Ausblicken auf das atlantischen Ozean. Mitt. Zool. Stat. Neapel, 14:1-38.

Chaetoceros subcoronatus
Krasske, G., 1941
Die Kieselalgen des chilenischen Küstenplankton (Aus dem südchilenischen Kusten gebiet IX). Arch. Hydrobiol. 38 (2):260-287.

Chaetoceros subsecundus
Braarud, T., 1945
A phytoplankton survey of the polluted waters of inner Oslo Fjord. Hvalrådets Skrifter, No.28, 142 pp., 19 text figs., 17 tables.

Chaetoceros subsecundus
Boden, B.P., 1950
Some marine plankton diatoms from the west coast of South Africa. Trans. R.Soc. S. Africa. 32:321-434, 100 text figs.

Chaetoceros subsecundus
Cupp, Easter E., 1943
Marine plankton diatoms of the west coast of North America. Bull. S.I.O. 5(1):1-238, 5 pls., 168 text figs.

Chaetoceros subsecundus
Cupp, E.E. and Allen, W.E., 1938
Plankton diatoms of the Gulf of California obtained by Allan Hancock Pacific Expedition of 1937. The Hancock Pacific Expeditions, The Univ. So. Calif. Publ. 3: 61-74, 1 map, pls.4-15.

Chaetoceros subsecundus

Ercegovic, A., 1936
Etudes qualitative et quantitatives du phytoplancton dans les eaux côtières de l'Adriatique oriental moyen au cours de l'année 1934. Acta Adriatica 1(9):1-126.

Chaetoceros subsecundus

Kokubo, S., 1952
Results of the observations on the plankton and oceanography of Mutsu Bay during 1950, reference being made also to the period 1946-1950. Bull Mar.Biol.Sta., Asamushi 5(1/4): 1-54, 3 tables,(fold-in), 1 fold-in.

Chaetoceros subsecundus

Lillick, L.C., 1940
Phytoplankton and planktonic protozoa of the offshore waters of the Gulf of Maine. Pt.II. Qualitative Composition of the Planktonic Flora. Trans. Am. Phil. Soc., n.s., 31(3):193-237, 13 text figs.

Chaetoceros subsecundus

Lillick, L.C., 1938
Preliminary report of the phytoplankton of the Gulf of Maine. Am. Mid. Nat. 20(3):624-640, 1 text fig, 37 tables.

Chaetoceros subsecundus

Uyeno, Fukuzo, 1961
Oceanographical and ecological studies on primary production of the sea, with special references to relationship between diatom production and temperature and chlorinity of water.
Rept., Fac. Fish., Pref. Univ., Mie, 4(1): 1-64.

Chaetoceros subtilis

Brunel, J., 1962
Le phytoplancton de la Baie de Chaleurs. Inst. Botan., Univ. Montréal, Contrib. No. 77: 365 pp., 66 pls.

Chaetoceros subtilis

Cleve-Euler, A., 1951
Die Diatomeen von Schweden und Finnland. Kungl. Svenska Vetenskaps Akad. Handl., Fjärde Ser. 2(1): 161 pp., 6 pls.

Chaetoceros subtile

Gran, H.H., 1908
Diatomeen. Nordisches Plankton, Botanischer Teil pp. XIX.1-XIX 146; 178 text figs.

Chaetoceros subtilis

Gran, H.H., and T. Braarud, 1935
A quantitative study of the phytoplankton in the Bay of Fundy and the Gulf of Maine (including observations on hydrography, chemistry, and turbidity). J. Biol. Bd., Canada, 1(5):279-467, 69 text figs.

Chaetoceros subtilis

Grøntved, J., 1960-61
Planktological contributions. IV. Taxonomical and productional investigations in shallow coastal waters.
Medd. Dansk Fisk. Havundersøgelser,n.s.,3(1): 1-17.

Chaetoceros subtile

Hendy, N. Ingram, 1964
An introductory account of the smaller algae of British coastal waters. V. Bacillariophyceae (Diatoms).
Her Majesty's Stationary Office, 317 pp., 45 pls.

Chaetoceros subtilis

Levander, K.M., 1947
Plankton gesammelt in den Jahren 1899-1910 an den Küsten Finnlands. Finnländische Hydrographisch-Biologische Untersuchungen (aus dem Wasserbiologischen Laboratorium der Societas Scientiarum Fennica) No.11:40 pp., 6 diagrams, 13 pls., tables.

Chaetoceros subtilis

Lillick, L.C., 1940
Phytoplankton and planktonic protozoa of the offshore waters of the Gulf of Maine. Pt.II. Qualitative Composition of the Planktonic Flora. Trans. Am. Phil. Soc., n.s., 31(3):193-237, 13 text figs.

Chaetoceros subtile

Meunier, A., 1913
Microplancton de la Mer Flamande. 1. Chaetoceros. Mem. Mus. Roy. Hist. Nat., Belgique 7(2):1-55, 7 pls.

Chaetoceros subtilis

Michailova, N.F., 1964.
On the spreading of the habitat into the Black Sea of species of the genus Chaetoceros of northern seas and their biogeography. (In Russian).
Trudy Sevastopol Biol. Sta., 7:231-248.

Chaetoceros subtilis

Morse, D.C., 1947
Some observations on seasonal variations in plankton population Patuxent River, Maryland 1943-1945. Bd. Nat. Res., Publ. No.65, Chesapeake Biol. Lab., 31, 3 figs.

Chaetoceros subtilis

Pettersson, H., 1934
Scattering and extinction of light in sea-water. Meddelanden från Göteborgs Högskolas Oceanografiska Institut. 9 (Göteborgs Kungl. vetenskaps-och vitterhetssamhälles Handlingar. Femte Foljden, Ser.B, 4(4)):1-16.

Chaetoceros subtilis

Steemann-Nielsen, Einar, 1951
The marine vegetation of the Isefjord. A study on ecology and production. Medd. Komm. Danmarks Fiskeri-og Havundersøgelser. Ser. Plankton. 5(4); 114pp., 46 text figs.

Chaetoceros sumatranum

Hendey, N.I., 1937
The plankton diatoms of the southern seas. Discovery Repts. 16:151-364, pls.6-13.

Chaetoceros tenuissimum n.sp.

Meunier, A., 1913
Microplancton de la Mer Flamande. 1. Chaetoceros. Mem. Mus. Roy. Hist. Nat., Belgique 7(2):1-55, 7 pls.

Chaetoceros teres

Aubert, M., et M. Gauthier 1966.
Origine et nature des substances antibiotiques présentes dans le milieu marin. 7. Note sur l'activité antibactérienne d'une diatomée marine: Chaetoceros teres (Cleve).
Rev. intern. Oceanogr. Med. 33-87.

Chaetoceros teres

Bigelow, H.B., and M. Leslie, 1930
Reconnaissance of the waters and plankton of Monterey Bay, July 1928. Bull. M.C.Z., 70(5):429-481, 43 text figs.

Chaetoceros teres

Boden, B.P., 1950
Some marine plankton diatoms from the west coast of South Africa. Trans. R.Soc. S. Africa. 32:321-434, 100 text figs.

Chaetoceros teres

Braarud, T., 1945
Experimental studies on marine plankton diatoms. Avhandlingar utgitt av Det Norske Videnskaps-Akademi i Oslo. 1. Mat.-Naturv. Klasse 1944. No.10, 1-16, 1 text fig.

Chaetoceros teres

Brunel, J., 1962
Le phytoplancton de la Baie de Chaleurs. Inst. Botan., Univ. Montréal, Contrib. No. 77: 365 pp., 66 pls.

Chaetoceros teres

Brunel, Jules, 1962
Le phytoplancton de la Baie des Chaleurs. Contrib. Ministère de la Chasse et des Pêcheries, Province de Québec, No. 91: 365 pp.

Chaetoceros teres

Cleve-Euler, A., 1951
Die Diatomeen von Schweden und Finnland. Kungl. Svenska Vetenskaps Akad. Handl., Fjärde Ser. 2(1): 161 pp., 6 pls.

Chaetoceros teres

Cupp, Easter E., 1943
Marine plankton diatoms of the west coast of North America. Bull. S.I.O. 5(1):1-238, 5 pls., 168 text figs.

Chaetoceros teres

Cupp, E.E., 1934
Analysis of marine diatom collections taken from the Canal Zone to California during March, 1933. Trans. Am. Micros. Soc. LIII (1):22-29, 1 map.

Chaetoceros teres

Gilson, H. C., 1937
Chemical and Physical Investigations. The nitrogen cycle. John Murray Exped., 1933-34, Sci. Repts., 2(2):21-81, 16 text figs.

Chaetoceros teres

Gran, H.H., 1908
Diatomeen. Nordisches Plankton, Botanischer Teil pp. XIX.1-XIX 146; 178 text figs.

Chaetoceros teres

Gran, H.H., 1897
Protophyta: Diatomaceae, Silico-flagellata and Cilioflagellata. Den Norske NorGhavs Expedition 1876-1878, h. 24, 36 pp., 4 pls.

Chaetoceros teres

Gran, H.H., and T. Braarud, 1935
A quantitative study of the phytoplankton in the Bay of Fundy and the Gulf of Maine (including observations on hydrography, chemistry, and turbidity). J. Biol. Bd., Canada, 1(5):279-467, 69 text figs.

Chaetoceros teres

Gran, H. H. and E. C. Angst, 1931
Plankton diatoms of Puget Sound. Publ. Puget Sound Biol. Sta. 7:417-519, 95 text figs.

Chaetoceros teres

Hendy, N. Ingram, 1964
An introductory account of the smaller algae of British coastal waters. V. Bacillariophyceae (Diatoms).
Her Majesty's Stationary Office, 317 pp., 45 pls.

Chaetoceros teres

Iselin, C., 1930
A report on the coastal waters of Labrador based on explorations of the "Chance" during the summer of 1926. Proc. Am. Acad. Arts Sci., 66(1):1-37, 14 text figs.

Chaetoceros teres

Jorgensen, E., 1900
Protophyten und Protozoen im Plankton aus der Norwegischen Westkerste. Bergens Mus. Aarb. 1899(6): 95 pp., 5 pls., 83 tables.

Chaetoceros teres

Lillick, L.C., 1940
Phytoplankton and planktonic protozoa of the offshore waters of the Gulf of Maine. Pt.II. Qualitative Composition of the Planktonic Flora. Trans. Am. Phil. Soc., n.s., 31(3):193-237, 13 text figs.

Chaetoceros teres
Lillick, L.C., 1938
Preliminary report of the phytoplankton of the Gulf of Maine. Am. Mid. Nat. 20(3):624-640, 1 text fig., 37 tables.

Chaetoceros teres
Lillick, L.C., 1937
Seasonal studies of the phytoplankton off Woods Hole, Massachusetts. Biol. Bull. LXXIII (3):488-503, 3 text figs.

Chaetoceros teres
Mangin, M. L., 1912
Phytoplancton de la croisière du "René" dans l'Atlantique (Septembre 1908). Ann. Inst. Ocean., n.s., 4(1):1-66, 2 pls., 41 text figs., 2 tables.

Chaetoceros teres
Meunier, A., 1913
Microplancton de la Mer Flamande. 1. Chaetoceros. Mem. Mus. Roy. Hist. Nat., Belgique 7(2):1-55, 7 pls.

Chaetoceros teres
Morse, D.C., 1947
Some observations on seasonal variations in plankton population Patuxant River, Maryland 1943-1945. Bd. Nat. Res., Publ. No.65, Chesapeake Biol. Lab., 31, 3 figs.

Chaetoceros teres
Phifer, L.D. (undated).
The occurrence and distribution of plankton diatoms in Bering Sea and Bering Strait, July 26-August 24, 1934. Report of Oceanographic cruise of U.S. Coast Guard Cutter Chelan 1934, Part II(A):1-44 (mimeographed) plus fig.1 (after Pt.B)

Chaetoceros tetras
Boden, B.P., 1950
Some marine plankton diatoms from the west coast of South Africa. Trans. R.Soc. S. Africa. 32:321-434, 100 text figs.

Chaetoceros tetrastichon
Allen, W.E., 1937
Plankton diatoms of the Gulf of California obtained by the G. Allan Hancock Expedition of 1936. The Hancock Pacific Expeditions, Univ. So. Calif. Publ. 3:47-59, 1 fig.

Chaetoceros tetrastichon
Brunel, J., 1962
Le phytoplancton de la Baie de Chaleurs. Inst. Botan., Univ. Montréal, Contrib. No. 77: 365 pp., 66 pls.

Chaetoceros tetrastichon
Allen, W. E., 1938
The Templeton Crocker Expedition to the Gulf of California in 1935 - The Phytoplankton. Amer. Microsc. Soc., Trans. 57:328-335.

Chaetoceros tetrastichon
Cupp, Easter E., 1943
Marine plankton diatoms of the west coast of North America. Bull. S.I.O. 5(1):1-238, 5 pls., 168 text figs.

Chaetoceros tetrastichon
Cupp, E.E., 1934
Analysis of marine diatom collections taken from the Canal Zone to California during March, 1933. Trans. Am. Micros. Soc. LIII (1):22-29, 1 map.

Chaetoceros tetrastichon
Ercegovic, A., 1936
Etudes qualitative et quantitatives du phytoplancton dans les eaux cotières de l'Adriatique oriental moyen au cours de l'année 1934. Acta Adriatica 1(9):1-126

Chaetoceros tetrastichon
Forti, A., 1922
Ricerche sulla flora pelagica (fitoplancton) di Quarto dei Mille. Mem. R. Com. Talass. Ital. 97:248 pp., 13 pls.

Chaetoceros tetrastichon
Gran, H.H., 1908
Diatomeen. Nordisches Plankton, Botanischer Teil pp. XIX.1-XIX 146; 178 text figs.

Chaetoceros tetrastichon
Hendy, N. Ingram, 1964
An introductory account of the smaller algae of British coastal waters. V. Bacillariophyceae (Diatoms). Her Majesty's Stationary Office, 317 pp., 45 pls.

Chaetoceros tetrastichon
Pavillard, J., 1925
Bacillariales. Rept. on the Danish Oceangr. Exped., 1908-10 to the Mediterranean and adj. seas. Vol.II., Biol. J4:72 pp., 116 text figs.

Chaetoceros tetrastichon
Pavillard, J., 1905
Recherches sur la flore pelagique (Phytoplankton) de l'Etang de Thau. Theses presentees a la Fac. Sci., Paris, 116 pp., 3 pls.

Chaetoceros tetrastichon
Rampi, L., 1942
Ricerche sul fitoplancton del Mare Ligure 6. Le diatomee delle acque di Sanremo. Nuovo Giornale Botanico Italiano, N.S., 49:252-268.

Chaetoceros tetrastichon
Schröder, B., 1900
Phytoplankton des Golfes von Neapel nebst vergleichenden Ausblicken auf das atlantischen Ozean. Mitt. Zool. Stat. Neapel, 14:1-38.

Chaetoceros tortissimus
Brunel, J., 1962
Le phytoplancton de la Baie de Chaleurs. Inst. Botan., Univ. Montréal, Contrib. No. 77: 365 pp., 66 pls.

Chaetoceros tortissimus
Cupp, Easter E., 1943
Marine plankton diatoms of the west coast of North America. Bull. S.I.O. 5(1):1-238, 5 pls., 168 text figs.

Chaetoceros tortissimus
Delegazione Italiana della Commissione Internazionale per l'Esplorazione Scientifica del Mediterraneo, 1941
Note sul plancton della Laguna veneta. [Memoria CCLXXXI], Arch. di Ocean. e Limn. Anno I, Fasc. I, 1941 XIX: 31-57 pp.

Chaetoceros tortissimus
Ercegovic, A., 1936
Etudes qualitative et quantitatives du phytoplancton dans les eaux cotières de l'Adriatique oriental moyen au cours de l'année 1934. Acta Adriatica 1(9):1-126

Chaetoceros tortissimum
Gran, H.H., 1908
Diatomeen. Nordisches Plankton, Botanischer Teil pp. XIX.1-XIX 146; 178 text figs.

Chaetoceros tortissimum
Hendy, N. Ingram, 1964
An introductory account of the smaller algae of British coastal waters. V. Bacillariophyceae (Diatoms). Her Majesty's Stationary Office, 317 pp., 45 pls.

Chaetoceros tortissimus
Mangin, L., 1915
Phytoplancton de L'Antartique. Deuxieme Exped. Ant. Francaise (1908-1910), 95 pp., 3 pls., 58 text figs.

Chaetoceros tortissimum
Pavillard, J., 1905
Recherches sur la flore pelagique (Phytoplankton) de l'Etang de Thau. Theses presentees a la Fac. Sci., Paris, 116 pp., 3 pls.

Chaetoceros uruguayensis
Brunel, J., 1962
Le phytoplancton de la Baie de Chaleurs. Inst. Botan., Univ. Montréal, Contrib. No. 77: 365 pp., 66 pls.

Chaetoceros uruguayensis n.sp.
Müller Melchers, F.C., 1953
New and little known diatoms from Uruguay and the South Atlantic coast. Com. Bot., Mus. Hist. Nat., Montevideo, 3(30):

Chaetoceros Van Heurckii
Allen, W.E., and E.E. Cupp, 1935
Plankton diatoms of the Java Sea. Annales du Jardin Botanique de Buitenzorg XLIV (2):101-174, figs.1-127.
(drawings of all species mentioned)

Chaetoceros vanheurckii
Boden, B.P., 1950
Some marine plankton diatoms from the west coast of South Africa. Trans. R.Soc. S. Africa. 32:321-434, 100 text figs.

Chaetoceros vanheurcki
Cupp, Easter E., 1943
Marine plankton diatoms of the west coast of North America. Bull. S.I.O. 5(1):1-238, 5 pls., 168 text figs.

Chaetoceros vanheurckii
Gilbert, J.Y., and W.E. Allen, 1943
The phytoplankton of the Gulf of California obtained by the "E.W. Scripps" in 1939 and 1940. J. Mar. Res. V(2):89-110, figs.30-31.

Chaetoceros vanheurckii
Gran, H. H. and E. C., Angst, 1931
Plankton diatoms of Puget Sound. Publ. Puget Sound Biol. Sta. 7:417-519, 95 text figs.

Chaetoceros Vanheurckii
Marukawa, H., 1921
Plankton lists and some new species of copepods, from the northern waters of Japan. Bull. Inst. Ocean., No.384, 15 pp., 3 pls., 1 chart. Monaco

Chaetoceros van Heurcki
Phifer, L.D. (undated)
The occurrence and distribution of plankton diatoms in Bering Sea and Bering Strait, July 26-August 24, 1934. Report of Oceanographic cruise of U.S. Coast Guard Cutter Chelan 1934, Part II(A):1-44 (mimeographed) plus fig.1 (after Pt.B)

Chaetoceros varians
Mann, A., 1893
List of Diatomaceae from a deep-sea dredging in the Atlantic Ocean off Delaware Bay by the U. S. Fish Commission Steamer Albatross. Proc. U. S. Nat. Mus. 16:303-312.

Chaetoceros vistulae
Cupp, Easter E., 1943
Marine plankton diatoms of the west coast of North America. Bull. S.I.O. 5(1):1-238, 5 pls., 168 text figs.

Chaetoceros vistulae

Brunel, J., 1962
Le phytoplancton de la Baie de Chaleurs. Inst. Botan., Univ. Montréal, Contrib. No. 77: 365 pp., 66 pls.

Chaetoceros vixvisibilis

Brunel, J., 1962
Le phytoplancton de la Baie de Chaleurs. Inst. Botan., Univ. Montréal, Contrib. No. 77: 365 pp., 66 pls.

Chaetoceros vixvisibilis

Ercegovic, A., 1940
Weitere Untersuchungen über einige hydrographische Verhältnisse und über die Phytoplanktonproduktion in den Gewässern der östlichen Mittelaria. Acta Adriatica 2(3):95-134, 8 text figs.

Chaetoceros Weissflogii

Bigelow, H.B., and M. Leslie, 1930
Reconnaissance of the waters and plankton of Monterey Bay, July 1928. Bull. M.C.Z., 70(5):429-481, 43 text figs.

Chaetoceros Weissflogii

Forti, A., 1922
Ricerche sulla flora pelagica (fitoplancton) di Quarto dei Mille. Mem. R. Com. Talass. Ital. 97:248 pp., 13 pls.

Chaetoceros Weissflogii

Gran, H.H., 1908
Diatomeen. Nordisches Plankton, Botanischer Teil pp. XIX.1-XIX 146; 178 text figs.

Chaetoceros Weissflogii

Mangin, M.L., 1912
Phytoplancton de la croisière du "René" dans l'Atlantique (Septembre 1908). Ann. Inst. Ocean., n.s., 4(1):1-66, 2 pls., 41 text figs., 2 tables.

Chaetoceros Weissflogii

Meunier, A., 1913
Microplancton de la Mer Flamande. 1. Chaetoceros. Mem. Mus. Roy. Hist. Nat., Belgique 7(2):1-55, 7 pls.

Chaetoceros Weisflogi

Pavillard, J., 1905
Recherches sur la flore pelagique (Phytoplankton) de l'Etang de Thau. Theses presentees a la Fac. Sci., Paris, 116 pp., 3 pls.

Chaetoceros Weissflogii

Schodduyn, M., 1926
Observations faites dans la baie d'Ambleteuse (Pas de Calais). Bull. Inst. Ocean., Monaco, No. 482: 64 pp.

Chaetoceros wighami

Brunel, J., 1962
Le phytoplancton de la Baie de Chaleurs. Inst. Botan., Univ. Montréal, Contrib. No. 77: 365 pp., 66 pls.

Chaetoceros wighami

Cleve-Euler, A., 1951
Die Diatomeen von Schweden und Finnland. Kungl. Svenska Vetenskaps Akad. Handl., Fjärde Ser. 2(1): 161 pp., 6 pls.

Chaetoceros wighami

Cupp, Easter E., 1943
Marine plankton diatoms of the west coast of North America. Bull. S.I.O. 5(1):1-238, 5 pls., 168 text figs.

Chaetoceros wighami

Ercegovic, A., 1936
Etudes qualitative et quantitatives du phytoplancton dans les eaux cotières de l'Adriatique oriental moyen au cours de l'année 1934. Acta Adriatica 1(9):1-126

Chaetoceros Wighami

Gran, H.H., 1908
Diatomeen. Nordisches Plankton, Botanischer Teil pp. XIX.1-XIX 146; 178 text figs.

Chaetoceros Wighami

Gran, H.H., 1897
Protophyta: Diatomaceae, Silico-flagellata and Cilioflagellata. Den Norske Nordhavs Expedition 1876-1878, h. 24, 36 pp., 4 pls.

Chaetoceros wighami

Hendy, N. Ingram, 1964
An introductory account of the smaller algae of British coastal waters. V. Bacillariophyceae (Diatoms). Her Majesty's Stationary Office, 317 pp., 45 pls.

Chaetoceros wighami

Levander, K.M., 1947
Plankton gesammelt in den Jahren 1899-1910 an den Küsten Finnlands. Finnländische Hydrographisch-Biologische Untersuchungen (aus dem Wasserbiologischen Laboratorin der Societas Scientiarum Fennica) No.11:40 pp., 6 diagrams, 13 pls., tables.

Chaetoceros Wighami

Lillick, L.C., 1940
Phytoplankton and planktonic protozoa of the offshore waters of the Gulf of Maine. Pt.II. Qualitative Composition of the Planktonic Flora. Trans. Am. Phil. Soc., n.s., 31(3):193-237, 13 text figs.

Chaetoceros Wighami

Meunier, A., 1913
Microplancton de la Mer Flamande. 1. Chaetoceros. Mem. Mus. Roy. Hist. Nat., Belgique 7(2):1-55, 7 pls.

Chaetoceros wighami

Michailova, N.F., 1964.
On the spreading of the habitat into the Black Sea of species of the genus Chaetoceros of northern seas and their biogeography. (In Russian). Trudy Sevastopol Biol. Sta., 7:231-248.

Chaetoceros wighami

Morse, D.C., 1947
Some observations on seasonal variations in plankton population Patuxant River, Maryland 1943-1945. Bd. Nat. Res., Publ. No.65, Chesapeake Biol. Lab., 31, 3 figs.

Chaetoceros wighami

Pavillard, J., 1925
Bacillariales. Rept. on the Danish Oceangr. Exped., 1908-10 to the Mediterranean and adj. seas. Vol.II., Biol. J4:72 pp., 116 text figs.

Chaetoceros Wighamii

Pavillard, J., 1905
Recherches sur la flore pelagique (Phytoplankton) de l'Etang de Thau. Theses presentees a la Fac. Sci., Paris, 116 pp., 3 pls.

Chaetoceros Wighami

Rampi, L., 1942
Ricerche sul fitoplancton del Mare Ligure 6. Le diatomee delle acque di Sanremo. Nuovo Giornale Botanico Italiano, N.S. 49:252-268.

Chaetoceros Wighami

Rumkówna, A., 1948
List of the phytoplankton species occurring in the superficial water layers in the Gulf of Gdańsk. Bull. Lab. mar., Gdynia, No. 4: 139-141 with tables in back.

Chaetoceros Willei

Bigelow, H.B., and M. Leslie, 1930
Reconnaissance of the waters and plankton of Monterey Bay, July 1928. Bull. M.C.Z., 70(5):429-481, 43 text figs.

Chaetoceros Willei

Braarud, T., 1945
A phytoplankton survey of the polluted waters of inner Oslo Fjord. Hvalrådets Skrifter, No.28, 142 pp., 19 text figs., 17 tables.

Chaetoceros willei

Cleve-Euler, A., 1951
Die Diatomeen von Schweden und Finnland. Kungl. Svenska Vetenskaps Akad. Handl., Fjärde Ser. 2(1): 161 pp., 6 pls.

Chaetoceros Willei

Forti, A., 1922
Ricerche sulla flora pelagica (fitoplancton) di Quarto dei Mille. Mem. R. Com. Talass. Ital. 97:248 pp., 13 pls.

Chaetoceros Willei

Gran, H.H., 1908
Diatomeen. Nordisches Plankton, Botanischer Teil pp. XIX.1-XIX 146; 178 text figs.

Chaetoceros Willei n.sp.

Gran, H.H., 1897
Protophyta: Diatomaceae, Silico-flagellata and Cilioflagellata. Den Norske Nordhavs Expedition 1876-1878, h. 24, 36 pp., 4 pls.

Chaetoceros willei

Hendy, N. Ingram, 1964
An introductory account of the smaller algae of British coastal waters. V. Bacillariophyceae (Diatoms). Her Majesty's Stationary Office, 317 pp., 45 pls.

Chaetoceros Willei

Jorgensen, E., 1900
Protophyten und Protozoen im Plankton aus der Norwegischen Westkerste. Bergens Mus. Aarb. 1899(6): 95 pp., 5 pls., 83 tables

Chaetoceros Willei

Schodduyn, M., 1926
Observations faites dans la baie d'Ambleteuse (Pas de Calais). Bull. Inst. Ocean., Monaco, No. 482: 64 pp.

Chaetoceros willei

Tratet, Gérard, 1964.
Variations du phytoplancton à Tanger. Trav. Inst. Sci., Cherifien, Rabat, Ser. Botan., (29):204 pp.

Chrysanthemodiscus floriatus

Takano, Hideaki, 1965.
New and rare diatoms from Japanese marine waters. 1. Bull. Tokai Reg. Fish. Res. Lab., No. 42:1-10.

Cistula lorenziana

Hendy, N. Ingram, 1964
An introductory account of the smaller algae of British coastal waters. V. Bacillariophyceae (Diatoms). Her Majesty's Stationary Office, 317 pp., 45 pls.

Cistula lorenziana

Takano, Hideaki, 1965.
New and rare diatoms from Japanese marine waters. 1. Bull. Tokai Reg. Fish. Res. Lab., No. 42:1-10.

Chuniella oceanica comb. nov.

Hendey, N.I., 1937
The plankton diatoms of the southern seas. Discovery Repts. 16:151-364, pls.6-13.

Cladogramma scandica
Cleve-Euler, A., 1951
Die Diatomeen von Schweden und Finnland. Kungl. Svenska Vetenskaps Akad. Handl., Fjärde Ser. 2(1): 161 pp., 6 pls.

Climacodium biconcavum
Bigelow, H. B., 1922
Exploration of the coastal water off the northeastern United States in 1916 by the U.S. Fisheries Schooner Grampus. Bull. M.C.Z. 65 (5):85-188, 53 text figs.

Climacodium biconcavum
Ghazzawi, F.M., 1939
Plankton of the Egyptian waters. A study of the Suez Canal Plankton. (A) Phytoplankton. Preliminary Report 83 pp. Notes and Memoires, Min. Commerce-Industry, Egypt, Hydrobiol. & Fish. 65 figs.

Climacodium biconcavum
Gran, H.H., 1908
Diatomeen. Nordisches Plankton, Botanischer Teil pp. XIX.1-XIX 146; 178 text figs.

Climacodium biconcavum
Hendey, N.I., 1937
The plankton diatoms of the southern seas. Discovery Repts. 16:151-364, pls.6-13.

Climacodium biconcavum
Kokubo, S., 1952
Results of the observations on the plankton and oceanography of Mutsu Bay during 1950, reference being made also to the period 1946-1950. Bull Mar.Biol.Sta., Asamushi 5(1/4): 1-54, 3 tables,(fold-in), 1 fold-in.

Climacodium biconcavum
Marshall, S. M., 1933
The production of microplankton in the Great Barrier Reef Region. Brit. Mus. (N.H.) Great Barrier Reef Exped. 1928-29, Sci. Repts. II(5):111-157, 14 text figs.

Climacodium biconcavum
Schodduyn, M., 1926
Observations faites dans la baie d'Ambleteuse (Pas de Calais). Bull. Inst. Ocean., Monaco, No. 482: 64 pp.

Climacodium Frauenfeldianum
Allen, W.E., and E.E. Cupp, 1935
Plankton diatoms of the Java Sea. Annales du Jardin Botanique de Buitenzorg XLIV (2):101-174, figs.1-127.
(drawings of all species mentioned)

Climacodium frauenfeldianum
Boden, B.P., 1950
Some marine plankton diatoms from the west coast of South Africa. Trans. R.Soc. S. Africa. 32:321-434, 100 text figs.

Climacodium frauenfeldianum
Cupp, Easter E., 1943
Marine plankton diatoms of the west coast of North America. Bull. S.I.O. 5(1):1-238, 5 pls., 168 text figs.

Climacodium frauenfeldianum
Cupp, E.E., 1934
Analysis of marine diatom collections taken from the Canal Zone to California during March, 1933. Trans. Am. Micros. Soc. LIII (1):22-29, 1 map.

Climacodium fraunfeldianum
Gilbert, J.Y., and W.E. Allen, 1943
The phytoplankton of the Gulf of California obtained by the "E.W. Scripps" in 1939 and 1940. J. Mar. Res. V(2):89-110, figs.30-31.

Climacodium Frauenfeldianum
Dangeard, P., 1927
Phytoplankton de la croisière du "Sylvana". Ann. Inst. Ocean., Monaco, n.s., 4(8):286-401, 54 text figs. (Feirer-Juin 1913).

Climacodium Frauenfeldianum
Gran, H.H., 1908
Diatomeen. Nordisches Plankton, Botanischer Teil pp. XIX.1-XIX 146; 178 text figs.

Climacodium Frauenfeldianum
Hendey, N.I., 1937
The plankton diatoms of the southern seas. Discovery Repts. 16:151-364, pls.6-13.

Climacodium frauenfeldianum
Kokubo, S., 1952
Results of the observations on the plankton and oceanography of Mutsu Bay during 1950, reference being made also to the period 1946-1950. Bull Mar.Biol.Sta., Asamushi 5(1/4): 1-54, 3 tables,(fold-in), 1 fold-in.

Climacodium frauenfeldianum
Kokubo, S., and S. Sato, 1947
Plankters in Jū-San Gata. Physiol. and Ecol. (Japan) 1(4):1-16, 3 text figs., tables.

Climacodium frauenfeldianum
Marshall, S. M., 1933
The production of microplankton in the Great Barrier Reef Region. Brit. Mus. (N.H.) Great Barrier Reef Exped. 1928-29, Sci. Repts. II(5):111-157, 14 text figs.

Climacodium frauenfeldium
Sousa e Silva, E., 1949
Diatomaceas e Dinoflagelados de Baía de Cascais. Portugaliae Acta Biol., Volume: Julio Henriques, Ser. B: 300-383, 9 pls, 2 fold-in tables.

Climacodium frauenfeldianum
Sukhanova, I.N., 1964.
The phytoplankton of the northeastern part of the Indian Ocean in the season of the northwest monsoon. Regularity of the distribution of oceanic plankton. (In Russian; English abstract). Trudy Inst. Okeanol., Akad. Nauk, SSSR, 65:24-31.

Climaconeis Lorenziana
Grunow, A., 1877
1. New Diatoms from Honduras. Monthly Micros. Jour., 18:165-186, pls. CXCIII-CXCVI.

Climacosphenia
Desikachary, T.V., 1956(1957).
Electron microscope studies on diatoms. J. R. Microsc. Soc. (3)76(1/2):9-36.

Climacosphenia littoralis n.sp.
Misra, J.N., 1956.
A systematic account of some littoral marine diatoms from the west coast of India. J. Bombay Nat. Hist. Soc., 53(4):537-568.

Climacosphenia moniligera
Cupp, Easter E., 1943
Marine plankton diatoms of the west coast of North America. Bull. S.I.O. 5(1):1-238, 5 pls., 168 text figs.

Climacosphenia moniligera
Eskinazi Enide e Shigekatsv Satô (1963-1964) 1966.
Contribuição ao estudo das diatomaceas da Praia de Piedade.
Trabhs Inst. Oceanogr., Univ. Recife, 5 (5/6): 73-114.

Climacosphenia moniligera
Grunow, A., 1863
Ueber einige neue und ungenügend bekannte Arten und Gattungen von Diatomaceen. Verhandl. d. K.K. Zool. Bot. Gesellsch., Vienna, 13: 137-162, pl.4-5 (Pl. 13-14).

Climacosphenia elongata
Mann, A., 1907
Report on the diatoms of the Albatross voyages in the Pacific Ocean, 1888-1904. Contrib. U. S. Nat. Herb. 10(5):221-419, Pls. XLIV-LIV.

Climacosphenia moniligera
Mann, A., 1907
Report on the diatoms of the Albatross voyages in the Pacific Ocean, 1888-1904. Contrib. U. S. Nat. Herb. 10(5):221-419, Pls. XLIV-LIV.

Climacosphenia moniligera
Rampi, L., 1942
Ricerche sul fitoplancton del Mare Ligure 6. Le diatomee delle acque di Sanremo. Nuovo Giornale Botanico Italiano, N.S., 49:252-258.

Climacosphenia moniligera
Rampi, L., 1940
Diatomee del Mare Adriatico. Nuovo Giornale Botanico Italiano, n.s., 47:559-608.

Climacosphenia moniligera
Takano, Hideaki, 1962
Notes on epiphytic diatoms upon sea-weeds from Japan.
J. Oceanogr. Soc., Japan, 18(1):29-33.

Climacosphenia moniligera
Takano, Hideaki, 1961.
Epiphytic diatoms upon Japanese agar sea-weeds. Bull. Tokai Reg. Fish. Res. Lab., No. 31:269-274.

Climacosphenia moniligera
Zanon, V., 1948
Diatomee marini di Sardegna e Pugillo di Alghe Marine della stressa. Boll. Pesca, Piscitutura e Idrobiologia, Anno 24, ns. 3(2): 202-244, 27 figs. on 1 pl.

Cocconeis sp.
Boden, B. P., 1949.
The diatoms collected by the U.S.S. CACO-PAN in the Antarctic in 1947. J. Mar. Res. 8(1):6-13, 3 textfigs.

Cocconeis sp.
Boden, Brian, 1948
Marine plankton diatoms on operation HIGHJUMP in: Some oceanographic observations on operation HIGHJUMP. By R.S. Dietz. USNEL Rept. No.55, 97 pp., 41 figs. 7 July 1948.

Cocconeis sp.
Braarud, T., 1945
A phytoplankton survey of the polluted waters of inner Oslo Fjord. Hvalrådets Skrifter, No.28, 142 pp., 19 text figs., 17 tables.

Cocconeis sp.
Brunel, J., 1962
Le phytoplancton de la Baie de Chaleurs. Inst. Botan. Univ. Montréal, Contrib. No. 77: 365 pp., 66 pls.

Cocconeis sp.
Brunel, Jules, 1962
Le phytoplancton de la Baie des Chaleurs. Contrib. Ministère de la Chasse et des Pêcheries, Province de Québec, No. 91: 365 pp.

cocconeis sp.
Bunt, John S, 1969.
Observations on photoheterotrophy in a marine diatom.
J. Phycol. 5(1):37-42

Cocconeis

Desikachary, T.V., 1956(1957).
Electron microscope studies on diatoms.
J.R. Microsc. Soc. (3)76(1/2):9-36.

Cocconeis sp.
Kokubo, S., and S. Sato., 1947
Plankters in Ju-San Gata. Physiol. and Ecol. (Japan) 1(4):1-16, 3 text figs., tables.

Cocconeis sp.
Lillick, L.C., 1940
Phytoplankton and planktonic protozoa of the offshore waters of the Gulf of Maine. Pt.II. Qualitative Composition of the Planktonic Flora. Trans. Am. Phil. Soc., n.s., 31(3):193-237, 13 text figs.

Cocconeis Adeliae n.sp.
Manguin, E., 1957.
Premier inventaire des diatomées de la Terre Adélie Antarctique. Espèces nouvelles. Rev. Algologique, n.s., 3(3):111-134.

Cocconeis antiqua
Hendey, N.I., 1937
The plankton diatoms of the southern seas. Discovery Repts. 16:151-364, pls.6-13.

Cocconeis antiqua
Mann, A., 1907
Report on the diatoms of the Albatross voyages in the Pacific Ocean, 1888-1904. Contrib. U. S. Nat. Herb. 10(5):221-419, Pls. XLIV-LIV.

Cocconeis baldjikiana
Mann, A., 1907
Report on the diatoms of the Albatross voyages in the Pacific Ocean, 1888-1904. Contrib. U. S. Nat. Herb. 10(5):221-419, Pls. XLIV-LIV.

Cocconeis binotata
Grunow, A., 1863
Ueber einige neue und ungenügend bekannte Arten und Gattungen von Diatomaceen. Verhandl. d. K.K. Zool. Bot. Gesellsch., Vienna, 13: 137-162, pl.4-5 (Pl. 13-14).

Coscinodiscus biplicatus
Zanon, V., 1948
Diatomee marini di Sardegna e Pugillo di Alghe Marine della stressa. Boll. Pesca, Piscitutura e Idrobiologia, Anno 24, ns. 3(2): 202-244, 27 figs. on 1 pl.

Cocconeis britannica
Hendy, N. Ingram, 1964
An introductory account of the smaller algae of British coastal waters. V. Bacillariophyceae (Diatoms).
Her Majesty's Stationary Office, 317 pp., 45 pls.

Cocconeis britannica
Rampi, L., 1942
Ricerche sul fitoplancton del Mare Ligure 6. Le diatomee delle acque di Sanremo. Nuovo Giornale Botanico Italiano, N.S., 49:252-268.

Cocconeis britannica
Rampi, L., 1940
Diatomee del Mare Adriatico. Nuovo Giornale Botanico Italiano, n.s., 47:559-608.

Cocconeis britannica
Zanon, V., 1948
Diatomee marini di Sardegna e Pugillo di Alghe Marine della stressa. Boll. Pesca, Piscitutura e Idrobiologia, Anno 24, ns. 3(2): 202-244, 27 figs. on 1 pl.

Cocconeis californica
Misra, J.N., 1956.
A systematic account of some littoral marine diatoms from the west coast of India. J. Bombay Nat. Hist. Soc., 53(4):537-568.

Cocconeis ceticola
Hendey, N.I., 1937
The plankton diatoms of the southern seas. Discovery Repts. 16:151-364, pls.6-13.

Cocconeis ceticola
Van der Werff, A., 1950.
On a characteristic diatom from the skin film of whales. Amsterdam Naturhist. 1(3):91-93, 2 figs.

Cocconeis clandestina
Misra, J.N., 1956.
A systematic account of some littoral marine diatoms from the west coast of India. J. Bombay Nat. Hist. Soc., 53(4):537-568.

Cocconeis clandestina
Hendy, N. Ingram, 1964
An introductory account of the smaller algae of British coastal waters. V. Bacillariophyceae (Diatoms).
Her Majesty's Stationary Office, 317 pp., 45 pls.

Cocconeis clandestina n. var. capensis
Cholnoky, B.J., 1963.
Beiträge zur Kenntnis des marinen Litorals von Südafrika.
Botanica Marina, 5(2/3):38-83.

Cocconeis costata
Bigelow, H.B., and M. Leslie, 1930
Reconnaissance of the waters and plankton of Monterey Bay, July 1928. Bull. M.C.Z., 70(5):429-481, 43 text figs.

Cocconeis costata
Boden, B.P., 1950
Some marine plankton diatoms from the west coast of South Africa. Trans. R.Soc. S. Africa. 32:321-434, 100 text figs.

Cocconeis costata
Hendy, N. Ingram, 1964
An introductory account of the smaller algae of British coastal waters. V. Bacillariophyceae (Diatoms).
Her Majesty's Stationary Office, 317 pp., 45 pls.

Cocconeis costata
Jørgensen, E., 1905
B. Protistplankton and the diatoms in bottom samples. Hydrographical and biological investigations in Norwegian fiords. Bergens Mus. Skr. 7: 49-225.

Cocconeis costata
Mann, A., 1907
Report on the diatoms of the Albatross voyages in the Pacific Ocean, 1888-1904. Contrib. U. S. Nat. Herb. 10(5):221-419, Pls. XLIV-LIV.

Cocconeis costata
Misra, J.N., 1956.
A systematic account of some littoral marine diatoms from the west coast of India. J. Bombay Nat. Hist. Soc., 53(4):537-568.

Cocconeis cunoniae n.sp.
Cholnoky, B.J., 1963.
Beiträge zur Kenntnis des marinen Litorals von Südafrika.
Botanica Marina, 5(2/3):38-83.

Cocconeis curvirotunda
Bigelow, H.B., and M. Leslie, 1930
Reconnaissance of the waters and plankton of Monterey Bay, July 1928. Bull. M.C.Z., 70(5):429-481, 43 text figs.

Cocconeis De Benedettii, n.sp.
Frenguelli, Joaquin, y Hector A. Orlando, 1958.
Diatomeas y silicoflagelados del sector Antartico Sudamericano.
Inst. Antartico Argentino, Publ., No. 5:191 pp.

Cocconeis Debesi
Rampi, L., 1940
Diatomee del Mare Adriatico. Nuovo Giornale Botanico Italiano, n.s., 47:559-608.

Cocconeis Bebesi
Zanon, V., 1948
Diatomee marini di Sardegna e Pugillo di Alghe Marine della stressa. Boll. Pesca, Piscitutura e Idrobiologia, Anno 24, ns. 3(2): 202-244, 27 figs. on 1 pl.

Cocconeis decipiens
Mann, A., 1907
Report on the diatoms of the Albatross voyages in the Pacific Ocean, 1888-1904. Contrib. U. S. Nat. Herb. 10(5):221-419, Pls. XLIV-LIV.

Cocconeis diminuta
Rumkówna, A., 1948
[List of the phytoplankton species occurring in the superficial water layers in the Gulf of Gdańsk] Bull. Lab. mar., Gdynia, No. 4: 139-141 with tables in back.

Cocconeis disrupta
Bigelow, H.B., and M. Leslie, 1930
Reconnaissance of the waters and plankton of Monterey Bay, July 1928. Bull. M.C.Z., 70(5):429-481, 43 text figs.

Cocconeis dirupta
Hendy, N. Ingram, 1964
An introductory account of the smaller algae of British coastal waters. V. Bacillariophyceae (Diatoms).
Her Majesty's Stationary Office, 317 pp., 45 pls.

Cocconeis dirupta
Hendey, N.I., 1951
Littoral diatoms of Chicester Harbour with special reference to fouling. J.Roy. Microsc. Soc. 71(1): 1-86, 18 pls.

Cocconeis dirupta
Hustedt, F. and A.A. Aleem, 1951
Littoral diatoms from the Salstone near Plymouth. JMBA 30(1): 177.196.

Cocconeis dirupta
Mann, A., 1907
Report on the diatoms of the Albatross voyages in the Pacific Ocean, 1888-1904. Contrib. U. S. Nat. Herb. 10(5):221-419, Pls. XLIV-LIV.

Cocconeis dirupta
Zanon, V., 1948
Diatomee marini di Sardegna e Pugillo di Alghe Marine della stressa. Boll. Pesca, Piscitutura e Idrobiologia, Anno 24, ns. 3(2): 202-244, 27 figs. on 1 pl.

Cocconeis dirupta
Politis, J., 1949
Diatomees marines de Bosphores et des ibes de la mer de Marmara. II Practica tou Hellenikou Hidrobiologikou Institutoutou 1929, Etoz 1929, 3(1):11-31.

Cocconeis dirupta var. flexella
Rampi, L., 1940
Diatomee del Mare Adriatico. Nuovo Giornale Botanico Italiano, n.s., 47:559-608.

Oceanographic Index: Marine Organisms Cumulation, 1946-1973

Cocconeis disculoides
Hendy, N. Ingram, 1964
An introductory account of the smaller algae of British coastal waters. V. Bacillariophyceae (Diatoms).
Her Majesty's Stationary Office, 317 pp., 45 pls.

Cocconeis disculus
Hendy, N. Ingram, 1964
An introductory account of the smaller algae of British coastal waters. V. Bacillariophyceae (Diatoms).
Her Majesty's Stationary Office, 317 pp., 45 pls.

Cocconeis disculus
Rumkówna, A., 1948
[List of the phytoplankton species occurring in the superficial water layers in the Gulf of Gdańsk] Bull. Lab. mar., Gdynia, No. 4: 139-141 with tables in back.

Cocconeis disculus
Zanon, D. V., 1949
Diatomee di Buenos Aires (Argentina) Atti Accad. Naz. Lincei, Memorie, Cl. Sci. fis., mat. e. nat., ser. 7, 11(3):59-151, 2 pls.

Cocconeis dirupta flexella
Misra, J.N., 1956.
A systematic account of some littoral marine diatoms from the west coast of India.
J. Bombay Nat. Hist. Soc., 53(4):537-568.

Cocconeis distans
Hendy, N. Ingram, 1964
An introductory account of the smaller algae of British coastal waters. V. Bacillariophyceae (Diatoms).
Her Majesty's Stationary Office, 317 pp., 45 pls.

Cocconeis distans
Jorgensen, E., 1900
Protophyten und Protozoen im Plankton aus der Norwegischen Westkerste. Bergens Mus. Aarb. 1899(6): 95 pp., 5 pls., 83 tables.

Cocconeis distans
Mann, A., 1907
Report on the diatoms of the Albatross voyages in the Pacific Ocean, 1888-1904.
Contrib. U. S. Nat. Herb. 10(5):221-419, Pls. XLIV-LIV.

Cocconeis distans
Mann, A., 1893
List of Diatomaceae from a deep-sea dredging in the Atlantic Ocean off Delaware Bay by the U. S. Fish Commission Steamer Albatross.
Proc. U. S. Nat. Mus. 16:303-312.

Cocconeis distans
Politis, J., 1949
Diatomees marines de Bosphores et des ibes de la mer de Marmara. II Practica tou Hellenikou Hidrobiologikou Institutoutou 1929, Etoz 1929, 3(1):11-31.

Cocconeis distans
Zanon, V., 1948
Diatomee marini di Sardegna e Pugillo di Alghe Marine della stressa. Boll. Pesca, Piscitutura e Idrobiologia, Anno 24, ns. 3(2): 202-244, 27 figs. on 1 pl.

Cocconeis divisa
Takano, Hideaki, 1963.
Notes on marine littoral diatoms of Japan. Bull. Tokai Reg. Fish. Res. Lab., No. 36:1-8.

Cocconeis fimbriata
Forti, A., 1922
Ricerche sulla flora pelagica (fitoplancton) di Quarto dei Mille. Mem. R. Com. Talass. Ital. 97:248 pp., 13 pls.

Cocconeis fluminensis
Rampi, L., 1940
Diatomee del Mare Adriatico. Nuovo Giornale Botanico Italiano, n.s., 47:559-608.

Cocconeis granulifera
Politis, J., 1949
Diatomees marines de Bosphores et des ibes de la mer de Marmara. II Practica tou Hellenikou Hidrobiologikou Institutoutou 1929, Etoz 1929, 3(1):11-31.

Cocconeis grevillei
Mann, A., 1907
Report on the diatoms of the Albatross voyages in the Pacific Ocean, 1888-1904.
Contrib. U. S. Nat. Herb. 10(5):221-419, Pls. XLIV-LIV.

Cocconeis Grevillei
Politis, J., 1949
Diatomees marines de Bosphores et des ibes de la mer de Marmara. II Practica tou Hellenikou Hidrobiologikou Institutoutou 1929, Etoz 1929, 3(1):11-31.

Cocconeis guttata
Hendy, N. Ingram, 1964
An introductory account of the smaller algae of British coastal waters. V. Bacillariophyceae (Diatoms).
Her Majesty's Stationary Office, 317 pp., 45 pls.

Cocconeis guttata n.sp.
Hustedt, F. and A.A. Aleem, 1951
Littoral diatoms from the Salstone near Plymouth. JMBA 30(1): 177,196.

Cocconeis heteroides
Misra, J.N., 1956.
A systematic account of some littoral marine diatoms from the west coast of India.
J. Bombay Nat. Hist. Soc., 53(4):537-568.

Cocconeis hirsuta n.sp.
Cholnoky, B.J., 1963.
Beiträge zur Kenntnis des marinen Litorals von Südafrika.
Botanica Marina, 5(2/3):38-83.

Cocconeis Hustedtii
Tsumura, K., 1956.
Diatomoj el la cirkaufoso de la restajo de la kastele de Odawara. J. Yokohama Municipal Univ., (C-14) No. 47:23 pp.

Cocconeis imperatrix
Balech, E., 1947
Contribution al conocimiento del plancton antartico. Plankton del Mar de Bellinghausen. Physis 20:75-91, 76 figs. on 8 pls.

Cocconeis imperatrix
Hendey, N.I., 1937
The plankton diatoms of the southern seas. Discovery Repts. 16:151-364, pls.6-13.

Cocconeis infirmata n.sp.
Manguin, E., 1957.
Premier inventaire des diatomées de la Terre Adélie Antarctique. Espèces nouvelles. Rev. Algologiques, n.s., 3(3):111-134.

Cocconeis interrupta
Grunow, A., 1863
Ueber einige neue und ungenügend bekannte Arten und Gattungen von Diatomaceen. Verhandl. d. K.K. Zool. Bot. Gesellsch., Vienna, 13: 137-162, pl.4-5 (Pl. 13-14).

Cocconeis lauriensis, n.sp.
Frenguelli, Joaquin, y Hector A. Orlando, 1958.
Diatomeas y silicoflagelados del sector Antartico Sudamericano.
Inst. Antartico Argentino, Publ., No. 5:191 pp.

Cocconeis lineatus
Rampi, L., 1940
Diatomee del Mare Adriatico. Nuovo Giornale Botanico Italiano, n.s., 47:559-608.

Cocconeis littoralis n.sp.
Subrahmanyan, R., 1946.
A systematic account of the marine plankton diatoms of the Madras coast. Proc. Indian Acad. Sci 24(4):85-197, Pl. 2, 440 textfigs.

Cocconeis maxima
Politis, J., 1949
Diatomees marines de Bosphores et des ibes de la mer de Marmara. II Practica tou Hellenikou Hidrobiologikou Institutoutou 1929, Etoz 1929, 3(1):11-31.

Cocconeis maxima
Rampi, L., 1940
Diatomee del Mare Adriatico. Nuovo Giornale Botanico Italiano, n.s., 47:559-608.

Cocconeis maxima
Zanon, V., 1948
Diatomee marini di Sardegna e Pugillo di Alghe Marine della stressa. Boll. Pesca, Piscitutura e Idrobiologia, Anno 24, ns. 3(2): 202-244, 27 figs. on 1 pl.

Cocconeis melchiori, n.sp.
Frenguelli, Joaquin, y Hector A. Orlando, 1958.
Diatomeas y silicoflagelados del sector Antartico Sudamericano.
Inst. Antartico Argentino, Publ., No. 5:191 pp.

Cocconeis molesta
Politis, J., 1949
Diatomees marines de Bosphores et des ibes de la mer de Marmara. II Practica tou Hellenikou Hidrobiologikou Institutoutou 1929, Etoz 1929, 3(1):11-31.

Cocconeis molesta
Rampi, L., 1940
Diatomee del Mare Adriatico. Nuovo Giornale Botanico Italiano, n.s., 47:559-608.

Cocconeis molesta (figs.)
Sousa e Silva, E., 1949
Diatomaceas e Dinoflagelados de Baia de Cascais. Portugaliae Acta Biol., Volume: Julio Henriques, Ser. B: 300-383, 9 pls, 2 fold-in tables.

Cocconeis molesta
Zanon, V., 1948
Diatomee marini di Sardegna e Pugillo di Alghe Marine della stressa. Boll. Pesca, Piscitutura e Idrobiologia, Anno 24, ns. 3(2): 202-244, 27 figs. on 1 pl.

Cocconeis monodii n.sp.
Amossé, A. 1970.
Diatomées marines et saumâtres du Sénégal et de la Côte d'Ivoire.
Bull. Inst. fond. Afr. Noire (A) 32(2):289-311.

Cocconeis nitidus
Rampi, L., 1940
Diatomee del Mare Adriatico. Nuovo Giornale Botanico Italiano, n.s., 47:559-608.

Cocconeis orbicularis, n.sp.
Frenguelli, Joaquin, y Hector A. Orlando, 1958.
Diatomeas y silicoflagelados del sector
Antartico Sudamericano.
Inst. Antartico Argentino, Publ., No. 5:191 pp.

Cocconeis ornata
Politis, J., 1949
Diatomees marines de Bosphores et des
ibes de la mer de Marmara. II Practica tou
Hellenikou Hidrobiologikou Institutoutou
1929, Etoz 1929, 3(1):11-31.

Cocconeis panniformis
Bigelow, H.B., and M. Leslie, 1930
Reconnaissance of the waters and
plankton of Monterey Bay, July 1928.
Bull. M.C.Z., 70(5):429-481, 43 text
figs.

Cocconeis pediculus
Rumkówna, A., 1948
[List of the phytoplankton species occurring in the superficial water layers in the Gulf of Gdańsk] Bull. Lab. mar., Gdynia, No. 4: 139-141 with tables in back.

Cocconeis pediculus
Zanon, V., 1948
Diatomee marini di Sardegna e Pugillo
di Alghe Marine della stressa. Boll. Pesca,
Piscitutura e Idrobiologia, Anno 24, ns. 3(2):
202-244, 27 figs. on 1 pl.

Cocconeis pellucida
de Sousa e Silva, E., 1956.
Contribution à l'étude du microplancton de Dakar
et des regions maritimes voisines.
Bull. I.F.A.N., 8(2):335-371, 7 pls.

Cocconeis pellucida
Grunow, A., 1863
Ueber einige neue und ungenügend bekannte
Arten und Gattungen von Diatomaceen. Verhandl.
d. K.K. Zool. Bot. Gesellsch., Vienna, 13:
137-162, pl.4-5 (Pl. 13-14).

Cocconeis pellucida
Mann, A., 1907
Report on the diatoms of the Albatross
voyages in the Pacific Ocean, 1888-1904.
Contrib. U. S. Nat. Herb. 10(5):221-419, Pls.
XLIV-LIV.

Cocconeis pellucida
Rampi, L., 1940
Diatomee del Mare Adriatico. Nuovo
Giornale Botanico Italiano, n.s., 47:559-608.

Cocconeis pellucida
Takano, Hideaki, 1962
Notes on epiphytic diatoms upon sea-weeds
from Japan.
J. Oceanogr. Soc., Japan, 18(1):29-33.

Cocconeis pellucida
Zanon, V., 1948
Diatomee marini di Sardegna e Pugillo
di Alghe Marine della stressa. Boll. Pesca,
Piscitutura e Idrobiologia, Anno 24, ns. 3(2):
202-244, 27 figs. on 1 pl.

Cocconeis pelta
Hustedt, F. and A.A. Aleem,1951
Littoral diatoms from the Salstone
near Plymouth. JMBA 30(1): 177.196.

Cocconeis peltoides
Hendy, N. Ingram, 1964
An introductory account of the smaller algae
of British coastal waters. V. Bacillariophyceae (Diatoms).
Her Majesty's Stationary Office, 317 pp.,
45 pls.

Cocconeis pinnata
Hendey, N.I. 1937
The plankton diatoms of the southern seas.
Discovery Repts. 16:151-364, pls.6-13.

Cocconeis pinnata
Jørgensen, E., 1905
B.Protistplankton and the diatoms
in bottom samples. Hydrographical and
biological investigations in Norwegian
fjords. Bergens Mus. Skr. 7: 49-225.

Cocconeis pinnata
Jorgensen, E., 1900
Protophyten und Protozoen im Plankton aus der Norwegischen Westkerste. Bergens Mus. Aarb. 1899(6): 95 pp., 5 pls., 83 tables

Cocconeis pinnata
Manguin, E., 1954
Diatomees marines provenant de l'ile
Heard (Australian National Antarctic
Expedition). Rev. Algol., n.s., 1:
14-24.

Cocconeis pinnata
Politis, J., 1949
Diatomees marines de Bosphores et des
ibes de la mer de Marmara. II Practica tou
Hellenikou Hidrobiologikou Institutoutou
1929, Etoz 1929, 3(1):11-31.

Cocconeis pinnata
Rampi, L., 1942
Ricerche sul fitoplancton del Mare Ligure
6. Le diatomee delle acque di Sanremo. Nuovo
Giornale Botanico Italiano, N.S., 49:252-268.

Cocconeis placentula
de Sousa e Silva, E., 1956.
Contribution à l'étude du microplancton de Dakar
et des regions maritimes voisines.
Bull. I.F.A.N., 8(2):335-371, 7 pls.

Cocconeis placentula
Lillick, L.C., 1940
Phytoplankton and planktonic
protozoa of the offshore waters
of the Gulf of Maine. Pt.II.
Qualitative Composition of the
Planktonic Flora. Trans. Am.
Phil. Soc., n.s., 31(3):193-237,
13 text figs.

Cocconeis placentula
Lillick, L.C., 1937
Seasonal studies of the phytoplankton
off Woods Hole, Massachusetts. Biol. Bull.
LXXIII (3):488-503, 3 text figs.

Cocconeis placentula
Mann, A., 1893
List of Diatomaceae from a deep-sea dredging in the Atlantic Ocean off Delaware Bay by the U. S. Fish Commission Steamer Albatross.
Proc. U. S. Nat. Mus. 16:303-312.

Cocconeis placentula
Takano, Hideaki, 1962
Notes on epiphytic diatoms upon sea-weeds
from Japan.
J. Oceanogr. Soc., Japan, 18(1):29-33.

Cocconeis placentula
Zanon, D. V., 1949
Diatomee di Buenos Aires (Argentina)
Atti Accad. Naz. Lincei, Memorie, Cl. Sci.
fis., mat. e. nat., ser. 7, 11(3):59-151,
2 pls.

Cocconeis placentula
Zanon, V., 1948
Diatomee marini di Sardegna e Pugillo
di Alghe Marine della stressa. Boll. Pesca,
Piscitutura e Idrobiologia, Anno 24, ns. 3(2):
202-244, 27 figs. on 1 pl.

Cocconeis placentula euglypta
Misra, J.N., 1956.
A systematic account of some littoral marine
diatoms from the west coast of India.
J. Bombay Nat. Hist. Soc., 53(4):537-568.

Cocconeis placentula euglypta
Rumkówna, A., 1948
[List of the phytoplankton species occurring in the superficial water layers in the Gulf of Gdańsk] Bull. Lab. mar., Gdynia, No. 4: 139-141 with tables in back.

Cocconeis pseudograta
Rampi, L., 1940
Diatomee del Mare Adriatico. Nuovo
Giornale Botanico Italiano, n.s., 47:559-608.

Cocconeis pseudomarginata
Hendy, N. Ingram, 1964
An introductory account of the smaller algae
of British coastal waters. V. Bacillariophyceae (Diatoms).
Her Majesty's Stationary Office, 317 pp.,
45 pls.

Cocconeis pseudo-marginata
Politis, J., 1949
Diatomees marines de Bosphores et des
ibes de la mer de Marmara. II Practica tou
Hellenikou Hidrobiologikou Institutoutou
1929, Etoz 1929, 3(1):11-31.

Cocconeis presudomarginata
Rampi, L., 1940
Diatomee del Mare Adriatico. Nuovo
Giornale Botanico Italiano, n.s., 47:559-608.

Cocconeis pseudomarginata
Takano, Hideaki, 1962
Notes on epiphytic diatoms upon sea-weeds
from Japan.
J. Oceanogr. Soc., Japan, 18(1):29-33.

Cocconeis pseudomarginata
Zanon, V., 1948
Diatomee marini di Sardegna e Pugillo
di Alghe Marine della stressa. Boll. Pesca,
Piscitutura e Idrobiologia, Anno 24, ns. 3(2):
202-244, 27 figs. on 1 pl.

Cocconeis quarnerensis
Hendy, N. Ingram, 1964
An introductory account of the smaller algae
of British coastal waters. V. Bacillariophyceae (Diatoms).
Her Majesty's Stationary Office, 317 pp.,
45 pls.

Cocconeis quarnerensis
Politis, J., 1949
Diatomees marines de Bosphores et des
ibes de la mer de Marmara. II Practica tou
Hellenikou Hidrobiologikou Institutoutou
1929, Etoz 1929, 3(1):11-31.

Cocconeis quarnerensis
Rampi, L., 1940
Diatomee del Mare Adriatico. Nuovo
Giornale Botanico Italiano, n.s., 47:559-608.

Cocconeis Quarnerensis
Zanon, V., 1948
Diatomee marini di Sardegna e Pugillo
di Alghe Marine della stressa. Boll. Pesca,
Piscitutura e Idrobiologia, Anno 24, ns. 3(2):
202-244, 27 figs. on 1 pl.

Cocconeis scutellum
Bigelow, H.B., and M. Leslie, 1930
Reconnaissance of the waters and
plankton of Monterey Bay, July 1928.
Bull. M.C.Z., 70(5):429-481, 43 text
figs.

Cocconeis scutellum
Ercegovic, A., 1936
Etudes qualitative et quantitatives du phytoplancton dans les eaux cotières de l'Adriatique oriental moyen au cours de l'année 1934. Acta Adriatica 1(9):1-126

Cocconeis scutellum
Eskinazi Enide e Shigekatsv Satô (1963-1964) 1966.
Contribuição ao estudo das diatomaceas da Praia de Piedade.
Trabhs Inst. Oceanogr., Univ. Recife, 5 (5/6): 73-114.

Cocconeis scutellum
Forti, A., 1922
Ricerche sulla flora pelagica (fitoplancton) di Quarto dei Mille. Mem. R. Com. Talass. Ital. 97:248 pp., 13 pls.

Cocconeis scutellum & varieties
Hendy, N. Ingram, 1964
An introductory account of the smaller algae of British coastal waters. V. Bacillariophyceae (Diatoms).
Her Majesty's Stationary Office, 317 pp., 45 pls.

Cocconeis scutellum
Hendey, N.I., 1951
Littoral diatoms of Chicester Harbour with special reference to fouling. J.Roy. Microscop. Soc. 71(1): 1-86, 18 pls.

Cocconeis scutellum
Hendey, N.I., 1937
The plankton diatoms of the southern seas. Discovery Repts. 16:151-364, pls.6-13.

Cocconeis scutellum
Hustedt, F. and A.A. Aleem, 1951
Littoral diatoms from the Salstone near Plymouth. JMBA 30(1): 177.196.

Cocconeis scutellum
Jørgensen, E., 1905
B. Protistplankton and the diatoms in bottom samples. Hydrographical and biological investigations in Norwegian fiords. Bergens Mus. Skr. 7: 49-225.

Cocconeis scutellum
Jorgensen, E., 1900
Protophyten und Protozoën im Plankton aus der Norwegischen Westkerste. Bergens Mus. Aarb. 1899(6): 95 pp., 5 pls., 83 tables.

Cocconeis Scutellum
Lillick, L.C., 1940
Phytoplankton and planktonic protozoa of the offshore waters of the Gulf of Maine. Pt.II. Qualitative Composition of the Planktonic Flora. Trans. Am. Phil. Soc., n.s., 31(3):193-237, 13 text figs.

Cocconeis Scutellum
Lillick, L.C., 1938
Preliminary report of the phytoplankton of the Gulf of Maine. Am. Mid. Nat. 20(3):624-640, 1 text fig, 37 tables.

Cocconeis scutellum
Mann, A., 1893
List of Diatomaceae from a deep-sea dredging in the Atlantic Ocean off Delaware Bay by the U. S. Fish Commission Steamer Albatross. Proc. U. S. Nat. Mus. 16:303-312.

Cocconeis scutellum
Morse, D.C., 1947
Some observations on seasonal variations in plankton population Patuxent River, Maryland 1943-1945. Bd. Nat. Res., Publ. No.65, Chesapeake Biol. Lab., 31, 3 figs.

Cocconeis scutellum
Politis, J., 1949
Diatomees marines de Bosphores et des ibes de la mer de Marmara. II Practica tou Hellenikou Hidrobiologikou Institutoutou 1929, Etoz 1929, 3(1):11-31.

Cocconeis scutellum
Rampi, L., 1942
Ricerche sul fitoplancton del Mare Ligure 6. Le diatomee delle acque di Sanremo. Nuovo Giornale Botanico Italiano, N.S., 49:252-268.

Cocconeis scutellum
Rampi, L., 1940
Diatomee del Mare Adriatico. Nuovo Giornale Botanico Italiano, n.s., 47:559-608.

Cocconeis scutellum
Sieburth, John McN. and Cynthia D. Thomas 1973.
Fouling on eelgrass (Zostera marina L.). J. Phycol. 9(1): 46-50.

Cocconeis scutellum
Takano, Hideaki, 1963.
Notes on marine littoral diatoms of Japan. Bull. Tokai Reg. Fish. Res. Lab., No. 36:1-8.

Cocconeis scutellum
Takano, Hideaki, 1962
Notes on epiphytic diatoms upon sea-weeds from Japan.
J. Oceanogr. Soc., Japan, 18(1):29-33.

Cocconeis scutellatum
Takano, Hideaki, 1961.
Epiphytic diatoms upon Japanese agar sea-weeds.
Bull. Tokai Reg. Fish. Res. Lab., No. 31:269-274.

Cocconeis scutellum
Zanon, D. V., 1949
Diatomee di Buenos Aires (Argentina) Atti Accad. Naz. Lincei, Memorie, Cl. Sci. fis., mat. e. nat., ser. 7, 11(3):59-151, 2 pls.

Cocconeis scutellum
Zanon, V., 1948
Diatomee marini di Sardegna e Pugillo di Alghe Marine della stressa. Boll. Pesca, Piscitutura e Idrobiologia, Anno 24, ns. 3(2): 202-244, 27 figs. on 1 pl.

Cocconeis senegalensis
Mann, A., 1907
Report on the diatoms of the Albatross voyages in the Pacific Ocean, 1888-1904. Contrib. U. S. Nat. Herb. 10(5):221-419, Pls. XLIV-LIV.

Cocconeis sigmoides n.sp.
Subrahmanyan, R., 1946.
A systematic account of the marine plankton diatoms of the Madras coast. Proc. Indian Acad. Sci. 24(4):85-197, Pl. 2, 440 textfigs.

Cocconeis speciosa
Hendy, N. Ingram, 1964
An introductory account of the smaller algae of British coastal waters. V. Bacillariophyceae (Diatoms).
Her Majesty's Stationary Office, 317 pp., 45 pls.

Cocconeis splendida
Mann, A., 1907
Report on the diatoms of the Albatross voyages in the Pacific Ocean, 1888-1904. Contrib. U. S. Nat. Herb. 10(5):221-419, Pls. XLIV-LIV.

Cocconeis stauriformis
Hendy, N. Ingram, 1964
An introductory account of the smaller algae of British coastal waters. V. Bacillariophyceae (Diatoms).
Her Majesty's Stationary Office, 317 pp., 45 pls.

Cocconeis stauroneiformis
Takano, Hideaki, 1963.
Notes on marine littoral diatoms of Japan. Bull. Tokai Reg. Fish. Res. Lab., No. 36:1-8.

Cocconeis Stompsii n.sp.
Cholnoky, B.J., 1963.
Beiträge zur Kenntnis des marinen Litorals von Südafrika.
Botanica Marina, 5(2/3):38-83.

Cocconeis sublittoralis
Hendy, N. Ingram, 1964
An introductory account of the smaller algae of British coastal waters. V. Bacillariophyceae (Diatoms).
Her Majesty's Stationary Office, 317 pp., 45 pls.

Cocconeis sublittoralis n.sp.
Hendey, N.I., 1951
Littoral diatoms of Chicester Harbour with special reference to fouling. J.Roy. Microscop. Soc. 71(1): 1-86, 18 pls.

Cocconema inaequale
Mann, A., 1907
Report on the diatoms of the Albatross voyages in the Pacific Ocean, 1888-1904. Contrib. U. S. Nat. Herb. 10(5):221-419, Pls. XLIV-LIV.

Cocconema kamtochatica
Mann, A., 1907
Report on the diatoms of the Albatross voyages in the Pacific Ocean, 1888-1904. Contrib. U. S. Nat. Herb. 10(5):221-419, Pls. XLIV-LIV.

Cocconema lanceolatum
Mann, A., 1907
Report on the diatoms of the Albatross voyages in the Pacific Ocean, 1888-1904. Contrib. U. S. Nat. Herb. 10(5):221-419, Pls. XLIV-LIV.

Coenobiodiscus muriformis n.gen., n.sp.
Loeblich, Alfred R. III, William W. Wright and W. Marshall Darley 1968.
A unique colonial marine centric diatom Coenobiodiscus muriformis gen. et sp. nov. J. Phycol. 4(1):23-29.

Coenobiodiscus muriformis
Round, F.E. 1972.
Some observations on colonies and ultrastructure of the frustule of Coenobiodiscus muriformis and its transfer to Planktoniella.
J. Phycol. 8(3):222-231

Corethron
Brunel, J., 1962
Le phytoplancton de la Baie de Chaleurs. Inst. Botan., Univ. Montréal, Contrib. No. 77: 365 pp., 66 pls.

Corethron
Desikachary, T.V., 1956(1957).
Electron microscope studies on diatoms. J.R. Microsc. Soc. (3)76(1/2):9-30.

Corethron

Neaverson, E., 1934
The sea-floor deposits. 1. General characteristics and distribution. Discovery Repts. 9: 297-349, Plates 17-22.

Corethron atlanticus

Jørgensen, E., 1905
B. Protistplankton and the diatoms in bottom samples. Hydrographical and biological investigations in Norwegian fjords. Bergens Mus. Skr. 7: 49-225.

Corethron borealis

Jørgensen, E., 1905
B. Protistplankton and the diatoms in bottom samples. Hydrographical and biological investigations in Norwegian fjords. Bergens Mus. Skr. 7: 49-225.

Corethron brevis

Jørgensen, E., 1905
B. Protistplankton and the diatoms in bottom samples. Hydrographical and biological investigations in Norwegian fjords. Bergens Mus. Skr. 7: 49-225.

Corethron constrictus

Jørgensen, E., 1905
B. Protistplankton and the diatoms in bottom samples. Hydrographical and biological investigations in Norwegian fjords. Bergens Mus. Skr. 7: 49-225.

Corethron contortus

Jørgensen, E., 1905
B. Protistplankton and the diatoms in bottom samples. Hydrographical and biological investigations in Norwegian fjords. Bergens Mus. Skr. 7: 49-225.

Corethron convolutus

Jørgensen, E., 1905
B. Protistplankton and the diatoms in bottom samples. Hydrographical and biological investigations in Norwegian fjords. Bergens Mus. Skr. 7: 49-225.

Corethron criophilum

Allen, W. E., 1938
The Templeton Crocker Expedition to the Gulf of California in 1935 - The Phytoplankton. Amer. Microsc. Soc., Trans. 57:328-335.

Corethron criophilum

Allen, W.E., 1937
Plankton diatoms of the Gulf of California obtained by the G. Allan Hancock Expedition of 1936. The Hancock Pacific Expeditions, Univ. So. Calif. Publ. 3:47-59, 1 fig.

Corethron criophilum

Allen, W.E., and E.E. Cupp, 1935
Plankton diatoms of the Java Sea. Annales du Jardin Botanique de Buitenzorg XLIV (2):101-174, figs.1-127.
(drawings of all species mentioned)

Corethron criophilus

Balech, E., 1947
Contribution al conocimiento del plancton antartico. Plankton del Mar de Bellinghausen. Physis 20:75-91, 76 figs. on 8 pls.

Corethron criophilum

Boden, B.P., 1950
Some marine plankton diatoms from the west coast of South Africa. Trans. R.Soc. S. Africa. 32:321-434, 100 text figs.

Corethron criophilum

Boden, B. P., 1949.
The diatoms collected by the U.S.S. CACOPAN in the Antarctic in 1947. J. Mar. Res. 8(1):6-13, 3 textfigs.

Corethron criophilum

Boden, Brian, 1948
Marine plankton diatoms on operation HIGHJUMP in: Some oceanographic observations on operation HIGHJUMP. By R.S. Dietz. USNEL Rept. No.55, 97 pp., 41 figs. 7 July 1948.

Corethron criophilum n.sp.

Castracane degli Antelminelli, F., 1886
1. Report on the Diatomaceae collected by H.M.S. Challenger during the years 1873-1876. Rept. Sci. Results, H.M.S. Challenger, Botany Vol. II, 178 pp., 30 pls.

Corethron criophilum

Cupp, E.E., 1934
Analysis of marine diatom collections taken from the Canal Zone to California during March, 1933. Trans. Am. Micros. Soc. LIII (1):22-29, 1 map.

Corethron criophilum

Cupp, E.E. and Allen, W.E., 1938
Plankton diatoms of the Gulf of California obtained by Allan Hancock Pacific Expedition of 1937. The Hancock Pacific Expeditions, The Univ. So. Calif. Publ. 3: 61-74, 1 map, pls.4-15.

Corethron criophilum

Gilbert, J.Y., and W.E. Allen, 1943
The phytoplankton of the Gulf of California obtained by the "E.W. Scripps" in 1939 and 1940. J. Mar. Res. V(2):89-110, figs.30-31.

Corethron criophilum

Dangeard, P., 1927
Phytoplankton de la croisière du "Sylvana". Ann. Inst. Ocean., Monaco, n.s., 4(8):286-401, 54 text figs. (Feirer-Juin 1913).

Corethron criophilum

Frenguelli, Joaquin, and Hector Antonio Orlando, 1959.
Operacion MERLUZA. Diatomeas y silicoflagelados del plancton del "VI Crucero". Servicio Hidrogr. Naval, Argentina, Publ. No. H. 619; 5-62.

Corethron criophilum

Gran, H.H., 1908
Diatomeen. Nordisches Plankton, Botanischer Teil pp. XIX 1-XIX 146; 178 text figs.

Corethron criophilum

Hendey, N.I., 1937
The plankton diatoms of the southern seas. Discovery Repts. 16:151-364, pls.6-13.

Corethron criophilus

Jørgensen, E., 1905
B. Protistplankton and the diatoms in bottom samples. Hydrographical and biological investigations in Norwegian fjords. Bergens Mus. Skr. 7: 49-225.

Corethron criophilum

Mangin, M. L., 1912
Phytoplancton de la croisière du "René" dans l'Atlantique (Septembre 1908). Ann. Inst. Ocean., n.s., 4(1):1-66, 2 pls., 41 text figs., 2 tables.

Corethron criophilum

Manguin, E., 1954
Diatomees marines provenant de l'ile Heard (Australian National Antarctic Expedition). Rev. Algol., n.s., 1: 14-24.

Corethron criophilum

Marukawa, H., 1921
Plankton lists and some new species of copepods, from the northern waters of Japan. Bull. Inst. Ocean., No.384, 15 pp., 3 pls., 1 chart. Monaco

Corethron criophilum

Pavillard, J., 1925
Bacillariales. Rept. on the Danish Oceangr. Exped., 1908-10 to the Mediterranean and adj. seas. Vol.II., Biol. J4:72 pp., 116 text figs.

Corethron criophilum

Sousa e Silva, E., 1949
Diatomaceas e Dinoflagelados de Baía de Cascais. Portugaliae Acta Biol. Volume: Julio Henriques, Ser. B: 300-383, 9 pls, 2 fold-in tables.

Corethron curvisetus

Jørgensen, E., 1905
B. Protistplankton and the diatoms in bottom samples. Hydrographical and biological investigations in Norwegian fjords. Bergens Mus. Skr. 7: 49-225.

Corethron danicus

Jørgensen, E., 1905
B. Protistplankton and the diatoms in bottom samples. Hydrographical and biological investigations in Norwegian fjords. Bergens Mus. Skr. 7: 49-225.

Corethron debilis

Jørgensen, E., 1905
B. Protistplankton and the diatoms in bottom samples. Hydrographical and biological investigations in Norwegian fjords. Bergens Mus. Skr. 7: 49-225.

Corethron decipiens

Jørgensen, E., 1905
B. Protistplankton and the diatoms in bottom samples. Hydrographical and biological investigations in Norwegian fjords. Bergens Mus. Skr. 7: 49-225.

Corethron densus

Jørgensen, E., 1905
B. Protistplankton and the diatoms in bottom samples. Hydrographical and biological investigations in Norwegian fjords. Bergens Mus. Skr. 7: 49-225.

Corethron diadema

Jørgensen, E., 1905
B. Protistplankton and the diatoms in bottom samples. Hydrographical and biological investigations in Norwegian fjords. Bergens Mus. Skr. 7: 49-225.

Corethron furcellatus

Jørgensen, E., 1905
B. Protistplankton and the diatoms in bottom samples. Hydrographical and biological investigations in Norwegian fjords. Bergens Mus. Skr. 7: 49-225.

Corethron hispidum n.sp.

Castracane degli Antelminelli, F., 1886
1. Report on the Diatomaceae collected by H.M.S. Challenger during the years 1873-1876. Rept. Sci. Results, H.M.S. Challenger, Botany Vol. II, 178 pp., 30 pls.

Corethron hystrix

Braarud, T., 1934
A note on the phytoplankton of the Gulf of Maine in the summer of 1933. Biol. Bull. 67(1):76-82. (Contribution No.46 of the Woods Hole Oceanographic Institution)

Corethron hystrix

Clarke, G.L., 1937.
On securing large quantities of diatoms from the sea for chemical analysis. Science 86(2243):593-594.

Corethron hystrix

Cleve-Euler, A., 1951.
Die Diatomeen von Schweden und Finnland. Kungl. Svenska Vetenskaps Akad. Handl., Fjärde Ser. 2(1): 161 pp., 6 pls.

Corethron hystrix
Cupp, Easter E., 1943
Marine plankton diatoms of the west coast of North America. Bull. S.I.O. 5(1):1-238, 5 pls., 168 text figs.

Corethron hystrix
Cupp, E.E., 1937
Seasonal distribution and occurrence of marine diatoms and dinoflagellates at Scotch Cap, Alaska. Bull. S.I.O. Tech. ser.4(3):71-100, 7 textfigs.

Corethron hystrix
Cupp, E.E. and Allen, W.E., 1938
Plankton diatoms of the Gulf of California obtained by Allan Hancock Pacific Expedition of 1937. The Hancock Pacific Expeditions, The Univ. So. Calif. Publ. 3: 61-74, 1 map, pls.4-15.

Corethron hystrix
Gilbert, J.Y., and W.E. Allen, 1943
The phytoplankton of the Gulf of California obtained by the "E.W. Scripps" in 1939 and 1940. J. Mar. Res. V(2):89-110, figs.30-31.

Corethron hystrix
Gran, H.H., and T. Braarud, 1935
A quantitative study of the phytoplankton in the Bay of Fundy and the Gulf of Maine (including observations on hydrography, chemistry, and turbidity). J. Biol. Bd., Canada, 1(5):279-467, 69 text figs.

Corethron hystrix
Gran, H. H. and E. C. Angst, 1931
Plankton diatoms of Puget Sound. Publ. Puget Sound Biol. Sta. 7:417-519, 95 text figs.

Corethron hystrix
Jorgensen, E., 1900
Protophyten und Protozoën im Plankton aus der Norwegischen Westkerste. Bergens Mus. Aarb. 1899(6): 95 pp., 5 pls., 83 tables.

Corethron hystrix
Lillick, L.C., 1940
Phytoplankton and planktonic protozoa of the offshore waters of the Gulf of Maine. Pt.II. Qualitative Composition of the Planktonic Flora. Trans. Am. Phil. Soc., n.s., 31(3):193-237, 13 text figs.

Corethron hystrix
Lillick, L.C., 1938
Preliminary report of the phytoplankton of the Gulf of Maine. Am. Mid. Nat. 20(3):624-640, 1 text fig, 37 tables.

Corethron hystrix
Lillick, L.C., 1937
Seasonal studies of the phytoplankton off Woods Hole, Massachusetts. Biol. Bull. LXXIII (3):488-503, 3 text figs.

Corethron hystrix
Paiva Carvalho, J., 1950
O plancton do Rio Maria Rodriques (Cananeis). 1. Diatomaceas e Dinoflagelados. Bol. Inst. Paulista Oceanogr. 1(1): 27-43, 2 fold-in tables, 2 figs.

Corethron hystrix
Phifer, L.D. (undated)
The occurrence and distribution of plankton diatoms in Bering Sea and Bering Strait, July 26-August 24, 1934. Report of Oceanographic cruise of U.S. Coast Guard Cutter Chelan 1934, Part II(A):1-44 (mimeographed) plus fig.1 (after Pt.B)

Corethron hystrix
Schulz, B., and A. Wulff, 1929
Hydrographie und Oberflächen plankton des westlichen Barentsmeeres im Sommer 1927. Ber. deutschen wissensch. Komm. F. Meeresforsch. n.s. 4(5):232-372, 13 tables, 25 text figs.

Corethron laciniosus
Jørgensen, E., 1905
B.Protistplankton and the diatoms in bottom samples. Hydrographical and biological investigations in Norwegian fjords. Bergens Mus. Skr. 7: 49-225.

Corethron lusitanicum n.sp.
Dangeard, P., 1927
Phytoplankton de la croisière du "Sylvana". Ann. Inst. Ocean., Monaco, n.s., 4(8):286-401, 54 text figs. (Feirer-Juin 1913).

Corethron lusitanicum
Sousa e Silva, E., 1949
Diatomaceas e Dinoflagelados de Baia de Cascais. Portugaliae Acta Biol. Volume: Julio Henriques, Ser. B: 300-383, 9 pls, 2 fold-in tables.

Corethron murrayanum n.sp.
Castracane degli Antelminelli, F., 1886
1. Report on the Diatomaceae collected by H.M.S. Challenger during the years 1873-1876. Rept. Sci. Results, H.M.S. Challenger, Botany Vol. II, 178 pp., 30 pls.

Corethron schüttii
Jørgensen, E., 1905
B.Protistplankton and the diatoms in bottom samples. Hydrographical and biological investigations in Norwegian fjords. Bergens Mus. Skr. 7: 49-225.

Corethron similis
Jørgensen, E., 1905
B.Protistplankton and the diatoms in bottom samples. Hydrographical and biological investigations in Norwegian fjords. Bergens Mus. Skr. 7: 49-225.

Corethron socialis
Jørgensen, E., 1905
B.Protistplankton and the diatoms in bottom samples. Hydrographical and biological investigations in Norwegian fjords. Bergens Mus. Skr. 7: 49-225.

Corethron teres
Jørgensen, E., 1905
B.Protistplankton and the diatoms in bottom samples. Hydrographical and biological investigations in Norwegian fjords. Bergens Mus. Skr. 7: 49-225.

Corethron willei
Jørgensen, E., 1905
B.Protistplankton and the diatoms in bottom samples. Hydrographical and biological investigations in Norwegian fjords. Bergens Mus. Skr. 7: 49-225.

Corethron valdiviae
Bigelow, H.B., and M. Leslie, 1930
Reconnaissance of the waters and plankton of Monterey Bay, July 1928. Bull. M.C.Z. 70(5):429-481, 43 text figs.

Corethron valdiviae
Cupp, E.E., 1937
Seasonal distribution and occurrence of marine diatoms and dinoflagellates at Scotch Cap, Alaska. Bull. S.I.O. Tech. ser.4(3):71-100, 7 textfigs.

Corethron valdiviae
Cupp, E.E., 1934
Analysis of marine diatom collections taken from the Canal Zone to California during March, 1933. Trans. Am. Micros. Soc LIII (1):22-29, 1 map.

Corethron Valdiviae
Mangin, L., 1915
Phytoplancton de L'Antartique. Deuxieme Exped. Ant. Francaise (1908-1910), 95 pp., 3 pls., 58 text figs.

Coscinodiscus sp.
Allen, W. E., 1938
The Templeton Crocker Expedition to the Gulf of California in 1935 - The Phytoplankton. Amer. Microsc. Soc. Trans. 57:328-335.

Coscinodiscus sp.
Allen, W.E., 1937
Plankton diatoms of the Gulf of California obtained by the G. Allan Hancock Expedition of 1936. The Hancock Pacific Expeditions, Univ. So. Calif. Publ. 3:47-59, 1 fig.

Coscinodiscus
Bens, Everett M., and Charles M. Drew, 1967.
Diatomaceous earth: scanning electron microscope of "chromosorb P". Nature, Lond. 216(5119):1046-1048.

Coscinodiscus sp.
Boden, Brian, 1948
Marine plankton diatoms on operation HIGHJUMP in: Some oceanographic observations on operation HIGHJUMP. By R.S. Dietz. USNEL Rept. No.55, 97 pp., 41 figs. 7 July 1948.

Coscinodiscus sp.
Braarud, T., 1945
A phytoplankton survey of the polluted waters of inner Oslo Fjord. Hvalrådets Skrifter, No.28, 142 pp., 19 text figs., 17 tables.

Coscinodiscus sp.
Braarud, T., and Adam Bursa, 1939
On the phytoplankton of the Oslo Fjord, 1933-1934. Hvalrådets Skr. No.19:1-63; 9 text figs. Reviewed. J. du. Cons. 14(3): 418-420. A.C. Gardiner.

Coscinodiscus sp.
Brunel, J., 1962
Le phytoplancton de la Baie de Chaleurs. Inst. Botan., Univ. Montréal, Contrib. No. 77: 365 pp., 66 pls.

Coscinodiscus sp.
Brunel, Jules, 1962
Le phytoplancton de la Baie des Chaleurs. Contrib. Ministère de la Chasse et des Pêcheries, Province de Québec, No. 91: 365 pp.

Coscinodiscus sp.
Cassie, Vivienne, 1961
Marine phytoplankton in New Zealand waters. Botanica Marina, 2(Suppl.):54 pp., 8 pls.

Coscinodiscus sp.
Central Meteorological Observatory, 1949
Report on sea and weather observation on Antarctic Whaling Ground (1947-48). Ocean. Mag., Japan, 1(1):49-88, 17 text figs.

Coscinodiscus sp.
Cleve-Euler, A., 1951
Die Diatomeen von Schweden und Finnland. Fungl. Svenska Vetenskaps Akad. Handl., Fjärde Ser. 2(1): 161 pp., 6 pls.

Coscinodiscus sp.
Cupp, E.E., 1934
Analysis of marine diatom collections taken from the Canal Zone to California during March, 1933. Trans. Am. Micros. Soc. LIII (1):22-29, 1 map.

Coscinodiscus sp.
Cupp, E.E. and Allen, W.E., 1938
Plankton diatoms of the Gulf of California obtained by Allan Hancock Pacific Expedition of 1937. The Hancock Pacific Expeditions, The Univ. So. Calif. Publ. 3: 61-74, 1 map, pls.4-15.

Oceanographic Index: Marine Organisms Cumulation, 1946-1973

Coscinodiscus
Desikachary, T.V., 1956(1957).
Electron microscope studies on diatoms.
J. R. Microsc. Soc. (3)76(1/2):9-36.

Coscinodiscus
Drebes, Gerhard 1968
Lagenisma coscinodisci gen.nov, spec. nov., ein Vertreter der Lagenidales in der marinen Diatomee Coscinodiscus.
Mar. Mykologie, Veröff. Inst. Meeresforsch. Bremerh. Sonderband 3:67-69.

Coscinodiscus sp.
Galtsoff, P.S., 1948
Red Tide. Progress Report on the investigations of the cause of the mortality of fish along the west coast of Florida conducted by the U.S. Fish and Wildlife Service and Cooperating Organizations. Fish and Wildlife Service, Special Scientific Rept. No.46, 44 pp. (mimeographed), 9 figs.

Coscinodiscus
Gilbert, J.Y., and W.E. Allen, 1943
The phytoplankton of the Gulf of California obtained by the "E.W. Scripps" in 1939 and 1940. J. Mar. Res. V(2):89-110, figs.30-31.

Coscinodiscus
Gunter G., F.G. Walton Smith, and R.H. Roberts, 1947
Mass mortality of marine animals on the lower west coast of Florida, November 1946 - January 1947. Science, 105(2723): 256-257.

Coscinodiscus sp.
Kucherova, Z.S., 1961
Vertical distribution of diatoms from Sevastopol Bay.
Trudy Sevastopol Biol. Sta., (14):64-78.

Coscinodiscus sp.
Levander, K.M., 1947
Plankton gesammelt in den Jahren 1899-1910 an den Küsten Finnlands. Finnländische Hydrographisch-Biologische Untersuchungen (aus dem Wasserbiologischen Laboratorin der Societas Scientiarum Fennica) No.11: 40 pp., 6 diagrams, 13 pls., tables.

Coscinodiscus sp.
Margalef, R., 1949
Fitoplancton nerítico de la Costa Brava en 1947-48. Publ. Inst. Biol. Aplicada, 5: 41-51, 3 text figs.

Coscinodiscus sp
Murphy, R.C., 1923
The oceanography of the Peruvian littoral with reference to the abundance and distribution of marine life. Geogr. Rev. 13:64-85.

Coscinodiscus
Neaverson, E., 1934
The sea-floor deposits. 1. General characteristics and distribution. Discovery Repts. 9: 297-349, Plates 17-22.

Coscinodiscus sp.
Ogino, C., 1963
Studies on the chemical composition of some natural foods of aquatic animals. In Japanese, English summary Bull. Jap. Soc. Sci. Fish, 29(5): 459 - 462

Coscinodiscus sp.
Parsons, T.R., 1961
On the pigment composition of eleven species of marine phytoplankton.
J. Fish. Res. Bd., Canada, 18(6):1017-1025.

Coscinodiscus sp.
Parsons, T.R., K. Stephens and J.D.H. Strickland, 1961
On the chemical composition of eleven species of marine phytoplankters.
J. Fish. Res. Bd., Canada, 18(6):1001-1016.

Coscinodiscus sp.
Pettersson, H., F. Gross, and F. Koczy, 1939
Large scale plankton cultures. Medd. från Oceanografiska Institutet i Göteborg No.3 (Göteborgs Kungl. vetenskaps-och Vitterhets-Samhälles Handlinger Femte Följden. Ser. B) Vol.6(13):1-24.
1. The plankton shaft. H. Pettersson
2. Experiment with phytoplankton. F. Gross and F. Koczy
3. Experiments with zooplankton.

Coscinodiscus
Rattray, J., 1890
A revision of the genus Coscinodiscus Ehrb. and some allied genera. Proc. Roy. Soc. Edinburgh, 16:

Coscinodiscus africanus rotunda
Castracane degli Antelminelli, F., 1886 n.sp.
1. Report on the Diatomaceae collected by H.M.S. Challenger during the years 1873-1876. Rept. Sci. Results, H.M.S. Challenger, Botany Vol. II, 178 pp., 30 pls.

Coscinodiscus alboranii n.sp.
Pavillard, J., 1925
Bacillariales. Rept. on the Danish Oceangr. Exped., 1908-10 to the Mediterranean and adj. seas. Vol.II., Biol. J4:72 pp., 116 text figs.

Coscinodiscus angstii n.sp.
Gran, H. H. and E. C. Angst, 1931
Plankton diatoms of Puget Sound. Publ. Puget Sound Biol. Sta. 7:417-519, 95 text figs.

Coscinodiscus angstii var. granulomarginata n.sp.
Gran, H. H. and E. C. Angst, 1931
Plankton diatoms of Puget Sound. Publ. Puget Sound Biol. Sta. 7:417-519, 95 text figs.

Coscinodiscus angstii
Kokubo, S., 1952
Results of the observations on the plankton and oceanography of Mutsu Bay during 1950, reference being made also to the period 1946-1950. Bull Mar.Biol.Sta., Asamushi 5(1/4): 1-54, 3 tables,(fold-in), 1 fold-in.

Coscinodiscus angustalineatus
Cleve-Euler, A., 1951
Die Diatomeen von Schweden und Finnland. Kungl. Svenska Vetenskaps Akad. Handl., Fjärde Ser. 2(1): 161 pp., 6 pls.

Coscinodiscus angustelineatus
Mangin, L., 1915
Phytoplancton de L'Antartique. Deuxieme Exped. Ant. Francaise (1908-1910), 95 pp., 3 pls., 58 text figs.

Coscinodiscus antarcticus
Castracane degli Antelminelli, F., 1886 n.sp.
1. Report on the Diatomaceae collected by H.M.S. Challenger during the years 1873-1876. Rept. Sci. Results, H.M.S. Challenger, Botany Vol. II, 178 pp., 30 pls.

Coscinodiscus antarcticus
Frenguelli, Joaquin, and Hector Antonio Orlando, 1959.
Operacion MERLUZA. Diatomeas y silicoflagelados del plancton del "VI Crucero". Servicio Hidrogr. Naval., Argentina, Publ. No. H. 619: 5-62.

Coscinodiscus antarctica n.sp.
Mangin, L., 1915
Phytoplancton de L'Antartique. Deuxieme Exped. Ant. Francaise (1908-1910), 95 pp., 3 pls., 58 text figs.

Coscinodiscus (Thalassiosira) antiquus
Cleve-Euler, A., 1951
Die Diatomeen von Schweden und Finnland. Kungl. Svenska Vetenskaps Akad. Handl., Fjärde Ser. 2(1): 161 pp., 6 pls.

Coscinodiscus apiculatus
Cleve-Euler, A., 1951
Die Diatomeen von Schweden und Finnland. Kungl. Svenska Vetenskaps Akad. Handl., Fjärde Ser. 2(1): 161 pp., 6 pls.

Coscinodiscus arafurensis
Castracane degli Antelminelli, F., 1886
1. Report on the Diatomaceae collected by H.M.S. Challenger during the years 1873-1876. Rept. Sci. Results, H.M.S. Challenger, Botany Vol. II, 178 pp., 30 pls.

Coscinodiscus aralensis
Cleve-Euler, A., 1951
Die Diatomeen von Schweden und Finnland. Kungl. Svenska Vetenskaps Akad. Handl., Fjärde Ser. 2(1): 161 pp., 6 pls.

Coscinodiscus Arentii
Cleve-Euler, A., 1951
Die Diatomeen von Schweden und Finnland. Kungl. Svenska Vetenskaps Akad. Handl., Fjärde Ser. 2(1): 161 pp., 6 pls.

Coscinodiscus argus
Cleve-Euler, A., 1951
Die Diatomeen von Schweden und Finnland. Kungl. Svenska Vetenskaps Akad. Handl., Fjärde Ser. 2(1): 161 pp., 6 pls.

Coscinodiscus argus
Florin, M-B., 1948
9. Diatomeae in submarine cores from the Tyrrhenian Sea. Medd. Ocean. Inst., Göteborg, 15 (Göteborgs Kungl. Vetenskaps-och Viterrhets Samhälles Handlingar, Sjätte Foljden, Ser. B 5(13):80-88.

Coscinodiscus argus
Lillick, L.C., 1940
Phytoplankton and planktonic protozoa of the offshore waters of the Gulf of Maine. Pt.II. Qualitative Composition of the Planktonic Flora. Trans. Am. Phil. Soc., n.s., 31(3):193-237, 13 text figs.

Coscinodiscus argus
Zanon, V., 1948
Diatomee marini di Sardegna e Pugillo di Alghe Marine della stressa. Boll. Pesca, Piscitutura e Idrobiologia, Anno 24, ns. 3(2): 202-244, 27 figs. on 1 pl.

Coscinodiscus argus heteroporus
Florin, M-B., 1948
9. Diatomeae in submarine cores from the Tyrrhenian Sea. Medd. Ocean. Inst., Göteborg, 15 (Göteborgs Kungl. Vetenskaps-och Viterrhets Samhälles Handlingar, Sjätte Foljden, Ser. B 5(13):80-88.

Coscinodiscus ? asperulus
Cleve-Euler, A., 1951
Die Diatomeen von Schweden und Finnland. Kungl. Svenska Vetenskaps Akad. Handl., Fjärde Ser. 2(1): 161 pp., 6 pls.

Coscinodiscus asteromphalus
Allen, W.E., and E.E. Cupp, 1935
Plankton diatoms of the Java Sea. Annales du Jardin Botanique de Buitenzorg XLIV (2):101-174, figs.1-127.
(drawings of all species mentioned)

Coscinodiscus asteromphalus

Bigelow, H.B., and M. Leslie, 1930
Reconnaissance of the waters and plankton of Monterey Bay, July 1928. Bull. M.C.Z., 70(5):429-481, 43 text figs.

Coscinodiscus asteromphalus

Chiba, T., 1949
On the distribution of the plankton in the eastern China Sea and Yellow Sea. 1. Plankton composition in the spring. J. Shimonoseki Coll. Fisheries, 1(1):57-63, 1 fig.

Coscinodiscus asteromphalus

Cleve-Euler, A., 1951
Die Diatomeen von Schweden und Finnland. Kungl. Svenska Vetenskaps Akad. Handl., Fjärde Ser. 2(1): 161 pp., 6 pls.

Coscinodiscus asteromphalus

Fox, M., 1957.
A first list of marine algae from Nigeria. J. Linn. Soc., London, 55g362):615-631.

Coscinodiscus asteromphalus

Frenguelli, Joaquin, and Hector Antonio Orlando, 1959.
Operacion MERLUZA. Diatomeas y silicoflagelados del plancton del "VI Crucero". Servicio Hidrogr. Naval., Argentina, Publ. No. H. 619: 5-62.

Coscinodiscus asteromphalus

Hendy, N. Ingram, 1964
An introductory account of the smaller algae of British coastal waters. V. Bacillariophyceae (Diatoms). Her Majesty's Stationary Office, 317 pp., 45 pls.

Coscinodiscus asteromphalus

Hendey, N-Ingram, 1958 [1957(Publ. 1958)]
Marine diatoms from some West African Ports-J. R.Microsc. Soc. (3) 77(1/2): 28-85.

Coscinodiscus Asteromphalus

Hendey, N.I., 1937
The plankton diatoms of the southern seas. Discovery Repts. 16:151-364, pls.6-13.

Coscinodiscus asteromphalus

Iselin, C., 1930
A report on the coastal waters of Labrador based on explorations of the "Chance" during the summer of 1926. Proc. Am. Acad. Arts Sci., 66(1):1-37, 14 text figs.

Coscinodiscus asteromphalus

Kokubo, S., 1952
Results of the observations on the plankton and oceanography of Mutsu Bay during 1950, reference being made also to the period 1946-1950. Bull Mar.Biol.Sta., Asamushi 5(1/4): 1-54, 3 tables,(fold-in), 1 fold-in.

Coscinodiscus asteromphalus

Lillick, L.C., 1940
Phytoplankton and planktonic protozoa of the offshore waters of the Gulf of Maine. Pt.II. Qualitative Composition of the Planktonic Flora. Trans. Am. Phil. Soc., n.s., 31(3):193-237, 13 text figs.

Coscinodiscus asteromphalus

Mann, A., 1907
Report on the diatoms of the Albatross voyages in the Pacific Ocean, 1888-1904. Contrib. U. S. Nat. Herb. 10(5):221-419, Pls. XLIV-LIV.

Coscinodiscus asteromphalus

Mann, A., 1893
List of Diatomaceae from a deep-sea dredging in the Atlantic Ocean off Delaware Bay by the U. S. Fish Commission Steamer Albatross. Proc. U. S. Nat. Mus. 16:303-312.

Coscinodiscus Asteromphalus

Marukawa, H., 1921
Plankton lists and some new species of copepods, from the northern waters of Japan. Bull. Inst. Ocean., No.384, 15 pp., 3 pls., 1 chart. Monaco

Coscinodiscus asteromphalus

Morse, D.C., 1947
Some observations on seasonal variations in plankton population Patuxant River, Maryland 1943-1945. Bd. Nat. Res., Publ. No.65, Chesapeake Biol. Lab., 31, 3 figs.

Coscinodiscus asteromphalus

Takano, Hideaki, 1963.
Notes on marine littoral diatoms of Japan. Bull. Tokai Reg. Fish. Res. Lab., No. 36:1-8.

Coscinodiscus asteromphalus

Zanon, D. V., 1949
Diatomee di Buenos Aires (Argentina) Atti Accad. Naz. Lincei, Memorie, Cl. Sci. fis., mat. e. nat., ser. 7, 11(3):59-151, 2 pls.

Coscinodiscus asteromphalus a

Werner, Dietrich 1971.
Der Entwicklungscyclus mit Sexualphase bei der marinen Diatomee Coscinodiscus asteromphalus. II. Oberflächenabhängige Differenzierung während der vegetativen Zellverkleinerung. III. Differenzierung und Spermatogenese.
Arch. Mikrobiol. 80 (2): 115-133; 134-146.

Coscinodiscus asteromphalus a

Werner, Dietrich, 1971.
Der Entwicklungscyclus mit Sexualphase bei der marinen Diatomee Coscinodiscus asteromphalus. 1. Kultur und Synchronisation von Entwicklungsstadien.
Arch. Mikrobiol. 80(1): 43-49.

Coscinodiscus australis

Boden, B. P., 1949.
The diatoms collected by the U.S.S. CACOPAN in the Antarctic in 1947. J. Mar. Res. 8(1):6-13, 3 textfigs.

Coscinodiscus australis

Boden, Brian, 1948
Marine plankton diatoms on observation HIGHJUMP in: Some oceanographic observations on operation HIGHJUMP. By R.S. Dietz. USNEL Rept. No.55, 97 pp., 41 figs. 7 July 1948.

Coscinodiscus australis

Mangin, L., 1915
Phytoplancton de L'Antartiqua. Deuxieme Exped. Ant. Francaise (1908-1910), 95 pp., 3 pls., 58 text figs.

Coscinodiscus (Thalassiosira) balticus

Cleve-Euler, A., 1951
Die Diatomeen von Schweden und Finnland. Kungl. Svenska Vetenskaps Akad. Handl., Fjärde Ser. 2(1): 161 pp., 6 pls.

Coscinodiscus bathyomphalus

Cleve-Euler, A., 1951
Die Diatomeen von Schweden und Finnland. Kungl. Svenska Vetenskaps Akad. Handl., Fjärde Ser. 2(1): 161 pp., 6 pls.

Coscinodiscus bathyomphalus

Misra, J.N., 1956.
A systematic account of some littoral marine diatoms from the west coast of India. J. Bombay Nat. Hist. Soc., 53(4):537-568.

Coscinodiscus belgicae

Frenguelli, Joaquin, and Hector Antonio Orlando, 1959.
Operacion MERLUZA. Diatomeas y silicoflagelados del plancton del "VI Crucero". Servicio Hidrogr. Naval., Argentina, Publ. No. H. 619: 5-62.

Coscinodiscus bergii

Cleve-Euler, A., 1951
Die Diatomeen von Schweden und Finnland. Kungl. Svenska Vetenskaps Akad. Handl., Fjärde Ser. 2(1): 161 pp., 6 pls.

Coscinodiscus beta

Hendey, N.I., 1937
The plankton diatoms of the southern seas. Discovery Repts. 16:151-364, pls.6-13.

Coscinodiscus (?) bifrons n.sp.

Castracane degli Antelminelli, F., 1886
1. Report on the Diatomaceae collected by H.M.S. Challenger during the years 1873-1876. Rept. Sci. Results, H.M.S. Challenger, Botany Vol. II, 178 pp., 30 pls.

Coscinodiscus bifrons

Mangin, L., 1915
Phytoplancton de L'Antartiqua. Deuxieme Exped. Ant. Francaise (1908-1910), 95 pp., 3 pls. 58 text figs.

Coscinodiscus biocularis

Jørgensen, E., 1905
B.Protistplankton and the diatoms in bottom samples. Hydrographical and biological investigations in Norwegian fjords. Bergens Mus. Skr. 7: 49-225.

Coscinodiscus bipartitas

Rampi, L., 1950.
Su di una rara diatomea plantonica il Coscinodiscus bipartitus Rattray 1889. Bull. Inst. Océan., Monaco, No. 981: 7 pp., 8 textfigs.

Coscinodiscus borealis

Mann, A., 1907
Report on the diatoms of the Albatross voyages in the Pacific Ocean, 1888-1904. Contrib. U. S. Nat. Herb. 10(5):221-419, Pls. XLIV-LIV.

Coscinodiscus bouvet

Hendey, N.I., 1937
The plankton diatoms of the southern seas. Discovery Repts. 16:151-364, pls.6-13.

Coscinodiscus bouvet

Jouse, A.P., G.S. Koroleva, G.A. Nagaeva, 1962
Diatoms in the surface layer of sediment in the Indian sector of the Antarctic. Investigations of marine bottom sediments. (In Russian; English summary). Trudy Inst. Okeanol., Akad. Nauk, SSSR, 61:19-92.

Coscinodiscus Bouvet

Mangin, L., 1915
Phytoplancton de L'Antartiqua. Deuxieme Exped. Ant. Francaise (1908-1910), 95 pp., 3 pls., 58 text figs.

Coscinodiscus bullatus

Jouse, A.P., G.S. Koroleva, G.A. Nagaeva, 1962
Diatoms in the surface layer of sediment in the Indian sector of the Antarctic. Investigations of marine bottom sediments. (In Russian; English summary). Trudy Inst. Okeanol., Akad. Nauk, SSSR, 61:19-92.

Coscinodiscus centralis

Atkins, W.R.G., and M. Parke, 1951.
Seasonal changes in the phytoplankton as indicated by chlorophyll estimation. J.M.B.A. 29(3):609-618.

Braarud, T., 1945
Experimental studies on marine plankton diatoms. Avhandlingar utgitt av Det Norske Videnskaps-Akademi i Oslo. 1. Mat.-Naturv. Klasse 1944. No.10, 1-16, 1 text fig.

Brunel, J., 1962
Le phytoplancton de la Baie de Chaleurs. Inst. Botan., Univ. Montréal, Contrib. No. 77: 365 pp., 66 pls.

Coscinodiscus centralis pacifica

Brunel, Jules, 1962
Le phytoplancton de la Baie des Chaleurs. Contrib. Ministère de la Chasse et des Pêcheries, Province de Québec, No. 91: 365 pp.

Coscinodiscus centralis

Castracane degli Antelminelli, F., 1886
1. Report on the Diatomaceae collected by H.M.S. Challenger during the years 1873-1876. Rept. Sci. Results, H.M.S. Challenger, Botany Vol. II, 178 pp., 30 pls.

Cleve-Euler, A., 1951
Die Diatomeen von Schweden und Finnland. Kungl. Svenska Vetenskaps Akad. Handl., Fjärde Ser. 2(1): 161 pp., 6 pls.

Cupp, Easter E., 1943
Marine plankton diatoms of the west coast of North America. Bull. S.I.O. 5(1):1-238, 5 pls., 168 text figs.

Cupp, E.E., 1937
Seasonal distribution and occurrence of marine diatoms and dinoflagellates at Scotch Cap, Alaska. Bull. S.I.O. Tech. ser.4(3):71-100, 7 textfigs.

Dangeard, P., 1927
Phytoplankton de la croisière du "Sylvana". Ann. Inst. Océan., Monaco, n.s., 4(8):286-401, 54 text figs. (Feirer-Juin 1913).

Delegazione Italiana della Commissione Internazionale per l'Esplorazione Scientifica del Mediterraneo, 1941
Note sul plancton della Laguna veneta. [Memoria CCLXXIX], Arch. di Ocean. e Limn. Anno I, Fasc. I, 1941 XIX: 31-57 pp.

Ercegovic, A., 1936
Etudes qualitative et quantitatives du phytoplancton dans les eaux cotières de l'Adriatique oriental moyen au cours de l'année 1934. Acta Adriatica 1(9):1-126

Gran, H.H., 1908
Diatomeen. Nordisches Plankton, Botanischer Teil pp. XIX.1-XIX 146; 178 text figs.

Forti, A., 1922
Ricerche sulla flora pelagica (fitoplancton) di Quarto dei Mille. Mem. R. Com. Talass. Ital. 97:248 pp., 13 pls.

Hendy, N. Ingram, 1964
An introductory account of the smaller algae of British coastal waters. V. Bacillariophyceae (Diatoms).
Her Majesty's Stationary Office, 317 pp., 45 pls.

Hendey, N. Ingram, 1958 [1957(Publ. 1958)]
Marine diatoms from some West African Ports. J. R. Microsc. Soc. (3) 77(1/2): 28-85.

Hendey, N.I., 1937
The plankton diatoms of the southern seas. Discovery Repts. 16:151-364, pls.6-13.

Hustedt, F. and A.A. Aleem, 1951
Littoral diatoms from the Salstone near Plymouth. JMBA 30(1): 177-196.

Johnson, T.W., and Jr., 1966
A Lagenidium in the marine diatom Coscinodiscus centralis. Mycologia, 58(1):131-135.

Jørgensen, E., 1905
B. Protistplankton and the diatoms in bottom samples. Hydrographical and biological investigations in Norwegian fjords. Bergens Mus. Skr. 7: 49-225.

Jorgensen, E., 1900
Protophyten und Protozoën im Plankton aus der Norwegischen Westkerste. Bergens Mus. Aarb. 1899(6): 95 pp., 5 pls., 83 tables.

Kokubo, S., 1952
Results of the observations on the plankton and oceanography of Mutsu Bay during 1950, reference being made also to the period 1946-1950. Bull Mar.Biol.Sta., Asamushi 5(1/4): 1-54, 3 tables,(fold-in), 1 fold-in.

Lillick, L.C., 1940
Phytoplankton and planktonic protozoa of the offshore waters of the Gulf of Maine. Pt.II. Qualitative Composition of the Planktonic Flora. Trans. Am. Phil. Soc., n.s., 31(3):193-237, 13 text figs.

Lillick, L.C., 1938
Preliminary report of the phytoplankton of the Gulf of Maine. Am. Mid. Nat. 20(3):624-640, 1 text fig., 37 tables.

Lillick, L.C., 1937
Seasonal studies of the phytoplankton off Woods Hole, Massachusetts. Biol. Bull. LXXIII (3):488-503, 3 text figs.

Manguin, E., 1954
Diatomees marines provenant de l'ile Heard (Australian National Antarctic Expedition). Rev. Algol., n.s., 1: 14-24.

Mann, A., 1907
Report on the diatoms of the Albatross voyages in the Pacific Ocean, 1888-1904. Contrib. U. S. Nat. Herb. 10(5):221-419, Pls. XLIV-LIV.

Marukawa, H., 1921
Plankton lists and some new species of copepods, from the northern waters of Japan. Bull. Inst. Océan., No.384, 15 pp., 3 pls., 1 chart. Monaco

Morse, D.C., 1947
Some observations on seasonal variations in plankton population Patuxant River, Maryland 1943-1945. Bd. Nat. Res., Publ. No.65, Chesapeake Biol. Lab., 31, 3 figs.

Pavillard, J., 1925
Bacillariales. Rept. on the Danish Oceangr. Exped., 1908-10 to the Mediterranean and adj. seas. Vol.II., Biol. J4:72 pp., 116 text figs.

Phifer, L.D. (undated)
The occurrence and distribution of plankton diatoms in Bering Sea and Bering Strait, July 26-August 24, 1934. Report of Oceanographic cruise of U.S. Coast Guard Cutter Chelan 1934, Part II(A):1-44 (mimeographed) plus fig.1 (after Pt.B)

Rampi, L., 1942
Ricerche sul fitoplancton del Mare Ligure 6. Le diatomee delle acque di Sanremo. Nuovo Giornale Botanico Italiano, N.S., 49:252-268.

Rumkówna, A., 1948
[List of the phytoplankton species occurring in the superficial water layers in the Gulf of Gdańsk] Bull. Lab. mar., Gdynia, No. 4: 139-141 with tables in back.

Zanon, D. V., 1949
Diatomee di Buenos Aires (Argentina). Atti Accad. Naz. Lincei, Memorie, Cl. Sci. fis., mat. e. nat., ser. 7, 11(3):59-151, 2 pls.

Coscinodiscus centralis pacifica

Cupp, Easter E., 1943
Marine plankton diatoms of the west coast of North America. Bull. S.I.O. 5(1):1-238, 5 pls., 168 text figs.

Coscinodiscus charcotii

Hendey, N.I., 1937
The plankton diatoms of the southern seas. Discovery Repts. 16:151-364, pls.6-13.

Coscinodiscus chromoradiatus

Mangin, L., 1915
Phytoplancton de L'Antartique. Deuxieme Exped. Ant. Francaise (1908-1910), 95 pp., 3 pls., 58 text figs.

Coscinodiscus chunii

Hendey, N.I., 1937
The plankton diatoms of the southern seas. Discovery Repts. 16:151-364, pls.6-13.

Coscinodiscus cinctus

Lillick, L.C., 1940
Phytoplankton and planktonic protozoa of the offshore waters of the Gulf of Maine. Pt.II. Qualitative Composition of the Planktonic Flora. Trans. Am. Phil. Soc., n.s., 31(3):193-237, 13 text figs.

Coscinodiscus cocconeiformis

Mann, A., 1907
Report on the diatoms of the Albatross voyages in the Pacific Ocean, 1888-1904. Contrib. U. S. Nat. Herb. 10(5):221-419, Pls. XLIV-LIV.

Coscinodiscus commutatus

Cleve-Euler, A., 1951
Die Diatomeen von Schweden und Finnland. Kungl. Svenska Vetenskaps Akad. Handl., Fjärde Ser. 2(1): 161 pp., 6 pls.

Coscinodiscus commutatus

Hendy, N. Ingram, 1964
An introductory account of the smaller algae of British coastal waters. V. Bacillariophyceae (Diatoms).
Her Majesty's Stationary Office, 317 pp., 45 pls.

Coscinodiscus comptus n.sp.

Castracane degli Antelminelli, F., 1886
1. Report on the Diatomaceae collected by H.M.S. Challenger during the years 1873-1876. Rept. Sci. Results, H.M.S. Challenger, Botany Vol. II, 178 pp., 30 pls.

Coscinodiscus concavus

Mann, A., 1907
Report on the diatoms of the Albatross voyages in the Pacific Ocean, 1888-1904. Contrib. U. S. Nat. Herb. 10(5):221-419, Pls. XLIV-LIV.

Coscinodiscus concinnus

Allen, W.E., and E.E. Cupp, 1935
Plankton diatoms of the Java Sea. Annales du Jardin Botanique de Buitenzorg XLIV (2):101-174, figs.1-127.
(drawings of all species mentioned)

Coscinodiscus concinnus

Bigelow, H.B., and M. Leslie, 1930
Reconnaissance of the waters and plankton of Monterey Bay, July 1928. Bull. M.C.Z., 70(5):429-481, 43 text figs.

Coscinodiscus concinnus

Boalch, G.T., 1971.
The typification of the diatom species Coscinodiscus concinnus WM.Smith and Coscinodiscus granii Gough. J. mar. biol. Ass. U.K. 51(3): 685-695.

Coscinodiscus concinnus?

Boden, Brian, 1948
Marine plankton diatoms on operation HIGHJUMP in: Some oceanographic observations on operation HIGHJUMP. By R.S. Dietz. USNEL Rept. No.55, 97 pp., 41 figs. 7 July 1948.

Coscinodiscus concinnus

Chiba, T., 1949
On the distribution of the plankton in the eastern China Sea and Yellow Sea. 1. Plankton composition in the spring. J. Shimonoseki Coll. Fisheries, 1(1):57-63, 1 fig.

Coscinodiscus concinnus

Cleve-Euler, A., 1951
Die Diatomeen von Schweden und Finnland. Kungl. Svenska Vetenskaps Akad. Handl., Fjärde Ser. 2(1): 161 pp., 6 pls.

Coscinodiscus concinnus

Cupp, Easter E., 1943
Marine plankton diatoms of the west coast of North America. Bull. S.I.O. 5(1):1-238, 5 pls., 168 text figs.

Coscinodiscus concinnus

Davidson, V.M., 1931.
Biological and oceanographic conditions in Hudson Bay. Contr. Canadian Biol. Fish., n.s., 6(26): 497-509, 7 textfigs.

Coscinodiscus concinnum

Drebes, Gerhard 1966.
Ein parasitischer Phycomycet (Lagenidiales) in Coscinodiscus. Helgoländer wiss. Meeresunters. 13(4): 426-435.

Coscinodiscus concinnus

Fraser, J. H., 1949
Plankton investigations from the Scottish Research Vessel. Ann. Biol., Int. Cons., 4: 66-67.

Coscinodiscus concinnus

Gran, H.H., 1908
Diatomeen. Nordisches Plankton, Botanischer Teil pp. XIX 1-XIX 146; 178 text figs.

Coscinodiscus concinnus

Gran, H. H. and E. C. Angst, 1931
Plankton diatoms of Puget Sound. Publ. Puget Sound Biol. Sta. 7:417-519, 95 text figs.

Coscinodiscus concinnus (oil pollution caused by)

Grøntved, J., 1952.
Investigations on the phytoplankton in the southern North Sea in May 1947. Medd. Komm. Danmarks Fisk.- og Havundersøgelser, Plankton ser., 5(5):1-49, 1 pl., 21 tables, 24 textfigs.

Coscinodiscus concinnus

Hendy, N. Ingram, 1964
An introductory account of the smaller algae of British coastal waters. V. Bacillariophyceae (Diatoms).
Her Majesty's Stationary Office, 317 pp., 45 pls.

Coscinodiscus concinnus

Hendey, N.I., 1937
The plankton diatoms of the southern seas. Discovery Repts. 16:151-364, pls.6-13.

Coscinodiscus concinnus

Holmes, Robert W., 1966.
Short-term temperature and light conditions associated with auxospore formation in the marine centric diatom Coscinodiscus Concinnus W. Smith.
Nature, 209 (5019):217-218.

Coscinodiscus concinnus

Holmes, Robert W., and Bernhard E. F. Reimann, 1966.
Variation in valve morphology during the life cycle of the marine diatom Coscinodiscus concinnus.
Phycologia 5(4):233-244.

Coscinodiscus concinnus

Jørgensen, E., 1905
B. Protistplankton and the diatoms in bottom samples. Hydrographical and biological investigations in Norwegian fjords. Bergens Mus. Skr. 7: 49-225.

Coscinodiscus concinnus

Jorgensen, E., 1900
Protophyten und Protozoën im Plankton aus der Norwegischen Westkerste. Bergens Mus. Aarb. 1899(6): 95 pp., 5 pls., 83 tables.

Coscinodiscus concinnus

Kokubo, S., 1952
Results of the observations on the plankton and oceanography of Mutsu Bay during 1950, reference being made also to the period 1946-1950. Bull Mar.Biol.Sta., Asamushi 5(1/4): 1-54, 3 tables,(fold-in), 1 fold-in.

Coscinodiscus concinnus

Lillick, L.C., 1940
Phytoplankton and planktonic protozoa of the offshore waters of the Gulf of Maine. Pt.II. Qualitative Composition of the Planktonic Flora. Trans. Am. Phil. Soc., n.s., 31(3):193-237, 13 text figs.

Coscinodiscus concinnus

Manguin, E., 1954
Diatomées marines provenant de l'île Heard (Australian National Antarctic Expedition). Rev. Algol., n.s., 1: 14-24.

Coscinodiscus concinnus

Mann, A., 1907
Report on the diatoms of the Albatross voyages in the Pacific Ocean, 1888-1904. Contrib. U. S. Nat. Herb. 10(5):221-419, Pls. XLIV-LIV.

Coscinodiscus concinnus

Meunier, A., 1915
Microplancton de la Mer Flamande. 2. Diatomées (excepté le genre Chaetoceros). Mem. Mus. Roy. Hist. Nat., Belgique, 7(3):1-118, Pls. VIII-XIV.

Coscinodiscus concinnus

Rumkówna, A., 1948
[List of the phytoplankton species occurring in the superficial water layers in the Gulf of Gdańsk] Bull. Lab. mar., Gdynia, No. 4: 139-141 with tables in back.

Coscinodiscus concinnus

Savage, R.E., 1937
The food of North Sea herring 1930-1934. Minstry of Agriculture and Fisheries. Fish. Invest. Ser. II, 15(5):1-60; 16 text figs.

Coscinodiscus concinnus

Schodduyn, M., 1926
Observations faites dans la baie d'Ambleteuse (Pas de Calais). Bull. Inst. Ocean., Monaco, No. 482: 64 pp.

Coscinodiscus concinnus

Sousa e Silva, E., 1949
Diatomaceas e Dinoflagelados de Baía de Cascais. Portugaliae Acta Biol., Volume: Julio Henriques, Ser. B: 300-383, 9 pls, 2 fold-in tables.

Coscinodiscus concinnus

Sousa e Silva, E., and J. Dos Santos-Pinto, 1948
O Plancton da Baía de S. Martinho do Porto. 1. Diatomaceas e Dinoflagelados. Bol. Soc. Portuguese de Ciencias Naturais, 16(2):134-187, 6 pls. (Trav. Sta. Biol. Mar. de Lisbonne No. 52).

Coscinodiscus concinnus

von Sydow, Burkard, und Robert Christenhuss 1972.
Rasterelektronmikroskopische Untersuchungen der Hohlräume in der Schalenwand einiger centrischer Kieselalgen. Arch. Protistenk 114 (3): 256-271.

Coscinodiscus concinnus

Wimpenny, R. S., 1949.
The dry weight and fat content of plankton. Ann. Biol. 5:89.

Coscinodiscus concinnus = 30-40%

Coscinodiscus concinnus

Zanon, D. V., 1949
Diatomee di Buenos Aires (Argentina) Atti Accad. Naz. Lincei, Memorie, Cl. Sci. fis., mat. e. nat., ser. 7, 11(3):59-151, 2 pls.

Coscinodiscus (Thalassiosira) condensatus

Cleve-Euler, A., 1951
Die Diatomeen von Schweden und Finnland. Kungl. Svenska Vetenskaps Akad. Handl., Fjärde Ser. 2(1): 161 pp., 6 pls.

Coscinodiscus confusus
Mann, A., 1893
List of Diatomaceae from a deep-sea dredging in the Atlantic Ocean off Delaware Bay by the U. S. Fish Commission Steamer Albatross. Proc. U. S. Nat. Mus. 16:303-312.

Coscinodiscus convexus
Mann, A., 1893
List of Diatomaceae from a deep-sea dredging in the Atlantic Ocean off Delaware Bay by the U. S. Fish Commission Steamer Albatross. Proc. U. S. Nat. Mus. 16:303-312.

Coscinodiscus coronula, n.sp.
Frenguelli, Joaquin, y Hector A. Orlando, 1958.
Diatomeas y silicoflagelados del sector Antartico Sudamericano.
Inst. Antartico Argentino, Publ., No. 5:191 pp.

Coscinodiscus crassus
Cleve-Euler, A., 1951
Die Diatomeen von Schweden und Finnland. Kungl. Svenska Vetenskaps Akad. Handl., Fjärde Ser. 2(1): 161 pp., 6 pls.

Coscinodiscus crenulatus
Hasle, Grethe Rytter, 1960
Phytoplankton and ciliate species from the Tropical Pacific.
Skr. Norske Videnskaps-Akad., Oslo, 1. Mat.-Nat. Kl. 1960(2): 1-50.

Coscinodiscus curvulatus
Bigelow, H.B., and M. Leslie, 1930
Reconnaissance of the waters and plankton of Monterey Bay, July 1928.
Bull. M.C.Z., 70(5):429-481, 43 text figs.

Coscinodiscus curvatulus
Boden, B.P., 1950
Some marine plankton diatoms from the west coast of South Africa. Trans. R.Soc. S. Africa. 32:321-434, 100 text figs.

Coscinodiscus curvatulus
Castracane degli Antelminelli, F., 1886
1. Report on the Diatomaceae collected by H.M.S. Challenger during the years 1873-1876. Rept. Sci. Results, H.M.S. Challenger, Botany Vol. II, 178 pp., 30 pls.

Coscinodiscus curvatulus
Cleve-Euler, A., 1951
Die Diatomeen von Schweden und Finnland. Kungl. Svenska Vetenskaps Akad. Handl., Fjärde Ser. 2(1): 161 pp., 6 pls.

Coscinodiscus curvatulus
Cupp, Easter E., 1943
Marine plankton diatoms of the west coast of North America. Bull. S.I.O. 5(1):1-238, 5 pls., 168 text figs.

Coscinodiscus curvatulus
Forti, A., 1922
Ricerche sulla flora pelagica (fitoplancton) di Quarto dei Mille. Mem. R. Com. Talass. Ital. 97:248 pp., 13 pls.

Coscinodiscus curvatulus
Gran, H.H., 1908
Diatomeen. Nordisches Plankton, Botanischer Teil pp. XIX.1-XIX.146; 178 text figs.

Coscinodiscus curvatulus
Gran, H. H. and E. C. Angst, 1931
Plankton diatoms of Puget Sound. Publ. Puget Sound Biol. Sta. 7:417-519, 95 text figs.

Coscinodiscus curvatulus
Hendy, N. Ingram, 1964
An introductory account of the smaller algae of British coastal waters. V. Bacillariophyceae (Diatoms).
Her Majesty's Stationary Office, 317 pp., 45 pls.

Coscinodiscus curvatulus
Hendey, N.I., 1937
The plankton diatoms of the southern seas. Discovery Repts. 16:151-364, pls.6-13.

Coscinodiscus curvatulus
Jørgensen, E., 1905
B.Protistplankton and the diatoms in bottom samples. Hydrographical and biological investigations in Norwegian fiords. Bergens Mus. Skr. 7: 49-225.

Coscinodiscus curvatulus
Jorgensen, E., 1900
Protophyten und Protozoën im Plankton aus der Norwegischen Westkerste. Bergens Mus. Aarb. 1899(6): 95 pp., 5 pls., 83 tables.

Coscinodiscus curvatulus
Lillick, L.C., 1940
Phytoplankton and planktonic protozoa of the offshore waters of the Gulf of Maine. Pt.II. Qualitative Composition of the Planktonic Flora. Trans. Am. Phil. Soc., n.s., 31(3):193-237, 13 text figs.

Coscinodiscus curvatulus
Mann, A., 1907
Report on the diatoms of the Albatross voyages in the Pacific Ocean, 1888-1904. Contrib. U. S. Nat. Herb. 10(5):221-419, Pls. XLIV-LIV.

Coscinodiscus curvatulus
Pavillard, J., 1925
Bacillariales. Rept. on the Danish Oceangr. Exped., 1908-10 to the Mediterranean and adj. seas. Vol.II., Biol. J4:72 pp., 116 text figs.

Coscinodiscus curvatulus
Phifer, L.D. (undated)
The occurrence and distribution of plankton diatoms in Bering Sea and Bering Strait, July 26-August 24, 1934. Report of Oceanographic cruise of U.S. Coast Guard Cutter Chelan 1934, Part II(A):1-44 (mimeographed) plus fig.1 (after Pt.B)

Coscinodiscus curvatulus
Rumkówna, A., 1948
List of the phytoplankton species occurring in the superficial water layers in the Gulf of Gdańsk. Bull. Lab. mar., Gdynia, No. 4: 139-141 with tables in back.

Coscinodiscus curvatulus
Zanon, D. V., 1949
Diatomee di Buenos Aires (Argentina) Atti Accad. Naz. Lincei, Memorie, Cl. Sci. fis., mat. e. nat., ser. 7, 11(3):59-151, 2 pls.

Coscinodiscus curvatulus
Zanon, V., 1948
Diatomee marini di Sardegna e Pugillo di Alghe Marine della stressa. Boll. Pesca, Piscitutura e Idrobiologia, Anno 24, ns. 3(2): 202-244, 27 figs. on 1 pl.

Coscinodiscus cycloteres n.sp.
Castracane degli Antelminelli, F., 1886
1. Report on the Diatomaceae collected by H.M.S. Challenger during the years 1873-1876. Rept. Sci. Results, H.M.S. Challenger, Botany Vol. II, 178 pp., 30 pls.

Coscinodiscus (Thalassiosira) decipiens
Cleve-Euler, A., 1951
Die Diatomeen von Schweden und Finnland. Kungl. Svenska Vetenskaps Akad. Handl., Fjärde Ser. 2(1): 161 pp., 6 pls.

Coscinodiscus decipiens
Jørgensen, E., 1905
B.Protistplankton and the diatoms in bottom samples. Hydrographical and biological investigations in Norwegian fiords. Bergens Mus. Skr. 7: 49-225.

Coscinodiscus decipiens
Jorgensen, E., 1900
Protophyten und Protozoën im Plankton aus der Norwegischen Westkerste. Bergens Mus. Aarb. 1899(6): 95 pp., 5 pls., 83 tables.

Coscinodiscus (Thalassiosira) decoratus
Cleve-Euler, A., 1951
Die Diatomeen von Schweden und Finnland. Kungl. Svenska Vetenskaps Akad. Handl., Fjärde Ser. 2(1): 161 pp., 6 pls.

Coscinodiscus decrescens
Castracane degli Antelminelli, F., 1886 n.sp.
1. Report on the Diatomaceae collected by H.M.S. Challenger during the years 1873-1876. Rept. Sci. Results, H.M.S. Challenger, Botany Vol. II, 178 pp., 30 pls.

Coscinodiscus decrescens
Cleve-Euler, A., 1951
Die Diatomeen von Schweden und Finnland. Kungl. Svenska Vetenskaps Akad. Handl., Fjärde Ser. 2(1): 161 pp., 6 pls.

Coscinodiscus decrescens
Florin, M-B., 1948
9. Diatomeae in submarine cores from the Tyrrhenian Sea. Medd. Ocean. Inst., Göteborg, 15 (Göteborgs Kungl. Vetenskaps-och Viterrhets Samhälles Handlingar, Sjätte Foljden, Ser. B 5(13):80-88.

Coscinodiscus decrescens
Hendy, N. Ingram, 1964
An introductory account of the smaller algae of British coastal waters. V. Bacillariophyceae (Diatoms).
Her Majesty's Stationary Office, 317 pp., 45 pls.

Coscinodiscus decrescens
Hendey, N.I., 1937
The plankton diatoms of the southern seas. Discovery Repts. 16:151-364, pls.6-13.

Coscinodiscus decrescens
Mann, A., 1907
Report on the diatoms of the Albatross voyages in the Pacific Ocean, 1888-1904. Contrib. U. S. Nat. Herb. 10(5):221-419, Pls. XLIV-LIV.

Coscinodiscus decrescens
Mann, A., 1893
List of Diatomaceae from a deep-sea dredging in the Atlantic Ocean off Delaware Bay by the U. S. Fish Commission Steamer Albatross. Proc. U. S. Nat. Mus. 16:303-312.

Coscinodiscus decrescens
Zanon, D. V., 1949
Diatomee di Buenos Aires (Argentina) Atti Accad. Naz. Lincei, Memorie, Cl. Sci. fis., mat. e. nat., ser. 7, 11(3):59-151, 2 pls.

Coscinodiscus deformatus n.sp.
Mann, A., 1907
Report on the diatoms of the Albatross voyages in the Pacific Ocean, 1888-1904. Contrib. U. S. Nat. Herb. 10(5):221-419, Pls. XLIV-LIV.

Coscinodiscus denarius
Bigelow, H.B., and M. Leslie, 1930
Reconnaissance of the waters and plankton of Monterey Bay, July 1928. Bull. M.C.Z., 70(5):429-481, 43 text figs.

Coscinodiscus denarius

Mangin, L., 1915
Phytoplankton de L'Antartique. Deuxieme Exped. Ant. Francaise (1908-1910), 95 pp., 3 pls., 58 text figs.

Coscinodiscus denarius

Mann, A., 1907
Report on the diatoms of the Albatross voyages in the Pacific Ocean, 1888-1904. Contrib. U. S. Nat. Herb. 10(5):221-419, Pls. XLIV-LIV.

Coscinodiscus denarius var. sinensis

Hendey, N.Ingram, 1958 [1957(Publ. 1958)]
Marine diatoms from some West African Ports. J. R. Microsc. Soc. (3) 77(1/2): 28-85.

Coscinodiscus denticulatus

Castracane degli Antelminelli, F., 1886 n.sp.
1. Report on the Diatomaceae collected by H.M.S. Challenger during the years 1873-1876. Rept. Sci. Results, H.M.S. Challenger, Botany Vol. II, 178 pp., 30 pls.

Coscinodiscus devius

Sousa e Silva, E., 1949
Diatomaceas e Dinoflagelados de Baia de Cascais. Portugaliae Acta Biol., Volume: Julio Henriques, Ser. B: 300-383, 9 pls, 2 fold-in tables.

Coscinodiscus (?) dimorphus

Castracane degli Antelminelli, F., 1886 n.sp.
1. Report on the Diatomaceae collected by H.M.S. Challenger during the years 1873-1876. Rept. Sci. Results, H.M.S. Challenger, Botany Vol. II, 178 pp., 30 pls.

Coscinodiscus diophthalmus

Castracane degli Antelminelli, F., 1886 n.sp.
1. Report on the Diatomaceae collected by H.M.S. Challenger during the years 1873-1876. Rept. Sci. Results, H.M.S. Challenger, Botany Vol. II, 178 pp., 30 pls.

Coscinodiscus diophthalmus monophthalma

Castracane degli Antelminelli, F., 1886 n.var.
1. Report on the Diatomaceae collected by H.M.S. Challenger during the years 1873-1876. Rept. Sci. Results, H.M.S. Challenger, Botany Vol. II, 178 pp., 30 pls.

Coscinodiscus diorema

Cleve-Euler, A., 1951
Die Diatomeen von Schweden und Finnland. Kungl. Svenska Vetenskaps Akad. Handl., Fjärde Ser. 2(1): 161 pp., 6 pls.

Coscinodiscus diorama

Florin, M-B., 1948
9. Diatomeae in submarine cores from the Tyrrhenian Sea. Medd. Ocean. Inst., Göteborg, 15 (Göteborgs Kungl. Vetenskaps-och Viterrhets Samhälles Handlingar, Sjätte Poljden, Ser. B 5(13):80-88.

Coscinodiscus divisus

Cleve-Euler, A., 1951
Die Diatomeen von Schweden und Finnland. Kungl. Svenska Vetenskaps Akad. Handl., Fjärde Ser. 2(1): 161 pp., 6 pls.

Coscinodiscus divisus

Frenguelli, Joaquin, and Hector Antonio Orlando, 1959.
Operacion MERLUZA. Diatomeas y silico-flagelados del plancton del "VI Crucero". Servicio Hidrogr. Naval., Argentina, Publ. No. H. 619: 5-62.

Coscinodiscus divisus

von Sydow, Burkard, und Robert Christenhuss 1972. Rasterelektronmikroskopische Untersuchungen der Hohlräume in der Schalenwand einiger centrischer Kieselalgen. Arch. Protistenk 114 (3): 256-271.

Coscinodiscus divisus

Zanon, D. V., 1949
Diatomee di Buenos Aires (Argentina) Atti Accad. Naz. Lincei, Memorie, Cl. Sci. fis., mat. e. nat., ser. 7, 11(3):59-151, 2 pls.

Coscinodiscus domifactus n.sp.

Hendey, N.Ingram, 1958 [1957(Publ. 1958)]
Marine diatoms from some West African Ports. J. R. Microsc. Soc. (3) 77(1/2): 28-85.

Coscinodiscus ebulliens

Castracane degli Antelminelli, F., 1886
1. Report on the Diatomaceae collected by H.M.S. Challenger during the years 1873-1876. Rept. Sci. Results, H.M.S. Challenger, Botany Vol. II, 178 pp., 30 pls.

Coscinodiscus elegans

Mann, A., 1907
Report on the diatoms of the Albatross voyages in the Pacific Ocean, 1888-1904. Contrib. U. S. Nat. Herb. 10(5):221-419, Pls. XLIV-LIV.

Coscinodiscus eta

Hendey, N.I., 1937
The plankton diatoms of the southern seas. Discovery Repts. 16:151-364, pls.6-13.

Coscinodiscus excentricus

Allen, W.E., and E.E. Cupp, 1935
Plankton diatoms of the Java Sea. Annales du Jardin Botanique de Buitenzorg XLIV (2):101-174, figs.1-127.
(drawings of all species mentioned)

Coscinodiscus excentricus

Bigelow, H.B., and M. Leslie, 1930
Reconnaissance of the waters and plankton of Monterey Bay, July 1928. Bull. M.C.Z., 70(5):429-481, 43 text figs.

Coscinodiscus excentricus

Boden, B.P., 1950
Some marine plankton diatoms from the west coast of South Africa. Trans. R.Soc. S. Africa. 32:321-434, 100 text figs.

Coscinodiscus excentricus

Boden, B. P., 1949.
The diatoms collected by the U.S.S. CACOPAN in the Antarctic in 1947. J. Mar. Res. 8(1):6-13, 3 textfigs.

Coscinodiscus excentricus

Boden, Brian, 1948
Marine plankton diatoms on operation HIGHJUMP in: Some oceanographic observations on operation HIGHJUMP. By R.S. Dietz. USNEL Rept. No.55, 97 pp., 41 figs. 7 July 1948.

Coscinodiscus excentricus

Braarud, T., 1945
Experimental studies on marine plankton diatoms. Avhandlingar utgitt av Det Norske Videnskaps-Akademi i Oslo. 1. Mat.-Naturv. Klasse 1944. No.10, 1-16, 1 text fig.

Coscinodiscus excentricus

Chiba, T., 1949
On the distribution of the plankton in the eastern China Sea and Yellow Sea. 1. Plankton composition in the spring. J. Shimonoseki Coll. Fisheries, 1(1):57-63, 1 fig.

Coscinodiscus excentricus

Chu, S. P., 1949
Experimental studies on the environmental factors influencing the growth of phytoplankton. Sci. & Tech. in China 2(3):37-52.

Coscinodiscus (Thalassiosira) excentricus

Cleve-Euler, A., 1951
Die Diatomeen von Schweden und Finnland. Kungl. Svenska Vetenskaps Akad. Handl., Fjärde Ser. 2(1): 161 pp., 6 pls.

Coscinodiscus excentricus

Cupp, Easter E., 1943
Marine plankton diatoms of the west coast of North America. Bull. S.I.O. 5(1):1-238, 5 pls., 168 text figs.

Coscinodiscus excentricus

Cupp, E.E., 1937
Seasonal distribution and occurrence of marine diatoms and dinoflagellates at Scotch Cap, Alaska. Bull. S.I.O. Tech. ser.4(3):71-100, 7 textfigs.

Coscinodiscus excentricus

Dangeard, P., 1927
Phytoplankton de la croisière du "Sylvana". Ann. Inst. Ocean., Monaco, n.s., 4(8):286-401, 54 text figs. (Feirer-Juin 1913).

Coscinodiscus excentricus

Ercegovic, A., 1936
Etudes qualitative et quantitatives du phytoplancton dans les eaux cotières de l'Adriatique oriental moyen au cours de l'année 1934. Acta Adriatica 1(9):1-126

Coscinodiscus excentricus

Florin, M-B., 1948
9. Diatomeae in submarine cores from the Tyrrhenian Sea. Medd. Ocean. Inst., Göteborg, 15 (Göteborgs Kungl. Vetenskaps-och Viterrhets Samhälles Handlingar, Sjätte Poljden, Ser. B 5(13):80-88.

Coscinodiscus excentricus

Forti, A., 1922
Ricerche sulla flora pelagica (fitoplancton) di Quarto dei Mille. Mem. R. Com. Talass. Ital. 97:248 pp., 13 pls.

Coscinodiscus excentricus

Frenguelli, Joaquin, and Hector Antonio Orlando, 1959.
Operacion MERLUZA. Diatomeas y silico-flagelados del plancton del "VI Crucero". Servicio Hidrogr. Naval., Argentina, Publ. No. H. 619: 5-62.

Coscinodiscus excentricus

Gaarder, Karen Ringdal, and Grethe Rytter Hasle, 1961(1962).
On the assumed symbiosis between diatoms and coccolithophorids in Branneckella. Nytt Mag. Botanikk, 9:145-149.

Coscinodiscus excentricus

Gran, H.H., 1908
Diatomeen. Nordisches Plankton, Botanischer Teil pp. XIX.1-XIX.146; 178 text figs.

Coscinodiscus excentricus

Gran, H.H. and T. Braarud, 1935
A quantitative study of the phytoplancton in the Bay of Fundy and the Gulf of Maine (including observations on hydrography, chemistry, and turbidity). J. Biol. Bd., Canada, 1(5):279-467, 69 text figs.

Coscinodiscus excentricus

Gran, H. H. and E. C. Angst, 1931
Plankton diatoms of Puget Sound. Publ. Puget Sound Biol. Sta. 7:417-519, 95 text figs.

Coscinodiscus excentricus

Harvey, H.W., 1947
Manganese and the growth of phytoplankton. JMBA 26(4): 562-579, 2 text figs.

Hendy, N. Ingram, 1964
An introductory account of the smaller algae of British coastal waters. V. Bacillariophyceae (Diatoms).
Her Majesty's Stationary Office, 317 pp., 45 pls.

Hendey, N-Ingram, 1958 [1957(Publ. 1958)]
Marine diatoms from some West African Ports. J. R.Microsc. Soc. (3) 77(1/2): 28-85.

Hendey, N.I., 1937
The plankton diatoms of the southern seas. Discovery Repts. 16:151-364, pls.6-13.

Hustedt, F. and A.A. Aleem, 1951
Littoral diatoms from the Salstone near Plymouth. JMBA 30(1): 177-196.

Jenkin, P.M., 1937
Oxygen production by the diatom Coscinodiscus excentricus Ehr. in relation to submarine illumination in the English Channel. J.M.B.A. (ns) 22:301-342.

Jørgensen, E., 1905
B.Protistplankton and the diatoms in bottom samples. Hydrographical and biological investigations in Norwegian fiords. Bergens Mus. Skr. 7: 49-225.

Jorgensen, E., 1900
Protophyten und Protozoën im Plankton aus der Norwegischen Westkerste. Bergens Mus. Aarb. 1899(6): 95 pp., 5 pls., 83 tables.

Kokubo, S., 1952
Results of the observations on the plankton and oceanography of Mutsu Bay during 1950, reference being made also to the period 1946-1950. Bull. Mar.Biol.Sta., Asamushi 5(1/4): 1-54, 3 tables,(fold-in), 1 fold-in.

Lillick, L.C., 1940
Phytoplankton and planktonic protozoa of the offshore waters of the Gulf of Maine. Pt.II. Qualitative Composition of the Planktonic Flora. Trans. Am. Phil. Soc., n.s., 31(3):193-237, 13 text figs.

Lillick, L.C., 1938
Preliminary report of the phytoplankton of the Gulf of Maine. Am. Mid. Nat. 20(3):624-640, 1 text fig. 37 tables.

Lillick, L.C., 1937
Seasonal studies of the phytoplankton off Woods Hole, Massachusetts. Biol. Bull. LXXIII (3):488-503, 3 text figs.

Macdonald, R., 1933
An examination of plankton hauls made in the Suez Canal during the year 1928. Fish. Res. Dis., Notes & Mem. No.3, 11 pp., 1 chart.

Manguin, E., 1954
Diatomees marines provenant de l'ile Heard (Australian National Antarctic Expedition). Rev. Algol., n.s., 1: 14-24.

Mann, A., 1907
Report on the diatoms of the Albatross voyages in the Pacific Ocean, 1888-1904. Contrib. U. S. Nat. Herb. 10(5):221-419, Pls. XLIV-LIV.

Mann, A., 1893
List of Diatomaceae from a deep-sea dredging in the Atlantic Ocean off Delaware Bay by the U. S. Fish Commission Steamer Albatross. Proc. U. S. Nat. Mus. 16:303-312.

Meunier, A., 1915
Microplancton de la Mer Flamande. 2. Diatomées (excepté le genre Chaetoceros). Mem. Mus. Roy. Hist. Nat., Belgique, 7(3):1-118, Pls. VIII-XIV.

Morse, D.C., 1947
Some observations on seasonal variations in plankton population Patuxant River, Maryland 1943-1945. Bd. Nat. Res., Publ. No.65, Chesapeake Biol. Lab., 31, 3 figs.

Paiva Carvalho, J., 1950
O plancton do Rio Maria Rodriques (Cananeis). 1. Diatomaceas e Dinoflagelados. Bol. Inst. Paulista Oceanogr. 1(1); 27-43, 2 fold-in tables, 2 figs.

Pavillard, J., 1925
Bacillariales. Rept. on the Danish Oceangr. Exped., 1908-10 to the Mediterranean and adj. seas. Vol.II., Biol. J4:72 pp., 116 text figs.

Pavillard, J., 1905
Recherches sur la flore pelagique (Phytoplankton) de l'Etang de Thau. Theses presentees a la Fac. Sci., Paris, 116 pp., 3 pls.

Phifer, L.D. (undated)
The occurrence and distribution of plankton diatoms in Bering Sea and Bering Strait, July 26-August 24, 1934. Report of Oceanographic cruise of U.S. Coast Guard Cutter Chelan 1934, Part II(A):1-44 (mimeographed) plus fig.1 (after Pt.B)

Coscinodiscus eccentricus

Pugh, P.R., 1971.
Changes in the fatty acid composition of Coscinodiscus eccentricus with culture-age and salinity. Mar. Biol. 11(2): 118-124.

Coscinodiscus excentricus

Rampi, L., 1942
Ricerche sul fitoplancton del Mare Ligure 6. Le diatomee delle acque di Sanremo. Nuovo Giornale Botanico Italiano, N.S., 49:252-268.

Rampi, L., 1940
Diatomee del Mare Adriatico. Nuovo Giornale Botanico Italiano, n.s., 47:559-608.

Rumkówna, A., 1948
List of the phytoplankton species occurring in the superficial water layers in the Gulf of Gdańsk. Bull. Lab. mar., Gdynia, No. 4: 139-141 with tables in back.

Schröder, B., 1900
Phytoplankton des Golfes von Neapel nebst vergleichenden Ausblicken auf das atlantischen Ozean. Mitt. Zool. Stat. Neapel, 14:1-38.

Coscinodiscus eccentricus [a]

Somers, D., 1972
Scanning electron microscope studies on some species of the centric diatom genera Thalassiosira and Coscinodiscus. Biol. Jaarb. Dodonaea 40:317-322. Also in: Coll. Repr. Inst. Zeewetenschap. Onderzoek Belg. 3(1973).

Coscinodiscus excentricus

Sousa e Silva, E., 1949
Diatomaceas e Dinoflagelados de Baía de Cascais. Portugaliae Acta Biol., Volume: Julio Henriques, Ser. B: 300-383, 9 pls, 2 fold-in tables.

Sousa a Silva, E., and J. Dos Santos-Pinto, 1948
O Plancton da Baía de S. Martinho do Porto. 1. Diatomaceas e Dinoflagelados. Bol. Soc. Portuguese de Ciencias Naturais, 16(2):134-187, 6 pls. (Trav. Sta. Biol. Mar. de Lisbonne No. 52).

Zanon, D. V., 1949
Diatomee di Buenos Aires (Argentina) Atti Accad. Naz. Lincei, Memorie, Cl. Sci. fis., mat. e. nat., ser. 7, 11(3):59-151, 2 pls.

Zanon, V., 1948
Diatomee marini di Sardegna e Pugillo di Alghe Marine della stressa. Boll. Pesca, Piscitutura e Idrobiologia, Anno 24, ns. 3(2): 202-244, 27 figs. on 1 pl.

Coscinodiscus excentricus var. catenata

Gran, H.H., 1897
Protophyta: Diatomaceae, Silico-flagellata and Cilioflagellata. Den Norske Nordhavs Expedition 1876-1878, h. 24, 36 pp., 4 pls.

Coscinodiscus excentricus var. minor

Ghazzawi, F.M., 1939
Plankton of the Egyptian waters. A study of the Suez Canal Plankton. (A) Phytoplankton. Preliminary Report 83 pp. Notes and Memoires, Min. Commerce-Industry, Egypt, Hydrobiol. & Fish. 65 figs.

Coscinodiscus (Thalassiosira) fallax

Cleve-Euler, A., 1951
Die Diatomeen von Schweden und Finnland. Kungl. Svenska Vetenskaps Akad. Handl., Fjärde Ser. 2(1): 161 pp., 6 pls.

Coscinodiscus fimbriatus

Schröder, B., 1900
Phytoplankton des Golfes von Neapel nebst vergleichenden Ausblicken auf das atlantischen Ozean. Mitt. Zool. Stat. Neapel, 14:1-38.

Coscinodiscus finicus n.sp.

Misra, J.N., 1956.
A systematic account of some littoral marine diatoms from the west coast of India. J. Bombay Nat. Hist. Soc., 53(4):537-568.

Coscinodiscus (Thalassiosira) fluviatilis

Cleve-Euler, A., 1951
Die Diatomeen von Schweden und Finnland. Kungl. Svenska Vetenskaps Akad. Handl., Fjärde Ser. 2(1): 161 pp., 6 pls.

Coscinodiscus (Thalassiosira) frigidus

Cleve-Euler, A., 1951
Die Diatomeen von Schweden und Finnland. Kungl. Svenska Vetenskaps Akad. Handl., Fjärde Ser. 2(1): 161 pp., 6 pls.

Coscinodiscus gigas
Rampi, L., 1942
Ricerche sul fitoplancton del Mare Ligure 6. Le diatomee delle acque di Sanremo. Nuovo Giornale Botanico Italiano, N.S., 49:252-268.

Coscinodiscus furcatus
Jouse, A.P., G.S. Koroleva, G.A. Nagaeva, 1962
Diatoms in the surface layer of sediment in the Indian sector of the Antarctic. Investigations of marine bottom sediments. (In Russian; English summary). Trudy Inst. Okeanol., Akad. Nauk. SSSR, 61:19-92.

Coscinodiscus galapagensis
Mann, A., 1907
Report on the diatoms of the Albatross voyages in the Pacific Ocean, 1888-1904. Contrib. U. S. Nat. Herb. 10(5):221-419, Pls. XLIV-LIV.

Coscinodiscus gemmatulus
Castracane degli Antelminelli, F., 1886 n.sp.
1. Report on the Diatomaceae collected by H.M.S. Challenger during the years 1873-1876. Rept. Sci. Results, H.M.S. Challenger, Botany Vol. II, 178 pp., 30 pls.

Coscinodiscus gigas
Boden, B.P., 1950
Some marine plankton diatoms from the west coast of South Africa. Trans. R.Soc. S. Africa. 32:321-434, 100 text figs.

Coscinodiscus gigas
Delsman, H. C., 1939
Preliminary plankton investigations in the Java Sea. Treubia, 17:139-181, 41 figs., 8 maps.

Coscinodiscus gigas
Fox, M., 1957.
A first list of marine algae from Nigeria. J. Linn. Soc., London, 55(362):615-631.

Coscinodiscus gigas
Frenguelli, Joaquin, and Hector Antonio Orlando, 1959.
Operacion MERLUZA. Diatomeas y silico-flagelados del plancton del "VI Crucero". Servicio Hidrogr. Naval., Argentina, Publ. No. H. 619: 5-62.

Coscinodiscus gigas
Ghazzawi, F.M., 1939
Plankton of the Egyptian waters. A study of the Suez Canal Plankton. (A) Phytoplankton. Preliminary Report 83 pp. Notes and Memoires. Min. Commerce-Industry, Egypt, Hydrobiol. & Fish. 65 figs.

Coscinodiscus gigas
Hendey, N.I., 1937
The plankton diatoms of the southern seas. Discovery Repts. 16:151-364, pls.6-13.

Coscinodiscus gigas
Proshkina-Lavrenko, A.I., 1961.
Variability of some Black Sea diatoms. Botan. Zhurn., Akad. Nauk, SSSR, 46(12):1794-1797.

Coscinodiscus gigas
Zanon, D. V., 1949
Diatomee di Buenos Aires (Argentina) Atti Accad. Naz. Lincei, Memorie, Cl. Sci. fis., mat. e. nat., ser. 7, 11(3):59-151, 2 pls.

Coscinodiscus gigas var. praetexta
Allen, W.E., and E.E. Cupp, 1935
Plankton diatoms of the Java Sea. Annales du Jardin Botanique de Buitenzorg XLIV (2):101-174, figs.1-127.
(drawings of all species mentioned)

Coscinodiscus gracilis
Boden, B. P., 1949.
The diatoms collected by the U.S.S. CACOPAN in the Antarctic in 1947. J. Mar. Res. 8(1):6-13, 3 textfigs.

Coscinodiscus gracilis
Boden, Brian, 1948
Marine plankton diatoms on operation HIGHJUMP in: Some oceanographic observations on operation HIGHJUMP. By R.S. Dietz. USNEL Rept. No.55, 97 pp., 41 figs. 7 July 1948.

Coscinodiscus gracilis
Hendey, N.I., 1937
The plankton diatoms of the southern seas. Discovery Repts. 16:151-364, pls.6-13.

Coscinodiscus grandenucleatus
Hendey, N.I., 1937
The plankton diatoms of the southern seas. Discovery Repts. 16:151-364, pls.6-13.

coscinodiscus granii
Boalch, G.T., 1971.
The typification of the diatom species Coscindiscus concinnus WM.Smith and Coscinodiscus granii Gough. J. mar. biol. Ass. U.K. 51(3): 685-695.

Coscinodiscus granii
Boden, B.P., 1950
Some marine plankton diatoms from the west coast of South Africa. Trans. R.Soc. S. Africa. 32:321-434, 100 text figs.

Coscinodiscus granii
Cleve-Euler, A., 1951
Die Diatomeen von Schweden und Finnland. Kungl. Svenska Vetenskaps Akad. Handl., Fjärde Ser. 2(1): 161 pp., 6 pls.

Coscinodiscus granii
Cupp, Easter E., 1943
Marine plankton diatoms of the west coast of North America. Bull. S.I.O. 5(1):1-238, 5 pls., 168 text figs.

Coscinodiscus granii
Drebes, Gerhard 1966.
Ein parasilischer Phycomycet (Lagenidiales) in Coscinodiscus. Helgoländer wiss. Meeresunters. 13(4), 426-435.

Coscinodiscus granii
Follman, Gerhard, 1958.
Plasmolyse-Verhalten und Vitalfaerbungs-Eigenschaften von Coscinodiscus granii. Gough. Bull. Inst. Océan., Monaco, No. 1116:22 pp.

Coscinodiscus Granii
Ghazzawi, F.M., 1939
Plankton of the Egyptian waters. A study of the Suez Canal Plankton. (A) Phytoplankton. Preliminary Report 83 pp. Notes and Memoires, Min. Commerce-Industry, Egypt, Hydrobiol. & Fish. 65 figs.

Coscinodiscus Granii
Gran, H.H., 1908
Diatomeen. Nordisches Plankton, Botanischer Teil pp. XIX.1-XIX 146; 178 text figs.

Coscinodiscus granii
Gran, H. H. and E. C. Angst, 1931
Plankton diatoms of Puget Sound. Publ. Puget Sound Biol. Sta. 7:417-519, 95 text figs.

Coscinodiscus grani
Hendy, N. Ingram, 1964
An introductory account of the smaller algae of British coastal waters. V. Bacillariophyceae (Diatoms). Her Majesty's Stationary Office, 317 pp., 45 pls.

Coscinodiscus grani
Hendey, N.I., 1937
The plankton diatoms of the southern seas. Discovery Repts. 16:151-364, pls.6-13.

Coscinodiscus granii
Iyengar, M.O.P. and G.Venkataraman,1951.
The ecology and seasonal succession of the algae flora of the River Cooum at Madras with special reference to the Diatomaceae. J. Madras Univ. 21, Sect. B(1): 140-192, 1 pl of 4 figs., 11 text figs.

Coscinodiscus granii
Kokubo, S., 1952
Results of the observations on the plankton and oceanography of Mutsu Bay during 1950, reference being made also to the period 1946-1950. Bull Mar.Biol.Sta., Asamushi 5(1/4): 1-54, 3 tables, (fold-in), 1 fold-in.

Coscinodiscus granii
Mangin, M. L., 1912
Phytoplancton de la croisière du "René" dans l'Atlantique (Septembre 1908). Ann. Inst. Ocean., n.s., 4(1):1-66, 2 pls., 41 text figs., 2 tables.

Coscinodiscus granii
Marukawa, H., 1921
Plankton lists and some new species of copepods, from the northern waters of Japan. Bull. Inst. Ocean., No.384, 15 pp., 3 pls., 1 chart. Monaco

Coscinodiscus granii
Meunier, A., 1915
Microplancton de la Mer Flamande. 2. Diatomées (excepté le genre Chaetoceros). Mem. Mus. Roy. Hist. Nat., Belgique, 7(3):1-118, Pls. VIII-XIV.

Coscinodiscus granii
Paiva Carvalho, J., 1950
O plancton do Rio Maria Rodrigues (Cananeis). 1. Diatomaceas e Dinoflagelados. Bol. Inst. Paulista Oceanogr. 1(1); 27-43, 2 fold-in tables, 2 figs.

Coscinodiscus Granii
Rumkówna, A., 1948
List of the phytoplankton species occurring in the superficial water layers in the Gulf of Gdańsk. Bull. Lab. mar., Gdynia, No. 4: 139-141 with tables in back.

Coscinodiscus granii
Schodduyn, M., 1926
Observations faites dans la baie d'Ambleteuse (Pas de Calais). Bull. Inst. Ocean., Monaco, No. 482: 64 pp.

Coscinodiscus Granii
Sousa a Silva, E., and J. Dos Santos-Pinto, 1948
O Plancton da Baia de S. Martinho do Porto. 1. Diatomaceas e Dinoflagelados. Bol. Soc. Portuguesa de Ciencias Naturais, 16(2):134-187, 6 pls. (Trav. Sta. Biol. Mar. de Lisbonne No. 52).

Coscinodiscus granulosus
Cleve-Euler, A., 1951
Die Diatomeen von Schweden und Finnland. Kungl. Svenska Vetenskaps Akad. Handl., Fjärde Ser. 2(1): 161 pp., 6 pls.

Coscinodiscus granulosus
Schröder, B., 1900
Phytoplankton des Golfes von Neapel nebst vergleichenden Ausblicken auf das atlantischen Ozean. Mitt. Zool. Stat. Neapel, 14:1-38.

Coscinodiscus (Thalassiosira) gravidus

Cleve-Euler, A., 1951
Die Diatomeen von Schweden und
Finnland. Kungl. Svenska Vetenskaps
Akad. Handl., Fjärde Ser. 2(1): 161 pp.,
6 pls.

Coscinodiscus griseus

Cleve-Euler, A., 1951
Die Diatomeen von Schweden und
Finnland. Kungl. Svenska Vetenskaps
Akad. Handl., Fjärde Ser. 2(1): 161 pp.,
6 pls.

Coscinodiscus gyratus

Jouse, A.P., G.S. Koroleva, G.A. Nagaeva,
1962
Diatoms in the surface layer of sediment
in the Indian sector of the Antarctic.
Investigations of marine bottom sediments.
(In Russian; English summary).
Trudy Inst. Okeanol., Akad. Nauk. SSSR.
61:19-92.

Coscinodiscus hartingii

Cleve-Euler, A., 1951
Die Diatomeen von Schweden und
Finnland. Kungl. Svenska Vetenskaps
Akad. Handl., Fjärde Ser. 2(1): 161 pp.,
6 pls.

Coscinodiscus hauckii

Cleve-Euler, A., 1951
Die Diatomeen von Schweden und
Finnland. Kungl. Svenska Vetenskaps
Akad. Handl., Fjärde Ser. 2(1): 161 pp.,
6 pls.

Coscinodiscus heteroporus

Cleve-Euler, A., 1951
Die Diatomeen von Schweden und
Finnland. Kungl. Svenska Vetenskaps
Akad. Handl., Fjärde Ser. 2(1): 161 pp.,
6 pls.

Coscinodiscus heteroporus

Mann, A., 1907
Report on the diatoms of the Albatross
voyages in the Pacific Ocean, 1888-1904.
Contrib. U. S. Nat. Herb. 10(5):221-419, Pls.
XLIV-LIV.

Coscinodiscus hexagonalis

Hendey, N.I., 1937
The plankton diatoms of the southern seas.
Discovery Repts. 16:151-364, pls.6-13.

Coscinodiscus hustedtii

Hendey, N-Ingram, 1958 [1957(Publ. 1958)]
Marine diatoms from some West African
Ports- J. R. Microsc. Soc. (3) 77(1/2):
28-85.

Coscinodiscus Hustedti n.sp.

Müller Melchers, F.C., 1953.
New and little known diatoms from Uruguay and
the South Atlantic coast.
Com. Bot., Mus. Hist. Nat., Montevideo, 3(30):
1-11, 8 pls.

Coscinodiscus (Thalassiosira) hyalinus

Cleve-Euler, A., 1951
Die Diatomeen von Schweden und
Finnland. Kungl. Svenska Vetenskaps
Akad. Handl., Fjärde Ser. 2(1): 161 pp.,
6 pls.

Coscinodiscus Imberti n.sp.

Manguin, E., 1957.
Premier inventaire des Diatomées de la Terre
Adélie Antarctique. Espèces nouvelles.
Rev. Algologique, n.s., 3(3):111-134.

Coscinodiscus incurvus

Hendey, N.I., 1937
The plankton diatoms of the southern seas.
Discovery Repts. 16:151-364, pls.6-13.

Coscinodiscus inflatus

Hendey, N.I., 1937
The plankton diatoms of the southern seas.
Discovery Repts. 16:151-364, pls.6-13.

Coscinodiscus inflatus

Jouse, A.P., G.S. Koroleva, G.A. Nagaeva,
1962
Diatoms in the surface layer of sediment
in the Indian sector of the Antarctic.
Investigations of marine bottom sediments.
(In Russian; English summary).
Trudy Inst. Okeanol., Akad. Nauk. SSSR.
61:19-92.

Coscinodiscus inflatus

Mangin, L., 1915
Phytoplancton de L'Antartique. Deuxieme
Exped. Ant. Francaise (1908-1910), 95 pp., 3 pls.,
58 text figs.

Coscinodiscus intermittens

Hendey, N.I., 1937
The plankton diatoms of the southern seas.
Discovery Repts. 16:151-364, pls.6-13.

Coscinodiscus Janischii var. arafurensie

Allen, W.E., and E.E. Cupp, 1935
Plankton diatoms of the Java Sea.
Annales du Jardin Botanique de Buitenzorg
XLIV (2):101-174, figs.1-127.

(drawings of all species mentioned)

Coscinodiscus janischii

Boden, B.P., 1950
Some marine plankton diatoms from the
west coast of South Africa. Trans. R.Soc.
S. Africa. 32:321-434, 100 text figs.

Coscinodiscus janischii

Nakai, Z., 1955.
The chemical composition, volume weight, and size
of the important marine plankton.
Tokai Reg. Fish. Res. Lab., Spec. Publ. 5:12-24.

Coscinodiscus Janischi

Zanon, D. V., 1949
Diatomee di Buenos Aires (Argentina)
Atti Accad. Naz. Lincei, Memorie, Cl. Sci.
fis., mat. e. nat., ser. 7, 11(3):59-151,
2 pls.

Coscinodiscus (?) Janus n.sp.

Castracane degli Antelminelli, F., 1886
1. Report on the Diatomaceae collected
by H.M.S. Challenger during the years 1873-
1876. Rept. Sci. Results, H.M.S. Challenger,
Botany Vol. II, 178 pp., 30 pls.

Coscinodiscus jonesianus

Alfimov, N.N., 1966
On the biology and bio chemistry of two mass
marine diatoms, Coscinodiscus jonesianus
(Grev.) Ostf. and Rhizosolenia calcar-avis
M. Schultze from the Azov and Caspain seas.
(In Russian)
Bot. Zh., 51(9): 1276-1283.

Coscinodiscus Jonesianus

Allen, W.E., and E.E. Cupp, 1935
Plankton diatoms of the Java Sea.
Annales du Jardin Botanique de Buitenzorg
XLIV (2):101-174, figs.1-127.

(drawings of all species mentioned)

Coscinodiscus jonesianus

Delsman, H. C., 1939.
Preliminary plankton investigations in the Java
Sea. Treubia, 17:139-181, 8 maps, 41 figs.

Coscinodiscus jonesianus

Fox, M., 1957.
A first list of marine algae from Nigeria.
J. Linn. Soc., London., 55,(362):615-631.

Coscinodiscus jonesianus

Hendy, N. Ingram, 1964
An introductory account of the smaller algae
of British coastal waters. V. Bacillariophyceae (Diatoms).
Her Majesty's Stationary Office, 317 pp.,
45 pls.

Coscinodiscus jonesianus

Hendey, N-Ingram, 1958 [1957(Publ. 1958)]
Marine diatoms from some West African
Ports- J. R. Microsc. Soc. (3) 77(1/2):
28-85.

Coscinodiscus Jonesianus var. commutata

Allen, W.E., and E.E. Cupp, 1935
Plankton diatoms of the Java Sea.
Annales du Jardin Botanique de Buitenzorg
XLIV (2):101-174, figs.1-127.

(drawings of all species mentioned)

Coscinodiscus kerguelensis

Hendey, N.I., 1937
The plankton diatoms of the southern seas.
Discovery Repts. 16:151-364, pls.6-13.

Coscinodiscus kerguelensis

Mangin, L., 1915
Phytoplancton de L'Antartique. Deuxieme
Exped. Ant. Francaise (1908-1910), 95 pp., 3 pls.,
58 text figs.

Coscinodiscus kolbei n.sp.

Jouse, A.P., G.S. Koroleva, G.A. Nagaeva,
1962
Diatoms in the surface layer of sediment
in the Indian sector of the Antarctic.
Investigations of marine bottom sediments.
(In Russian; English summary).
Trudy Inst. Okeanol., Akad. Nauk. SSSR.
61:19-92.

Coscinodiscus (Thalassiosira) kryophilus

Cleve-Euler, A., 1951
Die Diatomeen von Schweden und
Finnland. Kungl. Svenska Vetenskaps
Akad. Handl., Fjärde Ser. 2(1): 161 pp.,
6 pls.

Coscinodiscus Kryophilus

Hendey, N.I., 1937
The plankton diatoms of the southern seas.
Discovery Repts. 16:151-364, pls.6-13.

Coscinodiscus kurzii

Hendey, N-Ingram, 1958 [1957(Publ. 1958)]
Marine diatoms from some West African
Ports- J. R. Microsc. Soc. (3) 77(1/2):
28-85.

Coscinodiscus Kutzingii

Bigelow, H.B., and M. Leslie, 1930
Reconnaissance of the waters and
plankton of Monterey Bay, July 1928.
Bull. M.C.Z., 70(5):429-481, 43 text
figs.

Coscinodiscus kütsingii

Cleve-Euler, A., 1951
Die Diatomeen von Schweden und
Finnland. Kungl. Svenska Vetenskaps
Akad. Handl., Fjärde Ser. 2(1): 161 pp.,
6 pls.

Coscinodiscus Kützingii

Gran, H.H., 1908
Diatomeen. Nordisches Plankton, Botanischer Teil pp. XIX.1-XIX 146; 178 text figs.

Coscinodiscus kutzingii

Hendy, N. Ingram, 1964
An introductory account of the smaller algae of British coastal waters. V. Bacillariophyceae (Diatoms).
Her Majesty's Stationary Office, 317 pp., 45 pls.

Coscinodiscus kuetzingii

Hendey, N. Ingram, 1958 [1957(Publ. 1958)]
Marine diatoms from some West African Ports. J. R. Microsc. Soc. (3) 77(1/2): 28-85.

Coscinodiscus kutzingi

Somers, D., 1972
Scanning electron microscope studies on some species of the centric diatom genera Thalassiosira and Coscinodiscus. Biol. Jaarb. Dodonaea 40:317-322. Also in: Coll. Repr. Inst. Zeewetenschap. Onderzoek Belg. 3(1973).

Ditylum brightwelli

Mullin, Michael M. and Elaine R. Brooks, 1970
Growth and metabolism of two planktonic, marine copepods as influenced by temperature and type of food.
In: Marine Food Chains, J.H. Steele, editor, Oliver and Boyd, 74-95.

Coscinodiscus lacustris

Cleve-Euler, A., 1951
Die Diatomeen von Schweden und Finnland. Kungl. Svenska Vetenskaps Akad. Handl., Fjärde Ser. 2(1): 161 pp., 6 pls.

Coscinodiscus lacustris

Kokubo, S., and S. Sato., 1947
Plankters in Jū-San Gata. Physiol. and Ecol. (Japan) 1(4):1-16, 3 text figs., tables.

Coscinodiscus lacustris

Meunier, A., 1915
Microplancton de la Mer Flamande. 2. Diatomées (excepté le genre Chaetoceros). Mem. Mus. Roy. Hist. Nat., Belgique, 7(3):1-118, Pls. VIII-XIV.

Coscinodiscus lanceolatus n.sp.

Castracane degli Antelminelli, F., 1886
1. Report on the Diatomaceae collected by H.M.S. Challenger during the years 1873-1876. Rept. Sci. Results, H.M.S. Challenger, Botany Vol. II, 178 pp., 30 pls.

Coscinodiscus lentiginosus

Boden, B. P., 1949.
The diatoms collected by the U.S.S. CACOPAN in the Antarctic in 1947. J. Mar. Res. 8(1):6-13, 3 textfigs.

Coscinodiscus lentiginosus

Boden, Brian, 1948
Marine plankton diatoms on operation HIGHJUMP in: Some oceanographic observations on operation HIGHJUMP. By R.S. Dietz. USNEL Rept. No.55, 97 pp., 41 figs. 7 July 1948.

Coscinodiscus lentiginosus

Castracane degli Antelminelli, F., 1886
1. Report on the Diatomaceae collected by H.M.S. Challenger during the years 1873-1876. Rept. Sci. Results, H.M.S. Challenger, Botany Vol. II, 178 pp., 30 pls.

Coscinodiscus lentigenosus

Hendey, N.I., 1937
The plankton diatoms of the southern seas. Discovery Repts. 16:151-364, pls.6-13.

Coscinodiscus lentiginosus

Jouse, A.P., G.S. Koroleva, G.A. Nagaeva, 1962
Diatoms in the surface layer of sediment in the Indian sector of the Antarctic. Investigations of marine bottom sediments. (In Russian; English summary).
Trudy Inst. Okeanol., Akad. Nauk, SSSR, 61: 19-92.

Coscinodiscus lentiginosus

Manguin, E., 1954
Diatomees marines provenant de l'île Heard (Australian National Antarctic Expedition). Rev. Algol., n.s., 1: 14-24.

Coscinodiscus lentiginosus

Mann, A., 1907
Report on the diatoms of the Albatross voyages in the Pacific Ocean, 1888-1904. Contrib. U. S. Nat. Herb. 10(5):221-419, Pls. XLIV-LIV.

Coscinodiscus leptopus

Forti, A., 1922
Ricerche sulla flora pelagica (fitoplancton) di Quarto dei Mille. Mem. R. Com. Talass. Ital. 97:248 pp., 13 pls.

Coscinodiscus leptopus

Pavillard, J., 1925
Bacillariales. Rept. on the Danish Oceangr. Exped., 1908-10 to the Mediterranean and adj. seas. Vol.II., Biol. J4:72 pp., 116 text figs.

Coscinodiscus (Thalassiosira) levanderi

Cleve-Euler, A., 1951
Die Diatomeen von Schweden und Finnland. Kungl. Svenska Vetenskaps Akad. Handl., Fjärde Ser. 2(1): 161 pp., 6 pls.

Coscinodiscus lineatus

Allen, W.E., and E.E. Cupp, 1935
Plankton diatoms of the Java Sea. Annales du Jardin Botanique de Buitenzorg XLIV (2):101-174, figs.1-127.
(drawings of all species mentioned)

Coscinodiscus lineatus

Bigelow, H.B., and M. Leslie, 1930
Reconnaissance of the waters and plankton of Monterey Bay, July 1928. Bull. M.C.Z., 70(5):429-481, 43 text figs.

Coscinodiscus lineatus

Boden, B.P., 1950
Some marine plankton diatoms from the west coast of South Africa. Trans. R.Soc. S. Africa. 32:321-434, 100 text figs.

Coscinodiscus lineatus

Boden, B. P., 1949.
The diatoms collected by the U.S.S. CACOPAN in the Antarctic in 1947. J. Mar. Res. 8(1):6-13, 3 textfigs.

Coscinodiscus lineatus

Boden, Brian, 1948
Marine plankton diatoms on operation HIGHJUMP in: Some oceanographic observations on operation HIGHJUMP. By R.S. Dietz. USNEL Rept. No.55, 97 pp., 41 figs. 7 July 1948.

Coscinodiscus lineatus

Chiba, T., 1949
On the distribution of the plankton in the eastern China Sea and Yellow Sea. 1. Plankton composition in the spring. J. Shimonoseki Coll. Fisheries, 1(1):57-63, 1 fig.

Coscinodiscus lineatus

Cleve-Euler, A., 1951
Die Diatomeen von Schweden und Finnland. Kungl. Svenska Vetenskaps Akad. Handl., Fjärde Ser. 2(1): 161 pp., 6 pls.

Coscinodiscus lineatus

Cupp, Easter E., 1943
Marine plankton diatoms of the west coast of North America. Bull. S.I.O. 5(1):1-238, 5 pls., 168 text figs.

Coscinodiscus lineatus

Ercegovic, A., 1936
Etudes qualitative et quantitatives du phytoplancton dans les eaux cotières de l'Adriatique oriental moyen au cours de l'année 1934. Acta Adriatica 1(9):1-126

Coscinodiscus lineatus

Florin, M-B., 1948
9. Diatomeae in submarine cores from the Tyrrhenian Sea. Medd. Ocean. Inst., Göteborg, 15 (Göteborgs Kungl. Vetenskaps-och Viterrhets Samhälles Handlingar, Sjätte Foljden, Ser. B 5(13):80-88.

Coscinodiscus lineatus

Frenguelli, Joaquin, and Hector Antonio Orlando, 1959.
Operacion MERLUZA. Diatomeas y silicoflagelados del plancton del "VI Crucero". Servicio Hidrogr. Naval., Argentina, Publ. No. H. 619: 5-62.

Coscinodiscus lineatus

Gran, H.H., 1908
Diatomeen. Nordisches Plankton, Botanischer Teil pp. XIX.1-XIX 146; 178 text figs.

Coscinodiscus lineatus

Hasle, Grethe Rytter, 1960
Phytoplankton and ciliate species from the Tropical Pacific.
Skr. Norske Videnskaps-Akad., Oslo, 1. Mat.-Nat. Kl., 1960(2): 1-50.

Coscinodiscus lineatus

Hendey, N. Ingram, 1958 [1957(Publ. 1958)]
Marine diatoms from some West African Ports. J. R. Microsc. Soc. (3) 77(1/2): 28-85.

Coscinodiscus lineatus

Hendey, N.I., 1937
The plankton diatoms of the southern seas. Discovery Repts. 16:151-364, pls.6-13.

Coscinodiscus lineatus

Jørgensen, E., 1905
B. Protistplankton and the diatoms in bottom samples. Hydrographical and biological investigations in Norwegian fjords. Bergens Mus. Skr. 7: 49-225.

Coscinodiscus lineatus

Jorgensen, E., 1900
Protophyten und Protozoën im Plankton aus der Norwegischen Westkerste. Bergens Mus. Aarb. 1899(6): 95 pp., 5 pls., 83 tables.

Coscinodiscus lineatus

Kokubo, S., 1952
Results of the observations on the plankton and oceanography of Mutsu Bay during 1950, reference being made also to the period 1946-1950. Bull Mar.Biol.Sta., Asamushi 5(1/4): 1-54, 3 tables,(fold-in), 1 fold-in.

Coscinodiscus lineatus

Lillick, L.C., 1940
Phytoplankton and planktonic protozoa of the offshore waters of the Gulf of Maine. Pt.II. Qualitative Composition of the Planktonic Flora. Trans. Am. Phil. Soc., n.s., 31(3):193-237, 13 text figs.

Coscinodiscus lineatus
Lillick, L.C., 1937
Seasonal studies of the phytoplankton off Woods Hole, Massachusetts. Biol. Bull. LXXIII (3):488-503, 3 text figs.

Coscinodiscus lineatus
Mangin, L., 1915
Phytoplancton de L'Antartique. Deuxieme Exped. Ant. Francaise (1908-1910), 95 pp., 3 pls., 58 text figs.

Coscinodiscus lineatus
Mangin, M. L., 1912
Phytoplancton de la croisière du "René" dans l'Atlantique (Septembre 1908). Ann. Inst. Ocean., n.s., 4(1):1-66, 2 pls., 41 text figs., 2 tables.

Coscinodiscus lineatus
Manguin, E., 1954
Diatomees marines provenant de l'ile Heard (Australian National Antarctic Expedition). Rev. Algol., n.s., 1: 14-24.

Coscinodiscus lineatus
Mann, A., 1907
Report on the diatoms of the Albatross voyages in the Pacific Ocean, 1888-1904. Contrib. U. S. Nat. Herb. 10(5):221-419, Pls. XLIV-LIV.

Coscinodiscus lineatus
Mann, A., 1893
List of Diatomaceae from a deep-sea dredging in the Atlantic Ocean off Delaware Bay by the U. S. Fish Commission Steamer Albatross. Proc. U. S. Nat. Mus. 16:303-312.

Coscinodiscus lineatus
Politis, J., 1949
Diatomees marines de Bosphores et des ibes de la mer de Marmara. II Practica tou Hellenikou Hidrobiologikou Institutoutou 1929, Etoz 1929, 3(1):11-31.

Coscinodiscus lineatus
Rampi, L., 1942
Ricerche sul fitoplancton del Mare Ligure 6. Le diatomee delle acque di Sanremo. Nuovo Giornale Botanico Italiano, N.S., 49:252-268.

Coscinodiscus lineatus
Rumkówna, A., 1948
List of the phytoplankton species occurring in the superficial water layers in the Gulf of Gdańsk. Bull. Lab. mar., Gdynia, No. 4: 139-141 with tables in back.

Coscinodiscus lineatus
Zanon, D. V., 1949
Diatomee di Buenos Aires (Argentina) Atti Accad. Naz. Lincei, Memorie, Cl. Sci. fis., mat. e. nat., ser. 7, 11(3):59-151, 2 pls.

Coscinodiscus lineatus
Zanon, V., 1948
Diatomee marini di Sardegna e Pugillo di Alghe Marine della stressa. Boll. Pesca, Piscitutura e Idrobiologia, Anno 24, ns. 3(2): 202-244, 27 figs. on 1 pl.

Coscinodiscus margaritaceus
Castracane degli Antelminelli, F., 1886 n.sp.
1. Report on the Diatomaceae collected by H.M.S. Challenger during the years 1873-1876. Rept. Sci. Results, H.M.S. Challenger, Botany Vol. II, 178 pp., 30 pls.

Coscinodiscus magaritaceus
Jouse, A.P., G.S. Koroleva, G.A. Nagaeva, 1962
Diatoms in the surface layer of sediment in the Indian sector of the Antarctic. Investigations of marine bottom sediments. (In Russian; English summary).
Trudy Inst. Okeanol., Akad. Nauk, SSSR, 61: 19-92.

Coscinodiscus margaritaceus
Manguin, E., 1954
Diatomees marines provenant de l'ile Heard (Australian National Antarctic Expedition). Rev. Algol., n.s., 1: 14-24.

Coscinodiscus margaritae, n.sp.
Frenguelli, Joaquin, y Hector A. Orlando, 1958.
Diatomeas y silicoflagelados del sector Antartico Sudamericano.
Inst. Antartico Argentino, Publ., No. 5:191 pp.

Coscinodiscus marginatus
Allen, W.E., 1936.
Surface plankton diatoms in the North Pacific Ocean in 1934. Madroño 3(6):3 pp.

Coscinodiscus marginatus
Allen, W.E., and E.E. Cupp, 1935
Plankton diatoms of the Java Sea. Annales du Jardin Botanique de Buitenzorg XLIV (2):101-174, figs.1-127.
(drawings of all species mentioned)

Coscinodiscus marginatus
Brunel, J., 1962
Le phytoplancton de la Baie de Chaleurs. Inst. Botan., Univ. Montréal, Contrib. No. 77: 365 pp., 66 pls.

Coscinodiscus marginatus
Cleve-Euler, A., 1951
Die Diatomeen von Schweden und Finnland. Kungl. Svenska Vetenskaps Akad. Handl., Fjärde Ser. 2(1): 161 pp., 6 pls.

Coscinodiscus marginatus
Cupp, Easter E., 1943
Marine plankton diatoms of the west coast of North America. Bull. S.I.O. 5(1):1-238, 5 pls., 168 text figs.

Coscinodiscus marginatus
Florin, M-B., 1948
9. Diatomeae in submarine cores from the Tyrrhenian Sea. Medd. Ocean. Inst., Göteborg, 15 (Göteborgs Kungl. Vetenskaps-och Viterrhets Samhälles Handlingar, Sjätte Foljden, Ser. B 5(13):80-88.

Coscinodiscus marginatus
Gran, H.H., 1908
Diatomeen. Nordisches Plankton, Botanischer Teil pp. XIX.1-XIX 146; 178 text figs.

Coscinodiscus marginatus
Hendy, N. Ingram, 1964
An introductory account of the smaller algae of British coastal waters. V. Bacillariophyceae (Diatoms).
Her Majesty's Stationary Office, 317 pp., 45 pls.

Coscinodiscus marginatus
Hendey, N.I., 1937
The plankton diatoms of the southern seas. Discovery Repts. 16:151-364, pls.6-13.

Coscinodiscus marginatus
Kokubo, S., 1952
Results of the observations on the plankton and oceanography of Mutsu Bay during 1950, reference being made also to the period 1946-1950. Bull Mar.Biol.Sta., Asamushi 5(1/4):1-54, 3 tables,(fold-in), 1 fold-in.

Coscinodiscus marginatus
Kokubo, S., and S. Sato., 1947
Plankters in Ju-San Gata. Physiol. and Ecol. (Japan) 1(4):1-16, 3 text figs., tables.

Coscinodiscus marginatus
Macdonald, R., 1933
An examination of plankton hauls made in the Suez Canal during the year 1928. Fish. Res. Dis., Notes & Mem. No.3, 11 pp., 1 chart.

Coscinodiscus marginatus
Mann, A., 1907
Report on the diatoms of the Albatross voyages in the Pacific Ocean, 1888-1904. Contrib. U. S. Nat. Herb. 10(5):221-419, Pls. XLIV-LIV.

Coscinodiscus marginatus
Zanon, D. V., 1949
Diatomee di Buenos Aires (Argentina) Atti Accad. Naz. Lincei, Memorie, Cl. Sci. fis., mat. e. nat., ser. 7, 11(3):59-151, 2 pls.

Coscinodiscus megacoccus
Castracane degli Antelminelli, F., 1886 n.sp.
1. Report on the Diatomaceae collected by H.M.S. Challenger during the years 1873-1876. Rept. Sci. Results, H.M.S. Challenger, Botany Vol. II, 178 pp., 30 pls.

Coscinodiscus minimus
Boden, B. P., 1949.
The diatoms collected by the U.S.S. CACOPAN in the Antarctic in 1947. J. Mar. Res. 8(1):6-13, 3 textfigs.

Coscinodiscus minimus
Boden, Brian, 1948
Marine plankton diatoms on operation HIGHJUMP in: Some oceanographic observations on operation HIGHJUMP. By R.S. Dietz. USNEL Rept. No.55, 97 pp., 41 figs. 7 July 1948.

Coscinodiscus miocenicus n.sp.
Schrader, Hans-Joachim, 1973
Cenozoic diatoms from the northeast Pacific, leg 18. Initial Repts Deep Sea Drilling Proj. 18:673-797.

Coscinodiscus mirificus
Castracane degli Antelminelli, F., 1886 n.sp.
1. Report on the Diatomaceae collected by H.M.S. Challenger during the years 1873-1876. Rept. Sci. Results, H.M.S. Challenger, Botany Vol. II, 178 pp., 30 pls.

Coscinodiscus neoradiatus
Brunel, J., 1962
Le phytoplancton de la Baie de Chaleurs. Inst. Botan., Univ. Montréal, Contrib. No. 77: 365 pp., 66 pls.

Coscinodiscus neoradiatus
Cleve-Euler, A., 1951
Die Diatomeen von Schweden und Finnland. Kungl. Svenska Vetenskaps Akad. Handl., Fjärde Ser. 2(1): 161 pp., 6 pls.

Coscinodiscus nitidulus
Mann, A., 1907
Report on the diatoms of the Albatross voyages in the Pacific Ocean, 1888-1904. Contrib. U. S. Nat. Herb. 10(5):221-419, Pls. XLIV-LIV.

Coscinodiscus nitidus
Cleve-Euler, A., 1951
Die Diatomeen von Schweden und Finnland. Kungl. Svenska Vetenskaps Akad. Handl., Fjärde Ser. 2(1): 161 pp., 6 pls.

Coscinodiscus nitidus
Cupp, Easter E., 1943
Marine plankton diatoms of the west coast of North America. Bull. S.I.O. 5(1):1-238, 5 pls., 168 text figs.

Coscinodiscus nitidus
Gran, H.H., 1908
Diatomeen. Nordisches Plankton, Botanischer Teil pp. XIX.1-XIX 146; 178 text figs.

Coscinodiscus nitidus
Hendy, N. Ingram, 1964
An introductory account of the smaller algae of British coastal waters. V. Bacillariophyceae (Diatoms).
Her Majesty's Stationary Office, 317 pp., 45 pls.

Coscinodiscus nitidus
Hendey, N-Ingram, 1958 [1957(Publ. 1958)]
Marine diatoms from some West African Ports-J. R.Microsc. Soc. (3) 77(1/2): 28-85.

Coscinodiscus nitidus
Hendey, N.I., 1937
The plankton diatoms of the southern seas. Discovery Repts. 16:151-364, pls.6-13.

Coscinodiscus nitidus
Jørgensen, E., 1905
B.Protistplankton and the diatoms in bottom samples. Hydrographical and biological investigations in Norwegian fiords. Bergens Mus. Skr. 7: 49-225.

Coscinodiscus nitidus
Mann, A., 1907
Report on the diatoms of the Albatross voyages in the Pacific Ocean, 1888-1904. Contrib. U. S. Nat. Herb. 10(5):221-419, Pls. XLIV-LIV.

Coscinodiscus nitidus
Politis, J., 1949
Diatomees marines de Bosphores et des ibes de la mer de Marmara. II Practica tou Hellenikou Hidrobiologikou Institutoutou 1929, Etoz 1929, 3(1):11-31.

Coscinodiscus nitidus
Zanon, D. V., 1949
Diatomee di Buenos Aires (Argentina) Atti Accad. Naz. Lincei, Memorie, Cl. Sci. fis., mat. e. nat., ser. 7, 11(3):59-151, 2 pls.

Coscinodiscus nitidus
Zanon, V., 1948
Diatomee marini di Sardegna e Pugillo di Alghe Marine della stressa. Boll. Pesca, Piscitutura e Idrobiologia, Anno 24, ns. 3(2): 202-244, 27 figs. on 1 pl.

Coscinodiscus nobilis
Allen, W.E., and E.E. Cupp, 1935
Plankton diatoms of the Java Sea. Annales du Jardin Botanique de Buitenzorg XLIV (2):101-174, figs.1-127.
(drawings of all species mentioned)

Coscinodiscus nobilis
Delsman, H. C., 1939.
Preliminary plankton investigations in the Java Sea. Treubia, 17:139-181, 8 maps, 41 figs.

Coscinodiscus nobilis
Ghazzawi, F.M., 1939
Plankton of the Egyptian waters. A study of the Suez Canal Plankton. (A) Phytoplankton. Preliminary Report 83 pp. Notes and Memoires, Min. Commerce-Industry, Egypt, Hydrobiol. & Fish. 65 figs.

Coscinodiscus nobilis
Mann, A., 1907
Report on the diatoms of the Albatross voyages in the Pacific Ocean, 1888-1904. Contrib. U. S. Nat. Herb. 10(5):221-419, Pls. XLIV-LIV.

Coscinodiscus nodulifer
Allen, W.E., and E.E. Cupp, 1935
Plankton diatoms of the Java Sea. Annales du Jardin Botanique de Buitenzorg XLIV (2):101-174, figs.1-127.
(drawings of all species mentioned)

Coscinodiscus nodulifer
Florin, M-B., 1948
9. Diatomeae in submarine cores from the Tyrrhenian Sea. Medd. Ocean. Inst., Göteborg, 15 (Göteborgs Kungl. Vetenskaps-och Viterrhets Samhälles Handlingar, Sjätte Foljden, Ser. B 5(13):80-88.

Coscinodiscus nodulifer
Forti, A., 1922
Ricerche sulla flora pelagica (fitoplancton) di Quarto dei Mille. Mem. R. Com. Talass. Ital. 97:248 pp., 13 pls.

Coscinodiscus nodulifer
Hendy, N. Ingram, 1964
An introductory account of the smaller algae of British coastal waters. V. Bacillariophyceae (Diatoms).
Her Majesty's Stationary Office, 317 pp., 45 pls.

Coscinodiscus nodulifer
Hendey, N.I., 1937
The plankton diatoms of the southern seas. Discovery Repts. 16:151-364, pls.6-13.

Coscinodiscus nodulifer
Mann, A., 1907
Report on the diatoms of the Albatross voyages in the Pacific Ocean, 1888-1904. Contrib. U. S. Nat. Herb. 10(5):221-419, Pls. XLIV-LIV.

Coscinodiscus nodulifer
Pavillard, J., 1925
Bacillariales. Rept. on the Danish Oceangr. Exped., 1908-10 to the Mediterranean and adj. seas. Vol.II., Biol. J4:72 pp., 116 text figs.

Coscinodiscus nodulifer
Zanon, V., 1948
Diatomee marini di Sardegna e Pugillo di Alghe Marine della stressa. Boll. Pesca, Piscitutura e Idrobiologia, Anno 24, ns. 3(2): 202-244, 27 figs. on 1 pl.

Coscinodiscus nodulineatus n.sp.
Hendey, N-Ingram, 1958 [1957(Publ. 1958)]
Marine diatoms from some West African Ports-J. R.Microsc. Soc. (3) 77(1/2): 28-85.

Coscinodiscus (Thalassiosira) Nordenskiöldii
Cleve-Euler, A., 1951
Die Diatomeen von Schweden und Finnland. Kungl. Svenska Vetenskaps Akad. Handl., Fjärde Ser. 2(1): 161 pp., 6 pls.

Coscinodiscus Normanii
Bigelow, H.B., and M. Leslie, 1930
Reconnaissance of the waters and plankton of Monterey Bay, July 1928. Bull. M.C.Z. 70(5):429-481, 43 text figs.

Coscinodiscus normanii
Cleve-Euler, A., 1951
Die Diatomeen von Schweden und Finnland. Kungl. Svenska Vetenskaps Akad. Handl., Fjärde Ser. 2(1): 161 pp., 6 pls.

Coscinodiscus normani
Hendy, N. Ingram, 1964
An introductory account of the smaller algae of British coastal waters. V. Bacillariophyceae (Diatoms).
Her Majesty's Stationary Office, 317 pp., 45 pls.

Coscinodiscus normanni
Hendey, N-Ingram, 1958 [1957(Publ. 1958)]
Marine diatoms from some West African Ports-J. R.Microsc. Soc. (3) 77(1/2): 28-85.

Coscinodiscus normanni
Mann, A., 1907
Report on the diatoms of the Albatross voyages in the Pacific Ocean, 1888-1904. Contrib. U. S. Nat. Herb. 10(5):221-419, Pls. XLIV-LIV.

Coscinodiscus oblongus
Mann, A., 1893
List of Diatomaceae from a deep-sea dredging in the Atlantic Ocean off Delaware Bay by the U. S. Fish Commission Steamer Albatross. Proc. U. S. Nat. Mus. 16:303-312.

Coscinodiscus obovatus
Castracane degli Antelminelli, F., 1886 n.sp.
1. Report on the Diatomaceae collected by H.M.S. Challenger during the years 1873-1876. Rept. Sci. Results, H.M.S. Challenger, Botany Vol. II, 178 pp., 30 pls.

Coscinodiscus obscurus
Cleve-Euler, A., 1951
Die Diatomeen von Schweden und Finnland. Kungl. Svenska Vetenskaps Akad. Handl., Fjärde Ser. 2(1): 161 pp., 6 pls.

Coscinodiscus obscurus
Frenguelli, Joaquin, and Hector Antonio Orlando, 1959.
Operacion MERLUZA. Diatomeas y silicoflagelados del plancton del "VI Crucero". Servicio Hidrogr. Naval, Argentina, Publ. No. H. 619: 5-62.

Coscinodiscus obscurus
Mann, A., 1907
Report on the diatoms of the Albatross voyages in the Pacific Ocean, 1888-1904. Contrib. U. S. Nat. Herb. 10(5):221-419, Pls. XLIV-LIV.

Coscinodiscus obscurus
Zanon, D. V., 1949
Diatomee di Buenos Aires (Argentina) Atti Accad. Naz. Lincei, Memorie, Cl. Sci. fis., mat. e. nat., ser. 7, 11(3):59-151, 2 pls.

Coscinodiscus oculoides
Hendey, N.I., 1937
The plankton diatoms of the southern seas. Discovery Repts. 16:151-364, pls.6-13.

Coscinodiscus oculoides
Jouse, A.P., G.S. Koroleva, G.A. Nagaeva, 1962
Diatoms in the surface layer of sediment in the Indian sector of the Antarctic. Investigations of marine bottom sediments. (In Russian; English summary).
Trudy Inst. Okeanol., Akad. Nauk, SSSR, 61: 19-92.

Coscinodiscus oculus-iridis
Allen, W.E., and E.E. Cupp, 1935
Plankton diatoms of the Java Sea. Annales du Jardin Botanique de Buitenzorg XLIV (2):101-174, figs.1-127.
(drawings of all species mentioned)

Coscinodiscus oculus-iridis

Bigelow, H.B., and M. Leslie, 1930
Reconnaissance of the waters and plankton of Monterey Bay, July 1928. Bull. M.C.Z., 70(5):429-481, 43 text figs.

Coscinodiscus oculus iridis

Cleve-Euler, A., 1951
Die Diatomeen von Schweden und Finnland. Kungl. Svenska Vetenskaps Akad. Handl., Fjärde Ser. 2(1): 161 pp., 6 pls.

Coscinodiscus oculus iridis

Cupp, Easter E., 1943
Marine plankton diatoms of the west coast of North America. Bull. S.I.O. 5(1):1-238, 5 pls., 168 text figs.

Coscinodiscus oculus iridis

Florin, M-B., 1948
9. Diatomeae in submarine cores from the Tyrrhenian Sea. Medd. Ocean. Inst., Göteborg, 15 (Göteborgs Kungl. Vetenskaps-och Viterrhets Samhälles Handlingar, Sjätte Foljden, Ser. B 5(13):80-88.

Coscinodiscus oculis-iridis

Fox, M., 1957.
A first list of marine algae from Nigeria. J. Linn. Soc., London, 55(362):615-631.

Coscinodiscus oculus-iridis

Frenguelli, Joaquin, and Hector Antonio Orlando, 1959.
Operacion MERLUZA. Diatomeas y silico-flagelados del plancton del "VI Crucero". Servicio Hidrogr. Naval., Argentina, Publ. No. H. 619: 5-82.

Coscinodiscus oculus-iridis

Ghazzawi, F.M., 1939
Plankton of the Egyptian waters. A study of the Suez Canal Plankton. (A) Phytoplankton. Preliminary Report 83 pp. Notes and Memoires, Min. Commerce-Industry, Egypt, Hydrobiol. & Fish. 65 figs.

Coscinodiscus oculus-iridis

Hendy, N. Ingram, 1964
An introductory account of the smaller algae of British coastal waters. V. Bacillariophyceae (Diatoms). Her Majesty's Stationary Office, 317 pp., 45 pls.

Coscinodiscus oculus-iridis

Hendey, N-Ingram, 1958 [1957(Publ. 1958)]
Marine diatoms from some West African Ports-J. R.Microsc. Soc. (3) 77(1/2):28-85.

Coscinodiscus oculus-iridis

Hendey, N.I., 1937
The plankton diatoms of the southern seas. Discovery Repts. 16:151-364, pls.6-13.

Coscinodiscus oculus-iridis

Jorgensen, E., 1900
Protophyten und Protozoën im Plankton aus der Norwegischen Westkerste. Bergens Mus. Aarb. 1899(6): 95 pp., 5 pls., 83 tables.

Coscinodiscus osculis-iridis

Lillick, L.C., 1940
Phytoplankton and planktonic protozoa of the offshore waters of the Gulf of Maine. Pt.II. Qualitative Composition of the Planktonic Flora. Trans. Am. Phil. Soc., n.s., 31(3):193-237, 13 text figs.

Coscinodiscus oculus-iridis

Mangin, L., 1915
Phytoplancton de L'Antartique. Deuxieme Exped. Ant. Francaise (1908-1910), 95 pp., 3 pls., 58 text figs.

Coscinodiscus oculus-iridis

Mangin, M. L., 1912
Phytoplancton de la croisière du "René" dans l'Atlantique (Septembre 1908). Ann. Inst. Ocean., n.s., 4(1):1-66, 2 pls., 41 text figs., 2 tables.

Coscinodiscus oculus-iridis

Manguin, E., 1954
Diatomees marines provenant de l'ile Heard (Australian National Antarctic Expedition). Rev. Algol., n.s., 1: 14-24.

Coscinodiscus oculus-iridis

Mann, A., 1907
Report on the diatoms of the Albatross voyages in the Pacific Ocean, 1888-1904. Contrib. U. S. Nat. Herb. 10(5):221-419, Pls. XLIV-LIV.

Coscinodiscus oculus iridis

Meunier, A., 1915
Microplancton de la Mer Flamande. 2. Diatomées (excepté le genre Chaetoceros). Mem. Mus. Roy. Hist. Nat., Belgique, 7(3):1-118, Pls. VIII-XIV.

Coscinodiscus oculus-iridis

Morse, D.C., 1947
Some observations on seasonal variations in plankton population Patuxant River, Maryland 1943-1945. Bd. Nat. Res., Publ. No.65, Chesapeake Biol. Lab., 31, 3 figs.

Coscinodiscus oculus-iridis

Paiva Carvalho, J., 1950
O plancton do Rio Maria Rodriques (Cananeis). 1. Diatomaceas e Dinoflagelados. Bol. Inst. Paulista Oceanogr. 1(1); 27-43, 2 fold-in tables, 2 figs.

Coscinodiscus oculusiridis

Pavillard, J., 1905
Recherches sur la flore pelagique (Phytoplankton) de l'Etang de Thau. Theses presentees a la Fac. Sci., Paris, 116 pp., 3 pls.

Coscinodiscus oculus iridis

Rumkówna, A., 1948
List of the phytoplankton species occurring in the superficial water layers in the Gulf of Gdańsk. Bull. Lab. mar., Gdynia, No. 4: 139-141 with tables in back.

Coscinodiscus oculus iridis

Schodduyn, M., 1926
Observations faites dans la baie d'Ambleteuse (Pas de Calais). Bull. Inst. Ocean., Monaco, No. 482: 64 pp.

Coscinodiscus oculus-iridis

Schröder, B., 1900
Phytoplankton des Golfes von Neapel nebst vergleichenden Ausblicken auf das atlantischen Ozean. Mitt. Zool. Stat. Neapel, 14:1-38.

Coscinodiscus oculus-iridis

Sousa e Silva, E., 1949
Diatomaceas e Dinoflagelados de Baia de Cascais. Portugaliae Acta Biol., Volume: Julio Henriques, Ser. B: 300-383, 9 pls, 2 fold-in tables.

Coscinodiscus oculus iridis

Sousa e Silva, E., and J. Dos Santos-Pinto, 1948
O Plancton da Baía de S. Martinho do Porto. 1. Diatomaceas e Dinoflagelados. Bol. Soc. Portuguese de Ciencias Naturais, 16(2):134-187, 6 pls. (Trav. Sta. Biol. Mar. de Lisbonne No. 52).

Coscinodiscus oculus-iridis

Tavares de Lyra, Luiz, 1964.
Anomalia em Coscinodiscus oculus-iridis Ehrenberg, 1839 (Diatomacea). Memories Inst. Oswaldo Cruz, 62(1):19-23.

Coscinodiscus oculus-iridis

Zanon, D. V., 1949
Diatomee di Buenos Aires (Argentina) Atti Accad. Naz. Lincei, Memorie, Cl. Sci. fis., mat. e. nat., ser. 7, 11(3):59-151, 2 pls.

Coscinodiscus oculus-iridis

Zanon, V., 1948
Diatomee marini di Sardegna e Pugillo di Alghe Marine della stressa. Boll. Pesca, Piscitutura e Idrobiologia, Anno 24, ns. 3(2): 202-244, 27 figs. on 1 pl.

Coscinodiscus odontodiscus

Cleve-Euler, A., 1951
Die Diatomeen von Schweden und Finnland. Kungl. Svenska Vetenskaps Akad. Handl., Fjärde Ser. 2(1): 161 pp., 6 pls.

Coscinodiscus okunoi

Tsumura, K., 1956.
Diatomoj el la cirkaufoso de la restajo de la kastele de Odawara. J. Yokohama Municipal Univ., (C-14) No. 47:23 pp.

Coscinodiscus oppositus

Hendey, N.I., 1937
The plankton diatoms of the southern seas. Discovery Repts. 16:151-364, pls.6-13.

Coscinodiscus (Thalassiosira) ostenfeldii nom. nov.

Cleve-Euler, A., 1951
Die Diatomeen von Schweden und Finnland. Kungl. Svenska Vetenskaps Akad. Handl., Fjärde Ser. 2(1): 161 pp., 6 pls.

Coscinodiscus ovalis

Castracane degli Antelminelli, F., 1886
1. Réport on the Diatomaceae collected by H.M.S. Challenger during the years 1873-1876. Rept. Sci. Results, H.M.S. Challenger, Botany Vol. II, 178 pp., 30 pls.

Coscinodiscus pacificus

Bigelow, H.B., and M. Leslie, 1930
Reconnaissance of the waters and plankton of Monterey Bay, July 1928. Bull. M.C.Z., 70(5):429-481, 43 text figs.

Coscinodiscus praetextus

Bigelow, H.B., and M. Leslie, 1930
Reconnaissance of the waters and plankton of Monterey Bay, July 1928. Bull. M.C.Z., 70(5):429-481, 43 text figs.

Coscinodiscus (?) pacificus

Castracane degli Antelminelli, F., 1886 n.sp.
1. Réport on the Diatomaceae collected by H.M.S. Challenger during the years 1873-1876. Rept. Sci. Results, H.M.S. Challenger, Botany Vol. II, 178 pp., 30 pls.

Coscinodiscus papuanus n. sp.

Castracane degli Antelminelli, F., 1886
1. Réport on the Diatomaceae collected by H.M.S. Challenger during the years 1873-1876. Rept. Sci. Results, H.M.S. Challenger, Botany Vol. II, 178 pp., 30 pls.

Coscinodiscus parvulus

Boden, B.P., 1950
Some marine plankton diatoms from the west coast of South Africa. Trans. R.Soc. S. Africa. 32:321-434, 100 text figs.

Coscinodiscus patera n.sp.

Castracane degli Antelminelli, F., 1886
1. Réport on the Diatomaceae collected by H.M.S. Challenger during the years 1873-1876. Rept. Sci. Results, H.M.S. Challenger, Botany Vol. II, 178 pp., 30 pls.

Coscinodiscus pavillardi
Cupp, Easter E., 1943
Marine plankton diatoms of the west coast of North America. Bull. S.I.O. 5(1):1-238, 5 pls., 168 text figs.

Coscinodiscus pavillardii
Findlay, Ivan W. O. 1972.
Effect of external factors and cell size on the cell division rate of a marine diatom, Coscinodiscus pavillardii Forti.
Int. Revue ges. Hydrobiol. 57(4): 523-533

Coscinodiscus pavillardii
Findlay, I.W.O. 1969.
Cell size and spore formation in a clone of centric diatom, Coscinodiscus pavillardii Forti.
Phykos 8(1/2): 31-41

Coscinodiscus Pavillardii n.sp.
Forti, A., 1922
Ricerche sulla flora pelagica (fitoplancton) di Quarto dei Mille. Mem. R. Com. Talass. Ital. 97:248 pp., 13 pls.

Coscinodiscus pavillardii
Pavillard, J., 1925
Bacillariales. Rept. on the Danish Oceangr. Exped., 1908-10 to the Mediterranean and adj. seas. Vol.II., Biol. J4:72 pp., 116 text figs.

Coscinodiscus pavillardi
Sousa e Silva, E., 1949
Diatomacees e Dinoflagelados de Baia de Cascais. Portugalise Acta Biol., Volume: Julio Henriques, Ser. B: 300-383, 9 pls, 2 fold-in tables.

Coscinodiscus Pavillardii
Sousa e Silva, E., and J. Dos Santos-Pinto, 1948
O Plancton da Baia de S. Martinho do Porto. 1. Diatomaceas e Dinoflagelados. Bol. Soc. Portuguese de Ciencias Naturais, 16(2):134-187, 6 pls. (Trav. Sta. Biol. Mar. de Lisbonne No. 52).

Coscinodiscus payeri
Cleve-Euler, A., 1951
Die Diatomeen von Schweden und Finnland. Kungl. Svenska Vetenskaps Akad. Handl., Fjärde Ser. 2(1): 161 pp., 6 pls.

Coscinodiscus pentas
Mann, A., 1907
Report on the diatoms of the Albatross voyages in the Pacific Ocean, 1888-1904. Contrib. U. S. Nat. Herb. 10(5):221-419, Pls. XLIV-LIV.

Coscinodiscus perforatus
Cleve-Euler, A., 1951
Die Diatomeen von Schweden und Finnland. Kungl. Svenska Vetenskaps Akad. Handl., Fjärde Ser. 2(1): 161 pp., 6 pls.

Coscinodiscus perforatus
Cupp, Easter E., 1943
Marine plankton diatoms of the west coast of North America. Bull. S.I.O. 5(1):1-238, 5 pls., 168 text figs.

Coscinodiscus perforatus
Hendy, N. Ingram, 1964
An introductory account of the smaller algae of British coastal waters. V. Bacillariophyceae (Diatoms). Her Majesty's Stationary Office, 317 pp., 45 pls.

Coscinodiscus perforatus
Morse, D.C., 1947
Some observations on seasonal variations in plankton population Patuxent River, Maryland 1943-1945. Bd. Nat. Res., Publ. No.65, Chesapeake Biol. Lab., 31, 3 figs.

Coscinodiscus perforatus cellulosa
Cupp, Easter E., 1943
Marine plankton diatoms of the west coast of North America. Bull. S.I.O. 5(1):1-238, 5 pls., 168 text figs.

Coscinodiscus perforatus var. Pavillardi
Rampi, L., 1942
Ricerche sul fitoplancton del Mare Ligure 6. Le diatomee delle acque di Sanremo. Nuovo Giornale Botanico Italiano, N.S., 49:252-268.

Coscinodiscus plicatulus
Cleve-Euler, A., 1951
Die Diatomeen von Schweden und Finnland. Kungl. Svenska Vetenskaps Akad. Handl., Fjärde Ser. 2(1): 161 pp., 6 pls.

Coscinodiscus plicatus
Cleve-Euler, A., 1951
Die Diatomeen von Schweden und Finnland. Kungl. Svenska Vetenskaps Akad. Handl., Fjärde Ser. 2(1): 161 pp., 6 pls.

Coscinodiscus polyactis
Cleve-Euler, A., 1951
Die Diatomeen von Schweden und Finnland. Kungl. Svenska Vetenskaps Akad. Handl., Fjärde Ser. 2(1): 161 pp., 6 pls.

Coscinodiscus (Coscinosira) polychordus
Cleve-Euler, A., 1951
Die Diatomeen von Schweden und Finnland. Kungl. Svenska Vetenskaps Akad. Handl., Fjärde Ser. 2(1): 161 pp., 6 pls.

Coscinodiscus (?) polygonus
Castracane degli Antelminelli, F., 1886 n.sp.
1. Report on the Diatomaceae collected by H.M.S. Challenger during the years 1873-1876. Rept. Sci. Results, H.M.S. Challenger, Botany Vol. II, 178 pp., 30 pls.

Coscinodiscus polyradiatus
Castracane degli Antelminelli, F., 1886 n.sp.
1. Report on the Diatomaceae collected by H.M.S. Challenger during the years 1873-1876. Rept. Sci. Results, H.M.S. Challenger, Botany Vol. II, 178 pp., 30 pls.

Coscinodiscus praepaleaceus n.sp.
Schrader, Hans-Joachim, 1973
Cenozoic diatoms from the northeast Pacific, leg 18. Initial Repts Deep Sea Drilling Proj. 18:673-797.

Cosinodiscus praeyabei n.sp.
Schrader, Hans-Joachim, 1973
Cenozoic diatoms from the northeast Pacific, leg 18. Initial Repts Deep Sea Drilling Proj. 18:673-797.

Coscinodiscus punctiger n.comb.
Müller Melchers, F.C., 1953.
New and little known diatoms from Uruguay and the South Atlantic coast. Com. Bot., Mus. Hist. Nat., Montevideo, (30): 1-11, 2 pls.

Coscinodiscus pustulatus n.sp.
Mann, A., 1907
Report on the diatoms of the Albatross voyages in the Pacific Ocean, 1888-1904. Contrib. U. S. Nat. Herb. 10(5):221-419, Pls. XLIV-LIV.

Coscinodiscus pyrenoidophorus
Hendey, N.I., 1937
The plankton diatoms of the southern seas. Discovery Repts. 16:151-364, pls.6-13.

Coscinodiscus quadrifarius n.sp.
Manguin, E., 1957.
Premier inventaire des diatomées de la Terre Adélie Antarctique. Espèces nouvelles. Rev. Algologique, n.s., 3(3):111-134.

Coscinodiscus radiatus
Allen, W.E., and E.E. Cupp, 1935
Plankton diatoms of the Java Sea. Annales du Jardin Botanique de Buitenzorg XLIV (2):101-174, figs.1-127.
(drawings of all species mentioned)

Coscinodiscus radiatus
Bigelow, H.B., and M. Leslie, 1930
Reconnaissance of the waters and plankton of Monterey Bay, July 1928. Bull. M.C.Z., 70(5):429-481, 43 text figs.

Coscinodiscus radiatus
Boden, B.P., 1950
Some marine plankton diatoms from the west coast of South Africa. Trans. R.Soc. S. Africa. 32:321-434, 100 text figs.

Coscinodiscus radiatus
Brunel, J., 1962
Le phytoplancton de la Baie de Chaleurs. Inst. Botan., Univ. Montréal, Contrib. No. 77: 365 pp., 66 pls.

Coscinodiscus radiatus
Brunel, Jules, 1962
Le phytoplancton de la Baie des Chaleurs. Contrib. Ministère de la Chasse et des Pêcheries, Province de Québec, No. 91: 365 pp.

Coscinodiscus radiatus
Chiba, T., 1949
On the distribution of the plankton in the eastern China Sea and Yellow Sea. 1. Plankton composition in the spring. J. Shimonoseki Coll. Fisheries, 1(1):57-63, 1 fig.

Coscinodiscus radiatus
Cleve-Euler, A., 1951
Die Diatomeen von Schweden und Finnland. Kungl. Svenska Vetenskaps Akad. Handl., Fjärde Ser. 2(1): 161 pp., 6 pls.

Coscinodiscus radiatus
Copenhagen, W. J. and L. D., 1949
Variation in the phytoplankton of Table Bay, October 1934 to October 1935. With a note on the calorific value of Chaetoceros spp. Trans. Roy. Soc. S. Africa, 32(2):113-123, 2 text figs.

Coscinodiscus radiatus
Cupp, Easter E., 1943
Marine plankton diatoms of the west coast of North America. Bull. S.I.O. 5(1):1-238, 5 pls., 168 text figs.

Coscinodiscus radiatus
Cupp, E.E., 1937
Seasonal distribution and occurrence of marine diatoms and dinoflagellates at Scotch Cap, Alaska. Bull. S.I.O. Tech. ser.4(3):71-100, 7 textfigs.

Coscinodiscus radiatus
Dangeard, P., 1927
Phytoplankton de la croisière du "Sylvana". Ann. Inst. Ocean., Monaco, n.s., 4(8):286-401, 54 text figs. (Feirer-Juin 1913).

Coscinodiscus radiatus

Ercegovic, A., 1936
Etudes qualitative et quantitatives du phytoplancton dans les eaux cotières de l'Adriatique oriental moyen au cours de l'année 1934. Acta Adriatica 1(9):1-126.

Coscinodiscus radiatus

Frenguelli, Joaquin, and Hector Antonio Orlando, 1959.
Operacion MERLUZA. Diatomeas y silicoflagelados del plancton del "VI Crucero". Servicio Hidrogr. Naval., Argentina, Publ. No. H. 619: 5-62.

Coscinodiscus radiatus

Gran, H.H., 1908
Diatomeen. Nordisches Plankton, Botanischer Teil pp. XIX.1-XIX 146; 178 text figs.

Coscinodiscus radiatus

Gran, H. H. and E. C. Angst, 1931
Plankton diatoms of Puget Sound. Publ. Puget Sound Biol. Sta. 7:417-519, 95 text figs.

Coscinodiscus radiatus?

Gunter, G., R.H. Williams, C.C. Davis, and F.G. Walton Smith, 1948
Catastrophic mass mortality of marine animals and coincident phytoplankton bloom on the West Coast of Florida, November 1946 to August 1947. Ecol. Mon., 18:309-324, 2 text figs.

Coscinodiscus radiatus

Hendy, N. Ingram, 1964
An introductory account of the smaller algae of British coastal waters. V. Bacillariophyceae (Diatoms).
Her Majesty's Stationary Office, 317 pp., 45 pls.

Coscinodiscus radiatus

Hendey, N. Ingram, 1958 [1957(Publ. 1958)]
Marine diatoms from some West African Ports. J. R Microsc. Soc. (3) 77(1/2): 28-85.

Coscinodiscus radiatus

Hendey, N.I., 1937
The plankton diatoms of the southern seas. Discovery Repts. 16:151-364, pls.6-13.

Coscinodiscus radiatus

Hustedt, F. and A.A. Aleem, 1951
Littoral diatoms from the Salstone near Plymouth. JMBA 30(1): 177.196.

Coscinodiscus radiatus

Jørgensen, E., 1905
B. Protistplankton and the diatoms in bottom samples. Hydrographical and biological investigations in Norwegian fjords. Bergens Mus. Skr. 7: 49-225.

Coscinodiscus radiatus

Jorgensen, E., 1900
Protophyten und Protozoën im Plankton aus der Norwegischen Westkerste. Bergens Mus. Aarb. 1899(6): 95 pp., 5 pls., 83 tables.

Coscinodiscus radiatus

Kokubo, S., 1952
Results of the observations on the plankton and oceanography of Mutsu Bay during 1950, reference being made also to the period 1946-1950. Bull Mar.Biol.Sta., Asamushi 5(1/4): 1-54, 3 tables,(fold-in), 1 fold-in.

Coscinodiscus radiatus

Kokubo, S., and S. Sato., 1947
Plankters in JG-San Gata. Physiol. and Ecol. (Japan) 1(4):1-16, 3 text figs., tables.

Coscinodiscus radiatus

Lillick, L.C., 1940
Phytoplankton and planktonic protozoa of the offshore waters of the Gulf of Maine. Pt.II. Qualitative Composition of the Planktonic Flora. Trans. Am. Phil. Soc., n.s., 31(3):193-237, 13 text figs.

Coscinodiscus radiatus

Lillick, L.C., 1938
Preliminary report of the phytoplankton of the Gulf of Maine. Am. Mid. Nat. 20(3):624-640, 1 text figs 37 tables.

Coscinodiscus radiatus

Mangin, L., 1915
Phytoplancton de L'Antartique. Deuxieme Exped. Ant. Francaise (1908-1910), 95 pp., 3 pls., 58 text figs.

Coscinodiscus radiatus

Mangin, M. L., 1912
Phytoplancton de la croisière du "René" dans l'Atlantique (Septembre 1908). Ann. Inst. Ocean., n.s., 4(1):1-66, 2 pls., 41 text figs., 2 tables.

Coscinodiscus radiatus

Mann, A., 1907
Report on the diatoms of the Albatross voyages in the Pacific Ocean, 1888-1904. Contrib. U. S. Nat. Herb. 10(5):221-419, Pls. XLIV-LIV.

Coscinodiscus radiatus

Mann, A., 1893
List of Diatomaceae from a deep-sea dredging in the Atlantic Ocean off Delaware Bay by the U. S. Fish Commission Steamer Albatross. Proc. U. S. Nat. Mus. 16:303-312.

Coscinodiscus radiatus

Marukawa, H., 1921
Plankton lists and some new species of copepods, from the northern waters of Japan. Bull. Inst. Ocean., No.384, 15 pp., 3 pls., 1 chart. Monaco

Coscinodiscus radiatus

Meunier, A., 1915
Microplancton de la Mer Flamande. 2. Diatomées (excepté le genre Chaetoceros). Mem. Mus. Roy. Hist. Nat., Belgique, 7(3):1-118, Pls. VIII-XIV.

Coscinodiscus radiatus

Morse, D.C., 1947
Some observations on seasonal variations in plankton population Patuxant River, Maryland 1943-1945. Bd. Nat. Res., Publ. No.65, Chesapeake Biol. Lab., 31, 3 figs.

Coscinodiscus radiatus

Paiva Carvalho, J., 1950
O plancton do Rio Maria Rodrigues (Cananeis). 1. Diatomaceas e Dinoflagelados. Bol. Inst. Paulista Oceanogr. 1(1): 27-43, 2 fold-in tables, 2 figs.

Coscinodiscus radiatus

Pavillard, J., 1925
Bacillariales. Rept. on the Danish Oceangr. Exped., 1908-10 to the Mediterranean and adj. seas. Vol.II., Biol. J4:72 pp., 116 text figs.

Coscinodiscus radiatus

Pavillard, J., 1905
Recherches sur la flore pelagique (Phytoplankton) de l'Etang de Thau. Theses presentees a la Fac. Sci., Paris, 116 pp., 3 pls.

Coscinodiscus radiatus

Phifer, L.D. (undated)
The occurrence and distribution of plankton diatoms in Bering Sea and Bering Strait, July 26-August 24, 1934. Report of Oceanographic cruise of U.S. Coast Guard Cutter Chelan 1934, Part II(A):1-44 (mimeographed) plus fig.1 (after Pt.B)

Coscinodiscus radiatus

Rampi, L., 1942
Ricerche sul fitoplancton del Mare Ligure 6. Le diatomee delle acque di Sanremo. Nuovo Giornale Botanico Italiano, N.S., 49:252-268.

Coscinodiscus radiatus

Rumkówna, A., 1948
List of the phytoplankton species occurring in the superficial water layers in the Gulf of Gdańsk. Bull. Lab. mar., Gdynia, No. 4: 139-141 with tables in back.

Coscinodiscus radiatus

Schodduyn, M., 1926
Observations faites dans la baie d'Ambleteuse (Pas de Calais). Bull. Inst. Ocean., Monaco, No. 482: 64 pp.

Coscinodiscus radiatus

Sousa e Silva, E., 1949
Diatomaceas e Dinoflagelados de Baia de Cascais. Portugaliae Acta Biol., Volume: Julio Henriques, Ser. B: 300-383, 9 pls, 2 fold-in tables.

Coscinodiscus radiatus

Sousa a Silva, E., and J. Dos Santos-Pinto, 1948
O Plancton da Baía de S. Martinho do Porto. 1. Diatomaceas e Dinoflagelados. Bol. Soc. Portuguese de Ciencias Naturais, 16(2):134-187, 6 pls. (Trav. Sta. Biol. Mar. de Lisbonne No. 52).

Coscinodiscus radiatus

Zanon, D. V., 1949
Diatomee di Buenos Aires (Argentina) Atti Accad. Naz. Lincei, Memorie, Cl. Sci. fis., mat. e. nat., ser. 7, 11(3):59-151, 2 pls.

Coscinodiscus radiatus

Zanon, V., 1948
Diatomee marini di Sardegna e Pugillo di Alghe Marine della stressa. Boll. Pesca, Piscitutura e Idrobiologia, Anno 24, ns. 3(2): 202-244, 27 figs. on 1 pl.

Coscinodiscus radiatus abyssalis n.var.

Castracane degli Antelminelli, F., 1886
1. Report on the Diatomaceae collected by H.M.S. Challenger during the years 1873-1876. Rept. Sci. Results, H.M.S. Challenger, Botany Vol. II, 178 pp., 30 pls.

Coscinodiscus radiatus var pacifica n.var

Gran, H. H. and E. C. Angst, 1931
Plankton diatoms of Puget Sound. Publ. Puget Sound Biol. Sta. 7:417-519, 95 text figs.

Coscinodiscus reniformis

Castracane degli Antelminelli, F., 1886 n.sp.
1. Report on the Diatomaceae collected by H.M.S. Challenger during the years 1873-1876. Rept. Sci. Results, H.M.S. Challenger, Botany Vol. II, 178 pp., 30 pls.

Coscinodiscus rex

Bigelow, H.B., and M. Leslie, 1930
Reconnaissance of the waters and plankton of Monterey Bay, July 1928. Bull. M.C.Z., 70(5):429-481, 43 text figs.

Coscinodiscus rhombicus n.sp.

Castracane degli Antelminelli, F., 1886
1. Report on the Diatomaceae collected by H.M.S. Challenger during the years 1873-1876. Rept. Sci. Results, H.M.S. Challenger, Botany Vol. II, 178 pp., 30 pls.

Coscinodiscus ritscherii

Jouse, A.P., G.S. Koroleva, G.A. Nagaeva, 1962
Diatoms in the surface layer of sediment in the Indian sector of the Antarctic. Investigations of marine bottom sediments. (In Russian; English summary). Trudy Inst. Okeanol., Akad. Nauk, SSSR, 61: 19-92.

Coscinodiscus robustus

Mann, A., 1907
Report on the diatoms of the Albatross voyages in the Pacific Ocean, 1888-1904. Contrib. U. S. Nat. Herb. 10(5):221-419, Pls. XLIV-LIV.

Coscinodiscus robustus

Mann, A., 1893
List of Diatomaceae from a deep-sea dredging in the Atlantic Ocean off Delaware Bay by the U. S. Fish Commission Steamer Albatross. Proc. U. S. Nat. Mus. 16:303-312.

Coscinodiscus rothii

Cleve-Euler, A., 1951
Die Diatomeen von Schweden und Finnland. Kungl. Svenska Vetenskaps Akad. Handl., Fjärde Ser. 2(1): 161 pp., 6 pls.

Coscinodiscus rothii

Manguin, E., 1954
Diatomees marines provenant de l'ile Heard (Australian National Antarctic Expedition). Rev. Algol., n.s., 1: 14-24.

Coscinodiscus Rothii

Zanon, D. V., 1949
Diatomee di Buenos Aires (Argentina) Atti Accad. Naz. Lincei, Memorie, Cl. Sci. fis., mat. e. nat., ser. 7, 11(3):59-151, 2 pls.

Coscinodiscus (Thalassiosira) rotulus

Cleve-Euler, A., 1951
Die Diatomeen von Schweden und Finnland. Kungl. Svenska Vetenskaps Akad. Handl., Fjärde Ser. 2(1): 161 pp., 6 pls.

Coscinodiscus rotundus

Boden, B.P., 1950
Some marine plankton diatoms from the west coast of South Africa. Trans. R.Soc. S. Africa. 32:321-434, 100 text figs.

Coscinodiscus (?) rudis n.sp.

Castracane degli Antelminelli, F., 1886
1. Report on the Diatomaceae collected by H.M.S. Challenger during the years 1873-1876. Rept. Sci. Results, H.M.S. Challenger, Botany Vol. II, 178 pp., 30 pls.

Coscinodiscus simbirskianus

Boden, B. P., 1949.
The diatoms collected by the U.S.S. CACOPAN in the Antarctic in 1947. J. Mar. Res. 8(1):6-13, 3 textfigs.

Coscinodiscus simbirskianus

Boden, Brian, 1948
Marine plankton diatoms on operation HIGHJUMP in: Some oceanographic observations on operation HIGHJUMP. By R.S. Dietz. USNEL Rept. No.55, 97 pp., 41 figs. 7 July 1948.

Coscinodiscus simbirskianus

Hendey, N.I., 1937
The plankton diatoms of the southern seas. Discovery Repts. 16:151-364, pls.6-13.

Coscinodiscus simbirskianus

Mann, A., 1907
Report on the diatoms of the Albatross voyages in the Pacific Ocean, 1888-1904. Contrib. U. S. Nat. Herb. 10(5):221-419, Pls. XLIV-LIV.

Coscinodiscus stellaris

Castracane degli Antelminelli, F., 1886
1. Report on the Diatomaceae collected by H.M.S. Challenger during the years 1873-1876. Rept. Sci. Results, H.M.S. Challenger, Botany Vol. II, 178 pp., 30 pls.

Coscinodiscus stellaris

Cleve-Euler, A., 1951
Die Diatomeen von Schweden und Finnland. Kungl. Svenska Vetenskaps Akad. Handl., Fjärde Ser. 2(1): 161 pp., 6 pls.

Coscinodiscus stellaris

Cupp, Easter E., 1943
Marine plankton diatoms of the west coast of North America. Bull. S.I.O. 5(1):1-238, 5 pls., 168 text figs.

Coscinodiscus stellaris

Gran, H.H., 1908
Diatomeen. Nordisches Plankton, Botanischer Teil pp. XIX.1-XIX 146; 178 text figs.

Coscinodiscus stellaris

Gran, H. H. and E. C. Angst, 1931
Plankton diatoms of Puget Sound. Publ. Puget Sound Biol. Sta. 7:417-519, 95 text figs.

Coscinodiscus stellaris

Hendy, N. Ingram, 1964
An introductory account of the smaller algae of British coastal waters. V. Bacillariophyceae (Diatoms). Her Majesty's Stationary Office, 317 pp., 45 pls.

Coscinodiscus stellaris

Hendey, N.I., 1937
The plankton diatoms of the southern seas. Discovery Repts. 16:151-364, pls.6-13.

Coscinodiscus stellaris

Jørgensen, E., 1905
B. Protistplankton and the diatoms in bottom samples. Hydrographical and biological investigations in Norwegian fjords. Bergens Mus. Skr. 7: 49-225.

Coscinodiscus stellaris

Jorgensen, E., 1900
Protophyten und Protozoën, im Plankton aus der Norwegischen Westkerste. Bergens Mus. Aarb. 1899(6): 95 pp., 5 pls., 83 tables.

Coscinodiscus stellaris

Lillick, L.C., 1940
Phytoplankton and planktonic protozoa of the offshore waters of the Gulf of Maine. Pt.II. Qualitative Composition of the Planktonic Flora. Trans. Am. Phil. Soc., n.s., 31(3):193-237, 13 text figs.

Coscinodiscus stellaris

Mangin, L., 1915
Phytoplancton de L'Antartique. Deuxieme Exped. Ant. Francaise (1908-1910), 95 pp., 3 pls., 58 text figs.

Coscinodiscus stellaris

Pavillard, J., 1925
Bacillariales. Rept. on the Danish Oceangr. Exped., 1908-10 to the Mediterranean and adj. seas. Vol.II. Biol. J4:72 pp., 116 text figs.

Coscinodiscus stellaris

Castracane degli Antelminelli, F., 1886
1. Report on the Diatomaceae collected by H.M.S. Challenger during the years 1873-1876. Rept. Sci. Results, H.M.S. Challenger, Botany Vol. II, 178 pp., 30 pls.

fasciculata n.var.

Coscinodiscus stephanopixioides

Dangeard, P., 1927
Phytoplankton de la croisière du "Sylvana". Ann. Inst. Ocean., Monaco, n.s., 4(8):286-401, 54 text figs. (Feirer-Juin 1913).

Coscinodiscus subbulliens

Bigelow, H. B., 1922
Exploration of the coastal water off the northeastern United States in 1916 by the U.S. Fisheries Schooner Grampus. Bull. M.C.Z. 65 (5):85-188, 53 text figs.

Coscinodiscus subbulliens

Cleve-Euler, A., 1951
Die Diatomeen von Schweden und Finnland. Kungl. Svenska Vetenskaps Akad. Handl., Fjärde Ser. 2(1): 161 pp., 6 pls.

Coscinodiscus subbulliens

Gilson, H. C., 1937
Chemical and Physical Investigations. The nitrogen cycle. John Murray Exped., 1933-34, Sci. Repts., 2(2):21-81, 16 text figs.

Coscinodiscus subbulliens

Gran, H.H., 1908
Diatomeen. Nordisches Plankton, Botanischer Teil pp. XIX.1-XIX 146; 178 text figs.

Coscinodiscus subbulliens

Hendy, N. Ingram, 1964
An introductory account of the smaller algae of British coastal waters. V. Bacillariophyceae (Diatoms). Her Majesty's Stationary Office, 317 pp., 45 pls.

Coscinodiscus sub-bullians

Hendey, N.I., 1937
The plankton diatoms of the southern seas. Discovery Repts. 16:151-364, pls.6-13.

Coscinodiscus subbulliens n.sp.

Jørgensen, E., 1905
B. Protistplankton and the diatoms in bottom samples. Hydrographical and biological investigations in Norwegian fjords. Bergens Mus. Skr. 7: 49-225.

Coscinodiscus subbulliens

Mangin, L., 1915
Phytoplancton de L'Antartique. Deuxieme Exped. Ant. Francaise (1908-1910), 95 pp., 3 pls., 58 text figs.

Coscinodiscus subbullians

Rumkówna, A., 1948
List of the phytoplankton species occurring in the superficial water layers in the Gulf of Gdańsk] Bull. Lab. mar., Gdynia, No. 4: 139-141 with tables in back.

Coscinodiscus subbulliens

Schodduyn, M., 1926
Observations faites dans la baie d'Ambleteuse (Pas de Calais). Bull. Inst. Ocean., Monaco, No. 482: 64 pp.

Coscinodiscus subconcavus

Morse, D.C., 1947
Some observations on seasonal variations in plankton population Patuxant River, Maryland 1943-1945. Bd. Nat. Res., Publ. No.65, Chesapeake Biol. Lab., 31, 3 figs.

Coscinodiscus sublineatus

Cleve-Euler, A., 1951
Die Diatomeen von Schweden und Finnland. Kungl. Svenska Vetenskaps Akad. Handl., Fjärde Ser. 2(1): 161 pp., 6 pls.

Coscinodiscus sublineatus

Zanon, V., 1948
Diatomee marini di Sardegna e Pugillo di Alghe Marine della stressa. Boll. Pesca, Piscitutura e Idrobiologia, Anno 24, ns. 3(2): 202-244, 27 figs. on 1 pl.

Coscinodiscus subsalus

Cleve-Euler, A., 1951
Die Diatomeen von Schweden und
Finnland. Kungl. Svenska Vetenskaps
Akad. Handl., Fjärde Ser. 2(1): 161 pp.,
6 pls.

Coscinodiscus? subtabulatus n.sp.

Cleve-Euler, A., 1951
Die Diatomeen von Schweden und
Finnland. Kungl. Svenska Vetenskaps
Akad. Handl., Fjärde Ser. 2(1): 161 pp.,
6 pls.

Coscinodiscus subtilis

Allen, W.E., and E.E. Cupp, 1935
Plankton diatoms of the Java Sea.
Annales du Jardin Botanique de Buitenzorg
XLIV (2):101-174, figs.1-127.

(drawings of all species mentioned)

Coscinodiscus subtilis

Bigelow, H.B., and M. Leslie, 1930
Reconnaissance of the waters and
plankton of Monterey Bay, July 1928.
Bull. M.C.Z., 70(5):429-481, 43 text
figs.

Coscinodiscus subtilis

Boden, B.P., 1950
Some marine plankton diatoms from the
west coast of South Africa. Trans. R.Soc.
S. Africa. 32:321-434, 100 text figs.

Coscinodiscus subtilis

Cleve-Euler, A., 1951
Die Diatomeen von Schweden und
Finnland. Kungl. Svenska Vetenskaps
Akad. Handl., Fjärde Ser. 2(1): 161 pp.,
6 pls.

Coscinodiscus (Thalassiosira) subtilis

Cleve-Euler, A., 1951
Die Diatomeen von Schweden und
Finnland. Kungl. Svenska Vetenskaps
Akad. Handl., Fjärde Ser. 2(1): 161 pp.,
6 pls.

Coscinodiscus subtilis

Frenguelli, Joaquin, and Hector Antonio
Orlando, 1959.

Operacion MERLUZA. Diatomeas y silicoflagelados del plancton del "VI Crucero".
Servicio Hidrogr. Naval., Argentina, Publ.
No. H. 619: 5-62.

Coscinodiscus subtilis

Gran, H.H., 1908
Diatomeen. Nordisches Plankton, Botanischer Teil pp. XIX 1-XIX 146; 178 text figs.

Coscinodiscus subtilis

Hendey, N.I., 1937
The plankton diatoms of the southern seas.
Discovery Repts. 16:151-364, pls.6-13.

Coscinodiscus subtilis

Macdonald, R., 1933
An examination of plankton hauls made in
the Suez Canal during the year 1928. Fish.
Res. Dis., Notes & Mem. No.3, 11 pp., 1 chart.

Coscinodiscus subtilis

Mann, A., 1907
Report on the diatoms of the Albatross
voyages in the Pacific Ocean, 1888-1904.
Contrib. U. S. Nat. Herb. 10(5):221-419, Pls.
XLIV-LIV.

Coscinodiscus subtilis

Meunier, A., 1915
Microplancton de la Mer Flamande. 2.
Diatomées (excepté le genre Chaetoceros). Mem.
Mus. Roy. Hist. Nat., Belgique, 7(3):1-118,
Pls. VIII-XIV.

Coscinodiscus subtilis

Sousa e Silva, E., 1949
Diatomaceas e Dinoflagelados de
Baia de Cascais. Portugaliae Acta Biol.,
Volume: Julio Henriques, Ser. B: 300-383,
9 pls, 2 fold-in tables.

Coscinodiscus symbolophorus

Bigelow, H.B., and M. Leslie, 1930
Reconnaissance of the waters and
plankton of Monterey Bay, July 1928.
Bull. M.C.Z., 70(5):429-481, 43 text
figs.

Coscinodiscus symbolophorus

Jouse, A.P., G.S. Koroleva, G.A. Nagaeva,
1962
Diatoms in the surface layer of sediment in
the Indian sector of the Antarctic. Investigations of marine bottom sediments.
(In Russian; English summary).
Trudy Inst. Okeanol., Akad. Nauk, SSSR, 61:
19-92.

Coscinodiscus symbolophorus

Kozlova, O.G., 1962
Specific composition of diatoms in the waters
of the Indian sector of the Antarctic. Investigation of Marine Bottom Sediments. (In
Russian).
Trudy Inst. Okeanol., Akad. Nauk, SSSR, 61:
3-18.

English summary, p. 18.

Coscinodiscus symbolophorus

Mann, A., 1893
List of Diatomaceae from a deep-sea dredging in the Atlantic Ocean off Delaware Bay by
the U. S. Fish Commission Steamer Albatross.
Proc. U. S. Nat. Mus. 16:303-312.

Coscinodiscus symmetricus

Cleve-Euler, A., 1951
Die Diatomeen von Schweden und
Finnland. Kungl. Svenska Vetenskaps
Akad. Handl., Fjärde Ser. 2(1): 161 pp.,
6 pls.

Coscinodiscus symmetricus

Mann, A., 1893
List of Diatomaceae from a deep-sea dredging in the Atlantic Ocean off Delaware Bay by
the U. S. Fish Commission Steamer Albatross.
Proc. U. S. Nat. Mus. 16:303-312.

Coscinodiscus tabularis

Jouse, A.P., G.S. Koroleva, G.A. Nagaeva,
1962
Diatoms in the surface layer of sediment in
the Indian sector of the Antarctic. Investigations of marine bottom sediments.
(In Russian; English summary).
Trudy Inst. Okeanol., Akad. Nauk, SSSR, 61:
19-92.

Coscinodiscus tabularis

Manguin, E., 1954

Diatomees marines provenant de l'ile
Heard (Australian National Antarctic
Expedition). Rev. Algol., n.s., 1:
14-24.

Coscinodiscus termidus

Hendey, N.I., 1937
The plankton diatoms of the southern seas.
Discovery Repts. 16:151-364, pls.6-13.

Coscinodiscus thori n.sp.

Pavillard, J., 1925
Bacillariales. Rept. on the Danish Oceangr.
Exped., 1908-10 to the Mediterranean and adj.
seas. Vol.II., Biol. J4:72 pp., 116 text figs.

Coscinodiscus traduceus hispida

Mann, A., 1893
List of Diatomaceae from a deep-sea dredging in the Atlantic Ocean off Delaware Bay by
the U. S. Fish Commission Steamer Albatross.
Proc. U. S. Nat. Mus. 16:303-312.

Coscinodiscus trigonus

Hendey, N.I., 1937
The plankton diatoms of the southern seas.
Discovery Repts. 16:151-364, pls.6-13.

Coscinodiscus tumidus

Jouse, A.P., G.S. Koroleva, G.A. Nagaeva,
1962
Diatoms in the surface layer of sediment in
the Indian sector of the Antarctic. Investigations of marine bottom sediments.
(In Russian; English summary).
Trudy Inst. Okeanol., Akad. Nauk, SSSR, 61:
19-92.

Coscinodiscus umbonatus n.sp.

Castracane degli Antelminelli, F., 1886
1. Report on the Diatomaceae collected
by H.M.S. Challenger during the years 1873-
1876. Rept. Sci. Results, H.M.S. Challenger,
Botany Vol. II, 178 pp., 30 pls.

Coscinodiscus undulatus n.sp.

Castracane degli Antelminelli, F., 1886
1. Report on the Diatomaceae collected
by H.M.S. Challenger during the years 1873-
1876. Rept. Sci. Results, H.M.S. Challenger,
Botany Vol. II, 178 pp., 30 pls.

Coscinodiscus undulosus n.sp.

Mann, A., 1907
Report on the diatoms of the Albatross
voyages in the Pacific Ocean, 1888-1904.
Contrib. U. S. Nat. Herb. 10(5):221-419, Pls.
XLIV-LIV.

Coscinodiscus varians

Boden, B.P., 1950
Some marine plankton diatoms from the
west coast of South Africa. Trans. R.Soc.
S. Africa. 32:321-434, 100 text figs.

Coscinodiscus variolatus n.sp.

Castracane degli Antelminelli, F., 1886
1. Report on the Diatomaceae collected
by H.M.S. Challenger during the years 1873-
1876. Rept. Sci. Results, H.M.S. Challenger,
Botany Vol. II, 178 pp., 30 pls.

Coscinodiscus velatus

Cleve-Euler, A., 1951
Die Diatomeen von Schweden und
Finnland. Kungl. Svenska Vetenskaps
Akad. Handl., Fjärde Ser. 2(1): 161 pp.,
6 pls.

Coscinodiscus (?) venulosus

Castracane degli Antelminelli, F., 1886
1. Report on the Diatomaceae collected
by H.M.S. Challenger during the years 1873-
1876. Rept. Sci. Results, H.M.S. Challenger,
Botany Vol. II, 178 pp., 30 pls.

Coscinodiscus verecundus n.sp.

Mann, A., 1907
Report on the diatoms of the Albatross
voyages in the Pacific Ocean, 1888-1904.
Contrib. U. S. Nat. Herb. 10(5):221-419, Pls.
XLIV-LIV.

Coscinodiscus vidovichii

Hendey, N-Ingram, 1958 [1957(Publ. 1958)]

Marine diatoms from some West African
Ports- J. R. Microsc. Soc. (3) 77(1/2):
28-85.

Coscinodiscus Vidovichi, n.sp.

Müller Melchers, F.C., 1953.
New and little known diatoms from Uruguay and
the South Atlantic coast.
Com. Bot., Mus. Hist. Nat., Montevideo, 3(30):
1-11, 8 pls.

Coscinodiscus wailesii

Cassie, Vivienne, 1961
Marine phytoplankton in New Zealand waters.
Botanica Marina, 2(Suppl.):54 pp., 8 pls.

Coscinodiscus wailesii
Cupp, Easter E., 1943
Marine plankton diatoms of the west coast of North America. Bull. S.I.O. 5(1):1-238, 5 pls., 168 text figs.

Coscinodiscus wailesii (?)
Gilbert, J.Y., and W.E. Allen, 1943
The phytoplankton of the Gulf of California obtained by the "E.W. Scripps" in 1939 and 1940. J. Mar. Res. V(2):89-110, figs.30-31.

Coscinodiscus wailesii n.sp.
Gran, H. H. and E. C. Angst, 1931
Plankton diatoms of Puget Sound. Publ. Puget Sound Biol. Sta. 7:417-519, 95 text figs.

Coscinodiscus wailesii
Holmes, Robert W., 1966.
Light microscope observations on cytological manifestations of nitrate, phosphate, and silicate deficiency in four marine centric diatoms. J. Phycology, 2(4):136-140.

Coscinodiscus wailesii
Kesseler, Hanswerner, 1967.
Untersuchungen über die chemisch Zusammensetzung des Zellsaftes der Diatomee Coscinodiscus wailesii (Bacillariophyceae, Centrales). Helgoländer wiss. Meeresunters., 16(3):262-270.

Coscinodiscus wailesii
Kokubo, S., 1952
Results of the observations on the plankton and oceanography of Mutsu Bay during 1950, reference being made also to the period 1946-1950. Bull Mar. Biol. Sta., Asamushi 5(1/4): 1-54, 3 tables, (fold-in), 1 fold-in.

Coscinodiscus Woodwardii
Bigelow, H.B., and M. Leslie, 1930
Reconnaissance of the waters and plankton of Monterey Bay, July 1928. Bull. M.C.Z., 70(5):429-481, 43 text figs.

Coscinodiscus woodwardii
Mann, A., 1907
Report on the diatoms of the Albatross voyages in the Pacific Ocean, 1888-1904. Contrib. U. S. Nat. Herb. 10(5):221-419, Pls. XLIV-LIV.

Coscinosira antarctica
Jouse, A.P., G.S. Koroleva, G.A. Nagaeva, 1962
Diatoms in the surface layer of sediment in the Indian sector of the Antarctic. Investigations of marine bottom sediments. (In Russian; English summary).
Trudy Inst. Okeanol., Akad. Nauk. SSSR, 61:19-92.

Coscinosira antarctica nsp.
Kozlova, O.G., 1962
Specific composition of diatoms in the waters of the Indian sector of the Antarctic. Investigation of Marine Bottom Sediments. (In Russian).
Trudy Inst. Okeanol., Akad. Nauk. SSSR, 61: 3-18.

English summary, p. 18.

Coscinosira atlantica
Brunel, J., 1962
Le phytoplancton de la Baie de Chaleurs. Inst. Botan., Univ. Montréal, Contrib. No. 77: 365 pp., 66 pls.

Coscinosira atlantica n.sp.
Dangeard, P., 1927
Phytoplankton de la croisière du "Sylvana". Ann. Inst. Ocean., Monaco, n.s., 4(8):286-401, 54 text figs. (Feirer-Juin 1913).

Coscinosira mediterranea
Brunel, J., 1962
Le phytoplancton de la Baie de Chaleurs. Inst. Botan., Univ. Montréal, Contrib. No. 77: 365 pp., 66 pls.

Coscinosira Oestrupii
Allen, W.E., and E.E. Cupp, 1935
Plankton diatoms of the Java Sea. Annales du Jardin Botanique de Buitenzorg XLIV (2):101-174, figs.1-127.

(drawings of all species mentioned)

Coscinosira Oestrupi
Braarud, T., 1934
A note on the phytoplankton of the Gulf of Maine in the summer of 1933. Biol. Bull. 67(1):76-82.

Coscinosira Oestrupii
Gran, H.H., 1908
Diatomeen. Nordisches Plankton, Botanischer Teil pp. XIX 1-XIX 146; 178 text figs.

Coscinosira Oestrupi
Gran, H.H., and T. Braarud, 1935
A quantitative study of the phytoplankton in the Bay of Fundy and the Gulf of Maine (including observations on hydrography, chemistry, and turbidity). J. Biol. Bd., Canada, 1(5):279-467, 69 text figs.

Coscinosira oestrupii
Hendy, N. Ingram, 1964
An introductory account of the smaller algae of British coastal waters. V. Bacillariophyceae (Diatoms). Her Majesty's Stationary Office, 317 pp., 45 pls.

Coscinosira Oestrupi
Lillick, L.C., 1940
Phytoplankton and planktonic protozoa of the offshore waters of the Gulf of Maine. Pt.II. Qualitative Composition of the Planktonic Flora. Trans. Am. Phil. Soc., n.s., 31(3):193-237, 13 text figs.

Coscinosira oestrupii
Misra, J.N., 1956.
A systematic account of some littoral marine diatoms from the west coast of India. J. Bombay Nat. Hist. Soc., 53(4):537-568.

Coscinosira polychorda
Bigelow, H.B., and M. Leslie, 1930
Reconnaissance of the waters and plankton of Monterey Bay, July 1928. Bull. M.C.Z., 70(5):429-481, 43 text figs.

Coscinosira polychorda
Braarud, T., 1945
Experimental studies on marine plankton diatoms. Avhandlingar utgitt av Det Norske Videnskaps-Akademi i Oslo. 1. Mat.-Naturv. Klasse 1944. No.10, 1-16, 1 text fig.

Coscinosira polychorda
Braarud, T., 1945
A phytoplankton survey of the polluted waters of inner Oslo Fjord. Hvalrådets Skrifter, No.28, 142 pp., 19 text figs., 17 tables.

Coscinosira polychorda
Braarud, T., and Adam Burse, 1939
On the phytoplankton of the Oslo Fjord, 1933-1934. Hvalrådets Skr. No.19:1-63; 9 text figs. Reviewed. J. du. Cons. 14(3): 418-420. A.C. Gardiner.

Coscinosira polychorda
Brunel, J., 1962
Le phytoplancton de la Baie de Chaleurs. Inst. Botan., Univ. Montréal, Contrib. No. 77: 365 pp., 66 pls.

Coscinosira polychorda
Brunel, Jules, 1962
Le phytoplancton de la Baie des Chaleurs. Contrib. Ministère de la Chasse et des Pêcheries, Province de Québec, No. 91: 365 pp.

Coscinosira polychorda
Cupp, Easter E., 1943
Marine plankton diatoms of the west coast of North America. Bull. S.I.O. 5(1):1-238, 5 pls., 168 text figs.

Coscinosira polychorda
Davidson, V.M., 1931.
Biological and oceanographic conditions in Hudson Bay. Contr. Canadian Biol. Fish., n.s., 6(26):497-509, 7 textfigs.

Coscinosira polychorda
Gilbert, J.Y., and W.E. Allen, 1943
The phytoplankton of the Gulf of California obtained by the "E.W. Scripps" in 1939 and 1940. J. Mar. Res. V(2):89-110, figs.30-31.

Coscinosira polychorda
Gran, H.H., 1908
Diatomeen. Nordisches Plankton, Botanischer Teil pp. XIX 1-XIX 146; 178 text figs.

Coscinodiscus polychordus n.sp.
Gran, H.H., 1897
Protophyta: Diatomaceae, Silico-flagellata and Cilioflagellata. Den Norske Nordhavs Expedition 1876-1878, h. 24, 36 pp., 4 pls.

Coscinosira polychordus
Gran, H.H., 1897
Protophyta: Diatomaceae, Silico-flagellata and Cilioflagellata. Den Norske Nordhavs Expedition 1876-1878, h. 24, 36 pp., 4 pls.

Coscinosira polychorda
Gran, H.H., and T. Braarud, 1935
A quantitative study of the phytoplankton in the Bay of Fundy and the Gulf of Maine (including observations on hydrography, chemistry, and turbidity). J. Biol. Bd., Canada, 1(5):279-467, 69 text figs.

Coscinosira polychorda
Gran, H. H. and E. C. Angst, 1931
Plankton diatoms of Puget Sound. Publ. Puget Sound Biol. Sta. 7:417-519, 95 text figs.

Coscinosira polychorda
Hendy, N. Ingram, 1964
An introductory account of the smaller algae of British coastal waters. V. Bacillariophyceae (Diatoms). Her Majesty's Stationary Office, 317 pp., 45 pls.

Coscinosira polychorda
Jørgensen, E., 1905
B. Protistplankton and the diatoms in bottom samples. Hydrographical and biological investigations in Norwegian fjords. Bergens Mus. Skr. 7: 49-225.

Coscinosira polychorda
Lillick, L.C., 1940
Phytoplankton and planktonic protozoa of the offshore waters of the Gulf of Maine. Pt.II. Qualitative Composition of the Planktonic Flora. Trans. Am. Phil. Soc., n.s., 31(3):193-237, 13 text figs.

Coscinosira polychorda
Meunier, A., 1915
Microplancton de la Mer Flamande. 2. Diatomées (excepté le genre Chaetoceros). Mem. Mus. Roy. Hist. Nat., Belgique, 7(3):1-118, Pls. VIII-XIV.

Coscinosira polychorda
Morse, D.C., 1947
Some observations on seasonal variations in plankton population Patuxent River, Maryland 1943-1945. Bd. Nat. Res., Publ. No.65, Chesapeake Biol. Lab., 31, 3 figs.

Coscinosira polychorda
Pavillard, J., 1925
Bacillariales. Rept. on the Danish Oceanogr. Exped., 1908-10 to the Mediterranean and adj. seas. Vol.II., Biol. J4:72 pp., 116 text figs.

Coscinosira polycorda

Phifer, L.D. (undated)
The occurrence and distribution of plankton diatoms in Bering Sea and Bering Strait, July 26-August 24, 1934. Report of Oceanographic cruise of U.S. Coast Guard Cutter Chelan 1934, Part II(A):1-44 (mimeographed) plus fig.1 (after Pt.B)

Coscinosira poroseriata n.sp.

Ramsfjell, Einar, 1959
Two new phytoplankton species from the Norwegian Sea, the diatom Coscinosira poroseriata, and the dinoflagellate Goniaulax parva.
Nytt Magasin for Botanikk, 7: 175-177.

Craspidediscus sp.

Morse, D.C., 1947
Some observations on seasonal variations in plankton population Patuxant River, Maryland 1943-1945. Bd. Nat. Res., Publ. No.65, Chesapeake Biol. Lab., 31, 3 figs.

Craspedodiscus coscinodiscus

Mann, A., 1907
Report on the diatoms of the Albatross voyages in the Pacific Ocean, 1888-1904. Contrib. U.S. Nat. Herb. 10(5):221-419, Pls. XLIV-LIV.

Craspedodiscus mesotylus n.sp.

Barker, J.W. and S.H. Meakin, 1948
New and rare diatoms. J. Quakett Micros. Club, ser.4, 2(5):233-235, pl.25.

Craspelodiscus minor

Cleve-Euler, A., 1951
Die Diatomeen von Schweden und Finnland. Kungl. Svenska Vetenskaps Akad. Handl., Fjärde Ser. 2(1): 161 pp., 6 pls.

Craspelodiscus mölleri

Cleve-Euler, A., 1951
Die Diatomeen von Schweden und Finnland. Kungl. Svenska Vetenskaps Akad. Handl., Fjärde Ser. 2(1): 161 pp., 6 pls.

Craspedoporus elegans

Barker, J.W. and S.H. Meakin, 1948
New and rare diatoms. J. Quakett Micros. Club, ser.4, 2(5):233-235, pl.25.

Cyclophora n.gen.

Castracane degli Antelminelli, F., 1886
1. Report on the Diatomaceae collected by H.M.S. Challenger during the years 1873-1876. Rept. Sci. Results, H.M.S. Challenger, Botany Vol. II, 178 pp., 30 pls.

Cyclophora tenuis

Castracane degli Antelminelli, F., 1886
1. Report on the Diatomaceae collected by H.M.S. Challenger during the years 1873-1876. Rept. Sci. Results, H.M.S. Challenger, Botany Vol. II, 178 pp., 30 pls.

Cyclophora tenuis

Rampi, L., 1942
Ricerche sul fitoplancton del Mare Ligure 6. Le diatomee delle acque di Sanremo. Nuovo Giornale Botanico Italiano, N.S., 49:252-268.

Cyclophora tenuis

Rampi, L., 1940
Diatomee del Mare Adriatico. Nuovo Giornale Botanico Italiano, n.s., 47:559-608.

Cyclophora tenuis

Zanon, V., 1948
Diatomee marini di Sardegna e Pugillo di Alghe Marine della stressa. Boll. Pesca, Piscitutura e Idrobiologia, Anno 24, ns. 3(2): 202-244, 27 figs. on 1 pl.

Cyclotella

Brunel, J., 1962
Le phytoplancton de la Baie de Chaleurs. Inst. Botan., Univ. Montréal, Contrib. No. 77: 365 pp., 66 pls.

Cyclotella

Desikachary, T.V., 1956(1957).
Electron microscope studies on diatoms. J.R. Microsc. Soc. (3)76(1/2):9-36.

Cyclotella sp.

Lillick, L.C., 1940
Phytoplankton and planktonic protozoa of the offshore waters of the Gulf of Maine. Pt.II. Qualitative Composition of the Planktonic Flora. Trans. Am. Phil. Soc., n.s., 31(3):193-237, 13 text figs.

Cyclotella sp.

Lillick, L.C., 1938
Preliminary report of the phytoplankton of the Gulf of Maine. Am. Mid. Nat. 20(3):624-640, 1 text fig. 37 tables.

Cyclotella spp.

Round, E.E., 1970.
The delimitation of the genera Cyclotella and Stephanodiscus by light microscopy, transmission and reflecting electron microscopy.
Beihefte Nova Hedwigia 31: 591-604.

Cyclotella antiqua

Cleve-Euler, A., 1951
Die Diatomeen von Schweden und Finnland. Kungl. Svenska Vetenskaps Akad. Handl., Fjärde Ser. 2(1): 161 pp., 6 pls.

Cyclotella atomus

Hasle, Grethe Rytter, 1962
Three Cyclotella species from marine localities studied in the light and electron microscopes.
Nova Hedwigia, Zeits. für Kryptogamenkunde. 4(3/4):299-307, Pls. 57-63.

Cyclotella bodanica

Cleve-Euler, A., 1951
Die Diatomeen von Schweden und Finnland. Kungl. Svenska Vetenskaps Akad. Handl., Fjärde Ser. 2(1): 161 pp., 6 pls.

Cyclotella caspia

Braarud, T., and Bjørg Føyn, 1958.
Phytoplankton observations in a brackish water locality of south-east Norway.
Nytt Mag. Botan., 6:47-73.

Cyclotella caspia

Cleve-Euler, A., 1951
Die Diatomeen von Schweden und Finnland. Kungl. Svenska Vetenskaps Akad. Handl., Fjärde Ser. 2(1): 161 pp., 6 pls.

Cyclotella caspia

Hasle, Grethe Rytter, 1962
Three Cyclotella species from marine localities studied in the light and electron microscopes.
Nova Hedwigia, Zeits. für Kryptogemenkunde. 4(3/4):299-307, Pls. 57-63.

Cyclotella caspia

Hendy, N. Ingram, 1964
An introductory account of the smaller algae of British coastal waters. V. Bacillariophyceae (Diatoms).
Her Majesty's Stationary Office, 317 pp., 45 pls.

Cyclotella caspia

Wawrik, F., 1961.
Die horizontale Verteilung der Planktondiatomeen im Golf von Neapel.
Int. Revue Ges Hydrobiol., 46(3):460-479.

Cyclotella catenata

Cleve-Euler, A., 1951
Die Diatomeen von Schweden und Finnland. Kungl. Svenska Vetenskaps Akad. Handl., Fjärde Ser. 2(1): 161 pp., 6 pls.

Cyclotella comensis

Cleve-Euler, A., 1951
Die Diatomeen von Schweden und Finnland. Kungl. Svenska Vetenskaps Akad. Handl., Fjärde Ser. 2(1): 161 pp., 6 pls.

Cyclotella comte

Cleve-Euler, A., 1951
Die Diatomeen von Schweden und Finnland. Kungl. Svenska Vetenskaps Akad. Handl., Fjärde Ser. 2(1): 161 pp., 6 pls.

Cyclotella comta

Tsumura, K., 1956.
Diatomoj el la cirkaufoso de la restajo de la kastele de Odawara.
J. Yokohama Municipal Univ. (C-14) No. 47:23 pp.

Cyclotella cryptica

McLachlan, J., and J.S. Craigie, 1966.
Chitan fibres in Cyclotella cryptica and growth of C. cryptica and Thalassiosira fluviatilis.
In: Some contemporary studies in marine science H. Barnes, editor, George Allen & Unwin, Ltd., 511-517.

Cyclotella cryptica

Reimann, Bernhard E.F., Joyce M.C. Lewin and Robert R.L. Guillard, 1963
Cyclotella cryptica, a new brackish-water diatom species.
Phycologia 3(2):75-84.

Cyclotella cryptica

Schultz, Mary E., and Francis R. Trainor, 1968.
Production of male gametes and auxospores in the centric diatoms Cyclotella meneghiniana and C. cryptica.
J. Phycol. 4(2):85-88.

Cyclotella fimbriata n.sp.

Castracane degli Antelminelli, F., 1886
1. Report on the Diatomaceae collected by H.M.S. Challenger during the years 1873-1876. Rept. Sci. Results, H.M.S. Challenger, Botany Vol. II, 178 pp., 30 pls.

Cyclotella frigida n.sp.

Cleve-Euler, A., 1951
Die Diatomeen von Schweden und Finnland. Kungl. Svenska Vetenskaps Akad. Handl., Fjärde Ser. 2(1): 161 pp., 6 pls.

Cyclotella glomerata

Cleve-Euler, A., 1951
Die Diatomeen von Schweden und Finnland. Kungl. Svenska Vetenskaps Akad. Handl., Fjärde Ser. 2(1): 161 pp., 6 pls.

Cyclotella gothica

Cleve-Euler, A., 1951
Die Diatomeen von Schweden und Finnland. Kungl. Svenska Vetenskaps Akad. Handl., Fjärde Ser. 2(1): 161 pp., 6 pls.

Cyclotella iris

Cleve-Euler, A., 1951
Die Diatomeen von Schweden und Finnland. Kungl. Svenska Vetenskaps Akad. Handl., Fjärde Ser. 2(1): 161 pp., 6 pls.

Cyclotella Kützingiana

Cleve-Euler, A., 1951
Die Diatomeen von Schweden und Finnland. Kungl. Svenska Vetenskaps Akad. Handl., Fjärde Ser. 2(1): 161 pp., 6 pls.

Cyclotella kützingiana

de Sousa e Silva, E., 1956.
Contribution à l'étude du microplancton de Dakar et des regions maritimes voisines.
Bull. I.F.A.N., 8(2):335-371, 7 pls.

Cyclotella kützingiana

Iyengar, M.O.P. and G.Venkataraman, 1951.
The ecology and seasonal succession of the algae flora of the River Cooum at Madras with special reference to the Diatomaceae. J. Madras Univ. 21, Sect. B(1): 140-192, 1 pl of 4 figs., 11 text figs.

Cyclotella Kützingiana

Zanon, D. V., 1949
Diatomee di Buenos Aires (Argentina)
Atti Accad. Naz. Lincei, Memorie, Cl. Sci. fis., mat. e. nat., ser. 7, 11(3):59-151, 2 pls.

Cyclotella? ladogensis n.sp.

Cleve-Euler, A., 1951
Die Diatomeen von Schweden und Finnland. Kungl. Svenska Vetenskaps Akad. Handl., Fjärde Ser. 2(1): 161 pp., 6 pls.

Cyclotella melosiroides

Cleve-Euler, A., 1951
Die Diatomeen von Schweden und Finnland. Kungl. Svenska Vetenskaps Akad. Handl., Fjärde Ser. 2(1): 161 pp., 6 pls.

Cyclotella meneghiniana plana

Cleve-Euler, A., 1951
Die Diatomeen von Schweden und Finnland. Kungl. Svenska Vetenskaps Akad. Handl., Fjärde Ser. 2(1): 161 pp., 6 pls.

Cyclotella meneghiniana

Iyengar, M.O.P. and G.Venkataraman, 1951.
The ecology and seasonal succession of the algae flora of the River Cooum at Madras with special reference to the Diatomaceae. J. Madras Univ. 21, Sect. B(1): 140-192, 1 pl of 4 figs., 11 text figs.

Cyclotella meneghiniana

Misra, J.N., 1956.
A systematic account of some littoral marine diatoms from the west coast of India.
J. Bombay Nat. Hist. Soc., 53(4):537-568.

Rao, V.N.R. 1970. Cyclotella meneghiniana.
Studies on Cyclotella meneghiniana Kütz. 1. Sexual reproduction and auxospore formation.
Proc. Indian Acad. Sci. (B) 72 (6): 285-287.

Cyclotella Meneghiniana

Rumkówna, A., 1948
[List of the phytoplankton species occurring in the superficial water layers in the Gulf of Gdansk] Bull. Lab. mar. Gdynia, No. 4: 139-141 with tables in back.

Schultz, Mary E., and Francis R. Trainor 1968. Cyclotella meneghiniana
Production of male gametes and auxospores in the centric diatoms Cyclotella meneghiniana and C. cryptica.
J. Phycol. 4(2): 85-88

Cyclotella Meneghiniana

Steeman Nielsen, E., 1962.
On the maximum quantity of plankton chlorophyll per surface unit of a lake or the sea.
Int. Rev. Ges. Hydrobiol., 47(3):333-338.

Cyclotella meneghiniana

Tsumura, K., 1956. Diatomoj el la cirkaufoso de la restajo de la kastele de Odawara.
J. Yokohama Municipal Univ., (C-14) No. 47:23 pp.

Cyclotella Meneghiniana

Zanon, D. V., 1949
Diatomee di Buenos Aires (Argentina)
Atti Accad. Naz. Lincei, Memorie, Cl. Sci. fis., mat. e. nat., ser. 7, 11(3):59-151, 2 pls.

Cyclotella nana

Fuhs, G. Wolfgang 1969.
Phosphorus content and rate of growth in the diatoms Cyclotella nana and Thalassiosira fluviatilis.
J. Phycol. 5(4): 312-321

Cylotella nana

Guillard, Robert R.L., Peter Kilham and Togwell A. Jackson 1973.
Kinetics of silicon-limited growth in the marine diatom Thalassiosira pseudonana Hasle and Heimdal (= Cyclotella nana Hustedt).
J. Phycol. 9(3): 233-237.

Cyclotella nana

Guillard, R.R.L. and S. Myklestad, 1970.
Osmotic and ionic requirements of the marine centric diatom Cyclotella nana. Helgoländer wiss. Meeresunters, 20(1/4): 104-110.

Cyclotella nana

Guillard, Robert R.L., and John H. Ryther, 1962
Studies of marine planktonic diatoms. I. Cyclotella nana Hustedt and Detonula confervacea. (Cleve).
Canadian J. Microbiol., 8(2):229-240.

Cyclotella nana

Hobson, Louis A. and Robert J. Pariser, 1971.
The effect of inorganic nitrogen on macromolecular synthesis by Thalassiosira fluviatilis Hustedt and Cyclotella nana Hustedt grown in batch culture. J. exp. mar. Biol. Ecol., 6(1): 71-78.

Kanagawa, Akio, Mitsuki Yoshioka and Shin-ichi Teshima 1971. Cyclotella nana
The occurrence of brassicasterol in the diatoms, Cyclotella nana and Nitzschia closterium.
Bull. Jap. Soc. scient. Fish. 37(9): 899-903

Cyclotella nana

Paasche, E., 1973
Silicon and the ecology of marine plankton diatoms. 1. Thalassiosira pseudonana (Cyclotella nana) grown in a chemostat with silicate as limiting nutrient. Mar. Biol. 19(2): 117-126.

Cyclotella nana

Prakash, A., Liv Skoglund, Britt Rystad and Anne Jensen 1973
Growth and cell-size distribution of marine planktonic algae in batch and dialysis cultures.
J. Fish. Res. Bd. Can. 30(2):143-155

Cyclotella nana

Ryther, John H., and Dana D. Kramer, 1961
Relative iron requirement of some coastal and offshore algae.
Ecology, 42(2): 444-446.

Cyclotella nana [a]

Smayda, Theodore J., 1971.
Further enrichment experiments using the marine centric diatom Cyclotella nana (clone 13-1) as an assay organism. (Portuguese abstract). In: Fertility of the Sea, John D. Costlow, editor, Gordon Breach, 2: 493-509.

Cyclotella nana

Smayda, Theodore J., 1964.
Enrichment experiments using the marine centric diatom Cyclotella nana (clone 13-1) as an assay organism.
Narragansett Mar. Lab., Univ. Rhode Island, Occ. Publ., No. 2:25-32.

Cyclotella nana

Takano, Hideaki, 1967.
Some culture experiments of Cyclotella nana Hustedt. (In Japanese; English abstract).
Inf. Bull. Planktol. Japan, Comm. No. Dr. Y. Matsue, 231-

Cyclotella nana

Taylor, W. Rowland, 1964.
Inorganic nutrient requirements for marine phytoplankton organisms.
Narragansett Mar. Lab., Univ. Rhode Island, Occ. Publ., No. 2:17-24.

Cyclotella nana [a]

Wallen, D.G. and G.H. Geen, 1971.
Light quality and concentration of proteins, RNA, DNA and photosynthetic pigments in two species of marine plankton algae. Mar. Biol. 10(1): 44-51.

Cyclotella nana [a]

Wallen, D.G. and G.H. Geen, 1971.
Light quality in relation to growth, photosynthetic rates and carbon metabolism in two species of marine plankton algae. Mar. Biol. 10(1): 34-43.

Cyclotella nana

Yentsch, Charles S., and Carol A. Reichert, 1962
The interrelationship between water-soluble yellow pigment sybstances and chloroplastic pigments in marine algae.
Botanica Marina, 3(3/4):65-74.

Cyclotella nana

Yoo, Sung Kyoo, 1968.
Studies on the growth of algal food, Cyclotella nana, Chaetoceros calcitrans and Monochrysis lutheri. (In Korean; English abstract).
Bull. Pusan Fish. Coll. (Nat. Sci.), 8(2):123-126.

Cyclotella ocellata

Zanon, D. V., 1949
Diatomee di Buenos Aires (Argentina)
Atti Accad. Naz. Lincei, Memorie, Cl. Sci. fis., mat. e. nat., ser. 7, 11(3):59-151, 2 pls.

Cyclotella operculata

Cleve-Euler, A., 1951
Die Diatomeen von Schweden und Finnland. Kungl. Svenska Vetenskaps Akad. Handl., Fjärde Ser. 2(1): 161 pp., 6 pls.

Cyclotella physoplea

Mann, A., 1893
List of Diatomaceae from a deep-sea dredging in the Atlantic Ocean off Delaware Bay by the U.S. Fish Commission Steamer Albatross.
Proc. U.S. Nat. Mus. 16:303-312.

Cyclotella quadrijunata

Hyber-Pestalozzi, G., 1951.
Über Gallertbildung bei Cyclotella quadrijunata (Schröter) Husted. Rev. Suisse Hydrol. [Schweiz Zeits. Hydrol.] 13:291-299, 6 textfigs.

Cyclotella regina n.sp.
Mann, A., 1907
Report on the diatoms of the Albatross voyages in the Pacific Ocean, 1888-1904. Contrib. U. S. Nat. Herb. 10(5):221-419, Pls. XLIV-LIV.

Cyclotella socialis
Cleve-Euler, A., 1951
Die Diatomeen von Schweden und Finnland. Kungl. Svenska Vetenskaps Akad. Handl., Fjärde Ser. 2(1): 161 pp., 6 pls.

Cyclotella socialis
Rumkówna, A., 1948
[List of the phytoplankton species occurring in the superficial water layers in the Gulf of Gdańsk] Bull. Lab. mar., Gdynia, No. 4: 139-141 with tables in back.

Cyclotella stelligera
Cleve-Euler, A., 1951
Die Diatomeen von Schweden und Finnland. Kungl. Svenska Vetenskaps Akad. Handl., Fjärde Ser. 2(1): 161 pp., 6 pls.

Cyclotella striata
Bigelow, H.B., and M. Leslie, 1930
Reconnaissance of the waters and plankton of Monterey Bay, July 1928. Bull. M.C.Z., 70(5):429-481, 43 text figs.

Cyclotella striata
Cleve-Euler, A., 1951
Die Diatomeen von Schweden und Finnland. Kungl. Svenska Vetenskaps Akad. Handl., Fjärde Ser. 2(1): 161 pp., 6 pls.

Cyclotella striata
Hasle, Grethe Rytter, 1962
Three Cyclotella species from marine localities studied in the light and electron microscopes.
Nova Hedwigia, Zeits. für Kryptogamenkunde, 4(3/4):299-307, Pls. 57-63.

Cyclotella striata
Hendey, N. Ingram, 1964
An introductory account of the smaller algae of British coastal waters. V. Bacillariophyceae (Diatoms).
Her Majesty's Stationary Office, 317 pp., 45 pls.

Cyclotella striata
Hendey, N-Ingram, 1958 [1957(Publ. 1958)]
Marine diatoms from some West African Ports- J. R.Microsc. Soc. (3) 77(1/2): 28-85.

Cyclotella striata
Mann, A., 1907
Report on the diatoms of the Albatross voyages in the Pacific Ocean, 1888-1904. Contrib. U. S. Nat. Herb. 10(5):221-419, Pls. XLIV-LIV.

Cyclotella striata
Mann, A., 1893
List of Diatomaceae from a deep-sea dredging in the Atlantic Ocean off Delaware Bay by the U. S. Fish Commission Steamer Albatross. Proc. U. S. Nat. Mus. 16:303-312.

Cyclotella striata
Morse, D.C., 1947
Some observations on seasonal variations in plankton population Patuxant River, Maryland 1943-1945. Bd. Nat. Res., Publ. No.65, Chesapeake Biol. Lab. 31, 3 figs.

Cyclotella striata
Zanon, D. V., 1949
Diatomee di Buenos Aires (Argentina) Atti Accad. Naz. Lincei, Memorie, Cl. Sci. fis., mat. e. nat., ser. 7, 11(3):59-151, 2 pls.

Cyclotella stylorum minuta
Florin, M-B., 1948
9. Diatomeae in submarine cores from the Tyrrhenian Sea. Medd. Ocean. Inst., Göteborg, 15 (Göteborgs Kungl. Vetenskaps-och Viterrhets Samhälles Handlingar, Sjätte Földjen, Ser. B 5(13):80-88.

Cyclotella stylorum
Hendey, N-Ingram, 1958 [1957(Publ. 1958)]
Marine diatoms from some West African Ports- J. R.Microsc. Soc. (3) 77(1/2): 28-85.

Cyclotella stylorum
Mann, A., 1907
Report on the diatoms of the Albatross voyages in the Pacific Ocean, 1888-1904. Contrib. U. S. Nat. Herb. 10(5):221-419, Pls. XLIV-LIV.

Cyclotella stylorum
Zanon, D. V., 1949
Diatomee di Buenos Aires (Argentina) Atti Accad. Naz. Lincei, Memorie, Cl. Sci. fis., mat. e. nat., ser. 7, 11(3):59-151, 2 pls.

Cyclotella? vätteri n.sp.
Cleve-Euler, A., 1951
Die Diatomeen von Schweden und Finnland. Kungl. Svenska Vetenskaps Akad. Handl., Fjärde Ser. 2(1): 161 pp., 6 pls.

Cyclotella virihensis n.sp.
Cleve-Euler, A., 1951
Die Diatomeen von Schweden und Finnland. Kungl. Svenska Vetenskaps Akad. Handl., Fjärde Ser. 2(1): 161 pp., 6 pls.

Cyclotella worticose n.sp.
Cleve-Euler, A., 1951
Die Diatomeen von Schweden und Finnland. Kungl. Svenska Vetenskaps Akad. Handl., Fjärde Ser. 2(1): 161 pp., 6 pls.

Cyclotella woltereckii
Belcher, J.H., E.M.F. Swale and J. Heron 1966.
Ecological and morphological observations on a population of Cyclotella pseudostelligera Hustedt.
J. Ecol. 54(2): 335-340.

Cylindropyxis profunda n.sp.
Hendey, N. Ingram, 1964
An introductory account of the smaller algae of British coastal waters. V. Bacillariophyceae (Diatoms).
Her Majesty's Stationary Office, 317 pp., 45 pls.

Cylindropyxis tremulans n.gen., n.sp.
Hendey, N. Ingram, 1964
An introductory account of the smaller algae of British coastal waters. V. Bacillariophyceae (Diatoms).
Her Majesty's Stationary Office, 317 pp., 45 pls.

Cylindrotheca closterium
Humphrey, G.F., and D.V. Subba Rao, 1967.
Photosynthetic rate of the marine diatom Cylindrotheca closterium.
Aust. J. mar. Freshwat. Res., 18(2):123-127.

Cylindrotheca fusiformis a
Darley, W.M. and B.E. Volcani 1969.
Role of silicon in diatom metabolism.
Exp. Cell Res. 58: 334-342.
Also in Coll. Repr. Scripps Inst. Oceanogr. NO: 570-878.

Cylindrotheca fusiformis
Lewin, Joyce C., 1965.
The thiamine requirement of a marine diatom. Phycologia, 4(3):141-144.

Cylindrotheca fusiformis
Lewin, Joyce, and Johan A. Hellebust, 1970.
Heterotrophic nutrition of the marine pennate diatom, Cylindrotheca fusiformis.
Can. J. Microbiol. 16(11): 1123-1129.

Cylindrotheca gracilis
Hendey, N. Ingram, 1964
An introductory account of the smaller algae of British coastal waters. V. Bacillariophyceae (Diatoms).
Her Majesty's Stationary Office, 317 pp., 45 pls.

Cylindrotheca gracilis
Hustedt, F. and A.A. Aleem, 1951
Littoral diatoms from the Salstone near Plymouth. JMBA 30(1): 177.196.

Cymatodiscus gen. nov.
Hendey, N Ingram, 1958 [1957(Publ. 1958)]
Marine diatoms from some West African Ports- J. R.Microsc. Soc. (3) 77(1/2): 28-85.

Cymatodiscus planetophorus comb. nov.
Hendey, N-Ingram, 1958 [1957(Publ. 1958)]
Marine diatoms from some West African Ports- J. R.Microsc. Soc. (3) 77(1/2): 28-85.

Cymatoneis sulcata
Andrade e Clovis Teixeira, M.H. de, 1957.
Contribuição para o conhecimento das diatomáceas do Brasil.
Bol. Inst. Ocean., Univ. Sao Paulo, 8(1/2):171-225, 10 pls.

Cymatopleura
Desikachary, T.V., 1956(1957).
Electron microscope studies on diatoms. J.R. Microsc. Soc. (3)76(1/2):9-36.

Cymatopleura n.gen.
Smith, W., 1851
Notes on the Diatomaceae; with descriptions of British species included in the genera Campylodiscus, Surirella, and Cymatopleura. Ann. Mag. Nat. Hist., 2nd ser., 7:1-14, 3 pls.

Cymatopleura elliptica
Rumkówna, A., 1948
[List of the phytoplankton species occurring in the superficial water layers in the Gulf of Gdańsk] Bull. Lab. mar., Gdynia, No. 4: 139-141 with tables in back.

Cymatopleura elliptica
Smith, W., 1851
Notes on the Diatomaceae; with descriptions of British species included in the genera Campylodiscus, Surirella, and Cymatopleura. Ann. Mag. Nat. Hist., 2nd ser., 7:1-14, 3 pls.

Cymatopleura Hibernica
Smith, W., 1851
Notes on the Diatomaceae; with descriptions of British species included in the genera Campylodiscus, Surirella, and Cymatopleura. Ann. Mag. Nat. Hist., 2nd ser., 7:1-14, 3 pls.

Cymatopleura solea

Mann, A., 1893
List of Diatomaceae from a deep-sea dredging in the Atlantic Ocean off Delaware Bay by the U. S. Fish Commission Steamer Albatross. Proc. U. S. Nat. Mus. 16:303-312.

Cymatopleura solea

Smith, W., 1851
Notes on the Diatomaceae; with descriptions of British species included in the genera Campylodiscus, Surirella, and Cymatopleura. Ann. Mag. Nat. Hist., 2nd ser., 7:1-14, 3 pls.

Cymatopleura solea

Tsumura, K., 1956
Diatomoj el la cirkaufoso de la restajo de la kastele de Odawara. J. Yokohama Municipal Univ., (C-14), No. 47:23 pp.

Cymatosera

Desikachary, T.V., 1956(1957).
Electron microscope studies on diatoms. J.R. Microsc. Soc. (3)76(1/2):9-36.

Cymatosira belgica

Hendy, N. Ingram, 1964
An introductory account of the smaller algae of British coastal waters. V. Bacillariophyceae (Diatoms). Her Majesty's Stationary Office, 317 pp., 45 pls.

Cymatosira belgica

Meunier, A., 1915
Microplancton de la Mer Flamande. 2. Diatomées (excepté le genre Chaetoceros). Mem. Mus. Roy. Hist. Nat. Belgique, 7(3):1-118, Pls. VIII-XIV.

Cymatosira belgica

Zanon, V., 1948
Diatomee marini di Sardegna e Pugillo di Alghe Marine della stressa. Boll. Pesca, Piscitutura e Idrobiologia, Anno 24, ns. 3(2): 202-244, 27 figs. on 1 pl.

Cymatosira elliptica

Hendy, N. Ingram, 1964
An introductory account of the smaller algae of British coastal waters. V. Bacillariophyceae (Diatoms). Her Majesty's Stationary Office, 317 pp., 45 pls.

Cymatosira Laurenziana

Mann, A., 1893
List of Diatomaceae from a deep-sea dredging in the Atlantic Ocean off Delaware Bay by the U. S. Fish Commission Steamer Albatross. Proc. U. S. Nat. Mus. 16:303-312.

Cymatosira Lorenziana

Rampi, L., 1940
Diatomee del Mare Adriatico. Nuovo Giornale Botanico Italiano, n.s., 47:559-608.

Cymatosira Lorenziana

Zanon, V., 1948
Diatomee marini di Sardegna e Pugillo di Alghe Marine della stressa. Boll. Pesca, Piscitutura e Idrobiologia, Anno 24, ns. 3(2): 202-244, 27 figs. on 1 pl.

Cymatotheca gen. nov.

Hendey, N-Ingram, 1958 [1957(Publ. 1958)]
Marine diatoms from some West African Ports. J. R. Microsc. Soc. (3) 77(1/2): 28-85.

Cymatotheca weissflogii comb. nov.

Hendey, N-Ingram, 1958 [1957(Publ. 1958)]
Marine diatoms from some West African Ports. J.R. Microsc. Soc. (3) 77(1/2): 28-85.

Cymbasira minutula n.sp.

Grunow, A., 1863
Ueber einige neue und ungenügend bekannte Arten und Gattungen von Diatomaceen. Verhandl. d. K.K. Zool. Bot. Gesellsch., Vienna, 13: 137-162, pl.4-5 (Pl. 13-14).

Cymbella

Desikachary, T.V., 1956(1957).
Electron microscope studies on diatoms. J.R. Microsc. Soc. (3)76(1/2):9-36.

Cymbella sp.

Kokubo, S., and S. Sato, 1947
Planktera in Jū-San Gata. Physiol. and Ecol. (Japan) 1(4):1-16, 3 text figs., tables.

Cymbella sp.

Morse, D.C., 1947
Some observations on seasonal variations in plankton population Patuxant River, Maryland 1943-1945. Bd. Nat. Res., Publ. No.65, Chesapeake Biol. Lab., 31, 3 figs.

Cymbella affinis

Tsumura, K., 1956
Diatomoj el la cirkaufoso de la restajo de la kastele de Odawara. J. Yokohama Municipal Univ., (C-14), No. 47:23 pp.

Cymbella alpina

Grunow, A., 1863
Ueber einige neue und ungenügend bekannte Arten und Gattungen von Diatomaceen. Verhandl. d. K.K. Zool. Bot. Gesellsch., Vienna, 13: 137-162, pl.4-5 (Pl. 13-14).

Cymbella amphicephala

Rumkówna, A., 1948
[List of the phytoplankton species occurring in the superficial water layers in the Gulf of Gdańsk] Bull. Lab. mar., Gdynia, No. 4: 139-141 with tables in back.

Cymbella cistula

Mann, A., 1893
List of Diatomaceae from a deep-sea dredging in the Atlantic Ocean off Delaware Bay by the U. S. Fish Commission Steamer Albatross. Proc. U. S. Nat. Mus. 16:303-312.

Cymbella criophila n.sp.

Castracane degli Antelminelli, F., 1886
1. Report on the Diatomaceae collected by H.M.S. Challenger during the years 1873-1876. Rept. Sci. Results, H.M.S. Challenger, Botany Vol. II, 178 pp., 30 pls.

Cymbella cuspidata

Mann, A., 1893
List of Diatomaceae from a deep-sea dredging in the Atlantic Ocean off Delaware Bay by the U. S. Fish Commission Steamer Albatross. Proc. U. S. Nat. Mus. 16:303-312.

Cymbella Ehrenbergii

Zanon, D. V., 1949
Diatomee di Buenos Aires (Argentina) Atti Accad. Naz. Lincei, Memorie, Cl. Sci. fis., mat. e. nat., ser. 7, 11(3):59-151, 2 pls.

Cymbella exisa

Tsumura, K., 1956
Diatomoj el la cirkaufoso de la restajo de la kastele de Odawara. J. Yokohama Municipal Univ., (c-14), No. 47:23 pp.

Cymbella marina n.sp.

Castracane degli Antelminelli, F., 1886
1. Report on the Diatomaceae collected by H.M.S. Challenger during the years 1873-1876. Rept. Sci. Results, H.M.S. Challenger, Botany Vol. II, 178 pp., 30 pls.

Cymbella parva

Mann, A., 1893
List of Diatomaceae from a deep-sea dredging in the Atlantic Ocean off Delaware Bay by the U. S. Fish Commission Steamer Albatross. Proc. U. S. Nat. Mus. 16:303-312.

Cymbella pelagica n.sp.

Castracane degli Antelminelli, F., 1886
1. Report on the Diatomaceae collected by H.M.S. Challenger during the years 1873-1876. Rept. Sci. Results, H.M.S. Challenger, Botany Vol. II, 178 pp., 30 pls.

Cymbella sinuata

Tsumura, K., 1956
Diatomoj el la cirkaufoso de la restajo de la kastele de Odawara. J. Yokohama Municipal Univ., (C-14), No. 47:23 pp.

Cymbella tumida

Tsumura, K., 1956
Diatomoj el la cirkaufoso de la restajo de la kastele de Odawara. J. Yokohama Municipal Univ., (C-14), No. 47:23 pp.

Cymbella tumida

Zanon, D. V., 1949
Diatomee di Buenos Aires (Argentina) Atti Accad. Naz. Lincei, Memorie, Cl. Sci. fis., mat. e. nat., ser. 7, 11(3):59-151, 2 pls.

Cymbella tumidula

Tsumura, K., 1956
Diatomoj el la cirkaufoso de la restajo de la kastele de Odawara. J. Yokohama Municipal Univ., (C-14), No. 43:23 pp.

Cymbella turgida

Tsumura, K., 1956
Diatomoj el la cirkaufoso de la restajo de la kastele de Odawara. J. Yokohama Municipal Univ., (C-14), No. 47:23 pp.

Cymbella turgidula

Tsumura, K., 1957
Diatomoj el la cirkaufoso de la restajo de la kastele de Odawara. J. Yokohama Municipal Univ., (C-14), No. 47:23 pp.

Cymbella ventricosa & obtusa

Tsumura, K., 1956
Diatomoj el la cirkaufoso de la restajo de la kastele de Odawara. J. Yokohama Municipal Univ., (C-14), No. 47:23 pp.

Cymbella ventricosa

Zanon, D. V., 1949
Diatomee di Buenos Aires (Argentina) Atti Accad. Naz. Lincei, Memorie, Cl. Sci. fis., mat. e. nat., ser. 7, 11(3):59-151, 2 pls.

Cystopleura gibba

Mann, A., 1907
Report on the diatoms of the Albatross voyages in the Pacific Ocean, 1888-1904. Contrib. U. S. Nat. Herb. 10(5):221-419, Pls. XLIV-LIV.

Cystopleura turgida

Mann, A., 1907
Report on the diatoms of the Albatross voyages in the Pacific Ocean, 1888-1904. Contrib. U. S. Nat. Herb. 10(5):221-419, Pls. XLIV-LIV.

Dactyliosolen sp.

Allen, W. E., 1938
The Templeton Crocker Expedition to the Gulf of California in 1935 - The Phytoplankton. Amer. Microsc. Soc., Trans. 57:328-335.

Dactyliosolen sp.

Allen, W.E., 1937
Plankton diatoms of the Gulf of California obtained by the G. Allan Hancock Expedition of 1936. The Hancock Pacific Expeditions, Univ. So. Calif. Publ. 3:47-59, 1 fig.

Dactyliosolen

Brunel, J., 1962
Le phytoplancton de la Baie de Chaleurs. Inst. Botan., Univ. Montreal, Contrib. No. 77: 365 pp., 66 pls.

Dactyliosolen sp.
Cupp, E.E. and Allen, W.E., 1938
Plankton diatoms of the Gulf of California obtained by Allan Hancock Pacific Expedition of 1937. The Hancock Pacific Expeditions, The Univ. So. Calif. Publ. 3: 61-74, 1 map, pls.4-15.

Dactyliosolen
Desikachary, T.V., 1956(1957).
Electron microscope studies on diatoms. J. R. Microsc. Soc. (3)76(1/2):9-36.

Dactyliosolen antarcticus
Boden, B. P., 1949.
The diatoms collected by the U.S.S. CACOPAN in the Antarctic in 1947. J. Mar. Res. 8(1):6-13, 3 textfigs.

Dactyliosolen antarcticus
Boden, Brian, 1948
Marine plankton diatoms on operation HIGHJUMP in: Some oceanographic observations on operation HIGHJUMP. By R.S. Dietz. USNEL Rept. No.55, 97 pp., 41 figs. 7 July 1948.

Dactyliosolen antarcticus
Castracane degli Antelminelli, F., 1886 n.sp.
1. Report on the Diatomaceae collected by H.M.S. Challenger during the years 1873-1876. Rept. Sci. Results, H.M.S. Challenger, Botany Vol. II, 178 pp., 30 pls.

Dactyliosolen antarcticus
Central Meteorological Observatory, 1949
Report on sea and weather observation on Antarctic Whaling Ground (1947-48). Ocean. Mag., Japan, 1(1):49-88, 17 text figs.

Dactyliosolen antarcticus
Cupp, Easter E., 1943
Marine plankton diatoms of the west coast of North America. Bull. S.I.O. 5(1):1-238, 5 pls., 168 text figs.

Dactyliosolen antarcticus
Gilbert, J.Y., and W.E. Allen, 1943
The phytoplankton of the Gulf of California obtained by the "E.W. Scripps" in 1939 and 1940. J. Mar. Res. V(2):89-110, figs.30-31.

Dactyliosolen antarcticus
Gran, H.H., 1908
Diatomeen. Nordisches Plankton, Botanischer Teil pp. XIX 1-XIX 146; 178 text figs.

Dactyliosolen antarcticus
Hendy, N. Ingram, 1964
An introductory account of the smaller algae of British coastal waters. V. Bacillariophyceae (Diatoms). Her Majesty's Stationary Office, 317 pp., 45 pls.

Dactyliosolen antarcticus
Hendey, N.I., 1937
The plankton diatoms of the southern seas. Discovery Repts. 16:151-364, pls.6-13.

Dactyliosolen antarcticus
Jorgensen, E., 1900
Protophyten und Protozoën im Plankton aus der Norwegischen Westkerste. Bergens Mus. Aarb. 1899(6): 95 pp., 5 pls., 83 tables.

Dactyliosolen antarcticus
Manguin, E., 1954
Diatomees marines provenant de l'ile Heard (Australian National Antarctic Expedition). Rev. Algol., n.s., 1: 14-24.

Dactyliosolen antarcticus
Sargent, M. S. and T. J. Walker, 1948
Diatom populations associated with eddies off southern California in 1941. J. Mar. Res. 7(3):490-505, 15 text figs.

Dactyliosolen Bergonii
Forti, A., 1922
Ricerche sulla flora pelagica (fitoplancton) di Quarto dei Mille. Mem. R. Com. Talass. Ital. 97:248 pp., 13 pls.

Dactyliosolen Bergonii
Schröder, B., 1900
Phytoplankton des Golfes von Neapel nebst vergleichenden Ausblicken auf das atlantischen Ozean. Mitt. Zool. Stat. Neapel, 14:1-38.

Dactyliosolen (antarcticus var.?) curvatus n.sp.
Hasle, Grethe Rytter 1960
Phytoplankton and ciliate species from the Tropical Pacific.
Skr. Norske Videnskaps-Akad., Oslo, 1. Mat.-Nat. Kl., 1960(2): 1-50.

Dactyliosolen flexuosus n.sp.
Mangin, L., 1915
Phytoplancton de L'Antartique. Deuxieme Exped. Ant. Francaise (1908-1910), 95 pp., 3 pls., 58 text figs.

Dactyliosolen mediterraneus
Allen, W. E., 1936
Occurrence of marine plankton diatoms in a ten-year series of daily catches in southern California. Am. Jour. Bot. 23(1):60-63.

Dactyliosolen mediterraneus
Cupp, Easter E., 1943
Marine plankton diatoms of the west coast of North America. Bull. S.I.O. 5(1):1-238, 5 pls., 168 text figs.

Dactyliosolen mediterraneus
Cupp, E.E., 1934
Analysis of marine diatom collections taken from the Canal Zone to California during March, 1933. Trans. Am. Micros. Soc. LIII (1):22-29, 1 map.

Dactyliosolen mediterraneus
Cupp, E.E. and Allen, W.E., 1938
Plankton diatoms of the Gulf of California obtained by Allan Hancock Pacific Expedition of 1937. The Hancock Pacific Expeditions, The Univ. So. Calif. Publ. 3: 61-74, 1 map, pls.4-15.

Dactyliosolen mediterraneus
Delegazione Italiana della Commissione Internazionale per l'Esplorazione Scientifica del Mediterraneo, 1941
Note sul plancton della Laguna veneta. [Memoria CCLXXIX] , Arch. di Ocean. e Limn. Anno I, Fasc. I, 1941 XIX: 31-57 pp.

Dactyliosolen mediterraneus
Ercegovic, A., 1936
Etudes qualitative et quantitatives du phytoplancton dans les eaux cotières de l'Adriatique oriental moyen au cours de l'année 1934. Acta Adriatica 1(9):1-126

Dactyliosolen mediterraneus
Forti, A., 1922
Ricerche sulla flora pelagica (fitoplancton) di Quarto dei Mille. Mem. R. Com. Talass. Ital. 97:248 pp., 13 pls.

Dactyliosolen mediterraneus
Frenguelli, Joaquin, and Hector Antonio Orlando, 1959
Operacion MERLUZA. Diatomeas y silicoflagelados del plancton del "VI Crucero". Servicio Hidrogr. Naval., Argentina, Publ. No. H. 619: 5-62.

Dactyliosolen mediterranea
Gilbert, J.Y., and W.E. Allen, 1943
The phytoplankton of the Gulf of California obtained by the "E.W. Scripps" in 1939 and 1940. J. Mar. Res. V(2):89-110, figs.30-31.

Dactyliosolen mediterraneus
Gran, H.H. and T. Braarud, 1935
A quantitative study of the phytoplankton in the Bay of Fundy and the Gulf of Maine (including observations on hydrography, chemistry, and turbidity). J. Biol. Bd. Canada, 1(5):279-467, 69 text figs.

Dactyliosolen mediterraneus
Gran, H. H. and E. C. Angst, 1931
Plankton diatoms of Puget Sound. Publ. Puget Sound Biol. Sta. 7:417-519, 95 text figs.

Dactyliosolen mediterraneus
Hendy, N. Ingram, 1964
An introductory account of the smaller algae of British coastal waters. V. Bacillariophyceae (Diatoms). Her Majesty's Stationary Office, 317 pp., 45 pls.

Dactyliosolen mediterraneus
Hendey, N.I., 1937
The plankton diatoms of the southern seas. Discovery Repts. 16:151-364, pls.6-13.

Dactyliosolen mediterraneus
Jorgensen, E., 1900
Protophyten und Protozoën im Plankton aus der Norwegischen Westkerste. Bergens Mus. Aarb. 1899(6): 95 pp., 5 pls., 83 tables.

Dactyliosolen mediterraneus
Kokubo, S., 1952
Results of the observations on the plankton and oceanography of Mutsu Bay during 1950, reference being made also to the period 1946-1950. Bull Mar.Biol.Sta., Asamushi 5(1/4): 1-54, 3 tables, (fold-in), 1 fold-in.

Dactyliosolen mediterraneus
Lillick, L.C., 1940
Phytoplankton and planktonic protozoa of the offshore waters of the Gulf of Maine. Pt.II. Qualitative Composition of the Planktonic Flora. Trans. Am. Phil. Soc., n.s., 31(3):193-237, 13 text figs.

Dactyliosolen mediterraneus
Margalef, R., 1949
Fitoplancton nerítico de la Costa Brava en 1947-48. Publ. Inst. Biol. Aplicada, 5: 41-51, 3 text figs.

Dactyliosolen mediterraneus
Pavillard, J., 1925
Bacillariales. Rept. on the Danish Oceangr. Exped., 1908-10 to the Mediterranean and adj. seas. Vol.II., Biol. J4:72 pp., 116 text figs.

Dactyliosolen mediterraneus
Rampi, L., 1942
Ricerche sul fitoplancton del Mare Ligure 6. Le diatomee delle acque di Sanremo. Nuovo Giornale Botanico Italiano, N.S., 49:252-268.

Dactylisolen mediterraneus
Robinson, G.A., 1965.
Continuous plankton records: contribution towards a plankton atlas of the North Atlantic and the North Sea. IX. Seasonal cycles of phytoplankton.
Bulls. Mar. Ecol., Scottish Mar. Biol. Assoc., 6(4):104-122, pls. 26-61.

Dactyliosolen mediterraneus
Robinson, G.A., 1961
Contribution towards an atlas of the northeastern Atlantic and the North Sea. 1. Phytoplankton. Bulls. Mar. Ecology, 5(42): 81-89, Pls. 15-20.

Dactyliosolen mediterraneus
Schröder, B., 1900
Phytoplankton des Golfes von Neapel nebst vergleichenden Ausblicken auf das atlantischen Ozean. Mitt. Zool. Stat. Neapel, 14:1-38.

Dactyliosolen mediterraneus

Uyeno, Fukuzo, 1961
Oceanographical and ecological studies on primary production of the sea, with special references to relationship between diatom production and temperature and chlorinity of water.
Rept., Fac. Fish., Pref. Univ., Mie, 4(1): 1-64.

Dactyliosolen tenuis

Cleve-Euler, A., 1951
Die Diatomeen von Schweden und Finnland. Kungl. Svenska Vetenskaps Akad. Handl., Fjärde Ser. 2(1): 161 pp., 6 pls.

Dactyliosolen tenuis

Gran, H.H., 1908
Diatomeen. Nordisches Plankton, Botanischer Teil pp. XIX 1-XIX 146; 178 text figs.

Dactyliosolen tenuis

Mangin, M. L., 1912
Phytoplancton de la croisière du "René" dans l'Atlantique (Septembre 1908). Ann. Inst. Ocean., n.s., 4(1):1-66, 2 pls., 41 text figs., 2 tables.

Dactyliosolen tenuis

Sleggs., G.F., 1927.
Marine phytoplankton in the region of La Jolla, California during the summer of 1924. Bull. S.I.O., tech. ser., 1:93-117, 8 textfigs.

Dactyliosolen Voigtii n.sp.

Manguin, E., 1957.
Premier inventaire des diatomées de la Terre Adélie Antarctique. Espèces nouvelles.
Rev. Algologique, n.s., 3(3):111-134.

Debya insignis

Hendey, N.I., 1951
Littoral diatoms of Chicester Harbour with special reference to fouling. J.Roy. Microscop. Soc. 71(1): 1-86, 18 pls.

Denticula

Desikachary, T.V., 1956(1957).
Electron microscope studies on diatoms.
J.R. Microsc. Soc. (3)76(1/2):9-36.

Denticula dimorpha n.sp.

Schrader, Hans-Joachim, 1973
Cenozoic diatoms from the northeast Pacific, leg 18. Initial Repts Deep Sea Drilling Proj, 18:673-797.

Denticula elegans

Mann, A., 1893
List of Diatomaceae from a deep-sea dredging in the Atlantic Ocean off Delaware Bay by the U. S. Fish Commission Steamer Albatross. Proc. U. S. Natn. Mus. 16:303-312.

Denticula hyalina n.sp.

Schrader, Hans-Joachim, 1973
Cenozoic diatoms from the northeast Pacific, leg 18. Initial Repts Deep Sea Drilling Proj, 18:673-797.

Denticula miocenica n.sp.

Schrader, Hans-Joachim, 1973
Cenozoic diatoms from the northeast Pacific, leg 18. Initial Repts Deep Sea Drilling Proj, 18:673-797.

Denticula nicobarica

Mann, A., 1907
Report on the diatoms of the Albatross voyages in the Pacific Ocean, 1888-1904. Contrib. U. S. Nat. Herb. 10(5):221-419, Pls. XLIV-LIV.

Denticula punctata n.sp.

Schrader, Hans-Joachim, 1973
Cenozoic diatoms from the northeast Pacific, leg 18. Initial Repts Deep Sea Drilling Proj, 18:673-797.

Denticula tenuis

Zanon, D. V., 1949
Diatomee di Buenos Aires (Argentina) Atti Accad. Naz. Lincei, Memorie, Cl. Sci. fis., mat. e. nat., ser. 7, 11(3):59-151, 2 pls.

Desmogonium

Desikachary, T.V., 1956(1957).
Electron microscope studies on diatoms.
J.R. Microsc. Soc. (3)76(1/2):9-36.

Desmogonium guianense

Zanon, D. V., 1949
Diatomee di Buenos Aires (Argentina) Atti Accad. Naz. Lincei, Memorie, Cl. Sci. fis., mat. e. nat., ser. 7, 11(3):59-151, 2 pls.

Desmogonium Rabenhorstianum

Zanon, D. V., 1949
Diatomee di Buenos Aires (Argentina) Atti Accad. Naz. Lincei, Memorie, Cl. Sci. fis., mat. e. nat., ser. 7, 11(3):59-151, 2 pls.

Braarud, T., 1945
A phytoplankton survey of the polluted waters of inner Oslo Fjord. Hvalrådets Skrifter, No.28, 142 pp., 19 text figs., 17 tables.

Detonula confervacea

Braarud, T., 1934
A note on the phytoplankton of the Gulf of Maine in the summer of 1933. Biol. Bull. 67(1):76-82. (Contribution No.46 of the Woods Hole Oceanographic Institution)

Detonula confervacea

Brunel, J., 1962
Le phytoplancton de la Baie de Chaleurs. Inst. Botan., Univ. Montréal, Contrib. No. 77: 365 pp., 66 pls.

Detonula confervacea

Brunel, Jules, 1962
Le phytoplancton de la Baie des Chaleurs. Contrib. Ministère de la Chasse et des Pêcheries, Province de Québec, No. 91: 365 pp.

Detonula confervacea

Cleve-Euler, A., 1951
Die Diatomeen von Schweden und Finnland. Kungl. Svenska Vetenskaps Akad. Handl., Fjärde Ser. 2(1): 161 pp., 6 pls.

Detonula confervacea

Frenguelli, Joaquin, and Hector Antonio Orlando, 1959.
Operacion MERLUZA. Diatomeas y silicoflagelados del plancton del "VI Crucero". Servicio Hidrogr. Naval, Argentina, Publ. No. H. 619: 5-62.

Detonula confervacea

Gran, H.H., 1908
Diatomeen. Nordisches Plankton, Botanischer Teil pp. XIX 1-XIX 146; 178 text figs.

Detonula confervacea

Gran, H.H., and T. Braarud, 1935
A quantitative study of the phytoplankton in the Bay of Fundy and the Gulf of Maine (including observations on hydrography, chemistry, and turbidity). J. Biol. Bd., Canada, 1(5):279-467, 69 text figs.

Detonula confervacea

Grøntved, J., 1956.
Planktonological contributions. II. Taxonomical studies in some Danish coastal localities.
Medd. Denmarks Fiskeri- og Havundersøgelser, n.s. 1(12):13 pp.

Detonula confervacea

Guillard, Robert R.L., and John H. Ryther, 1962
Studies of marine planktonic diatoms. I. Cyclotella nana Hustedt and Detonula confervacea (Cleve).
Canadian J. Microbiol., 8(2):229-240.

Detonula confervacea

Hendy, N. Ingram, 1964
An introductory account of the smaller algae of British coastal waters. V. Bacillariophyceae (Diatoms).
Her Majesty's Stationary Office, 317 pp., 45 pls.

Detonula confervacea

Iselin, C., 1930
A report on the coastal waters of Labrador based on explorations of the "Chance" during the summer of 1926. Proc. Am. Acad. Arts Sci., 66(1):1-37, 14 text figs.

Detonula confervacea

Jørgensen, E., 1905
B.Protistplankton and the diatoms in bottom samples. Hydrographical and biological investigations in Norwegian fjords. Bergens Mus. Skr. 7: 49-225.

Detonula confervacea

Lillick, L.C., 1940
Phytoplankton and planktonic protozoa of the offshore waters of the Gulf of Maine. Pt.II. Qualitative Composition of the Planktonic Flora. Trans. Am. Phil. Soc., n.s., 31(3):193-237, 13 text figs.

Detonula confervacea

Lillick, L.C., 1938
Preliminary report of the phytoplankton of the Gulf of Maine. Am. Mid. Nat. 20(3):624-640, 1 text fig, 37 tables.

Detonula confervacea

Margalef, Ramón, y Dolores Blasco, 1970
Influencia del puerto de Barcelona sobre el fitoplancton de las áreas vecinas: una mancha de plancton de gran densidad, con dominancia de Thalassiosira, observada en agosto de 1969.
Inv. pesq. Barcelona, 34(2): 575-580

Detonula confervacea

Smayda, Theodor J., 1969.
Experimental observations on the influence of temperature, light and salinity on the cell division of the marine diatom, Detonula confervacea (Cleve) Gran.
J. Phycol. 5(2): 150-157.

Detonula confervacea

Steemann-Nielsen, Einar, 1951
The marine vegetation of the Isefjord. A study on ecology and production. Medd. Komm. Danmarks Fiskeri-og Havundersøgelser. Ser. Plankton. 5(4); 114 pp., 46 text figs.

Detonula cystifera

Gran, H.H., 1908
Diatomeen. Nordisches Plankton, Botanischer Teil pp. XIX 1-XIX 146; 178 text figs.

Detonula cystifera

Morse, D.C., 1947
Some observations on seasonal variations in plankton population Patuxant River, Maryland 1943-1945. Bd. Nat. Res., Publ. No.65, Chesapeake Biol. Lab., 31, 3 figs.

Gran, H.H., 1908
Diatomeen. Nordisches Plankton, Botanischer Teil pp. XIX.1-XIX.146; 178 text figs.

Detonula Schroederi

Schodduyn, M., 1926
Observations faites dans la baie d'Ambleteuse (Pas de Calais). Bull. Inst. Ocean., Monaco, No. 482: 64 pp.

Diatoma

Desikachary, T.V., 1956(1957).
Electron microscope studies on diatoms. J. R. Microsc. Soc. (3)76(1/2):9-36.

Diatoma anceps

Zanon, V., 1948
Diatomee marini di Sardegna e Pugillo di Alghe Marine della stressa. Boll. Pesca, Piscitutura e Idrobiologia, Anno 24, ns. 3(2): 202-244, 27 figs. on 1 pl.

Diatoma elongatum

Levander, K.M., 1947
Plankton gesammelt in den Jahren 1899-1910 an den Küsten Finnlands. Finnländische Hydrographisch-Biologische Untersuchungen (aus dem Wasserbiologischen Laboratorin der Societas Scientiarum Fennica) No.11:40 pp., 6 diagrams, 13 pls., tables.

Diatoma elongatum

Rumkówna, A., 1948
[List of the phytoplankton species occurring in the superficial water layers in the Gulf of Gdańsk] Bull. Lab. mar., Gdynia, No. 4: 139-141 with tables in back.

Diatoma elongatum var. tenuis

Levander, K.M., 1947
Plankton gesammelt in den Jahren 1899-1910 an den Küsten Finnlands. Finnländische Hydrographisch-Biologische Untersuchungen (aus dem Wasserbiologischen Laboratorin der Societas Scientiarum Fennica) No.11:40 pp., 6 diagrams, 13 pls., tables.

Diatoma hiemale

Zanon, V., 1948
Diatomee marini di Sardegna e Pugillo di Alghe Marine della stressa. Boll. Pesca, Piscitutura e Idrobiologia, Anno 24, ns. 3(2): 202-244, 27 figs. on 1 pl.

Diatoma rhombicum

Castracane degli Antelminelli, F., 1886
1. Rëport on the Diatomaceae collected by H.M.S. Challenger during the years 1873-1876. Rept. Sci. Results, H.M.S. Challenger, Botany Vol. II, 178 pp., 30 pls.

Diatoma vulgare

Rumkówna, A., 1948
[List of the phytoplankton species occurring in the superficial water layers in the Gulf of Gdańsk] Bull. Lab. mar., Gdynia, No. 4: 139-141 with tables in back.

Diatoma vulgare

Zanon, V., 1948
Diatomee marini di Sardegna e Pugillo di Alghe Marine della stressa. Boll. Pesca, Piscitutura e Idrobiologia, Anno 24, ns. 3(2): 202-244, 27 figs. on 1 pl.

Dictyoneis jamaicensis

Rampi, L., 1940
Diatomee del Mare Adriatico. Nuovo Giornale Botanico Italiano, n.s., 47:559-608.

Didymosphenia

Desikachary, T.V., 1956(1957).
Electron microscope studies on diatoms. J.R. Microsc. Soc. (3)76(1/2):9-36.

Dimerogramma antarcticum n.sp.

Frenguelli, Joaquin, y Hector A. Orlando, 1958.
Diatomeas y silicoflagelados del sector Antartico Sudamericano. Inst. Antartico Argentino, Publ., No. 5:191 pp.

Dimerogramma dubium

Rampi, L., 1940
Diatomee del Mare Adriatico. Nuovo Giornale Botanico Italiano, n.s., 47:559-608.

Dimerogramma fulvum

Hendy, N. Ingram, 1964
An introductory account of the smaller algae of British coastal waters. V. Bacillariophyceae (Diatoms). Her Majesty's Stationary Office, 317 pp., 45 pls.

Dimerogramma fulvum

Rampi, L., 1940
Diatomee del Mare Adriatico. Nuovo Giornale Botanico Italiano, n.s., 47:559-608.

Dimerogramma fulvum

Zanon, V., 1948
Diatomee marini di Sardegna e Pugillo di Alghe Marine della stressa. Boll. Pesca, Piscitutura e Idrobiologia, Anno 24, ns. 3(2): 202-244, 27 figs. on 1 pl.

Dimerogramma furcigerum

Rampi, L., 1940
Diatomee del Mare Adriatico. Nuovo Giornale Botanico Italiano, n.s., 47:559-608.

Dimerogramma inflatum n.sp.

Mann, A., 1907
Report on the diatoms of the Albatross voyages in the Pacific Ocean, 1888-1904. Contrib. U. S. Nat. Herb. 10(5):221-419, Pls. XLIV-LIV.

Dimerogramma marinum

Hendy, N. Ingram, 1964
An introductory account of the smaller algae of British coastal waters. V. Bacillariophyceae (Diatoms). Her Majesty's Stationary Office, 317 pp., 45 pls.

Dimerogramma marinum

Rampi, L., 1940
Diatomee del Mare Adriatico. Nuovo Giornale Botanico Italiano, n.s., 47:559-608.

Dimerogramma marinum

Zanon, D. V., 1949
Diatomee di Buenos Aires (Argentina) Atti Accad. Naz. Lincei, Memorie, Cl. Sci. fis., mat. e. nat., ser. 7, 11(3):59-151, 2 pls.

Dimerogramma minor & varieties

Hendy, N. Ingram, 1964
An introductory account of the smaller algae of British coastal waters. V. Bacillariophyceae (Diatoms). Her Majesty's Stationary Office, 317 pp., 45 pls.

Dimerogramma minor

Hendey, N.I., 1951
Littoral diatoms of Chicester Harbour with special reference to fouling. J.Roy. Microscop. Soc. 71(1): 1-86, 18 pls.

Dimerogramma minor

Hustedt, F. and A.A. Aleem, 1951
Littoral diatoms from the Salstone near Plymouth. JMBA 30(1): 177.196.

Dimerogramma minor

Politis, J., 1949
Diatomees marines de Bosphores et des ibes de la mer de Marmara. II Practica tou Hellenikou Hidrobiologikou Institutoutou 1929, Etoz 1929, 3(1):11-31.

Dimerogramma minor

Rampi, L., 1940
Diatomee del Mare Adriatico. Nuovo Giornale Botanico Italiano, n.s., 47:559-608.

Dimerogramma minor var. nana

Rampi, L., 1942
Ricerche sul fitoplancton del Mare Ligure 6. Le diatomee delle acque di Sanremo. Nuovo Giornale Botanico Italiano, N.S., 49:252-268.

Dimerogramma minus

Zanon, V., 1948
Diatomee marini di Sardegna e Pugillo di Alghe Marine della stressa. Boll. Pesca, Piscitutura e Idrobiologia, Anno 24, ns. 3(2): 202-244, 27 figs. on 1 pl.

Dimerogramma nanum var. thaitiensis

Castracane degli Antelminelli, F., 1886 s.nov.
1. Rëport on the Diatomaceae collected by H.M.S. Challenger during the years 1873-1876. Rept. Sci. Results, H.M.S. Challenger, Botany Vol. II, 178 pp., 30 pls.

Dimerogramma nanum

Jorgensen, E., 1900
Protophyten und Protozoën im Plankton aus der Norwegischen Westkerste. Bergens Mus. Aarb. 1899(6): 95 pp., 5 pls., 83 tables.

Diploneis

Desikachary, T.V., 1956(1957).
Electron microscope studies on diatoms. J.R. Microsc. Soc (3)76(1/2):9-36.

Diploneis Adeliae n.sp.

Manguin, E., 1957.
Premier inventaire des diatomées de la Terre Adélie Antarctique. Espèces nouvelles. Rev. Algologique, n.sp., 3(3):111-134.

Diploneis adonis

Zanon, V., 1948
Diatomee marini di Sardegna e Pugillo di Alghe Marine della stressa. Boll. Pesca, Piscitutura e Idrobiologia, Anno 24, ns. 3(2): 202-244, 27 figs. on 1 pl.

Diploneis advena var. sansegana

Rampi, L., 1942
Ricerche sul fitoplancton del Mare Ligure 6. Le diatomee delle acque di Sanremo. Nuovo Giornale Botanico Italiano, N.S., 49:252-268.

Diploneis aestuari

Hustedt, F. and A.A. Aleem, 1951
Littoral diatoms from the Salstone near Plymouth. JMBA 30(1): 177.196.

Diploneis Berrychiana

Rampi, L., 1940
Diatomee del Mare Adriatico. Nuovo Giornale Botanico Italiano, n.s., 47:559-608.

Diploneis bomboides

Rampi, L., 1940
Diatomee del Mare Adriatico. Nuovo Giornale Botanico Italiano, n.s., 47:559-608.

Diploneis bombus

Andrade e Clovis Teixeira, M.H. de, 1957.
Contribuição para o conhecimento das diatomáceas do Brasil. Bol. Inst. Ocean., Univ. Sao Paulo, 8(1/2):171-225, 10 pls.

Diploneis bombus
Hendy, N. Ingram, 1964
An introductory account of the smaller algae of British coastal waters. V. Bacillariophyceae (Diatoms).
Her Majesty's Stationary Office, 317 pp., 45 pls.

Diploneis bombus
Hendey, N.I., 1951
Littoral diatoms of Chicester Harbour with special reference to fouling. J.Roy. Microscop. Soc. 71(1): 1-86, 18 pls.

Diploneis bombus
Hustedt, F. and A.A. Aleem, 1951
Littoral diatoms from the Salstone near Plymouth. JMBA 30(1): 177-196.

Diploneis bombus
Rampi, L., 1940
Diatomee del Mare Adriatico. Nuovo Giornale Botanico Italiano, n.s., 47:559-608.

Diploneis bombus
Zanon, V., 1948
Diatomee marini di Sardegna e Pugillo di Alghe Marine della stressa. Boll. Pesca, Piscitutura e Idrobiologia, Anno 24, ns. 3(2): 202-244, 27 figs. on 1 pl.

Diploneis chersonensis
Andrade e Clovis Teixeira, M.H. de, 1957.
Contribuição para o conhecimento das diatomáceas do Brasil.
Bol. Inst. Ocean., Univ. Sao Paulo, 8(1/2):171-225, 10 pls.

Diploneis chersonensis
Hendy, N. Ingram, 1964
An introductory account of the smaller algae of British coastal waters. V. Bacillariophyceae (Diatoms).
Her Majesty's Stationary Office, 317 pp., 45 pls.

Diploneis Chersonensis
Zanon, V., 1948
Diatomee marini di Sardegna e Pugillo di Alghe Marine della stressa. Boll. Pesca, Piscitutura e Idrobiologia, Anno 24, ns. 3(2): 202-244, 27 figs. on 1 pl.

Diploneis chersonensis
Zanon, V., 1947
Diatomee delle Isole Ponziane - materiali per una florula diatomologica del Mare Tirreno.
Bol. Pesca, Piscol. Idrobiol., n.s., 2(1):36-53, 1 pl. of 10 figs.

Diploneis coffaeiformis
Andrade e Clovis Teixeira, M.H. de, 1957.
Contribuição para o conhecimento das diatomáceas do Brasil.
Bol. Inst. Ocean., Univ. Sao Paulo, 8(1/2):171-225, 10 pls.

Diploneis coffaeiformis
Rampi, L., 1940
Diatomee del Mare Adriatico. Nuovo Giornale Botanico Italiano, n.s., 47:559-608.

Diploneis constricta
Hendy, N. Ingram, 1964
An introductory account of the smaller algae of British coastal waters. V. Bacillariophyceae (Diatoms).
Her Majesty's Stationary Office, 317 pp., 45 pls.

Diploneis constricta
Rampi, L., 1940
Diatomee del Mare Adriatico. Nuovo Giornale Botanico Italiano, n.s., 47:559-608.

Diploneis crabo
Andrade e Clovis Teixeira, M.H. de, 1957.
Contribuição para o conhecimento das diatomáceas do Brasil.
Bol. Inst. Ocean., Univ. Sao Paulo, 8(1/2):171-225, 10 pls.

Diploneis crabro
Hendy, N. Ingram, 1964
An introductory account of the smaller algae of British coastal waters. V. Bacillariophyceae (Diatoms).
Her Majesty's Stationary Office, 317 pp., 45 pls.

Diploneis crabo
Hendey, N.I., 1951
Littoral diatoms of Chicester Harbour with special reference to fouling. J.Roy. Microscop. Soc. 71(1): 1-86, 18 pls.

Diploneis crabo
Rampi, L., 1942
Ricerche sul fitoplancton del Mare Ligure 6. Le diatomee delle acque di Sanremo. Nuovo Giornale Botanico Italiano, N.S., 49:252-268.

Diploneis crabro
Rampi, L., 1940
Diatomee del Mare Adriatico. Nuovo Giornale Botanico Italiano, n.s., 47:559-608.

Diploneis crabro
Schodduyn, M., 1926
Observations faites dans la baie d'Ambleteuse (Pas de Calais). Bull. Inst. Ocean., Monaco, No. 482: 64 pp.

Diploneis crabo
Zanon, V., 1948
Diatomee marini di Sardegna e Pugillo di Alghe Marine della stressa. Boll. Pesca, Piscitutura e Idrobiologia, Anno 24, ns. 3(2): 202-244, 27 figs. on 1 pl.

Diploneis dalmatica
Zanon, V., 1948
Diatomee marini di Sardegna e Pugillo di Alghe Marine della stressa. Boll. Pesca, Piscitutura e Idrobiologia, Anno 24, ns. 3(2): 202-244, 27 figs. on 1 pl.

Diploneis didyma
Hendy, N. Ingram, 1964
An introductory account of the smaller algae of British coastal waters. V. Bacillariophyceae (Diatoms).
Her Majesty's Stationary Office, 317 pp., 45 pls.

Diploneis didyma
Hendey, N.I., 1951
Littoral diatoms of Chicester Harbour with special reference to fouling. J.Roy. Microscop. Soc. 71(1): 1-86, 18 pls.

Diploneis didyma
Rumkówna, A., 1948
List of the phytoplankton species occurring in the superficial water layers in the Gulf of Gdańsk. Bull. Lab. mar., Gdynia, No. 4: 139-141 with tables in back.

Diploneis didyma
Zanon, V., 1948
Diatomee marini di Sardegna e Pugillo di Alghe Marine della stressa. Boll. Pesca, Piscitutura e Idrobiologia, Anno 24, ns. 3(2): 202-244, 27 figs. on 1 pl.

Diploneis divergens
Zanon, V., 1948
Diatomee marini di Sardegna e Pugillo di Alghe Marine della stressa. Boll. Pesca, Piscitutura e Idrobiologia, Anno 24, ns. 3(2): 202-244, 27 figs. on 1 pl.

Diploneis elliptica
Hendy, N. Ingram, 1964
An introductory account of the smaller algae of British coastal waters. V. Bacillariophyceae (Diatoms).
Her Majesty's Stationary Office, 317 pp., 45 pls.

Diploneis elliptica
Hendey, N.I., 1951
Littoral diatoms of Chicester Harbour with special reference to fouling. J.Roy. Microscop. Soc. 71(1): 1-86, 18 pls.

Diploneis elliptica
Kokubo, S., and S. Sato., 1947
Plankterx in Jū-San Gata. Physiol. and Ecol. (Japan) 1(4):1-16, 3 text figs., tables.

Diploneis elliptica
Zanon, D. V., 1949
Diatomee di Buenos Aires (Argentina) Atti Accad. Naz. Lincei, Memorie, Cl. Sci. fis., mat. e. nat., ser. 7, 11(3):59-151, 2 pls.

Diploneis elliptica
Zanon, V., 1948
Diatomee marini di Sardegna e Pugillo di Alghe Marine della stressa. Boll. Pesca, Piscitutura e Idrobiologia, Anno 24, ns. 3(2): 202-244, 27 figs. on 1 pl.

Diploneis finnica
Zanon, D. V., 1949
Diatomee di Buenos Aires (Argentina) Atti Accad. Naz. Lincei, Memorie, Cl. Sci. fis., mat. e. nat., ser. 7, 11(3):59-151, 2 pls.

Diploneis fusca
Hendy, N. Ingram, 1964
An introductory account of the smaller algae of British coastal waters. V. Bacillariophyceae (Diatoms).
Her Majesty's Stationary Office, 317 pp., 45 pls.

Diploneis fusca
Rampi, L., 1940
Diatomee del Mare Adriatico. Nuovo Giornale Botanico Italiano, n.s., 47:559-608.

Diploneis fusca var. delicata
Rampi, L., 1942
Ricerche sul fitoplancton del Mare Ligure 6. Le diatomee delle acque di Sanremo. Nuovo Giornale Botanico Italiano, N.S., 49:252-268.

Diploneis fusca
Zanon, V., 1948
Diatomee marini di Sardegna e Pugillo di Alghe Marine della stressa. Boll. Pesca, Piscitutura e Idrobiologia, Anno 24, ns. 3(2): 202-244, 27 figs. on 1 pl.

Diploneis fusca tyrrhenica n.var
Zanon, V., 1947
Diatomee delle Isole Ponziane - Materiali per una florula diatomologica del Mare Tirreno.
Bol. Pesca, Piscolt., Idrobiol., n.s., 2(1):36-53, 1 pl. of 10 figs.

Diploneis eudoxia
Florin, M-B., 1948
9. Diatomeae in submarine cores from the Tyrrhenian Sea. Medd. Ocean. Inst., Göteborg, 15 (Göteborgs Kungl. Vetenskaps-och Vitterhets Samhälles Handlingar, Sjätte Foljden, Ser. B 5(13):80-88.

Diploneis gemmata
Zanon, V., 1948
Diatomee marini di Sardegna e Pugillo di Alghe Marine della stressa. Boll. Pesca, Piscitutura e Idrobiologia, Anno 24, ns. 3(2): 202-244, 27 figs. on 1 pl.

Diploneis gemmatula
Rampi, L., 1940
Diatomee del Mare Adriatico. Nuovo Giornale Botanico Italiano, n.s., 47:559-608.

Diploneis Gründleri
Andrade e Clovis Teixeira, M.H. de, 1957.
Contribuição para o conhecimento das diatomáceas do Brasil.
Bol. Inst. Ocean., Univ. Sao Paulo, 8(1/2):171-225, 10 pls.

Diploneis incurvata
Rampi, L., 1940
Diatomee del Mare Adriatico. Nuovo Giornale Botanico Italiano, n.s., 47:559-608.

Diploneis incurvata
Zanon, V., 1948
Diatomee marini di Sardegna e Pugillo di Alghe Marine della stressa. Boll. Pesca, Piscitutura e Idrobiologia, Anno 24, ns. 3(2): 202-244, 27 figs. on 1 pl.

Diploneis interrupta & varieties
Hendy, N. Ingram, 1964
An introductory account of the smaller algae of British coastal waters. V. Bacillariophyceae (Diatoms).
Her Majesty's Stationary Office, 317 pp., 45 pls.

Diploneis interrupta
Iyengar, M.O.P. and G.Venkataraman, 1951.
The ecology and seasonal succession of the algae flora of the River Cooum at Madras with special reference to the Diatomaceae. J. Madras Univ. 21, Sect. B(1): 140-192, 1 pl of 4 figs., 11 text figs.

Diploneis interrupta
Rumkówna, A., 1948
[List of the phytoplankton species occurring in the superficial water layers in the Gulf of Gdańsk] Bull. Lab. mar., Gdynia, No. 4: 139-141 with tables in back.

Diploneis lineata
Hendy, N. Ingram, 1964
An introductory account of the smaller algae of British coastal waters. V. Bacillariophyceae (Diatoms).
Her Majesty's Stationary Office, 317 pp., 45 pls.

Diploneis littoralis
Hendy, N. Ingram, 1964
An introductory account of the smaller algae of British coastal waters. V. Bacillariophyceae (Diatoms).
Her Majesty's Stationary Office, 317 pp., 45 pls.

Diploneis littoralis
Hendey, N.I., 1951
Littoral diatoms of Chicester Harbour with special reference to fouling. J.Roy. Microscop. Soc. 71(1): 1-86, 18 pls.

Diploneis marginestriata
Rumkówna, A., 1948
[List of the phytoplankton species occurring in the superficial water layers in the Gulf of Gdańsk] Bull. Lab. mar., Gdynia, No. 4: 139-141 with tables in back.

Diploneis mediterranea
Rampi, L., 1940
Diatomee del Mare Adriatico. Nuovo Giornale Botanico Italiano, n.s., 47:559-608.

Diploneis mediterranea
Zanon, V., 1948
Diatomee marini di Sardegna e Pugillo di Alghe Marine della stressa. Boll. Pesca, Piscitutura e Idrobiologia, Anno 24, ns. 3(2): 202-244, 27 figs. on 1 pl.

Diploneis modicella n.sp.
Cholnoky, B.J., 1963.
Beiträge zur Kenntnis des marinen Litorals von Südafrika.
Botanica Marina, 5(2/3):38-83.

Diploneis nitescens
Andrade e Clovis Teixeira, M.H. de, 1957.
Contribuição para o conhecimento das diatomáceas do Brasil.
Bol. Inst. Ocean., Univ. Sao Paulo, 8(1/2):171-225, 10 pls.

Diploneis nitescens
Florin, M-B., 1948
9. Diatomeae in submarine cores from the Tyrrhenian Sea. Medd. Ocean. Inst., Göteborg, 15 (Göteborgs Kungl. Vetenskaps-och Viterrhets Samhälles Handlingar, Sjätte Foljden, Ser. B 5(13):80-88.

Diploneis nitescens
Rampi, L., 1940
Diatomee del Mare Adriatico. Nuovo Giornale Botanico Italiano, n.s., 47:559-608.

Diploneis notabilis
Andrade e Clovis Teixeira, M.H. de, 1957.
Contribuição para o conhecimento das diatomáceas do Brasil.
Bol. Inst. Ocean., Univ. Sao Paulo, 8(1/2):171-225, 10 pls.

Diploneis notabilis
Hendy, N. Ingram, 1964
An introductory account of the smaller algae of British coastal waters. V. Bacillariophyceae (Diatoms).
Her Majesty's Stationary Office, 317 pp., 45 pls.

Diploneis notabilis
Rampi, L., 1940
Diatomee del Mare Adriatico. Nuovo Giornale Botanico Italiano, n.s., 47:559-608.

Diploneis notabilis
Zanon, V., 1948
Diatomee marini di Sardegna e Pugillo di Alghe Marine della stressa. Boll. Pesca, Piscitutura e Idrobiologia, Anno 24, ns. 3(2): 202-244, 27 figs. on 1 pl.

Diploneis ovalis
Rumkówna, A., 1948
[List of the phytoplankton species occurring in the superficial water layers in the Gulf of Gdańsk] Bull. Lab. mar., Gdynia, No. 4: 139-141 with tables in back.

Diploneis ovalis
Zanon, D. V., 1949
Diatomee di Buenos Aires (Argentina) Atti Accad. Naz. Lincei, Memorie, Cl. Sci. fis., mat. e. nat., ser. 7, 11(3):59-151, 2 pls.

Diploneis ovalis
Zanon, V., 1948
Diatomee marini di Sardegna e Pugillo di Alghe Marine della stressa. Boll. Pesca, Piscitutura e Idrobiologia, Anno 24, ns. 3(2): 202-244, 27 figs. on 1 pl.

Diploneis papula
Salah, M.M., 1952(1953).
XII. Diatoms from Blakeney Point, Norfolk. New species and new records for Great Britain. J.R. Microsc. Soc., Ser. 3, 72(3):155-169, 3 pls.

Diploneis papula
Zanon, V., 1948
Diatomee marini di Sardegna e Pugillo di Alghe Marine della stressa. Boll. Pesca, Piscitutura e Idrobiologia, Anno 24, ns. 3(2): 202-244, 27 figs. on 1 pl.

Diploneis pseudopetersenii n.sp.
Cholnoky, B.J., 1963.
Beiträge zur Kenntnis des marinen Litorals von Südafrika.
Botanica Marina, 5(2/3):38-83.

Diploneis puella
Hendey, N.I., 1951
Littoral diatoms of Chicester Harbour with special reference to fouling. J.Roy. Microscop. Soc. 71(1): 1-86, 18 pls.

Diploneis puella
Rumkówna, A., 1948
[List of the phytoplankton species occurring in the superficial water layers in the Gulf of Gdańsk] Bull. Lab. mar., Gdynia, No. 4: 139-141 with tables in back.

Diploneis robustus n.sp.
Subrahmanyan, R., 1946.
A systematic account of the marine plankton diatoms of the Madras coast. Proc. Indian Acad. Sci. 24(4):85-197, Pl. 2, 440 textfigs.

Diploneis smithi
Andrade e Clovis Teixeira, M.H. de, 1957.
Contribuição para o conhecimento das diatomáceas do Brasil.
Bol. Inst. Ocean., Univ. Sao Paulo, 8(1/2):171-225, 10 pls.

Diploneis smithii
Boden, B.P., 1950
Some marine plankton diatoms from the west coast of South Africa. Trans. R.Soc. S. Africa. 32:321-434, 100 text figs.

Diploneis smithii & varieties
Hendy, N. Ingram, 1964
An introductory account of the smaller algae of British coastal waters. V. Bacillariophyceae (Diatoms).
Her Majesty's Stationary Office, 317 pp., 45 pls.

Diploneis Smithii
Hendey, N.I., 1951
Littoral diatoms of Chicester Harbour with special reference to fouling. J.Roy. Microscop. Soc. 71(1): 1-86, 18 pls.

Diploneis Smithi
Rampi, L., 1942
Ricerche sul fitoplancton del Mare Ligure 6. Le diatomee delle acque di Sanremo. Nuovo Giornale Botanico Italiano, N.S., 49:252-268.

Diploneis Smithi
Rampi, L., 1940
Diatomee del Mare Adriatico. Nuovo Giornale Botanico Italiano, n.s., 47:559-608.

Diploneis Smithii
Zanon, D. V., 1949
Diatomee di Buenos Aires (Argentina) Atti Accad. Naz. Lincei, Memorie, Cl. Sci. fis., mat. e. nat., ser. 7, 11(3):59-151, 2 pls.

Diploneis smithii
Zanon, V., 1948
Diatomee marini di Sardegna e Pugillo di Alghe Marine della stressa. Boll. Pesca, Piscitutura e Idrobiologia, Anno 24, ns. 3(2): 202-244, 27 figs. on 1 pl.

Diploneis splendida
Hendy, N. Ingram, 1964
An introductory account of the smaller algae of British coastal waters. V. Bacillariophyceae (Diatoms).
Her Majesty's Stationary Office, 317 pp., 45 pls.

Diploneis splendida
Rampi, L., 1942
Ricerche sul fitoplancton del Mare Ligure 6. Le diatomee delle acque di Sanremo. Nuovo Giornale Botanico Italiano, N.S., 49:252-268.

Diploneis stauroneiformis n.sp.[a]

Hendey, N. Ingram 1971.
Some marine diatoms from the
Galápagos Islands.
Nova Hedwigia 22 (1/2): 371-422.

Diploneis stroemi
Hendy, N. Ingram, 1964

An introductory account of the smaller algae of British coastal waters. V. Bacillario-phyceae (Diatoms).
Her Majesty's Stationary Office, 317 pp., 45 pls.

Diploneis subcincta
Rampi, L., 1940
Diatomee del Mare Adriatico. Nuovo Giornale Botanico Italiano, n.s., 47:559-608.

Diploneis suborbicularis
Hendy, N. Ingram, 1964

An introductory account of the smaller algae of British coastal waters. V. Bacillario-phyceae (Diatoms).
Her Majesty's Stationary Office, 317 pp., 45 pls.

Diploneis suborbicularis
Hendey, N.J., 1951
Littoral diatoms of Chicester Harbour with special reference to fouling. J. Roy. Microscop. Soc. 71(1): 1-86, 18 pls.

Diploneis suborbicularis
Rampi, L., 1940
Diatomee del Mare Adriatico. Nuovo Giornale Botanico Italiano, n.s., 47:559-608.

Diploneis suborbicularis
Zanon, V., 1948
Diatomee marini di Sardegna e Pugillo di Alghe Marine della stressa. Boll. Pesca, Piscitutura e Idrobiologia, Anno 24, ns. 3(2): 202-244, 27 figs. on 1 pl.

Diploneis subovalis
Tsumura, K., 1956.
Diatomoj el la cirkaufoso de la restajo de la kastele de Odawara. J. Yokohama Municipal Univ., (C-14), No. 47:23 pp.

Diploneis suezii n.sp.
Salah, M., and G. Tamás 1968.
Notes on new planktonic diatoms from Egypt.
Hydrobiologia 31 (2): 231-240.

Diploneis suspecta comb. nov.
Hendey, N-Ingram, 1958 [1957(Publ. 1958)]
Marine diatoms from some West African Ports. J. R. Microsc. Soc. (3) 77(1/2): 28-85.

Diploneis vacillans var. renitens
Rampi, L., 1942
Ricerche sul fitoplancton del Mare Ligure 6. Le diatomee delle acque di Sanremo. Nuovo Giornale Botanico Italiano, N.S., 49:252-268.

Diploneis vacillans
Zanon, D. V., 1949
Diatomee di Buenos Aires (Argentina) Atti Accad. Naz. Lincei, Memorie, Cl. Sci. fis., mat. e. nat., ser. 7, 11(3):59-151, 2 pls.

Diploneis vascillans
Zanon, V., 1948
Diatomee marini di Sardegna e Pugillo di Alghe Marine della stressa. Boll. Pesca, Piscitutura e Idrobiologia, Anno 24, ns. 3(2): 202-244, 27 figs. on 1 pl.

Diploneis vetula
Hendy, N. Ingram, 1964

An introductory account of the smaller algae of British coastal waters. V. Bacillario-phyceae (Diatoms).
Her Majesty's Stationary Office, 317 pp., 45 pls.

Diploneis vetula
Rampi, L., 1940
Diatomee del Mare Adriatico. Nuovo Giornale Botanico Italiano, n.s., 47:559-608.

Diploneis Weissflogi
Andrade e Clovis Teixeira, M.H. de, 1957.
Contribuição para o conhecimento das diatomáceas do Brasil.
Bol. Inst. Ocean., Univ. Sao Paulo, 8(1/2):171-225, 10 pls.

Diploneis Weisflogii
Zanon, V., 1948
Diatomee marini di Sardegna e Pugillo di Alghe Marine della stressa. Boll. Pesca, Piscitutura e Idrobiologia, Anno 24, ns. 3(2): 202-244, 27 figs. on 1 pl.

Ditylum
Desikachary, T.V., 1956(1957).

Electron microscope studies on diatoms. J.R. Microsc. Soc. (3)76(1/2):9-36.

Ditylum brightwelli
Allen, W. E., 1938
The Templeton Crocker Expedition to the Gulf of California in 1935 - The Phytoplankton. Amer. Microsc. Soc., Trans. 57:328-335.

Ditylum brightwelli
Allen, W.E., 1937
Plankton diatoms of the Gulf of California obtained by the G. Allan Hancock Expedition of 1936. The Hancock Pacific Expeditions, Univ. So. Calif. Publ. 3:47-59, 1 fig.

Ditylium Brightwelli
Bigelow, H.B., and M. Leslie, 1930
Reconnaissance of the waters and plankton of Monterey Bay, July 1928. Bull, M.C.Z., 70(5):429-481, 43 text figs.

Ditylum brightwelli
Boden, B.P., 1950
Some marine plankton diatoms from the west coast of South Africa. Trans. R.Soc. S. Africa. 32:321-434, 100 text figs.

Ditylum brightwelli

Boleyn, Brenda J. 1972.
Studies on the suspension of the marine centric diatom Ditylum brightwelli (West) Grunow.
Int. Revue ges. Hydrobiol. 57(4): 585-597.

Ditylum Brightwellii
Braarud, T., 1945
Experimental studies on marine plankton diatoms. Avhandlingar utgitt av Det Norske Videnskaps-Akademi i Oslo. 1. Mat.-Naturv. Klasse 1944. No.10, 1-16, 1 text fig.

Ditylum Brightwelli
Braarud, T., and Adam Bursa, 1939
On the phytoplankton of the Oslo Fjord, 1933-1934. Hvalrådets Skr. No.19:1-63; 9 text figs. Reviewed. J. du. Cons. 14(3): 418-420. A.C. Gardiner.

Ditylium Brightwelli
Chiba, T., 1949
On the distribution of the plankton in the eastern China Sea and Yellow Sea. 1. Plankton composition in the spring. J. Shimonoseki Coll. Fisheries, 1(1):57-63, 1 fig.

Ditylium Brightwelli
Chu, S. P., 1949
Experimental studies on the environmental factors influencing the growth of phytoplankton. Sci. & Tech. in China 2(3):37-52.

Ditylium brightwelli
Copenhagen, W. J. and L. D., 1949
Variation in the phytoplankton of Table Bay, October 1934 to October 1935. With a note on the calorific value of Chaetoceros spp. Trans. Roy. Soc. S. Africa, 32(2):113-123, 2 text figs.

Ditylum brightwelli
Cleve-Euler, A., 1951
Die Diatomeen von Schweden und Finnland. Kungl. Svenska Vetenskaps Akad. Handl., Fjärde Ser. 2(1): 161 pp., 6 pls.

Ditylum brightwelli
Cupp, Easter E., 1943
Marine plankton diatoms of the west coast of North America. Bull. S.I.O. 5(1):1-238, 5 pls., 168 text figs.

Ditylum brightwelli
Cupp, E.E., 1934
Analysis of marine diatom collections taken from the Canal Zone to California during March, 1933. Trans. Am. Micros. Soc. LIII (1):22-29, 1 map.

Ditylum brightwelli
Cupp, E.E. and Allen, W.E., 1938
Plankton diatoms of the Gulf of California obtained by Allan Hancock Pacific Expedition of 1937. The Hancock Pacific Expeditions, The Univ. So. Calif. Publ. 3: 61-74, 1 map, pls. 4-15.

Ditylum Brightwelli
Dangeard, P., 1927
Phytoplankton de la croisière du "Sylvana". Ann. Inst. Ocean. Monaco, n.s., 4(8):286-401, 54 text figs. (Feirer-Juin 1913).

Ditylum brightwelli
Eppley, Richard W., and James L. Coatsworth 1968.
Uptake of nitrate and nitrite by Ditylum brightwelli - Kinetics and mechanisms.
J. Phycol. 4 (2): 151-156.

Ditylum brightwellii
Eppley, Richard W., and Jane N. Rogers, 1970.
Inorganic nitrogen assimilation of Ditylum brightwellii, a marine plankton diatom.
J. Phycol. 6 (4): 344-351.

Ditylum Brightwelli
Ercegovic, A., 1936
Etudes qualitative et quantitatives du phytoplancton dans les eaux cotières de l'Adriatique oriental moyen au cours de l'année 1934. Acta Adriatica 1(9):1-126

Ditylum Brightwelli
Frenguelli, Joaquin, and Hector Antonio Orlando, 1959.
Operacion MERLUZA. Diatomeas y silico-flagelados del plancton del "VI Crucero". Servicio Hidrogr. Naval., Argentina, Publ. No. H. 619: 5-62.

Ditylium Brightwelli
Ghazzawi, F.M., 1939
Plankton of the Egyptian waters. A study of the Suez Canal Plankton. (A) Phytoplankton. Preliminary Report 83 pp. Notes and Memoires, Min. Commerce-Industry, Egypt, Hydrobiol. & Fish. 65 figs.

Ditylum brightwelli

Gilbert, J.Y., and W.E. Allen, 1943
The phytoplankton of the Gulf of California obtained by the "E.W. Scripps" in 1939 and 1940. J. Mar. Res. V(2):89-110, figs.30-31.

Ditylum Brightwelli

Gilson, H. C., 1937
Chemical and Physical Investigations. The nitrogen cycle. John Murray Exped., 1933-34, Sci. Repts., 2(2):21-81, 16 text figs.

Ditylum brightwelli

Gran, H.H., 1908
Diatomeen. Nordisches Plankton, Botanischer Teil pp. XIX.1-XIX 146; 178 text figs.

Ditylum Brightwelli

Gran, H.H., and T. Braarud, 1935
A quantitative study of the phytoplankton in the Bay of Fundy and the Gulf of Maine (including observations on hydrography, chemistry, and turbidity). J. Biol. Bd., Canada, 1(5):279-467, 69 text figs.

Ditylum Brightwelli

Gran, H. H. and E. C. Angst, 1931
Plankton diatoms of Puget Sound. Publ. Puget Sound Biol. Sta. 7:417-519, 95 text figs.

Ditylum brightwelli

Hendy, N. Ingram, 1964
An introductory account of the smaller algae of British coastal waters. V. Bacillariophyceae (Diatoms).
Her Majesty's Stationary Office, 317 pp., 45 pls.

Ditylum Brightwelli

Hendey, N.I., 1937
The plankton diatoms of the southern seas. Discovery Repts. 16:151-364, pls.6-13.

Ditylum brightwelli

Holmes, Robert W., 1966.
Light microscope observations on cytological manifestations of nitrate, phosphate, and silicate deficiency in four marine centric diatoms. J.Phycology,2(4):136-140.

Ditylum brightwelli

Jørgensen, E., 1905
B.Protistplankton and the diatoms in bottom samples. Hydrographical and biological investigations in Norwegian fjords. Bergens Mus. Skr. 7: 49-225.

Ditylum brightwelli

Jorgensen, E., 1900
Protophyten und Protozoen im Plankton aus der Norwegischen Westkerste. Bergens Mus. Aarb. 1899(6): 95 pp., 5 pls., 83 tables

Ditylum brightwelli

Lafon, M., M. Durchon and Y. Saudray, 1955.
Recherches sur les cycles saisonnières du plankton. Ann. Inst. Océan., 31(3):125-230.

Ditylum Brightwelli

Lillick, L.C. 1940
Phytoplankton and planktonic protozoa of the offshore waters of the Gulf of Maine. Pt.II. Qualitative Composition of the Planktonic Flora. Trans. Am. Phil. Soc., n.s., 31(3):193-237, 13 text figs.

Ditylum Brightwelli

Mangin, M. L., 1912
Phytoplancton de la croisière du "René" dans l'Atlantique (Septembre 1908). Ann. Inst. Océan., n.s., 4(1):1-66, 2 pls., 41 text figs., 2 tables.

Ditylum (Triceratium) Brightwelli

Mann, A., 1893
List of Diatomaceae from a deep-sea dredging in the Atlantic Ocean off Delaware Bay by the U. S. Fish Commission Steamer Albatross. Proc. U. S. Nat. Mus. 16:303-312.

Ditylium Brightwellii

Meunier, A., 1915
Microplancton de la Mer Flamande. 2. Diatomées (excepte le genre Chaetoceros). Mem. Mus. Roy. Hist. Nat., Belgique, 7(3):1-118, Pls. VIII-XIV.

Ditylum brightwelli

Morse, D.C., 1947
Some observations on seasonal variations in plankton population Patuxant River, Maryland 1943 1945. Bd. Nat. Res., Publ. No.65, Chesapeake Biol. Lab., 31, 3 figs.

Ditylum brightwelli

Paasche, E., 1973
The influence of cell size on growth rate, silica content and some other properties of four marine diatom species. Norwegian J. Bot. 20(2/3):199-204.

Ditylum brightwelli

Paasche, E. 1968.
Marine plankton algae grown with light-dark cycles II. Ditylum brightwelli and Nitzschia turgidula. Physiol. Plant. 21(1): 66-77.

Ditylum brightwelli

Paiva Carvalho, J., 1950
O plancton do Rio Maria Rodriques (Cananeis). 1. Diatomaceas e Dinoflagelados. Bol. Inst. Paulista Oceanogr. 1(1); 27-43, 2 fold-in tables, 2 figs.

Ditylum brightwelli

Pavillard, J., 1925
Bacillariales. Rept. on the Danish Oceangr. Exped., 1908-10 to the Mediterranean and adj. seas. Vol.II., Biol. J4:72 pp., 116 text figs.

Ditylum Brightwelli

Pettersson, H., F. Gross, and F. Koczy, 1939
Large scale plankton cultures. Medd. från Oceanografiska Institutet i Göteborg No.3 (Göteborgs Kungl. vetenskaps-och Vitterhets-Samhälles Handlingar Femte Följden. Ser. B) Vol.6(13):1-24.

1. The plankton shaft. H. Pettersson
2. Experiment with phytoplankton. F. Gross and F. Koczy
3. Experiments with zooplankton.

Ditylum Brightwelli

Phifer, L.D. (undated)
The occurrence and distribution of plankton diatoms in Bering Sea and Bering Strait, July 26-August 24, 1934. Report of Oceanographic cruise of U.S. Coast Guard Cutter Chelan 1934, Part II(A):1-44 (mimeographed) plus fig.1 (after Pt.B)

Ditylium Brightwelli

Schodduyn, M., 1926
Observations faites dans la baie d'Ambleteuse (Pas de Calais). Bull. Inst. Ocean., Monaco, No. 482: 64 pp.

Ditylum brightwelli

Sousa e Silva, E., 1949
Diatomaceas e Dinoflagelados de Baia de Cascais. Portugaliae Acta Biol. Volume; Julio Henriques, Ser. B: 300-383, 9 pls, 2 fold-in tables.

Ditylum Brightwelli

Sousa a Silva, E., and J. Dos Santos-Pinto, 1948
O Plancton da Baia de S. Martinho do Porto. 1. Diatomaceas e Dinoflagelados. Bol. Soc. Portuguese de Ciencias Naturais, 16(2):134-187, 6 pls. (Trav. Sta. Biol. Mar. de Lisbonne No. 52).

Ditylium intricatum

Forti, A., 1922
Ricerche sulla flora pelagica (fitoplancton) di Quarto dei Mille. Mem. R. Com. Talass. Ital. 97:248 pp., 13 pls.

Ditylium intricatum

Ghazzawi, F.M., 1939
Plankton of the Egyptian waters. A study of the Suez Canal Plankton. (A) Phytoplankton. Preliminary Report 83 pp. Notes and Memoires, Min. Cormerce-Industry, Egypt, Hydrobiol. & Fish. 65 figs.

Ditylium sol

Allen, W.E., and E.E. Cupp, 1935
Plankton diatoms of the Java Sea. Annales du Jardin Botanique de Buitenzorg XLIV (2):101-174, figs.1-127.

(drawings of all species mentioned)

Ditylium sol

Dangeard, P., 1927
Phytoplankton de la croisière du "Sylvana". Ann. Inst. Ocean., Monaco, n.s., 4(8):286-401, 54 text figs. (Feirer-Juin 1913).

Ditylium sol

Hendey, N.I., 1937
The plankton diatoms of the southern seas. Discovery Repts. 16:151-364, pls.6-13.

Ditylum sol

Mann, A., 1907
Report on the diatoms of the Albatross voyages in the Pacific Ocean, 1888-1904. Contrib. U. S. Nat. Herb. 10(5):221-419, Pls. XLIV-LIV.

Ditylium sol

Marshall, S. M., 1933
The production of microplankton in the Great Barrier Reef Region. Brit. Mus. (N.H.) Great Barrier Reef Exped. 1928-29, Sci. Repts. II(5):111-157, 14 text figs.

Ditylium sol

Marukawa, H., 1921
Plankton lists and some new species of copepods, from the northern waters of Japan. Bull. Inst. Ocean., No.384, 15 pp., 3 pls., 1 chart. Monaco

Ditylum undulatum

Mann, A., 1907
Report on the diatoms of the Albatross voyages in the Pacific Ocean, 1888-1904. Contrib. U. S. Nat. Herb. 10(5):221-419, Pls. XLIV-LIV.

Donkinia carinata

Hendy, N. Ingram, 1964
An introductory account of the smaller algae of British coastal waters. V. Bacillariophyceae (Diatoms).
Her Majesty's Stationary Office, 317 pp., 45 pls.

Donkinia recta

Hendy, N. Ingram, 1964
An introductory account of the smaller algae of British coastal waters. V. Bacillariophyceae (Diatoms).
Her Majesty's Stationary Office, 317 pp., 45 pls.

Donkinia recta

Rampi, L., 1940
Diatomee del Mare Adriatico. Nuovo Giornale Botanico Italiano, n.s., 47:559-608.

Donkinia recta

Schodduyn, M., 1926
Observations faites dans la baie d'Ambleteuse (Pas de Calais). Bull. Inst. Ocean., Monaco, No. 482: 64 pp.

Druridgea

Brunel, J., 1962
Le phytoplancton de la Baie de Chaleurs. Inst. Botan., Univ. Montréal, Contrib. No. 77: 365 pp., 66 pls.

Druridgea compressa

Hendy, N. Ingram, 1964
An introductory account of the smaller algae of British coastal waters. V. Bacillariophyceae (Diatoms). Her Majesty's Stationary Office, 317 pp., 45 pls.

Druridgea geminata

Schodduyn, M., 1926
Observations faites dans la baie d'Ambleteuse (Pas de Calais). Bull. Inst. Ocean., Monaco, No. 482: 64 pp.

Dunaliella

Götting, Klaus-Jurgen, 1963.
Zur Reincultur von Dunaliella. Helgoländer Wiss. Meeresuntersuch., 8(4):404-424.

Duneliella, spp.

Masyuk, N.P. and M.I. Radchenko, 1970.
Comparative chromatographic study of pigments in some species and strains of Dunaliella Teod. Gidrobiol. Zh., 6(3): 51-58.
(In Russian; English abstract)

Dunaliella bioculata

Saraiva, M.C., 1972.
Effet de l'irradiation gamma (cobalt 60) sur les cultures d'une chlorophycée, Dunaliella bioculata. Mar. Biol. 15(1): 74-80.

Dunaliella bioculata

Saraiva, M.C., R. Lo Go et J. Prudhomme 1971-1972.
L'influence de l'irradiation gamma sur la répartition des volumes cellulaires de l'algue Dunalielle bioculata Butcher 1959. Ann. Inst. Océanogr. Paris n.s. 48(2):179-185.

Dunaliella primolecta

Thomas, William H., 1964.
An experimental evaluation of the C14 method for measuring phytoplankton production, using cultures of Dunaliella primolecta Butcher. U.S.F.W.S. Fish. Bull., 63(2):273-292.

Dunaliella salina

Ehrhardt, J.P., R. Moncoulon et P. Niausset 1971.
Comportement in vitro de la Chlorophycée Dunaliella salina Dunal dans les milieux à salinité différente détermination d'un optimum de salinité. Vie Milieu Suppl. 22(1): 203-217.

Dunaliella salina

Mironyuk, V.I. and L.O. Einor, 1970.
Effect of phenol derivatives on oxygen exchange of Dunaliella salina Teod. Gidrobiol. Zh., 6(3): 91-95.
(In Russian)

Dunaliella salina

Parsons, T.R., 1961
On the pigment composition of eleven species of marine phytoplankton. J. Fish. Res. Bd., Canada, 18(6):1017-1025.

Dunaliella salina

Parsons, T.R., K. Stephens and J.D.H. Strickland, 1961
On the chemical composition of eleven species of marine phytoplankters. J. Fish. Res. Bd., Canada, 18(6):1001-1016.

Dunaliella salina

Thomas, Pierre, et Raoul Dumas 1970.
Contribution à l'étude de Dunaliella salina en cultures bactériennes, nutrition et composition. Téthys 2(1): 19-28.

Dunaliella tertiolecta

Davies, Anthony G., 1970.
Iron, chelation and the growth of marine phytoplankton. 1. Growth kinetics and chlorophyll production in cultures of the euryhaline flagellate Dunaliela tertiolecta under iron-limiting conditions. J. mar. biol. Ass., U.K. 50(1): 65-86.

Dunaliella tertiolecta

Eppley, Richard W., and James L. Coatsworth, 1966. Culture of the marine phytoplankter, Dunaliella tertiolecta, with light-dark cycles. Ark. Mikrobiol., 55:17-25.

Dunaliella tertiolecta

Huntsman Susan A. 1972.
Organic excretion by Dunaliella tertiolecta. J. Phycol. 8(1): 59-63.

Dunaliella tertiolecta

Jitts, H.R., C.D. McAllister, K. Stephens and J.D.H. Strickland, 1964.
The cell division rates of some marine phytoplanktens as a funtion of light and temperature. J. Fish. Res. Bd., Canada, 21(1):139-157.

Dunaliella tertiolecta

Hellebust, Johan A., and John Terborgh, 1967.
Effects of environmental conditions on the rate of photosynthesis and some photosynthetic enzymes in Dunaliella tertiolecta Butcher. Limnol. Oceanogr., 12(4):559-567.

Dunaliella tertiolecta

Latorella, A.H., and R.L. Vadas 1973.
Salinity adaptation by Dunaliella tertiolecta. 1. Increases in carbonic anhydrase activity and evidence for a light-dependent Na^+/H^+ exchange. J. Phycol. 9(5): 273-277.

Dunaliella tertiolecta

Luard Elizabeth J. 1973.
Sensitivity of Dunaliella and Scenedesmus (Chlorophyceae) to chlorinated hydrocarbons Phycologia 12 (1/2): 29-33.

Dunaliella tertiolecta

Wallen, D.G. and G.H. Geen, 1971.
Light quality and concentration of proteins, RNA, DNA and photosynthetic pigments in two species of marine plankton algae. Mar. Biol. 10(1): 44-51.

Dunaliella tertiolecta

Wallen, D.G. and G.H. Geen, 1971.
Light quality in relation to growth, photosynthetic rates and carbon metabolism in two species of marine plankton algae. Mar. Biol. 10(1): 34-43.

Endictya

Brunel, J., 1962
Le phytoplancton de la Baie de Chaleurs. Inst. Botan., Univ. Montréal, Contrib. No. 77: 365 pp., 66 pls.

Endictya lunyasekii

Cleve-Euler, A., 1951
Die Diatomeen von Schweden und Finnland. Kungl. Svenska Vetenskaps Akad. Handl., Fjärde Ser. 2(1): 161 pp., 6 pls.

Endictya oceanica

Cleve-Euler, A., 1951
Die Diatomeen von Schweden und Finnland. Kungl. Svenska Vetenskaps Akad. Handl., Fjärde Ser. 2(1): 161 pp., 6 pls.

Endictya oceanica

Gran, H.H. and T. Braarud, 1935
A quantitative study of the phytoplankton in the Bay of Fundy and the Gulf of Maine (including observations on hydrography, chemistry, and turbidity). J. Biol. Bd., Canada, 1(5):279-467, 69 text figs.

Endictya oceanica

Hendy, N. Ingram, 1964
An introductory account of the smaller algae of British coastal waters. V. Bacillariophyceae (Diatoms). Her Majesty's Stationary Office, 317 pp., 45 pls.

Endictya oceanica

Hendey, N. Ingram, 1958 [1957 (Publ. 1958)]
Marine diatoms from some West African Ports. J. R. Microsc. Soc, (3) 77(1/2): 28-85.

Endictya oceanica

Lillick, L.C., 1940
Phytoplankton and planktonic protozoa of the offshore waters of the Gulf of Maine. Pt. II. Qualitative Composition of the Planktonic Flora. Trans. Am. Phil. Soc., n.s., 31(3):193-237, 13 text figs.

Endictya oceanica

Paiva Carvalho, J., 1950
O plancton do Rio Maria Rodrigues (Cananeis). 1. Diatomaceas e Dinoflagelados. Bol. Inst. Paulista Oceanogr. 1(1): 27-43, 2 fold-in tables, 2 figs.

Endictya oceanica

Politis, J., 1949
Diatomees marines de Bosphores et des Ibes de la mer de Marmara. II Practica tou Hellenikou Hidrobiologikou Institutoutou 1929, Etoz 1929, 3(1):11-31.

Endictya oceanica

Zanon, V., 1948
Diatomee marini di Sardegna e Pugillo di Alghe Marine della stressa. Boll. Pesca, Piscitutura e Idrobiologia, Anno 24, ns. 3(2): 202-244, 27 figs. on 1 pl.

Endictya? Zabelinae

Cleve-Euler, A., 1951
Die Diatomeen von Schweden und Finnland. Kungl. Svenska Vetenskaps Akad. Handl., Fjärde Ser. 2(1): 161 pp., 6 pls.

Entopyla

Neaverson, E., 1934
The sea-floor deposits. 1. General characteristics and distribution. Discovery Repts. 9: 297-349, Plates 17-22.

Entopyla australis

Manguin, E., 1954
Diatomees marines provenant de l'Ile Heard (Australian National Antarctic Expedition). Rev. Algol., n.s., 1: 14-24.

Entopyla australis

Mann, A., 1907
Report on the diatoms of the Albatross voyages in the Pacific Ocean, 1888-1904. Contrib. U.S. Nat. Herb. 10(5):221-419, Pls. XLIV-LIV.

Entopyla kerguelensis
Hendey, N.I., 1937
The plankton diatoms of the southern seas. Discovery Repts. 16:151-364, pls.6-13.

Entopyla ocellata
Zanon, V., 1948
Diatomee marini di Sardegna e Pugillo di Alghe Marine della stressa. Boll. Pesca, Piscitutura e Idrobiologia, Anno 24, ns. 3(2): 202-244, 27 figs. on 1 pl.

Epithemia adnata proboscidea
Hendy, N. Ingram, 1964
An introductory account of the smaller algae of British coastal waters. V. Bacillariophyceae (Diatoms). Her Majesty's Stationary Office, 317 pp., 45 pls.

Ephithemia argus
Rumkówna, A., 1948
[List of the phytoplankton species occurring in the superficial water layers in the Gulf of Gdańsk] Bull. Lab. mar., Gdynia, No. 4: 139-141 with tables in back.

Ephithemia intermedia
Rumkówna, A., 1948
[List of the phytoplankton species occurring in the superficial water layers in the Gulf of Gdańsk] Bull. Lab. mar., Gdynia, No. 4: 139-141 with tables in back.

Epithemia musculus
Hendy, N. Ingram, 1964
An introductory account of the smaller algae of British coastal waters. V. Bacillariophyceae (Diatoms). Her Majesty's Stationary Office, 317 pp., 45 pls.

Epithemia musculus
Hendey, N.I., 1951
Littoral diatoms of Chicester Harbour with special reference to fouling. J. Roy. Microscop. Soc. 71(1): 1-86, 18 pls.

Ephithemia Reichelti
Rumkówna, A., 1948
[List of the phytoplankton species occurring in the superficial water layers in the Gulf of Gdańsk] Bull. Lab. mar., Gdynia, No. 4: 139-141 with tables in back.

Epithemia sorex
Hendy, N. Ingram, 1964
An introductory account of the smaller algae of British coastal waters. V. Bacillariophyceae (Diatoms). Her Majesty's Stationary Office, 317 pp., 45 pls.

Ephithemia sorex gracilis
Rumkówna, A., 1948
[List of the phytoplankton species occurring in the superficial water layers in the Gulf of Gdańsk] Bull. Lab. mar., Gdynia, No. 4: 139-141 with tables in back.

Epithemia turgida
Kokubo, S., and S. Sato, 1947
Plankters in Jū-San Gata. Physiol. and Ecol. (Japan) 1(4):1-16, 3 text figs., tables.

Epithemia turgida
Mann, A., 1893
List of Diatomaceae from a deep-sea dredging in the Atlantic Ocean off Delaware Bay by the U. S. Fish Commission Steamer Albatross. Proc. U. S. Nat. Mus. 16:303-312.

Epithemia turgida
Rumkówna, A., 1948
[List of the phytoplankton species occurring in the superficial water layers in the Gulf of Gdańsk] Bull. Lab. mar., Gdynia, No. 4: 139-141 with tables in back.

Epithemia turgida
Tsumura, K., 1956.
Diatomoj el la cirkaufoso de la restajo de la kastelo de Odawara. J. Yokohama Municipal Univ., (C-14); No. 47:23 pp.

Epithemia turgida
Zanon, D. V., 1949
Diatomee di Buenos Aires (Argentina) Atti Accad. Naz. Lincei, Memorie, Cl. Sci. fis., mat. e. nat., ser. 7, 11(3):59-151, 2 pls.

Epithemia Westermani
Mann, A., 1893
List of Diatomaceae from a deep-sea dredging in the Atlantic Ocean off Delaware Bay by the U. S. Fish Commission Steamer Albatross. Proc. U. S. Nat. Mus. 16:303-312.

Epithemia zebra
Kokubo, S., and S. Sato, 1947
Plankters in Jū-San Gata. Physiol. and Ecol. (Japan) 1(4):1-16, 3 text figs., tables.

Epithemia zebra
Zanon, D. V., 1949
Diatomee di Buenos Aires (Argentina) Atti Accad. Naz. Lincei, Memorie, Cl. Sci. fis., mat. e. nat., ser. 7, 11(3):59-151, 2 pls.

Epithemia zebra porcellus
Rumkówna, A., 1948
[List of the phytoplankton species occurring in the superficial water layers in the Gulf of Gdańsk] Bull. Lab. mar., Gdynia, No. 4: 139-141 with tables in back.

Erithemia
Desikachary, T.V., 1956(1957).
Electron microscope studies on diatoms. J. R. Microsc. Soc. (3)76(1/2):9-36.

Erithemia zebra
Zanon, V., 1948
Diatomee marini di Sardegna e Pugillo di Alghe Marine della stressa. Boll. Pesca, Piscitutura e Idrobiologia, Anno 24, ns. 3(2): 202-244, 27 figs. on 1 pl.

Ethmodiscus
Brunel, J., 1962
Le phytoplancton de la Baie de Chaleurs. Inst. Botan., Univ. Montréal, Contrib. No. 77: 365 pp., 66 pls.

Ethmodiscus
Desikachary, T.V., 1956(1957).
Electron microscope studies on diatoms. J.R. Microsc. Soc. (3)76(1/2):9-36.

Ethmosciscus appendiculatus n. comb.
Picard, Michel, 1970.
Observations sur les diatomées marines du genre Ethmodiscus Castr. Revue algol. no. 10(1):56-73

Ethmodiscus convexus n.sp.
Castracane degli Antelminelli, F., 1886
1. Report on the Diatomaceae collected by H.M.S. Challenger during the years 1873-1876. Rept. Sci. Results, H.M.S. Challenger, Botany Vol. II, 178 pp., 30 pls.

Ethmodiscus coronatus n.sp.
Castracane degli Antelminelli, F., 1886
1. Report on the Diatomaceae collected by H.M.S. Challenger during the years 1873-1876. Rept. Sci. Results, H.M.S. Challenger, Botany Vol. II, 178 pp., 30 pls.

Ethmodiscus diadema n.sp.
Castracane degli Antelminelli, F., 1886
1. Report on the Diatomaceae collected by H.M.S. Challenger during the years 1873-1876. Rept. Sci. Results, H.M.S. Challenger, Botany Vol. II, 178 pp., 30 pls.

Ethodiscus gasellae
Belyaeva, T.V., 1968.
Distribution and numbers of diatoms of the genus Ethmodiscus Castr. in the plankton and in bottom sediments of the Pacific Ocean. (In Russian; English abstract). Okeanologiia, Akad. Nauk. SSSR, 8(1): 102-110.

Ethmodiscus gazellae
Fraser, J.H., 1954.
Warm-water species in the plankton off the English Channel entrance. J.M.B.A., U.K., 33: 345-346.

Ethmodiscus gazellae
Hendey, N.I., 1937
The plankton diatoms of the southern seas. Discovery Repts. 16:151-364, pls.6-13.

Ethmodiscus gazellae
Picard, Michel, 1970.
Observations sur les diatomées marines du genre Ethmodiscus Castr. Revue algol. no. 10(1):56-73

Ethmodiscus gigas n.sp.
Castracane degli Antelminelli, F., 1886
1. Report on the Diatomaceae collected by H.M.S. Challenger during the years 1873-1876. Rept. Sci. Results, H.M.S. Challenger, Botany Vol. II, 178 pp., 30 pls.

Ethmodiscus humilis n.sp
Castracane degli Antelminelli, F., 1886
1. Report on the Diatomaceae collected by H.M.S. Challenger during the years 1873-1876. Rept. Sci. Results, H.M.S. Challenger, Botany Vol. II, 178 pp., 30 pls.

Ethmodiscus aponfeus n.sp.
Castracane degli Antelminelli, F., 1886
1. Report on the Diatomaceae collected by H.M.S. Challenger during the years 1873-1876. Rept. Sci. Results, H.M.S. Challenger, Botany Vol. II, 178 pp., 30 pls.

Ethmodiscus obovatus n.sp.
Castracane degli Antelminelli, F., 1886
1. Report on the Diatomaceae collected by H.M.S. Challenger during the years 1873-1876. Rept. Sci. Results, H.M.S. Challenger, Botany Vol. II, 178 pp., 30 pls.

Ethmodiscus perichantina n.sp.
Castracane degli Antelminelli, F., 1886
1. Report on the Diatomaceae collected by H.M.S. Challenger during the years 1873-1876. Rept. Sci. Results, H.M.S. Challenger, Botany Vol. II, 178 pp., 30 pls.

Ethmodiscus punctiger n.sp.
Castracane degli Antelminelli, F., 1886.
1. Report on the Diatomaceae collected by H.M.S. Challenger during the years 1873-1876. Rept. Sci. Results, H.M.S. Challenger, Botany Vol. II, 178 pp., 30 pls.

Ethmodiscus radiatus n.sp.
Castracane degli Antelminelli, F., 1886
1. Report on the Diatomaceae collected by H.M.S. Challenger during the years 1873-1876. Rept. Sci. Results, H.M.S. Challenger, Botany Vol. II, 178 pp., 30 pls.

Ethmodiscus rex
Anikouchine, William A., and Hsin-Yi Ling 1967
Evidence for turbidite accumulation in trenches in the Indo-Pacific region. Mar. Geol. 5(2):141-154.

Ethmodiscus rex
Beklemishev, C.W., M.N. Petrikova and H.J. Semina, 1961
[On the cause of the buoyancy of plankton diatoms.]
Trudy Inst. Okeanol., 51: 33-35.

Ethmodiscus rex
Belyaeva, T.V., 1968.
Distribution and numbers of diatoms of the genus Ethmodiscus Castr. in the plankton and in bottom sediments of the Pacific Ocean. (In Russian; English abstract). Okeanologiia. Akad. Nauk, SSSR, 8(1): 102-110.

Ethmodiscus rex
Kolbe, R.W., 1957.
Diatoms from Equatorial Indian Ocean cores. Rept. Swedish Deep-Sea Exped., 1947-48, 9(1):3-50.

Ethmodiscus rex
McHugh, J.L., 1954.
Distribution and abundance of the diatom Ethmodiscus rex off the west coast of North America.
Deep-Sea Res., 1(4):216-222.

Ethmodiscus rex
Riedel, W.R., 1954.
The age of the sediment collected at Challenger (1875) Station 225 and the distribution of Ethmodiscus rex (Rattray). Deep-Sea Res. 1(3): 170-175, 1 pl.

Ethmodiscus rex
Semina, H.G., 1959
[Distribution of the diatom Ethmodiscus rex (Wall) Hendey in the plankton.]
DAN, SSSR, 124(6):1309-1312.

Translation NIOT/37

Ethmodiscus rex
Wiseman, J.D.H., and N.I. Hendey, 1953.
The significance and diatom content of a deep-sea floor sample from the neighborhood of the greatest oceanic depth. Deep-sea Res. 1(1):47-59, 2 pls., 2 textfigs.

Ethmodiscus (?) sphaeroidalis n.sp.
Castracane degli Antelminelli, F., 1886
1. Report on the Diatomaceae collected by H.M.S. Challenger during the years 1873-1876. Rept. Sci. Results, H.M.S. Challenger, Botany Vol. II, 178 pp., 30 pls.

Ethmodiscus subtilis
Hendey, N.I., 1937
The plankton diatoms of the southern seas. Discovery Repts. 16:151-364, pls.6-13.

Ethmodiscus tympanum n.sp.
Castracane degli Antelminelli, F., 1886
1. Report on the Diatomaceae collected by H.M.S. Challenger during the years 1873-1876. Rept. Sci. Results, H.M.S. Challenger, Botany Vol. II, 178 pp., 30 pls.

Ethmodiscus wyvilleanus n.sp.
Castracane degli Antelminelli, F., 1886
1. Report on the Diatomaceae collected by H.M.S. Challenger during the years 1873-1876. Rept. Sci. Results, H.M.S. Challenger, Botany Vol. II, 178 pp., 30 pls.

Eucampia antarctica
Mangin, L., 1915
Phytoplankton de L'Antartique. Deuxieme Exped. Ant. Francaise (1908-1910), 95 pp., 3 pls., 58 text figs.

Eucampia balaustium
Balech, E., 1947
Contribution al conocimiento del plancton antarctico. Plankton del Mar de Bellinghausen. Physis 20:75-91, 76 figs. on 8 pls.

Eucampia balaustium
Boden, B. P., 1949.
The diatoms collected by the U.S.S. CACO-PAN in the Antarctic in 1947. J. Mar. Res. 8(1):6-13, 3 textfigs.

Eucampia baleustrum
Boden, Brian, 1948
Marine plankton diatoms on operation HIGHJUMP in: Some oceanographic observations on operation HIGHJUMP. By R.S. Dietz. USNEL Rept. No.55, 97 pp., 41 figs, 7 July 1948.

Eucampia balaustium n.sp.
Castracane degli Antelminelli, F., 1886
1. Report on the Diatomaceae collected by H.M.S. Challenger during the years 1873-1876. Rept. Sci. Results, H.M.S. Challenger, Botany Vol. II, 178 pp., 30 pls.

Eucampia balaustium
Frenguelli, Joaquin, and Hector Antonio Orlando, 1959.
Operacion MERLUZA. Diatomeas y silico-flagelados del plancton del "VI Crucero". Servicio Hidrogr. Naval., Argentina, Publ. No. H. 619: 5-62.

Eucampia balaustium
Hendey, N.I., 1937
The plankton diatoms of the southern seas. Discovery Repts. 16:151-364, pls.6-13.

Eucampia balaustium
Jouse, A.P., G.S. Koroleva, G.A. Nagaeva, 1962
Diatoms in the surface layer of sediment in the Indian sector of the Antarctic. Investigations of marine bottom sediments. (In Russian; English summary).
Trudy Inst. Okeanol., Akad. Nauk, SSSR, 61: 19-92.

Eucampia balaustium
Manguin, E., 1954
Diatomees marines provenant de l'ile Heard (Australian National Antarctic Expedition). Rev. Algol., n.s., 1: 14-24.

Eucampia balaustium minor n.var.
Castracane degli Antelminelli, F., 1886
1. Report on the Diatomaceae collected by H.M.S. Challenger during the years 1873-1876. Rept. Sci. Results, H.M.S. Challenger, Botany Vol. II, 178 pp., 30 pls.

Eucampia bioncavum
Marukawa, H., 1921
Plankton lists and some new species of copepods, from the northern waters of Japan. Bull. Inst. Ocean., No.384, 15 pp., 3 pls., 1 chart. Monaco

Eucampia braustium
Central Meteorological Observatory, 1949
Report on sea and weather observation on Antarctic Whaling Ground (1947-48). Ocean. Mag., Japan, 1(1):49-88, 17 text figs.

Eucampia cornuta
Allen, W.E., and E.E. Cupp, 1935
Plankton diatoms of the Java Sea. Annales du Jardin Botanique de Buitenzorg XLIV (2):101-174, figs.1-127.

(drawings of all species mentioned)

Eucampia cornuta
Boden, B.P., 1950
Some marine plankton diatoms from the west coast of South Africa. Trans. R.Soc. S. Africa. 32:321-434, 100 text figs.

Eucampia cornuta
Boden, B. P., 1949.
The diatoms collected by the U.S.S. CACO-PAN in the Antarctic in 1947. J. Mar. Res. 8(1):6-13, 3 textfigs.

Eucampia cornuta
Boden, Brian, 1948
Marine plankton diatoms on operation HIGHJUMP in: Some oceanographic observations on operation HIGHJUMP. By R.S. Dietz. USNEL Rept. No.55, 97 pp., 41 figs. 7 July 1948.

Eucampia cornuta
Cupp, Easter E., 1943
Marine plankton diatoms of the west coast of North America. Bull. S.I.O. 5(1):1-238, 5 pls., 168 text figs.

Eucampia cornuta
Cupp, E.E., 1934
Analysis of marine diatom collections taken from the Canal Zone to California during March, 1933. Trans. Am. Micros. Soc. LIII (1):22-29, 1 map.

Eucampia cornuta
Hendey, N.I., 1937
The plankton diatoms of the southern seas. Discovery Repts. 16:151-364, pls.6-13.

Eucampia groenlandica
Bigelow, H.B., and M. Leslie, 1930
Reconnaissance of the waters and plankton of Monterey Bay, July 1928. Bull. M.C.Z., 70(5):429-481, 43 text figs.

Eucampia groenlandica
Braarud, T., 1945
Experimental studies on marine plankton diatoms. Avhandlingar utgitt av Det Norske Videnskaps-Akademi i Oslo. 1. Mat.-Naturv. Klasse 1944. No.10, 1-16, 1 text fig.

Eucampia groenlandica
Cleve-Euler, A., 1951
Die Diatomeen von Schweden und Finnland. Kungl. Svenska Vetenskaps Akad. Handl., Fjärde Ser. 2(1): 161 pp., 6 pls.

Eucampia groenlandica
Gran, H.H., 1908
Diatomeen. Nordisches Plankton, Botanischer Teil pp. XIX.1-XIX 146; 178 text figs.

Eucampia groenlandica
Hendy, N. Ingram, 1964
An introductory account of the smaller algae of British coastal waters. V. Bacillariophyceae (Diatoms).
Her Majesty's Stationary Office, 317 pp., 45 pls.

Eucampia groenlandica
Jørgensen, E., 1905
B. Protistplankton and the diatoms in bottom samples. Hydrographical and biological investigations in Norwegian fjords. Bergens Mus. Skr. 7: 49-225.

Euodia orbicularis n.sp.
Castracane degli Antelminelli, F., 1886
1. Report on the Diatomaceae collected by H.M.S. Challenger during the years 1873-1876. Rept. Sci. Results, H.M.S. Challenger, Botany Vol. II, 178 pp., 30 pls.

Euodia radiata n.sp.
Castracane degli Antelminelli, F., 1886
1. Report on the Diatomaceae collected by H.M.S. Challenger during the years 1873-1876. Rept. Sci. Results, H.M.S. Challenger, Botany Vol. II, 178 pp., 30 pls.

Euodia recta n.sp.
Castracane degli Antelminelli, F., 1886
1. Report on the Diatomaceae collected by H.M.S. Challenger during the years 1873-1876. Rept. Sci. Results, H.M.S. Challenger, Botany Vol. II, 178 pp., 30 pls.

Eucampia recta n.sp.
Gran, H.H., and T. Braarud, 1935
A quantitative study of the phytoplankton in the Bay of Fundy and the Gulf of Maine (including observations on hydrography, chemistry, and turbidity). J. Biol. Bd., Canada, 1(5):279-467, 69 text figs.

Eucampia recta

Lillick, L.C., 1940
Phytoplankton and planktonic protozoa of the offshore waters of the Gulf of Maine. Pt.II. Qualitative Composition of the Planktonic Flora. Trans. Am. Phil. Soc., n.s., 31(3):193-237, 13 text figs.

Eucampia zodiacus

Lafon, M., M. Durchon and Y Saudray, 1955.
Recherches sur les cycles saisonnières du plankton Ann. Inst. Océan., 31(3):125-230.

Euodia ventricosa n.sp.

Castracane degli Antelminelli, F., 1886
1. Report on the Diatomaceae collected by H.M.S. Challenger during the years 1873-1876. Rept. Sci. Results, H.M.S. Challenger, Botany Vol. II, 178 pp., 30 pls.

Eucampia zodiacus

Allen, W. E., 1938
The Templeton Crocker Expedition to the Gulf of California in 1935 - The Phytoplankton. Amer. Microsc. Soc., Trans. 57:328-335.

Eucampia zoodiacus

Allen, W.E., 1937
Plankton diatoms of the Gulf of California obtained by the G. Allan Hancock Expedition of 1936. The Hancock Pacific Expeditions, Univ. So. Calif. Publ. 3:47-59, 1 fig.

Eucampia zoodiacus

Allen, W. E., 1936
Occurrence of marine plankton diatoms in a ten-year series of daily catches in southern California. Am. Jour. Bot. 23(1):60-63.

Eucampia zoodiacus

Allen, W.E., 1928.
Review of five years of studies of phytoplankton at Southern California piers, 1920-1924, inclusive. Bull. S.I.O., tech. ser., 1:357-401, 5 text-figs.

Eucampia zoodiacus

Allen, W.E., 1927.
Quantitative studies on inshore marine diatoms and dinoflagellates of Southern California in 1922. Bull. S.I.O., tech. ser., 1:31-38, 2 text-figs.

Eucampia zoodiacus

Allen, W.E., and E.E. Cupp, 1935
Plankton diatoms of the Java Sea. Annales du Jardin Botanique de Buitenzorg XLIV (2):101-174, figs.1-127.
(drawings of all species mentioned)

Eucampia zoodiacus

Bigelow, H.B., and M. Leslie, 1930
Reconnaissance of the waters and plankton of Monterey Bay, July 1928. Bull. M.C.Z. 70(5):429-481, 43 text figs.

Eucampia zoodiacus

Boden, B.P., 1950
Some marine plankton diatoms from the west coast of South Africa. Trans. R.Soc. S. Africa. 32:321-434, 100 text figs.

Eucampia zoodiacus

Braarud, T., 1945
Experimental studies on marine plankton diatoms. Avhandlingar utgitt av Det Norske Videnskaps-Akademi i Oslo. 1. Mat.-Naturv. Klasse 1944. No.10, 1-16, 1 text fig.

Eucampia zoodiacus

Brunel, J., 1962
Le phytoplancton de la Baie de Chaleurs. Inst. Botan., Univ. Montréal, Contrib. No. 77: 365 pp., 66 pls.

Eucampia zodiacus

Brunel, Jules, 1962
Le phytoplancton de la Baie des Chaleurs. Contrib. Ministère de la Chasse et des Pêcheries, Province de Québec, No. 91: 365 pp.

Eucampia zodiacus

Cleve-Euler, A., 1951
Die Diatomeen von Schweden und Finnland. Kungl. Svenska Vetenskaps Akad. Handl., Fjärde Ser. 2(1): 161 pp., 6 pls.

Eucampia zoodiacus

Cupp, Easter E., 1943
Marine plankton diatoms of the west coast of North America. Bull. S.I.O. 5(1):1-238, 5 pls., 168 text figs.

Eucampia zoodiacus

Cupp, E., 1930
Quantitative Studies of miscellaneous series of surface catches of marine diatoms and dinoflagellates taken between Seattle and the Canal Zone from 1924 to 1928. Trans. Am. Micro. Soc., XLIX (3):238-245.

Eucampia zodiacus

Cupp, E.E., 1934
Analysis of marine diatom collections taken from the Canal Zone to California during March, 1933. Trans. Am. Micros. Soc. LIII (1):22-29, 1 map.

Eucampia zodiacus

Cupp, E.E. and Allen, W.E., 1938
Plankton diatoms of the Gulf of California obtained by Allan Hancock Pacific Expedition of 1937. The Hancock Pacific Expeditions, The Univ. So. Calif. Publ. 3:61-74, 1 map, pls.4-15.

Eucampia zodiacus

Dangeard, P., 1927
Phytoplankton de la croisière du "Sylvana". Ann. Inst. Ocean., Monaco, n.s., 4(8):286-401, 54 text figs. (Feirer-Juin 1913).

Eucampia zodiacus

Frenguelli, Joaquin, and Hector Antonio Orlando. 1959.
Operacion MERLUZA. Diatomeas y silicoflagelados del plancton del "VI Crucero". Servicio Hidrogr. Naval, Argentina, Publ. No. H. 619: 5-62.

Eucampia zodiacus

Gilbert, J.Y., and W.E. Allen, 1943
The phytoplankton of the Gulf of California obtained by the "E.W. Scripps" in 1939 and 1940. J. Mar. Res. V(2):89-110, figs.30-31.

Eucampia zodiacus

Gran, H.H., 1908
Diatomeen. Nordisches Plankton, Botanischer Teil pp. XIX.1-XIX 146; 178 text figs.

Eucampia zodiacus

Gran, H.H., and T. Braarud, 1935
A quantitative study of the phytoplankton in the Bay of Fundy and the Gulf of Maine (including observations on hydrography, chemistry, and turbidity). J. Biol. Bd., Canada, 1(5):279-467, 69 text figs.

Eucampia zodiacus

Gran, H. H. and E. C. Angst, 1931
Plankton diatoms of Puget Sound. Publ. Puget Sound Biol. Sta. 7:417-519, 95 text figs.

Eucampia zodiacus

Hendy, N. Ingram, 1964
An introductory account of the smaller algae of British coastal waters. V. Bacillariophyceae (Diatoms). Her Majesty's Stationary Office, 317 pp., 45 pls.

Eucampia zoodiacus

Hendey, N.I., 1937
The plankton diatoms of the southern seas. Discovery Repts. 16:151-364, pls.6-13.

Eucampia zoodiacus

Hustedt, F. and A.A. Aleem, 1951
Littoral diatoms from the Salstone near Plymouth. JMBA 30(1): 177-196.

Eucampia zoodiacus

Jorgensen, E., 1900
Protophyten und Protozoën im Plankton aus der Norwegischen Westkerste. Bergens Mus. Aarb. 1899(6): 95 pp., 5 pls., 83 tables.

Eucampia zoodiacus

Kokubo, S., 1952
Results of the observations on the plankton and oceanography of Mutsu Bay during 1950, reference being made also to the period 1946-1950. Bull Mar.Biol.Sta., Asamushi 5(1/4): 1-54, 3 tables,(fold-in), 1 fold-in.

Eucampia zoodiacus

Lillick, L.C., 1940
Phytoplankton and planktonic protozoa of the offshore waters of the Gulf of Maine. Pt.II. Qualitative Composition of the Planktonic Flora. Trans. Am. Phil. Soc., n.s., 31(3):193-237, 13 text figs.

Eucampia zoodiacus

Lillick, L.C., 1938
Preliminary report of the phytoplankton of the Gulf of Maine. Am. Mid. Nat. 20(3):624-640, 1 text fig. 37 tables.

Eucampia zoodiacus

Mangin, M. L., 1912
Phytoplankton de la croisière du "René" dans l'Atlantique (Septembre 1908). Ann. Inst. Ocean., n.s., 4(1):1-66, 2 pls., 41 text figs., 2 tables.

Eucampia zoodiacus

Meunier, A., 1915
Microplancton de la Mer Flamande. 2. Diatomées (excepté le genre Chaetoceros). Mem. Mus. Roy. Hist. Nat., Belgique, 7(3):1-118, Pls. VIII-XIV.

Eucampia zoodiacus

Pavillard, J., 1925
Bacillariales. Rept. on the Danish Oceangr. Exped., 1908-10 to the Mediterranean and adj. seas. Vol.II., Biol. J4:72 pp., 116 text figs.

Eucampia zoodiacus

Pettersson, H., F. Gross, and F. Koczy, 1939
Large scale plankton cultures. Medd. från Oceanografiska Institutet i Göteborg No.3 (Göteborgs Kungl. vetenskaps-och Vitterhets-Samhälles Handlingar Femte Följden. Ser. B) Vol.6(13):1-24.

1. The plankton shaft. H. Pettersson
2. Experiment with phytoplankton. F. Gross and F. Koczy
3. Experiments with zooplankton.

Eucampia zoodiacus

Phifer, L.D. (undated)
The occurrence and distribution of plankton diatoms in Bering Sea and Bering Strait, July 26-August 24, 1934. Report of Oceanographic cruise of U.S. Coast Guard Cutter Chelan 1934, Part II(A):1-44 (mimeographed) plus fig.1 (after Pt.B)

Eucampia zodiacus

Schodduyn, M., 1926
Observations faites dans la baie d'Ambleteuse (Pas de Calais). Bull. Inst. Ocean., Monaco, No. 482: 64 pp.

Eucampia zoodiacus

Sleggs, G.F., 1927.
Marine phytoplankton in the region of La Jolla, California during the summer of 1924. Bull. S.I.O., tech. ser., 1:93-117, 8 textfigs.

Eucampia zodiacus

Sousa e Silva, E., 1949
  Diatomaceas e Dinoflagelados de Baía de Cascais. Portugaliae Acta Biol., Volume: Julio Henriques, Ser. B: 300-383, 9 pls, 2 fold-in tables.

Eucampia zodiacus

Sousa e Silva, E., and J. Dos Santos-Pinto, 1948
  O Plancton da Baía de S. Martinho do Porto. 1. Diatomaceas e Dinoflagelados. Bol. Soc. Portuguese de Ciencias Naturais, 16(2):134-187, 6 pls. (Trav. Sta. Biol. Mar. de Lisbonne No. 52).

Eucampia zoodiacus

Uyeno, Fukuzo, 1961
  Oceanographical and ecological studies on primary production of the sea, with special references to relationship between diatom production and temperature and chlorinity of water. Rept. Fac. Fish. Pref. Univ. Mie, 4(1):1-64.

Eucyonema prostratum

Mann, A., 1893
  List of Diatomaceae from a deep-sea dredging in the Atlantic Ocean off Delaware Bay by the U. S. Fish Commission Steamer Albatross. Proc. U. S. Nat. Mus. 16:303-312.

Eucyonema zebra

Mann, A., 1893
  List of Diatomaceae from a deep-sea dredging in the Atlantic Ocean off Delaware Bay by the U. S. Fish Commission Steamer Albatross. Proc. U. S. Nat. Mus. 16:303-312.

Eunotia

Desikachary, T. V., 1956(1957).
  Electron microscope studies on diatoms. J.R. Microsc. Soc. (3)76(1/2):9-36.

Eunotia sp.

Kokubo, S., and S. Sato., 1947
  Plankters in Jū-San Gata. Physiol. and Ecol. (Japan) 1(4):1-16, 3 text figs., tables.

Eunotia aequalis

Zanon, D. V., 1949
  Diatomee di Buenos Aires (Argentina) Atti Accad. Naz. Lincei, Memorie, Cl. Sci. fis., mat. e. nat., ser. 7, 11(3):59-151, 2 pls.

Eunotia arcus

Lillick, L.C., 1940
  Phytoplankton and planktonic protozoa of the offshore waters of the Gulf of Maine. Pt.II. Qualitative Composition of the Planktonic Flora. Trans. Am. Phil. Soc., n.s., 31(3):193-237, 13 text figs.

Eunotia arcus

Zanon, V., 1948
  Diatomee marini di Sardegna e Pugillo di Alghe Marine della stressa. Boll. Pesca, Piscitutura e Idrobiologia, Anno 24, ns. 3(2): 202-244, 27 figs. on 1 pl.

Eunotia bigibba

Zanon, D. V., 1949
  Diatomee di Buenos Aires (Argentina) Atti Accad. Naz. Lincei, Memorie, Cl. Sci. fis., mat. e. nat., ser. 7, 11(3):59-151, 2 pls.

Eunotia camelus

Zanon, D. V., 1949
  Diatomee di Buenos Aires (Argentina) Atti Accad. Naz. Lincei, Memorie, Cl. Sci. fis., mat. e. nat., ser. 7, 11(3):59-151, 2 pls.

Eunotia didyma

Zanon, D. V., 1949
  Diatomee di Buenos Aires (Argentina) Atti Accad. Naz. Lincei, Memorie, Cl. Sci. fis., mat. e. nat., ser. 7, 11(3):59-151, 2 pls.

Eunotia epithemioides

Zanon, D. V., 1949
  Diatomee di Buenos Aires (Argentina) Atti Accad. Naz. Lincei, Memorie, Cl. Sci. fis., mat. e. nat., ser. 7, 11(3):59-151, 2 pls.

Eunotia formica

Zanon, D. V., 1949
  Diatomee di Buenos Aires (Argentina) Atti Accad. Naz. Lincei, Memorie, Cl. Sci. fis., mat. e. nat., ser. 7, 11(3):59-151, 2 pls.

Eunotia gracilis

Zanon, D. V., 1949
  Diatomee di Buenos Aires (Argentina) Atti Accad. Naz. Lincei, Memorie, Cl. Sci. fis., mat. e. nat., ser. 7, 11(3):59-151, 2 pls.

Eunotia indica

Zanon, D. V., 1949
  Diatomee di Buenos Aires (Argentina) Atti Accad. Naz. Lincei, Memorie, Cl. Sci. fis., mat. e. nat., ser. 7, 11(3):59-151, 2 pls.

Eunotia lunaris

Zauer, L.M., 1950.
  [The movement of the diatom Eunotia lunaris (Ehr.) Grun. in connection with the question of the movements of diatoms generally.] Doklady Akad. Nauk, SSSR, 72(6):1131-

Eunotia monodon

Zanon, D. V., 1949
  Diatomee di Buenos Aires (Argentina) Atti Accad. Naz. Lincei, Memorie, Cl. Sci. fis., mat. e. nat., ser. 7, 11(3):59-151, 2 pls.

Eunotia monodon

Zanon, V., 1948
  Diatomee marini di Sardegna e Pugillo di Alghe Marine della stressa. Boll. Pesca, Piscitutura e Idrobiologia, Anno 24, ns. 3(2): 202-244, 27 figs. on 1 pl.

Eunotia pectinalis

Tsumura, K., 1956
  Diatomoj el la cirkaufoso de la restajo de la kastele de Odawara. J. Yokohama Municipal Univ., (C-14) No. 47:23 pp.

Eunotia pectinalis

Zanon, D. V., 1949
  Diatomee di Buenos Aires (Argentina) Atti Accad. Naz. Lincei, Memorie, Cl. Sci. fis., mat. e. nat., ser. 7, 11(3):59-151, 2 pls.

Eunotia pectinalis

Mann, A., 1893
  List of Diatomaceae from a deep-sea dredging in the Atlantic Ocean off Delaware Bay by the U. S. Fish Commission Steamer Albatross. Proc. U. S. Nat. Mus. 16:303-312.

Eunotia praerupta

Zanon, D. V., 1949
  Diatomee di Buenos Aires (Argentina) Atti Accad. Naz. Lincei, Memorie, Cl. Sci. fis., mat. e. nat., ser. 7, 11(3):59-151, 2 pls.

Eunotia praerupta

Zanon, V., 1948
  Diatomee marini di Sardegna e Pugillo di Alghe Marine della stressa. Boll. Pesca, Piscitutura e Idrobiologia, Anno 24, ns. 3(2): 202-244, 27 figs. on 1 pl.

Eunotia praerupta inflata

Tsumura, K., 1956
  Diatomoj el la cirkaufoso de la restajo de la kastele de odawara. J. Yokohama Municipal Univ., (C-14) No. 47:23 pp.

Eunotia sudetica

Zanon, D. V., 1949
  Diatomee di Buenos Aires (Argentina) Atti Accad. Naz. Lincei, Memorie, Cl. Sci. fis., mat. e. nat., ser. 7, 11(3):59-151, 2 pls.

Eunotia veneris

Zanon, D. V., 1949
  Diatomee di Buenos Aires (Argentina) Atti Accad. Naz. Lincei, Memorie, Cl. Sci. fis., mat. e. nat., ser. 7, 11(3):59-151, 2 pls.

Eunotia zygodon

Zanon, D. V., 1949
  Diatomee di Buenos Aires (Argentina) Atti Accad. Naz. Lincei, Memorie, Cl. Sci. fis., mat. e. nat., ser. 7, 11(3):59-151, 2 pls.

Eunotogramma producta

Cleve-Euler, A., 1951
  Die Diatomeen von Schweden und Finnland. Kungl. Svenska Vetenskaps Akad. Handl., Fjärde Ser. 2(1): 161 pp., 6 pls.

Eunotogramma rectum

Hendy, N. Ingram, 1964
  An introductory account of the smaller algae of British coastal waters. V. Bacillariophyceae (Diatoms). Her Majesty's Stationary Office, 317 pp., 45 pls.

Eunotogramma weissei

Cleve-Euler, A., 1951
  Die Diatomeen von Schweden und Finnland. Kungl. Svenska Vetenskaps Akad. Handl., Fjärde Ser. 2(1): 161 pp., 6 pls.

Euodia atlantica (figs.)

Sousa e Silva, E., 1949
  Diatomaceas e Dinoflagelados de Baía de Cascais. Portugaliae Acta Biol., Volume: Julio Henriques, Ser. B: 300-383, 9 pls, 2 fold-in tables.

Euodia cuneiformis

Gran, H.H., 1908
  Diatomeen. Nordisches Plankton, Botanischer Teil pp. XIX.1-XIX 146; 178 text figs.

Euodia (=Hemidiscus) cuneiformis

Mann, A., 1893
  List of Diatomaceae from a deep-sea dredging in the Atlantic Ocean off Delaware Bay by the U. S. Fish Commission Steamer Albatross. Proc. U. S. Nat. Mus. 16:303-312.

Euodia cuneiformis

Pavillard, J., 1925
  Bacillariales. Rept. on the Danish Oceangr. Exped., 1908-10 to the Mediterranean and adj. seas. Vol.II., Biol. J4:72 pp., 116 text figs.

Euodia Frauenfeldii n.sp.

Grunow, A., 1863
  Ueber einige neue und ungenügend bekannte Arten und Gattungen von Diatomaceen. Verhandl. d. K.K. Zool. Bot. Gesellsch., Vienna, 13: 137-162, pl.4-5 (Pl. 13-14).

Euodia gibba

Jørgensen, E., 1905
  B.Protistplankton and the diatoms in bottom samples. Hydrographical and biological investigations in Norwegian fjords. Bergens Mus. Skr. 7: 49-225.

Eupodiscus antiquus
Eskinazi Enide e Shigekatsv Satô (1963-1964) 1966.
Contribuição ao estudo das diatomaceas da Praia de Piedade.
Trabhs Inst. Oceanogr., Univ. Recife, 5 (5/6): 73-114.

Eupodiscus Argus
Gran, H.H., 1908
Diatomeen. Nordisches Plankton, Botanischer Teil pp. XIX.1-XIX 146; 178 text figs.

Eupodiscus argus
Meunier, A., 1915
Microplancton de la Mer Flamande. 2. Diatomées (excepté le genre Chaetoceros). Mem. Mus. Roy. Hist. Nat., Belgique, 7(3):1-118, Pls. VIII-XIV.

Eupodiscus argus
Schodduyn, M., 1926
Observations faites dans la baie d'Ambleteuse (Pas de Calais). Bull. Inst. Ocean., Monaco, No. 482: 64 pp.

Eupodiscus insutus n.sp.
Castracane degli Antelminelli, F., 1886
1. Report on the Diatomaceae collected by H.M.S. Challenger during the years 1873-1876. Rept. Sci. Results, H.M.S. Challenger, Botany Vol. II, 178 pp., 30 pls.

Eupodiscus radiatus
Hendy, N. Ingram, 1964
An introductory account of the smaller algae of British coastal waters. V. Bacillariophyceae (Diatoms).
Her Majesty's Stationary Office, 317 pp., 45 pls.

Eupodiscus radiatus
Mann, A., 1893
List of Diatomaceae from a deep-sea dredging in the Atlantic Ocean off Delaware Bay by the U. S. Fish Commission Steamer Albatross. Proc. U. S. Nat. Mus. 16:303-312.

Eupodiscus sparsus
Mann, A., 1893
List of Diatomaceae from a deep-sea dredging in the Atlantic Ocean off Delaware Bay by the U. S. Fish Commission Steamer Albatross. Proc. U. S. Nat. Mus. 16:303-312.

Eupodiscus tesselatus
Gran, H.H., 1908
Diatomeen. Nordisches Plankton, Botanischer Teil pp. XIX.1-XIX 146; 178 text figs.

Eupodiscus tesselatus
Mann, A., 1893
List of Diatomaceae from a deep-sea dredging in the Atlantic Ocean off Delaware Bay by the U. S. Fish Commission Steamer Albatross. Proc. U. S. Nat. Mus. 16:303-312.

Fenestrella antiqua
Cleve-Euler, A., 1951
Die Diatomeen von Schweden und Finnland. Kungl. Svenska Vetenskaps Akad. Handl., Fjärde Ser. 2(1): 161 pp., 6 pls.

Fragilaria sp.
Boden, B. P., 1949
The diatoms collected by the U.S.S. CACOPAN in the Antarctic in 1947. J. Mar. Res. 8(1):6-13, 3 textfigs.

Fragilaria sp.
Brunel, J., 1962
Le phytoplancton de la Baie de Chaleurs. Inst. Botan., Univ. Montréal, Contrib. No. 77: 365 pp., 66 pls.

Fragilaria sp.
Brunel, Jules, 1962
Le phytoplancton de la Baie des Chaleurs. Contrib. Ministère de la Chasse et des Pêcheries, Province de Québec, No. 91: 365 pp.

Fragilaria
Desikachary, T.V., 1956(1957).
Electron microscope studies on diatoms. J.R. Microsc. Soc. (3)76(1/2):9-36.

Fragilaria sp.
Gilbert, J.Y., and W.E. Allen, 1943
The phytoplankton of the Gulf of California obtained by the "E.W. Scripps" in 1939 and 1940. J. Mar. Res. V(2):89-110, figs.30-31.

Fragilaria sp.
Johnson, M. W., 1949.
Relation of plankton to hydrographic conditions in Sweetwater Lake. J. Am. Water Works Assoc. 41(4):347-356, 12 textfigs.

Fragilaria sp.
Kokubo, S., and S. Sato., 1947
Plankters in Jū-San Gata. Physiol. and Ecol. (Japan) 1(4):1-16, 3 text figs., tables.

Fragilaria sp.
Lillick, L.C., 1938
Preliminary report of the phytoplankton of the Gulf of Maine. Am. Mid. Nat. 20(3):624-640, 1 text figs 37 tables.

Fragillaria
Neaverson, E., 1934
The sea-floor deposits. 1. General characteristics and distribution. Discovery Repts. 9: 297-349, Plates 17-22.

Fragillaria sp.
Pettersson, H., 1934
Scattering and extinction of light in sea-water. Meddelanden från Göteborgs Högskolas Oceanografiska Institut: 9 (Göteborgs Kungl. vetenskaps-och vitterhetssamhälles Handlingar. Femte Foljden, Ser.B, 4(4)):1-16.

Fragilaria antarctica n.sp.
Castracane degli Antelminelli, F., 1886
1. Report on the Diatomaceae collected by H.M.S. Challenger during the years 1873-1876. Rept. Sci. Results, H.M.S. Challenger, Botany Vol. II, 178 pp., 30 pls.

Fragilaria antarctica
Central Meteorological Observatory, 1949
Report on sea and weather observation on Antarctic Whaling Ground (1947-48). Ocean. Mag., Japan, 1(1):49-88, 17 text figs.

Fragilaria atomus
Zanon, V., 1948
Diatomee marini di Sardegna e Pugillo di Alghe Marine della stressa. Boll. Pesca, Piscitutura e Idrobiologia, Anno 24, ns. 3(2): 202-244, 27 figs. on 1 pl.

Fragilaria brevistriata
Zanon, D. V., 1949
Diatomee di Buenos Aires (Argentina) Atti Accad. Naz. Lincei, Memorie, Cl. Sci. Fis., mat. e. nat., ser. 7, 11(3):59-151, 2 pls.

Fragilaria capensis n.sp.
Grunow, A., 1863
Ueber einige neue und ungenügend bekannte Arten und Gattungen von Diatomaceen. Verhandl. d. K.K. Zool. Bot. Gesellsch., Vienna, 13: 137-162, pl.4-5 (Pl. 13-14).

Fragilaria capucina
Hendy, N. Ingram, 1964
An introductory account of the smaller algae of British coastal waters. V. Bacillariophyceae (Diatoms).
Her Majesty's Stationary Office, 317 pp., 45 pls.

Fragilaria capucina
Mann, A., 1893
List of Diatomaceae from a deep-sea dredging in the Atlantic Ocean off Delaware Bay by the U. S. Fish Commission Steamer Albatross. Proc. U. S. Nat. Mus. 16:303-312.

Fragilaria capucina
Rumkówna, A., 1948
List of the phytoplankton species occurring in the superficial water layers in the Gulf of Gdańsk. Bull. Lab. mar., Gdynia, No. 4: 139-141 with tables in back.

Fragilaria Castracanei
Mangin, L., 1915
Phytoplancton de L'Antartique. Deuxieme Exped. Ant. Francaise (1908-1910), 95 pp., 3 pls. 58 text figs.

Fragilaria construens
Rumkówna, A., 1948
List of the phytoplankton species occurring in the superficial water layers in the Gulf of Gdańsk. Bull. Lab. mar., Gdynia, No. 4: 139-141 with tables in back.

Fragilaria construens
Zanon, V., 1948
Diatomee marini di Sardegna e Pugillo di Alghe Marine della stressa. Boll. Pesca, Piscitutura e Idrobiologia, Anno 24, ns. 3(2): 202-244, 27 figs. on 1 pl.

Fragilaria crotonensis
Brunel, J., 1962
Le phytoplancton de la Baie de Chaleurs. Inst. Botan., Univ. Montréal, Contrib. No. 77: 365 pp., 66 pls.

Fragilaria crotonensis
Brunel, Jules, 1962
Le phytoplancton de la Baie des Chaleurs. Contrib. Ministère de la Chasse et des Pêcheries, Province de Québec, No. 91: 365 pp.

Fragilaria crotonensis
Cupp, Easter E., 1943
Marine plankton diatoms of the west coast of North America. Bull. S.I.O. 5(1):1-238, 5 pls., 168 text figs.

Fragilaria crotonensis
Gran, H.H., 1908
Diatomeen. Nordisches Plankton, Botanischer Teil pp. XIX.1-XIX 146; 178 text figs.

Fragilaria crotonensis
Rumkówna, A., 1948
List of the phytoplankton species occurring in the superficial water layers in the Gulf of Gdańsk. Bull. Lab. mar., Gdynia, No. 4: 139-141 with tables in back.

Fragilaria cylindrus
Gran, H.H., 1908
Diatomeen. Nordisches Plankton, Botanischer Teil pp. XIX.1-XIX 146; 178 text figs.

Fragilaria cylindrus
Hendy, N. Ingram, 1964
An introductory account of the smaller algae of British coastal waters. V. Bacillariophyceae (Diatoms).
Her Majesty's Stationary Office, 317 pp., 45 pls.

Fragilaria cylindrus
Jørgensen, E., 1905
B. Protistplankton and the diatoms in bottom samples. Hydrographical and biological investigations in Norwegian fjords. Bergens Mus. Skr. 7: 49-225.

Fragilaria cylindrus
Lillick, L.C., 1940
Phytoplankton and planktonic protozoa of the offshore waters of the Gulf of Maine. Pt. II. Qualitative Composition of the Planktonic Flora. Trans. Am. Phil. Soc., n.s., 31(3):193-237, 13 text figs.

Fragilaria cylindrus
Misra, J.N., 1956.
A systematic account of some littoral marine diatoms from the west coast of India. J. Bombay Nat. Hist. Soc., 53(4):537-568.

Fragilaria cylindrus
Zanon, V., 1948
Diatomee marini di Sardegna e Pugillo di Alghe Marine della stressa. Boll. Pesca, Piscitutura e Idrobiologia, Anno 24, ns. 3(2): 202-244, 27 figs. on 1 pl.

Fragilaria exilis n.sp.
Grunow, A., 1863
Ueber einige neue und ungenügend bekannte Arten und Gattungen von Diatomaceen. Verhandl. d. K.K. Zool. Bot. Gesellsch., Vienna, 13: 137-162, pl.4-5 (Pl. 13-14).

Fragilaria granulata
Boden, B.P., 1950
Some marine plankton diatoms from the west coast of South Africa. Trans. R.Soc. S. Africa. 32:321-434, 100 text figs.

Fragilaria hyalina
Hendy, N. Ingram, 1964
An introductory account of the smaller algae of British coastal waters. V. Bacillariophyceae (Diatoms). Her Majesty's Stationary Office, 317 pp., 45 pls.

Fragilaria intermedia
Rumkówna, A., 1948
[List of the phytoplankton species occurring in the superficial water layers in the Gulf of Gdańsk] Bull. Lab. mar., Gdynia, No. 4: 139-141 with tables in back.

Fragilaria irrescens
Zanon, D. V., 1949
Diatomee di Buenos Aires (Argentina) Atti Accad. Naz. Lincei, Memorie, Cl. Sci. fis., mat. e. nat., ser. 7, 11(3):59-151, 2 pls.

Fragilaria islandica
Gran, H.H., 1908
Diatomeen. Nordisches Plankton, Botanischer Teil pp. XIX.1-XIX 146; 178 text figs.

Fragilaria islandica
Hendy, N. Ingram, 1964
An introductory account of the smaller algae of British coastal waters. V. Bacillariophyceae (Diatoms). Her Majesty's Stationary Office, 317 pp., 45 pls.

Fragilaria islandica
Iselin, C., 1930
A report on the coastal waters of Labrador based on explorations of the "Chance" during the summer of 1926. Proc. Am. Acad. Arts Sci., 66(1):1-37, 14 text figs.

Fragillaria islandica
Jørgensen, E., 1905
B. Protistplankton and the diatoms in bottom samples. Hydrographical and biological investigations in Norwegian fjords. Bergens Mus. Skr. 7: 49-225.

Fragilaria islandica
Rumkówna, A., 1948
[List of the phytoplankton species occurring in the superficial water layers in the Gulf of Gdańsk] Bull. Lab. mar., Gdynia, No. 4: 139-141 with tables in back.

Fragilaria karstenii
Boden, B.P., 1950
Some marine plankton diatoms from the west coast of South Africa. Trans. R.Soc. S. Africa. 32:321-434, 100 text figs.

Fragilaria linearis n.sp.
Castracane degli Antelminelli, F., 1886
1. Report on the Diatomaceae collected by H.M.S. Challenger during the years 1873-1876. Rept. Sci. Results, H.M.S. Challenger, Botany Vol. II, 178 pp., 30 pls.

Fragilaria linearis
Hendey, N.I., 1937
The plankton diatoms of the southern seas. Discovery Repts. 16:151-364, pls.6-13.

Fragilaria litoralis n.sp.
Cholnoky, B.J., 1963.
Beiträge zur Kenntnis des marinen Litorals von Südafrika. Botanica Marina, 5(2/3):38-83.

Fragilaria minima
Steemann-Nielsen, Einar, 1951
The marine vegetation of the Isefjord. A study on ecology and production. Medd. Komm. Danmarks Fiskeri-og Havundersøgelser. Ser. Plankton. 5(4); 114pp., 46 text figs.

Fragilaria nitzschioides
Rumkówna, A., 1948
[List of the phytoplankton species occurring in the superficial water layers in the Gulf of Gdańsk] Bull. Lab. mar., Gdynia, No. 4: 139-141 with tables in back.

Fragilaria oceanica
Brunel, J., 1962
Le phytoplancton de la Baie de Chaleurs. Inst. Botan., Univ. Montréal, Contrib. No. 77: 365 pp., 66 pls.

Fragilaria oceanica
Brunel, Jules, 1962
Le phytoplancton de la Baie des Chaleurs. Contrib. Ministère de la Chasse et des Pêcheries, Province de Québec, No. 91: 365 pp.

Fragilaria oceanica
Frenguelli, Joaquin, and Hector Antonio Orlando, 1959.
Operacion MERLUZA. Diatomeas y silicoflagelados del plankton del "VI Crucero". Servicio Hidrogr. Naval, Argentina, Publ. No. H. 619: 3-62.

Fragilaria oceanica
Gran, H.H., 1908
Diatomeen. Nordisches Plankton, Botanischer Teil pp. XIX.1-XIX 146; 178 text figs.

Fragillaria oceanica
Gran, H.H., 1897
Protophyta: Diatomaceae, Silico-flagellata and Cilioflagellata. Den Norske Nordhavs Expedition 1876-1878, h. 24, 36 pp., 4 pls.

Fragilaria oceanica
Gran, H.H., and T. Braarud, 1935
A quantitative study of the phytoplankton in the Bay of Fundy and the Gulf of Maine (including observations on hydrography, chemistry, and turbidity). J. Biol. Bd., Canada, 1(5):279-467, 69 text figs.

Fragilaria oceanica
Hendy, N. Ingram, 1964
An introductory account of the smaller algae of British coastal waters. V. Bacillariophyceae (Diatoms). Her Majesty's Stationary Office, 317 pp., 45 pls.

Fragillaria oceanica
Jørgensen, E., 1905
B. Protistplankton and the diatoms in bottom samples. Hydrographical and biological investigations in Norwegian fjords. Bergens Mus. Skr. 7: 49-225.

Fragilaria oceanica
Lillick, L.C., 1940
Phytoplankton and planktonic protozoa of the offshore waters of the Gulf of Maine. Pt. II. Qualitative Composition of the Planktonic Flora. Trans. Am. Phil. Soc., n.s., 31(3):193-237, 13 text figs.

Fragilaria oceanica
Nair, R. Velappan, and R. Subrahmanyan, 1955.
The diatom, Fragilaria oceanica Cleve an indicator of abundance of the Indian oil sardine, Sardinella longiceps Cuv. and Val. Current Sci. 24(2):41-42.

Fragilaria oceanica
Phifer, L.D. (undated)
The occurrence and distribution of plankton diatoms in Bering Sea and Bering Strait, July 26-August 24, 1934. Report of Oceanographic cruise of U.S. Coast Guard Cutter Chelan 1934, Part II(A):1-44 (mimeographed) plus fig.1 (after Pt.B)

Fragilaria pacifica
Grunow, A., 1863
Ueber einige neue und ungenügend bekannte Arten und Gattungen von Diatomaceen. Verhandl. d. K.K. Zool. Bot. Gesellsch., Vienna, 13: 137-162, pl.4-5 (Pl. 13-14).

Fragilaria pinnata
Hendy, N. Ingram, 1964
An introductory account of the smaller algae of British coastal waters. V. Bacillariophyceae (Diatoms). Her Majesty's Stationary Office, 317 pp., 45 pls.

Fragilaria pinnata
Hendey, N.I., 1951
Littoral diatoms of Chicester Harbour with special reference to fouling. J.Roy. Microsc. Soc. 71(1): 1-86, 18 pls.

Fragilaria pinnata
Zanon, D. V., 1949
Diatomee di Buenos Aires (Argentina) Atti Accad. Naz. Lincei, Memorie, Cl. Sci. fis., mat. e. nat., ser. 7, 11(3):59-151, 2 pls.

Fragilaria pinnata
Zanon, V., 1948
Diatomee marini di Sardegna e Pugillo di Alghe Marine della stressa. Boll. Pesca, Piscitutura e Idrobiologia, Anno 24, ns. 3(2): 202-244, 27 figs. on 1 pl.

Fragilaria pseudoatomus n.sp.
Manguin, E., 1957.
Premier inventaire des diatomées de la Terre Adélie Antarctique. Espèces nouvelles. Rev. Algologique, n.s., 3(3):111-134.

Fragilaria (rhombica?)
Boden, Brian, 1948
Marine plankton diatoms on operation HIGHJUMP in: Some oceanographic observations on operation HIGHJUMP. By R.S. Dietz. USNEL Rept. No.55, 97 pp., 41 figs. 7 July 1948.

Fragilaria schulzi
Hendy, N. Ingram, 1964
An introductory account of the smaller algae of British coastal waters. V. Bacillariophyceae (Diatoms). Her Majesty's Stationary Office, 317 pp., 45 pls.

Fragillaria striatella
Meunier, A., 1915
Microplancton de la Mer Flamande. 2. Diatomées (excepté le genre Chaetoceros). Mem. Mus. Roy. Hist. Nat., Belgique, 7(3):1-118, Pls. VIII-XIV.

Fragilaria striatula
Chu, S. P., 1949
Experimental studies on the environmental factors influencing the growth of phytoplankton. Sci. & Tech. in China 2(3):37-52.

Fragilaria striatula
Gran, H.H., 1908
Diatomeen. Nordisches Plankton, Botanischer Teil pp. XIX.1-XIX 146; 178 text figs.

Fragilaria striatula
Gran, H. H. and E. C. Angst, 1931
Plankton diatoms of Puget Sound. Publ. Puget Sound Biol. Sta. 7:417-519, 95 text figs.

Fragilaria striatula
Hendy, N. Ingram, 1964
An introductory account of the smaller algae of British coastal waters. V. Bacillariophyceae (Diatoms).
Her Majesty's Stationary Office, 317 pp., 45 pls.

Fragilaria striatula
Hendey, N.I., 1937
The plankton diatoms of the southern seas. Discovery Repts. 16:151-364, pls.6-13.

Fragilaria striatula
Iselin, C., 1930
A report on the coastal waters of Labrador based on explorations of the "Chance" during the summer of 1926. Proc. Am. Acad. Arts Sci., 66(1):1-37, 14 text figs.

Fragilaria striatula
Rumkówna, A., 1948
[List of the phytoplankton species occurring in the superficial water layers in the Gulf of Gdańsk] Bull. Lab. mar., Gdynia, No. 4: 139-141 with tables in back.

Fragilaria striatula var. california
Castenholz, Richard W., 1963.
An experimental study of the vertical distribution of littoral marine diatoms.
Limnology and Oceanography, 8(4):450-462.

Fragilaria sublinearis
Bunt, J., 1965.
Measurements of photosynthesis and respiration in a marine diatom with the mass spectrometer and with carbon-14.
Nature, 207(5007):1373-1375.

Fragilaria swartzii n.sp.
Grunow, A., 1863
Ueber einige neue und ungenügend bekannte Arten und Gattungen von Diatomaceen. Verhandl. d. K.K. Zool. Bot. Gesellsch., Vienna, 13: 137-162, pl.4-5 (Pl. 13-14).

Fragilaria Schwartzii
Mann, A., 1893
List of Diatomaceae from a deep-sea dredging in the Atlantic Ocean off Delaware Bay by the U. S. Fish Commission Steamer Albatross. Proc. U. S. Nat. Mus. 16:303-312.

Fragilaria ungeriana
Grunow, A., 1863
Ueber einige neue und ungenügend bekannte Arten und Gattungen von Diatomaceen. Verhandl. d. K.K. Zool. Bot. Gesellsch., Vienna, 13: 137-162, pl.4-5 (Pl. 13-14).

Fragilaria virescens oblonga
Grøntved, J., 1956.
Planktonological contributions. II. Taxonomical studies in some Danish coastal localities. Medd. Danmarks Fiskeri- og Havundersøgelser, n.s. 1(12):13 pp.

Fragilariella gen. nov.
Hendey, N-Ingram, 1958 [1957(Publ. 1958)]
Marine diatoms from some West African Ports. J. R. Microsc. Soc. (3) 77(1/2): 28-85.

Fragilariella dusenii comb. nov.
Hendey, N-Ingram, 1958 [1957(Publ. 1958)]
Marine diatoms from some West African Ports. J. R. Microsc. Soc. (3) 77(1/2): 28-85.

Fragilaropsis
Desikachary, T.V., 1956(1957).
Electron microscope studies on diatoms. J. R. Microsc. Soc. (3)76(1/2):9-36.

Fragilariopsis antarctica
Boden, B. P., 1949.
The diatoms collected by the U.S.S. CACOPAN in the Antarctic in 1947. J. Mar. Res. 8(1):6-13, 3 textfigs.

Fragilariopsis antarctica
Boden, Brian, 1948
Marine plankton diatoms on operation HIGHJUMP in: Some oceanographic observations on operation HIGHJUMP. By R.S. Dietz. USNEL Rept. No.55, 97 pp., 41 figs. 7 July 1948.

Fragilariopsis antarctica
Frenguelli, Joaquin, and Hector Antonio Orlando, 1959.
Operacion MERLUZA. Diatomeas y silicoflagelados del plancton del "VI Crucero". Servicio Hidrogr. Naval., Argentina, Publ. No. H. 619: 5-62.

Fregilieriopsis atlantica
Hasle, Grethe Rytter, 1965.
Nitzschia and Frafukuariopsis species studies in the light and electron microscopes. III. The genus Fragilieriopsis.
Norske Videnskaps- Akad., Oslo. I. Met.-Nature. Kl., N.S. No. 21: 49 pp., 17 pls.

Fragilariopsis antarctica
Hendey, N.I., 1937
The plankton diatoms of the southern seas. Discovery Repts. 16:151-364, pls.6-13.

Fragilariopsis antarctica
Jouse, A.P., G.S. Koroleva, G.A. Nagaeva, 1962
Diatoms in the surface layer of sediment in the Indian sector of the Antarctic. Investigations of marine bottom sediments. (In Russian; English summary).
Trudy Inst. Okeanol., Akad. Nauk, SSSR. 61: 19-92.

Fragilariopsis antarctica
Manguin, E., 1954
Diatomees marines provenant de l'ile Heard (Australian National Antarctic Expedition). Rev. Algol., n.s., 1: 14-24.

Fragilariopsis atlantica n.sp.
Paasche, E., 1961.
Notes on phytoplankton from the Norwegian Sea. Botanica Marina, 2(3/4):197-210.

Fragilariopsis curta
Hasle, Grethe Rytter, 1965.
Nitzschia and Fragilariopsis species studies in the light and electron microscopes. III. The genus Fragilieriopsis.
Norske Videnskaps- Akad., Oslo. I. Met.-Nature. Kl., N.S. No. 21: 49 pp. 17 pls.

Fragilariopsis curta
Jouse, A.P., G.S. Koroleva, G.A. Nagaeva, 1962
Diatoms in the surface layer of sediment in the Indian sector of the Antarctic. Investigations of marine bottom sediments. (In Russian; English summary).
Trudy Inst. Okeanol., Akad. Nauk. SSSR. 61: 19-92.

Fragilariopsis cylindrus
Hasle, Grethe Rytter, 1968.
Observations on the marine diatom Fragilariopsis Kerguelensis (O'Meara) Hust in the scanning electron microscope.
Nytt Mag. Bot. 15(3):205-208.

Fragilariopsis cylindrus
Hasle, Grethe Rytter, 1965.
Nitzschia and Fragilariopsis species studies in the light and electron microscopes. III. The genus Fragilieriopsis.
Norske Videnskaps- Akad., Oslo. I. Met.-Nature. Kl., N.S. No. 21: 49 pp., 17 pls.

Fragilariopsis cylindrus
Jouse, A.P., G.S. Koroleva, G.A. Nagaeva, 1962
Diatoms in the surface layer of sediment in the Indian sector of the Antarctic. Investigations of marine bottom sediments. (In Russian; English summary).
Trudy Inst. Okeanol., Akad. Nauk. SSSR. 61: 19-92.

Fragilariopsis Drakei n.sp.
Frenguelli, Joaquin, y Hector A. Orlando, 1958.
Diatomeas y silicoflagelados del sector Antartico Sudamericano.
Inst. Antartico Argentino, Publ., No. 5:191 pp.

Fragilariopsis Kerguelensis
Hasle, Grethe Rytter, 1968.
Observations on the marine diatom Fragilariopsis Kerguelensis (O'Meara)Hust. in the scanning electron microscope.
Nytt Mag. Bot., 15(3):205-208.

Fragilariopsis Kerguelensis
Hasle, Grethe Rytter, 1965.
Nitzschia and Fragilariopsis species studies in the light and electron microscopes. III. The genus Fragilieriopsis.
Norske Videnskaps- Akad., Oslo. I. Met.-Nature. Kl., N.S. No. 21: 49 pp., 17 pls.

Fragilariopsis linearis
Frenguelli, Joaquin, and Hector Antonio Orlando, 1959.
Operacion MERLUZA. Diatomeas y silicoflagelados del plancton del "VI Crucero". Servicio Hidrogr. Naval., Argentina, Publ. No. H. 619: 5-62.

Fragilariopsis linearis
Hasle, Grethe Rytter, 1965.
Nitzschia and Fragilariopsis species studies in the light and electron microscopes. III. The genus Fragilieriopsis.
Norske Videnskaps- Akad., Oslo. I. Met.-Nature. Kl., N.S. No. 21: 49 pp., 17 pls.

Fragilariopsis linearis
Jouse, A.P., G.S. Koroleva, G.A. Nagaeva, 1962
Diatoms in the surface layer of sediment in the Indian sector of the Antarctic. Investigations of marine bottom sediments. (In Russian; English summary).
Trudy Inst. Okeanol., Akad. Nauk. SSSR. 61: 19-92.

Fragilariopsis nana nov. comb.
Paasche, E., 1961.
Notes on phytoplankton from the Norwegian Sea. Botanica Marina, 2(3/4):197-210.

Fragilariopsis obliquecostata
Hasle, Grethe Ryther, 1965.
Nitzschia and Fragilariopsis species studies in the light and electron microscopes. III. The genus Fragilariopsis.
Norske Videnskaps- Akad., Oslo I. Met.- Nature. Kl., N.S. No. 21: 49 pp., 17 pls.

Fragilariopsis obliquecostata
Jouse, A.P., G.S. Koroleva, G.A. Nagaeva, 1962
Diatoms in the surface layer of sediment in the Indian sector of the Antarctic. Investigations of marine bottom sediments. (In Russian; English summary).
Trudy Inst. Okeanol., Akad. Nauk. SSSR. 61: 19-92.

Fragiliariopsis oceanica
Hasle, Grethe Ryther, 1965.
Nitzschia and Fragileriopsis species studies in the light and electron microscopes. III. The genus Fragileriopsis.
Norske Veidenskaps- Aked., Oslo I. Met.-Nature. Kl., N.S. No. 21: 49-pp., 17 pls.

Fragileriopsis pseudona nom. nov.
Hasle, Grethe Ryther, 1965.
Nitzschia and Fragileriopsis species studies in the light and electron microscopes. III. The genus Fragileriopsis.
Norske Videnskaps- Aked., Oslo I. Met.-Nature. Kl., N.S. No. 21: 49 pp., 17 pls.

Fragilariopsis rhombica
Frenguelli, Joaquin, and Hector Antonio Orlando, 1959.
Operacion MERLUZA. Diatomeas y silico-flagelados del plancton del "VI Crucero". Servicio Hidrogr. Naval., Argentina, Publ. No. H. 619: 5-62.

Fragilariopsis rhombica
Hasle, Grethe Ryther, 1965.
Nitzschia and Fragilariopsis species studies in the light and electron microscopes. III. The genus Fragilariopsis.
Norske Videnskaps-Aked., Oslo I. Met. -Nature. Kl., N.S. No. 21:49 pp., 17 pls.

Fragilariopsis rhombica
Manguin, E., 1954
Diatomees marines provenant de l'ile Heard (Australian National Antarctic Expedition). Rev. Algol., n.s., 1: 14-24.

Fragilariopsis ritscheri
Hasle, Grethe Rytter, 1968.
Observations on the marine diatom Fragilariopsis Kerguelensis (O'Meara) Hust in the scanning electron microscope.
Nytt Mag.Bot. 15(3):205-208.

Fragilariopsis ritscheri
Hasle, Grethe Ryther, 1965.
Nitzschia and Fragileriopsis species studies in the light and electron microscopes. III. The genus Fragileriopsis.
Norske Videnskaps - Aked., Oslo. I. Met. -Nature. Kl., N.S. No. 21: 49 pp. 17 pls.

Fragilariopsis Ritscherii
Jouse, A.P., G.S. Koroleva, G.A. Nagaeva, 1962
Diatoms in the surface layer of sediment in the Indian sector of the Antarctic. Investigations of marine bottom sediments. (In Russian; English summary).
Trudy Inst. Okeanol., Akad. Nauk. SSSR. 61: 19-92.

Fragilariopsis separanda
Hasle, Grethe Ryther, 1965.
Nitzschia and Fragilariopsis species studies in the light and eletron microscopes. III. The genus Fragilariopsis.
Norske Videnskaps- Aked., Oslo. I. Met.-Nature. Kl., N.S. No. 21: 49 pp., 17 pls.

Fragilariopsis separanda
Jouse, A.P., G.S. Koroleva, G.A. Nagaeva, 1962
Diatoms in the surface layer of sediment in the Indian sector of the Antarctic. Investigations of marine bottom sediments. (In Russian; English summary).
Trudy Inst. Okeanol., Akad. Nauk. SSSR. 61: 19-92.

Fragilariopsis sublinearis
Frenguelli, Joaquin, and Hector Antonio Orlando, 1959.
Operacion MERLUZA. Diatomeas y silico-flagelados del plancton del "VI Crucero". Servicio Hidrogr. Naval., Argentina, Publ. No. H. 619: 5-62.

Fragilariopsis sublinearis
Hasle, Grethe Ryther, 1965.
Nitzschia and Fragileriopsis species studies in the light and electron microscopes. III. The genus Fragileriopsis.
Norske Videnskaps-Aked., Oslo. I. Met. -Nature. Kl. N.S. No. 21: 49 pp. 17 pls.

Fragilariopsis sublinearis
Hendey, N.I., 1937
The plankton diatoms of the southern seas. Discovery Repts. 16:151-364, pls.6-13.

Fragilariopsis sublinearis
Jouse, A.P., G.S. Koroleva, G.A. Nagaeva, 1962
Diatoms in the surface layer of sediment in the Indian sector of the Antarctic. Investigations of marine bottom sediments. (In Russian; English summary).
Trudy Inst. Okeanol., Akad. Nauk. SSSR. 61: 19-92.

Fragilariopsis venheurckii
Hasle, Grethe Ryther, 1965.
Nitzschia and Fragilariopsis species studies in the light and electron microscopes. III. The genus Fragilariopsis.
Norske Videnskaps - Aked., Oslo. I. Met.-Nature. Kl., N.S. No. 21: 49 pp., 17 pls.

Fragilariopsis Vanheurckii
Jouse, A.P., G.S. Koroleva, G.A. Nagaeva, 1962
Diatoms in the surface layer of sediment in the Indian sector of the Antarctic. Investigations of marine bottom sediments. (In Russian; English summary).
Trudy Inst. Okeanol., Akad. Nauk. SSSR. 61: 19-92.

Frustulia
Desikachary, T.V., 1956(1957).
Electron microscope studies on diatoms. J.R. Microsc. Soc. (3)76(1/2):9-36.

Frustulia amphipleuroides
Hendy, N. Ingram, 1964
An introductory account of the smaller algae of British coastal waters. V. Bacillariophyceae (Diatoms).
Her Majesty's Stationary Office, 317 pp., 45 pls.

Frustulia antarctica n.sp.
Manguin, E., 1957.
Premier inventaire des diatomées de la Terre Adélie Antarctique. Especes nouvelles.
Rev. Algologique, n.s., 3(3):111-134.

Frustulia interposita
Tsumura, K., 1956.
Diatomoj el la cirkaufoso de la restajo de la kastele de Odawara. J. Yokohama Municipal Univ., (C-14), no. 47:23 pp.

Frustulia interposita
Zanon, D. V., 1949
Diatomee di Buenos Aires (Argentina) Atti Accad. Naz. Lincei, Memorie, Cl. Sci. fis., mat. e. nat., ser. 7, 11(3):59-151, 2 pls.

Frustulia linkei
Hendy, N. Ingram, 1964
An introductory account of the smaller algae of British coastal waters. V. Bacillariophyceae (Diatoms).
Her Majesty's Stationary Office, 317 pp., 45 pls.

Frustulia rhomboides
Hendy, N. Ingram, 1964
An introductory account of the smaller algae of British coastal waters. V. Bacillariophyceae (Diatoms).
Her Majesty's Stationary Office, 317 pp., 45 pls.

Frustulia rhomboides
Mann, A., 1907
Report on the diatoms of the Albatross voyages in the Pacific Ocean, 1888-1904. Contrib. U. S. Nat. Herb. 10(5):221-419, Pls. XLIV-LIV.

Frustulia vulgaris
Tsumura, K., 1956.
Diatomoj el la cirkaufoso de la restajo de la kastele de Odawara. J. Yokohama Municipal Univ., (C-14) No. 47:23 pp.

Frustulia vulgaris
Zanon, D. V., 1949
Diatomee di Buenos Aires (Argentina) Atti Accad. Naz. Lincei, Memorie, Cl. Sci. fis., mat. e. nat., ser. 7, 11(3):59-151, 2 pls.

Gallionella hyperborea
Jorgensen, E., 1900
Protophyten und Protozoën im Plankton aus der Norwegischen Westkerste. Bergens Mus. Aarb. 1899(6): 95 pp., 5 pls., 83 tables.

Gephyria
Desikachary, T.V., 1956(1957).
Electron microscope studies on diatoms. J. R. Microsc. Soc. (3)76(1/2):9-36.

Gephyria gigantea
Castracane degli Antelminelli, F., 1886
1. Rèport on the Diatomaceae collected by H.M.S. Challenger during the years 1873-1876. Rept. Sci. Results, H.M.S. Challenger, Botany Vol. II, 178 pp., 30 pls.

Gephyria media
Mann, A., 1907
Report on the diatoms of the Albatross voyages in the Pacific Ocean, 1888-1904. Contrib. U. S. Nat. Herb. 10(5):221-419, Pls. XLIV-LIV.

Glyphodesmis challengeriensis n.sp.
Castracane degli Antelminelli, F., 1886
1. Rèport on the Diatomaceae collected by H.M.S. Challenger during the years 1873-1876. Rept. Sci. Results, H.M.S. Challenger, Botany Vol. II, 178 pp., 30 pls.

Glyphodesmis distans
Hendy, N. Ingram, 1964
An introductory account of the smaller algae of British coastal waters. V. Bacillariophyceae (Diatoms).
Her Majesty's Stationary Office, 317 pp., 45 pls.

Glyphodesmis distans
Jorgensen, E., 1900
Protophyten und Protozoën im Plankton aus der Norwegischen Westkerste. Bergens Mus. Aarb. 1899(6): 95 pp., 5 pls., 83 tables.

Glyphodesmus distans
Rampi, L., 1940
Diatomee del Mare Adriatico. Nuovo Giornale Botanico Italiano, n.s., 47:559-608.

Glyphodesmis margaritacea
Castracane degli Antelminelli, F., 1886 n.sp.
1. Report on the Diatomaceae collected by H.M.S. Challenger during the years 1873-1876. Rept. Sci. Results, H.M.S. Challenger, Botany Vol. II, 178 pp., 30 pls.

Glyphodesmis murrayana n.sp.
Castracane degli Antelminelli, F., 1886
1. Report on the Diatomaceae collected by H.M.S. Challenger during the years 1873-1876. Rept. Sci. Results, H.M.S. Challenger, Botany Vol. II, 178 pp., 30 pls.

Glyphodesmis williamsonii
Hendy, N. Ingram, 1964
An introductory account of the smaller algae of British coastal waters. V. Bacillariophyceae (Diatoms).
Her Majesty's Stationary Office, 317 pp., 45 pls.

Glyphodesmis williamsonii
Jørgensen, E., 1905
B. Protistplankton and the diatoms in bottom samples. Hydrographical and biological investigations in Norwegian fjords. Bergens Mus. Skr. 7: 49-225.

Glyphodesmis williamsonii
Jorgensen, E., 1900
Protophyten und Protozoën im Plankton aus der Norwegischen Westkerste. Bergens Mus. Aarb. 1899(6): 95 pp., 5 pls., 83 tables.

Glyphodesmus Williamsonii
Rampi, L., 1940
Diatomee del Mare Adriatico. Nuovo Giornale Botanico Italiano, n.s., 47:559-608.

Gomphonema
Desikachary, T.V., 1956(1957).
Electron microscope studies on diatoms. J.R. Microsc. Soc. (3)76(1/2):9-36.

Gomphonema sp.
Levander, K.M., 1947
Plankton gesammelt in den Jahren 1899-1910 an den Küsten Finnlands. Finnländische Hydrographisch-Biologische Untersuchungen (aus dem Wasserbiologischen Laboratorin der Societas Scientiarum Fennica) No.11:40 pp., 6 diagrams, 13 pls., tables.

Gomphonema sp.
Schodduyn, M., 1926
Observations faites dans la baie d'Ambleteuse (Pas de Calais). Bull. Inst. Ocean., Monaco, No. 482: 64 pp.

Gomphonema acuminatum
Kokubo, S., and S. Sato., 1947
Plankters in Jū-San Gata. Physiol. and Ecol. (Japan) 1(4):1-16, 3 text figs., tables.

Gomphonema acuminatum
Tsumura, K., 1956.
Diatomoj el la cirkaufoso de la restajo de la kastele de Odawara. J. Yokohama Municipal Univ., (C-14), No. 47:23 pp.

Gomphonema acuminatum
Zanon, D. V., 1949
Diatomee di Buenos Aires (Argentina) Atti Accad. Naz. Lincei, Memorie, Cl. Sci. fis., mat. e. nat., ser. 7, 11(3):59-151, 2 pls.

Gomphonema acuminatum
Zanon, V., 1948
Diatomee marini di Sardegna e Pugillo di Alghe Marine della stressa. Boll. Pesca, Piscitutura e Idrobiologia, Anno 24, ns. 3(2): 202-244, 27 figs. on 1 pl.

Gomphonema constrictum
Kokubo, S., and S. Sato., 1947
Plankters in Jū-San Gata. Physiol. and Ecol. (Japan) 1(4):1-16, 3 text figs., tables.

Gomphonema constrictum
Zanon, D. V., 1949
Diatomee di Buenos Aires (Argentina) Atti Accad. Naz. Lincei, Memorie, Cl. Sci. fis., mat. e. nat., ser. 7, 11(3):59-151, 2 pls.

Gomphonema constrictum capitata
Tsumura, K., 1956.
Diatomoj el la cirkaufoso de la restajo de la kastele de Odawara. J. Yokohama Municipal Univ., (C-14), No. 47:23 pp.

Gomphonema cymbelloides n.sp.
Frenguelli, Joaquin, y Hector A. Orlando, 1958.
Diatomeas y silicoflagelados del sector Antartico Sudamericano.
Inst. Antartico Argentino, Publ., No. 5: 191 pp.

Gomphonema Demerarae
Zanon, D. V., 1949
Diatomee di Buenos Aires (Argentina) Atti Accad. Naz. Lincei, Memorie, Cl. Sci. fis., mat. e. nat., ser. 7, 11(3):59-151, 2 pls.

Gomphonema exiguum
Hustedt, F. and A.A. Aleem, 1951
Littoral diatoms from the Salstone near Plymouth. JMBA 30(1): 177.196.

Gomphonema gracile
Zanon, D. V., 1949
Diatomee di Buenos Aires (Argentina) Atti Accad. Naz. Lincei, Memorie, Cl. Sci. fis., mat. e. nat., ser. 7, 11(3):59-151, 2 pls.

Gomphonema gracile dichotoma
Tsumura, K., 1956.
Diatomoj el la cirkaufoso de la restajo de la kastele de Odawara. J. Yokohama Municipal Univ., (C-14), No. 47:23 pp.

Gomphonema herculeanum
Mann, A., 1907
Report on the diatoms of the Albatross voyages in the Pacific Ocean, 1888-1904. Contrib. U. S. Nat. Herb. 10(5):221-419, Pls. XLIV-LIV.

Gomphonema kamtschaticum
Hendy, N. Ingram, 1964
An introductory account of the smaller algae of British coastal waters. V. Bacillariophyceae (Diatoms).
Her Majesty's Stationary Office, 317 pp., 45 pls.

Gomphonema kamtschaticum var. californicum
Takano, Hideaki, 1961.
Epiphytic diatoms upon Japanese agar sea-weeds. Bull. Tokai Reg. Fish. Res. Lab., No. 31:269-274.

Gomphonema lanceolatum
Zanon, D. V., 1949
Diatomee di Buenos Aires (Argentina) Atti Accad. Naz. Lincei, Memorie, Cl. Sci. fis., mat. e. nat., ser. 7, 11(3):59-151, 2 pls.

Gomphonema longiceps
Zanon, D. V., 1949
Diatomee di Buenos Aires (Argentina) Atti Accad. Naz. Lincei, Memorie, Cl. Sci. fis., mat. e. nat., ser. 7, 11(3):59-151, 2 pls.

Gomphonema mammilla
Mann, A., 1907
Report on the diatoms of the Albatross voyages in the Pacific Ocean, 1888-1904. Contrib. U. S. Nat. Herb. 10(5):221-419, Pls. XLIV-LIV.

Gomphonema margaritae, n.sp.
Frenguelli, Joaquin, y Hector A. Orlando, 1958.
Diatomeas y silicoflagelados del sector Antartico Sudamericano.
Inst. Antartico Argentino, Publ. No. 5: 191 pp.

Gomphonema olivaceum
Kokubo, S., and S. Sato., 1947
Plankters in Jū-San Gata. Physiol. and Ecol. (Japan) 1(4):1-16, 3 text figs., tables.

Gomphonema olivaceum
Rumkówna, A., 1948
[List of the phytoplankton species occurring in the superficial water layers in the Gulf of Gdańsk.] Bull. Lab. mar., Gdynia, No. 4: 139-141 with tables in back.

Gomphonema olivaceum
Zanon, D. V., 1949
Diatomee di Buenos Aires (Argentina) Atti Accad. Naz. Lincei, Memorie, Cl. Sci. fis., mat. e. nat., ser. 7, 11(3):59-151, 2 pls.

Gomphonema parvulum
Low, E.M., 1955.
Studies on some chemical constituents of diatoms J. Mar. Res. 14(2):199-204.

Gomphonema parvulum
Tsumura, K., 1956.
Diatomoj el la cirkaufoso de la restajo de la kastele de Odawara. J. Yokohama Municipal Univ., (C-14), No. 47:23 pp.

Gomphonema parvulum
Zanon, D. V., 1949
Diatomee di Buenos Aires (Argentina) Atti Accad. Naz. Lincei, Memorie, Cl. Sci. fis., mat. e. nat., ser. 7, 11(3):59-151, 2 pls.

Gomphonema parvulum
Zanon, V., 1948
Diatomee marini di Sardegna e Pugillo di Alghe Marine della stressa. Boll. Pesca, Piscitutura e Idrobiologia, Anno 24, ns. 3(2): 202-244, 27 figs. on 1 pl.

Gomphonema pseudexiguum
Takano, Hideaki, 1963.
Notes on marine littoral diatoms of Japan. Bull. Tokai Reg. Fish. Res. Lab., No. 36:1-8.

Gomphonema sphaerophorum
Mann, A., 1893
List of Diatomaceae from a deep-sea dredging in the Atlantic Ocean off Delaware Bay by the U. S. Fish Commission Steamer Albatross. Proc. U. S. Nat. Mus. 16:303-312.

Gomphonema shaerophorum
Zanon, D. V., 1949
Diatomee di Buenos Aires (Argentina) Atti Accad. Naz. Lincei, Memorie, Cl. Sci. fis., mat. e. nat., ser. 7, 11(3):59-151, 2 pls.

Gossleriella punctata n.sp.
Wood, E. J. Ferguson, L.H. Crosby and Vivienne Cassie, 1959
Studies on Australian and New Zealand diatoms. III. Descriptions of further discoid species. Trans. R. Soc., N.Z., 87(3/4):211-219, pls. 15-17.

Gossleriella radiata
Schröder, B., 1900
Phytoplankton des Golfes von Neapel nebst vergleichenden Ausblicken auf das atlantischen Ozean. Mitt. Zool. Stat. Neapel, 14:1-38.

Gossleriella tropica
Gilbert, J.Y., and W.E. Allen, 1943
The phytoplankton of the Gulf of California obtained by the "E.W. Scripps" in 1939 and 1940. J. Mar. Res. V(2):89-110, figs.30-31.

Gossleriella tropica
Pavillard, J., 1925
Bacillariales. Rept. on the Danish Oceangr. Exped., 1908-10 to the Mediterranean and adj. seas. Vol.II., Biol. J4:72 pp., 116 text figs.

Glossleriella tropica
Rampi, L., 1942
Ricerche sul fitoplancton del Mare Ligure 6. Le diatomee delle acque di Sanremo. Nuovo Giornale Botanico Italiano, N.S., 49:252-268.

Gossleriella tropica
Vives, F., and A. Planas, 1952.
Plancton recogido por los laboratorios costeros. VI. Fitoplancton de las costas de Vinaroz, islas Columbretes y alrededores de la desembocadura del Elroa. Publ. Inst. Biol. Aplic. 11:141-156, 19 textfigs.

Grammatophora
Desikachary, T.V., 1956(1957).
Electron microscope studies on diatoms. J. R. Microsc. Soc. (3)76(1/2):9-36.

Grammatophora sp.
Levander, K.M., 1947
Plankton gesammelt in den Jahren 1899-1910 an den Küsten Finnlands. Finnländische Hydrographisch-Biologische Untersuchungen (aus dem Wasserbiologischen Laboratorin der Societas Scientiarum Fennica) No.11:40 pp., 6 diagrams, 13 pls., tables.

Grammatophora sp.
Meunier, A., 1915
Microplancton de la Mer Flamande. 2. Diatomées (excepté le genre Chaetoceros). Mem. Mus. Roy. Hist. Nat., Belgique, 7(3):1-118, Pls. VIII-XIV.

Grammatophora
Neaverson, E., 1934
The sea-floor deposits. 1. General characteristics and distribution. Discovery Repts. 9: 297-349, Plates 17-22.

Grammatophora anguina
Grunow, A., 1877
1. New Diatoms from Honduras. Monthly Micros. Jour., 18:165-186, pls. CXCIII-CXCVI.

Grammatophora angulosa
Cupp, Easter E., 1943
Marine plankton diatoms of the west coast of North America. Bull. S.I.O. 5(1):1-238, 5 pls., 168 text figs.

Grammatophora angulosa
Hendy, N. Ingram, 1964
An introductory account of the smaller algae of British coastal waters. V. Bacillariophyceae (Diatoms).
Her Majesty's Stationary Office, 317 pp., 45 pls.

Grammatophora angulosa
Misra, J.N., 1956.
A systematic account of some littoral marine diatoms from the west coast of India. J. Bombay Nat. Hist. Soc., 53(4):537-568.

Grammatophora angulosa
Rampi, L., 1942
Ricerche sul fitoplancton del Mare Ligure 6. Le diatomee delle acque di Sanremo. Nuovo Giornale Botanico Italiano, N.S., 49:252-268.

Grammatophora angulosa
Rampi, L., 1940
Diatomee del Mare Adriatico. Nuovo Giornale Botanico Italiano, n.s., 47:559-608.

Grammatophora angulosa
Zanon, V., 1948
Diatomee marini di Sardegna e Pugillo di Alghe Marine della stressa. Boll. Pesca, Piscitutura e Idrobiologia, Anno 24, ns. 3(2): 202-244, 27 figs. on 1 pl.

Grammatophora arcuata
Manguin, E., 1954
Diatomees marines provenant de l'ile Heard (Australian National Antarctic Expedition). Rev. Algol., n.s., 1: 14-24.

Grammatophora caulerpica n.sp.
Misra, J.N., 1956
A systematic account of some littoral marine diatoms from the west coast of India. J. Bombay Nat. Hist. Soc., 53(4):537-568.

Grammatophora diminutum
Rampi, L., 1942
Ricerche sul fitoplancton del Mare Ligure 6. Le diatomee delle acque di Sanremo. Nuovo Giornale Botanico Italiano, N.S., 49:252-268.

Grammatophora flexuosa
Mann, A., 1907
Report on the diatoms of the Albatross voyages in the Pacific Ocean, 1888-1904. Contrib. U. S. Nat. Herb. 10(5):221-419, Pls. XLIV-LIV.

Grammatophora gibberula
Zanon, D. V., 1949
Diatomee di Buenos Aires (Argentina) Atti Accad. Naz. Lincei, Memorie, Cl. Sci. fis., mat. e. nat., ser. 7, 11(3):59-151, 2 pls.

Grammatophora hamulifera
Eskinazi Enide e Shigekatsv Satô (1963-1964) 1966.
Contribuição ao estudo das diatomaceas da Praia de Piedade.
Trabhs Inst. Oceanogr., Univ. Recife, 5 (5/6): 73-114.

Grammatophora hamulifera
Hendy, N. Ingram, 1964
An introductory account of the smaller algae of British coastal waters. V. Bacillariophyceae (Diatoms).
Her Majesty's Stationary Office, 317 pp., 45 pls.

Grammatophora hamulifera
Misra, J.N., 1956.
A systematic account of some littoral marine diatoms from the west coast of India. J. Bombay Nat. Hist. Soc., 53(4):537-568.

Grammatophora hamulifera
Takano, Hideaki, 1962
Notes on epiphytic diatoms upon sea-weeds from Japan.
J. Oceanogr. Soc., Japan, 18(1):29-33.

Grammatophora hamulifera
Zanon, V., 1948
Diatomee marini di Sardegna e Pugillo di Alghe Marine della stressa. Boll. Pesca, Piscitutura e Idrobiologia, Anno 24, ns. 3(2): 202-244, 27 figs. on 1 pl.

Grammatophora islandica
Jørgensen, E., 1905
B.Protistplankton and the diatoms in bottom samples. Hydrographical and biological investigations in Norwegian fjords. Bergens Mus. Skr. 7: 49-225.

Grammatophora islandica
Jorgensen, E., 1900
Protophyten und Protozoën im Plankton aus der Norwegischen Westkerste. Bergens Mus. Aarb. 1899(6): 95 pp., 5 pls., 83 tables.

Grammatophora kerguelensis
Hendey, N.I., 1937
The plankton diatoms of the southern seas. Discovery Repts. 16:151-364, pls.6-13.

Grammatophora longissima
Rampi, L., 1940
Diatomee del Mare Adriatico. Nuovo Giornale Botanico Italiano, n.s., 47:559-608.

Grammatophora longissima
Zanon, V., 1948
Diatomee marini di Sardegna e Pugillo di Alghe Marine della stressa. Boll. Pesca, Piscitutura e Idrobiologia, Anno 24, ns. 3(2): 202-244, 27 figs. on 1 pl.

Grammatophora lyrata
Mann, A., 1907
Report on the diatoms of the Albatross voyages in the Pacific Ocean, 1888-1904. Contrib. U. S. Nat. Herb. 10(5):221-419, Pls. XLIV-LIV.

Grammatophora macilenta
Forti, A., 1922
Ricerche sulla flora pelagica (fitoplancton) di Quarto dei Mille. Mem. R. Com. Talass. Ital. 97:248 pp., 13 pls.

Grammatophora macilenta
Hendy, N. Ingram, 1964
An introductory account of the smaller algae of British coastal waters. V. Bacillariophyceae (Diatoms).
Her Majesty's Stationary Office, 317 pp., 45 pls.

Grammatophora macilenta
Mann, A., 1893
List of Diatomaceae from a deep-sea dredging in the Atlantic Ocean off Delaware Bay by the U. S. Fish Commission Steamer Albatross. Proc. U. S. Nat. Mus. 16:303-312.

Grammatophora marina
Bigelow, H.B., and M. Leslie, 1930
Reconnaissance of the waters and plankton of Monterey Bay, July 1928. Bull. M.C.Z., 70(5):429-481, 43 text figs.

Grammatophora marina
Aleem, A.A., 1949.
A quantitative method for estimating the periodicity of diatoms. J.M.B.A. 28(3):713-717, 1 textfig., 1 pl.

Grammatophora marina
Boden, B. P., 1949.
The diatoms collected by the U.S.S. CACOPAN in the Antarctic in 1947. J. Mar. Res. 8(1):6-13, 3 textfigs.

Grammatophora marina
Boden, Brian, 1948
Marine plankton diatoms on operation HIGHJUMP in: Some oceanographic observations on operation HIGHJUMP. By R.S. Dietz. USNEL Rept. No.55, 97 pp., 41 figs. 7 July 1948.

Grammatophora marina
Cupp, Easter E., 1943
Marine plankton diatoms of the west coast of North America. Bull. S.I.O. 5(1):1-238, 5 pls., 168 text figs.

Grammatophora marina
Eskinazi Enide e Shigekatsv Satô (1963-1964) 1966.
Contribuição ao estudo das diatomaceas da Praia de Piedade.
Trabhs Inst. Oceanogr., Univ. Recife, 5 (5/6): 73-114.

Grammatophora marina

Fox, M., 1957.
A first list of marine algae from Nigeria. J. Linn. Soc., London, 55(362):615-631.

Grammatophora marina

Grøntved, J., 1949
Investigations on the phytoplankton in the Danish Waddensea in July 1941. Medd. Komm. Danmarks Fiskeri og Havundersøgelser, ser. Plankton, 5(2):55 pp., 2 pls., 38 text figs.

Grammatophora marina

Forti, A., 1922
Ricerche sulla flora pelagica (fitoplancton) di Quarto dei Mille. Mem. R. Com. Talass. Ital. 97:248 pp., 13 pls.

Grammatophora marina

Hendy, N. Ingram, 1964
An introductory account of the smaller algae of British coastal waters. V. Bacillariophyceae (Diatoms). Her Majesty's Stationary Office, 317 pp., 45 pls.

Grammatophora marina

Hendey, N.J., 1951
Littoral diatoms of Chicester Harbour with special reference to fouling. J.Roy. Microscop. Soc. 71(1): 1-86, 18 pls.

Grammatophora marina

Hustedt, F. and A.A. Aleem, 1951
Littoral diatoms from the Salstone near Plymouth. JMBA 30(1): 177.196.

Grammatophora marina

Kucherova, Z.S., 1961.
Vertical distribution of diatoms from Sevastopol Bay. Trudy Sevastopol Biol. Sta., (14):64-78.

Grammatophora marina

Lillick, L.C., 1940
Phytoplankton and planktonic protozoa of the offshore waters of the Gulf of Maine. Pt.II. Qualitative Composition of the Planktonic Flora. Trans. Am. Phil. Soc., n.s., 31(3):193-237, 13 text figs.

Grammatophora marina

Lillick, L.C., 1938
Preliminary report of the phytoplankton of the Gulf of Maine. Am. Mid. Nat. 20(3):624-640, 1 text figs 37 tables.

Grammatophora marina

Lillick, L.C., 1937
Seasonal studies of the phytoplankton off Woods Hole, Massachusetts. Biol. Bull. LXXIII (3):488-503, 3 text figs.

Grammatophora marina

Mann, A., 1907
Report on the diatoms of the Albatross voyages in the Pacific Ocean, 1888-1904. Contrib. U. S. Nat. Herb. 10(5):221-419, Pls. XLIV-LIV.

Grammatophora marina

Morse, D.C., 1947
Some observations on seasonal variations in plankton population Patuxant River, Maryland 1943-1945. Bd. Nat. Res., Publ. No.65, Chesapeake Biol. Lab., 31, 3 figs.

Grammatophora marina

Murphy, R.C., 1923
The oceanography of the Peruvian littoral with reference to the abundance and distribution of marine life. Geogr. Rev. 13:64-85.

Grammatophora marina

Rampi, L., 1942
Ricerche sul fitoplancton del Mare Ligure 6. Le diatomee delle acque di Sanremo. Nuovo Giornale Botanico Italiano, N.S., 49:252-268.

Grammatophora marina

Rampi, L., 1940
Diatomee del Mare Adriatico. Nuovo Giornale Botanico Italiano, n.s., 47:559-608.

Grammatophora marina

Schodduyn, M., 1926
Observations faites dans la baie d'Ambleteuse (Pas de Calais). Bull. Inst. Ocean., Monaco, No. 482: 64 pp.

Grammatophora marina

Takano, Hideaki, 1962
Notes on epiphytic diatoms upon sea-weeds from Japan. J. Oceanogr. Soc., Japan, 18(1):29-33.

Grammatophora marina

Zanon, V., 1948
Diatomee marini di Sardegna e Pugillo di Alghe Marine della stressa. Boll. Pesca, Piscitutura e Idrobiologia, Anno 24, ns. 3(2): 202-244, 27 figs. on 1 pl.

Grammatophora marina adriatica

Cupp, Easter E., 1943
Marine plankton diatoms of the west coast of North America. Bull. S.I.O. 5(1):1-238, 5 pls., 168 text figs.

Grammatophora oceanica

Cupp, Easter E., 1943
Marine plankton diatoms of the west coast of North America. Bull. S.I.O. 5(1):1-238, 5 pls., 168 text figs.

Grammatophora maxima

Mann, A., 1907
Report on the diatoms of the Albatross voyages in the Pacific Ocean, 1888-1904. Contrib. U. S. Nat. Herb. 10(5):221-419, Pls. XLIV-LIV.

Grammatophora maxima

Misra, J.N., 1956.
A systematic account of some littoral marine diatoms from the west coast of India. J. Bombay Nat. Hist. Soc., 53(4):537-568.

Grammatophora maxima

Zanon, D. V., 1949
Diatomee di Buenos Aires (Argentina) Atti Accad. Naz. Lincei, Memorie, Cl. Sci. fis., mat. e. nat., ser. 7, 11(3):59-151, 2 pls.

Grammatophora maxima

Zanon, V., 1948
Diatomee marini di Sardegna e Pugillo di Alghe Marine della stressa. Boll. Pesca, Piscitutura e Idrobiologia, Anno 24, ns. 3(2): 202-244, 27 figs. on 1 pl.

Grammatophora oceanica macilenta

de Sousa e Silva, E., 1956.
Contribution à l'étude du microplancton de Dakar et des regions maritimes voisines. Bull. I.F.A.N., 8(2):335-371, 7 pls.

Grammatophora oceanica

Eskinazi Enide e Shigekatsv Satô (1963-1964) 1966.
Contribuição ao estudo das diatomaceas da Praia de Piedade.
Trabhs Inst. Oceanogr., Univ. Recife, 5 (5/6): 73-114.

Grammatophora oceanica

Grunow, A., 1877
1. New Diatoms from Honduras. Monthly Micros. Jour., 18:165-186, pls. CXCIII-CXCVI.

Grammatophora oceanica & varieties

Hendy, N. Ingram, 1964
An introductory account of the smaller algae of British coastal waters. V. Bacillariophyceae (Diatoms). Her Majesty's Stationary Office, 317 pp., 45 pls.

Grammatophora oceanica

Jørgensen, E., 1905
B.Protistplankton and the diatoms in bottom samples. Hydrographical and biological investigations in Norwegian fjords. Bergens Mus. Skr. 7: 49-225.

Grammatophora oceanica

Politis, J., 1949
Diatomees marines de Bosphores et des ibes de la mer de Marmara. II Practica tou Hellenikou Hidrobiologikou Institutoutou 1929, Etoz 1929, 3(1):11-31.

Grammatophora oceanica

Rampi, L., 1940
Diatomee del Mare Adriatico. Nuovo Giornale Botanico Italiano, n.s., 47:559-608.

Grammatophora oceanica

Sousa e Silva, E., 1949
Diatomaceas e Dinoflagelados de Baia de Cascais. Portugaliae Acta Biol., Volume: Julio Henriques, Ser. B: 300-383, 9 pls, 2 fold-in tables.

Grammatophora oceanica

Sousa a Silva, E., and J. Dos Santos-Pinto, 1948
O Plancton da Baia de S. Martinho do Porto. 1. Diatomaceas e Dinoflagelados. Bol. Soc. Portuguese de Ciencias Naturais, 16(2):134-187, 6 pls. (Trav. Sta. Biol. Mar. de Lisbonne No. 52).

Grammatophora oceanica

Zanon, D. V., 1949
Diatomee di Buenos Aires (Argentina) Atti Accad. Naz. Lincei, Memorie, Cl. Sci. fis., mat. e. nat., ser. 7, 11(3):59-151, 2 pls.

Grammatophora oceanica

Zanon, V., 1948
Diatomee marini di Sardegna e Pugillo di Alghe Marine della stressa. Boll. Pesca, Piscitutura e Idrobiologia, Anno 24, ns. 3(2): 202-244, 27 figs. on 1 pl.

Grammatophora oceanica macilenta

Florin, M-B., 1948
9. Diatomeae in submarine cores from the Tyrrhenian Sea. Medd. Ocean. Inst., Göteborg, 15 (Göteborgs Kungl. Vetenskaps-och Viterrhets Samhälles Handlingar, Sjätte Foljden, Ser. B 5(13):80-88.

Grammatophora oceanica var. macilenta

Rampi, L., 1942
Ricerche sul fitoplancton del Mare Ligure 6. Le diatomee delle acque di Sanremo. Nuovo Giornale Botanico Italiano, N.S., 49:252-268.

Grammatophora oceanica macilenta

Ercegovic, A., 1936
Etudes qualitative et quantitatives du phytoplancton dans les eaux cotières de l'Adriatique oriental moyen au cours de l'année 1934. Acta Adriatica 1(9):1-126

Grammatophora oceanica macilente

Misra, J.N., 1956.
A systematic account of some littoral marine diatoms from the west coast of India. J. Bombay Nat. Hist. Soc., 53(4):537-568.

Grammatophora oceanica subtilissima

Hendey, N.J., 1951
Littoral diatoms of Chicester Harbour with special reference to fouling. J.Roy. Microscop. Soc. 71(1): 1-86, 18 pls.

Grammatophora serpentina
Florin, M-B., 1948
9. Diatomeae in submarine cores from the Tyrrhenian Sea. Medd. Ocean. Inst., Göteborg, 15 (Göteborgs Kungl. Vetenskaps-och Viterrhets Samhälles Handlingar, Sjätte Foljden, Ser. B 5(13):80-88.

Grammatophora serpentina
Hendy, N. Ingram, 1964
An introductory account of the smaller algae of British coastal waters. V. Bacillariophyceae (Diatoms).
Her Majesty's Stationary Office, 317 pp., 45 pls.

Grammatophora serpintina
Hendey, N.I., 1951
Littoral diatoms of Chicester Harbour with special reference to fouling. J.Roy. Microscop. Soc. 71(1): 1-86, 18 pls.

Grammatophora serpentina
Hendey, N.I., 1937
The plankton diatoms of the southern seas. Discovery Repts. 16:151-364, pls.6-13.

Grammatophora serpentina
Hustedt, F. and A.A. Aleem, 1951
Littoral diatoms from the Salstone near Plymouth. JMBA 30(1): 177.196.

Grammatophora serpintina
Jorgensen, E., 1900
Protophyten und Protozoën im Plankton aus der Norwegischen Westkerste. Bergens Mus. Aarb. 1899(6): 95 pp., 5 pls., 83 tables.

Grammatophora serpentina
Morse, D.C., 1947
Some observations on seasonal variations in plankton population Patuxant River, Maryland 1943 1945. Bd. Nat. Res., Publ. No.65, Chesapeake Biol. Lab., 31, 3 figs.

Grammatophora serpentina
Politis, J., 1949
Diatomées marines de Bosphores et des ibes de la mer de Marmara. II Practica tou Hellenikou Hidrobiologikou Institutoutou 1929, Etoz 1929, 3(1):11-31.

Grammatophora serpentina
Rampi, L., 1942
Ricerche sul fitoplancton del Mare Ligure 6. Le diatomee delle acque di Sanremo. Nuovo Giornale Botanico Italiano, N.S., 49:252-268.

Grammatophora serpentina
Rampi, L., 1940
Diatomee del Mare Adriatico. Nuovo Giornale Botanico Italiano, n.s., 47:559-608.

Grammatophora serpentina
Sousa e Silva, E., 1949
Diatomaceas e Dinoflagelados de Baia de Cascais. Portugaliae Acta Biol., Volume: Julio Henriques, Ser. B: 300-383, 9 pls, 2 fold-in tables.

Grammatophora serpentina
Sousa e Silva, E., and J. Dos Santos-Pinto, 1948
O Plancton da Baia de S. Martinho do Porto. 1. Diatomaceas e Dinoflagelados. Bol. Soc. Portuguese de Ciencias Naturais, 16(2):134-187, 6 pls. (Trav. Sta. Biol. Mar. de Lisbonne No. 52).

Grammatophora serpentina
Steemann-Nielsen, Einar, 1951
The marine vegetation of the Isefjord. A study on ecology and production. Medd. Komm. Danmarks Fiskeri-og Havundersøgelser. Ser. Plankton. 5(4); 114pp., 46 text figs.

Grammatophora serpentina
Zanon, V., 1948
Diatomee marini di Sardegna e Pugillo di Alghe Marine della stressa. Boll. Pesca, Piscitutura e Idrobiologia, Anno 24, ns. 3(2): 202-244, 27 figs. on 1 pl.

Grammatophora souriei n.sp.
Amossé, A. 1970
Diatomées marines et saumâtres du Sénégal et de la Côte d'Ivoire. Bull. Inst. fond. Afr. Noire (A) 32 (2): 289-311.

Grammatophora stricta n.var.
Castracane degli Antelminelli, F., 1886
1. Report on the Diatomaceae collected by H.M.S. Challenger during the years 1873-1876. Rept. Sci. Results, H.M.S. Challenger, Botany Vol. II, 178 pp., 30 pls.

Grammatophora stricta
Mann, A., 1907
Report on the diatoms of the Albatross voyages in the Pacific Ocean, 1888-1904. Contrib. U. S. Nat. Herb. 10(5):221-419, Pls. XLIV-LIV.

Grammatophora undulata
Fox, M., 1957
A first list of marine algae from Nigeria. J. Linn. Soc., London, 55(362):615-631.

Grammatophora undulata
Rampi, L., 1940
Diatomee del Mare Adriatico. Nuovo Giornale Botanico Italiano, n.s., 47:559-608.

Grammatophora undulata
Zanon, V., 1948
Diatomee marini di Sardegna e Pugillo di Alghe Marine della stressa. Boll. Pesca, Piscitutura e Idrobiologia, Anno 24, ns. 3(2): 202-244, 27 figs. on 1 pl.

Groentvedia elliptica, n.gen., n.sp.
Hendy, N. Ingram, 1964
An introductory account of the smaller algae of British coastal waters. V. Bacillariophyceae (Diatoms).
Her Majesty's Stationary Office, 317 pp., 45 pls.

Guinardia
Brunel, J., 1962
Le phytoplancton de la Baie de Chaleurs. Inst. Botan., Univ. Montréal, Contrib. No. 77: 365 pp., 66 pls.

Guinardia
Desikachary, T.V., 1956(1957).
Electron microscope studies on diatoms. J.R. Microsc. Soc. (3)76(1/2):9-30.

Guinardia sp.
Mangin, L., 1915
Phytoplancton de L'Antartique. Deuxieme Exped. Ant. Francaise (1908-1910), 95 pp., 3 pls., 58 text figs.

Guinardia sp.
Wawrik, F., 1961.
Die horizontale Verteilung der Planktondiatomeen im Golf von Neapel. Int. Revue Ges. Hydrobiol., 46(3):460-479.

Guinardia blavyana
Boden, B.P., 1950
Some marine plankton diatoms from the west coast of South Africa. Trans. R.Soc. S. Africa. 32:321-434, 100 text figs.

Guinardia Blavyana
Ercegovic, A., 1936
Etudes qualitative et quantitatives du phytoplancton dans les eaux cotières de l'Adriatique oriental moyen au cours de l'année 1934. Acta Adriatica 1(9):1-126

Guinardia blavyana
Pavillard, J., 1925
Bacillariales. Rept. on the Danish Oceangr. Exped. 1908-10 to the Mediterranean and adj. seas. Vol.II., Biol. J4:72 pp., 116 text figs.

Guinardia Blavyana
Pavillard, J., 1905
Recherches sur la flore pelagique (Phytoplankton) de l'Etang de Thau. Theses presentees a la Fac. Sci., Paris, 116 pp., 3 pls.

Guinardia Blavyana
Rampi, L., 1942
Ricerche sul fitoplancton del Mare Ligure 6. Le diatomee delle acque di Sanremo. Nuovo Giornale Botanico Italiano, N.S., 49:252-268.

Guinardia Blavyana
Schröder, B., 1900
Phytoplankton des Golfes von Neapel nebst vergleichenden Ausblicken auf das atlantischen Ozean. Mitt. Zool. Stat. Neapel, 14:1-38.

Guinardia flaccida
Allen, W. E., 1938
The Templeton Crocker Expedition to the Gulf of California in 1935 - The Phytoplankton. Amer. Microsc. Soc., Trans. 57:328-335.

Guinardia flaccida
Allen, W.E., 1937
Plankton diatoms of the Gulf of California obtained by the G. Allan Hancock Expedition of 1936. The Hancock Pacific Expeditions, Univ. So. Calif. Publ. 3:47-59, 1 fig.

Guinardia flaccida
Allen, W. E., 1934
Marine plankton diatoms of Lower California in 1931. Bot. Gaz. 95(3):485-492, 1 fig.

Guinardia flaccida
Allen, W.E., and E.E. Cupp, 1935
Plankton diatoms of the Java Sea. Annales du Jardin Botanique de Buitenzorg XLIV (2):101-174, figs.1-127.
(drawings of all species mentioned)

Guinardia flaccida
Birnhax, Bruce I., Patricia V. Donnelly and Richard P. Saunders 1967.
Studies on Guinardia flaccida (Castracane) Peragallo.
Fla. Bd. Conserv. Mar. Lab. St. Petersburg, Leaflet Ser. Phytoplankton 1(3-3):23pp.

Guinardia flaccida
Boden, B.P., 1950
Some marine plankton diatoms from the west coast of South Africa. Trans. R.Soc. S. Africa. 32:321-434, 100 text figs.

Guinardia flaccida
Boden, B. P., 1949.
The diatoms collected by the U.S.S. CACOPAN in the Antarctic in 1947. J. Mar. Res. 8(1):6-13, 3 textfigs.

Guinardia flaccida
Boden, Brian, 1948
Marine plankton diatoms on operation HIGHJUMP in: Some oceanographic observations on operation HIGHJUMP. By R.S. Dietz. USNEL Rept. No.55, 97 pp., 41 figs. 7 July 1948.

Guinardia flaccida

Braarud, T., 1945
Experimental studies on marine plankton diatoms. Avhandlingar utgitt av Det Norske Videnskaps-Akademi i Oslo. 1. Mat.-Naturv. Klasse 1944. No.10, 1-16, 1 text fig.

Braarud, T., 1934
A note on the phytoplankton of the Gulf of Maine in the summer of 1933. Biol. Bull. 67(1):76-82. (Contribution No.46 of the Woods Hole Oceanographic Institution)

Braarud, T., and Adam Bursa, 1939
On the phytoplankton of the Oslo Fjord, 1933-1934. Hvalrådets Skr. No.19:1-63; 9 text figs. Reviewed. J. du. Cons. 14(3): 418-420. A.C. Gardiner.

Cleve-Euler, A., 1951
Die Diatomeen von Schweden und Finnland. Kungl. Svenska Vetenskaps Akad. Handl., Fjärde Ser. 2(1): 161 pp., 6 pls.

Cupp, Easter E., 1943
Marine plankton diatoms of the west coast of North America. Bull. S.I.O. 5(1):1-238, 5 pls., 168 text figs.

Cupp, E.E., 1934
Analysis of marine diatom collections taken from the Canal Zone to California during March, 1933. Trans. Am. Micros. Soc. LIII (1):22-29, 1 map.

Cupp, E.E. and Allen, W.E., 1938
Plankton diatoms of the Gulf of California obtained by Allan Hancock Pacific Expedition of 1937. The Hancock Pacific Expeditions, The Univ. So. Calif. Publ. 3: 61-74, 1 map, pls.4-15.

Dangeard, P., 1927
Phytoplankton de la croisière du "Sylvana". Ann. Inst. Ocean., Monaco, n.s., 4(8):286-401, 54 text figs. (Feirer-Juin 1913).

Delegazione Italiana della Commissione Internazionale per l'Esplorazione Scientifica del Mediterraneo, 1941
Note sul plancton della Laguna veneta. [Memoria CCLXXXIX], Arch. di Ocean. e Limn. Anno I, Fasc. I, 1941 XIX: 31-57 pp.

Delsman, H. C., 1939.
Preliminary plankton investigations in the Java Sea. Treubia, 17:139-181, 8 maps, 41 figs.

Ercegovic, A., 1936
Etudes qualitative et quantitatives du phytoplancton dans les eaux cotières de l'Adriatique oriental moyen au cours de l'année 1934. Acta Adriatica 1(9):1-126

Fox, M., 1957.
A first list of marine algae from Nigeria. J. Linn. Soc., London, 55(362):615-631.

Ghazzawi, F.M., 1939
Plankton of the Egyptian waters. A study of the Suez Canal Plankton. (A) Phytoplankton. Preliminary Report 83 pp. Notes and Memoires, Min. Commerce-Industry, Egypt, Hydrobiol. & Fish. 65 figs.

Gran, H.H., 1908
Diatomeen. Nordisches Plankton, Botanischer Teil pp. XIX.1-XIX 146; 178 text figs.

Grøntved, J., 1949(1950).
Investigations on the phytoplankton of the Danish Waddensee in July 1941. Medd. Komm. Danmarks Fiskeri- og Havundersøgelser, Serie Plankton, 5(2)(Medd. Skalling Lab. 10):55 pp., 2 pls., 38 textfigs.

Forti, A., 1922
Ricerche sulla flora pelagica (fitoplancton) di Quarto dei Mille. Mem. R. Com. Talass. Ital. 97:248 pp., 13 pls.

Gilbert, J.Y., and W.E. Allen, 1943
The phytoplankton of the Gulf of California obtained by the "E.W. Scripps" in 1939 and 1940. J. Mar. Res. V(2):89-110, figs.30-31.

Gran, H.H., and T. Braarud, 1935
A quantitative study of the phytoplankton in the Bay of Fundy and the Gulf of Maine (including observations on hydrography, chemistry, and turbidity). J. Biol. Bd., Canada, 1(5):279-467, 69 text figs.

Grøntved, J., 1949
Investigations on the phytoplankton in the Danish Waddensea in July 1941. Medd. Komm. Danmarks Fiskeri og Havundersøgelser, ser. Plankton, 5(2):55 pp., 2 pls., 38 text figs.

Hendy, N. Ingram, 1964
An introductory account of the smaller algae of British coastal waters. V. Bacillariophyceae (Diatoms). Her Majesty's Stationary Office, 317 pp., 45 pls.

Hendey, N.I., 1937
The plankton diatoms of the southern seas. Discovery Repts. 16:151-364, pls.6-13.

Höfler, Karl, und Luise Höfler, 1964.
Diatomeen des marine Planktons als Isotonobionten. Pubbl. Sta. Zool., Napoli, 33(3):315-330.

Jorgensen, E., 1900
Protophyten und Protozoën im Plankton aus der Norwegischen Westkerste. Bergens Mus. Aarb. 1899(6): 95 pp., 5 pls., 83 tables.

Kokubo, S., 1952
Results of the observations on the plankton and oceanography of Mutsu Bay during 1950, reference being made also to the period 1946-1950. Bull Mar.Biol.Sta., Asamushi 5(1/4): 1-54, 3 tables,(fold-in), 1 fold-in.

Lillick, L.C., 1940
Phytoplankton and planktonic protozoa of the offshore waters of the Gulf of Maine. Pt.II. Qualitative Composition of the Planktonic Flora. Trans. Am. Phil. Soc., n.s., 31(3):193-237, 13 text figs.

Lillick, L.C., 1938
Preliminary report of the phytoplankton of the Gulf of Maine. Am. Mid. Nat. 20(3):624-640, 1 text figs 37 tables.

Lillick, L.C., 1937
Seasonal studies of the phytoplankton off Woods Hole, Massachusetts. Biol. Bull. LXXIII (3):488-503, 3 text figs.

Mangin, M. L., 1912
Phytoplancton de la croisière du "René" dans l'Atlantique (Septembre 1908). Ann. Inst. Ocean, n.s., 4(1):1-66, 2 pls., 41 text figs., 2 tables.

Margalef, R., 1949
Fitoplancton nerítico de la Costa Brava en 1947-48. Publ. Inst. Biol. Aplicada, 5: 41-51, 3 text figs.

Meunier, A., 1915
Microplancton de la Mer Flamande. 2. Diatomées (excepté le genre Chaetoceros). Mem. Mus. Roy. Hist. Nat., Belgique, 7(3):1-118, Pls. VIII-XIV.

Pavillard, J., 1925
Bacillariales. Rept. on the Danish Oceangr. Exped., 1908-10 to the Mediterranean and adj. seas. Vol.II., Biol. J4:72 pp., 116 text figs.

Pavillard, J., 1905
Recherches sur la flore pelagique (Phytoplankton) de l'Etang de Thau. Theses presentees a la Fac. Sci., Paris, 116 pp., 3 pls.

Pettersson, H., 1934
Scattering and extinction of light in sea-water. Meddelanden från Göteborgs Högskolas Oceanografiska Institut - 9 (Göteborgs Kungl. vetenskaps-och vitterhetssamhälles Handlingar. Femte Foljden, Ser. B, 4(4)):1-16.

Rampi, L., 1942
Ricerche sul fitoplancton del Mare Ligure 6. Le diatomee delle acque di Sanremo. Nuovo Giornale Botanico Italiano, N.S., 49:252-268.

Rampi, L., 1940
Diatomee del Mare Adriatico. Nuovo Giornale Botanico Italiano, n.s., 47:559-608.

Robinson, G.A., 1961
Contributions towards an atlas of the northeastern Atlantic and the North Sea. 1. Phytoplankton. Bull. Mar. Ecology. 5(42): 81-89, Pls. 15-20.

Schodduyn, M., 1926
Observations faites dans la baie d'Ambleteuse (Pas de Calais). Bull. Inst. Ocean., Monaco, No. 482: 64 pp.

Schröder, B., 1900
Phytoplankton des Golfes von Neapel nebst vergleichenden Ausblicken auf das atlantischen Ozean. Mitt. Zool. Stat. Neapel, 14:1-38.

Guinardia flaccidus

Sousa e Silva, E., 1949
Diatomaceas e Dinoflagelados de Baia de Cascais. Portugaliae Acta Biol., Volume: Julio Henriques, Ser. B: 300-383, 9 pls, 2 fold-in tables.

Guinardia flaccida

Sousa a Silva, E., and J. Dos Santos-Pinto, 1948
O Plancton da Baia de S. Martinho do Porto.
1. Diatomaceas e Dinoflagelados. Bol. Soc.
Portuguese de Ciencias Naturais, 16(2):134-187,
6 pls. (Trav. Sta. Biol. Mar. de Lisbonne No. 52).

Guinardia flaccida

Stæmann-Nielsen, Einar, 1951
The marine vegetation of the Isefjord.
A study on ecology and production. Medd.
Komm. Danmarks Fiskeri-og Havundersøgelser.
Ser. Plankton. 5(4); 114 pp., 46 text figs.

Gyrosigma sp.

Brunel, J., 1962
Le phytoplancton de la Baie de Chaleurs.
Inst. Botan., Univ. Montréal, Contrib.
No. 77: 365 pp., 66 pls.

Gyrosigma sp.

Brunel, Jules, 1962
Le phytoplancton de la Baie des Chaleurs.
Contrib. Ministère de la Chasse et des
Pêcheries, Province de Québec, No. 91:
365 pp.

Gyrosygma

Desikachary, T.V., 1956(1957).
Electron microscope studies on diatoms.
J.R. Microsc. Soc. (3)76(1/2):9-30.

Gyrosigma spp.

Smayda, Theodore J., 1962.
Occurrence of unusual bodies in a marine pennate diatom.
Nature, 196(4850):191.

Gyrosigma acuminatum

Rumkówna, A., 1948
[List of the phytoplankton species occurring in the superficial water layers in the Gulf of Gdańsk] Bull. Lab. mar., Gdynia,
No. 4: 139-141 with tables in back.

Gyrosigma acuminatum

Zanon, D. V., 1949
Diatomee di Buenos Aires (Argentina)
Atti Accad. Naz. Lincei, Memorie, Cl. Sci.
fis., mat. e. nat., ser. 7, 11(3):59-151,
2 pls.

Gyrosigma aestuarii

Mann, A., 1907
Report on the diatoms of the Albatross
voyages in the Pacific Ocean, 1888-1904.
Contrib. U. S. Nat. Herb. 10(5):221-419, Pls.
XLIV-LIV.

Gyrosigma attenuatum

Zanon, D. V., 1949
Diatomee di Buenos Aires (Argentina)
Atti Accad. Naz. Lincei, Memorie, Cl. Sci.
fis., mat. e. nat., ser. 7, 11(3):59-151,
2 pls.

Gyrosigma attenuatum

Zanon, V., 1948
Diatomee marini di Sardegna e Pugillo
di Alghe Marine della stressa. Boll. Pesca,
Piscitutura e Idrobiologia, Anno 24, ns. 3(2):
202-244, 27 figs. on 1 pl.

Gyrosigma balticum

Andrade e Clovis Teixeira, M.H. de, 1957.
Contribuição para o conhecimento das diatomaceas
do Brasil.
Bol. Inst. Ocean., Univ. Sao Paulo, 8(1/2):171-225, 10 pls.

Gyrosigma balticum

Brunel, Jules, 1962
Le phytoplancton de la Baie des Chaleurs.
Contrib. Ministère de la Chasse et des
Pêcheries, Province de Québec, No. 91:
365 pp.

Gyrosigma balticum

Brunel, J., 1962
Le phytoplancton de la Baie de Chaleurs.
Inst. Botan., Univ. Montréal, Contrib.
No. 77: 365 pp., 66 pls.

Gyrosigma balticum

Eskinazi Enide e Shigekatsv Satô (1963-1964) 1966.
Contribuição ao estudo das diatomaceas da Praia de Piedade.
Trabhs Inst. Oceanogr., Univ. Recife, 5 (5/6):
73-114.

Gyrosigma balticum

Hendy, N. Ingram, 1964
An introductory account of the smaller algae
of British coastal waters. V. Bacillariophyceae (Diatoms).
Her Majesty's Stationary Office, 317 pp.,
45 pls.

Gyrosigma balticum

Hustedt, F. and A.A. Aleem, 1951
Littoral diatoms from the Salstone
near Plymouth. JMBA 30(1): 177.196.

Gyrosigma balticum

Iyengar, M.O.P. and G.Venkataraman, 1951.
The ecology and seasonal succession
of the algae flora of the River Cooum at
Madras with special reference to the Diatomaceae. J. Madras Univ. 21, Sect. B(1):
140-192, 1 pl of 4 figs., 11 text figs.

Gyrosigma balticum

Kokubo, S., and S. Sato., 1947
Plankters in Jū-San Gata. Physiol.
and Ecol. (Japan) 1(4):1-16, 3 text figs.,
tables.

Gyrosigma balticum

Rampi, L., 1942
Ricerche sul fitoplancton del Mare Ligure
6. Le diatomee delle acque di Sanremo. Nuovo
Giornale Botanico Italiano, N.S., 49:252-268.

Gyrosigma balticum

Rampi, L., 1940
Diatomee del Mare Adriatico. Nuovo
Giornale Botanico Italiano, n.s., 47:559-608.

Gyrosigma balticum

Zanon, D. V., 1949
Diatomee di Buenos Aires (Argentina)
Atti Accad. Naz. Lincei, Memorie, Cl. Sci.
fis., mat. e. nat., ser. 7, 11(3):59-151,
2 pls.

Gyrosigma balticum

Zanon, V., 1948
Diatomee marini di Sardegna e Pugillo
di Alghe Marine della stressa. Boll. Pesca,
Piscitutura e Idrobiologia, Anno 24, ns. 3(2):
202-244, 27 figs. on 1 pl.

Gyrosigma bigibbum n.sp.

Zanon, D. V., 1949
Diatomee di Buenos Aires (Argentina)
Atti Accad. Naz. Lincei, Memorie, Cl. Sci.
fis., mat. e. nat., ser. 7, 11(3):59-151,
2 pls.

Gyrosigma calcaritanum n.sp.

Zanon, V., 1948
Diatomee marini di Sardegna e Pugillo
di Alghe Marine della stressa. Boll. Pesca,
Piscitutura e Idrobiologia, Anno 24, ns. 3(2):
202-244, 27 figs. on 1 pl.

Gyrosigma compactum

Zanon, D. V., 1949
Diatomee di Buenos Aires (Argentina)
Atti Accad. Naz. Lincei, Memorie, Cl. Sci.
fis., mat. e. nat., ser. 7, 11(3):59-151,
2 pls.

Gyrosigma diminutum

Rampi, L., 1940
Diatomee del Mare Adriatico. Nuovo
Giornale Botanico Italiano, n.s., 47:559-608.

Gyrosigma diminutum

Zanon, V., 1948
Diatomee marini di Sardegna e Pugillo
di Alghe Marine della stressa. Boll. Pesca,
Piscitutura e Idrobiologia, Anno 24, ns. 3(2):
202-244, 27 figs. on 1 pl.

Gyrosigma distortum

Hustedt, F. and A.A. Aleem, 1951
Littoral diatoms from the Salstone
near Plymouth. JMBA 30(1): 177.196.

Gyrosigma distortum

Iyengar, M.O.P. and G.Venkataraman, 1951.
The ecology and seasonal succession
of the algae flora of the River Cooum at
Madras with special reference to the Diatomaceae. J. Madras Univ. 21, Sect. B(1):
140-192, 1 pl of 4 figs., 11 text figs.

Gyrosigma distortum

Kokubo, S., and S. Sato., 1947
Plankters in Jū-San Gata. Physiol.
and Ecol. (Japan) 1(4):1-16, 3 text figs.,
tables.

Gyrosigma distortum

Zanon, V., 1948
Diatomee marini di Sardegna e Pugillo
di Alghe Marine della stressa. Boll. Pesca,
Piscitutura e Idrobiologia, Anno 24, ns. 3(2):
202-244, 27 figs. on 1 pl.

Gyrosigma fasciola

Andrade e Clovis Teixeira, M.H. de, 1957.
Contribuição para o conhecimiento das diatomáceas do Brasil.
Bol. Inst. Ocean., Univ. Sao Paulo, 8(1/2):171-225, 10 pls.

Gyrosigma fasciola & varieties

Hendy, N. Ingram, 1964
An introductory account of the smaller algae
of British coastal waters. V. Bacillariophyceae (Diatoms).
Her Majesty's Stationary Office, 317 pp.,
45 pls.

Gyrosigma fasciola

Hustedt, F. and A.A. Aleem, 1951
Littoral diatoms from the Salstone
near Plymouth. JMBA 30(1): 177.196.

Gyrosigma formosum

Mann, A., 1907
Report on the diatoms of the Albatross
voyages in the Pacific Ocean, 1888-1904.
Contrib. U. S. Nat. Herb. 10(5):221-419, Pls.
XLIV-LIV.

Gyrosigma hippocampus

Hendy, N. Ingram, 1964
An introductory account of the smaller algae
of British coastal waters. V. Bacillariophyceae (Diatoms).
Her Majesty's Stationary Office, 317 pp.,
45 pls.

Gyrosigma inermedium

Mann, A., 1907
Report on the diatoms of the Albatross
voyages in the Pacific Ocean, 1888-1904.
Contrib. U. S. Nat. Herb. 10(5):221-419, Pls.
XLIV-LIV.

Gyrosigma itaparicanum
Andrade e Clovis Teixeira, M.H. de, 1957.
Contribuição para o conhecimento das diatomáceas do Brasil.
Bol. Inst. Ocean., Univ. Sao Paulo, 8(1/2):171-225, 10 pls.

Gyrosigma lineare
Zanon, D. V., 1949
Diatomee di Buenos Aires (Argentina)
Atti Accad. Naz. Lincei, Memorie, Cl. Sci. fis., mat. e. nat., ser. 7, 11(3):59-151, 2 pls.

Gyrosigma littorale
Hendy, N. Ingram, 1964

An introductory account of the smaller algae of British coastal waters. V. Bacillariophyceae (Diatoms).
Her Majesty's Stationary Office, 317 pp., 45 pls.

Gyrosigma littorale
Hustedt, F. and A.A. Aleem, 1951
Littoral diatoms from the Salstone near Plymouth. JMBA 30(1): 177.196.

Gyrosigma normanii
Mann, A., 1907
Report on the diatoms of the Albatross voyages in the Pacific Ocean, 1888-1904.
Contrib. U. S. Nat. Herb. 10(5):221-419, Pls. XLIV-LIV.

Gyrosigma prolongatum & varieties
Hendy, N. Ingram, 1964

An introductory account of the smaller algae of British coastal waters. V. Bacillariophyceae (Diatoms).
Her Majesty's Stationary Office, 317 pp., 45 pls.

Gyrosigma prolongatum closterioides
Brunel, Jules, 1962
Le phytoplancton de la Baie des Chaleurs.
Contrib. Ministère de la Chasse et des Pêcheries, Province de Québec, No. 91: 365 pp.

Gyrosigma prolongatum closterioides
Brunel, J., 1962
Le phytoplancton de la Baie de Chaleurs.
Inst. Botan., Univ. Montréal, Contrib. No. 77: 365 pp., 66 pls.

Gyrosigma rectum
Zanon, V., 1948
Diatomee marini di Sardegna e Pugillo di Alghe Marine della stressa. Boll. Pesca, Piscitutura e Idrobiologia, Anno 24, ns. 3(2): 202-244, 27 figs. on 1 pl.

Gyrosigma rigidum
Mann, A., 1907
Report on the diatoms of the Albatross voyages in the Pacific Ocean, 1888-1904.
Contrib. U. S. Nat. Herb. 10(5):221-419, Pls. XLIV-LIV.

Gyrosigma robustum
Zanon, V., 1948
Diatomee marini di Sardegna e Pugillo di Alghe Marine della stressa. Boll. Pesca, Piscitutura e Idrobiologia, Anno 24, ns. 3(2): 202-244, 27 figs. on 1 pl.

Gyrosigma sagitta
Mann, A., 1907
Report on the diatoms of the Albatross voyages in the Pacific Ocean, 1888-1904.
Contrib. U. S. Nat. Herb. 10(5):221-419, Pls. XLIV-LIV.

Gyrosigma speciosum
Mann, A., 1907
Report on the diatoms of the Albatross voyages in the Pacific Ocean, 1888-1904.
Contrib. U. S. Nat. Herb. 10(5):221-419, Pls. XLIV-LIV.

Gyrosigma spectabile
Zanon, D. V., 1949
Diatomee di Buenos Aires (Argentina)
Atti Accad. Naz. Lincei, Memorie, Cl. Sci. fis., mat. e. nat., ser. 7, 11(3):59-151, 2 pls.

Gyrosigma spenceri
Cupp, Easter E., 1943
Marine plankton diatoms of the west coast of North America. Bull. S.I.O. 5(1):1-238, 5 pls., 168 text figs.

Gyrosigma spenceri
Hustedt, F. and A.A. Aleem, 1951
Littoral diatoms from the Salstone near Plymouth. JMBA 30(1): 177.196.

Gyrosigma Spenceri
Zanon, D. V., 1949
Diatomee di Buenos Aires (Argentina)
Atti Accad. Naz. Lincei, Memorie, Cl. Sci. fis., mat. e. nat., ser. 7, 11(3):59-151, 2 pls.

Gyrosigma strigili
Paiva Carvalho, J., 1950
O plancton do Rio Maria Rodriques (Cananeis). 1. Diatomaceas e Dinoflagelados.
Bol. Inst. Paulista Oceanogr. 1(1); 27-43, 2 fold-in tables, 2 figs.

Gyrosigma strigilis
Zanon, V., 1948
Diatomee marini di Sardegna e Pugillo di Alghe Marine della stressa. Boll. Pesca, Piscitutura e Idrobiologia, Anno 24, ns. 3(2): 202-244, 27 figs. on 1 pl.

Gyrosigma tenuissimum
Hendy, N. Ingram, 1964

An introductory account of the smaller algae of British coastal waters. V. Bacillariophyceae (Diatoms).
Her Majesty's Stationary Office, 317 pp., 45 pls.

Gyrosigma tenuissimum
Hustedt, F. and A.A. Aleem, 1951
Littoral diatoms from the Salstone near Plymouth. JMBA 30(1): 177.196.

Gyrosigma thuringicum
Mann, A., 1907
Report on the diatoms of the Albatross voyages in the Pacific Ocean, 1888-1904.
Contrib. U. S. Nat. Herb. 10(5):221-419, Pls. XLIV-LIV.

Gyrosigma wansbeckii
Brunel, J., 1962
Le phytoplancton de la Baie de Chaleurs.
Inst. Botan., Univ. Montréal, Contrib. No. 77: 365 pp., 66 pls.

Gyrosigma wansbeckii
Hendy, N. Ingram, 1964

An introductory account of the smaller algae of British coastal waters. V. Bacillariophyceae (Diatoms).
Her Majesty's Stationary Office, 317 pp., 45 pls.

Gyrosigma Wansbeckii
Zanon, D. V., 1949
Diatomee di Buenos Aires (Argentina)
Atti Accad. Naz. Lincei, Memorie, Cl. Sci. fis., mat. e. nat., ser. 7, 11(3):59-151, 2 pls.

Gyrosigma Wansbeckii
Zanon, V., 1948
Diatomee marini di Sardegna e Pugillo di Alghe Marine della stressa. Boll. Pesca, Piscitutura e Idrobiologia, Anno 24, ns. 3(2): 202-244, 27 figs. on 1 pl.

Hantzschia
Desikachary, T.V., 1956(1957).
Electron microscope studies on diatoms. J.R. Microsc. Soc. (3)76(1/2):9-36.

Hantzschia amphioxys
Faure-Fremiet, E., 1951.
The tidal rhythm of the diatom Hantzschia amphioxys. Biol. Bull. 100(3):173-177, 1 textfig.

Hantzschia amphioxys
Iyengar, M.O.P. and G.Venkataraman, 1951.
The ecology and seasonal succession of the algae flora of the River Cooum at Madras with special reference to the Diatomaceae. J. Madras Univ. 21, Sect. B(1): 140-192, 1 pl of 4 figs., 11 text figs.

Hantzschia amphioxys
Tsumura, K., 1956.
Diatomoj el la cirkaufoso de la restaĵo de la kastele de Odawara. J. Yokohama Municipal Univ., (C-14), No. 47:23 pp.

Hantzschia amphioxys
Zanon, D. V., 1949
Diatomee di Buenos Aires (Argentina)
Atti Accad. Naz. Lincei, Memorie, Cl. Sci. fis., mat. e. nat., ser. 7, 11(3):59-151, 2 pls.

Hantzschia amphioxys
Zanon, V., 1948
Diatomee marini di Sardegna e Pugillo di Alghe Marine della stressa. Boll. Pesca, Piscitutura e Idrobiologia, Anno 24, ns. 3(2): 202-244, 27 figs. on 1 pl.

Hantzschia baltica n.sp.
Simonsen, Reimer, 1960.
Neue Diatomeen aus der Ostsee. II.
Kieler Meeresf., 16(1):126-130.

Hantzschia capitata
Tsumura, K., 1956.
Diatomoj el la cirkaufoso de la restaĵo de la kastele de Odawara. J. Yokohama Municipal Univ., (C-14), No. 47: 23 pp.

Hantzschia elongata
Zanon, D. V., 1949
Diatomee di Buenos Aires (Argentina)
Atti Accad. Naz. Lincei, Memorie, Cl. Sci. fis., mat. e. nat., ser. 7, 11(3):59-151, 2 pls.

Hantzschia hyalina
Zanon, V., 1948
Diatomee marini di Sardegna e Pugillo di Alghe Marine della stressa. Boll. Pesca, Piscitutura e Idrobiologia, Anno 24, ns. 3(2): 202-244, 27 figs. on 1 pl.

Hantzschia marina
Hendy, N. Ingram, 1964

An introductory account of the smaller algae of British coastal waters. V. Bacillariophyceae (Diatoms).
Her Majesty's Stationary Office, 317 pp., 45 pls.

Hantzschia marina
Hendey, N.I., 1951
Littoral diatoms of Chicester Harbour with special reference to fouling. J.Roy. Microscop. Soc. 71(1): 1-86, 18 pls.

Hantzschia marina
Zanon, V., 1948
Diatomee marini di Sardegna e Pugillo di Alghe Marine della stressa. Boll. Pesca, Piscitutura e Idrobiologia, Anno 24, ns. 3(2): 202-244, 27 figs. on 1 pl.

Hantzschia virgata & varieties
Hendy, N. Ingram, 1964

An introductory account of the smaller algae of British coastal waters. V. Bacillariophyceae (Diatoms).
Her Majesty's Stationary Office, 317 pp., 45 pls.

Hantzschia virgata
Palmer, John D., and Frank E. Round, 1967.
Persistent, vertical-migration rhythms in benthic microflora. VI. The tidal and diurnal nature of the rhythm in the diatom Hantzschia virgata. Biol. Bull. mar. biol. Lab., Woods Hole, 132(1):44-55.

Hantschia virgata
Schodduyn, M., 1926
Observations faites dans la baie d'Ambleteuse (Pas de Calais). Bull. Inst. Ocean., Monaco, No. 482: 64 pp.

Hantzschia virgata
Zanon, D. V., 1949
Diatomee di Buenos Aires (Argentina) Atti Accad. Naz. Lincei, Memorie, Cl. Sci. fis., mat. e. nat., ser. 7, 11(3):59-151, 2 pls.

Hemiaulus
Desikachary, T.V., 1956(1957).
Electron microscope studies on diatoms. J. R. Microsc. Soc. (3)76(1/2):9-36.

Hemiaulus ambiguus
Cleve-Euler, A., 1951
Die Diatomeen von Schweden und Finnland. Kungl. Svenska Vetenskaps Akad. Handl., Fjarde Ser. 2(1): 161 pp., 6 pls.

Hemiaulus arcticus
Cleve-Euler, A., 1951
Die Diatomeen von Schweden und Finnland. Kungl. Svenska Vetenskaps Akad. Handl., Fjarde Ser. 2(1): 161 pp., 6 pls.

Hemiaulus chinensis
Pavillard, J., 1905
Recherches sur la flore pelagique (Phytoplankton) de l'Etang de Thau. Theses presentees a la Fac. Sci., Paris, 116 pp., 3 pls.

Hemiaulus danicus
Cleve-Euler, A., 1951
Die Diatomeen von Schweden und Finnland. Kungl. Svenska Vetenskaps Akad. Handl., Fjarde Ser. 2(1): 161 pp., 6 pls.

Hemiaulus dubius
Cleve-Euler, A., 1951
Die Diatomeen von Schweden und Finnland. Kungl. Svenska Vetenskaps Akad. Handl., Fjarde Ser. 2(1): 161 pp., 6 pls.

Hemiaulus excavatus
Cleve-Euler, A., 1951
Die Diatomeen von Schweden und Finnland. Kungl. Svenska Vetenskaps Akad. Handl., Fjarde Ser. 2(1): 161 pp., 6 pls.

Hemiaulus exsculptus
Cleve-Euler, A., 1951
Die Diatomeen von Schweden und Finnland. Kungl. Svenska Vetenskaps Akad. Handl., Fjarde Ser. 2(1): 161 pp., 6 pls.

Hemiaulus februatus
Cleve-Euler, A., 1951
Die Diatomeen von Schweden und Finnland. Kungl. Svenska Vetenskaps Akad. Handl., Fjarde Ser. 2(1): 161 pp., 6 pls.

Hemiaulus glacialis n.sp.
Castracane degli Antelminelli, F., 1886
1. Report on the Diatomaceae collected by H.M.S. Challenger during the years 1873-1876. Rept. Sci. Results, H.M.S. Challenger, Botany Vol. II, 178 pp., 30 pls.

Hemiaulus hauckii
Allen, W. E., 1938
The Templeton Crocker Expedition to the Gulf of California in 1935 - The Phytoplankton. Amer. Microsc. Soc., Trans. 57:328-335.

Hemiaulus hauckii
Allen, W.E., 1937
Plankton diatoms of the Gulf of California obtained by the G. Allan Hancock Expedition of 1936. The Hancock Pacific Expeditions, Univ. So. Calif. Publ. 3:47-59, 1 fig.

Hemiaulus hauckii
Allen, W. E., 1936
Occurrence of marine plankton diatoms in a ten-year series of daily catches in southern California. Am. Jour. Bot. 23(1):60-63.

Hemiaulus hauckii
Allen, W.E., 1928.
Catches of marine diatoms and dinoflagellates taken by boat in Southern California waters in 1926. Bull. S.I.O., tech. ser., 1:201-246, 6 text figs.

Hemiaulus hauckii
Boden, B.P., 1950
Some marine plankton diatoms from the west coast of South Africa. Trans. R.Soc. S. Africa. 32:321-434, 100 text figs.

Hemiaulus hauckii
Cupp, Easter E., 1943
Marine plankton diatoms of the west coast of North America. Bull. S.I.O. 5(1):1-238, 5 pls., 168 text figs.

Hemiaulus hauckii
Cupp, E.E., 1934
Analysis of marine diatom collections taken from the Canal Zone to California during March, 1933. Trans. Am. Micros. Soc. LIII (1):22-29, 1 map.

Hemiaulus hauckii
Cupp, E.E. and Allen, W.E., 1938
Plankton diatoms of the Gulf of California obtained by Allan Hancock Pacific Expedition of 1937. The Hancock Pacific Expeditions, The Univ. So. Calif. Publ. 3:61-74, 1 map, pls. 4-15.

Hemiaulus hauckii
Delegazione Italiana della Commissione Internazionale per l'Esplorazione Scientifica del Mediterraneo, 1941
Note sul plancton della Laguna veneta. [Memoria CCLXXIX], Arch. di Ocean. e Limn. Anno I, Fasc. I, 1941 XIX: 31-57 pp.

Hemiaulus Haucki
Ercegovic, A., 1936
Etudes qualitative et quantitatives du phytoplancton dans les eaux cotières de l'Adriatique oriental moyen au cours de l'année 1934. Acta Adriatica 1(9):1-126

Hemiaulus Hauckii
Forti, A., 1922
Ricerche sulla flora pelagica (fitoplancton) di Quarto dei Mille. Mem. R. Com. Talass. Ital. 97:248 pp., 13 pls.

Hemiaulus hauchii
Gilbert, J.Y., and W.E. Allen, 1943
The phytoplankton of the Gulf of California obtained by the "E.W. Scripps" in 1939 and 1940. J. Mar. Res. V(2):89-110, figs. 30-31.

Hemiaulus Hauckii
Gran, H.H., 1908
Diatomeen. Nordisches Plankton, Botanischer Teil pp. XIX 1-XIX 146; 178 text figs.

Hemiaulus hauckii
Hendy, N. Ingram, 1964
An introductory account of the smaller algae of British coastal waters. V. Bacillariophyceae (Diatoms). Her Majesty's Stationary Office, 317 pp., 45 pls.

Hemiaulus Hauckii
Hendey, N.I., 1937
The plankton diatoms of the southern seas. Discovery Repts. 16:151-364, pls. 6-13.

Hemiaulus hauckii
Kokubo, S., 1952
Results of the observations on the plankton and oceanography of Mutsu Bay during 1950, reference being made also to the period 1946-1950. Bull Mar.Biol.Sta., Asamushi 5(1/4):1-54, 3 tables, (fold-in), 1 fold-in.

Hemiaulus Hauckii
Lillick, L.C., 1940
Phytoplankton and planktonic protozoa of the offshore waters of the Gulf of Maine. Pt.II. Qualitative Composition of the Planktonic Flora. Trans. Am. Phil. Soc., n.s., 31(3):193-237, 13 text figs.

Hemiaulus Hauckii
Lillick, L.C., 1937
Seasonal studies of the phytoplankton off Woods Hole, Massachusetts. Biol. Bull. LXXIII (3):488-503, 3 text figs.

Hemiaulus Hauckii
Mangin, M. L., 1912
Phytoplancton de la croisière du "René" dans l'Atlantique (Septembre 1908). Ann. Inst. Ocean., n.s., 4(1):1-66, 2 pls., 41 text figs., 2 tables.

Hemiaulus Hauckii
Margalef, R., 1949
Fitoplancton nerítico de la Costa Brava en 1947-48. Publ. Inst. Biol. Aplicada, 5:41-51, 3 text figs.

Hemiaulus hauckii
Marshall, S. M., 1933
The production of microplankton in the Great Barrier Reef Region. Brit. Mus. (N.H.) Great Barrier Reef Exped. 1928-29, Sci. Repts. II(5):111-157, 14 text figs.

Hemiaulus Haucki
Massutí Algamora, M., 1949
Estudio de diez y seis muestras de plancton del Golfo de Nápoles. Publ. Inst. Biol. Appl. 5:85-94, 1 fold-in table.

Hemiaulus hauckii
Pavillard, J., 1925
Bacillariales. Rept. on the Danish Oceangr. Exped., 1908-10 to the Mediterranean and adj. seas. Vol.II., Biol. J4:72 pp., 116 text figs.

Hemiaulus Hauckii
Pavillard, J., 1905
Recherches sur la flore pelagique (Phytoplankton) de l'Etang de Thau. Theses presentees a la Fac. Sci., Paris, 116 pp., 3 pls.

Hemiaulus Hauckii
Rampi, L., 1942
Ricerche sul fitoplancton del Mare Ligure 6. Le diatomee delle acque di Sanremo. Nuovo Giornale Botanico Italiano, N.S., 49:252-268.

Hemiaulus Hauckii
Schröder, B., 1900
Phytoplankton des Golfes von Neapel nebst vergleichenden Ausblicken auf das atlantischen Ozean. Mitt. Zool. Stat. Neapel, 14:1-38.

Hemiaulus haucki (figs.)
Sousa e Silva, E., 1949
Diatomaceas e Dinoflagelados de Baia de Cascais. Portugaliae Acta Biol., Volume: Julio Henriques, Ser. B: 300-383, 9 pls, 2 fold-in tables.

Hemiaulus hauckii
Uyeno, Fukuzo, 1961
Oceanographical and ecological studies on primary production of the sea, with special references to relationship between diatom production and temperature and chlorinity of water.
Rept. Fac. Fish., Pref. Univ., Mie, 4(1): 1-64.

Hemiaulus heibergii
Cupp, E.E., 1934
Analysis of marine diatom collections taken from the Canal Zone to California during March, 1933. Trans. Am. Micros. Soc. LIII (1):22-29, 1 map.

Hemiaulus Heirbergii
Ghazzawi, F.M., 1939
Plankton of the Egyptian waters. A study of the Suez Canal Plankton. (A) Phytoplankton. Preliminary Report 83 pp. Notes and Memoires, Min. Commerce-Industry, Egypt, Hydrobiol. & Fish. 65 figs.

Hemiaulus hostilis
Cleve-Euler, A., 1951
Die Diatomeen von Schweden und Finnland. Kungl. Svenska Vetenskaps Akad. Handl., Fjärde Ser. 2(1): 161 pp., 6 pls.

Hemiaulus hyperboreus
Cleve-Euler, A., 1951
Die Diatomeen von Schweden und Finnland. Kungl. Svenska Vetenskaps Akad. Handl., Fjärde Ser. 2(1): 161 pp., 6 pls.

Hemiaulus indicus
Allen, W.E., and E.E. Cupp, 1935
Plankton diatoms of the Java Sea. Annales du Jardin Botanique de Buitenzorg XLIV (2):101-174, figs.1-127.
(drawings of all species mentioned)

Hemiaulus indicus
Dangeard, P., 1927
Phytoplankton de la croisière du "Sylvana". Ann. Inst. Ocean., Monaco, n.s., 4(8):286-401, 54 text figs. (Feirer-Juin 1913).

Hemiaulus indicus
Marshall, S. M., 1933
The production of microplankton in the Great Barrier Reef Region. Brit. Mus. (N.H.) Great Barrier Reef Exped. 1928-29, Sci. Repts. II(5):111-157, 14 text figs.

Hemiaulus kittoni
Cleve-Euler, A., 1951
Die Diatomeen von Schweden und Finnland. Kungl. Svenska Vetenskaps Akad. Handl., Fjärde Ser. 2(1): 161 pp., 6 pls.

Hemiaulus membranaceus
Allen, W.E., and E.E. Cupp, 1935.
Plankton diatoms of the Java Sea. Annales du Jardin Botanique de Buitenzorg XLIV (2):101-174, figs.1-127.
(drawings of all species mentioned)

Hemiaulus membransceus
Cupp, Easter E., 1943
Marine plankton diatoms of the west coast of North America. Bull. S.I.O. 5(1):1-238, 5 pls., 168 text figs.

Hemiaulus membranaceus
Cupp, E.E., 1934
Analysis of marine diatom collections taken from the Canal Zone to California during March, 1933. Trans. Am. Micros. Soc. LIII (1):22-29, 1 map.

Hemiaulus mitra
Cleve-Euler, A., 1951
Die Diatomeen von Schweden und Finnland. Kungl. Svenska Vetenskaps Akad. Handl., Fjärde Ser. 2(1): 161 pp., 6 pls.

Hemiaulus pileolus
Cleve-Euler, A., 1951
Die Diatomeen von Schweden und Finnland. Kungl. Svenska Vetenskaps Akad. Handl., Fjärde Ser. 2(1): 161 pp., 6 pls.

Hemiaulus polycistinorum
Cleve-Euler, A., 1951
Die Diatomeen von Schweden und Finnland. Kungl. Svenska Vetenskaps Akad. Handl., Fjärde Ser. 2(1): 161 pp., 6 pls.

Hemiaulus polycistinorum
Mann, A., 1907
Report on the diatoms of the Albatross voyages in the Pacific Ocean, 1888-1904. Contrib. U. S. Nat. Herb. 10(5):221-419, Pls. XLIV-LIV.

Hemiaulus polycistinorum
Mann, A., 1893
List of Diatomaceae from a deep-sea dredging in the Atlantic Ocean off Delaware Bay by the U. S. Fish Commission Steamer Albatross. Proc. U. S. Nat. Mus. 16:303-312.

Hemiaulus polymorphus
Cleve-Euler, A., 1951
Die Diatomeen von Schweden und Finnland. Kungl. Svenska Vetenskaps Akad. Handl., Fjärde Ser. 2(1): 161 pp., 6 pls.

Hemiaulus proteus
Cleve-Euler, A., 1951
Die Diatomeen von Schweden und Finnland. Kungl. Svenska Vetenskaps Akad. Handl., Fjärde Ser. 2(1): 161 pp., 6 pls.

Hemiaulus pungens
Cleve-Euler, A., 1951
Die Diatomeen von Schweden und Finnland. Kungl. Svenska Vetenskaps Akad. Handl., Fjärde Ser. 2(1): 161 pp., 6 pls.

Hemiaulus regina
Cleve-Euler, A., 1951
Die Diatomeen von Schweden und Finnland. Kungl. Svenska Vetenskaps Akad. Handl., Fjärde Ser. 2(1): 161 pp., 6 pls.

Hemiaulus rusticus, n.sp.
Frenguelli, Joaquin, y Hector A. Orlando, 1958.
Diatomeas y silicoflagelados del sector Antartico Sudamericano.
Inst. Antartico Argentino, Publ., No. 5:191 pp.

Hemiaulus sinensis
Allen, W.E., and E.E. Cupp, 1935
Plankton diatoms of the Java Sea. Annales du Jardin Botanique de Buitenzorg XLIV (2):101-174, figs.1-127.
(drawings of all species mentioned)

Hemiaulus sinensis
Cupp, Easter E., 1943
Marine plankton diatoms of the west coast of North America. Bull. S.I.O. 5(1):1-238, 5 pls., 168 text figs.

Hemiaulus sinensis
Delsman, H. C., 1939.
Preliminary plankton investigations in the Java Sea. Treubia, 17:139-181, 8 maps, 41 figs.

Hemiaulus sinensis
Ercegovic, A., 1936
Etudes qualitative et quantitatives du phytoplancton dans les eaux cotières de l'Adriatique oriental moyen au cours de l'année 1934. Acta Adriatica 1(9):1-126

Hemiaulus sinensis
Müller Melchers, F.C., 1953.
New and little known diatoms from Uruguay and the South Atlantic coast.
Com. Bot., Mus. Hist. Nat., Montevideo, 3(30): 1-11, 8 pls.

Hemiaulus sinensis
Pavillard, J., 1925
Bacillariales. Rept. on the Danish Oceangr. Exped., 1908-10 to the Mediterranean and adj. seas. Vol.II., Biol. J4:72 pp., 116 text figs.

Hemiaulus sinensis
Rampi, L., 1942
Ricerche sul fitoplancton del Mare Ligure 6. Le diatomee delle acque di Sanremo. Nuovo Giornale Botanico Italiano, N.S., 49:252-268.

Hemidiscus
Desikachary, T.V., 1956(1957).
Electron microscope studies on diatoms. J. R. Microsc. Soc. (3)76(1/2):9-36

Hemidiscus cuneiformis
Boden, B.P., 1950
Some marine plankton diatoms from the west coast of South Africa. Trans. R.Soc. S. Africa. 32:321-434, 100 text figs.

Hemidiscus cuneiformis
Hendy, N. Ingram, 1964
An introductory account of the smaller algae of British coastal waters. V. Bacillariophyceae (Diatoms).
Her Majesty's Stationary Office, 317 pp., 45 pls.

Hemidiscus cuneiformis
Mann, A., 1907
Report on the diatoms of the Albatross voyages in the Pacific Ocean, 1888-1904. Contrib. U. S. Nat. Herb. 10(5):221-419, Pls. XLIV-LIV.

Hemidiscus cuneiformis
Rampi, L., 1942
Ricerche sul fitoplancton del Mare Ligure 6. Le diatomee delle acque di Sanremo. Nuovo Giornale Botanico Italiano, N.S., 49:252-268.

Hemidiscus cuneiformis
Rampi, L., 1940
Diatomee del Mare Adriatico. Nuovo Giornale Botanico Italiano, n.s., 47:559-608.

Hemidiscus cuneiformis
Simonsen, Reimer 1972.
Über die Diatomeengattung Hemidiscus Wallich und andere Angehörige der sogenannten "Hemidiscaceae".
Veröff. Inst. Meeresforsch. Bremerh. 13(2): 265-273.

Hemidiscus cuneiformis
Zanon, V., 1948
Diatomee marini di Sardegna e Pugillo di Alghe Marine della stressa. Boll. Pesca, Piscicoltura e Idrobiologia, Anno 24, ns. 3(2): 202-244, 27 figs. on 1 pl.

Hemidiscus cuneiformis ventricosa
Cupp, Easter E., 1943
Marine plankton diatoms of the west coast of North America. Bull. S.I.O. 5(1):1-238, 5 pls., 168 text figs.

Hemidiscus hardmanianus
Blasco, Dolores, 1970.
Ultraestructura de Hemidiscus hardmanianus Grev. y consideraciones sobre la filogenia de este género.
Inv. pesq. Barcelona, 34(2): 229-236

Hemidiscus Hardmanianus
Allen, W.E., and E.E. Cupp, 1935
Plankton diatoms of the Java Sea. Annales du Jardin Botanique de Buitenzorg XLIV (2):101-174, figs.1-127.
(drawings of all species mentioned)

Hemidiscus hardmannianus

Chua Thia Eng, 1970.
Notes on the abundance of the diatom
Hemidiscus hardmannianus (Grev.-Mann)
in the Singapore Straits.
Hydrobiologia, 36(1): 61-64.

Hemidiscus hardmannianus

Delsman, H. C., 1939.
Preliminary plankton investigations in the Java Sea. Treubia, 17:139-181, 8 maps, 41 figs.

Hemidiscus hardmannianus

Eng, Chua Thia, 1970.
Notes on the abundance of the diatom
Hemidiscus hardmannianus (Grev.-Mann)
in the Singapore Straits
Hydrobiologia 36(1):61-64

Hemidiscus hardmanianus

Eskinazi Enide e Shigekatsv Satô (1963-1964) 1966.
Contribuição ao estudo das diatomaceas da Praia de Piedade.

Trabhs Inst. Oceanogr., Univ. Recife, 5 (5/6): 73-114.

Hemidiscus hardmanianus

Kokubo, S., 1952
Results of the observations on the plankton and oceanography of Mutsu Bay during 1950, reference being made also to the period 1946-1950. Bull Mar.Biol.Sta., Asamushi 5(1/4): 1-54, 3 tables,(fold-in), 1 fold-in.

Hemidiscus Kanayanus n.sp.

Simonsen, Reimer 1972.
Über die Diatomeengattung Hemidiscus
Wallich und andere Angehörige der
sogenannten "Hemidiscaceae".
Veröff. Inst. Meeresforsch. Bremerh.
13(2): 265-273.

Hemidiscus Karsteni n.sp.

Jouse, A.P., G.S. Koroleva, G.A. Nagaeva, 1962
Diatoms in the surface layer of sediment in the Indian sector of the Antarctic. Investigations of marine bottom sediments. (In Russian; English summary).
Trudy Inst. Okeanol., Akad. Nauk, SSSR, 61: 19-92.

Hemidiscus rectus

Mann, A., 1907
Report on the diatoms of the Albatross voyages in the Pacific Ocean, 1888-1904. Contrib. U. S. Nat. Herb. 10(5):221-419, Pls. XLIV-LIV.

Hemidiscus suecicus

Cleve-Euler, A., 1951
Die Diatomeen von Schweden und Finnland. Kungl. Svenska Vetenskaps Akad. Handl., Fjärde Ser. 2(1): 161 pp., 6 pls.

Hemidiscus ventricosus

Mann, A., 1907
Report on the diatoms of the Albatross voyages in the Pacific Ocean, 1888-1904. Contrib. U. S. Nat. Herb. 10(5):221-419, Pls. XLIV-LIV.

Hemiptychus ehrenbergii

Mann, A., 1907
Report on the diatoms of the Albatross voyages in the Pacific Ocean, 1888-1904. Contrib. U. S. Nat. Herb. 10(5):221-419, Pls. XLIV-LIV.

Hemiptychus indicus

Mann, A., 1907
Report on the diatoms of the Albatross voyages in the Pacific Ocean, 1888-1904. Contrib. U. S. Nat. Herb. 10(5):221-419, Pls. XLIV-LIV.

Hemiptychus ornatus

Mann, A., 1907
Report on the diatoms of the Albatross voyages in the Pacific Ocean, 1888-1904. Contrib. U. S. Nat. Herb. 10(5):221-419, Pls. XLIV-LIV.

Hesslandia oceanica

Cleve-Euler, A., 1951
Die Diatomeen von Schweden und Finnland. Kungl. Svenska Vetenskaps Akad. Handl., Fjärde Ser. 2(1): 161 pp., 6 pls.

Heterodictyon jeffreysianum n.sp.

Castracane degli Antelminelli, F., 1886
1. Report on the Diatomaceae collected by H.M.S. Challenger during the years 1873-1876. Rept. Sci. Results, H.M.S. Challenger, Botany Vol. II, 178 pp., 30 pls.

Homoeocladia delicatissima

Meunier, A., 1915
Microplancton de la Mer Flamande. 2. Diatomées (excepté le genre Chaetoceros). Mem. Mus. Roy. Hist. Nat., Belgique, 7(3):1-118, Pls. VIII-XIV.

Hustedtiella baltica n.gen., n.sp.

Simonsen, Reimer, 1960.
Neue Diatomeen aus der Ostsee. II.
Kieler Meeref., 16(1):126-130.

Huttonia reichardtii

Hendy, N. Ingram, 1964
An introductory account of the smaller algae of British coastal waters. V. Bacillariophyceae (Diatoms).
Her Majesty's Stationary Office, 317 pp., 45 pls.

Huttonia Reichardtii

Zanon, V., 1948
Diatomee marini di Sardegna e Pugillo di Alghe Marine della stressa. Boll. Pesca, Piscitutura e Idrobiologia, Anno 24, ns. 3(2): 202-244, 27 figs. on 1 pl.

Huttonia reversa

Gosden, R., 1966.
Note on Biddulphia reversa Grove and Sturt.
J. Quekett Microsc. Club, 30(5):117.

Hyalodiscus

Brunel, J., 1962
Le phytoplankton de la Baie de Chaleurs.
Inst. Botan., Univ. Montréal, Contrib. No. 77: 365 pp., 66 pls.

Hyalodiscus ambiguus

Takano, Hideaki, 1962
Notes on epiphytic diatoms upon sea-weeds from Japan.
J. Oceanogr. Soc., Japan, 18(1):29-33.

Hyalodiscus chromatoaster

Hendey, N.I., 1937
The plankton diatoms of the southern seas.
Discovery Repts. 16:151-364, pls.6-13.

Hyalodiscus dubiosus

Manguin, E., 1954
Diatomees marines provenant de l'ile Heard (Australian National Antarctic Expedition). Rev. Algol., n.s., 1: 14-24.

Hyalodiscus kerguelensis

Hendey, N.I., 1937
The plankton diatoms of the southern seas.
Discovery Repts. 16:151-364, pls.6-13.

Hyalodiscus laevis

Hendey, N-Ingram, 1958 [1957(Publ. 1958)]
Marine diatoms from some West African Ports. J. R. Microsc. Soc. (3) 77(1/2): 28-85.

Hyalodiscus laevis

Zanon, D. V., 1949
Diatomee di Buenos Aires (Argentina)
Atti Accad. Naz. Lincei, Memorie, Cl. Sci. fis., mat. e. nat., ser. 7, 11(3):59-151, 2 pls.

Hyalodiscus (Pyxidicula) radiatus

Castracane degli Antelminelli, F., 1886
1. Report on the Diatomaceae collected by H.M.S. Challenger during the years 1873-1876. Rept. Sci. Results, H.M.S. Challenger, Botany Vol. II, 178 pp., 30 pls.

Hyalodiscus radiatus

Frenguelli, Joaquin, and Hector Antonio Orlando, 1959.

Operacion MERLUZA. Diatomeas y silicoflagelados del plancton del "VI Crucero". Servicio Hidrogr. Naval, Argentina, Publ. No. H. 619: 5-62.

Hyalodiscus radiatus

Manguin, E., 1954
Diatomees marines provenant de l'ile Heard (Australian National Antarctic Expedition). Rev. Algol., n.s., 1: 14-24.

Hyalodiscus radiatus

de Sousa e Silva E., 1956.
Contribution à l'étude du microplancton de Dakar et des regions maritimes voisines.
Bull. I.F.A.N., 8(2):335-371, 7 pls.

Hyalodiscus radiatus

Zanon, D. V., 1949
Diatomee di Buenos Aires (Argentina)
Atti Accad. Naz. Lincei, Memorie, Cl. Sci. fis., mat. e. nat., ser. 7, 11(3):59-151, 2 pls.

Hyalodiscus scoticus

Cleve-Euler, A., 1951
Die Diatomeen von Schweden und Finnland. Kungl. Svenska Vetenskaps Akad. Handl., Fjärde Ser. 2(1): 161 pp., 6 pls.

Hyalodiscus scoticus

Hendy, N. Ingram, 1964
An introductory account of the smaller algae of British coastal waters. V. Bacillariophyceae (Diatoms).
Her Majesty's Stationary Office, 317 pp., 45 pls.

Hyalodiscus scoticus

Hustedt, F. and A.A. Aleem,1951
Littoral diatoms from the Salstone near Plymouth. JMBA 30(1): 177.196.

Hyalodiscus scoticus

Jørgensen, E., 1905
B.Protistplankton and the diatoms in bottom samples. Hydrographical and biological investigations in Norwegian fjords. Bergens Mus. Skr, 7: 49-225.

Hyalodiscus scoticus

Jorgensen, E., 1900
Protophyten und Protozoën im Plankton aus der Norwegischen Westkerste. Bergens Mus. Aarb. 1899(6): 95 pp., 5 pls., 83 tables.

Hyalodiscus scoticus
Misra, J.N., 1956.
A systematic account of some littoral marine diatoms from the west coast of India.
J. Bombay Nat. Hist. Soc., 53(4):537-568.

Hyalodiscus scoticus
Takano, Hideaki, 1962
Notes on epiphytic diatoms upon sea-weeds from Japan.
J. Oceanogr. Soc., Japan, 18(1):29-33.

Hyalodiscus scoticus
von Sydow, Burkard, und Robert Christenhuss 1972.
Rasterelektronmikroskopische Untersuchungen der Hohlräume in der Schalenwand einiger centrischer Kieselalgen.
Arch. Protistenk. 114 (3): 256-271.

Hyalodiscus scoticus
Zanon, D. V., 1949
Diatomee di Buenos Aires (Argentina)
Atti Accad. Naz. Lincei, Memorie, Cl. Sci. fis., mat. e. nat., ser. 7, 11(3):59-151, 2 pls.

Hyalodiscus? spiniferus n.sp.
Manguin, E., 1957.
Premier inventaire des diatomées de la Terre Adélie Antarctique. Espèces nouvelles.
Rev. Algologique, n.s., 3(3):111-134.

Hyalodiscus stelliger
Cleve-Euler, A., 1951
Die Diatomeen von Schweden und Finnland. Kungl. Svenska Vetenskaps Akad. Handl., Fjärde Ser. 2(1): 161 pp., 6 pls.

Hyalodiscus stelliger
Dangeard, P., 1927
Phytoplankton de la croisière du "Sylvana". Ann. Inst. Ocean., Monaco, n.s., 4(8):286-401, 54 text figs. (Feirer-Juin 1913).

Hyalodiscus stelliger
Ghazzawi, F.M., 1939
Plankton of the Egyptian waters. A study of the Suez Canal Plankton. (A) Phytoplankton. Preliminary Report 83 pp. Notes and Memoires, Min. Commerce-Industry, Egypt, Hydrobiol. & Fish. 65 figs.

Hyalodiscus stelliger
Gran, H.H., 1908
Diatomeen. Nordisches Plankton, Botanischer Teil pp. XIX 1-XIX 146; 178 text figs.

Hyalodiscus stelliger
Hendey, N.I., 1937
The plankton diatoms of the southern seas.
Discovery Repts. 16:151-364, pls.6-13.

Hyalodiscus stelliger
Jørgensen, E., 1905
B.Protistplankton and the diatoms in bottom samples. Hydrographical and biological investigations in Norwegian fjords. Bergens Mus. Skr. 7: 49-225.

Hyalodiscus selliger
Jorgensen, E., 1900
Protophyten und Protozoën im Plankton aus der Norwegischen Westkerste. Bergens Mus. Aarb. 1899(6): 95 pp., 5 pls., 83 tables.

Hyalodiscus stelliger
Lafon, M., M. Durchon and Y. Saudray, 1955.
Recherches sur les cycles saisonnières du plankton. Ann. Inst. Océan., 31(3):125-230.

Hyalodiscus stelliger
Mangin, M. L., 1912
Phytoplancton de la croisière du "René" dans l'Atlantique (Septembre 1908). Ann. Inst. Ocean., n.s., 4(1):1-66, 2 pls., 41 text figs., 2 tables.

Hyalodiscus stelliger
Meunier, A., 1915
Microplancton de la Mer Flamande. 2. Diatomées (excepté le genre Chaetoceros). Mem. Mus. Roy. Hist. Nat., Belgique, 7(3):1-118, Pls. VIII-XIV.

Hyalodiscus stelliger
Pavillard, J., 1925
Bacillariales. Rept. on the Danish Oceangr. Exped., 1908-10 to the Mediterranean and adj. seas. Vol.II., Biol. J4:72 pp., 116 text figs.

Hyalodiscus stelliger
Schodduyn, M., 1926
Observations faites dans la baie d'Ambleteuse (Pas de Calais). Bull. Inst. Ocean., Monaco, No. 482: 64 pp.

Hyalodiscus stelliger
Sousa e Silva, E., 1949
Diatomaceas e Dinoflagelados de Baia de Cascais. Portugaliae Acta Biol., Volume: Julio Henriques, Ser. B: 300-383, 9 pls, 2 fold-in tables.

Hyalodiscus stelliger
Sousa a Silva, E., and J. Dos Santos-Pinto, 1948
O Plancton da Baia de S. Martinho do Porto.
1. Diatomaceas e Dinoflagelados. Bol. Soc. Portuguese de Ciencias Naturais, 16(2):134-187, 6 pls. (Trav. Sta. Biol. Mar. de Lisbonne No. 52).

Hyalodiscus subtilis
Bigelow, H.B., and M. Leslie, 1930
Reconnaissance of the waters and plankton of Monterey Bay, July 1928.
Bull. M.C.Z., 70(5):429-481, 43 text figs.

Hyalodiscus subtilis
Cleve-Euler, A., 1951
Die Diatomeen von Schweden und Finnland. Kungl. Svenska Vetenskaps Akad. Handl., Fjärde Ser. 2(1): 161 pp., 6 pls.

Hyalodiscus subtilis
Gran, H. H. and E. C. Angst, 1931
Plankton diatoms of Puget Sound. Publ. Puget Sound Biol. Sta. 7:417-519, 95 text figs.

Hyalodiscus subtilis
Hendy, N. Ingram, 1964
An introductory account of the smaller algae of British coastal waters. V. Bacillariophyceae (Diatoms).
Her Majesty's Stationary Office, 317 pp., 45 pls.

Hyalodiscus subtilis
Rampi, L., 1940
Diatomee del Mare Adriatico. Nuovo Giornale Botanico Italiano, n.s., 47:559-608.

Hyalodiscus subtilis
Jørgensen, E., 1905
B.Protistplankton and the diatoms in bottom samples. Hydrographical and biological investigations in Norwegian fjords. Bergens Mus. Skr. 7: 49-225.

Hyalodiscus subtilis
Manguin, E., 1954
Diatomees marines provenant de l'ile Heard (Australian National Antarctic Expedition). Rev. Algol., n.s., 1: 14-24.

Hyalodiscus subtilis
Meunier, A., 1915
Microplancton de la Mer Flamande. 2. Diatomées (excepté le genre Chaetoceros). Mem. Mus. Roy. Hist. Nat., Belgique, 7(3):1-118, Pls. VIII-XIV.

Hyalodiscus subtilis
Misra, J.N., 1956
A systematic account of some littoral marine diatoms from the west coast of India.
J. Bombay Nat. Hist. Soc., 53(4):537-568.

Hyalodiscus subtilis
Zanon, D. V., 1949
Diatomee di Buenos Aires (Argentina)
Atti Accad. Naz. Lincei, Memorie, Cl. Sci. fis., mat. e. nat., ser. 7, 11(3):59-151, 2 pls.

Hyalodiscus subtilis japonica
Castracane degli Antelminelli, F., 1886 n.var.
1. Report on the Diatomaceae collected by H.M.S. Challenger during the years 1873-1876. Rept. Sci. Results, H.M.S. Challenger, Botany Vol. II, 178 pp., 30 pls.

Hydrosera
Desikachary, T.V., 1956(1957).
Electron microscope studies on diatoms.
J.R. Microsc. Soc. (3)76(1/2):9-36.

Isthmia
Desikachary, T.V., 1956(1957).
Electron microscope studies on diatoms.
J.R. Microsc. Soc. (3)76(1/2):9-36.

Isthmia sp.
Wheeler, J.E.G., 1939
Plankton investigations. Bermuda Biological Station. Second Report. October 1939. 7 pp. (typed), 5 figs. Plymouth, Oct. 23, 1939.

Isthmia enervis
Cleve-Euler, A., 1951
Die Diatomeen von Schweden und Finnland. Kungl. Svenska Vetenskaps Akad. Handl., Fjärde Ser. 2(1): 161 pp., 6 pls.

Istmia enervis
Eskinazi Enide e Shigekatsv Satô (1963-1964) 1966.
Contribuição ao estudo das diatomaceas da Praia de Piedade.
Trabhs Inst. Oceanogr., Univ. Recife, 5 (5/6): 73-114.

Isthmia enervis
Hendy, N. Ingram, 1964
An introductory account of the smaller algae of British coastal waters. V. Bacillariophyceae (Diatoms).
Her Majesty's Stationary Office, 317 pp., 45 pls.

Isthmia enervis
Manguin, E., 1954
Diatomees marines provenant de l'ile Heard (Australian National Antarctic Expedition). Rev. Algol., n.s., 1: 14-24.

Isthmia enervis
Sousa a Silva, E., and J. Dos Santos-Pinto, 1948
O Plancton da Baia de S. Martinho do Porto.
1. Diatomaceas e Dinoflagelados. Bol. Soc. Portuguese de Ciencias Naturais, 16(2):134-187, 6 pls. (Trav. Sta. Biol. Mar. de Lisbonne No. 52).

Isthmia enervis japonica
Castracane degli Antelminelli, F., 1886 n.var.
1. Report on the Diatomaceae collected by H.M.S. Challenger during the years 1873-1876. Rept. Sci. Results, H.M.S. Challenger, Botany Vol. II, 178 pp., 30 pls.

Isthmia Lindigiana
Grunow, A., 1877
1. New Diatoms from Honduras. Monthly Micros. Jour., 18:165-186, pls. CXCIII-CXCVI.

Isthmia nervosa
Bigelow, H.B., and M. Leslie, 1930
Reconnaissance of the waters and plankton of Monterey Bay, July 1928. Bull. M.C.Z., 70(5):429-481, 43 text figs.

Isthmia nervosa
Cleve-Euler, A., 1951
Die Diatomeen von Schweden und Finnland. Kungl. Svenska Vetenskaps Akad. Handl., Fjärde Ser. 2(1): 161 pp., 6 pls.

Isthmia nervosa
Cupp, Easter E., 1943
Marine plankton diatoms of the west coast of North America. Bull. S.I.O. 5(1):1-238, 5 pls., 168 text figs.

Isthmia nervosa
Hendy, N. Ingram, 1964
An introductory account of the smaller algae of British coastal waters. V. Bacillariophyceae (Diatoms). Her Majesty's Stationary Office, 317 pp., 45 pls.

Isthmia nervosa
Schodduyn, M., 1926
Observations faites dans la baie d'Ambleteuse (Pas de Calais). Bull. Inst. Ocean., Monaco, No. 482: 64 pp.

Isthmia obliquata
Mann, A., 1907
Report on the diatoms of the Albatross voyages in the Pacific Ocean, 1888-1904. Contrib. U. S. Nat. Herb. 10(5):221-419, Pls. XLIV-LIV.

Isthmia sardoa
Zanon, V., 1948
Diatomee marini di Sardegna e Pugillo di Alghe Marine della stressa. Boll. Pesca, Piscitutura e Idrobiologia, Anno 24, ns. 3(2):202-244, 27 figs. on 1 pl.

Lauderia
Brunel, J., 1962
Le phytoplancton de la Baie de Chaleurs. Inst. Botan., Univ. Montréal, Contrib. No. 77: 365 pp., 66 pls.

Lauderia annulata
Allen, W.E., and E.E. Cupp, 1935
Plankton diatoms of the Java Sea. Annales du Jardin Botanique de Buitenzorg XLIV (2):101-174, figs.1-127.
(drawings of all species mentioned)

Lauderia annulata
Bigelow, H.B., and M. Leslie, 1930
Reconnaissance of the waters and plankton of Monterey Bay, July 1928. Bull. M.C.Z., 70(5):429-481, 43 text figs.

Lauderia annulata
Castracane degli Antelminelli, F., 1886
1. Report on the Diatomaceae collected by H.M.S. Challenger during the years 1873-1876. Rept. Sci. Results, H.M.S. Challenger, Botany Vol. II, 178 pp., 30 pls.

Lauderia annulata
Frenguelli, Joaquin, and Hector Antonio Orlando, 1959.
Operacion MERLUZA. Diatomeas y silicoflagelados del plancton del "VI Crucero". Servicio Hidrogr. Naval., Argentina, Publ. No. H. 619: 5-62.

Lauderia annulata
Gran, H. H. and E. C. Angst, 1931
Plankton diatoms of Puget Sound. Publ. Puget Sound Biol. Sta. 7:417-519, 95 text figs.

Lauderia annulata
Jorgensen, E., 1900
Protophyten und Protozoën im Plankton aus der Norwegischen Westkerste. Bergens Mus. Aarb. 1899(6): 95 pp., 5 pls., 83 tables.

Lauderia annulata
Mangin, M. L., 1912
Phytoplancton de la croisière du "René" dans l'Atlantique (Septembre 1908). Ann. Inst. Ocean., n.s., 4(1):1-66, 2 pls., 41 text figs., 2 tables.

Lauderia annulata
Pavillard, J., 1905
Recherches sur la flore pelagique (Phytoplankton) de l'Etang de Thau. Theses presentees a la Fac. Sci., Paris, 116 pp., 3 pls.

Lauderia annulata
Sousa a Silva, E., and J. Dos Santos-Pinto, 1948
O Plancton da Baía de S. Martinho do Porto. 1. Diatomaceas e Dinoflagelados. Bol. Soc. Portuguese de Ciencias Naturais, 16(2):134-187, 6 pls. (Trav. Sta. Biol. Mar. de Lisbonne No. 52).

Lauderia borealis
Allen, W. E., 1925
Statistical Studies of surface catches of marine diatoms and dinoflagellates made by the Yacht "Ohio" in tropical waters in 1924. Jan. Trans. Amer. Microscop. Soc.:24-30, 1 fig.

Lauderia borealis
Bigelow, H.B., and M. Leslie, 1930
Reconnaissance of the waters and plankton of Monterey Bay, July 1928. Bull. M.C.Z., 70(5):429-481, 43 text figs.

Lauderia borealis
Braarud, T., 1945
Experimental studies on marine plankton diatoms. Avhandlingar utgitt av Det Norske Videnskaps-Akademi i Oslo. 1. Mat.-Naturv. Klasse 1944. No.10, 1-16, 1 text fig.

Lauderia borealis
Cleve-Euler, A., 1951
Die Diatomeen von Schweden und Finnland. Kungl. Svenska Vetenskaps Akad. Handl., Fjärde Ser. 2(1): 161 pp., 6 pls.

Lauderia borealis
Cupp, Easter E., 1943
Marine plankton diatoms of the west coast of North America. Bull. S.I.O. 5(1):1-238, 5 pls., 168 text figs.

Lauderia borealis
Cupp, E.E., 1934
Analysis of marine diatom collections taken from the Canal Zone to California during March, 1933. Trans. Am. Micros. Soc. LIII (1):22-29, 1 map.

Lauderia borealis
Cupp, E.E. and Allen, W.E., 1938
Plankton diatoms of the Gulf of California obtained by Allan Hancock Pacific Expedition of 1937. The Hancock Pacific Expeditions, the Univ. So. Calif. Publ. 3: 61-74, 1 map, pls.4-15.

Lauderia borealis
Dangeard, P., 1927
Phytoplankton de la croisière du "Sylvana". Ann. Inst. Ocean., Monaco, n.s., 4(8):286-401, 54 text figs. (Feirer-Juin 1913).

Lauderia borealis
Gilbert, J.Y., and W.E. Allen, 1943
The phytoplankton of the Gulf of California obtained by the "E.W. Scripps" in 1939 and 1940. J. Mar. Res. V(2):89-110, figs.30-31.

Lauderia borealis
Gilson, H. C., 1937
Chemical and Physical Investigations. The nitrogen cycle. John Murray Exped., 1933-34, Sci. Repts., 2(2):21-81, 16 text figs.

Lauderia borealis
Gran, H.H., 1908
Diatomeen. Nordisches Plankton, Botanischer Teil pp. XIX.1-XIX 146; 178 text figs.

Lauderia borealis
Hasle, Grethe Rytter, 1968.
The valve Processes of the centric diatom genus Thalassiosira. Nytt Mag.Bot. 15(3):193-201.

Lauderia borealis
Hendy, N. Ingram, 1964
An introductory account of the smaller algae of British coastal waters. V. Bacillariophyceae (Diatoms). Her Majesty's Stationary Office, 317 pp., 45 pls.

Lauderia borealis
Hendey, N.I., 1937
The plankton diatoms of the southern seas. Discovery Repts. 16:151-364, pls.6-13.

Lauderia borealis
Kokubo, S., 1952
Results of the observations on the plankton and oceanography of Mutsu Bay during 1950, reference being made also to the period 1946-1950. Bull Mar.Biol.Sta., Asamushi 5(1/4):1-54, 3 tables,(fold-in), 1 fold-in.

Lauderia borealis
Lafon, M., M. Durchon and Y. Saudray, 1955.
Recherches sur les cycles saisonnières du plankton. Ann. Inst. Océan., 31(3):125-230.

Lauderia borealis
Meunier, A., 1915
Microplancton de la Mer Flamande. 2. Diatomées (excepté le genre Chaetoceros). Mem. Mus. Roy. Hist. Nat., Belgique, 7(3):1-118, Pls. VIII-XIV.

Lauderia borealis
Pavillard, J., 1925.
Bacillariales. Rept. on the Danish Oceangr. Exped., 1908-10 to the Mediterranean and adj. seas. Vol.II, Biol. J4:72 pp., 116 text figs.

Lauderia borealis
Rampi, L., 1942
Ricerche sul fitoplancton del Mare Ligure 6. Le diatomee delle acque di Sanremo. Nuovo Giornale Botanico Italiano, N.S., 49:252-268.

Lauderia borealis
Rumkówna, A., 1948
List of the phytoplankton species occurring in the superficial water layers in the Gulf of Gdańsk. Bull. Lab. mar., Gdynia, No. 4: 139-141 with tables in back.

Lauderia borealis
Schodduyn, M., 1926
Observations faites dans la baie d'Ambleteuse (Pas de Calais). Bull. Inst. Ocean., Monaco, No. 482: 64 pp.

Lauderia borealis
Sousa e Silva, E., 1949
Diatomaceas e Dinoflagelados de Baía de Cascais. Portugaliae Acta Biol., Volume: Julio Henriques, Ser. B: 300-383, 9 pls, 2 fold-in tables.

Lauderia borealis
Sousa a Silva, E., and J. Dos Santos-Pinto, 1948
O Plancton da Baía de S. Martinho do Porto. 1. Diatomaceas e Dinoflagelados. Bol. Soc. Portuguese de Ciencias Naturais, 16(2):134-187, 6 pls. (Trav. Sta. Biol. Mar. de Lisbonne No. 52).

Lauderia borealis
Wawrik, F., 1961.
Die horizontale Verteilung der Planktondiatomeen im Golf von Neapel. Int. Revue Ges. Hydrobiol., 46(3):460-479.

Lauderia delicatula
Bigelow, H.B., and M. Leslie, 1930
Reconnaissance of the waters and plankton of Monterey Bay, July 1928. Bull. M.C.Z., 70(5):429-481, 43 text figs.

Lauderia delicatula
Pavillard, J., 1905
Recherches sur la flore pelagique (Phytoplankton) de l'Etang de Thau. Theses presentees a la Fac. Sci., Paris, 116 pp., 3 pls.

Lauderia delicatula
Schröder, B., 1900
Phytoplankton des Golfes von Neapel nebst vergleichenden Ausblicken auf das atlantischen Ozean. Mitt. Zool. Stat. Neapel, 14:1-38.

Lauderia elongata n.sp.
Castracane degli Antelminelli, F., 1886
1. Report on the Diatomaceae collected by H.M.S. Challenger during the years 1873-1876. Rept. Sci. Results, H.M.S. Challenger, Botany Vol. II, 178 pp., 30 pls.

Lauderia glacialis
Bigelow, H.B., and M. Leslie, 1930
Reconnaissance of the waters and plankton of Monterey Bay, July 1928. Bull. M.C.Z., 70(5):429-481, 43 text figs.

Lauderia glacialis
Gran, H.H., 1908
Diatomeen. Nordisches Plankton, Botanischer Teil pp. XIX 1-XIX 146; 178 text figs.

Lauderia glacialis
Iselin, C., 1930
A report on the coastal waters of Labrador based on explorations of the "Chance" during the summer of 1926. Proc. Am. Acad. Arts Sci., 66(1):1-37, 14 text figs.

Lauderia glacialis
Meunier, A., 1915
Microplancton de la Mer Flamande. 2. Diatomées (excepté le genre Chaetoceros). Mem. Mus. Roy. Hist. Nat., Belgique, 7(3):1-118, Pls. VIII-XIV.

Lauderia (?) moseleyana n.sp.
Castracane degli Antelminelli, F., 1886
1. Report on the Diatomaceae collected by H.M.S. Challenger during the years 1873-1876. Rept. Sci. Results, H.M.S. Challenger, Botany Vol. II, 178 pp., 30 pls.

Lauderia moseleyana
Dangeard, P., 1927
Phytoplankton de la croisière du "Sylvana". Ann. Inst. Ocean., Monaco, n.s., 4(8):286-401, 54 text figs. (Feirer-Juin 1913).

Lauderia pumila n.sp.
Castracane degli Antelminelli, F., 1886
1. Report on the Diatomaceae collected by H.M.S. Challenger during the years 1873-1876. Rept. Sci. Results, H.M.S. Challenger, Botany Vol. II, 178 pp., 30 pls.

Lauderia punctata
Hendey, N.I., 1937
The plankton diatoms of the southern seas. Discovery Repts. 16:151-364, pls.6-13.

Lepidodiscus Hesslandia
Cleve-Euler, A., 1951
Die Diatomeen von Schweden und Finnland. Kungl. Svenska Vetenskaps Akad. Handl., Fjärde Ser. 2(1): 161 pp., 6 pls.

Leptocylindrus
Sears, M., 1941
Notes on the phytoplankton on Georges Bank in 1941. J. Mar. Res., IV(3):247-257; textfigs.

Leptocylindrus sp.
Wheeler, J.E.G., 1939
Plankton investigations. Bermuda Biological Station. Second Report. October 1939. 7 pp. (typed), 5 figs. Plymouth, Oct. 23, 1939.

Leptocylindrus adriaticus
Ercegovic, A., 1936
Etudes qualitative et quantitatives du phytoplancton dans les eaux cotières de l'Adriatique oriental moyen au cours de l'année 1934. Acta Adriatica 1(9):1-126

Leptocylindrus belgicus n.sp.
Meunier, A., 1915
Microplancton de la Mer Flamande. 2. Diatomées (excepté le genre Chaetoceros). Mem. Mus. Roy. Hist. Nat., Belgique, 7(3):1-118, Pls. VIII-XIV.

Leptocylindrus danicus
Allen, W. E., 1938
The Templeton Crocker Expedition to the Gulf of California in 1935 - The Phytoplankton. Amer. Microsc. Soc., Trans. 57:328-335.

Leptocylindrus danicus
Allen, W.E., 1937
Plankton diatoms of the Gulf of California obtained by the G. Allan Hancock Expedition of 1936. The Hancock Pacific Expeditions, Univ. So. Calif. Publ. 3:47-59, 1 fig.

Leptocylindrus danicus
Allen, W. E., 1936
Occurrence of marine plankton diatoms in a ten-year series of daily catches in southern California. Am. Jour. Bot. 23(1):60-63.

Leptocylindrus danicus
Allen, W.E., and E.E. Cupp, 1935
Plankton diatoms of the Java Sea. Annales du Jardin Botanique de Buitenzorg XLIV (2):101-174, figs.1-127.
(drawings of all species mentioned)

Leptocylindrus danicus
Bernhard, M., preparator, 1965.
Studies on the radioactive contamination of the sea, annual report 1964. Com. Naz. Energ. Nucleare, La Spezia, Rept. No. RT/BIC (65) 18:35 pp.

Leptocylindrus danicus
Bigelow, H.B., and M. Leslie, 1930
Reconnaissance of the waters and plankton of Monterey Bay, July 1928. Bull. M.C.Z., 70(5):429-481, 43 text figs.

Leptocylindrus danicus
Boden, B.P., 1950
Some marine plankton diatoms from the west coast of South Africa. Trans. R.Soc. S. Africa. 32:321-434, 100 text figs.

Leptocylindrus danicus
Braarud, T., 1945
Experimental studies on marine plankton diatoms. Avhandlingar utgitt av Det Norske Videnskaps-Akademi i Oslo. 1. Mat.-Naturv. Klasse 1944. No.10, 1-16, 1 text fig.

Leptocylindrus danicus
Braarud, T., 1945
A phytoplankton survey of the polluted waters of inner Oslo Fjord. Hvalrådets Skrifter, No.28, 142 pp., 19 text figs., 17 tables.

Leptocylindrus danicus
Braarud, T., 1934
A note on the phytoplankton of the Gulf of Maine in the summer of 1933. Biol. Bull. 67(1):76-82. (Contribution No.46 of the Woods Hole Oceanographic Institution)

Leptocylindrus danicus
Braarud, T., and Adam Bursa, 1939
On the phytoplankton of the Oslo Fjord, 1933-1934. Hvalrådets Skr. No.19:1-63; 9 text figs. Reviewed. J. du. Cons. 14(3): 418-420. A.C. Gardiner.

Leptocylindrus danicus
Brunel, J., 1962
Le phytoplancton de la Baie de Chaleurs. Inst. Botan., Univ. Montréal, Contrib. No. 77: 365 pp., 66 pls.

Leptocylindrus danicus
Brunel, Jules, 1962
Le phytoplancton de la Baie des Chaleurs. Contrib. Ministère de la Chasse et des Pêcheries, Province de Québec, No. 91: 365 pp.

Leptocylindrus danicus
Chiba, T., 1949
On the distribution of the plankton in the eastern China Sea and Yellow Sea. 1. Plankton composition in the spring. J. Shimonoseki Coll. Fisheries, 1(1):57-63, 1 fig.

Leptocylindrus danicus
Cleve-Euler, A., 1951
Die Diatomeen von Schweden und Finnland. Kungl. Svenska Vetenskaps Akad. Handl., Fjärde Ser. 2(1): 161 pp., 6 pls.

Leptocylindrus danicus
Cupp, Easter E., 1943
Marine plankton diatoms of the west coast of North America. Bull. S.I.O. 5(1):1-238, 5 pls., 168 text figs.

Leptocylindrus danicus
Cupp, E.E., 1934
Analysis of marine diatom collections taken from the Canal Zone to California during March, 1933. Trans. Am. Micros. Soc. LIII (1):22-29, 1 map.

Leptocylindrus danicus
Cupp, E.E. and Allen, W.E., 1938
Plankton diatoms of the Gulf of California obtained by Allan Hancock Pacific Expedition of 1937. The Hancock Pacific Expeditions, the Univ. So. Calif. Publ. 3: 61-74, 1 map, pls. 4-15.

Leptocylindrus danicus
Dangeard, P., 1927
Phytoplankton de la croisière du "Sylvana". Ann. Inst. Ocean., Monaco, n.s., 4(8):286-401, 54 text figs. (Feirer-Juin 1913).

Leptocylindrus danicus
Delegazione Italiana della Commissione Internazionale per l'Esplorazione Scientifica del Mediterraneo, 1941
Note sul plancton della Laguna veneta. [Memoria CCLXXXIX], Arch. di Ocean. e Limn. Anno I, Fasc. I, 1941 XIX: 31-57 pp.

Leptocylindrus danicus

Ercegovic, A., 1936
Etudes qualitative et quantitatives du phytoplancton dans les eaux cotières de l'Adriatique oriental moyen au cours de l'année 1934. Acta Adriatica 1(9):1-126

Leptocylindrus danicus

Forti, A., 1922
Ricerche sulla flora pelagica (fitoplancton) di Quarto dei Mille. Mem. R. Com. Talass. Ital. 97:248 pp., 13 pls.

Leptocylindrus danicus

Gilbert, J.Y., and W.E. Allen, 1943
The phytoplankton of the Gulf of California obtained by the "E.W. Scripps" in 1939 and 1940. J. Mar. Res. V(2):89-110, figs.30-31.

Leptocylindrus danicus

Gran, H.H., 1908
Diatomeen. Nordisches Plankton, Botanischer Teil pp. XIX.1-XIX 146; 178 text figs.

Leptocylindrus danicus

Gran, H.H., and T. Braarud, 1935
A quantitative study of the phytoplankton in the Bay of Fundy and the Gulf of Maine (including observations on hydrography, chemistry, and turbidity). J. Biol. Bd., Canada, 1(5):279-467, 69 text figs.

Leptocylindrus danicus

Gran, H. H. and E. C. Angst, 1931
Plankton diatoms of Puget Sound. Publ. Puget Sound Biol. Sta. 7:417-519, 95 text figs.

Leptocylindrus danicus

Hendy, N. Ingram, 1964
An introductory account of the smaller algae of British coastal waters. V. Bacillariophyceae (Diatoms). Her Majesty's Stationary Office, 317 pp., 45 pls.

Leptocylindrus danicus

Hendey, N.I., 1937
The plankton diatoms of the southern seas. Discovery Repts. 16:151-364, pls.6-13.

Leptocylindrus danicus

Jorgensen, E., 1900
Protophyten und Protozoën im Plankton aus der Norwegischen Westkerste. Bergens Mus. Aarb. 1899(6): 95 pp., 5 pls., 83 tables

Leptocylindrus danicus

Kokubo, S., 1952
Results of the observations on the plankton and oceanography of Mutsu Bay during 1950, reference being made also to the period 1946-1950. Bull Mar.Biol.Sta., Asamushi 5(1/4): 1-54, 3 tables,(fold-in), 1 fold-in.

Leptocylindrus danicus

Lillick, L.C., 1940
Phytoplankton and planktonic protozoa of the offshore waters of the Gulf of Maine. Pt.II. Qualitative Composition of the Planktonic Flora. Trans. Am. Phil. Soc., n.s., 31(3):193-237, 13 text figs.

Leptocylindrus danicus

Lillick, L.C., 1938
Preliminary report of the phytoplankton of the Gulf of Maine. Am. Mid. Nat. 20(3):624-640, 1 text fig, 37 tables.

Leptocylindrus danicus

Lillick, L.C., 1937
Seasonal studies of the phytoplankton off Woods Hole, Massachusetts. Biol. Bull. LXXIII (3):488-503, 3 text figs.

Leptocylindrus danicus

Mangin, M. L., 1912
Phytoplancton de la croisière du "René" dans l'Atlantique (Septembre 1908). Ann. Inst. Ocean., n.s., 4(1):1-66, 2 pls., 41 text figs., 2 tables.

Leptocylindrus danicus

Massutí Algamora, M., 1949
Estudio de diez y seis muestras de plancton del Golfo de Nápoles. Publ. Inst. Biol. Appl. 5:85-94, 1 fold-in table.

Leptocylindrus danicus

Meunier, A., 1915
Microplancton de la Mer Flamande. 2. Diatomées (excepté le genre Chaetoceros). Mem. Mus. Roy. Hist. Nat., Belgique, 7(3):1-118, Pls. VIII-XIV.

Leptocylindrus danicus

Morse, D.C., 1947
Some observations on seasonal variations in plankton population Patuxent River, Maryland 1943 1945. Bd. Nat. Res., Publ. No.65, Chesapeake Biol. Lab., 31, 3 figs.

Leptocylindrus danicus

Pavillard, J., 1925
Bacillariales. Rept. on the Danish Oceangr. Exped., 1908-10 to the Mediterranean and adj. seas. Vol.II, Biol. J4:72 pp., 116 text figs.

Leptocylindrus danicus

Pavillard, J., 1905
Recherches sur la flore pelagique (Phytoplankton) de l'Etang de Thau. Theses presentees a la Fac. Sci., Paris, 116 pp., 3 pls.

Leptocylindrus danicus

Pettersson, H., 1934
Scattering and extinction of light in sea-water. Meddelanden från Göteborgs Högskolas Oceanografiska Institut. 9 (Göteborgs Kungl. vetenskaps-och vitterhets-samhälles Handlingar. Femte Foljden, Ser.B, 4(4)):1-16.

Leptocylindricus danicus

Pettersson, H., F. Gross, and F. Koczy, 1939
Large scale plankton cultures. Medd. från Oceanografiska Institutet i Göteborg No.3 (Göteborgs Kungl. vetenskaps-och Vitterhets-Samhälles Handlinger Femte Följden. Ser. B) Vol.6(13):1-24.

1. The plankton shaft. H. Pettersson
2. Experiment with phytoplankton. F. Gross and F. Koczy
3. Experiments with zooplankton.

Leptocylindrus danicus

Phifer, L.D. (undated)
The occurrence and distribution of plankton diatoms in Bering Sea and Bering Strait, July 26-August 24, 1934. Report of Oceanographic cruise of U.S. Coast Guard Cutter Chelan 1934, Part II(A):1-44 (mimeographed) plus fig.1 (after Pt.B)

Leptocylindrus danicus

Proshkina-Lavrenko, A.I., 1961.
Variability of some Black Sea diatoms. Botan. Zhurn., 46(12):1794-1797.
Akad. Nauk, SSSR,

Leptocylindrus danicus

Rampi, L., 1942
Ricerche sul fitoplancton del Mare Ligure 6. Le diatomee delle acque di Sanremo. Nuovo Giornale Botanico Italiano, N.S., 49:252-268.

Leptocylindrus danicus

Schodduyn, M., 1926
Observations faites dans la baie d'Ambleteuse (Pas de Calais). Bull. Inst. Ocean., Monaco, No. 482: 64 pp.

Leptocylindrus danicus

Schröder, B., 1900
Phytoplankton des Golfes von Neapel nebst vergleichenden Ausblicken auf das atlantischen Ozean. Mitt. Zool. Stat. Neapel, 14:1-38.

Leptocylindrus danicus

Sleggs, G.F., 1927.
Marine phytoplankton in the region of La Jolla, California during the summer of 1924. Bull. S.I. O., tech. ser., 1:93-117, 8 textfigs.

Leptocylindrus danicus

Sousa e Silva, E., 1949
Diatomaceas e Dinoflagelados de Baía de Cascais. Portugaliae Acta Biol., Volume: Julio Henriques, Ser. B: 300-383, 9 pls, 2 fold-in tables.

Leptocylindrus danicus

Sousa a Silva, E., and J. Dos Santos-Pinto, 1948
O Plancton da Baía de S. Martinho do Porto. 1. Diatomaceas e Dinoflagelados. Bol. Soc. Portuguese de Ciencias Naturais, 16(2):134-187, 6 pls. (Trav. Sta. Biol. Mar. de Lisbonne No. 52).

Leptocylindrus danicus

Stemann-Nielsen, Einar, 1951
The marine vegetation of the Isefjord. A study on ecology and production. Medd. Komm. Danmarks Fiskeri-og Havundersøgelser. Ser. Plankton. 5(4); 114 pp., 46 text figs.

Leptocylindrus danicus

Uyeno, Fukuzo, 1961
Oceanographical and ecological studies on primary production of the sea, with special references to relationship between diatom production and temperature and chlorinity of water.
Rept. Fac. Fish., Pref. Univ., Mie, 4(1): 1-64.

Leptocylindrus danicus

Wawrik, F., 1961.
Die horizontale Verteilung der Planktondiatomeen im Golf von Neapel. Int. Revue Ges. Hydrobiol., 46(3):460-479.

Leptocylindrus minimus

Braarud, T., 1945
A phytoplankton survey of the polluted waters of inner Oslo Fjord. Hvalrådets Skrifter, No.28, 142 pp., 19 text figs., 17 tables.

Leptocylindrus minimus

Cleve-Euler, A., 1951.
Die Diatomeen von Schweden und Finnland. Kungl. Svenska Vetenskaps Akad. Handl., Fjärde Ser. 2(1): 161 pp., 6 pls.

Leptocylindrus minimus

Cupp, E.E., 1934
Analysis of marine diatom collections taken from the Canal Zone to California during March, 1933. Trans. Am. Micros. Soc. LIII (1):22-29, 1 map.

Leptocylindrus minimus

Gran, H.H., and T. Braarud, 1935
A quantitative study of the phytoplankton in the Bay of Fundy and the Gulf of Maine (including observations on hydrography, chemistry, and turbidity). J. Biol. Bd., Canada, 1(5):279-467, 69 text figs.

Leptocylindrus minimus

Gran, H. H. and E. C. Angst, 1931
Plankton diatoms of Puget Sound. Publ. Puget Sound Biol. Sta. 7:417-519, 95 text figs.

Leptocylindrus minimus

Hendy, N. Ingram, 1964
An introductory account of the smaller algae of British coastal waters. V. Bacillariophyceae (Diatoms). Her Majesty's Stationary Office, 317 pp., 45 pls.

Leptocylindrus minimus
Lillick, L.C., 1940
Phytoplankton and planktonic protozoa of the offshore waters of the Gulf of Maine. Pt. II. Qualitative Composition of the Planktonic Flora. Trans. Am. Phil. Soc., n.s., 31(3):193-237, 13 text figs.

Leptocylindrus minimus
Lillick, L.C., 1937
Seasonal studies of the phytoplankton off Woods Hole, Massachusetts. Biol. Bull. LXXIII (3):488-503, 3 text figs.

Leptocylindrus minimus
Pettersson, H., 1934
Scattering and extinction of light in sea-water. Meddelanden från Göteborgs Högskolas Oceanografiska Institut. 9 (Göteborgs Kungl. vetenskaps-och vitterhets-samhälles Handlingar. Femte Foljden, Ser.B, 4(4)):1-16.

Leptocylindrus minimus
Phifer, L.D. (undated)
The occurrence and distribution of plankton diatoms in Bering Sea and Bering Strait, July 26-August 24, 1934. Report of Oceanographic cruise of U.S. Coast Guard Cutter Chelan 1934, Part II(A):1-44 (mimeographed) plus fig.1 (after Pt.B)

Leptocylindrus minimus
Steemann-Nielsen, Einar, 1951
The marine vegetation of the Isefjord. A study on ecology and production. Medd. Komm. Danmarks Fiskeri-og Havundersøgelser. Ser. Plankton. 5(4); 114 pp., 46 text figs.

Licmophora sp.
Braarud, T., 1945
A phytoplankton survey of the polluted waters of inner Oslo Fjord. Hvalrådets Skrifter, No.28, 142 pp., 19 text figs., 17 tables.

Licmophora sp.
Braarud, T., and Adam Bursa, 1939
On the phytoplankton of the Oslo Fjord, 1933-1934. Hvalrådets Skr. No.19:1-63; 9 text figs. Reviewed. J. du. Cons. 14(3): 418-420. A.C. Gardiner.

Licmophora sp.
Brunel, J., 1962
Le phytoplancton de la Baie de Chaleurs. Inst. Botan., Univ. Montréal, Contrib. No. 77: 365 pp., 66 pls.

Licmophora sp.
Brunel, Jules, 1962
Le phytoplancton de la Baie des Chaleurs. Contrib. Ministere de la Chasse et des Pêcheries, Province de Québec, No. 91: 365 pp.

Licmophora sp.
Cupp, E.E. and Allen, W.E., 1938
Plankton diatoms of the Gulf of California obtained by Allan Hancock Pacific Expedition of 1937. The Hancock Pacific Expeditions, The Univ. So. Calif. Publ. 3: 61-74, 1 map, pls.4-15.

Licmophora
Desikachary, T.V., 1956(1957).
Electron microscope studies on diatoms. J.R. Microsc. Soc. (3)76(1/2):9-36.

Licmophora
Johnson, T.W., Jr. 1966.
Ectogella in marine species of Licmophora.
J. Elisha Mitchell Sci. Soc. 82(1):25-29.

Licmophora sp.
Kokubo, S., and S. Sato., 1947
Plankters in Ju-San Gata. Physiol. and Ecol. (Japan) 1(4):1-16, 3 text figs., tables.

Lichmophora sp.
Meunier, A., 1915
Microplancton de la Mer Flamande. 2. Diatomées (excepté le genre Chaetoceros). Mem. Mus. Roy. Hist. Nat., Belgique, 7(3):1-118, Pls. VIII-XIV.

Licomophora
Neaverson, E., 1934
The sea-floor deposits. 1. General characteristics and distribution. Discovery Repts. 9: 297-349, Plates 17-22.

Lichmophora sp.
Schodduyn, M., 1926
Observations faites dans la baie d'Ambleteuse (Pas de Calais). Bull. Inst. Ocean., Monaco, No. 482: 64 pp.

Licmophora abbreviata
Cupp, Easter E., 1943
Marine plankton diatoms of the west coast of North America. Bull. S.I.O. 5(1):1-238, 5 pls., 168 text figs.

Licmophora abbreviata
Cupp, E.E. and Allen, W.E., 1938
Plankton diatoms of the Gulf of California obtained by Allan Hancock Pacific Expedition of 1937. The Hancock Pacific Expeditions, The Univ. So. Calif. Publ. 3: 61-74, 1 map, pls.4-15.

Lichomophora abbreviata
Lillick, L.C., 1940
Phytoplankton and planktonic protozoa of the offshore waters of the Gulf of Maine. Pt. II. Qualitative Composition of the Planktonic Flora. Trans. Am. Phil. Soc., n.s., 31(3):193-237, 13 text figs.

Lichmophora abbreviata
Lillick, L.C., 1938
Preliminary report of the phytoplankton of the Gulf of Maine. Am. Mid. Nat. 20(3):624-640, 1 text figs 37 tables.

Lichmophora abbreviata
Lillick, L.C., 1937
Seasonal studies of the phytoplankton off Woods Hole, Massachusetts. Biol. Bull. LXXIII (3):488-503, 3 text figs.

Lichmophora abbreviata
Misra, J.N., 1956.
A systematic account of some littoral marine diatoms from the west coast of India. J. Bombay Nat. Hist. Soc., 53(4):537-568.

Licmophora abbreviata
Zanon, V., 1948
Diatomee marini di Sardegna e Pugillo di Alghe Marine della stressa. Boll. Pesca, Piscitutura e Idrobiologia, Anno 24, ns. 3(2): 202-244, 27 figs. on 1 pl.

Lichmophora bharadwajai n.sp.
Misra, J.N., 1956.
A systematic account of some littoral marine diatoms from the west coast of India. J. Bombay Nat. Hist. Soc., 53(4):537-568.

Licmophora californica
Bigelow, H.B., and M. Leslie, 1930
Reconnaissance of the waters and plankton of Monterey Bay, July 1928. Bull. M.C.Z. 70(5):429-481, 43 text figs.

Lichomophora capitata n.sp.
Zanon, V., 1947.
Diatomée delle Isole Ponziane - materiali per una florula diatomologica del Mare Tirreno. Bol. Pesca, Piscicol., Idrobiol., n.s., 2(1):36-53, 1 pl. of 10 figs.

Licmophora communis
Ercegovic, A., 1936
Etudes qualitative et quantitatives du phytoplankton dans les eaux cotières de l'Adriatique oriental moyen au cours de l'année 1934. Acta Adriatica 1(9):1-126

Licmophora communis
Forti, A., 1922
Ricerche sulla flora pelagica (fitoplancton) di Quarto dei Mille. Mem. R. Com. Talass. Ital. 97:248 pp., 13 pls.

Lichmophora communis
Misra, J.N., 1956.
A systematic account of some littoral marine diatoms from the west coast of India. J. Bombay Nat. Hist. Soc., 53(4):537-568.

Licmophora communis
Zanon, V., 1948
Diatomee marini di Sardegna e Pugillo di Alghe Marine della stressa. Boll. Pesca, Piscitutura e Idrobiologia, Anno 24, ns. 3(2): 202-244, 27 figs. on 1 pl.

Lichmophora ehrenbergii
Boden, B.P., 1950
Some marine plankton diatoms from the west coast of South Africa. Trans. R.Soc. S. Africa. 32:321-434, 100 text figs.

Licmophora ehrenbergii
Hendy, N. Ingram, 1964
An introductory account of the smaller algae of British coastal waters. V. Bacillariophyceae (Diatoms).
Her Majesty's Stationary Office, 317 pp., 45 pls.

Lichmophora Ehrenbergi
Hustedt, F. and A.A. Aleem, 1951
Littoral diatoms from the Salstone near Plymouth. JMBA 30(1): 177.196.

Licmophora Ehrenbergi
Kucherova, Z.S., 1961.
Vertical distribution of diatoms from Sevastopol Bay.
Trudy Sevastopol Biol. Sta., (14):64-78.

Licmophora Ehrenbergii
Politis, J., 1949
Diatomees marines de Bosphores et des ibes de la mer de Marmara. II Practica tou Hellenikou Hidrobiologikou Institutoutou 1929, Etoz 1929, 3(1):11-31.

Licmophora Ehrenbergi
Rampi, L., 1940
Diatomee del Mare Adriatico. Nuovo Giornale Botanico Italiano, n.s., 47:559-608.

Licmophora ehrenbergii
Takano, Hideaki, 1962
Notes on epiphytic diatoms upon sea-weeds from Japan.
J. Oceanogr. Soc., Japan, 18(1):29-33.

Licmophora Ehrenbergii
Zanon, V., 1948
Diatomee marini di Sardegna e Pugillo di Alghe Marine della stressa. Boll. Pesca, Piscitutura e Idrobiologia, Anno 24, ns. 3(2): 202-244, 27 figs. on 1 pl.

Licmophora flabellata
Ercegovic, A., 1936
Etudes qualitative et quantitatives du phytoplancton dans les eaux cotières de l'Adriatique oriental moyen au cours de l'année 1934. Acta Adriatica 1(9):1-126

Licmophora flabellata
Forti, A., 1922
Ricerche sulla flora pelagica (fitoplancton) di Quarto dei Mille. Mem. R. Com. Talass. Ital. 97:248 pp., 13 pls.

Licmophora flabellata
Hendy, N. Ingram, 1964
An introductory account of the smaller algae of British coastal waters. V. Bacillariophyceae (Diatoms). Her Majesty's Stationary Office, 317 pp., 45 pls.

Lichmophora flabellata
Hendey, N.I., 1951
Littoral diatoms of Chicester Harbour with special reference to fouling. J. Roy. Microscop. Soc. 71(1): 1-86, 18 pls.

Lichmophora flabellata
Hustedt, F. and A.A. Aleem, 1951
Littoral diatoms from the Salstone near Plymouth. JMBA 30(1): 177.196.

Lichmophora flabellata
Misra, J.N., 1956.
A systematic account of some littoral marine diatoms from the west coast of India. J. Bombay Nat. Hist. Soc., 53(4):537-568.

Licmophora flabellata
Morse, D.C., 1947
Some observations on seasonal variations in plankton population Patuxant River, Maryland 1943 1945. Bd. Nat. Res., Publ. No.65, Chesapeake Biol. Lab., 31, 3 figs.

Licmophora flabellata
Rampi, L., 1942
Ricerche sul fitoplancton del Mare Ligure 6. Le diatomee delle acque di Sanremo. Nuovo Giornale Botanico Italiano, N.S., 49:252-268.

Licmophora flabellata
Rampi, L., 1940
Diatomee del Mare Adriatico. Nuovo Giornale Botanico Italiano, n.s., 47:559-608.

Lychmophora flabellata
Sousa e Silva, E., 1949
Diatomacees e Dinoflagelados de Baia de Cascais. Portugaliae Acta Biol., Volume: Julio Henriques, Ser. B: 300-383, 9 pls, 2 fold-in tables.

Licmophora flabellata
Takano, Hideaki, 1962
Notes on epiphytic diatoms upon sea-weeds from Japan. J. Oceanogr. Soc., Japan, 18(1):29-33.

Licmophora gracilis & varieties
Hendy, N. Ingram, 1964
An introductory account of the smaller algae of British coastal waters. V. Bacillariophyceae (Diatoms). Her Majesty's Stationary Office, 317 pp., 45 pls.

Licmophora gracilis
Hendey, N.I., 1951
Littoral diatoms of Chicester Harbour with special reference to fouling. J. Roy. Microscop. Soc. 71(1): 1-86, 18 pls.

Lichmophora gracilis
Hustedt, F. and A.A. Aleem, 1951
Littoral diatoms from the Salstone near Plymouth. JMBA 30(1): 177.196.

Licmophora gracilis
Morse, D.C., 1947
Some observations on seasonal variations in plankton population Patuxant River, Maryland 1943 1945. Bd. Nat. Res., Publ. No.65, Chesapeake Biol. Lab., 31, 3 figs.

Licmophora gracilis
Zanon, V., 1948
Diatomee marini di Sardegna e Pugillo di Alghe Marine della stressa. Boll. Pesca, Piscitutura e Idrobiologia, Anno 24, ns. 3(2): 202-244, 27 figs. on 1 pl.

Licmophora gracilis anglica
Frenguelli, Joaquin, and Hector Antonio Orlando, 1959.
Operacion MERLUZA. Diatomeas y silicoflagelados del plancton del "VI Crucero". Servicio Hidrogr. Naval., Argentina, Publ. No. H. 619: 5-62.

Licmophora gracilis anglica
Takano, Hideaki, 1962
Notes on epiphytic diatoms upon sea-weeds from Japan. J. Oceanogr. Soc., Japan, 18(1):29-33.

Licmophora grandis
Zanon, V., 1948
Diatomee marini di Sardegna e Pugillo di Alghe Marine della stressa. Boll. Pesca, Piscitutura e Idrobiologia, Anno 24, ns. 3(2): 202-244, 27 figs. on 1 pl.

Lichmophora grandis somnathii n.var.
Misra, J.N., 1956.
A systematic account of some littoral marine diatoms from the west coast of India. J. Bombay Nat. Hist. Soc., 53(4):537-568.

Lichmophora hastata
Rampi, L., 1940
Diatomee del Mare Adriatico. Nuovo Giornale Botanico Italiano, n.s., 47:559-608.

Lichmophora hyalina
Hustedt, F. and A.A. Aleem, 1951
Littoral diatoms from the Salstone near Plymouth. JMBA 30(1): 177.196.

Licmophora hyalina
Paasche, E., 1973
The influence of cell size on growth rate, silica content and some other properties of four marine diatom species. Norwegian J. Bot. 20(2/3): 199-204.

Licmophora juergensii
Hendy, N. Ingram, 1964
An introductory account of the smaller algae of British coastal waters. V. Bacillariophyceae (Diatoms). Her Majesty's Stationary Office, 317 pp., 45 pls.

Lichmophora Juergensi
Hustedt, F. and A.A. Aleem, 1951
Littoral diatoms from the Salstone near Plymouth. JMBA 30(1): 177.196.

Lichmomophora Juergensii
Lillick, L.C., 1940
Phytoplankton and planktonic protozoa of the offshore waters of the Gulf of Maine. Pt.II. Qualitative Composition of the Planktonic Flora. Trans. Am. Phil. Soc., n.s., 31(3):193-237, 13 text figs.

Licmophora Juergensii
Zanon, V., 1948
Diatomee marini di Sardegna e Pugillo di Alghe Marine della stressa. Boll. Pesca, Piscitutura e Idrobiologia, Anno 24, ns. 3(2): 202-244, 27 figs. on 1 pl.

Licmophora juergensii elongata
Takano, Hideaki, 1961.
Epiphytic diatoms upon Japanese agar sea-weeds. Bull. Tokai Reg. Fish. Res. Lab., No. 31:269-274.

Licmophora Juergensii oedipus
Politis, J., 1949
Diatomees marines de Bosphores et des ibes de la mer de Marmara. II Practica tou Hellenikou Hidrobiologikou Institutoutou 1929, Etoz 1929, 3(1):11-31.

Licmophora luxuriosa
Hendey, N.I., 1937
The plankton diatoms of the southern seas. Discovery Repts. 16:151-364, pls.6-13.

Licmophora Lyngbyei
Allen, W.E., and E.E. Cupp, 1935
Plankton diatoms of the Java Sea. Annales du Jardin Botanique de Buitenzorg XLIV (2):101-174, figs.1-127.
(drawings of all species mentioned)

Licmophora Lyngbyei
Bigelow, H.B., and M. Leslie, 1930
Reconnaissance of the waters and plankton of Monterey Bay, July 1928. Bull. M.C.Z., 70(5):429-481, 43 text figs.

Lichmophora lyngbyei
Boden, B.P., 1950
Some marine plankton diatoms from the west coast of South Africa. Trans. R.Soc. S. Africa. 32:321-434, 100 text figs.

Licmophora Lyngbyei
Braarud, T., 1945
A phytoplankton survey of the polluted waters of inner Oslo Fjord. Hvalrådets Skrifter, No.28, 142 pp., 19 text figs., 17 tables.

Lichmophera lyngbyei
Gilbert, J.Y., and W.E. Allen, 1943
The phytoplankton of the Gulf of California obtained by the "E.W. Scripps" in 1939 and 1940. J. Mar. Res. V(2):89-110, figs.30-31.

Licmophora lyngbyei
Gran, H.N., 1908
Diatomeen. Nordisches Plankton, Botanischer Teil pp. XIX.1-XIX 146; 178 text figs.

Licmophora lyngbyei
Hendy, N. Ingram, 1964
An introductory account of the smaller algae of British coastal waters. V. Bacillariophyceae (Diatoms). Her Majesty's Stationary Office, 317 pp., 45 pls.

Lichmophora Lyngbyei
Hendey, N.I., 1951
Littoral diatoms of Chicester Harbour with special reference to fouling. J. Roy. Microscop. Soc. 71(1): 1-86, 18 pls.

Licmophora Lyngbyei
Hendey, N.I., 1937
The plankton diatoms of the southern seas. Discovery Repts. 16:151-364, pls.6-13.

Lichmophora lyngbyei
Jorgensen, E., 1900
Protophyten und Protozoën im Plankton aus der Norwegischen Westkerste. Bergens Mus. Aarb. 1899(6): 95 pp., 5 pls., 83 tables.

Lichmophora lyngbyei
Paiva Carvalho, J., 1950
O plancton do Rio Maria Rodriques (Cananeis). 1. Diatomaceas e Dinoflagelados. Bol. Inst. Paulista Oceanogr. 1(1): 27-43, 2 fold-in tables, 2 figs.

Licmophora Lyngbyei
Politis, J., 1949
Diatomees marines de Bosphores et des ibes de la mer de Marmara. II Practica tou Hellenikou Hidrobiologikou Institutoutou 1929, Etoz 1929, 3(1):11-31.

Licmophora Lyngbyei
Rumkówna, A., 1948
List of the phytoplankton species occurring in the superficial water layers in the Gulf of Gdańsk] Bull. Lab. mar., Gdynia, No. 4: 139-141 with tables in back.

Lycmophora lyngbyei
Sousa e Silva, E., 1949
Diatomaceas e Dinoflagelados de Baia de Cascais. Portugaliae Acta Biol., Volume: Julio Henriques, Ser. B: 300-383, 9 pls, 2 fold-in tables.

Lycmophora Lyngbyei
Sousa e Silva, E., and J. Dos Santos-Pinto, 1948
O Plancton da Baia de S. Martinho do Porto. 1. Diatomaceas e Dinoflagelados. Bol. Soc. Portuguese de Ciencias Naturais, 16(2):134-187, 6 pls. (Trav. Sta. Biol. Mar. de Lisbonne No. 52).

Licmophora nubecola
Zanon, V., 1948
Diatomee marini di Sardegna e Pugillo di Alghe Marine della stressa. Boll. Pesca, Piscitutura e Idrobiologia, Anno 24, ns. 3(2): 202-244, 27 figs. on 1 pl.

Licmophora paradoxa
Ercegovic, A., 1936
Etudes qualitative et quantitatives du phytoplancton dans les eaux cotières de l'Adriatique oriental moyen au cours de l'année 1934. Acta Adriatica 1(9):1-126

Licmophora paradoxa
Forti, A., 1922
Ricerche sulla flora pelagica (fitoplancton) di Quarto dei Mille. Mem. R. Com. Talass. Ital. 97:248 pp., 13 pls.

Licmophora paradoxa
Hendy, N. Ingram, 1964
An introductory account of the smaller algae of British coastal waters. V. Bacillariophyceae (Diatoms). Her Majesty's Stationary Office, 317 pp., 45 pls.

Lichmophora paradoxa
Hendey, N.I., 1951
Littoral diatoms of Chicester Harbour with special reference to fouling. J.Roy. Microscop. Soc. 71(1): 1-86, 18 pls.

Lichmophora paradoxa
Hustedt, F. and A.A. Aleem, 1951
Littoral diatoms from the Salstone near Plymouth. JMBA 30(1): 177.196.

Licmophora paradoxa
Rampi, L., 1942
Ricerche sul fitoplancton del Mare Ligure 6. Le diatomee delle acque di Sanremo. Nuovo Giornale Botanico Italiano, N.S. 49:252-268.

Licmophora paradoxa
Takano, Hideaki, 1963.
Notes on marine littoral diatoms of Japan. Bull. Tokai Reg. Fish. Res. Lab., No. 36:1-8.

Licmophora paradoxa
Zanon, V., 1948
Diatomee marini di Sardegna e Pugillo di Alghe Marine della stressa. Boll. Pesca, Piscitutura e Idrobiologia, Anno 24, ns. 3(2): 202-244, 27 figs. on 1 pl.

Lichmophora paradoxa media n.var.
Misra, J.N., 1956.
A systematic account of some littoral marine diatoms from the west coast of India. J. Bombay Nat. Hist. Soc., 53(4):537-568

Licmophora partita n.sp.
Giffen, Malcolm H. 1973.
Diatoms of the marine littoral of Steenberg's Cove in G. Helena Bay, Cape Province, South Africa. Botanica Marina 16(1): 32-48.

Licmophora proboscidea
Rampi, L., 1940
Diatomee del Mare Adriatico. Nuovo Giornale Botanico Italiano, n.s., 47:559-608.

Licmophora pseudohyalina, n.sp.
Brenguelli, Joaquin, y Hector A. Orlando, 1958.
Diatomeas y silicoflagelados del sector Antartico Sudamericano. Inst. Antartico Argentino, Publ., No. 5:191 pp.

Licmophora Reichardtii
Mangin, L., 1915
Phytoplancton de L'Antartique. Deuxieme Exped. Ant. Francaise (1908-1910), 95 pp., 3 pls., 58 text figs.

Licmophora Reichardtii
Zanon, V., 1948
Diatomee marini di Sardegna e Pugillo di Alghe Marine della stressa. Boll. Pesca, Piscitutura e Idrobiologia, Anno 24, ns. 3(2): 202-244, 27 figs. on 1 pl.

Licmophora Remulus
Grunow, A., 1877
1. New Diatoms from Honduras. Monthly Micros. Jour., 18:165-186, pls. CXCIII-CXCVI.

Licmophora tincta
Morse, D.C., 1947
Some observations on seasonal variations in plankton population Patuxant River, Maryland 1943 1945. Bd. Nat. Res., Publ. No.65, Chesapeake Biol. Lab., 31, 3 figs.

Licmosphenia hustedtii n.sp.
Amossé, A. 1970.
Diatomées marines et saumâtres du Sénégal et de la Côte d'Ivoire. Bull. Inst. fond. Afr. Noire (A) 32(2): 289-311.

Lithodesmum intricatum
Pavillard, J., 1925
Bacillariales. Rept. on the Danish Oceangr. Exped., 1908-10 to the Mediterranean and adj. seas. Vol.II., Biol. J4:72 pp., 116 text figs.

Lithodesmium intricatum (figs.)
Sousa e Silva, E., 1949
Diatomaceas e Dinoflagelados de Baia de Cascais. Portugaliae Acta Biol., Volume: Julio Henriques, Ser. B: 300-383, 9 pls, 2 fold-in tables.

Lithodesmium pliocenicum n.sp.
Schrader, Hans-Joachim, 1973
Cenozoic diatoms from the northeast Pacific, leg 18. Initial Repts Deep Sea Drilling Proj. 18:673-797.

Lithodesmium undulatum
Allen, W. E., 1938
The Templeton Crocker Expedition to the Gulf of California in 1935 - The Phytoplankton. Amer. Microsc. Soc., Trans. 57:328-335.

Lithodesmium undulatum
Allen, W.E., 1937
Plankton diatoms of the Gulf of California obtained by the G. Allan Hancock Expedition of 1936. The Hancock Pacific Expeditions, Univ. So. Calif. Publ. 3:47-59, 1 fig.

Lithodesmium undulatum
Bigelow, H.B., and M. Leslie, 1930
Reconnaissance of the waters and plankton of Monterey Bay, July 1928. Bull. M.C.Z. 70(5):429-481, 43 text figs.

Lithodesmium undulatum
Cleve-Euler, A., 1951
Die Diatomeen von Schweden und Finnland. Kungl. Svenska Vetenskaps Akad. Handl., Fjärde Ser. 2(1): 161 pp., 6 pls.

Lithodesmium undulatum
Cupp, Easter E., 1943
Marine plankton diatoms of the west coast of North America. Bull. S.I.O. 5(1):1-238, 5 pls., 168 text figs.

Lithodesmium undulatum
Cupp, E.E., 1934
Analysis of marine diatom collections taken from the Canal Zone to California during March, 1933. Trans. Am. Micros. Soc. LIII (1):22-29, 1 map.

Lithodesmium undulatum
Cupp, E.E. and Allen, W.E., 1938
Plankton diatoms of the Gulf of California obtained by Allan Hancock Pacific Expedition of 1937. The Hancock Pacific Expeditions, The Univ. So. Calif. Publ. 3: 61-74, 1 map, pls.4-15.

Lithodesmium undulatum
Forti, A., 1922
Ricerche sulla flora pelagica (fitoplancton) di Quarto dei Mille. Mem. R. Com. Talass. Ital. 97:248 pp., 13 pls.

Lithodesmium undulatum
Ghazzawi, F.M., 1939
Plankton of the Egyptian waters. A study of the Suez Canal Plankton. (A) Phytoplankton. Preliminary Report 83 pp. Notes and Memoires, Min. Commerce-Industry, Egypt, Hydrobiol. & Fish. 65 figs.

Lithodesmium undulatum
Gilbert, J.Y., and W.E. Allen, 1943
The phytoplankton of the Gulf of California obtained by the "E.W. Scripps" in 1939 and 1940. J. Mar. Res. V(2):89-110, figs.30-31.

Lithodesmium undulatum
Gran, H.H., 1908
Diatomeen. Nordisches Plankton, Botanischer Teil pp. XIX.1-XIX 146; 178 text figs.

Lithodesmium undulatum
Grøntved, J., 1949(1950).
Investigations on the phytoplankton in the Danish Waddensee in July 1941. Medd. Komm. Danmarks Fiskeri- og Havundersøgelser, Serie Plankton, 5(2)(Medd. Skalling Lab. 10):55 pp., 2 pls., 38 textfigs.

Lithodesmium undulatum
Grøntved, J., 1949
Investigations on the phytoplankton in the Danish Waddensea in July 1941. Medd. Komm. Danmarks Fiskeri og Havundersøgelser, ser. Plankton, 5(2):55 pp., 2 pls., 38 text figs.

Lithodesmium undulatum
Hendy, N. Ingram, 1964
An introductory account of the smaller algae of British coastal waters. V. Bacillariophyceae (Diatoms).
Her Majesty's Stationary Office, 317 pp., 45 pls.

Lithodesmium undulatum
Lafon, M., M. Durchon and Y. Saudray, 1955.
Recherches sur les cycles saisonnières du plankton. Ann. Inst. Océan., 31(3):125-230.

Lithodesmium undulatum
Mangin, M. L., 1912
Phytoplancton de la croisière du "René" dans l'Atlantique (Septembre 1908). Ann. Inst. Ocean., n.s., 4(1):1-66, 2 pls., 41 text figs., 2 tables.

Lithodesmium undulatum
Meunier, A., 1915
Microplancton de la Mer Flamande. 2. Diatomées (excepté le genre Chaetoceros). Mem. Mus. Roy. Hist. Nat., Belgique, 7(3):1-118, Pls. VIII-XIV.

Lithodesmium undulatum
Morse, D.C., 1947
Some observations on seasonal variations in plankton population Patuxant River, Maryland 1943-1945. Bd. Nat. Res., Publ. No.65, Chesapeake Biol. Lab., 31, 3 figs.

Lithodesmum undulatum
Pavillard, J., 1925
Bacillariales. Rept. on the Danish Oceangr. Exped., 1908-10 to the Mediterranean and adj. seas. Vol.II., Biol. J4:72 pp., 116 text figs.

Lithodesmium undulatum
Rampi, L., 1942
Ricerche sul fitoplancton del Mare Ligure 6. Le diatomee delle acque di Sanremo. Nuovo Giornale Botanico Italiano, N.S., 49:252-268.

Lithodesmium undulatum
Sousa e Silva, E., 1949
Diatomaceas e Dinoflagelados de Baia de Cascais. Portugalise Acta Biol., Volume: Julio Henriques, Ser. B: 300-383, 9 pls, 2 fold-in tables.

Lithodesmium Victoriae
Dangeard, P., 1927
Phytoplankton de la croisière du "Sylvana". Ann. Inst. Ocean., Monaco, n.s., 4(8):286-401, 54 text figs. (Feirer-Juin 1913).

Lysiogonium Juergensii
Jorgensen, E., 1900
Protophyten und Protozoën im Plankton aus der Norwegischen Westkerste. Bergens Mus. Aarb. 1899(6): 95 pp., 5 pls., 83 tables.

Mastogloia
Desikachary, T.V., 1956(1957).
Electron microscope studies on diatoms. J.R. Microsc. Soc. (3)76(1/2):9036.

Mastogloia sp.
Paiva Carvalho, J., 1950
O plancton do Rio Maria Rodriques (Cananeis). 1. Diatomaceas e Dinoflagelados. Bol. Inst. Paulista Oceanogr. 1(1); 27-43, 2 fold-in tables, 2 figs.

Mastogloia spp.
Voight, M., 1967.
Quelques Mastogloia de la Mélanésie. Cah. Pacif., (10):53-57.

Mastigloia acuta
Zanon, V., 1948
Diatomee marini di Sardegna e Pugillo di Alghe Marine della stressa. Boll. Pesca, Piscitutura e Idrobiologia, Anno 24, ns. 3(2): 202-244, 27 figs. on 1 pl.

Mastogloia acutuiscula
Zanon, V., 1948
Diatomee marini di Sardegna e Pugillo di Alghe Marine della stressa. Boll. Pesca, Piscitutura e Idrobiologia, Anno 24, ns. 3(2): 202-244, 27 figs. on 1 pl.

Mastogloia angulata
Grunow, A., 1877
1. New Diatoms from Honduras. Monthly Micros. Jour., 18:165-186, pls. CXCIII-CXCVI.

Mastogloia angulata
Politis, J., 1949
Diatomees marines de Bosphores et des ibes de la mer de Marmara. II Practica tou Hellenikou Hidrobiologikou Institutoutou 1929, Etoz 1929, 3(1):11-31.

Mastogloia angulata
Rampi, L., 1940
Diatomee del Mare Adriatico. Nuovo Giornale Botanico Italiano, n.s., 47:559-608.

Mastogloia angulata
Zanon, V., 1948
Diatomee marini di Sardegna e Pugillo di Alghe Marine della stressa. Boll. Pesca, Piscitutura e Idrobiologia, Anno 24, ns. 3(2): 202-244, 27 figs. on 1 pl.

Mastoglaia apiculata
Andrade e Clovis Teixeira, M. H. de, 1957.
Contribuição para o conhecimento das diatomáceas do Brasil.
Bol. Inst. Ocean., Univ. Sao Paulo, 8(1-2): 171 - 225, 10 pls.

Mastogloia apiculata
Mann, A., 1893
List of Diatomaceae from a deep-sea dredging in the Atlantic Ocean off Delaware Bay by the U. S. Fish Commission Steamer Albatross. Proc. U. S. Nat. Mus. 16:303-312.

Mastogloia apiculata
Politis, J., 1949
Diatomees marines de Bosphores et des ibes de la mer de Marmara. II Practica tou Hellenikou Hidrobiologikou Institutoutou 1929, Etoz 1929, 3(1):11-31.

Mastogloia apiculata
Zanon, V., 1948
Diatomee marini di Sardegna e Pugillo di Alghe Marine della stressa. Boll. Pesca, Piscitutura e Idrobiologia, Anno 24, ns. 3(2): 202-244, 27 figs. on 1 pl.

Mastogloia aquilegiae
Zanon, V., 1948
Diatomee marini di Sardegna e Pugillo di Alghe Marine della stressa. Boll. Pesca, Piscitutura e Idrobiologia, Anno 24, ns. 3(2): 202-244, 27 figs. on 1 pl.

Mastogloia Baldjikiana
Rampi, L., 1940
Diatomee del Mare Adriatico. Nuovo Giornale Botanico Italiano, n.s., 47:559-608.

Mastigloia Baldjikiana
Zanon, V., 1948
Diatomee marini di Sardegna e Pugillo di Alghe Marine della stressa. Boll. Pesca, Piscitutura e Idrobiologia, Anno 24, ns. 3(2): 202-244, 27 figs. on 1 pl.

Mastogloia binotata
Andrade e Clovis Teixeira, M.H. de, 1957.
Contribuição para o conhecimento das diatomáceas do Brasil.
Bol. Inst. Ocean., Univ. Sao Paulo, 8(1/2):171-225, 10 pls.

Mastogloia binotata
Eskinazi Enide e Shigekatsv Satô (1963-1964) 1966.
Contribuição ao estudo das diatomaceas da Praia de Piedade.
Trabhs Inst. Oceanogr., Univ. Recife, 5 (5/6): 73-114.

Mastogloia binotata
Hendy, N. Ingram, 1964
An introductory account of the smaller algae of British coastal waters. V. Bacillariophyceae (Diatoms).
Her Majesty's Stationary Office, 317 pp., 45 pls.

Mastogloia binotata
Rampi, L., 1942
Ricerche sul fitoplancton del Mare Ligure 6. Le diatomee delle acque di Sanremo. Nuovo Giornale Botanico Italiano, N.S., 49:252-268.

Mastogloia binotata
Rampi, L., 1940
Diatomee del Mare Adriatico. Nuovo Giornale Botanico Italiano, n.s., 47:559-608.

Mastogloia binotata
Zanon, V., 1948
Diatomee marini di Sardegna e Pugillo di Alghe Marine della stressa. Boll. Pesca, Piscitutura e Idrobiologia, Anno 24, ns. 3(2): 202-244, 27 figs. on 1 pl.

Mastogloia binotata oblonga n.var.
Zanon, V., 1947.
Diatomee delle Isole Ponziane - Materiali per una florula diatomologica del Mare Tirreno. Bol. Pesca, Piscicol., Idrobiol., n.s., 2(1):36-53, 1 pl., of 10 figs.

Mastogloia bisulcata
Grunow, A., 1877
1. New Diatoms from Honduras. Monthly Micros. Jour., 18:165-186, pls. CXCIII-CXCVI.

Mastogloia Braunii
Grunow, A., 1863
Ueber einige neue und ungenügend bekannte Arten und Gattungen von Diatomaceen. Verhandl. d. K.K. Zool. Bot. Gesellsch., Vienna, 13: 137-162, pl.4-5 (Pl. 13-14).

Mastogloia braunii
Hendy, N. Ingram, 1964
An introductory account of the smaller algae of British coastal waters. V. Bacillariophyceae (Diatoms).
Her Majesty's Stationary Office, 317 pp., 45 pls.

Mastogloia brauni
Paiva Carvalho, J., 1950
O plancton do Rio Maria Rodriques (Cananeis). 1. Diatomaceas e Dinoflagelados. Bol. Inst. Paulista Oceanogr. 1(1); 27-43, 2 fold-in tables, 2 figs.

Mastigloia brauni
Zanon, V., 1948
 Diatomee marini di Sardegna e Pugillo di Alghe Marine della stressa. Boll. Pesca, Piscitutura e Idrobiologia, Anno 24, ns. 3(2): 202-244, 27 figs. on 1 pl.

Mastigloia chersonensis
Rampi, L., 1940
 Diatomee del Mare Adriatico. Nuovo Giornale Botanico Italiano, n.s., 47:559-608.

Mastigloia citrus
Andrade e Clovis Teixeira, M.H. de, 1957.
Contribuição para o conhecimento das diatomáceas do Brasil.
Bol. Inst. Ocean., Univ. Sao Paulo, 8(1/2):171-225, 10 pls.

Mastigloia cocconeiformis
Andrade e Clovis Teixeira, M.H. de, 1957.
Contribuição para o conhecimento das diatomáceas do Brasil.
Bol. Inst. Ocean., Univ. Sao Paulo, 8(1/2):171-225, 10 pls.

Mastigloia corsicana
Politis, J., 1949
 Diatomees marines de Bosphores et des ibes de la mer de Marmara. II Practica tou Hellenikou Hidrobiologikou Institutoutou 1929, Etoz 1929, 3(1):11-31.

Mastigloia corsicana
Rampi, L., 1940
 Diatomee del Mare Adriatico. Nuovo Giornale Botanico Italiano, n.s., 47:559-608.

Mastigloia corsicana
Zanon, V., 1948
 Diatomee marini di Sardegna e Pugillo di Alghe Marine della stressa. Boll. Pesca, Piscitutura e Idrobiologia, Anno 24, ns. 3(2): 202-244, 27 figs. on 1 pl.

Mastigloia cribrosa
Andrade e Clovis Teixeira, M.H. de, 1957.
Contribuição para o conhecimento das diatomáceas do Brasil.
Bol. Inst. Ocean., Univ. Sao Paulo, 8(1/2):171-225, 10 pls.

Mastigloia crucicula
Rampi, L., 1942
 Ricerche sul fitoplancton del Mare Ligure 6. Le diatomee delle acque di Sanremo. Nuovo Giornale Botanico Italiano, N.S., 49:252-268.

Mastigloia crucicula
Rampi, L., 1940
 Diatomee del Mare Adriatico. Nuovo Giornale Botanico Italiano, n.s., 47:559-608.

Mastigloia crucicula
Zanon, V., 1948
 Diatomee marini di Sardegna e Pugillo di Alghe Marine della stressa. Boll. Pesca, Piscitutura e Idrobiologia, Anno 24, ns. 3(2): 202-244, 27 figs. on 1 pl.

Mastigloia decussata
Andrade, e Clovis Teixeira, M.H. de, 1957.
Contribuição para o conhecimento das diatomáceas do Brasil.
Bol. Inst. Ocean., Univ. Sao Paulo, 8(1/2):171-225, 10 pls.

Mastigloia delicatula
Zanon, V., 1948
 Diatomee marini di Sardegna e Pugillo di Alghe Marine della stressa. Boll. Pesca, Piscitutura e Idrobiologia, Anno 24, ns. 3(2): 202-244, 27 figs. on 1 pl.

Mastigloia depressa
Zanon, V., 1948
 Diatomee marini di Sardegna e Pugillo di Alghe Marine della stressa. Boll. Pesca, Piscitutura e Idrobiologia, Anno 24, ns. 3(2): 202-244, 27 figs. on 1 pl.

Mastigloia elliptica & var.
Hendy, N. Ingram, 1964

An introductory account of the smaller algae of British coastal waters. V. Bacillariophyceae (Diatoms).
Her Majesty's Stationary Office, 317 pp., 45 pls.

Mastigloia elliptica
Zanon, V., 1948
 Diatomee marini di Sardegna e Pugillo di Alghe Marine della stressa. Boll. Pesca, Piscitutura e Idrobiologia, Anno 24, ns. 3(2): 202-244, 27 figs. on 1 pl.

Mastigloia erythraca
Andrade e Clovis Teixeira, M.H. de, 1957.
Contribuição para o conhecimento das diatomáceas do Brasil.
Bol. Inst. Ocean., Univ. Sao Paulo, 8(1/2):171-225, 10 pls.

Mastigloia erythraea
Grunow, A., 1877
 1. New Diatoms from Honduras. Monthly Micros. Jour., 18:165-186, pls. CXCIII-CXCVI.

Mastigloia erythraea
Politis, J., 1949
 Diatomees marines de Bosphores et des ibes de la mer de Marmara. II Practica tou Hellenikou Hidrobiologikou Institutoutou 1929, Etoz 1929, 3(1):11-31.

Mastigloia erytraea
Rampi, L., 1940
 Diatomee del Mare Adriatico. Nuovo Giornale Botanico Italiano, n.s., 47:559-608.

Mastigloia Erythraea
Zanon, V., 1948
 Diatomee marini di Sardegna e Pugillo di Alghe Marine della stressa. Boll. Pesca, Piscitutura e Idrobiologia, Anno 24, ns. 3(2): 202-244, 27 figs. on 1 pl.

Mastigloia exigua
Andrade e Clovis Teixeira, M.H. de, 1957.
Contribuição para o conhecimento das diatomáceas do Brasil.
Bol. Inst. Ocean., Univ. Sao Paulo, 8(1/2):171-225, 10 pls.

Mastigloia exigua
Politis, J., 1949
 Diatomees marines de Bosphores et des ibes de la mer de Marmara. II Practica tou Hellenikou Hidrobiologikou Institutoutou 1929, Etoz 1929, 3(1):11-31.

Mastigloia exilis
Zanon, V., 1948
 Diatomee marini di Sardegna e Pugillo di Alghe Marine della stressa. Boll. Pesca, Piscitutura e Idrobiologia, Anno 24, ns. 3(2): 202-244, 27 figs. on 1 pl.

?Mastigloia fimbriata
Grunow, A., 1863
 Ueber einige neue und ungenügend bekannte Arten und Gattungen von Diatomaceen. Verhandl. d. K.K. Zool. Bot. Gesellsch., Vienna, 13: 137-162, pl.4-5 (Pl. 13-14).

Mastigloia fimbriata
Andrade e Clovis Teixeira, M.H. de, 1957.
Contribuição para o conhecimento das diatomáceas do Brasil.
Bol. Inst. Ocean., Univ. Sao Paulo, 8(1/2):171-225, 10 pls.

Mastigloia fimbrata
Rampi, L., 1942
 Ricerche sul fitoplancton del Mare Ligure 6. Le diatomee delle acque di Sanremo. Nuovo Giornale Botanico Italiano, N.S., 49:252-268.

Mastigloia fimbrata
Rampi, L., 1940
 Diatomee del Mare Adriatico. Nuovo Giornale Botanico Italiano, n.s., 47:559-608.

Mastigloia fimbriata
Zanon, V., 1948
 Diatomee marini di Sardegna e Pugillo di Alghe Marine della stressa. Boll. Pesca, Piscitutura e Idrobiologia, Anno 24, ns. 3(2): 202-244, 27 figs. on 1 pl.

Mastigloia gomphonemoides
Zanon, V., 1948
 Diatomee marini di Sardegna e Pugillo di Alghe Marine della stressa. Boll. Pesca, Piscitutura e Idrobiologia, Anno 24, ns. 3(2): 202-244, 27 figs. on 1 pl.

Mastigloia grunowi
Zanon, V., 1948
 Diatomee marini di Sardegna e Pugillo di Alghe Marine della stressa. Boll. Pesca, Piscitutura e Idrobiologia, Anno 24, ns. 3(2): 202-244, 27 figs. on 1 pl.

Mastigloia Jelineckii
Grunow, A., 1877
 1. New Diatoms from Honduras. Monthly Micros. Jour., 18:165-186, pls. CXCIII-CXCVI.

Mastigloia Jelinecki
Rampi, L., 1940
 Diatomee del Mare Adriatico. Nuovo Giornale Botanico Italiano, n.s., 47:559-608.

Mastigloia kerguelensis
Castracane degli Antelminelli, F., 1886   n.sp.
 1. Report on the Diatomaceae collected by H.M.S. Challenger during the years 1873-1876. Rept. Sci. Results, H.M.S. Challenger, Botany Vol. II, 178 pp., 30 pls.

Mastigloia labuensis
Zanon, V., 1948
 Diatomee marini di Sardegna e Pugillo di Alghe Marine della stressa. Boll. Pesca, Piscitutura e Idrobiologia, Anno 24, ns. 3(2): 202-244, 27 figs. on 1 pl.

Mastigloia laminaris
Zanon, V., 1948
 Diatomee marini di Sardegna e Pugillo di Alghe Marine della stressa. Boll. Pesca, Piscitutura e Idrobiologia, Anno 24, ns. 3(2): 202-244, 27 figs. on 1 pl.

Mastigloia lanceolata
Politis, J., 1949
 Diatomees marines de Bosphores et des ibes de la mer de Marmara. II Practica tou Hellenikou Hidrobiologikou Institutoutou 1929, Etoz 1929, 3(1):11-31.

Mastigloia lata
Andrade e Clovis Teixeira, M.H. de, 1957.
Contribuição para o conhecimento das diatomáceas do Brasil.
Bol. Inst. Ocean., Univ. Sao Paulo, 8(1/2):171-225, 10 pls.

Mastigloia lemnisca
Mann, A., 1907
 Report on the diatoms of the Albatross voyages in the Pacific Ocean, 1888-1904. Contrib. U.S. Nat. Herb. 10(5):221-419, Pls. XLIV-LIV.

Mastigloia marginulata
Grunow, A., 1877
 1. New Diatoms from Honduras. Monthly Micros. Jour., 18:165-186, pls. CXCIII-CXCVI.

Mastogloia maxima
Grunow, A., 1863
Ueber einige neue und ungenügend bekannte Arten und Gattungen von Diatomaceen. Verhandl. d. K.K. Zool. Bot. Gesellsch., Vienna, 13: 137-162, pl.4-5 (Pl. 13-14).

Mastogloia meleagris
Grunow, A., 1863
Ueber einige neue und ungenügend bekannte Arten und Gattungen von Diatomaceen. Verhandl. d. K.K. Zool. Bot. Gesellsch., Vienna, 13: 137-162, pl.4-5 (Pl. 13-14).

Mastogloia minuta
Allen, W.E., and E.E. Cupp, 1935
Plankton diatoms of the Java Sea. Annales du Jardin Botanique de Buitenzorg XLIV (2):101-174, figs.1-127.
(drawings of all species mentioned)

Mastogloia ovalis
Rampi, L., 1940
Diatomee del Mare Adriatico. Nuovo Giornale Botanico Italiano, n.s., 47:559-608.

Mastogloia ovalum
Rampi, L., 1940
Diatomee del Mare Adriatico. Nuovo Giornale Botanico Italiano, n.s., 47:559-608.

Mastogloia ovata
Andrade e Clovis Teixeira, M.H. de, 1957.
Contribuição para o conhecimento das diatomáceas do Brasil.
Bol. Inst. Ocean., Univ. Sao Paulo, 8(1/2):171-225, 10 pls.

Mastogloia ovata
Rampi, L., 1940
Diatomee del Mare Adriatico. Nuovo Giornale Botanico Italiano, n.s., 47:559-608.

Mastogloia ovata
Zanon, V., 1948
Diatomee marini di Sardegna e Pugillo di Alghe Marine della stressa. Boll. Pesca, Piscitutura e Idrobiologia, Anno 24, ns. 3(2): 202-244, 27 figs. on 1 pl.

Mastogloia ovulum
Zanon, V., 1948
Diatomee marini di Sardegna e Pugillo di Alghe Marine della stressa. Boll. Pesca, Piscitutura e Idrobiologia, Anno 24, ns. 3(2): 202-244, 27 figs. on 1 pl.

Mastogloia ovum paschae
Rampi, L., 1940
Diatomee del Mare Adriatico. Nuovo Giornale Botanico Italiano, n.s., 47:559-608.

Mastogloia ovum-paschale
Zanon, V., 1948
Diatomee marini di Sardegna e Pugillo di Alghe Marine della stressa. Boll. Pesca, Piscitutura e Idrobiologia, Anno 24, ns. 3(2): 202-244, 27 figs. on 1 pl.

Mastigloia paradoxa
Zanon, V., 1948
Diatomee marini di Sardegna e Pugillo di Alghe Marine della stressa. Boll. Pesca, Piscitutura e Idrobiologia, Anno 24, ns. 3(2): 202-244, 27 figs. on 1 pl.

Mastogloia Peragalli
Politis, J., 1949
Diatomees marines de Bosphores et des ibes de la mer de Marmara. II Practica tou Hellenikou Hidrobiologikou Institutoutou 1929, Etoz 1929, 3(1):11-31.

Mastigloia peragalli
Zanon, V., 1948
Diatomee marini di Sardegna e Pugillo di Alghe Marine della stressa. Boll. Pesca, Piscitutura e Idrobiologia, Anno 24, ns. 3(2): 202-244, 27 figs. on 1 pl.

Mastogloia portierana n.sp.
Grunow, A., 1863
Ueber einige neue und ungenügend bekannte Arten und Gattungen von Diatomaceen. Verhandl. d. K.K. Zool. Bot. Gesellsch., Vienna, 13: 137-162, pl.4-5 (Pl. 13-14).

Mastigloia pulchella
Zanon, V., 1948
Diatomee marini di Sardegna e Pugillo di Alghe Marine della stressa. Boll. Pesca, Piscitutura e Idrobiologia, Anno 24, ns. 3(2): 202-244, 27 figs. on 1 pl.

Mastogloia pumila
Hendy, N. Ingram, 1964
An introductory account of the smaller algae of British coastal waters. V. Bacillariophyceae (Diatoms).
Her Majesty's Stationary Office, 317 pp., 45 pls.

Mastogloia pumila
Takano, Hideaki, 1962
Notes on epiphytic diatoms upon sea-weeds from Japan.
J. Oceanogr.Soc., Japan, 18(1):29-33.

Mastigloia pumila
Zanon, V., 1948
Diatomee marini di Sardegna e Pugillo di Alghe Marine della stressa. Boll. Pesca, Piscitutura e Idrobiologia, Anno 24, ns. 3(2): 202-244, 27 figs. on 1 pl.

Mastigloia pusilla
Zanon, V., 1948
Diatomee marini di Sardegna e Pugillo di Alghe Marine della stressa. Boll. Pesca, Piscitutura e Idrobiologia, Anno 24, ns. 3(2): 202-244, 27 figs. on 1 pl.

Mastogloia quinquecostata
de Sousa e Silva, E., 1956.
Contribution à l'étude du microplancton de Dakar et des regions maritimes voisines.
Bull. I.F.A.N., 8(2):335-371, 7 pls.

Mastogloia quinquecostata
Rampi, L., 1942
Ricerche sul fitoplancton del Mare Ligure 6. Le diatomee delle acque di Sanremo. Nuovo Giornale Botanico Italiano, N.S., 49:252-268.

Mastogloia quinquecostata
Rampi, L., 1940
Diatomee del Mare Adriatico. Nuovo Giornale Botanico Italiano, n.s., 47:559-608.

Mastigloia quinquecostata
Zanon, V., 1948
Diatomee marini di Sardegna e Pugillo di Alghe Marine della stressa. Boll. Pesca, Piscitutura e Idrobiologia, Anno 24, ns. 3(2): 202-244, 27 figs. on 1 pl.

Mastogloia (?) reticulata
Grunow, A., 1877
1. New Diatoms from Honduras. Monthly Micros. Jour., 18:165-186, pls. CXCIII-CXCVI.

Mastogloia rostellata
Grunow, A., 1877
1. New Diatoms from Honduras. Monthly Micros. Jour., 18:165-186, pls. CXCIII-CXCVI.

Mastogloia Schmidti
Andrade e Clovis Teixeria, M.H. de, 1957.
Contribuição para o conhecimento das diatomáceas do Brasil.
Bol. Inst. Ocean., Univ. Sao Paulo, 8(1/2):171-225, 10 pls.

Mastogloia Schmidti
Rampi, L., 1940
Diatomee del Mare Adriatico. Nuovo Giornale Botanico Italiano, n.s., 47:559-608.

Mastigloia schmidtii
Zanon, V., 1948
Diatomee marini di Sardegna e Pugillo di Alghe Marine della stressa. Boll. Pesca, Piscitutura e Idrobiologia, Anno 24, ns. 3(2): 202-244, 27 figs. on 1 pl.

Mastogloia smithi
de Sousa e Silva, E., 1956.
Contribution à l'étude du microplancton de Dakar et des regions maritimes voisines.
Bull. I.F.A.N., 8(2):335-371, 7 pls.

Mastogloia smithii
Hendy, N. Ingram, 1964
An introductory account of the smaller algae of British coastal waters. V. Bacillariophyceae (Diatoms).
Her Majesty's Stationary Office, 317 pp., 45 pls.

Mastogloia Smithii
Politis, J., 1949
Diatomees marines de Bosphores et des ibes de la mer de Marmara. II Practica tou Hellenikou Hidrobiologikou Institutoutou 1929, Etoz 1929, 3(1):11-31.

Mastogloia Smithii
Zanon, V., 1948
Diatomee marini di Sardegna e Pugillo di Alghe Marine della stressa. Boll. Pesca, Piscitutura e Idrobiologia, Anno 24, ns. 3(2): 202-244, 27 figs. on 1 pl.

Mastogloia Smithi amphicephala
Rumkówna, A., 1948
[List of the phytoplankton species occurring in the superficial water layers in the Gulf of Gdańsk] Bull. Lab. mar., Gdynia, No. 4: 139-141 with tables in back.

Mastogloia splendida
Andrade e Clovis Teixeira, M.H. de, 1957.
Contribuição para o conhecimento das diatomáceas do Brasil.
Bol. Inst. Ocean., Univ. Sao Paulo, 8(1/2):171-225, 10 pls.

Mastogloia splendida
Eskinazi Enide e Shigekatsv Satô (1963-1964) 1966.
Contribuição ao estudo das diatomaceas da Praia de Piedade.
Trabhs Inst. Oceanogr., Univ. Recife, 5 (5/6): 73-114.

Mastogloia splendida
Hendy, N. Ingram, 1964
An introductory account of the smaller algae of British coastal waters. V. Bacillariophyceae (Diatoms).
Her Majesty's Stationary Office, 317 pp., 45 pls.

Mastogloia splendida
Rampi, L., 1942
Ricerche sul fitoplancton del Mare Ligure 6. Le diatomee delle acque di Sanremo. Nuovo Giornale Botanico Italiano, N.S., 49:252-268.

Mastogloia splendida
Rampi, L., 1940
Diatomee del Mare Adriatico. Nuovo Giornale Botanico Italiano, n.s., 47:559-608.

Mastogloia splendida
Zanon, V., 1948
Diatomee marini di Sardegna e Pugillo di Alghe Marine della stressa. Boll. Pesca, Piscitutura e Idrobiologia, Anno 24, ns. 3(2): 202-244, 27 figs. on 1 pl.

Mastigloia subaffirmata
Zanon, V., 1948
Diatomee marini di Sardegna e Pugillo di Alghe Marine della stressa. Boll. Pesca, Piscitutura e Idrobiologia, Anno 24, ns. 3(2): 202-244, 27 figs. on 1 pl.

Mastogloia tenera
Zanon, V., 1948
Diatomee marini di Sardegna e Pugillo di Alghe Marine della stressa. Boll. Pesca, Piscitutura e Idrobiologia, Anno 24, ns. 3(2): 202-244, 27 figs. on 1 pl.

Mastogloia tenuis
Takano, Hideaki, 1962
Notes on epiphytic diatoms upon sea-weeds from Japan. J. Oceanogr. Soc., Japan, 18(1):29-33.

Mastogloia thaitiana n.sp.
Castracane degli Antelminelli, F., 1886
1. Report on the Diatomaceae collected by H.M.S. Challenger during the years 1873-1876. Rept. Sci. Results, H.M.S. Challenger, Botany Vol. II, 178 pp., 30 pls.

Mastogloia undulata
Grunow, A., 1877
1. New Diatoms from Honduras. Monthly Micros. Jour., 18:165-186, pls. CXCIII-CXCVI.

Mastogloia undulata
Rampi, L., 1940
Diatomee del Mare Adriatico. Nuovo Giornale Botanico Italiano, n.s., 47:559-608.

Mastogloia varians
Zanon, V., 1948
Diatomee marini di Sardegna e Pugillo di Alghe Marine della stressa. Boll. Pesca, Piscitutura e Idrobiologia, Anno 24, ns. 3(2): 202-244, 27 figs. on 1 pl.

Mastogonia ?biaculeata n.sp.
Cleve-Euler, A., 1951
Die Diatomeen von Schweden und Finnland. Kungl. Svenska Vetenskaps Akad. Handl., Fjärde Ser. 2(1): 161 pp., 6 pls.

Melchersiela hexagolis n.gen., n.sp.
Teixeira, Clovis, 1958.
A new genus and a new species of diatom from Brazilian waters. Bol. Inst. Oceanogr., Univ. Sao Paulo, 9(1/2): 31-35, 2 pls.

Melosira sp.
Braarud, T., 1945
A phytoplankton survey of the polluted waters of inner Oslo Fjord. Hvalrådets Skrifter, No.28, 142 pp., 19 text figs., 17 tables.

Melosira
Desikachary, T.V., 1956 (1957).
Electron microscope studies on diatoms. J.R. Microsc. Soc. (3)76(1/2):9-36.

Melosira sp.
Gilbert, J.Y., and W.E. Allen, 1943
The phytoplankton of the Gulf of California obtained by the "E.W. Scripps" in 1939 and 1940. J. Mar. Res. V(2):89-110, figs. 30-31.

Melosira sp.
Kucherova, Z.S., 1961.
Vertical distribution of diatoms from Sevastopol Bay. Trudy Sevastopol Biol. Sta., (14):64-78.

Melosira sp.
Ukeles, R., 1961.
The effect of temperature on the growth and survival of several marine algal species. Biol. Bull., 120(2):255-264.

Melosira Adeliae n.sp.
Manguin, E., 1957.
Premier inventaire des diatomées de la Terre Adélie Antarctique. Espèces nouvelles. Rev. Algologique, n.s., 3(3):111-134.

Melosira ambigua
Zanon, D. V., 1949
Diatomee di Buenos Aires (Argentina) Atti Accad. Naz. Lincei, Memorie, Cl. Sci. fis., mat. e. nat., ser. 7, 11(3):59-151, 2 pls.

Melosira arctica
Braarud, T., 1945
A phytoplankton survey of the polluted waters of inner Oslo Fjord. Hvalrådets Skrifter, No.28, 142 pp., 19 text figs., 17 tables.

Melosira arctica
Brunel, J., 1962
Le phytoplancton de la Baie de Chaleurs. Inst. Botan. Univ. Montréal, Contrib. No. 77: 365 pp., 66 pls.

Melosira arctica
Cleve-Euler, A., 1951
Die Diatomeen von Schweden und Finnland. Kungl. Svenska Vetenskaps Akad. Handl., Fjärde Ser. 2(1): 161 pp., 6 pls.

Melosira arctica
Heimdal, Berit R., 1973
The fine structure of the frustules of Melosira nummuloides and M. arctica (Bacillariophyceae). Norwegian J. Bot. 20(2/3):139-149.

Melosira arctica
Levander, K.M., 1947
Plankton gesammelt in den Jahren 1899-1910 an den Küsten Finnlands. Finnländische Hydrographisch-Biologische Untersuchunger (aus dem Wasserbiologischen Laboratorin der Societas Scientiarum Fennica) No.11: 40 pp., 6 diagrams, 13 pls., tables.

Melosira arctica
Rothe F., 1942.
Quantitativen Untersuchungen über die Planktonverteilung in der östlichen Ostsee. Ber. Deutsch. Wiss. Komm. Meeresf., N.F.10:291-368, 33 text-figs.

Melosira arctica
Rothe, F., 1941
Quantitative Untersuchunger über die Planktonverteilung in der östlichen Ost see. Ber. Deut. Wiss. Komm. fur Meeresforschung. n.f. X(3):291-368, 33 text figs.

Melosira arenaria
Cleve-Euler, A., 1951
Die Diatomeen von Schweden und Finnland. Kungl. Svenska Vetenskaps Akad. Handl., Fjärde Ser. 2(1): 161 pp., 6 pls.

Melosira arenaria
Meunier, A., 1915
Microplancton de la Mer Flamande. 2. Diatomées (excepté le genre Chaetoceros). Mem. Mus. Roy. Hist. Nat., Belgique, 7(3):1-118, Pls. VIII-XIV.

Melossira Borreri
Chiba, T., 1949
On the distribution of the plankton in the eastern China Sea and Yellow Sea. 1. Plankton composition in the spring. J. Shimonoseki Coll. Fisheries, 1(1):57-63, 1 fig.

Melosira Borreri
Gran, H.H., 1908
Diatomeen. Nordisches Plankton, Botanischer Teil pp. XIX.1-XIX 146; 178 text figs.

Melosira borreri
Lafon, M., M. Durchon and Y. Saudray, 1955.
Recherches sur les cycles saisonnières du plankton. Ann. Inst. Océan., 31(3):125-230.

Melosira Borreri
Levander, K.M., 1947
Plankton gesammelt in den Jahren 1899-1910 an den Küsten Finnlands. Finnländische Hydrographisch-Biologische Untersuchunger (aus dem Wasserbiologischen Laboratorin der Societas Scientiarum Fennica) No.11: 40 pp., 6 diagrams, 13 pls., tables.

Melosira Borreri
Levring, T., 1945
Some culture experiments with marine plankton diatoms. Medd. Oceanografiska Institutet i Göteborg, No.9 (Göteborgs Kungl. Vetenskaps-och Vitterhets-Samhälles Handlingar, Sjätte Följden. Ser.B. Vol.3 (12), 1-17.

Melosira Borreri
Mangin, M. L., 1912
Phytoplancton de la croisière du "René" dans l'Atlantique (Septembre 1908). Ann. Inst. Ocean., n.s., 4(1):1-66, 2 pls., 41 text figs., 2 tables.

Melosira Borreri
Meunier, A., 1915
Microplancton de la Mer Flamande. 2. Diatomées (excepté le genre Chaetoceros). Mem. Mus. Roy. Hist. Nat., Belgique, 7(3):1-118, Pls. VIII-XIV.

Melosira borreri
Morse, D.C., 1947
Some observations on seasonal variations in plankton population Patuxent River, Maryland 1943-1945. Bd. Nat. Res., Publ. No.65, Chesapeake Biol. Lab., 31, 3 figs.

Melosira borreri
Paiva Carvalho, J., 1950
O plancton do Rio Maria Rodriques (Cananeis). 1. Diatomaceas e Dinoflagelados. Bol. Inst. Paulista Oceanogr. 1(1); 27-43, 2 fold-in tables, 2 figs.

Melosira borreri
Rothe, E. 1942.
Quantitativen Untersuchungen über die Planktonverteilung in der östlichen Ostsee. Ber. Deutsch. Wiss. Komm. Meeresf., N.F., 10:291-368, 33 text-figs.

Melosira Borreri
Rumkówna, A., 1948
List of the phytoplankton species occurring in the superficial water layers in the Gulf of Gdańsk. Bull. Lab. mar., Gdynia, No. 4: 139-141 with tables in back.

Melosira Borreri
Schodduyn, M., 1926
Observations faites dans la baie d'Ambleteuse (Pas de Calais). Bull. Inst. Ocean., Monaco, No. 482: 64 pp.

Melosira Borreri
Schröder, B., 1900
Phytoplankton des Golfes von Neapel nebst vergleichenden Ausblicken auf das atlantischen Ozean. Mitt. Zool. Stat. Neapel, 14:1-38.

Melosira borreri
Sousa e Silva, E., 1949
Diatomaceas e Dinoflagelados de Baia de Cascais. Portugaliae Acta Biol., Volume: Julio Henriques, Ser. B: 300-383, 9 pls, 2 fold-in tables.

Melosira Borreri

Sousa a Silva, E., and J. Dos Santos-Pinto, 1948
O Plancton da Baia de S. Martinho do Porto.
1. Diatomaceas e Dinoflagelados. Bol. Soc. Portuguese de Ciencias Naturais, 16(2):134-187, 6 pls. (Trav. Sta. Biol. Mar. de Lisbonne No. 52).

Melosira ? coronaria sp. nov.

Mann, A., 1907
Report on the diatoms of the Albatross voyages in the Pacific Ocean, 1888-1904.
Contrib. U. S. Nat. Herb. 10(5):221-419, Pls. XLIV-LIV.

Melosira costata

Castracane degli Antelminelli, F., 1886
1. Report on the Diatomaceae collected by H.M.S. Challenger during the years 1873-1876. Rept. Sci. Results, H.M.S. Challenger, Botany Vol. II, 178 pp., 30 pls.

Melosira crenulata

Meunier, A., 1915
Microplancton de la Mer Flamande. 2. Diatomées (excepté le genre Chaetoceros). Mem. Mus. Roy. Hist. Nat., Belgique, 7(3):1-118, Pls. VIII-XIV.

Melosira distans

Cleve-Euler, A., 1951
Die Diatomeen von Schweden und Finnland. Kungl. Svenska Vetenskaps Akad. Handl., Fjärde Ser. 2(1): 161 pp., 6 pls.

Melosira distans

Gran, H.H., 1908
Diatomeen. Nordisches Plankton, Botanischer Teil pp. XIX.1-XIX 146; 178 text figs.

Melosira distans

Zanon, D. V., 1949
Diatomee di Buenos Aires (Argentina)
Atti Accad. Naz. Lincei, Memorie, Cl. Sci. fis., mat. e. nat., ser. 7, 11(3):59-151, 2 pls.

Melosira Douguetti n.sp.

Manguin, E., 1957.
Premier inventaire des diatomées de la Terre Adélie Antarctique. Espèces nouvelles.
Rev. Algologique, n.s., 3(3):111-134.

Melosira dubia

Cleve-Euler, A., 1951
Die Diatomeen von Schweden und Finnland. Kungl. Svenska Vetenskaps Akad. Handl., Fjärde Ser. 2(1): 161 pp., 6 pls.

Melosira dubia

Misra, J.N., 1956.
A systematic account of some littoral marine diatoms from the west coast of India.
J. Bombay Nat. Hist. Soc., 53(4):537-568.

Melosira febigerii

Mann, A., 1907
Report on the diatoms of the Albatross voyages in the Pacific Ocean, 1888-1904.
Contrib. U. S. Nat. Herb. 10(5):221-419, Pls. XLIV-LIV.

Melosira fennoscandica n.sp.

Cleve-Euler, A., 1951
Die Diatomeen von Schweden und Finnland. Kungl. Svenska Vetenskaps Akad. Handl., Fjärde Ser. 2(1): 161 pp., 6 pls.

Melosira glacialis

Cleve-Euler, A., 1951
Die Diatomeen von Schweden und Finnland. Kungl. Svenska Vetenskaps Akad. Handl., Fjärde Ser. 2(1): 161 pp., 6 pls.

Melosira glomus n.sp.

Castracane degli Antelminelli, F., 1886
1. Report on the Diatomaceae collected by H.M.S. Challenger during the years 1873-1876. Rept. Sci. Results, H.M.S. Challenger, Botany Vol. II, 178 pp., 30 pls.

Melosira gothica n.sp.

Cleve-Euler, A., 1951
Die Diatomeen von Schweden und Finnland. Kungl. Svenska Vetenskaps Akad. Handl., Fjärde Ser. 2(1): 161 pp., 6 pls.

Melosira granulata

Cleve-Euler, A., 1951
Die Diatomeen von Schweden und Finnland. Kungl. Svenska Vetenskaps Akad. Handl., Fjärde Ser. 2(1): 161 pp., 6 pls.

Melosira granulata

Fox, M., 1957.
A first list of marine algae from Nigeria.
J. Linn. Soc., London, 55(362):615-631.

Melosira granulata

Gran, H.H., 1908
Diatomeen. Nordisches Plankton, Botanischer Teil pp. XIX.1-XIX 146; 178 text figs.

Melosira granulata

Tsumura, K., 1956.
Diatomoj el la cirkaufoso de la restajo de la kastele de Odawara. J. Yokohama Municipal Univ., (C-14) No. 47:23 pp.

Melosira granulata

Zanon, D. V., 1949
Diatomee di Buenos Aires (Argentina)
Atti Accad. Naz. Lincei, Memorie, Cl. Sci. fis., mat. e. nat., ser. 7, 11(3):59-151, 2 pls.

Melosira granulata

Zanon, V., 1948
Diatomee marini di Sardegna e Pugillo di Alghe Marine della stressa. Boll. Pesca, Piscitutura e Idrobiologia, Anno 24, ns. 3(2): 202-244, 27 figs. on 1 pl.

Melosira granulata angustissima

Cassie, Vivienne, 1961
Marine phytoplankton in New Zealand waters.
Botanica Marina, 2(Suppl.):54 pp., 8 pls.

Melosira granulata angustissima

Rumkówna, A., 1948
[List of the phytoplankton species occurring in the superficial water layers in the Gulf of Gdańsk] Bull. Lab. mar., Gdynia, No. 4: 139-141 with tables in back.

Melosira granulata v. augst.

Kokubo, S., and S. Sato., 1947
Plankters in Ju-San Gata. Physiol. and Ecol. (Japan) 1(4):1-16, 3 text figs., tables.

Melosira Herzogii

Cleve-Euler, A., 1951
Die Diatomeen von Schweden und Finnland. Kungl. Svenska Vetenskaps Akad. Handl., Fjärde Ser. 2(1): 161 pp., 6 pls.

Melosira Hustedti

Zanon, D. V., 1949
Diatomee di Buenos Aires (Argentina)
Atti Accad. Naz. Lincei, Memorie, Cl. Sci. fis., mat. e. nat., ser. 7, 11(3):59-151, 2 pls.

Melosira hormoides

Cleve-Euler, A., 1951
Die Diatomeen von Schweden und Finnland. Kungl. Svenska Vetenskaps Akad. Handl., Fjärde Ser. 2(1): 161 pp., 6 pls.

Melosira hyperborea

Gran, H.H., 1908
Diatomeen. Nordisches Plankton, Botanischer Teil pp. XIX.1-XIX 146; 178 text figs.

Melosira hyperborea

Levander, K.M., 1947
Plankton gesammelt in den Jahren 1899-1910 an den Küsten Finnlands.
Finnländische Hydrographisch-Biologishhe Untersuchunger (aus dem Wasserbiologischen Laboratorin der Societas Scientiarum Fennica) No.11: 40 pp., 6 diagrams, 13 pls., tables.

Melosira hyperborea

Rumkówna, A., 1948
[List of the phytoplankton species occurring in the superficial water layers in the Gulf of Gdańsk] Bull. Lab. mar., Gdynia, No. 4: 139-141 with tables in back.

Melosira hyalina n.sp.

Castracane degli Antelminelli, F., 1886
1. Report on the Diatomaceae collected by H.M.S. Challenger during the years 1873-1876. Rept. Sci. Results, H.M.S. Challenger, Botany Vol. II, 178 pp., 30 pls.

Melosira islandica

Cleve-Euler, A., 1951
Die Diatomeen von Schweden und Finnland. Kungl. Svenska Vetenskaps Akad. Handl., Fjärde Ser. 2(1): 161 pp., 6 pls.

Melosira islandica

Rumkówna, A., 1948
[List of the phytoplankton species occurring in the superficial water layers in the Gulf of Gdańsk] Bull. Lab. mar., Gdynia, No. 4: 139-141 with tables in back.

Melosira islandica

Zanon, D. V., 1949
Diatomee di Buenos Aires (Argentina)
Atti Accad. Naz. Lincei, Memorie, Cl. Sci. fis., mat. e. nat., ser. 7, 11(3):59-151, 2 pls.

Melosira islandica subsp. helvetica

Rodhe, W., 1948
Environmental requirements of fresh-water plankton algae. Experimental studies in the ecology of phytoplankton. Symbolae Botanicae Upsalienses X(1):149 pp., 30 figs.

Melosira italica

Cleve-Euler, A., 1951
Die Diatomeen von Schweden und Finnland. Kungl. Svenska Vetenskaps Akad. Handl., Fjärde Ser. 2(1): 161 pp., 6 pls.

Melosira italica

Lillick, L.C., 1940
Phytoplankton and planktonic protozoa of the offshore waters of the Gulf of Maine. Pt.II. Qualitative Composition of the Planktonic Flora. Trans. Am. Phil. Soc., n.s., 31(3):193-237, 13 text figs.

Melosira italica

Rumkówna, A., 1948
[List of the phytoplankton species occurring in the superficial water layers in the Gulf of Gdańsk] Bull. Lab. mar., Gdynia, No. 4: 139-141 with tables in back.

Melosira italica
Tsumura, K., 1956.
Diatomoj el la cirkaŭosa de la restajo de la kastele de Odawara. J. Yokohama Municipal Univ., (C-14) No. 47:23 pp.

Melosira italica
Zanon, D. V., 1949
Diatomee di Buenos Aires (Argentina) Atti. Accad. Naz. Lincei, Memorie, Cl. Sci. fis., mat. e. nat., ser. 7, 11(3):59-151, 2 pls.

Melosira Jürgensii
Cleve-Euler, A., 1951
Die Diatomeen von Schweden und Finnland. Kungl. Svenska Vetenskaps Akad. Handl., Fjärde Ser. 2(1): 161 pp., 6 pls.

Melosira Juergensii
Gran, H.H., 1908
Diatomeen. Nordisches Plankton, Botanischer Teil pp. XIX.1-XIX.146; 178 text figs.

Melosira juergensi
Gran, H. H. and E. C. Angst, 1931
Plankton diatoms of Puget Sound. Publ. Puget Sound Biol. Sta. 7:417-519, 95 text figs.

Melosira juergensii
Hendy, N. Ingram, 1964
An introductory account of the smaller algae of British coastal waters. V. Bacillariophyceae (Diatoms).
Her Majesty's Stationary Office, 317 pp., 45 pls.

Melosira juergensi
Kokubo, S., and S. Sato., 1947
Plankters in Jū-San Gata. Physiol. and Ecol. (Japan) 1(4):1-16, 3 text figs., tables.

Melosira Juergensi
Levander, K.M., 1947
Plankton gesammelt in den Jahren 1899-1910 an den Küsten Finnlands. Finnländische Hydrographisch-Biologische Untersuchungen (aus dem Wasserbiologischen Laboratorin der Societas Scientiarum Fennica) No.11: 40 pp., 6 diagrams, 13 pls., tables.

Melosira Jurgensii
Meunier, A., 1915
Microplancton de la Mer Flamande. 2. Diatomées (excepté le genre Chaetoceros). Mem. Mus. Roy. Hist. Nat., Belgique, 7(3):1-118, Pls. VIII-XIV.

Melosira juergensi
Misra, J.N., 1956.
A systematic account of some littoral marine diatoms from the west coast of India. J. Bombay Nat. Hist. Soc., 53(4):537-568.

Melosira Juergensi
Rampi, L., 1940
Diatomee del Mare Adriatico. Nuovo Giornale Botanico Italiano, n.s., 47:559-608.

Melosira gurgensi
Rumkówna, A., 1948
[List of the phytoplankton species occurring in the superficial water layers in the Gulf of Gdańsk] Bull. Lab. mar., Gdynia, No. 4: 139-141 with tables in back.

Melosira Jurgensi
Schodduyn, M., 1926
Observations faites dans la baie d'Ambleteuse (Pas de Calais). Bull. Inst. Ocean., Monaco, No. 482: 64 pp.

Melosira lineata
Brunel, J., 1962
Le phytoplancton de la Baie de Chaleurs. Inst. Botan., Univ. Montréal, Contrib. No. 77: 365 pp., 66 pls.

Melosira lineata
Cleve-Euler, A., 1951
Die Diatomeen von Schweden und Finnland. Kungl. Svenska Vetenskaps Akad. Handl., Fjärde Ser. 2(1): 161 pp., 6 pls.

Melosira lirata
Cleve-Euler, A., 1951
Die Diatomeen von Schweden und Finnland. Kungl. Svenska Vetenskaps Akad. Handl., Fjärde Ser. 2(1): 161 pp., 6 pls.

Melosira lucida
Cleve-Euler, A., 1951
Die Diatomeen von Schweden und Finnland. Kungl. Svenska Vetenskaps Akad. Handl., Fjärde Ser. 2(1): 161 pp., 6 pls.

Melosira medusa n.sp.
Mann, A., 1907
Report on the diatoms of the Albatross voyages in the Pacific Ocean, 1888-1904. Contrib. U. S. Nat. Herb. 10(5):221-419, Pls. XLIV-LIV.

Melosira moniliformis
Brunel, J., 1962
Le phytoplancton de la Baie de Chaleurs. Inst. Botan., Univ. Montréal, Contrib. No. 77: 365 pp., 66 pls.

Melosira moniliformis subglobosa

Melosira moniliformis
Brunel, Jules, 1962
Le phytoplancton de la Baie des Chaleurs. Contrib. Ministère de la Chasse et des Pêcheries, Province de Québec, No. 91: 365 pp.

Melosira moniliformis
Cleve-Euler, A., 1951
Die Diatomeen von Schweden und Finnland. Kungl. Svenska Vetenskaps Akad. Handl., Fjärde Ser. 2(1): 161 pp., 6 pls.

Melosira moniliformis
Cupp, Easter E., 1943
Marine plankton diatoms of the west coast of North America. Bull. S.I.O. 5(1):1-238, 5 pls., 168 text figs.

Melosira moniliformis
Gran, H. H. and E. C. Angst, 1931
Plankton diatoms of Puget Sound. Publ. Puget Sound Biol. Sta. 7:417-519, 95 text figs.

Melosira moniliformis
Hendy, N. Ingram, 1964
An introductory account of the smaller algae of British coastal waters. V. Bacillariophyceae (Diatoms).
Her Majesty's Stationary Office, 317 pp., 45 pls.

Melosira moniliformis
Hendey, N.I., 1951
Littoral diatoms of Chicester Harbour with special reference to fouling. J.Roy. Microscop. Soc., 71(1): 1-86, 18 pls.

Melosira moniliformis
Hustedt, F. and A.A. Aleem, 1951
Littoral diatoms from the Salstone near Plymouth. JMBA 30(1): 177-196.

Melosira moniliformis
Kokubo, S., and S. Sato., 1947
Plankters in Jū-San Gata. Physiol. and Ecol. (Japan) 1(4):1-16, 3 text figs., tables.

Melosira moniliformis
Levander, K.M., 1947
Plankton gesammelt in den Jahren 1899-1910 an den Küsten Finnlands. Finnländische Hydrographisch-Biologische Untersuchungen (aus dem Wasserbiologischen Laboratorin der Societas Scientiarum Fennica) No.11: 40 pp., 6 diagrams, 13 pls., tables.

Melosira moniliformis
Lillick, L.C., 1937
Seasonal studies of the phytoplankton off Woods Hole, Massachusetts. Biol. Bull. LXXIII (3):488-503, 3 text figs.

Melosira moniliformis
Migita, Seiji 1969.
Seasonal variation of cell size in Skeletonema costatum and Melosira moniliformis. (In Japanese; English abstract) Bull. Fac. Fish. Nagasaki Univ. 27: 9-17.

Melosira moniliformis
Takano, Hideaki, 1963.
Notes on marine littoral diatoms of Japan. Bull. Tokai Reg. Fish. Res. Lab., No. 36:1-8.

Melosira moniliformis
Takano, Hideaki, 1962
Notes on epiphytic diatoms upon sea-weeds from Japan.
J. Oceanogr. Soc., Japan, 18(1):29-33.

Melosira moniliformis
Zanon, V., 1948
Diatomee marini di Sardegna e Pugillo di Alghe Marine della stressa. Boll. Pesca, Piscicultura e Idrobiologia, Anno 24, ns. 3(2): 202-244, 27 figs. on 1 pl.

Melosira montagnei
Cleve-Euler, A., 1951
Die Diatomeen von Schweden und Finnland. Kungl. Svenska Vetenskaps Akad. Handl., Fjärde Ser. 2(1): 161 pp., 6 pls.

Melosira mucosa n.sp.
Mangin, L., 1915
Phytoplancton de L'Antartique. Deuxième Exped. Ant. Francaise (1908-1910), 95 pp., 3 pls., 58 text figs.

Melosira nummuloides
Brunel, J., 1962
Le phytoplancton de la Baie de Chaleurs. Inst. Botan., Univ. Montréal, Contrib. No. 77: 365 pp., 66 pls.

Melosira nummuloides
Brunel, Jules, 1962
Le phytoplancton de la Baie des Chaleurs. Contrib. Ministère de la Chasse et des Pêcheries, Province de Québec, No. 91: 365 pp.

Melosira nummuloides
Castenholz, Richard W., 1963.
An experimental study of the vertical distribution of littoral marine diatoms. Limnology and Oceanography, 8(4):450-462.

Melosira nummuloides
Cleve-Euler, A., 1951
Die Diatomeen von Schweden und Finnland. Kungl. Svenska Vetenskaps Akad. Handl., Fjärde Ser. 2(1): 161 pp., 6 pls.

Melosira nummuloides
Gran, H.H., 1908
Diatomeen. Nordisches Plankton, Botanischer Teil pp. XIX.1-XIX 146; 178 text figs.

Melosira numuloides
Gran, H. H. and E. C. Angst, 1931
Plankton diatoms of Puget Sound. Publ. Puget Sound Biol. Sta. 7:417-519, 95 text figs.

Melosira nummuloides
Heimdal, Berit R., 1973
The fine structure of the frustules of Melosira nummuloides and M. arctica (Bacillariophyceae). Norwegian J. Bot. 20(2/3):139-149.

Melosira nummuloides
Hellebust, Johan A., and Robert R.L. Guillard, 1967.
Uptake specificity for organic substrates by the marine diatom Melosira Nummuloides. J. Phycology, 3(3):132-136.

Melosira nummuloides
Hendy, N. Ingram, 1964
An introductory account of the smaller algae of British coastal waters. V. Bacillariophyceae (Diatoms). Her Majesty's Stationary Office, 317 pp., 45 pls.

Melosira nummuloides
Hendey, N.I., 1951
Littoral diatoms of Chicester Harbour with special reference to fouling. J.Roy. Microscop. Soc. 71(1): 1-86, 18 pls.

Melosira numuloides
Kokubo, S., and S. Sato., 1947
Plankters in Jû-San Gata. Physiol. and Ecol. (Japan) 1(4):1-16, 3 text figs., tables.

Melosira nummuloides
Meunier, A., 1915
Microplancton de la Mer Flamande. 2. Diatomées (excepté le genre Chaetoceros). Mem. Mus. Roy. Hist. Nat., Belgique, 7(3):1-118, Pls. VIII-XIV.

Melosira nummuloides
Rampi, L., 1940
Diatomee del Mare Adriatico. Nuovo Giornale Botanico Italiano, n.s., 47:559-608.

Melosira nummuloides
Rumkówna, A., 1948
List of the phytoplankton species occurring in the superficial water layers in the Gulf of Gdańsk. Bull. Lab. mar., Gdynia, No. 4: 139-141 with tables in back.

Melosira nummuloides
Takano, Hideaki, 1963.
Notes on marine littoral diatoms of Japan. Bull. Tokai Reg. Fish. Res. Lab., No. 36:1-8.

Melosira nummuloides
Takano, Hideaki, 1962
Notes on epiphytic diatoms upon sea-weeds from Japan. J. Oceanogr. Soc., Japan, 18(1):29-33.

Melosira nummuloides
Zanon, V., 1948
Diatomee marini di Sardegna e Pugillo di Alghe Marine della stressa. Boll. Pesca, Piscitutura e Idrobiologia, Anno 24, ns. 3(2): 202-244, 27 figs. on 1 pl.

Melosira nummulus
Iselin, C., 1930
A report on the coastal waters of Labrador based on explorations of the "Chance" during the summer of 1926. Proc. Am. Acad. Arts Sci., 66(1):1-37, 14 text figs.

Melosira nummulus n.sp.
Meunier, A., 1915
Microplancton de la Mer Flamande. 2. Diatomées (excepté le genre Chaetoceros). Mem. Mus. Roy. Hist. Nat., Belgique, 7(3):1-118, Pls. VIII-XIV.

Melosira ornata
Mann, A., 1893
List of Diatomaceae from a deep-sea dredging in the Atlantic Ocean off Delaware Bay by the U. S. Fish Commission Steamer Albatross. Proc. U. S. Nat. Mus. 16:303-312.

Melosira papillifera n.sp.
Hendey N. Ingram 1971.
Some marine diatoms from the Galapagos Islands.
Nova Hedwigia 22 (1/2):371-422.

Melosira polaris
Hendey, N.I., 1937
The plankton diatoms of the southern seas. Discovery Repts. 16:151-364, pls.6-13.

Melosira roeseana
Cleve-Euler, A., 1951
Die Diatomeen von Schweden und Finnland. Kungl. Svenska Vetenskaps Akad. Handl., Fjärde Ser. 2(1): 161 pp., 6 pls.

Melosira Roseana
Zanon, V., 1948
Diatomee marini di Sardegna e Pugillo di Alghe Marine della stressa. Boll. Pesca, Piscitutura e Idrobiologia, Anno 24, ns. 3(2): 202-244, 27 figs. on 1 pl.

Melosira (sulcata var.?) scopos n.sp.
Mann, A., 1907
Report on the diatoms of the Albatross voyages in the Pacific Ocean, 1888-1904. Contrib. U. S. Nat. Herb. 10(5):221-419, Pls. XLIV-LIV.

Melosira sculpta
Cleve-Euler, A., 1951
Die Diatomeen von Schweden und Finnland. Kungl. Svenska Vetenskaps Akad. Handl., Fjärde Ser. 2(1): 161 pp., 6 pls.

Melosira sol
Castracane degli Antelminelli, F., 1886
1. Report on the Diatomaceae collected by H.M.S. Challenger during the years 1873-1876. Rept. Sci. Results, H.M.S. Challenger, Botany Vol. II, 178 pp., 30 pls.

Melosira sol
Florin, M-B., 1948
9. Diatomeae in submarine cores from the Tyrrhenian Sea. Medd. Ocean. Inst., Göteborg, 15 (Göteborgs Kungl. Vetenskaps-och Viterrhets Samhälles Handlingar, Sjätte Foljden, Ser. B 5(13):80-88.

Melosira sol
Hendey, N.I., 1937
The plankton diatoms of the southern seas. Discovery Repts. 16:151-364, pls.6-13.

Melosira sol
Mangin, L., 1915
Phytoplancton de L'Antartique. Deuxieme Exped. Ant. Francaise (1908-1910), 95 pp., 3 pls, 58 text figs.

Melosira sol
Mann, A., 1907
Report on the diatoms of the Albatross voyages in the Pacific Ocean, 1888-1904. Contrib. U. S. Nat. Herb. 10(5):221-419, Pls. XLIV-LIV.

Melosira solida
Gran, H.H., 1897
Protophyta: Diatomaceae, Silico-flagellata and Cilioflagellata. Den Norske Nordhavs Expedition 1876-1878, h. 24, 36 pp., 4 pls.

Melosira solida
Morse, D.C., 1947
Some observations on seasonal variations in plankton population Patuxant River, Maryland 1943-1945. Bd. Nat. Res., Publ. No.65, Chesapeake Biol. Lab., 31, 3 figs.

Melosira sphaerica (fig)
Boden, B.P., 1950
Some marine plankton diatoms from the west coast of South Africa. Trans. R.Soc. S. Africa. 32:321-434, 100 text figs.

Melosira sphaerica
Hendey, N.I., 1937
The plankton diatoms of the southern seas. Discovery Repts. 16:151-364, pls.6-13.

Melosira sphaerica
Mangin, L., 1915
Phytoplancton de L'Antartique. Deuxieme Exped. Ant. Francaise (1908-1910), 95 pp., 3 pls, 58 text figs.

Melosira subsetosa n.sp.
Manguin, E., 1957.
Premier inventaire des diatomées de la Terre Adélie Antarctique. Espèces nouvelles. Rev. Algologique, n.s., 3(3):111-134.

Melosira sulcata
Braarud, T., 1934
A note on the phytoplankton of the Gulf of Maine in the summer of 1933. Biol. Bull. 67(1):76-82. (Contribution No.46 of the Woods Hole Oceanographic Institution)

Melosira sulcata
Brunel, J., 1962
Le phytoplancton de la Baie de Chaleurs. Inst. Botan., Univ. Montréal, Contrib. No. 77: 365 pp., 66 pls.

Melosira sulcata
Brunel, Jules, 1962
Le phytoplancton de la Baie des Chaleurs. Contrib. Ministère de la Chasse et des Pêcheries, Province de Québec, No. 91: 365 pp.

Melosira sulcata
Cupp, Easter E., 1943
Marine plankton diatoms of the west coast of North America. Bull. S.I.O. 5(1):1-238, 5 pls., 168 text figs.

Melosira sulcata
Fox, M., 1957.
A first list of Marine algae from Nigeria. J. Linn. Soc., London, 55(362):615-631.

Melosira (paralia) sulcata
Frenguelli, Joaquin, and Hector Antonio Orlando, 1959.
Operacion MaRLUZA. Diatomeas y silico-flagelados del plancton del "VI Crucero". Servicio Hidrogr. Naval., Argentina, Publ. No. H. 619: 5-62.

Melosira sulcata
Gran, H.H., and T. Braarud, 1935
A quantitative study of the phytoplankton in the Bay of Fundy and the Gulf of Maine (including observations on hydrography, chemistry, and turbidity). J. Biol. Bd., Canada, 1(5):279-467, 69 text figs.

Melosira sulcata
Grøntved, J., 1949(1950).
Investigations on the phytoplankton in the Danish Waddensee in July 1941. Medd. Komm. Danmarks Fiskeri- og Havundersøgelser, Serie Plankton, 5(2)(Medd. Skalling Lab. 10):55 pp., 2 pls., 38 textfigs.

Melosira sulcata
Grøntved, J., 1949
Investigations on the phytoplankton in the Danish Waddensea in July 1941. Medd. Komm. Danmarks Fiskeri og Havundersøgelser, ser. Plankton, 5(2):55 pp., 2 pls., 38 text figs.

Melosira sulcata
Hendey, N.I. 1951
Littoral diatoms of Chicester Harbour with special reference to fouling. J. Roy. Microscop. Soc. 71(1): 1-86, 18 pls.

Melosira sulcata
Hendey, N.I., 1937
The plankton diatoms of the southern seas. Discovery Repts. 16:151-364, pls.6-13.

Melosira sulcata
Hustedt, F. and A.A. Aleem, 1951
Littoral diatoms from the Salstone near Plymouth. JMBA 30(1): 177.196.

Melosira sulcata
Lillick, L.C., 1940
Phytoplankton and planktonic protozoa of the offshore waters of the Gulf of Maine. Pt.II. Qualitative Composition of the Planktonic Flora. Trans. Am. Phil. Soc., n.s., 31(3):193-237, 13 text figs.

Melosira sulcata
Lillick, L.C., 1938
Preliminary report of the phytoplankton of the Gulf of Maine. Am. Mid. Nat. 20(3):624-640, 1 text figs 37 tables.

Melosira sulcata
Lillick, L.C., 1937
Seasonal studies of the phytoplankton off Woods Hole, Massachusetts. Biol. Bull. LXXIII (3):488-503, 3 text figs.

Melosira sulcata
Mann, A., 1907
Report on the diatoms of the Albatross voyages in the Pacific Ocean, 1888-1904. Contrib. U. S. Nat. Herb. 10(5):221-419, Pls. XLIV-LIV.

Melosira sulcata
Mann, A., 1893
List of Diatomaceae from a deep-sea dredging in the Atlantic Ocean off Delaware Bay by the U. S. Fish Commission Steamer Albatross. Proc. U. S. Nat. Mus. 16:303-312.

Melosira sulcata
Misra, J.N., 1956.
A systematic account of some littoral marine diatoms from the west coast of India. J. Bombay Nat. Hist. Soc., 53(4):537-568.

Melosira sulcata
Morse, D.C., 1947
Some observations on seasonal variations in plankton population Patuxent River, Maryland 1943 1945. Bd. Nat. Res., Publ. No.65, Chesapeake Biol. Lab., 31, 3 figs.

Melosira sulcata
Politis, J., 1949
Diatomees marines de Bosphores et des ibes de la mer de Marmara. II Practica tou Hellenikou Hidrobiologikou Institutoutou 1929, Etoz 1929, 3(1):11-31.

Melosira sulcata
Ramamurthy, V.D., and K. Krishnamurthy, 1965.
On the culture of the diatom Melosira sulcata (Ehr.) Kutzing. Proc. Indian Acad. Sci., B, 42 (1):25-31.

Also in: Collected Reprints, Mar. Biol. Sta., Porto Novo, 1963/64.

Melosira sulcata
Ramamurthy, V.D., and K. Krishnamoorthy, 1965.
On the culture of the diatom Melosira sulcata (Ehr.) Kutzing. Proc. Indian Acad. Sci., (B), 62(1):25-31.

Melosira sulcata
Ramamurthy, V.D., and R. Seshadri, 1966.
Effects of gibberellic acid (GA) on laboratory cultures of Trichodesmium erythraeum (Ehr.) and Melosira sulcata (Ehr.). Proc. Indian Acad. Sci., (B), 64(3):146-151.

Melosira sulcata
Rampi, L., 1940
Diatomee del Mare Adriatico. Nuovo Giornale Botanico Italiano, n.s., 47:559-608.

Melosira sulcata
Zanon, D. V., 1949
Diatomee di Buenos Aires (Argentina) Atti Accad. Naz. Lincei, Memorie, Cl. Sci. fis., mat. e. nat., ser. 7, 11(3):59-151, 2 pls.

Melosira sulcata
Zanon, V., 1948
Diatomee marini di Sardegna e Pugillo di Alghe Marine della stressa. Boll. Pesca, Piscitutura e Idrobiologia, Anno 24, ns. 3(2): 202-244, 27 figs. on 1 pl.

Melosira sulcata radiata
Florin, M-B., 1948
9. Diatomeae in submarine cores from the Tyrrhenian Sea. Medd. Ocean. Inst., Göteborg, 15 (Göteborgs Kungl. Vetenskaps-och Viterrhets Samhälles Handlingar, Sjätte Foljden, Ser. B 5(13):80-88.

Melosira Tcherniai n.sp.
Manguin, E., 1957.
Premier inventaire des diatomées de la Terre Adélie, Antarctique. Espèces nouvelles. Rev. Algologique, n.s., 3(3):111-134.

Melosira thaitiensis n.sp.
Castracane degli Antelminelli, F., 1886
1. Report on the Diatomaceae collected by H.M.S. Challenger during the years 1873-1876. Rept. Sci. Results, H.M.S. Challenger, Botany Vol. II, 178 pp., 30 pls.

Melosira undulata
Cleve-Euler, A., 1951
Die Diatomeen von Schweden und Finnland. Kungl. Svenska Vetenskaps Akad. Handl., Fjärde Ser. 2(1): 161 pp., 6 pls.

Melosira undulata
Mann, A., 1907
Report on the diatoms of the Albatross voyages in the Pacific Ocean, 1888-1904. Contrib. U. S. Nat. Herb. 10(5):221-419, Pls. XLIV-LIV.

Melosira undulata
Morse, D.C., 1947
Some observations on seasonal variations in plankton population Patuxent River, Maryland 1943 1945. Bd. Nat. Res., Publ. No.65, Chesapeake Biol. Lab., 31, 3 figs.

Melosira varians
Cleve-Euler, A., 1951
Die Diatomeen von Schweden und Finnland. Kungl. Svenska Vetenskaps Akad. Handl., Fjärde Ser. 2(1): 161 pp., 6 pls.

Melosira varians
Crawford, Richard M. 1973.
The protoplasmic ultrastructure of The vegetative cell of Melosira varians C. A. Agardh. J. Phycol. 9(1): 50-61.

Melosira varians
Donchenko, N.S., 1970.
Effect of growing conditions of Melosira varians Ag. on its biochemical composition and fragrance. (In Russian). Gidrobiol. Zh., 6(6): 90-93.

Melosira varians
Kokubo, S., and S. Sato, 1947
Planktera in JO-San Gata. Physiol. and Ecol. (Japan) 1(4):1-16, 3 text figs., tables.

Melosira varians
Meunier, A., 1915
Microplancton de la Mer Flamande. 2. Diatomées (excepté le genre Chaetoceros). Mem. Mus. Roy. Hist. Nat., Belgique, 7(3):1-118, Pls. VIII-XIV.

Melosira varians
Mann, A., 1893
List of Diatomaceae from a deep-sea dredging in the Atlantic Ocean off Delaware Bay by the U. S. Fish Commission Steamer Albatross. Proc. U. S. Nat. Mus. 16:303-312.

Melosira varians
Rumkówna, A., 1948
List of the phytoplankton species occurring in the superficial water layers in the Gulf of Gdańsk. Bull. Lab. mar. Gdynia, No. 4: 139-141 with tables in back.

Melosira varians
von Stosch, H.-A., 1951.
Entwicklungsgeschichtliche Untersuchungen an zentrischen Diatomeen. 1. Die Auxosporenbildung von Melosira varians. Arch. f. Mikrobiol. 16(2): 102-135, 2 pls. (29 figs).

Melosira varians
von Stosch, 1950.
Oogamy in a centric diatom. Nature 165(4196): 531-532.

Melosira varians
Zanon, D. V., 1949
Diatomee di Buenos Aires (Argentina) Atti. Accad. Naz. Lincei, Memorie, Cl. Sci. fis., mat. e. nat., ser. 7, 11(3):59-151, 2 pls.

Melosira varians
Zanon, V., 1948
Diatomee marini di Sardegna e Pugillo di Alghe Marine della stressa. Boll. Pesca, Piscitutura e Idrobiologia, Anno 24, ns. 3(2): 202-244, 27 figs. on 1 pl.

Melosira westii
Castracane degli Antelminelli, F., 1886
1. Report on the Diatomaceae collected by H.M.S. Challenger during the years 1873-1876. Rept. Sci. Results, H.M.S. Challenger, Botany Vol. II, 178 pp., 30 pls.

Melosira westii
Cleve-Euler, A., 1951
Die Diatomeen von Schweden und Finnland. Kungl. Svenska Vetenskaps Akad. Handl., Fjärde Ser. 2(1): 161 pp., 6 pls.

Melosira westii

Gran, H.H., 1908
Diatomeen. Nordisches Plankton, Botanischer Teil pp. XIX.1-XIX 146; 178 text figs.

Melosira westii

Hendy, N. Ingram, 1964
An introductory account of the smaller algae of British coastal waters. V. Bacillariophyceae (Diatoms).
Her Majesty's Stationary Office, 317 pp., 45 pls.

Melosira Westii

Hustedt, F. and A.A. Aleem, 1951
Littoral diatoms from the Salstone near Plymouth. JMBA 30(1): 177.196.

Melosira Westii

Meunier, A., 1915
Microplancton de la Mer Flamande. 2. Diatomées (excepté le genre Chaetoceros). Mem. Mus. Roy. Hist. Nat., Belgique, 7(3):1-118, Pls. VIII-XIV.

Melosira Westii

Rampi, L., 1940
Diatomee del Mare Adriatico. Nuovo Giornale Botanico Italiano, n.s., 47:559-608.

Melosira Westii

Zanon, V., 1948
Diatomee marini di Sardegna e Pugillo di Alghe Marine della stressa. Boll. Pesca, Piscitutura e Idrobiologia, Anno 24, ns. 3(2): 202-244, 27 figs. on 1 pl.

Meridion

Desikachary, T.V., 1956(1957).
Electron microscope studies on diatoms. J.R.Microsc. Soc. (3)76(1/2):9-36.

Meridion circulare

Brunel, J., 1962
Le phytoplancton de la Baie de Chaleurs. Inst. Botan., Univ. Montréal, Contrib. No. 77: 365 pp., 66 pls.

Mölleria antarctica n.sp.

Castracane degli Antelminelli, F., 1886
1. Report on the Diatomaceae collected by H.M.S. Challenger during the years 1873-1876. Rept. Sci. Results, H.M.S. Challenger, Botany Vol. II, 178 pp., 30 pls.

Mölleria cornuta

Castracane degli Antelminelli, F., 1886
1. Report on the Diatomaceae collected by H.M.S. Challenger during the years 1873-1876. Rept. Sci. Results, H.M.S. Challenger, Botany Vol. II, 178 pp., 30 pls.

Navicula sp.

Allen, W.E., 1938
The Templeton Crocker Expedition to the Gulf of California in 1935 - The Phytoplankton. Amer. Microsc. Soc., Trans. 57:328-335.

Navicula sp.

Allen, W.E., 1937
Plankton diatoms of the Gulf of California obtained by the G. Allan Hancock Expedition of 1936. The Hancock Pacific Expeditions, Univ. So. Calif. Publ. 3:47-59, 1 fig.

Navicula sp.

Braarud, T., 1945
A phytoplankton survey of the polluted waters of inner Oslo Fjord. Hvalrådets Skrifter, No.28, 142 pp., 19 text figs., 17 tables.

Navicula sp.

Braarud, T., and Adam Bursa, 1939
On the phytoplankton of the Oslo Fjord, 1933-1934. Hvalrådets Skr. No.19:1-63; 9 text figs. Reviewed. J. du. Cons. 14(3): 418-420. A.C. Gardiner.

Navicula sp.

Brunel, J., 1962
Le phytoplancton de la Baie de Chaleurs. Inst. Botan., Univ. Montréal, Contrib. No. 77: 365 pp., 66 pls.

Navicula (Schizonema) sp.

Brunel, Jules, 1962
Le phytoplancton de la Baie des Chaleurs. Contrib. Ministère de la Chasse et des Pêcheries, Province de Québec, No. 91: 365 pp.

Navicula sp.

Cupp, E.E., 1934
Analysis of marine diatom collections taken from the Canal Zone to California during March, 1933. Trans. Am. Micros. Soc. LIII (1):22-29, 1 map.

Navicula sp.

Cupp, E.E. and Allen, W.E., 1938
Plankton diatoms of the Gulf of California obtained by Allan Hancock Pacific Expedition of 1937. The Hancock Pacific Expeditions, The Univ. So. Calif. Publ. 3: 61-74, 1 map, pls.4-15.

Navicula

Desikachary, T.V., 1956(1957).
Electron microscope studies on diatoms. J.R. Microsc. Soc. (3)76(1/2):9-36.

Navicula sp.

Galtsoff, P.S., 1948
Red Tide. Progress Report on the investigations of the cause of the mortality of fish along the west coast of Florida conducted by the U.S. Fish and Wildlife Service and Cooperating Organizations. Fish and Wildlife Service, Special Scientific Rept. No.46, 44 pp. (mimeographed), 9 figs.

Navicula sp.

Gilbert, J.Y., and W.E. Allen, 1943
The phytoplankton of the Gulf of California obtained by the "E.W. Scripps" in 1939 and 1940. J. Mar. Res. V(2):89-110, figs.30-31.

Navicula sp.

Kucherova, Z.S., 1961.
Vertical distribution of diatoms from Sevastopol Bay. Trudy Sevastopol Biol. Sta., (14):64-78.

Navicula spp.

Lillick, L.C., 1937
Seasonal studies of the phytoplankton off Woods Hole, Massachusetts. Biol. Bull. LXXIII (3):488-503, 3 text figs.

Navicula sp.

Mangin, M.L., 1912
Phytoplancton de la croisière du "René" dans l'Atlantique (Septembre 1908). Ann. Inst. Ocean., n.s., 4(1):1-66, 2 pls., 41 text figs., 2 tables.

Navicula

Neaverson, E., 1934
The sea-floor deposits. 1. General characteristics and distribution. Discovery Repts. 9: 297-349, Plates 17-22.

Navicula sp.

Pettersson, H., 1934
Scattering and extinction of light in sea-water. Meddelanden från Göteborgs Högskolas Oceanografiska Institut. 9 (Göteborgs Kungl. vetenskaps-och vitterhets-samhälles Handlingar. Femte Foljden, Ser.B, 4(4)):1-16.

Navicula sp.

Pettersson, H., F. Gross, and F. Koczy, 1939
Large scale plankton cultures. Medd. från Oceanografiska Institutet i Göteborg No.3 (Göteborgs Kungl. vetenskaps-och Vitterhets-Samhälles Handlingar Femte Följden. Ser. B) Vol.6(13):1-24.

1. The plankton shaft. H. Pettersson
2. Experiment with phytoplankton. F. Gross and F. Koczy
3. Experiments with zooplankton.

Navicula spp.

Sugawara, Ken, and Kikuo Terada, 1967.
Iodine assimilation by a marine Navicula sp. and the production of iodate accompanied by the growth of the algae.
Inf.Bull.Planktol.Japan,Comm.No.Dr.Y.Matsue, 213-218.

Navicula aberrans n.sp.

Simonsen, Reimer, 1960.
Neue Diatomeen aus der Ostsee. II. Kieler Meeresf., 16(1):126-130.

Navicula abnormis (?) n.sp.

Castracane degli Antelminelli, F., 1886
1. Report on the Diatomaceae collected by H.M.S. Challenger during the years 1873-1876. Rept. Sci. Results, H.M.S. Challenger, Botany Vol. II, 178 pp., 30 pls.

Navicula abnormis

Mann, A., 1893
List of Diatomaceae from a deep-sea dredging in the Atlantic Ocean off Delaware Bay by the U. S. Fish Commission Steamer Albatross. Proc. U. S. Nat. Mus. 16:303-312.

Navicula abrupta

Hendy, N. Ingram, 1964
An introductory account of the smaller algae of British coastal waters. V. Bacillariophyceae (Diatoms).
Her Majesty's Stationary Office, 317 pp., 45 pls.

Navicula abrupta

Hustedt, F. and A.A. Aleem, 1951
Littoral diatoms from the Salstone near Plymouth. JMBA 30(1): 177.196.

Navicula abrupta

Politis, J., 1949
Diatomees marines de Bosphores et des ibes de la mer de Marmara. II Practica tou Hellenikou Hidrobiologikou Institutoutou 1929, Etoz 1929, 3(1):11-31.

Navicula abrupta

Rampi, L., 1942
Ricerche sul fitoplancton del Mare Ligure 6. Le diatomee delle acque di Sanremo. Nuovo Giornale Botanico Italiano, N.S., 49:252-258.

Navicula abrupta

Rampi, L., 1940
Diatomee del Mare Adriatico. Nuovo Giornale Botanico Italiano, n.s., 47:559-608.

Navicula abrupta

Zanon, V., 1948
Diatomee marini di Sardegna e Pugillo di Alghe Marine della stressa. Boll. Pesca, Piscitutura e Idrobiologia, Anno 24, ns. 3(2): 202-244, 27 figs. on 1 pl.

Navicula accomoda

Jørgensen, E.G., 1952.
Notes on the ecology of the diatom Navicula accomoda Hustedt. Bot. Tidsskr. 49(2):189-191.

Navicula acus
Zanon, V., 1948
Diatomee marini di Sardegna e Pugillo di Alghe Marine della stressa. Boll. Pesca, Piscitutura e Idrobiologia, Anno 24, ns. 3(2): 202-244, 27 figs. on 1 pl.

Navicula adnatoides n.sp.
Cholnoky, B.J., 1963.
Beiträge zur Kenntnis des marinen Litorals von Südafrika.
Botanica Marina, 5(2/3):38-83.

Navicula aestiva
Mann, A., 1907
Report on the diatoms of the Albatross voyages in the Pacific Ocean, 1888-1904. Contrib. U. S. Nat. Herb. 10(5):221-419, Pls. XLIV-LIV.

Navicula aleemi
Hendy, N. Ingram, 1964
An introductory account of the smaller algae of British coastal waters. V. Bacillariophyceae (Diatoms).
Her Majesty's Stationary Office, 317 pp., 45 pls.

Navicula aleemi n.sp.
Hustedt, F. and A.A. Aleem, 1951
Littoral diatoms from the Salstone near Plymouth. JMBA 30(1): 177.196.

Navicula americana
Mann, A., 1893
List of Diatomaceae from a deep-sea dredging in the Atlantic Ocean off Delaware Bay by the U. S. Fish Commission Steamer Albatross.
Proc. U. S. Nat. Mus. 16:303-312.

Navicula ammophila
Hendy, N. Ingram, 1964
An introductory account of the smaller algae of British coastal waters. V. Bacillariophyceae (Diatoms).
Her Majesty's Stationary Office, 317 pp., 45 pls.

Navicula ammophila
Zanon, V., 1948
Diatomee marini di Sardegna e Pugillo di Alghe Marine della stressa. Boll. Pesca, Piscitutura e Idrobiologia, Anno 24, ns. 3(2): 202-244, 27 figs. on 1 pl.

Navicula amoena
Rampi, L., 1940
Diatomee del Mare Adriatico. Nuovo Giornale Botanico Italiano, n.s., 47:559-608.

Navicula anceps
Mann, A., 1907
Report on the diatoms of the Albatross voyages in the Pacific Ocean, 1888-1904. Contrib. U. S. Nat. Herb. 10(5):221-419, Pls. XLIV-LIV.

Navicula ancilla n.sp.
Hendy, N. Ingram, 1964
An introductory account of the smaller algae of British coastal waters. V. Bacillariophyceae (Diatoms).
Her Majesty's Stationary Office, 317 pp., 45 pls.

Navicula (Schizonema) antarctica n.sp.
Frenguelli, J., 1960.
Diatomeas y silicoflagelados recogidas en Tierra Adélia durante las Expediciones Polares Francesas de Paul-Emile VICTOR (1950-1952).
Revue Algologique, (1):1-47.

Navicula antillarum
Mann, A., 1907
Report on the diatoms of the Albatross voyages in the Pacific Ocean, 1888-1904. Contrib. U. S. Nat. Herb. 10(5):221-419, Pls. XLIV-LIV.

Navicula apiculata
Rampi, L., 1940
Diatomee del Mare Adriatico. Nuovo Giornale Botanico Italiano, n.s., 47:559-608.

Navicula apis
Politis, J., 1949
Diatomees marines de Bosphores et des ibes de la mer de Marmara. II Practica tou Hellenikou Hidrobiologikou Institutoutou 1929, Etoz 1929, 3(1):11-31.

Navicula approximata
Hendey, N-Ingram, 1958 [1957(Publ. 1958)]
Marine diatoms from some West African Ports—J. R Microsc. Soc. (3) 77(1/2): 28-85.

Navicula approximata
Zanon, V., 1948
Diatomee marini di Sardegna e Pugillo di Alghe Marine della stressa. Boll. Pesca, Piscitutura e Idrobiologia, Anno 24, ns. 3(2): 202-244, 27 figs. on 1 pl.

Navicula approximata var. niceoensis comb. nov.
Hendey, N-Ingram, 1958 [1957(Publ. 1958)]
Marine diatoms from some West African Ports—J. R Microsc. Soc. (3) 77(1/2): 28-85.

Navicula ardua n.sp.
Mann, A., 1907
Report on the diatoms of the Albatross voyages in the Pacific Ocean, 1888-1904. Contrib. U. S. Nat. Herb. 10(5):221-419, Pls. XLIV-LIV.

Navicula arenaria
Hendy, N. Ingram, 1964
An introductory account of the smaller algae of British coastal waters. V. Bacillariophyceae (Diatoms).
Her Majesty's Stationary Office, 317 pp., 45 pls.

Navicula arenaria
Hustedt, F. and A.A. Aleem, 1951
Littoral diatoms from the Salstone near Plymouth. JMBA 30(1): 177.196.

Navicula arenaria
Mann, A., 1907
Report on the diatoms of the Albatross voyages in the Pacific Ocean, 1888-1904. Contrib. U. S. Nat. Herb. 10(5):221-419, Pls. XLIV-LIV.

Navicula arenaria
Politis, J., 1949
Diatomees marines de Bosphores et des ibes de la mer de Marmara. II Practica tou Hellenikou Hidrobiologikou Institutoutou 1929, Etoz 1929, 3(1):11-31.

Navicula aspera
Mann, A., 1907
Report on the diatoms of the Albatross voyages in the Pacific Ocean, 1888-1904. Contrib. U. S. Nat. Herb. 10(5):221-419, Pls. XLIV-LIV.

Navicula aspera intermedia
Mann, A., 1893
List of Diatomaceae from a deep-sea dredging in the Atlantic Ocean off Delaware Bay by the U. S. Fish Commission Steamer Albatross.
Proc. U. S. Nat. Mus. 16:303-312.

Navicula assula n.sp.
Cholnoky, B.J., 1963.
Beiträge zur Kenntnis des marinen Litorals von Südafrika.
Botanica Marina, 5(2/3):38-83.

Navicula assuloides n.sp.
Giffen, Malcolm H. 1973.
Diatoms of the marine littoral of Steenberg's Cove in St. Helena Bay, Cape Province, South Africa.
Botanica Marina 16 (1): 32-48.

Navicula astrolabensis n.sp.
Hendey, N.I., 1937
The plankton diatoms of the southern seas.
Discovery Repts. 16:151-364, pls.6-13.

Navicula atlantica
Hendy, N. Ingram, 1964
An introductory account of the smaller algae of British coastal waters. V. Bacillariophyceae (Diatoms).
Her Majesty's Stationary Office, 317 pp., 45 pls.

Navicula auklandica n.sp.
Grunow, A., 1863
Ueber einige neue und ungenügend bekannte Arten und Gattungen von Diatomaceen. Verhandl. d. K.K. Zool. Bot. Gesellsch., Vienna, 13: 137-162, pl.4-5 (Pl. 13-14).

Navicula avenacea
Hendy, N. Ingram, 1964
An introductory account of the smaller algae of British coastal waters. V. Bacillariophyceae (Diatoms).
Her Majesty's Stationary Office, 317 pp., 45 pls.

Navicula avenacea
Hustedt, F. and A.A. Aleem, 1951
Littoral diatoms from the Salstone near Plymouth. JMBA 30(1): 177.196.

Navicula avenacea
Zanon, V., 1948
Diatomee marini di Sardegna e Pugillo di Alghe Marine della stressa. Boll. Pesca, Piscitutura e Idrobiologia, Anno 24, ns. 3(2): 202-244, 27 figs. on 1 pl.

Navicula bahusiensis & varieties
Hendy, N. Ingram, 1964
An introductory account of the smaller algae of British coastal waters. V. Bacillariophyceae (Diatoms).
Her Majesty's Stationary Office, 317 pp., 45 pls.

Navicula Baileyana
Zanon, V., 1948
Diatomee marini di Sardegna e Pugillo di Alghe Marine della stressa. Boll. Pesca, Piscitutura e Idrobiologia, Anno 24, ns. 3(2): 202-244, 27 figs. on 1 pl.

Navicula barberi n.sp.
Hendy, N. Ingram, 1964
An introductory account of the smaller algae of British coastal waters. V. Bacillariophyceae (Diatoms).
Her Majesty's Stationary Office, 317 pp., 45 pls.

Navicula Beyrichiana

Politis, J., 1949
Diatomees marines de Bosphores et des ibes de la mer de Marmara. II Practica tou Hellenikou Hidrobiologikou Institutoutou 1929, Etoz 1929, 3(1):11-31.

Navicula birostrata

Grunow, A., 1863
Ueber einige neue und ungenügend bekannte Arten und Gattungen von Diatomaceen. Verhandl. d. K.K. Zool. Bot. Gesellsch., Vienna, 13: 137-162, pl.4-5 (Pl. 13-14).

Navicula biskanteri

Hendy, N. Ingram, 1964
An introductory account of the smaller algae of British coastal waters. V. Bacillariophyceae (Diatoms).
Her Majesty's Stationary Office, 317 pp., 45 pls.

Navicula bisulcata

Mann, A., 1907
Report on the diatoms of the Albatross voyages in the Pacific Ocean, 1888-1904. Contrib. U. S. Nat. Herb. 10(5):221-419, Pls. XLIV-LIV.

Navicula bisulcata

Mann, A., 1893
List of Diatomaceae from a deep-sea dredging in the Atlantic Ocean off Delaware Bay by the U. S. Fish Commission Steamer Albatross. Proc. U. S. Nat. Mus. 16:303-312.

Navicula bombus

Mann, A., 1907
Report on the diatoms of the Albatross voyages in the Pacific Ocean, 1888-1904. Contrib. U. S. Nat. Herb. 10(5):221-419, Pls. XLIV-LIV.

Navicula bombus

Morse, D.C., 1947
Some observations on seasonal variations in plankton population Patuxant River, Maryland 1943 1945. Bd. Nat. Res., Publ. No.65, Chesapeake Biol. Lab., 31, 3 figs.

Navicula bombus

Politis, J., 1949
Diatomees marines de Bosphores et des ibes de la mer de Marmara. II Practica tou Hellenikou Hidrobiologikou Institutoutou 1929, Etoz 1929, 3(1):11-31.

Navicula bombus (figs.)

Sousa e Silva, E., 1949
Diatomacees e Dinoflagelados de Baia de Cascais. Portugaliae Acta Biol., Volume: Julio Henriques, Ser. B: 300-383, 9 pls, 2 fold-in tables.

Navicula borealis

Mann, A., 1893
List of Diatomaceae from a deep-sea dredging in the Atlantic Ocean off Delaware Bay by the U. S. Fish Commission Steamer Albatross. Proc. U. S. Nat. Mus. 16:303-312.

Navicula brasiliense

Castracane degli Antelminelli, F., 1886
1. Rèport on the Diatomaceae collected by H.M.S. Challenger during the years 1873-1876. Rept. Sci. Results, H.M.S. Challenger, Botany Vol. II, 178 pp., 30 pls.

Navicula brasiliensis n.sp.

Grunow, A., 1863
Ueber einige neue und ungenügend bekannte Arten und Gattungen von Diatomaceen. Verhandl. d. K.K. Zool. Bot. Gesellsch., Vienna, 13: 137-162, pl.4-5 (Pl. 13-14).

Navicula brasiliensis

Mann, A., 1907
Report on the diatoms of the Albatross voyages in the Pacific Ocean, 1888-1904. Contrib. U. S. Nat. Herb. 10(5):221-419, Pls. XLIV-LIV.

Navicula brasiliensis

Zanon, D. V., 1949
Diatomee di Buenos Aires (Argentina) Atti Accad. Naz. Lincei, Memorie, Cl. Sci. fis., mat. e. nat., ser. 7, 11(3):59-151, 2 pls.

Navicula bremeyeri

Hendy, N. Ingram, 1964
An introductory account of the smaller algae of British coastal waters. V. Bacillariophyceae (Diatoms).
Her Majesty's Stationary Office, 317 pp., 45 pls.

Navicula brevis

Mann, A., 1907
Report on the diatoms of the Albatross voyages in the Pacific Ocean, 1888-1904. Contrib. U. S. Nat. Herb. 10(5):221-419, Pls. XLIV-LIV.

Navicula britannica

Hendy, N. Ingram, 1964
An introductory account of the smaller algae of British coastal waters. V. Bacillariophyceae (Diatoms).
Her Majesty's Stationary Office, 317 pp., 45 pls.

Navicula britannica n.sp.

Hustedt, F. and A.A. Aleem, 1951
Littoral diatoms from the Salstone near Plymouth. JMBA 30(1): 177.196.

Navicula bullata var. carinata n.var.

Castracane degli Antelminelli, F., 1886
1. Rèport on the Diatomaceae collected by H.M.S. Challenger during the years 1873-1876. Rept. Sci. Results, H.M.S. Challenger, Botany Vol. II, 178 pp., 30 pls.

Navicula bullata var. obtusa n.var.

Castracane degli Antelminelli, F., 1886
1. Rèport on the Diatomaceae collected by H.M.S. Challenger during the years 1873-1876. Rept. Sci. Results, H.M.S. Challenger, Botany Vol. II, 178 pp., 30 pls.

Navicula bullata var. rhomboidea n.var.

Castracane degli Antelminelli, F., 1886
1. Rèport on the Diatomaceae collected by H.M.S. Challenger during the years 1873-1876. Rept. Sci. Results, H.M.S. Challenger, Botany Vol. II, 178 pp., 30 pls.

Navicula bulnheimii

Hendy, N. Ingram, 1964
An introductory account of the smaller algae of British coastal waters. V. Bacillariophyceae (Diatoms).
Her Majesty's Stationary Office, 317 pp., 45 pls.

Navicula calida n.sp.

Hendy, N. Ingram, 1964
An introductory account of the smaller algae of British coastal waters. V. Bacillariophyceae (Diatoms).
Her Majesty's Stationary Office, 317 pp., 45 pls.

Navicula cancellata

Hendy, N. Ingram, 1964
An introductory account of the smaller algae of British coastal waters. V. Bacillariophyceae (Diatoms).
Her Majesty's Stationary Office, 317 pp., 45 pls.

Navicula cancellata

Hustedt, F. and A.A. Aleem, 1951
Littoral diatoms from the Salstone near Plymouth. JMBA 30(1): 177.196.

Navicula cancelata

Politis, J., 1949
Diatomees marines de Bosphores et des ibes de la mer de Marmara. II Practica tou Hellenikou Hidrobiologikou Institutoutou 1929, Etoz 1929, 3(1):11-31.

Navicula cancellata

Rampi, L., 1940
Diatomee del Mare Adriatico. Nuovo Giornale Botanico Italiano, n.s., 47:559-608.

Navicula cancellata

Zanon, V., 1948
Diatomee marini di Sardegna e Pugillo di Alghe Marine della stressa. Boll. Pesca, Piscitutura e Idrobiologia, Anno 24, ns. 3(2): 202-244, 27 figs. on 1 pl.

Navicula cari

Zanon, V., 1948
Diatomee marini di Sardegna e Pugillo di Alghe Marine della stressa. Boll. Pesca, Piscitutura e Idrobiologia, Anno 24, ns. 3(2): 202-244, 27 figs. on 1 pl.

Navicula cari

Zanon, D. V., 1949
Diatomee di Buenos Aires (Argentina) Atti Accad. Naz. Lincei, Memorie, Cl. Sci. fis., mat. e. nat., ser. 7, 11(3):59-151, 2 pls.

Navicula caribaea

Mann, A., 1893
List of Diatomaceae from a deep-sea dredging in the Atlantic Ocean off Delaware Bay by the U. S. Fish Commission Steamer Albatross. Proc. U. S. Nat. Mus. 16:303-312.

Navicula carinifera

Hendy, N. Ingram, 1964
An introductory account of the smaller algae of British coastal waters. V. Bacillariophyceae (Diatoms).
Her Majesty's Stationary Office, 317 pp., 45 pls.

Navicula carinifera

Rampi, L., 1940
Diatomee del Mare Adriatico. Nuovo Giornale Botanico Italiano, n.s., 47:559-608.

Navicula cendronii n.sp.

Manguin, E., 1957
Premier inventaire des diatomées de la Terre Adélie Antarctique. Espèces nouvelles. Rev. Algologique, n.s., 3(3):111-134.

Navicula charlati

Zanon, D. V., 1949
Diatomee di Buenos Aires (Argentina) Atti Accad. Naz. Lincei, Memorie, Cl. Sci. fis., mat. e. nat., ser. 7, 11(3):59-151, 2 pls.

Navicula chi

Zanon, V., 1948
Diatomee marini di Sardegna e Pugillo di Alghe Marine della stressa. Boll. Pesca, Piscitutura e Idrobiologia, Anno 24, ns. 3(2): 202-244, 27 figs. on 1 pl.

Navicula cincta
Hendy, N. Ingram, 1964
An introductory account of the smaller algae of British coastal waters. V. Bacillariophyceae (Diatoms).
Her Majesty's Stationary Office, 317 pp., 45 pls.

Navicula cincta
Iyengar, M.O.P. and G.Venkataraman, 1951.
The ecology and seasonal succession of the algae flora of the River Cooum at Madras with special reference to the Diatomaceae. J. Madras Univ. 21, Sect. B(1): 140-192, 1 pl of 4 figs., 11 text figs.

Navicula cincta
Zanon, D. V., 1949
Diatomee di Buenos Aires (Argentina) Atti Accad. Naz. Lincei, Memorie, Cl. Sci. fis., mat. e. nat., ser. 7, 11(3):59-151, 2 pls.

Navicula cincta
Zanon, V., 1948
Diatomee marini di Sardegna e Pugillo di Alghe Marine della stressa. Boll. Pesca, Piscitutura e Idrobiologia, Anno 24, ns. 3(2): 202-244, 27 figs. on 1 pl.

Navicula clamans
Hendy, N. Ingram, 1964
An introductory account of the smaller algae of British coastal waters. V. Bacillariophyceae (Diatoms).
Her Majesty's Stationary Office, 317 pp., 45 pls.

Navicula clavata
Andrade e Clovis Teixeira, M.H. de, 1957.
Contribuição para o conhecimento das diatomaceas do Brasil.
Bol. Inst. Ocean., Univ. Sao Paulo, 8(1/2):171-225, 10 pls.

Navicula clavata
Hendy, N. Ingram, 1964
An introductory account of the smaller algae of British coastal waters. V. Bacillariophyceae (Diatoms).
Her Majesty's Stationary Office, 317 pp., 45 pls.

Navicula clavata
Mann, A., 1907
Report on the diatoms of the Albatross voyages in the Pacific Ocean, 1888-1904. Contrib. U. S. Nat. Herb. 10(5):221-419, Pls. XLIV-LIV.

Navicula clavata
Mann, A., 1893
List of Diatomaceae from a deep-sea dredging in the Atlantic Ocean off Delaware Bay by the U. S. Fish Commission Steamer Albatross. Proc. U. S. Nat. Mus. 16:303-312.

Navicula clavata
Rampi, L., 1940
Diatomee del Mare Adriatico. Nuovo Giornale Botanico Italiano, n.s., 47:559-608.

Navicula clavata
Takano, Hideaki, 1962
Notes on epiphytic diatoms upon sea-weeds from Japan.
J. Oceanogr. Soc., Japan, 18(1):29-33.

Navicula clavata
Zanon, V., 1948
Diatomee marini di Sardegna e Pugillo di Alghe Marine della stressa. Boll. Pesca, Piscitutura e Idrobiologia, Anno 24, ns. 3(2): 202-244, 27 figs. on 1 pl.

Navicula clavata caribaea
Takano, Hideaki, 1963.
Notes on marine littoral diatoms of Japan. Bull. Tokai Reg. Fish. Res. Lab., No. 36:1-8.

Navicula clementis
Hendy, N. Ingram, 1964
An introductory account of the smaller algae of British coastal waters. V. Bacillariophyceae (Diatoms).
Her Majesty's Stationary Office, 317 pp., 45 pls.

Navicula clepsydra
Schodduyn, M., 1926
Observations faites dans la baie d'Ambleteuse (Pas de Calais). Bull. Inst. Ocean., Monaco, No. 482: 64 pp.

Navicula cluthensis
Mann, A., 1893
List of Diatomaceae from a deep-sea dredging in the Atlantic Ocean off Delaware Bay by the U. S. Fish Commission Steamer Albatross. Proc. U. S. Nat. Mus. 16:303-312.

Navicula corymbosa
Hendey, N.I., 1937
The plankton diatoms of the southern seas. Discovery Repts. 16:151-364, pls.6-13.

Navicula corymbosa
Zanon, V., 1948
Diatomee marini di Sardegna e Pugillo di Alghe Marine della stressa. Boll. Pesca, Piscitutura e Idrobiologia, Anno 24, ns. 3(2): 202-244, 27 figs. on 1 pl.

Navicula crabo
Mann, A., 1907
Report on the diatoms of the Albatross voyages in the Pacific Ocean, 1888-1904. Contrib. U. S. Nat. Herb. 10(5):221-419, Pls. XLIV-LIV.

Navicula crabo
Meunier, A., 1915
Microplancton de la Mer Flamande. 2. Diatomées (excepté le genre Chaetoceros). Mem. Mus. Roy. Hist. Nat., Belgique, 7(3):1-118, Pls. VIII-XIV.

Navicula crabo (figs.)
Sousa e Silva, E., 1949
Diatomaceas e Dinoflagelados de Baia de Cascais. Portugaliae Acta Biol., Volume: Julio Henriques, Ser. B: 300-383, 9 pls, 2 fold-in tables.

Navicula criophila
Jouse, A.P., G.S. Koroleva, G.A. Nagaeva, 1962
Diatoms in the surface layer of sediment in the Indian sector of the Antarctic. Investigations of marine bottom sediments. (In Russian; English summary).
Trudy Inst. Okeanol., Akad. Nauk, SSSR, 61: 19-92.

Navicula cronullensis n.sp.
Wood, E.J. Ferguson, 1963.
Studies on Australian and New Zealand diatoms. VI. Tropical and subtropical species.
Trans. R. Soc., New Zealand, 2(15):189-218.

Navicula crucicula
Hendy, N. Ingram, 1964
An introductory account of the smaller algae of British coastal waters. V. Bacillariophyceae (Diatoms).
Her Majesty's Stationary Office, 317 pp., 45 pls.

Navicula crucicula
Hendey, N.I., 1951
Littoral diatoms of Chicester Harbour with special reference to fouling. J.Roy. Microscop. Soc. 71(1): 1-86, 18 pls.

Navicula crucifera
Hendy, N. Ingram, 1964
An introductory account of the smaller algae of British coastal waters. V. Bacillariophyceae (Diatoms).
Her Majesty's Stationary Office, 317 pp., 45 pls.

Navicula crucifera
Zanon, V., 1948
Diatomee marini di Sardegna e Pugillo di Alghe Marine della stressa. Boll. Pesca, Piscitutura e Idrobiologia, Anno 24, ns. 3(2): 202-244, 27 figs. on 1 pl.

Navicula crucigera
Hendy, N. Ingram, 1964
An introductory account of the smaller algae of British coastal waters. V. Bacillariophyceae (Diatoms).
Her Majesty's Stationary Office, 317 pp., 45 pls.

Navicula crucigera
Hendey, N.I., 1951
Littoral diatoms of Chicester Harbour with special reference to fouling. J.Roy. Microscop. Soc. 71(1): 1-86, 18 pls.

Navicula crucigera
Hustedt, F. and A.A. Aleem, 1951
Littoral diatoms from the Salstone near Plymouth. JMBA 30(1): 177.196.

Navicula crucigera
Schodduyn, M., 1926
Observations faites dans la baie d'Ambleteuse (Pas de Calais). Bull. Inst. Ocean., Monaco, No. 482: 64 pp.

Navicula cryophile n.sp.
Manguin, E., 1957.
Premier inventaire des diatomées de la Terre Adélie Antarctique. Espèces nouvelles.
Rev. Algologique, n.s., 3(3):111-134.

Navicula cryptocephala & varieties
Hendy, N. Ingram, 1964
An introductory account of the smaller algae of British coastal waters. V. Bacillariophyceae (Diatoms).
Her Majesty's Stationary Office, 317 pp., 45 pls.

Navicula cryptocephala
Hendey, N.I., 1951
Littoral diatoms of Chicester Harbour with special reference to fouling. J.Roy. Microscop. Soc. 71(1): 1-86, 18 pls.

Navicula cryptocephala
Hustedt, F. and A.A. Aleem, 1951
Littoral diatoms from the Salstone near Plymouth. JMBA 30(1): 177.196.

Navicula cryptocephala
Tsumura, K., 1956.
Diatomoj el la cirkaufoso de la restajo de la kastele de Odawara. J. Yokohama Municipal Univ., (C-14), No. 47:23 pp.

Navicula cryptocephala
Zanon, D. V., 1949
Diatomee di Buenos Aires (Argentina) Atti Accad. Naz. Lincei, Memoria, Cl. Sci. fis., mat. e. nat., ser. 7, 11(3):59-151, 2 pls.

Navicula cryptocephala
Zanon, V., 1948
Diatomee marini di Sardegna e Pugillo di Alghe Marine della stressa. Boll. Pesca, Piscitutura e Idrobiologia, Anno 24, ns. 3(2): 202-244, 27 figs. on 1 pl.

Navicula cryptostriata n.sp.
Salah, M.M., 1952(1953).
XII. Diatoms from Blakeney Point, Norfolk, New species and new records from Great Britain. J.R. Microsc. Soc., Ser. 3, 72(3):155-169, 3 pls.

Navicula cunoniae n.sp.
Cholnoky, B.J., 1963.
Beiträge zur Kenntnis des marinen Litorals von Südafrika. Botanica Marina, 5(2/3):38-83.

Navicula curvilineata n.sp.
Mann, A., 1907
Report on the diatoms of the Albatross voyages in the Pacific Ocean, 1888-1904. Contrib. U. S. Nat. Herb. 10(5):221-419, Pls. XLIV-LIV.

Navicula cuspidata
Mann, A., 1907
Report on the diatoms of the Albatross voyages in the Pacific Ocean, 1888-1904. Contrib. U. S. Nat. Herb. 10(5):221-419, Pls. XLIV-LIV.

Navicula cuspidata
Tsumura, K., 1956.
Diatomoj el la cirkaufoso de la restajo de la kastele de Odawara. J. Yokohama Municipal Univ., (C-14), No. 47:23 pp.

Navicula cuspidata
Zanon, D. V., 1949
Diatomee di Buenos Aires (Argentina) Atti Accad. Naz. Lincei, Memorie, Cl. Sci. fis., mat. e. nat., ser. 7, 11(3):59-151, 2 pls.

Navicula cuspidata
Zanon, V., 1948
Diatomee marini di Sardegna e Pugillo di Alghe Marine della stressa. Boll. Pesca, Piscitutura e Idrobiologia, Anno 24, ns. 3(2): 202-244, 27 figs. on 1 pl.

Navicula cuspidata ambigua
Rumkówna, A., 1948
List of the phytoplankton species occurring in the superficial water layers in the Gulf of Gdańsk] Bull. Lab. mar., Gdynia, No. 4: 139-141 with tables in back.

Navicula cyclophora n.sp.
Castracane degli Antelminelli, F., 1886
1. Report on the Diatomaceae collected by H.M.S. Challenger during the years 1873-1876. Rept. Sci. Results, H.M.S. Challenger, Botany Vol. II, 178 pp., 30 pls.

Navicula cyprinus
Hendey, N.J., 1951
Littoral diatoms of Chicester Harbour with special reference to fouling. J. Roy. Microscop. Soc. 71(1): 1-86, 18 pls.

Navicula decipiens n.sp.
Castracane degli Antelminelli, F., 1886
1. Report on the Diatomaceae collected by H.M.S. Challenger during the years 1873-1876. Rept. Sci. Results, H.M.S. Challenger, Botany Vol. II, 178 pp., 30 pls.

Navicula decussepunctata n.sp.
Simonsen, Reimer, 1960.
Neue Diatomeen aus der Ostsee. II. Kieler Meeresf., 16(1):126-130.

Navicula dehissa n.sp.
Giffen, Malcolm H. 1973. Diatoms of the marine littoral of Steenberg's Cove in St. Helena Bay, Cape Province, South Africa. Botanica Marina 16 (1): 32-48.

Navicula dicephala elginensis neglecta
Tsumura, K., 1956.
Diatomoj el la cirkaufoso de la restajo de la kastele de Odawara. J. Yokohama Municipal Univ., (C-14), No. 47:23 pp.

Navicula didyma
Mann, A., 1907
Report on the diatoms of the Albatross voyages in the Pacific Ocean, 1888-1904. Contrib. U. S. Nat. Herb. 10(5):221-419, Pls. XLIV-LIV.

Navicula digito-radiata
Hendy, N. Ingram, 1964
An introductory account of the smaller algae of British coastal waters. V. Bacillariophyceae (Diatoms). Her Majesty's Stationary Office, 317 pp., 45 pls.

Navicula digitoradiata
Hustedt, F. and A.A. Aleem, 1951
Littoral diatoms from the Salstone near Plymouth. JMBA 30(1): 177-196.

Navicula digitoradiata
Iyengar, M.O.P. and G.Venkataraman, 1951.
The ecology and seasonal succession of the algae flora of the River Cooum at Madras with special reference to the Diatomaceae. J. Madras Univ. 21, Sect. B(1): 140-192, 1 pl of 4 figs., 11 text figs.

Navicula digito-radiata rostrata
Salah, M.M., 1952(1953).
XII. Diatoms from Blakeney Point, Norfolk. New species and new records for Great Britain. J.R. Microsc. Soc., Ser. 3, 72(3):155-169, 3 pls.

Navicula digito-radiata
Zanon, V., 1948
Diatomee marini di Sardegna e Pugillo di Alghe Marine della stressa. Boll. Pesca, Piscitutura e Idrobiologia, Anno 24, ns. 3(2): 202-244, 27 figs. on 1 pl.

Navicula directa
Bigelow, H.B., and M. Leslie, 1930
Reconnaissance of the waters and plankton of Monterey Bay, July 1928. Bull. M.C.Z., 70(5):429-481, 43 text figs.

Navicula directa
Castenholz, Richard W., 1963.
An experimental study of the vertical distribution of littoral marine diatoms. Limnology and Oceanography, 8(4):450-462.

Navicula directa
Gran, H. H. and E. C. Angst, 1931
Plankton diatoms of Puget Sound. Publ. Puget Sound Biol. Sta. 7:417-519, 95 text figs.

Navicula directa
Jørgensen, E., 1905
B. Protistplankton and the diatoms in bottom samples. Hydrographical and biological investigations in Norwegian fjords. Bergens Mus. Skr. 7: 49-225.

Navicula directa
Politis, J., 1949
Diatomees marines de Bosphores et des ibes de la mer de Marmara. II Practica tou Hellenikou Hidrobiologikou Institutoutou 1929, Etoz 1929, 3(1):11-31.

Navicula directa
Rampi, L., 1940
Diatomee del Mare Adriatico. Nuovo Giornale Botanico Italiano, n.s., 47:559-608.

Navicula directa
Zanon, V., 1948
Diatomee marini di Sardegna e Pugillo di Alghe Marine della stressa. Boll. Pesca, Piscitutura e Idrobiologia, Anno 24, ns. 3(2): 202-244, 27 figs. on 1 pl.

Navicula directa-radiata and varieties
Hendy, N. Ingram, 1964
An introductory account of the smaller algae of British coastal waters. V. Bacillariophyceae (Diatoms). Her Majesty's Stationary Office, 317 pp., 45 pls.

Navicula diserta
Salah, M.M., 1952(1953).
XII. Diatoms from Blakeney Point, Norfolk. New species and new records fx for Great Britain. J.R. Microsc. Soc., Ser. 3, 72(3):155-169, 3 pls.

Navicula dissipata
Hendy, N. Ingram, 1964
An introductory account of the smaller algae of British coastal waters. V. Bacillariophyceae (Diatoms). Her Majesty's Stationary Office, 317 pp., 45 pls.

Navicula distans
Cupp, Easter E., 1943
Marine plankton diatoms of the west coast of North America. Bull. S.I.O. 5(1):1-238, 5 pls., 168 text figs.

Navicula distans
Gran, H. H. and E. C. Angst, 1931
Plankton diatoms of Puget Sound. Publ. Puget Sound Biol. Sta. 7:417-519, 95 text figs.

Navicula distans
Gran, H.H., and T. Braarud, 1935
A quantitative study of the phytoplankton in the Bay of Fundy and the Gulf of Maine (including observations on hydrography, chemistry, and turbidity). J. Biol. Bd., Canada, 1(5):279-467, 69 text figs.

Navicula distans
Hendy, N. Ingram, 1964
An introductory account of the smaller algae of British coastal waters. V. Bacillariophyceae (Diatoms). Her Majesty's Stationary Office, 317 pp., 45 pls.

Navicula distans
Lillick, L.C., 1938
Preliminary report of the phytoplankton of the Gulf of Maine. Am. Mid. Nat. 20(3):624-640, 1 text figs 37 tables.

Navicula distans
Lillick, L.C., 1940
Phytoplankton and planktonic protozoa of the offshore waters of the Gulf of Maine. Pt.II, Qualitative Composition of the Planktonic Flora. Trans. Am. Phil. Soc., n.s., 31(3):193-237, 13 text figs.

Navicula distans
Mann, A., 1907
Report on the diatoms of the Albatross voyages in the Pacific Ocean, 1888-1904. Contrib. U. S. Nat. Herb. 10(5):221-419, Pls. XLIV-LIV.

Navicula distans
Mann, A., 1893
List of Diatomaceae from a deep-sea dredging in the Atlantic Ocean off Delaware Bay by the U. S. Fish Commission Steamer Albatross. Proc. U. S. Nat. Mus. 16:303-312.

Navicula distans
Rampi, L., 1940
Diatomee del Mare Adriatico. Nuovo Giornale Botanico Italiano, n.s., 47:559-608.

Navicula dumontiae n.sp.
Baardseth, Egil and Jens Petter Taasen, 1973
Navicula dumontiae sp. nov., an endophytic diatom inhabiting the mucilage of Dumontia incrassata (Rhodophyceae). Norwegian J. Bot. 20(2/3):79-87.

Navicula dunstonii n.sp.
Salah, M.M., 1952(1953).
XII. Diatoms from Blakeney Point, Norfolk. New species and new records for Great Britain. J.R. Microsc. Soc., Ser. 3, 72(3):155-169, 3 pls.

Navicula elegans
Hendy, N. Ingram, 1964
An introductory account of the smaller algae of British coastal waters. V. Bacillariophyceae (Diatoms).
Her Majesty's Stationary Office, 317 pp., 45 pls.

Navicula elegans
Hendy, N.J., 1951
Littoral diatoms of Chicester Harbour with special reference to fouling. J.Roy. Microscop. Soc. 71(1): 1-86, 18 pls.

Navicula elegans
Kokubo, S., and S. Sato., 1947
Plankters in Jū-San Gata. Physiol. and Ecol. (Japan) 1(4):1-16, 3 text figs., tables.

Navicula elliptica
Allen, W.E., and E.E. Cupp, 1935
Plankton diatoms of the Java Sea. Annales du Jardin Botanique de Buitenzorg XLIV (2):101-174, figs.1-127.
(drawings of all species mentioned)

Navicula entomon
Castracane degli Antelminelli, F., 1886
1. Report on the Diatomaceae collected by H.M.S. Challenger during the years 1873-1876. Rept. Sci. Results, H.M.S. Challenger, Botany Vol. II, 178 pp., 30 pls.

Navicula entomon var. thaitiana
Castracane degli Antelminelli, F., 1886 n.var.
1. Report on the Diatomaceae collected by H.M.S. Challenger during the years 1873-1876. Rept. Sci. Results, H.M.S. Challenger, Botany Vol. II, 178 pp., 30 pls.

Navicula endophytica
Taasen, Jens Petter 1972.
Observations on Navicula endophytica Hasle (Bacillariophyceae).
Sarsia 51:67-82.

Navicula ergadensis minor & varieties
Hendy, N. Ingram, 1964
An introductory account of the smaller algae of British coastal waters. V. Bacillariophyceae (Diatoms).
Her Majesty's Stationary Office, 317 pp., 45 pls.

Navicula exemta
Mann, A., 1893
List of Diatomaceae from a deep-sea dredging in the Atlantic Ocean off Delaware Bay by the U. S. Fish Commission Steamer Albatross. Proc. U. S. Nat. Mus. 16:303-312.

Navicula exigua
Zanon, D. V., 1949
Diatomee di Buenos Aires (Argentina) Atti Accad. Naz. Lincei, Memorie, Cl. Sci. fis., mat. e. nat., ser. 7, 11(3):59-151, 2 pls.

Navicula expansa
Hendey, N-Ingram, 1958 [1957(Publ. 1958)]
Marine diatoms from some West African Ports-J. R Microsc. Soc. (3) 77(1/2): 28-85.

Navicula falaisiensis
Zanon, V., 1948
Diatomee marini di Sardegna e Pugillo di Alghe Marine della stressa. Boll. Pesca, Piscitutura e Idrobiologia, Anno 24, ns. 3(2): 202-244, 27 figs. on 1 pl.

Navicula favus n.sp.
Salah, M.M., 1952(1953).
XII. Diatoms from Blakeney Point, Norfolk. New species and new records from Great Britain. J.R. Microsc. Soc., Ser. 3, 72(3):155-169, 3 pls.

Navicula fenzlii
Grunow, A., 1863
Ueber einige neue und ungenügend bekannte Arten und Gattungen von Diatomaceen. Verhandl. d. K.K. Zool. Bot. Gesellsch., Vienna, 13: 137-162, pl.4-5 (Pl. 13-14).

Navicula finmarchica
Hendy, N. Ingram, 1964
An introductory account of the smaller algae of British coastal waters. V. Bacillariophyceae (Diatoms).
Her Majesty's Stationary Office, 317 pp., 45 pls.

Navicula firma tumescens
Mann, A., 1893
List of Diatomaceae from a deep-sea dredging in the Atlantic Ocean off Delaware Bay by the U. S. Fish Commission Steamer Albatross. Proc. U. S. Nat. Mus. 16:303-312.

Navicula flanatica
Hendy, N. Ingram, 1964
An introductory account of the smaller algae of British coastal waters. V. Bacillariophyceae (Diatoms).
Her Majesty's Stationary Office, 317 pp., 45 pls.

Navicula flanatica
Hustedt, F. and A.A. Aleem, 1951
Littoral diatoms from the Salstone near Plymouth. JMBA 30(1): 177.196.

Navicula flanatica
Salah, M.M., 1952(1953).
XII. Diatoms from Blakeney Point, Norfolk. New species and new records for Great Britain. J.R. Microsc. Soc., Ser. 3, 72(3):155-169, 3 pls.

Navicula flebilis n.sp.
Cholnoky, B.J., 1963.
Beiträge zur Kenntnis des marinen Litorals von Südafrika.
Botanica Marina, 5(2/3):38-83.

Navicula florinae
Hendy, N. Ingram, 1964
An introductory account of the smaller algae of British coastal waters. V. Bacillariophyceae (Diatoms).
Her Majesty's Stationary Office, 317 pp., 45 pls.

Navicula fluminensis
Mann, A., 1907
Report on the diatoms of the Albatross voyages in the Pacific Ocean, 1888-1904. Contrib. U. S. Nat. Herb. 10(5):221-419, Pls. XLIV-LIV.

Navicula forcipata
Andrade e Clovis Teixeira, M.H. de, 1957.
Contribuição para o conhecimento das diatomáceas do Brasil.
Bol. Inst. Ocean., Univ. Sao Paulo, 8(1/2):171-225, 10 pls.

Navicula forcipata & varieties
Hendy, N. Ingram, 1964
An introductory account of the smaller algae of British coastal waters. V. Bacillariophyceae (Diatoms).
Her Majesty's Stationary Office, 317 pp., 45 pls.

Navicula forcipata
Hendey, N.J., 1951
Littoral diatoms of Chicester Harbour with special reference to fouling. J.Roy. Microscop. Soc. 71(1): 1-86, 18 pls.

Navicula forcipata
Hustedt, F. and A.A. Aleem, 1951
Littoral diatoms from the Salstone near Plymouth. JMBA 30(1): 177.196.

Navicula forcipata
Politis, J., 1949
Diatomees marines de Bosphores et des ibes de la mer de Marmara. II Practica tou Hellenikou Hidrobiologikou Institutoutou 1929, Etoz 1929, 3(1):11-31.

Navicula forcipata
Rampi, L., 1942
Ricerche sul fitoplancton del Mare Ligure 6. Le diatomee delle acque di Sanremo. Nuovo Giornale Botanico Italiano, N.S., 49:252-268.

Navicula forcipata
Rampi, L., 1940
Diatomee del Mare Adriatico. Nuovo Giornale Botanico Italiano, n.s., 47:559-608.

Navicula forcipata
Zanon, V., 1948
Diatomee marini di Sardegna e Pugillo di Alghe Marine della stressa. Boll. Pesca, Piscitutura e Idrobiologia, Anno 24, ns. 3(2): 202-244, 27 figs. on 1 pl.

Navicula formosa
Bigelow, H.B., and M. Leslie, 1930
Reconnaissance of the waters and plankton of Monterey Bay, July 1928. Bull. M.C.Z., 70(5):429-481, 43 text figs.

Navicula formosa
Mann, A., 1907
Report on the diatoms of the Albatross voyages in the Pacific Ocean, 1888-1904. Contrib. U. S. Nat. Herb. 10(5):221-419, Pls. XLIV-LIV.

Navicula fortis
Hendy, N. Ingram, 1964
An introductory account of the smaller algae of British coastal waters. V. Bacillariophyceae (Diatoms).
Her Majesty's Stationary Office, 317 pp., 45 pls.

Navicula frigida n.sp.
Manguin, E., 1957.
Premier inventaire des diatomées de la Terre Adélie Antarctique. Espèces nouvelles. Rev. Algologique, n.s., 3(3):111-134.

Navicula frompii var.
Boden, Brian, 1948
Marine plankton diatoms on operation Highjump in: Some oceanographic observations on operation HIGHJUMP. By R.S. Dietz. USNEL Rept. No.55, 97 pp., 41 figs. 7 July 1948.

Navicula fusa
Politis, J., 1949
Diatomees marines de Bosphores et des ibes de la mer de Marmara. II Practica tou Hellenikou Hidrobiologikou Institutoutou 1929, Etoz 1929, 3(1):11-31.

Navicula fusca
Grunow, A., 1863
Ueber einige neue und ungenügend bekannte Arten und Gattungen von Diatomaceen. Verhandl. d. K.K. Zool. Bot. Gesellsch., Vienna, 13: 137-162, pl.4-5 (Pl. 13-14).

Navicula fusca
Politis, J., 1949
Diatomees marines de Bosphores et des ibes de la mer de Marmara. II Practica tou Hellenikou Hidrobiologikou Institutoutou 1929, Etoz 1929, 3(1):11-31.

Navicula fusca (figs.)
Sousa e Silva, E., 1949
Diatomaceas e Dinoflagelados de Baia de Cascais. Portugaliae Acta Biol., Volume: Julio Henriques, Ser. B: 300-383, 9 pls, 2 fold-in tables.

Navicula fusca delicata
Mann, A., 1893
List of Diatomaceae from a deep-sea dredging in the Atlantic Ocean off Delaware Bay by the U. S. Fish Commission Steamer Albatross. Proc. U. S. Nat. Mus. 16:303-312.

Navicula fusiformis
Grunow, A., 1877
1. New Diatoms from Honduras. Monthly Micros. Jour., 18:165-186, pls. CXCIII-CXCVI.

Navicula fusiformis
Ranson, M. Gilbert, 1943
Titres et travaux scientifiques. Paris: Masson et Cie, Editeurs, Libraries de l'Academie de Medecine, 120 Boulevard Saint-Germain. 88 pp.

Navicula fusiformis
Ransom, M.G., 1938
La Navicule bleue donc le Bassin d'Arcachon Observations complementaires sur sa biologie Bull. Stat. Biol. d'Arcachon XXXV:35-48, 1 fig.

Navicula fusiformis
Ranson, M. G., 1936
Sur la soi-disant dégénérescence de la Navicule bleue (Navicula fusiformis Grün. = N. ostrearia Bory). C. R. Acad. Sci., CCII:1702-1704.

Navicula fusiformis
Ranson, M. G., 1935.
Le déterminisme de la fixation saisonnière de Navicula fusiformis Grün (N. ostrearia Bory). Sa culture experimentale en ostréiculture. C.R. Acad. Sci. CCI:684-687.

Navicula fusiformis
Ranson, M.G., 1936
Essais de culture, dans la nature, de la Navicule bleue, cause du verdissement des Huîtres. Ostréiculture, Cultures marines, No.4

Navicula fusioides
Rampi, L., 1940
Diatomee del Mare Adriatico. Nuovo Giornale Botanico Italiano, n.s., 47:559-608.

Navicula fusioides
Zanon, V., 1948
Diatomee marini di Sardegna e Pugillo di Alghe Marine della stressa. Boll. Pesca, Piscitutura e Idrobiologia, Anno 24, ns. 3(2): 202-244, 27 figs. on 1 pl.

Navicula garkeana
Takano, Hideaki, 1962
Notes on epiphytic diatoms upon sea-weeds from Japan.
J. Oceanogr. Soc., Japan, 18(1):29-33.

Navicula gastrum placentula
Mann, A., 1893
List of Diatomaceae from a deep-sea dredging in the Atlantic Ocean off Delaware Bay by the U. S. Fish Commission Steamer Albatross. Proc. U. S. Nat. Mus. 16:303-312.

Navicula gemmata
Mann, A., 1907
Report on the diatoms of the Albatross voyages in the Pacific Ocean, 1888-1904. Contrib. U. S. Nat. Herb. 10(5):221-419, Pls. XLIV-LIV.

Navicula gemmifera n.sp.
Simonsen, Reimer, 1960.
Neue Diatomeen aus der Ostsee. II.
Kieler Meeresf., 16(1):126-130.

Navicula gotlandica
Hendy, N. Ingram, 1964
An introductory account of the smaller algae of British coastal waters. V. Bacillariophyceae (Diatoms).
Her Majesty's Stationary Office, 317 pp., 45 pls.

Navicula gracilis
Iyengar, M.O.P. and G.Venkataraman,1951.
The ecology and seasonal succession of the algae flora of the River Cooum at Madras with special reference to the Diatomaceae. J. Madras Univ. 21, Sect. B(1): 140-192, 1 pl of 4 figs., 11 text figs.

Navicula graeffii
Mann, A., 1907
Report on the diatoms of the Albatross voyages in the Pacific Ocean, 1888-1904. Contrib. U. S. Nat. Herb. 10(5):221-419, Pls. XLIV-LIV.

Navicula granii
Brunel, J., 1962
Le phytoplancton de la Baie de Chaleurs. Inst. Botan., Univ. Montréal, Contrib. No. 77: 365 pp., 66 pls.

Navicula granii
Gran, H.N., 1908
Diatomeen. Nordisches Plankton, Botanischer Teil pp. XIX.1-XIX 146; 178 text figs.

Navicula granulata
Hendy, N. Ingram, 1964
An introductory account of the smaller algae of British coastal waters. V. Bacillariophyceae (Diatoms).
Her Majesty's Stationary Office, 317 pp., 45 pls.

Navicula granulata
Hendey, N.J., 1951
Littoral diatoms of Chicester Harbour with special reference to fouling. J.Roy. Microscop. Soc. 71(1): 1-86, 18 pls.

Navicula granulata
Mann, A., 1893
List of Diatomaceae from a deep-sea dredging in the Atlantic Ocean off Delaware Bay by the U. S. Fish Commission Steamer Albatross. Proc. U. S. Nat. Mus. 16:303-312.

Navicula granulata
Zanon, D. V., 1949
Diatomee di Buenos Aires (Argentina)
Atti Accad. Naz. Lincei, Memorie, Cl. Sci. fis., mat. e. nat., ser. 7, 11(3):59-151, 2 pls.

Navicula granulata
Zanon, V., 1948
Diatomee marini di Sardegna e Pugillo di Alghe Marine della stressa. Boll. Pesca, Piscitutura e Idrobiologia, Anno 24, ns. 3(2): 202-244, 27 figs. on 1 pl.

Navicula gregalis n.sp.
Cholnoky, B.J., 1963.
Beiträge zur Kenntnis des marinen Litorals von Südafrika.
Botanica Marina, 5(2/3):38-83.

Navicula gregaria
Hustedt, F. and A.A. Aleem,1951
Littoral diatoms from the Salstone near Plymouth. JMBA 30(1): 177.196.

Navicula grevillei
Castenholz, Richard W., 1963.
An experimental study of the vertical distribution of littoral marine diatoms.
Limnology and Oceanography, 8(4):450-462.

Navicula grevilleana nom. nov.
Hendy, N. Ingram, 1964
An introductory account of the smaller algae of British coastal waters. V. Bacillariophyceae (Diatoms).
Her Majesty's Stationary Office, 317 pp., 45 pls.

Navicula Grevillei
Hendey, N.J., 1951
Littoral diatoms of Chicester Harbour with special reference to fouling. J.Roy. Microscop. Soc. 71(1): 1-86, 18 pls.

Navicula Grevillei
Hustedt, F. and A.A. Aleem,1951
Littoral diatoms from the Salstone near Plymouth. JMBA 30(1): 177.196.

Navicula grevillei
Takano, Hideaki, 1963.
Notes on marine littoral diatoms of Japan.
Bull. Tokai Reg. Fish. Res. Lab., No. 36:1-8.

Navicula grevillei
Zanon, V., 1948
Diatomee marini di Sardegna e Pugillo di Alghe Marine della stressa. Boll. Pesca, Piscitutura e Idrobiologia, Anno 24, ns. 3(2): 202-244, 27 figs. on 1 pl.

Navicula grimmioides n.sp.
Cholnoky, B.J., 1963.
Beiträge zur Kenntnis des marinen Litorals von Südafrika.
Botanica Marina, 5(2/3):38-83.

Navicula Groschopfi
Hustedt, F. and A.A. Aleem,1951
Littoral diatoms from the Salstone near Plymouth. JMBA 30(1): 177.196.

Navicula Gründleri
Schmidt, A., 18-
Ueber Navicula Weissflogii und Navicula Gründleri. Zeitschr. f. ges. Naturwiss. 41:403-410.

Navicula grunowi
Castracane degli Antelminelli, F., 1886
1. Report on the Diatomaceae collected by H.M.S. Challenger during the years 1873-1876. Rept. Sci. Results, H.M.S. Challenger, Botany Vol. II, 178 pp., 30 pls.

Navicula guttata
Rampi, L., 1940
Diatomee del Mare Adriatico. Nuovo Giornale Botanico Italiano, n.s., 47:559-608.

Navicula gyrinida n.sp.

Mann, A., 1907
Report on the diatoms of the Albatross voyages in the Pacific Ocean, 1888-1904. Contrib. U. S. Nat. Herb. 10(5):221-419, Pls. XLIV-LIV.

Navicula halophila

Hendy, N. Ingram, 1964
An introductory account of the smaller algae of British coastal waters. V. Bacillariophyceae (Diatoms).
Her Majesty's Stationary Office, 317 pp., 45 pls.

Navicula halophila

Hendey, N.I., 1951
Littoral diatoms of Chicester Harbour with special reference to fouling. J.Roy. Microscop. Soc. 71(1): 1-86, 18 pls.

Navicula halophila

Iyengar, M.O.P. and G.Venkataraman,1951.
The ecology and seasonal succession of the algae flora of the River Cooum at Madras with special reference to the Diatomaceae. J. Madras Univ. 21, Sect. B(1): 140-192, 1 pl of 4 figs., 11 text figs.

Navicula halophila

Subrahmanyan, R., 1945.
On somatic division, reduction division, auxospore-formation and sex differentiation in Navicula halophila (Grunow) Cleve. Curr. Sci. 14: 75-77.

Navicula halophila

Zanon, V., 1948
Diatomee marini di Sardegna e Pugillo di Alghe Marine della stressa. Boll. Pesca, Piscitutura e Idrobiologia, Anno 24, ns. 3(2): 202-244, 27 figs. on 1 pl.

Navicula hasta

Iyengar, M.O.P. and G.Venkataraman,1951.
The ecology and seasonal succession of the algae flora of the River Cooum at Madras with special reference to the Diatomaceae. J. Madras Univ. 21, Sect. B(1): 140-192, 1 pl of 4 figs., 11 text figs.

Navicula Hennedyi

Andrade, e Clovis Teixeira, M.H. de, 1957
Contribuição para o conhecimento das diatomáceas do Brasil.
Bol. Inst. Ocean., Univ. Sao Paulo, 8(1/2):171-225, 10 pls.

Navicula hennedyi

Hendy, N. Ingram, 1964
An introductory account of the smaller algae of British coastal waters. V. Bacillariophyceae (Diatoms).
Her Majesty's Stationary Office, 317 pp., 45 pls.

Navicula hennedyi

Mann, A., 1907
Report on the diatoms of the Albatross voyages in the Pacific Ocean, 1888-1904. Contrib. U. S. Nat. Herb. 10(5):221-419, Pls. XLIV-LIV.

Navicula Hennedyi

Mann, A., 1893
List of Diatomaceae from a deep-sea dredging in the Atlantic Ocean off Delaware Bay by the U. S. Fish Commission Steamer Albatross. Proc. U. S. Nat. Mus. 16:303-312.

Navicula Hennedyi

Politis, J., 1949
Diatomees marines de Bosphores et des ibes de la mer de Marmara. II Practica tou Hellenikou Hidrobiologikou Institutoutou 1929, Etoz 1929, 3(1):11-31.

Navicula Hennedyi

Rampi, L., 1940
Diatomee del Mare Adriatico. Nuovo Giornale Botanico Italiano, n.s., 47:559-608.

Navicula Hennedyi

Schodduyn, M., 1926
Observations faites dans la baie d'Ambleteuse (Pas de Calais). Bull. Inst. Ocean., Monaco, No. 482: 64 pp.

Navicula Hennedyi

Zanon, V., 1948
Diatomee marini di Sardegna e Pugillo di Alghe Marine della stressa. Boll. Pesca, Piscitutura e Idrobiologia, Anno 24, ns. 3(2): 202-244, 27 figs. on 1 pl.

Navicula hennedyi granulata

de Sousa e Silva, E., 1956.
Contribution à l'étude du microplancton de Dakar et des regions maritimes voisines. Bull. I.F.A.N. 8(2):335-371, 7 pls.

Navicula hennedyi nedra

de Sousa e Silva, E., 1956.
Contribution à l'étude du microplancton de Dakar et des regions maritimes voisines. Bull. I.F.A.N. 8(2):335-371, 7 pls.

Navicula hochstetteri n.sp.

Grunow, A., 1863
Ueber einige neue und ungenügend bekannt Arten und Gattungen von Diatomaceen. Verhandl. d. K.K. Zool. Bot. Gesellsch., Vienna, 13: 137-162, pl.4-5 (Pl. 13-14).

Navicula hudsonis

Hendy, N. Ingram, 1964
An introductory account of the smaller algae of British coastal waters. V. Bacillariophyceae (Diatoms).
Her Majesty's Stationary Office, 317 pp., 45 pls.

Navicula humerosa

Hendy, N. Ingram, 1964
An introductory account of the smaller algae of British coastal waters. V. Bacillariophyceae (Diatoms).
Her Majesty's Stationary Office, 317 pp., 45 pls.

Navicula humerosa

Hendey, N.I., 1951
Littoral diatoms of Chicester Harbour with special reference to fouling. J.Roy. Microscop. Soc. 71(1): 1-86, 18 pls.

Navicula humerosa

Mann, A., 1893
List of Diatomaceae from a deep-sea dredging in the Atlantic Ocean off Delaware Bay by the U. S. Fish Commission Steamer Albatross. Proc. U. S. Nat. Mus. 16:303-312.

Navicula humerosa

Rumkówna, A., 1948
[List of the phytoplankton species occurring in the superficial water layers in the Gulf of Gdańsk] Bull. Lab. mar., Gdynia, No. 4: 139-141 with tables in back.

Navicula humerosa minor

Salah, M.M., 1952(1953).
XII. Diatoms from Blakeney Point, Norfolk. New species and new records for Great Britain. J.R. Microsc. Soc., Ser. 3, 72(3):155-169, 3 pls.

Navicula humerosa

Schodduyn, M., 1926
Observations faites dans la baie d'Ambleteuse (Pas de Calais). Bull. Inst. Ocean., Monaco, No. 482: 64 pp.

Navicula humerosa

Sousa a Silva, E., and J. Dos Santos-Pinto, 1948
O Plancton da Baia de S. Martinho do Porto. 1. Diatomaceas e Dinoflagelados. Bol. Soc. Portuguese de Ciencias Naturais, 16(2):134-187, 6 pls. (Trav. Sta. Biol. Mar. de Lisbonne No. 52).

Navicula hungarica

Zanon, D. V., 1949
Diatomee di Buenos Aires (Argentina) Atti Accad. Naz. Lincei, Memorie, Cl. Sci. fis., mat. e. nat., ser. 7, 11(3):59-151, 2 pls.

Navicula hyalina

Hendy, N. Ingram, 1964
An introductory account of the smaller algae of British coastal waters. V. Bacillariophyceae (Diatoms).
Her Majesty's Stationary Office, 317 pp., 45 pls.

Navicula impressa

Mann, A., 1907
Report on the diatoms of the Albatross voyages in the Pacific Ocean, 1888-1904. Contrib. U. S. Nat. Herb. 10(5):221-419, Pls. XLIV-LIV.

Navicula inaurata n.sp.

Hendey, N-Ingram, 1958 [1957(Publ. 1958)]
Marine diatoms from some West African Ports. J.R. Microsc. Soc. (3) 77(1/2): 28-85.

Navicula inserta

Hendy, N. Ingram, 1964
An introductory account of the smaller algae of British coastal waters. V. Bacillariophyceae (Diatoms).
Her Majesty's Stationary Office, 317 pp., 45 pls.

Navicula incerta

Hendey, N.I., 1951
Littoral diatoms of Chicester Harbour with special reference to fouling. J.Roy. Microscop. Soc. 71(1): 1-86, 18 pls.

Navicula inclementis n.sp.

Hendy, N. Ingram, 1964
An introductory account of the smaller algae of British coastal waters. V. Bacillariophyceae (Diatoms).
Her Majesty's Stationary Office, 317 pp., 45 pls.

Navicula incurvata

Politis, J., 1949
Diatomees marines de Bosphores et des ibes de la mer de Marmara. II Practica tou Hellenikou Hidrobiologikou Institutoutou 1929, Etoz 1929, 3(1):11-31.

Navicula inflexa

Hendy, N. Ingram, 1964
An introductory account of the smaller algae of British coastal waters. V. Bacillariophyceae (Diatoms).
Her Majesty's Stationary Office, 317 pp., 45 pls.

Navicula insolubilis n.sp.

Cholnoky, B.J., 1963.
Beiträge zur Kenntnis des marinen Litorals von Südafrika.
Botanica Marina, 5(2/3):38-83.

Navicula insuta n.sp.

Manguin, E., 1957.
Premier inventaire des diatomées de la Terre Adélie Antarctique. Espèces nouvelles. Rev. Algologique, n.s., 3(3):111-134.

Navicula (aspera?) intermedia
Mann, A., 1907
Report on the diatoms of the Albatross voyages in the Pacific Ocean, 1888-1904. Contrib. U. S. Nat. Herb. 10(5):221-419, Pls. XLIV-LIV.

Navicula interrupta
Mann, A., 1893
List of Diatomaceae from a deep-sea dredging in the Atlantic Ocean off Delaware Bay by the U. S. Fish Commission Steamer Albatross. Proc. U. S. Nat. Mus. 16:303-312.

Navicula invenusta n.sp.
Mann, A., 1907
Report on the diatoms of the Albatross voyages in the Pacific Ocean, 1888-1904. Contrib. U. S. Nat. Herb. 10(5):221-419, Pls. XLIV-LIV.

Navicula irridula
Hendey, N.I., 1951
Littoral diatoms of Chicester Harbour with special reference to fouling. J.Roy. Microscop. Soc. 71(1): 1-86, 18 pls.

Navicula irrorata
Mann, A., 1907
Report on the diatoms of the Albatross voyages in the Pacific Ocean, 1888-1904. Contrib. U. S. Nat. Herb. 10(5):221-419, Pls. XLIV-LIV.

Navicula irrorata
Mann, A., 1893
List of Diatomaceae from a deep-sea dredging in the Atlantic Ocean off Delaware Bay by the U. S. Fish Commission Steamer Albatross. Proc. U. S. Nat. Mus. 16:303-312.

Navicula janischii n.sp.
Castracane degli Antelminelli, F., 1886
1. Report on the Diatomaceae collected by H.M.S. Challenger during the years 1873-1876. Rept. Sci. Results, H.M.S. Challenger, Botany Vol. II, 178 pp., 30 pls.

Navicula (?) jejuna
Castracane degli Antelminelli, F., 1886
1. Report on the Diatomaceae collected by H.M.S. Challenger during the years 1873-1876. Rept. Sci. Results, H.M.S. Challenger, Botany Vol. II, 178 pp., 30 pls.

Nitzschia Jelineckii n.sp.
Grunow, A., 1863
Ueber einige neue und ungenügend bekannte Arten und Gattungen von Diatomaceen. Verhandl. d. K.K. Zool. Bot. Gesellsch., Vienna, 13: 137-162, pl.4-5 (Pl. 13-14).

Navicula Jelineckii n.sp.
Grunow, A., 1863
Ueber einige neue und ungenügend bekannte Arten und Gattungen von Diatomaceen. Verhandl. d. K.K. Zool. Bot. Gesellsch., Vienna, 13: 137-162, pl.4-5 (Pl. 13-14).

Navicula kamorthensis n.sp.
Grunow, A., 1863
Ueber einige neue und ungenügend bekannte Arten und Gattungen von Diatomaceen. Verhandl. d. K.K. Zool. Bot. Gesellsch., Vienna, 13: 137-162, pl.4-5 (Pl. 13-14).

Navicula kariana
Jørgensen, E., 1905
B.Protistplankton and the diatoms in bottom samples. Hydrographical and biological investigations in Norwegian fjords. Bergens Mus. Skr. 7: 49-225.

Navicula kerguelensis n.sp.
Castracane degli Antelminelli, F., 1886
1. Report on the Diatomaceae collected by H.M.S. Challenger during the years 1873-1876. Rept. Sci. Results, H.M.S. Challenger, Botany Vol. II, 178 pp., 30 pls.

Navicula knysnensis n.sp.
Cholnoky, B.J., 1963.
Beiträge zur Kenntnis des marinen Litorals von Südafrika. Botanica Marina, 5(2/3):38-83.

Navicula lacrimans
Mann, A., 1907
Report on the diatoms of the Albatross voyages in the Pacific Ocean, 1888-1904. Contrib. U. S. Nat. Herb. 10(5):221-419, Pls. XLIV-LIV.

Navicula lacustris
Zanon, D. V., 1949
Diatomee di Buenos Aires (Argentina) Atti Accad. Naz. Lincei, Memorie, Cl. Sci. fis., mat. e. nat., ser. 7, 11(3):59-151, 2 pls.

Navicula Lagerheimi
Zanon, D. V., 1949
Diatomee di Buenos Aires (Argentina) Atti Accad. Naz. Lincei, Memorie, Cl. Sci. fis., mat. e. nat., ser. 7, 11(3):59-151, 2 pls.

Navicula lanceolata
Rampi, L., 1942
Ricerche sul fitoplancton del Mare Ligure 6. Le diatomee delle acque di Sanremo. Nuovo Giornale Botanico Italiano, N.S., 49:252-268.

Navicula lanceolata
Zanon, D. V., 1949
Diatomee di Buenos Aires (Argentina) Atti Accad. Naz. Lincei, Memorie, Cl. Sci. fis., mat. e. nat., ser. 7, 11(3):59-151, 2 pls.

Navicula lanceolata
Zanon, V., 1948
Diatomee marini di Sardegna e Pugillo di Alghe Marine della stressa. Boll. Pesca, Piscitutura e Idrobiologia, Anno 24, ns. 3(2): 202-244, 27 figs. on 1 pl.

Navicula lata
Mann, A., 1907
Report on the diatoms of the Albatross voyages in the Pacific Ocean, 1888-1904. Contrib. U. S. Nat. Herb. 10(5):221-419, Pls. XLIV-LIV.

Navicula laterostrata
Zanon, D. V., 1949
Diatomee di Buenos Aires (Argentina) Atti Accad. Naz. Lincei, Memorie, Cl. Sci. fis., mat. e. nat., ser. 7, 11(3):59-151, 2 pls.

Navicula latissima
Hendy, N. Ingram, 1964
An introductory account of the smaller algae of British coastal waters. V. Bacillariophyceae (Diatoms). Her Majesty's Stationary Office, 317 pp., 45 pls.

Navicula lauriensis, n.sp.
Frenguelli, Joaquin, y Hector A. Orlando, 1958. Diatomeas y silicoflagelados del sector Antartico Sudamericano. Inst. Antartico Argentino, Publ., No. 5:191 pp.

Navicula Leboimei n.sp.
Manguin, E., 1957.
Premier inventaire des diatomées de la Terre Adélie Antarctique. Espèces nouvelles. Rev. Algologique, n.s., 3(3):111-134.

Navicula libellus
Zanon, V., 1948
Diatomee marini di Sardegna e Pugillo di Alghe Marine della stressa. Boll. Pesca, Piscitutura e Idrobiologia, Anno 24, ns. 3(2): 202-244, 27 figs. on 1 pl.

Navicula liber
Grunow, A., 1863
Ueber einige neue und ungenügend bekannte Arten und Gattungen von Diatomaceen. Verhandl. d. K.K. Zool. Bot. Gesellsch., Vienna, 13: 137-162, pl.4-5 (Pl. 13-14).

Navicula liber
Politis, J., 1949
Diatomees marines de Bosphores et des ibes de la mer de Marmara. II Practica tou Hellenikou Hidrobiologikou Institutoutou 1929, Etoz 1929, 3(1):11-31.

Navicula lineata
Mann, A., 1893
List of Diatomaceae from a deep-sea dredging in the Atlantic Ocean off Delaware Bay by the U. S. Fish Commission Steamer Albatross. Proc. U. S. Nat. Mus. 16:303-312.

Navicula lithognatha n.sp.
Cholnoky, B.J., 1963.
Beiträge zur Kenntnis des marinen Litorals von Südafrika. Botanica Marina, 5(2/3):38-83.

Navicula lithognathoides n.sp.
Cholnoky, B.J., 1963.
Beiträge zur Kenntnis des marinen Litorals von Südafrika. Botanica Marina, 5(2/3):38-83.

Navicula litoris n.sp.
Salah, M.M., 1952(1953).
XII. Diatoms from Blakeney Point, Norfolk. New species and new records for Great Britain. J.R. Microsc. Soc., Ser. 3, 72(3):155-169, 3 pls.

Navicula lucens n.sp.
Salah, M.M., 1952(1953).
XII. Diatoms from Blakeney Point, Norfolk. New species and new records from Great Britain. J.R. Microsc. Soc., Ser. 3, 72(3):155-169, 3 pls.

Navicula lunatapicalis n.sp.
Salah, M.M., 1952(1953).
XII. Diatoms from Blakeney Point, Norfolk. New species and new records from Great Britain. J.R. Microsc. Soc., Ser. 3, 72(3):155-169, 3 pls.

Navicula lundstroemii
Hendy, N. Ingram, 1964
An introductory account of the smaller algae of British coastal waters. V. Bacillariophyceae (Diatoms). Her Majesty's Stationary Office, 317 pp., 45 pls.

Navicula lyra
Andrade e Clovis Teixeira, M.H. de, 1957.
Contribuição para o conhecimento das diatomeas do Brasil. Bol. Inst. Ocean., Univ. Sao Paulo, 8(1/2):171-225, 10 pls.

Navicula lyra
Hendy, N. Ingram, 1964
An introductory account of the smaller algae of British coastal waters. V. Bacillariophyceae (Diatoms). Her Majesty's Stationary Office, 317 pp., 45 pls.

Navicula lyra
Hendey, N-Ingram, 1958 [1957(Publ. 1958)]
Marine diatoms from some West African Ports. J. R.Microsc. Soc. (3) 77(1/2): 28-85.

Navicula lyra
Hendey, N.I., 1951
Littoral diatoms of Chicester Harbour with special reference to fouling. J.Roy. Microscop. Soc. 71(1): 1-86, 18 pls.

Navicula lyra
Hendey, N.I., 1937
The plankton diatoms of the southern seas. Discovery Repts. 16:151-364, pls.6-13.

Navicula lyra
Hendey, N.I., 1937
The plankton diatoms of the southern seas. Discovery Repts. 16:151-364, pls.6-13.

Navicula lyra
Hustedt, F. and A.A. Aleem, 1951
Littoral diatoms from the Salstone near Plymouth. JMBA 30(1): 177.196.

Navicula lyra
Mann, A., 1907
Report on the diatoms of the Albatross voyages in the Pacific Ocean, 1888-1904. Contrib. U. S. Nat. Herb. 10(5):221-419, Pls. XLIV-LIV.

Navicula lyra
Mann, A., 1893
List of Diatomaceae from a deep-sea dredging in the Atlantic Ocean off Delaware Bay by the U. S. Fish Commission Steamer Albatross. Proc. U. S. Nat. Mus. 16:303-312.

Navicula Lyra
Meunier, A., 1915
Microplancton de la Mer Flamande. 2. Diatomées (excepté le genre Chaetoceros). Mem. Mus. Roy. Hist. Nat., Belgique, 7(3):1-118, Pls. VIII-XIV.

Navicula lyra
Politis, J., 1949
Diatomees marines de Bosphores et des ibes de la mer de Marmara. II Practica tou Hellenikou Hidrobiologikou Institutoutou 1929, Etoz 1929, 3(1):11-31.

Navicula lyra
Rampi, L., 1942
Ricerche sul fitoplancton del Mare Ligure 6. Le diatomee delle acque di Sanremo. Nuovo Giornale Botanico Italiano, N.S., 49:252-268.

Navicula lyra
Rampi, L., 1940
Diatomee del Mare Adriatico. Nuovo Giornale Botanico Italiano, n.s., 47:559-608.

Navicula lyra (figs.)
Sousa e Silva, E., 1949
Diatomaceas e Dinoflagelados de Baia de Cascais. Portugaliae Acta Biol., Volume: Julio Henriques, Ser. B: 300-383, 9 pls, 2 fold-in tables.

Navicula lyra
Zanon, V., 1948
Diatomee marini di Sardegna e Pugillo di Alghe Marine della stressa. Boll. Pesca, Piscitutura e Idrobiologia, Anno 24, ns. 3(2): 202-244, 27 figs. on 1 pl.

Navicula lyra elliptica
Florin, M-B., 1948
9. Diatomeae in submarine cores from the Tyrrhenian Sea. Medd. Ocean. Inst., Göteborg, 15 (Göteborgs Kungl. Vetenskaps-och Vitterrhets Samhälles Handlingar, Sjätte Foljden, Ser. B 5(13):80-88.

Navicula lyra dilatata
Mann, A., 1893
List of Diatomaceae from a deep-sea dredging in the Atlantic Ocean off Delaware Bay by the U. S. Fish Commission Steamer Albatross. Proc. U. S. Nat. Mus. 16:303-312.

Navicula lyra elliptica
Mann, A., 1893
List of Diatomaceae from a deep-sea dredging in the Atlantic Ocean off Delaware Bay by the U. S. Fish Commission Steamer Albatross. Proc. U. S. Nat. Mus. 16:303-312.

Navicula lyra var. lyra
Hendey, N-Ingram, 1958 [1957(Publ. 1958)]
Marine diatoms from some West African Ports. J. R. Microsc. Soc. (3) 77(1/2): 28-85.

Navicula lyra magrinii
Zanon, V., 1947
Diatomee delle Isole Ponziane - materiali per una florula diatomologica del Mare Tirreno. Bol. Pesca, Piscicol., Idrobiol., n.s., 2(1):36-53, 1 pl., of 10 figs.

Navicula lyra var. signata
Castracane degli Antelminelli, F., 1886
1. Report on the Diatomaceae collected by H.M.S. Challenger during the years 1873-1876. Rept. Sci. Results, H.M.S. Challenger, Botany Vol. II, 178 pp., 30 pls.

Navicula lyroides
Hendy, N. Ingram, 1964
An introductory account of the smaller algae of British coastal waters. V. Bacillariophyceae (Diatoms). Her Majesty's Stationary Office, 317 pp., 45 pls.

Navicula lyroides stat. nov.
Hendey, N-Ingram, 1958 [1957(Publ. 1958)]
Marine diatoms from some West African Ports. J. R. Microsc. Soc. (3) 77(1/2): 28-85.

Navicula maculosa
Hendy, N. Ingram, 1964
An introductory account of the smaller algae of British coastal waters. V. Bacillariophyceae (Diatoms). Her Majesty's Stationary Office, 317 pp., 45 pls.

Navicula maculosa
Zanon, V., 1948
Diatomee marini di Sardegna e Pugillo di Alghe Marine della stressa. Boll. Pesca, Piscitutura e Idrobiologia, Anno 24, ns. 3(2): 202-244, 27 figs. on 1 pl.

Navicula major
Mann, A., 1907
Report on the diatoms of the Albatross voyages in the Pacific Ocean, 1888-1904. Contrib. U. S. Nat. Herb. 10(5):221-419, Pls. XLIV-LIV.

Navicula major
Mann, A., 1893
List of Diatomaceae from a deep-sea dredging in the Atlantic Ocean off Delaware Bay by the U. S. Fish Commission Steamer Albatross. Proc. U. S. Nat. Mus. 16:303-312.

Navicula mammalis n.sp.
Castracane degli Antelminelli, F., 1886
1. Report on the Diatomaceae collected by H.M.S. Challenger during the years 1873-1876. Rept. Sci. Results, H.M.S. Challenger, Botany Vol. II, 178 pp., 30 pls.

Navicula margino-nodularis n.sp.
Salah, M.M., 1952(1953).
XII. Diatoms from Blakeney Point, Norfolk. New species and new records for Great Britain. J.R. Microsc. Soc., Ser. 3, 72(3):155-169, 3 pls.

Navicula marginulata didyma
Mann, A., 1893
List of Diatomaceae from a deep-sea dredging in the Atlantic Ocean off Delaware Bay by the U. S. Fish Commission Steamer Albatross. Proc. U. S. Nat. Mus. 16:303-312.

Navicula marina
Hendy, N. Ingram, 1964
An introductory account of the smaller algae of British coastal waters. V. Bacillariophyceae (Diatoms). Her Majesty's Stationary Office, 317 pp., 45 pls.

Navicula Marnierii n.sp.
Manguin, E., 1957.
Premier inventaire des diatomées de la Terre Adélie Antarctique. Espèces nouvelles. Rev. Algologique, n.s., 3(3):111-134.

Navicula maxima
Politis, J., 1949
Diatomees marines de Bosphores et des ibes de la mer de Marmara. II Practica tou Hellenikou Hidrobiologikou Institutoutou 1929, Etoz 1929, 3(1):11-31.

Navicula mediterranea
Zanon, V., 1948
Diatomee marini di Sardegna e Pugillo di Alghe Marine della stressa. Boll. Pesca, Piscitutura e Idrobiologia, Anno 24, ns. 3(2): 202-244, 27 figs. on 1 pl.

Navicula membranacea
Cupp, Easter E., 1943
Marine plankton diatoms of the west coast of North America. Bull. S.I.O. 5(1):1-238, 5 pls., 168 text figs.

Navicula membranacea
Cupp, E.E., 1934
Analysis of marine diatom collections taken from the Canal Zone to California during March, 1933. Trans. Am. Micros. Soc. LIII (1):22-29, 1 map.

Navicula membranacea
Dangeard, P., 1927
Phytoplankton de la croisière du "Sylvana". Ann. Inst. Ocean. Monaco, n.s., 4(8):286-401, 54 text figs. (Feirer-Juin 1913).

Navicula membranacea
Gilbert, J.Y., and W.E. Allen, 1943
The phytoplankton of the Gulf of California obtained by the "E.W. Scripps" in 1939 and 1940. J. Mar. Res. V(2):89-110, figs.30-31.

Navicula membranacea
Gran, H.H., 1908
Diatomeen. Nordisches Plankton, Botanischer Teil pp. XIX.1-XIX 146; 178 text figs.

Navicula membranacea
Hendey, N.I., 1937
The plankton diatoms of the southern seas. Discovery Repts. 16:151-364, pls.6-13.

Navicula membranacea
Pavillard, J., 1925
Bacillariales. Rept. on the Danish Oceangr. Exped., 1908-10 to the Mediterranean and adj. seas. Vol.II., Biol. J4:72 pp., 116 text figs.

Navicula membranacea
Pavillard, J., 1905
Recherches sur la flore pelagique (Phytoplankton) de l'Etang de Thau. Theses presentees a la Fac. Sci., Paris, 116 pp., 3 pls.

Navicula membranacea

Schodduyn, M., 1926
Observations faites dans la baie d'Ambleteuse (Pas de Calais). Bull. Inst. Ocean., Monaco, No. 482: 64 pp.

Navicula membranacea

Sousa e Silva, E., 1949
Diatomaceas e Dinoflagelados de Baia de Cascais. Portugaliae Acta Biol., Volume: Julio Henriques, Ser. B: 300-383, 9 pls, 2 fold-in tables.

Navicula membranaceae

Sousa e Silva, E., and J. Dos Santos-Pinto, 1948
O Plancton da Baia de S. Martinho do Porto. 1. Diatomaceas e Dinoflagelados. Bol. Soc. Portuguese de Ciencias Naturais, 16(2):134-187, 6 pls. (Trav. Sta. Biol. Mar. de Lisbonne No. 52).

Navicula menaiana

Hendy, N. Ingram, 1964
An introductory account of the smaller algae of British coastal waters. V. Bacillariophyceae (Diatoms). Her Majesty's Stationary Office, 317 pp., 45 pls.

Navicula menisculus

Rumkówna, A., 1948
[List of the phytoplankton species occurring in the superficial water layers in the Gulf of Gdańsk.] Bull. Lab. mar., Gdynia, No. 4: 139-141 with tables in back.

Navicula meniscus

Hendy, N. Ingram, 1964
An introductory account of the smaller algae of British coastal waters. V. Bacillariophyceae (Diatoms). Her Majesty's Stationary Office, 317 pp., 45 pls.

Navicula mirabilis n.sp.

Castracane degli Antelminelli, F., 1886
1. Report on the Diatomaceae collected by H.M.S. Challenger during the years 1873-1876. Rept. Sci. Results, H.M.S. Challenger, Botany Vol. II, 178 pp., 30 pls.

Navicula misella

Hendy, N. Ingram, 1964
An introductory account of the smaller algae of British coastal waters. V. Bacillariophyceae (Diatoms). Her Majesty's Stationary Office, 317 pp., 45 pls.

Navicula molaris n.sp.

Grunow, A., 1863
Ueber einige neue und ungenügend bekannte Arten und Gattungen von Diatomaceen. Verhandl. d. K.K. Zool. Bot. Gesellsch., Vienna, 13: 137-162, pl.4-5 (Pl. 13-14).

Navicula mollis

Hendey, N.I., 1951
Littoral diatoms of Chicester Harbour with special reference to fouling. J. Roy. Microscop. Soc. 71(1): 1-86, 18 pls.

Navicula mollis

Zanon, V., 1948
Diatomee marini di Sardegna e Pugillo di Alghe Marine della stressa. Boll. Pesca, Piscitutura e Idrobiologia, Anno 24, ns. 3(2): 202-244, 27 figs. on 1 pl.

Navicula monilifera & varieties

Hendy, N. Ingram, 1964
An introductory account of the smaller algae of British coastal waters. V. Bacillariophyceae (Diatoms). Her Majesty's Stationary Office, 317 pp., 45 pls.

Navicula monilifera lacustris

Hendey, N.J., 1951
Littoral diatoms of Chicester Harbour with special reference to fouling. J. Roy. Microscop. Soc. 71(1): 1-86, 18 pls.

Navicula muralis

Zanon, D.V., 1949
Diatomee di Buenos Aires (Argentina) Atti Accad. Naz. Lincei, Memorie, Cl. Sci. fis., mat. e. nat., ser. 7, 11(3):59-151, 2 pls.

Navicula muscaeformis constricta

de Sousa e Silva, E., 1956.
Contribution à l'étude du microplancton de Dakar et des regions maritimes voisines. Bull. I.F.A.N., 8(2):335-371, 7 pls.

Navicula mutica

Hendy, N. Ingram, 1964
An introductory account of the smaller algae of British coastal waters. V. Bacillariophyceae (Diatoms). Her Majesty's Stationary Office, 317 pp., 45 pls.

Navicula mutica

Hendey, N.I., 1951
Littoral diatoms of Chicester Harbour with special reference to fouling. J. Roy. Microscop. Soc. 71(1): 1-86, 18 pls.

Navicula mutica

Zanon, V., 1948
Diatomee marini di Sardegna e Pugillo di Alghe Marine della stressa. Boll. Pesca, Piscitutura e Idrobiologia, Anno 24, ns. 3(2): 202-244, 27 figs. on 1 pl.

Navicula muticopsis muticopsis

Ko-bayashi, Tsuyako, 1963
Variations on some diatoms from Antarctica. 1.
Japan. Antarctic Res. Exped., 1956-1962. Sci. Repts., (E), No. 18:1-20, 16 pls.

Navicula nasuta n.sp.

Giffen, Malcolm H. 1973.
Diatoms of the marine littoral of Steenberg's Cove in St. Helena Bay, Cape Province, South Africa.
Botanica Marina 16 (1): 32-48.

Navicula nautica n.sp.

Cholnoky, B.J., 1963.
Beiträge zur Kenntnis des marinen Litorals von Südafrika.
Botanica Marina, 5(2/3):38-83.

Navicula Naveana n.sp.

Grunow, A., 1863
Ueber einige neue und ungenügend bekannte Arten und Gattungen von Diatomaceen. Verhandl. d. K.K. Zool. Bot. Gesellsch., Vienna, 13: 137-162, pl.4-5 (Pl. 13-14).

Navicula nebulosa

Hendy, N. Ingram, 1964
An introductory account of the smaller algae of British coastal waters. V. Bacillariophyceae (Diatoms). Her Majesty's Stationary Office, 317 pp., 45 pls.

Navicula nebulosa

Hendey, N.I., 1951
Littoral diatoms of Chicester Harbour with special reference to fouling. J. Roy. Microscop. Soc. 71(1): 1-86, 18 pls.

Navicula nicobarica n.sp.

Grunow, A., 1863
Ueber einige neue und ungenügend bekannte Arten und Gattungen von Diatomaceen. Verhandl. d. K.K. Zool. Bot. Gesellsch., Vienna, 13: 137-162, pl.4-5 (Pl. 13-14).

Navicula nitescens

Mann, A., 1907
Report on the diatoms of the Albatross voyages in the Pacific Ocean, 1888-1904. Contrib. U. S. Nat. Herb. 10(5):221-419, Pls. XLIV-LIV.

Navicula notabilis

Mann, A., 1907
Report on the diatoms of the Albatross voyages in the Pacific Ocean, 1888-1904. Contrib. U. S. Nat. Herb. 10(5):221-419, Pls. XLIV-LIV.

Navicula notabilis

Politis, J., 1949
Diatomees marines de Bosphores et des ibes de la mer de Marmara. II Practica tou Hellenikou Hidrobiologikou Institutoutou 1929, Etoz 1929, 3(1):11-31.

Navicula nummularia

Hendey, N-Ingram, 1958 [1957(Publ. 1958)]
Marine diatoms from some West African Ports. J. R. Microsc. Soc. (3) 77(1/2): 28-85.

Navicula oamaruensis

Mann, A., 1907
Report on the diatoms of the Albatross voyages in the Pacific Ocean, 1888-1904. Contrib. U. S. Nat. Herb. 10(5):221-419, Pls. XLIV-LIV.

Navicula octavosignata n.sp.

Salah, M.M., 1952(1953).
XII. Diatoms from Blakeney Point, Norfolk. New species and new records from Great Britain. J. R. Microsc. Soc., Ser. 3, 72(3):155-169, 3 pls

Navicula oris n.sp.

Zanon, D.V., 1949
Diatomee di Buenos Aires (Argentina) Atti Accad. Naz. Lincei, Memorie, Cl. Sci. fis., mat. e. nat., ser. 7, 11(3):59-151, 2 pls.

Navicula ostii n.sp.

Cholnoky, B.J., 1963.
Beiträge zur Kenntnis des marinen Litorals von Südafrika.
Botanica Marina, 5(2/3):38-83.

Navicula ostrearia

Hendy, N. Ingram, 1964
An introductory account of the smaller algae of British coastal waters. V. Bacillariophyceae (Diatoms). Her Majesty's Stationary Office, 317 pp., 45 pls.

Navicula ostrearia

Hustedt, F. and A.A. Aleem, 1951
Littoral diatoms from the Salstone near Plymouth. JMBA 30(1): 177-196.

Navicula ostrearia

Moreau, Jean, 1967.
Recherches préliminaires sur le verdissement en claires: l'évolution de leurs divers pigments liée au complexe pigmentaire de Navicula ostrearia Bory.
Rev.Trav.Ins.Peches marit., 31(4):373-382.

Navicula ostrearia

Neuville, Dominique, et Philippe Daste 1972.
Production de pigment bleu par la diatomée Navicula ostrearia (Gaillon) Bory maintenue en culture uni-algale sur un milieu synthétique carencé en azote nitrique.
C.r. hebd. Séanc. Acad. Sci. Paris (D) 274 (14): 2030-2033

Navicula ostrearia

Neuville, Dominique, Philippe Daste et Louis Genevès, 1971.
Premières données sur l'ultrastructure du frustule de la diatomée Navicula ostrearia (Gaillon) Bory.
C.r. hebd. Séanc. Acad. Sci. Paris (D) 273 (23): 2331-2334.

Navicula ostreanria

Neuville, Dominique, et Philippe Daste, 1970.
Observations concernant la production de pigment bleu par la diatomée Navicula ostrearia (Gaillon) Bory maintenue en culture uni-algale.
C.r. hebd. Séanc. Acad. Sci. Paris 271(25): 2389-2391

Navicula ostrearia

Neuville, Dominique, et Philippe Daste, 1971.
Observations concernant la production de pigment bleu par la diatomée Navicula ostrearia (Gaillon) Bory maintenue en culture unialgale sur un milieu synthétique.
C.r. hebd. Séanc. Acad. Sci. Paris (D) 272 (16): 2232-2234

Navicula ostrearia

Ranson, M. Gilbert, 1943
Titres et travaux scientifiques. Paris: Masson et Cie, Editeurs, Libraires de l'Academie de Medecine, 120 Boulevard Saint Germain. 88 pp.

Navicula ostrearia

Ransom, M.G., 1936.
Sur la soi-disant dégénérescence de la Navicule bleue (Navicula fusiformis Grün. – N. ostrearia Bory). C.R. Acad. Sci. CCII:1702-1704.

Navicula ostrearia

Ranson, G., 1927.
L'absorption de matières organiques dissoutes par la surface extérieure du corps chez les animaux aquatiques. Ann. Inst. Océan., n.s., 4(3):49-175, 1 pl. 17 textfigs.

Navicula ostrearia

Ranson, M. G., 1927.
Observations sur Navicula ostrearia Bory, origine du verdissement des huîtres. Rev. Algolog. 3:26-54, 1 pl.

Navicula ostrearia

Ranson, M.G., 1925.
Le déterminisme de la fixation saisonnière de Navicula fusiformis Grün (N. Ostrearia Bory). Sa culture experimentale en ostréiculture.
C.R. Acad. Sci. CCI:684-687.

Navicula ostrearia

Schodduyn, M., 1926
Observations faites dans la baie d'Ambleteuse (Pas de Calais). Bull. Inst. Ocean., Monaco, No. 482: 64 pp.

Navicula oxeia n.sp.

Castracane degli Antelminelli, F., 1886
1. Report on the Diatomaceae collected by H.M.S. Challenger during the years 1873-1876. Rept. Sci. Results, H.M.S. Challenger, Botany Vol. II, 178 pp., 30 pls.

Navicula paeninsulae n.sp.

Cholnoky, B.J., 1963.
Beiträge zur Kenntnis des marinen Litorals von Südafrika.
Botanica Marina, 5(2/3):38-83.

Navicula palpebralis

Hendy, N. Ingram, 1964
An introductory account of the smaller algae of British coastal waters. V. Bacillariophyceae (Diatoms).
Her Majesty's Stationary Office, 317 pp., 45 pls.

Navicula palpebralis

Hustedt, F. and A.A. Aleem, 1951
Littoral diatoms from the Salstone near Plymouth. JMBA 30(1): 177,196.

Navicula palpebralis

Paiva Carvalho, J., 1950
O plancton do Rio Maria Rodriques (Cananeis). 1. Diatomaceas e Dinoflagelados. Bol. Inst. Paulista Oceanogr. 1(1); 27-43, 2 fold-in tables, 2 figs.

Navicula palpebralis

Rampi, L., 1942
Ricerche sul fitoplancton del Mare Ligure 6. Le diatomee delle acque di Sanremo. Nuovo Giornale Botanico Italiano, N.S., 49:252-268.

Navicula palpebralis

Rampi, L., 1940
Diatomee del Mare Adriatico. Nuovo Giornale Botanico Italiano, n.s., 47:559-608.

Navicula palpebralis

Sousa e Silva, E., and J. Dos Santos-Pinto, 1948
O Plancton da Baia de S. Martinho do Porto. 1. Diatomaceas e Dinoflagelados. Bol. Soc. Portuguese de Ciencias Naturais, 16(2):134-187, 6 pls. (Trav. Sta. Biol. Mar. de Lisbonne No. 52).

Navicula palpebralis

Zanon, V., 1948
Diatomee marini di Sardegna e Pugillo di Alghe Marine della stressa. Boll. Pesca, Piscitutura e Idrobiologia, Anno 24, ns. 3(2): 202-244, 27 figs. on 1 pl.

Navicula pampeana oboesa n. var.

Zanon, D. V., 1949
Diatomee di Buenos Aires (Argentina) Atti Accad. Naz. Lincei, Memorie, Cl. Sci. fis., mat. e. nat., ser. 7, 11(3):59-151, 2 pls.

Navicula papula

Politis, J., 1949
Diatomees marines de Bosphores et des ibes de la mer de Marmara. II Practica tou Hellenikou Hidrobiologikou Institutoutou 1929, Etoz 1929, 3(1):11-31.

Navicula parallela n.sp.

Castracane degli Antelminelli, F., 1886
1. Report on the Diatomaceae collected by H.M.S. Challenger during the years 1873-1876. Rept. Sci. Results, H.M.S. Challenger, Botany Vol. II, 178 pp., 30 pls.

Navicula pavillardii

Hendy, N. Ingram, 1964
An introductory account of the smaller algae of British coastal waters. V. Bacillariophyceae (Diatoms).
Her Majesty's Stationary Office, 317 pp., 45 pls.

Navicula Pavillardi

Hendey, N.J., 1951
Littoral diatoms of Chicester Harbour with special reference to fouling. J. Roy. Microscop. Soc. 71(1): 1-86, 18 pls.

Navicula peculiaris n.sp.

Salah, M., and G. Tamás 1968.
Notes on new planktonic diatoms from Egypt.
Hydrobiologia 31(2): 231-240.

Navicula peisonis

Grunow, A., 1863
Ueber einige neue und ungenügend bekannte Arten und Gattungen von Diatomaceen. Verhandl. d. K.K. Zool. Bot. Gesellsch., Vienna, 13: 137-162, pl.4-5 (Pl. 13-14).

Navicula Pelagica

Allen, W.E., and E.E. Cupp, 1935
Plankton diatoms of the Java Sea. Annales du Jardin Botanique de Buitenzorg XLIV (2):101-174, figs.1-127.
(drawings of all species mentioned)

Navicula pelagica

Gilbert, J.Y., and W.E. Allen, 1943
The phytoplankton of the Gulf of California obtained by the "E.W. Scripps" in 1939 and 1940. J. Mar. Res. V(2):89-110, figs.30-31.

Navicula pelagica

Gran, H.H., 1908
Diatomeen. Nordisches Plankton, Botanischer Teil pp. XIX.1-XIX.146; 178 text figs.

Navicula pelagica

Hendy, N. Ingram, 1964
An introductory account of the smaller algae of British coastal waters. V. Bacillariophyceae (Diatoms).
Her Majesty's Stationary Office, 317 pp., 45 pls.

Navicula pelagica

Jørgensen, E., 1905
B. Protistplankton and the diatoms in bottom samples. Hydrographical and biological investigations in Norwegian fiords. Bergens Mus. Skr. 7: 49-225.

Navicula pelliculosa

Busby, William F., and Joyce Lewin, 1967.
Silicate uptake and silica shell formation by Synchronously dividing cells of the diatom Navicula pelliculosa (Bréb.) Hilse.
J. Phycology, 3(3):127-131.

Navicula pelliculosa

Healy, F.P., J. Coombs and B.E. Volcani, 1967.
Changes in pigment content of the diatom, Navicula pelliculosa (Bréb.) Hilse in silicon starvation synchrony.
Archiv. fur Mikrobiol., 59:131-142.

Navicula pelliculosa

Lewin, Joyce, 1966.
Silicon metabolism in diatoms. V. Germanium dioxide, a specific inhibitor of diatom growth. Phycologia, Int. Phycol. Soc., 6(1):1-12.

Navicula pelliculosa

Lewin, J.C., 1955.
The capsule of the diatom Navicula pelliculosa. J. Gen. Microbiol. 13(1):162-169.

Navicula pelliculosa

Low, E.M., 1955.
Studies on some chemical constituents of diatoms J. Mar. Res. 14(2):199-204.

Navicula pelliculosa

Reimann, Bernhard E.F., Joyce C. Lewin and Benjamin E. Volcani, 1966.
Studies on the biochemistry and fine structure of silica shell formation in diatoms. II The structure of the cell wall of Navicula pelliculosa (Bréb.) Hilse. J. Phycol. 2(2): 74-84.

Navicula pennata
Andrade e Clovis Teixeira, M.H. de, 1957.
Contribuição para o conhecimento das diatomaceas do Brasil.
Bol. Inst. Ocean., Univ. Sao Paulo, 8(1/2):171-225, 10 pls.

Navicula pennata
Hendy, N. Ingram, 1964
An introductory account of the smaller algae of British coastal waters. V. Bacillariophyceae (Diatoms).
Her Majesty's Stationary Office, 317 pp., 45 pls.

Navicula pennata
Mann, A., 1907
Report on the diatoms of the Albatross voyages in the Pacific Ocean, 1888-1904.
Contrib. U. S. Nat. Herb. 10(5):221-419, Pls. XLIV-LIV.

Navicula pennata
Mann, A., 1893
List of Diatomaceae from a deep-sea dredging in the Atlantic Ocean off Delaware Bay by the U. S. Fish Commission Steamer Albatross.
Proc. U. S. Nat. Mus. 16:303-312.

Navicula pennata
Margalef, R., 1949
Fitoplancton nerítico de la Costa Brava en 1947-48. Publ. Inst. Biol. Aplicada, 5: 41-51, 3 text figs.

Navicula pennata
Rampi, L., 1942
Ricerche sul fitoplancton del Mare Ligure 6. Le diatomee delle acque di Sanremo. Nuovo Giornale Botanico Italiano, N.S., 49:252-268.

Navicula pennata
Rampi, L., 1940
Diatomee del Mare Adriatico. Nuovo Giornale Botanico Italiano, n.s., 47:559-608.

Navicula pennata (figs.)
Sousa e Silva, E., 1949
Diatomaceas e Dinoflagelados de Baia de Cascais. Portugaliae Acta Biol., Volume: Julio Henriques, Ser. B: 300-383, 9 pls, 2 fold-in tables.

Navicula pennata
Zanon, V., 1948
Diatomee marini di Sardegna e Pugillo di Alghe Marine della stressa. Boll. Pesca, Piscitutura e Idrobiologia, Anno 24, ns. 3(2): 202-244, 27 figs. on 1 pl.

Navicula peregrina
Barber, H.G., 1961.
A note on unusual diatom deformities.
J. Quekett Microsc. Club, (4) 5(13):365.

Navicula peregrina
Hendy, N. Ingram, 1964
An introductory account of the smaller algae of British coastal waters. V. Bacillariophyceae (Diatoms).
Her Majesty's Stationary Office, 317 pp., 45 pls.

Navicula peregrina
Hendey, N.J., 1951
Littoral diatoms of Chicester Harbour with special reference to fouling. J.Roy. Microsc. Soc. 71(1): 1-86, 18 pls.

Navicula peregrina
Iyengar, M.O.P. and G.Venkataraman, 1951.
The ecology and seasonal succession of the algae flora of the River Cooum at Madras with special reference to the Diatomaceae. J. Madras Univ. 21, Sect. B(1): 140-192, 1 pl of 4 figs., 11 text figs.

Navicula peregrina
Rumkówna, A., 1948
[List of the phytoplankton species occurring in the superficial water layers in the Gulf of Gdańsk] Bull. Lab. mar., Gdynia, No. 4: 139-141 with tables in back.

Navicula peregrina
Zanon, D. V., 1949
Diatomee di Buenos Aires (Argentina)
Atti Accad. Naz. Lincei, Memorie, Cl. Sci. fis., mat. e. nat., ser. 7, 11(3):59-151, 2 pls.

Navicula perplexa
Hendy, N. Ingram, 1964
An introductory account of the smaller algae of British coastal waters. V. Bacillariophyceae (Diatoms).
Her Majesty's Stationary Office, 317 pp., 45 pls.

Navicula Perrotettii
Zanon, D. V., 1949
Diatomee di Buenos Aires (Argentina)
Atti Accad. Naz. Lincei, Memorie, Cl. Sci. fis., mat. e. nat., ser. 7, 11(3):59-151, 2 pls.

Navicula phoenicenteron
Mann, A., 1907
Report on the diatoms of the Albatross voyages in the Pacific Ocean, 1888-1904.
Contrib. U. S. Nat. Herb. 10(5):221-419, Pls. XLIV-LIV.

Navicula phyllepta
Hendy, N. Ingram, 1964
An introductory account of the smaller algae of British coastal waters. V. Bacillariophyceae (Diatoms).
Her Majesty's Stationary Office, 317 pp., 45 pls.

Navicula pinguis n.sp.
Mann, A., 1907
Report on the diatoms of the Albatross voyages in the Pacific Ocean, 1888-1904.
Contrib. U. S. Nat. Herb. 10(5):221-419, Pls. XLIV-LIV.

Navicula placentula
Rumkówna, A., 1948
[List of the phytoplankton species occurring in the superficial water layers in the Gulf of Gdańsk] Bull. Lab. mar., Gdynia, No. 4: 139-141 with tables in back.

Navicula placentula
Zanon, D. V., 1949
Diatomee di Buenos Aires (Argentina)
Atti Accad. Naz. Lincei, Memorie, Cl. Sci. fis., mat. e. nat., ser. 7, 11(3):59-151, 2 pls.

Navicula plagiostoma
Andrade e Clovis Teixeira, M.H. de, 1957.
Contribuição para o conhecimento das diatomaceas do Brasil.
Bol. Inst. Ocean., Univ. Sao Paulo, 8(1/2):171-225, 10 pls.

Navicula planamembranacea n.sp.
Hendy, N. Ingram, 1964
An introductory account of the smaller algae of British coastal waters. V. Bacillariophyceae (Diatoms).
Her Majesty's Stationary Office, 317 pp., 45 pls.

Navicula planamembranacea
Robinson, G.A., 1965.
Continuous plankton records: contribution towards a plankton atlas of the North Atlantic and the North Sea. X. Navicula planamembranacea, Hendey.
Bull. Mar. Ecol., 6(5): 141-145, Pls. 44-46.

Navicula platystoma
Rumkówna, A., 1948
[List of the phytoplankton species occurring in the superficial water layers in the Gulf of Gdańsk] Bull. Lab. mar., Gdynia, No. 4: 139-141 with tables in back.

Navicula pleurostaurum
Mann, A., 1907
Report on the diatoms of the Albatross voyages in the Pacific Ocean, 1888-1904.
Contrib. U. S. Nat. Herb. 10(5):221-419, Pls. XLIV-LIV.

Navicula plicata
Hendy, N. Ingram, 1964
An introductory account of the smaller algae of British coastal waters. V. Bacillariophyceae (Diatoms).
Her Majesty's Stationary Office, 317 pp., 45 pls.

Navicula plicata
Hustedt, F. and A.A. Aleem, 1951
Littoral diatoms from the Salstone near Plymouth. JMBA 30(1): 177-196.

Navicula polystricta
Rampi, L., 1940
Diatomee del Mare Adriatico. Nuovo Giornale Botanico Italiano, n.s., 47:559-608.

Navicula praestes
Politis, J., 1949
Diatomees marines de Bosphores et des ibes de la mer de Marmara. II Practica tou Hellenikou Hidrobiologikou Institutoutou 1929, Etoz 1929, 3(1):11-31.

Navicula praetexta
Andrade e Clovis Teixeira, M.H. de, 1957.
Contribuição para o conhecimento das diatomaceas do Brasil.
Bol. Inst. Ocean., Univ. Sao Paulo, 8(1/2):171-225, 10 pls.

Navicula praetexta
Hendy, N. Ingram, 1964
An introductory account of the smaller algae of British coastal waters. V. Bacillariophyceae (Diatoms).
Her Majesty's Stationary Office, 317 pp., 45 pls.

Navicula praetexta
Mann, A., 1907
Report on the diatoms of the Albatross voyages in the Pacific Ocean, 1888-1904.
Contrib. U. S. Nat. Herb. 10(5):221-419, Pls. XLIV-LIV.

Navicula praetexta
Mann, A., 1893
List of Diatomaceae from a deep-sea dredging in the Atlantic Ocean off Delaware Bay by the U. S. Fish Commission Steamer Albatross.
Proc. U. S. Nat. Mus. 16:303-312.

Navicula praetexta
Rampi, L., 1942
Ricerche sul fitoplancton del Mare Ligure 6. Le diatomee delle acque di Sanremo. Nuovo Giornale Botanico Italiano, N.S., 49:252-268.

Navicula praetexta
Rampi, L., 1940
Diatomee del Mare Adriatico. Nuovo Giornale Botanico Italiano, n.s., 47:559-608.

Navicula prodiga n.sp.
Mann, A., 1907
Report on the diatoms of the Albatross voyages in the Pacific Ocean, 1888-1904. Contrib. U. S. Nat. Herb. 10(5):221-419, Pls. XLIV-LIV.

Navicula protracta
Iyengar, M.O.P. and G.Venkataraman, 1951.
The ecology and seasonal succession of the algae flora of the River Cooum at Madras with special reference to the Diatomaceae. J. Madras Univ. 21, Sect. B(1): 140-192, 1 pl of 4 figs., 11 text figs.

Navicula protracta
Rumkówna, A., 1948
List of the phytoplankton species occurring in the superficial water layers in the Gulf of Gdańsk. Bull. Lab. mar., Gdynia, No. 4: 139-141 with tables in back.

Navicula protracta
Zanon, D. V., 1949
Diatomee di Buenos Aires (Argentina) Atti Accad. Naz. Lincei, Memorie, Cl. Sci. fis., mat. e. nat., ser. 7, 11(3):59-151, 2 pls.

Navicula protracta
Zanon, V., 1948
Diatomee marini di Sardegna e Pugillo di Alghe Marine della stressa. Boll. Pesca, Piscitutura e Idrobiologia, Anno 24, ns. 3(2): 202-244, 27 figs. on 1 pl.

Navicula pseudoapproximata n.sp.
Hendey, N-Ingram, 1958 [1957(Publ. 1958)]
Marine diatoms from some West African Ports. J. R. Microsc. Soc. (3) 77(1/2): 28-85.

Navicula pseudocarinifera n.sp.
Manguin, E., 1957.
Premier inventaire des diatomées de la Terre Adélie Antarctique. Espèces nouvelles. Rev. Algologique, n.s., 3(3):111-134.

Navicula pseudocomoides nom. nov.
Hendey, N. Ingram, 1964
An introductory account of the smaller algae of British coastal waters. V. Bacillariophyceae (Diatoms). Her Majesty's Stationary Office, 317 pp., 45 pls.

Navicula pseudopalpebralis n.sp.
Hendey, N. Ingram, 1964
An introductory account of the smaller algae of British coastal waters. V. Bacillariophyceae (Diatoms). Her Majesty's Stationary Office, 317 pp., 45 pls.

Navicula punctata
Hendey, N.I., 1953.
Taxonomic studies on some Naviculae punctatae. J.R. Microsc. Soc. (3) 73(3):156-161.

Navicula punctulata
Andrade e Clovis Teixeira, M.H. de, 1957.
Contribuição para o conhecimento das diatomaceas do Brasil. Bol. Inst. Ocean., Univ. Sao Paulo, 8(1/2):171-225, 10 pls.

Navicula pupula
Zanon, D. V., 1949
Diatomee di Buenos Aires (Argentina) Atti Accad. Naz. Lincei, Memorie, Cl. Sci. fis., mat. e. nat., ser. 7, 11(3):59-151, 2 pls.

Navicula pupula capitata
Tsumura, K., 1956.
Diatomoj el la cirkaufoso de la restajo de la kastele de Odawara. J. Yokohama Municipal Univ., (C-14), No. 47:23 pp.

Navicula pusilla
Grunow, A., 1863
Ueber einige neue und ungenügend bekannte Arten und Gattungen von Diatomaceen. Verhandl. d. K.K. Zool. Bot. Gesellsch., Vienna, 13: 137-162, pl.4-5 (Pl. 13-14).

Navicula pusilla
Hendy, N. Ingram, 1964
An introductory account of the smaller algae of British coastal waters. V. Bacillariophyceae (Diatoms). Her Majesty's Stationary Office, 317 pp., 45 pls.

Navicula pusila
Rumkówna, A., 1948
List of the phytoplankton species occurring in the superficial water layers in the Gulf of Gdańsk. Bull. Lab. mar., Gdynia, No. 4: 139-141 with tables in back.

Navicula pusilla
Zanon, D. V., 1949
Diatomee di Buenos Aires (Argentina) Atti Accad. Naz. Lincei, Memorie, Cl. Sci. fis., mat. e. nat., ser. 7, 11(3):59-151, 2 pls.

Navicula pusilla
Zanon, V., 1948
Diatomee marini di Sardegna e Pugillo di Alghe Marine della stressa. Boll. Pesca, Piscitutura e Idrobiologia, Anno 24, ns. 3(2): 202-244, 27 figs. on 1 pl.

Navicula pygmaea
Hendy, N. Ingram, 1964
An introductory account of the smaller algae of British coastal waters. V. Bacillariophyceae (Diatoms). Her Majesty's Stationary Office, 317 pp., 45 pls.

Navicula pygmaea
Hendey, N.J., 1951
Littoral diatoms of Chicester Harbour with special reference to fouling. J.Roy. Microscop. Soc. 71(1): 1-86, 18 pls.

Navicula pygmaea
Iyengar, M.O.P. and G.Venkataraman, 1951.
The ecology and seasonal succession of the algae flora of the River Cooum at Madras with special reference to the Diatomaceae. J. Madras Univ. 21, Sect. B(1): 140-192, 1 pl of 4 figs., 11 text figs.

Navicula pygmaea
Rumkówna, A., 1948
List of the phytoplankton species occurring in the superficial water layers in the Gulf of Gdańsk. Bull. Lab. mar., Gdynia, No. 4: 139-141 with tables in back.

Navicula pygmaea
Zanon, D. V., 1949
Diatomee di Buenos Aires (Argentina) Atti Accad. Naz. Lincei, Memorie, Cl. Sci. fis., mat. e. nat., ser. 7, 11(3):59-151, 2 pls.

Navicula quadripedis
Brunel, J., 1962
Le phytoplancton de la Baie de Chaleurs. Inst. Botan., Univ. Montréal, Contrib. No. 77: 365 pp., 66 pls.

Navicula quadripedis
Brunel, Jules, 1962
Le phytoplancton de la Baie des Chaleurs. Contrib. Ministère de la Chasse et des Pêcheries, Province de Québec, No. 91: 365 pp.

Navicula quadriseriata
Rampi, L., 1940
Diatomee del Mare Adriatico. Nuovo Giornale Botanico Italiano, n.s., 47:559-608.

Navicula quinquenodis
Grunow, A., 1863
Ueber einige neue und ungenügend bekannte Arten und Gattungen von Diatomaceen. Verhandl. d. K.K. Zool. Bot. Gesellsch., Vienna, 13: 137-162, pl.4-5 (Pl. 13-14).

Navicula radiosa
Zanon, V., 1948
Diatomee marini di Sardegna e Pugillo di Alghe Marine della stressa. Boll. Pesca, Piscitutura e Idrobiologia, Anno 24, ns. 3(2): 202-244, 27 figs. on 1 pl.

Navicula ramosissima
Castenholz, Richard W., 1963.
An experimental study of the vertical distribution of littoral marine diatoms. Limnology and Oceanography, 8(4):450-462.

Navicula ramosissima & varieties
Hendy, N. Ingram, 1964
An introductory account of the smaller algae of British coastal waters. V. Bacillariophyceae (Diatoms). Her Majesty's Stationary Office, 317 pp., 45 pls.

Navicula ramosissima
Hendey, N.J., 1951
Littoral diatoms of Chicester Harbour with special reference to fouling. J.Roy. Microscop. Soc. 71(1): 1-86, 18 pls.

Navicula ramosissima
Schodduyn, M., 1926
Observations faites dans la baie d'Ambleteuse (Pas de Calais). Bull. Inst. Ocean., Monaco, No. 482: 64 pp.

Navicula rancurellii n.sp
Amossé, A. 1970.
Diatomées marines et saumâtres du Sénégal et de la Côte d'Ivoire. Bull. Inst. fond Afr. Noire (A) 32(2), 289-311.

Navicula Reichardtii
Rampi, L., 1940
Diatomee del Mare Adriatico. Nuovo Giornale Botanico Italiano, n.s., 47:559-608.

Navicula Reinhardtii
Rumkówna, A., 1948
List of the phytoplankton species occurring in the superficial water layers in the Gulf of Gdańsk. Bull. Lab. mar., Gdynia, No. 4: 139-141 with tables in back.

Navicula rhombica
Hendy, N. Ingram, 1964
An introductory account of the smaller algae of British coastal waters. V. Bacillariophyceae (Diatoms). Her Majesty's Stationary Office, 317 pp., 45 pls.

Navicula rhombica
Manguin, E., 1954
Diatomees marines provenant de l'île Heard (Australian National Antarctic Expedition). Rev. Algol., n.s., 1: 14-24.

Navicula rhomboides
Mann, A., 1893
List of Diatomaceae from a deep-sea dredging in the Atlantic Ocean off Delaware Bay by the U. S. Fish Commission Steamer Albatross. Proc. U. S. Nat. Mus. 16:303-312.

Navicula rhyncocephala
Hendy, N. Ingram, 1964
An introductory account of the smaller algae of British coastal waters. V. Bacillariophyceae (Diatoms).
Her Majesty's Stationary Office, 317 pp., 45 pls.

Navicula rhyncocephala
Hendey, N.I., 1951
Littoral diatoms of Chicester Harbour with special reference to fouling. J.Roy. Microscop. Soc. 71(1): 1-86, 18 pls.

Navicula rhynchocephala
Rumkówna, A., 1948
[List of the phytoplankton species occurring in the superficial water layers in the Gulf of Gdańsk] Bull. Lab. mar., Gdynia, No. 4: 139-141 with tables in back.

Navicula robertsianan var. abnormis
Hendey, N-Ingram, 1958 [1957(Publ. 1958)]
Marine diatoms from some West African Ports.J. R.Microsc. Soc. (3) 77(1/2): 28-85.

Navicula robusta
Politis, J., 1949
Diatomees marines de Bosphores et des ibes de la mer de Marmara. II Practica tou Hellenikou Hidrobiologikou Institutoutou 1929, Etoz 1929, 3(1):11-31.

Navicula rossii
Hendy, N. Ingram, 1964
An introductory account of the smaller algae of British coastal waters. V. Bacillariophyceae (Diatoms).
Her Majesty's Stationary Office, 317 pp., 45 pls.

Navicula rostellata
Hendy, N. Ingram, 1964
An introductory account of the smaller algae of British coastal waters. V. Bacillariophyceae (Diatoms).
Her Majesty's Stationary Office, 317 pp., 45 pls.

Navicula rostellata
Mann, A., 1893
List of Diatomaceae from a deep-sea dredging in the Atlantic Ocean off Delaware Bay by the U. S. Fish Commission Steamer Albatross. Proc. U. S. Nat. Mus. 16:303-312.

Navicula rostellata
Rumkówna, A., 1948
[List of the phytoplankton species occurring in the superficial water layers in the Gulf of Gdańsk] Bull. Lab. mar., Gdynia, No. 4: 139-141 with tables in back.

Navicula salebrosa n.sp.
Cholnoky, B.J., 1963.
Beiträge zur Kenntnis des marinen Litorals von Südafrika.
Botanica Marina, 5(2/3):38-83.

Navicula salinaroides n.sp.
Cholnoky, B.J., 1963.
Beiträge zur Kenntnis des marinen Litorals von Südafrika.
Botanica Marina, 5(2/3):38-83.

Navicula salinarum
Hendy, N. Ingram, 1964
An introductory account of the smaller algae of British coastal waters. V. Bacillariophyceae (Diatoms).
Her Majesty's Stationary Office, 317 pp., 45 pls.

Navicula salinarum
Iyengar, M.O.P. and G.Venkataraman,1951.
The ecology and seasonal succession of the algae flora of the River Cooum at Madras with special reference to the Diatomaceae. J. Madras Univ. 21, Sect. B(1): 140-192, 1 pl of 4 figs., 11 text figs.

Navicula salinarum
Zanon, V., 1948
Diatomee marini di Sardegna e Pugillo di Alghe Marine della stressa. Boll. Pesca, Piscitutura e Idrobiologia, Anno 24, ns. 3(2): 202-244, 27 figs. on 1 pl.

Navicula salinicola
Hendy, N. Ingram, 1964
An introductory account of the smaller algae of British coastal waters. V. Bacillariophyceae (Diatoms).
Her Majesty's Stationary Office, 317 pp., 45 pls.

Navicula sandriana
Grunow, A., 1863
Ueber einige neue und ungenügend bekannte Arten und Gattungen von Diatomaceen. Verhandl. d. K.K. Zool. Bot. Gesellsch., Vienna, 13: 137-162, pl.4-5 (Pl. 13-14).

Navicula sandriana
Mann, A., 1907
Report on the diatoms of the Albatross voyages in the Pacific Ocean, 1888-1904. Contrib. U. S. Nat. Herb. 10(5):221-419, Pls. XLIV-LIV.

Navicula sandriana
Rampi, L., 1940
Diatomee del Mare Adriatico. Nuovo Giornale Botanico Italiano, n.s., 47:559-608.

Navicula Sandriana
Zanon, V., 1948
Diatomee marini di Sardegna e Pugillo di Alghe Marine della stressa. Boll. Pesca, Piscitutura e Idrobiologia, Anno 24, ns. 3(2): 202-244, 27 figs. on 1 pl.

Navicula Schuetti
Hendey, N.I., 1937
The plankton diatoms of the southern seas. Discovery Repts. 16:151-364, pls.6-13.

Navicula Schultzei
Mann, A., 1893
List of Diatomaceae from a deep-sea dredging in the Atlantic Ocean off Delaware Bay by the U. S. Fish Commission Steamer Albatross. Proc. U. S. Nat. Mus. 16:303-312.

Navicula scopulorum
Andrade e Clovis Teixeira, M.H. de, 1957.
Contribuicao para o conhecimento das diatomaceas do Brasil.
Bol. Inst. Ocean., Univ. Sao Paulo, 8(1/2):171-225, 10 pls.

Navicula scopulorum & varieties
Hendy, N. Ingram, 1964
An introductory account of the smaller algae of British coastal waters. V. Bacillariophyceae (Diatoms).
Her Majesty's Stationary Office, 317 pp., 45 pls.

Navicula scopulorum
Hendey, N.I., 1951
Littoral diatoms of Chicester Harbour with special reference to fouling. J.Roy. Microscop. Soc. 71(1): 1-86, 18 pls.

Navicula scopulorum
Rampi, L., 1940
Diatomee del Mare Adriatico. Nuovo Giornale Botanico Italiano, n.s., 47:559-608.

Navicula scopulorum
Zanon, V., 1948
Diatomee marini di Sardegna e Pugillo di Alghe Marine della stressa. Boll. Pesca, Piscitutura e Idrobiologia, Anno 24, ns. 3(2): 202-244, 27 figs. on 1 pl.

Navicula semen
Zanon, D. V., 1949
Diatomee di Buenos Aires (Argentina) Atti Accad. Naz. Lincei, Memorie, Cl. Sci. fis., mat. e. nat., ser. 7, 11(3):59-151, 2 pls.

Navicula semenoides
Tsumura, K., 1956.
Diatomoj el la cirkaufoso de la restajo de la kastele de Odawara. J. Yokohama Municipal Univ., (C-14), No. 47:23 pp.

Navicula septentrionalis
Gran, H.H., 1908
Diatomeen. Nordisches Plankton, Botanischer Teil pp. XIX.1-XIX 146; 178 text figs.

Navicula serena
Zanon, D. V., 1949
Diatomee di Buenos Aires (Argentina) Atti Accad. Naz. Lincei, Memorie, Cl. Sci. fis., mat. e. nat., ser. 7, 11(3):59-151, 2 pls.

Navicula serians
Mann, A., 1893
List of Diatomaceae from a deep-sea dredging in the Atlantic Ocean off Delaware Bay by the U. S. Fish Commission Steamer Albatross. Proc. U. S. Nat. Mus. 16:303-312.

Nitzschia seriata
Sargent, M. S. and T. J. Walker, 1948
Diatom populations associated with eddies off southern California in 1941. J. Mar. Res. 7(3):490-505, 15 text figs.

Navicula silicula
Mann, A., 1907
Report on the diatoms of the Albatross voyages in the Pacific Ocean, 1888-1904. Contrib. U. S. Nat. Herb. 10(5):221-419, Pls. XLIV-LIV.

Navicula Silvestrina n.sp.
Zanon, D. V., 1949
Diatomee di Buenos Aires (Argentina) Atti Accad. Naz. Lincei, Memorie, Cl. Sci. fis., mat. e. nat., ser. 7, 11(3):59-151, 2 pls.

Navicula simplex
Rumkówna, A., 1948
[List of the phytoplankton species occurring in the superficial water layers in the Gulf of Gdańsk] Bull. Lab. mar., Gdynia, No. 4: 139-141 with tables in back.

Navicula smithi
Mann, A., 1907
Report on the diatoms of the Albatross voyages in the Pacific Ocean, 1888-1904. Contrib. U. S. Nat. Herb. 10(5):221-419, Pls. XLIV-LIV.

Navicula smithii
Mann, A., 1893
List of Diatomaceae from a deep-sea dredging in the Atlantic Ocean off Delaware Bay by the U. S. Fish Commission Steamer Albatross. Proc. U. S. Nat. Mus. 16:303-312.

Navicula Smithii

Meunier, A., 1915
Microplancton de la Mer Flamande. 2. Diatomées (excepté le genre Chaetoceros). Mem. Mus. Roy. Hist. Nat., Belgique, 7(3):1-118, Pls. VIII-XIV.

Navicula smithi

Morse, D.C., 1947
Some observations on seasonal variations in plankton population Patuxent River, Maryland 1943-1945. Bd. Nat. Res., Publ. No.65, Chesapeake Biol. Lab., 31, 3 figs.

Navicula smithii

Politis, J., 1949
Diatomees marines de Bosphores et des ibes de la mer de Marmara. II Practica tou Hellenikou Hidrobiologikou Institutoutou 1929, Etoz 1929, 3(1):11-31.

Navicula sobria n.sp.

Cholnoky, B.J., 1963.
Beiträge zur Kenntnis des marinen Litorals von Südafrika.
Botanica Marina, 5(2/3):38-83.

Navicula solaris

Mann, A., 1907
Report on the diatoms of the Albatross voyages in the Pacific Ocean, 1888-1904. Contrib. U. S. Nat. Herb. 10(5):221-419, Pls. XLIV-LIV.

Navicula speciosa n.sp.

Mann, A., 1907
Report on the diatoms of the Albatross voyages in the Pacific Ocean, 1888-1904. Contrib. U. S. Nat. Herb. 10(5):221-419, Pls. XLIV-LIV.

Navicula spectabilis

Andrade e Clovis Teixeira, M.H. de, 1957.
Contribuição para o conhecimento das diatomáceas do Brasil.
Bol. Inst. Ocean., Univ. Sao Paulo, 8(1/2):171-225, 10 pls.

Navicula spectabilis

Hendy, N. Ingram, 1964
An introductory account of the smaller algae of British coastal waters. V. Bacillariophyceae (Diatoms).
Her Majesty's Stationary Office, 317 pp., 45 pls.

Navicula spectabilis

Mann, A., 1907
Report on the diatoms of the Albatross voyages in the Pacific Ocean, 1888-1904. Contrib. U. S. Nat. Herb. 10(5):221-419, Pls. XLIV-LIV.

Navicula spectabilis var. emarginata

Hendey, N-Ingram, 1958 [1957(Publ. 1958)]
Marine diatoms from some West African Ports- J.R Microsc. Soc. (3) 77(1/2): 28-85.

Navicula splendida

Mann, A., 1907
Report on the diatoms of the Albatross voyages in the Pacific Ocean, 1888-1904. Contrib. U. S. Nat. Herb. 10(5):221-419, Pls. XLIV-LIV.

Navicula splendida

Mann, A., 1893
List of Diatomaceae from a deep-sea dredging in the Atlantic Ocean off Delaware Bay by the U. S. Fish Commission Steamer Albatross. Proc. U. S. Nat. Mus. 16:303-312.

Navicula splendida

Politis, J., 1949
Diatomees marines de Bosphores et des ibes de la mer de Marmara. II Practica tou Hellenikou Hidrobiologikou Institutoutou 1929, Etoz 1929, 3(1):11-31.

Navicula spuma n.sp.

Mann, A., 1907
Report on the diatoms of the Albatross voyages in the Pacific Ocean, 1888-1904. Contrib. U. S. Nat. Herb. 10(5):221-419, Pls. XLIV-LIV.

Navicula Stompsii n.sp.

Cholnoky, B.J., 1963.
Beiträge zur Kenntnis des marinen Litorals von Südafrika.
Botanica Marina, 5(2/3):38-83.

Navicula Struggeri n.sp.

Cholnoky, B.J., 1963.
Beiträge zur Kenntnis des marinen Litorals von Südafrika.
Botanica Marina, 5(2/3):38-83.

Navicula subacuta

Mann, A., 1907
Report on the diatoms of the Albatross voyages in the Pacific Ocean, 1888-1904. Contrib. U. S. Nat. Herb. 10(5):221-419, Pls. XLIV-LIV.

Navicula subcarinata

Hendy, N. Ingram, 1964
An introductory account of the smaller algae of British coastal waters. V. Bacillariophyceae (Diatoms).
Her Majesty's Stationary Office, 317 pp., 45 pls.

Navicula subcarinata nom. nov.

Hendey, N.J., 1951
Littoral diatoms of Chicester Harbour with special reference to fouling. J.Roy. Microscop. Soc. 71(1): 1-86, 18 pls.

Navicula subcincta

Mann, A., 1907
Report on the diatoms of the Albatross voyages in the Pacific Ocean, 1888-1904. Contrib. U. S. Nat. Herb. 10(5):221-419, Pls. XLIV-LIV.

Navicula subcincta

Mann, A., 1893
List of Diatomaceae from a deep-sea dredging in the Atlantic Ocean off Delaware Bay by the U. S. Fish Commission Steamer Albatross. Proc. U. S. Nat. Mus. 16:303-312.

Navicula subcingulata n.sp.

Cholnoky, B.J., 1963.
Beiträge zur Kenntnis des marinen Litorals von Südafrika.
Botanica Marina, 5(2/3):38-83.

Navicula subdiffusa

Andrade e Clovis Teixeira, M.H. de, 1957.
Contribuição para o conhecimento das diatomáceas do Brasil.
Bol. Inst. Ocean., Univ. Sao Paulo, 8(1/2):171-225, 10 pls.

Navicula suborbicularis

Mann, A., 1893
List of Diatomaceae from a deep-sea dredging in the Atlantic Ocean off Delaware Bay by the U. S. Fish Commission Steamer Albatross. Proc. U. S. Nat. Mus. 16:303-312.

Navicula subpolaris nom. nov. (for N. cristata)

Hendey, N.I., 1937
The plankton diatoms of the southern seas. Discovery Repts. 16:151-364, pls.6-13.

Navicula subrhomboidea n.sp

Castracane degli Antelminelli, F., 1886
1. Report on the Diatomaceae collected by H.M.S. Challenger during the years 1873-1876. Rept. Sci. Results, H.M.S. Challenger, Botany Vol. II, 178 pp., 30 pls.

Navicula subtilyra n.sp.

Cholnoky, B.J., 1963.
Beiträge zur Kenntnis des marinen Litorals von Südafrika.
Botanica Marina, 5(2/3):38-83.

Navicula sulphurea, n.sp.

Frenguelli, Joaquin, y Hector A. Orlando, 1958.
Diatomeas y silicoflagelados del sector Antartico Sudamericano.
Inst. Antartico Argentino, Publ., No. 5:191 pp.

Navicula superciliaris n.sp.

Cholnoky, B.J., 1963.
Beiträge zur Kenntnis des marinen Litorals von Südafrika.
Botanica Marina, 5(2/3):38-83.

Navicula supralittoralis n.sp.

Aleem, A.A., and Fr. Hustedt, 1951.
Einige neue Diatomeen von der Südküste Englands. Botaniska Notiser 1951(1):13-20, 6 textfigs.

Navicula Surirellae n.sp.

Zanon, D. V., 1949
Diatomee di Buenos Aires (Argentina) Atti Accad. Naz. Lincei, Memorie, Cl. Sci. fis., mat. e. nat., ser. 7, 11(3):59-151, 2 pls.

Navicula tahitensis n.sp.

Grunow, A., 1863
Ueber einige neue und ungenügend bekannte Arten und Gattungen von Diatomaceen. Verhandl. d. K.K. Zool. Bot. Gesellsch., Vienna, 13: 137-162, pl.4-5 (Pl. 13-14).

Navicula takoradiensis nom. nov.

Hendey, N-Ingram, 1958 [1957(Publ. 1958)]
Marine diatoms from some West African Ports- J.R.Microsc. Soc. (3) 77(1/2): 28-85.

Navicula taylori

Hendy, N. Ingram, 1964
An introductory account of the smaller algae of British coastal waters. V. Bacillariophyceae (Diatoms).
Her Majesty's Stationary Office, 317 pp., 45 pls.

Navicula thaitiana n.sp.

Castracane degli Antelminelli, F., 1886
1. Report on the Diatomaceae collected by H.M.S. Challenger during the years 1873-1876. Rept. Sci. Results, H.M.S. Challenger, Botany Vol. II, 178 pp., 30 pls.

Navicula transfuga

Mann, A., 1893
List of Diatomaceae from a deep-sea dredging in the Atlantic Ocean off Delaware Bay by the U. S. Fish Commission Steamer Albatross. Proc. U. S. Nat. Mus. 16:303-312.

Navicula transitans

Iselin, C., 1930
A report on the coastal waters of Labrador based on explorations of the "Chance" during the summer of 1926. Proc. Am. Acad. Arts Sci., 66(1):1-37, 14 text figs.

Navicula transitans var. derasa

Heimdal, Berit R. 1970.
Morphology and distribution of two Navicula species in Norwegian coastal waters.
Nytt Mag. Bot. 17(2): 65-75

Navicula triundulata

Grunow, A., 1877
1. New Diatoms from Honduras. Monthly Micros. Jour., 18:165-186, pls. CXCIII-CXCVI.

Navicula trompii major

Boden, B. P., 1949.
The diatoms collected by the U.S.S. CACOPAN in the Antarctic in 1947. J. Mar. Res. 8(1):6-13, 3 textfigs.

Navicula tubulosa
Andrade e Clovis Teixeira, M.H. de, 1957.
Contribuição para o conhecimento das diatomáceas do Brasil.
Bol. Inst. Ocean., Univ. Sao Paulo, 8(1/2):171-225, 10 pls.

Navicula tuscula
Hendy, N. Ingram, 1964
An introductory account of the smaller algae of British coastal waters. V. Bacillariophyceae (Diatoms).
Her Majesty's Stationary Office, 317 pp., 45 pls.

Navicula tuscula
Rumkówna, A., 1948
[List of the phytoplankton species occurring in the superficial water layers in the Gulf of Gdańsk] Bull. Lab. mar., Gdynia, No. 4: 139-141 with tables in back.

Navicula unica n. sp.
Salah, M., and G. Tamás 1968.
Notes on new planktonic diatoms from Egypt.
Hydrobiologia 31 (2): 231-240.

Navicula unilatarea n.sp.
Salah, M.M., 1952(1953).
XII. Diatoms from Blakeney Point, Norfolk. New species and new records from Great Britain.
J.R. Microsc. Soc., Ser. 3, 72(3):155-169, 3 pls.

Navicula vacillans
Politis, J., 1949
Diatomees marines de Bosphores et des ibes de la mer de Marmara. II Practica tou Hellenikou Hidrobiologikou Institutoutou 1929, Etoz 1929, 3(1):11-31.

Navicula vagabunda
Mann, A., 1907
Report on the diatoms of the Albatross voyages in the Pacific Ocean, 1888-1904.
Contrib. U. S. Nat. Herb. 10(5):221-419, Pls. XLIV-LIV.

Navicula valens n.sp.
Cholnoky, B.J., 1963.
Beiträge zur Kenntnis des marinen Litorals von Südafrika.
Botanica Marina, 5(2/3):38-83.

Navicula vanhöffeni
Gran, H.W., 1908
Diatomeen. Nordisches Plankton, Botanischer Teil pp. XIX.I-XIX 146; 178 text figs.

Navicula Vanhöffeni
Gran, H.H., and T. Braarud, 1935
A quantitative study of the phytoplankton in the Bay of Fundy and the Gulf of Maine (including observations on hydrography, chemistry, and turbidity).
J. Biol. Bd., Canada, 1(5):279-467, 69 text figs.

Navicula vanhoeffenii [a]
Heimdal, Berit R. 1970.
Morphology and distribution of two Navicula species in Norwegian coastal waters.
Nytt Mag. Bot. 17(2): 65-75

Navicula vanhoffenii
Hendy, N. Ingram, 1964
An introductory account of the smaller algae of British coastal waters. V. Bacillariophyceae (Diatoms).
Her Majesty's Stationary Office, 317 pp., 45 pls.

Navicula vanhöffeni
Jørgensen, E., 1905
B.Protistplankton and the diatoms in bottom samples. Hydrographical and biological investigations in Norwegian fjords. Bergens Mus. Skr. 7: 49-225.

Navicula Vanhöffeni
Levander, K.M., 1947
Plankton gesammelt in den Jahren 1899-1910 an den Küsten Finnlands. Finnländische Hydrographisch-Biologische Untersuchunger (aus dem Wasserbiologischen Laboratorin der Societas Scientiarum Fennica) No.11:40 pp., 6 diagrams, 13 pls., tables.

Navicula Vanhoeffeni
Lillick, L.C., 1940
Phytoplankton and planktonic protozoa of the offshore waters of the Gulf of Maine. Pt.II. Qualitative Composition of the Planktonic Flora. Trans. Am. Phil. Soc., n.s., 31(3):193-237, 13 text figs.

Navicula Vanhoeffenii
Lillick, L.C., 1938
Preliminary report of the phytoplankton of the Gulf of Maine. Am. Mid. Nat. 20(3):624-640, 1 text figs 37 tables.

Navicula venusta
Rampi, L., 1942
Ricerche sul fitoplancton del Mare Ligure 6. Le diatomee delle acque di Sanremo. Nuovo Giornale Botanico Italiano, N.S., 49:252-288.

Navicula venusta
Rampi, L., 1940
Diatomee del Mare Adriatico. Nuovo Giornale Botanico Italiano, n.s., 47:559-608.

Navicula vidovichii
Grunow, A., 1863
Ueber einige neue und ungenügend bekannte Arten und Gattungen von Diatomaceen. Verhandl. d. K.K. Zool. Bot. Gesellsch., Vienna, 13: 137-162, pl.4-5 (Pl. 13-14).

Navicula vidovichii
Mann, A., 1907
Report on the diatoms of the Albatross voyages in the Pacific Ocean, 1888-1904.
Contrib. U. S. Nat. Herb. 10(5):221-419, Pls. XLIV-LIV.

Navicula viridis
Mann, A., 1907
Report on the diatoms of the Albatross voyages in the Pacific Ocean, 1888-1904.
Contrib. U. S. Nat. Herb. 10(5):221-419, Pls. XLIV-LIV.

Navicula viridula
Hendy, N. Ingram, 1964
An introductory account of the smaller algae of British coastal waters. V. Bacillariophyceae (Diatoms).
Her Majesty's Stationary Office, 317 pp., 45 pls.

Navicula viridula
Tsumura, K., 1956.
Diatomoj el la cirkaufoso de la restajo de la kastele de Odawara. J. Yokohama Municipal Univ., (c-14), No. 47:23 pp.

Navicula viridula
Zanon, D. V., 1949
Diatomee di Buenos Aires (Argentina) Atti Accad. Naz. Lincei, Memorie, Cl. Sci. fis., mat. e. nat., ser. 7, 11(3):59-151, 2 pls.

Navicula viridula
Zanon, V., 1948
Diatomee marini di Sardegna e Pugillo di Alghe Marine della stressa. Boll. Pesca, Piscitutura e Idrobiologia, Anno 24, ns. 3(2): 202-244, 27 figs. on 1 pl.

Navicula vulpina
Zanon, V., 1948
Diatomee marini di Sardegna e Pugillo di Alghe Marine della stressa. Boll. Pesca, Piscitutura e Idrobiologia, Anno 24, ns. 3(2): 202-244, 27 figs. on 1 pl.

Navicula Weissfalgii
Allen, W.E., and E.E. Cupp, 1935
Plankton diatoms of the Java Sea.
Annales du Jardin Botanique de Buitenzorg XLIV (2):101-174, figs.1-127.
(drawings of all species mentioned)

Navicula Weissflogii
Mann, A., 1893
List of Diatomaceae from a deep-sea dredging in the Atlantic Ocean off Delaware Bay by the U. S. Fish Commission Steamer Albatross.
Proc. U. S. Nat. Mus. 16:303-312.

Navicula Weissflogii
Schmidt, A., 18-
Ueber Navicula Weissflogii und Navicula Grundleri. Zeitsahr. f. ges. Naturwiss 41:403-410.

Navicula Zanardiniana
Rampi, L., 1940
Diatomee del Mare Adriatico. Nuovo Giornale Botanico Italiano, n.s., 47:559-608.

Navicula zanzibarica var. zebuana n.var.
Castracane degli Antelminelli, F., 1886
1. Report on the Diatomaceae collected by H.M.S. Challenger during the years 1873-1876. Rept. Sci. Results, H.M.S. Challenger, Botany Vol. II, 178 pp., 30 pls.

Navicula zohdyi n.sp.
Salah, M.M., 1952(1953).
XII Diatoms from Blakeney Point, Norfolk. New species and new records from Great Britain.
J.R. Microsc. Soc., Ser. 3, 72(3):155-169, 3 pls.

Navicula zostereti
Takano, Hideaki, 1962
Notes on epiphytic diatoms upon sea-weeds from Japan.
J. Oceanogr. Soc., Japan, 18(1):29-33.

Navicula zostereti
Zanon, V., 1948
Diatomee marini di Sardegna e Pugillo di Alghe Marine della stressa. Boll. Pesca, Piscitutura e Idrobiologia, Anno 24, ns. 3(2): 202-244, 27 figs. on 1 pl.

Neidium
Desikachary, T.V., 1956(1957).
Electron microscope studies on diatoms.
J.R. Microsc. Soc. (3)76(1/2):9-36.

Neidium, 2 spp.
Tsumura, K., 1956.
Diatomoj el la cirkaufoso de la restajo de la kastele de Odawara. J. Yokohama Municipal Univ., (C-14), No. 47:23 pp.

Neidium affine
Zanon, D. V., 1949
Diatomee di Buenos Aires (Argentina) Atti Accad. Naz. Lincei, Memorie, Cl. Sci. fis., mat. e. nat., ser. 7, 11(3):59-151, 2 pls.

Neidium affine longiceps
Rumkówna, A., 1948
[List of the phytoplankton species occurring in the superficial water layers in the Gulf of Gdańsk] Bull. Lab. mar., Gdynia, No. 4: 139-141 with tables in back.

Neidium bisulcatum
Tsumura, K., 1956.
Diatomoj el la cirkaufoso de la restajo de la kastele de Odawara. J. Yokohama Municipal Univ., (C-14), No. 47:23 pp.

Neidium dubium

Zanon, D. V., 1949
Diatomee di Buenos Aires (Argentina)
Atti Accad. Naz. Lincei, Memorie, Cl. Sci.
fis., mat. e. nat., ser. 7, 11(3):59-151,
2 pls.

Neidium dubium constricta

Rumkówna, A., 1948
List of the phytoplankton species occurring in the superficial water layers in the Gulf of Gdańsk. Bull. Lab. mar, Gdynia, No. 4: 139-141 with tables in back.

Neidium dubium cuneata

Tsumura, K., 1956.
Diatomoj el la cirkaufoso de la restajo de la kastele de Odawara. J. Yokohama Municipal Univ., (C-14), No. 47:23 pp.

Neidium iridis

Zanon, D. V., 1949
Diatomee di Buenos Aires (Argentina)
Atti Accad. Naz. Lincei, Memorie, Cl. Sci.
fis., mat. e. nat., ser. 7, 11(3):59-151,
2 pls.

Neidium magellanicum

Zanon, D. V., 1949
Diatomee di Buenos Aires (Argentina)
Atti Accad. Naz. Lincei, Memorie, Cl. Sci.
fis., mat. e. nat., ser. 7, 11(3):59-151,
2 pls.

Nitschia

Desikachary, T.V., 1956(1957).
Electron microscope studies on diatoms.
J. R. Microsc. Soc. (3)76(1/2):9-36.

Nitschia

Ketchum, B. H., 1939
The absorption of phosphate and nitrate by illuminated cultures of Nitschia closterium.
Am. J. Bot., 26:399-407

Nitschia sp.

Kolbe, R.W., 1951.
Elektronenmikroskopische Untersuchungen von Diatomeenmembranen II. Svenska Bot. Tidskr. 45(4):636-647, Pls. 1-4.

Nitzschia sp.

Meunier, A., 1915
Microplancton de la Mer Flamande. 2. Diatomées (excepté le genre Chaetoceros). Mem. Mus. Roy. Hist. Nat., Belgique, 7(3):1-118, Pls. VIII-XIV.

Nitschia

Riley, G. A., 1943
Physiological aspects of spring diatom flowerings. Bull. Bingham Oceanogr. Coll., VIII(4):1-53.

Nitschia

Stanbury, F. A., 1931
The effect of light of different intensities, reduced selectively and non-selectively, upon the rate of growth of Nitschia closterium.
JMBA, 17:633-653

Nitzschia

Wilson, D. P., and E. C. Lucas, 1942
Nitzschia cultures at Hull and at Plymouth.
Nature, 149:331

Nitzschia acicularis

Hutner, S.H., and L. Provasoli, 1953.
A pigmented marine diatom requiring Vitamin B12 and uracil. News Bull., Phycol. Soc., Amer., 6(18):7.

Nitzschia acicularis

Kokubo, S., and S. Sato., 1947
Plankterx in Jū-San Gata. Physiol. and Ecol. (Japan) 1(4):1-16, 3 text figs., tables.

Nitschia aciculariformis n.sp.

Manguin, E., 1957.
Premier inventaire des diatomées de la Terre Adélie Antarctique. Espèces nouvelles.
Rev. Algologique, n.s., 3(3):111-134.

Nitzschia acuminata

Hendy, N. Ingram, 1964
An introductory account of the smaller algae of British coastal waters. V. Bacillariophyceae (Diatoms).
Her Majesty's Stationary Office, 317 pp., 45 pls.

Nitzschia acuminata

Hustedt, F. and A.A. Aleem, 1951
Littoral diatoms from the Salstone near Plymouth. JMBA 30(1): 177.196.

Nitzschia acuminata

Kokubo, S., and S. Sato., 1947
Plankterx in Jū-San Gata. Physiol. and Ecol. (Japan) 1(4):1-16, 3 text figs., tables.

Nitzschia acuminata

Zanon, D. V., 1949
Diatomee di Buenos Aires (Argentina)
Atti Accad. Naz. Lincei, Memorie, Cl. Sci.
fis., mat. e. nat., ser. 7, 11(3):59-151,
2 pls.

Nitzschia acuminata

Zanon, V., 1948
Diatomee marini di Sardegna e Pugillo di Alghe Marine della stressa. Boll. Pesca, Piscitutura e Idrobiologia, Anno 24, ns. 3(2): 202-244, 27 figs. on 1 pl.

Nitzschia acuta

Zanon, D. V., 1949
Diatomee di Buenos Aires (Argentina)
Atti Accad. Naz. Lincei, Memorie, Cl. Sci.
fis., mat. e. nat., ser. 7, 11(3):59-151,
2 pls.

Nitschia adeliana n.sp.

Manguin, E., 1957.
Premier inventaire des diatomées de la Terre Adélie Antarctique. Espèces nouvelles.
Rev. Algologique, n.s., 3(3):111-134.

Nitzschia aequorea

Hustedt, F. and A.A. Aleem, 1951
Littoral diatoms from the Salstone near Plymouth. JMBA 30(1): 177.196.

Nitzschia aestatis n.sp.

Giffen, Malcolm H. 1973.
Diatoms of the marine littoral of Steenberg's Cove in St. Helena Bay, Cape Province, South Africa.
Botanica Marina 16 (1): 32-48.

Nitzschia alba

Lewin, Joyce, and Ching-hong Chen 1968.
Silicon metabolism in diatoms. VI. Silicic acid uptake by a colorless marine diatom, Nitzschia alba Lewin and Lewin.
J. Phycol. 4(2):161-166.

Nitzschia americana n.sp.

Hasle, Grethe Rytter, 1964.
Nitzschia and Fragilariopsis species studied in the light and electron microscopes. 1. Some marine species of the groups Nitzschiella and Lanceolatae.
Skrifter, Norske Videnskaps-Akad., Oslo, 1. Mat. Naturw. Kl., n.s., No. 16:48 pp.

Nitzschia amphibia

Mann, A., 1907
Report on the diatoms of the Albatross voyages in the Pacific Ocean, 1888-1904.
Contrib. U. S. Nat. Herb. 10(5):221-419, Pls. XLIV-LIV.

Nitzschia amphibia

Zanon, D. V., 1949
Diatomee di Buenos Aires (Argentina)
Atti Accad. Naz. Lincei, Memorie, Cl. Sci.
fis., mat. e. nat., ser. 7, 11(3):59-151,
2 pls.

Nitzschia angularis

Hendy, N. Ingram, 1964
An introductory account of the smaller algae of British coastal waters. V. Bacillariophyceae (Diatoms).
Her Majesty's Stationary Office, 317 pp., 45 pls.

Nitzschia angularis

Hendey, N.I., 1951
Littoral diatoms of Chicester Harbour with special reference to fouling. J.Roy. Microscop. Soc. 71(1): 1-86, 18 pls.

Nitzschia angularis

Hustedt, F. and A.A. Aleem, 1951
Littoral diatoms from the Salstone near Plymouth. JMBA 30(1): 177.196.

Nitzschia angularis

Jørgensen, E., 1905
B.Protistplankton and the diatoms in bottom samples. Hydrographical and biological investigations in Norwegian fjords. Bergens Mus. Skr. 7: 49-225.

Nitzschia angularis

Politis, J., 1949
Diatomees marines de Bosphores et des ibes de la mer de Marmara. II Practica tou Hellenikou Hidrobiologikou Institutoutou 1929, Etoz 1929, 3(1):11-31.

Nitzschia angularis

Zanon, V., 1948
Diatomee marini di Sardegna e Pugillo di Alghe Marine della stressa. Boll. Pesca, Piscitutura e Idrobiologia, Anno 24, ns. 3(2): 202-244, 27 figs. on 1 pl.

Nitzschia angustata

Mann, A., 1907
Report on the diatoms of the Albatross voyages in the Pacific Ocean, 1888-1904.
Contrib. U. S. Nat. Herb. 10(5):221-419, Pls. XLIV-LIV.

Nitzschia angustata

Zanon, D. V., 1949
Diatomee di Buenos Aires (Argentina)
Atti Accad. Naz. Lincei, Memorie, Cl. Sci.
fis., mat. e. nat., ser. 7, 11(3):59-151,
2 pls.

Nitzschia angustissima

Mangin, L., 1915
Phytoplancton de L'Antartique. Deuxieme Exped. Ant. Francaise (1908-1910), 95 pp., 3 pls., 58 text figs.

Nitzschia amphionys

Mann, A., 1893
List of Diatomaceae from a deep-sea dredging in the Atlantic Ocean off Delaware Bay by the U. S. Fish Commission Steamer Albatross.
Proc. U. S. Nat. Mus. 16:303-312.

Nitzschia apiculata

Hendy, N. Ingram, 1964
An introductory account of the smaller algae of British coastal waters. V. Bacillariophyceae (Diatoms).
Her Majesty's Stationary Office, 317 pp., 45 pls.

Nitzschia apiculata
Hendey, N.J., 1951
Littoral diatoms of Chicester Harbour with special reference to fouling. J.Roy. Microscop. Soc. 71(1): 1-86, 18 pls.

Nitzschia apiculata
Hustedt, F. and A.A. Aleem, 1951
Littoral diatoms from the Salstone near Plymouth. JMBA 30(1): 177-196.

Nitzschia apiculata
Rampi, L., 1940
Diatomee del Mare Adriatico. Nuovo Giornale Botanico Italiano, n.s., 47:559-608.

Nitzschia apiculata
Zanon, D. V., 1949
Diatomee di Buenos Aires (Argentina) Atti Accad. Naz. Lincei, Memorie, Cl. Sci. fis., mat. e. nat., ser. 7, 11(3):59-151, 2 pls.

Nitzschia apiculata
Zanon, V., 1948
Diatomee marini di Sardegna e Pugillo di Alghe Marine della stressa. Boll. Pesca, Piscitutura e Idrobiologia, Anno 24, ns. 3(2): 202-244, 27 figs. on 1 pl.

Nitzschia arctica
Jørgensen, E., 1905
B. Protistplankton and the diatoms in bottom samples. Hydrographical and biological investigations in Norwegian fjords. Bergens Mus. Skr. 7: 49-225.

Nitzschia baculum n.sp.
Frenguelli, Joaquin, and Hector Antonio Orlando, 1959.
Operacion MERLUZA. Diatomeas y silicoflagelados del plancton del "VI Crucero". Servicio Hidrogr. Naval., Argentina, Publ. No. H. 619: 5-62.

Nitzschia barbieri
Hasle, Grethe Rytter, 1965.
Nitzschia and Fragilariopsis species studies in the light and electron microscopes. III. The genus Fragilariopsis. Norske Videnskaps- Akad., Oslo. I. Met.- Nature. Kl., N.S. No. 21: 49 pp., 17 pls.

Nitzschia Barbieri
Hendey, N.I., 1937
The plankton diatoms of the southern seas. Discovery Repts. 16:151-364, pls.6-13.

Nitzschia barkleyi
Manguin, E., 1954
Diatomees marines provenant de l'ile Heard (Australian National Antarctic Expedition). Rev. Algol., n.s., 1: 14-24.

Nitzschia bicapitata
Frenguelli, Joaquin, and Hector Antonio Orlando, 1959.
Operacion MERLUZA. Diatomeas y silicoflagelados del plancton del "VI Crucero". Servicio Hidrogr. Naval., Argentina, Publ. No. H. 619: 5-62.

Nitzschia bicapitata
Hasle, Grethe Rytter, 1964.
Nitzschia and Fragilariopsis species studied in the light and electron microscopes. 1. Some marine species of the groups Nitzschiella and Lanceolatae. Skrifter, Det Norske Videnskaps-Akad., Oslo, 1. Mat.-Naturv. Kl., n.s., No. 16:48 pp.

Nitzschia bicapitata
Hasle, Grethe Rytter, 1960
Phytoplankton and ciliate species from the Tropical Pacific. Skr. Norske Videnskaps-Akad., Oslo, 1. Mat.-Nat. Kl., 1960(2): 1-50.

Nitzschia bilobata
Hendy, N. Ingram, 1964
An introductory account of the smaller algae of British coastal waters. V. Bacillariophyceae (Diatoms). Her Majesty's Stationary Office, 317 pp., 45 pls.

Nitzschia bilobata
Hendey, N.I., 1951
Littoral diatoms of Chicester Harbour with special reference to fouling. J.Roy. Microscop. Soc. 71(1): 1-86, 18 pls.

Nitzschia bilobata
Jørgensen, E., 1905
B. Protistplankton and the diatoms in bottom samples. Hydrographical and biological investigations in Norwegian fjords. Bergens Mus. Skr. 7: 49-225.

Nitzschia bilobata
Paiva Carvalho, J., 1950
O plancton do Rio Maria Rodrigues (Cananeia). 1. Diatomaceas e Dinoflagelados. Bol. Inst. Paulista Oceanogr. 1(1); 27-43, 2 fold-in tables, 2 figs.

Nitzschia bilobata
Sousa e Silva, E., and J. Dos Santos-Pinto, 1948
O Plancton da Baia de S. Martinho do Porto. 1. Diatomaceas e Dinoflagelados. Bol. Soc. Portuguese de Ciencias Naturais, 16(2):134-187, 6 pls. (Trav. Sta. Biol. Mar. de Lisbonne No. 52).

Nitzschia bilobata
Takano, Hideaki, 1962
Notes on epiphytic diatoms upon sea-weeds from Japan. J. Oceanogr. Soc., Japan, 18(1):29-33.

Nitzschia bilobata
Zanon, D. V., 1949
Diatomee di Buenos Aires (Argentina) Atti Accad. Naz. Lincei, Memorie, Cl. Sci. fis., mat. e. nat., ser. 7, 11(3):59-151, 2 pls.

Nitzschia bilobata adriatica
Ercegovic, A., 1936
Etudes qualitative et quantitatives du phytoplancton dans les eaux cotières de l'Adriatique oriental moyen au cours de l'année 1934. Acta Adriatica 1(9):1-126

Nitzschia bilobata minor
Cupp, Easter E., 1943
Marine plankton diatoms of the west coast of North America. Bull. S.I.O. 5(1):1-238, 5 pls., 168 text figs.

Nitzschia braarudii
Hasle, Grethe Rytter, 1964.
Nitzschia and Fragilariopsis species studied in the light and electron microscopes. 1. Some marine species of the groups Nitzschiella and Lanceolatae. Skrifter, Det Norske Videnskaps-Akad., Oslo, 1. Mat.-Naturv. Kl., n.s., No. 16:48 pp.

Nitzschia braarudii n.sp.
Hasle, Grethe Rytter, 1960
Phytoplankton and ciliate species from the Tropical Pacific. Skr. Norske Videnskaps-Akad., Oslo, 1. Mat.-Nat. Kl., 1960(2): 1-50.

Nitzschia cacumina n.sp.
Giffen, Malcolm H. 1973.
Diatoms of the marine littoral of Steenberg's Cove in St. Helena Bay, Cape Province, South Africa. Botanica Marina 16 (1): 32-48.

Nitzschia calcicola n.sp.
Aleem, A.A., and Fr. Hustedt, 1951.
Einige neue Diatomeen von der Südküste Englands. Botaniska Notiser 1951(1):13-20, 6 textfigs.

Nitzschia calida
Zanon, V., 1948
Diatomee marini di Sardegna e Pugillo di Alghe Marine della stressa. Boll. Pesca, Piscitutura e Idrobiologia, Anno 24, ns. 3(2): 202-244, 27 figs. on 1 pl.

Nitzschia californica n.sp.
Schrader, Hans-Joachim, 1973
Cenozoic diatoms from the northeast Pacific, leg 18. Initial Repts Deep Sea Drilling Proj. 18:673-797.

Nitzschia capitellata
Rumkówna, A., 1948
List of the phytoplankton species occurring in the superficial water layers in the Gulf of Gdańsk. Bull. Lab. mar., Gdynia, No. 4: 139-141 with tables in back.

Nitzschia challengeri n.sp.
Schrader, Hans-Joachim, 1973
Cenozoic diatoms from the northeast Pacific, leg 18. Initial Repts Deep Sea Drilling Proj. 18:673-797.

Nitzschia circumsuta
Hendy, N. Ingram, 1964
An introductory account of the smaller algae of British coastal waters. V. Bacillariophyceae (Diatoms). Her Majesty's Stationary Office, 317 pp., 45 pls.

Nitzschia circumsuta
Hendey, N.I., 1951
Littoral diatoms of Chicester Harbour with special reference to fouling. J.Roy. Microscop. Soc. 71(1): 1-86, 18 pls.

Nitzschia circumsuta
Zanon, D. V., 1949
Diatomee di Buenos Aires (Argentina) Atti Accad. Naz. Lincei, Memorie, Cl. Sci. fis., mat. e. nat., ser. 7, 11(3):59-151, 2 pls.

Nitzschia clausii
Zanon, D. V., 1949
Diatomee di Buenos Aires (Argentina) Atti Accad. Naz. Lincei, Memorie, Cl. Sci. fis., mat. e. nat., ser. 7, 11(3):59-151, 2 pls.

Nitzschia closterium
Allen, W.E., and E.E. Cupp, 1935
Plankton diatoms of the Java Sea. Annales du Jardin Botanique de Buitenzorg XLIV (2):101-174, figs.1-127.

(drawings of all species mentioned)

Nitzschia closterium
Atkins, W.R.G., and M. Parke, 1951.
Seasonal changes in the phytoplankton as indicated by chlorophyll estimation. J.M.B.A. 29(3): 609-618.

Nitzschia closterium
Bainbridge, R., 1949
Movement of zooplankton in diatom gradients
Nature 163(4154):910-911, 2 text figs.

Nitzschia closterium
Boden, B.P., 1950
Some marine plankton diatoms from the west coast of South Africa. Trans. R. Soc. S. Africa: 32:321-434, 100 text figs.

Nitzschia closterium
Boden, B. P., 1949.
The diatoms collected by the U.S.S. CACOPAN in the Antarctic in 1947. J. Mar. Res. 8(1):6-13, 3 textfigs.

Nitzschia closterium
Boden, Brian, 1948
Marine plankton diatoms on operation HIGHJUMP in: Some oceanographic observations on operation HIGHJUMP. By R.S. Dietz. USNEL Rept. No.55, 97 pp., 41 figs. 7 July 1948.

Nitschia closterium
Bonham, K., A.H. Seymour, L.R. Donaldson, and A.D. Welander, 1947.
Lethal effect of X-rays on marine plankton organisms. Science, 106(2750):245-246.

Nitzschia closterium
Braarud, T., 1945
A phytoplankton survey of the polluted waters of inner Oslo Fjord. Hvalrådets Skrifter, No.28, 142 pp., 19 text figs., 17 tables.

Nitzschia closterium
Braarud, T., 1934
A note on the phytoplankton of the Gulf of Maine in the summer of 1933. Biol. Bull. 67(1):76-82.

Nitzschia closterium
Braarud, T., and Adam Bursa, 1939
On the phytoplankton of the Oslo Fjord, 1933-1934. Hvaldrådets Skr. No.19:1-63; 9 text figs. Reviewed. J. du. Cons. 14(3): 418-420. A.C. Gardiner.

Nitzschia closterium
Brunel, J., 1962
Le phytoplancton de la Baie de Chaleurs. Inst. Botan., Univ. Montréal, Contrib. No. 77: 365 pp., 66 pls.

Nitzschia closterium
Brunel, Jules, 1962
Le phytoplancton de la Baie des Chaleurs. Contrib. Ministère de la Chasse et des Pêcheries, Province de Québec, No. 91: 365 pp.

Nitzschia closterium
Chu, S.P., 1947.
The utilization of organic phosphorus by phytoplankton. J.M.B.A. 26(3):285-295, 1 diagr.

Nitzschia closterium
Cupp, Easter E., 1943
Marine plankton diatoms of the west coast of North America. Bull. S.I.O. 5(1):1-238, 5 pls., 168 text figs.

Nitzschia closterium
Cupp, E.E., 1934
Analysis of marine diatom collections taken from the Canal Zone to California during March, 1933. Trans. Am. Micros. Soc. LIII (1):22-29, 1 map.

Nitzschia closterium
Dangeard, P., 1927
Phytoplankton de la croisière du "Sylvana". Ann. Inst. Ocean., Monaco, n.s., 4(8):286-401, 54 text figs. (Janvier-Juin 1913).

Nitzschia closterium
Davidson, V.M., 1931.
Biological and oceanographic conditions in Hudson Bay. Contr. Canadian Biol. Fish., n.s., 6(26): 497-509, 7 textfigs.

Nitzschia closterium
Gran, H.H., 1908
Diatomeen. Nordisches Plankton, Botanischer Teil pp. XIX.1-XIX 146; 178 text figs.

Nitzschia closterium
Gran, H.H., and T. Braarud, 1935
A quantitative study of the phytoplankton in the Bay of Fundy and the Gulf of Maine (including observations on hydrography, chemistry, and turbidity). J. Biol. Bd., Canada, 1(5):279-467, 69 text figs.

Nitzschia closterium
Gran, H. H. and E. C. Angst, 1931
Plankton diatoms of Puget Sound. Publ. Puget Sound Biol. Sta. 7:417-519, 95 text figs.

Nitzschia closterium
Grøntved, J., 1949(1950)
Investigations on the phytoplankton in the Danish Waddensee in July 1941. Medd. Komm. Danmarks Fiskeri- og Havundersøgelser, Serie Plankton, 5(2)(Medd. Skalling Lab. 10):55 pp., 2 pls., 38 textfigs.

Nitzschia closterium
Gunter, G., R.H. Williams, C.C. Davis, and F.G. Walton Smith, 1948
Catastrophic mass mortality of marine animals and coincident phytoplankton bloom on the West Coast of Florida, November 1946 to August 1947. Ecol. Mon., 18:309-324, 2 text figs.

Nitschia closterium
Hasle, Grethe Rytter, 1964
Nitschia and Fragilariopsis species studied in the light and electron microscope. 1. Some marine species of the groups Nitzschiella and Lanceolatae Skrifter, Det Norske Videnskaps-Akad., Oslo, 1. Mat.-Naturv. Kl., N.s., No. 16:48 pp.

Nitschia closterium
Harvey, H.W., 1953.
Synthesis of organic compounds and chlorophyll by Nitzschia closterium. J.M.B.A. 31(3):477-487, 4 textfigs.

Nitzschia closterium
Harvey, H.W., 1953.
Note on the absorption of organic phosphorus compounds by Nitzschia closterium in the dark. J.M.B.A. 31(3):475-476.

Nitzschia closterium
Hendy, N. Ingram, 1964
An introductory account of the smaller algae of British coastal waters. V. Bacillariophyceae (Diatoms). Her Majesty's Stationary Office, 317 pp., 45 pls.

Nitzschia Closterium
Hendey, N.I., 1951
Littoral diatoms of Chicester Harbour with special reference to fouling. J. Roy. Microscop. Soc. 71(1): 1-86, 18 pls.

Nitzschia closterium
Hendey, N.I., 1937
The plankton diatoms of the southern seas. Discovery Repts. 16:151-364, pls.6-13.

Nitzschia Costatum
Hustedt, F. and A.A. Aleem, 1951
Littoral diatoms from the Salstone near Plymouth. JMBA 30(1): 177.196.

Nitschia Closterium
Iyengar, M.O.P. and G. Venkataraman, 1951.
The ecology and seasonal succession of the algae flora of the River Cooum at Madras with special reference to the Diatomaceae. J. Madras Univ. 21, Sect. B(1): 140-192, 1 pl of 4 figs., 11 text figs.

Nitzschia closterium
Johannes, R.E., and Masako Satomi. 1966.
Composition and nutritive value of faecal pellets of a marine crustacean. Limnol. Oceanogr., 11(2):191-197.

Nitzschia closterium
Jørgensen, E., 1905
B. Protistplankton and the diatoms in bottom samples. Hydrographical and biological investigations in Norwegian fjords. Bergens Mus. Skr. 7: 49-225.

Nitzschia closterium
Jorgensen, E., 1900
Protophyten und Protozoën im Plankton aus der Norwegischen Westkerste. Bergens Mus. Aarb. 1899(6): 95 pp., 5 pls., 83 tables.

Nitzschia closterium
Kanagawa, Akio, Mitsuki Yoshioka and Shin-ichi Teshima 1971
The occurrence of brassicasterol in the diatoms Cyclotella nana and Nitzschia closterium.
Bull. Jap. Soc. scient. Fish. 37(9): 899-903

Nitzschia Closterium
Ketchum, B. H., and A.C. Redfield, 1949.
Some physical and chemical characteristics of algae grown in mass culture. J. Cell. & Comp. Physiol. 33(3):281-299, 2 textfigs.

Nitschia Closterium
Ketchum, B.H., L. Lillick, and A.C. Redfield, 1949.
The growth and optimum yields of unicellular algae in mass culture. J. Cell. & Comp. Physiol. 33(3):267-279, 3 textfigs.

Nitzschia closterium
Kokubo, S., and S. Sato., 1947
Planktern in Jū-San Gata. Physiol. and Ecol. (Japan) 1(4):1-16, 3 text figs., tables.

Nitzschia Closterium
Lillick, L.C., 1940
Phytoplankton and planktonic protozoa of the offshore waters of the Gulf of Maine. Pt.II. Qualitative Composition of the Planktonic Flora. Trans. Am. Phil. Soc., n.s., 31(3):193-237, 13 text figs.

Nitzschia closterium
Lillick, L.C. 1938
Preliminary report of the phytoplankton of the Gulf of Maine. Am. Mid. Nat. 20(3):624-640, 1 text figs 37 tables.

Nitzschia Closterium
Lillick, L.C., 1937
Seasonal studies of the phytoplankton off Woods Hole, Massachusetts. Biol. Bull. LXXIII (3):488-503, 3 text figs.

Nitzschia closterium
Low, E.M., 1955.
Studies on some chemical constituents of diatoms J. Mar. Res. 14(2):199-204.

Nitzschia closterium
Maddux, William S., and Raymond F. Jones, 1964.
Some interactions of temperature, light intensity and nutrient concentration during the continuous culture of Nitzschia closterium and Tetraselmis sp. Limnology and Oceanography, 9(1):79-86.

Nitzschia closterium
Margalef, Ramón, 1963
Modelos simplificados del ambiente marino para el estudio de la sucesión y distribución del fitoplancton y del valor indicador de sus pigmentos.
Invest. Pesquera. Barcelona, 23:11-52.

Nitzschia closterium
Margalef, R., 1954.
Consideraciones sobre la determinación cuantitativa del fitoplancton por la valoración de pigmentos solubles y los factores que afectan a la relación entre cantidad de pigmento y peso seco. Publ. Inst. Biol. Aplic. 16:71-84.

Nitzschia closterium
Marshall, S. M., 1933
The production of microplankton in the Great Barrier Reef Region. Brit. Mus. (N.H.) Great Barrier Reef Exped. 1928-29, Sci. Repts. II(5):111-157, 14 text figs.

Nitzschia Closterium
Mangin, L., 1915
Phytoplancton de L'Antartique. Deuxieme Exped. Ant. Francaise (1908-1910), 95 pp., 3 pls., 58 text figs.

Nitzschia closterium
Mangin, M. L., 1912
Phytoplancton de la croisière du "René" dans l'Atlantique (Septembre 1908). Ann. Inst. Ocean., n.s., 4(1):1-66, 2 pls., 41 text figs., 2 tables.

Nitzschia closterium
Marshall, S.M. and A.P. Orr, 1948
Further experiments on the fertilization of a sea loch (Loch Craiglin). The effect of different plant nutrients on the phytoplankton. J.M.B.A. 27(2):360-379, 10 text figs.

Nitzschia closterium
Morse, D.C., 1947
Some observations on seasonal variations in plankton population Patuxent River, Maryland 1943-1945. Bd. Nat. Res., Publ. No.65, Chesapeake Biol. Lab., 31, 3 figs.

Nitzschia closterium
Pettersson, H., 1934
Scattering and extinction of light in sea-water. Meddelanden från Göteborgs Högskolas Oceanografiska Institut. 9 (Göteborgs Kungl. vetenskaps-och vitterhetssamhälles Handlingar. Femte Foljden, Ser.B, 4(4)):1-16.

Nitzschia Closterium
Phifer, L.D. (undated)
The occurrence and distribution of plankton diatoms in Bering Sea and Bering Strait, July 26-August 24, 1934. Report of Oceanographic cruise of U.S. Coast Guard Cutter Chelan 1934, Part II(A):1-44 (mimeographed) plus fig.1 (after Pt.B)

Nitzschia closterium
Rampi, L., 1942
Ricerche sul fitoplancton del Mare Ligure 6. Le diatomee delle acque di Sanremo. Nuovo Giornale Botanico Italiano, N.S., 49:252-268.

Nitzschia closterium
Rice, T.R., 1953.
Phosphorus exchange in marine phytoplankton. Fish. Bull. 80, Vol. 54:77-89, 3 textfigs.

Nitzschia closterium
Rumkówna, A., 1948
List of the phytoplankton species occurring in the superficial water layers in the Gulf of Gdańsk] Bull. Lab. mar., Gdynia, No. 4: 139-141 with tables in back.

Nitzschia closterium
Sato, Tadao and Makoto Serikawa, 1968.
Mass culture of a marine diatom, Nitzschia closterium. (In Japanese; English abstract). Bull. Plankton Soc., Japan, 15(1): 13-16.

Nitzschia closterium
Schodduyn, M., 1926
Observations faites dans la baie d'Ambleteuse (Pas de Calais). Bull. Inst. Ocean., Monaco, No. 482: 64 pp.

Nitzschia closterium
Steemann-Nielsen, Einar, 1951
The marine vegetation of the Isefjord. A study on ecology and production. Medd. Komm. Danmarks Fiskeri-og Havundersøgelser. Ser. Plankton. 5(4); 114pp., 46 text figs.

Nitzschia closterium
Taylor, W. Rowland, 1964.
Inorganic nutrient requirements for marine phytoplankton organisms. Narragansett Mar. Lab., Univ. Rhode Island, Occ. Publ., No. 2:17-24.

Nitzschia closterium
Tokuda, Hiroshi, 1969.
Excretion of carbohydrate by a marine pennate diatom, Nitzschia closterium. Rec. oceanogr.Wks. Japan, 10(1):109-122.

Nitzschia closterium
Tokuda, Hiroshi, 1967.
Elimination of lag phase from the growth of Nitzschia closterium, a marine pennate diatom, with glycolic acid.
Inf.Bull.Planktol.Japan,Comm.No.Dr.Y.Matsue,261-269.

Nitzschia closterium
Tokuda, Hiroshi, 1966.
Studies on the growth of a marine diatom, Nitzschia closterium. 1. Its requirement for thiamine. Bull. Jap. Soc. scient. Fish., 32(7): 565-567.

Nitzschia closterium
Von Brand, and Th., and N.W. Rakestraw, 1940.
Decomposition and regeneration of nitrogenous organic matter in sea water. III. Influence of temperature and source and condition of water. Biol. Bull. 79(2):231-236, 2 textfigs.

Reviewed: J. du Cons. 16(1):113-116 by H. Barnes

Nitzschia closterium
von Brand, Th., N.W. Rakestraw, and C.E. Renn, 1939.
Further experiments on the decomposition and regeneration of nitrogenous organic matter in sea water. Biol. Bull. 77(2):285-296, 3 textfigs

Reviewed: J. du Cons. 16(1):113-116 by H. Barnes

Nitzschia closterium
Wilson, Douglas, P., 1946.
The triradiate and other forms of Nitzschia closterium (Ehrenberg) Wm. Smith forma Minutissima of Allen and Nelson. JMBA, XXVI (3):235-270.

Nitzschia closterium minutissima
Chu, S. P., 1949
Experimental studies on the environmental factors influencing the growth of phytoplankton. Sci. & Tech. in China 2(3):37-52.

Nitzschia closterium minutissima
Hutner, S.H., and L. Provasoli, 1953.
A pigmented marine diatom requiring Vitamin B12 and uracil. New Bull., Phycol. Soc., Amer., 6(18):7.

Nitzschia closterium f. minutissima
Levring, T., 1945
Some culture experiments with marine plankton diatoms. Medd. Oceanografiska Institutet i Göteborg, No.9 (Göteborgs Kungl. Vetenskpas-och Vitterhets-Samhälles Handlingar, Sjätte Följden. Ser. B. Vol.3 (12), 1-17.

Nitzschia closterium var minutissima
Pettersson, H., F. Gross, and F. Koczy, 1939
Large scale plankton cultures. Medd. från Oceanografiska Institutet i Göteborg No.3 (Göteborgs Kungl. vetenskaps-och vitterhets-Samhälles Handlingar Femte Följden. Ser. B) Vol.6(13):1-24.

1. The plankton shaft. H. Pettersson
2. Experiment with phytoplankton. F. Gross and F. Koczy
3. Experiments with zooplankton.

Nitzschia closterium minutissima
Spencer, C.P., 1954.
Studies on the culture of a marine diatom. J.M.B.A. 33:265-290, 16 textfigs.

Nitzschia closterium minutissima
Wilson, D. P., 1947.
The triradiate and other forms of Nitzschia closterium (Ehrenberg) W, Smith, forma minutissima of Allen and Nelson. J.M.B.A.26(3): 235-270, 9 textfigs.

Nitzschia closterium var. recta n.var.
Gran, H. H. and E. C. Angst, 1931
Plankton diatoms of Puget Sound. Publ. Puget Sound Biol. Sta. 7:417-519, 95 text figs.

Nitzschia coarctata
Jorgensen, E., 1900
Protophyten und Protozoen im Plankton aus der Norwegischen Westkerste. Bergens Mus. Aarb. 1899(6): 95 pp., 5 pls., 83 tables.

Nitzschia communis
Zanon, D. V., 1949
Diatomee di Buenos Aires (Argentina) Atti Accad. Naz. Lincei, Memorie, Cl. Sci. fis., mat. e. nat., ser. 7, 11(3):59-151, 2 pls.

Nitzschia commutata
Hustedt, F. and A.A. Aleem,1951
Littoral diatoms from the Salstone near Plymouth. JMBA 30(1): 177.196.

Nitzschia constricta
Hustedt, F. and A.A. Aleem,1951
Littoral diatoms from the Salstone near Plymouth. JMBA 30(1): 177.196.

Nitschia corpulenta n.sp.
Hendey, N.Ingram, 1958 [1957(Publ. 1958)]
Marine diatoms from some West African Ports. J. R.Microsc. Soc. (3) 77(1/2): 28-85.

Nitzschia decipiens
Hasle, Grethe Rytter, 1964.
Nitzschia and Fragilariopsis species studied in the light and electron microscopes. 1. Some marine species of the groups Nitzschiella and Lanceolatae. Skrifter, Det Norske Videnskaps-Akad. Oslo, 1. Mat.-Naturw. Kl., n.s., No. 16:48 pp.

Nitzschia delicatissima
Aleem, A.A., 1951.
Sur la présence de Nitzschia delicatissima Cleve dans le plancton méditerranéen. Vie et Milieu 2(4):441-447.

Nitzschia delicatissima
Boden, B.P., 1950
Some marine plankton diatoms from the west coast of South Africa. Trans. R.Soc. S. Africa. 32:321-434, 100 text figs.

Nitzschia delicatissima
Braarud, T., 1945
A phytoplankton survey of the polluted waters of inner Oslo Fjord. Hvalrådets Skrifter, No.28, 142 pp., 19 text figs., 17 tables.

Nitzschia delicatissima
Braarud, T., 1934
A note on the phytoplankton of the Gulf of Maine in the summer of 1933. Biol. Bull. 67(1):76-82.

Nitzschia delicatissima
Braarud, T., and Adam Bursa, 1939
On the phytoplankton of the Oslo Fjord, 1933-1934. Hvalrådets Skr. No.19:1-63; 9 text figs. Reviewed. J. du. Cons. 14(3): 418-420. A.C. Gardiner.

Nitzschia delicatissima
Braarud, T., K.R. Gaarder and J. Grøntved, 1953.
The phytoplankton of the North Sea and adjacent waters in May 1948. Rapp. Proc. Verb, Cons. Perm. Int. Expl. Mer. 133:1-87, 29 tables, Pls. A-B, 18 textfigs.

Nitzschia delicatissima
Brunel, J., 1962
Le phytoplancton de la Baie de Chaleurs. Inst. Botan., Univ. Montréal, Contrib. No. 77: 365 pp., 66 pls.

Nitzschia delicatissima
Cupp, Easter E., 1943
Marine plankton diatoms of the west coast of North America. Bull. S.I.O. 5(1):1-238, 5 pls., 168 text figs.

Nitzschia delicatissima
Cupp, E.E., 1934
Analysis of marine diatom collections taken from the Canal Zone to California during March, 1933. Trans. Am. Micros. Soc. LIII (1):22-29, 1 map.

Nitzschia delicatissima
Davidson, V.M., 1931.
Biological and oceanographic conditions in Hudson Bay. Contr. Canadian Biol. Fish., n.s., 6(26):

Nitzschia delicatissima
Gran, H.H., 1908
Diatomeen. Nordisches Plankton, Botanischer Teil pp. XIX.1-XIX 146; 178 text figs.

Nitzschia delicatissima
Gran, H.H., and T. Braarud, 1935
A quantitative study of the phytoplankton in the Bay of Fundy and the Gulf of Maine (including observations on hydrography, chemistry, and turbidity). J. Biol. Bd., Canada, 1(5):279-467, 69 text figs.

Nitzschia delicatissima
Gran, H.H. and E.C. Angst, 1931
Plankton diatoms of Puget Sound. Publ. Puget Sound Biol. Sta. 7:417-519, 95 text figs.

Nitzschia delicatissima?
Hasle, Grethe Rytter, 1960
Phytoplankton and ciliate species from the Tropical Pacific. Skr. Norske Videnskaps-Akad., Oslo, 1. Mat.-Nat. Kl., 1960(2): 1-50.

Nitzschia delicatissima
Hendy, N. Ingram, 1964
An introductory account of the smaller algae of British coastal waters. V. Bacillariophyceae (Diatoms). Her Majesty's Stationary Office, 317 pp., 45 pls.

Nitzschia delicatissima
Herrera, Juan, y Ramón Margalef, 1963
Hidrografía y fitoplancton de la costa comprendida entre Castellón y la desembocadura del Ebro, de julio de 1960 a junio de 1961. Inv. Pesq., Barcelona, 24:33-112.

Nitzschia delicatissima
Jørgensen, E., 1905
B. Protistplankton and the diatoms in bottom samples. Hydrographical and biological investigations in Norwegian fjords. Bergens Mus. Skr. 7: 49-225.

Nitzschia delicatissima
Lillick, L.C., 1940
Phytoplankton and planktonic protozoa of the offshore waters of the Gulf of Maine. Pt.II. Qualitative Composition of the Planktonic Flora. Trans. Am. Phil. Soc., n.s., 31(3):193-237, 13 text figs.

Lillick, L.C., 1938 Nitzschia delicatissima
Preliminary report of the phytoplankton of the Gulf of Maine. Am. Mid. Nat. 20(3):624-640, 1 text figs 37 tables.

Nitzschia delicatissima
MacFarlane, Robert B., Walter A. Glooschenko and Robert C. Harriss 1972.
The interaction of light intensity and DDT concentration upon the marine diatom, Nitzschia delicatissima Cleve.
Hydrobiologia 39 (3): 373-382

Nitzschia delicatissima
Pettersson, H., H. Höglund, S. Landberg, 1934.
Submarine daylight and the photosynthesis of phytoplankton. Medd. från Göteborgs Högskolas Oceanografiska Inst. 10 (Göteborgs Kungl. Vetenskaps-och Vitterhets-samhälles handlingar. Femte Följden, Ser. B 4(5):1-17.

Nitzschia delicatissima
Steemann-Nielsen, Einar, 1951
The marine vegetation of the Isefjord. A study on ecology and production. Medd. Komm. Danmarks Fiskeri-og Havundersøgelser. Ser. Plankton. 5(4); 114pp., 46 text figs.

Nitzschia delicatissima
Uyeno, Fukuzo, 1961
Oceanographical and ecological studies on primary production of the sea, with special references to relationship between diatom production and temperature and chlorinity of water.
Rept., Fac. Fish., Pref. Univ., Mie, 4(1): 1-64.

Nitzschia delicatula
Hasle, Grethe Rytter, and Blanca Rojas E. de Mendiola 1967.
The fine structure of some Thalassionema and Thalassiothrix species.
Phycologia 6 (2/3): 107-118.

Nitzschia denticula
Zanon, D. V., 1949
Diatomee di Buenos Aires (Argentina) Atti Accad. Naz. Lincei, Memorie, Cl. Sci. fis., mat. e. nat., ser. 7, 11(3):59-151, 2 pls.

Nitzschia denticulata
Zanon, V., 1948
Diatomee marini di Sardegna e Pugillo di Alghe Marine della stressa. Boll. Pesca, Piscicultura e Idrobiologia, Anno 24, ns. 3(2): 202-244, 27 figs. on 1 pl.

Nitzschia dissipata
Rumkówna, A., 1948
List of the phytoplankton species occurring in the superficial water layers in the Gulf of Gdańsk. Bull. Lab. mar., Gdynia, No. 4: 139-141 with tables in back.

Nitzschia dissipata
Tsumura, K., 1955.
Diatomoj el la cirkaufoso de la restajo de la kastelo de Odawara. J. Yokohama Municipal Univ., (C-14), No. 47:23 pp.

Nitzschia distans
Hendy, N. Ingram, 1964
An introductory account of the smaller algae of British coastal waters. V. Bacillariophyceae (Diatoms). Her Majesty's Stationary Office, 317 pp., 45 pls.

Nitzschia distans
Hustedt, F. and A.A. Aleem, 1951
Littoral diatoms from the Salstone near Plymouth. JMBA 30(1): 177.196.

Nitzschia distans
Rampi, L., 1940
Diatomee del Mare Adriatico. Nuovo Giornale Botanico Italiano, n.s., 47:559-608.

Nitzschia droebakensis n.sp.
Hasle, Grethe Rytter, 1964.
Nitzschia and Fragilariopsis species studied in the light and electron microscopes. 1. Some marine species of the groups Nitzschiella and Lanceolatae.
Skrifter, Det Norske Videnskaps-Akad., Oslo, 1. Mat.-Natur. Kl., n.s., No. 16:48 pp.

Nitzschia dubia
Hendy, N. Ingram, 1964
An introductory account of the smaller algae of British coastal waters. V. Bacillariophyceae (Diatoms). Her Majesty's Stationary Office, 317 pp., 45 pls.

Nitzschia dubia
Rampi, L., 1942
Ricerche sul fitoplancton del Mare Ligure 6. Le diatomee delle acque di Sanremo. Nuovo Giornale Botanico Italiano, N.S., 49:252-268.

Nitzschia dubiformis
Hendy, N. Ingram, 1964
An introductory account of the smaller algae of British coastal waters. V. Bacillariophyceae (Diatoms). Her Majesty's Stationary Office, 317 pp., 45 pls.

Nitzschia dubiiformis
Hustedt, F. and A.A. Aleem, 1951
Littoral diatoms from the Salstone near Plymouth. JMBA 30(1): 177.196.

Nitzschia fasciculata
Rumkówna, A., 1948
List of the phytoplankton species occurring in the superficial water layers in the Gulf of Gdańsk. Bull. Lab. mar., Gdynia, No. 4: 139-141 with tables in back.

Nitzschia fasciculata
Zanon, D. V., 1949
Diatomee di Buenos Aires (Argentina) Atti Accad. Naz. Lincei, Memorie, Cl. Sci. fis., mat. e. nat., ser. 7, 11(3):59-151, 2 pls.

Nitzschia fonticola
Tsumura, K., 1956.
Diatomoj el la cirkaŭfoso de la restaĵo de la kastele de Odawara. J. Yokohama Municipal Univ., (C-14), No. 47:23 pp.

Nitzschia fonticola
Zanon, D. V., 1949
Diatomee di Buenos Aires (Argentina)
Atti Accad. Naz. Lincei, Memorie, Cl. Sci. fis., mat. e. nat., ser. 7, 11(3):59-151, 2 pls.

Nitzschia fraudulenta
Jorgensen, E., 1900
Protophyten und Protozoën im Plankton aus der Norwegischen Westkerste. Bergens Mus. Aarb. 1899(6): 95 pp., 5 pls., 83 tables

Nitzschia fraudulenta
Schröder, B., 1900
Phytoplankton des Golfes von Neapel nebst vergleichenden Ausblicken auf das atlantischen Ozean. Mitt. Zool. Stat. Neapel, 14:1-38.

Nitzschia frigida
Gran, H.W., 1908
Diatomeen. Nordisches Plankton, Botanischer Teil pp. XIX.1-XIX 146; 178 text figs.

Nitzschia frigida
Grøntved, J., 1950.
Phytoplankton studies. 1. Nitzschia frigida Grun an Arctic-inner-Baltic diatom found in Danish waters. Kgl. Danske Nidensk. Selsk., Biol. Medd. 18(12):19 pp., 1pl., 6 textfigs.

Nitzschia frigida
Hendy, N. Ingram, 1964
An introductory account of the smaller algae of British coastal waters. V. Bacillariophyceae (Diatoms).
Her Majesty's Stationary Office, 317 pp., 45 pls.

Nitzschia frigida
Jørgensen, E., 1905
B.Protistplankton and the diatoms in bottom samples. Hydrographical and biological investigations in Norwegian fjords. Bergens Mus. Skr. 7: 49-225.

Nitzschia frigida
Levander, K.M., 1947
Plankton gesammelt in den Jahren 1899-1910 an den Küsten Finnlands. Finnländische Hydrographisch-Biologische Untersuchunger (aus dem Wasserbiologischen Laboratorin der Societas Scientiarum Fennica) No.11:40 pp., 6 diagrams, 13 pls., tables.

Nitzschia frigida
Rumkówna, A., 1948
[List of the phytoplankton species occurring in the superficial water layers in the Gulf of Gdańsk] Bull. Lab. mar., Gdynia, No. 4: 139-141 with tables in back.

Nitzschia frustulum
Hendy, N. Ingram, 1964
An introductory account of the smaller algae of British coastal waters. V. Bacillariophyceae (Diatoms).
Her Majesty's Stationary Office, 317 pp., 45 pls.

Nitzschia frustulum
Hendey, N.I., 1951
Littoral diatoms of Chicester Harbour with special reference to fouling. J.Roy. Microscop. Soc. 71(1): 1-86, 18 pls.

Nitzschia frustulum
Zanon, D. V., 1949
Diatomee di Buenos Aires (Argentina)
Atti Accad. Naz. Lincei, Memorie, Cl. Sci. fis., mat. e. nat., ser. 7, 11(3):59-151, 2 pls.

Nitzschia gaarderi n.sp.
Hasle, Grethe Rytter, 1960
Phytoplankton and ciliate species from the Tropical Pacific.
Skr. Norske Videnskaps-Akad., Oslo. 1. Mat.-Nat. Kl., 1960(2): 1-50.

Nitzschia gazellae
Bigelow, H.B., and M. Leslie, 1930
Reconnaissance of the waters and plankton of Monterey Bay, July 1928.
Bull. M.C.Z., 70(5):429-481, 43 text figs.

Nitzschia gazellae
Mangin, L., 1915
Phytoplancton de L'Antarctique. Deuxieme Exped. Ant. Francaise (1908-1910), 95 pp., 3 pls., 58 text figs.

Nitzschia gracilis
Mann, A., 1893
List of Diatomaceae from a deep-sea dredging in the Atlantic Ocean off Delaware Bay by the U. S. Fish Commission Steamer Albatross. Proc. U. S. Nat. Mus. 16:303-312.

Nitzschia gracilis
Tsumura, K., 1956.
Diatomoj el la cirkaŭfoso de la restaĵo de la kastele de Odawara. J. Yokohama Municipal Univ., (C-14), No. 47:23 pp.

Nitzschia intercedens
Rampi, L., 1942
Ricerche sul fitoplancton del Mare Ligure 6. Le diatomee delle acque di Sanremo. Nuovo Giornale Botanico Italiano, N.S., 49:252-268.

Nitzschia gracilis
Zanon, D. V., 1949
Diatomee di Buenos Aires (Argentina)
Atti Accad. Naz. Lincei, Memorie, Cl. Sci. fis., mat. e. nat., ser. 7, 11(3):59-151, 2 pls.

Nitzschia granii n.sp.
Hasle, Grethe Rytter, 1964.
Nitzschia and Fragilariopsis species studied in the light and electron microscopes. 1. Some marine species of the groups Nitzschiella and Lanceolatae.
Skrifter, Det Norske Videnskaps-Akad., Oslo. 1. Mat.-Naturv. Kl., n.s., No. 16:48 pp.

Nitzschia granulata
Hendy, N. Ingram, 1964
An introductory account of the smaller algae of British coastal waters. V. Bacillariophyceae (Diatoms).
Her Majesty's Stationary Office, 317 pp., 45 pls.

Nitzschia granulata
Hendey, N.I., 1951
Littoral diatoms of Chicester Harbour with special reference to fouling. J.Roy. Microscop. Soc. 71(1): 1-86, 18 pls.

Nitzschia habirshawii
Hendy, N. Ingram, 1964
An introductory account of the smaller algae of British coastal waters. V. Bacillariophyceae (Diatoms).
Her Majesty's Stationary Office, 317 pp., 45 pls.

Nitzschia heteropolica n.sp.
Schrader, Hans-Joachim, 1973
Cenozoic diatoms from the northeast Pacific, leg 18. Initial Repts Deep Sea Drilling Proj. 18:673-797.

Nitzschia holsatica
Rumkówna, A., 1948
[List of the phytoplankton species occurring in the superficial water layers in the Gulf of Gdańsk] Bull. Lab. mar., Gdynia, No. 4: 139-141 with tables in back.

Nitzschia hungarica
Hendy, N. Ingram, 1964
An introductory account of the smaller algae of British coastal waters. V. Bacillariophyceae (Diatoms).
Her Majesty's Stationary Office, 317 pp., 45 pls.

Nitzschia hungarica
Hendey, N.I., 1951
Littoral diatoms of Chicester Harbour with special reference to fouling. J.Roy. Microscop. Soc. 71(1): 1-86, 18 pls.

Nitzschia hungarica
Zanon, D. V., 1949
Diatomee di Buenos Aires (Argentina)
Atti Accad. Naz. Lincei, Memorie, Cl. Sci. fis., mat. e. nat., ser. 7, 11(3):59-151, 2 pls.

Nitzschia Hungarica
Zanon, V., 1948
Diatomee marini di Sardegna e Pugillo di Alghe Marine della stressa. Boll. Pesca, Piscitutura e Idrobiologia, Anno 24, ns. 3(2): 202-244, 27 figs. on 1 pl.

Nitschia hustedtiana n.sp.
Salah, M.M., 1952(1953).
XII. Diatoms from Blakeney Point, Norfolk. New species and new records for Great Britain.
J.R. Microsc. Soc., Ser. 3, 72(3):155-169, 3 pls

Nitzschia hybrida
Jørgensen, E., 1905
B.Protistplankton and the diatoms in bottom samples. Hydrographical and biological investigations in Norwegian fjords. Bergens Mus. Skr. 7: 49-225.

Nitzschia hybrida
Rumkówna, A., 1948
[List of the phytoplankton species occurring in the superficial water layers in the Gulf of Gdańsk] Bull. Lab. mar., Gdynia, No. 4: 139-141 with tables in back.

Nitzschia hybrida
Zanon, D. V., 1949
Diatomee di Buenos Aires (Argentina)
Atti Accad. Naz. Lincei, Memorie, Cl. Sci. fis., mat. e. nat., ser. 7, 11(3):59-151, 2 pls.

Nitschia improvisa n.sp.
Simonsen, Reimer, 1960.
Neue Diatomeen aus der Ostsee. II.
Kieler Meeresf., 16(1):126-130.

Nitzschia incerta
Rampi, L., 1940
Diatomee del Mare Adriatico. Nuovo Giornale Botanico Italiano, n.s., 47:559-608.

Nitzschia incognita
Krasske, G., 1941
Die Kieselalgen des chilenischen Küstenplankton (Aus dem südchilenischen Kusten gebiet IX). Arch. Hydrobiol. 38 (2):260-287.

Nitzschia insignis
Mann, A., 1907
Report on the diatoms of the Albatross voyages in the Pacific Ocean, 1888-1904.
Contrib. U. S. Nat. Herb. 10(5):221-419, Pls. XLIV-LIV.

Nitzschia insignis
Rampi, L., 1942
Ricerche sul fitoplancton del Mare Ligure 6. Le diatomee delle acque di Sanremo. Nuovo Giornale Botanico Italiano, N.S., 49:252-268.

Nitzschia insignis

Rampi, L., 1940
Diatomee del Mare Adriatico. Nuovo Giornale Botanico Italiano, n.s., 47:559-608.

Nitzschia insignes

Sousa e Silva, E., 1949
Diatomaceas e Dinoflagelados de Baia de Cascais. Portugaliae Acta Biol., Volume: Julio Henriques, Ser. B: 300-383, 9 pls, 2 fold-in tables.

Nitzschia insignis

Sousa e Silva, E., and J. Dos Santos-Pinto, 1948
O Plancton da Baia de S. Martinho do Porto. 1. Diatomaceas e Dinoflagelados. Bol. Soc. Portuguese de Ciencias Naturais, 16(2):134-187, 6 pls. (Trav. Sta. Biol. Mar. de Lisbonne No. 52).

Nitzschia insignis

Zanon, V., 1948
Diatomee marini di Sardegna e Pugillo di Alghe Marine della stressa. Boll. Pesca, Piscitutura e Idrobiologia, Anno 24, ns. 3(2): 202-244, 27 figs. on 1 pl.

Nitzschia intercedens

Rampi, L., 1940
Diatomee del Mare Adriatico. Nuovo Giornale Botanico Italiano, n.s., 47:559-608.

Nitzschia invisa n.sp.

Schrader, Hans-Joachim, 1973
Cenozoic diatoms from the northeast Pacific, leg 18. Initial Repts Deep Sea Drilling Proj, 18:673-797.

Nitzschia irregularis

Hendy, N. Ingram, 1964
An introductory account of the smaller algae of British coastal waters. V. Bacillariophyceae (Diatoms). Her Majesty's Stationary Office, 317 pp., 45 pls.

Nitzschia irregularis n.sp.

Ross, R., and G. Abdin, 1949.
Notes on some diatoms from Norfolk. J. Roy. Mic. Soc., ser. 3, 69(4):225-230, 4 figs. on 1 pl.

Nitzschia knysnensis n.sp.

Cholnoky, B.J., 1963.
Beiträge zur Kenntnis des marinen Litorals von Südafrika. Botanica Marina, 5(2/3):38-83.

Nitzschia Kolaizeckii

Grunow, A., 1877
1. New Diatoms from Honduras. Monthly Micros. Jour., 18:165-186, pls. CXCIII-CXCVI.

Nitzschia kolaczekii

Hasle, Grethe Rytter, 1960
Phytoplankton and ciliate species from the Tropical Pacific. Skr. Norske Videnskaps-Akad., Oslo, 1. Mat.-Nat. Kl., 1960(2): 1-50.

Nitzschia kutzingiana

Tsumura, K., 1956.
Diatomoj el la cirkaufoso de la restajo de la kastele de Odawara. J. Yokohama Municipal Univ., (C-14), No. 47:23 pp.

Nitzschia Kützingiana

Zanon, D. V., 1949
Diatomee di Buenos Aires (Argentina) Atti Accad. Naz. Lincei, Memorie, Cl. Sci. fis., mat. e. nat., ser. 7, 11(3):59-151, 2 pls.

Nitzschia lanceolata

Jørgensen, E., 1905
B: Protistplankton and the diatoms in bottom samples. Hydrographical and biological investigations in Norwegian fiords. Bergens Mus. Skr. 7: 49-225.

Nitzschia lanceola

Zanon, V., 1948
Diatomee marini di Sardegna e Pugillo di Alghe Marine della stressa. Boll. Pesca, Piscitutura e Idrobiologia, Anno 24, ns. 3(2): 202-244, 27 figs. on 1 pl.

Nitzschia lanceolata

Politis, J., 1949
Diatomees marines de Bosphores et des ibes de la mer de Marmara. II Practica tou Hellenikou Hidrobiologikou Institutoutou 1929, Etoz 1929, 3(1):11-31.

Nitzschia lanceolata

Zanon, V., 1948
Diatomee marini di Sardegna e Pugillo di Alghe Marine della stressa. Boll. Pesca, Piscitutura e Idrobiologia, Anno 24, ns. 3(2): 202-244, 27 figs. on 1 pl.

Nitzschia lecointrei

Hasle, Grethe Rytter, 1964.
Nitzschia and Fragilariopsis species studied in the light and electron microscopes. 1. Some Marine species of the groups Nitzschiella and Lanceolatae Skrifter, Det Norske Videnskaps-Akad., Oslo, 1. Mat.-Naturv. Kl., n.s., No. 16:48 pp.

Nitzschia lesbia n.sp.

Cholnoky, B.J., 1963.
Beiträge zur Kenntnis des marinen Litorals von Südafrika. Botanica Marina, 5(2/3):38-83.

Nitzschia levidensis

Hendy, N. Ingram, 1964
An introductory account of the smaller algae of British coastal waters. V. Bacillariophyceae (Diatoms). Her Majesty's Stationary Office, 317 pp., 45 pls.

Nitzschia linearis

Low, E.M., 1955.
Studies on some chemical constituents of diatoms J. Mar. Res. 14(2):199-204.

Nitzschia linearis

Tsumura, K., 1956.
Diatomoj el la cirkaufoso de la restajo de la kastele de Odawara. J. Yokohama Municipal Univ., (C-14), No. 47:23 pp.

Nitzschia linearis

Zanon, D. V., 1949
Diatomee di Buenos Aires (Argentina) Atti Accad. Naz. Lincei, Memorie, Cl. Sci. fis., mat. e. nat., ser. 7, 11(3):59-151, 2 pls.

Nitzschia littoralis

Hendy, N. Ingram, 1964
An introductory account of the smaller algae of British coastal waters. V. Bacillariophyceae (Diatoms). Her Majesty's Stationary Office, 317 pp., 45 pls.

Nitzschia littoralis

Jorgensen, E., 1900
Protophyten und Protozoen im Plankton aus der Norwegischen Westkerste. Bergens Mus. Aarb. 1899(6): 95 pp., 5 pls., 83 tables.

Nitzschia littoralis

Zanon, D. V., 1949
Diatomee di Buenos Aires (Argentina) Atti Accad. Naz. Lincei, Memorie, Cl. Sci. fis., mat. e. nat., ser. 7, 11(3):59-151, 2 pls.

Nitzschia littorea

Hustedt, F. and A.A. Aleem, 1951
Littoral diatoms from the Salstone near Plymouth. JMBA 30(1): 177-196.

Nitzschia longa

Zanon, V., 1948
Diatomee marini di Sardegna e Pugillo di Alghe Marine della stressa. Boll. Pesca, Piscitutura e Idrobiologia, Anno 24, ns. 3(2): 202-244, 27 figs. on 1 pl.

Nitzschia longicollum n.s

Hasle, Grethe Rytter, 1960
Phytoplankton and ciliate species from the Tropical Pacific. Skr. Norske Videnskaps-Akad., Oslo, 1. Mat.-Nat. Kl., 1960(2): 1-50.

Nitzschia longissima

Allen, W. E., 1938
The Templeton Crocker Expedition to the Gulf of California in 1935 - The Phytoplankton. Amer. Microsc. Soc., Trans. 57:328-335.

Nitzschia longissima

Allen, W.E., 1937
Plankton diatoms of the Gulf of California obtained by the G. Allan Hancock Expedition of 1936. The Hancock Pacific Expeditions, Univ. So. Calif. Publ. 3:47-59, 1 fig.

Nitzschia longissima

Allen, W.E., and E.E. Cupp, 1935
Plankton diatoms of the Java Sea. Annales du Jardin Botanique de Buitenzorg XLIV (2):101-174, figs.1-127.
(drawings of all species mentioned)

Nitzschia longissima

Boden, B.P., 1950
Some marine plankton diatoms from the west coast of South Africa. Trans. R.Soc. S. Africa. 32:321-434, 100 text figs.

Nitzschia longissima

Brunel, J., 1962
Le phytoplancton de la Baie de Chaleurs. Inst. Botan., Univ. Montréal, Contrib. No. 77: 365 pp., 66 pls.

Nitzschia longissima closterium

Nitzschia longissima

Cupp, Easter E., 1943
Marine plankton diatoms of the west coast of North America. Bull. S.I.O. 5(1):1-238, 5 pls., 168 text figs.

Nitzschia longissima

Cupp, E.E., 1934
Analysis of marine diatom collections taken from the Canal Zone to California during March, 1933. Trans. Am. Micros. Soc. LIII (1):22-29, 1 map.

Nitzschia longissima

Cupp, E.E. and Allen, W.E., 1938
Plankton diatoms of the Gulf of California obtained by Allan Hancock Pacific Expedition of 1937. The Hancock Pacific Expeditions, The Univ. So. Calif. Publ. 3: 61-74, 1 map, pls.4-15.

Nitzschia longissima

Gilbert, J.Y., and W.E. Allen, 1943
The phytoplankton of the Gulf of California obtained by the "E.W. Scripps" in 1939 and 1940. J. Mar. Res. V(2):89-110, figs.30-31.

Nitzschia longissima

Gran, H.H., 1897
Protophyta: Diatomaceae, Silico-flagellata and Cilioflagellata. Den Norske Nordhavs Expedition 1876-1878, h. 24, 36 pp., 4 pls.

Nitzschia longissima
Hasle, Grethe Rytter, 1964.
Nitzschia and Fragilariopsis species studied in light and electron microscopes. 1. Some marine species of the groups Nitzschiella and Lanceolatae.
Skrifter, Det Norske Videnskaps-Akad., Oslo, 1. Mat. Naturv. Kl., n.s., No. 18:48 pp.

Nitzschia longissima
Hendy, N. Ingram, 1964
An introductory account of the smaller algae of British coastal waters. V. Bacillariophyceae (Diatoms). Her Majesty's Stationary Office, 317 pp., 45 pls.

Nitzschia longissima
Hendey, N.I., 1951
Littoral diatoms of Chicester Harbour with special reference to fouling. J. Roy. Microscop. Soc. 71(1): 1-86, 18 pls.

Nitzschia longissimum
Hustedt, F. and A.A. Aleem, 1951
Littoral diatoms from the Salstone near Plymouth. JMBA 30(1): 177.196.

Nitzschia longissima
Jørgensen, E., 1905
B. Protistplankton and the diatoms in bottom samples. Hydrographical and biological investigations in Norwegian fjords. Bergens Mus. Skr. 7: 49-225.

Nitzschia longissima
Kokubo, S., and S. Sato, 1947
Plankters in Jū-San Gata. Physiol. and Ecol. (Japan) 1(4):1-16, 3 text figs., tables.

Nitzschia longissima
Levander, K.M., 1947
Plankton gesammelt in den Jahren 1899-1910 an den Küsten Finnlands. Finnländische Hydrographisch-Biologische Untersuchunger (aus dem Wasserbiologischen Laboratorium der Societas Scientiarum Fennica) No.11:40 pp., 6 diagrams, 13 pls., tables.

Nitzschia longissima
Lillick, L.C., 1940
Phytoplankton and planktonic protozoa of the offshore waters of the Gulf of Maine. Pt.II. Qualitative Composition of the Planktonic Flora. Trans. Am. Phil. Soc., n.s., 31(3):193-237, 13 text figs.

Nitzschia longissima
Lillick, L.C., 1937
Seasonal studies of the phytoplankton off Woods Hole, Massachusetts. Biol. Bull. LXXIII (3):488-503, 3 text figs.

Nitzschia longissima
Meunier, A., 1915
Microplancton de la Mer Flamande. 2. Diatomées (excepté le genre Chaetoceros). Mem. Mus. Roy. Hist. Nat., Belgique, 7(3):1-118, Pls. VIII-XIV.

Nitzschia longissima
Morse, D.C., 1947
Some observations on seasonal variations in plankton population Patuxant River, Maryland 1943-1945. Bd. Nat. Res., Publ. No.65, Chesapeake Biol. Lab., 31, 3 figs.

Nitzschia longissima
Pavillard, J., 1925
Bacillariales. Rept. on the Danish Oceanogr. Exped., 1908-10 to the Mediterranean and adj. seas. Vol.II., Biol. J4:72 pp., 116 text figs.

Nitzschia longissima
Pavillard, J., 1905
Recherches sur la flore pelagique (Phytoplankton) de l'Etang de Thau. Theses presentees a la Fac. Sci., Paris, 116 pp., 3 pls.

Nitzschiella longissima
Politis, J., 1949
Diatomees marines de Bosphores et des ibes de la mer de Marmara. II Practica tou Hellenikou Hidrobiologikou Institutoutou 1929, Etoz 1929, 3(1):11-31.

Nitzschia longissima
Rampi, L., 1942
Ricerche sul fitoplancton del Mare Ligure 6. Le diatomee delle acque di Sanremo. Nuovo Giornale Botanico Italiano, N.S., 49:252-268.

Nitzschia longissima
Rampi, L., 1940
Diatomee del Mare Adriatico. Nuovo Giornale Botanico Italiano, n.s., 47:559-608.

Nitzschia longissima
Schodduyn, M., 1926
Observations faites dans la baie d'Ambleteuse (Pas de Calais). Bull. Inst. Ocean., Monaco, No. 482: 64 pp.

Nitzschia longissima
Schröder, B., 1900
Phytoplankton des Golfes von Neapel nebst vergleichenden Ausblicken auf das atlantischen Ozean. Mitt. Zool. Stat. Neapel, 14:1-38.

Nitzschia longissima
Sousa e Silva, E., 1949
Diatomaceas e Dinoflagelados de Baía de Cascais. Portugaliae Acta Biol., Volume: Julio Henriques, Ser. B: 300-383, 9 pls, 2 fold-in tables.

Nitzschia longissima
Sousa a Silva, E., and J. Dos Santos-Pinto, 1948
O Plancton da Baía de S. Martinho do Porto. 1. Diatomaceas e Dinoflagelados. Bol. Soc. Portuguese de Ciencias Naturais, 16(2):134-187, 6 pls. (Trav. Sta. Biol. Mar. de Lisbonne No. 52).

Nitzschia longissima
Takano, Hideaki, 1963.
Notes on marine littoral diatoms of Japan. Bull. Tokai Reg. Fish. Res. Lab., No. 36:1-8.

Nitzschia longissima
Takano, Hideaki, 1962
Notes on epiphytic diatoms upon sea-weeds from Japan. J. Oceanogr. Soc., Japan, 18(1):29-33.

Nitzschia longissima
Zanon, V., 1948
Diatomee marini di Sardegna e Pugillo di Alghe Marine della stressa. Boll. Pesca, Piscitutura e Idrobiologia, Anno 24, ns. 3(2): 202-244, 27 figs. on 1 pl.

Nitzschia longissima var. closterioi
Bigelow, H.B., and M. Leslie, 1930 des
Reconnaissance of the waters and plankton of Monterey Bay, July 1928. Bull. M.C.Z., 70(5):429-481, 43 text figs.

Nitzschia longissima closterium
Margalef, R., 1949
Fitoplancton nerítico de la Costa Brava en 1947-48. Publ. Inst. Biol. Aplicada, 5: 41-51, 3 text figs.

Nitzschia lorenziana
Kokubo, S., and S. Sato, 1947
Plankters in Jū-San Gata. Physiol. and Ecol. (Japan) 1(4):1-16, 3 text figs., tables.

Nitzschiella Lorenziana
Politis, J., 1949
Diatomees marines de Bosphores et des ibes de la mer de Marmara. II Practica tou Hellenikou Hidrobiologikou Institutoutou 1929, Etoz 1929, 3(1):11-31.

Nitzschia Lorenziana
Rampi, L., 1940
Diatomee del Mare Adriatico. Nuovo Giornale Botanico Italiano, n.s., 47:559-608.

Nitzschia Lorenziana
Zanon, V., 1948
Diatomee marini di Sardegna e Pugillo di Alghe Marine della stressa. Boll. Pesca, Piscitutura e Idrobiologia, Anno 24, ns. 3(2): 202-244, 27 figs. on 1 pl.

Nitzschia Lorenziana var. incurva
Allen, W.E., and E.E. Cupp, 1935
Plankton diatoms of the Java Sea. Annales du Jardin Botanique de Buitenzorg XLIV (2):101-174, figs.1-127.
(drawings of all species mentioned)

Nitschia macilenta
Hustedt, F. and A.A. Aleem, 1951
Littoral diatoms from the Salstone near Plymouth. JMBA 30(1): 177.196.

Nitzschia macilenta
Politis, J., 1949
Diatomees marines de Bosphores et des ibes de la mer de Marmara. II Practica tou Hellenikou Hidrobiologikou Institutoutou 1929, Etoz 1929, 3(1):11-31.

Nitzschia macilenta
Rampi, L., 1940
Diatomee del Mare Adriatico. Nuovo Giornale Botanico Italiano, n.s., 47:559-608.

Nitzschia macilenta
Zanon, V., 1948
Diatomee marini di Sardegna e Pugillo di Alghe Marine della stressa. Boll. Pesca, Piscitutura e Idrobiologia, Anno 24, ns. 3(2): 202-244, 27 figs. on 1 pl.

Nitzschia maiuscula
Zanon, V., 1948
Diatomee marini di Sardegna e Pugillo di Alghe Marine della stressa. Boll. Pesca, Piscitutura e Idrobiologia, Anno 24, ns. 3(2): 202-244, 27 figs. on 1 pl.

Nitzschia mammalis n.sp.
Castracane degli Antelminelli, F., 1886
1. Report on the Diatomaceae collected by H.M.S. Challenger during the years 1873-1876. Rept. Sci. Results, H.M.S. Challenger, Botany Vol. II, 178 pp., 30 pls.

Nitzschia marginulata
Politis, J., 1949
Diatomees marines de Bosphores et des ibes de la mer de Marmara. II Practica tou Hellenikou Hidrobiologikou Institutoutou 1929, Etoz 1929, 3(1):11-31.

Nitzschia marginulata
Zanon, D. V., 1949
Diatomee di Buenos Aires (Argentina) Atti Accad. Naz. Lincei, Memorie, Cl. Sci. fis., mat. e. nat., ser. 7, 11(3):59-151, 2 pls.

Nitzschia marginulata
Zanon, V., 1948
Diatomee marini di Sardegna e Pugillo di Alghe Marine della stressa. Boll. Pesca, Piscitutura e Idrobiologia, Anno 24, ns. 3(2): 202-244, 27 figs. on 1 pl.

Nitzschia marina
Mann, A., 1893
List of Diatomaceae from a deep-sea dredging in the Atlantic Ocean off Delaware Bay by the U. S. Fish Commission Steamer Albatross. Proc. U. S. Nat. Mus. 16:303-312.

Nitzschia media
Zanon, V., 1948
Diatomee marini di Sardegna e Pugillo di Alghe Marine della stressa. Boll. Pesca, Piscitutura e Idrobiologia, Anno 24, ns. 3(2): 202-244, 27 figs. on 1 pl.

Nitzschia microcephala
Zanon, D. V., 1949
Diatomee di Buenos Aires (Argentina) Atti Accad. Naz. Lincei, Memorie, Cl. Sci. fis., mat. e. nat., ser. 7, 11(3):59-151, 2 pls.

Nitzschia microcephala
Zanon, V., 1948
Diatomee marini di Sardegna e Pugillo di Alghe Marine della stressa. Boll. Pesca, Piscitutura e Idrobiologia, Anno 24, ns. 3(2): 202-244, 27 figs. on 1 pl.

Nitschia migrans (fig.)
Vives, F., and A. Planas, 1952.
Plancton recogido por los laboratorios costeros. VI Fitoplancton de las costas de Vinaroz, islas Columbretes y alrededores de la desembocadura del Elroa. Publ. Inst. Biol. Aplic. 11:141-156, 19 textfigs.

Nitzschia miserabilis n.sp.
Cholnoky, B.J., 1963.
Beiträge zur Kenntnis des marinen Litorals von Südafrika.
Botanica Marina, 5(2/3):38-83.

Nitzschia mitchelliana
Jørgensen, E., 1905
B.Protistplankton and the diatoms in bottom samples. Hydrographical and biological investigations in Norwegian fjords. Bergens Mus. Skr. 7: 49-225.

Nitzschia morosa n.sp.
Cholnoky, B.J., 1963.
Beiträge zur Kenntnis des marinen Litorals von Südafrika.
Botanica Marina, 5(2/3):38-83.

Nitzschia navicularis
Hendy, N. Ingram, 1964
An introductory account of the smaller algae of British coastal waters. V. Bacillariophyceae (Diatoms). Her Majesty's Stationary Office, 317 pp., 45 pls.

Nitzschia navicularis
Hendey, N.I., 1951
Littoral diatoms of Chicester Harbour with special reference to fouling. J.Roy. Microscop. Soc. 71(1): 1-86, 18 pls.

Nitzschia navicularis
Zanon, D. V., 1949
Diatomee di Buenos Aires (Argentina) Atti Accad. Naz. Lincei, Memorie, Cl. Sci. fis., mat. e. nat., ser. 7, 11(3):59-151, 2 pls.

Nitzschia norvegica n.sp.
Hasle, Grethe Rytter, 1964.
Nitzschia and Fragilariopsis species studied in the light and electron microscopes. 1. Some marine species of the groups Nitzschiella and Lanceolatae.
Skrifter, Det Norske Videnskaps-Akad., Oslo. 1. Nat.-Maturv. Kl., n.s., No. 16:48 pp.

Nitzschia obesa n.sp.
Castracane degli Antelminelli, F., 1886
1. Report on the Diatomaceae collected by H.M.S. Challenger during the years 1873-1876. Rept. Sci. Results, H.M.S. Challenger, Botany Vol. II, 178 pp., 30 pls.

Nitzschia obtusa
Hendy, N. Ingram, 1964
An introductory account of the smaller algae of British coastal waters. V. Bacillariophyceae (Diatoms). Her Majesty's Stationary Office, 317 pp., 45 pls.

Nitzschia obtusa
Iyengar, M.O.P. and G.Venkataraman, 1951.
The ecology and seasonal succession of the algae flora of the River Cooum at Madras with special reference to the Diatomaceae. J. Madras Univ. 21, Sect. B(1): 140-192, 1 pl of 4 figs., 11 text figs.

Nitzschia obtusa
Kokubo, S., and S. Sato, 1947
Plankters in Jū-San Gata. Physiol. and Ecol. (Japan) 1(4):1-16, 3 text figs., tables.

Nitzschia obtusa
Zanon, D. V., 1949
Diatomee di Buenos Aires (Argentina) Atti Accad. Naz. Lincei, Memorie, Cl. Sci. fis., mat. e. nat., ser. 7, 11(3):59-151, 2 pls.

Nitzschia obtusa
Zanon, V., 1948
Diatomee marini di Sardegna e Pugillo di Alghe Marine della stressa. Boll. Pesca, Piscitutura e Idrobiologia, Anno 24, ns. 3(2): 202-244, 27 figs. on 1 pl.

Nitschia obtusa var. lata
Hendey, N-Ingram, 1958 [1957(Publ. 1958)]
Marine diatoms from some West African Ports. J. R Microsc. Soc. (3) 77(1/2): 28-85.

Nitzschia oceanica n.sp.
Hasle, Grethe Rytter, 1960
Phytoplankton and ciliate species from the Tropical Pacific.
Skr. Norske Videnskaps-Akad., Oslo, 1. Mat.-Nat. Kl., 1960(2): 1-50.

Nitzschia ovalis
North, Barbara B. and Grover C Stephens 1972.
Amino acid transport in Nitzschia ovalis Arnott.
J. Phycol. 8(1): 64-68

Nitzschia pacifica, n.sp.
Cupp, Easter E., 1943
Marine plankton diatoms of the west coast of North America. Bull. S.I.O. 5(1):1-238, 5 pls., 168 text figs.

Nitzschia pacifica
Manguin, E., 1954
Diatomees marines provenant de l'ile Heard (Australian National Antarctic Expedition). Rev. Algol., n.s., 1: 14-24.

Nitzschia palea (?)
Hutner, S.H., and L. Provasoli, 1953.
A pigmented marine diatom requiring Vitamin B12 and uracil. News Bull., Phycol. Soc., Amer., 6(18):7.

Nitschia palea
Iyengar, M.O.P. and G.Venkataraman, 1951.
The ecology and seasonal succession of the algae flora of the River Cooum at Madras with special reference to the Diatomaceae. J. Madras Univ. 21, Sect. B(1): 140-192, 1 pl of 4 figs., 11 text figs.

Nitzschia palea
Low, E.M., 1955.
Studies on some chemical constituents of diatoms J. Mar. Res. 14(2):199-204.

Nitzschia palea
Mann, A., 1893
List of Diatomaceae from a deep-sea dredging in the Atlantic Ocean off Delaware Bay by the U. S. Fish Commission Steamer Albatross. Proc. U. S. Nat. Mus. 16:303-312.

Nitzschia palea
Rumkówna, A., 1948
List of the phytoplankton species occurring in the superficial water layers in the Gulf of Gdansk. Bull. Lab. mar., Gdynia, No. 4: 139-141 with tables in back.

Nitzschia palea
Tsumura, K., 1956.
Diatomoj el la cirkaŭfoso de la restaĵo de la kastele de Odawara. J. Yokohama Municipal Univ., (C-14), No. 47:23 pp.

Nitzschia palea
Zanon, D. V., 1949
Diatomee di Buenos Aires (Argentina) Atti Accad. Naz. Lincei, Memorie, Cl. Sci. fis., mat. e. nat., ser. 7, 11(3):59-151, 2 pls.

Nitzschia panduriformis
Hendy, N. Ingram, 1964
An introductory account of the smaller algae of British coastal waters. V. Bacillariophyceae (Diatoms). Her Majesty's Stationary Office, 317 pp., 45 pls.

Nitschia panduriformis
Hustedt, F. and A.A. Aleem, 1951
Littoral diatoms from the Salstone near Plymouth. JMBA 30(1): 177.196.

Nitzschia panduriformis
Jorgensen, E., 1900
Protophyten und Protozoen im Plankton aus der Norwegischen Westkerste. Bergens Mus. Aarb. 1899(6): 95 pp., 5 pls., 83 tables.

Nitzschia panduriformis
Manguin, E., 1954
Diatomees marines provenant de l'ile Heard (Australian National Antarctic Expedition). Rev. Algol., n.s., 1: 14-24.

Nitzschia panduriformis
Mann, A., 1907
Report on the diatoms of the Albatross voyages in the Pacific Ocean, 1888-1904. Contrib. U. S. Nat. Herb. 10(5):221-419, Pls. XLIV-LIV.

Nitzschia panduriformis
Mann, A., 1893
List of Diatomaceae from a deep-sea dredging in the Atlantic Ocean off Delaware Bay by the U. S. Fish Commission Steamer Albatross. Proc. U. S. Nat. Mus. 16:303-312.

Nitzschia panduriformis
Politis, J., 1949
Diatomees marines de Bosphores et des ibes de la mer de Marmara. II Practica tou Hellenikou Hidrobiologikou Institutoutou 1929, Etoz 1929, 3(1):11-31.

Nitschia panduiformis
Rampi, L., 1942
Ricerche sul fitoplancton del Mare Ligure 6. Le diatomee delle acque di Sanremo. Nuovo Giornale Botanico Italiano, N.S., 49:252-268.

Nitzschia panduriformis
Rampi, L., 1940
Diatomee del Mare Adriatico. Nuovo Giornale Botanico Italiano, n.s., 47:559-608.

Nitzschia panduriformis
Zanon, V., 1948
Diatomee marini di Sardegna e Pugillo di Alghe Marine della stressa. Boll. Pesca, Piscitutura e Idrobiologia, Anno 24, ns. 3(2): 202-244, 27 figs. on 1 pl.

Nitzschia panduriformis var. continuua
Allen, W.E., and E.E. Cupp, 1935
Plankton diatoms of the Java Sea. Annales du Jardin Botanique de Buitenzorg XLIV (2):101-174, figs.1-127.

(drawings of all species mentioned)

Nitzschia paradoxa
Boden, B.P., 1950
Some marine plankton diatoms from the west coast of South Africa. Trans. R.Soc. S. Africa. 32:321-434, 100 text figs.

Nitzschia paradoxa
Cupp, Easter E., 1943
Marine plankton diatoms of the west coast of North America. Bull. S.I.O. 5(1):1-238, 5 pls., 168 text figs.

Nitzschia paradoxa
Eskinazi Enide e Shigekatsv Satô (1963-1964) 1966.
Contribuição ao estudo das diatomaceas da Praia de Piedade.
Trabhs Inst. Oceanogr., Univ. Recife, 5 (5/6): 73-114.

Nitzschia paradoxa
Gran, H. H. and E. C. Angst, 1931
Plankton diatoms of Puget Sound. Publ. Puget Sound Biol. Sta. 7:417-519, 95 text figs.

Nitzschia pelagica
Hendey, N.I., 1937
The plankton diatoms of the southern seas. Discovery Repts. 16:151-364, pls.6-13.

Nitzschia peragalli nom.nov.
Hasle, Grethe Ryther, 1965.
Nitzschia and Fragilariopsis species studies in the light and electron microscopes. III. The genus Fragilariopsis.
Norske Videnskaps- Akad., Oslo. I. Met.-Nature. Kl., N.S. No. 21: 49 pp., 17 pls.

Nitschia pertenuis n.sp.
Frenguelli, Joaquin, and Hector Antonio Orlando, 1959.
Operacion MERLUZA. Diatomeas y silicoflagelados del plancton del "VI Crucero". Servicio Hidrogr. Naval., Argentina, Publ. No. H. 619: 5-62.

Nitzschia plana
Hendy, N. Ingram, 1964
An introductory account of the smaller algae of British coastal waters. V. Bacillariophyceae (Diatoms).
Her Majesty's Stationary Office, 317 pp., 45 pls.

Nitzschia plana
Mann, A., 1907
Report on the diatoms of the Albatross voyages in the Pacific Ocean, 1888-1904. Contrib. U. S. Nat. Herb. 10(5):221-419, Pls. XLIV-LIV.

Nitzschia plana
Politis, J., 1949
Diatomees marines de Bosphores et des ibes de la mer de Marmara. II Practica tou Hellenikou Hidrobiologikou Institutoutou 1929, Etoz 1929, 3(1):11-31.

Nitzschia plana var. zebriana n. var.
Castracane degli Antelminelli, F., 1886
1. Report on the Diatomaceae collected by H.M.S. Challenger during the years 1873-1876. Rept. Sci. Results, H.M.S. Challenger, Botany Vol. II, 178 pp., 30 pls.

Nitzschia praefossilis n.sp.
Schrader, Hans-Joachim, 1973
Cenozoic diatoms from the northeast Pacific, leg 18. Initial Repts Deep Sea Drilling Proj. 18:673-797.

Nitschia proceroides n.sp.
Giffen, Malcolm H. 1973.
Diatoms of the marine littoral of Steenberg's Cove in St. Helena Bay, Cape Province, South Africa.
Botanica Marina 16 (1): 32-48.

Nitschia prolongata n.sp.
Manguin, E., 1957.
Premier inventaire des diatomées de la Terre Adélie Antarctique. Espèces nouvelles. Rev. Algologique, n.s., 3(3):111-134.

Nitzschia pseudodelicatissima nom. nov.
Hasle, Grethe Ryter, and Blanca Rojas E. de Mendiola 1967.
The fine structure of some Thalassionema and Thalassiothrix species.
Phycologia 6 (2/3): 107-118.

Nitzschia punctata & varieties
Hendy, N. Ingram, 1964
An introductory account of the smaller algae of British coastal waters. V. Bacillariophyceae (Diatoms).
Her Majesty's Stationary Office, 317 pp., 45 pls.

Nitzschia punctata
Hustedt, F. and A.A. Aleem, 1951
Littoral diatoms from the Salstone near Plymouth. JMBA 30(1): 177.196.

Nitzschia punctata
Mann, A., 1907
Report on the diatoms of the Albatross voyages in the Pacific Ocean, 1888-1904. Contrib. U. S. Nat. Herb. 10(5):221-419, Pls. XLIV-LIV.

Nitzschia punctata
Mann, A., 1893
List of Diatomaceae from a deep-sea dredging in the Atlantic Ocean off Delaware Bay by the U. S. Fish Commission Steamer Albatross. Proc. U. S. Nat. Mus. 16:303-312.

Nitzschia punctata
Politis, J., 1949
Diatomees marines de Bosphores et des ibes de la mer de Marmara. II Practica tou Hellenikou Hidrobiologikou Institutoutou 1929, Etoz 1929, 3(1):11-31.

Nitzschia punctata
Rampi, L., 1940
Diatomee del Mare Adriatico. Nuovo Giornale Botanico Italiano, n.s., 47:559-608.

Nitzschia punctata
Zanon, D. V., 1949
Diatomee di Buenos Aires (Argentina) Atti Accad. Naz. Lincei, Memorie, Cl. Sci. fis., mat. e. nat., ser. 7, 11(3):59-151, 2 pls.

Nitzschia punctata
Zanon, V., 1948
Diatomee marini di Sardegna e Pugillo di Alghe Marine della stressa. Boll. Pesca, Piscitutura e Idrobiologia, Anno 24, ns. 3(2): 202-244, 27 figs. on 1 pl.

Nitzschia pungens
Gilbert, J.Y., and W.E. Allen, 1943
The phytoplankton of the Gulf of California obtained by the "E.W. Scripps" in 1939 and 1940. J. Mar. Res. V(2):89-110, figs.30-31.

Nitzschia pungens
Gran, H.H., 1908
Diatomeen. Nordisches Plankton, Botanischer Teil pp. XIX.1-XIX 146; 178 text figs.

Nitzschia pungens
Hasle, Grethe Rytter, 1968.
Observations on the marine diatom Fragilariopsis Kerguelensis (O'Meara) Hust. in the scanning electron microscope. Nytt Mag. Bot.15(3):205-208.

Nitzschia pungens
Hustedt, F. and A.A. Aleem, 1951
Littoral diatoms from the Salstone near Plymouth. JMBA 30(1): 177.196.

Nitzschia pungens atlantica
Cupp, Easter E., 1943
Marine plankton diatoms of the west coast of North America. Bull. S.I.O. 5(1):1-238, 5 pls., 168 text figs.

Nitzschia pungens var. atlantica
Cupp, E.E. and Allen, W.E., 1938
Plankton diatoms of the Gulf of California obtained by Allan Hancock Pacific Expedition of 1937. The Hancock Pacific Expeditions, the Univ. So. Calif. Publ. 3: 61-74, 1 map, pls.4-15.

Nitzschia pungiformis
Hasle, Grethe Rytter 1971.
Nitzschia pungiformis (Bacillariophyceae), a new species of the Nitzschia seriata group.
Norw. J. Bot. 18 (3/4): 139-144

Nitzschia putrida
Hutner, S.H., and L. Provasoli, 1953.
A pigmented marine diatom requiring Vitamin B12 and uracil. News Bull., Phycol. Soc., Amer., 6(18):7.

Nitzschia putrida
Pringsheim, E.G., 1951.
Über farblose Diatomeen. Arch. f. Mikrobiol. 16: 18-27.

Nitschia putrida
Wagner, J., 1934.
Beitrage zur Kenntnis der Nitschia putrida Benecke, insbesondere ihrer Bewegung. Arch. Protistk. 82:86-113, 10 textfigs.

Nitschia recta
de Sousa e Silva, E., 1956.
Contribution à l'étude du microplancton de Dakar et des regions maritimes voisines. Bull. I.F.A.N., 8(2):335-371, 7 pls.

Nitzschia recta
Rampi, L., 1942
Ricerche sul fitoplancton del Mare Ligure 6. Le diatomee delle acque di Sanremo. Nuovo Giornale Botanico Italiano, N.S., 49:252-258.

Nitzschia rigida
Ercegovic, A., 1936
Etudes qualitative et quantitatives du phytoplancton dans les eaux cotières de l'Adriatique oriental moyen au cours de l'année 1934. Acta Adriatica 1(9):1-126

Nitzschia rigida
Rampi, L., 1940
Diatomee del Mare Adriatico. Nuovo Giornale Botanico Italiano, n.s., 47:559-608.

Nitzschia rigida (figs.)
Sousa e Silva, E., 1949
Diatomaceas e Dinoflagelados de Baia de Cascais. Portugalise Acta Biol., Volume: Julio Henriques, Ser. B: 300-383, 9 pls, 2 fold-in tables.

Nitzschia rigida
Zanon, V., 1948
Diatomee marini di Sardegna e Pugillo di Alghe Marine della stressa. Boll. Pesca, Piscitutura e Idrobiologia, Anno 24, ns. 3(2): 202-244, 27 figs. on 1 pl.

Nitzschia rolandii n.sp.
Schrader, Hans-Joachim, 1973
Cenozoic diatoms from the northeast Pacific, leg 18. Initial Repts Deep Sea Drilling Proj. 18:673-797.

Nitzschia romana
Zanon, D. V., 1949
Diatomee di Buenos Aires (Argentina) Atti Accad. Naz. Lincei, Memorie, Cl. Sci. fis., mat. e. nat., ser. 7, 11(3):59-151, 2 pls.

Nitzschia salinarum
Mann, A., 1893
List of Diatomaceae from a deep-sea dredging in the Atlantic Ocean off Delaware Bay by the U. S. Fish Commission Steamer Albatross. Proc. U. S. Nat. Mus. 16:303-312.

Nitzschia salinicola n.sp.
Aleem, A.A., and Fr. Hustedt, 1951.
Einige neue Diatomeen von der Südküste Englands. Botaniska Notiser 1951(1):13-20, 6 textfigs.

Nitzschia scabra
Mann, A., 1907
Report on the diatoms of the Albatross voyages in the Pacific Ocean, 1888-1904. Contrib. U. S. Nat. Herb. 10(5):221-419, Pls. XLIV-LIV.

Nitzschia scalaris
Zanon, D. V., 1949
Diatomee di Buenos Aires (Argentina) Atti Accad. Naz. Lincei, Memorie, Cl. Sci. fis., mat. e. nat., ser. 7, 11(3):59-151, 2 pls.

Nitzschia scalaris
Zanon, V., 1948
Diatomee marini di Sardegna e Pugillo di Alghe Marine della stressa. Boll. Pesca, Piscitutura e Idrobiologia, Anno 24, ns. 3(2): 202-244, 27 figs. on 1 pl.

Nitzschia schweinfurthii
Morse, D.C., 1947
Some observations on seasonal variations in plankton population Patuxant River, Maryland 1943-1945. Bd. Nat. Res., Publ. No.65, Chesapeake Biol. Lab., 31, 3 figs.

Nitzschia seriata
Allen, W. E., 1938
The Templeton Crocker Expedition to the Gulf of California in 1935 - The Phytoplankton. Amer. Microsc. Soc., Trans. 57:328-335.

Nitzschia seriata
Allen, W.E., 1937
Plankton diatoms of the Gulf of California obtained by the G. Allan Hancock Expedition of 1936. The Hancock Pacific Expeditions, Univ. So. Calif. Publ. 3:47-59, 1 fig.

Nitzschia seriata
Allen, W. E., 1936
Occurrence of marine plankton diatoms in a ten-year series of daily catches in southern California. Am. Jour. Bot. 23(1):60-63.

Nitzschia seriata
Allen, W. E., 1934
Marine plankton diatoms of Lower California in 1931. Bot. Gaz. 95(3):485-492, 1 fig.

Nitzschia seriata
Allen, W.E., 1928.
Catches of marine diatoms and dinoflagellates taken by boat in Southern California waters in 1926. Bull. S.I.O., tech. ser., 1:201-246, 6 textfigs.

Nitzschia seriata
Allen, W.E., 1928.
Review of five years of studies of phytoplankton at Southern California piers, 1920-1924, inclusive. Bull. S.I.O., tech. ser., 1:357-401, 5 textfigs.

Nitzschia seriata
Allen, W.E., 1927.
Quantitative studies on inshore marine diatoms and dinoflagellates of Southern California in 1922. Bull. S.I.O., tech. ser., 1:31-38, 2 textfigs.

Nitzschia seriata
Allen, W. E., 1925
Statistical Studies of surface catches of marine diatoms and dinoflagellates made by the Yacht "Ohio" in tropical waters in 1924. Jan. Trans. Amer. Microscop. Soc.:24-30, 1 fig.

Nitzschia seriata
Allen, W.E., and E.E. Cupp, 1935
Plankton diatoms of the Java Sea. Annales du Jardin Botanique de Buitenzorg XLIV (2):101-174, figs.1-127.
(drawings of all species mentioned)

Nitzschia seriata
Bigelow, H. B., 1922
Exploration of the coastal water off the northeastern United States in 1916 by the U.S. Fisheries Schooner Grampus. Bull. M.C.Z. 65 (5):85-188, 53 text figs.

Nitzschia seriata
Bigelow, H.B., and M. Leslie, 1930
Reconnaissance of the waters and plankton of Monterey Bay, July 1928. Bull. M.C.Z., 70(5):429-481, 43 text figs.

Nitzschia seriata
Boden, B.P., 1950
Some marine plankton diatoms from the west coast of South Africa. Trans. R.Soc. S. Africa. 32:321-434, 100 text figs.

Nitzschia seriata
Boden, B. P., 1949.
The diatoms collected by the U.S.S. CACOPAN in the Antarctic in 1947. J. Mar. Res. 8(1):6-13, 3 textfigs.

Nitzschia seriata
Boden, Brian, 1948
Marine plankton diatoms on operation HIGHJUMP in: Some oceanographic observations on operation HIGHJUMP. By R.S. Dietz. USNEL Rept. No.55, 97 pp., 41 figs. 7 July 1948.

Nitzschia seriata
Braarud, T., 1945
A phytoplankton survey of the polluted waters of inner Oslo Fjord. Hvalrådets Skrifter, No.28, 142 pp., 19 text figs., 17 tables.

Nitzschia seriata
Braarud, T., 1945
Experimental studies on marine plankton diatoms. Avhandlingar utgitt av Det Norske Videnskaps-Akademi i Oslo. 1. Mat.-Naturv. Klasse 1944. No.10, 1-16, 1 text fig.

Nitzschia seriata
Braarud, T., 1934
A note on the phytoplankton of the Gulf of Maine in the summer of 1933. Biol. Bull. 67(1):76-82.

Nitzschia seriata
Brunel, J., 1962
Le phytoplancton de la Baie de Chaleurs. Inst. Botan., Univ. Montréal, Contrib. No. 77: 365 pp., 66 pls.

Nitzschia seriata
Brunel, Jules, 1962
Le phytoplancton de la Baie des Chaleurs. Contrib. Ministère de la Chasse et des Pêcheries, Province de Québec, No. 91: 365 pp.

Nitzschia seriata
Central Meteorological Observatory, 1949
Report on sea and weather observation on Antarctic Whaling Ground (1947-48). Ocean. Mag., Japan, 1(1):49-88, 17 text figs.

Nitzschia seriata
Copenhagen, W. J. and L. D., 1949
Variation in the phytoplankton of Table Bay, October 1934 to October 1935. With a note on the calorific value of Chaetoceros spp. Trans. Roy. Soc. S. Africa, 32(2):113-123, 2 text figs.

Nitzschia seriata
Cupp, Easter E., 1943
Marine plankton diatoms of the west coast of North America. Bull. S.I.O. 5(1):1-238, 5 pls., 168 text figs.

Nitzschia seriata
Cupp, E.E., 1937
Seasonal distribution and occurrence of marine diatoms and dinoflagellates at Scotch Cap, Alaska. Bull. S.I.O. Tech. ser.4 (3):71-100, 7 textfigs.

Nitzschia seriata
Cupp, E.E., 1934
Analysis of marine diatom collections taken from the Canal Zone to California during March, 1933. Trans. Am. Micros. Soc. LIII (1):22-29, 1 map.

Nitzschia seriata
Cupp, E., 1930
Quantitative Studies of miscellaneous series of surface catches of marine diatoms and dinoflagellates taken between Seattle and the Canal Zone from 1924 to 1928. Trans. Am. Micro. Soc., XLIX (3):238-245.

Nitzschia Seriata
Cupp, E.E. and Allen, W.E., 1938
Plankton diatoms of the Gulf of California obtained by Allan Hancock Pacific Expedition of 1937. The Hancock Pacific Expeditions, The Univ. So. Calif. Publ. 3: 61-74, 1 map, pls.4-15.

Nitzschia seriata
Dangeard, P., 1927
Phytoplankton de la croisière du "Sylvana". Ann. Inst. Océan., Monaco, n.s., 4(8):286-401, 54 text figs. (Février-Juin 1913).

Nitzschia seriata
Davidson, V.M., 1931.
Biological and oceanographic conditions in Hudson Bay. Contr. Canadian Biol. Fish., n.s., 6(26): 497-509, 7 textfigs.

Nitzschia seriata
Delegazione Italiana della Commissione Internazionale per l'Esplorazione Scientifica del Mediterraneo, 1941
Note sul plancton della Laguna veneta. [Memoria CCLXXXIX], Arch. di Ocean. e Limn. Anno I, Fasc. I, 1941 XIX: 31-57 pp.

Nitzschia seriata
Ercegovic, A., 1936
Etudes qualitative et quantitatives du phytoplancton dans les eaux cotières de l'Adriatique oriental moyen au cours de l'année 1934. Acta Adriatica 1(9):1-126

Nitzschia seriata
Gilbert, J.Y., and W.E. Allen, 1943
The phytoplankton of the Gulf of California obtained by the "E.W. Scripps" in 1939 and 1940. J. Mar. Res. V(2):89-110, figs.30-31.

Nitzschia seriata
Forti, A., 1922
Ricerche sulla flora pelagica (fitoplancton) di Quarto dei Mille. Mem. R. Com. Talass. Ital. 97:248 pp., 13 pls.

Nitzschia seriata
Ghazzawi, F.M., 1939
Plankton of the Egyptian waters. A study of the Suez Canal Plankton. (A) Phytoplankton. Preliminary Report 83 pp. Notes and Memoires, Min. Commerce-Industry, Egypt, Hydrobiol. & Fish. 65 figs.

Nitzschia seriata
Gran, H.W., 1908
Diatomeen. Nordisches Plankton, Botanischer Teil pp. XIX.1-XIX 146; 178 text figs.

Nitzschia seriata
Gran, H.H., and T. Braarud, 1935
A quantitative study of the phytoplankton in the Bay of Fundy and the Gulf of Maine (including observations on hydrography, chemistry, and turbidity). J. Biol. Bd., Canada, 1(5):279-467, 69 text figs.

Nitzschia seriata
Gran, H. H. and E. C. Angst, 1931
Plankton diatoms of Puget Sound. Publ. Puget Sound Biol. Sta. 7:417-519, 95 text figs.

Nitzschia seriata
Hendy, N. Ingram, 1964
An introductory account of the smaller algae of British coastal waters. V. Bacillariophyceae (Diatoms). Her Majesty's Stationary Office, 317 pp., 45 pls.

Nitzschia seriata
Hendey, N.I., 1937
The plankton diatoms of the southern seas. Discovery Repts. 16:151-364, pls.6-13.

Nitzschia seriata
Herrera, Juan, y Ramón Margalef, 1963
Hidrografía y fitoplancton de la costa comprendida entre Castellón y la desembocadura del Ebro, de julio de 1960 a junio de 1961. Inv. Pesq., Barcelona, 24:33-112.

Nitzschia seriata
Iselin, C., 1930
A report on the coastal waters of Labrador based on explorations of the "Chance" during the summer of 1926. Proc. Am. Acad. Arts Sci., 66(1):1-37, 14 text figs.

Nitzschia seriata
Jørgensen, E., 1905
B.Protistplankton and the diatoms in bottom samples. Hydrographical and biological investigations in Norwegian fjords. Bergens Mus. Skr. 7: 49-225.

Nitzschia seriata
Kokubo, S., 1952
Results of the observations on the plankton and oceanography of Mutsu Bay during 1950, reference being made also to the period 1946-1950. Bull Mar.Biol.Sta., Asamushi 5(1/4): 1-54, 3 tables,(fold-in), 1 fold-in.

Nitzschia seriata
Lillick, L.C., 1940
Phytoplankton and planktonic protozoa of the offshore waters of the Gulf of Maine. Pt.II. Qualitative Composition of the Planktonic Flora. Trans. Am. Phil. Soc., n.s., 31(3):193-237, 13 text figs.

Nitzschia seriata
Lillick, L.C., 1938
Preliminary report of the phytoplankton of the Gulf of Maine. Am. Mid. Nat. 20(3):624-640, 1 text fig, 37 tables.

Nitzschia seriata
Lillick, L.C., 1937
Seasonal studies of the phytoplankton off Woods Hole, Massachusetts. Biol. Bull. LXXIII (3):488-503, 3 text figs.

Nitzschia seriata
Mangin, L., 1915
Phytoplancton de L'Antartique. Deuxieme Exped. Ant. Francaise (1908-1910), 95 pp., 3 pls., 58 text figs.

Nitzschia seriata
Mangin, M. L., 1912
Phytoplancton de la croisière du "René" dans l'Atlantique (Septembre 1908). Ann. Inst. Ocean, n.s., 4(1):1-66, 2 pls., 41 text figs., 2 tables.

Nitzschia seriata
Marukawa, H., 1921
Plankton lists and some new species of copepods, from the northern waters of Japan. Bull. Inst. Ocean., No.384, 15 pp., 3 pls., 1 chart. Monaco

Nitzschia seriata
Meunier, A., 1915
Microplancton de la Mer Flamande. 2. Diatomées (excepté le genre Chaetoceros). Mem. Mus. Roy. Hist. Nat., Belgique, 7(3):1-118, Pls. VIII-XIV.

Nitzschia seriata
Morse, D.C., 1947
Some observations on seasonal variations in plankton population Patuxent River, Maryland 1943-1945. Bd. Nat. Res., Publ. No.65, Chesapeake Biol. Lab., 31, 3 figs.

Nitzschia seriata
Paiva Carvalho, J., 1950
O plancton do Rio Maria Rodrigues (Cananeis). 1. Diatomaceas e Dinoflagelados. Bol. Inst. Paulista Oceanogr. 1(1); 27-43, 2 fold-in tables, 2 figs.

Nitzschia seriata
Pavillard, J., 1925
Bacillariales. Rept. on the Danish Oceanogr. Exped., 1908-10 to the Mediterranean and adj. seas. Vo..II., Biol. J4:72 pp., 116 text figs.

Nitzschia seriata
Pavillard, J., 1905
Recherches sur la flore pelagique (Phytoplankton) de l'Etang de Thau. Theses presentees a la Fac. Sci., Paris, 116 pp., 3 pls.

Nitzschia seriata
Petrova, V.I. and Ch. Skolka, 1964.
The mass development of Nitzschia seriata Cl. in the waters of the Black Sea. (In Russian). Rev. Roumaine Biol., Ser. Botan., 9(1):51-65.

Nitzschia seriata
Pettersson, H., 1934
Scattering and extinction of light in sea-water. Meddelanden från Göteborgs Högskolas Oceanografiska Institut. 9 (Göteborgs Kungl. vetenskaps-och vitterhets-samhälles Handlingar. Femte Foljden, Ser.B, 4(4)):1-16.

Nitzschia seriata
Phifer, L.D. (undated)
The occurrence and distribution of plankton diatoms in Bering Sea and Bering Strait, July 26-August 24, 1934. Report of Oceanographic cruise of U.S. Coast Guard Cutter Chelan 1934, Part II(A):1-44 (mimeographed) plus fig.1 (after Pt.B)

Nitzschia seriata
Robinson, G.A., 1965.
Continuous plankton records: contribution towards a plankton atlas of the North Atlantic and the North Sea. IX. Seasonal cycles of phytoplankton. Bulls. Mar. Ecol.,Scottish Mar. Biol.Assoc., 6(4):104-122, pls.26-61.

Summary p.120-121.

Nitzschia seriata
Schodduyn, M., 1926
Observations faites dans la baie d'Ambleteuse (Pas de Calais). Bull. Inst. Ocean., Monaco, No. 482: 64 pp.

Nitzschia seriata
Smayda, Theodore J., and Brenda J. Boleyn, 1965.
Experimental observations on the flotation of marine diatoms. 1. Thalassiosira cf. nana, Thalassiosira rotula and Nitzschia seriata. Limnol. Oceanogr., 10(4):499-509.

Nitzschia seriata
Sousa e Silva, E., 1949
Diatomaceas e Dinoflagelados de Baia de Cascais. Portugalise Acta Biol., Volume: Julio Henriques, Ser. B: 300-383, 9 pls, 2 fold-in tables.

Nitzschia seriata
Sousa e Silva, E., and J. Dos Santos-Pinto, 1948
O Plancton da Baia de S. Martinho do Porto. 1. Diatomaceas e Dinoflagelados. Bol. Soc. Portuguese de Ciencias Naturais, 16(2):134-187, 6 pls. (Trav. Sta. Biol. Mar. de Lisbonne No. 52).

Nitzschia seriata
Stemann-Nielsen, Einar, 1951
The marine vegetation of the Isefjord. A study on ecology and production. Medd. Komm. Danmarks Fiskeri-og Havundersøgelser. Ser. Plankton. 5(4); 114pp., 46 text figs.

Nitzschia seriata
Štirn Jože, 1965.
The importance of Nitzschia seriata Clev. in the northern Adriatic phytoplankton. Rapp. Proc.-verb. Réun. Comm. int explor. scient. Mer Méditerranée.
Rapp. Proc.-verb. Réun. Comm. int. Explor. scient. Mer Méditerranée, 19(3): 577-580

Nitzschia seriata
Uyeno, Fukuzo, 1961
Oceanographical and ecological studies on primary production of the sea, with special references to relationship between diatom production and temperature and chlorinity of water. Rept., Fac. Fish., Pref. Univ., Mie, 4(1): 1-64.

Nitzschia sibula n.sp.
Giffen, Malcolm H. 1973.
Diatoms of the marine littoral of Steenberg's Cove in St. Helena Bay, Cape Province, South Africa.
Botanica marina 16 (1): 32-48.

Nitzschia sicula
Hasle, Grethe Rytter, 1964.
Nitzschia and Fragilariopsis species studied in the light and electron microscopes. 1. Some marine species of the groups Nitzschiella and Lanceolatae. Skrifter, Det Norske Videnskaps-Akad., Oslo, 1. Math.-Naturv. Kl., n.s., No. 16:48 pp.

Nitzschia sicula
Hasle, Grethe Rytter, 1960
Phytoplankton and ciliate species from the Tropical Pacific. Skr. Norske Videnskaps-Akad., Oslo, 1, Mat.-Nat. Kl., 1960(2): 1-50.

Nitzschia sicula
Rampi, L., 1940
Diatomee del Mare Adriatico. Nuovo Giornale Botanico Italiano, n.s., 47:559-608.

Nitzschia sigma & varieties
Hendy, N. Ingram, 1964
An introductory account of the smaller algae of British coastal waters. V. Bacillariophyceae (Diatoms). Her Majesty's Stationary Office, 317 pp., 45 pls.

Nitzschia sigma
Hendey, N.I., 1951
Littoral diatoms of Chicester Harbour with special reference to fouling. J.Roy. Microscop. Soc. 71(1): 1-86, 18 pls.

Nitzschia sigma
Hustedt, F. and A.A. Aleem, 1951
Littoral diatoms from the Salstone near Plymouth. JMBA 30(1): 177.196.

Nitzschia sigma
Jorgensen, E., 1900
Protophyten und Protozoën im Plankton aus der Norwegischen Westkerste. Bergens Mus. Aarb. 1899(6): 95 pp., 5 pls., 83 tables.

Nitzschia sigma
Levander, K.M., 1947
Plankton gesammelt in den Jahren 1899-1910 an den Küsten Finnlands. Finnländische Hydrographisch-Biologische Untersuchunger (aus dem Wasserbiologischen Laboratorien der Societas Scientiarum Fennica) No.11:40 pp., 6 diagrams, 13 pls., tables.

Nitzschia sigma
Mann, A., 1907
Report on the diatoms of the Albatross voyages in the Pacific Ocean, 1888-1904. Contrib. U.S. Nat. Herb. 10(5):221-419, Pls. XLIV-LIV.

Nitzschia sigma
Mann, A., 1893
List of Diatomaceae from a deep-sea dredging in the Atlantic Ocean off Delaware Bay by the U.S. Fish Commission Steamer Albatross. Proc. U.S. Nat. Mus. 16:303-312.

Nitzschia sigma
Rampi, L., 1940
Diatomee del Mare Adriatico. Nuovo Giornale Botanico Italiano, n.s., 47:559-608.

Nitzschia sigma
Rumkówna, A., 1948
List of the phytoplankton species occurring in the superficial water layers in the Gulf of Gdańsk. Bull. Lab. mar., Gdynia, No. 4: 139-141 with tables in back.

Nitzschia sigma
Schodduyn, M., 1926
Observations faites dans la baie d'Ambleteuse (Pas de Calais). Bull. Inst. Ocean., Monaco, No. 482: 64 pp.

Nitzschia sigma
Zanon, V., 1948
Diatomee marini di Sardegna e Pugillo di Alghe Marine della stressa. Boll. Pesca, Piscitutura e Idrobiologia, Anno 24, ns. 3(2): 202-244, 27 figs. on 1 pl.

Nitzschia sigma
Zanon, D. V., 1949
Diatomee di Buenos Aires (Argentina) Atti Accad. Naz. Lincei, Memorie, Cl. Sci. fis., mat. e. nat., ser. 7, 11(3):59-151, 2 pls.

Nitzschia sigma curvula
Morse, D.C., 1947
Some observations on seasonal variations in plankton population Patuxant River, Maryland 1943-1945. Bd. Nat. Res., Publ. No.65, Chesapeake Biol. Lab., 31, 3 figs.

Nitzschia sigma var. indica
Allen, W.E., and E.E. Cupp, 1935
Plankton diatoms of the Java Sea. Annales du Jardin Botanique de Buitenzorg XLIV (2):101-174, figs.1-127.
(drawings of all species mentioned)

Nitzschia sigma var. intercedens
Allen, W.E., and E.E. Cupp, 1935
Plankton diatoms of the Java Sea. Annales du Jardin Botanique de Buitenzorg XLIV (2):101-174, figs.1-127.
(drawings of all species mentioned)

Nitzschia sigma intercedens
Takano, Hideaki, 1962
Notes on epiphytic diatoms upon sea-weeds from Japan. J. Oceanogr. Soc., Japan, 18(1):29-33.

Nitschia sigmoidea
Croasdell, G.C., 1964.
Nitzschia sigmoidea. J. Quekett Microsc. Club, 29(10):250.
freshwater parasitized by Amphora minutissima

Nitzschia sigmoidea
Geitler, L., 1949-1951.
Die Auxosporenbildung von Nitzschia sigmoidea und die Geschlechtsbestimmung bei den Diatomeen. Portugales Acta Biol., Vol. R. Goldschmidt, Ser. A:79-97.

Nitzschia sigmoidea
Rumkówna, A., 1948
List of the phytoplankton species occurring in the superficial water layers in the Gulf of Gdańsk. Bull. Lab. mar., Gdynia, No. 4: 139-141 with tables in back.

Nitzschia sigmoidea
Zanon, D. V., 1949
Diatomee di Buenos Aires (Argentina) Atti Accad. Naz. Lincei, Memorie, Cl. Sci. fis., mat. e. nat., ser. 7, 11(3):59-151, 2 pls.

Nitzschia socialis & varieties
Hendy, N. Ingram, 1964
An introductory account of the smaller algae of British coastal waters. V. Bacillariophyceae (Diatoms). Her Majesty's Stationary Office, 317 pp., 45 pls.

Nitzschia socialis
Hustedt, F. and A.A. Aleem, 1951
Littoral diatoms from the Salstone near Plymouth. JMBA 30(1): 177.196.

Nitzschia socialis
Rampi, L., 1942
Ricerche sul fitoplancton del Mare Ligure 6. Le diatomee delle acque di Sanremo. Nuovo Giornale Botanico Italiano, N.S., 49:252-268.

Nitzschia solida n.sp.
Frenguelli, Joaquin, y Hector A. Orlando, 1958. Diatomees y silicoflagelados del sector Antartico Sudamericano. Inst. Antartico Argentino, Publ., No. 5:191 pp.

Nitzschia spathulata
Hendy, N. Ingram, 1964
An introductory account of the smaller algae of British coastal waters. V. Bacillariophyceae (Diatoms). Her Majesty's Stationary Office, 317 pp., 45 pls.

Nitzschia spathulata
Jørgensen, E., 1905
B. Protistplankton and the diatoms in bottom samples. Hydrographical and biological investigations in Norwegian fjords. Bergens Mus. Skr. 7: 49-225.

Nitzschia spathulata
Sousa e Silva, E., 1949
Diatomaceas e Dinoflagelados de Baia de Cascais. Portugaliae Acta Biol. Volume: Julio Henriques, Ser. B: 300-383, 9 pls, 2 fold-in tables.

Nitzschia spathulata
Zanon, V., 1948
Diatomee marini di Sardegna e Pugillo di Alghe Marine della stressa. Boll. Pesca, Piscitutura e Idrobiologia, Anno 24, ns. 3(2): 202-244, 27 figs. on 1 pl.

Nitzschia spectabilis
Kokubo, S., and S. Sato, 1947
Plankterx in JG-San Gata. Physiol. and Ecol. (Japan) 1(4):1-16, 3 text figs., tables.

Nitzschia spectabilis
Zanon, D. V., 1949
Diatomee di Buenos Aires (Argentina) Atti Accad. Naz. Lincei, Memorie, Cl. Sci. fis., mat. e. nat., ser. 7, 11(3):59-151, 2 pls.

Nitzschia stagnorum
Zanon, D. V., 1949
Diatomee di Buenos Aires (Argentina) Atti Accad. Naz. Lincei, Memorie, Cl. Sci. fis., mat. e. nat., ser. 7, 11(3):59-151, 2 pls.

Nitzschia steenbergensis n.sp.
Giffen, Malcolm H. 1973.
Diatoms of the marine littoral of Steenberg's Cove in St. Helena Bay, Cape Province, South Africa.
Botanica Marina 16 (1): 32-48.

Nitschia stellata
Jouse, A.P., G.S. Koroleva, G.A. Nagaeva, 1962
Diatoms in the surface layer of sediment in the Indian sector of the Antarctic. Investigations of marine bottom sediments. (In Russian; English summary). Trudy Inst. Okeanol., Akad. Nauk. SSSR, 61: 19-92.

Nitschia stellata n.sp.
Manguin, E., 1957.
Premier inventaire des diatomees de la Terre Adélie Antarctique. Espèces nouvelles. Rev. Algologique, n.s., 3(3):111-134.

Nitzschia Stompsii n.sp.
Cholnoky, B.J., 1963.
Beiträge zur Kenntnis des marinen Litorals von Südafrika.
Botanica Marina, 5(2/3):38-83.

Nitzschia subarcuata n.sp.
Hasle, Grethe Rytter, 1964.
Nitzschia and Fragilariopsis species studied in the light and electron microscopes. 1. Some marine species of the groups Nitzschiella and Lanceolatae.
Skrifter, Det Norske Videnskaps-Akad., Oslo, 1. Mat.-Naturv. Kl., n.s., No. 16:48 pp.

Nitzschia subcurvata n.sp.
Hasle, Grethe Rytter, 1964.
Nitzschia and Fragilariopsis species studied in the light and electron microscopes. 1. Some marine species of the groups Nitzschiella and Lanceolatae.
Skrifter, Norske Videnskaps-Akad., Oslo, 1. Mat.-Naturw. Kl., n.s., No. 16:48 pp.

Nitschia subfrequens n.sp.
Simonsen, Reimer, 1960.
Neue Diatomeen aus der Ostsee. II.
Kieler Meeresf., 16(1):126-130.

Nitzschia subtilis
Zanon, D. V., 1949
Diatomee di Buenos Aires (Argentina)
Atti Accad. Naz. Lincei, Memorie, Cl. Sci. fis., mat. e. nat., ser. 7, 11(3):59-151, 2 pls.

Nitzschia tenuirostris
Hasle, Grethe Rytter, 1960
Phytoplankton and ciliate species from the Tropical Pacific.
Skr. Norske Videnskaps-Akad., Oslo, 1. Mat.-Nat. Kl., 1960(2): 1-50.

Nitzschia thermalis
Kokubo, S., and S. Sato., 1947
Plankters in Jū-San Gata. Physiol. and Ecol. (Japan) 1(4):1-16, 3 text figs., tables.

Nitzschia thermalis
Mann, A., 1893
List of Diatomaceae from a deep-sea dredging in the Atlantic Ocean off Delaware Bay by the U. S. Fish Commission Steamer Albatross.
Proc. U. S. Nat. Mus. 16:303-312.

Nitzschia themalis
Rumkówna, A., 1948
[List of the phytoplankton species occurring in the superficial water layers in the Gulf of Gdańsk] Bull. Lab. mar., Gdynia, No. 4: 139-141 with tables in back.

Nitzschia thermalis
Zanon, D. V., 1949
Diatomee di Buenos Aires (Argentina)
Atti Accad. Naz. Lincei, Memorie, Cl. Sci. fis., mat. e. nat., ser. 7, 11(3):59-151, 2 pls.

Nitzschia tryblionella
Hendy, N. Ingram, 1964
An introductory account of the smaller algae of British coastal waters. V. Bacillariophyceae (Diatoms).
Her Majesty's Stationary Office, 317 pp., 45 pls.

Nitzschia tryblionella
Zanon, D. V., 1949
Diatomee di Buenos Aires (Argentina)
Atti Accad. Naz. Lincei, Memorie, Cl. Sci. fis., mat. e. nat., ser. 7, 11(3):59-151, 2 pls.

Nitzschia tryblionella
Zanon, V., 1948
Diatomee marini di Sardegna e Pugillo di Alghe Marine della stressa. Boll. Pesca, Piscitutura e Idrobiologia, Anno 24, ns. 3(2): 202-244, 27 figs. on 1 pl.

Nitzschia tryblionella taxi levidensis
Iyengar, M.O.P. and G.Venkataraman, 1951.
The ecology and seasonal succession of the algae flora of the River Cooum at Madras with special reference to the Diatomaceae. J. Madras Univ. 21, Sect. B(1): 140-192, 1 pl of 4 figs., 11 text figs.

Nitzschia turgidula
Paasche, E. 1968.
Marine plankton algae grown with light-dark cycles. II. Ditylum brightwelli and Nitzschia turgidula.
Physiol. Plant. 21(4):66-77.

Nitzschia tryblionella & victoriae
Tsumura, K., 1956.
Diatomoj el la cirkaŭfoso de la restajo de la kastele de Odawara. J. Yokohama Municipal Univ., (C-14), No. 47:23 pp.

Nitzschia turgiduloides
Venrick, E.L. 1972.
Small-scale distributions of oceanic diatoms.
Fish. Bull. U.S. nat. mar. Fish. Serv. NOAA 70(2):363-372

Nitzschia valida
Rampi, L., 1940
Diatomee del Mare Adriatico. Nuovo Giornale Botanico Italiano, n.s., 47:559-608.

Nitzschia valida
Zanon, D. V., 1949
Diatomee di Buenos Aires (Argentina)
Atti Accad. Naz. Lincei, Memorie, Cl. Sci. fis., mat. e. nat., ser. 7, 11(3):59-151, 2 pls.

Nitzschia valida
Zanon, V., 1948
Diatomee marini di Sardegna e Pugillo di Alghe Marine della stressa. Boll. Pesca, Piscitutura e Idrobiologia, Anno 24, ns. 3(2): 202-244, 27 figs. on 1 pl.

Nitzschia vermicularis
Morse, D.C., 1947
Some observations on seasonal variations in plankton population Patuxent River, Maryland 1943-1945. Bd. Nat. Res., Publ. No.65, Chesapeake Biol. Lab., 31, 3 figs.

Nitzschia vermiculata
Castracane degli Antelminelli, F., 1886 n.sp.
1. Report on the Diatomaceae collected by H.M.S. Challenger during the years 1873-1876. Rept. Sci. Results, H.M.S. Challenger, Botany Vol. II, 178 pp., 30 pls.

Nitzschia vermicularis
Zanon, D. V., 1949
Diatomee di Buenos Aires (Argentina)
Atti Accad. Naz. Lincei, Memorie, Cl. Sci. fis., mat. e. nat., ser. 7, 11(3):59-151, 2 pls.

Nitzschia visurgis
Hendy, N. Ingram, 1964
An introductory account of the smaller algae of British coastal waters. V. Bacillariophyceae (Diatoms).
Her Majesty's Stationary Office, 317 pp., 45 pls.

Nitzschia vitrea
Iyengar, M.O.P. and G.Venkataraman, 1951.
The ecology and seasonal succession of the algae flora of the River Cooum at Madras with special reference to the Diatomaceae. J. Madras Univ. 21, Sect. B(1): 140-192, 1 pl of 4 figs., 11 text figs.

Nitzschia vitrea
Zanon, D. V., 1949
Diatomee di Buenos Aires (Argentina)
Atti Accad. Naz. Lincei, Memorie, Cl. Sci. fis., mat. e. nat., ser. 7, 11(3):59-151, 2 pls.

Nitzschia vitrea
Zanon, V., 1948
Diatomee marini di Sardegna e Pugillo di Alghe Marine della stressa. Boll. Pesca, Piscitutura e Idrobiologia, Anno 24, ns. 3(2): 202-244, 27 figs. on 1 pl.

Nitzschia vulpeculoides n.sp.
Giffen, Malcolm H. 1973.
Diatoms of the marine littoral of Steenberg's Cove in St. Helena Bay, Cape Province, South Africa.
Botanica Marina 16 (1): 32-48.

Nitzschia Weisflogii
Zanon, V., 1948
Diatomee marini di Sardegna e Pugillo di Alghe Marine della stressa. Boll. Pesca, Piscitutura e Idrobiologia, Anno 24, ns. 3(2): 202-244, 27 figs. on 1 pl.

Nitzschiella insignis
Politis, J., 1949
Diatomees marines de Bosphores et des ibes de la mer de Marmara. II Practica tou Hellenikou Hidrobiologikou Institutoutou 1929, Etoz 1929, 3(1):11-31.

Odontella expedita
Zanon, V., 1947.
Diatomée delle Isole Ponziane - materiali per una florula diatomologica del Mare Tirreno.
Bol. Pesca, Piscol. Idrobiol., n.s., 2(1):36-53, 1 pl., of 10 figs.

Odontotropis carinata
Cleve-Euler, A., 1951
Die Diatomeen von Schweden und Finnland. Kungl. Svenska Vetenskaps Akad. Handl., Fjärde Ser. 2(1): 161 pp., 6 pls.

Odontotropis cristata
Cleve-Euler, A., 1951
Die Diatomeen von Schweden und Finnland. Kungl. Svenska Vetenskaps Akad. Handl., Fjärde Ser. 2(1): 161 pp., 6 pls.

Oestrupia musca
Hendy, N. Ingram, 1964
An introductory account of the smaller algae of British coastal waters. V. Bacillariophyceae (Diatoms).
Her Majesty's Stationary Office, 317 pp., 45 pls.

Oestrupia Powelli
Andrade e Clovis Teixeira, M.H. de, 1957.
Contribuição para o conhecimento das diatomaceas do Brasil.
Bol. Inst. Ocean., Univ. Sao Paulo, 8(1/2):171-225, 10 pls.

Oestrupia powelli
Zanon, V., 1948
Diatomee marini di Sardegna e Pugillo di Alghe Marine della stressa. Boll. Pesca, Piscitutura e Idrobiologia, Anno 24, ns. 3(2): 202-244, 27 figs. on 1 pl.

Oestrupia supergradata
Hendy, N. Ingram, 1964
An introductory account of the smaller algae of British coastal waters. V. Bacillariophyceae (Diatoms).
Her Majesty's Stationary Office, 317 pp., 45 pls.

Okedenia inflexa
Hendy, N. Ingram, 1964
An introductory account of the smaller algae of British coastal waters. V. Bacillariophyceae (Diatoms).
Her Majesty's Stationary Office, 317 pp., 45 pls.

Omphalopelta japonica n.sp.
Castracane degli Antelminelli, F., 1886
1. Report on the Diatomaceae collected by H.M.S. Challenger during the years 1873-1876. Rept. Sci. Results, H.M.S. Challenger, Botany Vol. II, 178 pp., 30 pls.

Omphalopelta parada n.sp.
Castracane degli Antelminelli, F., 1886
1. Réport on the Diatomaceae collected by H.M.S. Challenger during the years 1873-1876. Rept. Sci. Results, H.M.S. Challenger, Botany Vol. II, 178 pp., 30 pls.

Omphalopelta shrubsoliana
Castracane degli Antelminelli, F., 1886 n.sp.
1. Réport on the Diatomaceae collected by H.M.S. Challenger during the years 1873-1876. Rept. Sci. Results, H.M.S. Challenger, Botany Vol. II, 178 pp., 30 pls.

Opephora
Desikachary, T.V., 1956(1957).
Electron microscope studies on diatoms. J.R. Microsc. Soc. (3)76(1/2):9-36.

Opephora marina
Hendy, N. Ingram, 1964
An introductory account of the smaller algae of British coastal waters. V. Bacillariophyceae (Diatoms).
Her Majesty's Stationary Office, 317 pp., 45 pls.

Opephora marina
Hustedt, F. and A.A. Aleem, 1951
Littoral diatoms from the Salstone near Plymouth. JMBA 30(1): 177.196.

Opephora marina
Zanon, V., 1948
Diatomee marini di Sardegna e Pugillo di Alghe Marine della stressa. Boll. Pesca, Piscicitura e Idrobiologia, Anno 24, ns. 3(2): 202-244, 27 figs. on 1 pl.

Opephora olseni
Hendy, N. Ingram, 1964
An introductory account of the smaller algae of British coastal waters. V. Bacillariophyceae (Diatoms).
Her Majesty's Stationary Office, 317 pp., 45 pls.

Opephora pacifica
Hendy, N. Ingram, 1964
An introductory account of the smaller algae of British coastal waters. V. Bacillariophyceae (Diatoms).
Her Majesty's Stationary Office, 317 pp., 45 pls.

Opephora schwartzii
Hendy, N. Ingram, 1964
An introductory account of the smaller algae of British coastal waters. V. Bacillariophyceae (Diatoms).
Her Majesty's Stationary Office, 317 pp., 45 pls.

Orthoneis aspera
Politis, J., 1949
Diatomees marines de Bosphores et des ibes de la mer de Marmara. II Practica tou Hellenikou Hidrobiologikou Institutoutou 1929, Etoz 1929, 3(1):11-31.

Orthoneis binotata
Politis, J., 1949
Diatomees marines de Bosphores et des ibes de la mer de Marmara. II Practica tou Hellenikou Hidrobiologikou Institutoutou 1929, Etoz 1929, 3(1):11-31.

Orthoneis crucicula
Grunow, A., 1877
1. New Diatoms from Honduras. Monthly Micros. Jour., 18:165-186, pls. CXCIII-CXCVI.

Orthoneis ovata
Politis, J., 1949
Diatomees marines de Bosphores et des ibes de la mer de Marmara. II Practica tou Hellenikou Hidrobiologikou Institutoutou 1929, Etoz 1929, 3(1):11-31.

Orthoneis splendida
Politis, J., 1949
Diatomees marines de Bosphores et des ibes de la mer de Marmara. II Practica tou Hellenikou Hidrobiologikou Institutoutou 1929, Etoz 1929, 3(1):11-31.

Palmeria herdmaniana
Marukawa, H., 1921
Plankton lists and some new species of copepods, from the northern waters of Japan. Bull. Inst. Ocean., No.384, 15 pp., 3 pls., 1 chart. Monaco

Paralia
Brunel, J., 1962
Le phytoplancton de la Baie de Chaleurs. Inst. Botan., Univ. Montréal, Contrib. No. 77: 365 pp., 66 pls.

Paralia ornata
Cleve-Euler, A., 1951
Die Diatomeen von Schweden und Finnland. Kungl. Svenska Vetenskaps Akad. Handl., Fjärde Ser. 2(1): 161 pp., 6 pls.

Paralia sulcata
Allen, W.E., and E.E. Cupp, 1935
Plankton diatoms of the Java Sea. Annales du Jardin Botanique de Buitenzorg XLIV (2):101-174, figs.1-127.
(drawings of all species mentioned)

Paralia sulcata
Cleve-Euler, A., 1951
Die Diatomeen von Schweden und Finnland. Kungl. Svenska Vetenskaps Akad. Handl., Fjärde Ser. 2(1): 161 pp., 6 pls.

Paralia sulcata
Dangeard, P., 1927
Phytoplankton de la croisière du "Sylvana". Ann. Inst. Ocean., Monaco, n.s., 4(8):286-401, 54 text figs. (fevrier-Juin 1913).

Paralia sulcata
Gilbert, J.Y., and W.E. Allen, 1943
The phytoplankton of the Gulf of California obtained by the "E.W. Scripps" in 1939 and 1940. J. Mar. Res. V(2):89-110, figs.30-31.

Paralia sulcata
Gran, H.H., 1908
Diatomeen. Nordisches Plankton, Botanischer Teil pp. XIX.1-XIX.146; 178 text figs.

Paralia sulcata
Gran, H. H. and E. C. Angst, 1931
Plankton diatoms of Puget Sound. Publ. Puget Sound Biol. Sta. 7:417-519, 95 text figs.

Paralia sulcata
Hendy, N. Ingram, 1964
An introductory account of the smaller algae of British coastal waters. V. Bacillariophyceae (Diatoms).
Her Majesty's Stationary Office, 317 pp., 45 pls.

Paralia sulcata
Hendey, N-Ingram, 1958 [1957(Publ. 1958)]
Marine diatoms from some West African Ports. J. R.Microsc. Soc. (3) 77(1/2): 28-85.

Paralia sulcata
Jørgensen, E., 1905
B.Protistplankton and the diatoms in bottom samples. Hydrographical and biological investigations in Norwegian fjords. Bergens Mus. Skr. 7: 49-225.

Paralia sulcata
Jorgensen, E., 1900
Protophyten und Protozoen im Plankton aus der Norwegischen Westkerste. Bergens Mus. Aarb. 1899(6): 95 pp., 5 pls., 83 tables.

Paralia sulcata
Meunier, A., 1915
Microplancton de la Mer Flamande. 2. Diatomées (exceptè le genre Chaetoceros). Mem. Mus. Roy. Hist. Nat., Belgique, 7(3):1-118, Pls. VIII-XIV.

Paralia sulcata
Paiva Carvalho, J., 1950
O plancton do Rio Maria Rodriques (Cananeis). 1. Diatomaceas e Dinoflagelados. Bol. Inst. Paulista Oceanogr. 1(1); 27-43, 2 fold-in tables, 2 figs.

Paralia sulcata
Pavillard, J., 1925
Bacillariales. Rept. on the Danish Oceangr. Exped., 1908-10 to the Mediterranean and adj. seas. Vol.II., Biol. J4:72 pp., 116 text figs.

Paralia sulcata
Pavillard, J., 1905
Recherches sur la flore pelagique (Phytoplankton) de l'Etang de Thau. Theses presentees a la Fac. Sci., Paris, 116 pp., 3 pls.

Paralia sulcata
Phifer, L.D. (undated)
The occurrence and distribution of plankton diatoms in Bering Sea and Bering Strait, July 26-August 24, 1934. Report of Oceanographic cruise of U.S. Coast Guard Cutter Chelan 1934, Part II(A):1-44 (mimeographed) plus fig.1 (after Pt.B)

Paralia sulcata
Schodduyn, M., 1926
Observations faites dans la baie d'Ambleteuse (Pas de Calais). Bull. Inst. Ocean., Monaco, No. 482: 64 pp.

Paralia sulcata
Sousa e Silva, E., 1949
Diatomaceas e Dinoflagelados de Baia de Cascais. Portugaliae Acta Biol., Volume: Julio Henriques, Ser. B: 300-383 9 pls, 2 fold-in tables.

Paralia sulcata
Sousa e Silva, E., and J. Dos Santos-Pinto, 1948
O Plancton da Baia de S. Martinho do Porto. 1. Diatomaceas e Dinoflagelados. Bol. Soc. Portuguese de Ciencias Naturais, 16(2):134-187, 6 pls. (Trav. Sta. Biol. Mar. de Lisbonne No. 52).

Paralia sulcata
Steemann-Nielsen, Einar, 1951
The marine vegetation of the Isefjord. A study on ecology and production. Medd. Komm. Danmarks Fiskeri-og Havundersøgelser. Ser. Plankton. 5(4); 114pp., 46 text figs.

Peragallia
Okamura, K., 1907
Some Chaetoceros and Peragallia of Japan. Bot. Mag., Tokyo, 21:

Peronia fibula
Ross, R., 1956. Notulae Diatomologicae. Ann. Mag. Nat. Hist., (12) 9:76-80.

Phaeodactylum spp.
Adams, M.N.E., and J.E.G. Raymont, 1958.
Studies on the mass culture of Phaeodactylum. Ann. Rept., Challenger Soc., 1958, 3(10):

Phaeodactylum tricornutum
Ansell, Alan D., J. Coughlan, K.F. Lander and F.A. Loosmore, 1964.
Studies on the mass culture of Phaeodactylum. IV. Production and nutrient utilization in outdoor mass culture.
Limnology and Oceanography, 9(3):334-342.

Phaeodactylum tricornutum
Ansell, A.D., J.E.G. Raymont and K.F. Lander, 1963
Studies on the mass culture of Phaeodactylum. III. Small-scale experiments.
Limnol. and Oceanogr., 8(2):207-213.

Phaeodactylum tricornutum
Ansell, A.D., J.E.G. Raymont, K.F. Lander, E. Crowley and P. Shackle, 1963
Studies on the mass culture of Phaeodactylum. II. The growth of Phaeodactylum and other species in outdoor tanks.
Limnol. and Oceanogr., 8(2):184-206.

Phaeodactylum tricornutum
Berland, Brigitte, 1966.
Contribution à l'étude des cultures de diatomees marines.
Recl. Trav.Stn.mar., Endoume, 40(56):3-82.

Phaeodactylum tricornutum
Bernhard, M., L. Rampi and A. Zattera 1971
First trophic level of the food chain.
CNEN Rept. RT/Bio (70)-11, M. Bernhard, editor: 23-40

Phaeodactylum tricorrutum
Besnier, Vincent, 1969.
Adion des ultrasous sur l'extraction des acides aminés libres d'une algae planctonique.
C.r.hebd. Séanc. Acad. Sci., Paris, (1) 268(11): 1505-1507.

Phaeodactylum tricornutum
Ceccaldi, Hubert J., et Brigitte Berland, 1964.
Contribution à l'étude de dosages quantitatifs du plancton. 2. Lyophilisation ou filtration sur filtres "millipore", comparaison de la solubilité des pigments photosynthetiques de la diatomée Phaeodactylum tricornutum (Bohlin) par quelques solvants organiques à diverses concentrations.
Rec. Trav. Sta. Mar., Endoume, 51(35):17-42.

Phaeodactylum tricornutum
Chapman, G. and A.C. Rae, 1969.
Excretion of photosynthate by a benthic diatom.
Marine Biol., 3(4): 341-351.

Phaeodactylum tricornutum
Coughlin, John, 1962
Chain formation by Phaeodactylum.
Nature, 195(4843):831-832.

Phaeodactylum tricornutum
Darley, W. Marshall, 1968.
Deoxyribonucleic acid content of the three cell types of Phaeodactylum tricornutum Bohlin.
J. Phycol., 4(3):219-220.

Phaeodactylum tricornutum
French, C. Stacy, 1967.
Changes with age in the absorption spectrum of chlorophyll a in a diatom.
Archiv für Mikrobiol., 59: 93-103.

Phaeodactylum tricornutum
Griffiths, D.J., 1973
Factors affecting the photosynthetic capacity of laboratory cultures of the diatom Phaeodactylum tricornutum.
Mar. Biol. 21(2):91-97.

Phaeodactylum tricornutum
Hayward, J., 1970.
Studies on the growth of Phaeodactylum tricornutum. VI. The relationship to sodium, potassium, calcium and magnesium.
J. mar. biol. Ass. U.K. 50(2): 293-299.

Phaeodactylum tricornutum
Hayward, J., 1969.
Studies on the growth of Phaeodactylum tricornutum. V. The relationship to iron, manganese and zinc.
J. mar. biol. Ass., U.K., 49(2): 439-446.

Phaeodactylum tricornutum
Hayward, J., 1968.
Studies on the growth of Phaeodactylum tricornutum. IV. Comparison of different isolates.
J.mar.biol.Ass.,U.K., 48(3):657-666.

Phaeodactylum tricornutum
Hayward, J., 1968.
Studies on the growth of Phaeodactylum tricornutum. III. The effect of iron on growth.
J. mar. biol. Ass., U.K., 48(2):295-302.

Phaeodactylum tricornutum
Hendy, N. Ingram, 1964
An introductory account of the smaller algae of British coastal waters. V. Bacillariophyceae (Diatoms).
Her Majesty's Stationary Office, 317 pp., 45 pls.

Phaeodactylum tricornutum
Kuenzler, Edward J., and Bostwick H. Ketchum, 1962
Rate of phosphorus uptake by Phaeodactylum tricornutum.
Biol. Bull., 123(1):134-145.

Phaeodactylum tricornutum
Lacaze, J.C., 1969.
Effets d'une pollution du type "Torrey Canyon" sur l'algue unicellulaire marine. Phaeodactylum tricomutum.
Rev. intern. Océanogr.Méd., 13-14:157-179.

Phaeodactylum tricornutum
Lacaze, Jean-Claude, 1967.
Etude de la croissance d'une algue planctonique en presence d'un detergent utilisé pour la destruction des nappes de petrole en mer.
C.r. hebd. Séanc. Acad. Sci., Paris, (D)265(20):1489-1491.

Phaeodactylum tricornutum
Mann, James E., and Jack Myers, 1968.
On pigments growth and photosynthesis of Phaeodactylum tricornutum.
J. Phycol., 4(4):349-355.

Phaeodactylum tricornutum
Lewin, Joyce, 1966.
Silicon metabolism in diatoms. V. Germanium dioxide, a specific inhibitor of diatom growth.
Phycologia, Int.Phycol.Soc., 6(1);1-12.

Phaeodactylum tricornutum
Lewin, J.C., 1958.
The taxonomic position of Phaeodactylum tricornutum. J. Gen. Microbiol., 18(2):427-432.

Phaeodactylum tricornutum
Lewin, J.C., R.A. Lewin and D.E. Philpott, 1958.
Observations on Phaeodactylum tricornutum.
J. Gen. Microbiol., 18(2):418-426.

Phaeodactylum tricornutum
Parsons, T.R., 1961
On the pigment composition of eleven species of marine phytoplankton.
J. Fish. Res. Bd., Canada, 18(6):1017-1025.

Phaeodactylum tricornutum
Parsons, T.R., K. Stephens and J.D.H. Strickland, 1961
On the chemical composition of eleven species of marine phytoplankters.
J. Fish. Res. Bd., Canada, 18(6):1001-1016.

Phaeodactylum tricornutum
Prakash, A., Liv Skoglund, Britt Rystad and Anne Jensen 1973
Growth and cell-size distribution of marine planktonic algae in batch and dialysis cultures.
J. Fish. Res. Bd. Can. 30(2):143-155

Phaeodactylum tricornutum
Takahashi, Masayuki, Sooji Shimura, Yukuya Yamaguchi and Yoshihiko Fujita 1971.
Photo-inhibition of phytoplankton photosynthesis as a function of exposure time.
J. oceanogr. Soc. Japan. 27(2):43-50

Phaeodactylum tricornutum
Ukeles, R., 1961.
The effect of temperature on the growth and survival of several marine algal species.
Biol. Bull., 120(2):255-264.

Phaeodactylum tricornutum
Uno, Shiroh, 1971.
Turbidometric continuous culture of phytoplankton constructions of the apparatus and experiments on the daily periodicity in photosynthetic activity of Phaeodactylum tricornutum and Skeletonema costatum. Bull. Plankt. Soc. Japan 18(1): 14-27.

Phaeodactylum tricornutum
Zattera, Antonio, e Michael Bernhard 1969.
L'importanza dello stato chimico-fisico degli elementi per l'accumulo negli organismi marini. II Accumulo di zinco stabile e radioattivo in Phaeodactylum tricornutum.
Pubbl. Staz. Zool. Napoli. 37 (2 Suppl.): 386-399.

Pinnularia sp.
Brunel, J., 1962
Le phytoplancton de la Baie de Chaleurs.
Inst. Botan., Univ. Montréal, Contrib. No. 77: 365 pp., 66 pls.

Pinnularia (?) sp.
Brunel, Jules, 1962
Le phytoplancton de la Baie des Chaleurs.
Contrib. Ministère de la Chasse et des Pêcheries, Province de Québec, No. 91: 365 pp.

Pinnularia
Desikachary, T.V., 1956(1957).
Electron microscope studies on diatoms.
J.R. Microsc. Soc. (3)76(1/2):9-36.

Pinnularia sp.
Lillick, L.C., 1940
Phytoplankton and planktonic protozoa of the offshore waters of the Gulf of Maine. Pt.II. Qualitative Composition of the Planktonic Flora. Trans. Am. Phil. Soc., n.s., 31(3):193-237, 13 text figs.

Lillick, L.C., 1938 Pinnularia sp.
Preliminary report of the phytoplankton of the Gulf of Maine. Am. Mid. Nat. 20(3):624-640, 1 text figs 37 tables.

Pinnularia acrosphaeria
Zanon, D. V., 1949
Diatomee di Buenos Aires (Argentina)
Atti Accad. Naz. Lincei, Memorie, Cl. Sci. fis., mat. e. nat., ser. 7, 11(3):59-151, 2 pls.

Pinnularia acrosphaerica var. turgidula
Tsumura, K., 1956.
Diatomoj el la cirkaŭfoso de la restaĵo de la kastelo de Odawara. J. Yokohama Municipal Univ., (C-14), No. 47:23 pp.

Pinnularia ambigua
Hendy, N. Ingram, 1964

An introductory account of the smaller algae of British coastal waters. V. Bacillariophyceae (Diatoms).
Her Majesty's Stationary Office, 317 pp., 45 pls.

Pinnularia ambigua
Hustedt, F. and A.A. Aleem, 1951
Littoral diatoms from the Salstone near Plymouth. JMBA 30(1): 177.196.

Pinnularia borealis
Tsumura, K., 1956.
Diatomoj el la cirkaŭfoso de la restaĵo de la kastelo de Odawara. J. Yokohama Municipal Univ., (C-14), No. 47:23 pp.

Pinnularia Borealis
Zanon, D. V., 1949
Diatomee di Buenos Aires (Argentina)
Atti Accad. Naz. Lincei, Memorie, Cl. Sci. fis., mat. e. nat., ser. 7, 11(3):59-151, 2 pls.

Pinnularia borealis
Zanon, V., 1948
Diatomee marini di Sardegna e Pugillo di Alghe Marine della stressa. Boll. Pesca, Piscitutura e Idrobiologia, Anno 24, ns. 3(2): 202-244, 27 figs. on 1 pl.

Pinnularia brevicostata
Zanon, D. V., 1949
Diatomee di Buenos Aires (Argentina)
Atti Accad. Naz. Lincei, Memorie, Cl. Sci. fis., mat. e. nat., ser. 7, 11(3):59-151, 2 pls.

Pinnularia cardinalis
Zanon, D. V., 1949
Diatomee di Buenos Aires (Argentina)
Atti Accad. Naz. Lincei, Memorie, Cl. Sci. fis., mat. e. nat., ser. 7, 11(3):59-151, 2 pls.

Pinnularia criophila n.sp.
Castracane degli Antelminelli, F., 1886
1. Report on the Diatomaceae collected by H.M.S. Challenger during the years 1873-1876. Rept. Sci. Results, H.M.S. Challenger, Botany Vol. II, 178 pp., 30 pls.

Pinnularia cruciformis
Hendy, N. Ingram, 1964

An introductory account of the smaller algae of British coastal waters. V. Bacillariophyceae (Diatoms).
Her Majesty's Stationary Office, 317 pp., 45 pls.

Pinnularia dactylus
Zanon, D. V., 1949
Diatomee di Buenos Aires (Argentina)
Atti Accad. Naz. Lincei, Memorie, Cl. Sci. fis., mat. e. nat., ser. 7, 11(3):59-151, 2 pls.

Pinnularia De Benedettii n.sp.
Frenguelli, Joaquin, y Hector A. Orlando, 1958.
Diatomeas y silicoflagelados del sector Antartico Sudamericano.
Inst. Antartico Argentino, Publ., No. 5:191 pp.

Pinnularia divergens
Zanon, D. V., 1949
Diatomee di Buenos Aires (Argentina)
Atti Accad. Naz. Lincei, Memorie, Cl. Sci. fis., mat. e. nat., ser. 7, 11(3):59-151, 2 pls.

Pinnularia Frenguellii n.sp.
Zanon, D. V., 1949
Diatomee di Buenos Aires (Argentina)
Atti Accad. Naz. Lincei, Memorie, Cl. Sci. fis., mat. e. nat., ser. 7, 11(3):59-151, 2 pls.

Pinnularia fritschii n.sp.
Salah, M.M., 1952(1953).
XII. Diatoms from Blakeney Point, Norfolk. New species and new records for Great Britain.
J.R. Microsc. Soc., Ser. 3, 72(3):155-169, 3 pls.

Pinnularia gentilis
Tsumura, K., 1956.
Diatomoj el la cirkaŭfoso de la restaĵo de la kastelo de Odawara. J. Yokohama Municipal Univ., (C-14), No. 47:23 pp.

Pinnularia gentilis
Zanon, D. V., 1949
Diatomee di Buenos Aires (Argentina)
Atti Accad. Naz. Lincei, Memorie, Cl. Sci. fis., mat. e. nat., ser. 7, 11(3):59-151, 2 pls.

Pinnularia gibba
Tsumura, K., 1956.
Diatomoj el la cirkaŭfoso de la restaĵo de la kastelo de Odawara. J. Yokohama Municipal Univ., (C-14), No. 47:23 pp.

Pinnularia gibba
Zanon, D. V., 1949
Diatomee di Buenos Aires (Argentina)
Atti Accad. Naz. Lincei, Memorie, Cl. Sci. fis., mat. e. nat., ser. 7, 11(3):59-151, 2 pls.

Pinnularia gibba
Zanon, V., 1948
Diatomee marini di Sardegna e Pugillo di Alghe Marine della stressa. Boll. Pesca, Piscitutura e Idrobiologia, Anno 24, ns. 3(2): 202-244, 27 figs. on 1 pl.

Pinnularia Hartleyana
Zanon, D. V., 1949
Diatomee di Buenos Aires (Argentina)
Atti Accad. Naz. Lincei, Memorie, Cl. Sci. fis., mat. e. nat., ser. 7, 11(3):59-151, 2 pls.

Pinnularia interrupta
Zanon, D. V., 1949
Diatomee di Buenos Aires (Argentina)
Atti Accad. Naz. Lincei, Memorie, Cl. Sci. fis., mat. e. nat., ser. 7, 11(3):59-151, 2 pls.

Pinnularia interrupta genuina
Iyengar, M.O.P. and G.Venkataraman, 1951.
The ecology and seasonal succession of the algae flora of the River Cooum at Madras with special reference to the Diatomaceae. J. Madras Univ. 21, Sect. B(1): 140-192, 1 pl of 4 figs., 11 text figs.

Pinnularia lanceolata
Boden, B. P., 1949.
The diatoms collected by the U.S.S. CACOPAN in the Antarctic in 1947. J. Mar. Res. 8(1):6-13, 3 textfigs.

Pinnularia lanceolata
Boden, Brian, 1948
Marine plankton diatoms on operation HIGHJUMP in: Some oceanographic observations on operation HIGHJUMP. By R.S. Dietz. USNEL Rept. No.55, 97 pp., 41 figs. 7 July 1948.

Pinnularia latevittata
Zanon, D. V., 1949
Diatomee di Buenos Aires (Argentina)
Atti Accad. Naz. Lincei, Memorie, Cl. Sci. fis., mat. e. nat., ser. 7, 11(3):59-151, 2 pls.

Pinnularia leptosoma
Tsumura, K., 1956.
Diatomoj el la cirkaŭfoso de la restaĵo de la kastelo de Odawara. J. Yokohama Municipal Univ., (C-14), No. 47:23 pp.

Pinnularia maior
Zanon, D. V., 1949
Diatomee di Buenos Aires (Argentina)
Atti Accad. Naz. Lincei, Memorie, Cl. Sci. fis., mat. e. nat., ser. 7, 11(3):59-151, 2 pls.

Pinnularia microstauron
Zanon, D. V., 1949
Diatomee di Buenos Aires (Argentina)
Atti Accad. Naz. Lincei, Memorie, Cl. Sci. fis., mat. e. nat., ser. 7, 11(3):59-151, 2 pls.

Pinnularia microstauron
Zanon, V., 1948
Diatomee marini di Sardegna e Pugillo di Alghe Marine della stressa. Boll. Pesca, Piscitutura e Idrobiologia, Anno 24, ns. 3(2): 202-244, 27 figs. on 1 pl.

Pinnularia nobilis
Zanon, D. V., 1949
Diatomee di Buenos Aires (Argentina)
Atti Accad. Naz. Lincei, Memorie, Cl. Sci. fis., mat. e. nat., ser. 7, 11(3):59-151, 2 pls.

Pinnularia rašana n.sp.
Castracane degli Antelminelli, F., 1886
1. Report on the Diatomaceae collected by H.M.S. Challenger during the years 1873-1876. Rept. Sci. Results, H.M.S. Challenger, Botany Vol. II, 178 pp., 30 pls.

Pinnularis quadratarea
Hendy, N. Ingram, 1964

An introductory account of the smaller algae of British coastal waters. V. Bacillariophyceae (Diatoms).
Her Majesty's Stationary Office, 317 pp., 45 pls.

Pinnularia rectangulata
Hendy, N. Ingram, 1964

An introductory account of the smaller algae of British coastal waters. V. Bacillariophyceae (Diatoms).
Her Majesty's Stationary Office, 317 pp., 45 pls.

Pinnularia Silvestrina n.sp.
Zanon, D. V., 1949
Diatomee di Buenos Aires (Argentina)
Atti Accad. Naz. Lincei, Memorie, Cl. Sci. fis., mat. e. nat., ser. 7, 11(3):59-151, 2 pls.

Pinnularia streptoraphe
Zanon, D. V., 1949
Diatomee di Buenos Aires (Argentina)
Atti Accad. Naz. Lincei, Memorie, Cl. Sci. fis., mat. e. nat., ser. 7, 11(3):59-151, 2 pls.

Pinnularia subcapitata
Ko-bayashi, Tsuyako, 1963
Variations on some pennate diatoms from Antarctica. 1.
Japan. Antarctic Res. Exped., 1956-1962. Sci. Repts., (E), No. 18:1-20. 16 pls.

Pinnularia trevelyana
Hendy, N. Ingram, 1964

An introductory account of the smaller algae of British coastal waters. V. Bacillariophyceae (Diatoms).
Her Majesty's Stationary Office, 317 pp., 45 pls.

Pinnularia Trevelyana
Zanon, V., 1948
Diatomee marini di Sardegna e Pugillo di Alghe Marine della stressa. Boll. Pesca, Piscitutura e Idrobiologia, Anno 24, ns. 3(2): 202-244, 27 figs. on 1 pl.

Pinnularia viridis
Barber, H.G., 1961.
A note on unusual diatom deformities.
J. Quekett Microsc. Club, (4) 5(13):365.

Pinnularia viridis?
Kokubo, S., and S. Sato, 1947
Plankters in Jū-San Gata. Physiol. and Ecol. (Japan) 1(4):1-16, 3 text figs., tables.

Pinnularia viridis
Tsumura, K., 1956.
Diatomoj el la cirkaŭfoso de la restajo de la kastelo de Odawara. J. Yokohama Municipal Univ., (C-14), No. 47:23 pp.

Pinnularia viridis
Zanon, D. V., 1949
Diatomee di Buenos Aires (Argentina) Atti Accad. Naz. Lincei, Memorie, Cl. Sci. fis., mat. e. nat., ser. 7, 11(3):59-151, 2 pls.

Plagiodiscus Martensianus
Grunow, A., 1877
1. New Diatoms from Honduras. Monthly Micros. Jour., 18:165-186, pls. CXCIII-CXCVI.

Plagiodiscus nervatus
Grunow, A., 1877
1. New Diatoms from Honduras. Monthly Micros. Jour., 18:165-186, pls. CXCIII-CXCVI.

Plagiogramma brockmanni
Hendy, N. Ingram, 1964
An introductory account of the smaller algae of British coastal waters. V. Bacillariophyceae (Diatoms).
Her Majesty's Stationary Office, 317 pp., 45 pls.

Plagiogramma elongatum
Mann, A., 1907
Report on the diatoms of the Albatross voyages in the Pacific Ocean, 1888-1904. Contrib. U. S. Nat. Herb. 10(5):221-419, Pls. XLIV-LIV.

Plagiogramma exiguum n.sp.
Hendey, N-Ingram, 1958 [1957(Publ. 1958)]
Marine diatoms from some West African Ports- J. R Microsc. Soc. (3) 77(1/2): 28-85.

Plagiogramma Grevilleanum n.sp.
Grunow, A., 1863
Ueber einige neue und ungenügend bekannte Arten und Gattungen von Diatomaceen. Verhandl. d. K.K. Zool. Bot. Gesellsch., Vienna, 13: 137-162, pl.4-5 (Pl. 13-14).

Plagiogramma interruptum
Rampi, L., 1940
Diatomee del Mare Adriatico. Nuovo Giornale Botanico Italiano, n.s., 47:559-608.

Plagiogramma interruptum
Zanon, V., 1948
Diatomee marini di Sardegna e Pugillo di Alghe Marine della stressa. Boll. Pesca, Piscitutura e Idrobiologia, Anno 24, ns. 3(2): 202-244, 27 figs. on 1 pl.

Plagiogramma margaritaceum n.sp
Castracane degli Antelminelli, F., 1886
1. Report on the Diatomaceae collected by H.M.S. Challenger during the years 1873-1876. Rept. Sci. Results, H.M.S. Challenger, Botany Vol. II, 178 pp., 30 pls.

Plagiogramma minimum
Hendy, N. Ingram, 1964
An introductory account of the smaller algae of British coastal waters. V. Bacillariophyceae (Diatoms).
Her Majesty's Stationary Office, 317 pp., 45 pls.

Plagiogramma parallelum
Hendy, N. Ingram, 1964
An introductory account of the smaller algae of British coastal waters. V. Bacillariophyceae (Diatoms).
Her Majesty's Stationary Office, 317 pp., 45 pls.

Plagiogramma pulchellum
Rampi, L., 1940
Diatomee del Mare Adriatico. Nuovo Giornale Botanico Italiano, n.s., 47:559-608.

Plagiogramma pulchellum
Zanon, V., 1948
Diatomee marini di Sardegna e Pugillo di Alghe Marine della stressa. Boll. Pesca, Piscitutura e Idrobiologia, Anno 24, ns. 3(2): 202-244, 27 figs. on 1 pl.

Plagiogramma pulchellum intermedia n.var.
Misra, J.N., 1956.
A systematic account of some littoral marine diatoms from the west coast of India. J. Bombay Nat. Hist. Soc., 53(4):537-568.

Plagiogramma sceptrum n.sp.
Mann, A., 1907
Report on the diatoms of the Albatross voyages in the Pacific Ocean, 1888-1904. Contrib. U. S. Nat. Herb. 10(5):221-419, Pls. XLIV-LIV.

Plagiogramma sigmoideum
Hendy, N. Ingram, 1964
An introductory account of the smaller algae of British coastal waters. V. Bacillariophyceae (Diatoms).
Her Majesty's Stationary Office, 317 pp., 45 pls.

Plagiogramma staurophorum
Hendy, N. Ingram, 1964
An introductory account of the smaller algae of British coastal waters. V. Bacillariophyceae (Diatoms).
Her Majesty's Stationary Office, 317 pp., 45 pls.

Plagiogramma staurophorum
Hendey, N-Ingram, 1958 [1957(Publ. 1958)]
Marine diatoms from some West African Ports- J. R Microsc. Soc. (3) 77(1/2): 28-85.

Plagiogramma staurophorum
Jorgensen, E., 1900
Protophyten und Protozoën im Plankton aus der Norwegischen Westkerste. Bergens Mus. Aarb. 1899(6): 95 pp., 5 pls., 83 tables.

Plagiogramma tesselatum
Mann, A., 1907
Report on the diatoms of the Albatross voyages in the Pacific Ocean, 1888-1904. Contrib. U. S. Nat. Herb. 10(5):221-419, Pls. XLIV-LIV.

Plagiogramma thaitiense n.sp.
Castracane degli Antelminelli, F., 1886
1. Report on the Diatomaceae collected by H.M.S. Challenger during the years 1873-1876. Rept. Sci. Results, H.M.S. Challenger, Botany Vol. II, 178 pp., 30 pls.

Plagiogramma vanheurckii
Cupp, Easter E., 1943
Marine plankton diatoms of the west coast of North America. Bull. S.I.O. 5(1):1-238, 5 pls., 168 text figs.

Plagiotropis van heurckii
Mann, A., 1907
Report on the diatoms of the Albatross voyages in the Pacific Ocean, 1888-1904. Contrib. U. S. Nat. Herb. 10(5):221-419, Pls. XLIV-LIV.

Plagiogramma Van Heurckii
Meunier, A., 1915
Microplancton de la Mer Flamande. 2. Diatomées (excepté le genre Chaetoceros). Mem. Mus. Roy. Hist. Nat., Belgique, 7(3):1-118, Pls. VIII-XIV.

Planktoniella
Brunel, J., 1962
Le phytoplancton de la Baie de Chaleurs. Inst. Botan., Univ. Montréal, Contrib. No. 77: 365 pp., 66 pls.

Planktoniella flora n.sp.
Wood, E.J. Ferguson, L.H. Crosby and Vivienne Cassie, 1959
Studies on Australian and New Zealand diatoms. III. Descriptions of further discoid species. Trans. R. Soc., N.Z., 87(3/4):211-219, pls. 15-17.

Planktoniella formosa
Hendey, N.I., 1937
The plankton diatoms of the southern seas. Discovery Repts. 16:151-364, pls.6-13.

Planktoniella muriformis
Round, F.E. 1972.
Some observations on colonies and ultrastructure of the frustule of Coenobiodiscus muriformis and its transfer to Planktoniella. J. Phycol. 8(3):222-231.

Planktoniella sol
Allen, W. E., 1938
The Templeton Crocker Expedition to the Gulf of California in 1935 - The Phytoplankton. Amer. Microsc. Soc., Trans. 57:328-335.

Planktoniella sol
Allen, W.E., 1937
Plankton diatoms of the Gulf of California obtained by the G. Allan Hancock Expedition of 1936. The Hancock Pacific Expeditions, Univ. So. Calif. Publ. 3:47-59, 1 fig.

Planktoniella sol
Allen, W.E., and E.E. Cupp, 1935
Plankton diatoms of the Java Sea. Annales du Jardin Botanique de Buitenzorg XLIV (2):101-174, figs.1-127.
(drawings of all species mentioned)

Planktoniella sol
Boden, B.P., 1950
Some marine plankton diatoms from the west coast of South Africa. Trans. R.Soc. S. Africa. 32:321-434, 100 text figs.

Planktoniella sol
Copenhagen, W. J. and L, D., 1949
Variation in the phytoplankton of Table Bay, October 1934 to October 1935. With a note on the calorific value of Chaetoceros spp. Trans. Roy. Soc. S. Africa, 32(2):113-123, 2 text figs.

Planktoniella sol
Cupp, Easter E., 1943
Marine plankton diatoms of the west coast of North America. Bull. S.I.O. 5(1):1-238, 5 pls., 168 text figs.

Planktoniella sol

Cupp, E.E., 1934
Analysis of marine diatom collections taken from the Canal Zone to California during March, 1933. Trans. Am. Micros. Soc. LIII (1):22-29, 1 map.

Planktoniella sol

Cupp, E.E. and Allen, W.E., 1938
Plankton diatoms of the Gulf of California obtained by Allan Hancock Pacific Expedition of 1937. The Hancock Pacific Expeditions, The Univ. So. Calif. Publ. 3: 61-74, 1 map, pls. 4-15.

Planktoniella sol

Dangeard, P., 1927
Phytoplankton de la croisière du "Sylvana". Ann. Inst. Ocean., Monaco, n.s., 4(8):286-401, 54 text figs. (Feirer-Juin 1913).

Planktoniella sol

Frenguelli, Joaquin, and Hector Antonio Orlando, 1959.
Operacion MERLUZA. Diatomeas y silicoflagelados del plancton del "VI Crucero". Servicio Hidrogr. Naval., Argentina, Publ. No. H. 619: 5-62.

Planktoniella sol

Enloff Johannes, 1970.
Elektronenmikroskopische Untersuchungen an Diatomeenschalen. VII. Der Bau der Schale von Planktoniella sol (Wallich) Schütt.
Beihefte Nova Hedwigia 31: 203-234

Planktonella sol

Gilbert, J.Y., and W.E. Allen, 1943
The phytoplankton of the Gulf of California obtained by the "E.W. Scripps" in 1939 and 1940. J. Mar. Res. V(2):89-110, figs. 30-31.

Planktoniella sol

Gran, H.H., 1908
Diatomeen. Nordisches Plankton, Botanischer Teil pp. XIX.1-XIX 146; 178 text figs.

Planktoniella sol

Gran, H. H. and E. C. Angst, 1931
Plankton diatoms of Puget Sound. Publ. Puget Sound Biol. Sta. 7:417-519, 95 text figs.

Planktoniella sol

Hasle, Grethe Rytter, 1960
Phytoplankton and ciliate species from the Tropical Pacific.
Skr. Norske Videnskaps-Akad., Oslo, 1. Mat.-Nat. Kl., 1960(2): 1-50.

Planktoniella sol

Hendy, N. Ingram, 1964
An introductory account of the smaller algae of British coastal waters. V. Bacillariophyceae (Diatoms).
Her Majesty's Stationary Office, 317 pp., 45 pls.

Planktoniella sol

Hendey, N.I., 1937
The plankton diatoms of the southern seas. Discovery Repts. 16:151-364, pls. 6-13.

Planktoniella sol

Pavillard, J., 1925
Bacillariales. Rept. on the Danish Oceangr. Exped., 1908-10 to the Mediterranean and adj. seas. Vol.II., Biol. J4:72 pp., 116 text figs.

Planktoniella sol

Rampi, L., 1942
Ricerche sul fitoplancton del Mare Ligure 6. Le diatomee delle acque di Sanremo. Nuovo Giornale Botanico Italiano, N.S., 49:252-268.

Planktoniella sol

Schröder, B., 1900
Phytoplankton des Golfes von Neapel nebst vergleichenden Ausblicken auf das atlantischen Ozean. Mitt. Zool. Stat. Neapel, 14:1-38.

Planktionella sol

Smayda, Theodore J., 1958
Biogeographical studies of marine phytoplankton. Oikos, 9(2): 158-191.

Pleurosigma sp.

Allen, W.E., 1937
Plankton diatoms of the Gulf of California obtained by the G. Allan Hancock Expedition of 1936. The Hancock Pacific Expeditions, Univ. So. Calif. Publ. 3:47-59, 1 fig.

Boden, Brian, 1948 Pleurosigma sp.
Marine plankton diatoms on operation HIGHJUMP in: Some oceanographic observations on operation HIGHJUMP. By R.S. Dietz. USNEL Rept. No.55, 97 pp., 41 figs. 7 July 1948.

Pleurosigma sp.

Braarud, T., 1945
A phytoplankton survey of the polluted waters of inner Oslo Fjord. Hvalrådets Skrifter, No.28, 142 pp., 19 text figs., 17 tables.

Pleurosigma acus

Bigelow, H.B., and M. Leslie, 1930
Reconnaissance of the waters and plankton of Monterey Bay, July 1928. Bull. M.C.Z., 70(5):429-481, 43 text figs.

Pleurosigma sp.

Cupp, E.E., 1934
Analysis of marine diatom collections taken from the Canal Zone to California during March, 1933. Trans. Am. Micros. Soc. LIII (1):22-29, 1 map.

Pleurosigma sp.

Cupp, E.E. and Allen, W.E., 1938
Plankton diatoms of the Gulf of California obtained by Allan Hancock Pacific Expedition of 1937. The Hancock Pacific Expeditions, The Univ. So. Calif. Publ. 3: 61-74, 1 map, pls. 4-15.

Pleurosigma

Desikachary, T.V., 1956(1957).
Electron microscope studies on diatoms. J.R. Microsc. Soc. (3)76(1/2):9-36.

Pleurosigma sp.

Gilbert, J.Y., and W.E. Allen, 1943
The phytoplankton of the Gulf of California obtained by the "E.W. Scripps" in 1939 and 1940. J. Mar. Res. V(2):89-110, figs. 30-31.

Pleurosigma sp.

Kokubo, S., and S. Sato., 1947
Plankters in Jū-San Gata. Physiol. and Ecol. (Japan) 1(4):1-16, 3 text figs., tables.

Pleurosigma sp.

Margalef, R., 1949
Fitoplancton nerítico de la Costa Brava en 1947-48. Publ. Inst. Biol. Aplicada, 5: 41-51, 3 text figs.

Pleurosigma

Neaverson, E., 1934
The sea-floor deposits. 1. General characteristics and distribution. Discovery Repts. 9: 297-349, Plates 17-22.

Pleurosigma

Oliveira, Lejeune P.H. de, and H. Muth, 1960.
Microscopia electronica de seis diatomaceas Pleurosigma, com uma critica do genero (Naviculaceae, Bacillariophyceae).
Mem. Inst. Oswaldo Cruz, Brasil, 58(1):1-38.

Pleurosigma sp.

Peiva Carvalho, J., 1950
O plancton do Rio Maria Rodrigues (Cananeis). 1. Diatomaceas e Dinoflagelados. Bol. Inst. Paulista Oceanogr. 1(1); 27-43, 2 fold-in tables, 2 figs.

Pleurosigma

Péragallo, H., 1891
Monographie du genre Pleurosigma et des genres alliés. Le Diatomiste 1: 1-35.

Pleurosigma (Gyrosigma) acuminatum

Oliveira, Lejeune P.H. de, and H. Muth, 1960.
Microscopia electronica de seis diatomaceas Pleurosigma, com uma critica do genero (Naviculaceae, Bacillariophyceae).
Mem. Inst. Oswaldo Cruz, Brasil, 58(1):1-38.

Pleurosigma acumenatum

Smith, W., 1852
Notes on the Diatomaceae; with descriptions of British species included in the genus Pleurosigma. (cont. Vol. 7, p. 14). Ann. Mag. Nat. Hist. II Vol. 9:1-12, 2 pls.

Pleurosigma acuminatum

Zanon, V., 1948
Diatomee marini di Sardegna e Pugillo di Alghe Marine della stressa. Boll. Pesca. Piscitutura e Idrobiologia, Anno 24, ns. 3(2): 202-244, 27 figs. on 1 pl.

Pleurosigma acutum

Zanon, V., 1948
Diatomee marini di Sardegna e Pugillo di Alghe Marine della stressa. Boll. Pesca. Piscitutura e Idrobiologia, Anno 24, ns. 3(2): 202-244, 27 figs. on 1 pl.

Pleurosigma aestuarii

Allen, W.E., and E.E. Cupp, 1935
Plankton diatoms of the Java Sea. Annales du Jardin Botanique de Buitenzorg XLIV (2):101-174, figs.1-127.
(drawings of all species mentioned)

Pleurosigma aestuarii

Hendy, N. Ingram, 1964.
An introductory account of the smaller algae of British coastal waters. V. Bacillariophyceae (Diatoms).
Her Majesty's Stationary Office, 317 pp., 45 pls.

Pleurosigma aestuarii

Hendey, N.I., 1951
Littoral diatoms of Chicester Harbour with special reference to fouling. J. Roy. Microscop. Soc. 71(1): 1-86, 18 pls.

Pleurosigma aestuari

Hustedt, F. and A.A. Aleem, 1951
Littoral diatoms from the Salstone near Plymouth. JMBA 30(1): 177.196.

Pleurosigma aestuarii

Kolbe, R.W., 1951.
Elektronenmikroskopische Untersuchungen von Diatomeenmembranen II. Svenska Bot. Tidskr. 45(4):637-647, Pls. 1-4.

Pleurosigma (Gyrosigma) aestuarii

Oliveira, Lejeune P.H. de, and H. Muth, 1960.
Microscopia electronica de seis diatomaceas Pleurosigma, com uma critica do genero (Naviculaceae, Bacillariophyceae).
Mem. Inst. Oswaldo Cruz, Brasil, 58(1):1-38.

Pleurosigma aestuarii

Zanon, V., 1948
Diatomee marini di Sardegna e Pugillo di Alghe Marine della stressa. Boll. Pesca, Piscitutura e Idrobiologia, Anno 24, ns. 3(2): 202-244, 27 figs. on 1 pl.

Pleurosigma affine

Mann, A., 1893
List of Diatomaceae from a deep-sea dredging in the Atlantic Ocean off Delaware Bay by the U. S. Fish Commission Steamer Albatross. Proc. U. S. Nat. Mus. 16:303-312.

Pleurosigma affine

Politis, J., 1949
Diatomees marines de Bosphores et des ibes de la mer de Marmara. II Practica tou Hellenikou Hidrobiologikou Institutoutou 1929, Etoz 1929, 3(1):11-31.

Pleurosigma angulatum

Andrade e Clovis Teixeira, M.H. de, 1957.
Contribução para o conhecimento das diatomáceas do Brasil.
Bol. Inst. Ocean., Univ. Sao Paulo, 8(1/2):171-225, 10 pls.

Pleurosigma angulatum

Forti, A., 1922
Ricerche sulla flora pelagica (fitoplancton) di Quarto dei Mille. Mem. R. Com. Talass. Ital. 97:248 pp., 13 pls.

Pleurosigma angulatum

Hendy, N. Ingram, 1964
An introductory account of the smaller algae of British coastal waters. V. Bacillariophyceae (Diatoms).
Her Majesty's Stationary Office, 317 pp., 45 pls.

Pleurosigma angulatum

Hendey, N.I., 1951
Littoral diatoms of Chicester Harbour with special reference to fouling. J.Roy. Microscop. Soc. 71(1): 1-86, 18 pls.

Pleurosigma angulatum

Hustedt, F. and A.A. Aleem, 1951
Littoral diatoms from the Salstone near Plymouth. JMBA 30(1): 177.196.

Pleurosigma angulatum

Iyengar, M.O.P. and G.Venkataraman, 1951.
The ecology and seasonal succession of the algae flora of the River Cooum at Madras with special reference to the Diatomaceae. J. Madras Univ. 21, Sect. B(1): 140-192, 1 pl of 4 figs., 11 text figs.

Pleurosigma angulatum

Jørgensen, E., 1905
B. Protistplankton and the diatoms in bottom samples. Hydrographical and biological investigations in Norwegian fjords. Bergens Mus. Skr. 7: 49-225.

Pleurosigma angulatum

Lillick, L.C., 1940
Phytoplankton and planktonic protozoa of the offshore waters of the Gulf of Maine. Pt.II. Qualitative Composition of the Planktonic Flora. Trans. Am. Phil. Soc., n.s., 31(3):193-237, 13 text figs.

Pleurosigma angulatum

Meunier, A., 1915
Microplancton de la Mer Flamande. 2. Diatomées (excepté le genre Chaetoceros). Mem. Mus. Roy. Hist. Nat., Belgique, 7(3):1-118, Pls. VIII-XIV.

Pleurosigma angulatum

Morse, D.C., 1947
Some observations on seasonal variations in plankton population Patuxant River, Maryland 1943 1945. Bd. Nat. Res., Publ. No.65, Chesapeake Biol. Lab., 31, 3 figs.

Pleurosigma angulatum

Politis, J., 1949
Diatomees marines de Bosphores et des ibes de la mer de Marmara. II Practica tou Hellenikou Hidrobiologikou Institutoutou 1929, Etoz 1929, 3(1):11-31.

Pleurosigma angulatum

Rampi, L., 1942
Ricerche sul fitoplancton del Mare Ligure 6. Le diatomee delle acque di Sanremo. Nuovo Giornale Botanico Italiano, N.S. 49:252-268.

Pleurosigma angulatum

Rampi, L., 1940
Diatomee del Mare Adriatico. Nuovo Giornale Botanico Italiano, n.s., 47:559-608.

Pleurosigma angulatum

Schodduyn, M., 1926
Observations faites dans la baie d'Ambleteuse (Pas de Calais). Bull. Inst. Ocean., Monaco, No. 482: 64 pp.

Pleurosigma angulatum

Smith, W., 1852
Notes on the Diatomaceae; with descriptions of British species included in the genus Pleurosigma. (cont. Vol. 7, p. 14). Ann. Mag. Nat. Hist. II Vol. 9:1-12, 2 pls.

Pleurosigma angulatum

Subrahmanyan, R., 1945.
On the cell division and mitosis in some South Indian diatoms. Proc. Indian Acad. Sci., Sect. B, 22(6):331-354, Pl. 39, 88 textfigs.

Pleurosigma angulatum

Zanon, V., 1948
Diatomee marini di Sardegna e Pugillo di Alghe Marine della stressa. Boll. Pesca, Piscitutura e Idrobiologia, Anno 24, ns. 3(2): 202-244, 27 figs. on 1 pl.

Pleurosigma angulatum var. strigosa

Allen, W.E., and E.E. Cupp, 1935
Plankton diatoms of the Java Sea. Annales du Jardin Botanique de Buitenzorg XLIV (2):101-174, figs.1-127.

(drawings of all species mentioned)

Pleurosigma angulatum strigosum

Ercegovic, A., 1936
Etudes qualitative et quantitatives du phytoplancton dans les eaux cotières de l'Adriatique oriental moyen au cours de l'année 1934. Acta Adriatica 1(9):1-126

Pleurosigma angulatum strigosa

Forti, A., 1922
Ricerche sulla flora pelagica (fitoplancton) di Quarto dei Mille. Mem. R. Com. Talass. Ital. 97:248 pp., 13 pls.

Pleurosigma arafurense n.sp.

Castracane degli Antelminelli, F., 1886
1. Report on the Diatomaceae collected by H.M.S. Challenger during the years 1873-1876. Rept. Sci. Results, H.M.S. Challenger, Botany Vol. II, 178 pp., 30 pls.

Pleurosigma attenuatum

Morse, D.C., 1947
Some observations on seasonal variations in plankton population Patuxant River, Maryland 1943. 1945. Bd. Nat. Res., Publ. No.65, Chesapeake Biol. Lab., 31, 3 figs.

Pleurosigma attenuatum

Smith, W., 1852
Notes on the Diatomaceae; with descriptions of British species included in the genus Pleurosigma. (cont. Vol. 7, p. 14). Ann. Mag. Nat. Hist. II Vol. 9:1-12, 2 pls.

Pleurosigma australe

Zanon, D. V., 1949
Diatomee di Buenos Aires (Argentina) Atti Accad. Naz. Lincei, Memorie, Cl. Sci. fis., mat. e. nat., ser. 7, 11(3):59-151, 2 pls.

Pleurosigma balticum

Hendey, N.I., 1951
Littoral diatoms of Chicester Harbour with special reference to fouling. J.Roy. Microscop. Soc. 71(1): 1-86, 18 pls.

Pleurosigma balticum

Lillick, L.C., 1940
Phytoplankton and planktonic protozoa of the offshore waters of the Gulf of Maine. Pt.II. Qualitative Composition of the Planktonic Flora. Trans. Am. Phil. Soc., n.s., 31(3):193-237, 13 text figs.

Pleurosigma balticum

Meunier, A., 1915
Microplancton de la Mer Flamande. 2. Diatomées (excepté le genre Chaetoceros). Mem. Mus. Roy. Hist. Nat., Belgique, 7(3):1-118, Pls. VIII-XIV.

Pleurosigma (Gyrosigma) balticum

Oliveira, Lejeune P.H. de, and H. Muth, 1960.
Microscopia electronica de seis diatomaceas Pleurosigma com uma critica do genero (Naviculaceae, Bacillariophyceae).
Mem. Inst. Oswaldo Cruz, Brasil, 58(1):1-38.

Pleurosigma balticum

Schodduyn, M., 1926
Observations faites dans la baie d'Ambleteuse (Pas de Calais). Bull. Inst. Ocean., Monaco, No. 482: 64 pp.

Pleurosigma Balticum

Smith, W., 1852
Notes on the Diatomaceae; with descriptions of British species included in the genus Pleurosigma. (cont. Vol. 7, p. 14). Ann. Mag. Nat. Hist. II Vol. 9:1-12, 2 pls.

Pleurosigma capense

Boden, B.P., 1950
Some marine plankton diatoms from the west coast of South Africa. Trans. R.Soc. S. Africa. 32:321-434, 100 text figs.

Pleurosigma chilensis

Krasske, G., 1941
Die Kieselalgen des chilenischen Küstenplankton (Aus dem südchilenischen Kusten gebiet IX). Arch. Hydrobiol. 38 (2):260-287.

Pleurosigma compactum

Allen, W.E., and E.E. Cupp, 1935
Plankton diatoms of the Java Sea. Annales du Jardin Botanique de Buitenzorg XLIV (2):101-174, figs.1-127.

(drawings of all species mentioned)

Pleurosigma cuspidatum

Hendy, N. Ingram, 1964
An introductory account of the smaller algae of British coastal waters. V. Bacillariophyceae (Diatoms).
Her Majesty's Stationary Office, 317 pp., 45 pls.

Pleurosigma decorum
Hendy, N. Ingram, 1964
An introductory account of the smaller algae of British coastal waters. V. Bacillariophyceae (Diatoms).
Her Majesty's Stationary Office, 317 pp., 45 pls.

Pleurosigma decorum
Hendey, N.J., 1951
Littoral diatoms of Chicester Harbour with special reference to fouling. J.Roy. Microscop. Soc. 71(1): 1-86, 18 pls.

Pleurosigma decorum
Lillick, L.C., 1940
Phytoplankton and planktonic protozoa of the offshore waters of the Gulf of Maine. Pt.II. Qualitative Composition of the Planktonic Flora. Trans. Am. Phil. Soc., n.s., 31(3):193-237, 13 text figs.

Pleurosigma decorum
Meunier, A., 1915
Microplancton de la Mer Flamande. 2. Diatomées (excepté le genre Chaetoceros). Mem. Mus. Roy. Hist. Nat., Belgique, 7(3):1-118, Pls. VIII-XIV.

Pleurosigma decorum
Rampi, L., 1942
Ricerche sul fitoplancton del Mare Ligure 6. Le diatomee delle acque di Sanremo. Nuovo Giornale Botanico Italiano, N.S., 49:252-268.

Pleurosigma decorum
Rampi, L., 1940
Diatomee del Mare Adriatico. Nuovo Giornale Botanico Italiano, n.s., 47:559-608.

Pleurosigma delicatulum
Bigelow, H.B., and M. Leslie, 1930
Reconnaissance of the waters and plankton of Monterey Bay, July 1928. Bull. M.C.Z., 70(5):429-481, 43 text figs.

Pleurosigma delicatulum
Hendy, N. Ingram, 1964
An introductory account of the smaller algae of British coastal waters. V. Bacillariophyceae (Diatoms).
Her Majesty's Stationary Office, 317 pp., 45 pls.

Pleurosigma delicatulum
Jørgensen, E., 1905
B.Protistplankton and the diatoms in bottom samples. Hydrographical and biological investigations in Norwegian fjords. Bergens Mus. Skr. 7: 49-225.

Pleurosigma delicatulum
Rampi, L., 1942
Ricerche sul fitoplancton del Mare Ligure 6. Le diatomee delle acque di Sanremo. Nuovo Giornale Botanico Italiano, N.S., 49:252-268.

Pleurosigma delicatulum
Rampi, L., 1940
Diatomee del Mare Adriatico. Nuovo Giornale Botanico Italiano, n.s., 47:559-608.

Pleurosigma delicatulum n.sp.
Smith, W., 1852
Notes on the Diatomaceae; with descriptions of British species included in the genus Pleurosigma. (cont. Vol. 7, p. 14). Ann. Mag. Nat. Hist. II Vol. 9:1-12, 2 pls.

Pleurosigma directum
Boden, B. P., 1949.
The diatoms collected by the U.S.S. CACOPAN in the Antarctic in 1947. J. Mar. Res. 8(1):6-13, 3 textfigs.

Pleurosigma directum
Boden, Brian, 1948
Marine plankton diatoms on operation HIGHJUMP in: Some oceanographic observations on operation HIGHJUMP. By R.S. Dietz. USNEL Rept. No.55, 97 pp., 41 figs. 7 July 1948.

Pleurosigma directum
Hendey, N.I., 1937
The plankton diatoms of the southern seas. Discovery Repts. 16:151-364, pls.6-13.

Pleurosigma directum-secundum
Hendey, N.I., 1937
The plankton diatoms of the southern seas. Discovery Repts. 16:151-364, pls.6-13.

Pleurosigma distortum n.sp.
Smith, W., 1852
Notes on the Diatomaceae; with descriptions of British species included in the genus Pleurosigma. (cont. Vol. 7, p. 14). Ann. Mag. Nat. Hist. II Vol. 9:1-12, 2 pls.

Pleurosigma elegantissimum n.sp.
Castracane degli Antelminelli, F., 1886
1. Report on the Diatomaceae collected by H.M.S. Challenger during the years 1873-1876. Rept. Sci. Results, H.M.S. Challenger, Botany Vol. II, 178 pp., 30 pls.

Pleurosigma elongatum
Allen, W.E., and E.E. Cupp, 1935
Plankton diatoms of the Java Sea. Annales du Jardin Botanique de Buitenzorg XLIV (2):101-174, figs.1-127.
(drawings of all species mentioned)

Pleurosigma elongatum
Andrade e Clovis Teixeira, M.H. de, 1957.
Contribuição para o conhecimento das diatomaceas do Brasil.
Bol. Inst. Ocean., Univ. Sao Paulo, 8(1/2):171-225, 10 pls.

Pleurosigma elongatum
Cupp, Easter E., 1943
Marine plankton diatoms of the west coast of North America. Bull. S.I.O. 5(1):1-238, 5 pls., 168 text figs.

Pleurosigma elongatum
Hendy, N. Ingram, 1964
An introductory account of the smaller algae of British coastal waters. V. Bacillariophyceae (Diatoms).
Her Majesty's Stationary Office, 317 pp., 45 pls.

Pleurosigma elongatum
Hendey, N.J., 1951
Littoral diatoms of Chicester Harbour with special reference to fouling. J.Roy. Microscop. Soc. 71(1): 1-86, 18 pls.

Pleurosigma elongatum
Kokubo, S., and S. Sato., 1947
Plankters in Jū-San Gata. Physiol. and Ecol. (Japan) 1(4):1-16, 3 text figs., tables.

Pleurosigma elongatum
Kucherova, Z.S., 1961.
Vertical distribution of diatoms from Sevastopol Bay.
Trudy Sevastopol Biol. Sta., (14):64-78.

Pleurosigma elongatum
Lillick, L.C., 1940
Phytoplankton and planktonic protozoa of the offshore waters of the Gulf of Maine. Pt.II. Qualitative Composition of the Planktonic Flora. Trans. Am. Phil. Soc., n.s., 31(3):193-237, 13 text figs.

Pleurosigma elongatum
Meunier, A., 1915
Microplancton de la Mer Flamande. 2. Diatomées (excepté le genre Chaetoceros). Mem. Mus. Roy. Hist. Nat., Belgique, 7(3):1-118, Pls. VIII-XIV.

Pleurosigma elongatum
Morse, D.C., 1947
Some observations on seasonal variations in plankton population Patuxant River, Maryland 1943-1945. Bd. Nat. Res., Publ. No.65, Chesapeake Biol. Lab., 31, 3 figs.

Pleurosigma elongatum
Rampi, L., 1942
Ricerche sul fitoplancton del Mare Ligure 6. Le diatomee delle acque di Sanremo. Nuovo Giornale Botanico Italiano, N.S., 49:252-268.

Pleurosigma elongatum
Rumkówna, A., 1948
List of the phytoplankton species occurring in the superficial water layers in the Gulf of Gdańsk. Bull. Lab. mar., Gdynia, No. 4: 139-141 with tables in back.

Pleurosigma elongatum n.sp.
Smith, W., 1852
Notes on the Diatomaceae; with descriptions of British species included in the genus Pleurosigma. (cont. Vol. 7, p. 14). Ann. Mag. Nat. Hist. II Vol. 9:1-12, 2 pls.

Pleurosigma elongatum
Zanon, V., 1948
Diatomee marini di Sardegna e Pugillo di Alghe Marine della stressa. Boll. Pesca, Piscitutura e Idrobiologia, Anno 24, ns. 3(2): 202-244, 27 figs. on 1 pl.

Pleurosigma fasciola
Gran, H. H. and E. C. Angst, 1931
Plankton diatoms of Puget Sound. Publ. Puget Sound Biol. Sta. 7:417-519, 95 text figs.

Pleurosigma fasciola
Hendey, N.J., 1951
Littoral diatoms of Chicester Harbour with special reference to fouling. J.Roy. Microscop. Soc. 71(1): 1-86, 18 pls.

Pleurosigma fasciola
Jørgensen, E., 1905
B.Protistplankton and the diatoms in bottom samples. Hydrographical and biological investigations in Norwegian fjords. Bergens Mus. Skr. 7: 49-225.

Pleurosigma fasciola
Lillick, L.C., 1940
Phytoplankton and planktonic protozoa of the offshore waters of the Gulf of Maine. Pt.II. Qualitative Composition of the Planktonic Flora. Trans. Am. Phil. Soc., n.s., 31(3):193-237, 13 text figs.

Pleurosigma Fasciola
Meunier, A., 1915
Microplancton de la Mer Flamande. 2. Diatomées (excepté le genre Chaetoceros). Mem. Mus. Roy. Hist. Nat., Belgique, 7(3):1-118, Pls. VIII-XIV.

Pleurosigma fascicola
Morse, D.C., 1947
Some observations on seasonal variations in plankton population Patuxant River, Maryland 1943-1945. Bd. Nat. Res., Publ. No.65, Chesapeake Biol. Lab., 31, 3 figs.

Pleurosigma Fasciola

Smith, W., 1852
Notes on the Diatomaceae; with descriptions of British species included in the genus Pleurosigma. (cont. Vol. 7, p. 14). Ann. Mag. Nat. Hist. II Vol. 9:1-12, 2 pls.

Pleurosigma fasciola (figs.)

Sousa e Silva, E., 1949
Diatomaceas e Dinoflagelados de Baia de Cascais. Portugaliae Acta Biol., Volume: Julio Henriques, Ser. B: 300-383, 9 pls, 2 fold-in tables.

Pleurosigma finnmarchicum

Hendy, N. Ingram, 1964
An introductory account of the smaller algae of British coastal waters. V. Bacillariophyceae (Diatoms). Her Majesty's Stationary Office, 317 pp., 45 pls.

Pleurosigma formosum

Brunel, J., 1962
Le phytoplancton de la Baie de Chaleurs. Inst. Botan., Univ. Montréal, Contrib. No. 77: 365 pp., 66 pls.

Pleurosigma formosum

Brunel, Jules, 1962
Le phytoplancton de la Baie des Chaleurs. Contrib. Ministère de la Chasse et des Pêcheries, Province de Québec, No. 91: 365 pp.

Pleurosigma formosum

Gran, H. H. and E. C. Angst, 1931
Plankton diatoms of Puget Sound. Publ. Puget Sound Biol. Sta. 7:417-519, 95 text figs.

Pleurosigma formosum

Hendy, N. Ingram, 1964
An introductory account of the smaller algae of British coastal waters. V. Bacillariophyceae (Diatoms). Her Majesty's Stationary Office, 317 pp., 45 pls.

Pleurosigma formosum

Hendey, N.I., 1951
Littoral diatoms of Chicester Harbour with special reference to fouling. J.Roy. Microscop. Soc. 71(1): 1-86, 18 pls.

Pleurosigma formosum

Hustedt, F. and A.A. Aleem, 1951
Littoral diatoms from the Salstone near Plymouth. JMBA 30(1): 177-196.

Pleurosigma (Eupleurosigma) formosum

Oliveira, Lejeune P.H.de, and H. Muth, 1960.
Microscopia electronica de seis diatomaceas Pleurosigma, com uma critica do genero (Naviculaceae, Bacillariophyceae). Mem. Inst. Oswaldo Cruz, Brasil, 58(1):1-38.

Pleurosigma formosum

Rampi, L., 1942
Ricerche sul fitoplancton del Mare Ligure 6. Le diatomee delle acque di Sanremo. Nuovo Giornale Botanico Italiano, N.S., 49:252-268.

Pleurosigma formosum

Rampi, L., 1940
Diatomee del Mare Adriatico. Nuovo Giornale Botanico Italiano, n.s., 47:559-608.

Pleurosigma formosum n.sp.

Smith, W., 1852
Notes on the Diatomaceae; with descriptions of British species included in the genus Pleurosigma. (cont. Vol. 7, p. 14). Ann. Mag. Nat. Hist. II Vol. 9:1-12, 2 pls.

Pleurosigma formosum

Zanon, D. V., 1949
Diatomee di Buenos Aires (Argentina) Atti Accad. Naz. Lincei, Memorie, Cl. Sci. fis., mat. e. nat., ser. 7, 11(3):59-151, 2 pls.

Pleurosigma formosum

Zanon, V., 1948
Diatomee marini di Sardegna e Pugillo di Alghe Marine della stressa. Boll. Pesca, Piscitutura e Idrobiologia, Anno 24, ns. 3(2): 202-244, 27 figs. on 1 pl.

Pleurosigma giganteum

Grunow, A., 1863
Ueber einige neue und ungenügend bekannte Arten und Gattungen von Diatomaceen. Verhandl. d. K.K. Zool. Bot. Gesellsch., Vienna, 13: 137-162, pl.4-5 (Pl. 13-14).

Pleurosigma hamuliferum

Cupp, Easter E., 1943
Marine plankton diatoms of the west coast of North America. Bull. S.I.O. 5(1):1-238, 5 pls., 168 text figs.

Pleurosigma hippocampus

Meunier, A., 1915
Microplancton de la Mer Flamande. 2. Diatomées (excepté le genre Chaetoceros). Mem. Mus. Roy. Hist. Nat., Belgique, 7(3):1-118, Pls. VIII-XIV.

Pleurosigma Hippocampus

Smith, W., 1852
Notes on the Diatomaceae; with descriptions of British species included in the genus Pleurosigma. (cont. Vol. 7, p. 14). Ann. Mag. Nat. Hist. II Vol. 9:1-12, 2 pls.

Pleurosigma inflatum

Mann, A., 1893
List of Diatomaceae from a deep-sea dredging in the Atlantic Ocean off Delaware Bay by the U. S. Fish Commission Steamer Albatross. Proc. U. S. Nat. Mus. 16:303-312.

Pleurosigma intermedium

Andrade e Clovis Teixeira, M.H. de, 1957.
Contribuição para o conhecimento das diatomáceas do Brasil.
Bol. Inst. Ocean., Univ. Sao Paulo, 8(1/2):171-225, 10 pls.

Pleurosigma intermedium

Hendy, N. Ingram, 1964
An introductory account of the smaller algae of British coastal waters. V. Bacillariophyceae (Diatoms). Her Majesty's Stationary Office, 317 pp., 45 pls.

Pleurosigma intermedium

Hendey, N.I., 1951
Littoral diatoms of Chicester Harbour with special reference to fouling. J.Roy. Microscop. Soc. 71(1): 1-86, 18 pls.

Pleurosigma japonicum n.sp.

Castracane degli Antelminelli, F., 1886
1. Report on the Diatomaceae collected by H.M.S. Challenger during the years 1873-1876. Rept. Sci. Results, H.M.S. Challenger, Botany Vol. II, 178 pp., 30 pls.

Pleurosigma Kützingii

Mann, A., 1893
List of Diatomaceae from a deep-sea dredging in the Atlantic Ocean off Delaware Bay by the U. S. Fish Commission Steamer Albatross. Proc. U. S. Nat. Mus. 16:303-312.

Pleurosigma lacustre n.sp.

Smith, W., 1852
Notes on the Diatomaceae; with descriptions of British species included in the genus Pleurosigma. (cont. Vol. 7, p. 14). Ann. Mag. Nat. Hist. II Vol. 9:1-12, 2 pls.

Pleurosigma latum

Rampi, L., 1942
Ricerche sul fitoplancton del Mare Ligure 6. Le diatomee delle acque di Sanremo. Nuovo Giornale Botanico Italiano, N.S., 49:252-268.

Pleurosigma latum

Zanon, V., 1948
Diatomee marini di Sardegna e Pugillo di Alghe Marine della stressa. Boll. Pesca, Piscitutura e Idrobiologia, Anno 24, ns. 3(2): 202-244, 27 figs. on 1 pl.

Pleurosigma littorale n.sp.

Smith, W., 1852
Notes on the Diatomaceae; with descriptions of British species included in the genus Pleurosigma. (cont. Vol. 7, p. 14). Ann. Mag. Nat. Hist. II Vol. 9:1-12, 2 pls.

Pleurosigma longum

Hendy, N. Ingram, 1964
An introductory account of the smaller algae of British coastal waters. V. Bacillariophyceae (Diatoms). Her Majesty's Stationary Office, 317 pp., 45 pls.

Pleurosigma longum

Zanon, V., 1948
Diatomee marini di Sardegna e Pugillo di Alghe Marine della stressa. Boll. Pesca, Piscitutura e Idrobiologia, Anno 24, ns. 3(2): 202-244, 27 figs. on 1 pl.

Pleurosigma marinum

Hendy, N. Ingram, 1964
An introductory account of the smaller algae of British coastal waters. V. Bacillariophyceae (Diatoms). Her Majesty's Stationary Office, 317 pp., 45 pls.

Pleurosigma marinum

Rampi, L., 1940
Diatomee del Mare Adriatico. Nuovo Giornale Botanico Italiano, n.s., 47:559-608.

Pleurosigma maroccanum

Hustedt, F. and A.A. Aleem, 1951
Littoral diatoms from the Salstone near Plymouth. JMBA 30(1): 177-196.

Pleurosigma naviculaceum

Allen, W.E., and E.E. Cupp, 1935
Plankton diatoms of the Java Sea. Annales du Jardin Botanique de Buitenzorg XLIV (2):101-174, figs.1-127.
(drawings of all species mentioned)

Pleurosigma naviculaceum

Andrade e Clovis Teixeira, M.H. de, 1957.
Contribuição para o conhecimento das diatomáceas do Brasil.
Bol. Inst. Ocean., Univ. Sao Paulo, 8(1/2):171-225, 10 pls.

Pleurosigma naviculaceum

Castracane degli Antelminelli, F., 1886
1. Report on the Diatomaceae collected by H.M.S. Challenger during the years 1873-1876. Rept. Sci. Results, H.M.S. Challenger, Botany Vol. II, 178 pp., 30 pls.

Pleurosigma naviculaceum
Hendy, N. Ingram, 1964
An introductory account of the smaller algae of British coastal waters. V. Bacillariophyceae (Diatoms).
Her Majesty's Stationary Office, 317 pp., 45 pls.

Pleurosigma naviculaceum
Hustedt, F. and A.A. Aleem, 1951
Littoral diatoms from the Salstone near Plymouth. JMBA 30(1): 177.196.

Pleurosigma naviculaceum
Jørgensen, E., 1905
B. Protistplankton and the diatoms in bottom samples. Hydrographical and biological investigations in Norwegian fiords. Bergens Mus. Skr. 7: 49-225.

Pleurosigma nicobaricum
Cupp, Easter E., 1943
Marine plankton diatoms of the west coast of North America. Bull. S.I.O. 5(1):1-238, 5 pls., 168 text figs.

Pleurosigma nicobaricum
Ercegovic, A., 1936
Etudes qualitative et quantitatives du phytoplancton dans les eaux cotières de l'Adriatique oriental moyen au cours de l'année 1934. Acta Adriatica 1(9):1-126

Pleurosigma nicobaricum
Forti, A., 1922
Ricerche sulla flora pelagica (fitoplancton) di Quarto dei Mille. Mem. R. Com. Talass. Ital. 97:248 pp., 13 pls.

Pleurosigma nicobaricum
Rampi, L., 1942
Ricerche sul fitoplancton del Mare Ligure 6. Le diatomee delle acque di Sanremo. Nuovo Giornale Botanico Italiano, N.S., 49:252-268.

Pleurosigma nicobaricum
Rampi, L., 1940
Diatomee del Mare Adriatico. Nuovo Giornale Botanico Italiano, n.s., 47:559-608.

Pleurosigma Nicobaricum
Zanon, V., 1948
Diatomee marini di Sardegna e Pugillo di Alghe Marine della stressa. Boll. Pesca, Piscitutura e Idrobiologia, Anno 24, ns. 3(2): 202-244, 27 figs. on 1 pl.

Pleurosigma Normanii
Allen, W.E., and E.E. Cupp, 1935
Plankton diatoms of the Java Sea. Annales du Jardin Botanique de Buitenzorg XLIV (2):101-174, figs.1-127.
(drawings of all species mentioned)

Pleurosigma Normani
Andrade e Clovis Teixeira, M.H. de, 1957
Contribuição para o conhecimento das diatomáceas do Brasil.
Bol. Inst. Ocean., Univ. Sao Paulo, 8(1/2):171-225, 10 pls.

Pleurosigma normani
Cupp, Easter E., 1943
Marine plankton diatoms of the west coast of North America. Bull. S.I.O. 5(1):1-238, 5 pls., 168 text figs.

Pleurosigma Normani
Gran, H.H., and T. Braarud, 1935
A quantitative study of the phytoplankton in the Bay of Fundy and the Gulf of Maine (including observations on hydrography, chemistry, and turbidity). J. Biol. Bd., Canada, 1(5):279-467, 69 text figs.

Pleurosigma normanii
Hendy, N. Ingram, 1964
An introductory account of the smaller algae of British coastal waters. V. Bacillariophyceae (Diatoms).
Her Majesty's Stationary Office, 317 pp., 45 pls.

Pleurosigma normani
Hustedt, F. and A.A. Aleem, 1951
Littoral diatoms from the Salstone near Plymouth. JMBA 30(1): 177.196.

Pleurosigma normanni
Jørgensen, E., 1905
B. Protistplankton and the diatoms in bottom samples. Hydrographical and biological investigations in Norwegian fiords. Bergens Mus. Skr. 7: 49-225.

Pleurosigma Normani
Lillick, L.C., 1940
Phytoplankton and planktonic protozoa of the offshore waters of the Gulf of Maine. Pt.II. Qualitative Composition of the Planktonic Flora. Trans. Am. Phil. Soc., n.s., 31(3):193-237, 13 text figs.

Pleurosigma Normani
Lillick, L.C., 1938
Preliminary report of the phytoplankton of the Gulf of Maine. Am. Mid. Nat. 20(3):624-640, 1 text fig. 37 tables.

Pleurosigma Normanii
Lillick, L.C., 1937
Seasonal studies of the phytoplankton off Woods Hole, Massachusetts. Biol. Bull. LXXIII (3):488-503, 3 text figs.

Pleurosigma Normani
Zanon, V., 1948
Diatomee marini di Sardegna e Pugillo di Alghe Marine della stressa. Boll. Pesca, Piscitutura e Idrobiologia, Anno 24, ns. 3(2): 202-244, 27 figs. on 1 pl.

Pleurosigma nubecola
Zanon, V., 1948
Diatomee marini di Sardegna e Pugillo di Alghe Marine della stressa. Boll. Pesca, Piscitutura e Idrobiologia, Anno 24, ns. 3(2): 202-244, 27 figs. on 1 pl.

Pleurosigma obliquum
Hendey, N.J., 1951
Littoral diatoms of Chicester Harbour with special reference to fouling. J. Roy. Microscop. Soc. 71(1): 1-86, 18 pls.

Pleurosigma obscurum
Hendy, N. Ingram, 1964
An introductory account of the smaller algae of British coastal waters. V. Bacillariophyceae (Diatoms).
Her Majesty's Stationary Office, 317 pp., 45 pls.

Pleurosigma obscurum
Hustedt, F. and A.A. Aleem, 1951
Littoral diatoms from the Salstone near Plymouth. JMBA 30(1): 177.196.

Pleurosigma obscurum
Rampi, L., 1942
Ricerche sul fitoplancton del Mare Ligure 6. Le diatomee delle acque di Sanremo. Nuovo Giornale Botanico Italiano, N.S., 49:252-268.

Pleurosigma obscurum n.sp.
Smith, W., 1852
Notes on the Diatomaceae; with descriptions of British species included in the genus Pleurosigma. (cont. Vol. 7, p. 14). Ann. Mag. Nat. Hist. II Vol. 9:1-12, 2 pls.

Pleurosigma pelagicum
Allen, W.E., and E.E. Cupp, 1935
Plankton diatoms of the Java Sea. Annales du Jardin Botanique de Buitenzorg XLIV (2):101-174, figs.1-127.
(drawings of all species mentioned)

Pleurosigma prolongatum
Hendey, N.J., 1951
Littoral diatoms of Chicester Harbour with special reference to fouling. J. Roy. Microscop. Soc. 71(1): 1-86, 18 pls.

Pleurosigma prolongatum n.sp.
Smith, W., 1852
Notes on the Diatomaceae; with descriptions of British species included in the genus Pleurosigma. (cont. Vol. 7, p. 14). Ann. Mag. Nat. Hist. II Vol. 9:1-12, 2 pls.

Pleurosigma rectum
Allen, W.E., and E.E. Cupp, 1935
Plankton diatoms of the Java Sea. Annales du Jardin Botanique de Buitenzorg XLIV (2):101-174, figs.1-127.
(drawings of all species mentioned)

Pleurosigma rigidum
Hendy, N. Ingram, 1964
An introductory account of the smaller algae of British coastal waters. V. Bacillariophyceae (Diatoms).
Her Majesty's Stationary Office, 317 pp., 45 pls.

Pleurosigma rigidum
Politis, J., 1949
Diatomees marines de Bosphores et des iles de la mer de Marmara. II Practica tou Hellenikou Hidrobiologikou Institutoutou 1929, Etoz 1929, 3(1):11-31.

Pleurosigma rigidum
Rampi, L., 1942
Ricerche sul fitoplancton del Mare Ligure 6. Le diatomee delle acque di Sanremo. Nuovo Giornale Botanico Italiano, N.S., 49:252-268.

Pleurosigma rigidum
Rampi, L., 1940
Diatomee del Mare Adriatico. Nuovo Giornale Botanico Italiano, n.s., 47:559-608.

Pleurosigma rigidum
Zanon, V., 1948
Diatomee marini di Sardegna e Pugillo di Alghe Marine della stressa. Boll. Pesca, Piscitutura e Idrobiologia, Anno 24, ns. 3(2): 202-244, 27 figs. on 1 pl.

Pleurosigma salinarum
Iyengar, M.O.P. and G. Venkataraman, 1951
The ecology and seasonal succession of the algae flora of the River Cooum at Madras with special reference to the Diatomaceae. J. Madras Univ. 21, Sect. B(1); 140-192, 1 pl of 4 figs., 11 text figs.

Pleurosigma salinarum
Rumkówna, A., 1948
List of the phytoplankton species occurring in the superficial water layers in the Gulf of Gdańsk. Bull. Lab. mar., Gdynia, No. 4: 139-141 with tables in back.

Pleurosigma Sardoum n.sp.
Zanon, V., 1948
Diatomee marini di Sardegna e Pugillo di Alghe Marine della stressa. Boll. Pesca, Piscitutura e Idrobiologia, Anno 24, ns. 3(2): 202-244, 27 figs. on 1 pl.

Pleurosigma smithianum n.sp.
Castracane degli Antelminelli, F., 1886
1. Report on the Diatomaceae collected by H.M.S. Challenger during the years 1873-1876. Rept. Sci. Results, H.M.S. Challenger, Botany Vol. II, 178 pp., 30 pls.

Pleurosigma Smithianum
Hendey, N.I., 1937
The plankton diatoms of the southern seas. Discovery Repts. 16:151-364, pls.6-13.

Pleurosigma speciosum
Castracane degli Antelminelli, F., 1886
1. Report on the Diatomaceae collected by H.M.S. Challenger during the years 1873-1876. Rept. Sci. Results, H.M.S. Challenger, Botany Vol. II, 178 pp., 30 pls.

Pleurosigma speciosum n.sp.
Smith, W., 1852
Notes on the Diatomaceae; with descriptions of British species included in the genus Pleurosigma. (cont. Vol. 7, p. 14). Ann. Mag. Nat. Hist. II Vol. 9:1-12, 2 pls.

Pleurosigma speciosum
Zanon, V., 1948
Diatomee marini di Sardegna e Pugillo di Alghe Marine della stressa. Boll. Pesca, Piscitutura e Idrobiologia, Anno 24, ns. 3(2): 202-244, 27 figs. on 1 pl.

Pleurosigma Spencerii
Allen, W.E., and E.E. Cupp, 1935
Plankton diatoms of the Java Sea. Annales du Jardin Botanique de Buitenzorg XLIV (2):101-174, figs.1-127.
(drawings of all species mentioned)

Pleurosigma spenceri
Morse, D.C., 1947
Some observations on seasonal variations in plankton population Patuxant River, Maryland 1943-1945. Bd. Nat. Res., Publ. No.65, Chesapeake Biol. Lab., 31, 3 figs.

Pleurosigma Spencerii
Smith, W., 1852
Notes on the Diatomaceae; with descriptions of British species included in the genus Pleurosigma. (cont. Vol. 7, p. 14). Ann. Mag. Nat. Hist. II Vol. 9:1-12, 2 pls.

Pleurosigma strigilis
Smith, W., 1852
Notes on the Diatomaceae; with descriptions of British species included in the genus Pleurosigma. (cont. Vol. 7, p. 14). Ann. Mag. Nat. Hist. II Vol. 9:1-12, 2 pls.

Pleurosigma strigosum
Hendy, N. Ingram, 1964
An introductory account of the smaller algae of British coastal waters. V. Bacillariophyceae (Diatoms). Her Majesty's Stationary Office, 317 pp., 45 pls.

Pleurosigma strigosum
Hendey, N.I., 1951
Littoral diatoms of Chicester Harbour with special reference to fouling. J.Roy. Microscop. Soc. 71(1): 1-86, 18 pls.

Pleurosigma (Eupleurosigma) strigosum
Oliveira, Lejeune P.H. de, and H. Muth, 1960. Microscopia electronica de seis diatomaceas Pleurosigma, com uma critica do genero (Naviculaceae, Bacillariophyceae). Mem. Inst. Oswaldo Cruz, Brasil, 58(1):1-38.

Pleurosigma strigosum
Rampi, L., 1942
Ricerche sul fitoplancton del Mare Ligure 6. Le diatomee delle acque di Sanremo. Nuovo Giornale Botanico Italiano, N.S., 49:252-268.

Pleurosigma strigosum n.sp.
Smith, W., 1852
Notes on the Diatomaceae; with descriptions of British species included in the genus Pleurosigma. (cont. Vol. 7, p. 14). Ann. Mag. Nat. Hist. II Vol. 9:1-12, 2 pls.

Pleurosigma strigosum
Takano, Hideaki, 1963.
Notes on marine littoral diatoms of Japan. Bull. Tokai Reg. Fish. Res. Lab., No. 36:1-8.

Pleurosigma Stuxbergii
Zanon, V., 1948
Diatomee marini di Sardegna e Pugillo di Alghe Marine della stressa. Boll. Pesca, Piscitutura e Idrobiologia, Anno 24, ns. 3(2): 202-244, 27 figs. on 1 pl.

Pleurosigma subhyalinum n.sp.
Hustedt, F. and A.A. Aleem, 1951
Littoral diatoms from the Salstone near Plymouth. JMBA 30(1): 177-196.

Pleurosigma tenerum n.sp.
Jørgensen, E., 1905
B. Protistplankton and the diatoms in bottom samples. Hydrographical and biological investigations in Norwegian fjords. Bergens Mus. Skr. 7: 49-225.

Pleurosigma tenuissimum hyperborea
Jørgensen, E., 1905
B. Protistplankton and the diatoms in bottom samples. Hydrographical and biological investigations in Norwegian fjords. Bergens Mus. Skr. 7: 49-225.

Pleurosigma tenuirostre
Jørgensen, E., 1905
B. Protistplankton and the diatoms in bottom samples. Hydrographical and biological investigations in Norwegian fjords. Bergens Mus. Skr. 7: 49-225.

Pleurosigma thaitiense n.sp.
Castracane degli Antelminelli, F., 1886
1. Report on the Diatomaceae collected by H.M.S. Challenger during the years 1873-1876. Rept. Sci. Results, H.M.S. Challenger, Botany Vol. II, 178 pp., 30 pls.

Pleurosigma (Eupleurosigma) thuringicum
Oliveira, Lejeune P.H. de, and H. Muth, 1960.
Microscopia electronica de seis diatomaceas Pleurosigma, com uma critica do genero (Naviculaceae, Bacillariophyceae). Mem. Inst. Oswaldo Cruz, Brasil, 58(1):1-38.

Pleurosigma Wansbeckii
Hendey, N.I., 1951
Littoral diatoms of Chicester Harbour with special reference to fouling. J.Roy. Microscop. Soc. 71(1): 1-86, 18 pls.

Pleurosigma wigginsianum n.sp.
Hendey, N. Ingram 1971.
Some marine diatoms from the Galapagos Islands.
Nova Hedwigia 22(1/2):371-422.

Ploiaria petasiformis
Mann, A., 1907
Report on the diatoms of the Albatross voyages in the Pacific Ocean, 1888-1904. Contrib. U. S. Nat. Herb. 10(5):221-419, Pls. XLIV-LIV.

Podocystis
Desikachary, T. V., 1956(1957).
Electron microscope studies on diatoms. J. R. Microsc. Soc. (3)76(1/2):9-36.

Podocystis adriatica
Eskinazi Enide e Shigekatsv Satô (1963-1964) 1966.
Contribuição ao estudo das diatomaceas da Praia de Piedade.
Trabhs Inst. Oceanogr., Univ. Recife, 5 (5/6): 73-114.

Podocystis adriatica
Hendy, N. Ingram, 1964
An introductory account of the smaller algae of British coastal waters. V. Bacillariophyceae (Diatoms).
Her Majesty's Stationary Office, 317 pp., 45 pls.

Podocystis adriatica
Rampi, L., 1942
Ricerche sul fitoplancton del Mare Ligure 6. Le diatomee delle acque di Sanremo. Nuovo Giornale Botanico Italiano, N.S., 49:252-268.

Podocystis adriatica
Zanon, V., 1948
Diatomee marini di Sardegna e Pugillo di Alghe Marine della stressa. Boll. Pesca, Piscitutura e Idrobiologia, Anno 24, ns. 3(2): 202-244, 27 figs. on 1 pl.

Podocystis ovalis n.sp.
Misra, J.N., 1956.
A systematic account of some littoral marine diatoms from the west coast of India. J. Bombay Nat. Hist. Soc., 53(4):537-568.

Podosira
Brunel, J., 1962
Le phytoplancton de la Baie de Chaleurs. Inst. Botan., Univ. Montréal, Contrib. No. 77: 365 pp., 66 pls.

Podosira
Desikachary, T.V., 1956,(1957).
Electron microscope studies on diatoms. J.R. Microsc. Soc. (3)76(1/2):9-36.

Podosira sp.
Hasle, Grethe Rytter, 1960
Phytoplankton and ciliate species from the Tropical Pacific.
Skr. Norske Videnskaps-Akad., Oslo, 1. Mat.-Nat. Kl., 1960(2): 1-50.

Podosira sp.
Manguin, E., 1954
Diatomees marines provenant de l'ile Heard (Australian National Antarctic Expedition). Rev. Algol., n.s., 1: 14-24.

Podosira Adeliae n.sp.
Manguin, E., 1957.
Premier inventaire des diatomées de la Terre Adélie Antarctique. Espèces nouvelles. Rev. Algologique, n.s., 3(3):111-134.

Podosira argus
Mann, A., 1907
Report on the diatoms of the Albatross voyages in the Pacific Ocean, 1888-1904. Contrib. U. S. Nat. Herb. 10(5):221-419, Pls. XLIV-LIV.

Podosira compressa
Mann, A., 1893
List of Diatomaceae from a deep-sea dredging in the Atlantic Ocean off Delaware Bay by the U. S. Fish Commission Steamer Albatross. Proc. U. S. Nat. Mus. 16:303-312.

Podosira dubia
Politis, J., 1949
Diatomees marines de Bosphores et des ibes de la mer de Marmara. II Practica tou Hellenikou Hidrobiologikou Institutoutou 1929, Etoz 1929, 3(1):11-31.

Porosira glacialis
Braarud, T., 1945
A phytoplankton survey of the polluted waters of inner Oslo Fjord. Hvalrådets Skrifter, No.28, 142 pp., 19 text figs., 17 tables.

Porosira glacialis
Braarud, T., 1945
Experimental studies on marine plankton diatoms. Avhandlingar utgitt av Det Norske Videnskaps-Akademi i Oslo. 1. Mat.-Naturv. Klasse 1944. No.10, 1-16, 1 text fig.

Porosira glacialis
Braarud, T., and Adam Bursa, 1939
On the phytoplankton of the Oslo Fjord, 1933-1934. Hvalrådets Skr. No.19:1-63; 9 text figs. Reviewed. J. du. Cons. 14(3): 418-420. A.C. Gardiner.

Porosira glacialis
Hendy, N. Ingram, 1964
An introductory account of the smaller algae of British coastal waters. V. Bacillariophyceae (Diatoms). Her Majesty's Stationary Office, 317 pp., 45 pls.

Podosira hormoides
Frenguelli, Joaquin, and Hector Antonio Orlando, 1959.
Operacion MERLUZA. Diatomeas y silicoflagelades del plancton del "VI Crucero". Servicio Hidrogr. Naval., Argentina, Publ. No. H. 619: 5-62.

Podosira hormoides
Zanon, V., 1948
Diatomee marini di Sardegna e Pugillo di Alghe Marine della stressa. Boll. Pesca, Piscitutura e Idrobiologia, Anno 24, ns. 3(2): 202-244, 27 figs. on 1 pl.

Podosira Listardii n.sp.
Manguin, E., 1957.
Premier inventaire des diatomées de la Terre Adélie Antarctique. Espèces nouvelles. Rev. Algologique, n.s., 3(3):111-134.

Podosira maculata
Mann, A., 1893
List of Diatomaceae from a deep-sea dredging in the Atlantic Ocean off Delaware Bay by the U. S. Fish Commission Steamer Albatross. Proc. U. S. Nat. Mus. 16:303-312.

Podosira montagnei
de Sousa e Silva E., 1956.
Contribution à l'étude du microplancton de Dakar et des regions maritimes voisines. Bull. I.F.A.N., 8(2):335-371, 7 pls.

Podosira montagnei
Fox, M., 1957.
A first list of marine algae from Nigeria. J. Linn. Soc., London, 55(362):615-631.

Podosira montagnei
Hendy, N. Ingram, 1964
An introductory account of the smaller algae of British coastal waters. V. Bacillariophyceae (Diatoms). Her Majesty's Stationary Office, 317 pp., 45 pls.

Podosira montagnei
Hendey, N-Ingram, 1958 [1957(Publ. 1958)]
Marine diatoms from some West African Ports-J. R. Microsc. Soc. (3) 77(1/2): 28-85.

Podosira montagnei
Hustedt, F. and A.A. Aleem, 1951
Littoral diatoms from the Salstone near Plymouth. JMBA 30(1): 177.196.

Podosira montagnei
Misra, J.N., 1956.
A systematic account of some littoral marine diatoms from the west coast of India. J. Bombay Nat. Hist. Soc., 53(4):537-568.

Podosira montagnei
Zanon, V., 1948
Diatomee marini di Sardegna e Pugillo di Alghe Marine della stressa. Boll. Pesca, Piscitutura e Idrobiologia, Anno 24, ns. 3(2): 202-244, 27 figs. on 1 pl.

Podosira stelliger
Hendy, N. Ingram, 1964
An introductory account of the smaller algae of British coastal waters. V. Bacillariophyceae (Diatoms). Her Majesty's Stationary Office, 317 pp., 45 pls.

Podosira stelliger
Hendey, N.I., 1951
Littoral diatoms of Chicester Harbour with special reference to fouling. J. Roy. Microscop. Soc. 71(1): 1-86, 18 pls.

Podosira stelliger
Hustedt, F. and A.A. Aleem, 1951
Littoral diatoms from the Salstone near Plymouth. JMBA 30(1): 177.196.

Podosira stilliger
Mann, A., 1907
Report on the diatoms of the Albatross voyages in the Pacific Ocean, 1888-1904. Contrib. U. S. Nat. Herb. 10(5):221-419, Pls. XLIV-LIV.

Podosira stelliger
Rampi, L., 1940
Diatomee del Mare Adriatico. Nuovo Giornale Botanico Italiano, n.s., 47:559-608.

Podosira stelliger
Zanon, D. V., 1949
Diatomee di Buenos Aires (Argentina) Atti. Accad. Naz. Lincei, Memorie, Cl. Sci. fis., mat. e. nat., ser. 7, 11(3):59-151, 2 pls.

Podosira stelliger
Zanon, V., 1948
Diatomee marini di Sardegna e Pugillo di Alghe Marine della stressa. Boll. Pesca, Piscitutura e Idrobiologia, Anno 24, ns. 3(2): 202-244, 27 figs. on 1 pl.

Podosira subtilis
Mann, A., 1907
Report on the diatoms of the Albatross voyages in the Pacific Ocean, 1888-1904. Contrib. U. S. Nat. Herb. 10(5):221-419, Pls. XLIV-LIV.

Podosira tenebra
Hendey, N-Ingram, 1958 [1957(Publ. 1958)]
Marine diatoms from some West African Ports-J. R. Microsc. Soc. (3) 77(1/2): 28-85.

Podosphenia Pappeana n.sp.
Grunow, A., 1863
Ueber einige neue und ungenügend bekannte Arten und Gattungen von Diatomaceen. Verhandl. d. K.K. Zool. Bot. Gesellsch., Vienna, 13: 137-162, pl.4-5 (Pl. 13-14).

Porodiscus interruptus
Barker, J.W. and S.H. Meakin, 1948
New and rare diatoms. J. Quakett Micros. Club, ser.4, 2(5):233-235, pl.25.

Porodiscus stolterfothii n.sp.
Castracane degli Antelminelli, F., 1886
1. Report on the Diatomaceae collected by H.M.S. Challenger during the years 1873-1876. Rept. Sci. Results, H.M.S. Challenger, Botany Vol. II, 178 pp., 30 pls.

Porosira n.gen.
Jørgensen, E., 1905
B.Protistplankton and the diatoms in bottom samples. Hydrographical and biological investigations in Norwegian fiords. Bergens Mus. Skr. 7: 49-225.

Porosira antarctica n. sp.
Kozlova, O.G., 1962
Specific composition of diatoms in the waters of the Indian sector of the Antarctic. Investigation of Marine Bottom Sediments. (In Russian). Trudy Inst. Okeanol., Akad. Nauk. SSSR, 61: 3-18.
English summary, p. 18.

Porosira dichotomica nsp.
Kozlova, O.G., 1962
Specific composition of diatoms in the waters of the Indian sector of the Antarctic. Investigation of Marine Bottom Sediments. (In Russian). Trudy Inst. Okeanol., Akad. Nauk. SSSR, 61: 3-18.
English summary, p. 18

Porosira glacialis
Brunel, J., 1962
Le phytoplankton de la Baie de Chaleurs. Inst. Botan., Univ. Montréal, Contrib. No. 77: 365 pp., 66 pls.

Porosira glacialis
Brunel, Jules, 1962
Le phytoplancton de la Baie des Chaleurs. Contrib. Ministère de la Chasse et des Pêcheries, Province de Québec, No. 91: 365 pp.

Porosira glacialis
Gran, H.H., and T. Braarud, 1935
A quantitative study of the phytoplankton in the Bay of Fundy and the Gulf of Maine (including observations on hydrography, chemistry, and turbidity). J. Biol. Bd., Canada, 1(5):279-467, 69 text figs.

Porosira glacialis
Jørgensen, E., 1905
B.Protistplankton and the diatoms in bottom samples. Hydrographical and biological investigations in Norwegian fiords. Bergens Mus. Skr. 7: 49-225.

Porosira glacialis
Lillick, L.C., 1940
Phytoplankton and planktonic protozoa of the offshore waters of the Gulf of Maine. Pt.II. Qualitative Composition of the Planktonic Flora. Trans. Am. Phil. Soc., n.s., 31(3):193-237, 13 text figs.

Porosira glacialis
Lillick, L.C., 1938
Preliminary report of the phytoplankton of the Gulf of Maine. Am. Mid. Nat. 20(3):624-640, 1 text figs. 37 tables.

Porosira pseudodenticulata comb nov
Jouse, A.P., G.S. Koroleva, G.A. Nagaeva, 1962
Diatoms in the surface layer of sediment in the Indian sector of the Antarctic. Investigations of marine bottom sediments. (In Russian; English summary). Trudy Inst. Okeanol., Akad. Nauk. SSSR, 61:19-92.

Porosira pseudodenticulata
Kozlova, O.G., 1962
Specific composition of diatoms in the waters of the Indian sector of the Antarctic. Investigation of Marine Bottom Sediments. (In Russian). Trudy Inst. Okeanol., Akad. Nauk. SSSR, 61: 3-18.
English summary, p. 18.

Porpeia quadriceps
Mann, A., 1907
Report on the diatoms of the Albatross voyages in the Pacific Ocean, 1888-1904. Contrib. U. S. Nat. Herb. 10(5):221-419, Pls. XLIV-LIV.

Protoraphis hustediana n.gen., n.sp.
Protoraphidaceae, eine neue Familie der Diatomeen
Beiheft Nova Hedwigia 31: 383-394.

Pseudoamphiprora Manginii n.sp.
Manguin, E., 1957.
Premier inventaire des diatomées de la Terre Adélie Antarctique. Espèces nouvelles. Rev. Algologique, n.s., 3(3):111-134.

Pseudoamphiprora monodii n.sp.
Amossé, A. 1970.
Diatomées marines et saumâtres du Sénégal et de la Côte d'Ivoire. Bull. Inst. fond. Afr. Noire (A) 32 (2): 289-311.

Pseudoamphiprora stauropter-a
Hendy, N. Ingram, 1964
An introductory account of the smaller algae of British coastal waters. V. Bacillariophyceae (Diatoms).
Her Majesty's Stationary Office, 317 pp., 45 pls.

Pseudoeunotia doliolus
Allen, W.E., 1937
Plankton diatoms of the Gulf of California obtained by the G. Allan Hancock Expedition of 1936. The Hancock Pacific Expeditions, Univ. So. Calif. Publ. 3:47-59, 1 fig.

Pseudoeunotia doliolus
Cupp, Easter E., 1943
Marine plankton diatoms of the west coast of North America. Bull. S.I.O. 5(1):1-238, 5 pls., 168 text figs.

Pseudoeunotia doliolus
Cupp, E.E., 1934
Analysis of marine diatom collections taken from the Canal Zone to California during March, 1933. Trans. Am. Micros. Soc. LIII (1):22-29, 1 map.

Pseudoeunotia doliolus
Cupp, E.E. and Allen, W.E., 1938
Plankton diatoms of the Gulf of California obtained by Allan Hancock Pacific Expedition of 1937. The Hancock Pacific Expeditions, The Univ. So. Calif. Publ. 3: 61-74, 1 map, pls.4-15.

Pseudoeunotia doliolus
Gilbert, J.Y., and W.E. Allen, 1943
The phytoplankton of the Gulf of California obtained by the "E.W. Scripps" in 1939 and 1940. J. Mar. Res. V(2):89-110, figs.30-31.

Pseudoeunotia doliolus
Hasle, Grethe Rytter, 1960
Phytoplankton and ciliate species from the Tropical Pacific.
Skr. Norske Videnskaps-Akad., Oslo. 1. Mat.-Nat. Kl., 1960(2): 1-50.

Pseudohimantidium adriaticum
Simonsen, Reimer 1970
Protoraphidaceae, eine neue Familie der Diatomeen
Beiheft Nova Hedwigia 31: 383-394.

Pseudohimantidium pacificum
Krasske, G., 1941
Die Kieselalgen des chilenischen Küstenplankton (Aus dem südchilenischen Kusten gebiet IX). Arch. Hydrobiol. 38 (2):260-287.

Pseudohimantidium pacificum
Simonsen, Reimer 1970
Protoraphidaceae, eine neue Familie der Diatomeen.
Beiheft Nova Hedwigia 31: 383-394.

Pseudonitzschia
Desikachary, T.V., 1956(1957).
Electron microscope studies on diatoms. J.R. Microsc. Soc. (3)76(1/2):9-30.

Pseudonitzschia antarctica n.sp.
Manguin, E., 1957.
Premier inventaire des diatomées de la Terre Adélie Antarctique. Espèces nouvelles. Rev. Algologique, n.s., 3(3):111-134.

Pseudonitschia Heimii n.sp.
Manguin, E., 1957.
Premier inventaire des diatomées de la Terre Adélie Antarctique. Espèces nouvelles. Rev. Algologique, n.s., 3(3):111-134.

Pseudonitzschia seriata
Frenguelli, Joaquin, and Hector Antonio Orlando, 1959.
Operacion MERLUZA. Diatomeas y silicoflagelados del plancton del "VI Crucero". Servicio Hidrogr. Naval., Argentina, Publ. No. H. 619: 5-62.

Pseudonitschia seriata
Rampi, L., 1942
Ricerche sul fitoplancton del Mare Ligure 6. Le diatomee delle acque di Sanremo. Nuovo Giornale Botanico Italiano, N.S., 49:252-268.

Pseudonitschia sicula
Rampi, L., 1942
Ricerche sul fitoplancton del Mare Ligure 6. Le diatomee delle acque di Sanremo. Nuovo Giornale Botanico Italiano, N.S., 49:252-268.

Pseudostictodiscus angulatus
Cleve-Euler, A., 1951
Die Diatomeen von Schweden und Finnland. Kungl. Svenska Vetenskaps Akad. Handl., Fjärde Ser. 2(1): 161 pp., 6 pls.

Pseudo-triceratium cinnamomeum
Hendey, N.I., 1937
The plankton diatoms of the southern seas. Discovery Repts. 16:151-364, pls.6-13.

Pterotheca spada
Cleve-Euler, A., 1951
Die Diatomeen von Schweden und Finnland. Kungl. Svenska Vetenskaps Akad. Handl., Fjärde Ser. 2(1): 161 pp., 6 pls.

Pyrgupyxis escena, n-gen., n.sp.
Hendey, N. Ingram 1969.
Pyrgupyxis, a new genus of diatoms from a South Atlantic Eocen core.
Occ. Pap. Calif. Acad. Sci. (72) 6pp.

Pyxidicula
Brunel, J., 1962
Le phytoplancton de la Baie de Chaleurs. Inst. Botan., Univ. Montréal, Contrib. No. 77: 365 pp., 66 pls.

Pyxidicula mediterranea
Zanon, V., 1948
Diatomee marini di Sardegna e Pugillo di Alghe Marine della stressa. Boll. Pesca, Piscitutura e Idrobiologia, Anno 24, ns. 3(2): 202-244, 27 figs. on 1 pl.

Pyxidicula minuta
Manguin, E., 1954
Diatomées marines provenant de l'ile Heard (Australian National Antarctic Expedition). Rev. Algol., n.s., 1: 14-24.

Pyxilla aculeifera
Cleve-Euler, A., 1951
Die Diatomeen von Schweden und Finnland. Kungl. Svenska Vetenskaps Akad. Handl., Fjärde Ser. 2(1): 161 pp., 6 pls.

Pyxilla antiqua n.sp.
Cleve-Euler, A., 1951
Die Diatomeen von Schweden und Finnland. Kungl. Svenska Vetenskaps Akad. Handl., Fjärde Ser. 2(1): 161 pp., 6 pls.

Pyxilla Baltica
Mann, A., 1893
List of Diatomaceae from a deep-sea dredging in the Atlantic Ocean off Delaware Bay by the U. S. Fish Commission Steamer Albatross. Proc. U. S. Nat. Mus. 16:303-312.

Pyxilla capitata n.sp.
Barker, J.W. and S.H. Meakin, 1948 (fossile)
New and rare diatoms. J. Quakett Micros. Club, ser.4, 2(5):233-235, pl.25.

Pyxilla carinifera
Cleve-Euler, A., 1951
Die Diatomeen von Schweden und Finnland. Kungl. Svenska Vetenskaps Akad. Handl., Fjärde Ser. 2(1): 161 pp., 6 pls.

Pyxilla dubia
Cleve-Euler, A., 1951
Die Diatomeen von Schweden und Finnland. Kungl. Svenska Vetenskaps Akad. Handl., Fjärde Ser. 2(1): 161 pp., 6 pls.

Pyxilla Johnsoniana
Barker, J.W. and S.H. Meakin, 1948
New and rare diatoms. J. Quakett Micros. Club, ser.4, 2(5):233-235, pl.25.

Radiodiscus chafferaci
Fox, M., 1957.
A first list of marine algae from Nigeria. J. Linn. Soc., London, 55(362):615-631.

Radiodiscus hispidus
Fox, M., 1957.
A first list of marine algae from Nigeria. J. Linn. Soc., London, 55(362):615-631.

Rhaphoneis
Desikachary, T.V., 1956(1957).
Electron microscope studies on diatoms. J.R. Microsc. Soc. (3)76(1/2):9-36.

Raphoneis sp.
Morse, D.C., 1947
Some observations on seasonal variations in plankton population Patuxant River, Maryland 1943-1945. Bd. Nat. Res., Publ. No.65, Chesapeake Biol. Lab., 31, 3 figs.

Raphoneis amphiceros
Allen, W.E., and E.E. Cupp, 1935
Plankton diatoms of the Java Sea.
Annales du Jardin Botanique de Buitenzorg
XLIV (2):101-174, figs.1-127.

(drawings of all species mentioned)

Raphoneis amphiceros
Hendy, N. Ingram, 1964

An introductory account of the smaller algae of British coastal waters. V. Bacillariophyceae (Diatoms).
Her Majesty's Stationary Office, 317 pp., 45 pls.

Raphoneis amphiceros
Hendey, N.I., 1951
Littoral diatoms of Chicester Harbour with special reference to fouling. J.Roy. Microscop. Soc. 71(1): 1-86, 18 pls.

Raphoneis amphiceros
Hustedt, F. and A.A. Aleem, 1951
Littoral diatoms from the Salstone near Plymouth. JMBA 30(1): 177.196.

Raphoneis amphiceros
Lillick, L.C., 1940
Phytoplankton and planktonic protozoa of the offshore waters of the Gulf of Maine. Pt.II. Qualitative Composition of the Planktonic Flora. Trans. Am. Phil. Soc., n.s., 31(3):193-237, 13 text figs.

Raphoneis amphiceros
Mann, A., 1893
List of Diatomaceae from a deep-sea dredging in the Atlantic Ocean off Delaware Bay by the U. S. Fish Commission Steamer Albatross. Proc. U. S. Nat. Mus. 16:303-312.

Raphoneis amphiceros
Meunier, A., 1915
Microplancton de la Mer Flamande. 2. Diatomées (excepté le genre Chaetoceros). Mem. Mus. Roy. Hist. Nat., Belgique, 7(3):1-118, Pls. VIII-XIV.

Raphoneis amphiceros
Politis, J., 1949
Diatomees marines de Bosphores et des ibes de la mer de Marmara. II Practica tou Hellenikou Hidrobiologikou Institutoutou 1929, Etoz 1929, 3(1):11-31.

Raphoneis amphiceros
Schodduyn, M., 1926
Observations faites dans la baie d'Ambleteuse (Pas de Calais). Bull. Inst. Ocean., Monaco, No. 482: 64 pp.

Raphoneis amphiceros
Takano, Hideaki, 1962
Notes on epiphytic diatoms upon sea-weeds from Japan.
J. Oceanogr. Soc., Japan, 18(1):29-33.

Raphoneis amphiceros
Zanon, V., 1948
Diatomee marini di Sardegna e Pugillo di Alghe Marine della stressa. Boll. Pesca, Piscitutura e Idrobiologia, Anno 24, ns. 3(2): 202-244, 27 figs. on 1 pl.

Raphoneis amphiceros rhombica
Mann, A., 1893
List of Diatomaceae from a deep-sea dredging in the Atlantic Ocean off Delaware Bay by the U. S. Fish Commission Steamer Albatross. Proc. U. S. Nat. Mus. 16:303-312.

Raphoneis belgica
Meunier, A., 1915
Microplancton de la Mer Flamande. 2. Diatomées (excepté le genre Chaetoceros). Mem. Mus. Roy. Hist. Nat., Belgique, 7(3):1-118, Pls. VIII-XIV.

Raphoneis belgica
Zanon, V., 1948
Diatomee marini di Sardegna e Pugillo di Alghe Marine della stressa. Boll. Pesca, Piscitutura e Idrobiologia, Anno 24, ns. 3(2): 202-244, 27 figs. on 1 pl.

Raphoneis bilineata lancettula
Takano, Hideaki, 1965.
New and rare diatoms from Japanese marine waters. 1.
Bull. Tokai Reg. Fish. Res. Lab., No. 42:1-10.

Raphoneis bilineata protracta
Takano, Hideaki, 1965.
New and rare diatoms from Japanese marine waters. 1.
Bull. Tokai Reg. Fish. Res. Lab., No. 42:1-10.

Raphoneis cocconeides n.sp.
Schrader, Hans-Joachim, 1973
Cenozoic diatoms from the northeast Pacific, leg 18. Initial Repts Deep Sea Drilling Proj. 18:673-797.

Raphoneis discoides n.sp.
Subrahmanyan, R., 1946.
A systematic account of the marine plankton diatoms of the Madras coast. Proc. Indian Acad. Sci. 24(4):85-197, Pl. 2, 440 textfigs.

Raphoneis gemmifera
Mann, A., 1893
List of Diatomaceae from a deep-sea dredging in the Atlantic Ocean off Delaware Bay by the U. S. Fish Commission Steamer Albatross. Proc. U. S. Nat. Mus. 16:303-312.

Raphoneis japonica n.sp.
Castracane degli Antelminelli, F., 1886
1. Report on the Diatomaceae collected by H.M.S. Challenger during the years 1873-1876. Rept. Sci. Results, H.M.S. Challenger, Botany Vol. II, 178 pp., 30 pls.

Raphoneis mammalis n.sp.
Castracane degli Antelminelli, F., 1886
1. Report on the Diatomaceae collected by H.M.S. Challenger during the years 1873-1876. Rept. Sci. Results, H.M.S. Challenger, Botany Vol. II, 178 pp., 30 pls.

Raphoneis miocenica n.sp.
Schrader, Hans-Joachim, 1973
Cenozoic diatoms from the northeast Pacific, leg 18. Initial Repts Deep Sea Drilling Proj. 18:673-797.

Raphoneis nitida
Rampi, L., 1940
Diatomee del Mare Adriatico. Nuovo Giornale Botanico Italiano, n.s., 47:559-608.

Raphoneis nitida
Sousa a Silva, E., and J. Dos Santos-Pinto, 1948
O Plancton da Baia de S. Martinho do Porto. 1. Diatomaceas e Dinoflagelados. Bol. Soc. Portuguese de Ciencias Naturais, 16(2):134-187, 6 pls. (Trav. Sta. Biol. Mar. de Lisbonne No. 52).

Raphoneis nitida
Zanon, V., 1948
Diatomee marini di Sardegna e Pugillo di Alghe Marine della stressa. Boll. Pesca, Piscitutura e Idrobiologia, Anno 24, ns. 3(2): 202-244, 27 figs. on 1 pl.

Raphoneis surirella
Allen, W.E., and E.E. Cupp, 1935
Plankton diatoms of the Java Sea.
Annales du Jardin Botanique de Buitenzorg
XLIV (2):101-174, figs.1-127.

(drawings of all species mentioned)

Raphoneis surirella
Hendy, N. Ingram, 1964

An introductory account of the smaller algae of British coastal waters. V. Bacillariophyceae (Diatoms).
Her Majesty's Stationary Office, 317 pp., 45 pls.

Raphoneis surirella
Hustedt, F. and A.A. Aleem, 1951
Littoral diatoms from the Salstone near Plymouth. JMBA 30(1): 177.196.

Raphoneis surirella
Lillick, L.C., 1940
Phytoplankton and planktonic protozoa of the offshore waters of the Gulf of Maine. Pt.II. Qualitative Composition of the Planktonic Flora. Trans. Am. Phil. Soc., n.s., 31(3):193-237, 13 text figs.

Raphoneis surirella
Mann, A., 1893
List of Diatomaceae from a deep-sea dredging in the Atlantic Ocean off Delaware Bay by the U. S. Fish Commission Steamer Albatross. Proc. U. S. Nat. Mus. 16:303-312.

Raphoneis Surirella
Meunier, A., 1915
Microplancton de la Mer Flamande. 2. Diatomées (excepté le genre Chaetoceros). Mem. Mus. Roy. Hist. Nat., Belgique, 7(3):1-118, Pls. VIII-XIV.

Raphoneis surirella
Rampi, L., 1942
Ricerche sul fitoplancton del Mare Ligure 6. Le diatomee delle acque di Sanremo. Nuovo Giornale Botanico Italiano, N.S., 49:252-268.

Raphoneis surirella
Zanon, D. V., 1949
Diatomee di Buenos Aires (Argentina)
Atti Accad. Naz. Lincei, Memorie, Cl. Sci. fis., mat. e. nat., ser. 7, 11(3):59-151, 2 pls.

Raphoneis surirella
Zanon, V., 1948
Diatomee marini di Sardegna e Pugillo di Alghe Marine della stressa. Boll. Pesca, Piscitutura e Idrobiologia, Anno 24, ns. 3(2): 202-244, 27 figs. on 1 pl.

Raphoneis surirella var. australis
Rampi, L., 1940
Diatomee del Mare Adriatico. Nuovo Giornale Botanico Italiano, n.s., 47:559-608.

Raphoneis surirella australis
Zanon, V., 1947.
Diatomée delle Isole Ponziane - Materiale per una florula diatomologica del Mare Tirreno. Bol. Pesca, Piscicol., Idrobiol., n.s., 2(1):36-53, 1 pl. of 10 figs.

Rhabdonema sp.
Braarud, T., 1945
A phytoplankton survey of the polluted waters of inner Oslo Fjord. Hvalrådets Skrifter, No.28, 142 pp., 19 text figs., 17 tables.

Rhabdonema sp.
Meunier, A., 1915
Microplancton de la Mer Flamande. 2. Diatomées (excepté le genre Chaetoceros). Mem. Mus. Roy. Hist. Nat., Belgique, 7(3):1-118, Pls. VIII-XIV.

Rhabdonema adriaticum
Ercegovic, A., 1936
Etudes qualitative et quantitatives du phytoplancton dans les eaux cotières de l'Adriatique oriental moyen au cours de l'année 1934. Acta Adriatica 1(9):1-126

Rhabdonema adriaticum
Eskinazi Enide e Shigekatsv Satô (1963-1964) 1966.
Contribuição ao estudo das diatomaceas da Praia de Piedade.
Trabhs Inst. Oceanogr., Univ. Recife, 5 (5/6): 73-114.

Rhabdonema adriaticum
Forti, A., 1922
Ricerche sulla flora pelagica (fitoplancton) di Quarto dei Mille. Mem. R. Com. Talass. Ital. 97:248 pp., 13 pls.

Rhabdonema adriaticum
Ghazzawi, F.M., 1939
Plankton of the Egyptian waters. A study of the Suez Canal Plankton. (A) Phytoplankton. Preliminary Report 83 pp. Notes and Memoires, Min. Commerce-Industry, Egypt, Hydrobiol. & Fish. 65 figs.

Rhabdonema adriaticum
Hendy, N. Ingram, 1964
An introductory account of the smaller algae of British coastal waters. V. Bacillariophyceae (Diatoms).
Her Majesty's Stationary Office, 317 pp., 45 pls.

Rhabdonema adriaticum
Hendey, N.I., 1951
Littoral diatoms of Chicester Harbour with special reference to fouling. J.Roy. Microscop. Soc. 71(1): 1-86, 18 pls.

Rhabdonema adriaticum
Hendey, N.I., 1937
The plankton diatoms of the southern seas. Discovery Repts. 16:151-364, pls.6-13.

Rhabdonema adriaticum
Jørgensen, E., 1905
B.Protistplankton and the diatoms in bottom samples. Hydrographical and biological investigations in Norwegian fjords. Bergens Mus. Skr. 7: 49-225.

Rhabdonema adriaticum
Jorgensen, E., 1900
Protophyten und Protozoën im Plankton aus der Norwegischen Westkerste. Bergens Mus. Aarb. 1899(6): 95 pp., 5 pls., 83 tables.

Rhabdonema adriaticum
Lillick, L.C., 1940
Phytoplankton and planktonic protozoa of the offshore waters of the Gulf of Maine. Pt.II. Qualitative Composition of the Planktonic Flora. Trans. Am. Phil. Soc., n.s., 31(3):193-237, 13 text figs.

Rhabdonema adriaticum
Lillick, L.C., 1937
Seasonal studies of the phytoplankton off Woods Hole, Massachusetts. Biol. Bull. LXXIII (3):488-503, 3 text figs.

Rhabdonema adriaticum
Morse, D.C., 1947
Some observations on seasonal variations in plankton population Patuxant River, Maryland 1943-1945. Bd. Nat. Res., Publ. No.65, Chesapeake Biol. Lab., 31, 3 figs.

Rhabdonema adriaticum
Pavillard, J., 1925
Bacillariales. Rept. on the Danish Oceangr. Exped., 1908-10 to the Mediterranean and adj. seas. Vol.II., Biol. J4:72 pp., 116 text figs.

Rhabdonema adriaticum
Politis, J., 1949
Diatomees marines de Bosphores et des ibes de la mer de Marmara. II Practica tou Hellenikou Hidrobiologikou Institutoutou 1929, Etoz 1929, 3(1):11-31.

Rhabdonema adriaticum
Rampi, L., 1942
Ricerche sul fitoplancton del Mare Ligure 6. Le diatomee delle acque di Sanremo. Nuovo Giornale Botanico Italiano, N.S., 49:252-268.

Rhabdonema adriaticum
Rampi, L., 1940
Diatomee del Mare Adriatico. Nuovo Giornale Botanico Italiano, n.s., 47:559-608.

Rhabdonema adriaticum
Sousa e Silva, E., 1949
Diatomaceas e Dinoflagelados de Baia de Cascais. Portugaliae Acta Biol., Volume: Julio Henriques, Ser. B: 300-383, 9 pls, 2 fold-in tables.

Rhabdonema Adriaticum
Sousa a Silva, E., and J. Dos Santos-Pinto, 1948
O Plancton da Baia de S. Martinho do Porto. 1. Diatomaceas e Dinoflagelados. Bol. Soc. Portuguese de Ciencias Naturais, 16(2):134-187, 6 pls. (Trav. Sta. Biol. Mar. de Lisbonne No. 52).

Rhabdonema adriaticum
Takano, Hideaki, 1962
Notes on epiphytic diatoms upon sea-weeds from Japan.
J. Oceanogr. Soc., Japan, 18(1):29-33.

Rhabdonema adriaticum
Zanon, V., 1948
Diatomee marini di Sardegna e Pugillo di Alghe Marine della stressa. Boll. Pesca, Piscitutura e Idrobiologia, Anno 24, ns. 3(2): 202-244, 27 figs. on 1 pl.

Rhabdonema arcuatum
Brunel, J., 1962
Le phytoplancton de la Baie de Chaleurs. Inst. Botan., Univ. Montréal, Contrib. No. 77: 365 pp., 66 pls.

Rhabdonema arcuatum ventricosum

Rhabdonema arcuatum
Brunel, Jules, 1962
Le phytoplancton de la Baie des Chaleurs. Contrib. Ministère de la Chasse et des Pêcheries, Province de Québec, No. 91: 365 pp.

Rhabdonema arcuatum
Florin, M-B., 1948
9. Diatomeae in submarine cores from the Tyrrhenian Sea. Medd. Ocean. Inst., Göteborg, 15 (Göteborgs Kungl. Vetenskaps-och Viterrhets Samhälles Handlingar, Sjätte Foljden, Ser. B 5(13):80-88.

Rhabdonema arcuatum
Hendy, N. Ingram, 1964
An introductory account of the smaller algae of British coastal waters. V. Bacillariophyceae (Diatoms).
Her Majesty's Stationary Office, 317 pp., 45 pls.

Rhabdonema arcuatum
Hendey, N.I., 1951
Littoral diatoms of Chicester Harbour with special reference to fouling. J.Roy. Microscop. Soc. 71(1): 1-86, 18 pls.

Rhabdonema arcuatum
Hustedt, F. and A.A. Aleem, 1951
Littoral diatoms from the Salstone near Plymouth. JMBA 30(1): 177.196.

Rhabdonema arcuatum
Jørgensen, E., 1905
B.Protistplankton and the diatoms in bottom samples. Hydrographical and biological investigations in Norwegian fjords. Bergens Mus. Skr. 7: 49-225.

Rhabdonema arcuatum
Jorgensen, E., 1900
Protophyten und Protozoën im Plankton aus der Norwegischen Westkerste. Bergens Mus. Aarb. 1899(6): 95 pp., 5 pls., 83 tables.

Rhabdonema arcuatum
Lillick, L.C., 1940
Phytoplankton and planktonic protozoa of the offshore waters of the Gulf of Maine. Pt.II. Qualitative Composition of the Planktonic Flora. Trans. Am. Phil. Soc., n.s., 31(3):193-237, 13 text figs.

Rhabdonema arcuatum
Takano, Hideaki, 1963
Notes on marine littoral diatoms of Japan. Bull. Tokai Reg. Fish. Res. Lab., No. 36:1-8.

Rhabdonema crassum n.sp.
Hendy, N. Ingram, 1964
An introductory account of the smaller algae of British coastal waters. V. Bacillariophyceae (Diatoms).
Her Majesty's Stationary Office, 317 pp., 45 pls.

Rhabdonema indicum n.sp.
Misra, J.N., 1956.
A systematic account of some littoral marine diatoms from the west coast of India.
J. Bombay Nat. Hist. Soc., 53(4):537-568.

Rhabdonema minutum
Hendy, N. Ingram, 1964
An introductory account of the smaller algae of British coastal waters. V. Bacillariophyceae (Diatoms).
Her Majesty's Stationary Office, 317 pp., 45 pls.

Rhabdonema minutum
Hustedt, F. and A.A. Aleem, 1951
Littoral diatoms from the Salstone near Plymouth. JMBA 30(1): 177.196.

Rhabdonema minutum
Jørgensen, E., 1905
B.Protistplankton and the diatoms in bottom samples. Hydrographical and biological investigations in Norwegian fjords. Bergens Mus. Skr. 7: 49-225.

Rhabdonema minutum
Jorgensen, E., 1900
Protophyten und Protozoën im Plankton aus der Norwegischen Westkerste. Bergens Mus. Aarb. 1899(6): 95 pp., 5 pls., 83 tables.

Rhabdonema minutum
Mann, A., 1893
List of Diatomaceae from a deep-sea dredging in the Atlantic Ocean off Delaware Bay by the U. S. Fish Commission Steamer Albatross. Proc. U. S. Nat. Mus. 16:303-312.

Rhabdonema minutum
Rampi, L., 1940
Diatomee del Mare Adriatico. Nuovo Giornale Botanico Italiano, n.s., 47:559-608.

Rhabdonema minutum (figs.
Sousa e Silva, E., 1949
Diatomaceas e Dinoflagelados de Baia de Cascais. Portugaliae Acta Biol., Volume: Julio Henriques, Ser. B: 300-383, 9 pls, 2 fold-in tables.

Rhizosolenia sp.
Boden, Brian, 1948
Marine plankton diatoms on operation HIGHJUMP in: Some oceanographic observations on operation HIGHJUMP. By R.S. Dietz. USNEL Rept. No.55, 97 pp., 41 figs. 7 July 1948.

Rhizosolenia sp.
Cupp, E.E., 1934
Analysis of marine diatom collections taken from the Canal Zone to California during March, 1933. Trans. Am. Micros. Soc. LIII (1):22-29, 1 map.

Rhizosolenia sp.
Cupp, E.E. and Allen, W.E., 1938
Plankton diatoms of the Gulf of California obtained by Allan Hancock Pacific Expedition of 1937. The Hancock Pacific Expeditions, The Univ. So. Calif. Publ. 3: 61-74, 1 map, pls.4-15.

Rhizosolenia
Desikachary, T.V., 1956(1957)
Electron microscope studies on diatoms. J.R. Microsc. Soc. (3)76(1/2):9-36.

Rhizosolenia sp.
Galtsoff, P.S., 1948
Red Tide. Progress Report on the investigations of the cause of the mortality of fish along the west coast of Florida conducted by the U.S. Fish and Wildlife Service and Cooperating Organizations. Fish and Wildlife Service, Special Scientific Rept. No.46, 44 pp. (mimeographed), 9 figs.

Rhizosolenia
Gunter G., F.G. Walton Smith, and R.H. Roberts, 1947
Mass mortality of marine animals on the lower west coast of Florida, November 1946 - January 1947. Science, 105(2723): 256-257.

Rhizosolenia
Höfler, Karl, und Luise Höfler, 1964.
Diatomeen des marinen Planktons als Isotonobionten. Pubbl. Staz. Zool., Napoli, 33(3):315-330.

Rhizosolenia
Lucas, C.E., 1940.
Ecological investigations with the continuous plankton recorder: the phytoplankton in the southern North Sea, 1932-37. Hull Bull. Mar. Ecol., 1(3):73-170, 17 textfigs., 21 pls.

Rhizosolenia
Neaverson, E., 1934
The sea-floor deposits. 1. General characteristics and distribution. Discovery Repts. 9: 297-349, Plates 17-22.

Rhizosolenia sp.
Paiva Carvalho, J., 1950
O plancton do Rio Maria Rodrigues (Cananeis). 1. Diatomaceas e Dinoflagelados. Bol. Inst. Paulista Oceanogr. 1(1); 27-43, 2 fold-in tables, 2 figs.

Rhizosolenia
Peragallo, H., 1892
Monographie de Rhizosolenia et de quelques genres voisins. Le Diatomiste,1:

Rhizosolenia acuminata
Cupp, Easter E., 1943
Marine plankton diatoms of the west coast of North America. Bull. S.I.O. 5(1):1-238, 5 pls., 168 text figs.

Rhizosolenia acuminata
Cupp, E.E., 1934
Analysis of marine diatom collections taken from the Canal Zone to California during March, 1933. Trans. Am. Micros. Soc. LIII (1):22-29, 1 map.

Rhizosolenia acuminata
Cupp, E.E. and Allen, W.E., 1938
Plankton diatoms of the Gulf of California obtained by Allan Hancock Pacific Expeditions, The Univ. So. Calif. Publ. 3: 61-74, 1 map, pls.4-15.

Rhizosolenia acuminata
Dangeard, P., 1927
Phytoplankton de la croisière du "Sylvana". Ann. Inst. Ocean., Monaco, n.s., 4(8):286-401, 54 text figs. (Février-Juin 1913).

Rhizosolenia acuminata
Eskinazi Enide e Shigekatsu Sato (1963-1964) 1966.
Contribuição ao estudo das diatomaceas da Praia de Piedade. Trabhs Inst. Oceanogr., Univ. Recife, 5 (5/6): 73-114.

Rhizosolenia acuminata
Forti, A., 1922
Ricerche sulla flora pelagica (fitoplancton) di Quarto dei Mille. Mem. R. Com. Talass. Ital. 97:248 pp., 13 pls.

Rhizosolenia acuminata
Gilbert, J.Y., and W.E. Allen, 1943
The phytoplankton of the Gulf of California obtained by the "E.W. Scripps" in 1939 and 1940. J. Mar. Res. V(2):89-110, figs.30-31.

Rhizosolenia acuminata
Gran, H.H., 1908
Diatomeen. Nordisches Plankton, Botanischer Teil pp. XIX 1-XIX 146; 178 text figs.

Rhizosolenia acuminata
Hendy, N. Ingram, 1964
An introductory account of the smaller algae of British coastal waters. V. Bacillariophyceae (Diatoms). Her Majesty's Stationary Office, 317 pp., 45 pls.

Rhizosolenia acuminata
Macdonald, R., 1933
An examination of plankton hauls made in the Suez Canal during the year 1928. Fish. Res. Dis., Notes & Mem. No.3, 11 pp., 1 chart.

Rhizosolenia acuminata
Okuno, H., 1957.
Electron-microscopical study on fine structures of diatom frustules. 15. Observation on genus Rhizosolenia. Bot. Mag., Tokyo, 71(826):101-107.

Rhizosolenia acuminata
Pavillard, J., 1925
Bacillariales. Rept. on the Danish Oceangr. Exped., 1908-10 to the Mediterranean and adj. seas. Vol.II., Biol. J4:72 pp., 116 text figs.

Rhizosolenia acuminata
Rampi, L., 1942
Ricerche sul fitoplancton del Mare Ligure 6. Le diatomee delle acque di Sanremo. Nuovo Giornale Botanico Italiano, N.S., 49:252-268.

Rhizosolenia alata
Allen, W.E., 1938
The Templeton Crocker Expedition to the Gulf of California in 1935 - The Phytoplankton. Amer. Microsc. Soc., Trans. 57:328-335.

Rhizosolenia alata
Allen, W.E., 1937
Plankton diatoms of the Gulf of California obtained by the G. Allan Hancock Expedition of 1936. The Hancock Pacific Expeditions, Univ. So. Calif. Publ. 3:47-59, 1 fig.

Rhizosolenia alata
Bigelow, H.B., 1922
Exploration of the coastal water off the northeastern United States in 1916 by the U.S. Fisheries Schooner Grampus. Bull. M.C.Z. 65 (5):85-188, 53 text figs.

Rhizosolenia alata
Boden, B.P., 1950
Some marine plankton diatoms from the west coast of South Africa. Trans. R.Soc. S. Africa. 32:321-434, 100 text figs.

Rhizosolenia alata
Boden, Brian, 1948
Marine plankton diatoms on operation HIGHJUMP in: Some oceanographic observations on operation HIGHJUMP. By R.S. Dietz. USNEL Rept. No.55, 97 pp., 41 figs. 7 July 1948.

Rhizosolenia alata
Braarud, T., 1945
A phytoplankton survey of the polluted waters of inner Oslo Fjord. Hvalrådets Skrifter, No.28, 142 pp., 19 text figs., 17 tables.

Rhizosolenia alata
Braarud, T., 1934
A note on the phytoplankton of the Gulf of Maine in the summer of 1933. Biol. Bull. 67(1):76-82.

Rhizosolenia alata
Braarud, T., and Adam Bursa, 1939
On the phytoplankton of the Oslo Fjord, 1933-1934. Hvalrådets Skr. No.19:1-63; 9 text figs. Reviewed. J. du. Cons. 14(3): 418-420. A.C. Gardiner.

Rhizosolenia alata
Central Meteorological Observatory, 1949
Report on sea and weather observation on Antarctic Whaling Ground (1947-48). Ocean. Mag., Japan, 1(1):49-88, 17 text figs.

Rhizosolenia alata
Clarke, G.L., 1937.
On securing large quantities of diatoms from the sea for chemical analysis. Science 86(2243):593-594.

Rhizosolenia alata
Cleve-Euler, A., 1951
Die Diatomeen von Schweden und Finnland. Kungl. Svenska Vetenskaps Akad. Handl., Fjärde Ser. 2(1): 161 pp., 6 pls.

Rhizosolenia alata
Copenhagen, W.J. and L.D., 1949
Variation in the phytoplankton of Table Bay, October 1934 to October 1935. With a note on the calorific value of Chaetoceros spp. Trans. Roy. Soc. S. Africa, 32(2):113-123, 2 text figs.

Rhizosolenia alata
Cupp, Easter E., 1943
Marine plankton diatoms of the west coast of North America. Bull. S.I.O. 5(1):1-238, 5 pls., 168 text figs.

Rhizosolenia alata
Cupp, E.E., 1937
Seasonal distribution and occurrence of marine diatoms and dinoflagellates at Scotch Cap, Alaska. Bull. S.I.O. Tech. ser.4(3):71-100, 7 textfigs.

Rhizosolenia alata
Cupp, E.E., 1934
Analysis of marine diatom collections taken from the Canal Zone to California during March, 1933. Trans. Am. Micros. Soc. LIII (1):22-29, 1 map.

Rhizosolenia alata
Cupp, E.E. and Allen, W.E., 1938
Plankton diatoms of the Gulf of California obtained by Allan Hancock Pacific Expedition of 1937. The Hancock Pacific Expeditions, The Univ. So. Calif. Publ. 3: 61-74, 1 map, pls.4-15.

Oceanographic Index: Marine Organisms Cumulation, 1946-1973

Rhizosolenia alata
Dangeard, P., 1927
Phytoplankton de la croisière du "Sylvana". Ann. Inst. Ocean., Monaco, n.s., 4(8):286-401, 54 text figs. (Février-Juin 1913).

Rhizosolenia alata
Davidson, V.M., 1931.
Biological and oceanographic conditions in Hudson Bay. Contr. Canadian Biol. Fish., n.s., 6(26):497-509, 7 textfigs.

Rhizosolenia alata
Delegazione Italiana della Commissione Internazionale per l'Esplorazione Scientifica del Mediterraneo, 1941
Note sul plancton della Laguna veneta. [Memoria CCLXXIX], Arch. di Ocean. e Limn. Anno I, Fasc. I, 1941 XIX: 31-57 pp.

Rhizosolenia alata
Delsman, H. C., 1939.
Preliminary plankton investigations in the Java Sea. Treubia, 17:139-181, 8 maps, 41 figs.

Rhizosolenia alata
Ercegovic, A., 1936
Etudes qualitative et quantitatives du phytoplancton dans les eaux cotières de l'Adriatique oriental moyen au cours de l'année 1934. Acta Adriatica 1(9):1-126

Rhizosolenia alata
Forti, A., 1922
Ricerche sulla flora pelagica (fitoplancton) di Quarto dei Mille. Mem. R. Com. Talass. Ital. 97:248 pp., 13 pls.

Rhizosolenia alata
Frenguelli, Joaquin, and Hector Antonio Orlando, 1959.
Operacion MERLUZA. Diatomeas y silicoflagelados del plancton del "VI Crucero". Servicio Hidrogr. Naval., Argentina, Publ. No. H. 619: 5-62.

Rhizosolenia alata
Ghazzawi, F.M., 1939
Plankton of the Egyptian waters. A study of the Suez Canal Plankton. (A) Phytoplankton. Preliminary Report 83 pp. Notes and Memoires, Min. Commerce-Industry, Egypt, Hydrobiol. & Fish. 65 figs.

Rhizosolenia alata
Gilbert, J.Y., and W.E. Allen, 1943
The phytoplankton of the Gulf of California obtained by the "E.W. Scripps" in 1939 and 1940. J. Mar. Res. V(2):89-110, figs.30-31.

Rhizosolenia alata
Gran, H.H., 1908
Diatomeen. Nordisches Plankton, Botanischer Teil pp. XIX.1-XIX 146; 178 text figs.

Rhizosolenia alata
Gran, H.H., and T. Braarud, 1935
A quantitative study of the phytoplankton in the Bay of Fundy and the Gulf of Maine (including observations on hydrography, chemistry, and turbidity). J. Biol. Bd., Canada, 1(5):279-467, 69 text figs.

Rhizosolenia alata & forma
Hendy, N. Ingram, 1964
An introductory account of the smaller algae of British coastal waters. V. Bacillariophyceae (Diatoms).
Her Majesty's Stationary Office, 317 pp., 45 pls.

Rhizosolenia alata
Hendey, N.I., 1937
The plankton diatoms of the southern seas. Discovery Repts. 16:151-364, pls.6-13.

Rhizosolenia alata
Jørgensen, E., 1905
B.Protistplankton and the diatoms in bottom samples. Hydrographical and biological investigations in Norwegian fjords. Bergens Mus. Skr. 7: 49-225.

Rhizosolenia alata
Jorgensen, E., 1900
Protophyten und Protozoen im Plankton aus der Norwegischen Westkerste. Bergens Mus. Aarb. 1899(6): 95 pp., 5 pls., 83 tables.

Rhizosolenia alata
Kokubo, S., 1952
Results of the observations on the plankton and oceanography of Mutsu Bay during 1950, reference being made also to the period 1946-1950. Bull Mar.Biol.Sta., Asamushi 5(1/4):1-54, 3 tables,(fold-in), 1 fold-in.

Rhizosolenia alata
Lillick, L.C., 1940
Phytoplankton and planktonic protozoa of the offshore waters of the Gulf of Maine. Pt.II. Qualitative Composition of the Planktonic Flora. Trans. Am. Phil. Soc., n.s., 31(3):193-237, 13 text figs.

Rhizosolenia alata
Lillick, L.C., 1938
Preliminary report of the phytoplankton of the Gulf of Maine. Am. Mid. Nat. 20(3):624-640, 1 text figs 37 tables.

Rhizosolenia alata
Lillick, L.C., 1937
Seasonal studies of the phytoplankton off Woods Hole, Massachusetts. Biol. Bull. LXXIII (3):488-503, 3 text figs.

Rhizosolenia alata
Mangin, M. L., 1912
Phytoplancton de la croisière du "René" dans l'Atlantique (Septembre 1908). Ann. Inst. Ocean., n.s., 4(1):1-66, 2 pls., 41 text figs., 2 tables.

Rhizosolenia alata
Manguin, E., 1954
Diatomees marines provenant de l'île Heard (Australian National Antarctic Expedition). Rev. Algol., n.s., 1: 14-24.

Rhizosolenia alata
Marukawa, H., 1921
Plankton lists and some new species of copepods, from the northern waters of Japan. Bull. Inst. Ocean., No.384, 15 pp., 3 pls., 1 chart. Monaco

Rhizosolenia alata
Massutí Algamora, M., 1949
Estudio de diez y seis muestras de plancton del Golfo de Nápoles. Publ. Inst. Biol. Appl. 5:85-94, 1 fold-in table.

Rhizosolenia alata
Pavillard, J., 1925
Bacillariales. Rept. on the Danish Oceangr. Exped., 1908-10 to the Mediterranean and adj. seas. Vol.II, Biol. J4:72 pp., 116 text figs.

Rhizosolenia alata
Pavillard, J., 1905
Recherches sur la flore pelagique (Phytoplankton) de l'Etang de Thau. Theses presentees a la Fac. Sci., Paris, 116 pp., 3 pls.

Rhizosolenia alata
Pettersson, H., 1934
Scattering and extinction of light in sea-water. Meddelanden från Göteborgs Högskolas Oceanografiska Institut 9 (Göteborgs Kungl. vetenskaps-och vitterhets-samhälles Handlingar. Femte Foljden, Ser.B, 4(4)):1-16.

Rhizosolenia alata
Pettersson, H., H. Höglund, S. Landberg, 1934.
Submarine daylight and the photosynthesis of phytoplankton. Medd. fran Göteborgs Högskolas Oceanografiska Inst. 10 (Göteborgs Kungl. Vetenskaps-och Vitterhets-samhälles handlingar. Femte Földjen, Ser. B 4(5)):1-17.

Rhizosolenia alata
Rampi, L., 1942
Ricerche sul fitoplancton del Mare Ligure 6. Le diatomee delle acque di Sanremo. Nuovo Giornale Botanico Italiano, N.S., 49:252-268.

Rhizosolenia alata
Rampi, L., 1940
Diatomee del Mare Adriatico. Nuovo Giornale Botanico Italiano, n.s., 47:559-608.

Rhizosolenia alata
Robinson, G.A., 1957.
The forms of Rhizosolenia alata Brightwell. Bull. Mar. Ecol., 4(36):203-209.

Rhizosolenia alata
Sargent, M. S. and T. J. Walker, 1948
Diatom populations associated with eddies off southern California in 1941. J. Mar. Res. 7(3):490-505, 15 text figs.

Rhizosolenia alata
Saunders, Richard P., 1967.
Rhizosolenia alata Brightwell. Leeflet Ser., Fis Bd. Conserv., Mar.Lab. I. (Phytoplankton) (2-Diatoms) (4): 7 pp.

Rhizosolenia alata
Schröder, B., 1900
Phytoplankton des Golfes von Neapel nebst vergleichenden Ausblicken auf das atlantischen Ozean. Mitt. Zool. Stat. Neapel, 14:1-38.

Rhizosolenia alata
Schütt, F., 1886.
Auxosporenbildung von Rhizosolenia alata Ber. Deutschen Bot. Gesellsch. 4:8-14.

Rhizosolenia alata
Smirnova, L.I., 1958.
[The mode of propagation in the diatom Rhizosolenia alata Bright.] Doklady Akad. Nauk, SSSR, 118(1):192-

Rhizosolenia alata
Sousa e Silva, E., 1949
Diatomaceas e Dinoflagelados de Baia de Cascais. Portugaliae Acta Biol., Volume: Julio Henriques, Ser.B: 300-383, 9 pls, 2 fold-in tables.

Rhizosolenia alata
Sousa a Silva, E., and J. Dos Santos-Pinto, 1948
O Plancton da Baía de S. Martinho do Porto. 1. Diatomaceas e Dinoflagelados. Bol. Soc. Portuguese de Ciencias Naturais, 16(2):134-187, 6 pls. (Trav. Sta. Biol. Mar. de Lisbonne No. 52).

Rhizosolenia alata
Stæmann-Nielsen, Einar, 1951
The marine vegetation of the Isefjord. A study on ecology and production. Medd. Komm. Danmarks Fiskeri-og Havundersøgelser. Ser. Plankton. 5(4); 114 pp., 46 text figs.

Rhizosolenia alata
Wheeler, J.E.G., 1939
Plankton investigations. Bermuda Biological Station. Second Report. October 1939, 7 pp. (typed), 5 pls. Plymouth, Oct. 23, 1939.

Rhizosolenia alata
Wimpenny, R. S., 1936.
The size of diatoms. I. The diameter of Rhizosolenia styliformis Brightw. and R. alata Brightw. in particular and of pelagic marine diatoms in general. JMBA, XXI:29-60

Rhizosolenia alata
Woodmansee, Robert A., 1963.
Cell-diameter frequency distributions of the planktonic diatom Rhizosolenia alata.
Publ. Inst. Mar. Sci., Port Aransas, 9:117-131.

Rhizosolenia alata alata
Robinson, G.A., 1961
Contribution towards an atlas of the north-eastern Atlantic and the North Sea. 1. Phytoplankton.
Bulls. Mar. Ecology, 5(42): 81-89, Pls. 15-20.

Rhizosolenia alata curvirostris
Cupp, Easter E., 1943
Marine plankton diatoms of the west coast of North America. Bull. S.I.O. 5(1):1-238, 5 pls., 168 text figs.

Rhizosolenia alata form. genuina
Allen, W.E., and E.E. Cupp, 1935
Plankton diatoms of the Java Sea.
Annales du Jardin Botanique de Buitenzorg XLIV (2):101-174, figs.1-127.
(drawings of all species mentioned)

Rhizosolenia alata gracile
Forti, A., 1922
Ricerche sulla flora pelagica (fitoplancton) di Quarto dei Mille. Mem. R. Com. Talass. Ital. 97:248 pp., 13 pls.

Rhizosolenia alata form. gracillima
Allen, W.E., and E.E. Cupp, 1935
Plankton diatoms of the Java Sea.
Annales du Jardin Botanique de Buitenzorg XLIV (2):101-174, figs.1-127.
(drawings of all species mentioned)

Rhizosolenia alata/gracillima
Cupp, Easter E., 1943
Marine plankton diatoms of the west coast of North America. Bull. S.I.O. 5(1):1-238, 5 pls., 168 text figs.

Rhizosolenia alata gracillima
Margalef, R., 1949
Fitoplancton nerítico de la Costa Brava en 1947-48. Publ. Inst. Biol. Aplicada, 5: 41-51, 3 text figs.

Rhizosolenia alata gracillima
Pavillard, J., 1905
Recherches sur la flore pelagique (Phytoplankton) de l'Etang de Thau. Theses presentees a la Fac. Sci., Paris, 116 pp., 3 pls.

Rhizosolenia alata f. gracillima
Pettersson, H., F. Gross, and F. Koczy, 1939
Large scale plankton cultures. Medd. fran Oceanografiska Institutet i Göteborg No.3 (Göteborgs Kungl. vetenskaps-och Vitterhets-Samhälles Handlingar Femte Följden. Ser. B) Vol.6(13):1-24.
1. The plankton shaft. H. Pettersson
2. Experiment with phytoplankton. F. Gross and F. Koczy
3. Experiments with zooplankton.

Rhizosolenia alata form. indica
Allen, W.E., and E.E. Cupp, 1935
Plankton diatoms of the Java Sea.
Annales du Jardin Botanique de Buitenzorg XLIV (2):101-174, figs.1-127.
(drawings of all species mentioned)

Rhizosolenia alata indica
Boden, B. P., 1949.
The diatoms collected by the U.S.S. CACOPAN in the Antarctic in 1947. J. Mar. Res. 8(1):6-13, 3 textfigs.

Rhizosolenia f. indica
Boden, Brian, 1948
Marine plankton diatoms on operation HIGHJUMP in: Some oceanographic observations on operation HIGHJUMP. By R.S. Dietz. USNEL Rept. No.55, 97 pp., 41 figs. 7 July 1948.

Rhizosolenia alata indica
Cupp, Easter E., 1943
Marine plankton diatoms of the west coast of North America. Bull. S.I.O. 5(1):1-238, 5 pls., 168 text figs.

Rhizosolenia alata f. indica
Cupp, E.E. and Allen, W.E., 1938
Plankton diatoms of the Gulf of California obtained by Allan Hancock Pacific Expedition of 1937. The Hancock Pacific Expeditions, The Univ. So. Calif. Publ. 3: 61-74, 1 map, pls.4-15.

Rhizosolenia alata indica
Eskinazi Enide e Shigekatsv Sato (1963-1964) 1966.
Contribuição ao estudo das diatomaceas da Praia de Piedade.
Trabhs Inst. Oceanogr., Univ. Recife, 5 (5/6): 73-114.

Rhizosolenia alata indica
Robinson, G.A., 1965.
Continuous plankton records: contribution towards a plankton atlas of the North Atlantic and the North Sea.IX. Seasonal cycles of phytoplankton.
Bulls. Mar. Ecol., Scottish Mar. Biol. Assoc., 5(4):104-122, pls.26-61.

Rhizosolenia alata indica
Robinson, G.A., 1961
Contribution towards an atlas of the north-eastern Atlantic and the North Sea. 1. Phytoplankton.
Bulls. Mar. Ecology, 5(42): 81-89, Pls. 15-20.

Rhizosolenia alata inermis
Boden, B. P., 1949.
The diatoms collected by the U.S.S. CACOPAN in the Antarctic in 1947. J. Mar. Res. 8(1):6-13, 3 textfigs.

Rhizosolenia f. inermis
Boden, Brian, 1948
Marine plankton diatoms on operation HIGHJUMP in: Some oceanographic observations on operation HIGHJUMP. By R.S. Dietz. USNEL Rept. No.55, 97 pp., 41 figs. 7 July 1948.

Rhizosolenia alata inermis
Mangin, L., 1915
Phytoplancton de L'Antartique. Deuxieme Exped. Ant. Francaise (1908-1910), 95 pp., 3 pls., 58 text figs.

Rhizosolenia alata f. inermis
Phifer, L.D. (undated)
The occurrence and distribution of plankton diatoms in Bering Sea and Bering Strait, July 26-August 24, 1934. Report of Oceanographic cruise of U.S. Coast Guard Cutter Chelan 1934, Part II(A):1-44 (mimeographed) plus fig.1 (after Pt.B)

Rhizosolenia alata var. obtusa
Bigelow, H.B., and M. Leslie, 1930
Reconnaissance of the waters and plankton of Monterey Bay, July 1928.
Bull. M.C.Z., 70(5):429-481, 43 text figs.

Rhizosolenia alata var. truncata
Gran, H.H., 1897
Protophyta: Diatomaceae, Silicoflagellata and Cilioflagellata. Den Norske Nordhavs Expedition 1876-1878, h. 24, 36 pp., 4 pls.

Rhizosolenia alata truncata
Schröder, B., 1900
Phytoplankton des Golfes von Neapel nebst vergleichenden Ausblicken auf das atlantischen Ozean. Mitt. Zool. Stat. Neapel, 14:1-38.

Rhizosolenia amputata
Pavillard, J., 1905
Recherches sur la flore pelagique (Phytoplankton) de l'Etang de Thau. Theses presentees a la Fac. Sci., Paris, 116 pp., 3 pls.

Rhizosolenia annulata
Hendey, N.I., 1937
The plankton diatoms of the southern seas.
Discovery Repts. 16:151-364, pls.6-13.

Rhizosolenia antarctica
Boden, B. P., 1949.
The diatoms collected by the U.S.S. CACOPAN in the Antarctic in 1947. J. Mar. Res. 8(1):6-13, 3 textfigs.

Rhizosolenia antarctica
Boden, Brian, 1948
Marine plankton diatoms on operation HIGHJUMP in: Some oceanographic observations on operation HIGHJUMP. By R.S. Dietz. USNEL Rept. No.55, 97 pp., 41 figs. 7 July 1948.

Rhizosolenia antarctica
Mangin, L., 1915
Phytoplancton de L'Antartique. Deuxieme Exped. Ant. Francaise (1908-1910), 95 pp., 3 pls., 58 text figs.

Rhizosolenia antarctica
Manguin, E., 1954
Diatomees marines provenant de l'ile Heard (Australian National Antarctic Expedition). Rev. Algol., n.s., 1: 14-24.

Rhizosolenia arafurensis
Allen, W.E., and E.E. Cupp, 1935
Plankton diatoms of the Java Sea.
Annales du Jardin Botanique de Buitenzorg XLIV (2):101-174, figs.1-127.
(drawings of all species mentioned)

Rhizosolenia arafurensis
Castracane degli Antelminelli, F., 1886
1. Report on the Diatomaceae collected by H.M.S. Challenger during the years 1873-1876. Rept. Sci. Results, H.M.S. Challenger, Botany Vol. II, 178 pp., 30 pls.

Rhizosolenia arafurensis
Delsman, H. C., 1939.
Preliminary plankton investigations in the Java Sea. Treubia, 17:139-181, 8 maps, 41 figs.

Rhizosolenia arafurensis
Gran, H.H., 1908
Diatomeen. Nordisches Plankton, Botanischer Teil pp. XIX.1-XIX 146; 178 text figs.

Rhizosolenia arafurensis
Hendy, N. Ingram, 1964
An introductory account of the smaller algae of British coastal waters. V. Bacillariophyceae (Diatoms).
Her Majesty's Stationary Office, 317 pp., 45 pls.

Rhizosolenia arafurensis n. sp.
Wood, E.J. Ferguson, 1963.
Studies on Australian and New Zealand diatoms. VI. Tropical and subtropical species.
Trans. R. Soc., New Zealand, 2(15):189-218.

Rhizosolenia Bergonii
Allen, W.E., and E.E. Cupp, 1935
Plankton diatoms of the Java Sea.
Annales du Jardin Botanique de Buitenzorg XLIV (2):101-174, figs.1-127.
(drawings of all species mentioned)

Rhizosolenia bergonii
Cupp, Easter E., 1943
Marine plankton diatoms of the west coast of North America. Bull. S.I.O. 5(1):1-238, 5 pls., 168 text figs.

Rhizosolenia bergonii
Cupp, E.E., 1934
Analysis of marine diatom collections taken from the Canal Zone to California during March, 1933. Trans. Am. Micros. Soc. LIII (1):22-29, 1 map.

Rhizosolenia bergonii
Cupp, E.E. and Allen, W.E., 1938
Plankton diatoms of the Gulf of California obtained by Allan Hancock Pacific Expedition of 1937. The Hancock Pacific Expeditions, The Univ. So. Calif. Publ. 3: 61-74, 1 map, pls.4-15.

Rhizosolenia bergonii
Dangeard, P., 1927
Phytoplankton de la croisière du "Sylvana". Ann. Inst. Ocean., Monaco, n.s., 4(8):286-401, 54 text figs. (Feirer-Juin 1913).

Rhizosolenia Bergoni
Ercegovic, A., 1936
Etudes qualitative et quantitatives du phytoplancton dans les eaux cotières de l'Adriatique oriental moyen au cours de l'année 1934. Acta Adriatica 1(9):1-126

Rhizosolenia bergonii
Gilbert, J.Y., and W.E. Allen, 1943
The phytoplankton of the Gulf of California obtained by the "E.W. Scripps" in 1939 and 1940. J. Mar. Res. V(2):89-110, figs.30-31.

Rhizosolenia Bergonii
Gran, H.H., 1908
Diatomeen. Nordisches Plankton, Botanischer Teil pp. XIX.1-XIX 146; 178 text figs.

Rhizosolenia bergonii
Hendy, N. Ingram, 1964
An introductory account of the smaller algae of British coastal waters. V. Bacillariophyceae (Diatoms). Her Majesty's Stationary Office, 317 pp., 45 pls.

Rhizosolenia Bergonii
Hendey, N.I., 1937
The plankton diatoms of the southern seas. Discovery Repts. 16:151-364, pls.6-13.

Rhizosolenia Bergonii
Lillick, L.C., 1940
Phytoplankton and planktonic protozoa of the offshore waters of the Gulf of Maine. Pt.II. Qualitative Composition of the Planktonic Flora. Trans. Am. Phil. Soc., n.s., 31(3):193-237, 13 text figs.

Rhizosolenia Bergoni
Lillick, L.C., 1938
Preliminary report of the phytoplankton of the Gulf of Maine. Am. Mid. Nat. 20(3):624-640, 1 text figs 37 tables.

Rhizosolenia Bergoni
Massutí Algamora, M., 1949
Estudio de diez y seis muestras de plancton del Golfo de Nápoles. Publ. Inst. Biol. Appl. 5:85-94, 1 fold-in table.

Rhizosolenia bergonii
Pavillard, J., 1925
Bacillariales. Rept. on the Danish Oceangr. Exped., 1908-10 to the Mediterranean and adj. seas. Vol.II., Biol. J4:72 pp., 116 text figs.

Rhizosolenia Bergoni
Rampi, L., 1942
Ricerche sul fitoplancton del Mare Ligure 6. Le diatomee delle acque di Sanremo. Nuovo Giornale Botanico Italiano, N.S., 49:252-268.

Rhizosolenia Bergonii
Schröder, B., 1900
Phytoplankton des Golfes von Neapel nebst vergleichenden Ausblicken auf das atlantischen Ozean. Mitt. Zool. Stat. Neapel, 14:1-38.

Rhizosolenia bergonii
Sousa e Silva, E., 1949
Diatomaceas e Dinoflagelados de Baía de Cascais. Portugalise Acta Biol., Volume: Julio Henriques, Ser. B: 300-383, 9 pls, 2 fold-in tables.

Rhizosolenia bergonii
Uyeno, Fukuzo, 1961
Oceanographical and ecological studies on primary production of the sea, with special references to relationship between diatom production and temperature and chlorinity of water. Rept. Fac. Fish., Pref. Univ., Mie, 4(1): 1-64.

Rhizosolenia bidens
Boden, B. P., 1949
The diatoms collected by the U.S.S. CACOPAN in the Antarctic in 1947. J. Mar. Res. 8(1):6-13, 3 textfigs.

Rhizosolenia bidens
Boden, Brian, 1948
Marine plankton diatoms on operation HIGHJUMP in: Some oceanographic observations on operation HIGHJUMP. By R.S. Dietz. USNEL Rept. No.55, 97 pp., 41 figs. 7 July 1948.

Rhizosolenia bidens (taxonomic note)
Takano, Hideaki, 1964
Diatom culture in artificial sea water. II. Cultures without using soil extract. Bull. Tokai Reg. Fish. Res. Lab., No. 38:45-56.

Rhizosolenia bidens
Hendey, N.I., 1937
The plankton diatoms of the southern seas. Discovery Repts. 16:151-364, pls.6-13.

Rhizosolenia calcar-avis
Alfimov, N.N., 1966
On the biology and bio chemistry of two mass marine diatoms, Coscinodiscus jonesianus (Grev.) Ostf. and Rhizosolenia calcar-avis M. Schultze from the Azov and Caspian seas. (In Russian) Bot. Zh., 51(9): 1276-1283.

Rhizosolenia calcar avis
Allen, W. E., 1938
The Templeton Crocker Expedition to the Gulf of California in 1935 - The Phytoplankton. Amer. Microsc. Soc., Trans. 57:328-335.

Rhizosolenia calcar avis
Allen, W.E., 1937
Plankton diatoms of the Gulf of California obtained by the G. Allan Hancock Expedition of 1936. The Hancock Pacific Expeditions, Univ. So. Calif. Publ. 3:47-59, 1 fig.

Rhizosolenia calcar-avis
Allen, W.E., and E.E. Cupp, 1935
Plankton diatoms of the Java Sea. Annales du Jardin Botanique de Buitenzorg XLIV (2):101-174, figs.1-127.
(drawings of all species mentioned)

Rhizosolenia calcar-avis
Bigelow, H. B., 1922
Exploration of the coastal water off the northeastern United States in 1916 by the U.S. Fisheries Schooner Grampus. Bull. M.C.Z. 65 (5):85-188, 53 text figs.

Rhizosolenia calcar-avis
Cleve-Euler, A., 1951.
Die Diatomeen von Schweden und Finnland. Kungl. Svenska Vetenskaps Akad. Handl., Fjärde Ser. 2(1): 161 pp., 6 pls.

Rhizosolenia calcar avis
Cupp, Easter E., 1943
Marine plankton diatoms of the west coast of North America. Bull. S.I.O. 5(1):1-238, 5 pls., 168 text figs.

Rhizosolenia calcar avis
Cupp, E.E., 1934
Analysis of marine diatom collections taken from the Canal Zone to California during March, 1933. Trans. Am. Micros. Soc. LIII (1):22-29, 1 map.

Rhizosolenia calcar avis
Cupp, E.E. and Allen, W.E., 1938
Plankton diatoms of the Gulf of California obtained by Allan Hancock Pacific Expedition of 1937. The Hancock Pacific Expeditions, The Univ. So. Calif. Publ. 3: 61-74, 1 map, pls.4-15.

Rhizosolenia calcaravis
Dangeard, P., 1927
Phytoplankton de la croisière du "Sylvana". Ann. Inst. Ocean., Monaco, n.s., 4(8):286-401, 54 text figs. (Feirer-Juin 1913).

Rhizosolenia calcar avis
Delegazione Italiana della Commissione Internazionale per l'Esplorazione Scientifica del Mediterraneo, 1941
Note sul plancton della Laguna veneta. [Memoria CCLXXXI], Arch. di Ocean. e Limn. Anno I, Fasc. I, 1941 XIX: 31-57 pp.

Rhizosolenia calcaravis
Delsman, H. C., 1939.
Preliminary plankton investigations in the Java Sea. Treubia, 17:139-181, 8 maps, 41 figs.

Rhizosolenia calcar-avis
Ercegovic, A., 1936
Etudes qualitative et quantitatives du phytoplancton dans les eaux cotières de l'Adriatique oriental moyen au cours de l'année 1934. Acta Adriatica 1(9):1-126

Rhizosolenia calcar-avis
Eskinazi Enide e Shigekatsv Satô (1963-1964) 1966.
Contribuição ao estudo das diatomaceas da Praia de Piedade.
Trabhs Inst. Oceanogr., Univ. Recife, 5 (5/6): 73-114.

Rhizosolenia calcar avis
Forti, A., 1922
Ricerche sulla flora pelagica (fitoplancton) di Quarto dei Mille. Mem. R. Com. Talass. Ital. 97:248 pp., 13 pls.

Rhizosolenia calcar-avis
Frenguelli, Joaquin, and Hector Antonio Orlando, 1959.
Operacion MERLUZA. Diatomeas y silicoflagelados del plancton del "VI Crucero". Servicio Hidrogr. Naval., Argentina, Publ. No. H. 619: 5-62.

Rhizosolenia calcar-avis
Ghazzawi, F.M., 1939
Plankton of the Egyptian waters. A study of the Suez Canal Plankton. (A) Phytoplankton. Preliminary Report 83 pp. Notes and Memoires, Min. Commerce-Industry, Egypt, Hydrobiol. & Fish. 65 figs.

Rhizosolenia calcar-avis
Gilbert, J.Y., and W.E. Allen, 1943
The phytoplankton of the Gulf of California obtained by the "E.W. Scripps" in 1939 and 1940. J. Mar. Res. V(2):89-110, figs.30-31.

Rhizosolenia calcar-avis
Gran, H.H., 1908
Diatomeen. Nordisches Plankton, Botanischer Teil pp. XIX.1-XIX 146; 178 text figs.

Rhizosolenia calcar-avis
Hendy, N. Ingram, 1964
An introductory account of the smaller algae of British coastal waters. V. Bacillariophyceae (Diatoms). Her Majesty's Stationary Office, 317 pp., 45 pls.

Rhizosolenia calcar-avis
Hendey, N.I., 1937
The plankton diatoms of the southern seas. Discovery Repts. 16:151-364, pls.6-13.

Rhizosolenia calcar-avis
Jorgensen, E., 1900
Protophyten und Protozoën im Plankton aus der Norwegischen Westkerste. Bergens Mus. Aarb. 1899(6): 95 pp., 5 pls., 83 tables.

Rhizosolenia calcar-avis
Lillick, L.C., 1940
Phytoplankton and planktonic protozoa of the offshore waters of the Gulf of Maine. Pt.II. Qualitative Composition of the Planktonic Flora. Trans. Am. Phil. Soc., n.s., 31(3):193-237, 13 text figs.

Rhizosolenia calcar-avis
Lillick, L.C., 1938
Preliminary report of the phytoplankton of the Gulf of Maine. Am. Mid. Nat. 20(3):624-640, 1 text fig, 37 tables.

Rhizosolenia calcar-avis
Lillick, L.C., 1937
Seasonal studies of the phytoplankton off Woods Hole, Massachusetts. Biol. Bull. LXXIII (3):488-503, 3 text figs.

Rhizosolenia calcar-avis
Macdonald, R., 1933
An examination of plankton hauls made in the Suez Canal during the year 1928. Fish. Res. Dis., Notes & Mem. No.3, 11 pp., 1 chart.

Rhizosolenia calcaravis
Margalef, R., 1949
Fitoplancton nerítico de la Costa Brava en 1947-48. Publ. Inst. Biol. Aplicada, 5: 41-51, 3 text figs.

Rhizosolenia calcar-avis
Marukawa, H., 1921
Plankton lists and some new species of copepods, from the northern waters of Japan. Bull. Inst. Ocean., No.384, 15 pp., 3 pls., 1 chart. Monaco

Rhizosolenia calcarvis
Massuti Algamora, M., 1949
Estudio de diez y seis muestras de plancton del Golfo de Nápoles. Publ. Inst. Biol. Appl. 5:85-94, 1 fold-in table.

Rhizosolenia calcar-avis
Morse, D.C., 1947
Some observations on seasonal variations in plankton population Patuxent River, Maryland 1943-1945. Bd. Nat. Res., Publ. No.65, Chesapeake Biol. Lab., 31, 3 figs.

Rhizosolenia calcar-avis
Pavillard, J., 1925
Bacillariales. Rept. on the Danish Oceangr. Exped., 1908-10 to the Mediterranean and adj. seas. Vol.II, Biol. J4:72 pp., 116 text figs.

Rhizosolenia calcar avis
Pavillard, J., 1905
Recherches sur la flore pelagique (Phytoplankton) de l'Etang de Thau. Theses presentees a la Fac. Sci., Paris, 116 pp., 3 pls.

Rhizosolenia calcar-avis
Proshkina-Lavrenko, A.I., 1961
Variability of some Black Sea diatoms. Botan. Zhurn., Akad. Nauk, SSSR, 46(12):1794-1794.

Rhizosolenia calcaravis
Rampi, L., 1942
Ricerche sul fitoplancton del Mare Ligure 6. Le diatomee delle acque di Sanremo. Nuovo Giornale Botanico Italiano, N.S., 49:252-268.

Rhizosolenia calcar avis
Rampi, L., 1940
Diatomee del Mare Adriatico. Nuovo Giornale Botanico Italiano, n.s., 47:559-608.

Rhizosolenia calcar-avis
Schröder, B., 1900
Phytoplankton des Golfes von Neapel nebst vergleichenden Ausblicken auf das atlantischen Ozean. Mitt. Zool. Stat. Neapel, 14:1-38.

Rhizosolenia calcar-avis
Sousa e Silva, E., 1949
Diatomaceas e Dinoflagelados de Baia de Cascais. Portugaliae Acta Biol. Volume: Julio Henriques, Ser. B: 300-383, 9 pls, 2 fold-in tables.

Rhizosolenia castracanei
Cupp, Easter E., 1943
Marine plankton diatoms of the west coast of North America. Bull. S.I.O. 5(1):1-238, 5 pls., 168 text figs.

Rhizosolenia castracanei
Dangeard, P., 1927
Phytoplankton de la croisière du "Sylvana". Ann. Inst. Ocean., Monaco, n.s., 4(8):286-401, 54 text figs. (Feirer-Juin 1913).

Rhizosolenia Castracanei
Ercegovic, A., 1936
Etudes qualitative et quantitatives du phytoplankton dans les eaux cotières de l'Adriatique oriental moyen au cours de l'année 1934. Acta Adriatica 1(9):1-126

Rhizosolenia Castracanei
Forti, A., 1922
Ricerche sulla flora pelagica (fitoplancton) di Quarto dei Mille. Mem. R. Com. Talass. Ital. 97:248 pp., 13 pls.

Rhizosolenia Castracanei
Gilbert, J.Y., and W.E. Allen, 1943
The phytoplankton of the Gulf of California obtained by the "E.W. Scripps" in 1939 and 1940. J. Mar. Res. V(2):89-110, figs.30-31.

Rhizosolenia castracanei
Gran, H.H., 1908
Diatomeen. Nordisches Plankton, Botanischer Teil pp. XIX.1-XIX 146; 178 text figs.

Rhizosolenia castracanei
Hendy, N. Ingram, 1964
An introductory account of the smaller algae of British coastal waters. V. Bacillariophyceae (Diatoms). Her Majesty's Stationary Office, 317 pp., 45 pls.

Rhizosolenia Castracanei
Hendey, N.I., 1937
The plankton diatoms of the southern seas. Discovery Repts. 16:151-364, pls.6-13.

Rhizosolenia Castracanei
Marukawa, H., 1921
Plankton lists and some new species of copepods, from the northern waters of Japan. Bull. Inst. Ocean., No.384, 15 pp., 3 pls., 1 chart. Monaco

Rhizosolenia castracanei
Pavillard, J., 1925
Bacillariales. Rept. on the Danish Oceangr. Exped., 1908-10 to the Mediterranean and adj. seas. Vol.II, Biol. J4:72 pp., 116 text figs.

Rhizosolenia castracanei
Pavillard, J., 1905
Recherches sur la flore pelagique (Phytoplankton) de l'Etang de Thau. Theses presentees a la Fac. Sci., Paris, 116 pp., 3 pls.

Rhizosolenia castracanei
Rampi, L., 1942
Ricerche sul fitoplancton del Mare Ligure 6. Le diatomee delle acque di Sanremo. Nuovo Giornale Botanico Italiano, N.S., 49:252-268.

Rhizosolenia Castracanei
Schröder, B., 1900
Phytoplankton des Golfes von Neapel nebst vergleichenden Ausblicken auf das atlantischen Ozean. Mitt. Zool. Stat. Neapel, 14:1-38.

Rhizosolenia chunii
Hendey, N.I., 1937
The plankton diatoms of the southern seas. Discovery Repts. 16:151-364, pls.6-13.

Rhizosolenia chunii
Manguin, E., 1954
Diatomees marines provenant de l'ile Heard (Australian National Antarctic Expedition). Rev. Algol., n.s., 1: 14-24.

Rhizosolenia Clevei
Allen, W.E., and E.E. Cupp, 1935
Plankton diatoms of the Java Sea. Annales du Jardin Botanique de Buitenzorg XLIV (2):101-174, figs.1-127.
(drawings of all species mentioned)

Rhizosolenia clevei
Delsman, H. C., 1939.
Preliminary plankton investigations in the Java Sea. Treubia, 17:139-181, 8 maps, 41 figs.

Rhizosolenia clevei
Marukawa, H., 1921
Plankton lists and some new species of copepods, from the northern waters of Japan. Bull. Inst. Ocean., No.384, 15 pp., 3 pls., 1 chart. Monaco

Rhizosolenia crassa
Hendey, N.I., 1937
The plankton diatoms of the southern seas. Discovery Repts. 16:151-364, pls.6-13.

Rhizosolenia crassispina
Dangeard, P., 1927
Phytoplankton de la croisière du "Sylvana". Ann. Inst. Ocean., Monaco, n.s., 4(8):286-401, 54 text figs. (Feirer-Juin 1913).

Rhizosolenia curvata
Hart, T.J., 1937.
Rhizosolenia curvata Zacharias, an indicator species in the Southern Ocean. Discovery Repts., 16:415-446, Pl. 14.

Rhizosolenia curvata
Hendey, N.I., 1937
The plankton diatoms of the southern seas. Discovery Repts. 16:151-364, pls.6-13.

Rhizosolenia curvatulus n.sp
Wood, E.J. Ferguson, 1963.
Studies on Australian and New Zealand diatoms.
VI. Tropical and subtropical species.
Trans. R.Soc., New Zealand, 2(15):189-218.

Rhizosolenia cylindrus
Allen, W.E., and E.E. Cupp, 1935
Plankton diatoms of the Java Sea.
Annales du Jardin Botanique de Buitenzorg
XLIV (2):101-174, figs.1-127.

(drawings of all species mentioned)

Rhizosolenia cylindrus
Cupp, Easter E., 1943
Marine plankton diatoms of the west coast of North America. Bull. S.I.O. 5(1):1-238, 5 pls., 168 text figs.

Rhizosolenia cylindrus
Dangeard, P., 1927
Phytoplankton de la croisière du "Sylvana". Ann. Inst. Ocean., Monaco, n.s., 4(8):286-401, 54 text figs. (Feirer-Juin 1913).

Rhizosolenia cylindrus
Gran, H.H., 1908
Diatomeen. Nordisches Plankton, Botanischer Teil pp. XIX.1-XIX 146; 178 text figs.

Rhizosolenia cylindrus
Hendy, N. Ingram, 1964
An introductory account of the smaller algae of British coastal waters. V. Bacillariophyceae (Diatoms).
Her Majesty's Stationary Office, 317 pp., 45 pls.

Rhizosolenia cylindrus
Schröder, B., 1900
Phytoplankton des Golfes von Neapel nebst vergleichenden Ausblicken auf das atlantischen Ozean. Mitt. Zool. Stat. Neapel, 14:1-38.

Rhizosolenia delicatula
Allen, W. E., 1938
The Templeton Crocker Expedition to the Gulf of California in 1935 - The Phytoplankton. Amer. Microsc. Soc., Trans. 57:328-335.

Rhizosolenia delicatula
Allen, W.E., 1937
Plankton diatoms of the Gulf of California obtained by the G. Allan Hancock Expedition of 1936. The Hancock Pacific Expeditions, Univ. So. Calif. Publ. 3:47-59, 1 fig.

Rhizosolenia delicatula
Boden, B.P., 1950
Some marine plankton diatoms from the west coast of South Africa. Trans. R.Soc. S. Africa. 32:321-434, 100 text figs.

Rhizosolenia delicatula
Boden, B. P., 1949.
The diatoms collected by the U.S.S. CACOPAN in the Antarctic in 1947. J. Mar. Res. 8(1):6-13, 3 textfigs.

Rhizosolenia delicatula
Boden, Brian, 1948
Marine plankton diatoms on operation HIGHJUMP in: Some oceanographic observations on operation HIGHJUMP. By R.S. Dietz. USNEL Rept. No.55, 97 pp., 41 figs. 7 July 1948.

Rhizosolenia delicatula
Cleve-Euler, A., 1951
Die Diatomeen von Schweden und Finnland. Kungl. Svenska Vetenskaps Akad. Handl., Fjärde Ser. 2(1): 161 pp., 6 pls.

Rhizosolenia delicatula
Cupp, Easter E., 1943
Marine plankton diatoms of the west coast of North America. Bull. S.I.O. 5(1):1-238, 5 pls., 168 text figs.

Rhizosolenia delicatula
Cupp, E.E., 1934
Analysis of marine diatom collections taken from the Canal Zone to California during March, 1933. Trans. Am. Micros. Soc. LIII (1):22-29, 1 map.

Rhizosolenia delicatula
Cupp, E.E. and Allen, W.E., 1938
Plankton diatoms of the Gulf of California obtained by Allan Hancock Pacific Expedition of 1937. The Hancock Pacific Expeditions, The Univ. So. Calif. Publ. 3: 61-74, 1 map, pls.4-15.

Rhizosolenia delicatula
Dangeard, P., 1927
Phytoplankton de la croisière du "Sylvana". Ann. Inst. Ocean., Monaco, n.s., 4(8):286-401, 54 text figs. (Feirer-Juin 1913).

Rhizosolenia delicatula
Gilbert, J.Y., and W.E. Allen, 1943
The phytoplankton of the Gulf of California by the "E.W. Scripps" in 1939 and 1940. J. Mar. Res. V(2):89-110, figs.30-31.

Rhizosolenia delicatula
Grall, J.-R., 1972
Développement "printanier" de la diatomée Rhizosolenia delicatula près de Roscoff. Mar. Biol. 16(1): 41-48.

Rhizosolenia delicatula
Gran, H.H., 1908
Diatomeen. Nordisches Plankton, Botanischer Teil pp. XIX.1-XIX 146; 178 text figs.

Rhizosolenia delicatula
Gran, H. H. and E. C. Angst, 1931
Plankton diatoms of Puget Sound. Publ. Puget Sound Biol. Sta. 7:417-519, 95 text figs.

Rhizosolenia delicatula
Hendy, N. Ingram, 1964
An introductory account of the smaller algae of British coastal waters. V. Bacillariophyceae (Diatoms).
Her Majesty's Stationary Office, 317 pp., 45 pls.

Rhizosolenia delicatula
Hendey, N.I., 1937
The plankton diatoms of the southern seas. Discovery Repts. 16:151-364, pls.6-13.

Rhizosolenia delicatula
Mangin, M. L., 1912
Phytoplancton de la croisière du "René" dans l'Atlantique (Septembre 1908). Ann. Inst. Ocean., n.s., 4(1):1-66, 2 pls., 41 text figs., 2 tables.

Rhizosolenia delicatula
Meunier, A., 1915
Microplancton de la Mer Flamande. 2. Diatomées (excepté le genre Chaetoceros). Mem. Mus. Roy. Hist. Nat., Belgique, 7(3):1-118, Pls. VIII-XIV.

Rhizosolenia delicatula
Pavillard, J., 1925
Bacillariales. Rept. on the Danish Oceangr. Exped., 1908-10 to the Mediterranean and adj. seas. Vol.II., Biol. J4:72 pp., 116 text figs.

Rhizosolenia delicatula
Phifer, L.D. (undated)
The occurrence and distribution of plankton diatoms in Bering Sea and Bering Strait, July 26-August 24, 1934. Report of Oceanographic cruise of U.S. Coast Guard Cutter Chelan 1934, Part II(A):1-44 (mimeographed) plus fig.1 (after Pt.B)

Rhizosolenia delicatula
Schodduyn, M., 1926
Observations faites dans la baie d'Ambleteuse (Pas de Calais). Bull. Inst. Ocean., Monaco, No. 482: 64 pp.

Rhizosolenia eriensis
Cleve-Euler, A., 1951
Die Diatomeen von Schweden und Finnland. Kungl. Svenska Vetenskaps Akad. Handl., Fjärde Ser. 2(1): 161 pp., 6 pls.

Rhizosolenia firma
Ercegovic, A., 1940
Weitere Untersuchungen über einige hydrographische Verhältnisse und über die Phytoplanktonproduktion in den Gewässern der östlichen Mitteladria. Acta Adriatica 2(3):95-134, 8 text figs.

Rhizosolenia firma
Gangemi, Giuseppe 1969.
Sulla presenza di Rhizosolenia firma Karsten nell'area idrografica dello Stretto di Messina.
Boll. Pesca Piscic. Idrobiol. 24(2): 245-264

Rhizosolenia faeröensis
Gran, H.H., 1908
Diatomeen. Nordisches Plankton, Botanischer Teil pp. XIX.1-XIX 146; 178 text figs.

Rhizosolenia firma
Pavillard, J., 1925
Bacillariales. Rept. on the Danish Oceangr. Exped., 1908-10 to the Mediterranean and adj. seas. Vol.II., Biol. J4:72 pp., 116 text figs.

Rhizosolenia (?) flaccida n.sp
Castracane degli Antelminelli, F., 1886
1. Report on the Diatomaceae collected by H.M.S. Challenger during the years 1873-1876. Rept. Sci. Results, H.M.S. Challenger. Botany Vol. II, 178 pp., 30 pls.

Rhizosolenia formosa
Dangeard, P., 1927
Phytoplankton de la croisière du "Sylvana". Ann. Inst. Ocean., Monaco, n.s., 4(8):286-401, 54 text figs. (Feirer-Juin 1913).

Rhizosolenia formosa
Forti, A., 1922
Ricerche sulla flora pelagica (fitoplancton) di Quarto dei Mille. Mem. R. Com. Talass. Ital. 97:248 pp., 13 pls.

Rhizosolenia formosa
Pavillard, J., 1925
Bacillariales. Rept. on the Danish Oceangr. Exped., 1908-10 to the Mediterranean and adj. seas. Vol.II., Biol. J4:72 pp., 116 text figs.

Rhizosolenia formosa
Schröder, B., 1900
Phytoplankton des Golfes von Neapel nebst vergleichenden Ausblicken auf das atlantischen Ozean. Mitt. Zool. Stat. Neapel, 14:1-38.

Rhizosolenia formosa
Sousa e Silva, E., 1949
Diatomaceas e Dinoflagelados de Baia de Cascais. Portugaliae Acta Biol. Volume: Julio Henriques, Ser. B: 300-383, 9 pls, 2 fold-in tables.

Rhizosolenia fragillima
Forti, A., 1922
Ricerche sulla flora pelagica (fitoplancton) di Quarto dei Mille. *Mem. R. Com. Talass. Ital.* 97:248 pp., 13 pls.

Rhizosolenia fragillima
Ghazzawi, F.M., 1939
Plankton of the Egyptian waters. A study of the Suez Canal Plankton. (A) Phytoplankton. Preliminary Report 83 pp. *Notes and Memoires, Min. Commerce-Industry, Egypt, Hydrobiol. & Fish.* 65 figs.

Rhizosolenia fragillima
Gran, H.H., 1908
Diatomeen. Nordisches Plankton, Botanischer Teil pp. XIX.1-XIX 146; 178 text figs.

Rhizosolenia fragillima
Macdonald, R., 1933
An examination of plankton hauls made in the Suez Canal during the year 1928. *Fish. Res. Dis., Notes & Mem.* No.3, 11 pp., 1 chart.

Rhizosolenia fragillima
Schodduyn, M., 1926
Observations faites dans la baie d'Ambleteuse (Pas de Calais). *Bull. Inst. Ocean.*, Monaco, No. 482: 64 pp.

Rhizosolenia fragillima
Steemann-Nielsen, Einar, 1951
The marine vegetation of the Isefjord. A study on ecology and production. *Medd. Komm. Danmarks Fiskeri-og Havundersøgelser. Ser. Plankton.* 5(4); 114pp., 46 text figs.

Rhizosolenia fragillissima
Allen, W. E., 1938
The Templeton Crocker Expedition to the Gulf of California in 1935 - The Phytoplankton. *Amer. Microsc. Soc., Trans.* 57:328-335.

Rhizosolenia fragillissima
Allen, W.E., 1937
Plankton diatoms of the Gulf of California obtained by the G. Allan Hancock Expedition of 1936. *The Hancock Pacific Expeditions, Univ. So. Calif. Publ.* 3:47-59, 1 fig.

Rhizosolenia fragillissima
Boden, B.P., 1950
Some marine plankton diatoms from the west coast of South Africa. *Trans. R.Soc. S. Africa.* 32:321-434, 100 text figs.

Rhizosolenia fragillissima
Braarud, T., 1945
A phytoplankton survey of the polluted waters of inner Oslo Fjord. *Hvalrådets Skrifter*, No.28, 142 pp., 19 text figs., 17 tables.

Rhizosolenia fragillissima
Braarud, T., 1934
A note on the phytoplankton of the Gulf of Maine in the summer of 1933. *Biol. Bull.* 67(1):76-82.

Rhizosolenia fragillissima
Brunel, J., 1962
Le phytoplancton de la Baie de Chaleurs. *Inst. Botan., Univ. Montréal*, Contrib. No. 77: 365 pp., 66 pls.
Rhizosolenia fragillissima bergonii
Rhizosolenia fragillissima faeroënsis

Rhizosolenia fragillissima
Brunel, Jules, 1962
Le phytoplancton de la Baie des Chaleurs. *Contrib. Ministère de la Chasse et des Pêcheries, Province de Québec*, No. 91: 365 pp.

Rhizosolenia fragilissima
Cleve-Euler, A., 1951
Die Diatomeen von Schweden und Finnland. *Kungl. Svenska Vetenskaps Akad. Handl., Fjärde Ser.* 2(1): 161 pp., 6 pls.

Rhizosolenia fragilissima
Cupp, Easter E., 1943
Marine plankton diatoms of the west coast of North America. *Bull. S.I.O.* 5(1):1-238, 5 pls., 168 text figs.

Rhizosolenia fragilissima
Cupp, E.E. and Allen, W.E., 1938
Plankton diatoms of the Gulf of California obtained by Allan Hancock Pacific Expedition of 1937. *The Hancock Pacific Expeditions, the Univ. So. Calif. Publ.* 3: 61-74, 1 map, pls.4-15.

Rhizosolenia fragilissima
Delegazione Italiana della Commissione Internazionale per l'Esplorazione Scientifica del Mediterraneo, 1941
Note sul plancton della Laguna veneta. Memoria CCLXXXI, *Arch. di Ocean. e Limn.* Anno I, Fasc. I, 1941 XIX: 31-57 pp.

Rhizosolenia fragilissima
Gilbert, J.Y., and W.E. Allen, 1943
The phytoplankton of the Gulf of California obtained by the "E.W. Scripps" in 1939 and 1940. *J. Mar. Res.* V(2):89-110, figs.30-31.

Rhizosolenia fragilissima
Gran, H.H. and T. Braarud, 1935
A quantitative study of the phytoplankton in the Bay of Fundy and the Gulf of Maine (including observations on hydrography, chemistry, and turbidity). *J. Biol. Bd., Canada*, 1(5):279-467, 69 text figs.

Rhizosolenia fragilissima
Gran, H. H. and E. C. Angst, 1931
Plankton diatoms of Puget Sound. *Publ. Puget Sound Biol. Sta.* 7:417-519, 95 text figs.

Rhizosolenia fragilissima
Hendy, N. Ingram, 1964
An introductory account of the smaller algae of British coastal waters. V. Bacillariophyceae (Diatoms). Her Majesty's Stationary Office, 317 pp., 45 pls.

Rhizosolenia fragilissima
Hendey, N.I., 1937
The plankton diatoms of the southern seas. *Discovery Repts.* 16:151-364, pls.6-13.

Rhizosolenia fragilissima
Ignatiades, Lydia, and Theodore J. Smayda, 1970.
Autecological studies on the marine diatom Rhizosolenia fragilissima Bergon. II Enrichment and dark variability experiments. *J. Phycol.* 6(4):357-364

Rhizosolenia fragilissima
Ignatiades, Lydia, and Theodore J. Smayda, 1970.
Autecological studies on the marine diatom Rhizosolenia fragilissima Bergon. I. The influence of light, temperature and salinity. *J. Phycol.* 6(4):332-339

Rhizosolenia fragilissima
Lillick, L.C., 1940
Phytoplankton and planktonic protozoa of the offshore waters of the Gulf of Maine. Pt.II. Qualitative Composition of the Planktonic Flora. *Trans. Am. Phil. Soc.*, n.s., 31(3):193-237, 13 text figs.

Rhizosolenia fragilissima
Lillick, L.C., 1938
Preliminary report of the phytoplankton of the Gulf of Maine. *Am. Mid. Nat.* 20(3):624-640, 1 text fig, 37 tables.

Rhizosolenia fragilissima
Lillick, L.C., 1937
Seasonal studies of the phytoplankton off Woods Hole, Massachusetts. *Biol. Bull.* LXXIII (3):488-503, 3 text figs.

Rhizosolenia fragilissima
Meunier, A., 1915
Microplancton de la Mer Flamande. 2. Diatomées (excepté le genre Chaetoceros). *Mem. Mus. Roy. Hist. Nat., Belgique*, 7(3):1-118, Pls. VIII-XIV.

Rhizosolenia fragilissima
Pavillard, J., 1925
Bacillariales. Rept. on the Danish Oceangr. Exped., 1908-10 to the Mediterranean and adj. seas. Vol.II., *Biol.* J4:72 pp., 116 text figs.

Rhizosolenia fragilissima
Pavillard, J., 1905
Recherches sur la flore pelagique (Phytoplankton) de l'Etang de Thau. *Theses presentees a la Fac. Sci.*, Paris, 116 pp., 3 pls.

Rhizosolenia fragilissima
Phifer, L.D. (undated)
The occurrence and distribution of plankton diatoms in Bering Sea and Bering Strait, July 26-August 24, 1934. *Report of Oceanographic cruise of U.S. Coast Guard Cutter Chelan 1934*, Part II(A):1-44 (mimeographed) plus fig.1 (after Pt.B)

Rhizosolenia fragilissima
Rampi, L., 1942
Ricerche sul fitoplancton del Mare Ligure 6. Le diatomee delle acque di Sanremo. *Nuovo Giornale Botanico Italiano*, N.S., 49:252-268.

Rhizosolenia gracillima
Mangin, M. L., 1912
Phytoplancton de la croisière du "René" dans l'Atlantique (Septembre 1908). *Ann. Inst. Ocean.*, n.s., 4(1):1-66, 2 pls., 41 text figs., 2 tables.

Rhizosolenia gracillima
Schröder, B., 1900
Phytoplankton des Golfes von Neapel nebst vergleichenden Ausblicken auf das atlantischen Ozean. *Mitt. Zool. Stat. Neapel*, 14:1-38.

Rhizosolenia hebetata
Boden, B.P., 1950
Some marine plankton diatoms from the west coast of South Africa. *Trans. R.Soc. S. Africa.* 32:321-434, 100 text figs.

Rhizosolenia hebetata
Brunel, J., 1962
Le phytoplancton de la Baie de Chaleurs. *Inst. Botan., Univ. Montréal*, Contrib. No. 77: 365 pp., 66 pls.

Rhizosolenia hebetata hiemalis
Rhizosolenia hebetata semispina

Rhizosolenia hebetata
Cupp, Easter E., 1943
Marine plankton diatoms of the west coast of North America. *Bull. S.I.O.* 5(1):1-238, 5 pls., 168 text figs.

Rhizosolenia hebetata
Delsman, H. C., 1939.
Preliminary plankton investigations in the Java Sea. *Treubia*, 17:139-181, 8 maps, 41 figs.

Rhizosolenia hebetata

Gran, H.H., 1908
Diatomeen. Nordisches Plankton, Botanischer Teil pp. XIX.1-XIX 146; 178 text figs.

Rhizosolenia hebetata & forma

Hendy, N. Ingram, 1964
An introductory account of the smaller algae of British coastal waters. V. Bacillariophyceae (Diatoms).
Her Majesty's Stationary Office, 317 pp., 45 pls.

Rhizosolenia hebetata

Hendey, N.I., 1937
The plankton diatoms of the southern seas. Discovery Repts. 16:151-364, pls.6-13.

Rhizosolenia hebetata

Manguin, E., 1954
Diatomees marines provenant de l'ile Heard (Australian National Antarctic Expedition). Rev. Algol., n.s., 1: 14-24.

Rhizosolenia hebetata

Mann, A., 1907
Report on the diatoms of the Albatross voyages in the Pacific Ocean, 1888-1904. Contrib. U. S. Nat. Herb. 10(5):221-419, Pls. XLIV-LIV.

Rhizosolenia hebetata

Marukawa, H., 1921
Plankton lists and some new species of copepods, from the northern waters of Japan. Bull. Inst. Ocean., No.384, 15 pp., 3 pls., 1 chart. Monaco

Rhizosolenia hebetata

Morse, D.C., 1947
Some observations on seasonal variations in plankton population Patuxant River, Maryland 1943-1945. Bd. Nat. Res., Publ. No.65, Chesapeake Biol. Lab., 31, 3 figs.

Rhizosolenia hebetata

Steemann-Nielsen, Einar, 1951
The marine vegetation of the Isefjord. A study on ecology and production. Medd. Komm. Danmarks Fiskeri-og Havundersøgelser. Ser. Plankton. 5(4); 114pp., 46 text figs.

Rhizosolenia hebetata hiemalis

Cleve-Euler, A., 1951
Die Diatomeen von Schweden und Finnland. Kungl. Svenska Vetenskaps Akad. Handl., Fjärde Ser. 2(1): 161 pp., 6 pls.

Rhizosolenia hebetata hiemalis

Cupp, Easter E., 1943
Marine plankton diatoms of the west coast of North America. Bull. S.I.O. 5(1):1-238, 5 pls., 168 text figs.

Rhizosolenia hebetata hiemalis

Okuno, H., 1957.
Electron-microscopical study on fine structures of diatom frustules. 15. Observation on the genus Rhizosolenia. Bot. Mag., Tokyo, 71(826):101-107.

Rhizosolenia hebetata f. hiemalis

Phifer, L.D. (undated)
The occurrence and distribution of plankton diatoms in Bering Sea and Bering Strait, July 26-August 24, 1934. Report of Oceanographic cruise of U.S. Coast Guard Cutter Chelan 1934, Part II(A):1-44 (mimeographed) plus fig.1 (after Pt.B)

Rhizosolenia hebetata form. semispina

Allen, W.E., and E.E. Cupp, 1935
Plankton diatoms of the Java Sea. Annales du Jardin Botanique de Buitenzorg XLIV (2):101-174, figs.1-127.
(drawings of all species mentioned)

Rhizosolenia hebetata semispina

Boden, B. P., 1949.
The diatoms collected by the U.S.S. CACOPAN in the Antarctic in 1947. J. Mar. Res. 8(1):6-13, 3 textfigs.

Rhizosolenia hebetata f. semispina

Boden, Brian, 1948
Marine plankton diatoms on operation HIGHJUMP in: Some oceanographic observations on operation HIGHJUMP. By R.S. Dietz. USNEL Rept. No.55, 97 pp., 41 figs. 7 July 1948.

Rhizosolenia hebetata, f. semispina

Braarud, T., 1945
A phytoplankton survey of the polluted waters of inner Oslo Fjord. Hvalrådets Skrifter, No.28, 142 pp., 19 text figs., 17 tables.

Rhizosolenia hebetata f. semispina

Braarud, T., 1945
Experimental studies on marine plankton diatoms. Avhandlingar utgitt av Det Norske Videnskaps-Akademi i Oslo. 1. Mat.-Naturv. Klasse 1944. No.10, 1-16, 1 text fig.

Rhizosolenia hebetata semispina

Brunel, Jules, 1962
Le phytoplancton de la Baie des Chaleurs. Contrib. Ministère de la Chasse et des Pêcheries, Province de Québec, No. 91: 365 pp.

Rhizosolenia hebetata f. semispina

Central Meteorological Observatory, 1949
Report on sea and weather observation on Antarctic Whaling Ground (1947-48). Ocean. Mag., Japan, 1(1):49-88, 17 text figs.

Rhizosolenia hebetata semispina

Cleve-Euler, A., 1951
Die Diatomeen von Schweden und Finnland. Kungl. Svenska Vetenskaps Akad. Handl., Fjärde Ser. 2(1): 161 pp., 6 pls.

Rhizosolenia hebetata semispina

Cupp, Easter E., 1943
Marine plankton diatoms of the west coast of North America. Bull. S.I.O. 5(1):1-238, 5 pls., 168 text figs.

Rhizosolenia hebetata semispina

Eskinazi Enide e Shigekatsv Satô (1963-1964) 1966.
Contribuição ao estudo das diatomaceas da Praia de Piedade.
Trabhs Inst. Oceanogr., Univ. Recife, 5 (5/6): 73-114.

Rhizosolenia hebetata semispina

Fraser, J. H., and A. Saville, 1949
Plankton distribution in Scottish and adjacent waters in 1948. Ann. Biol. 5:61-62.

Rhizosolenia hebetata semispina

Frenguelli, Joaquin, and Hector Antonio Orlando, 1959.
Operacion MERLUZA. Diatomeas y silicoflagelados del plancton del "VI Crucero". Servicio Hidrogr. Naval, Argentina, Publ. No. H. 619: 5-62.

Rhizosolenia hebetata f. semispina

Ghazzawi, F.M., 1939
Plankton of the Egyptian waters. A study of the Suez Canal Plankton. (A) Phytoplankton. Preliminary Report 83 pp. Notes and Memoires, Min. Commerce-Industry, Egypt, Hydrobiol. & Fish. 65 figs.

Rhizosolenia hebetata f. semispina

Gilbert, J.Y., and W.E. Allen, 1943
The phytoplankton of the Gulf of California obtained by the "E.W. Scripps" in 1939 and 1940. J. Mar. Res. V(2):89-110, figs.30-31.

Rhizosolenia hebetata semispina

Gran, H.H., 1908
Diatomeen. Nordisches Plankton, Botanischer Teil pp. XIX.1-XIX 146; 178 text figs.

Rhizosolenia hebetata f. semispina

Gran, H.H., and T. Braarud, 1935
A quantitative study of the phytoplankton in the Bay of Fundy and the Gulf of Maine (including observations on hydrography, chemistry, and turbidity). J. Biol. Bd., Canada, 1(5):279-467, 69 text figs.

Rhizosolenia hebetata semispina

Holmes, Robert W., 1966.
Light microscope observations on cytological manifestations of nitrate, phosphate, and silicate deficiency in four marine centric diatoms. J.Phycology,2(4):136-140.

Rhizosolenia hebetata semispina

Kamshilov, M.M., 1950.
The peculiarity of the division of the diatom Rhizosolenia hebetata semispina. Doklady Akad. Nauk, SSSR, 75(5):747-748.

Rhizosolenia hebetata semispina

Kokubo, S., 1952
Results of the observations on the plankton and oceanography of Mutsu Bay during 1950, reference being made also to the period 1946-1950. Bull Mar.Biol.Sta., Asamushi 5(1/4): 1-54, 3 tables,(fold-in), 1 fold-in.

Rhizosolenia hebetata var. semispina

Lillick, L.C., 1940
Phytoplankton and planktonic protozoa of the offshore waters of the Gulf of Maine. Pt.II. Qualitative Composition of the Planktonic Flora. Trans. Am. Phil. Soc., n.s., 31(3):193-237, 13 text figs.

Rhizosolenia hebetata var. semispina

Lillick, L.C., 1938
Preliminary report of the phytoplankton of the Gulf of Maine. Am. Mid. Nat. 20(3):624-640, 1 text figs 37 tables.

Rhizosolenia hebetata var. semispina

Lillick, L.C., 1937
Seasonal studies of the phytoplankton off Woods Hole, Massachusetts. Biol. Bull. LXXIII (3):488-503, 3 text figs.

Rhizosolenia hebetata f. semispina

Rampi, L., 1942
Ricerche sul fitoplanton del Mare Ligure 6. Le diatomee delle acque di Sanremo. Nuovo Giornale Botanico Italiano, N.S., 49:252-268.

Rhizosolenia hebetata f. semispina

Ramsfjell, Einar, 1959
Dimorphism and the simultaneous occurrence of auxospores and microspores in the diatom Rhizosolenia hebetata f. semispina (Hensen) Gran.
Nytt Magasin for Botanikk, 7: 169-173.

Rhizosolenia hebetata semispina

Robinson, G.A., 1965.
Continuous plankton records: contribution towards a plankton atlas of the North Atlantic and the North Sea. IX. Seasonal cycles of phytoplankton.
Bulls. Mar. Ecol.,Scottish Mar. Biol. Assoc. 6(4):104-122, pls. 26-61.

Rhizosolenia hebetata semispina

Robinson, G.A., 1961
Contribution towards an atlas of the northeastern Atlantic and the North Sea. 1. Phytoplankton.
Bulls. Mar. Ecology, 5(42): 81-89, Pls. 15-20.

Rhizosolenia hebetata

Seaton, D.D., 1970.
Reproduction in Rhizosolenia hebetata and its linkage with Rhizosolenia styliformis. J. mar. biol. Ass. U.K., 50(1): 97-106.

Rhizosolenia hebetata f. semispina

Takano, Hideaki, 1972
Remarks on the morphology of the diatom Rhizosolenia hebetata forma semispina occuring in cold and warm waters. In: Biological oceanography of the northern North Pacific Ocean, A.Y. Takenouti, Chief Editor, Idemitsu Shoten, Tokyo, 165-172.

Rhizosolenia hyalina

Dangeard, P., 1927
Phytoplankton de la croisière du "Sylvana". Ann. Inst. Ocean., Monaco, n.s., 4(8):286-401, 54 text figs. (Feirer-Juin 1913).

Rhizosolenia imbricata

Allen, W. E., 1938
The Templeton Crocker Expedition to the Gulf of California in 1935 - The Phytoplankton. Amer. Microsc. Soc., Trans. 57:328-335.

Rhizosolenia imbricata

Allen, W.E., and E.E. Cupp, 1935
Plankton diatoms of the Java Sea. Annales du Jardin Botanique de Buitenzorg XLIV (2):101-174, figs.1-127.

(drawings of all species mentioned)

Rhizosolenia imbricata

Castracane degli Antelminelli, F., 1886
1. Report on the Diatomaceae collected by H.M.S. Challenger during the years 1873-1876. Rept. Sci. Results, H.M.S. Challenger, Botany Vol. II, 178 pp., 30 pls.

Rhizosolenia imbricata

Cupp, Easter E., 1943
Marine plankton diatoms of the west coast of North America. Bull. S.I.O. 5(1):1-238, 5 pls., 168 text figs.

Rhizosolenia imbricata

Cupp, E.E., 1934
Analysis of marine diatom collections taken from the Canal Zone to California during March, 1933. Trans. Am. Micros. Soc. LIII (1):22-29, 1 map.

Rhizosolenia imbricata

Dangeard, P., 1927
Phytoplankton de la croisière du "Sylvana". Ann. Inst. Ocean., Monaco, n.s., 4(8):286-401, 54 text figs, (Février-Juin 1913).

Rhizosolenia imbricata

Delsman, H. C., 1939.
Preliminary plankton investigations in the Java Sea. Treubia, 17:139-181, 8 maps, 41 figs.

Rhizosolenia imbricata

Ercegovic, A., 1936
Etudes qualitative et quantitatives du phytoplancton dans les eaux cotières de l'Adriatique oriental moyen au cours de l'année 1934. Acta Adriatica 1(9):1-126

Rhizosolenia imbricata

Forti, A., 1922
Ricerche sulla flora pelagica (fitoplancton) di Quarto dei Mille. Mem. R. Com. Talass. Ital. 97:248 pp., 13 pls.

Rhizosolenia imbricata

Frenguelli, Joaquin, and Hector Antonio Orlando, 1959.
Operacion MERLUZA. Diatomeas y silicoflagelados del plancton del "VI Crucero". Servicio Hidrogr. Naval., Argentina, Publ. No. H. 619: 5-62.

Rhizosolenia imbricata

Gilbert, J.Y., and W.E. Allen, 1943
The phytoplankton of the Gulf of California obtained by the "E.W. Scripps" in 1939 and 1940. J. Mar. Res. V(2):89-110, figs.30-31.

Rhizosolenia imbricata

Gran, H.H., and T. Braarud, 1935
A quantitative study of the phytoplankton in the Bay of Fundy and the Gulf of Maine (including observations on hydrography, chemistry, and turbidity). J. Biol. Bd., Canada, 1(5):279-467, 69 text figs.

Rhizosolenia imbricata

Hendy, N. Ingram, 1964
An introductory account of the smaller algae of British coastal waters. V. Bacillariophyceae (Diatoms). Her Majesty's Stationary Office, 317 pp., 45 pls.

Rhizosolenia imbricata

Hendey, N.I., 1937
The plankton diatoms of the southern seas. Discovery Repts. 16:151-364, pls.6-13.

Rhizosolenia imbricata

Kokubo, S., 1952
Results of the observations on the plankton and oceanography of Mutsu Bay during 1950, reference being made also to the period 1946-1950. Bull Mar.Biol.Sta., Asamushi 5(1/4): 1-54, 3 tables,(fold-in), 1 fold-in.

Rhizosolenia imbricata

Kokubo, S., and S. Sato., 1947
Plankters in JG-San Gata. Physiol. and Ecol. (Japan) 1(4):1-16, 3 text figs., tables.

Rhizosolenia imbricata

Margalef, R., 1949
Fitoplancton nerítico de la Costa Brava en 1947-48. Publ. Inst. Biol. Aplicada, 5: 41-51, 3 text figs.

Rhizosolenia imbricata

Pavillard, J., 1905
Recherches sur la flore pelagique (Phytoplankton) de l'Etang de Thau. Theses presentees a la Fac. Sci., Paris, 116 pp., 3 pls

Rhizosolenia imbricata

Rampi, L., 1942
Ricerche sul fitoplancton del Mare Ligure 6. Le diatomee delle acque di Sanremo. Nuovo Giornale Botanico Italiano, N.S., 49:252-268.

Rhizosolenia imbricata

Schröder, B., 1900
Phytoplankton des Golfes von Neapel nebst vergleichenden Ausblicken auf das atlantischen Ozean. Mitt. Zool. Stat. Neapel, 14:1-38.

Rhizosolenia imbricata var. Shrubsolei

Allen, W.E., and E.E. Cupp, 1935
Plankton diatoms of the Java Sea. Annales du Jardin Botanique de Buitenzorg XLIV (2):101-174, figs.1-127.

(drawings of all species mentioned)

Rhizosolenia imbricata-shrubsole

Boden, B.P., 1950
Some marine plankton diatoms from the west coast of South Africa. Trans. R.Soc. S. Africa. 32:321-434, 100 text figs.

Rhizosolenia imbricata shrubsolei

Braarud, T., K.R. Gaarder, and J. Grøntved, 1953.
The phytoplankton of the North Sea and adjacent waters in May 1948. Rapp. Proc. Verb., Cons. Perm. Int. Expl. Mer, 133:1-87, 29 tables, Pls. A-B, 18 textfigs.

Rhizosolenia imbricata shrubsolei

Cupp, Easter E., 1943
Marine plankton diatoms of the west coast of North America. Bull. S.I.O. 5(1):1-238, 5 pls., 168 text figs.

Rhizosolenia imbricata var. shrubsolei

Cupp, E.E., 1934
Analysis of marine diatom collections taken from the Canal Zone to California during March, 1933. Trans. Am. Micros. Soc. LIII (1):22-29, 1 map.

Rhizosolenia imbricata var. shrubsolei

Cupp, E.E. and Allen, W.E., 1938
Plankton diatoms of the Gulf of California obtained by Allan Hancock Pacific Expedition of 1937. The Hancock Pacific Expeditions, The Univ. So. Calif. Publ. 3: 61-74, 1 map, pls.4-15.

Rhizosolenia imbricata Shrubsolei

Delegazione Italiana della Commissione Internazionale per l'Esplorazione Scientifica del Mediterraneo, 1941
Note sul plancton della Laguna veneta. [Memoria CCLXXXI], Arch. di Ocean. e Limn. Anno I, Fasc. I, 1941 XIX: 31-57 pp.

Rhizosolenia imbricata shrubsolei

Eskinazi Enide e Shigekatsv Sato (1963-1964) 1966.
Contribuição ao estudo das diatomaceas da Praia de Piedade.

Trabhs Inst. Oceanogr., Univ. Recife, 5 (5/6): 73-114.

Rhizosolenia imbricata shrubsolei

Grøntved, J., 1949(1950).
Investigations on the phytoplankton in the Danish Waddensee in July 1941. Medd. Komm. Danmarks Fiskeri- og Havundersøgelser, Serie Plankton, 5(2)(Medd. Skalling Lab. 10):55 pp., 2 pls., 38 textfigs.

Rhizosolenia imbricata shrubsolei

Herrera, Juan, y Ramón Margalef, 1963
Hidrografía y fitoplancton de la costa comprendida entre Castellón y la desembocadura del Ebro, de julio de 1960 a junio de 1961. Inv. Pesq., Barcelona, 24:33-112.

Rhizosolenia imbricata var. Shrubsolei

Lillick, L.C., 1940
Phytoplankton and planktonic protozoa of the offshore waters of the Gulf of Maine. Pt.II. Qualitative Composition of the Planktonic Flora. Trans. Am. Phil. Soc., n.s., 31(3):193-237, 13 text figs.

Rhizosolenia imbricata var Shrubsolei

Lillick, L.C., 1938
Preliminary report of the phytoplankton of the Gulf of Maine. Am. Mid. Nat. 20(3):624-640, 1 text figs 37 tables.

Rhizosolenia imbricata var. Shrubsolei

Lillick, L.C., 1937
Seasonal studies of the phytoplankton off Woods Hole, Massachusetts. Biol. Bull. LXXIII (3):488-503, 3 text figs.

Rhizosolenia imbricata shrubsolei

Okuno, H., 1957.
Electron-microscopical study on fine structures of diatom frustules. 15. Observation on the genus Rhizosolenia. Bot. Mag., Tokyo, 71(826):101-107.

Rhizosolenia imbricata shrubsolei

Robinson, G.A., 1961
Contribution towards an atlas of the northeastern Atlantic and the North Sea. 1. Phytoplankton. Bulls. Mar. Ecology, 5(42): 81-89, Pls. 15-20.

Rhizosolenia indica

Dangeard, P., 1927
Phytoplankton de la croisière du "Sylvana". Ann. Inst. Ocean., Monaco, n.s., 4(8):286-401, 54 text figs. (Feirer-Juin 1913).

Rhizosolenia indica

Pavillard, J., 1925
Bacillariales. Rept. on the Danish Oceangr. Exped., 1908-10 to the Mediterranean and adj. seas. Vol.II., Biol. J4:72 pp., 116 text figs.

Rhizosolenia inaequalis n.sp.

Castracane degli Antelminelli, F., 1886
1. Report on the Diatomaceae collected by H.M.S. Challenger during the years 1873-1876. Rept. Sci. Results, H.M.S. Challenger, Botany Vol. II, 178 pp., 30 pls.

Rhizosolenia inermis n.sp.

Castracane degli Antelminelli, F., 1886
1. Report on the Diatomaceae collected by H.M.S. Challenger during the years 1873-1876. Rept. Sci. Results, H.M.S. Challenger, Botany Vol. II, 178 pp., 30 pls.

Rhizosolenia inermis

Cupp, Easter E., 1943
Marine plankton diatoms of the west coast of North America. Bull. S.I.O. 5(1):1-238, 5 pls., 168 text figs.

Rhizosolenia japonica n.sp.

Castracane degli Antelminelli, F., 1886
1. Report on the Diatomaceae collected by H.M.S. Challenger during the years 1873-1876. Rept. Sci. Results, H.M.S. Challenger, Botany Vol. II, 178 pp., 30 pls.

Rhizosolenia longiseta

Asmund, B., 1955.
Five Danish waters and their population of Rhizosolenia longiseta. Dansk Bot. Ark., 15(5): 68 pp.

Rhizosolenia longiseta

Cleve-Euler, A., 1951.
Die Diatomeen von Schweden und Finnland. Kungl. Svenska Vetenskaps Akad. Handl., Fjärde Ser. 2(1): 161 pp., 6 pls.

Rhizosolenia longiseta

Meunier, A., 1915
Microplancton de la Mer Flamande. 2. Diatomées (excepté le genre Chaetoceros). Mem. Mus. Roy. Hist. Nat., Belgique, 7(3):1-118, Pls. VIII-XIV.

Rhizosolenia longiseta

Okuno, H., 1957.
Electron-microscopical study on fine structures of diatom frustules. 15. Observation on the genus Rhizosolenia. Bot. Mag., Tokyo, 71(826):101-107.

Rhizosolenia minima

Cleve-Euler, A., 1951.
Die Diatomeen von Schweden und Finnland. Kungl. Svenska Vetenskaps Akad. Handl., Fjärde Ser. 2(1): 161 pp., 6 pls.

Rhizosolenia murrayana n.sp.

Castracane degli Antelminelli, F., 1886
1. Report on the Diatomaceae collected by H.M.S. Challenger during the years 1873-1876. Rept. Sci. Results, H.M.S. Challenger, Botany Vol. II, 178 pp., 30 pls.

Rhizosolenia obtusa

Bigelow, H. B., 1922
Exploration of the coastal water off the northeastern United States in 1916 by the U.S. Fisheries Schooner Grampus. Bull. M.C.Z. 65 (5):85-188, 53 text figs.

Rhizosolenia obtusa

Central Meteorological Observatory, 1949
Report on sea and weather observation on Antarctic Whaling Ground (1947-48). Ocean. Mag., Japan, 1(1):49-88, 17 text figs.

Rhizosolenia obtusa

Cupp, E.E., 1937
Seasonal distribution and occurrence of marine diatoms and dinoflagellates at Scotch Cap, Alaska. Bull. S.I.O. Tech ser.4(3):71-100, 7 textfigs.

Rhizosolenia obtusa

Gran, H.H., 1908
Diatomeen. Nordisches Plankton, Botanischer Teil pp. XIX.1-XIX 146; 178 text figs.

Rhizosolenia obtusa

Hendy, N. Ingram, 1964
An introductory account of the smaller algae of British coastal waters. V. Bacillariophyceae (Diatoms). Her Majesty's Stationary Office, 317 pp., 45 pls.

Rhizosolenia obtusa

Schröder, B., 1900
Phytoplankton des Golfes von Neapel nebst vergleichenden Ausblicken auf das atlantischen Ozean. Mitt. Zool. Stat. Neapel, 14:1-38.

Rhizosolenia polydactyla n.sp.

Castracane degli Antelminelli, F., 1886
1. Report on the Diatomaceae collected by H.M.S. Challenger during the years 1873-1876. Rept. Sci. Results, H.M.S. Challenger, Botany Vol. II, 178 pp., 30 pls.

Rhizosolenia polydactyla

Hendey, N.I., 1937
The plankton diatoms of the southern seas. Discovery Repts. 16:151-364, pls.6-13.

Rhizosolenia polydactyla

Mangin, L., 1915
Phytoplancton de L'Antartique. Deuxieme Exped. Ant. Francaise (1908-1910), 95 pp., 3 pls., 58 text figs.

Rhizosolenia praealata n.sp. [a]

Schrader, Hans-Joachim, 1973
Cenozoic diatoms from the northeast Pacific, leg 18. Initial Repts Deep Sea Drilling Proj. 18:673-797.

Rhizosolenia praebarboi n.sp. [a]

Schrader, Hans-Joachim, 1973
Cenozoic diatoms from the northeast Pacific, leg 18. Initial Repts Deep Sea Drilling Proj. 18:673-797.

Rhizosolenia pungens

Brunel, J., 1962
Le phytoplancton de la Baie de Chaleurs. Inst. Botan., Univ. Montréal, Contrib. No. 77: 365 pp., 66 pls.

Rhizosolenia pungens

Brunel, Jules, 1962
Le phytoplancton de la Baie des Chaleurs. Contrib. Ministere de la Chasse et des Pecheries. Province de Québec, No. 91: 365 pp.

Rhizosolenia pungens

Cleve-Euler, A., 1951.
Die Diatomeen von Schweden und Finnland. Kungl. Svenska Vetenskaps Akad. Handl., Fjärde Ser. 2(1): 161 pp., 6 pls.

Rhizosolenia rhombus

Hendey, N.I., 1937
The plankton diatoms of the southern seas. Discovery Repts. 16:151-364, pls.6-13.

Rhizosolenia rhombus

Mangin, L., 1915
Phytoplancton de L'Antartique. Deuxieme Exped. Ant. Francaise (1908-1910), 95 pp., 3 pls., 58 text figs.

Rhizosolenia robusta

Allen, W.E., and E.E. Cupp, 1935
Plankton diatoms of the Java Sea. Annales du Jardin Botanique de Buitenzorg XLIV (2):101-174, figs.1-127.
(drawings of all species mentioned)

Rhizosolenia robusta

Boden, B.P., 1950
Some marine plankton diatoms from the west coast of South Africa. Trans. R.Soc. S. Africa. 32:321-434, 100 text figs.

Rhizosolenia robusta

Cleve-Euler, A., 1951.
Die Diatomeen von Schweden und Finnland. Kungl. Svenska Vetenskaps Akad. Handl., Fjärde Ser. 2(1): 161 pp., 6 pls.

Rhizosolenia robusta

Cupp, Easter E., 1943
Marine plankton diatoms of the west coast of North America. Bull. S.I.O. 5(1):1-238, 5 pls., 168 text figs.

Rhizosolenia robusta

Dangeard, P., 1927
Phytoplankton de la croisière du "Sylvana". Ann. Inst. Ocean., Monaco, n.s., 4(8):286-401, 54 text figs. (Février-Juin 1913).

Rhizosolenia robusta

Delegazione Italiana della Commissione Internazionale per l'Esplorazione Scientifica del Mediterraneo, 1941
Note sul plancton della Laguna veneta. [Memoria CCLXXIX] Arch. di Ocean. e Limn. Anno I, Fasc. I, 1941 XIX: 31-57 pp.

Rhizosolenia robusta

Delsman, H. C., 1939.
Preliminary plankton investigations in the Java Sea. Treubia, 17:139-181, 8 maps, 41 figs.

Rhizosolenia robusta

Ercegovic, A., 1936
Etudes qualitative et quantitatives du phytoplancton dans les eaux cotières de l'Adriatique oriental moyen au cours de l'année 1934. Acta Adriatica 1(9):1-126

Rhizosolenia robusta

Eskinazi Enide e Shigekatsv Satô (1963-1964) 1966.
Contribuição ao estudo das diatomaceas da Praia de Piedade.
Trabhs Inst. Oceanogr., Univ. Recife, 5 (5/6): 73-114.

Rhizosolenia robusta

Forti, A., 1922
Ricerche sulla flora pelagica (fitoplancton) di Quarto dei Mille. Mem. R. Com. Talass. Ital. 97:248 pp., 13 pls.

Rhizosolenia robusta

Gilbert, J.Y., and W.E. Allen, 1943
The phytoplankton of the Gulf of California obtained by the "E.W. Scripps" in 1939 and 1940. J. Mar. Res. V(2):89-110, figs.30-31.

Gran, H.H. 1908
Diatomeen. Nordisches Plankton, Botanischer Teil pp. XIX 1-XIX 146; 178 text figs.

Gunter, G., R.H. Williams, C.C. Davis, and F.G. Walton Smith, 1948
Catastrophic mass mortality of marine animals and coincident phytoplankton bloom on the West Coast of Florida, November 1946 to August 1947. Ecol. Mon., 18:309-324, 2 text figs.

Hendy, N. Ingram, 1964
An introductory account of the smaller algae of British coastal waters. V. Bacillariophyceae (Diatoms). Her Majesty's Stationary Office, 317 pp., 45 pls.

Hendey, N.I., 1937
The plankton diatoms of the southern seas. Discovery Repts. 16:151-364, pls.6-13.

Mangin, M. L., 1912
Phytoplancton de la croisière du "René" dans l'Atlantique (Septembre 1908). Ann. Inst. Ocean., n.s., 4(1):1-66, 2 pls., 41 text figs., 2 tables.

Mann, A., 1907
Report on the diatoms of the Albatross voyages in the Pacific Ocean, 1888-1904. Contrib. U. S. Nat. Herb. 10(5):221-419, Pls. XLIV-LIV.

Margalef, R., 1949
Fitoplancton nerítico de la Costa Brava en 1947-48. Publ. Inst. Biol. Aplicada, 5: 41-51, 3 text figs.

Marshall, S. M., 1933
The production of microplankton in the Great Barrier Reef Region. Brit. Mus. (N.H.) Great Barrier Reef Exped. 1928-29, Sci. Repts. II(5):111-157, 14 text figs.

Meunier, A., 1915
Microplancton de la Mer Flamande. 2. Diatomées (excepté le genre Chaetoceros). Mem. Mus. Roy. Hist. Nat., Belgique, 7(3):1-118, Pls. VIII-XIV.

Okuno, H., 1957.
Electron-microscopical study on fine structures of diatom frustules. 15. Observation on the genus Rhizosolenia. Bot. Mag., Tokyo, 71(826):101-107.

Pavillard, J., 1925
Bacillariales. Rept. on the Danish Oceangr. Exped., 1908-10 to the Mediterranean and adj. seas. Vol.II., Biol. J4:72 pp., 116 text figs.

Pavillard, J., 1905
Recherches sur la flore pelagique (Phytoplankton) de l'Etang de Thau. Theses presentees a la Fac. Sci., Paris, 116 pp., 3 pls.

Rampi, L., 1942
Ricerche sul fitoplancton del Mare Ligure 6. Le diatomee delle acque di Sanremo. Nuovo Giornale Botanico Italiano, N.S., 49:252-268.

Schröder, B., 1900
Phytoplankton des Golfes von Neapel nebst vergleichenden Ausblicken auf das atlantischen Ozean. Mitt. Zool. Stat. Neapel, 14:1-38.

Sousa e Silva, E., 1949
Diatomaceas e Dinoflagelados de Baia de Cascais. Portugaliae Acta Biol. Volume: Julio Henriques, Ser. B: 300-383, 9 pls, 2 fold-in tables.

Sousa e Silva, E., and J. Dos Santos-Pinto, 1948
O Plancton da Baía de S. Martinho do Porto. 1. Diatomaceas e Dinoflagelados. Bol. Soc. Portuguese de Ciencias Naturais, 16(2):134-187, 6 pls. (Trav. Sta. Biol. Mar. de Lisbonne No. 52).

Rhizosolenia semispina

Allen, W. E., 1938
The Templeton Crocker Expedition to the Gulf of California in 1935 - The Phytoplankton. Amer. Microsc. Soc., Trans. 57:328-335.

Allen, W.E., 1937
Plankton diatoms of the Gulf of California obtained by the G. Allan Hancock Expedition of 1936. The Hancock Pacific Expeditions, Univ. So. Calif. Publ. 3:47-59, 1 fig.

Bigelow, H. B., 1922
Exploration of the coastal water off the northeastern United States in 1916 by the U.S. Fisheries Schooner Grampus. Bull. M.C.Z. 65 (5):85-188, 53 text figs.

Bigelow, H.B., and M. Leslie, 1930
Reconnaissance of the waters and plankton of Monterey Bay, July 1928. Bull. M.C.Z., 70(5):429-481, 43 text figs.

Braarud, T., 1945
A phytoplankton survey of the polluted waters of inner Oslo Fjord. Hvalrådets Skrifter, No.28, 142 pp., 19 text figs., 17 tables.

Braarud, T., 1934
A note on the phytoplankton of the Gulf of Maine in the summer of 1933. Biol. Bull. 67(1):76-82.

Braarud, T., and Adam Bursa, 1939
On the phytoplankton of the Oslo Fjord, 1933-1934. Hvaldrådets Skr. No.19:1-63; 9 text figs. Reviewed. J. du. Cons. 14(3): 418-420. A.C. Gardiner.

Cupp, E.E., 1937
Seasonal distribution and occurrence of marine diatoms and dinoflagellates at Scotch Cap, Alaska. Bull. S.I.O. Tech. ser.4(3):71-100, 7 textfigs.

Cupp, E.E., 1934
Analysis of marine diatom collections taken from the Canal Zone to California during March, 1933. Trans. Am. Micros. Soc. LIII (1):22-29, 1 map.

Dangeard, P., 1927
Phytoplankton de la croisière du "Sylvana". Ann. Inst. Ocean. Monaco, n.s., 4(8):286-401, 54 text figs. (Feirer-Juin 1913).

Forti, A., 1922
Ricerche sulla flora pelagica (fitoplancton) di Quarto dei Mille. Mem. R. Com. Talass. Ital. 97:248 pp., 13 pls.

Gran, H.H. 1897
Protophyta: Diatomaceae, Silico-flagellata and Cilioflagellata. Den Norske Nordhavs Expedition 1876-1878, h. 24, 36 pp., 4 pls.

Gran, H. H. and E. C. Angst, 1931
Plankton diatoms of Puget Sound. Publ. Puget Sound Biol. Sta. 7:417-519, 95 text figs.

Jørgensen, E., 1905
B. Protistplankton and the diatoms in bottom samples. Hydrographical and biological investigations in Norwegian fjords. Bergens Mus. Skr. 7: 49-225.

Jorgensen, E., 1900
Protophyten und Protozoen im Plankton aus der Norwegischen Westkerste. Bergens Mus. Aarb. 1899(6): 95 pp., 5 pls., 83 tables.

Macdonald, R., 1933
An examination of plankton hauls made in the Suez Canal during the year 1928. Fish. Res. Dis., Notes & Mem. No.3, 11 pp., 1 chart.

Mangin, L., 1915
Phytoplancton de L'Antartique. Deuxieme Exped. Ant. Francaise (1908-1910), 95 pp., 3 pls., 58 text figs.

Mangin, M. L., 1912
Phytoplancton de la croisière du "René" dans l'Atlantique (Septembre 1908). Ann. Inst. Ocean., n.s., 4(1):1-66, 2 pls., 41 text figs., 2 tables.

Paiva Carvalho, J., 1950
O plancton do Rio Maria Rodrigues (Cananeis). 1. Diatomaceas e Dinoflagelados. Bol. Inst. Paulista Oceanogr. 1(1); 27-43, 2 fold-in tables, 2 figs.

Pavillard, J., 1925
Bacillariales. Rept. on the Danish Oceangr. Exped., 1908-10 to the Mediterranean and adj. seas. Vol.II., Biol. J4:72 pp., 116 text figs.

Pettersson, H., 1934
Scattering and extinction of light in sea-water. Meddelanden från Göteborgs Högskolas Oceanografiska Institut. 9 (Göteborgs Kungl. vetenskaps-och vitterhets-samhälles Handlingar. Femte Foljden, Ser.B, 4(4)):1-16.

Phifer, L.D. (undated)
The occurrence and distribution of plankton diatoms in Bering Sea and Bering Strait, July 26-August 24, 1934. Report of Oceanographic cruise of U.S. Coast Guard Cutter Chelan 1934, Part II(A):1-44 (mimeographed) plus fig.1 (after Pt.B)

Rhizosolenia semispina
Schulz, B., and A. Wulff, 1929
Hydrographie und Oberflächen plankton des westlichen Barentsmeeres im Sommer 1927. Ber. deutschen wissensch. Komm. F. Meeresforsch. n.s. 4(5):232-372, 13 tables, 25 text figs.

Rhizosolenia semispina
Sleggs, G.F., 1927.
Marine phytoplankton in the region of La Jolla, California during the summer of 1924. Bull. S.I.O., tech. ser., 1:93-117, 8 textfigs.

Rhizosolenia semispina
Sousa e Silva, E., 1949
Diatomaceas e Dinoflagelados de Baia de Cascais. Portugaliae Acta Biol., Volume: Julio Henriques, Ser. B: 300-383, 9 pls, 2 fold-in tables.

Rhizosolenia semispina
Sousa e Silva, E., and J. Dos Santos-Pinto, 1948
O Plancton da Baia de S. Martinho do Porto. 1. Diatomaceas e Dinoflagelados. Bol. Soc. Portuguese de Ciencias Naturais, 16(2):134-167, 6 pls. (Trav. Sta. Biol. Mar. de Lisbonne No. 52).

Rhizosolenia setigera
Allen, W.E., 1938
The Templeton Crocker Expedition to the Gulf of California in 1935 - The Phytoplankton. Amer. Microsc. Soc., Trans. 57:328-335.

Rhizosolenia setigera
Allen, W.E., 1937
Plankton diatoms of the Gulf of California obtained by the G. Allan Hancock Expedition of 1936. The Hancock Pacific Expeditions, Univ. So. Calif. Publ. 3:47-59, 1 fig.

Rhizosolenia setigera
Allen, W.E., and E.E. Cupp, 1935
Plankton diatoms of the Java Sea. Annales du Jardin Botanique de Buitenzorg XLIV (2):101-174, figs.1-127.
(drawings of all species mentioned)

Rhizosolenia setigera
Aurich, H.J., 1949.
Die Verbreitung des Nannoplanktons im Oberflächenwasser vor der Nordfriesischen Kuste. Ber. Deutschen Wiss. Komm. f. Meeresf., n. f., II(4):403-405, 2 figs.

Rhizosolenia setigera
Bigelow, H.B., and M. Leslie, 1930
Reconnaissance of the waters and plankton of Monterey Bay, July 1928. Bull. M.C.Z., 70(5):429-481, 43 text figs.

Rhizosolenia setigera
Braarud, T., and Adam Bursa, 1939
On the phytoplankton of the Oslo Fjord, 1933-1934. Hvalrådets Skr. No.19:1-63; 9 text figs. Reviewed. J. du. Cons. 14(3): 418-420. A.C. Gardiner.

Rhizosolenia setigera
Brunel, J., 1962
Le phytoplancton de la Baie de Chaleurs. Inst. Botan., Univ. Montréal, Contrib. No. 77: 365 pp., 66 pls.
Rhizosolenia/pungens (setigera)

Rhizosolenia setigera
Brunel, Jules, 1962
Le phytoplancton de la Baie des Chaleurs. Contrib. Ministère de la Chasse et des Pêcheries, Province de Québec, No. 91: 365 pp.

Rhizosolenia setigera
Cleve-Euler, A., 1951
Die Diatomeen von Schweden und Finnland. Kungl. Svenska Vetenskaps Akad. Handl., Fjärde Ser. 2(1): 161 pp., 6 pls.

Rhizosolenia setigera
Cupp, Easter E., 1943
Marine plankton diatoms of the west coast of North America. Bull. S.I.O. 5(1):1-238, 5 pls., 168 text figs.

Rhizosolenia setigera
Cupp, E.E., 1934
Analysis of marine diatom collections taken from the Canal Zone to California during March, 1933. Trans. Am. Micros. Soc. LIII (1):22-29, 1 map.

Rhizosolenia setigera
Cupp, E.E. and Allen, W.E., 1938
Plankton diatoms of the Gulf of California obtained by Allan Hancock Pacific Expedition of 1937. The Hancock Pacific Expeditions, The Univ. So. Calif. Publ. 3: 61-74, 1 map, pls.4-15.

Rhizosolenia setigera
Dangeard, P., 1927
Phytoplankton de la croisière du "Sylvana". Ann. Inst. Ocean., Monaco, n.s., 4(8):286-401, 54 text figs. (Février-Juin 1913).

Rhizosolenia setigera
Davidson, V.M., 1931.
Biological and oceanographic conditions in Hudson Bay. Contr. Canadian Biol. Fish., n.s., 497-509, 7 textfigs.

Rhizosolenia setigera
Forti, A., 1922
Ricerche sulla flora pelagica (fitoplancton) di Quarto dei Mille. Mem. R. Com. Talass. Ital. 97:248 pp., 13 pls.

Rhizosolenia setigera
Egusa, S., 1949.
The seasonal variation of size in phytoplankton and the reliability of net collection. Bull. Jap. Soc. Sci. Fish. 15(2):78-82.

Rhizosolenia setigera
Eskinazi Enide e Shigekatsv Sato (1963-1964) 1966.
Contribuição ao estudo das diatomaceas da Praia de Piedade.
Trabhs Inst. Oceanogr., Univ. Recife, 5 (5/6): 73-114.

Rhizosolenia setigera
Gilbert, J.Y., and W.E. Allen, 1943
The phytoplankton of the Gulf of California obtained by the "E.W. Scripps" in 1939 and 1940. J. Mar. Res. V(2):89-110, figs.30-31.

Rhizosolenia setigera
Gran, H.H., 1908
Diatomeen. Nordisches Plankton, Botanischer Teil pp. XIX 1-XIX 146; 178 text figs.

Rhizosolenia setigera
Gran, H.H., and T. Braarud, 1935
A quantitative study of the phytoplankton in the Bay of Fundy and the Gulf of Maine (including observations on hydrography, chemistry, and turbidity). J. Biol. Bd., Canada, 1(5):279-467, 69 text figs.

Rhizosolenia setigera
Gran, H.H. and E.C. Angst, 1931
Plankton diatoms of Puget Sound. Publ. Puget Sound Biol. Sta. 7:417-519, 95 text figs.

Rhizosolenia setigera
Gunter, G., R.H. Williams, C.C. Davis, and F.G. Walton Smith, 1948
Catastrophic mass mortality of marine animals and coincident phytoplankton bloom on the West Coast of Florida, November 1946 to August 1947. Ecol. Mon., 18:309-324, 2 text figs.

Rhizosolenia setigera
Hendy, N. Ingram, 1964
An introductory account of the smaller algae of British coastal waters. V. Bacillariophyceae (Diatoms). Her Majesty's Stationary Office, 317 pp., 45 pls.

Rhizosolenia setigera
Hendey, N.I., 1937
The plankton diatoms of the southern seas. Discovery Repts. 16:151-364, pls.6-13.

Rhizosolenia setigera
Iselin, C., 1930
A report on the coastal waters of Labrador based on explorations of the "Chance" during the summer of 1926. Proc. Am. Acad. Arts Sci., 66(1):1-37, 14 text figs.

Rhizosolenia setigera
Jorgensen, E., 1900
Protophyten und Protozoën im Plankton aus der Norwegischen Westkerste. Bergens Mus. Aarb. 1899(6): 95 pp., 5 pls., 83 tables.

Rhizosolenia setigera
Jørgensen, E., 1905
B. Protistplankton and the diatoms in bottom samples. Hydrographical and biological investigations in Norwegian fjords. Bergens Mus. Skr. 7: 49-225.

Rhizosolenia setigera
Kokubo, S., 1952
Results of the observations on the plankton and oceanography of Mutsu Bay during 1950, reference being made also to the period 1946-1950. Bull Mar.Biol.Sta., Asamushi 5(1/4): 1-54, 3 tables,(fold-in), 1 fold-in.

Rhizosolenia setigera
Lillick, L.C., 1940
Phytoplankton and planktonic protozoa of the offshore waters of the Gulf of Maine. Pt.II. Qualitative Composition of the Planktonic Flora. Trans. Am. Phil. Soc., n.s., 31(3):193-237, 13 text figs.

Rhizosolenia setigera
Lillick, L.C., 1938
Preliminary report of the phytoplankton of the Gulf of Maine. Am. Mid. Nat. 20(3):624-640, 1 text figs 37 tables.

Rhizosolenia setigera
Lillick, L.C., 1937
Seasonal studies of the phytoplankton off Woods Hole, Massachusetts. Biol. Bull. LXXIII (3):488-503, 3 text figs.

Rhizosolenia setigera
Macdonald, R., 1933
An examination of plankton hauls made in the Suez Canal during the year 1928. Fish. Res. Dis., Notes & Mem. No.3, 11 pp., 1 chart.

Rhizosolenia setigera
Mangin, M.L., 1912
Phytoplankton de la croisière du "René" dans l'Atlantique (Septembre 1908). Ann. Inst. Ocean., n.s., 4(1):1-66, 2 pls., 41 text figs., 2 tables.

Rhizosolenia setigera
Mann, A., 1921.
The dependence of fishes on diatoms. Ecology, 2(2):79-83.

Rhizosolenia setigera
Marukawa, H., 1921
Plankton lists and some new species of copepods, from the northern waters of Japan. Bull. Inst. Ocean., No.384, 15 pp., 3 pls., 1 chart. Monaco

Rhizosolenia setigera
Meunier, A., 1915
Microplancton de la Mer Flamande. 2. Diatomées (excepte le genre Chaetoceros). Mem. Mus. Roy. Hist. Nat., Belgique, 7(3):1-118, Pls. VIII-XIV.

Rhizosolenia setigera
Morse, D.C., 1947
Some observations on seasonal variations in plankton population Patuxant River, Maryland 1943 1945. Bd. Nat. Res., Publ. No.65, Chesapeake Biol. Lab., 31, 3 figs.

Rhizosolenia setigera
Paasche, E., 1961.
Notes on phytoplankton from the Norwegian Sea. Botanica Marina, 2(3/4):197-210.

Rhizosolenia setigera
Pavillard, J., 1925
Bacillariales. Rept. on the Danish Oceangr. Exped., 1908-10 to the Mediterranean and adj. seas. Vol.II., Biol. J4:72 pp., 116 text figs.

Rhizosolenia setigera
Pavillard, J., 1905
Recherches sur la flore pelagique (Phytoplankton) de l'Etang de Thau. Theses presentees a la Fac. Sci., Paris, 116 pp., 3 pls.

Rhizosolenia setigera
Phifer, L.D. (undated)
The occurrence and distribution of plankton diatoms in Bering Sea and Bering Strait, July 26-August 24, 1934. Report of Oceanographic cruise of U.S. Coast Guard Cutter Chelan 1934, Part II(A):1-44 (mimeographed) plus fig.1 (after Pt.B)

Rhizosolenia setigera
Rampi, L., 1942
Ricerche sul fitoplancton del Mare Ligure 6. Le diatomee delle acque di Sanremo. Nuovo Giornale Botanico Italiano, N.S., 49:252-268.

Rhizosolenia setigera
Schodduyn, M., 1926
Observations faites dans la baie d'Ambleteuse (Pas de Calais). Bull. Inst. Ocean., Monaco, No. 482: 64 pp.

Rhizosolenia setigera
Schröder, B., 1900
Phytoplankton des Golfes von Neapel nebst vergleichenden Ausblicken auf das atlantischen Ozean. Mitt. Zool. Stat. Neapel, 14:1-38.

Rhizosolenia setigera
Smayda, Theodore J., and Brenda J. Boleyn, 1966
Experimental observations on the flotation of marine diatoms. II. Skeletonema costatum and Rhizosolenia setigera. Limnol. and Oceanogr., 11(1):18-34.

Rhizosolenia setigera
Sousa e Silva, E., 1949
Diatomaceas e Dinoflagelados de Baia de Cascais. Portugaliae Acta Biol., Volume: Julio Henriques, Ser. B: 300-383, 9 pls, 2 fold-in tables.

Rhizosolenia setigera
Sousa a Silva, E., and J. Dos Santos-Pinto, 1948
O Plancton da Baia de S. Martinho do Porto. 1. Diatomaceas e Dinoflagelados. Bol. Soc. Portuguese de Ciencias Naturais, 16(2):134-187, 6 pls. (Trav. Sta. Biol. Mar. de Lisbonne No. 52).

Rhizosolenia setigera
Steemann-Nielsen, Einar, 1951
The marine vegetation of the Isefjord. A study on ecology and production. Medd. Komm. Danmarks Fiskeri-og Havundersøgelser. Ser. Plankton. 5(4); 114pp., 46 text figs.

Rhizosolenia setigera
Uyeno, Fukuzo, 1961
Oceanographical and ecological studies on primary production of the sea, with special references to relationship between diatom production and temperature and chlorinity of water. Rept., Fac. Fish., Pref. Univ., Mie, 4(1): 1-64.

Rhizosolenia setigera
Wheeler, J.E.G., 1939
Plankton investigations. Bermuda Biological Station. Second Report. October 1939. 7 pp. (typed), 5 figs. Plymouth, Oct. 23, 1939.

Rhizosolenia setigera
Woodmansee, Robert A., 1963.
Cell-diameter frequency distributions of the planktonic diatom Rhizosolenia alata. Publ. Inst. Mar. Sci., Port Aransas, 9:117-131.

Rhizosolenia shrubsolei
Bigelow, H. B., 1922
Exploration of the coastal water off the northeastern United States in 1916 by the U.S. Fisheries Schooner Grampus. Bull. M.C.Z. 65 (5):85-188, 53 text figs.

Rhizosolenia shrubsolei
Braarud, T., and Adam Bursa, 1939
On the phytoplankton of the Oslo Fjord, 1933-1934. Hvalrådets Skr. No.19:1-63; 9 text figs. Reviewed. J. du Cons. 14(3): 418-420. A.C. Gardiner.

Rhizosolenia shrubsolei
Cleve-Euler, A., 1951
Die Diatomeen von Schweden und Finnland. Kungl. Svenska Vetenskaps Akad. Handl., Fjärde Ser. 2(1): 161 pp., 6 pls.

Rhizosolenia Schrubsolei
Dangeard, P., 1927
Phytoplankton de la croisière du "Sylvana". Ann. Inst. Ocean., Monaco, n.s., 4(8):286-401, 54 text figs. (Février-Juin 1913).

Rhizosolenia shrubsolei
Davidson, V.M., 1931.
Biological and oceanographic conditions in Hudson Bay. Contr. Canadian Biol. Fish., n.s., 6(26):497-509, 7 textfigs.

Rhizosolenia Shrubsolei
Forti, A., 1922
Ricerche sulla flora pelagica (fitoplancton) di Quarto dei Mille. Mem. R. Com. Talass. Ital. 97:248 pp., 13 pls.

Rhizosolenia shrubsolei
Fraser, J. H., and A. Saville, 1949
Plankton distribution in Scottish and adjacent waters in 1948. Ann. Biol. 5:61-62.

Rhizosolenia Shrubsolei
Ghazzawi, F.M., 1939
Plankton of the Egyptian waters. A study of the Suez Canal Plankton. (A) Phytoplankton. Preliminary Report 83 pp. Notes and Memoires, Min. Commerce-Industry, Egypt, Hydrobiol. & Fish. 65 figs.

Rhizosolenia Shrubsolei
Gran, H.H., 1908
Diatomeen. Nordisches Plankton, Botanischer Teil pp. XIX.1-XIX 146; 178 text figs.

Rhizosolenia schrubsolei
Grøntved, J., 1949
Investigations on the phytoplankton in the Danish Waddensea in July 1941. Medd. Komm. Danmarks Fiskeri og Havundersøgelser, ser. Plankton, 5(2):55 pp., 2 pls., 38 text figs.

Rhizosolenia shrubsolei
Hendy, N. Ingram, 1964
An introductory account of the smaller algae of British coastal waters. V. Bacillariophyceae (Diatoms). Her Majesty's Stationary Office, 317 pp., 45 pls.

Rhizosolenia Shrubsolii
Hendey, N.I., 1937
The plankton diatoms of the southern seas. Discovery Repts. 16:151-364, pls.6-13.

Rhizosolenia shrubsolei
Jørgensen, E., 1905
B.Protistplankton and the diatoms in bottom samples. Hydrographical and biological investigations in Norwegian fjords. Bergens Mus. Skr. 7: 49-225.

Rhizosolenia shrubsolei
Jorgensen, E., 1900
Protophyten und Protozoen im Plankton aus der Norwegischen Westkerste. Bergens Mus. Aarb. 1899(6): 95 pp., 5 pls., 83 tables.

Rhizosolenia shrubsolei
Lafon, M., M. Durchon and Y. Saudray, 1953.
Recherches sur les cycles saisonnières du plankton. Ann. Inst. Océan., 31(3):125-230.

Rhizosolenia shrubsolii
Macdonald, R., 1933
An examination of plankton hauls made in the Suez Canal during the year 1928. Fish. Res. Dis., Notes & Mem. No.3, 11 pp., 1 chart.

Rhizosolenia Shrubsolei
Meunier, A., 1915
Microplancton de la Mer Flamande. 2. Diatomées (excepté le genre Chaetoceros). Mem. Mus. Roy. Hist. Nat., Belgique, 7(3):1-118, Pls. VIII-XIV.

Rhizosolenia shrubsolei
Paiva Carvalho, J., 1950
O plancton do Rio Maria Rodriques (Cananeis). 1. Diatomaceas e Dinoflagelados. Bol. Inst. Paulista Oceanogr. 1(1); 27-43, 2 fold-in tables, 2 figs.

Rhizosolenia shrubsolei
Pavillard, J., 1925
Bacillariales. Rept. on the Danish Oceanogr. Exped., 1908-10 to the Mediterranean and adj. seas. Vol.II., Biol. J4:72 pp., 116 text figs.

Rhizosolenia shrubsolei
Pavillard, J., 1905
Recherches sur la flore pelagique (Phytoplankton) de l'Etang de Thau. Theses presentees a la Fac. Sci., Paris, 116 pp., 3 pls.

Rhizosolenia Shrubsolii
Schodduyn, M., 1926
Observations faites dans la baie d'Ambleteuse (Pas de Calais). Bull. Inst. Ocean., Monaco, No. 482: 64 pp.

Rhizosolenia Shrubsolei
Schröder, B., 1900
Phytoplankton des Golfes von Neapel nebst vergleichenden Ausblicken auf das atlantischen Ozean. Mitt. Zool. Stat. Neapel, 14:1-38.

Rhizosolenia shrubsolei
Sousa e Silva, E., 1949
Diatomaceas e Dinoflagelados de Baia de Cascais. Portugaliae Acta Biol., Volume: Julio Henriques, Ser. B: 300-383, 9 pls, 2 fold-in tables.

Rhizosolenia Shrubsolei
Sousa a Silva, E., and J. Dos Santos-Pinto, 1948
O Plancton da Baia de S. Martinho do Porto. 1. Diatomaceas e Dinoflagelados. Bol. Soc. Portuguese de Ciencias Naturais, 16(2):134-187, 6 pls. (Trav. Sta. Biol. Mar. de Lisbonne No. 52).

Rhizosolenia shrubsolei

Steemann-Nielsen, Einar, 1951
The marine vegetation of the Isefjord. A study on ecology and production. Medd. Komm. Danmarks Fiskeri-og Havundersøgelser. Ser. Plankton. 5(4); 114pp., 46 text figs.

Rhizosolenia sigma

Schröder, B., 1900
Phytoplankton des Golfes von Neapel nebst vergleichenden Ausblicken auf das atlantischen Ozean. Mitt. Zool. Stat. Neapel, 14:1-38.

Rhizosolenia sima n. sp.

Castracane degli Antelminelli, F., 1886
1. Report on the Diatomaceae collected by H.M.S. Challenger during the years 1873-1876. Rept. Sci. Results, H.M.S. Challenger, Botany Vol. II, 178 pp., 30 pls.

Rhizosolenia similoides nom. nov.

Cleve-Euler, A., 1953
Die Diatomeen von Schweden und Finnland. Kungl. Svenska Vetenskaps Akad. Handl., Fjärde Ser. 2(1): 161 pp., 6 pls.

Rhizosolenia simplex

Boden, B.P., 1950
Some marine plankton diatoms from the west coast of South Africa. Trans. R. Soc. S. Africa. 32:321-434, 100 text figs.

Rhizosolenia simplex

Dangeard, P., 1927
Phytoplankton de la croisière du "Sylvana". Ann. Inst. Ocean., Monaco, n.s., 4(8):286-401, 54 text figs. (Feirer-Juin 1913).

Rhizosolenia simplex

Hendey, N.I., 1937
The plankton diatoms of the southern seas. Discovery Repts. 16:151-364, pls.6-13.

Rhizosolenia simplex

Manguin, E., 1954
Diatomees marines provenant de l'ile Heard (Australian National Antarctic Expedition). Rev. Algol., n.s., 1: 14-24.

Rhizosolenia stolterfothii

Allen, W. E., 1938
The Templeton Crocker Expedition to the Gulf of California in 1935 - The Phytoplankton. Amer. Microsc. Soc., Trans. 57:328-335.

Rhizosolenia stolterfothii

Allen, W.E., 1937
Plankton diatoms of the Gulf of California obtained by the G. Allan Hancock Expedition of 1936. The Hancock Pacific Expeditions, Univ. So. Calif. Publ. 3:47-59, 1 fig.

Rhizosolenia Stolterfothii

Allen, W.E., and E.E. Cupp, 1935
Plankton diatoms of the Java Sea. Annales du Jardin Botanique de Buitenzorg XLIV (2):101-174, figs.1-127.
(drawings of all species mentioned)

Rhizosolenia Stolforthii

Bigelow, H.B., and M. Leslie, 1930
Reconnaissance of the waters and plankton of Monterey Bay, July 1928. Bull. M.C.Z., 70(5):429-481, 43 text figs.

Rhizosolenia stolterfothii

Boden, B.P., 1950
Some marine plankton diatoms from the west coast of South Africa. Trans. R. Soc. S. Africa. 32:321-434, 100 text figs.

Rhizosolenia Stolterfothii

Braarud, T., 1945
A phytoplankton survey of the polluted waters of inner Oslo Fjord. Hvalrådets Skrifter, No.28, 142 pp., 19 text figs., 17 tables.

Rhizosolenia stolterfothii

Cleve-Euler, A., 1953
Die Diatomeen von Schweden und Finnland. Kungl. Svenska Vetenskaps Akad. Handl., Fjärde Ser. 2(1): 161 pp., 6 pls.

Rhizosolenia stoltifothii

Cupp, Easter E., 1943
Marine plankton diatoms of the west coast of North America. Bull. S.I.O. 5(1):1-238, 5 pls., 168 text figs.

Rhizosolenia stolterfothii

Cupp, E.E., 1934
Analysis of marine diatom collections taken from the Canal Zone to California during March, 1933. Trans. Am. Micros. Soc. LIII (1):22-29, 1 map.

Rhizosolenia stolterfothii

Cupp, E.E. and Allen, W.E., 1938
Plankton diatoms of the Gulf of California obtained by Allan Hancock Pacific Expedition of 1937. The Hancock Pacific Expeditions, The Univ. So. Calif. Publ. 3: 61-74, 1 map, pls.4-15.

Rhizosolenia stolterfothii

Dangeard, P., 1927
Phytoplankton de la croisière du "Sylvana". Ann. Inst. Ocean., Monaco, n.s., 4(8):286-401, 54 text figs. (Février-Juin 1913).

Rhizosolenia stolterfothii

Davidson, V.M., 1931.
Biological and oceanographic conditions in Hudson Bay. Contr. Canadian Biol. Fish., n.s., 6(26):497-509, 7 textfigs.

Rhizosolenia stoltifothii

Delegazione Italiana della Commissione Internazionale per l'Esplorazione Scientifica del Mediterraneo, 1941
Note sul plancton della Laguna veneta. [Memoria CCLXXXI], Arch. di Ocean. e Limn. Anno I, Fasc. I, 1941 XIX: 31-57 pp.

Rhizosolenia Stolterfothi

Ercegovic, A., 1936
Etudes qualitative et quantitatives du phytoplancton dans les eaux cotières de l'Adriatique oriental moyen au cours de l'année 1934. Acta Adriatica 1(9):1-126

Rhizosolenia stolterfothii

Eskinazi Enide e Shigekatsv Satô (1963-1964) 1966.
Contribuição ao estudo das diatomaceas da Praia de Piedade.
Trabhs Inst. Oceanogr., Univ. Recife, 5 (5/6): 73-114.

Rhizosolenia Stoltifothii

Forti, A., 1922
Ricerche sulla flora pelagica (fitoplancton) di Quarto dei Mille. Mem. R. Com. Talass. Ital. 97:248 pp., 13 pls.

Rhizosolenia stolterfothii

Ghazzawi, F.M., 1939
Plankton of the Egyptian waters. A study of the Suez Canal Plankton. (A) Phytoplankton. Preliminary Report 83 pp. Notes and Memoires, Min. Commerce-Industry, Egypt, Hydrobiol. & Fish. 65 figs.

Rhizosolenia stolterfothii

Gilbert, J.Y., and W.E. Allen, 1943
The phytoplankton of the Gulf of California obtained by the "E.W. Scripps" in 1939 and 1940. J. Mar. Res. V(2):89-110, figs.30-31.

Rhizosolenia Stoltifothii

Gran, H.H., 1908
Diatomeen. Nordisches Plankton, Botanischer Teil pp. XIX.1-XIX 146; 178 text figs.

Rhizosolenia stolterfothii

Gran, H. H. and E. C. Angst, 1931
Plankton diatoms of Puget Sound. Publ. Puget Sound Biol. Sta. 7:417-519, 95 text figs.

Rhizosolenia stolterfothii

Hendy, N. Ingram, 1964
An introductory account of the smaller algae of British coastal waters. V. Bacillariophyceae (Diatoms). Her Majesty's Stationary Office, 317 pp., 45 pls.

Rhizosolenia Stolterfothii

Hendey, N.I., 1937
The plankton diatoms of the southern seas. Discovery Repts. 16:151-364, pls.6-13.

Rhizosolenia stolterfothii

Jorgensen, E., 1900
Protophyten und Protozoën im Plankton aus der Norwegischen Westküste. Bergens Mus. Aarb. 1899(6): 95 pp., 5 pls., 83 tables

Rhizosolenia stolterfothii

Kokubo, S., 1952
Results of the observations on the plankton and oceanography of Mutsu Bay during 1950, reference being made also to the period 1946-1950. Bull. Mar.Biol.Sta., Asamushi 5(1/4): 1-54, 3 tables,(fold-in), 1 fold-in.

Rhizosolenia Stolterfothi

Macdonald, R., 1933
An examination of plankton hauls made in the Suez Canal during the year 1928. Fish. Res. Dis., Notes & Mem. No.3, 11 pp., 1 chart.

Rhizosolenia stolterfothii

Lafon, M., M. Durchon and Y. Saudray, 1955. Recherches sur les cycles saisonnières du plankton. Ann. Inst. Océan., 31(3):125-230.

Rhizosolenia Stolterfothi

Mangin, M. L., 1912
Phytoplancton de la croisière du "René" dans l'Atlantique (Septembre 1908). Ann. Inst. Ocean., n.s., 4(1):1-66, 2 pls., 41 text figs., 2 tables.

Rhizosolenia Stolterfothii

Margalef, R., 1949
Fitoplancton nerítico de la Costa Brava en 1947-48. Publ. Inst. Biol. Aplicada, 5: 41-51, 3 text figs.

Rhizosolenia stolterfothii

Marshall, S. M., 1933
The production of microplankton in the Great Barrier Reef Region. Brit. Mus. (N.H.) Great Barrier Reef Exped. 1928-29, Sci. Repts. II(5):111-157, 14 text figs.

Rhizosolenia stoltifothi

Massutí Algamora, M., 1949
Estudio de diez y seis muestras de plancton del Golfo de Nápoles. Publ. Inst. Biol. Appl. 5:85-94, 1 fold-in table.

Rhizosolenia Stolterfothii

Meunier, A., 1915
Microplancton de la Mer Flamande. 2. Diatomées (excepté le genre Chaetoceros). Mem. Mus. Roy. Hist. Nat., Belgique, 7(3):1-118, Pls. VIII-XIV.

Rhizosolenia stoltifothii

Pavillard, J., 1925
Bacillariales. Rept. on the Danish Oceangr. Exped., 1908-10 to the Mediterranean and adj. seas. Vol.II., Biol. J4:72 pp., 116 text figs.

Rhizosolenia Stolterfothii
Pavillard, J., 1905
Recherches sur la flore pelagique (Phytoplankton) de l'Etang de Thau. Theses presentees a la Fac. Sci., Paris, 116 pp., 3 pls.

Rhizosolenia Stolterfothii
Rampi, L., 1942
Ricerche sul fitoplancton del Mare Ligure 6. Le diatomee delle acque di Sanremo. Nuovo Giornale Botanico Italiano, N.S., 49:252-268.

Rhizosolenia Stolterfothii
Schodduyn, M., 1926
Observations faites dans la baie d'Ambleteuse (Pas de Calais). Bull. Inst. Ocean., Monaco, No. 482: 64 pp.

Rhizosolenia Stolterfothii
Schröder, B., 1900
Phytoplankton des Golfes von Neapel nebst vergleichenden Ausblicken auf das atlantischen Ozean. Mitt. Zool. Stat. Neapel, 14:1-38.

Rhizosolenia stolterforthii
Sousa e Silva, E., 1949
Diatomaceas e Dinoflagelados de Baia de Cascais. Portugaliae Acta Biol., Volume: Julio Henriques, Ser. B: 300-383, 9 pls, 2 fold-in tables.

Rhizosolenia stolterfothii
Sousa e Silva, E., and J. Dos Santos-Pinto, 1948
O Plancton da Baia de S. Martinho do Porto. 1. Diatomaceas e Dinoflagelados. Bol. Soc. Portuguese de Ciencias Naturais, 16(2):134-187, 6 pls. (Trav. Sta. Biol. Mar. de Lisbonne No. 52).

Rhizosolenia stolterfothii
Uyeno, Fukuzo, 1961
Oceanographical and ecological studies on primary production of the sea, with special references to relationship between diatom production and temperature and chlorinity of water.
Rept., Fac. Fish., Pref. Univ., Mie, 4(1): 1-64.

Rhizosolenia stolterfothii
Woodmansee, Robert A., 1963.
Cell-diameter frequency distributions of the planktonic diatom Rhizosolenia alata. Publ. Inst. Mar. Sci., Port Aransas, 9:117-131.

Rhizosolenia styliformis
Allen, W. E., 1938
The Templeton Crocker Expedition to the Gulf of California in 1935 - The Phytoplankton. Amer Microsc. Soc., Trans. 57:328-335.

Rhizosolenia styliformis
Allen, W.E., 1937
Plankton diatoms of the Gulf of California obtained by the G. Allan Hancock Expedition of 1936. The Hancock Pacific Expeditions, Univ. So. Calif. Publ. 3:47-59, 1 fig.

Rhizosolenia styliformis
Allen, W.E., and E.E. Cupp, 1935
Plankton diatoms of the Java Sea. Annales du Jardin Botanique de Buitenzorg XLIV (2):101-174, figs.1-127.

(drawings of all species mentioned)

? Rhizosolenia styliformis
Balech, E., 1947
Contribution al conocimiento del plancton antartico. Plancton del Mar de Bellinghausen. Physis 20:75-91, 76 figs. on 8 pls.

Rhizosolenia styliformis
Bigelow, H. B., 1922
Exploration of the coastal water off the northeastern United States in 1916 by the U.S. Fisheries Schooner Grampus. Bull. M.C.Z. 65 (5):85-188, 53 text figs.

Rhizosolenia styliformis
Boden, B.P., 1950
Some marine plankton diatoms from the west coast of South Africa. Trans. R.Soc. S. Africa. 32:321-434, 100 text figs.

Rhizosolenia styliformis
Boden, B. P., 1949.
The diatoms collected by the U.S.S. CACOPAN in the Antarctic in 1947. J. Mar. Res. 8(1):6-13, 3 textfigs.

Boden, Brian, 1948 Rhizosolenia styliformis
Marine plankton diatoms on operation HIGHJUMP in: Some oceanographic observations on operation HIGHJUMP. By R.S. Dietz. USNEL Rept. No.55, 97 pp., 41 figs. 7 July 1948.

Rhizosolenia styliformis
Braarud, T., 1962
Species distribution in marine phytoplankton.
J. Oceanogr. Soc., Japan. 20th Ann. Vol., 628-649.

Rhizosolenia styliformis
Braarud, T., 1934
A note on the phytoplankton of the Gulf of Maine in the summer of 1933. Biol. Bull. 67(1):76-82.

Rhizosolenia styliformis
Braarud, T., and Adam Bursa, 1939
On the phytoplankton of the Oslo Fjord, 1933-1934. Hvalrådets Skr. No.19:1-63; 9 text figs. Reviewed. J. du. Cons. 14(3): 418-420. A.C. Gardiner.

Rhizosolenia styliformis
Central Meteorological Observatory, 1949
Report on sea and weather observation on Antarctic Whaling Ground (1947-48). Ocean. Mag., Japan, 1(1):49-88, 17 text figs.

Rhizosolenia styliformis
Cleve-Euler, A., 1951
Die Diatomeen von Schweden und Finnland. Kungl. Svenska Vetenskaps Akad. Handl., Fjärde Ser. 2(1): 161 pp., 6 pls.

Rhizosolenia styliformis
Cupp, Easter E., 1943
Marine plankton diatoms of the west coast of North America. Bull. S.I.O. 5(1):1-238, 5 pls., 168 text figs.

Rhizosolenia styliformis
Cupp, E.E., 1937
Seasonal distribution and occurrence of marine diatoms and dinoflagellates at Scotch Cap, Alaska. Bull. S.I.O. Tech. ser.4(3):71-100, 7 textfigs.

Rhizosolenia styliformis
Cupp, E.E. and Allen, W.E., 1938
Plankton diatoms of the Gulf of California obtained by Allan Hancock Pacific Expedition of 1937. The Hancock Pacific Expeditions, The Univ. So. Calif. Publ. 3: 61-74, 1 map, pls.4-15.

Rhizosolenia styliformis
Dangeard, P., 1927
Phytoplankton de la croisière du "Sylvana". Ann. Inst. Ocean., Monaco, n.s., 4(8):286-401, 54 text figs. (Janvier-Juin 1913).

Rhizosolenia styliformis
Delsman, H. C., 1939.
Preliminary plankton investigations in the Java Sea. Treubia, 17:139-181, 8 maps, 41 figs.

Rhizosolenia styliformis
Ercegovic, A., 1936
Etudes qualitative et quantitatives du phytoplancton dans les eaux cotières de l'Adriatique oriental moyen au cours de l'année 1934. Acta Adriatica 1(9):1-126

Rhizosolenia styliformis
Forti, A., 1922
Ricerche sulla flora pelagica (fitoplancton) di Quarto dei Mille. Mem. R. Com. Talass. Ital. 97:248 pp., 13 pls.

Rhizosolenia styliformis
Fraser, J. H., and A. Saville, 1949
Plankton distribution in Scottish and adjacent waters in 1948. Ann. Biol. 5:61-62.

Rhizosolenia styliformis
Frenguelli, Joaquin, and Hector Antonio Orlando, 1959.
Operacion MERLUZA. Diatomeas y silicoflagelados del plancton del "VI Crucero". Servicio Hidrogr. Naval., Argentina, Publ. No. H. 619: 5-62.

Rhizosolenia styliformis
Gran, H.H., 1908
Diatomeen. Nordisches Plankton, Botanischer Teil pp. XIX 1-XIX 146; 178 text figs.

Rhizosolenia styliformis
Gran, H.H., 1897
Protophyta: Diatomaceae, Silico-flagellata and Cilioflagellata. Den Norske Nordhavs Expedition 1876-1878, h. 24, 36 pp., 4 pls.

Rhizosolenia styliformis
Gran, H.H., and T. Braarud, 1935
A quantitative study of the phytoplankton in the Bay of Fundy and the Gulf of Maine (including observations on hydrography, chemistry, and turbidity). J. Biol. Bd., Canada, 1(5):279-467, 69 text figs.

Rhizosolenia styliformis
Gran, H. H. and E. C. Angst, 1931
Plankton diatoms of Puget Sound. Publ. Puget Sound Biol. Sta. 7:417-519, 95 text figs.

Rhizosolenia styliformis
Hendy, N. Ingram, 1964
An introductory account of the smaller algae of British coastal waters. V. Bacillariophyceae (Diatoms).
Her Majesty's Stationary Office, 317 pp., 45 pls.

Rhizosolenia styliformis
Hendey, N.I., 1937
The plankton diatoms of the southern seas. Discovery Repts. 16:151-364, pls.6-13.

Rhizosolenia styliformis
Iselin, C., 1930
A report on the coastal waters of Labrador based on explorations of the "Chance" during the summer of 1926. Proc. Am. Acad. Arts Sci., 66(1):1-37, 14 text figs.

Rhizosolenia styliformis
Jørgensen, E., 1905
B.Protistplankton and the diatoms in bottom samples. Hydrographical and biological investigations in Norwegian fjords. Bergens Mus. Skr. 7: 49-225.

Rhizosolenia styliformis
Jorgensen, E., 1900
Protophyten und Protozoen im Plankton aus der Norwegischen Westkerste. Bergens Mus. Aarb. 1899(6): 95 pp., 5 pls., 83 tables.

Rhizosolenia styliformis
Kokubo, S., and S. Sato, 1947
Plankters in Jū-San Gata. Physiol. and Ecol. (Japan) 1(4):1-16, 3 text figs., tables.

Rhizosolenia styliformis
Lucas, C.E. and H.G. Stubbings, 1948
Size variations in diatoms and their ecological significance. Hull Bull. Mar. Ecol., 2(12):133-171, 14 text figs.

Rhizosolenia styliformis
Lillick, L.C., 1940
Phytoplankton and planktonic protozoa of the offshore waters of the Gulf of Maine. Pt.II. Qualitative Composition of the Planktonic Flora. Trans. Am. Phil. Soc., n.s., 31(3):193-237, 13 text figs.

Rhizosolenia styliformis
Lillick, L.C., 1938
Preliminary report of the phytoplankton of the Gulf of Maine. Am. Mid. Nat. 20(3):624-640, 1 text fig. 37 tables.

Rhizosolenia styliformis
Lillick, L.C., 1937
Seasonal studies of the phytoplankton off Woods Hole, Massachusetts. Biol. Bull. LXXIII (3):488-503, 3 text figs.

Rhizosolenia styliformis
Lucas, C. E., 1949
Notes on continuous plankton records at 10 m depth in the North Sea and Northeastern Atlantic during 1946-1947. Ann. Biol., Int. Cons., 4:63-66, text fig. 4.

Rhizosolenia styliformis
Mangin, L., 1915
Phytoplancton de L'Antartique. Deuxieme Exped. Ant. Francaise (1908-1910), 95 pp., 3 pls., 58 text figs.

Rhizosolenia styliformis
Mangin, M. L., 1912
Phytoplancton de la croisière du "René" dans l'Atlantique (Septembre 1908). Ann. Inst. Ocean., n.s., 4(1):1-66, 2 pls., 41 text figs., 2 tables.

Rhizosolenia styliformis
Mann, A., 1893
List of Diatomaceae from a deep-sea dredging in the Atlantic Ocean off Delaware Bay by the U. S. Fish Commission Steamer Albatross. Proc. U. S. Nat. Mus. 16:303-312.

Rhizosolenia styliformis
Marshall, S. M., 1933
The production of microplankton in the Great Barrier Reef Region. Brit. Mus. (N.H.) Great Barrier Reef Exped. 1928-29, Sci. Repts. II(5):111-157, 14 text figs.

Rhizosolenia styliformis
Marukawa, H., 1921
Plankton lists and some new species of copepods, from the northern waters of Japan. Bull. Inst. Ocean. No.384, 15 pp., 3 pls., 1 chart. Monaco

Rhizosolenia styliformis
Morse, D.C., 1947
Some observations on seasonal variations in plankton population Patuxent River, Maryland 1943-1945. Bd. Nat. Res., Publ. No.65, Chesapeake Biol. Lab., 31, 3 figs.

Rhizosolenia styliformis
Pavillard, J., 1925
Bacillariales. Rept. on the Danish Oceangr. Exped., 1908-10 to the Mediterranean and adj. seas. Vol.II., Biol. J4:72 pp., 116 text figs.

Rhizosolenia styliformis
Phifer, L.D. (undated)
The occurrence and distribution of plankton diatoms in Bering Sea and Bering Strait, July 26-August 24, 1934. Report of Oceanographic cruise of U.S. Coast Guard Cutter Chelan 1934, Part II(A):1-44 (mimeographed) plus fig.1 (after Pt.B)

Rhizosolenia styliformis
Qasim, S.Z., P.M.A. Bhattathiri and V.P. Devassy, 1972
The effect of intensity and quality of illumination on the photosynthesis of some tropical marine phytoplankton. Mar. Biol. 16(1): 22-27.

Rhizosolenia styliformis
Rampi, L., 1940
Diatomee del Mare Adriatico. Nuovo Giornale Botanico Italiano, n.s., 47:559-608.

Rhizosolenia styliformis
Robinson, G.A., 1965.
Continuous plankton records: contribution towards a plankton atlas of the North Atlantic and the North Sea. IX. Seasonal cycles of phytoplankton. Bulls. Mar. Ecol., Scottish Mar. Biol. Assoc., 6(4):104-122, pls. 26-61.

Rhizosolenia styliformis
Robinson, G.A., 1961
Contribution towards an atlas of the northeastern Atlantic and the North Sea. 1. Phytoplankton. Bulls. Mar. Ecology, 5(42): 81-89, Pls. 15-20.

Rhizosolenia styliformis
Robinson, G.A., and D.J. Colbourn, 1970.
Continuous plankton records: further studies on the distribution of Rhizosolenia styliformis Brightwell. Bull. mar. Ecol. 6 (9): 303-33.

Rhizosolenia styliformis
Robinson, G.A., and O.R. Waller, 1966.
The distribution of Rhizosolenia styliformis Brightwell and its varieties. In: Some contemporary studies in marine science H. Barnes, editor, George Allen & Unwin, Ltd., 645-663.

Rhizosolenia styliformis
Schröder, B., 1900
Phytoplankton des Golfes von Neapel nebst vergleichenden Ausblicken auf das atlantischen Ozean. Mitt. Zool. Stat. Neapel, 14:1-38.

Rhizosolenia styliformis
Schulz, B., and A. Wulff, 1929
Hydrographie und Oberflächen plankton des westlichen Barentsmeeres im Sommer 1927. Ber. deutschen wissensch. Komm. F. Meeresforsch. n.s. 4(5):232-372, 13 tables, 25 text figs.

Rhizosolenia styliformis
Seaton, D.D., 1970.
Reproduction in Rhizosolenia hebetata and its linkage with Rhizosolenia styliformis. J. mar. biol. Ass., U.K., 50(1): 97-106.

Rhizosolenia styliformis
Sousa e Silva, E., 1949
Diatomaceas e Dinoflagelados de Baia de Cascais. Portugaliae Acta Biol., Volume: Julio Henriques, Ser. B: 300-383, 9 pls, 2 fold-in tables.

Rhizosolenia styliformis
Sousa e Silva, E., and J. Dos Santos-Pinto, 1948
O Plancton da Baia de S. Martinho do Porto. 1. Diatomaceas e Dinoflagelados. Bol. Soc. Portuguese de Ciencias Naturais, 16(2):134-187, 6 pls. (Trav. Sta. Biol. Mar. de Lisbonne No. 52).

Rhizosolenia styliformis
Steemann-Nielsen, Einar, 1951
The marine vegetation of the Isefjord. A study on ecology and production. Medd. Komm. Danmarks Fiskeri-og Havundersøgelser. Ser. Plankton. 5(4); 114pp., 46 text figs.

Rhizosolenia styliformis
Uyeno, Fukuzo, 1961
Oceanographical and ecological studies on primary production of the sea, with special references to relationship between diatom production and temperature and chlorinity of water. Rept., Fac. Fish., Pref. Univ., Mie, 4(1): 1-64.

Rhizosolenia styliformis
Wheeler, J.E.G., 1939
Plankton investigations. Bermuda Biological Station. Second Report. October 1939. 7 pp. (typed), 5 figs. Plymouth, Oct. 23, 1939.

Rhizosolenia styliformis
Wimpenny, R.S., 1947
The size of diatoms. II. Further observations, on Rhizosolenia styliformis (Brightwell). JMBA 26 (3):271-284, pl.5 2 textfigs.

Rhizosolenia styliformis
Wimpenny, R. R., 1936.
The size of diatoms. I. The diameter of Rhizosolenia styliformis Brightw. and R. alata Brightw. in particular and of pelagic marine diatoms in general. JMBA, XXI:29-60

Rhizosolenia styliformis var. latissima
Allen, W.E., and E.E. Cupp, 1935
Plankton diatoms of the Java Sea. Annales du Jardin Botanique de Buitenzorg XLIV (2):101-174, figs.1-127.
(drawings of all species mentioned)

Rhizosolenia styliformis var. latissima
Cupp, E.E., 1934
Analysis of marine diatom collections taken from the Canal Zone to California during March, 1933. Trans. Am. Micros. Soc. LIII (1):22-29, 1 map.

Rhizosolenia styliformis latissima
Okuno, H., 1957.
Electron-microscopical study on fine structures of diatom frustules. 15. Observation on the genus Rhizosolenia. Bot. Mag., Tokyo, 71(826):101-107.

Rhizosolenia styliformis var. longispina
Allen, W.E., and E.E. Cupp, 1935
Plankton diatoms of the Java Sea. Annales du Jardin Botanique de Buitenzorg XLIV (2):101-174, figs.1-127.
(drawings of all species mentioned)

Rhizosolenia styliformis longispina
Boden, B. P., 1949.
The diatoms collected by the U.S.S. CACAPAN in the Antarctic in 1947. J. Mar. Res. 8(1):6-13, 3 textfigs.

Rhizosolenia styliformis var. longispina
Boden, Brian, 1948
Marine plankton diatoms on operation HIGHJUMP in: Some oceanographic observations on operation HIGHJUMP. By R.S. Dietz. USNEL Rept. No.55, 97 pp., 41 figs. 7 July 1948.

Rhizosolenia styliformis longispina
Cupp, Easter E., 1943
Marine plankton diatoms of the west coast of North America. Bull. S.I.O. 5(1):1-238, 5 pls., 168 text figs.

Rhizosolenia styliformis var. longispina
Cupp, E.E. and Allen, W.E., 1938
Plankton diatoms of the Gulf of California obtained by Allan Hancock Pacific Expedition of 1937. The Hancock Pacific Expeditions, The Univ. So. Calif. Publ. 3: 61-74, 1 map, pls.4-15.

Rhizosolenia styliformis var. longispina
Delsman, H. C., 1939.
Preliminary plankton investigations in the Java Sea. Treubia, 17:139-181, 8 maps, 41 figs.

Rhizosolenia styliformis oceanica
Wimpenny, R.S., 1966.
The size of diatoms. IV. The cell diameter in Rhizosolenia styliformis var. oceanica. J. mar. biol. Assoc. U.K., 46(3):541-546.

Rhizosolenia temperei
Margalef, R., 1949
Fitoplancton nerítico de la Costa Brava en 1947-48. Publ. Inst. Biol. Aplicada, 5: 41-51, 3 text figs.

Rhizosolenia Temperei
Massutí Algamora, M., 1949
Estudio de diez y seis muestras de plancton del Golfo de Nápoles. Publ. Inst. Biol. Appl. 5:85-94, 1 fold-in table.

Rhizosolenia temperi
Pavillard, J., 1925
Bacillariales. Rept. on the Danish Oceangr. Exped., 1908-10 to the Mediterranean and adj. seas. Vol. II., Biol. J4:72 pp., 116 text figs.

Rhizosolenia temperei
Rampi, L., 1942
Ricerche sul fitoplancton del Mare Ligure 6. Le diatomee delle acque di Sanremo. Nuovo Giornale Botanico Italiano, N.S., 49:252-269.

Rhizosolenia temperei acuminata
Pavillard, J., 1905
Recherches sur la flore pelagique (Phytoplankton) de l'Etang de Thau. Theses presentees a la Fac. Sci., Paris, 116 pp., 3 pls.

Rhizosolenia Temperei acuminata
Schröder, B., 1900
Phytoplankton des Golfes von Neapel nebst vergleichenden Ausblicken auf das atlantischen Ozean. Mitt. Zool. Stat. Neapel, 14:1-38.

Rhizosolenia tenuijuncta n.sp.
Manguin, E., 1957.
Premier inventaire des diatomées de la Terre Adélie Antarctique. Espèces nouvelles. Rev. Algologique, n.s., 3(3):111-134.

Rhizosolenia truncata
Hendey, N.I., 1937
The plankton diatoms of the southern seas. Discovery Repts. 16:151-364, pls. 6-13.

Rhizosolenia truncata
Mangin, L., 1915
Phytoplancton de L'Antarctique. Deuxieme Exped. Ant. Francaise (1908-1910), 95 pp., 3 pls., 58 text figs.

Rhoicosigma arcticum
Jørgensen, E., 1905
B. Protistplankton and the diatoms in bottom samples. Hydrographical and biological investigations in Norwegian fjords. Bergens Mus. Skr. 7: 49-225.

Rhoicosigma compactum
Hendy, N. Ingram, 1964
An introductory account of the smaller algae of British coastal waters. V. Bacillariophyceae (Diatoms).
Her Majesty's Stationary Office, 317 pp., 45 pls.

Rhoicosigma mediterraneum
Rampi, L., 1940
Diatomee del Mare Adriatico. Nuovo Giornale Botanico Italiano, n.s., 47:559-608.

Rhoicosigma Reichardtii
Grunow, A., 1877
1. New Diatoms from Honduras. Monthly Micros. Jour., 18:165-186, pls. CXCIII-CXCVI.

Rhoicosphenia
Desikachary, T.V., 1956(1957).
Electron microscope studies on diatoms. J.R. Microsc. Soc. (3)76(1/2):9-36.

Rhoicosphenia curvata
Hendy, N. Ingram, 1964
An introductory account of the smaller algae of British coastal waters. V. Bacillariophyceae (Diatoms).
Her Majesty's Stationary Office, 317 pp., 45 pls.

Rhoicosphenia curvata
Rumkówna, A., 1948
[List of the phytoplankton species occurring in the superficial water layers in the Gulf of Gdańsk] Bull. Lab. mar., Gdynia, No. 4: 139-141 with tables in back.

Rhoicosphaenia curvata
Tsumura, K., 1956.
Diatomoj el la cirkaufoso de la restajo de la kastele de Odawara. J. Yokohama Municipal Univ., (C-14) No. 47: 23 pp.

Rhoicosphenia marina
Rampi, L., 1940
Diatomee del Mare Adriatico. Nuovo Giornale Botanico Italiano, n.s., 47:559-608.

Rhoicosphaenia curvata marina
Politis, J., 1949
Diatomees marines de Bosphores et des ibes de la mer de Marmara. II Practica tou Hellenikou Hidrobiologikou Institutoutou 1929, Etoz 1929, 3(1):11-31.

Rhoicosphenia stauroneiformis
Hendy, N. Ingram, 1964
An introductory account of the smaller algae of British coastal waters. V. Bacillariophyceae (Diatoms).
Her Majesty's Stationary Office, 317 pp., 45 pls.

Rhoikoneis Bolleana n.gen. & n.sp.
Grunow, A., 1863
Ueber einige neue und ungenügend bekannte Arten und Gattungen von Diatomaceen. Verhandl. d. K.K. Zool. Bot. Gesellsch., Vienna, 13: 137-162, pl. 4-5 (Pl. 13-14).

Rhoikoneis Garkeana n.sp.
Grunow, A., 1863
Ueber einige neue und ungenügend bekannte Arten und Gattungen von Diatomaceen. Verhandl. d. K.K. Zool. Bot. Gesellsch., Vienna, 13: 137-162, pl. 4-5 (Pl. 13-14).

Rhopalodia
Brunel, J., 1962
Le phytoplancton de la Baie de Chaleurs. Inst. Botan., Univ. Montréal, Contrib. No. 77: 365 pp., 66 pls.

Rhopalodia
Desikachary, T.V., 1956(1957).
Electron microscope studies on diatoms. J. R. Microsc. Soc. (3)76(1/2):9-36.

Rhopalodia gibba & varieties
Hendy, N. Ingram, 1964
An introductory account of the smaller algae of British coastal waters. V. Bacillariophyceae (Diatoms).
Her Majesty's Stationary Office, 317 pp., 45 pls.

Rhopalodia gibba
Politis, J., 1949
Diatomees marines de Bosphores et des ibes de la mer de Marmara. II Practica tou Hellenikou Hidrobiologikou Institutoutou 1929, Etoz 1929, 3(1):11-31.

Rhopalodia gibba
Zanon, D. V., 1949
Diatomee di Buenos Aires (Argentina) Atti Accad. Naz. Lincei, Memorie, Cl. Sci. fis., mat. e. nat., ser. 7, 11(3):59-151, 2 pls.

Rhopalodia gibba
Zanon, V., 1948
Diatomee marini di Sardegna e Pugillo di Alghe Marine della stressa. Boll. Pesca, Piscitutura e Idrobiologia, Anno 24, ns. 3(2): 202-244, 27 figs. on 1 pl.

Rhopalodia gibberula producta
Hendy, N. Ingram, 1964
An introductory account of the smaller algae of British coastal waters. V. Bacillariophyceae (Diatoms).
Her Majesty's Stationary Office, 317 pp., 45 pls.

Rhopalodia gibberula
Zanon, D. V., 1949
Diatomee di Buenos Aires (Argentina) Atti Accad. Naz. Lincei, Memorie, Cl. Sci. fis., mat. e. nat., ser. 7, 11(3):59-151, 2 pls.

Rhopalodia gibberula
Zanon, V., 1948
Diatomee marini di Sardegna e Pugillo di Alghe Marine della stressa. Boll. Pesca, Piscitutura e Idrobiologia, Anno 24, ns. 3(2): 202-244, 27 figs. on 1 pl.

Rhopalodia minusculus
Tsumura, K., 1956.
Diatomoj el la cirkaufoso de la restajo de la kastele de Odawara. J. Yokohama Municipal Univ., (C-14), No. 47:23 pp.

Rhopalodia musculus
Hustedt, F. and A.A. Aleem, 1951
Littoral diatoms from the Salstone near Plymouth. JMBA 30(1): 177.196.

Rhopalodia musculus
Politis, J., 1949
Diatomees marines de Bosphores et des ibes de la mer de Marmara. II Practica tou Hellenikou Hidrobiologikou Institutoutou 1929, Etoz 1929, 3(1):11-31.

Rhopalodia musculus
Rampi, L., 1940
Diatomee del Mare Adriatico. Nuovo Giornale Botanico Italiano, n.s., 47:559-608.

Rhopalodia musculus
Tsumura, K., 1956.
Diatomoj el la cirkaufoso de la restajo de la kastele de Odawara. J. Yokohama Municipal Univ., (C-14), No. 47:23 pp.

Rhopalodia musculus
Zanon, D. V., 1949
Diatomee di Buenos Aires (Argentina) Atti Accad. Naz. Lincei, Memorie, Cl. Sci. fis., mat. e. nat., ser. 7, 11(3):59-151, 2 pls.

Rhopalodia musculus
Zanon, V., 1948
Diatomee marini di Sardegna e Pugillo di Alghe Marine della stressa. Boll. Pesca, Piscitutura e Idrobiologia, Anno 24, ns. 3(2): 202-244, 27 figs. on 1 pl.

Rhopalodia parallela
Kokubo, S., and S. Sato, 1947
Plankters in Jū-San Gata. Physiol. and Ecol. (Japan) 1(4):1-16, 3 text figs., tables.

Rhopalopia contorta
Tsumura, K., 1955.
Diatomoj el la cirkaŭfoso de la restajo de la kastelo de Odawara. J. Yokohama Municipal Univ., (C-14), No. 47:23 pp.

Roperia tesselata
Cleve-Euler, A., 1951
Die Diatomeen von Schweden und Finnland. Kungl. Svenska Vetenskaps Akad. Handl., Fjärde Ser. 2(1): 161 pp., 6 pls.

Roperia tessellata
Frenguelli, Joaquin, and Hector Antonio Orlando, 1959.
Operacion MERLUZA. Diatomeas y silico-flagelados del plancton del "VI Crucero". Servicio Hidrogr. Naval., Argentina, Publ. No. H. 619: 5-62.

Roperia tesselata
Hendy, N. Ingram, 1964
An introductory account of the smaller algae of British coastal waters. V. Bacillariophyceae (Diatoms). Her Majesty's Stationary Office, 317 pp., 45 pls.

Roperia tessellata
Hustedt, F. and A.A. Aleem, 1951
Littoral diatoms from the Salstone near Plymouth. JMBA 30(1): 177.196.

Roperia tesselata
Jørgensen, E., 1905
B.Protistplankton and the diatoms in bottom samples. Hydrographical and biological investigations in Norwegian fjords. Bergens Mus. Skr. 7: 49-225.

Rouxia californica
Florin, M-B., 1948
9. Diatomeae in submarine cores from the Tyrrhenian Sea. Medd. Ocean. Inst., Göteborg, 15 (Göteborgs Kungl. Vetenskaps-och Viterrhets Samhälles Handlingar, Sjätte Foljden, Ser. B 5(13):80-88.

Rouxia diploneides n.sp.
Schrader, Hans-Joachim, 1973
Cenozoic diatoms from the northeast Pacific, leg 18. Initial Repts Deep Sea Drilling Proj. 18:673-797.

Rouxia naviculoides n.sp.
Schrader, Hans-Joachim, 1973
Cenozoic diatoms from the northeast Pacific, leg 18. Initial Repts Deep Sea Drilling Proj. 18:673-797.

Rutilaria edentula n.sp.
Castracane degli Antelminelli, F., 1886
1. Report on the Diatomaceae collected by H.M.S. Challenger during the years 1873-1876. Rept. Sci. Results, H.M.S. Challenger, Botany Vol. II, 178 pp., 30 pls.

Rutilaria epsilon
Mann, A., 1907
Report on the diatoms of the Albatross voyages in the Pacific Ocean, 1888-1904. Contrib. U. S. Nat. Herb. 10(5):221-419, Pls. XLIV-LIV.

Rutilaria tulkii n.sp.
Castracane degli Antelminelli, F., 1886
1. Report on the Diatomaceae collected by H.M.S. Challenger during the years 1873-1876. Rept. Sci. Results, H.M.S. Challenger, Botany Vol. II, 178 pp., 30 pls.

Sceptroneis cuneata
Grunow, A., 1877
1. New Diatoms from Honduras. Monthly Micros. Jour., 18:165-186, pls. CXCIII-CXCVI.

Sceptroneis dubia
Grunow, A., 1877
1. New Diatoms from Honduras. Monthly Micros. Jour., 18:165-186, pls. CXCIII-CXCVI.

Schimperella antarctica
Boden, B. P., 1949
The diatoms collected by the U.S.S. CACO-PAN in the Antarctic in 1947. J. Mar. Res. 8(1):6-13, 3 textfigs.

Schimperiella antarctica
Boden, Brian, 1948
Marine plankton diatoms on operation HIGHJUMP in: Some oceanographic observations on operation HIGHJUMP. By R.S. Dietz. USNEL Rept. No.55, 97 pp., 41 figs. 7 July 1948.

Schimperiella antarctica
Hendey, N.I., 1937
The plankton diatoms of the southern seas. Discovery Repts. 16:151-364, pls.6-13.

Schimperiella antarctica
Jouse, A.P., G.S. Koroleva, G.A. Nagaeva, 1962
Diatoms in the surface layer of sediment in the Indian sector of the Antarctic. Investigations of marine bottom sediments. (In Russian; English summary). Trudy Inst. Okeanol., Akad. Nauk, SSSR, 61: 19-92.

Schimperiella valdiviae
Hendey, N.I., 1937
The plankton diatoms of the southern seas. Discovery Repts. 16:151-364, pls.6-13.

Schizonema sp.
Brunel, J., 1962
Le phytoplancton de la Baie de Chaleurs. Inst. Botan., Univ. Montréal, Contrib. No. 77: 365 pp., 66 pls.

Schizonema grevillei
Jørgensen, E., 1905
B.Protistplankton and the diatoms in bottom samples. Hydrographical and biological investigations in Norwegian fjords. Bergens Mus. Skr. 7: 49-225.

Schizonema mucosa n.sp.
Meunier, A., 1915
Microplancton de la Mer Flamande. 2. Diatomées (excepté le genre Chaetoceros). Mem. Mus. Roy. Hist. Nat., Belgique, 7(3):1-118, Pls. VIII-XIV.

Schizonema vulgare
Mann, A., 1893
List of Diatomaceae from a deep-sea dredging in the Atlantic Ocean off Delaware Bay by the U. S. Fish Commission Steamer Albatross. Proc. U. S. Nat. Mus. 16:303-312.

Schizostauron n.g.
Grunow, A., 1877
1. New Diatoms from Honduras. Monthly Micros. Jour., 18:165-186, pls. CXCIII-CXCVI.

Schizostauron Lindigii
Grunow, A., 1877
1. New Diatoms from Honduras. Monthly Micros. Jour., 18:165-186, pls. CXCIII-CXCVI.

Schizostauron Reichardtii
Grunow, A., 1877
1. New Diatoms from Honduras. Monthly Micros. Jour., 18:165-186, pls. CXCIII-CXCVI.

Schroederella
Brunel, J., 1962
Le phytoplancton de la Baie de Chaleurs. Inst. Botan., Univ. Montréal, Contrib. No. 77: 365 pp., 66 pls.

Schröderella delicatula
Allen, W.E., and E.E. Cupp, 1935
Plankton diatoms of the Java Sea. Annales du Jardin Botanique de Buitenzorg XLIV (2):101-174, figs.1-127.
(drawings of all species mentioned)

Schröderella delicatula
Boden, B.P., 1950
Some marine plankton diatoms from the west coast of South Africa. Trans. R.Soc. S. Africa. 32:321-434, 100 text figs.

Schröderella delicatula
Cupp, Easter E., 1943
Marine plankton diatoms of the west coast of North America. Bull. S.I.O. 5(1):1-238, 5 pls., 168 text figs.

Schröderella delicatula
Cupp, E.E. and Allen, W.E., 1938
Plankton diatoms of the Gulf of California obtained by Allan Hancock Pacific Expedition of 1937. The Hancock Pacific Expeditions, The Univ. So. Calif. Publ. 3: 61-74, 1 map, pls.4-15.

Schroederella delicatula
Ercegovic, A., 1936
Etudes qualitative et quantitatives du phytoplancton dans les eaux cotières de l'Adriatique oriental moyen au cours de l'année 1934. Acta Adriatica 1(9):1-126

Schroederella delicatula
Forti, A., 1922
Ricerche sulla flora pelagica (fitoplancton) di Quarto dei Mille. Mem. R. Com. Talass. Ital. 97:248 pp., 13 pls.

Schroederella delicatula
Gilbert, J.Y., and W.E. Allen, 1943
The phytoplankton of the Gulf of California obtained by the "E.W. Scripps" in 1939 and 1940. J. Mar. Res. V(2):89-110, figs.30-31.

Schroederella delicatula
Hendy, N. Ingram, 1964
An introductory account of the smaller algae of British coastal waters. V. Bacillariophyceae (Diatoms). Her Majesty's Stationary Office, 317 pp., 45 pls.

Schroederella delicatula
Hendey, N.I., 1937
The plankton diatoms of the southern seas. Discovery Repts. 16:151-364, pls.6-13.

Schröderella delicatula
Morse, D.C., 1947
Some observations on seasonal variations in plankton population Patuxent River, Maryland 1943-1945. Bd. Nat. Res., Publ. No.65, Chesapeake Biol. Lab., 31, 3 figs.

Schröderella delicatula
Pavillard, J., 1925
Bacillariales. Rept. on the Danish Oceanogr. Exped., 1908-10 to the Mediterranean and adj. seas. Vol.II., Biol. J4:72 pp., 116 text figs.

Schroederella delicatula

Rampi, L., 1942
Ricerche sul fitoplancton del Mare Ligure 6. Le diatomee delle acque di Sanremo. Nuovo Giornale Botanico Italiano, N.S., 49:252-268.

Schroederella schroderi

Dangeard, P., 1927
Phytoplankton de la croisière du "Sylvana". Ann. Inst. Ocean., Monaco, n.s., 4(8):286-401, 54 text figs. (Feirer-Juin 1913).

Schroederella schroderi

Hendey, N.I., 1937
The plankton diatoms of the southern seas. Discovery Repts. 16:151-364, pls.6-13.

Schröderella schröderi

Pavillard, J., 1925
Bacillariales. Rept. on the Danish Oceangr. Exped., 1908-10 to the Mediterranean and adj. seas. Vol.II., Biol. J4:72 pp., 116 text figs.

Schröderella schröderi

Sousa e Silva, E., 1949
Diatomaceas e Dinoflagelados de Baía de Cascais. Portugaliae Acta Biol., Volume: Julio Henriques, Ser. B: 300-383, 9 pls, 2 fold-in tables.

Schroederella Schröderi

Sousa a Silva, E., and J. Dos Santos-Pinto, 1948
O Plancton da Baía de S. Martinho do Porto. 1. Diatomaceas e Dinoflagelados. Bol. Soc. Portuguesa de Ciencias Naturais, 16(2):134-187, 6 pls. (Trav. Sta. Biol. Mar. de Lisbonne No. 52).

Schuettia annulata

Hendey, N-Ingram, 1958 [1957(Publ. 1958)]
Marine diatoms from some West African Ports. J. R.Microsc. Soc. (3) 77(1/2): 28-85.

Scoliopleura

Desikachary, T.V., 1956(1957).
Electron microscope studies on diatoms. J.R. Microsc. Soc. (3)76(1/2):9-36.

Scoliopleura tumida

Hendy, N. Ingram, 1964
An introductory account of the smaller algae of British coastal waters. V. Bacillariophyceae (Diatoms). Her Majesty's Stationary Office, 317 pp., 45 pls.

Scoliopleura tumida

Hendey, N.I., 1951
Littoral diatoms of Chicester Harbour with special reference to fouling. J.Roy. Microscop. Soc. 71(1): 1-86, 18 pls.

Scoliopleura tumida

Hustedt, F. and A.A. Aleem, 1951
Littoral diatoms from the Salstone near Plymouth. JMBA 30(1): 177.196.

Scoliopleura tumida

Rampi, L., 1942
Ricerche sul fitoplancton del Mare Ligure 6. Le diatomee delle acque di Sanremo. Nuovo Giornale Botanico Italiano, N.S., 49:252-268.

Scoliopleura Westii

Hendey, N.I., 1951
Littoral diatoms of Chicester Harbour with special reference to fouling. J.Roy. Microscop. Soc. 71(1): 1-86, 18 pls.

Scoliotropis latestriata

Hendy, N. Ingram, 1964
An introductory account of the smaller algae of British coastal waters. V. Bacillariophyceae (Diatoms). Her Majesty's Stationary Office, 317 pp., 45 pls.

Scoliotropis latestriata

Hendey, N.I., 1951
Littoral diatoms of Chicester Harbour with special reference to fouling. J.Roy. Microscop. Soc. 71(1): 1-86, 18 pls.

Scoliotropis latestriata

Hustedt, F. and A.A. Aleem, 1951
Littoral diatoms from the Salstone near Plymouth. JMBA 30(1): 177.196.

Scoresbya Kempii n.gen. n.sp.

Hendey, N.I., 1937
The plankton diatoms of the southern seas. Discovery Repts. 16:151-364, pls.6-13.

Skeletonema sp.

Chin, T.G., C.F.Chen, S.C.Liu and S.S. Wu, 1965.
Influence of temperature and salinity on the growth of three plankton diatom species. (In Chinese; English abstract). (Not seen). Oceanol. et Limnol.Sinica. 7(4):373-384.

Skeletonema

Desikachary, T.V., 1956 (1957).
Electron microscope studies on diatoms. J.R. Microsc. Soc. (3)76(1/2):9-36.

Skeletonema

Wheeler, J.E.G., 1939
Plankton investigations. Bermuda Biological Station. Second Report. October 1939. 7 pp. (typed), 5 figs. Plymouth, Oct. 23, 1939.

Skeletonema costatum

Ackman, R.G., P.M. Jangaard, R.J. Hoyle and H. Brockerhoff, 1964.
Origin of marine fatty acids. 1. Analyses of the fatty acids produced by the diatom Skeletonema costatum.
J. Fish. Res. Bd., Canada, 21(4):747-756.

Skeletonema costatum

Allen, W. E., 1938
The Templeton Crocker Expedition to the Gulf of California in 1935 - The Phytoplankton. Amer. Microsc. Soc., Trans. 57:328-335.

Skeletonema costatum

Allen, W.E., 1937
Plankton diatoms of the Gulf of California obtained by the G. Allan Hancock Expedition of 1936. The Hancock Pacific Expeditions, Univ. So. Calif. Publ. 3:47-59, 1 fig.

Skeletonema costatum

Allen, W. E., 1936
Occurrence of marine plankton diatoms in a ten-year series of daily catches in southern California. Am. Jour. Bot. 23(1):60-63.

Skeletonema costatum

Allen, W. E., 1934
Marine plankton diatoms of Lower California in 1931. Bot. Gaz. 95(3):485-492, 1 fig.

Skeletonema costatum

Allen, W.E., 1928.
Review of five years of studies of phytoplankton at Southern California piers, 1920-1924, inclusive. Bull. S.I.O., tech. ser., 1:357-401, 5 text-figs.

Sceletonema costatum

Allen, W.E., and E.E. Cupp, 1935
Plankton diatoms of the Java Sea. Annales du Jardin Botanique de Buitenzorg XLIV (2):101-174, figs.1-127.

(drawings of all species mentioned)

Skeletonema costatum

Bigelow, H.B., and M. Leslie, 1930
Reconnaissance of the waters and plankton of Monterey Bay, July 1928. Bull. M.C.Z., 70(5):429-481, 43 text figs.

Skeletonema costatum

Boden, B.P., 1950
Some marine plankton diatoms from the west coast of South Africa. Trans. R.Soc. S. Africa. 32:321-434, 100 text figs.

Skeletonema costatum

Braarud, T., 1962
Species distribution in marine phytoplankton.
J. Oceanogr. Soc. Japan. 20th Ann. Vol., 628-649.

Skeletonema costatum

Braarud, T., 1945
A phytoplankton survey of the polluted waters of inner Oslo Fjord. Hvalrådets Skrifter, No.28, 142 pp., 19 text figs., 17 tables.

Sceletonema costatum

Braarud, T., 1945
Experimental studies on marine plankton diatoms. Avhandlingar utgitt av Det Norske Videnskaps-Akademi i Oslo. 1. Mat.-Naturv. Klasse 1944. No.10, 1-16, 1 text fig.

Sceletonema costatum

Braarud, T., 1939
Observations on the phytoplankton of the Oslo Fjord, March-April, 1937. Nytt Magasin for Naturvidenskapene, 80:211-218, 1 text fig.

Sceletonema costatum

Braarud, T., and Adam Bursa, 1939
On the phytoplankton of the Oslo Fjord, 1933-1934. Hvalrådets Skr. No.19:1-63; 9 text figs. Reviewed. J. du. Cons. 14(3): 418-420. A.C. Gardiner.

Skeletonema costatum

Brunel, J., 1962
Le phytoplancton de la Baie de Chaleurs. Inst. Botan., Univ. Montréal, Contrib. No. 77: 365 pp., 66 pls.

Skeletonema costatum

Brunel, Jules, 1962
Le phytoplancton de la Baie des Chaleurs. Contrib. Ministère de la Chasse et des Pêcheries, Province de Québec, No. 91: 365 pp.

Skeletonema costatum

Carlucci, A.F., and Peggy M. Bowes, 1970.
Production of vitamin B_{12}, thiamine and biotin by phytoplankton.
J. Phycol. 6(4): 351-357.

Skeletonema costatum

Cassie, R. Morrison, 1963.
Relationship between plant pigments and gross primary production in Skeletonema costatum. Limnology and Oceanography, 8(4):433-439.

Skeletonema costatum

Castellvi, Josefina, 1971.
Contribución a la biología de Skeletonema costatum (Grev.) Cleve. Investigación pesq. 35(2): 365-520.

Oceanographic Index: Marine Organisms Cumulation, 1946-1973

Skeletonema costatum
Castellví, Josefina, 1963
Pigmentos de la diatomea marina Skeletonema costatum (Grev.) en su dependencia de los factores ambientales y de la dinamica de las poblaciones.
Inv. Pesq., Barcelona. 24:129-137.

Skeletonema costatum
Chu, S. P., 1949
Experimental studies on the environmental factors influencing the growth of phytoplankton. Sci. & Tech. in China 2(3):37-52.

Skeletonema costatum
Chu, S.P., 1947.
The utilization of organic phosphorus by phytoplankton. J.M.B.A. 26(3):285-295, 1 diag.

Skeletonema costatum
Cleve-Euler, A., 1951
Die Diatomeen von Schweden und Finnland. Kungl. Svenska Vetenskaps Akad. Handl., Fjärde Ser. 2(1): 161 pp., 6 pls.

Skeletonema costatum
Conover, Shirley A. McMillan, 1965.
Measurement of the Photosynthetic quotient in Skeletonema costatum. (Abstract).
Ocean Sci. and Ocean Eng., Mar. Techn. Soc., Amer. Soc. Limnol. Oceanogr., 1:293.

Skeletonema costatum
Cupp, Easter E., 1943
Marine plankton diatoms of the west coast of North America. Bull. S.I.O. 5(1):1-238, 5 pls., 168 text figs.

Sceletonema costatum
Cupp, E.E., 1934
Analysis of marine diatom collections taken from the Canal Zone to California during March, 1933. Trans. Am. Micros. Soc. LIII (1):22-29, 1 map.

Sceletonema costatum
Cupp, E., 1930
Quantitative Studies of miscellaneous series of surface catches of marine diatoms and dinoflagellates taken between Seattle and the Canal Zone from 1924 to 1928. Trans. Am. Micro. Soc. XLIX (3):238-245.

Sceletonema costatum
Cupp, E.E. and Allen, W.E., 1938
Plankton diatoms of the Gulf of California obtained by Allan Hancock Pacific Expedition of 1937. The Hancock Pacific Expeditions, The Univ. So. Calif. Publ. 3: 61-74, 1 map, pls.4-15.

Skeletonema costatum
Curl, Herbert, Jr., 1962
Effect of divalent sulfur and vitamin B12 in controlling the distribution of Skeletonema costatum.
Limnol. and Oceanogr., 7(3):422-424.

Skeletonema costatum
Curl, Herbert, Jr., and G.C. McLeod, 1961
The physiological ecology of a marine diatom, Skeletonema costatum (Grev.) Cleve.
J. Mar. Res., 19(2): 70-88.

Skeletonema costatum
Cushing, D.H., 1963
Studies on a Calanus patch. II. The estimation of algal productive rates.
J. Mar. Biol. Assoc., U.K., 43(2):339-347.

Skeletonema costatum
Davis, Curtiss O., Paul J. Harrison and Richard C. Dugdale 1973.
Continuous culture of marine diatoms under silicate limitation. I. Synchronized life cycle of Skeletonema costatum.
J. Phycol. 9(2): 175-180.

Skeletonema costatum
Delegazione Italiana della Commissione Internazionale per l'Esplorazione Scientifica del Mediterraneo, 1941
Note sul plancton della Laguna veneta. [Memoria CCLXXIX], Arch. di Ocean. e Limn. Anno I, Fasc. I, 1941 XIX: 31-57 pp.

Skeletonema costatum
Droop, M.R., 1955.
A pelagic marine diatom requiring cobalamin.
J.M.B.A. 34(2):229-231.

Skeletonema costatum
Egusa, S., 1949.
[Size variations in planktonic diatoms and some considerations of their ecological significance. 1. Skeletonema costatum and Biddulphia sinensis.]
Bull. Japan. Soc. Sci. Fish. 15(7):332-336, 3 textfigs. (In Japanese).

Skeletonema costatum
Eppley, Richard W., Jan N. Rogers, James J. McCarthy and Alain Sournia, 1971.
Light/dark periodicity in nitrogen assimilation of the marine phytoplankters Skeletonema costatum and Coccolithus huxleyi in N-limited chemostat cultures.
J. Phycol. 7(2): 150-154.

Sceletonema costatum
Ercegović, A., 1940
Weitere Untersuchungen über einige hydrographische Verhältnisse und über die Phytoplanktonproduktion in den Gewässern der Östlichen Mitteladria. Acta Adriatica 2(3):95-134, 8 text figs.

Skeletonema costatum
Eskinazi Enide e Shigekatsu Satô (1963-1964) 1966.
Contribuição ao estudo das diatomaceas da Praia de Piedade.
Trabhs Inst. Oceanogr., Univ. Recife, 5 (5/6): 73-114.

Skeletonema costatum
Forti, A., 1922
Ricerche sulla flora pelagica (fitoplancton) di Quarto dei Mille. Mem. R. Com. Talass. Ital. 97:248 pp., 13 pls.

Skeletonema costatum
Frenguelli, Joaquin, and Hector Antonio Orlando, 1959.
Operacion MERLUZA. Diatomeas y silicoflagelados del plancton del "VI Crucero".
Servicio Hidrogr. Naval., Argentina, Publ. No. H. 619: 5-62.

Skeletonema costatum
Fudinami, M., and H. Kasahara, 1942.
[Rearing and metamorphosis of Balanus amphitrite hawaiiensis Bloch.] Zool. Mag., Tokyo, 54(3):108-118.

Skeletonema costatum
Ghazzawi, F.M., 1939
Plankton of the Egyptian waters. A study of the Suez Canal Plankton. (A) Phytoplankton. Preliminary Report 83 pp. Notes and Memoires, Min. Commerce-Industry, Egypt, Hydrobiol. & Fish. 65 figs.

Sceletonema costatum
Gilbert, J.Y., and W.E. Allen, 1943
The phytoplankton of the Gulf of California obtained by the "E.W. Scripps" in 1939 and 1940. J. Mar. Res. V(2):89-110, figs.30-31.

Skeletonema costatum
Gran, H.H., 1908
Diatomeen. Nordisches Plankton, Botanischer Teil pp. XIX.1-XIX.146; 178 text figs.

Skeletonema costatum
Gran, H.H. and T. Braarud, 1935
A quantitative study of the phytoplankton in the Bay of Fundy and the Gulf of Maine (including observations on hydrography, chemistry, and turbidity). J. Biol. Bd., Canada, 1(5):279-467, 69 text figs.

Skeletonema costatum
Gran, H. H. and E. C. Angst, 1931
Plankton diatoms of Puget Sound. Publ. Puget Sound Biol. Sta. 7:417-519, 95 text figs.

Skeletonema costatum
Grøntved, J., 1949(1950).
Investigations on the phytoplankton in the Danish Waddensee in July 1941. Medd. Komm. Danmarks Fiskeri- og Havundersøgelser, Serie Plankton, 5(2)(Medd. Skalling Lab. 10):55 pp., 2 pls., 38 textfigs.

Skeletonema costatum
Grøntved, J., 1949
Investigations on the phytoplankton in the Danish Waddensea in July 1941. Medd. Komm. Danmarks Fiskeri og Havundersøgelser, ser. Plankton, 5(2):55 pp., 2 pls., 38 text figs.

Skeletonema costatum (effect of)
Gross, M. Grant, Sevket M. Gucluer, Joe S. Creager and William A. Dawson, 1963.
Varved marine sediments in a stagnant fjord.
Science, 141(3584):918-919.

Skeletonema costatum ?
Gunter, G., R.H. Williams, C.C. Davis, and F.G. Walton Smith, 1948
Catastrophic mass mortality of marine animals and coincident phytoplankton bloom on the West Coast of Florida, November 1946 to August 1947. Ecol. Mon., 18:309-324, 2 text figs.

Skeletonema costatum
Hasle, Grethe R., 1973
Morphology and taxonomy of Skeletonema costatum (Bacillariophyceae). Norwegian J. Bot. 20(2/3):109-137.

Skeletonema costatum
Hendy, N. Ingram, 1964
An introductory account of the smaller algae of British coastal waters. V. Bacillariophyceae (Diatoms).
Her Majesty's Stationary Office, 317 pp., 45 pls.

Skeletonema costatum
Hendey, N.I., 1937
The plankton diatoms of the southern seas.
Discovery Repts. 16:151-364, pls.6-13.

Skeletonema costatum
Hirano, Reijiro, 1957.
Studies on the red tide. 1. Summer flowering of Skeletonema costatum and Aoshiro (green tide) in the northern region of Tokyo Bay.
Coll. Wks., Fish. Sci., Jubilee Publ., Prof. I. Amemiya, Univ. Tokyo Press, 407-411.
Abstr. in:
Rec. Res., Fac. Agric., Univ. Tokyo, 7(1958):49.

Skeletonema costatum
Hogetsu, Kinji, Mitsuru Sakamoto, and Hiroshi Sumikawa, 1959.
On the high photosynthetic activity of Skeletonema costatum under the strong light intensity. Botan. Mag., Tokyo, 72:431-422.
Also in:
Collected Papers on Science of Atmosphere and Hydrosphere, 1959-1963, Water Res. Lab., Nagoya Univ., 1 (4).

Skeletonema costatum
Holmes, Robert W., 1966.
Light microscope observations on cytological manifestations of nitrate, phosphate, and silicate deficiency in four marine centric diatoms.
J. Phycology, 2(4):136-140.

Skeletonema costatum

Hulburt, E.M., 1963.
The occurrence of Skeletonema costatum (Bacillariophyceae) in the Gulf Stream and Sargasso Sea. Bull. Mar. Sci., Gulf and Caribbean, 13(2):219-223.

Hustedt, F. and A.A. Aleem, 1951
Littoral diatoms from the Salstone near Plymouth. JMBA 30(1): 177.196.

Iizuka, Shoji, and Haruhiko Irie, 1964
Electron micrographic study on the marine diatoms especially Skeletonema costatum (Grev.) Cleve. Bull. Fac. Fish., Nagasaki Univ., 15:92-99.

Iselin, C., 1930
A report on the coastal waters of Labrador based on explorations of the "Chance" during the summer of 1926. Proc. Am. Acad. Arts Sci., 66(1):1-37, 14 text figs.

Japan, Kobe Marine Observatory, 1963.
Report of the oceanographic observations in the sea south of Honshu from July to August, and from the cold water region south of Enshu Nada October to November 1960. (In Japanese). Bull. Kobe Mar. Obs., 171(3):36-52.

Jitts, H.R., C.D. McAllister, K. Stephens and J.D.H. Strickland, 1964.
The cell division of some marine phytoplankters as a function of light and temperature. J. Fish. Res. Bd., Canada, 21(1):139-157.

Jørgensen, E., 1905
B. Protistplankton and the diatoms in bottom samples. Hydrographical and biological investigations in Norwegian fjords. Bergens Mus. Skr. 7: 49-225.

Jorgensen, E., 1900
Protophyten und Protozoën im Plankton aus der Norwegischen Westkerste. Bergens Mus. Aarb. 1899(6): 95 pp., 5 pls., 83 tables.

Jørgensen, Erik G., 1970.
The adaptation of plankton algae. V. Variation in the photosynthetic characteristics of Skeletonema costatum cells grown at low light intensity. Physiol. Plant. 23 (1): 11-17

Jørgensen, Erik G. 1968.
The adaptation of plankton algae. II Aspects of the temperature adaptation of Skeletonema costatum. Physiologia Plant. 21(2): 423-427.

Jørgensen, Erik G. 1967.
Photosynthetic activity during the life cycle of synchronous Skeletonema cells. Physiol. Plantarum 19 (3): 789-799

Jørgensen, Erik G., 1966.
Photosynthetic activity during the life cycle of synchronous Skeletenema cell. Physiol. Plant., 19:789-799.

Kimura, Tomohiro, Akio Mizokami and Toshimasa Hashimoto, 1972
The red tide that caused severe damage to the fishery resources in Hiroshima Bay: outline of its occurrence and the environmental conditions. (In Japanese; English abstract). Bull. Plankt. Soc. Japan 19(2):24-38 (82-112).

Kokubo, S., 1952
Results of the observations on the plankton and oceanography of Mutsu Bay during 1950, reference being made back to the period 1946-1950. Bull Mar.Biol.Sta., Asamushi 5(1/4): 1-54, 3 tables,(fold-in), 1 fold-in.

Levander, K.M., 1947
Plankton gesammelt in den Jahren 1899-1910 an den Küsten Finnlands. Finnländische Hydrographisch-Biologische Untersuchungen (aus dem Wasserbiologischen Laboratorin der Societas Scientiarum Fennica) No.11: 40 pp., 6 diagrams, 13 pls., tables.

Levring, T., 1945
Some culture experiments with marine plankton diatoms. Medd. Oceanografiska Institutet i Göteborg, No.9 (Göteborg Kungl. Vetenskaps-och Vitterhets-Samhälles Handlingar, Sjätte Följden. Ser.B. Vol.3 (12). 1-17.

Lillick, L.C., 1940
Phytoplankton and planktonic protozoa of the offshore waters of the Gulf of Maine. Pt.II. Qualitative Composition of the Planktonic Flora. Trans. Am. Phil. Soc., n.s., 31(3):193-237, 13 text figs.

Lillick, L.C., 1938
Preliminary report of the phytoplankton of the Gulf of Maine. Am. Mid. Nat. 20(3):624-640, 1 text figs 37 tables.

Lillick, L.C., 1937
Seasonal studies of the phytoplankton off Woods Hole, Massachusetts. Biol. Bull. LXXIII (3):488-503, 3 text figs.

Mangin, M. L., 1912
Phytoplancton de la croisière du "René" dans l'Atlantique (Septembre 1908). Ann. Inst. Ocean., n.s., 4(1):1-66, 2 pls., 41 text figs., 2 tables.

Margalef, Ramón, 1964.
Modelos experimentales de poblaciones de fitoplancton: nuevas observaciones sobre pigmentos y fijación de carbono inorgánico. Inv. Pesq., Barcelona, 26:195-203.

Margalef, Ramón, 1963
Modelos simplificados del ambiente marino para el estudio de la sucesión y distribución del fitoplancton y del valor indicador de sus pigmentos. Invest. Pesquera, Barcelona, 23:11-52.

Marshall, S.M., 1947
An experiment in marine fish cultivation: III. The plankton of a fertilized loch. Proc. Roy. Soc., Edinburgh, Sect.B., 63, Pt.I(3):21-33, 7 text figs.

Marukawa, H., 1921
Plankton lists and some new species of copepods, from the northern waters of Japan. Bull. Inst. Ocean., No.384, 15 pp., 3 pls., 1 chart. Monaco

Matsue, Y., 1954.
On the culture of the marine plankton diatom, Skeletonema costatum (Grev.) Cleve. Rev. Fish. Sci., Japan, (Suisangaku no Gaikan):1-4

Matsue, Y., 1950.
(Phytoplankton and its oxidisability by permanganate.) Bull. Jap. Soc. Sci. Fish. 15(12): 813-817.

Matsue, Y., 1950.
(The variation of titratable base in sea water by the growth of diatoms.) J. Ocean. Soc., Tokyo, 6(1):32-38, 2 textfigs. (In Japanese with English abstract).

Matsue, Y., 1950.
(Phytoplankton and its oxidisability by permanganate.) Bull. Japan. Soc. Sci. Fish. 15(12): 813-817, 2 textfigs. (In Japanese, with English summary).

Matsue, Y., 1949.
The physiological analysis of brine and the preservation of phosphoric acid in Skeletonema costatum. J. Fish. Res. Inst. 2:34-49.

Matsue, Yoshiyuki, 1957.
On the absorption of nitrogen compounds in different forms in sea water by a marine plankton diatom, Skeletonema costatum (Grev.) Cleve. Coll. Wks. Fish. Sci. Jubilee Publ. Prof. I. Amemiya, Tokyo Univ. Press; 249-257. Abstr. in: Rec. Res., Fac. Agric., Tokyo Univ., Mar. 1958, 7(79):57.

Meunier, A., 1915
Microplancton de la Mer Flamande. 2. Diatomées (excepté le genre Chaetoceros). Mem. Mus. Roy. Hist. Nat., Belgique, 7(3):1-118, Pls. VIII-XIV.

Migita, Seiji 1969.
Seasonal variation of cell size in Skeletonema costatum and Melosira moniliformis. (In Japanese; English abstract) Bull. Fac. Fish. Nagasaki Univ. 27:9-17.

Morse, D.C., 1947
Some observations on seasonal variations in plankton population Patuxant River, Maryland 1943-1945. Bd. Nat. Res., Publ. No.65, Chesapeake Biol. Lab., 31, 3 figs.

Murphy, R.C., 1923
The oceanography of the Peruvian littoral with reference to the abundance and distribution of marine life. Geogr. Rev. 13:64-85.

Ogiino, C., 1963.
Studies on the chemical composition of some natural foods of aquatic animals. (In Japanese; English abstract). Bull. Jap. Soc. Sci. Fish., 29(5):459-462.

Paasche, E., 1973
The influence of cell size on growth rate, silica content and some other properties of four marine diatom species. Norwegian J. Bot. 20(2/3): 199-204.

Parsons, T.R., 1961
On the pigment composition of eleven species of marine phytoplankton. J. Fish. Res. Bd., Canada, 18(6):1017-1025.

Skeletonema costatum
Parsons, T.R., K. Stephens and J.D.H. Strickland, 1961
On the chemical composition of eleven species of marine phytoplankters.
J. Fish. Res. Bd. Canada, 18(6):1001-1016.

Skeletonema costata
Pavillard, J., 1925
Bacillariales. Rept. on the Danish Oceanogr. Exped., 1908-10 to the Mediterranean and adj. seas. Vol.II., Biol. J4:72 pp., 116 text figs.

Skeletonema costatum
Pavillard, J., 1905
Recherches sur la flore pelagique (Phytoplankton) de l'Etang de Thau. Theses presentees a la Fac. Sci., Paris, 116 pp., 3 pls.

Skeletonema costatum
Pettersson, H., 1934
Scattering and extinction of light in sea-water. Meddelanden från Göteborgs Högskolas Oceanografiska Institut. 9 (Göteborgs Kungl. vetenskaps-och vitterhets-samhälles Handlingar. Femte Foljden, Ser.B, 4(4)):1-16.

Skeletonema costatum
Pettersson, H., F. Gross, and F. Koczy, 1939
Large scale plankton cultures. Medd. från Oceanografiska Institutet i Göteborg No.3 (Göteborgs Kungl. vetenskaps-och Vitterhets-Samhälles Handlingar Femte Földjen. Ser. B) Vol.6(13):1-24.
1. The plankton shaft. H. Pettersson
2. Experiment with phytoplankton. F. Gross and F. Koczy
3. Experiments with zooplankton.

Sceletonema costatum
Pettersson, H., H. Höglund, S. Landberg, 1934.
Submarine daylight and the photosynthesis of phytoplankton. Medd. fran Göteborgs Högskolas Oceanografiska Inst. 10 (Göteborgs Kungl. Vetenskaps-och Vitterhets-samhälles handlingar. Femte Földjen, Ser. B 4(5):1-17.

Skeletonema costatum [a]
Prakash, A., Liv Skoglund, Britt Rystad and Anne Jensen 1973
Growth and cell-size distribution of marine planktonic algae in batch and dialysis cultures.
J. Fish. Res. Bd. Can. 30(2):143-155

Skeletonema costatum
Pratt, David M., 1966.
Competition between Skeletonema costatum and Olisthodiscus luteus in Narragensett Bay and in culture.
Limnol. Oceanogr., 11(4):447-455.

Sceletonema costatum
Proshkina-Lavrenko, A.I., 1961.
[Variability of some Black Sea diatoms.]
Botan. Zhurn., Akad. Nauk, SSSR, 46(12):1794-1797.

Sceletonema costatum
Rampi, L., 1942
Ricerche sul fitoplancton del Mare Ligure 6. Le diatomee delle acque di Sanremo. Nuovo Giornale Botanico Italiano, N.S., 49:252-268.

Skeletonema costatum
Rothe, F., 1942.
Quantitativen Untersuchungen über die Planktonverteilung in der östlichen Ostsee. Ber. Deutsch Wiss. Komm. Meeresf., N.B., 10:291-368, 33 text-figs.

Skeletonema costatum
Rothe, F., 1941
Quantitative Untersuchunger über die Plankton verteilung in der östlichen Ost see. Ber. Deut. Wiss. Komm. fur Meeresforschung. n.f. X(3):291-368, 33 text figs.

Sceletonema costatum
Rumkówna, A., 1948
[List of the phytoplankton species occurring in the superficial water layers in the Gulf of Gdańsk.] Bull. Lab. mar., Gdynia, No. 4: 139-141 with tables in back.

Skeletonema costatum
Rustad, E., 1946.
Experiments on photosynthesis and respiration at different depths in the Oslo Fjord. Nytt Mag. Naturvidensk. 85:223-229, 4 figs.

Skeletonema costatum
Ryther, John H., and Dana D. Kramer, 1961
Relative iron requirement of some coastal and offshore algae.
Ecology, 42(2):444-446.

Skeletonema costatum
Sakshaug, Egil 1972
Quantitative phytoplankton investigations in near-shore water masses
Skr. K. norske Vidensk. Selsk. (3):1-8.

Skeletonema costatum
Schodduyn, M., 1926
Observations faites dans la baie d'Ambleteuse (Pas de Calais). Bull. Inst. Ocean., Monaco, No. 482: 64 pp.

Skeletonema costatum
Sieburth, John McN. and David M. Pratt, 1962.
Anticoliform activity of sea water associated with the termination of Skeletonema costatum blooms.
Trans. N.Y. Acad. Sci., (2), 24(5):498-501.

Skeletonema costatum [a]
Smayda, Theodore J., 1973
The growth of Skeletonema costatum during a winter-spring bloom in Narragansett Bay, Rhode Island.
Norwegian J. Bot. 20(2/3):219-247.

Skeletonema costatum
Smayda, Theodore J., and Brenda J. Boleyn, 1966
Experimental observations on the flotation of marine diatoms. II. Skeletonema costatum and Rhizosolenia setigera.
Limnol. and Oceanogr., 11(1):18-34.

Skeletonema costatum (figs.)
Sousa e Silva, E., 1949
Diatomaceas e Dinoflagelados de Baia de Cascais. Portugaliae Acta Biol., Volume: Julio Henriques, Ser. B: 300-383, 9 pls, 2 fold-in tables.

Skeletonema costatum
Steemann-Nielsen, Einar, 1951
The marine vegetation of the Isefjord. A study on ecology and production. Medd. Komm. Danmarks Fiskeri-og Havundersøgelser. Ser. Plankton. 5(4); 114pp., 46 text figs.

Skeletonema costatum
Steemann Nielsen, E., and E.G. Jørgensen 1968
The adaptation of plankton algae. I. General part
Physiol. Plant. 21(2):401-413

Skeletonema costatum
Takano, Hideaki, 1961.
Epiphytic diatoms upon Japanese agar sea-weeds.
Bull. Tokai Reg. Fish. Res. Lab., No. 31:269-274.

Skeletonema costatum [a]
Uno, Shiroh, 1971.
Turbidometric continuous culture of phytoplankton constructions of the apparatus and experiments on the daily periodicity in photosynthetic activity of Phaeodactylum tricornutum and Skeletonema costatum. Bull. Plankt. Soc. Japan 18(1): 14-27.

Skeletonema costatum
Uyeno, Fukuzo, 1961
Oceanographical and ecological studies on primary production of the sea, with special references to relationship between diatom production and temperature and chlorinity of water.
Rept., Fac. Fish., Pref. Univ., Mie, 4(1): 1-64.

Sceletonema costatum (fig.)
Vives, F., and A. Planas, 1952.
Plancton recogido por los laboratorios costeros. VI. Fitoplancton de las costas de Vinaroz, islas Columbretes y alrededores de la desembocadura del Elroa. Publ. Inst. Biol. Aplic. 11:141-156, 19 textfigs.

Skeletonema costatum
Zgurovskaya, L.N., and N.G. Kustenko, 1968.
The influence of different nitrite nitrogen concentrations on photosynthesis, pigment accumulation and cell division of Skeletonema costatum (Grev.) Cl. (In Russian; English abstract).
Okeanologiia, Akad. Nauk, SSSR, 8(6):1053-1058.

Skeletonema costatum [a]
Zgurovskaya, L.N., and N.G. Kustenko, 1968.
The action of ammonia nitrogen on cell division, photosynthesis and the accumulation of pigments in Skeletonema costatum (Grev.), Chaetoceros sp. and Prorcentrum micans. (In Russian; English abstract).
Okeanologiia, Akad. Nauk, SSSR, 8(1): 116-125.

Skeletonema mediterraneum
Wawrik, F., 1961.
Die horizontale Verteilung der Planktondiatomeen im Golf von Neapel.
Int. Revue Ges. Hydrobiol. 46(3):460-479.

Skeletonema mirabile
Cleve-Euler, A., 1951
Die Diatomeen von Schweden und Finnland. Kungl. Svenska Vetenskaps Akad. Handl., Fjärde Ser. 2(1): 161 pp., 6 pls.

Sceletonema mirabile
Jorgensen, E., 1900
Protophyten und Protozoën im Plankton aus der Norwegischen Westkerste. Bergens Mus. Aarb. 1899(6): 95 pp., 5 pls., 83 tables.

Skeletonema mirabile
Gran, H.H., 1908
Diatomeen. Nordisches Plankton, Botanischer Teil pp. XIX.1-XIX 146; 178 text figs.

Skeletonema peniculus
Cleve-Euler, A., 1951
Die Diatomeen von Schweden und Finnland. Kungl. Svenska Vetenskaps Akad. Handl., Fjärde Ser. 2(1): 161 pp., 6 pls.

Skeletonema subsalum
Brunel, J., 1962
Le phytoplancton de la Baie de Chaleurs.
Inst. Botan., Univ. Montréal, Contrib. No. 77: 365 pp., 66 pls.

Skeletonema subsalum?
Brunel, Jules, 1962
Le phytoplancton de la Baie des Chaleurs.
Contrib. Ministère de la Chasse et des Pêcheries, Province de Québec, No. 91: 365 pp.

Skeletonema subsalum

Cleve-Euler, A., 1951
Die Diatomeen von Schweden und Finnland. Kungl. Svenska Vetenskaps Akad. Handl., Fjärde Ser. 2(1): 161 pp., 6 pls.

Skeletonema tropicum

Hulburt, Edward M., and Robert R. L. Guillard, 1968.
The relationship of the distribution of the diatom Skeletonema tropicum to temperature. Ecology, 49(2):337-339.

Spatangidium arachne

Pavillard, J., 1925
Bacillariales. Rept. on the Danish Oceangr. Exped., 1908-10 to the Mediterranean and adj. seas. Vol.II., Biol. J4:72 pp., 116 text figs.

Sphinctocystis librile

Mann, A., 1907
Report on the diatoms of the Albatross voyages in the Pacific Ocean, 1888-1904. Contrib. U. S. Nat. Herb. 10(5):221-419, Pls. XLIV-LIV.

Sphinctocystis undulata

Mann, A., 1907
Report on the diatoms of the Albatross voyages in the Pacific Ocean, 1888-1904. Contrib. U. S. Nat. Herb. 10(5):221-419, Pls. XLIV-LIV.

Stauroneis

Desikachary, T.V., 1956(1957).
Electron microscope studies on diatoms. J.R. Microsc. Soc. (3)76(1/2):9-36.

Stauroneis africana

Hendy, N. Ingram, 1964
An introductory account of the smaller algae of British coastal waters. V. Bacillariophyceae (Diatoms).
Her Majesty's Stationary Office, 317 pp., 45 pls.

Stauroneis Alabamae

Zanon, D. V., 1949
Diatomee di Buenos Aires (Argentina) Atti Accad. Naz. Lincei, Memorie, Cl. Sci. fis., mat. e. nat., ser. 7, 11(3):59-151, 2 pls.

Stauroneis amphioxys & varieties

Hendy, N. Ingram, 1964
An introductory account of the smaller algae of British coastal waters. V. Bacillariophyceae (Diatoms).
Her Majesty's Stationary Office, 317 pp., 45 pls.

Stauroneis anceps

Mann, A., 1893
List of Diatomaceae from a deep-sea dredging in the Atlantic Ocean off Delaware Bay by the U. S. Fish Commission Steamer Albatross. Proc. U. S. Nat. Mus. 16:303-312.

Stauroneis anceps leiostauron

Tsumura, K., 1956.
Diatomoj el la cirkaufoso de la restajo de la kastele de Odawara. J. Yokohama Municipal Univ., (C-14), No. 47:23 pp.

Stauroneis Bacillum n.sp.

Grunow, A., 1863
Ueber einige neue und ungenügend bekannte Arten und Gattungen von Diatomaceen. Verhandl. d. K.K. Zool. Bot. Gesellsch., Vienna, 13: 137-162, pl.4-5 (Pl. 13-14).

Stauroneis biformis n.sp.

Grunow, A., 1863
Ueber einige neue und ungenügend bekannte Arten und Gattungen von Diatomaceen. Verhandl. d. K.K. Zool. Bot. Gesellsch., Vienna, 13: 137-162, pl.4-5 (Pl. 13-14).

Stauroneis brebissonii n.sp.

Castracan degli Antelminelli, F., 1886
1. Report on the Diatomaceae collected by H.M.S. Challenger during the years 1873-1876. Rept. Sci. Results, H.M.S. Challenger, Botany Vol. II, 178 pp., 30 pls.

Stauroneis constricta

Hendey, N.I., 1951
Littoral diatoms of Chicester Harbour with special reference to fouling. J.Roy. Microscop. Soc. 71(1): 1-86, 18 pls.

Stauroneis (Libellus) constricta minor n.var.

Zanon, V., 1947.
Diatomee delle Isole Ponziane - materiali per una florula diatomologica del Mare Tirreno. Bol. Pesca, Piscicol., Idrobiol., n.s., 2(1):36-53, 1 pl. of 10 figs.

Stauroneis decipiens

Hendy, N. Ingram, 1964
An introductory account of the smaller algae of British coastal waters. V. Bacillariophyceae (Diatoms).
Her Majesty's Stationary Office, 317 pp., 45 pls.

Stauroneis glacialis n.sp.

Castracan degli Antelminelli, F., 1886
1. Report on the Diatomaceae collected by H.M.S. Challenger during the years 1873-1876. Rept. Sci. Results, H.M.S. Challenger, Botany Vol. II, 178 pp., 30 pls.

Stauroneis gracilis

Mann, A., 1893
List of Diatomaceae from a deep-sea dredging in the Atlantic Ocean off Delaware Bay by the U. S. Fish Commission Steamer Albatross. Proc. U. S. Nat. Mus. 16:303-312.

Stauroneis grani n.sp.

Jørgensen, E., 1905
B.Protistplankton and the diatoms in bottom samples. Hydrographical and biological investigations in Norwegian fjords. Bergens Mus. Skr. 7: 49-225.

Stauroneis Gregorii

Hendey, N.I., 1951
Littoral diatoms of Chicester Harbour with special reference to fouling. J.Roy. Microscop. Soc. 71(1): 1-86, 18 pls.

Stauroneis Gregorii

Zanon, D. V., 1949
Diatomee di Buenos Aires (Argentina) Atti Accad. Naz. Lincei, Memorie, Cl. Sci. fis., mat. e. nat., ser. 7, 11(3):59-151, 2 pls.

Stauroneis gregorii diminuta

Tsumura, K., 1956.
Diatomoj el la cirkaufoso de la restajo de la kastele de Odawara. J. Yokohama Municipal Univ., (C-14), No. 47:23 pp.

Stauroneis Heufleriana

Grunow, A., 1863
Ueber einige neue und ungenügend bekannte Arten und Gattungen von Diatomaceen. Verhandl. d. K.K. Zool. Bot. Gesellsch., Vienna, 13: 137-162, pl.4-5 (Pl. 13-14).

Stauroneis indistincta n.sp.

Cholnoky, B.J., 1963.
Beiträge zur Kenntnis des marinen Litorals von Südafrika. Botanica Marina, 5(2/3):38-83.

Stauroneis melchiori, n.sp.

Frenguelli, Joaquin, y Hector A. Orlando, 1958.
Diatomeas y silicoflagelados del sector Antartico Sudamericano. Inst. Antartico Argentino, Publ., No. 5:191 pp.

Stauroneis membranacea

Hendy, N. Ingram, 1964
An introductory account of the smaller algae of British coastal waters. V. Bacillariophyceae (Diatoms).
Her Majesty's Stationary Office, 317 pp., 45 pls.

Stauroneis oblonga

Castracan degli Antelminelli, F., 1886
1. Report on the Diatomaceae collected by H.M.S. Challenger during the years 1873-1876. Rept. Sci. Results, H.M.S. Challenger, Botany Vol. II, 178 pp., 30 pls.

Stauroneis pacifica n.sp.

Castracan degli Antelminelli, F., 1886
1. Report on the Diatomaceae collected by H.M.S. Challenger during the years 1873-1876. Rept. Sci. Results, H.M.S. Challenger, Botany Vol. II, 178 pp., 30 pls.

Stauroneis Phoecicenteron gracilis

Mann, A., 1893
List of Diatomaceae from a deep-sea dredging in the Atlantic Ocean off Delaware Bay by the U. S. Fish Commission Steamer Albatross. Proc. U. S. Nat. Mus. 16:303-312.

Stauroneis phoenicentron & lanceolata

Tsumura, K., 1956.
Diatomoj el la cirkaufoso de la restajo de la kastele de Odawara. J. Yokohama Municipal Univ., (C-14), No. 47:23 pp.

Stauroneis phoenicentron

Zanon, D. V., 1949
Diatomee di Buenos Aires (Argentina) Atti Accad. Naz. Lincei, Memorie, Cl. Sci. fis., mat. e. nat., ser. 7, 11(3):59-151, 2 pls.

Stauroneis prominula

Hendy, N. Ingram, 1964
An introductory account of the smaller algae of British coastal waters. V. Bacillariophyceae (Diatoms).
Her Majesty's Stationary Office, 317 pp., 45 pls.

Stauroneis pseudothermicola n.sp.

Cholnoky, B.J., 1963.
Beiträge zur Kenntnis des marinen Litorals von Südafrika. Botanica Marina, 5(2/3):38-83.

Stauroneis pygmaea n.sp.

Castracan degli Antelminelli, F., 1886
1. Report on the Diatomaceae collected by H.M.S. Challenger during the years 1873-1876. Rept. Sci. Results, H.M.S. Challenger, Botany Vol. II, 178 pp., 30 pls.

Stauroneis quadripedis

Hendy, N. Ingram, 1964
An introductory account of the smaller algae of British coastal waters. V. Bacillariophyceae (Diatoms).
Her Majesty's Stationary Office, 317 pp., 45 pls.

Stauroneis salina

Castracan degli Antelminelli, F., 1886
1. Report on the Diatomaceae collected by H.M.S. Challenger during the years 1873-1876. Rept. Sci. Results, H.M.S. Challenger, Botany Vol. II, 178 pp., 30 pls.

Stauroneis salina

Hendy, N. Ingram, 1964
An introductory account of the smaller algae of British coastal waters. V. Bacillariophyceae (Diatoms).
Her Majesty's Stationary Office, 317 pp., 45 pls.

Stauroneis salina

Hendey, N.I., 1951
Littoral diatoms of Chicester Harbour with special reference to fouling. J.Roy. Microscop. Soc. 71(1): 1-86, 18 pls.

Stauroneis salina

Hustedt, F. and A.A. Aleem, 1951
Littoral diatoms from the Salstone near Plymouth. JMBA 30(1): 177.196.

Stauroneis salina

Zanon, V., 1948
Diatomee marini di Sardegna e Pugillo di Alghe Marine della stressa. Boll. Pesca, Piscitutura e Idrobiologia, Anno 24, ns. 3(2): 202-244, 27 figs. on 1 pl.

Stauroneis septentrionalis

Jørgensen, E., 1905
B. Protistplankton and the diatoms in bottom samples. Hydrographical and biological investigations in Norwegian fjords. Bergens Mus. Skr. 7: 49-225.

Stauroneis Smithii

Mann, A., 1893
List of Diatomaceae from a deep-sea dredging in the Atlantic Ocean off Delaware Bay by the U. S. Fish Commission Steamer Albatross. Proc. U. S. Nat. Mus. 16:303-312.

Stauroneis thaitiana n.sp

Castracane degli Antelminelli, F., 1886
1. Report on the Diatomaceae collected by H.M.S. Challenger during the years 1873-1876. Rept. Sci. Results, H.M.S. Challenger, Botany Vol. II, 178 pp., 30 pls.

Stauropsis membranacea

Meunier, A., 1915
Microplancton de la Mer Flamande. 2. Diatomées (excepté le genre Chaetoceros). Mem. Mus. Roy. Hist. Nat., Belgique, 7(3):1-118, Pls. VIII-XIV.

Stauropsis septentrionalis

Brunel, J., 1962
Le phytoplancton de la Baie de Chaleurs. Inst. Botan., Univ. Montréal, Contrib. No. 77: 365 pp., 66 pls.

Stauroptera pedkii

Grunow, A., 1863
Ueber einige neue und ungenügend bekannte Arten und Gattungen von Diatomaceen. Verhandl. d. K.K. Zool. Bot. Gesellsch., Vienna, 13: 137-162, pl.4-5 (Pl. 13-14).

Stenoneis inconspicua

Hendy, N. Ingram, 1964
An introductory account of the smaller algae of British coastal waters. V. Bacillariophyceae (Diatoms). Her Majesty's Stationary Office, 317 pp., 45 pls.

Stenopterobia

Desikachary, T.V., 1956(1957).
Electron microscope studies on diatoms. J.R. Microsc. Soc. (3)76(1/2):9-36.

Stephanodiscus

Brunel, J., 1962
Le phytoplancton de la Baie de Chaleurs. Inst. Botan., Univ. Montréal, Contrib. No. 77: 365 pp., 66 pls.

Stephanodiscus

Desikachary, T.V., 1956(1957).
Electron microscope studies on diatoms. J.R. Microsc. Soc. (3)76(1/2):9-36.

Stepanodiscus sp.

Kokubo, S., and S. Sato, 1947
Plankters in Jū-San Gata. Physiol. and Ecol. (Japan) 1(4):1-16, 3 text figs., tables.

Stephanodiscus sps

Round, F.E., 1970.
The delineation of the genera Cyclotella and Stephanodiscus by light microscopy transmission and reflecting electron microscopy. Beihefte Nova Hedwigie 31: 591-604.

Stephanodiscus astraea

Cleve-Euler, A., 1951
Die Diatomeen von Schweden und Finnland. Kungl. Svenska Vetenskaps Akad. Handl., Fjärde Ser. 2(1): 161 pp., 6 pls.

Stephanodiscus astrea

Lillick, L.C., 1940
Phytoplankton and planktonic protozoa of the offshore waters of the Gulf of Maine. Pt.II. Qualitative Composition of the Planktonic Flora. Trans. Am. Phil. Soc., n.s., 31(3):193-237, 13 text figs.

Stephanodiscus astraea

Zanon, D. V., 1949
Diatomee di Buenos Aires (Argentina) Atti Accad. Naz. Lincei, Memorie, Cl. Sci. fis., mat. e. nat., ser. 7, 11(3):59-151, 2 pls.

Stephanodiscus astraea

Zanon, V., 1948
Diatomee marini di Sardegna e Pugillo di Alghe Marine della stressa. Boll. Pesca, Piscitutura e Idrobiologia, Anno 24, ns. 3(2): 202-244, 27 figs. on 1 pl.

Stephanodiscus binderanus

Cleve-Euler, A., 1951
Die Diatomeen von Schweden und Finnland. Kungl. Svenska Vetenskaps Akad. Handl., Fjärde Ser. 2(1): 161 pp., 6 pls.

Stephanodiscus dubius

Cleve-Euler, A., 1951
Die Diatomeen von Schweden und Finnland. Kungl. Svenska Vetenskaps Akad. Handl., Fjärde Ser. 2(1): 161 pp., 6 pls.

Stephanodiscus dubius

Zanon, V., 1948
Diatomee marini di Sardegna e Pugillo di Alghe Marine della stressa. Boll. Pesca, Piscitutura e Idrobiologia, Anno 24, ns. 3(2): 202-244, 27 figs. on 1 pl.

Stephanodiscus Hantzschii

Cleve-Euler, A., 1951
Die Diatomeen von Schweden und Finnland. Kungl. Svenska Vetenskaps Akad. Handl., Fjärde Ser. 2(1): 161 pp., 6 pls.

Stephanodiscus Hantzschianus

Mann, A., 1893
List of Diatomaceae from a deep-sea dredging in the Atlantic Ocean off Delaware Bay by the U. S. Fish Commission Steamer Albatross. Proc. U. S. Nat. Mus. 16:303-312.

Stephanodiscus hantzschii

Sakevich, A.I., 1970.
Detection of methyl amines in the culture of Stephanodiscus hantzschii Grun. Gidrobiol. Zh., 6(3): 98-100. (In Russian)

Stephanodiscus subtilis

Cleve-Euler, A., 1951
Die Diatomeen von Schweden und Finnland. Kungl. Svenska Vetenskaps Akad. Handl., Fjärde Ser. 2(1): 161 pp., 6 pls.

Stephanodiscus torula

Hendy, N. Ingram, 1964
An introductory account of the smaller algae of British coastal waters. V. Bacillariophyceae (Diatoms). Her Majesty's Stationary Office, 317 pp., 45 pls.

Stephanogonia sp.?

Florin, M-B., 1948
9. Diatomeae in submarine cores from the Tyrrhenian Sea. Medd. Ocean. Inst., Göteborg, 15 (Göteborgs Kungl. Vetenskaps-och Viterrhets Samhälles Handlingar, Sjätte Följden, Ser. B 5(13):80-88.

Stephanogonia actinoptychus

Cleve-Euler, A., 1951
Die Diatomeen von Schweden und Finnland. Kungl. Svenska Vetenskaps Akad. Handl., Fjärde Ser. 2(1): 161 pp., 6 pls.

Stephanogonia danica

Cleve-Euler, A., 1951
Die Diatomeen von Schweden und Finnland. Kungl. Svenska Vetenskaps Akad. Handl., Fjärde Ser. 2(1): 161 pp., 6 pls.

Stephanogonia Danica

Mann, A., 1893
List of Diatomaceae from a deep-sea dredging in the Atlantic Ocean off Delaware Bay by the U. S. Fish Commission Steamer Albatross. Proc. U. S. Nat. Mus. 16:303-312.

Stephanogonia polygona

Cleve-Euler, A., 1951
Die Diatomeen von Schweden und Finnland. Kungl. Svenska Vetenskaps Akad. Handl., Fjärde Ser. 2(1): 161 pp., 6 pls.

Stephanopyxis sp.

Allen, W. E., 1938
The Templeton Crocker Expedition to the Gulf of California in 1935 - The Phytoplankton. Amer. Microsc. Soc., Trans. 57:328-335.

Stephanopyxis sp.

Allen, W.E., 1937
Plankton diatoms of the Gulf of California obtained by the G. Allan Hancock Expedition of 1936. The Hancock Pacific Expeditions, Univ. So. Calif. Publ. 3:47-59, 1 fig.

Stephanopyxis

Brunel, J., 1962
Le phytoplancton de la Baie de Chaleurs. Inst. Botan., Univ. Montréal, Contrib. No. 77: 365 pp., 66 pls.

Stephanopyxis sp.

Cupp, E.E. and Allen, W.E., 1938
Plankton diatoms of the Gulf of California obtained by Allan Hancock Pacific Expedition of 1937. The Hancock Pacific Expeditions, The Univ. So. Calif. Publ. 3: 61-74, 1 map, pls. 4-15.

Stephanopyxis

Desikachary, T.V., 1956 (1957).
Electron microscope studies on diatoms. J.R. Microsc. Soc. (3)76(1/2):9-36.

Stephanopyxis appendicula

Mann, A., 1907
Report on the diatoms of the Albatross voyages in the Pacific Ocean, 1888-1904. Contrib. U. S. Nat. Herb. 10(5):221-419, Pls. XLIV-LIV.

Stephanopyxis broschii

Cleve-Euler, A., 1951
Die Diatomeen von Schweden und Finnland. Kungl. Svenska Vetenskaps Akad. Handl., Fjärde Ser. 2(1): 161 pp., 6 pls.

Stephanopyxis californica n.sp.

Schrader, Hans-Joachim, 1973
Cenozoic diatoms from the northeast Pacific, leg 18. Initial Repts Deep Sea Drilling Proj. 18:673-797.

Stephanopyxis campana n.sp.

Castracane degli Antelminelli, F., 1886
1. Report on the Diatomaceae collected by H.M.S. Challenger during the years 1873-1876. Rept. Sci. Results, H.M.S. Challenger, Botany Vol. II, 178 pp., 30 pls.

Stephanopyxis corona

Bigelow, H.B., and M. Leslie, 1930
Reconnaissance of the waters and plankton of Monterey Bay, July 1928. Bull. M.C.Z., 70(5):429-481, 43 text figs.

Stephanopyxis corona

Mann, A., 1907
Report on the diatoms of the Albatross voyages in the Pacific Ocean, 1888-1904. Contrib. U. S. Nat. Herb. 10(5):221-419, Pls. XLIV-LIV.

Stephanopyxis costata

Braarud, T., 1962
Species distribution in marine phytoplankton.
J. Oceanogr. Soc., Japan, 20th Ann. Vol., 628-649.

Stephanopyxis costata

Carpenter, Edward J., Charles G. Remsen and Brian W. Schroeder, 1972.
Comparison of laboratory and in situ measurements of urea decomposition by a marine diatom. J. exp. mar. Biol. Ecol. 8(3): 259-264.

Stephanopyxis costata

Wawrik, F., 1961.
Die horizontale Verteilung der Planktondiatomeen im Golf von Neapel.
Int. Rev. Ges. Hydrobiol., 46(3):460-479.

Stephanopyxis cruciata

Cleve-Euler, A., 1951
Die Diatomeen von Schweden und Finnland. Kungl. Svenska Vetenskaps Akad. Handl., Fjärde Ser. 2(1): 161 pp., 6 pls.

Stephanopyxis dimorpha n.sp.

Schrader, Hans-Joachim, 1973
Cenozoic diatoms from the northeast Pacific, leg 18. Initial Repts Deep Sea Drilling Proj. 18:673-797.

Stephanopyxis ferox

Cleve-Euler, A., 1951
Die Diatomeen von Schweden und Finnland. Kungl. Svenska Vetenskaps Akad. Handl., Fjärde Ser. 2(1): 161 pp., 6 pls.

Stephanopyxis kittoniana n.sp.

Castracane degli Antelminelli, F., 1886
1. Report on the Diatomaceae collected by H.M.S. Challenger during the years 1873-1876. Rept. Sci. Results, H.M.S. Challenger, Botany Vol. II, 178 pp., 30 pls.

Stephanopyxis kulmii n.sp.

Schrader, Hans-Joachim, 1973.
Cenozoic diatoms from the northeast Pacific, leg 18. Initial Repts Deep Sea Drilling Proj. 18: 673-797.

Stephanopyxis marginata

Cleve-Euler, A., 1951
Die Diatomeen von Schweden und Finnland. Kungl. Svenska Vetenskaps Akad. Handl., Fjärde Ser. 2(1): 161 pp., 6 pls.

Stephanopyxis megapora

Cleve-Euler, A., 1951
Die Diatomeen von Schweden und Finnland. Kungl. Svenska Vetenskaps Akad. Handl., Fjärde Ser. 2(1): 161 pp., 6 pls.

Stephanopyxis melosiroides n.sp.

Cleve-Euler, A., 1951
Die Diatomeen von Schweden und Finnland. Kungl. Svenska Vetenskaps Akad. Handl., Fjärde Ser. 2(1): 161 pp., 6 pls.

Stephanopyxis minuta

Cleve-Euler, A., 1951
Die Diatomeen von Schweden und Finnland. Kungl. Svenska Vetenskaps Akad. Handl., Fjärde Ser. 2(1): 161 pp., 6 pls.

Stephanopyxis mölleri

Cleve-Euler, A., 1951
Die Diatomeen von Schweden und Finnland. Kungl. Svenska Vetenskaps Akad. Handl., Fjärde Ser. 2(1): 161 pp., 6 pls.

Stephanopyxis nipponica

Cupp, Easter E., 1943
Marine plankton diatoms of the west coast of North America. Bull. S.I.O. 5(1):1-238, 5 pls., 168 text figs.

Stephanopyxis nipponica

Gran, H. H. and E. C. Angst, 1931
Plankton diatoms of Puget Sound. Publ. Puget Sound Biol. Sta. 7:417-519, 95 text figs.

Stephanopyxis nipponica

Marukawa, H., 1921
Plankton lists and some new species of copepods, from the northern waters of Japan. Bull. Inst. Ocean., No.384, 15 pp., 3 pls., 1 chart. Monaco

Stephanopyxis nipponica

Phifer, L.D. (undated)
The occurrence and distribution of plankton diatoms in Bering Sea and Bering Strait, July 26-August 24, 1934. Report of Oceanographic cruise of U.S. Coast Guard Cutter Chelan 1934, Part II(A):1-44 (mimeographed) plus fig.1 (after Pt.B)

Stephanopyxis orbicularis

Cassie, Vivienne 1961.
Marine phytoplankton in New England waters.
Botanica marina 2 (Suppl.):54pp. 8pls.

Stephanopyxis orbicularis n.sp.

Wood, E.J. Ferguson, L.H. Crosby and Vivienne Cassie, 1959
Studies on Australian and New Zealand diatoms. III. Descriptions of further discoid species. Trans. R. Soc., N.Z., 87(3/4):211-219, pls. 15-17

Stephanopyxis Palmeriana

Allen, W.E., and E.E. Cupp, 1935
Plankton diatoms of the Java Sea. Annales du Jardin Botanique de Buitenzorg XLIV (2):101-174, figs.1-127.
(drawings of all species mentioned)

Stephanopyxis palmeriana

Boden, B.P., 1950
Some marine plankton diatoms from the west coast of South Africa. Trans. R.Soc. S. Africa. 32:321-434, 100 text figs.

Stephanopyxis Palmeriana

Chiba, T., 1949
On the distribution of the plankton in the eastern China Sea and Yellow Sea. 1. Plankton composition in the spring. J. Shimonoseki Coll. Fisheries, 1(1):57-63, 1 fig.

Stephanopyxis palmeriana

Cupp, Easter E., 1943
Marine plankton diatoms of the west coast of North America. Bull. S.I.O. 5(1):1-238, 5 pls., 168 text figs.

Stephanopyxis Palmeriana

Dangeard, P., 1927
Phytoplankton de la croisière du "Sylvana". Ann. Inst. Ocean., Monaco, n.s., 4(8):286-401, 54 text figs. (Janvier-Juin 1913).

Stephanopyxis palmeriana

Delsman, H. C., 1939.
Preliminary plankton investigations in the Java Sea. Treubia, 17:139-181, 8 maps, 41 figs.

Stephanopyxis palmeriana

Gilbert, J.Y., and W.E. Allen, 1943
The phytoplankton of the Gulf of California obtained by the "E.W. Scripps" in 1939 and 1940. J. Mar. Res. V(2):89-110, figs.30-31.

Stephanopyxis palmeriana

Gran, H. H. and E. C. Angst, 1931
Plankton diatoms of Puget Sound. Publ. Puget Sound Biol. Sta. 7:417-519, 95 text figs.

Stephanopyxis Palmeriana

Hendey, N.I., 1937
The plankton diatoms of the southern seas. Discovery Repts. 16:151-364, pls.6-13.

Stephanopyxis palmeriana

Marukawa, H., 1921
Plankton lists and some new species of copepods, from the northern waters of Japan. Bull. Inst. Ocean., No.384, 15 pp., 3 pls., 1 chart. Monaco

Stephanopyxis palmeriana

Pavillard, J., 1925
Bacillariales. Rept. on the Danish Oceangr. Exped., 1908-10 to the Mediterranean and adj. seas. Vol.II., Biol. J4:72 pp., 116 text figs.

Stephanopyxis palmeriana

Sousa e Silva, E., 1949
Diatomaceas e Dinoflagelados de Baia de Cascais. Portugaliae Acta Biol. Volume: Julio Henriques, Ser. B: 300-383, 9 pls, 2 fold-in tables.

Stephanopyxis palmeriana

Sousa e Silva, E., and J. Dos Santos-Pinto, 1948
O Plancton da Baia de S. Martinho do Porto. 1. Diatomaceas e Dinoflagelados. Bol. Soc. Portuguese de Ciencias Naturais, 16(2):134-187, 6 pls. (Trav. Sta. Biol. Mar. de Lisbonne No. 52).

Stephanopyxis rapax n.sp.

Castracane degli Antelminelli, F., 1886
1. Report on the Diatomaceae collected by H.M.S. Challenger during the years 1873-1876. Rept. Sci. Results, H.M.S. Challenger, Botany Vol. II, 178 pp., 30 pls.

Stephanopyxis trisculpta n.sp.

Mann, A., 1907
Report on the diatoms of the Albatross voyages in the Pacific Ocean, 1888-1904. Contrib. U. S. Nat. Herb. 10(5):221-419, Pls. XLIV-LIV.

Stephanopyxis tunis

Copenhagen, W. J. and L. D., 1949
Variation in the phytoplankton of Table Bay, October 1934 to October 1935. With a note on the calorific value of Chaetoceros spp. Trans. Roy. Soc. S. Africa, 32(2):113-123, 2 text figs.

Stephanopyxis turris

Bigelow, H.B., and M. Leslie, 1930
Reconnaissance of the waters and plankton of Monterey Bay, July 1928. Bull. M.C.Z., 70(5):429-481, 43 text figs.

Stephanopyxis turris (fig)

Boden, B.P., 1950
Some marine plankton diatoms from the west coast of South Africa. Trans. R. Soc. S. Africa. 32:321-434, 100 text figs.

Stephanopyxis turris

Carlucci, A.F., and Peggy M. Bowes, 1970.
Production of vitamin B_{12}, thiamine and biotin by phytoplankton. J. Phycol. 6(4): 351-357.

Stephanopyxis turris

Castracane degli Antelminelli, F., 1886
1. Report on the Diatomaceae collected by H.M.S. Challenger during the years 1873-1876. Rept. Sci. Results, H.M.S. Challenger, Botany Vol. II, 178 pp., 30 pls.

Stephanopyxis turris

Cleve-Euler, A., 1951
Die Diatomeen von Schweden und Finnland. Kungl. Svenska Vetenskaps Akad. Handl., Fjärde Ser. 2(1): 161 pp., 6 pls.

Stephanopyxis turris

Copenhagen, W. J. and L. D., 1949
Variation in the phytoplankton of Table Bay, October 1934 to October 1935. With a note on the calorific value of Chaetoceros spp. Trans. Roy. Soc. S. Africa, 32(2):113-123, 2 text figs.

Stephanopyxis turris

Cupp, Easter E., 1943
Marine plankton diatoms of the west coast of North America. Bull. S.I.O. 5(1):1-238, 5 pls., 168 text figs.

Stephanopyxis turris

Cupp, E.E., 1934
Analysis of marine diatom collections taken from the Canal Zone to California during March, 1933. Trans. Am. Micros. Soc. LIII (1):22-29, 1 map.

Stephanopyxis turris

Cupp, E.E. and Allen, W.E., 1938
Plankton diatoms of the Gulf of California obtained by Allan Hancock Pacific Expedition of 1937. The Hancock Pacific Expeditions, the Univ. So. Calif. Publ. 3: 61-74, 1 map, pls. 4-15.

Stephanopyxis turris

Drebes, G., 1964.
Über den Lebenszyklus der marinen Planktondiatomee Stephanopyxis turris (Centrales) und seine Steuerung im Experiment. Helgoländer Wiss. Meeresuntersuch., 10(1/4):153-154.

Stephanopyxis turris

Florin, M-B., 1948
9. Diatomeae in submarine cores from the Tyrrhenian Sea. Medd. Ocean. Inst., Göteborg, 15 (Göteborgs Kungl. Vetenskaps-och Viterrhets Samhälles Handlingar, Sjätte Foljden, Ser. B 5(13):80-88.

Stephanopyxis turris

Frenguelli, Joaquin, and Hector Antonio Orlando, 1959.
Operacion MERLUZA. Diatomeas y silicoflagelados del plancton del "VI Crucero". Servicio Hidrogr. Naval, Argentina, Publ. No. H. 619: 5-62.

Stephanopyxis turris

Gilbert, J.Y., and W.E. Allen, 1943
The phytoplankton of the Gulf of California obtained by the "E.W. Scripps" in 1939 and 1940. J. Mar. Res. V(2):89-110, figs. 30-31.

Stephanopyxis turris

Gran, H.H., 1908
Diatomeen. Nordisches Plankton, Botanischer Teil pp. XIX.1-XIX 146; 178 text figs.

Stephanopyxis turris

Hendy, N. Ingram, 1964
An introductory account of the smaller algae of British coastal waters. V. Bacillariophyceae (Diatoms). Her Majesty's Stationary Office, 317 pp., 45 pls.

Stephanopyxis turris

Hendey, N.I., 1937
The plankton diatoms of the southern seas. Discovery Repts. 16:151-364, pls.6-13.

Stephenopyxis turris

Holmes, Robert W., 1966.
Light microscope observations on cytological manifestations of nitrate, phosphate, and silicate deficiency in four marine centric diatoms. J. Phycology, 2(4):136-140.

Stephopyxis turris

Jorgensen, E., 1900
Protophyten und Protozoen im Plankton aus der Norwegischen Westkerste. Bergens Mus. Aarb. 1899(6): 95 pp., 5 pls., 83 tables

Stephanopyxis turris

Lillick, L.C., 1940
Phytoplankton and planktonic protozoa of the offshore waters of the Gulf of Maine. Pt.II. Qualitative Composition of the Planktonic Flora. Trans. Am. Phil. Soc., n.s., 31(3):193-237, 13 text figs.

Stephanopyxis turris

Mangin, M. L., 1912
Phytoplancton de la croisière du "René" dans l'Atlantique (Septembre 1908). Ann. Inst. Ocean., n.s., 4(1):1-66, 2 pls., 41 text figs., 2 tables.

Stephanopyxis turris

Manguin, E., 1954
Diatomees marines provenant de l'ile Heard (Australian National Antarctic Expedition). Rev. Algol., n.s., 1: 14-24.

Stephanopyxis turris

Mann, A., 1893
List of Diatomaceae from a deep-sea dredging in the Atlantic Ocean off Delaware Bay by the U. S. Fish Commission Steamer Albatross. Proc. U. S. Nat. Mus. 16:303-312.

Stephanopyxis turris

Paiva Carvalho, J., 1950
O plancton do Rio Maria Rodriques (Cananeis). 1. Diatomaceas e Dinoflagelados. Bol. Inst. Paulista Oceanogr. 1(1); 27-43, 2 fold-in tables, 2 figs.

Stephanopyxis turris

Rampi, L., 1940
Diatomee del Mare Adriatico. Nuovo Giornale Botanico Italiano, n.s., 47:559-608.

Stephanopyxis turris

Sousa e Silva, E., 1949
Diatomaceas e Dinoflagelados de Baía de Cascais. Portugaliae Acta Biol., Volume: Julio Henriques, Ser. B: 300-383, 9 pls, 2 fold-in tables.

Stephanopyxis turris

Sousa e Silva, E., and J. Dos Santos-Pinto, 1948
O Plancton da Baía de S. Martinho do Porto. 1. Diatomaceas e Dinoflagelados. Bol. Soc. Portuguese de Ciencias Naturais, 16(2):134-187, 6 pls. (Trav. Sta. Biol. Mar. de Lisbonne No. 52).

Stephanopyxis turris

Von Stosch, Hans A., und Gerhard Drebes, 1964. Entwicklungsgeschtchtliche Untersuchungen an zentrischen Diatomeen. IV. Die Planktondiatomee Stephanopyxis turris - ihre Behandlung und Entwicklungsgeschichte. Helgoländer Wiss. Meeresuntersuchungen, 11(3/4): 209-257.

Stephanopyxis vasta

Cleve-Euler, A., 1951
Die Diatomeen von Schweden und Finnland. Kungl. Svenska Vetenskaps Akad. Handl., Fjärde Ser. 2(1): 161 pp., 6 pls.

Stephanosira decussata (figs.)

Sousa e Silva, E., 1949
Diatomaceas e Dinoflagelados de Baía de Cascais. Portugaliae Acta Biol., Volume: Julio Henriques, Ser. B: 300-383, 9 pls, 2 fold-in tables.

Stictodiscus affinis n.sp.

Castracane degli Antelminelli, F., 1886
1. Report on the Diatomaceae collected by H.M.S. Challenger during the years 1873-1876. Rept. Sci. Results, H.M.S. Challenger, Botany Vol. II, 178 pp., 30 pls.

Stictodiscus affinis late-zonata n.var.

Castracane degli Antelminelli, F., 1886
1. Report on the Diatomaceae collected by H.M.S. Challenger during the years 1873-1876. Rept. Sci. Results, H.M.S. Challenger, Botany Vol. II, 178 pp., 30 pls.

Stictodiscus anceps n.sp.

Castracane degli Antelminelli, F., 1886
1. Report on the Diatomaceae collected by H.M.S. Challenger during the years 1873-1876. Rept. Sci. Results, H.M.S. Challenger, Botany Vol. II, 178 pp., 30 pls.

Stictodiscus bicoronatus n.sp.

Castracane degli Antelminelli, F., 1886
1. Report on the Diatomaceae collected by H.M.S. Challenger during the years 1873-1876. Rept. Sci. Results, H.M.S. Challenger, Botany Vol. II, 178 pp., 30 pls.

Stictodiscus bicoronatus punctigera

Castracane degli Antelminelli, F., 1886 n.var.
1. Report on the Diatomaceae collected by H.M.S. Challenger during the years 1873-1876. Rept. Sci. Results, H.M.S. Challenger, Botany Vol. II, 178 pp., 30 pls.

Stictodiscus buryanus

Mann, A., 1907
Report on the diatoms of the Albatross voyages in the Pacific Ocean, 1888-1904. Contrib. U. S. Nat. Herb. 10(5):221-419, Pls. XLIV-LIV.

Stictodiscus elegans n.sp.

Castracane degli Antelminelli, F., 1886
1. Report on the Diatomaceae collected by H.M.S. Challenger during the years 1873-1876. Rept. Sci. Results, H.M.S. Challenger, Botany Vol. II, 178 pp., 30 pls.

Stictodiscus eulensteinii

Castracane degli Antelminelli, F., 1886
1. Report on the Diatomaceae collected by H.M.S. Challenger during the years 1873-1876. Rept. Sci. Results, H.M.S. Challenger, Botany Vol. II, 178 pp., 30 pls.

Stictodiscus gelidus n.sp.

Mann, A., 1907
Report on the diatoms of the Albatross voyages in the Pacific Ocean, 1888-1904. Contrib. U. S. Nat. Herb. 10(5):221-419, Pls. XLIV-LIV.

Stictodiscus hexagonus n.sp.

Castracane degli Antelminelli, F., 1886.
1. Report on the Diatomaceae collected by H.M.S. Challenger during the years 1873-1876. Rept. Sci. Results, H.M.S. Challenger, Botany Vol. II, 178 pp., 30 pls.

Stictodiscus japonicus n.sp.
Castracane degli Antelminelli, F., 1886
  1. Report on the Diatomaceae collected by H.M.S. Challenger during the years 1873-1876. Rept. Sci. Results, H.M.S. Challenger, Botany Vol. II, 178 pp., 30 pls.

Stictodiscus johnsonianus
Cleve-Euler, A., 1951
  Die Diatomeen von Schweden und Finnland. Kungl. Svenska Vetenskaps Akad. Handl., Fjärde Ser. 2(1): 161 pp., 6 pls.

Stictodiscus johnsonianus
Mann, A., 1907
  Report on the diatoms of the Albatross voyages in the Pacific Ocean, 1888-1904. Contrib. U. S. Nat. Herb. 10(5):221-419, Pls. XLIV-LIV.

Stictodiscus kittonianus
Cleve-Euler, A., 1951
  Die Diatomeen von Schweden und Finnland. Kungl. Svenska Vetenskaps Akad. Handl., Fjärde Ser. 2(1): 161 pp., 6 pls.

Stictodiscus kittonianus
Mann, A., 1907
  Report on the diatoms of the Albatross voyages in the Pacific Ocean, 1888-1904. Contrib. U. S. Nat. Herb. 10(5):221-419, Pls. XLIV-LIV.

Stictodiscus margaritaceus
Castracane degli Antelminelli, F., 1886
  1. Report on the Diatomaceae collected by H.M.S. Challenger during the years 1873-1876. Rept. Sci. Results, H.M.S. Challenger, Botany Vol. II, 178 pp., 30 pls.

Stictodiscus morsianus
Cleve-Euler, A., 1951
  Die Diatomeen von Schweden und Finnland. Kungl. Svenska Vetenskaps Akad. Handl., Fjärde Ser. 2(1): 161 pp., 6 pls.

Stictodiscus radiondianus n.sp.
Castracane degli Antelminelli, F., 1886
  1. Report on the Diatomaceae collected by H.M.S. Challenger during the years 1873-1876. Rept. Sci. Results, H.M.S. Challenger, Botany Vol. II, 178 pp., 30 pls.

Stictodiscus radiatus n.sp.
Castracane degli Antelminelli, F., 1886
  1. Report on the Diatomaceae collected by H.M.S. Challenger during the years 1873-1876. Rept. Sci. Results, H.M.S. Challenger, Botany Vol. II, 178 pp., 30 pls.

Stictodiscus reticulatus n.sp.
Castracane degli Antelminelli, F., 1886
  1. Report on the Diatomaceae collected by H.M.S. Challenger during the years 1873-1876. Rept. Sci. Results, H.M.S. Challenger, Botany Vol. II, 178 pp., 30 pls.

Stictodiscus trigonus n.sp.
Castracane degli Antelminelli, F., 1886
  1. Report on the Diatomaceae collected by H.M.S. Challenger during the years 1873-1876. Rept. Sci. Results, H.M.S. Challenger, Botany Vol. II, 178 pp., 30 pls.

Stictodiscus trigonus
Cleve-Euler, A., 1951
  Die Diatomeen von Schweden und Finnland. Kungl. Svenska Vetenskaps Akad. Handl., Fjärde Ser. 2(1): 161 pp., 6 pls.

Stictodiscus varians n.sp.
Castracane degli Antelminelli, F., 1886
  1. Report on the Diatomaceae collected by H.M.S. Challenger during the years 1873-1876. Rept. Sci. Results, H.M.S. Challenger, Botany Vol. II, 178 pp., 30 pls.

Streptotheca sp
Chin, T.G., C.F. Chen, S.C. Liu and S.S. Wu, 1965.
  Influence of temperature and salinity on the growth of three plankton diatom species. (In Chinese; English abstract). (Not seen). Oceanol. et Limnol. Sinica, 7(4):373-384.

Streptotheca indica
Allen, W.E., and E.E. Cupp, 1935
  Plankton diatoms of the Java Sea. Annales du Jardin Botanique de Buitenzorg XLIV (2):101-174, figs.1-127.
  (drawings of all species mentioned)

Streptotheca indica
Delsman, H. C., 1939.
  Preliminary plankton investigations in the Java Sea. Treubia, 17:139-181, 8 maps, 41 figs.

Streptotheca thamesis
Boden, B.P., 1950
  Some marine plankton diatoms from the west coast of South Africa. Trans. R.Soc. S. Africa. 32:321-434, 100 text figs.

Streptotheca thamesis
Cupp, Easter E., 1943
  Marine plankton diatoms of the west coast of North America. Bull. S.I.O. 5(1):1-238, 5 pls., 168 text figs.

Streptotheca thamesis
Gran, H.H., 1908
  Diatomeen. Nordisches Plankton, Botanischer Teil pp. XIX.1-XIX 146; 178 text figs.

Streptotheca thamesis
Gran, H.H., and T. Braarud, 1935
  A quantitative study of the phytoplankton in the Bay of Fundy and the Gulf of Maine (including observations on hydrography, chemistry, and turbidity). J. Biol. Bd., Canada, 1(5):279-467, 69 text figs.

Streptotheca tamesis
Hendy, N. Ingram, 1964
  An introductory account of the smaller algae of British coastal waters. V. Bacillariophyceae (Diatoms). Her Majesty's Stationary Office, 317 pp., 45 pls.

Streptotheca thamesis
Hendey, N.I., 1937
  The plankton diatoms of the southern seas. Discovery Repts. 16:151-364, pls.6-13.

Streptotheca thamesis
Lafon, M., M. Durchon and Y. Saudray, 1955.
  Recherches sur les cycles saisonnières du plankton. Ann. Inst. Océan., 31(3):125-230.

Streptotheca thamesis
Lillick, L.C., 1940
  Phytoplankton and planktonic protozoa of the offshore waters of the Gulf of Maine. Pt.II. Qualitative Composition of the Planktonic Flora. Trans. Am. Phil. Soc., n.s., 31(3):193-237, 13 text figs.

Streptotheca tamesis
Meunier, A., 1915
  Microplancton de la Mer Flamande. 2. Diatomées (excepté le genre Chaetoceros). Mem. Mus. Roy. Hist. Nat., Belgique, 7(3):1-118, Pls. VIII-XIV.

Streptotheca thamesis
Pavillard, J., 1925
  Bacillariales. Rept. on the Danish Oceangr. Exped., 1908-10 to the Mediterranean and adj. seas. Vol.II, Biol. J4:72 pp., 116 text figs.

Streptotheca thamesis
Schodduyn, M., 1926
  Observations faites dans la baie d'Ambleteuse (Pas de Calais). Bull. Inst. Ocean., Monaco, No. 482: 64 pp.

Streptotheca thamesis
Sousa e Silva, E., and J. Dos Santos-Pinto, 1948
  O Plancton da Baía de S. Martinho do Porto. 1. Diatomaceas e Dinoflagelados. Bol. Soc. Portuguese de Ciencias Naturais, 16(2):134-187, 6 pls. (Trav. Sta. Biol. Mar. de Lisbonne No. 52).

Streptotheca thamesis
Sousa e Silva, E., 1949
  Diatomaceas e Dinoflagelados de Baía de Cascais. Portugaliae Acta Biol., Volume: Julio Henriques, Ser. B: 300-383, 9 pls, 2 fold-in tables.

Streptotheca thamesis
Sukhanova, I.N., 1964.
  The phytoplankton of the northeastern part of the Indian ocean in the season of the southwest monsoon. Regularity of the distribution of the oceanic plankton. Trudy Oceanol. Inst., Akad. Nauk, SSSR, 65:24-31.

Striatella
Desikachary, T.V., 1956(1957).
  Electron microscope studies on diatoms. J. R. Microsc. Soc. (3)76(1/2):9-36.

Striatella sp.
Meunier, A., 1915
  Microplancton de la Mer Flamande. 2. Diatomées (excepté le genre Chaetoceros). Mem. Mus. Roy. Hist. Nat., Belgique, 7(3):1-118, Pls. VIII-XIV.

Striatella crumena
Hendey, N.I., 1951
  Littoral diatoms of Chicester Harbour with special reference to fouling. J.Roy. Microscop. Soc. 71(1): 1-86, 18 pls.

Striatella delicatula
Cupp, Easter E., 1943
  Marine plankton diatoms of the west coast of North America. Bull. S.I.O. 5(1):1-238, 5 pls., 168 text figs.

Striatella delicatula
Hendy, N. Ingram, 1964
  An introductory account of the smaller algae of British coastal waters. V. Bacillariophyceae (Diatoms). Her Majesty's Stationary Office, 317 pp., 45 pls.

Striatella delicatula
Hendey, N.I., 1951
  Littoral diatoms of Chicester Harbour with special reference to fouling. J.Roy. Microscop. Soc. 71(1): 1-86, 18 pls.

Striatella delicatula
Zanon, V., 1948
  Diatomee marini di Sardegna e Pugillo di Alghe Marine della stressa. Boll. Pesca, Piscitutura e Idrobiologia, Anno 24, ns. 3(2): 202-244, 27 figs. on 1 pl.

Striatella intermedia
Grunow, A., 1877
  1. New Diatoms from Honduras. Monthly Micros. Jour., 18:165-186, pls. CXCIII-CXCVI.

Striatella interrupta
Ercegovic, A., 1936
  Etudes qualitative et quantitatives du phytoplancton dans les eaux cotières de l'Adriatique oriental moyen au cours de l'année 1934. Acta Adriatica 1(9):1-126

Striatella interrupta
Pavillard, J., 1925
  Bacillariales. Rept. on the Danish Oceangr. Exped., 1908-10 to the Mediterranean and adj. seas. Vol.II, Biol. J4:72 pp., 116 text figs.

Striatella interrupta

Politis, J., 1949
Diatomees marines de Bosphores et des ibes de la mer de Marmara. II Practica tou Hellenikou Hidrobiologikou Institutoutou 1929, Etoz 1929, 3(1):11-31.

Striatella interrupta

Rampi, L., 1942
Ricerche sul fitoplancton del Mare Ligure 6. Le diatomee delle acque di Sanremo. Nuovo Giornale Botanico Italiano, N.S., 49:252-268.

Striatella interrupta

Rampi, L., 1940
Diatomee del Mare Adriatico. Nuovo Giornale Botanico Italiano, n.s., 47:559-608.

Striatella interrupta

Zanon, V., 1948
Diatomee marini di Sardegna e Pugillo di Alghe Marine della stressa. Boll. Pesca, Piscitutura e Idrobiologia, Anno 24, ns. 3(2): 202-244, 27 figs. on 1 pl.

Striatella Lindigiana

Grunow, A., 1877
1. New Diatoms from Honduras. Monthly Micros. Jour., 18:165-186, pls. CXCIII-CXCVI.

Striatella ovata

Hendey, N.I., 1951
Littoral diatoms of Chicester Harbour with special reference to fouling. J.Roy. Microscop. Soc. 71(1): 1-86, 18 pls.

Striatella salina

Hendey, N.I., 1951
Littoral diatoms of Chicester Harbour with special reference to fouling. J.Roy. Microscop. Soc. 71(1): 1-86, 18 pls.

Striatella unipunctata

Brunel, J., 1962
Le phytoplancton de la Baie de Chaleurs. Inst. Botan., Univ. Montréal, Contrib. No. 77: 365 pp., 66 pls.

Striatella unipunctata

Brunel, Jules, 1962
Le phytoplancton de la Baie des Chaleurs. Contrib. Ministère de la Chasse et des Pêcheries, Province de Québec, No. 91: 365 pp.

Striatella unipunctata

Cupp, Easter E., 1943
Marine plankton diatoms of the west coast of North America. Bull. S.I.O. 5(1):1-238, 5 pls., 168 text figs.

Striatella unipunctata

Ercegovic, A., 1936
Etudes qualitative et quantitatives du phytoplancton dans les eaux cotières de l'Adriatique oriental moyen au cours de l'année 1934. Acta Adriatica 1(9):1-126

Striatella unipunctata

Forti, A., 1922
Ricerche sulla flora pelagica (fitoplancton) di Quarto dei Mille. Mem. R. Com. Talass. Ital. 97:248 pp., 13 pls.

Striatella unipunctata

Hendy, N. Ingram, 1964
An introductory account of the smaller algae of British coastal waters. V. Bacillariophyceae (Diatoms). Her Majesty's Stationary Office, 317 pp., 45 pls.

Striatella unipunctata

Hustedt, F. and A.A. Aleem, 1951
Littoral diatoms from the Salstone near Plymouth. JMBA 30(1): 177.196.

Striatella unipunctata

Jørgensen, E., 1905
B.Protistplankton and the diatoms in bottom samples. Hydrographical and biological investigations in Norwegian fjords. Bergens Mus. Skr. 7: 49-225.

Striatella unipunctata

Jorgensen, E., 1900
Protophyten und Protozoen im Plankton aus der Norwegischen Westkerste. Bergens Mus. Aarb. 1899(6): 95 pp., 5 pls., 83 tables.

Striatella unipunctata

Lillick, L.C., 1937
Seasonal studies of the phytoplankton off Woods Hole, Massachusetts. Biol. Bull. LXXIII (3):488-503, 3 text figs.

Striatella unipunctata

Massuti Algamora, M., 1949
Estudio de diez y seis muestras de plancton del Golfo de Nápoles. Publ. Inst. Biol. Appl. 5:85-94, 1 fold-in table.

Striatella unipunctata

Pavillard, J., 1925
Bacillariales. Rept. on the Danish Oceangr. Exped., 1908-10 to the Mediterranean and adj. seas, Vol.II., Biol. J4:72 pp., 116 text figs.

Striatella unipunctata

Pavillard, J., 1905
Recherches sur la flore pelagique (Phytoplankton) de l'Etang de Thau. Theses presentees a la Fac. Sci., Paris, 116 pp., 3 pls.

Striatella unipunctata

Politis, J., 1949
Diatomees marines de Bosphores et des ibes de la mer de Marmara. II Practica tou Hellenikou Hidrobiologikou Institutoutou 1929, Etoz 1929, 3(1):11-31.

Striatella unipunctata

Rampi, L., 1942
Ricerche sul fitoplancton del Mare Ligure 6. Le diatomee delle acque di Sanremo. Nuovo Giornale Botanico Italiano, N.S., 49:252-268.

Striatella unipunctata

Rampi, L., 1940
Diatomee del Mare Adriatico. Nuovo Giornale Botanico Italiano, n.s., 47:559-608.

Striatella unipunctata (figs.)

Sousa e Silva, E., 1949
Diatomaceas e Dinoflagelados de Baia de Cascais. Portugaliae Acta Biol., Volume: Julio Henriques, Ser. B: 300-383, 9 pls, 2 fold-in tables.

Striatella unipunctata

Takano, Hideaki, 1963.
Notes on marine littoral diatoms of Japan. Bull. Tokai Reg. Fish. Res. Lab., No. 36:1-8.

Striatella unipunctata

Zanon, V., 1948
Diatomee marini di Sardegna e Pugillo di Alghe Marine della stressa. Boll. Pesca, Piscitutura e Idrobiologia, Anno 24, ns. 3(2): 202-244, 27 figs. on 1 pl.

Surirella sp.

Cupp, E.E. and Allen, W.E., 1938
Plankton diatoms of the Gulf of California obtained by Allan Hancock Pacific Expedition of 1937. The Hancock Pacific Expeditions, The Univ. So. Calif. Publ. 3: 61-74, 1 map, pls.4-15.

Surirella

Desikachary, T.V., 1956(1957).
Electron microscope studies on diatoms. J. R. Microsc. Soc. (3)76(1/2):9-36.

Surirella sp.

Gilbert, J.Y., and W.E. Allen, 1943
The phytoplankton of the Gulf of California obtained by the "E.W. Scripps" in 1939 and 1940. J. Mar. Res. V(2):89-110, figs.30-31.

Surirella sp.

Levander, K.M., 1947
Plankton gesammelt in den Jahren 1899-1910 an den Küsten Finnlands. Finnländische Hydrographisch-Biologische Untersuchunger (aus dem Wasserbiologischen Laboratorin der Societas Scientiarum Fennica) No.11:40 pp., 6 diagrams, 13 pls., tables.

Surirella (Podocystis) adriatica

de Sousa e Silva, E., 1956.
Contribution à l'étude du microplancton de Dakar et des regions maritimes voisines. Bull. I.F.A.N., 8(2):335-371, 7 pls.

Surirella amoricana

Hendy, N. Ingram, 1964
An introductory account of the smaller algae of British coastal waters. V. Bacillariophyceae (Diatoms). Her Majesty's Stationary Office, 317 pp., 45 pls.

Surirella anceps

Morse, D.C., 1947
Some observations on seasonal variations in plankton population Patuxent River, Maryland 1943 1945. Bd. Nat. Res., Publ. No.65, Chesapeake Biol. Lab., 31, 3 figs.

Surirella angusta

Tsumura, K., 1956.
Diatomoj el la cirkaufoso de la restajo de la kastele de Odawara. J. Yokohama Municipal Univ., (C-14), No. 47:23 pp.

Surirella angusta

Zanon, D. V., 1949
Diatomee di Buenos Aires (Argentina) Atti Accad. Naz, Lincei, Memorie, Cl. Sci. fis., mat. e. nat., ser. 7, 11(3):59-151, 2 pls.

Surirella apiculata

Tsumura, K., 1956.
Diatomoj el la cirkaufoso de la restajo de la kastele de Odawara. J. Yokohama Municipal Univ., (C-14), No. 47:23 pp.

Surirella arachnoidea n.sp.

Wood, E.J. Ferguson, 1963.
Studies on Australian and New Zealand diatoms. VI. Tropical and subtropical species. Trans. R. Soc., New Zealand, 2(15):189-218.

Surirella argus n.sp.

Castracane degli Antelminelli, F., 1886
1. Report on the Diatomaceae collected by H.M.S. Challenger during the years 1873-1876. Rept. Sci. Results, H.M.S. Challenger, Botany Vol. II, 178 pp., 30 pls.

Surirella armoricana

de Sousa e Silva, E., 1956.
Contribution à l'étude du microplancton de Dakar et des regions maritimes voisines. Bull. I.F.A.N., 8(2):335-371, 7 pls.

Surirella bifrons

Mann, A., 1907
Report on the diatoms of the Albatross voyages in the Pacific Ocean, 1888-1904. Contrib. U. S. Nat. Herb. 10(5):221-419, Pls. XLIV-LIV.

Surirella biseriata

de Sousa e Silva, E., 1956.
Contribution à l'étude du microplancton de Dakar et des regions maritimes voisines. Bull. I.F.A.N., 8(2):335-371, 7 pls.

Surirella biseriata

Smith, W., 1851
Notes on the Diatomaceae; with descriptions of British species included in the genera Campylodiscus, Surirella, and Cymatopleura. Ann. Mag. Nat. Hist., 2nd ser., 7:1-14, 3 pls.

Surirella biseriata

Zanon, D. V., 1949
Diatomee di Buenos Aires (Argentina) Atti Accad. Naz. Lincei, Memorie, Cl. Sci. fis., mat. e. nat., ser. 7, 11(3):59-151, 2 pls.

Surirella biseriata bifrons

Tsumura, K., 1956.
Diatomoj el la cirkaufoso de la restajo de la kastele de Odawara. J. Yokohama Municipal Univ., (C-14), No. 47:23 pp.

Surirella capronii

Kokubo, S., and S. Sato, 1947
Plankton in Jū-San Gata. Physiol. and Ecol. (Japan) 1(4):1-16, 3 text figs., tables.

Surirella capronii

Zanon, D. V., 1949
Diatomee di Buenos Aires (Argentina) Atti Accad. Naz. Lincei, Memorie, Cl. Sci. fis., mat. e. nat., ser. 7, 11(3):59-151, 2 pls.

Surirella caspia

Hustedt, F. and A.A. Aleem, 1951
Littoral diatoms from the Salstone near Plymouth. JMBA 30(1): 177.196.

Surirella comis

Hendy, N. Ingram, 1964
An introductory account of the smaller algae of British coastal waters. V. Bacillariophyceae (Diatoms). Her Majesty's Stationary Office, 317 pp., 45 pls.

Surirella comis

Rampi, L., 1940
Diatomee del Mare Adriatico. Nuovo Giornale Botanico Italiano, n.s., 47:559-608.

Surirella craticula

Smith, W., 1851
Notes on the Diatomaceae; with descriptions of British species included in the genera Campylodiscus, Surirella, and Cymatopleura. Ann. Mag. Nat. Hist., 2nd ser., 7:1-14, 3 pls.

Surirella crumena

Hendy, N. Ingram, 1964
An introductory account of the smaller algae of British coastal waters. V. Bacillariophyceae (Diatoms). Her Majesty's Stationary Office, 317 pp., 45 pls.

Surirella delicatissima

Zanon, D. V., 1949
Diatomee di Buenos Aires (Argentina) Atti Accad. Naz. Lincei, Memorie, Cl. Sci. fis., mat. e. nat., ser. 7, 11(3):59-151, 2 pls.

Surirella dives n.sp.

Castracane degli Antelminelli, F., 1886
1. Report on the Diatomaceae collected by H.M.S. Challenger during the years 1873-1876. Rept. Sci. Results, H.M.S. Challenger, Botany Vol. II, 178 pp., 30 pls.

Surirella elegans

Iyengar, M.O.P. and G.Venkataraman,1951.
The ecology and seasonal succession of the algae flora of the River Cooum at Madras with special reference to the Diatomaceae. J. Madras Univ. 21, Sect. B(1): 140-192; 1 pl of 4 figs., 11 text figs.

Surirella elegans

Rumkówna, A., 1948
List of the phytoplankton species occurring in the superficial water layers in the Gulf of Gdańsk. Bull. Lab. mar., Gdynia, No. 4: 139-141 with tables in back.

Surirella elegans

Zanon, D. V., 1949
Diatomee di Buenos Aires (Argentina) Atti Accad. Naz. Lincei, Memorie, Cl. Sci. fis., mat. e. nat., ser. 7, 11(3):59-151, 2 pls.

Surirella fastuosa

Eskinazi Enide e Shigekatsv Satô (1963-1964) 1966.
Contribuição ao estudo das diatomaceas da Praia de Piedade.
Trabhs Inst. Oceanogr., Univ. Recife, 5 (5/6): 73-114.

Surirella fastuosa

Hendy, N. Ingram, 1964
An introductory account of the smaller algae of British coastal waters. V. Bacillariophyceae (Diatoms). Her Majesty's Stationary Office, 317 pp., 45 pls.

Surirella fastuosa

Hendey, N.I., 1951
Littoral diatoms of Chicester Harbour with special reference to fouling. J. Roy. Microscop. Soc. 71(1): 1-86, 18 pls.

Surirella fastuosa

Hustedt, F. and A.A. Aleem, 1951
Littoral diatoms from the Salstone near Plymouth. JMBA 30(1): 177.196.

Surirella fastuosa

Jørgensen, E., 1905
B.Protistplankton and the diatoms in bottom samples. Hydrographical and biological investigations in Norwegian fiords. Bergens Mus. Skr. 7: 49-225.

Surirella fastuosa

Jorgensen, E., 1900
Protophyten und Protozoën im Plankton aus der Norwegischen Westkerste. Bergens Mus. Aarb. 1899(6): 95 pp., 5 pls., 83 tables.

Surirella fastuosa

Mann, A., 1907
Report on the diatoms of the Albatross voyages in the Pacific Ocean, 1888-1904. Contrib. U. S. Nat. Herb. 10(5):221-419, Pls. XLIV-LIV.

Surirella fastuosa

Politis, J., 1949
Diatomees marines de Bosphores et des ibes de la mer de Marmara. II Practica tou Hellenikou Hidrobiologikou Institutoutou 1929, Etoz 1929, 3(1):11-31.

Surirella fastuosa

Rampi, L., 1942
Ricerche sul fitoplancton del Mare Ligure 6. Le diatomee delle acque di Sanremo. Nuovo Giornale Botanico Italiano, N.S., 49:252-268.

Surirella fastuosa

Morse, D.C., 1947
Some observations on seasonal variations in plankton population Patuxant River, Maryland 1943-1945. Bd. Nat. Res., Publ. No.65, Chesapeake Biol. Lab., 31, 3 figs.

Surirella fastuosa

Rampi, L., 1940
Diatomee del Mare Adriatico. Nuovo Giornale Botanico Italiano, n.s., 47:559-608.

Surirella fastuosa

Schodduyn, M., 1926
Observations faites dans la baie d'Ambleteuse (Pas de Calais). Bull. Inst. Ocean., Monaco, No. 482: 64 pp.

Surirella fastuosa

Smith, W., 1851
Notes on the Diatomaceae; with descriptions of British species included in the genera Campylodiscus, Surirella, and Cymatopleura. Ann. Mag. Nat. Hist., 2nd ser., 7:1-14, 3 pls.

Surirella fastuosa

Zanon, V., 1948
Diatomee marini di Sardegna e Pugillo di Alghe Marine della stressa. Boll. Pesca, Piscitutura e Idrobiologia, Anno 24, ns. 3(2): 202-244, 27 figs. on 1 pl.

Surirella fastuosa recedens

Cupp, Easter E., 1943
Marine plankton diatoms of the west coast of North America. Bull. S.I.O. 5(1):1-238, 5 pls., 168 text figs.

Surirella febigerii

Eskinazi Enide e Shigekatsv Satô (1963-1964) 1966.
Contribuição ao estudo das diatomaceas da Praia de Piedade.
Trabhs Inst. Oceanogr., Univ. Recife, 5 (5/6): 73-114.

Surirella fluminensis

Allen, W.E., and E.E. Cupp, 1935
Plankton diatoms of the Java Sea. Annales du Jardin Botanique de Buitenzorg XLIV (2):101-174, figs.1-127.
(drawings of all species mentioned)

Surirella fluminensis

Morse, D.C., 1947
Some observations on seasonal variations in plankton population Patuxant River, Maryland 1943-1945. Bd. Nat. Res., Publ. No.65, Chesapeake Biol. Lab., 31, 3 figs.

Surirella fluminensis

Rampi, L., 1942
Ricerche sul fitoplancton del Mare Ligure 6. Le diatomee delle acque di Sanremo. Nuovo Giornale Botanico Italiano, N.S., 49:252-268.

Surirella fluminensis

Rampi, L., 1940
Diatomee del Mare Adriatico. Nuovo Giornale Botanico Italiano, n.s., 47:559-608.

Surirella Fluminensis

Zanon, V., 1948
Diatomee marini di Sardegna e Pugillo di Alghe Marine della stressa. Boll. Pesca, Piscitutura e Idrobiologia, Anno 24, ns. 3(2): 202-244, 27 figs. on 1 pl.

Surirella formosa

Mann, A., 1907
Report on the diatoms of the Albatross voyages in the Pacific Ocean, 1888-1904. Contrib. U. S. Nat. Herb. 10(5):221-419, Pls. XLIV-LIV.

Surirella gemma
Allen, W.E., and E.E. Cupp, 1935
Plankton diatoms of the Java Sea.
Annales du Jardin Botanique de Buitenzorg
XLIV (2):101-174, figs.1-127.
(drawings of all species mentioned)

Surirella gemma
Hendy, N. Ingram, 1964
An introductory account of the smaller algae of British coastal waters. V. Bacillariophyceae (Diatoms).
Her Majesty's Stationary Office, 317 pp., 45 pls.

Surirella gemma
Hendey, N.I., 1951
Littoral diatoms of Chicester Harbour with special reference to fouling. J.Roy. Microscop. Soc. 71(1): 1-86, 18 pls.

Surirella gemma
Hopkins, J.T., 1966.
The role of water in the behaviour of an estuarine mud-flat diatom.
Jour. mar. biol. Assoc., U.K. 46(3):617-626.

Surirella gemma
Hustedt, F. and A.A. Aleem, 1951
Littoral diatoms from the Salstone near Plymouth. JMBA 30(1): 177.196.

Surirella gemma
Jorgensen, E., 1900
Protophyten und Protozoën im Plankton aus der Norwegischen Westkerste. Bergens Mus. Aarb. 1899(6): 95 pp., 5 pls., 83 tables.

Surirrella Gemma?
Lillick, L.C., 1937
Seasonal studies of the phytoplankton off Woods Hole, Massachusetts. Biol. Bull. LXXIII (3):488-503, 3 text figs.

Surirella gemma
Meunier, A., 1915
Microplancton de la Mer Flamande. 2. Diatomées (excepté le genre Chaetoceros). Mem. Mus. Roy. Hist. Nat., Belgique, 7(3):1-118, Pls. VIII-XIV.

Surirella gemma
Morse, D.C., 1947
Some observations on seasonal variations in plankton population Patuxent River, Maryland 1943-1945. Bd. Nat. Res., Publ. No.65, Chesapeake Biol. Lab., 31, 3 figs.

Surirella gemma
Paiva Carvalho, J., 1950
O plancton do Rio Maria Rodriques (Cananeis). 1. Diatomaceas e Dinoflagelados. Bol. Inst. Paulista Oceanogr. 1(1): 27-43, 2 fold-in tables, 2 figs.

Surirella gemma
Pavillard, J., 1925
Bacillariales. Rept. on the Danish Oceanogr. Exped., 1908-10 to the Mediterranean and adj. seas. Vol.II., Biol. J4:72 pp., 116 text figs.

Surirella gemma
Pavillard, J., 1905
Recherches sur la flore pelagique (Phytoplankton) de l'Etang de Thau. Theses presentees a la Fac. Sci., Paris, 116 pp., 3 pls.

Surirella gemma
Rampi, L., 1940
Diatomee del Mare Adriatico. Nuovo Giornale Botanico Italiano, n.s., 47:559-608.

Surirella gemma
Smith, W., 1851
Notes on the Diatomaceae; with descriptions of British species included in the genera Campylodiscus, Surirella, and Cymatopleura. Ann. Mag. Nat. Hist., 2nd ser., 7:1-14, 3 pls.

Surirella gemma
Sousa a Silva, E., and J. Dos Santos-Pinto, 1948
O Plancton da Baia de S. Martinho do Porto. 1. Diatomaceas e Dinoflagelados. Bol. Soc. Portuguese de Ciencias Naturais, 16(2):134-187, 6 pls. (Trav. Sta. Biol. Mar. de Lisbonne No. 52).

Surirella gracilis
Zanon, D. V., 1949
Diatomee di Buenos Aires (Argentina) Atti Accad. Naz. Lincei, Memorie, Cl. Sci. fis., mat. e. nat., ser. 7, 11(3):59-151, 2 pls.

Surirella granduiscula n.sp.
Castracane degli Antelminelli, F., 1886
1. Report on the Diatomaceae collected by H.M.S. Challenger during the years 1873-1876. Rept. Sci. Results, H.M.S. Challenger, Botany Vol. II, 178 pp., 30 pls.

Surirella guatimalensis
Zanon, D. V., 1949
Diatomee di Buenos Aires (Argentina) Atti Accad. Naz. Lincei, Memorie, Cl. Sci. fis., mat. e. nat., ser. 7, 11(3):59-151, 2 pls.

Surirella linearis
Zanon, D. V., 1949
Diatomee di Buenos Aires (Argentina) Atti Accad. Naz. Lincei, Memorie, Cl. Sci. fis., mat. e. nat., ser. 7, 11(3):59-151, 2 pls.

Surirella guinardii
de Sousa e Silva, E., 1956.
Contribution à l'étude du microplancton de Dakar et des regions maritimes voisines. Bull. I.F.A.N., 8(2):335-371, 7 pls.

Surirella hispida
Hendy, N. Ingram, 1964
An introductory account of the smaller algae of British coastal waters. V. Bacillariophyceae (Diatoms).
Her Majesty's Stationary Office, 317 pp., 45 pls.

Surirella hispida, n. sp.
Ross, R., and G. Abdin, 1949.
Notes on some diatoms from Norfolk. J. Roy. Micr. Soc., ser. 3, 69(4):225-230, 4 figs. on 1 pl.

Surirella hybrida
Rampi, L., 1940
Diatomee del Mare Adriatico. Nuovo Giornale Botanico Italiano, n.s., 47:559-608.

Surirella hybrida (figs.)
Sousa e Silva, E., 1949
Diatomaceas e Dinoflagelados de Baia de Cascais. Portugaliae Acta Biol., Volume: Julio Henriques, Ser. B: 300-383, 9 pls, 2 fold-in tables.

Surirella hybrida
Zanon, V., 1948
Diatomee marini di Sardegna e Pugillo di Alghe Marine della stressa. Boll. Pesca, Piscitutura e Idrobiologia, Anno 24, ns. 3(2): 202-244, 27 figs. on 1 pl.

Surirella intercedens
Rampi, L., 1942
Ricerche sul fitoplancton del Mare Ligure 6. Le diatomee delle acque di Sanremo. Nuovo Giornale Botanico Italiano, N.S., 49:252-268.

Surirella intercedens
Rampi, L., 1940
Diatomee del Mare Adriatico. Nuovo Giornale Botanico Italiano, n.s., 47:559-608.

Surirella japonica n.sp.
Castracane degli Antelminelli, F., 1886
1. Report on the Diatomaceae collected by H.M.S. Challenger during the years 1873-1876. Rept. Sci. Results, H.M.S. Challenger, Botany Vol. II, 178 pp., 30 pls.

Surirella lata
Jørgensen, E., 1905
B.Protistplankton and the diatoms in bottom samples. Hydrographical and biological investigations in Norwegian fiords. Bergens Mus. Skr. 7: 49-225.

Surirella lata
Rampi, L., 1942
Ricerche sul fitoplancton del Mare Ligure 6. Le diatomee delle acque di Sanremo. Nuovo Giornale Botanico Italiano, N.S., 49:252-268.

Surirella lata var. robusta
Rampi, L., 1940
Diatomee del Mare Adriatico. Nuovo Giornale Botanico Italiano, n.s., 47:559-608.

Surirella linearis
Rumkówna, A., 1948
[List of the phytoplankton species occurring in the superficial water layers in the Gulf of Gdańsk] Bull. Lab. mar., Gdynia, No. 4: 139-141 with tables in back.

Surirella linearis
Tsumura, K., 1956.
Diatomoj el la cirkaufoso de la restajo de la kastele de odawara. J. Yokohama Municipal Univ., (C-14), No. 47:23 pp.

Surirella minima
Hendy, N. Ingram, 1964
An introductory account of the smaller algae of British coastal waters. V. Bacillariophyceae (Diatoms).
Her Majesty's Stationary Office, 317 pp., 45 pls.

Surirella minima, n. sp.
Ross, R., and G. Abdin, 1949.
Notes on some diatoms from Norfolk. J. Roy. Micr. Soc., ser. 3, 69(4):225-230, 4 figs. on 1 pl.

Surirella minuta
Mann, A., 1893
List of Diatomaceae from a deep-sea dredging in the Atlantic Ocean off Delaware Bay by the U. S. Fish Commission Steamer Albatross. Proc. U. S. Nat. Mus. 16:303-312.

Surirella minuta
Smith, W., 1851
Notes on the Diatomaceae; with descriptions of British species included in the genera Campylodiscus, Surirella, and Cymatopleura. Ann. Mag. Nat. Hist., 2nd ser., 7:1-14, 3 pls.

Surirella Mölleriana
Zanon, D. V., 1949
Diatomee di Buenos Aires (Argentina) Atti Accad. Naz. Lincei, Memorie, Cl. Sci. fis., mat. e. nat., ser. 7, 11(3):59-151, 2 pls.

Surirella multicostata n.sp.
Castracane degli Antelminelli, F., 1886
1. Report on the Diatomaceae collected by H.M.S. Challenger during the years 1873-1876. Rept. Sci. Results, H.M.S. Challenger, Botany Vol. II, 178 pp., 30 pls.

Surirella Neumeyeri
Rampi, L., 1940
 Diatomee del Mare Adriatico. Nuovo Giornale Botanico Italiano, n.s., 47:559-608.

Surirella ocellata n.sp.
Castracane degli Antelminelli, F., 1886
 l. Report on the Diatomaceae collected by H.M.S. Challenger during the years 1873-1876. Rept. Sci. Results, H.M.S. Challenger, Botany Vol. II, 178 pp., 30 pls.

Surirella ovalis
Hendy, N. Ingram, 1964
 An introductory account of the smaller algae of British coastal waters. V. Bacillariophyceae (Diatoms). Her Majesty's Stationary Office, 317 pp., 45 pls.

Surirella ovalis
Hustedt, F. and A.A. Aleem, 1951
 Littoral diatoms from the Salstone near Plymouth. JMBA 30(1): 177.196.

Surirella ovalis
Mann, A., 1893
 List of Diatomaceae from a deep-sea dredging in the Atlantic Ocean off Delaware Bay by the U. S. Fish Commission Steamer Albatross. Proc. U. S. Nat. Mus. 16:303-312.

Surirella ovalis
Meunier, A., 1915
 Microplancton de la Mer Flamande. 2. Diatomées (excepté le genre Chaetoceros). Mem. Mus. Roy. Hist. Nat., Belgique, 7(3):1-118, Pls. VIII-XIV.

Surirella ovalis
Zanon, D. V., 1949
 Diatomee di Buenos Aires (Argentina) Atti Accad. Naz. Lincei, Memorie, Cl. Sci. fis., mat. e. nat., ser. 7, 11(3):59-151, 2 pls.

Surirella ovalis
Zanon, V., 1948
 Diatomee marini di Sardegna e Pugillo di Alghe Marine della stressa. Boll. Pesca, Piscitutura e Idrobiologia, Anno 24, ns. 3(2): 202-244, 27 figs. on 1 pl.

Surirella ovata
Hendy, N. Ingram, 1964
 An introductory account of the smaller algae of British coastal waters. V. Bacillariophyceae (Diatoms). Her Majesty's Stationary Office, 317 pp., 45 pls.

Surirella ovata
Rumkówna, A., 1948
 [List of the phytoplankton species occurring in the superficial water layers in the Gulf of Gdańsk] Bull. Lab. mar., Gdynia, No. 4: 139-141 with tables in back.

Surirella ovata
Zanon, D. V., 1949
 Diatomee di Buenos Aires (Argentina) Atti Accad. Naz. Lincei, Memorie, Cl. Sci. fis., mat. e. nat., ser. 7, 11(3):59-151, 2 pls.

Surirella ovata
Zanon, V., 1948
 Diatomee marini di Sardegna e Pugillo di Alghe Marine della stressa. Boll. Pesca, Piscitutura e Idrobiologia, Anno 24, ns. 3(2): 202-244, 27 figs. on 1 pl.

Surirella ovata pinnuata
Tsumura, K., 1956.
Diatomoj el la cirkaufoso de la restajo de la kastelo de Odawara. J. Yokohama Municipal Univ., (C-14), No. 47:23 pp.

?Surirella ovulum
Ross, R., and G. Abdin, 1949.
Notes on some diatoms from Norfolk. J. Roy. Micr. Soc., ser. 3, 69(4):225-230, 4 figs. on 1 pl.

Surirella Pandura
Zanon, V., 1948
 Diatomee marini di Sardegna e Pugillo di Alghe Marine della stressa. Boll. Pesca, Piscitutura e Idrobiologia, Anno 24, ns. 3(2): 202-244, 27 figs. on 1 pl.

Surirella patens
Mann, A., 1907
 Report on the diatoms of the Albatross voyages in the Pacific Ocean, 1888-1904. Contrib. U. S. Nat. Herb. 10(5):221-419, Pls. XLIV-LIV.

Surirella recedens
Mann, A., 1893
 List of Diatomaceae from a deep-sea dredging in the Atlantic Ocean off Delaware Bay by the U. S. Fish Commission Steamer Albatross. Proc. U. S. Nat. Mus. 16:303-312.

Surirella recedens
Zanon, V., 1948
 Diatomee marini di Sardegna e Pugillo di Alghe Marine della stressa. Boll. Pesca, Piscitutura e Idrobiologia, Anno 24, ns. 3(2): 202-244, 27 figs. on 1 pl.

Surirella reniformis
Eskinazi Enide e Shigekatsv Satô (1963-1964) 1966.
Contribuição ao estudo das diatomaceas da Praia de Piedade.
Trabhs Inst. Oceanogr., Univ. Recife, 5 (5/6): 73-114.

Surirella reniformis
Politis, J., 1949
 Diatomees marines de Bosphores et des ibes de la mer de Marmara. II Practica tou Hellenikou Hidrobiologikou Institutoutou 1929, Etoz 1929, 3(1):11-31.

Surirella reniformis
Zanon, V., 1948
 Diatomee marini di Sardegna e Pugillo di Alghe Marine della stressa. Boll. Pesca, Piscitutura e Idrobiologia, Anno 24, ns. 3(2): 202-244, 27 figs. on 1 pl.

Surirella robusta
de Sousa e Silva, E., 1956.
Contribution à l'étude du microplancton de Dakar et des regions maritimes voisines. Bull. I.F.A.N., 8(2):335-371, 7 pls.

Surirella robusta
Kokubo, S., and S. Sato., 1947
 Plankterx in Jū-San Gata. Physiol. and Ecol. (Japan) 1(4):1-16, 3 text figs. tables.

Surirella robusta
Mann, A., 1907
 Report on the diatoms of the Albatross voyages in the Pacific Ocean, 1888-1904. Contrib. U. S. Nat. Herb. 10(5):221-419, Pls. XLIV-LIV.

Surirella robusta
Zanon, D. V., 1949
 Diatomee di Buenos Aires (Argentina) Atti Accad. Naz. Lincei, Memorie, Cl. Sci. fis., mat. e. nat., ser. 7, 11(3):59-151, 2 pls.

Surirella rorata
Zanon, D. V., 1949
 Diatomee di Buenos Aires (Argentina) Atti Accad. Naz. Lincei, Memorie, Cl. Sci. fis., mat. e. nat., ser. 7, 11(3):59-151, 2 pls.

Surirella rugosa n.sp.
Salah, M.M., 1952(1953).
XII. Diatoms from Blakeney Point, Norfolk. New species and new records for Great Britain. J.R. Microsc. Soc., Ser. 3, 72(3):155-169, 3 pls

Surirella salina
Hendy, N. Ingram, 1964
 An introductory account of the smaller algae of British coastal waters. V. Bacillariophyceae (Diatoms). Her Majesty's Stationary Office, 317 pp., 45 pls.

Surirella salina
Smith, W., 1851
 Notes on the Diatomaceae; with descriptions of British species included in the genera Campylodiscus, Surirella, and Cymatopleura. Ann. Mag. Nat. Hist., 2nd ser., 7:1-14, 3 pls.

Surirella saxonica
de Sousa e Silva, E., 1956.
Contribution à l'étude du microplancton de Dakar et des regions maritimes voisines. Bull. I.F.A.N., 8(2):335-371, 7 pls.

Surirella senta n.sp.
Hendey, N. Ingram, 1958 [1957(Publ. 1958)]
Marine diatoms from some West African Ports. J. R Microsc. Soc. (3) 77(1/2): 28-85.

Surirella smithii
Hendy, N. Ingram, 1964
 An introductory account of the smaller algae of British coastal waters. V. Bacillariophyceae (Diatoms). Her Majesty's Stationary Office, 317 pp., 45 pls.

Surirella splendida
Smith, W., 1851
 Notes on the Diatomaceae; with descriptions of British species included in the genera Campylodiscus, Surirella, and Cymatopleura. Ann. Mag. Nat. Hist., 2nd ser., 7:1-14, 3 pls.

Surirella striatula
Hendy, N. Ingram, 1964
 An introductory account of the smaller algae of British coastal waters. V. Bacillariophyceae (Diatoms). Her Majesty's Stationary Office, 317 pp., 45 pls.

Surirella striatula
Hendey, N.J., 1951
 Littoral diatoms of Chicester Harbour with special reference to fouling. J.Roy. Microscop. Soc. 71(1): 1-86, 18 pls.

Surirella striatula
Rampi, L., 1940
 Diatomee del Mare Adriatico. Nuovo Giornale Botanico Italiano, n.s., 47:559-608.

Surirella striatula
Rumkówna, A., 1948
 [List of the phytoplankton species occurring in the superficial water layers in the Gulf of Gdańsk] Bull. Lab. mar., Gdynia, No. 4: 139-141 with tables in back.

Surirella striatula
Smith, W., 1851
 Notes on the Diatomaceae; with descriptions of British species included in the genera Campylodiscus, Surirella, and Cymatopleura. Ann. Mag. Nat. Hist., 2nd ser., 7:1-14, 3 pls.

Surirella striatula
Zanon, D. V., 1949
Diatomee di Buenos Aires (Argentina)
Atti Accad. Naz. Lincei, Memorie, Cl. Sci. fis., mat. e. nat., ser. 7, 11(3):59-151, 2 pls.

Surirella tenera
Mann, A., 1893
List of Diatomaceae from a deep-sea dredging in the Atlantic Ocean off Delaware Bay by the U. S. Fish Commission Steamer Albatross. Proc. U. S. Nat. Mus. 16:303-312.

Surirella tenera
Tsumura, K., 1956.
Diatomoj el la cirkaŭfoso de la restajo de la kastelo de Odawara. J. Yokohama Municipal Univ., (C-14), No. 47:23 pp.

Surirella tenera
Zanon, D. V., 1949
Diatomee di Buenos Aires (Argentina)
Atti Accad. Naz. Lincei, Memorie, Cl. Sci. fis., mat. e. nat., ser. 7, 11(3):59-151, 2 pls.

Surirella thaitiana n.sp.
Castracane degli Antelminelli, F., 1886
1. Report on the Diatomaceae collected by H.M.S. Challenger during the years 1873-1876. Rept. Sci. Results, H.M.S. Challenger, Botany Vol. II, 178 pp., 30 pls.

Surirella turgida
Rampi, L., 1940
Diatomee del Mare Adriatico. Nuovo Giornale Botanico Italiano, n.s., 47:559-608.

Surirella turgida
Zanon, D. V., 1949
Diatomee di Buenos Aires (Argentina)
Atti Accad. Naz. Lincei, Memorie, Cl. Sci. fis., mat. e. nat., ser. 7, 11(3):59-151, 2 pls.

Synedra sp.
Braarud, T., 1945
A phytoplankton survey of the polluted waters of inner Oslo Fjord. Hvalrådets Skrifter, No.28, 142 pp., 19 text figs., 17 tables.

Syndendrium diadema
Mann, A., 1893
List of Diatomaceae from a deep-sea dredging in the Atlantic Ocean off Delaware Bay by the U. S. Fish Commission Steamer Albatross. Proc. U. S. Nat. Mus. 16:303-312.

Synedra
Desikachary, T.V., 1956(1957).
Electron microscope studies on diatoms.
J. R. Microsc. Soc. (3)76(1/2):9-36.

Synedra actinastroides
Kokubo, S., and S. Sato., 1947
Plankters in Ju-San Gata. Physiol. and Ecol. (Japan) 1(4):1-16, 3 text figs., tables.

Synedra acus
Rumkówna, A., 1948
[List of the phytoplankton species occurring in the superficial water layers in the Gulf of Gdańsk] Bull. Lab. mar., Gdynia, No. 4: 139-141 with tables in back.

Synedra acus
Zanon, D. V., 1949
Diatomee di Buenos Aires (Argentina)
Atti Accad. Naz. Lincei, Memorie, Cl. Sci. fis., mat. e. nat., ser. 7, 11(3):59-151, 2 pls.

Synedra Adeliae n.sp.
Manguin, E., 1957.
Premier inventaire des diatomées de la Terre Adélie Antarctique. Espèces nouvelles.
Rev. Antarctique, n.s., 3(3):111-134.

Synedra affinis
Hendy, N. Ingram, 1964
An introductory account of the smaller algae of British coastal waters. V. Bacillariophyceae (Diatoms).
Her Majesty's Stationary Office, 317 pp., 45 pls.

Synedra affinis
Hendey, N.I., 1951
Littoral diatoms of Chicester Harbour with special reference to fouling. J.Roy. Microscop. Soc. 71(1): 1-86, 18 pls.

Synedra affinis
Jorgensen, E., 1900
Protophyten und Protozoën im Plankton aus der Norwegischen Westkerste. Bergens Mus. Aarb. 1899(6): 95 pp., 5 pls., 83 tables.

Synedra affinis
Levander, K.M., 1947
Plankton gesammelt in den Jahren 1899-1910 an den Küsten Finnlands. Finnländische Hydrographisch-Biologische Untersuchunger (aus dem Wasserbiologischen Laboratorin der Societas Scientiarum Fennica) No.11:40 pp., 6 diagrams, 13 pls., tables.

Synedra affinis
Politis, J., 1949
Diatomees marines de Bosphores et des ibes de la mer de Marmara. II Practica tou Hellenikou Hidrobiologikou Institutoutou 1929, Etoz 1929, 3(1):11-31.

Synedra affinis
Rumkówna, A., 1948
[List of the phytoplankton species occurring in the superficial water layers in the Gulf of Gdańsk] Bull. Lab. mar., Gdynia, No. 4: 139-141 with tables in back.

Synedra affinis acuminata
Takano, Hideaki, 1962
Notes on epiphytic diatoms upon sea-weeds from Japan.
J. Oceanogr. Soc., Japan, 18(1):29-33.

Synedra affinis fasciculata
Rumkówna, A., 1948
[List of the phytoplankton species occurring in the superficial water layers in the Gulf of Gdańsk] Bull. Lab. mar., Gdynia, No. 4: 139-141 with tables in back.

Synedra affinis obtusa
Rumkówna, A., 1948
[List of the phytoplankton species occurring in the superficial water layers in the Gulf of Gdańsk] Bull. Lab. mar., Gdynia, No. 4: 139-141 with tables in back.

Synedra amphicephala
Rumkówna, A., 1948
[List of the phytoplankton species occurring in the superficial water layers in the Gulf of Gdańsk] Bull. Lab. mar., Gdynia, No. 4: 139-141 with tables in back.

Synedra amphicephala
Zanon, D. V., 1949
Diatomee di Buenos Aires (Argentina)
Atti Accad. Naz. Lincei, Memorie, Cl. Sci. fis., mat. e. nat., ser. 7, 11(3):59-151, 2 pls.

Synedra atlantica n.sp.
Castracane degli Antelminelli, F., 1886
1. Report on the Diatomaceae collected by H.M.S. Challenger during the years 1873-1876. Rept. Sci. Results, H.M.S. Challenger, Botany Vol. II, 178 pp., 30 pls.

Synedra auriculata
Hendey, N.I., 1937
The plankton diatoms of the southern seas. Discovery Repts. 16:151-364, pls.6-13.

Synedra Baculus
Grunow, A., 1877
1. New Diatoms from Honduras. Monthly Micros. Jour., 18:165-186, pls. CXCIII-CXCVI.

Synedra baculus
Hendy, N. Ingram, 1964
An introductory account of the smaller algae of British coastal waters. V. Bacillariophyceae (Diatoms).
Her Majesty's Stationary Office, 317 pp., 45 pls.

Synedra baculus
Politis, J., 1949
Diatomees marines de Bosphores et des ibes de la mer de Marmara. II Practica tou Hellenikou Hidrobiologikou Institutoutou 1929, Etoz 1929, 3(1):11-31.

Synedra baculus
Zanon, V., 1948
Diatomee marini di Sardegna e Pugillo di Alghe Marine della stressa. Boll. Pesca, Piscitutura e Idrobiologia, Anno 24, ns. 3(2):202-244, 27 figs. on 1 pl.

Synedra barbatula
Zanon, V., 1948
Diatomee marini di Sardegna e Pugillo di Alghe Marine della stressa. Boll. Pesca, Piscitutura e Idrobiologia, Anno 24, ns. 3(2):202-244, 27 figs. on 1 pl.

Synedra berolinensis
Rumkówna, A., 1948
[List of the phytoplankton species occurring in the superficial water layers in the Gulf of Gdańsk] Bull. Lab. mar., Gdynia, No. 4: 139-141 with tables in back.

Synedra capitata
Zanon, D. V., 1949
Diatomee di Buenos Aires (Argentina)
Atti Accad. Naz. Lincei, Memorie, Cl. Sci. fis., mat. e. nat., ser. 7, 11(3):59-151, 2 pls.

Synedra capitata
Zanon, V., 1948
Diatomee marini di Sardegna e Pugillo di Alghe Marine della stressa. Boll. Pesca, Piscitutura e Idrobiologia, Anno 24, ns. 3(2):202-244, 27 figs. on 1 pl.

Synedra capetulata n.sp.
Castracane degli Antelminelli, F., 1886
1. Report on the Diatomaceae collected by H.M.S. Challenger during the years 1873-1876. Rept. Sci. Results, H.M.S. Challenger, Botany Vol. II, 178 pp., 30 pls.

Synedra capillaris n.sp.
Grunow, A., 1877
1. New Diatoms from Honduras. Monthly Micros. Jour., 18:165-186, pls. CXCIII-CXCVI.

Synedra cloisterioides
Misra, J.N., 1956
A systematic account of some littoral marine diatoms from the west coast of India.
J. Bombay Nat. Hist. Soc., 53(4):537-568.

Synedra crystallena
Grunow, A., 1877
1. New Diatoms from Honduras. Monthly Micros. Jour., 18:165-186, pls. CXCIII-CXCVI.

Synedra crystallina

Hasle, Grethe Rytter, 1960
Phytoplankton and ciliate species from the Tropical Pacific.
Skr. Norske Videnskaps-Akad., Oslo, 1. Mat.-Nat. Kl., 1960(2): 1-50.

Synedra crystallina

Hendy, N. Ingram, 1964
An introductory account of the smaller algae of British coastal waters. V. Bacillariophyceae (Diatoms).
Her Majesty's Stationary Office, 317 pp., 45 pls.

Synedra crystallina

Hustedt, F. and A.A. Aleem, 1951
Littoral diatoms from the Salstone near Plymouth. JMBA 30(1): 177.196.

Synedra crystallina

Politis, J., 1949
Diatomees marines de Bosphores et des ibes de la mer de Marmara. II Practica tou Hellenikou Hidrobiologikou Institutoutou 1929, Etoz 1929, 3(1):11-31.

Synedra cristallina

Zanon, D. V., 1949
Diatomee di Buenos Aires (Argentina)
Atti Accad. Naz. Lincei, Memorie, Cl. Sci. fis., mat. e. nat., ser. 7, 11(3):59-151, 2 pls.

Synedra crystallina

Zanon, V., 1948
Diatomee marini di Sardegna e Pugillo di Alghe Marine della stressa. Boll. Pesca, Piscitutura e Idrobiologia, Anno 24, ns. 3(2): 202-244, 27 figs. on 1 pl.

Synedra decloitrei n.sp.

Amossé, A. 1970.
Diatomées marines et saumâtres du Sénégal et de la Côte d'Ivoire.
Bull. Inst. fond. Afr. Noire (A) 32(2): 289-311.

Synedra delicatissima mesoleia

Mann, A., 1893
List of Diatomaceae from a deep-sea dredging in the Atlantic Ocean off Delaware Bay by the U. S. Fish Commission Steamer Albatross.
Proc. U. S. Nat. Mus. 16:303-312.

Synedra famelica

Misra, J.N., 1956.
A systematic account of some littoral marine diatoms from the west coast of India.
J. Bombay Nat. Hist. Soc., 53(4):537-568.

Synedra fimbriata n.sp.

Castracane degli Antelminelli, F., 1886
1. Report on the Diatomaceae collected by H.M.S. Challenger during the years 1873-1876. Rept. Sci. Results, H.M.S. Challenger, Botany Vol. II, 178 pp., 30 pls.

Synedra formosa

Takano, Hideaki, 1963.
Notes on marine littoral diatoms of Japan.
Bull. Tokai Reg. Fish. Res. Sta., No. 36:1-8.

Synedra formosa

Zanon, V., 1948
Diatomee marini di Sardegna e Pugillo di Alghe Marine della stressa. Boll. Pesca, Piscitutura e Idrobiologia, Anno 24, ns. 3(2): 202-244, 27 figs. on 1 pl.

Synedra fragilis n.sp.

Manguin, E., 1957.
Premier inventaire des diatomées de la Terre Adélie Antarctique. Espèces nouvelles.
Rev. Algologique, n.s., 3(3):111-134.

Synedra fulgens

Hendy, N. Ingram, 1964
An introductory account of the smaller algae of British coastal waters. V. Bacillariophyceae (Diatoms).
Her Majesty's Stationary Office, 317 pp., 45 pls.

Synedra fulgans

Hendey, N.I., 1951
Littoral diatoms of Chicester Harbour with special reference to fouling. J.Roy. Microscop. Soc. 71(1): 1-86, 18 pls.

Synedra fulgens

Politis, J., 1949
Diatomees marines de Bosphores et des ibes de la mer de Marmara. II Practica tou Hellenikou Hidrobiologikou Institutoutou 1929, Etoz 1929, 3(1):11-31.

Synedra fulgens

Zanon, V., 1948
Diatomee marini di Sardegna e Pugillo di Alghe Marine della stressa. Boll. Pesca, Piscitutura e Idrobiologia, Anno 24, ns. 3(2): 202-244, 27 figs. on 1 pl.

Synedra Gaillonii

Florin, M-B., 1948
9. Diatomeae in submarine cores from the Tyrrhenian Sea. Medd. Ocean. Inst., Göteborg, 15 (Göteborgs Kungl. Vetenskaps-och Viterrhets Samhälles Handlingar, Sjätte Foljden, Ser. B 5(13):80-88.

Synedra gaillonii

Hendy, N. Ingram, 1964
An introductory account of the smaller algae of British coastal waters. V. Bacillariophyceae (Diatoms).
Her Majesty's Stationary Office, 317 pp., 45 pls.

Synedra Gaillionii

Hendey, N.I., 1951
Littoral diatoms of Chicester Harbour with special reference to fouling. J.Roy. Microscop. Soc. 71(1): 1-86, 18 pls.

Synedra Gailloni

Hustedt, F. and A.A. Aleem, 1951
Littoral diatoms from the Salstone near Plymouth. JMBA 30(1): 177.196.

Synedra gallionii

Jorgensen, E., 1900
Protophyten und Protozoën im Plankton aus der Norwegischen Westkerste. Bergens Mus. Aarb. 1899(6): 95 pp., 5 pls., 83 tables.

Synedra Gaillonii

Lillick, L.C., 1940
Phytoplankton and planktonic protozoa of the offshore waters of the Gulf of Maine. Pt.II. Qualitative Composition of the Planktonic Flora. Trans. Am. Phil. Soc., n.s., 31(3):193-237, 13 text figs.

Synedra Gallionii?

Lillick, L.C., 1937
Seasonal studies of the phytoplankton off Woods Hole, Massachusetts. Biol. Bull. LXXIII (3):488-503, 3 text figs.

Synedra gaillonii

Paiva Carvalho, J., 1950
O plancton do Rio Maria Rodriques (Cananeis). 1. Diatomaceas e Dinoflagelados. Bol. Inst. Paulista Oceanogr. 1(1): 27-43, 2 fold-in tables, 2 figs.

Synedra Gaillonii

Politis, J., 1949
Diatomees marines de Bosphores et des ibes de la mer de Marmara. II Practica tou Hellenikou Hidrobiologikou Institutoutou 1929, Etoz 1929, 3(1):11-31.

Synedra Gaillonii

Rampi, L., 1940
Diatomee del Mare Adriatico. Nuovo Giornale Botanico Italiano, n.s., 47:559-608.

Synedra Gaillonii

Rumkówna, A., 1948
List of the phytoplankton species occurring in the superficial water layers in the Gulf of Gdansk. Bull. Lab. mar., Gdynia, No. 4: 139-141 with tables in back.

Synedra gallionii

Schodduyn, M., 1926
Observations faites dans la baie d'Ambleteuse (Pas de Calais). Bull. Inst. Ocean., Monaco, No. 482: 64 pp.

Synedra gailonii

Sousa e Silva, E., 1949
Diatomaceas e Dinoflagelados de Baia de Cascais. Portugaliae Acta Biol., Volume: Julio Henriques, Ser. B: 300-383, 9 pls, 2 fold-in tables.

Synedra Gaillonii

Sousa a Silva, E., and J. Dos Santos-Pinto, 1948
O Plancton da Baia de S. Martinho do Porto. 1. Diatomaceas e Dinoflagelados. Bol. Soc. Portuguese de Ciencias Naturais, 16(2):134-187, 6 pls. (Trav. Sta. Biol. Mar. de Lisbonne No. 52).

Synedra Gailloni

Zanon, V., 1948
Diatomee marini di Sardegna e Pugillo di Alghe Marine della stressa. Boll. Pesca, Piscitutura e Idrobiologia, Anno 24, ns. 3(2): 202-244, 27 figs. on 1 pl.

Synedra goulardi

Zanon, D. V., 1949
Diatomee di Buenos Aires (Argentina)
Atti Accad. Naz. Lincei, Memorie, Cl. Sci. fis., mat. e. nat., ser. 7, 11(3):59-151, 2 pls.

Synedra hennedyana

Hendy, N. Ingram, 1964
An introductory account of the smaller algae of British coastal waters. V. Bacillariophyceae (Diatoms).
Her Majesty's Stationary Office, 317 pp., 45 pls.

Synedra Hennedyana

Politis, J., 1949
Diatomees marines de Bosphores et des ibes de la mer de Marmara. II Practica tou Hellenikou Hidrobiologikou Institutoutou 1929, Etoz 1929, 3(1):11-31.

Synedra Hennedyana

Rampi, L., 1942
Ricerche sul fitoplancton del Mare Ligure 6. Le diatomee delle acque di Sanremo. Nuovo Giornale Botanico Italiano, N.S., 49:252-268.

Synedra Hennedeyana

Rampi, L., 1940
Diatomee del Mare Adriatico. Nuovo Giornale Botanico Italiano, n.s., 47:559-608.

Synedra hennedyana

Zanon, V., 1948
Diatomee marini di Sardegna e Pugillo di Alghe Marine della stressa. Boll. Pesca, Piscitutura e Idrobiologia, Anno 24, ns. 3(2): 202-244, 27 figs. on 1 pl.

Synedra hyperborea rostellata

Misra, J.N., 1956.
A systematic account of some littoral marine diatoms from the west coast of India.
J. Bombay Nat. Hist. Soc., 53(4):537-568.

Synedra investiens
Hendy, N. Ingram, 1964

An introductory account of the smaller algae of British coastal waters. V. Bacillariophyceae (Diatoms).
Her Majesty's Stationary Office, 317 pp., 45 pls.

Synedra investiens
Hustedt, F. and A.A. Aleem, 1951
Littoral diatoms from the Salstone near Plymouth. JMBA 30(1): 177.196.

Synedra japonica
Tsumura, K., 1956.
Diatomoj el la kirkaufoso de la restajo de la kastelo de Odawara. J. Yokohama Municipal Univ. (C-14) No. 47:23 pp.

Synedra kashyapiens n.sp.
Misra, J.N., 1956.
A systematic account of some littoral marine diatoms from the west coast of India.
J. Bombay Nat. Hist. Soc., 53(4):537-568.

Synedra laevigata n.sp.
Grunow, A., 1877
1. New Diatoms from Honduras. Monthly Micros. Jour., 18:165-186, pls. CXCIII-CXCVI.

Synedra laevigata
Rampi, L., 1940
Diatomee del Mare Adriatico. Nuovo Giornale Botanico Italiano, n.s., 47:559-608.

Synedra laevigata
Zanon, V., 1948
Diatomee marini di Sardegna e Pugillo di Alghe Marine della stressa. Boll. Pesca, Piscitutura e Idrobiologia, Anno 24, ns. 3(2): 202-244, 27 figs. on 1 pl.

Synedra lanceolata n.sp.
Castracane degli Antelminelli, F., 1886
1. Report on the Diatomaceae collected by H.M.S. Challenger during the years 1873-1876. Rept. Sci. Results, H.M.S. Challenger, Botany Vol. II, 178 pp., 30 pls.

Synedra lanceolata n.sp. var. thailandiae
Castracane degli Antelminelli, F., 1886
1. Report on the Diatomaceae collected by H.M.S. Challenger during the years 1873-1876. Rept. Sci. Results, H.M.S. Challenger, Botany Vol. II, 178 pp., 30 pls.

Synedra merluzae n.sp.
Frenguelli, Joaquin, and Hector Antonio Orlando, 1959.
Operacion MERLUZA. Diatomeas y silicoflagelados del plancton del "VI Crucero". Servicio Hidrogr. Naval., Argentina, Publ. No. H. 619: 5-62.

Synedra minuscula
Zanon, V., 1948
Diatomee marini di Sardegna e Pugillo di Alghe Marine della stressa. Boll. Pesca, Piscitutura e Idrobiologia, Anno 24, ns. 3(2): 202-244, 27 figs. on 1 pl.

Synedra nitzschioides
Bigelow, H.B., and M. Leslie, 1930
Reconnaissance of the waters and plankton of Monterey Bay, July 1928.
Bull. M.C.Z., 70(5):429-481, 43 text figs.

Synedra Nitzschioides
Meunier, A., 1915
Microplancton de la Mer Flamande. 2. Diatomées (excepté le genre Chaetoceros). Mem. Mus. Roy. Hist. Nat., Belgique, 7(3):1-118, Pls. VIII-XIV.

Synedra parva
Takano, Hideaki, 1962
Notes on epiphytic diatoms upon sea-weeds from Japan.
J. Oceanogr. Soc., Japan, 18(1):29-33.

Synedra pelagica
Boden, B. P., 1949.
The diatoms collected by the U.S.S. CACOPAN in the Antarctic in 1947. J. Mar. Res. 8(1):6-13, 3 textfigs.

Synedra pelagica
Boden, Brian, 1948
Marine plankton diatoms on operation HIGHJUMP in: Some oceanographic observations on operation HIGHJUMP. By R.S. Dietz. USNEL Rept. No.55, 97 pp., 41 figs. 7 July 1948.

Synedra pelagica nom.nov.
Hendey, N.I., 1937
The plankton diatoms of the southern seas.
Discovery Repts. 16:151-364, pls.6-13.

Synedra philippinarum n.sp
Castracane degli Antelminelli, F., 1886
1. Report on the Diatomaceae collected by H.M.S. Challenger during the years 1873-1876. Rept. Sci. Results, H.M.S. Challenger, Botany Vol. II, 178 pp., 30 pls.

Synedra provincialis n.sp.
Grunow, A., 1877
1. New Diatoms from Honduras. Monthly Micros. Jour., 18:165-186, pls. CXCIII-CXCVI.

Synedra pulchella
Hendy, N. Ingram, 1964

An introductory account of the smaller algae of British coastal waters. V. Bacillariophyceae (Diatoms).
Her Majesty's Stationary Office, 317 pp., 45 pls.

Synedra pulchella
Iselin, C., 1930
A report on the coastal waters of Labrador based on explorations of the "Chance" during the summer of 1926. Proc. Am. Acad. Arts Sci. 66(1):1-37, 14 text figs.

Synedra pulchella
Levander, K.M., 1947
Plankton gesammelt in den Jahren 1899-1910 an den Küsten Finnlands. Finnländische Hydrographisch-Biologische Untersuchungen (aus dem Wasserbiologischen Laboratorin der Societas Scientiarum Fennica) No.11:40 pp., 6 diagrams, 13 pls., tables.

Synedra pulchella
Mann, A., 1893
List of Diatomaceae from a deep-sea dredging in the Atlantic Ocean off Delaware Bay by the U. S. Fish Commission Steamer Albatross. Proc. U. S. Nat. Mus. 16:303-312.

Synedra pulchella
Zanon, V., 1948
Diatomee marini di Sardegna e Pugillo di Alghe Marine della stressa. Boll. Pesca, Piscitutura e Idrobiologia, Anno 24, ns. 3(2): 202-244, 27 figs. on 1 pl.

Synedra pulcherrina
Zanon, V., 1948
Diatomee marini di Sardegna e Pugillo di Alghe Marine della stressa. Boll. Pesca, Piscitutura e Idrobiologia, Anno 24, ns. 3(2): 202-244, 27 figs. on 1 pl.

Synedra reinboldi
Balech, E., 1947
Contribution al conocimiento del plancton antarctico. Plankton del Mar de Bellinghausen. Physis 20:75-91, 76 figs. on 8 pls.

Synedra Reinboldi
Frenguelli, Joaquin, and Hector Antonio Orlando, 1959.

Operacion MERLUZA. Diatomeas y silicoflagelados del plancton del "VI Crucero". Servicio Hidrogr. Naval., Argentina, Publ. No. H. 619: 5-62.

Synedra reinboldii
Jouse, A.P., G.S. Koroleva, G.A. Nagaeva, 1962
Diatoms in the surface layer of sediment in the Indian sector of the Antarctic. Investigations of marine bottom sediments. (In Russian; English summary).
Trudy Inst. Okeanol., Akad. Nauk, SSSR, 61: 19-92.

Synedra Reinboldii
Mangin, L., 1915
Phytoplancton de L'Antartique. Deuxieme Exped. Ant. Francaise (1908-1910), 95 pp., 3 pls., 58 text figs.

Synedra robusta
Rampi, L., 1942
Ricerche sul fitoplancton del Mare Ligure 6. Le diatomee delle acque di Sanremo. Nuovo Giornale Botanico Italiano, N.S., 49:252-268.

Synedra robusta
Zanon, V., 1948
Diatomee marini di Sardegna e Pugillo di Alghe Marine della stressa. Boll. Pesca, Piscitutura e Idrobiologia, Anno 24, ns. 3(2): 202-244, 27 figs. on 1 pl.

Synedra rumpens
Kokubo, S., and S. Sato, 1947
Plankters in Jū-San Gata. Physiol. and Ecol. (Japan) 1(4):1-16, 3 text figs., tables.

Synedra rupens
Tsumura, K., 1956.
Diatomoj el la cirkaufoso de la restajo de la kastelo de Odawara. J. Yokohama Municipal Univ., (C-14) No. 47:23 pp.

S. rumpens fragilarioides
S. rumpens scotica
S. rumpens familiaris

Synedra stricta
Hendey, N.I., 1937
The plankton diatoms of the southern seas.
Discovery Repts. 16:151-364, pls.6-13.

Synedra superba
Hendy, N. Ingram, 1964

An introductory account of the smaller algae of British coastal waters. V. Bacillariophyceae (Diatoms).
Her Majesty's Stationary Office, 317 pp., 45 pls.

Synedra tabulata
Hendy, N. Ingram, 1964

An introductory account of the smaller algae of British coastal waters. V. Bacillariophyceae (Diatoms).
Her Majesty's Stationary Office, 317 pp., 45 pls.

Synedra tabulata
Hendey, N.I., 1951
Littoral diatoms of Chicester Harbour with special reference to fouling. J. Roy. Microscop. Soc. 71(1): 1-86, 18 pls.

Synedra tabulata
Kucherova, Z.S., 1961.
Vertical distribution of diatoms from Sevastopol Bay.
Trudy Sevastopol Biol. Sta., (14):64-78.

Synedra tabulata
Levander, K.M., 1947
Plankton gesammelt in den Jahren 1899-1910 an den Küsten Finnlands. Finnländische Hydrographisch-Biologische Untersuchungen (aus dem Wasserbiologischen Laboratorin der Societas Scientiarum Fennica) No.11:40 pp., 6 diagrams, 13 pls., tables.

Synedra tabulata
Rampi, L., 1942
Ricerche sul fitoplancton del Mare Ligure 6. Le diatomee delle acque di Sanremo. Nuovo Giornale Botanico Italiano, N.S., 49:252-268.

Synedra tabulata
Rampi, L., 1940
Diatomee del Mare Adriatico. Nuovo Giornale Botanico Italiano, n.s., 47:559-608.

Synedra tabulata
Zanon, D.V., 1949
Diatomee di Buenos Aires (Argentina) Atti Accad. Naz. Lincei, Memorie, Cl. Sci. fis., mat. e. nat., ser. 7, 11(3):59-151, 2 pls.

Synedra tabulata
Zanon, V., 1948
Diatomee marini di Sardegna e Pugillo di Alghe Marine della stressa. Boll. Pesca, Piscitutura e Idrobiologia, Anno 24, ns. 3(2): 202-244, 27 figs. on 1 pl.

Synedra tabulata fasciculata
Misra, J.N., 1956.
A systematic account of some littoral marine diatoms from the west coast of India. J. Bombay Nat. Hist. Soc., 53(4):537-568.

Synedra toxoneides
Rampi, L., 1940
Diatomee del Mare Adriatico. Nuovo Giornale Botanico Italiano, n.s., 47:559-608.

Synedra ulna
Brunel, J., 1962
Le phytoplancton de la Baie de Chaleurs. Inst. Botan., Univ. Montréal, Contrib. No. 77: 365 pp., 66 pls.

Synedra ulna
Brunel, Jules, 1962
Le phytoplancton de la Baie des Chaleurs. Contrib. Ministère de la Chasse et des Pêcheries, Province de Québec, No. 91: 365 pp.

Synedra ulna
Iyengar, M.O.P. and G.Venkataraman,1951.
The ecology and seasonal succession of the algae flora of the River Cooum at Madras with special reference to the Diatomaceae. J. Madras Univ. 21, Sect. B(1): 140-192, 1 pl of 4 figs., 11 text figs.

Synedra ulna
Kokubo, S., and S. Sato, 1947
Plankters in Jū-San Gata. Physiol. and Ecol. (Japan) 1(4):1-16, 3 text figs., tables.

Synedra ulna
Mann, A., 1893
List of Diatomaceae from a deep-sea dredging in the Atlantic Ocean off Delaware Bay by the U. S. Fish Commission Steamer Albatross. Proc. U. S. Nat. Mus. 16:303-312.

Synedra ulna
Rumkówna, A., 1948
[List of the phytoplankton species occurring in the superficial water layers in the Gulf of Gdańsk] Bull. Lab. mar., Gdynia, No. 4: 139-141 with tables in back.

Synedra ulna
Tsumura, K., 1956.
Diatomoj el la kirkaufoso de la restajo de la kastelo de Odawara. J. Yokohama Municipal Univ., (C-14) No. 47:23 pp.

Synedra ulna
Zanon, D.V., 1949
Diatomee di Buenos Aires (Argentina) Atti Accad. Naz. Lincei, Memorie, Cl. Sci. fis., mat. e. nat., ser. 7, 11(3):59-151, 2 pls.

Synedra ulna
Zanon, V., 1948
Diatomee marini di Sardegna e Pugillo di Alghe Marine della stressa. Boll. Pesca, Piscitutura e Idrobiologia, Anno 24, ns. 3(2): 202-244, 27 figs. on 1 pl.

Synedra ulna danica
Iyengar, M.O.P. and G.Venkataraman,1951.
The ecology and seasonal succession of the algae flora of the River Cooum at Madras with special reference to the Diatomaceae. J. Madras Univ. 21, Sect. B(1): 140-192, 1 pl of 4 figs., 11 text figs.

Synedra ulna spathulifera
Mann, A., 1893
List of Diatomaceae from a deep-sea dredging in the Atlantic Ocean off Delaware Bay by the U. S. Fish Commission Steamer Albatross. Proc. U. S. Nat. Mus. 16:303-312.

Synedra subaequalis ulna
Mann, A., 1893
List of Diatomaceae from a deep-sea dredging in the Atlantic Ocean off Delaware Bay by the U. S. Fish Commission Steamer Albatross. Proc. U. S. Nat. Mus. 16:303-312.

Synedra undosa
Grunow, A., 1877
1. New Diatoms from Honduras. Monthly Micros. Jour., 18:165-186, pls. CXCIII-CXCVI.

Synedra undulata
Cupp, Easter E., 1943
Marine plankton diatoms of the west coast of North America. Bull. S.I.O. 5(1):1-238, 5 pls., 168 text figs.

Synedra undulata
Hendy, N. Ingram, 1964
An introductory account of the smaller algae of British coastal waters. V. Bacillariophyceae (Diatoms). Her Majesty's Stationary Office, 317 pp., 45 pls.

Synedra undulata
Margalef, R., 1949
Fitoplancton nerítico de la Costa Brava en 1947-48. Publ. Inst. Biol. Aplicada, 5: 41-51, 3 text figs.

Synedra undulata
Politis, J., 1949
Diatomees marines de Bosphores et des ibes de la mer de Marmara. II Practica tou Hellenikou Hidrobiologikou Institutoutou 1929, Etoz 1929, 3(1):11-31.

Synedra undulata
Rampi, L., 1942
Ricerche sul fitoplancton del Mare Ligure 6. Le diatomee delle acque di Sanremo. Nuovo Giornale Botanico Italiano, N.S., 49:252-268.

Synedra undulata
Rampi, L., 1940
Diatomee del Mare Adriatico. Nuovo Giornale Botanico Italiano, n.s., 47:559-608.

Synedra undulata
Schröder, B., 1900
Phytoplankton des Golfes von Neapel nebst vergleichenden Ausblicken auf das atlantischen Ozean. Mitt. Zool. Stat. Neapel, 14:1-38.

Synedra undulata
Zanon, V., 1948
Diatomee marini di Sardegna e Pugillo di Alghe Marine della stressa. Boll. Pesca, Piscitutura e Idrobiologia, Anno 24, ns. 3(2): 202-244, 27 figs. on 1 pl.

Synedra vaucheriae capitellata
Tsumura, K., 1956.
Diatomoj el la kirkaufoso de la restajo de la kastele de Odawara. J. Yokohama Municipal Univ., (C-14) No. 47:23 pp.

Systephania aculeata
Castracane degli Antelminelli, F., 1886
1. Report on the Diatomaceae collected by H.M.S. Challenger during the years 1873-1876. Rept. Sci. Results, H.M.S. Challenger, Botany Vol. II, 178 pp., 30 pls.

Systephania rašana n.sp.
Castracane degli Antelminelli, F., 1886
1. Report on the Diatomaceae collected by H.M.S. Challenger during the years 1873-1876. Rept. Sci. Results, H.M.S. Challenger, Botany Vol. II, 178 pp., 30 pls.

Tabellaria
Desikachary, T.V., 1956(1957).
Electron microscope studies on diatoms. J. R. Microsc. Soc. (3)76(1/2):9-36.

Tabellaria fenestrata
Braarud, T., 1945
A phytoplankton survey of the polluted waters of inner Oslo Fjord. Hvalrådets Skrifter, No.28, 142 pp., 19 text figs., 17 tables.

Tabellaria fenestrata
Mann, A., 1893
List of Diatomaceae from a deep-sea dredging in the Atlantic Ocean off Delaware Bay by the U. S. Fish Commission Steamer Albatross. Proc. U. S. Nat. Mus. 16:303-312.

Tabellaria fenestrata
Levander, K.M., 1947
Plankton gesammelt in den Jahren 1899-1910 an den Küsten Finnlands. Finnländische Hydrographisch-Biologische Untersuchungen (aus dem Wasserbiologischen Laboratorin der Societas Scientiarum Fennica) No.11:40 pp., 6 diagrams, 13 pls., tables.

Tabellaria fenestrata
Lillick, L.C., 1940
Phytoplankton and planktonic protozoa of the offshore waters of the Gulf of Maine. Pt.II. Qualitative Composition of the Planktonic Flora. Trans. Am. Phil. Soc., n.s., 31(3):193-237, 13 text figs.

Tabellaria fenestrata asterionelloides
Rumkówna, A. 1948
[List of the phytoplankton species occurring in the superficial water layers in the Gulf of Gdańsk] Bull. Lab. mar., Gdynia, No. 4: 139-141 with tables in back.

Tabellaria flocculosa
Braarud, T., 1945
A phytoplankton survey of the polluted waters of inner Oslo Fjord. Hvalrådets Skrifter, No.28, 142 pp., 19 text figs., 17 tables.

Tabellaria flocculosa
Jorgensen, E., 1900
Protophyten und Protozoën im Plankton aus der Norwegischen Westkerste. Bergens Mus. Aarb. 1899(6): 95 pp., 5 pls. 83 tables.

Tabellaria flocculosa
Levander, K.M., 1947
Plankton gesammelt in den Jahren 1899-1910 an den Küsten Finnlands. Finnländische Hydrographisch-Biologische Untersuchunger (aus dem Wasserbiologischen Laboratorin der Societas Scientiarum Fennica) No.11:40 pp., 6 diagrams, 13 pls., tables.

Tabellaria flocculosa
Rumkówna, A., 1948
List of the phytoplankton species occurring in the superficial water layers in the Gulf of Gdańsk. Bull. Lab. mar., Gdynia, No. 4: 139-141 with tables in back.

Terpinsoë americana
Cleve-Euler, A., 1951
Die Diatomeen von Schweden und Finnland. Kungl. Svenska Vetenskaps Akad. Handl., Fjärde Ser. 2(1): 161 pp., 6 pls.

Terpsinoë americana
Paiva Carvalho, J., 1950
O plancton do Rio Maria Rodrigues (Cananeia). 1. Diatomaceas e Dinoflagelados. Bol. Inst. Paulista Oceanogr. 1(1); 27-43, 2 fold-in tables, 2 figs.

Terpsinoë americana
Sousa a Silva, E., and J. Dos Santos-Pinto, 1948
O Plancton da Baia de S. Martinho do Porto. 1. Diatomaceas e Dinoflagelados. Bol. Soc. Portuguese de Ciencias Naturais, 16(2):134-187, 6 pls. (Trav. Sta. Biol. Mar. de Lisbonne No. 52).

Terpsinoe americana
Zanon, D. V., 1949
Diatomee di Buenos Aires (Argentina) Atti Accad. Naz. Lincei, Memorie, Cl. Sci. fis., mat. e. nat., ser. 7, 11(3):59-151, 2 pls.

Terpsinoë intermedia
de Sousa e Silva, E., 1956.
Contribution à l'étude du microplancton de Dakar et des regions maritimes voisines. Bull. I.F.A.N., 8(2):335-371, 7 pls.

Terpsinoe minima
Zanon, D. V., 1949
Diatomee di Buenos Aires (Argentina) Atti Accad. Naz. Lincei, Memorie, Cl. Sci. fis., mat. e. nat., ser. 7, 11(3):59-151, 2 pls.

Terpsinoe musica
de Sousa e Silva, E., 1956.
Contribution à l'étude du microplancton de Dakar et des regions maritimes voisines. Bull. I.F.A.N., 8(2):335-371, 7 pls.

Terpsinoe musica
Mann, A., 1907
Report on the diatoms of the Albatross voyages in the Pacific Ocean, 1888-1904. Contrib. U. S. Nat. Herb. 10(5):221-419, Pls. XLIV-LIV.

Terpsinoë musica
Subrahmanyan, R., 1945.
On the cell division and mitosis in some South Indian diatoms. Proc. Indian Acad. Sci., Sect. B 22(6):331-354, Pl. 39, 88 textfigs.

Terpsinoe musica
Zanon, D. V., 1949
Diatomee di Buenos Aires (Argentina) Atti Accad. Naz. Lincei, Memorie, Cl. Sci. fis., mat. e. nat., ser. 7, 11(3):59-151, 2 pls.

Tessella adriatica
Mann, A., 1907
Report on the diatoms of the Albatross voyages in the Pacific Ocean, 1888-1904. Contrib. U. S. Nat. Herb. 10(5):221-419, Pls. XLIV-LIV.

Tessella catena
Mann, A., 1907
Report on the diatoms of the Albatross voyages in the Pacific Ocean, 1888-1904. Contrib. U. S. Nat. Herb. 10(5):221-419, Pls. XLIV-LIV.

Tessella japonica
Mann, A., 1907
Report on the diatoms of the Albatross voyages in the Pacific Ocean, 1888-1904. Contrib. U. S. Nat. Herb. 10(5):221-419, Pls. XLIV-LIV.

Thalassionema antiqua n.sp.
Schrader, Hans-Joachim, 1973.
Cenozoic diatoms from the northeast Pacific, leg 18. Initial Repts Deep Sea Drilling Proj. 18: 673-797.

Thalassionema bacillaris
Hasle, Grethe Rytter, and Blanca Rojas E. de Mendiola, 1967.
The fine structure of some Thalassionema and Thalassiothrix species. Phycologia 6 (2/3): 107-118.

Thalassionema capitulata
Hasle, Grethe Rytter, 1960
Phytoplankton and ciliate species from the Tropical Pacific. Skr. Norske Videnskaps-Akad., Oslo, 1. Mat.-Nat. Kl., 1960(2): 1-50.

Thalassionema claviformis n.sp.
Schrader, Hans-Joachim, 1973
Cenozoic diatoms from the northeast Pacific, leg 18. Initial Repts Deep Sea Drilling Proj. 18:673-797.

Thalassionema elegans
Hasle, Grethe Rytter, 1960
Phytoplankton and ciliate species from the Tropical Pacific. Skr. Norske Videnskaps-Akad., Oslo, 1. Mat.-Nat. Kl., 1960(2): 1-50.

Thalassionema nitzschioides
Allen, W. E., 1938
The Templeton Crocker Expedition to the Gulf of California in 1935 - The Phytoplankton. Amer. Microsc. Soc., Trans. 57:328-335.

Thalassionema nitzschioides
Allen, W.E., 1937
Plankton diatoms of the Gulf of California obtained by the G. Allan Hancock Expedition of 1936. The Hancock Pacific Expeditions, Univ. So. Calif. Publ. 3:47-59, 1 fig.

Thalassionema nitzschioides
Boden, B.P., 1950
Some marine plankton diatoms from the west coast of South Africa. Trans. R.Soc. S. Africa. 32:321-434, 100 text figs.

Thalassionema nitzschioides
Braarud, T., 1945
Experimental studies on marine plankton diatoms. Avhandlingar utgitt av Det Norske Videnskaps-Akademi i Oslo. 1. Mat.-Naturv. Klasse 1944. No.10, 1-16, 1 text fig.

Thalassionema nitzschioides
Braarud, T., and Adam Bursa, 1939
On the phytoplankton of the Oslo Fjord, 1933-1934. Hvalrådets Skr. No.19:1-63; 9 text figs. Reviewed. J. du. Cons. 14(3): 418-420. A.C. Gardiner.

Thalassionema nitzschioides
Colton, John B., Jr., and Robert R. Marak, 1962.
Use of the Hardy continuous plankton recorder in a fishery research program. Bull. Mar. Ecology, 5(49):231-246.

Thalassionema nitzschioides
Copenhagen, W. J. and L. D., 1949.
Variation in the phytoplankton of Table Bay, October 1934 to October 1935. With a note on the calorific value of Chaetoceros spp. Trans. Roy. Soc. S. Africa, 32(2):113-123, 2 text figs.

Thalassionema nitzschioides
Cupp, Easter E., 1943
Marine plankton diatoms of the west coast of North America. Bull. S.I.O. 5(1):1-238, 5 pls., 168 text figs.

Thalassionema nitzschioides
Cupp, E.E. and Allen, W.E., 1938
Plankton diatoms of the Gulf of California obtained by Allan Hancock Pacific Expedition of 1937. The Hancock Pacific Expeditions, The Univ. So. Calif. Publ. 3: 61-74, 1 map, pls.4-15.

Thalassionema nitzschioides
Florin, M-B., 1948
9. Diatomeae in submarine cores from the Tyrrhenian Sea. Medd. Ocean. Inst., Göteborg, 15 (Göteborgs Kungl. Vetenskaps-och Viterrhets Samhälles Handlingar, Sjätte Följden, Ser. B 5(13):80-88.

Thalassionema nitzschioides
Frenguelli, Joaquin, and Hector Antonio Orlando, 1959.
Operacion MERLUZA. Diatomeas y silicoflagelados del plancton del "VI Crucero". Servicio Hidrogr. Naval, Argentina, Publ. No. H. 619: 5-62.

Thalassionema nitzschioides
Gilbert, J.Y., and W.E. Allen, 1943
The phytoplankton of the Gulf of California obtained by the "E.W. Scripps" in 1939 and 1940. J. Mar. Res. V(2):89-110, figs.30-31.

Thalassionema nitzschioides
Gran, H.H., 1908
Diatomeen. Nordisches Plankton, Botanischer Teil pp. XIX.1-XIX 146; 178 text figs.

Thalassionema nitzschioides
Gran, H.H., and T. Braarud, 1935
A quantitative study of the phytoplankton in the Bay of Fundy and the Gulf of Maine (including observations on hydrography, chemistry, and turbidity). J. Biol. Bd., Canada, 1(5):279-467, 69 text figs.

Thalassionema nitzschioides
Hasle, Grethe Rytter, 1960
Phytoplankton and ciliate species from the Tropical Pacific. Skr. Norske Videnskaps-Akad., Oslo, 1. Mat.-Nat. Kl., 1960(2): 1-50.

Thalassionema nitzschioides
Hasle, Grethe Rytter, and Blanca Rojas E. de Mendiola, 1967.
The fine structure of some Thalassionema and Thalassiothrix species. Phycologia 6 (2/3): 107-118.

Thalassionema nitzschioides
Hendy, N. Ingram, 1964
An introductory account of the smaller algae of British coastal waters. V. Bacillariophyceae (Diatoms).
Her Majesty's Stationary Office, 317 pp., 45 pls.

Thalassionema nitzschioides
Hendey, N.I., 1937
The plankton diatoms of the southern seas. Discovery Repts. 16:151-364, pls.6-13.

Thalassionema nitzschioides
Hustedt, F. and A.A. Aleem, 1951
Littoral diatoms from the Salstone near Plymouth. JMBA 30(1): 177.196.

Oceanographic Index: Marine Organisms Cumulation, 1946-1973

Thalassionema nitzschioides

Kokubo, S., 1952
Results of the observations on the plankton and oceanography of Mutsu Bay during 1950, reference being made also to the period 1946-1950. Bull Mar.Biol.Sta., Asamushi 5(1/4): 1-54, 3 tables,(fold-in), 1 fold-in.

Thalassionema nitzschioides

Lillick, L.C., 1940
Phytoplankton and planktonic protozoa of the offshore waters of the Gulf of Maine. Pt.II. Qualitative Composition of the Planktonic Flora. Trans. Am. Phil. Soc., n.s., 31(3):193-237, 13 text figs.

Thalassionema nitzschioides

Lillick, L.C., 1938
Preliminary report of the phytoplankton of the Gulf of Maine. Am. Mid. Nat. 20(3):624-640, 1 text fig, 37 tables.

Thalassionema nitzschioides

Lillick, L.C., 1937
Seasonal studies of the phytoplankton off Woods Hole, Massachusetts. Biol. Bull. LXXIII (3):488-503, 3 text figs.

Thalassionema nitzschioides

Manguin, E., 1954
Diatomees marines provenant de l'ile Heard (Australian National Antarctic Expedition). Rev. Algol., n.s., 1: 14-24.

Thalassionema nitzschioides

Margalef, R., 1949
Fitoplancton nerítico de la Costa Brava en 1947-48. Publ. Inst. Biol. Aplicada, 5: 41-51, 3 text figs.

Thalassionema nitzschioides

Paiva Carvalho, J., 1950
O plancton do Rio Maria Rodriques (Cananeis). 1. Diatomaceas e Dinoflagelados. Bol. Inst. Paulista Oceanogr. 1(1); 27-43, 2 fold-in tables, 2 figs.

Thalassionema nitzschioides

Phifer, L.D. (undated)
The occurrence and distribution of plankton diatoms in Bering Sea and Bering Strait, July 26-August 24, 1934. Report of Oceanographic cruise of U.S. Coast Guard Cutter Chelan 1934, Part II(A):1-44 (mimeographed) plus fig.1 (after Pt.B)

Thalassionema nitzschioides

Rampi, L., 1942
Ricerche sul fitoplancton del Mare Ligure 6. Le diatomee delle acque di Sanremo. Nuovo Giornale Botanico Italiano, N.S., 49:252-268.

Thalassionema nitzschioides

Robinson, G.A., 1965.
Continuous plankton records: contribution towards a plankton atlas of the North Atlantic and the North Sea. IX. Seasonal cycles of phytoplankton. Bulls. Mar. Ecol.,Scottish Mar. Biol. Assoc., 6(4):104-122, pls. 26-61.

Thalassionema nitzschioides

Robinson, G.A., 1961
Contribution towards an atlas of the northeastern Atlantic and the North Sea. 1. Phytoplankton. Bulls. Mar. Ecology, 5(42): 81-89, Pls. 15-20.

Thalassionema nitzschioides

Sakshaug, Egil 1970.
Quantitative phytoplankton investigations in near-shore water masses.
Skr. K. Norske Vidensk. Selsk. (3):1-8.

Thalassionema nitzschioides

Smayda, Theodore J., 1958
Biogeographical studies of marine phytoplankton Oikos, 9(2): 158-191.

Thalassionema nitzschioides

Uyeno, Fukuzo, 1961
Oceanographical and ecological studies on primary production of the sea, with special references to relationship between diatom production and temperature and chlorinity of water.
Rept.. Fac. Fish.. Pref. Univ., Mie, 4(1): 1-64.

Thalassionema nitzschioides

Zanon, D. V., 1949
Diatomee di Buenos Aires (Argentina) Atti Accad. Naz. Lincei, Memorie, Cl. Sci. fis., mat. e. nat., ser. 7, 11(3):59-151, 2 pls.

Thalassionema nitzschioides lanceolata

Ercegovic, A., 1936
Etudes qualitative et quantitatives du phytoplancton dans les eaux cotières de l'Adriatique oriental moyen au cours de l'année 1934. Acta Adriatica 1(9):1-126

Thalassionema robusta n.sp.

Schrader, Hans-Joachim, 1973
Cenozoic diatoms from the northeast Pacific, leg 18. Initial Repts Deep Sea Drilling Proj. 18:673-797.

Thalassiophysa rhipidis n.gen., n. sp.

Conger, P.S., 1954.
A new genus and species of plankton diatom from the Florida Straits.
(Publ. 4171) Smithsonian Misc. Coll. 122(14): 1-8, 4 pls.

Thalassiosira sp.

Braarud, T., 1945
A phytoplankton survey of the polluted waters of inner Oslo Fjord. Hvalrådets Skrifter, No.28, 142 pp., 19 text figs., 17 tables.

Thalassiosira sp.

Braarud, T., and Adam Bursa, 1939
On the phytoplankton of the Oslo Fjord, 1933-1934. Hvalrådets Skr. No.19:1-63; 9 text figs. Reviewed. J. du Cons. 14(3): 418-420. A.C. Gardiner.

Thalassiosira

Desikachary, T.V., 1956 (1957).
Electron microscope studies on diatoms. J. R. Microsc. Soc. (3)76(1/2):9-36.

Thalassiosira sp.

Frenguelli, Joaquin, and Hector Antonio Orlando, 1959.
Operacion MERLUZA. Diatomeas y silicoflagelados del plancton del "VI Crucero". Servicio Hidrogr. Naval., Argentina, Publ. No. H. 619: 5-62.

Thalassiosira spp

Hasle, G.R. and B.R. Heimdal 1970.
Some species of the centric diatom genus Thalassiosira studied in the light and electron microscopes.
Beiheft Nova Hedwigia 31: 559-589

Thalassiosira

Marshall, S.M., 1947
An experiment in marine fish cultivation: III. The plankton of a fertilized loch. Proc. Roy. Soc., Edinburgh, Sect.B., 63, Pt.I(3):21-33, 7 text figs.

Thalassiosira

Neaverson, E., 1934
The sea-floor deposits. 1. General characteristics and distribution. Discovery Repts. 9: 297-349, Plates 17-22.

Thalassiosira sp.

Pettersson, H., H. Höglund, S. Landberg, 1934.
Submarine daylight and the photosynthesis of phytoplankton. Medd. fran Göteborgs Högskolas Oceanografiska Inst. 10 (Göteborgs Kungl. Vetenskaps-och Vitterhets-samhälles handlingar. Femte Följden, Ser. B 4(5):1-17.

Thalassiosira

Sears, M., 1941
Notes on the phytoplankton on Georges Bank in 1941. J. Mar. Res., IV(3):247-257; textfigs. 54-58.

Thalassiosira Adeliae n.sp.

Manguin, E., 1957.
Premier inventaire des diatomées de la Terre Adélie Antarctique. Espèces nouvelles. Rev. Algologique, n.s., 3(3):111-134.

Thalassiosira aestivalis

Cupp, Easter E., 1943
Marine plankton diatoms of the west coast of North America. Bull. S.I.O. 5(1):1-238, 5 pls., 168 text figs.

Thalassiosira aestivalis n.sp.

Gran, H. H. and E. C. Angst, 1931
Plankton diatoms of Puget Sound. Publ. Puget Sound Biol. Sta. 7:417-519, 95 text figs.

Thalassiosira allenii n.sp.

Takano, Hideaki, 1965.
New and rare diatoms from Japanese marine waters. 1.
Bull. Tokai Reg. Fish. Res. Lab., No. 42:1-10.

Thalassiosira ambique n. sp.

Kozlova, O.G., 1962
Specific composition of diatoms in the waters of the Indian sector of the Antarctic. Investigation of Marine Bottom Sediments. (In Russian).
Trudy Inst. Okeanol., Akad. Nauk, SSSR, 61: 3-18.
English summary, p. 18

Thalassiosira antarctica

Boden, B. P., 1949.
The diatoms collected by the U.S.S. CACOPAN in the Antarctic in 1947. J. Mar. Res. 8(1):6-13, 3 textfigs.

Thalassiosira antarctica

Boden, Brian, 1948
Marine plankton diatoms on operation HIGHJUMP in: Some oceanographic observations on operation HIGHJUMP. By R.S. Dietz. USNEL Rept. No.55, 97 pp., 41 figs. 7 July 1948.

Thalassiosira antarctica

Central Meteorological Observatory, 1949
Report on sea and weather observation on Antarctic Whaling Ground (1947-48). Ocean. Mag., Japan, 1(1):49-88, 17 text figs.

Thalassiosira antarctica

Frenguelli, Joaquin, and Hector Antonio Orlando, 1959.
Operacion MERLUZA. Diatomeas y silicoflagelados del plancton del "VI Crucero". Servicio Hidrogr. Naval., Argentina, Publ. No. H. 619: 5-62.

Thalassiosira antarctica

Hasle, Grethe Rytter, 1960
Phytoplankton and ciliate species from the Tropical Pacific.
Skr. Norske Videnskaps-Akad., Oslo, 1. Mat.-Nat. Kl., 1960(2): 1-50.

Thalassiosira antarctica
Hasle, G.R., and B.R. Heimdal, 1968
Morphology and distribution of the marine centric diatom Thalassiosira antarctica Comber.
Jl R. microsc. Soc. 88(3): 357-369.

Thalassiosira antarctica
Hendey, N.I., 1937
The plankton diatoms of the southern seas. Discovery Repts. 16:151-364, pls.6-13.

Thalassiosira antarctica
Jouse, A.P., G.S. Koroleva, G.A. Nagaeva, 1962
Diatoms in the surface layer of sediment in the Indian sector of the Antarctic. Investigations of marine bottom sediments. (In Russian; English summary). Trudy Inst. Okeanol. Akad. Nauk. SSSR. 61:19-92.

Thalassiosira antarctica
Mangin, L., 1915
Phytoplankton de L'Antartique. Deuxieme Exped. Ant. Francaise (1908-1910), 95 pp., 3 pls., 58 text figs.

Thalassiosira antarctica
Manguin, E., 1954
Diatomees marines provenant de l'ile Heard (Australian National Antarctic Expedition). Rev. Algol., n.s., 1: 14-24.

Thalassiosira antarctica
Smayda, Theodore J., 1958
Biogeographical studies of marine phytoplankton Oikos, 9(2): 158-191.

Thalassiosira baltica
Bigelow, H.B., and M. Leslie, 1930
Reconnaissance of the waters and plankton of Monterey Bay, July 1928. Bull. M.C.Z., 70(5):429-481, 43 text figs.

Thalassiosira baltica
Braarud, T., 1962
Species distribution in marine phytoplankton.
J. Oceanogr. Soc., Japan. 20th Ann. Vol., 628-649.

Thalassiosira baltica
Brunel, J., 1962
Le phytoplancton de la Baie de Chaleurs. Inst. Botan., Univ. Montréal, Contrib. No. 77: 365 pp., 66 pls.

Thalassiosira baltica
Gran, H.H., 1908
Diatomeen. Nordisches Plankton, Botanischer Teil pp. XIX 1-XIX 146; 178 text figs.

Thalassiosira baltica
Hendy, N. Ingram, 1964
An introductory account of the smaller algae of British coastal waters. V. Bacillariophyceae (Diatoms).
Her Majesty's Stationary Office, 317 pp., 45 pls.

Thalassiosira baltica
Levander, K.M., 1947
Plankton gesammelt in den Jahren 1899-1910 an den Küsten Finnlands. Finnländische Hydrographisch-Biologische Untersuchungen (aus dem Wasserbiologischen Laboratorin der Societas Scientiarum Fennica) No.11: 40 pp., 6 diagrams, 13 pls., tables.

Thalassiosira baltica
Lillick, L.C., 1940
Phytoplankton and planktonic protozoa of the offshore waters of the Gulf of Maine. Pt.II. Qualitative Composition of the Planktonic Flora. Trans. Am. Phil. Soc., n.s., 31(3):193-237, 13 text figs.

Thalassiosira baltica
Rothe, F., 1942
Quantitativen Untersuchungen über die Planktonverteilung in der östlichen Ostsee. Ber. Deutsch Wiss. Komm. Meeresf., N.F., 10:291-368, 33 text figs.

Thalassoisira baltica
Rothe, F., 1941
Quantitative Untersuchungen über die Plankton verteilung in der östlichen Ost see. Ber. Deut. Wiss. Komm. fur Meeresforschung. n.f. X(3):291-368, 33 text figs.

Thalassiosira baltica
Rumkówna, A., 1948
List of the phytoplankton species occurring in the superficial water layers in the Gulf of Gdańsk. Bull. Lab. mar., Gdynia, No. 4: 139-141 with tables in back.

Thalassiosira baltica
Schodduyn, M., 1926
Observations faites dans la baie d'Ambleteuse (Pas de Calais). Bull. Inst. Ocean., Monaco, No. 482: 64 pp.

Thalassiosira aff. baltica
Somers, D., 1972
Scanning electron microscope studies on some species of the centric diatom genera Thalassiosira and Coscinodiscus. Biol. Jaarb. Dodonaea 40:317-322. Also in: Coll. Repr. Inst. Zeewetenschap. Onderzoek Belg. 3(1973).

Thalassiosira bioculata
Brunel, J., 1962
Le phytoplancton de la Baie de Chaleurs. Inst. Botan., Univ. Montréal, Contrib. No. 77: 365 pp., 66 pls.

Thalassiosira bioculata?
Brunel, Jules, 1962
Le phytoplancton de la Baie des Chaleurs. Contrib. Ministère de la Chasse et des Pêcheries, Province de Québec, No. 91: 365 pp.

Thalassiosira bioculata
Gran, H.H., 1908
Diatomeen. Nordisches Plankton, Botanischer Teil pp. XIX 1-XIX 146; 178 text figs.

Thalassiosira bioculata
Gran, H. H. and E. C. Angst, 1931
Plankton diatoms of Puget Sound. Publ. Puget Sound Biol. Sta. 7:417-519, 95 text figs.

Thalassiosira bioculata
Gran, H.H., and T. Braarud, 1935
A quantitative study of the phytoplankton in the Bay of Fundy and the Gulf of Maine (including observations on hydrography, chemistry, and turbidity). J. Biol. Bd., Canada, 1(5):279-467, 69 text figs.

Thalassiosira bioculata
Lillick, L.C., 1940
Phytoplankton and planktonic protozoa of the offshore waters of the Gulf of Maine. Pt.II. Qualitative Composition of the Planktonic Flora. Trans. Am. Phil. Soc., n.s., 31(3):193-237, 13 text figs.

Thalassiosira bioculata
Lillick, L.C., 1938
Preliminary report of the phytoplankton of the Gulf of Maine. Am. Mid. Nat. 20(3):624-640, 1 text fig. 37 tables.

Thalassiosira bioculata
Phifer, L.D. (undated)
The occurrence and distribution of plankton diatoms in Bering Sea and Bering Strait, July 26-August 24, 1934. Report of Oceanographic cruise of U.S. Coast Guard Cutter Chelan 1934, Part II(A):1-44 (mimeographed) plus fig.1 (after Pt.B)

Thalassiosira bioculata raripora
Paasche, E., 1961.
Notes on phytoplankton from the Norwegian Sea. Botanica Marina, 2(3/4):197-210.

Thalassiosira Clevei n.sp.
Gran, H.H., 1897
Protophyta: Diatomaceae, Silico-flagellata and Cilioflagellata. Den Norske Nordhavs Expedition 1876-1878, h. 24, 36 pp., 4 pls.

Thalassiosira chilensis
Krasske, G., 1941
Die Kieselalgen des chilenischen Küstenplankton (Aus dem südchilenischen Kusten gebiet IX). Arch. Hydrobiol. 38(2):260-287.

Thalassiosira condensata
Allen, W. E., 1938
The Templeton Crocker Expedition to the Gulf of California in 1935 - The Phytoplankton. Amer. Microsc. Soc., Trans. 57:328-335.

Thalassiosira condensata
Allen, W.E., 1937
Plankton diatoms of the Gulf of California obtained by the G. Allan Hancock Expedition of 1936. The Hancock Pacific Expeditions, Univ. So. Calif. Publ. 3:47-59, 1 fig.

Thalassiosira condensata
Boden, B.P., 1950
Some marine plankton diatoms from the west coast of South Africa. Trans. R.Soc. S. Africa. 32:321-434, 100 text figs.

Thalassiosira condensata
Brunel, J., 1962
Le phytoplancton de la Baie de Chaleurs. Inst. Botan., Univ. Montréal, Contrib. No. 77: 365 pp., 66 pls.

Thalassiosira condensata
Cupp, E.E. and Allen, W.E., 1938
Plankton diatoms of the Gulf of California obtained by Allan Hancock Pacific Expedition of 1937. The Hancock Pacific Expeditions, The Univ. So. Calif. Publ. 3: 61-74, 1 map, pls.4-15.

Thalassiosira condensata
Fraser, J. H., and A. Saville, 1949
Plankton distribution in Scottish and adjacent waters in 1948. Ann. Biol. 5:61-62.

Thalassiosira condensata
Gran, H. H. and E. C. Angst, 1931
Plankton diatoms of Puget Sound. Publ. Puget Sound Biol. Sta. 7:417-519, 95 text figs.

Thalassiosira condensata
Hendy, N. Ingram, 1964
An introductory account of the smaller algae of British coastal waters. V. Bacillariophyceae (Diatoms).
Her Majesty's Stationary Office, 317 pp., 45 pls.

Thalassiosira condensata
Hendey, N.I., 1937
The plankton diatoms of the southern seas. Discovery Repts. 16:151-364, pls.6-13.

Thalassiosira constricta n.sp.
Gaarder, Karen Ringdal.
Phytoplankton studies from the Tromsø district, 1930-31. Tromsø Mus. Årshefter, Naturhist. Avd. 11, 55(1):159 pp., 4 fold-in pls., 12 textfigs.

Thalassiosira constricta

Heimdal, Berit Riddervold, 1971.
Vegetative cells and resting spores of *Thalassiosira constricta* Gaarder (Bacillariophyceae)
Norw. J. Bot. 18(3/4): 153-159.

Thalassiosira coramandeliana n.sp.

Subrahmanyan, R., 1946.
A systematic account of the marine plankton diatoms of the Madras coast. Proc. Indian Acad. Sci.

Thalassiosira coronata

Hasle, Grethe Rytter, 1968.
The valve processes of the centric diatom genus *Thalassiosira*.
Nytt Mag. Bot. 15(3)193-201.

Thalassiosira decipiens

Allen, W. E., 1938
The Templeton Crocker Expedition to the Gulf of California in 1935 - The Phytoplankton. Amer. Microsc. Soc., Trans. 57:328-335.

Thalassiosira decipiens

Allen, W.E., 1937
Plankton diatoms of the Gulf of California obtained by the G. Allan Hancock Expedition of 1936. The Hancock Pacific Expeditions, Univ. So. Calif. Publ. 3:47-59, 1 fig.

Thalassiosira decipiens

Allen, W. E., 1936
Occurrence of marine plankton diatoms in a ten-year series of daily catches in southern California. Am. Jour. Bot. 23(1):60-63.

Thalassiosira decipiens

Bigelow, H.B., and M. Leslie, 1930
Reconnaissance of the waters and plankton of Monterey Bay, July 1928. Bull. M.C.Z., 70(5):429-481, 43 text figs.

Thalassiosira decipiens

Boden, B.P., 1950
Some marine plankton diatoms from the west coast of South Africa. Trans. R. Soc. S. Africa. 32:321-434, 100 text figs.

Thalassiosira decipiens

Boden, B. P., 1949.
The diatoms collected by the U.S.S. CACOPAN in the Antarctic in 1947. J. Mar. Res. 8(1):6-13, 3 textfigs.

Thalassiosira decipiens

Boden, Brian, 1948
Marine plankton diatoms on operation HIGHJUMP in: Some oceanographic observations on operation HIGHJUMP. By R.S. Dietz. USNEL Rept. No.55, 97 pp., 41 figs. 7 July 1948.

Thalassiosira decipiens

Braarud, T., 1945
Experimental studies on marine plankton diatoms. Avhandlingar utgitt av Det Norske Videnskaps-Akademi i Oslo. 1. Mat.-Naturv. Klasse 1944. No.10, 1-16, 1 text fig.

Thalassiosira decipiens

Braarud, T., 1945
A phytoplankton survey of the polluted waters of inner Oslo Fjord. Hvalrådets Skrifter, No.28, 142 pp., 19 text figs., 17 tables.

Thalassiosira decipiens

Braarud, T., 1934
A note on the phytoplankton of the Gulf of Maine in the summer of 1933. Biol. Bull. 67(1):76-82.

Thalassiosira decipiens

Braarud, T., and Adam Bursa, 1939
On the phytoplankton of the Oslo Fjord, 1933-1934. Hvalrådets Skr. No.19:1-63; 9 text figs. Reviewed. J. du. Cons. 14(3): 418-420. A.C. Gardiner.

Thalassiosira decipiens

Brunel, J., 1962
Le phytoplancton de la Baie de Chaleurs. Inst. Botan., Univ. Montréal, Contrib. No. 77: 365 pp., 66 pls.

Thalassiosira decipiens

Chu, S. P., 1949
Experimental studies on the environmental factors influencing the growth of phytoplankton. Sci. & Tech. in China 2(3):37-52.

Thalassiosira decipiens

Cupp, Easter E., 1943
Marine plankton diatoms of the west coast of North America. Bull. S.I.O. 5(1):1-238, 5 pls., 168 text figs.

Thalassiosira decipiens

Cupp, E.E. and Allen, W.E., 1938
Plankton diatoms of the Gulf of California obtained by Allan Hancock Pacific Expedition of 1937. The Hancock Pacific Expeditions, The Univ. So. Calif. Publ. 3: 61-74, 1 map, pls. 4-15.

Thalassiosira decipiens

Dangeard, P., 1927
Phytoplancton de la croisière du "Sylvana". Ann. Inst. Ocean., Monaco, n.s., 4(8):286-401, 54 text figs. (Feirer-Juin 1913).

Thalassiosira decipiens

Forti, A., 1922
Ricerche sulla flora pelagica (fitoplancton) di Quarto dei Mille. Mem. R. Com. Talass. Ital. 97:248 pp., 13 pls.

Thalassiosira decipiens

Frenguelli, Joaquin, and Hector Antonio Orlando, 1959.
Operacion MERLUZA. Diatomeas y silicoflagelados del plancton del "VI Crucero". Servicio Hidrogr. Naval., Argentina, Publ. No. H. 619: 5-62.

Thalassiosira decipiens

Gilbert, J.Y., and W.E. Allen, 1943
The phytoplankton of the Gulf of California obtained by the "E.W. Scripps" in 1939 and 1940. J. Mar. Res. V(2):89-110, figs. 30-31.

Thalassiosira decipiens

Gran, H. H. and E. C. Angst, 1931
Plankton diatoms of Puget Sound. Publ. Puget Sound Biol. Sta. 7:417-519, 95 text figs.

Thalassiosira decipiens

Gran, H.H., and T. Braarud, 1935
A quantitative study of the phytoplankton in the Bay of Fundy and the Gulf of Maine (including observations on hydrography, chemistry, and turbidity). J. Biol. Bd., Canada, 1(5):279-467, 69 text figs.

Thalassiosira decipiens

Hasle, Grethe Rytter, 1960
Phytoplankton and ciliate species from the Tropical Pacific. Skr. Norske Videnskaps-Akad., Oslo, 1. Mat.-Nat. Kl., 1960(2): 1-50.

Thalassiosira decipiens

Hendy, N. Ingram, 1964
An introductory account of the smaller algae of British coastal waters. V. Bacillariophyceae (Diatoms).
Her Majesty's Stationary Office, 317 pp., 45 pls.

Thalassiosira decipiens

Hendey, N.I., 1937
The plankton diatoms of the southern seas. Discovery Repts. 16:151-364, pls. 6-13.

Thalassiosira decipiens

Hustedt, F. and A.A. Aleem, 1951
Littoral diatoms from the Salstone near Plymouth. JMBA 30(1): 177.196.

Thalassiosira decipiens

Iselin, C., 1930
A report on the coastal waters of Labrador based on explorations of the "Chance" during the summer of 1926. Proc. Am. Acad. Arts Sci., 66(1):1-37, 14 text figs.

Thalassiosira decipiens

Jørgensen, E., 1905
B. Protistplankton and the diatoms in bottom samples. Hydrographical and biological investigations in Norwegian fjords. Bergens Mus. Skr. 7: 49-225.

Thalassiosira decipiens

Krasske, G., 1941
Die Kieselalgen des chilenischen Küstenplankton (Aus dem südchilenischen Kusten gebiet IX). Arcl. Hydrobiol. 38 (2):260-287.

Thalassiosira decipiens

Lillick, L.C., 1940
Phytoplankton and planktonic protozoa of the offshore waters of the Gulf of Maine. Pt. II. Qualitative Composition of the Planktonic Flora. Trans. Am. Phil. Soc., n.s., 31(3):193-237, 13 text figs.

Thalassiosira decipiens

Lillick, L.C., 1938
Preliminary report of the phytoplankton of the Gulf of Maine. Am. Mid. Nat. 20(3):624-640, 1 text fig. 37 tables.

Thalassiosira decipiens

Lillick, L.C., 1937
Seasonal studies of the phytoplankton off Woods Hole, Massachusetts. Biol. Bull. LXXIII (3):488-503, 3 text figs.

Thalassiosira decipiens

Manguin, E., 1954
Diatomées marines provenant de l'île Heard (Australian National Antarctic Expedition). Rev. Algol., n.s., 1: 14-24.

Thalassiosira decipiens

Marukawa, H., 1921
Plankton lists and some new species of copepods, from the northern waters of Japan. Bull. Inst. Ocean., No.384, 15 pp., 3 pls., 1 chart. Monaco

Thalassiosira decipiens

Meunier, A., 1915
Microplancton de la Mer Flamande. 2. Diatomées (excepté le genre Chaetoceros). Mem. Mus. Roy. Hist. Nat., Belgique, 7(3):1-118, Pls. VIII-XIV.

Thalassiosira decipiens

Morse, D.C., 1947
Some observations on seasonal variations in plankton population Patuxant River, Maryland 1943-1945. Bd. Nat. Res., Publ. No.65, Chesapeake Biol. Lab. 31, 3 figs.

Thalassiosira decipens

Paasche, E., 1973
The influence of cell size on growth rate, silica content and some other properties of four marine diatom species. Norwegian J. Bot. 20(2/3): 199-204.

Thalassiosira decipiens

Pavillard, J., 1925
Bacillariales. Rept. on the Danish Oceangr. Exped., 1908-10 to the Mediterranean and adj. seas. Vol.II., Biol. J4:72 pp., 116 text figs.

Thalassiosira decipiens

Phifer, L.D. (undated)
The occurrence and distribution of plankton diatoms in Bering Sea and Bering Strait, July 26-August 24, 1934. Report of Oceanographic cruise of U.S. Coast Guard Cutter Chelan 1934, Part II(A):1-44 (mimeographed) plus fig.1 (after Pt.B)

Thalassiosira decipiens

Rampi, L., 1942
Ricerche sul fitoplancton del Mare Ligure 6. Le diatomee delle acque di Sanremo. Nuovo Giornale Botanico Italiano, N.S., 49:252-268.

Thalassiosira decipens

Somers, D., 1972
Scanning electron microscope studies on some species of the centric diatom genera Thalassiosira and Coscinodiscus. Biol. Jaarb. Dodonaea 40:317-322. Also in: Coll. Repr. Inst. Zeewetenschap. Onderzoek Belg. 3(1973).

Thalassiosira decipiens

Sousa e Silva, E., 1949
Diatomaceas e Dinoflagelados de Baia de Cascais. Portugaliae Acta Biol., Volume: Julio Henriques, Ser. B: 300-383, 9 pls, 2 fold-in tables.

Thalassiosira decipiens

Takano, H., 1956.
Harmful blooming of minute cells of Thalassiosira decipiens in coastal water in Tokyo Bay. J. Ocean. Soc., Japan, 12(2):63-68.

Thalassiosira decipiens

Takano, H., 1946.
Harmful blooming of minute cells of Thalassiosira decipiens in coastal water in Tokyo Bay. J. Ocean. Soc., Japan, 12(2):63-67.

Thalassiosira delicatula

Jouse, A.P., G.S. Koroleva, G.A. Nagaeva, 1962
Diatoms in the surface layer of sediment in the Indian sector of the Antarctic. Investigations of marine bottom sediments. (In Russian; English summary). Trudy Inst. Okeanol., Akad. Nauk, SSSR, 61:19-92.

Thalassiosira diporocyclus n.sp.

Hasle, Grethe Rytter 1972.
Thalassiosira subtilis (Bacillariophyceae) and two allied species. Norw. J. Bot. 19(2): 111-137.

Thalassiosira dubia

Jouse, A.P., G.S. Koroleva, G.A. Nagaeva, 1962
Diatoms in the surface layer of sediment in the Indian sector of the Antarctic. Investigations of marine bottom sediments. (In Russian; English summary). Trudy Inst. Okeanol., Akad. Nauk, SSSR, 61:19-92.

Thalassiosira dubia n.sp.

Kozlova, O.G., 1962
Specific composition of diatoms in the waters of the Indian sector of the Antarctic. Investigation of Marine Bottom Sediments. (In Russian). Trudy Inst. Okeanol., Akad. Nauk, SSSR, 61: 3-18.
English summary, p. 18.

Thalassiosira excentrica

Boden, B.P., 1950
Some marine plankton diatoms from the west coast of South Africa. Trans. R.Soc. S. Africa. 32:321-434, 100 text figs.

Thalassiosira eccentrica

Fryxell, Greta A., and Grethe R. Hasle 1972.
Thalassiosira eccentrica (Ehrenb.) Cleve, T. symmetrica sp.nov., and some related centric diatoms. J. Phycol. 8(4): 297-317.

Thalassiosira fallax

Brunel, J., 1962
Le phytoplancton de la Baie de Chaleurs. Inst. Botan., Univ. Montréal, Contrib. No. 77: 365 pp., 66 pls.

Thalassiosira fallax

Hendy, N. Ingram, 1964
An introductory account of the smaller algae of British coastal waters. V. Bacillariophyceae (Diatoms). Her Majesty's Stationary Office, 317 pp., 45 pls.

Thalassiosira fallax

Jouse, A.P., G.S. Koroleva, G.A. Nagaeva, 1962
Diatoms in the surface layer of sediment in the Indian sector of the Antarctic. Investigations of marine bottom sediments. (In Russian; English summary). Trudy Inst. Okeanol., Akad. Nauk, SSSR, 61:19-92.

Thalassiosira fallax, n.sp.

Kozlova, O.G., 1962
Specific composition of diatoms in the waters of the Indian sector of the Antarctic. Investigation of Marine Bottom Sediments. (In Russian). Trudy Inst. Okeanol., Akad. Nauk, SSSR, 61: 3-18.
English summary, p. 18

Thalassiosira fluviatilis

Brunel, J., 1962
Le phytoplancton de la Baie de Chaleurs. Inst. Botan., Univ. Montréal, Contrib. No. 77: 365 pp., 66 pls.

Thalassiosira fluviatilis

Dweltz, N.E., and J. Ross Colvin, 1968.
The structure of the diatom Thalassiosira fluviatilis. Can. J. Microbiol., 14(10):1049-1052.

Thalassiosira fluviatilis

Fuhs, G. Wolfgang, 1969.
Phosphorus content and rate of growth in the diatoms Cyclotella nana and Thalassiosira fluviatilis. J. Phycol., 5(6): 312-321

Thalassiosira fluviatilis

Hasle, Grethe Rytter, 1961(1962).
The morphology of Thalassiosira fluviatilis from the polluted Inner Oslofjord. Nytt Mag. Botanikk, 9:151-154.

Thalassiosira fluviatilis

Hendy, N. Ingram, 1964
An introductory account of the smaller algae of British coastal waters. V. Bacillariophyceae (Diatoms). Her Majesty's Stationary Office, 317 pp., 45 pls.

Thalassiosira fluviatilis

Hobson, Louis A. and Robert J. Pariser, 1971.
The effect of inorganic nitrogen on macromolecular synthesis by Thalassiosira fluviatilis Hustedt and Cyclotella nana Hustedt grown in batch culture. J. exp. mar. Biol. Ecol., 6(1): 71-78.

Thalassiosira fluviatilis

McLachlan, J., and J.S. Craigie, 1966.
Chitan fibres in Cyclotella cryptica and growth of C. cryptica and Thalassiosira fluviatilis. In: Some contemporary studies in marine science, H. Barnes, editor, George Allen & Unwin, Ltd., 511-517.

Thalassiosira fluviatilis

Mullin, Michael M. and Elaine R. Brooks, 1970
Growth and metabolism of two planktonic, marine copepods as influenced by temperature and type of food. In: Marine Food Chains, J.H. Steele, editor, Oliver and Boyd, 74-95.

Thalassiosira fluviatilus

Takano, Hideaki, 1963.
Notes on marine littoral diatoms of Japan. Bull. Tokai Reg. Fish. Res. Lab., No. 36:1-8.

Thalassiosira gelatinosa

Jorgensen, E., 1900
Protophyten und Protozoën im Plankton aus der Norwegischen Westkerste. Bergens Mus. Aarb. 1899(6): 95 pp., 5 pls., 83 tables.

Thalassiosira gracilis

Jouse, A.P., G.S. Koroleva, G.A. Nagaeva, 1962
Diatoms in the surface layer of sediment in the Indian sector of the Antarctic. Investigations of marine bottom sediments. (In Russian; English summary). Trudy Inst. Okeanol., Akad. Nauk, SSSR, 61:19-92.

Thalassiosira gravida

Allen, E. J., 1914.
On the culture of the plankton diatom, Thalassiosira gravida Cleve, in artificial sea water. J.M.B.A., n.s., 10:417-439.

Thalassiosira gravida

Bigelow, H.B., and M. Leslie, 1930
Reconnaissance of the waters and plankton of Monterey Bay, July 1928. Bull. M.C.Z., 70(5):429-481, 43 text figs.

Thalassiosira gravida

Braarud, T., 1945
Experimental studies on marine plankton diatoms. Avhandlingar utgitt av Det Norske Videnskaps-Akademi i Oslo. 1. Mat.-Naturv. Klasse 1944. No.10, 1-16, 1 text fig.

Thalassiosira gravida

Braarud, T., 1945
A phytoplankton survey of the polluted waters of inner Oslo Fjord. Hvalrådets Skrifter, No.28, 142 pp., 19 text figs., 17 tables.

Thalassiosira gravida

Braarud, T., 1934
A note on the phytoplankton of the Gulf of Maine in the summer of 1933. Biol. Bull. 67(1):76-82.

Thalassiosira gravida

Braarud, T., and Adam Bursa, 1939
On the phytoplankton of the Oslo Fjord, 1933-1934. Hvalrådets Skr. No.19:1-63; 9 text figs. Reviewed. J. du Cons. 14(3): 418-420. A.C. Gardiner.

Thalassiosira gravida

Brunel, J., 1962
Le phytoplancton de la Baie de Chaleurs. Inst. Botan., Univ. Montréal, Contrib. No. 77: 365 pp., 66 pls.

Thalassiosira gravida

Brunel, Jules, 1962
Le phytoplancton de la Baie des Chaleurs. Contrib. Ministère de la Chasse et des Pêcheries, Province de Québec, No. 91: 365 pp.

Thalassiosira gravida

Chu, S. P., 1949
Experimental studies on the environmental factors influencing the growth of phytoplankton. Sci. & Tech. in China 2(3):37-52.

Thalassiosira gravida

Cupp, Easter E., 1943
Marine plankton diatoms of the west coast of North America. Bull. S.I.O. 5(1):1-238, 5 pls., 168 text figs.

Thalassiosira gravida

Cupp, E.E., 1934
Analysis of marine diatom collections taken from the Canal Zone to California during March, 1933. Trans. Am. Micros. Soc. LIII (1):22-29, 1 map.

Thalassiosira gravida

Fraser, J.H., and A. Saville, 1949
Plankton distribution in Scottish and adjacent waters in 1948. Ann. Biol. 5:61-62.

Thalassiosira gravida
Gilbert, J.Y., and W.E. Allen, 1943
The phytoplankton of the Gulf of California obtained by the "E.W. Scripps" in 1939 and 1940. J. Mar. Res. V(2):89-110, figs.30-31.

Thalassiosira gravida
Gran, H.H., 1908
Diatomeen. Nordisches Plankton, Botanischer Teil pp. XIX.1-XIX 146; 178 text figs.

Thalassiosira gravida
Gran, H.H., 1897
Protophyta: Diatomaceae, Silico-flagellata and Cilioflagellata. Den Norske Nordhavs Expedition 1876-1878, h. 24, 36 pp., 4 pls.

Thalassiosira gravida
Gran, H.H., and T. Braarud, 1935
A quantitative study of the phytoplankton in the Bay of Fundy and the Gulf of Maine (including observations on hydrography, chemistry, and turbidity). J. Biol. Bd., Canada, 1(5):279-467, 69 text figs.

Thalassiosira gravida
Hasle, Grethe Rytter, 1968.
The valve Processes of the centric diatom genus Thalassiosira. Nytt Mag.Bot.15(3):193-201.

Thalassiosira gravida
Hendy, N. Ingram, 1964
An introductory account of the smaller algae of British coastal waters. V. Bacillariophyceae (Diatoms). Her Majesty's Stationary Office, 317 pp., 45 pls.

Thalassiosira gravida
Hendey, N.I., 1937
The plankton diatoms of the southern seas. Discovery Repts. 16:151-364, pls.6-13.

Thalassiosira gravida
Iselin, C., 1930
A report on the coastal waters of Labrador based on explorations of the "Chance" during the summer of 1926. Proc. Am. Acad. Arts Sci., 66(1):1-37, 14 text figs.

Thalassiosira gravida
Jørgensen, E., 1905
B.Protistplankton and the diatoms in bottom samples. Hydrographical and biological investigations in Norwegian fiords. Bergens Mus. Skr. 7: 49-225.

Thalassiosira gravida
Jorgensen, E., 1900
Protophyten und Protozoen im Plankton aus der Norwegischen Westkerste. Bergens Mus. Aarb. 1899(6): 95 pp., 5 pls., 83 tables.

Thalassiosira gravida
Lefon, M., M. Durchon and Y. Saudray, 1955.
Recherches sur les cycles saisonnières du plankton Ann. Inst. Océan., 31(3):125-230.

Thalassiosira gravida
Lillick, L.C., 1940
Phytoplankton and planktonic protozoa of the offshore waters of the Gulf of Maine. Pt.II. Qualitative Composition of the Planktonic Flora. Trans. Am. Phil. Soc., n.s., 31(3):193-237, 13 text figs.

Thalassiosira gravida
Lillick, L.C., 1938
Preliminary report of the phytoplankton of the Gulf of Maine. Am. Mid. Nat. 20(3):624-640, 1 text figs 37 tables.

Thalassiosira gravida
Lillick, L.C., 1937
Seasonal studies of the phytoplankton off Woods Hole, Massachusetts. Biol. Bull. LXXIII (3):488-503, 3 text figs.

Thalassiosira gravida
Mangin, M. L., 1912
Phytoplancton de la croisière du "René" dans l'Atlantique (Septembre 1908). Ann. Inst. Ocean., n.s., 4(1):1-66, 2 pls., 41 text figs., 2 tables.

Thalassiosira gravida
Marukawa, H., 1921
Plankton lists and some new species of copepods, from the northern waters of Japan. Bull. Inst. Ocean., No.384, 15 pp., 3 pls., 1 chart. Monaco

Thalassiosira gravida
Morse, D.C., 1947
Some observations on seasonal variations in plankton population Patuxant River, Maryland 1943-1945. Bd. Nat. Res., Publ. No.65, Chesapeake Biol. Lab., 31, 3 figs.

Thalassiosira gravida
Paasche, E., 1961.
Notes on phytoplankton from the Norwegian Sea. Botanica Marina, 2(3/4):197-210.

Thalassiosira gravida
Pettersson, H., F. Gross, and F. Koczy, 1939
Large scale plankton cultures. Medd. fran Oceanografiska Institutet i Göteborg No.3 (Göteborgs Kungl. vetenskaps-och Vitterhets-Samhälles Handlingar Femte Följden. Ser. B) Vol.6(13):1-24.
1. The plankton shaft. H. Pettersson
2. Experiment with phytoplankton. F. Gross and F. Koczy
3. Experiments with zooplankton.

Thalassiosira gravida
Phifer, L.D. (undated)
The occurrence and distribution of plankton diatoms in Bering Sea and Bering Strait, July 26-August 24, 1934. Report of Oceanographic cruise of U.S. Coast Guard Cutter Chelan 1934, Part II(A):1-44 (mimeographed) plus fig.1 (after Pt.B)

Thalassiosira gravida
Sousa e Silva, E., 1949
Diatomaceas e Dinoflagelados de Baia de Cascais. Portugaliae Acta Biol., Volume: Julio Henriques, Ser. B: 300-383.

Thalassiosira gravida
Stemann-Nielsen, Einar, 1951
The marine vegetation of the Isefjord. A study on ecology and production. Medd. Komm. Danmarks Fiskeri-og Havundersøgelser. Ser. Plankton. 5(4); 114pp., 46 text figs.

Thalassiosira hispanica
Paulsen, O., 1930.
Études sur le microplancton de la mer d'Alboran. Trab. Inst. Esp. Ocean. No. 4:1-108, 61 textfigs.

Thalassiosira hyalina
Bigelow, H.B., and M. Leslie, 1930
Reconnaissance of the waters and plankton of Monterey Bay, July 1928. Bull. M.C.Z. 70(5):429-481, 43 text figs.

Thalassiosira hyalina
Boden, B.P., 1950
Some marine plankton diatoms from the west coast of South Africa. Trans. R.Soc. S. Africa. 32:321-434, 100 text figs.

Thalassiosira hyalina
Brunel, J., 1962
Le phytoplancton de la Baie de Chaleurs. Inst. Botan., Univ. Montréal, Contrib. No. 77: 365 pp., 66 pls.

Thalassiosira hyalina
Cassie Vivienne, 1961
Marine phytoplankton in New Zealand waters. Botanica Marina, 2(Suppl.): 54 pp., 8 pls.

Thalassiosira hyalina
Davidson, V.M., 1931.
Biological and oceanographic conditions in Hudson Bay. Contr. Canadian Biol. Fish., n.s., 6(26):497-509, 7 textfigs.

Thalassiosira hyalina
Gran, H.H., 1908
Diatomeen. Nordisches Plankton, Botanischer Teil pp. XIX.1-XIX 146; 178 text figs.

Thalassiosira hyalina
Gran, H.H., and T. Braarud, 1935
A quantitative study of the phytoplankton in the Bay of Fundy and the Gulf of Maine (including observations on hydrography, chemistry, and turbidity). J. Biol. Bd., Canada, 1(5):279-467, 69 text figs.

Thalassiosira hyalina
Hendy, N. Ingram, 1964
An introductory account of the smaller algae of British coastal waters. V. Bacillariophyceae (Diatoms). Her Majesty's Stationary Office, 317 pp., 45 pls.

Thalassiosira hyalina
Hendey, N.I., 1937
The plankton diatoms of the southern seas. Discovery Repts. 16:151-364, pls.6-13.

Thalassiosira hyalina
Jørgensen, E., 1905
B.Protistplankton and the diatoms in bottom samples. Hydrographical and biological investigations in Norwegian fiords. Bergens Mus. Skr. 7: 49-225.

Thalassiosira hyalina
Krasske, G., 1941
Die Kieselalgen des chilenischen Küstenplankton (Aus dem südchilenischen Kusten gebiet IX). Arcl. Hydrobiol. 38 (2):260-287.

Thalassiosira hyalina
Lillick, L.C., 1940
Phytoplankton and planktonic protozoa of the offshore waters of the Gulf of Maine. Pt.II. Qualitative Composition of the Planktonic Flora. Trans. Am. Phil. Soc., n.s., 31(3):193-237, 13 text figs.

Thalassiosira hyalina
Lillick, L.C., 1938
Preliminary report of the phytoplankton of the Gulf of Maine. Am. Mid. Nat. 20(3):624-640, 1 text figs 37 tables.

Thalassiosira hyalina
Motoda, Sigeru, Teruyoshi Kawamura, Tsuneyoshi Suzuki and Takashi Minoda, 1963.
Photosynthesis of a natural phytoplankton population mainly composed of a cold diatom, Thalassiosira hyalina, in Hakodate harbor, March 1962. Bull. Fac. Fish., Hokkaido Univ., 14(3):127-130.

Thalassiosira hyalina
Phifer, L.D. (undated)
The occurrence and distribution of plankton diatoms in Bering Sea and Bering Strait, July 26-August 24, 1934. Report of Oceanographic cruise of U.S. Coast Guard Cutter Chelan 1934, Part II(A):1-44 (mimeographed) plus fig.1 (after Pt.B)

Thalassiosira hyalina
Smayda, Theodore J., 1958
Biogeographical studies of marine phytoplankton. Oikos, 9(2): 158-191.

Thalassiosira hyalina
Somers, D., 1972
Scanning electron microscope studies on some species of the centric diatom genera Thalassiosira and Coscinodiscus. Biol. Jaarb. Dodonaea 40:317-322. Also in: Coll. Repr. Inst. Zeewetenschap. Onderzoek Belg. 3(1973).

Thalassiosira kryophila
Brunel, J., 1962
Le phytoplancton de la Baie de Chaleurs.
Inst. Botan., Univ. Montréal, Contrib. No. 77: 365 pp., 66 pls.

Thalassiosira kryoptula
Krasske, G., 1941
Die Kieselalgen des chilenischen Küstenplankton (Aus dem südchilenischen Kusten gebiet IX). Arcl. Hydrobiol. 38(2):260-287.

Thalassiosira levanderi
Brunel, J., 1962
Le phytoplancton de la Baie de Chaleurs.
Inst. Botan., Univ. Montréal, Contrib. No. 77: 365 pp., 66 pls.

Thalassiosira mala n.sp.
Takano, Hideaki, 1965.
New and rare diatoms from Japanese marine waters. 1.
Bull. Tokai Reg. Fish. Res. Lab., No. 42:1-10.

Thalassiosira margaritae
Jouse, A.P., G.S. Koroleva, G.A. Nagaeva, 1962
Diatoms in the surface layer of sediment in the Indian sector of the Antarctic. Investigations of marine bottom sediments. (In Russian; English summary).
Trudy Inst. Okeanol., Akad. Nauk. SSSR. 61:19-92.

Thalassiosira margaritae n. comb.
Kozlova, O.G., 1962
Specific composition of diatoms in the waters of the Indian sector of the Antarctic. Investigation of Marine Bottom Sediments. (In Russian).
Trudy Inst. Okeanol., Akad. Nauk. SSSR, 61:3-18.

English summary, p. 18

Thalassiosira marginata
Iyengar, M.O.P. and G.Venkataraman,1951.
The ecology and seasonal succession of the algae flora of the River Cooum at Madras with special reference to the Diatomaceae. J. Madras Univ, 21, Sect. B(1): 140-192, 1 pl of 4 figs., 11 text figs.

Thalassiosira mendiolana n. sp.
Hasle, G.R., and B.R. Heimdal 1970.
Some Species of the centric diatom genus Thalassiosira studied in the light and electron microscopes.
Beiheft Nova Hedwigia 31: 559-589.

Thalassiosira minuscula
Krasske, G., 1941
Die Kieselalgen des chilenischen Küstenplankton (Aus dem südchilenischen Kusten gebiet IX). Arck. Hydrobiol. 38(2):260-287.

Thalassiosira mirniae n.sp.
Jouse, A.P., G.S. Koroleva, G.A. Nagaeva, 1962
Diatoms in the surface layer of sediment in the Indian sector of the Antarctic. Investigations of marine bottom sediments. (In Russian; English summary).
Trudy Inst. Okeanol., Akad. Nauk. SSSR. 61:19-92.

Thalassiosira monoporocyclus n.sp.
Hasle, Grethe Rytter 1972.
Thalassiosira subtilis (Bacillariophyceae) and two allied species.
Norw. J. Bot. 19(2): 111-137.

Thalassiosira nana
Brunel, J., 1962
Le phytoplancton de la Baie de Chaleurs.
Inst. Botan., Univ. Montréal, Contrib. No. 77: 365 pp., 66 pls.

Thalassiosira cf. nana
Smayda, Theodore J., and Brenda J. Boleyn, 1965.
Experimental observations on the flotation of marine diatoms. 1. Thalassiosira cf. nana, Thalassiosira rotula and Nitzschia seriata.
Limnol. Oceanogr., 10(4):499-509.

Thalassiosira nana
Steemann-Nielsen, Einar, 1951
The marine vegetation of the Isefjord. A study on ecology and production. Medd. Komm. Danmarks Fiskeri-og Havundersøgelser. Ser. Plankton. 5(4); 114pp., 46 text figs.

Thalassiosira nitzschioides
Davidson, V.M., 1931.
Biological and oceanographic conditions in Hudson Bay. Contr. Canadian Biol. Fish., n.s., 6(26):497-509, 7 textfigs.

Thalassiosira Nordenskiöldii
Bigelow, H.B., and M. Leslie, 1930
Reconnaissance of the waters and plankton of Monterey Bay, July 1928.
Bull, M.C.Z., 70(5):429-481, 43 text figs.

Thalassiosira nordenskioeldii
Braarud, T., 1962
Species distribution in marine phytoplankton.
J. Oceanogr. Soc., Japan. 20th Ann. Vol., 628-649.

Thalassiosira Nordenskiöldii
Braarud, T., 1945
Experimental studies on marine plankton diatoms. Avhandlingar utgitt av Det Norske Videnskaps-Akademi i Oslo. 1. Mat.-Naturv. Klasse 1944. No.10, 1-16, 1 text fig.

Thalassiosira Nordenskiöldii
Braarud, T., 1945
A phytoplankton survey of the polluted waters of inner Oslo Fjord. Hvalrådets Skrifter, No.28, 142 pp., 19 text figs., 17 tables.

Thalassiosira Nordenskiöldi
Braarud, T., 1939
Observations on the phytoplankton of the Oslo Fjord, March-April, 1937. Nytt Magasin for Naturvidenskapene, 80:211-218, 1 text fig.

Thalassiosira Nordenskiöldi
Braarud, T., and Adam Bursa, 1939
On the phytoplankton of the Oslo Fjord, 1933-1934. Hvalrådets Skr. No.19:1-63; 9 text figs. Reviewed. J. du. Cons. 14(3): 418-420. A.C. Gardiner.

Thalassiosira nordenskiöld-ii
Brunel, J., 1962
Le phytoplancton de la Baie de Chaleurs.
Inst. Botan., Univ. Montréal, Contrib. No. 77: 365 pp., 66 pls.

Thalassiosira nordenskiöld-ii
Brunel, Jules, 1962
Le phytoplancton de la Baie des Chaleurs. Contrib. Ministère de la Chasse et des Pêcheries, Province de Québec, No. 91: 365 pp.

Thalassiosira nordenskiöl-dii
Castracane degli Antelminelli, F., 1886
1. Report on the Diatomaceae collected by H.M.S. Challenger during the years 1873-1876. Rept. Sci. Results, H.M.S. Challenger, Botany Vol. II, 178 pp., 30 pls.

Thalassiosira nordenskioldii
Colton, John B., Jr., and Robert R. Marak, 1962.
Use of the Hardy continuous plankton recorder in a fishery research program.
Bull. Mar. Ecology, 5(49):231-246.

Thalassiosira nordenskioldii
Cupp, Easter E., 1943
Marine plankton diatoms of the west coast of North America. Bull. S.I.O. 5(1):1-238, 5 pls., 168 text figs.

Thalassiosira Nordenskioldii
Dangeard, P., 1927
Phytoplankton de la croisière du "Sylvana". Ann. Inst. Ocean., Monaco, n.s., 4(8):286-401, 54 text figs. (Feirer-Juin 1913).

Thalassiosira nordenskioldii
Davidson, V.M., 1931
Biological and oceanographic conditions in Hudson Bay. Contr. Canadian Biol. Fish., n.s., 6(26):497-509, 7 textfigs.

Thalassiosira Nordenskioldii
Gran, H.H., 1908
Diatomeen. Nordisches Plankton, Botanischer Teil pp. XIX.1-XIX 146; 178 text figs.

Thalassiosira Nordenskiöldii
Gran, H.H., 1897
Protophyta: Diatomaceae, Silico-flagellata and Cilioflagellata. Den Norske Nordhavs Expedition 1876-1878, h. 24, 36 pp., 4 pls.

Thalassiosira Nordenskioeldi
Gran, H.H., and T. Braarud, 1935
A quantitative study of the phytoplankton in the Bay of Fundy and the Gulf of Maine (including observations on hydrography, chemistry, and turbidity). J. Biol. Bd., Canada, 1(5):279-467, 69 text figs.

Thalassiosira nordenskioeldii
Gran, H. H. and E. C. Angst, 1931
Plankton diatoms of Puget Sound. Publ. Puget Sound Biol. Sta. 7:417-519, 95 text figs.

Thalassiosira nordenskioeldi
Hasle, Gretha Rytter, 1968.
The valve Processes of the centric diatom genus Thalassiosira.
Nytt Mag. bot.15(3):193-201.

Thalassiosira nordenskioldii
Hendy, N. Ingram, 1964
An introductory account of the smaller algae of British coastal waters. V. Bacillariophyceae (Diatoms).
Her Majesty's Stationary Office, 317 pp., 45 pls.

Thalassiosira Nordenskjoldi
Iselin, C., 1930
A report on the coastal waters of Labrador based on explorations of the "Chance" during the summer of 1926. Proc. Am. Acad. Arts Sci., 66(1):1-37, 14 text figs.

Thalassiodira nordenskiöldi
Jørgensen, E., 1905
B.Protistplankton and the diatoms in bottom samples. Hydrographical and biological investigations in Norwegian fjords. Bergens Mus. Skr. 7: 49-225.

Thalassiosira nordenskioldii
Jitts, H.R., C.D. McAllister, K. Stephens and J.D.H. Strickland, 1964.
The cell division rates of some marine phytoplankters as a function of light and temperature. J. Fish. Res. Bd., Canada, 21(1):139-157.

Thalassiosira nordenski-ldii
Jorgensen, E., 1900
Protophyten und Protozoën im Plankton aus der Norwegischen Westkerste. Bergens Mus. Aarb. 1899(6): 95 pp., 5 pls., 83 tables

Thalassiosira Nordenskioeldi
Lillick, L.C., 1940
Phytoplankton and planktonic protozoa of the offshore waters of the Gulf of Maine. Pt.II. Qualitative Composition of the Planktonic Flora. Trans. Am. Phil. Soc., n.s., 31(3):193-237, 13 text figs.

Thalassiosira Nordenskioeldii
Lillick, L.C., 1938
Preliminary report of the phytoplankton of the Gulf of Maine. Am. Mid. Nat. 20(3):624-640, 1 text figs 37 tables.

Thalassiosira Nordenskioeldii

Lillick, L.C., 1937
Seasonal studies of the phytoplankton off Woods Hole, Massachusetts. Biol. Bull. LXXIII (3):488-503, 3 text figs.

Thalassiosira Nordenskioildii

Marukawa, H., 1921
Plankton lists and some new species of copepods, from the northern waters of Japan. Bull. Inst. Ocean., No.384, 15 pp., 3 pls., 1 chart. Monaco

Thalassiosira Nordenskiöldii

Meunier, A., 1915
Microplancton de la Mer Flamande. 2. Diatomées (excepté le genre Chaetoceros). Mem. Mus. Roy. Hist. Nat., Belgique, 7(3):1-118, Pls. VIII-XIV.

Thalassiosira nordenskiöldii

Pavillard, J., 1925
Bacillariales. Rept. on the Danish Oceangr. Exped., 1908-10 to the Mediterranean and adj. seas. Vol.II., Biol. J4:72 pp., 116 text figs.

Thalassiosira Nordenskioeldi

Phifer, L.D. (undated)
The occurrence and distribution of plankton diatoms in Bering Sea and Bering Strait, July 26-August 24, 1934. Report of Oceanographic cruise of U.S. Coast Guard Cutter Chelan 1934, Part II(A):1-44 (mimeographed) plus fig.1 (after Pt.B)

Thalassiosira Nordenskioeldii

Rampi, L., 1940
Diatomee del Mare Adriatico. Nuovo Giornale Botanico Italiano, n.s., 47:559-608.

Thalassiosira nordenskioldi

Redfield, A.C., 1934.
On the proportions of organic derivatives in sea water and their relation to the composition of plankton. In : James Johnstone Memorial Volume:176-192, 5 textfigs.

Thalassiosira Nordenskioildii

Rumkówna, A., 1948
[List of the phytoplankton species occurring in the superficial water layers in the Gulf of Gdańsk] Bull. Lab. mar., Gdynia, No. 4: 139-141 with tables in back.

Thalassiosira nordenskioldii [a]

Somers, D., 1972
Scanning electron microscope studies on some species of the centric diatom genera Thalassiosira and Coscinodiscus. Biol. Jaarb. Dodonaea 40:317-322. Also in: Coll. Repr. Inst. Zeewetenschap. Onderzoek Belg. 3(1973).

Thalassiosira nordenskjöldii

Steemann-Nielsen, Einar, 1951
The marine vegetation of the Isefjord. A study on ecology and production. Medd. Komm. Danmarks Fiskeri-og Havundersøgelser. Ser. Plankton. 5(4); 114pp., 46 text figs.

Thalassiosira Nordenskiøldii

Steenstrup, K.J.V. 1893
Beretning om Undersøgelsesrejserne i Nord-Grønland i Aårene 1878-1880 Medd. Grønland, 5:1-41.

Thalassiosira obica

Jouse, A.P., G.S. Koroleva, G.A. Nagaeva, 1962
Diatoms in the surface layer of sediment in the Indian sector of the Antarctic. Investigations of marine bottom sediments. (In Russian; English summary). Trudy Inst. Okeanol., Akad. Nauk. SSSR. 61:19-92.

Thalassiosira obica nsp.

Kozlova, O.G., 1962
Specific composition of diatoms in the waters of the Indian sector of the Antarctic. Investigation of Marine Bottom Sediments. (In Russian). Trudy Inst. Okeanol., Akad. Nauk. SSSR, 61: 3-18.

Thalassiosira oestrupii

Hasle, Grethe Rytter, 1960
Phytoplankton and ciliate species from the Tropical Pacific. Skr. Norske Videnskaps-Akad., Oslo, 1. Mat.-Nat. Kl., 1960(2): 1-50.

Thalassiosira pacifica

Gilbert, J.Y., and W.E. Allen, 1943
The phytoplankton of the Gulf of California obtained by the "E.W. Scripps" in 1939 and 1940. J. Mar. Res. V(2):89-110, figs.30-31.

Thalassiosira pacifica n.sp.

Gran, H.H. and E.C. Angst, 1931
Plankton diatoms of Puget Sound. Publ. Puget Sound Biol. Sta. 7:417-519, 95 text figs.

Thalassiosira parthenia n.sp.

Schrader, Hans-Joachim 1972.
Thalassiosira partheneis, eine neue Gallert-lager bildende zentrale Diatomee. Meteor Forsch. Ergebn. (D): 58-64

Thalassiosira perpusilla n.sp.

Kozlova, O.G., 1962
Specific composition of diatoms in the waters of the Indian sector of the Antarctic. Investigation of Marine Bottom Sediments. (In Russian). Trudy Inst. Okeanol., Akad. Nauk, SSSR, 61: 3-18.

English summary, p. 18

Thalassiosira polychorda

Jorgensen, E., 1900
Protophyten und Protozoën im Plankton aus der Norwegischen Westkerste. Bergens Mus. Aarb. 1899(6): 95 pp., 5 pls., 83 tables.

Thalassiosira poro-irregulata n.sp.

Hasle, G.R. and B.R. Heimdal 1970.
Some species of the centric diatom genus Thalassiosira studied in the light and electron microscopes. Beiheft Nova Hedwigia 31: 559-589.

Thalassiosira poroseriata

Hasle, Grethe Rytter, 1968.
The valve Processes of the centric diatom genus Thalassiosira. Nytt Mag. Bot. 15(3):193-201.

Thalassiosira pseudonana

Erickson, Stanton J. 1972.
Toxicity of copper to Thalassiosira pseudonana. J. Phycol. 8(4): 318-323.

Thalassiosira pseudonana [a]

Guillard, Robert R.L., Peter Kilham and Togwell A. Jackson 1973.
Kinetics of Silicon-limited growth in the marine diatom Thalassiosira pseudonana Hasle and Heimdal (= Cyclotella nana Hustedt). J. Phycol. 9(3): 233-237.

Thalassiosira pseudonana

Paasche, E., 1973
Silicon and the ecology of marine plankton diatoms. 1. Thalassiosira pseudonana (Cyclotella nana) grown in a chemostat with silicate as limiting nutrient. Mar. Biol. 19(2): 117-126.

Thalassiosira pulchella n.sp.

Takano, Hideaki, 1963.
Notes on marine littoral diatoms of Japan. Bull. Tokai Reg. Fish. Res. Lab., No. 36:1-8.

Thalassiosira rotula

Allen, W.E., 1937
Plankton diatoms of the Gulf of California obtained by the G. Allan Hancock Expedition of 1936. The Hancock Pacific Expeditions, Univ. So. Calif. Publ. 3:47-59, 1 fig.

Thalassiosira rotula

Boden, B.P., 1950
Some marine plankton diatoms from the west coast of South Africa. Trans. R.Soc. S. Africa. 32:321-434, 100 text figs.

Thalassiosira rotula

Brunel, J., 1962
Le phytoplancton de la Baie de Chaleurs. Inst. Botan., Univ. Montréal, Contrib. No. 77: 365 pp., 66 pls.

Thalassiosira rotula

Cupp, Easter E., 1943
Marine plankton diatoms of the west coast of North America. Bull. S.I.O. 5(1):1-238, 5 pls., 168 text figs.

Thalassiosira rotula

Cupp, E.E. and Allen, W.E., 1938
Plankton diatoms of the Gulf of California obtained by Allan Hancock Pacific Expedition of 1937. The Hancock Pacific Expeditions, The Univ. So. Calif. Publ. 3: 61-74, 1 map, pls.4-15.

Thalassiosira rotula

Dangeard, P., 1927
Phytoplankton de la croisière du "Sylvana". Ann. Inst. Ocean., Monaco, n.s., 4(8):286-401, 54 text figs. (février-Juin 1913).

Thalassiosira rotula

Gilbert, J.Y., and W.E. Allen, 1943
The phytoplankton of the Gulf of California obtained by the "E.W. Scripps" in 1939 and 1940. J. Mar. Res. V(2):89-110, figs. 30-31.

Thalassiosira rotula

Gran, H.H. and E.C. Angst, 1931
Plankton diatoms of Puget Sound. Publ. Puget Sound Biol. Sta. 7:417-519, 95 text figs.

Thalassiosira rotula

Hasle, Grethe Rytter, 1968.
The valve Processes of the centric diatom genus Thalassiosira. Nytt Mag. Bot. 15(3):193-201.

Thalassiosira rotula

Hendy, N. Ingram, 1964
An introductory account of the smaller algae of British coastal waters. V. Bacillariophyceae (Diatoms). Her Majesty's Stationary Office, 317 pp., 45 pls.

Thalassiosira rotula

Meunier, A., 1915
Microplancton de la Mer Flamande. 2. Diatomées (excepté le genre Chaetoceros). Mem. Mus. Roy. Hist. Nat., Belgique, 7(3):1-118, Pls. VIII-XIV.

Thalassiosira rotula

Pavillard, J., 1925
Bacillariales. Rept. on the Danish Oceangr. Exped., 1908-10 to the Mediterranean and adj. seas. Vol.II., Biol. J4:72 pp., 116 text figs.

Thalassiosira rotula [a]

Schöne, H.K., 1972.
Experimentelle Untersuchungen zur Ökologie der marinen Kieselalge Thalassiosira rotula. I. Temperatur und Licht. Mar. Biol. 13(4): 284-291.

Thalassiosira rotula
Smayda, Theodore J., and Brenda J. Boleyn, 1965.
Experimental observations on the flotation of marine diatoms. 1. Thalassiosira cf. nana, Thalassiosira rotula and Nitzschia seriata. Limnol. Oceanogr., 10(4):499-509.

Thalassiosira salvadoriana
Margalef, Ramón, y Dolores Blasco, 1970. Influencia del puerto de Barcelona sobre el fitoplancton de las áreas vecinas: una mancha de plancton de gran densidad, con dominancia de Thalassiosira, observada en agosto de 1969. Inv. pesq. Barcelona, 34(2): 575-580

Thalassiosira saturni
Brunel, J., 1962
Le phytoplancton de la Baie de Chaleurs. Inst. Botan., Univ. Montréal, Contrib. No. 77: 365 pp., 66 pls.

Thalassiosira spicula n. sp.
Kozlova, O.G., 1962
Specific composition of diatoms in the waters of the Indian sector of the Antarctic. Investigation of Marine Bottom Sediments. (In Russian).
Trudy Inst. Okeanol., Akad. Nauk. SSSR, 61: 3-18.

English summary, p. 18.

Thalassiosira subtilis
Bigelow, H.B., and M. Leslie, 1930
Reconnaissance of the waters and plankton of Monterey Bay, July 1928. Bull. M.C.Z., 70(5):429-481, 43 text figs.

Thalassiosira subtilis
Boden, B.P., 1950
Some marine plankton diatoms from the west coast of South Africa. Trans. R.Soc. S. Africa. 32:321-434, 100 text figs.

Thalassiosira subtilis (?)
Braarud, T., 1945
A phytoplankton survey of the polluted waters of inner Oslo Fjord. Hvalrådets Skrifter, No.28, 142 pp., 19 text figs., 17 tables.

Thalassiosira subtilis
Brunel, J., 1962
Le phytoplancton de la Baie de Chaleurs. Inst. Botan., Univ. Montréal, Contrib. No. 77: 365 pp., 66 pls.

Thalassiosira subtilis
Cupp, Easter E., 1943
Marine plankton diatoms of the west coast of North America. Bull. S.I.O. 5(1):1-238, 5 pls., 168 text figs.

Thalassiosira subtilis
Cupp, E.E., 1937
Seasonal distribution and occurrence of marine diatoms and dinoflagellates at Scotch Cap, Alaska. Bull. S.I.O. Tech. ser.4(3):71-100, 7 textfigs.

Thalassiosira subtilis
Cupp, E.E., 1934
Analysis of marine diatom collections taken from the Canal Zone to California during March, 1933. Trans. Am. Micros. Soc. LIII (1):22-29, 1 map.

Thalassiosira subtilis
Gilbert, J.Y., and W.E. Allen, 1943
The phytoplankton of the Gulf of California obtained by the "E.W. Scripps" in 1939 and 1940. J. Mar. Res. V(2):89-110, figs.30-31.

Thalassiosira subtilis
Gran, H.H., 1908
Diatomeen. Nordisches Plankton, Botanischer Teil pp. XIX.1-XIX.146; 178 text figs.

Thalassiosira subtilis
Hasle, Grethe Rytter 1972. Thalassiosira subtilis (Bacillariophyceae) and two allied species. Norw. J. Bot. 19(2): 111-137.

Thalassiosira subtilis
Hendy, N. Ingram, 1964
An introductory account of the smaller algae of British coastal waters. V. Bacillariophyceae (Diatoms).
Her Majesty's Stationary Office, 317 pp., 45 pls.

Thalassiosira subtilis
Hendey, N.I., 1937
The plankton diatoms of the southern seas. Discovery Repts. 16:151-364, pls.6-13.

Thalassiosira subtilis
Iselin, C., 1930
A report on the coastal waters of Labrador based on explorations of the "Chance" during the summer of 1926. Proc. Am. Acad. Arts Sci., 66(1):1-37, 14 text figs.

Thalassiosira subtilis
Lillick, L.C., 1940
Phytoplankton and planktonic protozoa of the offshore waters of the Gulf of Maine. Pt.II. Qualitative Composition of the Planktonic Flora. Trans. Am. Phil. Soc., n.s., 31(3):193-237, 13 text figs.

Thalassiosira Tcherniae n.sp. Adélie
Manguin, E., 1957.
Premier inventaire des diatomées de la Terre Antarctique. Espèces nouvelles.
Rev. Algologique, n.s., 3(3):111-134.

Thalassiosira subtilis
Rumkówna, A., 1948
List of the phytoplankton species occurring in the superficial water layers in the Gulf of Gdańsk. Bull. Lab. mar., Gdynia, No. 4: 139-141 with tables in back.

Thalassiosira subtilis
Sousa e Silva, E., 1949
Diatomaceas e Dinoflagelados de Baia de Cascais. Portugaliae Acta Biol., Volume: Julio Henriques, Ser. B: 300-383, 9 pls, 2 fold-in tables.

Thalassiosira symmetrica n.sp.
Fryxell, Greta A., and Grethe R. Hasle 1972. Thalassiosira eccentrica (Ehrenb.) Cleve, T. symmetrica sp.nov., and some related centric diatoms.
J. Phycol. 8 (4): 297-317.

Thalassiosira tcherniae
Jouse, A.P., G.S. Koroleva, G.A. Nagaeva, 1962
Diatoms in the surface layer of sediment in the Indian sector of the Antarctic. Investigations of marine bottom sediments. (In Russian; English summary).
Trudy Inst. Okeanol., Akad. Nauk. SSSR. 61:19-92.

Thalassiosira tropica n.sp. littoral
Misra, J.N., 1956.
A systematic account of some/marine diatoms from the west coast of India. J. Bombay Nat. Hist. Soc. 53(4): 537-568.

Thalassiosira tumida
El-Sayed, Sayed Z. 1971. Observations on phytoplankton bloom in the Weddell Sea. Biology of the Antarctic Seas IV, George A. Llano and I. Eugene Wallen, editors. Antarct. Res. Ser. Am. Geophys. Un. 17: 301-312.

Thalassiothrix
Bajkov, A.D. and T.W. Robinson, 1947
The distribution and occurrence of marine plankton. Aero Medical Laboratory Army Air Forces. Air Material Command. Engineering Division, Memo. Rept., Ser. No. TSEAA-691-3F-1, Expenditure Order No.691-12, ADB/ml, 7 pp. (ozalid) 30 Apr. 1947.

Thalassiothrix
Desikachary, T.V., 1956(1957).
Electron microscope studies on diatoms. J. R. Microsc. Soc. (3)76(1/2):9-36.

Thalassiothrix
Wheeler, J.E.G., 1939
Plankton investigations. Bermuda Biological Station. Second Report. October 1939. 7 pp. (typed), 5 figs. Plymouth, Oct. 23, 1939.

Thalassiothrix acuta (?)
Allen, W. E., 1938
The Templeton Crocker Expedition to the Gulf of California in 1935 - The Phytoplankton. Amer. Microsc. Soc., Trans. 57:328-335.

Thalassiothrix acuta
Cupp, E.E., 1934
Analysis of marine diatom collections taken from the Canal Zone to California during March, 1933. Trans. Am. Micros. Soc. LIII (1):22-29, 1 map.

Thalassiothrix acuta
Hendey, N.I., 1937
The plankton diatoms of the southern seas. Discovery Repts. 16:151-364, pls.6-13.

Thalassiothrix antarctica
Frenguelli, Joaquin, and Hector Antonio Orlando, 1959.
Operación MERLUZA. Diatomeas y silicoflagelados del plancton del "VI Crucero". Servicio Hidrogr. Naval., Argentina, Publ. No. H. 619: 5-62.

Thalassiothrix antarctica
Hendey, N.I., 1937
The plankton diatoms of the southern seas. Discovery Repts. 16:151-364, pls.6-13.

Thalassiothrix antarctica
Jouse, A.P., G.S. Koroleva, G.A. Nagaeva, 1962
Diatoms in the surface layer of sediment in the Indian sector of the Antarctic. Investigations of marine bottom sediments. (In Russian; English summary).
Trudy Inst. Okeanol., Akad. Nauk. SSSR, 61: 19-92.

Thalassiothrix curvata n.sp
Castracane degli Antelminelli, F., 1886
1. Report on the Diatomaceae collected by H.M.S. Challenger during the years 1873-1876. Rept. Sci. Results, H.M.S. Challenger, Botany Vol. II, 178 pp., 30 pls.

Thalassiothrix curvata
Schröder, B., 1900
Phytoplankton des Golfes von Neapel nebst vergleichenden Ausblicken auf das atlantischen Ozean. Mitt. Zool. Stat. Neapel, 14:1-38.

Thalassiothrix delicatula n.sp.
Cupp, Easter E., 1943
Marine plankton diatoms of the west coast of North America. Bull. S.I.O. 5(1):1-238, 5 pls., 168 text figs.

Thalassiothrix elongata
Allen, W.E., and E.E. Cupp, 1935
Plankton diatoms of the Java Sea. Annales du Jardin Botanique de Buitenzorg XLIV (2):101-174, figs.1-127.

(drawings of all species mentioned)

Thalassiothrix frauenfeldii

Allen, W. E., 1938
The Templeton Crocker Expedition to the Gulf of California in 1935 - The Phytoplankton. Amer. Microsc. Soc., Trans. 57:328-335.

Allen, W.E., 1937
Plankton diatoms of the Gulf of California obtained by the G. Allan Hancock Expedition of 1936. The Hancock Pacific Expeditions, Univ. So. Calif. Publ. 3:47-59, 1 fig.

Allen, W. E., 1936
Occurrence of marine plankton diatoms in a ten-year series of daily catches in southern California. Am. Jour. Bot. 23(1):60-63.

Allen, W.E., 1928
Catches of marine diatoms and dinoflagellates taken by boat in Southern California waters in 1926. Bull. S.I.O., tech. ser., 1:201-246, 6 textfigs.

Allen, W.E., and E.E. Cupp, 1935
Plankton diatoms of the Java Sea. Annales du Jardin Botanique de Buitenzorg XLIV (2):101-174, figs.1-127.
(drawings of all species mentioned)

Bigelow, H.B., and M. Leslie, 1930
Reconnaissance of the waters and plankton of Monterey Bay, July 1928. Bull. M.C.Z., 70(5):429-481, 43 text figs.

Castracane degli Antelminelli, F., 1886
1. Report on the Diatomaceae collected by H.M.S. Challenger during the years 1873-1876. Rept. Sci. Results, H.M.S. Challenger, Botany Vol. II, 178 pp., 30 pls.

Cupp, Easter E., 1943
Marine plankton diatoms of the west coast of North America. Bull. S.I.O. 5(1):1-238, 5 pls., 168 text figs.

Cupp, E.E., 1937
Seasonal distribution and occurrence of marine diatoms and dinoflagellates at Scotch Cap, Alaska. Bull. S.I.O. Tech. ser.4(3):71-100, 7 textfigs.

Cupp, E.E. and Allen, W.E., 1938
Plankton diatoms of the Gulf of California obtained by Allan Hancock Pacific Expedition of 1937. The Hancock Pacific Expeditions, The Univ. So. Calif. Publ. 3: 61-74, 1 map, pls.4-15.

Dangeard, P., 1927
Phytoplankton de la croisière du "Sylvana". Ann. Inst. Ocean., Monaco, n.s., 4(8):286-401, 54 text figs. (Février-Juin 1913).

Delegazione Italiana della Commissione Internazionale per l'Esplorazione Scientifica del Mediterraneo, 1941
Note sul plancton della Laguna veneta. [Memoria CCLXXXIX], Arch. di Ocean. e Limn. Anno I, Fasc. I, 1941 XIX: 31-57 pp.

Delsman, H. C., 1939.
Preliminary plankton investigations in the Java Sea. Treubia, 17:139-181, 8 maps, 41 figs.

Ercegovic, A., 1936
Etudes qualitative et quantitatives du phytoplancton dans les eaux cotières de l'Adriatique oriental moyen au cours de l'année 1934. Acta Adriatica 1(9):1-126

Forti, A., 1922
Ricerche sulla flora pelagica (fitoplancton) di Quarto dei Mille. Mem. R. Com. Talass. Ital. 97:248 pp., 13 pls.

Frenguelli, Joaquin, and Hector Antonio Orlando, 1959.
Operacion MERLUZA. Diatomeas y silico-flagelados del plancton del "VI Crucero". Servicio Hidrogr. Naval., Argentina, Publ. No. H. 619: 5-62.

Ghazzawi, F.M., 1939
Plankton of the Egyptian waters. A study of the Suez Canal Plankton. (A) Phytoplankton. Preliminary Report 83 pp., and Memoires, Min. Commerce-Industry, Egypt, Hydrobiol. & Fish. 65 figs.

Gilbert, J.Y., and W.E. Allen, 1943
The phytoplankton of the Gulf of California obtained by the "E.W. Scripps" in 1939 and 1940. J. Mar. Res. V(2):89-110, figs.30-31.

Gran, H.H., 1908
Diatomeen. Nordisches Plankton, Botanischer Teil pp. XIX.1-XIX 146; 178 text figs.

Hasle, Grethe Rytter, and Blanca Rojas E. de Mendiola 196?
The fine structure of some Thalassionema and Thalassiothrix species. Phycologia 6 (2/3): 107-118.

Hendy, N. Ingram, 1964
An introductory account of the smaller algae of British coastal waters. V. Bacillariophyceae (Diatoms).
Her Majesty's Stationary Office, 317 pp., 45 pls.

Jorgensen, E., 1900
Protophyten und Protozoën im Plankton aus der Norwegischen Westkerste. Bergens Mus. Aarb. 1899(6): 95 pp., 5 pls., 83 tables.

Kokubo, S., 1952
Results of the observations on the plankton and oceanography of Mutsu Bay during 1950, reference being made also to the period 1946-1950. Bull Mar.Biol.Sta., Asamushi 5(1/4): 1-54, 3 tables,(fold-in), 1 fold-in.

Lillick, L.C., 1940
Phytoplankton and planktonic protozoa of the offshore waters of the Gulf of Maine. Pt.II. Qualitative Composition of the Planktonic Flora. Trans. Am. Phil. Soc., n.s., 31(3):193-237, 13 text figs.

Lillick, L.C., 1937
Seasonal studies of the phytoplankton off Woods Hole, Massachusetts. Biol. Bull. LXXIII (3):488-503, 3 text figs.

Massutí Algamora, M., 1949
Estudio de diez y seis muestras de plancton del Golfo de Nápoles. Publ. Inst. Biol. Appl. 5:85-94, 1 fold-in table.

Paiva Carvalho, J., 1950
O plancton do Rio Maria Rodriques (Cananeia). 1. Diatomaceas e Dinoflagelados. Bol. Inst. Paulista Oceanogr. 1(1); 27-43, 2 fold-in tables, 2 figs.

Pavillard, J., 1925
Bacillariales. Rept. on the Danish Oceangr. Exped., 1908-10 to the Mediterranean and adj. seas. Vol.II., Biol. J4:72 pp., 116 text figs.

Pavillard, J., 1905
Recherches sur la flore pelagique (Phytoplankton) de l'Etang de Thau. Theses presentees a la Fac. Sci., Paris, 116 pp., 3 pls.

Politis, J., 1949
Diatomees marines de Bosphores et des ibes de la mer de Marmara. II Practica tou Hellenikou Hidrobiologikou Institutoutou 1929, Etoz 1929, 3(1):11-31.

Rampi, L., 1942
Ricerche sul fitoplancton del Mare Ligure 6. Le diatomee delle acque di Sanremo. Nuovo Giornale Botanico Italiano, N.S., 49:252-268.

Rampi, L., 1940
Diatomee del Mare Adriatico. Nuovo Giornale Botanico Italiano, n.s., 47:559-608.

Schröder, B., 1900
Phytoplankton des Golfes von Neapel nebst vergleichenden Ausblicken auf das atlantischen Ozean. Mitt. Zool. Stat. Neapel, 14:1-38.

Sousa e Silva, E., 1949
Diatomaceas e Dinoflagelados de Baia de Cascais. Portugaliae Acta Biol., Volume: Julio Henriques, Ser. B: 300-383, 9 pls, 2 fold-in tables.

Sousa e Silva, E., and J. Dos Santos-Pinto, 1948
O Plancton da Baía de S. Martinho do Porto. 1. Diatomaceas e Dinoflagelados. Bol. Soc. Portuguese de Ciencias Naturais, 16(2):134-187, 6 pls. (Trav. Sta. Biol. Mar. de Lisbonne No. 52).

Uyeno, Fukuzo, 1961
Oceanographical and ecological studies on primary production of the sea, with special references to relationship between diatom production and temperature and chlorinity of water.
Rept., Fac. Fish., Pref. Univ., Mie, 4(1): 1-64.

Thalassiothrix gibberula n.sp.

Hasle, Grethe Rytter, 1960
Phytoplankton and ciliate species from the Tropical Pacific.
Skr. Norske Videnskaps-Akad., Oslo, 1. Mat.-Nat. Kl., 1960(2): 1-50.

Thalassiothrix heteromorpha

Allen, W.E., 1937
Plankton diatoms of the Gulf of California obtained by the G. Allan Hancock Expedition of 1936. The Hancock Pacific Expeditions, Univ. So. Calif. Publ. 3:47-59, 1 fig.

Thalassiothrix hetermorpha

Gilbert, J.Y., and W.E. Allen, 1943
The phytoplankton of the Gulf of California obtained by the "E.W. Scripps" in 1939 and 1940. J. Mar. Res. V(2):89-110, figs.30-31.

Thalassiothrix longissima

Allen, W. E., 1938
The Templeton Crocker Expedition to the Gulf of California in 1935 - The Phytoplankton. Amer. Microsc. Soc., Trans. 57:328-335.

Thalassiothrix longissima
Allen, W.E., 1937
Plankton diatoms of the Gulf of California obtained by the G. Allan Hancock Expedition of 1936. The Hancock Pacific Expeditions, Univ. So. Calif. Publ. 3:47-59, 1 fig.

Thalassiothrix longissima
Bigelow, H. B., 1922
Exploration of the coastal water off the northeastern United States in 1916 by the U.S. Fisheries Schooner Grampus. Bull. M.C.Z. 65 (5):85-188, 53 text figs.

Thalassiothrix longissima
Bigelow, H.B., and M. Leslie, 1930
Reconnaissance of the waters and plankton of Monterey Bay, July 1928. Bull. M.C.Z., 70(5):429-481, 43 text figs.

Thalassiothrix longissima
Braarud, T., 1962
Species distribution in marine phytoplankton. J. Oceanogr. Soc., Japan, 20th Ann. Vol., 628-649.

Thalassiothrix longissima
Braarud, T., 1945
A phytoplankton survey of the polluted waters of inner Oslo Fjord. Hvalrådets Skrifter, No.28, 142 pp., 19 text figs., 17 tables.

Thalassiothrix longissima
Boden, B.P., 1950
Some marine plankton diatoms from the west coast of South Africa. Trans. R.Soc. S. Africa. 32:321-434, 100 text figs.

Thalassiothrix longissima
Chiba, T., 1949
On the distribution of the plankton in the eastern China Sea and Yellow Sea. 1. Plankton composition in the spring. J. Shimonoseki Coll. Fisheries, 1(1):57-63, 1 fig.

Thalassiothrix longissima
Cupp, Easter E., 1943
Marine plankton diatoms of the west coast of North America. Bull. S.I.O. 5(1):1-238, 5 pls., 168 text figs.

Thalassiothrix longissima
Cupp, E.E., 1937
Seasonal distribution and occurrence of marine diatoms and dinoflagellates at Scotch Cap, Alaska, Bull. S.I.O. Tech. ser.4(3):71-100, 7 textfigs.

Thalassiothrix longissima
Cupp, E.E, and Allen, W.E., 1938
Plankton diatoms of the Gulf of California obtained by Allan Hancock Pacific Expedition of 1937. The Hancock Pacific Expeditions, The Univ. So. Calif. Publ. 3: 61-74, 1 map, pls.4-15.

Thalassiothrix longissima
Dangeard, P., 1927
Phytoplankton de la croisière du "Sylvana". Ann. Inst. Ocean., Monaco, n.s., 4(8):286-401, 54 text figs. (Feirer-Juin 1913).

Thalassiothrix longissima
Ercegovic, A., 1936
Etudes qualitative et quantitatives du phytoplancton dans les eaux cotières de l'Adriatique oriental moyen au cours de l'année 1934. Acta Adriatica 1(9):1-126

Thalassiothrix longissima
Forti, A., 1922
Ricerche sulla flora pelagica (fitoplancton) di Quarto dei Mille. Mem. R. Com. Talass. Ital. 97:248 pp., 13 pls.

Thalassiothrix longissima
Ghazzawi, F.M., 1939
Plankton of the Egyptian waters. A study of the Suez Canal Plankton. (A) Phytoplankton. Preliminary Report 83 pp., Notes and Memoires, Min. Commerce-Industry, Egypt, Hydrobiol. & Fish. 65 figs.

Thalassiothrix longissima
Gilbert, J.Y., and W.E. Allen, 1943
The phytoplankton of the Gulf of California obtained by the "E.W. Scripps" in 1939 and 1940. J. Mar. Res. V(2):89-110, figs.30-31.

Thalassiothrix longissima
Gran, H.H., 1908
Diatomeen. Nordisches Plankton, Botanischer Teil pp. XIX.1-XIX 146; 178 text figs.

Thalassiothrix longissima
Gran, H.H., 1897
Protophyta: Diatomaceae, Silico-flagellata and Cilioflagellata. Den Norske Nordhavs Expedition 1876-1878, h. 24, 36 pp., 4 pls.

Thalassiothrix longissima
Gran, H. H. and E. C. Angst, 1931
Plankton diatoms of Puget Sound. Publ. Puget Sound Biol. Sta. 7:417-519, 95 text figs.

Thalassiothrix longissima
Gran, H.H., and T. Braarud, 1935
A quantitative study of the phytoplankton in the Bay of Fundy and the Gulf of Maine (including observations on hydrography, chemistry, and turbidity). J. Biol. Bd., Canada, 1(5):279-467, 69 text figs.

Thalassiothrix longissima
Hendy, N. Ingram, 1964
An introductory account of the smaller algae of British coastal waters. V. Bacillariophyceae (Diatoms). Her Majesty's Stationary Office, 317 pp., 45 pls.

Thalassiothrix longissima
Hendey, N.I., 1937
The plankton diatoms of the southern seas. Discovery Repts. 16:151-364, pls.6-13,

Thalassiothrix longissima
Hustedt, F. and A.A. Aleem, 1951
Littoral diatoms from the Salstone near Plymouth. JMBA 30(1): 177.196.

Thalassiothrix longissima
Iselin, C., 1930
A report on the coastal waters of Labrador based on explorations of the "Chance" during the summer of 1926. Proc. Am. Acad. Arts Sci., 66(1):1-37, 14 text figs.

Thalassiothrix longissima
Jørgensen, E., 1905
B. Protistplankton and the diatoms in bottom samples. Hydrographical and biological investigations in Norwegian fiords. Bergens Mus. Skr. 7: 49-225.

Thalassiothrix longissima
Jorgensen, E., 1900
Protophyten und Protozoën im Plankton aus der Norwegischen Westkerste. Bergens Mus. Aarb. 1899(6): 95 pp., 5 pls., 83 tables.

Thalassiothrix longissima
Lillick, L.C., 1940
Phytoplankton and planktonic protozoa of the offshore waters of the Gulf of Maine. Pt.II. Qualitative Composition of the Planktonic Flora. Trans. Am. Phil. Soc., n.s., 31(3):193-237, 13 text figs.

Thalassiothrix longissima
Lillick, L.C., 1938
Preliminary report of the phytoplankton of the Gulf of Maine. Am. Mid. Nat. 20(3):624-640, 1 text figs 37 tables.

Thalassiothrix longissima
Lillick, L.C., 1937
Seasonal studies of the phytoplankton off Woods Hole, Massachusetts. Biol. Bull. LXXIII (3):488-503, 3 text figs.

Thalassiothrix longissima
Manguin, E., 1954
Diatomees marines provenant de l'île Heard (Australian National Antarctic Expedition). Rev. Algol., n.s., 1: 14-24.

Thalassiothrix longissima
Marukawa, H., 1921
Plankton lists and some new species of copepods, from the northern waters of Japan. Bull. Inst. Ocean. Monaco, No.384, 15 pp., 3 pls., 1 chart.

Thalassiothrix longissima
Pavillard, J., 1925
Bacillariales. Rept. on the Danish Oceanogr. Exped., 1908-10 to the Mediterranean and adj. seas. Vol.II., Biol. J4:72 pp., 116 text figs.

Thalassiothrix longissima
Pavillard, J., 1905
Recherches sur la flore pelagique (Phytoplankton) de l'Etang de Thau. Theses presentees a la Fac. Sci., Paris, 116 pp., 3 pls.

Thalassiothrix longissima
Phifer, L.D. (undated)
The occurrence and distribution of plankton diatoms in Bering Sea and Bering Strait, July 26-August 24, 1934. Report of Oceanographic cruise of U.S. Coast Guard Cutter Chelan 1934, Part II(A):1-44 (mimeographed) plus fig.1 (after Pt.B)

Thalassiothrix longissima
Rampi, L., 1942
Ricerche sul fitoplancton del Mare Ligure 6. Le diatomee delle acque di Sanremo. Nuovo Giornale Botanico Italiano, N.S., 49:252-268.

Thalassiothrix longissima
Robinson, G.A., 1965.
Continuous plankton records: contribution towards a plankton atlas of the North Atlantic and the North Sea. IX. Seasonal cycles of phytoplankton. Bulls. Mar. Ecol., Scottish Mar. Biol. Assoc., 6(4):104-122, pls. 26-61.

Thalassiothrix longissima
Robinson, G.A., 1961
Contribution towards an atlas of the northeastern Atlantic and the North Sea. 1. Phytoplankton. Bulls. Mar. Ecology. 5(42): 81-89, Pls. 15-20.

Thalassiothrix longissima antarctica
Boden, B. P., 1949.
The diatoms collected by the U.S.S. CACOPAN in the Antarctic in 1947. J. Mar. Res. 8(1):6-13, 3 textfigs.

Thalassiothrix longissima var. antarctica
Boden, Brian, 1948
Marine plankton diatoms on operation HIGHJUMP in: Some oceanographic observations on operation HIGHJUMP. By R.S. Dietz. USNEL Rept. No.55, 97 pp., 41 figs. 7 July 1948.

Thalassiothrix mediterranea
Cupp, E.E. and Allen, W.E., 1938
Plankton diatoms of the Gulf of California obtained by Allan Hancock Pacific Expedition of 1937. The Hancock Pacific Expeditions, The Univ. So. Calif. Publ. 3: 61-74, 1 map, pls.4-15.

Thalassiothrix mediterranea var. pacifica n.var.
Cupp, Easter E., 1943
Marine plankton diatoms of the west coast of North America. Bull. S.I.O. 5(1):1-238, 5 pls., 168 text figs.

Thalassiothrix mediterranea
Ercegovic, A., 1936
Etudes qualitative et quantitatives du phytoplancton dans les eaux cotières de l'Adriatique oriental moyen au cours de l'année 1934. Acta Adriatica 1(9):1-126

Thalassiothrix mediterranea var pacifica, n. var.
Cupp, E. E., 1943.
Marine diatoms of the west coast of North America. Bull. S. I. O., 5(1):1-238, 5 pls., 168 textfigs.

Thalassiothrix miocenica n.sp.
Schrader, Hans-Joachim, 1973
Cenozoic diatoms from the northeast Pacific, leg 18. Initial Repts Deep Sea Drilling Proj. 18:673-797.

Thalassiothrix nitzschioides
Allen, W. E., 1936
Occurrence of marine plankton diatoms in a ten-year series of daily catches in southern California. Am. Jour. Bot. 23(1):60-63.

Thalassiothrix nitzschioides
Allen, W.E., 1928.
Catches of marine diatoms and dinoflagellates taken by boat in Southern California waters in 1926. Bull. S.I.O., tech. ser., 1:201-246, 6 textfigs.

Thalassiothrix nitzschioides
Allen, W.E., and E.E. Cupp, 1935
Plankton diatoms of the Java Sea. Annales du Jardin Botanique de Buitenzorg XLIV (2):101-174, figs.1-127.
(drawings of all species mentioned)

Thalassiothrix nitschioides
Bigelow, H. B., 1922
Exploration of the coastal water off the northeastern United States in 1916 by the U.S. Fisheries Schooner Grampus. Bull. M.C.Z. 65 (5):85-188, 53 text figs.

Thalassiothrix nitzschioides
Brunel, J., 1962
Le phytoplancton de la Baie de Chaleurs. Inst. Botan., Univ. Montréal, Contrib. No. 77: 365 pp., 66 pls.

Thalassiothrix nitzschioides
Brunel, Jules, 1962
Le phytoplancton de la Baie des Chaleurs. Contrib. Ministère de la Chasse et des Pêcheries, Province de Québec, No. 91: 365 pp.

Thalassiothrix nitzschioides
Chiba, T., 1949
On the distribution of the plankton in the eastern China Sea and Yellow Sea. 1. Plankton composition in the spring. J. Shimonoseki Coll. Fisheries, 1(1):57-63, 1 fig.

Thalassiothrix nitzschioides
Copenhagen, W. J. and L. D., 1949
Variation in the phytoplankton of Table Bay, 1934 to October 1935. With a note on the calorific value of Chaetoceros spp. Trans. Roy. Soc. S. Africa, 32(2):113-123, 2 text figs.

Thalassiothrix nitzschioides
Dangeard, P., 1927
Phytoplankton de la croisière du "Sylvana". Ann. Inst. Ocean., Monaco, n.s., 4(8):286-401, 54 text figs. (février-Juin 1913).

Thalassiothrix nitzschioides
Delsman, H. C., 1939.
Preliminary plankton investigations in the Java Sea. Treubia, 17:139-181, 8 maps, 41 figs.

Thalassionema nitzschioides
Ercegovic, A., 1936
Etudes qualitative et quantitatives du phytoplancton dans les eaux cotières de l'Adriatique oriental moyen au cours de l'année 1934. Acta Adriatica 1(9):1-126

Thalassiothrix nitzschioides
Forti, A., 1922
Ricerche sulla flora pelagica (fitoplancton) di Quarto dei Mille. Mem. R. Com. Talass. Ital. 97:248 pp., 13 pls.

Thalassiothrix nitzschioides
Cupp, E.E., 1934
Analysis of marine diatom collections taken from the Canal Zone to California during March, 1933. Trans. Am. Micros. Soc. LIII (1):22-29, 1 map.

Thalassiothrix nitzschioides
Gran, H. H. and E. C. Angst, 1931
Plankton diatoms of Puget Sound. Publ. Puget Sound Biol. Sta. 7:417-519, 95 text figs.

Thalassiothrix nitzschioides
Iselin, C., 1930
A report on the coastal waters of Labrador based on explorations of the "Chance" during the summer of 1926. Proc. Am. Acad. Arts Sci., 66(1):1-37, 14 text figs.

Thalassiothrix nitzschioides
Jørgensen, E., 1905
B. Protistplankton and the diatoms in bottom samples. Hydrographical and biological investigations in Norwegian fjords. Bergens Mus. Skr. 7: 49-225.

Thalassiothrix nitzschioides
Lafon, M., M. Durchon and Y. Saudray, 1955.
Recherches sur les cycles saisonnières du plankton. Ann. Inst. Océan., 31(3): 125-230.

Thalassiothrix nitzschioides
Lovegrove, T., 1958(1960).
Plankton investigations from Aberdeen in 1958. Ann. Biol., Cons. Perm. Int. Expl. Mer, 15:55.

Thalassiothrix nitzschioides
Macdonald, R., 1933
An examination of plankton hauls made in the Suez Canal during the year 1928. Fish. Res. Dis., Notes & Mem. No.3, 11 pp., 1 chart.

Thalassionema nitzschioides
Mangin, M. L., 1912
Phytoplankton de la croisière du "René" dans l'Atlantique (Septembre 1908). Ann. Inst. Ocean., n.s., 4(1):1-66, 2 pls., 41 text figs., 2 tables.

Thalassiothrix nitzschioides
Marukawa, H., 1921
Plankton lists and some new species of copepods, from the northern waters of Japan. Bull. Inst. Ocean., No.384, 15 pp., 3 pls., 1 chart. Monaco

Thalassiothrix nitzschioides
Morse, D.C., 1947
Some observations on seasonal variations in plankton population Patuxant River, Maryland 1943-1945. Bd. Nat. Res., Publ. No.65, Chesapeake Biol. Lab., 31, 3 figs.

Thalassiothrix nitzschioides
Pavillard, J., 1925
Bacillariales. Rept. on the Danish Oceangr. Exped., 1908-10 to the Mediterranean and adj. seas. Vol.II., Biol. J4:72 pp., 116 text figs.

Thalassiothrix nitzschioides
Pavillard, J., 1905
Recherches sur la flore pelagique (Phytoplankton) de l'Etang de Thau. Theses presentees a la Fac. Sci., Paris, 116 pp., 3 pls

Thalassiothrix nitzschioides
Pettersson, H., 1934
Scattering and extinction of light in sea-water. Meddelanden från Göteborgs Högskolas Oceanografiska Institut 9 (Göteborgs Kungl. vetenskaps-och vitterhets-samhälles Handlingar. Femte Följden, Ser.B, 4(4)):1-16.

Thalassiothrix nitzschioides
Pettersson, H., F. Gross, and F. Koczy, 1939
Large scale plankton cultures. Medd. från Oceanografiska Institutet i Göteborg No.3 (Göteborgs Kungl. vetenskaps-och Vitterhets-Samhälles Handlingar Femte Följden. Ser. B) Vol.6(13):1-24.
1. The plankton shaft. H. Pettersson
2. Experiment with phytoplankton. F. Gross and F. Koczy
3. Experiments with zooplankton.

Thalassionema nitzschioides
Politis, J., 1949
Diatomees marines de Bosphores et des ibes de la mer de Marmara. II Practica tou Hellenikou Hidrobiologikou Institutoutou 1929, Etoz 1929, 3(1):11-31.

Thalassiothrix nitzschioides
Rumkówna, A., 1948
List of the phytoplankton species occurring in the superficial water layers in the Gulf of Gdańsk. Bull. Lab. mar., Gdynia, No. 4: 139-141 with tables in back.

Thalassiothrix Nitzschioides
Schodduyn, M., 1926
Observations faites dans la baie d'Ambleteuse (Pas de Calais). Bull. Inst. Ocean., Monaco, No. 482: 64 pp.

Thalassiothrix nitzschioides
Smith, F.G.W., R.H. Williams, and C.C. Davis, 1950.
An ecological survey of the subtropical inshore waters adjacent to Miami. Ecol. 31(1):119-146, 7 textfigs.

Thalassiothrix nitzschioides
Sousa e Silva, E., 1949
Diatomaceas e Dinoflagelados de Baía de Cascais. Portugaliae Acta Biol., Volume: Julio Henriques, Ser. B: 300-383, 9 pls, 2 fold-in tables.

Thalassiothrix nitzschioides
Sousa a Silva, E., and J. Dos Santos-Pinto, 1948
O Plancton da Baía de S. Martinho do Porto.
1. Diatomaceas e Dinoflagelados. Bol. Soc. Portuguesa de Ciencias Naturais, 16(2):134-187, 6 pls. (Trav. Sta. Biol. Mar. de Lisbonne No. 52).

Thalassiothrix nitzschioides
Steemann-Nielsen, Einar, 1951
The marine vegetation of the Isefjord. A study on ecology and production. Medd. Komm. Danmarks Fiskeri-og Havundersøgelser. Ser. Plankton. 5(4); 114pp., 46 text figs.

Thalassiothrix vanhöffenii
Hasle, Grethe Rytter, 1960
Phytoplankton and ciliate species from the Tropical Pacific. Skr. Norske Videnskaps-Akad., Oslo. 1. Mat.-Nat. Kl., 1960(2): 1-50.

Tortilaria Briggerii n.sp. (fossil)
Barker, J.W. and S.H. Meakin, 1948
New and rare diatoms. J. Quakett Micros. Club, ser.4, 2(5):233-235, pl.25.

Toxarium undulatum
Forti, A., 1922
Ricerche sulla flora pelagica (fitoplancton) di Quarto dei Mille. Mem. R. Com. Talass. Ital. 97:248 pp., 13 pls.

Toxarium undulatum
Jorgensen, E., 1900
Protophyten und Protozoën im Plankton aus der Norwegischen Westkerste. Bergens Mus. Aarb. 1899(6): 95 pp., 5 pls., 83 tables.

Toxarium undulatum
Pavillard, J., 1925
Bacillariales. Rept. on the Danish Oceangr. Exped., 1908-10 to the Mediterranean and adj. seas. Vol.II., Biol. J4:72 pp., 116 text figs.

Toxonidea challengeriensis n.sp.
Castracane degli Antelminelli, F., 1886
1. on the Diatomaceae collected by H.M.S. Challenger during the years 1873-1876. Rept. Sci. Results, H.M.S. Challenger, Botany Vol. II, 178 pp., 30 pls.

Toxonidea gregoryana
Hendy, N. Ingram, 1964
An introductory account of the smaller algae of British coastal waters. V. Bacillariophyceae (Diatoms). Her Majesty's Stationary Office, 317 pp., 45 pls.

Toxonidia Gregoriana
Schodduyn, M., 1926
Observations faites dans la baie d'Ambleteuse (Pas de Calais). Bull. Inst. Ocean., Monaco, No. 482: 64 pp.

Toxonidea insignis
Hendy, N. Ingram, 1964
An introductory account of the smaller algae of British coastal waters. V. Bacillariophyceae (Diatoms). Her Majesty's Stationary Office, 317 pp., 45 pls.

Toxonidea insignis
Meunier, A., 1915
Microplancton de la Mer Flamande. 2. Diatomées (excepté le genre Chaetoceros). Mem. Mus. Roy. Hist. Nat., Belgique, 7(3):1-118, Pls. VIII-XIV.

Toxonidea insignis
Rampi, L., 1942
Ricerche sul fitoplancton del Mare Ligure 6. Le diatomee delle acque di Sanremo. Nuovo Giornale Botanico Italiano, N.S., 49:252-268.

Toxonidea insignis
Zanon, V., 1948
Diatomee marini di Sardegna e Pugillo di Alghe Marine della stressa. Boll. Pesca, Piscitutura e Idrobiologia, Anno 24, ns. 3(2): 202-244, 27 figs. on 1 pl.

Trachyneis
Desikachary, T.V., 1956(1957).
Electron microscope studies on diatoms. J.R. Microsc. Soc. (3)76(1/2):9-36.

Trachyneis aspera
Hendey, N-Ingram, 1958 [1957(Publ. 1958)]
Marine diatoms from some West African Ports. J. R. Microsc. Soc. (3) 77(1/2): 28-85.

Trachyneis aspera
Hendey, N.I., 1951
Littoral diatoms of Chicester Harbour with special reference to fouling. J.Roy. Microscop. Soc. 71(1): 1-86, 18 pls.

Trachyneis aspera
Hendey, N.I. 1937
The plankton diatoms of the southern seas. Discovery Repts. 16:151-364, pls.6-13.

Trachyneis aspera
Hustedt, F. and A.A. Aleem,1951
Littoral diatoms from the Salstone near Plymouth. JMBA 30(1): 177.196.

Trachyneis aspera
Politis, J., 1949
Diatomees marines de Bosphores et des ibes de la mer de Marmara. II Practica tou Hellenikou Hidrobiologikou Institutoutou 1929, Etoz 1929, 3(1):11-31.

Trachyneis aspera
Rampi, L., 1942
Ricerche sul fitoplancton del Mare Ligure 6. Le diatomee delle acque di Sanremo. Nuovo Giornale Botanico Italiano, N.S., 49:252-268.

Trachyneis aspera
Rampi, L., 1940
Diatomee del Mare Adriatico. Nuovo Giornale Botanico Italiano, n.s., 47:559-608.

Trachyneis aspera (figs.)
Sousa e Silva, E., 1949
Diatomaceas e Dinoflagelados de Baia de Cascais. Portugaliae Acta Biol., Volume: Julio Henriques, Ser. B: 300-383, 9 pls, 2 fold-in tables.

Trachyneis aspera
Takano, Hideaki, 1961.
Epiphytic diatoms upon Japanese agar sea-weeds. Bull. Tokai Reg. Fish. Res. Lab., No. 31:269-274.

Trachyneis aspera
Zanon, V., 1948
Diatomee marini di Sardegna e Pugillo di Alghe Marine della stressa. Boll. Pesca, Piscitutura e Idrobiologia, Anno 24, ns. 3(2): 202-244, 27 figs. on 1 pl.

Tranchyneis aspera & var.
Hendy, N. Ingram, 1964
An introductory account of the smaller algae of British coastal waters. V. Bacillariophyceae (Diatoms). Her Majesty's Stationary Office, 317 pp., 45 pls.

Trachyneis aspera intermedia
Takano, Hideaki, 1962
Notes on epiphytic diatoms upon sea-weeds from Japan. J. Oceanogr. Soc., Japan, 18(1):29-33.

Trachyneis clepsydra
Politis, J., 1949
Diatomees marines de Bosphores et des ibes de la mer de Marmara. II Practica tou Hellenikou Hidrobiologikou Institutoutou 1929, Etoz 1929, 3(1):11-31.

Trachyneis Schmidtiana
Rampi, L., 1940
Diatomee del Mare Adriatico. Nuovo Giornale Botanico Italiano, n.s., 47:559-608.

Trachysphenia australis
Hendy, N. Ingram, 1964
An introductory account of the smaller algae of British coastal waters. V. Bacillariophyceae (Diatoms). Her Majesty's Stationary Office, 317 pp., 45 pls.

Trachysphaenia australis
Salah, M.M., 1952(1953).
XII. Diatoms from Blakeney Point, Norfolk. New species and new records for Great Britain. J.R. Microsc. Soc., Ser. 3, 72(3):155-169, 3 pls

Triceratium sp.
Cupp, E.E. and Allen, W.E., 1938
Plankton diatoms of the Gulf of California obtained by Allan Hancock Pacific Expedition of 1937. The Hancock Pacific Expeditions, the Univ. So. Calif. Publ. 3: 61-74, 1 map, pls.4-15.

Triceratium
Desikachary, T.V., 1956(1957).
Electron microscope studies on diatoms. J. R. Microsc. Soc. (3)76(1/2):9-36.

Triceratium
Neaverson, E., 1934
The sea-floor deposits. 1. General characteristics and distribution. Discovery Repts. 9: 297-349, Plates 17-22.

Triceratium abyssale n.sp.
Castracane degli Antelminelli, F., 1886
1. Report on the Diatomaceae collected by H.M.S. Challenger during the years 1873-1876. Rept. Sci. Results, H.M.S. Challenger, Botany Vol. II, 178 pp., 30 pls.

Triceratium abyssorum
Cleve-Euler, A., 1951
Die Diatomeen von Schweden und Finnland. Kungl. Svenska Vetenskaps Akad. Handl., Fjärde Ser. 2(1): 161 pp., 6 pls.

Triceratium acutum
Mann, A., 1893
List of Diatomaceae from a deep-sea dredging in the Atlantic Ocean off Delaware Bay by the U. S. Fish Commission Steamer Albatross. Proc. U. S. Nat. Mus. 16:303-312.

Triceratium alternans
Cleve-Euler, A., 1951
Die Diatomeen von Schweden und Finnland. Kungl. Svenska Vetenskaps Akad. Handl., Fjärde Ser. 2(1): 161 pp., 6 pls.

Triceratium alternans
Dangeard, P., 1927
Phytoplankton de la croisière du "Sylvana". Ann. Inst. Ocean., Monaco, n.s., 4(8):286-401, 54 text figs. (Feirer-Juin 1913).

Triceratium alternans
Hendey, N.I. 1951
Littoral diatoms of Chicester Harbour with special reference to fouling. J.Roy. Microscop. Soc. 71(1): 1-86, 18 pls.

Triceratium alternans
Hustedt, F. and A.A. Aleem,1951
Littoral diatoms from the Salstone near Plymouth. JMBA 30(1): 177.196.

Triceratium alternans
Lillick, L.C., 1940
Phytoplankton and planktonic protozoa of the offshore waters of the Gulf of Maine. Pt.II. Qualitative Composition of the Planktonic Flora. Trans. Am. Phil. Soc., n.s., 31(3):193-237, 13 text figs.

Oceanographic Index: Marine Organisms Cumulation, 1946-1973

Lillick, L.C., 1938 Triceratium alternans
Preliminary report of the phytoplankton of the Gulf of Maine. Am. Mid. Nat. 20(3):624-640, 1 text figs 37 tables.

Triceratium alternans
Mann, A., 1893
List of Diatomaceae from a deep-sea dredging in the Atlantic Ocean off Delaware Bay by the U. S. Fish Commission Steamer Albatross. Proc. U. S. Nat. Mus. 16:303-312.

Triceratium alternans
Pavillard, J., 1925
Bacillariales. Rept. on the Danish Oceangr. Exped., 1908-10 to the Mediterranean and adj. seas. Vol.II., Biol. J4:72 pp., 116 text figs.

Triceratium alternans
Sousa e Silva, E., 1949
Diatomaceas e Dinoflagelados de Baia de Cascais. Portugaliae Acta Biol., Volume: Julio Henriques, Ser. B: 300-383, 9 pls, 2 fold-in tables.

Triceratium alternans
Zanon, V., 1948
Diatomee marini di Sardegna e Pugillo di Alghe Marine della stressa. Boll. Pesca, Piscitutura e Idrobiologia, Anno 24, ns. 3(2): 202-244, 27 figs. on 1 pl.

Triceratium americanum
Cleve-Euler, A., 1951
Die Diatomeen von Schweden und Finnland. Kungl. Svenska Vetenskaps Akad. Handl., Fjärde Ser. 2(1): 161 pp., 6 pls.

Triceratium antediluvianum
Cleve-Euler, A., 1951
Die Diatomeen von Schweden und Finnland. Kungl. Svenska Vetenskaps Akad. Handl., Fjärde Ser. 2(1): 161 pp., 6 pls.

Triceratium antediluvianum
Eskinazi Enide e Shigekatsv Satô (1963-1964) 1966.
Contribuição ao estudo das diatomaceas da Praia de Piedade.
Trabhs Inst. Oceanogr., Univ. Recife, 5 (5/6): 73-114.

Triceratium antediluvianum
Politis, J., 1949
Diatomees marines de Bosphores et des ibes de la mer de Marmara. II Practica tou Hellenikou Hidrobiologikou Institutoutou 1929, Etoz 1929, 3(1):11-31.

Triceratium antediluvianum
Rampi, L., 1942
Ricerche sul fitoplancton del Mare Ligure 6. Le diatomee delle acque di Sanremo. Nuovo Giornale Botanico Italiano, N.S., 49:252-268.

Triceratium antediluvianum
Rampi, L., 1940
Diatomee del Mare Adriatico. Nuovo Giornale Botanico Italiano, n.s., 47:559-608.

Triceratium antediluvianum
von Sydow, Burkard, und Robert Christenhuss 1972.
Rasterelektronmikroskopische Untersuchungen der Hohlräume in der Schalenwand einiger centrischer Kieselalgen.
Arch. Protistenk 114 (3): 256-271.

Triceratium antediluvianum
Zanon, V., 1948
Diatomee marini di Sardegna e Pugillo di Alghe Marine della stressa. Boll. Pesca, Piscitutura e Idrobiologia, Anno 24, ns. 3(2): 202-244, 27 figs. on 1 pl.

Triceratium arcticum
Cleve-Euler, A., 1951
Die Diatomeen von Schweden und Finnland. Kungl. Svenska Vetenskaps Akad. Handl., Fjärde Ser. 2(1): 161 pp., 6 pls.

Triceratium arcticum
Forti, A., 1922
Ricerche sulla flora pelagica (fitoplancton) di Quarto dei Mille. Mem. R. Com. Talass. Ital. 97:248 pp., 13 pls.

Triceratium arcticum
Florin, M-B., 1948
9. Diatomeae in submarine cores from the Tyrrhenian Sea. Medd. Ocean. Inst., Göteborg. 15 (Göteborgs Kungl. Vetenskaps-och Viterrhets Samhälles Handlingar, Sjätte Foljden, Ser. B 5(13):80-88.

Triceratium arcticum
Manguin, E., 1954
Diatomees marines provenant de l'ile Heard (Australian National Antarctic Expedition). Rev. Algol., n.s., 1: 14-24.

Triceratium arcticum
Zanon, V., 1948
Diatomee marini di Sardegna e Pugillo di Alghe Marine della stressa. Boll. Pesca, Piscitutura e Idrobiologia, Anno 24, ns. 3(2): 202-244, 27 figs. on 1 pl.

Triceratium arcticum kerguelensis n.var.
Castracane degli Antelminelli, F., 1886
1. Report on the Diatomaceae collected by H.M.S. Challenger during the years 1873-1876. Rept. Sci. Results, H.M.S. Challenger, Botany Vol. II, 178 pp., 30 pls.

Triceratium arcticum v. kerguelenensis
Mangin, L., 1915
Phytoplancton de L'Antartique. Deuxieme Exped. Ant. Francaise (1908-1910), 95 pp., 3 pls., 58 text figs.

Triceratium armatum
Castracane degli Antelminelli, F., 1886
1. Report on the Diatomaceae collected by H.M.S. Challenger during the years 1873-1876. Rept. Sci. Results, H.M.S. Challenger, Botany Vol. II, 178 pp., 30 pls.

Triceratium atlanticum n.sp.
Castracane degli Antelminelli, F., 1886
1. Report on the Diatomaceae collected by H.M.S. Challenger during the years 1873-1876. Rept. Sci. Results, H.M.S. Challenger, Botany Vol. II, 178 pp., 30 pls.

Triceratium balearicum
Misra, J.N., 1956.
A systematic account of some littoral marine diatoms from the west coast of India.
J. Bombay Nat. Hist. Soc., 53(4):537-568.

Triceratium balearicum
Zanon, V., 1948
Diatomee marini di Sardegna e Pugillo di Alghe Marine della stressa. Boll. Pesca, Piscitutura e Idrobiologia, Anno 24, ns. 3(2): 202-244, 27 figs. on 1 pl.

Triceratium bicorne
Mann, A., 1893
List of Diatomaceae from a deep-sea dredging in the Atlantic Ocean off Delaware Bay by the U. S. Fish Commission Steamer Albatross. Proc. U. S. Nat. Mus. 16:303-312.

Triceratium calvescens n.sp.
Castracane degli Antelminelli, F., 1886
1. Report on the Diatomaceae collected by H.M.S. Challenger during the years 1873-1876. Rept. Sci. Results, H.M.S. Challenger, Botany Vol. II, 178 pp., 30 pls.

Triceratium cariosum n.sp.
Castracane degli Antelminelli, F., 1886
1. Report on the Diatomaceae collected by H.M.S. Challenger during the years 1873-1876. Rept. Sci. Results, H.M.S. Challenger, Botany Vol. II, 178 pp., 30 pls.

Triceratium cinnamomeum
Mann, A., 1893
List of Diatomaceae from a deep-sea dredging in the Atlantic Ocean off Delaware Bay by the U. S. Fish Commission Steamer Albatross. Proc. U. S. Nat. Mus. 16:303-312.

Triceratium cinnamomeum var. minor
Hasle, Grethe Rytter, 1960
Phytoplankton and ciliate species from the Tropical Pacific.
Skr. Norske Videnskaps-Akad., Oslo, 1. Mat.-Nat. Kl., 1960(2): 1-50.

Triceratium Contortum
Eskinazi Enide e Shigekatsv Satô (1963-1964) 1966.
Contribuição ao estudo das diatomaceas da Praia de Piedade.
Trabhs Inst. Oceanogr., Univ. Recife, 5 (5/6): 73-114.

Triceratium contortum
Hendey, N. Ingram, 1958 [1957(Publ. 1958)]
Marine diatoms from some West African Ports- J. R. Microsc. Soc. (3) 77(1/2): 28-85.

Triceratium coronatum n.sp.
Castracane degli Antelminelli, F., 1886
1. Report on the Diatomaceae collected by H.M.S. Challenger during the years 1873-1876. Rept. Sci. Results, H.M.S. Challenger, Botany Vol. II, 178 pp., 30 pls.

Triceratium distinctum
Paiva Carvalho, J., 1950
O plancton do Rio Maria Rodriques (Cananeis). 1. Diatomaceas e Dinoflagelados. Bol. Inst. Paulista Oceanogr. 1(1): 27-43, 2 fold-in tables, 2 figs.

Triceratium dubium
de Sousa e Silva, E., 1956.
Contribution à l'étude du microplancton de Dakar et des regions maritimes voisines.
Bull. I.F.A.N., 8(2):335-371, 7 pls.

Triceratium dubium
Hendey, N. Ingram, 1958 [1957(Publ. 1958)]
Marine diatoms from some West African Ports- J. R. Microsc. Soc. (3) 77(1/2): 28-85.

Triceratium dubium
Misra, J.N., 1956.
A systematic account of some littoral marine diatoms from the west ocast of India.
J. Bombay Nat. Hist. Soc., 53(4):537-568.

Triceratium dubium
Subrahmanyan, R., 1945.
On the cell division and mitosis of some South Indian diatoms. Proc. Indian Acad. Sci., Sect. B, 22(6):331-354, Pl. 39, 88 textfigs.

Triceratium dubium
Takano, Hideaki, 1963.
Notes on marine littoral diatoms of Japan. Bull. Tokai Reg. Fish. Res. Lab., No. 36:1-8.

Triceratium elongatum
Grunow, A., 1877
1. New Diatoms from Honduras. Monthly Micros. Jour., 18:165-186, pls. CXCIII-CXCVI.

Triceratium elongatum
Pavillard, J., 1925
Bacillariales. Rept. on the Danish Oceangr. Exped., 1908-10 to the Mediterranean and adj. seas. Vol.II., Biol. J4:72 pp., 116 text figs.

Triceratium favus

Brunel, J., 1962
Le phytoplancton de la Baie de Chaleurs. Inst. Botan., Univ. Montréal, Contrib. No. 77: 365 pp., 66 pls.

Triceratium favus

Cleve-Euler, A., 1951
Die Diatomeen von Schweden und Finnland. Kungl. Svenska Vetenskaps Akad. Handl., Fjärde Ser. 2(1): 161 pp., 6 pls.

Triceratium Favus

Dangeard, P., 1927
Phytoplankton de la croisière du "Sylvana". Ann. Inst. Ocean., Monaco, n.s., 4(8):286-401, 54 text figs. (Février-Juin 1913).

Triceratium favus

Eskinazi Enide e Shigekatsv Satô (1963-1964) 1966.
Contribuição ao estudo das diatomaceas da Praia de Piedade.
Trabhs Inst. Oceanogr., Univ. Recife, 5 (5/6): 73-114.

Triceratium favus

Fox, M., 1957.
A first list of marine algae from Nigeria. J. Linn. Soc., London, 55(362):615-631.

Triceratium favus

Grøntved, J., 1949
Investigations on the phytoplankton in the Danish Waddensea in July 1941. Medd. Komm. Danmarks Fiskeri og Havundersøgelser, ser. Plankton, 5(2):55 pp., 2 pls., 38 text figs.

Triceratium favus

Hendy, N. Ingram, 1964
An introductory account of the smaller algae of British coastal waters. V. Bacillariophyceae (Diatoms). Her Majesty's Stationary Office, 317 pp., 45 pls.

Triceratium favus

Hendey, N. Ingram, 1958 [1957(Publ. 1958)]
Marine diatoms from some West African Ports. J. R. Microsc. Soc. (3) 77(1/2): 28-85.

Triceratium favus

Hendey, N.I., 1937
The plankton diatoms of the southern seas. Discovery Repts. 16:151-364, pls.6-13.

Triceratium favus

Rampi, L., 1940
Diatomee del Mare Adriatico. Nuovo Giornale Botanico Italiano, n.s., 47:559-608.

Triceratium favus

Sousa e Silva, E., 1949
Diatomaceas e Dinoflagelados de Baía de Cascais. Portugaliae Acta Biol., Volume: Julio Henriques, Ser. B: 300-383, 9 pls., 2 fold-in tables.

Triceratium favus

Zanon, D. V., 1949
Diatomee di Buenos Aires (Argentina) Atti Accad. Naz. Lincei, Memorie, Cl. Sci. fis., mat. e. nat., ser. 7, 11(3):59-151, 2 pls.

Triceratium favus

Zanon, V., 1948
Diatomee marini di Sardegna e Pugillo di Alghe Marine della stressa. Boll. Pesca, Piscitutura e Idrobiologia, Anno 24, ns. 3(2): 202-244, 27 figs. on 1 pl.

Triceratium favus pacifica n.var.

Castracane degli Antelminelli, F., 1886
1. Rēport on the Diatomaceae collected by H.M.S. Challenger during the years 1873-1876. Rept. Sci. Results, H.M.S. Challenger, Botany Vol. II, 178 pp., 30 pls.

Triceratium favus quadrata

de Sousa e Silva, E., 1956.
Contribution à l'étude du microplancton de Dakar et des regions maritimes voisines. Bull. I.F.A.N., 8(2):335-371, 7 pls.

Triceratium favus late-areo-lata n.var.

Castracane degli Antelminelli, F., 1886
1. Rēport on the Diatomaceae collected by H.M.S. Challenger during the years 1873-1876. Rept. Sci. Results, H.M.S. Challenger, Botany Vol. II, 178 pp., 30 pls.

Triceratium ferox n.sp.

Castracane degli Antelminelli, F., 1886
1. Rēport on the Diatomaceae collected by H.M.S. Challenger during the years 1873-1876. Rept. Sci. Results, H.M.S. Challenger, Botany Vol. II, 178 pp., 30 pls.

Triceratium fimbriatum

Castracane degli Antelminelli, F., 1886
1. Rēport on the Diatomaceae collected by H.M.S. Challenger during the years 1873-1876. Rept. Sci. Results, H.M.S. Challenger, Botany Vol. II, 178 pp., 30 pls.

Triceratium flos

Cleve-Euler, A., 1951
Die Diatomeen von Schweden und Finnland. Kungl. Svenska Vetenskaps Akad. Handl., Fjärde Ser. 2(1): 161 pp., 6 pls.

Triceratium formosum

Rampi, L., 1942
Ricerche sul fitoplancton del Mare Ligure 6. Lo diatomee delle acque di Sanremo. Nuovo Giornale Botanico Italiano, N.S., 49:252-268.

Triceratium formosum

Zanon, V., 1948
Diatomee marini di Sardegna e Pugillo di Alghe Marine della stressa. Boll. Pesca, Piscitutura e Idrobiologia, Anno 24, ns. 3(2): 202-244, 27 figs. on 1 pl.

Triceratium grunowianum n.sp.

Castracane degli Antelminelli, F., 1886
1. Rēport on the Diatomaceae collected by H.M.S. Challenger during the years 1873-1876. Rept. Sci. Results, H.M.S. Challenger, Botany Vol. II, 178 pp., 30 pls.

Triceratium Hardmanianum

Barker, J.W. and S.H. Meakin, 1948
New and rare diatoms. J. Quakett Micros. Club, ser.4, 2(5):233-235, pl.25.

Triceratium heibergii

Cleve-Euler, A., 1951
Die Diatomeen von Schweden und Finnland. Kungl. Svenska Vetenskaps Akad. Handl., Fjärde Ser. 2(1): 161 pp., 6 pls.

Triceratium incrassatum n.sp.

Castracane degli Antelminelli, F., 1886
1. Rēport on the Diatomaceae collected by H.M.S. Challenger during the years 1873-1876. Rept. Sci. Results, H.M.S. Challenger, Botany Vol. II, 178 pp., 30 pls.

Triceratium inelegans

Mann, A., 1893
List of Diatomaceae from a deep-sea dredging in the Atlantic Ocean off Delaware Bay by the U. S. Fish Commission Steamer Albatross. Proc. U. S. Nat. Mus. 16:303-312.

Triceratium inautum n.sp.

Castracane degli Antelminelli, F., 1886
1. Rēport on the Diatomaceae collected by H.M.S. Challenger during the years 1873-1876. Rept. Sci. Results, H.M.S. Challenger, Botany Vol. II, 178 pp., 30 pls.

Tropidoneis lepidoptera

Hendey, N.I., 1951
Littoral diatoms of Chicester Harbour with special reference to fouling. J. Roy. Microscop. Soc. 71(1): 1-86, 18 pls.

Triceratium mesoleium

Cleve-Euler, A., 1951
Die Diatomeen von Schweden und Finnland. Kungl. Svenska Vetenskaps Akad. Handl., Fjärde Ser. 2(1): 161 pp., 6 pls.

Triceratium orbiculatum

Schröder, B., 1900
Phytoplankton des Golfes von Neapel nebst vergleichenden Ausblicken auf das atlantischen Ozean. Mitt. Zool. Stat. Neapel, 14:1-38.

Triceratium ornatum

Mann, A., 1893
List of Diatomaceae from a deep-sea dredging in the Atlantic Ocean off Delaware Bay by the U. S. Fish Commission Steamer Albatross. Proc. U. S. Nat. Mus. 16:303-312.

Triceratium parallelum

Cleve-Euler, A., 1951
Die Diatomeen von Schweden und Finnland. Kungl. Svenska Vetenskaps Akad. Handl., Fjärde Ser. 2(1): 161 pp., 6 pls.

Triceratium pavimentosum n.sp.

Castracane degli Antelminelli, F., 1886
1. Rēport on the Diatomaceae collected by H.M.S. Challenger during the years 1873-1876. Rept. Sci. Results, H.M.S. Challenger, Botany Vol. II, 178 pp., 30 pls.

Triceratium pentacrinus

de Sousa e Silva, E., 1956.
Contribution à l'étude du microplancton de Dakar et des regions maritimes voisines. Bull. I.F.A.N., 8(2):335-371, 7 pls.

Triceratium pentacrinus

Eskinazi Enide e Shigekatsv Satô (1963-1964) 1966.
Contribuição ao estudo das diatomaceas da Praia de Piedade.
Trabhs Inst. Oceanogr., Univ. Recife, 5 (5/6): 73-114.

Triceratium pentacrinus

Misra, J.N., 1956.
A systematic account of some littoral marine diatoms from the west coast of India. J. Bombay Nat. Hist. Soc., 53(4):537-568.

Triceratium pentacrinus

Rampi, L., 1940
Diatomee del Mare Adriatico. Nuovo Giornale Botanico Italiano, n.s., 47:559-608.

Triceratium pentacrinus

Takano, Hideaki, 1963.
Notes on marine littoral diatoms of Japan. Bull. Tokai Reg. Fish. Res. Lab., No. 36:1-8.

Triceratium pulvillus n.sp.

Castracane degli Antelminelli, F., 1886
1. Rēport on the Diatomaceae collected by H.M.S. Challenger during the years 1873-1876. Rept. Sci. Results, H.M.S. Challenger, Botany Vol. II, 178 pp., 30 pls.

Triceratium punctatum

Mann, A., 1893
List of Diatomaceae from a deep-sea dredging in the Atlantic Ocean off Delaware Bay by the U. S. Fish Commission Steamer Albatross. Proc. U. S. Nat. Mus. 16:303-312.

Triceratium punctigerum n.sp.

Castracane degli Antelminelli, F., 1886
1. Rēport on the Diatomaceae collected by H.M.S. Challenger during the years 1873-1876. Rept. Sci. Results, H.M.S. Challenger, Botany Vol. II, 178 pp., 30 pls.

Triceratium radiatum
Barker, J.W. and S.H. Meakin, 1948
New and rare diatoms. J. Quakett Micros. Club, ser.4, 2(5):233-235, pl.25.

Triceratium reticulum
Cleve-Euler, A., 1951
Die Diatomeen von Schweden und Finnland. Kungl. Svenska Vetenskaps Akad. Handl., Fjärde Ser. 2(1): 161 pp., 6 pls.

Triceratium reticulatum
Rampi, L., 1940
Diatomee del Mare Adriatico. Nuovo Giornale Botanico Italiano, n.s., 47:559-608.

Triceratium reticulum
Zanon, D. V., 1949
Diatomee di Buenos Aires (Argentina) Atti Accad. Naz. Lincei, Memorie, Cl. Sci. fis., mat. e. nat., ser. 7, 11(3):59-151, 2 pls.

Triceratium reticulum
Zanon, V., 1948
Diatomee marini di Sardegna e Pugillo di Alghe Marine della stressa. Boll. Pesca, Piscitutura e Idrobiologia, Anno 24, ns. 3(2): 202-244, 27 figs. on 1 pl.

Triceratium Robertsianum
Rampi, L., 1940
Diatomee del Mare Adriatico. Nuovo Giornale Botanico Italiano, n.s., 47:559-608.

Triceratium robertsianum dwardensum n.var.
Misra, J.N., 1956.
A systematic account of some littoral marine diatoms from the west coast of India. J. Bombay Nat. Hist. Soc., 53(4):537-568.

Triceratium sarcophagus n.sp.
Castracane degli Antelminelli, F., 1886
1. Report on the Diatomaceae collected by H.M.S. Challenger during the years 1873-1876. Rept. Sci. Results, H.M.S. Challenger, Botany Vol. II, 178 pp., 30 pls.

Triceratium sculptum
Rampi, L., 1940
Diatomee del Mare Adriatico. Nuovo Giornale Botanico Italiano, n.s., 47:559-608.

Triceratium sculptum
de Sousa e Silva, E., 1956
Contribution à l'étude du microplancton de Dakar et des regions maritimes voisines. Bull. I.F.A.N., 8(2):335-371, 7 pls.

Triceratium Shadboltianum
Rampi, L., 1940
Diatomee del Mare Adriatico. Nuovo Giornale Botanico Italiano, n.s., 47:559-608.

Triceratium Shadboltianum
Zanon, V., 1948
Diatomee marini di Sardegna e Pugillo di Alghe Marine della stressa. Boll. Pesca, Piscitutura e Idrobiologia, Anno 24, ns. 3(2): 202-244, 27 figs. on 1 pl.

Triceratium spinosum
Hendy, N. Ingram, 1964
An introductory account of the smaller algae of British coastal waters. V. Bacillariophyceae (Diatoms). Her Majesty's Stationary Office, 317 pp., 45 pls.

Triceratium spinosum
Rampi, L., 1940
Diatomee del Mare Adriatico. Nuovo Giornale Botanico Italiano, n.s., 47:559-608.

Tricertaium spinosum tetragona
Misra, J.N., 1956.
A systematic account of some littoral marine diatoms from the west coast of India. J. Bombay Nat. Hist. Soc., 53(4):537-568.

Triceratium thaitiense n.sp.
Castracane degli Antelminelli, F., 1886
1. Report on the Diatomaceae collected by H.M.S. Challenger during the years 1873-1876. Rept. Sci. Results, H.M.S. Challenger, Botany Vol. II, 178 pp., 30 pls.

Triceratium trifoliatum
Cleve-Euler, A., 1951
Die Diatomeen von Schweden und Finnland. Kungl. Svenska Vetenskaps Akad. Handl., Fjärde Ser. 2(1): 161 pp., 6 pls.

Triceratium tumescens n.sp.
Castracane degli Antelminelli, F., 1886
1. Report on the Diatomaceae collected by H.M.S. Challenger during the years 1873-1876. Rept. Sci. Results, H.M.S. Challenger, Botany Vol. II, 178 pp., 30 pls.

Triceratium umbilicatum
Cleve-Euler, A., 1951
Die Diatomeen von Schweden und Finnland. Kungl. Svenska Vetenskaps Akad. Handl., Fjärde Ser. 2(1): 161 pp., 6 pls.

Triceratium Weissii
Mann, A., 1893
List of Diatomaceae from a deep-sea dredging in the Atlantic Ocean off Delaware Bay by the U. S. Fish Commission Steamer Albatross. Proc. U. S. Nat. Mus. 16:303-312.

Trigonium arcticum
Bigelow, H.B., and M. Leslie, 1930
Reconnaissance of the waters and plankton of Monterey Bay, July 1928. Bull. M.C.Z., 70(5):429-481, 43 text figs.

Trigonum arcticum
Hendy, N. Ingram, 1964
An introductory account of the smaller algae of British coastal waters. V. Bacillariophyceae (Diatoms). Her Majesty's Stationary Office, 317 pp., 45 pls.

Trigonium arcticum
Hendey, N.I., 1937
The plankton diatoms of the southern seas. Discovery Repts. 16:151-364, pls.6-13.

Trigonium arcticum
Mann, A., 1907
Report on the diatoms of the Albatross voyages in the Pacific Ocean, 1888-1904. Contrib. U. S. Nat. Herb. 10(5):221-419, Pls. XLIV-LIV.

Trigonium adspersum n.sp.
Mann, A., 1907
Report on the diatoms of the Albatross voyages in the Pacific Ocean, 1888-1904. Contrib. U. S. Nat. Herb. 10(5):221-419, Pls. XLIV-LIV.

Trigonium alternans
Mann, A., 1907
Report on the diatoms of the Albatross voyages in the Pacific Ocean, 1888-1904. Contrib. U. S. Nat. Herb. 10(5):221-419, Pls. XLIV-LIV.

Trigonium cinnamomeum
Mann, A., 1907
Report on the diatoms of the Albatross voyages in the Pacific Ocean, 1888-1904. Contrib. U. S. Nat. Herb. 10(5):221-419, Pls. XLIV-LIV.

Trigonium coscinoides
Mann, A., 1907
Report on the diatoms of the Albatross voyages in the Pacific Ocean, 1888-1904. Contrib. U. S. Nat. Herb. 10(5):221-419, Pls. XLIV-LIV.

Trigonium montereyi
Bigelow, H.B., and M. Leslie, 1930
Reconnaissance of the waters and plankton of Monterey Bay, July 1928. Bull. M.C.Z., 70(5):429-481, 43 text figs.

Trigonium parallelum
Mann, A., 1907
Report on the diatoms of the Albatross voyages in the Pacific Ocean, 1888-1904. Contrib. U. S. Nat. Herb. 10(5):221-419, Pls. XLIV-LIV.

Trigonium plano-woncavum
Mann, A., 1907
Report on the diatoms of the Albatross voyages in the Pacific Ocean, 1888-1904. Contrib. U. S. Nat. Herb. 10(5):221-419, Pls. XLIV-LIV.

Trigonium rusticum n.sp.
Mann, A., 1907
Report on the diatoms of the Albatross voyages in the Pacific Ocean, 1888-1904. Contrib. U. S. Nat. Herb. 10(5):221-419, Pls. XLIV-LIV.

Trigonium sculptum
Mann, A., 1907
Report on the diatoms of the Albatross voyages in the Pacific Ocean, 1888-1904. Contrib. U. S. Nat. Herb. 10(5):221-419, Pls. XLIV-LIV.

Trigonium striolatum
Mann, A., 1907
Report on the diatoms of the Albatross voyages in the Pacific Ocean, 1888-1904. Contrib. U. S. Nat. Herb. 10(5):221-419, Pls. XLIV-LIV.

Trigonium tabellarium
Mann, A., 1907
Report on the diatoms of the Albatross voyages in the Pacific Ocean, 1888-1904. Contrib. U. S. Nat. Herb. 10(5):221-419, Pls. XLIV-LIV.

Trigonium trinitas
Mann, A., 1907
Report on the diatoms of the Albatross voyages in the Pacific Ocean, 1888-1904. Contrib. U. S. Nat. Herb. 10(5):221-419, Pls. XLIV-LIV.

Trigonium zonulatum
Mann, A., 1907
Report on the diatoms of the Albatross voyages in the Pacific Ocean, 1888-1904. Contrib. U. S. Nat. Herb. 10(5):221-419, Pls. XLIV-LIV.

Trinacria excavata
Mann, A., 1893
List of Diatomaceae from a deep-sea dredging in the Atlantic Ocean off Delaware Bay by the U. S. Fish Commission Steamer Albatross. Proc. U. S. Nat. Mus. 16:303-312.

Trinacria ventricosa
Barker, J.W. and S.H. Meakin, 1948
New and rare diatoms. J. Quakett Micros. Club, ser.4, 2(5):233-235, pl.25.

Tripodiscus affinis
Mann, A., 1907
Report on the diatoms of the Albatross voyages in the Pacific Ocean, 1888-1904. Contrib. U. S. Nat. Herb. 10(5):221-419, Pls. XLIV-LIV.

Tripodiscus beringensis
Mann, A., 1907
Report on the diatoms of the Albatross voyages in the Pacific Ocean, 1888-1904. Contrib. U. S. Nat. Herb. 10(5):221-419, Pls. XLIV-LIV.

Tripodiscus concentricus
Mann, A., 1907
Report on the diatoms of the Albatross voyages in the Pacific Ocean, 1888-1904. Contrib. U. S. Nat. Herb. 10(5):221-419, Pls. XLIV-LIV.

Tripodiscus cosmiodiscus n.sp.
Mann, A., 1907
Report on the diatoms of the Albatross voyages in the Pacific Ocean, 1888-1904. Contrib. U. S. Nat. Herb. 10(5):221-419, Pls. XLIV-LIV.

Tripodiscus kinderi
Mann, A., 1907
Report on the diatoms of the Albatross voyages in the Pacific Ocean, 1888-1904. Contrib. U. S. Nat. Herb. 10(5):221-419, Pls. XLIV-LIV.

Tripodiscus margaritaceus
Mann, A., 1907
Report on the diatoms of the Albatross voyages in the Pacific Ocean, 1888-1904. Contrib. U. S. Nat. Herb. 10(5):221-419, Pls. XLIV-LIV.

Tripodiscus oregonus
Mann, A., 1907
Report on the diatoms of the Albatross voyages in the Pacific Ocean, 1888-1904. Contrib. U. S. Nat. Herb. 10(5):221-419, Pls. XLIV-LIV.

Tripodiscus orientalis
Mann, A., 1907
Report on the diatoms of the Albatross voyages in the Pacific Ocean, 1888-1904. Contrib. U. S. Nat. Herb. 10(5):221-419, Pls. XLIV-LIV.

Tripodiscus radiosus
Mann, A., 1907
Report on the diatoms of the Albatross voyages in the Pacific Ocean, 1888-1904. Contrib. U. S. Nat. Herb. 10(5):221-419, Pls. XLIV-LIV.

Tripodiscus rogersii
Mann, A., 1907
Report on the diatoms of the Albatross voyages in the Pacific Ocean, 1888-1904. Contrib. U. S. Nat. Herb. 10(5):221-419, Pls. XLIV-LIV.

Tripodiscus scaber
Mann, A., 1907
Report on the diatoms of the Albatross voyages in the Pacific Ocean, 1888-1904. Contrib. U. S. Nat. Herb. 10(5):221-419, Pls. XLIV-LIV.

Tripodiscus tripartitus
Mann, A., 1907
Report on the diatoms of the Albatross voyages in the Pacific Ocean, 1888-1904. Contrib. U. S. Nat. Herb. 10(5):221-419, Pls. XLIV-LIV.

Trochosira mirabilis
Cleve-Euler, A., 1951
Die Diatomeen von Schweden und Finnland. Kungl. Svenska Vetenskaps Akad. Handl., Fjärde Ser. 2(1): 161 pp., 6 pls.

Trochosira spinosa
Cleve-Euler, A., 1951
Die Diatomeen von Schweden und Finnland. Kungl. Svenska Vetenskaps Akad. Handl., Fjärde Ser. 2(1): 161 pp., 6 pls.

Tropidoneis sp.
Gilbert, J.Y., and W.E. Allen, 1943
The phytoplankton of the Gulf of California obtained by the "E.W. Scripps" in 1939 and 1940. J. Mar. Res. V(2):89-110, figs.30-31.

Tropidoneis Adeliae n.sp.
Manguin, E., 1957.
Premier inventaire des diatomées de la Terre Adélie Antarctique. Espèces nouvelles. Rev. Algologique, n.s., 3(3):111-134.

Tropidoneis antarctica
Bigelow, H.B., and M. Leslie, 1930
Reconnaissance of the waters and plankton of Monterey Bay, July 1928. Bull. M.C.Z., 70(5):429-481, 43 text figs.

Tropidoneis antarctica
Boden, Brian, 1948
Marine plankton diatoms on operation HIGHJUMP in: Some oceanographic observations on operation HIGHJUMP. By R.S. Dietz. USNEL Rept. No.55, 97 pp., 41 figs. 7 July 1948.

Tropidoneis antarctica
Hendey, N.I., 1937
The plankton diatoms of the southern seas. Discovery Repts. 16:151-364, pls.6-13.

Tropidoneis antarctica
Manguin, E., 1954
Diatomées marines provenant de l'ile Heard (Australian National Antarctic Expedition). Rev. Algol., n.s., 1: 14-24.

Tropidoneis antarctica polyplasta
Cupp, Easter E., 1943
Marine plankton diatoms of the west coast of North America. Bull. S.I.O. 5(1):1-238, 5 pls., 168 text figs.

Tropidoneis arctica var. polyplasta
Gran, H. H. and E. C. Angst, 1931
Plankton diatoms of Puget Sound. Publ. Puget Sound Biol. Sta. 7:417-519, 95 text figs.

Tropidoneis belgicae
Hendey, N.I., 1937
The plankton diatoms of the southern seas. Discovery Repts. 16:151-364, pls.6-13.

Tropidoneis confusa n.sp.
Hendy, N. Ingram, 1964
An introductory account of the smaller algae of British coastal waters. V. Bacillariophyceae (Diatoms). Her Majesty's Stationary Office, 317 pp., 45 pls.

Tropidoneis elegans
Hendy, N. Ingram, 1964
An introductory account of the smaller algae of British coastal waters. V. Bacillariophyceae (Diatoms). Her Majesty's Stationary Office, 317 pp., 45 pls.

Tropidoneis elegans
Rampi, L., 1940
Diatomee del Mare Adriatico. Nuovo Giornale Botanico Italiano, n.s., 47:559-608.

Tropidoneis elegans
Zanon, V., 1948
Diatomee marini di Sardegna e Pugillo di Alghe Marine della stressa. Boll. Pesca, Piscitutura e Idrobiologia, Anno 24, ns. 3(2): 202-244, 27 figs. on 1 pl.

Tropidoneis elegans glacialis n.sp.
Frenguelli, Joaquin, y Hector A. Orlando, 1958.
Diatomeas y silicoflagelados del sector Antartico Sudamericano. Inst. Antartico Argentino, Publ., No. 5:191 pp.

Tropidoneis fusiformis
Manguin, E., 1957.
Premier inventaire des diatomées de la Terre Adélie Antarctique. Espèces nouvelles. Rev. Algologique, n.s., 3(3):111-134.

Tropidoneis gibberula
Zanon, V., 1948
Diatomee marini di Sardegna e Pugillo di Alghe Marine della stressa. Boll. Pesca, Piscitutura e Idrobiologia, Anno 24, ns. 3(2): 202-244, 27 figs. on 1 pl.

Tropidoneis glacialis contricta
Boden, B. P., 1949
The diatoms collected by the U.S.S. CACOPAN in the Antarctic in 1947. J. Mar. Res. 8(1):6-13, 3 textfigs.

Tropidoneis glacialis var.
Boden, Brian, 1948
Marine plankton diatoms on operation HIGHJUMP in: Some oceanographic observations on operation HIGHJUMP. By R.S. Dietz. USNEL Rept. No.55, 97 pp., 41 figs. 7 July 1948.

Tropidoneis Hustedtii n.sp.
Manguin, E., 1957.
Premier inventaire des diatomées de la Terre Adélie Antarctique. Espèces nouvelles. Rev. Algologiques, n.s., 3(3):111-134.

Tropidoneis laevissima
Ko-bayashi, Tsuyako, 1963
Variations on some pennate diatoms from Antarctica. 1. Japan. Antarctic Res. Exped., 1956-1962, Sci Repts., (E), No. 18:1-20. 16 pls.

Tropidoneis lepidoptera
Cupp, Easter E., 1943
Marine plankton diatoms of the west coast of North America. Bull. S.I.O. 5(1):1-238, 5 pls., 168 text figs.

Tropidoneis lepidoptera lepidoptera
Hendy, N. Ingram, 1964
An introductory account of the smaller algae of British coastal waters. V. Bacillariophyceae (Diatoms). Her Majesty's Stationary Office, 317 pp., 45 pls.

Tropidoneis lepidoptera
Iyengar, M.O.P. and G. Venkataraman, 1951.
The ecology and seasonal succession of the algae flora of the River Cooum at Madras with special reference to the Diatomaceae. J. Madras Univ. 21, Sect. B(1): 140-192, 1 pl of 4 figs., 11 text figs.

Tropidoneis lepidoptera
Rampi, L., 1942
Ricerche sul fitoplancton del Mare Ligure 6. Le diatomee delle acque di Sanremo. Nuovo Giornale Botanico Italiano, N.S., 49:252-258.

Tropidoneis lepidoptera
Rampi, L., 1940
Diatomee del Mare Adriatico. Nuovo Giornale Botanico Italiano, n.s., 47:559-608.

Tropidoneis lepidoptera
Zanon, V., 1948
Diatomee marini di Sardegna e Pugillo di Alghe Marine della stressa. Boll. Pesca, Piscitutura e Idrobiologia, Anno 24, ns. 3(2): 202-244, 27 figs. on 1 pl.

Tropidoneis maxima
Hendy, N. Ingram, 1964
An introductory account of the smaller algae of British coastal waters. V. Bacillariophyceae (Diatoms). Her Majesty's Stationary Office, 317 pp., 45 pls.

Tropidoneis maxima
Zanon, V., 1948
Diatomee marini di Sardegna e Pugillo di Alghe Marine della stressa. Boll. Pesca, Piscitutura e Idrobiologia, Anno 24, ns. 3(2): 202-244, 27 figs. on 1 pl.

Tropidoneis membranacea
Bigelow, H.B., and M. Leslie, 1930
Reconnaissance of the waters and plankton of Monterey Bay, July 1928. Bull. M.C.Z., 70(5):429-481, 43 text figs.

Tropidoneis Pergalloi nom. nov.
Frenguelli, Joaquin, y Hector A. Orlando, 1958.
Diatomeas y silicoflagelados del sector Antartico Sudamericano.
Inst. Antartico Argentino, Publ., No. 5:191 pp.

Tripodoneis proteus
Hendey, N.I., 1937
The plankton diatoms of the southern seas.
Discovery Repts. 16:151-364, pls.6-13.

Tropidoneis pusilla
Hendy, N. Ingram, 1964
An introductory account of the smaller algae of British coastal waters. V. Bacillariophyceae (Diatoms).
Her Majesty's Stationary Office, 317 pp., 45 pls.

Tropidoneis vanheurckii
Hendy, N. Ingram, 1964
An introductory account of the smaller algae of British coastal waters. V. Bacillariophyceae (Diatoms).
Her Majesty's Stationary Office, 317 pp., 45 pls.

Tropidoneis Vanheurcki
Hustedt, F. and A.A. Aleem, 1951
Littoral diatoms from the Salstone near Plymouth. JMBA 30(1): 177.196.

Tropidoneis vitrea
Hendy, N. Ingram, 1964
An introductory account of the smaller algae of British coastal waters. V. Bacillariophyceae (Diatoms).
Her Majesty's Stationary Office, 317 pp., 45 pls.

Tropidoneis vitra
Hustedt, F. and A.A. Aleem, 1951
Littoral diatoms from the Salstone near Plymouth. JMBA 30(1): 177.196.

Tropidoneis vitrea
Rampi, L., 1940
Diatomee del Mare Adriatico. Nuovo Giornale Botanico Italiano, n.s., 47:559-608.

Tropidoneis vitrea
Zanon, V., 1948
Diatomee marini di Sardegna e Pugillo di Alghe Marine della stressa. Boll. Pesca, Piscitutura e Idrobiologia, Anno 24, ns. 3(2): 202-244, 27 figs. on 1 pl.

Tryblioptychus gen. nov.
Hendey, N-Ingram, 1958 [1957(Publ. 1958)]
Marine diatoms from some west African Ports- J. R Microsc. Soc. (3) 77(1/2): 28-85.

Tryblioptychus cocconeiformis comb. nov.
Hendey, N-Ingram, 1958 [1957(Publ. 1958)]
Marine diatoms from some west African Ports- J. R Microsc. Soc. (3) 77(1/2): 28-85.

Willemoësia n.gen.
Castracane degli Antelminelli, F., 1886
1. Report on the Diatomaceae collected by H.M.S. Challenger during the years 1873-1876. Rept. Sci. Results, H.M.S. Challenger, Botany Vol. II, 178 pp., 30 pls.

Xanthiopyxis oblonga
Mann, A., 1907
Report on the diatoms of the Albatross voyages in the Pacific Ocean, 1888-1904.
Contrib. U. S. Nat. Herb. 10(5):221-419, Pls. XLIV-LIV.

---

protozoa
Abe, T.H., 1941.
Studies on protozoan fauna of Shimoda Bay, the Diplopsalis group. Rec. Ocean. Wks. Japan 12(2):121-144, 45 figs.

protozoans
Bernard, Francis, 1967.
Research on phytoplankton and pelagic Protozoa in the Mediterranean Sea from 1953-1966.
Oceanogr. Mar. Biol., Ann. Rev. H. Barnes, editor, George Allen and Unwin, Ltd., 5:205-229.

protists
Bernstein, T., 1931.
[Pelagic protists of the northwest part of the Kara Sea] Trans. Arctic Inst. USSR 3(1):1-23.

protozoa
Bigelow, H.B., L.C. Lillick, and M. Sears, 1940.
Phytoplankton and planktonic protozoa of the offshore waters of the Gulf of Maine. Pt. 1. Numerical distribution. Trans. Am. Phil. Soc., n.s., 31(3):149-191, 10 textfigs.

Protozoa, ciliates
Burkovsky, I.V. 1971
Ecology of psammophilous ciliates in the White Sea (In Russian).
Zool. Zh. 50(9): 1285-1302

protozoa
Calkins, G.N., 1926
The biology of the protozoa. 623 pp., 238 textfigs. Lea and Febiger, Phila. and N.Y.

protozoa
Campbell, A. S., 1942.
Scientific Results of cruise VII of the Carnegie during 1928-1929 under command of Captain J. P. Ault. Biology II. The oceanic tintinnoina of the plankton gathered during the last Cruise of the Carnegie. Carnegie Inst., Washington, Publ. No. 537:1-163

protozoa
Hasle, G.R., 1960
Phytoplankton and ciliate species from the tropical Pacific.
Skrif. Norske Vidensk.-Akad., Oslo, 1(2): 50 pp.

protozoa
Hornell, J. 1917
A new Protozoan Cause of Widespread Mortality among Marine Fishes. Madras Fish Bull. No.11:53-66.

protozoa
Kamshilov, M.M., 1963.
The effect produced by ultra-violet rays on microbiocoenoses of marine protozoa. (In Russian Doklady, Akad. Nauk, SSSR, 150(6):1363-1365.

protozoa
Lackey, J. B., 1936.
Occurrence and distribution of the marine protozoan species in the Woods Hole area.
Biol. Bull., 70:264-278.

Protozoa
Lighthart, Bruce, 1969.
Planktonic and benthic bacteriovorous Protozoa at eleven stations in Puget Sound and adjacent Pacific Ocean.
J. Fish. Res. Bd. Can. 26(2): 299-304.

planktonic protozoa
Lillick, L. C., 1940
Phytoplankton and planktonic protozoa of the offshore waters of the Gulf of Maine. Pt. II. Qualitative composition of the planktonic flora. Trans. Am. Phil. Soc., n.s., 31:193-237

Protozoa
Ostenfeld, C.H., 1910
Protozoa. Danmark Exped. 1906-08. Medd. om Grønland, 43(11):287-300.

Protozoa
Sawyer, Thomas K., 1971.
Isolation and identification of free-living marine amoebae from upper Chesapeake Bay, Maryland.
Trans. Am. Micros. Soc., 90(1): 43-51

Protozoa
Small, Eugene B. 1972.
Free-living Protozoa of the Chesapeake Bay exclusive of Foraminifera and the flagellates.
Chesapeake Sci. 13 (Suppl.): S.96-597.

protozoa
Wang, C. C., and D. Nie, 1932
A survey of the marine protozoa of Amoy.
Contr. Biol. Lab. Sci. Soc., China, Zool. Ser. 8:285-385, 89 text figs.

protozoa, lists of spp.
Biernacka, I., 1962
Die Protozoenfauna in der Danziger Bucht. 1. Die Protozoen in einigen Biotopen der Seeküste
Polskie Arch. Hydrobiol. 10(23):39-109.

Protozoa, lists of spp.
Biernacka, I., 1962
Die Protozoenfauna in Danziger Bucht. II. Die Charakteristik der Protozoen in Untersuchten Biotopen der Seekuste.
Polskie Arch. Hydrobiol., 11(24) (1):17-75.

Protozoa, lists of spp.
Tibbs, John F., 1967.
On some planktonic Protozoa taken from the track of drift Station ARLIS I, 1960-61.
Arctic, 20(4):247-254.

protozoa
Turner, H.J., jr., 1954.
An improved method of staining the external organelles of hypotrichs. J. Protozool. 1(1): 18-19.

protozoa (physiol-ecol.)
Provosoli, Luigi, 1958
Nutrition and ecology of protozoa and algae.
Ann. Rev. Microbiol., 12:279-308.

protozoa, classification of
Honigberg, B.M., W. Balamuth, E.C. Boree, J.O. Corlis, M. Gojdics, R.P. Hall, R.R. Kudo, N.D. Levine, A.R. Loeblich, Jr., J. Weiser, and D.H. Wenrich, 1964
A revised classification of the phylum Protozoa.
J. Protozool., 11:7-20.

protozoa, classification of
The Committee on Taxonomy and Taxonomic Problems of the Society of Protozoologists, 1964.
A revised classification of the phylum Protozoa.
J. Protozoology, 11(1):7-19.

Protozoa, effect of
Johannes, R.E., 1965.
Influence of marine Protozoa on nutrient regeneration.
Limnol. Oceanogr., 10(3):434-442.

Acantharia, lists of spp.
Bottazzi-Massera, E., and G. Andreoli 1972.
Acantharia collected in the Tyrrhenian and northern Adriatic seas during three oceanographic cruises of the R/V Bannock. The problem of the upper and lower Adriatic Sea.
Archo Oceanogr. Limnol. 17(3): 191-207

acantharians

Beers, John R. and Gene L. Stewart, 1970.
The preservation of acantharians in fixed plankton samples. Limnol. Oceanogr., 15(5): 825-827.

Acantharia

Bottazzi Massera, E., G. Nencini and A. Vannucci, 1965.
Ulteriori ricerche sulla sistematica e sulla ecologia degli acantari (Protozoa) del Mar Tirreno.
Boll. Pesca, Piscicolt. e Idrobiol., n.s., 20(1):9-39.

Acantharia

Bottazzi, Elsa Massera, Bruno Schreiber and Vaughan T. Bowen, 1971.
Acantharia in the Atlantic Ocean, their abundance and preservation. Limnol. Oceanogr. 16(4): 677-684.

Acantharia

Bottazzi, E. Massera, and A. Vannucci, 1965.
Acantharia in the Atlantic Ocean: a systematic and ecological analysis of plankton collections made during Cruise 25 of R.V. Chain, of the Woods Hole Oceanographic Institution. 2nd Contrib.
Arch. Oceanogr. Limnol., 14(1):1-68.

Acantharia

Massera Bottazzi, E., and A. Vannucci, 1964.
Acantharia in the Atlantic Ocean - a systematic and ecological analysis of plankton collections made during cruises "Chain" 17 and "Chain" 21 of the Woods Hole Oceanographic Institution. 1st contribution.
Arch. Oceanogr. e Limnol., 13(3):315-385.

Acantharia

Nencini, Giuliano, e Luigina Saglia 1968
Ricerche biometriche comparative fra Acantari della stessa specie del Mediterraneo e dell'Atlantico.
Boll. Zool. 35(1/2):9-18.

Acantharia

Schreiber, B., 1963.
Acantharia as "scavengers" for strontium and their role in the sedimentation of radioactive debris.
In: Nuclear detonations and marine radioactivity the report of a symposium held at the Norwegian Defence Research Establishment, 16-20 September, 1963. Forsvarets Forskningsinstitutt, P.O. Box, 25, Kjeller, Norge, 113-126.
(multilithed)

acantharians, anat. physiol.

Febvre, Jean 1973.
Le cortex des acanthaires. II. Ultrastructure des zones de jonction entre les pièces corticales.
Protistologica 9(1):87-94

Acantharia, anat.

Febvre, Jean, 1971
Le myonème d'Acanthaire: essai d'interprétation ultrastructurale et cinétique.
Protistologica 7(3):379-391.

Acantharia, anat.

Hollande, André, Jean Cachon et Monique Cachon Enjumet, 1965.
Les modalités de l'enkystement présporogénétique chez les acanthaires.
Protistologica, 1(2):91-104.
Also in:
Trav. Sta. Zool., Villefranche-sur-Mer, 25.

Acantharia, anat.

Acantharia, life history of

Hollande, André, Jean Cachon et Monique Cachon-Enjumet 1965.
Les modalités de l'enkystement présporogénétique chez les acanthaires.
Protistologica 1(2):91-104.

Acantharia, lists of spp.

Bottazzi Massera, E., e Gabriella Andreoli 1974.
Ulteriori ricerche sugli Acantari (Protozoa) del mar Tirreno.
Boll. Pesca Piscic. Idrobiol. 26(1/2): 87-107.

Acantharia, lists of spp.

Massera Bottazzi, E., K. Vijayakrishnan Nair and M.C. Balani, 1968
On the occurrence of Acantharia in the Arabian Sea.
Arch. Oceanogr. Limnol., 15(1):63-67.

Acantharia, lists of spp.

Kimor, Baruch, 1971.
Some considerations on the distribution of Acantharia and Radiolaria in the Eastern Mediterranean.
Rapp. P.-v. Comm. int. Explor. scient. mer Medit. 20(3): 349-351.

Diaphanoeca grandis

Throndsen, J., 1970.
Marine planktonic acanthoecaceans (Craspedophyceae) from Arctic waters.
Nytt Mag. Bot. 17(2): 103-111

Parvicorbicula quadricostata

Throndsen, J., 1970.
Marine planktonic acanthoecaceans (Craspedophyceae) from Arctic waters.
Nytt Mag. Bot. 17(2): 103-111

Parvicorbicula socialis

Throndsen, J., 1970.
Marine planktonic acanthoecaceans (Craspedophyceae) from Arctic waters.
Nytt Mag. Bot. 17(2): 103-111

Pleurasiga minima n. sp.

Throndsen, J., 1970.
Marine planktonic acanthoecaceans (Craspedophyceae) from Arctic waters.
Nytt Mag. Bot. 17(2): 103-111

Pleurasiga reynoldsii n. sp.

Throndsen, J., 1970.
Marine planktonic acanthoecaceans (Craspedophyceae) from Arctic waters.
Nytt Mag. Bot. 17(2): 103-111

flagellates

Atkins, W. R. G., 1945
Autrophic flagellates as the major constituent of the oceanic phytoplankton. Nature 156(3963): 446-447.

flagellates

Bandel, W., 1940.
Phytoplankton- und Nahrstoffgehalt der Ostsee im Gebiet der Darsser Schwelle. Internat. Rev. ges. Hydrobiol. u. Hydrogr., 40:249-304.

flagellates

Bernard, Francis, 1965,
Production de flagellés en zone aphotique Mediterraneenne.
Rapp. Proc. Verb. Réunions, Comm. Int. Expl. Sci. Mer Mediterranée, Monaco, 18(2):341-344.

flagellates

Bernard, F., & B. Elkaim, 1962
Importance de la chute des Flagellés calcaires pour la fertilité profonde des mers chaudes.
Comptes Rendus Acad. Sci., Paris, 254(24): 4208-4210.

phytoflagellate

Blasco, Dolores 1973.
Étude cytologique de Chattonella subsala Biecheler. (Abstract). Rapp. Proc.-v. Reun. Comm. int. Explor. scient. Mer Medit. Monaco 21(8):423.

monads

Braarud, T., and Adam Bursa, 1939
On the phytoplankton of the Oslo Fjord, 1933-1934. Hvalrådets Skr. No.19:1-63; 9 text figs. Reviewed. J. du. Cons. 14(3): 418-420. A.C. Gardiner.

protozoa

Gemeinhardt, K, 1930
Silico flagellatae. Rabenhorst's Kryptogamen-flora Deutschlands, Österreichs, und der Schweiz 10(2):1-87. Leipzig

pseudomonad

Gow, John A., I.W. DeVoe and Robert A. MacLeod 1973.
Dissociation in a marine pseudomonad.
Can. J. Microbiol. 19(6): 695-701.

flagellates, physiol.

Halldal, Per, 1963
Zur Frage des Photoreceptors bei der Topophototaxis der Flagellaten.
Ber. Deutsch. Botan. Gesellschaft, 76(8): 323-327.

flagellates

Hartmann, M., and C. Chagas, 1910.
Flagellaten-Studien. Mem. de Inst. Osw.-Cruz 2:64-125

flagellates

Imai, T., and M. Hatanaka, 1950.
Studies on marine non-colored flagellates, Monas sp., favorite food of larvae of various marine animals. 1. Preliminary research on cultural requirements. Sci. Repts., Tohoku Univ., 4th ser., 18(3):304-315.

flagellates

Lackey, J.B., 1940.
Some new flagellates from the Woods Hole area.
Amer. Midl. Nat., 23(2):463-471. 21 text figs.

choanoflagellates

Leadbeater, B.S.C., 1972.
Fine-structural observations on some marine choanoflagellates from the coast of Norway.
J. mar. biol. Ass. U.K. 52(1): 67-79.

flagellates

Marshall, S.M., 1947
An experiment in marine fish cultivation: III. The plankton of a fertilized loch. Proc. Roy. Soc., Edinburgh, Sect.B., 63, Pt.I(3):21-33, 7 text figs.

choanoflagellates

Norris, Richard E., 1965.
Neustonic marine craspedomonadales (choanoflagellates) from Washington and California. J. Protozool., 12(4): 589-602.

flagellates

Parke, M., 1949.
Studies on marine flagellates. J. M. B. A. 28(1): 255-285.

flagellates

Roukiyainen, M.I. 1971.
Small flagellates in the Black Sea.
Rapp. P.-v. Comm. int. Explor. scient. mer
Medit. 20(3): 323-326.

monads

Smayda, Theodore J., 1966.
A quantitative analysis of the phytoplankton
of the Gulf of Panama. III General ecological
conditions and the phytoplankton dynamics at
8o 45'N, 79o 23'W from November 1954 to May
1957.
Inter-Amer. Trop. Tuna Comm., Bull., 11(5):
355-612.

flagellates, anat. physiol.

Allen, Mary Belle 1968.
Physiological studies on marine
chrysomonads.
Bull. Misaki mar. biol. Inst. Kyoto Univ.
12: 33-34.

flagellates, physiol.

Iwasaki, Hideo, 1972
The physiological characteristics of
neritic red tide flagellates. (In
Japanese; English abstract). Bull.
Plankt. Soc. Japan 19(2):46-56(104-114).

flagellates, anat.-physiol.

Jahn, T.L., M.D. Landman and J.R. Fonseca. 1964.
The mechanism of locomotion of flagellates. II.
Function of the Mastigonemes of Ochromonas.
J. Protozool., 11(3):291-296.

flagellates

Leadbeater, B.S.C., 1971.
Observations by means of ciné photography on
the behaviour of the haptonema in plankton
flagellates of the class Haptophyceae. J. mar.
biol. Ass., U.K., 51(1): 207-217.

flagellates, anat.-physiol.

Manton, I., and M. Parke, 1965.
Observations on the fine structure of two
species of Platymonas with special reference to
flagellar scales and the mode of origin of the
theca.
J. mar. biol. Ass., U.K., 45(3):743-754.

flagellates, anat.-physiol

Ronkin, R.R., 1959.
Motility and power dissipation in flagellated
cells, especially Chlamydomonas.
Biol. Bull., 116(2):285-293.

flagellates, anat., physiol.

Ryther, J.H., 1956.
Interrelation between photosynthesis and respiration in the marine flagellate, Dunaliella
euchlora. Nature 178(4538):861-863.

flagellates, chemistry of

Marker, A.F.H., 1965.
Extracellular carbohydrate liberation in the
flagellates Isochrysis galbana and Prymnesium
parvum.
J. mar. biol. Ass., U.K., 45(3):755-772.

flagellates

Roukhiyainen, M.I., 1970.
On the quantitative development of flagellate
algae in the southern seas. (In Russian; English
abstract). Okeanologiia, 1066-1070.
10(4):

flagellates, misc., lists of spp.

Lackey, James B., and Elsie W. Lackey, 1963
Microscopic algae and protozoa in the waters
near Plymouth in August 1962.
J. Mar. Biol. Assoc., U.K. 43(3):797-805.

choanoflagellates, lists of spp.

Leadbeater, Barry S.C. 1973.
External morphology of some marine
choanoflagellates from the coast
of Jugoslavia.
Arch. Protistenk. 115 (2/3): 234-252

flagellates, lists of spp.

Leadbeater B.S.C. 1972
Identification, by means of electron
microscopy, of flagellate nannoplankton
from the coast of Norway.
Sarsia 49: 107-124.

flagellates, lists of spp.

Throndsen J., 1970.
Flagellates from Arctic waters.
Nytt Mag Bot., 17(1): 49-57.

flagellates, lists of spp.

Throndsen, Jahn 1969.
Flagellates of Norwegian coastal
waters.
Nytt Mag. Bot. 16 (3/4): 161-216.

Anisomonas astigmatica n.sp.

Scagel, Robert F., and Janet R. Stein, 1961.
Marine nannoplankton from a British Columbia
fjord.
Canadian J. Botany, 39:1205-1213.

Bodo sp.

Quayle, D.B.
Paralytic shellfish poisoning in British Colombia.
Bull. Fish Res. Bd. Can., 168: 68 pp.

Bodo bacillariophagus n.sp.

Bursa, Adam, 1963.
Phytoplankton in coastal waters of the Arctic
Ocean at Point Barrow, Alaska.
Arctic, 16(4):239-262.

Boda marina

Braarud, T., 1945
A phytoplankton survey of the polluted
waters of inner Oslo Fjord. Hvalrådets
Skrifter, No.28, 142 pp., 19 text figs.,
17 tables.

? Bodo marina

Braarud, T., 1934
A note on the phytoplankton of the Gulf
of Maine in the summer of 1933. Biol. Bull.
67(1):76-82.

Bodo marina

Braarud, T., and Adam Bursa, 1939
On the phytoplankton of the Oslo Fjord,
1933-1934. Hvalrådets Skr. No.19:1-63;
9 text figs. Reviewed. J. du. Cons. 14(3):
418-420. A.C. Gardiner.

Bodo marina

Gran, H.H., and T. Braarud, 1935
A quantitative study of the phytoplankton in the Bay of Fundy and the
Gulf of Maine (including observations
on hydrography, chemistry, and turbidity).
J. Biol. Bd., Canada, 1(5):279-467, 69
text figs.

Bodopsis platyformis n.s

Lackey, J.B., 1940.
Some new flagellates from the Woods Hole area.
Amer. Midl. Nat., 23(2):463-471.

Calkinsia aureus n.gen., n.sp.

Lackey, James B., 1960.
Calkinsia aureus gen. et sp. nov., a new marine
euglenid.
Trans. Amer. Microsc. Soc., 79(1):105-107.

Carteria sp.

Braarud, T., 1945
A phytoplankton survey of the polluted
waters of inner Oslo Fjord. Hvalrådets
Skrifter, No.28, 142 pp., 19 text figs.,
17 tables.

Carteria sp.

Braarud, T., and Adam Bursa, 1939
On the phytoplankton of the Oslo Fjord,
1933-1934. Hvalrådets Skr. No.19:1-63;
9 text figs. Reviewed. J. du. Cons. 14(3):
418-420. A.C. Gardiner.

Carteria? sp.

Ronkin, R.R., 1959
Motility and power dissipation in flagellated cells, especially Chlamydomonas.
Biol. Bull., 116(2):285-293.

Chilomonas marina

Hasle, Grethe Rytter, 1960
Phytoplankton and ciliate species from the
Tropical Pacific.
Skr. Norske Videnskaps-Akad., Oslo, 1.
Mat.-Nat. Kl., 1960(2): 1-50.

Chilomonas marina

Norris, R.E., 1961.
Observations on phytoplankton organisms collected on the N.Z.O.I. Pacific Cruise, September
1958.
N.Z.J. Sci., 4(1):162-188.

Chlamydomonas sp.

Braarud, T., 1945
A phytoplankton survey of the polluted
waters of inner Oslo Fjord. Hvalrådets
Skrifter, No.28, 142 pp., 19 text figs.,
17 tables.

Chlamydomonas (3 strains)

Ronkin, R.R., 1959
Motility and power dissipation in flagellated cells, especially Chlamydomonas.
Biol. Bull., 116(2):285-293.

Chlamydomonas parkeae

Ettl, H., 1967.
Chlamydomonas parkeae, eine neue marine
Chlamydomonade.
Int. Rev. ges. Hydrobiol., 52(3):437-440.

Chlamydomonas reinhardi

Levine, R.P., 1960.
Genetic control of photosynthesis in
Chlamydomonas reinhardi.
Proc. Nat. Acad. Sci., 46(7):972-977.

Chromulina pleiades n.sp.

Parke, M., 1949.
Studies on marine flagellates. J.M.B.A. 28(1):
255-285.

Chromulina sp.

Scagel, Robert F., and Janet R. Stein, 1961.
Marine nannoplankton from a British Columbia
fjord.
Canadian J. Botany, 39:1205-1213.

Chryptochrysis atlantica n.sp.

Lackey, J.B., 1940.
Some new flagellates from the Woods Hole area.
Amer. Midl. Nat., 23(2):463-471.

Cryptomonas sp.

Braarud, T., 1951.
Salinity as an ecological factor in marine phytoplankton. Physiol. Plant. 4:28-34, 3 textfigs.

Chrysamoeba nana n.sp.

Scagel, Robert F., and Janet R. Stein, 1961.
Marine nannoplankton from a British Columbia
fjord.
Canadian J. Botany, 39:1205-1213.

Chrysochromulina sp.

Norris, R.E., 1961.
Observations on phytoplankton organisms collected on the N.Z.O.I. Pacific Cruise, September
1958.
N.Z. J. Sci., 4(1):162-188.

Chrysochromulina sp.
Scagel, Robert F., and Janet R. Stein, 1961.
Marine nannoplankton from a British Columbia fjord.
Canadian J. Botany, 39:1205-1213.

Chrysochromulina alifera n.sp.
Parke, M., I. Manton and B. Clarke, 1956.
Studies on marine flagellates. III. Three further species of Chrysochromulina. J.M.B.A., 35(2):387-414.

Chrysochromulina bergenensis n.sp.
Leadbeater, B.S.C. 1972.
Fine structural observations of six new species of Chrysochromulina (Haptophyceae) from Norway with preliminary observations on scale production in C. microcylindra sp. nov.
Sarsia 49:65-80.

Chrysochromulina brevifilum n.sp.
Parke, M., I. Manton and B. Parke, 1955.
Studies on marine flagellates. II. Three new species of Chrysochromulina. J.M.B.A. 34:579-609, Pls. 1-9.

Chrysochromulina ephippium n.sp.
Parke, M., I. Manton and B. Clarke, 1956.
Studies on marine flagellates. III. Three further species of Chrysochromulina. J.M.B.A., 35(2):387-414.

Chrysochromulina ericina
Manton, I., and G.F. Leedale, 1961.
Further observations on the fine structure of Chrysochromulina ericina Parke & Manton.
J.M.B.A., U.K., 41(1):145-155.

Chrysochromulina ericina n.sp.
Parke, M., I. Manton and B. Clarke, 1956.
Studies on marine flagellates. III. Three further species of Chrysochromulina. J.M.B.A., 35(2):387-414.

Chrysochromulina fragilis n. sp.
Leadbeater, B.S.C. 1972.
Fine structural observations of six new species of Chrysochromulina (Haptophyceae) from Norway with preliminary observations on scale production in C. microcylindra sp. nov.
Sarsia 49:65-80.

Chrysochromulina herdlensis n.sp.
Leadbeater, B.S.C. 1972.
Fine structural observations of six new species of Chrysochromulina (Haptophyceae), from Norway with preliminary observations on scale production in C. microcylindra sp. nov.
Sarsia 49:65-80.

Chrysochromulina kappa n.sp.
Parke, M., I. Manton and B. Clarke, 1955.
Studies on marine flagellates. II. Three new species of Chrysochromulina. J.M.B.A. 34:579-609, 9 pls.

Chrysochromulina minor n.sp.
Parke, M., I. Manton and B. Clark, 1955.
Studies on marine flagellates. II. Three new species of Chrysochromulina. J.M.B.A. 34:579-609, 9 pls.

Chrysochromulina parkeae
Green, J.C. and B.S.C. Leadbeater, 1972.
Chrysochromulina parkeae sp. nov. (Haptophyceae) a new species recorded from S.W. England and Norway. J. mar. biol. Ass. U.K. 52(2): 469-474.

Chrysochromulina microcylinda n.sp
Leadbeater B.S.C. 1972.
Fine structural observations of six new species of Chrysochromulina (Haptophyceae), from Norway with preliminary observations on scale production in C. microcylindra sp. nov.
Sarsia 49:65-80.

Chrysochromulina mantoniae n.sp.
Leadbeater B.S.C. 1972.
Fine structural observations of six new species of Chrysochromulina (Haptophyceae), from Norway with preliminary observations on scale production in C. microcylindra sp. nov.
Sarsia 49:65-80.

Chrysochromulina parva
Parke, M., J.W.G. Lund and I. Manton, 1962
Observations on the biology and fine structure of the type species of Chrysochromulina. C. parva Lackey) in the English Lake District.
Arch. Mikrobiol., 42:333-352.

Abstr. in:
J.M.B.A., U.K., 42(3):705-706.

Chrysochromulina polylepis n.sp.
Manton, Irene and Mary Parke, 1962
Preliminary observations on scales and their mode of origin in Chrysochromulina polylepis sp. nov.
J. Mar. Biol. Lab., U.K., 42(3):565-578.

Chrysochromulina pringsheimii n.sp.
Parke, Mary, and Irene Manton, 1962
Studies on marine flagellates. VI. Chrosochromulina pringsheimii sp. nov.
J. Mar. Biol. Assoc., U.K., 42(2):391-404.

Chrysochromulina strobilus n.sp.
Parke, Mary, Irene Manton and B. Clarke, 1959.
Studies on marine flagellates. V. Morphology and microanatomy of Chrysochromulina strobilus n.sp.
J.M.B.A., U.K., 38(1):169-188.

Chrysococcus cinctus n.sp.
Lackey, J.B., 1940.
Some new flagellates from the Woods Hole area.
Amer. Midl. Nat., 23(2):463-471.

Chrysastrella deceptionis n.sp.
Frenguelli, Joaquin, y Hector A. Orlando, 1958.
Diatomeas y silicoflagelados del sector Antartico Sudamericano.
Inst. Antartico Argentino, Publ., No. 5:191 pp.

Chlamydomonas globosa
Kroes, H.W., 1972
Growth interactions between Chlamydomonas globosa Snow and Chlorococcum ellipsoideum Deason and Bold: the role of extracellular products. Limnol. Oceanogr. 17(3): 423-432.

Clericia antarctica n.sp.
Frenguelli, Joaquin, y Hector A. Orlando, 1958.
Diatomeas y silicoflagelados del sector Antartico Sudamericano.
Inst. Antartico Argentino, Publ., No. 5:191 pp.

Clericia fasciciculata n.sp.
Frenguelli, Joaquin, y Hector A. Orlando, 1958.
Diatomeas y silicoflagelados del sector Antartico Sudamericano.
Inst. Antartico Argentino, Publ., No. 5:191 pp.

Corbicula socialis
Gran, H.H., and T. Braarud, 1935
A quantitative study of the phytoplankton in the Bay of Fundy and the Gulf of Maine (including observations on hydrography, chemistry, and turbidity).
J. Biol. Bd., Canada, 1(5):279-467, 69 text figs.

Dicrateria gilva n.sp.
Parke, M., 1949.
Studies on marine flagellates. J.M.B.A. 28(1): 255-285.

Dicrateria inornata n.gen., n. sp.
Parke, M., 1949.
Studies on marine flagellates. J.M.B.A. 28(1): 255-285.

Dinobryon sp.
Braarud, T., and Adam Bursa, 1939
On the phytoplankton of the Oslo Fjord, 1933-1934. Hvalrådets Skr. No.19:1-63; 9 text figs. Reviewed. J. du. Cons. 14(3): 418-420. A.C. Gardiner.

Dinobryon cylindricum
Braarud, T., 1945
A phytoplankton survey of the polluted waters of inner Oslo Fjord. Hvalrådets Skrifter, No.28, 142 pp., 19 text figs., 17 tables.

Dinobryon pellucidum
Jorgensen, E., 1900
Protophyten und Protozoen im Plankton aus der Norwegischen Westkerste. Bergens Mus. Aarb. 1899(6): 95 pp., 5 pls., 83 tables.

Dinobryon pellucidum
Stæmann-Nielsen, Einar, 1951
The marine vegetation of the Isefjord. A study on ecology and production. Medd. Komm. Danmarks Fiskeri-og Havundersøgelser. Ser. Plankton. 5(4); 114pp., 46 text figs.

Dinobryon stipidatum
Braarud, T., 1945
A phytoplankton survey of the polluted waters of inner Oslo Fjord. Hvalrådets Skrifter, No.28, 142 pp., 19 text figs., 17 tables.

Dunaliella euchlora
McLachlan, Jack, and Charles S. Yentsch, 1959
Observations on the growth of Dunaliella euchlora in culture. Biol. Bull. 116(3): 461-483.

Dunaliella sp.
Ronkin, R.R., 1959
Motility and power dissipation in flagellated cells, especially Chlamydomonas.
Biol. Bull., 116(2):285-293.

Dunaliella euchlora
Ryther, J., 1956.
Interrelation between photosynthesis and respiration in the marine flagellate, Dunaliella euchlora. Nature 178(4538):861-863.

Dunaliella euchlora
Ukeles, R., 1961.
The effect of temperature on the growth and survival of several marine algal species.
Biol. Bull., 120(2):255-264.

Euglena spp
Butcher, R.W., 1961
An introductory account of the smaller algae of British coastal waters. Pt. VIII. Euglenophyceae.
Min. Agric. Fish. Food, Fish. Invest., (4): 17 pp.

Individual species of genus are described

Euglena
Pringsheim, E.G., 1953.
Salzwasser Euglena. Arch. f. Mikrobiol. 18:149-164, 11 textfigs.

Euglena gracilis
Gross, J. A., 1965.
Effects of high hydrostatic pressures on Euglena gracilis.
Ocean Sci. and Ocean Eng., Mar. Techn. Soc., Amer. Soc. Limnol. Oceanogr., 1:81.

Euglena obtusa
Palmer, John D., and Frank B. Round, 1965.
Persistent, vertical-migration rhythms in benthic microflora. 1. The effect of light and temperature on the rhythmic behaviour of Euglena obtusa.
J. mar. biol. Ass., U.K., 45(3):567-582.

Euglena viridis
Dragesco, Jean, 1965.
Etude cytologique de quelques flagellés mesopsammiques.
Cahiers Biol. Mar., Roscoff, 6(1):83-115.

Euglenidae
Pringsheim, E.G., 1948.
Taxonomic problems in the Euglenidae. Biol. Rev. 23:46-61, 4 textfigs.

Euglenopsis zabra n.sp.
Norris, R.E., 1961.
Observations on phytoplankton organisms collected on the N.Z.O.I. Pacific Cruise, September 1958.
N.Z.J. Sci., 4(1):162-188.

Eutreptia spp
Butcher, R.W., 1961
An introductory account of the smaller algae of British coastal waters. Pt. VIII. Euglenophyceae.
Min. Agric. Fish. Food. Fish. Invest., (4): 17 pp.

Individual species of genus are described.

Eutreptia Lanowii
Braarud, T., 1945
A phytoplankton survey of the polluted waters of inner Oslo Fjord. Hvalrådets Skrifter, No.28, 142 pp., 19 text figs., 17 tables.

Eutreptia Lanowi
Braarud, T., and Adam Bursa, 1939
On the phytoplankton of the Oslo Fjord, 1933-1934. Hvalrådets Skr. No.19:1-63; 9 text figs. Reviewed. J. du. Cons. 14(3): 418-420. A.C. Gardiner.

Eutreptia Lanowi
Gran, H.H., and T. Braarud, 1935
A quantitative study of the phytoplankton in the Bay of Fundy and the Gulf of Maine (including observations on hydrography, chemistry, and turbidity).
J. Biol. Bd., Canada, 1(5):279-467, 69 text figs.

Eutreptia viridis
Braarud, T., 1934
A note on the phytoplankton of the Gulf of Maine in the summer of 1933. Biol. Bull. 67(1):76-82.

Eutreptiella spp
Butcher, R.W., 1961
An introductory account of the smaller algae of British coastal waters. Pt. VIII. Euglenophyceae.
Min. Agric. Fish. Food. Fish. Invest., (4): 17 pp.

Individual species of genus are described.

Hemieutreptia antigua
Kimura, Tomohiro, Akio Mizokami and Toshimasa Hashimoto, 1972
The red tide that caused severe damage to the fishery resources in Hiroshima Bay: outline of its occurrence and the environmental conditions. (In Japanese; English abstract). Bull. Plankt. Soc. Japan 19(2):24-38 (81-112).

Hemieutreptia antiqua
Takayama, Haruyoshi, 1972.
Observations on an interesting marine euglenoid flagellate blooming in Hiroshima Bay 1969 and 1970. Bull. Hiroshima Fish. Exp. Stn, 3: 1-7. (In Japanese; English abstract).

Hemiselmis rufescens h. gen., n. sp.
Parke, M., 1949.
Studies on marine flagellates. J. M. B. A. 28(1): 255-285.

Heteromastix angulata
Manton, I., D.G. Rayns and H. Ettl, 1965.
Further observations on green flagellates with scaly flagella: the genus Heteromastix Korshikov.
Jour. Mar. Biol. Assoc., U.K., 45(1):241-255.

Heteromastix longifilis
Manton, I., D.G. Rayns and H. Ettl, 1965.
Further observations on green flagellates with scaly flagella: the genus Heteromastix Korshikov.
Jour. Mar. Biol. Assoc., U.K., 45(1):241-255.

Heteromastix rotunda
Manton, I., D.G. Rayns and H. Ettl, 1965.
Further observations on green flagellates with scaly flagella: the genus Heteromastix Korshikov.
Jour. Mar. Biol. Assoc., U.K., 45(1):241-255.

Isochrysis galbana
Kain, J.A. and G.E. Fogg, 1958
Studies on the growth of marine phytoplankton II Isochrysis galbana Parke. J. Mar. Biol. Ass. 37(3): pp.781-788.

Isochrysis galbana
Marker, A.F.H., 1965.
Extracellular carbohydrate liberation in the flagellates Isochrysis galbana and Prymnesium parvum.
J. mar. biol. Ass., U.K., 45(3):755-772.

Isochrysis galbana n. sp.
Parke, M., 1949.
Studies on marine flagellates. J.M.B.A 28(1): 255-285.

Isochyris galbana
Ryther, John H., and Dana D. Kramer, 1961
Relative iron requirement of some coastal and offshore algae.
Ecology, 42(2): 444-446.

Isochrysis galbana
Ukeles, R., 1961.
The effect of temperature on the growth and survival of several marine algal species.
Biol. Bull., 120(2):255-264.

Isonema nigricans
Schuster, F.L., S. Goldstein and B. Hershenov, 1968.
Ultrastructure of a flagellate Isonema nigrigricans nov.gen.Nov. sp., from a polluted marine habitat.
Protistologica, 4(1):141-149.

Micromonas (Chromulina) pusilla comb. nov.
Manton, I., 1960
Further observations on small green flagellates with special reference to possible relatives of Chromulina pusilla Butcher.
J. Mar. Biol. Assoc., U.K., 39(2): 275-298.

Micromonas squamata sp.nov
Manton, I., 1960
Further observations on small green flagellates with special reference to possible relatives of Chromulina pusilla Butcher.
J. Mar. Biol. Assoc., U.K., 39(2): 275-298.

Micromonas squamata
Parke, M., and D.G. Rayns, 1964.
Studies on marine flagellates. VII. Nephroselmis gilva sp. nov. and some allied forms.
J. Mar. Biol. Assoc., U.K., 44(1):209-217.

Microsportella fimbriadensis n.gen., n.sp.
Scagel, Robert F., and Janet R. Stein, 1961.
Marine nannoplankton from a British Columbia fjord.
Canadian J. Botany, 39:1205-1213.

Monochrysis lutheri
Droop, M.R., 1968.
Vitamin B_{12} and marine ecology. IV. The kinetics of uptake, growth and inhibition in Monochrysis lutheri.
J. mar. biol. Ass., 48(3):689-733.

Monochrysis lutheri
Jitts, H.R., C.D. McAllister, K. Stephens and J.D.H. Strickland, 1964.
The cell division rates of some marine phytoplankters as a function of light and temperature.
J. Fish. Res. Bd., Canada, 21(1):139-157.

Monochrysis lutheri
Droop, M.R., 1961.
Vitamin B12 and marine ecology: the response of Monochrysis lutheri.
J.M.B.A., U.K., 41(1):69-76.

Monochrysis lutheri
Ukeles, R., 1961.
The effect of temperature on the growth and survival of several marine algal species.
Biol. Bull., 120(2):255-264.

Monochrysis lutheri
Yoo, Sung Kyoo, 1968.
Studies on the growth of algal food, Cyclotella nana, Chaetoceros calcitrans and Monochrysis lutheri. (In Korean; English abstract).
Bull. Pusan Fish. Coll. (Nat. Sci.), 8(2):123-126.

Monosiga sp.
Bursa, Adam S., 1961
The annual oceanographic cycle at Igloolik in the Canadian Arctic. II. The phytoplankton.
J. Fish. Res. Bd., Canada, 18(4):563-615.

Monosiga marina n.sp. (flagellate)
Grøntved, J., 1952.
Investigations on the phytoplankton in the southern North Sea in May 1947. Medd. Komm. Danmarks Fisk.- og Havundersøgelser, Plankton Ser., 5(5): 1-49, 1 pl., 21 tables, 24 textfigs.

Monosiga marina
Hasle, Grethe Rytter, 1960
Phytoplankton and ciliate species from the Tropical Pacific.
Skr. Norske Videnskaps-Akad., Oslo, 1. Mat.-Nat. Kl., 1960(2): 1-50.

Monosiga marina minima n.var.
Paasche, E., 1961.
Notes on phytoplankton from the Norwegian Sea.
Botanica Marina, 2(3/4):197-210.

a colorless flagellate

Ochromonas glacialis n.sp.
Bursa, Adam, 1963.
Phytoplankton in coastal waters of the Arctic Ocean at Point Barrow, Alaska.
Arctic, 16(4):239-262.

Ochromonas malhamensis
Droop, M.R., 1961.
Vitamin B12 and marine ecology: the response of Monochrysis lutheri.
J.M.B.A., U.K., 41(1):69-76.

Ochromonas marina n.sp.
Lackey, J.B., 1940.
Some new flagellates from the Woods Hole area.
Amer. Midl. Nat., 23(2):463-471.

Ochromonas pearyi n.sp.
Bursa, Adam, 1963.
Phytoplankton in coastal waters of the Arctic Ocean at Point Barrow, Alaska.
Arctic, 16(4):239-262.

Ochromonas ?vallesiaca
Scagel, Robert F., and Janet R. Stein, 1961.
Marine nannoplankton from a British Columbia fjord.
Canadian J. Botany, 39:1205-1213.

Olisthodiscus sp.
Iizuka, Shoji, and Haruhiko Irie, 1968.
Discoloration phenomena by microalgae in Nagasaki Pref. in 1956 and ecology of causative organisms, Olisthodiscus. (In Japanese; English abstract).
Bull. Fac. Fish., Nagasaki Univ., 26: 25-35.

Olisthodiscus luteus
Pratt, David M., 1966.
Competition between Skeletonema costatum and Olisthodiscus luteus in Narragansett Bay and in culture.
Limnol. Oceanogr., 11(4):447-455.

Pachysphaera marshalliae n. sp.
Parke, M., 1966.
The genus Pachysphaera (Prasinophyceae).
In: Some contemporary studies in marine science
H. Barnes, editor, George Allen & Unwin, Ltd., 555-563.

Pachysphaera pelagica
Parke, M., 1966.
The genus Pachysphaera (Prasinophyceae).
In: Some contemporary studies in marine science
H. Barnes, editor, George Allen & Unwin, Ltd., 555-563.

Paraphysomonas cylicophora n.sp.
Leadbeater, B.S.C. 1972.
Paraphysomonas cylicophora sp. nov., a marine species from the coast of Norway.
Norw. J. Bot. 19(3/4): 179-185.

Paraphysomonas cribosa n.sp
Lucas, I.A.N., 1968.
A new member of the Chrysophyceae, bearing polymorphic scales.
J. mar. biol. Ass., U.K., 48(2):437-441.

Paraphysomonas foraminifera n. sp.
*Lucas, I.A.N., 1967.
Two new marine species of Paraphysomonas.
J. mar. biol. Ass., U.K., 47(2):329-334.

Paraphysomonas imperforatus n. sp.
Lucas, I.A.N., 1967.
Two new marine species of Paraphysomonas.
J. mar. biol. Ass., U.K., 47(2):329-334.

Parvicorbicula campaniformis n.sp.
Leadbeater, Barry S.C. 1973.
External morphology of some marine choanoflagellates from the coast of Jugoslavia.
Arch. Protistenk. 115 (2/3): 234-252

Parvicorbicula pedicellata n.sp.
Leadbeater, Barry S.C. 1973.
External morphology of some marine choanoflagellates from the coast of Jugoslavia.
Arch. Protistenk. 115 (2/3): 234-252

Parvicorsicula spinifera n.sp.
Leadbeater, Barry S.C. 1973.
External morphology of some marine choanoflagellates from the coast of Jugoslavia.
Arch. Protistenk. 115 (2/3): 234-252

Pavlova gyrans
Green, J.C., and I. Manton, 1970.
Studies in the fine structure and taxonomy of flagellates in the genus Pavlova. 1. A revision of Pavlova gyrans, the type species.
J. mar. biol. Ass., U.K., 50(4): 1113-1130.

Phacus spp
Butcher, R.W., 1961
An introductory account of the smaller algae of British coastal waters. Pt. VIII. Euglenophyceae.
Min. Agric. Fish. Food. Fish. Invest., (4): 17 pp.

Individual species of genus are described

Phaeocystis
Brunel, J., 1962
Le phytoplancton de la Baie de Chaleurs.
Inst. Botan., Univ. Montréal, Contrib. No. 77: 365 pp., 66 pls.

Phaeocystis
Lucas, C.E. 1942
Continuous plankton records: Phytoplankton in the North Sea 1938-39. II. Dinoflagellates, Phaeocystis, etc. Hull Bull. Mar. Ecol., II(9):47-70, 6 textfigs., 24 pls.

Phaeocystis
Lucas, C. E., 1940.
Ecological investigations with the continuous plankton recorder: the phytoplankton in the southern North Sea, 1932-37. Hull Bull. Mar. Ecol., 1(3):73-170, 17 textfigs., 21 pls.

phaeocystis
Savage, R. E., 1930
The influence of Phaeocystis on the migrations of the Herring. Min. Agric. Fish., Fish. Invest. ser. ii, XII(2)

Phaeocystis sp.
Sieburth, J. McN. 1960
Acrylic acid an "antibiotic" principle in Phaeocystis blooms in Antarctic waters.
Science 132: 676-

Phaeocystis
Wulf, A., 1934.
Über Hydrographie und Oberflachenplankton nebst Verbreitung von Phaeocystis in der Deutschen Buch im Mai 1933. Ber. d. Deutsch. Wiss. Komm. Meeresforsch., N. F., VII(3):343-350

Phaeocystis globosa
Galtsoff, P.S., 1949
The mystery of the red tide. Sci. Mon. LXVIII (2):108-117.

Phaeocystis globosa
Lemmermann, E., 1908.
XXI. Flagellatae, Chlorophyceae, Coccosphaerales, und Silicoflagellatae. Nordisches Plankton, Bot. Teil:1-40, 135 textfigs.

Phaeocystis globosa
Schodduyn, M., 1926
Observations faites dans la baie d'Ambleteuse (Pas de Calais). Bull. Inst. Ocean., Monaco, No. 482: 64 pp.

Phaeocystis Poucheti
Braarud, T., and Adam Bursa, 1939
On the phytoplankton of the Oslo Fjord, 1933-1934. Hvalrådets Skr. No.19:1-63; 9 text figs. Reviewed. J. du. Cons. 14(3): 418-420. A.C. Gardiner.

Phaeocystis Poucheti
Chu, S. P., 1947.
The utilization of organic phosphorus by phytoplankton. J.M.B.A. 26(3):285-295, 1 diagr.

Phaeocystis poucheti
Galtsoff, P.S., 1949
The mystery of the red tide. Sci. Mon. LXVIII (2):108-117.

Phaeocystis poucheti
Guillard, R.R.L. and Johan A. Hellebust 1971
Growth and the production of extracellular substances by two strains of Phaeocystis poucheti.
J. Phycol. 7 (4): 330-338.

Phaeocystis poucheti
Jørgensen, E., 1905
B. Protistplankton and the diatoms in bottom samples. Hydrographical and biological investigations in Norwegian fjords. Bergens Mus. Skr. 7: 49-225.

Phaeocystis poucheti
Jorgensen, E., 1900
Protophyten und Protozoen im Plankton aus der Norwegischen Westkerste. Bergens Mus. Aarb. 1899(6): 95 pp., 5 pls., 83 tables

Phaeocystis pouchetii (flagellate)
Kashkin, N.I., 1963
Material on the ecology of Phaeocystis pouchetii (Hariot) Lagerheim, 1893 (Chrysophyceae). (In Russian).
Okeanologiia, Akad. Nauk. SSSR, 3(4):697-705.

Phaeocystis poucheti
Kayser, H. 1970.
Experimental-ecological investigations on Phaeocystis poucheti (Haptophyceae): cultivation and waste water test.
Meeresunters. 20(1/4): 195-212.

Phaeocystis poucheti
Lemmermann, E., 1908.
XXI. Flagellatae, Chlorophyceae, Coccosphaerales, und Silicoflagellatae. Nordisches Plankton, Bot. Teil:1-40, 135 textfigs.

Phaeocystis Poucheti
Meunier, A., 1919
Microplankton de la Mer Flamande. 4. Les Tintinnides et Coetera. Mem. Mus. Roy. Hist. Nat., Belgique, 8(2):59pp., Pls. 22-23.

Phaeocystis pouchetii
Parke, M., J.C. Green and I. Manton, 1971.
Observations on the fine structure of zoids of the genus Phaeocystis (Haptophyceae).
J. mar. biol. Ass. U.K. 51(4): 927-941.

Phaeocystis Poucheti
Schodduyn, M., 1926
Observations faites dans la baie d'Ambleteuse (Pas de Calais). Bull. Inst. Ocean., Monaco, No. 482: 64 pp.

Phaeocystis pouchetii
Sieburth, John McNeill, 1963.
Bacterial habitats in the Antarctic environment Ch. 49 in: Symposium on Marine Microbiology, C.H. Oppenheimer, Editor, C.C. Thomas, Springfield, Illinois, 533-548.

Phyllomitus salinus n.sp.
Lackey, J.B., 1940.
Some new flagellates from the Woods Hole.
Amer. Midl. Nat., 23(2):463-471.

Platymonas
Manton, I., and M. Parke, 1965.
Observations on the fine structure of two species of Platymonas with special reference to flagellar scales and the mode of origin of the theca.
J. mar. biol. Ass., U.K., 45(3):743-754.

Platymonas convolutae n.sp.
Parke, Mary, and Irene Manton, 1967.
The specific identity of the algal symbiont in Convoluta roscoffensis.
J. mar. biol. Ass., U.K., 47(2):445-464.

Platymonas impellucide n.sp.
Mclachlan, J., and M. Parke, 1967.
Platymonas impellucide sp. nov. from Puerto Rico.
J. mar. biol. Ass., U.K., 47(3):723-733.

Platymonas impellucida

Mihnea, Pia Elena 1971.
Notes on Platymonas impellucida (McLachlan + Parke) species in monoalgal cultures.
Cercetări mar. Constanta (1): 83-94.

Platymonas viridis

Spektorova, L.V., 1972
Content of Na, K and Ca in the unicellular alga Platymonas viridis Rouch sp. nov. cultivated on water from different seas. (In Russian; English abstract). Okeanologiia 12(4): 695-700.

Pleurasiga orculaeformis

Leadbeater, Barry S.C. 1973.
External morphology of some marine choanoflagellates from the coast of Jugoslavia.
Arch. Protistenk. 115 (2|3): 234-252

Pleurasiga reynoldsii

Leadbeater, Barry S.C. 1973.
External morphology of some marine choanoflagellates from the coast of Jugoslavia.
Arch. Protistenk. 115 (2|3): 234-252

Protoeuglena noctilucae n.gen., n.sp.

Subrahmanyan, R., 1954.
A new member of the Euglenineae, Protoeuglena noctilucae gen. et sp. nov., occurring in the sea off Calicut.
Proc. Indian Acad. Sci., 39:118-127.

Protospongia dybsoeënsis n.sp.

Grøntved, J., 1956.
Planktonological contributions. II. Taxonomical studies in some Danish coastal localities.
Medd. Danmarks Fiskeri- og Havundersøgelser, n.s. 1(12):13 pp.

Pyramimonas

Parke, M., and D.G. Rayns, 1964.
Studies on marine flagellates. VII. Nephroselmis gilva sp. nov. and some allied forms.
J. Mar. Biol. Assoc., U.K., 44(1):209-217.

Pyraminomas sp.

Yentsch, Charles S., and Carol A. Reichert, 1962.
The interrelationship between water-soluble yellow substances and chloroplastic pigments in marine algae.
Botanica Marina, 3(3/4):65-74.

Prymnesium parvum

Marker, A.F.H., 1965.
Extracellular carbohydrate liberation in the flagellates Isochrysis galbana and Prymnesium parvum.
J. mar. biol. Ass., U.K., 45(3):755-772.

Prymnesium parvum

Martin, Dean F., George M. Padilla, Michael G. Heyl and Priscilla A. Brown 1972
Effect of Gymnodinium breve toxin on hemolysis induced by Prymnesium parvum toxin.
Toxicon 10(3): 285-290.

Pseudomicrosportella ornata n.gen., n. sp.

Scagel, Robert F., and Janet R. Stein, 1961.
Marine nannoplankton in a British Columbia fjord.
Canadian J. Botany, 39:1205-1213.

Pyramimonas grossii n.sp.

Parke, M., 1949.
Studies on marine flagellates. J.M.B.A. 28(1): 255-285.

Rhynchobodo agilis n.sp.

Lackey, J.B., 1940.
Some new flagellates from the Woods Hole area.
Amer. Midl. Nat., 23(2):463-471.

Salpingoeca sp.

Leadbeater, Barry S.C. 1973.
External morphology of some marine choanoflagellates from the coast of Jugoslavia.
Arch. Protistenk. 115 (2|3): 234-252

Salpingoeca natans n.sp.

Grøntved, J., 1956.
Planktonological contributions. II. Taxonomical studies in some Danish coastal localities.
Medd. Danmarks Fiskeri- og Havundersøgelser, n.s. 1(12):13 pp.

Sticholonche zanclea

Chiba, T., 1949
On the distribution of the plankton in the eastern China Sea and Yellow Sea. 1. Plankton composition in the spring. J. Shimonoseki Coll. Fisheries, 1(1):57-63, 1 fig.

Stylochromonas minuta n.gen., n.sp.

Lackey, J.B., 1940.
Some new flagellates from the Woods Hole area.
Amer. Midl. Nat., 23(2):463-471.

Thalassomonas exurgens

Scagel, Robert R., and Janet R. Stein, 1961.
Marine nannoplankton from a British Columbia fjord.
Canadian J. Botany, 39:1205-1213.

Thekadinium kofoidi

Dragesco, Jean, 1965.
Étude cytologique de quelques flagellés mésopsammiques.
Cahiers Biol. Mar., 6(1):83-115.
Roscoff.

Trachelomonas spp

Butcher, R.W., 1961
An introductory account of the smaller algae of British coastal waters. Pt. VIII. Euglenophyceae.
Min. Agric. Fish. Food. Fish. Invest., (4): 17 pp.

Trachelomonas abrupta

Dragesco, Jean, 1965.
Étude cytologique de quelques flagellés mésopsammiques.
Cahiers Biol. Mar., Roscoff, 6(1):83-115.

Trachelomonas bernardiensis

Pringsheim, E.G., 1952.
Observations on some species of Trachelomonas grown in culture. The New Phytologist 52(2):93-113, 5 textfigs.

Trachelomonas hispida

Pringsheim, E.G., 1952.
Observations on some species of Trachelomonas grown in culture. The New Phytologist 52(2):93-113, 5 textfigs.

Trachelomonas lefevrei

Pringsheim, E.G., 1952.
Observations on some species of Trachelomonas grown in culture. The New Phytologist 52(2):93-113, 5 textfigs.

Trachelomonas volvocinopsis

Pringsheim, E.G., 1952.
Observations on some species of Trachelomonas grown in culture. The New Phytologist 52(2):93-113, 5 textfigs.

Trachelomonas zorensis

Pringsheim, E.G., 1952.
Observations on some species of Trachelomonas grown in culture. The New Phytologist 52(2):93-113, 5 textfigs.

Triangulomonas rigida n.gen., n.s

Lackey, J.B., 1940.
Some new flagellates from the Woods Hole area.
Amer. Midl. Nat., 23(2):463-471.

Xanthomonas thalassoides n.gen., n. sp.

Scagel, Robert F., and Janet R. Stein, 1961.
Marine nannoplankton from a British Columbia fjord.
Canadian J. Botany, 39:1205-1213.

dinoflagellates

Abé, Tohru H. 1966.
The armoured Dinoflagellata I. Podolampodae.
Publs. Seto mar. biol. Lab., 14(2):129-154.

dinoflagellates

Abe, T.H., 1941.
Studies on protozoan fauna of Shimoda Bay, the Diplopsalis group. Rec. Ocean. Wks., Japan, 12(2):121-144, 45 figs.

dinoflagellates

Allen, W.E., 1928.
Quantitative studies on inshore marine diatoms and dinoflagellates collected in Southern California in 1924. Bull. S.I.O., tech. ser., 1:347-356, 1 textfig.

dinoflagellates

Allen, W.E., 1925
Statistical studies of surface catches of marine diatoms and dinoflagellates made by the yacht 'Ohio' in tropical waters in 1924. Trans. Amer. Micros. Soc.: 24-30, 1 fig.

dinoflagellates

Allen, W. E., 1925
Statistical Studies of surface catches of marine diatoms and dinoflagellates made by the Yacht "Ohio" in tropical waters in 1924. Jan. Trans. Amer. Microscop. Soc.:24-30, 1 fig.

dinoflagellates

Allen, W.E., 1927.
Surface catches of marine diatoms and dinoflagellates made by the U.S.S."Pioneer" in Alaskan waters in 1923. Bull. S.I.O., tech. ser., 1:39-48, 2 textfigs.

dinoflagellates

Allen, W.E., 1927.
Quantitative studies on inshore marine diatoms and dinoflagellates of Southern California in 1921. Bull. S.I.O., tech. ser., 1:19-29, 2 textfigs.

dinoflagellates

Allen, W.E., 1927.
Quantitative studies on inshore marine diatoms and dinoflagellates of Southern California in 1922. Bull. S.I.O., tech. ser., 1:31-38, 2 textfigs.

dinoflagellates

Allen, W.E., 1928.
Catches of marine diatoms and dinoflagellates taken by boat in Southern California waters in 1926. Bull. S.I.O., tech. ser., 1:201-246, 6 text

dinoflagellates

Allen, W. E. 1941
Twenty years' statistical studies of marine plankton dinoflagellates of Southern California.
Amer. Mid. Nat., Vol. 26, pp. 603-635

dinoflagellates

Allen, W.E., and R. Lewis, 1927.
Surface catches of marine diatoms and dinoflagellates from Pacific high seas in 1925 and 1926.
Bull. S.I.O., tech. ser., 1:197-200.

dinoflagellates

Asakura, K., 1910.
"Akashiwo" in the sea bordering Kanagawa Prefecture. J. Met. Soc., Japan, 29:227-235, (In Japanese).

dinoflagellates

Balech, Enrique, 1962
Tintinnoinea y Dinoflagellata del Pacifico segun material de las Expediciones NORPAC y DOWNWIND del Instituto Scripps de Oceanografia.
Revista, Mus. Argentino Ciencias Nat. "Bernardino Rivadavia", Ciencias Zool., 7(1):1-253.

dinoflagellates

Bal, D.V., and L.B. Pradhan 1945.
A preliminary note on the plankton of Bombay harbor.
Current Sci. 14(8): 211-212.

Peridinea

Bandel, W., 1940.
Phytoplankton- und Nahrstoffgehalt der Ostsee im Gebiet der Darsser Schwelle. Internat. Rev. ges. Hydrobiol. u. Hydrogr., 40:249-304.

dinoflagellates

Barnes, H., and G.R. Hasle, 1956.
A statistical examination of the distribution of some species of dinoflagellates in the polluted Inner Oslo Fjord. Nytt Mag. Bot., 5:113-124.

peridinians

Biecheler, B., 1952.
Recherches sur les Peridiniens.
Bull. Biol., France Belg., Suppl. 36:1-149, 75 textfigs.

dinoflagellates, anat.

Braarud, T., 1945.
Morphological observations on marine dinoflagellate cultures (Porella perforata, Goniaulax tamarensis, Protoceratium reticulatum). Avhandl. Norske Vidensk.-Akad., Oslo, Mat. Natur. Kl., 1944(11):1-18, 4 pls., 6 textfigs.

dinoflagellates, anatomy

Braarud, T., 1945.
Morphological observations on marine dinoflagellates (Porella perforata, Goniaulax tamarensis, Protoceratium reticulatum). Avhandl. Norske Vidensk.-Akad., Oslo, Matem-Naturvidenskap. Kl. 1944(11):1-18.

dinoflagellates

Brunel, J., 1962
Le phytoplancton de la Baie de Chaleurs.
Inst. Botan., Univ. Montréal, Contrib. No. 77: 365 pp., 66 pls.

dinoflagellates

Cachon, Jean, 1964
Contribution a l'étude des péridiniens parasites Cytologie, cycles evolutifs.
Ann. Sci. Nat. Zool., Paris, (12). 6:1-158.
Also: Trav. Sta. Zool., Villefranche-sur-Mer, 24 bis.

peridinians

Cachon, Jean, et Monique Cachon 1970
Ultrastructure des Amoebophryidae (péridiniens Duboscquodinida). II. Systèmes atractophoriens et microtubulaires; leur intervention dans la mitose.
Protistologia 6(1): 57-70.

dinoflagellates

Cachon, Jean, et Monique Cachon-Enjumet, 1965.
Atlanticellodinium tregouboffi nov. gen. nov. sp., péridinien Blastuloidae Neresheimer, parasite de Planktonetta atlantica Borgert, Phaeodarie Atlanticellide. Cytologie, cycle biologique, évolution nucléaire au cours de la sporogenèse.
Arch. Zool. exp. gen., 105:369-380.
Also in:
Trav. Sta. Zool., Villefranche-sur-Mer, 25.

dinoflagellates

Cachon, Jean, Monique Cachon, et Françoise Bouquaheux, 1965.
Stylodinium gastrophilum Cachon péridinien dinococcide parasite de siphonophores.
Bull. Inst. Oceanogr., Monaco, 65(1359):8 pp.
also in:
Trav. Sta. Zool., Villefranche-sur-Mer, 25.

dinoflagellates

Cupp, E., 1930
Quantitative Studies of miscellaneous series of surface catches of marine diatoms and dinoflagellates taken between Seattle and the Canal Zone from 1924 to 1928. Trans. Am. Micro. Soc., XLIX (3):238-245.

dinoflagellates

Cupp, E.E., 1937
Seasonal distribution and occurrence of marine diatoms and dinoflagellates at Scotch Cap. Alaska Bull. S.I.O. Tech. ser. 4(3):71-100, 7 textfigs.

peridinians

Dangeard, P., 1927.
Péridiniens nouveaux ou peu connus de la croisière du "Sylvana". Bull. Inst. Océan., Monaco, No. 491:16 pp., 9 textfigs.

dinoflagellates

Dorman, H.P., 1927.
Quantitative studies on marine diatoms and dinoflagellates at four inshore stations on the coast of California in 1923. Bull. S.I.O., tech. ser., 1:73-89, 4 textfigs.

dinoflagellates

Dorman, H.P., 1927.
Studies on marine diatoms and dinoflagellates caught with the Kofoid bucket in 1923. Bull. S.I.O., tech. ser., 1:49-61, 4 textfigs.

dinoflagellates

Dumitrică, Paulian, 1973
Cenozoic endoskeletal dinoflagellates in southwestern Pacific sediments cored during leg 21 of the DSDP.
Initial Repts, Deep Sea Drilling Project, 21:819-835.

Peridiniales

Graham, H. W., 1942
Studies in the morphology, taxonymy, and ecology of the Peridiniales. Sci. Res. Cruise VII of the Carnegie, 1928-1929---Biol. III(542): 129 pp., 67 figs.

peridiniens

Grenet, Claude 1972.
Les critères de détermination chez les péridiniens Warnowiidae Lindemann.
Protistologica 8(4): 461-469

dinoflagellates

Hardy, A.C., and R.H. Kay, 1964
Experimental studies of plankton luminescence.
Jour. Mar. Biol. Assoc., U.K. 44(2):435-484.

Dinoflagellates

Hulburt, E.M., 1957.
The taxonomy of unarmoured Dinophyceae of shallow embayments on Cape Cod, Massachusetts.
Biol. Bull., 112(2):196-219.

Peridiniales

Jörgensen, E., 1910
Peridiniales, Ceratium. Bull. Trim. Cons. Int. Explor. Mer:205-250.

peridinians

Jörgensen, E., 1900
Protophyten und Protozoen im Plankton aus der Norwegischen Westkerste. Bergens Mus. Aarb. 1899(6): 95 pp., 5 pls., 83 tables.

Dinophysiaceae

Jörgensen, E., 1923
Mediterranean Dinophysiaceae. Rept. Danish Oceanogr. Expeds. 1908-10, to the Mediterranean and adjacent seas, Vol.II, Biol. J 2, 48 pp., 64 text figs.

peridinians

Käsler, R., 1938
Die Verbreitung der Dinophysiales im Sudatlantischen Ozean. Wiss. Ergeb. Deutschen Atlantischen Expedition----"Meteor" 1925-1927, 12(2):162-237, text figs. 85-118.

dinoflagellate

Kimball, J.F., Jr., and E.J. Ferguson Wood, 1965.
A dinoflagellate with characters of Gymnodinium and Gyrodinium.
J. Protozool., 12(4):577-580.

dinoflagellata

Kofoid, C. A., 1907
Dinoflagellata of the San Diego region. III. Descriptions of new species. Univ. Calif. Publ., Zool. 3:299-340, Pls. 22-33.

dinoflagellata

Kofoid, A. C., 1911
Dinoflagellata of the San Diego region. IV The genus Gonyaulax. Univ. Calif. Publ. Zool. 8(4):187-286, Pls. 9-17.

dinoflagellates

Kofoid, C.A., and A.M. Adamson, 1933.
36. The Dinoflagellata: the Family Heterodinidea of the Peridiniodiae. Mem. Mus. Comp.Zool. 54(1):1-136, 22pls.

dinoflagellates

Kofoid, C. A. and T. Skogsberg, 1928
XXXV. The Dinoflagellata: The Dinophysiodae. Reports on the scientific results of the expedition to the Eastern Tropical Pacific, in charge of Alexander Agassiz, by the U. S. Fish Commission Steamer "Albatross" from October 1904 to March 1905----. Mem. M. C. Z. 51:766 pp., 31 pls.

dinoflagellates

Kon, H., 1953.
On the distribution of phytoplankton in the north-eastern part of the Pacific (Tohoku District) in autumn. (Oct.-Nov., 1951).
J. Ocean. Soc., Japan, 9(2):109-114, 4 textfigs.

dinoflagellates

Kruger, D., 1950.
Variations quantitatives des protistes marins au voisinage du Port d'Alger durant l'hiver 1949-1950. Bull. Inst., Monaco, No. 978:20 pp., 5 textfigs.

dinoflagellates -systematic

Lebour, M.V., 1935
The dinoflagellates of Northern Seas. The Marine Biological Association of the United Kingdom. Plymouth.

dinoflagellates

LeBour, M.V., 1925
The dinoflagellates of Northern Seas. The Marine Biological Association of the United Kingdom, Plymouth, 250 pp., 35 pls., 53 text figs.

dinoflagellates

Lewis, R., 1927.
Surface catches of marine diatoms and dinoflagellates pff the coast of Oregon by U.S.S. "Guide" in 1924. Bull. S.I.O., tech. ser. 1:189-196, 3 textfigs.

peridiniens

Lindemann, E., 1924
Peridineen aus dem goldenen Horn und dem Bosphorus. Bot. Arch. 5:216-233, 98 text figs.

Oceanographic Index: Marine Organisms Cumulation, 1946-1973

peridinians

Lindemann, E., 1925
Neubeobachtungen an den Winter peridineen des Golfes von Neapel. Bot. Arch. 9:95-102, 19 text figs.

dinoflagellates

Lucas, C. E. 1942
Continuous plankton records: Phytoplankton in the North Sea 1938-1939. II. Dinoflagellates, Phaeocystis, etc. Hull Bull. Mar. Ecol., II(5):47-70, 6 textfigs., 24 pls.

dinoflagellates

Lucas, C. E., 1940
Ecological investigations with the continuous plankton recorder: the phytoplankton in the southern North Sea, 1932-37. Hull Bull. Mar. Ecol., 1(3):73-170, 17 textfigs., 21 pls.

dinoflagellates

Lucas, C. E., N.B. Marshall, and C.B. Rees 1942
Continuous plankton records: The Faeroe-Shetland Channel 1939. Hull Bull. Mar. Ecol.,

dinoflagellates

Margalef, Ramón, 1967.
Las algas inferiores.
In: Ecología marina. Monogr. Fundación La Salle de Ciencias Naturales, Caracas, 14:230-272.

dinoflagellates

Marshall, S.M., 1947
An experiment in marine fish cultivation: III. The plankton of a fertilized loch. Proc. Roy. Soc., Edinburgh, Sect.B., 63, Pt.I(3):21-33, 7 text figs.

dinoflagellates

Matzenauer, L., 1933
Die Dinoflagellaten des indischen Ozeans (mit Ausnahme der Gattung Ceratium.) Bot. Arch. 35:437-510, 77 text figs., 2 charts.

dinoflagellates

Menon, M.A.S., 1945.
Observations on the seasonal distribution of the plankton, Trivandrum Coast. Proc. Indian Acad. Sci., Sect. B, 22(2):31-62, 1 textfig.

peridinians

Meunier, A., 1919
Microplancton de la Mer Flamande 3. Les Péridiniens. Mem. Mus. Roy. Hist. Nat., Belgique 8(1):1-116, Pls. XV-XXI.

dinoflagellates

Navarro, F. de P., and L. Bellon Uriarte, 1945.
Catálogo de la flora del Mar de Baleares (con exclusión de las diatomeas). Notas y Res., Inst. Español Ocean., 2nd ser., No. 124:160-295.

Norris, Richard E., 1966. *dinoflagellates*
Unarmoured marine dinoflagellates.
Endeavour, 25(96):124-128.

dinoflagellates

Paiva Carvalho, J., 1950
O plancton do Rio Maria Rodriques (Cananeis). 1. Diatomaceas e Dinoflagelados. Bol. Inst. Paulista Oceanogr. 1(1): 27-43, 2 fold-in tables, 2 figs.

dinoflagellates

Paulsen, O., 1949
Observations on dinoflagellates. (Ed. J. Grøntved) Kongl. Dansk. Videnskab. Selsk., Biol. Skr. 6(4):67 pp., 30 text figs.

peridinians

Pavillard, J., 1953.
Remarques systématiques sur quelques péridiniens. Bull. Inst. Ocean., Monaco, No. 1022:3pp.

peridinians

Pavillard, J., 1923
A propos de la systématique des Péridiniens. Bull. Soc. Bot. de France 70:876-882; 914-918.

peridinians

Pavillard, J., 1916
Recherches sur les Peridiniens du Golfe du Lion. Mem. Univ. Montpellier. Trav. Inst. Bot., Univ. Montpellier. Serie mixte No.4, 70 pp., 3 pls., 15 text figs.

peridinians

Pavillard, J., 1905
Recherches sur la flore pelagique (Phytoplankton) de l'Etang de Thau. Theses presentees a la Fac. Sci., Paris, 116 pp., 3 pls.

dinoflagellates

Pincemin, Jean-Marc, 1966.
Note preliminaire à l'étude écologique des dinoflagellés de la baie d'Alger et comparaison avec les diatomés. Pelagos, (6):9-47.

dinoflagellates

Pomeroy, L.R., H.H. Haskins and R.A. Ragotzkie, 1956.
Observations on dinoflagellate blooms. Limnol. & Oceanogr. 1(1):54-60.

peridinians

Rampi, L., 1951.
Ricerche sul fitoplancton del Mare Ligure. 10) Peridiniale delle acque di Sanremo. Pubbl. Centro Talassografico Tirreno No. 6: [Atti Acad. Ligure Sci. Lett.] 7(1): 8 pp., Pls. 3-4.

dinoflagellates

Rampi, L. 1940.
Ricerche sul Fitoplancton del mare Ligure. II Le tecatali e le dinofisiali delle acque di Sanremo. Boll. Pesca, Piscicolt., Idrobiol. (18) 16(2): 243-274, 56 figs.

peridinians

Rampi, L., 1943
Richerche sul fitoplancton del Mare Ligure. F. Le Goniaulacee delle acque di Sanremo. Ahi della Soc. Ital. di Scienze Naturali, 82:1-12, figs.1-16.

peridinians

Rampi, L., 1950.
Péridiniens rares ou nouveaux pour le Pacifique Sud-Equatorial. Bull. Inst. Océan., Monaco, No. 974, 11 pp., 26 textfigs.

dinoflagellates, physiol. etc.

Richards, F.A., 1952.
The estimation and characterization of plankton populations by pigment analyses. I. The absorption spectra of some pigments occuring in diatoms, dinoflagellates and brown algae. J. Mar. Res. 11(2):147-155.

peridinians

Roxas, H.A., 1941.
Marine protozoa of the Philippines. Philippine J. Sci. 74:91-136, 17 pls., 2 textfigs.

dinoflagellates

Silva, E. de S.E., and J.D. Santos Pinto, 1948.
O plancton de Baia de S. Martinho de Porto. 1. Diatomáceas e Dinoflagelados. Bol. Soc. Portuguesa Ciencias Natur. 16(2):134-187.

dinoflagellates

Sousa e Silva, E., 1949
Diatomaceas e Dinoflagelados de Baia de Cascais. Portugaliae Acta Biol., Volume: Julio Henriques, Ser. B: 300-383, 9 pls, 2 fold-in tables.

dinoflagellates

Steidinger, Karen A., Joanne T. Davis and Jean Williams, 1966.
Observations of Gymnodinium breve Davis and other dinoflagellates. Prof. Pap. Ser. Fla Bd Conserv.8: 8-15.

dinoflagellates

Tai, Si-Sun, & T. Skogsberg, 1934
Studies on the Dinophysoidae, marine armored dinoflagellates of Monterey Bay, California. Arch. Protistenk. 82:380-482, 14 text figs. Pls. 11-12.

dinoflagellates

Toriumi, Saburo, 1966.
Cultivation of marine dinoflagellates. (In Japanese; English abstract). Inf. Bull. Planktol. Japan, No. 13:41-49.

dinoflagellata

Torrey, H. B. 1902
An unusual occurrence of dinoflagellata on the California coast. Amer. Nat. Vol. 36, pp. 187-192.

dinoflagellates

Tregouboff, G., 1956.
Rapport sur les travaux concernant le plancton Mediterranéen publiés entre Novembre 1952 et Novembre 1954. Rapp. Proc. Verb., Comm. Int. Expl. Sci., Mer Mediterranee, 13:65-100

dinoflagellates

Wall, David, 1971.
Biological problems concerning fossilizable dinoflagellates.
In: Science and Man, Louisiana State Univ., 3: 1-15.

dinoflagellates (fossil)

Wall, D. 1967.
Fossil microplankton in deep-sea cores from the Caribbean Sea.
Palaeontology 10(1): 95-123

dinoflagellates

Wall, David and Barrie Dale, 1968.
Quaternary calcareous dinoflagellates (Calciodinellideae) and their natural affinities. J. Paleont., 42(6):1395-1408.

peridiniens

Wauthy, B., R. Desrosières et J. Le Bourhis, 1967.
Importance presumée de l'ultraplancton dans les eaux tropicales oligotrophes du Pacifique central sud. Cah. ORSTOM, Sér. Océanogr., 5(2):109-116.

dinoflagellates

Williams, David B., and William A.S. Sarjeant, 1967.
Organic-walled microfossils as depth and shoreline indicators. Marine Geol., 5(5/6):389-412.

dinoflagellates

Wood, E.J. Ferguson 1968.
Dinoflagellates of the Caribbean Sea and adjacent areas.
Univ. Miami Press 143 pp.

dinoflagellates

Wood, E.J.F., 1963
Dinoflagellates in the Australian region. II. Recent collections. Comm. Sci. and Industr. Res. Org., Div. Fish. and Oceanogr., Techn. Paper, No. 14: 55 pp.

dinoflagellates

Wood, E.J.F., 1954.
Dinoflagellates in the Australian region. Australian J. Mar. Freshw. Res., 5(2):171-351.

dinoflagellates, anat.-physiol
Abe, Tohru H., 1967
The armoured Dinoflagellata: II. Brorocentridae and Dinophysidae (A).
Publs Seto Mar. biol. Lab., 14(5): 369-389.

peridiniens, anat.-phys.
Bècheler, B., 1952.
Recherches sur les Peridiniens.
Bull. Biol. France Belg., Suppl., 36:1-149, 75 textfigs.

dinoflagellates, physiol.
Braarud, T., 1951.
Taxonomical studies of marine dinoflagellates.
Nytt Mag. Naturvidensk 88:43-48.

dinoflagellates, anat. physiol. [a]
Boltovskoy, Andrés 1973.
Formación del arqueopilo en tecas de dinoflagelados
Revista esp. Micropaleontol. 5(1): 81-98.

peridiniens, anat.-physiol.
Burkholder, Paul R., Lillian M. Burkholder and Luis R. Almodóvar, 1967.
Carbon assimilation of marine flagellate blooms in neritic waters of southern Puerto Rico.
Bull. mar. Sci. Miami, 17(1):1-15.

dinoflagellates, anat.physiol.
Cachon, Jean, Monique Cachon et Claude Greuet 1970.
Le système pusulaire de quelques péridiniens libres ou parasites.
Protistologica 6 (4): 467-476.

dinoflagellates, anat.-physiol
Dodge, J.D., 1968.
The fine structure of chloroplasts and pyrenoids in some marine dinoflagellates.
J. Cell Sci., 3(1):41-48.

dinoflagellates anat. physiol.
Dodge, John D., 1965.
Thecal fine-structure in the dinoflagellate genera Prorocentrum and Exuviaella.
J. mar. biol. Ass., U.K., 45(3):607-614.

dinoflagellates, anat. physiol. [a]
Droop, M.R. and J.F. Pennock, 1971.
Terpenoid quinones and sterols in the nutrition of Oxyrrhis marina. J. mar. biol. Ass. U.K. 51(2): 455-470.

protozoa, anat.-physiol.
Entz, G., Jr., 1935.
Ueber das Problem der Kerne und kemähnlichen Einschlüsse bei Petalotricha ampulla.
Fol. Biol. General. 11(1):15-26.

Evitt, William R., 1967 **dinoflagellates, anat. physiol.**
Dinoflagellate studies II. The archeopyle.
Publ. Stanford Univ., Geol. Sci., 10(3):1-82

dinoflagellates, anat.-physiol. [a]
Forward, Richard, and Demorest Davenport, 1968.
Red and far-red light effects on a short-term behavioral response of a dinoflagellate.
Science, 161 (3845): 1028-1029.

dinoflagellates, anat.-phy
Gold, K., 1964.
Aspects of marine dinoflagellate nutrition measures by C14 assimiliation.
J. Protozool., 11(1):85-89.

dinoflagellates, anat. physiol.
Gold, Kenneth, and Kathryn Stein Pokorny 1973.
Effects of carbon, nitrogen and allopurinol on the the abundance of particulate inclusions in a marine dinoflagellate.
J. Phycol. 9(2): 225-229.

dinoflagellates, anat.
Grasso, Pierre-P., André Hollande, Jean Cachon et Monique Cachon-Enjumet, 1965.
Interprétation de quelques aspects infrastructuraux des chromosomes de péridiniens en division.
C.R. Séanc. Hebd., Acad. Sci., Paris, 260:6975-6978.
Also in:
Trav. Sta. Zool., Villefranche-sur-Mer, 25.

dinoflagellates, anat.
Greuet Claude 1972.
Intervention de lamelles annelées dans la formation de couches squelettiques au niveau de la capsule périnucléaire de péridiniens Warnowiidae.
Protistologica 8(2): 155-168

peridinians, anat. [a]
Greuet, Claude, 1971
Etude ultrastructurale et évolution des cnidocystes de Nematodinium, Péridinien Warnowiidae Lindemann. Protistologica 7(3):345-355.

dinoflagellates, physiol.
Halldal, P., 1958.
Action spectra of phototaxis and related problems in Volvocales, Ulva-gametes and Dinophyceae.
Physiol. Plant. 11:118-153.

dinoflagellates anat. physiol
Hand, William G., Richard Forward and Demorest Davenport, 1967.
Short-term photic regulation of a receptor mechanism in a dinoflagellate.
Biol. Bull., mar. biol. Lab., Woods Hole, 133(1):150-165.

dinoflagellates, physiol.
Hasle, G.R., 1954.
More on phototactic diurnal migration in marine dinoflagellates. Nytt Mag. Bot. 2:139-147.

dinoflagellates,anat-physiol
Hastings, J.W., Marcie Vergin, and R.De Sa, 1966.
Scintillons; the biochemistry of dinoflagellate bioluminescence.
In: Bioluminescence in Progress, F.H. Johnson and Y. Haneda, editors, Princeton Univ. Press, 301-329.

dinoflagellates, anat.-phys.
Hochachka, Peter W., and John M. Teal, 1964.
Respiratory metabolism in a marine dinoflagellate
Biol. Bull., 126(2):274-281.

dinoflagellates, anat. physiol.
Jahn, T.L., W.M. Harmon and M. Landman, 1963.
Mechanisms of locomotion in flagellates. 1. Ceratium.
J. Protozoology, 10(3):358-363.

dinoflagellates,anat.physiol
Leadbeater, B., and J.D. Dodge, 1967.
Fine structure of the dinoflagellate transverse flagellum.
Nature, Lond., 213(5074):421-422.

dinoflagellates, anat.-physiol.
Mendiola, Leticia R., C.A. Price and R.R.L. Guillard, 1966.
Isolation of nuclei from a marine dinoflagellate
Science, 153(3744):1661-1663.

dinoflagellates, anat.-physiol.
Nie, D., 1947.
Thecal morphology of some dinoflagellates of Woods Hole - with special reference to the "ventral area". Biol. Bull. 93(2):210.

dinoflagellates, anat. [a]
Pokorny, Kathryn Stein, and Kenneth Gold 1973
Two morphological types of particulate inclusions in marine dinoflagellates.
J. Phycol. 9(2): 218-224.

dinoflagellates, anat.-physiol
Prakash, A. and M.A. Rashid, 1968.
Influence of humic substances on the growth of marine phytoplankton: dinoflagellates.
Limnol. Oceanogr., 13(4):598-606.

dinoflagellates, anat.-phys.
Sweeney, Beatrice M., 1963.
Bioluminescent dinoflagellates.
Biol. Bull., 125(1):177-181.

dinoflagellates, anat. physiol.
Vien, Cao 1967.
Sur l'existence de phénomènes sexuels chez un péridinien libre, l'Amphidinium carteri.
C.r. Hebd. Séanc. Acad. Sci Paris (D) 264 (5): 1006-1008.

dinoflagellates, anat.
Wall, David, and Barrie Dale, 1968.
Modern dinoflagellate cysts and evolution of the Peridiniales.
Micropaleontology, 14(3):265-304.

dinoflagellate bioluminescence [a]
Esaias, Wayne E. and Herbert C. Curl, Jr., 1972.
Effect of dinoflagellate bioluminescence on copepod ingestion rates. Limnol. Oceanogr. 17(6): 901-906.

dinoflagellates, luminescent forms
Kelly, Mahlon G., and Steven Katona, 1966.
An endogenous diurnal rhythm of bioluminescence in a natural population of dinoflagellates.
Biol. Bull., 131(1):115-126.

dinoflagellate bloom, effect of [a]
Harrison, William G., 1973
Nitrate reductase activity during a dinoflagellate bloom. Limnol. Oceanogr, 18(3):457-465.

dinoflagellate bloom
Odum, William E., 1968.
Mullet grazing on a dinoflagellate bloom.
Chesapeake Sci., 9(3):202-204.

peridinians, chemical composition
Bordovskiy, O.K., 1965.
Accumulation and transformation of organic substance in marine sediments. 1. Summary and introduction. 2. Sources of organic matter in marine basins. 3. Accumulation of organic matter in bottom sediments. 4. Transformation of organic matter in marine sediments.
Marine Geology, Elsevir Publ. Co. 3(½):3-4; 5-31; 33-82; 83-114.

dinoflagellates, chemistry of
Haidek, David J., Christopher K. Mathews and Beatrice M. Sweeney, 1966.
Pigment protein complex from Gonyaulax.
Science, 152 (3719):212-213.

Oceanographic Index: Marine Organisms Cumulation, 1946-1973

dinoflagellates, chemistry

Harrington, Glenn W., David H. Beach, Joyce E. Dunham and George G. Holz, Jr., 1970.
The polyunsaturated fatty acids of marine dinoflagellates.
J. Protozool. 17(2): 213-219.

dinoflagellates, chemistry

Sakshaug, E., S. Myklestad, T. Krogh and G. Westin, 1973
Production of protein and carbohydrate in the dinoflagellate Amphidinium carteri. Some preliminary results.
Norwegian J. Bot. 20(2/3):211-218.

dinoflagellates, cytology of

Dodge, John D., 1963.
Chromosome numbers in some marine dinoflagellates.
Botanica Marina, 5(4):121-127.

dinoflagellates, anat.-physiol

Sweeney, Beatrice M., and G.B. Bouck, 1966.
Crystal-like particles in luminous and non-luminous dinoflagellates.
In: Bioluminescence in progress, F.H. Johnson and Y. Haneda, editors, Princeton Univ. Press, 331-348.

dinoflagellate distribution

Williams, D.B., 1971.
The distribution of marine dinoflagellates in relation to physical and chemical conditions.
In: Micropaleontology of oceans, B.M. Funnell and W.R. Riedel, editors, Cambridge Univ. Press, 91-95.

dinoflagellates, ecology

Nordli, E., 1957.
Experimental studies on the ecology of Ceratia.
Oikos 8(2):200-265.

dinoflagellates, effect of

Aubert, M., J. Aubert, M. Gauthier et D. Pesando, 1967.
Etude de phénomènes antibiotiques liés à une efflorescence de péridiniens.
Rev. intern Océanogr., Méd. 6/7:43-52.

dinoflagellates, fossil

De Coninck, Jan, 1965.
Microfossiles planctoniques du sable Y presien à Merelbeke. Dinophyceae et Acritarchs.
Acad. Roy. Belg., Cl. Sci., Mem., 36(2):54pp. 14 pls.

dinoflagellates (fossil)

Downie, Charles, and William Antony S. Sarjent, 1964.
Bibliography and index of fossil dinoflagellates and acritarchs.
Geol. Soc., Amer., Mem., 94:180 pp.

dinoflagellates, fossil

Wall, David, Barrie Dale and Kenichi Harada 1973.
Description of new fossil dinoflagellates from the Late Quaternary of the Black Sea.
Micropaleontology 19(1): 18-31

dinoflagellates, growth of

Sergeeva, L.M., D.K. Krupatkina Akinina, 1971.
Growth of Dinoflagellata with absence of some elements of mineral nutrition. (In Russian).
Gidrobiol. Zh. 7(5): 82-87.

dinoflagellates, hystrichosphaerids

Wall, David and Barrie Dale 1970.
Living hystrichosphaerid dinoflagellate spores from Bermuda and Puerto Rico.
Micropaleontology 16(1): 47-58

dinoflagellates

Wall, David, and Barrie Dale, 1966.
"Living fossils" in western Atlantic plankton.
Nature 211(5053):1025-1026.

peridinians, lists of spp.

Angot, Michel, et Robert Gerard, 1966.
Hydrologie et phytoplancton de l'eau de surface en avril 1965 à Nosy Be.
Cah. ORSTOM, Ser. Oceanogr., 4(1):95-136.

dinoflagellates (data only)

Australia, Commonwealth Scientific and Industrial Research Organization, 1964.
Oceanographical observations in the Indian Ocean, in 1961, H.M.A.S. Diamantina, Cruise Dm 3/61.
Div. Fish. and Oceanogr., Oceanogr. Cruise Rept., No. 11:215 pp.

dinoflagellates, lists of spp

Australia, Commonwealth Scientific and Industrial Research Organization, Division of Fisheries and Oceanography, 1963.
Oceanographical observations in the Pacific Ocean in 1960, H.M.A.S. Gascoyne, Cruise G 3/
Oceanographical Cruise Report, No. 6: 115 pp.

dinoflagellates (lists of spp)

Australia, Commonwealth Scientific and Industrial Research Organization, Division of Fisheries and Oceanography, 1963.
Oceanographical observations in the Indian Ocean in 1960, H.M.A.S. Diamantina, Cruise Dm 2/60.
Oceanographical Cruise Report No. 3: 347 pp.

dinoflagellates, lists of spp.

Australia, Commonwealth Scientific and Industrial Organization, 1962.
Oceanographical observations in the Indian Ocean in 1960, H.M.A.S. Diamantina, Cruise Dm 1/60.
Oceanogr. Cruise Rept., Div. Fish. and Oceanogr., No. 2:128 pp.

dinoflagellates, lists of spp.

Australia, Commonwealth Scientific and Industrial Organization, 1962.
Oceanographical observations in the Indian Ocean in 1959, H.M.A.S. Diamantina, Cruises Dm 1/59 and 2/59.
Oceanogr. Cruise Rept., Div. Fish. and Oceanogr., No. 1:134 pp.

dinoflagellates, lists of spp.

Australia, Commonwealth Scientific and Industrial Research Organization, Division of Fisheries and Oceanography, 1962.
Oceanographic observations in the Pacific Ocean in 1960, H.M.A.S. Gascoyne, Cruises G 1/60 and G 2/60.
Oceanographical Cruise Report No. 5:255 pp.

dinoflagellates, lists of spp.

Australia, Commonwealth Scientific and Industrial Research Organization, 1961.
F.R.V. "Derwent Hunter".
C.S.I.R.O., Div. Fish. and Oceanogr., Rept., No. 32:56 pp.

dinoflagellates, lists of spp.

Australia, Marine Biological Laboratory, Cronulla, 1960.
F.R.V. "Derwent Hunter", scientific report of cruises 10-20/58 —
C.S.I.R.O., Div. Fish. & Oceanogr., Rept., 30: 53 pp., numerous figs. (mimeographed).
For complete "title", see author card.

dinoflagellates, lists of spp.

Avaria P., Sergio 1970.
Fitoplancton de la expedición del Doña Berta en la zona Puerto Montt-Aysen.
Rev. Biol. mar. Valparaiso 14(2): 1-17.

dinoflagellates, lists of spp.

Balech, Enrique, 1971.
Microplancton de la campaña oceanographica: Productividad III.
Revta Mus. argent. Cienc. Nat. Bernadina Rivadavia, Hydrobiol. 3(1):1-202, 39 pls.

dinoflagellates, lists of spp.

Balech, Enrique, 1971
Dinoflagelados y tintinnidos del Golfo de México y Caribe: sus relaciones con el Atlántico Ecuatorial. Symp. Investigations and resources of the Caribbean Sea and adjacent regions, UNESCO, 18-26 Nov. 1968, Curaçao: 297-301.

dinoflagellates, lists of spp.

Balech, Enrique, 1967.
Dinoflagellates and tintinnids in the northeastern Gulf of Mexico.
Bull. mar. Sci., Miami, 17(2):280-298.

dinoflagellates, lists of spp.

Balech, Enrique, 1959
Operacion Oceanografica Merluza, V Crucero. Plancton.
Servicio Hidrografia Naval, Publ. H 618: 1-43.

dinoflagellates, lists of spp.

Bernard, Francis, 1967.
Contribution à l'étude du nannoplancton, de 0 à 3000 m, dans les zones atlantiques lusitanienne et mauritanienne (Campagnes de la Calypso, 1960, et du Coriolus, 1964).
Pelagos, Alger, 7:1-81.

dinoflagellates, lists of spp.

Bernhard, M., L. Rampi e A. Zattera 1969.
La distribuzione del fitoplancton nel mar Ligure.
Pubbl. Staz. Zool. Napoli 37 (2 Suppl.): 73-114

dinoflagellates, lists of spp.

Boyd, R.J. 1972.
The zooplankton of Cork Harbour.
Ir. Nat. J. 17(8): 256-262.

dinoflagellates, lists of spp.

Bursa, Adam, 1961.
Phytoplankton of the "Calanus" Expeditions in Hudson Bay, 1953 and 1954.
J. Fish. Res. Bd., Canada, 18(1):51-83.

dinoflagellates, lists of spp.

Bursa, Adam S., 1961
The annual oceanographic cycle at Igloolik in the Canadian Arctic. II. The phytoplankton
J. Fish. Res. Bd., Canada, 18(4):563-615.

dinoflagellates, lists of spp

Cassie, Vivienne, 1966.
Diatoms, dinoflagellates and hydrology in the Hauraki Gulf, 1964-1965.
New Zealand J. Sci. 9(3):569-585.

dinoflagellates, lists of spp.

Cassie, Vivienne, 1960
Seasonal changes in diatoms and dinoflagellates off the east coast of New Zealand during 1957 and 1958. N.Z.J. Sci. 3(1): 137-172.

---

dinoflagellates, lists of spp.

El-Sayed, S.Z., W.M. Sackett, L.M. Jeffrey, A.D. Fredericks, R.P. Saunders, P.S. Conger, G.A. Fryxell, K.A. Steidinger and S.A. Earle 1972.
Chemistry, primary productivity and benthic algae of the Gulf of Mexico.
Ser. Atlas Mar. Environm. Am. Geogr. Soc. 22: 29 pp., 6 pls. (quarto)

---

dinoflagellates (lists of spp)

Establier, R., y R. Margalef, 1964
Fitoplancton e hidrografía de las costas de Cádiz (Barbate) de junio de 1961 a agosto de 1962.
Inv. Pesq., Barcelona. 25:5-31.

---

dinoflagellates, lists of spp.

Furnestin, Marie-Louise 1972.
Phytoplancton et production primaire dans le secteur sud-occidental de la Méditerranée.
Rev. Trav. Inst. Pêches marit. 37(1): 19-68.

---

peridinians, lists of spp.

Genovese, S., G. Gangemi e F. De Domenio 1972.
Campagna estiva 1970 della n/o Bannock nel Mar Tirreno – misure di produzione primaria lungo la trasversale Palermo-Cagliari.
Boll. Pesca Piscic. Idrobiol. 27(1):139-157.

---

dinoflagellates, lists of spp. (1970)

Hada, Yoshine, 1970.
The protozoan plankton of the Antarctic and Subantarctic seas.
Scient. Repts, Japan. Antarct. Res. Exped., (E) 31:51 pp.

---

dinoflagellates, lists of spp. (photographs)

Halim, Youssef, 1967.
Dinoflagellates of the south-east Caribbean Sea (east Venezuela). Int. Revue ges. Hydrobiol., 52(5): 701-755.

---

dinoflagellates, lists of spp.

Halim, Youssef, 1960
Étude quantitative et qualitative du cycle écologique des dinoflagellés dans les eaux de Villefranche-sur-Mer. (1953-1955).
Ann. Inst. Océanogr., Monaco, 38:123-232.

---

dinoflagellates, lists of spp. [a]

Heimdal, Berit R., Grethe R. Hasle and Jahn Throndsen 1973
An annotated check-list of plankton algae from the Oslofjord, Norway (1951-1972).
Norw. J. Bot. 20(1): 13-19.

---

dinoflagellates, lists of spp.

Herrera, Juan, y Ramón Margalef, 1963
Hidrografía y fitoplancton de la costa comprendida entre Castellón y la desembocadura del Ebro, de julio de 1960 a junio de 1961.
Inv. Pesq., Barcelona, 24:33-112.

---

dinoflagellates, lists of spp.

Hulburt, Edward M., 1964.
Succession and diversity in the plankton flora of the western North Atlantic.
Bull. Mar. Sci., Gulf and Caribbean, 14(1):33-44.

---

dinoflagellates, lists of spp

Jacques, G. 1969.
Aspects quantitatifs du phytoplancton de Banyuls-sur-Mer (Golfe du Lion). III. Diatomées et dinoflagellés de juin 1965 à juin 1968.
Vie Milieu (B) 20(1): 91-126

---

dinoflagellates, lists of spp.

Japan, Hokkaido University, Faculty of Fisheries, 1967.
Data record of oceanographic observations and exploratory fishing, 11: 383 pp.

---

dinoflagellates, lists of spp.

Klement, Karl W., 1964.
Armored dinoflagellates of the Gulf of California.
Bull. Scripps Inst. Oceanogr., 8(5):347-372.

---

dinoflagellates, lists of spp.

Lackey, James B., 1967.
The microbiota of estuaries and their roles.
In: Estuaries, G.H. Lauff, editor, Publs Am. Ass. Advmt Sci., 83:291-302.

---

dinoflagellates, lists of spp.

Lackey, James B., and Elsie W. Lackey, 1963
Microscopic algae and protozoa in the waters near Plymouth in August 1962.
J. Mar. Biol. Assoc., U.K., 43(3):797-805.

---

dinoflagellates, lists of spp. [a]

Lapshina, V.I. 1971.
Characteristic plankton of the north tropical and equatorial zones in the eastern Pacific Ocean. (In Russian).
Izv. Tichookean. nauchno-issled. Inst. ribn. Choz. Okean, 79: 100-126.
(TINRO)

---

dinoflagellates, lists of spp.

Lecal, J. 1967.
Le nannoplancton des côtes d'Israel.
Hydrobiologia, 29(3/4): 305-357

---

dinoflagellates, lists of spp.

Magazzù, Giuseppe, e Carlo Andreoli 1971.
Trasferimenti fitoplanctonici attraverso lo Stretto di Messina in relazione alle condizioni idrologiche.
Boll. Pesca Piscic. Idrobiol. 26(1/2): 125-193

---

dinoflagellates, lists of spp. [a]

Margalef, Ramón, 1973
Fitoplancton marino de la región de afloramiento del NW de África. II. Composicion y distribucion del fitoplancton (Campaña Sahara II del Cornide de Saavedra. Result. Exped. cient. Cornide de Saavedra, Madrid 2:65-94.

---

dinoflagellates, lists of spp.

Margalef, Ramón, 1964.
Fitoplancton de las costas de Blanes (provincia de Gerona, Mediterráneo Occidental), de julio de 1959 a junio de 1963.
Inv. Pesq., Barcelona, 26:131-164.

---

dinoflagellates, lists of spp.

Margalef, Ramón, y Juan Herrera, 1964.
Hidrografía y fitoplancton de la costa comprendida entre Castellón y la desembocadura del Ebro, de julio de 1961 a julio de 1962.
Inv. Pesq., Barcelona, 26:49-90.

---

dinoflagellates, lists of spp.

Markina, N.P. 1971.
(COCTAB) Composition and distribution of plankton along the west and southern coasts of Australia in October-January 1962-63. (In Russian).
Izv. Tichookean. nauchno-issled. Inst. ribn. Choz. Okean, (TINRO) 79:127-140.

---

dinoflagellates, lists of spp. [a]

Mulford, R.A. 1972
An annual plankton cycle on the Chesapeake Bay in the vicinity of Calvert Cliffs, Maryland, June 1969-May 1970.
Proc. Acad. nat. Sci. Phila. 124(3):17-40.

---

dinoflagellates, lists of spp.

Paredes, J.F., 1962.
34. On an occurrence of red waters in the coast of Angola.
Trab. Cent. Biol. Pisc., (32-35):89-114.
Also:
Mem. Junta Invest. Ultram., (2), MO. 33.

---

dinoflagellates, lists of sp

Pieterse, F., and D.C. van der Post, 1967.
The pilchard of South West Africa (Sardinops ocellata): oceanographic conditions associated with red-tides and fish mortalities in the Walvis Bay region.
Investl Rept., Mar. Res. Lab., SWest Africa, 14: 1s5 pp.

---

dinoflagellates, lists of spp. [a]

Portugal Instituto Hidrográfico 1973
CAPEC-II Janeiro/Fevereiro-1971: resultados preliminares 9: 149 pp.

---

dinoflagellates, lists of spp.

Portugal Instituto Hidrográfico 1973
Companha oceanográfica para Apoio às Pescas do continente. CAPEC1 (12 Outubro a 15 de Novembro de 1970. Resultados Preliminares, 6:123pp

---

dinoflagellates, lists of spp.

Rampi, Leopoldo 1969.
Péridiniens Heterococcales et Pterospermales rares, intéressants ou nouveaux récoltés dans la Mer Ligurienne (Méditerranée occidentale).
Natura, Milano 60(4):313-333.

---

dinoflagellates, lists of spp.

Reyssac, Josette. 1972.
Premières observations sur le cycle annuel des diatomées et dinoflagellés dans la baie du Lévrier (Mauritanie).
Bull. Inst. fond. Afrique Noire 34(2):278-291.

---

dinoflagellates, lists of spp.

Ricard, M. 1970.
Premier inventaire des diatomées et des dinoflagellés du plancton côtier de Tahiti.
Cah. Pacifique 14: 245-254.

dinoflagellates, lists of spp.

Reyssac, Josette, 1972.
Phytoplancton récolté par le navire Ombango au large d'Angola (10-27 novembre 1965)
Bull. Inst. fond. Afr. Noire (A) 34(4): 796-808.

dinoflagellates, lists of spp.

Riley, Gordon A., and Shirley M. Conover 1967
Phytoplankton of Long Island Sound, 1954-1955.
Bull. Bingham oceanogr. Coll. 19(2):5-34.

dinoflagellates, lists of spp.

Skolka, V.H., 1961.
Données sur le phytoplancton des parages prébosphoriques de la Mer Noire.
Rapp. Proc. Verb., Réunions, Comm. Int. Expl. Sci. Mer. Méditerranée, Monaco, 16(2):129-132.

dinoflagellates, lists of spp

Smayda, Theodore J., 1966.
A quantitative analysis of the phytoplankton of the Gulf of Panama. III General ecological conditions and the phytoplankton dynamics at 8o 45'N, 79o 23'W from November 1954 to May 1957.
Inter-Amer. Trop. Tuna Comm., Bull.,11(5): 355-612.

dinoflagellates lists of spp.

Sournia, Alain, 1972
Une période de poussées phytoplanctoniques près de Nosy-Bé (Madagascar) en 1971. 1. Espèces rares ou nouvelles du phytoplancton
Cah. ORSTOM sér. Océanogr. 10(2): 151-159.

dinoflagellates, lists of spp.

Sournia, A., 1970.
A checklist of planktonic diatoms and dinoflagellates from the Mozambique Channel.
Bull. mar. Sci., 20(3): 678-696.

dinoflagellates, lists of spp.

Sournia, A. 1968.
Quelques nouvelles données sur le phytoplancton marin et la production primaire à Tuléar (Madagascar).
Hydrobiologia 31 (3/4): 545-560.

peridiniens, lists of spp.

Sournia, A., 1967.
Contribution à la connaissance des péridiniens microplanctoniques du Canal de Mozambique.
Bull.Mus.natn.Hist.nat.,Paris,(2)39(2):417-438.

dinoflagellates, lists of spp.

Steidinger, Karen A., Joanne T. Davis and Jean Williams, 1967.
Dinoflagellate studies on the inshore waters of the west coast of Florida. Prof. Pap. Ser. Fla Bd Conserv. 9: 4- 47

dinoflagellates, lists of spp.

Steidinger, Karen A., Joanne T. Davis and Jean Williams 1967.
A Key to the marine dinoflagellate genera of the west coast of Florida.
Techn. Ser. Fla. Bd Conserv. 52: 45 pp.

dinoflagellates, lists of spp.

Steidinger,Karen A., Joanne T. Davis and Jean Williams, 1966.
Observations of Gymnodinium breve Davis and other dinoflagellates.
Florida Bd., Conserv., St. Petersburg,Mar. Lab. Prof. Papers Ser., No. 8:8-15.

dinoflagellates, lists of spp.

Steidinger, Karen A., and Jean Williams 1970
Dinoflagellates.
Mem. Hourglass Cruises, Mar. Res. Lab., Fla. Dept. Nat. Res. 2:1-251.

dinoflagellates, lists of spp

Stroukuna, V.G., 1950.
[Phytoplankton of the Black Sea in the vicinity of Karadaga and its seasonal dynamics.]
Trudy Karadagsk Biol. Sta., 10:38-52.

dinoflagellates, lists of spp.

Tett, P.B., 1971.
The relation between dinoflagellates and the bioluminescence of sea water. J. mar. biol. Ass., U.K., 51(1): 183-206.

dinoflagellates, lists of spp.

Travers, A., 1962.
Recherches sur le phytoplancton du Golfe de Marseille. 1. Etude qualitative des Diatomées et des Dinoflagellés du Golfe de Marseille.
Rec. Trav., Sta. Mar., Endoume, Bull., 26(41):7-69.

dinoflagellates, lists of spp.

Tu, Hu-Kung and Young-Meng Chiang, 1972
Dinoflagellates collected from the north-eastern part of the South China Sea. Acta oceanogr. Taiwanica (2):134-146.

Dinoflagellates, lists of spp.

Wood, E.J. Ferguson, 1965.
Protoplankton of the Benguela-Guinea current region.
Bull. Mar. Sci.,15(2):475-479.

dinoflagellates, lists of spp.

Wood, E.J.F., 1963.
Check-list of dinoflagellates recorded from the Indian Ocean.
C.S.I.R.O., Div. Fish. Oceanogr., Australia, Rept., No. 28:58 pp. (multilithed).

dinoflagellates, parasitic

Cachon,Jean et Monique, 1968.
Cytologie et cycle évolutif des Chytriodinium (Chatton).
Protistologica, 4(2):249-261.

dinoflagellates, parasitic

Cachon, Jean, et Monique Cachon-Enjumet, 1964.
Cycle évolutif et cytologie de Neresheimeria catenata Neresheimer, péridinien parasite d'appendiculaires.
Ann. Sci. Nat., Zool. et Biol. Animale (12), 4: 779-800.

dinoflagellate populations

Swift, Elijah and Edward G. Durbin, 1972.
The phased division and cytological characteristics of Pyrocystis spp. can be used to estimate doubling times of their populations in the sea. Deep-Sea Res. 19(3): 189-198.

dinoflagellates, quantitative

Hirota,Keiichiro, and Takuo Endo,1965.
On primary production in the Seto Inland Sea. II. Primary production and plankton. (In Japanese;English abstract).
J. Fac. Fish., Animal Husbandry,Hiroshima,Univ. 6(1):101-132.

dinoflagellates, quantitative (data only)

Krauel, David P., 1969
Bedford Basin data report, 1967.
Techn. Rept. Fish. Res. Bd., Can., 120:84 pp (multilithed).

dinoflagellates, vertical migration

Eppley,R.W., O. Holm-Hansen and J.D.H. Strickland,1968.
Some observations on the vertical migration of dinoflagellates.
J.Phycol., 4(4):333-340.

dinoflagellates, resting spores

Wall,David, and Barrie Dale,1967.
The resting cysts of modern marine dinoflagellates and their palaeontological significance.
Palaeobot. Palynol., 2(1/4):349-354.

Acanthogonyaulax spinifera

Graham, H. W., 1942
Studies in the morphology, taxonymy, and ecology of the Peridiniales. Sci. Res. Cruise VII of the Carnegie, 1928-1929---Biol. III(542): 129 pp., 67 figs.

Achradina pulchra

Nivel, P. 1969
Données écologiques sur quelques protozoaires planctoniques rares en Méditerranée.
Protistologica 5(2): 215-225.

Achradina pulchra

Paulsen, O., 1908
XVIII Peridiniales. Nordisches Plankton. Bot. Teil: 1-124, 155 text figs.

Actinodinium apsteini n.g., n.sp.

Chatton, E., and H. Hovasse, 1937.
Actinodinium apsteini n.g., n.sp., peridinien parasite enterocoelomique des Acartia (Copepodes)
Arch. Zool. Exp. Gen. 76(N&R):24-29.

Alexandrium minutum

Halim, Youssef, 1960
Alexandrium minutum nov. g., nov. sp., Dinoflagellé provocant des "eaux rouges".
Vie et Milieu, 11(1): 102-105.

Amphidiniopsis kofoidi

Bursa, Adam, 1963.
Phytoplankton in coastal waters of the Arctic Ocean at Point Barrow, Alaska.
Arctic, 16(4):239-262.

Amphidiniopsis kofoidi

Rumkówna, A., 1948
[List of the phytoplankton species occurring in the superficial water layers in the Gulf of Gdańsk] Bull. Lab. mar., Gdynia, No. 4: 139-141 with tables in back.

Amphidinium sp.

Braarud, T., 1951.
Salinity as an ecological factor in marine phytoplankton. Physiol. Plant. 4:28-34, 3 textfigs.

Amphidinium sp.

Braarud, T., 1945
A phytoplankton survey of the polluted waters of inner Oslo Fjord. Hvalrådets Skrifter, No.28, 142 pp., 19 text figs., 17 tables.

Amphidinium sp.

Braarud, T., and Adam Bursa, 1939
On the phytoplankton of the Oslo Fjord, 1933-1934. Hvalrådets Skr. No.19:1-63; 9 text figs. Reviewed. J. du. Cons. 14(3): 418-420. A.C. Gardiner.

Amphidinium sp.

Gran, H.H., and T. Braarud, 1935
A quantitative study of the phytoplankton in the Bay of Fundy and the Gulf of Maine (including observations on hydrography, chemistry, and turbidity).
J. Biol. Bd., Canada, 1(5):279-467, 69 text figs.

Amphidinium spp.

Jeffrey, S.W., and F.T. Haxo, 1968.
Photosynthetic pigments of symbiotic dinoflagellates (zooxanthellae) from corals and clams. Biol. Bull., mar. biol. Lab., Woods Hole, 135(1): 149-165.

Amphidinium aculeatum
Schröder, B., 1900
Phytoplankton des Golfes von Neapel nebst vergleichenden Ausblicken auf das atlantischen Ozean. Mitt. Zool. Stat. Neapel, 14:1-38.

Amphidinium acutissimum
Ercegovic, A., 1936
Etudes qualitative et quantitatives du. phytoplancton dans les eaux cotières de l'Adriatique oriental moyen au cours de l'année 1934. Acta Adriatica 1(9):1-126

Amphidinium acutissimum
Hada, Yoshina, 1970. (1970)
The protozoan plankton of the Antarctic and Subantarctic seas.
Scient. Repts, Japan. Antarct. Res. Exped., (E)31:51 pp.

Amphidinium acutissimum
Wood, E.J.F., 1963
Dinoflagellates in the Australian region. II. Recent collections.
Comm. Sci. and Industr. Res. Org., Div. Fish. and Oceanogr., Techn. Paper, No. 14: 55 pp.

Amphidinium acutum.
Norrise, R.E., 1961.
Observations on phytoplankton organisms collected on the N.Z.O.I. Pacific Cruise, September 1958.
N.Z.J. Sci., 4(1):162-188.

Amphidinium aloxalocium n.sp.
Norris, R.E., 1961.
Observations on phytoplankton organisms collected on the N.Z.O.I. Pacific Cruise, September 1958.
N.Z.J.Sci., 4(1):162-188.

Amphidinium amphidinioides
Wood, E.J.F., 1963
Dinoflagellates in the Australian region. II. Recent collections.
Comm. Sci. and Industr. Res. Org., Div. Fish. and Oceanogr., Techn. Paper, No. 14: 55 pp.

Amphidinium bipes
LeBour, M.V., 1925
The dinoflagellates of Northern Seas. The Marine Biological Association of the United Kingdom, Plymouth, 250 pp., 35 pls., 53 text figs.

Amphidinium bipes
Wood, E.J.F., 1963
Dinoflagellates in the Australian region. II. Recent collections.
Comm. Sci. and Industr. Res. Org., Div. Fish. and Oceanogr., Techn. Paper, No. 14: 55 pp.

Amphidinium britannicum
LeBour, M.V., 1925
The dinoflagellates of Northern Seas. The Marine Biological Association of the United Kingdom, Plymouth, 250 pp., 35 pls., 53 text figs.

Amphidinium carteri
Carlucci, A.F., and S.B. Silbernagel 1967.
Bioassay of seawater. IV. The determination of dissolved biotin in seawater using ^{14}C uptake by cells of Amphidinium carteri.
Can. J. Microbiol. 13(8):979-986.

Amphidinium carteri nom.nov.
Hulburt, E.M., 1957.
The taxonomy of unarmored Dinophyceae of shallow embayments on Cape Cod, Massachusetts.
Biol. Bull., 112(2):196-219.

Amphidinium carteri
Jitts, H.R., C.D. McAllister, K. Stephens and J.D.H. Strickland, 1964.
The cell division rates of some marine phytoplankters as a function of light and temperature.
J. Fish. Res. Bd., Canada, 21(1):139-157.

Amphidinium carteri
Parsons, T.R., 1961
On the pigment composition of eleven species of marine phytoplankton.
J. Fish. Res. Bd., Canada, 18(6):1017-1025.

Amphidinium carteri
Parsons, T.R., K. Stephens and J.D.H. Strickland, 1961
On the chemical composition of eleven species of marine phytoplankters.
J. Fish. Res. Bd., Canada, 18(6):1001-1016.

Amphidinium carteri
Ryther, John H., and Dana D. Kramer, 1961
Relative iron requirement of some coastal and offshore algae.
Ecology, 42(2): 444-446.

Amphidinium carteri
Sakshaug, E., S. Myklestad, T. Krogh and G. Westin, 1973
Production of protein and carbohydrate in the dinoflagellate Amphidinium carteri. Some preliminary results.
Norwegian J. Bot. 20(2/3):211-218.

Amphidinium carteri
Taylor, W. Rowland, 1964.
Inorganic nutrient requirements for marine phytoplankton organisms.
Narragansett Mar. Lab., Univ. Rhode Island, Occ. Publ., No. 2:17-24.

Amphidinium carteri
Thurberg, Frederick P., and John J. Sasner, Jr. 1973
Biological activity of a cell extract from the dinoflagellate Amphidinium carteri.
Chesapeake Sci. 14(6):48-51.

Amphidinium carteri
Vien, Cao, 1968.
Sur la germination du zygote et sur un mode particulier de multiplication végétative chez le péridinien libre Amphidinium carteri.
C.r.hebd.Seanc.Acad.Sci., Paris, 267(7):701-703.

Amphidinium carteri
Vien, Cao 1967.
Sur l'existence de phénomènes sexuels chez un péridinien libre, l'Amphidinium carteri.
C.r. hebd. Seanc. Acad. Sci. Paris (D) 264 (8): 1006-1008.

Amphidinium conus
Ercegovic, A., 1936
Etudes qualitative et quantitatives du. phytoplancton dans les eaux cotières de l'Adriatique oriental moyen au cours de l'année 1934. Acta Adriatica 1(9):1-126

Amphidinium crassum
Hulburt, E.M., 1957.
The taxonomy of unarmored Dinophyceae of shallow embayments on Cape Cod, Massachusetts.
Biol. Bull., 112(2):196-219.

Amphidinium crassum
LeBour, M.V., 1925
The dinoflagellates of Northern Seas. The Marine Biological Association of the United Kingdom, Plymouth, 250 pp., 35 pls., 53 text figs.

Amphidinium crassum
Paulsen, O., 1908
XVIII Peridiniales. Nordisches Plankton, Bot. Teil: 1-124, 155 text figs.

Amphidinium cucurbita
Wood, E.J.F., 1963
Dinoflagellates in the Australian region. II. Recent collections.
Comm. Sci. and Industr. Res. Org., Div. Fish. and Oceanogr., Techn. Paper, No. 14: 55 pp.

Amphidinium curvatum
Ercegovic, A., 1936
Etudes qualitative et quantitatives du. phytoplancton dans les eaux cotières de l'Adriatique oriental moyen au cours de l'année 1934. Acta Adriatica 1(9):1-126

Amphidinium discoidalis
LeBour, M.V., 1925
The dinoflagellates of Northern Seas. The Marine Biological Association of the United Kingdom, Plymouth, 250 pp., 35 pls., 53 text figs.

Amphidinium eludens
LeBour, M.V., 1925
The dinoflagellates of Northern Seas. The Marine Biological Association of the United Kingdom, Plymouth, 250 pp., 35 pls., 53 text figs.

Amphidinium emarginatum
LeBour, M.V., 1925
The dinoflagellates of Northern Seas. The Marine Biological Association of the United Kingdom, Plymouth, 250 pp., 35 pls., 53 text figs.

Amphidinium extensum
LeBour, M.V., 1925
The dinoflagellates of Northern Seas. The Marine Biological Association of the United Kingdom, Plymouth, 250 pp., 35 pls., 53 text figs.

Amphidinium flagellans
Wood, E.J.F., 1963
Dinoflagellates in the Australian region. II. Recent collections.
Comm. Sci. and Industr. Res. Org., Div. Fish. and Oceanogr., Techn. Paper, No. 14: 55 pp.

Amphidinium flexum
LeBour, M.V., 1925
The dinoflagellates of Northern Seas. The Marine Biological Association of the United Kingdom, Plymouth, 250 pp., 35 pls., 53 text figs.

Amphidinium flexum
Morse, D.C., 1947
Some observations on seasonal variations in plankton population Patuxant River, Maryland 1943-1945. Bd. Nat. Res., Publ. No.65, Chesapeake Biol. Lab., 31, 3 figs.

Amphidinium globosum
Ercegovic, A., 1936
Etudes qualitative et quantitatives du. phytoplancton dans les eaux cotières de l'Adriatique oriental moyen au cours de l'année 1934. Acta Adriatica 1(9):1-126

Amphidinium globosum
Matzenauer, L., 1933
Die Dinoflagellaten des indischen Ozeans (mit Ausnahme der Gattung Ceratium.) Bot. Arch. 35:437-510, 77 text figs., 2 charts.

Amphidinium herdmani
Dragesco, Jean, 1965.
Étude cytologique de quelques flagellés mésopsammiques.
Cahiers Biol. Mar., Roscoff, 6(1):83-115.

Amphidinium herdmanni
LeBour, M.V., 1925
The dinoflagellates of Northern Seas. The Marine Biological Association of the United Kingdom, Plymouth, 250 pp., 35 pls., 53 text figs.

Amphidinium höfleri
Elbrächter, Malte 1972
Begrenzte Heterotrophie bei Amphidinium (Dinoflagellata).
Kieler Meeresf. 28(1): 84-91.

Amphidinium inflatum
Wood, E.J.F., 1954.
Dinoflagellates in the Australian region.
Australian J. Mar. Freshwater Res., 5(2):171-351.

Amphidinium kesslitzi
Wood, E.J.F., 1954.
Dinoflagellates in the Australian region.
Australian J. Mar. Freshwater Res., 5(4):171-351.

Amphidinium klebsi
Dragesco, Jean, 1965.
Étude cytologique de quelques flagellés mesopsammiques.
Cahiers Biol. Mar., Roscoff, 6(1):83-115.

Amphidinium klebsi
LeBour, M.V., 1925
The dinoflagellates of Northern Seas. The Marine Biological Association of the United Kingdom, Plymouth, 250 pp., 35 pls., 53 text figs.

Amphidinium klebsii
Mandelli, Enrique F. 1969.
Carotenoid interconversion in light dark cultures of the dinoflagellate Amphidinium klebsii.
J. Phycol. 5(4): 382-384.

Amphidinium klebsi
Ronkin, R.R., 1959
Motility and power dissipation in flagellated cells, especially Chlamydomonas.
Biol. Bull., 116(2):285-293.

Amphidinium klebsi
Wood, E.J.F., 1954.
Dinoflagellates in the Australian region.
Australian J. Mar. Freshwater Res., 5(2):171-351.

Amphidinium kofoidi
LeBour, M.V., 1925
The dinoflagellates of Northern Seas. The Marine Biological Association of the United Kingdom, Plymouth, 250 pp., 35 pls., 53 text figs.

Amphidinium lacustriforme
Norris, R.E., 1961.
Observations on phytoplankton organisms collected on the N.Z.O.I. Pacific Cruise, September 1958.
N.Z.J. Sci., 4(1):162-188.

Amphidinium lanceolatum
Ercegovic, A., 1936
Etudes qualitative et quantitatives du phytoplancton dans les eaux cotières de l'Adriatique oriental moyen au cours de l'année 1934. Acta Adriatica 1(9):1-126

Amphidinium latum n.sp.
LeBour, M.V., 1925
The dinoflagellates of Northern Seas. The Marine Biological Association of the United Kingdom, Plymouth, 250 pp., 35 pls., 53 text figs.

Amphidinium lissae
Ercegovic, A., 1936
Etudes qualitative et quantitatives du phytoplancton dans les eaux cotières de l'Adriatique oriental moyen au cours de l'année 1934. Acta Adriatica 1(9):1-126

Amphidinium longum
Ercegovic, A., 1936
Etudes qualitative et quantitatives du phytoplancton dans les eaux cotières de l'Adriatique oriental moyen au cours de l'année 1934. Acta Adriatica 1(9):1-126

Amphidinium longum
LeBour, M.V., 1925
The dinoflagellates of Northern Seas. The Marine Biological Association of the United Kingdom, Plymouth, 250 pp., 35 pls., 53 text figs.

Amphidinium longum
Paulsen, O., 1908
XVIII Peridiniales. Nordisches Plankton, Bot. Teil: 1-124, 155 text figs.

Amphidinium manannini
LeBour, M.V., 1925
The dinoflagellates of Northern Seas. The Marine Biological Association of the United Kingdom, Plymouth, 250 pp., 35 pls., 53 text figs.

Amphidinium massarti n.sp
Biecheler, B., 1952.
Recherches sur les Peridiniens.
Bull. Biol. France Belg., Suppl., 36:1-149, 75 textfigs.

Amphidinium microcephalum n.sp.
Norris, R.E., 1961.
Observations on phytoplankton organisms collected on the N.Z.O.I. Pacific Cruise, September 1958.
N.Z.J. Sci., 4(1):162-188.

Amphidinium oceanicum
Lillick, L.C., 1940
Phytoplankton and planktonic protozoa of the offshore waters of the Gulf of Maine. Pt.II. Qualitative Composition of the Planktonic Flora. Trans. Am. Phil. Soc., n.s., 31(3):193-237, 13 text figs.

Amphidinium oceanicum
Lillick, L.C., 1938
Preliminary report of the phytoplankton of the Gulf of Maine. Am. Mid. Nat. 20(3):624-640, 1 text figs 37 tables.

Amphidinium operculatum
Calkins, G.N., 1902
Marine protozoa from Woods Hole. U.S. Fish Comm. Bull. for 1901, pp. 413-468, 69 text figs.

Amphidinium operculatum
LeBour, M.V., 1925
The dinoflagellates of Northern Seas. The Marine Biological Association of the United Kingdom, Plymouth, 250 pp., 35 pls., 53 text figs.

Amphidinium operculatum
Levander, K.M., 1947
Plankton gesammelt in den Jahren 1899-1910 an den Küsten Finnlands. Finnländische Hydrographisch-Biologische Untersuchungen (aus dem Wasserbiologischen Laboratorin der Societas Scientiarum Fennica) No.11: 40 pp., 6 diagrams, 13 pls., tables.

Amphidinium operculatum
Paulsen, O., 1908
XVIII Peridiniales. Nordisches Plankton, Bot. Teil: 1-124, 155 text figs.

Amphidinium rotundatum
Paulsen, O., 1908
XVIII Peridiniales. Nordisches Plankton, Bot. Teil: 1-124, 155 text figs.

Amphidinium operculatum
Rumkówna, A., 1948
List of the phytoplankton species occurring in the superficial water layers in the Gulf of Gdańsk. Bull. Lab. mar., Gdynia, No. 4: 139-141 with tables in back.

Amphidinium operculatum
Schröder, B., 1900
Phytoplankton des Golfes von Neapel nebst vergleichenden Ausblicken auf das atlantischen Ozean. Mitt. Zool. Stat. Neapel, 14:1-38.

Amphidinium ovum
LeBour, M.V., 1925
The dinoflagellates of Northern Seas. The Marine Biological Association of the United Kingdom, Plymouth, 250 pp., 35 pls., 53 text figs.

Amphidinium pelagicum n.sp.
LeBour, M.V., 1925
The dinoflagellates of Northern Seas. The Marine Biological Association of the United Kingdom, Plymouth, 250 pp., 35 pls., 53 text figs.

Amphidinium pellucidum
Dragesco, Jean, 1965.
Étude cytologique de quelques flagellés mesopsammiques.
Cahiers Biol. Mar., Roscoff, 6(1):83-115.

Amphidinium pellucidum
LeBour, M.V., 1925
The dinoflagellates of Northern Seas. The Marine Biological Association of the United Kingdom, Plymouth, 250 pp., 35 pls., 53 text figs.

Amphidinium phaeocysticola n.sp.
LeBour, M.V., 1925
The dinoflagellates of Northern Seas. The Marine Biological Association of the United Kingdom, Plymouth, 250 pp., 35 pls., 53 text figs.

Amphidinium rhynchocephalum
Aleem, A.A., 1952.
Données écologiques sur deux espèces de peridiniens des eaux saumatres. Vie et Milieu, Bull. Lab. Arago, Univ. Paris, 3(3):281-287, 2 textfigs.

Amphidinium scissoides n.sp.
LeBour, M.V., 1925
The dinoflagellates of Northern Seas. The Marine Biological Association of the United Kingdom, Plymouth, 250 pp., 35 pls., 53 text figs.

Amphidinium scissum
LeBour, M.V., 1925
The dinoflagellates of Northern Seas. The Marine Biological Association of the United Kingdom, Plymouth, 250 pp., 35 pls., 53 text figs.

Amphidinium semilunatum
LeBour, M.V., 1925
The dinoflagellates of Northern Seas. The Marine Biological Association of the United Kingdom, Plymouth, 250 pp., 35 pls., 53 text figs.

Amphidinium semilunatum
Rumkówna, A., 1948
List of the phytoplankton species occurring in the superficial water layers in the Gulf of Gdańsk. Bull. Lab. mar., Gdynia, No. 4: 139-141 with tables in back.

Oceanographic Index: Marine Organisms Cumulation, 1946-1973

Amphidinium sphenoides
Hulburt, E.M., 1957.
The taxonomy of unarmored Dinophyceae of shallow embayments on Cape Cod, Massachusetts.
Biol. Bull., 112(2):196-219.

Amphidinium sphenoides
LeBour, M.V., 1925
The dinoflagellates of Northern Seas. The Marine Biological Association of the United Kingdom, Plymouth, 250 pp., 35 pls., 53 text figs.

Amphidinium sphenoides
Wood, E.J.F., 1963
Dinoflagellates in the Australian region. II. Recent collections.
Comm. Sci. and Industr. Res. Org., Div. Fish. and Oceanogr., Techn. Paper, No. 14: 55 pp.

Amphidinium steini
LeBour, M.V., 1925
The dinoflagellates of Northern Seas. The Marine Biological Association of the United Kingdom, Plymouth, 250 pp., 35 pls., 53 text figs.

Amphidinium stigmatum
Ercegovic, A., 1936
Etudes qualitative et quantitatives du phytoplancton dans les eaux cotières de l'Adriatique oriental moyen au cours de l'année 1934. Acta Adriatica 1(9):1-126

Amphidinium subsalum n.sp
Biecheler, B., 1952.
Recherches sur les Peridiniens.
Bull. Biol., France Belg., Suppl. 36:1-149, 75 textfigs.

Amphidinium sulcatum n.sp.
Kofoid, C. A., 1907
Dinoflagellata of the San Diego region. III. Descriptions of new species. Univ. Calif. Publ., Zool. 3:299-340, Pls. 22-33.

Amphidinium sulcatum
Wood, E.J.F., 1954.
Dinoflagellates in the Australian region.
Australian J. Mar. Freshwater Res., 5(2):171-351.

Amphidinium testudo
LeBour, M.V., 1925
The dinoflagellates of Northern Seas. The Marine Biological Association of the United Kingdom, Plymouth, 250 pp., 35 pls., 53 text figs.

Amphidinium turbo
Matzenauer, L., 1933
Die Dinoflagellaten des indischen Ozeans (mit Ausnahme der Gattung Ceratium.) Bot. Arch. 35:437-510, 77 text figs., 2 charts.

Amphidinium turbo
Wood, E.J.F., 1963
Dinoflagellates in the Australian region. II. Recent collections.
Comm. Sci. and Industr. Res. Org., Div. Fish. and Oceanogr., Techn. Paper, No. 14: 55 pp.

Amphidinium vasculum
Wood, E.J.F., 1963
Dinoflagellates in the Australian region. II. Recent collections.
Comm. Sci. and Industr. Res. Org., Div. Fish. and Oceanogr., Techn. Paper, No. 14: 55 pp.

Amphidinium vitreum
LeBour, M.V., 1925
The dinoflagellates of Northern Seas. The Marine Biological Association of the United Kingdom, Plymouth, 250 pp., 35 pls., 53 text figs.

Amphidinium wislouchi, n.sp.
Hulburt, E.M., 1957.
The taxonomy of unarmoured Dinophyceae of shallow embayments on Cape Cod, Massachusetts.
Biol. Bull., 112(2):196-219.

Amphidoma sp.
Wood, E.J.F., 1963
Dinoflagellates in the Australian region. II. Recent collections.
Comm. Sci. and Industr. Res. Org., Div. Fish. and Oceanogr., Techn. Paper, No. 14: 55 pp.

Amphidoma biconica n. sp.
Kofoid, C. A., 1907
Dinoflagellata of the San Diego region. III. Descriptions of new species. Univ. Calif. Publ., Zool. 3:299-340, Pls. 22-33.

Amphidoma caudata n.sp.
Halldal, P., 1953.
Phytoplankton investigations from Weather Ship M in the Norwegian Sea, 1948-49 (including observations during the "Armauer Hansen" cruise, July 1949). Hvalrådets Skrifter No. 38:91 pp., 20 tables, 21 textfigs.

Amphidoma nucula
Balech, Enrique, 1971
Microplancton del Atlantico ecuatorial oeste (Equalant 1)
Publ. Serv. Hidrograf. Naval, Argentina H. 654: 103 pp., 122 figs.

Amphidoma nucula
Murray, G., and F. G. Whitting, 1899
New Peridiniaceae from the Atlantic. Trans. Linn. Soc., London, Bot., ser 2, 5: 321-342, Pls. 27-33, 9 tables.

Amphisolenia sp.
Lindemann, E., 1925
Neubeobachtungen an den Winter peridineen des Golfes von Neapel. Bot. Arch. 9:95-102, 19 text figs.

Amphisolenia astragalus
Kofoid, C. A. and T. Skogsberg, 1928
XXV. The Dinoflagellata: The Dinophysiodae. Reports on the scientific results of the expedition to the Eastern Tropical Pacific, in charge of Alexander Agassiz, by the U. S. Fish Commission Steamer "Albatross" from October 1904 to March 1905----. Mem. M. C. Z. 51:766 pp., 31 pls.

Amphisolenia astragalus
Wood, E.J.F., 1963
Dinoflagellates in the Australian region. II. Recent collections.
Comm. Sci. and Industr. Res. Org., Div. Fish. and Oceanogr., Techn. Paper, No. 14: 55 pp.

Amphisolenia asymmetrica
Käsler, R., 1938
Die Verbreitung der Dinophysiales im Sud-atlantischen Ozean. Wiss. Ergeb. Deutschen Atlantischen Expedition----"Meteor" 1925-1927, 12(2):162-237, text figs. 85-118.

Amphisolenia asymmetrica
Kofoid, C. A. and T. Skogsberg, 1928
XXV. The Dinoflagellata: The Dinophysiodae. Reports on the scientific results of the expedition to the Eastern Tropical Pacific, in charge of Alexander Agassiz, by the U. S. Fish Commission Steamer "Albatross" from October 1904 to March 1905----. Mem. M. C. Z. 51:766 pp., 31 pls.

Amphisolenia bidentata
Abe, Tohru H., 1967.
The armoured Dinoflagellata:II. Prorocentridae and Dinophysidae (C) - Ornithocercus, Histioneis, Amphisolenia and others.
Publs Seto mar. biol.Lab.,15(2):79-116.

Amphisolenia bidentata
Balech, Enrique, 1962
Tintinnoinea y Dinoflagellata del Pacifico segun material de las Expediciones NORPAC y DOWNWIND del Instituto Scripps de Oceanografia.
Revista, Mus. Argentino Ciencias Nat. "Bernardino Rivadavia", Ciencias Zool., 7(1):1-253.

Amphisolenia bidentata
Dangeard, P., 1927
Phytoplankton de la croisière du "Sylvana". Ann. Inst. Ocean., Monaco, n.s., 4(8):286-401, 54 text figs. (Avrier-Juin 1913).

Amphisolenia bidentata
Ercegovic, A., 1936
Etudes qualitative et quantitatives du phytoplancton dans les eaux cotières de l'Adriatique oriental moyen au cours de l'année 1934. Acta Adriatica 1(9):1-126

Amphisolenia bidentata
Forti, A., 1922
Ricerche sulla flora pelagica (fitoplancton) di Quarto dei Mille. Mem. R. Com. Talass. Ital. 97:248 pp., 13 pls.

Amphisolenia bidentata
Halim, Y., 1965.
Microplancton des eaux Egyptiennes. II. Chrysomonadines; Ebriediens et dinoflagellés nouveaux on d'interêt biogeographique.
Rapp. Proc. Verb. Reunions. Comm. Int. Expl. Sci., Mer Méditerranée, Monaco, 18(2):373-379.

Amphisolenia bidentata
Jörgensen, E., 1923
Mediterranean Dinophysiaceae. Rept. Danish Oceanogr. Exped. 1908-10, to the Mediterranean and adjacent seas, Vol.II, Biol. J 2, 48 pp., 64 text figs.

Amphisolenia bidentata
Käsler, R., 1938
Die Verbreitung der Dinophysiales im Sud-atlantischen Ozean. Wiss. Ergeb. Deutschen Atlantischen Expedition----"Meteor" 1925-1927, 12(2):162-237, text figs. 85-118.

Amphisolenia bidentata
Kofoid, C. A. and T. Skogsberg, 1928
XXV. The Dinoflagellata: The Dinophysiodae. Reports on the scientific results of the expedition to the Eastern Tropical Pacific, in charge of Alexander Agassiz, by the U. S. Fish Commission Steamer "Albatross" from October 1904 to March 1905----. Mem. M. C. Z. 51:766 pp., 31 pls.

Amphisolenia bidentata
Margalef, R., 1949
Fitoplancton nerítico de la Costa Brava en 1947-48. Publ. Inst. Biol. Aplicada, 5: 41-51, 3 text figs.

Amphisolenia bidentata
Marshall, S. M., 1933
The production of microplankton in the Great Barrier Reef Region. Brit. Mus. (N.H.) Great Barrier Reef Exped. 1928-29, Sci. Repts. II(5):111-157, 14 text figs.

Amphisolenia bidentata
Matzenauer, L., 1933
Die Dinoflagellaten des indischen Ozeans (mit Ausnahme der Gattung Ceratium.) Bot. Arch. 35:437-510, 77 text figs., 2 charts.

Amphisolenia bidentata
Pavillard, J., 1916
Recherches sur les Peridiniens du Golfe du Lion. Mem. Univ. Montpellier. Trav. Inst. Bot., Univ. Montpellier. Serie mixte No.4, 70 pp., 3 pls., 15 text figs.

Amphisolenia bidentata
Rampi, L., 1940
Ricerche sul Fitoplancton del mare Ligure. Boll. di Pesca, di Piscicoltura e di Idrobiologa, 18(2):1-34, 56 text figs.

Amphisolenia bidentata
Rampi, L. 1940.
Ricerche sul Fitoplancton del mare Ligure. II Le tecatali e le dinofisiali delle acque di Sanremo. Boll. Pesca, Piscicolt., Idrobiol. (18) 16(2): 243-274, 56 figs.

Amphisolenia bidentata n.sp.
Schröder, B., 1900
Phytoplankton des Golfes von Neapel nebst vergleichenden Ausblicken auf das atlantischen Ozean. Mitt. Zool. Stat. Neapel, 14:1-38.

Amphisolenia bidentata
Wood, E.J.F., 1954
Dinoflagellates in the Australian region. Australian J. Mar. Freshwater Res., 5(2):171-351.

Amphisolenia bifurcata
Balech, Enrique, 1962
Tintinnoinea y Dinoflagellata del Pacifico segun material de las Expediciones NORPAC y DOWNWIND del Instituto Scripps de Oceanografia. Revista, Mus. Argentino Ciencias Nat. "Bernardino Rivadavia", Ciencias Zool., 7(1):1-253.

Amphisolenia bifurcata
Käsler, R., 1938
Die Verbreitung der Dinophysiales im Sud-atlantischen Ozean. Wiss. Ergeb. Deutschen Atlantischen Expedition----"Meteor" 1925-1927, 12(2):162-237, text figs. 85-118.

Amphisolenia bifurcata
Kofoid, C. A. and T. Skogsberg, 1928
XXV. The Dinoflagellata: The Dinophysiodae. Reports on the scientific results of the expedition to the Eastern Tropical Pacific, in charge of Alexander Agassiz, by the U. S. Fish Commission Steamer "Albatross" from October 1904 to March 1905----. Mem. M. C. Z. 51:766 pp., 31 pls.

Amphisolenia bifurcata sp. n.
Murray, G., and F. G. Whitting, 1899
New Peridiniaceae from the Atlantic. Trans. Linn. Soc., London, Bot., ser 2, 5: 321-342, Pls. 27-33, 9 tables.

Amphisolenia bifurcata
Wood, E.J.F., 1963
Dinoflagellates in the Australian region. II. Recent collections. Comm. Sci. and Industr. Res. Org., Div. Fish. and Oceanogr., Techn. Paper, No. 14: 55 pp.

Amphisolenia bispinosa
Käsler, R., 1938
Die Verbreitung der Dinophysiales im Sud-atlantischen Ozean. Wiss. Ergeb. Deutschen Atlantischen Expedition----"Meteor" 1925-1927, 12(2):162-237, text figs. 85-118.

Amphisolenia bispinosa
Kofoid, C. A. and T. Skogsberg, 1928
XXV. The Dinoflagellata: The Dinophysiodae. Reports on the scientific results of the expedition to the Eastern Tropical Pacific, in charge of Alexander Agassiz, by the U. S. Fish Commission Steamer "Albatross" from October 1904 to March 1905----. Mem. M. C. Z. 51:766 pp., 31 pls.

Amphisolenia bispinosa
Wood, E.J.F., 1954
Dinoflagellates in the Australian region. Australian J. Mar. Freshwater Res., 5(2):171-351.

Amphisolenia brevicauda
Käsler, R., 1938
Die Verbreitung der Dinophysiales im Sud-atlantischen Ozean. Wiss. Ergeb. Deutschen Atlantischen Expedition----"Meteor" 1925-1927, 12(2):162-237, text figs. 85-118.

Amphisolenia brevicauda
Kofoid, C. A. and T. Skogsberg, 1928
XXV. The Dinoflagellata: The Dinophysiodae. Reports on the scientific results of the expedition to the Eastern Tropical Pacific, in charge of Alexander Agassiz, by the U. S. Fish Commission Steamer "Albatross" from October 1904 to March 1905----. Mem. M. C. Z. 51:766 pp., 31 pls.

Amphisolenia brevicauda
Wood, E.J.F., 1963
Dinoflagellates in the Australian region. II. Recent collections. Comm. Sci. and Industr. Res. Org., Div. Fish. and Oceanogr., Techn. Paper, No. 14: 55 pp.

Amphisolenia clavipes
Käsler, R., 1938
Die Verbreitung der Dinophysiales im Sud-atlantischen Ozean. Wiss. Ergeb. Deutschen Atlantischen Expedition----"Meteor" 1925-1927, 12(2):162-237, text figs. 85-118.

Amphisolenia clavipes
Kofoid, C. A. and T. Skogsberg, 1928
XXV. The Dinoflagellata: The Dinophysiodae. Reports on the scientific results of the expedition to the Eastern Tropical Pacific, in charge of Alexander Agassiz, by the U. S. Fish Commission Steamer "Albatross" from October 1904 to March 1905----. Mem. M. C. Z. 51:766 pp., 31 pls.

Amphisolenia clavipes
Wood, E.J.F., 1963
Dinoflagellates in the Australian region. II. Recent collections. Comm. Sci. and Industr. Res. Org., Div. Fish. and Oceanogr., Techn. Paper, No. 14: 55 pp.

Amphisolenia complanata n.sp.
Kofoid, C. A. and T. Skogsberg, 1928
XXV. The Dinoflagellata: The Dinophysiodae. Reports on the scientific results of the expedition to the Eastern Tropical Pacific, in charge of Alexander Agassiz, by the U. S. Fish Commission Steamer "Albatross" from October 1904 to March 1905----. Mem. M. C. Z. 51:766 pp., 31 pls.

Amphisolenia curvata
Kofoid, C. A. and T. Skogsberg, 1928
XXV. The Dinoflagellata: The Dinophysiodae. Reports on the scientific results of the expedition to the Eastern Tropical Pacific, in charge of Alexander Agassiz, by the U. S. Fish Commission Steamer "Albatross" from October 1904 to March 1905----. Mem. M. C. Z. 51:766 pp., 31 pls.

Amphisolenia curvata
Wood, E.J.F., 1954.
Dinoflagellates in the Australian region. Australian J. Mar. Freshwater Res., 5(2):171-351.

Amphisolenia elegans
Käsler, R., 1938
Die Verbreitung der Dinophysiales im Sud-atlantischen Ozean. Wiss. Ergeb. Deutschen Atlantischen Expedition----"Meteor" 1925-1927, 12(2):162-237, text figs. 85-118.

Amphisolenia elongata
Käsler, R., 1938
Die Verbreitung der Dinophysiales im Sud-atlantischen Ozean. Wiss. Ergeb. Deutschen Atlantischen Expedition----"Meteor" 1925-1927, 12(2):162-237, text figs. 85-118.

Amphisolenia elongata n.sp.
Kofoid, C. A. and T. Skogsberg, 1928
XXV. The Dinoflagellata: The Dinophysiodae. Reports on the scientific results of the expedition to the Eastern Tropical Pacific, in charge of Alexander Agassiz, by the U. S. Fish Commission Steamer "Albatross" from October 1904 to March 1905----. Mem. M. C. Z. 51:766 pp., 31 pls.

Amphisolenia extensa
Jörgensen, E., 1923
Mediterranean Dinophysiaceae. Rept. Danish Oceanogr. Expeds. 1908-10, to the Mediterranean and adjacent seas, Vol.II, Biol. J 2, 48 pp., 64 text figs.

Amphisolenia extensa
Käsler, R., 1938
Die Verbreitung der Dinophysiales im Sud-atlantischen Ozean. Wiss. Ergeb. Deutschen Atlantischen Expedition----"Meteor" 1925-1927, 12(2):162-237, text figs. 85-118.

Amphisolenia extensa
Kofoid, C. A. and T. Skogsberg, 1928
XXV. The Dinoflagellata: The Dinophysiodae. Reports on the scientific results of the expedition to the Eastern Tropical Pacific, in charge of Alexander Agassiz, by the U. S. Fish Commission Steamer "Albatross" from October 1904 to March 1905----. Mem. M. C. Z. 51:766 pp., 31 pls.

Amphisolenia extensa
Rampi, L., 1945
Osservazioni sulla distribuzione qualitativa del fitoplancton nel mare Mediterraneo. Atti della Soc. Ital. di Sci. Nat. 84:105-113.

Amphisolenia extensa
Rampi, L., 1942
II Fitoplancton mediterraneo: Problemi ed affinita interoceaniche. Boll. di Pesca di Piscicoltura e di Idrobiologia, Anno 18, Fasc. 4:7-19.

Amphisolenia globifera
Balech, Enrique, 1962
Tintinnoinea y Dinoflagellata del Pacifico segun material de las Expediciones NORPAC y DOWNWIND del Instituto Scripps de Oceanografia. Revista, Mus. Argentino Ciencias Nat. "Bernardino Rivadavia", Ciencias Zool., 7(1):1-253.

Amphisolenia globifera
Halim, Youssef, 1960
Étude quantitative et qualitative du cycle écologique des dinoflagellés dans les eaux de Villefranche-sur-Mer. (1953-1955). Ann. Inst. Océanogr., Monaco, 38:123-232.

Amphisolenia globifera
Jörgensen, E., 1923
Mediterranean Dinophysiaceae. Rept. Danish Oceanogr. Expeds. 1908-10, to the Mediterranean and adjacent seas, Vol.II, Biol. J 2, 48 pp., 64 text figs.

Amphisolenia globifera
Käsler, R., 1938
Die Verbreitung der Dinophysiales im Sud-atlantischen Ozean. Wiss. Ergeb. Deutschen Atlantischen Expedition----"Meteor" 1925-1927, 12(2):162-237, text figs. 85-118.

Amphisolenia globifera
Kofoid, C. A. and T. Skogsberg, 1928
XXV. The Dinoflagellata: The Dinophysiodae. Reports on the scientific results of the expedition to the Eastern Tropical Pacific, in charge of Alexander Agassiz, by the U. S. Fish Commission Steamer "Albatross" from October 1904 to March 1905----. Mem. M. C. Z. 51:766 pp., 31 pls.

Amphisolenia globifera
LeBour, M.V., 1925
The dinoflagellates of Northern Seas. The Marine Biological Association of the United Kingdom, Plymouth, 250 pp., 35 pls., 53 text figs.

Amphisolenia globifera
Paulsen, O., 1908
XVIII Peridiniales. Nordisches Plankton, Bot. Teil: 1-124, 155 text figs.

Amphisolenia globifera
Wood, E.J.F., 1963
Dinoflagellates in the Australian region. II. Recent collections.
Comm. Sci. and Industr. Res. Org., Div. Fish. and Oceanogr., Techn. Paper, No. 14: 55 pp.

Amphisolenia inflata
Kofoid, C. A. and T. Skogsberg, 1928
XXXV. The Dinoflagellata: The Dinophysiodae. Reports on the scientific results of the expedition to the Eastern Tropical Pacific, in charge of Alexander Agassiz, by the U. S. Fish Commission Steamer "Albatross" from October 1904 to March 1905----. Mem. M. C. Z. 51:766 pp., 31 pls.

Amphisolenia inflata
LeBour, M.V., 1925
The dinoflagellates of Northern Seas. The Marine Biological Association of the United Kingdom, Plymouth, 250 pp., 35 pls., 53 text figs.

Amphisolenia inflata n.sp.
Murray, G., and F. G. Whitting, 1899
New Peridiniaceae from the Atlantic.
Trans. Linn. Soc., London, Bot., ser 2, 5: 321-342, Pls. 27-33, 9 tables.

Amphisolenia inflata
Paulsen, O., 1908
XVIII Peridiniales. Nordisches Plankton, Bot. Teil: 1-124, 155 text figs.

Amphisolenia laticincta
Kofoid, C. A. and T. Skogsberg, 1928
XXXV. The Dinoflagellata: The Dinophysiodae. Reports on the scientific results of the expedition to the Eastern Tropical Pacific, in charge of Alexander Agassiz, by the U. S. Fish Commission Steamer "Albatross" from October 1904 to March 1905----. Mem. M. C. Z. 51:766 pp., 31 pls.

Amphisolenia lemmermanni
Balech, Enrique, 1962
Tintinnoinea y Dinoflagellata del Pacifico segun material de las Expediciones NORPAC y DOWNWIND del Instituto Scripps de Oceanografia.
Revista, Mus. Argentino Ciencias Nat. "Bernardino Rivadavia", Ciencias Zool., 7(1):1-253.

Amphisolenia Lemmermanni
Käsler, R., 1938
Die Verbreitung der Dinophysiales im Sudatlantischen Ozean. Wiss. Ergeb. Deutschen Atlantischen Expedition----"Meteor" 1925-1927, 12(2):162-237, text figs. 85-118.

Amphisolenia lemmermanni
Kofoid, C. A. and T. Skogsberg, 1928
XXXV. The Dinoflagellata: The Dinophysiodae. Reports on the scientific results of the expedition to the Eastern Tropical Pacific, in charge of Alexander Agassiz, by the U. S. Fish Commission Steamer "Albatross" from October 1904 to March 1905----. Mem. M. C. Z. 51:766 pp., 31 pls.

Amphisolenia lemmermanni
Matzenauer, L., 1933
Die Dinoflagellaten des indischen Ozeans (mit Ausnahme der Gattung Ceratium.) Bot. Arch. 35:437-510, 77 text figs., 2 charts.

Amphisolenia lemmermanni
Wood, E.J.F., 1963
Dinoflagellates in the Australian region. II. Recent collections.
Comm. Sci. and Industr. Res. Org., Div. Fish. and Oceanogr., Techn. Paper, No. 14: 55 pp.

Amphisolenia microcephalus n.sp.
Abe, Tohru H., 1967
The armoured Dinoflagellata: II. Prorocentridae and Dinophusidae (C)- Ornithocercus, Histioneis, Amphisilenia and others.
Publs Seto mar. Biol. Lab., 15(2):79-116.

Amphisolenia mozembica n. sp.
Sournie, A., 1967
Contribution à la connaissance des péridiniens microplanctoniques du Canal de Mozambique.
Bull. Mus. natn. Hist. nat., Paris, (2)39(2):417-438.

Amphisolenia palaeotheroides
Balech, Enrique, 1962
Tintinnoinea y Dinoflagellata del Pacifico segun material de las Expediciones NORPAC y DOWNWIND del Instituto Scripps de Oceanografia.
Revista, Mus. Argentino Ciencias Nat. "Bernardino Rivadavia", Ciencias Zool., 7(1):1-253.

Amphisolenia palaeotheroides
Käsler, R., 1938
Die Verbreitung der Dinophysiales im Sudatlantischen Ozean. Wiss. Ergeb. Deutschen Atlantischen Expedition----"Meteor" 1925-1927, 12(2):162-237, text figs. 85-118.

Amphisolenia palaeotheroides
Kofoid, C. A. and T. Skogsberg, 1928
XXXV. The Dinoflagellata: The Dinophysiodae. Reports on the scientific results of the expedition to the Eastern Tropical Pacific, in charge of Alexander Agassiz, by the U. S. Fish Commission Steamer "Albatross" from October 1904 to March 1905----. Mem. M. C. Z. 51:766 pp., 31 pls.

Amphisolenia palaeotheroides
Wood, E.J.F., 1963
Dinoflagellates in the Australian region. II. Recent collections.
Comm. Sci. and Industr. Res. Org., Div. Fish. and Oceanogr., Techn. Paper, No. 14: 55 pp.

Amphisolenia palmata
Abe, Tohru H., 1967
The armoured Dinoflagellata: II. Prorocentridae and Dinophysidae (C) - Ornithocercus, Histioneis, Amphisolenia and others.
Publs Seto mar. biol. Lab., 15(2):79-116.

Amphisolenia palmata
Balech, Enrique, 1962
Tintinnoinea y Dinoflagellata del Pacifico segun material de las Expediciones NORPAC y DOWNWIND del Instituto Scripps de Oceanografia.
Revista, Mus. Argentino Ciencias Nat. "Bernardino Rivadavia", Ciencias Zool., 7(1):1-253.

Amphisolenia palmata
Jörgensen, E., 1923
Mediterranean Dinophysiaceae. Rept. Danish Oceanogr. Expeds. 1908-10, to the Mediterranean and adjacent seas, Vol.II, Biol J 2, 48 pp., 64 text figs.

Amphisolenia palmata
Käsler, R., 1938
Die Verbreitung der Dinophysiales im Sudatlantischen Ozean. Wiss. Ergeb. Deutschen Atlantischen Expedition----"Meteor" 1925-1927, 12(2):162-237, text figs. 85-118.

Amphisolenia palmata
Kofoid, C. A. and T. Skogsberg, 1928
XXXV. The Dinoflagellata: The Dinophysiodae. Reports on the scientific results of the expedition to the Eastern Tropical Pacific, in charge of Alexander Agassiz, by the U. S. Fish Commission Steamer "Albatross" from October 1904 to March 1905----. Mem. M. C. Z. 51:766 pp., 31 pls.

Amphisolenia palmata
Murray, G., and F. G. Whitting, 1899
New Peridiniaceae from the Atlantic.
Trans. Linn. Soc., London, Bot., ser 2, 5: 321-342, Pls. 27-33, 9 tables.

Amphisolenia palmata
Wood, E.J.F., 1954
Dinoflagellates in the Australian region.
Australian J. Mar. Freshwater Res., 5(2):171-351

Amphisolenia projecta
Kofoid, C. A. and T. Skogsberg, 1928
XXXV. The Dinoflagellata: The Dinophysiodae. Reports on the scientific results of the expedition to the Eastern Tropical Pacific, in charge of Alexander Agassiz, by the U. S. Fish Commission Steamer "Albatross" from October 1904 to March 1905----. Mem. M. C. Z. 51:766 pp., 31 pls.

Amphisolenia quadricauda
Kofoid, C. A. and T. Skogsberg, 1928
XXXV. The Dinoflagellata: The Dinophysiodae. Reports on the scientific results of the expedition to the Eastern Tropical Pacific, in charge of Alexander Agassiz, by the U. S. Fish Commission Steamer "Albatross" from October 1904 to March 1905----. Mem. M. C. Z. 51:766 pp., 31 pls.

Amphisolenia quadrispina
Käsler, R., 1938
Die Verbreitung der Dinophysiales im Sudatlantischen Ozean. Wiss. Ergeb. Deutschen Atlantischen Expedition----"Meteor" 1925-1927, 12(2):162-237, text figs. 85-118.

Amphisolenia quadrispina
Kofoid, C. A. and T. Skogsberg, 1928
XXXV. The Dinoflagellata: The Dinophysiodae. Reports on the scientific results of the expedition to the Eastern Tropical Pacific, in charge of Alexander Agassiz, by the U. S. Fish Commission Steamer "Albatross" from October 1904 to March 1905----. Mem. M. C. Z. 51:766 pp., 31 pls.

Amphisolenia quinquecauda
Kofoid, C. A. and T. Skogsberg, 1928
XXXV. The Dinoflagellata: The Dinophysiodae. Reports on the scientific results of the expedition to the Eastern Tropical Pacific, in charge of Alexander Agassiz, by the U. S. Fish Commission Steamer "Albatross" from October 1904 to March 1905----. Mem. M. C. Z. 51:766 pp., 31 pls.

Amphisolenia rectangulata
Abe, Tohru H., 1967
The armoured Dinoflagellata:II. Prorocentridae and Dinophusidae (C) - Ornithocercus, Histioneis, Amphisolenia and others.
Publs Seto mar.biol.Lab., 15(2):79-116.

Amphisolenia rectangulata
Käsler, R., 1938
Die Verbreitung der Dinophysiales im Sudatlantischen Ozean. Wiss. Ergeb. Deutschen Atlantischen Expedition----"Meteor" 1925-1927, 12(2):162-237, text figs. 85-118.

Amphisolenia rectangulata
Kofoid, C. A. and T. Skogsberg, 1928
XXXV. The Dinoflagellata: The Dinophysiodae. Reports on the scientific results of the expedition to the Eastern Tropical Pacific, in charge of Alexander Agassiz, by the U. S. Fish Commission Steamer "Albatross" from October 1904 to March 1905----. Mem. M. C. Z. 51:766 pp., 31 pls.

Amphisolenia rectangulata
Wood, E.J.F., 1963
Dinoflagellates in the Australian region. II. Recent collections.
Comm. Sci. and Industr. Res. Org., Div. Fish. and Oceanogr., Techn. Paper, No. 14: 55 pp.

Amphisolenia Schauinslandi
Käsler, R., 1938
Die Verbreitung der Dinophysiales im Sud-atlantischen Ozean. Wiss. Ergeb. Deutschen Atlantischen Expedition----"Meteor" 1925-1927, 12(2):162-237, text figs. 85-118.

Amphisolenia schauinslandi
Kofoid, C. A. and T. Skogsberg, 1928
XXV. The Dinoflagellata: The Dinophysiodae. Reports on the scientific results of the expedition to the Eastern Tropical Pacific, in charge of Alexander Agassiz, by the U. S. Fish Commission Steamer "Albatross" from October 1904 to March 1905----. Mem. M. C. Z. 51:766 pp., 31 pls.

Amphisolenia Schauinslandi
Matzenauer, L., 1933
Die Dinoflagellaten des indischen Ozeans (mit Ausnahme der Gattung Ceratium.) Bot. Arch. 35:437-510, 77 text figs., 2 charts.

Amphisolenia schauinslandi
Wood, E.J.F., 1963
Dinoflagellates in the Australian region. II. Recent collections.
Comm. Sci. and Industr. Res. Org., Div. Fish. and Oceanogr., Techn. Paper, No. 14: 55 pp.

Amphisolenia schröderi
Balech, Enrique, 1962
Tintinnoinea y Dinoflagellata del Pacifico segun material de las Expediciones NORPAC y DOWNWIND del Instituto Scripps de Oceanografia.
Revista. Mus. Argentino Ciencias Nat. "Bernardino Rivadavia", Ciencias Zool., 7(1):1-253.

Amphisolenia Schröderi
Käsler, R., 1938
Die Verbreitung der Dinophysiales im Sud-atlantischen Ozean. Wiss. Ergeb. Deutschen Atlantischen Expedition----"Meteor" 1925-1927, 12(2):162-237, text figs. 85-118.

Amphisolenia schröderi
Kofoid, C. A. and T. Skogsberg, 1928
XXV. The Dinoflagellata: The Dinophysiodae. Reports on the scientific results of the expedition to the Eastern Tropical Pacific, in charge of Alexander Agassiz, by the U. S. Fish Commission Steamer "Albatross" from October 1904 to March 1905----. Mem. M. C. Z. 51:766 pp., 31 pls.

Amphisolenia Schröderi
Rampi, L., 1945
Osservazioni sulla distribuzione qualitativa del fitoplancton nel mare Mediterraneo. Atti della Soc. Ital. di Sci. Nat. 84:105-113.

Amphisolenia Schröderi
Rampi, L., 1942
II Fitoplancton mediterraneo: Problemi ed affinita interoceaniche. Boll. di Pesca di Piscicoltura e di Idrobiologia, Anno 18, Fasc. 4:7-19.

Amphisolenia schroederi
Wood, E.J.F., 1963
Dinoflagellates in the Australian region. II. Recent collections.
Comm. Sci. and Industr. Res. Org., Div. Fish. and Oceanogr., Techn. Paper, No. 14: 55 pp.

Amphisolenia sigma nsp.
Halim, Y., 1965.
Microplancton des eaux Egyptiennes. II. Chrysomonedines; Ebriediens et dinoflagellés nouveaux on d'intérêt biogeographique. Rapp. Proc. Verb., Reunions. Comm. Int. Expl. Sci., Mer Mediterranée, Monaco, 18(2):373-379.

Amphisolenia spinulosa
Jörgensen, E., 1923
Mediterranean Dinophysiaceae. Rept. Danish Oceanogr. Expeds. 1908-10, to the Mediterranean and adjacent seas, Vol.II, Biol. J 2, 48 pp., 64 text figs.

Amphisolenia spinulosa n. sp.
Kofoid, C. A., 1907
Dinoflagellata of the San Diego region. III. Descriptions of new species. Univ. Calif. Publ., Zool. 3:299-340, Pls. 22-33.

Amphisolenia spinulosa
Rampi, L., 1945
Osservazioni sulla distribuzione qualitativa del fitoplancton nel mare Mediterraneo. Atti della Soc. Ital. di Sci. Nat. 84:105-113.

Amphisolenia spinulosa
Rampi, L., 1942
II Fitoplancton mediterraneo: Problemi ed affinita interoceaniche. Boll. di Pesca di Piscicoltura e di Idrobiologia, Anno 18, Fasc. 4:7-19.

Amphisolenia testa n.sp.
Balech, Enrique, 1962
Tintinnoinea y Dinoflagellata del Pacifico segun material de las Expediciones NORPAC y DOWNWIND del Instituto Scripps de Oceanografia.
Revista. Mus. Argentino Ciencias Nat. "Bernardino Rivadavia", Ciencias Zool., 7(1):1-253.

Amphisolenia thrinax
Abe, Tohru H., 1967.
The armoured Dinoflagellata: II. Prorocentridae and Dinophysidae (C) - Ornithocercus, Histioneis, Amphisolenia and others.
Publs Seto mar. biol. Lab., 15(2):79-116.

Amphisolenia thrinax
Balech, Enrique, 1962
Tintinnoinea y Dinoflagellata del Pacifico segun material de las Expediciones NORPAC y DOWNWIND del Instituto Scripps de Oceanografia.
Revista. Mus. Argentino Ciencias Nat. "Bernardino Rivadavia", Ciencias Zool., 7(1):1-253.

Amphisolenia thrinax
Käsler, R., 1938
Die Verbreitung der Dinophysiales im Sud-atlantischen Ozean. Wiss. Ergeb. Deutschen Atlantischen Expedition----"Meteor" 1925-1927, 12(2):162-237, text figs. 85-118.

Amphisolenia thrinax
Kofoid, C. A. and T. Skogsberg, 1928
XXV. The Dinoflagellata: The Dinophysiodae. Reports on the scientific results of the expedition to the Eastern Tropical Pacific, in charge of Alexander Agassiz, by the U. S. Fish Commission Steamer "Albatross" from October 1904 to March 1905----. Mem. M. C. Z. 51:766 pp., 31 pls.

Amphisolenia thrinax
Matzenauer, L., 1933
Die Dinoflagellaten des indischen Ozeans (mit Ausnahme der Gattung Ceratium.) Bot. Arch. 35:437-510, 77 text figs., 2 charts.

Amphisolema thrinax
Murray, G., and F. G. Whitting, 1899
New Peridiniaceae from the Atlantic.
Trans. Linn. Soc., London, Bot., ser 2, 5: 321-342, Pls. 27-33, 9 tables.

Amphisolenia thrinax
Wood, E.J.F., 1954.
Dinoflagellates in the Australian region.
Australian J. Mar. Freshwater Res., 5(2):171-351.

Amphisolenia truncata
Halim, Youssef, 1960
Étude quantitative et qualitative du cycle écologique des dinoflagellés dans les eaux de Villefranche-sur-Mer. (1953-1955). Ann. Inst. Océanogr., Monaco, 38:123-232.

Amphisolenia truncata
Jörgensen, E., 1923
Mediterranean Dinophysiaceae. Rept. Danish Oceanogr. Expeds. 1908-10, to the Mediterranean and adjacent seas, Vol.II, Biol. J 2, 48 pp., 64 text figs.

Amphisolenia truncata
Käsler, R., 1938
Die Verbreitung der Dinophysiales im Sud-atlantischen Ozean. Wiss. Ergeb. Deutschen Atlantischen Expedition----"Meteor" 1925-1927, 12(2):162-237, text figs. 85-118.

Amphisolenia truncata
Kofoid, C. A. and T. Skogsberg, 1928
XXV. The Dinoflagellata: The Dinophysiodae. Reports on the scientific results of the expedition to the Eastern Tropical Pacific, in charge of Alexander Agassiz, by the U. S. Fish Commission Steamer "Albatross" from October 1904 to March 1905----. Mem. M. C. Z. 51:766 pp., 31 pls.

Amphisolenia truncata
Rampi, L., 1942
II Fitoplancton mediterraneo: Problemi ed affinita interoceaniche. Boll. di Pesca di Piscicoltura e di Idrobiologia, Anno 18, Fasc. 4:7-19.

Amphorellopsis sp.
Balech, Enrique, and Sayed Z. El-Sayed, 1965.
The microplankton of the Weddell Sea.
In: Biology of Antarctic seas. II.
Antarctic Res. Ser., Amer. Geophys. Union, 5:107-124.

Amphorides amphora
Balech, Enrique, 1971
Microplancton del Atlantico ecuatorial oeste (Equalant 1)
Publ. Serv. Hidrograf. Naval, Argentina H. 654: 103 pp., 122 figs.

Amylax diacantha
Meunier, A., 1919
Microplancton de la Mer Flamande 3. Les Péridiniens. Mem. Mus. Roy. Hist. Nat., Belgique 8(1):1-116, Pls. XV-XXI.

Angulochrysis erratica n.gen., n.sp.
Lackey, J.B., 1940.
Some new flagellates from the Woods Hole area.
Amer. Midl. Nat., 23(2):463-471.

Anisonema orbiculatum n.sp.
Lackey, J.B., 1940.
Some new flagellates from the Woods Hole area.
Amer. Midl. Nat., 23(2):463-471.

Archaeosphaerodiniopsis verrucosum, n. gen., n. sp.
Rampi, L., 1943.
Su qualche altra Peridinea nuova o rara delle acque di Sanremo. Atti della Soc. Ital. Sci. Nat. 82:151-157, 9 textfigs.

Arnithocircus heteroporus
Käsler, R., 1938
Die Verbreitung der Dinophysiales im Sud-atlantischen Ozean. Wiss. Ergeb. Deutschen Atlantischen Expedition----"Meteor" 1925-1927, 12(2):162-237, text figs. 85-118.

Arnithocircus magnificus
Käsler, R., 1938
Die Verbreitung der Dinophysiales im Sud-atlantischen Ozean. Wiss. Ergeb. Deutschen Atlantischen Expedition----"Meteor" 1925-1927, 12(2):162-237, text figs. 85-118.

Arnithocircus splendidus
Käsler, R., 1938
Die Verbreitung der Dinophysiales im Sud-atlantischen Ozean. Wiss. Ergeb. Deutschen Atlantischen Expedition----"Meteor" 1925-1927, 12(2):162-237, text figs. 85-118.

Arnithocircus thurnii
Käsler, R., 1938
Die Verbreitung der Dinophysiales im Sudatlantischen Ozean. Wiss. Ergeb. Deutschen Atlantischen Expedition----"Meteor" 1925-1927, 12(2):162-237, text figs. 85-118.

Ascampbelliella armilla
Balech, Enrique, 1971
Microplancton del Atlantico ecuatorial oeste (Equalant 1)
Publ. Serv. Hidrograf. Naval, Argentina H. 654: 103 pp., 122 figs.

Ascampselliella urceolata
Balech, Enrique, 1971
Microplancton del Atlantico ecuatorial oeste (Equalant 1)
Publ. Serv. Hidrograf. Naval, Argentina H. 654: 103 pp., 122 figs.

Asterodinium gracile n. gen. n. sp.
Sournia, Alain 1972.
Quatre nouveaux dinoflagellés du plancton marin.
Phycologia 11(1):71-74.

Asterodinium spinosum n.sp.
Sournia, Alain, 1972
Une période de poussées phytoplanctoniques près de Nosy-Bé (Madagascar) en 1971. 1. Espèces rares ou nouvelles du phytoplancton
Cah. ORSTOM sér. Océanogr. 10(2): 151-159.

Aureodinium pigmentosum
Dodge, J.D., 1968.
The fine structure of chloroplasts and pyrenoids in some marine dinoflagellates.
J. Cell. Sci., 3(1):41-48.

Bernardinium salinum
Javornicky, Pavel, 1962.
Two scarcely known genera of the class Dinophyceae: Bernardinium Chodat and Crypthecodinium Biecheler.
Preslia (Czechoslovakia), 34:98-113.

Bernardinium thiophilum
Javornicky, Pavel, 1962.
Two scarcely known genera of the class Dinophyceae: Bernardinium Chodat and Crypthecodinium Biecheler.
Preslia (Czechoslovakia), 34:98-113.

Blepharocysta denticulada
Balech, Enrique, 1963.
La familia Podolampacea (Dinoflagellata).
Bol. Inst. Biol. Mar., Mar del Plata, Argentina, No. 2: 30 pp.

Blephalocysta okamurai n. sp.
Abé, Tohru H., 1966.
The armoured Dinoflagellata I. Podolampidae.
Publs. Seto mar. biol. Lab., 14(2):129-154.

Blephalocysta Paulseni
Rampi, L., 1941
Ricerche sul fitoplancton del Mare Ligure. 5. Le Podolampacer delle acque di Sanremo. Annali del Mus. Civico di Storia Naturale di Genova, 51:141-152, Pl. 5.

Blepharocysta splendiformis
Marshall, S. M., 1933
The production of microplankton in the Great Barrier Reef Region. Brit. Mus. (N.H.) Great Barrier Reef Exped. 1928-29, Sci. Repts. II(5):111-157, 14 text figs.

Blephalocysta splendor-maris
Abé, Tohru H., 1966.
The armoured Dinoflagellata I. Podolampidae.
Publs. Seto mar. biol. Lab., 14(2):129-154.

Blepharocysta splendor maris
Balech, Enrique, 1971.
Microplancton de la campaña oceanographica: Productividad III.
Revta Mus. argent. Cienc. Nat. Bernadina Rivadavia, Hydrobiol. 3(1):1-202, 39 pls.

Blepharocysta splendor-maris
Balech, Enrique, 1963.
La familia Podolampacea (Dinoflagellata).
Bol. Inst. Biol. Mar., Mar del Plata, Argentina, No. 2:30 pp.

Blepharocysta splendor-maris
Dangeard, P., 1927
Phytoplankton de la croisière du "Sylvana". Ann. Inst. Ocean., Monaco, n.s., 4(8):286-401, 54 text figs. (février-Juin 1913).

Blepharocysta splendor-maris
LeBour, M.V., 1925
The dinoflagellates of Northern Seas. The Marine Biological Association of the United Kingdom. Plymouth, 250 pp., 35 pls. 53 text figs.

Blepharocysta splendormais
Lindemann, E., 1925
Neubeobachtungen an den Winter peridineen des Golfes von Neapel. Bot. Arch. 9:95-102, 19 text figs.

Blepharocysta Splendor-maris
Mangin, M. L., 1912
Phytoplancton de la croisière du "René" dans l'Atlantique (Septembre 1908). Ann. Inst. Ocean., n.s., 4(1):1-66, 2 pls., 41 text figs., 2 tables.

Blepharocysta splendormaris
Matzenauer, L., 1933
Die Dinoflagellaten des indischen Ozeans (mit Ausnahme der Gattung Ceratium.) Bot. Arch. 35:437-510, 77 text figs., 2 charts.

Blepharocysta splendor maris
Paulsen, O., 1908
XVIII Peridiniales. Nordisches Plankton, Bot. Teil: 1-124, 155 text figs.

Blepharocysta splendor maris
Rampi, L., 1948
Sur quelques Peridiniens rares ou interessants du Pacifique subtropical (Recoltes Alain Gerbault). Bull. l'Inst. Ocean., Monaco, No.937: 7 pp., 8 text figs.

Blepharocysta splendor maris
Rampi, L., 1941
Ricerche sul fitoplancton del Mare Ligure. 5. Le Podolampacer delle acque di Sanremo. Annali del Mus. Civico di Storia Naturale di Genova, 51:141-152, Pl. 5.

Blepharocysta splendor maris
Schröder, B., 1900
Phytoplankton des Golfes von Neapel nebst vergleichenden Ausblicken auf das atlantischen Ozean. Mitt. Zool. Stat. Neapel, 14:1-38.

Blepharocysta splendomaris
Wood, E.J.F., 1963
Dinoflagellates in the Australian region. II. Recent collections.
Comm. Sci. and Industr. Res. Org., Div. Fish. and Oceanogr., Techn. Paper, No. 14: 55 pp.

Blepharocysta striata
Rampi, L., 1941
Ricerche sul fitoplancton del Mare Ligure. 5. Le Podolampacer delle acque di Sanremo. Annali del Mus. Civico di Storia Naturale di Genova, 51:141-152, Pl. 5.

Brachydinium brevipes n.sp. [a]
Sournia, Alain, 1972
Une période de poussées phytoplanctoniques près de Nosy-Bé (Madagascar) en 1971. 1. Espèces rares ou nouvelles du phytoplancton
Cah. ORSTOM sér. Océanogr. 10(2): 151-159.

Brachydinium capitatum [a]
Sournia, Alain, 1972
Une période de poussées phytoplanctoniques près de Nosy-Bé (Madagascar) en 1971. 1. Espèces rares ou nouvelles du phytoplancton
Cah. ORSTOM sér. Océanogr. 10(2): 151-159.

Brachydinium taylorii n.sp. [a]
Sournia, Alain, 1972
Une période de poussées phytoplanctoniques près de Nosy-Bé (Madagascar) en 1971. 1. Espèces rares ou nouvelles du phytoplancton
Cah. ORSTOM sér. Océanogr. 10(2): 151-159.

Brandtiella palliata
Balech, Enrique, 1971
Microplancton del Atlantico ecuatorial oeste (Equalant 1)
Publ. Serv. Hidrograf. Naval, Argentina H. 654: 103 pp., 122 figs.

Cachonina niei n.gen., n.sp.
Loeblich, Alfred R. III, 1968.
A new marine dinoflagellate genus, Cachonina, in axenic culture, from the Salton Sea, California, with remarks on the genus, Peridinium.
Proc. Biol. Soc. Wash. 81:91-96.

Cachonina niei
von Stosch, H.A., 1969.
Dinoflagellaten aus der Nordsee I. Über Cachonina niei Loeblich (1968), Gonyaulax grindleyi Reinecke (1967) und eine Methode zur Darstellung von Peridineenpanzern.
Helgoländer wiss. Meeresunters. 1969:558-565.

Cenchridium globosum
Bursa, Adam S., 1961
The annual oceanographic cycle at Igloolik in the Canadian Arctic. II. The phytoplankton.
J. Fish. Res. Bd., Canada, 18(4):563-615.

Cenchridinium globosum
Pavillard, J., 1905
Recherches sur la flore pelagique (Phytoplankton) de l'Etang de Thau. Theses presentees a la Fac. Sci., Paris, 116 pp., 3 pls.

Cenchridium spherula
Bursa, Adam S., 1961
The annual oceanographic cycle at Igloolik in the Canadian Arctic. II. The phytoplankton.
J. Fish. Res. Bd., Canada, 18(4):563-615.

Centrodinium sp.
Balech, Enrique, 1962
Tintinnoinea y Dinoflagellata del Pacifico segun material de las Expediciones NORPAC y DOWNWIND del Instituto Scripps de Oceanografia.
Revista. Mus. Argentino Ciencias Nat. "Bernardino Rivadavia", Ciencias Zool., 7(1):1-253.

Centrodinium complanatum

Pavillard, J., 1916
Recherches sur les Peridiniens du Golfe du Lion. Mem. Univ. Montpellier. Trav. Inst. Bot., Univ. Montpellier. Serie mixte No. 4, 70 pp., 3 pls., 15 text figs.

Centrodinium complanatum

Rampi, L., 1951.
Ricerche sul fitoplancton del Mare Ligure. 10) Peridiniale delle acque di Sanremo. Pubbl. Centro Talassografico Tirreno No. 6: [Atti Acad. Ligure Sci. Lett.] 7(1): 8 pp., Pls. 3-4.

Centrodinium complanatum

Vives, F., and A. Planas, 1952.
Plancton recogido por los laboratorios costeros. VI. Fitoplancton de las costas de Vinaroz, islas Columbretes, y alrededores de la desembocadura del Elroa. Publ. Inst. Biol. Aplic. 11:141-156, 19 textfigs.

Centrodinium complanatum

Wood, E.J.F., 1963
Dinoflagellates in the Australian region. II. Recent collections. Comm. Sci. and Industr. Res. Org., Div. Fish. and Oceanogr., Techn. Paper, No. 14: 55 pp.

Centrodinium deflexoides n.sp.

Balech, Enrique, 1962
Tintinnoinea y Dinoflagellata del Pacifico segun material de las Expediciones NORPAC y DOWNWIND del Instituto Scripps de Oceanografia. Revista. Mus. Argentino Ciencias Nat. "Bernardino Rivadavia", Ciencias Zool., 7(1):1-253.

Centrodinium eminens pulchrum

Balech, Enrique, 1962
Tintinnoinea y Dinoflagellata del Pacifico segun material de las Expediciones NORPAC y DOWNWIND del Instituto Scripps de Oceanografia. Revista. Mus. Argentino Ciencias Nat. "Bernardino Rivadavia", Ciencias Zool., 7(1):1-253.

Centrodinium eminens pulchrum

Rampi, L., 1951.
Ricerche sul fitoplancton del Mare Ligure. 10) Peridiniale dell acque di Sanremo. Pubbl. Centro Talassografico Tirreno No. 6: [Atti Acad. Ligure Sci. Lett.] 7(1): 8 pp., Pls. 3-4.

Centrodinium intermedium

Balech, Enrique, 1962
Tintinnoinea y Dinoflagellata del Pacifico segun material de las Expediciones NORPAC y DOWNWIND del Instituto Scripps de Oceanografia. Revista. Mus. Argentino Ciencias Nat. "Bernardino Rivadavia", Ciencias Zool., 7(1):1-253.

Centrodinium intermedium

Rampi, L., 1951.
Ricerche sul fitoplancton del Mare Ligure. 10) Peridiniale delle acque di Sanremo. Pubbl. Centro Talassografico Tirreno No. 6: [Atti Acad. Ligure Sci. Lett.] 7(1): 8 pp., Pls. 3-4.

Centrodinium maximum

Rampi, L., 1951.
Ricerche sul fitoplancton del Mare Ligure. 10) Peridiniale delle acque di Sanremo. Pubbl. Contro Talassografico Tirreno No. 6: [Atti Acad. Ligure Sci. Lett.] 7(1): 8 pp., Pls. 3-4.

Ceratium

Bandel, W., 1940.
Phytoplankton- und Nahrstoffgehalt der Ostsee im Gebiet der Darsser Schwelle. Internat. Rev. ges. Hydrobiol. u. Hydrogr., 40:249-304

Ceratium sp.

Braarud, T., and Adam Bursa, 1939
On the phytoplankton of the Oslo Fjord, 1933-1934. Hvalrådets Skr. No.19:1-63; 9 text figs. Reviewed. J. du. Cons. 14(3): 418-420. A.C. Gardiner.

Ceratium spp.

Braarud, T., K.R. Gaarder and J. Grøntved, 1953.
The phytoplankton of the North Sea and adjacent waters in May 1948. Rapp. Proc. Verb., Cons. Perm. Int. Expl. Mer, 133:1-87, 29 tables, Pls. A-B, 18 textfigs.

Ceratium

Frost, N., 1938
The genus Ceratium and its use as an indicator of hydrographic conditions in Newfoundland waters. Res. Bull. No. 5. Dept. Nat Res., St. John's Newfoundland.

Ceratium

Graham, H. W. and N. Bronikovsky, 1944
The genus Ceratium in the Pacific and North Atlantic Oceans. Sci. Res. Cruise VII of the Carnegie, 1928-1929 ----- Biol. V (565):209 pp., 54 charts, 27 figs., 54 tables.

Ceratium sp.

Johnson, M. W., 1949.
Relation of plankton to hydrographic conditions in Sweetwater Lake. J. Am. Water Works Assoc. 41(4):347-356, 12 textfigs.

Ceratium

Jörgensen, E. 1911
Die Ceratien. Klinkhardt. Leipzig

Ceratium

Jörgensen, E., 1910
Peridiniales, Ceratium. Bull. Trim. Cons. Int. Explor. Mer, 205-250.

CERATIA

Jörgensen, E. 1920
Mediterranean Ceratia. Report on the Danish Oceanographical Expeditions to the Mediterranean and adjacent seas, 1908-1910, II (Biology).

Ceratium

Kofoid, C.A., 1907.
The plates of Ceratium with a note on the unity of the genus. Zool. Anz. 32:177-183, 8 figs.

Ceratium, lists of spp.

López, J., 1966.
Variación y regulación de la forma en el género Ceratium. Inv. Pesq., Barcelona, 30:325-427.

Ceratia

Nielsen, J., 1956.
Temporary variations in certain marine Ceratia. Oikos, 7(2):256-272.

Ceratia

Nordli, E., 1957.
Experimental studies on the ecology of Ceratia. Oikos 8(2):200-265.

Ceratium

Okitsu, T., 1954.
On the seasonal change of Ceratium in Aomori Bay. Bull. Mar. Biol. Sta., Asamushi, 7(1):17-20, 2 textfigs.

lists of species.

Ceratium

Peters, Nicolaus, 1934
Die Bevolkerung des Sudatlantischen Ozeans mit Ceratien. Biol. Sonderuntersuchungen 1. Wiss Ergeb. Deutschen Atlantischen Exped.--- "Meteor" 1925-1927, 12(1): 1-69, 28 text figs.

Ceratium, lists of spp.

Sournia, A., 1967.
Le genre Ceratium (péridinien planctonique) dans le Canal de Mozambique. Contribution à une révision mondiale. Vie Milieu (A) 18(2): 375-440

Ceratium (lists of spp.)

Sournia, Alain, 1966.
Sur la variabilité infraspécifique du genre Ceratium (Péridinien planctonique) en milieu marin. C.r.hebd.Séanc., Acad.Sci., Paris, (D)263(25):1980-1983.

Ceratium sp.

Sousa e Silva, E., 1949
Diatomaceas e Dinoflagelados de Baia de Cascais. Portugaliae Acta Biol., Volume: Julio Henriques, Ser. B: 300-383, 9 pls., 2 fold-in tables.

Ceratia

Steemann Nielsen, E., 1934
Untersuchungen über die Verbreitung, Biologie und Variation der Ceratien im südlichen Stillen Ozean. Dana Rept. No. 4: 67 pp., 73 figs. and 11 charts in text.

Ceratium

Rampi, L., 1942.
Ricerche sul fitoplancton del Mare Ligure. 4. I Ceratium delle acque di Sanremo. Nuovo Giornale Botanico Italiano, N.S., 49: 221-236, 19 text figs.

Ceratia

Steemann Nielsen, E., 1934
Untersuchungen über die Verbreitung, Biologie und Variation der Ceratien im südlichen Stillen Ozean. Dana Rept. No.4:70 pp.

Ceratium

Toriumi, Saburo, 1968.
Cultivation of marine Ceratia using natural sea water culture media. I. Bull. Plankton Soc., Japan, 15(1): 1-6. (In Japanese; English abstract)

Ceratium, distribution of

Williams, D.M. 1971.
The distribution of marine dinoflagellates in relation to physical and chemical conditions. In: Micropalaeontology of oceans, B.M. Funnell and W.R. Riedel, editors, Cambridge Univ. Press, 91-95.

Ceratium Allieri

Gourret, P., 1883
Sur les Peridiniens du Golfe de Marseille. Ann. du Musee d'hist. Nat., Marseille, Zool., 1 (Mme. 8):1-114, 4 pls.

Ceratium sp. ecology

Nordli, E., 1957.
Experimental studies on the ecology of Ceratia. Oikos, 8(2):201-265.

Ceratium arcticum

Bigelow, H. B., 1922
Exploration of the coastal water off the northeastern United States in 1916 by the U.S. Fisheries Schooner Grampus. Bull. M.C.Z. 65 (5):85-188, 53 text figs.

Ceratium arcticum

Brunel, J., 1962
Le phytoplancton de la Baie de Chaleurs. Inst. Botan., Univ. Montréal, Contrib. No. 77: 365 pp., 66 pls.

Ceratium arcticum ventricosum

Ceratium arcticum

Graham, H. W. and N. Bronikovsky, 1944
The genus Ceratium in the Pacific and North Atlantic Oceans. Sci. Res. Cruise VII of the Carnegie, 1928-1929 ----- Biol. V (565):209 pp., 54 charts, 27 figs., 54 tables.

Ceratium arcticum

Gran, H.H., and T. Braarud, 1935
A quantitative study of the phytoplankton in the Bay of Fundy and the Gulf of Maine (including observations on hydrography, chemistry, and turbidity). J. Biol. Bd., Canada, 1(5):279-467, 69 text figs.

Ceratium arcticum
Iselin, C., 1930
A report on the coastal waters of Labrador based on explorations of the "Chance" during the summer of 1926. Proc. Am. Acad. Arts Sci., 66(1):1-37, 14 text figs.

Ceratium arcticum
Jørgensen, E., 1905
B. Protistplankton and the diatoms in bottom samples. Hydrographical and biological investigations in Norwegian fjords. Bergens Mus. Skr. 7: 49-225.

Ceratium arcticum
LeBour, M.V., 1925
The dinoflagellates of Northern Seas. The Marine Biological Association of the United Kingdom. Plymouth, 250 pp., 35 pls. 53 text figs.

Ceratium arcticum
Lillick, L.C., 1940
Phytoplankton and planktonic protozoa of the offshore waters of the Gulf of Maine. Pt.II. Qualitative Composition of the Planktonic Flora. Trans. Am. Phil. Soc., n.s., 31(3):193-237, 13 text figs.

Ceratium arcticum
Marukawa, H., 1921
Plankton lists and some new species of copepods, from the northern waters of Japan. Bull. Inst. Ocean., No.384, 15 pp., 3 pls., 1 chart. Monaco

Ceratium arcticum
Mulford, Richard A., 1963
Distribution of the dinoflagellate genus Ceratium in the tidal and offshore waters of Virginia. Chesapeake Science, 4(2):84-89.

Ceratium arcticum
Paulsen, O., 1908
XVIII Peridiniales. Nordisches Plankton, Bot. Teil: 1-124, 155 text figs.

Ceratium arcticum
Robinson, G.A., 1961
Contribution towards an atlas of the north-eastern Atlantic and the North Sea. 1. Phytoplankton. Bulls. Mar. Ecology, 5(42): 81-89, Pls. 15-20.

Ceratium arcticum
Schulz, B., and A. Wulff, 1929
Hydrographie und Oberflächen plankton des westlichen Barentsmeeres im Sommer 1927. Ber. deutschen wissensch. Komm. F. Meeresforsch. n.s. 4(5):232-372, 13 tables, 25 text figs.

Ceratium arcticum
Yarranton, G.A. 1967.
Parameters for use in distinguishing populations of Euceratium Gran. Bull. mar. Ecol. 6(6): 147-158

Ceratium arcticum longipes
Graham, H. W. and N. Bronikovsky, 1944
The genus Ceratium in the Pacific and North Atlantic Oceans. Sci. Res. Cruise VII of the Carnegie, 1928-1929 ----- Biol. V (565):209 pp., 54 charts, 27 figs., 54 tables.

Ceratium arcticum ventricosum
Graham, H. W. and N. Bronikovsky, 1944
The genus Ceratium in the Pacific and North Atlantic Oceans. Sci. Res. Cruise VII of the Carnegie, 1928-1929 ----- Biol. V (565):209 pp., 54 charts, 27 figs., 54 tables.

Ceratium arcuatum
Brunel, Jules, 1962
Le phytoplancton de la Baie des Chaleurs. Contrib. Ministère de la Chasse et des Pêcheries, Province de Québec, No. 91: 365 pp.

Ceratium arcuatum
Ercegovic, A., 1936
Etudes qualitative et quantitatives du phytoplancton dans les eaux cotières de l'Adriatique oriental moyen au cours de l'année 1934. Acta Adriatica 1(9):1-126

Ceratium arcuatum
Forti, A., 1922
Ricerche sulla flora pelagica (fitoplancton) di Quarto dei Mille. Mem. R. Com. Talass. Ital. 97:248 pp., 13 pls.

Ceratium arcuatum
Lindemann, E., 1925
Neubeobachtungen an den Winter peridineen des Golfes von Neapel. Bot. Arch. 9:95-102, 19 text figs.

Ceratium arcuatum
Pavillard, J., 1923
A propos de la systématique des Péridiniens Bull. Soc. Bot. de France 70:876-882; 914-918.

Ceratium arcuatum
Pavillard, J., 1916
Recherches sur les Peridiniens du Golfe du Lion. Mem. Univ. Montpellier. Trav. Inst. Bot., Univ. Montpellier. Serie mixte No.4, 70 pp., 3 pls., 15 text figs.

Ceratium arcuatum
Pavillard, J., 1905
Recherches sur la flore pelagique (Phytoplankton) de l'Etang de Thau. Theses presentees a la Fac. Sci., Paris, 116 pp., 3 pls.

Ceratium arcuatum
Peters, Nicolaus, 1934
Die Bevolkerung des Sudatlantischen Ozeans mit Ceratien. Biol. Sonderuntersuchungen 1. Wiss Ergeb. Deutschen Atlantischen Exped.--- "Meteor" 1925-1927, 12(1): 1-69, 28 text figs.

Ceratium arcuatum
Sousa e Silva, E., 1949
Diatomaceas e Dinoflagelados de Baia de Cascais. Portugaliae Acta Biol., Volume: Julio Henriques, Ser. B: 300-383, 9 pls, 2 fold-in tables.

Ceratium arietinum
Dangeard, P., 1927
Phytoplankton de la croisière du "Sylvana". Ann. Inst. Ocean., Monaco, n.s., 4(8):286-401, 54 text figs. (février-Juin 1913).

Ceratium arietinum
Ercegovic, A., 1936
Etudes qualitative et quantitatives du phytoplancton dans les eaux cotières de l'Adriatique oriental moyen au cours de l'année 1934. Acta Adriatica 1(9):1-126

Ceratium arietinum
Forti, A., 1922
Ricerche sulla flora pelagica (fitoplancton) di Quarto dei Mille. Mem. R. Com. Talass. Ital. 97:248 pp., 13 pls.

Ceratium arietinum
Graham, H. W. and N. Bronikovsky, 1944
The genus Ceratium in the Pacific and North Atlantic Oceans. Sci. Res. Cruise VII of the Carnegie, 1928-1929 ----- Biol. V (565):209 pp., 54 charts, 27 figs., 54 tables.

Ceratium arietinum
Jørgensen, E., 1920
Mediterranean Ceratia. Rept. Danish Oceanogr. Exped. 1908-10 to the Mediterranean and adjacent seas, Vol.II, Biol. J 1:110 pp., 94 text figs., 26 charts.

Ceratium arietinum
Lindemann, E., 1925
Neubeobachtungen an den Winter peridineen des Golfes von Neapel. Bot. Arch. 9:95-102, 19 text figs.

Ceratium arietinum
Massutí Algamora, M., 1949
Estudio de diez y seis muestras de plancton del Golfo de Nápoles. Publ. Inst. Biol. Appl. 5:85-94, 1 fold-in table.

Ceratium arietinum
Pavillard, J., 1916
Recherches sur les Peridiniens du Golfe du Lion. Mem. Univ. Montpellier. Trav. Inst. Bot., Univ. Montpellier. Serie mixte No.4, 70 pp., 3 pls., 15 text figs.

Ceratium arietinum
Peters, Nicolaus, 1934
Die Bevolkerung des Sudatlantischen Ozeans mit Ceratien. Biol. Sonderuntersuchungen 1. Wiss Ergeb. Deutschen Atlantischen Exped.--- "Meteor" 1925-1927, 12(1): 1-69, 28 text figs.

Ceratium arietinum
Sousa e Silva, E., 1949
Diatomaceas e Dinoflagelados de Baia de Cascais. Portugaliae Acta Biol., Volume: Julio Henriques, Ser. B: 300-383, 9 pls, 2 fold-in tables.

Ceratium arietinum
Steemann-Nielsen, E., 1939
Die Ceratien des Indischen Ozeans und der Ostasiatischen Gewässer mit einer allgemeinen zusammenfassung über die Verbreitung der Ceratien in den Weltmeeren. Dana Rept. No. 17: 33 pp., 8 maps in text.

Ceratium arietinum
Steemann Nielsen, E., 1934
Untersuchungen über die Verbreitung, Biologie und Variation der Ceratien im südlichen Stillen Ozean. Dana Rept. No. 4: 67 pp., 73 figs. and 11 charts in text.

Ceratium arietinum arietinum
Graham, H. W. and N. Bronikovsky, 1944
The genus Ceratium in the Pacific and North Atlantic Oceans. Sci. Res. Cruise VII of the Carnegie, 1928-1929 ----- Biol. V (565):209 pp., 54 charts, 27 figs., 54 tables.

Ceratium arietinum bucephalum
Graham, H. W. and N. Bronikovsky, 1944
The genus Ceratium in the Pacific and North Atlantic Oceans. Sci. Res. Cruise VII of the Carnegie, 1928-1929 ----- Biol. V (565):209 pp., 54 charts, 27 figs., 54 tables.

Ceratium arietinum var. detortum
Ghazzawi, F.M., 1939
Plankton of the Egyptian waters. A study of the Suez Canal Plankton. (A) Phytoplankton. Preliminary Report 83 pp. Notes and Memoires, Min. Commerce-Industry, Egypt, Hydrobiol. & Fish. 65 figs.

Ceratium arietinum var. gracilentum
Ghazzawi, F.M., 1939
Plankton of the Egyptian waters. A study of the Suez Canal Plankton. (A) Phytoplankton. Preliminary Report 83 pp. Notes and Memoires, Min. Commerce-Industry, Egypt, Hydrobiol. & Fish. 65 figs.

Ceratium arietinum gracilentum
Graham, H. W. and N. Bronikovsky, 1944
The genus Ceratium in the Pacific and North Atlantic Oceans. Sci. Res. Cruise VII of the Carnegie, 1928-1929 ----- Biol. V (565):209 pp., 54 charts, 27 figs., 54 tables.

Ceratium aultii n.sp.
Graham, H. W. and N. Bronikovsky, 1944
The genus Ceratium in the Pacific and North Atlantic Oceans. Sci. Res. Cruise VII of the Carnegie, 1928-1929 ----- Biol. V (565):209 pp., 54 charts, 27 figs., 54 tables.

Ceratium axiale
Graham, H. W. and N. Bronikovsky, 1944
The genus Ceratium in the Pacific and North Atlantic Oceans. Sci. Res. Cruise VII of the Carnegie, 1928-1929 ----- Biol. V (565):209 pp., 54 charts, 27 figs., 54 tables.

Ceratium axiale
Peters, Nicolaus, 1934
Die Bevolkerung des Sudatlantischen Ozeans mit Ceratien. Biol. Sonderuntersuchungen 1. Wiss Ergeb. Deutschen Atlantischen Exped.---"Meteor" 1925-1927, 12(1): 1-69, 28 text figs.

Ceratium axiale
Steemann-Nielsen, E., 1939
Die Ceratien des Indischen Ozeans und der Ostasiatischen Gewässer mit einer allgemeinen zusammenfassung über die Verbreitung der Ceratien in den Weltmeeren. Dana Rept. No. 17: 33 pp., 8 maps in text.

Ceratium axiale
Steemann Nielsen, E., 1934
Untersuchungen über die Verbreitung, Biologie und Variation der Ceratien im südlichen Stillen Ozean. Dana Rept. No. 4: 67 pp., 73 figs. and 11 charts in text.

Ceratium azoricum
Dangeard, P., 1927
Phytoplankton de la croisière du "Sylvana". Ann. Inst. Ocean., Monaco, n.s., 4(8):286-401, 54 text figs. (Feirer-Juin 1913).

Ceratium azoricum
Ercegovic, A., 1936
Etudes qualitative et quantitatives du phytoplancton dans les eaux cotières de l'Adriatique oriental moyen au cours de l'année 1934. Acta Adriatica 1(9):1-126

Ceratium azoricum
Forti, A., 1922
Ricerche sulla flora pelagica (fitoplancton) di Quarto dei Mille. Mem. R. Com. Talass. Ital. 97:248 pp., 13 pls.

Ceratium azoricum
Graham, H. W. and N. Bronikovsky, 1944
The genus Ceratium in the Pacific and North Atlantic Oceans. Sci. Res. Cruise VII of the Carnegie, 1928-1929 ----- Biol. V (565):209 pp., 54 charts, 27 figs., 54 tables.

Ceratium azoricum
Jörgensen, E., 1920
Mediterranean Ceratia. Rept. Danish Oceanogr. Exped. 1908-10 to the Mediterranean and adjacent seas, Vol.II, Biol. J 1:110 pp., 94 text figs., 26 charts.

Ceratium azoricum
LeBour, M.V., 1925
The dinoflagellates of Northern Seas. The Marine Biological Association of the United Kingdom. Plymouth, 250 pp., 35 pls. 53 text figs.

Ceratium azoricum
Paulsen, O., 1908
XVIII Peridiniales. Nordisches Plankton, Bot. Teil: 1-124, 155 text figs.

Ceratium azoricum
Pavillard, J., 1916
Recherches sur les Peridiniens du Golfe du Lion. Mem. Univ. Montpellier. Trav. Inst. Bot., Univ. Montpellier. Serie mixte No.4, 70 pp., 3 pls., 15 text figs.

Ceratium azoricum
Pavillard, J., 1905
Recherches sur la flore pelagique (Phytoplankton) de l'Etang de Thau. Theses presentees a la Fac. Sci., Paris, 116 pp., 3 pls.

Ceratium azoricum
Peters, Nicolaus, 1934
Die Bevolkerung des Sudatlantischen Ozeans mit Ceratien. Biol. Sonderuntersuchungen 1. Wiss Ergeb. Deutschen Atlantischen Exped.---"Meteor" 1925-1927, 12(1): 1-69, 28 text figs.

Ceratium azoricum
Robinson, G.A., 1961
Contribution towards an atlas of the northeastern Atlantic and the North Sea. 1. Phytoplankton. Bulls. Mar. Ecology, 5(42): 81-89, Pls. 15-20.

Ceratium azoricum
Sousa e Silva, E., 1949
Diatomaceas e Dinoflagelados de Baia de Cascais. Portugaliae Acta Biol., Volume: Julio Henriques, Ser. B: 300-383, 9 pls, 2 fold-in tables.

Ceratium azoricum
Steemann-Nielsen, E., 1939
Die Ceratien des Indischen Ozeans und der Ostasiatischen Gewässer mit einer allgemeinen zusammenfassung über die Verbreitung der Ceratien in den Weltmeeren. Dana Rept. No. 17: 33 pp., 8 maps in text.

Ceratium azoricum
Steemann Nielsen, E., 1934
Untersuchungen über die Verbreitung, Biologie und Variation der Ceratien im südlichen Stillen Ozean. Dana Rept. No. 4: 67 pp., 73 figs. and 11 charts in text.

Ceratium balticum
Balech, Enrique, 1944
Contribucion al conocimiento del Plancton de Lennox y Cabo de Hornos. Physis XIX:423-446, 6 pls. with 67 figs.

Ceratium batavum
Meunier, A., 1919
Microplancton de la Mer Flamande 3. Les Péridiniens. Mem. Mus. Roy. Hist. Nat., Belgique 8(1):1-116, Pls. XV-XXI.

Ceratium batavum
Paulsen, O., 1908
XVIII Peridiniales. Nordisches Plankton, Bot. Teil: 1-124, 155 text figs.

Ceratium belone
Dangeard, P., 1927
Phytoplankton de la croisière du "Sylvana". Ann. Inst. Ocean., Monaco, n.s., 4(8):286-401, 54 text figs. (Feirer-Juin 1913).

Ceratium belone
Graham, H. W. and N. Bronikovsky, 1944
The genus Ceratium in the Pacific and North Atlantic Oceans. Sci. Res. Cruise VII of the Carnegie, 1928-1929 ----- Biol. V (565):209 pp., 54 charts, 27 figs., 54 tables.

Ceratium belone
Halim, Youssef, 1960
Etude quantitative et qualitative du cycle écologique des dinoflagellés dans les eaux de Villefranche-sur-Mer. (1953-1955). Ann. Inst. Océanogr. Monaco, 38:123-232.

Ceratium belone
Jörgensen, E., 1920
Mediterranean Ceratia. Rept. Danish Oceanogr. Exped. 1908-10 to the Mediterranean and adjacent seas, Vol.II, Biol. J 1:110 pp., 94 text figs., 26 charts.

Ceratium belone
Margalef, R., 1949
Fitoplancton nerítico de la Costa Brava en 1947-48. Publ. Inst. Biol. Aplicada, 5: 41-51, 3 text figs.

Ceratium belone
Massuti Algamora, M., 1949
Estudio de diez y seis muestras de plancton del Golfo de Nápoles. Publ. Inst. Biol. Appl. 5:85-94, 1 fold-in table.

Ceratium belone
Pavillard, J., 1916
Recherches sur les Peridiniens du Golfe du Lion. Mem. Univ. Montpellier. Trav. Inst. Bot., Univ. Montpellier. Serie mixte No.4, 70 pp., 3 pls., 15 text figs.

Ceratium belone
Peters, Nicolaus, 1934
Die Bevolkerung des Sudatlantischen Ozeans mit Ceratien. Biol. Sonderuntersuchungen 1. Wiss Ergeb. Deutschen Atlantischen Exped.---"Meteor" 1925-1927, 12(1): 1-69, 28 text figs.

Ceratium belone
Rampi, L., 1942.
Ricerche sul fitoplancton del Mare Ligure. 4. I Ceratium delle acque di Sanremo. Nuovo Giornale Botanico Italiano, N.S., 49: 221-236, 19 text figs.

Ceratium belone
Sousa e Silva, E., 1949
Diatomaceas e Dinoflagelados de Baia de Cascais. Portugaliae Acta Biol., Volume: Julio Henriques, Ser. B: 300-383, 9 pls, 2 fold-in tables.

Ceratium belone
Steemann-Nielsen, E., 1939
Die Ceratien des Indischen Ozeans und der Ostasiatischen Gewässer mit einer allgemeinen zusammenfassung über die Verbreitung der Ceratien in den Weltmeeren. Dana Rept. No. 17: 33 pp., 8 maps in text.

Ceratium belone
Steemann Nielsen, E., 1934
Untersuchungen über die Verbreitung, Biologie und Variation der Ceratien im südlichen Stillen Ozean. Dana Rept. No. 4: 67 pp., 73 figs. and 11 charts in text.

Ceratium Berghi
Gourret, P., 1883
Sur les Peridiniens du Golfe de Marseille. Ann. du Musee d'hist. Nat., Marseille, Zool., 1 (Mme. 8):1-114, 4 pls.

Ceratium bicorne
Gourret, P., 1883
Sur les Peridiniens du Golfe de Marseille. Ann. du Musee d'hist. Nat., Marseille, Zool., 1 (Mme. 8):1-114, 4 pls.

Ceratium biconicum n.sp.
Murray, G., and F. G. Whitting, 1899
New Peridiniaceae from the Atlantic. Trans. Linn. Soc., London, Bot., ser 2, 5: 321-342, Pls. 27-33, 9 tables.

Ceratium bigelowi
Balech, Enrique, 1962
Tintinnoinea y Dinoflagellata del Pacifico segun material de las Expediciones NORPAC y DOWNWIND del Instituto Scripps de Oceanografia. Revista, Mus. Argentino Ciencias Nat. "Bernardino Rivadavia", Ciencias Zool., 7(1):1-253.

Ceratium bigelowii
Graham, H. W. and N. Bronikovsky, 1944
The genus Ceratium in the Pacific and North Atlantic Oceans. Sci. Res. Cruise VII of the Carnegie, 1928-1929 ----- Biol. V (565):209 pp., 54 charts, 27 figs., 54 tables.

Oceanographic Index: Marine Organisms Cumulation, 1946-1973

Ceratium Bigelowi
Steemann-Nielsen, E., 1939
Die Ceratien des Indischen Ozeans und der Ostasiatischen Gewässer mit einer allgemeinen zusammenfassung über die Verbreitung der Ceratien in den Weltmeeren. Dana Rept. No. 17: 33 pp., 8 maps in text.

Ceratium Bigelowi
Steemann Nielsen, E., 1934
Untersuchungen über die Verbreitung, Biologie und Variation der Ceratien im südlichen Stillen Ozean. Dana Rept. No. 4: 67 pp., 73 figs. and 11 charts in text.

Ceratium bigelowi
Wood, E.J.F., 1963
Dinoflagellates in the Australian region. II. Recent collections. Comm. Sci. and Industr. Res. Org., Div. Fish. and Oceanogr., Techn. Paper, No. 14: 55 pp.

Ceratium böhmii n.sp.
Graham, H. W. and N. Bronikovsky, 1944
The genus Ceratium in the Pacific and North Atlantic Oceans. Sci. Res. Cruise VII of the Carnegie, 1928-1929 ----- Biol. V (565):209 pp., 54 charts, 27 figs., 54 tables.

Ceratium breve
Chiba, T., 1949
On the distribution of the plankton in the eastern China Sea and Yellow Sea. 1. Plankton composition in the spring. J. Shimonoseki Coll. Fisheries, 1(1):57-63, 1 fig.

Ceratium breve
Dangeard, P., 1927
Phytoplankton de la croisière du "Sylvana". Ann. Inst. Ocean., Monaco, n.s., 4(8):286-401, 54 text figs. (Feirer-Juin 1913).

Ceratium breve
Graham, H. W. and N. Bronikovsky, 1944
The genus Ceratium in the Pacific and North Atlantic Oceans. Sci. Res. Cruise VII of the Carnegie, 1928-1929 ----- Biol. V (565):209 pp., 54 charts, 27 figs., 54 tables.

Ceratium breve var. curvulum
Kokubo, S., and S. Sato., 1947
Plankters in Jū-San Gata. Physiol. and Ecol. (Japan) 1(4):1-16, 3 text figs., tables.

Ceratium breve
Sousa e Silva, E., 1949
Diatomaceas e Dinoflagelados de Baia de Cascais. Portugaliae Acta Biol., Volume: Julio Henriques, Ser. B: 300-383, 9 pls, 2 fold-in tables.

Ceratium breve
Steemann-Nielsen, E., 1939
Die Ceratien des Indischen Ozeans und der Ostasiatischen Gewässer mit einer allgemeinen zusammenfassung über die Verbreitung der Ceratien in den Weltmeeren. Dana Rept. No. 17: 33 pp., 8 maps in text.

Ceratium breve
Steemann Nielsen, E., 1934
Untersuchungen über die Verbreitung, Biologie und Variation der Ceratien im südlichen Stillen Ozean. Dana Rept. No. 4: 67 pp., 73 figs. and 11 charts in text.

Ceratium breve
Wang, C. C., and D. Nie, 1932
A survey of the marine protozoa of Amoy. Contr. Biol. Lab. Sci. Soc., China, Zool. Ser., 8:285-385, 89 text figs.

Ceratium breve curvulum
Peters, Nicolaus, 1934
Die Bevolkerung des Sudatlantischen Ozeans mit Ceratien. Biol. Sonderuntersuchungen 1. Wiss Ergeb. Deutschen Atlantischen Exped.--- "Meteor" 1925-1927, 12(1): 1-69, 28 text figs.

Ceratium brunelli n.sp.
Rampi, L., 1942.
Ricerche sul fitoplancton del Mare Ligure. 4. I Ceratium delle acque di Sanremo. Nuovo Giornale Botanico Italiano, N.S., 49: 221-236, 19 text figs.

Ceratium bucephalum
Bigelow, H. B., 1922
Exploration of the coastal water off the northeastern United States in 1916 by the U.S. Fisheries Schooner Grampus. Bull. M.C.Z. 65 (5):85-188, 53 text figs.

Ceratium bucephalum
Dangeard, P., 1926
Description des Péridiniens Testacés recueillis para la Mission Charcot pendent le mois d'Aout 1924. Ann. Inst. Ocean. n.s. 3(7):307-334, 15 text figs.

Ceratium bucephalum
Gran, H.H., and T. Braarud, 1935
A quantitative study of the phytoplankton in the Bay of Fundy and the Gulf of Maine (including observations on hydrography, chemistry, and turbidity). J. Biol. Bd., Canada, 1(5):279-467, 69 text figs.

Ceratium bucephalum
Jørgensen, E., 1920
Mediterranean Ceratia. Rept. Danish Oceanogr. Exped. 1908-10 to the Mediterranean and adjacent seas, Vol.II, Biol. J 1:110 pp., 94 text figs., 26 charts.

Ceratium bucephalum
Jørgensen, E., 1905.
B.Protistplankton and the diatoms in bottom samples. Hydrographical and biological investigations in Norwegian fjords. Bergens Mus. Skr. 7: 49-225.

Ceratium bucephalum
LeBour, M.V., 1925
The dinoflagellates of Northern Seas. The Marine Biological Association of the United Kingdom. Plymouth, 250 pp., 35 pls. 53 text figs.

Ceratium bucephalum
Lillick, L.C., 1940
Phytoplankton and planktonic protozoa of the offshore waters of the Gulf of Maine. Pt.II. Qualitative Composition of the Planktonic Flora. Trans. Am. Phil. Soc., n.s., 31(3):193-237, 13 text figs.

Ceratium bucephalum
Lillick, L.C., 1938
Preliminary report of the phytoplankton of the Gulf of Maine. Am. Mid. Nat. 20(3):624-640, 1 text fig., 37 tables.

Ceratium bucephalum
López, J., 1966.
Variación y regulación de la forma en el género Ceratium. Inv. Pesq., Barcelona, 30:325-427.

Ceratium bucephalum
Mulford, Richard A., 1963
Distribution of the dinoflagellate genus Ceratium in the tidal and offshore waters of Virginia. Chesapeake Science, 4(2):84-89.

Ceratium bucephalum
Paulsen, O., 1908
XVIII Peridiniales. Nordisches Plankton, Bot. Teil: 1-124, 155 text figs.

Ceratium bucephalum
Peters, Nicolaus, 1934
Die Bevolkerung des Sudatlantischen Ozeans mit Ceratien. Biol. Sonderuntersuchungen 1. Wiss Ergeb. Deutschen Atlantischen Exped.--- "Meteor" 1925-1927, 12(1): 1-69, 28 text figs.

Ceratium bucephalum
Sousa e Silva, E., 1949
Diatomaceas e Dinoflagelados de Baia de Cascais. Portugaliae Acta Biol., Volume: Julio Henriques, Ser. B: 300-383, 9 pls, 2 fold-in tables.

Ceratium bucephalum
Steemann-Nielsen, Einar, 1951
The marine vegetation of the Isefjord. A study on ecology and production. Medd. Komm. Danmarks Fiskeri-og Havundersøgelser. Ser. Plankton. 5(4); 114pp., 46 text figs.

Ceratium buceros
Pavillard, J., 1923
A propos de la systématique des Péridiniens Bull. Soc. Bot. de France 70:876-882; 914-918.

Ceratium buceros
Rampi, L., 1942.
Ricerche sul fitoplancton del Mare Ligure. 4. I Ceratium delle acque di Sanremo. Nuovo Giornale Botanico Italiano, N.S., 49: 221-236, 19 text figs.

Ceratium buceros
Vives, F., and A. Planas, 1952.
Plancton recogido por los laboratorios costeros. VI. Fitoplanton de las costas de Vinaroz, islas Columbretes y alrededores de la desembocadura del Elroa. Publ. Inst. Biol. Aplic. 11:141-156, 19 textfigs.

Ceratium californiense n.sp.
Kofoid, C. A., 1907
Dinoflagellata of the San Diego region. III. Descriptions of new species. Univ. Calif. Publ., Zool. 3:299-340, Pls. 22-33.

Ceratium candelabrum
Chiba, T., 1949
On the distribution of the plankton in the eastern China Sea and Yellow Sea. 1. Plankton composition in the spring. J. Shimonoseki Coll. Fisheries, 1(1):57-63, 1 fig.

Ceratium candelabrum
Dangeard, P., 1927
Phytoplankton de la croisière du "Sylvana". Ann. Inst. Ocean., Monaco, n.s., 4(8):286-401, 54 text figs. (Feirer-Juin 1913).

Ceratium candelabrum
Delegazione Italiana della Commissione Internazionale per l'Esplorazione Scientifica del Mediterraneo, 1941
Note sul plancton della Laguna veneta. [Memoria CCLXXXIX] , Arch. di Ocean. e Limn. Anno I, Fasc. I, 1941 XIX: 31-57 pp.

Ceratium candelabrum
Ercegovic, A., 1936
Etudes qualitative et quantitatives du phytoplancton dans les eaux cotières de l'Adriatique oriental moyen au cours de l'année 1934. Acta Adriatica 1(9):1-126

Ceratium candelabrum
Forti, A., 1922
Ricerche sulla flora pelagica (fitoplancton) di Quarto dei Mille. Mem. R. Com. Talass. Ital. 97:248 pp., 13 pls.

Ceratium candelabrum
Ghazzawi, F.M., 1939
Plankton of the Egyptian waters. A study of the Suez Canal Plankton. (A) Phytoplankton. Preliminary Report 83 pp. Notes and Memoires, Min. Cormerce-Industry, Egypt, Hydrobiol. & Fish. 65 figs.

Ceratium candelabrum
Gilbert, J.Y., and W.E. Allen, 1943
The phytoplankton of the Gulf of California obtained by the "E.W. Scripps" in 1939 and 1940. J. Mar. Res. V(2):89-110, figs.30-31.

Ceratium candelabrum
Graham, H. W. and N. Bronikovsky, 1944
The genus Ceratium in the Pacific and North Atlantic Oceans. Sci. Res. Cruise VII of the Carnegie, 1928-1929 ----- Biol. V (565):209 pp., 54 charts, 27 figs., 54 tables.

Ceratium candelabrum

Halim, Y., 1963.
Microplancton des eaux égyptiennes. Le genre Ceratium Schrank (Dinoflagellés).
Rapp. Proc. Verb. Réunions, Comm. Int. Expl. Sci., Mer Méditerranée, Monaco, 17(2):495-502.

Ceratium candelabrum

Jörgensen, E., 1920
Mediterranean Ceratia. Rept. Danish Oceanogr. Exped. 1908-10 to the Mediterranean and adjacent seas, Vol. II, Biol. J 1:110 pp., 94 text figs., 26 charts.

Ceratium candelabrum

LeBour, M.V., 1925
The dinoflagellates of Northern Seas. The Marine Biological Association of the United Kingdom. Plymouth, 250 pp., 35 pls. 53 text figs.

Ceratium candelabrum

Lindemann, E., 1925
Neubeobachtungen an den Winter peridineen des Golfes von Neapel. Bot. Arch. 9:95-102, 19 text figs.

Ceratium candelabrum

López, J., 1966.
Variación y regulación de la forma en el género Ceratium.
Inv. Pesq., Barcelona, 30:325-427.

Ceratium candelabrum

Mangin, M. L., 1912
Phytoplankton de la croisière du "René" dans l'Atlantique (Septembre 1908). Ann. Inst. Ocean., n.s., 4(1):1-66, 2 pls., 41 text figs., 2 tables.

Ceratium candelabrum

Margalef, R., 1949
Fitoplancton nerítico de la Costa Brava en 1947-48. Publ. Inst. Biol. Aplicada, 5: 41-51, 3 text figs.

Ceratium candelabrum

Marshall, S. M., 1933
The production of microplankton in the Great Barrier Reef Region. Brit. Mus. (N.H.) Great Barrier Reef Exped. 1928-29, Sci. Repts. II(5):111-157, 14 text figs.

Ceratium candelabrum

Massutí Algamora, M., 1949
Estudio de diez y seis muestras de plancton del Golfo de Nápoles. Publ. Inst. Biol. Appl. 5:85-94, 1 fold-in table.

Ceratium candelabrum

Mulford, Richard A., 1963
Distribution of the dinoflagellate genus Ceratium in the tidal and offshore waters of Virginia.
Chesapeake Science, 4(2):84-89.

Ceratium candelabrum

Murray, G., and F. G. Whitting, 1899
New Peridiniaceae from the Atlantic.
Trans. Linn. Soc., London, Bot., ser 2, 5: 321-342, Pls. 27-33, 9 tables.

Ceratium candelabrum

Paulsen, O., 1908
XVIII Peridiniales. Nordisches Plankton, Bot. Teil: 1-124, 155 text figs.

Ceratium candelabrum

Pavillard, J., 1916
Recherches sur les Peridiniens du Golfe du Lion. Mem. Univ. Montpellier. Trav. Inst. Bot., Univ. Montpellier. Serie mixte No. 4, 70 pp., 3 pls., 15 text figs.

Ceratium candelabrum

Pavillard, J., 1905
Recherches sur la flore pelagique (Phytoplankton) de l'Etang de Thau. Theses presentees a la Fac. Sci., Paris, 116 pp., 3 pls.

Ceratium candelabrum

Peters, Nicolaus, 1934
Die Bevolkerung des Sudatlantischen Ozeans mit Ceratien. Biol. Sonderuntersuchungen 1. Wiss Ergeb. Deutschen Atlantischen Exped.--- "Meteor" 1925-1927, 12(1): 1-69, 28 text figs.

Ceratium candelabrum

Schröder, B., 1900
Phytoplankton des Golfes von Neapel nebst vergleichenden Ausblicken auf das atlantischen Ozean. Mitt. Zool. Stat. Neapel, 14:1-38.

Ceratium candelabrum

Sousa e Silva, E., 1949
Diatomaceas e Dinoflagelados de Baia de Cascais. Portugaliae Acta Biol., Volume: Julio Henriques, Ser. B: 300-383, 9 pls, 2 fold-in tables.

Ceratium candelabrum

Steemann-Nielsen, E., 1939
Die Ceratien des Indischen Ozeans und der Ostasiatischen Gewässer mit einer allgemeinen zusammenfassung über die Verbreitung der Ceratien in den Weltmeeren. Dana Rept. No. 17: 33 pp., 8 maps in text.

Ceratium candelabrum

Steemann Nielsen, E., 1934
Untersuchungen über die Verbreitung, Biologie und Variation der Ceratien im südlichen Stillen Ozean. Dana Rept. No. 4: 67 pp., 73 figs. and 11 charts in text.

Ceratium candelabrum

Wheeler, J.E.G., 1939
Plankton investigations. Bermuda Biological Station. Second Report. October 1939. 7 pp. (typed), 5 figs. Plymouth, Oct. 23, 1939.

Ceratium candelabrum depressum

Forti, A., 1922
Ricerche sulla flora pelagica (fitoplancton) di Quarto dei Mille. Mem. R. Com. Talass. Ital. 97:248 pp., 13 pls.

Ceratium candelabrum dilatatum

Lindemann, E., 1925
Neubeobachtungen an den Winter peridineen des Golfes von Neapel. Bot. Arch. 9:95-102, 19 text figs.

Ceratium candelabrum dilatatum

Pavillard, J., 1916
Recherches sur les Peridiniens du Golfe du Lion. Mem. Univ. Montpellier. Trav. Inst. Bot., Univ. Montpellier. Serie mixte No. 4, 70 pp., 3 pls., 15 text figs.

Ceratium candelabrum depressum

Steemann Nielsen, E., 1934
Untersuchungen über die Verbreitung, Biologie und Variation der Ceratien im südlichen Stillen Ozean. Dana Rept. No. 4: 67 pp., 73 figs. and 11 charts in text.

Ceratium carnegiei n.sp.

Graham, H. W. and N. Bronikovsky, 1944
The genus Ceratium in the Pacific and North Atlantic Oceans. Sci. Res. Cruise VII of the Carnegie, 1928-1929 ----- Biol. V (565):209 pp., 54 charts, 27 figs., 54 tables.

Ceratium cannegiei

Sournia, A., 1967.
Le genre Ceratium (peridinien planctonique) dans le canal de Mozambique. Contribution a une revision mondiale (fin).
Vie Milieu (A) 18(3):441-449.

Ceratium carriense

Dangeard, P., 1927
Phytoplankton de la croisière du "Sylvana". Ann. Inst. Ocean., Monaco, n.s., 4(8):286-401, 54 text figs. (Feirer-Juin 1913).

Ceratium carriense

Ercegovic, A., 1936
Etudes qualitative et quantitatives du phytoplancton dans les eaux cotières de l'Adriatique oriental moyen au cours de l'année 1934. Acta Adriatica 1(9):1-126

Ceratium carriense

Forti, A., 1922
Ricerche sulla flora pelagica (fitoplancton) di Quarto dei Mille. Mem. R. Com. Talass. Ital. 97:248 pp., 13 pls.

Ceratium carriense

Gourret, P., 1883
Sur les Peridiniens du Golfe de Marseille. Ann. du Musee d'hist. Nat., Marseille, Zool. 1 (Mme. 8):1-114, 4 pls.

Ceratium carriense

Graham, H. W. and N. Bronikovsky, 1944
The genus Ceratium in the Pacific and North Atlantic Oceans. Sci. Res. Cruise VII of the Carnegie, 1928-1929 ----- Biol. V (565):209 pp., 54 charts, 27 figs., 54 tables.

Ceratium carriense

Jörgensen, E., 1920
Mediterranean Ceratia. Rept. Danish Oceanogr. Exped. 1908-10 to the Mediterranean and adjacent seas, Vol. II, Biol. J 1:110 pp., 94 text figs., 26 charts.

Ceratium carriense

Margalef, R., 1949
Fitoplancton nerítico de la Costa Brava en 1947-48. Publ. Inst. Biol. Aplicada, 5: 41-51, 3 text figs.

Ceratium carriense

Margalef, R., 1948.
Le phytoplancton estival de la "Costa Brava" catalane en 1946. Hydrobiol. 1(1):15-21.

Ceratium carriense

Pavillard, J., 1923
A propos de la systématique des Péridiniens Bull. Soc. Bot. de France 70:876-882; 914-918.

Ceratium carriense

Pavillard, J., 1916
Recherches sur les Peridiniens du Golfe du Lion. Mem. Univ. Montpellier. Trav. Inst. Bot., Univ. Montpellier. Serie mixte No. 4, 70 pp., 3 pls., 15 text figs.

Ceratium carriense

Peters, Nicolaus, 1934
Die Bevolkerung des Sudatlantischen Ozeans mit Ceratien. Biol. Sonderuntersuchungen 1. Wiss Ergeb. Deutschen Atlantischen Exped.--- "Meteor" 1925-1927, 12(1): 1-69, 28 text figs.

Ceratium carriense

Robinson, G.A., 1961
Contribution towards an atlas of the northeastern Atlantic and the North Sea. 1. Phytoplankton.
Bulls. Mar. Ecology. 5(42): 81-89, Pls. 15-20.

Ceratium carriense

Steemann-Nielsen, E., 1939
Die Ceratien des Indischen Ozeans und der Ostasiatischen Gewässer mit einer allgemeinen zusammenfassung über die Verbreitung der Ceratien in den Weltmeeren. Dana Rept. No. 17: 33 pp., 8 maps in text.

Ceratium carriense

Sournia, A., 1967.
Le genre Ceratium (peridinien planctonique) dans le canal de Mozambique. Contribution a une revision mondiale (fin).
Vie Milieu (A) 18(3):441-449.

Ceratium carriense
Sukhanova, I.N., 1964.
The phytoplankton of the northeastern part of the Indian Ocean in the season of the southwest monsoon. Regularity of the distribution of oceanic plankton. (In Russian; English abstract).
Trudy Inst. Okeanol., Akad. Nauk, SSSR, 65:24-31.

Ceratium carriense ceylanicum
Lindemann, E., 1925
Neubeobachtungen an den Winter peridineen des Golfes von Neapel. Bot. Arch. 9:95-102, 19 text figs.

Ceratium carriense volans
Delegazione Italiana della Commissione Internazionale per l'Esplorazione Scientifica del Mediterraneo, 1941
Note sul plancton della Laguna veneta. [Memoria CCLXXXIX], Arch. di Ocean. e Limn. Anno I, Fasc. I, 1941 XIX: 31-57 pp.

Ceratium carriense volans
Forti, A., 1922
Ricerche sulla flora pelagica (fitoplancton) di Quarto dei Mille. Mem. R. Com. Talass. Ital. 97:248 pp., 13 pls.

Ceratium carriense volans
Lindemann, E., 1925
Neubeobachtungen an den Winter peridineen des Golfes von Neapel. Bot. Arch. 9:95-102, 19 text figs.

Ceratium carriense volans
Pavillard, J., 1916
Recherches sur les Peridiniens du Golfe du Lion. Mem. Univ. Montpellier. Trav. Inst. Bot., Univ. Montpellier. Serie mixte No.4, 70 pp., 3 pls., 15 text figs.

Ceratium carsteni
Ercegovic, A., 1936
Etudes qualitative et quantitatives du phytoplancton dans les eaux cotières de l'Adriatique oriental moyen au cours de l'année 1934. Acta Adriatica 1(9):1-126

Ceratium cephalotum
Graham, H. W. and N. Bronikovsky, 1944
The genus Ceratium in the Pacific and North Atlantic Oceans. Sci. Res. Cruise VII of the Carnegie, 1928-1929 ----- Biol. V (565):209 pp., 54 charts, 27 figs., 54 tables.

Ceratium cephalotum
Peters, Nicolaus, 1934
Die Bevolkerung des Sudatiantischen Ozeans mit Ceratien. Biol. Sonderuntersuchungen 1. Wiss Ergeb. Deutschen Atlantischen Exped.--- "Meteor" 1925-1927, 12(1): 1-69, 28 text figs.

Ceratium cephalotum
Steemann-Nielsen, E., 1939
Die Ceratien des Indischen Ozeans und der Ostasiatischen Gewässer mit einer allgemeinen zusammenfassung über die Verbreitung der Ceratien in den Weltmeeren. Dana Rept. No. 17: 33 pp., 8 maps in text.

Ceratium cephalotum
Steemann Nielsen, E., 1934
Untersuchungen über die Verbreitung, Biologie und Variation der Ceratien im südlichen Stillen Ozean. Dana Rept. No. 4: 67 pp., 73 figs. and 11 charts in text.

Ceratium claviger
Pavillard, J., 1923
A propos de la systématique des Péridiniens Bull. Soc. Bot. de France 70:876-882; 914-918.

?Ceratium claviger
Pavillard, J., 1916
Recherches sur les Peridiniens du Golfe du Lion. Mem. Univ. Montpellier. Trav. Inst. Bot., Univ. Montpellier. Serie mixte No.4, 70 pp., 3 pls., 15 text figs.

Ceratium claviger
Steemann Nielsen, E. 1934
Untersuchungen über die Verbreitung, Biologie und Variation der Ceratien im südlichen Stillen Ozean. Dana Rept. No. 4: 67 pp., 73 figs. and 11 charts in text.

Ceratium coarctatum n.sp.
Pavillard, J., 1905
Recherches sur la flore pelagique (Phytoplankton) de l'Etang de Thau. Theses presentees a la Fac. Sci., Paris, 116 pp., 3 pls.

Ceratium compressum
Graham, H. W. and N. Bronikovsky, 1944
The genus Ceratium in the Pacific and North Atlantic Oceans. Sci. Res. Cruise VII of the Carnegie, 1928-1929 ----- Biol. V (565):209 pp., 54 charts, 27 figs., 54 tables.

Ceratium compressum
LeBour, M.V., 1925
The dinoflagellates of Northern Seas. The Marine Biological Association of the United Kingdom. Plymouth, 250 pp., 35 pls. 53 text figs.

Ceratium compressum
Paulsen, O., 1908
XVIII Peridiniales. Nordisches Plankton, Bot. Teil: 1-124, 155 text figs.

Ceratium concilians
Dangeard, P., 1927
Phytoplankton de la croisière du "Sylvana". Ann. Inst. Ocean., Monaco, n.s., 4(8):286-401, 54 text figs. (Février-Juin 1913).

Ceratium concilians
Forti, A., 1922
Ricerche sulla flora pelagica (fitoplancton) di Quarto dei Mille. Mem. R. Com. Talass. Ital. 97:248 pp., 13 pls.

Ceratium concilians
Graham, H. W. and N. Bronikovsky, 1944
The genus Ceratium in the Pacific and North Atlantic Oceans. Sci. Res. Cruise VII of the Carnegie, 1928-1929 ----- Biol. V (565):209 pp., 54 charts, 27 figs., 54 tables.

Ceratium concilians n.sp.
Jörgensen, E., 1920
Mediterranean Ceratia. Rept. Danish Oceanogr. Exped. 1908-10 to the Mediterranean and adjacent seas, Vol.II, Biol. J 1:110 pp., 94 text figs., 26 charts.

Ceratium concilians
Margalef, R., 1949
Fitoplancton nerítico de la Costa Brava en 1947-48. Publ. Inst. Biol. Aplicada, 5: 41-51, 3 text figs.

Ceratium concilians
Margalef, R., 1948.
Le phytoplancton estival de la "Costa Brava" catalane en 1946. Hydrobiol. 1(1):15-21.

Ceratium concilians
Peters, Nicolaus, 1934
Die Bevolkerung des Sudatiantischen Ozeans mit Ceratien. Biol. Sonderuntersuchungen 1. Wiss Ergeb. Deutschen Atlantischen Exped.--- "Meteor" 1925-1927, 12(1): 1-69, 28 text figs.

Ceratium concilians
Sournia, A., 1967.
Le genre Ceratium (peridinien planctonique) dans le canal de Mozambique. Contribution a une revision mondiale (fin).
Vie Milieu (A) 18(3):441-449.

Ceratium concilians
Steemann-Nielsen, E., 1939
Die Ceratien des Indischen Ozeans und der Ostasiatischen Gewässer mit einer allgemeinen zusammenfassung über die Verbreitung der Ceratien in den Weltmeeren. Dana Rept. No. 17: 33 pp., 8 maps in text.

Ceratium concilians
Steemann Nielsen, E. 1934
Untersuchungen über die Verbreitung, Biologie und Variation der Ceratien im südlichen Stillen Ozean. Dana Rept. No. 4: 67 pp., 73 figs. and 11 charts in text.

Centrodinium conplanatum
Ercegovic, A., 1936
Etudes qualitative et quantitatives du phytoplancton dans les eaux cotières de l'Adriatique oriental moyen au cours de l'année 1934. Acta Adriatica 1(9):1-126

Ceratium contortum
Dangeard, P., 1927
Phytoplankton de la croisière du "Sylvana". Ann. Inst. Ocean., Monaco, n.s., 4(8):286-401, 54 text figs. (Février-Juin 1913).

Ceratium contortum
Ercegovic, A., 1936
Etudes qualitative et quantitatives du phytoplancton dans les eaux cotières de l'Adriatique oriental moyen au cours de l'année 1934. Acta Adriatica 1(9):1-126

Ceratium contortum
Graham, H. W. and N. Bronikovsky, 1944
The genus Ceratium in the Pacific and North Atlantic Oceans. Sci. Res. Cruise VII of the Carnegie, 1928-1929 ----- Biol. V (565):209 pp., 54 charts, 27 figs., 54 tables.

Ceratium contortum
Halim, Y., 1965.
Microplancton des eaux Egyptiennes. II. Chrysomonadines, Ebriediens et dinoflagellés nouveaux ou d'intérêt biogeographique.
Rapp. Proc. Verb. Réunions, Comm. Int. Expl. Sci., Mer. Mediterranee, Monaco, 18(2):373-379.

Ceratium contortum
Halim, Y., 1963.
Microplancton des eaux égyptiennes. Le genre Ceratium Schrank (Dinoflagellés).
Rapp. Proc. Verb. Réunions, Comm. Int. Expl. Sci., Mer Méditerranée, Monaco, 17(2):495-502.

Ceratium contortum
Mulford, Richard A., 1963
Distribution of the dinoflagellate genus Ceratium in the tidal and offshore waters of Virginia.
Chesapeake Science, 4(2):84-89.

Ceratium contortum
Peters, Nicolaus, 1934
Die Bevolkerung des Sudatiantischen Ozeans mit Ceratien. Biol. Sonderuntersuchungen 1. Wiss Ergeb. Deutschen Atlantischen Exped.--- "Meteor" 1925-1927, 12(1): 1-69, 28 text figs.

Ceratium contortum & var.
Sournia, A., 1967.
Le genre Ceratium (peridinien planctonique) dans le canal de Mozambique. Contribution a une revision mondiale (fin).
Vie Milieu (A) 18(3):441-449.

Ceratium contortum
Steemann-Nielsen, E., 1939
Die Ceratien des Indischen Ozeans und der Ostasiatischen Gewässer mit einer allgemeinen zusammenfassung über die Verbreitung der Ceratien in den Weltmeeren. Dana Rept. No. 17: 33 pp., 8 maps in text.

Ceratium contortum
Steemann Nielsen, E. 1934
Untersuchungen über die Verbreitung, Biologie und Variation der Ceratien im südlichen Stillen Ozean. Dana Rept. No. 4: 67 pp., 73 figs. and 11 charts in text.

Ceratium contrarium
Dangeard, P., 1927
Phytoplankton de la croisière du "Sylvana". Ann. Inst. Ocean., Monaco, n.s., 4(8):286-401, 54 text figs. (Février-Juin 1913).

Ceratium contrarium

Delegazione Italiana della Commissione Internazionale per l'Esplorazione Scientifica del Mediterraneo, 1941
Note sul plancton della Laguna veneta. [Memoria CCLXXXIX], Arch. di Ocean. e Limn. Anno I, Fasc. I, 1941 XIX: 31-57 pp.

Ceratium contrarium

Graham, H. W. and N. Bronikovsky, 1944
The genus Ceratium in the Pacific and North Atlantic Oceans. Sci. Res. Cruise VII of the Carnegie, 1928-1929 ----- Biol. V (565):209 pp., 54 charts, 27 figs., 54 tables.

Ceratium contrarium

Jörgenson, E., 1920
Mediterranean Ceratia. Rept. Danish Oceanogr. Exped. 1908-10 to the Mediterranean and adjacent seas, Vol. II, Biol. J 1:110 pp., 94 text figs., 26 charts.

Ceratium contrarium

Margalef, R., 1949
Fitoplancton nerítico de la Costa Brava en 1947-48. Publ. Inst. Biol. Aplicada, 5: 41-51, 3 text figs.

Ceratium contrarium

Pavillard, J., 1923
A propos de la systématique des Péridiniens Bull. Soc. Bot. de France 70:876-882; 914-918.

Ceratium contrarium

Pavillard, J., 1923
A propos de la systématique des Péridiniens Bull. Soc. Bot. de France 70:876-882; 914-918.

Ceratium contrarium nom. nov.

Pavillard, J., 1905
Recherches sur la flore pelagique (Phytoplankton) de l'Etang de Thau. Theses presentees a la Fac. Sci., Paris, 116 pp., 3 pls.

Ceratium contrarium

Peters, Nicolaus, 1934
Die Bevolkerung des Sudatlantischen Ozeans mit Ceratien. Biol. Sonderuntersuchungen 1. Wiss Ergeb. Deutschen Atlantischen Exped.---"Meteor" 1925-1927, 12(1): 1-69, 28 text figs.

Ceratium contrarium

Sournia, A., 1967.
Le genre Ceratium (peridinien planctonique) dans le canal de Mozambique. Contribution a une revision mondiale (fin). Vie Milieu (A) 18(3):441-449.

Ceratium contrarium

Sousa e Silva, E., 1949
Diatomaceas e Dinoflagelados de Baia de Cascais. Portugaliae Acta Biol., Volume: Julio Henriques, Ser. B: 300-383, 9 pls, 2 fold-in tables.

Ceratium contrarium

Steemann-Nielsen, E., 1939
Die Ceratien des Indischen Ozeans und der Ostasiatischen Gewässer mit einer allgemeinen zusammenfassung über die Verbreitung der Ceratien in den Weltmeeren. Dana Rept. No. 17: 33 pp., 8 maps in text.

Ceratium contrarium

Steemann Nielsen, E., 1934
Untersuchungen über die Verbreitung, Biologie und Variation der Ceratien im südlichen Stillen Ozean. Dana Rept. No. 4: 67 pp., 73 figs. and 11 charts in text.

Ceratium cornutum

Meunier, A., 1919
Microplancton de la Mer Flamande 3. Les Péridiniens. Mem. Mus. Roy. Hist. Nat., Belgique 8(1):1-116, Pls. XV-XXI.

Ceratium curvicorne

López, J., 1966.
Variación y regulación de la forma en el género Ceratium. Inv. Pesq., Barcelona, 30:325-427.

Ceratium curvicorne

Pavillard, J., 1905
Recherches sur la flore pelagique (Phytoplankton) de l'Etang de Thau. Theses presentees a la Fac. Sci., Paris, 116 pp., 3 pls.

Ceratium dalmaticum

Ercegovic, A., 1936
Etudes qualitative et quantitatives du phytoplancton dans les eaux cotières de l'Adriatique oriental moyen au cours de l'année 1934. Acta Adriatica 1(9):1-126

Ceratium declinatum

Dangeard, P., 1927
Phytoplankton de la croisière du "Sylvana". Ann. Inst. Ocean., Monaco, n.s., 4(8):286-401, 54 text figs. (Feirer-Juin 1913).

Ceratium declinatum

Delegazione Italiana della Commissione Internazionale per l'Esplorazione Scientifica del Mediterraneo, 1941
Note sul plancton della Laguna veneta. [Memoria CCLXXXIX], Arch. di Ocean. e Limn. Anno I, Fasc. I, 1941 XIX: 31-57 pp.

Ceratium declinatum

Ercegovic, A., 1936
Etudes qualitative et quantitatives du phytoplancton dans les eaux cotières de l'Adriatique oriental moyen au cours de l'année 1934. Acta Adriatica 1(9):1-126

Ceratium declinatum

Forti, A., 1922
Ricerche sulla flora pelagica (fitoplancton) di Quarto dei Mille. Mem. R. Com. Talass. Ital. 97:248 pp., 13 pls.

Ceratium declinatum

Graham, H. W. and N. Bronikovsky, 1944
The genus Ceratium in the Pacific and North Atlantic Oceans. Sci. Res. Cruise VII of the Carnegie, 1928-1929 ----- Biol. V (565):209 pp., 54 charts, 27 figs., 54 tables.

Ceratium declinatum

Halim, Youssef, 1960
Etude quantitative et qualitative du cycle écologique des dinoflagellés dans les eaux de Villefranche-sur-Mer. (1953-1955). Ann. Inst. Océanogr. Monaco, 38:123-232.

Ceratium declinatum

Jörgensen, E., 1920
Mediterranean Ceratia. Rept. Danish Oceanogr. Exped. 1908-10 to the Mediterranean and adjacent seas, Vol. II, Biol. J 1:110 pp., 94 text figs., 26 charts.

Ceratium declinatum

Lindemann, E., 1925
Neubeobachtungen an den Winter peridineen des Golfes von Neapel. Bot. Arch. 9:95-102, 19 text figs.

Ceratium declinatum?

Mangin, M. L., 1912
Phytoplancton de la croisière du "René" dans l'Atlantique (Septembre 1908). Ann. Inst. Ocean., n.s., 4(1):1-66, 2 pls., 41 text figs., 2 tables.

Ceratium declinatum

Margalef, R., 1949
Fitoplancton nerítico de la Costa Brava en 1947-48. Publ. Inst. Biol. Aplicada, 5: 41-51, 3 text figs.

Ceratium declinatum

Pavillard, J., 1916
Recherches sur les Peridiniens du Golfe du Lion. Mem. Univ. Montpellier. Trav. Inst. Bot., Univ. Montpellier. Serie mixte No.4, 70 pp., 3 pls., 15 text figs.

Ceratium declinatum

Peters, Nicolaus, 1934
Die Bevolkerung des Sudatlantischen Ozeans mit Ceratien. Biol. Sonderuntersuchungen 1. Wiss Ergeb. Deutschen Atlantischen Exped.---"Meteor" 1925-1927, 12(1): 1-69, 28 text figs.

Ceratium declinatum majus

Forti, A., 1922
Ricerche sulla flora pelagica (fitoplancton) di Quarto dei Mille. Mem. R. Com. Talass. Ital. 97:248 pp., 13 pls.

Ceratium declinatum

Steemann-Nielsen, E., 1939
Die Ceratien des Indischen Ozeans und der Ostasiatischen Gewässer mit einer allgemeinen zusammenfassung über die Verbreitung der Ceratien in den Weltmeeren. Dana Rept. No. 17: 33 pp., 8 maps in text.

Ceratium declinatum

Steemann Nielsen, E., 1934
Untersuchungen über die Verbreitung, Biologie und Variation der Ceratien im südlichen Stillen Ozean. Dana Rept. No. 4: 67 pp., 73 figs. and 11 charts in text.

Ceratium declinatum

Steemann Nielsen, E., 1934
Untersuchungen über die Verbreitung, Biologie und Variation der Ceratien im südlichen Stillen Ozean. Dana Rept. No. 4: 67 pp., 73 figs. and 11 charts in text.

Ceratium declinatum angusticornum

Steemann Nielsen, E., 1934
Untersuchungen über die Verbreitung, Biologie und Variation der Ceratien im südlichen Stillen Ozean. Dana Rept. No. 4: 67 pp., 73 figs. and 11 charts in text.

Ceratium deflexum

Dangeard, P., 1927
Phytoplankton de la croisière du "Sylvana". Ann. Inst. Ocean., Monaco, n.s., 4(8):286-401, 54 text figs. (Février-Juin 1913).

Ceratium deflexum

Forti, A., 1922
Ricerche sulla flora pelagica (fitoplancton) di Quarto dei Mille. Mem. R. Com. Talass. Ital. 97:248 pp., 13 pls.

Ceratium deflexum

Graham, H. W. and N. Bronikovsky, 1944f
The genus Ceratium in the Pacific and North Atlantic Oceans. Sci. Res. Cruise VII of the Carnegie, 1928-1929 ----- Biol. V (565):209 pp., 54 charts, 27 figs., 54 tables.

Ceratium deflexum

Marukawa, H., 1921
Plankton lists and some new species of copepods, from the northern waters of Japan. Bull. Inst. Ocean., No.384, 15 pp., 3 pls., 1 chart. Monaco

Ceratium deflexum

Sournia, A., 1967.
Le genre Ceratium (peridinien planctonique) dans le canal de Mozambique. Contribution a une revision mondiale (fin). Vie Milieu (A) 18(3):441-449.

Ceratium deflexum

Steemann-Nielsen, E., 1939
Die Ceratien des Indischen Ozeans und der Ostasiatischen Gewässer mit einer allgemeinen zusammenfassung über die Verbreitung der Ceratien in den Weltmeeren. Dana Rept. No. 17: 33 pp., 8 maps in text.

Ceratium deflexum
Steemann Nielsen, E., 1934
Untersuchungen über die Verbreitung, Biologie und Variation der Ceratien im südlichen Stillen Ozean. Dana Rept. No. 4: 67 pp., 73 figs. and 11 charts in text.

Ceratium dens
Sournia, A., 1967.
Le genre Ceratium (peridinien planctonique) dans le canal de Mozambique. Contribution a une revision mondiale (fin).
Vie Milieu (A) 18(3):441-449.

Ceratium dens
Steemann-Nielsen, E., 1939
Die Ceratien des Indischen Ozeans und der Ostasiatischen Gewässer mit einer allgemeinen zusammenfassung über die Verbreitung der Ceratien in den Weltmeeren. Dana Rept. No. 17: 33 pp., 8 maps in text.

Ceratium dens
Steemann Nielsen, E., 1934
Untersuchungen über die Verbreitung, Biologie und Variation der Ceratien im südlichen Stillen Ozean. Dana Rept. No. 4: 67 pp., 73 figs. and 11 charts in text.

Ceratium depressum
Gourret, P., 1883
Sur les Peridiniens du Golfe de Marseille. Ann. du Musee d'hist. Nat., Marseille, Zool., 1 (Mme. 8):1-114, 4 pls.

Ceratium depressum
Pavillard, J., 1923
A propos de la systématique des Péridiniens Bull. Soc. Bot. de France 70:876-882; 914-918.

Ceratium digitatum
Forti, A., 1922
Ricerche sulla flora pelagica (fitoplancton) di Quarto dei Mille. Mem. R. Com. Talass. Ital. 97:248 pp., 13 pls.

Ceratium digitatum
Graham, H. W. and N. Bronikovsky, 1944
The genus Ceratium in the Pacific and North Atlantic Oceans. Sci. Res. Cruise VII of the Carnegie, 1928-1929 ----- Biol. V (565):209 pp., 54 charts, 27 figs., 54 tables.

Ceratium digitatum
Jörgensen, E., 1920
Mediterranean Ceratia. Rept. Danish Oceanogr. Exped. 1908-10 to the Mediterranean and adjacent seas, Vol.II, Biol. J 1:110 pp., 94 text figs., 26 charts.

Ceratium digitatum
Murray, G., and F. G. Whitting, 1899
New Peridiniaceae from the Atlantic. Trans. Linn. Soc., London, Bot., ser 2, 5: 321-342, Pls. 27-33, 9 tables.

Ceratium digitatum
Pavillard, J., 1916
Recherches sur les Peridiniens du Golfe du Lion. Mem. Univ. Montpellier. Trav. Inst. Bot. Univ. Montpellier. Serie mixte No.4, 70 pp., 3 pls., 15 text figs.

Ceratium digitatum
Peters, Nicolaus, 1934
Die Bevolkerung des Sudatlantischen Ozeans mit Ceratien. Biol. Sonderuntersuchungen 1. Wiss Ergeb. Deutschen Atlantischen Exped.--- "Meteor" 1925-1927, 12(1): 1-69, 28 text figs.

Ceratium digitatum
Steemann-Nielsen, E., 1939
Die Ceratien des Indischen Ozeans und der Ostasiatischen Gewässer mit einer allgemeinen zusammenfassung über die Verbreitung der Ceratien in den Weltmeeren. Dana Rept. No. 17: 33 pp., 8 maps in text.

Ceratium digitatum
Steemann Nielsen, E., 1934
Untersuchungen über die Verbreitung, Biologie und Variation der Ceratien im südlichen Stillen Ozean. Dana Rept. No. 4: 67 pp., 73 figs. and 11 charts in text.

Ceratium digitatum
Wood, E.J.F., 1963
Dinoflagellates in the Australian region. II. Recent collections.
Comm. Sci. and Industr. Res. Org., Div. Fish. and Oceanogr., Techn. Paper, No. 14: 55 pp.

Ceratium dilatatum
Gourret, P., 1883
Sur les Peridiniens du Golfe de Marseille. Ann. du Musee d'hist. Nat., Marseille, Zool., 1 (Mme. 8):1-114, 4 pls.

Ceratium dilatatum
Pavillard, J., 1923
A propos de la systématique des Péridiniens Bull. Soc. Bot. de France 70:876-882; 914-918.

Ceratium dilatatum parvum
Gourret, P., 1883
Sur les Peridiniens du Golfe de Marseille. Ann. du Musee d'hist. Nat., Marseille, Zool., 1 (Mme. 8):1-114, 4 pls.

Ceratium divaricatum?
Brokaw, C.J., and Leigh Wright, 1963.
Bending waves of the posterior flagellum of Ceratium.
Science, 142(3596):1169-1170.

Ceratium egyptiacum
Dowidar, N.M., 1972
Morphological variations in Ceratium egyptiacum in different natural habitats. Mar. Biol. 16(2): 138-149.

Ceratium egyptiacum
Dowidar, Naim M. 1971.
Distribution and ecology of Ceratium egyptiacum Halim and its validity as indicator of the current regime in the Suez Canal.
Int. Revue ges. Hydrobiol. 56(6): 957-966.

Ceratium egypticum suez ensus f. non.
Halim, Y., 1965.
Microplancton des eaux Egyptiennes. II. Chrysomonedines; Ebriediens et dinoflagellés nouveaux on d'intérêt biogeographique.
Rapp. Proc. Verb., Réunions, Comm. Int. Expl. Sci., Mer Mediterranee, Monaco, 18(2):373-379.

Ceratium euarcuatum
Dangeard, P., 1927
Phytoplankton de la croisière du "Sylvana". Ann. Inst. Ocean., Monaco, n.s., 4(8):286-401, 54 text figs. (Feirer-Juin 1913).

Ceratium euarcuatum
Forti, A., 1922
Ricerche sulla flora pelagica (fitoplancton) di Quarto dei Mille. Mem. R. Com. Talass. Ital. 97:248 pp., 13 pls.

Ceratium euarcuatum
Graham, H. W. and N. Bronikovsky, 1944
The genus Ceratium in the Pacific and North Atlantic Oceans. Sci. Res. Cruise VII of the Carnegie, 1928-1929 ----- Biol. V (565):209 pp., 54 charts, 27 figs., 54 tables.

Ceratium euarcuatum
Halim, Youssef, 1960
Étude quantitative et qualitative du cycle écologique des dinoflagellés dans les eaux de Villefranche-sur-Mer. (1953-1955). Ann. Inst. Océanogr. Monaco, 38:123-232.

Ceratium euarcuatum n. nom.
Jörgensen, E., 1920
Mediterranean Ceratia. Rept. Danish Oceanogr. Exped. 1908-10 to the Mediterranean and adjacent seas, Vol.II, Biol. J 1:110 pp., 94 text figs., 26 charts.

Ceratium euarcuatum
Massutí Algamora, M., 1949
Estudio de diez y seis muestras de plancton del Golfo de Nápoles. Publ. Inst. Biol. Appl. 5:85-94, 1 fold-in table.

Ceratium euarcuatum
Pavillard, J., 1923
A propos de la systématique des Péridiniens Bull. Soc. Bot. de France 70:876-882; 914-918.

Ceratium euarcuatum
Peters, Nicolaus, 1934
Die Bevolkerung des Sudatlantischen Ozeans mit Ceratien. Biol. Sonderuntersuchungen 1. Wiss Ergeb. Deutschen Atlantischen Exped.--- "Meteor" 1925-1927, 12(1): 1-69, 28 text figs.

Ceratium euarcuatum
Steemann-Nielsen, E., 1939
Die Ceratien des Indischen Ozeans und der Ostasiatischen Gewässer mit einer allgemeinen zusammenfassung über die Verbreitung der Ceratien in den Weltmeeren. Dana Rept. No. 17: 33 pp., 8 maps in text.

Ceratium euarcuatum
Steemann Nielsen, E., 1934
Untersuchungen über die Verbreitung, Biologie und Variation der Ceratien im südlichen Stillen Ozean. Dana Rept. No. 4: 67 pp., 73 figs. and 11 charts in text.

Ceratium eupulchellum
Vives, F., and A. Planas, 1952.
Plancton recogido por los laboratorios costeros. VI. Fitoplancton de las costas de Vinaroz, islas Columbretes y alrededores de la desembocadura del Elroa. Publ. Inst. Biol. Aplic. 11:141-156, 19 textfigs.

Ceratium extensum
Dangeard, P., 1927
Phytoplankton de la croisière du "Sylvana". Ann. Inst. Ocean., Monaco, n.s., 4(8):286-401, 54 text figs. (Feirer-Juin 1913).

Ceratium extensum
Delegazione Italiana della Commissione Internazionale per l'Esplorazione Scientifica del Mediterraneo, 1941
Note sul plancton della Laguna veneta. [Memoria CCLXXIX] , Arch. di Ocean. e Limn. Anno I, Fasc. I, 1941 XIX: 31-57 pp.

Ceratium extensum
Ercegovic, A., 1936
Etudes qualitative et quantitatives du phytoplancton dans les eaux cotières de l'Adriatique oriental moyen au cours de l'année 1934. Acta Adriatica 1(9):1-126

Ceratium extensum
Graham, H. W. and N. Bronikovsky, 1944
The genus Ceratium in the Pacific and North Atlantic Oceans. Sci. Res. Cruise VII of the Carnegie, 1928-1929 ----- Biol. V (565):209 pp., 54 charts, 27 figs., 54 tables.

Ceratium extensum
Halim, Y., 1963.
Microplancton des eaux égyptiennes. Le genre Ceratium Schrank (Dinoflagellés).
Rapp. Proc. Verb., Réunions, Comm. Int. Expl. Sci. Mer Méditerranée, Monaco, 17(2):495-502.

Ceratium extensum
Jörgensen, E., 1920
Mediterranean Ceratia. Rept. Danish Oceanogr. Exped. 1908-10 to the Mediterranean and adjacent seas, Vol.II, Biol. J 1:110 pp., 94 text figs., 26 charts.

Ceratium extensum
LeBour, M.V., 1925
The dinoflagellates of Northern Seas. The Marine Biological Association of the United Kingdom. Plymouth, 250 pp., 35 pls. 53 text figs.

Ceratium extensum

Lindemann, E., 1925
Neubeobachtungen an den Winter peridineen des Golfes von Neapel. Bot. Arch. 9:95-102, 19 text figs.

Ceratium extensum

Margalef, R., 1949
Fitoplancton nerítico de la Costa Brava en 1947-48. Publ. Inst. Biol. Aplicada, 5: 41-51, 3 text figs.

Ceratium extensum

Massutí Algamora, M., 1949
Estudio de diez y seis muestras de plancton del Golfo de Nápoles. Publ. Inst. Biol. Appl. 5:85-94, 1 fold-in table.

Ceratium extensum

Mulford, Richard A., 1963
Distribution of the dinoflagellate genus Ceratium in the tidal and offshore waters of Virginia. Chesapeake Science, 4(2):84-89.

Ceratium extensum

Paulsen, O., 1908
XVIII Peridiniales. Nordisches Plankton, Bot. Teil: 1-124, 155 text figs.

Ceratium extensum

Pavillard, J., 1916
Recherches sur les Peridiniens du Golfe du Lion. Mem. Univ. Montpellier. Trav. Inst. Bot., Univ. Montpellier. Serie mixte No.4, 70 pp., 3 pls., 15 text figs.

Ceratium extensum

Pavillard, J., 1905
Recherches sur la flore pelagique (Phytoplankton) de l'Etang de Thau. Theses presentees a la Fac. Sci., Paris, 116 pp., 3 pls.

Ceratium extensum

Peters, Nicolaus, 1934
Die Bevolkerung des Sudatlantischen Ozeans mit Ceratien. Biol. Sonderuntersuchungen 1. Wiss Ergeb. Deutschen Atlantischen Exped. "Meteor" 1925-1927, 12(1): 1-69, 28 text figs.

Ceratium extensum

Sousa e Silva, E., 1949
Diatomaceas e Dinoflagelados de Baia de Cascais. Portugaliae Acta Biol., Volume: Julio Henriques, Ser. B: 300-383, 9 pls, 2 fold-in tables.

Ceratium extensum

Steemann-Nielsen, E., 1939
Die Ceratien des Indischen Ozeans und der Ostasiatischen Gewässer mit einer allgemeinen zusammenfassung über die Verbreitung der Ceratien in den Weltmeeren. Dana Rept. No. 17: 33 pp., 8 maps in text.

Ceratium extensum

Steemann Nielsen, E., 1934
Untersuchungen über die Verbreitung, Biologie und Variation der Ceratien im südlichen Stillen Ozean. Dana Rept. No. 4: 67 pp., 73 figs. and 11 charts in text.

Ceratium falcatiforme

Balech, Enrique, 1962
Tintinnoinea y Dinoflagellata del Pacifico segun material de las Expediciones NORPAC y DOWNWIND del Instituto Scripps de Oceanografia. Revista, Mus. Argentino Ciencias Nat. "Bernardino Rivadavia", Ciencias Zool., 7(1):1-253.

Ceratium falcatiforme

Halim, Youssef, 1960
Etude quantitative et qualitative du cycle écologique des dinoflagellés dans les eaux de Villefranche-sur-Mer. (1953-1955). Ann. Inst. Océanogr., Monaco, 38:123-232.

Ceratium falcatiforme n. sp.

Jörgensen, E., 1920
Mediterranean Ceratia. Rept. Danish Oceanogr. Exped. 1908-10 to the Mediterranean and adjacent seas, Vol.II, Biol. J 1:110 pp., 94 text figs., 26 charts.

Ceratium falcatiforme

Steemann Nielsen, E., 1934
Untersuchungen über die Verbreitung, Biologie und Variation der Ceratien im südlichen Stillen Ozean. Dana Rept. No. 4: 67 pp., 73 figs. and 11 charts in text.

Ceratium falcatum

Dangeard, P., 1927
Phytoplankton de la croisière du "Sylvana". Ann. Inst. Ocean., Monaco, n.s., 4(8):286-401, 54 text figs. (Feirer-Juin 1913).

Ceratium falcatum

Delegazione Italiana della Commissione Internazionale per l'Esplorazione Scientifica del Mediterraneo, 1941
Note sul plancton della Laguna veneta. [Memoria CCLXXIX], Arch. di Ocean. e Limn. Anno I, Fasc. I, 1941 XIX: 31-57 pp.

Ceratium falcatum

Ercegovic, A., 1940
Weitere Untersuchungen über einige hydrographische Verhältnisse und über die Phytoplanktonproduktion in den Gewässern der Östlichen Mitteladria. Acta Adriatica 2(3):95-134, 8 text figs.

Ceratium falcatum

Ghazzawi, F.M., 1939
Plankton of the Egyptian waters. A study of the Suez Canal Plankton. (A) Phytoplankton. Preliminary Report 83 pp. Notes and Memoires, Min. Commerce-Industry, Egypt, Hydrobiol. & Fish. 65 figs.

Ceratium falcatum

Graham, H. W. and N. Bronikovsky, 1944
The genus Ceratium in the Pacific and North Atlantic Oceans. Sci. Res. Cruise VII of the Carnegie, 1928-1929 ----- Biol. V (565):209 pp., 54 charts, 27 figs., 54 tables.

Ceratium falcatum

Jörgensen, E., 1920
Mediterranean Ceratia. Rept. Danish Oceanogr. Exped. 1908-10 to the Mediterranean and adjacent seas, Vol.II, Biol. J 1:110 pp., 94 text figs., 26 charts.

Ceratium falcatum

Margalef, R., 1949
Fitoplancton nerítico de la Costa Brava en 1947-48. Publ. Inst. Biol. Aplicada, 5: 41-51, 3 text figs.

Ceratium falcatum

Massutí Algamora, M., 1949
Estudio de diez y seis muestras de plancton del Golfo de Nápoles. Publ. Inst. Biol. Appl. 5:85-94, 1 fold-in table.

Ceratium falcatum

Steemann-Nielsen, E., 1939
Die Ceratien des Indischen Ozeans und der Ostasiatischen Gewässer mit einer allgemeinen zusammenfassung über die Verbreitung der Ceratien in den Weltmeeren. Dana Rept. No. 17: 33 pp., 8 maps in text.

Ceratium falcatum

Steemann Nielsen, E., 1934
Untersuchungen über die Verbreitung, Biologie und Variation der Ceratien im südlichen Stillen Ozean. Dana Rept. No. 4: 67 pp., 73 figs. and 11 charts in text.

Ceratium filicorne

Graham, H. W. and N. Bronikovsky, 1944
The genus Ceratium in the Pacific and North Atlantic Oceans. Sci. Res. Cruise VII of the Carnegie, 1928-1929 ----- Biol. V (565):209 pp., 54 charts, 27 figs., 54 tables.

Ceratium filicorne

Steemann-Nielsen, E., 1939
Die Ceratien des Indischen Ozeans und der Ostasiatischen Gewässer mit einer allgemeinen zusammenfassung über die Verbreitung der Ceratien in den Weltmeeren. Dana Rept. No. 17: 33 pp., 8 maps in text.

Ceratium filicorne n. sp.

Steemann Nielsen, E., 1934
Untersuchungen über die Verbreitung, Biologie und Variation der Ceratien im südlichen Stillen Ozean. Dana Rept. No. 4: 67 pp., 73 figs. and 11 charts in text.

Ceratium furca

Allen, W.E., 1927.
Quantitative studies on inshore marine diatoms and dinoflagellates of Southern California in 1922 Bull. S.I.O., tech. ser., 1:31-38, 2 textfigs.

Ceratium furca

Braarud, T., 1945
A phytoplankton survey of the polluted waters of inner Oslo Fjord. Hvalrådets Skrifter, No.28, 142 pp., 19 text figs., 17 tables.

Ceratium furca

Braarud, T., and Adam Bursa, 1939
On the phytoplankton of the Oslo Fjord, 1933-1934. Hvalrådets Skr. No.19:1-63; 9 text figs. Reviewed. J. du Cons. 14(3): 418-420. A.C. Gardiner.

Ceratium furca

Copenhagen, W. J. and L. D., 1949
Variation in the phytoplankton of Table Bay, October 1934 to October 1935. With a note on the calorific value of Chaetoceros spp. Trans. Roy. Soc. S. Africa, 32(2):113-123, 2 text figs.

Ceratium furca

Cupp, E., 1930
Quantitative Studies of miscellaneous series of surface catches of marine diatoms and dinoflagellates taken between Seattle and the Canal Zone from 1924 to 1928. Trans. Am. Micro. Soc., XLIX (3):238-245.

Ceratium furca

Dangeard, P., 1927
Phytoplankton de la croisière du "Sylvana". Ann. Inst. Ocean., Monaco, n.s., 4(8):286-401, 54 text figs. (Fevrier -Juin 1913).

Ceratium furca

Dangeard, P., 1926
Description des Péridiniens Testacés recueillis para la Mission Charcot pendent le mois d'Aout 1924. Ann. Inst. Ocean. n.s. 3(7):307-334, 15 text figs.

Ceratium furca

de Azevedo Souza, J., 1950.
Nota sôbre variação específica em Ceratium furca Dujardin, do plâncton do litoral Paulista. Bol. Inst. Paulista Oceanogr. 1(2):93-97, 1 textfig.

Ceratium furca

Delegazione Italiana della Commissione Internazionale per l'Esplorazione Scientifica del Mediterraneo, 1941
Note sul plancton della Laguna veneta. [Memoria CCLXXIX], Arch. di Ocean. e Limn. Anno I, Fasc. I, 1941 XIX: 31-57 pp.

Ceratium furca

Ercegovic, A., 1936
Etudes qualitative et quantitatives du phytoplancton dans les eaux côtières de l'Adriatique oriental moyen au cours de l'année 1934. Acta Adriatica 1(9):1-126

Ceratium furca

Forti, A., 1922
Ricerche sulla flora pelagica (fitoplancton) di Quarto dei Mille. Mem. R. Com. Talass. Ital. 97:248 pp., 13 pls.

Ceratium furca
Ghazzawi, F.M., 1939
  Plankton of the Egyptian waters. A study of the Suez Canal Plankton. (A) Phytoplankton. Preliminary Report 83 pp. Notes and Memoires, Min. Commerce-Industry, Egypt, Hydrobiol. & Fish. 65 figs.

Ceratium furca
Gilbert, J.Y., and W.E. Allen, 1943
  The phytoplankton of the Gulf of California obtained by the "E.W. Scripps" in 1939 and 1940. J. Mar. Res. V(2):89-110, figs.30-31.

Ceratium furca
Gourret, P., 1883
  Sur les Péridiniens du Golfe de Marseille. Ann. du Musee d'hist. Nat., Marseille, Zool., 1 (Mme. 8):1-114, 4 pls.

Ceratium furca
Graham, H. W. and N. Bronikovsky, 1944
  The genus Ceratium in the Pacific and North Atlantic Oceans. Sci. Res. Cruise VII of the Carnegie, 1928-1929 ----- Biol. V (565):209 pp., 54 charts, 27 figs., 54 tables.

Ceratium furca
Halim, Y., 1963.
  Microplancton des eaux égyptiennes. Le genre Ceratium Schrank (Dinoflagellés). Rapp. Proc. Verb., Réunions, Comm. Int. Expl. Sci. Mer Méditerranée, Monaco, 17(2):495-502.

Ceratium furca
Halim, Youssef, 1960
  Étude quantitative et qualitative du cycle écologique des dinoflagellés dans les eaux de Villefranche-sur-Mer. (1953-1955). Ann. Inst. Océanogr., Monaco, 38:123-232.

Ceratium furca
Jahn, T.L., W.M. Harmon and M. Landman, 1963.
  Mechanisms of locomotion in flagellates. 1. Ceratium.
  J. Protozoology, 10(3):358-363.

Ceratium furca
Jørgensen, E., 1920
  Mediterranean Ceratia. Rept. Danish Oceanogr. Exped. 1908-10 to the Mediterranean and adjacent seas, Vol.II, Biol. J 1:110 pp., 94 text figs., 26 charts.

Ceratium furca
Jørgensen, E., 1905
  B.Protistplankton and the diatoms in bottom samples. Hydrographical and biological investigations in Norwegian fjords. Bergens Mus. Skr. 7: 49-225.

Ceratium furca
Jorgensen, E., 1900
  Protophyten und Protozoën im Plankton aus der Norwegischen Westkerste. Bergens Mus. Aarb. 1899(6): 95 pp., 5 pls., 83 tables.

Ceratium furca
Kayser, H., 1969.
Züchtungsexperimente an zwei Marinen Flagellaten (Dinophyta) und ihre Anwendung im toxikologischen Abwassertest.
Helgoländer wiss. Meeresunters. 19(1): 21-44

Ceratium furca
LeBour, M.V., 1925
  The dinoflagellates of Northern Seas. The Marine Biological Association of the United Kingdom. Plymouth, 250 pp., 35 pls. 53 text figs.

Ceratium furca
Lindemann, E., 1924
  Peridineen aus dem goldenen Horn und dem Bosphorus. Bot. Arch. 5:216-233, 98 text figs.

Ceratium furca
López, J., 1966.
  Variación y regulación de la forma en el género Ceratium.
  Inv. Pesq., Barcelona, 30:325-427.

Ceratium furca
Macdonald, R., 1933
  An examination of plankton hauls made in the Suez Canal during the year 1928. Fish. Res. Dis., Notes & Mem. No.3, 11 pp., 1 chart.

Ceratium furca
Mangin, M. L., 1912
  Phytoplankton de la croisière du "René" dans l'Atlantique (Septembre 1908). Ann. Inst. Ocean., n.s., 4(1):1-66, 2 pls., 41 text figs., 2 tables.

Ceratium furca
Margalef, R., 1949
  Fitoplancton nerítico de la Costa Brava en 1947-48. Publ. Inst. Biol. Aplicada, 5: 41-51, 3 text figs.

Ceratium furca
Margalef, R., 1948.
  Le phytoplancton estival de la "Costa Brava" catalane en 1946. Hydrobiol. 1(1):15-21.

Ceratium furca
Massutí Algamora, M., 1949
  Estudio de diez y seis muestras de plancton del Golfo de Nápoles. Publ. Inst. Biol. Appl. 5:85-94, 1 fold-in table.

Ceratium furca
Meunier, A., 1919
  Microplancton de la Mer Flamande 3. Les Péridiniens. Mem. Mus. Roy. Hist. Nat., Belgique 8(1):1-116, Pls. XV-XXI.

Ceratium furca
Moberg, E.G., and W.E. Allen, 1927.
  Effect of tidal changes on physical, chemical, and biological conditions in the sea water of the San Diego region. 1. Observations on the effect of tidal changes on physical and chemical conditions of sea water in the San Diego region. 2. Half-hourly collections of marine microplankton taken at the Scripps Institution pier in 1923. Bull. S.I.O., tech. ser., 1:1-17, 4 textfigs.

Ceratium furca
Morse, D.C., 1947
  Some observations on seasonal variations in plankton population Patuxent River, Maryland 1943 1945. Bd. Nat. Res., Publ. No.65, Chesapeake Biol. Lab., 31, 3 figs.

Ceratium furca
Mulford, Richard A., 1963
  Distribution of the dinoflagellate genus Ceratium in the tidal and offshore waters of Virginia.
  Chesapeake Science, 4(2):84-89.

Ceratium furca
Murray, G., and F. G. Whitting, 1899
  New Peridiniaceae from the Atlantic. Trans. Linn. Soc., London, Bot., ser 2, 5: 321-342, Pls. 27-33, 9 tables.

Ceratium furca
Nielsen, J., 1956.
  Temporary variations in certain marine Ceratia. Oikos, 7(2):256-272.

Ceratium furca
Nordli, E., 1957.
  Experimental studies on the ecology of Ceratia. Oikos 8(2):200-265.

Ceratium furca
Nordli, E., 1953.
  Salinity and temperature as controlling factors for distribution and mass occurrence of Ceratia. Blyttia 11:16-18, 4 textfigs.

Ceratium furca
Paulsen, O., 1908
  XVIII Peridiniales. Nordisches Plankton, Bot. Teil: 1-124, 155 text figs.

Ceratium furca
Paiva Carvalho, J., 1950
  O plancton do Rio Maria Rodrigues (Cananeis). 1. Diatomaceas e Dinoflagelados. Bol. Inst. Paulista Oceanogr. 1(1); 27-43, 2 fold-in tables, 2 figs.

Ceratium furca
Pavillard, J., 1905
  Recherches sur la flore pelagique (Phytoplankton) de l'Etang de Thau. Theses presentees a la Fac. Sci., Paris, 116 pp., 3 pls.

Ceratium furca
Peters, Nicolaus, 1934
  Die Bevolkerung des Sudatlantischen Ozeans mit Ceratien. Biol. Sonderuntersuchungen 1. Wiss Ergeb. Deutschen Atlantischen Exped.---"Meteor" 1925-1927, 12(1): 1-69, 28 text figs.

Ceratium furca
Qasim, S.Z., P.M.A. Bhattathiri and V.P. Devassy, 1973
  Growth kinetics and nutrient requirements of two tropical marine phytoplankters. Mar. Biol. 21(4):299-304.

Ceratium furca
Robinson, G.A., 1961
  Contribution towards an atlas of the northeastern Atlantic and the North Sea. 1. Phytoplankton.
  Bulls. Mar. Ecology, 5(42): 81-89, Pls. 15-20.

Ceratium furca
Rodriguez Villar, Luis, 1966.
  Promera cita de las especias componentes del "Huirihue o marea roja".
  Est.Oceanol., Chile, 2:91-93.

Ceratium furca
Schröder, B., 1900
  Phytoplankton des Golfes von Neapel nebst vergleichenden Ausblicken auf das atlantischen Ozean. Mitt. Zool. Stat. Neapel, 14:1-38.

Ceratium furca
Smith, F.G.W., R.H. Williams, and C.C. Davis, 1950.
  An ecological survey of the subtropical inshore waters adjacent to Miami. Ecol. 31(1):119-146, 7 textfigs.

Ceratium furca
Sousa e Silva, E., 1949
  Diatomaceas e Dinoflagelados de Baía de Cascais. Portugaliae Acta Biol., Volume: Julio Henriques, Ser. B: 300-383, 9 pls, 2 fold-in tables.

Ceratium furca
Steemann-Nielsen, E., 1939
  Die Ceratien des Indischen Ozeans und der Ostasiatischen Gewässer mit einer allgemeinen zusammenfassung über die Verbreitung der Ceratien in den Weltmeeren. Dana Rept. No. 17: 33 pp., 8 maps in text.

Ceratium furca
Steemann Nielsen, E., 1934
  Untersuchungen über die Verbreitung, Biologie und Variation der Ceratien im südlichen Stillen Ozean. Dana Rept. No. 4: 67 pp., 73 figs. and 11 charts in text.

Ceratium furca
Wang, C. C., and D. Nie, 1932
  A survey of the marine protozoa of Amoy. Contr. Biol. Lab. Sci. Soc., China, Zool. Ser., 8:285-385, 89 text figs.

Ceratium furca

Woodmansee, R.A., 1958.
The seasonal distribution of the zooplankton off Chicken Key in Biscayne Bay, Florida. Ecology, 39(2):247-261.

Ceratium furca Berghii

Steemann Nielsen, E., 1934
Untersuchungen über die Verbreitung, Biologie und Variation der Ceratien im südlichen Stillen Ozean. Dana Rept. No. 4: 67 pp., 73 figs. and 11 charts in text.

Ceratium furca eugrammum

Lindemann, E., 1925
Neubeobachtungen an den Winter peridineen des Golfes von Neapel. Bot. Arch. 9:95-102, 19 text figs.

Ceratium furca eugrammum

Marshall, S. M., 1933
The production of microplankton in the Great Barrier Reef Region. Brit. Mus. (N.H.) Great Barrier Reef Exped. 1928-29, Sci. Repts. II(5):111-157, 14 text figs.

Ceratium furca eugrammum

Pavillard, J., 1916
Recherches sur les Peridiniens du Golfe du Lion. Mem. Univ. Montpellier. Trav. Inst. Bot., Univ. Montpellier. Serie mixte No.4, 70 pp., 3 pls., 15 text figs.

Ceratium furca

Robinson, G.A., 1965.
Continuous plankton records: contribution towards a plankton atlas of the North Atlantic and the North Sea. IX. Seasonal cycles of phytoplankton.
Bulls. Mar. Ecol., Scottish Mar. Biol. Assoc., 6(4):104-122, pls. 26-61.

Ceratium furca eugrammum

Steemann Nielsen, E., 1934
Untersuchungen über die Verbreitung, Biologie und Variation der Ceratien im südlichen Stillen Ozean. Dana Rept. No. 4: 67 pp., 73 figs. and 11 charts in text.

Ceratium furca mediterraneum

Gourret, P., 1883
Sur les Peridiniens du Golfe de Marseille. Ann. du Musee d'hist. Nat., Marseille, Zool., 1 (Mme. 8):1-114, 4 pls.

Ceratium furca medium

Gourret, P., 1883
Sur les Peridiniens du Golfe de Marseille. Ann. du Musee d'hist. Nat., Marseille, Zool., 1 (Mme. 8):1-114, 4 pls.

Ceratium furca singulare

Gourret, P., 1883
Sur les Peridiniens du Golfe de Marseille. Ann. du Musee d'hist. Nat., Marseille, Zool., 1 (Mme. 8):1-114, 4 pls.

Ceratium furca tertium

Gourret, P., 1883
Sur les Peridiniens du Golfe de Marseille. Ann. du Musee d'hist. Nat., Marseille, Zool., 1 (Mme. 8):1-114, 4 pls.

Ceratium fusus

Balech, Enrique, 1944
Contribucion al conocimiento del Plancton de Lennox y Cabo de Hornos. Physis XIX:423-446, 6 pls. with 67 figs.

Ceratium fusus

Bigelow, H. B., 1922
Exploration of the coastal water off the northeastern United States in 1916 by the U.S. Fisheries Schooner Grampus. Bull. M.C.Z. 65 (5):85-188, 53 text figs.

Ceratium fusus

Braarud, T., 1945
A phytoplankton survey of the polluted waters of inner Oslo Fjord. Hvalrådets Skrifter, No.28, 142 pp., 19 text figs., 17 tables.

Ceratium fusus

Braarud, T., and Adam Bursa, 1939
On the phytoplankton of the Oslo Fjord, 1933-1934. Hvalrådets Skr. No.19:1-63; 9 text figs. Reviewed. J. du. Cons. 14(3): 418-420. A.C. Gardiner.

Ceratium fusus

Brunel, J., 1962
Le phytoplancton de la Baie de Chaleurs. Inst. Botan., Univ. Montréal, Contrib. No. 77: 365 pp., 66 pls.

Ceratium fusus

Brunel, Jules, 1962
Le phytoplancton de la Baie des Chaleurs. Contrib. Ministère de la Chasse et des Pêcheries, Province de Québec, No. 91: 365 pp.

Ceratium fusus

Calkins, G. N., 1902
Marine protozoa from Woods Hole. U.S. Fish Comm. Bull. for 1901, pp. 413-468, 69 text figs.

Ceratium fusus

Chiba, T., 1949
On the distribution of the plankton in the eastern China Sea and Yellow Sea. 1. Plankton composition in the spring. J. Shimonoseki Coll. Fisheries, 1(1):57-63, 1 fig.

Ceratium fusus

Cupp, E., 1930
Quantitative Studies of miscellaneous series of surface catches of marine diatoms and dinoflagellates taken between Seattle and the Canal Zone from 1924 to 1928. Trans. Am. Micro. Soc., XLIX (3):238-245.

Ceratium fusus

Dangeard, P., 1927
Phytoplankton de la croisière du "Sylvana". Ann. Inst. Ocean., Monaco, n.s., 4(8):286-401, 54 text figs. (février-Juin 1913).

Ceratium fusus

Dangeard, P., 1926
Description des Péridiniens Testacés recueillis para la Mission Charcot pendent le mois d'Aout 1924. Ann. Inst. Ocean. n.s. 3(7):307-334, 15 text figs.

Ceratium fusus

Ercegovic, A., 1936
Etudes qualitative et quantitatives du phytoplankton dans les eaux cotières de l'Adriatique oriental moyen au cours de l'année 1934. Acta Adriatica 1(9):1-126

Ceratium fusus

Forti, A., 1922
Ricerche sulla flora pelagica (fitoplancton) di Quarto dei Mille. Mem. R. Com. Talass. Ital. 97:248 pp., 13 pls.

Ceratium fusus

Ghazzawi, F.M., 1939
Plankton of the Egyptian waters. A study of the Suez Canal Plankton. (A) Phytoplankton. Preliminary Report 83 pp. Notes and Memoires, Min. Commerce-Industry, Egypt, Hydrobiol. & Fish. 65 figs.

Ceratium fusus

Gilbert, J.Y., and W.E. Allen, 1943
The phytoplankton of the Gulf of California obtained by the "E.W. Scripps" in 1939 and 1940. J. Mar. Res. V(2):89-110, figs.30-31.

Ceratium fusus

Graham, H. W. and N. Bronikovsky, 1944
The genus Ceratium in the Pacific and North Atlantic Oceans. Sci. Res. Cruise VII of the Carnegie, 1928-1929 ----- Biol. V (565):209 pp., 54 charts, 27 figs., 54 tables.

Ceratium fusus

Gran, H.H., 1897
Protophyta: Diatomaceae, Silico-flagellata and Cilioflagellata. Den Norske Nordhavs Expedition 1876-1878, h. 24, 36 pp., 4 pls.

Ceratium fusus

Gran, H.H., and T. Braarud, 1935
A quantitative study of the phytoplankton in the Bay of Fundy and the Gulf of Maine (including observations on hydrography, chemistry, and turbidity). J. Biol. Bd., Canada, 1(5):279-467, 69 text figs.

Ceratium fusus

Grøntved, J., 1949
Investigations on the phytoplankton in the Danish Waddensea in July 1941. Medd. Komm. Danmarks Fiskeri og Havundersøgelser, ser. Plankton, 5(2):55 pp., 2 pls., 38 text figs.

Ceratium fusum

Hada, Yoshina, 1970. (1970)
The protozoan plankton of the Antarctic and Subantarctic seas.
Scient. Repts., Japan. Antarct. Res. Exped., (E)31:51 pp.

Ceratium fusus

Halim, Y., 1963.
Microplancton des eaux égyptiennes. Le genre Ceratium Schrank (Dinoflagellés). Rapp. Proc. Verb., Réunion, Comm. Int. Expl. Sci., Mer Méditerranée, Monaco, 17(2):495-502.

Ceratium fusus

Hasle, G.R., and E. Nordli, 1952.
Form variations in Ceratium fusus and tripos populations in cultures and from the sea. Abhandl. Norske Videnskaps-Akad., Oslo, 1. Math.-Naturv. Kl., 1951(4):1-25, 8 textfigs.

ceratium fusus

Hickel, Wolfgang, 1967.
Untersuchungen über die Phytoplanktonblüte in der westlichen Ostsee.
Helgoländer wiss. Meeresunters., 16(½):1-66.

Ceratium fusus

Jörgensen, E., 1920
Mediterranean Ceratia. Rept. Danish Oceanogr. Exped. 1908-10 to the Mediterranean and adjacent seas, Vol.II, Biol. J 1:110 pp., 94 text figs., 26 charts.

Ceratium fusus

Jørgensen, E., 1905
B. Protistplankton and the diatoms in bottom samples. Hydrographical and biological investigations in Norwegian fjords. Bergens Mus. Skr. 7: 49-225.

Ceratium fusus

Jorgensen, E., 1900
Protophyten und Protozoën im Plankton aus der Norwegischen Westkerste. Bergens Mus. Aarb. 1899(6): 95 pp., 5 pls., 83 tables

Ceratium fusus

Kokubo, S., and S. Sato, 1947
Planktere in Jū-San Gata. Physiol. and Ecol. (Japan) 1(4):1-16, 3 text figs., tables.

Ceratium fusus

LeBour, M.V., 1925
The dinoflagellates of Northern Seas. The Marine Biological Association of the United Kingdom. Plymouth, 250 pp., 35 pls. 53 text figs.

Ceratium Fusus
Lillick, L.C., 1940
Phytoplankton and planktonic protozoa of the offshore waters of the Gulf of Maine. Pt. II. Qualitative Composition of the Planktonic Flora. Trans. Am. Phil. Soc., n.s., 31(3):193-237, 13 text figs.

Ceratium Fusus
Lillick, L.C., 1938
Preliminary report of the phytoplankton of the Gulf of Maine. Am. Mid. Nat. 20(3):624-640, 1 text fig. 37 tables.

Ceratium Fusus
Lillick, L.C., 1937
Seasonal studies of the phytoplankton off Woods Hole, Massachusetts. Biol. Bull. LXXIII (3):488-503, 3 text figs.

Ceratium fusus
Lindemann, E., 1925
Neubeobachtungen an den Winter peridineen des Golfes von Neapel. Bot. Arch. 9:95-102, 19 text figs.

Ceratium fusus
Lindemann, E., 1924
Peridineen aus dem goldenen Horn und dem Bosphorus. Bot. Arch. 5:216-233, 98 text figs.

Ceratium fusus
Macdonald, R., 1933
An examination of plankton hauls made in the Suez Canal during the year 1928. Fish. Res. Dis., Notes & Mem. No.3, 11 pp., 1 chart.

Ceratium fusus
Margalef, R., 1949
Fitoplancton nerítico de la Costa Brava en 1947-48. Publ. Inst. Biol. Aplicada, 5: 41-51, 3 text figs.

Ceratium fusus
Marshall, S. M., 1933
The production of microplankton in the Great Barrier Reef Region. Brit. Mus. (N.H.) Great Barrier Reef Exped. 1928-29, Sci. Repts. II(5):111-157, 14 text figs.

Ceratium fusus
Marukawa, H., 1921
Plankton lists and some new species of copepods, from the northern waters of Japan. Bull. Inst. Ocean., No.384, 15 pp., 3 pls., 1 chart. Monaco

Ceratium fusus
Massutí Algamora, M., 1949
Estudio de diez y seis muestras de plancton del Golfo de Nápoles. Publ. Inst. Biol. Appl. 5:85-94, 1 fold-in table.

Ceratium fusus
Meunier, A., 1919
Microplancton de la Mer Flamande 3. Les Péridiniens. Mem. Mus. Roy. Hist. Nat., Belgique 8(1):1-116, Pls. XV-XXI.

Ceratium fusus
Morse, D.C., 1947
Some observations on seasonal variations in plankton population Patuxant River, Maryland 1943 1945. Bd. Nat. Res., Publ. No.65, Chesapeake Biol. Lab., 31, 3 figs.

Ceratium fusus
Mulford, Richard A., 1963
Distribution of the dinoflagellate genus Ceratium in the tidal and offshore waters of Virginia. Chesapeake Science, 4(2):84-89.

Ceratium fusus
Murray, G., and F. G. Whitting, 1899
New Peridiniaceae from the Atlantic. Trans. Linn. Soc., London, Bot., ser 2, 5: 321-342, Pls. 27-33, 9 tables.

Ceratium fusus
Nielsen, J., 1956.
Temporary variations in certain marine Ceratia. Oikos, 7(2):256-272.

Ceratium fusus
Nordli, E., 1957.
Experimental studies on the ecology of Ceratia. Oikos, 8(2):201-265.

Ceratium fusus
Nordli, E., 1953.
Salinity and temperature as controlling factors for distribution and mass occurrence of Ceratia. Blyttia 11:16-18, 4 textfigs.

Ceratium fusus
Paiva Carvalho, J., 1950
O plancton do Río María Rodriques (Cananeis). 1. Diatomaceas e Dinoflagelados. Bol. Inst. Paulista Oceanogr. 1(1); 27-43, 2 fold-in tables, 2 figs.

Ceratium fusus
Paulsen, O., 1908
XVIII Peridiniales. Nordisches Plankton, Bot. Teil: 1-124, 155 text figs.

Ceratium fusus
Pavillard, J., 1916
Recherches sur les Peridiniens du Golfe du Lion. Mem. Univ. Montpellier. Trav. Inst. Bot., Univ. Montpellier. Serie mixte No.4, 70 pp., 3 pls., 15 text figs.

Ceratium fusus
Pavillard, J., 1905
Recherches sur la flore pelagique (Phytoplankton) de l'Etang de Thau. Theses presentees a la Fac. Sci., Paris, 116 pp., 3 pls.

Ceratium fusus
Peters, Nicolaus, 1934
Die Bevolkerung des Sudatlantischen Ozeans mit Ceratien. Biol. Sonderuntersuchungen 1. Wiss Ergeb. Deutschen Atlantischen Exped.--- "Meteor" 1925-1927, 12(1): 1-69, 28 text figs.

Ceratium fusus
Robinson, G.A., 1965.
Continuous plankton records: contribution towards a plankton atlas of the North Atlantic and the North Sea. IX. Seasonal cycles of phytoplankton. Bulls. Mar. Ecol., Scottish Mar.Biol. Assoc., 6(4):104-122, pls. 26-61.

Ceratium fusus
Robinson, G.A., 1961
Contribution towards an atlas of the north-eastern Atlantic and the North Sea. 1. Phytoplankton. Bulls. Mar. Ecology, 5(42): 81-89, Pls. 15-20.

Ceratium fusus
Schodduyn, M., 1926
Observations faites dans la baie d'Ambleteuse (Pas de Calais). Bull. Inst. Ocean., Monaco, No. 482: 64 pp.

Ceratium fusus
Schröder, B., 1900
Phytoplankton des Golfes von Neapel nebst vergleichenden Ausblicken auf das atlantischen Ozean. Mitt. Zool. Stat. Neapel, 14:1-38.

Ceratium fusus
Sousa e Silva, E., 1949
Diatomaceas e Dinoflagelados de Baía de Cascais. Portugaliae Acta Biol., Volume: Julio Henriques, Ser. B: 300-383, 9 pls, 2 fold-in tables.

Ceratium fusus
Steemann-Nielsen, Einar, 1951
The marine vegetation of the Isefjord. A study on ecology and production. Medd. Komm. Danmarks Fiskeri-og Havundersøgelser. Ser. Plankton. 5(4); 114pp., 46 text figs.

Ceratium fusus
Steemann-Nielsen, E., 1939
Die Ceratien des Indischen Ozeans und der Ostasiatischen Gewässer mit einer allgemeinen zusammenfassung über die Verbreitung der Ceratien in den Weltmeeren. Dana Rept. No. 17: 33 pp., 8 maps in text.

Ceratium fusus
Steemann Nielsen, E., 1934
Untersuchungen über die Verbreitung, Biologie und Variation der Ceratien im südlichen Stillen Ozean. Dana Rept. No. 4: 67 pp., 73 figs. and 11 charts in text.

Ceratium fusus
Wheeler, J.E.G., 1939
Plankton investigations. Bermuda Biological Station. Second Report. October 1939. 7 pp. (typed), 5 figs. Plymouth, Oct. 23, 1939.

Ceratium fusus
Woodmansee, R.A., 1958.
The seasonal distribution of the zooplankton off Chicken Key in Biscayne Bay, Florida. Ecology, 39(2):247-261.

Ceratium fusus concavum
Gourret, P., 1883
Sur les Peridiniens du Golfe de Marseille. Ann. du Musee d'hist. Nat., Marseille, Zool., 1 (Mme. 8):1-114, 4 pls.

Ceratium fusus concave
Pavillard, J., 1905
Recherches sur la flore pelagique (Phytoplankton) de l'Etang de Thau. Theses presentees a la Fac. Sci., Paris, 116 pp., 3 pls.

Ceratium fusus extensum
Gourret, P., 1883
Sur les Peridiniens du Golfe de Marseille. Ann. du Musee d'hist. Nat., Marseille, Zool., 1 (Mme. 8):1-114, 4 pls.

Ceratium fusus seta
Nordli, E., 1957.
Experimental studies on the ecology of Ceratia. Oikos 8(2):200-265.

Ceratium gallicum n.sp.
Kofoid, C. A., 1907
Dinoflagellata of the San Diego region. III. Descriptions of new species. Univ. Calif. Publ., Zool. 3:299-340, Pls. 22-33.

Ceratium geniculatum
Ercegovic, A., 1940
Weitere Untersuchungen über einige hydrographische Verhältnisse und über die Phytoplanktonproduktion in den Gewässern der Östlichen Mittelradria. Acta Adriatica 2(3):95-134, 8 text figs.

Ceratium geniculatum
Graham, H. W. and N. Bronikovsky, 1944
The genus Ceratium in the Pacific and North Atlantic Oceans. Sci. Res. Cruise VII of the Carnegie, 1928-1929 ----- Biol. V (565):209 pp., 54 charts, 27 figs., 54 tables.

Ceratium geniculatum
Jörgensen, E., 1920
Mediterranean Ceratia. Rept. Danish Oceanogr. Exped. 1908-10 to the Mediterranean and adjacent seas, Vol.II, Biol. J 1:110 pp., 94 text figs., 26 charts.

Ceratium geniculatum
Peters, Nicolaus, 1934
Die Bevolkerung des Sudatlantischen Ozeans mit Ceratien. Biol. Sonderuntersuchungen 1. Wiss Ergeb. Deutschen Atlantischen Exped.--- "Meteor" 1925-1927, 12(1): 1-69, 28 text figs.

Ceratium geniculatum
Steemann-Nielsen, E., 1939
Die Ceratien des Indischen Ozeans und der Ostasiatischen Gewässer mit einer allgemeinen zusammenfassung über die Verbreitung der Ceratien in den Weltmeeren. Dana Rept. No. 17: 33 pp., 8 maps in text.

Ceratium geniculatum
Sousa e Silva, E., 1949
Diatomaceas e Dinoflagelados de Baia de Cascais. Portugaliae Acta Biol., Volume: Julio Henriques, Ser. B: 300-383, 9 pls, 2 fold-in tables.

Ceratium geniculatum
Steemann Nielsen, E., 1934
Untersuchungen über die Verbreitung, Biologie und Variation der Ceratien im südlichen Stillen Ozean. Dana Rept. No. 4: 67 pp., 73 figs. and 11 charts in text.

Ceratium geniculatum
Steemann Nielsen, E., 1934
Untersuchungen über die Verbreitung, Biologie und Variation der Ceratien im südlichen Stillen Ozean. Dana Rept. No. 4: 67 pp., 73 figs. and 11 charts in text.

Ceratium geniculatum
Wood, E.J.F., 1963
Dinoflagellates in the Australian region. II. Recent collections. Comm. Sci. and Industr. Res. Org., Div. Fish. and Oceanogr., Techn. Paper, No. 14: 55 pp.

Ceratium gibberum
Chiba, T., 1949
On the distribution of the plankton in the eastern China Sea and Yellow Sea. 1. Plankton composition in the spring. J. Shimonoseki Coll. Fisheries, 1(1):57-63, 1 fig.

Ceratium gibberum
Dangeard, P., 1927
Phytoplankton de la croisière du "Sylvana". Ann. Inst. Ocean., Monaco, n.s., 4(8):286-401, 54 text figs. (Fevrier-Juin 1913).

Ceratium gibberum
Ercegovic, A., 1936
Etudes qualitative et quantitatives du phytoplancton dans les eaux cotières de l'Adriatique oriental moyen au cours de l'année 1934. Acta Adriatica 1(9):1-126

Ceratium gibberum
Forti, A., 1922
Ricerche sulla flora pelagica (fitoplancton) di Quarto dei Mille. Mem. R. Com. Talass. Ital. 97:248 pp., 13 pls.

Ceratium gibberum
Gourret, P., 1883
Sur les Peridiniens du Golfe de Marseille. Ann. du Musee d'hist. Nat., Marseille, Zool., 1 (Mme. 8):1-114, 4 pls.

Ceratium gibberum
Graham, H. W. and N. Bronikovsky, 1944
The genus Ceratium in the Pacific and North Atlantic Oceans. Sci. Res. Cruise VII of the Carnegie, 1928-1929 ----- Biol. V (565):209 pp., 54 charts, 27 figs., 54 tables.

Ceratium gibberum
Jörgensen, E., 1920
Mediterranean Ceratia. Rept. Danish Oceanogr. Exped. 1908-10 to the Mediterranean and adjacent seas, Vol.II, Biol. J 1:110 pp., 94 text figs., 26 charts.

Ceratium gibberum
LeBour, M.V., 1925
The dinoflagellates of Northern Seas. The Marine Biological Association of the United Kingdom, Plymouth, 250 pp., 35 pls. 53 text figs.

Ceratium gibberum
Macdonald, R., 1933
An examination of plankton hauls made in the Suez Canal during the year 1928. Fish. Res. Dis., Notes & Mem. No.3, 11 pp., 1 chart.

Ceratium gibberum
Mangin, M. L., 1912
Phytoplancton de la croisière du "René" dans l'Atlantique (Septembre 1908). Ann. Inst. Ocean., n.s., 4(1):1-66, 2 pls., 41 text figs., 2 tables.

Ceratium gibberum
Margalef, R., 1949
Fitoplancton nerítico de la Costa Brava en 1947-48. Publ. Inst. Biol. Aplicada, 5: 41-51, 3 text figs.

Ceratium gibberum
Marukawa, H., 1921
Plankton lists and some new species of copepods, from the northern waters of Japan. Bull. Inst. Ocean. No.384, 15 pp., 3 pls., 1 chart. Monaco

Ceratium gibberum
Massutí Algamora, M., 1949
Estudio de diez y seis muestras de plancton del Golfo de Nápoles. Publ. Inst. Biol. Appl. 5:85-94, 1 fold-in table.

Ceratium gibberum
Mulford, Richard A., 1963
Distribution of the dinoflagellate genus Ceratium in the tidal and offshore waters of Virginia. Chesapeake Science, 4(2):84-89.

Ceratium gibberum
Paulsen, O., 1908
XVIII Peridiniales. Nordisches Plankton, Bot. Teil: 1-124, 155 text figs.

Ceratium gibberum
Pavillard, J., 1916
Recherches sur les Peridiniens du Golfe du Lion. Mem. Univ. Montpellier. Trav. Inst. Bot., Univ. Montpellier. Serie mixte No.4, 70 pp., 3 pls., 15 text figs.

Ceratium gibberum
Peters, Nicolaus, 1934
Die Bevolkerung des Sudatlantischen Ozeans mit Ceratien. Biol. Sonderuntersuchungen 1. Wiss Ergeb. Deutschen Atlantischen Exped.--- "Meteor" 1925-1927, 12(1): 1-69, 28 text figs.

Ceratium gibberum
Sournia, A., 1967.
Le genre Ceratium (peridinien planctonique) dans le canal de Mozambique. Contribution a une revision mondiale (fin). Vie Milieu (A) 18(3):441-449.

Ceratium gibberum
Sousa e Silva, E., 1949
Diatomaceas e Dinoflagelados de Baia de Cascais. Portugaliae Acta Biol., Volume: Julio Henriques, Ser. B: 300-383, 9 pls, 2 fold-in tables.

Ceratium gibberum
Steemann-Nielsen, E., 1939
Die Ceratien des Indischen Ozeans und der Ostasiatischen Gewässer mit einer allgemeinen zusammenfassung über die Verbreitung der Ceratien in den Weltmeeren. Dana Rept. No. 17: 33 pp., 8 maps in text.

Ceratium gibberum
Steemann Nielsen, E., 1934
Untersuchungen über die Verbreitung, Biologie und Variation der Ceratien im südlichen Stillen Ozean. Dana Rept. No. 4: 67 pp., 73 figs. and 11 charts in text.

Ceratium gibberum
Wheeler, J.E.G., 1939
Plankton investigations. Bermuda Biological Station. Second Report. October 1939. 7 pp. (typed), 5 figs. Plymouth, Oct. 23, 1939.

Ceratium gibberum contortum
Gourret, P., 1883
Sur les Peridiniens du Golfe de Marseille. Ann. du Musee d'hist. Nat., Marseille, Zool., 1 (Mme. 8):1-114, 4 pls.

Ceratium gibberum dispar
Forti, A., 1922
Ricerche sulla flora pelagica (fitoplancton) di Quarto dei Mille. Mem. R. Com. Talass. Ital. 97:248 pp., 13 pls.

Ceratium gibberum sinistra
Ercegovic, A., 1936
Etudes qualitative et quantitatives du phytoplancton dans les eaux cotières de l'Adriatique oriental moyen au cours de l'année 1934. Acta Adriatica 1(9):1-126

Ceratium gibberum sinistrum
Gourret, P., 1883
Sur les Peridiniens du Golfe de Marseille. Ann. du Musee d'hist. Nat., Marseille, Zool., 1 (Mme. 8):1-114, 4 pls.

Ceratium gibberum sinistrum
Lindemann, E., 1925
Neubeobachtungen an den Winter peridineen des Golfes von Neapel. Bot. Arch. 9:95-102, 19 text figs.

Ceratium gibberum sinistrum
Pavillard, J., 1916
Recherches sur les Peridiniens du Golfe du Lion. Mem. Univ. Montpellier. Trav. Inst. Bot., Univ. Montpellier. Serie mixte No.4, 70 pp., 3 pls., 15 text figs.

Ceratium gibberum subaequale
Graham, H. W. and N. Bronikovsky, 1944
The genus Ceratium in the Pacific and North Atlantic Oceans. Sci. Res. Cruise VII of the Carnegie, 1928-1929 ----- Biol. V (565):209 pp., 54 charts, 27 figs., 54 tables.

Ceratium globatum
Gourret, P., 1883
Sur les Peridiniens du Golfe de Marseille. Ann. du Musee d'hist. Nat., Marseille, Zool., 1 (Mme. 8):1-114, 4 pls.

Ceratium globosum
Gourret, P., 1883
Sur les Peridiniens du Golfe de Marseille. Ann. du Musee d'hist. Nat., Marseille, Zool., 1 (Mme. 8):1-114, 4 pls.

Ceratium gracile
Dangeard, P., 1927
Phytoplankton de la croisière du "Sylvana". Ann. Inst. Ocean., Monaco, n.s., 4(8):286-401, 54 text figs. (février-Juin 1913).

Ceratium gracile
Ercegovic, A., 1936
Etudes qualitative et quantitatives du phytoplancton dans les eaux côtières de l'Adriatique oriental moyen au cours de l'année 1934. Acta Adriatica 1(9):1-126

Ceratium gracile
Forti, A., 1922
Ricerche sulla flora pelagica (fitoplancton) di Quarto dei Mille. Mem. R. Com. Talass. Ital. 97:248 pp., 13 pls.

Ceratium gracile
Jörgensen, E., 1920
Mediterranean Ceratia. Rept. Danish Oceanogr. Exped. 1908-10 to the Mediterranean and adjacent seas, Vol.II, Biol. J 1:110 pp., 94 text figs., 26 charts.

Ceratium gracile
Pavillard, J., 1923
A propos de la systématique des Péridiniens Bull. Soc. Bot. de France 70:876-882; 914-918.

Ceratium gracile
Pavillard, J., 1916
Recherches sur les Peridiniens du Golfe du Lion. Mem. Univ. Montpellier. Trav. Inst. Bot., Univ. Montpellier. Serie mixte No.4, 70 pp., 3 pls., 15 text figs.

Ceratium gracile
Pavillard, J., 1905
Recherches sur la flore pelagique (Phytoplankton) de l'Etang de Thau. Theses presentees a la Fac. Sci., Paris, 116 pp., 3 pls.

Ceratium gracile
Peters, Nicolaus, 1934
Die Bevolkerung des Sudatlantischen Ozeans mit Ceratien. Biol. Sonderuntersuchungen 1. Wiss Ergeb. Deutschen Atlantischen Exped.--- "Meteor" 1925-1927, 12(1): 1-69, 28 text figs.

Ceratium gracile
Sousa e Silva, E., 1949
Diatomaceas e Dinoflagelados de Baia de Cascais. Portugaliae Acta Biol. Volume: Julio Henriques, Ser. B: 300-383, 9 pls, 2 fold-in tables.

Ceratium gracile symmetricum
Ercegovic, A., 1936
Etudes qualitative et quantitatives du phytoplancton dans les eaux côtières de l'Adriatique oriental moyen au cours de l'année 1934. Acta Adriatica 1(9):1-126

Ceratium gracile var. symmetricum
Ghazzawi, F.M., 1939
Plankton of the Egyptian waters. A study of the Suez Canal Plankton. (A) Phytoplankton. Preliminary Report 83 pp. Notes and Memoires, Min. Commerce-Industry, Egypt, Hydrobiol. & Fish. 65 figs.

Ceratium gravidum
Dangeard, P., 1927
Phytoplankton de la croisière du "Sylvana". Ann. Inst. Ocean., Monaco, n.s., 4(8):286-401, 54 text figs. (février-Juin 1913).

Ceratium gravidum
Ercegovic, A., 1940
Weitere Untersuchungen über einige hydrographische Verhältnisse und über die Phytoplanktonproduktion in den Gewässern der Östlichen Mitteladria. Acta Adriatica 2(3):95-134, 8 text figs.

Ceratium gravidum
Forti, A., 1922
Ricerche sulla flora pelagica (fitoplancton) di Quarto dei Mille. Mem. R. Com. Talass. Ital. 97:248 pp., 13 pls.

Ceratium gravidum
Gourret, P., 1883
Sur les Peridiniens du Golfe de Marseille. Ann. du Musee d'hist. Nat., Marseille, Zool., 1 (Mme. 8):1-114, 4 pls.

Ceratium gravidum
Graham, H. W. and N. Bronikovsky, 1944
The genus Ceratium in the Pacific and North Atlantic Oceans. Sci. Res. Cruise VII of the Carnegie, 1928-1929 ----- Biol. V (565):209 pp., 54 charts, 27 figs., 54 tables.

Ceratium gravidum
Jörgensen, E., 1920
Mediterranean Ceratia. Rept. Danish Oceanogr. Exped. 1908-10 to the Mediterranean and adjacent seas, Vol.II, Biol. J 1:110 pp., 94 text figs., 26 charts.

Ceratium gravidum
Marukawa, H., 1921
Plankton lists and some new species of copepods, from the northern waters of Japan. Bull. Inst. Ocean., No.384, 15 pp., 3 pls., 1 chart. Monaco

Ceratium gravidum
Murray, G., and F. G. Whitting, 1899
New Peridiniaceae from the Atlantic. Trans. Linn. Soc., London, Bot., ser 2, 5: 321-342, Pls. 27-33, 9 tables.

Ceratium gravidum
Pavillard, J., 1916
Recherches sur les Peridiniens du Golfe du Lion. Mem. Univ. Montpellier. Trav. Inst. Bot., Univ. Montpellier. Serie mixte No.4, 70 pp., 3 pls., 15 text figs.

Ceratium gravidum
Peters, Nicolaus, 1934
Die Bevolkerung des Sudatlantischen Ozeans mit Ceratien. Biol. Sonderuntersuchungen 1. Wiss Ergeb. Deutschen Atlantischen Exped.--- "Meteor" 1925-1927, 12(1): 1-69, 28 text figs.

Ceratium gravidum
Schröder, B., 1900
Phytoplankton des Golfes von Neapel nebst vergleichenden Ausblicken auf das atlantischen Ozean. Mitt. Zool. Stat. Neapel, 14:1-38.

Ceratium gravidum
Steemann-Nielsen, E., 1939
Die Ceratien des Indischen Ozeans und der Ostasiatischen Gewässer mit einer allgemeinen zusammenfassung über die Verbreitung der Ceratien in den Weltmeeren. Dana Rept. No. 17: 33 pp., 8 maps in text.

Ceratium gravidum
Steemann Nielsen, E., 1934
Untersuchungen über die Verbreitung, Biologie und Variation der Ceratien im südlichen Stillen Ozean. Dana Rept. No. 4: 67 pp., 73 figs. and 11 charts in text.

Ceratium gravidum elongatum n.var.
Wood, E.J.F., 1963
Dinoflagellates in the Australian region. II. Recent collections. Comm. Sci. and Industr. Res. Org., Div. Fish. and Oceanogr., Techn. Paper, No. 14: 55 pp.

Ceratium heterocamptum
Bigelow, H. B., 1922
Exploration of the coastal water off the northeastern United States in 1916 by the U.S. Fisheries Schooner Grampus. Bull. M.C.Z. 65 (5):85-188, 53 text figs.

Ceratium heterocamptum
Paulsen, O., 1908
XVIII Peridiniales. Nordisches Plankton, Bot. Teil: 1-124, 155 text figs.

Ceratium heterocamptum
Pavillard, J., 1905
Recherches sur la flore pelagique (Phytoplankton) de l'Etang de Thau. Theses presentees a la Fac. Sci., Paris, 116 pp., 3 pls.

Ceratium hexacanthum
Dangeard, P., 1927
Phytoplankton de la croisière du "Sylvana". Ann. Inst. Ocean., Monaco, n.s., 4(8):286-401, 54 text figs. (février-Juin 1913).

Ceratium hexacanthum
Davis, Joanne T., 1965
Leaflet series: Plankton, Florida Bd., Conservation, Mar. Lab., St. Petersburg, 1(9):4 pp.

no title!

Ceratium hexacanthum
Delegazione Italiana della Commissione Internazionale per l'Esplorazione Scientifica del Mediterraneo, 1941
Note sul plancton della Laguna veneta. [Memoria CCLXXIX], Arch. di Ocean. e Limn. Anno I, Fasc. I, 1941 XIX: 31-57 pp.

Ceratium hexacanthum
Gourret, P., 1883
Sur les Peridiniens du Golfe de Marseille. Ann. du Musee d'hist. Nat., Marseille, Zool., 1 (Mme. 8):1-114, 4 pls.

Ceratium hexacanthum
Graham, H. W. and N. Bronikovsky, 1944
The genus Ceratium in the Pacific and North Atlantic Oceans. Sci. Res. Cruise VII of the Carnegie, 1928-1929 ----- Biol. V (565):209 pp., 54 charts, 27 figs., 54 tables.

Ceratium hexacanthum
Jörgensen, E., 1920
Mediterranean Ceratia. Rept. Danish Oceanogr. Exped. 1908-10 to the Mediterranean and adjacent seas, Vol.II, Biol. J 1:110 pp., 94 text figs., 26 charts.

Ceratium hexacanthum
Margalef, R., 1949
Fitoplancton nerítico de la Costa Brava en 1947-48. Publ. Inst. Biol. Aplicada, 5: 41-51, 3 text figs.

Ceratium hexacanthum
Peters, Nicolaus, 1934
Die Bevolkerung des Sudatlantischen Ozeans mit Ceratien. Biol. Sonderuntersuchungen 1. Wiss Ergeb. Deutschen Atlantischen Exped.--- "Meteor" 1925-1927, 12(1): 1-69, 28 text figs.

Ceratium hexacanthum
Rampi, L., 1942
Ricerche sul fitoplancton del Mare Ligure. 4. I Ceratium delle acque di Sanremo. Nuovo Giornale Botanico Italiano, N.S., 49: 221-236, 19 text figs.

Ceratium hexacanthum
Robinson, G.A., 1961
Contribution towards an atlas of the northeastern Atlantic and the North Sea. 1. Phytoplankton. Bulls. Mar. Ecology, 5(42): 81-89, Pls. 15-20.

Ceratium hexacanthum
Sournia, A., 1967.
Le genre Ceratium (peridinien planctonique) dans le canal de Mozambique. Contribution a une revision mondiale (fin). Vie Milieu (A) 18(3):441-449.

Ceratium hexacanthum
Sousa e Silva, E., 1949
Diatomaceas e Dinoflagelados de Baia de Cascais. Portugaliae Acta Biol., Volume: Julio Henriques, Ser. B: 300-383, 9 pls, 2 fold-in tables.

Ceratium hexacanthum
Steemann-Nielsen, E., 1939
Die Ceratien des Indischen Ozeans und der Ostasiatischen Gewässer mit einer allgemeinen zusammenfassung über die Verbreitung der Ceratien in den Weltmeeren. Dana Rept. No. 17: 33 pp., 8 maps in text.

Ceratium hexacanthum
Steemann Nielsen, E., 1934
Untersuchungen über die Verbreitung, Biologie und Variation der Ceratien im südlichen Stillen Ozean. Dana Rept. No. 4: 67 pp., 73 figs. and 11 charts in text.

Ceratium hexacanthum var. hiemale
Rampi, L., 1939
Péridiniens rares ou interessants recoltes dans la mer Ligure. Bull. Soc. Fran. de Microscopie, 8(2/3):106-112, 13 text figs.

Ceratium hirundella
Becheler, B., 1952
Recherches sur les Peridiniens. Bull. Biol., France Belg., Suppl., 36:1-149, 75 textfigs.

Ceratium hirundella
Birte, M., 1955.
Variabilite e differenzione del Ceratium hirundinella nei lagi Italiani. Arch. Ocean. e Limnol. 10(1/2):47-65.

Ceratium hirundella
De Angelis, C.M., 1962
Distribuzione ed ecologia di alcune specie di Dinoflagellati di acque salmastre. Pubbl. Staz. Zool., Napoli, 32 (Suppl.):301-314.

Ceratium hirundinella
Braarud, T., 1945
A phytoplankton survey of the polluted waters of inner Oslo Fjord. Hvalrådets Skrifter, No.28, 142 pp., 19 text figs., 17 tables.

Ceratium hirundinella
Brunel, J., 1962
Le phytoplancton de la Baie de Chaleurs. Inst. Botan., Univ. Montréal, Contrib. No. 77: 365 pp., 66 pls.

Ceratium hirundinella
Dodge, John D., and Richard M. Crawford 1970.
The morphology and fine structure of Ceratium hirundinella (Dinophyceae) J. Phycol. 6(2): 137-149.

Ceratium hirundinella
LeBour, M.V., 1925
The dinoflagellates of Northern Seas. The Marine Biological Association of the United Kingdom. Plymouth, 250 pp., 35 pls. 53 text figs.

Ceratium hirundinella
Levander, K.M., 1947
Plankton gesammelt in den Jahren 1899-1910 an den Küsten Finnlands. Finnländische Hydrographisch-Biologische Untersuchunger (aus dem Wasserbiologischen Laboratorin der Societas Scientiarum Fenniea) No.11: 40 pp., 6 diagrams, 13 pls., tables.

Ceratium hirundinella
Meunier, A., 1919
Microplancton de la Mer Flamande 3. Les Péridiniens. Mém. Mus. Roy. Hist. Nat. Belgique 8(1):1-116, Pls. XV-XXI.

Ceratium hirundinella
Poulsen, O., 1908
XVIII Peridiniales. Nordisches Plankton, Bot. Teil: 1-124, 155 text figs.

Ceratium hirundinella
Sebestyen, O., 1959.
The ecological niche of Ceratium hirundinella Schrank in the plankton community and in lacustrine life in general. Acta Biologica, Acad. Sci. Hung., 10(2):235-244.

Ceratium hirundinella
Shoji, S., 1952.
Synecological notes on the variation of the number of spines in Ceratium hirundinella O.F. Müller Ecol. Rev. Mt. Hakkoda Bot. Lab., Sendai, 13(2): 95-98, 2 textfigs.

Ceratium horridum
Brunel, J., 1962
Le phytoplancton de la Baie de Chaleurs. Inst. Botan., Univ. Montréal, Contrib. No. 77: 365 pp., 66 pls.

Ceratium horridum
Colebrook, J.M., and G.A. Robinson, 1964.
Continuous plankton records: annual variations of abundance of plankton 1948-1960. Bull. Mar. Ecol., 6(3):52-69.

Ceratium horridum
Dangeard, P., 1927
Phytoplankton de la croisière du "Sylvana". Ann. Inst. Ocean., Monaco, n.s., 4(8):286-401, 54 text figs. (Feirer-Juin 1913).

Ceratocorys horrida
Delegazione Italiana della Commissione Internazionale per l'Esplorazione Scientifica del Mediterraneo, 1941
Note sul plancton della Laguna veneta. [Memoria CCLXXIX], Arch. di Ocean. e Limn. Anno I, Fasc. I, 1941 XIX: 31-57 pp.

Ceratium horridum
Fraser, J. H., and A. Saville, 1949
Plankton distribution in Scottish and adjacent waters in 1948. Ann. Biol. 5:61-62.

Ceratium horridum
Graham, H. W. and N. Bronikovsky, 1944
The genus Ceratium in the Pacific and North Atlantic Oceans. Sci. Res. Cruise VII of the Carnegie, 1928-1929 ----- Biol. V (565):209 pp., 54 charts, 27 figs., 54 tables.

Ceratium horridum
Jörgensen, E., 1920
Mediterranean Ceratia. Rept. Danish Oceanogr. Exped. 1908-10 to the Mediterranean and adjacent seas, Vol.II, Biol. J 1:110 pp., 94 text figs., 26 charts.

Ceratium horridum
Halim, Y., 1963.
Microplancton des eaux égyptiennes. Le genre Ceratium Schrank (Dinoflagellés). Rapp. Proc. Verb. Réunions, Comm. Int. Expl. Sci. Mer Méditerranée, Monaco, 17(2):495-502.

Ceratium horridum
LeBour, M.V., 1925
The dinoflagellates of Northern Seas. The Marine Biological Association of the United Kingdom. Plymouth, 250 pp., 35 pls. 53 text figs.

Ceratium horridum
Margalef, R., 1949
Fitoplancton nerítico de la Costa Brava en 1947-48. Publ. Inst. Biol. Aplicada, 5:41-51, 3 text figs.

Ceratium horridum
Massutí Algamora, M., 1949
Estudio de diez y seis muestras de plancton del Golfo de Nápoles. Publ. Inst. Biol. Appl. 5:85-94, 1 fold-in table.

Ceratium horridum
Pavillard, J., 1923
A propos de la systématique des Péridiniens Bull. Soc. Bot. de France 70:876-882; 914-918.

Ceratium horridum
Peters, Nicolaus, 1934
Die Bevolkerung des Sudatlantischen Ozeans mit Ceratien. Biol. Sonderuntersuchungen 1. Wiss Ergeb. Deutschen Atlantischen Exped.---- "Meteor" 1925-1927, 12(1): 1-69, 28 text figs.

Ceratium horridum
Robinson, G.A., 1965.
Continuous plankton records: contribution towards a plankton atlas of the North Atlantic and the North Sea. IX. Seasonal cycles of phytoplankton. Bulls. Mar. Ecol., Scottish Mar. Biol. Assoc., 6(4):104-122, pls. 26-61.

Ceratium horridum
Robinson, G.A., 1961
Contribution towards an atlas of the northeastern Atlantic and the North Sea. 1. Phytoplankton. Bulls. Mar. Ecology, 5(42): 81-89, Pls. 15-20.

Ceratium horridum
Sournia, A., 1967.
Le genre Ceratium (peridinien planctonique) dans le canal de Mozambique. Contribution a une revision mondiale (fin). Vie Milieu (A) 18(3):441-449.

Ceratium horridum
Sousa e Silva, E., 1949
Diatomaceas e Dinoflagelados de Baia de Cascais. Portugaliae Acta Biol., Volume: Julio Henriques, Ser. B: 300-383, 9 pls, 2 fold-in tables.

Ceratium horridum
Von Stosch, H.A., 1964.
Zum Problem der sexuallen Fortpflanzung in der Peridineengattung Ceratium. Helgoländer Wiss. Meeresuntersuch., 10(1/4):140-152.

Ceratium horridum
Yarranton, G.A. 1967.
Parameters for use in distinguishing populations of Euceratium Gran. Bull. mar. Ecol. 6(6):147-158.

Ceratium horridum claviger
Graham, H. W. and N. Bronikovsky, 1944
The genus Ceratium in the Pacific and North Atlantic Oceans. Sci. Res. Cruise VII of the Carnegie, 1928-1929 ----- Biol. V (565):209 pp., 54 charts, 27 figs., 54 tables.

Ceratium horridum horridum
Graham, H. W. and N. Bronikovsky, 1944
The genus Ceratium in the Pacific and North Atlantic Oceans. Sci. Res. Cruise VII of the Carnegie, 1928-1929 ----- Biol. V (565):209 pp., 54 charts, 27 figs., 54 tables.

Ceratium horridum molle
Graham, H. W. and N. Bronikovsky, 1944
The genus Ceratium in the Pacific and North Atlantic Oceans. Sci. Res. Cruise VII of the Carnegie, 1928-1929 ----- Biol. V (565):209 pp., 54 charts, 27 figs., 54 tables.

Ceratium humile
Graham, H. W. and N. Bronikovsky, 1944
The genus Ceratium in the Pacific and North Atlantic Oceans. Sci. Res. Cruise VII of the Carnegie, 1928-1929 ----- Biol. V (565):209 pp., 54 charts, 27 figs., 54 tables.

Ceratium humile
Steemann-Nielsen, E., 1939
Die Ceratien des Indischen Ozeans und der Ostasiatischen Gewässer mit einer allgemeinen zusammenfassung über die Verbreitung der Ceratien in den Weltmeeren. Dana Rept. No. 17: 33 pp., 8 maps in text.

Ceratium humile
Steemann Nielsen, E., 1934
Untersuchungen über die Verbreitung, Biologie und Variation der Ceratien im südlichen Stillen Ozean. Dana Rept. No. 4: 67 pp., 73 figs. and 11 charts in text.

Ceratium incisum
Graham, H. W. and N. Bronikovsky, 1944
The genus Ceratium in the Pacific and North Atlantic Oceans. Sci. Res. Cruise VII of the Carnegie, 1928-1929 ----- Biol. V (565):209 pp., 54 charts, 27 figs., 54 tables.

Ceratium incisum
Jörgensen, E., 1920
Mediterranean Ceratia. Rept. Danish Oceanogr. Exped. 1908-10 to the Mediterranean and adjacent seas, Vol.II, Biol. J 1:110 pp., 94 text figs., 26 charts.

Ceratium incisum
Peters, Nicolaus, 1934
Die Bevolkerung des Sudatlantischen Ozeans mit Ceratien. Biol. Sonderuntersuchungen 1. Wiss Ergeb. Deutschen Atlantischen Exped.---"Meteor" 1925-1927, 12(1): 1-69, 28 text figs.

Ceratium incisum
Steemann-Nielsen, E., 1939
Die Ceratien des Indischen Ozeans und der Ostasiatischen Gewässer mit einer allgemeinen zusammenfassung über die Verbreitung der Ceratien in den Weltmeeren. Dana Rept. No. 17: 33 pp., 8 maps in text.

Ceratium incisum
Steemann Nielsen, E., 1934
Untersuchungen über die Verbreitung, Biologie und Variation der Ceratien im südlichen Stillen Ozean. Dana Rept. No. 4: 67 pp., 73 figs. and 11 charts in text.

Ceratium inclinatum n.sp.
Kofoid, C. A., 1907
Dinoflagellata of the San Diego region. III. Descriptions of new species. Univ. Calif. Publ., Zool. 3:299-340, Pls. 22-33.

Ceratium inclinatum
Pavillard, J., 1923
A propos de la systematique des Péridiniens Bull. Soc. Bot. de France 70:876-882; 914-918.

Ceratium inflatum
Dangeard, P., 1927
Phytoplankton de la croisière du "Sylvana". Ann. Inst. Ocean., Monaco, n.s., 4(8):286-401, 54 text figs. (Feirer-Juin 1913).

Ceratium inflatum
Delegazione Italiana della Commissione Internazionale per l'Esplorazione Scientifica del Mediterraneo, 1941
Note sul plancton della Laguna veneta. [Memoria CCLXXXI], Arch. di Ocean. e Limn. Anno I, Fasc. I, 1941 XIX: 31-57 pp.

Ceratium inflatum
Ercegovic, A., 1936
Etudes qualitative et quantitatives du phytoplancton dans les eaux cotières de l'Adriatique oriental moyen au cours de l'année 1934. Acta Adriatica 1(9):1-126

Ceratium inflatum
Forti, A., 1922
Ricerche sulla flora pelagica (fitoplancton) di Quarto dei Mille. Mem. R. Com. Talass. Ital. 97:248 pp., 13 pls.

Ceratium inflatum
Graham, H. W. and N. Bronikovsky, 1944
The genus Ceratium in the Pacific and North Atlantic Oceans. Sci. Res. Cruise VII of the Carnegie, 1928-1929 ----- Biol. V (565):209 pp., 54 charts, 27 figs., 54 tables.

Ceratium inflatum
Jörgensen, E., 1920
Mediterranean Ceratia. Rept. Danish Oceanogr. Exped. 1908-10 to the Mediterranean and adjacent seas, Vol.II, Biol. J 1:110 pp., 94 text figs., 26 charts.

Ceratium inflatum
Pavillard, J., 1916
Recherches sur les Peridiniens du Golfe du Lion. Mem. Univ. Montpellier. Trav. Inst. Bot., Univ. Montpellier. Serie mixte No.4, 70 pp., 3 pls., 15 text figs.

Ceratium inflatum
Peters, Nicolaus, 1934
Die Bevolkerung des Sudatlantischen Ozeans mit Ceratien. Biol. Sonderuntersuchungen 1. Wiss Ergeb. Deutschen Atlantischen Exped.---"Meteor" 1925-1927, 12(1): 1-69, 28 text figs.

Ceratium inflatum
Steemann-Nielsen, E., 1939
Die Ceratien des Indischen Ozeans und der Ostasiatischen Gewässer mit einer allgemeinen zusammenfassung über die Verbreitung der Ceratien in den Weltmeeren. Dana Rept. No. 17: 33 pp., 8 maps in text.

Ceratium inflatum
Steemann-Nielsen, E., 1934
Untersuchungen über die Verbreitung, Biologie und Variation der Ceratien im südlichen Stillen Ozean. Dana Rept. No. 4: 67 pp., 73 figs. and 11 charts in text.

Ceratium inflexum
Ercegovic, A., 1936
Etudes qualitative et quantitatives du phytoplancton dans les eaux cotières de l'Adriatique oriental moyen au cours de l'année 1934. Acta Adriatica 1(9):1-126

Ceratium inflexum
Forti, A., 1922
Ricerche sulla flora pelagica (fitoplancton) di Quarto dei Mille. Mem. R. Com. Talass. Ital. 97:248 pp., 13 pls.

Ceratium inflexum
Kokubo, S., and S. Sato, 1947
Plankters in JG-San Gata. Physiol. and Ecol. (Japan) 1(4):1-16, 3 text figs., tables.

Ceratium inflexum
Lindemann, E., 1925
Neubeobachtungen an den Winter peridineen des Golfes von Neapel. Bot. Arch. 9:95-102, 19 text figs.

Ceratium inflexum
Lindemann, E., 1924
Peridineen aus dem goldenen Horn und dem Bosphorus. Bot. Arch. 5:216-233, 98 text figs.

Ceratium inflexum
Marukawa, H., 1921
Plankton lists and some new species of copepods, from the northern waters of Japan. Bull. Inst. Ocean., No.384, 15 pp., 3 pls., 1 chart. Monaco

Ceratium inflexum
Pavillard, J., 1923
A propos de la systematique des Péridiniens Bull. Soc. Bot. de France 70:876-882; 914-918.

Ceratium inflexum
Pavillard, J., 1916
Recherches sur les Peridiniens du Golfe du Lion. Mem. Univ. Montpellier. Trav. Inst. Bot., Univ. Montpellier. Serie mixte No.4, 70 pp., 3 pls., 15 text figs.

Ceratium inflexum claviceps
Forti, A., 1922
Ricerche sulla flora pelagica (fitoplancton) di Quarto dei Mille. Mem. R. Com. Talass. Ital. 97:248 pp., 13 pls.

Ceratium intermedium
Braarud, T., and Adam Bursa, 1939
On the phytoplankton of the Oslo Fjord, 1933-1934. Hvalrådets Skr. No.19:1-63; 9 text figs. Reviewed. J. du. Cons. 14(3): 418-420. A.C. Gardiner.

Ceratium intermedium
Dangeard, P., 1926
Description des Péridiniens Testacés recueillis para la Mission Charcot pendent le mois d'Aout 1924. Ann. Inst. Ocean. n.s. 3(7):307-334, 15 text figs.

Ceratium intermedium
Jörgensen, E., 1905
B.Protistplankton and the diatoms in bottom samples. Hydrographical and biological investigations in Norwegian fjords. Bergens Mus. Skr. 7: 49-225.

Ceratium intermedium
Paulsen, O., 1908
XVIII Peridiniales. Nordisches Plankton, Bot. Teil: 1-124, 155 text figs.

Ceratium intermedium
Pavillard, J., 1923
A propos de la systematique des Péridiniens Bull. Soc. Bot. de France 70:876-882; 914-918.

Ceratium intermedium
Pavillard, J., 1905
Recherches sur la flore pelagique (Phytoplankton) de l'Etang de Thau. Theses presentees a la Fac. Sci., Paris, 116 pp., 3 pls.

Ceratium intermedium
Wang, C. C., and D. Nie, 1932
A survey of the marine protozoa of Amoy. Contr. Biol. Lab. Sci. Soc. China, Zool. Ser. 8:285-385, 89 text figs.

Ceratium karsteni
Balech, Enrique, 1962
Tintinnoinea y Dinoflagellata del Pacifico segun material de las Expediciones NORPAC y DOWNWIND del Instituto Scripps de Oceanografia. Revista. Mus. Argentino Ciencias Nat. "Bernardino Rivadavia", Ciencias Zool., 7(1):1-253.

Ceratium Karstenii
Dangeard, P., 1927
Phytoplankton de la croisière du "Sylvana". Ann. Inst. Ocean., Monaco, n.s., 4(8):286-401, 54 text figs. (février-Juin 1913).

Ceratium karsteni
Delegazione Italiana della Commissione Internazionale per l'Esplorazione Scientifica del Mediterraneo, 1941
Note sul plancton della Laguna veneta. [Memoria CCLXXXI], Arch. di Ocean. e Limn. Anno I, Fasc. I, 1941 XIX: 31-57 pp.

Ceratium karsteni
Halim, Y., 1963.
Microplancton des eaux égyptiennes. Le genre Ceratium Schrank (Dinoflagellés). Rapp. Proc. Verb. Réunions, Comm. Int. Expl Sci. Mer Méditerranée, Monaco, 17(2):495-502.

Ceratium karsteni

Lindemann, E., 1925
 Neubeobachtungen an den Winter peridineen des Golfes von Neapel. Bot. Arch. 9:95-102, 19 text figs.

Ceratium Karsteni

Margalef, R., 1949
 Fitoplancton nerítico de la Costa Brava en 1947-48. Publ. Inst. Biol. Aplicada, 5: 41-51, 3 text figs.

Ceratium Karstenii

Pavillard, J., 1923
 A propos de la systématique des Péridiniens Bull. Soc. Bot. de France 70:876-882; 914-918.

Ceratium Karsteni

Pavillard, J., 1916
 Recherches sur les Peridiniens du Golfe du Lion. Mem. Univ. Montpellier. Trav. Inst. Bot., Univ. Montpellier. Serie mixte No.4, 70 pp., 3 pls., 15 text figs.

Ceratium Karstenii

Steemann-Nielsen, E., 1939
 Die Ceratien des Indischen Ozeans und der Ostasiatischen Gewässer mit einer allgemeinen zusammenfassung über die Verbreitung der Ceratien in den Weltmeeren. Dana Rept. No. 17: 33 pp., 8 maps in text.

Ceratium karstenii

Steemann Nielsen, E., 1934
 Untersuchungen über die Verbreitung, Biologie und Variation der Ceratien im südlichen Stillen Ozean. Dana Rept. No. 4: 67 pp., 73 figs. and 11 charts in text.

Ceratium Kofoidii

Dangeard, P., 1927
 Phytoplankton de la croisière du "Sylvana". Ann. Inst. Ocean., Monaco, n.s., 4(8):286-401, 54 text figs. (février-Juin 1913).

Ceratium kofoidii

Gilbert, J.Y., and W.E. Allen, 1943
 The phytoplankton of the Gulf of California obtained by the "E.W. Scripps" in 1939 and 1940. J. Mar. Res. V(2):89-110, figs.30-31.

Ceratium kofoidii

Graham, H. W. and N. Bronikovsky, 1944
 The genus Ceratium in the Pacific and North Atlantic Oceans. Sci. Res. Cruise VII of the Carnegie, 1928-1929 ----- Biol. V (565):209 pp., 54 charts, 27 figs., 54 tables.

Ceratium kofoidi

Hada, Yoshina, 1970. (1970)
 The protozoan plankton of the Antarctic and Subantarctic seas. Scient. Repts, Japan. Antarct. Res. Exped., (E)31:51 pp.

Ceratium Kofoidii

Jörgensen, E., 1920
 Mediterranean Ceratia. Rept. Danish Oceanogr. Exped. 1908-10 to the Mediterranean and adjacent seas, Vol.II, Biol. J 1:110 pp., 94 text figs., 26 charts.

Ceratium kofoidii

Marshall, S. M., 1933
 The production of microplankton in the Great Barrier Reef Region. Brit. Mus. (N.H.) Great Barrier Reef Exped. 1928-29, Sci. Repts. II(5):111-157, 14 text figs.

Ceratium kofoidi

Peters, Nicolaus, 1934
 Die Bevolkerung des Sudatlantischen Ozeans mit Ceratien. Biol. Sonderuntersuchungen 1. Wiss Ergeb. Deutschen Atlantischen Exped.---- "Meteor" 1925-1927, 12(1): 1-69, 28 text figs.

Ceratium kofoidi

Rampi, L., 1942.
 Ricerche sul fitoplancton del Mare Ligure. 4. I Ceratium delle acque di Sanremo. Nuovo Giornale Botanico Italiano, N.S., 49: 221-236, 19 text figs.

Ceratium kofoidi

Rampi, L., 1939
 Péridiniens rares ou intéressants recoltés dans la mer Ligure. Bull. Soc. Fran. de Microscopie, 8(2/3):106-112, 13 text figs.

Ceratium kofoidii

Steemann-Nielsen, E., 1939
 Die Ceratien des Indischen Ozeans und der Ostasiatischen Gewässer mit einer allgemeinen zusammenfassung über die Verbreitung der Ceratien in den Weltmeeren. Dana Rept. No. 17: 33 pp., 8 maps in text.

Ceratium kofoidii

Steemann Nielsen, E., 1934
 Untersuchungen über die Verbreitung, Biologie und Variation der Ceratien im südlichen Stillen Ozean. Dana Rept. No. 4: 67 pp., 73 figs. and 11 charts in text.

Ceratium lamellicorne

LeBour, M.V., 1925
 The dinoflagellates of Northern Seas. The Marine Biological Association of the United Kingdom, Plymouth, 250 pp., 35 pls. 53 text figs.

Ceratium leptosomum

Dangeard, P., 1927
 Phytoplankton de la croisière du "Sylvana". Ann. Inst. Ocean., Monaco, n.s., 4(8):286-401, 54 text figs. (février-Juin 1913).

Ceratium limulus

Dangeard, P., 1927
 Phytoplankton de la croisière du "Sylvana". Ann. Inst. Ocean., Monaco, n.s., 4(8):286-401, 54 text figs. (février-Juin 1913).

Ceratium limulus

Forti, A., 1922
 Ricerche sulla flora pelagica (fitoplancton) di Quarto dei Mille. Mem. R. Com. Talass. Ital. 97:248 pp., 13 pls.

Ceratium limulus

Gourret, P., 1883
 Sur les Peridiniens du Golfe de Marseille. Ann. du Musee d'hist. Nat., Marseille, Zool., 1 (Mme. 8):1-114, 4 pls.

Ceratium limulus

Graham, H. W. and N. Bronikovsky, 1944
 The genus Ceratium in the Pacific and North Atlantic Oceans. Sci. Res. Cruise VII of the Carnegie, 1928-1929 ----- Biol. V (565):209 pp., 54 charts, 27 figs., 54 tables.

Ceratium limulus

Pavillard, J., 1916
 Recherches sur les Peridiniens du Golfe du Lion. Mem. Univ. Montpellier. Trav. Inst. Bot., Univ. Montpellier. Serie mixte No.4, 70 pp., 3 pls., 15 text figs.

Ceratium limulus

Pavillard, J., 1905
 Recherches sur la flore pelagique (Phytoplankton) de l'Etang de Thau. Theses presentees a la Fac. Sci., Paris, 116 pp., 3 pls.

Ceratium limulus

Peters, Nicolaus, 1934
 Die Bevolkerung des Sudatlantischen Ozeans mit Ceratien. Biol. Sonderuntersuchungen 1. Wiss Ergeb. Deutschen Atlantischen Exped.---- "Meteor" 1925-1927, 12(1): 1-69, 28 text figs.

Ceratium limulus

Sournia, A., 1967.
 Le genre Ceratium (peridinien planctonique) dans le canal de Mozambique. Contribution a une revision mondiale (fin). Vie Milieu (A) 18(3):441-449.

Ceratium limulus

Sousa e Silva, E., 1949
 Diatomaceas e Dinoflagelados de Baia de Cascais. Portugaliae Acta Biol. Volume: Julio Henriques, Ser. B: 300-383, 9 pls, 2 fold-in tables.

Ceratium limulus

Steemann-Nielsen, E., 1939
 Die Ceratien des Indischen Ozeans und der Ostasiatischen Gewässer mit einer allgemeinen zusammenfassung über die Verbreitung der Ceratien in den Weltmeeren. Dana Rept. No. 17: 33 pp., 8 maps in text.

Ceratium limulus

Steemann Nielsen, E., 1934
 Untersuchungen über die Verbreitung, Biologie und Variation der Ceratien im südlichen Stillen Ozean. Dana Rept. No. 4: 67 pp., 73 figs. and 11 charts in text.

Ceratium lineatum

Balech, Enrique, 1944
 Contribucion al conocimiento del Plancton de Lennox y Cabo de Hornos. Physis XIX:423-446, 6 pls. with 67 figs.

Ceratium lineatus

Braarud, T., 1945
 A phytoplankton survey of the polluted waters of inner Oslo Fjord. Hvalrådets Skrifter, No.28, 142 pp., 19 text figs., 17 tables.

Ceratium lineatum

Braarud, T., and Adam Bursa, 1939
 On the phytoplankton of the Oslo Fjord, 1933-1934. Hvalrådets Skr. No.19:1-63; 9 text figs. Reviewed. J. du. Cons. 14(3): 418-420. A.C. Gardiner.

Ceratium lineatum

Chiba, T., 1949
 On the distribution of the plankton in the eastern China Sea and Yellow Sea. 1. Plankton composition in the spring. J. Shimonoseki Coll. Fisheries, 1(1):57-63, 1 fig.

Ceratium lineatum

Dangeard, P., 1927
 Phytoplankton de la croisière du "Sylvana". Ann. Inst. Ocean., Monaco, n.s., 4(8):286-401, 54 text figs. (février-Juin 1913).

Ceratium lineatum

Dangeard, P., 1926
 Description des Péridiniens Testacés recueillis par la Mission Charcot pendent le mois d'Aout 1924. Ann. Inst. Ocean. n.s. 3(7):307-334, 15 text figs.

Ceratium lineatum

Ercegovic, A., 1936
 Etudes qualitative et quantitatives du phytoplankton dans les eaux cotières de l'Adriatique oriental moyen au cours de l'année 1934. Acta Adriatica 1(9):1-126

Ceratium lineatum

Graham, H. W. and N. Bronikovsky, 1944
 The genus Ceratium in the Pacific and North Atlantic Oceans. Sci. Res. Cruise VII of the Carnegie, 1928-1929 ----- Biol. V (565):209 pp., 54 charts, 27 figs., 54 tables.

Ceratium lineatum

Gran, H.H., and T. Braarud, 1935
 A quantitative study of the phytoplankton in the Bay of Fundy and the Gulf of Maine (including observations on hydrography, chemistry, and turbidity). J. Biol. Bd., Canada, 1(5):279-467, 69 text figs.

Ceratium lineatum
Hada, Yoshina, 1970. (1970)
The protozoan plankton of the Antarctic and Subantarctic seas.
Scient. Repts, Japan. Antarct. Res. Exped., (E) 31:51 pp.

Ceratium lineatum
Jörgensen, E., 1920
Mediterranean Ceratia. Rept. Danish Oceanogr. Exped. 1908-10 to the Mediterranean and adjacent seas. Vol. II, Biol. J 1:110 pp., 94 text figs., 26 charts.

Ceratium lineatum
Jörgensen, E., 1905
B. Protistplankton and the diatoms in bottom samples. Hydrographical and biological investigations in Norwegian fiords. Bergens Mus. Skr. 7: 49-225.

Ceratium lineatum
Kokubo, S., and S. Sato, 1947
Plankters in Jū-San Gata. Physiol. and Ecol. (Japan) 1(4):1-16, 3 text figs., tables.

Ceratium lineatum
LeBour, M.V., 1925
The dinoflagellates of Northern Seas. The Marine Biological Association of the United Kingdom. Plymouth, 250 pp., 35 pls. 53 text figs.

Ceratium lineatum
Lillick, L.C., 1940
Phytoplankton and planktonic protozoa of the offshore waters of the Gulf of Maine. Pt. II. Qualitative Composition of the Planktonic Flora. Trans. Am. Phil. Soc., n.s., 31(3):193-237, 13 text figs.

Ceratium lineatum
Lillick, L.C., 1938
Preliminary report of the phytoplankton of the Gulf of Maine. Am. Mid. Nat. 20(3):624-640, 1 text fig. 37 tables.

Ceratium lineatum
Lillick, L.C., 1937
Seasonal studies of the phytoplankton off Woods Hole, Massachusetts. Biol. Bull. LXXIII (3):488-503, 3 text figs.

Ceratium lineatum
Meunier, A., 1919
Microplancton de la Mer Flamande 3. Les Péridiniens. Mem. Mus. Roy. Hist. Nat., Belgique 8(1):1-116, Pls. XV-XXI.

Ceratium lineatum
Mulford, Richard A., 1963
Distribution of the dinoflagellate genus Ceratium in the tidal and offshore waters of Virginia.
Chesapeake Science, 4(2):84-89.

Ceratium lineatum
Nordli, E., 1957.
Experimental studies on the ecology of Ceratia.
Oikos, 8(2):201-265.

Ceratium lineatum
Pavillard, J., 1905
Recherches sur la flore pelagique (Phytoplankton) de l'Etang de Thau. Theses presentees a la Fac. Sci., Paris, 116 pp., 3 pls.

Ceratium lineatum
Robinson, G.A., 1965.
Continuous plankton records: contribution towards a plankton atlas of the North Atlantic and the North Sea. IX. Seasonal cycles of phytoplankton.
Bulls. Mar. Ecol., Scottish Mar. Biol. Assoc., 6(4):104-122, pls. 26-61.

Ceratium lineatum
Robinson, G.A., 1961
Contribution towards an atlas of the north-eastern Atlantic and the North Sea. 1. Phytoplankton.
Bulls. Mar. Ecology, 5(42): 81-89, Pls. 15-20.

Ceratium lineatum
Sousa e Silva, E., 1949
Diatomaceas e Dinoflagelados de Baía de Cascais. Portugaliae Acta Biol., Volume: Julio Henriques, Ser. B: 300-383, 9 pls, 2 fold-in tables.

Ceratium lineatum
Stæmann-Nielsen, Einar, 1951
The marine vegetation of the Isefjord. A study on ecology and production. Medd. Komm. Danmarks Fiskeri-og Havundersøgelser. Ser. Plankton. 5(4); 114pp., 46 text figs.

Ceratium lineatum
Wheeler, J.E.G., 1939
Plankton investigations. Bermuda Biological Station. Second Report. October 1939. 7 pp. (typed), 5 figs. Plymouth, Oct. 23, 1939.

Ceratium longinum
Dangeard, P., 1927
Phytoplankton de la croisière du "Sylvana". Ann. Inst. Ocean., Monaco, n.s., 4(8):286-401, 54 text figs. (Feirer-Juin 1913).

Ceratium longinum
Ercegovic, A., 1936
Etudes qualitative et quantitatives du phytoplancton dans les eaux cotières de l'Adriatique oriental moyen au cours de l'année 1934. Acta Adriatica 1(9):1-126

Ceratium longinum
Marukawa, H., 1921
Plankton lists and some new species of copepods, from the northern waters of Japan. Bull. Inst. Ocean., No. 384, 15 pp., 3 pls., 1 chart. Monaco

Ceratium longipes
Braarud, T., 1934
A note on the phytoplankton of the Gulf of Maine in the summer of 1933. Biol. Bull. 67(1):76-82. (Contribution No. 46 of the Woods Hole Oceanographic Institution)

Ceratium longipes
Braarud, T., and Adam Bursa, 1939
On the phytoplankton of the Oslo Fjord, 1933-1934. Hvalrådets Skr. No. 19:1-63; 9 text figs. Reviewed. J. du. Cons. 14(3): 418-420. A.C. Gardiner.

Ceratium longipes
Brunel, J., 1962
Le phytoplancton de la Baie de Chaleurs. Inst. Botan. Univ. Montréal, Contrib. No. 77: 365 pp., 66 pls.

Ceratium longipes baltica
Ceratium longipes ventricosa

Ceratium longipes
Brunel, Jules, 1962
Le phytoplancton de la Baie des Chaleurs. Contrib. Ministère de la Chasse et des Pêcheries, Province de Québec, No. 91: 365 pp.

Ceratium longipes
Chiba, T., 1949
On the distribution of the plankton in the eastern China Sea and Yellow Sea. 1. Plankton composition in the spring. J. Shimonoseki Coll. Fisheries, 1(1):57-83, 1 fig.

Ceratium longipes
Dangeard, P., 1926
Description des Péridiniens Testacés recueillis para la Mission Charcot pendent le mois d'Aout 1924. Ann. Inst. Ocean. n.s. 3(7):307-334, 15 text figs.

Ceratium longipes
Gran, H.H., and T. Braarud, 1935
A quantitative study of the phytoplankton in the Bay of Fundy and the Gulf of Maine (including observations on hydrography, chemistry, and turbidity). J. Biol. Bd., Canada, 1(5):279-467, 69 text figs.

Ceratium longipes
Jörgensen, E., 1905
B. Protistplankton and the diatoms in bottom samples. Hydrographical and biological investigations in Norwegian fiords. Bergens Mus. Skr. 7: 49-225.

Ceratium longipes
LeBour, M.V., 1925
The dinoflagellates of Northern Seas. The Marine Biological Association of the United Kingdom. Plymouth, 250 pp., 35 pls. 53 text figs.

Ceratium longipes
Lillick, L.C., 1940
Phytoplankton and planktonic protozoa of the offshore waters of the Gulf of Maine. Pt. II. Qualitative Composition of the Planktonic Flora. Trans. Am. Phil. Soc., n.s., 31(3):193-237, 13 text figs.

Ceratium longipes
Lillick, L.C., 1938
Preliminary report of the phytoplankton of the Gulf of Maine. Am. Mid. Nat. 20(3):624-640, 1 text fig. 37 tables.

Ceratium longipes
Lillick, L.C., 1937
Seasonal studies of the phytoplankton off Woods Hole, Massachusetts. Biol. Bull. LXXIII (3):488-503, 3 text figs.

Ceratium longipes
Meunier, A., 1919
Microplancton de la Mer Flamande 3. Les Péridiniens. Mem. Mus. Roy. Hist. Nat., Belgique 8(1):1-116, Pls. XV-XXI.

Ceratium longipes
Paulsen, O., 1908
XVIII Peridiniales. Nordisches Plankton, Bot. Teil: 1-124, 155 text figs.

Ceratium longipes
Robinson, G.A., 1961
Contribution towards an atlas of the north-eastern Atlantic and the North Sea. 1. Phytoplankton.
Bulls. Mar. Ecology, 5(42): 81-89, Pls. 15-20.

Ceratium longipes
Schulz, B., and A. Wulff, 1929
Hydrographie und Oberflächen plankton des westlichen Barentsmeeres im Sommer 1927. Ber. deutschen wissensch. Komm. F. Meeresforsch. n.s. 4(5):232-372, 13 tables, 25 text figs.

Ceratium longipes
Stæmann-Nielsen, Einar, 1951
The marine vegetation of the Isefjord. A study on ecology and production. Medd. Komm. Danmarks Fiskeri-og Havundersøgelser. Ser. Plankton. 5(4); 114pp., 46 text figs.

Ceratium longipes
Yarranton, G.A. 1967
Parameters for use in distinguishing populations of Euceratium Gran.
Bull. mar. Ecol. 6 (6): 147-158.

Ceratium longipes atlanticum
Bigelow, H. B., 1922
Exploration of the coastal water off the northeastern United States in 1916 by the U.S. Fisheries Schooner Grampus. Bull. M.C.Z. 65 (5):85-188, 53 text figs.

Ceratium longipes var. oceanicum
Lillick, L.C., 1937
Seasonal studies of the phytoplankton off Woods Hole, Massachusetts. Biol. Bull. LXXIII (3):488-503, 3 text figs.

Ceratium longirostrum
Dangeard, P., 1927
Phytoplankton de la croisière du "Sylvana". Ann. Inst. Ocean., Monaco, n.s., 4(8):286-401, 54 text figs. (février-Juin 1913).

Ceratium longirostrum
Delegazione Italiana della Commissione Internazionale per l'Esplorazione Scientifica del Mediterraneo, 1941
Note sul plancton della Laguna veneta. [Memoria CCLXXXIX], Arch. di Ocean. e Limn. Anno I, Fasc. I, 1941 XIX: 31-57 pp.

Ceratium longirostrum
Ercegovic, A., 1936
Etudes qualitative et quantitatives du phytoplancton dans les eaux cotières de l'Adriatique oriental moyen au cours de l'année 1934. Acta Adriatica 1(9):1-126

Ceratium longirostrum
Gourret, P., 1883
Sur les Peridiniens du Golfe de Marseille. Ann. du Musee d'hist. Nat., Marseille, Zool., 1 (Mme. 8):1-114, 4 pls.

Ceratium longirostrum
Graham, H. W. and N. Bronikovsky, 1944
The genus Ceratium in the Pacific and North Atlantic Oceans. Sci. Res. Cruise VII of the Carnegie, 1928-1929 ----- Biol. V (565):209 pp., 54 charts, 27 figs., 54 tables.

Ceratium longirostrum
Jörgensen, E., 1920
Mediterranean Ceratia. Rept. Danish Oceanogr. Exped. 1908-10 to the Mediterranean and adjacent seas, Vol.II, Biol. J 1:110 pp., 94 text figs., 26 charts.

Ceratium longirostrum
Macdonald, R., 1933
An examination of plankton hauls made in the Suez Canal during the year 1928. Fish. Res. Dis., Notes & Mem. No.3, 11 pp., 1 chart.

Ceratium longirostrum
Margalef, R., 1949
Fitoplancton nerítico de la Costa Brava en 1947-48. Publ. Inst. Biol. Aplicada, 5: 41-51, 3 text figs.

Ceratium longirostrum
Paiva Carvalho, J., 1950
O plancton do Rio Maria Rodrigues (Cananeis). 1. Diatomaceas e Dinoflagelados. Bol. Inst. Paulista Oceanogr. 1(1); 27-43, 2 fold-in tables, 2 figs.

Ceratium longirostrum
Steemann-Nielsen, E., 1939
Die Ceratien des Indischen Ozeans und der Ostasiatischen Gewässer mit einer allgemeinen zusammenfassung über die Verbreitung der Ceratien in den Weltmeeren. Dana Rept. No. 17: 33 pp., 8 maps in text.

Ceratium longirostrum
Steemann Nielsen, E., 1934
Untersuchungen über die Verbreitung, Biologie und Variation der Ceratien im südlichen Stillen Ozean. Dana Rept. No. 4: 67 pp., 73 figs. and 11 charts in text.

Ceratium longissimum
Graham, H. W. and N. Bronikovsky, 1944
The genus Ceratium in the Pacific and North Atlantic Oceans. Sci. Res. Cruise VII of the Carnegie, 1928-1929 ----- Biol. V (565):209 pp., 54 charts, 27 figs., 54 tables.

Ceratium longissimum
Jörgensen, E., 1920
Mediterranean Ceratia. Rept. Danish Oceanogr. Exped. 1908-10 to the Mediterranean and adjacent seas, Vol.II, Biol. J 1:110 pp., 94 text figs., 26 charts.

Ceratium longissimum
Pavillard, J., 1916
Recherches sur les Peridiniens du Golfe du Lion. Mem. Univ. Montpellier. Trav. Inst. Bot., Univ. Montpellier. Serie mixte No.4, 70 pp., 3 pls., 15 text figs.

Ceratium longissimum
Rampi, L., 1942
Ricerche sul fitoplancton del Mare Ligure. 4. I Ceratium delle acque di Sanremo. Nuovo Giornale Botanico Italiano, N.S., 49: 221-236, 19 text figs.

Ceratium longissimum
Rampi, L., 1939
Péridiniens rares ou intéressants rocoltes dans la mer Ligure. Bull. Soc. Fran. de Microscopie, 8(2/3):106-112, 13 text figs.

Ceratium longissimum
Steemann-Nielsen, E., 1939
Die Ceratien des Indischen Ozeans und der Ostasiatischen Gewässer mit einer allgemeinen zusammenfassung über die Verbreitung der Ceratien in den Weltmeeren. Dana Rept. No. 17: 33 pp., 8 maps in text.

Ceratium longissimum
Steemann Nielsen, E., 1934
Untersuchungen über die Verbreitung, Biologie und Variation der Ceratien im südlichen Stillen Ozean. Dana Rept. No. 4: 67 pp., 73 figs. and 11 charts in text.

Ceratium lunula
Dangeard, P., 1927
Phytoplankton de la croisière du "Sylvana". Ann. Inst. Ocean., Monaco, n.s., 4(8):286-401, 54 text figs. (février-Juin 1913).

Ceratium lunula
Graham, H. W. and N. Bronikovsky, 1944
The genus Ceratium in the Pacific and North Atlantic Oceans. Sci. Res. Cruise VII of the Carnegie, 1928-1929 ----- Biol. V (565):209 pp., 54 charts, 27 figs., 54 tables.

Ceratium lunula
Jörgensen, E., 1920
Mediterranean Ceratia. Rept. Danish Oceanogr. Exped. 1908-10 to the Mediterranean and adjacent seas, Vol.II, Biol. J 1:110 pp., 94 text figs., 26 charts.

Ceratium lunula
Norris, Dean R., 1969.
Possible phagotrophic feeding in Ceratium lunula Schimper. Limnol. Oceanogr. 14(3): 448-449.

Ceratium lunula
Peters, Nicolaus, 1934
Die Bevolkerung des Sudatlantischen Ozeans mit Ceratien. Biol. Sonderuntersuchungen 1. Wiss Ergeb. Deutschen Atlantischen Exped.--- "Meteor" 1925-1927, 12(1): 1-69, 28 text figs.

Ceratium lunula
Sournia, A., 1967.
Le genre Ceratium (peridinien planctonique) dans le canal de Mozambique. Contribution a une revision mondiale (fin).
Vie Milieu (A) 18(3):441-449.

Ceratium lunula
Steemann-Nielsen, E., 1939
Die Ceratien des Indischen Ozeans und der Ostasiatischen Gewässer mit einer allgemeinen zusammenfassung über die Verbreitung der Ceratien in den Weltmeeren. Dana Rept. No. 17: 33 pp., 8 maps in text.

Ceratium lunula
Steemann Nielsen, E., 1934
Untersuchungen über die Verbreitung, Biologie und Variation der Ceratien im südlichen Stillen Ozean. Dana Rept. No. 4: 67 pp., 73 figs. and 11 charts in text.

Ceratium lunula
Wang, C. C., and D. Nie, 1932
A survey of the marine protozoa of Amoy. Contr. Biol. Lab. Sci. Soc., China, Zool. Ser., 8:285-385, 89 text figs.

Ceratium macroceros
Bigelow, H. B., 1922
Exploration of the coastal water off the northeastern United States in 1916 by the U.S. Fisheries Schooner Grampus. Bull. M.C.Z. 65 (5):85-188, 53 text figs.

Ceratium macroceros
Braarud, T., 1945
A phytoplankton survey of the polluted waters of inner Oslo Fjord. Hvalrådets Skrifter, No.28, 142 pp., 19 text figs., 17 tables.

Ceratium macroceros
Braarud, T., and Adam Bursa, 1939
On the phytoplankton of the Oslo Fjord, 1933-1934. Hvalrådets Skr. No.19:1-63; 9 text figs. Reviewed. J. du. Cons. 14(3): 418-420. A.C. Gardiner.

Ceratium macroceros
Dangeard, P., 1927
Phytoplankton de la croisière du "Sylvana". Ann. Inst. Ocean., Monaco, n.s., 4(8):286-401, 54 text figs. (février-Juin 1913).

Ceratium macroceros
Dangeard, P., 1926
Description des Péridiniens Testacés recueillis para la Mission Charcot pendent le mois d'Aout 1924. Ann. Inst. Ocean. n.s. 3(7):307-334, 15 text figs.

Ceratium macroceros
Forti, A., 1922
Ricerche sulla flora pelagica (fitoplancton) di Quarto dei Mille. Mem. R. Com. Talass. Ital. 97:248 pp., 13 pls.

Ceratium macroceros
Fraser, J. H., 1949
Plankton investigations from the Scottish Research Vessel. Ann. Biol., Int. Cons., 4: 66-67.

Ceratium macroceros
Ghazzawi, F.M., 1939
Plankton of the Egyptian waters. A study of the Suez Canal Plankton. (A) Phytoplankton. Preliminary Report 83 pp. Notes and Memoires, Min. Commerce-Industry. Egypt, Hydrobiol. & Fish. 65 figs.

Ceratium macroceros
Gilbert, J.Y., and W.E. Allen, 1943
The phytoplankton of the Gulf of California obtained by the "E.W. Scripps" in 1939 and 1940. J. Mar. Res. V(2):89-110, figs.30-31.

Ceratium macroceros
Graham, H. W. and N. Bronikovsky, 1944
The genus Ceratium in the Pacific and North Atlantic Oceans. Sci. Res. Cruise VII of the Carnegie, 1928-1929 ----- Biol. V (565):209 pp., 54 charts, 27 figs., 54 tables.

Ceratium macroceros
Jörgensen, E., 1920
Mediterranean Ceratia. Rept. Danish Oceanogr. Exped. 1908-10 to the Mediterranean and adjacent seas, Vol.II, Biol. J 1:110 pp., 94 text figs., 26 charts.

Ceratium macroceros
Jørgensen, E., 1905
B. Protistplankton and the diatoms in bottom samples. Hydrographical and biological investigations in Norwegian fiords. Bergens Mus. Skr. 7: 49-225.

Ceratium macroceros
Kokubo, S., and S. Sato., 1947
Plankters in JQ-San Gata. Physiol. and Ecol. (Japan) 1(4):1-16, 3 text figs., tables.

Ceratium macroceros
LeBour, M.V., 1925
The dinoflagellates of Northern Seas. The Marine Biological Association of the United Kingdom. Plymouth, 250 pp., 35 pls. 53 text figs.

Ceratium macroceros
Lillick, L.C., 1940
Phytoplankton and planktonic protozoa of the offshore waters of the Gulf of Maine. Pt. II. Qualitative Composition of the Planktonic Flora. Trans. Am. Phil. Soc., n.s., 31(3):193-237, 13 text figs.

Ceratium macroceros
Lillick, L.C., 1938
Preliminary report of the phytoplankton of the Gulf of Maine. Am. Mid. Nat. 20(3):624-640, 1 text figs 37 tables.

Ceratium macroceros
Lillick, L.C., 1937
Seasonal studies of the phytoplankton off Woods Hole, Massachusetts. Biol. Bull. LXXIII (3):488-503, 3 text figs.

Ceratium macroceros
Macdonald, R., 1933
An examination of plankton hauls made in the Suez Canal during the year 1928. Fish. Res. Dis., Notes & Mem. No.3, 11 pp., 1 chart.

Ceratium macroceros
Mangin, M. L., 1912
Phytoplancton de la croisière du "René" dans l'Atlantique (Septembre 1908). Ann. Inst. Ocean., n.s., 4(1):1-66, 2 pls., 41 text figs., 2 tables.

Ceratium macroceros
Marshall, S. M., 1933
The production of microplankton in the Great Barrier Reef Region. Brit. Mus. (N.H.) Great Barrier Reef Exped. 1928-29, Sci. Repts. II(5):111-157, 14 text figs.

Ceratium macroceras
Marukawa, H., 1921
Plankton lists and some new species of copepods, from the northern waters of Japan. Bull. Inst. Ocean., No.384, 15 pp., 3 pls., 1 chart. Monaco

Ceratium macroceros
Mulford, Richard A., 1963
Distribution of the dinoflagellate genus Ceratium in the tidal and offshore waters of Virginia.
Chesapeake Science, 4(2):84-89.

Ceratium macroceros
Paulsen, O., 1908
XVIII Peridiniales. Nordisches Plankton, Bot. Teil: 1-124, 155 text figs.

Ceratium macroceros
Pavillard, J., 1905
Recherches sur la flore pelagique (Phytoplankton) de l'Etang de Thau. Theses presentees a la Fac. Sci., Paris, 116 pp., 3 pls.

Ceratium macroceros
Peters, Nicolaus, 1934
Die Bevolkerung des Sudatlantischen Ozeans mit Ceratien. Biol. Sonderuntersuchungen 1. Wiss Ergeb. Deutschen Atlantischen Exped. "Meteor" 1925-1927, 12(1): 1-69, 28 text figs.

Ceratium macroceros
Robinson, G.A., 1961
Contribution towards an atlas of the northeastern Atlantic and the North Sea. 1. Phytoplankton.
Bulls. Mar. Ecology, 5(42): 81-89, Pls. 15-20.

Ceratium macroceros
Sournia, A., 1967.
Le genre Ceratium (peridinien planctonique) dans le canal de Mozambique. Contribution a une revision mondiale (fin).
Vie Milieu (A) 18(3):441-449.

Ceratium macroceros
Sousa e Silva, E., 1949
Diatomaceas e Dinoflagelados de Baia de Cascais. Portugaliae Acta Biol. Volume: Julio Henriques, Ser. B: 300-383, 9 pls, 2 fold-in tables.

Ceratium macroceros
Steemann-Nielsen, Einar, 1951
The marine vegetation of the Isefjord. A study on ecology and production. Medd. Komm. Danmarks Fiskeri-og Havundersøgelser. Ser. Plankton. 5(4); 114pp., 46 text figs.

Ceratium macroceros
Steemann-Nielsen, E., 1939
Die Ceratien des Indischen Ozeans und der Ostasiatischen Gewässer mit einer allgemeinen zusammenfassung über die Verbreitung der Ceratien in den Weltmeeren. Dana Rept. No. 17: 33 pp., 8 maps in text.

Ceratium macroceros
Steemann Nielsen, E., 1934
Untersuchungen über die Verbreitung, Biologie und Variation der Ceratien im südlichen Stillen Ozean. Dana Rept. No. 4: 67 pp., 73 figs. and 11 charts in text.

Ceratium macroceros
Wheeler, J.E.G., 1939
Plankton investigations. Bermuda Biological Station. Second Report. October 1939. 7 pp. (typed), 5 figs. Plymouth, Oct. 23, 1939.

Ceratium macroceros
Woodmansee, R.A., 1958.
The seasonal distribution of the zooplankton off Chicken Key in Biscayne Bay, Florida. Ecology 39(2):247-261.

Ceratium macroceros
Yarranton, G.A. 1967.
Parameters for use in distinguishing populations of Euceratium Gran. Bull. mar. Ecol. 6(6): 147-158.

Ceratium macroceros californiense
Forti, A., 1922
Ricerche sulla flora pelagica (fitoplancton) di Quarto dei Mille. Mem. R. Com. Talass. Ital. 97:248 pp., 13 pls.

Ceratium macroceros deflexum n. subsp.
Kofoid, C. A., 1907
Dinoflagellata of the San Diego region. III. Descriptions of new species. Univ. Calif. Publ., Zool. 3:299-340, Pls. 22-33.

Ceratium macroceros gallicum
Delegazione Italiana della Commissione Internazionale per l'Esplorazione Scientifica del Mediterraneo, 1941
Note sul plancton della Laguna veneta. [Memoria CCLXXXIX], Arch. di Ocean. e Limn. Anno I, Fasc. I, 1941 XIX: 31-57 pp.

Ceratium macroceros gallicum
Ercegovic, A., 1936
Etudes qualitative et quantitatives du phytoplancton dans les eaux cotières de l'Adriatique oriental moyen au cours de l'année 1934. Acta Adriatica 1(9):1-126

Ceratium macroceros subsp. gallicum
Ghazzawi, F.M., 1939
Plankton of the Egyptian waters. A study of the Suez Canal Plankton. (A) Phytoplankton. Preliminary Report 83 pp. Notes and Memoires, Min. Commerce-Industry, Egypt, Hydrobiol. & Fish. 65 figs.

Ceratium macroceros gallicum
Graham, H. W. and N. Bronikovsky, 1944
The genus Ceratium in the Pacific and North Atlantic Oceans. Sci. Res. Cruise VII of the Carnegie, 1928-1929 ----- Biol. V (565):209 pp., 54 charts, 27 figs., 54 tables.

Ceratium macroceros gallicum
Lindemann, E., 1925
Neubeobachtungen an den Winter peridineen des Golfes von Neapel. Bot. Arch. 9:95-102, 19 text figs.

Ceratium macroceros gallicum
Margalef, R., 1949
Fitoplancton nerítico de la Costa Brava en 1947-48. Publ. Inst. Biol. Aplicada, 5: 41-51, 3 text figs.

Ceratium macroceros gallicum
Massutí Algamora, M., 1949
Estudio de diez y seis muestras de plancton del Golfo de Nápoles. Publ. Inst. Biol. Appl. 5:85-94, 1 fold-in table.

Ceratium macroceros gallicum
Pavillard, J., 1916
Recherches sur les Peridiniens du Golfe du Lion. Mem. Univ. Montpellier. Trav. Inst. Bot., Univ. Montpellier. Serie mixte No.4, 70 pp., 3 pls., 15 text figs.

Ceratium macroceros macroceros
Graham, H. W. and N. Bronikovsky, 1944
The genus Ceratium in the Pacific and North Atlantic Oceans. Sci. Res. Cruise VII of the Carnegie, 1928-1929 ----- Biol. V (565):209 pp., 54 charts, 27 figs., 54 tables.

Ceratium massiliens
Chiba, T., 1949
On the distribution of the plankton in the eastern China Sea and Yellow Sea. 1. Plankton composition in the spring. J. Shimonoseki Coll. Fisheries, 1(1):57-83, 1 fig.

Ceratium massiliense
Dangeard, P., 1927
Phytoplankton de la croisière du "Sylvana". Ann. Inst. Ocean., Monaco, n.s., 4(8):286-401, 54 text figs. (février-Juin 1913).

Ceratium massiliense
Delegazione Italiana della Commissione Internazionale per l'Esplorazione Scientifica del Mediterraneo, 1941
Note sul plancton della Laguna veneta. [Memoria CCLXXXIX], Arch. di Ocean. e Limn. Anno I, Fasc. I, 1941 XIX: 31-57 pp.

Ceratium massiliense
Forti, A., 1922
Ricerche sulla flora pelagica (fitoplancton) di Quarto dei Mille. Mem. R. Com. Talass. Ital. 97:248 pp., 13 pls.

Ceratium massiliense
Ghazzawi, F.M., 1939
Plankton of the Egyptian waters. A study of the Suez Canal Plankton. (A) Phytoplankton. Preliminary Report 83 pp. Notes and Memoires, Min. Commerce-Industry, Egypt, Hydrobiol. & Fish. 65 figs.

Ceratium massiliense
Graham, H. W. and N. Bronikovsky, 1944
The genus Ceratium in the Pacific and North Atlantic Oceans. Sci. Res. Cruise VII of the Carnegie, 1928-1929 ----- Biol. V (565):209 pp., 54 charts, 27 figs., 54 tables.

Ceratium massiliense
Halim, Y., 1963.
Microplancton des eaux égyptiennes. Le genre Ceratium Schrank (Dinoflagellés). Rapp. Proc. Verb., Réunions, Comm. Int. Expl. Sci., Mer Méditerranée, Monaco, 17(2):495-502.

Ceratium massiliense
Lindemann, E., 1925
Neubeobachtungen an den Winter peridineen des Golfes von Neapel. Bot. Arch. 9:95-102, 19 text figs.

Ceratium massiliense
Jörgensen, E., 1920
Mediterranean Ceratia. Rept. Danish Oceanogr. Exped. 1908-10 to the Mediterranean and adjacent seas, Vol.II, Biol. J 1:110 pp., 94 text figs., 26 charts.

Ceratium massiliense
Macdonald, R., 1933
An examination of plankton hauls made in the Suez Canal during the year 1928. Fish. Res. Dis., Notes & Mem. No.3, 11 pp., 1 chart.

Ceratium massiliense
Mangin, M. L., 1912
Phytoplancton de la croisière du "René" dans l'Atlantique (Septembre 1908). Ann. Inst. Ocean., n.s., 4(1):1-66, 2 pls., 41 text figs., 2 tables.

Ceratium massiliense
Margalef, R., 1949
Fitoplancton nerítico de la Costa Brava en 1947-48. Publ. Inst. Biol. Aplicada, 5: 41-51, 3 text figs.

Ceratium massiliense
Mulford, Richard A., 1963
Distribution of the dinoflagellate genus Ceratium in the tidal and offshore waters of Virginia. Chesapeake Science, 4(2):84-89.

Ceratium massiliense
Pavillard, J., 1923
A propos de la systématique des Péridiniens Bull. Soc. Bot. de France 70:876-882; 914-918.

Ceratium massiliense
Pavillard, J., 1916
Recherches sur les Peridiniens du Golfe du Lion. Mem. Univ. Montpellier. Trav. Inst. Bot., Univ. Montpellier. Serie mixte No.4, 70 pp., 3 pls., 15 text figs.

Ceratium massiliense
Peters, Nicolaus, 1934
Die Bevölkerung des Sudatlantischen Ozeans mit Ceratien. Biol. Sonderuntersuchungen 1. Wiss Ergeb. Deutschen Atlantischen Exped.--- "Meteor" 1925-1927, 12(1): 1-69, 28 text figs.

Ceratium massiliense
Roubault, A., 1946
Observations sur la Répartition du Plancton. Bull. Mus. Inst. Ocean., No.902: 4 pp.

Ceratium massiliense
Sournia, A., 1967.
Le genre Ceratium (peridinien planctonique) dans le canal de Mozambique. Contribution a une revision mondiale (fin). Vie Milieu (A) 18(3):441-449.

Ceratium massiliense
Sousa e Silva, E., 1949
Diatomaceas e Dinoflagelados de Baía de Cascais. Portugaliae Acta Biol., Volume: Julio Henriques, Ser. B: 300-383, 9 pls, 2 fold-in tables.

Ceratium massiliense
Steemann-Nielsen, E., 1939
Die Ceratien des Indischen Ozeans und der Ostasiatischen Gewässer mit einer allgemeinen zusammenfassung über die Verbreitung der Ceratien in den Weltmeeren. Dana Rept. No. 17: 33 pp., 8 maps in text.

Ceratium massiliense
Steemann Nielsen, E., 1934
Untersuchungen über die Verbreitung, Biologie und Variation der Ceratien im südlichen Stillen Ozean. Dana Rept. No. 4: 67 pp., 73 figs. and 11 charts in text.

Ceratium massiliense
Sukhanova, I.N., 1964.
The phytoplankton of the northeastern part of the Indian Ocean in the season of the southwest monsoon. Regularity of the distribution of oceanic plankton. (In Russian; English abstract). Trudy Inst. Okeanol., Akad. Nauk, SSSR, 65: 24-31.

Ceratium massiliense
Vives, F., and A. Planas, 1952.
Plancton recogido por los laboratorios costeros. VI. Fitoplancton de las costas de Vinaroz, islas Columbretes y alrededores de la desembocadura del Ebro. Publ. Inst. Biol. Aplic. 11:141-156, 19 textfigs.

Ceratium massiliense
Wang, C. C., and D. Nie, 1932
A survey of the marine protozoa of Amoy. Contr. Biol. Lab. Sci. Soc., China, Zool. Ser., 8:285-385, 89 text figs.

Ceratium massiliense armatum
Forti, A., 1922
Ricerche sulla flora pelagica (fitoplancton) di Quarto dei Mille. Mem. R. Com. Talass. Ital. 97:248 pp., 13 pls.

Ceratium massiliense protuberans
Delegazione Italiana della Commissione Internazionale per l'Esplorazione Scientifica del Mediterraneo, 1941
Note sul plancton della Laguna veneta. [Memoria CCLXXXIX], Arch. di Ocean. e Limn. Anno I, Fasc. I, 1941 XIX: 31-57 pp.

Ceratium massiliense protuberans
Forti, A., 1922
Ricerche sulla flora pelagica (fitoplancton) di Quarto dei Mille. Mem. R. Com. Talass. Ital. 97:248 pp., 13 pls.

Ceratium massiliense protuberans
Pavillard, J., 1916
Recherches sur les Peridiniens du Golfe du Lion. Mem. Univ. Montpellier. Trav. Inst. Bot., Univ. Montpellier. Serie mixte No.4, 70 pp., 3 pls., 15 text figs.

Ceratium microceros
Halim, Y., 1963.
Microplancton des eaux égyptiennes. Le genre Ceratium (Schrank (Dinoflagellés). Rapp. Proc. Verb., Réunions, Comm. Int. Expl. Sci., Mer Méditerranée, Monaco, 17(2):495-502.

Ceratium minus
Gourret, P., 1883
Sur les Peridiniens du Golfe de Marseille. Ann. du Musee d'hist. Nat., Marseille, Zool., 1 (Mme. 8):1-114, 4 pls.

Ceratium minutum
Dangeard, P., 1927
Phytoplankton de la croisière du "Sylvana". Ann. Inst. Ocean. Monaco, n.s., 4(8):286-401, 54 text figs. (Février-Juin 1913).

Ceratium minutum
Dangeard, P., 1926
Description des Péridiniens Testacés recueillis para la Mission Charcot pendent le mois d'Aout 1924. Ann. Inst. Ocean. n.s. 3(7):307-334, 15 text figs.

Ceratium minutum n.nom.
Jörgensen, E., 1920
Mediterranean Ceratia. Rept. Danish Oceanogr. Exped. 1908-10 to the Mediterranean and adjacent seas, Vol.II, Biol. J 1:110 pp., 94 text figs., 26 charts.

Ceratium minutum
LeBour, M.V., 1925
The dinoflagellates of Northern Seas. The Marine Biological Association of the United Kingdom. Plymouth, 250 pp., 35 pls. 53 text figs.

Ceratium minutum
Peters, Nicolaus, 1934
Die Bevölkerung des Sudatlantischen Ozeans mit Ceratien. Biol. Sonderuntersuchungen 1. Wiss Ergeb. Deutschen Atlantischen Exped.--- "Meteor" 1925-1927, 12(1): 1-69, 28 text figs.

Ceratium minutum
Rampi, L., 1942.
Ricerche sul fitoplancton del Mare Ligure. 4. I Ceratium delle acque di Sanremo. Nuovo Giornale Botanico Italiano, N.S., 49: 221-236, 19 text figs.

Ceratium minutum
Rampi, L., 1939
Péridiniens rares ou intéressants récoltés dans la mer Ligure. Bull. Soc. Fran. de Microscopie, 8(2/3):106-112, 13 text figs.

Ceratium molle
Delegazione Italiana della Commissione Internazionale per l'Esplorazione Scientifica del Mediterraneo, 1941
Note sul plancton della Laguna veneta. [Memoria CCLXXXIX], Arch. di Ocean. e Limn. Anno I, Fasc. I, 1941 XIX: 31-57 pp.

Ceratium molle
Ercegovic, A., 1936
Etudes qualitative et quantitatives du phytoplancton dans les eaux cotières de l'Adriatique oriental moyen au cours de l'année 1934. Acta Adriatica 1(9):1-126

Ceratium molle
Forti, A., 1922
Ricerche sulla flora pelagica (fitoplancton) di Quarto dei Mille. Mem. R. Com. Talass. Ital. 97:248 pp., 13 pls.

Ceratium mollis n. sp.
Kofoid, C. A., 1907
Dinoflagellata of the San Diego region. III. Descriptions of new species. Univ. Calif. Publ. Zool. 3:299-340, Pls. 22-33.

Ceratium molle
Margalef, R., 1949
Fitoplancton nerítico de la Costa Brava en 1947-48. Publ. Inst. Biol. Aplicada, 5: 41-51, 3 text figs.

Ceratium molle
Pavillard, J., 1923
A propos de la systématique des Péridiniens Bull. Soc. Bot. de France 70:876-882; 914-918.

Ceratium molle
Steemann-Nielsen, E., 1939
Die Ceratien des Indischen Ozeans und der Ostasiatischen Gewässer mit einer allgemeinen zusammenfassung über die Verbreitung der Ceratien in den Weltmeeren. Dana Rept. No. 17: 33 pp., 8 maps in text.

Ceratium molle
Steemann Nielsen, E., 1934
Untersuchungen über die Verbreitung, Biologie und Variation der Ceratien im südlichen Stillen Ozean. Dana Rept. No. 4: 67 pp., 73 figs. and 11 charts in text.

Ceratium obliquum
Gourret, P., 1883
Sur les Peridiniens du Golfe de Marseille. Ann. du Musee d'hist. Nat., Marseille, Zool., 1 (Mme. 8):1-114, 4 pls.

Ceratium obtusum
Gourret, P., 1883
Sur les Peridiniens du Golfe de Marseille. Ann. du Musee d'hist. Nat., Marseille, Zool., 1 (Mme. 8):1-114, 4 pls.

Ceratium obtusum
Pavillard, J., 1923
A propos de la systématique des Péridiniens Bull. Soc. Bot. de France 70:876-882; 914-918.

Ceratium ostenfeldi n. sp.
Kofoid, C. A., 1907
Dinoflagellata of the San Diego region. III. Descriptions of new species. Univ. Calif. Publ., Zool. 3:299-340, Pls. 22-33.

Ceratium pacificum n.sp.
Wood, E.J.F., 1963
Dinoflagellates in the Australian region. II. Recent collections. Comm. Sci. and Industr. Res. Org., Div. Fish. and Oceanogr., Techn. Paper, No. 14: 55 pp.

Ceratium palmatum
Ercegovic, A., 1936
Etudes qualitative et quantitatives du phytoplancton dans les eaux cotières de l'Adriatique oriental moyen au cours de l'année 1934. Acta Adriatica 1(9):1-126

Ceratium palmatum
Forti, A., 1922
Ricerche sulla flora pelagica (fitoplancton) di Quarto dei Mille. Mem. R. Com. Talass. Ital. 97:248 pp., 13 pls.

Ceratium palmatum ranipes
Ercegovic, A., 1936
Etudes qualitative et quantitatives du phytoplancton dans les eaux cotières de l'Adriatique oriental moyen au cours de l'année 1934. Acta Adriatica 1(9):1-126

Ceratium palmatum ranipes
Forti, A., 1922
Ricerche sulla flora pelagica (fitoplancton) di Quarto dei Mille. Mem. R. Com. Talass. Ital. 97:248 pp., 13 pls.

Ceratium palmatum
Lindemann, E., 1925
Neubeobachtungen an den Winter peridineen des Golfes von Neapel. Bot. Arch. 9:95-102, 19 text figs.

Ceratium palmatum ranipes
Pavillard, J., 1916
Recherches sur les Peridiniens du Golfe du Lion. Mem. Univ. Montpellier. Trav. Inst. Bot., Univ. Montpellier. Serie mixte No.4, 70 pp., 3 pls., 15 text figs.

Ceratium paradoxydes
Dangeard, P., 1927
Phytoplankton de la croisière du "Sylvana". Ann. Inst. Ocean. Monaco, n.s., 4(8):286-401, 54 text figs. (février-Juin 1913).

Ceratium paradoxides
Graham, H. W. and N. Bronikovsky, 1944
The genus Ceratium in the Pacific and North Atlantic Oceans. Sci. Res. Cruise VII of the Carnegie, 1928-1929 ----- Biol. V (565):209 pp., 54 charts, 27 figs., 54 tables.

Ceratium pardoxides
Jörgensen, E., 1920
Mediterranean Ceratia. Rept. Danish Oceanogr. Exped. 1908-10 to the Mediterranean and adjacent seas, Vol.II, Biol. J 1:110 pp., 94 text figs., 26 charts.

Ceratium paradoxoides
Peters, Nicolaus, 1934
Die Bevolkerung des Sudatlantischen Ozeans mit Ceratien. Biol. Sonderuntersuchungen 1. Wiss Ergeb. Deutschen Atlantischen Exped.--- "Meteor" 1925-1927, 12(1): 1-69, 28 text figs.

Ceratium paradoxoides
Sournia, A., 1967.
Le genre Ceratium (peridinien planctonique) dans le canal de Mozambique. Contribution a une revision mondiale (fin). Vie Milieu (A) 18(3):441-449.

Ceratium paradoxoides
Steemann-Nielsen, E., 1939
Die Ceratien des Indischen Ozeans und der Ostasiatischen Gewässer mit einer allgemeinen zusammenfassung über die Verbreitung der Ceratien in den Weltmeeren. Dana Rept. No. 17: 33 pp., 8 maps in text.

Ceratium paradoxoides
Steemann Nielsen, E., 1934
Untersuchungen über die Verbreitung, Biologie und Variation der Ceratien im südlichen Stillen Ozean. Dana Rept. No. 4: 67 pp., 73 figs. and 11 charts in text.

Ceratium paradoxoides
Wood, E.J.F., 1963
Dinoflagellates in the Australian region. II. Recent collections. Comm. Sci. and Industr. Res. Org., Div. Fish. and Oceanogr., Techn. Paper, No. 14: 55 pp.

Ceratium parvum
Gourret, P., 1883
Sur les Peridiniens du Golfe de Marseille. Ann. du Musee d'hist. Nat., Marseille, Zool., 1 (Mme. 8):1-114, 4 pls.

Ceratium patentissimum
Pavillard, J., 1923
A propos de la systématique des Péridiniens Bull. Soc. Bot. de France 70:876-882; 914-918.

Ceratium pavillardii
Dangeard, P., 1927
Phytoplankton de la croisière du "Sylvana". Ann. Inst. Ocean. Monaco, n.s., 4(8):286-401, 54 text figs. (février-Juin 1913).

Ceratium Pavillardi
Ercegovic, A., 1936
Etudes qualitative et quantitatives du phytoplancton dans les eaux cotières de l'Adriatique oriental moyen au cours de l'année 1934. Acta Adriatica 1(9):1-126

Ceratium Pavillardii
Forti, A., 1922
Ricerche sulla flora pelagica (fitoplancton) di Quarto dei Mille. Mem. R. Com. Talass. Ital. 97:248 pp., 13 pls.

Ceratium pavillardii
Graham, H. W., 1942
Studies in the morphology, taxonymy, and ecology of the Peridiniales. Sci. Res. Cruise VII of the Carnegie, 1928-1929---Biol. III(542) 129 pp., 67 figs.

Ceratium Pavillardii
Jörgensen, E., 1920
Mediterranean Ceratia. Rept. Danish Oceanogr. Exped. 1908-10 to the Mediterranean and adjacent seas, Vol.II, Biol. J 1:110 pp., 94 text figs., 26 charts.

Ceratium Pavillardii
Lindemann, E., 1925
Neubeobachtungen an den Winter peridineen des Golfes von Neapel. Bot. Arch. 9:95-102, 19 text figs.

Ceratium Pavillardi
Margalef, R., 1949
Fitoplancton nerítico de la Costa Brava en 1947-48. Publ. Inst. Biol. Aplicada, 5: 41-51, 3 text figs.

Ceratium Pavillardii
Pavillard, J., 1916
Recherches sur les Peridiniens du Golfe du Lion. Mem. Univ. Montpellier. Trav. Inst. Bot., Univ. Montpellier. Serie mixte No.4, 70 pp., 3 pls., 15 text figs.

Ceratium pavillardii
Sournia, A., 1967.
Le genre Ceratium (peridinien planctonique) dans le canal de Mozambique. Contribution a une revision mondiale (fin). Vie Milieu (A) 18(3):441-449.

Ceratium pavillardii
Steemann-Nielsen, E., 1939
Die Ceratien des Indischen Ozeans und der Ostasiatischen Gewässer mit einer allgemeinen zusammenfassung über die Verbreitung der Ceratien in den Weltmeeren. Dana Rept. No. 17: 33 pp., 8 maps in text.

Ceratium pellucidum
Gourret, P., 1883
Sur les Peridiniens du Golfe de Marseille. Ann. du Musee d'hist. Nat., Marseille, Zool., 1 (Mme. 8):1-114, 4 pls.

Ceratium pennatum
Chiba, T., 1949
On the distribution of the plankton in the eastern China Sea and Yellow Sea. 1. Plankton composition in the spring. J. Shimonoseki Coll. Fisheries, 1(1):57-63, 1 fig.

Ceratium pennatum
Ercegovic, A., 1936
Etudes qualitative et quantitatives du phytoplancton dans les eaux cotières de l'Adriatique oriental moyen au cours de l'année 1934. Acta Adriatica 1(9):1-126

Ceratium pennatum
Marshall, S. M., 1933
The production of microplankton in the Great Barrier Reef Region. Brit. Mus. (N.H.) Great Barrier Reef Exped. 1928-29, Sci. Repts. II(5):111-157, 14 text figs.

Ceratium pennatum
Pavillard, J., 1916
Recherches sur les Peridiniens du Golfe du Lion. Mem. Univ. Montpellier. Trav. Inst. Bot., Univ. Montpellier. Serie mixte No.4, 70 pp., 3 pls., 15 text figs.

Ceratium pentagonum
Balech, Enrique, 1962
Tintinnoinea y Dinoflagellata del Pacifico segun material de las Expediciones NORPAC y DOWNWIND del Instituto Scripps de Oceanografia. Revista, Mus. Argentino Ciencias Nat. "Bernardino Rivadavia", Ciencias Zool., 7(1):1-253.

Ceratium pentagonum
Dangeard, P., 1927
Phytoplankton de la croisière du "Sylvana". Ann. Inst. Ocean., Monaco, n.s., 4(8):286-401, 54 text figs. (février-Juin 1913).

Ceratium pentagonum
Delegazione Italiana della Commissione Internazionale per l'Esplorazione Scientifica del Mediterraneo, 1941
Note sul plancton della Laguna veneta. [Memoria CCLXXXIX], Arch. di Ocean. e Limn. Anno I, Fasc. I, 1941 XIX: 31-57 pp.

Ceratium pentagonum
Ercegovic, A., 1936
Etudes qualitative et quantitatives du phytoplankton dans les eaux cotières de l'Adriatique oriental moyen au cours de l'année 1934. Acta Adriatica 1(9):1-126

Ceratium pentagonum
Forti, A., 1922
Ricerche sulla flora pelagica (fitoplancton) di Quarto dei Mille. Mem. R. Com. Talass. Ital. 97:248 pp., 13 pls.

Ceratium pentagonum
Gilbert, J.Y., and W.E. Allen, 1943
The phytoplankton of the Gulf of California obtained by the "E.W. Scripps" in 1939 and 1940. J. Mar. Res. V(2):89-110, figs.30-31.

Ceratium pentagonum
Gourret, P., 1883
Sur les Peridiniens du Golfe de Marseille. Ann. du Musee d'hist. Nat., Marseille, Zool., 1 (Mme. 8):1-114, 4 pls.

Ceratium pentagonum
Graham, H. W. and N. Bronikovsky, 1944
The genus Ceratium in the Pacific and North Atlantic Oceans. Sci. Res. Cruise VII of the Carnegie, 1928-1929 ----- Biol. V (565):209 pp., 54 charts, 27 figs., 54 tables.

Ceratium pentagonum
Jörgensen, E., 1920
Mediterranean Ceratia. Rept. Danish Oceanogr. Exped. 1908-10 to the Mediterranean and adjacent seas, Vol.II, Biol. J 1:110 pp., 94 text figs., 26 charts.

Ceratium pentagonum
Lindemann, E., 1925
Neubeobachtungen an den Winter peridineen des Golfes von Neapel. Bot. Arch. 9:95-102, 19 text figs.

Ceratium pentagonum
Lindemann, E., 1924
Peridineen aus dem goldenen Horn und dem Bosphorus. Bot. Arch. 5:216-233, 98 text figs.

Ceratium pentagonum
López, J., 1966
Variación y regulación de la forma en el género Ceratium. Inv. Pesq., Barcelona, 30:325-427.

English summary

Coratium
Ceratium candelabrum
Ceratium pentagonum
Ceratium tres
Ceratium furca
Ceratium types
Ceratium gutterii
Ceratium lingshinium
Ceratium schmidtii
Ceratium curvicorne
Keys Ceratium
Mediterranean, west

Ceratium pentagonum
Margalef, R., 1949
Fitoplancton nerítico de la Costa Brava en 1947-48. Publ. Inst. Biol. Aplicada, 5: 41-51, 3 text figs.

Ceratium pentagonum
Margalef, R., 1948
Le phytoplancton estival de la "Costa Brava" catalane en 1946. Hydrobiol. 1(1):15-21.

Ceratium pentagonum
Pavillard, J., 1923
A propos de la systématique des Péridiniens Bull. Soc. Bot. de France 70:876-882; 914-918.

Ceratium pentagonum
Pavillard, J., 1916
Recherches sur les Peridiniens du Golfe du Lion. Mem. Univ. Montpellier. Trav. Inst. Bot., Univ. Montpellier. Serie mixte No.4, 70 pp., 3 pls., 15 text figs.

Ceratium pentagonum
Peters, Nicolaus, 1934
Die Bevolkerung des Sudatlantischen Ozeans mit Ceratien. Biol. Sonderuntersuchungen 1. Wiss Ergeb. Deutschen Atlantischen Exped. "Meteor" 1925-1927, 12(1): 1-69, 28 text figs.

Ceratium pentagonum
Roubault, A., 1946
Observations sur la Répartition du plancton. Bull. Mus. Inst. Océan, No.902:4 pp.

Ceratium pentagonum
Steemann-Nielsen, E., 1939
Die Ceratien des Indischen Ozeans und der Ostasiatischen Gewässer mit einer allgemeinen zusammenfassung über die Verbreitung der Ceratien in den Weltmeeren. Dana Rept. No. 17: 33 pp., 8 maps in text.

Ceratium pentagonum
Steemann Nielsen, E., 1934
Untersuchungen über die Verbreitung, Biologie und Variation der Ceratien im südlichen Stillen Ozean. Dana Rept. No. 4: 67 pp., 73 figs. and 11 charts in text.

Ceratium pentagonum longisetum
Forti, A., 1922
Ricerche sulla flora pelagica (fitoplancton) di Quarto dei Mille. Mem. R. Com. Talass. Ital. 97:248 pp., 13 pls.

Ceratium pentagonum pacificum
Graham, H. W. and N. Bronikovsky, 1944
The genus Ceratium in the Pacific and North Atlantic Oceans. Sci. Res. Cruise VII of the Carnegie, 1928-1929 ----- Biol. V (565):209 pp., 54 charts, 27 figs., 54 tables.

Ceratium pentagonum rectum
Gourret, P., 1883
Sur les Peridiniens du Golfe de Marseille. Ann. du Musee d'hist. Nat., Marseille, Zool., 1 (Mme. 8):1-114, 4 pls.

Ceratium pentagonum subrobustum
Forti, A., 1922
Ricerche sulla flora pelagica (fitoplancton) di Quarto dei Mille. Mem. R. Com. Talass. Ital. 97:248 pp., 13 pls.

Ceratium pentagonum v. subrobustum
Ghazzawi, F.M., 1939
Plankton of the Egyptian waters. A study of the Suez Canal Plankton. (A) Phytoplankton. Preliminary Report 83 pp. Notes and Memoires, Min. Commerce-Industry, Egypt, Hydrobiol. & Fish. 65 figs.

Ceratium pentagonum tenerum
Graham, H. W. and N. Bronikovsky, 1944
The genus Ceratium in the Pacific and North Atlantic Oceans. Sci. Res. Cruise VII of the Carnegie, 1928-1929 ----- Biol. V (565):209 pp., 54 charts, 27 figs., 54 tables.

Ceratium pentagonum turgidum
Pavillard, J., 1916
Recherches sur les Peridiniens du Golfe du Lion. Mem. Univ. Montpellier. Trav. Inst. Bot., Univ. Montpellier. Serie mixte No.4, 70 pp., 3 pls., 15 text figs.

Ceratium petersi
Balech, Enrique, 1962
Tintinnoinea y Dinoflagellata del Pacifico segun material de las Expediciones NORPAC y DOWNWIND del Instituto Scripps de Oceanografia. Revista, Mus. Argentino Ciencias Nat. "Bernardino Rivadavia", Ciencias Zool., 7(1):1-253.

Ceratium petersii
Balech, Enrique, 1944
Contribucion al conocimiento del Plancton de Lennox y Cabo de Hornos. Physis XIX:423-446, 6 pls. with 67 figs.

Ceratium petersii
Graham, H. W. and N. Bronikovsky, 1944
The genus Ceratium in the Pacific and North Atlantic Oceans. Sci. Res. Cruise VII of the Carnegie, 1928-1929 ----- Biol. V (565):209 pp., 54 charts, 27 figs., 54 tables.

Ceratium petersii
López, J., 1966
Variación y regulación de la forma en el género Ceratium. Inv. Pesq., Barcelona, 30:325-427.

Ceratium Petersii n. sp.
Steemann Nielsen, E., 1934
Untersuchungen über die Verbreitung, Biologie und Variation der Ceratien im südlichen Stillen Ozean. Dana Rept. No. 4: 67 pp., 73 figs. and 11 charts in text.

Ceratium platycorne
Dangeard, P., 1927
Phytoplankton de la croisière du "Sylvana". Ann. Inst. Ocean., Monaco, n.s., 4(8):286-401, 54 text figs. (février-Juin 1913).

Ceratium platycorne
Forti, A., 1922
Ricerche sulla flora pelagica (fitoplancton) di Quarto dei Mille. Mem. R. Com. Talass. Ital. 97:248 pp., 13 pls.

Ceratium platycorne
Graham, H. W. and N. Bronikovsky, 1944
The genus Ceratium in the Pacific and North Atlantic Oceans. Sci. Res. Cruise VII of the Carnegie, 1928-1929 ----- Biol. V (565):209 pp., 54 charts, 27 figs., 54 tables.

Ceratium platycorne
Jörgensen, E., 1920
Mediterranean Ceratia. Rept. Danish Oceanogr. Exped. 1908-10 to the Mediterranean and adjacent seas, Vol.II, Biol. J 1:110 pp., 94 text figs., 26 charts.

Ceratium platycorne
LeBour, M.V., 1925
The dinoflagellates of Northern Seas. The Marine Biological Association of the United Kingdom. Plymouth, 250 pp., 35 pls. 53 text figs.

Ceratium platycorne
Paulsen, O., 1908
XVIII Peridiniales. Nordisches Plankton, Bot. Teil: 1-124, 155 text figs.

Ceratium platycorne
Pavillard, J., 1916
Recherches sur les Peridiniens du Golfe du Lion. Mem. Univ. Montpellier. Trav. Inst. Bot., Univ. Montpellier. Serie mixte No.4, 70 pp., 3 pls., 15 text figs.

Ceratium platycorne
Pavillard, J., 1905
Recherches sur la flore pelagique (Phytoplankton) de l'Etang de Thau. Theses presentees a la Fac. Sci., Paris, 116 pp., 3 pls

Ceratium platycorne
Peters, Nicolaus, 1934
Die Bevolkerung des Sudatlantischen Ozeans mit Ceratien. Biol. Sonderuntersuchungen 1. Wiss Ergeb. Deutschen Atlantischen Exped.--- "Meteor" 1925-1927, 12(1): 1-69, 28 text figs.

Ceratium platycorne
Sousa, A., 1967.
Le genre Ceratium (peridinien planctonique) dans le canal de Mozambique. Contribution a une revision mondiale (fin).
Vie Milieu (A) 18(3):441-449.

Ceratium platycorne
Sousa e Silva, E., 1949
Diatomaceas e Dinoflagelados de Baia de Cascais. Portugaliae Acta Biol., Volume: Julio Henriques, Ser. B: 300-383, 9 pls, 2 fold-in tables.

Ceratium platycorne
Steemann-Nielsen, E., 1939
Die Ceratien des Indischen Ozeans und der Ostasiatischen Gewässer mit einer allgemeinen zusammenfassung über die Verbreitung der Ceratien in den Weltmeeren. Dana Rept. No. 17: 33 pp., 8 maps in text.

Ceratium platycorne
Steemann Nielsen, E. 1934
Untersuchungen über die Verbreitung, Biologie und Variation der Ceratien im südlichen Stillen Ozean. Dana Rept. No. 4: 67 pp., 73 figs. and 11 charts in text.

Ceratium platycorne v. Daday f. cuneatum
Rampi, L., 1942.
Ricerche sul fitoplancton del Mare Ligure. 4. I Ceratium delle acque di Sanremo. Nuovo Giornale Botanico Italiano, N.S., 49: 221-236, 19 text figs.

Ceratium platycorne dilatatum
Steemann Nielsen, E., 1934
Untersuchungen über die Verbreitung, Biologie und Variation der Ceratien im südlichen Stillen Ozean. Dana Rept. No. 4: 67 pp., 73 figs. and 11 charts in text.

Ceratium platycorne v. Daday fa. incisum
Rampi, L., 1942.
Ricerche sul fitoplancton del Mare Ligure. 4. I Ceratium delle acque di Sanremo. Nuovo Giornale Botanico Italiano, N.S., 49: 221-236, 19 text figs.

Ceratium porrectum
Dangeard, P., 1927
Phytoplankton de la croisière du "Sylvana". Ann. Inst. Ocean., Monaco, n.s., 4(8):286-401, 54 text figs. (Feirer-Juin 1913).

Ceratium porrectum
Sousa e Silva, E., 1949
Diatomaceas e Dinoflagelados de Baia de Cascais. Portugaliae Acta Biol., Volume: Julio Henriques, Ser. B: 300-383, 9 pls, 2 fold-in tables.

Ceratium praelongum
Graham, H. W. and N. Bronikovsky, 1944
The genus Ceratium in the Pacific and North Atlantic Oceans. Sci. Res. Cruise VII of the Carnegie, 1928-1929 ----- Biol. V (565):209 pp., 54 charts, 27 figs., 54 tables.

Ceratium praelongum
Peters, Nicolaus, 1934
Die Bevolkerung des Sudatlantischen Ozeans mit Ceratien. Biol. Sonderuntersuchungen 1. Wiss Ergeb. Deutschen Atlantischen Exped.--- "Meteor" 1925-1927, 12(1): 1-69, 28 text figs.

Ceratium praelongum
Steemann-Nielsen, E., 1939
Die Ceratien des Indischen Ozeans und der Ostasiatischen Gewässer mit einer allgemeinen zusammenfassung über die Verbreitung der Ceratien in den Weltmeeren. Dana Rept. No. 17: 33 pp., 8 maps in text.

Ceratium praelongum
Steemann Nielsen, E., 1934
Untersuchungen über die Verbreitung, Biologie und Variation der Ceratien im südlichen Stillen Ozean. Dana Rept. No. 4: 67 pp., 73 figs. and 11 charts in text.

Ceratium praelongum
Wood, E.J.F., 1963
Dinoflagellates in the Australian region. II. Recent collections.
Comm. Sci. and Industr. Res. Org., Div. Fish. and Oceanogr., Techn. Paper, No. 14: 55 pp.

Ceratium procerum
Gourret, P., 1883
Sur les Peridiniens du Golfe de Marseille. Ann. du Musee d'hist. Nat., Marseille, Zool., I (Mme. 8):1-114, 4 pls.

Ceratium procerum divergens
Gourret, P., 1883
Sur les Peridiniens du Golfe de Marseille. Ann. du Musee d'hist. Nat., Marseille, Zool., I (Mme. 8):1-114, 4 pls.

Ceratium pulchellum
Chiba, T., 1949
On the distribution of the plankton in the eastern China Sea and Yellow Sea. 1. Plankton composition in the spring. J. Shimonoseki Coll. Fisheries, 1(1):57-63, 1 fig.

Ceratium pulchellum
Dangeard, P., 1927
Phytoplankton de la croisière du "Sylvana". Ann. Inst. Ocean., Monaco, n.s., 4(8):286-401, 54 text figs. (Feirer-Juin 1913).

Ceratium pulchellum
Delegazione Italiana della Commissione Internazionale per l'Esplorazione Scientifica del Mediterraneo, 1941
Note sul plancton della Laguna veneta. [Memoria CCLXXXI], Arch. di Ocean. e Limn. Anno I, Fasc. I, 1941 XIX: 31-57 pp.

Ceratium pulchellum
Ercegovic, A., 1936
Etudes qualitative et quantitatives du phytoplancton dans les eaux cotieres de l'Adriatique oriental moyen au cours de l'année 1934. Acta Adriatica 1(9):1-126

Ceratium pulchellum
Forti, A., 1922
Ricerche sulla flora pelagica (fitoplancton) di Quarto dei Mille. Mem. R. Com. Talass. Ital. 97:248 pp., 13 pls.

Ceratium pulchellum
Ghazzawi, F.M., 1939
Plankton of the Egyptian waters. A study of the Suez Canal Plankton. (A) Phytoplankton. Preliminary Report 83 pp. Notes and Memoires, Min. Commerce-Industry, Egypt, Hydrobiol. & Fish. 65 figs.

Ceratium pulchellum
Graham, H. W. and N. Bronikovsky, 1944
The genus Ceratium in the Pacific and North Atlantic Oceans. Sci. Res. Cruise VII of the Carnegie, 1928-1929 ----- Biol. V (565):209 pp., 54 charts, 27 figs., 54 tables.

Ceratium pulchellum
Halim, Y., 1963.
Microplancton des eaux égyptiennes. Le genre Ceratium Schrank (Dinoflagellés). Rapp. Proc. Verb., Réunions, Comm. Int. Expl. Sci., Mer Méditerranée, Monaco, 17(2):495-502.

Ceratium pulchellum
Jörgensen, E., 1920
Mediterranean Ceratia. Rept. Danish Oceanogr. Exped. 1908-10 to the Mediterranean and adjacent seas, Vol.II, Biol. J 1:110 pp., 94 text figs., 26 charts.

Ceratium pulchellum
Lindemann, E., 1925
Neubeobachtungen an den Winter peridineen des Golfes von Neapel. Bot. Arch. 9:95-102, 19 text figs.

Ceratium pulchellum
Macdonald, R., 1933
An examination of plankton hauls made in the Suez Canal during the year 1928. Fish. Res. Dis., Notes & Mem. No.3, 11 pp., 1 chart.

Ceratium pulchellum
Pavillard, J., 1916
Recherches sur les Peridiniens du Golfe du Lion. Mem. Univ. Montpellier. Trav. Inst. Bot., Univ. Montpellier. Serie mixte No.4, 70 pp., 3 pls., 15 text figs.

Ceratium pulchellum
Sousa e Silva, E., 1949
Diatomaceas e Dinoflagelados de Baia de Cascais. Portugaliae Acta Biol., Volume: Julio Henriques, Ser. B: 300-383, 9 pls, 2 fold-in tables.

Ceratium pulchellum
Steemann Nielsen, E., 1934
Untersuchungen über die Verbreitung, Biologie und Variation der Ceratien im südlichen Stillen Ozean. Dana Rept. No. 4: 67 pp., 73 figs. and 11 charts in text.

Ceratium pulchellum s. eupulchellum
Ghazzawi, F.M., 1939
Plankton of the Egyptian waters. A study of the Suez Canal Plankton. (A) Phytoplankton. Preliminary Report 83 pp. Notes and Memoires, Min. Commerce-Industry, Egypt, Hydrobiol. & Fish. 65 figs.

Ceratium pulchellum f. semipulchellum
Ghazzawi, F.M., 1939
Plankton of the Egyptian waters. A study of the Suez Canal Plankton. (A) Phytoplankton. Preliminary Report 83 pp. Notes and Memoires, Min. Commerce-Industry, Egypt, Hydrobiol. & Fish. 65 figs.

Ceratium quinquecorne
Gourret, P., 1883
Sur les Peridiniens du Golfe de Marseille. Ann. du Musee d'hist. Nat., Marseille, Zool., I (Mme. 8):1-114, 4 pls.

Ceratium ranipes
Balech, Enrique, 1962
Tintinnoinea y Dinoflagellata del Pacifico segun material de las Expediciones NORPAC y DOWNWIND del Instituto Scripps de Oceanografia.
Revista, Mus. Argentino Ciencias Nat. "Bernardino Rivadavia", Ciencias Zool., 7(1):1-253.

Ceratium ranipes
Dangeard, P., 1927
Phytoplankton de la croisière du "Sylvana". Ann. Inst. Ocean., Monaco, n.s., 4(8):286-401, 54 text figs. (février-Juin 1913).

Ceratium ranipes
Graham, H. W. and N. Bronikovsky, 1944
The genus Ceratium in the Pacific and North Atlantic Oceans. Sci. Res. Cruise VII of the Carnegie, 1928-1929 ----- Biol. V (565):209 pp., 54 charts, 27 figs., 54 tables.

Ceratium ranipes
Jörgensen, E., 1920
Mediterranean Ceratia. Rept. Danish Oceanogr. Exped. 1908-10 to the Mediterranean and adjacent seas, Vol.II, Biol. J 1:110 pp., 94 text figs., 26 charts.

Ceratium ranipes
Peters, Nicolaus, 1934
Die Bevolkerung des Sudatlantischen Ozeans mit Ceratien. Biol. Sonderuntersuchungen 1. Wiss Ergeb. Deutschen Atlantischen Exped.— "Meteor" 1925-1927, 12(1): 1-69, 28 text figs.

Ceratium ranipes
Sournia, A., 1967.
Le genre Ceratium (peridinien planctonique) dans le canal de Mozambique. Contribution a une revision mondiale (fin). Vie Milieu (A) 18(3):441-449.

Ceratium ranipes
Steemann-Nielsen, E., 1939
Die Ceratien des Indischen Ozeans und der Ostasiatischen Gewässer mit einer allgemeinen zusammenfassung über die Verbreitung der Ceration in den Weltmeeren. Dana Rept. No. 17: 33 pp., 8 maps in text.

Ceratium ranipes
Steemann Nielsen, E., 1934
Untersuchungen über die Verbreitung, Biologie und Variation der Ceratien im südlichen Stillen Ozean. Dana Rept. No. 4: 67 pp., 73 figs. and 11 charts in text.

Ceratium ranipes
Vives, F., and A. Planas, 1952.
Plancton recogido por los laboratorios costeros. VI. Fitoplancton de las costas de Vinaroz, islas Columbretes y alrededores de la desembocadura del Ebro. Publ. Inst. Biol. Aplic. 11:141-156, 19 textfigs.

Ceratium reflexum
Graham, H. W. and N. Bronikovsky, 1944
The genus Ceratium in the Pacific and North Atlantic Oceans. Sci. Res. Cruise VII of the Carnegie, 1928-1929 ----- Biol. V (565):209 pp., 54 charts, 27 figs., 54 tables.

ceratium reflexum
Sournia, A., 1967.
Le genre Ceratium (peridinien planctonique) dans le canal de Mozambique. Contribution a une revision mondiale (fin). Vie Milieu (A) 18(3):441-449.

Ceratium reflexum
Steemann-Nielsen, E., 1939
Die Ceratien des Indischen Ozeans und der Ostasiatischen Gewässer mit einer allgemeinen zusammenfassung über die Verbreitung der Ceration in den Weltmeeren. Dana Rept. No. 17: 33 pp., 8 maps in text.

Ceratium reflexum
Steemann Nielsen, E., 1934
Untersuchungen über die Verbreitung, Biologie und Variation der Ceratien im südlichen Stillen Ozean. Dana Rept. No. 4: 67 pp., 73 figs. and 11 charts in text.

Ceratium reflexum
Wood, E.J.F., 1963
Dinoflagellates in the Australian region. II. Recent collections. Comm. Sci. and Industr. Res. Org., Div. Fish. and Oceanogr., Techn. Paper, No. 14: 55 pp.

Ceratium reticulatum
Ercegovic, A., 1936
Etudes qualitative et quantitatives du phytoplancton dans les eaux cotières de l'Adriatique oriental moyen au cours de l'année 1934. Acta Adriatica 1(9):1-126

Ceratium reticulatum
Forti, A., 1922
Ricerche sulla flora pelagica (fitoplancton) di Quarto dei Mille. Mem. R. Com. Talass. Ital. 97:248 pp., 13 pls.

Ceratium reticulatum
Fraser, J. H., and A. Saville, 1949
Plankton distribution in Scottish and adjacent waters in 1948. Ann. Biol. 5:61-62.

Ceratium reticulatum
LeBour, M.V., 1925
The dinoflagellates of Northern Seas. The Marine Biological Association of the United Kingdom. Plymouth, 250 pp., 35 pls. 53 text figs.

Ceratium reticulatum
Lindemann, E., 1925
Neubeobachtungen an den Winter peridineen des Golfes von Neapel. Bot. Arch. 9:95-102, 19 text figs.

Ceratium reticulatum
Mangin, M. L., 1912
Phytoplancton de la croisière du "René" dans l'Atlantique (Septembre 1908). Ann. Inst. Ocean., n.s., 4(1):1-66, 2 pls., 41 text figs., 2 tables.

Ceratium reticulatum
Paulsen, O., 1908
XVIII Peridiniales. Nordisches Plankton, Bot. Teil: 1-124, 155 text figs.

Ceratium reticulatum
Pavillard, J., 1916
Recherches sur les Peridiniens du Golfe du Lion. Mem. Univ. Montpellier. Trav. Inst. Bot. Univ. Montpellier. Serie mixte No.4, 70 pp., 3 pls., 15 text figs.

Ceratium reticulatum
Pavillard, J., 1905
Recherches sur la flore pelagique (Phytoplankton) de l'Etang de Thau. Theses presentees a la Fac. Sci., Paris, 116 pp., 3 pls

Ceratium reticulatum spirale n. subsp.
Kofoid, C. A., 1907
Dinoflagellata of the San Diego region. III. Descriptions of new species. Univ. Calif. Publ., Zool. 3:299-340, Pls. 22-33.

Ceratium rostellum
Gourret, P., 1883
Sur les Peridiniens du Golfe de Marseille. Ann. du Musee d'hist. Nat., Marseille, Zool., 1 (Mme. 8):1-114, 4 pls.

Ceratium scapiforme
Steemann Nielsen, E., 1934
Untersuchungen über die Verbreitung, Biologie und Variation der Ceratien im südlichen Stillen Ozean. Dana Rept. No. 4: 67 pp., 73 figs. and 11 charts in text.

Ceratium schmidti
Balech, Enrique, 1962
Tintinnoinea y Dinoflagellata del Pacifico segun material de las Expediciones NORPAC y DOWNWIND del Instituto Scripps de Oceanografia. Revista, Mus. Argentino Ciencias Nat. "Bernardino Rivadavia", Ciencias Zool., 7(1):1-253.

Ceratium schmidtii
López, J., 1966.
Variación y regulación de la forma en el género Ceratium. Inv. Pesq., Barcelona, 30:325-427.

Ceratium schmidtii
Steemann-Nielsen, E., 1939
Die Ceratien des Indischen Ozeans und der Ostasiatischen Gewässer mit einer allgemeinen zusammenfassung über die Verbreitung der Ceratien in den Weltmeeren. Dana Rept. No. 17: 33 pp., 8 maps in text.

Ceratium Schmidtii
Steemann Nielsen, E., 1934
Untersuchungen über die Verbreitung, Biologie und Variation der Ceratien im südlichen Stillen Ozean. Dana Rept. No. 4: 67 pp., 73 figs. and 11 charts in text.

Ceratium schmidti
Wang, C. C., and D. Nie, 1932
A survey of the marine protozoa of Amoy. Contr. Biol. Lab. Sci. Soc., China, Zool. Ser., 8:285-385, 89 text figs.

Ceratium Schroeteri
Jörgensen, E., 1920
Mediterranean Ceratia. Rept. Danish Oceanogr. Exped. 1908-10 to the Mediterranean and adjacent seas, Vol.II, Biol. J 1:110 pp., 94 text figs., 26 charts.

Ceratium schroeteri
Wood, E.J.F., 1963
Dinoflagellates in the Australian region. II. Recent collections. Comm. Sci. and Industr. Res. Org., Div. Fish. and Oceanogr., Techn. Paper, No. 14: 55 pp.

Ceratium semipulchellum
Steemann-Nielsen, E., 1939
Die Ceratien des Indischen Ozeans und der Ostasiatischen Gewässer mit einer allgemeinen zusammenfassung über die Verbreitung der Ceratien in den Weltmeeren. Dana Rept. No. 17: 33 pp., 8 maps in text.

Ceratium semipulchellum n.sp.
Steemann Nielsen, E., 1934
Untersuchungen über die Verbreitung, Biologie und Variation der Ceratien im südlichen Stillen Ozean. Dana Rept. No. 4: 67 pp., 73 figs. and 11 charts in text.

Ceratium seta
Forti, A., 1922
Ricerche sulla flora pelagica (fitoplancton) di Quarto dei Mille. Mem. R. Com. Talass. Ital. 97:248 pp., 13 pls.

Ceratium seta
Wang, C. C., and D. Nie, 1932
A survey of the marine protozoa of Amoy. Contr. Biol. Lab. Sci. Soc., China, Zool. Ser., 8:285-385, 89 text figs.

Ceratium setaceum
Ercegovic, A., 1936
Etudes qualitative et quantitatives du phytoplancton dans les eaux cotières de l'Adriatique oriental moyen au cours de l'année 1934. Acta Adriatica 1(9):1-126

Ceratium setaceum
Graham, H. W. and N. Bronikovsky, 1944
The genus Ceratium in the Pacific and North Atlantic Oceans. Sci. Res. Cruise VII of the Carnegie, 1928-1929 ----- Biol. V (565):209 pp., 54 charts, 27 figs., 54 tables.

Ceratium setaceum
Jörgensen, E., 1920
Mediterranean Ceratia. Rept. Danish Oceanogr. Exped. 1908-10 to the Mediterranean and adjacent seas, Vol.II, Biol. J 1:110 pp., 94 text figs., 26 charts.

Ceratium setaceum
Mangin, M. L., 1912
Phytoplancton de la croisière du "René" dans l'Atlantique (Septembre 1908). Ann. Inst. Ocean., n.s., 4(1):1-66, 2 pls., 41 text figs., 2 tables.

Ceratium setaceum
Peters, Nicolaus, 1934
Die Bevölkerung des Südatlantischen Ozeans mit Ceratien. Biol. Sonderuntersuchungen 1. Wiss Ergeb. Deutschen Atlantischen Exped. "Meteor" 1925-1927, 12(1): 1-69, 28 text figs.

Ceratium setaceum
Rampi, L., 1942
Ricerche sul fitoplancton del Mare Ligure. 4. I Ceratium delle acque di Sanremo. Nuovo Giornale Botanico Italiano, N.S., 49: 221-236, 19 text figs.

Ceratium setaceum
Steemann-Nielsen, E., 1939
Die Ceratien des Indischen Ozeans und der Ostasiatischen Gewässer mit einer allgemeinen zusammenfassung über die Verbreitung der Ceratien in den Weltmeeren. Dana Rept. No. 17: 33 pp., 8 maps in text.

Ceratium setaceum
Steemann Nielsen, E., 1934
Untersuchungen über die Verbreitung, Biologie und Variation der Ceratien im südlichen Stillen Ozean. Dana Rept. No. 4: 67 pp., 73 figs. and 11 charts in text.

Ceratium schranki n. sp.
Kofoid, C. A., 1907
Dinoflagellata of the San Diego region. III. Descriptions of new species. Univ. Calif. Publ., Zool. 3:299-340, Pls. 22-33.

Ceratium strictum
Balech, Enrique, 1962
Tintinnoinea y Dinoflagellata del Pacifico segun material de las Expediciones NORPAC y DOWNWIND del Instituto Scripps de Oceanografia. Revista, Mus. Argentino Ciencias Nat. "Bernardino Rivadavia", Ciencias Zool., 7(1):1-253.

Ceratium strictum
Dangeard, P., 1927
Phytoplankton de la croisière du "Sylvana". Ann. Inst. Ocean., Monaco, n.s., 4(8):286-401, 54 text figs. (Feirer-Juin 1913).

Ceratium strictum
Ercegovic, A., 1936
Etudes qualitative et quantitatives du phytoplancton dans les eaux cotières de l'Adriatique oriental moyen au cours de l'année 1934. Acta Adriatica 1(9):1-126

Ceratium strictum
Jörgensen, E., 1920
Mediterranean Ceratia. Rept. Danish Oceanogr. Exped. 1908-10 to the Mediterranean and adjacent seas, Vol.II, Biol. J 1:110 pp., 94 text figs., 26 charts.

Ceratium strictum
Forti, A., 1922
Ricerche sulla flora pelagica (fitoplancton) di Quarto dei Mille. Mem. R. Com. Talass. Ital. 97:248 pp., 13 pls.

Ceratium strictum
Pavillard, J., 1916
Recherches sur les Peridiniens du Golfe du Lion. Mem. Univ. Montpellier. Trav. Inst. Bot., Univ. Montpellier. Serie mixte No.4, 70 pp., 3 pls., 15 text figs.

Ceratium subrobustum
Graham, H. W. and N. Bronikovsky, 1944
The genus Ceratium in the Pacific and North Atlantic Oceans. Sci. Res. Cruise VII of the Carnegie, 1928-1929 ----- Biol. V (565):209 pp., 54 charts, 27 figs., 54 tables.

Ceratium subrobustum
Steemann-Nielsen, E., 1939
Die Ceratien des Indischen Ozeans und der Ostasiatischen Gewässer mit einer allgemeinen zusammenfassung über die Verbreitung der Ceratien in den Weltmeeren. Dana Rept. No. 17: 33 pp., 8 maps in text.

Ceratium subrobustum n. sp.
Steemann Nielsen, E., 1934
Untersuchungen über die Verbreitung, Biologie und Variation der Ceratien im südlichen Stillen Ozean. Dana Rept. No. 4: 67 pp., 73 figs. and 11 charts in text.

Ceratium sumatranum
Dangeard, P., 1927
Phytoplankton de la croisière du "Sylvana". Ann. Inst. Ocean., Monaco, n.s., 4(8):286-401, 54 text figs. (Feirer-Juin 1913).

Ceratium sumatranum
Steemann-Nielsen, E., 1939
Die Ceratien des Indischen Ozeans und der Ostasiatischen Gewässer mit einer allgemeinen zusammenfassung über die Verbreitung der Ceratien in den Weltmeeren. Dana Rept. No. 17: 33 pp., 8 maps in text.

Ceratium symmetricum
Delegazione Italiana della Commissione Internazionale per l'Esplorazione Scientifica del Mediterraneo, 1941
Note sul plancton della Laguna veneta. [Memoria CCLXXIX], Arch. di Ocean. e Limn. Anno I, Fasc. I, 1941 XIX: 31-57 pp.

Ceratium symmetricum
Forti, A., 1922
Ricerche sulla flora pelagica (fitoplancton) di Quarto dei Mille. Mem. R. Com. Talass. Ital. 97:248 pp., 13 pls.

Ceratium symmetricum
Graham, H. W. and N. Bronikovsky, 1944
The genus Ceratium in the Pacific and North Atlantic Oceans. Sci. Res. Cruise VII of the Carnegie, 1928-1929 ----- Biol. V (565):209 pp., 54 charts, 27 figs., 54 tables.

Ceratium symmetricum
Pavillard, J., 1923
A propos de la systématique des Péridiniens Bull. Soc. Bot. de France 70:876-882; 914-918.

Ceratium symmetricum
Pavillard, J., 1916
Recherches sur les Peridiniens du Golfe du Lion. Mem. Univ. Montpellier. Trav. Inst. Bot., Univ. Montpellier. Serie mixte No.4, 70 pp., 3 pls., 15 text figs.

Ceratium symmetricum n.sp.
Pavillard, J., 1905
Recherches sur la flore pelagique (Phytoplankton) de l'Etang de Thau. Theses presentees a la Fac. Sci., Paris, 116 pp., 3 pls.

Ceratium symmetricum
Steemann-Nielsen, E., 1939
Die Ceratien des Indischen Ozeans und der Ostasiatischen Gewässer mit einer allgemeinen zusammenfassung über die Verbreitung der Ceratien in den Weltmeeren. Dana Rept. No. 17: 33 pp., 8 maps in text.

Ceratium symmetricum
Steemann Nielsen, E., 1934
Untersuchungen über die Verbreitung, Biologie und Variation der Ceratien im südlichen Stillen Ozean. Dana Rept. No. 4: 67 pp., 73 figs. and 11 charts in text.

Ceratium symmetricum coarctatum
Graham, H. W. and N. Bronikovsky, 1944
The genus Ceratium in the Pacific and North Atlantic Oceans. Sci. Res. Cruise VII of the Carnegie, 1928-1929 ----- Biol. V (565):209 pp., 54 charts, 27 figs., 54 tables.

Ceratium symmetricum orthoceros
Graham, H. W. and N. Bronikovsky, 1944
The genus Ceratium in the Pacific and North Atlantic Oceans. Sci. Res. Cruise VII of the Carnegie, 1928-1929 ----- Biol. V (565):209 pp., 54 charts, 27 figs., 54 tables.

Ceratium symmetricum symmetricum
Graham, H. W. and N. Bronikovsky, 1944
The genus Ceratium in the Pacific and North Atlantic Oceans. Sci. Res. Cruise VII of the Carnegie, 1928-1929 ----- Biol. V (565):209 pp., 54 charts, 27 figs., 54 tables.

Ceratium tasmaniae n.sp.
Wood, E.J.F., 1963
Dinoflagellates in the Australian region. II. Recent collections. Comm. Sci. and Industr. Res. Org., Div. Fish. and Oceanogr., Techn. Paper, No. 14: 55 pp.

Ceratium tenue
Ercegovic, A., 1936
Etudes qualitative et quantitatives du phytoplancton dans les eaux cotières de l'Adriatique oriental moyen au cours de l'année 1934. Acta Adriatica 1(9):1-126

Ceratium tenue
Forti, A., 1922
Ricerche sulla flora pelagica (fitoplancton) di Quarto dei Mille. Mem. R. Com. Talass. Ital. 97:248 pp., 13 pls.

Ceratium tenua
Graham, H. W. and N. Bronikovsky, 1944
The genus Ceratium in the Pacific and North Atlantic Oceans. Sci. Res. Cruise VII of the Carnegie, 1928-1929 ----- Biol. V (565):209 pp., 54 charts, 27 figs., 54 tables.

Ceratium tenue
Mangin, M. L., 1912
Phytoplancton de la croisière du "René" dans l'Atlantique (Septembre 1908). Ann. Inst. Ocean., n.s., 4(1):1-66, 2 pls., 41 text figs., 2 tables.

Ceratium tenue
Pavillard, J., 1923
A propos de la systématique des Péridiniens Bull. Soc. Bot. de France 70:876-882; 914-918.

Ceratium tenue
Pavillard, J., 1916
Recherches sur les Peridiniens du Golfe du Lion. Mem. Univ. Montpellier. Trav. Inst. Bot., Univ. Montpellier. Serie mixte No.4, 70 pp., 3 pls., 15 text figs.

Ceratium tenue
Steemann-Nielsen, E., 1939
Die Ceratien des Indischen Ozeans und der Ostasiatischen Gewässer mit einer allgemeinen zusammenfassung über die Verbreitung der Ceratien in den Weltmeeren. Dana Rept. No. 17: 33 pp., 8 maps in text.

Ceratium tenue
Steemann Nielsen, E., 1934
Untersuchungen über die Verbreitung, Biologie und Variation der Ceratien im südlichen Stillen Ozean. Dana Rept. No. 4: 67 pp., 73 figs. and 11 charts in text.

Ceratium tenue buceros
Forti, A., 1922
Ricerche sulla flora pelagica (fitoplancton) di Quarto dei Mille. Mem. R. Com. Talass. Ital. 97:248 pp., 13 pls.

Ceratium tenua inclinatum
Graham, H. W. and N. Bronikovsky, 1944
The genus Ceratium in the Pacific and North Atlantic Oceans. Sci. Res. Cruise VII of the Carnegie, 1928-1929 ----- Biol. V (565):209 pp., 54 charts, 27 figs., 54 tables.

Oceanographic Index: Marine Organisms Cumulation, 1946-1973

Ceratium tenua tenuissimum
Graham, H. W. and N. Bronikovsky, 1944
 The genus Ceratium in the Pacific and North Atlantic Oceans. Sci. Res. Cruise VII of the Carnegie, 1928-1929 ----- Biol. V (565):209 pp., 54 charts, 27 figs., 54 tables.

Ceratium tenuissimum n. sp.
Kofoid, C. A., 1907
 Dinoflagellata of the San Diego region. III. Descriptions of new species. Univ. Calif. Publ., Zool. 3:299-340, Pls. 22-33.

Ceratium tenuissimum
Pavillard, J., 1923
 A propos de la systématique des Péridiniens Bull. Soc. Bot. de France 70:876-882; 914-918.

Ceratium teres
Dangeard, P., 1927
 Phytoplankton de la croisière du "Sylvana". Ann. Inst. Ocean., Monaco, n.s., 4(8):286-401, 54 text figs. (février -Juin 1913).

Ceratium teres
Delegazione Italiana della Commissione Internazionale per l'Esplorazione Scientifica del Mediterraneo, 1941
 Note sul plancton della Laguna veneta. [Memoria CCLXXIX], Arch. di Ocean. e Limn. Anno I, Fasc. I, 1941 XIX: 31-57 pp.

Ceratium teres
Ercegovic, A., 1936
 Etudes qualitative et quantitatives du phytoplankton dans les eaux cotières de l'Adriatique oriental moyen au cours de l'année 1934. Acta Adriatica 1(9):1-126

Ceratium teres
Forti, A., 1922
 Ricerche sulla flora pelagica (fitoplancton) di Quarto dei Mille. Mem. R. Com. Talass. Ital. 97:248 pp., 13 pls.

Ceratium teres
Graham, H. W. and N. Bronikovsky, 1944
 The genus Ceratium in the Pacific and North Atlantic Oceans. Sci. Res. Cruise VII of the Carnegie, 1928-1929 ----- Biol. V (565):209 pp., 54 charts, 27 figs., 54 tables.

Ceratium teres
Jörgensen, E., 1920
 Mediterranean Ceratia. Rept. Danish Oceanogr. Exped. 1908-10 to the Mediterranean and adjacent seas, Vol.II, Biol. J 1:110 pp., 94 text figs., 26 charts.

Ceratium teres n. sp.
Kofoid, C. A., 1907
 Dinoflagellata of the San Diego region. III. Descriptions of new species. Univ. Calif. Publ., Zool. 3:299-340, Pls. 22-33.

Ceratium teres
López, J., 1966
 Variación y regulación de la forma en el género Ceratium. Inv. Pesq., Barcelona, 30:325-427.

Ceratium teres
Margalef, R., 1949
 Fitoplancton nerítico de la Costa Brava en 1947-48. Publ. Inst. Biol. Aplicada, 5: 41-51, 3 text figs.

Ceratium teres
Marshall, S. M., 1933
 The production of microplankton in the Great Barrier Reef Region. Brit. Mus. (N.H.) Great Barrier Reef Exped. 1928-29, Sci. Repts. II(5):111-157, 14 text figs.

Ceratium teres
Pavillard, J., 1916
 Recherches sur les Peridiniens du Golfe du Lion. Mem. Univ. Montpellier. Trav. Inst. Bot., Univ. Montpellier. Serie mixte No.4, 70 pp., 3 pls., 15 text figs.

Ceratium teres
Peters, Nicolaus, 1934
 Die Bevolkerung des Sudatlantischen Ozeans mit Ceratien. Biol. Sonderuntersuchungen 1. Wiss Ergeb. Deutschen Atlantischen Exped. "Meteor" 1925-1927, 12(1): 1-69, 28 text figs.

Ceratium teres (figs.)
Sousa e Silva, E., 1949
 Diatomaceas e Dinoflagelados de Baia de Cascais. Portugaliae Acta Biol., Volume: Julio Henriques, Ser. B: 300-383, 9 pls, 2 fold-in tables.

Ceratium teres
Steemann-Nielsen, E., 1939
 Die Ceratien des Indischen Ozeans und der Ostasiatischen Gewässer mit einer allgemeinen zusammenfassung über die Verbreitung der Ceratien in den Weltmeeren. Dana Rept. No. 17: 33 pp., 8 maps in text.

Ceratium teres
Steemann Nielsen, E., 1934
 Untersuchungen über die Verbreitung, Biologie und Variation der Ceratien im südlichen Stillen Ozean. Dana Rept. No. 4: 67 pp., 73 figs. and 11 charts in text.

Ceratium trichoceros
Davis, Joanne, and Karen A. Steidinger
 Ceratium trichoceros.
 Florida Bd. Conserv. Mar. Lab. Leaflet Ser.
 1 (Phytoplankton) (1) Dinoflagellates (1). 3pp.

Ceratium trichoceros
Delegazione Italiana della Commissione Internazionale per l'Esplorazione Scientifica del Mediterraneo, 1941
 Note sul plancton della Laguna veneta. [Memoria CCLXXIX], Arch. di Ocean. e Limn. Anno I, Fasc. I, 1941 XIX: 31-57 pp.

Ceratium trichoceros
Ercegovic, A., 1936
 Etudes qualitative et quantitatives du phytoplankton dans les eaux cotières de l'Adriatique oriental moyen au cours de l'année 1934. Acta Adriatica 1(9):1-126

Ceratium trichoceros
Forti, A., 1922
 Ricerche sulla flora pelagica (fitoplancton) di Quarto dei Mille. Mem. R. Com. Talass. Ital. 97:248 pp., 13 pls.

Ceratium trichoceros
Ghazzawi, F.M., 1939
 Plankton of the Egyptian waters. A study of the Suez Canal Plankton. (A) Phytoplankton. Preliminary Report 83 pp. Notes and Memoires, Min. Commerce-Industry, Egypt, Hydrobiol. & Fish. 65 figs.

Ceratium trichoceros
Graham, H. W. and N. Bronikovsky, 1944
 The genus Ceratium in the Pacific and North Atlantic Oceans. Sci. Res. Cruise VII of the Carnegie, 1928-1929 ----- Biol. V (565):209 pp., 54 charts, 27 figs., 54 tables.

Ceratium trichoceros
Jörgensen, E., 1920
 Mediterranean Ceratia. Rept. Danish Oceanogr. Exped. 1908-10 to the Mediterranean and adjacent seas, Vol.II, Biol. J 1:110 pp., 94 text figs., 26 charts.

Ceratium trichoceros
Lindemann, E., 1925
 Neubeobachtungen an den Winter peridineen des Golfes von Neapel. Bot. Arch. 9:95-102, 19 text figs.

Ceratium trichoceros
Macdonald, R., 1933
 An examination of plankton hauls made in the Suez Canal during the year 1928. Fish. Res. Dis., Notes & Mem. No.3, 11 pp., 1 chart.

Ceratium trichoceros
Margalef, R., 1949
 Fitoplancton nerítico de la Costa Brava en 1947-48. Publ. Inst. Biol. Aplicada, 5: 41-51, 3 text figs.

Ceratium trichoceros
Marukawa, H., 1921
 Plankton lists and some new species of copepods, from the northern waters of Japan. Bull. Inst. Ocean., No.384, 15 pp., 3 pls., 1 chart. Monaco

Ceratium trichoceros
Mulford, Richard A., 1963
 Distribution of the dinoflagellate genus Ceratium in the tidal and offshore waters of Virginia. Chesapeake Science, 4(2):84-89.

Ceratium trichoceros
Peters, Nicolaus, 1934
 Die Bevolkerung des Sudatlantischen Ozeans mit Ceratien. Biol. Sonderuntersuchungen 1. Wiss Ergeb. Deutschen Atlantischen Exped. "Meteor" 1925-1927, 12(1): 1-69, 28 text figs.

Ceratium trichoceros
Sournia A., 1967.
 Le genre Ceratium (peridinien planctonique) dans le canal de Mozambique. Contribution a une revision mondiale (fin). Vie Milieu (A) 18(3):441-449.

Ceratium trichoceros
Sousa e Silva, E., 1949
 Diatomaceas e Dinoflagelados de Baia de Cascais. Portugaliae Acta Biol., Volume: Julio Henriques, Ser. B: 300-383, 9 pls, 2 fold-in tables.

Ceratium trichoceros
Steemann-Nielsen, E., 1939
 Die Ceratien des Indischen Ozeans und der Ostasiatischen Gewässer mit einer allgemeinen zusammenfassung über die Verbreitung der Ceratien in den Weltmeeren. Dana Rept. No. 17: 33 pp., 8 maps in text.

Ceratium trichoceros
Steemann Nielsen, E., 1934
 Untersuchungen über die Verbreitung, Biologie und Variation der Ceratien im südlichen Stillen Ozean. Dana Rept. No. 4: 67 pp., 73 figs. and 11 charts in text.

Ceratium trichoceros
Sukhanova, I.N., 1964.
 The phytoplankton of the northeastern part of the Indian Ocean in the season of the southwest monsoon. Regularity of the distribution of oceanic plankton. (In Russian; English abstract). Trudy Inst. Okeanol., Akad. Nauk, SSSR, 65:24-31.

Ceratium trichoceros
Wang, C. C., and D. Nie, 1932
 A survey of the marine protozoa of Amoy. Contr. Biol. Lab. Sci. Soc., China, Zool. Ser., 8:285-385, 89 text figs.

Ceratium tripodoides n. sp.
Steemann Nielsen, E., 1934
 Untersuchungen über die Verbreitung, Biologie und Variation der Ceratien im südlichen Stillen Ozean. Dana Rept. No. 4: 67 pp., 73 figs. and 11 charts in text.

Ceratium tripos
Allen, W.E., 1927.
 Quantitative studies on inshore marine diatoms and dinoflagellates of Southern California in 1922. Bull. S.I.O., tech. ser., 1:31-38, 2 textfigs.

Ceratium tripos
Balech, Enrique, 1944
Contribucion al conocimiento del Plancton de Lennox y Cabo de Hornos. Physis XIX:423-446, 6 pls. with 67 figs.

Ceratium tripos
Bigelow, H. B., 1922
Exploration of the coastal water off the northeastern United States in 1916 by the U.S. Fisheries Schooner Grampus. Bull. M.C.Z. 65(5):85-188, 53 text figs.

Ceratium tripos
Brandes, C. -H., 1939(1951).
Über die räumlichen und zeitlichen Unterschiede in der Zusammensetzunk des Ostseeplanktons. Mitt. Hamburg Zool. Mus. u. Inst. 48:1-47, 23 textfigs.

Ceratium tripos
Braarud, T., 1945
A phytoplankton survey of the polluted waters of inner Oslo Fjord. Hvalrådets Skrifter, No.28, 142 pp., 19 text figs., 17 tables.

Ceratium tripos
Braarud, T., 1934
A note on the phytoplankton of the Gulf of Maine in the summer of 1933. Biol. Bull. 67(1):76-82.

Ceratium tripos
Braarud, T., and Adam Bursa, 1939
On the phytoplankton of the Oslo Fjord, 1933-1934. Hvalrådets Skr. No.19:1-63; 9 text figs. Reviewed. J. du. Cons. 14(3):418-420. A.C. Gardiner.

Ceratium tripos
Brunel, J., 1962
Le phytoplancton de la Baie de Chaleurs. Inst. Botan., Univ. Montréal, Contrib. No. 77: 365 pp., 66 pls.

Ceratium tripos
Calkins, G. N., 1902
Marine protozoa from Woods Hole. U.S. Fish Comm. Bull. for 1901, pp. 413-468, 69 text figs.

Ceratium tripos
Chiba, T., 1949
On the distribution of the plankton in the eastern China Sea and Yellow Sea. 1. Plankton composition in the spring. J. Shimonoseki Coll. Fisheries, 1(1):57-63, 1 fig.

Ceratium tripos
Copenhagen, W. J. and L. D., 1949
Variation in the phytoplankton of Table Bay, October 1934 to October 1935. With a note on the calorific value of Chaetoceros spp. Trans. Roy. Soc. S. Africa, 32(2):113-123, 2 text figs.

Ceratium tripos
Cupp, E.E., 1934
Analysis of marine diatom collections taken from the Canal Zone to California during March, 1933. Trans. Am. Micros. Soc. LIII (1):22-29, 1 map.

Ceratium tripos
Cupp, E., 1930
Quantitative Studies of miscellaneous series of surface catches of marine diatoms and dinoflagellates taken between Seattle and the Canal Zone from 1924 to 1928. Trans. Am. Micro. Soc. XLIX (3):238-245.

Ceratium tripos
Dangeard, P., 1927
Phytoplankton de la croisière du "Sylvana". Ann. Inst. Ocean., Monaco, n.s., 4(8):286-401, 54 text figs. (Février-Juin 1913).

Ceratium tripos
Fraser, J. H., and A. Saville, 1949
Plankton distribution in Scottish and adjacent waters in 1948. Ann. Biol. 5:61-62.

Ceratium tripos
Ghazzawi, F.M., 1939
Plankton of the Egyptian waters. A study of the Suez Canal Plankton. (A) Phytoplankton. Preliminary Report 83 pp. Notes and Memoires, Min. Commerce-Industry, Egypt, Hydrobiol. & Fish. 65 figs.

Ceratium tripos
Gilbert, J.Y., and W.E. Allen, 1943
The phytoplankton of the Gulf of California obtained by the "E.W. Scripps" in 1939 and 1940. J. Mar. Res. V(2):89-110, figs.30-31.

Ceratium tripos
Gourret, P., 1883
Sur les Peridiniens du Golfe de Marseille. Ann. du Musee d'hist. Nat., Marseille, Zool., 1 (Mme. 8):1-114, 4 pls.

Ceratium tripos
Graham, H. W. and N. Bronikovsky, 1944
The genus Ceratium in the Pacific and North Atlantic Oceans. Sci. Res. Cruise VII of the Carnegie, 1928-1929 ----- Biol. V (565):209 pp., 54 charts, 27 figs., 54 tables.

Ceratium tripos
Gran, H.H., 1897
Protophyta: Diatomaceae, Silico-flagellata and Cilioflagellata. Den Norske Nordhavs Expedition 1876-1878, h. 24, 36 pp., 4 pls.

Ceratium tripos
Gran, H.H., and T. Braarud, 1935
A quantitative study of the phytoplankton in the Bay of Fundy and the Gulf of Maine (including observations on hydrography, chemistry, and turbidity). J. Biol. Bd., Canada, 1(5):279-467, 69 text figs.

Ceratium tripos
Hasle, G.R., and E. Nordli, 1952.
Form variation in Ceratium fusus and tripos populations in cultures and from the sea. Abhandl Norske Videnskaps-Akad., Oslo, 1. Math.-Naturv. Kl., 1951(4)1-25, 8 textfigs.

ceratium tripos
Hickel, Wolfgang, 1967.
Untersuchungen über die Phytoplanktonblüte in der westlichen Ostsee. Helgoländer wiss. Meeresunters., 16(½):1-66.

Ceratium tripos
Jahn, T.L., W.M. Harmon and M. Landman, 1963. Mechanisms of locomotion in flagellates. L. Ceratium. J. Protozoology, 10(3):358-363.

Ceratium tripos
Jørgensen, E., 1920
Mediterranean Ceratia. Rept. Danish Oceanogr. Exped. 1908-10 to the Mediterranean and adjacent seas, Vol.II, Biol. J 1:110 pp., 94 text figs., 26 charts.

Ceratium tripos
Jørgensen, E., 1905
B.Protistplankton and the diatoms in bottom samples. Hydrographical and biological investigations in Norwegian fjords. Bergens Mus. Skr. 7: 49-225.

Ceratium tripos
Jorgensen, E., 1900
Protophyten und Protozoën im Plankton aus der Norwegischen Westkerste. Bergens Mus. Aarb. 1899(6): 95 pp., 5 pls., 83 tables

Ceratium tripos
LeBour, M.V., 1925
The dinoflagellates of Northern Seas. The Marine Biological Association of the United Kingdom. Plymouth, 250 pp., 35 pls. 53 text figs.

Ceratium tripos
Lillick, L.C. 1940
Phytoplankton and planktonic protozoa of the offshore waters of the Gulf of Maine. Pt.II. Qualitative Composition of the Planktonic Flora. Trans. Am. Phil. Soc., n.s., 31(3):193-237, 13 text figs.

Ceratium tripos
Lillick, L.C. 1938
Preliminary report of the phytoplankton of the Gulf of Maine. Am. Mid. Nat. 20(3):624-640, 1 text fig. 37 tables.

Ceratium Tripos
Lillick, L.C. 1937
Seasonal studies of the phytoplankton off Woods Hole, Massachusetts. Biol. Bull. LXXIII (3):488-503, 3 text figs.

Ceratium tripos
Lindemann, E., 1924
Peridineen aus dem goldenen Horn und dem Bosphorus. Bot. Arch. 5:216-233, 98 text figs.

Ceratium typos
López, J., 1966.
Variación y regulación de la forma en el género Ceratium. Inv. Pesq., Barcelona, 30:325-427.

Ceratium tripos
López, J., 1955.
Variación alométrica en Ceratium tripos. Invest. Pesq. 2:131-159.

Ceratium tripos
Mangin, M. L., 1912
Phytoplancton de la croisière du "René" dans l'Atlantique (Septembre 1908). Ann. Inst. Ocean., n.s., 4(1):1-66, 2 pls., 41 text figs., 2 tables.

Ceratium tripos
Marshall, S. M., 1933
The production of microplankton in the Great Barrier Reef Region. Brit. Mus. (N.H.) Great Barrier Reef Exped. 1928-29, Sci. Repts. II(5):111-157, 14 text figs.

Ceratium tripos
Marukawa, H., 1921
Plankton lists and some new species of copepods, from the northern waters of Japan. Bull. Inst. Ocean., No.384, 15 pp., 3 pls., 1 chart. Monaco

Ceratium tripos
Massuti Algamora, M., 1949
Estudio de diez y seis muestras de plancton del Golfo de Nápoles. Publ. Inst. Biol. Appl. 5:85-94, 1 fold-in table.

Ceratium tripos
Meunier, A., 1919
Microplancton de la Mer Flamande 3. Les Péridiniens. Mem. Mus. Roy. Hist. Nat., Belgique 8(1):1-116, Pls. XV-XXI.

Ceratium tripos
Mulford, Richard A., 1963
Distribution of the dinoflagellate genus Ceratium in the tidal and offshore waters of Virginia.
Chesapeake Science, 4(2):84-89.

Ceratium tripos
Murray, G., and F. G. Whitting, 1899
New Peridiniaceae from the Atlantic. Trans. Linn. Soc., London, Bot., ser 2, 5:321-342, Pls. 27-33, 9 tables.

Ceratium tripos
Nielsen, J., 1956.
Temporary variations in certain marine Ceratia. Oikos, 7(2):256-272.

Ceratium tripos
Nordli, E., 1957.
Experimental studies on the ecology of Ceratia. Oikos, 8(2):201-265.

Ceratium tripos
Nordli, E., 1953.
Salinity and temperature as controlling factors for distribution and mass occurrence of Ceratia. Blyttia 11:16-18, 4 textfigs.

Ceratium tripos
Paiva Carvalho, J., 1950
O plancton do Rio Maria Rodrigues (Cananeis). 1. Diatomaceas e Dinoflagelados. Bol. Inst. Paulista Oceanogr. 1(1); 27-43, 2 fold-in tables, 2 figs.

Ceratium tripos
Paulsen, O., 1908
XVIII Peridiniales. Nordisches Plankton, Bot. Teil: 1-124, 155 text figs.

Ceratium tripos
Pavillard, J., 1905
Recherches sur la flore pelagique (Phytoplankton) de l'Etang de Thau. Theses presentees a la Fac. Sci., Paris, 116 pp., 3 pls.

Ceratium tripos
Peters, Nicolaus, 1934
Die Bevolkerung des Sudatlantischen Ozeans mit Ceratien. Biol. Sonderuntersuchungen 1. Wiss Ergebn Deutschen Atlantischen Exped. "Meteor" 1925-1927, 12(1): 1-69, 28 text figs.

Ceratium tripos
Pettersson, H., 1934
Scattering and extinction of light in sea-water. Meddelanden från Göteborgs Högskolas Oceanografiska Institut. 9 (Göteborgs Kungl. vetenskaps-och vitterhets-samhälles Handlingar. Femte Följden, Ser.B, 4(4)):1-16.

Ceratium tripos
Pettersson, H., F. Gross, and F. Koczy, 1939
Large scale plankton cultures. Medd. från Oceanografiska Institutet i Göteborg No.3 (Göteborgs Kungl. vetenskaps-och Vitter-hets-Samhälles Handlinger Femte Följden. Ser. B) Vol.6(13):1-24.
1. The plankton shaft. H. Pettersson
2. Experiment with phytoplankton. F. Gross and F. Koczy
3. Experiments with zooplankton.

Ceratium tripos
Robinson, G.A., 1965.
Continuous plankton records: contribution towards a plankton atlas of the North Atlantic and the North Sea. IX. Seasonal cycles of phytoplankton. Bulls. Mar. Ecol. Scottish Mar. Biol. Assoc., 6(4):104-122, pls. 26-61.

Ceratium tripos
Robinson, G.A., 1961
Contribution towards an atlas of the north-eastern Atlantic and the North Sea. 1. Phytoplankton. Bulls. Mar. Ecology, 5(42): 81-89, Pls. 15-20.

Ceratium tripos
Roubault, A., 1946
Observations sur la Répartition du Plancton. Bull. Mus. Inst. Océan., No.902: 4 pp.

Ceratium tripos
Rumkówna, A., 1948
List of the phytoplankton species occurring in the superficial water layers in the Gulf of Gdańsk. Bull. Lab. mar., Gdynia, No. 4: 139-141 with tables in back.

Ceratium tripos
Schodduyn, M., 1926
Observations faites dans la baie d'Ambleteuse (Pas de Calais). Bull. Inst. Ocean., Monaco, No. 482: 64 pp.

Ceratium tripos (with numerous varieties)
Schröder, B., 1900
Phytoplankton des Golfes von Neapel nebst vergleichenden Ausblicken auf das atlantischen Ozean. Mitt. Zool. Stat. Neapel, 14:1-38.

Ceratium tripos
Sommer, H., and F. N. Clarke, 1946.
Effect of red water on marine life in Santa Monica Bay, California. California Fish and Game, 32(2):100-101.

Ceratium tripos
Sousa e Silva, E., 1949
Diatomaceas e Dinoflagelados de Baia de Cascais. Portugaliae Acta Biol., Volume: Julio Henriques, Ser. B: 300-383, 9 pls, 2 fold-in tables.

Ceratium tripos
Steemann-Nielsen, Einar, 1951
The marine vegetation of the Isefjord. A study on ecology and production. Medd. Komm. Danmarks Fiskeri-og Havundersøgelser. Ser. Plankton. 5(4); 114pp., 46 text figs.

Ceratium tripos
Steemann-Nielsen, E., 1939
Die Ceratien des Indischen Ozeans und der Ostasiatischen Gewässer mit einer allgemeinen zusammenfassung über die Verbreitung der Ceratien in den Weltmeeren. Dana Rept. No. 17: 33 pp., 8 maps in text.

Ceratium tripos
Steemann Nielsen, E., 1934
Untersuchungen über die Verbreitung, Biologie und Variation der Ceratien im südlichen Stillen Ozean. Dana Rept. No. 4: 67 pp., 73 figs. and 11 charts in text.

Ceratium tripos
Wang, C. C., and D. Nie, 1932
A survey of the marine protozoa of Amoy. Contr. Biol. Lab. Sci. Soc., China, Zool. Ser., 8:285-385, 89 text figs.

Ceratium tripos
Wheeler, J.E.G., 1939
Plankton investigations. Bermuda Biological Station. Second Report. October 1939. 7 pp. (typed), 5 figs. Plymouth, Oct. 23, 1939.

Ceratium tripos
Yarranton G.A. 1967
Parameters for use in distinguishing populations of Euceratium Gran. Bull. mar. Ecol. 6(6): 147-158.

Ceratium/arcuatum tripos
Gourret, P., 1883
Sur les Peridiniens du Golfe de Marseille. Ann. du Musee d'hist. Nat., Marseille, Zool., 1 (Mme. 8):1-114, 4 pls.

Ceratium tripos arcuatum
Pavillard, J., 1923
A propos de la systématique des Péridiniens Bull. Soc. Bot. de France 70:876-882; 914-918.

Ceratium tripos atlanticum
Dangeard, P., 1926
Description des Péridiniens Testacés recueillis para la Mission Charcot pendent le mois d'Aout 1924. Ann. Inst. Ocean. n.s. 3(7):307-334, 15 text figs.

Ceratium tripos atlanticum
Graham, H. W. and N. Bronikovsky, 1944
The genus Ceratium in the Pacific and North Atlantic Oceans. Sci. Res. Cruise VII of the Carnegie, 1928-1929 ----- Biol. V (565):209 pp., 54 charts, 27 figs., 54 tables.

Ceratium tripos var. atlanticum
Lillick, L.C., 1940
Phytoplankton and planktonic protozoa of the offshore waters of the Gulf of Maine. Pt.II. Qualitative Composition of the Planktonic Flora. Trans. Am. Phil. Soc., n.s., 31(3):193-237, 13 text figs.

Ceratium tripos var. atlanticum
Lillick, L.C., 1938
Preliminary report of the phytoplankton of the Gulf of Maine. Am. Mid. Nat. 20(3):624-640, 1 text figs 37 tables.

Ceratium tripos atlanticum
Nordli, E., 1957.
Experimental studies on the ecology of Ceratia. Oikos 8(2):200-265.

Ceratium tripos contrarium
Gourret, P., 1883
Sur les Peridiniens du Golfe de Marseille. Ann. du Musee d'hist. Nat., Marseille, Zool., 1 (Mme. 8):1-114, 4 pls.

Ceratium/graciale tripos
Gourret, P., 1883
Sur les Peridiniens du Golfe de Marseille. Ann. du Musee d'hist. Nat., Marseille, Zool., 1 (Mme. 8):1-114, 4 pls.

Ceratium tripos gracile
Pavillard, J., 1923
A propos de la systématique des Péridiniens Bull. Soc. Bot. de France 70:876-882; 914-918.

Ceratium tripos inaequale
Gourret, P., 1883
Sur les Peridiniens du Golfe de Marseille. Ann. du Musee d'hist. Nat., Marseille, Zool., 1 (Mme. 8):1-114, 4 pls.

Ceratium tripos inflexum
Gourret, P., 1883
Sur les Peridiniens du Golfe de Marseille. Ann. du Musee d'hist. Nat., Marseille, Zool., 1 (Mme. 8):1-114, 4 pls.

Ceratium tripos lata
Paulsen, O., 1908
XVIII Peridiniales. Nordisches Plankton, Bot. Teil: 1-124, 155 text figs.

Ceratium tripos lineatum
Dangeard, P., 1926
Description des Péridiniens Testacés recueillis para la Mission Charcot pendent le mois d'Aout 1924. Ann. Inst. Ocean. n.s. 3(7):307-334, 15 text figs.

Ceratium tripos lineata
Paulsen, O., 1908
XVIII Peridiniales. Nordisches Plankton, Bot. Teil: 1-124, 155 text figs.

Ceratium tripos macroceros
Gourret, P., 1883
Sur les Peridiniens du Golfe de Marseille. Ann. du Musee d'hist. Nat., Marseille, Zool., 1 (Mme. 8):1-114, 4 pls.

Ceratium tripos massiliense
Gourret, P., 1883
Sur les Peridiniens du Golfe de Marseille. Ann. du Musee d'hist. Nat., Marseille, Zool., 1 (Mme. 8):1-114, 4 pls.

Ceratium tripos mediterraneum

Delegazione Italiana della Commissione Internazionale per l'Esplorazione Scientifica del Mediterraneo, 1941
Note sul plancton della Laguna veneta. Memoria CCLXXXIX, Arch. di Ocean. e Limn. Anno I, Fasc. I, 1941 XIX: 31-57 pp.

Ceratium tripos mediterraneum

Margalef, R., 1949
Fitoplancton nerítico de la Costa Brava en 1947-48. Publ. Inst. Biol. Aplicada, 5: 41-51, 3 text figs.

Ceratium tripos semipulchellum

Graham, H. W. and N. Bronikovsky, 1944
The genus Ceratium in the Pacific and North Atlantic Oceans. Sci. Res. Cruise VII of the Carnegie, 1928-1929 ----- Biol. V (565):209 pp., 54 charts, 27 figs., 54 tables.

Ceratium tripos subsalsum

Ercegovic, A., 1936
Etudes qualitative et quantitatives du phytoplancton dans les eaux cotières de l'Adriatique oriental moyen au cours de l'année 1934. Acta Adriatica 1(9):1-126

Ceratium tripos var. subsalsum

Lillick, L.C., 1940
Phytoplankton and planktonic protozoa of the offshore waters of the Gulf of Maine. Pt. II. Qualitative Composition of the Planktonic Flora. Trans. Am. Phil. Soc., n.s., 31(3):193-237, 13 text figs.

Ceratium tripos var. subsalsum

Lillick, L.C., 1938
Preliminary report of the phytoplankton of the Gulf of Maine. Am. Mid. Nat. 20(3):624-640, 1 text figs 37 tables.

Ceratium tripos typicum

Gourret, P., 1883
Sur les Peridiniens du Golfe de Marseille. Ann. du Musee d'hist. Nat., Marseille, Zool., 1 (Mme. 8):1-114, 4 pls.

Ceratium uncinus n. sp.

Sournia, Alain 1972.
Quatre nouveaux dinoflagellés du plancton marin. Phycologia 11(1):71-74.

Ceratium volans

Margalef, R., 1949
Fitoplancton nerítico de la Costa Brava en 1947-48. Publ. Inst. Biol. Aplicada, 5: 41-51, 3 text figs.

Ceratium volans

Pavillard, J., 1905
Recherches sur la flore pelagique (Phytoplankton) de l'Etang de Thau. Theses presentees a la Fac. Sci., Paris, 116 pp., 3 pls.

Ceratium vultur

Balech, Enrique, 1962
Tintinnoinea y Dinoflagellata del Pacifico segun material de las Expediciones NORPAC y DOWNWIND del Instituto Scripps de Oceanografia. Revista, Mus. Argentino Ciencias Nat. "Bernardino Rivadavia", Ciencias Zool., 7(1):1-253.

Ceratium vultur

Dangeard, P., 1927
Phytoplankton de la croisière du "Sylvana". Ann. Inst. Ocean., Monaco, n.s., 4(8):286-401, 54 text figs, (février-Juin 1913).

Ceratium vultur

Graham, H. W. and N. Bronikovsky, 1944
The genus Ceratium in the Pacific and North Atlantic Oceans. Sci. Res. Cruise VII of the Carnegie, 1928-1929 ----- Biol. V (565):209 pp., 54 charts, 27 figs., 54 tables.

Ceratium vultur

Pavillard, J., 1905
Recherches sur la flore pelagique (Phytoplankton) de l'Etang de Thau. Theses presentees a la Fac. Sci., Paris, 116 pp., 3 pls.

Ceratium vultur

Peters, Nicolaus, 1934
Die Bevolkerung des Sudatlantischen Ozeans mit Ceratien. Biol. Sonderuntersuchungen 1. Wiss Ergeb. Deutschen Atlantischen Exped. "Meteor" 1925-1927, 12(1): 1-69, 28 text figs.

Ceratium vultur

Steemann-Nielsen, E., 1939
Die Ceratien des Indischen Ozeans und der Ostasiatischen Gewässer mit einer allgemeinen zusammenfassung über die Verbreitung der Ceratien in den Weltmeeren. Dana Rept. No. 17: 33 pp., 8 maps in text.

Ceratium vulture

Steemann Nielsen, E., 1934
Untersuchungen über die Verbreitung, Biologie und Variation der Ceratien im südlichen Stillen Ozean. Dana Rept. No. 4: 67 pp., 73 figs. and 11 charts in text.

Ceratium japonicum vultur

Graham, H. W. and N. Bronikovsky, 1944
The genus Ceratium in the Pacific and North Atlantic Oceans. Sci. Res. Cruise VII of the Carnegie, 1928-1929 ----- Biol. V (565):209 pp., 54 charts, 27 figs., 54 tables.

Ceratium vultur pavillardii

Graham, H. W. and N. Bronikovsky, 1944
The genus Ceratium in the Pacific and North Atlantic Oceans. Sci. Res. Cruise VII of the Carnegie, 1928-1929 ----- Biol. V (565):209 pp., 54 charts, 27 figs., 54 tables.

Ceratium vultur recurvum

Graham, H. W. and N. Bronikovsky, 1944
The genus Ceratium in the Pacific and North Atlantic Oceans. Sci. Res. Cruise VII of the Carnegie, 1928-1929 ----- Biol. V (565):209 pp., 54 charts, 27 figs., 54 tables.

Ceratium vultur regulare n.var.

Graham, H. W. and N. Bronikovsky, 1944
The genus Ceratium in the Pacific and North Atlantic Oceans. Sci. Res. Cruise VII of the Carnegie, 1928-1929 ----- Biol. V (565):209 pp., 54 charts, 27 figs., 54 tables.

Ceratium vultur reversum n.var.

Graham, H. W. and N. Bronikovsky, 1944
The genus Ceratium in the Pacific and North Atlantic Oceans. Sci. Res. Cruise VII of the Carnegie, 1928-1929 ----- Biol. V (565):209 pp., 54 charts, 27 figs., 54 tables.

Ceratium vultur sumatranum

Graham, H. W. and N. Bronikovsky, 1944
The genus Ceratium in the Pacific and North Atlantic Oceans. Sci. Res. Cruise VII of the Carnegie, 1928-1929 ----- Biol. V (565):209 pp., 54 charts, 27 figs., 54 tables.

Ceratium vultur sumatranum

Steemann Nielsen, E., 1934
Untersuchungen über die Verbreitung, Biologie und Variation der Ceratien im südlichen Stillen Ozean. Dana Rept. No. 4: 67 pp., 73 figs. and 11 charts in text.

Ceratium vultur vultur

Graham, H. W. and N. Bronikovsky, 1944
The genus Ceratium in the Pacific and North Atlantic Oceans. Sci. Res. Cruise VII of the Carnegie, 1928-1929 ----- Biol. V (565):209 pp., 54 charts, 27 figs., 54 tables.

Ceratium vultus

Sournia, A., 1967.
Le genre Ceratium (peridinien planctonique) dans le canal de Mozambique. Contribution a une revision mondiale (fin). Vie Milieu (A) 18(3):441-449.

Ceratocorys

Kofoid, C.A., 1910.
A revision of the genus Ceratocorys based on skeletal morphology. Univ. Calif. Publ., Zool. 6(8):177-187.

Ceratocorys sp.

Wheeler, J.E.G., 1939
Plankton investigations. Bermuda Biological Station. Second Report. October 1939. 7 pp. (typed), 5 figs. Plymouth, Oct. 23, 1939.

Ceratocorys sp.

Wood, E.J.F., 1963
Dinoflagellates in the Australian region. II. Recent collections. Comm. Sci. and Industr. Res. Org., Div. Fish. and Oceanogr., Techn. Paper, No. 14: 55 pp.

Ceratocorys armata

Balech, Enrique, 1962
Tintinnoinea y Dinoflagellata del Pacifico segun material de las Expediciones NORPAC y DOWNWIND del Instituto Scripps de Oceanografia. Revista, Mus. Argentino Ciencias Nat. "Bernardino Rivadavia", Ciencias Zool., 7(1):1-253.

Ceratocorys armatum

Dangeard, P., 1927
Phytoplankton de la croisière du "Sylvana". Ann. Inst. Ocean., Monaco, n.s., 4(8):286-401, 54 text figs, (février-Juin 1913).

Ceratocorys armata

Ercegovic, A., 1936
Etudes qualitative et quantitatives du phytoplancton dans les eaux cotières de l'Adriatique oriental moyen au cours de l'année 1934. Acta Adriatica 1(9):1-126

Ceratocorys armata

Forti, A., 1922
Ricerche sulla flora pelagica (fitoplancton) di Quarto dei Mille. Mem. R. Com. Talass. Ital. 97:248 pp., 13 pls.

Ceratocorys armata

Graham, H. W., 1942
Studies in the morphology, taxonymy, and ecology of the Peridiniales. Sci. Res. Cruise VII of the Carnegie, 1928-1929---Biol. III(542): 129 pp., 67 figs.

Ceratocorys armatum

Kofoid, C.A., 1910.
A revision of the genus Ceratocorys based on skeletal morphology. Univ. Calif. Publ., Zool. 6(8):177-187.

Ceratocorys armata

Margalef, R., 1949
Fitoplancton nerítico de la Costa Brava en 1947-48. Publ. Inst. Biol. Aplicada, 5: 41-51, 3 text figs.

Ceratocorys armata

Matzenauer, L., 1933
Die Dinoflagellaten des indischen Ozeans (mit Ausnahme der Gattung Ceratium). Bot. Arch. 35:437-510, 77 text figs., 2 charts.

Ceratocorys armatum
Pavillard, J., 1916
Recherches sur les Peridiniens du Golfe du Lion. Mem. Univ. Montpellier. Trav. Inst. Bot., Univ. Montpellier. Serie mixte No.4, 70 pp., 3 pls., 15 text figs.

Cerratocorys armata
Rampi, L., 1951.
Ricerche sul fitoplancton del Mare Ligure. 10) Peridiniale delle acque di Sanremo. Pubbl. Centro Talassografico Tirreno No. 6: [Atti Acad. Ligure Sci. Lett.] 7(1): 8 pp., Pls. 3-4.

Ceratocorys aultii n.sp.
Graham, H. W., 1942
Studies in the morphology, taxonymy, and ecology of the Peridiniales. Sci. Res. Cruise VII of the Carnegie, 1928-1929---Biol. III(542): 129 pp., 67 figs.

Ceratocorys bipes
Dangeard, P., 1927
Phytoplankton de la croisière du "Sylvana". Ann. Inst. Ocean., Monaco, n.s., 4(8):286-401, 54 text figs. (février-Juin 1913).

Ceratocorys bipes
Graham, H. W., 1942
Studies in the morphology, taxonymy, and ecology of the Peridiniales. Sci. Res. Cruise VII of the Carnegie, 1928-1929---Biol. III(542): 129 pp., 67 figs.

Ceratocorys bipes
Kofoid, C.A., 1910.
A revision of the genus Ceratocorys based on skeletal morphology. Univ. Calif. Publ., Zool. 6(8):177-187.

Ceratocorys gouretti
Balech, Enrique, 1962
Tintinnoinea y Dinoflagellata del Pacifico segun material de las Expediciones NORPAC y DOWNWIND del Instituto Scripps de Oceanografia. Revista, Mus. Argentino Ciencias Nat. "Bernardino Rivadavia", Ciencias Zool., 7(1):1-253.

Ceratocorys Gourreti
Ercegovic, A., 1940
Weitere Untersuchungen über einige hydrographische Verhältnisse und über die Phytoplanktonproduktion in den Gewässern der östlichen Mitteladria. Acta Adriatica 2(3):95-134, 8 text figs.

Ceratocorys gourretii
Graham, H. W., 1942
Studies in the morphology, taxonymy, and ecology of the Peridiniales. Sci. Res. Cruise VII of the Carnegie, 1928-1929---Biol. III(542): 129 pp., 67 figs.

Cerratocorys Gourreti
Rampi, L., 1951.
Ricerche sul fitoplancton del Mare Ligure. 10) Peridiniale delle acque di Sanremo. Pubbl. Centro Talassografico Tirreno No. 6: [Atti Acad. Ligure Sci. Lett.] 7(1): 8 pp., Pls. 3-4.

Ceratocorys hirsuta n.sp.
Matzenauer, L., 1933
Die Dinoflagellaten des indischen Ozeans (mit Ausnahme der Gattung Ceratium.) Bot. Arch. 35:437-510, 77 text figs., 2 charts.

Ceratocorys horrida
Balech, Enrique, 1962
Tintinnoinea y Dinoflagellata del Pacifico segun material de las Expediciones NORPAC y DOWNWIND del Instituto Scripps de Oceanografia. Revista, Mus. Argentino Ciencias Nat. "Bernardino Rivadavia", Ciencias Zool., 7(1):1-253.

Ceratocorys horrida
Dangeard, P., 1927
Phytoplankton de la croisière du "Sylvana". Ann. Inst. Ocean., Monaco, n.s., 4(8):286-401, 54 text figs. (Feirer-Juin 1913).

Ceratocorys horrida
Ercegovic, A., 1936
Etudes qualitative et quantitatives du phytoplancton dans les eaux cotières de l'Adriatique oriental moyen au cours de l'année 1934. Acta Adriatica 1(9):1-126

Ceratocorys horrida
Forti, A., 1922
Ricerche sulla flora pelagica (fitoplancton) di Quarto dei Mille. Mem. R. Com. Talass. Ital. 97:248 pp., 13 pls.

Ceratocorys horrida
Graham, H. W., 1942
Studies in the morphology, taxonymy, and ecology of the Peridiniales. Sci. Res. Cruise VII of the Carnegie, 1928-1929---Biol. III(542): 129 pp., 67 figs.

Ceratocorys horrida
Kofoid, C.A., 1910.
A revision of the genus Ceratocorys based on skeletal morphology. Univ. Calif. Publ., Zool. 6(8):177-187.

Ceratocorys horrida
Lindemann, E., 1925
Neubeobachtungen an den Winter peridineen des Golfes von Neapel. Bot. Arch. 9:95-102, 19 text figs.

Ceratocorys horrida
Margalef, R., 1949
Fitoplancton nerítico de la Costa Brava en 1947-48. Publ. Inst. Biol. Aplicada, 5: 41-51, 3 text figs.

Ceratocorys horrida
Marukawa, H., 1921
Plankton lists and some new species of copepods, from the northern waters of Japan. Bull. Inst. Ocean., No.384, 15 pp., 3 pls., 1 chart. Monaco

Ceratocorys horrida
Matzenauer, L., 1933
Die Dinoflagellaten des indischen Ozeans (mit Ausnahme der Gattung Ceratium.) Bot. Arch. 35:437-510, 77 text figs., 2 charts.

Ceratocorys horrida
Murray, G., and F. G. Whitting, 1899
New Peridiniaceae from the Atlantic. Trans. Linn. Soc., London, Bot., ser 2, 5: 321-342, Pls. 27-33, 9 tables.

Ceratocorys horrida
Pavillard, J., 1916
Recherches sur les Peridiniens du Golfe du Lion. Mem. Univ. Montpellier. Trav. Inst. Bot., Univ. Montpellier. Serie mixte No.4, 70 pp., 3 pls., 15 text figs.

Ceratocorys horrida
Pavillard, J., 1905
Recherches sur la flore pelagique (Phytoplankton) de l'Etang de Thau. Theses presentees a la Fac. Sci., Paris, 116 pp., 3 pls.

Cerratocorys horrida
Rampi, L., 1951.
Ricerche sul fitoplancton del Mare Ligure. 10) Peridiniale delle acque di Sanremo. Pubbl. Centro Talassografico Tirreno No. 6: [Atti Acad. Ligure Sci. Lett.] 7(1): 8 pp., Pls. 3-4.

Ceratocorys horrida
Schröder, B., 1900
Phytoplankton des Golfes von Neapel nebst vergleichenden Ausblicken auf das atlantischen Ozean. Mitt. Zool. Stat. Neapel, 14:1-38.

Ceratocorys horrida
Sousa e Silva, E., 1949
Diatomaceas e Dinoflagelados de Baia de Cascais. Portugaliae Acta Biol., Volume: Julio Henriques, Ser. B: 300-383, 9 pls, 2 fold-in tables.

Ceratocorys horrida extensa
Balech, E., 1949
Estudio de "Ceratocorys horrida" Stein var. "extensa" Pavillard. Physis 20(57):165-173, 29 figs. on 1 pl.

Ceratocorys horrida extensa
Margalef, R., 1949
Fitoplancton nerítico de la Costa Brava en 1947-48. Publ. Inst. Biol. Aplicada, 5: 41-51, 3 text figs.

Ceratocorys Jourdani
Dangeard, P., 1927
Phytoplankton de la croisière du "Sylvana". Ann. Inst. Ocean., Monaco, n.s., 4(8):286-401, 54 text figs. (Feirer-Juin 1913).

Ceratocorys Jourdani
Ercegovic, A., 1936
Etudes qualitative et quantitatives du phytoplancton dans les eaux cotières de l'Adriatique oriental moyen au cours de l'année 1934. Acta Adriatica 1(9):1-126

Ceratocorys Jourdani
Forti, A., 1922
Ricerche sulla flora pelagica (fitoplancton di Quarto dei Mille. Mem. R. Com. Talass. Ital. 97:248 pp., 13 pls.

Ceratocorys jourdani
Kofoid, C.A., 1910.
A revision of the genus Ceratocorys based on skeletal morphology. Univ. Calif. Publ., Zool. 6(8):177-187.

Ceratocorys Jourdani
Matzenauer, L., 1933
Die Dinoflagellaten des indischen Ozeans (mit Ausnahme der Gattung Ceratium.) Bot. Arch. 35:437-510, 77 text figs., 2 charts.

Ceratocorys Jourdani
Pavillard, J., 1916
Recherches sur les Peridiniens du Golfe du Lion. Mem. Univ. Montpellier. Trav. Inst. Bot., Univ. Montpellier. Serie mixte No.4, 70 pp., 3 pls., 15 text figs.

Ceratocorys jourdani
Sousa e Silva, E., 1949
Diatomaceas e Dinoflagelados de Baia de Cascais. Portugaliae Acta Biol., Volume: Julio Henriques, Ser. B: 300-383, 9 pls, 2 fold-in tables.

Ceratocorys kofoidi n.sp. ad int.
Paulsen, O., 1930.
Etudes sur le microplancton de la mer d'Alboran. Trab. Inst. Esp. Ocean. No. 4:1-108, 61 textfigs.

Ceratocorys magna n.sp.
Kofoid, C.A., 1910.
A revision of the genus Ceratocorys based on skeletal morphology. Univ. Calif. Publ., Zool. 6(8):177-187.

Ceratocorys reticulata
Balech, Enrique, 1962
Tintinnoinea y Dinoflagellata del Pacifico segun material de las Expediciones NORPAC y DOWNWIND del Instituto Scripps de Oceanografia. Revista, Mus. Argentino Ciencias Nat. "Bernardino Rivadavia", Ciencias Zool., 7(1):1-253.

Ceratocorys reticulata n.sp.
Graham, H. W., 1942
Studies in the morphology, taxonymy, and ecology of the Peridiniales. Sci. Res. Cruise VII of the Carnegie, 1928-1929---Biol. III(542): 129 pp., 67 figs.

Ceratocorys skogsbergii n.sp.
Graham, H. W., 1942
Studies in the morphology, taxonymy, and ecology of the Peridiniales. Sci. Res. Cruise VII of the Carnegie, 1928-1929---Biol. III(542): 129 pp., 67 figs.

Ceratocorys spinifera
Murray, G., and F. G. Whitting, 1899
New Peridiniaceae from the Atlantic. Trans. Linn. Soc., London, Bot., ser 2, 5: 321-342, Pls. 27-33, 9 tables.

Ceratocorys tridentata
Schröder, B., 1900
Phytoplankton des Golfes von Neapel nebst vergleichenden Ausblicken auf das atlantischen Ozean. Mitt. Zool. Stat. Neapel, 14:1-38.

Citharistes
Abe, Tohru H., 1967.
The armoured Dinoflagellata: II. Prorocentridae and Dinophusidae (C)- Ornithocercus, Histioneis, Amphisolenia and others. Publs Seto mar.biol.Lab.,15(2):79-116.

Citharistes spp.
Balech, Enrique, 1971
Microplancton del Atlantico ecuatorial oeste (Equalant 1) Publ. Serv. Hidrograf. Naval, Argentina H. 654: 103 pp., 122 figs.

Citharistes sp.
Schröder, B., 1900
Phytoplankton des Golfes von Neapel nebst vergleichenden Ausblicken auf das atlantischen Ozean. Mitt. Zool. Stat. Neapel, 14:1-38.

Citharistes apsteinii
Balech, Enrique, 1962
Tintinnoinea y Dinoflagellata del Pacifico segun material de las Expediciones NORPAC y DOWNWIND del Instituto Scripps de Oceanografia. Revista, Mus. Argentino Ciencias Nat. "Bernardino Rivadavia", Ciencias Zool., 7(1):1-253.

Citharistes Apsteinii
Käsler, R., 1938
Die Verbreitung der Dinophysiales im Sudatlantischen Ozean. Wiss. Ergeb. Deutschen Atlantischen Expedition----"Meteor" 1925-1927, 12(2):162-237, text figs. 85-118.

Citharistes apsteini
Kofoid, C. A. and T. Skogsberg, 1928
XXXV. The Dinoflagellata: The Dinophysiodae. Reports on the scientific results of the expedition to the Eastern Tropical Pacific, in charge of Alexander Agassiz, by the U. S. Fish Commission Steamer "Albatross" from October 1904 to March 1905----. Mem. M. C. Z. 51:766 pp., 31 pls.

Citharistes Apsteinii
Murray, G., and F. G. Whitting, 1899
New Peridiniaceae from the Atlantic. Trans. Linn. Soc., London, Bot., ser 2, 5: 321-342, Pls. 27-33, 9 tables.

Citharistes apsteini
Rao, D.V. Subba 1973.
Occurrence of Citharistes apsteini Schutt, a rare dinophysoid dinoflagellate, in the Bay of Bengal. Phycologia 12(1/2): 89-90.

Citharistes apsteini
Wood, E.J.F., 1963
Dinoflagellates in the Australian region. II. Recent collections. Comm. Sci. and Industr. Res. Org., Div. Fish. and Oceanogr., Techn. Paper, No. 14: 55 pp.

Citharistes regius
Balech, Enrique, 1962
Tintinnoinea y Dinoflagellata del Pacifico segun material de las Expediciones NORPAC y DOWNWIND del Instituto Scripps de Oceanografia. Revista, Mus. Argentino Ciencias Nat. "Bernardino Rivadavia", Ciencias Zool., 7(1):1-253.

Citharistes regius
Halim, Youssef, 1960
Étude quantitative et qualitative du cycle écologique des dinoflagellés dans les eaux de Villefranche-sur-Mer. (1953-1955). Ann. Inst. Océanogr., Monaco, 38:123-232.

Citharistes regius
Kofoid, C. A. and T. Skogsberg, 1928
XXXV. The Dinoflagellata: The Dinophysiodae. Reports on the scientific results of the expedition to the Eastern Tropical Pacific, in charge of Alexander Agassiz, by the U. S. Fish Commission Steamer "Albatross" from October 1904 to March 1905----. Mem. M. C. Z. 51:766 pp., 31 pls.

Citharistes regius
Murray, G., and F. G. Whitting, 1899
New Peridiniaceae from the Atlantic. Trans. Linn. Soc., London, Bot., ser 2, 5: 321-342, Pls. 27-33, 9 tables.

Citharistes regius
Rampi, L., 1940
Ricerche sul Fitoplancton del mare Ligure. II Le tecatali e le dinofisiali delle acque di Sanremo. Boll. Pesca, Piscicolt., Idrobiol. (18) 16(2): 243-274, 56 figs.

Citharistes regius
Rampi, L., 1940
Ricerche sul Fitoplancton del mare Ligure. Boll. di Pesca, di Piscicoltura e di Idrobiologa, 18(2):1-34, 56 text figs.

Cladopyxis brachiolata
Balech, Enrique, 1964.
El genero "Cladopyxis"(Dinoflagellata). Comunicaciones, Museo Argentina, Ciencias Nat. e Inst. Nacional, Invest. Ciencias Naturales, Hidrobiol., 1(4):27-39.

Cladopyxis brachiolata
Balech, Enrique, 1962
Tintinnoinea y Dinoflagellata del Pacifico segun material de las Expediciones NORPAC y DOWNWIND del Instituto Scripps de Oceanografia. Revista, Mus. Argentino Ciencias Nat. "Bernardino Rivadavia", Ciencias Zool., 7(1):1-253.

Cladopyxis brachiolata
Murray, G., and F. G. Whitting, 1899
New Peridiniaceae from the Atlantic. Trans. Linn. Soc., London, Bot., ser 2, 5: 321-342, Pls. 27-33, 9 tables.

Cladopyxis brachiolata
Rampi, L., 1951.
Ricerche sul fitoplancton del Mare Ligure. 10) Peridiniale delle acque di Sanremo. Pubbl. Centro Talassografico Tirreno No. 6: Atti Acad. Ligure Sci. Lett. 7(1): 8 pp., Pls. 3-4.

Cladopyxis brachiolata
Wood, E.J.F., 1963
Dinoflagellates in the Australian region. II. Recent collections. Comm. Sci. and Industr. Res. Org., Div. Fish. and Oceanogr., Techn. Paper, No. 14: 55 pp.

Cladopyxis caryophyllum
Rampi, L., 1945
Osservazioni sulla distribuzione qualitativa del fitoplancton nel mare Mediterraneo. Atti della Soc. Ital. di Sci. Nat. 84:105-113.

Cladopyxis caryophyllum
Rampi, L., 1942
Il Fitoplancton mediterraneo: Problemi ed affinita interoceaniche. Boll. di Pesca di Piscicoltura e di Idrobiologia, Anno 18, Fasc. 4:7-19.

Cladopyxis caryophyllum
Wood, E.J.F., 1963
Dinoflagellates in the Australian region. II. Recent collections. Comm. Sci. and Industr. Res. Org., Div. Fish. and Oceanogr., Techn. Paper, No. 14: 55 pp.

Cladopyxis claytoni n.sp.
Holmes, R.W., 1956.
The annual cycle of phytoplankton in the Labrador Sea, 1950-51. Bull. Bingham Oceanogr. Coll., 1-74.

Cladopyxis hemibrachiata n. sp.
Balech, Enrique, 1964.
El genero "Cladopyxis"(Dinoflagellata). Comunicaciones, Museo Argentina, Ciencias Nat. e Inst. Nacional, Invest. Ciencias Naturales, Hidrobiol., 1(4):27-39.

Cladopyxis quadrispina
Rampi, L., 1951.
Ricerche sul fitoplancton del Mare Ligure. 10) Peridiniale delle acque di Sanremo. Pubbl. Centro Talassografico Tirreno No. 6: Atti Acad. Ligure Sci. Lett. 7(1): 8 pp., Pls. 3-4.

Cladopyxis spinosa
Rampi, L., 1950.
Peridiniens rares ou nouveaux pour le Pacifique Sud-Equatorial. Bull. Inst. Océan., Monaco, No. 974:11pp., 28 textfigs.

Cochlodinium spp.
Holmes, R.W., P.M. Williams and R.W. Eppley, 1967. Red water in La Jolla Bay, 1964-1965. Limnol. Oceanogr., 12(3):503-512.

Cochlodinium heterolobatum n.sp.
de Sousa e Silva, Estela, 1967.
Cochlodinium heterolobatum n.sp.: Structure and cytophysiological aspects. J. Protozool., 14(4):745-754.

Cochlodinium sp.
Hope, B., 1954.
Floristic and taxonomic observations on marine phytoplankton from Nordavåtn, near Bergen. Nytt Mag. f. Botanikk 2:149-153, 1 fig.

Cochlodinium achromaticum n.sp.
LeBour, M.V., 1925
The dinoflagellates of Northern Seas. The Marine Biological Association of the United Kingdom, Plymouth, 250 pp., 35 pls., 53 text figs.

Cochlodinium archimedes
LeBour, M.V., 1925
The dinoflagellates of Northern Seas. The Marine Biological Association of the United Kingdom, Plymouth, 250 pp., 35 pls., 53 text figs.

Cochlodinium Archimedis
Paulsen, O., 1908
XVIII Peridiniales. Nordisches Plankton, Bot. Teil: 1-124, 155 text figs.

Cochlodinium archimedes
Wood, E.J.F., 1954.
Dinoflagellates in the Australian region. Australian J. Mar. Freshwater Res., 5(2):171-351.

Cochlodinium brandti
LeBour, M.V., 1925
The dinoflagellates of Northern Seas. The Marine Biological Association of the United Kingdom, Plymouth, 250 pp., 35 pls., 53 text figs.

Cochlodinium cnidophorum
Biecheler, B., 1952.
Recherches sur les Péridiniens.
Bull. Biol., France Belg., Suppl., 36:1-149, 75 textfigs.

Cochlodinium flavum
Wood, E.J.F., 1963
Dinoflagellates in the Australian region. II. Recent collections.
Comm. Sci. and Industr. Res. Org., Div. Fish. and Oceanogr., Techn. Paper, No. 14: 55 pp.

Cochlodinium helicoides nom. nov.
LeBour, M.V., 1925
The dinoflagellates of Northern Seas. The Marine Biological Association of the United Kingdom, Plymouth, 250 pp., 35 pls., 53 text figs.

Cochlodinium helix
LeBour, M.V., 1925
The dinoflagellates of Northern Seas. The Marine Biological Association of the United Kingdom, Plymouth, 250 pp., 35 pls., 53 text figs.

Cochlodinium helix
Paulsen, O., 1908
XVIII Peridiniales. Nordisches Plankton, Bot. Teil: 1-124, 155 text figs.

Cochlodinium helix
Schodduyn, M., 1926
Observations faites dans la baie d'Ambleteuse (Pas de Calais). Bull. Inst. Ocean., Monaco, No. 482: 64 pp.

Cochlodinium helix
Wood, E.J.F., 1954.
Dinoflagellates in the Australian region. Australian J. Mar. Freshwater Res., 5(2):171-351.

Cochlodinium longum
Paulsen, O., 1908
XVIII Peridiniales. Nordisches Plankton, Bot. Teil: 1-124, 155 text figs.

Cochlodinium pellucidum
Paulsen, O., 1908
XVIII Peridiniales. Nordisches Plankton, Bot. Teil: 1-124, 155 text figs.

Cochlodinium polykrikoides n.sp.
Margalef, Ramón, 1961
Hidrografía y fitoplancton de un área marina de la costa meridional de Puerto Rico.
Inv. Pesq., Barcelona, 18:38-96.

Cochlodinium pulchellum
LeBour, M.V., 1925
The dinoflagellates of Northern Seas. The Marine Biological Association of the United Kingdom, Plymouth, 250 pp., 35 pls., 53 text figs.

Cochlodinum pulchellum
Marshall, S.M., 1933
The production of microplankton in the Great Barrier Reef Region. Brit. Mus. (N.H.) Great Barrier Reef Exped. 1928-29, Sci. Repts. II(5):111-157, 14 text figs.

Cochlodinium pupa n.sp.
LeBour, M.V., 1925
The dinoflagellates of Northern Seas. The Marine Biological Association of the United Kingdom, Plymouth, 250 pp., 35 pls., 53 text figs.

Cochlodinium pupa
Wood, E.J.F., 1963
Dinoflagellates in the Australian region. II. Recent collections.
Comm. Sci. and Industr. Res. Org., Div. Fish. and Oceanogr., Techn. Paper, No. 14: 55 pp.

Cochlodinium rosaceum
Wood, E.J.F., 1963
Dinoflagellates in the Australian region. II. Recent collections.
Comm. Sci. and Industr. Res. Org., Div. Fish. and Oceanogr., Techn. Paper, No. 14: 55 pp.

Cochlodinium schuetti
LeBour, M.V., 1925
The dinoflagellates of Northern Seas. The Marine Biological Association of the United Kingdom, Plymouth, 250 pp., 35 pls., 53 text figs.

Cochlodinium schuetti
Morse, D.C., 1947
Some observations on seasonal variations in plankton population Patuxant River, Maryland 1943-1945. Bd. Nat. Res., Publ. No.65, Chesapeake Biol. Lab., 31, 3 figs.

Cochlodinium vinctum
LeBour, M.V., 1925
The dinoflagellates of Northern Seas. The Marine Biological Association of the United Kingdom, Plymouth, 250 pp., 35 pls., 53 text figs.

Cochlodinium virescens
Wood, E.J.F., 1963
Dinoflagellates in the Australian region. II. Recent collections.
Comm. Sci. and Industr. Res. Org., Div. Fish. and Oceanogr., Techn. Paper, No. 14: 55 pp.

Congruentidium compressum
Matzenauer, L., 1933
Die Dinoflagellaten des indischen Ozeans (mit Ausnahme der Gattung Ceratium.) Bot. Arch. 35:437-510, 77 text figs., 2 charts.

Congruentidium compressum
Rampi, L., 1950.
Péridiniens rares ou nouveaux pour le Pacifique Sud-Equatorial. Bull. Inst. Océan., Monaco, No. 974:11 pp., 28 textfigs.

Coolia n.gen.
Meunier, A., 1919
Microplancton de la Mer Flamande 3. Les Péridiniens. Mem. Mus. Roy. Hist. Nat., Belgique 8(1):1-116, Pls. XV-XXI.

Coolia monotis
de Sousa e Silva, E., 1956.
Contribution à l'étude du microplancton de Dakar et des regions maritimes voisines.
Bull. I.F.A.N., 8(2):335-371, 7 pls.

Coolia monotis
LeBour, M.V., 1925
The dinoflagellates of Northern Seas. The Marine Biological Association of the United Kingdom, Plymouth, 250 pp., 35 pls. 53 text figs.

Coolia monotis n.sp.
Meunier, A., 1919
Microplancton de la Mer Flamande 3. Les Péridiniens. Mem. Mus. Roy. Hist. Nat., Belgique 8(1):1-116, Pls. XV-XXI.

Craspedotella pileolus
Cachon, Jean, et Monique Cachon 1969. Contribution à l'étude des Noctilucidae Saville-Kent. Evolution morphologique, cytologie, systématique. II. Les Leptodiscinae Cachon J. et M. Protistologica 5(1):11-33

Crypthecodinium cohnii comb. nov.
Javornicky, Pavel. 1962.
Two scarcely known genera of the class Dinophyceae: Bernardinium Chodat and Crypthecodinium Biecheler.
Preslia (Czechoslovakia), 34:98-113.

Crypthecodinium cohnii
Keller, Steven E., S.H. Hutner and Dolores E. Keller, 1968.
Rearing the colorless marine dinoflagellate Cryptothecodinium cohnii for use as a biochemical tool.
J. Protozool., 15(4):792-795.

Crypthecodinium setense n.gen., n.sp.
Biecheler, B., 1952.
Recherches sur les Péridiniens.
Bull. Biol., France Belg., Suppl., 36:1-149, 75 textfigs.

Cymbodinium elegans
Cachon, Jean, et Monique Cachon 1969. Contribution à l'étude des Noctilucidae Saville-Kent. Evolution morphologique, cytologie, systématique. II. Les Leptodiscinae Cachon J. et M. Protistologica 5(1):11-33

Cymbodinium elegans n.gen., n.sp.
Cachon, Jean et Monique, 1967.
Cymbodinium elegans nov.gen.nov.sp. péridinien Noctilucidae Saville-Kent.
Protistologica 3(3):313-318.

cystodinium sp?
Balech, Enrique, 1971.
Microplancton de la campaña oceanographica: Productividad III.
Revta Mus. argent. Cienc. Nat. Bernadina Rivadavia, Hydrobiol. 3(1):1-202, 39 pls.

Didinium balbianii nanum
*Hada, Yoshina, 1970.   (1970)
The protozoan plankton of the Antarctic and Subantarctic seas.
Scient. Repts, Japan. Antarct. Res. Exped., (E)31:51 pp.

Didinium gargantum
*Hada, Yoshina, 1970.   (1970)
The protozoan plankton of the Antarctic and Subantarctic seas.
Scient. Repts, Japan. Antarct. Res. Exped., (E)31:51 pp.

Didinium nasutum
Braarud, T., 1945
A phytoplankton survey of the polluted waters of inner Oslo Fjord. Hvalrådets Skrifter, No.28, 142 pp., 19 text figs., 17 tables.

Didinium nasutum
Levander, K.M., 1947
Plankton gesammelt in den Jahren 1899-1910 an den Küsten Finnlands. Finnländische Hydrographisch-Biologische Untersuchungen (aus dem Wasserbiologischen Laboratorin der Societas Scientiarum Fennica) No.11:40 pp., 6 diagrams, 13 pls., tables.

Didinium parvulum n.sp.
Gaarder, Karen Ringdal, 1938.
Phytoplankton studies from the Tromsø district, 1930-31. Tromsø Mus. Årshefter, Naturhist. Avd., 11, 55(1):159 pp., 4 fold-in pls., 12 textfigs.

Didinium parvulum
Gran, H.H., and T. Braarud, 1935
A quantitative study of the phytoplankton in the Bay of Fundy and the Gulf of Maine (including observations on hydrography, chemistry, and turbidity).
J. Biol. Bd., Canada, 1(5):279-467, 69 text figs.

Dinofurcula
Abe, Tohru H., 1967.
The armoured Dinoflagellates: II. Prorocentridae and Dinophysidae (B)- Dinophysis and its allied genera.
Publs. Seto mar.biol. Lab., 15(1):37-78.

Dinofurcula ultima n.gen.
Kofoid, C. A. and T. Skogsberg, 1928
XXXV. The Dinoflagellata: The Dinophysiodae. Reports on the scientific results of the expedition to the Eastern Tropical Pacific, in charge of Alexander Agassiz, by the U. S. Fish Commission Steamer "Albatross" from October 1904 to March 1905----. Mem. M. C. Z. 51:766 pp., 31 pls.

Dinofurcula ventralis n.sp.
Kofoid, C. A. and T. Skogsberg, 1928
XXXV. The Dinoflagellata: The Dinophysiodae. Reports on the scientific results of the expedition to the Eastern Tropical Pacific, in charge of Alexander Agassiz, by the U. S. Fish Commission Steamer "Albatross" from October 1904 to March 1905----. Mem. M. C. Z. 51:766 pp., 31 pls.

Dinogymnium sp.
Evitt, William R., Robin F.A. Clarke and Jean-Pierre Verdier, 1967.
Dinoflagellate studies III. Dinogymnium acuminatum n.gen., n.sp. (Maastrichtian) and other fossils referable to Gymnodinium Stein.
Publ. Stanford Univ., Geol. Sci., 10(4):1-27.

Dinogymnium acuminatum
Evitt, William R., Robin F.A. Clarke and Jean-Pierre Verdier 1967.
Dinoflagellate studies III. Dinogymnium acuminatum n.gen., n.sp. (Maastrichtian) and other fossils referable to Gymnodinium Stein.
Publ. Stanford Univ. Geol. Sci. 10(4):1-27.

Dinogymnium avellana
Evitt, William R., Robin F.A. Clarke and Jean-Pierre Verdier, 1967.
Dinoflagellate studies III. Dinogymnium acuminatum n.gen., n.sp. (Maastrichtian) and other fossils referable to Gymnodinium Stein.
Publ. Stanford Univ., Geol. Sci., 10(4):1-27.

Dinogymnium cretaceum
Evitt, William R., Robin F.A. Clarke and Jean-Pierre Verdier, 1967.
Dinoflagellate studies III. Dinogymnium acuminatum n.gen., n.sp. (Maastrichtian) and other fossils referable to Gymnodinium Stein.
Publ. Stanford Univ., Geol. Sci., 10(4):1-27.

Dinogymnium decorum
Evitt, William R., Robin F.A. Clarke and Jean-Pierre Verdier, 1967.
Dinoflagellate studies III. Dinogymnium acuminatum n.gen., n.sp. (Maastrichtian) and other fossils referable to Gymnodinium Stein.
Publ. Stanford Univ., Geol. Sci., 10(4):1-27.

Dinogymnium denticulatum
Evitt, William R., Robin F.A. Clarke and Jean-Pierre Verdier, 1967.
Dinoflagellate studies III. Dinogymnium acuminatum n.gen., n.sp. (Maastrichtian) and other fossils referable to Gymnodinium Stein.
Publ. Stanford Univ., Geol. Sci., 10(4):1-27.

Dinogymnium digitus
Evitt, William R., Robin F.A. Clarke and Jean-Pierre Verdier, 1967.
Dinoflagellate studies III. Dinogymnium acuminatum n.gen., n.sp. (Maastrichtian) and other fossils referable to Gymnodinium Stein.
Publ. Stanford Univ., Geol. Sci., 10(4):1-27.

Dinogymnium heterocostetum
Evitt, William R., Robin F.A. Clarke and Jean-Pierre Verdier, 1967.
Dinoflagellate studies III. Dinogymnium acuminatum n.gen., n.sp. (Maastrichtian) and other fossils referable to Gymnodinium Stein.
Publ. Stanford Univ., Geol. Sci., 10(4):1-27.

?Dinogymnium hexagonum
Evitt, William R., Robin F.A. Clarke and Jean-Pierre Verdier 1967.
Dinoflagellate studies III. Dinogymnium acuminatum n.gen., n.sp. (Maastrichtian) and other fossils referable to Gymnodinium Stein.
Publ. Stanford Univ. Geol. Sci. 10(4):1-27.

Dinogymnium laticinctum
Evitt, William R., Robin F.A. Clarke and Jean-Pierre Verdier 1967.
Dinoflagellate studies III. Dinogymnium acuminatum n.gen., n.sp. (Maastrichtian) and other fossils referable to Gymnodinium Stein.
Publ. Stanford Univ., Geol. Sci. 10(4):1-27.

Dinogymnium marthae
Evitt, William R., Robin F.A. Clarke and Jean-Pierre Verdier 1967.
Dinoflagellate studies III. Dinogymnium acuminatum n.gen., n.sp. (Maastrichtian) and other fossils referable to Gymnodinium Stein.
Publ. Stanford Univ., Geol. Sci. 10(4):1-27.

Dinogymnium nelsonense
Evitt, William R., Robin F.A. Clarke and Jean-Pierre Verdier, 1967.
Dinoflagellate studies III. Dinogymnium acuminatum n.gen., n.sp. (Maastrichtian) and other fossils referable to Gymnodinium Stein.
Publ. Stanford Univ., Geol. Sci., 10(4):1-27.

?Dinogymnium sibiricum
Evitt, William R., Robin F.A. Clarke and Jean-Pierre Verdier, 1967.
Dinoflagellate studies III. Dinogymnium acuminatum n.gen., n.sp. (Maastrichtian) and other fossils referable to Gymnodinium Stein.
Publ. Stanford Univ., Geol. Sci., 10(4):1-27.

Dinogymnium strombomorphum
Evitt, William R., Robin F.A. Clarke and Jean-Pierre Verdier, 1967.
Dinoflagellate studies III. Dinogymnium acuminatum n.gen., n.sp. (Maastrichtian) and other fossils referable to Gymnodinium Stein.
Publ. Stanford Univ., Geol. Sci., 10(4):1-27.

Dinogymnium westralium
Evitt, William R., Robin F.A. Clarke and Jean-Pierre Verdier, 1967.
Dinoflagellate studies III. Dinogymnium acuminatum n.gen., n.sp. (Maastrichtian) and other fossils referable to Gymnodinium Stein.
Publ. Stanford Univ., Geol. Sci., 10(4):1-27.

Dinophysis
Abe, Tohru H., 1967.
The armoured Dinoflagellata: II. Prorocentridae and Dinophysidae (B) - Dinophysis and its allied genera.
Publs. Seto mar. biol. Lab., 15(1):37-78.

Dinophysis sp.
Balech, Enrique, 1967.
Dinoflagelados nuevos o interesantes del Golfo de Mexico y Caribe.
Revta Mus. argent. Cienc. nat. Bernardino Rivadavia Inst. nac. Invest. Cienc. nat., Hidrobiol. 2(3):77-126.

Dinophysis sp.
Brunel, J., 1962
Le phytoplancton de la Baie de Chaleurs.
Inst. Botan., Univ. Montréal, Contrib. No. 77: 365 pp., 66 pls.

Dinophysis sp.
Chiba, T., 1949
On the distribution of the plankton in the eastern China Sea and Yellow Sea. 1. Plankton composition in the spring. J. Shimonoseki Coll. Fisheries, 1(1):57-63, 1 fig.

Dinophysis sp.
Wheeler, J.E.G., 1939
Plankton investigations. Bermuda Biological Station. Second Report. October 1939. 7 pp. (typed), 5 figs. Plymouth, Oct. 23, 1939.

Dinophysis acuminata
Abe, Tohru H., 1967.
The armoured Dinoflagellata: II. Prorocentridae and Dinophysidae (B) - Dinophysis and its allied genera.
Publs. Seto mar. biol. Lab., 15(1):37-78.

Dinophysis acuminata
Balech, Enrique, 1944
Contribucion al conocimiento del Planeton de Lennox y Cabo de Hornos. Physis XIX:423-446, 6 pls. with 67 figs.

Dinophysis acuminata
Braarud, T., 1945
A phytoplankton survey of the polluted waters of inner Oslo Fjord. Hvalrådets Skrifter, No.28, 142 pp., 19 text figs., 17 tables.

Dinophysis acuminata
Braarud, T., and Adam Bursa, 1939
On the phytoplankton of the Oslo Fjord, 1933-1934. Hvalrådets Skr. No.19:1-63; 9 text figs. Reviewed. J. du. Cons. 14(3): 418-420. A.C. Gardiner.

Dinophysis acuminata
Cupp, E., 1930
Quantitative Studies of miscellaneous series of surface catches of marine diatoms and dinoflagellates taken between Seattle and the Canal Zone from 1924 to 1928. Trans. Am. Micro. Soc., XLIX (3):238-245.

Dinophysis acuminata
Dangeard, P., 1926
Description des Péridiniens Testacés recueillis para la Mission Charcot pendent le mois d'Aout 1924. Ann. Inst. Ocean. n.s. 3(7):307-334, 15 text figs.

Dinophysis acuminata
De Angelis, C.M., 1962.
Distribuzione ed ecologia di alcune specie di Dinoflagellati di acque salmastre.
Pubbl. Staz. Zool., Napoli, 32 (Suppl.):301-314.

Dinophysis acuminata
Gran, H.H., and T. Braarud, 1935
A quantitative study of the phytoplankton in the Bay of Fundy and the Gulf of Maine (including observations on hydrography, chemistry, and turbidity). J. Biol. Bd., Canada, 1(5):279-467, 69 text figs.

Dinophysis acuminata
Jörgensen, E., 1923
Mediterranean Dinophysiaceae. Rept. Danish Oceanogr. Exped. 1908-10, to the Mediterranean and adjacent seas, Vol.II, Biol. J 2, 48 pp., 64 text figs.

Dinophysis acuminata

Jørgensen, E., 1905
B. Protistplankton and the diatoms in bottom samples. Hydrographical and biological investigations in Norwegian fjords. Bergens Mus. Skr. 7: 49-225.

Dinophysis acuminata

Jorgensen, E., 1900
Protophyten und Protozoën im Plankton aus der Norwegischen Westkerste. Bergens Mus. Aarb. 1899(6): 95 pp., 5 pls., 83 tables.

Dinophysis acuminata

Käsler, R., 1938
Die Verbreitung der Dinophysiales im Sud-atlantischen Ozean. Wiss. Ergeb. Deutschen Atlantischen Expedition----"Meteor" 1925-1927, 12(2):162-237, text figs. 85-118.

Dinophysis acuminata

LeBour, M.V., 1925
The dinoflagellates of Northern Seas. The Marine Biological Association of the United Kingdom, Plymouth, 250 pp., 35 pls., 53 text figs.

Dinophysis acuminata

Lillick, L.C., 1938
Preliminary report of the phytoplankton of the Gulf of Maine. Am. Mid. Nat. 20(3):624-640, 1 text figs 37 tables.

Dinophysis acuminata

Levander, K.M., 1947
Plankton gesammelt in den Jahren 1899-1910 an den Küsten Finnlands. Finnländische Hydrographisch-Biologische Untersuchungen (aus dem Wasserbiologischen Laboratorin der Societas Scientiarum Fennica) No.11: 40 pp., 6 diagrams, 13 pls., tables.

Dinophysis acuminata

Lillick, L.C., 1940
Phytoplankton and planktonic protozoa of the offshore waters of the Gulf of Maine. Pt.II. Qualitative Composition of the Planktonic Flora. Trans. Am. Phil. Soc., n.s., 31(3):193-237, 13 text figs.

Dinophysis acuminata

Lillick, L.C., 1937
Seasonal studies of the phytoplankton off Woods Hole, Massachusetts. Biol. Bull. LXXIII (3):488-503, 3 text figs.

Dinophysis acuminata

Lindemann, E., 1924
Peridineen aus dem goldenen Horn und dem Bosphorus. Bot. Arch. 5:216-233, 98 text figs.

Dinophysis acuminata

Morse, D.C., 1947
Some observations on seasonal variations in plankton population Patuxant River, Maryland 1943-1945. Bd. Nat. Res., Publ. No.65, Chesapeake Biol. Lab., 31, 3 figs.

Dinophysis acuminata

Paulsen, O., 1949
Observations on dinoflagellates. (Ed. J. Grøntoed) Kongl. Dansk. Videnskab. Selsk., Biol. Skr. 6(4):67 pp., 30 text figs.

Dinophysis acuminata

Paulsen, O., 1908
XVIII Peridiniales. Nordisches Plankton, Bot. Teil: 1-124, 155 text figs.

Dinophysis acuminata

Pavillard, J., 1923
A propos de la systématique des Péridiniens Bull. Soc. Bot. de France 70:876-882; 914-918.

Dinophysis acuminata

Pavillard, J., 1905
Recherches sur la flore pelagique (Phytoplankton) de l'Etang de Thau. Theses presentees a la Fac. Sci., Paris, 116 pp., 3 pls.

Dinophysis acuminata

Rampi, L., 1940
Ricerche sul Fitoplancton del mare Ligure. II Le tecatali e le dinofisiali delle acque di Sanremo. Boll. Pesca. Piscicolt., Idrobiol. (18) 16(2): 243-274, 56 figs.

Dinophysis acuminata

Rampi, L., 1940
Ricerche sul Fitoplancton del mare Ligure. Boll. di Pesca, di Piscicoltura e di Idrobiologa, 18(2):1-34, 56 text figs.

Dinophysis acuminata

Rumkówna, A., 1948
List of the phytoplankton species occurring in the superficial water layers in the Gulf of Gdańsk. Bull. Lab. mar., Gdynia, No. 4: 139-141 with tables in back.

Dinophysis acuminata

Solum, Ingrid, 1962.
The taxonomy of Dinophysis populations in Norwegian waters in view of biometric observations. Nytt Mag. Botanikk, 10:5-32.

Dinophysis acuminata

Steemann-Nielsen, Einar, 1951
The marine vegetation of the Isefjord. A study on ecology and production. Medd. Komm. Danmarks Fiskeri-og Havundersøgelser. Ser. Plankton. 5(4); 114pp., 46 text figs.

Dinophysis acuminata

Tai, Si-Sun, & T. Skogsberg, 1934
Studies on the Dinophysordae, marine armored dinoflagellates of Monterey Bay, California. Arch. Protistenk. 82:380-482, 14 text figs. Pls. 11-12.

Dinophysis acuminata

Wood, E.J.F., 1954.
Dinoflagellates in the Australian region. Australian J. Mar. Freshwater Sci., 5(2):171-351.

Dinophysis acutum

Abe, Tohru H., 1967.
The armoured Dinoflagellata: II. Prorocentridae and Dinophysidae (B)- Dinophysis and its allied genera. Publs. Seto mar.biol. Lab., 15(1):37-78.

Dinophysis acuta

Braarud, T., 1945
A phytoplankton survey of the polluted waters of inner Oslo Fjord. Hvalrådets Skrifter, No.28, 142 pp., 19 text figs., 17 tables.

Dinophysis acuta

Braarud, T., and Adam Bursa, 1939
On the phytoplankton of the Oslo Fjord, 1933-1934. Hvalrådets Skr. No.19:1-63; 9 text figs. Reviewed. J. du. Cons. 14(3): 418-420. A.C. Gardiner.

Dinophysis acuta

Dangeard, P., 1927
Phytoplankton de la croisière du "Sylvana". Ann. Inst. Ocean. Monaco, n.s., 4(8):286-401, 54 text figs. (Jenvier -Juin 1913).

Dinophysis acuta

Dangeard, P., 1926
Description des Péridiniens Testacés recueillis para la Mission Charcot pendent le mois d'Aout 1924. Ann. Inst. Ocean. n.s. 3(7):307-334, 15 text figs.

Dinophysis acuta

Forti, A., 1922
Ricerche sulla flora pelagica (fitoplancton) di Quarto dei Mille. Mem. R. Com. Talass. Ital. 97:248 pp., 13 pls.

Dinophysis acuta

Gran, H.H., and T. Braarud, 1935
A quantitative study of the phytoplankton in the Bay of Fundy and the Gulf of Maine (including observations on hydrography, chemistry, and turbidity). J. Biol. Bd., Canada, 1(5):279-467, 69 text figs.

Dinophysis acuta

Jørgensen, E., 1923
Mediterranean Dinophysiaceae. Rept. Danish Oceanogr. Expeds. 1908-10, to the Mediterranean and adjacent seas, Vol.II, Biol. J 2, 48 pp., 64 text figs.

Dinophysis acuta

Jørgensen, E., 1905
B. Protistplankton and the diatoms in bottom samples. Hydrographical and biological investigations in Norwegian fjords. Bergens Mus. Skr. 7: 49-225.

Dinophysis acuta

Jorgensen, E., 1900
Protophyten und Protozoën im Plankton aus der Norwegischen Westkerste. Bergens Mus. Aarb. 1899(6): 95 pp., 5 pls., 83 tables.

Dinophysis acuta

LeBour, M.V., 1925
The dinoflagellates of Northern Seas. The Marine Biological Association of the United Kingdom, Plymouth, 250 pp., 35 pls., 53 text figs.

Dinophysis acuta

Lillick, L.C., 1940
Phytoplankton and planktonic protozoa of the offshore waters of the Gulf of Maine. Pt.II. Qualitative Composition of the Planktonic Flora. Trans. Am. Phil. Soc., n.s. 31(3):193-237, 13 text figs.

Dinophysis acuta

Lindemann, E., 1925
Neubeobachtungen an den Winter peridineen des Golfes von Neapel. Bot. Arch. 9:95-102, 19 text figs.

Dinophysis acuta

Lindemann, E., 1924
Peridineen aus dem goldenen Horn und dem Bosphorus. Bot. Arch. 5:216-233, 98 text figs.

Dinophysis acuta

Mangin, M. L., 1912
Phytoplancton de la croisière du "René" dans l'Atlantique (Septembre 1908). Ann. Inst. Ocean., n.s., 4(1):1-66, 2 pls., 41 text figs., 2 tables.

Dinophysis acuta

Murray, G., and F. G. Whitting, 1899
New Peridiniaceae from the Atlantic. Trans. Linn. Soc., London, Bot., ser 2, 5: 321-342, Pls. 27-33, 9 tables.

Dinophysis acuta

Nordli, O., 1951.
Dinoflagellates from Lofoten. Nytt Mag. Naturvidensk 88:49-55, 7 textfigs.

Dinophysis acuta

Paulsen, O., 1949
Observations on dinoflagellates. (Ed. J. Grøntoed) Kongl. Dansk. Videnskab. Selsk., Biol. Skr. 6(4):67 pp., 30 text figs.

Dinophysis acuta

Paulsen, O., 1908
XVIII Peridiniales. Nordisches Plankton, Bot. Teil: 1-124, 155 text figs.

Dinophysis acuta
Pavillard, J., 1916
Recherches sur les Peridiniens du Golfe du Lion. Mem. Univ. Montpellier. Trav. Inst. Bot., Univ. Montpellier. Serie mixte No. 4, 70 pp., 3 pls., 15 text figs.

Dinophysis acuta
Pavillard, J., 1905
Rechorches sur la flore pelagique (Phytoplankton) de l'Etang de Thau. Theses presentees a la Fac. Sci., Paris, 116 pp., 3 pls.

Dinophysis acuta
Rampi, L. 1940
Ricerche sul Fitoplancton del mare Ligure. II Le tecatali e le dinofisiali delle acque di Sanremo.. Boll. Pesca, Piscicolt..Idrobiol. (18) 16(2): 243-274, 56 figs.

Dinophysis acuta
Rampi, L., 1940
Ricerche sul Fitoplancton del mare Ligure. Boll. di Pesca, di Piscicoltura e di Idrobiologa, 18(2):1-34, 56 text figs.

Dinophysis acuta
Schröder, B., 1900
Phytoplankton des Golfes von Neapel nebst vergleichenden Ausblicken auf das atlantischen Ozean. Mitt. Zool. Stat. Neapel, 14:1-38.

Dinophysis acuta
Solum, Ingrid, 1962
The taxonomy of Dinophysis populations in Norwegian waters in view of biometric observations.
Nytt Mag. Botanikk, 10:5-32.

Dinophysis acuta
Sousa e Silva, E., 1949
Diatomaceas e Dinoflagelados de Baia de Cascais. Portugaliae Acta Biol., Volume: Julio Henriques, Ser. B: 300-383, 9 pls., 2 fold-in tables.

Dinophysis acuta
Sousa a Silva, E., and J. Dos Santos-Pinto, 1948
O Plancton da Baia de S. Martinho do Porto. 1. Diatomaceas e Dinoflagelados. Bol. Soc. Portuguese de Ciencias Naturais, 16(2):134-187, 6 pls. (Trav. Sta. Biol. Mar. de Lisbonne No. 52).

Dinophysis acuta
Tai, Si-Sun, & T. Skogsberg, 1934
Studies on the Dinophysordae, marine armored dinoflagellates of Monterey Bay, California. Arch. Protistenk. 82:380-482, 14 text figs. Pls. 11-12.

Dinophysis acuta
Wood, E.J.F., 1954.
Dinoflagellates in the Australian region. Australian J. Mar. Freshwater Res., 5(2):171-351.

Dinophysis alata n.sp.
Jörgensen, E., 1923
Mediterranean Dinophysiaceae. Rept. Danish Oceanogr. Exped. 1908-10, to the Mediterranean and adjacent seas, Vol. II, Biol. J 2, 48 pp., 64 text figs.

Dinophysis Allieri
Gourret, P., 1883
Sur les Peridiniens du Golfe de Marseille. Ann. du Musee d'hist. Nat., Marseille, Zool., 1 (Mme. 8):1-114, 4 pls.

Dinophysis amphora n. sp.
Balech, Enrique, 1971.
Microplancton de la campaña oceanographica: Productividad III.
Revta Mus. argent. Cienc. Nat. Bernadina Rivadavia, Hydrobiol. 3(1):1-202, 39 pls.

Dinophysis amygdala
Balech, Enrique, 1971
Microplancton del Atlantico ecuatorial oeste (Equalant 1)
Publ. Serv. Hidrograf. Naval, Argentina H. 654: 103 pp., 122 figs.

Dinophysis anabilis n. sp.
Abe, Tohru H., 1967.
The armoured Dinoflagellata: ii. Prorocentridae and Dinophysidae (B) - Dinophysis and its allied genera.
Publs. Seto mar.biol.Lab., 15(1):37-78.

Dinophysis antarcticum n.sp.
Balech, Enrique, 1958.
Plancton de la Campana Antartica Argentina, 1954-1955.
Physis, 21(60):75-108.

Dinophysis antarcticum n.sp.
Hada, Yoshina, 1970. (1970)
The protozoan plankton of the Antarctic and Subantarctic seas.
Scient. Repts, Japan. Antarct. Res. Exped., (E)31:51 pp.

Dinophysis apicatum
Abe, Tohru H.,1967.
The armoured Dinoflagellata: II. Prorocentridae and Dinophysidae (B)- Dinophysis and its allied genera.
Publs. Seto mar.biol. Lab., 15(1):37-78.

Dinophysis arctica
Abe,Tohru.,1967.
The armoured Dinoflagellata: II. Prorocentridae and Dinophysidae (B) - Dinophysis and its allied genera.
Publs. Seto mar. biol.Lab., 15(1):37-78.

Dinophysis arctica
Gran, H.H., and T. Braarud, 1935
A quantitative study of the phytoplankton in the Bay of Fundy and the Gulf of Maine (including observations on hydrography, chemistry, and turbidity). J. Biol. Bd., Canada, 1(5):279-467, 69 text figs.

Dinophysis arctica
LeBour, M.V., 1925
The dinoflagellates of Northern Seas. The Marine Biological Association of the United Kingdom, Plymouth, 250 pp., 35 pls,;53 text figs.

Dinophysis arctica
Lillick, L.C., 1940
Phytoplankton and planktonic protozoa of the offshore waters of the Gulf of Maine. Pt.II. Qualitative Composition of the Planktonic Flora. Trans. Am. Phil. Soc., n.s., 31(3):193-237, 13 text figs.

Dinophysis arctica
Paulsen, O., 1949
Observations on dinoflagellates. (Ed. J. Grøntoed) Kongl. Dansk. Videnskab. Selsk., Biol. Skr. 6(4):67 pp., 30 text figs.

Dinophysis arctica
Paulsen, O., 1908
XVIII Peridiniales. Nordisches Plankton, Bot. Teil: 1-124, 155 text figs.

Dinophysis arctica
Rumkówna, A., 1948
[List of the phytoplankton species occurring in the superficial water layers in the Gulf of Gdańsk] Bull. Lab. mar., Gdynia, No. 4: 139-141 with tables in back.

Dinophysis arctica
Wood, E.J.P., 1954.
Dinoflagellates in the Australian region. Australian J. Mar. Freshwater Res., 5(2):171-351.

Dinophysis argus
Abe,Tohru H., 1967.
The armoured Dinoflagellata: II. Prorocentridae and Dinophysidae (B)- Dinophysis and its allied genera.
Publs. Seto mar biol. Lab., 15(1):37-78.

Dinophysis armata
Schröder, B., 1900
Phytoplankton des Golfes von Neapel nebst vergleichenden Ausblicken auf das atlantischen Ozean. Mitt. Zool. Stat. Neapel, 14:1-38.

Dinophysis balechi n. sp.
Norris, Dean R., and Leo D. Berner Jr. 1970.
Thecal morphology of selected species of Dinophysis (Dinoflagellata) from the Gulf of Mexico.
Contrib. mar. Sci. Port Aransas 15:145-192

Dinophysis baltica
Rumkówna, A., 1948
[List of the phytoplankton species occurring in the superficial water layers in the Gulf of Gdańsk] Bull. Lab. mar., Gdynia, No. 4: 139-141 with tables in back.

Dinophysis bibulbus n. sp
Balech, Enrique, 1971.
Microplancton de la campaña oceanographica: Productividad III.
Revta Mus. argent. Cienc. Nat. Bernadina Rivadavia, Hydrobiol. 3(1):1-202, 39 pls.

Dinophysis boehmii
Balech, Enrique, 1971.
Microplancton de la campaña oceanographica: Productividad III.
Revta Mus. argent. Cienc. Nat. Bernadina Rivadavia, Hydrobiol. 3(1):1-202, 39 pls.

Dinophysis borealis
Brunel, J., 1962
Le phytoplancton de la Baie de Chaleurs. Inst. Botan., Univ. Montréal, Contrib. No. 77: 365 pp., 66 pls.

Dinophysis borealis
Brunel, Jules, 1962
Le phytoplancton de la Baie des Chaleurs. Contrib. Ministère de la Chasse et des Pêcheries, Province de Québec, No. 91: 365 pp.

Dinophysis borealis n.sp.
Paulsen, O., 1949
Observations on dinoflagellates. (Ed. J. Grøntoed) Kongl. Dansk. Videnskab. Selsk., Biol. Skr. 6(4):67 pp., 30 text figs.

Dinophysis brevisulcus n.sp.
Tai, Si-Sun, & T. Skogsberg, 1934
Studies on the Dinophysordae, marine armored dinoflagellates of Monterey Bay, California. Arch. Protistenk. 82:380-482, 14 text figs. Pls. 11-12.

Dinophysis capitulata n.sp.
Balech, Enrique, 1967.
Dinoflagelados nuevos o interesantes del Golfo de Mexico y Caribe.
Revta Mus. argent.Cienc.nat. Bernardino Rivadavia Inst.nac.Invest.Cienc.nat., Hidrobiol. 2(3):77-126.

Dinophysis caudata
Abe, Tohru H., 1967.
The armoured Dinoflagellata: II. Prorocentridae and Dinophysidae (B)-Dinophysis and its allied genera.
Publs. Seto mar. biol. Lab., 15(1):37-78.

Dinophysis caudata
Allen, W.E., 1927.
Quantitative studies on inshore marine diatoms and dinoflagellates of Southern California in 1922. Bull. S.I.O., tech. ser., 1:31-38, 2 textfigs.

Dinophysis caudata
Balech, E., 1951.
Sobre dos variedades de Dinophysis caudata Kent. Comun. Zool. Mus. Hist. Nat., Montevideo, 3(60): 1-9, 1 textfig., 4 pls.

Dinophysis caudata
Delegazione Italiana della Commissione Internazionale per l'Esplorazione Scientifica del Mediterraneo, 1941
Note sul plancton della Laguna veneta.
[Memoria CCLXXIX], Arch. di Ocean. e Limn. Anno I, Fasc. I, 1941 XIX: 31-57 pp.

Dinophysis caudata
Ercegovic, A., 1936
Etudes qualitative et quantitatives du phytoplancton dans les eaux cotières de l'Adriatique oriental moyen au cours de l'année 1934. Acta Adriatica 1(9):1-126

Dinophysis caudata
Ghazzawi, F.M., 1939
Plankton of the Egyptian waters. A study of the Suez Canal Plankton. (A) Phytoplankton. Preliminary Report 83 pp. Notes and Memoires, Min. Commerce-Industry, Egypt, Hydrobiol. & Fish. 65 figs.

Dinophysis caudata
Gilbert, J.Y., and W.E. Allen, 1943
The phytoplankton of the Gulf of California obtained by the "E.W. Scripps" in 1939 and 1940. J. Mar. Res. V(2):89-110, figs. 30-31.

Dinophysis caudata
Jörgensen, E., 1923
Mediterranean Dinophysiaceae. Rept. Danish Oceanogr. Expeds. 1908-10, to the Mediterranean and adjacent seas, Vol. II, Biol. J 2, 48 pp., 64 text figs.

Dinophysis caudata
Käsler, R., 1938
Die Verbreitung der Dinophysiales im Sudatlantischen Ozean. Wiss. Ergeb. Deutschen Atlantischen Expedition----"Meteor" 1925-1927, 12(2):162-237, text figs. 85-118.

Dinophysis caudata
Kofoid, C. A. and T. Skogsberg, 1928
XXXV. The Dinoflagellata: The Dinophysiodae. Reports on the scientific results of the expedition to the Eastern Tropical Pacific, in charge of Alexander Agassiz, by the U. S. Fish Commission Steamer "Albatross" from October 1904 to March 1905----. Mem. M. C. Z. 51:766 pp., 31 pls.

Dinophysis caudata
LeBour, M.V., 1925
The dinoflagellates of Northern Seas. The Marine Biological Association of the United Kingdom, Plymouth, 250 pp., 35 pls., 53 text figs.

Dinophysis caudata
Macdonald, R., 1933
An examination of plankton hauls made in the Suez Canal during the year 1928. Fish. Res. Dis., Notes & Mem. No. 3, 11 pp., 1 chart.

Dinophysis caudata
Margalef, R., 1949
Fitoplancton nerítico de la Costa Brava en 1947-48. Publ. Inst. Biol. Aplicada, 5: 41-51, 3 text figs.

Dinophysis caudata
Marshall, S. M., 1933
The production of microplankton in the Great Barrier Reef Region. Brit. Mus. (N.H.) Great Barrier Reef Exped. 1928-29, Sci. Repts. II(5):111-157, 14 text figs.

Dinophysis caudata
Massutí Algamora, M., 1949
Estudio de diez y seis muestras de plancton del Golfo de Nápoles. Publ. Inst. Biol. Appl. 5:85-94, 1 fold-in table.

Dinophysis caudata
Matzenauer, L., 1933
Die Dinoflagellaten des indischen Ozeans (mit Ausnahme der Gattung Ceratium.) Bot. Arch. 35:437-510, 77 text figs., 2 charts.

Dinophysis caudata
Paiva Carvalho, J., 1950
O plancton do Rio Maria Rodrigues (Cananeis). 1. Diatomaceas e Dinoflagelados. Bol. Inst. Paulista Oceanogr. 1(1); 27-43, 2 fold-in tables, 2 figs.

Dinophysis caudata
Pavillard, J., 1923
A propos de la systématique des Péridiniens Bull. Soc. Bot. de France 70:876-882; 914-918.

Dinophysis caudata
Rampi, L., 1940
Ricerche sul Fitoplancton del mare Ligure. II Le tecatali e le dinofisiali delle acque di Sanremo. Boll. Pesca, Piscicolt., Idrobiol. (18) 16(2): 243-274, 56 figs.

Dinophysis caudata
Rampi, L., 1940
Ricerche sul Fitoplancton del mare Ligure. Boll. di Pesca, di Piscicoltura e di Idrobiologa, 18(2):1-34, 56 text figs.

Dinophysis caudata
Sousa e Silva, E., 1949
Diatomaceas e Dinoflagelados de Baia de Cascais. Portugaliae Acta Biol. Volume: Julio Henriques, Ser. B: 300-383, 9 pls, 2 fold-in tables.

Dinophysis caudata
Sousa a Silva, E., and J. Dos Santos-Pinto, 1948
O Plancton da Baia de S. Martinho do Porto. 1. Diatomaceas e Dinoflagelados. Bol. Soc. Portuguese de Ciencias Naturais, 16(2):134-187, 6 pls. (Trav. Sta. Biol. Mar. de Lisbonne No. 52).

Dinophysis caudata
Tai, Si-Sun, & T. Skogsberg, 1934
Studies on the Dinophysordae, marine armored dinoflagellates of Monterey Bay, California. Arch. Protistenk. 82:380-482, 14 text figs. Pls. 11-12.

Dinophysis caudata
Wang, C. C., and D. Nie, 1932
A survey of the marine protozoa of Amoy. Contr. Biol. Lab. Sci. Soc., China, Zool. Ser., 8:285-385, 89 text figs.

Dinophysis caudata
Wood, E.J.P., 1954.
Dinoflagellates in the Australian region. Australian J. Mar. Freshwater Res., 5(2):171-351.

Dinophysis caudata acutiformis
Ercegovic, A., 1936
Etudes qualitative et quantitatives du phytoplancton dans les eaux cotières de l'Adriatique oriental moyen au cours de l'année 1934. Acta Adriatica 1(9):1-126

Dinophysis caudata acutiformis n.f.
Kofoid, C. A. and T. Skogsberg, 1928
XXXV. The Dinoflagellata: The Dinophysiodae. Reports on the scientific results of the expedition to the Eastern Tropical Pacific, in charge of Alexander Agassiz, by the U. S. Fish Commission Steamer "Albatross" from October 1904 to March 1905----. Mem. M. C. Z. 51:766 pp., 31 pls.

Dinophysis caudata f. acutiformis
Rampi, L., 1940
Ricerche sul Fitoplancton del mare Ligure. Boll. di Pesca, di Piscicoltura e di Idrobiologa, 18(2):1-34, 56 text figs.

Dinophysis caudata maris rubri nf.
Matzenauer, L., 1933
Die Dinoflagellaten des indischen Ozeans (mit Ausnahme der Gattung Ceratium.) Bot. Arch. 35:437-510, 77 text figs., 2 charts.

Dinophysis collaris
Kofoid, C. A. and T. Skogsberg, 1928
XXXV. The Dinoflagellata: The Dinophysiodae. Reports on the scientific results of the expedition to the Eastern Tropical Pacific, in charge of Alexander Agassiz, by the U. S. Fish Commission Steamer "Albatross" from October 1904 to March 1905----. Mem. M. C. Z. 51:766 pp., 31 pls.

Dinophysis circumsuta
Norris, Dean R., and Leo D. Berner Jr. 1970.
Thecal morphology of selected species of Dinophysis (Dinoflagellata) from the Gulf of Mexico.
Contrib. mar. Sci. Port Aransas 15:145-192

Dinophysis cornuta
Balech, Enrique, 1971.
Microplancton de la campaña oceanographica: Productividad III.
Revta Mus. argent. Cienc. Nat. Bernadina Rivadavia, Hydrobiol. 3(1):1-202, 39 pls.

Dinophysis cumeus
Abe, Tohru H., 1967.
The armoured Dinoflagellata : II. Prorocentridae and Dinophysidae (B)-Dinophysis and its allied genera.
Publs. Seto mar. biol. Lab., 15(1):37-78.

Dinophysis debilior
Paulsen, O., 1949
Observations on dinoflagellates. (Ed. J. Grøntoed) Kongl. Dansk. Videnskab. Selsk., Biol. Skr. 6(4):67 pp., 30 text figs.

Dinophysis dens
Dangeard, P., 1926
Description des Péridiniens Testacés recueillis para la Mission Charcot pendent le mois d'Aout 1924. Ann. Inst. Ocean. n.s. 3(7):307-334, 15 text figs.

Dinophysis dens
Forti, A., 1922
Ricerche sulla flora pelagica (fitoplancton) di Quarto dei Mille. Mem. R. Com. Talass. Ital. 97:248 pp., 13 pls.

Dinophysis dens
Jörgensen, E., 1923
Mediterranean Dinophysiaceae. Rept. Danish Oceanogr. Expeds. 1908-10, to the Mediterranean and adjacent seas, Vol. II, Biol. J 2, 48 pp., 64 text figs.

Dinophysis dens
Paulsen, O., 1949
Observations on dinoflagellates. (Ed. J. Grøntoed) Kongl. Dansk. Videnskab. Selsk., Biol. Skr. 6(4):67 pp., 30 text figs.

Dinophysis dens
Pavillard, J., 1923
A propos de la systématique des Péridiniens Bull. Soc. Bot. de France 70:876-882; 914-918.

Dinophysis dens
Pavillard, J., 1916
Recherches sur les Péridiniens du Golfe du Lion. Mem. Univ. Montpellier. Trav. Inst. Bot., Univ. Montpellier. Serie mixte No. 4, 70 pp., 3 pls., 15 text figs.

Dinophysis dens
Solum, Ingrid, 1962
The taxonomy of Dinophysis populations in Norwegian waters in view of biometric observations.
Nytt Mag. Botanikk, 10:5-32.

Dinophysis diegensis n. sp.
Kofoid, C. A., 1907
Dinoflagellata of the San Diego region. III. Descriptions of new species. Univ. Calif. Publ., Zool. 3:299-340, Pls. 22-33.

Dinophysis diegensis
Pavillard, J., 1923
A propos de la systématique des Péridiniens Bull. Soc. Bot. de France 70:876-882; 914-918.

Dinophysis diegensis caudata
Pavillard, J., 1916
Recherches sur les Péridiniens du Golfe du Lion. Mem. Univ. Montpellier. Trav. Inst. Bot., Univ. Montpellier. Serie mixte No. 4, 70 pp., 3 pls., 15 text figs.

Dinophysis diegensis
Sousa e Silva, E., 1949
Diatomaceas e Dinoflagelados de Baia de Cascais. Portugaliae Acta Biol., Volume: Julio Henriques, Ser. B: 300-383, 9 pls, 2 fold-in tables.

Dinophysis diegensis caudata
Sousa a Silva, E., and J. Dos Santos-Pinto, 1948
O Plancton da Baía de S. Martinho do Porto. 1. Diatomaceas e Dinoflagelados. Bol. Soc. Portuguese de Ciencias Naturais, 16(2):134-187, 6 pls. (Trav. Sta. Biol. Mar. de Lisbonne No. 52).

Dinophysis doryphorum
Abe, Tohru H., 1967.
The armoured Dinoflagellata: II. Prorocentridae and Dinophysidae (B) - Dinophysis and its allied genera.
Publs. Seto mar. biol. Lab., 15(1):37-78.

Dinophysis doryphora
Norris, Dean R., and Leo D. Berner Jr. 1970.
Thecal morphology of selected species of Dinophysis (Dinoflagellata) from the Gulf of Mexico.
Contrib. mar. Sci. Port Aransas 15:145-192

Dinophysis ellipsoides
Brunel, J., 1962
Le phytoplancton de la Baie de Chaleurs. Inst. Botan., Univ. Montréal, Contrib. No. 77: 365 pp., 66 pls.

Dinophysis ellipsoides
Brunel, Jules, 1962
Le phytoplancton de la Baie des Chaleurs. Contrib. Ministère de la Chasse et des Pêcheries, Province de Québec, No. 91: 365 pp.

Dinophysis ellipsoides
Gilbert, J.Y., and W.E. Allen, 1943
The phytoplankton of the Gulf of California obtained by the "E.W. Scripps" in 1939 and 1940. J. Mar. Res. V(2):89-110, figs. 30-31.

Dinophysis ellipsoides n. sp.
Kofoid, C. A., 1907
Dinoflagellata of the San Diego region. III. Descriptions of new species. Univ. Calif. Publ., Zool. 3:299-340, Pls. 22-33.

Dinophysis ellipsoides
Pavillard, J., 1923
A propos de la systématique des Péridiniens Bull. Soc. Bot. de France 70:876-882; 914-918.

Dinophysis elongatum
Abe, Tohru H., 1967.
The armoured Dinoflagellata: II. Prorocentridae and Dinophysidae (B) - Dinophysis and its allied genera.
Publs. Seto mar. biol. Lab., 15(1):37-78.

Dinophysis equalanti n. sp.
Balech, Enrique, 1971
Microplancton del Atlantico ecuatorial oeste (Equalant 1)
Publ. Serv. Hidrograf. Naval, Argentina H. 654: 103 pp., 122 figs.

Dinophysis exigua
Balech, Enrique, 1967.
Dinoflagelados nuevos o interesantes del Golfo de Mexico y Caribe.
Revta Mus. argent. Cienc. nat. Bernardino Rivadavia Inst. nac. Invest. Cienc. nat., Hidrobiol. 2(3):77-126.

Dinophysis exigua
Käsler, R., 1938
Die Verbreitung der Dinophysiales im Sud-atlantischen Ozean. Wiss. Ergeb. Deutschen Atlantischen Expedition----"Meteor" 1925-1927, 12(2):162-237, text figs. 85-118.

Dinophysis exigua n.sp.
Kofoid, C. A. and T. Skogsberg, 1928
XXXV. The Dinoflagellata: The Dinophysio-dae. Reports on the scientific results of the expedition to the Eastern Tropical Pacific, in charge of Alexander Agassiz, by the U. S. Fish Commission Steamer "Albatross" from October 1904 to March 1905----. Mem. M. C. Z. 51:766 pp., 31 pls.

Dinophysis exigua
Wood, E.J.F., 1963
Dinoflagellates in the Australian region. II. Recent collections.
Comm. Sci. and Industr. Res. Org., Div. Fish. and Oceanogr., Techn. Paper, No. 14: 55 pp.

Dinophysis favus
Abe, Tohru H., 1967.
The armoured Dinoflagellata: II. Prorocentridae and Dinophysidae (B) - Dinophysis and its allied genera.
Publs. Seto mar. biol. Lab., 15(1):37-78.

Dinophysis forti
Abe, Tohru H., 1967.
The armoured Dinoflagellata: II. Prorocentridae and Dinophysidae (B) - Dinophysis and its allied genera.
Publs. Seto mar. biol. Lab., 15(1):37-78.

Dinophysis fortii
Balech, Enrique, 1962
Tintinnoinea y Dinoflagellata del Pacifico segun material de las Expediciones NORPAC y DOWNWIND del Instituto Scripps de Oceanografia.
Revista. Mus. Argentino Ciencias Nat. "Bernardino Rivadavia", Ciencias Zool., 7(1):1-253.

Dinophysis fortii
De Angelis, C.M., 1962.
Distribuzione ed ecologia di alcune specie di Dinoflagellati di acque salmastre.
Pubbl. Staz. Zool., Napoli, 32 Suppl.):301-314.

Dinophysis fortii
Käsler, R., 1938
Die Verbreitung der Dinophysiales im Sud-atlantischen Ozean. Wiss. Ergeb. Deutschen Atlantischen Expedition----"Meteor" 1925-1927, 12(2):162-237, text figs. 85-118.

Dinophysis fortii
Kofoid, C. A. and T. Skogsberg, 1928
XXXV. The Dinoflagellata: The Dinophysio-dae. Reports on the scientific results of the expedition to the Eastern Tropical Pacific, in charge of Alexander Agassiz, by the U. S. Fish Commission Steamer "Albatross" from October 1904 to March 1905----. Mem. M. C. Z. 51:766 pp., 31 pls.

Dinophysis Fortii
Matzenauer, L., 1933
Die Dinoflagellaten des indischen Ozeans (mit Ausnahme der Gattung Ceratium.) Bot. Arch. 35:437-510, 77 text figs., 2 charts.

Dinophysis Fortii nom. nov.
Pavillard, J., 1923
A propos de la systématique des Péridiniens Bull. Soc. Bot. de France 70:876-882; 914-918.

Dinophysis fortii
Rampi, L. 1940.
Ricerche sul Fitoplancton del mare Ligure. II Le tecatali e le dinofisiali delle acque di Sanremo.. Boll. Pesca, Piscicolt., Idrobiol. (18) 16(2): 243-274, 56 figs.

Dinophysis Forti
Rampi, L., 1940
Ricerche sul Fitoplancton del mare Ligure. Boll. di Pesca, di Piscicoltura e di Idrobiologa, 18(2):1-34, 56 text figs.

Dinophysis fortii
Tai, Si-Sun, & T. Skogsberg, 1934
Studies on the Dinophysordae, marine armored dinoflagellates of Monterey Bay, California. Arch. Protistenk. 82:380-482, 14 text figs. Pls. 11-12.

Dinophysis fortii
Wood, E.J.P., 1954.
Dinoflagellates in the Australian region. Australian J. Mar. Freshwater Res., 5(2):171-351.

Dinophysis Granii n.nom.
Paulsen, O., 1949
Observations on dinoflagellates. (Ed. J. Grøntoed) Kongl. Dansk. Videnskab. Selsk., Biol. Skr. 6(4):67 pp., 30 text figs.

Dinophysis hastata
Abe, Tohru H., 1967.
The armoured Dinoflagellata: II. Prorocentridae and Dinophysidae (B) - Dinophysis and its allied genera.
Publs. Seto mar. biol. Lab., 15(1):37-78.

Dinophysis hastata
Dangeard, P., 1927
Phytoplankton de la croisière du "Sylvana". Ann. Inst. Océan., Monaco, n.s., 4(8):286-401, 54 text figs. (Février-Juin 1913).

Dinophysis hastata
Delegazione Italiana della Commissione Internazionale per l'Esplorazione Scientifica del Mediterraneo, 1941
Note sul plancton della Laguna veneta. Memoria CCLXXXI, Arch. di Ocean. e Limn. Anno I, Fasc. I, 1941 XIX: 31-57 pp.

Dinophysis hastata
Ercegovic, A., 1936
Etudes qualitative et quantitatives du phytoplancton dans les eaux côtières de l'Adriatique oriental moyen au cours de l'année 1934. Acta Adriatica 1(9):1-126

Dinophysis hastata
Jörgensen, E., 1923
Mediterranean Dinophysiaceae. Rept. Danish Oceanogr. Expeds. 1908-10, to the Mediterranean and adjacent seas, Vol.II, Biol. J 2, 48 pp., 64 text figs.

Dinophysis hastata
Jorgensen, E., 1900
Protophyten und Protozoën im Plankton aus der Norwegischen Westkerste. Bergens Mus. Aarb. 1899(6): 95 pp., 5 pls., 83 tables.

Dinophysis hastata
Käsler, R., 1938
Die Verbreitung der Dinophysiales im Sudatlantischen Ozean. Wiss. Ergeb. Deutschen Atlantischen Expedition----"Meteor" 1925-1927, 12(2):162-237, text figs. 85-118.

Dinophysis hastata
Kofoid, C. A. and T. Skogsberg, 1928
XXXV. The Dinoflagellata: The Dinophysiodae. Reports on the scientific results of the expedition to the Eastern Tropical Pacific, in charge of Alexander Agassiz, by the U. S. Fish Commission Steamer "Albatross" from October 1904 to March 1905----. Mem. M. C. Z. 51:766 pp., 31 pls.

Dinophysis hastata
LeBour, M.V., 1925
The dinoflagellates of Northern Seas. The Marine Biological Association of the United Kingdom, Plymouth, 250 pp., 35 pls., 53 text figs.

Dinophysis hastata
Mangin, M. L., 1912
Phytoplancton de la croisière du "René" dans l'Atlantique (Septembre 1908). Ann. Inst. Ocean., n.s., 4(1):1-66, 2 pls., 41 text figs., 2 tables.

Dinophysis hastata
Matzenauer, L., 1933
Die Dinoflagellaten des indischen Ozeans (mit Ausnahme der Gattung Ceratium.) Bot. Arch. 35:437-510, 77 text figs., 2 charts.

Dinophysis hastata
Murray, G., and F. G. Whitting, 1899
New Peridiniaceae from the Atlantic. Trans. Linn. Soc., London, Bot., ser 2, 5: 321-342, Pls. 27-33, 9 tables.

Dinophysis hastata
Norris, Dean R., and Leo D. Berner Jr. 1970.
Thecal morphology of selected species of Dinophysis (Dinoflagellata) from the Gulf of Mexico. Contrib. mar. Sci. Port Aransas 15:145-192

Dinophysis hastata
Paulsen, O., 1908
XVIII Peridiniales. Nordisches Plankton, Bot. Teil: 1-124, 155 text figs.

Dinophysis hastata
Pavillard, J., 1923
A propos de la systématique des Péridiniens Bull. Soc. Bot. de France 70:876-882; 914-918.

Dinophysis hastata
Pavillard, J., 1916
Recherches sur les Peridiniens du Golfe du Lion. Mem. Univ. Montpellier. Trav. Inst. Bot., Univ. Montpellier. Serie mixte No.4, 70 pp., 3 pls., 15 text figs.

Dinophysis hastata
Rampi, L. 1940
Ricerche sul Fitoplancton del mare Ligure. II Le tecatali e le dinofisiali delle acque di Sanremo. Boll. Pesca, Piscicolt., Idrobiol. (18) 16(2): 243-274, 56 figs.

Dinophysis hastata
Rampi, L., 1940
Ricerche sul Fitoplancton del mare Ligure. Boll. di Pesca, di Piscicoltura e di Idrobiologa. 18(2):1-34, 56 text figs.

Dinophysis hastata
Wood, E.J.P., 1954
Dinoflagellates in the Australian region. Australian J. Mar. Freshwater Res., 5(2):171-351.

Dinophysis hastata parvula n.var.
Lindemann, E., 1924
Peridineen aus dem goldenen Horn und dem Bosphorus. Bot. Arch. 5:216-233, 98 text figs.

Dinophysis homunculus
Bigelow, H. B., 1922
Exploration of the coastal water off the northeastern United States in 1916 by the U.S. Fisheries Schooner Grampus. Bull. M.C.Z. 65 (5):85-188, 53 text figs.

Dinophysis homunculus
Dangeard, P., 1927
Phytoplankton de la croisière du "Sylvana". Ann. Inst. Ocean., Monaco, n.s., 4(8):286-401, 54 text figs. (Feirer-Juin 1913).

Dinophysis homunculus
Forti, A., 1922
Ricerche sulla flora pelagica (fitoplancton) di Quarto dei Mille. Mem. R. Com. Talass. Ital. 97:248 pp., 13 pls.

Dinophysis homunculus
Jørgensen, E., 1905
B. Protistplankton and the diatoms in bottom samples. Hydrographical and biological investigations in Norwegian fjords. Bergens Mus. Skr. 7: 49-225.

Dinophysis homunculus
Lindemann, E., 1925
Neubeobachtungen an den Winter peridineen des Golfes von Neapel. Bot. Arch. 9:95-102, 19 text figs.

Dinophysis homunculus
Lindemann, E., 1924
Peridineen aus dem goldenen Horn und dem Bosphorus. Bot. Arch. 5:216-233, 98 text figs.

Dinophysis homunculus
Mangin, M. L., 1912
Phytoplancton de la croisière du "René" dans l'Atlantique (Septembre 1908). Ann. Inst. Ocean., n.s., 4(1):1-66, 2 pls., 41 text figs., 2 tables.

Dinophysis homunculus
Moberg, E.G., and W.E. Allen, 1927.
Effect of tidal changes on physical, chemical, and biological conditions in the sea water of the San Diego region. 1. Observations on the effect of tidal changes on physical and chemical conditions of sea water in the San Diego region. 2. Half-hourly collections of marine microplankton taken at the Scripps Institution pier in 1923. Bull. S.I.O., tech. ser., 1:1-17, 4 textfigs.

Dinophysis homunculus
Murray, G., and F. G. Whitting, 1899
New Peridiniaceae from the Atlantic. Trans. Linn. Soc., London, Bot., ser 2, 5: 321-342, Pls. 27-33, 9 tables.

Dinophysis homunculus
Paulsen, O., 1908
XVIII Peridiniales. Nordisches Plankton, Bot. Teil: 1-124, 155 text figs.

Dinophysis homunculus
Pavillard, J., 1923
A propos de la systématique des Péridiniens Bull. Soc. Bot. de France 70:876-882; 914-918.

Dinophysis homunculus
Pavillard, J., 1916
Recherches sur les Peridiniens du Golfe du Lion. Mem. Univ. Montpellier. Trav. Inst. Bot., Univ. Montpellier. Serie mixte No.4, 70 pp., 3 pls., 15 text figs.

Dinophysis homunculus
Pavillard, J., 1905
Recherches sur la flore pelagique (Phytoplankton) de l'Etang de Thau. Theses presentees a la Fac. Sci., Paris, 116 pp., 3 pls.

Dinophysis homunculus tripos
Marukawa, H., 1921
Plankton lists and some new species of copepods, from the northern waters of Japan. Bull. Inst. Ocean. No.384, 15 pp., 3 pls., 1 chart. Monaco

Dinophysis homunculus
Schröder, B., 1900
Phytoplankton des Golfes von Neapel nebst vergleichenden Ausblicken auf das atlantischen Ozean. Mitt. Zool. Stat. Neapel, 14:1-38.

Dinophysis homunculus tripos
Pavillard, J., 1905
Recherches sur la flore pelagique (Phytoplankton) de l'Etang de Thau. Theses presentees a la Fac. Sci., Paris, 116 pp., 3 pls.

Dinophysis homunculus ventricosa
Dangeard, P., 1927.
Péridiniens nouveaux ou peu connus de la croisière du "Sylvana". Bull. Inst. Ocean., Monaco, No. 491:16 pp., 9 textfigs.

Dinophysis homunculus ventricosa
Forti, A., 1922
Ricerche sulla flora pelagica (fitoplancton) di Quarto dei Mille. Mem. R. Com. Talass. Ital. 97:248 pp., 13 pls.

Dinophysis homunculus ventricosa
Pavillard, J., 1916
Recherches sur les Peridiniens du Golfe du Lion. Mem. Univ. Montpellier. Trav. Inst. Bot., Univ. Montpellier. Serie mixte No.4, 70 pp., 3 pls., 15 text figs.

Dinophysis inaequalis
Gourret, P., 1883
Sur les Peridiniens du Golfe de Marseille. Ann. du Musee d'hist. Nat., Marseille, Zool., I (Mme. 8):1-114, 4 pls.

Dinophysis infundibulus
Abe, Tohru H., 1967.
The armoured Dinoflagellata: II. Prorocentridae and Dinophysidae (B) Dinophysis and its allied genera. Publs. Seto mar. biol. Lab., 15(1):37-78.

Dinophysis intermedia
Dangeard, P., 1927
Phytoplankton de la croisière du "Sylvana". Ann. Inst. Ocean., Monaco, n.s., 4(8):286-401, 54 text figs. (Feirer-Juin 1913).

Dinophysis intermedia
Forti, A., 1922
Ricerche sulla flora pelagica (fitoplancton) di Quarto dei Mille. Mem. R. Com. Talass. Ital. 97:248 pp., 13 pls.

Dinophysis intermedia
Jörgensen, E., 1923
Mediterranean Dinophysiaceae. Rept. Danish Oceanogr. Expeds. 1908-10, to the Mediterranean and adjacent seas, Vol.II, Biol. J 2, 48 pp., 64 text figs.

Dinophysis intermedia
Pavillard, J., 1923
A propos de la systématique des Péridiniens Bull. Soc. Bot. de France 70:876-882; 914-918.

Dinophysis intermedia n.sp.
Pavillard, J., 1916
Recherches sur les Peridiniens du Golfe du Lion. Mem. Univ. Montpellier. Trav. Inst. Bot., Univ. Montpellier. Serie mixte No.4, 70 pp., 3 pls., 15 text figs.

Dinophysis intermedia
Sousa e Silva, E., 1949
Diatomaceas e Dinoflagelados de Baia de Cascais. Portugaliae Acta Biol., Volume: Julio Henriques, Ser. B: 300-383, 9 pls, 2 fold-in tables.

Dinophysis islandica n.sp.
Paulsen, O., 1949
Observations on dinoflagellates. (Ed. J. Grøntoed) Kongl. Dansk. Videnskab. Selsk., Biol. Skr. 6(4):67 pp., 30 text figs.

Dinophysis islandica
Solum, Ingrid, 1962.
The taxonomy of Dinophysis populations in Norwegian waters in view of biometric observations.
Nytt Mag. Botanikk, 10:5-32.

Dinophysis Jörgenseni
Käsler, R., 1938
Die Verbreitung der Dinophysiales im Sudatlantischen Ozean. Wiss. Ergeb. Deutschen Atlantischen Expedition----"Meteor" 1925-1927, 12(2):162-237, text figs. 85-118.

Dinophysis jörgenseni n.sp.
Kofoid, C. A. and T. Skogsberg, 1928
XXXV. The Dinoflagellata: The Dinophysiodae. Reports on the scientific results of the expedition to the Eastern Tropical Pacific, in charge of Alexander Agassiz, by the U. S. Fish Commission Steamer "Albatross" from October 1904 to March 1905----. Mem. M. C. Z. 51:766 pp., 31 pls.

Dinophysis Jourdani
Gourret, P., 1883
Sur les Peridiniens du Golfe de Marseille. Ann. du Musee d'hist. Nat., Marseille, Zool., 1 (Mme. 8):1-114, 4 pls.

Dinophysis kofoidii n.sp.
Jörgensen, E., 1923
Mediterranean Dinophysiaceae. Rept. Danish Oceanogr. Expeds. 1908-10, to the Mediterranean and adjacent seas, Vol.II, Biol. J 2, 48 pp., 64 text figs.

Dinophysis Kofoidi
Pavillard, J., 1923
A propos de la systématique des Peridiniens Bull. Soc. Bot. de France 70:876-882; 914-918.

Dinophysis lachmanni n.nom.
Paulsen, O., 1949
Observations on dinoflagellates. (Ed. J. Grøntoed) Kongl. Dansk. Videnskab. Selsk., Biol. Skr. 6(4):67 pp., 30 text figs.

Dinophysis lachmanni (= D. borealis)
Solum, Ingrid, 1962.
The taxonomy of Dinophysis populations in Norwegian waters in view of biometric observations.
Nytt Mag. Botanikk, 10:5-32.

Dinophysis lata n. sp.
Balech, Enrique, 1971.
Microplancton de la campaña oceanographica: Productividad III.
Revta Mus. argent. Cienc. Nat. Bernadina Rivadavia, Hydrobiol. 3(1):1-202, 39 pls.

Dinophysis lepidistrigiliformis n.sp.
Abe, Tohru H., 1967.
The armoured Dinoflagellata: II. Prorocentridae and Dinophysidae (B)-Dinophysis and its allied genera.
Publs. Seto mar. biol. Lab., 15(1):37-78.

Dinophysis lenticula
Abe, Tohru H., 1967.
The armoured Dinoflagellata: II. Prorocentridae and Dinophysidae (B) - Dinophysis and its allied genera.
Publs. Seto mar. biol. Lab., 15(1):37-78.

Dinophysis lenticula
Jörgensen, E., 1923
Mediterranean Dinophysiaceae. Rept. Danish Oceanogr. Expeds. 1908-10, to the Mediterranean and adjacent seas, Vol.II, Biol. J 2, 48 pp., 64 text figs.

Dinophysis lenticula
Käsler, R., 1938
Die Verbreitung der Dinophysiales im Sudatlantischen Ozean. Wiss. Ergeb. Deutschen Atlantischen Expedition----"Meteor" 1925-1927, 12(2):162-237, text figs. 85-118.

Dinophysis lenticula
LeBour, M.V., 1925
The dinoflagellates of Northern Seas. The Marine Biological Association of the United Kingdom, Plymouth, 250 pp., 35 pls., 53 text figs.

Dinophysis lenticula
Morse, D.C., 1947
Some observations on seasonal variations in plankton population Patuxent River, Maryland 1943-1945. Bd. Nat. Res., Publ. No.65, Chesapeake Biol. Lab., 31, 3 figs.

Dinophysis lenticula n.sp.
Pavillard, J., 1916
Recherches sur les Peridiniens du Golfe du Lion. Mem. Univ. Montpellier. Trav. Inst. Bot., Univ. Montpellier. Serie mixte No.4, 70 pp., 3 pls., 15 text figs.

Dinophysis lenticula
Sousa e Silva, E., 1949
Diatomaceas e Dinoflagelados de Baia de Cascais. Portugaliae Acta Biol., Volume: Julio Henriques, Ser. B: 300-383, 9 pls, 2 fold-in tables.

Dinophysis lindemanni
Käsler, R., 1938
Die Verbreitung der Dinophysiales im Sudatlantischen Ozean. Wiss. Ergeb. Deutschen Atlantischen Expedition----"Meteor" 1925-1927, 12(2):162-237, text figs. 85-118.

Dinophysis longi-alata n.sp.
Gran, H.H., and T. Braarud, 1935
A quantitative study of the phytoplankton in the Bay of Fundy and the Gulf of Maine (including observations on hydrography, chemistry, and turbidity). J. Biol. Bd., Canada, 1(5):279-467, 69 text figs.

Dinophysis longi-alata
Lillick, L.C., 1940
Phytoplankton and planktonic protozoa of the offshore waters of the Gulf of Maine. Pt.II. Qualitative Composition of the Planktonic Flora. Trans. Am. Phil. Soc., n.s., 31(3):193-237, 13 text figs.

Dinophysis mawsonii
Balech, Enrique, 1971.
Microplancton de la campaña oceanographica: Productividad III.
Revta Mus. argent. Cienc. Nat. Bernadina Rivadavia, Hydrobiol. 3(1):1-202, 39 pls.

Dinophysis meteori
Balech, Enrique, 1971.
Microplancton de la campaña oceanographica: Productividad III.
Revta Mus. argent. Cienc. Nat. Bernadina Rivadavia, Hydrobiol. 3(1):1-202, 39 pls.

Dinophysis meteori
Käsler, R., 1938
Die Verbreitung der Dinophysiales im Sudatlantischen Ozean. Wiss. Ergeb. Deutschen Atlantischen Expedition----"Meteor" 1925-1927, 12(2):162-237, text figs. 85-118.

Dinophysis micropleura n. sp.
Balech, Enrique, 1971.
Microplancton de la campaña oceanographica: Productividad III.
Revta Mus. argent. Cienc. Nat. Bernadina Rivadavia, Hydrobiol. 3(1):1-202, 39 pls.

Dinophysis micropterygia n.sp.
Dangeard, P., 1927.
Peridiens nouveaux ou peu connus de la croisière du "Sylvana". Bull. Inst. Ocean., Monaco, No. 491:16 pp., 9 textfigs.

Dinophysis micropterygia
Dangeard, P., 1927
Phytoplankton de la croisière du "Sylvana". Ann. Inst. Ocean., Monaco, n.s., 4(8):286-401, 54 text figs. (Janvier-Juin 1913).

Dinophysis micropterygia
Wood, E.J.F., 1963
Dinoflagellates in the Australian region. II. Recent collections.
Comm. Sci. and Industr. Res. Org., Div. Fish. and Oceanogr., Techn. Paper, No. 14: 55 pp.

Dinophysis microstrigiliformis n.sp.
Abe, Tohru H., 1967.
The armoured Dinoflagellata: II. Prorocentridae and Dino physidae (B)-Dinophysis and its allied genera.
Publs. Seto mar. biol. Lab., 15(1):37-78.

Dinophysis miles
Matzenauer, L., 1933
Die Dinoflagellaten des indischen Ozeans (mit Ausnahme der Gattung Ceratium.) Bot. Arch. 35:437-510, 77 text figs., 2 charts.

Dinophysis miles
Qasim, S.Z., P.M.A. Bhattathiri and V.P. Devassy, 1972
The effect of intensity and quality of illumination on the photosynthesis of some tropical marine phytoplankton. Mar. Biol. 16(1): 22-27.

Dinophysis miles
Wood, E.J.F., 1954.
Dinoflagellates in the Australian region. Australian J. Mar. Freshwater Res., 5(2):171-351.

Dinophysis miles arabica nf.
Matzenauer, L., 1933
Die Dinoflagellaten des indischen Ozeans (mit Ausnahme der Gattung Ceratium.) Bot. Arch. 35:437-510, 77 text figs., 2 charts.

Dinophysis miles indica
Matzenauer, L., 1933
Die Dinoflagellaten des indischen Ozeans (mit Ausnahme der Gattung Ceratium.) Bot. Arch. 35:437-510, 77 text figs., 2 charts.

Dinophysis miles maris ubri
Matzenauer, L., 1933
Die Dinoflagellaten des indischen Ozeans (mit Ausnahme der Gattung Ceratium.) Bot. Arch. 35:437-510, 77 text figs., 2 charts.

Dinophysis miles Schröderi
Matzenauer, L., 1933
Die Dinoflagellaten des indischen Ozeans (mit Ausnahme der Gattung Ceratium.) Bot. Arch. 35:437-510, 77 text figs., 2 charts.

Dinophysis miles tripsoidea nf.
Matzenauer, L., 1933
Die Dinoflagellaten des indischen Ozeans (mit Ausnahme der Gattung Ceratium.) Bot. Arch. 35:437-510, 77 text figs., 2 charts.

Dinophysis mitra
Abe, Tohru H., 1967.
The armoured Dinoflagellata: II. Prorocentridae and Dinophysidae (B)- Dinophysis and its allied genera.
Publs. Seto mar. biol. Lab., 15(1):37-78.

Dinophysis monoacantha
Kästler, R., 1938
Die Verbreitung der Dinophysiales im Sud-atlantischen Ozean. Wiss. Ergeb. Deutschen Atlantischen Expedition----"Meteor" 1925-1927, 12(2):162-237, text figs. 85-118.

Dinophysis monacantha n.sp.
Kofoid, C. A. and T. Skogsberg, 1928
XXXV. The Dinoflagellata: The Dinophysiodae. Reports on the scientific results of the expedition to the Eastern Tropical Pacific, in charge of Alexander Agassiz, by the U. S. Fish Commission Steamer "Albatross" from October 1904 to March 1905----. Mem. M. C. Z. 51:766 pp., 31 pls.

Dinophysis moresbyensis n.sp.
Wood, E.J.F., 1963
Dinoflagellates in the Australian region. II. Recent collections. Comm. Sci. and Industr. Res. Org., Div. Fish. and Oceanogr., Techn. Paper, No. 14: 55 pp.

Dinophysis nias
Kästler, R., 1938
Die Verbreitung der Dinophysiales im Sud-atlantischen Ozean. Wiss. Ergeb. Deutschen Atlantischen Expedition----"Meteor" 1925-1927, 12(2):162-237, text figs. 85-118.

Dinophysis nias
Kofoid, C. A. and T. Skogsberg, 1928
XXXV. The Dinoflagellata: The Dinophysiodae. Reports on the scientific results of the expedition to the Eastern Tropical Pacific, in charge of Alexander Agassiz, by the U. S. Fish Commission Steamer "Albatross" from October 1904 to March 1905----. Mem. M. C. Z. 51:766 pp., 31 pls.

Dinophysis norvegica
Braarud, T., 1945
A phytoplankton survey of the polluted waters of inner Oslo Fjord. Hvalrådets Skrifter, No.28, 142 pp., 19 text figs., 17 tables.

Dinophysis norvegica
Braarud, T., and Adam Bursa, 1939
On the phytoplankton of the Oslo Fjord, 1933-1934. Hvalrådets Skr. No.19:1-63; 9 text figs. Reviewed. J. du. Cons. 14(3): 418-420. A.C. Gardiner.

Dinophysis norvegica
Brunel, Jules, 1962
Le phytoplancton de la Baie des Chaleurs. Contrib. Ministère de la Chasse et des Pêcheries, Province de Québec, No. 91: 365 pp.

Dinophysis norvegica
Brunel, J., 1962
Le phytoplancton de la Baie de Chaleurs. Inst. Botan., Univ. Montréal, Contrib. No. 77: 365 pp., 66 pls.

Dinophysis norvegica crassior
Dinophysis norvegica debilior

Dinophysis norvegica
Dangeard, P., 1926
Description des Péridiniens Testacés recueillis para la Mission Charcot pendent le mois d'Aout 1924. Ann. Inst. Ocean. n.s. 3(7):307-334, 15 text figs.

Dinophysis norvegica
Gran, H.H. and T. Braarud, 1935
A quantitative study of the phytoplankton in the Bay of Fundy and the Gulf of Maine (including observations on hydrography, chemistry, and turbidity). J. Biol. Bd., Canada, 1(5):279-467, 69 text figs.

Dinophysis norvegica
Iselin, C., 1930
A report on the coastal waters of Labrador based on explorations of the "Chance" during the summer of 1926. Proc. Am. Acad. Arts Sci., 66(1):1-37, 14 text figs.

Dinophysis norvegica
Jørgensen, E., 1905
B.Protistplankton and the diatoms in bottom samples. Hydrographical and biological investigations in Norwegian fjords. Bergens Mus. Skr. 7: 49-225.

Dinophysis norvegica
Jorgensen, E., 1900
Protophyten und Protozoën im Plankton aus der Norwegischen Westkerste. Bergens Mus. Aarb. 1899(6): 95 pp., 5 pls., 83 tables.

Dinophysis norvegica
Kofoid, C. A. and T. Skogsberg, 1928
XXXV. The Dinoflagellata: The Dinophysiodae. Reports on the scientific results of the expedition to the Eastern Tropical Pacific, in charge of Alexander Agassiz, by the U. S. Fish Commission Steamer "Albatross" from October 1904 to March 1905----. Mem. M. C. Z. 51:766 pp., 31 pls.

Dinophysis norvegica
LeBour, M.V., 1925
The dinoflagellates of Northern Seas. The Marine Biological Association of the United Kingdom, Plymouth, 250 pp., 35 pls., 53 text figs.

Dinophysis norvegica
Lillick, L.C., 1940
Phytoplankton and planktonic protozoa of the offshore waters of the Gulf of Maine. Pt.II. Qualitative Composition of the Planktonic Flora. Trans. Am. Phil. Soc., n.s., 31(3):193-237, 13 text figs.

Dinophysis norvegica
Levander, K.M., 1947
Plankton gesammelt in den Jahren 1899-1910 an den Küsten Finnlands. Finnländische Hydrographisch-Biologische Untersuchungen (aus dem Wasserbiologischen Laboratorin der Societas Scientiarum Fennica) No.11: 40 pp., 6 diagrams, 13 pls., tables.

Dinophysis norvegica
Lillick, L.C., 1938
Preliminary report of the phytoplankton of the Gulf of Maine. Am. Mid. Nat. 20(3):624-640, 1 text figs 37 tables.

Dinophysis norvegica
Lillick, L.C., 1937
Seasonal studies of the phytoplankton off Woods Hole, Massachusetts. Biol. Bull. LXXIII (3):488-503, 3 text figs.

Dinophysis norvegica
Marukawa, H., 1921
Plankton lists and some new species of copepods, from the northern waters of Japan. Bull. Inst. Ocean., No.384, 15 pp., 3 pls., 1 chart. Monaco

Dinophysis norvegica
Paulsen, O., 1949
Observations on dinoflagellates. (Ed. J. Grøntoed) Kongl. Dansk. Videnskab. Selsk., Biol. Skr. 6(4):67 pp., 30 text figs.

Dinophysis norvegica
Paulsen, O., 1908
XVIII Peridiniales. Nordisches Plankton, Bot. Teil: 1-124, 155 text figs.

Dinophysis norvegica debilios
Rumkówna, A., 1948
List of the phytoplankton species occurring in the superficial water layers in the Gulf of Gdańsk. Bull. Lab. mar., Gdynia, No. 4: 139-141 with tables in back.

Dinophysis norvegica (= D. debilior)
Solum, Ingrid, 1962.
The taxonomy of Dinophysis populations in Norwegian waters in view of biometric observations. Nytt Mag. Botanikk, 10:5-32.

Dinophysis norvegica
Stemann-Nielsen, Einar, 1951
The marine vegetation of the Isefjord. A study on ecology and production. Medd. Komm. Danmarks Fiskeri-og Havundersøgelser. Ser. Plankton. 5(4); 114pp., 46 text figs.

Dinophysis odiosa
Balech, Enrique, 1962
Tintinnoinea y Dinoflagellata del Pacifico segun material de las Expediciones NORPAC y DOWNWIND del Instituto Scripps de Oceanografia. Revista. Mus. Argentino Ciencias Nat. "Bernardino Rivadavia", Ciencias Zool., 7(1):1-253.

Dinophysis odiosa
Norris, Dean R., and Leo D. Berner Jr. 1970.
Thecal morphology of selected species of Dinophysis (Dinoflagellata) from the Gulf of Mexico. Contrib. mar. Sci. Port Aransas 15:145-192

Dinophysis odiosa
Tai, Si-Sun, & T. Skogsberg, 1934
Studies on the Dinophysordae, marine armored dinoflagellates of Monterey Bay, California. Arch. Protistenk. 82:380-482, 14 text figs. Pls. 11-12.

Dinophysis okamurai
Abe, Tohru H., 1967.
The armoured Dinoflagellata: II. Prorocentridae and Dinophysidae (B): - Dinophysis and its allied genera. Publs. Seto mar. biol. Lab., 15(1):37-78.

Dinophysis okamurai
Balech, Enrique, 1971.
Microplancton de la campaña oceanographica: Productividad III. Revta Mus. argent. Cienc. Nat. Bernadina Rivadavia, Hydrobiol. 3(1):1-202, 39 pls.

Dinophysis okamurai n.sp.
Kofoid, C. A. and T. Skogsberg, 1928
XXXV. The Dinoflagellata: The Dinophysiodae. Reports on the scientific results of the expedition to the Eastern Tropical Pacific, in charge of Alexander Agassiz, by the U. S. Fish Commission Steamer "Albatross" from October 1904 to March 1905----. Mem. M. C. Z. 51:766 pp., 31 pls.

Dinophysis okamurai
Wood, E.J.F., 1954.
Dinoflagellates in the Australian region. Australian J. Mar. Freshwater Res., 5(2):171-351.

Dinophysis operculata
Balech, Enrique, 1971.
Microplancton de la campaña oceanographica: Productividad III. Revta Mus. argent. Cienc. Nat. Bernadina Rivadavia, Hydrobiol. 3(1):1-202, 39 pls.

Dinophysis ovatum
Marukawa, H., 1921
Plankton lists and some new species of copepods, from the northern waters of Japan. Bull. Inst. Ocean., No.384, 15 pp., 3 pls., 1 chart. Monaco

Dinophysis ovum
Abe, Tohru H., 1967.
The armoured Dinoflagellata: II. Prorocentridae and Dinophysidae (B) - Dinophysis and its allied genera. Publs. Seto mar. biol. Lab., 15(1):37-78.

Dinophysis ovum
Dangeard, P., 1927
Phytoplankton de la croisière du "Sylvana". Ann. Inst. Ocean., Monaco, n.s., 4(8):286-401, 54 text figs. (Janvier-Juin 1913).

Dinophysis ovum

Dangeard, P., 1926
Description des Péridiniens Testacés recueillis para la Mission Charcot pendent le mois d'Aout 1924. Ann. Inst. Ocean. n.s. 3(7):307-334, 15 text figs.

Dinophysis ovum

Ercegovic, A., 1940
Weitere Untersuchungen über einige hydrographische Verhältnisse und über die Phytoplanktonproduktion in den Gewässern der Östlichen Mitteladria. Acta Adriatica 2(3):95-134, 8 text figs.

Dinophysis ovum

Forti, A., 1922
Ricerche sulla flora pelagica (fitoplancton) di Quarto dei Mille. Mem. R. Com. Talass. Ital. 97:248 pp., 13 pls.

Dinophysis ovum

Gran, H.H., and T. Braarud, 1935
A quantitative study of the phytoplankton in the Bay of Fundy and the Gulf of Maine (including observations on hydrography, chemistry, and turbidity). J. Biol. Bd., Canada, 1(5):279-467, 69 text figs.

Dinophysis ovum

Jörgensen, E., 1923
Mediterranean Dinophysiaceae. Rept. Danish Oceanogr. Expeds. 1908-10, to the Mediterranean and adjacent seas, Vol.II, Biol. J 2, 48 pp., 64 text figs.

Dinophysis ovum

Käsler, R., 1938
Die Verbreitung der Dinophysiales im Sudatlantischen Ozean. Wiss. Ergeb. Deutschen Atlantischen Expedition----"Meteor" 1925-1927, 12(2):162-237, text figs. 85-118.

Dinophysis ovum

LeBour, M.V., 1925
The dinoflagellates of Northern Seas. The Marine Biological Association of the United Kingdom, Plymouth, 250 pp., 35 pls., 53 text figs.

Dinophysis ovum

Lillick, L.C., 1940
Phytoplankton and planktonic protozoa of the offshore waters of the Gulf of Maine. Pt.II. Qualitative Composition of the Planktonic Flora. Trans. Am. Phil. Soc., n.s., 31(3):193-237, 13 text figs.

Dinophysis ovum

Lillick, L.C., 1938
Preliminary report of the phytoplankton of the Gulf of Maine. Am. Mid. Nat. 20(3):624-640, 1 text figs 37 tables.

Dinophysis ovum

Lillick, L.C., 1937
Seasonal studies of the phytoplankton off Woods Hole, Massachusetts. Biol. Bull. LXXIII (3):488-503, 3 text figs.

Dinophysis ovum

Mangin, M. L., 1912
Phytoplancton de la croisière du "René" dans l'Atlantique (Septembre 1908). Ann. Inst. Ocean., n.s., 4(1):1-66, 2 pls., 41 text figs., 2 tables.

Dinophysis ovum

Morse, D.C., 1947
Some observations on seasonal variations in plankton population Patuxent River, Maryland 1943-1945. Bd. Nat. Res., Publ. No.65, Chesapeake Biol. Lab., 31, 3 figs.

Dinophysis ovum

Paulsen, O., 1908
XVIII Peridiniales. Nordisches Plankton Bot. Teil: 1-124, 155 text figs.

Dinophysis ovum

Pavillard, J., 1916
Recherches sur les Peridiniens du Golfe du Lion. Mem. Univ. Montpellier. Trav. Inst. Bot., Univ. Montpellier. Serie mixte No.4, 70 pp., 3 pls., 15 text figs.

Dinophysis ovum

Pavillard, J., 1905
Recherches sur la flore pelagique (Phytoplankton) de l'Etang de Thau. Theses presentees a la Fac. Sci., Paris, 116 pp., 3 pls.

Dinophysis ovum

Rampi, L. 1940
Ricerche sul Fitoplancton del mare Ligure. II Le tecatali e le dinofisiali delle acque di Sanremo. Boll. Pesca, Piscicolt.,Idrobiol. (18) 16(2): 243-274, 56 figs.

Dinophysis ovum

Rampi, L., 1940
Ricerche sul Fitoplancton del mare Ligure. Boll. di Pesca, di Piscicoltura e di Idrobiologa, 18(2):1-34, 56 text figs.

Dinophysis ovum

Sousa e Silva, E., 1949
Diatomaceas e Dinoflagelados de Baia de Cascais. Portugaliae Acta Biol., Volume: Julio Henriques, Ser. B: 300-383, 9 pls, 2 fold-in tables.

Dinophysis ovum

Wood, E.J.F., 1954.
Dinoflagellates in the Australian region. Australian J. Mar. Freshwater Res., 5(2):171-351.

Dinophysis pacifica n.sp.

Wood, E.J.F., 1963
Dinoflagellates in the Australian region. II. Recent collections. Comm. Sci. and Industr. Res. Org., Div. Fish. and Oceanogr., Techn. Paper, No. 14: 55 pp.

Dinophysis parva

Rampi, L. 1940
Ricerche sul Fitoplancton del mare Ligure. II Le tecatali e le dinofisiali delle acque di Sanremo. Boll. Pesca, Piscicolt.,Idrobiol. (18) 16(2): 243-274, 56 figs.

Dinophysis parva

Rampi, L., 1940
Ricerche sul Fitoplancton del mare Ligure. Boll. di Pesca, di Piscicoltura e di Idrobiologa, 18(2):1-34, 56 text figs.

Dinophysis parva

Wood, E.J.F., 1963
Dinoflagellates in the Australian region. II. Recent collections. Comm. Sci. and Industr. Res. Org., Div. Fish. and Oceanogr., Techn. Paper, No. 14: 55 pp.

Dinophysis parvula

Balech, Enrique, 1971.
Microplancton de la campaña oceanographica: Productividad III. Revta Mus. argent. Cienc. Nat. Bernardina Rivadavia, Hydrobiol. 3(1):1-202, 39 pls.

Dinophysis paulseni

Balech,Enrique,1967.
Dinoflagelados nuevos o interesantes del Golfo de Mexico y Caribe. Revta Mus.argent. Cienc.nat.Bernardino Rivadavia Inst.nac.Invest.Cienc.nat., Hidrobiol. 2(3):77-126.

Dinophysis paulseni

Norris, Dean R., and Leo D. Berner Jr. 1970.
Thecal morphology of selected species of Dinophysis (Dinoflagellata) from the Gulf of Mexico. Contrib. mar. Sci. Port Aransas 15:145-192

Dinophysis Pavillardii

Pavillard, J., 1923
A propos de la systématique des Péridiniens Bull. Soc. Bot. de France 70:876-882; 914-918.

Dinophysis Pavillardi

Pavillard, J., 1916
Recherches sur les Peridiniens du Golfe du Lion. Mem. Univ. Montpellier. Trav. Inst. Bot., Univ. Montpellier. Serie mixte No.4, 70 pp., 3 pls., 15 text figs.

Dinophysis porodictyum

Abe,Tohru H.,1967.
The armoured Dinoflagellata: II. Prorocentridae and Dinophysidae (B)- Dinophysis and its allied genera. Publs. Seto mar. biol. Lab., 15(1):37-78.

Dinophysis punctata

Balech, Enrique, 1971.
Microplancton de la campaña oceanographica: Productividad III. Revta Mus. argent. Cienc. Nat. Bernardina Rivadavia, Hydrobiol. 3(1):1-202, 39 pls.

Dinophysis punctata n.sp.

Jörgensen, E., 1923
Mediterranean Dinophysiaceae. Rept. Danish Oceanogr. Expeds. 1908-10, to the Mediterranean and adjacent seas, Vol.II, Biol. J 2, 48 pp., 64 text figs.

Dinophysis punctata

LeBour, M.V., 1925
The dinoflagellates of Northern Seas. The Marine Biological Association of the United Kingdom, Plymouth, 250 pp., 35 pls., 53 text figs.

Dinophysis punctata

Matzenauer, L., 1933
Die Dinoflagellaten des indischen Ozeans (mit Ausnahme der Gattung Ceratium.) Bot. Arch. 35:437-510, 77 text figs., 2 charts.

Dinophysis punctata

Rampi, L. 1940
Ricerche sul Fitoplancton del mare Ligure. II Le tecatali e le dinofisiali delle acque di Sanremo. Boll. Pesca, Piscicolt.,Idrobiol. (18) 16(2): 243-274, 56 figs.

Dinophysis punctata

Rampi, L., 1940
Ricerche sul Fitoplancton del mare Ligure. Boll. di Pesca, di Piscicoltura e di Idrobiologa, 18(2):1-34, 56 text figs.

Dinophysis punctata

Rampi, L., 1939
Peridiniens rares ou interessants recoltes dans la mer Ligure. Bull. Soc. Fran. de Microscopie, 8(2/3):106-112, 13 text figs.

Dinophysis pusilla

Balech,Enrique,1967.
Dinoflagelados nuevos o interesantes del Golfo de Mexico y Caribe. Revta Mus.argent.Cienc.nat.Bernardino Rivadavia Inst.nac.Invest.Cienc. nat., Hidrobiol. 2(3):77-126.

Dinophysis pusilla n.sp.

Jörgensen, E., 1923
Mediterranean Dinophysiaceae. Rept. Danish Oceanogr. Expeds. 1908-10, to the Mediterranean and adjacent seas, Vol.II, Biol. J 2, 48 pp., 64 text figs.

Dinophysis pusilla

Norris, Dean R., and Leo D. Berner Jr. 1970.
Thecal morphology of selected species of Dinophysis (Dinoflagellata) from the Gulf of Mexico. Contrib. mar. Sci. Port Aransas 15:145-192

Dinophysis rapa

Abe, Tohru H., 1967.
The armoured Dinoflagellata: II. Prorocentridae and Dinophysidae (B)- Dinophysis and its allied genera.
Publs. Seto mar. biol. Lab., 15(1):37-78.

Dinophysis recurva

Ercegovic, A., 1936
Etudes qualitative et quantitatives du phytoplancton dans les eaux cotières de l'Adriatique oriental moyen au cours de l'année 1934. Acta Adriatica 1(9):1-126

Dinophysis recurva

Rampi, L., 1940
Ricerche sul Fitoplancton del mare Ligure. II Le tecatali e le dinofisiali delle acque di Sanremo. Boll. Pesca, Piscicolt., Idrobiol. (18) 16(2): 243-274, 56 figs.

Dinophysis recurva

Rampi, L., 1940
Ricerche sul Fitoplancton del mare Ligure. Boll. di Pesca, di Piscicoltura e di Idrobiologa, 18(2):1-34, 56 text figs.

Dinophysis recurva

Wood, E.J.F., 1963
Dinoflagellates in the Australian region. II. Recent collections.
Comm. Sci. and Industr. Res. Org., Div. Fish. and Oceanogr., Techn. Paper, No. 14: 55 pp.

Dinophysis robusta n.sp.

Gran, H.H., and T. Braarud, 1935
A quantitative study of the phytoplankton in the Bay of Fundy and the Gulf of Maine (including observations on hydrography, chemistry, and turbidity).
J. Biol. Bd., Canada, 1(5):279-467, 69 text figs.

Dinophysis robusta

Lillick, L.C., 1940
Phytoplankton and planktonic protozoa of the offshore waters of the Gulf of Maine. Pt.II, Qualitative Composition of the Planktonic Flora. Trans. Am. Phil. Soc., n.s., 31(3):193-237, 13 text figs.

Dinophysis robusta

Lillick, L.C., 1938
Preliminary report of the phytoplankton of the Gulf of Maine. Am. Mid. Nat. 20(3):624-640, 1 text fig, 37 tables.

Dinophysis rotundata

Abe, Tohru H., 1967.
The armoured Dinoflagellata: II. Prorocentridae and Dinophysidae (B)- Dinophysis and its allied genera.
Publs. Seto mar. biol. Lab., 15(1):37-78.

Dinophysis rotundata

Dangeard, P., 1926
Description des Péridiniens Testacés recueillis para la Mission Charcot pendent le mois d'Aout 1924. Ann. Inst. Ocean. n.s. 3(7):307-334, 15 text figs.

Dinophysis rotundata

Jørgensen, E., 1905
B. Protistplankton and the diatoms in bottom samples. Hydrographical and biological investigations in Norwegian fjords. Bergens Mus. Skr. 7: 49-225.

Dinophysis rotundata

Jorgensen, E., 1900
Protophyten und Protozoën im Plankton aus der Norwegischen Westkerste. Bergens Mus. Aarb. 1899(6): 95 pp., 5 pls., 83 tables.

Dinophysis rotundata

Levander, K.M., 1947
Plankton gesammelt in den Jahren 1899-1910 an den Küsten Finnlands. Finnländische Hydrographisch-Biologische Untersuchungen (aus den Wasserbiologischen Laboratorin der Societas Scientiarum Fennica) No.11: 40 pp., 6 diagrams, 13 pls., tables.

Dinophysis rotundata

Lindemann, E., 1924
Peridineen aus dem goldenen Horn und dem Bosphorus. Bot. Arch. 5:216-233, 98 text figs.

Dinophysis rotundata

Mangin, M. L., 1912
Phytoplancton de la croisière du "René" dans l'Atlantique (Septembre 1908). Ann. Inst. Ocean., n.s., 4(1):1-66, 2 pls., 41 text figs., 2 tables.

Dinophysis rotundata

Meunier, A., 1919
Microplancton de la Mer Flamande 3. Les Péridiniens. Mem. Mus. Roy. Hist. Nat. Belgique 8(1):1-116, Pls. XV-XXI.

Dinophysis rotundata

Murray, G., and F. G. Whitting, 1899
New Peridiniaceae from the Atlantic. Trans. Linn. Soc., London, Bot., ser 2, 5: 321-342, Pls. 27-33, 9 tables.

Dinophysis rotundata

Paulsen, O., 1908
XVIII Peridiniales. Nordisches Plankton, Bot. Teil: 1-124, 155 text figs.

Dinophysis rotundata

Pavillard, J., 1916
Recherches sur les Peridiniens du Golfe du Lion. Mem. Univ. Montpellier. Trav. Inst. Bot., Univ. Montpellier. Serie mixte No.4, 70 pp., 3 pls., 15 text figs.

Dinophysis rotundata

Pettersson, H., 1934
Scattering and extinction of light in sea-water. Meddelanden från Göteborgs Högskolas Oceanografiska Institut: 9 (Göteborgs Kungl. vetenskaps-och vitterhetssamhälles Handlingar. Femte Foljden, Ser.B, 4(4)):1-16.

Dinophysis rotundata

Schodduyn, M., 1926
Observations faites dans la baie d'Ambleteuse (Pas de Calais). Bull. Inst. Ocean., Monaco, No. 482: 64 pp.

Dinophysis rotundata

Schröder, B., 1900
Phytoplankton des Golfes von Neapel nebst vergleichenden Ausblicken auf das atlantischen Ozean. Mitt. Zool. Stat. Neapel, 14:1-38.

Dinophysis rotundata

Tai, Si-Sun, & T. Skogsberg, 1934
Studies on the Dinophysordae, marine armored dinoflagellates of Monterey Bay, California. Arch. Protistenk. 82:380-482, 14 text figs. Pls. 11-12.

Dinophysis rudgei

Abe, Tohru H., 1967.
The armoured Dinoflagellata: II. Prorocentridae and Dinophysidae (B)- Dinophysis and its allied genera.
Publs. Seto mar.biol. Lab., 15(1):37-78.

Dinophysis Rudgei n.sp.

Murray, G., and F. G. Whitting, 1899
New Peridiniaceae from the Atlantic. Trans. Linn. Soc., London, Bot., ser 2, 5: 321-342, Pls. 27-33, 9 tables.

Dinophysis sacculus

Dangeard, P., 1927
Phytoplankton de la croisière du "Sylvana". Ann. Inst. Ocean., Monaco, n.s., 4(8):286-401, 54 text figs. (Feirer-Juin 1913).

Dinophysis sacculus

Dangeard, P., 1926
Description des Péridiniens Testacés recueillis para la Mission Charcot pendent le mois d'Aout 1924. Ann. Inst. Ocean. n.s. 3(7):307-334, 15 text figs.

Dinophysis sacculus

Ercegovic, A., 1936
Etudes qualitative et quantitatives du phytoplancton dans les eaux cotières de l'Adriatique oriental moyen au cours de l'année 1934. Acta Adriatica 1(9):1-126

Dinophysis sacculus

Forti, A., 1922
Ricerche sulla flora pelagica (fitoplancton) di Quarto dei Mille. Mem. R. Com. Talass. Ital. 97:248 pp., 13 pls.

Dinophysis sacculus

Jörgensen, E., 1923
Mediterranean Dinophysiaceae. Rept. Danish Oceanogr. Expeds. 1908-10, to the Mediterranean and adjacent seas, Vol.II, Biol. J 2, 48 pp., 64 text figs.

Dinophysis sacculus

Käsler, R., 1938
Die Verbreitung der Dinophysiales im Sudatlantischen Ozean. Wiss. Ergeb. Deutschen Atlantischen Expedition---- "Meteor" 1925-1927, 12(2):162-237, text figs. 85-118.

Dinophysis sacculus

Margalef, R., 1949
Fitoplancton nerítico de la Costa Brava en 1947-48. Publ. Inst. Biol. Aplicada, 5: 41-51, 3 text figs.

Dinophysis sacculus

Murray, G., and F. G. Whitting, 1899
New Peridiniaceae from the Atlantic. Trans. Linn. Soc., London, Bot., ser 2, 5: 321-342, Pls. 27-33, 9 tables.

Dinophysis sacculus

Pavillard, J., 1923
A propos de la systématique des Péridiniens Bull. Soc. Bot. de France 70:876-882; 914-918.

Dinophysis sacculus

Pavillard, J., 1916
Recherches sur les Peridiniens du Golfe du Lion. Mem. Univ. Montpellier. Trav. Inst. Bot., Univ. Montpellier. Serie mixte No.4, 70 pp., 3 pls., 15 text figs.

Dinophysis sacculus

Rampi, L., 1940
Ricerche sul Fitoplancton del mare Ligure. II Le tecatali e le dinofisiali delle acque di Sanremo. Boll. Pesca, Piscicolt., Idrobiol. (18) 16(2): 243-274, 56 figs.

Dinophysis sacculus

Rampi, L., 1940
Ricerche sul Fitoplancton del mare Ligure. Boll. di Pesca, di Piscicoltura e di Idrobiologa, 18(2):1-34, 56 text figs.

Dinophysis sacculus

Schröder, B., 1900
Phytoplankton des Golfes von Neapel nebst vergleichenden Ausblicken auf das atlantischen Ozean. Mitt. Zool. Stat. Neapel, 14:1-38.

Binophysis sacculus

Sousa e Silva, E., 1949
Diatomaceas e Dinoflagelados de Baia de Cascais. Portugaliae Acta Biol., Volume: Julio Henriques, Ser. B: 300-383, 9 pls, 2 fold-in tables.

Dinophysis sacculus

Sousa e Silva, E., and J. Dos Santos-Pinto, 1948
O Plancton da Baia de S. Martinho do Porto. 1. Diatomaceas e Dinoflagelados. Bol. Soc. Portuguese de Ciencias Naturais, 16(2):134-187, 6 pls. (Trav. Sta. Biol. Mar. de Lisbonne No. 52).

Dinophysis sacculus

Wood, E.J.F., 1954.
Dinoflagellates in the Australian region.
Australian J. Mar. Freshwater Res., 5(2):171-351.

Dinoshysis schroederi

Balech, Enrique, 1971.
Microplancton de la campaña oceanographica:
Productividad III.
Revta Mus. argent. Cienc. Nat. Bernardina
Rivadavia, Hydrobiol. 3(1):1-202, 39 pls.

Dinophysis Schröderi

Delegazione Italiana della Commissione
Internazionale per l'Esplorazione Scientifica del Mediterraneo, 1941
Note sul plancton della Laguna veneta.
[Memoria CCLXXXIX], Arch. di Ocean. e Limn.
Anno I, Fasc. I, 1941 XIX: 31-57 pp.

Dinophysis Schröderi

Ercegovic, A., 1936
Etudes qualitative et quantitatives du phytoplancton dans les eaux cotières de l'Adriatique oriental moyen au cours de l'année 1934. Acta Adriatica 1(9):1-126

Dinophysis Schroederi

Forti, A., 1922
Ricerche sulla flora pelagica (fitoplancton) di Quarto dei Mille. Mem. R. Com. Talass. Ital. 97:248 pp., 13 pls.

Dinophysis schroederi

Jörgensen, E., 1923
Mediterranean Dinophysiaceae. Rept.
Danish Oceanogr. Expeds. 1908-10, to the Mediterranean and adjacent seas, Vol.II, Biol. J 2, 48 pp., 64 text figs.

Dinophysis Schröderi

Käsler, R., 1938
Die Verbreitung der Dinophysiales im Sud-atlantischen Ozean. Wiss. Ergeb. Deutschen Atlantischen Expedition----"Meteor" 1925-1927, 12(2):162-237, text figs. 85-118.

Dinophysis schröderi

Kofoid, C. A. and T. Skogsberg, 1928
XXXV. The Dinoflagellata: The Dinophysiodae. Reports on the scientific results of the expedition to the Eastern Tropical Pacific, in charge of Alexander Agassiz, by the U. S. Fish Commission Steamer "Albatross" from October 1904 to March 1905----. Mem. M. C. Z. 51:766 pp., 31 pls.

Dinophysis Schroederi

Margalef, R., 1949
Fitoplancton nerítico de la Costa Brava en 1947-48. Publ. Inst. Biol. Aplicada, 5: 41-51, 3 text figs.

Dinophysis schuetti

Norris, Dean R., and Leo D. Berner Jr.
1970.
Thecal morphology of selected species of Dinophysis (Dinoflagellata) from the Gulf of Mexico.
Contrib. mar. Sci. Port Aransas 15:145-192

Dinophysis Schröderi

Pavillard, J., 1916
Recherches sur les Peridiniens du Golfe du Lion. Mem. Univ. Montpellier. Trav. Inst. Bot., Univ. Montpellier. Serie mixte No.4, 70 pp., 3 pls., 15 text figs.

Dinophysis schröderi

Rampi, L. 1940.
Ricerche sul Fitoplancton del mare Ligure. II Le tecatali e le dinofisiali delle acque di Sanremo. Boll. Pesca, Piscicolt.,Idrobiol. (18) 16(2): 243-274, 56 figs.

Dinophysis Schröderi

Rampi, L., 1940
Ricerche sul Fitoplancton del mare Ligure. Boll. di Pesca, di Piscicoltura e di Idrobiologa. 18(2):1-34, 56 text figs.

Dinophysis schroederi

Sousa e Silva, E., 1949
Diatomaceas e Dinoflagelados de Baia de Cascais. Portugaliae Acta Biol., Volume: Julio Henriques, Ser. B: 300-383, 9 pls, 2 fold-in tables.

Dinophysis schroederi

Wood, E.J.F., 1954.
Dinoflagellates in the Australian region.
Australian J. Mar. Freshwater Res., 5(2):171-351.

Dinophysis Schuthii

Dangeard, P., 1927
Phytoplankton de la croisière du "Sylvana". Ann. Inst. Ocean., Monaco, n.s., 4(8):286-401, 54 text figs. (Feirer-Juin 1913).

Dinophysis schuetti

Jörgensen, E., 1923
Mediterranean Dinophysiaceae. Rept.
Danish Oceanogr. Expeds. 1908-10, to the Mediterranean and adjacent seas, Vol.II, Biol. J 2, 48 pp., 64 text figs.

Dinophysis Schüttü

Käsler, R., 1938
Die Verbreitung der Dinophysiales im Sud-atlantischen Ozean. Wiss. Ergeb. Deutschen Atlantischen Expedition----"Meteor" 1925-1927, 12(2):162-237, text figs. 85-118.

Dinophysis schütti

Kofoid, C. A. and T. Skogsberg, 1928
XXXV. The Dinoflagellata: The Dinophysiodae. Reports on the scientific results of the expedition to the Eastern Tropical Pacific, in charge of Alexander Agassiz, by the U. S. Fish Commission Steamer "Albatross" from October 1904 to March 1905----. Mem. M. C. Z. 51:766 pp., 31 pls.

Dinophysis shuettii

LeBour, M.V., 1925
The dinoflagellates of Northern Seas. The Marine Biological Association of the United Kingdom, Plymouth, 250 pp., 35 pls, 53 text figs.

Dinophysis Schuettii

Murray, G., and F. G. Whitting, 1899
New Peridiniaceae from the Atlantic.
Trans. Linn. Soc., London, Bot., ser 2, 5: 321-342, Pls. 27-33, 9 tables.

Dinophysis Schutti

Paulsen, O., 1908
XVIII Peridiniales. Nordisches Plankton.
Bot. Teil: 1-124, 155 text figs.

Dinophysis Schüttii

Pavillard, J., 1916
Recherches sur les Peridiniens du Golfe du Lion. Mem. Univ. Montpellier. Trav. Inst. Bot., Univ. Montpellier. Serie mixte No.4, 70 pp., 3 pls., 15 text figs.

Dinophysis schüttii

Rampi, L. 1940
Ricerche sul Fitoplancton del mare Ligure. II Le tecatali e le dinofisiali delle acque di Sanremo. Boll. Pesca, Piscicolt.,Idrobiol. (18) 16(2): 243-274, 56 figs.

Dinophysis Schüttii

Rampi, L., 1940
Ricerche sul Fitoplancton del mare Ligure. Boll. di Pesca, di Piscicoltura e di Idrobiologa. 18(2):1-34, 56 text figs.

Dinophysis schuetti

Wood, E.J.F., 1963
Dinoflagellates in the Australian region.
II. Recent collections. Comm. Sci. and Industr. Res. Org., Div. Fish. and Oceanogr., Techn. Paper, No. 14: 55 pp.

Dinophysis scrobiculata n. sp.

Balech, Enrique, 1971.
Microplancton de la campaña oceanographica:
Productividad III.
Revta Mus. argent. Cienc. Nat. Bernardina
Rivadavia, Hydrobiol. 3(1):1-202, 39 pls.

Dinophysis similis

Käsler, R., 1938
Die Verbreitung der Dinophysiales im Sud-atlantischen Ozean. Wiss. Ergeb. Deutschen Atlantischen Expedition----"Meteor" 1925-1927, 12(2):162-237, text figs. 85-118.

Dinophysis similis n.sp.

Kofoid, C. A. and T. Skogsberg, 1928
XXXV. The Dinoflagellata: The Dinophysiodae. Reports on the scientific results of the expedition to the Eastern Tropical Pacific, in charge of Alexander Agassiz, by the U. S. Fish Commission Steamer "Albatross" from October 1904 to March 1905----. Mem. M. C. Z. 51:766 pp., 31 pls.

Dinophysis similis

Matzenauer, L., 1933
Die Dinoflagellaten des indischen Ozeans (mit Ausnahme der Gattung Ceratium.) Bot. Arch. 35:437-510, 77 text figs., 2 charts.

Dinophysis similis

Wood, E.J.F., 1954.
Dinoflagellates in the Australian region.
Australian J. Mar. Freshwater Res., 5(2):171-351.

Dinophysis simplex

Balech, Enrique, 1971.
Microplancton de la campaña oceanographica:
Productividad III.
Revta Mus. argent. Cienc. Nat. Bernardina
Rivadavia, Hydrobiol. 3(1):1-202, 39 pls.

Dinophysis simplex n.sp.

Balech, Enrique, 1962
Tintinnoinea y Dinoflagellata del Pacifico segun material de las Expediciones NORPAC y DOWNWIND del Instituto Scripps de Oceanografia.
Revista. Mus. Argentino Ciencias Nat. "Bernardino Rivadavia", Ciencias Zool., 7(1):1-253.

Dinophysis simplex

Käsler, R., 1938
Die Verbreitung der Dinophysiales im Sud-atlantischen Ozean. Wiss. Ergeb. Deutschen Atlantischen Expedition----"Meteor" 1925-1927, 12(2):162-237, text figs. 85-118.

Dinophysis skagii

Brunel, J., 1962
Le phytoplancton de la Baie de Chaleurs.
Inst. Botan., Univ. Montréal, Contrib. No. 77: 365 pp., 66 pls.

Dinophysis Skagi n.sp.

Paulsen, O., 1949
Observations on dinoflagellates. (Ed. J. Grøntoed) Kongl. Dansk. Videnskab. Selsk., Biol. Skr. 6(4):67 pp., 30 text figs.

Dinophysis sp. cf. D. sphaerica

Balech, Enrique, 1971
Microplancton del Atlantico ecuatorial oeste (Equalant 1)
Publ. Serv. Hidrograf. Naval, Argentina H. 654: 103 pp., 122 figs.

Dinophysis sphaerica

Gran, H.H., and T. Braarud, 1935
A quantitative study of the phytoplankton in the Bay of Fundy and the Gulf of Maine (including observations on hydrography, chemistry, and turbidity). J. Biol. Bd., Canada, 1(5):279-467, 69 text figs.

Dinophysis sphaerica
Jörgensen, E., 1923
Mediterranean Dinophysiaceae. Rept. Danish Oceanogr. Expeds. 1908-10, to the Mediterranean and adjacent seas, Vol.II, Biol. J 2, 48 pp., 64 text figs.

Dinophysis sphaerica
Käsler, R., 1938
Die Verbreitung der Dinophysiales im Sud-atlantischen Ozean. Wiss. Ergeb. Deutschen Atlantischen Expedition----"Meteor" 1925-1927, 12(2):162-237, text figs. 85-118.

Dinophysis sphaerica
Kofoid, C. A. and T. Skogsberg, 1928
XXXV. The Dinoflagellata: The Dinophysiodae. Reports on the scientific results of the expedition to the Eastern Tropical Pacific, in charge of Alexander Agassiz, by the U. S. Fish Commission Steamer "Albatross" from October 1904 to March 1905----. Mem. M. C. Z. 51:766 pp., 31 pls.

Dinophysis sphaerica
LeBour, M.V., 1925
The dinoflagellates of Northern Seas. The Marine Biological Association of the United Kingdom, Plymouth, 250 pp., 35 pls., 53 text figs.

Dinophysis sphaerica
Lillick, L.C., 1940
Phytoplankton and planktonic protozoa of the offshore waters of the Gulf of Maine. Pt.II. Qualitative Composition of the Planktonic Flora. Trans. Am. Phil. Soc., n.s., 31(3):193-237, 13 text figs.

Dinophysis sphaerica
Lillick, L.C., 1938
Preliminary report of the phytoplankton of the Gulf of Maine. Am. Mid. Nat. 20(3):624-640, 1 text figs 37 tables.

Dinophysis sphaerica
Matzenauer, L., 1933
Die Dinoflagellaten des indischen Ozeans (mit Ausnahme der Gattung Ceratium). Bot. Arch. 35:437-510, 77 text figs., 2 charts.

Dinophysis sphaerica
Murray, G., and F. G. Whitting, 1899
New Peridiniaceae from the Atlantic. Trans. Linn. Soc., London, Bot., ser 2, 5: 321-342, Pls. 27-33, 9 tables.

Dinophysis sphaerica
Paulsen, O., 1908
XVIII Peridiniales. Nordisches Plankton, Bot. Teil: 1-124, 155 text figs.

Dinophysis sphaerica
Pavillard, J., 1916
Recherches sur les Peridiniens du Golfe du Lion. Mem. Univ. Montpellier. Trav. Inst. Bot., Univ. Montpellier. Serie mixte No.4, 70 pp., 3 pls., 15 text figs.

Dinophysis sphaerica
Schröder, B., 1900
Phytoplankton des Golfes von Neapel nebst vergleichenden Ausblicken auf das atlantische Ozean. Mitt. Zool. Stat. Neapel, 14:1-38.

Dinophysis sphaerica
Wood, E.J.F., 1954
Dinoflagellates in the Australian region. Australian J. Mar. Freshwater Res., 5(2):171-351.

Dinophysis spinosa n.sp.
Rampi, L., 1950.
Péridiniens rares ou nouveaux pour le Pacifique Sud-Equatorial. Bull. Inst. Océan., Monaco, No. 974:11 pp., 26 textfigs.

Dinophysis subcircularis n.nom.
Paulsen, O., 1949
Observations on dinoflagellates. (Ed. J. Grøntoed) Kongl. Dansk. Videnskab. Selsk., Biol. Skr. 6(4):67 pp., 30 text figs.

Dinophysis swezyi
Käsler, R., 1938
Die Verbreitung der Dinophysiales im Sud-atlantischen Ozean. Wiss. Ergeb. Deutschen Atlantischen Expedition----"Meteor" 1925-1927, 12(2):162-237, text figs. 85-118.

Dinophysis swezyi n.sp.
Kofoid, C. A. and T. Skogsberg, 1928
XXXV. The Dinoflagellata: The Dinophysiodae. Reports on the scientific results of the expedition to the Eastern Tropical Pacific, in charge of Alexander Agassiz, by the U. S. Fish Commission Steamer "Albatross" from October 1904 to March 1905----. Mem. M. C. Z. 51:766 pp., 31 pls.

dinophysis swezyi
Norris, Dean R., and Leo D. Berner Jr. 1970.
Thecal morphology of selected species of Dinophysis (Dinoflagellata) from the Gulf of Mexico.
Contrib. mar. Sci. Port Aransas 15:145-192

Dinophysis taii
Balech, Enrique, 1971
Microplancton del Atlantico ecuatorial oeste (Equalant 1)
Publ. Serv. Hidrograf. Naval, Argentina H. 654: 103 pp., 122 figs.

Dinophysis taii nom. nov.
Balech, Enrique, 1971.
Microplancton de la campaña oceanographica: Productividad III.
Revta Mus. argent. Cienc. Nat. Bernadina Rivadavia, Hydrobiol. 3(1):1-202, 39 pls.

Dinophysis trapezium n.sp.
Kofoid, C. A. and T. Skogsberg, 1928
XXXV. The Dinoflagellata: The Dinophysiodae. Reports on the scientific results of the expedition to the Eastern Tropical Pacific, in charge of Alexander Agassiz, by the U. S. Fish Commission Steamer "Albatross" from October 1904 to March 1905----. Mem. M. C. Z. 51:766 pp., 31 pls.

Dinophysis triacantha
Jörgensen, E., 1923
Mediterranean Dinophysiaceae. Rept. Danish Oceanogr. Expeds. 1908-10, to the Mediterranean and adjacent seas, Vol.II, Biol. J 2, 48 pp., 64 text figs.

Dinophysis triacantha
Käsler, R., 1938
Die Verbreitung der Dinophysiales im Sud-atlantischen Ozean. Wiss. Ergeb. Deutschen Atlantischen Expedition----"Meteor" 1925-1927, 12(2):162-237, text figs. 85-118.

Dinophysis triacantha
Kofoid, C. A. and T. Skogsberg, 1928
XXXV. The Dinoflagellata: The Dinophysiodae. Reports on the scientific results of the expedition to the Eastern Tropical Pacific, in charge of Alexander Agassiz, by the U. S. Fish Commission Steamer "Albatross" from October 1904 to March 1905----. Mem. M. C. Z. 51:766 pp., 31 pls.

Dinophysis triacantha
Rampi, L., 1942
II Fitoplancton mediterraneo: Problemi ed affinita interoceaniche. Boll. di Pesca di Piscicoltura e di Idrobiologia, Anno 18, Fasc. 4:7-19.

Dinophysis triacantha
Rampi, L., 1945
Osservazioni sulla distribuzione qualitativa del fitoplancton nel mare Mediterraneo. Atti della Soc. Ital. di Sci. Nat. 84:105-113.

Dinophysis tripos
Balech, Enrique, 1944
Contribucion al conocimiento del Plancton de Lennox y Cabo de Hornos. Physis XIX:423-446, 6 pls. with 67 figs.

Dinophysis tripos
Dangeard, P., 1927
Phytoplankton de la croisière du "Sylvana". Ann. Inst. Océan. Monaco, n.s., 4(8):286-401, 54 text figs. (Feirer-Juin 1913).

Dinophysis tripos
Ercegovic, A., 1936
Etudes qualitative et quantitatives du phytoplancton dans les eaux cotières de l'Adriatique oriental moyen au cours de l'année 1934. Acta Adriatica 1(9):1-126

Dinophysis tripos
Ghazzawi, F.M., 1939
Plankton of the Egyptian waters. A study of the Suez Canal Plankton. (A) Phytoplankton. Preliminary Report 83 pp. Notes and Memoires, Min. Commerce-Industry, Egypt, Hydrobiol. & Fish. 65 figs.

Dinophysis tripos
Gourret, P., 1883
Sur les Peridiniens du Golfe de Marseille. Ann. du Musee d'hist. Nat. Marseille, Zool., 1 (Mme. 8):1-114, 4 pls.

Dinophysis tripos
Jörgensen, E., 1923
Mediterranean Dinophysiaceae. Rept. Danish Oceanogr. Expeds. 1908-10, to the Mediterranean and adjacent seas, Vol.II, Biol. J 2, 48 pp., 64 text figs.

Dinophysis tripos
Käsler, R., 1938
Die Verbreitung der Dinophysiales im Sud-atlantischen Ozean. Wiss. Ergeb. Deutschen Atlantischen Expedition----"Meteor" 1925-1927, 12(2):162-237, text figs. 85-118.

Dinophysis tripos
LeBour, M.V., 1925
The dinoflagellates of Northern Seas. The Marine Biological Association of the United Kingdom, Plymouth, 250 pp., 35 pls., 53 text figs.

Dinophysis tripos
Margalef, R., 1949
Fitoplancton nerítico de la Costa Brava en 1947-48. Publ. Inst. Biol. Aplicada, 5: 41-51, 3 text figs.

Dinophysis tripos
Paiva Carvalho, J., 1950
O plancton do Rio Maria Rodrigues (Cananeis). 1. Diatomaceas e Dinoflagelados. Bol. Inst. Paulista Oceanogr. 1(1): 27-43, 2 fold-in tables, 2 figs.

Dinophysis tripos
Pavillard, J., 1916
Recherches sur les Peridiniens du Golfe du Lion. Mem. Univ. Montpellier. Trav. Inst. Bot., Univ. Montpellier. Serie mixte No.4, 70 pp., 3 pls., 15 text figs.

Dinophysis tripos
Rampi, L., 1940
Ricerche sul Fitoplancton del mare Ligure. II Le tecatali e le dinofisiali delle acque di Sanremo.. Boll. Pesca, Piscicolt. Idrobiol. (18) 16(2): 243-274, 56 figs.

Dinophysis tripos
Rampi, L., 1940
Ricerche sul Fitoplancton del mare Ligure. Boll. di Pesca, di Piscicoltura e di Idrobiologa, 18(2):1-34, 56 text figs.

Dinophysis tripos
Sousa e Silva, E., 1949
Diatomaceas e Dinoflagelados de Baia de Cascais. Portugaliae Acta Biol., Volume: Julio Henriques, Ser. B: 300-383, 9 pls, 2 fold-in tables.

Dinophysis tripos
Tai, Si-Sun, & T. Skogsberg, 1934
Studies on the Dinophysoidae, marine armored dinoflagellates of Monterey Bay, California. Arch. Protistenk. 82:380-482, 14 text figs. Pls. 11-12.

Dinophysis tripos
Wood, E.J.P., 1954.
Dinoflagellates in the Australian region.
Australian J. Mar. Freshwater Res., 5(2):171-351

Dinophysis truncata
Balech, Enrique, 1944
Contribucion al conocimiento del Plancton de Lennox y Cabo de Hornos. Physis XIX:423-446, 6 pls. with 67 figs.

Dinophysis truncata
Käsler, R., 1938
Die Verbreitung der Dinophysiales im Sud-atlantischen Ozean. Wiss. Ergeb. Deutschen Atlantischen Expedition----"Meteor" 1925-1927, 12(2):162-237, text figs. 85-118.

Dinophysis truncata
Pavillard, J., 1923
A propos de la systematique des Péridiniens Bull. Soc. Bot. de France 70:876-882; 914-918.

Dinophysis truncata
Rampi, L., 1950.
Péridiniens rares ou nouveaux pour le Pacifique Sud-Equatorial. Bull. Inst. Océan., Monaco, No. 974:11 pp., 26 textfigs.

Dinophysis truncata
Wood, E.J.P., 1954.
Dinoflagellates in the Australian region.
Australian J. Mar. Freshwater Res., 5(2):171-351.

Dinophysis tuberculata
Argentina, Secretaria de Marina, Servicio de Hidrografia Naval, 1962.
Plancton de las campanas oceanograficas DRAKE I y II.
Publico. H. 627:57.

Dinophysis tuberculata
Käsler, R., 1938
Die Verbreitung der Dinophysiales im Sud-atlantischen Ozean. Wiss. Ergeb. Deutschen Atlantischen Expedition----"Meteor" 1925-1927, 12(2):162-237, text figs. 85-118.

Dinophysis tuberculata
Wood, E.J.P., 1954.
Dinoflagellates in the Australian region.
Australian J. Mar. Freshwater Res., 5(2):171-351.

Dinophysis uracantha
Dangeard, P., 1927
Phytoplankton de la croisière du "Sylvana". Ann. Inst. Océan., Monaco, n.s., 4(8):286-401, 54 text figs. (février-Juin 1913).

Dinophysis uracantha
de Sousa e Silva, E., 1956.
Contribution à l'étude du microplancton de Dakar et des regions maritimes voisines.
Bull. I.F.A.N., 8(2):335-371, 7 pls.

Dinophysis uracantha
Ercegovic, A., 1940
Weitere Untersuchungen über einige hydrographische Verhältnisse und über die Phytoplanktonproduktion in den Gewässern der Östlichen Mitteladria. Acta Adriatica 2(3):95-134, 8 text figs.

Dinophysis uracantha
Jörgensen, E., 1923
Mediterranean Dinophysiaceae. Rept. Danish Oceanogr. Expeds. 1908-10, to the Mediterranean and adjacent seas, Vol.II, Biol. J 2, 48 pp., 64 text figs.

Dinophysis urocantha
Käsler, R., 1938
Die Verbreitung der Dinophysiales im Sud-atlantischen Ozean. Wiss. Ergeb. Deutschen Atlantischen Expedition----"Meteor" 1925-1927, 12(2):162-237, text figs. 85-118.

Dinophysis uracantha
Kofoid, C. A. and T. Skogsberg, 1928
XXV. The Dinoflagellata: The Dinophysiodae. Reports on the scientific results of the expedition to the Eastern Tropical Pacific, in charge of Alexander Agassiz, by the U. S. Fish Commission Steamer "Albatross" from October 1904 to March 1905----. Mem. M. C. Z. 51:766 pp., 31 pls.

Dinophysis uracantha
LeBour, M.V., 1925
The dinoflagellates of Northern Seas. The Marine Biological Association of the United Kingdom, Plymouth. 250 pp., 35 pls., 53 text figs.

Dinophysis uracantha
Matzenauer, L., 1933
Die Dinoflagellaten des indischen Ozeans (mit Ausnahme der Gattung Ceratium.) Bot. Arch. 35:437-510, 77 text figs., 2 charts.

Dinophysis uracantha
Murray, G., and F. G. Whitting, 1899
New Peridiniaceae from the Atlantic. Trans. Linn. Soc., London, Bot., ser 2, 5: 321-342, Pls. 27-33, 9 tables.

Dinophysis uracantha
Norris, Dean R., and Leo D. Berner Jr. 1970.
Thecal morphology of selected species of Dinophysis (Dinoflagellata) from the Gulf of Mexico.
Contrib. mar. Sci. Port Aransas 15:145-192

Dinophysis uracantha
Pavillard, J., 1923
A propos de la systematique des Péridiniens Bull. Soc. Bot. de France 70:876-882; 914-918.

Dinophysis uracantha
Pavillard, J., 1916
Recherches sur les Peridiniens du Golfe du Lion. Mem. Univ. Montpellier. Trav. Inst. Bot., Univ. Montpellier. Serie mixte No.4, 70 pp., 3 pls., 15 text figs.

Dinophysis uracantha
Rampi, L. 1940
Ricerche sul Fitoplancton del mare Ligure. II Le tecatali e le dinofisiali delle acque di Sanremo. Boll. Pesca, Piscicolt.,Idrobiol. (18) 16(2): 243-274, 56 figs.

Dinophysis uracantha
Rampi, L., 1940
Ricerche sul Fitoplancton del mare Ligure. Boll. di Pesca, di Piscicoltura e di Idrobiologa, 18(2):1-34, 56 text figs.

Dinophysis uracantha
Wood, E.J.P., 1954.
Dinoflagellates in the Australian region.
Australian J. Mar. Freshwater Sci., 5(2):171-351

Dinophysis urceolus
Käsler, R., 1938
Die Verbreitung der Dinophysiales im Sud-atlantischen Ozean. Wiss. Ergeb. Deutschen Atlantischen Expedition----"Meteor" 1925-1927, 12(2):162-237, text figs. 85-118.

Dinophysis urceolus n.sp.
Kofoid, C. A. and T. Skogsberg, 1928
XXV. The Dinoflagellata: The Dinophysiodae. Reports on the scientific results of the expedition to the Eastern Tropical Pacific, in charge of Alexander Agassiz, by the U. S. Fish Commission Steamer "Albatross" from October 1904 to March 1905----. Mem. M. C. Z. 51:766 pp., 31 pls.

Dinophysis vanhöffeni
Abe, Tohru H. 1967.
The armoured Dinoflagellata: II.Prorocentridae and Dinophysidae (B)- Dinophysis and its allied genera.
Publs. Seto mar.biol.Lab., 15(1):37-78.

Dinophysis Vanhoffeni
Marukawa, H., 1921
Plankton lists and some new species of copepods, from the northern waters of Japan. Bull. Inst. Ocean., No.384, 15 pp., 3 pls., 1 chart.
Monaco

Dinophysis Vanhöffenii
Paulsen, O., 1949
Observations on dinoflagellates. (Ed. J. Grøntoed) Kongl. Dansk. Videnskab. Selsk., Biol. Skr. 6(4):67 pp., 30 text figs.

Dinophysis ventrecta
Wood, E.J.F., 1963
Dinoflagellates in the Australian region. II. Recent collections.
Comm. Sci. and Industr. Res. Org., Div. Fish. and Oceanogr., Techn. Paper, No. 14: 55 pp.

Dinophysis ventricosa
Dangeard, P., 1927
Phytoplankton de la croisière du "Sylvana". Ann. Inst. Ocean., Monaco, n.s., 4(8):286-401, 54 text figs. (Feirer-Juin 1913).

Dinophysis vertex
Pavillard, J., 1916
Recherches sur les Peridiniens du Golfe du Lion. Mem. Univ. Montpellier. Trav. Inst. Bot., Univ. Montpellier. Serie mixte No.4, 70 pp., 3 pls., 15 text figs.

Dinophysis whittingae
Balech, Enrique, 1971.
Microplancton et de la campaña oceanographica: Productividad III.
Revta Mus. argent. Cienc. Nat. Bernadina Rivadavia, Hydrobiol. 3(1):1-202, 39 pls.

Dinoporella perforata = Porella perforata
Halim, Youssef, 1960
Etude quantitative et qualitative du cycle écologique des dinoflagellés dans les eaux de Villefranche-sur-Mer. (1953-1955). Ann. Inst. Océanogr., Monaco, 38:123-232.

Diplomorpha paradoxa
Cachon, J., 1953.
Morphologie et cycle évolutif de Diplomorpha paradoxa (Rose et Cachon) péridinien parasite des siphonophores. Bull. Soc. Zool., France, 78(5/6): 408-414, 5 textfigs.

Diplomorpha paradoxa
Rose, M., et J. Cachon, 1952.
L'émission des bras du Diplomorpha paradoxa. C.R. Acad. Sci., Paris, 234:2306-2308.

Diplomorpha paradoxa
Rose, M., et J. Cachon, 1952.
Le mouvement chez Diplomorpha paradoxa, parasite des siphonophores. C.R. Acad. Sci., Paris, 234:669-671.

Diplomorpha paradoxa n.gen., n.sp.
Rose, M., and J. Cachon, 1951.
Diplomorpha paradoxa n. g., nov. sp., protiste parasite de l'ectoderm des siphonophores. C.R. Acad. Sci., Paris, 233:451-452.

Diplopelta bomba
Pavillard, J., 1916
Recherches sur les Peridiniens du Golfe du Lion. Mem. Univ. Montpellier. Trav. Inst. Bot., Univ. Montpellier. Serie mixte No.4, 70 pp., 3 pls., 15 text figs.

Diplopelta symmetrica
Forti, A., 1922
Ricerche sulla flora pelagica (fitoplancton) di Quarto dei Mille. Mem. R. Com. Talass. Ital. 97:248 pp., 13 pls.

Diplopelta symmetrica
Pavillard, J., 1916
Recherches sur les Peridiniens du Golfe du Lion. Mem. Univ. Montpellier. Trav. Inst. Bot., Univ. Montpellier. Serie mixte No.4, 70 pp., 3 pls., 15 text figs.

Diplopeltopsis granulosa n.sp.
Balech, Enrique, 1958.
Plancton de la Campaña Antartica Argentina, 1954-1955.
Physis, 21(60):75-108.

Diplopeltopsis minor
Dangeard, P., 1927
Phytoplankton de la croisière du "Sylvana". Ann. Inst. Ocean., Monaco, n.s., 4(8):286-401, 54 text figs. (Février-Juin 1913).

Diplopeltopsis minor
Hada, Yoshine, 1970. (1970)
The protozoan plankton of the Antarctic and Subantarctic seas.
Scient. Repts, Japan. Antarct. Res. Exped., (E)31:51 pp.

Diplopeltopsis minor
Balech, Enrique, 1959.
Operacion Oceanografica Merluza V Crucero. Plancton.
Publico, Servicio de Hidrografia Naval, Argentina, H. 618:43 pp.

Diplopeltopsis minor
Balech, Enrique, 1958.
Plancton de la Campaña Antartica Argentina, 1954-1955.
Physis, 21(60):75-108.

Diplopeltopsis minor
Dangeard, P., 1926
Description des Péridiniens Testacés recueillis para la Mission Charcot pendent le mois d'Aout 1924. Ann. Inst. Ocean. n.s. 3(7):307-334, 15 text figs.

Diplopeltopsis minor
LeBour, M.V., 1925
The dinoflagellates of Northern Seas. The Marine Biological Association of the United Kingdom, Plymouth, 250 pp., 35 pls., 53 text figs.

Diplopeltopsis orbicularis
Dangeard, P., 1927
Phytoplankton de la croisière du "Sylvana". Ann. Inst. Ocean., Monaco, n.s., 4(8):286-401, 54 text figs. (Feirer-Juin 1913).

Diplopeltopsis perlata n. sp.
Balech, Enrique, 1971.
Microplancton de la campaña oceanographica: Productividad III.
Revta Mus. argent. Cienc. Nat. Bernadina Rivadavia, Hydrobiol. 3(1):1-202, 39 pls.

Diplosalis sp.
Braarud, T., and Adam Bursa, 1939
On the phytoplankton of the Oslo Fjord, 1933-1934. Hvalrådets Skr. No.19:1-63; 9 text figs. Reviewed. J. du. Cons. 14(3): 418-420. A.C. Gardiner.

Diplosalis excentrica n. sp.
Nie, Dashu, 1943.
Dinoflagellata of the Hainan region. VI. On the genus Diplosalis. Sinensia, 14:1-21.

Diplosalis hainanensis n.sp.
Nie, Dashu, 1943.
Dinoflagellata of the Hainan region. VI. On the genus Diplosalis. Sinensia, 14:1-21.

Diplosalis lebourae nov.comb.
Balech, Enrique, 1967.
Dinoflagelados nuevos o interesantes del Golfo de Mexico y Caribe.
Revta Mus. argent.Cienc.nat.Bernardino Rivadavis Inst.nac.Invest.Cienc.nat.,Hidrobiol. 2(3):77-126.

Diplosalis lenticula
Dangeard, P., 1927
Phytoplankton de la croisière du "Sylvana". Ann. Inst. Ocean., Monaco, n.s., 4(8):286-401, 54 text figs. (Février-Juin 1913).

Diplopsalis lenticula
Forti, A., 1922
Ricerche sulla flora pelagica (fitoplancton) di Quarto dei Mille. Mem. R. Com. Talass. Ital. 97:248 pp., 13 pls.

Diplopsalis lenticula
Gran, H.H., and T. Braarud, 1935
A quantitative study of the phytoplankton in the Bay of Fundy and the Gulf of Maine (including observations on hydrography, chemistry, and turbidity). J. Biol. Bd., Canada, 1(5):279-467, 69 text figs.

Diplosalis lenticulata
Jørgensen, E., 1905
B.Protistplankton and the diatoms in bottom samples. Hydrographical and biological investigations in Norwegian fiords. Bergens Mus. Skr. 7: 49-225.

Diplopsalis lenticula
Jorgensen, E., 1900
Protophyten und Protozoën im Plankton aus der Norwegischen Westkerste. Bergens Mus. Aarb. 1899(6): 95 pp., 5 pls., 83 tables.

Diplopsalis lenticula
LeBour, M.V., 1925
The dinoflagellates of Northern Seas. The Marine Biological Association of the United Kingdom, Plymouth, 250 pp., 35 pls., 53 text figs.

Diplosalis lenticula
Levander, K.M., 1947
Plankton gesammelt in den Jahren 1899-1910 an den Küsten Finnlands. Finnländische Hydrographisch-Biologische Untersuchungen (aus der Wasserbiologischen Laboratorin der Societas Scientiarum Fennica) No.11: 40 pp., 6 diagrams, 13 pls.; tables.

Diplopsalis lenticula
Lindemann, E., 1924
Peridineen aus dem goldenen Horn und dem Bosphorus. Bot. Arch. 5:216-233, 98 text figs.

Diplopsalis lenticula
Matzenauer, L., 1933
Die Dinoflagellaten des indischen Ozeans (mit Ausnahme der Gattung Ceratium.) Bot. Arch. 35:437-510, 77 text figs., 2 charts.

Diplopsalis lenticula
Meunier, A., 1919
Microplancton de la Mer Flamande 3. Les Péridiniens. Mem. Mus. Roy. Hist. Nat., Belgique 8(1):1-116, Pls. XV-XXI.

Diplopsalis lenticula
Morse, D.C., 1947
Some observations on seasonal variations in plankton population Patuxent River, Maryland 1943 1945. Bd. Nat. Res., Publ. No.65, Chesapeake Biol. Lab., 31, 3 figs.

Diplopsalis lenticula
Murray, G., and F. G. Whitting, 1899
New Peridiniaceae from the Atlantic. Trans. Linn. Soc., London, Bot., ser 2, 5: 321-342, Pls. 27-33, 9 tables.

Diplopsalis lenticulata
Paulsen, O., 1908
XVIII Peridiniales. Nordisches Plankton, Bot. Teil: 1-124, 155 text figs.

Diplopsalis lenticula
Pavillard, J., 1916
Recherches sur les Peridiniens du Golfe du Lion. Mem. Univ. Montpellier. Trav. Inst. Bot., Univ. Montpellier. Serie mixte No.4, 70 pp., 3 pls., 15 text figs.

Diplosalis lenticula
Pavillard, J., 1905
Recherches sur la flore pelagique (Phytoplankton) de l'Etang de Thau. Theses presentees a la Fac. Sci., Paris, 116 pp., 3 pls.

Diplopsalis lenticula
Schodduyn, M., 1926
Observations faites dans la baie d'Ambleteuse (Pas de Calais). Bull. Inst. Ocean., Monaco, No. 482: 64 pp.

Diplopsalis lenticula
Schröder, B., 1900
Phytoplankton des Golfes von Neapel nebst vergleichenden Ausblicken auf das atlantischen Ozean. Mitt. Zool. Stat. Neapel, 14:1-38.

Diplopsalis lenticula
Wall, David, and Barrie Dale, 1968.
Modern dinoflagellate cysts and evolution of the Peridiniales.
Micropaleontology, 14(3):265-304.

Diplopsalis lenticula
Wang, C. C., and D. Nie, 1932
A survey of the marine protozoa of Amoy. Contr. Biol. Lab. Sci. Soc., China, Zool. Ser., 8:285-385, 89 text figs.

Diplopsalis lenticula
Wood, E.J.P., 1954.
Dinoflagellates in the Australian region. Australian J. Mar. Freshwater Res., 5(2):171-351.

Diplosalis lenticula var lebourii n. var.
Nie Dashu, 1943.
Dinoflagellata of the Hainan region. VI. On the genus Diplosalis. Sinensia, 14:1-21.

Diplosalis minima n.sp.
Mangin, M. L., 1912
Phytoplancton de la croisière du "René" dans l'Atlantique (Septembre 1908). Ann. Inst. Ocean., n.s., 4(1):1-66, 2 pls., 41 text figs., 2 tables.

Diplopeltopsis minor
Wall, David, and Barrie Dale, 1968.
Modern dinoflagellate cysts and evolution of the Peridiniales.
Micropaleontology, 14(3):265-304.

Diplosalis orbiculare
de Sousa e Silva, E., 1956.
Contribution à l'étude du microplancton de Dakar et des regions maritimes voisines. Bull. I.F.A.N., 8(2):335-371, 7 pls.

Diplopsalopsis orbicularis
Wall, David, and Barrie Dale, 1968.
Modern dinoflagellate cysts and evolution of the Peridiniales.
Micropaleontology, 14(3):265-304.

Diplopsalis orbicularis

Wood, E.J.P., 1954.
Dinoflagellates in the Australian region.
Australian J. Mar. Freshwater Res., 5(2):171-351.

Diplosalis pillula

Levander, K.M., 1947
Plankton gesammelt in den Jahren 1899-1910 an den Küsten Finnlands. Finnländische Hydrographisch-Biologische Untersuchungen (aus dem Wasserbiologischen Laboratorin der Societas Scientiarum Fennica) No.11: 40 pp., 6 diagrams, 13 pls., tables.

Diplosalis Pillula

Paulsen, O., 1908
XVIII Peridiniales. Nordisches Plankton, Bot. Teil: 1-124, 155 text figs.

Diplosalis pingii n.sp.

Nie, Dashu, 1943.
Dinoflagellata of the Hainan region. VI. On the genus Diplosalis.
Sinensia, 14:1-21.

Diplopsalis rotundata

Wood, E.J.P., 1954.
Dinoflagellates in the Australian region.
Australian J. Mar. Freshwater Res., 5(2):171-351.

Diplopsalis saecularis n.sp.

Murray, G., and F. G. Whitting, 1899
New Peridiniaceae from the Atlantic.
Trans. Linn. Soc., London, Bot., ser 2, 5: 321-342, Pls. 27-33, 9 tables.

Diplopsalis saecularis

Paulsen, O., 1908
XVIII Peridiniales. Nordisches Plankton, Bot. Teil: 1-124, 155 text figs.

Diplosalis saecularis

Schröder, B., 1900
Phytoplankton des Golfes von Neapel nebst vergleichenden Ausblicken auf das atlantischen Ozean. Mitt. Zool. Stat. Neapel, 14:1-38.

Diplopsalopsis orbicularis

LeBour, M.V., 1925
The dinoflagellates of Northern Seas. The Marine Biological Association of the United Kingdom, Plymouth, 250 pp., 35 pls., 53 text figs.

Diplopsalopsis orbicularis

Matzenauer, L., 1933
Die Dinoflagellaten des indischen Ozeans (mit Ausnahme der Gattung Ceratium.) Bot. Arch. 35:437-510, 77 text figs., 2 charts.

Diplopsalopsis orbicularis (figs.)

Sousa e Silva, E., 1949
Diatomaceas e Dinoflagelados de Baia de Cascais. Portugaliae Acta Biol. Volume: Julio Henriques, Ser. B: 300-383, 9 pls, 2 fold-in tables.

Diplopsalopsis sphaerica

Balech, Enrique, 1962
Tintinnoinea y Dinoflagellata del Pacifico segun material de las Expediciones NORPAC y DOWNWIND del Instituto Scripps de Oceanografia.
Revista, Mus. Argentino Ciencias Nat. "Bernardino Rivadavia", Ciencias Zool., 7(1):1-253.

Discoerisma psilonereiella n. gen., n. sp.

Taylor, F.J.R., and J. Allen Cattell 1969.
Discoerisma psilonereiella gen. et sp. n., a new dinoflagellate from British Columbia waters.
Protistologica 5(2):169-172.

Dissodinium (Pyrocystis) elegans

Matzenauer, L., 1933
Die Dinoflagellaten des indischen Ozeans (mit Ausnahme der Gattung Ceratium.) Bot. Arch. 35:437-510, 77 text figs., 2 charts.

Dissodinium (Pyrocystis) fusiformis

Matzenauer, L., 1933
Die Dinoflagellaten des indischen Ozeans (mit Ausnahme der Gattung Ceratium.) Bot. Arch. 35:437-510, 77 text figs., 2 charts.

Dissodinium (Pyrocystis) fusiformis detruncata n.f.

Matzenauer, L., 1933
Die Dinoflagellaten des indischen Ozeans (mit Ausnahme der Gattung Ceratium.) Bot. Arch. 35:437-510, 77 text figs., 2 charts.

Dissodinium (Pyrocystis) hamulus

Matzenauer, L., 1933
Die Dinoflagellaten des indischen Ozeans (mit Ausnahme der Gattung Ceratium.) Bot. Arch. 35:437-510, 77 text figs., 2 charts.

Dissodinium (Pyrocystis) lanceolata

Matzenauer, L., 1933
Die Dinoflagellaten des indischen Ozeans (mit Ausnahme der Gattung Ceratium.) Bot. Arch. 35:437-510, 77 text figs., 2 charts.

Dissodinium lunula

Aurich, H. J., 1949.
Die Verbreitung des Nannoplanktons im Oberflächenwasser vor der Nordfriesischen Küste. Ber. Deutschen Wiss. Komm. f. Meeresf., n.f., 11(4): 403-405, 2 figs.

Dissodinium minima n.sp.

Matzenauer, L., 1933
Die Dinoflagellaten des indischen Ozeans (mit Ausnahme der Gattung Ceratium.) Bot. Arch. 35:437-510, 77 text figs., 2 charts.

Dissodinium (Pyrocystis) obtusa

Matzenauer, L., 1933
Die Dinoflagellaten des indischen Ozeans (mit Ausnahme der Gattung Ceratium.) Bot. Arch. 35:437-510, 77 text figs., 2 charts.

Dissodinium pseudocalani

Drebes, G., 1969.
Dissodinium pseudocalani sp. nov., ein parasitisch er Dinoflagellat auf Copepodeneiern. Helgoländer wiss. Meersunters. 19(1): 58-67.

Dissodinium pseudolunula n.sp.

Swift, Elijah 1973.
Dissodinium pseudolunula n.sp.
Phycologia 12(1/2):90-91.

Dissodinium (Pyrocystis) pseudonoctiluca

Matzenauer, L., 1933
Die Dinoflagellaten des indischen Ozeans (mit Ausnahme der Gattung Ceratium.) Bot. Arch. 35:437-510, 77 text figs., 2 charts.

Dissodinium rhomboides n.sp

Matzenauer, L., 1933
Die Dinoflagellaten des indischen Ozeans (mit Ausnahme der Gattung Ceratium.) Bot. Arch. 35:437-510, 77 text figs., 2 charts.

Dissodinium (Pyrocystis) robusta

Matzenauer, L., 1933
Die Dinoflagellaten des indischen Ozeans (mit Ausnahme der Gattung Ceratium.) Bot. Arch. 35:437-510, 77 text figs., 2 charts.

Dissodinium (Pyrocystis) semicircularis

Matzenauer, L., 1933
Die Dinoflagellaten des indischen Ozeans (mit Ausnahme der Gattung Ceratium.) Bot. Arch. 35:437-510, 77 text figs., 2 charts.

Endodinium chattonii

Taylor, D.L., 1971.
Ultrastructure of the 'zooxanthella' Endodinium chattonii in situ. J. mar. biol. Ass., U.K. 51(1): 227-234.

Ensiculifera mexicana n.gen.n.sp.

Balech, Enrique, 1967.
Dinoflagelados nuevos o interesantes del Golfo de Mexico y Caribe.
Revta Mus. argent. Cienc. nat. Bernardino Rivadavia Inst. nac. Invest. Cienc. nat., Hidrobiol. 2(3):77-126.

Entzia acuta

LeBour, M.V., 1925
The dinoflagellates of Northern Seas. The Marine Biological Association of the United Kingdom, Plymouth, 250 pp., 35 pls., 53 text figs.

Epiperidinium michaelsarsii

Hasle, Grethe Rytter, 1960
Phytoplankton and ciliate species from the Tropical Pacific.
Skr. Norske Videnskaps-Akad., Oslo, 1. Mat.-Nat. Kl., 1960(2): 1-50.

Erythropsis agilis

Pavillard, J., 1905
Recherches sur la flore pelagique (Phytoplankton) de l'Etang de Thau. Theses presentees a la Fac. Sci., Paris, 116 pp., 3 pls.

Erythropsis pavillardi

Greuet, Claude, 1968.
Organisation ultrastructurale de l'ocelle de deux peridiniens Warmowiidae, Erythropsis pavillardi Kofoid et Swezy et Warnowia pulchra Schiller.
Protistologica, 4(2):209-228.

Erthropsis pavillardi

Greuet, Claude, 1967.
Organisation ultrastructurale du tentacule d'Erythropsis pavillardi Kofoid et Swezy peridinien warnowiidae Lindemann.
Protistologica 3(3):335-344.

Exuviaella sp.

Balech, Enrique and Sayed Z. El-Sayed, 1965.
Microplankton of the Weddell Sea.
In: Biology of Antarctic seas. II.
Antarctic Res. Ser., Amer. Geophys. Union, 5:107-124.

Exuviella

Bernard, F., 1957.
Présence des flagellés marins Coccolithus et Exuviella dans le plancton de la Mer Morte. C.R. Acad. Sci., Paris, 245(20):1754-1756.

Exuviaella sp.

Braarud, T., 1945
A phytoplankton survey of the polluted waters of inner Oslo Fjord. Hvalrådets Skrifter, No.28, 142 pp., 19 text figs., 17 tables.

Exuviaella

Bucalossi, G., 1960.
Etude quantitative des variations du phytoplancton dans la baie d'Alger en fonction du milieu (novembre 1959 à mai 1960).
Bull. Inst. Océanogr., Monaco, 57(1189):1-40.

Exuviella sp.

Parsons, T.R., 1961
On the pigment composition of eleven species of marine phytoplankters.
J. Fish. Res. Bd., Canada, 18(6):1017-1025.

Exuviella sp.
Parsons, T.R., K. Stephens and J.D.H. Strickland, 1961
On the chemical composition of eleven species of marine phytoplankters.
J. Fish. Res. Bd., Canada, 18(6):1001-1016.

Exuviella sp.
Steven, D.M., 1966.
Characteristics of a red-water bloom in Kingston Harbor, Jamaica, W.I.
J. Mar. Res., 24(2):113-123.

Exuviaella aequatorialis n.sp.
Hasle, Grethe Rytter, 1960
Phytoplankton and ciliate species from the Tropical Pacific.
Skr. Norske Videnskaps-Akad., Oslo, 1. Mat.-Nat. Kl., 1960(2): 1-50.

Exuviella antarctica
Hada, Yoshina, 1970. (1970)
The protozoan plankton of the Antarctic and Subantarctic seas.
Scient. Repts, Japan. Antarct. Res. Exped., (E)31:51 pp.

Exuviaella apora
Balech, Enrique, 1962
Tintinnoinea y Dinoflagellata del Pacifico segun material de las Expediciones NORPAC y DOWNWIND del Instituto Scripps de Oceanografia.
Revista, Mus. Argentino Ciencias Nat. "Bernardino Rivadavia", Ciencias Zool., 7(1):1-253.

Exuviella apora
LeBour, M.V., 1925
The dinoflagellates of Northern Seas. The Marine Biological Association of the United Kingdom, Plymouth, 250 pp., 35 pls.,53 text figs.

Exuviella apora
Morse, D.C., 1947
Some observations on seasonal variations in plankton population Patuxant River, Maryland 1943-1945. Bd. Nat. Res., Publ. No.65, Chesapeake Biol. Lab., 31, 3 figs.

Exuviaella baltica
Adachi, Rokuro, 1972
Surface nannoplankton collected from the northwestern North Pacific Ocean in summer 1967. In: Biological oceanography of the northern North Pacific Ocean, A.Y. Takenouti, Chief Editor, Idemitsu Shoten, Tokyo, 139-144.

Exuviaella baltica
Balech, Enrique, 1971.
Microplancton de la campaña oceanographica: Productividad III.
Revta Mus. argent. Cienc. Nat. Bernadina Rivadavia, Hydrobiol. 3(1):1-202, 39 pls.

Exuviaella baltica
Braarud, T., 1962
Species distribution in marine phytoplankton.
J. Oceanogr. Soc., Japan, 20th Ann. Vol., 628-649.

Exuviaella baltica
Braarud, T., 1951. marine
Salinity as an ecological factor in/phytoplankton. Physiol. Plant. 4:28-34, 3 textfigs.

Exuviaella baltica
Braarud, T., 1945
A phytoplankton survey of the polluted waters of inner Oslo Fjord. Hvalrådets Skrifter, No.28, 142 pp., 19 text figs., 17 tables.

Exuviaella baltica
Braarud, T., 1934
A note on the phytoplankton of the Gulf of Maine in the summer of 1933. Biol. Bull. 67(1):76-82.

Exuviaella baltica
Braarud, T., and Adam Bursa, 1939
On the phytoplankton of the Oslo Fjord, 1933-1934. Hvalrådets Skr. No.19:1-63; 9 text figs. Reviewed. J. du. Cons. 14(3): 418-420. A.C. Gardiner.

Exuviaella baltica
Braarud, T., J. Markali and E. Nordli, 1958.
A note on the thecal structure of Exuviaella baltica Lohm.
Nytt Mag. Botan., 6:43-45.

Exuviaella baltica
Dodge, John D., 1965.
Thecal fine-structure in the dinoflagellate genera Prorocentrum and Exuviaella.
J. mar. biol. Ass., U.K., 45(3):607-614.

Exuviaella baltica
Gran, H.H., and T. Braarud, 1935
A quantitative study of the phytoplankton in the Bay of Fundy and the Gulf of Maine (including observations on hydrography, chemistry, and turbidity).
J. Biol. Bd., Canada, 1(5):279-467, 69 text figs.

Exuviella baltica
Halim, Youssef, 1960
Étude quantitative et qualitative du cycle écologique des dinoflagellés dans les eaux de Villefranche-sur-Mer. (1953-1955).
Ann. Inst. Océanogr., Monaco, 38:123-232.

Exuviella baltica
Hasle, Grethe Rytter, 1960
Phytoplankton and ciliate species from the Tropical Pacific.
Skr. Norske Videnskaps-Akad., Oslo, 1. Mat.-Nat. Kl., 1960(2): 1-50.

Exuviella baltica
LeBour, M.V., 1925
The dinoflagellates of Northern Seas. The Marine Biological Association of the United Kingdom, Plymouth, 250 pp., 35 pls.,53 text figs.

Exuviaella baltica
Lillick, L.C., 1940
Phytoplankton and planktonic protozoa of the offshore waters of the Gulf of Maine. Pt.II. Qualitative Composition of the Planktonic Flora. Trans. Am. Phil. Soc., n.s. 31(3):193-237, 13 text figs.

Exuviaella baltica
Lillick, L.C., 1938
Preliminary report of the phytoplankton of the Gulf of Maine. Am. Mid. Nat. 20(3):624-640, 1 text figs 37 tables.

Exuviaella baltica
Lillick, L.C., 1937
Seasonal studies of the phytoplankton off Woods Hole, Massachusetts. Biol. Bull. LXXIII (3):488-503, 3 text figs.

Exuviaella baltica
Paredes, J.F., 1962.
35. A brief comment on Peridinium nudum.
Trab. Cent. Biol. Pisc., 35:115-120.
Also in:
Mem. Junta Invest. Ultram., (2), No. 33.

Exuviaella baltica
Paulsen, O., 1908
XVIII Peridiniales. Nordisches Plankton, Bot. Teil: 1-124, 155 text figs.

Exuviaella baltica
Rampi, L. 1948
Ricerche sul Fitoplancton del mare Ligure. II Le tecatali e le dinofisiali delle acque di Sanremo. Boll. Pesca, Piscicolt.,Idrobiol. (18) 16(2): 243-274, 56 figs.

Exuviaella baltica
Rampi, L., 1940
Ricerche sul Fitoplancton del mare Ligure. Boll. di Pesca, di Piscicoltura e di Idrobiologa, 18(2):1-34, 56 text figs.

Exuviaella baltica
Rampi, L., 1939
Péridiniens rares ou intéressants récoltés dans la mer Ligure. Bull. Soc. Fran. de Microscopie, 8(2/3):106-112, 13 text figs.

Exuviaella baltica
Rumkówna, A., 1948
List of the phytoplankton species occurring in the superficial water layers in the Gulf of Gdańsk. Bull. Lab. mar., Gdynia, No. 4: 139-141 with tables in back.

Exuviaella baltica
Steemann-Nielsen, Einar, 1951
The marine vegetation of the Isefjord. A study on ecology and production. Medd. Komm. Danmarks Fiskeri-og Havundersøgelser. Ser. Plankton. 5(4); 114pp., 46 text figs.

Exuviaella baltica
Wheeler, Bernice, 1966.
Phototactic vertical migration in Exuviaella baltica.
Botanica Marina, 9 (1/2):15-17.

Exuviaella baltica
Wood, E.J.F., 1963
Dinoflagellates in the Australian region. II. Recent collections.
Comm. Sci. and Industr. Res. Org., Div. Fish. and Oceanogr., Techn. Paper, No. 14: 55 pp.

Exuviaella cassubica
Dodge, John D., 1965.
Thecal fine-structure in the dinoflagellate genera Prorocentrum and Exuviaella.
J. mar. biol. Ass., U.K., 45(3):607-614.

Exuviella cassubica
Rumkówna, A., 1948
List of the phytoplankton species occurring in the superficial water layers in the Gulf of Gdańsk. Bull. Lab. mar., Gdynia, No. 4: 139-141 with tables in back.

Exuviella cincta
Halim, Youssef, 1960
Étude quantitative et qualitative du cycle écologique des dinoflagellés dans les eaux de Villefranche-sur-Mer. (1953-1955).
Ann. Inst. Océanogr., Monaco, 38:123-232.

Exuviella compressa
Balech, Enrique, 1971.
Microplancton de la campaña oceanographica: Productividad III.
Revta Mus. argent. Cienc. Nat. Bernadina Rivadavia, Hydrobiol. 3(1):1-202, 39 pls.

Exuviella compressa
Balech, Enrique, 1962
Tintinnoinea y Dinoflagellata del Pacifico segun material de las Expediciones NORPAC y DOWNWIND del Instituto Scripps de Oceanografia.
Revista, Mus. Argentino Ciencias Nat. "Bernardino Rivadavia", Ciencias Zool., 7(1):1-253.

Exuviella compressa
Dangeard, P., 1927
Phytoplankton de la croisière du "Sylvana". Ann. Inst. Ocean., Monaco, n.s., 4(8):286-401, 54 text figs. (février-Juin 1913).

Exuviella compressa
Ercegovic, A., 1936
Etudes qualitative et quantitatives du phytoplancton dans les eaux cotières de l'Adriatique oriental moyen au cours de l'année 1934. Acta Adriatica 1(9):1-126

Exuviella compressa
Forti, A., 1922
Ricerche sulla flora pelagica (fitoplancton) di Quarto dei Mille. Mem. R. Com. Talass. Ital. 97:248 pp., 13 pls.

Exuviella compressa
Hasle, Grethe Rytter, 1960
Phytoplankton and ciliate species from the Tropical Pacific. Skr. Norske Videnskaps-Akad., Oslo, 1. Mat.-Nat. Kl., 1960(2): 1-50.

Exuviella compressa
LeBour, M.V., 1925
The dinoflagellates of Northern Seas. The Marine Biological Association of the United Kingdom, Plymouth, 250 pp., 35 pls., 53 text figs.

Exuviella compressa
Mangin, M. L., 1912
Phytoplancton de la croisière du "René" dans l'Atlantique (Septembre 1908). Ann. Inst. Ocean., n.s., 4(1):1-66, 2 pls., 41 text figs., 2 tables.

Exuviella compressa
Margalef, R., 1949
Fitoplancton nerítico de la Costa Brava en 1947-48. Publ. Inst. Biol. Aplicada, 5: 41-51, 3 text figs.

Exuviella compressa
Matzenauer, L., 1933
Die Dinoflagellaten des indischen Ozeans (mit Ausnahme der Gattung Ceratium). Bot. Arch. 35:437-510, 77 text figs., 2 charts.

Exuviella compressa
Paulsen, O., 1908
XVIII Peridiniales. Nordisches Plankton, Bot. Teil: 1-124, 155 text figs.

Exuviella compressa
Pavillard, J., 1916
Recherches sur les Peridiniens du Golfe du Lion. Mem. Univ. Montpellier. Trav. Inst. Bot., Univ. Montpellier. Serie mixte No.4, 70 pp., 3 pls., 15 text figs.

Exuviella compressa
Rampi, L. 1940
Ricerche sul Fitoplancton del mare Ligure. II Le tecatali e le dinofisiali delle acque di Sanremo. Boll. Pesca, Piscicolt., Idrobiol. (18) 16(2): 243-274, 56 figs.

Exuviella compressa
Rampi, L., 1940
Ricerche sul Fitoplancton del mare Ligure. Boll. di Pesca, di Piscicoltura e di Idrobiologa, 18(2):1-34, 56 text figs.

Exuviella compressa
Wood, E.J.F., 1954
Dinoflagellates in the Australian region. Australian J. Mar. Freshw. Res. 5(2):171-351.

Exuviella cordata
Halim, Y., 1965.
Microplancton des eaux Egyptiennes. II. Chrysomonadines; Ebriediens et dinoflagellés nouveaux ou d'intérêt biogeographique. Rapp. Proc. Verb. Reunions, Comm. Int. Expl. Sci., Mer Méditerranée, Monaco, 18(2):373-379.

Exuviella cordata
Ionesco, Al, et H. Skolka 1968.
Notes sur la physiologie et l'écologie du dinoflagellé Exuviella cordata. Trav. Mus. Hist. nat. "Grigore Antipa" 8(1): 207-215.

Exuviella cordata
Skolka, Vidor Hilarius, et Ileana Cautis 1971.
Floraison d'Exuviella cordata Ostenf. et ses consequences sur la pêche maritime en Roumanie au cours de l'année 1969. Cercetări marine, Constanta, (1). 59-82

Exuviella grani n.sp.
Gaarder, Karen Ringdal, 1938.
Phytoplankton studies from the Tromsø district, 1930-31. Tromsø Mus. Årshefter, Naturhist. Avd., 11, 55(1):159 pp., 4 fold-in pls., 12 textfigs.

Exuviella laevis
Schröder, B., 1900
Phytoplankton des Golfes von Neapel nebst vergleichenden Ausblicken auf das atlantischen Ozean. Mitt. Zool. Stat. Neapel, 14:1-38.

Exuviella lenticulata n.sp.
Matzenauer, L., 1933
Die Dinoflagellaten des indischen Ozeans (mit Ausnahme der Gattung Ceratium.) Bot. Arch. 35:437-510, 77 text figs., 2 charts.

Exuviella lenticulata
Rampi, L., 1948
Sur quelques Peridiniens rares ou interessants du Pacifique subtropical (Recoltes Alain Gerbault). Bull. l'Inst. Ocean, Monaco, No.937: 7 pp., 8 text figs.

Exuviella lima
Calkins, G. N., 1902
Marine protozoa from Woods Hole. U.S. Fish Comm. Bull. for 1901, pp. 413-468, 69 text figs.

Exuviella Lima
Paulsen, O., 1908
XVIII Peridiniales. Nordisches Plankton, Bot. Teil: 1-124, 155 text figs.

Exuviella lima
Schodduyn, M., 1926
Observations faites dans la baie d'Ambleteuse (Pas de Calais). Bull. Inst. Ocean., Monaco, No. 482: 64 pp.

Exuviella mariae-lebouriae
Dodge, John D., 1965.
Thecal fine-structure in the dinoflagellate genera Prorocentrum and Exuviella. J. mar. biol. Ass., U.K., 45(3):607-614.

Exuviella mariae-lebourae
Nakazima, Masao 1968.
Studies on the source of shellfish poison in Lake Hamana. 4. Identification and collection of the noxious dinoflagellate. Bull. Jap. Soc. scient. Fish. 34(2):131-132

Exuviella mariae-lebouriae n.sp.
Parke, M., and D. Ballentine, 1957.
A new marine dinoflagellate: Exuviella mariae-lebouriae n.sp. J.M.B.A., 36(3):643-650.

Exuviella marina
Aleem, A.A., 1952.
Données écologiques sur deux espèces de peridiniens des eaux saumâtres. Vie et Milieu, Bull. Lab. Arago, Univ. Paris, 3(3):281-287, 2 textfigs

Exuviella marina
Biecheler, B., 1952.
Recherches sur les Peridiniens. Bull. Biol. France Belg., Suppl. 36:1-149, 75 textfigs.

Exuviella marina
Dodge, John D., 1965.
Thecal fine-structure in the dinoflagellate genera Prorocentrum and Exuviella. J. mar. biol. Ass., U.K., 45(3):607-614.

Exuviella marina
Dragesco, Jean, 1965.
Étude cytologique de quelques flagellés mésopsammiques. Cahiers Biol. Mar., Roscoff, 6(1):83-115.

Exuviella marina
Ercegovic, A., 1936
Etudes qualitative et quantitatives du phytoplancton dans les eaux cotières de l'Adriatique oriental moyen au cours de l'année 1934. Acta Adriatica 1(9):1-126

Exuviella marina
Calkins, G. N., 1902
Marine protozoa from Woods Hole. U.S. Fish Comm. Bull. for 1901, pp. 413-468, 69 text figs.

Exuviella marina
Hada, Yoshina, 1970. (1970)
The protozoan plankton of the Antarctic and Subantarctic seas. Scient. Repts, Japan. Antarct. Res. Exped., (E) 31:51 pp.

Exuviella marina
LeBour, M.V., 1925
The dinoflagellates of Northern Seas. The Marine Biological Association of the United Kingdom, Plymouth, 250 pp., 35 pls., 53 text figs.

Exuviella marina
Lillick, L.C., 1940
Phytoplankton and planktonic protozoa of the offshore waters of the Gulf of Maine. Pt.II. Qualitative Composition of the Planktonic Flora. Trans. Am. Phil. Soc., n.s., 31(3):193-237, 13 text figs.

Exuviella marina
Lillick, L.C., 1938
Preliminary report of the phytoplankton of the Gulf of Maine. Am. Mid. Nat. 20(3):624-640, 1 text fig, 37 tables.

Exuviella marina
Pavillard, J., 1916
Recherches sur les Peridiniens du Golfe du Lion. Mem. Univ. Montpellier. Trav. Inst. Bot., Univ. Montpellier. Serie mixte No.4, 70 pp., 3 pls., 15 text figs.

Exuviella marina
Rampi, L. 1940
Ricerche sul Fitoplancton del mare Ligure. II Le tecatali e le dinofisiali delle acque di Sanremo. Boll. Pesca, Piscicolt., Idrobiol. (18) 16(2): 243-274, 56 figs.

Exuviella marina
Rampi, L., 1940
Ricerche sul Fitoplancton del mare Ligure. Boll. di Pesca, di Piscicoltura e di Idrobiologa, 18(2):1-34, 56 text figs.

Oceanographic Index: Marine Organisms Cumulation, 1946-1973

Exuviaella marina var. lima
Rampi, L., 1939
Péridiniens rares ou intéressants recoltés dans la mer Ligure. Bull. Soc. Fran. de Microscopie, 8(2/3):106-112, 13 text figs.

Exuviella marina
Tratet, Gérard, 1964.
Variations du phytoplancton à Tanger. Trav. Inst. Sci., Cherifien, Rabat, Ser. Botan., (29):204 pp.

Exuviaella marina
Wang, C. C., and D. Nie, 1932
A survey of the marine protozoa of Amoy. Contr. Biol. Lab. Sci. Soc., China, Zool. Ser., 8:285-385, 89 text figs.

Exuviaella marina
Schröder, B., 1900
Phytoplankton des Golfes von Neapel nebst vergleichenden Ausblicken auf das atlantischen Ozean. Mitt. Zool. Stat. Neapel, 14:1-38.

Exuviaella marina
Wood, E.J.F., 1954
Dinoflagellates in the Australian region. Australian J. Mar. Freshw. Res., 5(2):171-351.

Exuviaella minima
Forti, A., 1922
Ricerche sulla flora pelagica (fitoplancton) di Quarto dei Mille. Mem. R. Com. Talass. Ital. 97:248 pp., 13 pls.

Exuviaella minima n.sp.
Pavillard, J., 1916
Recherches sur les Péridiniens du Golfe du Lion. Mem. Univ. Montpellier. Trav. Inst. Bot., Univ. Montpellier. Serie mixte No.4, 70 pp., 3 pls., 15 text figs.

Exuviella oblonga
Halim, Youssef, 1960
Étude quantitative et qualitative du cycle écologique des dinoflagellés dans les eaux de Villefranche-sur-Mer. (1953-1955). Ann. Inst. Océanogr., Monaco, 38:123-232.

Exuviaella oblonga
Matzenauer, L., 1933
Die Dinoflagellaten des indischen Ozeans (mit Ausnahme der Gattung Ceratium.) Bot. Arch. 35:437-510, 77 text figs., 2 charts.

Exuviaella perforata
Gran, H.H., and T. Braarud, 1935
A quantitative study of the phytoplankton in the Bay of Fundy and the Gulf of Maine (including observations on hydrography, chemistry, and turbidity). J. Biol. Bd., Canada, 1(5):279-467, 69 text figs.

Exuviella perforata
LeBour, M.V., 1925
The dinoflagellates of Northern Seas. The Marine Biological Association of the United Kingdom, Plymouth, 250 pp., 35 pls., 53 text figs.

Exuviaella perforata
Lillick, L.C., 1940
Phytoplankton and planktonic protozoa of the offshore waters of the Gulf of Maine. Pt.II. Qualitative Composition of the Planktonic Flora. Trans. Am. Phil. Soc., n.s., 31(3):193-237, 13 text figs.

Exuviaella pusilla
Dodge, John D., 1965.
Thecal fine-structure in the dinoflagellate genera Prorocentrum and Exuviaella. J. mar. biol. Ass., U.K., 45(3):607-614.

Exuviella pusilla
Ercegovic, A., 1936
Etudes qualitative et quantitatives du phytoplancton dans les eaux cotières de l'Adriatique oriental moyen au cours de l'année 1934. Acta Adriatica 1(9):1-126

Filodinium hovassei N.gen.,N.sp.
Cachon, Jean et Monique, 1968.
Filodinium hovassei nov.gen.nov.sp., péridinien phoretique d'appendiculaires. Protistologica, 4(1):15-18.

Fragilidium heterolobum, n. gen., n. sp.
Balech, Enrique, 1959
Two new genera of dinoflagellates from California. Biol. Bull., 116(2): 195-203.

Gessnerium mochimaensis n.gen., n.sp.
Halim, Youssef, 1967
Dinoflagellates of the south-east Caribbean Sea (east Venezuela). Int. Revue ges. Hydrobiol., 52(5): 701-755.

Glenodinium
Gran, H.H., and T. Braarud, 1935
A quantitative study of the phytoplankton in the Bay of Fundy and the Gulf of Maine (including observations on hydrography, chemistry, and turbidity). J. Biol. Bd., Canada, 1(5):279-467, 69 text figs.

Glenodinium sp.
Kokubo, S., and S. Sato., 1947
Plankterr in Jū-San Gata. Physiol. and Ecol. (Japan) 1(4):1-16, 3 text figs., tables.

Glenodinium sp.
Lindemann, E., 1924
Peridineen aus dem goldenen Horn und dem Bosphorus. Bot. Arch. 5:216-233, 98 text figs.

Glenodinium sp.
Mangin, M. L., 1912
Phytoplancton de la croisière du "René" dans l'Atlantique (Septembre 1908). Ann. Inst. Ocean., n.s., 4(1):1-66, 2 pls., 41 text figs., 2 tables.

Glenodinium (?) sp.
Paasche, E., 1961.
Notes on phytoplankton from the Norwegian Sea. Botanica Marina, 2(3/4):197-210.

Glenodinium sp.
Stæmann-Nielsen, Einar, 1951
The marine vegetation of the Isefjord. A study on ecology and production. Medd. Komm. Danmarks Fiskeri-og Havundersøgelser. Ser. Plankton. 5(4); 114pp., 46 text figs.

Glenodinium acuminatum
Jorgensen, E., 1900
Protophyten und Protozoën im Plankton aus der Norwegischen Westkerste. Bergens Mus. Aarb. 1899(6): 95 pp., 5 pls., 83 tables.

Glenodinium bacilliferum n.sp.
Biecheler, B., 1952.
Recherches sur les Péridiniens. Bull. Biol., France Belg., Suppl., 36:1-149, 75 textfigs.

Glenodinium bipes
Meunier, A., 1919
Microplancton de la Mer Flamande 3. Les Péridiniens. Mem. Mus. Roy. Hist. Nat., Belgique 8(1):1-116, Pls. XV-XXI.

Glenodinium cinctum
Calkins, G. N., 1902
Marine protozoa from Woods Hole. U.S. Fish Comm. Bull. for 1901, pp. 413-468, 69 text figs.

Glenodinium compressa n.sp.
Calkins, G. N., 1902
Marine protozoa from Woods Hole. U.S. Fish Comm. Bull. for 1901, pp. 413-468, 69 text figs.

Glenodinum cristatum n. sp.
Balech, Enrique, 1961
Glenodinium cristatum, sp. nov. (Dinoflagellatae). Neotropica, 7:47-51.

Glenodinium danicum
Ercegovic, A., 1936
Etudes qualitative et quantitatives du phytoplancton dans les eaux cotières de l'Adriatique oriental moyen au cours de l'année 1934. Acta Adriatica 1(9):1-126

Glenodinium danicum
LeBour, M.V., 1925
The dinoflagellates of Northern Seas. The Marine Biological Association of the United Kingdom, Plymouth, 250 pp., 35 pls., 53 text figs.

Glenodinium danicum
Lillick, L.C., 1940
Phytoplankton and planktonic protozoa of the offshore waters of the Gulf of Maine. Pt.II. Qualitative Composition of the Planktonic Flora. Trans. Am. Phil. Soc., n.s., 31(3):193-237, 13 text figs.

Glenodinium foliaceum
De Sousa e Silva Estela, 1962.
Some observations on marine dinoflagellate cultures. II. Glenodinium foliaceum Stein and Goniaulax diacantha (Meunier) Schiller. Botanica Marina, 3(3/4):75-100.

Glenodinium foliaceum
de Sousa e Silva, E., 1962.
Some observations on marine Dinoflagellata cultures. II. Glenodinium foliaceum Stein and Goniaulax diacantha (Meunier) Schiller. Notas e Estudos, Inst. Biol. Marit., Lisbon, No. 24:75-100.

Glenodinium foliaceum
Grøntved, J., 1956.
Planktonological contributions. II. Taxonomical studies in some Danish coastal waters. Medd. Danmarks Fiskeri- og Havundersøgelser, n. 1(12):13 pp.

Glenodinium foliaceum
Mandelli, Enrique F., 1968.
Carotenoid pigments of the dinoflagellate Glenodinium foliaceum Stein. J. Phycol., 4(4):347-348.

Glenodinium foliaceum
Paulsen, O., 1908
XVIII Peridiniales. Nordisches Plankton, Bot. Teil: 1-124, 155 text figs.

Glenodinium foliaceum
Schröder, B., 1900
Phytoplankton des Golfes von Neapel nebst vergleichenden Ausblicken auf das atlantischen Ozean. Mitt. Zool. Stat. Neapel, 14:1-38.

Gymnodinium gracile var. sphaerica
Calkins, G. N., 1902
Marine protozoa from Woods Hole. U.S. Fish Comm. Bull. for 1901, pp. 413-468, 69 text figs.

Glenodinium gymnodinium
LeBour, M.V., 1925
The dinoflagellates of Northern Seas. The Marine Biological Association of the United Kingdom, Plymouth, 250 pp., 35 pls., 53 text figs.

Glenodinium Gymnodinium
Paulsen, O., 1908
XVIII Peridiniales. Nordisches Plankton, Bot. Teil: 1-124, 155 text figs.

Glenodinium halli n.sp.
Freudenthal, H.D., and J.J. Lee, 1963.
Glenodinium halli n.sp. and Gyrodinium instriatum n.s., dinoflagellates from New York waters.
J. Protozoology, 10(2):182-189.

Glenodinium halli
Gold, K., 1964.
Aspects of marine dinoflagellate nutrition measures by C14 assimilation.
J. Protozool., 11(1):85-89.

Glenodinium lenticulum
Braarud, T., 1945
A phytoplankton survey of the polluted waters of inner Oslo Fjord. Hvalrådets Skrifter, No.28, 142 pp., 19 text figs., 17 tables.

Glenodinium lenticula
De Angelis, C.M., 1962.
Distribuzione ed ecologia di alcune specie di Dinoflagellati di acque di acque salmastre.
Pubbl. Staz. Zool., Napoli, 32 Suppl.):301-314.

Glenodinium lenticula
Ercegovic, A., 1936
Etudes qualitative et quantitatives du phytoplancton dans les eaux cotières de l'Adriatique oriental moyen au cours de l'année 1934. Acta Adriatica 1(9):1-126

Glenodinium lenticula
Kelly, Mahlon G., and Steven Katona, 1966.
An endogenous diurnal rhythm of bioluminescence in a natural population of dinoflagellates.
Biol. Bull., 131(1):115-126.

Glenodinium lenticula
Levander, K.M., 1947
Plankton gesammelt in den Jahren 1899-1910 an den Küsten Finnlands. Finnländische Hydrographisch-Biologische Untersuchunger (aus dem Wasserbiologischen Laboratorin der Societas Scientiarum Fennica) No.11: 40 pp., 6 diagrams, 13 pls., tables.

Glenodinium lenticulata
Lillick, L.C., 1940
Phytoplankton and planktonic protozoa of the offshore waters of the Gulf of Maine. Pt.II. Qualitative Composition of the Planktonic Flora. Trans. Am. Phil. Soc., n.s., 31(3):193-237, 13 text figs.

Glenodinium lenticula
Rumkówna, A., 1948
[List of the phytoplankton species occurring in the superficial water layers in the Gulf of Gdańsk] Bull. Lab. mar., Gdynia, No. 4: 139-141 with tables in back.

Glenodinium lenticula asymmetrica
Rampi, L., 1951.
Ricerche sul fitoplancton del Mare Ligure. 10) Peridináale delle acque di Sanremo.
Pubbl. Centro Talassografico Tirreno No. 6: Atti Acad. Ligure Sci. Lett., 7(1): 8 pp., Pls. 3-4.

Glenodinium monensis
LeBour, M.V., 1925
The dinoflagellates of Northern Seas. The Marine Biological Association of the United Kingdom, Plymouth, 250 pp., 35 pls.,53 text figs.

Glenodinium monotis
Biecheler, B., 1952.
Recherches sur les Peridiniens.
Bull. Biol., France Belg., Suppl., 36:1-149, 75 textfigs.

Glenodinium monotis
Pincemin, J.-M. 1972.
Besoins en vitamines de trois organismes phytoplanctoniques, Asterionella japonica Prorocentrum micans Glenodinium monotis. Recherche du taux optimal de B12 pour Glenodinium monotis.
Rev. int. Oceanogr. Médic. 26:85-97.

Glenodinium monotis
Pincemin, J.-M. 1971.
Télémediateurs chimiques et équilibre biologique océanique. 3. Étude in vitro de relations entre populations phytoplanctoniques.
Rev. int. Oceanogr. Méd. 22-23:165-196

Glenodinium monotis
Pincemin, J.M. 1962.
Influence de la salinité sur le dinoflagellé Glenodinium monotis.
Rev. int. Oceanogr. Méd. 25:71-87.

Glenodinium obliquum
LeBour, M.V., 1925
The dinoflagellates of Northern Seas. The Marine Biological Association of the United Kingdom, Plymouth, 250 pp., 35 pls.,53 text figs.

Glenodinium obliquum
Schröder, B., 1900
Phytoplankton des Golfes von Neapel nebst vergleichenden Ausblicken auf das atlantischen Ozean. Mitt. Zool. Stat. Neapel, 14:1-38.

Glendonium pillula
Levander, K.M., 1947
Plankton gesammelt in den Jahren 1899-1910 an den Küsten Finnlands. Finnländische Hydrographisch-Biologische Untersuchunger (aus dem Wasserbiologischen Laboratorin der Societas Scientiarum Fennica) No.11: 40 pp., 6 diagrams, 13 pls., tables.

Glenodinium rotundum
Braarud, T., 1945
A phytoplankton survey of the polluted waters of inner Oslo Fjord. Hvalrådets Skrifter, No.28, 142 pp., 19 text figs., 17 tables.

Glenodinium sphaera
Paulsen, O., 1908
XVIII Peridiniales. Nordisches Plankton, Bot. Teil: 1-124, 155 text figs.

Glenodinium sphaera
Schröder, B., 1900
Phytoplankton des Golfes von Neapel nebst vergleichenden Ausblicken auf das atlantischen Ozean. Mitt. Zool. Stat. Neapel, 14:1-38.

Glenodinium trochoideum
Lillick, L.C., 1937
Seasonal studies of the phytoplankton off Woods Hole, Massachusetts. Biol. Bull. LXXIII (3):488-503, 3 text figs.

Glenodinium trochoideum
Murray, G., and F. G. Whitting, 1899
New Peridiniaceae from the Atlantic.
Trans. Linn. Soc., London, Bot., ser 2, 5: 321-342, Pls. 27-33, 9 tables.

Glenodinium trochoideum
Schröder, B., 1900
Phytoplankton des Golfes von Neapel nebst vergleichenden Ausblicken auf das atlantischen Ozean. Mitt. Zool. Stat. Neapel, 14:1-38.

Glenodinium turbo
Paulsen, O., 1908
XVIII Peridiniales. Nordisches Plankton, Bot. Teil: 1-124, 155 text figs.

Glenodinium warmingii
LeBour, M.V., 1925
The dinoflagellates of Northern Seas. The Marine Biological Association of the United Kingdom, Plymouth, 250 pp., 35 pls.,53 text figs.

Gloedinium marinum n.sp.
Bouqhaheux, Françoise, 1971
Gloedinium marinum nov. sp. Peridinien Dinocapsale. Arch. Protistenk Bd. 113: 314-321.

Gonioderma polyedricum
Graham, H. W., 1942
Studies in the morphology, taxonymy, and ecology of the Peridiniales. Sci. Res. Cruise VII of the Carnegie, 1928-1929---Biol. III(542): 129 pp., 67 figs.

Goniodinium ?
Balech, Enrique, 1962
Tintinnoinea y Dinoflagellata del Pacifico segun material de las Expediciones NORPAC y DOWNWIND del Instituto Scripps de Oceanografia.
Revista, Mus. Argentino Ciencias Nat. "Bernardino Rivadavia", Ciencias Zool., 7(1):1-253.

Goniodinium spiniferum n.sp.
Dangeard, P., 1927
Phytoplankton de la croisière du "Sylvana". Ann. Inst. Ocean. Monaco, n.s., 4(8):286-401, 54 text figs. (Feirer-Juin 1913).

Goniodinium cristatum n.sp.
Dangeard, P., 1927
Phytoplankton de la croisière du "Sylvana". Ann. Inst. Ocean. Monaco, n.s., 4(8):286-401, 54 text figs. (Feirer-Juin 1913).

Goniodoma spp.
Sousa e Silva Estela 1969.
Cytological aspects on multiplication of Goniodoma sp.
Botanica mar. 12 (1/2):233-243.

Goniodoma sp.
de Sousa e Silva, E., 1956.
Contribution à l'étude du microplancton de Dakar et des regions maritimes voisines.
Bull. I.F.A.N., 8(2):335-371, 7 pls.

Goniodoma
Murray, G., and F. G. Whitting, 1899
New Peridiniaceae from the Atlantic.
Trans. Linn. Soc., London, Bot., ser 2, 5: 321-342, Pls. 27-33, 9 tables.

Goniodoma acuminatum
Ercegovic, A., 1936
Etudes qualitative et quantitatives du phytoplancton dans les eaux cotières de l'Adriatique oriental moyen au cours de l'année 1934. Acta Adriatica 1(9):1-126

Goniodoma acuminatum
Murray, G., and F. G. Whitting, 1899
New Peridiniaceae from the Atlantic.
Trans. Linn. Soc., London, Bot., ser 2, 5: 321-342, Pls. 27-33, 9 tables.

Goniodoma acuminatum
Pavillard, J., 1916
Recherches sur les Peridiniens du Golfe du Lion. Mem. Univ. Montpellier. Trav. Inst. Bot., Univ. Montpellier. Serie mixte No.4, 70 pp., 3 pls., 15 text figs.

Goniodoma acuminatum
Pavillard, J., 1905
Recherches sur la flore pelagique (Phytoplankton) de l'Etang de Thau. Theses presentees a la Fsc. Sci., Paris, 116 pp., 3 pls.

Goniodoma acuminatum
Schröder, B., 1900
Phytoplankton des Golfes von Neapel nebst vergleichenden Ausblicken auf das atlantischen Ozean. Mitt. Zool. Stat. Neapel, 14:1-38.

Goniodoma acuminatum
Wang, C. C., and D. Nie, 1932
A survey of the marine protozoa of Amoy. Contr. Biol. Lab. Sci. Soc., China, Zool. Ser., 8:285-385, 89 text figs.

Goniodoma acuminatum armatum
Lindemann, E., 1925
Neubeobachtungen an den Winter peridineen des Golfes von Neapel. Bot. Arch. 9:95-102, 19 text figs.

Goniodoma armatum
Pavillard, J., 1905
Recherches sur la flore pelagique (Phytoplankton) de l'Etang de Thau. Theses presentees a la Fac. Sci., Paris, 116 pp., 3 pls.

Goniodoma concavum
Balech, Enrique, 1962
Tintinnoinea y Dinoflagellata del Pacifico segun material de las Expediciones NORPAC y DOWNWIND del Instituto Scripps de Oceanografia. Revista, Mus. Argentino Ciencias Nat. "Bernardino Rivadavia", Ciencias Zool., 7(1):1-253.

Goniodoma fimbriatum
Murray, G., and F. G. Whitting, 1899
New Peridiniaceae from the Atlantic. Trans. Linn. Soc., London, Bot., ser 2, 5: 321-342, Pls. 27-33, 9 tables.

Goniodoma lacustris n.sp.
Lindemann, E., 1924
Peridineen aus dem goldenen Horn und dem Bosphorus. Bot. Arch. 5:216-233, 98 text figs.

Goniodoma Milneri n.sp.
Murray, G., and F. G. Whitting, 1899
New Peridiniaceae from the Atlantic. Trans. Linn. Soc., London, Bot., ser 2, 5: 321-342, Pls. 27-33, 9 tables.

Goniodoma ostenfeldi
LeBour, M.V., 1925
The dinoflagellates of Northern Seas. The Marine Biological Association of the United Kingdom, Plymouth, 250 pp., 35 pls.,53 text figs.

Goniodoma Ostenfeldii
Paulsen, O., 1908
XVIII Peridiniales. Nordisches Plankton, Bot. Teil: 1-124, 155 text figs.

Goniodoma Ostenfeldi
Rumkówna, A., 1948
[List of the phytoplankton species occurring in the superficial water layers in the Gulf of Gdańsk] Bull. Lab. mar. Gdynia, No. 4: 139-141 with tables in back.

Goniodoma polyedricum
Dangeard, P., 1927
Phytoplankton de la croisière du "Sylvana". Ann. Inst. Ocean., Monaco, n.s., 4(8):286-401, 54 text figs. (Feirer-Juin 1913).

Goniodoma polyhedricum
Ercegovic, A., 1936
Etudes qualitative et quantitatives du phytoplancton dans les eaux cotières de l'Adriatique oriental moyen au cours de l'année 1934. Acta Adriatica 1(9):1-126

Goniodoma polyhedricum
Forti, A., 1922
Ricerche sulla flora pelagica (fitoplankton) di Quarto dei Mille. Mem. R. Com. Talass. Ital. 97:248 pp., 13 pls.

Goniodoma polyedricum
Lindemann, E., 1925
Neubeobachtungen an den Winter peridineen des Golfes von Neapel. Bot. Arch. 9:95-102, 19 text figs.

Goniodoma polyedricum
LeBour, M.V., 1925
The dinoflagellates of Northern Seas. The Marine Biological Association of the United Kingdom, Plymouth, 250 pp., 35 pls.,53 text figs.

Goniodoma polyedricum
Mangin, M. L., 1912
Phytoplancton de la croisière du "René" dans l'Atlantique (Septembre 1908). Ann. Inst. Ocean., n.s., 4(1):1-66, 2 pls., 41 text figs., 2 tables.

Gonodioma polyedricum
Margalef, R., 1949
Fitoplancton nerítico de la Costa Brava en 1947-48. Publ. Inst. Biol. Aplicada, 5: 41-51, 3 text figs.

Goniodoma polyedricum
Margalef, R., 1948.
Le phytoplancton estival de la "Costa Brava" catalane en 1946. Hydrobiol. 1(1):15-21.

Goniodoma polyedricum maior
Margalef, R., 1949
Fitoplancton nerítico de la Costa Brava en 1947-48. Publ. Inst. Biol. Aplicada, 5: 41-51, 3 text figs.

Goniodoma polyedricum
Matzenauer, L., 1933
Die Dinoflagellaten des indischen Ozeans (mit Ausnahme der Gattung Ceratium.) Bot. Arch. 35:437-510, 77 text figs., 2 charts.

Goniodoma polyedricum
Paulsen, O., 1908
XVIII Peridiniales. Nordisches Plankton, Bot. Teil: 1-124, 155 text figs.

Goniodoma polyedricum
Rampi, L., 1951.
Ricerche sul fitoplancton del Mare Ligure. 10) Peridiniale delle acque di Sanremo. Pubbl. Centro Talassografico Tirreno No. 6; [Atti Acad. Ligure Sci. Lett.] 7(1): 8 pp., Pls. 3-4.

Goniodoma polyedricum
Wood, E.J.F., 1963
Dinoflagellates in the Australian region. II. Recent collections. Comm. Sci. and Industr. Res. Org., Div. Fish. and Oceanogr., Techn. Paper, No. 14: 55 pp.

Goniodoma pseudogoniaulax n.sp.
Biecheler, B., 1952.
Recherches sur les Peridiniens. Bull. Biol., France Belg., Suppl., 36:1-149, 75 textfigs.

Goniodoma pseudogoniaulax
de Sousa e Silva, Estela, 1965.
Note on some cytophysiological aspects in Prorocentrum micans Ehr. and Goniodoma pseudogoniaulax Biech. from cultures. Notas e Estudos do Inst. Biol. Marit., Lisboa, No. 30:30 pp., 26 pls.

Goniodoma sphaericum
Dangeard, P., 1927
Phytoplankton de la croisière du "Sylvana". Ann. Inst. Ocean., Monaco, n.s., 4(8):286-401, 54 text figs, (Février-Juin 1913).

Goniodoma sphaericum
Dangeard, P., 1927
Péridiniens nouveaux ou peu connus de la croisière du "Sylvana". Bull. Inst. Océan., Monaco, No. 491:16 pp., 9 textfigs.

Goniodoma sphaerica
Hada, Yoshina, 1970. (1970)
The protozoan plankton of the Antarctic and Subantarctic seas. Scient. Repts, Japan. Antarct. Res. Exped., (E)31:51 pp.

Gonodioma sphaericum
Margalef, R., 1949
Fitoplancton nerítico de la Costa Brava en 1947-48. Publ. Inst. Biol. Aplicada, 5: 41-51, 3 text figs.

Goniodoma sphaericum
Matzenauer, L., 1933
Die Dinoflagellaten des indischen Ozeans (mit Ausnahme der Gattung Ceratium.) Bot. Arch. 35:437-510, 77 text figs., 2 charts.

Goniodoma sphaericum n.sp.
Murray, G., and F. G. Whitting, 1899
New Peridiniaceae from the Atlantic. Trans. Linn. Soc., London, Bot., ser 2, 5: 321-342, Pls. 27-33, 9 tables.

Goniodema sphaericum
Rampi, L., 1951.
Ricerche dul fitoplancton del Mare Ligure. 10) Peridiniale delle acque di Sanremo. Pubbl. Centro Talassografico Tirreno. No. 6; [Atti Acad. Ligure Sci. Lett.] 7(1): 8 pp., Pls. 3-4.

Goniodoma sphaericum
Schröder, B., 1900
Phytoplankton des Golfes von Neapel nebst vergleichenden Ausblicken auf das atlantischen Ozean. Mitt. Zool. Stat. Neapel, 14:1-38.

Gonyaulax sp.
Balech, Enrique, 1962
Tintinnoinea y Dinoflagellata del Pacifico segun material de las Expediciones NORPAC y DOWNWIND del Instituto Scripps de Oceanografia. Revista, Mus. Argentino Ciencias Nat. "Bernardino Rivadavia", Ciencias Zool., 7(1):1-253.

Goniaulax sp.
Braarud, T., 1945
A phytoplankton survey of the polluted waters of inner Oslo Fjord. Hvalrådets Skrifter, No.28, 142 pp., 19 text figs., 17 tables.

Goniaulax sp.
Gilbert, J.Y., and W.E. Allen, 1943
The phytoplankton of the Gulf of California obtained by the "E.W. Scripps" in 1939 and 1940. J. Mar. Res. V(2):89-110, figs.30-31.

Gonyaulax
Kofoid, A. C., 1911
Dinoflagellata of the San Diego region. IV The genus Gonyaulax. Univ. Calif. Publ. Zool. 8(4):187-286, Pls. 9-17.

Gonyaulax
Nishikawa, T., 1901
Gonyaulax and the discolored water in the Bay of Agu. Annotationes Zool. Japan, IV(1):31-34.

Gonyaulax sp.
Sommer, H., Whedon, W.F., Kofoid, C.A., and A. Stohler, 1937.
Relation of paralytic shellfish poison to certain plankton organisms of the genus Gonyaulax. Arch. Path., Vol. 24, pp. 537-559

Oceanographic Index: Marine Organisms Cumulation, 1946-1973

Gonyaulax spp.
Wall, David, and Barrie Dale, 1968.
Modern dinoflagellate cysts and evolution of the peridiniales.
Micropaleontology. 14(3):265-304.

Gonyaulax acatenella
Prakash, A., and F.J.R. Taylor, 1966.
A "red water" bloom of Gonyaulex acatenella in the Strait of Georgia and its relation to paralytic shellfish toxicity.
J. Fish. Res. Bd., Canada. 23(8):1265-1270.

Gonyaulax acatenella
Quayle, D.B.
Paralytic shellfish poisoning in British Colombia.
Bull. Fish Res. Bd. Can., 168: 68 pp.

Gonyiaulax acatanella
Whedon, W.F., and Kofoid, C.A., 1936
Dinoflagellata of the San Francisco Region. 1. On the skeletal morphology of two new species, Gonyiaulex catanella and G. acatanella. Univ. Calif. Publ. Zool. 41: 25-34.

Goniaulax africana?
Conover, S.A.M., 1954.
Observation on the structure of red tides in New Haven, Connecticut. J. Mar. Res. 13(1):145-155, 6 textfigs.

Gonyaulax alaskensis
Dangeard, P., 1927
Phytoplankton de la croisière du "Sylvana". Ann. Inst. Ocean., Monaco, n.s., 4(8):286-401, 54 text figs. (Feirer-Juin 1913).

Gonyaulax alaskensis
Dangeard, P., 1926
Description des Péridiniens Testacés recueillis para la Mission Charcot pendent le mois d'Aout 1924. Ann. Inst. Ocean. n.s. 3(7):307-334, 15 text figs.

Gonyaulax alaskensis n.sp.
Kofoid, A. C., 1911
Dinoflagellata of the San Diego region. IV The genus Gonyaulax. Univ. Calif. Publ. Zool. 8(4):187-286, Pls. 9-17.

Gonyaulax (Steiniella) alaskensis
Pavillard, J., 1916
Recherches sur les Peridiniens du Golfe du Lion. Mem. Univ. Montpellier. Trav. Inst. Bot., Univ. Montpellier. Serie mixte No.4, 70 pp., 3 pls., 15 text figs.

Goniaulax alaskensis
Rampi, L., 1945
Osservazioni sulla distribuzione qualitativa del fitoplancton nel mare Mediterraneo. Atti della Soc. Ital. di Sci. Nat. 84:105-113.

Goniaulax alaskensis
Rampi, L., 1943
Richerche sul fitoplancton del Mare Ligure. F. Le Goniaulacee delle acque di Sanremo. Atti della Soc. Ital. di Scienze Naturali, 82:1-12, figs.1-16.

Gonyaulax alaskensis
Rampi, L., 1942
Il Fitoplancton mediterraneo: Problemi ed affinita interoceaniche. Boll. di Pesca di Piscicoltura e di Idrobiologia, Anno 18, Fasc. 4:7-19.

Goniaulax apiculata
LeBour, M.V., 1925
The dinoflagellates of Northern Seas. The Marine Biological Association of the United Kingdom, Plymouth, 250 pp., 35 pls.,53 text figs.

Gonyaulax apiculata
Paulsen, O., 1908
XVIII Peridiniales. Nordisches Plankton, Bot. Teil: 1-124, 155 text figs.

Gonyaulax balechii n.sp.
Steidinger, Karen A. 1971.
Gonyaulax balechii sp.nov. (Dinophyceae) with a discussion of the genera Gonyaulax and Heteraulus.
Phycologia 10(2/3):183-187.

Gonyaulax birostris
Balech, Enrique, 1962
Tintinnoinea y Dinoflagellata del Pacifico segun material de las Expediciones NORPAC y DOWNWIND del Instituto Scripps de Oceanografia.
Revista. Mus. Argentino Ciencias Nat. "Bernardino Rivadavia", Ciencias Zool. 7(1):1-253.

Gonyaulax birostris
Dangeard, P., 1926
Description des Péridiniens Testacés recueillis para la Mission Charcot pendent le mois d'Aout 1924. Ann. Inst. Ocean. n.s. 3(7):307-334, 15 text figs.

Gonyaulax birostris
Kofoid, A. C., 1911
Dinoflagellata of the San Diego region. IV The genus Gonyaulax. Univ. Calif. Publ. Zool. 8(4):187-286, Pls. 9-17.

Gonyaulax birostris
Matzenauer, L., 1933
Die Dinoflagellaten des indischen Ozeans (mit Ausnahme der Gattung Ceratium.) Bot. Arch. 35:437-510, 77 text figs., 2 charts.

Gonyaulax birostris
Murray, G., and F. G. Whitting, 1899
New Peridiniaceae from the Atlantic.
Trans. Linn. Soc., London, Bot., ser 2, 5: 321-342, Pls. 27-33, 9 tables.

Goniaulax birostris
Rampi, L., 1943
Richerche sul fitoplancton del Mare Ligure. F. Le Goniaulacee delle acque di Sanremo. Atti della Soc. Ital. di Scienze Naturali, 82:1-12, figs.1-16.

Goniaulax borealis n.sp.
Nordli, O., 1951.
Dinoflagellates from Lofoten. Nytt Mag. Naturvidensk 88:49-55, 7 textfigs.

Goniaulax braarudii nom. nov.
Hasle, Grethe Rytter, 1960
Phytoplankton and ciliate species from the Tropical Pacific.
Skr. Norske Videnskaps-Akad., Oslo, 1. Mat.-Nat. Kl., 1960(2): 1-50.

Gonyaulax brevisulcatum n.sp.
Dangeard, P., 1927
Phytoplankton de la croisière du "Sylvana". Ann. Inst. Ocean., Monaco, n.s., 4(8):286-401, 54 text figs. (Fevier -Juin 1913).

Gonyaulax buxus n.sp.
Balech,Enrique,1967.
Dinoflagelados nuevos o interesantes del Golfo de Mexico y Caribe.
Revta Mus.argent.Cienc.nat.Bernardino Rivadavia Inst.nac.Invest.Cienc.nat., Hidrobiol. 2(3):77-126.

Gonyaulax catenata
Brunel, J., 1962
Le phytoplancton de la Baie de Chaleurs. Inst. Botan., Univ. Montreal, Contrib. No. 77: 365 pp., 66 pls.

Gonyaulax catenata
Connell, C.H., and J.B., 1950.
Mass mortality of fish associated with the protozoan Gonyaulax in the Gulf of Mexico. Science 112(2909):359-363, 5 textfigs.

Goniaulax catenata
Ghazzawi, F.M., 1939
Plankton of the Egyptian waters. A study of the Suez Canal Plankton. (A) Phytoplankton. Preliminary Report 83 pp. Notes and Memoires, Min. Commerce-Industry, Egypt, Hydrobiol. & Fish. 65 figs.

Goniaulax catenata
LeBour, M.V., 1925
The dinoflagellates of Northern Seas. The Marine Biological Association of the United Kingdom, Plymouth, 250 pp., 35 pls.,53 text figs.

Goniaulax cantenata
Levander, K.M., 1947
Plankton gesammelt in den Jahren 1899-1910 an den Küsten Finnlands. Finnländische Hydrographisch-Biologishbe Untersuchunger (aus dem Wasserbiologischen Laboratorin der Societas Scientiarum Fenniea) No.11: 40 pp., 6 diagrams, 13 pls., tables.

Goniaulax catenata
Rumkówna, A., 1948
List of the phytoplankton species occurring in the superficial water layers in the Gulf of Gdańsk. Bull. Lab. mar., Gdynia, No. 4: 139-141 with tables in back.

Goniaulax catenella
Fowler, H.L., 1943.
Shellfish poison. Nat. Hist. 51(5):228-229.

Goniaulax catennella
Gilbert, J.Y., and W.E. Allen, 1943
The phytoplankton of the Gulf of California obtained by the "E.W. Scripps" in 1939 and 1940. J. Mar. Res. V(2):89-110, figs.30-31.

Goniaulax catenella
Halldal, P., 1958.
Action spectra of phototaxis and related problems in Volvocales, Ulva-gametes and Dinophyceae. Physiol. Plant., 11:118-153.

Goniaulax catenella
Morse, D.C., 1947
Some observations on seasonal variations in plankton population Patuxant River, Maryland 1943-1945. Bd. Nat. Res., Publ. No.65, Chesapeake Biol. Lab., 31, 3 figs.

Gonyaulax catenella
Taylor,F.J.R.,1968.
Parasitism of the toxin-producing dinoflagellate Gonyaulax catenella by the endo parasitic dinoflagellate Amoebophyra ceratii.
J. Fish.Res.Bd.,Can. 25(10):2241-2245.

Gonyiaulax catenella
Whedon, W.F., and Kofoid, C.A., 1936
Dinoflagellata of the San Francisco Region. 1. On the skeletal morphology of two new species, Gonyiaulax catenella and G. acatanella. Univ. Calif. Publ. Zool. 41: 25-34.

Gonyaulax claudus n.sp.
Marukawa, H., 1928.
Ueber 4 neue Arten der Peridinialen. Annot. Oceanogr. Res., 2(1):1-2.

Gonisulax cochlea?
Conover, S.A.M., 1954.
Observation on the structure of red tides in New Haven Harbor, Connecticut. J. Mar. Res. 13(1):145-155, 6 textfigs.

Goniaulax cochlea n.sp.
Meunier, A., 1919
Microplancton de la Mer Flamande 3. Les Péridiniens. Mem. Mus. Roy. Hist. Nat., Belgique 8(1):1-116, Pls. XV-XXI.

Gonyaulax cohorticula n.sp.
Balech,Enrique,1967.
Dinoflagelados nuevos o interesantes del Golfo de Mexico y Caribe.
Revta Mus.argent.Cienc.nat.Bernardino Rivadavia Inst.nac.Invest.Cienc.nat., Hidrobiol. 2(3):77-126.

Gonyaulax concava nov. comb.

Balech, Enrique, 1967.
Dinoflagelados nuevos o interesantes del Golfo de Mexico y Caribe.
Revta Mus. argent. Cienc. nat. Bernardino Rivadavia Inst. nac. Invest. Cienc. nat. Hidrobiol. 2(3):77-126.

Goniaulax diacantha

de Sousa e Silva, E., 1962.
Some observations on marine Dinoflagellata cultures. II. Glenodinium foliaceum Stein and
Notas e Estudos, Inst. Biol. Marit., Lisbon, No. 24:75-100.

Goniaulax diacantha

De Sousa e Silva, Estela, 1962.
Some observations on marine dinoflagellate cultures. II. Glenodinium foliaceum Stein and Goniaulax diacantha (Meunier) Schiller.
Botanica Marina, 3(3/4):75-100.

Goniaulax diacantha

Ercegovic, A., 1936
Etudes qualitative et quantitatives du phytoplancton dans les eaux cotières de l'Adriatique oriental moyen au cours de l'année 1934. Acta Adriatica 1(9):1-126

Gonyaulax diegensis

Dangeard, P., 1927
Phytoplankton de la croisière du "Sylvana". Ann. Inst. Ocean., Monaco, n.s., 4(8):286-401, 54 text figs. (Feirer-Juin 1913).

Gonyaulax diegensis n.sp.

Kofoid, A. C., 1911
Dinoflagellata of the San Diego region. IV The genus Gonyaulax. Univ. Calif. Publ. Zool. 8(4):187-286, Pls. 9-17.

Goniaulax diegensis

LeBour, M.V., 1925
The dinoflagellates of Northern Seas. The Marine Biological Association of the United Kingdom, Plymouth, 250 pp., 35 pls., 53 text figs.

Gonyaulax diegensis

Pavillard, J., 1916
Recherches sur les Peridiniens du Golfe du Lion. Mem. Univ. Montpellier. Trav. Inst. Bot., Univ. Montpellier. Serie mixte No.4, 70 pp., 3 pls., 15 text figs.

Goniaulax diegensis

Rampi, L., 1943
Richerche sul fitoplancton del Mare Ligure. F. Le Goniaulacee delle acque di Sanremo. AH1 della Soc. Ital. di Scienze Naturali, 82:1-12, figs.1-16.

Goniaulax diegensis (figs.)

Sousa e Silva, E., 1949
Diatomaceas e Dinoflagelados de Baia de Cascais. Portugaliae Acta Biol., Volume: Julio Henriques, Ser. B: 300-383, 9 pls, 2 fold-in tables.

Gonyaulax digitale

Dangeard, P., 1927
Phytoplankton de la croisière du "Sylvana". Ann. Inst. Ocean., Monaco, n.s., 4(8):286-401, 54 text figs. (Février-Juin 1913).

Goniaulax digitale

Dangeard, P., 1926
Description des Péridiniens Testacés recueillis para la Mission Charcot pendent le mois d'Aout 1924. Ann. Inst. Ocean. n.s. 3(7):307-334, 15 text figs.

Goniaulax digitale

Ercegovic, A., 1936
Etudes qualitative et quantitatives du phytoplancton dans les eaux cotières de l'Adriatique oriental moyen au cours de l'année 1934. Acta Adriatica 1(9):1-126

Gonyaulax digitale

Evitt, William R., and Susan E. Davidson, 1964.
Dinoflagellate studies. 1. Dinoflagellate cysts and thecae.
Stanford Univ. Publ., Geol. Sci., 10(1):1-12.

Gonyaulax digitale

Forti, A., 1922
Ricerche sulla flora pelagica (fitoplancton) di Quarto dei Mille. Mem. R. Com. Talass. Ital. 97:248 pp., 13 pls.

Gonyaulax digitale

Kofoid, A. C., 1911
Dinoflagellata of the San Diego region. IV The genus Gonyaulax. Univ. Calif. Publ. Zool. 8(4):187-286, Pls. 9-17.

Goniaulax digitale

LeBour, M.V., 1925
The dinoflagellates of Northern Seas. The Marine Biological Association of the United Kingdom, Plymouth, 250 pp., 35 pls., 53 text figs.

Goniaulax digitale

Lillick, L.C., 1940
Phytoplankton and planktonic protozoa of the offshore waters of the Gulf of Maine. Pt.II. Qualitative Composition of the Planktonic Flora. Trans. Am. Phil. Soc., n.s., 31(3):193-237, 13 text figs.

Gonyaulax digitale

Margalef, R., 1949
Fitoplancton nerítico de la Costa Brava en 1947-48. Publ. Inst. Biol. Aplicada, 5: 41-51, 3 text figs.

Goniaulax digitale

Morse, D.C., 1947
Some observations on seasonal variations in plankton population Patuxant River, Maryland 1943-1945. Bd. Nat. Res., Publ. No.65, Chesapeake Biol. Lab., 31, 3 figs.

Gonyaulax digitale

Pavillard, J., 1916
Recherches sur les Peridiniens du Golfe du Lion. Mem. Univ. Montpellier. Trav. Inst. Bot., Univ. Montpellier. Serie mixte No.4, 70 pp., 3 pls., 15 text figs.

Goniaulax digitale

Rampi, L., 1943
Richerche sul fitoplancton del Mare Ligure. F. Le Goniaulacee delle acque di Sanremo. AH1 della Soc. Ital. di Scienze Naturali, 82:1-12, figs.1-16.

Gonyaulax digitalis

Wall, David, and Barrie Dale 1970.
Living hystrichosphaerid dinoflagellate spores from Bermuda and Puerto Rico.
Micropaleontology 16(1):47-58

Gonyaulax digitalis

Wall, David, and Barrie Dale 1968.
Modern dinoflagellate cysts and evolution of the Peridiniales.
Micropaleontology, 14(3):265-304.

Goniaulax dimorpha n.sp.

Biecheler, B., 1952.
Recherches sur les Peridiniens.
Bull. Biol., France Belg., Suppl., 36:1-149, 75 textfigs.

Gonyaulax excavata n. comb.

Balech, Enrique, 1971
Microplancton del Atlantico ecuatorial oeste (Equalant 1)
Publ. Serv. Hidrograf. Naval, Argentina H. 654: 103 pp., 122 figs.

Gonyaulax fragilis

Balech, Enrique, 1962
Tintinnoinea y Dinoflagellata del Pacifico segun material de las Expediciones NORPAC y DOWNWIND del Instituto Scripps de Oceanografia.
Revista. Mus. Argentino Ciencias Nat. "Bernardino Rivadavia". Ciencias Zool., 7(1):1-253.

Gonyaulax fragilis

Dangeard, P., 1926
Description des Péridiniens Testacés recueillis para la Mission Charcot pendent le mois d'Aout 1924. Ann. Inst. Ocean. n.s. 3(7):307-334, 15 text figs.

Gonyaulax fragilis

Kofoid, A. C., 1911
Dinoflagellata of the San Diego region. IV The genus Gonyaulax. Univ. Calif. Publ. Zool. 8(4):187-286, Pls. 9-17.

Goniaulax fragilis

LeBour, M.V., 1925
The dinoflagellates of Northern Seas. The Marine Biological Association of the United Kingdom, Plymouth, 250 pp., 35 pls., 53 text figs.

Gonyaulax (Steiniella) fragilis

Pavillard, J., 1916
Recherches sur les Peridiniens du Golfe du Lion. Mem. Univ. Montpellier. Trav. Inst. Bot., Univ. Montpellier. Serie mixte No.4, 70 pp., 3 pls., 15 text figs.

Goniaulax fragilis

Rampi, L., 1943
Richerche sul fitoplancton del Mare Ligure. F. Le Goniaulacee delle acque di Sanremo. AH1 della Soc. Ital. di Scienze Naturali, 82:1-12, figs.1-16.

Gonyaulax fratercula n.sp.

Balech, Enrique, 1964.
El plancton de Mar del Plata durante el periodo 1961-1962.
Bol. Inst. Biol. Mar., Buenos Aires, (4):56 pp.
49 pp. + plates

Gonyaulax fusiformis

Balech, Enrique, 1962
Tintinnoinea y Dinoflagellata del Pacifico segun material de las Expediciones NORPAC y DOWNWIND del Instituto Scripps de Oceanografia.
Revista. Mus. Argentino Ciencias Nat. "Bernardino Rivadavia". Ciencias Zool., 7(1):1-253.

Gonyaulax fusiformis n.sp.

Graham, H. W., 1942
Studies in the morphology, taxonymy, and ecology of the Peridiniales. Sci. Res. Cruise VII of the Carnegie, 1928-1929---Biol. III(542): 129 pp., 67 figs.

Gonyaualx glyptorhynchus n.sp.

Murray, G., and F. G. Whitting, 1899
New Peridiniaceae from the Atlantic.
Trans. Linn. Soc., London, Bot., ser 2, 5: 321-342, Pls. 27-33, 9 tables.

Gonyaulax golliffei

Schröder, B., 1900
Phytoplankton des Golfes von Neapel nebst vergleichenden Ausblicken auf das atlantischen Ozean. Mitt. Zool. Stat. Neapel, 14:1-38.

Gonyaulax grindleyi

Balech, Enrique, 1971.
Microplancton de la campaña oceanographica: Productividad III.
Revta Mus. argent. Cienc. Nat. Bernadina Rivadavia, Hydrobiol. 3(1):1-202, 39 pls.

Gonyaulax grindleyi = Protoceratium reticulatum
Grindley, J.R. and E.A. Nel, 1970.
Redwater and mussel poisoning at
Elands Bay, December 1966.
Fish. Bull. S.Afr. 6: 36-55.

Gonyaulax grindleyi
Von Stosch, H.A. 1969.
Dinoflagellaten aus der Nordsee. I.
Über Cachonina niei Loeblich (1968),
Gonyaulax grindleyi Reinecke (1967)
und eine Methode zur Darstellung
von Peridineenpanzern.
Helgoländer wiss. Meeresunters. 19(4):
558-568.

Goniaulax helensis

Rumkówna, A., 1948
[List of the phytoplankton species occurring in the superficial water layers in the Gulf of Gdańsk] Bull. Lab. mar., Gdynia, No. 4: 139-141 with tables in back.

Gonyaualx higleii n.sp.

Murray, G., and F. G. Whitting, 1899
New Peridiniaceae from the Atlantic.
Trans. Linn. Soc., London, Bot., ser 2, 5: 321-342, Pls. 27-33, 9 tables.

Gonyaulax hyalina

Matzenauer, L., 1933
Die Dinoflagellaten des indischen Ozeans (mit Ausnahme der Gattung Ceratium.) Bot. Arch. 35:437-510, 77 text figs., 2 charts.

Gonyaulax inflata

Balech, Enrique, 1962
Tintinnoinea y Dinoflagellata del Pacifico segun material de las Expediciones NORPAC y DOWNWIND del Instituto Scripps de Oceanografia.
Revista, Mus. Argentino Ciencias Nat. "Bernardino Rivadavia", Ciencias Zool., 7(1):1-253.

Goniaulax inflata

Ercegovic, A., 1936
Etudes qualitative et quantitatives du phytoplancton dans les eaux cotières de l'Adriatique oriental moyen au cours de l'année 1934. Acta Adriatica 1(9):1-126

Goniaulax inflata

Rampi, L., 1950.
Péridiniens rares ou nouveaux pour le Pacifique Sud-Equatorial. Bull. Inst. Océan., Monaco, No. 974:11 pp., 26 textfigs.

Gonyaulax japonicus n.sp.

Marukawa, H., 1928.
Ueber 4 neue Arten der Peridinialen.
Annot. Oceanogr. Res., 2(1):1-2.

Gonyaulax Jolliffei n.sp.

Murray, G., and F. G. Whitting, 1899
New Peridiniaceae from the Atlantic.
Trans. Linn. Soc., London, Bot., ser 2, 5: 321-342, Pls. 27-33, 9 tables.

Gonyaulax kofoidi

Balech, Enrique, 1962
Tintinnoinea y Dinoflagellata del Pacifico segun material de las Expediciones NORPAC y DOWNWIND del Instituto Scripps de Oceanografia.
Revista, Mus. Argentino Ciencias Nat. "Bernardino Rivadavia", Ciencias Zool., 7(1):1-253.

Gonyaulax kofoidii

Dangeard, P., 1927
Phytoplankton de la croisière du "Sylvana". Ann. Inst. Ocean., Monaco, n.s., 4(8):286-401, 54 text figs. (Feirer-Juin 1913).

Gonyaulax kofoidi

Kofoid, A. C., 1911
Dinoflagellata of the San Diego region. IV The genus Gonyaulax. Univ. Calif. Publ. Zool. 8(4):187-286, Pls. 9-17.

Gonyaulax kofoidi

Lindemann, E., 1925
Neubeobachtungen an den Winter peridineen des Golfes von Neapel. Bot. Arch. 9:95-102, 19 text figs.

Gonyaulax kofoidi

Pavillard, J., 1916
Recherches sur les Peridiniens du Golfe du Lion. Mem. Univ. Montpellier. Trav. Inst. Bot., Univ. Montpellier. Serie mixte No.4, 70 pp., 3 pls., 15 text figs.

Goniaulax kofoidi

Rampi, L., 1943
Richerche sul fitoplancton del Mare Ligure. F. Le Goniaulacee delle acque di Sanremo. AHi della Soc. Ital. di Scienze Naturali, 82:1-12, figs.1-16.

Goniaulax kofoidi

Sousa e Silva, E., 1949
Diatomaceas e Dinoflagelados de Baia de Cascais. Portugaliae Acta Biol., Volume: Julio Henriques, Ser. B: 300-383, 9 pls, 2 fold-in tables.

Goniaulax Levanderi

Levander, K.M., 1947
Plankton gesammelt in den Jahren 1899-1910 an den Küsten Finnlands. Finnländische Hydrographisch-Biologische Untersuchungen (aus dem Wasserbiologischen Laboratorin der Societas Scientiarum Fenniea) No.11: 40 pp., 6 diagrams, 13 pls., tables.

Gonyaulax levanderi

Lindemann, E., 1924
Peridineen aus dem goldenen Horn und dem Bosphorus. Bot. Arch. 5:216-233, 98 text figs.

Gonyaulax Levanderi

Marukawa, H., 1921
Plankton lists and some new species of copepods, from the northern waters of Japan. Bull. Inst. Ocean., No.384, 15 pp., 3 pls., 1 chart. Monaco

Gonyaulax Levanderi

Paulsen, O., 1908
XVIII Peridiniales. Nordisches Plankton, Bot., Teil: 1-124, 155 text figs.

Goniaulax loculatum n.sp.

Meunier, A., 1919
Microplankton de la Mer Flamande 3. Les Péridiniens. Mem. Mus. Roy. Hist. Nat., Belgique 8(1):1-116, Pls. XV-XXI.

Goniaulax longispina n.sp.

LeBour, M.V., 1925
The dinoflagellates of Northern Seas. The Marine Biological Association of the United Kingdom, Plymouth, 250 pp., 35 pls.,53 text figs.

Gonyaulax longispina

Matzenauer, L., 1933
Die Dinoflagellaten des indischen Ozeans (mit Ausnahme der Gattung Ceratium.) Bot. Arch. 35:437-510, 77 text figs., 2 charts.

Gonyaulax macroporus ?

Balech, Enrique, 1971
Microplancton de la campaña oceanographica: Productividad III.
Revta Mus. argent. Cienc. Nat. Bernadina Rivadavia, Hydrobiol. 3(1):1-202, 39 pls.

Goniaulax milneri

Wood, E.J.F., 1963
Dinoflagellates in the Australian region. II. Recent collections.
Comm. Sci. and Industr. Res. Org., Div. Fish. and Oceanogr., Techn. Paper, No. 14: 55 pp.

Gonyaulax minima n.sp.

Matzenauer, L., 1933
Die Dinoflagellaten des indischen Ozeans (mit Ausnahme der Gattung Ceratium,) Bot. Arch. 35:437-510, 77 text figs., 2 charts.

Goniaulax mitra

Ercegovic, A., 1936
Etudes qualitative et quantitatives du phytoplancton dans les eaux cotières de l'Adriatique oriental moyen au cours de l'année 1934. Acta Adriatica 1(9):1-126

Goniaulax (Steiniella) mitra

Pavillard, J., 1916
Recherches sur les Peridiniens du Golfe du Lion. Mem. Univ. Montpellier. Trav. Inst. Bot., Univ. Montpellier. Serie mixte No.4, 70 pp., 3 pls., 15 text figs.

Goniaulax monacantha

De Angelis, C.M., 1962.
Distribuzione ed ecologia di alcune specie di Dinoflagellati di acque salmastre.
Pubbl. Staz. Zool., Napoli, 32 (Suppl.):301-314.

Goniaulax monacantha

Ercegovic, A., 1936
Etudes qualitative et quantitatives du phytoplancton dans les eaux cotières de l'Adriatique oriental moyen au cours de l'année 1934. Acta Adriatica 1(9):1-126

Gonyaulax monacantha

Forti, A., 1922
Ricerche sulla flora pelagica (fitoplancton) di Quarto dei Mille. Mem. R. Com. Talass. Ital. 97:248 pp., 13 pls.

Gonyaulax monacantha n.sp.

Pavillard, J., 1916
Recherches sur les Peridiniens du Golfe du Lion. Mem. Univ. Montpellier. Trav. Inst. Bot., Univ. Montpellier. Serie mixte No.4, 70 pp., 3 pls., 15 text figs.

Goniaulax monacantha

Rampi, L., 1943
Richerche sul fitoplancton del Mare Ligure. F. Le Goniaulacee delle acque di Sanremo. AHi della Soc. Ital. di Scienze Naturali, 82:1-12, figs.1-16.

Gonyaulax monacantha major

Forti, A., 1922
Ricerche sulla flora pelagica (fitoplancton) di Quarto dei Mille. Mem. R. Com. Talass. Ital. 97:248 pp., 13 pls.

Gonyaulax monacantha maur

Margalef, R., 1949
Fitoplancton nerítico de la Costa Brava en 1947-48. Publ. Inst. Biol. Aplicada, 5: 41-51, 3 text figs.

Gonyaulax monacantha minor

Balech, Enrique, 1971
Microplancton del Atlantico ecuatorial oeste (Equalant 1)
Publ. Serv. Hidrograf. Naval, Argentina H. 654: 103 pp., 122 figs.

Gonyaulax monacantha minor
Forti, A., 1922
Ricerche sulla flora pelagica (fitoplancton) di Quarto dei Mille. Mem. R. Com. Talass. Ital. 97:248 pp., 13 pls.

Gonyaulax manocantha minor
Margalef, R., 1949
Fitoplancton nerítico de la Costa Brava en 1947-48. Publ. Inst. Biol. Aplicada, 5: 41-51, 3 text figs.

Gonyaulax monilata
Aldrich, David V., Sammy M. Ray and William B. Wilson, 1967.
Gonyaulax monilata: population growth and development of toxicity in cultures.
J. Protozool. 14(4):636-639.

Gonyaulax monilata
Chunosoff, Laura, and H.I. Hirshfield, 1968.
The effects of chloramphenicol and actinomycin D on the nucleus of the dinoflagellate Gonyaulax monilata.
J. gen. Microbiol., 50(2):281-283.

Gonyaulax monilata
Chunosoff, Laura, and H.I. Hirshfield, 1968.
Giant carbohydrate-rich cells of the dinoflagellate Gonyaulax Monilata.
J. gen. Microbiol., 50(2):277-279.

Gonyaulax monilata
Gates, Jean A., and William B. Wilson, 1960
The toxicity of Gonyaulax monilata Howell to Mugil cephalus. Limnol. & Oceanogr. 5(2): 171-174.

Gonyaulax monilata
Gaudsmith, Joanne T., and Clinton J. Dawes 1972
The ultrastructure of several dinoflagellates with emphasis on Gonyaulax polyedra Stein and Gonyaulax monilata Davis
Phycologia 11(2): 123-132

Gonyaulax monilata sp. nov
Howell, J.F., 1953.
Gonyaulax monilata sp. nov., the causative dinoflagellate of a red tide on the east coast of Florida in August-September 1951. Trans. Amer. Microsc. Soc. 72(2):153-156, 5 textfigs.

Gonyaulax monilata
Sievers, Anita M., 1969.
Comparative toxicity of Gonyaulax monilata and Gymnodinium breve to annelids, crustaceans, molluscs and a fish.
J. Protozool., 16(3): 401-404

Gonyaulax monilata
Williams, Jean and Robert M. Ingle 1972.
Ecological notes on Gonyaulax monilata (Dinophyceae) blooms along the west coast of Florida.
Leaflet Ser. Mar. Res. Lab. St. Petersberg, Fla. Dept. Nat. Res. 1 (Phytop.) (1-Dinoflagellates) No. 5: 12 pp.

Goniaulax orientalis
LeBour, M.V., 1925
The dinoflagellates of Northern Seas. The Marine Biological Association of the United Kingdom, Plymouth, 250 pp., 35 pls., 53 text figs.

Goniaulax orientalis
Lillick, L.C., 1940
Phytoplankton and planktonic protozoa of the offshore waters of the Gulf of Maine. Pt.II. Qualitative Composition of the Planktonic Flora. Trans. Am. Phil. Soc., n.s. 31(3):193-237, 13 text figs.

Gonyaulax orientalis n.sp.
Lindemann, E., 1924
Peridineen aus dem goldenen Horn und dem Bosphorus. Bot. Arch. 5:216-233, 98 text figs.

Gonyaulax orientalis
Matzenauer, L., 1933
Die Dinoflagellaten des indischen Ozeans (mit Ausnahme der Gattung Ceratium). Bot. Arch. 35:437-510, 77 text figs., 2 charts.

Goniaulax Ostenfeldii
Paulsen, O., 1949
Observations on dinoflagellates. (Ed. J. Grøntoed) Kongl. Dansk. Videnskab. Selsk. Biol. Skr. 6(4):67 pp., 30 text figs.

Gonyaulax ovata n.sp.
Matzenauer, L., 1933
Die Dinoflagellaten des indischen Ozeans (mit Ausnahme der Gattung Ceratium) Bot. Arch. 35:437-510, 77 text figs., 2 charts.

Gonyaulax pacifica
Balech, Enrique, 1962
Tintinnoinea y Dinoflagellata del Pacifico segun material de las Expediciones NORPAC y DOWNWIND del Instituto Scripps de Oceanografia.
Revista, Mus. Argentino Ciencias Nat. "Bernardino Rivadavia", Ciencias Zool., 7(1):1-253.

Gonyaulax pacifica
Dangeard, P., 1927
Phytoplankton de la croisière du "Sylvana". Ann. Inst. Ocean., Monaco, n.s., 4(8):286-401, 54 text figs. (février-Juin 1913).

Gonyaulax pacifica
Graham, H. W., 1942
Studies in the morphology, taxonymy, and ecology of the Peridiniales. Sci. Res. Cruise VII of the Carnegie, 1928-1929---Biol. III(542): 129 pp., 67 figs.

Gonyaulax pacifica
Kofoid, A. C., 1911
Dinoflagellata of the San Diego region. IV The genus Gonyaulax. Univ. Calif. Publ. Zool. 8(4):187-286, Pls. 9-17.

Gonyaulax pacifica n. sp.
Kofoid, C. A., 1907
Dinoflagellata of the San Diego region. III. Descriptions of new species. Univ. Calif. Publ., Zool. 3:299-340, Pls. 22-33.

Gonyaulax pacifica
Pavillard, J., 1916
Recherches sur les Peridiniens du Golfe du Lion. Mem. Univ. Montpellier. Trav. Inst. Bot., Univ. Montpellier. Serie mixte No.4, 70 pp., 3 pls., 15 text figs.

Goniaulax pacifica
Rampi, L., 1945
Osservazioni sulla distribuzione qualitativa del fitoplancton nel mare Mediterraneo. Atti della Soc. Ital. di Sci. Nat. 84:105-113.

Goniaulax pacifica
Rampi, L., 1943
Richerche sul fitoplancton del Mare Ligure. F. Le Goniaulacee delle acque di Sanremo. AH1 della Soc. Ital. di Scienze Naturali, 82:1-12, figs.1-16.

Gonyaulax pacifica
Rampi, L., 1942
II Fitoplancton mediterraneo: Problemi ed affinita interoceaniche. Boll. di Pesca di Piscicoltura e di Idrobiologia, Anno 18, Fasc. 4:7-19.

Goniaulax pacifica
Rampi, L., 1941
Ricerche sul microplancton del Mare Ligure. 3. Le Haterodiniacee e le Oxytoxacee dell acque di Sanremo. Annali del Mus. Civico di Storia Naturale di Genova, 81:50-70, 2 pls.

Goniaulax pacifica
Sousa e Silva, E., 1949
Diatomaceas e Dinoflagelados de Baia de Cascais. Portugaliae Acta Biol., Volume: Julio Henriques, Ser. B: 300-383, 9 pls, 2 fold-in tables.

Gonyaulax pacifica
Taylor, F.J.R., 1969.
Peri-nuclear structural elements formed in the dinoflagellate Gonyaulax pacifica Kofoid.
Protistologica, 5(2): 165-167

Goniaulax parva n. sp.
Ramsfjell, Einar, 1959
Two new phytoplankton species from the Norwegian Sea, the diatom Coscinosira poroseriata, and the dinoflagellate Goniaulax parva.
Nytt Magasin for Botanikk, 7: 175-177.

Gonyaulax Pavillardii n.sp.
Dangeard, P., 1927
Phytoplankton de la croisière du "Sylvana". Ann. Inst. Ocean., Monaco, n.s., 4(8):286-401, 54 text figs. (février-Juin 1913).

Gonyaulax polyedra
Allen, W.S., 1946
"Red water" in La Jolla Bay in 1945.
Trans. Amer. Micros. Soc. 65(2): 149-153

Gonyaulax polyedra Stein
Allen, W.E., 1943
"Red water" in La Jolla Bay in 1942.
Trans. Am. Micr. Soc. LXII (3):262-264.

Goniaulax polyedra
Allen, W. E., 1942.
Occurrences of "red water" near San Diego. Science, 96(2499):471.

Gonyaulax polyedra
Allen, W.E., 1938.
"Red water" along the west coast of the United States in 1938. Science 88:55-56.

Gonyaulax polyedra
Bode, V.C., R. DeSa and J.W. Hastings, 1963
Daily rhythm of luciferin activity in Gonyaulax polyedra.
Science, 141(3584):913-915.

Goniaulax polyedra
Braarud, T., 1945
A phytoplankton survey of the polluted waters of inner Oslo Fjord. Hvalrådets Skrifter, No.28, 142 pp., 19 text figs., 17 tables.

Gonyaulax polyedra
Braarud, T., and Adam Bursa, 1939
On the phytoplankton of the Oslo Fjord, 1933-1934. Hvalrådets Skr. No.19:1-63; 9 text figs. Reviewed. J. du. Cons. 14(3): 418-420. A.C. Gardiner.

Gonyaulax polyedra
Carlucci, A.F., and Peggy M. Bowes, 1970
Production of vitamin B12, thiamine and biotin by phytoplankton.
J. Phycol. 6(4): 351-357.

Gonyaulax polyedra
Dangeard, P., 1927
Phytoplankton de la croisière du "Sylvana". Ann. Inst. Ocean., Monaco, n.s., 4(8):286-401, 54 text figs. (février-Juin 1913).

Gonyaulax polyedra

Dangeard, P., 1926
Description des Péridiniens Testacés recueillis para la Mission Charcot pendant le mois d'Aout 1924. Ann. Inst. Ocean. n.s. 3(7):307-334, 15 text figs.

Goniaulax poliedra

dos Santos-Pinto, J., 1949.
Un caso de "red water" motivato por abundancia anormal de Goniaulax poliedra Stein. Bol. Sci. Port. Cienc. Natur. 17(2):94-96.

Goniaulax polyhedra

Ercegovic, A., 1936
Etudes qualitative et quantitatives du phytoplancton dans les eaux cotières de l'Adriatique oriental moyen au cours de l'année 1934. Acta Adriatica 1(9):1-126

Gonyaulax polyedra

Evitt, William R., and Susan E. Davidson, 1964.
Dinoflagellate studies. 1. Dinoflagellate cysts and thecae. Stanford Univ. Publ., Geol. Sci., 10(1):1-12.

Gonyaulax polyhedra

Forti, A., 1922
Ricerche sulla flora pelagica (fitoplancton) di Quarto dei Mille. Mem. R. Com. Talass. Ital. 97:248 pp., 13 pls.

Gonyaulax polyedra

Fuller, C.W., Paul Kreiss and H.H. Seliger, 1972.
Particulate bioluminescence in dinoflagellates: dissociation and partial reconstitution. Science 177(4052): 884-885.

Gonyaulax polyedra

Goudsmith, Joanne T., and Clinton J. Dawes 1972.
The ultrastructure of several dinoflagellates with emphasis on Gonyaulax polyedra Stein and Gonyaulax monilata Davis. Phycologia 11(2):123-132.

Goniaulax polyhedra

Gilbert, J.Y., and W.E. Allen, 1943
The phytoplankton of the Gulf of California obtained by the "E.W. Scripps" in 1939 and 1940. J. Mar. Res. V(2):89-110, figs.30-31.

Gonyaulax polyedra

Haidak, David J., Christopher K. Mathews and Beatrice M. Sweeney, 1966.
Pigment protein complex from Gonyaulax. Science, 152 (3719):212-213.

Gonyaulax polyedra

Hand, William G., Patricia A. Collard and Demorest Davenport, 1965.
The effects of temperature and salinity change on swimming rate in the dinoflagellates, Gonyaulax and Gyrodinium. Biol. Bull., 128(1):90-101.

Gonyaulax polyedra

Hansen, Palle G., and Eric G. Barham, 1962.
Resonant cavity measurements of the effects of "red water" plankton on the attenuation of underwater sound. Limnology & Oceanography, 7(1):8-13.

Gonyaulax polyedra

Hastings, J. Woodland, and Beatrice M. Sweeney, 1958.
A persistent diurnal rhythm of luminescence in Gonyaulax polyedra. Biol. Bull., 115(3):440-458.

Gonyaulax polyedra

Hastings, J.W., and B.M. Sweeney, 1957.
On the mechanism of temperature independence in a biological clock. Proc. U.S. Nat. Acad., 43(8):804-811.

Gonyaulax polyedra

Hastings, J.W., and B.M. Sweeney, 1957.
The luminescent reaction in extracts of the marine dinoflagellate, Gonyaulax polyedra. J. Cell. Comp. Physiol., 49(2):209-226.

Gonyaulax polyedra

Hastings, J. Woodland, and Vernon C. Bode, 1961.
Ionic extracts upon bioluminescence in Gonyaulax extracts. Light and Life, W.D. McElroy and Bentley Glass, Edits., Johns Hopkins Univ. Press, 294-

Gonyaulax polyedra

Holmes, R.W., P.M. Williams and R.W. Eppley, 1967.
Red water in La Jolla Bay, 1964-1965. Limnol. Oceanogr., 12(3):503-512.

Gonyaulax polyedra

Jorgensen, E., 1900
Protophyten und Protozoën im Plankton aus der Norwegischen Westkerste. Bergens Mus. Aarb. 1899(6): 95 pp., 5 pls., 83 tables.

Gonyaulax polyedra

Kofoid, A. C., 1911
Dinoflagellata of the San Diego region. IV The genus Gonyaulax. Univ. Calif. Publ. Zool. 8(4):187-286, Pls. 9-17.

Gonyaulax polyedra

Krieger, Neil, and J.W. Hastings, 1968.
Bioluminescence: pH activity profiles of related luciferase fractions. Science, 161(3841): 586-589.

Goniaulax polyedra

LeBour, M.V., 1925
The dinoflagellates of Northern Seas. The Marine Biological Association of the United Kingdom, Plymouth, 250 pp., 35 pls., 53 text figs.

Gonyaulax polyedra

Lindemann, E., 1924
Peridineen aus dem goldenen Horn und dem Bosphorus. Bot. Arch. 5:216-233, 98 text figs.

Gonyaulax polyedra

Margalef, R., 1949
Fitoplancton nerítico de la Costa Brava en 1947-48. Publ. Inst. Biol. Aplicada, 5: 41-51, 3 text figs.

Goniaulax polyedra

Marshall, S.M., 1947
An experiment in marine fish cultivation: III. The plankton of a fertilized loch. Proc. Roy. Soc., Edinburgh, Sect.B., 63, Pt.1(3):21-33, 7 text figs.

Gonyaulax polyedra

Matzenauer, L., 1933
Die Dinoflagellaten des indischen Ozeans (mit Ausnahme der Gattung Ceratium.) Bot. Arch. 35:437-510, 77 text figs., 2 charts.

Gonyaulax polyedra

McMurry, Laura and J.W. Hastings, 1972
Circadian rhythms: mechanism of luciferase activity changes in Gonyaulax. Biol. Bull. mar. biol. Lab, Woods Hole 143(1): 196-206.

Goniaulax polyedra

Meunier, A., 1919
Microplancton de la Mer Flamande 3. Les Péridiniens. Mem. Mus. Roy. Hist. Nat., Belgique 8(1):1-116, Pls. XV-XXI.

Goniaulax polyedra

Morse, D.C., 1947
Some observations on seasonal variations in plankton population Patuxant River, Maryland 1943 1945. Bd. Nat. Res., Publ. No.65, Chesapeake Biol. Lab., 31, 3 figs.

Gonyaulax polyedra

Nordli, E., 1951.
Resting spore of Goniaulax polyedra. Nytt Mag. Naturvidensk. 88:207-212, 3 textfigs.

Gonyaulax polyedra

Patton, Stuart, P.T. Chandler, E.B. Kalen, A.R. Loeblich III, G. Fuller and A.A. Benson,1967.
Food value of red tide (Gonyaulax polyedra). Science. 158(3802):789-790.

Gonyaulax polyedra

Patton, S., G. Fuller, A.R. Loeblich, III and A.A. Benson,1966.
Fatty acids of the "red tide" organism, Gonyaulax polyedra. Biochimica et Biophysica Acta, 116:577-579.

Gonyaulax polyedra

Paulsen, O., 1908
XVIII Peridiniales. Nordisches Plankton. Bot. Teil: 1-124, 155 text figs.

Gonyaulax polyedra

Pavillard, J., 1916
Recherches sur les Peridiniens du Golfe du Lion. Mem. Univ. Montpellier. Trav. Inst. Bot., Univ. Montpellier. Serie mixte No.4, 70 pp., 3 pls., 15 text figs.

Gonyaulax polyedra

Pavillard, J., 1905
Recherches sur la flore pelagique (Phytoplankton) de l'Etang de Thau. Theses presentees a la Fac. Sci., Paris, 116 pp., 3 pls.

Gonyaulax polyedra

Poliarpov, G.G. and A.V. Tokareva 1970.
On the cellular cycle of the dinoflagellates Peridinium trochoideum (Stein) and Goniaulax polyedra (Stein) (Microautoradiographic investigation). (In Russian; English abstract) Gidrobiol. Zh. 6(5):66-69.

Goniaulax polyedra

Rampi, L., 1943
Richerche sul fitoplancton del Mare Ligure. F. Le Goniaulacee delle acque di Sanremo. Ahi della Soc. Ital. di Scienze Naturali, 82:1-12, figs.1-16.

Gonyaulax polyedra

Reynolds, George T., J.W. Hastings, Hidemi Sato and A. Randolph Sweeney, 1966.
The identity and photon yield of scintillons of Gonyaulax polyedra. (Abstract only). Biol. Bull., 131(2):403.

Goniaulax poliedra

Santos-Pinto, J. dos, 1949.
Um caso de "red water" motivado por abundancia anormal de Goniaulax poliedra Stein. Bol. Soc. Portuguesa Ciênc. Nat. 17(2):94-96, figs.

Gonyaulax polyedra

Schröder, B., 1900
Phytoplankton des Golfes von Neapel nebst vergleichenden Ausblicken auf das atlantischen Ozean. Mitt. Zool. Stat. Neapel, 14:1-38.

Gonyaulax polyhedra

Soli, Giorgio, 1966.
Bioluminescent cycle of photosynthetic dino-flagellates. Limnol. Oceanogr., 11(3):355-363.

Gonyaulax polyedra

Sweeney, Beatrice M., 1969.
Transducing mechanisms between circadian clock and overt rhythms in Gonyaulax. Can. J. Bot., 47(2):299-308.

Gonyaulax polyedra

Thomas, William H., Donald W. Lear, Jr., and Francis T. Haxo, 1962.
Oceanographic studies during Operation "Wigwam". Uptake of the marine dinoflagellate, Gonyaulax polyedra, of radioactivity formed during an underwater nuclear test. Limnol. and Oceanogr., Suppl. to Vol. 7:lxvi-lxxi.

Gonyaulax polyedra
Wall, David, and Barrie Dale, 1968.
Modern dinoflagellate cysts and evolution of the Peridiniales.
Micropaleontology, 14(3):265-304.

Goniodoma polyedricum
Sousa e Silva, E., 1949
Diatomaceas e Dinoflagelados de Baia de Cascais. Portugaliae Acta Biol., Volume: Julio Henriques, Ser. B: 300-383, 9 pls, 2 fold-in tables.

Gonyaulax polygramma
Dangeard, P., 1927
Phytoplankton de la croisière du "Sylvana". Ann. Inst. Ocean., Monaco, n.s., 4(8):286-401, 54 text figs. (Fevier-Juin 1913).

Gonyaulax polygramma
Delegazione Italiana della Commissione Internazionale per l'Esplorazione Scientifica del Mediterraneo, 1941
Note sul plancton della Laguna veneta. [Memoria CCLXXXIX], Arch. di Ocean. e Limn. Anno I, Fasc. I, 1941 XIX: 31-57 pp.

Goniaulax polygramma
Ercegovic, A., 1936
Etudes qualitative et quantitatives du phytoplancton dans les eaux cotières de l'Adriatique oriental moyen au cours de l'année 1934. Acta Adriatica 1(9):1-126

Gonyaulax polygramma
Forti, A., 1922
Ricerche sulla flora pelagica (fitoplancton) di Quarto dei Mille. Mem. R. Com. Talass. Ital. 97:248 pp., 13 pls.

Gonyaulax polygramma
Grindley, J.R., and F.J.R. Taylor, 1970.
Factors affecting plankton blooms in False Bay. Trans. roy. Soc. SAfr. 39(2): 201-210.

Gonyaulax polygramma
Grindley, J.R., and F.J.R. Taylor, 1964.
Red water and marine fauna mortality near Cape Town.
Trans. R. Soc. S. Africa, 37(2):110-130.

Gonyaulax polygramma
Kofoid, A. C., 1911
Dinoflagellata of the San Diego region. IV The genus Gonyaulax. Univ. Calif. Publ. Zool. 8(4):187-286, Pls. 9-17.

Goniaulax polygramma
LeBour, M.V., 1925
The dinoflagellates of Northern Seas. The Marine Biological Association of the United Kingdom, Plymouth, 250 pp., 35 pls., 53 text figs.

Gonyaulax polygramma
Lewis, E.J. 1967.
On a Gonyaulax bloom off Mt. Dalley in the Arabian Sea. In: Proc. Seminar, Sea, Salt and Plants, V. Krishnamurthy, editor, Bhavnagar, India, 224-226.

Gonyaulax polygramma
Lindemann, E., 1925
Neubeobachtungen an den Winter peridineen des Golfes von Neapel. Bot. Arch. 9:95-102, 19 text figs.

Gonyaulax polygramma
Mangin, M. L., 1912
Phytoplancton de la croisière du "René" dans l'Atlantique (Septembre 1908). Ann. Inst. Ocean., n.s., 4(1):1-66, 2 pls., 41 text figs., 2 tables.

Gonyaulax polygramma
Margalef, R., 1949
Fitoplancton nerítico de la Costa Brava en 1947-48. Publ. Inst. Biol. Aplicada, 5: 41-51, 3 text figs.

Gonyaulax polygramma
Matzenauer, L., 1933
Die Dinoflagellaten des indischen Ozeans (mit Ausnahme der Gattung Ceratium.) Bot. Arch. 35:437-510, 77 text figs., 2 charts.

Goniaulax polygramma
Morse, D.C., 1947
Some observations on seasonal variations in plankton population Patuxant River, Maryland 1943-1945. Bd. Nat. Res., Publ. No.65, Chesapeake Biol. Lab., 31, 3 figs.

Gonyaualx polygramma
Murray, G., and F. G. Whitting, 1899
New Peridiniaceae from the Atlantic. Trans. Linn. Soc., London, Bot., ser 2, 5: 321-342, Pls. 27-33, 9 tables.

Gonyaulax polygramma
Paulsen, O., 1908
XVIII Peridiniales. Nordisches Plankton, Bot. Teil: 1-124, 155 text figs.

Gonyaulax polygramma
Pavillard, J., 1916
Recherches sur les Peridiniens du Golfe du Lion. Mem. Univ. Montpellier. Trav. Inst. Bot., Univ. Montpellier. Serie mixte No.4, 70 pp., 3 pls., 15 text figs.

Gonyaulax polygramma
Pavillard, J., 1905
Recherches sur la flore pelagique (Phytoplankton) de l'Etang de Thau. Theses presentees a la Fac. Sci., Paris, 116 pp., 3 pls.

Gonyaulax polygramma
Schröder, B., 1900
Phytoplankton des Golfes von Neapel nebst vergleichenden Ausblicken auf das atlantischen Ozean. Mitt. Zool. Stat. Neapel, 14:1-38.

Goniaulax polygramma
Shiokawa, Tsukasa, Masaru Tateishi, Shoji Iizuka and Haruhiko Irie, 1966.
On the mass mortality of benthic animals occurred in the season of red water, 1962. (In Japanese; English abstract).
Bull. Fac. Fish. Nagasaki Univ., No. 21:45-58.

Goniaulax polygramma
Sousa e Silva, E., 1949
Diatomaceas e Dinoflagelados de Baia de Cascais. Portugaliae Acta Biol., Volume: Julio Henriques, Ser. B: 300-383, 9 pls, 2 fold-in tables.

Gonyaulax polygramma
Steidinger, Karen A., 1968.
The genus Gonyaulax in Florida waters. 1. Morphology and thecal development in Gonyaulax polygramma Stein, 1883. (Phytoplankton) (Dinoflagellates)
Leaflet Ser., Florida Bd.Conserv.1 (1-4): 5 pp.

Gonyaulax polygramma
Taylor, F.J.R., 1962
Gonyaulax polygramma Stein in Cape waters: a taxonomic problem related to developmental morphology.
The Journal of South African Botany, 28(3): 237-242.

Also in:
Univ. Cape Town Dept. Oceanogr., Collected Reprints, Vol. 1. (1962)

Gonyaulax polygramma pulchra n. subsp.
Margalef, R., and M. Duran, 1953.
Microplancton de Vigo de octubre de 1951 a septiembre de 1952. Publ. Inst. Biol. Aplic. 13: 5-78, 22 textfigs.

Gonyaulax rostratum n.sp.
Dangeard, P., 1927
Phytoplankton de la croisière du "Sylvana". Ann. Inst. Ocean., Monaco, n.s., 4(8):286-401, 54 text figs. (Feirer-Juin 1913).

Goniaulax Rouchi
Rampi, L., 1948
Sur quelques Peridiniens rares ou interessants du Pacifique subtropical (Recoltes Alain Gerbault). Bull. l'Inst. Ocean., Monaco, No.937: 7 pp., 8 text figs.

Goniaulax polygramma
Rampi, L., 1943
Richerche sul fitoplancton del Mare Ligure. F. Le Goniaulacee delle acque di Sanremo. AHi della Soc. Ital. di Scienze Naturali, 82:1-12, figs.1-16.

Gonyaulax schilleri n.sp.
Matzenauer, L., 1933
Die Dinoflagellaten des indischen Ozeans (mit Ausnahme der Gattung Ceratium.) Bot. Arch. 35:437-510, 77 text figs., 2 charts.

Gonyaulax aff. scrippsae
Balech, Enrique, 1962
Tintinnoinea y Dinoflagellata del Pacifico segun material de las Expediciones NORPAC y DOWNWIND del Instituto Scripps de Oceanografia.
Revista. Mus. Argentino Ciencias Nat. "Bernardino Rivadavia", Ciencias Zool., 7(1):1-253.

Gonyaulax scrippsae
Dangeard, P., 1927
Phytoplankton de la croisière du "Sylvana". Ann. Inst. Ocean., Monaco, n.s., 4(8):286-401, 54 text figs. (Fevier-Juin 1913).

Gonyaulax scrippsae n.sp.
Kofoid, A. C., 1911
Dinoflagellata of the San Diego region. IV The genus Gonyaulax. Univ. Calif. Publ. Zool. 8(4):187-286, Pls. 9-17.

Goniaulax scrippsae
LeBour, M.V., 1925
The dinoflagellates of Northern Seas. The Marine Biological Association of the United Kingdom, Plymouth, 250 pp., 35 pls., 53 text figs.

Goniaulax scrippsae
Morse, D.C., 1947
Some observations on seasonal variations in plankton population Patuxant River, Maryland 1943-1945. Bd. Nat. Res., Publ. No.65, Chesapeake Biol. Lab., 31, 3 figs.

Goniaulax scrippsae
Rampi, L., 1943
Richerche sul fitoplancton del Mare Ligure. F. Le Goniaulacee delle acque di Sanremo. AHi della Soc. Ital. di Scienze Naturali, 82:1-12, figs.1-16.

Gonyaulax scrippsae
Wall, David, and Barrie Dale, 1968.
Modern dinoflagellate cysts and evolution of the Peridiniales.
Micropaleontology, 14(3):265-304.

Goniaulax scupsae
Macdonald, R., 1933
An examination of plankton hauls made in the Suez Canal during the year 1928. Fish. Res. Dis., Notes & Mem. No.3, 11 pp., 1 chart.

Gonyaulax sousae
Balech, Enrique, 1971.
Microplancton de la campaña oceanographica: Productividad III.
Revta Mus. argent. Cienc. Nat. Bernadina Rivadavia, Hydrobiol. 3(1):1-202, 39 pls.

Goniaulax sousi n.sp.
Balech, Enrique, 1959.
Operacion Oceanografica Merluza V Crucero. Plancton.
Publico, Servicio de Hidrografia Naval, Argentina, H. 618:43 pp.

Gonyaulax sphaeroidea
Balech, Enrique, 1962
Tintinnoinea y Dinoflagellata del Pacifico segun material de las Expediciones NORPAC y DOWNWIND del Instituto Scripps de Oceanografia.
Revista, Mus. Argentino Ciencias Nat. "Bernardino Rivadavia", Ciencias Zool., 7(1):1-253.

Gonyaualx spheroidea
Dangeard, P., 1927
Phytoplankton de la croisière du "Sylvana". Ann. Inst. Ocean., Monaco, n.s., 4(8):286-401, 54 text figs. (Feirer-Juin 1913).

Gonyaulax sphaeroidea n.sp.
Kofoid, A. C., 1911
Dinoflagellata of the San Diego region. IV The genus Gonyaulax. Univ. Calif. Publ. Zool. 8(4):187-286, Pls. 9-17.

Gonyaulax sphaeroidea
Matzenauer, L., 1933
Die Dinoflagellaten des indischen Ozeans (mit Ausnahme der Gattung Ceratium.) Bot. Arch. 35:437-510, 77 text figs., 2 charts.

Goniaulax sphaeroidea
Rampi, L., 1943
Richerche sul fitoplancton del Mare Ligure. F. Le Goniaulacee delle acque di Sanremo. AHi della Soc. Ital. di Scienze Naturali, 82:1-12, figs.1-16.

Gonyaulax spinifera
Balech, Enrique, 1962
Tintinnoinea y Dinoflagellata del Pacifico segun material de las Expediciones NORPAC y DOWNWIND del Instituto Scripps de Oceanografia.
Revista, Mus. Argentino Ciencias Nat. "Bernardino Rivadavia", Ciencias Zool., 7(1):1-253.

Goniaulax spinifera
Braarud, T., 1945
A phytoplankton survey of the polluted waters of inner Oslo Fjord. Hvalrådets Skrifter, No.28, 142 pp., 19 text figs., 17 tables.

Goniaulax spinifera
Dangeard, P., 1927
Phytoplankton de la croisière du "Sylvana". Ann. Inst. Ocean., Monaco, n.s., 4(8):286-401, 54 text figs. (Fevier-Juin 1913).

Gonyaulax spinifera
Dangeard, P., 1926
Description des Péridiniens Testacés recueillis par la Mission Charcot pendent lo mois d'Aout 1924. Ann. Inst. Ocean. n.s. 3(7):307-334, 15 text figs.

Goniaulax spinifera
De Angelis, C.M., 1962.
Distribuzione ed ecologia di alcune specie di Dinoflagellati di acque salmastre.
Pubbl. Staz. Zool., Napoli, 32 (Suppl.):301-314.

Goniaulax spinifera
Ercegovic, A., 1936
Etudes qualitative et quantitatives du phytoplancton dans les eaux cotières de l'Adriatique oriental moyen au cours de l'année 1934. Acta Adriatica 1(9):1-126

Gonyaulax spinifera
Forti, A., 1922
Ricerche sulla flora pelagica (fitoplancton) di Quarto dei Mille. Mem. R. Com. Talass. Ital. 97:248 pp., 13 pls.

Gonyaulax spinifera
Gran, H.H., and T. Braarud, 1935
A quantitative study of the phytoplankton in the Bay of Fundy and the Gulf of Maine (including observations on hydrography, chemistry, and turbidity). J. Biol. Bd., Canada, 1(5):279-467, 69 text figs.

Gonyaulax spinifera
Hada, Yoshina, 1970. (1970)
The protozoan plankton of the Antarctic and Subantarctic seas.
Scient. Repts, Japan. Antarct. Res. Exped., (E)31:51 pp.

Gonyaulax spinifera
Jørgensen, E., 1905
B.Protistplankton and the diatoms in bottom samples. Hydrographical and biological investigations in Norwegian fjords. Bergens Mus. Skr. 7: 49-225.

Gonyaulax spinifera
Jorgensen, E., 1900
Protophyten und Protozoën im Plankton aus der Norwegischen Westkerste. Bergens Mus. Aarb. 1899(6): 95 pp., 5 pls., 83 tables.

Gonyaulax spinifera
Kofoid, A. C., 1911
Dinoflagellata of the San Diego region. IV The genus Gonyaulax. Univ. Calif. Publ. Zool. 8(4):187-286, Pls. 9-17.

Goniaulax spinifera
LeBour, M.V., 1925
The dinoflagellates of Northern Seas. The Marine Biological Association of the United Kingdom, Plymouth, 250 pp., 35 pls.,53 text figs.

Goniaulax spinifera
Levander, K.M., 1947
Plankton gesammelt in den Jahren 1899-1910 an den Küsten Finnlands. Finnländische Hydrographisch-Biologische Untersuchungen (aus dem Wasserbiologischen Laboratorin der Societas Scientiarum Fennica) No.11: 40 pp., 6 diagrams, 13 pls., tables.

Goniaulax spinifera
Lillick, L.C., 1940
Phytoplankton and planktonic protozoa of the offshore waters of the Gulf of Maine. Pt.II. Qualitative Composition of the Planktonic Flora. Trans. Am. Phil. Soc., n.s., 31(3):193-237, 13 text figs.

Gonyaulax spinifera
Lindemann, E., 1924
Peridineen aus dem goldenen Horn und dem Bosphorus. Bot. Arch. 5:216-233, 98 text figs.

Gonyaulax spinifera
Mangin, M. L., 1912
Phytoplankton de la croisière du "René" dans l'Atlantique (Septembre 1908). Ann. Inst. Ocean., n.s., 4(1):1-66, 2 pls., 41 text figs., 2 tables.

Gonyaulax spinifera
Matzenauer, L., 1933
Die Dinoflagellaten des indischen Ozeans (mit Ausnahme der Gattung Ceratium.) Bot. Arch. 35:437-510, 77 text figs., 2 charts.

Goniaulax spinifera
Morse, D.C., 1947
Some observations on seasonal variations in plankton population Patuxant River, Maryland 1943 1945. Bd. Nat. Res., Publ. No.65, Chesapeake Biol. Lab., 31, 3 figs.

Gonyaulax spinifera
Paulsen, O., 1908
XVIII Peridiniales. Nordisches Plankton, Bot. Teil: 1-124, 155 text figs.

Gonyaulax spinifera
Pavillard, J., 1905
Recherches sur la flore pelagique (Phytoplankton) de l'Etang de Thau. Theses presentees a la Fac. Sci., Paris, 116 pp., 3 pls.

Goniaulax spinifera
Rumkówna, A., 1948
[List of the phytoplankton species occurring in the superficial water layers in the Gulf of Gdańsk] Bull. Lab. mar., Gdynia, No. 4: 139-141 with tables in back.

Gonyaulax spinifera
Schröder, B., 1900
Phytoplankton des Golfes von Neapel nebst vergleichenden Ausblicken auf das atlantischen Ozean. Mitt. Zool. Stat. Neapel, 14:1-38.

Goniaulax spinifera
Sousa e Silva, E., 1949
Diatomaceas e Dinoflagelados de Baia de Cascais. Portugaliae Acta Biol., Volume: Julio Henriques, Ser. B: 300-383, 9 pls, 2 fold-in tables.

Gonyaulax spinifera
Wall, David, and Barrie Dale 1970.
Living hystrichosphaerid dinoflagellate spores from Bermuda and Puerto Rico. Micropaleontology 16(1): 47-58.

Gonyaulax spinifera
Wall, David, and Barrie Dale, 1968.
Modern dinoflagellate cysts and evolution of the Peridiniales.
Micropaleontology, 14(3):265-304.

Gonyaulax spinifera estelae
Margalef, R., and M. Duran, 1953. n. subsp.
Microplancton de Vigo de octubre 1951 a septiembre 1952. Publ. Inst. Biol. Aplic. 13:5-78, 22 textfigs.

Goniaulax striata
Balech, Enrique, 1959.
Operacion Oceanografica Merluza V Crucero. Plancton.
Publico, Servicio de Hidrografia Naval, Argentina, H. 618:43 pp.

Gonyaulax tamarensis
Anon., 1968.
Mussels not for eating.
Nature, Lond., 220(5162):13.

Goniaulax tamarensis
Braarud, T., 1951.
Taxinomical studies of marine dinoflagellates. Nytt Mag. Naturvidensk 88:43-48.

Goniaulax tamarensis
Braarud, T., 1945.
Morphological observations on marine dinoflagellate cultures (Porella perforata, Goniaulax tamarensis, Protoceratium reticulatum). Avhandl. Norske Videnskaps-Akademi, Oslo. I. Mat.-Naturv. Kl. 1944, No. 11:1-18, 6 textfigs., 4 pls.

Gonyaulax Tamarensis
Braarud, T., 1934
A note on the phytoplankton of the Gulf of Maine in the summer of 1933. Biol. Bull. 67(1):76-82.

Gonyaulax tamarensis
Gran, H.H., and T. Braarud, 1935
A quantitative study of the phytoplankton in the Bay of Fundy and the Gulf of Maine (including observations on hydrography, chemistry, and turbidity). J. Biol. Bd., Canada, 1(5):279-467, 69 text figs.

Goniaulax tamarensis n.sp.
LeBour, M.V., 1925
The dinoflagellates of Northern Seas. The Marine Biological Association of the United Kingdom, Plymouth, 250 pp., 35 pls.,53 text figs.

Goniaulax tamarensis
Lillick, L.C., 1940
Phytoplankton and planktonic protozoa of the offshore waters of the Gulf of Maine. Pt. II. Qualitative Composition of the Planktonic Flora. Trans. Am. Phil. Soc., n.s., 31(3):193-237, 13 text figs.

Goniaulax tamarensis
Lillick, L.C., 1938
Preliminary report of the phytoplankton of the Gulf of Maine. Am. Mid. Nat. 20(3):624-640, 1 text figs. 37 tables.

Gonyaulax tamarensis
Lillick, L.C., 1937
Seasonal studies of the phytoplankton off Woods Hole, Massachusetts. Biol. Bull. LXXIII (3):488-503, 3 text figs.

Goniaulax tamarensis
Needler, A. B., 1949.
Paralytic shellfish poisoning and Goniaulax tamarensis. J. Fish. Res. Bd., Canada, 7(8):490-504, 3 textfigs.

Goniaulax tamarensis
Nordli, O., 1951.
Dinoflagellates from Lofoten. Nytt Mag. Naturvidensk 88:49-55, 7 textfigs.

Gonyaulax tamarensis
Prakash, A., 1967.
Growth and toxicity of a marine dinoflagellate, Gonyaulax tamarensis
J. Fish. Res. Bd. Can. 24 (7): 1589-1606

Gonyaulax tamarensis
Prakash, A., J.C. Medcof and A.D. Tennant, 1971.
Paralytic shellfish poisoning in eastern Canada.
Bull. Fish. Res. Bd. Can. 177: 87pp.

Gonyaulax tamarensis
Prakash, A., Liv Skoglund, Britt Rystad and Anne Jensen 1973
Growth and cell-size distribution of marine planktonic algae in batch and dialysis cultures.
J. Fish. Res. Bd. Can. 30(2):143-155

Gonyaulax tamarensis
Robinson, G.A., 1968.
Distribution of Gonyaulax tamarensis Lebour in the western North Sea in April, May and June 1968. Nature, Lond., 220(5162):22-23.

Gonyaulax tamarensis
Dakshaug, Egil, and Anne Jensen 1971.
Gonyaulax tamarensis and paralytic mussel toxicity in Trondheimsfjorden, 1963-1969.
Skr. K. Norske Vidensk. Selskab. (15): 15pp.

Goniaulax tamarensis
Wood, E.J.F., 1963
Dinoflagellates in the Australian region. II. Recent collections.
Comm. Sci. and Industr. Res. Org., Div. Fish. and Oceanogr., Techn. Paper, No. 14: 55 pp.

Gonyaulax tamarensis
Wood, P.C., 1968.
Dinoflagellate crop in the North Sea. Nature, Lond., 220(5162):21.

Gonyaulax turbynei
Balech, Enrique, 1971.
Microplancton de la campaña oceanographica: Productividad III.
Revta Mus. argent. Cienc. Nat. Bernadina Rivadavia, Hydrobiol. 3(1):1-202, 39 pls.

Goniaulax triacantha
Braarud, T., and Adam Bursa, 1939
On the phytoplankton of the Oslo Fjord, 1933-1934. Hvalrådets Skr. No.19:1-63; 9 text figs. Reviewed. J. du. Cons. 14(3): 418-420. A.C. Gardiner.

Goniaulax triacantha
Gran, H.H., and T. Braarud, 1935
A quantitative study of the phytoplankton in the Bay of Fundy and the Gulf of Maine (including observations on hydrography, chemistry, and turbidity). J. Biol. Bd., Canada, 1(5):279-467, 69 text figs.

Gonyaulax ? triacantha n.sp.
Jorgensen, E., 1900
Protophyten und Protozoën im Plankton aus der Norwegischen Westkerste. Bergens Mus. Aarb. 1899(6): 95 pp., 5 pls., 83 tables.

Gonyaulax triacantha
Kofoid, A. C., 1911
Dinoflagellata of the San Diego region. IV The genus Gonyaulax. Univ. Calif. Publ. Zool. 8(4):187-286, Pls. 9-17.

Goniaulax triacantha
LeBour, M.V., 1925
The dinoflagellates of Northern Seas. The Marine Biological Association of the United Kingdom, Plymouth, 250 pp., 35 pls., 53 text figs.

Goniaulax triacantha
Levander, K.M., 1947
Plankton gesammelt in den Jahren 1899-1910 an den Küsten Finnlands. Finnländische Hydrographisch-Biologische Untersuchungen (aus dem Wasserbiologischen Laboratorin der Societas Scientiarum Fennica) No.11: 40 pp., 6 diagrams, 13 pls., tables.

Goniaulax triacantha
Lillick, L.C., 1940
Phytoplankton and planktonic protozoa of the offshore waters of the Gulf of Maine. Pt. II. Qualitative Composition of the Planktonic Flora. Trans. Am. Phil. Soc., n.s., 31(3):193-237, 13 text figs.

Gonyaulax triacantha
Paulsen, O., 1908
XVIII Peridiniales. Nordisches Plankton. Bot. Teil: 1-124, 155 text figs.

Gonyaulax triacanta
Rumkówna, A., 1948
List of the phytoplankton species occurring in the superficial water layers in the Gulf of Gdańsk. Bull. Lab. mar., Gdynia, No. 4: 139-141 with tables in back.

Gonyaulax turbynei
Balech, Enrique, 1962
Tintinnoinea y Dinoflagellata del Pacifico segun material de las Expediciones NORPAC y DOWNWIND del Instituto Scripps de Oceanografia.
Revista. Mus. Argentino Ciencias Nat. "Bernardino Rivadavia", Ciencias Zool., 7(1):1-253.

Gonyaulax turbynei
Dangeard, P., 1927
Phytoplankton de la croisière du "Sylvana". Ann. Inst. Ocean. Monaco, n.s., 4(8):286-401, 54 text figs. (février-Juin 1913).

Goniaulax turbynei
Ercegovic, A., 1936
Etudes qualitative et quantitatives du phytoplancton dans les eaux cotières de l'Adriatique oriental moyen au cours de l'année 1934. Acta Adriatica 1(9):1-126

Gonyaulax Turbynei
Forti, A., 1922
Ricerche sulla flora pelagica (fitoplancton) di Quarto dei Mille. Mem. R. Com. Talass. Ital. 97:248 pp., 13 pls.

Gonyaulax turbynei
Hada, Yoshina, 1970. (1970)
The protozoan plankton of the Antarctic and Subantarctic seas.
Scient. Repts, Japan. Antarct. Res. Exped., (E) 31:51 pp.

Gonyaulax turbynei
Kofoid, A. C., 1911
Dinoflagellata of the San Diego region. IV The genus Gonyaulax. Univ. Calif. Publ. Zool. 8(4):187-286, Pls. 9-17.

Goniaulax turbynei
LeBour, M.V., 1925
The dinoflagellates of Northern Seas. The Marine Biological Association of the United Kingdom, Plymouth, 250 pp., 35 pls., 53 text figs.

Gonyaulax turbynei
Matzenauer, L., 1933
Die Dinoflagellaten des indischen Ozeans (mit Ausnahme der Gattung Ceratium.) Bot. Arch. 35:437-510, 77 text figs., 2 charts.

Gonyaulax turbynei n.sp.
Murray, G., and F. G. Whitting, 1899
New Peridiniaceae from the Atlantic. Trans. Linn. Soc., London, Bot., ser 2, 5: 321-342, Pls. 27-33, 9 tables.

Gonyaulax turbynei
Pavillard, J., 1916
Recherches sur les Peridiniens du Golfe du Lion. Mem. Univ. Montpellier. Trav. Inst. Bot., Univ. Montpellier. Serie mixte No.4, 70 pp., 3 pls., 15 text figs.

Gonyaulax turbynei
Rampi, L., 1943
Richerche sul fitoplancton del Mare Ligure. F. Le Goniaulacee delle acque di Sanremo. AH1 della Soc. Ital. di Scienze Naturali, 82:1-12, figs.1-16.

Goniaulax unicornis n.sp.
LeBour, M.V., 1925
The dinoflagellates of Northern Seas. The Marine Biological Association of the United Kingdom, Plymouth, 250 pp., 35 pls., 53 text figs.

Goniaulax unicornis
Morse, D.C., 1947
Some observations on seasonal variations in plankton population Patuxant River, Maryland 1943-1945. Bd. Nat. Res., Publ. No.65, Chesapeake Biol. Lab., 31, 3 figs.

Gymnaster pentasterias
Hovasse, R., 1946.
Flagellés à squelette siliceux: Silicoflagellés et Ebriidés provenant du plancton recueilli au cours des campagnes scientifiques du Prince Albert Ier de Monaco (1885-1912). Res. Camp. Sci Monaco, 107:19 pp., 1 pl.

Gymnaster pentasterias
Jørgensen, E. 1905
B. Protistplankton and the diatoms in bottom samples. Hydrographical and biological investigations in Norwegian fjords. Bergens Mus. Skr. 7: 49-225.

Gymnaster pentasterias
Jorgensen, E., 1900
Protophyten und Protozoën im Plankton aus der Norwegischen Westkerste. Bergens Mus. Aarb. 1899(6): 95 pp., 5 pls., 83 tables.

Gymnaster pentasterias
Nival, P. 1969.
Données écologiques sur quelques protozoaires planctoniques en Méditerranée.
Protistologica 5(2): 215-225.

Gymnaster pentasterias
Paulsen, O., 1908
XVIII Peridiniales. Nordisches Plankton, Bot. Teil: 1-124, 155 text figs.

Gymnodinium sp.
Atkins, W.R.G., and M. Parke, 1951.
Seasonal changes in the phytoplankton as indicated by chlorophyll estimation. J.M.B.A. 29(3): 609-618.

Gymnodinium sp.
Biecheler, B., 1952.
Recherches sur les Péridiniens.
Bull. Biol. France Belg., Suppl., 36:1-149, 75 textfigs.

Gymnodinium sp.
Braarud, T., 1945
A phytoplankton survey of the polluted waters of inner Oslo Fjord. Hvalrådets Skrifter, No.28, 142 pp., 19 text figs., 17 tables.

Gymnodinium sp.
Braarud, T., and Adam Bursa, 1939
On the phytoplankton of the Oslo Fjord, 1933-1934. Hvalrådets Skr. No.19:1-63; 9 text figs. Reviewed. J. du Cons. 14(3): 418-420. A.C. Gardiner.

Gymnodinium
Brunel, J., 1962
Le phytoplancton de la Baie de Chaleurs.
Inst. Botan., Univ. Montréal, Contrib. No. 77: 365 pp., 66 pls.

Gymnodinium
Evitt, William R., Robin F.A. Clarke, and Jean-Pierre Verdier, 1967.
Dinoflagellate studies III. Dinogymnium acuminatum n.gen., n.sp. (Maastrichtian) and other fossils referable to Gymnodinium Stein.
Publ. Stanford Univ. Geol. Sci., 10(4): 1-27.

Gymnodinium sp.
Galtsoff, P.S., 1948
Emergency survey report on Florida's "Red Tide". Fishing Gaz. 65(1):66-67.

Gymnodinium sp.
Galtsoff, P.S., 1948
Red Tide. Progress Report on the investigations of the cause of the mortality of fish along the west coast of Florida conducted by the U.S. Fish and Wildlife Service and Cooperating Organizations. Fish and Wildlife Service, Special Scientific Rept. No.46, 44 pp. (mimeographed), 9 figs.

Gymnodinium
Gran, H.H., and T. Braarud, 1935
A quantitative study of the phytoplankton in the Bay of Fundy and the Gulf of Maine (including observations on hydrography, chemistry, and turbidity). J. Biol. Bd., Canada, 1(5):279-467, 69 text figs.

Gymnodinium
Gunter G., F.G. Walton Smith, and R.H. Roberts, 1947
Mass mortality of marine animals on the lower west coast of Florida, November 1946 - January 1947. Science, 105(2723): 256-257.

Gymnodinium spp.
Hickel, W., E. Hagmeier and G. Drebes, 1971.
Gymnodinium blooms in the Helgoland Bight (North Sea) during August, 1968. Helgoländer wiss. Meeresunters 22(3/4): 401-416.

Gymnodinium sp.
Hirayama, Kazutsugu, Shoji Iizuka and Takashi Yoneji, 1972
On culture of Gymnodinium type-'65 in the sea water sampled in Omura Bay during summer 1971. Bull. Fac. Fish. Nagasaki Univ. 33: 11-20. (In Japanese; English abstract).

Gymnodinium spp.
Holmes, R.W., P.M. Williams and R.W. Eppley, 1967.
Red water in La Jolla Bay, 1964-1965.
Limnol. Oceanogr., 12(3):503-512.

Gymnodinium sp.?
Iizuka, Shoji and Haruhiko Irie, 1970.
Anoxic status of bottom waters and occurrences of Gymnodinium red water in Omura Bay. (In Japanese; English abstract). Bull. Plankt. Soc. Japan, 16(2):99-115.

Gymnodinium sp.
Iizuka, Shoji, and Haruhiko Irie, 1966.
The hydrographic conditions and the fisheries damages by the red water occurred in Omura Bay in summer 1965 - II. The biological aspects of a dominant species in the red water. (In Japanese; English abstract).
Bull. Fac. Fish. Nagasaki Univ., No. 21:67-101.

Gymnodinium spp.
Jeffrey, S.W., and F.T. Haxo, 1968.
Photosynthetic pigments of symbiotic dinoflagellates (zooxanthellae) from corals and clams. Biol. Bull., mar. biol. Lab., Woods Hole, 135(1): 149-165.

Gymnodinium sp.
Ketchum, B.H., and J. Keen, 1948
Unusual phosphorus concentrations in the Florida "Red Tide" sea water. J. Mar. Res. 7(1): 17-21, fig.2.

Gymnodinium
Kimball, J.F., Jr., and E.J. Ferguson Wood, 1965.
A dinoflagellate with characters of Gymnodinium and Gyrodinium.
J. Protozool. 12(4):577-580.

Gymnodinium sp. (new)
Marine Laboratory, University of Miami, 1947
Red Tide and Fish Mortality on the Florida West Coast. Special Service Bulletin, (#195), 5 pp. (mimeographed). Coral Gables. July 1947

Gymnodinium sp.
Marshall, S.M. and A.P. Orr, 1948
Further experiments on the fertilization of a sea loch (Loch Craiglin). The effect of different plant nutrients on the phytoplankton. J.M.B.A. 27(2):360-379, 10 text figs.

Gymnodinium sp.
Meunier, A., 1919
Microplancton de la Mer Flamande 3. Les Péridiniens. Mem. Mus. Roy. Hist. Nat., Belgique 8(1):1-116, Pls. XV-XXI.

Gymnodinium sp.
Numaguchi, Katsuyuki and Kazutsugu Hirayama, 1972
On the suitable pH and chlorinity to the growth of Gymnodinium type - '65, causative organism of the red tide in Omura Bay. Bull. Fac. Fish. Nagasaki Univ. 33: 7-10. (In Japanese; English abstract).

Gymnodinium sp.
Pettersson, H., 1934
Scattering and extinction of light in sea-water. Meddelanden från Göteborgs Högskolas Oceanografiska Institut: 9 (Göteborgs Kungl. vetenskaps-och vitterhetssamhälles Handlingar. Femte Följden, Ser.B, 4(4)):1-16.

Gymnodinium sp.
Shiokawa, Tsukasa, and Haruhiko Irie, 1966.
The hydrographic conditions and the fisheries damages by the red water occurred in Omura Bay in summer 1965 - IV. Mass-mortality of fishes resultant in the red water. (In Japanese; English abstract).
Bull. Fac. Fish. Nagasaki Univ., No. 21:115-129.

Gymnodinium sp.
Wood, E.J.F., 1963
Dinoflagellates in the Australian region. II. Recent collections.
Comm. Sci. and Industr. Res. Org., Div. Fish. and Oceanogr., Techn. Paper, No. 14: 55 pp.

Gymnodinium abbreviatum
LeBour, M.V., 1925
The dinoflagellates of Northern Seas. The Marine Biological Association of the United Kingdom, Plymouth, 250 pp., 35 pls., 53 text figs.

Gymnodinium achromaticum
LeBour, M.V., 1925
The dinoflagellates of Northern Seas. The Marine Biological Association of the United Kingdom, Plymouth, 250 pp., 35 pls., 53 text figs.

Gymnodinium aequatoriale n.sp.
Hasle, Grethe Rytter, 1960
Phytoplankton and ciliate species from the Tropical Pacific.
Skr. Norske Videnskaps-Akad., Oslo, 1. Mat.-Nat. Kl., 1960(2): 1-50.

Gymnodinium aeruginosum
Paulsen, O., 1908
XVIII Peridiniales. Nordisches Plankton, Bot. Teil: 1-124, 155 text figs.

Gymnodinium agile
LeBour, M.V., 1925
The dinoflagellates of Northern Seas. The Marine Biological Association of the United Kingdom, Plymouth, 250 pp., 35 pls., 53 text figs.

Gymnodinium agiliforme
Ercegovic, A., 1936
Etudes qualitative et quantitatives du phytoplancton dans les eaux cotières de l'Adriatique oriental moyen au cours de l'année 1934. Acta Adriatica 1(9):1-126

Gymnodinium alaskensis n.sp.
Bursa, Adam, 1963.
Phytoplankton in coastal waters of the Arctic Ocean at Point Barrow, Alaska.
Arctic, 16(4):239-262.

Gymnodinium arcticum
LeBour, M.V., 1925
The dinoflagellates of Northern Seas. The Marine Biological Association of the United Kingdom, Plymouth, 250 pp., 35 pls., 53 text figs.

Gymnodinium arenicolus n.sp.
Dragesco, Jean, 1965.
Étude cytologique de quelques flagellés mésopsammiques.
Cahiers Biol. Mar., Roscoff, 6(1):83-115.

Gymnodinium baccatum n.sp.
Balesh, Enrique and Sayed Z. El-Sayed, 1965.
Microplankton of the Weddell Sea.
In: Biology of Antarctic seas. II.
Antarctic Res. Ser., Amer. Geophys. Union, 5:107-124.

Gymnodinium baccatum
Hada, Yoshina, 1970.   (1970)
The protozoan plankton of the Antarctic and Subantarctic seas.
Scient. Repts. Japan. Antarct. Res. Exped., (E) 31:51 pp.

Gymnodinium bicaudatum n.sp.
Pavillard, J., 1905
Recherches sur la flore pelagique (Phytoplankton) de l'Etang de Thau. Theses presentees a la Fac. Sci., Paris, 116 pp., 3 pls.

Gymnodinium biconicum
Wood, E.J.F., 1963
Dinoflagellates in the Australian region. II. Recent collections.
Comm. Sci. and Industr. Res. Org., Div. Fish. and Oceanogr., Techn. Paper, No. 14: 55 pp.

Gymnodinium bipes
Paulsen, O., 1908
XVIII Peridiniales. Nordisches Plankton, Bot. Teil: 1-124, 155 text figs.

Gymnodinium bogoriense
Wood, E.J.F., 1963
Dinoflagellates in the Australian region. II. Recent collections.
Comm. Sci. and Industr. Res. Org., Div. Fish. and Oceanogr., Techn. Paper, No. 14: 55 pp.

Gymnodinium breve
Abbott, B.C. and Paster Z. 1970
Actions of toxins from Gymnodinium breve.
Toxicon 8: 120.

Gymnodinium breve
Alam, M., J.J. Sasner, Jr. and M. Ikawa 1973
Isolation of Gymnodinium breve toxin from Florida red tide water.
Toxicon 11 (2): 201-202.

Gymnodinium breve
Aldrich, David V., 1962
Photoautotrophy in Gymnodinium breve Davis. Science, 137(3534):988-990.

Gymnodinium breve
Aldrich, David V., and William B. Wilson, 1960.
The effect of salinity on growth of Gymnodinium breve Davis. Biol. Bull., 119(1):57-64.

Gymnodinium brevis
Chew, F., 1953.
Results of hydrographic and chemical investigations in the region of the "red tide" bloom on the west coast of Florida in November 1952. Bull. Mar. Sci., Gulf and Caribbean 2(4):610-625, 10 textfigs.

Gymnodinium brevis
Collier, A., 1953.
Titanium and zirconium in bloom of Gymnodinium brevis Davis. Science 118(3064):329.

Gymnodinium breve
Collier, Albert, W.B. Wilson and Marilynn Borkowski, 1969.
Responses of Gymnodinium breve Davis to natural waters of diverse origin.
J. Phycol. 5(2): 168-172

Gymnodinium breve
Cummins, J.M., A.C. Jones and A.A. Stevens 1971
Occurrence of toxic bivalve molluscs during a Gymnodinium breve 'red tide'.
Trans. Am. Fish. Soc. 100(1): 112-116.

Gymnodinium breve
Doig, M.T. III, and D.F. Martin 1973.
Anticoagulant properties of a red tide toxin.
Toxicon 11 (4): 351-355.

Gymnodinium breve
Doig, Marion T., and Dean F. Martin 1972.
Physical and chemical stability of ichthyotoxins produced by Gymnodinium breve.
Environmental Letts 3(4): 279-288.

Gymnodinium breve
Donnelly, P.V., J. Vuille, M.C. Jayaswal, R.A. Overstreet, J. Williams, M.A. Burklew and R.M. Ingle, 1966.
A study of contributory chemical parameters to red tide in Apalachee Bay.
Florida Bd., Conserv. St. Petersburg, Mar. Lab. Prof. Papers Ser. No. 8:43-83.

Gymnodinium breve
Dragovich, Alexander, 1968.
Morphological variations of Gymnodinium breve Davis. Quarterly Journal of the Florida Academy of Sciences, 30(1967)(4): 245-249, 1968. Also in: Collected Repr. Div. Biol. Res., Bur. Comm. Fish. U.S. Fish Wildl. Serv., 1968, 1.

Gymnodinium breve
Dragovich, A., 1963.
Hydrology and plankton of coastal waters at Naples, Florida.
Q.J. Florida Acad. Sci., 26(1):22-47.

Gymnodinium breve
Dragovich, Alexander, John H. Finucane, John A. Kelly, Jr., and Billie Z. May, 1963.
Counts of red-tide organisms, Gymnodinium breve, west coast, 1960-61.
U.S. Fish and Wildlife Service, Spec. Sci. Repts. Fish. No. 455:1-40.
(and associated oceanographic data from Florida)

Gymnodinium breve
Dragovich, Alexander, John H. Finucane and Billie Z. May, 1961.
Counts of red tide organisms, Gymnodinium breve, and associated oceanographic data from Florida west coast, 1957-1959.
U.S.F.W.S. Spec. Sci. Rept., Fish., No. 369:175 pp.

counts only!

Gymnodinium breve
Finucane, John H., 1964.
Distribution and seasonal occurrence of Gymnodinium breve on the west coast of Florida, 1954-1957.
U.S.F.W.S. Spec. Sci. Rept., Fish., No. 487:14 pp.

Gymnodinium brevis n.sp.
Davis, C.C., 1948.
Gymnodinium brevis sp. nov., a cause of discolored water and animal mortality in the Gulf of Mexico. Bot. Gaz. 109(3):358-360, 2 figs.

60,000,000 per liter

Gymnodinium brevis
Finucane, John H., and Alexander Dragovich, 1959
Counts of red tide organisms, Gymnodinium brevis and associated oceanographic data from Florida west coast, 1954-1957. USFWS Spec. Sci. Rept., Fish. No. 289:220 pp.

Gymnodinium brevis
Galtsoff, P.S., 1949.
The mystery of the red tide. Sci. Mon. LXVIII (2):108-117, 6 figs.

Gymnodinium brevis
Gunter, G., 1949.
The "red tide" and the Florida fisheries. Proc. Gulf and Caribbean Fish. Inst., Inaugural Sess., Aug. 1948:31-32.

Gymnodinium brevis
Gunter, G., R.H. Williams, C.C. Davis, and F.G. Walton Smith, 1948
Catastrophic mass mortality of marine animals and coincident phytoplankton bloom on the West Coast of Florida, November 1946 to August 1947. Ecol. Mon., 18:309-324, 2 text figs.

Gymnodinium brevis
Hela, I., 1955.
Ecological observations on a locally limited red tide bloom. Bull. Mar. Sci., Gulf and Caribbean, 5(4):269-291, 16 textfigs.

Gymnodinium brevis
Hutton, Robert F., 1960.
Notes on the causes of discolored water along the southwestern coast of Florida.
Q.J. Florida Acad. Sci., 23(2):163-164.

Gymnodinium brevis
Long, E.J., 1953.
The red tide hits and runs. Nature Mag. 46(3): 125-128.

Gymnodinium breve
Martin, Dean F., and Ashwin B. Chatterjee, 1971.
Some chemical and physical properties of two toxins from the red-tide organism, Gymnodinium breve.
Fish. Bull. U.S. Nat. Ocean. Atmos. Adm. 68(3): 433-443

Gymnodinium breve
Martin, Dean F., George M. Padilla, Michael Heyl and Priscilla A. Brown 1972
Effect of Gymnodinium breve toxin on hemolysis induced by Prymnesium parvum toxin.
Toxicon 10 (3): 285-290.

Gymnodinium breve
Marvin, Kenneth T., and Raphael R. Proctor, Jr., 1964.
Preliminary results of the systematic screening of 4,306 compounds as 'red tide' toxicants.
U.S. Dept. Interior, Fish Wildl. Serv., Bur. Comm. Fish., Data Rept., 2:3 cards (microfiche).

Gymnodinium breve
McFarren E.F., H. Tanabe, F.J. Silva, W.B. Wilson, J.E. Campbell and K.H. Lewis 1965.
The occurrence of a ciguatera-like poison in oysters, clams and Gymnodinium breve cultures.
Toxicon 3: 111-

Gymnodinium brevis
Odum, H.T., J.B. Lackey, J. Hynes and Nelson Marshall, 1955.
Some red tide characteristics during 1952-1954.
Bull. Mar. Sci., Gulf and Caribbean, 5(4):247-258

Gymnodinium breve
Paster, Zvi and Bernard C. Abbott, 1970.
Gibberellic acid: a growth factor in the unicellular alga Gymnodinium breve. Science, 169(3945): 600-601.

Gymnodinium breve
Ray, Sammy M., and David V. Aldrich, 1965.
Gymnodinium breve: induction of shellfish poisoning in chicks.
Science, 148(3678):1748-1749.

Gymnodinium brevis
Ray, S.M., and W.B. Wilson, 1957.
Effects of unialgal and bacteria-free cultures of Gymnodinium brevis on fish. USFWS Fish. Bull. 123:469-496.

Gymnodinium brevis
Ray, S.M., and W.B. Wilson, 1957.
The effects of unialgal and bacteria-free cultures of *Gymnodinium brevis* on fish and notes on related studies with bacteria. U.S.F.W.S. Fish. Bull. (in press).
Issued first as Special Sci. Rept., Fish., No. 211.

Gymnodinium breve
Rounsefell, George A., and Alexander Dragovich, 1966.
Correlation between oceanographic factors and abundance of the Florida red-tide (*Gymnodinium breve* Davis), 1954-61.
Bull. Mar. Sci., 16(3):404-422.

Gymnodinium breve
Rounsefell, George A., and Walter R. Nelson, 1966.
Red-tide research summarized to 1964 including an annotated bibliography.
Spec. Scient. Rep. U.S. Fish Wildl. Serv., Fish., 535: 85pp. (multilithed).

Gymnodinium breve
Sasner, J.J. Jr., M. Ikawa, F. Thurberg and M. Alam 1972.
Physiological and chemical studies on *Gymnodinium breve* Davis toxins.
Toxicon 10(2): 163-172.

Gymnodinium breve
Saunders, Richard P., and Carol L. Wahlquist, 1966.
Diatoms and their relationship to *Gymnodinium breve* Davis.
Florida Bd. Conserv., St. Petersburg, Mar. Lab. Prof. Papers Ser., No. 8:16-33.

Gymnodinium breve
Sievers, Anita M., 1969.
Comparative toxicity of *Gonyaulax monilata* and *Gymnodinium breve* to annelids, crustaceans, molluscs and a fish.
J. Protozool., 16(3): 401-404

Gymnodinium brevis
Smith, F.G. Walton, 1948(1949).
Probable fundamental causes of red tide off the west coast of Florida. Quart. J. Florida Acad. Sci. 11(1):1-6.

Gymnodinium breve
Spikes, John J., Sammy M. Ray, David V. Aldrich and Joe B. Nash, 1968.
Toxicity variations of *Gymnodinium breve* cultures.
Toxicon 5: 171-174.
Also in Contr. Oceanogr. Texas A&M Univ. 12 (Contr. 378)

Gymnodinium breve
Spikes, J.J., S.M. Ray, D.V. Aldrich and J.B. Nash 1968.
Toxicity variations of *Gymnodinium breve* cultures.
Toxicon 5:171-

Gymnodinium breve
Steidinger, Karen A., Joanne T. Davis and Jean Williams, 1966.
Observations of *Gymnodinium breve* Davis and other dinoflagellates.
Florida Bd. Conserv., St. Petersburg, Mar. Lab. Prof. Papers Ser., No. 8:8-15.

Gymnodinium breve
Steidinger, Karen A. and Edwin A. Joyce, Jr. 1973.
Florida red tides.
Education. Ser. Dept. Nat. Resources, Florida, 17: 26pp.

Gymnodinium breve
Stewart V.N., H. Wahlquist, R. Burket and C. Wahlquist, 1966.
Observations of vitamin B_{12} distribution in Apalachee Bay, Florida. Prof. Pap. Ser. Fla Bd Conserv. 8: 34-38.

Gymnodinium breve
Trieff, N.M., J.J. Spikes, S.M. Ray and J.B. Nash 1972
Isolation of *Gymnodinium breve* toxin
Toxicon 8: 157-158

Gymnodinium breve
Wilson, William B. 1967.
Forms of the dinoflagellate *Gymnodinium breve* Davis in cultures.
Contrib. mar. Sci. Port Aransas 12: 120-134.

Gymnodinium breve
Wilson, William B., 1966.
The suitability of sea-water for the survival and growth of *Gymnodinium breve* Davis; and some effects of phosphorus and nitrogen on its growth.
Florida State Univ., Prof. Papers Ser. No. 7: 42 pp.

Gymnodinium brevis
Wilson, W.B., and A. Collier, 1955.
Preliminary notes on the culturing of *Gymnodinium brevis* Davis. Science 121(3142):394-395.

Gymnodinium brevis
Wilson, W.B., and S.M. Ray, 1956.
The occurrence of *Gymnodinium brevis* in the western Gulf of Mexico. Ecology 37(2):388.

Gymnodinium cassiei n.sp.
Norris, R.E., 1961.
Observations on phytoplankton organisms collected on the N.Z.O.I. Pacific Cruise, September, 1958.
N.Z.J. Sci., 4(1):162-188.

Gymnodinium catenatum
Balech, Enrique, 1964.
El plancton de Mar del Plata durante el periodo 1961-1962.
Bol. Inst. Biol. Mar., Buenos Aires, (4):56 pp.
49 pp. + plates

Gymnodinium catenatum n.sp.
Graham, H.W., 1943.
Gymnodinium catenatum, a new dinoflagellate from the Gulf of California. Trans. Amer. Microsc. Soc 62(3):259-261, 2 textfigs.

Gymnodinium cinctum
Hada, Yoshine, 1970. (1970)
The protozoan plankton of the Antarctic and Subantarctic seas.
Scient. Repts, Japan. Antarct. Res. Exped., (E)31;51 pp.

Gymnodinium cinctum
Wood, E.J.F., 1963
Dinoflagellates in the Australian region.
II. Recent collections.
Comm. Sci. and Industr. Res. Org., Div. Fish. and Oceanogr., Techn. Paper, No. 14: 55 pp.

Gymnodinium coerulatum
Wood, E.J.F., 1963
Dinoflagellates in the Australian region.
II. Recent collections.
Comm. Sci. and Industr. Res. Org., Div. Fish. and Oceanogr., Techn. Paper, No. 14: 55 pp.

Gymnodinium cohni
Becheler, B., 1952.
Recherches sur les Péridiniens.
Bull. Biol. France Belg., Suppl., 36:1-149, 75 textfigs.

Gymnodinium conicum
LeBour, M.V., 1925
The dinoflagellates of Northern Seas. The Marine Biological Association of the United Kingdom, Plymouth, 250 pp., 35 pls., 53 text figs.

Gymnodinium Cori
Ercegovic, A., 1936
Etudes qualitative et quantitatives du phytoplancton dans les eaux cotières de l'Adriatique oriental moyen au cours de l'année 1934. Acta Adriatica 1(9):1-126

Gymnodinium costatum
Wood, E.J.F., 1963
Dinoflagellates in the Australian region.
II. Recent collections.
Comm. Sci. and Industr. Res. Org., Div. Fish. and Oceanogr., Techn. Paper, No. 14: 55 pp.

Gymnodinium danicum
Paulsen, O., 1908
XVIII Peridiniales. Nordisches Plankton, Bot. Teil: 1-124, 155 text figs.

Gymnodinium diamphidium n.sp.
Norris, R.E., 1961.
Observations on phytoplankton organisms collected on the N.Z.O.I. Pacific Cruise, September, 1958.
N.Z.J. Sci., 4(1):162-188.

Gymnodinium diploconus
Ercegovic, A., 1936
Etudes qualitative et quantitatives du phytoplancton dans les eaux cotières de l'Adriatique oriental moyen au cours de l'année 1934. Acta Adriatica 1(9):1-126

Gymnodinium diploconus
Wood, E.J.F., 1963
Dinoflagellates in the Australian region.
II. Recent collections.
Comm. Sci. and Industr. Res. Org., Div. Fish. and Oceanogr., Techn. Paper, No. 14: 55 pp.

Gymnodinium elongatum n.sp.
Hope, B., 1954.
Floristic and taxonomic observations on marine phytoplankton from Nordavatn, near Bergen.
Nytt Mag. f. Botanikk 2:149-153, 1 fig.

Gymnodinium exechegloutum n.sp.
Norris, R.E., 1961.
Observations on phytoplankton organisms collected on the N.Z.O.I. Pacific Cruise, September, 1958.
N.Z.J. Sci., 4(1):162-188.

Gymnodinium filum
LeBour, M.V., 1925
The dinoflagellates of Northern Seas. The Marine Biological Association of the United Kingdom, Plymouth, 250 pp., 35 pls., 53 text figs.

Gymnodinium flavum
Hada, Yoshine, 1970. (1970)
The protozoan plankton of the Antarctic and Subantarctic seas.
Scient. Repts, Japan. Antarct. Res. Exped., (E)31;51 pp.

Gymnodinium flavum
Lackey, James B., and K.A. Clendenning, 1963.
A possible fish-killing yellow tide in California waters.
Q.J. Florida Acad. Sci., 26(3):263-268.

Gymnodinium flavum
Wilton, J.W., and E.G. Barham, 1968.
A yellow-water bloom of Gymnodinium flavum Kofoid and Swezy.
J. exp. mar. Biol. Ecol., 2(2):167-173.

Gymnodinium flavum
Wood, E.J.F., 1963
Dinoflagellates in the Australian region. II. Recent collections.
Comm. Sci. and Industr. Res. Org., Div. Fish. and Oceanogr., Techn. Paper, No. 14: 55 pp.

Gymnodinium frigidum n.sp.
Balech, Enrique and Sayed Z. El-Sayed, 1965.
Microplankton of the Weddell Sea.
In: Biology of Antarctic seas. II.
Antarctic Res. Ser., Amer. Geophys. Union, 5:107-124.

Gymnodinium frigidum
Hada, Yoshina, 1970. (1970)
The protozoan plankton of the Antarctic and Subantarctic seas.
Scient. Repts, Japan. Antarct. Res. Exped., (E)31:51 pp.

Gymnodinium fungiforme
Biecheler, B., 1952.
Recherches sur les Peridiniens.
Bull. Biol., France Belg., Suppl., 36:1-149, 75 textfigs.

Gymnodinium fusus
Wood, E.J.F., 1963
Dinoflagellates in the Australian region. II. Recent collections.
Comm. Sci. and Industr. Res. Org., Div. Fish. and Oceanogr., Techn. Paper, No. 14: 55 pp.

Gymnodinium galatheae n.sp.
Braarud Trygve, 1957.
A red water organism from Walvis Bay (Gymnodinium galatheae n.sp.) Galathea Rept. No. 1:137-138.

Gymnodinium galeaeforme
Wood, E.J.F., 1963
Dinoflagellates in the Australian region. II. Recent collections.
Comm. Sci. and Industr. Res. Org., Div. Fish. and Oceanogr., Techn. Paper, No. 14: 55 pp.

Gymnodinium gelbum
Wood, E.J.F., 1963
Dinoflagellates in the Australian region. II. Recent collections.
Comm. Sci. and Industr. Res. Org., Div. Fish. and Oceanogr., Techn. Paper, No. 14: 55 pp.

Gymnodinium glandula
LeBour, M.V., 1925
The dinoflagellates of Northern Seas. The Marine Biological Association of the United Kingdom, Plymouth, 250 pp., 35 pls., 53 text figs.

Gymnodinium gracile
LeBour, M.V., 1925
The dinoflagellates of Northern Seas. The Marine Biological Association of the United Kingdom, Plymouth, 250 pp., 35 pls., 53 text figs.

Gymnodinium gracile
Paulsen, O., 1908
XVIII Peridiniales. Nordisches Plankton, Bot. Teil: 1-124, 155 text figs.

Gymnodinium gracile exiguum
Paulsen, O., 1908
XVIII Peridiniales. Nordisches Plankton, Bot. Teil: 1-124, 155 text figs.

Gymnodinium grammaticum
LeBour, M.V., 1925
The dinoflagellates of Northern Seas. The Marine Biological Association of the United Kingdom, Plymouth, 250 pp., 35 pls., 53 text figs.

Gymnodinium grammaticum
Norris, R.E., 1961.
Observations on phytoplankton organisms collected on the N.Z.O.I. Pacific Cruise, September 1958.
N.Z. J. Sci., 4(1):162-188.

Gymnodinium grammaticum
Wood, E.J.F., 1963
Dinoflagellates in the Australian region. II. Recent collections.
Comm. Sci. and Industr. Res. Org., Div. Fish. and Oceanogr., Techn. Paper, No. 14: 55 pp.

Gymnodinium halophilum n.sp.
Biecheler, B., 1952.
Recherches sur les Peridiniens.
Bull. Biol., France Belg., Suppl., 36:1-149, 75 textfigs.

Gymnodinium heterostriatum
Ercegovic, A., 1936
Etudes qualitative et quantitatives du phytoplancton dans les eaux cotières de l'Adriatique oriental moyen au cours de l'année 1934. Acta Adriatica 1(9):1-126

Gymnodinium heterostriatum
LeBour, M.V., 1925
The dinoflagellates of Northern Seas. The Marine Biological Association of the United Kingdom, Plymouth, 250 pp., 35 pls., 53 text figs.

Gymnodinium heterostriatum
Wood, E.J.F., 1963
Dinoflagellates in the Australian region. II. Recent collections.
Comm. Sci. and Industr. Res. Org., Div. Fish. and Oceanogr., Techn. Paper, No. 14: 55 pp.

Gymnodinium hyalinum n.sp.
LeBour, M.V., 1925
The dinoflagellates of Northern Seas. The Marine Biological Association of the United Kingdom, Plymouth, 250 pp., 35 pls., 53 text figs.

Gymnodinium incertum
LeBour, M.V., 1925
The dinoflagellates of Northern Seas. The Marine Biological Association of the United Kingdom, Plymouth, 250 pp., 35 pls., 53 text figs.

Gymnodinium intercalaris nsp.
Bursa, Adam S., 1961
The annual oceanographic cycle at Igloolik in the Canadian Arctic. II. The phytoplankton
J. Fish. Res. Bd., Canada, 18(4):563-615.

Gymnodinium irregulare n.sp.
Hope, B., 1954.
Floristic and taxonomic observations on marine phytoplankton from Nordavåtn, near Bergen.
Nytt Mag. f. Botanikk 2:149-153, 1 fig.

Gymnodinium kovalevskii
Akinina, D.K., 1969.
Relative velocity of settling of dinoflagellata as dependent on their division rates. (In Russi; English abstract). Okeanologiia, 9(2): 301-305.

Gymnodinium kovalevskii
Akinina, D.K., 1966.
Dependence of light saturation of two mass species of dinoflagellates on a number of factors. (In Russian; English Abstract).
Okeanologiia, Akad.Nauk,SSSR,6(5)L861-868.

Gymnodinium lazulum n.sp.
Hulburt, E.M., 1957.
The taxonomy of unarmored Dinophyceae of shallow embayments on Cape Cod, Massachusetts.
Biol. Bull., 112(2):196-219.

Gymnodinium Lebourii
LeBour, M.V., 1925
The dinoflagellates of Northern Seas. The Marine Biological Association of the United Kingdom, Plymouth, 250 pp., 35 pls., 53 text figs.

Gymnodinium letum n.sp.
Norris, R.E., 1961.
Observations on phytoplankton organisms collected on the N.Z.O.I. Pacific Cruise, September 1958.
N.Z.J. Sci., 4(1):162-188.

Gymnodinium Lohmannii
Braarud, T., 1945
A phytoplankton survey of the polluted waters of inner Oslo Fjord. Hvalrådets Skrifter, No.28, 142 pp., 19 text figs., 17 tables.

Gymnodinium Lohmannii
Braarud, T., 1934
A note on the phytoplankton of the Gulf of Maine in the summer of 1933. Biol. Bull. 67(1):76-82.

Gymnodinium Lohmanni
Braarud, T., and Adam Bursa, 1939
On the phytoplankton of the Oslo Fjord, 1933-1934. Hvalrådets Skr. No.19:1-63; 9 text figs. Reviewed. J. du. Cons. 14(3): 418-420. A.C. Gardiner.

Gymnodinium lohmanni
LeBour, M.V., 1925
The dinoflagellates of Northern Seas. The Marine Biological Association of the United Kingdom, Plymouth, 250 pp., 35 pls., 53 text figs.

Gymnodinium Lohmanni
Lillick, L.C., 1940
Phytoplankton and planktonic protozoa of the offshore waters of the Gulf of Maine. Pt.II. Qualitative Composition of the Planktonic Flora. Trans. Am. Phil. Soc., n.s., 31(3):193-237, 13 text figs.

Gymnodinium lohmanni
Paulsen, O., 1908
XVIII Peridiniales. Nordisches Plankton, Bot. Teil: 1-124, 155 text figs.

Gymnodinium lunula
LeBour, M.V., 1925
The dinoflagellates of Northern Seas. The Marine Biological Association of the United Kingdom, Plymouth, 250 pp., 35 pls., 53 text figs.

Gymnodinium lunula
Morse, D.C., 1947
Some observations on seasonal variations in plankton population Patuxant River, Maryland 1943-1945. Bd. Nat. Res., Publ. No.65, Chesapeake Biol. Lab., 31, 3 figs.

Gimnodinium lunula
Sousa e Silva, E., 1949
Diatomaceas e Dinoflagelados de Baía de Cascais. Portugaliae Acta Biol., Volume: Julio Henriques, Ser. B: 300-383, 9 pls, 2 fold-in tables.

Gymnodinium maguelonnense
Biecheler, B., 1952.
Recherches sur les Peridiniens.
Bull. Biol., France Belg., Suppl., 36:1-149.

Gymnodinium maximum nom. prov.
Nordli, O., 1951.
Dinoflagellates from Lofoten. Nytt Mag. Naturvidensk 88:49-55, 7 textfigs.

Gymnodinium marinum
LeBour, M.V., 1925
The dinoflagellates of Northern Seas. The Marine Biological Association of the United Kingdom, Plymouth, 250 pp., 35 pls., 53 text figs.

Gymnodinium marinum
Wood, E.J.F., 1963
Dinoflagellates in the Australian region. II. Recent collections.
Comm. Sci. and Industr. Res. Org., Div. Fish. and Oceanogr., Techn. Paper, No. 14: 55 pp.

Gymnodinium mikimotoi
Takano, Hideaki, 1971.
The original description of Gymnodinium mikimotoi. Bull. Plankt. Soc. Japan 18(1): 103-104. (In Japanese and English).

Gymnodinium minor
Ercegovic, A., 1936
Etudes qualitative et quantitatives du. phytoplancton dans les eaux cotières de l'Adriatique oriental moyen au cours de l'année 1934. Acta Adriatica 1(9):1-126

Gymnodinium minor
Hada, Yoshina, 1970. (1970)
The protozoan plankton of the Antarctic and Subantarctic seas.
Scient. Repts, Japan. Antarct. Res. Exped., (E)31:51 pp.

Gymnodinium minor
LeBour, M.V., 1925
The dinoflagellates of Northern Seas. The Marine Biological Association of the United Kingdom, Plymouth, 250 pp., 35 pls., 53 text figs.

Gymnodinium minor
Norris, R.E., 1961.
Observations on phytoplankton organisms collected on the N.Z.O.I. Pacific Cruise, September 1958.
N.Z. J. Sci., 4(1):162-188.

Gymnodinium minor
Wood, E.J.F., 1963
Dinoflagellates in the Australian region. II. Recent collections.
Comm. Sci. and Industr. Res. Org., Div. Fish. and Oceanogr., Techn. Paper, No. 14: 55 pp.

Gymnodinium minutum nom. nov.
LeBour, M.V., 1925
The dinoflagellates of Northern Seas. The Marine Biological Association of the United Kingdom, Plymouth, 250 pp., 35 pls., 53 text figs.

Gymnodinium multistriatum
Wood, E.J.F., 1963
Dinoflagellates in the Australian region. II. Recent collections.
Comm. Sci. and Industr. Res. Org., Div. Fish. and Oceanogr., Techn. Paper, No. 14: 55 pp.

Gymnodinium najadeum
Ercegovic, A., 1936
Etudes qualitative et quantitatives du. phytoplancton dans les eaux cotières de l'Adriatique oriental moyen au cours de l'année 1934. Acta Adriatica 1(9):1-126

Gymnodinium nanum
Wood, E.J.F., 1963
Dinoflagellates in the Australian region. II. Recent collections.
Comm. Sci. and Industr. Res. Org., Div. Fish. and Oceanogr., Techn. Paper, No. 14: 55 pp.

Gymnodinium nelsoni
Hochachka, Peter W., and John M. Teal, 1964.
Respiratory metabolism in a marine dinoflagellate
Biol. Bull., 126(2):274-281.

Gymnodinium nelsoni
Hulburt, E.M., 1957.
The taxonomy of unarmored Dinophyceae of shallow embayments on Cape Cod, Massachusetts.
Biol. Bull., 112(2):196-219.

Gymnodinium nelsoni
Mendiola, Leticia R., C. A. Price and R.R.L. Guillard, 1966.
Isolation of nuclei from a marine dinoflagellate
Science, 153(3744):1661-1663.

Gymnodinium nelsoni
Morse, D.C., 1947
Some observations on seasonal variations in plankton population Patuxant River, Maryland 1943-1945. Bd. Nat. Res., Publ. No.65, Chesapeake Biol. Lab., 31, 3 figs.

Gymnodinium obesum
Wood, E.J.F., 1963
Dinoflagellates in the Australian region. II. Recent collections.
Comm. Sci. and Industr. Res. Org., Div. Fish. and Oceanogr., Techn. Paper, No. 14: 55 pp.

Gymnodinium obliquum
Paulsen, O., 1908
XVIII Peridiniales. Nordisches Plankton, Bot. Teil: 1-124, 155 text figs.

Gymnodinium oceanicum n.sp.
Hasle, Grethe Rytter, 1960
Phytoplankton and ciliate species from the Tropical Pacific.
Skr. Norske Videnskaps-Akad., Oslo, 1. Mat.-Nat. Kl., 1960(2): 1-50.

Gymnodinium ochraceum
Wood, E.J.F., 1963
Dinoflagellates in the Australian region. II. Recent collections.
Comm. Sci. and Industr. Res. Org., Div. Fish. and Oceanogr., Techn. Paper, No. 14: 55 pp.

Gymnodinium ovatum
Gourret, P., 1883
Sur les Peridiniens du Golfe de Marseille. Ann. du Musee d'hist. Nat., Marseille, Zool., 1 (Mme. 8):1-114, 4 pls.

Gymnodinium Paulseni
Ercegovic, A., 1936
Etudes qualitative et quantitatives du. phytoplancton dans les eaux cotières de l'Adriatique oriental moyen au cours de l'année 1934. Acta Adriatica 1(9):1-126

Gymnodinium patagoniam n. sp.
Balech, Enrique, 1971.
Microplancton de la campaña oceanographica: Productividad III.
Revta Mus. argent. Cienc. Nat. Bernadina Rivadavia, Hydrobiol. 3(1):1-202, 39 pls.

Gymnodinium pavillardi
Biecheler, B., 1952.
Recherches sur les Peridiniens.
Bull. Biol., France Belg., Suppl., 36:1-149, 75 textfigs.

Gymnodinium pellucidum
LeBour, M.V., 1925
The dinoflagellates of Northern Seas. The Marine Biological Association of the United Kingdom, Plymouth, 250 pp., 35 pls., 53 text figs.

Gymnodinium placidum
LeBour, M.V., 1925
The dinoflagellates of Northern Seas. The Marine Biological Association of the United Kingdom, Plymouth, 250 pp., 35 pls., 53 text figs.

Gymnodinium Pouchetii
Paulsen, O., 1908
XVIII Peridiniales. Nordisches Plankton, Bot. Teil: 1-124, 155 text figs.

Gymnodinium Pouchetii
Pavillard, J., 1905
Recherches sur la flore pelagique (Phytoplankton) de l'Etang de Thau. Theses presentees a la Fac. Sci., Paris, 116 pp., 3 pls.

Gymnodinium Pouchetii
Schröder, B., 1900
Phytoplankton des Golfes von Neapel nebst vergleichenden Ausblicken auf das atlantischen Ozean. Mitt. Zool. Stat. Neapel, 14:1-38.

Gymnodinium pseudonoctiluca
LeBour, M.V., 1925
The dinoflagellates of Northern Seas. The Marine Biological Association of the United Kingdom, Plymouth, 250 pp., 35 pls., 53 text figs.

Gymnodinium pseudonoctiluca
Meunier, A., 1919
Microplancton de la Mer Flamande 3. Les Péridiniens. Mem. Mus. Roy. Hist. Nat., Belgique 8(1):1-116, Pls. XV-XXI.

Gymnodinium pseudonoctiluca
Paulsen, O., 1908
XVIII Peridiniales. Nordisches Plankton, Bot. Teil: 1-124, 155 text figs.

Gymnodinium punctatum
LeBour, M.V., 1925
The dinoflagellates of Northern Seas. The Marine Biological Association of the United Kingdom, Plymouth, 250 pp., 35 pls., 53 text figs.

Gymnodinium punctatum
Paulsen, O., 1908
XVIII Peridiniales. Nordisches Plankton, Bot. Teil: 1-124, 155 text figs.

Gymnodinium punctatum grammatica
Schröder, B., 1900
Phytoplankton des Golfes von Neapel nebst vergleichenden Ausblicken auf das atlantischen Ozean. Mitt. Zool. Stat. Neapel, 14:1-38.

Gymnodinium punctatum
Wood, E.J.F., 1963
Dinoflagellates in the Australian region. II. Recent collections.
Comm. Sci. and Industr. Res. Org., Div. Fish. and Oceanogr., Techn. Paper, No. 14: 55 pp.

Gymnodinium pygmaeum n.sp.
LeBour, M.V., 1925
The dinoflagellates of Northern Seas. The Marine Biological Association of the United Kingdom, Plymouth, 250 pp., 35 pls., 53 text figs.

Gymnodinium pygmaeum

Wood, E.J.F., 1963
Dinoflagellates in the Australian region.
II. Recent collections.
Comm. Sci. and Industr. Res. Org., Div.
Fish. and Oceanogr., Techn. Paper, No. 14:
55 pp.

Gymnodinium pyrocystis

LeBour, M.V., 1925
The dinoflagellates of Northern Seas. The
Marine Biological Association of the United
Kingdom, Plymouth, 250 pp., 35 pls., 53 text
figs.

Gymnodinium rhomboides

Ercegovic, A., 1936
Etudes qualitative et quantitatives du
phytoplancton dans les eaux cotières de
l'Adriatique oriental moyen au cours de l'année
1934. Acta Adriatica 1(9):1-126

Gymnodinium rhomboides

LeBour, M.V., 1925
The dinoflagellates of Northern Seas. The
Marine Biological Association of the United
Kingdom, Plymouth, 250 pp., 35 pls., 53 text
figs.

Gymnodinium rhomboides

Paulsen, O., 1908
XVIII Peridiniales. Nordisches Plankton,
Bot. Teil: 1-124, 155 text figs.

Gymnodinium rhomboides

Rumkówna, A., 1948
List of the phytoplankton species occurring in the superficial water layers in the
Gulf of Gdańsk] Bull. Lab. mar., Gdynia,
No. 4: 139-141 with tables in back.

Gymnodinium rotundatum

Wood, E.J.F., 1963
Dinoflagellates in the Australian region.
II. Recent collections.
Comm. Sci. and Industr. Res. Org., Div.
Fish. and Oceanogr., Techn. Paper, No. 14:
55 pp.

Gymnodinium rubricinetum n.sp.

LeBour, M.V., 1925
The dinoflagellates of Northern Seas. The
Marine Biological Association of the United
Kingdom, Plymouth, 250 pp., 35 pls., 53 text
figs.

Gymnodinium rubrum

Wood, E.J.F., 1963
Dinoflagellates in the Australian region.
II. Recent collections.
Comm. Sci. and Industr. Res. Org., Div.
Fish. and Oceanogr., Techn. Paper, No. 14:
55 pp.

Gymnodinium sanguineum n.sp.

Hirasaka, K., 1922.
On a case of discolored water. Annotat. Zool.
Japon. 10(5):161-164.

Gymnodinium scopulosum

Wood, E.J.F., 1963
Dinoflagellates in the Australian region.
II. Recent collections.
Comm. Sci. and Industr. Res. Org., Div.
Fish. and Oceanogr., Techn. Paper, No. 14:
55 pp.

Gymnodinium simplex

Ercegovic, A., 1936
Etudes qualitative et quantitatives du
phytoplancton dans les eaux cotières de
l'Adriatique oriental moyen au cours de l'année
1934. Acta Adriatica 1(9):1-126

Gymnodinium simplex

LeBour, M.V., 1925
The dinoflagellates of Northern Seas. The
Marine Biological Association of the United
Kingdom, Plymouth, 250 pp., 35 pls., 53 text
figs.

Gymnodinium simplex

Morse, D.C., 1947
Some observations on seasonal variations in
plankton population Patuxant River, Maryland 1943-
1945. Bd. Nat. Res. Publ. No.65, Chesapeake
Biol. Lab., 31, 3 figs.

Gymnodinium simplex

Norris, R.E., 1961.
Observations on phytoplankton organisms collected on the N.Z.O.I. Pacific Cruise, September 1958.
N.Z.J. Sci., 4(1):162-188.

Gymnodinium simplex

Wood, E.J.F., 1963
Dinoflagellates in the Australian region.
II. Recent collections.
Comm. Sci. and Industr. Res. Org., Div.
Fish. and Oceanogr., Techn. Paper, No. 14:
55 pp.

Gymnodinium situla

Wood, E.J.F., 1963
Dinoflagellates in the Australian region.
II. Recent collections.
Comm. Sci. and Industr. Res. Org., Div.
Fish. and Oceanogr., Techn. Paper, No. 14:
55 pp.

Gymnodinium soyai n. sp.

Hada, Yoshina, 1970.   (1970)
The protozoan plankton of the Antarctic and
Subantarctic seas.
Scient. Repts, Japan. Antarct. Res. Exped.,
(E)31:51 pp.

Gymnodinium sphaericum

Wood, E.J.F., 1963
Dinoflagellates in the Australian region.
II. Recent collections.
Comm. Sci. and Industr. Res. Org., Div.
Fish. and Oceanogr., Techn. Paper, No. 14:
55 pp.

Gymnodinium spiralis

Jorgensen, E., 1900
Protophyten und Protozoen im Plankton aus der Norwegischen Westkerste. Bergens
Mus. Aarb. 1899(6): 95 pp., 5 pls., 83 tables

Gymnodinium splendens

Biecheler, B., 1952.
Recherches sur les Peridiniens.
Bull. Biol., France Belg., Suppl., 36:1-149,
75 textfigs.

Gymnodinium splendens

Dandonneau, Y., 1970.
Un phénomène d'eaux rouges au large de la
Côte d'Ivoire causé par Gymnodinium splendens Lebour. Doc. scient. Centre Rech.
Oceanogr. Abidjan, 1(1): 11-19.

Gymnodinium splendens

Grøntved, J., 1960-61
Planktological contributions. IV. Taxonomical
and productional investigations in shallow
coastal waters.
Medd. Dansk Fisk. Havundersøgelser, n.s., 3(1):
1-17.

Gymnodinium splendens

Hutton, Robert, F., 1960.
Notes on the causes of discolored water along the
southwestern coast of Florida.
Q.J. Florida Acad. Sci., 23(2):163-164.

Gymnodinium splendens n.sp.

LeBour, M.V., 1925
The dinoflagellates of Northern Seas. The
Marine Biological Association of the United
Kingdom, Plymouth, 250 pp., 35 pls., 53 text
figs.

Gymnodinium splendens

Quayle, D.B.
Paralytic shellfish poisoning in British Colombia.
Bull. Fish Res. Bd. Can., 168: 68 pp.

Gymnodinium splendens

Sweeney, B.M., 1954.
Gymnodinium splendens, a marine dinoflagellate
requiring Vitamin B12. Amer. J. Bot., 41:821-824.

Gymnodinium splendens

Wood, E.J.F., 1963
Dinoflagellates in the Australian region.
II. Recent collections.
Comm. Sci. and Industr. Res. Org., Div.
Fish. and Oceanogr., Techn. Paper, No. 14:
55 pp.

Gymnodinium stellatum n.sp.

Hulburt, E.M., 1957.
The taxonomy of unarmored Dinophyceae of shallow
embayments on Cape Cod, Massachusetts.
Biol. Bull., 112(2):196-219.

Gymnodinium striatissimum n.sp.

Hulburt, E.M., 1957.
The taxonomy of unarmored Dinophyceae of shallow
embayments on Cape Cod, Massachusetts.
Biol. Bull., 112(2):196-219.

Gymnodinium sulcatum

Wood, E.J.F., 1963
Dinoflagellates in the Australian region.
II. Recent collections.
Comm. Sci. and Industr. Res. Org., Div.
Fish. and Oceanogr., Techn. Paper, No. 14:
55 pp.

Gymnodinium teredo

Paulsen, O., 1908
XVIII Peridiniales. Nordisches Plankton,
Bot. Teil: 1-124, 155 text figs.

Gymnodinium teredo

Schröder, B., 1900
Phytoplankton des Golfes von Neapel
nebst vergleichenden Ausblicken auf das
atlantischen Ozean. Mitt. Zool. Stat.
Neapel, 14:1-38.

Gymnodinium tintinnicula

LeBour, M.V., 1925
The dinoflagellates of Northern Seas. The
Marine Biological Association of the United
Kingdom, Plymouth, 250 pp., 35 pls., 53 text
figs.

Gymnodinium triangularis

LeBour, M.V., 1925
The dinoflagellates of Northern Seas. The
Marine Biological Association of the United
Kingdom, Plymouth, 250 pp., 35 pls., 53 text
figs.

Gymnodinium trochoideum

Paulsen, O., 1908
XVIII Peridiniales. Nordisches Plankton,
Bot. Teil: 1-124, 155 text figs.

Gymnodinium uberrimum

Wood, E.J.F., 1963
Dinoflagellates in the Australian region.
II. Recent collections.
Comm. Sci. and Industr. Res. Org., Div.
Fish. and Oceanogr., Techn. Paper, No. 14:
55 pp.

Gymnodinium variabile
Dragesco, Jean, 1965.
Etude cytologique de quelques flagellés mésopsammiques.
Cahiers Biol. Mar., Roscoff, 6(1):83-115.

Gymnodinium variabile
LeBour, M.V., 1925
The dinoflagellates of Northern Seas. The Marine Biological Association of the United Kingdom, Plymouth, 250 pp., 35 pls., 53 text figs.

Gymnodinium varians
Wood, E.J.F., 1963
Dinoflagellates in the Australian region. II. Recent collections.
Comm. Sci. and Industr. Res. Org., Div. Fish. and Oceanogr., Techn. Paper, No. 14: 55 pp.

Gymnodinium veneficium
Abbott, B.C., and D. Ballentine, 1957.
The toxin from Gymnodinium veneficium Ballentine.
J.M.B.A., 36(1):169-190.

not ordinarily poisonous

Gymnodinium veneficium n.sp.
Ballentine, D., 1956.
Two new marine species of Gymnodinium isolated from the Plymouth area. J.M.B.A., 35(3):467-474.

Gymnodinium vestifici
LeBour, M.V., 1925
The dinoflagellates of Northern Seas. The Marine Biological Association of the United Kingdom, Plymouth, 250 pp., 35 pls., 53 text figs.

Gymnodinium vestifici
Paulsen, O., 1908
XVIII Peridiniales. Nordisches Plankton, Bot. Teil: 1-124, 155 text figs.

Gymnodinium vitiligo n.sp.
Ballentine, Dorothy, 1956.
Two new marine species of Gymnodinium isolated from the Plymouth area. J.M.B.A., 35(3):467-474.

Gymnodinium vorax n.sp.
Biecheler, B., 1952.
Recherches sur les Peridiniens.
Bull. Biol. France Belg., Suppl., 36:1-149, 75 textfigs.

Gymnodinium Warmingii
Paulsen, O., 1908
XVIII Peridiniales. Nordisches Plankton, Bot. Teil: 1-124, 155 text figs.

Gymnodinium wilczeki
LeBour, M.V., 1925
The dinoflagellates of Northern Seas. The Marine Biological Association of the United Kingdom, Plymouth, 250 pp., 35 pls., 53 text figs.

Gymnodinium Wilczecki
Paulsen, O., 1908
XVIII Peridiniales. Nordisches Plankton, Bot. Teil: 1-124, 155 text figs.

Gyrodinium sp.
Balech, Enrique and Sayed Z. El-Sayed, 1965.
Microplankton of the Weddell Sea.
In: Biology of Antarctic seas. II. Antarctic Res. Ser., Amer. Geophys. Union, 5:107-124.

Gyrodinium sp.
Braarud, T., 1945
A phytoplankton survey of the polluted waters of inner Oslo Fjord. Hvalrådets Skrifter, No.28, 142 pp., 19 text figs., 17 tables.

Gyrodinium
Kimball, J.F., Jr., and E.J. Ferguson Wood, 1965
A dinoflagellate with characters of Gymnodinium and Gyrodinium.
J. Protozool., 12(4):577-580.

Gyrodinium sp?
Provasoli, L., and J.F. Howell, 1952.
Culture of a marine Gyrodinium in a synthetic medium. Proc. Am. Soc. Protozool. 3:6.

Gyrodinium sp.
Hand, William G., Patricia A. Collard and Demarest Davenport, 1965.
The effects of temperature and salinity change on swimming rate in the dinoflagellates, Gonyaulax and Gyrodinium.
Biol. Bull., 128(1):90-101.

Gyrodinium adriaticum
Ercegovic, A., 1936
Etudes qualitative et quantitatives du phytoplancton dans les eaux cotières de l'Adriatique oriental moyen au cours de l'année 1934. Acta Adriatica 1(9):1-126

Gyrodinium apidiomorphum n.sp.
Norris, R.E., 1961.
Observations on phytoplankton organisms collected on the N.Z.O.I. Pacific Cruise, September 1958.
N.Z.J. Sci., 4(1):162-188.

Gyrodinium arcticum n.sp.
Bursa, Adam S., 1961
The annual oceanographic cycle at Igloolik in the Canadian Arctic. II. The phytoplankton
J. Fish. Res. Bd., Canada, 18(4):563-615.

Gyrodinium aureolum
Braarud, Trygve, and Berit R. Heimdal
Brown water on the Norwegian coast in autumn 1966
Nytt Mag. Bot. 17(2):91-97

Gyrodinium aureolum n.sp.
Hulburt, E.M., 1957.
The taxonomy of unarmored Dinophyceae of shallow embayments on Cape Cod, Massachusetts.
Biol. Bull., 112(2):196-219.

Gyrodinium bepo
LeBour, M.V., 1925
The dinoflagellates of Northern Seas. The Marine Biological Association of the United Kingdom, Plymouth, 250 pp., 35 pls., 53 text figs.

Gyrodinium britannia
LeBour, M.V., 1925
The dinoflagellates of Northern Seas. The Marine Biological Association of the United Kingdom, Plymouth, 250 pp., 35 pls., 53 text figs.

Gyrodinium calyptoglyphe n.sp.
LeBour, M.V., 1925
The dinoflagellates of Northern Seas. The Marine Biological Association of the United Kingdom, Plymouth, 250 pp., 35 pls., 53 text figs.

Gyrodinium calyptoglyphe
Morse, D.C., 1947
Some observations on seasonal variations in plankton population Patuxent River, Maryland 1943-1945. Bd. Nat. Res., Publ. No.65, Chesapeake Biol. Lab., 31, 3 figs.

Gyrodinium caudatum
Wood, E.J.F., 1963
Dinoflagellates in the Australian region. II. Recent collections.
Comm. Sci. and Industr. Res. Org., Div. Fish. and Oceanogr., Techn. Paper, No. 14: 55 pp.

Gyrodinium chiasmonetrium n.sp.
Norris, R.E., 1961.
Observations on phytoplankton organisms collected on the N.Z.O.I. Pacific Cruise, September 1958.
N.Z.J. Sci., 4(1):162-188.

Gyrodinium cochlea n.sp.
LeBour, M.V., 1925
The dinoflagellates of Northern Seas. The Marine Biological Association of the United Kingdom, Plymouth, 250 pp., 35 pls., 53 text figs.

Gyrodinium concentricum
LeBour, M.V., 1925
The dinoflagellates of Northern Seas. The Marine Biological Association of the United Kingdom, Plymouth, 250 pp., 35 pls., 53 text figs.

Gyrodinium cornutum
LeBour, M.V., 1925
The dinoflagellates of Northern Seas. The Marine Biological Association of the United Kingdom, Plymouth, 250 pp., 35 pls., 53 text figs.

Gyrodinium cornutum
Wood, E.J.F., 1963
Dinoflagellates in the Australian region. II. Recent collections.
Comm. Sci. and Industr. Res. Org., Div. Fish. and Oceanogr., Techn. Paper, No. 14: 55 pp.

Gyrodinium crassum
LeBour, M.V., 1925
The dinoflagellates of Northern Seas. The Marine Biological Association of the United Kingdom, Plymouth, 250 pp., 35 pls., 53 text figs.

Gyrodinium cuneatum
LeBour, M.V., 1925
The dinoflagellates of Northern Seas. The Marine Biological Association of the United Kingdom, Plymouth, 250 pp., 35 pls., 53 text figs.

Gyrodinium dominans n.sp.
Hulburt, E.M., 1957.
The taxonomy of unarmored Dinophyceae of shallow embayments on Cape Cod, Massachusetts.
Biol. Bull., 112(2):196-219.

Gyrodinium dorsum
Forward, Richard B. Jr. 1973.
Phototaxis in a dinoflagellate: action spectra as evidence for a two-pigment system.
Planta 111:167-178

Gyrodinium dorsum
Forward, Richard, and Demorest Davenport, 1968.
Red and far-red light effects on a short-term behavioral response of a dinoflagellate.
Science, 161 (3845): 1028-1029.

Gyrodinium dorsum
Hand, William G., Richard Forward and Demorest Davenport, 1967.
Short-term photic regulation of a receptor mechanism in a dinoflagellate.
Biol. Bull., mar. biol. Lab., Woods Hole, 133(1):150-165.

Gyrodinium estuariale n.sp.
Hulburt, E.M., 1957.
The taxonomy of unarmored Dinophyceae of shallow embayments on Cape Cod, Massachusetts
Biol. Bull., 112(2):196-219.

Gyrodinium falcatum
LeBour, M.V., 1925
The dinoflagellates of Northern Seas. The Marine Biological Association of the United Kingdom, Plymouth, 250 pp., 35 pls., 53 text figs.

Gyrodinium fissum
LeBour, M.V., 1925
The dinoflagellates of Northern Seas. The Marine Biological Association of the United Kingdom, Plymouth, 250 pp., 35 pls., 53 text figs.

Gyrodinium fissum
Levander, K.M., 1947
Plankton gesammelt in den Jahren 1899-1910 an den Küsten Finnlands. Finnländische Hydrographisch-Biologische Untersuchunger (aus dem Wasserbiologischen Laboratorin der Societas Scientiarum Fennica) No.11: 40 pp., 6 diagrams, 13 pls., tables.

Gyrodinium fucorum
LeBour, M.V., 1925
The dinoflagellates of Northern Seas. The Marine Biological Association of the United Kingdom, Plymouth, 250 pp., 35 pls., 53 text figs.

Gyrodinium fusiforme
LeBour, M.V., 1925
The dinoflagellates of Northern Seas. The Marine Biological Association of the United Kingdom, Plymouth, 250 pp., 35 pls., 53 text figs.

Gyrodinium glacilis n. sp.
Hada, Yoshina, 1970.   (1970)
The protozoan plankton of the Antarctic and Subantarctic seas.
Scient. Repts, Japan. Antarct. Res. Exped., (E)31:51 pp.

Gyrodinium glaebum n.sp.
Hulburt, E.M., 1957.
The taxonomy of unarmored Dinophyceae of shallow embayments on Cape Cod, Massachusetts.
Biol. Bull., 112(2):196-219.

Gyrodinium glaucum
LeBour, M.V., 1925
The dinoflagellates of Northern Seas. The Marine Biological Association of the United Kingdom, Plymouth, 250 pp., 35 pls., 53 text figs.

Gyrodinium grave
LeBour, M.V., 1925
The dinoflagellates of Northern Seas. The Marine Biological Association of the United Kingdom, Plymouth, 250 pp., 35 pls., 53 text figs.

Gyrodinium grenlandicum
Paasche, E., 1961.
Notes on phytoplankton from the Norwegian Sea.
Botanica Marina, 2(3/4):197-210.

Gyrodinium instriatum
Freudenthal, H.D., and J.J. Lee, 1963. n.sp.
Glenodinium halli, n.sp., and Gyrodinium instriatum n.sp., dinoflagellates from New York waters.
J. Protozoology, 10(2):182-189.

Gyrodinium kofoidii n.sp.
Norris, R.E., 1961.
Observations on phytoplankton organisms collected on the N.Z.O.I. Pacific Cruise, September 1958.
N.Z.J. Sci., 4(1):162-188.

Gyrodinium lachryma
Balech, Enrique and Sayed Z. El-Sayed, 1965.
Microplankton of the Weddell Sea.
In: Biology of Antarctic seas. II.
Antarctic Res. Ser., Amer. Geophys. Union, 5:107-124.

Gyrodinium lachrymum
Hada, Yoshina, 1970.   (1970)
The protozoan plankton of the Antarctic and Subantarctic seas.
Scient. Repts, Japan. Antarct. Res. Exped., (E)31:51 pp.

Gyrodinium lachryma
LeBour, M.V., 1925
The dinoflagellates of Northern Seas. The Marine Biological Association of the United Kingdom, Plymouth, 250 pp., 35 pls., 53 text figs.

Gyrodinium lebourae
LeBour, M.V., 1925
The dinoflagellates of Northern Seas. The Marine Biological Association of the United Kingdom, Plymouth, 250 pp., 35 pls., 53 text figs.

Gyrodinium lingulifera n.sp.
LeBour, M.V., 1925
The dinoflagellates of Northern Seas. The Marine Biological Association of the United Kingdom, Plymouth, 250 pp., 35 pls., 53 text figs.

Gyrodinium longum
LeBour, M.V., 1925
The dinoflagellates of Northern Seas. The Marine Biological Association of the United Kingdom, Plymouth, 250 pp., 35 pls., 53 text figs.

Gyrodinium metum n.sp.
Hulburt, E.M., 1957. unarmored
The taxonomy of the Dinophyceae of shallow embayments on Cape Cod, Massachusetts. Biol. Bull. 112(2):196-219.

Gyrodinium nasutum
Wood, E.J.F., 1963
Dinoflagellates in the Australian region. II. Recent collections.
Comm. Sci. and Industr. Res. Org., Div. Fish. and Oceanogr., Techn. Paper, No. 14: 55 pp.

Gyrodinium norvegica debilis
Aurich, H.J., 1949.
Die Verbreitung des Nannoplanktons in Oberflächenwasser vor der Nordfriesischen Kuste. Ber. Deutschen Wiss. Komm. f. Meeresf., n.f., II(4): 403-405, 2 figs.

Gyrodinium obtusum
LeBour, M.V., 1925
The dinoflagellates of Northern Seas. The Marine Biological Association of the United Kingdom, Plymouth, 250 pp., 35 pls., 53 text figs.

Gyrodinium ochraceum
Wood, E.J.F., 1963
Dinoflagellates in the Australian region. II. Recent collections.
Comm. Sci. and Industr. Res. Org., Div. Fish. and Oceanogr., Techn. Paper, No. 14: 55 pp.

Gyrodinium optimum
LeBour, M.V., 1925
The dinoflagellates of Northern Seas. The Marine Biological Association of the United Kingdom, Plymouth, 250 pp., 35 pls., 53 text figs.

Gyrodinium ovatum
LeBour, M.V., 1925
The dinoflagellates of Northern Seas. The Marine Biological Association of the United Kingdom, Plymouth, 250 pp., 35 pls., 53 text figs.

Gyrodinium phorkorium n.sp.
Norris, R.E., 1961.
Observations on phytoplankton organisms collected on the N.Z.O.I. Pacific Cruise, September 1958
N.Z. J. Sci., 4(1):162-188.

Gyrodinium pingue
LeBour, M.V., 1925
The dinoflagellates of Northern Seas. The Marine Biological Association of the United Kingdom, Plymouth, 250 pp., 35 pls., 53 text figs.

Gyrodinium pingue
Wood, E.J.F., 1963
Dinoflagellates in the Australian region. II. Recent collections.
Comm. Sci. and Industr. Res. Org., Div. Fish. and Oceanogr., Techn. Paper, No. 14: 55 pp.

Gyrodinium prunus
LeBour, M.V., 1925
The dinoflagellates of Northern Seas. The Marine Biological Association of the United Kingdom, Plymouth, 250 pp., 35 pls., 53 text figs.

Gyrodinium prunus
Wood, E.J.F., 1963
Dinoflagellates in the Australian region. II. Recent collections.
Comm. Sci. and Industr. Res. Org., Div. Fish. and Oceanogr., Techn. Paper, No. 14: 55 pp.

Gyrodinium resplendens
Hulburt, E.M., 1957.
The taxonomy of unarmored Dinophyceae of shallow embayments on Cape Cod, Massachusetts.
Biol. Bull., 112(2):196-219.

Gyrodinium spirale
Hulburt, E.M., 1957.
The taxonomy of unarmored Dinophyceae of shallow embayments on Cape Cod, Massachusetts.
Biol. Bull., 112(2):196-219.

Gyrodinium spirale
LeBour, M.V., 1925
The dinoflagellates of Northern Seas. The Marine Biological Association of the United Kingdom, Plymouth, 250 pp., 35 pls., 53 text figs.

Gyrodinium spirale
Wood, E.J.F., 1963
Dinoflagellates in the Australian region. II. Recent collections.
Comm. Sci. and Industr. Res. Org., Div. Fish. and Oceanogr., Techn. Paper, No. 14: 55 pp.

Gyrodinium submarinum
Wood, E.J.F., 1963
Dinoflagellates in the Australian region. II. Recent collections.
Comm. Sci. and Industr. Res. Org., Div. Fish. and Oceanogr., Techn. Paper, No. 14: 55 pp.

Gyrodinium uncatenum n.sp.
Hulburt, E.M., 1957.
The taxonomy of unarmored Dinophyceae of shallow embayments on Cape Cod, Massachusetts.
Biol. Bull., 112(2):196-219.

Gyrodinium undulans n.sp.
Hulburt, E.M., 1957.
The taxonomy of unarmored Dinophyceae of shallow embayments on Cape Cod, Massachusetts.
Biol. Bull., 112(2):196-219.

Halosphaera
Lovegrove, T., 1958(1960).
Plankton investigations from Aberdeen in 1958.
Ann. Biol., Cons. Perm. Int. Expl. Mer, 15:55.

Halosphaera
Manton, I., K. Oates and M. Parke, 1963
Observations on the fine structure of the Pyramimonas stage of Halosphaera and preliminary observations on three species of Pyramimonas.
J. Mar. Biol. Assoc., U.K., 43(1):225-238.

Halosphaera
Parke, M., and D.G. Rayns, 1964.
Studies on marine flagellates. VII. Nephroselmis gilva sp. nov. and some allied forms.
J. Mar. Biol. Assoc., U.K., 44(1):209-217.

Halosphaera

Travers, Anne, et Marc Travers 1973.
Le genre *Halosphaera* Schmitz dans le golfe de Marseille. Rapp. Proc.-v. Reun. Commn int. Explor. scient. Mer Medit. Monaco 21(8):425-428.

Halosphaera minor

Jorgensen, E., 1900
Protophyten und Protozoën im Plankton aus der Norwegischen Westkerste. Bergens Mus. Aarb. 1899(6): 95 pp., 5 pls., 83 tables

Halosphaera minor

Parke, Mary, and I. den Hartog-Adams, 1965.
Three species of *Halosphaera*. Jour. Mar. Biol. Assoc., U.K.,45(2):537-557.

Halosphaera minor

Wall, D., 1962
Evidence from Recent plankton regarding the biological affinities of *Tasmanites* Newton 1875 and *Leiosphaeridia* Eisenack 1958. Geol. Mag., 99(4):353-362.

Halosphaera parkeae n. sp.

Boalch, G.T. and J.P. Mommaerts, 1969.
A new punctate species of *Halosphaera*. J. mar. biol. Ass., U.K., 49(1):129-139.

Halosphaera russellii n. sp.

Parke, Mary I. and I. den Hartog-Adams, 1965.
Three species of *Halosphaera*. Jour. Mar. Biol. Assoc.,U.K., 45(2):537-557.

Halosphaera viridis

Braarud, T., 1962
Species distribution in marine phytoplankton.
J. Oceanogr. Soc., Japan, 20th Ann. Vol., 628-649.

Halosphaera viridis

Braarud, T., 1945
A phytoplankton survey of the polluted waters of inner Oslo Fjord. Hvalrådets Skrifter, No.28, 142 pp., 19 text figs., 17 tables.

Halosphaera viridis

Braarud, Trygve, and Erling Nordli, 1963.
Reproduction and size variation in *Halosphaera viridis* of northern waters.
Nytt Mag. Botanikk, 10:131-136.

Halosphaera viridis

Brunel, J., 1962
Le phytoplancton de la Baie de Chaleurs. Inst. Botan., Univ. Montréal, Contrib. No. 77: 365 pp., 66 pls.

Halosphaera viridis

Brunel, Jules, 1962.
Le phytoplancton de la Baie des Chaleurs. Contrib. Ministère de la Chasse et des Pêcheries, Province de Québec. No. 91: 365 pp.

Halosphaera viridis

Forti, A., 1922
Ricerche sulla flora pelagica (fitoplancton) di Quarto dei Mille. Mem. R. Com. Talass. Ital. 97:248 pp., 13 pls.

Halosphaera viridis

Jørgensen, E., 1905
B. Protistplankton and the diatoms in bottom samples. Hydrographical and biological investigations in Norwegian fjords. Bergens Mus. Skr. 7: 49-225.

Halosphaera viridis

Margalef, R., 1949
Fitoplancton nerítico de la Costa Brava en 1947-48. Publ. Inst. Biol. Aplicada, 5: 41-51, 3 text figs.

Halosphaera viridis

Marukawa, H., 1921
Plankton lists and some new species of copepods, from the northern waters of Japan. Bull. Inst. Ocean., No.384, 15 pp., 3 pls., 1 chart. Monaco

Halosphaera viridis

Massuti Algamora, M., 1949
Estudio de diez y seis muestras de planctor del Golfo de Nápoles. Publ. Inst. Biol. Appl. 5:85-94, 1 fold-in table.

Halosphaera viridis

Murray, G., and F. G. Whitting, 1899
New Peridiniaceae from the Atlantic. Trans. Linn. Soc., London, Bot., ser 2, 5: 321-342, Pls. 27-33, 9 tables.

Halosphaera viridis

Chiba, T., 1949
On the distribution of the plankton in the eastern China Sea and Yellow Sea. 1. Plankton composition in the spring. J. Shimonoseki Coll. Fisheries, 1(1):57-83, 1 fig.

Halosphaera viridis

Parke, Mary, and I. den Hartog-Adams, 1965.
Three species of *Halosphaera*. Jour. Mar. Biol. Assoc., U.K. 45(2): 537-557.

Halosphaera viridis

Savage, R.E., 1937
The food of North Sea herring 1930-1934. Ministry of Agriculture and Fisheries. Fish. Invest. Ser. II, 15(5):1-60; 16 text figs.

Halosphaera viridis

Schröder, B., 1900
Phytoplankton des Golfes von Neapel nebst vergleichenden Ausblicken auf das atlantischen Ozean. Mitt. Zool. Stat. Neapel, 14:1-38.

Halosphaera viridis

Schulz, B., and A. Wulff, 1929
Hydrographie und Oberflächen plankton des westlichen Barentsmeeres im Sommer 1927. Ber. deutschen wissensch. Komm. F. Meeresforsch. n.s. 4(5):232-372, 13 tables, 25 text figs.

Haplodinium indicum n.sp.

Subrahmanyan, R., 1966(1967).
New species of Dinophyceae from Indian waters. I. The genera Haplodinium Klebs emend. Subrahmanyan and Mesoporos Lillick. Phykos, 5:175-180.

Haplodinium iyengaricum n.sp.

Subrahmanyan, R., 1966(1967).
New species of Dinophyceae from Indian waters. I. The genera Haplodinium Klebs emend. Subrahmanyan and Mesoporos Lillick. Phykos., 5:175-180.

Haplodinium jonesicum n.sp.

Subrahmanyan, R., 1966(1967).
New species of Dinophyceae from Indian waters. I. The genera Haplodinium Klebs emend. Subrahmanyan and Mesoporos Lillick. Phykos, 5:175-180.

Helgolandinium subglobosum

von Stosch, H.A., 1969.
Dinoflagellaten aus der Nordsee II. Helgolandinium subglobosum gen. et. spec. nov. Helgoländer wiss. Meeresunters. 19(4): 569-577

Helgolandinium subglobosum, n.gen. nspp.

von Stosch, H.A., 1969.
Dinoflagellaten aus der Nordsee II. Helgolandinium subglobosum gen. et. spec. nov. Helgoländer wiss. Meeresuntersuch., 19(4): 569-577.

Hemidinium mediterraneum

Ercegovic, A., 1936
Etudes qualitative et quantitatives du phytoplancton dans les eaux cotières de l'Adriatique oriental moyen au cours de l'année 1934. Acta Adriatica 1(9):1-126

Hemidinium nasutum

LeBour, M.V., 1925
The dinoflagellates of Northern Seas. The Marine Biological Association of the United Kingdom, Plymouth, 250 pp., 35 pls.,53 text figs.

Hemidinium nasutum

Lindemann, E., 1924
Peridineen aus dem goldenen Horn und dem Bosphorus. Bot. Arch. 5:216-233, 98 text figs.

Hemidinium nasutum

Paulsen, O., 1908
XVIII Peridiniales. Nordisches Plankton, Bot. Teil: 1-124, 155 text figs.

Hemidinium nasutum

Rumkówna, A., 1948
List of the phytoplankton species occurring in the superficial water layers in the Gulf of Gdańsk. Bull. Lab. mar., Gdynia, No. 4: 139-141 with tables in back.

Hemidinium nasutum

Wood, E.J.F., 1954.
Dinoflagellates in the Australian region. Australian J. Mar. Freshwater Res., 5(2):171-351.

Heterocapsa pacifica n. sp.

Kofoid, C. A., 1907
Dinoflagellata of the San Diego region. III. Descriptions of new species. Univ. Calif. Publ., Zool. 3:299-340, Pls. 22-33.

Heterocapsa triqueter

Jorgensen, E., 1900
Protophyten und Protozoën im Plankton aus der Norwegischen Westkerste. Bergens Mus. Aarb. 1899(6): 95 pp., 5 pls., 83 tables.

Heterocapsa triquetra

Lindemann, E., 1925
Neubeobachtungen an den Winter peridineen des Golfes von Neapel. Bot. Arch. 9:95-102, 19 text figs.

Heterocapsa triquetra

Paulsen, O., 1908
XVIII Peridiniales. Nordisches Plankton, Bot. Teil: 1-124, 155 text figs.

Heterocapsa triquetra

Schröder, B., 1900
Phytoplankton des Golfes von Neapel nebst vergleichenden Ausblicken auf das atlantischen Ozean. Mitt. Zool. Stat. Neapel, 14:1-38.

Heterocapsa triquetra apiculata

Lindemann, E., 1924
Peridineen aus dem goldenen Horn und dem Bosphorus. Bot. Arch. 5:216-233, 98 text figs.

Heterocapsa triquetra littoralis

Lindemann, E., 1924
Peridineen aus dem goldenen Horn und dem Bosphorus. Bot. Arch. 5:216-233, 98 text figs.

Heterodinium

Kofoid, C. A., 1906
Contributions from the Laboratory of the Marine Biological Association of San Diego. VIII. Dinoflagellata of the San Diego Region. 1. On Heterodinium, a new genus of the Peridinidae. Univ. Calif. Publ. Zool. 2(8):341-368, Pls. 17-19.

Heterodinium agassizi
Kofoid, C.A., and A.M. Adamson, 1933.
36. The Dinoflagellata: the Family Heterodiniidea of the Peridiniodae.
Mem. Mus. Comp. Zool., 54(1):1-136, 22 pls.

Heterodinium agassizi
Rampi, L., 1945
Osservazioni sulla distribuzione qualitativa del fitoplancton nel mare Mediterraneo. Atti della Soc. Ital. di Sci. Nat. 84:105-113.

Heterodinium agassizi
Rampi, L., 1942
II Fitoplancton mediterraneo: Problemi ed affinita interoceaniche. Boll. di Pesca di Piscicoltura e di Idrobiologia, Anno 18, Fasc. 4:7-19.

Heterodinium angulatum
Kofoid, C.A., and A.M. Adamson, 1933.
36. The Dinoflagellata: the Family heterodiniidea of the Peridiniodae.
Mem. Mus. Com. Zool., 54(1):1-136, 22 pls.

Heterodinium asymmetricum nom. nov.
Kofoid, C.A., and A.M. Adamson, 1933.
36. The Dinoflagellata: the Family Heterodiniidea of the Peridiniodae.
Mem. Mus. Comp. Zool., 54(1):1-136, 22 pls.

Heterodinium australiae n.sp.
Wood, E.J.F., 1963
Dinoflagellates in the Australian region. II. Recent collections.
Comm. Sci. and Industr. Res. Org., Div. Fish. and Oceanogr., Techn. Paper, No. 14: 55 pp.

Heterodinium blackmani
Balech, Enrique, 1962
Tintinnoinea y Dinoflagellata del Pacifico segun material de las Expediciones NORPAC y DOWNWIND del Instituto Scripps de Oceanografia.
Revista, Mus. Argentino Ciencias Nat. "Bernardino Rivadavia", Ciencias Zool., 7(1):1-253.

Heterodinium blackmani
Kofoid, C. A., 1906
Contributions from the Laboratory of the Marine Biological Association of San Diego. VIII. Dinoflagellata of the San Diego Region. 1. On Heterodinium, a new genus of the Peridinidae. Univ. Calif. Publ. Zool. 2(8):341-368, Pls. 17-19.

Heterodinium blackmani
Kofoid, C.A., and A.M. Adamson, 1933.
36. The Dinoflagellata: the Family Heterodiniidea of the Peridiniodae.
Mem. Mus. Comp. Zool., 54(1):1-136, 22 pls.

Heterodinium calvum
Kofoid, C.A., and A.M. Adamson, 1933.
36. The Dinoflagellata: the Family Heterodiniidea of the Peridiniodae.
Mem. Mus. Comp. Zool., 54(1):1-136, 22 pls.

Heterodinium crassipes
Rampi, L., 1945
Osservazioni sulla distribuzione qualitativa del fitoplancton nel mare Mediterraneo. Atti della Soc. Ital. di Sci. Nat. 84:105-113.

Heterodinium crassipes
Wood, E.J.F., 1963
Dinoflagellates in the Australian region. II. Recent collections.
Comm. Sci. and Industr. Res. Org., Div. Fish. and Oceanogr., Techn. Paper, No. 14: 55 pp.

Heterodinium curvatum
Balech, Enrique, 1962
Tintinnoinea y Dinoflagellata del Pacifico segun material de las Expediciones NORPAC y DOWNWIND del Instituto Scripps de Oceanografia.
Revista, Mus. Argentino Ciencias Nat. "Bernardino Rivadavia", Ciencias Zool., 7(1):1-253.

Heterodinium curvatum
Dangeard, P., 1927
Phytoplankton de la croisière du "Sylvana". Ann. Inst. Ocean., Monaco, n.s., 4(8):286-401, 54 text figs. (Feirer-Juin 1913).

Heterodinium curvatum
Kofoid, C.A., and A.M. Adamson, 1933.
36. The Dinoflagellata: the Family Heterodiniidea of the Peridiniodae.
Mem. Mus. Comp. Zool., 54(1):1-136, 22 pls.

Heterodinium Debeauxi
Rampi, L., 1945
Osservazioni sulla distribuzione qualitativa del fitoplancton nel mare Mediterraneo. Atti della Soc. Ital. di Sci. Nat. 84:105-113.

Heterodinium Debrauxi n.sp.
Rampi, L., 1941
Ricerche sul microplancton del Mare Ligure. 3. Le Haterodiniacee e le Oxytoxacee dell acque di Sanremo. Annali del Mus. Civico di Storia Naturale di Genova, 61:50-70, 2 pls.

Heterodinium deformatum
Kofoid, C.A., and A.M. Adamson, 1933.
36. The Dinoflagellata: the Family Heterodiniidea of the Peridiniodae.
Mem. Mus. Comp. Zool., 54(1):1-136, 22 pls.

Heterodinium Detonii
Halim, Youssef, 1960
Etude quantitative et qualitative du cycle écologique des dinoflagellés dans les eaux de Villefranche-sur-Mer. (1953-1955). Ann. Inst. Oceanogr., Monaco, 38:123-232.

Heterodinium detonii
Rampi, L., 1950.
Péridiniens rares ou nouveaux pour le Pacifique Sud-Equatorial. Bull. Inst. Ocean., Monaco, No. 974:11 pp., 26 textfigs.

Heterodinium De Tonii
Rampi, L., 1948
Sur quelques Peridiniens rares ou interessants du Pacifique subtropical (Recoltes Alain Gerbault). Bull. l'Inst. Ocean., Monaco, No.937: 7 pp., 8 text figs.

Heterodinium detonii
Rampi, L., 1945
Osservazioni sulla distribuzione qualitativa del fitoplancton nel mare Mediterraneo. Atti della Soc. Ital. di Sci. Nat. 84:105-113.

Heterodinium Detonii
Rampi, L., 1943.
Su qualche altra Peridinea nuova o rara delle acque di Sanremo. Atti Soc. Ital. Sci. Nat. 82: 151-157, 9 textfigs.

Heterodinium dispar, n.sp.
Kofoid, C.A., and A.M. Adamson, 1933.
36. The Dinoflagellata: the Family Heterodiniidea of the Peridiniodae.
Mem. Mus. Comp. Zool., 54(1):1-136, 22 pls.

Heterodinium doma
Balech, Enrique, 1962
Tintinnoinea y Dinoflagellata del Pacifico segun material de las Expediciones NORPAC y DOWNWIND del Instituto Scripps de Oceanografia.
Revista, Mus. Argentino Ciencias Nat. "Bernardino Rivadavia", Ciencias Zool., 7(1):1-253.

Heterodinium doma
Kofoid, C. A., 1906
Contributions from the Laboratory of the Marine Biological Association of San Diego. VIII. Dinoflagellata of the San Diego Region. 1. On Heterodinium, a new genus of the Peridinidae. Univ. Calif. Publ. Zool. 2(8):341-368, Pls. 17-19.

Heterodinium doma
Kofoid, C.A., and A.M. Adamson, 1933.
36. The Dinoflagellata: the Family Heterodiniidea of the Peridiniodae.
Mem. Mus. Comp. Zool., 54(1):1-136, 22 pls.

Heterodinium doma
Rampi, L., 1945
Osservazioni sulla distribuzione qualitativa del fitoplancton nel mare Mediterraneo. Atti della Soc. Ital. di Sci. Nat. 84:105-113.

Heterodinium doma
Rampi, L., 1942
II Fitoplancton mediterraneo: Problemi ed affinita interoceaniche. Boll. di Pesca di Piscicoltura e di Idrobiologia, Anno 18, Fasc. 4:7-19.

Heterodinium doma
Rampi, L., 1941
Ricerche sul microplancton del Mare Ligure. 3. Le Haterodiniacee e le Oxytoxacee dell acque di Sanremo. Annali del Mus. Civico di Storia Naturale di Genova, 61:50-70, 2 pls.

Heterodinium dubium
Rampi, L., 1950.
Péridiniens rares ou nouveaux pour le Pacifique Sud-Equatorial. Bull. Inst. Ocean., Monaco, No. 974:11 pp., 26 textfigs.

Heterodinium dubium
Rampi, L., 1945
Osservazioni sulla distribuzione qualitativa del fitoplancton nel mare Mediterraneo. Atti della Soc. Ital. di Sci. Nat. 84:105-113.

Heterodinium dubium n.sp.
Rampi, L., 1941
Ricerche sul microplancton del Mare Ligure. 3. Le Haterodiniacee e le Oxytoxacee dell acque di Sanremo. Annali del Mus. Civico di Storia Naturale di Genova, 61:50-70, 2 pls.

Heterodinium elongatum
Kofoid, C.A., and A.M. Adamson, 1933.
36. The Dinoflagellata: the Family Heterodiniidea of the Peridiniodae.
Mem. Mus. Comp. Zool., 54(1):1-136, 22 pls.

Heterodinium expansum
Kofoid, C.A., and A.M. Adamson, 1933.
36. The Dinoflagellata: the Family Heterodiniidea of the Peridiniodae.
Mem. Mus. Comp. Zool., 54(1):1-136, 22 pls.

Heterodinium extremum
Kofoid, C.A., and A.M. Adamson, 1933.
36. The Dinoflagellata: the Family Heterodiniidea of the Peridiniodae.
Mem. Mus. Comp. Zool., 54(1):1-136, 22 pls.

Heterodinium fenestratum
Kofoid, C.A., and A.M. Adamson, 1933.
36. The Dinoflagellata: the Family Heterodiniidea of the Peridiniodae.
Mem. Mus. Comp. Zool., 54(1):1-136, 22 pls.

Heterodinium fenestratum
Wood, E.J.F., 1963
Dinoflagellates in the Australian region. II. Recent collections.
Comm. Sci. and Industr. Res. Org., Div. Fish. and Oceanogr., Techn. Paper, No. 14: 55 pp.

Heterodinium fides
Kofoid, C.A., and A.M. Adamson, 1933.
36. The Dinoflagellata: the Family Heterodiniidea of the Peridiniodiae.
Mem. Mus. Comp. Zool., 54(1):1-136, 22 pls.

Heterodinium gesticulatum
Kofoid, C.A., and A.M. Adamson, 1933.
36. The Dinoflagellata: the Family Heterodiniidea of the Peridiniodiae.
Mem. Mus. Comp. Zool., 54(1):1-136, 22 pls.

Heterodinium globosum
Balech, Enrique, 1962
Tintinnoinea y Dinoflagellata del Pacifico segun material de las Expediciones NORPAC y DOWNWIND del Instituto Scripps de Oceanografia.
Revista, Mus. Argentino Ciencias Nat. "Bernardino Rivadavia", Ciencias Zool., 7(1):1-253.

Heterodinium globosum
Kofoid, C.A., and A.M. Adamson, 1933.
36. The Dinoflagellata: the Family Heterodiniidea of the Peridiniodiae.
Mem. Mus. Comp. Zool., 54(1):1-136, 22 pls.

Heterodinium globosum
Rampi, L., 1945
Osservazioni sulla distribuzione qualitativa del fitoplancton nel mare Mediterraneo. Atti della Soc. Ital. di Sci. Nat. 84:105-113.

Heterodinium globosum
Rampi, L., 1942
Il Fitoplanton mediterraneo: Problemi ed affinita interoceaniche. Boll. di Pesca di Piscicoltura e di Idrobiologia, Anno 18, Fasc. 4:7-19.

Heterodinium globosum
Rampi, L., 1941
Ricercho sul microplancton del Mare Ligure. 3. Le Hatorodiniacee e le Oxytoxacee dell acque di Sanremo. Annali del Mus. Civico di Storia Naturale di Genova, 61:50-70, 2 pls.

Heterodinium grahami
Halim, Youssef, 1960
Étude quantitative et qualitative du cycle écologique des dinoflagellés dans les eaux de Villefranche-sur-Mer. (1953-1955).
Ann. Inst. Océanogr., Monaco, 38:123-232.

Heterodinium hindmarchii
Balech, Enrique, 1962
Tintinnoinea y Dinoflagellata del Pacifico segun material de las Expediciones NORPAC y DOWNWIND del Instituto Scripps de Oceanografia.
Revista, Mus. Argentino Ciencias Nat. "Bernardino Rivadavia", Ciencias Zool., 7(1):1-253.

Heterodinium hindmarchi
Kofoid, C. A., 1906
Contributions from the Laboratory of the Marine Biological Association of San Diego.
VIII. Dinoflagellata of the San Diego Region. 1. On Heterodinium, a new genus of the Peridinidae. Univ. Calif. Publ. Zool. 2(8):341-368, Pls. 17-19.

Heterodinium hindmarchii
Kofoid, C.A., and AM. Adamson, 1933.
36. The Dinoflagellata: the Family Heterodiniidea of the Peridiniodiae.
Mem. Mus. Comp. Zool. 54(1):1-136, 22 pls.

Heterodinium hindmarchi
Wood, E.J.F., 1963
Dinoflagellates in the Australian region. II. Recent collections.
Comm. Sci. and Industr. Res. Org., Div. Fish. and Oceanogr., Techn. Paper, No. 14: 55 pp.

Heterodinium inaequale n. sp.
Kofoid, C. A., 1906
Contributions from the Laboratory of the Marine Biological Association of San Diego.
VIII. Dinoflagellata of the San Diego Region. 1. On Heterodinium, a new genus of the Peridinidae. Univ. Calif. Publ. Zool. 2(8):341-368, Pls. 17-19.

Heterodinium inaequale
Rampi, L., 1950
Péridiniens rares ou nouveaux pour le Pacifique Sud-Equatorial. Bull. Inst. Océan., Monaco, No. 974:11 pp., 26 textfigs.

Heterodinium Kofoidi
Pavillard, J., 1916
Recherches sur les Peridiniens du Golfe du Lion. Mem. Univ. Montpellier. Trav. Inst. Bot. Univ. Montpellier. Serie mixte No.4, 70 pp., 3 pls., 15 text figs.

Heterodinium Kofoidi
Rampi, L., 1945
Osservazioni sulla distribuzione qualitativa del fitoplancton nel mare Mediterraneo. Atti della Soc. Ital. di Sci. Nat. 84:105-113.

Heterodinium laeve
Kofoid, C.A., and A.M. Adamson, 1933.
36. The Dinoflagellata: the Family Heterodiniidea of the Peridiniodiae.
Mem. Mus. Comp. Zool., 54(1):1-136, 22 pls.

Heterodinium laticinctum
Kofoid, C.A., and A.M. Adamson, 1933.
36. The Dinoflagellata: the Family Heterodiniidea of the Peridiniodiae.
Mem. Mus. Comp. Zool., 54(1):1-136, 22 pls.

Heterodinium leiorynchum
Halim, Youssef, 1960
Étude quantitative et qualitative du cycle écologique des dinoflagellés dans les eaux de Villefranche-sur-Mer. (1953-1955).
Ann. Inst. Océanogr., Monaco, 38:123-232.

Heterodinium leiorhynchum
Kofoid, C. A., 1906
Contributions from the Laboratory of the Marine Biological Association of San Diego.
VIII. Dinoflagellata of the San Diego Region. 1. On Heterodinium, a new genus of the Peridinidae. Univ. Calif. Publ. Zool. 2(8):341-368, Pls. 17-19.

Heterodinium leiorhynchum
Kofoid, C.A., and A.M. Adamson, 1933.
36. The Dinoflagellata: the Family Heterodiniidea of the Peridiniodiae.
Mem. Mus. Comp. Zool., 54(1):1-136, 22 pls.

Heterodinium leiorhynchum
Rampi, L., 1945
Osservazioni sulla distribuzione qualitativa del fitoplancton nel mare Mediterraneo. Atti della Soc. Ital. di Sci. Nat. 84:105-113.

Heterodinium leiorhynchum
Rampi, L., 1941
Ricercho sul microplancton del Mare Ligure. 3. Le Hatorodiniacee e le Oxytoxacee dell acque di Sanremo. Annali del Mus. Civico di Storia Naturale di Genova, 61:50-70, 2 pls.

Heterodinium longicollis
Rampi, L., 1942
Il Fitoplancton mediterraneo: Problemi ed affinita interoceaniche. Boll. di Pesca di Piscicoltura e di Idrobiologia, Anno 18, Fasc. 4:7-19.

Heterodinium mediocre
Kofoid, C.A., and A.M. Adamson, 1933.
36. The Dinoflagellata: the Family Heterodiniidea of the Peridiniodiae.
Mem. Mus. Comp. Zool., 54(1):1-136, 22 pls.

Heterodinium mediocre
Rampi, L., 1945
Osservazioni sulla distribuzione qualitativa del fitoplancton nel mare Mediterraneo. Atti della Soc. Ital. di Sci. Nat. 84:105-113.

Heterodinium mediocre
Rampi, L., 1942
Il Fitoplancton mediterraneo: Problemi ed affinita interoceaniche. Boll. di Pesca di Piscicoltura e di Idrobiologia, Anno 18, Fasc. 4:7-19.

Heterodinium mediterraneum
Balech, Enrique, 1962
Tintinnoinea y Dinoflagellata del Pacifico segun material de las Expediciones NORPAC y DOWNWIND del Instituto Scripps de Oceanografia.
Revista, Mus. Argentino Ciencias Nat. "Bernardino Rivadavia", Ciencias Zool., 7(1):1-253.

Heterodinium mediterraneum
Halim, Youssef, 1960
Étude quantitative et qualitative du cycle écologique des dinoflagellés dans les eaux de Villefranche-sur-Mer. (1953-1955).
Ann. Inst. Océanogr., Monaco, 38:123-232.

Heterodinium mediterraneum
Rampi, L., 1945
Osservazioni sulla distribuzione qualitativa del fitoplancton nel mare Mediterraneo. Atti della Soc. Ital. di Sci. Nat. 84:105-113.

Heterodinium mediterraneum
Rampi, L., 1941
Ricercho sul microplancton del Mare Ligure. 3. Le Hatorodiniacee e le Oxytoxacee dell acque di Sanremo. Annali del Mus. Civico di Storia Naturale di Genova, 61:50-70, 2 pls.

Heterodinium mediterraneum
Wood, E.J.F., 1963
Dinoflagellates in the Australian region. II. Recent collections.
Comm. Sci. and Industr. Res. Org., Div. Fish. and Oceanogr., Techn. Paper, No. 14: 55 pp.

Heterodinium milneri
Balech, Enrique, 1971.
Microplancton de la campaña oceanographica: Productividad III.
Revta Mus. argent. Cienc. Nat. Bernadina Rivadavia, Hydrobiol. 3(1):1-202, 39 pls.

Heterodinium milneri
Balech, Enrique, 1962
Tintinnoinea y Dinoflagellata del Pacifico segun material de las Expediciones NORPAC y DOWNWIND del Instituto Scripps de Oceanografia.
Revista, Mus. Argentino Ciencias Nat. "Bernardino Rivadavia", Ciencias Zool., 7(1):1-253.

Heterodinium milneri
Halim, Youssef, 1960
Étude quantitative et qualitative du cycle écologique des dinoflagellés dans les eaux de Villefranche-sur-Mer. (1953-1955).
Ann. Inst. Océanogr., Monaco, 38:123-232.

Heterodinium milneri
Kofoid, C. A., 1906
Contributions from the Laboratory of the Marine Biological Association of San Diego.
VIII. Dinoflagellata of the San Diego Region. 1. On Heterodinium, a new genus of the Peridinidae. Univ. Calif. Publ. Zool. 2(8):341-368, Pls. 17-19.

Heterodinium milneri
Kofoid, C.A., and A.M. Adamson, 1933.
36. The Dinoflagellata: the Family Heterodiniidea of the Peridiniodiae.
Mem. Mus. Comp. Zool., 54(1):1-136, 22 pls.

Heterodinium Milneri
Rampi, L., 1948
Sur quelques Peridiniens rares ou intéressants du Pacifique subtropical (Recoltes Alain Gerbault). Bull. l'Inst. Ocean., Monaco, No. 937: 7 pp., 8 text figs.

Heterodinium milneri
Rampi, L., 1945
Osservazioni sulla distribuzione qualitativa del fitoplancton nel mare Mediterraneo. Atti della Soc. Ital. di Sci. Nat. 84:105-113.

Heterodinium milneri
Rampi, L., 1941
Ricerche sul microplancton del Mare Ligure. 3. Le Heterodiniacee e le Oxytoxacee dell acque di Sanremo. Annali del Mus. Civico di Storia Naturale di Genova, 61:50-70, 2 pls.

Heterodinium milneri
Wood, E.J.F., 1963
Dinoflagellates in the Australian region. II. Recent collections.
Comm. Sci. and Industr. Res. Org., Div. Fish. and Oceanogr., Techn. Paper, No. 14: 55 pp.

Heterodinium minutum
Kofoid, C.A., and A.M. Adamson, 1933.
36. The Dinoflagellata: the Family Heterodiniidea of the Peridiniodiae.
Mem. Mus. Comp. Zool. 54(1):1-136, 22 pls.

Heterodinium murrayi
Halim, Youssef, 1960
Étude quantitative et qualitative du cycle écologique des dinoflagellés dans les eaux de Villefranche-sur-Mer. (1953-1955).
Ann. Inst. Océanogr., Monaco, 38:123-232.

Heterodinium murrayi nom. nov.
Kofoid, C. A., 1906
Contributions from the Laboratory of the Marine Biological Association of San Diego. VIII. Dinoflagellata of the San Diego Region. 1. On Heterodinium, a new genus of the Peridinidae. Univ. Calif. Publ. Zool. 2(8):341-368, Pls. 17-19.

Heterodinium murrayi
Kofoid, C.A., and A.M. Adamson, 1933.
36. The Dinoflagellata: the Family Heterodiniidea of the Peridiniodiae.
Mem. Mus. Comp. Zool. 54(1):1-136, 22 pls.

Heterodinium Murrayi
Rampi, L., 1948
Sur quelques Peridiniens rares ou interessants du Pacifique subtropical (Recoltes Alain Gerbault). Bull. l'Inst. Ocean., Monaco, No.937: 7 pp., 8 text figs.

Heterodinium murrayi
Rampi, L., 1945
Osservazioni sulla distribuzione qualitativa del fitoplancton nel mare Mediterraneo. Atti della Soc. Ital. di Sci. Nat. 84:105-113.

Heterodinium murrayi
Rampi, L., 1942
Il Fitoplancton mediterraneo: Problemi ed affinita interoceaniche. Boll. di Pesca di Piscicoltura e di Idrobiologia, Anno 18, Fasc. 4:7-19.

Heterodinium murrayi
Rampi, L., 1941
Ricerche sul microplancton del Mare Ligure. 3. Le Heterodiniacee e le Oxytoxacee dell acque di Sanremo. Annali del Mus. Civico di Storia Naturale di Genova, 61:50-70, 2 pls.

Heterodinium obesum
Kofoid, C.A. and A.M. Adamson, 1933.
36. The Dinoflagellata: the Family Heterodiniidea of the Peridiniodiae.
Mem. Mus. Comp. Zool., 54(1): 1-136, 22 pls.

Heterodinium origdenae n.sp
Kofoid, C. A., 1906
Contributions from the Laboratory of the Marine Biological Association of San Diego. VIII. Dinoflagellata of the San Diego Region. 1. On Heterodinium, a new genus of the Peridinidae. Univ. Calif. Publ. Zool. 2(8):341-368, Pls. 17-19.

Heterodinium praetextum
Kofoid, C.A., and A.M. Adamson, 1933.
36. The Dinoflagellata: the Family Heterodiniidea of the Peridiniodiae.
Mem. Mus. Comp. Zool. 54(1):1-136, 22 pls.

Heterodinium richardi
Halim, Youssef, 1960
Étude quantitative et qualitative du cycle écologique des dinoflagellés dans les eaux de Villefranche-sur-Mer. (1953-1955).
Ann. Inst. Océanogr., Monaco, 38:123-232.

Heterodinium Richardi
Rampi, L., 1945
Osservazioni sulla distribuzione qualitativa del fitoplancton nel mare Mediterraneo. Atti della Soc. Ital. di Sci. Nat. 84:105-113.

Heterodinium ridgenae
Balech, Enrique, 1962
Tintinnoinea y Dinoflagellata del Pacifico segun material de las Expediciones NORPAC y DOWNWIND del Instituto Scripps de Oceanografia.
Revista, Mus. Argentino Ciencias Nat. "Bernardino Rivadavia", Ciencias Zool., 7(1):1-253.

Heterodinium rigdeni
Kofoid, C.A., and A.M. Adamson, 1933.
36. The Dinoflagellata: the Family Heterodiniidea of the Peridiniodiae.
Mem. Mus. Comp. Zool., 54(1):1-136, 22 pls.

Heterodinium Schilleri
Rampi, L., 1945
Osservazioni sulla distribuzione qualitativa del fitoplancton nel mare Mediterraneo. Atti della Soc. Ital. di Sci. Nat. 84:105-113.

Heterodinium scotti n.sp.
Kofoid, C.A., and A.M. Adamson, 1933.
36. The Dinoflagellata: the Family Heterodiniidea of the Peridiniodiae.
Mem. Mus. Comp. Zool., 54(1):1-136, 22 pls.

Heterodinium scrippsi n.sp.
Kofoid, C. A., 1906
Contributions from the Laboratory of the Marine Biological Association of San Diego. VIII. Dinoflagellata of the San Diego Region. 1. On Heterodinium, a new genus of the Peridinidae. Univ. Calif. Publ. Zool. 2(8):341-368, Pls. 17-19.

Heterodinium scrippsi
Kofoid, C.A., and A.M. Adamson, 1933.
36. The Dinoflagellata: the Family Heterodiniidea of the Peridiniodiae.
Mem. Mus. Comp. Zool., 54(1):1-136, 22 pls.

Heterodinium scrippsi
Rampi, L., 1945
Osservazioni sulla distribuzione qualitativa del fitoplancton nel mare Mediterraneo. Atti della Soc. Ital. di Sci. Nat. 84:105-113.

Heterodinium scrippsi
Rampi, L., 1942
Il Fitoplancton mediterraneo: Problemi ed affinita interoceaniche. Boll. di Pesca di Piscicoltura e di Idrobiologia, Anno 18, Fasc. 4:7-19.

Heterodinium sinistrum n.sp.
Kofoid, C.A., and A.M. Adamson, 1933.
36. The Dinoflagellata: the Family Heterodiniidea of the Peridiniodiae.
Mem. Mus. Comp. Zool., 54(1):1-136, 22 pls.

Heterodinium sphaeroideum n. sp.
Kofoid, C. A., 1906
Contributions from the Laboratory of the Marine Biological Association of San Diego. VIII. Dinoflagellata of the San Diego Region. 1. On Heterodinium, a new genus of the Peridinidae. Univ. Calif. Publ. Zool. 2(8):341-368, Pls. 17-19.

Heterodinium spiniferum
Kofoid, C.A., and A.M. Adamson, 1933.
36. The Dinoflagellata: the Family Heterodiniidea of the Peridiniodiae.
Mem. Mus. Comp. Zool., 54(1):1-136, 22 pls.

Heterodinium superbum
Kofoid, C.A., and A.M. Adamson, 1933.
36. The Dinoflagellata: the Family Heterodiniidea of the Peridiniodiae.
Mem. Mus. Comp. Zool., 54(1):1-136, 22 pls.

Heterodinium superbum
Rampi, L., 1950.
Péridiniens rares ou nouveaux pour le Pacifique Sud-Equatorial. Bull. Inst. Océan., Monaco, No. 974:11 pp., 28 textfigs.

Heterodinium triacantha
Kofoid, C. A., 1906
Contributions from the Laboratory of the Marine Biological Association of San Diego. VIII. Dinoflagellata of the San Diego Region. 1. On Heterodinium, a new genus of the Peridinidae. Univ. Calif. Publ. Zool. 2(8):341-368, Pls. 17-19.

Heterodinium trirostre
Kofoid, C. A., 1906
Contributions from the Laboratory of the Marine Biological Association of San Diego. VIII. Dinoflagellata of the San Diego Region. 1. On Heterodinium, a new genus of the Peridinidae. Univ. Calif. Publ. Zool. 2(8):341-368, Pls. 17-19.

Heterodinium varicator n.sp.
Kofoid, C.A., and A.M. Adamson, 1933.
36. The Dinoflagellata: the Family Heterodiniidea of the Peridiniodiae.
Mem. Mus. Comp. Zool., 54(1):1-136, 22 pls.

Heterodinium whittingae n. sp.
Kofoid, C. A., 1906
Contributions from the Laboratory of the Marine Biological Association of San Diego. VIII. Dinoflagellata of the San Diego Region. 1. On Heterodinium, a new genus of the Peridinidae. Univ. Calif. Publ. Zool. 2(8):341-368, Pls. 17-19.

Heterodinium whittingae
Kofoid, C.A., and A.M. Adamson, 1933.
36. The Dinoflagellata: the Family Heterodiniidea of the Peridiniodiae.
Mem. Mus. Comp. Zool., 54(1):1-136.

Heterodinium Whittingae
Rampi, L., 1945
Osservazioni sulla distribuzione qualitativa del fitoplancton nel mare Mediterraneo. Atti della Soc. Ital. di Sci. Nat. 84:105-113.

Heterodinium whittingae
Rampi, L., 1942
Il Fitoplancton mediterraneo: Problemi ed affinita interoceaniche. Boll. di Pesca di Piscicoltura e di Idrobiologia, Anno 18, Fasc. 4:7-19.

Heteromastix fissa n.sp.
Lackey, J.B., 1940.
Some new flagellates from the Woods Hole area.
Amer. Mid. Nat. 23(2):463-471.

Heteromastix longifils
Ettl, H., 1967.
Der Verlauf der Zellteilung bei zwei marinen Heteromastix-Arten.
Int. Revue ges. Hydrobiol. 52(3):441-445.

Heteromastix rotunda
Ettl, H., 1967.
Der Verlauf der Zellteilung bei zwei marinen Heteromastix-Arten.
Int. Revue ges. Hydrobiol. 52(3):441-445.

Heteroschisma
Abe, Tohru H., 1967.
The armoured Dinoflagellata: II. Prorocentridae and Dinophysidae (C) - Ornthocercus, Histioneis, Amphisolenia and others.
Publs Seto mar.biol. Lab. 15(2):79-116.

Heteroschisma inaequale
Balech, Enrique, 1967.
Dinoflagelados nuevos o interesantes del Golfo de Mexico y Caribe.
Revta Mus.argent.Cienc.nat.Bernardino Rivadavia Inst.nac.Invest.Cienc.nat., Hidrobiol. 2(3):77-126.

Heteroschisma inaequale
Balech, Enrique, 1971.
Microplancton de la campaña oceanographica: Productividad III.
Revta Mus. argent. Cienc. Nat. Bernardina Rivadavia, Hydrobiol. 3(1):1-202, 39 pls.

Heteroschisma inaequale
Käsler, R., 1938
Die Verbreitung der Dinophysiales im Sudatlantischen Ozean. Wiss. Ergeb. Deutschen Atlantischen Expedition----"Meteor" 1925-1927, 12(2):162-237, text figs. 85-118.

Heteroschisma longialata
Balech, Enrique, 1967.
Dinoflagelados nuevos o interesantes del Golfo de Mexico y Caribe.
Revta Mus.argent.Cienc.nat.Bernardino Rivadavia Inst.nac.Invest.Cienc.nat., Hidrobiol. 2(3):77-126.

Heteroschisma pirum
Balech, Enrique, 1967.
Dinoflagelados nuevos o interesantes del Golfo de Mexico y Caribe.
Revta Mus.argent.Cienc.nat.Bernardino Rivadavia Inst.nac.Invest.Cienc.nat., Hidrobiol. 2(3):77-126.

Heterosigma
Honjo, Tsuneo and Tasuku Hanaoka, 1972
Studies on the mechanisms of red tide occurrence in Hakata Bay. II. General features of red tide flagellate, Heterosigma sp. (In Japanese; English abstract). Bull. Plankt. Soc. Japan 19(2):17-23 (75-81).

Heterosigma inlandica
Iwasaki, Hideo, and Ken Sasada, 1969.
Studies on the red tide Dinoflagellates. II. On Heterosigma inlandica appeared in Gokasho Bay, Shima Peninsula. (In Japanese; English abstract).
Bull. Jap. Soc. scient. Fish. 35(10): 943-947

Heterosigma inlandica
Kimura, Tomohiro, Akio Mizokami and Toshimasa Hashimoto, 1972
The red tide that caused severe damage to the fishery resources in Hiroshima Bay: outline of its occurrence and the environmental conditions. (In Japanese; English abstract). Bull. Plankt. Soc. Japan 19(2):24-38 (82-112).

Heterosigma inlandica
Uyeno, Fukuzo and Ko Nagai, 1972
Seasonal change of growth promoting effect of mud extracts and sea water collected during various seasons at the Ise Bay on a red tide flagellate Heterosigma inlandica Hada. Bull. Plankt. Soc. Japan 19(2):39-45 (97-103) (In Japanese; English abstract).

Histioneis sp.? (2)
Böhm, A., 1934
Zur Verbreitung einiger Peridineen. Arch. Protistenk. 75:498-501, 6 text figs.

Histioneis aequatorialis n.sp.
Wood, E.J.F., 1963
Dinoflagellates in the Australian region. II. Recent collections.
Comm. Sci. and Industr. Res. Org., Div. Fish. and Oceanogr., Techn. Paper, No. 14: 55 pp.

Histioneis alata n.sp.
Rampi, L., 1947
Osservazioni sulle Histioneis (Peridinea) raccolte nel Mare Ligure presso Sanremo. Bull. Inst. Océan. Monaco, No.920:1-16, 17 text figs.

Histioneis australiae n.sp.
Wood, E.J.F., 1963
Dinoflagellates in the Australian region. II. Recent collections.
Comm. Sci. and Industr. Res. Org., Div. Fish. and Oceanogr., Techn. Paper, No. 14: 55 pp.

Histioneis bernhardi n. sp.
Rampi, Leopoldo, 1969.
Peridiniens, Heterococcales et Pterospermales rares, intéressants ou nouveaux récoltés dans la Mer Ligurienne (Méditerranée occidentale).
Natura, Milano 60(4): 313-333

Histioneis biremis
Balech, Enrique, 1962
Tintinnoinea y Dinoflagellata del Pacifico segun material de las Expediciones NORPAC y DOWNWIND del Instituto Scripps de Oceanografia.
Revista, Mus. Argentino Ciencias Nat. "Bernardino Rivadavia", Ciencias Zool., 7(1):1-253.

Histioneis biremis
Kofoid, C. A. and T. Skogsberg, 1928
XXXV. The Dinoflagellata: The Dinophysiodae. Reports on the scientific results of the expedition to the Eastern Tropical Pacific, in charge of Alexander Agassiz, by the U. S. Fish Commission Steamer "Albatross" from October 1904 to March 1905----. Mem. M. C. Z. 51:766 pp., 31 pls.

Histioneis biremis
Murray, G., and F. G. Whitting, 1899
New Peridiniaceae from the Atlantic.
Trans. Linn. Soc., London, Bot., ser 2, 5: 321-342, Pls. 27-33, 9 tables.

Histioneis bougainvilleae n.sp.
Wood, E.J.F., 1963
Dinoflagellates in the Australian region. II. Recent collections.
Comm. Sci. and Industr. Res. Org., Div. Fish. and Oceanogr., Techn. Paper, No. 14: 55 pp.

Histioneis carinata
Böhm, A., 1934
Zur Verbreitung einiger Peridineen. Arch. Protistenk. 75:498-501, 6 text figs.

Histioneis carinata
Kofoid, C. A. and T. Skogsberg, 1928
XXXV. The Dinoflagellata: The Dinophysiodae. Reports on the scientific results of the expedition to the Eastern Tropical Pacific, in charge of Alexander Agassiz, by the U. S. Fish Commission Steamer "Albatross" from October 1904 to March 1905----. Mem. M. C. Z. 51:766 pp., 31 pls.

Histioneis carinata
Rampi, L., 1945
Osservazioni sulla distribuzione qualitativa del fitoplancton nel mare Mediterraneo. Atti della Soc. Ital. di Sci. Nat. 84:105-113.

Histioneis carinata
Rampi, L., 1942
Il Fitoplancton mediterraneo: Problemi ed affinita interoceaniche. Boll. di Pesca di Piscicoltura e di Idrobiologia, Anno 18, Fasc. 4:7-19.

Histioneis carinata
Rampi, L., 1940.
Ricerche sul Fitoplancton del mare Ligure. II Le tecatali e le dinofisiali delle acque di Sanremo. Boll. Pesca, Piscicolt., Idrobiol. (18) 16(2): 243-274, 56 figs.

Histioneis carinata
Rampi, L., 1940
Ricerche sul Fitoplancton del mare Ligure. Boll. di Pesca, di Piscicoltura e di Idrobiologa. 18(2):1-34, 56 text figs.

Histioneis carinata
Wood, E.J.F., 1954
Dinoflagellates in the Australian region. Australian J. Mar. Freshwater Res., 5(2):171-351

Histioneis cerasus
Balech, Enrique, 1971
Microplancton del Atlantico ecuatorial oeste (Equalant 1)
Publ. Serv. Hidrograf. Naval, Argentina H. 654: 103 pp., 122 figs.

Histioneis cerasus n.sp.
Böhm, A., 1934
Zur Verbreitung einiger Peridineen. Arch. Protistenk. 75:498-501, 6 text figs.

Histioneis cerasus
Wood, E.J.F., 1963
Dinoflagellates in the Australian region. II. Recent collections.
Comm. Sci. and Industr. Res. Org., Div. Fish. and Oceanogr., Techn. Paper, No. 14: 55 pp.

Histioneis cleaveri n.sp.
Rampi, L., 1952.
Ricerche sul microplancton di superficie del Pacifico tropicale. Bull. Inst. Océan., Monaco, No. 1014:16 pp., 5 textfigs.

Histioneis costata
Böhm, A., 1934
Zur Verbreitung einiger Peridineen. Arch. Protistenk. 75:498-501, 6 text figs.

Histioneis costata
Kofoid, C. A. and T. Skogsberg, 1928
XXXV. The Dinoflagellata: The Dinophysiodae. Reports on the scientific results of the expedition to the Eastern Tropical Pacific, in charge of Alexander Agassiz, by the U. S. Fish Commission Steamer "Albatross" from October 1904 to March 1905----. Mem. M. C. Z. 51:766 pp., 31 pls.

Histioneis costata
Wood, E.J.F., 1963
Dinoflagellates in the Australian region. II. Recent collections.
Comm. Sci. and Industr. Res. Org., Div. Fish. and Oceanogr., Techn. Paper, No. 14: 55 pp.

Histioneis crateriformis
Balech, Enrique, 1971
Microplancton del Atlantico ecuatorial oeste (Equalant 1)
Publ. Serv. Hidrograf. Naval, Argentina H. 654: 103 pp., 122 figs.

Histioneis crateriformis
Schröder, B., 1900
Phytoplankton des Golfes von Neapel nebst vergleichenden Ausblicken auf das atlantischen Ozean. Mitt. Zool. Stat. Neapel, 14:1-38.

Histioneis cymbalaria?
Balech, Enrique, 1971.
Microplancton de la campaña oceanographica: Productividad III.
Revta Mus. argent. Cienc. Nat. Bernardina Rivadavia, Hydrobiol. 3(1):1-202, 39 pls.

Histioneis cymbalaria
Rampi, L., 1948
Sur quelques Peridiniens rares ou intéressants du Pacifique subtropical (Recoltes Alain Gerbault). Bull. l'Inst. Ocean., Monaco, No.937: 7 pp., 8 text figs.

Histioneis depressa
Balech, Enrique, 1971
Microplancton del Atlantico ecuatorial oeste (Equalant 1)
Publ. Serv. Hidrograf. Naval, Argentina H. 654: 103 pp., 122 figs.

Histioneis depressa
Böhm, A., 1934
Zur Verbreitung einiger Peridineen. Arch. Protistenk. 75:498-501, 6 text figs.

Histioneis depressa
Rampi, L., 1947
Osservazioni sulle Histioneis (Peridinea) raccolte nel Mare Ligure presso Sanremo. Bull. Inst. Océan., Monaco, No.920:1-16, 17 text figs.

Histioneis depressa
Rampi, L., 1945
Osservazioni sulla distribuzione qualitativa del fitoplancton nel mare Mediterraneo. Atti della Soc. Ital. di Sci. Nat. 84:105-113.

Histioneis depressa
Wood, E.J.F., 1963
Dinoflagellates in the Australian region. II. Recent collections.
Comm. Sci. and Industr. Res. Org. Div. Fish. and Oceanogr., Techn. Paper, No. 14: 55 pp.

Histioneis depressa
Wood, E.J.F., 1954.
Dinoflagellates in the Australian region.
Australian J. Mar. Freshwater Res., 5(2):171-351.

Histioneis dentata n.sp.
Murray, G., and F. G. Whitting, 1899
New Peridiniaceae from the Atlantic.
Trans. Linn. Soc., London, Bot., ser 2, 5: 321-342, Pls. 27-33, 9 tables.

Histioneis detonii
Rampi, L., 1947
Osservazioni sulle Histioneis (Peridinea) raccolte nel Mare Ligure presso Sanremo. Bull. Inst. Océan., Monaco, No.920:1-16, 17 text figs.

Histioneis dolon
Balech, Enrique, 1962
Tintinnoinea y Dinoflagellata del Pacifico segun material de las Expediciones NORPAC y DOWNWIND del Instituto Scripps de Oceanografia.
Revista, Mus. Argentino Ciencias Nat. "Bernardino Rivadavia", Ciencias Zool., 7(1):1-253.

Histioneis dolon
Böhm, A., 1934
Zur Verbreitung einiger Peridineen. Arch. Protistenk. 75:498-501, 6 text figs.

Histioneis dolon
Käsler, R., 1938
Die Verbreitung der Dinophysiales im Sudatlantischen Ozean. Wiss. Ergeb. Deutschen Atlantischen Expedition----"Meteor" 1925-1927, 12(2):162-237, text figs. 85-118.

Histioneis dolon
Kofoid, C. A. and T. Skogsberg, 1928
XXXV. The Dinoflagellata: The Dinophysiodae. Reports on the scientific results of the expedition to the Eastern Tropical Pacific, in charge of Alexander Agassiz, by the U. S. Fish Commission Steamer "Albatross" from October 1904 to March 1905----. Mem. M. C. Z. 51:766 pp., 31 pls.

Histioneis Dolon n.sp.
Murray, G., and F. G. Whitting, 1899
New Peridiniaceae from the Atlantic.
Trans. Linn. Soc., London, Bot., ser 2, 5: 321-342, Pls. 27-33, 9 tables.

Histioneis dolon
Rampi, L., 1945
Osservazioni sulla distribuzione qualitativa del fitoplancton nel mare Mediterraneo. Atti della Soc. Ital. di Sci. Nat. 84:105-113.

Histioneis dolon
Wood, E.J.F., 1963
Dinoflagellates in the Australian region. II. Recent collections.
Comm. Sci. and Industr. Res. Org. Div. Fish. and Oceanogr., Techn. Paper, No. 14: 55 pp.

Histioneis dolon
Wood, E.J.F., 1954.
Dinoflagellates in the Australian region.
Australian J. Mar. Freshwater Res., 5(2):171-351.

Histioneis elegans n.sp.
Halim, Youssef, 1960
Étude quantitative et qualitative du cycle écologique des dinoflagellés dans les eaux de Villefranche-sur-Mer. (1953-1955).
Ann. Inst. Océanogr., Monaco, 38:123-232.

Histioneis elongata
Böhm, A., 1934
Zur Verbreitung einiger Peridineen. Arch. Protistenk. 75:498-501, 6 text figs.

Histioneis elongata
Kofoid, C. A. and T. Skogsberg, 1928
XXXV. The Dinoflagellata: The Dinophysiodae. Reports on the scientific results of the expedition to the Eastern Tropical Pacific, by the U. S. Fish Commission Steamer "Albatross" from October 1904 to March 1905----. Mem. M. C. Z. 51:766 pp., 31 pls.

Histioneis elongata
Rampi, L., 1945
Osservazioni sulla distribuzione qualitativa del fitoplancton nel mare Mediterraneo. Atti della Soc. Ital. di Sci. Nat. 84:105-113.

Histioneis elongata
Rampi, L., 1942
Il Fitoplancton mediterraneo: Problemi ed affinita interoceaniche. Boll. di Pesca di Piscicoltura e di Idrobiologia, Anno 18, Fasc. 4:7-19.

Histioneis elongata
Wood, E.J.F., 1963
Dinoflagellates in the Australian region. II. Recent collections.
Comm. Sci. and Industr. Res. Org. Div. Fish. and Oceanogr., Techn. Paper, No. 14: 55 pp.

Histioneis elongata curvata n.var.
Wood, E.J.F., 1963
Dinoflagellates in the Australian region. II. Recent collections.
Comm. Sci. and Industr. Res. Org. Div. Fish. and Oceanogr., Techn. Paper, No. 14: 55 pp.

Histioneis expansa n.sp.
Rampi, L., 1947
Osservazioni sulle Histioneis (Peridinea) raccolte nel Mare Ligure presso Sanremo. Bull. Inst. Océan., Monaco, No.920:1-16, 17 text figs.

Histioneis faouzii n.sp.
Halim, Youssef, 1960
Étude quantitative et qualitative du cycle écologique des dinoflagellés dans les eaux de Villefranche-sur-Mer. (1953-1955).
Ann. Inst. Océanogr., Monaco, 38:123-232.

Histioneis fragilis n.sp.
Böhm, A., 1934
Zur Verbreitung einiger Peridineen. Arch. Protistenk. 75:498-501, 6 text figs.

Histioneis Francescae n.sp.
Murray, G., and F. G. Whitting, 1899
New Peridiniaceae from the Atlantic.
Trans. Linn. Soc., London, Bot., ser 2, 5: 321-342, Pls. 27-33, 9 tables.

Histioneis gubernans
Kofoid, C. A. and T. Skogsberg, 1928
XXXV. The Dinoflagellata: The Dinophysiodae. Reports on the scientific results of the expedition to the Eastern Tropical Pacific, in charge of Alexander Agassiz, by the U. S. Fish Commission Steamer "Albatross" from October 1904 to March 1905----. Mem. M. C. Z. 51:766 pp., 31 pls.

Histioneis gubernans
Rampi, L., 1947
Osservazioni sulle Histioneis (Peridinea) raccolte nel Mare Ligure presso Sanremo. Bull. Inst. Océan., Monaco, No.920:1-16, 17 text figs.

Histioneis gubernans
Rampi, L., 1945
Osservazioni sulla distribuzione qualitativa del fitoplancton nel mare Mediterraneo. Atti della Soc. Ital. di Sci. Nat. 84:105-113.

Histioneis gubernans
Rampi, L., 1940
Ricerche sul Fitoplancton del mare Ligure. II Le tecatali e le dinofisiali delle acque di Sanremo. Boll. Pesca, Piscicolt., Idrobiol. (18) 16(2): 243-274, 56 figs.

Histioneis gubernans
Rampi, L., 1940
Ricerche sul Fitoplancton del mare Ligure. Boll. di Pesca, di Piscicoltura e di Idrobiologa, 18(2):1-34, 56 text figs.

Histioneis helenae
Balech, Enrique, 1962
Tintinnoinea y Dinoflagellata del Pacifico segun material de las Expediciones NORPAC y DOWNWIND del Instituto Scripps de Oceanografia.
Revista, Mus. Argentino Ciencias Nat. "Bernardino Rivadavia", Ciencias Zool., 7(1):1-253.

Histioneis helenae
Kofoid, C. A. and T. Skogsberg, 1928
XXXV. The Dinoflagellata: The Dinophysiodae. Reports on the scientific results of the expedition to the Eastern Tropical Pacific, in charge of Alexander Agassiz, by the U. S. Fish Commission Steamer "Albatross" from October 1904 to March 1905----. Mem. M. C. Z. 51:766 pp., 31 pls.

Histioneis Helenae n.sp.
Murray, G., and F. G. Whitting, 1899
New Peridiniaceae from the Atlantic.
Trans. Linn. Soc., London, Bot., ser 2, 5: 321-342, Pls. 27-33, 9 tables.

Histioneis Helenae
Rampi, L., 1948
Sur quelques Peridiniens rares ou intéressants du Pacifique subtropical (Recoltes Alain Gerbault). Bull. l'Inst. Ocean., Monaco, No.937: 7 pp., 8 text figs.

Histioneis helenae
Wood, E.J.F., 1963
Dinoflagellates in the Australian region. II. Recent collections.
Comm. Sci. and Industr. Res. Org. Div. Fish. and Oceanogr., Techn. Paper, No. 14: 55 pp.

Histioneis highleyi
Balech, Enrique, 1971
Microplancton del Atlantico ecuatorial oeste (Equalant 1)
Publ. Serv. Hidrograf. Naval, Argentina H. 654: 103 pp., 122 figs.

Histioneis highleyi
Käsler, R., 1938
Die Verbreitung der Dinophysiales im Sudatlantischen Ozean. Wiss. Ergeb. Deutschen Atlantischen Expedition----"Meteor" 1925-1927, 12(2):162-237, text figs. 85-118.

Histioneis highleyi
Kofoid, C. A. and T. Skogsberg, 1928
XXXV. The Dinoflagellata: The Dinophysiodae. Reports on the scientific results of the expedition to the Eastern Tropical Pacific, in charge of Alexander Agassiz, by the U. S. Fish Commission Steamer "Albatross" from October 1904 to March 1905----. Mem. M. C. Z. 51:766 pp., 31 pls.

Histioneis Highleii n.sp.
Murray, G., and F. G. Whitting, 1899
New Peridiniaceae from the Atlantic. Trans. Linn. Soc., London, Bot., ser 2, 5: 321-342, Pls. 27-33, 9 tables.

Histioneis highleyi
Wood, E.J.F., 1963
Dinoflagellates in the Australian region. II. Recent collections. Comm. Sci. and Industr. Res. Org.,Div. Fish. and Oceanogr., Techn. Paper, No. 14: 55 pp.

Histioneis hippoperoides
Abe, Tohru H., 1967.
The armoured Dinoflagellata: II. Prorocentridae and Dinophysidae (C)- Ornithocercus, Histioneis, Amphisolenia and others. Publs Seto mar.biol.Lab., 15(2):79-116.

Histioneis hipperoides
Wood, E.J.F., 1954.
Dinoflagellates in the Australian region. Australian J. Mar. Freshwater Res., 5(2):171-351.

Histioneis hippoperoides
Kofoid, C. A. and T. Skogsberg, 1928
XXXV. The Dinoflagellata: The Dinophysiodae. Reports on the scientific results of the expedition to the Eastern Tropical Pacific, in charge of Alexander Agassiz, by the U. S. Fish Commission Steamer "Albatross" from October 1904 to March 1905----. Mem. M. C. Z. 51:766 pp., 31 pls.

Histioneis Houckii
Rampi, L., 1945
Osservazioni sulla distribuzione qualitativa del fitoplancton nel mare Mediterraneo. Atti della Soc. Ital. di Sci. Nat. 84:105-113.

Histioneis hyalina
Balech, Enrique, 1971
Microplancton del Atlantico ecuatorial oeste (Equalant 1) Publ. Serv. Hidrograf. Naval, Argentina H. 654: 103 pp., 122 figs.

Histioneis hyalina
Böhm, A., 1934
Zur Verbreitung einiger Peridineen. Arch. Protistenk. 75:498-501, 6 text figs.

Histioneis hyalina
Kofoid, C. A. and T. Skogsberg, 1928
XXXV. The Dinoflagellata: The Dinophysiodae. Reports on the scientific results of the expedition to the Eastern Tropical Pacific, in charge of Alexander Agassiz, by the U. S. Fish Commission Steamer "Albatross" from October 1904 to March 1905----. Mem. M. C. Z. 51:766 pp., 31 pls.

Histioneis hyalina
Wood, E.J.F., 1963
Dinoflagellates in the Australian region. II. Recent collections. Comm. Sci. and Industr. Res. Org.,Div. Fish. and Oceanogr., Techn. Paper, No. 14: 55 pp.

Histioneis imbricata n.sp.
Halim, Youssef, 1960
Etude quantitative et qualitative du cycle écologique des dinoflagellés dans les eaux de Villefranche-sur-Mer. (1953-1955). Ann. Inst. Océanogr., Monaco, 38:123-232.

Histioneis inclinata
Balech, Enrique, 1971
Microplancton del Atlantico ecuatorial oeste (Equalant 1) Publ. Serv. Hidrograf. Naval, Argentina H. 654: 103 pp., 122 figs.

Histioneis inclinata
Böhm, A., 1934
Zur Verbreitung einiger Peridineen. Arch. Protistenk. 75:498-501, 6 text figs.

Histioneis inclinata
Kofoid, C. A. and T. Skogsberg, 1928
XXXV. The Dinoflagellata: The Dinophysiodae. Reports on the scientific results of the expedition to the Eastern Tropical Pacific, in charge of Alexander Agassiz, by the U. S. Fish Commission Steamer "Albatross" from October 1904 to March 1905----. Mem. M. C. Z. 51:766 pp., 31 pls.

Histioneis inclinata
Rampi, L., 1947
Osservazioni sulle Histioneis (Peridinea) raccolte nel Mare Ligure presso Sanremo. Bull. Inst. Océan., Monaco, No.920:1-16, 17 text figs.

Histioneis inclinata
Rampi, L., 1945
Osservazioni sulla distribuzione qualitativa del fitoplancton nel mare Mediterraneo. Atti della Soc. Ital. di Sci. Nat. 84:105-113.

Histioneis inclinata
Rampi, L., 1943.
Su qualche altra Peridinea nuova o rara delle acque di Sanremo. Atti della Soc. Ital. di Sci. Nat. 82:151-157, 9 textfigs.

Histioneis inclinata
Rampi, L., 1942
II Fitoplancton mediterraneo: Problemi ed affinita interoceaniche. Boll. di Pesca di Piscicoltura e di Idrobiologia, Anno 18, Fasc. 4:7-19.

Histioneis inclinata
Wood, E.J.F., 1963
Dinoflagellates in the Australian region. II. Recent collections. Comm. Sci. and Industr. Res. Org.,Div. Fish. and Oceanogr., Techn. Paper, No. 14: 55 pp.

Histioneis inornata
Kofoid, C. A. and T. Skogsberg, 1928
XXXV. The Dinoflagellata: The Dinophysiodae. Reports on the scientific results of the expedition to the Eastern Tropical Pacific, in charge of Alexander Agassiz, by the U. S. Fish Commission Steamer "Albatross" from October 1904 to March 1905----. Mem. M. C. Z. 51:766 pp., 31 pls.

Histioneis Isseli
Rampi, L., 1945
Osservazioni sulla distribuzione qualitativa del fitoplancton nel mare Mediterraneo. Atti della Soc. Ital. di Sci. Nat. 84:105-113.

Histioneis Jörgerseni
Ercegovic, A., 1936
Etudes qualitative et quantitatives du phytoplancton dans les eaux cotières de l'Adriatique oriental moyen au cours de l'année 1934. Acta Adriatica 1(9):1-126

Histioneis jorgenseni
Rampi, L., 1947
Osservazioni sulle Histioneis (Peridinea) raccolte nel Mare Ligure presso Sanremo. Bull. Inst. Océan., Monaco, No.920:1-16, 17 text figs.

Histioneis Jörgenseni
Rampi, L., 1945
Osservazioni sulla distribuzione qualitativa del fitoplancton nel mare Mediterraneo. Atti della Soc. Ital. di Sci. Nat. 84:105-113.

Histioneis jorgenseni
Wood, E.J.F., 1963
Dinoflagellates in the Australian region. II. Recent collections. Comm. Sci. and Industr. Res. Org.,Div. Fish. and Oceanogr., Techn. Paper, No. 14: 55 pp.

Histioneis josephinae
Kofoid, C. A. and T. Skogsberg, 1928
XXXV. The Dinoflagellata: The Dinophysiodae. Reports on the scientific results of the expedition to the Eastern Tropical Pacific, in charge of Alexander Agassiz, by the U. S. Fish Commission Steamer "Albatross" from October 1904 to March 1905----. Mem. M. C. Z. 51:766 pp., 31 pls.

Histioneis Karsteni
Balech, Enrique, 1971
Microplancton del Atlantico ecuatorial oeste (Equalant 1) Publ. Serv. Hidrograf. Naval, Argentina H. 654: 103 pp., 122 figs.

Histioneis Kofoidi
Rampi, L., 1945
Osservazioni sulla distribuzione qualitativa del fitoplancton nel mare Mediterraneo. Atti della Soc. Ital. di Sci. Nat. 84:105-113.

Histioneis lanceolata n.sp.
Wood, E.J.F., 1963
Dinoflagellates in the Australian region. II. Recent collections. Comm. Sci. and Industr. Res. Org.,Div. Fish. and Oceanogr., Techn. Paper, No. 14: 55 pp.

Histioneis ligustica
Rampi, L., 1947
Osservazioni sulle Histioneis (Peridinea) raccolte nel Mare Ligure presso Sanremo. Bull. Inst. Océan., Monaco, No.920:1-16, 17 text figs.

Histioneis ligustica
Rampi, L., 1945
Osservazioni sulla distribuzione qualitativa del fitoplancton nel mare Mediterraneo. Atti della Soc. Ital. di Sci. Nat. 84:105-113.

Histioneis ligustica
Rampi, L., 1940
Ricerche sul Fitoplancton del mare Ligure. Boll. di Pesca, di Piscicoltura e di Idrobiologa 18(2):1-34, 56 text figs.

Histineis lingustica n.sp.
Rampi, L. 1940.
Ricerche sul Fitoplancton del mare Ligure. II Le tecatali e le dinofisiali delle acque di Sanremo. Boll. Pesca, Piscicolt.,Idrobiol. (18) 16(2): 243-274, 56 figs.

Histioneis longicollis
Böhm, A., 1934
Zur Verbreitung einiger Peridineen. Arch. Protistenk. 75:498-501, 6 text figs.

Histioneis longicollis
Halim, Youssef, 1960
Etude quantitative et qualitative du cycle écologique des dinoflagellés dans les eaux de Villefranche-sur-Mer. (1953-1955). Ann. Inst. Océanogr., Monaco, 38:123-232.

Histioneis longicollis
Kofoid, C. A. and T. Skogsberg, 1928
XXXV. The Dinoflagellata: The Dinophysiodae. Reports on the scientific results of the expedition to the Eastern Tropical Pacific, in charge of Alexander Agassiz, by the U. S. Fish Commission Steamer "Albatross" from October 1904 to March 1905----. Mem. M. C. Z. 51:766 pp., 31 pls.

Histioneis longicollis
Rampi, L., 1948
Sur quelques Péridiniens rares ou intéressants du Pacifique subtropical (Récoltes Alain Gerbault). Bull. l'Inst. Ocean., Monaco, No.937: 7 pp., 8 text figs.

Histioneis longicollis
Rampi, L., 1947
Osservazioni sulle Histioneis (Peridinea) raccolte nel Mare Ligure presso Sanremo. Bull. Inst. Océan., Monaco, No.920:1-16, 17 text figs.

Histioneis longicollis
Rampi, L., 1945
Osservazioni sulla distribuzione qualitativa del fitoplancton nel mare Mediterraneo. Atti della Soc. Ital. di Sci. Nat. 84:105-113.

Histioneis longicollis
Rampi, L., 1940
Ricerche sul Fitoplancton del mare Ligure. II Le tecatali e le dinofisiali delle acque di Sanremo. Boll. Pesca, Piscicolt., Idrobiol. (18) 16(2): 243-274, 56 figs.

Histioneis longicollis
Rampi, L., 1940
Ricerche sul Fitoplancton del mare Ligure. Boll. di Pesca, di Piscicoltura e di Idrobiologa, 18(2):1-34, 56 text figs.

Histioneis longicollis
Wood, E.J.F., 1963
Dinoflagellates in the Australian region. II. Recent collections. Comm. Sci. and Industr. Res. Org., Div. Fish. and Oceanogr., Techn. Paper, No. 14: 55 pp.

Histioneis magnifica
Forti, A., 1922
Ricerche sulla flora pelagica (fitoplancton) di Quarto dei Mille. Mem. R. Com. Talass. Ital. 97:248 pp., 13 pls.

Histioneis magnifica
Murray, G., and F. G. Whitting, 1899
New Peridiniaceae from the Atlantic. Trans. Linn. Soc., London, Bot., ser 2, 5: 321-342, Pls. 27-33, 9 tables.

Histioneis magnifica
Pavillard, J., 1916
Recherches sur les Péridiniens du Golfe du Lion. Mem. Univ. Montpellier. Trav. Inst. Bot., Univ. Montpellier. Serie mixte No.4, 70 pp., 3 pls., 15 text figs.

Histioneis magnifica
Schröder, B., 1900
Phytoplankton des Golfes von Neapel nebst vergleichenden Ausblicken auf das atlantischen Ozean. Mitt. Zool. Stat. Neapel, 14:1-38.

Histioneis Marchesonii
Rampi, L., 1945
Osservazioni sulla distribuzione qualitativa del fitoplancton nel mare Mediterraneo. Atti della Soc. Ital. di Sci. Nat. 84:105-113.

Histioneis marchesonii
Rampi, L., 1947
Osservazioni sulle Histioneis (Peridinea) raccolte nel Mare Ligure presso Sanremo. Bull. Inst. Océan., Monaco, No.920:1-16, 17 text figs.

Histioneis megalocopa
Böhm, A., 1934
Zur Verbreitung einiger Peridineen. Arch. Protistenk. 75:498-501, 6 text figs.

Histioneis milneri
Balech, Enrique, 1962
Tintinnoinea y Dinoflagellata del Pacifico segun material de las Expediciones NORPAC y DOWNWIND del Instituto Scripps de Oceanografia. Revista, Mus. Argentino Ciencias Nat. "Bernardino Rivadavia", Ciencias Zool., 7(1):1-253.

Histioneis milneri
Kofoid, C. A. and T. Skogsberg, 1928
XXV. The Dinoflagellata: The Dinophysiodae. Reports on the scientific results of the expedition to the Eastern Tropical Pacific, in charge of Alexander Agassiz, by the U. S. Fish Commission Steamer "Albatross" from October 1904 to March 1905----. Mem. M. C. Z. 51:766 pp., 31 pls.

Histioneis Milneri n.sp.
Murray, G., and F. G. Whitting, 1899
New Peridiniaceae from the Atlantic. Trans. Linn. Soc., London, Bot., ser 2, 5: 321-342, Pls. 27-33, 9 tables.

Histioneis minuscula n.sp.
Rampi, L., 1950
Péridiniens rares ou nouveaux pour le Pacifique Sud-Equatorial. Bull. Inst. Océan., Monaco, No. 974:11 pp., 26 textfigs.

Histioneis mitchellana
Kofoid, C. A. and T. Skogsberg, 1928
XXV. The Dinoflagellata: The Dinophysiodae. Reports on the scientific results of the expedition to the Eastern Tropical Pacific, in charge of Alexander Agassiz, by the U. S. Fish Commission Steamer "Albatross" from October 1904 to March 1905----. Mem. M. C. Z. 51:766 pp., 31 pls.

Histioneis mitchellana
Abe, Tohru H., 1967.
The armoured Dinoflagellata: II. Prorocentridae and Dinophysidae (C) - Ornithocercus, Histioneis, Amphisolenia and others. Publs Seto mar.biol.Lab., 15(2):79-116.

Histioneis Mitchellana n.sp.
Murray, G., and F. G. Whitting, 1899
New Peridiniaceae from the Atlantic. Trans. Linn. Soc., London, Bot., ser 2, 5: 321-342, Pls. 27-33, 9 tables.

Histioneis mitchellana
Wood, E.J.F., 1963
Dinoflagellates in the Australian region. II. Recent collections. Comm. Sci. and Industr. Res. Org., Div. Fish. and Oceanogr., Techn. Paper, No. 14: 55 pp.

Histioneis moresbyensis n.sp.
Wood, E.J.F., 1963
Dinoflagellates in the Australian region. II. Recent collections. Comm. Sci. and Industr. Res. Org., Div. Fish. and Oceanogr., Techn. Paper, No. 14: 55 pp.

Histioneis navicula
Kofoid, C. A. and T. Skogsberg, 1928
XXV. The Dinoflagellata: The Dinophysiodae. Reports on the scientific results of the expedition to the Eastern Tropical Pacific, in charge of Alexander Agassiz, by the U. S. Fish Commission Steamer "Albatross" from October 1904 to March 1905----. Mem. M. C. Z. 51:766 pp., 31 pls.

Histioneis oceanica n.sp.
Rampi, L., 1950
Péridiniens rares ou nouveaux pour le Pacifique Sud-Equatorial. Bull. Inst. Océan., Monaco, No. 974:11 pp., 26 textfigs.

Histioneis oxypteris
Balech, Enrique, 1971
Microplancton del Atlantico ecuatorial oeste (Equalant 1) Publ. Serv. Hidrograf. Naval, Argentina H. 654: 103 pp., 122 figs.

Histioneis oxypteris
Rampi, L., 1945
Osservazioni sulla distribuzione qualitativa del fitoplancton nel mare Mediterraneo. Atti della Soc. Ital. di Sci. Nat. 84:105-113.

Histioneis oxypteris
Wood, E.J.F., 1963
Dinoflagellates in the Australian region. II. Recent collections. Comm. Sci. and Industr. Res. Org., Div. Fish. and Oceanogr., Techn. Paper, No. 14: 55 pp.

Histioneis pacifica
Balech, Enrique, 1971
Microplancton del Atlantico ecuatorial oeste (Equalant 1) Publ. Serv. Hidrograf. Naval, Argentina H. 654: 103 pp., 122 figs.

Histioneis pacifica
Böhm, A., 1934
Zur Verbreitung einiger Peridineen. Arch. Protistenk. 75:498-501, 6 text figs.

Histioneis pacifica n.sp.
Kofoid, C. A. and T. Skogsberg, 1928
XXV. The Dinoflagellata: The Dinophysiodae. Reports on the scientific results of the expedition to the Eastern Tropical Pacific, in charge of Alexander Agassiz, by the U. S. Fish Commission Steamer "Albatross" from October 1904 to March 1905----. Mem. M. C. Z. 51:766 pp., 31 pls.

Histioneis panaria
Balech, Enrique, 1971
Microplancton del Atlantico ecuatorial oeste (Equalant 1) Publ. Serv. Hidrograf. Naval, Argentina H. 654: 103 pp., 122 figs.

Histioneis panaria n.sp.
Kofoid, C. A. and T. Skogsberg, 1928
XXV. The Dinoflagellata: The Dinophysiodae. Reports on the scientific results of the expedition to the Eastern Tropical Pacific, in charge of Alexander Agassiz, by the U. S. Fish Commission Steamer "Albatross" from October 1904 to March 1905----. Mem. M. C. Z. 51:766 pp., 31 pls.

Histioneis panda
Kofoid, C. A. and T. Skogsberg, 1928
XXV. The Dinoflagellata: The Dinophysiodae. Reports on the scientific results of the expedition to the Eastern Tropical Pacific, in charge of Alexander Agassiz, by the U. S. Fish Commission Steamer "Albatross" from October 1904 to March 1905----. Mem. M. C. Z. 51:766 pp., 31 pls.

Histioneis Para n.sp.
Murray, G., and F. G. Whitting, 1899
New Peridiniaceae from the Atlantic. Trans. Linn. Soc., London, Bot., ser 2, 5: 321-342, Pls. 27-33, 9 tables.

Histioneis paraformis n. comb.
Balech, Enrique, 1971
Microplancton del Atlantico ecuatorial oeste (Equalant 1) Publ. Serv. Hidrograf. Naval, Argentina H. 654: 103 pp., 122 figs.

Histioneis Paulseni
Böhm, A., 1934
Zur Verbreitung einiger Peridineen. Arch. Protistenk. 75:498-501, 6 text figs.

Histioneis paulseni
Kofoid, C. A. and T. Skogsberg, 1928
XXV. The Dinoflagellata: The Dinophysiodae. Reports on the scientific results of the expedition to the Eastern Tropical Pacific, in charge of Alexander Agassiz, by the U. S. Fish Commission Steamer "Albatross" from October 1904 to March 1905----. Mem. M. C. Z. 51:766 pp., 31 pls.

Histioneis paulseni

Wood, E.J.F., 1963
Dinoflagellates in the Australian region. II. Recent collections.
Comm. Sci. and Industr. Res. Org.,Div. Fish. and Oceanogr., Techn. Paper, No. 14: 55 pp.

Histioneis pavillardi

Rampi, L., 1947
Osservazioni sulle Histioneis (Peridinea) raccolte nel Mare Ligure presso Sanremo. Bull. Inst. Océan., Monaco, No.920:1-16, 17 text figs.

Histioneis Pavillardi

Rampi, L., 1945
Osservazioni sulla distribuzione qualitativa del fitoplancton nel mare Mediterraneo. Atti della Soc. Ital. di Sci. Nat. 84:105-113.

Histioneis pavillardii

Rampi, L. 1940
Ricerche sul Fitoplancton del mare Ligure. II Le tecatali e le dinofisiali delle acque di Sanremo.. Boll. Pesca, Piscicolt.,Idrobiol. (18) 16(2): 243-274, 56 figs.

Histioneis Pavillardii

Rampi, L., 1940
Ricerche sul Fitoplancton del mare Ligure. Boll. di Pesca, di Piscicoltura e di Idrobiologa, 18(2):1-34, 56 text figs.

Histioneis peroides

Böhm, A., 1934
Zur Verbreitung einiger Peridineen. Arch. Protistenk. 75:498-501, 6 text figs.

Histioneis pietschmani

Abe,Tohru H.,1967.
The armoured Dinoflagellata:II.Prorocentridae, and Dinophysidae (C) - Ornithocercus,Histioneis, Amphisolenia and others.
Publs Seto mar.biol.Lab., 15(2):79-116.

Histioneis pietschmanr -1

Balech, Enrique, 1962
Tintinnoinea y Dinoflagellata del Pacifico segun material de las Expediciones NORPAC y DOWNWIND del Instituto Scripps de Oceanografia.
Revista, Mus. Argentino Ciencias Nat. "Bernardino Rivadavia", Ciencias Zool., 7(1):1-253.

Histioneis Pietschmanni n.sp.

Böhm, A., 1934
Zur Verbreitung einiger Peridineen. Arch. Protistenk. 75:498-501, 6 text figs.

Histioneis pietschmanni

Rampi, L., 1950.
Péridiniens rares ou nouveaux pour le Pacifique Sud-Equatorial. Bull. Inst. Océan., Monaco, No. 974:11 pp., 26 textfigs.

Histioneis pietschmanni

Wood, E.J.F., 1963
Dinoflagellates in the Australian region. II. Recent collections.
Comm. Sci. and Industr. Res. Org.,Div. Fish. and Oceanogr., Techn. Paper, No. 14: 55 pp.

Histioneis planeta n.sp.

Wood, E.J.F., 1963
Dinoflagellates in the Australian region. II. Recent collections.
Comm. Sci. and Industr. Res. Org.,Div. Fish. and Oceanogr., Techn. Paper, No. 14: 55 pp.

Histioneis pulchra

Kälser, R., 1938
Die Verbreitung der Dinophysiales im Sudatlantischen Ozean. Wiss. Ergeb. Deutschen Atlantischen Expedition----"Meteor" 1925-1927, 12(2):162-237, text figs. 85-118.

Histioneis pulchra

Kofoid, C. A. and T. Skogsberg, 1928
XXXV. The Dinoflagellata: The Dinophysiodae. Reports on the scientific results of the expedition to the Eastern Tropical Pacific, in charge of Alexander Agassiz, by the U. S. Fish Commission Steamer "Albatross" from October 1904 to March 1905----. Mem. M. C. Z. 51:766 pp., 31 pls.

Histioneis pulchra

Wood, E.J.F., 1963
Dinoflagellates in the Australian region. II. Recent collections.
Comm. Sci. and Industr. Res. Org.,Div. Fish. and Oceanogr., Techn. Paper, No. 14: 55 pp.

Histioneis rampii n.sp.

Halim, Youssef, 1960
Etude quantitative et qualitative du cycle écologique des dinoflagellés dans les eaux de Villefranche-sur-Mer. (1953-1955). Ann. Inst. Océanogr., Monaco, 38:123-232.

Histioneis reginella

Kofoid, C. A. and T. Skogsberg, 1928
XXXV. The Dinoflagellata: The Dinophysiodae. Reports on the scientific results of the expedition to the Eastern Tropical Pacific, in charge of Alexander Agassiz, by the U. S. Fish Commission Steamer "Albatross" from October 1904 to March 1905----. Mem. M. C. Z. 51:766 pp., 31 pls.

Histioneis remora

Jörgensen, E., 1923
Mediterranean Dinophysiaceae. Rept. Danish Oceanogr. Expeds. 1908-10, to the Mediterranean and adjacent seas, Vol.II, Biol. J 2, 48 pp., 64 text figs.

Histioneis remora

Murray, G., and F. G. Whitting, 1899
New Peridiniaceae from the Atlantic. Trans. Linn. Soc., London, Bot., ser 2, 5: 321-342, Pls. 27-33, 9 tables.

Histioneis remora

Pavillard, J., 1916
Recherches sur les Peridiniens du Golfe du Lion. Mem. Univ. Montpellier. Trav. Inst. Bot., Univ. Montpellier. Serie mixte No.4, 70 pp., 3 pls., 15 text figs.

Histioneis remora

Rampi, L., 1945
Osservazioni sulla distribuzione qualitativa del fitoplancton nel mare Mediterraneo. Atti della Soc. Ital. di Sci. Nat. 84:105-113.

Histioneis remora

Schröder, B., 1900
Phytoplankton des Golfes von Neapel nebst vergleichenden Ausblicken auf das atlantischen Ozean. Mitt. Zool. Stat... Neapel, 14:1-38.

Histioneis sp. aff. H. reticulata

Balech, Enrique, 1971
Microplancton del Atlantico ecuatorial oeste (Equalant 1)
Publ. Serv. Hidrograf. Naval, Argentina H. 654: 103 pp., 122 figs.

Histioneis Schilleri n.sp.

Böhm, A., 1934
Zur Verbreitung einiger Peridineen. Arch. Protistenk. 75:498-501, 6 text figs.

Histioneis schilleri

Wood, E.J.F., 1963
Dinoflagellates in the Australian region. II. Recent collections.
Comm. Sci. and Industr. Res. Org.,Div. Fish. and Oceanogr., Techn. Paper, No. 14: 55 pp.

Histioneis simplex n.sp.

Wood, E.J.F., 1963
Dinoflagellates in the Australian region. II. Recent collections.
Comm. Sci. and Industr. Res. Org.,Div. Fish. and Oceanogr., Techn. Paper, No. 14: 55 pp.

Histioneis speciosa n. sp.

Rampi, Leopoldo, 1969.
Péridiniens, Heterococcales et Pterospermales rares, intéressants ou nouveaux récoltés dans la Mer Ligurienne (Méditerranée Occidentale).
Natura, Milano 60(4):313-333

Histioneis splendida

Murray, G., and F. G. Whitting, 1899
New Peridiniaceae from the Atlantic. Trans. Linn. Soc., London, Bot., ser 2, 5: 321-342, Pls. 27-33, 9 tables.

Histioneis striata

Balech, Enrique, 1971
Microplancton del Atlantico ecuatorial oeste (Equalant 1)
Publ. Serv. Hidrograf. Naval, Argentina H. 654: 103 pp., 122 figs.

Histioneis striata

Kofoid, C. A. and T. Skogsberg, 1928
XXXV. The Dinoflagellata: The Dinophysiodae. Reports on the scientific results of the expedition to the Eastern Tropical Pacific, in charge of Alexander Agassiz, by the U. S. Fish Commission Steamer "Albatross" from October 1904 to March 1905----. Mem. M. C. Z. 51:766 pp., 31 pls.

Histioneis striata

Rampi, L., 1948
Sur quelques Peridiniens rares ou interessants du Pacifique subtropical (Recoltes Alain Gerbault). Bull. l'Inst. Ocean., Monaco, No.937: 7 pp., 8 text figs.

Histioneis subcarinata

Balech, Enrique, 1971
Microplancton del Atlantico ecuatorial oeste (Equalant 1)
Publ. Serv. Hidrograf. Naval, Argentina H. 654: 103 pp., 122 figs.

Histioneis subcarinata

Rampi, L., 1948
Sur quelques Peridiniens rares ou interessants du Pacifique subtropical (Recoltes Alain Gerbault). Bull. l'Inst. Ocean., Monaco, No.937: 7 pp., 8 text figs.

Histioneis subcarinata n.sp.

Rampi, L., 1947
Osservazioni sulle Histioneis (Peridinea) raccolte nel Mare Ligure presso Sanremo. Bull. Inst. Océan., Monaco, No.920:1-16, 17 text figs.

Histioneis sublongicollis n.sp.

Halim, Youssef, 1960
Etude quantitative et qualitative du cycle écologique des dinoflagellés dans les eaux de Villefranche-sur-Mer. (1953-1955). Ann. Inst. Océanogr., Monaco, 38:123-232.

Histioneis tubifera

Wood, E.J.F., 1963
Dinoflagellates in the Australian region. II. Recent collections.
Comm. Sci. and Industr. Res. Org.,Div. Fish. and Oceanogr., Techn. Paper, No. 14: 55 pp.

Histioneis variabilis
Halim, Youssef, 1960
Étude quantitative et qualitative du cycle écologique des dinoflagellés dans les eaux de Villefranche-sur-Mer. (1953-1955).
Ann. Inst. Océanogr., Monaco, 38:123-232.

Histioneis variabilis
Rampi, L., 1947
Osservazioni sulle Histioneis (Peridinea) raccolte nel Mare Ligure presso Sanremo. Bull. Inst. Océan., Monaco, No.920:1-16, 17 text figs.

Histioneis variabilis
Rampi, L., 1945
Osservazioni sulla distribuzione qualitativa del fitoplancton nel mare Mediterraneo. Atti della Soc. Ital. di Sci. Nat. 84:105-113.

Histioneis variabilis
Rampi, L., 1940
Ricerche sul Fitoplancton del mare Ligure. II Le tecatali e le dinofisiali delle acque di Sanremo. Boll. Pesca, Piscicolt., Idrobiol. (18) 16(2): 243-274, 56 figs.

Histioneis variabilis
Rampi, L., 1940
Ricerche sul Fitoplancton del mare Ligure. Boll. di Pesca, di Piscicoltura e di Idrobiologa, 18(2):1-34, 56 text figs.

Histioneis variabilis
Wood, E.J.F., 1963
Dinoflagellates in the Australian region. II. Recent collections.
Comm. Sci. and Industr. Res. Org., Div. Fish. and Oceanogr., Techn. Paper, No. 14: 55 pp.

Histioneis villafranca n.sp.
Halim, Youssef, 1960
Étude quantitative et qualitative du cycle écologique des dinoflagellés dans les eaux de Villefranche-sur-Mer. (1953-1955).
Ann. Inst. Océanogr., Monaco, 38:123-232.

Histioneis Vouki
Böhm, A., 1934
Zur Verbreitung einiger Peridineen. Arch. Protistenk. 75:498-501, 6 text figs.

Histioneis Voukii
Ercegovic, A., 1936
Études qualitative et quantitatives du phytoplancton dans les eaux côtières de l'Adriatique oriental moyen au cours de l'année 1934. Acta Adriatica 1(9):1-126

Histioneis voukii
Halim, Youssef, 1960
Étude quantitative et qualitative du cycle écologique des dinoflagellés dans les eaux de Villefranche-sur-Mer. (1953-1955).
Ann. Inst. Océanogr., Monaco, 38:123-232.

Histioneis voucki
Wood, E.J.F., 1963
Dinoflagellates in the Australian region. II. Recent collections.
Comm. Sci. and Industr. Res. Org., Div. Fish. and Oceanogr., Techn. Paper, No. 14: 55 pp.

Histiophysis
Abe, Tohru H., 1967.
The armoured Dinoflagellates: II. Prorocentridae and Dinophydsae (B)- Dinophysis and its allied genera.
Publs. Seto mar. biol. Lab., 15(1):37-78.

Histiophysis rugosa n.gen.
Kofoid, C.A. and T. Skogsberg, 1928
XXXV. The Dinoflagellata: The Dinophysiodae. Reports on the scientific results of the expedition to the Eastern Tropical Pacific, in charge of Alexander Agassiz, by the U.S. Fish Commission Steamer "Albatross" from October 1904 to March 1905----. Mem. M. C. Z. 51:766 pp., 31 pls.

Katodinium dorsalisulcum n. sp.
Hulburt, E.M., J.J.A. McLaughlin and P.A. Zahl, 1960
Katodinium dorsalisulcum, a new species of unarmored Dinophyceae. J. Protozool, 7(4): 323-326.

Kofoidinium sp.
Balech, Enrique, 1971
Microplancton del Atlantico ecuatorial oeste (Equalant 1)
Publ. Serv. Hidrograf. Naval, Argentina H. 654: 103 pp., 122 figs.

Kofoidinium pavillardi n.sp.
Cachon, Jean et Monique avec Odette et Gérard Ferru, 1967.
Contribution a l'étude des Noctilucidae Saville-Kent. 1. Les Kofoidininae Cachon J. et M., évolution morphologique et Systématique.
Protistologica 3(4):427-443.

Kofoidinium splendens n.sp.
Cachon, Jean et Monique avec Odette et Gérard Ferru, 1967.
Contribution a l'étude des Noctilucidae Saville-Kent. 1. Les Kofoidininae Cachon J. et M., évolution morphologique et Systématique.
Protistologica 3(4):427-443.

Kofoidinium velelloides
Balech, Enrique, 1971.
Microplancton de la campaña oceanographica: Productividad III.
Revta Mus. argent. Cienc. Nat. Bernardina Rivadavia, Hydrobiol. 3(1):1-202, 39 pls.

Kofoidium velelloides
Balech, Enrique, 1962.
Notulas de la Estacion Hidrobiologica de Puerto Quequen.
Revista, Mus. Argentino, Ciencias Nat., "Bernardino Rivadavia", Ciencias Zool., 8(6):81-87.

Kofoidinium velelloides
Balech, Enrique, 1962
Tintinnoinea y Dinoflagellata del Pacifico segun material de las Expediciones NORPAC y DOWNWIND del Instituto Scripps de Oceanografia.
Revista, Mus. Argentino Ciencias Nat. "Bernardino Rivadavia", Ciencias Zool., 7(1):1-253.

Kofoidinium velelloides=Gymnodinium pseudonoctiluca
Cachon, Jean et Monique avec Odette et Gérard Ferru, 1967.
Contribution a l'étude des Noctilucidae Saville-Kent. 1. Les Kofoidininae Cachon J. et M., évolution morphologique et Systématique.
Protistologica 3(4):427-443.

Kofoidinium velelloides
Fenaux, R., 1958.
Contribution a l'etude de Kofoidinium velelloides Pavillard.
Bull. Inst. Oceanogr., Monaco, No. 1118:11pp.
Also in: Trav. Sta. Zool., Villefranche-sur-Mer. 17(3)

Kryptoperidinium foliaceum
LeBour, M.V., 1925
The dinoflagellates of Northern Seas. The Marine Biological Association of the United Kingdom, Plymouth, 250 pp., 35 pls., 53 text figs.

Kryptoperidinium foliaceum
Morse, D.C., 1947
Some observations on seasonal variations in plankton population Patuxent River, Maryland 1943 1945. Bd. Nat. Res., Publ. No.65, Chesapeake Biol. Lab., 31, 3 figs.

Kryptopodinium sp.
Odum, William E., 1968.
Mullet grazing on a dinoflagellate bloom.
Chesapeake Sci., 9(3):202-204.

Leptodinium caudatum n. gen. n. sp.
Cachon, Jean, et Monique Cachon, 1969.
Contribution à l'étude des Noctilucidae Saville-Kent. Évolution morphologique, cytologie, systématique. II. Les Leptodiscinae Cachon J. et M.
Protistologica 5(1):11-33

Leptodiscus medusoides
Cachon, Jean, et Monique Cachon, 1969.
Contribution à l'étude des Noctilucidae Saville-Kent. Évolution morphologique, cytologie, systématique. II. Les Leptodiscinae Cachon J. et M.
Protistologica 5(1):11-33

Leptophyllus dasypus
Cachon, Jean, et Monique Cachon, 1969.
Contribution à l'étude des Noctilucidae Saville-Kent. Évolution morphologique, cytologie, systématique. II. Les Leptodiscinae Cachon J. et M.
Protistologica 5(1):11-33

Leptophyllus dasypus n.gen., n.sp
Cachon, Jean, et Monique Cachon-Enjumet, 1964.
Leptospathium navicula, nov. gen. nov. sp. et Leptophyllus dasypus nov. gen. nov. sp., péridiniens Noctilucidae (Hertwig) du plancton néritique de Villefranche-sur-Mer.
Bull. Inst. Océanogr., Monaco, 62(1292):12 pp.

Leptospathium navicula n.gen. n.sp.
Cachon, Jean, et Monique Cachon-Enjumat, 1964.
Leptospathium navicula nov. gen. nov. sp. et Leptophyllus dasypus nov. gen. nov. sp., péridiens Noctilucidae (Hertwig) du plancton néritique de Villefranche-sur-Mer.
Bull. Inst. Océanogr., Monaco, 62(1292):12 pp.

Lissodinium Schilleri n.gen. n.sp.
Matzenauer, L., 1933
Die Dinoflagellaten des indischen Ozeans (mit Ausnahme der Gattung Ceratium.) Bot. Arch. 35:437-510, 77 text figs., 2 charts.

Massartia asymmetrica
Biecheler, B., 1952.
Recherches sur les Peridiniens.
Bull. Biol., France Belg., Suppl., 36:1-149.

Massartia asymmetrica
Hulburt, E.M., 1957.
The taxonomy of unarmored Dinophyceae of shallow embayments on Cape Cod, Massachusetts.
Biol. Bull., 112(2):196-219.

Massartia glauca
Hulburt, E.M., 1957.
The taxonomy of unarmored Dinophyceae of shallow embayments on Cape Cod, Massachusetts.
Biol. Bull., 112(2):196-219.

Massartia rotundata
Hulburt, E.M., 1957.
The taxonomy of unarmored Dinophyceae of shallow embayments on Cape Cod, Massachusetts.
Biol. Bull., 112(2):196-219.

Massartia rotundata
Wood, E.J.F., 1963
Dinoflagellates in the Australian region. II. Recent collections.
Comm. Sci. and Industr. Res. Org., Div. Fish. and Oceanogr., Techn. Paper, No. 14: 55 pp.

Melanodinium nigricans
Wood, E.J.F., 1963
Dinoflagellates in the Australian region. II. Recent collections.
Comm. Sci. and Industr. Res. Org., Div. Fish. and Oceanogr., Techn. Paper, No. 14: 55 pp.

Mesoporus (Porella) adriaticus
Rampi, L., 1950.
Péridiniens rares ou nouveaux pour le Pacifique Sud-Equatorial. Bull. Inst. Océan., Monaco, No. 974:11 pp., 26 textfigs.

Mesoporos asymmetricus
Lillick, L.C. 1940
Phytoplankton and planktonic protozoa of the offshore waters of the Gulf of Maine. Pt.II. Qualitative Composition of the Planktonic Flora. Trans. Am. Phil. Soc., n.s., 31(3):193-237, 13 text figs.

Mesoporos asymmetricus
Lillick, L.C., 1937
Seasonal studies of the phytoplankton off Woods Hole, Massachusetts. Biol. Bull. LXXIII (3):488-503, 3 text figs.

Mesoporos parthasarathicus n.sp.
Subrahmanyan, R., 1966(1967).
New species of Dinophyceae from Indian waters. I. The genera Haplodinium Klebs emend. Subrahmanyan and Mesoporos Lillick. Phykos, 5:175-180.

Mesoporos perforatus
Lillick, L.C., 1940
Phytoplankton and planktonic protozoa of the offshore waters of the Gulf of Maine. Pt.II. Qualitative Composition of the Planktonic Flora. Trans. Am. Phil. Soc., n.s., 31(3):193-237, 13 text figs.

Mesoporos perforata
Lillick, L.C., 1938
Preliminary report of the phytoplankton of the Gulf of Maine. Am. Mid. Nat. 20(3):624-640, 1 text figs 37 tables.

Metaphalacroma
Abe, Tohru H., 1967.
The armoured Dinoflagellata: II. Procentridae and Dinophysidae (C) - Ornithocercus, Histioneis, Amphisolenia and others. Publs Seto mar.biol.Lab., 15(2):79-116.

Metaphalacroma skogsbergi
Balech, Enrique, 1971.
Microplancton de la campaña oceanographica: Productividad III. Revta Mus. argent. Cienc. Nat. Bernardina Rivadavia, Hydrobiol. 3(1):1-202, 39 pls.

Metaphalacroma skogsbergi n.gen. n.sp.
Tai, Si-Sun, & T. Skogsberg, 1934
Studies on the Dinophysordae, marine armored dinoflagellates of Monterey Bay, California. Arch. Protistenk. 82:380-482, 14 text figs. Pls. 11-12.

Microceratium orstomii n. gen. n.sp.
Sournia, Alain 1972.
Quatre nouveaux dinoflagellés du plancton marin. Phycologia 11(1):71-74.

Minuscula bipes
Braarud, T., 1934
A note on the phytoplankton of the Gulf of Maine in the summer of 1933. Biol. Bull. 67(1):76-82. (Contribution No.46 of the Woods Hole Oceanographic Institution)

Minuscula bipes
Gran, H.H., and T. Braarud, 1935
A quantitative study of the phytoplankton in the Bay of Fundy and the Gulf of Maine (including observations on hydrography, chemistry, and turbidity). J. Biol. Bd., Canada, 1(5):279-467, 69 text figs.

Minuscula bipes n.gen.
Lebour, M.V., 1925.
The dinoflagellates of northern seas. Marine Biological Association, United Kingdom, 250 pp., 35 pls., 53 textfigs.

Minusculus bipes
Lillick, L.C., 1937
Seasonal studies of the phytoplankton off Woods Hole, Massachusetts. Biol. Bull. LXXIII (3):488-503, 3 text figs.

Minuscula bipes
Steemann-Nielsen, Einar, 1951
The marine vegetation of the Isefjord. A study on ecology and production. Medd. Komm. Danmarks Fiskeri-og Havundersøgelser. Ser. Plankton. 5(4); 114pp., 46 text figs.

Murrayella sp?
Balech, Enrique, 1962
Tintinnoinea y Dinoflagellata del Pacifico segun material de las Expediciones NORPAC y DOWNWIND del Instituto Scripps de Oceanografia. Revista, Mus. Argentino Ciencias Nat. "Bernardino Rivadavia", Ciencias Zool., 7(1):1-253.

Murrayella biconica
Wood, E.J.F., 1963
Dinoflagellates in the Australian region. II. Recent collections. Comm. Sci. and Industr. Res. Org., Div. Fish. and Oceanogr., Techn. Paper, No. 14: 55 pp.

Murrayella intermedia
Forti, A., 1922
Ricerche sulla flora pelagica (fitoplancton) di Quarto dei Mille. Mem. R. Com. Talass. Ital. 97:248 pp., 13 pls.

Murrayella intermedia
Pavillard, J., 1916
Recherches sur les Peridiniens du Golfe du Lion. Mem. Univ. Montpellier. Trav. Inst. Bot., Univ. Montpellier. Serie mixte No.4, 70 pp., 3 pls., 15 text figs.

Murrayella intermedia
Rampi, L., 1941
Ricerche sul microplancton del Mare Ligure. 3. Le Hatorodiniacee o le Oxytoxaee dell acque di Sanremo. Annali del Mus. Civico di Storia Naturale di Genova, 61:50-70, 2 pls.

Murrayella intermedia
Wood, E.J.F., 1963
Dinoflagellates in the Australian region. II. Recent collections. Comm. Sci. and Industr. Res. Org., Div. Fish. and Oceanogr., Techn. Paper, No. 14: 55 pp.

Murrayella mimetica n.sp.
Balech, Enrique, 1967.
Dinoflagelados nuevos o interesantes del Golfo de Mexico y Caribe. Revta Mus.argent.Cienc.nat.Bernardino Rivadavia Inst.nac.Invest.Cienc.nat., Hidrobiol. 2(3):77-126.

Murrayella punctata
Balech, Enrique, 1971
Microplancton del Atlantico ecuatorial oeste (Equalant 1) Publ. Serv. Hidrograf. Naval, Argentina H. 654: 103 pp., 122 figs.

Murrayella spinosa
Wood, E.J.F., 1963
Dinoflagellates in the Australian region. II. Recent collections. Comm. Sci. and Industr. Res. Org., Div. Fish. and Oceanogr., Techn. Paper, No. 14: 55 pp.

Murrayella splendida
Balech, Enrique, 1967.
Dinoflagelados nuevos o interesantes del Golfo de Mexico y Caribe. Revta Mus.argent.Cienc.nat.Bernardino Rivadavia Inst.nac.Invest.Cienc.nat., Hidrobiol. 2(3):77-126.

Murrayella splendida n.sp.
Rampi, L., 1941
Ricerche sul microplancton del Mare Ligure. 3. Le Hatorodiniacee o le Oxytoxaee dell acque di Sanremo. Annali del Mus. Civico di Storia Naturale di Genova, 61:50-70, 2 pls.

Nematodinium
Greuet, Claude, 1971
Etude ultrastructurale et évolution des cnidocystes de Nematodinium, Péridinien Warnowiidae Lindemann. Protistologica 7(3):345-355.

Nematodinium armatum
Hulburt, E.M., 1957.
The taxonomy of unarmored Dinophyceae of shallow embayments on Cape Cod, Massachusetts. Biol. Bull. 112(2):196-219.

Nematodinium armatum
Lebour, M.V., 1925
The dinoflagellates of Northern Seas. The Marine Biological Association of the United Kingdom, Plymouth, 250 pp., 35 pls., 53 text figs.

Nematodinium torpedo
Wood, E.J.F., 1963
Dinoflagellates in the Australian region. II. Recent collections. Comm. Sci. and Industr. Res. Org., Div. Fish. and Oceanogr., Techn. Paper, No. 14: 55 pp.

Neresheimeria catenata
Cachon, Jean, et Monique Cachon, 1966.
Ultrastructure d'un peridinien parasite d'appendiculaires, Neresheimeria catenata (Neresheimer) Protistologica, 2(4):17-25.

Also in: Trav. Stn zool., Villefranche, 26.

Neresheimeria catenata
Cachon, Jean, et Monique Cachon-Enjumet, 1964.
Cycle évolutif et cytologie de Neresheimeria catenata Neresheimer, péridinien parasite d'appendiculaires. Ann. Sci. Nat., Zool. et Biol. Animale (12), 4: 779-800.

Noctiluca
Brongersma-Sanders, M., 1957.
On the desirability of a research into certain phenomena in the region of upwelling along the coast of southwest Africa. Kon. Neder. Akad. Wetens., Proc. 50(6):659-665.

Noctiluca
Brunel, J., 1962
Le phytoplancton de la Baie de Chaleurs. Inst. Botan., Univ. Montréal, Contrib. No. 77: 365 pp., 66 pls.

Noctiluca
Charernphol, LCDR Swarng, 1958
Preliminary study of discoloration of sea water in the Gulf of Thailand. Proc. Ninth Pacific Sci. Congr., Pacific Sci. Assoc., 1957, 16(Oceanogr.):131-134.

Noctiluca
Delsman, H. C., 1939.
Preliminary plankton investigations in the Java Sea. Treubia, 17:139-181, 8 maps, 41 figs.

Noctiluca
Devanesen, D. W., 1942.
Plankton studies in the fisheries branch of the Department of Industries and Commerce, Madras. Current Sci., 11(4):142-143.

Noctiluca
Jayaraman, R., 1954.
Seasonal variations in salinity, dissolved oxygen and nutrient salts in the inshore waters of the Gulf of Manner and Palk Bay near Mandapam (S. India). Indian J. Fish. 1:345-364.

Oceanographic Index: Marine Organisms Cumulation, 1946-1973

Noctiluca sp.
Johnson, M.W. (undated)
The production and distribution of Zooplankton in the surface waters of Bering Sea and Bering Strait, with special reference to copepods, echinoderms, mollusks and annelids. Report of oceanographic cruise of U.S. Coast Guard Cutter Chelan 1934 Part II (B):45-85, fig.1, table 1.

Noctiluca
Koshtojantz, H.S., 1953.
The relation of biological luminosity of Noctiluca to the composition of the reactive group of albumens and the gaseous exchange. Dok. Akad. Nauk. SSSR, 91(5):1229.

Noctiluca sp.
Lebour, M.V., 1947
Notes on the inshore plankton of Plymouth. JMBA 26(4):527-547.

Noctiluca
Nicol, J.A.C. 1958.
Observations on luminescence in Noctiluca. J.M.B.A., U.K. 37(3):535-550.

Noctiluca sp.
Nozawa, K., 1943.
Specific gravity and its adaptation to environmental sea water in Noctiluca. (In Japanese). Zool. Mag., 55(9/10):305-314.

Noctiluca
Puissegur, C., 1946
Les Etres vivants lumineux. Séd. et Vie, Vol.LXX (348), translated by R. Widmer, David Taylor Model Basin Translation 221, dated May 1927 (as Luminous Living Organisms), 17 pp., 9 text figs.

Noctiluca
Ramamurthy, S., 1965.
Swarming of Noctiluca in the North Kanara coast with observations on some oceanographic factors. Proc. Symp. Oceanogr. cent. Adv. Study Mar. Biol. Annomalai Univ., Porto Novo, 1965: 32-40.

Noctiluca
Ritchie, G.S., 1952.
H.M.S. Challenger's investigations in the Pacific. J. Inst. Navigation, 5(3):251-261, 1 textfig

Noctiluca sp.
Seguin, Gerard, 1966.
Contribution à l'étude de la biologie du plancton de surface de la baie de Dakar (Sénégal). Etude quantitative et observations écologiques au cours d'un cycle annuel. Inst. Francais d'Afrique Noire, (A), 28(1):1-90. Bull.

Noctiluca sp.
Smidt, E.L.B., 1944
Biological Studies of the Invertebrate Fauna of the Harbor of Copehagen. Vidensk. Medd. fra Dansk naturh. Foren. 107:235-316, 23 text figs.

Noctiluca
Thompson, J.V., 1830.
On the luminosity of the ocean with descriptions of some remarkable species of luminous animals (Pyrosoma pigmaea and Sapphirina indica), and particularly of the new genera, Noctiluca (sic), Cynthis, Lucifer, and Podopsis of the Schizopoda (sic). (Addendum to Memoir 1. Addendum to Memoir 2) Zoological researches and illustrations on natural history of nondescript or imperfectly known animals in a series of memoirs. Vol. 1, Mem. 3:110 pp., 14 pls.

Noctiluca, effect of
Holmes, R.W., P.M. Williams and R.W. Eppley, 1967.
Red water in La Jolla Bay, 1964-1965. Limnol. Oceanogr., 12(3):503-512.

Noctiluca miliaris
Bityukov, E.P., 1968.
Characteristics of the daily rhythm of bioluminescence of Noctiluca miliaria (Flagellata, Peridinea). (In Russian; English abstract). Zool. Zh., 47(1):36-40.

Noctiluca miliaris
Aiyar, R.G., 1936.
Mortality of fish on the Madras coast in June, 1935. Current Science 4:488-489.

Noctiluca miliaris
Bhimachar, B.S., and P.C. George, 1950.
Abrupt set-backs in the fisheries of the Malabar and Kanara coasts and "red water" phenomenon as their probable cause. Proc. Indian Acad. Sci. 31:339-350, 1 graph.

Noctiluca miliaris
Bigelow, H.B., 1922
Exploration of the coastal water off the northeastern United States in 1916 by the U.S. Fisheries Schooner Grampus. Bull. M.C.Z. 65 (5):85-188, 53 text figs.

Noctiluca miliaris
Cooper, L.H.N. 1968.
Scientific consequences of the wreck of the Torrey Canyon. Helgoländer wiss. Meeresunters. 17: 340-355.

Noctiluca miliaris
Drews, R., 1960.
Der Erreger des Meeresleuchtens: das Geisseltierchen Noctiluca. Mikrokosmos, 49(12):353-355.

Noctiluca miliaris
Eckert, Roger, 1966.
Subcellular sources of luminescence in Noctiluca. Science, 151(3708):349-352.

Noctiluca miliaris
Eckert, Roger, 1965.
Bioelectric control of bioluminescence in the dinoflagellate Noctiluca. II. Asynchronous flash initiation by a propagated triggering potential. Science, 147(3662):1142-1145.

Noctiluca miliaris
Eckert, Roger, 1965.
Bioelectric control of bioluminescence in the dinoflagellate Noctiluca. I. Specific nature of triggering events. Science, 147(3662):1140-1142.

Noctiluca miliaris
Gilbert, J.Y., and W.E. Allen, 1943
The phytoplankton of the Gulf of California obtained by the "E.W. Scripps" in 1939 and 1940. J. Mar. Res. V(2):89-110, figs.30-31.

Noctiluca miliaris
Gross, F., 1934.
Zur Biologie und Entwickelungsgeschichte von Noctiluca miliaris. Arch. f. Protistenkunde 83: 178-196, 6 textfigs.

Noctiluca miliaris
Kesseler, Hanswerner, 1966.
Beitrag zur Kenntnis der chemischen und physikalischen Eigenschaften des Zellsaftes von Noctiluca miliaris. Veröff. Inst. Meeresforsch., Bremerh., Sonderband II:357-368.

Noctiluca Miliaris
Lillick, L.C. 1940
Phytoplankton and planktonic protozoa of the offshore waters of the Gulf of Maine. Pt.II. Qualitative Composition of the Planktonic Flora. Trans. Am. Phil. Soc., n.s., 31(3):193-237, 13 text figs.

Noctiluca miliaris
Lillick, L.C. 1938
Preliminary report of the phytoplankton of the Gulf of Maine. Am. Mid. Nat. 20(3):624-640, 1 text figs 37 tables.

Noctiluca miliaris
Lillick, L.C. 1937
Seasonal studies of the phytoplankton off Woods Hole, Massachusetts. Biol. Bull. LXXIII (3):488-503, 3 text figs.

Noctiluca miliaris
Marukawa, H., 1921
Plankton lists and some new species of copepods, from the northern waters of Japan. Bull. Inst. Ocean. No.384, 15 pp., 3 pls., 1 chart. Monaco

Noctiluca miliaris
Meunier, A., 1919
Microplankton de la Mer Flamande. 4. Les Tintinnides et Coetera. Mem. Mus. Roy. Hist. Nat., Belgique, 8(2):59pp., Pls. 22-23.

Noctiluca miliaris
Mironov G.N. 1966.
On the digestion of Black Sea copepod eggs by Noctiluca miliaris Sur. (In Russian; English abstract)
Zool. Zh. 45(7):1093.

Noctiluca miliaris
Niaussal, Pierre-Marie, and Roland Bourcart, 1963.
Contribution à l'étude du plancton dans les eaux de l'embouchure de la Gironde. Prédominance du dino-flagellé "Notula miliaris". Cahiers Oceanogr., C.C.O.E.C., 15(10):722-725.

Noctiluca miliaris
Plate, L., 1889.
Observations on Noctiluca miliaris Suriray, and the sea luminosity produced by it. Ann. Mag. Nat Hist., 6th ser, 3(13):22-28.

Noctiluca miliaris
Prasad, R.R., 1953.
Swarming of Noctiluca in the Palk Bay and its effect on the "Choodai" fishery, with a note on the possible use of Noctiluca as an indicator species. Proc. Indian Acad. Sci., B, 38(1):40-47.

Noctiluca miliaris
Rajagopal, P.K., 1962
Respiration of some marine planktonic organisms. Proc. Indian Acad. Sci., (B) 55(2):76-81.

Noctiluca miliaris
Schodduyn, M., 1926
Observations faites dans la baie d'Ambleteuse (Pas de Calais). Bull. Inst. Ocean., Monaco, No. 482: 64 pp.

Noctiluca miliaris
Soyer, Marie-Odile, 1969.
Etude ultrastructurale des inclusions paracristallines intra-mitochondriates et intra-vacuolaires chez Noctiluca miliaris S. dinoflagelle, Noctilucidae et observations concernant leur role dans la genese des trichocystes fibreux et muqueux. Protistologica 5(3): 327-334.

Noctiluca miliaris
Soyer, Marie-Odile, 1969.
Etude cytologique ultrastructurale d'un dinoflagellé libre, Noctiluca miliaris Suriray: trichocystes et inclusions paracristallines.
Vie Milieu 20(1B): 305-314

Noctiluca miliaris
Soyer, Marie-Odile, 1968.
Présence de formations fibrillaires complexes chez Noctiluca miliaris Suriray et discussion de leur rôle dans la motilité de ce dinoflagellé.
C. r. hebd. Séanc. Acad. Sci. Paris (D) 266(26): 2425-2430.

Noctiluca miliaris

Sweeney, Beatrice M. 1971.
Laboratory studies of a green Noctiluca from New Guinea.
J. Phycol. 7(1):53-58.

Noctiluca miliaris

Uhlig, G., 1972
Entwicklung von Noctiluca miliaris.
Publ. wiss. Film (Sect.Biol) 5:387-399.

Noctiluca miliaris

Wood, E.J.F., 1954
Dinoflagellates in the Australian region.
Australian J. Mar. Freshwater Res., 5(2):171-351.

Noctiluca miliaris

Zingmark, Richard G., 1970.
Sexual reproduction in the dinoflagellate Noctiluca miliaris Suriray.
J. Phycol. 6(2): 122-126.

Noctiluca scintillans

Akatsuka, K. (deceased), F. Uyeno, K. Mitani and M. Miyamura, 1960.
On the relation between the distribution of plankton and the annual changes of sea conditions in Ise Bay.
J. Oceanogr. Soc., Japan, 16(2):83-91.

Noctiluca scintillans

Allen, W.E., 1937.
A large catch of Noctiluca. Science 86(2226): 197-198.

Noctiluca scintillans

Allen, W.E., 1927.
Quantitative studies on inshore marine diatoms and dinoflagellates of Southern California in 1922. Bull. S.I.O., tech. ser., 1:31-38, 2 text-figs.

Noctiluca scintillans

Chiba, T., 1949
On the distribution of the plankton in the eastern China Sea and Yellow Sea. 1. Plankton composition in the spring. J. Shimonoseki Coll. Fisheries, 1(1):57-63, 1 fig.

Noctiluca scintillans

Fung, Y.C. and L.B. Trott, 1973
The occurrence of a Noctiluca scintillans (Macartney) induced red tide in Hong Kong. Limnol. Oceanogr. 18(3):472-476.

Noctiluca scintillens

Grindley, J.R., and A.E.F. Heydorn 1970.
Red water and associated phenomena in St. Lucia.
S.Afr. Jl. Sci. 66 (7): 210-213.

Noctiluca scintillans

Grindley, J.R., and F.J.R. Taylor, 1970.
Factors affecting plankton blooms in False Bay. Trans. roy. Soc. SAfr. 39(2): 201-210.

Noctiluca scintillans

Hada, Yoshine, 1970.   (1970)
The protozoan plankton of the Antarctic and Subantarctic seas.
Scient. Repts, Japan. Antarct. Res. Exped., (E)31:51 pp.

Noctiluca scintillans

Hattori, S., 1962
Predatory activity of Noctiluca on anchovy eggs.
Bull. Tokai Reg. Fish. Res. Lab., Tokyo, No.9 211-220.

(In English; Japanese summary).

Noctiluca scintillans

LeBour, M.V., 1925
The dinoflagellates of Northern Seas. The Marine Biological Association of the United Kingdom, Plymouth, 250 pp., 35 pls.,53 text figs.

Noctiluca scintillans

Le Fèvre, J. and J.R. Grall, 1970.
On the relationships of Noctiluca swarming off the western coast of Brittany with hydrological features and plankton characteristics of the environment. J. exp. mar. Biol. Ecol., 4(3): 287-308.

Noctiluca scintillans

López, J. y P. Arté, 1971.
Aguas rojas en las costas catalanas. Investigación pesq. 35(2): 699-708.

Noctiluca scintillans

Morse, D.C., 1947
Some observations on seasonal variations in plankton population Patuxent River, Maryland 1943-1945. Bd. Nat. Res., Publ. No.65, Chesapeake Biol. Lab., 31, 3 figs.

Noctilueca scintillans

Quayle, D.B.
Paralytic shellfish poisoning in British Colombia.
Bull. Fish Res. Bd. Can., 168: 68 pp.

Oblea baculifera

Balech, Enrique, 1971.
Microplancton de la campaña oceanographica: Productividad III.
Revta Mus. argent. Cienc. Nat. Bernadina Rivadavia, Hydrobiol. 3(1):1-202, 39 pls.

Oblea bacculifera n. gen. n. sp.

Balech, Enrique, 1964
El plancton de Mar del Plata durante el periodo 1961 - 1962 (Buenos Aires, Argentina).
Bol. Inst. Biol. Mar., Mar del Plata, Argentina, No. 4: 49pp. & plates

Oodinium cyprinodontum

Lawler, Adrian R., 1968.
New host record for the parasitic dinoflagellate Oodinium cyprinodontum Lawler,1967.
Chesapeake Sci., 9(4):263.

Oodinium dogieli

Cachon, Jean et Monique Cachon, 1971
Ultrastructures du genre Oodinium Chatton. Différenciations cellulaires en rapport avec la vie parasitaire.
Protistologica 7(2):153-169.

Oodinium fritillariae

Cachon, Jean et Monique Cachon, 1971
Ultrastructures du genre Oodinium Chatton. Différenciations cellulaires en rapport avec la vie parasitaire.
Protistologica 7(2):153-169.

Oodinium poucheti

Cachon, Jean et Monique Cachon, 1971
Ultrastructures du genre Oodinium Chatton. Différenciations cellulaires en rapport avec la vie parasitaire.
Protistologica 7(2):153-169.

Ornithocercus sp.

Jörgensen, E., 1923
Mediterranean Dinophysiaceae. Rept. Danish Oceanogr. Expeds. 1908-10, to the Mediterranean and adjacent seas, Vol.II, Biol. J 2, 48 pp., 64 text figs.

Ornithocercus

Matzenauer, L., 1933
Die Dinoflagellaten des indischen Ozeans (mit Ausnahme der Gattung Ceratium.) Bot. Arch. 35:437-510, 77 text figs., 2 charts.

Ornithocercus spp.

Taylor, F.J.R., 1973.
Topography of cell division in the structurally complex dinoflagellate genus Ornithocercus.
J. Phycol. 9(1): 1-10.

Ornithocercus spp.

Taylor, F.J.R., 1971.
Scanning electron microscopy of thecae of the dinoflagellate genus Ornithocercus.
J. Phycol. 7(3): 249-258.

Ornithocercus sp.

Wheeler, J.E.G., 1939
Plankton investigations. Bermuda Biological Station. Second Report. October 1939. 7 pp. (typed), 5 figs. Plymouth, Oct. 23, 1939.

Ornithocercus sp.

Wood, E.J.F., 1963
Dinoflagellates in the Australian region. II. Recent collections.
Comm. Sci. and Industr. Res. Org., Div. Fish. and Oceanogr., Techn. Paper, No. 14: 55 pp.

Ornithocercus assimilis n.sp.

Jörgensen, E., 1923
Mediterranean Dinophysiaceae. Rept. Danish Oceanogr. Expeds. 1908-10, to the Mediterranean and adjacent seas, Vol.II, Biol. J 2, 48 pp., 64 text figs.

Ornithocercus australis n.sp.

Wood, E.J.F., 1963
Dinoflagellates in the Australian region. II. Recent collections.
Comm. Sci. and Industr. Res. Org., Div. Fish. and Oceanogr., Techn. Paper, No. 14: 55 pp.

Ornithocercus biclavatus n.sp.

Wood, E.J., 1954
Dinoflagellates in the Australian region.
Australian J. Mar. Freshwater Res., 5(2):171-351.

Ornithocercus bilobatus n. sp.

Rampi, L., 1950.
Péridiniens rares ou nouveaux pour le Pacifique Sud-Equatorial. Bull. Inst. Ocean., Monaco, No. 974:11 pp., 26 textfigs.

Ornithocercus carolinae

Balech, Enrique, 1962
Tintinnoinea y Dinoflagellata del Pacifico segun material de las Expediciones NORPAC y DOWNWIND del Instituto Scripps de Oceanografia.
Revista, Mus. Argentino Ciencias Nat. "Bernardino Rivadavia", Ciencias Zool., 7(1):1-253.

Ornithocercus carolinae

Dangeard, P., 1927
Phytoplankton de la croisière du "Sylvana". Ann. Inst. Ocean., Monaco, n.s., 4(8):286-401, 54 text figs. (février-Juin 1913).

Ornithocercus Carolinae

Jörgensen, E., 1923
Mediterranean Dinophysiaceae. Rept. Danish Oceanogr. Expeds. 1908-10, to the Mediterranean and adjacent seas, Vol.II, Biol. J 2, 48 pp., 64 text figs.

Ornithocircus carolinae

Käsler, R., 1938
Die Verbreitung der Dinophysiales im Sud-atlantischen Ozean. Wiss. Ergeb. Deutschen Atlantischen Expedition---- "Meteor" 1925-1927, 12(2):162-237, text figs. 85-118.

Ornithocercus carolinae

Kofoid, C. A. and T. Skogsberg, 1928
XXXV. The Dinoflagellata: The Dinophysiodae. Reports on the scientific results of the expedition to the Eastern Tropical Pacific, in charge of Alexander Agassiz, by the U. S. Fish Commission Steamer "Albatross" from October 1904 to March 1905----. Mem. M. C. Z. 51:766 pp., 31 pls.

Ornithocercus carolinae

Rampi, L., 1945
Osservazioni sulla distribuzione qualitativa del fitoplancton nel mare Mediterraneo. Atti della Soc. Ital. di Sci. Nat. 84:105-113.

Ornithocercus carolinae

Rampi, L., 1942
II Fitoplancton mediterraneo: Problemi ed affinita interoceaniche. Boll. di Pesca di Piscicoltura e di Idrobiologia, Anno 18, Fasc. 4:7-19.

Ornithocercus carolinae

Wood, E.J.F., 1954.
Dinoflagellates in the Australian region. Australian J. Mar. Freshwater Res., 5(2):171-351.

Ornithocercus cristatus

Balech, Enrique, 1967.
Dinoflagelados nuevos o interesantes del Golfo de Mexico y Caribe.
Revta Mus. argent. Cienc. nat. Bernardino Rivadavia Inst. nac. Invest. Cienc. nat., Hidrobiol. 2(3):77-126.

Ornithocercus cristatus n.sp.

Matzenauer, L., 1933
Die Dinoflagellaten des indischen Ozeans (mit Ausnahme der Gattung Ceratium.) Bot. Arch. 35:437-510, 77 text figs., 2 charts.

Ornithocercus fimbriatus n.sp.

Marukawa, H., 1928.
Ueber 4 neue Arten der Peridinialen. Annot. Oceanogr. Res., 2(1):1-2.

Ornithocercus formosus

Balech, Enrique, 1962
Tintinnoinea y Dinoflagellata del Pacifico segun material de las Expediciones NORPAC y DOWNWIND del Instituto Scripps de Oceanografia.
Revista. Mus. Argentino Ciencias Nat. "Bernardino Rivadavia". Ciencias Zool., 7(1):1-253.

Ornithocercus formosus

Kofoid, C. A. and T. Skogsberg, 1928
XXXV. The Dinoflagellata: The Dinophysiodae. Reports on the scientific results of the expedition to the Eastern Tropical Pacific, in charge of Alexander Agassiz, by the U. S. Fish Commission Steamer "Albatross" from October 1904 to March 1905----. Mem. M. C. Z. 51:766 pp., 31 pls.

Ornithocercus formosus

Wood, E.J.F., 1963
Dinoflagellates in the Australian region. II. Recent collections.
Comm. Sci. and Industr. Res. Org., Div. Fish. and Oceanogr., Techn. Paper, No. 14: 55 pp.

Ornithocercus francescae

Abe, Tohru H., 1967.
The armoured Dinoflagellata: II. Prorocentridae and Dinophysidae (C)- Ornithocercus, Histioneis, Amphisolenia and others.
Publs Seto mar. biol. Lab., 15(2):79-116.

Ornithocercus francescae

Balech, Enrique, 1962
Tintinnoinea y Dinoflagellata del Pacifico segun material de las Expediciones NORPAC y DOWNWIND del Instituto Scripps de Oceanografia.
Revista. Mus. Argentino Ciencias Nat. "Bernardino Rivadavia". Ciencias Zool., 7(1):1-253.

Ornithocercus galea

Abe, Tohru H., 1967.
The armoured Dinoflagellata: II. Prorocentridae and Dinophusidae (C) - Ornithocercus, Histioneis, Amphisolenia and others.
Publs Seto mar. biol. Lab., 15(2):79-116.

Ornithocercus geniculatus n.sp.

Dangeard, P., 1927
Phytoplankton de la croisière du "Sylvana". Ann. Inst. Ocean., Monaco, n.s., 4(8):286-401, 54 text figs. (février-Juin 1913).

Ornithocercus geniculatus

Wood, E.J.F., 1963
Dinoflagellates in the Australian region. II. Recent collections.
Comm. Sci. and Industr. Res. Org., Div. Fish. and Oceanogr., Techn. Paper, No. 14: 55 pp.

Ornithocercus heteropoides n.sp.

Abe, Tohru H., 1967.
The armoured Dinoflagellata: II. Prorocentridae and Dinophysidae (C) - Ornithocercus, Histioneis, Amphisolenia and others.
Publs Seto mar. biol. Lab., 15(2):79-116.

Ornithocercus heteroporus

Abe, Tohru H., 1967.
The armoured Dinoflagellata: II. Prorocentridae and Dinophysidae (C)- Ornithocercus, Histioneis, Amphisolenia and others.
Publs Seto mar. biol. Lab., 15(2):79-116.

Ornithocercus heteroporus

Jörgensen, E., 1923
Mediterranean Dinophysiaceae. Rept. Danish Oceanogr. Expeds. 1908-10, to the Mediterranean and adjacent seas, Vol. II, Biol. J 2, 48 pp., 64 text figs.

Ornithocercus heteroporus

Kofoid, C. A. and T. Skogsberg, 1928
XXXV. The Dinoflagellata: The Dinophysiodae. Reports on the scientific results of the expedition to the Eastern Tropical Pacific, in charge of Alexander Agassiz, by the U. S. Fish Commission Steamer "Albatross" from October 1904 to March 1905----. Mem. M. C. Z. 51:766 pp., 31 pls.

Ornithocercus heteroporus

Rampi, L., 1940
Ricerche sul Fitoplancton del mare Ligure. II Le tecatali e le dinofisiali delle acque di Sanremo.. Boll. Pesca, Piscicolt., Idrobiol. (18) 16(2): 243-274, 56 figs.

Ornithocercus heteroporus

Rampi, L., 1940
Ricerche sul Fitoplancton del mare Ligure. Boll. di Pesca, di Piscicoltura e di Idrobiologa, 18(2):1-34, 56 text figs.

Ornithocercus heteroporus

Wood, E.J.F., 1954.
Dinoflagellates in the Australian region. Australian J. Mar. Freshwater Res., 5(2):171-351.

Ornithocercus magnificus

Abe, Tohru H., 1967.
The armoured Dinoflagellata: II. Prorocentridae and Dinophysidae (C)- Ornithocercus, Histioneis, Amphisolenia and others.
Publs Seto mar. biol. Lab., 15(2):79-116.

Ornithocercus magnificus

Chiba, T., 1949
On the distribution of the plankton in the eastern China Sea and Yellow Sea. 1. Plankton composition in the spring. J. Shimonoseki Coll. Fisheries, 1(1):57-63, 1 fig.

Ornithocercus magnificus

Dangeard, P., 1927
Phytoplankton de la croisière du "Sylvana". Ann. Inst. Ocean., Monaco, n.s., 4(8):286-401, 54 text figs. (février-Juin 1913).

Ornithocercus magnificus

Delegazione Italiana della Commissione Internazionale per l'Esplorazione Scientifica del Mediterraneo, 1941
Note sul plancton della Laguna veneta. [Memoria CCLXXXIX], Arch. di Ocean. e Limn. Anno I, Fasc. I, 1941 XIX: 31-57 pp.

Ornithocercus magnificus

Ercegovic, A., 1936
Etudes qualitative et quantitatives du phytoplancton dans les eaux cotières de l'Adriatique oriental moyen au cours de l'année 1934. Acta Adriatica 1(9):1-126

Ornithocercus magnificus

Kofoid, C. A. and T. Skogsberg, 1928
XXXV. The Dinoflagellata: The Dinophysiodae. Reports on the scientific results of the expedition to the Eastern Tropical Pacific, in charge of Alexander Agassiz, by the U. S. Fish Commission Steamer "Albatross" from October 1904 to March 1905----. Mem. M. C. Z. 51:766 pp., 31 pls.

Ornithocercus magnificus

Margalef, R., 1949
Fitoplancton nerítico de la Costa Brava en 1947-48. Publ. Inst. Biol. Aplicada, 5: 41-51, 3 text figs.

Ornithocercus magnificus

Matzenauer, L., 1933
Die Dinoflagellaten des indischen Ozeans (mit Ausnahme der Gattung Ceratium.) Bot. Arch. 35:437-510, 77 text figs., 2 charts.

Ornithocercus magnificus

Norris, Dean R., 1969.
Thecal morphology of Ornithocercus magnificus (Dinoflagellata) with notes on related species. Bull. Mar. Sci., 19(1): 175-193.

Ornithocercus magnificus

Rampi, L., 1940
Ricerche sul Fitoplancton del mare Ligure. II Le tecatali e le dinofisiali delle acque di Sanremo.. Boll. Pesca, Piscicolt., Idrobiol. (18) 16(2): 243-274, 56 figs.

Ornithocercus magnificus

Rampi, L., 1940
Ricerche sul Fitoplancton del mare Ligure. Boll. di Pesca, di Piscicoltura e di Idrobiologa 18(2):1-34, 56 text figs.

Ornithocercus magnificus (fig).

Sousa e Silva, E., 1949
Diatomaceas e Dinoflagelados de Baia de Cascais. Portugaliae Acta Biol., Volume: Julio Henriques, Ser. B: 300-383, 9 pls, 2 fold-in tables.

Ornithocercus magnificus

Wood, E.J.F., 1963
Dinoflagellates in the Australian region. II. Recent collections.
Comm. Sci. and Industr. Res. Org., Div. Fish. and Oceanogr., Techn. Paper, No. 14: 55 pp.

Ornithocercus magnificus

Wood, E.J.F., 1954.
Dinoflagellates in the Australian region. Australian J. Mar. Freshwater Res., 5(2):171-351.

Ornithocercus orbiculatus

Kofoid, C. A. and T. Skogsberg, 1928
XXXV. The Dinoflagellata: The Dinophysiodae. Reports on the scientific results of the expedition to the Eastern Tropical Pacific, in charge of Alexander Agassiz, by the U. S. Fish Commission Steamer "Albatross" from October 1904 to March 1905----. Mem. M. C. Z. 51:766 pp., 31 pls.

Ornithocercus orbiculatus

Matzenauer, L., 1933
Die Dinoflagellaten des indischen Ozeans (mit Ausnahme der Gattung Ceratium.) Bot. Arch. 35:437-510, 77 text figs., 2 charts.

Ornithocercus quadratus

Abe, Tohru H., 1967.
The armoured Dinoflagellata:II.Prorocentridae and Dinophysidae (C) - Ornithocercus, Histioneis, Amphisolenia and others.
Publs Seto mar.biol.Lab., 15(2):79-116.

Ornithocercus quadratus

Dangeard, P., 1927
Phytoplankton de la croisière du "Sylvana". Ann. Inst. Ocean., Monaco, n.s., 4(8):286-401, 54 text figs. (Février-Juin 1913).

Ornithocercus quadratus

Ercegovic, A., 1936
Etudes qualitative et quantitatives du phytoplancton dans les eaux cotières de l'Adriatique oriental moyen au cours de l'année 1934. Acta Adriatica 1(9):1-126

Ornithocercus quadratus

Jörgensen, E., 1923
Mediterranean Dinophysiaceae. Rept. Danish Oceanogr. Expeds. 1908-10, to the Mediterranean and adjacent seas, Vol.II, Biol. J 2, 48 pp., 64 text figs.

Ornithocircus quadratus

Käsler, R., 1938
Die Verbreitung der Dinophysiales im Sud-atlantischen Ozean. Wiss. Ergeb. Deutschen Atlantischen Expedition----"Meteor" 1925-1927, 12(2):162-237, text figs. 85-118.

Ornithocercus quadratus

Kofoid, C. A. and T. Skogsberg, 1928
XXXV. The Dinoflagellata: The Dinophysiodae. Reports on the scientific results of the expedition to the Eastern Tropical Pacific, in charge of Alexander Agassiz, by the U. S. Fish Commission Steamer "Albatross" from October 1904 to March 1905----. Mem. M. C. Z. 51:766 pp., 31 pls.

Ornithocercus quadratus

Matzenauer, L., 1933
Die Dinoflagellaten des indischen Ozeans (mit Ausnahme der Gattung Ceratium.) Bot. Arch. 35:437-510, 77 text figs., 2 charts.

Ornithocercus quadratus

Rampi, L. 1940
Ricerche sul Fitoplancton del mare Ligure. II Le tecatali e le dinofisiali delle acque di Sanremo.. Boll. Pesca, Piscicolt.,Idrobiol. (18) 16(2): 243-274, 56 figs.

Ornithocercus quadratus

Rampi, L., 1940
Ricerche sul Fitoplancton del mare Ligure. Boll. di Pesca, di Piscicoltura e di Idrobiologa, 18(2):1-34, 56 text figs.

Ornithocercus quadratus

Sournia, A.,1967.
Contribution à la connaissance des péridiniens microplanctoniques du Canal de Mozambique.
Bull.Mus.natn.Hist.nat.,Paris, (2)39(2):417-438.

Ornithocercus quadratus (figs.)

Sousa e Silva, E., 1949
Diatomaceas e Dinoflagelados de Baia de Cascais. Portugaliae Acta Biol., Volume: Julio Henriques, Ser. B: 300-383, 9 pls, 2 fold-in tables.

Ornithocercus quadratus

Wood, E.J.P., 1954.
Dinoflagellates in the Australian region.
Australian J. Mar. Freshwater Res., 5(2):171-351.

Ornithocercus serratus

Dangeard, P., 1927
Phytoplankton de la croisière du "Sylvana". Ann. Inst. Ocean., Monaco, n.s., 4(8):286-401, 54 text figs. (Février-Juin 1913).

Ornithocercus serratus

Jörgensen, E., 1923
Mediterranean Dinophysiaceae. Rept. Danish Oceanogr. Expeds. 1908-10, to the Mediterranean and adjacent seas, Vol.II, Biol. J 2, 48 pp., 64 text figs.

Ornithocercus serratus

Rampi, L., 1945
Osservazioni sulla distribuzione qualitativa del fitoplancton nel mare Mediterraneo. Atti della Soc. Ital. di Sci. Nat. 84:105-113.

Ornithocercus serratus

Rampi, L., 1942
Il Fitoplancton mediterraneo: Problemi ed affinita interoceaniche. Boll. di Pesca di Piscicoltura e di Idrobiologia, Anno 18, Fasc. 4:7-19.

Ornithocercus skogsbergi n.sp.

Abe, Tohru H.,1967.
The armoured Dinoflagellata: II.Prorocentridae and Dinophysidae (C) - Ornithocercus, Histioneis, Amphisolenia and others.
Publs Seto mar.biol.Lab., 15(2):79-116.

Ornithocercus splendidus

Abe, Tohru H.,1967.
The armoured Dinoflagellata: II.Prorocentridae and Dinophysidae (C) - Ornithocercus, Histioneis, Amphisolenia and others.
Publs Seto mar.biol.Lab., 15(2):79-116.

Ornithocercas splendidus

Chiba, T., 1949
On the distribution of the plankton in the eastern China Sea and Yellow Sea. 1. Plankton composition in the spring. J. Shimonoseki Coll. Fisheries, 1(1):57-63, 1 fig.

Ornithocercus splendidus

Kofoid, C. A. and T. Skogsberg, 1928
XXXV. The Dinoflagellata: The Dinophysiodae. Reports on the scientific results of the expedition to the Eastern Tropical Pacific, in charge of Alexander Agassiz, by the U. S. Fish Commission Steamer "Albatross" from October 1904 to March 1905----. Mem. M. C. Z. 51:766 pp., 31 pls.

Ornithocercus splendidus

Lindemann, E., 1925
Neubeobachtungen an den Winter peridineen des Golfes von Neapel. Bot. Arch. 9:95-102, 19 text figs.

Ornithocercius splendidus

Matzenauer, L., 1933
Die Dinoflagellaten des indischen Ozeans (mit Ausnahme der Gattung Ceratium.) Bot. Arch. 35:437-510, 77 text figs., 2 charts.

Ornithocercus splendidus

Wood, E.J.P., 1954.
Dinoflagellates in the Australian region.
Australian J. Mar. Freshwater Sci., 5(2):171-351.

Ornithocercus steini

Abe, Tohru H.,1967.
The armoured Dinoflagellata:II.Prorocentridae and Dinophysidae (C)- Ornithocercus, Histioneis, Amphisolenia and others.
Publs Seto mar.biol.Lab., 15(2):79-116.

Ornithocercus Steinii

Dangeard, P., 1927
Phytoplankton de la croisière du "Sylvana". Ann. Inst. Ocean., Monaco, n.s., 4(8):286-401, 54 text figs. (Février-Juin 1913).

Ornithocercus Steini

Jörgensen, E., 1923
Mediterranean Dinophysiaceae. Rept. Danish Oceanogr. Expeds. 1908-10, to the Mediterranean and adjacent seas, Vol.II, Biol. J 2, 48 pp., 64 text figs.

Ornithocircus Steinii

Käsler, R., 1938
Die Verbreitung der Dinophysiales im Sud-atlantischen Ozean. Wiss. Ergeb. Deutschen Atlantischen Expedition----"Meteor" 1925-1927, 12(2):162-237, text figs. 85-118.

Ornithocercus steini

Kofoid, C. A. and T. Skogsberg, 1928
XXXV. The Dinoflagellata: The Dinophysiodae. Reports on the scientific results of the expedition to the Eastern Tropical Pacific, in charge of Alexander Agassiz, by the U. S. Fish Commission Steamer "Albatross" from October 1904 to March 1905----. Mem. M. C. Z. 51:766 pp., 31 pls.

Ornithocercus Steinii

Matzenauer, L., 1933
Die Dinoflagellaten des indischen Ozeans (mit Ausnahme der Gattung Ceratium.) Bot. Arch. 35:437-510, 77 text figs., 2 charts.

Ornithocercus steinii

Sousa e Silva, E., 1949
Diatomaceas e Dinoflagelados de Baia de Cascais. Portugaliae Acta Biol., Volume: Julio Henriques, Ser. B: 300-383, 9 pls, 2 fold-in tables.

Ornithocercus steini

Wood, E.J.P., 1954.
Dinoflagellates in the Australian region.
Australian J. Mar. Freshwater Res., 5(2):171-351.

Ornithocercus thurnii

Abe, Tohru H.,1967.
The armoured Dinoflagellata:II.Prorocentridae and Dinophysidae (C)- Ornithocercus, Histioneis, Amphisolenia and others.
Publs Seto mar.biol.Lab., 15(2):79-116.

Ornithocercus thurnii

Establier, R., and R. Margalef, 1964
Fitoplancton e hidrografía de las costas de Cádiz (Barbate), de junio de 1961 a agosto de 1962.
Invest. Pesquera, Barcelona, 25:5-31.

Ornithocercus thurni

Kofoid, C. A. and T. Skogsberg, 1928
XXXV. The Dinoflagellata: The Dinophysiodae. Reports on the scientific results of the expedition to the Eastern Tropical Pacific, in charge of Alexander Agassiz, by the U. S. Fish Commission Steamer "Albatross" from October 1904 to March 1905----. Mem. M. C. Z. 51:766 pp., 31 pls.

Ornithocercus Thurnii

Matzenauer, L., 1933
Die Dinoflagellaten des indischen Ozeans (mit Ausnahme der Gattung Ceratium.) Bot. Arch. 35:437-510, 77 text figs., 2 charts.

Ornithocercus thurni

Nie, Dashu, 1943
Dinoflagellata of the Hainan region. VII. On the thecal morphology of Ornithocercus thurni (Schmidt) Kofoid and Skogsberg.
Sinensia, 14:23-28.

Ornithocercus thurni

Wood, E.J.P., 1954.
Dinoflagellates in the Australian region.
Australian J. Mar. Freshwater Res., 5(2):171-351.

Ornithocercus triclavatus n.sp.

Wood, E.J.P., 1954.
Dinoflagellates in the Australian region.
Australian J. Mar. Freshwater Res., 5(2):171-351.

Ostreopsis monotis

Halim, Youssef, 1960
Étude quantitative et qualitative du cycle écologique des dinoflagellés dans les eaux de Villefranche-sur-Mer. (1953-1955).
Ann. Inst. Océanogr., Monaco, 38:123-232.

Oxynotus centrina

Della Croce, Norberto, 1964.
Ritrovamenti di Oxynotus centrina (L.) nel Mar Ligure (Pisces).
Atti Soc. ital. Scil. nat., 103(3):205-211.

Oxyphysis

Abe, Tohru H., 1967.
The armoured Dinoflagellata: II. Prorocentridae and Dinophysidae (C)- Ornithocercus, Histioneis, Amphisolenia and others.
Publs Seto mar. biol. Lab., 15(2):79-116.

Oxyphysis oxytoxoides n.gen., n.sp.

Kofoid, C.A., 1926.
On Oxyphysis oxytosoides gen. nov., sp. nov., a dinophysid dinoflagellate convergent toward the peridinioid type. Univ. Calif. Publ. Zool. 28(10):203-216, Pl. 18.

Oxyphysis oxytoxoides

Tai, Si-Sun, & T. Skogsberg, 1934
Studies on the Dinophysoidae, marine armored dinoflagellates of Monterey Bay, California. Arch. Protistenk. 82:380-482, 14 text figs. Pls. 11-12.

Oxyrrhis marina

Biecheler, B., 1952.
Recherches sur les Peridiniens.
Bull. Biol., France Belg., Suppl., 36:1-149, 75 textfigs.

Oxyrrhis marina

Dodge, J.D., and R.M. Crawford 1971.
Fine structure of the dinoflagellate Oxyrrhis marina 1. The general structure of the cell.
Protistologica 7(2): 295-304.

Oxyrrhis marina

Droop, M.R., 1966.
The role of algae in the nutrition of Heteramoeba clara Droop, with notes on Oxyrrhis marina Dujardin and Philodina roseola Ehrenberg.
In: Some contemporary studies in marine biology, H. Barnes, editor, George Allen & Unwin, Ltd., 269-282.

Oxyrrhis marina

Droop, M.R. and J.F. Pennock 1971.
Terpenoid quinones and steroids in the nutrition of Oxyrrhis marina. J. mar. biol. Ass. U.K. 51(2): 455-470.

Oxyrrhis marina

Ercegovic, A., 1936
Etudes qualitative et quantitatives du. phytoplancton dans les eaux cotières de l'Adriatique oriental moyen au cours de l'année 1934. Acta Adriatica 1(9):1-126

Oxyrhis marina

Grøntved, J., 1960-61
Planktological contributions. IV. Taxonomical and productional investigations in shallow coastal waters.
Medd. Dansk Fisk. Havundersøgelser, n.s., 3(1):1-17.

Oxyrrhis marina

Hausmann, K., 1973
Cytologische Studien an Trichocysten VI. Feinstruktur und Funktionsmodus der Trichocysten des Flagellaten Oxyrrhis marina und des Ciliaten Pleuronema marinum. Helgoländer wiss. Meeresunters, 25(1):39-62.

Oxyrrhis marina

Hulburt, E.M., 1957.
The taxonomy of unarmored Dinophyceae of shallow embayments on Cape Cod, Massachusetts.
Biol. Bull., 112(2):196-219.

Oxyrrhis marina

LeBour, M.V., 1925
The dinoflagellates of Northern Seas. The Marine Biological Association of the United Kingdom, Plymouth, 250 pp., 35 pls., 53 text figs.

Oxyrrhis marina

Nozawa, K., 1940.
Problem in the diurnal rhythm in the cell division of the dinoflagellate, Oxyrrhis marina.
Annot. Zool. Japonensis, 19(3):170-174.
Contrib. No. 90, Seto Mar. Biol. Lab.

Oxyrrhis marina

Osorio Tafall, B.F., 1946.
Nuevos datos sobre la distribución del dinoflagelado Oxyrrhis marina Duj.
Rev. Soc. Mexicana Hist. Nat., 7(1/4):41-48.

Oxyrrhis marina

Wood, E.J.F., 1954.
Dinoflagellates in the Australian region.
Australian J. Mar. Freshwater Res., 5(2):171-351.

Oxytoxa obliquum

Wood, E.J.F., 1963
Dinoflagellates in the Australian region. II. Recent collections.
Comm. Sci. and Industr. Res. Org. Div. Fish. and Oceanogr., Techn. Paper, No. 14: 55 pp.

Oxytoxum sp.

Balech, Enrique, 1971.
Microplancton de la campaña oceanographica: Productividad III.
Revta Mus. argent. Cienc. Nat. Bernadina Rivadavia, Hydrobiol. 3(1):1-202, 39 pls.

Oxytoxum sp.

Wood, E.J.F., 1963
Dinoflagellates in the Australian region. II. Recent collections.
Comm. Sci. and Industr. Res. Org. Div. Fish. and Oceanogr., Techn. Paper, No. 14: 55 pp.

Oxytoxum areolatum n.sp.

Rampi, L., 1941
Ricerche sul microplancton del Mare Ligure. 3. Le Hatorodiniacee e le Oxytoxacee dell'acque di Sanremo. Annali del Mus. Civico di Storia Naturale di Genova, 61:50-70, 2 pls.

Oxytoxum belgicae

Balech, Enrique, 1971.
Microplancton de la campaña oceanographica: Productividad III.
Revta Mus. argent. Cienc. Nat. Bernadina Rivadavia, Hydrobiol. 3(1):1-202, 39 pls.

Oxytoxum belgicae

LeBour, M.V., 1925
The dinoflagellates of Northern Seas. The Marine Biological Association of the United Kingdom. Plymouth, 250 pp., 35 pls. 53 text figs.

Oxytoxum belgicae

Wood, E.J.F., 1963
Dinoflagellates in the Australian region. II. Recent collections.
Comm. Sci. and Industr. Res. Org. Div. Fish. and Oceanogr., Techn. Paper, No. 14: 55 pp.

Oxytoxum Brunelli

Halim, Youssef, 1960
Étude quantitative et qualitative du cycle écologique des dinoflagellés dans les eaux de Villefranche-sur-Mer. (1953-1955).
Ann. Inst. Océanogr., Monaco, 38:123-232.

Oxytoxum Brunelli

Rampi, L., 1941
Ricerche sul microplancton del Mare Ligure. 3. Le Hatorodiniacee e le Oxytoxacee dell'acque di Sanremo. Annali del Mus. Civico di Storia Naturale di Genova, 61:50-70, 2 pls.

Oxytoxum caudatum

Hasle, Grethe Rytter, 1960
Phytoplankton and ciliate species from the Tropical Pacific.
Skr. Norske Videnskaps-Akad., Oslo, 1. Mat.-Nat. Kl., 1960(2): 1-50.

Oxytoxum caudatum

Wood, E.J.F., 1963
Dinoflagellates in the Australian region. II. Recent collections.
Comm. Sci. and Industr. Res. Org. Div. Fish. and Oceanogr., Techn. Paper, No. 14: 55 pp.

Oxytoxum challengeroides

Balech, Enrique, 1971
Microplancton del Atlantico ecuatorial oeste (Equalant 1)
Publ. Serv. Hidrograf. Naval, Argentina H. 654: 103 pp., 122 figs.

Oxytoxum challengeroides

Wood, E.J.F., 1963
Dinoflagellates in the Australian region. II. Recent collections.
Comm. Sci. and Industr. Res. Org. Div. Fish. and Oceanogr., Techn. Paper, No. 14: 55 pp.

Oxytoxum compressum

Wood, E.J.F., 1963
Dinoflagellates in the Australian region. II. Recent collections.
Comm. Sci. and Industr. Res. Org. Div. Fish. and Oceanogr., Techn. Paper, No. 14: 55 pp.

Oxytoxum constrictum

Ercegovic, A., 1936
Etudes qualitative et quantitatives du. phytoplancton dans les eaux cotières de l'Adriatique oriental moyen au cours de l'année 1934. Acta Adriatica 1(9):1-126

Oxytoxum constrictum

Halim, Youssef, 1960
Étude quantitative et qualitative du cycle écologique des dinoflagellés dans les eaux de Villefranche-sur-Mer. (1953-1955).
Ann. Inst. Océanogr., Monaco, 38:123-232.

Oxytoxum constrictum

Margalef, R., 1949
Fitoplancton nerítico de la Costa Brava en 1947-48. Publ. Inst. Biol. Aplicada, 5: 41-51, 3 text figs.

Oxytoxum constrictum

Murray, G., and F.G. Whitting, 1899
New Peridiniaceae from the Atlantic.
Trans. Linn. Soc., London, Bot., ser 2, 5: 321-342, Pls. 27-33, 9 tables.

Oxytoxum constrictum
Pavillard, J., 1916
Recherches sur les Peridiniens du Golfe du Lion. Mem. Univ. Montpellier. Trav. Inst. Bot., Univ. Montpellier. Serie mixte No.4, 70 pp., 3 pls., 15 text figs.

Oxytoxum constrictum
Rampi, L., 1948
Sur quelques Peridiniens rares ou interessants du Pacifique subtropical (Recoltes Alain Gerbault). Bull. l'Inst. Ocean., Monaco, No.937: 7 pp., 8 text figs.

Oxytoxum constrictum
Rampi, L., 1941
Ricerche sul microplancton del Mare Ligure. 3. Le Haterodiniacee o le Oxytoxacee dell acque di Sanremo. Annali del Mus. Civico di Storia Naturale di Genova, 61:50-70, 2 pls.

Oxytoxum constrictum
Schröder, B., 1900
Phytoplankton des Golfes von Neapel nebst vergleichenden Ausblicken auf das atlantischen Ozean. Mitt. Zool. Stat. Neapel, 14:1-38.

Oxytoxum constrictum
Wood, E.J.F., 1963
Dinoflagellates in the Australian region. II. Recent collections. Comm. Sci. and Industr. Res. Org., Div. Fish. and Oceanogr., Techn. Paper, No. 14: 55 pp.

Oxytoxum coronatum
Wood, E.J.F., 1963
Dinoflagellates in the Australian region. II. Recent collections. Comm. Sci. and Industr. Res. Org., Div. Fish. and Oceanogr., Techn. Paper, No. 14: 55 pp.

Oxytoxum cribosum
Wood, E.J.F., 1963
Dinoflagellates in the Australian region. II. Recent collections. Comm. Sci. and Industr. Res. Org., Div. Fish. and Oceanogr., Techn. Paper, No. 14: 55 pp.

Oxytoxum criophilum n.sp.
Balech, Enrique, and Sayed Z. El-Sayed, 1965.
Microplankton of the Weddell Sea. In: Biology of Antarctic seas. II. Antarctic Res. Ser., Amer. Geophys. Union, 5:107-124.

Oxytoxum cristatum
Balech, Enrique, 1962
Tintinnoinea y Dinoflagellata del Pacifico segun material de las Expediciones NORPAC y DOWNWIND del Instituto Scripps de Oceanografia. Revista, Mus. Argentino Ciencias Nat. "Bernardino Rivadavia", Ciencias Zool., 7(1):1-253.

Oxytoxum cristatum
Rampi, L., 1950.
Peridiniens rares ou nouveaux pour le Pacifique Sud-Equatorial. Bull. Inst. Océan., Monaco, No. 974:11 pp., 28 textfigs.

Oxytoxum cristatum
Rampi, L., 1945
Osservazioni sulla distribuzione qualitativa del fitoplancton nel mare Mediterraneo. Atti della Soc. Ital. di Sci. Nat. 84:105-113.

Oxytoxum cristatum
Rampi, L., 1942
Il Fitoplancton mediterraneo: Problemi ed affinita interoceaniche. Boll. di Pesca di Piscicoltura e di Idrobiologia, Anno 18, Fasc. 4:7-19.

Oxytoxum curvatum
Hasle, Grethe Rytter, 1960
Phytoplankton and ciliate species from the Tropical Pacific. Skr. Norske Videnskaps-Akad., Oslo, 1. Mat.-Nat. Kl., 1960(2): 1-50.

Oxytoxum curvatum
Wood, E.J.F., 1963
Dinoflagellates in the Australian region. II. Recent collections. Comm. Sci. and Industr. Res. Org., Div. Fish. and Oceanogr., Techn. Paper, No. 14: 55 pp.

Oxlytoxum diploconus
Balech, Enrique, 1971.
Microplancton de la campaña oceanographica: Productividad III. Revta Mus. argent. Cienc. Nat. Bernardina Rivadavia, Hydrobiol. 3(1):1-202, 39 pls.

Oxytoxum diploconus
Dangeard, P., 1927
Phytoplankton de la croisière du "Sylvana". Ann. Inst. Ocean., Monaco, n.s., 4(8):286-401, 54 text figs. (février-Juin 1913).

Oxytoxum diploconus
Ercegovic, A., 1936
Etudes qualitative et quantitatives du phytoplancton dans les eaux cotières de l'Adriatique oriental moyen au cours de l'année 1934. Acta Adriatica 1(9):1-126

Oxytoxum diploconus
Jørgensen, E., 1905
B. Protistplankton and the diatoms in bottom samples: Hydrographical and biological investigations in Norwegian fiords. Bergens Mus. Skr. 7: 49-225.

Oxytoxum diploconus
Jorgensen, E., 1900
Protophyten und Protozoën im Plankton aus der Norwegischen Westkerste. Bergens Mus. Aarb. 1899(6): 95 pp., 5 pls., 83 tables

Oxytoxum diploconus
LeBour, M.V., 1925
The dinoflagellates of Northern Seas. The Marine Biological Association of the United Kingdom. Plymouth, 250 pp., 35 pls. 53 text figs.

Oxytoxum diploconus
Murray, G., and F. G. Whitting, 1899
New Peridiniaceae from the Atlantic. Trans. Linn. Soc., London, Bot., ser 2, 5: 321-342, Pls. 27-33, 9 tables.

Oxytoxum diploconus
Paulsen, O., 1908
XVIII Peridiniales. Nordisches Plankton, Bot. Teil: 1-124, 155 text figs.

Oxytoxum diploconus
Rampi, L., 1941
Ricerche sul microplancton del Mare Ligure. 3. Le Haterodiniacee o le Oxytoxacee dell acque di Sanremo. Annali del Mus. Civico di Storia Naturale di Genova, 61:50-70, 2 pls.

Oxytoxum elegans
Balech, Enrique, 1971
Microplancton del Atlantico ecuatorial oeste (Equalant 1) Publ. Serv. Hidrograf. Naval, Argentina H. 654: 103 pp., 122 figs.

Oxytoxum elegans
Balech, Enrique, 1962
Tintinnoinea y Dinoflagellata del Pacifico segun material de las Expediciones NORPAC y DOWNWIND del Instituto Scripps de Oceanografia. Revista, Mus. Argentino Ciencias Nat. "Bernardino Rivadavia", Ciencias Zool., 7(1):1-253.

Oxytoxum elegans
Halim, Youssef, 1960
Etude quantitative et qualitative du cycle écologique des dinoflagellés dans les eaux de Villefranche-sur-Mer. (1953-1955). Ann. Inst. Océanogr., Monaco, 38:123-232.

Oxytoxum elegans
Matzenauer, L., 1933
Die Dinoflagellaten des indischen Ozeans (mit Ausnahme der Gattung Ceratium.) Bot. Arch. 35:437-510, 77 text figs., 2 charts.

Oxytoxum elegans n.sp.
Pavillard, J., 1916
Recherches sur les Peridiniens du Golfe du Lion. Mem. Univ. Montpellier. Trav. Inst. Bot., Univ. Montpellier. Serie mixte No.4, 70 pp., 3 pls., 15 text figs.

Oxytoxum elegans
Rampi, L., 1941
Ricerche sul microplancton del Mare Ligure. 3. Le Haterodiniacee o le Oxytoxacee dell acque di Sanremo. Annali del Mus. Civico di Storia Naturale di Genova, 61:50-70, 2 pls.

Oxytoxum elegans
Ercegovic, A., 1940
Weitere Untersuchungen über einige hydrographische Verhältnisse und über die Phytoplanktonproduktion in den Gewässern der Östlichen Mitteladria. Acta Adriatica 2(3):95-134, 8 text figs.

Oxytoxum elegans
Wood, E.J.F., 1963
Dinoflagellates in the Australian region. II. Recent collections. Comm. Sci. and Industr. Res. Org., Div. Fish. and Oceanogr., Techn. Paper, No. 14: 55 pp.

Oxytoxum elongatum n.sp.
Wood, E.J.F., 1963
Dinoflagellates in the Australian region. II. Recent collections. Comm. Sci. and Industr. Res. Org., Div. Fish. and Oceanogr., Techn. Paper, No. 14: 55 pp.

Oxytoxum Frenguellii n.sp.
Rampi, L., 1941
Ricerche sul microplancton del Mare Ligure. 3. Le Haterodiniacee o le Oxytoxacee dell acque di Sanremo. Annali del Mus. Civico di Storia Naturale di Genova, 61:50-70, 2 pls.

Oxytoxum gigas
Balech, Enrique, 1962
Tintinnoinea y Dinoflagellata del Pacifico segun material de las Expediciones NORPAC y DOWNWIND del Instituto Scripps de Oceanografia. Revista, Mus. Argentino Ciencias Nat. "Bernardino Rivadavia", Ciencias Zool., 7(1):1-253.

Oxytoxum gigas
Ercegovic, A., 1940
Weitere Untersuchungen über einige hydrographische Verhältnisse und über die Phytoplanktonproduktion in den Gewässern der Östlichen Mitteladria. Acta Adriatica 2(3):95-134, 8 text figs.

Oxytoxum gladiolus
LeBour, M.V., 1925
The dinoflagellates of Northern Seas. The Marine Biological Association of the United Kingdom. Plymouth, 250 pp., 35 pls. 53 text figs.

Oxytoxum gladiolus
Paulsen, O., 1908
XVIII Peridiniales. Nordisches Plankton, Bot. Teil: 1-124, 155 text figs.

Oxytoxum gracile

Braarud, T., 1934
A note on the phytoplankton of the Gulf of Maine in the summer of 1933. Biol. Bull. 67(1):76-82. (Contribution No.46 of the Woods Hole Oceanographic Institution)

Oxytoxum gracile

Lillick, L.C., 1940
Phytoplankton and planktonic protozoa of the offshore waters of the Gulf of Maine. Pt.II. Qualitative Composition of the Planktonic Flora. Trans. Am. Phil. Soc., n.s., 31(3):193-237, 13 text figs.

Oxytoxum gracile

Wood, E.J.F., 1963
Dinoflagellates in the Australian region. II. Recent collections. Comm. Sci. and Industr. Res. Org., Div. Fish. and Oceanogr., Techn. Paper, No. 14: 55 pp.

Oxytoxum laticeps

Wood, E.J.F., 1963
Dinoflagellates in the Australian region. II. Recent collections. Comm. Sci. and Industr. Res. Org., Div. Fish. and Oceanogr., Techn. Paper, No. 14: 55 pp.

Oxytoxum latum

Balech, Enrique, 1962
Tintinnoinea y Dinoflagellata del Pacifico segun material de las Expediciones NORPAC y DOWNWIND del Instituto Scripps de Oceanografia. Revista, Mus. Argentino Ciencias Nat. "Bernardino Rivadavia", Ciencias Zool., 7(1):1-253.

Oxytoxum ligusticum n. sp.

Rampi, Leopoldo, 1969.
Péridiniens, Hétérococcales et Pterospermales rares, intéressants ou nouveaux récoltés dans la Mer Ligurienne (Méditerranée occidentale).
Natura, Milano 60(4): 313-333

Oxytoxum ligusticum n.sp.

Rampi, L., 1941
Ricerche sul microplancton del Mare Ligure. 3. Le Heterodiniaceae e le Oxytoxaceae dell acque di Sanremo. Annali del Mus. Civico di Storia Naturale di Genova, 61:50-70, 2 pls.

Oxytoxum longiceps

Rampi, L., 1950
Péridiniens rares ou nouveaux pour le Pacifique Sud-Equatorial. Bull. Inst. Océan., Monaco, No. 974:11 pp., 26 textfigs.

Oxytoxum longiceps

Wood, E.J.F., 1963
Dinoflagellates in the Australian region. II. Recent collections. Comm. Sci. and Industr. Res. Org., Div. Fish. and Oceanogr., Techn. Paper, No. 14: 55 pp.

Oxytoxum longipes

Balech, Enrique, 1962
Tintinnoinea y Dinoflagellata del Pacifico segun material de las Expediciones NORPAC y DOWNWIND del Instituto Scripps de Oceanografia. Revista, Mus. Argentino Ciencias Nat. "Bernardino Rivadavia", Ciencias Zool., 7(1):1-253.

Oxytoxum longiceps

Margalef, R., 1949
Fitoplancton nerítico de la Costa Brava en 1947-48. Publ. Inst. Biol. Aplicada, 5: 41-51, 3 text figs.

Oxytoxum longiceps

de Sousa e Silva, E., 1956
Contribution à l'étude du microplancton de Dakar et des regions maritimes voisines. Bull. I.F.A.N., 8(2):335-371, 7 pls.

Oxytoxum longum

Wood, E.J.F., 1963
Dinoflagellates in the Australian region. II. Recent collections. Comm. Sci. and Industr. Res. Org., Div. Fish. and Oceanogr., Techn. Paper, No. 14: 55 pp.

Oxytoxum mediterraneum

Balech, Enrique, 1971
Microplancton del Atlantico ecuatorial oeste (Equalant 1) Publ. Serv. Hidrograf. Naval, Argentina H. 654: 103 pp., 122 figs.

Oxytoxum milneri

Balech, Enrique, 1971
Microplancton del Atlantico ecuatorial oeste (Equalant 1) Publ. Serv. Hidrograf. Naval, Argentina H. 654: 103 pp., 122 figs.

Oxytoxum milneri

Balech, Enrique, 1962
Tintinnoinea y Dinoflagellata del Pacifico segun material de las Expediciones NORPAC y DOWNWIND del Instituto Scripps de Oceanografia. Revista, Mus. Argentino Ciencias Nat. "Bernardino Rivadavia", Ciencias Zool., 7(1):1-253.

Oxytoxum milneri

Ercegovic, A., 1936
Etudes qualitative et quantitatives du phytoplancton dans les eaux cotières de l'Adriatique oriental moyen au cours de l'année 1934. Acta Adriatica 1(9):1-126

Oxytoxum Milneri

Halim, Youssef, 1960
Étude quantitative et qualitative du cycle écologique des dinoflagellés dans les eaux de Villefranche-sur-Mer. (1953-1955). Ann. Inst. Océanogr., Monaco, 38:123-232.

Oxytoxum milnerii

Hasle, Grethe Rytter, 1960
Phytoplankton and ciliate species from the Tropical Pacific. Skr. Norske Videnskaps-Akad., Oslo, 1. Mat.-Nat. Kl., 1960(2): 1-50.

Oxytoxum milneri

LeBour, M.V., 1925
The dinoflagellates of Northern Seas. The Marine Biological Association of the United Kingdom. Plymouth, 250 pp., 35 pls. 53 text figs.

Oxytoxum milneri

Murray, G., and F. G. Whitting, 1899
New Peridiniaceae from the Atlantic. Trans. Linn. Soc., London, Bot., ser 2, 5: 321-342, Pls. 27-33, 9 tables.

Oxytoxum milneri

Pavillard, J., 1916
Recherches sur les Peridiniens du Golfe du Lion. Mem. Univ. Montpellier. Trav. Inst. Bot., Univ. Montpellier. Serie mixte No.4, 70 pp., 3 pls., 15 text figs.

Oxytoxum milneri

Rampi, L., 1941
Ricerche sul microplancton del Mare Ligure. 3. Le Heterodiniaceae e le Oxytoxaceae dell acque di Sanremo. Annali del Mus. Civico di Storia Naturale di Genova, 61:50-70, 2 pls.

Oxytoxum milneri

Schröder, B., 1900
Phytoplankton des Golfes von Neapel nebst vergleichenden Ausblicken auf das atlantischen Ozean. Mitt. Zool. Stat. Neapel, 14:1-38.

Oxytoxum milneri

Wood, E.J.F., 1963
Dinoflagellates in the Australian region. II. Recent collections. Comm. Sci. and Industr. Res. Org., Div. Fish. and Oceanogr., Techn. Paper, No. 14: 55 pp.

Oxytoxum minutum

Rampi, L., 1948
Sur quelques Peridiniens rares ou intéressants du Pacifique subtropical (Recoltes Alain Gerbault). Bull. l'Inst. Ocean., Monaco, No.937: 7 pp., 8 text figs.

Oxytoxum minutum n.sp.

Rampi, L., 1941
Ricerche sul microplancton del Mare Ligure. 3. Le Heterodiniaceae e le Oxytoxaceae dell acque di Sanremo. Annali del Mus. Civico di Storia Naturale di Genova, 61:50-70, 2 pls.

Oxytoxum mitra

Wood, E.J.F., 1963
Dinoflagellates in the Australian region. II. Recent collections. Comm. Sci. and Industr. Res. Org., Div. Fish. and Oceanogr., Techn. Paper, No. 14: 55 pp.

Oxytoxum mucronatum n.sp.

Hope, B., 1954.
Floristic and taxonomic observations on marine phytoplankton from Nordavatn, near Bergen. Nytt Mag. f. Botanikk 2:149-153, 1 fig.

Oxytoxum nanum n.sp.

Halldal, P., 1953.
Phytoplankton investigations from Weather Ship M in the Norwegian Sea, 1948-49 (including observations during the "Armauer Hansen" cruise, July 1949). Hvalrådets Skrifter No. 38:91 pp., 20 tables, 21 textfigs.

Oxytoxum obesum n. sp.

Rampi, Leopoldo, 1969.
Péridiniens, Hétérococcales et Pterospermales rares, intéressants ou nouveaux récoltés dans la Mer Ligurienne (Méditerranée occidentale).
Natura, Milano 60(4): 313-333

Oxytoxum pachyderme

Wood, E.J.F., 1963
Dinoflagellates in the Australian region. II. Recent collections. Comm. Sci. and Industr. Res. Org., Div. Fish. and Oceanogr., Techn. Paper, No. 14: 55 pp.

Oxytoxum parvum

Wood, E.J.F., 1963
Dinoflagellates in the Australian region. II. Recent collections. Comm. Sci. and Industr. Res. Org., Div. Fish. and Oceanogr., Techn. Paper, No. 14: 55 pp.

Oxytoxum punctulatum

Hada, Yoshina, 1970. (1970)
The protozoan plankton of the Antarctic and Subantarctic seas.
Scient. Repts, Japan. Antarct. Res. Exped., (E)31:51 pp.

Oxytoxum radiosum n.sp.

Rampi, L., 1941
Ricerche sul microplancton del Mare Ligure. 3. Le Hatorodiniacee o le Oxytoxacee dell acque di Sanremo. Annali del Mus. Civico di Storia Naturale di Genova, 61:50-70, 2 pls.

Oxytoxum reticulatum

Balech, Enrique, 1971
Microplancton del Atlantico ecuatorial oeste (Equalant 1)
Publ. Serv. Hidrograf. Naval, Argentina H. 654: 103 pp., 122 figs.

Oxytoxum reticulatum

LeBour, M.V., 1925
The dinoflagellates of Northern Seas. The Marine Biological Association of the United Kingdom. Plymouth, 250 pp., 35 pls. 53 text figs.

Oxytoxum reticulatum

Lillick, L.C., 1940
Phytoplankton and planktonic protozoa of the offshore waters of the Gulf of Maine. Pt.II. Qualitative Composition of the Planktonic Flora. Trans. Am. Phil. Soc., n.s. 31(3):193-237, 13 text figs.

Oxytoxum reticulatum

Murray, G., and F. G. Whitting, 1899
New Peridiniaceae from the Atlantic.
Trans. Linn. Soc., London, Bot., ser 2, 5: 321-342, Pls. 27-33, 9 tables.

Oxytoxum reticulatum

Paulsen, O., 1908
XVIII Peridiniales. Nordisches Plankton, Bot. Teil: 1-124, 155 text figs.

Oxytoxum reticulatum

Pavillard, J., 1916
Recherches sur les Peridiniens du Golfe du Lion. Mem. Univ. Montpellier. Trav. Inst. Bot., Univ. Montpellier. Serie mixte No.4, 70 pp., 3 pls., 15 text figs.

Oxytoxum reticulatum

Pavillard, J., 1905
Recherches sur la flore pelagique (Phytoplankton) de l'Etang de Thau. Theses presentees a la Fac. Sci., Paris, 116 pp., 3 pls.

Oxytoxum robustum

Wood, E.J.F., 1963
Dinoflagellates in the Australian region. II. Recent collections.
Comm. Sci. and Industr. Res. Org.,Div. Fish. and Oceanogr., Techn. Paper, No. 14: 55 pp.

Oxytoxum sceptrum

Balech, Enrique, 1971
Microplancton del Atlantico ecuatorial oeste (Equalant 1)
Publ. Serv. Hidrograf. Naval, Argentina H. 654: 103 pp., 122 figs.

Oxytoxum sceptrum

Hasle, Grethe Rytter, 1960
Phytoplankton and ciliate species from the Tropical Pacific.
Skr. Norske Videnskaps-Akad., Oslo, 1. Mat.-Nat. Kl., 1960(2): 1-50.

Oxytoxum sceptrum

Pavillard, J., 1916
Recherches sur les Peridiniens du Golfe du Lion. Mem. Univ. Montpellier. Trav. Inst. Bot., Univ. Montpellier. Serie mixte No.4, 70 pp., 3 pls., 15 text figs.

Oxytoxum sceptrum

Schröder, B., 1900
Phytoplankton des Golfes von Neapel nebst vergleichenden Ausblicken auf das atlantischen Ozean. Mitt. Zool. Stat. Neapel, 14:1-38.

Oxytoxum sceptrum

Wood, E.J.F., 1963
Dinoflagellates in the Australian region. II. Recent collections.
Comm. Sci. and Industr. Res. Org.,Div. Fish. and Oceanogr., Techn. Paper, No. 14: 55 pp.

Oxytoxum scolopax

Balech, Enrique, 1971.
Microplancton de la campaña oceanographica: Productividad III.
Revta Mus. argent. Cienc. Nat. Bernardina Rivadavia, Hydrobiol. 3(1):1-202, 39 pls.

Oxytoxum scolopax

Balech, Enrique, 1962
Tintinnoinea y Dinoflagellata del Pacifico segun material de las Expediciones NORPAC y DOWNWIND del Instituto Scripps de Oceanografia.
Revista, Mus. Argentino Ciencias Nat. "Bernardino Rivadavia", Ciencias Zool., 7(1):1-253.

Oxytoxum scolopax

Dangeard, P., 1927
Phytoplankton de la croisière du "Sylvana". Ann. Inst. Ocean., Monaco, n.s., 4(8):286-401, 54 text figs. (fevrier-Juin 1913).

Oxytoxum scolopax

Ercegovic, A., 1936
Etudes qualitative et quantitatives du phytoplancton dans les eaux cotières de l'Adriatique oriental moyen au cours de l'année 1934. Acta Adriatica 1(9):1-126

Oxytoxum scolopax

LeBour, M.V., 1925
The dinoflagellates of Northern Seas. The Marine Biological Association of the United Kingdom. Plymouth, 250 pp., 35 pls. 53 text figs.

Oxytoxum scolopax

Lindemann, E., 1925
Neubeobachtungen an den Winter peridineen des Golfes von Neapel. Bot. Arch. 9:95-102, 19 text figs.

Oxytoxum scolopax

Marshall, S. M., 1933
The production of microplankton in the Great Barrier Reef Region. Brit. Mus. (N.H.) Great Barrier Reef Exped. 1928-29, Sci. Repts. II(5):111-157, 14 text figs.

Oxytoxum scolopax

Murray, G., and F. G. Whitting, 1899
New Peridiniaceae from the Atlantic.
Trans. Linn. Soc., London, Bot., ser 2, 5: 321-342, Pls. 27-33, 9 tables.

Oxytoxum scolopax

Paulsen, O., 1908
XVIII Peridiniales. Nordisches Plankton, Bot. Teil: 1-124, 155 text figs.

Oxytoxum scolopax

Pavillard, J., 1916
Recherches sur les Peridiniens du Golfe du Lion. Mem. Univ. Montpellier. Trav. Inst. Bot., Univ. Montpellier. Serie mixte No.4, 70 pp., 3 pls., 15 text figs.

Oxytoxum scolopax

Pavillard, J., 1905
Recherches sur la flore pelagique (Phytoplankton) de l'Etang de Thau. Theses presentees a la Fac. Sci., Paris, 116 pp., 3 pls.

Oxytoxum scolopax

Rampi, L., 1941
Ricerche sul microplancton del Mare Ligure. 3. Le Hatorodiniacee o le Oxytoxacee dell acque di Sanremo. Annali del Mus. Civico di Storia Naturale di Genova, 61:50-70, 2 pls.

Oxytoxum scolopax

Schröder, B., 1900
Phytoplankton des Golfes von Neapel nebst vergleichenden Ausblicken auf das atlantischen Ozean. Mitt. Zool. Stat. Neapel, 14:1-38.

Oxytoxum sphaeroides

Wood, E.J.F., 1963
Dinoflagellates in the Australian region. II. Recent collections.
Comm. Sci. and Industr. Res. Org.,Div. Fish. and Oceanogr., Techn. Paper, No. 14: 55 pp.

Oxytoxum sphaeroideum

Ercegovic, A., 1936
Etudes qualitative et quantitatives du phytoplancton dans les eaux cotières de l'Adriatique oriental moyen au cours de l'année 1934. Acta Adriatica 1(9):1-126

Oxytoxum sphaeroideum

LeBour, M.V., 1925
The dinoflagellates of Northern Seas. The Marine Biological Association of the United Kingdom. Plymouth, 250 pp., 35 pls. 53 text figs.

Oxytoxum sphaeroideum

Paulsen, O., 1908
XVIII Peridiniales. Nordisches Plankton, Bot. Teil: 1-124, 155 text figs.

Oxytoxum sphaeroideum

Pavillard, J., 1916
Recherches sur les Peridiniens du Golfe du Lion. Mem. Univ. Montpellier. Trav. Inst. Bot., Univ. Montpellier. Serie mixte No.4, 70 pp., 3 pls., 15 text figs.

Oxytoxum sphaeroideum

Rampi, L., 1948
Sur quelques Peridiniens rares ou intéressants du Pacifique subtropical (Recoltes Alain Gerbault). Bull. l'Inst. Ocean., Monaco, No.937: 7 pp., 8 text figs.

Oxytoxum sphaeroideum

Rampi, L., 1941
Ricerche sul microplancton del Mare Ligure. 3. Le Hatorodiniacee o le Oxytoxacee dell acque di Sanremo. Annali del Mus. Civico di Storia Naturale di Genova, 61:50-70, 2 pls.

Oxytoxum sphaeroideum

Schröder, B., 1900
Phytoplankton des Golfes von Neapel nebst vergleichenden Ausblicken auf das atlantischen Ozean. Mitt. Zool. Stat. Neapel, 14:1-38.

Oxytoxum spinosum

Halim, Youssef, 1960
Etude quantitative et qualitative du cycle écologique des dinoflagellés dans les eaux de Villefranche-sur-Mer. (1953-1955). Ann. Inst. Océanogr., Monaco, 38:123-232.

Oxytoxum spinosum n.sp.

Rampi, L., 1941
Ricerche sul microplancton del Mare Ligure. 3. Le Heterodiniaceae e le Oxytoxaceae dell'acque di Sanremo. Annali del Mus. Civico di Storia Naturale di Genova, 61:50-70, 2 pls.

Oxytoxum subulatum

Balech, Enrique, 1962
Tintinnoinea y Dinoflagellata del Pacifico segun material de las Expediciones NORPAC y DOWNWIND del Instituto Scripps de Oceanografia. Revista. Mus. Argentino Ciencias Nat. "Bernardino Rivadavia", Ciencias Zool., 7(1):1-253.

Oxytoxum tenuistriatum n.sp.

Rampi, L., 1941
Ricerche sul microplancton del Mare Ligure. 3. Le Heterodiniaceae e le Oxytoxaceae dell acque di Sanremo. Annali del Mus. Civico di Storia Naturale di Genova, 61:50-70, 2 pls.

Oxytoxum tesselatum

Balech, Enrique, 1971
Microplancton del Atlantico ecuatorial oeste (Equalant 1)
Publ. Serv. Hidrograf. Naval, Argentina H. 654: 103 pp., 122 figs.

Oxytoxum tesselatum

Dangeard, P., 1927
Phytoplankton de la croisière du "Sylvana". Ann. Inst. Ocean., Monaco, n.s., 4(8):286-401, 54 text figs. (février-juin 1913).

Oxytoxum tesselatum

Halim, Youssef, 1960
Étude quantitative et qualitative du cycle écologique des dinoflagellés dans les eaux de Villefranche-sur-Mer. (1953-1955). Ann. Inst. Océanogr., Monaco, 38:123-232.

Oxytoxum tessalatum

Matzenauer, L., 1933
Die Dinoflagellaten des indischen Ozeans (mit Ausnahme der Gattung Ceratium.) Bot. Arch. 35:437-510, 77 text figs., 2 charts.

Oxytoxum tesselatum

Pavillard, J., 1916
Recherches sur les Peridiniens du Golfe du Lion. Mem. Univ. Montpellier. Trav. Inst. Bot., Univ. Montpellier. Serie mixte No.4, 70 pp., 3 pls., 15 text figs.

Oxytoxum tessallatum

Murray, G., and F. G. Whitting, 1899
New Peridiniaceae from the Atlantic. Trans. Linn. Soc., London, Bot., ser 2, 5: 321-342, Pls. 27-33, 9 tables.

Oxytoxum tesselatum

Pavillard, J., 1905
Recherches sur la flore pelagique (Phytoplankton) de l'Etang de Thau. Theses presentees a la Fac. Sci., Paris, 116 pp., 3 pls.

Oxytoxum tesselatum

Rampi, L., 1941
Ricerche sul microplancton del Mare Ligure. 3. Le Heterodiniaceae e le Oxytoxaceae dell acque di Sanremo. Annali del Mus. Civico di Storia Naturale di Genova, 61:50-70, 2 pls.

Oxytoxum tesselatum

Schröder, B., 1900
Phytoplankton des Golfes von Neapel nebst vergleichenden Ausblicken auf das atlantischen Ozean. Mitt. Zool. Stat. Neapel, 14:1-38.

Oxytoxum turbo

Rampi, L., 1945
Osservazioni sulla distribuzione qualitativa del fitoplancton nel mare Mediterraneo. Atti della Soc. Ital. di Sci. Nat. 84:105-113.

Oxytoxum turbo

Rampi, L., 1942
Il Fitoplancton mediterraneo: Problemi ed affinita interoceaniche. Boll. di Pesca di Piscicoltura e di Idrobiologia, Anno 18, Fasc. 4:7-19.

Oxytoxum turbo var.

Wood, E.J.F., 1963
Dinoflagellates in the Australian region. II. Recent collections. Comm. Sci. and Industr. Res. Org., Div. Fish. and Oceanogr., Techn. Paper, No. 14: 55 pp.

Oxytoxum variabile

Hasle, Grethe Rytter, 1960
Phytoplankton and ciliate species from the Tropical Pacific. Skr. Norske Videnskaps-Akad., Oslo, 1. Mat.-Nat. Kl., 1960(2):1-50.

Oxytoxum variabile

Wood, E.J.F., 1963
Dinoflagellates in the Australian region. II. Recent collections. Comm. Sci. and Industr. Res. Org., Div. Fish. and Oceanogr., Techn. Paper, No. 14: 55 pp.

Pachydinium indicum n.sp.

Matzenauer, L., 1933
Die Dinoflagellaten des indischen Ozeans (mit Ausnahme der Gattung Ceratium.) Bot. Arch. 35:437-510, 77 text figs., 2 charts.

Pachydinium mediterraneum

Pavillard, J., 1916
Recherches sur les Peridiniens du Golfe du Lion. Mem. Univ. Montpellier. Trav. Inst. Bot., Univ. Montpellier. Serie mixte No.4, 70 pp., 3 pls., 15 text figs.

Pachydinium mediterraneum

Rampi, L., 1951.
Ricerche dul fitoplancton del Mare Ligure. 10) Peridiniale dalle acque di Sanremo. Pubbl. Centro Talassografico Terreno No. 6: Atti Acad. Ligure Sci. Lett. 7(1): 8 pp., Pls. 3-4.

Palaeophalacroma

Abe, Tohru H., 1967.
The armoured Dinoflagellata: II. Prorocentridae and Dinophysidae (B) - Dinophysis and its allied genera. Publs. Seto mar.biol.Lab., 15(1):37-78.

Palaeophalacroma unicinctum

Balech, Enrique, 1967.
Palaeophalacroma Schiller, otro miembro de la familia Cladopyxidae (Dinoflagellata). Neotropica, 13(42):105-112.

Paleophalacroma unicinctum

Ercegovic, A., 1936
Etudes qualitatives et quantitatives du phytoplancton dans les eaux cotières de l'Adriatique oriental moyen au cours de l'année 1934. Acta Adriatica 1(9):1-126

Palaeophalacroma unicinctum

Rampi, L. 1940.
Ricerche sul Fitoplancton del mare Ligure. II Le tecatali e le dinofisjali delle acque di Sanremo. Boll. Pesca, Piscicolt.,Idrobiol. (18) 16(2): 243-274, 56 figs.

Palaeophalacroma unicinctum

Rampi, L., 1940
Ricerche sul Fitoplancton del mare Ligure. Boll. di Pesca, di Piscicoltura e di Idrobiologa, 18(2):1-34, 56 text figs.

Palaeophalacroma unicinctum

Rampi, L., 1939
Peridiniens rares ou interessants recoltés dans la mer Ligure. Bull. Soc. Fran. de Microscopie, 8(2/3):106-112, 13 text figs.

Paradinium poucheti

Cachon, Jean, Monique Cachon et Chandra K. Pyne, 1968.
Structure et ultrastructure de Paradinium poucheti Chatton 1910, et position systématique des paradinides. Protistologica 4 (3) 303-311. Also in: Trav. Sta. zool. Villefranche-sur-mer. 28.1968

Parahistioneis

Abe, Tohru H., 1967.
The armoured Dinoflagellata: II. Prorocentridae and Dinophysidae (C)- Ornithocercus, Histioneis, Amphisolenia and others. Publs Seto mar.biol.Lab., 15(2):79-116.

Parahistioneis acuta

Böhm, A., 1934
Zur Verbreitung einiger Peridineen. Arch. Protistenk. 75:498-501, 6 text figs.

Parahistioneis acutiformis

Rampi, L., 1950.
Peridiniens rares ou nouveaux pour le Pacifique Sud-Equatorial. Bull. Inst. Ocean., Monaco, No. 974, 11 pp., 26 textfigs.

Parahistioneis acutiformis n.sp.

Rampi, L., 1947
Osservazioni sulle Histioneis (Peridinea) raccolte nel Mare Ligure presso Sanremo. Bull. Inst. Océan., Monaco, No.920:1-16, 17 text figs.

Parahistioneis conica n.sp.

Böhm, A., 1934
Zur Verbreitung einiger Peridineen. Arch. Protistenk. 75:498-501, 6 text figs.

Parahistioneis crateriformis

Wood, E.J.F., 1963
Dinoflagellates in the Australian region. II. Recent collections. Comm. Sci. and Industr. Res. Org., Div. Fish. and Oceanogr., Techn. Paper, No. 14: 55 pp.

Parahistioneis diomedeae

Kofoid, C. A. and T. Skogsberg, 1928
XXXV. The Dinoflagellata: The Dinophysiodae. Reports on the scientific results of the expedition to the Eastern Tropical Pacific, in charge of Alexander Agassiz, by the U. S. Fish Commission Steamer "Albatross" from October 1904 to March 1905----. Mem. M. C. Z. 51:766 pp., 31 pls.

Parahistioneis garretti

Kofoid, C. A. and T. Skogsberg, 1928
XXXV. The Dinoflagellata: The Dinophysiodae. Reports on the scientific results of the expedition to the Eastern Tropical Pacific, in charge of Alexander Agassiz, by the U. S. Fish Commission Steamer "Albatross" from October 1904 to March 1905----. Mem. M. C. Z. 51:766 pp., 31 pls.

Parahistioneis garretti

Wood, E.J.F., 1963
Dinoflagellates in the Australian region. II. Recent collections. Comm. Sci. and Industr. Res. Org., Div. Fish. and Oceanogr., Techn. Paper, No. 14: 55 pp.

Parahistioneis gascoynensis n.sp.

Wood, E.J.F., 1963
Dinoflagellates in the Australian region. II. Recent collections. Comm. Sci. and Industr. Res. Org., Div. Fish. and Oceanogr., Techn. Paper, No. 14: 55 pp.

Parahistioneis karsteni

Kofoid, C. A. and T. Skogsberg, 1928
XXXV. The Dinoflagellata: The Dinophysiodae. Reports on the scientific results of the expedition to the Eastern Tropical Pacific, in charge of Alexander Agassiz, by the U. S. Fish Commission Steamer "Albatross" from October 1904 to March 1905----. Mem. M. C. Z. 51:766 pp., 31 pls.

Parahistioneis Karsteni
Rampi, L., 1947
Osservazioni sulle Histioneis (Peridinea) raccolte nel Mare Ligure presso Sanremo. Bull. Inst. Océan., Monaco, No.920:1-16, 17 text figs.

Parahistioneis Karsteni
Rampi, L., 1945
Osservazioni sulla distribuzione qualitativa del fitoplancton nel mare Mediterraneo. Atti della Soc. Ital. di Sci. Nat. 84:105-113.

Parahistioneis karsteni
Rampi, L., 1942
Il Fitoplancton mediterraneo: Problemi ed affinita interoceaniche. Boll. di Pesca di Piscicoltura e di Idrobiologia, Anno 18, Fasc. 4:7-19.

Parahistioneis karsteni
Rampi, L. 1940
Ricerche sul Fitoplancton del mare Ligure. II Le tecatali e le dinofisiali delle acque di Sanremo. Boll. Pesca, Piscicolt., Idrobiol. (18) 16(2): 243-274, 56 figs.

Parahistioneis karsteni
Rampi, L., 1940
Ricerche sul Fitoplancton del mare Ligure. Boll. di Pesca, di Piscicoltura e di Idrobiologa. 18(2):1-34, 56 text figs.

Parahistioneis mediterranea
Rampi, L., 1945
Osservazioni sulla distribuzione qualitativa del fitoplancton nel mare Mediterraneo. Atti della Soc. Ital. di Sci. Nat. 84:105-113.

Parahistioneis pachypus n.sp.
Böhm, A., 1934
Zur Verbreitung einiger Peridineen. Arch. Protistenk. 75:498-501, 6 text figs.

Parahistioneis para
Balech, Enrique, 1962
Tintinnoinea y Dinoflagellata del Pacifico segun material de las Expediciones NORPAC y DOWNWIND del Instituto Scripps de Oceanografia. Revista, Mus. Argentino Ciencias Nat. "Bernardino Rivadavia", Ciencias Zool., 7(1):1-253.

Parahistioneis para
Kofoid, C. A. and T. Skogsberg, 1928
XXV. The Dinoflagellata: The Dinophysiodae. Reports on the scientific results of the expedition to the Eastern Tropical Pacific, in charge of Alexander Agassiz, by the U. S. Fish Commission Steamer "Albatross" from October 1904 to March 1905----. Mem. M. C. Z. 51:766 pp., 31 pls.

Parahistioneis paraformis n.sp.
Kofoid, C. A. and T. Skogsberg, 1928
XXV. The Dinoflagellata: The Dinophysiodae. Reports on the scientific results of the expedition to the Eastern Tropical Pacific, in charge of Alexander Agassiz, by the U. S. Fish Commission Steamer "Albatross" from October 1904 to March 1905----. Mem. M. C. Z. 51:766 pp., 31 pls.

Parahistioneis paraformis
Rampi, L., 1950.
Péridiniens rares ou nouveaux pour le Pacifique Sud-Equatorial. Bull. Inst. Océan., Monaco, No. 974, 11 pp., 26 textfigs.

Parahistioneis paraformis
Rampi, L., 1947
Osservazioni sulle Histioneis (Peridinea) raccolte nel Mare Ligure presso Sanremo. Bull. Inst. Océan., Monaco, No.920:1-16, 17 text figs.

Parahistioneis paraformis
Wood, E.J.F., 1963
Dinoflagellates in the Australian region. II. Recent collections. Comm. Sci. and Industr. Res. Org., Div. Fish. and Oceanogr., Techn. Paper, No. 14: 55 pp.

Parahistioneis reticulata
Kofoid, C. A. and T. Skogsberg, 1928
XXV. The Dinoflagellata: The Dinophysiodae. Reports on the scientific results of the expedition to the Eastern Tropical Pacific, in charge of Alexander Agassiz, by the U. S. Fish Commission Steamer "Albatross" from October 1904 to March 1905----. Mem. M. C. Z. 51:766 pp., 31 pls.

Parahistioneis reticulata
Wood, E.J.P., 1954.
Dinoflagellates in the Australian region. Australian J. Mar. Freshwater Res., 5(2):171-351.

Parahistioneis rotundata
Käsler, R., 1938
Die Verbreitung der Dinophysiales im Sudatlantischen Ozean. Wiss. Ergeb. Deutschen Atlantischen Expedition----"Meteor" 1925-1927, 12(2):162-237, text figs. 85-118.

Parahistioneis rotundata
Kofoid, C. A. and T. Skogsberg, 1928
XXV. The Dinoflagellata: The Dinophysiodae. Reports on the scientific results of the expedition to the Eastern Tropical Pacific, in charge of Alexander Agassiz, by the U. S. Fish Commission Steamer "Albatross" from October 190 to March 1905----. Mem. M. C. Z. 51:766 pp., 31 pls.

Parahistioneis rotundata
Rampi, L., 1948
Sur quelques Peridiniens rares ou interessants du Pacifique subtropical (Recoltes Alain Gerbault). Bull. l'Inst. Ocean., Monaco, No.937: 7 pp., 8 text figs.

Parahistioneis sphaeroidea n.sp.
Rampi, L., 1947
Osservazioni sulle Histioneis (Peridinea) raccolte nel Mare Ligure presso Sanremo. Bull. Inst. Océan., Monaco, No.920:1-16, 17 text figs.

Parahistioneis rotundata
Wood, E.J.F., 1954.
Dinoflagellates in the Australian region. Australian J. Mar. Freshwater Res., 5(2):171-351.

Parahistioneis rotundata
Wood, E.J.F., 1963
Dinoflagellates in the Australian region. II. Recent collections. Comm. Sci. and Industr. Res. Org., Div. Fish. and Oceanogr., Techn. Paper, No. 14: 55 pp.

Paraperidinium Sectioaredata
Matzenauer, L., 1933
Die Dinoflagellaten des indischen Ozeans (mit Ausnahme der Gattung Ceratium.) Bot. Arch. 35:437-510, 77 text figs., 2 charts.

Parrocelia n.gen.
Gourret, P., 1883
Sur les Peridiniens du Golfe de Marseille. Ann. du Musee d'hist. Nat., Marseille, Zool., 1 (Mme. 8):1-114, 4 pls.

Pavillardinium (Murrayella) biconicum
Rampi, L., 1948
Sur quelques Peridiniens rares ou interessants du Pacifique subtropical (Recoltes Alain Gerbault). Bull. l'Inst. Ocean., Monaco, No.937: 7 pp., 8 text figs.

Pavillardinium (Murrayella) globosum
Rampi, L., 1950.
Péridiniens rares ou nouveaux pour le Pacifique Sud-Equatorial. Bull. Inst. Océan., Monaco, No. 974:11 pp., 26 textfigs.

Pavillardinium (Murrayella) pacificum n.sp.
Rampi, L., 1950.
Péridiniens rares ou nouveaux pour le Pacifique Sud-Equatorial. Bull. Inst. Océan., Monaco, No. 974:11 pp., 26 textfigs.

Pavillardinium (Murrayella) splendidum
Rampi, L., 1950.
Péridiniens rares ou Nouveaux pour le Pacifique Sud-Equatorial. Bull. Inst. Océan., Monaco, No. 974:11 pp., 26 textfigs.

Peranemopsis striata n.gen., n.sp.
Lackey, J.B., 1940.
Some new flagellates from the Woods Hole area. Amer. Midl. Nat., 23(2):463-471.

Peridiniopsis asymmetrica
Dangeard, P., 1927
Phytoplankton de la croisière du "Sylvana". Ann. Inst. Ocean., Monaco, n.s., 4(8):286-401, 54 text figs. (Feirer-Juin 1913).

Peridiniopsis asymmetrica
Dangeard, P., 1926
Description des Péridiniens Testacés recueillis par la Mission Charcot pendent le mois d'Aout 1924. Ann. Inst. Ocean. n.s. 3(7):307-334, 15 text figs.

Peridinopsis asymmetrica
LeBour, M.V., 1925
The dinoflagellates of Northern Seas. The Marine Biological Association of the United Kingdom, Plymouth, 250 pp., 35 pls.,53 text figs.

Peridiniopsis asymmetrica
Mangin, M. L., 1912
Phytoplancton de la croisière du "René" dans l'Atlantique (Septembre 1908). Ann. Inst. Ocean., n.s., 4(1):1-66, 2 pls., 41 text figs., 2 tables.

Peridiniopsis asymmetrica
Margalef, R., 1949
Fitoplancton nerítico de la Costa Brava en 1947-48. Publ. Inst. Biol. Aplicada, 5: 41-51, 3 text figs.

Peridiniopsis asymmetrica
Matzenauer, L., 1933
Die Dinoflagellaten des indischen Ozeans (mit Ausnahme der Gattung Ceratium.) Bot. Arch. 35:437-510, 77 text figs., 2 charts.

Peridiniopsis asymmetrica?
Vives, F., and A. Planas, 1952.
Plancton recogido por los laboratorios costeros. VI. Fitoplancton de las costas de Vinaroz, islas Columbretes y alrededores de la desembocadura del Elroa. Publ. Inst. Biol. Aplic. 11:141-156, 19 textfigs.

Peridiniopsis rotunda
Gran, H.H., and T. Braarud, 1935
A quantitative study of the phytoplankton in the Bay of Fundy and the Gulf of Maine (including observations on hydrography, chemistry, and turbidity). J. Biol. Bd., Canada, 1(5):279-467, 69 text figs.

Peridinopsis rotunda
LeBour, M.V., 1925
The dinoflagellates of Northern Seas. The Marine Biological Association of the United Kingdom, Plymouth, 250 pp., 35 pls.,53 text figs.

Peridiniopsis rotunda
Morse, D.C., 1947
Some observations on seasonal variations in plankton population Patuxent River, Maryland 1943-1945. Bd. Nat. Res., Publ. No.65, Chesapeake Biol. Lab., 31, 3 figs.

Peridinium sp.
Balech, Enrique, 1971.
Microplancton de la campaña oceanographica: Productividad III. Revta Mus. argent. Cienc. Nat. Bernadina Rivadavia, Hydrobiol. 3(1):1-202, 39 pls.

Peridinium

Barrows, A. L., 1918.
The significance of skeletal variation in the genus Peridinium. Univ. Calif. Publ. Zool. 18(15): 397-478, 19 textfigs., Pls. 17-20.

Peridinium sp.

Braarud, T., 1945
A phytoplankton survey of the polluted waters of inner Oslo Fjord. Hvalrådets Skrifter, No.28, 142 pp., 19 text figs., 17 tables.

Peridinium spp.

Braarud, T., 1939
Observations on the phytoplankton of the Oslo Fjord, March-April, 1937. Nytt Magasin for Naturvidenskapene, 80:211-218, 1 text fig.

Peridinium sp.

de Sousa e Silva, E., 1956.
Contribution à l'étude du microplancton de Dakar et des regions maritimes voisines. Bull. I.F.A.N., 8(2):335-371, 7 pls.

Peridinium

Diwald, K., 1939.
Ein Beitrag zur Variabilität und Systematik der Gattung Peridinium. Arch. f. Protistenk. 93:121-184.

Peridinium sp.

Gilbert, J.Y., and W.E. Allen, 1943
The phytoplankton of the Gulf of California obtained by the "E.W. Scripps" in 1939 and 1940. J. Mar. Res. V(2):89-110, figs.30-31.

Peridinium sp.

Kokubo, S., and S. Sato, 1947
Plankton in Jū-San Gata. Physiol. and Ecol. (Japan) 1(4):1-16, 3 text figs., tables.

Peridinium spp.

Lillick, L.C., 1937
Seasonal studies of the phytoplankton off Woods Hole, Massachusetts. Biol. Bull. LXXIII (3):488-503, 3 text figs.

Peridinium

Loeblich, Alfred R. III, 1968.
A new marine dinoflagellate genus, Cachonina, in axenic culture from the Salton Sea, California, with remarks on the genus Peridinium.
Proc. Biol. Soc. Wash. 81:91-96.

Peridinium sp.

Massutí Algamora, M., 1949
Estudio de diez y seis muestras de plancton del Golfo de Nápoles. Publ. Inst. Biol. Appl. 5:85-94, 1 fold-in table.

Peridinium

Mead, A.D. 1898
Peridinium and the "redwater" in Narragansett Bay. Science, Vol. 8, pp. 707-709.

Peridinium sp.

Meunier, A., 1919
Microplancton de la Mer Flamande 3. Les Péridiniens. Mem. Mus. Roy. Hist. Nat., Belgique 8(1):1-116, Pls. XV-XXI.

Peridinium sp.

Pettersson, H., F. Gross, and F. Koczy, 1939
Large scale plankton cultures. Medd. från Oceanografiska Institutet i Göteborg No.3 (Göteborgs Kungl. vetenskaps-och Vitterhets-Samhälles Handlingar Femte Följden. Ser. B) Vol.6(13):1-24.
1. The plankton shaft. H. Pettersson.
2. Experiment with phytoplankton. F. Gross and F. Koczy
3. Experiments with zooplankton.

Peridinium

Smith, H.M., 1908
Peridinium (editorial) Philippine J. Sci., A, 3(3):187-188.

poisonous organisms

Peridinium spp.

Wall, David, and Barrie Dale, 1968.
Modern dinoflagellate cysts and evolution of the Peridiniales. Micropaleontology, 14(3):265-304.

Peridinium abei

Wood, E.J.F., 1954
Dinoflagellates in the Australian region. Australian J. Mar. Freshwater Res., 5(2):171-351.

Peridinium acanthophorum n. spp.

Argentina, Secretaria de Marina, Servicio de Hidrografía Naval, 1962.
Plancton de las campanas oceanograficas DRAKE I y II. Publico, H. 627:57.

Peridinium achromaticum

Becheler, B., 1952.
Recherches sur les Peridiniens. Bull. Biol., France Belg., Suppl., 36:1-149.

Peridinium achromaticum

Gran, H.H., and T. Braarud, 1935
A quantitative study of the phytoplankton in the Bay of Fundy and the Gulf of Maine (including observations on hydrography, chemistry, and turbidity). J. Biol. Bd. Canada, 1(5):279-467, 69 text figs.

Peridinium achromaticum

LeBour, M.V., 1925
The dinoflagellates of Northern Seas. The Marine Biological Association of the United Kingdom, Plymouth, 250 pp., 35 pls., 53 text figs.

Peridinium achromaticum

Levander, K.M., 1947
Plankton gesammelt in den Jahren 1899-1910 an den Küsten Finnlands. Finnländische Hydrographisch-Biologische Untersuchungen (aus dem Wasserbiologischen Laboratorin der Societas Scientiarum Fennica) No.11: 40 pp., 6 diagrams, 13 pls., tables.

Peridinium achromaticum

Lillick, L.C., 1940
Phytoplankton and planktonic protozoa of the offshore waters of the Gulf of Maine. Pt.II. Qualitative Composition of the Planktonic Flora. Trans. Am. Phil. Soc., n.s., 31(3):193-237, 13 text figs.

Peridinium achromaticum

Matzenauer, L., 1933
Die Dinoflagellaten des indischen Ozeans (mit Ausnahme der Gattung Ceratium.) Bot. Arch. 35:437-510, 77 text figs., 2 charts.

Peridinium achromaticum

Morse, D.C., 1947
Some observations on seasonal variations in plankton population Patuxant River, Maryland 1943-1945. Bd. Nat. Res., Publ. No.65, Chesapeake Biol. Lab., 31, 3 figs.

Peridinium achromaticum

Paulsen, O., 1908
XVIII Peridiniales. Nordisches Plankton, Bot. Teil: 1-124, 155 text figs.

Peridinium achromatum

Rumkówna, A., 1948
List of the phytoplankton species occurring in the superficial water layers in the Gulf of Gdańsk. Bull. Lab. mar., Gdynia, No. 4: 139-141 with tables in back.

Peridinium aciculiferum

Rumkówna, A., 1948
List of the phytoplankton species occurring in the superficial water layers in the Gulf of Gdańsk. Bull. Lab. mar., Gdynia, No. 4: 139-141 with tables in back.

Peridinium acutipes n.sp.

Dangeard, P., 1927
Phytoplankton de la croisière du "Sylvana". Ann. Inst. Océan. Monaco, n.s., 4(8):286-401, 54 text figs. (Feirer-Juin 1913).

Peridinium acutipes

Matzenauer, L., 1933
Die Dinoflagellaten des indischen Ozeans (mit Ausnahme der Gattung Ceratium.) Bot. Arch. 35:437-510, 77 text figs., 2 charts.

Peridinium adeliense

Argentina, Secretaria de Marina, Servicio de Hidrografía Naval, 1962.
Plancton de las campanas oceanograficas DRAKE I y II. Publico, H. 627:57.

Peridinium adeliense

Balech, Enrique, and Sayed Z. El-Sayed, 1965.
Microplankton of the Weddell Sea.
In: Biology of Antarctic seas. II.
Antarctic Res. Ser., Amer. Geophys. Union, 5:107-124.

Peridinium adense n.sp.

Matzenauer, L., 1933
Die Dinoflagellaten des indischen Ozeans (mit Ausnahme der Gattung Ceratium.) Bot. Arch. 35:437-510, 77 text figs., 2 charts.

Peridinium adriaticum

Dangeard, P., 1927
Phytoplankton de la croisière du "Sylvana". Ann. Inst. Océan. Monaco, n.s., 4(8):286-401, 54 text figs. (Feirer-Juin 1913).

Peridinium adriaticum

Dangeard, P., 1926
Description des Péridiniens Testacés recueillis para la Mission Charcot pendent le mois d'Aout 1924. Ann. Inst. Océan. n.s. 3(7):307-334, 15 text figs.

Peridinium adriaticum

Forti, A., 1922
Ricerche sulla flora pelagica (fitoplancton) di Quarto dei Mille. Mem. R. Com. Talass. Ital. 97:248 pp., 13 pls.

Peridinium adriaticum

Pavillard, J., 1916
Recherches sur les Peridiniens du Golfe du Lion. Mem. Univ. Montpellier. Trav. Inst. Bot., Univ. Montpellier. Serie mixte No.4, 70 pp., 3 pls., 15 text figs.

Peridinium adultum n. sp.

Balech, Enrique, 1971.
Microplancton de la campaña oceanographica: Productividad III.
Revta Mus. argent. Cienc. Nat. Bernadina Rivadavia, Hydrobiol. 3(1):1-202, 39 pls.

Peridinium aequalis

Paulsen, O., 1908
XVIII Peridiniales. Nordisches Plankton, Bot. Teil: 1-124, 155 text figs.

Peridinium aequatoriale n. sp.

Balech, Enrique, 1971
Microplancton del Atlantico ecuatorial oeste (Equalant 1)
Publ. Serv. Hidrograf. Naval, Argentina H. 654: 103 pp., 122 figs.

Peridinium affine

Argentina, Secretaria de Marina, Servicio de Hidrografía Naval, 1962.
Plancton de las campanas oceanograficas DRAKE I y II. Publico, H. 627:57.

Peridinium affine
Balech, Enrique, 1971.
Microplancton de la campaña oceanographica:
Productividad III.
Revta Mus. argent. Cienc. Nat. Bernadina
Rivadavia, Hydrobiol. 3(1):1-202, 39 pls.

Peridinium africanoides n.sp.
Dangeard, P., 1927
Phytoplankton de la croisière du
"Sylvana". Ann. Inst. Ocean., Monaco, n.s.,
4(8):286-401, 54 text figs, (février-Juin 1913).

Peridinium africanoides
Matzenauer, L., 1933
Die Dinoflagellaten des indischen Ozeans
(mit Ausnahme der Gattung Ceratium.) Bot.
Arch. 35:437-510, 77 text figs., 2 charts.

Peridinium africanum
Wood, E.J.F., 1954.
Dinoflagellates in the Australian region.
Australian J. Mar. Freshwater Res., 5(2):171-351.

Peridinium aliferum?
Balech, Enrique, 1971.
Microplancton de la campaña oceanographica:
Productividad III.
Revta Mus. argent. Cienc. Nat. Bernadina
Rivadavia, Hydrobiol. 3(1):1-202, 39 pls.

Peridinium americanum n.sp.
Gran, H.H., and T. Braarud, 1935
A quantitative study of the phyto-
plankton in the Bay of Fundy and the
Gulf of Maine (including observations
on hydrography, chemistry, and turbidity).
J. Biol. Bd., Canada, 1(5):279-467, 69
text figs.

Peridinium americanum
Lillick, L.C., 1940
Phytoplankton and planktonic
protozoa of the offshore waters
of the Gulf of Maine. Pt.II.
Qualitative Composition of the
Planktonic Flora. Trans. Am.
Phil. Soc., n.s., 31(3):193-237,
13 text figs.

Peridinium americanum
Lillick, L.C., 1938
Preliminary report of the phyto-
plankton of the Gulf of Maine. Am.
Mid. Nat. 20(3):624-640, 1 text figs
37 tables.

Peridinium amplum n.sp.
Matzenauer, L., 1933
Die Dinoflagellaten des indischen Ozeans
(mit Ausnahme der Gattung Ceratium.) Bot.
Arch. 35:437-510, 77 text figs., 2 charts.

Peridinium ampulla nov. nom.
Balech, Enrique, 1971.
Microplancton de la campaña oceanographica:
Productividad III.
Revta Mus. argent. Cienc. Nat. Bernadina
Rivadavia, Hydrobiol. 3(1):1-202, 39 pls.

Peridinium ampulliforme n.sp.
Wood, E.J.F., 1954.
Dinoflagellates in the Australian region.
Australian J. Mar. Freshwater Res., 5(2):171-351.

Peridinium anguipes n.sp.
Balech, Enrique, 1967.
Dinoflagelados nuevos o interesantes del Golfo
de Mexico y Caribe.
Revta Mus.argent.Cienc.nat.Bernardino
Rivadavia Inst.nac.Invest.Cienc.nat.,Hidrobiol.
2(3):77-126.

Peridinium angustatum n.sp.
Dangeard, P., 1927
Phytoplankton de la croisière du
"Sylvana". Ann. Inst. Ocean., Monaco, n.s.,
4(8):286-401, 54 text figs, (février-Juin 1913).

Peridinium anomaloplacum n.sp.
Balech, Enrique, 1964
El plancton de Mar del Plata durante el periodo
1961-1962 (Buenos Aires, Argentina).
Bol. Inst. Biol. Mar., Mar del Plata, Argentina, No. 4:49 pp.

Peridinium anomaloplaxum n.sp.
Balech, Enrique, 1964.
El plancton de Mar del Plata durante el periodo
1961-1962.
Bol. Inst. Biol. Mar., Buenos Aires, (4):56 pp.
49 pp. + plates

Peridinium antarcticum
Balech, Enrique, 1958.
Plancton de la Campaña Antartica Argentina,
1954-1955.
Physis 21(60):75-108.

Peridinium antarcticum
Balech, E., 1947
Contribution al conocimiento del plancton
antartico. Plankton del Mar de Bellinghausen.
Physis 20:75-91, 76 figs. on 8 pls.

Peridinium antarcticum
Balech, Enrique, 1947
Contribucion al conocimiento del Plancton
antartico Plancton del Mar de Bellingshausen.
Physis 22(56):75-91, 7 pls. with 76 figs.

Peridinium anthonyi
LeBour, M.V., 1925
The dinoflagellates of Northern Seas. The
Marine Biological Association of the United
Kingdom, Plymouth, 250 pp., 35 pls., 53 text
figs.

Peridinium applanatum
Balech, Enrique, 1971.
Microplancton de la campaña oceanographica:
Productividad III.
Revta Mus. argent. Cienc. Nat. Bernadina
Rivadavia, Hydrobiol. 3(1):1-202, 39 pls.

Peridinium applanatum
Balech, Enrique, 1958.
Plancton de la Campaña Antartica Argentina,
1954-1955.
Physis, 21(60):75-108.

Peridinium applanatum
Hada, Yoshina, 1970.   (1970)
The protozoan plankton of the Antarctic and
Subantarctic seas.
Scient. Repts, Japan. Antarct. Res. Exped.,
(E)31:51 pp.

Peridinium applanatum n.sp.
Mangin, L., 1915
Phytoplancton de L'Antartiqua. Deuxieme
Exped. Ant. Francaise (1908-1910), 95 pp., 3 pls.,
58 text figs.

Peridinium archiovatum
Argentina, Secretaria de Marina, Servicio de
Hidrografia Naval, 1962.
Plancton de las campanas oceanograficas DRAKE I
y II.
Publico, H. 627:57.

Peridinium archiovatum, n.sp.
Balech, Enrique, 1958.
Plancton de la Campana Antartica Argentina,
1954-1955.
Physis, 21(60):75-108.

Peridinium archiovatum
Balech, Enrique, and Sayed Z. El-Sayed, 1965.
Microplankton of the Weddell Sea.
In: Biology of Antarctic seas. II.
Antarctic Res. Ser., Amer. Geophys. Union, 5:107-124.

Peridinium aspidiotum n. sp.
Balech, Enrique, 1964
El plancton de Mar del Plata durante el
periodo 1961-1962 (Buenos Aires, Argentina).
Bol. Inst. Biol. Mar., Mar del Plata,
Argentina, No. 4:49 pp.

Peridinium assymetrica n.sp.
Mangin, M. L., 1912
Phytoplankton de la croisière du "René"
dans l'Atlantique (Septembre 1908). Ann.
Inst. Ocean., n.s., 4(1):1-66, 2 pls., 41
text figs., 2 tables.

Peridinium asymmetricum
Matzenauer, L., 1933
Die Dinoflagellaten des indischen Ozeans
(mit Ausnahme der Gattung Ceratium.) Bot.
Arch. 35:437-510, 77 text figs., 2 charts.

Peridinium avellana
LeBour, M.V., 1925
The dinoflagellates of Northern Seas. The
Marine Biological Association of the United
Kingdom, Plymouth, 250 pp., 35 pls.,53 text
figs.

Peridinium avellana
Wall, David, and Barrie Dale, 1968.
Modern dinoflagellate cysts and evolution of the
Peridiniales.
Micropaleontology, 14(3):265-304.

Peridinium avellana
Wood, E.J.F., 1963
Dinoflagellates in the Australian region.
II. Recent collections.
Comm. Sci. and Industr. Res. Org., Div.
Fish. and Oceanogr., Techn. Paper, No. 14:
55 pp.

Peridinium bahamense (bioluminescense)
Margalef, Ramon, 1957.
Fitoplancton de las costas de Puerto Rico.
Inv. Pesq., Barcelona, 6:39-52.

Peridinium balticum
Becheler, B., 1952.
Recherches sur les Peridiniens.
Bull. Biol., France Belg., Suppl., 36:1-149,
75 textfigs.

Peridinium balticum
Lebour, M.V., 1925.
The dinoflagellates of northern seas. Marine
Biological Association, United Kingdom, 250 pp.,
35 pls., 53 textfigs.

Peridinium balticum
Levander, K.M., 1947
Plankton gesammelt in den Jahren
1899-1910 an den Küsten Finnlands.
Finnländische Hydrographisch-Biologishhe
Untersuchungen (aus dem Wasserbiologis-
chen Laboratorin der Societas Scientiarum
Fennica) No.11: 40 pp., 6 diagrams, 13 pls.,
tables.

Peridinium balticum
Paulsen, O., 1908
XVIII Peridiniales. Nordisches Plankton,
Bot. Teil: 1-124, 155 text figs.

Peridinium balticum
Rumkówna, A., 1948
[List of the phytoplankton species occur-
ring in the superficial water layers in the
Gulf of Gdańsk] Bull. Lab. mar., Gdynia,
No. 4: 139-141 with tables in back.

Peridinium balticum
Tomas, Ronald N., and Eleanor R. Cox, 1973.
Observations on the symbiosis of
Peridinium balticum and its intracellular
alga. I. Ultrastructure.
J. Phycol. 9(3): 304-323

Peridinium balticum

Tomas, Ronald N., Glenn R. Cox and Karen A. Steidinger 1973
Peridinium balticum (Levander) Lemmermann, an unusual dinoflagellate with a mesocaryotic and an eucaryotic nucleus.
J. Phycol. 9(1): 91-98.

Peridinium bellulum n. sp.

Balech, Enrique, 1971.
Microplancton de la campaña oceanographica: Productividad III.
Revta Mus. argent. Cienc. Nat. Bernadina Rivadavia, Hydrobiol. 3(1):1-202, 39 pls.

Peridinium biconicum n.sp.

Dangeard, P., 1927.
Péridiniens nouveaux ou peu connus de la croisière du "Sylvana". Bull. Inst. Océan., Monaco, No. 491:16 pp., 9 textfigs.

Peridinium biconicum

Dangeard, P., 1927
Phytoplankton de la croisière du "Sylvana". Ann. Inst. Ocean., Monaco, n.s., 4(8):286-401, 54 text figs. (Feirer-Juin 1913).

Peridinium biconicum

Matzenauer, L., 1933
Die Dinoflagellaten des indischen Ozeans (mit Ausnahme der Gattung Ceratium.) Bot. Arch. 35:437-510, 77 text figs., 2 charts.

Peridinium bipes

Braarud, T., and Adam Bursa, 1939
On the phytoplankton of the Oslo Fjord, 1933-1934. Hvalrådets Skr. No.19:1-63; 9 text figs. Reviewed. J. du. Cons. 14(3): 418-420. A.C. Gardiner.

Peridinium bipes

Lindemann, E., 1924
Peridineen aus dem goldenen Horn und dem Bosphorus. Bot. Arch. 5:216-233, 98 text figs.

Peridinium bisintercalares n.f.

Graham, H. W., 1942
Studies in the morphology, taxonymy, and ecology of the Peridiniales. Sci. Res. Cruise VII of the Carnegie, 1928-1929---Biol. III(542): 129 pp., 67 figs.

Peridinium Blackmani n.sp.

Murray, G., and F. G. Whitting, 1899
New Peridiniaceae from the Atlantic. Trans. Linn. Soc., London, Bot., ser 2, 5: 321-342, Pls. 27-33, 9 tables.

Peridinium breve

Braarud, T., 1945
A phytoplankton survey of the polluted waters of inner Oslo Fjord. Hvalrådets Skrifter, No.28, 142 pp., 19 text figs., 17 tables.

Peridinium breve

Dangeard, P., 1927
Phytoplankton de la croisière du "Sylvana". Ann. Inst. Ocean., Monaco, n.s., 4(8):286-401, 54 text figs. (février-Juin 1913).

Peridinium breve

Gran, H.H., and T. Braarud, 1935
A quantitative study of the phytoplankton in the Bay of Fundy and the Gulf of Maine (including observations on hydrography, chemistry, and turbidity). J. Biol. Bd. Canada, 1(5):279-467, 69 text figs.

Peridinium breve

LeBour, M.V., 1925.
The dinoflagellates of northern seas. Marine Biological Association, United Kingdom, 250 pp., 35 pls., 53 textfigs.

Peridinium breve

Lillick, L.C., 1940
Phytoplankton and planktonic protozoa of the offshore waters of the Gulf of Maine. Pt.II. Qualitative Composition of the Planktonic Flora. Trans. Am. Phil. Soc., n.s., 31(3):193-237, 13 text figs.

Peridinium breve

Lillick, L.C., 1938
Preliminary report of the phytoplankton of the Gulf of Maine. Am. Mid. Nat. 20(3):624-640, 1 text figs 37 tables.

Peridinium breve

Lillick, L.C., 1937
Seasonal studies of the phytoplankton off Woods Hole, Massachusetts. Biol. Bull. LXXIII (3):488-503, 3 text figs.

Peridinium breve

Mangin, M. L., 1912
Phytoplancton de la croisière du "René" dans l'Atlantique (Septembre 1908). Ann. Inst. Ocean., n.s., 4(1):1-66, 2 pls., 41 text figs., 2 tables.

Peridium breve

Matzenauer, L., 1933
Die Dinoflagellaten des indischen Ozeans (mit Ausnahme der Gattung Ceratium.) Bot. Arch. 35:437-510, 77 text figs., 2 charts.

Peridinium breve

Paulsen, O., 1908
XVIII Peridiniales. Nordisches Plankton, Bot. Teil: 1-124, 155 text figs.

Peridinium breve

Wood, E.J.F., 1954.
Dinoflagellates in the Australian region. Australian J. Mar. Freshwater Res., 5(2):171-351.

Peridinium brevipes

Balech, Enrique, 1971.
Microplancton de la campaña oceanographica: Productividad III.
Revta Mus. argent. Cienc. Nat. Bernadina Rivadavia, Hydrobiol. 3(1):1-202, 39 pls.

Peridinium brevipes

Braarud, T., 1945
A phytoplankton survey of the polluted waters of inner Oslo Fjord. Hvalrådets Skrifter, No.28, 142 pp., 19 text figs., 17 tables.

Peridinium brevipes

Braarud, T., and Adam Bursa, 1939
On the phytoplankton of the Oslo Fjord, 1933-1934. Hvalrådets Skr. No.19:1-63; 9 text figs. Reviewed. J. du. Cons. 14(3): 418-420. A.C. Gardiner.

Peridinium brevipes

Lebour, M.V., 1925.
The dinoflagellates of northern seas. The Marine Biological Association, United Kingdom, 250 pp., 35 pls., 53 textfigs.

Peridinium brevipes

Lillick, L.C., 1940
Phytoplankton and planktonic protozoa of the offshore waters of the Gulf of Maine. Pt.II. Qualitative Composition of the Planktonic Flora. Trans. Am. Phil. Soc., n.s., 31(3):193-237, 13 text figs.

Peridinium brevipes

Morse, D.C., 1947
Some observations on seasonal variations in plankton population Patuxent River, Maryland 1943-1945. Bd. Nat. Res., Publ. No.65, Chesapeake Biol. Lab., 31, 3 figs.

Peridinium brevipes n. nom.

Paulsen, O., 1908
XVIII Peridiniales. Nordisches Plankton, Bot. Teil: 1-124, 155 text figs.

Peridinium brevipes

Rumkówna, A., 1948
List of the phytoplankton species occurring in the superficial water layers in the Gulf of Gdansk. Bull. Lab. mar. Gdynia, No. 4: 139-141 with tables in back.

Peridinium brevipes

Schodduyn, M., 1926
Observations faites dans la baie d'Ambleteuse (Pas de Calais). Bull. Inst. Ocean., Monaco, No. 482: 64 pp.

Peridinium brevipes

Wood, E.J.P., 1954.
Dinoflagellates in the Australian region. Australian J. Mar. Freshwater Res., 5(2):171-351.

Peridinium brevipes and Peridinium africans

Gran, H.H., and T. Braarud, 1935
A quantitative study of the phytoplankton in the Bay of Fundy and the Gulf of Maine (including observations on hydrography, chemistry, and turbidity). J. Biol. Bd. Canada, 1(5):279-467, 69 text figs.

Peridinium brintoni n.sp.

Balech, Enrique, 1962
Tintinnoinea y Dinoflagellata del Pacifico segun material de las Expediciones NORPAC y DOWNWIND del Instituto Scripps de Oceanografia.
Revista, Mus. Argentino Ciencias Nat. "Bernardino Rivadavia", Ciencias Zool., 7(1):1-253.

Peridinium brochi

Balech, E., 1951.
Deuxième contribution à la connaissance des Peridinium. Hydrobiol., Acta Hydrobiol., Limnol., Protistol., 3(4):305-330, 7 pls., of 137 figs.

Peridinium Brochi

Ercegovic, A., 1936
Etudes qualitative et quantitatives du phytoplancton dans les eaux cotières de l'Adriatique oriental moyen au cours de l'année 1934. Acta Adriatica 1(9):1-126

Peridinium Brochii

Margalef, R., 1949
Fitoplancton nerítico de la Costa Brava én 1947-48. Publ. Inst. Biol. Aplicada, 5: 41-51, 3 text figs.

Peridinium brochii

Margalef, R., 1948.
Le phytoplancton estival de la "Costa Brava" catalane en 1946. Hydrobiol. 1(1):15-21.

Peridinium brochi

Rampi, L., 1950.
Ricerche sul fitoplancton del Mare Ligure. 9) I Peridinium delle acque di Sanremo. Pubbl. Centro Talassografico Tirreno No. 5:10 pp. 2 pls.

Atti Acad. Ligure Sci. & Lett. 7(1):

Peridinium brochii

Wood, E.J.P., 1954.
Dinoflagellates in the Australian region. Australian J. Mar. Freshwater Res., 5(2):171-351

Peridinium bulla

LeBour, M.V., 1925
The dinoflagellates of Northern Seas. The Marine Biological Association of the United Kingdom, Plymouth, 250 pp., 35 pls., 53 text figs.

Peridinium bulla

Morse, D.C., 1947
Some observations on seasonal variations in plankton population Patuxant River, Maryland 1943 1945. Bd. Nat. Res., Publ. No.65, Chesapeake Biol. Lab., 31, 3 figs.

Perididium capderillei

Balech, Enrique, 1971.
Microplancton de la campaña oceanographica: Productividad III.
Revta Mus. argent. Cienc. Nat. Bernadina Rivadavia, Hydrobiol. 3(1):1-202, 39 pls.

Peridinium capdevillei n. sp.

Balech, Enrique, 1959.
Operacion Oceanografica Merluza V Crucero. Plancton.
Publico, Servicio de Hidrografia Naval, Argentina, H. 618:43 pp.

Peridinium capurroi n. sp.

Balech, Enrique, 1959.
Operacion Oceanografica Merluza V Crucero. Plancton.
Publico, Servicio de Hidrografia Naval, Argentina, H. 618:43 pp.

Peridinium capurroi capurroi

Balech, Enrique, 1971.
Microplancton de la campaña oceanographica: Productividad III.
Revta Mus. argent. Cienc. Nat. Bernadina Rivadavia, Hydrobiol. 3(1):1-202, 39 pls.

Peridinium capurroi subpellucidum n. subsp.

Balech, Enrique, 1971.
Microplancton de la campaña oceanographica: Productividad III.
Revta Mus. argent. Cienc. Nat. Bernadina Rivadavia, Hydrobiol. 3(1):1-202, 39 pls.

Peridinium cassum n. sp.

Balech, Enrique, 1971.
Microplancton de la campaña oceanographica: Productividad III.
Revta Mus. argent. Cienc. Nat. Bernadina Rivadavia, Hydrobiol. 3(1):1-202, 39 pls.

Peridinium catenatum

Levander, K.M., 1947
Plankton gesammelt in den Jahren 1899-1910 an den Küsten Finnlands. Finnländische Hydrographisch-Biologische Untersuchunger (aus dem Wasserbiologischen Laboratorin der Societas Scientiarum Fennica) No.11: 40 pp., 6 diagrams, 13 pls., tables.

Peridinium catenatum

Paulsen, O., 1908
XVIII Peridiniales. Nordisches Plankton, Bot. Teil: 1-124, 155 text figs.

Peridinium catenatum

Rothe, F., 1942
Quantitativen Untersuchungen über die Planktonverteilung in der östlichen Ostsee. Ber. Deutsch. Wiss. Komm. Meeresf., N.F., 10:291-368, 33 text-figs.

Peridinium catenatum

Rothe, F., 1941
Quantitative Untersuchunger über die Plankton verteilung in der östlichen Ost see. Ber. Deut. Wiss. Komm. fur Meeresforschung. n.f. X(3):291-368, 33 text figs.

Peridinium centenniale

Wood, E.J.P., 1954.
Dinoflagellates in the Australian region. Australian J. Mar. Freshwater Res., 5(2):171-351.

Peridinium cepa n. sp.

Balech, Enrique, 1971.
Microplancton de la campaña oceanographica: Productividad III.
Revta Mus. argent. Cienc. Nat. Bernadina Rivadavia, Hydrobiol. 3(1):1-202, 39 pls.

Peridinium cerasus

Braarud, T., and Adam Bursa, 1939
On the phytoplankton of the Oslo Fjord, 1933-1934. Hvalrådets Skr. No.19:1-63; 9 text figs. Reviewed. J. du. Cons. 14(3): 418-420. A.C. Gardiner.

Peridinium cerasus

Dangeard, P., 1927
Phytoplankton de la croisière du "Sylvana". Ann. Inst. Ocean., Monaco, n.s., 4(8):286-401, 54 text figs. (février-Juin 1913).

Peridinium cerasus

Forti, A., 1922
Ricerche sulla flora pelagica (fitoplancton) di Quarto dei Mille. Mem. R. Com. Talass. Ital. 97:248 pp., 13 pls.

Peridinium cerasus

Ghazzawi, F.M. 1939
Plankton of the Egyptian waters. A study of the Suez Canal Plankton. (A) Phytoplankton. Preliminary Report 83 pp. Notes and Memoires, Min. Commerce-Industry, Egypt. Hydrobiol. & Fish. 65 figs.

Peridinium cerasus

Gran, H.H., and T. Braarud, 1935
A quantitative study of the phytoplankton in the Bay of Fundy and the Gulf of Maine (including observations on hydrography, chemistry, and turbidity). J. Biol. Bd., Canada, 1(5):279-467, 69 text figs.

Peridinium ~~Gyrodinium~~ cerasus

LeBour, M.V., 1925
The dinoflagellates of Northern Seas. The Marine Biological Association of the United Kingdom, Plymouth, 250 pp., 35 pls., 53 text figs.

Peridinium Cerasus

Lillick, L.C., 1937
Seasonal studies of the phytoplankton off Woods Hole, Massachusetts. Biol. Bull. LXXIII (3):488-503, 3 text figs.

Peridinium cerasus

Mangin, M. L., 1912
Phytoplancton de la croisière du "René" dans l'Atlantique (Septembre 1908). Ann. Inst. Ocean., n.s., 4(1):1-66, 2 pls., 41 text figs., 2 tables.

Peridinium cerasus

Margalef, R., 1949
Fitoplancton nerítico de la Costa Brava en 1947-48. Publ. Inst. Biol. Aplicada, 5: 41-51, 3 text figs.

Peridinium cerasus

Paulsen, O., 1908
XVIII Peridiniales. Nordisches Plankton, Bot. Teil: 1-124, 155 text figs.

Peridinium cerasus

Pavillard, J., 1916
Recherches sur les Peridiniens du Golfe du Lion. Mem. Univ. Montpellier. Trav. Inst. Bot., Univ. Montpellier. Serie mixte No.4, 70 pp., 3 pls., 15 text figs.

Peridinium cerasus

Rampi, L., 1950.
Ricerche sul fitoplancton del Mare Ligure. 9) I Peridinium delle acque di Sanremo. Pubbl. Centro Talassografico Tirreno No. 5:10 pp. 2 pls.

Atti Acad. Ligure Sci. e Lett. 7(1):

Peridinium cerasus

Sousa e Silva, E., 1949
Diatomaceas e Dinoflagelados de Baia de Cascais. Portugaliae Acta Biol., Volume: Julio Henriques, Ser. B: 300-383, 9 pls, 2 fold-in tables.

Peridinium cerasus

Wood, E.J.P., 1954.
Dinoflagellates in the Australian region. Australian J. Mar. Freshwater Res., 5(2):171-351.

Peridinium chattoni n.sp.

Becheler, B., 1952.
Recherches sur les Peridinians. Bull. Biol., France Belg., Suppl., 36:1-149, 75 textfigs.

Peridinium cinctum

De Angelis, C.M., 1962.
Distribuzione ed ecologia di alcune specie di Dinoflagellati di aoque salmastre. Pubbl. Staz. Zool., Napoli, 32(Suppl.):301-314.

Peridinium cinctum westii

Messer, Glenda, and Yehuda Ben-Shaul 1972
Changes in chloroplast structure during culture growth of Peridinium cinctum Fa. westii (Dinophyceae). Phycologia 11 (3/4): 291-299.

Peridinium claudicanoides n.sp.

Graham, H. W., 1942
Studies in the morphology, taxonymy, and ecology of the Peridiniales. Sci. Res. Cruise VII of the Carnegie, 1928-1929---Biol. III(542): 129 pp., 67 figs.

Peridinium claudicans

Balech, E., 1951.
Deuxième contribution à la connaissance des Peridinium. Hydrobiol., Acta Hydrobiol., Limnol. Protistol. 3(4):305-330, 7 pls., of 137 figs.

Peridinium claudicans

LeBour, M.V., 1925
The dinoflagellates of Northern Seas. The Marine Biological Association of the United Kingdom, Plymouth, 250 pp., 35 pls., 53 text figs.

Peridinium claudicans

Lindemann, E., 1924
Peridineen aus dem goldenen Horn und dem Bosphorus. Bot. Arch. 5:216-233, 98 text figs.

Peridinium claudicans

Matzenauer, L., 1933
Die Dinoflagellaten des indischen Ozeans (mit Ausnahme der Gattung Ceratium.) Bot. Arch. 35:437-510, 77 text figs., 2 charts.

Peridinium claudicans

Paulsen, O., 1908
XVIII Peridiniales. Nordisches Plankton, Bot. Teil: 1-124, 155 text figs.

Peridinium claudicans

Rampi, L., 1950.
Ricerche sul fitoplancton del Mare Ligure. 9) I Peridinium delle acque di Sanremo. Pubbl. Centro Talassografico Tirreno No. 5:10 pp. 2 pls.

Atti Acad. Ligure Sci. e Lett. 7(1):

Peridinium claudicans

Sousa e Silva, E., 1949
Diatomaceas e Dinoflagelados de Baia de Cascais. Portugaliae Acta Biol., Volume: Julio Henriques, Ser. B: 300-383, 9 pls, 2 fold-in tables.

Peridinium claudicans

Wall, David, and Barrie Dale, 1968.
Modern dinoflagellate cysts and evolution of the Peridiniales. Micropaleontology, 14(3):265-304.

Peridinium colombonense n.sp.

Matzenauer, L., 1933
Die Dinoflagellaten des indischen Ozeans (mit Ausnahme der Gattung Ceratium.) Bot. Arch. 35:437-510, 77 text figs., 2 charts.

Peridinium concavum concavum n. subsp.

Balech, Enrique, 1971.
Microplancton de la campaña oceanographica: Productividad III.
Revta Mus. argent. Cienc. Nat. Bernadina Rivadavia, Hydrobiol. 3(1):1-202, 39 pls.

Peridinium concavum radius n. subsp.

Balech, Enrique, 1971.
Microplancton de la campaña oceanographica: Productividad III.
Revta Mus. argent. Cienc. Nat. Bernadina Rivadavia, Hydrobiol. 3(1):1-202, 39 pls.

Peridinium conicoides

Gran, H.H., and T. Braarud, 1935
A quantitative study of the phytoplankton in the Bay of Fundy and the Gulf of Maine (including observations on hydrography, chemistry, and turbidity). J. Biol. Bd. Canada, 1(5):279-467, 69 text figs.

Peridinium conicoides

LeBour, M.V., 1925
The dinoflagellates of Northern Seas. The Marine Biological Association of the United Kingdom, Plymouth, 250 pp., 35 pls., 53 text figs.

Peridinium conicoides

Lillick, L.C., 1940
Phytoplankton and planktonic protozoa of the offshore waters of the Gulf of Maine. Pt.II. Qualitative Composition of the Planktonic Flora. Trans. Am. Phil. Soc., n.s., 31(3):193-237, 13 text figs.

Peridinium conicoides

Lillick, L.C., 1938
Preliminary report of the phytoplankton of the Gulf of Maine. Am. Mid. Nat. 20(3):624-640, 1 text figs 37 tables.

Peridinium conicoides

Lillick, L.C., 1937
Seasonal studies of the phytoplankton off Woods Hole, Massachusetts. Biol. Bull. LXXIII (3):488-503, 3 text figs.

Peridinium conicoides

Marukawa, H., 1921
Plankton lists and some new species of copepods, from the northern waters of Japan. Bull. Inst. Ocean., No.384, 15 pp., 3 pls., 1 chart. Monaco

Peridinium conicoides

Meunier, A., 1919
Microplancton de la Mer Flamande 3. Les Péridiniens. Mem. Mus. Roy. Hist. Nat., Belgique 8(1):1-116, Pls. XV-XXI.

Peridinium conicoides

Paulsen, O., 1908
XVIII Peridiniales. Nordisches Plankton, Bot. Teil: 1-124, 155 text figs.

Peridinium (Euperidinium) conicoides n.sp.

Paulsen, O., 1905.
On some Peridinese and plankton diatoms. Medd. Komm. Havundersøgelser, Ser. Plankton, 1(3): 7 pp.

Peridinium conicoides

Sousa e Silva, E., 1949
Diatomaceas e Dinoflagelados de Baia de Cascais. Portugaliae Acta Biol., Volume: Julio Henriques, Ser. B: 300-383, 9 pls, 2 fold-in tables.

Peridinium conicum

Balech, E., 1949.
Etude de quelques especes de Peridinium, souvent confondues. Hydrobiol., Acta Hydrobiol., Limnol. et Protistol., 1(4):390-409, 6 pls.

Peridinium conicum

Braarud, T., 1945
A phytoplankton survey of the polluted waters of inner Oslo Fjord. Hvalrådets Skrifter, No.28, 142 pp., 19 text figs., 17 tables.

Peridinium conicum

Braarud, T., and Adam Bursa, 1939
On the phytoplankton of the Oslo Fjord, 1933-1934. Hvalrådets Skr. No.19:1-63; 9 text figs. Reviewed. J. du Cons. 14(3): 418-420. A.C. Gardiner.

Peridinium conicum

Brunel, J., 1962
Le phytoplancton de la Baie de Chaleurs. Inst. Botan., Univ. Montréal, Contrib. No. 77: 365 pp., 66 pls.

Peridinium conicum

Brunel, Jules, 1962
Le phytoplancton de la Baie des Chaleurs. Contrib. Ministère de la Chasse et des Pêcheries, Province de Québec, No. 91: 365 pp.

Peridinium conicum

Dangeard, P., 1927
Phytoplankton de la croisière du "Sylvana". Ann. Inst. Ocean., Monaco, n.s., 4(8):286-401, 54 text figs. (Février-Juin 1913).

Peridinium conicum

Dangeard, P., 1927.
Notes sur la variation de la genre Peridinium. Bull. Inst. Ocean., Monaco, No. 507:16 pp., 9 textfigs.

Peridinium conicum

Dangeard, P., 1926
Description des Péridiniens Testacés recueillis para la Mission Charcot pendent le mois d'Aout 1924. Ann. Inst. Ocean. n.s. 3(7):307-334, 15 text figs.

Peridinium conicum

Delegazione Italiana della Commissione Internazionale per l'Esplorazione Scientifica del Mediterraneo, 1941
Note sul plancton della Laguna veneta. Memoria CCLXXXI, Arch. di Ocean. e Limn. Anno I, Fasc. I, 1941 XIX: 31-57 pp.

Peridinium conicum

Ercegovic, A., 1936
Etudes qualitative et quantitatives du phytoplancton dans les eaux côtières de l'Adriatique oriental moyen au cours de l'année 1934. Acta Adriatica 1(9):1-126

Peridinium conicum

Forti, A., 1922
Ricerche sulla flora pelagica (fitoplancton) di Quarto dei Mille. Mem. R. Com. Talass. Ital. 97:248 pp., 13 pls.

Peridinium conicum

Gran, H.H., and T. Braarud, 1935
A quantitative study of the phytoplankton in the Bay of Fundy and the Gulf of Maine (including observations on hydrography, chemistry, and turbidity). J. Biol. Bd. Canada, 1(5):279-467, 69 text figs.

Peridinium conicum

Jørgensen, E., 1905
B. Protistplankton and the diatoms in bottom samples. Hydrographical and biological investigations in Norwegian fjords. Bergens Mus. Skr. 7: 49-225.

Peridinium conicum

Kelly, Mahon G., and Steven Katona, 1966.
An endogenous diurnal rhythm of bioluminescence in a natural population of dinoflagellates. Biol. Bull. 131(1):115-126.

Peridinium conicum

LeBour, M.V., 1925
The dinoflagellates of Northern Seas. The Marine Biological Association of the United Kingdom, Plymouth, 250 pp., 35 pls., 53 text figs.

Peridinium conicum

Lillick, L.C., 1940
Phytoplankton and planktonic protozoa of the offshore waters of the Gulf of Maine. Pt.II. Qualitative Composition of the Planktonic Flora. Trans. Am. Phil. Soc., n.s., 31(3):193-237, 13 text figs.

Peridinium conicum

Lillick, L.C., 1938
Preliminary report of the phytoplankton of the Gulf of Maine. Am. Mid. Nat. 20(3):624-640, 1 text figs 37 tables.

Peridinium conicum

Lillick, L.C., 1937
Seasonal studies of the phytoplankton off Woods Hole, Massachusetts. Biol. Bull. LXXIII (3):488-503, 3 text figs.

Peridinium conicum

Lindemann, E., 1924
Peridineen aus dem goldenen Horn und dem Bosphorus. Bot. Arch. 5:216-233, 98 text figs.

Peridinium conicum

Marukawa, H., 1921
Plankton lists and some new species of copepods, from the northern waters of Japan. Bull. Inst. Ocean., No.384, 15 pp., 3 pls., 1 chart. Monaco

Peridinium conicum

Matzenauer, L., 1933
Die Dinoflagellaten des indischen Ozeans (mit Ausnahme der Gattung Ceratium.) Bot. Arch. 35:437-510, 77 text figs., 2 charts.

Peridinium conicum

Meunier, A., 1919
Microplancton de la Mer Flamande 3. Les Péridiniens. Mem. Mus. Roy. Hist. Nat., Belgique 8(1):1-116, Pls. XV-XXI.

Peridinium conicum

Paulsen, O., 1908
XVIII Peridiniales. Nordisches Plankton, Bot. Teil: 1-124, 155 text figs.

Peridinium conicum

Pavillard, J., 1916
Recherches sur les Peridiniens du Golfe du Lion. Mem. Univ. Montpellier. Trav. Inst. Bot., Univ. Montpellier. Serie mixte No.4, 70 pp., 3 pls., 15 text figs.

Peridinium conicum

Rampi, L., 1950.
Ricerche sul fitoplancton del Mare Ligure. 9) I Peridinium delle acque di Sanremo. Pubbl. Centro Talassografico Tirreno No. 5:10 pp. 2 pls.

Atti Acad. Ligure Sci. e Lett. 7(1):

Peridinium conicum

Sousa e Silva, E., 1949
Diatomaceas e Dinoflagelados de Baia de Cascais. Portugaliae Acta Biol., Volume: Julio Henriques, Ser. B: 300-383, 9 pls, 2 fold-in tables.

Peridinium conicum

Sousa a Silva, E., and J. Dos Santos-Pinto, 1948
O Plancton da Baía de S. Martinho do Porto.
1. Diatomaceas e Dinoflagelados. Bol. Soc. Portuguese de Ciencias Naturais, 16(2):134-187, 6 pls. (Trav. Sta. Biol. Mar. de Lisbonne No. 52).

Peridinium conicum

Wall, David, and Barrie Dale, 1968.
Modern dinoflagellate cysts and evolution of the Peridiniales.
Micropaleontology, 14(3):265-304.

Peridinium conicum var. Asamushi

Lillick, L.C., 1940
Phytoplankton and planktonic protozoa of the offshore waters of the Gulf of Maine. Pt.II. Qualitative Composition of the Planktonic Flora. Trans. Am. Phil. Soc., n.s., 31(3):193-237, 13 text figs.

Peridinium conicum var. Asamushi

Lillick, L.C., 1938
Preliminary report of the phytoplankton of the Gulf of Maine. Am. Mid. Nat. 20(3):624-640, 1 text figs 37 tables.

Peridinium convexius n. var.

Graham, H. W., 1942
Studies in the morphology, taxonymy, and ecology of the Peridiniales. Sci. Res. Cruise VII of the Carnegie, 1928-1929---Biol. III(542): 129 pp., 67 figs.

Peridinium crassipes

Argentina, Secretaria de Marina, Servicio de Hidrografia Naval, 1962.
Plancton de las campanas oceanograficas DRAKE I y II.
Publico, H. 627:57.

Peridinium sp cf. P. crassipes

Balech, Enrique, 1971.
Microplancton de la campaña oceanográfica: Productividad III.
Revta Mus. argent. Cienc. Nat. Bernadina Rivadavia, Hydrobiol. 3(1):1-202, 39 pls.

Peridinium crassipes

Braarud, T., 1945
A phytoplankton survey of the polluted waters of inner Oslo Fjord. Hvalrådets Skrifter, No.28, 142 pp., 19 text figs., 17 tables.

Peridinium crassipes

Braarud, T., and Adam Bursa, 1939
On the phytoplankton of the Oslo Fjord, 1933-1934. Hvalrådets Skr. No.19:1-63; 9 text figs. Reviewed. J. du. Cons. 14(3): 418-420. A.C. Gardiner.

Peridinium crassipes

Brunel, J., 1962
Le phytoplancton de la Baie de Chaleurs. Inst. Botan., Univ. Montréal, Contrib. No. 77: 365 pp., 66 pls.

Peridinium crassipes

Brunel, Jules, 1962
Le phytoplancton de la Baie des Chaleurs. Contrib. Ministère de la Chasse et des Pêcheries, Province de Québec, No. 91: 365 pp.

Peridinium crassipes

Dangeard, P., 1927
Phytoplankton de la croisière du "Sylvana". Ann. Inst. Ocean., Monaco, n.s., 4(8):286-401, 54 text figs. (février-Juin 1913).

Peridinium crassipes

Dangeard, P., 1926
Description des Péridiniens Testacés recueillis para la Mission Charcot pendent le mois d'Aout 1924. Ann. Inst. Ocean. n.s. 3(7):307-334, 15 text figs.

Peridinium crassipes

Delegazione Italiana della Commissione Internazionale per l'Esplorazione Scientifica del Mediterraneo, 1941
Note sul plancton della Laguna veneta. [Memoria CCLXXIX], Arch. di Ocean. e Limn. Anno I, Fasc. I, 1941 XIX: 31-57 pp.

Peridinium crassipes

Ercegovic, A., 1936
Etudes qualitative et quantitatives du phytoplancton dans les eaux cotières de l'Adriatique oriental moyen au cours de l'année 1934. Acta Adriatica 1(9):1-126

Peridinium crassipes

Forti, A., 1922
Ricerche sulla flora pelagica (fitoplancton) di Quarto dei Mille. Mem. R. Com. Talass. Ital. 97:248 pp., 13 pls.

Peridinium crassipes

Gilbert, J.Y., and W.E. Allen, 1943
The phytoplankton of the Gulf of California obtained by the "E.W. Scripps" in 1939 and 1940. J. Mar. Res. V(2):89-110, figs.30-31.

Peridinium crassipes

Graham, H. W., 1942
Studies in the morphology, taxonymy, and ecology of the Peridiniales. Sci. Res. Cruise VII of the Carnegie, 1928-1929---Biol. III(542): 129 pp., 67 figs.

Peridinium crassipes n. sp.

Kofoid, C. A., 1907
Dinoflagellata of the San Diego region. III. Descriptions of new species. Univ. Calif. Publ. Zool. 3:299-340, Pls. 22-33.

Peridinium crassipes

Lillick, L.C., 1940
Phytoplankton and planktonic protozoa of the offshore waters of the Gulf of Maine. Pt.II. Qualitative Composition of the Planktonic Flora. Trans. Am. Phil. Soc., n.s., 31(3):193-237, 13 text figs.

Peridinium crassipes

Lillick, L.C., 1938
Preliminary report of the phytoplankton of the Gulf of Maine. Am. Mid. Nat. 20(3):624-640, 1 text figs 37 tables.

Peridinium crassipes

Lindemann, E., 1924
Peridineen aus dem goldenen Horn und dem Bosphorus. Bot. Arch. 5:216-233, 98 text figs.

Peridinium crassipes

Mangin, M. L., 1912
Phytoplancton de la croisière du "René" dans l'Atlantique (Septembre 1908). Ann. Inst. Ocean. n.s., 4(1):1-66, 2 pls., 41 text figs., 2 tables.

Peridinium crassipes

Margalef, R., 1949
Fitoplancton nerítico de la Costa Brava en 1947-48. Publ. Inst. Biol. Aplicada, 5: 41-51, 3 text figs.

Peridinium crassipes

Matzenauer, L., 1933
Die Dinoflagellaten des indischen Ozeans (mit Ausnahme der Gattung Ceratium.) Bot. Arch. 35:437-510, 77 text figs., 2 charts.

Peridinium crassipes

Margalef, R., 1948.
Le phytoplancton estival de la "Costa Brava" catalane en 1946. Hydrobiol. 1(1):15-21.

Peridinium crassipes

Marukawa, H., 1921
Plankton lists and some new species of copepods, from the northern waters of Japan. Bull. Inst. Ocean. No.384, 15 pp., 3 pls., 1 chart. Monaco

Peridinium crassipes

Paiva Carvalho, J., 1950
O plancton do Rio Maria Rodriques (Cananeis). 1. Diatomaceas e Dinoflagelados. Bol. Inst. Paulista Oceanogr. 1(1): 27-43, 2 fold-in tables, 2 figs.

Peridinium crassipes

Paulsen, O., 1908
XVIII Peridiniales. Nordisches Plankton, Bot. Teil: 1-124, 155 text figs.

Peridinium crassipes

Pavillard, J., 1916
Recherches sur les Peridiniens du Golfe du Lion. Mem. Univ. Montpellier. Trav. Inst. Bot., Univ. Montpellier. Serie mixte No.4, 70 pp., 3 pls., 15 text figs.

Peridinium crassipes

Sousa e Silva, E., 1949
Diatomaceas e Dinoflagelados de Baia de Cascais. Portugaliae Acta Biol. Volume: Julio Henriques, Ser. B: 300-383, 9 pls, 2 fold-in tables.

Peridinium crassipes

Rampi, L., 1950.
Ricerche sul fitoplancton del Mare Ligure. 9) I Peridinium delle acque di Sanremo. Pubbl. Centro Talassografico Tirreno No. 5:10 pp. 2 pls.

Atti Acad. Ligure Sci. e Lett. 7(1):

Peridinium crassipes

Sousa a Silva, E., and J. Dos Santos-Pinto, 1948
O Plancton da Baía de S. Martinho do Porto.
1. Diatomaceas e Dinoflagelados. Bol. Soc. Portuguese de Ciencias Naturais, 16(2):134-187, 6 pls. (Trav. Sta. Biol. Mar. de Lisbonne No. 52).

Peridinium crassipes

Wood, E.J.F., 1954.
Dinoflagellates in the Australian region. Australian J. Mar. Freshwater Res., 5(2):171-351.

Peridinium crassipes autumnalis

Forti, A., 1922
Ricerche sulla flora pelagica (fitoplancton) di Quarto dei Mille. Mem. R. Com. Talass. Ital. 97:248 pp., 13 pls.

Peridinium crassipyrum nom. nov.

Balech, Enrique, 1961
Notula sobre Peridinium (Dinoflagellatae). Neotropica, 7:29-32.

Peridinium crassipyrum n. sp.

Balech, Enrique, 1959.
Operacion Oceanografica Merluza V Crucero. Plancton.
Publico, Servicio de Hidrografia Naval, Argentina, H. 618:43 pp.

Peridinium crassum n.sp.

Balech, Enrique, 1959
Operacion Oceanografica Merluza, V Crucero. Plancton.
Servicio Hidrografia Naval, Publ. H 618: 1-43.

Peridinium crassum n.sp.

Dangeard, P., 1927
Phytoplankton de la croisière du "Sylvana". Ann. Inst. Ocean., Monaco, n.s., 4(8):286-401, 54 text figs. (février-Juin 1913).

Peridinium cruciferum n. sp.
Balech, Enrique, 1971.
Microplancton de la campaña oceanographica:
Productividad III.
Revta Mus. argent. Cienc. Nat. Bernadina
Rivadavia, Hydrobiol. 3(1):1-202, 39 pls.

Peridinium curtipes
Dangeard, P., 1927
Phytoplankton de la croisière du "Sylvana". Ann. Inst. Ocean., Monaco, n.s., 4(8):286-401, 54 text figs. (Février-Juin 1913).

Peridinium curtipes
LeBour, M.V., 1925
The dinoflagellates of Northern Seas, The Marine Biological Association of the United Kingdom, Plymouth. 250 pp., 35 pls., 53 text figs.

Peridinium curtipes
Wood, E.J.P., 1954.
Dinoflagellates in the Australian region.
Australian J. Mar. Freshwater Res., 5(2):171-351.

Peridinium curtum
Argentina, Secretaria de Marina, Servicio de Hidrografia Naval, 1962.
Plancton de las campanas oceanograficas DRAKE I y II.
Publico. H. 627:57.

Peridinium curtum
Hada, Yoshina, 1970.   (1970)
The protozoan plankton of the Antarctic and Subantarctic seas.
Scient. Repts, Japan. Antarct. Res. Exped., (E)31:51 pp.

Peridinium aff. curvipes
Balech, Enrique, 1959.
Operacion Oceanografica Merluza V Crucero. Plancton.
Publico, Servicio de Hidrografia Naval, Argentina, H. 618:43 pp.

Peridinium curvipes
Braarud, T., 1945
A phytoplankton survey of the polluted waters of inner Oslo Fjord. Hvalrådets Skrifter, No.28, 142 pp., 19 text figs., 17 tables.

Peridinium curvipes
Braarud, T., and Adam Bursa, 1939
On the phytoplankton of the Oslo Fjord, 1933-1934. Hvalrådets Skr. No.19:1-63; 9 text figs. Reviewed. J. du. Cons. 14(3): 418-420. A.C. Gardiner.

Peridinium curvipes
Dangeard, P., 1927
Phytoplankton de la croisière du "Sylvana". Ann. Inst. Ocean., Monaco, n.s., 4(8):286-401, 54 text figs. (Février-Juin 1913).

Peridinium curvipes
Dangeard, P., 1926
Description des Péridiniens Testacés recueillis par la Mission Charcot pendent le mois d'Aout 1924. Ann. Inst. Ocean. n.s. 3(7):307-334, 15 text figs.

Peridinium curvipes
Forti, A., 1922
Ricerche sulla flora pelagica (fitoplancton) di Quarto dei Mille. Mem. R. Com. Talass. Ital. 97:248 pp., 13 pls.

Peridinium curvipes
Gran, H.H., and T. Braarud, 1935
A quantitative study of the phytoplankton in the Bay of Fundy and the Gulf of Maine (including observations on hydrography, chemistry, and turbidity). J. Biol. Bd., Canada, 1(5):279-467, 69 text figs.

Peridinium curvipes
Lebour, M.V., 1925.
The dinoflagellates of northern seas. Marine Biological Association, United Kingdom, 250 pp., 35 pls., 53 textfigs.

Peridinium curvipes
Lillick, L.C., 1940
Phytoplankton and planktonic protozoa of the offshore waters of the Gulf of Maine. Pt.II. Qualitative Composition of the Planktonic Flora. Trans. Am. Phil. Soc., n.s., 31(3):193-237, 13 text figs.

Peridinium curvipes
Macdonald, R., 1933
An examination of plankton hauls made in the Suez Canal during the year 1928. Fish. Res. Dir., Notes & Mem. No.3, 11 pp., 1 chart.

Peridinium curvipes
Matzenauer, L., 1933
Die Dinoflagellaten des indischen Ozeans (mit Ausnahme der Gattung Ceratium.) Bot. Arch. 35:437-510, 77 text figs., 2 charts.

Peridinium curvipes
Paulsen, O., 1908
XVIII Peridiniales. Nordisches Plankton, Bot. Teil: 1-124, 155 text figs.

Peridinium curvipes
Pavillard, J., 1916
Recherches sur les Peridiniens du Golfe du Lion. Mem. Univ. Montpellier. Trav. Inst. Bot., Univ. Montpellier. Serie mixte No.4, 70 pp., 3 pls., 15 text figs.

Peridinium curvipes
Rampi, L., 1950.
Ricerche sul fitoplancton del Mare Ligure. 9) I Peridinium delle acque di Sanremo. Pubbl. Centro Talassografico Tirreno No. 5:10 pp. 2 pls.

Atti Acad. Ligure Sci. e Lett. 7(1):

Peridinium curvipes
Rumkówna, A., 1948
List of the phytoplankton species occurring in the superficial water layers in the Gulf of Gdańsk. Bull. Lab. mar., Gdynia, No. 4: 139-141 with tables in back.

Peridinium curvipes
Sousa e Silva, E., 1949
Diatomaceas e Dinoflagelados de Baia de Cascais. Portugaliae Acta Biol., Volume: Julio Henriques, Ser. B: 300-383, 9 pls, 2 fold-in tables.

Peridinium curvipes
Wood, E.J.P., 1954.
Dinoflagellates in the Australian region.
Australian J. Mar. Freshwater Res., 5(2):171-351.

Peridinium cystiferum n.sp.
Pavillard, J., 1916
Recherches sur les Peridiniens du Golfe du Lion. Mem. Univ. Montpellier. Trav. Inst. Bot., Univ. Montpellier. Serie mixte No.4, 70 pp., 3 pls., 15 text figs.

Peridinium dakariensis n. sp.
Dangeard, P., 1927
Phytoplankton de la croisière du "Sylvana". Ann. Inst. Ocean., Monaco, n.s., 4(8):286-401, 54 text figs. (Février-Juin 1913).

Peridinium dakariensis
Wood, E.J.P., 1954.
Dinoflagellates in the Australian region.
Australian J. Mar. Freshwater Res., 5(2):171-351.

Peridinium decens n. sp.
Balech, Enrique, 1971.
Microplancton de la campaña oceanographica:
Productividad III.
Revta Mus. argent. Cienc. Nat. Bernadina
Rivadavia, Hydrobiol. 3(1):1-202, 39 pls.

Peridinium decipiens n.sp.
Jorgensen, E., 1900
Protophyten und Protozoen im Plankton aus der Norwegischen Westkerste. Bergens Mus. Aarb. 1899(6): 95 pp., 5 pls., 83 tables.

Peridinium decipiens
Lebour, M.V., 1925.
The dinoflagellates of northern seas. Marine Biological Association, United Kingdom, 250 pp., 35 pls., 53 textfigs.

Peridinium decipiens
Matzenauer, L., 1933
Die Dinoflagellaten des indischen Ozeans (mit Ausnahme der Gattung Ceratium.) Bot. Arch. 35:437-510, 77 text figs., 2 charts.

Peridinium decipiens
Paulsen, O., 1908
XVIII Peridiniales. Nordisches Plankton, Bot. Teil: 1-124, 155 text figs.

Peridinium decipiens
Wood, E.J.P., 1954.
Dinoflagellates in the Australian region.
Australian J. Mar. Freshwater Res., 5(2):171-351.

Peridinium decollatum n. sp.
Balech, Enrique, 1971.
Microplancton de la campaña oceanographica:
Productividad III.
Revta Mus. argent. Cienc. Nat. Bernadina
Rivadavia, Hydrobiol. 3(1):1-202, 39 pls.

Peridinium defectum n.sp.
Balech, Enrique, and Sayed Z. El-Sayed, 1965.
Microplankton of the Weddell Sea.
In: Biology of Antarctic seas. II.
Antarctic Res. Ser., Amer. Geophys. Union, 5:107-124.

Peridinium deficiens n.sp.
Meunier, A., 1919
Microplancton de la Mer Flamande 3. Les Péridiniens. Mem. Mus. Roy. Hist. Nat., Belgique 8(1):1-116, Pls. XV-XXI.

Peridinium deficiens
Rumkówna, A., 1948
List of the phytoplankton species occurring in the superficial water layers in the Gulf of Gdańsk. Bull. Lab. mar., Gdynia, No. 4: 139-141 with tables in back.

Peridinium delicatissimum
Halim, Y., 1965.
Microplancton des eaux Egyptiennes. II. Chrysomonadines; Ebriediens et dinoflagellés nouveaux ou d'intérêt biogeographique. Rapp. Proc. Verb. Reunions, Comm. Int. Expl. Sci. Mer Méditerranée, Monaco, 18(2):373-379.

Peridinium denticulatum
Brunel, J., 1962
Le phytoplancton de la Baie de Chaleurs. Inst. Botan., Univ. Montréal, Contrib. No. 77: 365 pp., 66 pls.

Peridinium denticulatum
Brunel, Jules, 1962
Le phytoplancton de la Baie des Chaleurs. Contrib. Ministère de la Chasse et des Pêcheries, Province de Québec, No. 91: 365 pp.

Peridinium denticulatum n. sp.
Gran, H.H., and T. Braarud, 1935
A quantitative study of the phytoplankton in the Bay of Fundy and the Gulf of Maine (including observations on hydrography, chemistry, and turbidity). J. Biol. Bd., Canada, 1(5):279-467, 69 text figs.

Peridinium denticulatum
Lillick, L.C., 1940
Phytoplankton and planktonic protozoa of the offshore waters of the Gulf of Maine. Pt.II. Qualitative Composition of the Planktonic Flora. Trans. Am. Phil. Soc., n.s., 31(3):193-237, 13 text figs.

Peridinium denticulatum
Lillick, L.C., 1938
Preliminary report of the phytoplankton of the Gulf of Maine. Am. Mid. Nat. 20(3):624-640, 1 text figs 37 tables.

Peridinium denticulatum
Lillick, L.C., 1937
Seasonal studies of the phytoplankton off Woods Hole, Massachusetts. Biol. Bull. LXXIII (3):488-503, 3 text figs.

Peridinium ? denticulatum
Wall, David, and Barrie Dale, 1968.
Modern dinoflagellate cysts and evolution of the Peridiniales. Micropaleontology, 14(3):265-304.

Peridinium depressum
Braarud, T., 1945
A phytoplankton survey of the polluted waters of inner Oslo Fjord. Hvalrådets Skrifter, No.28, 142 pp., 19 text figs., 17 tables.

Peridinium depressum
Braarud, T., and Adam Bursa, 1939
On the phytoplankton of the Oslo Fjord, 1933-1934. Hvalrådets Skr. No.19:1-63; 9 text figs. Reviewed. J. du. Cons. 14(3): 418-420. A.C. Gardiner.

Peridinium depressum
Brunel, J., 1962
Le phytoplancton de la Baie de Chaleurs. Inst. Botan., Univ. Montréal, Contrib. No. 77: 365 pp., 66 pls.

Peridinium depressum
Brunel, Jules, 1962
Le phytoplancton de la Baie des Chaleurs. Contrib. Ministère de la Chasse et des Pêcheries, Province de Québec, No. 91: 365 pp.

Peridinium depressum
Dangeard, P., 1927
Phytoplankton de la croisière du "Sylvana". Ann. Inst. Ocean., Monaco, n.s., 4(8):286-401, 54 text figs. (février-Juin 1913).

Peridinium depressum
Dangeard, P., 1927
Notes sur la variation de la genre Peridinium. Bull. Inst. Ocean., Monaco, No. 507: 16 pp., 9 textfigs.

Peridinium depressum
Dangeard, P., 1926
Description des Péridiniens Testacés recueillis par la Mission Charcot pendant le mois d'Aout 1924. Ann. Inst. Ocean. n.s. 3(7):307-334, 15 text figs.

Peridinium depressum
Delegazione Italiana della Commissione Internazionale per l'Esplorazione Scientifica del Mediterraneo, 1941
Note sul plancton della Laguna veneta. [Memoria CCLXXXIX], Arch. di Ocean. e Limn. Anno I, Fasc. I, 1941 XIX: 31-57 pp.

Peridinium depressum
Ercegovic, A., 1936
Etudes qualitative et quantitatives du phytoplancton dans les eaux cotières de l'Adriatique oriental moyen au cours de l'année 1934. Acta Adriatica 1(9):1-126

Peridinium depressum
Forti, A., 1922
Ricerche sulla flora pelagica (fitoplancton) di Quarto dei Mille. Mem. R. Com. Talass. Ital. 97:248 pp., 13 pls.

Peridinium depressum
Gilbert, J.Y., and W.E. Allen, 1943
The phytoplankton of the Gulf of California obtained by the "E.W. Scripps" in 1939 and 1940. J. Mar. Res. V(2):89-110, figs.30-31.

Peridinium depressum
Graham, H.W., 1942
Studies in the morphology, taxonymy, and ecology of the Peridiniales. Sci. Res. Cruise VII of the Carnegie, 1928-1929---Biol. III(542): 129 pp., 67 figs.

Peridinium depressum
Gran, H.H., and T. Braarud, 1935
A quantitative study of the phytoplankton in the Bay of Fundy and the Gulf of Maine (including observations on hydrography, chemistry, and turbidity). J. Biol. Bd., Canada, 1(5):279-467, 69 text figs.

Peridinium depressum
Iselin, C., 1930
A report on the coastal waters of Labrador based on explorations of the "Chance" during the summer of 1926. Proc. Am. Acad. Arts Sci., 66(1):1-37, 14 text figs.

Peridinium depressum
Jørgensen, E., 1905
B. Protistplankton and the diatoms in bottom samples. Hydrographical and biological investigations in Norwegian fjords. Bergens Mus. Skr. 7: 49-225.

Peridinium depressum
Jorgensen, E., 1900
Protophyten und Protozoën im Plankton aus der Norwegischen Westkerste. Bergens Mus. Aarb. 1899(6): 95 pp., 5 pls., 83 tables

Peridinium depressum
LeBour, M.V., 1925
The dinoflagellates of Northern Seas. The Marine Biological Association of the United Kingdom, Plymouth, 250 pp., 35 pls., 53 text figs.

Peridinium depressum
Lillick, L.C., 1940
Phytoplankton and planktonic protozoa of the offshore waters of the Gulf of Maine. Pt.II. Qualitative Composition of the Planktonic Flora. Trans. Am. Phil. Soc., n.s., 31(3):193-237, 13 text figs.

Peridinium depressum
Lillick, L.C., 1938
Preliminary report of the phytoplankton of the Gulf of Maine. Am. Mid. Nat. 20(3):624-640, 1 text figs 37 tables.

Peridinium depressum
Lillick, L.C., 1937
Seasonal studies of the phytoplankton off Woods Hole, Massachusetts. Biol. Bull. LXXIII (3):488-503, 3 text figs.

Peridinium depressum
Margalef, R., 1949
Fitoplancton nerítico de la Costa Brava en 1947-48. Publ. Inst. Biol. Aplicada, 5: 41-51, 3 text figs.

Peridinium depressum
Marukawa, H., 1921
Plankton lists and some new species of copepods, from the northern waters of Japan. Bull. Inst. Ocean. No.384, 15 pp., 3 pls., 1 chart. Monaco

Peridinium depressum
Matzenauer, L., 1933
Die Dinoflagellaten des indischen Ozeans (mit Ausnahme der Gattung Ceratium.) Bot. Arch. 35:437-510, 77 text figs., 2 charts.

Peridineum depressum
Paulsen, O., 1949
Observations on dinoflagellates. (Ed. J. Grøntoed) Kongl. Dansk. Videnskab. Selsk., Biol. Skr. 6(4):67 pp., 30 text figs.

Peridinium depressum
Paulsen, O., 1908
XVIII Peridiniales. Nordisches Plankton, Bot. Teil: 1-124, 155 text figs.

Peridinium depressum
Pavillard, J., 1916
Recherches sur les Peridiniens du Golfe du Lion. Mem. Univ. Montpellier. Trav. Inst. Bot., Univ. Montpellier. Serie mixte No.4, 70 pp., 3 pls., 15 text figs.

PERIDINIUM DEPRESSUM
Piccinetti, C., e G. Manfrin, 1969.
Osservazioni sulla mortalità di pesci e di altri organismi verificatasi nel 1969 in Adriatico. Note Lab. Biol. mar. pesca - Fano, Ist. Zool. Univ. Bologna, 3(4):73-92.

Peridinium depressum
Rampi, L., 1950.
Ricerche sul fitoplancton del Mare Ligure. 9) I Peridinium delle acque di Sanremo. Pubbl. Centro Talassografico Tirreno No. 5:10 pp., 2 pls.

Atti Acad. Ligure Sci. e Lett. 7(1):

Peridinium depressum
Schulz, B., and A. Wulff, 1929
Hydrographie und Oberflächen plankton des westlichen Barentsmeeres im Sommer 1927. Ber. deutschen wissensch. Komm. F. Meeresforsch. n.s. 4(5):232-372, 13 tables, 25 text figs.

Peridinium depressum
Sousa e Silva, E., 1949
Diatomaceas e Dinoflagelados de Baia de Cascais. Portugaliae Acta Biol., Volume: Julio Henriques, Ser. B: 300-383, 9 pls, 2 fold-in tables.

Peridinium depressum
Sousa a Silva, E., and J. Dos Santos-Pinto, 1948
O Plancton da Baia de S. Martinho do Porto. 1. Diatomaceas e Dinoflagelados. Bol. Soc. Portuguese de Ciencias Naturais, 16(2):134-187, 6 pls. (Trav. Sta. Biol. Mar. de Lisbonne No. 52).

Peridinium depressum
Wheeler, J.E.G., 1939
Plankton investigations. Bermuda Biological Station. Second Report. October 1939. 7 pp. (typed), 5 figs. Plymouth, Oct. 23, 1939.

Peridinium diabolus
Dangeard, P., 1927
Phytoplankton de la croisière du "Sylvana". Ann. Inst. Ocean. Monaco, n.s., 4(8):286-401, 54 text figs. (février-Juin 1913).

Peridinium diabolus
Ercegovic, A., 1936
Etudes qualitative et quantitatives du phytoplancton dans les eaux cotières de l'Adriatique oriental moyen au cours de l'année 1934. Acta Adriatica 1(9):1-126

Peridinium diabolus
Forti, A., 1922
Ricerche sulla flora pelagica (fitoplancton) di Quarto dei Mille. Mem. R. Com. Talass. Ital. 97:248 pp., 13 pls.

Peridinium diabolus
Lebour, M.V., 1925.
The dinoflagellates of northern seas. Marine Biological Association, United Kingdom, 250 pp., 53 textfigs., 35 pls.

Peridinium diabolus
Macdonald, R., 1933
An examination of plankton hauls made in the Suez Canal during the year 1928. Fish. Res. Dis., Notes & Mem. No.3, 11 pp., 1 chart.

Peridinium diabolus
Margalef, R., 1948.
Le phytoplancton estival de la "Costa Brava" catalane en 1946. Hydrobiol. 1(1):15-21.

Peridinium diabolus
Matzenauer, L., 1933
Die Dinoflagellaten des indischen Ozeans (mit Ausnahme der Gattung Ceratium.) Bot. Arch. 35:437-510, 77 text figs., 2 charts.

Peridinium diabolus
Pavillard, J., 1916
Recherches sur les Peridiniens du Golfe du Lion. Mem. Univ. Montpellier. Trav. Inst. Bot., Univ. Montpellier. Serie mixte No.4, 70 pp., 3 pls., 15 text figs.

Peridinium diabolus
Rampi, L., 1950.
Ricerche sul fitoplancton del Mare Ligure. 9) I Peridinium delle acque di Sanremo. Pubbl. Centro Talassografico Tirreno No. 5:10 pp 2 pls.

Atti Acad. Ligure Sci. e Lett. 7(1):

Peridinium diabolus
Sousa e Silva, E., 1949
Diatomaceas e Dinoflagelados de Baia de Cascais. Portugaliae Acta Biol., Volume: Julio Henriques, Ser. B: 300-383, 9 pls., 2 fold-in tables.

Peridinium diabolus
Sousa e Silva, E., and J. Dos Santos-Pinto, 1948
O Plancton da Baía de S. Martinho do Porto. 1. Diatomaceas e Dinoflagelados. Bol. Soc. Portuguese de Ciencias Naturais, 16(2):134-187, 6 pls. (Trav. Sta. Biol. Mar. de Lisbonne No. 52).

Peridinium diabolus
Wood, E.J.F., 1954.
Dinoflagellates in the Australian region. Australian J. Mar. Freshwater Res., 5(2):171-351.

Peridinium digitale
Calkins, G. N., 1902
Marine protozoa from Woods Hole. U.S. Fish Comm. Bull. for 1901, pp. 413-468, 69 text figs.

Peridinium divaricatum
Dangeard, P., 1927
Phytoplankton de la croisière du "Sylvana". Ann. Inst. Ocean., Monaco, n.s., 4(8):286-401, 54 text figs. (février-Juin 1913).

Peridinium divaricatum
LeBour, M.V., 1925
The dinoflagellates of Northern Seas. The Marine Biological Association of the United Kingdom, Plymouth, 250 pp., 35 pls., 53 text figs.

Peridinium divaricatum n.sp.
Meunier, A., 1919
Microplancton de la Mer Flamande 3. Les Péridiniens. Mem. Mus. Roy. Hist. Nat., Belgique 8(1):1-116, Pls. XV-XXI.

Peridinium divergens
Allen, W.E., 1927.
Quantitative studies on inshore marine diatoms and dinoflagellates of Southern California in 1922. Bull. S.I.O., tech. ser., 1:31-38, 2 text-figs.

Peridinium divergens
Braarud, T., 1945
A phytoplankton survey of the polluted waters of inner Oslo Fjord. Hvalrådets Skrifter, No.28, 142 pp., 19 text figs., 17 tables.

Peridinium divergens
Braarud, T., and Adam Bursa, 1939
On the phytoplankton of the Oslo Fjord, 1933-1934. Hvalrådets Skr. No.19:1-63; 9 text figs. Reviewed. J. du. Cons. 14(3): 418-420. A.C. Gardiner.

Peridinium divergens
Brunel, J., 1962
Le phytoplancton de la Baie de Chaleurs. Inst. Botan., Univ. Montréal, Contrib. No. 77: 365 pp., 66 pls.

Peridinium divergens
Calkins, G. N., 1902
Marine protozoa from Woods Hole. U.S. Fish Comm. Bull. for 1901, pp. 413-468, 69 text figs.

Peridinium divergens
Chiba, T., 1949
On the distribution of the plankton in the eastern China Sea and Yellow Sea. 1. Plankton composition in the spring. J. Shimonoseki Coll. Fisheries, 1(1):57-63, 1 fig.

Peridinium divergens
Copenhagen, W. J. and L. D., 1949
Variation in the phytoplankton of Table Bay, October 1934 to October 1935. With a note on the calorific value of Chaetoceros spp. Trans. Roy. Soc. S. Africa, 32(2):113-123, 2 text figs.

Peridinium divergens
Cupp, E., 1930
Quantitative Studies of miscellaneous series of surface catches of marine diatoms and dinoflagellates taken between Seattle and the Canal Zone from 1924 to 1928. Trans. Am. Micro. Soc., XLIX (3):238-245.

Peridinium divergens
Dangeard, P., 1927
Phytoplankton de la croisière du "Sylvana". Ann. Inst. Ocean., Monaco, n.s., 4(8):286-401, 54 text figs. (février-Juin 1913).

Peridinium divergens
Delegazione Italiana della Commissione Internazionale per l'Esplorazione Scientifica del Mediterraneo, 1941
Note sul plancton della Laguna veneta. [Memoria CCLXXXIX], Arch. di Ocean. e Limn. Anno I, Fasc. I, 1941 XIX: 31-57 pp.

Peridinium divergens
Ercegovic, A., 1936
Etudes qualitative et quantitatives du phytoplancton dans les eaux cotières de l'Adriatique oriental moyen au cours de l'année 1934. Acta Adriatica 1(9):1-126

Peridinium divergens
Forti, A., 1922
Ricerche sulla flora pelagica (fitoplancton) di Quarto dei Mille. Mem. R. Com. Talass. Ital. 97:248 pp., 13 pls.

Peridinium divergens
Gilbert, J.Y., and W.E. Allen, 1943
The phytoplankton of the Gulf of California obtained by the "E.W. Scripps" in 1939 and 1940. J. Mar. Res. V(2):89-110, figs.30-31.

Peridinium divergens
Gourret, P., 1883
Sur les Peridiniens du Golfe de Marseille. Ann. du Musee d'hist. Nat., Marseille, Zool. 1 (Mme. 8):1-114, 4 pls.

Peridinium divergens
Gran, H.H., and T. Braarud, 1935
A quantitative study of the phytoplankton in the Bay of Fundy and the Gulf of Maine (including observations on hydrography, chemistry, and turbidity). J. Biol. Bd., Canada, 1(5):279-467, 69 text figs.

Peridinium divergens
Jørgensen, E., 1905
B. Protistplankton and the diatoms in bottom samples. Hydrographical and biological investigations in Norwegian fiords. Bergens Mus. Skr. 7: 49-225.

Peridinium divergens
LeBour, M.V., 1925
The dinoflagellates of Northern Seas. The Marine Biological Association of the United Kingdom, Plymouth, 250 pp., 35 pls., 53 text figs.

Peridinium divergens
Lillick, L.C., 1940
Phytoplankton and planktonic protozoa of the offshore waters of the Gulf of Maine. Pt. II. Qualitative Composition of the Planktonic Flora. Trans. Am. Phil. Soc., n.s., 31(3):193-237, 13 text figs.

Peridinium divergens
Lindemann, E., 1924
Peridineen aus dem goldenen Horn und dem Bosphorus. Bot. Arch. 5:216-233, 98 text figs.

Peridinium divergens
Mangin, M. L., 1912
Phytoplancton de la croisière du "René" dans l'Atlantique (Septembre 1908). Ann. Inst. Ocean., n.s., 4(1):1-66, 2 pls., 41 text figs., 2 tables.

Peridinium divergens
Matzenauer, L., 1933
Die Dinoflagellaten des indischen Ozeans (mit Ausnahme der Gattung Ceratium.) Bot. Arch. 35:437-510, 77 text figs., 2 charts.

Peridinium divergens
Meunier, A., 1919
Microplancton de la Mer Flamande 3. Les Péridiniens. Mem. Mus. Roy. Hist. Nat., Belgique 8(1):1-116, Pls. XV-XXI.

Peridinium divergens
Moberg, E.G., and W.E. Allen, 1927.
Effect of tidal changes on physical, chemical, and biological conditions of the sea water of the San Diego region. 1. Observations on the effect of tidal changes on physical and chemical conditions of sea water in the San Diego region. 2. Half-hourly collections of marine microplankton taken at the Scripps Institution pier in 1923 Bull. S.I.O., tech. ser., 1:1-17, 4 textfigs.

Peridinium divergens
Murray, G., and F. G. Whitting, 1899
New Peridiniaceae from the Atlantic. Trans. Linn. Soc., London, Bot., ser 2, 5: 321-342, Pls. 27-33, 9 tables.

Peridinium divergens
Paiva Carvalho, J., 1950
O plancton do Rio Maria Rodriques (Cananeis). 1. Diatomaceas e Dinoflagelados. Bol. Inst. Paulista Oceanogr. 1(1); 27-43, 2 fold-in tables, 2 figs.

Peridinium divergens
Pavillard, J., 1905
Recherches sur la flore pelagique (Phytoplankton) de l'Etang de Thau. Theses presentees a la Fac. Sci., Paris, 116 pp., 3 pls.

Peridinium divergens
Rampi, L., 1950.
Ricerche sul fitoplancton del Mare Ligure. 9) I Peridinium delle acque di Sanremo. Pubbl. Centro Talassografico Tirreno No. 5:10 pp. 2 pls.

Atti Acad. Ligure Sci. e Lett. 7(1):

Peridinium divergens
Schodduyn, M., 1926
Observations faites dans la baie d'Ambleteuse (Pas de Calais). Bull. Inst. Ocean., Monaco, No. 482: 64 pp.

Peridinium divergens
Schröder, B., 1900
Phytoplankton des Golfes von Neapel nebst vergleichenden Ausblicken auf das atlantischen Ozean. Mitt. Zool. Stat. Neapel, 14:1-38.

Peridinium divergens
Sousa e Silva, E., 1949
Diatomaceas e Dinoflagelados de Baia de Cascais. Portugaliae Acta Biol., Volume: Julio Henriques, Ser. B: 300-383, 9 pls, 2 fold-in tables.

Peridinium divergens
Sousa a Silva, E., and J. Dos Santos-Pinto, 1948
O Plancton da Baia de S. Martinho do Porto. 1. Diatomaceas e Dinoflagelados. Bol. Soc. Portuguese de Ciencias Naturais, 16(2):134-187, 5 pls. (Trav. Sta. Biol. Mar. de Lisbonne No. 52).

Peridinium divergens
Steemann-Nielsen, Einar, 1951
The marine vegetation of the Isefjord. A study on ecology and production. Medd. Komm. Danmarks Fiskeri-og Havundersøgelser. Ser. Plankton. 5(4); 114pp., 46 text figs.

Peridinium divergens
Wang, C. C., and D. Nie, 1932
A survey of the marine protozoa of Amoy. Contr. Biol. Lab. Sci. Soc., China, Zool. Ser., 8:285-385, 89 text figs.

Peridinium divergens
Wood, E.J.F., 1954.
Dinoflagellates in the Australian regions Australian J. Mar. Freshwater Res., 5(2):171-351.

Peridinium doma n.sp.
Murray, G., and F. G. Whitting, 1899
New Peridiniaceae from the Atlantic. Trans. Linn. Soc., London, Bot., ser 2, 5: 321-342, Pls. 27-33, 9 tables.

Peridinium elegans
Matzenauer, L., 1933
Die Dinoflagellaten des indischen Ozeans (mit Ausnahme der Gattung Ceratium.) Bot. Arch. 35:437-510, 77 text figs., 2 charts.

Peridinium elegans
Pavillard, J., 1916
Recherches sur les Peridiniens du Golfe du Lion. Mem. Univ. Montpellier. Trav. Inst. Bot., Univ. Montpellier. Serie mixte No.4, 70 pp., 3 pls., 15 text figs.

Peridinium elegans
Wood, E.J.F., 1954.
Dinoflagellates in the Australian region. Australian J. Mar. Freshwater Res., 5(2):171-351.

Peridinium elegantissimum n.sp.
Balech, Enrique, 1958.
Plancton de la Campana Antartica Argentina, 1954-1955. Physis, 21(60):75-108.

Peridinium ellipsoides
Dangeard, P., 1927
Phytoplankton de la croisière du "Sylvana". Ann. Inst. Ocean., Monaco, n.s., 4(8):286-401, 54 text figs. (février-Juin 1913).

Peridinium ellipsoideum n.sp.
Dangeard, P., 1927
Péridinium nouveaux ou peu connus de la croisière du "Sylvana". Bull. Inst. Océan., Monaco, No. 491:16 pp., 9 text figs.

Peridinium elongatum
LeBour, M.V., 1925
The dinoflagellates of Northern Seas. The Marine Biological Association of the United Kingdom, Plymouth, 250 pp., 35 pls., 53 text figs.

Peridinium excentricum
Balech, E., 1951.
Deuxième contribution à la connaissance des Peridinium. Hydrobiol., Acta Hydrobiol., Limnol. Protistol. 3(4):305-330, 7 pls. of 137 figs.

Peridinium excentricum
Dangeard, P., 1927
Phytoplankton de la croisière du "Sylvana". Ann. Inst. Ocean., Monaco, n.s., 4(8):286-401, 54 text figs. (février-Juin 1913).

Peridinium excentricum
Dangeard, P., 1926
Description des Péridiniens Testacés recueillis para la Mission Charcot pendent le mois d'Aout 1924. Ann. Inst. Ocean. n.s. 3(7):307-334, 15 text figs.

Peridinium excentricum
Gran, H.H., and T. Braarud, 1935
A quantitative study of the phytoplankton in the Bay of Fundy and the Gulf of Maine (including observations on hydrography, chemistry, and turbidity). J. Biol. Bd., Canada, 1(5):279-467, 69 text figs.

Peridinium excentricum
LeBour, M.V., 1925
The dinoflagellates of Northern Seas. The Marine Biological Association of the United Kingdom, Plymouth, 250 pp., 35 pls., 53 text figs.

Peridinium excentricum
Lillick, L.C., 1940
Phytoplankton and planktonic protozoa of the offshore waters of the Gulf of Maine. Pt.II. Qualitative Composition of the Planktonic Flora. Trans. Am. Phil. Soc., n.s., 31(3):193-237, 13 text figs.

Peridinium excentricum
Mangin, M. L., 1912
Phytoplancton de la croisière du "René" dans l'Atlantique (Septembre 1908). Ann. Inst. Ocean., n.s., 4(1):1-66, 2 pls., 41 text figs., 2 tables.

Peridinium excentricum
Matzenauer, L., 1933
Die Dinoflagellaten des indischen Ozeans (mit Ausnahme der Gattung Ceratium.) Bot. Arch. 35:437-510, 77 text figs., 2 charts.

Peridinium excentricum
Meunier, A., 1919
Microplancton de la Mer Flamande 3. Les Péridiniens. Mem. Mus. Roy. Hist. Nat., Belgique 8(1):1-116, Pls. XV-XXI.

Peridinium excentricum
Paulsen, O., 1908
XVIII Peridiniales. Nordisches Plankton, Bot. Teil: 1-124, 155 text figs.

Peridinium excentricum
Pavillard, J., 1916
Recherches sur les Peridiniens du Golfe du Lion. Mem. Univ. Montpellier. Trav. Inst. Bot., Univ. Montpellier. Serie mixte No.4, 70 pp., 3 pls., 15 text figs.

Peridinium excentricum
Sousa a Silva, E., and J. Dos Santos-Pinto, 1948
O Plancton da Baia de S. Martinho do Porto. 1. Diatomaceas e Dinoflagelados. Bol. Soc. Portuguese de Ciencias Naturais, 16(2):134-187, 5 pls. (Trav. Sta. Biol. Mar. de Lisbonne No. 52).

Peridinium excentricum
Wall, David, and Barrie Dale, 1968.
Modern dinoflagellate cysts and evolution of the Peridiniales. Micropaleontology, 14(3):265-304.

Peridinium excentricum
Wood, E.J.F., 1954.
Dinoflagellates in the Australian region. Australian J. Mar. Freshwater Res., 5(2):171-351.

Peridinium exiguum
Paulsen, O., 1908
XVIII Peridiniales. Nordisches Plankton, Bot. Teil: 1-124, 155 text figs.

Peridinium faeroense see P. thorianum
Gran, H.H., and T. Braarud, 1935
A quantitative study of the phytoplankton in the Bay of Fundy and the Gulf of Maine (including observations on hydrography, chemistry, and turbidity). J. Biol. Bd., Canada, 1(5):279-467, 69 text figs.

Peridinium faeroense
LeBour, M.V., 1925
The dinoflagellates of Northern Seas. The Marine Biological Association of the United Kingdom, Plymouth, 250 pp., 35 pls., 53 text figs.

Peridinium faeroense
Morse, D.C., 1947
Some observations on seasonal variations in plankton population Patuxent River, Maryland 1943 1945. Bd. Nat. Res., Publ. No.65, Chesapeake Biol. Lab. 31, 3 figs.

Peridinium faeroense
Paulsen, O., 1908
XVIII Peridiniales. Nordisches Plankton, Bot. Teil: 1-124, 155 text figs.

Peridinium faroense n.sp.
Paulsen, O., 1905.
On some Peridineae and plankton diatoms. Medd. Komm. Havundersøgelser, Ser. Plankton, 1(3):7 pp.

Peridinium fatulipes
Matzenauer, L., 1933
Die Dinoflagellaten des indischen Ozeans (mit Ausnahme der Gattung Ceratium.) Bot. Arch. 35:437-510, 77 text figs., 2 charts.

Peridinium fatulipes
Wood, E.J.F., 1963
Dinoflagellates in the Australian region. II. Recent collections. Comm. Sci. and Industr. Res. Org., Div. Fish. and Oceanogr., Techn. Paper, No. 14: 55 pp.

Peridinium fimbriatum
LeBour, M.V., 1925
The dinoflagellates of Northern Seas. The Marine Biological Association of the United Kingdom, Plymouth, 250 pp., 35 pls., 53 text figs.

Peridinium fimbriatum n.sp.
Meunier, A., 1919
Microplancton de la Mer Flamande 3. Les Péridiniens. Mem. Mus. Roy. Hist. Nat. Belgique 8(1):1-116, Pls. XV-XXI.

Peridinium finlandicum
Levander, K.M., 1947
Plankton gesammelt in den Jahren 1899-1910 an den Küsten Finnlands. Finnländische Hydrographisch-Biologische Untersuchunger (aus dem Wasserbiologischen Laboratorium der Societas Scientiarum Fennica) No.11: 40 pp., 6 diagrams, 13 pls., tables.

Peridinium finlandicum
Paulsen, O., 1908
XVIII Peridiniales. Nordisches Plankton, Bot. Teil: 1-124, 155 text figs.

Peridinium foliaceum
Becheler, B., 1952.
Recherches sur les Peridiniens. Bull. Biol., France Belg., Suppl., 36:1-149, 75 textfigs.

Peridinium foliaceum
Rae, B.B., R. Johnston and J.A. Adams, 1965.
The incidence of dead and dying fish in the Moray Firth, September 1963.
Jour. Mar. Biol. Assoc., U.K., 45(1):29-47.

Peridinium formosum
Dangeard, P., 1927
Phytoplankton de la croisière du "Sylvana". Ann. Inst. Ocean., Monaco, n.s., 4(8):286-401, 54 text figs. (février-Juin 1913).

Peridinium formosum
Forti, A., 1922
Ricerche sulla flora pelagica (fitoplancton) di Quarto dei Mille. Mem. R. Com. Talass. Ital. 97:248 pp., 13 pls.

Peridinium formosum
Pavillard, J., 1916
Recherches sur les Peridiniens du Golfe du Lion. Mem. Univ. Montpellier. Trav. Inst. Bot. Univ. Montpellier. Serie mixte No.4, 70 pp., 3 pls., 15 text figs.

Peridinium gainii
Argentina, Secretaria de Marina, Servicio de Hidrografia Naval, 1962.
Plancton de las campanas oceanograficas DRAKE I y II.
Publico, H. 627:57.

Peridinium garderae n.sp.
Balech, Enrique, 1967.
Dinoflagelados nuevos o interesantes del Golfo de Mexico y Caribe.
Revta Mus. argent. Cienc. nat. Bernardino Rivadavia Inst. nac. Invest. Cienc. nat., Hidrobiol. 2(3):77-126.

Peridinium Gainii n.sp.
Dangeard, P., 1927
Phytoplankton de la croisière du "Sylvana". Ann. Inst. Ocean., Monaco, n.s., 4(8):286-401, 54 text figs. (février-Juin 1913).

Peridinium gargantua
Becheler, B., 1952.
Recherches sur les Peridiniens. Bull. Biol., France Belg., Suppl., 36:1-149, 75 textfigs.

Peridinium gatunense
Wood, E.J.F., 1954.
Dinoflagellates in the Australian region.
Australian J. Mar. Freshwater Res., 5(2):171-351.

Peridinium gibbosum
Balech, Enrique, 1971.
Microplancton de la campaña oceanographica: Productividad III.
Revta Mus. argent. Cienc. Nat. Bernadina Rivadavia, Hydrobiol. 3(1):1-202, 39 pls.

Peridinium gibbosum n.sp.
Matzenauer, L., 1933
Die Dinoflagellaten des indischen Ozeans (mit Ausnahme der Gattung Ceratium.) Bot. Arch. 35:437-510, 77 text figs., 2 charts.

Peridinium globulus
Braarud, T., 1945
A phytoplankton survey of the polluted waters of inner Oslo Fjord. Hvalrådets Skrifter, No.28, 142 pp., 19 text figs., 17 tables.

Peridinium globulus
Braarud, T., and Adam Bursa, 1939
On the phytoplankton of the Oslo Fjord, 1933-1934. Hvalrådets Skr. No.19:1-63; 9 text figs. Reviewed. J. du. Cons. 14(3): 418-420. A.C. Gardiner.

Peridinium globulus
Dangeard, P., 1927
Phytoplankton de la croisière du "Sylvana". Ann. Inst. Ocean., Monaco, n.s., 4(8):286-401, 54 text figs. (février-Juin 1913).

Peridinium globulus
Dangeard, P., 1927
Notes sur la variation de la genre Peridinium. Bull. Inst. Ocean., Monaco, No. 507:16 pp., 9 textfigs.

Peridinium globulus
Ercegovic, A., 1936
Etudes qualitative et quantitatives du phytoplancton dans les eaux cotières de l'Adriatique oriental moyen au cours de l'année 1934. Acta Adriatica 1(9):1-126

Peridinium globulus
Forti, A., 1922
Ricerche sulla flora pelagica (fitoplancton) di Quarto dei Mille. Mem. R. Com. Talass. Ital. 97:248 pp., 13 pls.

Peridinium globulus
Jorgensen, E., 1900
Protophyten und Protozoën im Plankton aus der Norwegischen Westkerste. Bergens Mus. Aarb. 1899(6): 95 pp., 5 pls., 83 tables.

Peridinium globulus
LeBour, M.V., 1925
The dinoflagellates of Northern Seas. The Marine Biological Association of the United Kingdom, Plymouth, 250 pp., 35 pls., 53 text figs.

Peridinium globulus
Lillick, L.C., 1940
Phytoplankton and planktonic protozoa of the offshore waters of the Gulf of Maine. Pt.II. Qualitative Composition of the Planktonic Flora. Trans. Am. Phil. Soc., n.s., 31(3):193-237, 13 text figs.

Peridinium globulus
Lindemann, E., 1925
Neubeobachtungen an den Winter peridineen des Golfes von Neapel. Bot. Arch. 9:95-102, 19 text figs.

Peridinium globulus
Lindemann, E., 1924
Peridineen aus dem goldenen Horn und dem Bosphorus. Bot. Arch. 5:216-233, 98 text figs.

Peridinium globulus
Mangin, M. L., 1912
Phytoplankton de la croisière du "René" dans l'Atlantique (Septembre 1908). Ann. Inst. Ocean., n.s., 4(1):1-66, 2 pls., 41 text figs., 2 tables.

Peridinium globulus
Matzenauer, L., 1933
Die Dinoflagellaten des indischen Ozeans (mit Ausnahme der Gattung Ceratium.) Bot. Arch. 35:437-510, 77 text figs., 2 charts.

Peridinium globulus
Murray, G. and F. G. Whitting, 1899
New Peridiniaceae from the Atlantic. Trans. Linn. Soc., London, Bot., ser 2, 5: 321-342, Pls. 27-33, 9 tables.

Peridinium globulus
Paulsen, O., 1908
XVIII Peridiniales. Nordisches Plankton, Bot. Teil: 1-124, 155 text figs.

Peridinium globulus
Meunier, A., 1919
Microplancton de la Mer Flamande 3. Les Péridiniens. Mem. Mus. Roy. Hist. Nat. Belgique 8(1):1-116, Pls. XV-XXI.

Peridinium globulus
Pavillard, J., 1916
Recherches sur les Peridiniens du Golfe du Lion. Mem. Univ. Montpellier. Trav. Inst. Bot. Univ. Montpellier. Serie mixte No.4, 70 pp., 3 pls., 15 text figs.

Peridinium globulus
Pavillard, J., 1905
Recherches sur la flore pelagique (Phytoplankton) de l'Etang de Thau. Theses presentees a la Fac. Sci., Paris, 116 pp., 3 pls.

Peridinium globulus & var.
Rampi, L., 1950.
Ricerche sul fitoplancton del Mare Ligure. 9) I Peridinium dell'acque di Sanremo. Pubbl. Centro Talassografico Tirreno No. 5:20 pp. 2 pls.

Atti Acad. Ligure Sci. e Lett. 7(1):

Peridinium globulus
Sousa e Silva, E., 1949
Diatomaceas e Dinoflagelados de Baia de Cascais. Portugaliae Acta Biol., Volume: Julio Henriques, Ser. B: 300-383, 9 pls, 2 fold-in tables.

Peridinium globulus
Wood, E.J.F., 1954.
Dinoflagellates in the Australian region.
Australian J. Mar. Freshwater Res., 5(2):171-351.

Peridinium globulus Quarnerense
Ercegovic, A., 1936
Etudes qualitative et quantitatives du phytoplancton dans les eaux cotières de l'Adriatique oriental moyen au cours de l'année 1934. Acta Adriatica 1(9):1-126

Peridinium globulus quarnerensis n. var.
Schröder, B., 1900
Phytoplankton des Golfes von Neapel nebst vergleichenden Ausblicken auf das atlantischen Ozean. Mitt. Zool. Stat. Neapel, 14:1-38.

Peridinium globulus ovatum
Ercegovic, A., 1940
Weitere Untersuchungen über einige hydrographische Verhältnisse und über die Phytoplanktonproduktion in den Gewässern der Östlichen Mittelladria. Acta Adriatica 2(3):95-134, 8 text figs.

Peridinium globulus var. ovatum
Lillick, L.C., 1940
Phytoplankton and planktonic protozoa of the offshore waters of the Gulf of Maine. Pt.II. Qualitative Composition of the Planktonic Flora. Trans. Am. Phil. Soc., n.s., 31(3):193-237, 13 text figs.

Peridinium globulus var. quarnerensis
Lillick, L.C., 1940
Phytoplankton and planktonic protozoa of the offshore waters of the Gulf of Maine. Pt.II. Qualitative Composition of the Planktonic Flora. Trans. Am. Phil. Soc., n.s., 31(3):193-237, 13 text figs.

Peridinium globosum n.sp.
Dangeard, P., 1927
Phytoplankton de la croisière du "Sylvana". Ann. Inst. Ocean., Monaco, n.s., 4(8):286-401, 54 text figs. (février-Juin 1913).

Peridinium gracile n.sp.
Gran, H.H., and T. Braarud, 1935
A quantitative study of the phytoplankton in the Bay of Fundy and the Gulf of Maine (including observations on hydrography, chemistry, and turbidity). J. Biol. Bd., Canada, 1(5):279-467, 69 text figs.

Peridinium gracile
Lillick, L.C., 1937
Seasonal studies of the phytoplankton off Woods Hole, Massachusetts. Biol. Bull. LXXIII (3):488-503, 3 text figs.

Peridinium grandi
Dangeard, P., 1927
Phytoplankton de la croisière du "Sylvana". Ann. Inst. Ocean., Monaco, n.s., 4(8):286-401, 54 text figs. (février-Juin 1913).

Peridinium grande
Matzenauer, L., 1933
Die Dinoflagellaten des indischen Ozeans (mit Ausnahme der Gattung Ceratium.) Bot. Arch. 35:437-510, 77 text figs., 2 charts.

Peridinium grande
Pavillard, J., 1916
Recherches sur les Peridiniens du Golfe du Lion. Mem. Univ. Montpellier. Trav. Inst. Bot., Univ. Montpellier. Serie mixte No.4, 70 pp., 3 pls., 15 text figs.

Peridinium grande
Rampi, L., 1950.
Ricerche sul fitoplancton del Mare Ligure. 9) I Peridinium delle acque di Sanremo. Pubbl. Centro Talassografico Tirreno No. 5:10 pp. 2 pls.
Ligure
Atti Acad. Sci. e Lett. 7(1):

Peridinium grande
Wood, E.J.P., 1954.
Dinoflagellates in the Australian region. Australian J. Mar. Freshwater Res., 5(2):171-351.

Peridinium granii
Balech, Enrique, 1971.
Microplancton de la campaña oceanographica: Productividad III.
Revta Mus. argent. Cienc. Nat. Bernadina Rivadavia, Hydrobiol. 3(1):1-202, 39 pls.

Peridinium Granii
Braarud, T., 1945
A phytoplankton survey of the polluted waters of inner Oslo Fjord. Hvalrådets Skrifter, No.28, 142 pp., 19 text figs., 17 tables.

Peridinium Grani
Braarud, T., and Adam Bursa, 1939
On the phytoplankton of the Oslo Fjord, 1933-1934. Hvalrådets Skr. No.19:1-63; 9 text figs. Reviewed. J. du. Cons. 14(3): 418-420. A.C. Gardiner.

Peridinium Granii
Dangeard, P., 1927
Notes sur la variation de la genre Peridinium. Bull. Inst. Ocean., Monaco, No. 507:16 pp., 9 textfigs.

Peridinium granii
Ganapati, P.N., D.G.V. Prasada Rao and M.V. Lakshmana Rao, 1959.
Bioluminescence in Visakha Patnam Harbour. Current Sci., 28:246-427.

Peridinium Grani
Gran, H.H., and T. Braarud, 1935
A quantitative study of the phytoplankton in the Bay of Fundy and the Gulf of Maine (including observations on hydrography, chemistry, and turbidity). J. Biol. Bd., Canada, 1(5):279-467, 69 text figs.

Peridinium granii
Kelly, Mahlon G., and Steven Katona, 1966.
An endogenous diurnal rhythm of bioluminescence in a natural population of dinoflagellates. Biol. Bull., 131(1):115-126.

Peridinium granii
LeBour, M.V., 1925
The dinoflagellates of Northern Seas. The Marine Biological Association of the United Kingdom, Plymouth, 250 pp., 35 pls.,53 text figs.

Peridinium Granii
Lillick, L.C., 1940
Phytoplankton and planktonic protozoa of the offshore waters of the Gulf of Maine. Pt.II. Qualitative Composition of the Planktonic Flora. Trans. Am. Phil. Soc., n.s., 31(3):193-237, 13 text figs.

Peridinium Granii
Lillick, L.C., 1938
Preliminary report of the phytoplankton of the Gulf of Maine. Am. Mid. Nat. 20(3):624-640, 1 text figs 37 tables.

Peridinium Granii
Lillick, L.C., 1937
Seasonal studies of the phytoplankton off Woods Hole, Massachusetts. Biol. Bull. LXXIII (3):488-503, 3 text figs.

Peridinium granii
Lindemann, E., 1924
Peridineen aus dem goldenen Horn und dem Bosphorus. Bot. Arch. 5:216-233, 98 text figs.

Peridinium Granii
Matzenauer, L., 1933
Die Dinoflagellaten des indischen Ozeans (mit Ausnahme der Gattung Ceratium.) Bot. Arch. 35:437-510, 77 text figs., 2 charts.

Peridinium Granii
Meunier, A., 1919
Microplancton de la Mer Flamande 3. Les Péridiniens. Mem. Mus. Roy. Hist. Nat., Belgique 8(1):1-116, Pls. XV-XXI.

Peridinium granii
Nordli, O., 1951.
Dinoflagellates from Lofoten. Nytt Mag. Naturvidensk 88:49-55, 7 textfigs.

Peridinium Granii
Paulsen, O., 1908
XVIII Peridiniales. Nordisches Plankton, Bot. Teil: 1-124, 155 text figs.

Peridinium Granii
Rumkówna, A., 1948
List of the phytoplankton species occurring in the superficial water layers in the Gulf of Gdańsk. Bull. Lab. mar., Gdynia, No. 4: 139-141 with tables in back.

Peridinium Grani
Schodduyn, M., 1926
Observations faites dans la baie d'Ambleteuse (Pas de Calais). Bull. Inst. Ocean., Monaco, No. 482: 64 pp.

Peridinium grani
Schulz, B., and A. Wulff, 1929
Hydrographie und Oberflächenplankton des westlichen Barentsmeeres im Sommer 1927. Ber. deutschen wissensch. Komm. F. Meeresforsch. n.s. 4(5):232-372, 13 tables, 25 text figs.

Peridinium granii & var.
Rampi, L., 1950.
Ricerche sul fitoplancton del Mare Ligure. 9) I Peridinium delle acque di Sanremo. Pubbl. Centro Talassografico Tirreno No. 5:10 pp. 2 pls.

Atti Acad. Ligure Sci. e Lett. 7(1):

Peridinium granii
Wood, E.J.P., 1954.
Dinoflagellates in the Australian region. Australian J. Mar. Freshwater Res., 5(2):171-351.

Peridinium gregarium
Lombard, Eugene H., and Brian Capon 1971.
Observations on the tidepool ecology and behavior of Peridinium gregarium. J. Phycol. 7(3):158-194.

Peridinium gregarium n.sp.
Lombard, Eugene H., and Brian Capon 1971.
Peridinium gregarium, a new species of dinoflagellate.
J. Phycol. 7(3):184-187.

Peridinium grenlandicum
Rumkówna, A., 1948
List of the phytoplankton species occurring in the superficial water layers in the Gulf of Gdańsk. Bull. Lab. mar., Gdynia, No. 4: 139-141 with tables in back.

Peridinium hangoei
Lillick, L.C., 1940
Phytoplankton and planktonic protozoa of the offshore waters of the Gulf of Maine. Pt.II. Qualitative Composition of the Planktonic Flora. Trans. Am. Phil. Soc., n.s., 31(3):193-237, 13 text figs.

Peridinium helix n.sp
Balech, Enrique, 1962
Tintinnoinea y Dinoflagellata del Pacifico segun material de las Expediciones NORPAC y DOWNWIND del Instituto Scripps de Oceanografia.
Revista. Mus. Argentino Ciencias Nat. "Bernardino Rivadavia", Ciencias Zool., 7(1):1-253.

Peridinium heteracanthum
Dangeard, P., 1927
Phytoplankton de la croisière du "Sylvana". Ann. Inst. Ocean., Monaco, n.s., 4(8):286-401, 54 text figs. (février-Juin 1913).

Peridinium heteracanthum n.sp.
Dangeard, P., 1927.
Péridinium nouveaux ou peu connus de la croisière du "Sylvana". Bull. Inst. Ocean., Monaco, No. 491:16 pp., 9 textfigs.

Peridinium heteracanthum
Matzenauer, L., 1933
Die Dinoflagellaten des indischen Ozeans (mit Ausnahme der Gattung Ceratium.) Bot. Arch. 35:437-510, 77 text figs., 2 charts.

Peridinium heteroconicum n.sp.
Matzenauer, L., 1933
Die Dinoflagellaten des indischen Ozeans (mit Ausnahme der Gattung Ceratium.) Bot. Arch. 35:437-510, 77 text figs., 2 charts.

Peridinium Hindmarchii n.sp.
Murray, G., and F. G. Whitting, 1899
New Peridiniaceae from the Atlantic.
Trans. Linn. Soc., London, Bot., ser 2, 5: 321-342, Pls. 27-33, 9 tables.

Peridinium hirobis?
Balech, Enrique, 1971.
Microplancton de la campaña oceanographica: Productividad III.
Revta Mus. argent. Cienc. Nat. Bernadina Rivadavia, Hydrobiol. 3(1):1-202, 39 pls.

Peridinium hirobis
Wood, E.J.P., 1954.
Dinoflagellates in the Australian region.
Australian J. Mar. Freshwater Res., 5(2):171-351.

Peridinium incertum n.sp.
Balech, Enrique, 1958.
Plancton de la Campaña Antartica Argentina, 1954-1955.
Physis, 21(60)75-108.

Peridinium incertum
Hada, Yoshina, 1970. (1970)
The protozoan plankton of the Antarctic and Subantarctic seas.
Scient. Repts, Japan. Antarct. Res. Exped., (E)31:51 pp.

Peridinium inclinatum n. sp.
Balech, Enrique, 1964.
Tercera contribucion al conocimiento del genero "Peridinium".
Revista, Mus. Argentino, Ciencias Nat. "Bernardino Rivadavia" e Inst. Nacional, Invest. Ciencias Naturales, Hidrobiol., 1(6):179-195.

Peridinium incognitum
Argentina, Secretaria de Marina, Servicio de Hidrografia Naval, 1962.
Plancton de las campanas oceanograficas DRAKE I y II.
Publico, H. 627:57.

Peridinium incognitum
Balech, Enrique, 1971.
Microplancton de la campaña oceanographica: Productividad III.
Revta Mus. argent. Cienc. Nat. Bernadina Rivadavia, Hydrobiol. 3(1):1-202, 39 pls.

Peridinium incognitum n.sp.
Balech, Enrique, 1959.
Operacion Oceanografica Merluza V Crucero. Plancton.
Publico, Servicio de Hidrografia Naval, Argentina, H. 618:43 pp.

Peridinium inconspicuum
Rumkówna, A., 1948
[List of the phytoplankton species occurring in the superficial water layers in the Gulf of Gdańsk] Bull. Lab. mar., Gdynia, No. 4: 139-141 with tables in back.

Peridinium inconspicuum
Wood, E.J.P., 1954.
Dinoflagellates in the Australian region.
Australian J. Mar. Freshwater Res., 5(2):171-351.

Peridinium inflatum
Margalef, R., 1949
Fitoplanton nerítico de la Costa Brava en 1947-48. Publ. Inst. Biol. Aplicada, 5: 41-51, 3 text figs.

Peridinium inflatum
Matzenauer, L., 1933
Die Dinoflagellaten des indischen Ozeans (mit Ausnahme der Gattung Ceratium.) Bot. Arch. 35:437-510, 77 text figs., 2 charts.

Peridinium inflexum
Marukawa, H., 1921
Plankton lists and some new species of copepods, from the northern waters of Japan.
Bull. Inst. Ocean., No.384, 15 pp., 3 pls., 1 chart. Monaco

Peridinium islandicum
Lebour, M.V., 1925.
The dinoflagellates of northern seas. Marine Biological Association, United Kingdom, 250 pp., 35 pls., 53 textfigs.

Peridinium islandicum
Marukawa, H., 1921
Plankton lists and some new species of copepods, from the northern waters of Japan.
Bull. Inst. Ocean., No.384, 15 pp., 3 pls., 1 chart. Monaco

Peridineum islandicum
Paulsen, O., 1949
Observations on dinoflagellates. (Ed. J. Grøntoed) Kongl. Dansk. Videnskab. Selsk. Biol. Skr. 6(4):67 pp., 30 text figs.

Peridinium islandicum
Paulsen, O., 1908
XVIII Peridiniales. Nordisches Plankton, Bot. Teil: 1-124, 155 text figs.

Peridinium islandicum
Schulz, B., and A. Wulff, 1929
Hydrographie und Oberflächen plankton des westlichen Barentsmeeres im Sommer 1927.
Ber. deutschen wissensch. Komm. F. Meeres-Forsch. n.s. 4(5):232-372, 13 tables, 25 text figs.

Peridinium joergenseni n. sp.
Balech, Enrique, 1971.
Microplancton de la campaña oceanographica: Productividad III.
Revta Mus. argent. Cienc. Nat. Bernadina Rivadavia, Hydrobiol. 3(1):1-202, 39 pls.

Peridinium joubini n.sp.
Dangeard, P., 1927
Phytoplankton de la croisière du "Sylvana". Ann. Inst. Ocean., Monaco, n.s., 4(8):286-401, 54 text figs. (Feirer-Juin 1913).

Peridinium latidorsale nov. comb
Balech, E., 1951.
Deuxième contribution à la connaissance des Peridinium. Hydrobiol., Acta, Hydrobiol., Limnol. Protistol. 3(4):305-330, 7 pls. of 137 figs.

Peridinium latipyrum n.sp.
Balech, Enrique, 1959.
Operacion Oceanografica Merluza V Crucero. Plancton.
Publico, Servicio de Hidrografica Naval, Argentina, H. 618:43 pp.

Peridinium latispinum
Wood, E.J.P., 1954.
Dinoflagellates in the Australian region.
Australian J. Mar. Freshwater Res., 5(2):171-351.

Peridinium latissimum
Argentina, Secretaria, de Marina, Servicio de Hidrografia Naval, 1962.
Plancton de las campanas oceanograficas DRAKE I y II.
Publico, H. 627:57.

Peridinium latissimum
Matzenauer, L., 1933
Die Dinoflagellaten des indischen Ozeans (mit Ausnahme der Gattung Ceratium.) Bot. Arch. 35:437-510, 77 text figs., 2 charts.

Peridinium latissimum
Wall, David, and Barrie Dale, 1968.
Modern dinoflagellate cysts and evolution of the Peridiniales.
Micropaleontology, 14(3):265-304.

Peridinium latistriatum n.sp.
Balech, Enrique, 1958.
Plancton de la Campaña Antartica Argentina, 1954-1955.
Physis, 21(60):75-108.

Peridinium latum nom. nov.
Paulsen, O., 1908
XVIII Peridiniales. Nordisches Plankton, Bot. Teil: 1-124, 155 text figs.

Peridinium latum
Rumkówna, A., 1948
[List of the phytoplankton species occurring in the superficial water layers in the Gulf of Gdańsk] Bull. Lab. mar., Gdynia, No. 4: 139-141 with tables in back.

Peridinium latum
Wood, E.J.P., 1954.
Dinoflagellates in the Australian region.
Australian J. Mar. Freshwater Res., 5(2):171-351.

Peridinium leiorhynchum n.sp.
Murray, G., and F. G. Whitting, 1899
New Peridiniaceae from the Atlantic.
Trans. Linn. Soc., London, Bot., ser 2, 5: 321-342, Pls. 27-33, 9 tables.

Peridinium Leonis
Dangeard, P., 1927
Phytoplankton de la croisière du "Sylvana". Ann. Inst. Ocean., Monaco, n.s., 4(8):286-401, 54 text figs. (Février-Juin 1913).

Peridinium leonis
Dangeard, P., 1927.
Notes sur la variation de la genre Peridinium. Bull. Inst. Ocean., Monaco, No. 507: 16 pp., 9 textfigs.

Peridinium leonis
Evitt, William R., and Susan E. Davidson, 1964.
Dinoflagellate studies. 1. Dinoflagellate cysts and thecae.
Stanford Univ. Publ., Geol. Sci., 10(1):1-12.

Peridinium leonis
Forti, A., 1922
Ricerche sulla flora pelagica (fitoplancton) di Quarto dei Mille. Mem. R. Com. Talass. Ital. 97:248 pp., 13 pls.

Peridinium leonis
Kelly, Mahlon G., and Steven Katona, 1966.
An endogenous diurnal rhythm of bioluminescence in a natural population of dinoflagellates.
Biol. Bull., 131(1):115-126.

Peridinium leonis
LeBour, M.V., 1925.
The dinoflagellates of Northern Seas. The Marine Biological Association of the United Kingdom, Plymouth, 250 pp., 35 pls., 53 text figs.

Peridinium leonis
Matzenauer, L., 1933
Die Dinoflagellaten des indischen Ozeans (mit Ausnahme der Gattung Ceratium.) Bot. Arch. 35:437-510, 77 text figs., 2 charts.

Peridinium leonis
Morse, D.C., 1947
Some observations on seasonal variations in plankton population Patuxant River, Maryland 1943-1945. Bd. Nat. Res., Publ. No.65, Chesapeake Biol. Lab., 31, 3 figs.

Peridinium leonis nom.nov.
Pavillard, J., 1916
Recherches sur les Peridiniens du Golfe du Lion. Mem. Univ. Montpellier. Trav. Inst. Bot., Univ. Montpellier. Serie mixte No.4, 70 pp., 3 pls., 15 text figs.

Peridinium leonis
Rampi, L., 1950.
Ricerche sul fitoplancton del Mare Ligure. 9) I Peridinium delle acque di Sanremo. Pubbl. Centro Talassografico Tirreno No. 5:10 pp. 2 pls.

Atti Acad. Ligure Sci. e Lett. 7(1):

Peridinium leonis
Sousa e Silva, E., 1949
Diatomaceas e Dinoflagelados de Baia de Cascais. Portugaliae Acta Biol., Volume: Julio Henriques, Ser. B: 300-383, 9 pls, 2 fold-in tables.

Peridinium leonis
Sousa a Silva, E., and J. Dos Santos-Pinto, 1948
O Plancton da Baia de S. Martinho do Porto. 1. Diatomaceas e Dinoflagelados. Bol. Soc. Portuguese de Ciencias Naturais, 16(2):134-187, 6 pls. (Trav. Sta. Biol. Mar. de Lisbonne No. 52).

Peridium leonis
Wall, David, and Barrie Dale, 1968.
Modern dinoflagellate cysts and evolution of the Peridiniales. Micropaleontology, 14(3):265-304.

Peridinium Levanderi
Matzenauer, L., 1933
Die Dinoflagellaten des indischen Ozeans (mit Ausnahme der Gattung Ceratium.) Bot. Arch. 35:437-510, 77 text figs., 2 charts.

Peridinium limbatum
Wall, David, and Barrie Dale, 1968.
Modern dinoflagellate cysts and evolution of the Peridiniales. Micropaleontology, 14(3):265-304.

Peridinium lipopodium n.sp.
Balech, Enrique, 1964.
El plancton de Mar del Plate durante el periodo 1961-1962. Bol. Inst. Biol. Mar., Buenos Aires, (4):56 pp.
49 pp. + plates

Peridinium longicollum
Dangeard, P., 1927
Phytoplankton de la croisière du "Sylvana". Ann. Inst. Ocean., Monaco, n.s., 4(8):286-401, 54 text figs. (février-Juin 1913).

Peridinium longicollum
Forti, A., 1922
Ricerche sulla flora pelagica (fitoplancton) di Quarto dei Mille. Mem. R. Com. Talass. Ital. 97:248 pp., 13 pls.

Peridinium longicollum
Pavillard, J., 1916
Recherches sur les Peridiniens du Golfe du Lion. Mem. Univ. Montpellier. Trav. Inst. Bot., Univ. Montpellier. Serie mixte No.4, 70 pp., 3 pls., 15 text figs.

Peridinium longicollum
Sousa e Silva, E., 1949
Diatomaceas e Dinoflagelados de Baia de Cascais. Portugaliae Acta Biol., Volume: Julio Henriques, Ser. B: 300-383, 9 pls, 2 fold-in tables.

Peridinium longipes
Balech, Enrique, 1964.
Tercera contribucion al conocimiento del genero "Peridinium". Revista, Mus. Argentino, Ciencias Nat. "Bernardino Rivadavia" e Inst. Nacional. Invest. Ciencias Naturales, Hidrobiol.,1(6):179-195.

Peridinium longipes
Balech, Enrique, 1959.
Operacion Oceanografica Merluza V Crucero. Plancton. Publico, Servicio de Hidrografia Naval, Argentina, H. 618:43 pp.

Peridinium longipes
Matzenauer, L., 1933
Die Dinoflagellaten des indischen Ozeans (mit Ausnahme der Gattung Ceratium.) Bot. Arch. 35:437-510, 77 text figs., 2 charts.

Peridinium longipes
Rampi, L., 1950.
Ricerche sul fitoplancton del Mare Ligure. 9) I Peridinium delle acque di Sanremo. Pubbl. Centro Talassografico Tirreno No. 5:10 pp. 2 pls.

Atti Acad. Ligure Sci. e Lett. 7(1):

Peridinium macrapicatum n. nom.
Balech, Enrique, 1971.
Microplancton de la campaña oceanographica: Productividad III. Revta Mus. argent. Cienc. Nat. Bernardina Rivadavia, Hydrobiol. 3(1):1-202, 39 pls.

Peridinium macrospinum
Lebour, M.V., 1925.
The dinoflagellates of northern seas. Marine Biological Association, United Kingdom, 250 pp., 35 pls., 53 textfigs.

Peridinium macrospinum n.sp.
Mangin, M. L., 1912
Phytoplancton de la croisière du "René" dans l'Atlantique (Septembre 1908). Ann. Inst. Ocean., n.s., 4(1):1-66, 2 pls., 41 text figs., 2 tables.

Peridinium majus n. sp.
Dangeard, P., 1927
Phytoplankton de la croisière du "Sylvana". Ann. Inst. Ocean., Monaco, n.s., 4(8):286-401, 54 text figs. (février-Juin 1913).

Peridinium mangini n. sp.
Balech, Enrique, 1971.
Microplancton de la campaña oceanographica: Productividad III. Revta Mus. argent. Cienc. Nat. Bernardina Rivadavia, Hydrobiol. 3(1):1-202, 39 pls.

Peridinium marchicum
Iyengar, M.O.P. and G.Venkataraman,1951.
The ecology and seasonal succession of the algae flora of the River Cooum at Madras with special reference to the Diatomaceae. J. Madras Univ. 21, Sect. B(1): 140-192, 1 pl of 4 figs., 11 text figs.

Peridinium marielebourae nom. nov
Paulsen, O., 1930.
Etudes sur le microplancton de la mer d'Alboran. Trab. Inst. Esp. Ocean. No. 4:1-108, 61 textfigs.

Peridinium marinum n.sp.
Lindemann, E., 1925
Neubeobachtungen an den Winter peridineen des Golfes von Neapel. Bot. Arch. 9:95-102, 19 text figs.

Peridinium marinum travectum
Lindemann, E., 1925
Neubeobachtungen an den Winter peridineen des Golfes von Neapel. Bot. Arch. 9:95-102, 19 text figs.

Peridinium mastophorum n. sp.
Balech, Enrique, 1971.
Microplancton de la campaña oceanographica: Productividad III. Revta Mus. argent. Cienc. Nat. Bernardina Rivadavia, Hydrobiol. 3(1):1-202, 39 pls.

Peridinium mediocre
Balech, Enrique, 1971.
Microplancton de la campaña oceanographica: Productividad III. Revta Mus. argent. Cienc. Nat. Bernardina Rivadavia, Hydrobiol. 3(1):1-202, 39 pls.

Peridinium mediterraneum
Balech, Enrique, 1964.
Tercera contribucion al conocimiento del genero "Peridinium". Revista, Mus. Argentino, Ciencias Nat. "Bernardino Rivadavia" e Inst. Nacional. Invest. Ciencias Naturales, Hidrobiol.,1(6):179-195.

Peridinium melo n. sp.
Balech, Enrique, 1971.
Microplancton de la campaña oceanographica: Productividad III. Revta Mus. argent. Cienc. Nat. Bernardina Rivadavia, Hydrobiol. 3(1):1-202, 39 pls.

Peridinium mediocre
Hada, Yoshina, 1970.   (1970)
The protozoan plankton of the Antarctic and Subantarctic seas. Scient. Repts., Japan. Antarct. Res. Exped., (E)31:51 pp.

Peridinium metananum
Balech, Enrique, 1971.
Microplancton de la campaña oceanographica: Productividad III. Revta Mus. argent. Cienc. Nat. Bernardina Rivadavia, Hydrobiol. 3(1):1-202, 39 pls.

Peridinium metananum n.sp.
Balech, Enrique, and Sayed Z. El-Sayed, 1965.
Microplankton of the Weddell Sea. In: Biology of Antarctic seas. II. Antarctic Res. Ser., Amer. Geophys. Union, 5:107-124.

Peridinium micans
Gran, H.H., and T. Braarud, 1935
A quantitative study of the phytoplankton in the Bay of Fundy and the Gulf of Maine (including observations on hydrography, chemistry, and turbidity). J. Biol. Bd., Canada, 1(5):279-467, 69 text figs.

Peridinium Michaelis
Murray, G., and F. G. Whitting, 1899
New Peridiniaceae from the Atlantic. Trans. Linn. Soc., London, Bot., ser 2, 5: 321-342, Pls. 27-33, 9 tables.

Peridinium michaelis
Schröder, B., 1900
Phytoplankton des Golfes von Neapel nebst vergleichenden Ausblicken auf das atlantischen Ozean. Mitt. Zool. Stat. Neapel, 14:1-38.

Peridinium michaelis
Wang, C. C., and D. Nie, 1932
A survey of the marine protozoa of Amoy. Contr. Biol. Lab. Sci. Soc., China, Zool. Ser., 8:285-385, 89 text figs.

Peridinium micrapium n.sp.
Meunier, A., 1919
Microplancton de la Mer Flamande 3. Les Péridiniens. Mem. Mus. Roy. Hist. Nat., Belgique 8(1):1-116, Pls. XV-XXI.

Peridinium Milneri n.sp.
Murray, G., and F. G. Whitting, 1899
New Peridiniaceae from the Atlantic. Trans. Linn. Soc., London, Bot., ser 2, 5: 321-342, Pls. 27-33, 9 tables.

Peridinium minusculum
Braarud, T., 1945
A phytoplankton survey of the polluted waters of inner Oslo Fjord. Hvalrådets Skrifter, No.28, 142 pp., 19 text figs., 17 tables.

Oceanographic Index: Marine Organisms Cumulation, 1946-1973

Peridinium minusculum
Ercegovic, A., 1936
Etudes qualitative et quantitatives du phytoplancton dans les eaux cotières de l'Adriatique oriental moyen au cours de l'année 1934. Acta Adriatica 1(9):1-126

Peridinium minusculum
Lillick, L.C., 1940
Phytoplankton and planktonic protozoa of the offshore waters of the Gulf of Maine. Pt.II. Qualitative Composition of the Planktonic Flora. Trans. Am. Phil. Soc., n.s., 31(3):193-237, 13 text figs.

Peridinium minusculum
Lillick, L.C., 1938
Preliminary report of the phytoplankton of the Gulf of Maine. Am. Mid. Nat. 20(3):624-640, 1 text fig, 37 tables.

Peridinium minusculum
Pavillard, J., 1916
Recherches sur les Peridiniens du Golfe du Lion. Mem. Univ. Montpellier. Trav. Inst. Bot., Univ. Montpellier. Serie mixte No.4, 70 pp., 3 pls., 15 text figs.

Peridinium minusculum n.sp.
Pavillard, J., 1905
Recherches sur la flore pelagique (Phytoplankton) de l'Etang de Thau. Theses presentees a la Fac. Sci., Paris, 116 pp., 3 pls.

Peridinium minusculum
Rumkówna, A., 1948
[List of the phytoplankton species occurring in the superficial water layers in the Gulf of Gdańsk]. Bull. Lab. mar., Gdynia, No. 4: 139-141 with tables in back.

Peridinium minutum
Balech, Enrique, 1964.
Tercera contribucion al conocimiento del genero "Peridinium". Revista. Mus. Argentino, Ciencias Nat. "Bernardino Rivadavia" e Inst. Nacional, Inv est. Ciencias Naturales, Hidrobiol.,1(6):179-195.

Peridinium minutum n. sp.
Kofoid, C. A., 1907
Dinoflagellata of the San Diego region. III. Descriptions of new species. Univ. Calif. Publ. Zool. 3:299-340, Pls. 22-33.

Peridinium minutum
Wall, David, and Barrie Dale, 1968.
Modern dinoflagellate cysts and evolution of the Peridiniales. Micropaleontology, 14(3):265-304.

Peridinium mite
Balech, Enrique, 1971.
Microplancton de la campaña oceanographica Productividad III. Revta Mus. argent. Cienc. Nat. Bernadina Rivadavia, Hydrobiol. 3(1):1-202, 39 pls.

Peridinium mite
Dangeard, P., 1927
Phytoplankton de la croisière du "Sylvana". Ann. Inst. Ocean., Monaco, n.s., 4(8):286-401, 54 text figs. (février-Juin 1913).

Peridinium mite
Forti, A., 1922
Ricerche sulla flora pelagica (fitoplancton) di Quarto dei Mille. Mem. R. Com. Talass. Ital. 97:248 pp., 13 pls.

Peridinium mite
LeBour, M.V., 1925
The dinoflagellates of Northern Seas. The Marine Biological Association of the United Kingdom, Plymouth, 250 pp., 35 pls., 53 text figs.

Peridinium mite
Margalef, R., 1949
Fitoplancton nerítico de la Costa Brava en 1947-48. Publ. Inst. Biol. Aplicada, 5: 41-51, 3 text figs.

Peridinium mite
Matzenauer, L., 1933
Die Dinoflagellaten des indischen Ozeans (mit Ausnahme der Gattung Ceratium.) Bot. Arch. 35:437-510, 77 text figs., 2 charts.

Peridinium mite
Pavillard, J., 1916
Recherches sur les Peridiniens du Golfe du Lion. Mem. Univ. Montpellier. Trav. Inst. Bot., Univ. Montpellier. Serie mixte No.4, 70 pp., 3 pls., 15 text figs.

Peridinium mite
Sousa e Silva, E., 1949
Diatomaceas e Dinoflagelados de Baia de Cascais. Portugaliae Acta Biol., Volume: Julio Henriques, Ser. B: 300-383, 9 pls, 2 fold-in tables.

Peridinium mite
Sousa e Silva, E., and J. Dos Santos-Pinto, 1948
O Plancton da Baia de S. Martinho do Porto. 1. Diatomaceas e Dinoflagelados. Bol. Soc. Portuguesa de Ciencias Naturais, 16(2):134-187, 6 pls. (Trav. Sta. Biol. Mar. de Lisbonne No. 52).

Peridinium mite
Wood, E.J.P., 1954.
Dinoflagellates in the Australian region. Australian J. Mar. Freshwater Res., 5(2):171-351.

Peridinium monacanthus
Lebour, M.V., 1925.
The dinoflagellates of northern seas. Marine Biological Association, United Kingdom, 250 pp., 35 pls., 53 textfigs.

Peridinium monacanthum
Wood, E.J.P., 1954.
Dinoflagellates of the Australian region. Australian J. Mar. Freshwater Res., 5(2):171-351.

Peridinium monacanthus
Gran, H.H., and T. Braarud, 1935
A quantitative study of the phytoplankton in the Bay of Fundy and the Gulf of Maine (including observations on hydrography, chemistry, and turbidity). J. Biol. Bd., Canada, 1(5):279-467, 69 text figs.

Peridinium monacanthus
Lillick, L.C., 1940
Phytoplankton and planktonic protozoa of the offshore waters of the Gulf of Maine. Pt.II. Qualitative Composition of the Planktonic Flora. Trans. Am. Phil. Soc., n.s., 31(3):193-237, 13 text figs.

Peridinium monospinum
Dangeard, P., 1927
Phytoplankton de la croisière du "Sylvana". Ann. Inst. Ocean., Monaco, n.s., 4(8):286-401, 54 text figs. (février-Juin 1913).

Peridinium monospinum
LeBour, M.V., 1925
The dinoflagellates of Northern Seas. The Marine Biological Association of the United Kingdom, Plymouth, 250 pp., 35 pls.,53 text figs.

Peridinium monospinum
Mangin, M. L., 1912
Phytoplancton de la croisière du "René" dans l'Atlantique (Septembre 1908). Ann. Inst. Ocean., n.s., 4(1):1-66, 2 pls., 41 text figs., 2 tables.

Peridinium monospinum
Paulsen, O., 1908
XVIII Peridiniales. Nordisches Plankton, Bot. Teil: 1-124, 155 text figs.

Peridinium monospinum
Sousa e Silva, E., 1949
Diatomaceas e Dinoflagelados de Baia de Cascais. Portugaliae Acta Biol., Volume: Julio Henriques, Ser. B: 300-383, 9 pls, 2 fold-in tables.

Peridinium multistriatum n. sp.
Kofoid, C. A., 1907
Dinoflagellata of the San Diego region. III. Descriptions of new species. Univ. Calif. Publ. Zool. 3:299-340, Pls. 22-33.

Peridinium multitabulatum n.f.
Graham, H. W., 1942
Studies in the morphology, taxonymy, and ecology of the Peridiniales. Sci. Res. Cruise VII of the Carnegie, 1928-1929---Biol. III(542): 129 pp., 67 figs.

Peridinium Murrayi
Margalef, R., 1949
Fitoplancton nerítico de la Costa Brava en 1947-48. Publ. Inst. Biol. Aplicada, 5: 41-51, 3 text figs.

Peridinium murrayi
Matzenauer, L., 1933
Die Dinoflagellaten des indischen Ozeans (mit Ausnahme der Gattung Ceratium.) Bot. Arch. 35:437-510, 77 text figs., 2 charts.

Peridinium manum n. sp.
Argentina, Secretaria de Marina, Servicio de Hidrografia Naval, 1962.
Plancton de las campanas oceanograficas DRAKE I y II. Publico, H. 627:57.

Peridinium nanum
Hada, Yoshina, 1970. (1970)
The protozoan plankton of the Antarctic and Subantarctic seas. Scient. Repts, Japan. Antarct. Res. Exped., (E)31:51 pp.

Peridinium nipponica
Matzenauer, L., 1933
Die Dinoflagellaten des indischen Ozeans (mit Ausnahme der Gattung Ceratium.) Bot. Arch. 35:437-510, 77 text figs., 2 charts.

Peridinium norpacense n.sp
Balech, Enrique, 1962
Tintinnoinea y Dinoflagellata del Pacifico segun material de las Expediciones NORPAC y DOWNWIND del Instituto Scripps de Oceanografia. Revista. Mus. Argentino Ciencias Nat. "Bernardino Rivadavia". Ciencias Zool., 7(1):1-253.

Peridinium novascotiense n. sp.
Gran, H.H., and T. Braarud, 1935
A quantitative study of the phytoplankton in the Bay of Fundy and the Gulf of Maine (including observations on hydrography, chemistry, and turbidity). J. Biol. Bd., Canada, 1(5):279-467, 69 text figs.

Peridinium novascotiense
Lillick, L.C., 1938
Preliminary report of the phytoplankton of the Gulf of Maine. Am. Mid. Nat. 20(3):624-640, 1 text fig, 37 tables.

Peridinium nudum
LeBour, M.V., 1925
The dinoflagellates of Northern Seas. The Marine Biological Association of the United Kingdom, Plymouth, 250 pp., 35 pls., 53 text figs.

Peridinium nudum n.sp.
Meunier, A., 1919
Microplancton de la Mer Flamande 3. Les Péridiniens. Mem. Mus. Roy. Hist. Nat., Belgique 8(1):1-116, Pls. XV-XXI.

Peridinium nudum
Paredes, J.F., 1962.
35. A brief comment on Peridinium nudum. Trab. Cent. Biol. Pisc., 38: 115-120.
Also in:
Mem. Junta Invest. Ultram., (2)No. 33.

Peridinium nudum
Sousa e Silva, E., 1949
Diatomaceas e Dinoflagelados de Baia de Cascais. Portugaliae Acta Biol., Volume: Julio Henriques, Ser. B: 300-383, 9 pls, 2 fold-in tables.

Peridinium ? nudum
Wall, David, and Barrie Dale, 1968.
Modern dinoflagellate cysts and evolution of the Peridiniales. Micropaleontology, 14(3):265-304.

Peridinium nudum
Wood, E.J.P., 1954.
Dinoflagellates in the Australian region. Australian J. Mar. Freshwater Res., 5(2):171-351.

Peridinium obesum n.sp.
Matzenauer, L., 1933
Die Dinoflagellaten des indischen Ozeans (mit Ausnahme der Gattung Ceratium.) Bot. Arch. 35:437-510, 77 text figs., 2 charts.

Peridinium obliquum n.sp.
Dangeard, P., 1927
Phytoplankton de la croisière du "Sylvana". Ann. Inst. Ocean., Monaco, n.s., 4(8):286-401, 54 text figs. (Janvier-Juin 1913).

Peridinium obliquum
Matzenauer, L., 1933
Die Dinoflagellaten des indischen Ozeans (mit Ausnahme der Gattung Ceratium.) Bot. Arch. 35:437-510, 77 text figs., 2 charts.

Peridinium oblongum
Braarud, T., and Adam Bursa, 1939
On the phytoplankton of the Oslo Fjord, 1933-1934. Hvalrådets Skr. No.19:1-63; 9 text figs. Reviewed. J. du. Cons. 14(3): 418-420. A.C. Gardiner.

Peridinium oblongum
Dangeard, P., 1927
Notes sur la variation de la genre Peridinium. Bull. Inst. Ocean., Monaco, No. 507:16 pp., 9 textfigs.

Peridinium oblongum
LeBour, M.V., 1925
The dinoflagellates of Northern Seas. The Marine Biological Association of the United Kingdom, Plymouth, 250 pp., 35 pls., 53 text figs.

Peridinium oblongum
Matzenauer, L., 1933
Die Dinoflagellaten des indischen Ozeans (mit Ausnahme der Gattung Ceratium.) Bot. Arch. 35:437-510, 77 text figs., 2 charts.

Peridinium oblongum
Rampi, L., 1950.
Ricerche sul fitoplancton del Mare Ligure. 9) I Peridinium delle acque di Sanremo. Pubbl. Centro Talassografico Tirreno No. 5:10 pp. 2 pls.

Atti Acad. Ligure Sci. e Lett. 7(1):

Peridinium oblongum
Sousa e Silva, E., 1949
Diatomaceas e Dinoflagelados de Baia de Cascais. Portugaliae Acta Biol., Volume: Julio Henriques, Ser. B: 300-383, 9 pls, 2 fold-in tables.

Peridinium oblongum
Wall, David, and Barrie Dale, 1968.
Modern dinoflagellate cysts and evolution of the Peridiniales. Micropaleontology, 14(3):265-304.

Peridinium obovatum
Argentina, Secretaria de Marina, Servicio de Hidrografía Naval, 1962.
Plancton de las campanas oceanograficas DRAKE I y II. Publico, H. 627:57.

Peridinium obovatum
Balech, Enrique, and Sayed Z. El-Sayed, 1965.
Microplankton of the Weddell Sea.
In: Biology of Antarctic seas. II.
Antarctic Res. Ser., Amer. Geophys. Union, 5:107-124.

Peridinium obovatum n.sp.
Wood, E.J.P., 1954.
Dinoflagellates in the Australian region. Australian J. Mar. Freshwater Res., 5(2):171-351.

Peridinium obtusum
Balech, E., 1949.
Etude de quelques especes de Peridinium souvent confonues. Hydrobiol., Acta Hydrobiol., Limnol., et Protistol. 1(4):390-409, 6 pls.

Peridinium obtusum
LeBour, M.V., 1925
The dinoflagellates of Northern Seas. The Marine Biological Association of the United Kingdom, Plymouth, 250 pp., 35 pls., 53 text figs.

Peridinium obtusum
Lillick, L.C., 1940
Phytoplankton and planktonic protozoa of the offshore waters of the Gulf of Maine. Pt.II. Qualitative Composition of the Planktonic Flora. Trans. Am. Phil. Soc., n.s., 31(3):193-237, 13 text figs.

Peridinium obtusum
Morse, D.C., 1947
Some observations on seasonal variations in plankton population Patuxant River, Maryland 1943-1945. Bd. Nat. Res., Publ. No.65, Chesapeake Biol. Lab., 31, 3 figs.

Peridinium oceanicum
Balech, E., 1951.
Deuxième contribution à la connaissance des Peridinium. Hydrobiol., Acta Hydrobiol., Limnol. Protistol. 3(4):305-330, 7 pls., of 137 figs.

Peridinium oceanicum
Brunel, J., 1962
Le phytoplancton de la Baie de Chaleurs. Inst. Botan., Univ. Montréal, Contrib. No. 77: 365 pp., 66 pls.

Peridinium oceanicum
Dangeard, P., 1927
Phytoplankton de la croisière du "Sylvana". Ann. Inst. Ocean., Monaco, n.s., 4(8):286-401, 54 text figs. (Janvier-Juin 1913).

Peridinium oceanicum
Dangeard, P., 1927.
Notes sur la variation de la genre Peridinium. Bull. Inst. Ocean., Monaco, No. 507:16 pp., 9 textfigs.

Peridinium oceanicum
Dangeard, P., 1926
Description des Péridiniens Testacés recueillis para la Mission Charcot pendent le mois d'Aout 1924. Ann. Inst. Ocean. n.s. 3(7):307-334, 15 text figs.

Peridinium oceanicum
Delegazione Italiana della Commissione Internazionale per l'Esplorazione Scientifica del Mediterraneo, 1941
Note sul plancton della Laguna veneta. [Memoria CCLXXIX], Arch. di Ocean. e Limn. Anno I, Fasc. I, 1941 XIX: 31-57 pp.

Peridinium oceanicum
Ercegovic, A., 1936
Etudes qualitative et quantitatives du phytoplancton dans les eaux cotières de l'Adriatique oriental moyen au cours de l'année 1934. Acta Adriatica 1(9):1-126

Peridinium oceanicum
Forti, A., 1922
Ricerche sulla flora pelagica (fitoplancton) di Quarto dei Mille. Mem. R. Com. Talass. Ital. 97:248 pp., 13 pls.

Peridinium oceanicum
Gilbert, J.Y., and W.E. Allen, 1943
The phytoplankton of the Gulf of California obtained by the "E.W. Scripps" in 1939 and 1940. J. Mar. Res. V(2):89-110, figs.30-31.

Peridinium oceanicum
Graham, H. W., 1942
Studies in the morphology, taxonymy, and ecology of the Peridiniales. Sci. Res. Cruise VII of the Carnegie, 1928-1929---Biol. III(542): 129 pp., 67 figs.

Peridinium oceanicum
Jørgensen, E., 1905
B. Protistplankton and the diatoms in bottom samples. Hydrographical and biological investigations in Norwegian fjords. Bergens Mus. Skr. 7: 49-225.

Peridinium oceanicum
LeBour, M.V., 1925
The dinoflagellates of Northern Seas. The Marine Biological Association of the United Kingdom, Plymouth, 250 pp., 35 pls., 53 text figs.

Peridinium oceanicum
Mangin, M. L., 1912
Phytoplankton de la croisière du "René" dans l'Atlantique (Septembre 1908). Ann. Inst. Ocean., n.s., 4(1):1-66, 2 pls., 41 text figs., 2 tables.

Peridinium oceanicum
Margalef, R., 1949
Fitoplancton nerítico de la Costa Brava en 1947-48. Publ. Inst. Biol. Aplicada, 5: 41-51, 3 text figs.

Peridinium oceanicum
Marukawa, H., 1921
Plankton lists and some new species of copepods, from the northern waters of Japan. Bull. Inst. Ocean., No.384, 15 pp., 3 pls., 1 chart. Monaco

Peridinium oceanicum
Matzenauer, L., 1933
Die Dinoflagellaten des indischen Ozeans (mit Ausnahme der Gattung Ceratium.) Bot. Arch. 35:437-510, 77 text figs., 2 charts.

Peridinium oceanicum
Meunier, A., 1919
Microplancton de la Mer Flamande 3. Les Péridiniens. Mem. Mus. Roy. Hist. Nat., Belgique 8(1):1-116, Pls. XV-XXI.

Peridinium oceanicum
Paulsen, O., 1908
XVIII Peridiniales. Nordisches Plankton, Bot. Teil: 1-124, 155 text figs.

Peridinium oceanicum
Pavillard, J., 1916
Recherches sur les Peridiniens du Golfe du Lion. Mem. Univ. Montpellier. Trav. Inst. Bot., Univ. Montpellier. Serie mixte No.4, 70 pp., 3 pls., 15 text figs.

Peridinium oceanicum
Rampi, L., 1950
Ricerche sul fitoplancton del Mare Ligure. 9) I Peridinium delle acque di Sanremo. Pubbl. Centro Talassografico Tirreno No. 5:10 pp. 2 pls.

Atti Acad. Ligure Sci. e Lett. 7(1):

Peridinium oceanicum
Sousa e Silva, E., 1949
Diatomaceas e Dinoflagelados de Baia de Cascais. Portugaliae Acta Biol., Volume: Julio Henriques, Ser. B: 300-383, 9 pls, 2 fold-in tables.

Peridinium oceanicum bisintercalares n.f.
Graham, H. W., 1942
Studies in the morphology, taxonymy, and ecology of the Peridiniales. Sci. Res. Cruise VII of the Carnegie, 1928-1929---Biol. III(542): 129 pp., 67 figs.

Peridinium oceanicum oblongum
Forti, A., 1922
Ricerche sulla flora pelagica (fitoplancton) di Quarto dei Mille. Mem. R. Com. Talass. Ital. 97:248 pp., 13 pls.

Peridinium oceanicum oblongum
Lindemann, E., 1924
Peridineen aus dem goldenen Horn und dem Bosphorus. Bot. Arch. 5:216-233, 98 text figs.

Peridinium oceanicum oblongum
Paulsen, O., 1908
XVIII Peridiniales. Nordisches Plankton, Bot. Teil: 1-124, 155 text figs.

Peridinium oceanicum spiniferum n.f.
Graham, H. W., 1942
Studies in the morphology, taxonymy, and ecology of the Peridiniales. Sci. Res. Cruise VII of the Carnegie, 1928-1929---Biol. III(542): 129 pp., 67 figs.

Peridinium oceanicum var. tenellum n.var.
Graham, H. W., 1942
Studies in the morphology, taxonymy, and ecology of the Peridiniales. Sci. Res. Cruise VII of the Carnegie, 1928-1929---Biol. III(542): 129 pp., 67 figs.

Peridinium oceanicum tricornutum n.f.
Graham, H. W., 1942
Studies in the morphology, taxonymy, and ecology of the Peridiniales. Sci. Res. Cruise VII of the Carnegie, 1928-1929---Biol. III(542): 129 pp., 67 figs.

Peridinium oceanicum typica
Paulsen, O., 1908
XVIII Peridiniales. Nordisches Plankton, Bot. Teil: 1-124, 155 text figs.

Peridinium okamurai n.sp.
Marukawa, H., 1928.
Ueber 4 neue Arten der Peridinialen. Annot. Oceanogr. Res., 2(1):1-2.

Peridinium okamurai
Wood, E.J.F., 1954.
Dinoflagellates in the Australian region. Australian J. Mar. Freshwater Res., 5(2):171-351

Peridinium orbiculare
Gilbert, J.Y., and W.E. Allen, 1943
The phytoplankton of the Gulf of California obtained by the "E.W. Scripps" in 1939 and 1940. J. Mar. Res. V(2):89-110, figs.30-31.

Peridinium orbiculare
Paulsen, O., 1908
XVIII Peridiniales. Nordisches Plankton, Bot. Teil: 1-124, 155 text figs.

Peridinium orientale n.sp.
Matzenauer, L., 1933
Die Dinoflagellaten des indischen Ozeans (mit Ausnahme der Gattung Ceratium.) Bot. Arch. 35:437-510, 77 text figs., 2 charts.

Peridinium ovatum?
Balech, Enrique, 1971.
Microplancton de la campaña oceanographica: Productividad III. Revta Mus. argent. Cienc. Nat. Bernadina Rivadavia, Hydrobiol. 3(1):1-202, 39 pls.

Peridinium aff. ovatum
Balech, Enrique, 1959.
Operacion Oceanografica Merluza V Crucero. Plancton. Publico, Servicio de Hidrografia Naval, Argentina, H. 618:43 pp.

Peridinium ovatum
Brunel, J., 1962
Le phytoplancton de la Baie de Chaleurs. Inst. Botan., Univ. Montréal, Contrib. No. 77: 365 pp., 66 pls.

Peridinium ovatum minor

Peridinium ovatum
Brunel, Jules, 1962
Le phytoplancton de la Baie des Chaleurs. Contrib. Ministère de la Chasse et des Pêcheries, Province de Québec, No. 91: 365 pp.

Peridinium ovatum
Dangeard, P., 1927
Phytoplankton de la croisière du "Sylvana". Ann. Inst. Ocean., Monaco, n.s., 4(8):286-401, 54 text figs. (Février-Juin 1913).

Peridinium ovatum
Dangeard, P., 1927
Notes sur la variation de la genre Peridinium. Bull. Inst. Ocean., Monaco, No. 507: 16 pp., 9 textfigs.

Peridinium ovatum
Dangeard, P., 1926
Description des Péridiniens Testacés recueillis pendant la Mission Charcot pendent le mois d'Aout 1924. Ann. Inst. Ocean. n.s. 3(7):307-334, 15 text figs.

Peridinium ovatum
Forti, A., 1922
Ricerche sulla flora pelagica (fitoplancton) di Quarto dei Mille. Mem. R. Com. Talass. Ital. 97:248 pp., 13 pls.

Peridinium ovatum
Gran, H.H., and T. Braarud, 1935
A quantitative study of the phytoplankton in the Bay of Fundy and the Gulf of Maine (including observations on hydrography, chemistry, and turbidity). J. Biol. Bd., Canada, 1(5):279-467, 69 text figs.

Peridinium ovatum
Iselin, C., 1930
A report on the coastal waters of Labrador based on explorations of the "Chance" during the summer of 1926. Proc. Am. Acad. Arts Sci., 66(1):1-37, 14 text figs.

Peridinium ovatum
Jørgensen, E., 1905
B.Protistplankton and the diatoms in bottom samples. Hydrographical and biological investigations in Norwegian fjords. Bergens Mus. Skr. 7: 49-225.

Peridinium ovatum
Jorgensen, E., 1900
Protophyten und Protozoën im Plankton aus der Norwegischen Westkerste. Bergens Mus. Aarb. 1899(6): 95 pp., 5 pls., 83 tables.

Peridinium ovatum
LeBour, M.V., 1925
The dinoflagellates of Northern Seas. The Marine Biological Association of the United Kingdom, Plymouth, 250 pp., 35 pls., 53 text figs.

Peridinium ovatum
Mangin, M. L., 1912
Phytoplancton de la croisière du "René" dans l'Atlantique (Septembre 1908). Ann. Inst. Ocean., n.s., 4(1):1-66, 2 pls., 41 text figs., 2 tables.

Peridinium ovatum
Meunier, A., 1919
Microplancton de la Mer Flamande 3. Les Péridiniens. Mem. Mus. Roy. Hist. Nat., Belgique 8(1):1-116, Pls. XV-XXI.

Peridinium ovatum
Paulsen, O., 1908
XVIII Peridiniales. Nordisches Plankton, Bot. Teil: 1-124, 155 text figs.

Peridinium ovatum
Pavillard, J., 1916
Recherches sur les Peridiniens du Golfe du Lion. Mem. Univ. Montpellier. Trav. Inst. Bot., Univ. Montpellier. Serie mixte No.4, 70 pp., 3 pls., 15 text figs.

Peridinium ovatum
Schulz, B., and A. Wulff, 1929
Hydrographie und Oberflächen plankton des westlichen Barentsmeeres im Sommer 1927. Ber. deutschen wissensch. Komm. F. Meeresforsch. n.s. 4(5):232-372, 13 tables, 25 text figs.

Peridinium ovatum
Sousa e Silva, E., 1949
Diatomaceas e Dinoflagelados de Baia de Cascais. Portugaliae Acta Biol., Volume: Julio Henriques, Ser. B: 300-383, 9 pls, 2 fold-in tables.

Peridinium ovatum
Sousa e Silva, E., and J. Dos Santos-Pinto, 1948
O Plancton da Baia de S. Martinho do Porto. 1. Diatomaceas e Dinoflagelados. Bol. Soc. Portuguese de Ciencias Naturais, 16(2):134-187, 6 pls. (Trav. Sta. Biol. Mar. de Lisbonne No. 52).

Peridinium ovatum

Wood, E.J.P., 1954.
Dinoflagellates in the Australian region.
Australian J. Mar. Freshwater Res., 5(2):171-351.

Peridinium ovatum inarmata nv.

Matzenauer, L., 1933
Die Dinoflagellaten des indischen Ozeans (mit Ausnahme der Gattung Ceratium.) Bot. Arch. 35:437-510, 77 text figs., 2 charts.

Peridinium aff. oviforme

Balech, Enrique, 1959.
Operacion Oceanografia Merluza V Crucero. Plancton.
Publico, Servicio de Hidrografia Naval, Argentina, H. 618:43 pp.

Peridinium oviforme

Dangeard, P., 1927
Phytoplankton de la croisière du "Sylvana". Ann. Inst. Ocean., Monaco, n.s., 4(8):286-401, 54 text figs. (février -Juin 1913).

Peridinium oviforme n.sp.

Dangeard, P., 1927.
Péridiniens nouveaux ou peu connus de la croisière du "Sylvana". Bull. Inst. Océan., Monaco, No. 491:16 pp., 9 textfigs.

Peridinium oviforme

Margalef, R., 1949
Fitoplancton nerítico de la Costa Brava en 1947-48. Publ. Inst. Biol. Aplicada, 5: 41-51, 3 text figs.

Peridinium oviforme

Matzenauer, L., 1933
Die Dinoflagellaten des indischen Ozeans (mit Ausnahme der Gattung Ceratium.) Bot. Arch. 35:437-510, 77 text figs., 2 charts.

Peridinium ovum

Balech, Enrique, 1971.
Microplancton de la campaña oceanographica: Productividad III.
Revta Mus. argent. Cienc. Nat. Bernadina Rivadavia, Hydrobiol. 3(1):1-202, 39 pls.

Peridinium ovum

Ercegovic, A., 1936
Etudes qualitative et quantitatives du phytoplancton dans les eaux cotières de l'Adriatique oriental moyen au cours de l'année 1934. Acta Adriatica 1(9):1-126

Peridinium ovum

Gilbert, J.Y., and W.E. Allen, 1943
The phytoplankton of the Gulf of California obtained by the "E.W. Scripps" in 1939 and 1940. J. Mar. Res. V(2):89-110, figs.30-31.

Peridinium ovum

Rampi, L., 1950.
Ricerche sul fitoplancton del Mare Ligure. 9) I Peridinium delle acque di Sanremo.
Pubbl. Centro Talassografico Tirreno No. 5:10 pp., 2 pls.

Atti Acad. Ligure Sci. e Lett. 7(1):

Peridinium ovum

Wood, E.J.P., 1954.
Dinoflagellates in the Australian region.
Australian J. Mar. Freshwater Res., 5(2):171-351.

Peridinium pallidum

Braarud, T., and Adam Bursa, 1939
On the phytoplankton of the Oslo Fjord, 1933-1934. Hvalrådets Skr. No.19:1-63; 9 text figs. Reviewed. J. du. Cons. 14(3): 418-420. A.C. Gardiner.

Peridinium pallidum

Brunel, Jules, 1962
Le phytoplancton de la Baie des Chaleurs. Contrib. Ministère de la Chasse et des Pêcheries, Province de Québec, No. 91: 365 pp.

Peridinium pallidum

Brunel, J., 1962
Le phytoplancton de la Baie de Chaleurs. Inst. Botan., Univ. Montréal, Contrib. No. 77: 365 pp., 66 pls.

Peridinium pallidum

Dangeard, P., 1927
Phytoplankton de la croisière du "Sylvana". Ann. Inst. Ocean., Monaco, n.s., 4(8):286-401, 54 text figs. (février -Juin 1913).

Peridinium pallidum

Dangeard, P., 1926
Description des Péridiniens Testacés recueillis para la Mission Charcot pendent le mois d'Aout 1924. Ann. Inst. Océan. n.s. 3(7):307-334, 15 text figs.

Peridinium pallidum

Ercegovic, A., 1936
Etudes qualitative et quantitatives du phytoplancton dans les eaux cotières de l'Adriatique oriental moyen au cours de l'année 1934. Acta Adriatica 1(9):1-126

Peridinium pallidum

Forti, A., 1922
Ricerche sulla flora pelagica (fitoplancton) di Quarto dei Mille. Mem. R. Com. Talass. Ital. 97:248 pp., 13 pls.

Peridinium pallidum

Graham, H. W., 1942
Studies in the morphology, taxonymy, and ecology of the Peridiniales. Sci. Res. Cruise VII of the Carnegie, 1928-1929---Biol. III(542): 129 pp., 67 figs.

Peridinium pallidum

Gran, H.H., and T. Braarud, 1935
A quantitative study of the phytoplankton in the Bay of Fundy and the Gulf of Maine (including observations on hydrography, chemistry, and turbidity). J. Biol. Bd., Canada, 1(5):279-467, 69 text figs.

Peridinium pallidum

Jørgensen, E., 1905
B.Protistplankton and the diatoms in bottom samples. Hydrographical and biological investigations in Norwegian fjords. Bergens Mus. Skr. 7: 49-225.

Peridinium pallidum

Lebour, M.V., 1925.
The dinoflagellates of northern seas. Marine Biological Association, United Kingdom, 250 pp., 35 pls., 53 textfigs.

Peridinium pallidum

Lillick, L.C., 1940
Phytoplankton and planktonic protozoa of the offshore waters of the Gulf of Maine. Pt.II. Qualitative Composition of the Planktonic Flora. Trans. Am. Phil. Soc., n.s., 31(3):193-237, 13 text figs.

Peridinium pallidum

Lindemann, E., 1924
Peridineen aus dem goldenen Horn und dem Bosphorus. Bot. Arch. 5:216-233, 98 text figs.

Peridinium pallidum

Paulsen, O., 1908
XVIII Peridiniales. Nordisches Plankton, Bot. Teil: 1-124, 155 text figs.

Peridinium pallidum

Margalef, R., 1949
Fitoplancton nerítico de la Costa Brava en 1947-48. Publ. Inst. Biol. Aplicada, 5: 41-51, 3 text figs.

Peridinium pallidum

Marukawa, H., 1921
Plankton lists and some new species of copepods, from the northern waters of Japan. Bull. Inst. Ocean., No.384, 15 pp., 3 pls., 1 chart. Monaco

Peridinium pallidum

Meunier, A., 1919
Microplancton de la Mer Flamande 3. Les Péridiniens. Mem. Mus. Roy. Hist. Nat., Belgique 8(1):1-116, Pls. XV-XXI.

Peridinium pallidum

Pavillard, J., 1916
Recherches sur les Peridiniens du Golfe du Lion. Mem. Univ. Montpellier. Trav. Inst. Bot., Univ. Montpellier. Serie mixte No.4, 70 pp., 3 pls., 15 text figs.

Peridinium pallidum

Schulz, B., and A. Wulff, 1929
Hydrographie und Oberflächen plankton des westlichen Barentsmeeres im Sommer 1927. Ber. deutschen wissensch. Komm. F. Meeresforsch. n.s. 4(5):232-372, 13 tables, 25 text figs.

Peridinium pallidum

Sousa e Silva, E., 1949
Diatomaceas e Dinoflagelados de Baia de Cascais. Portugaliae Acta Biol., Volume: Julio Henriques, Ser. B: 300-383, 9 pls, 2 fold-in tables.

Peridinium pallidum

Wood, E.J.P., 1954.
Dinoflagellates in the Australian region.
Australian J. Mar. Freshwater Res., 5(2):171-351.

Peridinium pallidum & var.

Rampi, L., 1950.
Ricerche sul fitoplancton del Mare Ligure. 9) I Peridinium delle acque di Sanremo.
Pubbl. Centro Talassografico Tirreno No. 5:10 pp., 2 pls.

Atti. Acad. Ligure Sci. e Lett. 7(1):

Peridinium palmipes

Pavillard, J., 1916
Recherches sur les Peridiniens du Golfe du Lion. Mem. Univ. Montpellier. Trav. Inst. Bot., Univ. Montpellier. Serie mixte No.4, 70 pp., 3 pls., 15 text figs.

Peridinium parallelum

Graham, H. W., 1942
Studies in the morphology, taxonymy, and ecology of the Peridiniales. Sci. Res. Cruise VII of the Carnegie, 1928-1929---Biol. III(542): 129 pp., 67 figs.

Peridinium parallelum

Marukawa, H., 1921
Plankton lists and some new species of copepods, from the northern waters of Japan. Bull. Inst. Ocean., No.384, 15 pp., 3 pls., 1 chart. Monaco

Peridinium parallelum

Paulsen, O., 1908
XVIII Peridiniales. Nordisches Plankton, Bot. Teil: 1-124, 155 text figs.

peridinium parapyriforme n.sp

Hermosilla S., Jorge 1965-1966
Peridinium parapyriforme, nueva especie de dinoflagellata.
Bol. Soc. Biol. Concepción, 40:125-130

Peridinium parcum n. sp.
Balech, Enrique, 1971.
Microplancton de la campaña oceanographica:
Productividad III.
Revta Mus. argent. Cienc. Nat. Bernadina
Rivadavia, Hydrobiol. 3(1):1-202, 39 pls.

Peridinium parvicollum
Argentina, Secretaria de Marina, Servicio de
Hidrografia Naval, 1962.
Plancton de las campanas oceanograficas DRAKE I
y II.
Publico, H. 627:57 pp.

Peridinium parvicollum
Balech, Enrique, 1971.
Microplancton de la campaña oceanographica:
Productividad III.
Revta Mus. argent. Cienc. Nat. Bernadina
Rivadavia, Hydrobiol. 3(1):1-202, 39 pls.

Peridinium parvicollum
Balech, Enrique, 1959.
Operacion Oceanografica Merluza V Crucero.
Plancton.
Publico, Servicio de Hidrografia Naval,
Argentina, H. 618:43 pp.

Peridinium parvicollum n.sp.
Balech, Enrique, 1958.
Plancton de la Campaña Antartica Argentina,
1954-1955.
Physis, 21(60):75-108.

Peridinium parvicollum
Hada, Yoshina, 1970. (1970)
The protozoan plankton of the Antarctic and
Subantarctic seas.
Scient. Repts, Japan. Antarct. Res. Exped.,
(E)31:51 pp.

Peridinium parvispinum
Balech, Enrique, 1959.
Operacion Oceanografica Merluza V Crucero.
Plancton.
Publico, Servicio de Hidrografia Naval,
Argentina, H. 618:43 pp.

Peridinium patagonicum n.sp.
Balech, Enrique, 1959.
Operacion Oceanografica Merluza V Crucero.
Plancton.
Publico, Servicio de Hidrografia Naval,
Argentina, H. 618:43 pp.

Peridinium patens n.sp.
Dangeard, P., 1927
Phytoplankton de la croisière du
"Sylvana". Ann. Inst. Ocean., Monaco, n.s.,
4(8):286-401, 54 text figs. (fevrier-Juin 1913).

Peridinium patens
Matzenauer, L., 1933
Die Dinoflagellaten des indischen Ozeans
(mit Ausnahme der Gattung Ceratium.) Bot.
Arch. 35:437-510, 77 text figs., 2 charts.

Peridinium Paulsenii
Forti, A., 1922
Ricerche sulla flora pelagica (fitoplancton)
di Quarto dei Mille. Mem. R. Com. Talass.
Ital. 97:248 pp., 13 pls.

Peridinium Paulseni
Pavillard, J., 1916
Recherches sur les Peridiniens du Golfe
du Lion. Mem. Univ. Montpellier. Trav. Inst.
Bot., Univ. Montpellier. Serie mixte No.4,
70 pp., 3 pls., 15 text figs.

Peridinium pedunculatum
Ercegovic, A., 1940
Weitere Untersuchungen über einige hydrographische Verhältnisse und über die Phytoplanktonproduktion in den Gewässern der Östlichen Mitteladria. Acta Adriatica 2(3):95-134, 8 text figs.

Peridinium pedunculatum
Jørgensen, E., 1905
B. Protistplankton and the diatoms
in bottom samples. Hydrographical and
biological investigations in Norwegian
fjords. Bergens Mus. Skr. 7: 49-225.

Peridinium pedunculatum
Jorgensen, E., 1900
Protophyten und Protozoën im Plank-
ton aus der Norwegischen Westkerste. Bergens
Mus. Aarb. 1899(6): 95 pp., 5 pls., 83 tables.

Peridinium pedunculatum
Lindemann, E., 1924
Peridineen aus dem goldenen Horn und
dem Bosphorus. Bot. Arch. 5:216-233, 98 text figs.

Peridinium pedunculatum
Matzenauer, L., 1933
Die Dinoflagellaten des indischen Ozeans
(mit Ausnahme der Gattung Ceratium.) Bot.
Arch. 35:437-510, 77 text figs., 2 charts.

Peridinium pedunculatum
Paulsen, O., 1908
XVIII Peridiniales. Nordisches Plankton,
Bot. Teil: 1-124, 155 text figs.

Peridinium pedunculatum
Schröder, B., 1900
Phytoplankton des Golfes von Neapel
nebst vergleichenden Ausblicken auf das
atlantischen Ozean. Mitt. Zool. Stat.
Neapel, 14:1-38.

Peridinium pedunculatum
Wood, E.J.P., 1954.
Dinoflagellates in the Australian region.
Australian J. Mar. Freshwater Res., 5(2):171-351.

Peridinium pellucidum
Balech, Enrique, 1964.
Tercera contribucion al conocimiento del
genero "Peridinium".
Revista, Mus. Argentino, Ciencias Nat.
"Bernardino Rivadavia" e Inst. Nacional. Invest.
Ciencias Naturales, Hidrobiol.,1(6):179-195.

Peridinium pellucidum
Braarud, T., 1945
A phytoplankton survey of the polluted
waters of inner Oslo Fjord. Hvalrådets
Skrifter, No.28, 142 pp., 19 text figs.,
17 tables.

Peridinium pellucidum
Braarud, T., and Adam Bursa, 1939
On the phytoplankton of the Oslo Fjord,
1933-1934. Hvalrådets Skr. No.19:1-63;
9 text figs. Reviewed. J. du. Cons. 14(3):
418-420. A.C. Gardiner.

Peridinium pellucidum
Brunel, J., 1962
Le phytoplancton de la Baie de Chaleurs.
Inst. Botan., Univ. Montréal, Contrib.
No. 77: 365 pp., 66 pls.

Peridinium pellucidum?
Brunel, Jules, 1962
Le phytoplancton de la Baie des Chaleurs.
Contrib. Ministère de la Chasse et des
Pêcheries, Province de Québec, No. 91:
365 pp.

Peridinium pellucidum
Dangeard, P., 1927
Phytoplankton de la croisière du
"Sylvana". Ann. Inst. Ocean., Monaco, n.s.,
4(8):286-401, 54 text figs. (fevrier-Juin 1913).

Peridinium pellucidum
Dangeard, P., 1926
Description des Péridiniens Testacés
recueillis para la Mission Charcot pendent
le mois d'Aout 1924. Ann. Inst. Ocean. n.s.
3(7):307-334, 15 text figs.

Peridinium pellucidum
Forti, A., 1922
Ricerche sulla flora pelagica (fitoplancton)
di Quarto dei Mille. Mem. R. Com. Talass.
Ital. 97:248 pp., 13 pls.

Peridinium pellucidum
Gilbert, J.Y., and W.E. Allen, 1943
The phytoplankton of the Gulf of
California obtained by the "E.W. Scripps"
in 1939 and 1940. J. Mar. Res. V(2):89-110, figs.30-31.

Peridinium pellucidum
Gran, H.H., and T. Braarud, 1935
A quantitative study of the phyto-
plankton in the Bay of Fundy and the
Gulf of Maine (including observations
on hydrography, chemistry, and turbidity).
J. Biol. Bd., Canada, 1(5):279-467, 69 text figs.

Peridinium pellucidum
Iselin, C., 1930
A report on the coastal waters of Labrador
based on explorations of the "Chance" during
the summer of 1926. Proc. Am. Acad. Arts Sci.,
66(1):1-37, 14 text figs.

Peridinium pellucidum
Jørgensen, E., 1905
B. Protistplankton and the diatoms
in bottom samples. Hydrographical and
biological investigations in Norwegian
fjords. Bergens Mus. Skr. 7: 49-225.

Peridinium pellucidum
Jorgensen, E., 1900
Protophyten und Protozoën im Plank-
ton aus der Norwegischen Westkerste. Bergens
Mus. Aarb. 1899(6): 95 pp., 5 pls., 83 tables.

Peridinium pellucidum
Levander, K.M., 1947
Plankton gesammelt in den Jahren
1899-1910 an den Küsten Finnlands.
Finnländische Hydrographisch-Biologische
Untersuchungen (aus dem Wasserbiologis-
chen Laboratorin der Societas Scientiarum
Fennica) No.11: 40 pp., 6 diagrams, 13 pls.,
tables.

Peridinium pellucidum
Lillick, L.C., 1940
Phytoplankton and planktonic
protozoa of the offshore waters
of the Gulf of Maine. Pt.II.
Qualitative Composition of the
Planktonic Flora. Trans. Am.
Phil. Soc., n.s., 31(3):193-237,
13 text figs.

Peridinium pellucidum
Lindemann, E., 1924
Peridineen aus dem goldenen Horn und
dem Bosphorus. Bot. Arch. 5:216-233, 98 text figs.

Peridinium pellucidum
Mangin, M. L., 1912
Phytoplancton de la croisière du "René"
dans l'Atlantique (Septembre 1908). Ann.
Inst. Ocean., n.s., 4(1):1-66, 2 pls., 41
text figs., 2 tables.

Peridinium pellucidum
Matzenauer, L., 1933
Die Dinoflagellaten des indischen Ozeans
(mit Ausnahme der Gattung Ceratium.) Bot.
Arch. 35:437-510, 77 text figs., 2 charts.

Peridinium pellucidum
Meunier, A., 1919
Microplancton de la Mer Flamande 3. Les
Péridiniens. Mem. Mus. Roy. Hist. Nat.,
Belgique 8(1):1-116, Pls. XV-XXI.

Peridinium pellucidum
Nordli, O., 1951.
Dinoflagellates from Lofoten. Nytt Mag. Naturvidensk 88:49-55, 7 textfigs.

Peridinium pellucidum
Paulsen, O., 1908
XVIII Peridiniales. Nordisches Plankton, Bot. Teil: 1-124, 155 text figs.

Peridinium pellucidum
Pavillard, J., 1916
Recherches sur les Peridiniens du Golfe du Lion. Mem. Univ. Montpellier. Trav. Inst. Bot., Univ. Montpellier. Serie mixte No. 4, 70 pp., 3 pls., 15 text figs.

Peridinium pellucidum
Pavillard, J., 1905
Recherches sur la flore pelagique (Phytoplankton) de l'Etang de Thau. Theses presentees a la Fac. Sci., Paris, 116 pp., 3 pls.

Peridinium pellucidum
Rothe, F., 1942.
Quantitativen Untersuchungen über die Planktonverteilung in der östlichen Ostsee. Ber. Deutsch. Wiss. Komm. Meeresf., N.F., 10:291-368, 33 textfigs.

Peridinium pellucidum
Rothe, F., 1941
Quantitative Untersuchunger über die Plankton verteilung in der östlichen Ost see. Ber. Deut. Wiss. Komm. fur Meeresforschung, n.f. X(3):291-368, 33 text figs.

Peridinium pellucidum
Rumkówna, A., 1948
[List of the phytoplankton species occurring in the superficial water layers in the Gulf of Gdańsk] Bull. Lab. mar., Gdynia, No. 4: 139-141 with tables in back.

Peridinium pellucidum
Schulz, B., and A. Wulff, 1929
Hydrographie und Oberflächen plankton des westlichen Barentsmeeres im Sommer 1927. Ber. deutschen wissensch. Komm. F. Meeresforsch. n.s. 4(5):232-372, 13 tables, 25 text figs.

Peridinium pellucidum
Sousa e Silva, E., 1949
Diatomaceas e Dinoflagelados de Baia de Cascais. Portugaliae Acta Biol., Volume: Julio Henriques, Ser. B: 300-383, 9 pls, 2 fold-in tables.

Peridinium pellucidum
Stæmann-Nielsen, Einar, 1951
The marine vegetation of the Isefjord. A study on ecology and production. Medd. Komm. Danmarks Fiskeri-og Havundersøgelser. Ser. Plankton. 5(4); 114pp., 46 text figs.

Peridinium pellucidum
Vives, F., and A. Planas, 1952.
Plancton recogido por los laboratorios costeros. VI. Fitoplancton de las costas de Vinaroz, islas Columbretes y alrededores de la desembocadura del Elroa. Publ. Inst. Biol. Aplic. 11:141-156, 19 textfigs.

Peridinium pellucidum
Wood, E.J.P., 1954.
Dinoflagellates in the Australian region. Australian J. Mar. Freshwater Res., 5(2):171-351.

Peridinium pendunculatum
Mangin, M. L., 1912
Phytoplancton de la croisière du "René" dans l'Atlantique (Septembre 1908). Ann. Inst. Ocean., n.s., 4(1):1-66, 2 pls., 41 text figs., 2 tables.

Peridinium penitum n. sp.
Balech, Enrique, 1971.
Microplancton de la campaña oceanographica: Productividad III.
Revta Mus. argent. Cienc. Nat. Bernadina Rivadavia, Hydrobiol. 3(1):1-202, 39 pls.

Peridinium pendulatum
Mangin, M. L., 1912
Phytoplancton de la croisière du "René" dans l'Atlantique (Septembre 1908). Ann. Inst. Ocean., n.s., 4(1):1-66, 2 pls., 41 text figs., 2 tables.

Peridinium pedunculatum
Rampi, L., 1950.
Ricerche sul fitoplancton del Mare Ligure. 9) I Peridinium delle acque di Sanremo. Pubbl. Centro Talassografico Tirreno No. 5:10 pp. 2 pls.

Atti Acad. Ligure Sci. e Lett. 7(1):

Peridinium pentagonoides n.sp.
Balech, E., 1949.
Etude de quelques especes de Peridinium, souvent confondues. Hydrobiol., Acta Hydrobiol., Limnol. et Protistol. 1(4):390-409, 6 pls.

Peridinium pentagonoides
de Sousa e Silva, E., 1956.
Contribution à l'étude du microplancton de Dakar et des regions maritimes voisines. Bull. I.F.A.N., 8(2):335-371, 7 pls.

Peridinium pentagonum
Brunel, J., 1962
Le phytoplancton de la Baie de Chaleurs. Inst. Botan., Univ. Montréal, Contrib. No. 77: 365 pp., 66 pls.

Peridinium pentagonum
Dangeard, P., 1927
Phytoplankton de la croisière du "Sylvana". Ann. Inst. Ocean., Monaco, n.s., 4(8):286-401, 54 text figs. (Février-Juin 1913).

Peridinium pentagona
Dangeard, P., 1926
Description des Péridiniens Testacés recueillis para la Mission Charcot pendent le mois d'Aout 1924. Ann. Inst. Ocean. n.s. 3(7):307-334, 15 text figs.

Peridinium pentagonum
Gran, H.H., and T. Braerud, 1935
A quantitative study of the phytoplankton in the Bay of Fundy and the Gulf of Maine (including observations on hydrography, chemistry, and turbidity). J. Biol. Bd., Canada, 1(5):279-467, 69 text figs.

Peridinium pentagonum depressum
Halim, Y., 1965.
Microplancton des eaux Egyptiennes. II. Chrysomonadines; Ebriediens et dinoflagellés nouveaux on d'intéret biogeographique. Rapp. Proc.Verb. Réunions. Comm. Int. Expl. Sci., Mer. Mediterranée, Monaco. 18(2):373-379.

Peridinium pentagonum
Jørgensen, E., 1905
B.Protistplankton and the diatoms in bottom samples. Hydrographical and biological investigations in Norwegian fjords. Bergens Mus. Skr. 7: 49-225.

Peridinium pentagonum
LeBour, M.V., 1925
The dinoflagellates of Northern Seas. The Marine Biological Association of the United Kingdom, Plymouth. 250 pp., 35 pls., 53 text figs.

Peridinium pentagonum
Lillick, L.C., 1940
Phytoplankton and planktonic protozoa of the offshore waters of the Gulf of Maine. Pt.II. Qualitative Composition of the Planktonic Flora. Trans. Am. Phil. Soc., n.s., 31(3):193-237, 13 text figs.

Peridinium pentagonum
Mangin, M. L., 1912
Phytoplancton de la croisière du "René" dans l'Atlantique (Septembre 1908). Ann. Inst. Ocean., n.s., 4(1):1-66, 2 pls., 41 text figs., 2 tables.

Peridinium pentagonum
Meunier, A., 1919
Microplancton de la Mer Flamande 3. Les Péridiniens. Mem. Mus. Roy. Hist. Nat., Belgique 8(1):1-116, Pls. XV-XXI.

Peridinium pentagonum
Morse, D.C., 1947
Some observations on seasonal variations in plankton population Patuxant River, Maryland 1943-1945. Bd. Nat. Res., Publ. No.65, Chesapeake Biol. Lab., 31, 3 figs.

Peridinium pentagonum
Paulsen, O., 1908
XVIII Peridiniales. Nordisches Plankton, Bot. Teil: 1-124, 155 text figs.

Peridinium pellucidum
Rampi, L., 1950.
Ricerche sul fitoplancton del Mare Ligure. 9) I Peridinium delle acque di Sanremo. Pubbl. Centro Talassografico Tirreno No. 5:10 pp. 2 pls.

Atti Acad. Ligure Sci. e Lett. 7(1):

Peridinium pentagonum
Sousa e Silva, E., 1949
Diatomaceas e Dinoflagelados de Baia de Cascais. Portugaliae Acta Biol., Volume: Julio Henriques, Ser. B: 300-383, 9 pls, 2 fold-in tables.

Peridinium pentagonum
Wall, David, and Barrie Dale, 1968.
Modern dinoflagellate cysts and evolution of the Peridiniales. Micropaleontology, 14(3):265-304.

Peridium sp. cf. P. pentagonum
Wall, David, and Barrie Dale, 1968.
Modern dinoflagellate cysts and evolution of the Peridiniales. Micropaleontology, 14(3):265-304.

Peridinium pentagonum
Wang, C. C., and D. Nie, 1932
A survey of the marine protozoa of Amoy. Contr. Biol. Lab. Sci. Soc., China, Zool. Ser., 8:285-385, 89 text figs.

Peridinium perbreve n. sp.
Balech, Enrique, y Leo de Oliveira Soares, 1966.
Dos dinoflagelados de la Bahia de Guanabara y proximades (Brasil). Neotropica, 12(39):103-109.

Peridinium perplexum n. sp.
Balech, Enrique, 1971.
Microplancton de la campaña oceanographica: Productividad III.
Revta Mus. argent. Cienc. Nat. Bernadina Rivadavia, Hydrobiol. 3(1):1-202, 39 pls.

Peridinium peruvianum sp. nov.
Balech, Enrique, 1961
Notula sobre Peridinium (Dinoflagellatae). Neotropica, 7:29-32.

Peridinium petersi n.sp.
Balech, Enrique, 1958.
Plancton de la Campaña Antartica Argentina, 1954-1955. Physis, 21(60):75-108.

Peridinium piriforme
Wood, E.J.P., 1954.
Dinoflagellates in the Australian Region. Australian J. Mar. Freshwater Res., 5(2):171-351.

Peridinium polonicum
Adachi, Rokuro, 1965.
Studies on a dinoflagellate Peridinium polonicum Woloszynska. I. The structure of the skeleton. (In Japanese; English summary). J. Fac. Fish. prefect. Univ. Mie. 6(3):317-326.

Peridinium polonicum
Hashimoto, Yoshiro, Tomotoshi Okaichi, Le Dung Dang, 2nd Tamas Noguchi, 1968.
Glenodinine an ichthyotoxic substance produced by a dinoflagellate Peridinium polonicum.
Bull. Jap. Soc. Sci. Fish, 34(6): 528-534

Peridinium polymorphum n.sp.
Lindemann, E., 1924
Peridineen aus dem goldenen Horn und dem Bosphorus. Bot. Arch. 5:216-233, 98 text figs.

Peridinium pseudoantarcticum n.sp.
Balech, Enrique, 1958.
Plancton de la Campaña Antartica Argentina, 1954-1955. Physis, 21(60):75-108.

Pericinium punctulatum
Balech, Enrique, 1971.
Microplancton de la campaña oceanographica: Productividad III. Revta Mus. argent. Cienc. Nat. Bernadina Rivadavia, Hydrobiol. 3(1):1-202, 39 pls.

Peridinium punctulatum
Balech, E., 1949.
Etude de quelques especes de Peridinium souvent confondues. Hydrobiol., Acta Hydrobiol., Limnol. et Protistol. 1(4):390-409, 6 pls.

Peridinium punctulatum
Dangeard, P., 1927
Phytoplankton de la croisière du "Sylvana". Ann. Inst. Ocean., Monaco, n.s., 4(8):286-401, 54 text figs. (février-Juin 1913).

Peridinium punctulatum
LeBour, M.V., 1925
The dinoflagellates of Northern Seas. The Marine Biological Association of the United Kingdom, Plymouth, 250 pp., 35 pls., 53 text figs.

Peridinium punctulatum
Lindemann, E., 1924
Peridineen aus dem goldenen Horn und dem Bosphorus. Bot. Arch. 5:216-233, 98 text figs.

Peridinium punctulatum
Marukawa, H., 1921
Plankton lists and some new species of copepods, from the northern waters of Japan. Bull. Inst. Ocean., No.384, 15 pp., 3 pls., 1 chart. Monaco

Peridinium punctulatum
Matzenauer, L., 1933
Die Dinoflagellaten des indischen Ozeans (mit Ausnahme der Gattung Ceratium.) Bot. Arch. 35:437-510, 77 text figs., 2 charts.

Peridinium punctulatum
Meunier, A., 1919
Microplancton de la Mer Flamande 3. Les Péridiniens. Mem. Mus. Roy. Hist. Nat., Belgique 8(1):1-116, Pls. XV-XXI.

Peridineum punctulatum
Paulsen, O., 1949
Observations on dinoflagellates. (Ed. J. Grøntoed) Kongl. Dansk. Videnskab. Selsk., Biol. Skr. 6(4):67 pp., 30 text figs.

Peridinium punctulatum
Paulsen, O., 1908
XVIII Peridiniales. Nordisches Plankton, Bot. Teil: 1-124, 155 text figs.

Peridinium punctulatum
Pavillard, J., 1916
Recherches sur les Peridiniens du Golfe du Lion. Mem. Univ. Montpellier. Trav. Inst. Bot., Univ. Montpellier. Serie mixte No.4, 70 pp., 3 pls., 15 text figs.

Peridinium punctulatum
Sousa e Silva, E., 1949
Diatomaceas e Dinoflagelados de Baia de Cascais. Portugaliae Acta Biol., Volume: Julio Henriques, Ser. B: 300-383, 9 pls, 2 fold-in tables.

Peridinium punctulatum
Sousa a Silva, E., and J. Dos Santos-Pinto, 1948
O Plancton da Baia de S. Martinho do Porto. 1. Diatomaceas e Dinoflagelados. Bol. Soc. Portuguese de Ciencias Naturais, 16(2):134-187, 6 pls. (Trav. Sta. Biol. Mar. de Lisbonne No. 52).

Peridinium punctulatum
Vives, F., and A. Planas, 1952
Plancton recogido por los laboratorios costeros. VI. Fitoplancton de las costas de Vinaroz, islas Columbretes y alrededores de la desembocadura del Elroa. Publ. Inst. Biol. Aplic. 11:141-156, 19 textfigs.

Peridinium punctulatum
Wall, David, and Barrie Dale, 1968.
Modern dinoflagellate cysts and evolution of the Peridiniales. Micropaleontology, 14(3):265-304.

Peridinium pusillum
Klottler, H.E., 1951.
Eine Wasserblüte von Peridinium pusillum (Penard) Lemmermann. Arch. Hydrobiol., Suppl.-Band. 20:144-156, 9 textfigs.

Peridinium pusillum
Wood, E.J.F., 1954.
Dinoflagellates in the Australian region. Australian J. Mar. Freshwater Res., 5(2):171-351.

Peridinium pyriforme
Gilbert, J.Y., and W.E. Allen, 1943
The phytoplankton of the Gulf of California obtained by the "E.W. Scripps" in 1939 and 1940. J. Mar. Res. V(2):89-110, figs.30-31.

Peridinium pyriforme
Gran, H.H., and T. Braarud, 1935
A quantitative study of the phytoplankton in the Bay of Fundy and the Gulf of Maine (including observations on hydrography, chemistry, and turbidity). J. Biol. Bd., Canada, 1(5):279-467, 69 text figs.

Peridinium pyriforme
Lillick, L.C., 1940
Phytoplankton and planktonic protozoa of the offshore waters of the Gulf of Maine. Pt.II. Qualitative Composition of the Planktonic Flora. Trans. Am. Phil. Soc., n.s., 31(3):193-237, 13 text figs.

Peridinium pyriforme
Moberg, E.G., and W.E. Allen, 1927.
Effect of tidal changes on physical, chemical, and biological conditions in the sea water of the San Diego region. 1. Observations on the effect of tidal changes on physical and chemical conditions of sea water in the San Diego region. 2. Half-hourly collections of marine microplankton taken at the Scripps Institution pier in 1923. Bull. S.I.O., tech. ser., 1:1-17, 4 textfigs.

Peridinium pyriforme
Paulsen, O., 1908
XVIII Peridiniales. Nordisches Plankton, Bot. Teil: 1-124, 155 text figs.

Peridineum piriforme
Paulsen, O., 1949
Observations on dinoflagellates. (Ed. J. Grøntoed) Kongl. Dansk. Videnskab. Selsk., Biol. Skr. 6(4):67 pp., 30 text figs.

Peridinium pyriforme
Rampi, L., 1950.
Ricerche sul fitoplancton del Mare Ligure. 9) I Peridinium delle acque di Sanremo. Pubbl. Centro Talassografico Tirreno No. 5:10 pp 2 pls.

Atti Acad. Ligure Sci. e Lett. 7(1):

Peridinium pyriforme
Sousa e Silva, E., 1949
Diatomaceas e Dinoflagelados de Baia de Cascais. Portugaliae Acta Biol., Volume: Julio Henriques, Ser. B: 300-383, 9 pls, 2 fold-in tables.

Peridinium pyrum
Balech, Enrique, 1971.
Microplancton de la campaña oceanographica: Productividad III. Revta Mus. argent. Cienc. Nat. Bernadina Rivadavia, Hydrobiol. 3(1):1-202, 39 pls.

Peridinium pyrum n.sp.
Balech, Enrique, 1959.
Operacion Oceanografice Merluza V Crucero. Plancton. Publico Servico de Hidrografia Naval, Argentina, H. 618:43 pp.

Peridinium quadratum n.sp.
Matzenauer, L., 1933
Die Dinoflagellaten des indischen Ozeans (mit Ausnahme der Gattung Ceratium.) Bot. Arch. 35:437-510, 77 text figs., 2 charts.

Peridinium quarnerense
Dangeard, P., 1927
Phytoplankton de la croisière du "Sylvana". Ann. Inst. Ocean., Monaco, n.s., 4(8):286-401, 54 text figs. (février-Juin 1913).

Peridinium quarerense
Wood, E.J.F., 1954.
Dinoflagellates in the Australian region. Australian J. Mar. Freshwater Res., 5(2):171-351.

Peridinium quarnerense
Dangeard, P., 1927.
Notes sur la variation de la genre Peridinium. Bull. Inst. Ocean., Monaco, No. 507:16 pp., 9 textfigs.

Peridinium quarnerense
Delegazione Italiana della Commissione Internazionale per l'Esplorazione Scientifica del Mediterraneo, 1941
Note sul plancton della Laguna veneta. [Memoria CCLXXIX], Arch. di Ocean. e Limn. Anno I, Fasc. I, 1941 XIX: 31-57 pp.

Peridinium quarnerense
Forti, A., 1922
Ricerche sulla flora pelagica (fitoplancton) di Quarto dei Mille. Mem. R. Com. Talass. Ital. 97:248 pp., 13 pls.

Peridinium quarnerense
Matzenauer, L., 1933
Die Dinoflagellaten des indischen Ozeans (mit Ausnahme der Gattung Ceratium.) Bot. Arch. 35:437-510, 77 text figs., 2 charts.

Peridinium quernerense
Pavillard, J., 1916
Recherches sur les Peridiniens du Golfe du Lion. Mem. Univ. Montpellier. Trav. Inst. Bot., Univ. Montpellier. Serie mixte No.4, 70 pp., 3 pls., 15 text figs.

Peridinium quinquecorne
Helim, Y., 1965.
Microplancton des eaux Egyptiennes. II. Chrysomonadines; Ebriediens et dinoglagellés nouveaux ou d'intérêt biogeographique. Rapp. Proc. Verb. Reunions. Comm. Int. Expl. Sci., Mer Mediterranée, Monaco. 18(2):373-379.

Peridinium rampii
Balech, Enrique, 1971.
Microplancton de la campaña oceanographica: Productividad III.
Revta Mus. argent. Cienc. Nat. Bernadina Rivadavia, Hydrobiol. 3(1):1-202, 39 pls.

Peridinium rampii n.sp.
Balech, Enrique, 1959.
Operacion Oceanografica Merluza V Crucero. Plancton.
Publico, Servicio de Hidrografia Naval, Argentina, H. 618:43 pp.

Peridinium raphanum n.sp.
Balech, Enrique, 1958.
Plancton de la Campaña Antartica Argentina, 1954-1955.
Physis, 21(60):75-108.

Peridinium rectius n.var.
Graham, H. W., 1942
Studies in the morphology, taxonymy, and ecology of the Peridiniales. Sci. Res. Cruise VII of the Carnegie, 1928-1929---Biol. III(542): 129 pp., 67 figs.

Peridinium rectum
Forti, A., 1922
Ricerche sulla flora pelagica (fitoplancton) di Quarto dei Mille. Mem. R. Com. Talass. Ital. 97:248 pp., 13 pls.

Peridinium rectum n. sp.
Kofoid, C. A., 1907
Dinoflagellata of the San Diego region. III. Descriptions of new species. Univ. Calif. Publ., Zool. 3:299-340, Pls. 22-33.

Peridinium rectum
Pavillard, J., 1916
Recherches sur les Peridiniens du Golfe du Lion. Mem. Univ. Montpellier. Trav. Inst. Bot., Univ. Montpellier. Serie mixte No.4, 70 pp., 3 pls., 15 text figs.

Peridinium rectum
Sousa e Silva, E., 1949
Diatomaceas e Dinoflagelados de Baía de Cascais. Portugaliae Acta Biol., Volume: Julio Henriques, Ser. B: 300-383, 9 pls, 2 fold-in tables.

Peridinium remotum
Matzenauer, L., 1933
Die Dinoflagellaten des indischen Ozeans (mit Ausnahme der Gattung Ceratium.) Bot. Arch. 35:437-510, 77 text figs., 2 charts.

Peridinium remotum
Wood, E.J.P., 1954.
Dinoflagellates in the Australian region. Australian J. Mar. Freshwater Res., 5(2):171-351.

Peridinium retiferum n.sp.
Matzenauer, L., 1933
Die Dinoflagellaten des indischen Ozeans (mit Ausnahme der Gattung Ceratium.) Bot. Arch. 35:437-510, 77 text figs., 2 charts.

Peridinium robustum
LeBour, M.V., 1925
The dinoflagellates of Northern Seas. The Marine Biological Association of the United Kingdom, Plymouth, 250 pp., 35 pls.,53 text figs.

Peridinium rosaceum
Argentina, Secretaria de Marina, Servicio de Hidrografia Naval, 1962.
Plancton de las campanas oceanograficas DRAKE I y II.
Publico, H. 627:57.

Peridinium rosaceum
Balech, Enrique, 1971.
Microplancton de la campaña oceanographica: Productividad III.
Revta Mus. argent. Cienc. Nat. Bernadina Rivadavia, Hydrobiol. 3(1):1-202, 39 pls.

Peridinium rosaceum n.sp.
Balech, Enrique, 1958.
Plancton de la Campaña Antartica Argentina, 1954-1955.
Physis, 21(60):75-108.

Peridinium roscoffiense n.sp.
Balech, Enrique, 1962.
Notulas de la Estacion Hidrobiologica de Puerto Quequen.
Revista. Mus. Argentino. Ciencias Nat., "Bernardino Rivadavia", Ciencias Zool., 8(6):81-87.

Peridinium roseum
Lebour, M.V., 1925.
The dinoflagellates of Northern seas. The Marine Biological Association, United Kingdom, 250 pp., 35 pls., 53 textfigs.

Peridinium roseum
Lillick, L.C., 1940
Phytoplankton and planktonic protozoa of the offshore waters of the Gulf of Maine. Pt.II. Qualitative Composition of the Planktonic Flora. Trans. Am. Phil. Soc., n.s., 31(3):193-237, 13 text figs.

Peridineum roseum
Paulsen, O., 1949
Observations on dinoflagellates. (Ed. J. Grøntoed) Kongl. Dansk. Videnskab. Selsk., Biol. Skr. 6(4):67 pp., 30 text figs.

Peridinium roseum
Paulsen, O., 1908
XVIII Peridiniales. Nordisches Plankton, Bot. Teil: 1-124, 155 text figs.

Peridinium roseum
Schulz, B., and A. Wulff, 1929
Hydrographie und Oberflächen plankton des westlichen Barentsmeeres im Sommer 1927. Ber. deutschen wissensch. Komm. F. Meeresforsch. n.s. 4(5):232-372, 13 tables, 25 text figs.

Peridinium roseum
Wood, E.J.P., 1954.
Dinoflagellates in the Australian region. Australian J. Mar. Freshwater Res., 5(2):171-351.

Peridinium rotundum
Lillick, L.C., 1940
Phytoplankton and planktonic protozoa of the offshore waters of the Gulf of Maine. Pt.II. Qualitative Composition of the Planktonic Flora. Trans. Am. Phil. Soc., n.s., 31(3):193-237, 13 text figs.

Peridinium (?) rubrum
Paulsen, O., 1908
XVIII Peridiniales. Nordisches Plankton, Bot. Teil: 1-124, 155 text figs.

Peridinium saltans
LeBour, M.V., 1925
The dinoflagellates of Northern Seas. The Marine Biological Association of the United Kingdom, Plymouth, 250 pp., 35 pls.,53 text figs.

Peridinium saltans Pavillard 1915 non Meunier
Pavillard, J., 1916
Recherches sur les Peridiniens du Golfe du Lion. Mem. Univ. Montpellier. Trav. Inst. Bot., Univ. Montpellier. Serie mixte No.4, 70 pp., 3 pls., 15 text figs.

Peridinium schilleri n.sp.
Paulsen, O., 1930.
Etudes sur le microplancton de la mer d'Alboran. Trab. Inst. Esp. Ocean. No. 4:1-108, 61 textfigs.

Peridinium simplex
Brunel, J., 1962
Le phytoplancton de la Baie de Chaleurs. Inst. Botan., Univ. Montréal, Contrib. No. 77: 365 pp., 66 pls.

Peridinium simplex?
Brunel, Jules, 1962.
Le phytoplancton de la Baie des Chaleurs. Contrib. Ministère de la Chasse et des Pêcheries. Province de Québec. No. 91: 365 pp.

Peridinium simplex n.sp.
Gran, H.H., and T. Braarud, 1935
A quantitative study of the phytoplankton in the Bay of Fundy and the Gulf of Maine (including observations on hydrography, chemistry, and turbidity). J. Biol. Bd., Canada, 1(5):279-467, 69 text figs.

Peridinium simplex
Lillick, L.C., 1940
Phytoplankton and planktonic protozoa of the offshore waters of the Gulf of Maine. Pt.II. Qualitative Composition of the Planktonic Flora. Trans. Am. Phil. Soc., n.s., 31(3):193-237, 13 text figs.

Peridinium simplex
Lillick, L.C., 1938
Preliminary report of the phytoplankton of the Gulf of Maine. Am. Mid. Nat. 20(3):624-640, 1 text figs 37 tables.

Peridinium simplex
Lillick, L.C., 1937
Seasonal studies of the phytoplankton off Woods Hole, Massachusetts. Biol. Bull. LXXIII (3):488-503, 3 text figs.

Peridinium simulum
Balech, Enrique, 1959.
Operacion Oceanografica Merluza V Crucero. Plancton.
Publico, Servicio.Hidrografia Naval, Argentina, H. 618:43 pp.

Peridinium simulum nom. nov.
Paulsen, O., 1930.
Etudes sur le microplancton de la mer d'Alboran. Trab. Inst. Esp. Ocean. No. 4:1-108, 61 textfigs.

Peridinium sinaicum n.sp.
Matzenauer, L., 1933
Die Dinoflagellaten des indischen Ozeans (mit Ausnahme der Gattung Ceratium.) Bot. Arch. 35:437-510, 77 text figs., 2 charts.

Peridinium sinaicum
Rampi, L., 1950.
Peridiniens rares ou nouveaux pour le Pacifique Sud-Equatorial. Bull. Inst. Ocean., Monaco, No. 974:11 pp., 26 textfigs.

Peridinium sociale
Biecheler, B., 1952.
Recherches sur les Peridiniens. Bull. Biol., France Belg., Suppl., 36:1-149, 75 textfigs.

Peridinium solidicorne
Balech, Enrique, 1971
Microplancton del Atlantico ecuatorial oeste (Equalant 1)
Publ. Serv. Hidrograf. Naval, Argentina H. 654: 103 pp., 122 figs.

Peridinium solidicorne
Dangeard, P., 1927
Phytoplankton de la croisière du "Sylvana". Ann. Inst. Ocean., Monaco, n.s., 4(8):286-401, 54 text figs. (Février-Juin 1913).

Peridinium solidicorne
Dangeard, P., 1927.
Péridiniens nouveaux ou peu connus de la croisière du "Sylvana". Bull. Inst. Océan., Monaco, No. 491:16 pp., 9 textfigs.

Peridinium solidicorne
Matzenauer, L., 1933
Die Dinoflagellaten des indischen Ozeans (mit Ausnahme der Gattung Ceratium.) Bot. Arch. 35:437-510, 77 text figs., 2 charts.

Peridinium solidicorne
Rampi, L., 1950.
Ricerche sul fitoplancton del Mare Ligure. 9) I Peridinium delle acque di Sanremo. Pubbl. Centro Talassografico Tirreno No. 5: 10 pp. 2 pls.

Atti Acad. Ligure Sci. e Lett. 7(1):

Peridinium solidicorne
Wood, E.J.P., 1954.
Dinoflagellates in the Australian region. Australian J. Mar. Freshwater Res., 5(2):171-351.

Peridinium solitarium
Balech, Enrique, 1971.
Microplancton de la campaña oceanographica: Productividad III. Revta Mus. argent. Cienc. Nat. Bernadina Rivadavia, Hydrobiol. 3(1):1-202, 39 pls.

Peridinium somma n.sp.
Matzenauer, L., 1933
Die Dinoflagellaten des indischen Ozeans (mit Ausnahme der Gattung Ceratium.) Bot. Arch. 35:437-510, 77 text figs., 2 charts.

Peridinium sphaeroideum
Balech, Enrique, 1971.
Microplancton de la campaña oceanographica: Productividad III. Revta Mus. argent. Cienc. Nat. Bernadina Rivadavia, Hydrobiol. 3(1):1-202, 39 pls.

Peridium sphaericum
Matzenauer, L., 1933
Die Dinoflagellaten des indischen Ozeans (mit Ausnahme der Gattung Ceratium.) Bot. Arch. 35:437-510, 77 text figs., 2 charts.

Peridinium sphaericum
Pavillard, J., 1916 Okamura 1911, non P. sphaericum Murray & Whitting 1899
Recherches sur les Peridiniens du Golfe du Lion. Mem. Univ. Montpellier. Trav. Inst. Bot. Univ. Montpellier. Serie mixte No.4, 70 pp., 3 pls., 15 text figs.

Peridinium sphaericum
Wood, E.J.P., 1954.
Dinoflagellates in the Australian region. Australian J. Mar. Freshwater Res., 5(2):171-351.

Peridinium sphaeroides n.sp.
Dangeard, P., 1927.
Péridiniens nouveaux ou peu connus de la croisière du "Sylvana". Bull. Inst. Océan., Monaco, No. 491:16 pp., 9 textfigs.

Peridinium spheroides
Dangeard, P., 1927
Phytoplankton de la croisière du "Sylvana". Ann. Inst. Ocean., Monaco, n.s., 4(8):286-401, 54 text figs. (Février-Juin 1913).

Peridinium spheroides
Margalef, R., 1949
Fitoplancton nerítico de la Costa Brava en 1947-48. Publ. Inst. Biol. Aplicada, 5: 41-51, 3 text figs.

Peridinium sphaeroides
Matzenauer, L., 1933
Die Dinoflagellaten des indischen Ozeans (mit Ausnahme der Gattung Ceratium.) Bot. Arch. 35:437-510, 77 text figs., 2 charts.

Peridinium sphaeroides
Matzenauer, L., 1933
Die Dinoflagellaten des indischen Ozeans (mit Ausnahme der Gattung Ceratium.) Bot. Arch. 35:437-510, 77 text figs., 2 charts.

Peridinium Spheroides
Sousa e Silva, E., 1949
Diatomaceas e Dinoflagelados de Baia de Cascais. Portugaliae Acta Biol., Volume: Julio Henriques, Ser. B: 300-383, 9 pls, 2 fold-in tables.

Peridinium spinifera
Pavillard, J., 1916
Recherches sur les Peridiniens du Golfe du Lion. Mem. Univ. Montpellier. Trav. Inst. Bot. Univ. Montpellier. Serie mixte No.4, 70 pp., 3 pls., 15 text figs.

Peridinium spirale nov. comb.
Balech, Enrique, 1971.
Microplancton de la campaña oceanographica: Productividad III. Revta Mus. argent. Cienc. Nat. Bernadina Rivadavia, Hydrobiol. 3(1):1-202, 39 pls.

Peridinium stagnale n.sp.
Meunier, A., 1919
Microplancton de la Mer Flamande 3. Les Péridiniens. Mem. Mus. Roy. Hist. Nat., Belgique 8(1):1-116, Pls. XV-XXI.

Peridinium Steinii
Braarud, T., 1945
A phytoplankton survey of the polluted waters of inner Oslo Fjord. Hvalrådets Skrifter, No.28, 142 pp., 19 text figs., 17 tables.

Peridinium Steini
Braarud, T., and Adam Bursa, 1939
On the phytoplankton of the Oslo Fjord, 1933-1934. Hvalrådets Skr. No.19:1-63; 9 text figs. Reviewed. J. du. Cons. 14(3): 418-420. A.C. Gardiner.

Peridinium Steinii
Dangeard, P., 1927
Phytoplankton de la croisière du "Sylvana". Ann. Inst. Ocean., Monaco, n.s., 4(8):286-401, 54 text figs. (Février-Juin 1913).

Peridinium Steinii
Dangeard, P., 1926
Description des Péridiniens Testacés recueillis para la Mission Charcot pendent le mois d'Aout 1924. Ann. Inst. Ocean. n.s. 3(7):307-334, 15 text figs.

Peridinium Steinii
Delegazione Italiana della Commissione Internazionale per l'Esplorazione Scientifica del Mediterraneo, 1941
Note sul plancton della Laguna veneta. [Memoria CCLXXXIX], Arch. di Ocean. e Limn. Anno I, Fasc. I, 1941 XIX: 31-57 pp.

Peridinium Steinii
Ercegovic, A., 1936
Etudes qualitative et quantitatives du phytoplancton dans les eaux cotières de l'Adriatique oriental moyen au cours de l'année 1934. Acta Adriatica 1(9):1-126

Peridinium Steinii
Forti, A., 1922
Ricerche sulla flora pelagica (fitoplancton) di Quarto dei Mille. Mem. R. Com. Talass. Ital. 97:248 pp., 13 pls.

Peridinium steinii
Gilbert, J.Y., and W.E. Allen, 1943
The phytoplankton of the Gulf of California obtained by the "E.W. Scripps" in 1939 and 1940. J. Mar. Res. V(2):89-110, figs.30-31.

Peridinium Steini
Gran, H.H., and T. Braarud, 1935
A quantitative study of the phytoplankton in the Bay of Fundy and the Gulf of Maine (including observations on hydrography, chemistry, and turbidity). J. Biol. Bd., Canada, 1(5):279-467, 69 text figs.

Peridium steinii
Jørgensen, E., 1905
B.Protistplankton and the diatoms in bottom samples. Hydrographical and biological investigations in Norwegian fjords. Bergens Mus. Skr. 7: 49-225.

Peridinium steinii nom. nov.
Jorgensen, E., 1900
Protophyten und Protozoën im Plankton aus der Norwegischen Westkurste. Bergens Mus. Aarb. 1899(6): 95 pp., 5 pls., 83 tables.

Peridinium steinii
LeBour, M.V., 1925
The dinoflagellates of Northern Seas. The Marine Biological Association of the United Kingdom, Plymouth, 250 pp., 35 pls., 53 text figs.

Peridinium Steinii
Lillick, L.C., 1940
Phytoplankton and planktonic protozoa of the offshore waters of the Gulf of Maine. Pt.II. Qualitative Composition of the Planktonic Flora. Trans. Am. Phil. Soc., n.s., 31(3):193-237, 13 text figs.

Peridinium steinii
Lindemann, E., 1924
Peridineen aus dem goldenen Horn und dem Bosphorus. Bot. Arch. 5:216-233, 98 text figs.

Peridinium Steinii
Mangin, M. L., 1912
Phytoplancton de la croisière du "René" dans l'Atlantique (Septembre 1908). Ann. Inst. Ocean., n.s., 4(1):1-66, 2 pls., 41 text figs., 2 tables.

Peridinium Steinii
Margalef, R., 1949
Fitoplancton nerítico de la Costa Brava en 1947-48. Publ. Inst. Biol. Aplicada, 5: 41-51, 3 text figs.

Peridinium Steini
Marukawa, H., 1921
Plankton lists and some new species of copepods, from the northern waters of Japan. Bull. Inst. Ocean., No.384, 15 pp., 3 pls., 1 chart. Monaco

Peridinium Steinii
Matzenauer, L., 1933
Die Dinoflagellaten des indischen Ozeans (mit Ausnahme der Gattung Ceratium.) Bot. Arch. 35:437-510, 77 text figs., 2 charts.

Peridinium Steinii
Paulsen, O., 1908
XVIII Peridiniales. Nordisches Plankton, Bot. Teil: 1-124, 155 text figs.

Peridinium steinii
Paulsen, O., 1905.
On some Peridinese and plankton diatoms.
Medd. Komm. Havundersøgelser, Ser. Plankton, 1(3):7 pp.

Peridinium Steinii
Pavillard, J., 1916
Recherches sur les Peridiniens du Golfe du Lion. Mem. Univ. Montpellier. Trav. Inst. Bot., Univ. Montpellier. Serie mixte No.4, 70 pp., 3 pls., 15 text figs.

Peridinium Steinii
Pavillard, J., 1905
Recherches sur la flore pelagique (Phytoplankton) de l'Etang de Thau. Theses presentees a la Fac. Sci., Paris, 116 pp., 3 pls.

Peridinium steinii
Sousa e Silva, E., 1949
Diatomaceas e Dinoflagelados de Baia de Cascais. Portugaliae Acta Biol., Volume: Julio Henriques, Ser. B: 300-383, 9 pls, 2 fold-in tables.

Peridinium steinii & var.
Rampi, L., 1950.
Ricerche sul fitoplancton del Mare Ligure. 9) I Peridinium delle acque di Sanremo.
Pubbl. Centro Talassografico Tirreno No. 5:10 pp. 2 pls.

Atti Acad. Ligure Sci. e Lett. 7(1):

Peridinium Steinii mediterraneum Kofoid 1909
Pavillard, J., 1916
Recherches sur les Peridiniens du Golfe du Lion. Mem. Univ. Montpellier. Trav. Inst. Bot., Univ. Montpellier. Serie mixte No.4, 70 pp., 3 pls., 15 text figs.

Peridinium Steinii
Rumkówna, A., 1948
[List of the phytoplankton species occurring in the superficial water layers in the Gulf of Gdańsk] Bull. Lab. mar., Gdynia, No. 4: 139-141 with tables in back.

Peridinium steinii
Wood, E.J.F., 1954.
Dinoflagellates in the Australian region.
Australian J. Mar. Freshwater Res., 5(2):171-351.

Peridinium Steinii var. africanum n.var.
Dangeard, P., 1927.
Péridiniens nouveau ou peu connus de la croisière du "Sylvana". Bull. Inst. Océan., Monaco, No. 491:16 pp., 9 textfigs.

Peridinium Steinii mediterraneum
Ercegovic, A., 1936
Etudes qualitative et quantitatives du phytoplancton dans les eaux cotières de l'Adriatique oriental moyen au cours de l'année 1934. Acta Adriatica 1(9):1-126

Peridinium stellatum n. sp.
Wall, David, and Barrie Dale, 1968.
Modern dinoflagellate cysts and evolution of the Peridiniales. Micropaleontology, 14(3):265-304.

Peridinium striolatum
Wood, E.J.P., 1954.
Dinoflagellates in the Australian region.
Australian J. Mar. Freshwater Res., 5(2):171-351.

Peridinium sub-curvipes
Balech, Enrique, 1959.
Operacion Oceanografica Merluza V Crucero. Plancton.
Publico, Servicio de Hidrografia Naval, Argentina, H. 618:43 pp.

Peridinium subcurvipes
Dangeard, P., 1927
Phytoplankton de la croisière du "Sylvana". Ann. Inst. Ocean. Monaco, n.s., 4(8):286-401, 54 text figs. (février-Juin 1913).

Peridinium sub-curvipes
Gran, H.H., and T. Braarud, 1935
A quantitative study of the phytoplankton in the Bay of Fundy and the Gulf of Maine (including observations on hydrography, chemistry, and turbidity). J. Biol. Bd., Canada, 1(5):279-467, 69 text figs.

Peridinium sub-curvipes
Lebour, M.V., 1925.
The dinoflagellates of northern seas. Marine Biological Association, United Kingdom, 250 pp., 35 pls., 53 textfigs.

Peridinium subinerme?
Balech, Enrique, 1971.
Microplancton de la campaña oceanographica: Productividad III.
Revta Mus. argent. Cienc. Nat. Bernadina Rivadavia, Hydrobiol. 3(1):1-202, 39 pls.

Peridinium subinerme
Dangeard, P., 1927
Phytoplankton de la croisière du "Sylvana". Ann. Inst. Ocean., Monaco, n.s., 4(8):286-401, 54 text figs. (février-Juin 1913).

Peridinium subimerme
Dangeard, P., 1926
Description des Péridiniens Testacés recueillis para la Mission Charcot pendent le mois d'Aout 1924. Ann. Inst. Ocean. n.s. 3(7):307-334, 15 text figs.

Peridinium subinerme
Gran, H.H., and T. Braarud, 1935
A quantitative study of the phytoplankton in the Bay of Fundy and the Gulf of Maine (including observations on hydrography, chemistry, and turbidity). J. Biol. Bd., Canada, 1(5):279-467, 69 text figs.

Peridinium subinerme
LeBour, M.V., 1925
The dinoflagellates of Northern Seas. The Marine Biological Association of the United Kingdom, Plymouth, 250 pp., 35 pls., 53 text figs.

Peridinium subinerme
Lillick, L.C., 1940
Phytoplankton and planktonic protozoa of the offshore waters of the Gulf of Maine. Pt.II. Qualitative Composition of the Planktonic Flora. Trans. Am. Phil. Soc., n.s. 31(3):193-237, 13 text figs.

Peridinium subinerme
Mangin, M. L., 1912
Phytoplancton de la croisière du "René" dans l'Atlantique (Septembre 1908). Ann. Inst. Ocean., n.s., 4(1):1-66, 2 pls., 41 text figs., 2 tables.

Peridinium subinerme
Margalef, R., 1949
Fitoplancton nerítico de la Costa Brava en 1947-48. Publ. Inst. Biol. Aplicada, 5: 41-51, 3 text figs.

Peridinium subinermis
Marukawa, H., 1921
Plankton lists and some new species of copepods, from the northern waters of Japan. Bull. Inst. Ocean, No.384, 15 pp., 3 pls., 1 chart. Monaco

Peridinium subinerme
Matzenauer, L., 1933
Die Dinoflagellaten des indischen Ozeans (mit Ausnahme der Gattung Ceratium.) Bot. Arch. 35:437-510, 77 text figs., 2 charts.

Peridinium subinerme
Meunier, A., 1919
Microplancton de la Mer Flamande 3. Les Péridiniens. Mem. Mus. Roy. Hist. Nat., Belgique 8(1):1-116, Pls. XV-XXI.

Peridineum subinerme
Paulsen, O., 1949
Observations on dinoflagellates. (Ed. J. Grøntoed) Kongl. Dansk. Videnskab. Selsk., Biol. Skr. 6(4):67 pp., 30 text figs.

Peridinium subinerme
Wall, David, and Barrie Dale, 1968.
Modern dinoflagellate cysts and evolution of the Peridiniales.
Micropaleontology, 14(3):265-304.

Peridinium subinermis
Paulsen, O., 1908
XVIII Peridiniales. Nordisches Plankton, Bot. Teil: 1-124, 155 text figs.

Peridinium subinerme
Rampi, L., 1950.
Ricerche sul fitoplancton del Mare Ligure. 9) I Peridinium delle acque di Sanremo.
Pubbl. Centro Talassografico Tirreno No. 5:10 pp. 2 pls.

Atti Acad. Ligure Sci. e Lett. 7(1):

Peridinium subinerme
Schulz, B., and A. Wulff, 1929
Hydrographie und Oberflächen plankton des westlichen Barentsmeeres im Sommer 1927.
Ber. deutschen wissensch. Komm. F. Meeresforsch. n.s. 4(5):232-372, 13 tables, 25 text figs.

Peridinium subinerme
Sousa e Silva, E., 1949
Diatomaceas e Dinoflagelados de Baia de Cascais. Portugaliae Acta Biol., Volume: Julio Henriques, Ser. B: 300-383, 9 pls, 2 fold-in tables.

Peridinium subpyriforme
Balech, Enrique, 1962
Tintinnoinea y Dinoflagellata del Pacifico segun material de las Expediciones NORPAC y DOWNWIND del Instituto Scripps de Oceanografia.
Revista, Mus. Argentino Ciencias Nat. "Bernardino Rivadavia", Ciencias Zool., 7(1):1-253.

Peridinium subpyriforme n.sp.
Dangeard, P., 1927
Phytoplankton de la croisière du "Sylvana". Ann. Inst. Ocean., Monaco, n.s., 4(8):286-401, 54 text figs. (février-Juin 1913).

Peridinium subpyriforme
Matzenauer, L., 1933
Die Dinoflagellaten des indischen Ozeans (mit Ausnahme der Gattung Ceratium.) Bot. Arch. 35:437-510, 77 text figs., 2 charts.

Peridinium subpyriforme
Sousa e Silva, E., 1949
Diatomaceas e Dinoflagelados de Baia de Cascais. Portugaliae Acta Biol., Volume: Julio Henriques, Ser. B: 300-383, 9 pls, 2 fold-in tables.

Peridinium subsalsum
Balech, Enrique, 1964.
Tercera contribucion al conocimiento del genero "Peridinium".
Revista, Mus. Argentino, Ciencias Nat. "Bernardino Rivadavia" e Inst. Nacional. Invest. Ciencias Naturales, Hidrobiol. 1(6):179-195.

Peridinium subsphaericum
Balech, Enrique, 1971.
Microplancton de la campaña oceanographica:
Productividad III.
Revta Mus. argent. Cienc. Nat. Bernadina
Rivadavia, Hydrobiol. 3(1):1-202, 39 pls.

Peridinium subsphaericum n.sp.
Balech, Enrique, 1959.
Operacion Oceanografia Merluza V Crucero.
Plancton.
Publico, Servicio de Hidrografia Naval,
Argentina, H. 618:43 pp.

Peridinium sylvanae
Dangeard, P., 1927
Phytoplankton de la croisière du
"Sylvana". Ann. Inst. Ocean., Monaco, n.s.,
4(8):286-401, 54 text figs. (février-Juin 1913).

Peridinium sylvanae n.sp.
Dangeard, P., 1927.
Péridiniens nouveaux ou peu connus, de la croisière du "Sylvana". Bull. Inst. Océan., Monaco,
No. 491:16 pp., 2 textfigs.

Peridinium tabulatum
LeBour, M.V., 1925
The dinoflagellates of Northern Seas. The
Marine Biological Association of the United
Kingdom, Plymouth, 250 pp., 35 pls., 53 text figs.

Peridinium tabulatum
Meunier, A., 1919
Microplancton de la Mer Flamande 3. Les
Péridiniens. Mem. Mus. Roy. Hist. Nat.,
Belgique 8(1):1-116, Pls. XV-XXI.

Peridinium tabulatum
Paulsen, O., 1908
XVIII Peridiniales. Nordisches Plankton,
Bot. Teil: 1-124, 155 text figs.

Peridinium täningi n.sp.
Rampi, L., 1950.
Péridiniens rares ou nouveaux pour le Pacifique
Sud-Equatorial. Bull. Inst. Océan., Monaco,
No. 974:11 pp., 26 textfigs.

Peridinium tenuissimum
Dangeard, P., 1927
Phytoplankton de la croisière du
"Sylvana". Ann. Inst. Ocean., Monaco, n.s.,
4(8):286-401, 54 text figs. (février-Juin 1913).

Peridinium tenuissimum
Matzenauer, L., 1933
Die Dinoflagellaten des indischen Ozeans
(mit Ausnahme der Gattung Ceratium.) Bot.
Arch. 35:437-510, 77 text figs., 2 charts.

Peridinium sp. cf P. thorianum
Balech, Enrique, 1971.
Microplancton de la campaña oceanographica:
Productividad III.
Revta Mus. argent. Cienc. Nat. Bernadina
Rivadavia, Hydrobiol. 3(1):1-202, 39 pls.

Peridinium Thorianum
Dangeard, P., 1927
Phytoplankton de la croisière du
"Sylvana". Ann. Inst. Ocean., Monaco, n.s.,
4(8):286-401, 54 text figs. (février-Juin 1913).

Peridinium Thorianum
Gran, H.H., and T. Braarud, 1935
A quantitative study of the phytoplankton in the Bay of Fundy and the
Gulf of Maine (including observations
on hydrography, chemistry, and turbidity).
J. Biol. Bd., Canada, 1(5):279-467, 69 text figs.

Peridinium thorianum
LeBour, M.V., 1925
The dinoflagellates of Northern Seas. The
Marine Biological Association of the United
Kingdom, Plymouth, 250 pp., 35 pls., 53 text figs.

Peridinium Thorianum
Lillick, L.C., 1940
Phytoplankton and planktonic
protozoa of the offshore waters
of the Gulf of Maine. Pt.II.
Qualitative Composition of the
Planktonic Flora. Trans. Am.
Phil. Soc., n.s., 31(3):193-237,
13 text figs.

Peridinium Thorianum
Marukawa, H., 1921
Plankton lists and some new species of
copepods, from the northern waters of Japan.
Bull. Inst. Ocean., No.384, 15 pp., 3 pls.,
1 chart. Monaco

Peridinium Thoranium
Matzenauer, L., 1933
Die Dinoflagellaten des indischen Ozeans
(mit Ausnahme der Gattung Ceratium.) Bot.
Arch. 35:437-510, 77 text figs., 2 charts.

Peridinium thorianum
Paulsen, O., 1908
XVIII Peridiniales. Nordisches Plankton,
Bot. Teil: 1-124, 155 text figs.

Peridinium (Euperidinium) thorianum n.sp.
Paulsen, O., 1905.
On some Peridineae and plankton diatoms.
Medd. Komm. Havundersøgelser, Ser. Plankton,
1(3):7 pp.

Peridinium thorianum
Wood, E.J.F., 1954.
Dinoflagellates in the Australian region.
Australian J. Mar. Freshwater Res., 5(2):171-351.

Peridinium thulesense n.sp.
Balech, Enrique, 1958.
Plancton de la Campaña Antartica Argentina,
1954-1955.
Physis, 21(60):75-108.

Peridinium Tregouboffi
Halim, Youssef, 1960
Étude quantitative et qualitative du cycle
écologique des dinoflagellés dans les
eaux de Villefranche-sur-Mer. (1953-1955).
Ann. Inst. Océanogr., Monaco, 38:123-232.

Peridinium Tregouboffi n.sp.
Halim, Y., 1955.
Note sur Peridinium Tregouboffi n.sp.
(Dinoflagellé). Bull. Inst. Océan., Monaco, No.
1056:1-5, 2 pls.

Peridinium trochoideum
Braarud, T., 1945
A phytoplankton survey of the polluted
waters of inner Oslo Fjord. Hvalrådets
Skrifter, No.28, 142 pp., 19 text figs.,
17 tables.

Peridinium trochoideum
Braarud, T., 1934
A note on the phytoplankton of the Gulf
of Maine in the summer of 1933. Biol. Bull.
67(1):76-82.

Peridinium trochoideum
Braarud, T., and Adam Bursa, 1939
On the phytoplankton of the Oslo Fjord,
1933-1934. Hvalrådets Skr. No.19:1-63;
9 text figs. Reviewed. J. du. Cons. 14(3):
418-420. A.C. Gardiner.

Peridinium trochoideum
Polikarpov, G.G. and A.V. Tokareva, 1970.
On the cellular cycle of the dinoflagellates
Peridinium trochoideum (Stein.) and
Goniaulax polyedra (Stein.) (Microautoradiographic investigation). (In Russian;
English abstract).
Gidrobiol. Zh. 6(5):66-69.

Peridinium trochoideum
Wall, D., R.R.L. Guillard, B. Dale and
E. Swift, 1970
Calcitic resting cysts in Peridinium
trochoideum (Stein) Lemmermann, an
autotrophic marine dinoflagellate.
Phycologia 9(2): 151-156

Peridinium trochoideum
Braarud, T., 1962
Species distribution in marine phytoplankton.
J. Oceanogr. Soc., Japan, 20th Ann. Vol.,
628-649.

Peridinium trichoideum
De Angelis, C.M., 1962.
Distribuzione ed ecologia di alcune specie di
Dinoflagellati di acque salmastre.
Pubbl. Staz. Zool., Napoli, 32(Suppl.):301-314.

Peridinium triquetrum
Braarud, T., 1962
Species distribution in marine phytoplankton.
J. Oceanogr. Soc., Japan, 20th Ann. Vol.,
628-649.

Peridinium Triquetrum
Braarud, T., 1945
A phytoplankton survey of the polluted
waters of inner Oslo Fjord. Hvalrådets
Skrifter, No.28, 142 pp., 19 text figs.,
17 tables.

Peridinium triquetrum
Braarud, T., 1934
A note on the phytoplankton of the Gulf
of Maine in the summer of 1933. Biol. Bull.
67(1):76-82.

Peridinium triquetrum
Braarud, T., and Adam Bursa, 1939
On the phytoplankton of the Oslo Fjord,
1933-1934. Hvalrådets Skr. No.19:1-63;
9 text figs. Reviewed. J. du. Cons. 14(3):
418-420. A.C. Gardiner.

Peridinium triquetrum
Braarud, T., and Irene Pappas, 1951.
Experimental studies on the dinoflagellate
Peridinium triquetrum (Ehrb.) Lebour.
Norske Videnskaps-Akad., Oslo, 1. Mat.-Naturvid.
Kl., 1951, No. 2:1-23, 17 textfigs.

Peridinium triquetrum
Gran, H.H., and T. Braarud, 1935
A quantitative study of the phytoplankton in the Bay of Fundy and the
Gulf of Maine (including observations
on hydrography, chemistry, and turbidity).
J. Biol. Bd., Canada, 1(5):279-467, 69 text figs.

Peridinium triquetrum
Harrison, William G., 1973
Nitrate reductase activity during a
dinoflagellate bloom. Limnol. Oceanogr.
18(3):457-465.

Peridinium triqueta
LeBour, M.V., 1925
The dinoflagellates of Northern Seas. The
Marine Biological Association of the United
Kingdom, Plymouth, 250 pp., 35 pls., 53 text figs.

Peridinium triquetum
Lillick, L.C., 1940
Phytoplankton and planktonic protozoa of the offshore waters of the Gulf of Maine. Pt.II. Qualitative Composition of the Planktonic Flora. Trans. Am. Phil. Soc., n.s., 31(3):193-237, 13 text figs.

Peridinium triquetra
Marshall, S.M., 1947
An experiment in marine fish cultivation: III. The plankton of a fertilized loch. Proc. Roy. Soc. Edinburgh, Sect.B., 63, Pt.I(3):21-33, 7 text figs.

Peridinium triquetra
Marshall, S.M. and A.P. Orr, 1948
Further experiments on the fertilization of a sea loch (Loch Craiglin). The effect of different plant nutrients on the phytoplankton. J.M.B.A. 27(2):360-379, 10 text figs.

Peridinium triqueta
Morse, D.C., 1947
Some observations on seasonal variations in plankton population Patuxant River, Maryland 1943-1945. Bd. Nat. Res., Publ. No.65, Chesapeake Biol. Lab., 31, 3 figs.

Peridinium triquetrum
Steemann-Nielsen, Einar, 1951
The marine vegetation of the Isefjord. A study on ecology and production. Medd. Komm. Danmarks Fiskeri-og Havundersøgelser. Ser. Plankton. 5(4); 114pp., 46 text figs.

Peridinium tripos
Dangeard, P., 1926
Description des Péridiniens Testacés recueillis para la Mission Charcot pendant le mois d'Aout 1924. Ann. Inst. Ocean. n.s. 3(7):307-334, 15 text figs.

Peridinium tripos
Forti, A., 1922
Ricerche sulla flora pelagica (fitoplancton) di Quarto dei Mille. Mem. R. Com. Talass. Ital. 97:248 pp., 13 pls.

Peridinium tripos n.sp.
Murray, G., and F. G. Whitting, 1899
New Peridiniaceae from the Atlantic. Trans. Linn. Soc., London, Bot., ser 2, 5: 321-342, Pls. 27-33, 9 tables.

Peridinium tripos
Paulsen, O., 1908
XVIII Peridiniales. Nordisches Plankton, Bot. Teil: 1-124, 155 text figs.

Peridinium tripos
Pavillard, J., 1916
Recherches sur les Peridiniens du Golfe du Lion. Mem. Univ. Montpellier. Trav. Inst. Bot., Univ. Montpellier. Serie mixte No.4, 70 pp., 3 pls., 15 text figs.

Peridinium trirostre
Murray, G., and F. G. Whitting, 1899
New Peridiniaceae from the Atlantic. Trans. Linn. Soc., London, Bot., ser 2, 5: 321-342, Pls. 27-33, 9 tables.

Peridinium tristylum
Balech, E., 1951
Deuxième contribution à la connaissance des Peridinium. Hydrobiol., Acta Hydrobiol., Limnol., Protistol., 3(4):305-330, 7 pls. of 137 figs.

Peridinium tristylum
Dangeard, P., 1927
Phytoplankton de la croisière du "Sylvana". Ann. Inst. Ocean., Monaco, n.s., 4(8):286-401, 54 text figs. (Février-Juin 1913).

Peridinium tristylum
Ercegovic, A., 1936
Etudes qualitative et quantitatives du phytoplancton dans les eaux cotières de l'Adriatique oriental moyen au cours de l'année 1934. Acta Adriatica 1(9):1-126

Peridinium tristylum
Forti, A., 1922
Ricerche sulla flora pelagica (fitoplancton) di Quarto dei Mille. Mem. R. Com. Talass. Ital. 97:248 pp., 13 pls.

Peridinium tristylum
Murray, G., and F. G. Whitting, 1899
New Peridiniaceae from the Atlantic. Trans. Linn. Soc., London, Bot., ser 2, 5: 321-342, Pls. 27-33, 9 tables.

Peridinium tristylum ovata n. var.
Schröder, B., 1900
Phytoplankton des Golfes von Neapel nebst vergleichenden Ausblicken auf das atlantischen Ozean. Mitt. Zool. Stat. Neapel, 14:1-38.

Peridinium trochoideum
Becheler, B., 1952.
Recherches sur les Peridiniens. Bull. Biol. France Belg., Suppl., 36:1-149, 75 textfigs.

Peridinium trochoideum
Braarud, Trygve, 1958.
Observations on Peridinium trochoideum (Stein) Lemm. in culture. Nytt Mag. Botan., 6:39-42.

Peridinium trochoideum
Braarud, T., 1951.
Salinity as an ecological factor in marine phytoplankton. Physiol. Plant. 4:28-34, 3 textfigs.

Peridinium trochoideum
Braarud, T., 1951.
Taxinomical studies of marine dinoflagellates. Nytt Mag. Natudvidensk. 88:43-48.

Peridinium trochoideum
Gran, H.H., and T. Braarud, 1935
A quantitative study of the phytoplankton in the Bay of Fundy and the Gulf of Maine (including observations on hydrography, chemistry, and turbidity). J. Biol. Bd., Canada, 1(5):279-467, 69 text figs.

Peridinium trochoideum
Halldal, P., 1958.
Action spectra of phototaxis and related problems in Volvocales, Ulva-gametes and Dinophyceae. Physiol. Plant. 11:118-153.

Peridinium trochoideum
LeBour, M.V., 1925
The dinoflagellates of Northern Seas. The Marine Biological Association of the United Kingdom, Plymouth, 250 pp., 35 pls.,53 text figs.

Peridinium trochoideum
Lillick, L.C., 1940
Phytoplankton and planktonic protozoa of the offshore waters of the Gulf of Maine. Pt.II. Qualitative Composition of the Planktonic Flora. Trans. Am. Phil. Soc., n.s., 31(3):193-237, 13 text figs.

Peridinium trochoideum
Lillick, L.C., 1938
Preliminary report of the phytoplankton of the Gulf of Maine. Am. Mid. Nat. 20(3):624-640, 1 text fig. 37 tables.

Peridinium trochoideum
Morse, D.C., 1947
Some observations on seasonal variations in plankton population Patuxant River, Maryland 1943-1945. Bd. Nat. Res., Publ. No.65, Chesapeake Biol. Lab., 31, 3 figs.

Peridinium trochoideum
Paredes, J.F., 1962
35. A brief comment on Peridinium nudum. Trab. Cent. Biol. Pisc., 35:115-120.
Also in:
Mem. Junta Invest. Ultram., (2), No. 33.

Peridinium trochoideum
Steemann-Nielsen, Einar, 1951
The marine vegetation of the Isefjord. A study on ecology and production. Medd. Komm. Danmarks Fiskeri-og Havundersøgelser. Ser. Plankton. 5(4); 114pp., 46 text figs.

Peridinium truncatum n.sp.
Graham, H. W., 1942
Studies in the morphology, taxonymy, and ecology of the Peridiniales. Sci. Res. Cruise VII of the Carnegie, 1928-1929---Biol. III(542): 129 pp., 67 figs.

Peridinium truncatum acutum
Graham, H. W., 1942
Studies in the morphology, taxonymy, and ecology of the Peridiniales. Sci. Res. Cruise VII of the Carnegie, 1928-1929---Biol. III(542): 129 pp., 67 figs.

Peridinium tuberosum n.sp.
Meunier, A., 1919
Microplancton de la Mer Flamande 3. Les Péridiniens. Mem. Mus. Roy. Hist. Nat., Belgique 8(1):1-116, Pls. XV-XXI.

Peridinium tumidum
Marukawa, H., 1921
Plankton lists and some new species of copepods, from the northern waters of Japan. Bull. Inst. Ocean., No.384, 15 pp., 3 pls., 1 chart. Monaco

Peridinium tumidum
Matzenauer, L., 1933
Die Dinoflagellaten des indischen Ozeans (mit Ausnahme der Gattung Ceratium.) Bot. Arch. 35:437-510, 77 text figs., 2 charts.

Peridinium turbinatum
Argentina, Secretaria de Marina, Servicio de Hidrografia Naval, 1962.
Plancton de las campanas oceanograficas DRAKE I y II. Publico, H. 627:57.

Peridinium turbinatum
Balech, Enrique, 1959.
Operacion Oceanografica Merluza V Crucero. Plancton. Publico, Servicio Hidrografia Naval, Argentina, H. 618:43 pp.

Peridinium turbinatum
Balech, Enrique, 1958.
Plancton de la Campaña Antartica Argentina, 1954-1955. Physis, 21(60):75-108.

Peridineum tychodiscus inflatus
Paulsen, O., 1949
Observations on dinoflagellates. (Ed. J. Grøntoed) Kongl. Dansk. Videnskab. Selsk. Biol. Skr. 6(4):67 pp., 30 text figs.

Peridinium umbonatum
Wood, E.J.F., 1954.
Dinoflagellates in the Australian region. Australian J. Mar. Freshwater Res., 5(2):171-351.

Perinidinium unipes n. sp.
Argentina, Secretaria de Marina, Servicio de Hidrografía Naval, 1962.
Plancton de las campanas oceanograficas DRAKE I y II.
Publico. H. 627:57.

Peridinium varicans
Braarud, T., and Adam Bursa, 1939
On the phytoplankton of the Oslo Fjord, 1933-1934. Hvalrådets Skr. No.19:1-63; 9 text figs. Reviewed. J. du. Cons. 14(3): 418-420. A.C. Gardiner.

Peridinium varicans
Lebour, M.V., 1925.
The dinoflagellates of northern seas. Marine Biological Association, United Kingdom, 250 pp., 35 pls., 53 textfigs.

Peridinium variegatum
Balech, Enrique, 1958.
Plancton de la Campaña Antartica Argentina, 1954-1955.
Physis, 21(60):75-108.

Peridinium variegatum
Gran, H.H., and T. Braarud, 1935
A quantitative study of the phytoplankton in the Bay of Fundy and the Gulf of Maine (including observations on hydrography, chemistry, and turbidity).
J. Biol. Bd., Canada, 1(5):279-467, 69 text figs.

Peridinium variegatum
Lillick, L.C., 1940
Phytoplankton and planktonic protozoa of the offshore waters of the Gulf of Maine. Pt.II. Qualitative Composition of the Planktonic Flora. Trans. Am. Phil. Soc., n.s., 31(3):193-237, 13 text figs.

Peridinium ventricum
Wood, E.J.P., 1954.
Dinoflagellates in the Australian region.
Australian J. Mar. Freshwater Res., 5(2):171-351.

Peridinium venustum n.sp.
Matzenauer, L., 1933
Die Dinoflagellaten des indischen Ozeans (mit Ausnahme der Gattung Ceratium.) Bot. Arch. 35:437-510, 77 text figs., 2 charts.

Peridinium variegatum
Wood, E.J.P., 1954.
Dinoflagellates in the Australian region.
Australian J. Mar. Freshwater Res., 5(2):171-351.

Peridinium verrucosum
LeBour, M.V., 1925
The dinoflagellates of Northern Seas. The Marine Biological Association of the United Kingdom, Plymouth, 250 pp., 35 pls.,53 text figs.

Peridinium vexans n.sp.
Murray, G., and F. G. Whitting, 1899
New Peridiniaceae from the Atlantic.
Trans. Linn. Soc., London, Bot., ser 2, 5: 321-342, Pls. 27-33, 9 tables.

Peridinium vexans
Paulsen, O., 1908
XVIII Peridiniales. Nordisches Plankton, Bot. Teil: 1-124, 155 text figs.

Peridinium volzii
Wood, E.J.P., 1954.
Dinoflagellates in the Australian region.
Australian J. Mar. Freshwater Res., 5(2):171-351.

Peridinium wiesneri
Balech, Enrique, 1971
Microplancton del Atlantico ecuatorial oeste (Equalant 1)
Publ. Serv. Hidrograf. Naval, Argentina H. 654: 103 pp., 122 figs.

Peridinium wiesneri
Wood, E.J.P., 1954.
Dinoflagellates in the Australian region.
Australian J. Mar. Freshwater Res., 5(2):171-351.

Peridinium willei
LeBour, M.V., 1925
The dinoflagellates of Northern Seas. The Marine Biological Association of the United Kingdom, Plymouth, 250 pp., 35 pls.,53 text figs.

Peridinium willei stagnale
Lindemann, E., 1924
Peridineen aus dem goldenen Horn und dem Bosphorus. Bot. Arch. 5:216-233, 98 text figs.

Peridinium Willei
Meunier, A., 1919
Microplancton de la Mer Flamande 3. Les Péridiniens. Mem. Mus. Roy. Hist. Nat., Belgique 8(1):1-116, Pls. XV-XXI.

Peridinium willei
Morse, D.C., 1947
Some observations on seasonal variations in plankton population Patuxant River, Maryland 1943-1945. Bd. Nat. Res., Publ. No.65, Chesapeake Biol. Lab., 31, 3 figs.

Peridinium Willei
Paulsen, O., 1908
XVIII Peridiniales. Nordisches Plankton, Bot. Teil: 1-124, 155 text figs.

Peridinium wisconsinense
Wall, David, and Barrie Dale, 1968.
Modern dinoflagellate cysts and evolution of the Peridiniales.
Micropaleontology, 14(3):265-304.

Peridinium yserense
LeBour, M.V., 1925
The dinoflagellates of Northern Seas. The Marine Biological Association of the United Kingdom, Plymouth, 250 pp., 35 pls.,53 text figs.

Peridinium Yserense n.sp.
Meunier, A., 1919
Microplancton de la Mer Flamande 3. Les Péridiniens. Mem. Mus. Roy. Hist. Nat., Belgique 8(1):1-116, Pls. XV-XXI.

Petalodinium porcello n.gen. n. sp.
Cachon, Jean, et Monique Cachon, 1969.
Contribution à l'étude des Noctilucidae Saville-Kent. Evolution morphologique, cytologie, systématique. V. Les Leptodiscinae Cachon J. et M.
Protistologica 5(1):11-33

Petalotricha ampulla
Entz, G., jr., 1935.
Ueber das Problem der Kerne und kernähnlichen Einschlüsse bei Petalotricha ampulla.
Fol. Biol. General. 11(1):15-26.

Phalacroma sp.
Balech, Enrique and Sayed Z. El-Sayed, 1964.
Microplankton of the Weddell Sea.
In: Biology of Antarctic seas. II.
Antarctic Res. Ser., Amer. Geophys. Union, 5:107-124.

Phalacroma sp.
Balech, Enrique, 1944
Contribucion al conocimiento del Plancton de Lennox y Cabo de Hornos. Physis XIX:423-446, 6 pls. with 67 figs.

Phalacroma Stein non corde
Balech, Enrique, 1944
Contribucion al conocimiento del Plancton de Lennox y Cabo de Hornos. Physis XIX:423-446, 6 pls. with 67 figs.

Phalacroma sp.
Lindemann, E., 1924
Peridineen aus dem goldenen Horn und dem Bosphorus. Bot. Arch. 5:216-233, 98 text figs.

Phalacroma sp.
Massuti Algamora, M., 1949
Estudio de diez y seis muestras de plancton del Golfo de Nápoles. Publ. Inst. Biol. Appl. 5:85-94, 1 fold-in table.

Phalacroma acutum
Ercegovic, A., 1936
Etudes qualitative et quantitatives du phytoplancton dans les eaux cotières de l'Adriatique oriental moyen au cours de l'année 1934. Acta Adriatica 1(9):1-126

Phalachroma acutum
Jörgensen, E., 1923
Mediterranean Dinophysiaceae. Rept. Danish Oceanogr. Expeds. 1908-10, to the Mediterranean and adjacent seas, Vol.II, Biol. J 2, 48 pp., 64 text figs.

Phalacroma acutum
Matzenauer, L., 1933
Die Dinoflagellaten des indischen Ozeans (mit Ausnahme der Gattung Ceratium.) Bot. Arch. 35:437-510, 77 text figs., 2 charts.

Phalacroma acutum
Pavillard, J., 1923
A propos de la systématique des Péridiniens Bull. Soc. Bot. de France 70:876-882; 914-918.

Phalacroma acutum n.sp.
Pavillard, J., 1916
Recherches sur les Peridiniens du Golfe du Lion. Mem. Univ. Montpellier. Trav. Inst. Bot., Univ. Montpellier. Serie mixte No.4, 70 pp., 3 pls., 15 text figs.

Phalacroma acutum
Rampi, L. 1940
Ricerche sul Fitoplancton del mare Ligure. II Le tecatali e le dinofisiali delle acque di Sanremo.. Boll. Pesca. Piscicolt.,Idrobiol. (18) 16(2): 243-274, 56 figs.

Phalacroma acutum
Rampi, L., 1940
Ricerche sul Fitoplancton del mare Ligure. Boll. di Pesca, di Piscicoltura e di Idrobiologa, 18(2):1-34, 56 text figs.

Phalacroma acutum
Vives, F., and A. Planas, 1952.
Plancton recogido por los laboratorios costeros. VI. Fitoplancton de las costas de Vinaroz, islas Columbretes y alrededores de las desembocadura del Ebro. Publ. Inst. Biol. Aplic. 11:141-156, 19 textfigs.

Phalacroma acutum
Wood, E.J.P., 1954.
Dinoflagellates in the Australian region.
Australian J. Mar. Freshwater Res., 5(2):171-351

Phalacroma alata n.sp.
Wood, E.J.P., 1954.
Dinoflagellates in the Australian region.
Australian J. Mar. Freshwater Res., 5(2):171-351

Phalacroma apicatum

Balech, Enrique, 1962
Tintinnoinea y Dinoflagellata del Pacifico segun material de las Expediciones NORPAC y DOWNWIND del Instituto Scripps de Oceanografia. Revista, Mus. Argentino Ciencias Nat. "Bernardino Rivadavia", Ciencias Zool., 7(1):1-253.

Phalacroma apicatum

Kässler, R., 1938
Die Verbreitung der Dinophysiales im Sudatlantischen Ozean. Wiss. Ergeb. Deutschen Atlantischen Expedition----"Meteor" 1925-1927, 12(2):162-237, text figs. 85-118.

Phalacroma apicatum n.sp.

Kofoid, C. A. and T. Skogsberg, 1928
XXXV. The Dinoflagellata: The Dinophysiodae. Reports on the scientific results of the expedition to the Eastern Tropical Pacific, in charge of Alexander Agassiz, by the U. S. Fish Commission Steamer "Albatross" from October 1904 to March 1905----. Mem. M. C. Z. 51:766 pp., 31 pls.

Phalacroma apicatum

Rampi, L., 1945
Osservazioni sulla distribuzione qualitativa del fitoplancton nel mare Mediterraneo. Atti della Soc. Ital. di Sci. Nat. 84:105-113.

Phalacroma apicatum

Rampi, L., 1942
Il Fitoplancton mediterraneo: Problemi ed affinita interoceaniche. Boll. di Pesca di Piscicoltura e di Idrobiologia, Anno 18, Fasc. 4:7-19.

Phalacroma apicatum

Wood, E.J.F., 1954.
Dinoflagellates in the Australian region. Australian J. Mar. Freshwater Res., 5(2):171-351.

Phalacroma argus

Balech, Enrique, 1962
Tintinnoinea y Dinoflagellata del Pacifico segun material de las Expediciones NORPAC y DOWNWIND del Instituto Scripps de Oceanografia. Revista, Mus. Argentino Ciencias Nat. "Bernardino Rivadavia", Ciencias Zool., 7(1):1-253.

Phalacroma argus

Dangeard, P., 1927
Phytoplankton de la croisière du "Sylvana". Ann. Inst. Ocean., Monaco, n.s., 4(8):286-401, 54 text figs. (février-Juin 1913).

Phalacroma argus

Ercegovic, A., 1936
Etudes qualitatives et quantitatives du phytoplancton dans les eaux cotières de l'Adriatique oriental moyen au cours de l'année 1934. Acta Adriatica 1(9):1-126

Phalachroma argus

Jörgensen, E., 1923
Mediterranean Dinophysiaceae. Rept. Danish Oceanogr. Expeds. 1908-10, to the Mediterranean and adjacent seas, Vol.II, Biol. J 2, 48 pp., 64 text figs.

Phalacroma argus

Kässler, R., 1938
Die Verbreitung der Dinophysiales im Sudatlantischen Ozean. Wiss. Ergeb. Deutschen Atlantischen Expedition----"Meteor" 1925-1927, 12(2):162-237, text figs. 85-118.

Phalacroma argus

Kofoid, C. A. and T. Skogsberg, 1928
XXXV. The Dinoflagellata: The Dinophysiodae. Reports on the scientific results of the expedition to the Eastern Tropical Pacific, in charge of Alexander Agassiz, by the U. S. Fish Commission Steamer "Albatross" from October 1904 to March 1905----. Mem. M. C. Z. 51:766 pp., 31 pls.

Phalacroma argus

Margalef, R., 1949
Fitoplancton nerítico de la Costa Brava en 1947-48. Publ. Inst. Biol. Aplicada, 5:41-51, 3 text figs.

Phalacroma argus

Matzenauer, L., 1933
Die Dinoflagellaten des indischen Ozeans (mit Ausnahme der Gattung Ceratium.) Bot. Arch. 35:437-510, 77 text figs., 2 charts.

Phalacroma argus

Pavillard, J., 1916
Recherches sur les Peridiniens du Golfe du Lion. Mem. Univ. Montpellier. Trav. Inst. Bot., Univ. Montpellier. Serie mixte No.4, 70 pp., 3 pls., 15 text figs.

Phalacroma argus

Rampi, L., 1940
Ricerche sul Fitoplancton del mare Ligure. II Le tecatali e le dinofisiali delle acque di Sanremo. Boll. Pesca, Piscicolt..Idrobiol. (18) 16(2): 243-274, 56 figs.

Phalacroma argus

Rampi, L., 1940
Ricerche sul Fitoplancton del mare Ligure. Boll. di Pesca, di Piscicoltura e di Idrobiologa, 18(2):1-34, 56 text figs.

Phalacroma argus

Sousa e Silva, E., 1949
Diatomaceas e Dinoflagelados de Baía de Cascais. Portugaliae Acta Biol., Volume: Julio Henriques, Ser. B: 300-383, 9 pls, 2 fold-in tables.

Phalacroma argus

Wood, E.J.F., 1954.
Dinoflagellates in the Australian region. Australian J. Mar. Freshwater Res., 5(2):171-351.

Phalacroma bipartitum n.sp.

Kofoid, C. A. and T. Skogsberg, 1928
XXXV. The Dinoflagellata: The Dinophysiodae. Reports on the scientific results of the expedition to the Eastern Tropical Pacific, in charge of Alexander Agassiz, by the U. S. Fish Commission Steamer "Albatross" from October 1904 to March 1905----. Mem. M. C. Z. 51:766 pp., 31 pls.

Phalacroma Blackmani n.sp.

Murray, G., and F. G. Whitting, 1899
New Peridiniaceae from the Atlantic. Trans. Linn. Soc., London, Bot., ser 2, 5:321-342, Pls. 27-33, 9 tables.

Phalacroma braarudi n.sp.

Nordli, O., 1951.
Dinoflagellates from Lofoten. Nytt Mag. Naturvidensk 88:49-55, 7 textfigs.

Phalacroma circumcinctum

Kofoid, C. A. and T. Skogsberg, 1928
XXXV. The Dinoflagellata: The Dinophysiodae. Reports on the scientific results of the expedition to the Eastern Tropical Pacific, in charge of Alexander Agassiz, by the U. S. Fish Commission Steamer "Albatross" from October 1904 to March 1905----. Mem. M. C. Z. 51:766 pp., 31 pls.

Phalachroma circumsutum

Jörgensen, E., 1923
Mediterranean Dinophysiaceae. Rept. Danish Oceanogr. Expeds. 1908-10, to the Mediterranean and adjacent seas, Vol.II, Biol. J 2, 48 pp., 64 text figs.

Phalacroma circumsutum

Kässler, R., 1938
Die Verbreitung der Dinophysiales im Sudatlantischen Ozean. Wiss. Ergeb. Deutschen Atlantischen Expedition----"Meteor" 1925-1927, 12(2):162-237, text figs. 85-118.

Phalacroma circumsutum

Kofoid, C. A. and T. Skogsberg, 1928
XXXV. The Dinoflagellata: The Dinophysiodae. Reports on the scientific results of the expedition to the Eastern Tropical Pacific, in charge of Alexander Agassiz, by the U. S. Fish Commission Steamer "Albatross" from October 1904 to March 1905----. Mem. M. C. Z. 51:766 pp., 31 pls.

Phalacroma circumsutum

Matzenauer, L., 1933
Die Dinoflagellaten des indischen Ozeans (mit Ausnahme der Gattung Ceratium.) Bot. Arch. 35:437-510, 77 text figs., 2 charts.

Phalacroma circumsutum

Wood, E.J.F., 1963
Dinoflagellates in the Australian region. II. Recent collections. Comm. Sci. and Industr. Res. Org., Div. Fish. and Oceanogr., Techn. Paper, No. 14: 55 pp.

Phalacroma contractum

Kässler, R., 1938
Die Verbreitung der Dinophysiales im Sudatlantischen Ozean. Wiss. Ergeb. Deutschen Atlantischen Expedition----"Meteor" 1925-1927, 12(2):162-237, text figs. 85-118.

Phalacroma contractum n.sp.

Kofoid, C. A. and T. Skogsberg, 1928
XXXV. The Dinoflagellata: The Dinophysiodae. Reports on the scientific results of the expedition to the Eastern Tropical Pacific, in charge of Alexander Agassiz, by the U. S. Fish Commission Steamer "Albatross" from October 1904 to March 1905----. Mem. M. C. Z. 51:766 pp., 31 pls.

Phalacroma contractum

Wood, E.J.F., 1963
Dinoflagellates in the Australian region. II. Recent collections. Comm. Sci. and Industr. Res. Org., Div. Fish. and Oceanogr., Techn. Paper, No. 14: 55 pp.

Phalacroma cornuta

Argentina, Secretaria de Marina, Servicio de Hidrografia Naval, 1962.
Plancton de las campanas oceanograficas DRAKE I y II. Publico, H. 627:57.

Phalacroma cornutum inerme n. subsp.

Balech, Enrique and Sayed Z. El-Sayed, 1965.
Microplankton of the Weddell Sea. In: Biology of Antarctic seas. II. Antarctic Res. Ser., Amer. Geophys. Union, 5:107-124.

Phalacroma cuneolus

Kässler, R., 1938
Die Verbreitung der Dinophysiales im Sudatlantischen Ozean. Wiss. Ergeb. Deutschen Atlantischen Expedition----"Meteor" 1925-1927, 12(2):162-237, text figs. 85-118.

Phalacroma cuneolus n.sp.

Kofoid, C. A. and T. Skogsberg, 1928
XXXV. The Dinoflagellata: The Dinophysiodae. Reports on the scientific results of the expedition to the Eastern Tropical Pacific, in charge of Alexander Agassiz, by the U. S. Fish Commission Steamer "Albatross" from October 1904 to March 1905----. Mem. M. C. Z. 51:766 pp., 31 pls.

Phalacroma cuneus

Balech, Enrique, 1962
Tintinnoinea y Dinoflagellata del Pacifico segun material de las Expediciones NORPAC y DOWNWIND del Instituto Scripps de Oceanografia. Revista, Mus. Argentino Ciencias Nat. "Bernardino Rivadavia", Ciencias Zool., 7(1):1-253.

Phalacroma cuneus
Dangeard, P., 1927
Phytoplankton de la croisière du "Sylvana". Ann. Inst. Ocean., Monaco, n.s., 4(8):286-401, 54 text figs. (Feirer-Juin 1913).

Phalachroma cuneus
Jörgensen, E., 1923
Mediterranean Dinophysiaceae. Rept. Danish Oceanogr. Expeds. 1908-10, to the Mediterranean and adjacent seas, Vol.II, Biol. J 2, 48 pp., 64 text figs.

Phalacroma cuneus
Käsler, R., 1938
Die Verbreitung der Dinophysiales im Sudatlantischen Ozean. Wiss. Ergeb. Deutschen Atlantischen Expedition----"Meteor" 1925-1927, 12(2):162-237, text figs. 85-118.

Phalacroma cuneus
Kofoid, C. A. and T. Skogsberg, 1928
XXXV. The Dinoflagellata: The Dinophysiodae. Reports on the scientific results of the expedition to the Eastern Tropical Pacific, in charge of Alexander Agassiz, by the U. S. Fish Commission Steamer "Albatross" from October 1904 to March 1905----. Mem. M. C. Z. 51:766 pp., 31 pls.

Phalacroma cuneus
Matzenauer, L., 1933
Die Dinoflagellaten des indischen Ozeans (mit Ausnahme der Gattung Ceratium.) Bot. Arch. 35:437-510, 77 text figs., 2 charts.

Phalacroma cuneus
Murray, G., and F. G. Whitting, 1899
New Peridiniaceae from the Atlantic. Trans. Linn. Soc., London, Bot., ser 2, 5: 321-342, Pls. 27-33, 9 tables.

Phalacroma cuneus
Pavillard, J., 1916
Recherches sur les Peridiniens du Golfe du Lion. Mem. Univ. Montpellier. Trav. Inst. Bot., Univ. Montpellier. Serie mixte No.4, 70 pp., 3 pls., 15 text figs.

Phalacroma cuneus
Rampi, L., 1940
Ricerche sul Fitoplancton del mare Ligure. II Le tecatali e le dinofisiali delle acque di Sanremo.. Boll. Pesca, Piscicolt.,Idrobiol. (18) 16(2): 243-274, 56 figs.

Phalacroma cuneus
Rampi, L., 1940
Ricerche sul Fitoplancton del mare Ligure. Boll. di Pesca, di Piscicoltura e di Idrobiologa, 18(2):1-34, 56 text figs.

Phalacroma cuneus
Schröder, B., 1900
Phytoplankton des Golfes von Neapel nebst vergleichenden Ausblicken auf das atlantischen Ozean. Mitt. Zool. Stat. Neapel, 14:1-38.

Phalacroma cuneus
Sousa e Silva, E., 1949
Diatomaceas e Dinoflagelados de Baia de Cascais. Portugaliae Acta Biol., Volume: Julio Henriques, Ser. B: 300-383, 9 pls, 2 fold-in tables.

Phalacroma cuneus
Wood, E.J.P., 1954.
Dinoflagellates in the Australian region. Australian J. Mar. Freshwater Res., 5(2):171-351.

Phalacroma dolichopterygium
Dangeard, P., 1927
Phytoplankton de la croisière du "Sylvana". Ann. Inst. Ocean., Monaco, n.s., 4(8):286-401, 54 text figs. (Feirer-Juin 1913).

Phalacroma dolichopterygium
Delegazione Italiana della Commissione Internazionale per l'Esplorazione Scientifica del Mediterraneo, 1941
Note sul plancton della Laguna veneta. [Memoria CCLXXXIX], Arch. di Ocean. e Limn. Anno I, Fasc. I, 1941 XIX: 31-57 pp.

Phalachroma dolichopterygium
Jörgensen, E., 1923
Mediterranean Dinophysiaceae. Rept. Danish Oceanogr. Expeds. 1908-10, to the Mediterranean and adjacent seas, Vol.II, Biol. J 2, 48 pp., 64 text figs.

Phalacroma dolichopteryguim n.sp.
Murray, G., and F. G. Whitting, 1899
New Peridiniaceae from the Atlantic. Trans. Linn. Soc., London, Bot., ser 2, 5: 321-342, Pls. 27-33, 9 tables.

Phalacroma dolichoptergium
Pavillard, J., 1923
A propos de la systématique des Péridiniens Bull. Soc. Bot. de France 70:876-882; 914-918.

Phalacroma dolichoptergium
Wood, E.J.P., 1954.
Dinoflagellates in the Australian region. Australian J. Mar. Freshwater Res., 5(2):171-351.

Phalacroma doryphoroides n.sp
Dangeard, P., 1927
Phytoplankton de la croisière du "Sylvana". Ann. Inst. Ocean., Monaco, n.s., 4(8):286-401, 54 text figs. (Feirer-Juin 1913).

Phalacroma doryphorum
Dangeard, P., 1927
Phytoplankton de la croisière du "Sylvana". Ann. Inst. Ocean., Monaco, n.s., 4(8):286-401, 54 text figs. (février-Juin 1913).

Phalacroma doryphorum
Ercegovic, A., 1936
Etudes qualitative et quantitatives du phytoplancton dans les eaux cotières de l'Adriatique oriental moyen au cours de l'année 1934. Acta Adriatica 1(9):1-126

Phalacroma doryphorum
Forti, A., 1922
Ricerche sulla flora pelagica (fitoplancton) di Quarto dei Mille. Mem. R. Com. Talass. Ital. 97:248 pp., 13 pls.

Phalacroma doryphorum
Jörgensen, E., 1923
Mediterranean Dinophysiaceae. Rept. Danish Oceanogr. Expeds. 1908-10, to the Mediterranean and adjacent seas, Vol.II, Biol. J 2, 48 pp., 64 text figs.

Phalacroma doryphorum
Käsler, R., 1938
Die Verbreitung der Dinophysiales im Sudatlantischen Ozean. Wiss. Ergeb. Deutschen Atlantischen Expedition----"Meteor" 1925-1927, 12(2):162-237, text figs. 85-118.

Phalacroma doryphorum
Kofoid, C. A. and T. Skogsberg, 1928
XXXV. The Dinoflagellata: The Dinophysiodae. Reports on the scientific results of the expedition to the Eastern Tropical Pacific, in charge of Alexander Agassiz, by the U. S. Fish Commission Steamer "Albatross" from October 1904 to March 1905----. Mem. M. C. Z. 51:766 pp., 31 pls.

Phalacroma doryphorum
Margalef, R., 1949
Fitoplancton nerítico de la Costa Brava en 1947-48. Publ. Inst. Biol. Aplicada, 5: 41-51, 3 text figs.

Phalacroma doryphorum
Matzenauer, L., 1933
Die Dinoflagellaten des indischen Ozeans (mit Ausnahme der Gattung Ceratium.) Bot. Arch. 35:437-510, 77 text figs., 2 charts.

Phalacroma doryphorum
Murray, G., and F. G. Whitting, 1899
New Peridiniaceae from the Atlantic. Trans. Linn. Soc., London, Bot., ser 2, 5: 321-342, Pls. 27-33, 9 tables.

Phalacroma doryphorum
Pavillard, J., 1916
Recherches sur les Peridiniens du Golfe du Lion. Mem. Univ. Montpellier. Trav. Inst. Bot., Univ. Montpellier. Serie mixte No.4, 70 pp., 3 pls., 15 text figs.

Phalacroma doryphorum
Pavillard, J., 1905
Recherches sur la flore pelagique (Phytoplankton) de l'Etang de Thau. Theses presentees a la Fac. Sci., Paris, 116 pp., 3 pls.

Phalacroma doryphorum
Rampi, L., 1940
Ricerche sul Fitoplancton del mare Ligure. II Le tecatali e le dinofisiali delle acque di Sanremo. Boll. Pesca, Piscicolt.,Idrobiol. (18) 16(2): 243-274, 56 figs.

Phalacroma doryphorum
Rampi, L., 1940
Ricerche sul Fitoplancton del mare Ligure. Boll. di Pesca, di Piscicoltura e di Idrobiologa, 18(2):1-34, 56 text figs.

Phalacroma doryphorum
Schröder, B., 1900
Phytoplankton des Golfes von Neapel nebst vergleichenden Ausblicken auf das atlantischen Ozean. Mitt. Zool. Stat. Neapel, 14:1-38.

Phalacroma doryphorum
Sousa e Silva, E., 1949
Diatomaceas e Dinoflagelados de Baia de Cascais. Portugaliae Acta Biol., Volume: Julio Henriques, Ser. B: 300-383, 9 pls, 2 fold-in tables.

Phalacroma doryphorum
Wood, E.J.P., 1954.
Dinoflagellates in the Australian region. Australian J. Mar. Freshwater Res., 5(2):171-351.

Phalacroma ebriolum
LeBour, M.V., 1925
The dinoflagellates of Northern Seas. The Marine Biological Association of the United Kingdom, Plymouth, 250 pp., 35 pls.,53 text figs.

Phalachroma elongatum n.sp.
Jörgensen, E., 1923
Mediterranean Dinophysiaceae. Rept. Danish Oceanogr. Expeds. 1908-10, to the Mediterranean and adjacent seas, Vol.II, Biol. J 2, 48 pp., 64 text figs.

Phalacroma elongatum
Pavillard, J., 1923
A propos de la systématique des Péridiniens Bull. Soc. Bot. de France 70:876-882; 914-918.

Phalacroma elongatum
Wood, E.J.P., 1954.
Dinoflagellates in the Australian region. Australian J. Mar. Freshwater Res., 5(2):171-351

Phalacroma expulsum

Balech, Enrique, 1962
Tintinnoinea y Dinoflagellata del Pacifico segun material de las Expediciones NORPAC y DOWNWIND del Instituto Scripps de Oceanografia. Revista. Mus. Argentino Ciencias Nat. "Bernardino Rivadavia", Ciencias Zool., 7(1):1-253.

Phalacroma expulsum

Käsler, R., 1938
Die Verbreitung der Dinophysiales im Sudatlantischen Ozean. Wiss. Ergeb. Deutschen Atlantischen Expedition----"Meteor" 1925-1927, 12(2):162-237, text figs. 85-118.

Phalacroma expulsum

Kofoid, C. A. and T. Skogsberg, 1928
XXXV. The Dinoflagellata: The Dinophysiodae. Reports on the scientific results of the expedition to the Eastern Tropical Pacific, in charge of Alexander Agassiz, by the U. S. Fish Commission Steamer "Albatross" from October 1904 to March 1905----. Mem. M. C. Z. 51:766 pp., 31 pls.

Phalacroma expulsum

Rampi, L., 1945
Osservazioni sulla distribuzione qualitativa del fitoplancton nel mare Mediterraneo. Atti della Soc. Ital. di Sci. Nat. 84:105-113.

Phalacroma expulsum

Rampi, L., 1942
II Fitoplancton mediterraneo: Problemi ed affinita interoceaniche. Boll. di Pesca di Piscicoltura e di Idrobiologia, Anno 18, Fasc. 4:7-19.

Phalacroma favus

Dangeard, P., 1927
Phytoplankton de la croisière du "Sylvana". Ann. Inst. Ocean., Monaco, n.s., 4(8):286-401, 54 text figs. (Janvier-Juin 1913).

Phalacroma favus

Ercegovic, A., 1936
Etudes qualitative et quantitatives du phytoplancton dans les eaux cotières de l'Adriatique oriental moyen au cours de l'année 1934. Acta Adriatica 1(9):1-126

Phalachroma favus

Jörgensen, E., 1923
Mediterranean Dinophysiaceae. Rept. Danish Oceanogr. Expeds. 1908-10, to the Mediterranean and adjacent seas, Vol.II, Biol. J 2, 48 pp., 64 text figs.

Phalacroma favus

Käsler, R., 1938
Die Verbreitung der Dinophysiales im Sudatlantischen Ozean. Wiss. Ergeb. Deutschen Atlantischen Expedition----"Meteor" 1925-1927, 12(2):162-237, text figs. 85-118.

Phalacroma favus

Matzenauer, L., 1933
Die Dinoflagellaten des indischen Ozeans (mit Ausnahme der Gattung Ceratium.) Bot. Arch. 35:437-510, 77 text figs., 2 charts.

Phalacroma favus

Pavillard, J., 1923
A propos de la systématique des Péridiniens Bull. Soc. Bot. de France 70:876-882; 914-918.

Phalacroma favus

Rampi, L. 1940
Ricerche sul Fitoplancton del mare Ligure. II Le tecatali e le dinofisiali delle acque di Sanremo. Boll. Pesca, Piscicolt., Idrobiol. (18) 16(2): 243-274, 56 figs.

Phalacroma favus

Rampi, L., 1940
Ricerche sul Fitoplancton del mare Ligure. Boll. di Pesca, di Piscicoltura e di Idrobiologa, 18(2):1-34, 56 text figs.

Phalacroma favus

Sousa e Silva, E., 1949
Diatomaceas e Dinoflagelados de Baia de Cascais. Portugaliae Acta Biol., Volume: Julio Henriques, Ser. B: 300-383, 9 pls, 2 fold-in tables.

Phalacroma favus

Wood, E.J.P., 1954.
Dinoflagellates in the Australian region. Australian J. Mar. Freshwater Res., 5(2):171-351.

Phalacroma fimbriatum

Kofoid, C. A. and T. Skogsberg, 1928
XXXV. The Dinoflagellata: The Dinophysiodae. Reports on the scientific results of the expedition to the Eastern Tropical Pacific, in charge of Alexander Agassiz, by the U. S. Fish Commission Steamer "Albatross" from October 1904 to March 1905----. Mem. M. C. Z. 51:766 pp., 31 pls.

Phalacroma giganteum

Käsler, R., 1938
Die Verbreitung der Dinophysiales im Sudatlantischen Ozean. Wiss. Ergeb. Deutschen Atlantischen Expedition----"Meteor" 1925-1927, 12(2):162-237, text figs. 85-118.

Phalacroma giganteum

Kofoid, C. A. and T. Skogsberg, 1928
XXXV. The Dinoflagellata: The Dinophysiodae. Reports on the scientific results of the expedition to the Eastern Tropical Pacific, in charge of Alexander Agassiz, by the U. S. Fish Commission Steamer "Albatross" from October 1904 to March 1905----. Mem. M. C. Z. 51:766 pp., 31 pls.

Phalacroma giganteum

Rampi, L., 1945
Osservazioni sulla distribuzione qualitativa del fitoplancton nel mare Mediterraneo. Atti della Soc. Ital. di Sci. Nat. 84:105-113.

Phalacroma giganteum

Rampi, L., 1942
II Fitoplancton mediterraneo: Problemi ed affinita interoceaniche. Boll. di Pesca di Piscicoltura e di Idrobiologia, Anno 18, Fasc. 4:7-19.

Phalacroma globulus

Murray, G., and F. G. Whitting, 1899
New Peridiniaceae from the Atlantic. Trans. Linn. Soc., London, Bot., ser 2, 5: 321-342, Pls. 27-33, 9 tables.

Phalacroma globulus

Schröder, B., 1900
Phytoplankton des Golfes von Neapel nebst vergleichenden Ausblicken auf das atlantischen Ozean. Mitt. Zool. Stat. Neapel, 14:1-38.

Phalacroma hastatum

Forti, A., 1922
Ricerche sulla flora pelagica (fitoplancton) di Quarto dei Mille. Mem. R. Com. Talass. Ital. 97:248 pp., 13 pls.

Phalacroma hastatum

Pavillard, J., 1923
A propos de la systématique des Péridiniens Bull. Soc. Bot. de France 70:876-882; 914-918.

Phalacroma hastatum

Pavillard, J., 1916
Recherches sur les Peridiniens du Golfe du Lion. Mem. Univ. Montpellier. Trav. Inst. Bot., Univ. Montpellier. Serie mixte No.4, 70 pp., 3 pls., 15 text figs.

Phalacroma hindmarchii

Balech, Enrique, 1962
Tintinnoinea y Dinoflagellata del Pacifico segun material de las Expediciones NORPAC y DOWNWIND del Instituto Scripps de Oceanografia. Revista. Mus. Argentino Ciencias Nat. "Bernardino Rivadavia", Ciencias Zool., 7(1):1-253.

Phalacroma Hindmarchii

Käsler, R., 1938
Die Verbreitung der Dinophysiales im Sudatlantischen Ozean. Wiss. Ergeb. Deutschen Atlantischen Expedition----"Meteor" 1925-1927, 12(2):162-237, text figs. 85-118.

Phalacroma hindmarchi

Kofoid, C. A. and T. Skogsberg, 1928
XXXV. The Dinoflagellata: The Dinophysiodae. Reports on the scientific results of the expedition to the Eastern Tropical Pacific, in charge of Alexander Agassiz, by the U. S. Fish Commission Steamer "Albatross" from October 1904 to March 1905----. Mem. M. C. Z. 51:766 pp., 31 pls.

Phalacroma Hindmarchii n.sp.

Murray, G., and F. G. Whitting, 1899
New Peridiniaceae from the Atlantic. Trans. Linn. Soc., London, Bot., ser 2, 5: 321-342, Pls. 27-33, 9 tables.

Phalacroma Hindmarchii

Pavillard, J., 1923
A propos de la systématique des Péridiniens Bull. Soc. Bot. de France 70:876-882; 914-918.

Phalacroma Hindmarchii

Pavillard, J., 1916
Recherches sur les Peridiniens du Golfe du Lion. Mem. Univ. Montpellier. Trav. Inst. Bot., Univ. Montpellier. Serie mixte No.4, 70 pp., 3 pls., 15 text figs.

Phalacroma hindmarchii

Wood, E.J.P 1954.
Dinoflagellates in the Australian region. Australian J. Mar. Freshwater Res., 5(2):171-351.

Phalacroma irregulare

Brunel, J., 1962
Le phytoplancton de la Baie de Chaleurs. Inst. Botan., Univ. Montréal, Contrib. No. 77: 365 pp., 66 pls.

Phalacroma irregularis n.sp.

LeBour, M.V., 1925
The dinoflagellates of Northern Seas. The Marine Biological Association of the United Kingdom, Plymouth, 250 pp., 35 pls.,53 text figs.

Phalacroma irregulare?

Lillick, L.C., 1937
Seasonal studies of the phytoplankton off Woods Hole, Massachusetts. Biol. Bull. LXXIII (3):488-503, 3 text figs.

Phalacroma irregulare

Paulsen, O., 1949
Observations on dinoflagellates. (Ed. J. Grøntoed) Kongl. Dansk. Videnskab. Selsk., Biol. Skr. 6(4):67 pp., 30 text figs.

Phalacroma irregulare

Woods, E.J.P., 1954.
Dinoflagellates in the Australian region. Australian J. Mar. Freshwater Res., 5(2):171-351.

Phalacroma jibbonse n.sp.

Wood, E.J.P. 1954.
Dinoflagellates in the Australian region. Australian J. Mar. Freshwater Res. 5(2):171-351.

Phalacroma Jourdani
Murray, G., and F. G. Whitting, 1899
New Peridiniaceae from the Atlantic.
Trans. Linn. Soc., London, Bot., ser 2, 5: 321-342, Pls. 27-33, 9 tables.

Phalacroma Jourdani
Pavillard, J., 1905
Recherches sur la flore pelagique (Phytoplankton) de l'Etang de Thau. Theses presentees a la Fac. Sci., Paris, 116 pp., 3 pls.

Phalacroma Jourdani
Schröder, B., 1900
Phytoplankten des Golfes von Neapel nebst vergleichenden Ausblicken auf das atlantischen Ozean. Mitt. Zool. Stat. Neapel, 14:1-38.

Phalacroma kofoidi
LeBour, M.V., 1925
The dinoflagellates of Northern Seas. The Marine Biological Association of the United Kingdom, Plymouth, 250 pp., 35 pls.,53 text figs.

Phalacroma lativelatum
Käsler, R., 1938
Die Verbreitung der Dinophysiales im Sudatlantischen Ozean. Wiss. Ergeb. Deutschen Atlantischen Expedition----"Meteor" 1925-1927, 12(2):162-237, text figs. 85-118.

Phalacroma lativelatum n.sp.
Kofoid, C. A. and T. Skogsberg, 1928
XXXV. The Dinoflagellata: The Dinophysiodae. Reports on the scientific results of the expedition to the Eastern Tropical Pacific, in charge of Alexander Agassiz, by the U. S. Fish Commission Steamer "Albatross" from October 1904 to March 1905----. Mem. M. C. Z. 51:766 pp., 31 pls.

Phalacroma lativelatum
Matzenauer, L., 1933
Die Dinoflagellaten des indischen Ozeans (mit Ausnahme der Gattung Ceratium.) Bot. Arch. 35:437-510, 77 text figs., 2 charts.

Phalacroma lens
Käsler, R., 1938
Die Verbreitung der Dinophysiales im Sudatlantischen Ozean. Wiss. Ergeb. Deutschen Atlantischen Expedition----"Meteor" 1925-1927, 12(2):162-237, text figs. 85-118.

Phalacroma lens n.sp.
Kofoid, C. A. and T. Skogsberg, 1928
XXXV. The Dinoflagellata: The Dinophysiodae. Reports on the scientific results of the expedition to the Eastern Tropical Pacific, in charge of Alexander Agassiz, by the U. S. Fish Commission Steamer "Albatross" from October 1904 to March 1905----. Mem. M. C. Z. 51:766 pp., 31 pls.

Phalacroma lens
Wood, E.J.F., 1954.
Dinoflagellates in the Australian region. Australian J. Mar. Freshwater Res., 5(2):171-351.

Phalacroma lenticula
Kofoid, C. A. and T. Skogsberg, 1928
XXXV. The Dinoflagellata: The Dinophysiodae. Reports on the scientific results of the expedition to the Eastern Tropical Pacific, in charge of Alexander Agassiz, by the U. S. Fish Commission Steamer "Albatross" from October 1904 to March 1905----. Mem. M. C. Z. 51:766 pp., 31 pls.

Phalacroma lenticula
Wood, E.J.F., 1963
Dinoflagellates in the Australian region. II. Recent collections.
Comm. Sci. and Industr. Res. Org., Div. Fish. and Oceanogr., Techn. Paper, No. 14: 55 pp.

Phalacroma limbatum
Käsler, R., 1938
Die Verbreitung der Dinophysiales im Sudatlantischen Ozean. Wiss. Ergeb. Deutschen Atlantischen Expedition----"Meteor" 1925-1927, 12(2):162-237, text figs. 85-118.

Phalacroma limbatum
Kofoid, C. A. and T. Skogsberg, 1928
XXXV. The Dinoflagellata: The Dinophysiodae. Reports on the scientific results of the expedition to the Eastern Tropical Pacific, in charge of Alexander Agassiz, by the U. S. Fish Commission Steamer "Albatross" from October 1904 to March 1905----. Mem. M. C. Z. 51:766 pp., 31 pls.

Phalacroma mawsoni n.sp.
Wood, E.J.P., 1954.
Dinoflagellates in the Australian region. Australian J. Mar. Freshwater Res. 5(2):171-351.

Phalacroma minutum
Käsler, R., 1938
Die Verbreitung der Dinophysiales im Sudatlantischen Ozean. Wiss. Ergeb. Deutschen Atlantischen Expedition----"Meteor" 1925-1927, 12(2):162-237, text figs. 85-118.

Phalacroma minutum
LeBour, M.V., 1925
The dinoflagellates of Northern Seas. The Marine Biological Association of the United Kingdom, Plymouth, 250 pp., 35 pls.,53 text figs.

Phalacroma minutum
Mangin, M. L., 1912
Phytoplancton de la croisière du "René" dans l'Atlantique (Septembre 1908). Ann. Inst. Ocean., n.s., 4(1):1-66, 2 pls., 41 text figs., 2 tables.

Phalacroma minutum
Paulsen, O., 1908
XVIII Peridiniales. Nordisches Plankton, Bot. Teil: 1-124, 155 text figs.

Phalacroma minutum
Pavillard, J., 1923
A propos de la systématique des Péridiniens Bull. Soc. Bot. de France 70:876-882; 914-918.

Phalacroma minutum
Wood, E.J.P., 1954.
Dinoflagellates in the Australian region. Australian J. Mar. Freshwater Res., 5(2):171-351.

Phalacroma mitra
Balech, Enrique, 1962
Tintinnoinea y Dinoflagellata del Pacifico segun material de las Expediciones NORPAC y DOWNWIND del Instituto Scripps de Oceanografia.
Revista. Mus. Argentino Ciencias Nat. "Bernardino Rivadavia", Ciencias Zool., 7(1):1-253.

Phalacroma mitra
Ercegovic, A., 1936
Etudes qualitative et quantitatives du phytoplancton dans les eaux cotières de l'Adriatique oriental moyen au cours de l'année 1934. Acta Adriatica 1(9):1-126

Phalacroma mitra
Forti, A., 1922
Ricerche sulla flora pelagica (fitoplancton) di Quarto dei Mille. Mem. R. Com. Talass. Ital. 97:248 pp., 13 pls.

Phalacroma mitra
Käsler, R., 1938
Die Verbreitung der Dinophysiales im Sudatlantischen Ozean. Wiss. Ergeb. Deutschen Atlantischen Expedition----"Meteor" 1925-1927, 12(2):162-237, text figs. 85-118.

Phalacroma mitra
Murray, G., and F. G. Whitting, 1899
New Peridiniaceae from the Atlantic.
Trans. Linn. Soc., London, Bot., ser 2, 5: 321-342, Pls. 27-33, 9 tables.

Phalacroma mitra
Pavillard, J., 1923
A propos de la systématique des Péridiniens Bull. Soc. Bot. de France 70:876-882; 914-918.

Phalacroma mitra
Pavillard, J., 1916
Recherches sur les Peridiniens du Golfe du Lion. Mem. Univ. Montpellier. Trav. Inst. Bot., Univ. Montpellier. Serie mixte No.4, 70 pp., 3 pls., 15 text figs.

Phalacroma mitra
Pavillard, J., 1905
Recherches sur la flore pelagique (Phytoplankton) de l'Etang de Thau. Theses presentees a la Fac. Sci., Paris, 116 pp., 3 pls.

Phalacroma mitra
Rampi, L. 1940
Ricerche sul Fitoplancton del mare Ligure. II Le tecatali e le dinofisiali delle acque di Sanremo.. Boll. Pesca, Piscicolt.,Idrobiol. (18) 16(2): 243-274, 56 figs.

Phalacroma mitra
Rampi, L., 1940
Ricerche sul Fitoplancton del mare Ligure. Boll. di Pesca, di Piscicoltura e di Idrobiologa, 18(2):1-34, 56 text figs.

Phalacroma mitra
Wood, E.J.P., 1954.
Dinoflagellates in the Australian region. Australian J. Mar. Freshwater Res., 5(2):171-351.

Phalacroma mucronatum
Hasle, Grethe Rytter, 1960
Phytoplankton and ciliate species from the Tropical Pacific.
Skr. Norske Videnskaps-Akad., Oslo, 1. Mat.-Nat. Kl., 1960(2): 1-50.

Phalacroma mucronatum
Käsler, R., 1938
Die Verbreitung der Dinophysiales im Sudatlantischen Ozean. Wiss. Ergeb. Deutschen Atlantischen Expedition----"Meteor" 1925-1927, 12(2):162-237, text figs. 85-118.

Phalacroma mucronatum n.sp.
Kofoid, C. A. and T. Skogsberg, 1928
XXXV. The Dinoflagellata: The Dinophysiodae. Reports on the scientific results of the expedition to the Eastern Tropical Pacific, in charge of Alexander Agassiz, by the U. S. Fish Commission Steamer "Albatross" from October 1904 to March 1905----. Mem. M. C. Z. 51:766 pp., 31 pls.

Phalacroma mucronatum
Wood, E.J.F., 1963
Dinoflagellates in the Australian region. II. Recent collections.
Comm. Sci. and Industr. Res. Org., Div. Fish. and Oceanogr., Techn. Paper, No. 14: 55 pp.

Phalacroma odiosum
Rampi, L. 1940
Ricerche sul Fitoplancton del mare Ligure. II Le tecatali e le dinofisiali delle acque di Sanremo.. Boll. Pesca, Piscicolt.,Idrobiol. (18) 16(2): 243-274, 56 figs.

Phalacroma odiosum
Rampi, L., 1940
Ricerche sul Fitoplancton del mare Ligure. Boll. di Pesca, di Piscicoltura e di Idrobiologa, 18(2):1-34, 56 text figs.

Phalacroma operculatum

Balech, Enrique, 1962
Tintinnoinea y Dinoflagellata del Pacifico segun material de las Expediciones NORPAC y DOWNWIND del Instituto Scripps de Oceanografia. Revista, Mus. Argentino Ciencias Nat. "Bernardino Rivadavia", Ciencias Zool., 7(1):1-253.

Phalacroma operculatum

Forti, A., 1922
Ricerche sulla flora pelagica (fitoplancton) di Quarto dei Mille. Mem. R. Com. Talass. Ital. 97:248 pp., 13 pls.

Phalachroma operculatum

Jörgensen, E., 1923
Mediterranean Dinophysiaceae. Rept. Danish Oceanogr. Expeds. 1908-10, to the Mediterranean and adjacent seas, Vol.II, Biol. J 2, 48 pp., 64 text figs.

Phalacroma operculatum

Lindemann, E., 1925
Neubeobachtungen an den Winter peridineen des Golfes von Neapel. Bot. Arch. 9:95-102, 19 text figs.

Phalacroma operculatum

Murray, G., and F. G. Whitting, 1899
New Peridiniaceae from the Atlantic. Trans. Linn. Soc., London, Bot., ser 2, 5: 321-342, Pls. 27-33, 9 tables.

Phalacroma operculatum

Pavillard, J., 1916
Recherches sur les Peridiniens du Golfe du Lion. Mem. Univ. Montpellier. Trav. Inst. Bot., Univ. Montpellier. Serie mixte No.4, 70 pp., 3 pls., 15 text figs.

Phalacroma operculatum

Pavillard, J., 1905
Recherches sur la flore pelagique (Phytoplankton) de l'Etang de Thau. Theses presentees a la Fac. Sci., Paris, 116 pp., 3 pls.

Phalacroma operculatum

Rampi, L. 1940
Ricerche sul Fitoplancton del mare Ligure. II Le tecatali e le dinofisiali delle acque di Sanremo. Boll. Pesca, Piscicolt.,Idrobiol. (18) 16(2): 243-274, 56 figs.

Phalacroma operculatum

Rampi, L., 1940
Ricerche sul Fitoplancton del mare Ligure. Boll. di Pesca, di Piscicoltura e di Idrobiologa, 18(2):1-34, 56 text figs.

Phalacroma operculatum

Schröder, B., 1900
Phytoplankton des Golfes von Neapel nebst vergleichenden Ausblicken auf das atlantischen Ozean. Mitt. Zool. Stat. Neapel, 14:1-38.

Phalacroma operculatum

Sousa e Silva, E., 1949
Diatomaceas e Dinoflagelados de Baia de Cascais. Portugaliae Acta Biol., Volume: Julio Henriques, Ser. B: 300-383, 9 pls, 2 fold-in tables.

Phalacroma operculatum

Wood, E.J.P., 1954.
Dinoflagellates in the Australian region. Australian J. Mar. Freshwater Res., 5(2):171-351.

Phalacroma operculoides

Balech, Enrique, 1962
Tintinnoinea y Dinoflagellata del Pacifico segun material de las Expediciones NORPAC y DOWNWIND del Instituto Scripps de Oceanografia. Revista, Mus. Argentino Ciencias Nat. "Bernardino Rivadavia", Ciencias Zool., 7(1):1-253.

Phalacroma operculoides

Dangeard, P., 1927
Phytoplankton de la croisière du "Sylvana". Ann. Inst. Ocean., Monaco, n.s., 4(8):286-401, 54 text figs. (Feirer-Juin 1913).

Phalacroma operculoides

Forti, A., 1922
Ricerche sulla flora pelagica (fitoplancton) di Quarto dei Mille. Mem. R. Com. Talass. Ital. 97:248 pp., 13 pls.

Phalachroma operculoides

Jörgensen, E., 1923
Mediterranean Dinophysiaceae. Rept. Danish Oceanogr. Expeds. 1908-10, to the Mediterranean and adjacent seas, Vol.II, Biol. J 2, 48 pp., 64 text figs.

Phalacroma operculoides

Margalef, R., 1949
Fitoplancton nerítico de la Costa Brava en 1947-48. Publ. Inst. Biol. Aplicada, 5: 41-51, 3 text figs.

Phalacroma operculoides

Pavillard, J., 1923
A propos de la systématique des Péridiniens Bull. Soc. Bot. de France 70:876-882; 914-918.

Phalacroma operculoides

Pavillard, J., 1916
Recherches sur les Peridiniens du Golfe du Lion. Mem. Univ. Montpellier. Trav. Inst. Bot., Univ. Montpellier. Serie mixte No.4, 70 pp., 3 pls., 15 text figs.

Phalacroma operculoides

Rampi, L. 1940.
Ricerche sul Fitoplancton del mare Ligure. II Le tecatali e le dinofisiali delle acque di Sanremo. Boll. Pesca, Piscicolt.,Idrobiol. (18) 16(2): 243-274, 56 figs.

Phalacroma operculoides

Rampi, L., 1940
Ricerche sul Fitoplancton del mare Ligure. Boll. di Pesca, di Piscicoltura e di Idrobiologa, 18(2):1-34, 56 text figs.

Phalacroma ovatum

de Sousa e Silva, E., 1956.
Contribution à l'étude du microplancton de Dakar et des regions maritimes voisines. Bull. I.F.A.N., 8(2):335-371, 7 pls.

Phalachroma ovatum

Jörgensen, E., 1923
Mediterranean Dinophysiaceae. Rept. Danish Oceanogr. Expeds. 1908-10, to the Mediterranean and adjacent seas, Vol.II, Biol. J 2, 48 pp., 64 text figs.

Phalacroma ovum

Balech, Enrique, 1962
Tintinnoinea y Dinoflagellata del Pacifico segun material de las Expediciones NORPAC y DOWNWIND del Instituto Scripps de Oceanografia. Revista, Mus. Argentino Ciencias Nat. "Bernardino Rivadavia", Ciencias Zool., 7(1):1-253.

Phalacroma ovum

Käsler, R., 1938
Die Verbreitung der Dinophysiales im Sudatlantischen Ozean. Wiss. Ergeb. Deutschen Atlantischen Expedition----"Meteor" 1925-1927, 12(2):162-237, text figs. 85-118.

Phalacroma ovum

Kofoid, C. A. and T. Skogsberg, 1928
XXXV. The Dinoflagellata: The Dinophysiodae. Reports on the scientific results of the expedition to the Eastern Tropical Pacific, in charge of Alexander Agassiz, by the U. S. Fish Commission Steamer "Albatross" from October 1904 to March 1905----. Mem. M. C. Z. 51:766 pp., 31 pls.

Phalacroma ovum

Rampi, L. 1940
Ricerche sul Fitoplancton del mare Ligure. II Le tecatali e le dinofisiali delle acque di Sanremo. Boll. Pesca, Piscicolt.,Idrobiol. (18) 16(2): 243-274, 56 figs.

Phalacroma ovum

Rampi, L., 1940
Ricerche sul Fitoplancton del mare Ligure. Boll. di Pesca, di Piscicoltura e di Idrobiologa, 18(2):1-34, 56 text figs.

Phalacroma ovum

Wood, E.J.P., 1954.
Dinoflagellates in the Australian region. Australian J. Mar. Freshwater Res., 5(2):171-351

Phalacroma parvulum

Balech, Enrique, 1962
Tintinnoinea y Dinoflagellata del Pacifico segun material de las Expediciones NORPAC y DOWNWIND del Instituto Scripps de Oceanografia. Revista, Mus. Argentino Ciencias Nat. "Bernardino Rivadavia", Ciencias Zool., 7(1):1-253.

Phalacroma parvulum

Dangeard, P., 1927
Phytoplankton de la croisière du "Sylvana". Ann. Inst. Ocean., Monaco, n.s., 4(8):286-401, 54 text figs. (Feirer-Juin 1913).

Phalacroma parvulum

Ercegovic, A., 1936
Etudes qualitative et quantitatives du. phytoplancton dans les eaux cotières de l'Adriatique oriental moyen au cours de l'année 1934. Acta Adriatica 1(9):1-186

Phalachroma parvulum

Jörgensen, E., 1923
Mediterranean Dinophysiaceae. Rept. Danish Oceanogr. Expeds. 1908-10, to the Mediterranean and adjacent seas, Vol.II, Biol. J 2, 48 pp., 64 text figs.

Phalacroma parvulum

Käsler, R., 1938
Die Verbreitung der Dinophysiales im Sudatlantischen Ozean. Wiss. Ergeb. Deutschen Atlantischen Expedition----"Meteor" 1925-1927, 12(2):162-237, text figs. 85-118.

Phalacroma parvulum

Kofoid, C. A. and T. Skogsberg, 1928
XXXV. The Dinoflagellata: The Dinophysiodae. Reports on the scientific results of the expedition to the Eastern Tropical Pacific, in charge of Alexander Agassiz, by the U. S. Fish Commission Steamer "Albatross" from October 1904 to March 1905----. Mem. M. C. Z. 51:766 pp., 31 pls.

Phalacroma parvulum

Lillick, L.C., 1940
Phytoplankton and planktonic protozoa of the offshore waters of the Gulf of Maine. Pt.II. Qualitative Composition of the Planktonic Flora. Trans. Am. Phil. Soc., n.s., 31(3):193-237, 13 text figs.

Phalacroma parvulum

Lillick, L.C., 1938
Preliminary report of the phytoplankton of the Gulf of Maine. Am. Mid. Nat. 20(3):624-640, 1 text figs 37 tables.

Phalacroma parvulum

Margalef, R., 1949
Fitoplancton nerítico de la Costa Brava en 1947-48. Publ. Inst. Biol. Aplicada, 5: 41-51, 3 text figs.

Phalacroma parvulum

Matzenauer, L., 1933
Die Dinoflagellaten des indischen Ozeans (mit Ausnahme der Gattung Ceratium.) Bot. Arch. 35:437-510, 77 text figs., 2 charts.

Phalacroma parvulum

Rampi, L., 1940
Ricerche sul Fitoplancton del mare Ligure. Boll. di Pesca, di Piscicoltura e di Idrobiologa, 18(2):1-34, 56 text figs.

Phalacroma parvulum

Rampi, L. 1940.
Ricerche sul Fitoplancton del mare Ligure. II Le tecatali e le dinofisiali delle acque di Sanremo. Boll. Pesca, Piscicolt.,Idrobiol. (18) 16(2): 243-274, 56 figs.

Phalacroma parvulum

Wood, E.J.F., 1963
Dinoflagellates in the Australian region. II. Recent collections. Comm. Sci. and Industr. Res. Org., Div. Fish. and Oceanogr., Techn. Paper, No. 14: 55 pp.

Phalacroma porodictyum

Balech, Enrique, 1962
Tintinnoinea y Dinoflagellata del Pacifico segun material de las Expediciones NORPAC y DOWNWIND del Instituto Scripps de Oceanografia. Revista, Mus. Argentino Ciencias Nat. "Bernardino Rivadavia", Ciencias Zool., 7(1):1-253.

Phalacroma porodictyum

Dangeard, P., 1927
Phytoplankton de la croisière du "Sylvana". Ann. Inst. Ocean., Monaco, n.s., 4(8):286-401, 54 text figs. (février-Juin 1913).

Phalacroma porodictyum

Ercegovic, A., 1936
Etudes qualitative et quantitatives du phytoplancton dans les eaux cotières de l'Adriatique oriental moyen au cours de l'année 1934. Acta Adriatica 1(9):1-126

Phalacroma porodictyum

Forti, A., 1922
Ricerche sulla flora pelagica (fitoplancton) di Quarto dei Mille. Mem. R. Com. Talass. Ital. 97:248 pp., 13 pls.

Phalachroma porodictyum

Jörgensen, E., 1923
Mediterranean Dinophysiaceae. Rept. Danish Oceanogr. Expeds. 1908-10, to the Mediterranean and adjacent seas, Vol.II, Biol. J 2, 48 pp., 64 text figs.

Phalacroma porodictyum

Käsler, R., 1938
Die Verbreitung der Dinophysiales im Sud-atlantischen Ozean. Wiss. Ergeb. Deutschen Atlantischen Expedition----"Meteor" 1925-1927, 12(2):162-237, text figs. 85-118.

Phalacroma porodictyum

Kofoid, C. A. and T. Skogsberg, 1928
XXXV. The Dinoflagellata: The Dinophysiodae. Reports on the scientific results of the expedition to the Eastern Tropical Pacific, in charge of Alexander Agassiz, by the U. S. Fish Commission Steamer "Albatross" from October 1904 to March 1905----. Mem. M. C. Z. 51:766 pp., 31 pls.

Phalacroma porodictyum

Matzenauer, L., 1933
Die Dinoflagellaten des indischen Ozeans (mit Ausnahme der Gattung Ceratium.) Bot. Arch. 35:437-510, 77 text figs., 2 charts.

Phalacroma porodictyum

Murray, G., and F. G. Whitting, 1899
New Peridiniaceae from the Atlantic. Trans. Linn. Soc., London, Bot., ser 2, 5: 321-342, Pls. 27-33, 9 tables.

Phalacroma porodictyum

Pavillard, J., 1916
Recherches sur les Peridiniens du Golfe du Lion. Mem. Univ. Montpellier. Trav. Inst. Bot., Univ. Montpellier. Serie mixte No.4, 70 pp., 3 pls., 15 text figs.

Phalacroma porodictyum

Pavillard, J., 1905
Recherches sur la flore pelagique (Phytoplankton) de l'Etang de Thau. Theses presentees a la Fac. Sci., Paris, 116 pp., 3 pls.

Phalacroma porodictyum

Rampi, L. 1940.
Ricerche sul Fitoplancton del mare Ligure. II Le tecatali e le dinofisiali delle acque di Sanremo. Boll. Pesca, Piscicolt.,Idrobiol. (18) 16(2): 243-274, 56 figs.

Phalacroma porodictyum

Rampi, L., 1940
Ricerche sul Fitoplancton del mare Ligure. Boll. di Pesca, di Piscicoltura e di Idrobiologa, 18(2):1-34, 56 text figs.

Phalacroma porodictum

Schröder, B., 1900
Phytoplankton des Golfes von Neapel nebst vergleichenden Ausblicken auf das atlantischen Ozean. Mitt. Zool. Stat. Neapel, 14:1-38.

Phalacroma pulchellum

Wood, E.J.P., 1954.
Dinoflagellaets in the Australian region. Australian J. Mar. Freshwater Res., 5(2):171-351.

Phalacroma porosum

Brunel, J., 1962
Le phytoplancton de la Baie de Chaleurs. Inst. Botan., Univ. Montreal, Contrib. No. 77: 365 pp., 66 pls.

Phalacroma porosum

Käsler, R., 1938
Die Verbreitung der Dinophysiales im Sud-atlantischen Ozean. Wiss. Ergeb. Deutschen Atlantischen Expedition----"Meteor" 1925-1927, 12(2):162-237, text figs. 85-118.

Phalacroma porosum

Kofoid, C. A. and T. Skogsberg, 1928
XXXV. The Dinoflagellata: The Dinophysiodae. Reports on the scientific results of the expedition to the Eastern Tropical Pacific, in charge of Alexander Agassiz, by the U. S. Fish Commission Steamer "Albatross" from October 1904 to March 1905----. Mem. M. C. Z. 51:766 pp., 31 pls.

Phalacroma porosum

Wood, E.J.P., 1954.
Dinoflagellates in the Australian region. Australian J. Mar. Freshwater Res., 5(2):171-351.

Phalacroma praetextum

Kofoid, C. A. and T. Skogsberg, 1928
XXXV. The Dinoflagellata: The Dinophysiodae. Reports on the scientific results of the expedition to the Eastern Tropical Pacific, in charge of Alexander Agassiz, by the U. S. Fish Commission Steamer "Albatross" from October 1904 to March 1905----. Mem. M. C. Z. 51:766 pp., 31 pls.

Phalacroma protuberans n.sp.

Kofoid, C. A. and T. Skogsberg, 1928
XXXV. The Dinoflagellata: The Dinophysiodae. Reports on the scientific results of the expedition to the Eastern Tropical Pacific, in charge of Alexander Agassiz, by the U. S. Fish Commission Steamer "Albatross" from October 1904 to March 1905----. Mem. M. C. Z. 51:766 pp., 31 pls.

Phalacroma pugiunculus

Establier, R., and R. Margalef, 1964
Fitoplancton e hidrografía de las costas de Cádiz (Barbate), de junio de 1961 a agosto de 1962. Invest. Pesquera, Barcelona, 25:5-31.

Phalachroma pugiunculus n.sp.

Jörgensen, E., 1923
Mediterranean Dinophysiaceae. Rept. Danish Oceanogr. Expeds. 1908-10, to the Mediterranean and adjacent seas, Vol.II, Biol. J 2, 48 pp., 64 text figs.

Phalacroma pulchellum

Halim, Youssef, 1960
Etude quantitative et qualitative du cycle écologique des dinoflagellés dans les eaux de Villefranche-sur-Mer. (1953-1955). Ann. Inst. Oceanogr., Monaco, 38:123-232.

Phalacroma pulchellum

LeBour, M.V., 1925
The dinoflagellates of Northern Seas. The Marine Biological Association of the United Kingdom, Plymouth, 250 pp., 35 pls.,53 text figs.

Phalacroma pulchellum

Matzenauer, L., 1933
Die Dinoflagellaten des indischen Ozeans (mit Ausnahme der Gattung Ceratium.) Bot. Arch. 35:437-510, 77 text figs., 2 charts.

Phalacroma pulchrum

Käsler, R., 1938
Die Verbreitung der Dinophysiales im Sud-atlantischen Ozean. Wiss. Ergeb. Deutschen Atlantischen Expedition----"Meteor" 1925-1927, 12(2):162-237, text figs. 85-118.

Phalacroma pulchrum

Kofoid, C. A. and T. Skogsberg, 1928
XXXV. The Dinoflagellata: The Dinophysiodae. Reports on the scientific results of the expedition to the Eastern Tropical Pacific, in charge of Alexander Agassiz, by the U. S. Fish Commission Steamer "Albatross" from October 1904 to March 1905----. Mem. M. C. Z. 51:766 pp., 31 pls.

Phalacroma pulchrum

Wood, E.J.P., 1954.
Dinoflagellates in the Australian region. Australian J. Mar. Freshwater Res., 5(2):171-351.

Phalacroma pyriforme n.sp.

Kofoid, C. A. and T. Skogsberg, 1928
XXXV. The Dinoflagellata: The Dinophysiodae. Reports on the scientific results of the expedition to the Eastern Tropical Pacific, in charge of Alexander Agassiz, by the U. S. Fish Commission Steamer "Albatross" from October 1904 to March 1905----. Mem. M. C. Z. 51:766 pp., 31 pls.

Phalacroma rapa

Dangeard, P., 1927
Phytoplankton de la croisière du "Sylvana". Ann. Inst. Ocean., Monaco, n.s., 4(8):286-401, 54 text figs. (Feirer-Juin 1913).

Phalacroma rapa

Delegazione Italiana della Commissione Internazionale per l'Esplorazione Scientifica del Mediterraneo, 1941
Note sul plancton della Laguna veneta. [Memoria CCLXXXIX] Arch. di Ocean. e Limn. Anno I, Fasc. I, 1941 XIX: 31-57 pp.

Phalacroma rapa

de Sousa e Silva, E., 1956.
Contribution à l'étude du microplancton de Dakar et des regions maritimes voisines. Bull. I.F.A.N., 8(2):335-371, 7 pls.

Phalacroma rapa

Ercegovic, A., 1936
Etudes qualitative et quantitatives du phytoplancton dans les eaux cotières de l'Adriatique oriental moyen au cours de l'année 1934. Acta Adriatica 1(9):1-126

Phalachroma rapa
Jörgensen, E., 1923
Mediterranean Dinophysiaceae. Rept. Danish Oceanogr. Expeds. 1908-10, to the Mediterranean and adjacent seas, Vol.II, Biol. J 2, 48 pp., 64 text figs.

Phalacroma rapa
Käsler, R., 1938
Die Verbreitung der Dinophysiales im Sudatlantischen Ozean. Wiss. Ergeb. Deutschen Atlantischen Expedition----"Meteor" 1925-1927, 12(2):162-237, text figs. 85-118.

Phalacroma rapa
Kofoid, C. A. and T. Skogsberg, 1928
XXXV. The Dinoflagellata: The Dinophysiodae. Reports on the scientific results of the expedition to the Eastern Tropical Pacific, in charge of Alexander Agassiz, by the U. S. Fish Commission Steamer "Albatross" from October 1904 to March 1905----. Mem. M. C. Z. 51:766 pp., 31 pls.

Phalacroma rapa
Margalef, R., 1949
Fitoplancton nerítico de la Costa Brava en 1947-48. Publ. Inst. Biol. Aplicada, 5: 41-51, 3 text figs.

Phalacroma rapa
Matzenauer, L., 1933
Die Dinoflagellaten des indischen Ozeans (mit Ausnahme der Gattung Ceratium.) Bot. Arch. 35:437-510, 77 text figs., 2 charts.

Phalacroma rapa
Pavillard, J., 1923
A propos de la systématique des Péridiniens Bull. Soc. Bot. de France 70:876-882; 914-918.

Phalacroma rapa
Pavillard, J., 1916
Recherches sur les Peridiniens du Golfe du Lion. Mem. Univ. Montpellier. Trav. Inst. Bot., Univ. Montpellier. Serie mixte No.4, 70 pp., 3 pls., 15 text figs.

Phalacroma rapa
Rampi, L. 1940.
Ricerche sul Fitoplancton del mare Ligure. II Le tecatali e le dinofisiali delle acque di Sanremo.. Boll. Pesca, Piscicolt.,Idrobiol. (18) 16(2): 243-274, 56 figs.

Phalacroma rapa
Rampi, L., 1940
Ricerche sul Fitoplancton del mare Ligure. Boll. di Pesca, di Piscicoltura e di Idrobiologa, 18(2):1-34, 56 text figs.

Phalacroma rapa
Schröder, B., 1900
Phytoplankton des Golfes von Neapel nebst vergleichenden Ausblicken auf das atlantischen Ozean. Mitt. Zool. Stat. Neapel, 14:1-38.

Phalacroma rapa
Wood, E.J.F., 1954.
Dinoflagellates in the Australian region. Australian J. Mar. Freshwater Res., 5(2):171-351.

Phalacroma reticulatum
Käsler, R., 1938
Die Verbreitung der Dinophysiales im Sudatlantischen Ozean. Wiss. Ergeb. Deutschen Atlantischen Expedition----"Meteor" 1925-1927, 12(2):162-237, text figs. 85-118.

Phalacroma reticulatum
Kofoid, C. A. and T. Skogsberg, 1928
XXXV. The Dinoflagellata: The Dinophysiodae. Reports on the scientific results of the expedition to the Eastern Tropical Pacific, in charge of Alexander Agassiz, by the U. S. Fish Commission Steamer "Albatross" from October 1904 to March 1905----. Mem. M. C. Z. 51:766 pp., 31 pls.

Phalacroma reticulatum
Rampi, L., 1945
Osservazioni sulla distribuzione qualitativa del fitoplancton nel mare Mediterraneo. Atti della Soc. Ital. di Sci. Nat. 84:105-113.

Phalacroma reticulatum
Pavillard, J., 1923
A propos de la systématique des Péridiniens Bull. Soc. Bot. de France 70:876-882; 914-918.

Phalacroma reticulatum
Rampi, L., 1942
Il Fitoplancton mediterraneo: Problemi ed affinita interoceaniche. Boll. di Pesca di Piscicoltura e di Idrobiologia, Anno 18, Fasc. 4:7-19.

Phalacroma rotundatum
Balech, Enrique, 1962
Tintinnoinea y Dinoflagellata del Pacifico segun material de las Expediciones NORPAC y DOWNWIND del Instituto Scripps de Oceanografia. Revista, Mus. Argentino Ciencias Nat. "Bernardino Rivadavia", Ciencias Zool., 7(1):1-253.

Phalacroma rotundata
Braarud, T., 1945
A phytoplankton survey of the polluted waters of inner Oslo Fjord. Hvalrådets Skrifter, No.28, 142 pp., 19 text figs., 17 tables.

Phalacroma rotundata
Braarud, T., and Adam Bursa, 1939
On the phytoplankton of the Oslo Fjord, 1933-1934. Hvalrådets Skr. No.19:1-63; 9 text figs. Reviewed. J. du. Cons. 14(3): 418-420. A.C. Gardiner.

Phalacroma rotundatum
Brunel, J., 1962
Le phytoplancton de la Baie de Chaleurs. Inst. Botan., Univ. Montréal, Contrib. No. 77: 365 pp., 66 pls.

Phalacroma rotundatum
Brunel, Jules, 1962.
Le phytoplancton de la Baie des Chaleurs. Contrib. Ministère de la Chasse et des Pêcheries, Province de Québec, No. 91: 365 pp.

Phalacroma rotundatum
Dangeard, P., 1927
Phytoplankton de la croisière du "Sylvana". Ann. Inst. Ocean., Monaco, n.s., 4(8):286-401, 54 text figs. (fevrier-Juin 1913).

Phalacroma rotundatum
Ercegovic, A., 1940
Weitere Untersuchungen über einige hydrographische Verhältnisse und über die Phytoplanktonproduktion in den Gewässern der östlichen Mitteladria. Acta Adriatica 2(3):95-134, 8 text figs.

Phalacroma rotundatum
Gilbert, J.Y., and W.E. Allen, 1943
The phytoplankton of the Gulf of California obtained by the "E.W. Scripps" in 1939 and 1940. J. Mar. Res. V(2):89-110, figs.30-31.

Phalachroma rotundatum
Jörgensen, E., 1923
Mediterranean Dinophysiaceae. Rept. Danish Oceanogr. Expeds. 1908-10, to the Mediterranean and adjacent seas, Vol.II, Biol. J 2, 48 pp., 64 text figs.

Phalacroma rotundatum
Käsler, R., 1938
Die Verbreitung der Dinophysiales im Sudatlantischen Ozean. Wiss. Ergeb. Deutschen Atlantischen Expedition----"Meteor" 1925-1927, 12(2):162-237, text figs. 85-118.

Phalacroma rotundatum
LeBour, M.V., 1925
The dinoflagellates of Northern Seas. The Marine Biological Association of the United Kingdom, Plymouth, 250 pp., 35 pls.,53 text figs.

Phalacroma rotundatum
Margalef, R., 1949
Fitoplancton nerítico de la Costa Brava en 1947-48. Publ. Inst. Biol. Aplicada, 5: 41-51, 3 text figs.

Phalacroma rotundatum
Matzenauer, L., 1933
Die Dinoflagellaten des indischen Ozeans (mit Ausnahme der Gattung Ceratium.) Bot. Arch. 35:437-510, 77 text figs., 2 charts.

Phalacroma rotundatum
Rampi, L. 1940.
Ricerche sul Fitoplancton del mare Ligure. II Le tecatali e le dinofisiali delle acque di Sanremo.. Boll. Pesca, Piscicolt.,Idrobiol. (18) 16(2): 243-274, 56 figs.

Phalacroma rotundatum
Rampi, L., 1940
Ricerche sul Fitoplancton del mare Ligure. Boll. di Pesca, di Piscicoltura e di Idrobiologa, 18(2):1-34, 56 text figs.

Phalacroma rotundatum
Rumkówna, A., 1948
[List of the phytoplankton species occurring in the superficial water layers in the Gulf of Gdańsk] Bull. Lab. mar., Gdynia, No. 4: 139-141 with tables in back.

Phalacroma rotundata
Sousa e Silva, E., 1949
Diatomaceas e Dinoflagelados de Baia de Cascais. Portugaliae Acta Biol., Volume; Julio Henriques, Ser. B: 300-383, 9 pls, 2 fold-in tables.

Phalacroma rotundata
Sousa e Silva, E., and J. Dos Santos-Pinto, 1948
O Plancton da Baia de S. Martinho do Porto. 1. Diatomaceas e Dinoflagelados. Bol. Soc. Portuguese de Ciencias Naturais, 16(2):134-187, 6 pls. (Trav. Sta. Biol. Mar. de Lisbonne No. 52).

Phalacroma Rudgei
Käsler, R., 1938
Die Verbreitung der Dinophysiales im Sudatlantischen Ozean. Wiss. Ergeb. Deutschen Atlantischen Expedition----"Meteor" 1925-1927, 12(2):162-237, text figs. 85-118.

Phalacroma rudgei
Matzenauer, L., 1933
Die Dinoflagellaten des indischen Ozeans (mit Ausnahme der Gattung Ceratium.) Bot. Arch. 35:437-510, 77 text figs., 2 charts.

Phalacroma Rudgei n.sp.
Murray, G., and F. G. Whitting, 1899
New Peridiniaceae from the Atlantic. Trans. Linn. Soc., London, Bot., ser 2, 5: 321-342, Pls. 27-33, 9 tables.

Phalacroma Rudgei
Paulsen, O., 1908
XVIII Peridiniales. Nordisches Plankton, Bot. Teil: 1-124, 155 text figs.

Phalacroma Rudgei
Schröder, B., 1900
Phytoplankton des Golfes von Neapel nebst vergleichenden Ausblicken auf das atlantischen Ozean. Mitt. Zool. Stat. Neapel, 14:1-38.

Phalacroma rudgei

Wood, E.J.P., 1954.
Dinoflagellates in the Australian region.
Australian J. Mar. Freshwater Res., 5(2):171-351.

Phalacroma Ruudii

Braarud, T., 1945
A phytoplankton survey of the polluted waters of inner Oslo Fjord. Hvalrådets Skrifter, No.28, 142 pp., 19 text figs., 17 tables.

Phalacroma ruudii

Hasle, Grethe Rytter, 1960
Phytoplankton and ciliate species from the Tropical Pacific.
Skr. Norske Videnskaps-Akad., Oslo, 1.
Mat.-Nat. Kl., 1960(2): 1-50.

Phalacroma spinatum

Käsler, R., 1938
Die Verbreitung der Dinophysiales im Sud-atlantischen Ozean. Wiss. Ergeb. Deutschen Atlantischen Expedition----"Meteor" 1925-1927, 12(2):162-237, text figs. 85-118.

Phalachroma stenopterygium n.sp.

Jörgensen, E., 1923
Mediterranean Dinophysiaceae. Rept. Danish Oceanogr. Expeds. 1908-10, to the Mediterranean and adjacent seas, Vol.II, Biol. J 2, 48 pp., 64 text figs.

Phalachroma striatum

Jörgensen, E., 1923
Mediterranean Dinophysiaceae. Rept. Danish Oceanogr. Expeds. 1908-10, to the Mediterranean and adjacent seas, Vol.II, Biol. J 2, 48 pp., 64 text figs.

Phalacroma striatum

Käsler, R., 1938
Die Verbreitung der Dinophysiales im Sud-atlantischen Ozean. Wiss. Ergeb. Deutschen Atlantischen Expedition----"Meteor" 1925-1927, 12(2):162-237, text figs. 85-118.

Phalacroma striatum

Kofoid, C. A. and T. Skogsberg, 1928
XXXV. The Dinoflagellata: The Dinophysiodae. Reports on the scientific results of the expedition to the Eastern Tropical Pacific, in charge of Alexander Agassiz, by the U. S. Fish Commission Steamer "Albatross" from October 1904 to March 1905----. Mem. M. C. Z. 51:766 pp., 31 pls.

Phalacroma striatum

Rampi, L., 1945
Osservazioni sulla distribuzione qualitativa del fitoplancton nel mare Mediterraneo. Atti della Soc. Ital. di Sci. Nat. 84:105-113.

Phalacroma striatum

Rampi, L., 1942
II Fitoplancton mediterraneo: Problemi ed affinita interoceaniche. Boll. di Pesca di Piscicoltura e di Idrobiologia, Anno 18, Fasc. 4:7-19.

Phalacroma thompsonii n.sp.

Wood, E.J.P., 1954.
Dinoflagellates in the Australian region.
Australian J. Mar. Freshwater Res., 5(2):171-351.

Phalacroma triangulare n.sp.

Wood, E.J.P., 1954.
Dinoflagellates in the Australian region.
Australian J. Mar. Freshwater Res., 5(2):171-351.

Phalacroma triata

Wood, E.J.F., 1963
Dinoflagellates in the Australian region.
II. Recent collections.
Comm. Sci. and Industr. Res. Org., Div. Fish. and Oceanogr., Techn. Paper, No. 14: 55 pp.

Phalacroma turbineum

Kofoid, C. A. and T. Skogsberg, 1928
XXXV. The Dinoflagellata: The Dinophysiodae. Reports on the scientific results of the expedition to the Eastern Tropical Pacific, in charge of Alexander Agassiz, by the U. S. Fish Commission Steamer "Albatross" from October 1904 to March 1905----. Mem. M. C. Z. 51:766 pp., 31 pls.

Phalacroma vastiforme n.sp.

Tai, Si-Sun, & T. Skogsberg, 1934
Studies on the Dinophysordae, marine armored dinoflagellates of Monterey Bay, California. Arch. Protistenk. 82:380-482, 14 text figs. Pls. 11-12.

Phalacroma vastum acuta

Pavillard, J., 1905
Recherches sur la flore pelagique (Phytoplankton) de l'Etang de Thau. Theses presentees a la Fac. Sci., Paris, 116 pp., 3 pls.

Phalacroma vastum

Schröder, B., 1900
Phytoplankton des Golfes von Neapel nebst vergleichenden Ausblicken auf das atlantischen Ozean. Mitt. Zool. Stat. Neapel, 14:1-38.

Phalacroma whiteleggei n.sp.

Wood, E.J.P., 1954.
Dinoflagellates in the Australian region.
Australian J. Mar. Freshwater Res., 5(2):171-351.

Plectodinium nucleovolvatum

Becheler, B., 1952.
Recherches sur les Peridiniens.
Bull. Biol. France Belg., Suppl., 36:1-149, 75 textfigs.

Podolampace

Rampi, L., 1941
Ricerche sul fitoplancton del Mare Ligure. 5. Le Podolampacer delle acque di Sanremo. Annali del Mus. Civico di Storia Naturale di Genova, 51:141-152, Pl. 5.

Podolampus antarctica n.sp.

Balech, Enrique, and Sayed Z. El-Sayed, 1965.
Microplankton of the Weddell Sea.
In: Biology of Antarctic seas. II.
Antarctic Res. Ser., Amer. Geophys. Union, 5:107-124.

Podolampas bipes

Abé, Tohru H., 1966.
The armoured Dinoflagellata I. Podolampidae.
Publs Seto mar. biol. Lab., 14(2):129-154.

Podolampas bipes

Balech, Enrique, 1963.
La familia Podolampacea (Dinoflagellata).
Bol. Inst. Biol. Mar. Mar del Plata, Argentina, No.2: 30 pp.

Podolampas elegans

Balech, Enrique, 1963.
La familia Podolampacea (Dinoflagellata).
Bol. Inst. Biol. Mer. Mar. del Plata, Argentina, No. 2:30 pp.

Podolampas bipes

Dangeard, P., 1927
Phytoplankton de la croisière du "Sylvana". Ann. Inst. Ocean., Monaco, n.s., 4(8):286-401, 54 text figs. (Janvier-Juin 1913).

Podolampas bipes

Delegazione Italiana della Commissione Internazionale per l'Esplorazione Scientifica del Mediterraneo, 1941
Note sul plancton della Laguna veneta.
[Memoria CCLXXIX], Arch. di Ocean. e Limn. Anno I, Fasc. I, 1941 XIX: 31-57 pp.

Podolampas bipes

Ercegovic, A., 1936
Etudes qualitative et quantitatives du phytoplancton dans les eaux cotières de l'Adriatique oriental moyen au cours de l'année 1934. Acta Adriatica 1(9):1-126

Podolampas bipes

Forti, A., 1922
Ricerche sulla flora pelagica (fitoplancton) di Quarto dei Mille. Mem. R. Com. Talass. Ital. 97:248 pp., 13 pls.

Podolampas bipes

LeBour, M.V., 1925
The dinoflagellates of Northern Seas. The Marine Biological Association of the United Kingdom. Plymouth, 250 pp., 35 pls. 53 text figs.

Podolampas bipes

Lindemann, E., 1925
Neubeobachtungen an den Winter peridineen des Golfes von Neapel. Bot. Arch. 9:95-102, 19 text figs.

Podolampas bipes

Matzenauer, L., 1933
Die Dinoflagellaten des indischen Ozeans (mit Ausnahme der Gattung Ceratium.) Bot. Arch. 35:437-510, 77 text figs., 2 charts.

Podolampas bipes

Murray, G., and F. G. Whitting, 1899
New Peridiniaceae from the Atlantic.
Trans. Linn. Soc., London, Bot., ser 2, 5: 321-342, Pls. 27-33, 9 tables.

Podolampas bipes

Paulsen, O., 1908
XVIII Peridiniales. Nordisches Plankton, Bot. Teil: 1-124, 155 text figs.

Podolampas bipes

Pavillard, J., 1916
Recherches sur les Peridiniens du Golfe du Lion. Mem. Univ. Montpellier. Trav. Inst. Bot., Univ. Montpellier. Serie mixte No.4, 70 pp., 3 pls., 15 text figs.

Podolampas bipes

Pavillard, J., 1905
Recherches sur la flore pelagique (Phytoplankton) de l'Etang de Thau. Theses presentees a la Fac. Sci., Paris, 116 pp., 3 pls.

Podolampas bipes

Rampi, L., 1941
Ricerche sul fitoplancton del Mare Ligure. 5. Le Podolampacer delle acque di Sanremo. Annali del Mus. Civico di Storia Naturale di Genova, 51:141-152, Pl. 5.

Podolampas bipes

Schröder, B., 1900
Phytoplankton des Golfes von Neapel nebst vergleichenden Ausblicken auf das atlantischen Ozean. Mitt. Zool. Stat. Neapel, 14:1-38.

Podolampas bipes

Sousa e Silva, E., 1949
Diatomaceas e Dinoflagelados de Baia de Cascais. Portugaliae Acta Biol. Volume: Julio Henriques, Ser. B: 300-383, 9 pls, 2 fold-in tables.

Podolampas curvatus

Wood, E.J.F., 1963
Dinoflagellates in the Australian region.
II. Recent collections.
Comm. Sci. and Industr. Res. Org., Div. Fish. and Oceanogr., Techn. Paper, No. 14: 55 pp.

Podolampas elegans
Abé, Tohru H., 1966.
The armoured Dinoflagellata I. Podolampidae.
Publs Seto mar. biol. Lab., 14(2):129-154.

Podolampas elegans
Matzenauer, L., 1933
Die Dinoflagellaten des indischen Ozeans (mit Ausnahme der Gattung Ceratium.) Bot. Arch. 35:437-510, 77 text figs., 2 charts.

Podolampas elegans
Pavillard, J., 1905
Recherches sur la flore pelagique (Phytoplankton) de l'Etang de Thau. Theses presentees a la Fac. Sci., Paris, 116 pp., 3 pls.

Podolampas elegans
Rampi, L., 1941
Ricerche sul fitoplancton del Mare Ligure. 5. Le Podolampacer delle acque di Sanremo. Annali del Mus. Civico di Storia Naturale di Genova, 51:141-152, Pl. 5.

Podolampas elegans
Wood, E.J.F., 1963
Dinoflagellates in the Australian region. II. Recent collections.
Comm. Sci. and Industr. Res. Org., Div. Fish. and Oceanogr., Techn. Paper, No. 14: 55 pp.

Podolampas palmipes
Abé, Tohru H., 1966.
The armoured Dinoflagellata I. Podolampidae.
Publs Seto mar. biol. Lab., 14(2):129-154.

Podolampas palmipes
Balech, Enrique, 1963.
La familia Podolampacea (Dinoflagellata).
Bol. Inst.Biol. Mar., Mar del Plata, Argentina, No. 2:30 pp.

Podolampas palmipes
Dangeard, P., 1927
Phytoplankton de la croisière du "Sylvana". Ann. Inst. Ocean., Monaco, n.s., 4(8):286-401, 54 text figs. (fevrier-Juin 1913).

Podolampas palmipes
Ercegovic, A., 1936
Etudes qualitative et quantitatives du phytoplancton dans les eaux cotières de l'Adriatique oriental moyen au cours de l'année 1934. Acta Adriatica 1(9):1-126

Podolampas palmipes
Gilbert, J.Y., and W.E. Allen, 1943
The phytoplankton of the Gulf of California obtained by the "E.W. Scripps" in 1939 and 1940. J. Mar. Res. V(2):89-110, figs.30-31.

Podolampas palmipes
Jørgensen, E., 1905
B.Protistplankton and the diatoms in bottom samples. Hydrographical and biological investigations in Norwegian fjords. Bergens Mus. Skr. 7: 49-225.

Podolampas palmipes
Jorgensen, E., 1900
Protophyten und Protozoën im Plankton aus der Norwegischen Westkerste. Bergens Mus. Aarb. 1899(6): 95 pp., 5 pls., 83 tables

Podolampas palmipes
LeBour, M.V., 1925
The dinoflagellates of Northern Seas. The Marine Biological Association of the United Kingdom. Plymouth, 250 pp., 35 pls. 53 text figs.

Podolampas palmipes
Lindemann, E., 1925
Neubeobachtungen an den Winter peridineen des Golfes von Neapel. Bot. Arch. 9:95-102, 19 text figs.

Podolampas palmipes
Margalef, R., 1949
Fitoplancton nerítico de la Costa Brava en 1947-48. Publ. Inst. Biol. Aplicada, 5: 41-51, 3 text figs.

Podolampas palmipes
Massuti Algamora, M., 1949
Estudio de diez y seis muestras de plancton del Golfo de Nápoles. Publ. Inst. Biol. Appl. 5:85-94, 1 fold-in table.

Podolampas palmipes
Matzenauer, L., 1933
Die Dinoflagellaten des indischen Ozeans (mit Ausnahme der Gattung Ceratium.) Bot. Arch. 35:437-510, 77 text figs., 2 charts.

Podolampas palmipes
Murray, G., and F. G. Whitting, 1899
New Peridiniaceae from the Atlantic.
Trans. Linn. Soc., London, Bot., ser 2, 5: 321-342, Pls. 27-33, 9 tables.

Podolampas palmipes
Paulsen, O., 1908
XVIII Peridiniales. Nordisches Plankton, Bot. Teil: 1-124, 155 text figs.

Podolampas palmipes
Pavillard, J., 1905
Recherches sur la flore pelagique (Phytoplankton) de l'Etang de Thau. Theses presentees a la Fac. Sci., Paris, 116 pp., 3 pls.

Podolampas palmipes
Rampi, L., 1941
Ricerche sul fitoplancton del Mare Ligure. 5. Le Podolampacer delle acque di Sanremo. Annali del Mus. Civico di Storia Naturale di Genova, 51:141-152, Pl. 5.

Podolampas palmipes
Schröder, B., 1900
Phytoplankton des Golfes von Neapel nebst vergleichenden Ausblicken auf das atlantischen Ozean. Mitt. Zool. Stat. Neapel, 14:1-38.

Podolampas reticulata
Balech, Enrique, 1963.
La familia Podolampacea (Dinoflagellata).
Bol. Inst.Biol.Mar., Mar del Plata, Argentina, No. 2 30 pp.

Podolampas reticulata
Matzenauer, L., 1933
Die Dinoflagellaten des indischen Ozeans (mit Ausnahme der Gattung Ceratium.) Bot. Arch. 35:437-510, 77 text figs., 2 charts.

Podolampas spinifera
Abé, Tohru H., 1966.
The armoured Dinoflagellata I. Podolampidae.
Publs Seto mar. biol. Lab., 14(2):129-154.

Podolampas spinifer
Balech, Enrique, 1963.
La familia Podolampacea (Dinoflagellata).
Bol.Inst.Biol.Mar., Mar del Plata, Argentina, No.2: 30 pp.

Podolampas spinifera
Ercegovic, A., 1936
Etudes qualitative et quantitatives du phytoplancton dans les eaux cotières de l'Adriatique oriental moyen au cours de l'année 1934. Acta Adriatica 1(9):1-126

Podolampas spinifera
Forti, A., 1922
Ricerche sulla flora pelagica (fitoplancton) di Quarto dei Mille. Mem. R. Com. Talass. Ital. 97:248 pp., 13 pls.

Podolampas spinifera
Margalef, R., 1949
Fitoplancton nerítico de la Costa Brava en 1947-48. Publ. Inst. Biol. Aplicada, 5: 41-51, 3 text figs.

Podolampas spinifer
Rampi, L., 1941
Ricerche sul fitoplancton del Mare Ligure. 5. Le Podolampacer delle acque di Sanremo. Annali del Mus. Civico di Storia Naturale di Genova, 51:141-152, Pl. 5.

Podolampas spinifer
Wood, E.J.F., 1963
Dinoflagellates in the Australian region. II. Recent collections.
Comm. Sci. and Industr. Res. Org., Div. Fish. and Oceanogr., Techn. Paper, No. 14: 55 pp.

Polykrikos sp.
Meunier, A., 1919
Microplancton de la Mer Flamande 3. Les Péridiniens. Mem. Mus. Roy. Hist. Nat., Belgique 8(1):1-116, Pls. XV-XXI.

Polykrikos sp., effect of
Holmes, R.W., P.M. Williams and R.W. Eppley, 1967.
Red water in La Jolla Bay, 1964-1965.
Limnol. Oceanogr., 12(3):503-512.

Polykrikos auricularia
Forti, A., 1922
Ricerche sulla flora pelagica (fitoplancton) di Quarto dei Mille. Mem. R. Com. Talass. Ital. 97:248 pp., 13 pls.

Polykrikos auricularia
Jorgensen, E., 1900
Protophyten und Protozoën im Plankton aus der Norwegischen Westkerste. Bergens Mus. Aarb. 1899(6): 95 pp., 5 pls., 83 tables.

Polykrikos auricularia
Pavillard, J., 1905
Recherches sur la flore pelagique (Phytoplankton) de l'Etang de Thau. Theses presentees a la Fac. Sci., Paris, 116 pp., 3 pls.

Polykrikos hartmanni
Hulburt, E.M., 1957.
The taxonomy of unarmored Dinophyceae of shallow embayments on Cape Cod, Massachusetts.
Biol. Bull., 112(2):196-219.

Polykrikos kofoidi
Morse, D.C., 1947
Some observations on seasonal variations in plankton population Patuxent River, Maryland 1943-1945. Bd. Nat. Res., Publ. No.65, Chesapeake Biol. Lab., 31, 3 figs.

Polykrikos lebourae
Dragesco, Jean, 1965.
Étude cytologique de quelques flagellés mésopsammiques.
Cahiers Biol. Mar., Roscoff, 6(1):83-115.

Polykrikos lebourae
Hulburt, E.M., 1957.
The taxonomy of unarmored Dinophyceae of shallow embayments on Cape Cod, Massachusetts.
Biol. Bull., 112(2):196-219.

Polykrikos lebourae
LeBour, M.V., 1925
The dinoflagellates of Northern Seas. The Marine Biological Association of the United Kingdom, Plymouth, 250 pp., 35 pls., 53 text figs.

Polykrikos schwartzi

Becheler, B., 1952.
Recherches sur les Peridiniens.
Bull. Biol., France Belg., Suppl., 36:1-149, 75 textfigs.

Polykrikos schwartzi

Hulburt, E.M., 1957.
The taxonomy of unarmored Dinophyceae of shallow embayments on Cape Cod, Massachusetts.
Biol. Bull., 112(2):196-219.

Polykrikos schwarzi

Iwasaki, Hideo 1971.
Studies on the red tide dinoflagellates
I On Polykrikos schwarzi Bütschli.
(In Japanese; English abstract)
Bull. Jap. Soc. scient. Fish. 37(7):606-609.

Polykrikos schwarzi

LeBour, M.V., 1925
The dinoflagellates of Northern Seas. The Marine Biological Association of the United Kingdom, Plymouth, 250 pp., 35 pls., 53 text figs.

Polykrikos schwartzii

Paulsen, O., 1908
XVIII Peridiniales. Nordisches Plankton, Bot. Teil: 1-124, 155 text figs.

Polykrikos Schwartzii

Schodduyn, M., 1926
Observations faites dans la baie d'Ambleteuse (Pas de Calais). Bull. Inst. Ocean., Monaco, No. 482: 64 pp.

Polykrikos

Sommer, H., and F. N. Clarke, 1946.
Effect of red water on marine life in Santa Monica Bay, California. California Fish and Game, 32(2):100-101.
P. schwartzi.

Polykrikos schwarzi

Sousa e Silva, E., 1949
Diatomaceas e Dinoflagelados de Baía de Cascais. Portugaliae Acta Biol., Volume: Julio Henriques, Ser. B: 300-383, 9 pls, 2 fold-in tables.

Porella adriatica

Ercegovic, A., 1936
Etudes qualitative et quantitatives du phytoplancton dans les eaux cotières de l'Adriatique oriental moyen au cours de l'année 1934. Acta Adriatica 1(9):1-126

Porella adriatica

Rampi, L. 1940
Ricerche sul Fitoplancton del mare Ligure. II Le tecatali e le dinofisiali delle acque di Sanremo.. Boll. Pesca, Piscicolt. Idrobiol. (18) 16(2): 243-274, 56 figs.

Porella adriatica

Rampi, L., 1940
Ricerche sul Fitoplancton del mare Ligure. Boll. di Pesca, di Piscicoltura e di Idrobiologa, 18(2):1-34, 56 text figs.

Porella adriatica

Rampi, L., 1939
Péridiniens rares ou intéressants recoltés dans la mer Ligure. Bull. Soc. Fran. de Microscopie, 8(2/3):106-112, 13 text figs.

Porella globulus

Rampi, L. 1940.
Ricerche sul Fitoplancton del mare Ligure. II Le tecatali e le dinofisiali delle acque di Sanremo.. Boll. Pesca, Piscicolt. Idrobiol. (18) 16(2): 243-274, 56 figs.

Porella globulus

Rampi, L., 1940
Ricerche sul Fitoplancton del mare Ligure. Boll. di Pesca, di Piscicoltura e di Idrobiologa, 18(2):1-34, 56 text figs.

Porella globulus

Rampi, L., 1939
Péridiniens rares ou intéressants recoltés dans la mer Ligure. Bull. Soc. Fran. de Microscopie, 8(2/3):106-112, 13 text figs.

Porella perforata

Braarud, T., 1945.
Morphological observations on marine dinoflagellates cultures (Porella perforata, Goniaulax tamarensis, Protoceratium reticulatum). Avhandl. Norske Videnskaps-Akademi i Oslo, I Mat.-Naturv. Kl. 1944, No. 11:1-18, 6 textfigs., 4 pls.

Porella perforata

Braarud, T., 1945.
Morphological observations on marine dinoflagellate cultures (Porella perforata, Goniaulax tamarensis, Protoceratium reticulatum). Avhandl. Norske Vidensk.-Akad., Oslo, Mat.-Naturh. Kl., 1944(11):1-18, 4 pls., 6 textfigs.

Porella perforata

Woods, E.J.P., 1954.
Dinoflagellates in the Australian region.
Australian J. Mar. Freshwater Res., 5(2):751-351.

Pomatodinium impatiens

Cachon, Jean, et Monique Cachon-Enjumet, 1966.
Pomatodinium impatiens nov. gen. nov. sp. péridinien Noctilucidae Kent.
Protistilogica, 2(1):23-29
Also in: Trav. Stn. zool., Villefranche, 26.

Pomatodinium impatiens

Cachon, Jean et Monique avec Odette et Gérard Ferru, 1967.
Contribution a l'étude des Noctilucidae Saville-Kent. 1. Les Kofoidininae Cachon J. et M., évolution morphologique et Systématique.
Protistologica 3(4):427-443.

Pomatodinium impatiens

Travers, Anne, et Marc Travers, 1971.
Observation en Atlantique nord-est de Pomatodinium impatiens Cachon et Cachon Enjumet, dinoflagellé Noctilucidae, décrit de Méditerranée.
Rapp. p.-v. Comm. int. Explor. scient. mer Médit. 20(3): 315-316.

Postprorocentrum n. gen.

Gourret, P., 1883
Sur les Peridiniens du Golfe de Marseille. Ann. du Musee d'hist. Nat., Marseille, Zool., 1 (Mme. 8):1-114, 4 pls.

Postprorocentrum maximum

Gourret, P., 1883
Sur les Peridiniens du Golfe de Marseille. Ann. du Musee d'hist. Nat., Marseille, Zool., 1 (Mme. 8):1-114, 4 pls.

Postprorocentrum ovale

Gourret, P., 1883
Sur les Peridiniens du Golfe de Marseille. Ann. du Musee d'hist. Nat., Marseille, Zool., 1 (Mme. 8):1-114, 4 pls.

Pouchetia fusus

LeBour, M.V., 1925
The dinoflagellates of Northern Seas. The Marine Biological Association of the United Kingdom, Plymouth, 250 pp., 35 pls., 53 text figs.

Pouchetia nigra

Pavillard, J., 1905
Recherches sur la flore pelagique (Phytoplankton) de l'Etang de Thau. Theses presentees a la Fac. Sci., Paris, 116 pp., 3 pls.

Pouchetia parva

LeBour, M.V., 1925
The dinoflagellates of Northern Seas. The Marine Biological Association of the United Kingdom, Plymouth, 250 pp., 35 pls., 53 text figs.

Pouchetia parva

Paulsen, O., 1908
XVIII Peridiniales. Nordisches Plankton, Bot. Teil: 1-124, 155 text figs.

Pouchetia polyphemus

LeBour, M.V., 1925
The dinoflagellates of Northern Seas. The Marine Biological Association of the United Kingdom, Plymouth, 250 pp., 35 pls., 53 text figs.

Pouchetia rosea

Forti, A., 1922
Ricerche sulla flora pelagica (fitoplankton) di Quarto dei Mille. Mem. R. Com. Talass. Ital. 97:248 pp., 13 pls.

Pouchetia rosea

LeBour, M.V., 1925
The dinoflagellates of Northern Seas. The Marine Biological Association of the United Kingdom, Plymouth, 250 pp., 35 pls., 53 text figs.

Pouchetia rosea

Paulsen, O., 1908
XVIII Peridiniales. Nordisches Plankton, Bot. Teil: 1-124, 155 text figs.

Pouchetia rosea

Pavillard, J., 1905
Recherches sur la flore pelagique (Phytoplankton) de l'Etang de Thau. Theses presentees a la Fac. Sci., Paris, 116 pp., 3 pls.

Pouchetia rosea

Schröder, B., 1900
Phytoplankton des Golfes von Neapel nebst vergleichenden Ausblicken auf das atlantischen Ozean. Mitt. Zool. Stat. Neapel, 14:1-38.

Prodinophysis n.nov

Balech, Enrique, 1944
Contribucion al conocimiento del Plancton de Lennox y Cabo de Hornos. Physis XIX:423-446, 6 pls. with 67 figs.

Prodinophysis rotundata

Balech, Enrique, 1944
Contribucion al conocimiento del Plancton de Lennox y Cabo de Hornos. Physis XIX:423-446, 6 pls. with 67 figs.

Proheteroschisma

Abe, Tohru H., 1967.
The armoured Dinoflagellata: II. Prorocentridae and Dinophysidae (B)- Dinophysis and its allied genera.
Publs. Seto mar. biol. Lab., 15(1):37-78.

Proheteroschisma connectens n.gen., n.sp.

Tai, Si-Sun, & T. Skogsberg, 1934
Studies on the Dinophysordae, marine armored dinoflagellates of Monterey Bay, California. Arch. Protistenk. 82:380-482, 14 text figs. Pls. 11-12.

Pronoctiluca sp.

Braarud, T., 1945
A phytoplankton survey of the polluted waters of inner Oslo Fjord. Hvalrådets Skrifter, No.28, 142 pp., 19 text figs., 17 tables.

Pronoctiluca acuta

Wood, E.J.P., 1954.
Dinoflagellates in the Australian region.
Australian J. Mar. Freshwater Res., 5(2):171-351.

Pronoctiluca pelagia
LeBour, M.V., 1925
The dinoflagellates of Northern Seas. The Marine Biological Association of the United Kingdom, Plymouth, 250 pp., 35 pls.,53 text figs.

Pronoctiluca pelagica
Wood, E.J.F., 1954.
Dinoflagellates in the Australian region.
Australian J. Mar. Freshwater Res., 5(2):171-351.

Pronoctiluca spinifera
Wood, E.J.F., 1954.
Dinoflagellates in the Australian region.
Australian J. Mar. Freshwater Res., 5(2):171-351.

Properidinium n.gen.
Meunier, A., 1919
Microplancton de la Mer Flamande 3. Les Péridiniens. Mem. Mus. Roy. Hist. Nat., Belgique 8(1):1-116, Pls. XV-XXI.

Properidinium aspinum
Meunier, A., 1919
Microplancton de la Mer Flamande 3. Les Péridiniens. Mem. Mus. Roy. Hist. Nat., Belgique 8(1):1-116, Pls. XV-XXI.

Properidinium avellana n.sp.
Meunier, A., 1919
Microplancton de la Mer Flamande 3. Les Péridiniens. Mem. Mus. Roy. Hist. Nat., Belgique 8(1):1-116, Pls. XV-XXI.

Properidinium apiculatum
Meunier, A., 1919
Microplancton de la Mer Flamande 3. Les Péridiniens. Mem. Mus. Roy. Hist. Nat., Belgique 8(1):1-116, Pls. XV-XXI.

Properidinium Heterocapsa
Meunier, A., 1919
Microplancton de la Mer Flamande 3. Les Péridiniens. Mem. Mus. Roy. Hist. Nat., Belgique 8(1):1-116, Pls. XV-XXI.

Properidinium inaequale
Meunier, A., 1919
Microplancton de la Mer Flamande 3. Les Péridiniens. Mem. Mus. Roy. Hist. Nat., Belgique 8(1):1-116, Pls. XV-XXI.

Properidinium Thorianum
Meunier, A., 1919
Microplancton de la Mer Flamande 3. Les Péridiniens. Mem. Mus. Roy. Hist. Nat., Belgique 8(1):1-116, Pls. XV-XXI.

Properidinium umbonatum
Meunier, A., 1919
Microplancton de la Mer Flamande 3. Les Péridiniens. Mem. Mus. Roy. Hist. Nat., Belgique 8(1):1-116, Pls. XV-XXI.

Proplectella ellipsoida
Hasle, Grethe Rytter, 1960
Phytoplankton and ciliate species from the Tropical Pacific.
Skr. Norske Videnskaps-Akad., Oslo, 1. Mat.-Nat. Kl., 1960(2): 1-50.

Proplectella junei n.sp.
Rampi, L., 1952.
Ricerche sul microplancton di superficie del Pacifico tropicale. Bull. Inst. Ocean., Monaco, No. 1014:16 pp., 5 textfigs.

Prorocentrum
Allen, W. E. 1946
"Red water" in La Jolla Bay in 1946.
Trans Am. Micro. Soc., Vol. LXV, No. 3, pp. 149-153.

Prorocentrum sp.
Balech, Enrique, 1971
Microplancton del Atlantico ecuatorial oeste (Equalant 1)
Publ. Serv. Hidrograf. Naval, Argentina H. 654: 103 pp., 122 figs.

Prorocentrum sp.
Nakazima, Masao, 1965.
Studies on the source of shellfish poison in Lake Hamana. 1. Relation of the abundance of a species of Dinoflagellata, Prorocentrum sp. to shellfish toxicity. 2. Shellfish toxicity during the red tide.
Bull. Jap. Soc. Sci. Fish., 31(3):198-203,204-207

Prorocentrum adriaticum
Halim, Youssef, 1960
Étude quantitative et qualitative du cycle écologique des dinoflagellés dans les eaux de Villefranche-sur-Mer. (1953-1955).
Ann. Inst. Océanogr., Monaco, 38:123-232.

Prorocentrum arcuatum
Wood, E.J.F., 1963
Dinoflagellates in the Australian region. II. Recent collections.
Comm. Sci. and Industr. Res. Org., Div. Fish. and Oceanogr., Techn. Paper, No. 14: 55 pp.

Prorocentrum caudatum
Massuti Algamora, M., 1949
Estudio de diez y seis muestras de plancton del Golfo de Nápoles. Publ. Inst. Biol. Appl. 5:85-94, 1 fold-in table.

Prorocentrum compressum
Abe, Tohru H., 1967.
The armoured Dinoflagellata: II. Brorocentridae and Dinophysidae (A)°
Publs. Seto mar. biol. Lab., 14(5):369-389.

Prorocentrum dentatum
Halim, Youssef, 1960
Étude quantitative et qualitative du cycle écologique des dinoflagellés dans les eaux de Villefranche-sur-Mer. (1953-1955).
Ann. Inst. Océanogr., Monaco, 38:123-232.

Prorocentrum dentatum
LeBour, M.V., 1925
The dinoflagellates of Northern Seas. The Marine Biological Association of the United Kingdom, Plymouth, 250 pp., 35 pls.,53 text figs.

Prorocentrum dentatum
Paulsen, O., 1908
XVIII Peridiniales. Nordisches Plankton, Bot. Teil: 1-124, 155 text figs.

Prorocentrum dentatum
Schröder, B., 1900
Phytoplankton des Golfes von Neapel nebst vergleichenden Ausblicken auf das atlantischen Ozean. Mitt. Zool. Stat. Neapel, 14:1-38.

Prorocentrum gibbosum
Ercegovic, A., 1936
Etudes qualitative et quantitatives du phytoplancton dans les eaux cotières de l'Adriatique oriental moyen au cours de l'année 1934. Acta Adriatica 1(9):1-126

Prorocentrum gibbosum
Matzenauer, L., 1933
Die Dinoflagellaten des indischen Ozeans (mit Ausnahme der Gattung Ceratium.) Bot. Arch. 35:437-510, 77 text figs., 2 charts.

Prorocentrum graciale
Dangeard, P., 1927
Phytoplankton de la croisière du "Sylvana". Ann. Inst. Ocean., Monaco, n.s., 4(8):286-401, 54 text figs. (Feirer-Juin 1913).

Prorocentrum maximum
Halim, Youssef, 1960
Étude quantitative et qualitative du cycle écologique des dinoflagellés dans les eaux de Villefranche-sur-Mer. (1953-1955).
Ann. Inst. Océanogr., Monaco, 38:123-232.

Prorocentrum maximum
Hasle, Grethe Rytter, 1960
Phytoplankton and ciliate species from the Tropical Pacific.
Skr. Norske Videnskaps-Akad., Oslo, 1. Mat.-Nat. Kl., 1960(2): 1-50.

Prorocentrum maximum
Matzenauer, L., 1933
Die Dinoflagellaten des indischen Ozeans. (mit Ausnahme der Gattung Ceratium.) Bot. Arch. 35:437-510, 77 text figs., 2 charts.

Prorocentrum micans
Abe, Tohru H., 1967.
The armoured Dinoflagellata: II. Prorocentridae and Dinophysidae (A)°
Publs Seto mar. biol. Lab., 14(5):369-389.

Prorocentrum micans
Akinina, D.K., 1969.
Relative velocity of settling of dinoflagellata as dependent on their division rates. (In Russian, English abstract). Okeanologiia, 9(2): 301-305.

Prorocentrum micans
Akinina, D.K., 1966.
Dependence of light saturation of two mass species of dinoflagellates on a number of factors. (In Russian; English abstract).
Okeanologiia, Akad. Nauk, SSSR, 6(5)L861-868.

Prorocentrum micans
Allen, W.S., 1946
"Red water" in La Jolla Bay in 1945.
Trans. Amer. Micros. Soc. 65(2): 149-153

Prorocentrum micans Ehrenberg
Allen, W.E., 1943
"Red water" in La Jolla Bay in 1942.
Trans. Am. Micro. Soc. LXII (3):262-264.

Prorocentrum micans
Allen, W. E., 1942
Occurrences of "red water" near San Diego
Science, 96(2499):471.

Prorocentrum micans
Allen, W.E., 1927.
Quantitative studies on inshore marine diatoms and dinoflagellates of Southern California in 1922
Bull. S.I.O., tech. ser., 1:31-38, 2 textfigs.

Prorocentrum micans
Biecheler, B., 1952.
Recherches sur les Peridiniens.
Bull. Biol. France Belg., Suppl. 36:1-149, 75 textfg.

Prorocentrum micans
Braarud, T., 1951.
Salinity as an ecological factor in marine phytoplankton. Physiol. Plant. 4:28-34, 3 textfigs.

Prorocentrum micans

Braarud, T., 1945
A phytoplankton survey of the polluted waters of inner Oslo Fjord. Hvalrådets Skrifter, No.28, 142 pp., 19 text figs., 17 tables.

Prorocentrum micans

Braarud, T., and Adam Bursa, 1939
On the phytoplankton of the Oslo Fjord, 1933-1934. Hvalrådets Skr. No.19:1-63; 9 text figs. Reviewed. J. du. Cons. 14(3): 418-420. A.C. Gardiner.

Prorocentrum micans

Braarud, T., and Ellen Rossavik, 1952.
Observations on the marine dinoflagellate, Prorocentrum micans Ehrenb. in culture. Avhandl. Norske Videnskaps-Akad., Oslo 1951, Mat.-Naturvidens. Kl., No. 1:1-18, 8 textfigs.

Prorocentrum micans

Cupp, E., 1930
Quantitative Studies of miscellaneous series of surface catches of marine diatoms and dinoflagellates taken between Seattle and the Canal Zone from 1924 to 1928. Trans. Am. Micro. Soc., XLIX (3):238-245.

Prorocentrum micans

Dangeard, P., 1927
Phytoplankton de la croisière du "Sylvana". Ann. Inst. Ocean., Monaco, n.s., 4(8):286-401, 54 text figs. (Feirer-Juin 1913).

Prorocentrum micans

Dangeard, P., 1926
Description des Péridiniens Testacés recueillis para la Mission Charcot pendent le mois d'Aout 1924. Ann. Inst. Ocean. n.s. 3(7):307-334, 15 text figs.

Prorocentrum micans

De Angelis, C.M., 1962
Distribuzione ed ecologia di alcune specie di Dinoflagellati di acque salmastre. Pubbl. Staz. Zool., Napoli, 32 (Suppl.):301-314.

Prorocentrum micans

Delegazione Italiana della Commissione Internazionale per l'Esplorazione Scientifica del Mediterraneo, 1941
Note sul plancton della Laguna veneta. [Memoria CCLXXIX], Arch. di Ocean. e Limn. Anno I, Fasc. I, 1941 XIX: 31-57 pp.

Prorocentrum micans

de Sousa e Silva, Estela, 1965.
Note on some cytophysiological aspects in Prorocentrum micans Ehr. and Goniodoma pseudogoniaulax Biech. from cultures. Notas e Estudos do Inst. Biol. Marit., Lisboa, No. 30:30 pp., 26 pls.

Prorocentrum micans

Dodge, John D., 1965.
Thecal fine-structure in the dinoflagellate genera Prorocentrum and Exuviaella. J. mar. biol. Ass., U.K. 45(3):607-614.

Prorocentrum micans

Dos Santos Pinto, J., and E. de Sousa e Silva, 1956.
The toxicity of Cardium edule L. and its possible relation to the dinoflagellate Prorocentrum micans Ehr. Notas e Estudo, Inst. Biol. Marit., Lisboa, No. 12:20 pp.

Prorocentrum micans

Ercegovic, A., 1936
Etudes qualitative et quantitatives du phytoplancton dans les eaux cotières de l'Adriatique oriental moyen au cours de l'année 1934. Acta Adriatica 1(9):1-126

Prorocentrum micans

Forti, A., 1922
Ricerche sulla flora pelagica (fitoplancton) di Quarto dei Mille. Mem. R. Com. Talass. Ital. 97:248 pp., 13 pls.

Prorocentrum micans

Ghazzawi, F.M., 1939
Plankton of the Egyptian waters. A study of the Suez Canal Plankton. (A) Phytoplankton. Preliminary Report 83 pp. Notes and Memoires, Min. Commerce-Industry, Egypt, Hydrobiol. & Fish. 65 figs.

Prorocentrum micans

Gilbert, J.Y., and W.E. Allen, 1943
The phytoplankton of the Gulf of California obtained by the "E.W. Scripps" in 1939 and 1940. J. Mar. Res. V(2):89-110, figs.30-31.

Prorocentrum micans

Grindley, J.R., and F.J.R. Taylor, 1964.
Red water and marine fauna mortality near Cape Town. Trans. R. Soc., S. Africa, 37(2):110-130.

Prorocentrum micans

Halim, Youssef, 1960
Etude quantitative et qualitative du cycle écologique des dinoflagellés dans les eaux de Villefranche-sur-Mer. (1953-1955). Ann. Inst. Océanogr., Monaco, 38:123-232.

Prorocentrum micans

Halldal, P., 1958.
Action spectra of phototaxis and related problems in Volvocales, Ulva-gametes and Dinophyceae. Physiol. Plant., 11:118-153.

Prorocentrum micans

Holmes, R.W., P.M. Williams and R.W. Eppley, 1967.
Red water in La Jolla Bay, 1964-1965. Limnol. Oceanogr., 12(3):503-512.

Prorocentrum micans

Hope, B., 1954.
Floristic and taxonomic observations on marine phytoplankton from Nordåsvatn, near Bergen. Nytt Mag. f. Botanikk 2:149-153, 1 fig.

Prorocentrum micans

Jørgensen, E., 1905
B. Protistplankton and the diatoms in bottom samples. Hydrographical and biological investigations in Norwegian fjords. Bergens Mus. Skr. 7: 49-225.

Prorocentrum micans

Jorgensen, E., 1900
Protophyten und Protozoën im Plankton aus der Norwegischen Westkerste. Bergens Mus. Aarb. 1899(6): 95 pp., 5 pls., 83 tables.

Prorocentrum micans

Kain, Joanna M., and G.E. Fogg, 1960.
Studies on the growth of marine plankton. III. Prorocentrum micans Ehrenberg. J.M.B.A., U.K., 39:33-50.

Prorocentrum micans

Kayser, H., 1969.
Züchtungsexperimente an zwei marinen Flagellaten (Dinophyta) und ihre Anwendung im toxikologischen Abwassertest. Helgoländer wiss. Meeresunters. 19(1): 21-44

Prorocentrum micans

Kowallik, K.V. 1971.
The use of proteases for improved presentation of DNA in chromosomes and chloroplasts of Prorocentrum micans (Dinophyceae). Arch. Mikrobiol. 80(2): 154-165

Prorocentrum micans

LeBour, M.V. 1925
The dinoflagellates of Northern Seas. The Marine Biological Association of the United Kingdom, Plymouth, 250 pp., 35 pls., 53 text figs.

Prorocentrum micans

Lillick, L.C., 1940
Phytoplankton and planktonic protozoa of the offshore waters of the Gulf of Maine. Pt.II. Qualitative Composition of the Planktonic Flora. Trans. Am. Phil. Soc., n.s., 31(3):193-237, 13 text figs.

Prorocentrum micans

Lillick, L.C., 1938
Preliminary report of the phytoplankton of the Gulf of Maine. Am. Mid. Nat. 20(3):624-640, 1 text fig; 37 tables.

Prorocentrum micans

Lillick, L.C., 1937
Seasonal studies of the phytoplankton off Woods Hole, Massachusetts. Biol. Bull. LXXIII (3):488-503, 3 text figs.

Prorocentrum micans

Lindemann, E., 1925
Neubeobachtungen an den Winter peridineen des Golfes von Neapel. Bot. Arch. 9:95-102, 19 text figs.

Prorocentrum micans

Lindemann, E., 1924
Peridineen aus dem goldenen Horn und dem Bosphorus. Bot. Arch. 5:216-233, 98 text figs.

Prorocentrum micans

Macdonald, R., 1933
An examination of plankton hauls made in the Suez Canal during the year 1928. Fish. Res. Dis., Notes & Mem. No.3, 11 pp., 1 chart.

Prorocentrum micans

Mangin, M.L., 1912
Phytoplancton de la croisière du "René" dans l'Atlantique (Septembre 1908). Ann. Inst. Ocean., n.s., 4(1):1-66, 2 pls., 41 text figs., 2 tables.

Prorocentrum micans

Margalef, R., 1949
Fitoplancton nerítico de la Costa Brava en 1947-48. Publ. Inst. Biol. Aplicada, 5: 41-51, 3 text figs.

Prorocentrum micans

Marshall, S.M., 1947
An experiment in marine fish cultivation: III. The plankton of a fertilized loch. Proc. Roy. Soc., Edinburgh, Sect.B., 63, Pt.I(3):21-33, 7 text figs.

Prorocentrum micans

Matzenauer, L., 1933
Die Dinoflagellaten des indischen Ozeans (mit Ausnahme der Gattung Ceratium.) Bot. Arch. 35:437-510, 77 text figs., 2 charts.

Prorocentrum micans

Meunier, A., 1919
Micropiancton de la Mer Flamande 3. Les Péridiniens. Mem. Mus. Roy. Hist. Nat., Belgique 8(1):1-116, Pls. XV-XXI.

Prorocentrum micans

Morse, D.C., 1947
Some observations on seasonal variations in plankton population Patuxent River, Maryland 1943-1945. Bd. Nat. Res., Publ. No.65, Chesapeake Biol. Lab., 31, 3 figs.

Prorocentrum micans

Paulsen, O., 1908
XVIII Peridiniales. Nordisches Plankton, Bot. Teil: 1-124, 155 text figs.

Prorocentrum micans
Pavillard, J., 1916
Recherches sur les Péridiniens du Golfe du Lion. Mem. Univ. Montpellier. Trav. Inst. Bot., Univ. Montpellier. Serie mixte No. 4, 70 pp., 3 pls., 15 text figs.

Prorocentrum micans
Pavillard, J., 1905
Recherches sur la flore pélagique (Phytoplankton) de l'Etang de Thau. Theses presentees a la Fac. Sci., Paris, 116 pp., 3 pls.

Prorocentrum micans
Pincemin, J.-M. 1972.
Besoins en vitamines de trois organismes phytoplanctoniques, Asterionella japonica, Prorocentrum micans, Glenodinium monotis. Recherch du taux optimal de B12 pour Glenodinium monotis. Rev. int. Oceanogr. Medic. 26:85-97.

Prorocentrum micans
Polikarpov, G.G., and L.A. Lanskaia, 1961.
Culture of unicellular algae Prorocentrum micans Ehr. in bulk in the presence of S35.
Trudy Sevastopol Biol. Sta., (14):329-333.

Prorocentrum micans
Rampi, L. 1942
Ricerche sul Fitoplancton del mare Ligure. II Le tecatali e le dinofisiali delle acque di Sanremo.. Boll. Pesca, Piscicolt.,Idrobiol. (18) 16(2): 243-274, 56 figs.

Prorocentrum micans
Rampi, L., 1940
Ricerche sul Fitoplancton del mare Ligure. Boll. di Pesca, di Piscicoltura e di Idrobiologa, 18(2):1-34, 56 text figs.

Prorocentrum micans
Rodriguez Villar, Luis, 1966.
Promera cita de las especias componentes del "Huirihue o marea roja".
Est. Oceanol., Chile, 2 91-93.

Prorocentrum micans
Schodduyn, M., 1926
Observations faites dans la baie d'Ambleteuse (Pas de Calais). Bull. Inst. Ocean., Monaco, No. 482: 64 pp.

Prorocentrum micans
Schröder, B., 1900
Phytoplankton des Golfes von Neapel nebst vergleichenden Ausblicken auf das atlantischen Ozean. Mitt. Zool. Stat. Neapel, 14:1-38.

Prorocentrum micans
Sousa e Silva, E., 1949
Diatomaceas e Dinoflagelados de Baia de Cascais. Portugalise Acta Biol., Volume: Julio Henriques, Ser. B: 300-383, 9 pls, 2 fold-in tables.

Prorocentrum micans
Sousa e Silva, E., and J. Dos Santos-Pinto, 1948
O Plancton da Baia de S. Martinho do Porto. 1. Diatomaceas e Dinoflagelados. Bol. Soc. Portuguese de Ciencias Naturais, 16(2):134-187, 6 pls. (Trav. Sta. Biol. Mar. de Lisbonne No. 52).

Prorocentrum micans
Stæmann-Nielsen, Einar, 1951
The marine vegetation of the Isefjord. A study on ecology and production. Medd. Komm. Danmarks Fiskeri-og Havundersøgelser. Ser. Plankton. 5(4); 114pp., 46 text figs.

Prorocentrum micans
Vien Cao 1967.
Un mode particulier de multiplication végétative chez un péridinien libre, le Prorocentrum micans Ehrenberg.
C.r. hebd. Séanc. Acad. Sci. Paris (D) 265(2): 108-110.

Prorocentrum micans
Wang, C. C., and D. Nie, 1932
A survey of the marine protozoa of Amoy. Contr. Biol. Lab. Sci. Soc., China, Zool. Ser., 8:285-385, 89 text figs.

Prorocentrum micans
Wood, E.J.F., 1954.
Dinoflagellates in the Australian region. Australian J. Mar. Freshwater Res., 5(2):171-351

Prorocentrum micans
Zgurovskaya, L.N., and N.G. Kustenko, 1968.
The action of ammonia nitrogen on cell division, photosynthesis and the accumulation of pigments in Skeletonema costatum (Grev.), Chaetoceros sp. and Prorocentrum micans. (In Russian; English Abstract).
Okeanologiia, Akad. Nauk, SSSR, 8(1):116-125.

Prorocentrum minimum
Lillick, L.C., 1940
Phytoplankton and planktonic protozoa of the offshore waters of the Gulf of Maine. Pt.II. Qualitative Composition of the Planktonic Flora. Trans. Am. Phil. Soc., n.s., 31(3):193-237, 13 text figs.

Prorocentrum minimum
Lillick, L.C., 1938
Preliminary report of the phytoplankton of the Gulf of Maine. Am. Mid. Nat. 20(3):624-640, 1 text figs 37 tables.

Prorocentrum minimum
Lillick, L.C., 1937
Seasonal studies of the phytoplankton off Woods Hole, Massachusetts. Biol. Bull. LXXIII (3):488-503, 3 text figs.

Prorocentrum obtusidens
Hasle, Grethe Rytter, 1960
Phytoplankton and ciliate species from the Tropical Pacific.
Skr. Norske Videnskaps-Akad., Oslo, 1. Mat.-Nat. Kl., 1960(2): 1-50.

Prorocentrum obtusidens
Wood, E.J.F., 1963
Dinoflagellates in the Australian region. II. Recent collections.
Comm. Sci. and Industr. Res. Org., Div. Fish. and Oceanogr., Techn. Paper, No. 14: 55 pp.

Prorocentrum ovalis n.sp.
Rampi, L., 1940.
Ricerche sul Fitoplancton del mare Ligure. II Le tecatali e le dinofisiali delle acque di Sanremo.. Boll. Pesca, Piscicolt.,Idrobiol. (18) 16(2): 243-274, 56 figs.

Prorocentrum ovalis
Rampi, L., 1940
Ricerche sul Fitoplancton del mare Ligure. Boll. di Pesca, di Piscicoltura e di Idrobiologa, 18(2):1-34, 56 text figs.

Prorocentrum pacificum n. sp.
Wood, E.J.F., 1963
Dinoflagellates in the Australian region. II. Recent collections.
Comm. Sci. and Industr. Res. Org., Div. Fish. and Oceanogr., Techn. Paper, No. 14: 55 pp.

Prorocentrum reticulatum
Evitt, William R., and Susan E. Davidson, 1964.
Dinoflagellate studies. 1. Dinoflagellate cysts and thecae.
Stanford Univ. Publ., Geol. Sci., 10(1):1-12.

Prorocentrum rostratum
Hasle, Grethe Rytter, 1960
Phytoplankton and ciliate species from the Tropical Pacific.
Skr. Norske Videnskaps-Akad., Oslo, 1. Mat.-Nat. Kl., 1960(2): 1-50.

Prorocentrum rostratum
Rampi, L., 1950.
Péridiniens rares ou nouveaux pour le Pacifique Sud-Equatorial. Bull. Inst. Ocean., Monaco, No. 974:11 pp., 26 textfigs.

Prorocentrum rostratum
Wood, E.J.P., 1954.
Dinoflagellates in the Australian region. Australian J. Mar. Freshwater Res., 5(2):171-351.

Prorocentrum rotundatum
Rampi, L., 1950.
Péridiniens rares ou nouveaux pour le Pacifique Sud-Equatorial. Bull. Inst. Ocean., Monaco, No. 974:11 pp., 26 textfigs.

Prorocentrum rotundatum
Rampi, L. 1942.
Ricerche sul Fitoplancton del mare Ligure. II Le tecatali e le dinofisiali delle acque di Sanremo.. Boll. Pesca, Piscicolt.,Idrobiol. (18) 16(2): 243-274, 56 figs.

Prorocentrum rotundatum
Rampi, L., 1940
Ricerche sul Fitoplancton del mare Ligure. Boll. di Pesca, di Piscicoltura e di Idrobiologa, 18(2):1-34, 56 text figs.

Prorocentrum rotundatum
Rampi, L., 1939
Péridiniens rares ou intéressants recoltés dans la mer Ligure. Bull. Soc. Fran. de Microscopie, 8(2/3):106-112, 13 text figs.

Prorocentrum Schilleri
Ercegovic, A., 1936
Etudes qualitative et quantitatives du phytoplancton dans les eaux cotières de l'Adriatique oriental moyen au cours de l'année 1934. Acta Adriatica 1(9):1-126

Prorocentrum schilleri
Rampi, L. 1940.
Ricerche sul Fitoplancton del mare Ligure. II Le tecatali e le dinofisiali delle acque di Sanremo.. Boll. Pesca, Piscicolt.,Idrobiol. (18) 16(2): 243-274, 56 figs.

Prorocentrum Schilleri
Rampi, L., 1940
Ricerche sul Fitoplancton del mare Ligure. Boll. di Pesca, di Piscicoltura e di Idrobiologa, 18(2):1-34, 56 text figs.

Prorocentrum Schilleri
Rampi, L., 1939
Péridiniens rares ou intéressants recoltés dans la mer Ligure. Bull. Soc. Fran. de Microscopie, 8(2/3):106-112, 13 text figs.

Prorocentrum schilleri
Wood, E.J.F., 1963
Dinoflagellates in the Australian region. II. Recent collections.
Comm. Sci. and Industr. Res. Org., Div. Fish. and Oceanogr., Techn. Paper, No. 14: 55 pp.

Prorocentrum scutellum
De Angelis, C.M., 1962.
Distribuzione ed ecologia di alcune specie di Dinoflagellati di acque salmastre.
Pubbl. Staz. Zool., Napoli, 32 (Suppl.):301-314.

Prorocentrum scutellum

Delegazione Italiana della Commissione Internazionale per l'Esplorazione Scientifica del Mediterraneo, 1941
Note sul plancton della Laguna veneta. Memoria CCLXXXIX , Arch. di Ocean. e Limn. Anno I, Fasc. I, 1941 XIX: 31-57 pp.

Prorocentrum scutellum

Ercegovic, A., 1936
Etudes qualitative et quantitatives du phytoplancton dans les eaux cotières de l'Adriatique oriental moyen au cours de l'année 1934. Acta Adriatica 1(9):1-126

Prorocentrum scutellum

Forti, A., 1922
Ricerche sulla flora pelagica (fitoplancton) di Quarto dei Mille. Mem. R. Com. Talass. Ital. 97:248 pp., 13 pls.

Prorocentrum scutellum

LeBour, M.V., 1925
The dinoflagellates of Northern Seas. The Marine Biological Association of the United Kingdom, Plymouth. 250 pp., 35 pls., 53 text figs.

Prorocentrum Scutellum

Lillick, L.C., 1940
Phytoplankton and planktonic protozoa of the offshore waters of the Gulf of Maine. Pt.II. Qualitative Composition of the Planktonic Flora. Trans. Am. Phil. Soc., n.s. 31(3):193-237, 13 text figs.

Prorocentrum Scutellum

Lillick, L.C., 1938
Preliminary report of the phytoplankton of the Gulf of Maine. Am. Mid. Nat. 20(3):624-640, 1 text figs 37 tables.

Prorocentrum Scutellum

Lillick, L.C., 1937
Seasonal studies of the phytoplankton off Woods Hole, Massachusetts. Biol. Bull. LXXIII (3):488-503, 3 text figs.

Prorocentrum scutellum

Lindemann, E., 1924
Peridineen aus dem goldenen Horn und dem Bosphorus. Bot. Arch. 5:216-233, 98 text figs.

Prorocentrum scutellum

Margalef, R., 1949
Fitoplancton nerítico de la Costa Brava en 1947-48. Publ. Inst. Biol. Aplicada, 5: 41-51, 3 text figs.

Prorocentrum scutellum

Rampi, L., 1940
Ricerche sul Fitoplancton del mare Ligure. Boll. di Pesca, di Piscicoltura e di Idrobiologa 18(2):1-34, 56 text figs.

Prorocentrum scutellum

Paulsen, O., 1908
XVIII Peridiniales. Nordisches Plankton, Bot. Teil: 1-124, 155 text figs.

Prorocentrum scutellum

Pavillard, J., 1916
Recherches sur les Peridiniens du Golfe du Lion. Mem. Univ. Montpellier. Trav. Inst. Bot., Univ. Montpellier. Serie mixte No.4, 70 pp., 3 pls., 15 text figs.

Prorocentrum scutellum

Rampi, L. 1940
Ricerche sul Fitoplancton del mare Ligure. II Le tecatali e le dinofisiali delle acque di Sanremo.. Boll. Pesca, Piscicolt. Idrobiol. (18) 16(2): 243-274, 56 figs.

Prorocentrum scutellum

Schröder, B., 1900
Phytoplankton des Golfes von Neapel nebst vergleichenden Ausblicken auf das atlantischen Ozean. Mitt. Zool. Stat. Neapel, 14:1-38.

Prorocentrum scutellum n.sp.

Sousa a Silva, E., and J. Dos Santos-Pinto, 1948
O Plancton da Baia de S. Martinho do Porto. l. Diatomaceas e Dinoflagelados. Bol. Soc. Portuguese de Ciencias Naturais, 16(2):134-187, 6 pls. (Trav. Sta. Biol. Mar. de Lisbonne No. 52).

Prorocentrum scutellum

Wood, E.J.F., 1954
Dinoflagellates in the Australian region. Australian J. Mar. Freshwater Res. 5(2):171-351.

Prorocentrum triangulatum

Morse, D.C., 1947
Some observations on seasonal variations in plankton population Patuxant River, Maryland 1943-1945. Bd. Nat. Res., Publ. No.65, Chesapeake Biol. Lab. 31, 3 figs.

Prorocentrum triestinum

Dodge, John D., 1965
Thecal fine-structure in the dinoflagellate genera Prorocentrum and Exuviaella. J. mar. biol. Ass., U.K., 45(3):607-614.

Prorocentrum triestinum

Ercegovic, A., 1936
Etudes qualitative et quantitatives du phytoplancton dans les eaux cotières de l'Adriatique oriental moyen au cours de l'année 1934. Acta Adriatica 1(9):1-126

Prorocentrum triestinum

Halim, Youssef, 1960
Etude quantitative et qualitative du cycle écologique des dinoflagellés dans les eaux de Villefranche-sur-Mer. (1953-1955). Ann. Inst. Océanogr., Monaco, 38:123-232.

Pronoctiluca pelagica

Marshall, S. M., 1933
The production of microplankton in the Great Barrier Reef Region. Brit. Mus. (N.H.) Great Barrier Reef Exped. 1928-29, Sci. Repts. II(5):111-157, 14 text figs.

Protoceratium sp.

Balech, Enrique, 1962
Tintinnoinea y Dinoflagellata del Pacifico segun material de las Expediciones NORPAC y DOWNWIND del Instituto Scripps de Oceanografia. Revista, Mus. Argentino Ciencias Nat. "Bernardino Rivadavia", Ciencias Zool., 7(1):1-253.

Protoceratium areolatum

Balech, Enrique, 1962
Tintinnoinea y Dinoflagellata del Pacifico segun material de las Expediciones NORPAC y DOWNWIND del Instituto Scripps de Oceanografia. Revista, Mus. Argentino Ciencias Nat. "Bernardino Rivadavia", Ciencias Zool., 7(1):1-253.

Protoceratium areolatum

Matzenauer, L., 1933
Die Dinoflagellaten des indischen Ozeans (mit Ausnahme der Gattung Ceratium.) Bot. Arch. 35:437-510, 77 text figs., 2 charts.

Protoceratium areolatum

Rampi, L., 1951
Ricerche sul fitoplancton del Mare Ligure. 10) Peridiniale dell acque di Sanremo. Pubbl. Centro Talassografico Tirreno No. 6: Atti Acad. Ligure Sci. Lett. 7(1): 8 pp., Pls. 3-4.

Protoceratium areolatum

Rampi, L., 1948
Sur quelques Peridiniens rares ou interessants du Pacifique subtropical (Recoltes Alain Gerbault). Bull. l'Inst. Ocean., Monaco, No.937: 7 pp., 8 text figs.

Protoceratium densum

Gourret, P., 1883
Sur les Peridiniens du Golfe de Marseille. Ann. du Musee d'hist. Nat. Marseille, Zool., 1 (Mme. 8):1-114, 4 pls.

Protoceratium pepo

Rampi, L., 1948
Sur quelques Peridiniens rares ou interessants du Pacifique subtropical (Recoltes Alain Gerbault). Bull. l'Inst. Ocean., Monaco, No.937: 7 pp., 8 text figs.

Protoceratium reticulatum

Braarud, T., 1945
A phytoplankton survey of the polluted waters of inner Oslo Fjord. Hvalrådets Skrifter, No.28, 142 pp., 19 text figs., 17 tables.

Protoceratium reticulatum

Braarud, T., 1945
Morphological observations on marine dinoflagellate cultures (Porella perforata, Goniaulax tamarensis, Protoceratium reticulatum). Avhandl. Norske Vidensk.-Akad., Oslo, Math-Naturh. Kl., 1944(11):1-18, 4 pls., 6 textfigs.

Protoceratium reticulatum

Braarud, T., 1945
Morphological observations on marine dinoflagellate cultures. (Porella perforata, Goniaulax tamarensis, Protoceratium reticulatum). Avhandl. Norske Vidensk.-Akad., Oslo, Matem.-Naturvidensk. Kl. 1944(11):1-18.

Protoceratium reticulatum

Braarud, T., and Adam Bursa, 1939
On the phytoplankton of the Oslo Fjord, 1933-1934. Hvalrådets Skr. No.19:1-63; 9 text figs. Reviewed. J. du. Cons. 14(3): 418-420. A.C. Gardiner.

Protoceratium reticulatum

Dangeard, P., 1927
Phytoplankton de la croisière du "Sylvana". Ann. Inst. Ocean., Monaco, n.s., 4(8):286-401, 54 text figs, (-Juin 1913).

Protoceratium reticulatum

Dangeard, P., 1926
Description des Péridiniens Testacés recueillis para la Mission Charcot pendent le mois d'Aout 1924. Ann. Inst. Ocean. n.s. 3(7):307-334, 15 text figs.

Protoceratium reticulatum

Forti, A., 1922
Ricerche sulla flora pelagica (fitoplancton) di Quarto dei Mille. Mem. R. Com. Talass. Ital. 97:248 pp., 13 pls.

Protoceratium reticulatum

Jørgensen, E., 1905
B. Protistplankton and the diatoms in bottom samples. Hydrographical and biological investigations in Norwegian fjords. Bergens Mus. Skr. 7: 49-225.

Protoceratium reticulatum

Jorgensen, E., 1900
Protophyten und Protozoen im Plankton aus der Norwegischen Westküste. Bergens Mus. Aarb. 1899(6): 95 pp., 5 pls., 83 tables

Protoceratium reticulatum

LeBour, M.V., 1925
The dinoflagellates of Northern Seas. The Marine Biological Association of the United Kingdom, Plymouth. 250 pp., 35 pls., 53 text figs.

Protoceratium reticulatum

Levander, K.M., 1947
Plankton gesammelt in den Jahren 1899-1910 an den Küsten Finnlands. Finnländische Hydrographisch-Biologische Untersuchunger (aus dem Wasserbiologischen Laboratorin der Societas Scientiarum Fenniea) No.11: 40 pp., 6 diagrams, 13 pls., tables.

Protoceratium reticulatum
Lillick, L.C., 1940
Phytoplankton and planktonic protozoa of the offshore waters of the Gulf of Maine. Pt.II. Qualitative Composition of the Planktonic Flora. Trans. Am. Phil. Soc., n.s. 31(3):193-237, 13 text figs.

Protoceratium reticulatum
Lindemann, E., 1924
Peridineen aus dem goldenen Horn und dem Bosphorus. Bot. Arch. 5:216-233, 98 text figs.

Protoceratium reticulatum
Matzenauer, L., 1933
Die Dinoflagellaten des indischen Ozeans (mit Ausnahme der Gattung Ceratium.) Bot. Arch. 35:437-510, 77 text figs., 2 charts.

Protoceratium reticulatum
Meunier, A., 1919
Microplancton de la Mer Flamande 3. Les Péridiniens. Mem. Mus. Roy. Hist. Nat., Belgique 8(1):1-116, Pls. XV-XXI.

Protoceratium reticulatum
Paulsen, O., 1908
XVIII Peridiniales. Nordisches Plankton, Bot. Teil: 1-124, 155 text figs.

Protoceratium reticulatum
Pavillard, J., 1916
Recherches sur les Peridiniens du Golfe du Lion. Mem. Univ. Montpellier. Trav. Inst. Bot., Univ. Montpellier. Serie mixte No.4, 70 pp., 3 pls., 15 text figs.

Protoceratium reticulatum
Rumkówna, A., 1948
[List of the phytoplankton species occurring in the superficial water layers in the Gulf of Gdańsk] Bull. Lab. mar., Gdynia, No. 4: 139-141 with tables in back.

Protoceratium reticulatum
Schröder, B., 1900
Phytoplankton des Golfes von Neapel nebst vergleichenden Ausblicken auf das atlantischen Ozean. Mitt. Zool. Stat. Neapel, 14:1-38.

Protoceratium reticulatum
Stæmann-Nielsen, Einar, 1951
The marine vegetation of the Isefjord. A study on ecology and production. Medd. Komm. Danmarks Fiskeri-og Havundersøgelser. Ser. Plankton. 5(4); 114pp., 46 text figs.

Protoceratium reticulatum
von Stosch, H.A., 1969.
Dinoglagellatin aus der Nordsee I. Über Cachonina niei Loeblich (1968), Gonyaulax grindleyi Reinecke (1967) und eine Methode zur Darstellung von Peridineenpanzern. Helgolander wiss. Meeresuntersuch., 19(4): 558-568.

Protoceratium reticulatum
Wall, David, and Barrie Dale, 1968.
Modern dinoflagellate cysts and evolution of the Peridiniales. Micropaleontology, 14(3):265-304.

Protoceratium spinulosum
Balech, Enrique, 1962
Tintinnoinea y Dinoflagellata del Pacifico segun material de las Expediciones NORPAC y DOWNWIND del Instituto Scripps de Oceanografia. Revista, Mus. Argentino Ciencias Nat. "Bernardino Rivadavia", Ciencias Zool., 7(1):1-253.

Protodinium chattoni
Cachon, Jean et Monique Cachon, 1971
Protoodinium chattoni Hovasse. Manifestations ultrastructurales des rapports entre le Péridinien et la Méduse-hôte: fixation, phagocytose. Arch. Protistenk, Bd.113:293-305.

Protodinium simplex
Paulsen, O., 1908
XVIII Peridiniales. Nordisches Plankton, Bot. Teil: 1-124, 155 text figs.

Protoerythropsis crassicauda
Wood, E.J.F., 1963
Dinoflagellates in the Australian region. II. Recent collections. Comm. Sci. and Industr. Res. Org., Div. Fish. and Oceanogr., Techn. Paper, No. 14: 55 pp.

Protoerythropsis vigilans
LeBour, M.V., 1925
The dinoflagellates of Northern Seas. The Marine Biological Association of the United Kingdom, Plymouth, 250 pp., 35 pls., 53 text figs.

Protoperidinium irrida
Paulsen, O., 1908
XVIII Peridiniales. Nordisches Plankton, Bot. Teil: 1-124, 155 text figs.

Protopsis elongata
Wood, E.J.F., 1963
Dinoflagellates in the Australian region. II. Recent collections. Comm. Sci. and Industr. Res. Org., Div. Fish. and Oceanogr., Techn. Paper, No. 14: 55 pp.

Protopsis nigra
LeBour, M.V., 1925
The dinoflagellates of Northern Seas. The Marine Biological Association of the United Kingdom, Plymouth, 250 pp., 35 pls., 53 text figs.

Protopsis simplex n.sp.
LeBour, M.V., 1925
The dinoflagellates of Northern Seas. The Marine Biological Association of the United Kingdom, Plymouth, 250 pp., 35 pls., 53 text figs.

Protopsis simplex
Wood, E.J.F., 1963
Dinoflagellates in the Australian region. II. Recent collections. Comm. Sci. and Industr. Res. Org., Div. Fish. and Oceanogr., Techn. Paper, No. 14: 55 pp.

Pseliodinium vaubanii n.sp.
Sournia, Alain, 1972
Une période de poussées phytoplanctoniques près de Nosy-Bé (Madagascar) en 1971. 1. Espèces rares ou nouvelles du phytoplancton Cah. ORSTOM ser. Océanogr. 10(2): 151-159.

Pseudophalacroma nasutum n.gen.
Jörgensen, E., 1923
Mediterranean Dinophysiaceae. Rept. Danish Oceanogr. Expeds. 1908-10, to the Mediterranean and adjacent seas, Vol.II, Biol. J 2, 48 pp., 64 text figs.

Pseudophalacroma nasutum
LeBour, M.V., 1925
The dinoflagellates of Northern Seas. The Marine Biological Association of the United Kingdom, Plymouth, 250 pp., 35 pls., 53 text figs.

Pseudophalacroma nasutum
Tai, Si-Sun, & T. Skogsberg, 1934
Studies on the Dinophysidae, marine armored dinoflagellates of Monterey Bay, California. Arch. Protistenk. 82:380-482, 14 text figs. Pls. 11-12.

?Pseudophalacroma nasutum
Wood, E.J.F., 1954.
Dinoflagellates in the Australian region. Australian J. Mar. Freshwater Res., 5(2):171-351.

Pterosphaera möbii n.gen., n.sp.
Jorgensen, E., 1900
Protophyten und Protozoën im Plankton aus der Norwegischen Westkerste. Bergens Mus. Aarb. 1899(6): 95 pp., 5 pls., 83 tables.

Pterosphaera vanhöffeni n.sp.
Jorgensen, E., 1900
Protophyten und Protozoën im Plankton aus der Norwegischen Westkerste. Bergens Mus. Aarb. 1899(6): 95 pp., 5 pls., 83 tables.

Ptychodiscus carinatus
Matzenauer, L., 1933
Die Dinoflagellaten des indischen Ozeans (mit Ausnahme der Gattung Ceratium.) Bot. Arch. 35:437-510, 77 text figs., 2 charts.

Ptychodiscus carinatus
Rampi, L., 1945
Osservazioni sulla distribuzione qualitativa del fitoplancton nel mare Mediterraneo. Atti della Soc. Ital. di Sci. Nat. 84:105-113.

Ptychodiscus carinatus
Rampi, L., 1942
Il Fitoplancton mediterraneo: Problemi ed affinita interoceaniche. Boll. di Pesca di Piscicoltura e di Idrobiologia, Anno 18, Fasc. 4:7-19.

Ptychodiscus inflatus
Balech, Enrique, 1971.
Microplancton de la campaña oceanographica: Productividad III. Revta Mus. argent. Cienc. Nat. Bernadina Rivadavia, Hydrobiol. 3(1):1-202, 39 pls.

Ptychodiscus inflatus
Balech, Enrique, 1962
Tintinnoinea y Dinoflagellata del Pacifico segun material de las Expediciones NORPAC y DOWNWIND del Instituto Scripps de Oceanografia. Revista, Mus. Argentino Ciencias Nat. "Bernardino Rivadavia", Ciencias Zool., 7(1):1-253.

Ptychodiscus inflatus
Margalef, R., 1949
Fitoplancton nerítico de la Costa Brava en 1947-48. Publ. Inst. Biol. Aplicada, 5: 41-51, 3 text figs.

Ptychodiscus inflatus
Matzenauer, L., 1933
Die Dinoflagellaten des indischen Ozeans (mit Ausnahme der Gattung Ceratium.) Bot. Arch. 35:437-510, 77 text figs., 2 charts.

Ptychodiscus inflatus
Rampi, L., 1950.
Péridiniens rares ou nouveaux pour le Pacifique Sud-Equatorial. Bull. Inst. Ocean., Monaco, No. 974:11 pp., 26 textfigs.

Pyrocystis robusta
Sousa e Silva, E., 1949
Diatomaceas e Dinoflagelados de Baia de Cascais. Portugaliae Acta Biol. Volume: Julio Henriques, Ser. B: 300-383, 9 pls, 2 fold-in tables.

Ptychodiscus noctiluca
Balech, Enrique, 1967.
Dinoflagelados nuevos o interesantes del Golfo de Mexico y Caribe.
Revta Mus. argent. Cienc. nat. Bernardino Rivadavia Inst. nac. Invest. Cienc. nat., Hidrobiol. 2(3):77-126.

Ptychodiscus noctiluca
Murray, G., and F. G. Whitting, 1899
New Peridiniaceae from the Atlantic.
Trans. Linn. Soc., London, Bot., ser 2, 5: 321-342, Pls. 27-33, 9 tables.

Pyrocystis sp.
Taylor, F. J. R. 1972.
Unpublished observations on the thecate stage of the dinoflagellate genus Pyrocystis by the late C. A. Kofoid and Josephine Michener.
Phycologia 11 (1): 47-55.

Pyrocystis sp.
Wheeler, J.E.G., 1939
Plankton investigations. Bermuda Biological Station. Second Report. October 1939. 7 pp. (typed), 5 figs. Plymouth, Oct. 23, 1939.

Pyrocystis acuta
Rampi, L., 1945
Osservazioni sulla distribuzione qualitativa del fitoplancton nel mare Mediterraneo. Atti della Soc. Ital. di Sci. Nat. 84:105-113.

Pyrocystis acuta
Rampi, L., 1942
II Fitoplancton mediterraneo: Problemi ed affinita interoceaniche. Boll. di Pesca di Piscicoltura e di Idrobiologia, Anno 18, Fasc. 4:7-19.

Pyrocystis acuta
Swift, Elijah and David Wall 1972.
Asexual reproduction through a thecate stage in Pyrocystis acuta Kofoid, 1907 (Dinophyceae).
Phycologia 11 (1): 57-65.

Pyrocystis bicornis
Murray, G., and F. G. Whitting, 1899
New Peridiniaceae from the Atlantic.
Trans. Linn. Soc., London, Bot., ser 2, 5: 321-342, Pls. 27-33, 9 tables.

Pyrocystis elegans
Bouquaheux, Françoise 1972.
Variations morphologiques de Pyrocystis fusiformis Murray 1876 et Pyrocystis elegans Pavillard 1931.
Cah. Biol. mar. 13 (1): 1-8.

Pyrocystis elegans
Margalef, R., 1949
Fitoplancton nerítico de la Costa Brava en 1947-48. Publ. Inst. Biol. Aplicada, 5: 41-51, 3 text figs.

Pyrocystis (Dissodinium) elegans
Rampi, L., 1951.
Ricerche sul fitoplancton del Mare Ligure. 10) Peridiniale della acque di Sanremo.
Pubbl. Centro Talassografico Tirreno No. 6: Atti Acad. Ligure Sci. Lett. 7(1): 8 pp., Pls.

Pyrocystis fusiformis
Bouquaheux, Françoise 1972.
Variations morphologiques de Pyrocystis fusiformis Murray 1876 et Pyrocystis elegans Pavillard 1931.
Cah. Biol. mar. 13 (1): 1-8.

Pyrocystis fusiformis
Marukawa, H., 1921
Plankton lists and some new species of copepods, from the northern waters of Japan. Bull. Inst. Ocean., No.384, 15 pp., 3 pls., 1 chart. Monaco

Pyrocystis fusiformis
Murray, G., and F. G. Whitting, 1899
New Peridiniaceae from the Atlantic.
Trans. Linn. Soc., London, Bot., ser 2, 5: 321-342, Pls. 27-33, 9 tables.

Pyrocystis (Dissodinium) fusiformis
Rampi, L., 1951.
Ricerche sul fitoplancton del Mare Ligure. 10) Peridiniale delle acque di Sanremo.
Pubbl. Centro Talassografico Tirreno No. 6: Atti Acad. Ligure Sci. Lett. 7(1): 8 pp., Pls. 3-4.

Pyrocystis fusiformis
Swift, Elijah and Edward G. Durbin, 1972.
The phased division and cytological characteristics of Pyrocystis spp. can be used to estimate doubling times of their populations in the sea. Deep-Sea Res. 19(3): 189-198.

Pyrocystis fusiformis
Taylor, F. J. R. 1972.
Unpublished observations on the thecate stage of the dinoflagellate genus Pyrocystis by the late C. A. Kofoid and Josephine Michener.
Phycologia 11 (1): 47-55.

Pyrocystis lanceolata
Forti, A., 1922
Ricerche sulla flora pelagica (fitoplancton) di Quarto dei Mille. Mem. R. Com. Talass. Ital. 97:248 pp., 13 pls.

Pyrocystis (Dissodinium) lanceolata
Rampi, L., 1951.
Ricerche sul fitoplancton del Mare Ligure. 10) Peridiniale delle acque di Sanremo.
Pubbl. Centro Talassografico Tirreno No. 6: Atti Acad Ligure Sci. Lett. 7(1): 8 pp., Pls. 3-4.

Pyrocystis lanceolata n.sp.
Schröder, B., 1900
Phytoplankton des Golfes von Neapel nebst vergleichenden Ausblicken auf das atlantischen Ozean. Mitt. Zool. Stat. Neapel, 14:1-38.

Pyrocystis lunula
Forti, A., 1922
Ricerche sulla flora pelagica (fitoplancton) di Quarto dei Mille. Mem. R. Com. Talass. Ital. 97:248 pp., 13 pls.

Pyrocystis lunula
Fuller, C.W., Paul Kreiss and H.H. Seliger, 1972.
Particulate bioluminescence in dinoflagellates: dissociation and partial reconstitution.
Science 177(4052): 884-885.

Pyrocystis lunula
Jorgensen, E., 1900
Protophyten und Protozoën im Plankton aus der Norwegischen Westkerste. Bergens Mus. Aarb. 1899(6): 95 pp., 5 pls., 83 tables.

Pyrocystis lunula
Meunier, A., 1919
Microplancton de la Mer Flamande 3. Les Péridiniens. Mem. Mus. Roy. Hist. Nat., Belgique 8(1):1-116, Pls. XV-XXI.

Pyrocystis lunula
Murray, G., and F. G. Whitting, 1899
New Peridiniaceae from the Atlantic.
Trans. Linn. Soc., London, Bot., ser 2, 5: 321-342, Pls. 27-33, 9 tables.

Pyrocystis lunula
Paulsen, O., 1908
XVIII Peridiniales. Nordisches Plankton, Bot. Teil: 1-124, 155 text figs.

Pyrocystis lunula
Pavillard, J., 1905
Recherches sur la flore pelagique (Phytoplankton) de l'Etang de Thau. Theses presentees a la Fac. Sci., Paris, 116 pp., 3 pls.

Pyrocystis lunula globosa
Paulsen, O., 1908
XVIII Peridiniales. Nordisches Plankton, Bot. Teil: 1-124, 155 text figs.

Pyrocystis lunula lunula
Paulsen, O., 1908
XVIII Peridiniales. Nordisches Plankton, Bot. Teil: 1-124, 155 text figs.

Pyrocystis (Dissodinium) lunula
Rampi, L., 1951.
Ricerche sul fitoplancton del Mare Ligure. 10) Peridiniale delle acque di Sanremo.
Pubbl. Centro Talassografico Tirreno No. 6: Atti Acad. Ligure Sci. Lett. 7(1): 8 pp., Pls. 3-4.

Pyrocystis lunula
Rumkówna, A., 1948
List of the phytoplankton species occurring in the superficial water layers in the Gulf of Gdańsk. Bull. Lab. mar., Gdynia, No. 4: 139-141 with tables in back.

Pyrocystis lunula
Schodduyn, M., 1926
Observations faites dans la baie d'Ambleteuse (Pas de Calais). Bull. Inst. Ocean., Monaco, No. 482: 64 pp.

Pyrocystis lunula
Schröder, B., 1900
Phytoplankton des Golfes von Neapel nebst vergleichenden Ausblicken auf das atlantischen Ozean. Mitt. Zool. Stat. Neapel, 14:1-38.

Pyrocystis lunula
Swift, Elijah and W. Rowland Taylor, 1967.
Bioluminescence and chloroplast movement in the dinoflagellate Pyrocystis lunula.
J. Phycology, 3(2):77-81.

Pyrocystis noctiluca
Murray, G., and F. G. Whitting, 1899
New Peridiniaceae from the Atlantic.
Trans. Linn. Soc., London, Bot., ser 2, 5: 321-342, Pls. 27-33, 9 tables.

Pyrocystis noctiluca
Schröder, B., 1900
Phytoplankton des Golfes von Neapel nebst vergleichenden Ausblicken auf das atlantischen Ozean. Mitt. Zool. Stat. Neapel, 14:1-38.

Pyrocystis noctiluca
Swift, Elijah and Edward G. Durbin, 1972.
The phased division and cytological characteristics of Pyrocystis spp. can be used to estimate doubling times of their populations in the sea. Deep-Sea Res. 19(3): 189-198.

Pyrocystis pseudonoctiluca
Forti, A., 1922
Ricerche sulla flora pelagica (fitoplancton) di Quarto dei Mille. Mem. R. Com. Talass. Ital. 97:248 pp., 13 pls.

Pyrocystis pseudonoctiluca
Marukawa, H., 1921
Plankton lists and some new species of copepods, from the northern waters of Japan. Bull. Inst. Ocean., No.384, 15 pp., 3 pls., 1 chart. Monaco

Pyrocystis pseudonoctiluca
Pavillard, J., 1905
Recherches sur la flore pelagique (Phytoplankton) de l'Etang de Thau. Theses presentees a la Fac. Sci., Paris, 116 pp., 3 pls.

Pyrocystis pseudonoctiluca
Sukhanova, I.N., 1964.
The phytoplankton of the northeastern part of the Indian Ocean in the season of the southwest monsoon. Regularity of the distribution of oceanic plankton. (In Russian; English abstract). Trudy Inst. Okeanol., Akad. Nauk, SSSR, 65:24-31.

Pyrocystis pseudonoctiluca
Taylor F.J.R. 1972.
Unpublished observations on the thecate stage of the dinoflagellate genus Pyrocystis by the late C.A. Kofoid and Josephine Michener.
Phycologia 11(6): 47-55.

Pyrocystis (Dissodinium) robusta
Rampi, L., 1951.
Ricerche sul fitoplancton del Mare Ligure. 10) Peridiniale dell acque di Sanremo. Pubbl. Centro Talassografico Tirreno No. 6: Atti Acad. Ligure Sci. Lett. 7(1): 8 pp., Pls. 3-4.

Pyrodinium bahamense
Buchanan, R.J., 1968.
Studies at Oyster Bay in Jamaica, West Indies, IV. Observations on the morphology and a sexual cycle of Pyrodinium bahamense Plate. J. Phycol. 4(4):272-277.

Pyrodinium bahamense
Carpenter, J.H., and H.H. Seliger, 1968.
Studies at Oyster Bay in Jamaica, West Indies. 2. Effects of flow patterns and exchange on bioluminescent distributions.
J.mar.Res., 26(3):256-272.

Pyrodinium bahamense
Fuller, C.W., Paul Kreiss and H.H. Seliger, 1972.
Particulate bioluminescence in dinoflagellates: dissociation and partial reconstitution.
Science 177(4052): 884-885.

Pyrodinium bahamense
Gold, Kenneth, 1965.
A note on the distribution of luminescent dinoflagellates and water constituents in Phosphorescent Bay, Puerto Rico.
Ocean Sci. and Ocean Eng., Mar. Techn. Soc.,- Amer. Soc. Limnol. Oceanogr., 1:77-80.

Pyrodinium bahamense
Margalef, Ramón, 1961
Hidrografía y fitoplancton de un área marina de la costa meridional de Puerto Rico.
Inv. Pesq., Barcelona, 18:38-96.

Pyrodinium bahamense
McLaughlin, John J.A., and Paul A. Zahl, 1961.
In vitro culture of Pyrodinium.
Science, 134(3493):1878.
Also in:
Contrib., Inst. Mar. Sci., Univ. Puerto Rico, 3.

Pyrodinium bahamense
Plate, L., 1906. n.gen., n.sp.
Pyrodinium bahamense n.g., n.sp., die leucht-Peridinee des "Feuersees" von Nassau, Bahamas. Arch. Protistenk. 7:411-429, Pl. 19.

Pyrodinium bahamense
Seliger, H.H., J.H. Carpenter, M. Loftus and W.D. McElroy, 1970.
Mechanisms for the accumulation of high concentrations of dinoflagellates in a bioluminescent bay. Limnol. Oceanogr., 15(2): 234-245.

Pyrodinium bahamense
Seliger, H.H., and W.G. Fastie, 1968.
Studies at Oyster Bay in Jamaica, West Indies. 3. Measurement of underwater-sunlight spectra. J. mar. Res., 26(3):273-280.

Pyrodinium bahamense
Seliger, H.H., and W.D. McElroy, 1968.
Studies at Oyster Bay in Jamaica, West Indies. 1. Intensity patterns of bioluminescence in a natural environment.
J. mar. Res., 26(3):244-255.

Pyrodinium bahamense
Soli, Giogio, 1966.
Bioluminescent cycle of photosynthetic dinoflagellates. Limnol. Oceanogr., 11(3):355-363.

Pyrodinium bahamense
Taylor W. Rowland, H.H. Selinger, W.G. Fastie and W.D. McElroy, 1966.
Biological and physical observations on a phosphorescent bay in Falmouth Harbor, Jamaica, W.I.
J. Mar. Res., 24(1):28-43.

Pyrodinium bahamense
Wall, David and Barrie Dale, 1969.
The "hystrichosphaerid" resting spore of the dinoflagellate Pyrodinium bahamense, Plate 1906.
J. Phycol., 5(2): 140-149

Pyridinium pyriforme
LeBour, M.V., 1925.
The dinoflagellates of Northern Seas. The Marine Biological Association of the United Kingdom, Plymouth, 250 pp., 35 pls., 53 text figs.

Pyrophacus spp.
Steidinger, Karen A., and Joanne T. Davis, 1967.
The genus Pyrophacus, with a description of a new form.
Leaflet Ser. Fla. Bd. Conserv. 1(Phytoplankton) 1-Dinoflagettates) (3):8pp.

Pyrophacus horologicum
Dangeard, P., 1927
Phytoplankton de la croisière du "Sylvana". Ann. Inst. Ocean., Monaco, n.s., 4(8):286-401, 54 text figs. (Février-Juin 1913).

Pyrophacus horologicus
Braarud, T., 1945
A phytoplankton survey of the polluted waters of inner Oslo Fjord. Hvalrådets Skrifter, No.28, 142 pp., 19 text figs., 17 tables.

Pyrophacus horologicum
Dangeard, P., 1926
Description des Péridiniens Testacés recueillis para la Mission Charcot pendent le mois d'Aout 1924. Ann. Inst. Ocean. n.s. 3(7):307-334, 15 text figs.

Pyrophacus horologium
Forti, A., 1922
Ricerche sulla flora pelagica (fitoplancton) di Quarto dei Mille. Mem. R. Com. Talass. Ital. 97:248 pp., 13 pls.

Pyrophacus horologicum
Jørgensen, E., 1905
B. Protistplankton and the diatoms in bottom samples. Hydrographical and biological investigations in Norwegian fjords. Bergens Mus. Skr. 7: 49-225.

Pyrophacus horologicum
Kokubo, S., and S. Sato., 1947
Plankterz in JG-San Gata. Physiol. and Ecol. (Japan) 1(4):1-16, 3 text figs., tables.

Pyrophacus horologicum
Jorgensen, E., 1900
Protophyten und Protozoën im Plankton aus der Norwegischen Westkerste. Bergens Mus. Aarb. 1899(6): 95 pp., 5 pls., 83 tables.

Pyrophacus horologicum
LeBour, M.V., 1925
The dinoflagellates of Northern Seas, The Marine Biological Association of the United Kingdom. Plymouth, 250 pp., 35 pls. 53 text figs.

Pyrophacus horologium
Lillick, L.C., 1940
Phytoplankton and planktonic protozoa of the offshore waters of the Gulf of Maine. Pt.II. Qualitative Composition of the Planktonic Flora. Trans. Am. Phil. Soc., n.s., 31(3):193-237, 13 text figs.

Pyrophycus horologicum
Lindemann, E., 1925
Neubeobachtungen an den Winter peridineen des Golfes von Neapel. Bot. Arch. 9:95-102, 19 text figs.

Pyrophacus horologicum
Lindemann, E., 1924
Peridineen aus dem goldenen Horn und dem Bosphorus. Bot. Arch. 5:216-233, 98 text figs.

Pyrophacus horologicum
Mangin, M. L., 1912
Phytoplancton de la croisière du "René" dans l'Atlantique (Septembre 1908). Ann. Inst. Ocean., n.s., 4(1):1-66, 2 pls., 41 text figs., 2 tables.

Pyrophacus horologicum
Margalef, R., 1948.
Le phytoplancton estival de la "Costa Brava" catalane en 1946. Hydrobiol. 1(1):15-21.

Pyrophacus horologicum
Marshall, S. M., 1933
The production of microplankton in the Great Barrier Reef Region. Brit. Mus. (N.H.) Great Barrier Reef Exped. 1928-29, Sci. Repts. II(5):111-157, 14 text figs.

Pyrophacus horologicum
Margalef, R., 1949
Fitoplancton nerítico de la Costa Brava en 1947-48. Publ. Inst. Biol. Aplicada, 5: 41-51, 3 text figs.

Pyrophacus horologicum
Matzenauer, L., 1933
Die Dinoflagellaten des indischen Ozeans (mit Ausnahme der Gattung Ceratium.) Bot. Arch. 35:437-510, 77 text figs., 2 charts.

Pyrophacus horologicum
Marukawa, H., 1921
Plankton lists and some new species of copepods, from the northern waters of Japan. Bull. Inst. Ocean., No.384, 15 pp., 3 pls., 1 chart. Monaco

Pyrophaecus horologicum
Massutí Algamora, M., 1949
Estudio de diez y seis muestras de plancton del Golfo de Nápoles. Publ. Inst. Biol. Appl. 5:85-94, 1 fold-in table.

Pyrophacus horologicum
Meunier, A., 1919
Microplancton de la Mer Flamande 3. Les Péridiniens. Mem. Mus. Roy. Hist. Nat., Belgique 8(1):1-116, Pls. XV-XXI.

Pyrophacus horologicum
Paulsen, O., 1908
XVIII Peridiniales. Nordisches Plankton, Bot. Teil: 1-124, 155 text figs.

Pyrophacus horologicum

Pavillard, J., 1916
Recherches sur les Peridiniens du Golfe du Lion. Mem. Univ. Montpellier. Trav. Inst. Bot., Univ. Montpellier. Serie mixte No.4, 70 pp., 3 pls., 15 text figs.

Pyrophacus horologicum

Pavillard, J., 1905
Recherches sur la flore pelagique (Phytoplankton) de l'Etang de Thau. Theses presentees a la Fac. Sci., Paris, 116 pp., 3 pls.

Pyrophacus horologicum with var.

Rampi, L., 1951
Ricerche sul fitoplancton del Mare Ligure. 10) Peridiniale delle acque di Sanremo. Pubbl. Centro Talassografico Tirreno No. 6: Atti Acad. Ligure Sci. Lett, 7(1):8 pp., Pls. 3-4.

Pyrophacus horologicum

Schröder, B., 1900
Phytoplankton des Golfes von Neapel nebst vergleichenden Ausblicken auf das atlantischen Ozean. Mitt. Zool. Stat. Neapel, 14:1-38.

Pyrophacus horologicum (figs.)

Sousa e Silva, E., 1949
Diatomaceas e Dinoflagelados de Baia de Cascais. Portugaliae Acta Biol., Volume: Julio Henriques, Ser. B: 300-383, 9 pls, 2 fold-in tables.

Pyrophacus horologicum

Steidinger, Karen A., and Joanne T. Davis, 1967.
The genus Pyrophacus, with a description of a new form. Leaflet Ser. Fla. Bd. Conerv. 1(Phytoplankton)(1-Dinoflagellates) (3):8 pp.

Pyrophacus horologicum

Wall, David, and Barrie Dale 1971.
A reconsideration of living and fossil Pyrophacus Stein, 1883 (Dinophyceae). J. Phycol. 7(3): 221-235.

Pyrophacus horologicum

Wood, E.J.F., 1954.
Dinoflagellates in the Australian region. Australian J. Mar. Freshwater Res., 5(2):171-351.

Pyrophacus steinii nov. comb.

Wall, David, and Barrie Dale 1971.
A reconsideration of living and fossil Pyrophacus Stein, 1883 (Dinophyceae). J. Phycol. 7(3): 221-235.

Pyrophacus vancampoae nov. comb.

Wall, David, and Barrie Dale 1971.
A reconsideration of living and fossil Pyrophacus Stein, 1883 (Dinophyceae). J. Phycol. 7(3): 221-235.

Ptychodiscus carinatus

Pavillard, J., 1916
Recherches sur les Peridiniens du Golfe du Lion. Mem. Univ. Montpellier. Trav. Inst. Bot., Univ. Montpellier. Serie mixte No.4, 70 pp., 3 pls., 15 text figs.

Ptychodiscus inflatus n.sp.

Pavillard, J., 1916
Recherches sur les Peridiniens du Golfe du Lion. Mem. Univ. Montpellier. Trav. Inst. Bot., Univ. Montpellier. Serie mixte No.4, 70 pp., 3 pls., 15 text figs.

Roulea n. gen.

Gourret, P., 1883
Sur les Peridiniens du Golfe de Marseille. Ann. du Musee d'hist. Nat., Marseille, Zool., 1 (Mme. 8):1-114, 4 pls.

Roulea obliqua

Gourret, P., 1883
Sur les Peridiniens du Golfe de Marseille. Ann. du Musee d'hist. Nat., Marseille, Zool., 1 (Mme. 8):1-114, 4 pls.

Roulea spinifera

Gourret, P., 1883
Sur les Peridiniens du Golfe de Marseille. Ann. du Musee d'hist. Nat., Marseille, Zool., 1 (Mme. 8):1-114, 4 pls.

Scaphodinium mirabile

Cachon, Jean, et Monique Cachon 1969.
Contribution à l'étude des Noctilucidae Saville-Kent. Evolution morphologique, cytologie, systématique. II. Les Leptodiscinae Cachon J. et M. Protistologica 5(1): 11-33

Scaphodinium mirabile n.gen., n.sp.

Margalef, R., 1958?
Scaphodinium mirabile nov. gen., nov. sp., un nuevo dinoflagelado aberrante del plancton marino. Miscelánea Zoológica, Mus. Zool., Barcelona, 1(5):1-2.

Scrippsiella spp.

San Feliu, J.M., F. Muñoz y P. Suau, 1971.
Sobre la aparición de una "Purga de mar" en el puerto de Castellón. Investigación pesq. 35(2): 681-685.

Scrippsiella faeroense nov.comb.

Balech, Enrique, y Leo de Oliveira Soares, 1966.
Dos dinoflegelados de la Bahia de Guanabara y proximades (Brasil). Neotropica, 12(39):103-109.

Scrippsiella sweeneyi, n. gen., n.sp.

Balech, Enrique, 1959
Two new genera of dinoflagellates from California. Biol. Bull., 116(2): 195-203.

Sinophysis microcephalus

Nie, D., and C.C. Wang, 1944.
Dinoflagellata of the Hainan region. VIII. On Sinophysis microcephalus, a new genus and species of Dinophysidae. Sinensia 15(1/6):145-151.

Spatulodinium pseudonoctiluca n.gen.(=Gymnodinium pseudonoctiluca

Cachon, Jean et Monique avec Odette et Gérard Ferru, 1967.
Contribution à l'étude des Noctilucidae Saville-Kent. 1. Les Kofoidininae Cachon J. et M., évolution morphologique et Systématique. Protistologica 3(4):427-443.

Spiniferites bentori

Wall, David, and Barrie Dale 1970.
Living hystrichosphaerid dinoflagellate spores from Bermuda and Puerto Rico. Micropaleontology 16(1): 47-58.

Spiraulax Jollifei

Dangeard, P., 1927
Phytoplankton de la croisière du "Sylvana". Ann. Inst. Ocean., Monaco, n.s., 4(8):286-401, 54 text figs. (Janvier-Juin 1913).

Spiraulax Jollifei

Ercegovic, A., 1936
Etudes qualitative et quantitatives du phytoplancton dans les eaux cotières de l'Adriatique oriental moyen au cours de l'année 1934. Acta Adriatica 1(9):1-126

Spiraulax Jolliffei

Forti, A., 1922
Ricerche sulla flora pelagica (fitoplancton) di Quarto dei Mille. Mem. R. Com. Talass. Ital. 97:248 pp., 13 pls.

Spiraulax Jollifei

Margalef, R., 1949
Fitoplancton nerítico de la Costa Brava en 1947-48. Publ. Inst. Biol. Aplicada, 5: 41-51, 3 text figs.

Spiraulax Jolliffei

Matzenauer, L., 1933
Die Dinoflagellaten des indischen Ozeans (mit Ausnahme der Gattung Ceratium.) Bot. Arch. 35:437-510, 77 text figs., 2 charts.

Spiraulax Jolliffei

Pavillard, J., 1916
Recherches sur les Peridiniens du Golfe du Lion. Mem. Univ. Montpellier. Trav. Inst. Bot., Univ. Montpellier. Serie mixte No.4, 70 pp., 3 pls., 15 text figs.

Spiraulax jollifei

Rampi, L., 1943
Richerche sul fitoplancton del Mare Ligure. F. Le Goniaulacee delle acque di Sanremo. AHi della Soc. Ital. di Scienze Naturali, 82:1-12, figs.1-16.

Spiraulax kofoidii new name

Graham, H.W., 1942
Studies in the morphology, taxonymy, and ecology of the Peridiniales. Sci. Res. Cruise VII of the Carnegie, 1928-1929---Biol. III(542): 129 pp., 67 figs.

Spirodinium crassum

Paulsen, O., 1908
XVIII Peridiniales. Nordisches Plankton, Bot. Teil: 1-124, 155 text figs.

Spirodinium crassum

Pavillard, J., 1905
Recherches sur la flore pelagique (Phytoplankton) de l'Etang de Thau. Theses presentees a la Fac. Sci., Paris, 116 pp., 3 pls.

Spirodinium fissum

Levander, K.M., 1947
Plankton gesammelt in den Jahren 1899-1910 an den Küsten Finnlands. Finnländische Hydrographisch-Biologische Untersuchunger (aus dem Wasserbiologischen Laboratorin der Societas Scientiarum Fennica) No.11: 40 pp., 6 diagrams, 13 pls., tables.

Spirodinium fissum

Paulsen, O., 1908
XVIII Peridiniales. Nordisches Plankton, Bot. Teil: 1-124, 155 text figs.

Spirodinium fusus

Meunier, A., 1919
Microplancton de la Mer Flamande 3. Les Péridiniens. Mem. Mus. Roy. Hist. Nat., Belgique 8(1):1-116, Pls. XV-XXI.

Spirodinium spirale

Paulsen, O., 1908
XVIII Peridiniales. Nordisches Plankton, Bot. Teil: 1-124, 155 text figs.

Spirodinium spirale

Pavillard, J., 1905
Recherches sur la flore pelagique (Phytoplankton) de l'Etang de Thau. Theses presentees a la Fac. Sci., Paris, 116 pp., 3 pls.

Spirodinium spirale

Schodduyn, M., 1926
Observations faites dans la baie d'Ambleteuse (Pas de Calais). Bull. Inst. Ocean., Monaco, No. 482: 64 pp.

Spirodinium spirale

Schröder, B., 1900
Phytoplankton des Golfes von Neapel nebst vergleichenden Ausblicken auf das atlantischen Ozean. Mitt. Zool. Stat. Neapel, 14:1-38.

Steiniella fragilis

Forti, A., 1922
Ricerche sulla flora pelagica (fitoplancton) di Quarto dei Mille. Mem. R. Com. Talass. Ital. 97:248 pp., 13 pls.

Steiniella fragilis

Schröder, B., 1900
Phytoplankton des Golfes von Neapel nebst vergleichenden Ausblicken auf das atlantischen Ozean. Mitt. Zool. Stat. Neapel, 14:1-38.

Steiniella mitra

Lindemann, E., 1925
Neubeobachtungen an den Winter peridineen des Golfes von Neapel. Bot. Arch. 9:95-102, 19 text figs.

Steiniella mitra

Schröder, B., 1900
Phytoplankton des Golfes von Neapel nebst vergleichenden Ausblicken auf das atlantischen Ozean. Mitt. Zool. Stat. Neapel, 14:1-38.

Steiniella fragillis

Paulsen, O., 1908
XVIII Peridiniales. Nordisches Plankton, Bot. Teil: 1-124, 155 text figs.

Stichodiscus (Triceratium) trigonus

de Sousa e Silva, E., 1956.
Contribution à l'étude du microplancton de Dakar et des regions maritimes voisines. Bull. I.F.A.N., 8(2):335-371, 7 pls.

Torodinium robustum

LeBour, M.V., 1925
The dinoflagellates of Northern Seas. The Marine Biological Association of the United Kingdom, Plymouth, 250 pp., 35 pls., 53 text figs.

Torodinium teredo

LeBour, M.V., 1925
The dinoflagellates of Northern Seas. The Marine Biological Association of the United Kingdom, Plymouth, 250 pp., 35 pls., 53 text figs.

Triposolenia

Kofoid, C. A., 1906.
On the significance of the asymmetry in Triposolenia. Univ. Calif. Publ. Zool. 3(8):127-133.

Triposolenia

Kofoid, C. A., 1906.
A discussion of species characters in Triposolenia. 1. Nature of species characters. II. The adaptive significance of species characters. III The coincident distribution of related species. Univ. Calif. Publ. Zool. 3(7):117-126.

Triposolenia

Kofoid, C. A., 1906.
XIII. Dinoflagellata of the San Diego Region. II. On Triposolenia, a new genus of the Dinophysidae. Contributions from the Laboratory of the Marine Biological Association of San Diego. Univ. Calif. Publ. Zool. 3(6):93-116, Pls. 15-17.

Triposolenia ambulacris

Rampi, L., 1945
Osservazioni sulla distribuzione qualitativa del fitoplancton nel mare Mediterraneo. Atti della Soc. Ital. di Sci. Nat. 84:105-113.

Triposolenia ambulatrix

Jörgensen, E., 1923
Mediterranean Dinophysiaceae. Rept. Danish Oceanogr. Expeds. 1908-10, to the Mediterranean and adjacent seas, Vol.II, Biol. J 2, 48 pp., 64 text figs.

Triposolenia ambulatrix

Kofoid, C. A. and T. Skogsberg, 1928
XXXV. The Dinoflagellata: The Dinophysiodae. Reports on the scientific results of the expedition to the Eastern Tropical Pacific, in charge of Alexander Agassiz, by the U. S. Fish Commission Steamer "Albatross" from October 1904 to March 1905----. Mem. M. C. Z. 51:766 pp., 31 pls.

Triposolenia ambulatrix

Rampi, L., 1945
Osservazioni sulla distribuzione qualitativa del fitoplancton nel mare Mediterraneo. Atti della Soc. Ital. di Sci. Nat. 84:105-113.

Triposolenia ambulatrix

Rampi, L., 1942
Il Fitoplancton mediterraneo: Problemi ed affinita interoceaniche. Boll. di Pesca di Piscicoltura e di Idrobiologia, Anno 18, Fasc. 4:7-19.

Triposolenia bicornis

Abe, Tohru H., 1967.
The armoured Dinoflagellata: II. Prorocentridae and Dinophysidae (C) - Ornithocercus, Histioneis, Amphisolenia and others. Publs Seto mar. biol. Lab., 15(2):79-116.

Triposolenia bicornis

Jörgensen, E., 1923
Mediterranean Dinophysiaceae. Rept. Danish Oceanogr. Expeds. 1908-10, to the Mediterranean and adjacent seas, Vol.II, Biol. J 2, 48 pp., 64 text figs.

Triposolenia bicornis n. sp.

Kofoid, C. A., 1906.
XIII. Dinoflagellata of the San Diego Region. II. On Triposolenia, a new genus of the Dinophysidae. Contributions from the Laboratory of the Marine Biological Association of San Diego. Univ. Calif. Publ. Zool. 3(6):93-116, Pls. 15-17.

Triposolenia bicornis

Kofoid, C. A. and T. Skogsberg, 1928
XXXV. The Dinoflagellata: The Dinophysiodae. Reports on the scientific results of the expedition to the Eastern Tropical Pacific, in charge of Alexander Agassiz, by the U. S. Fish Commission Steamer "Albatross" from October 1904 to March 1905----. Mem. M. C. Z. 51:766 pp., 31 pls.

Triposolenia bicornis

Rampi, L., 1945
Osservazioni sulla distribuzione qualitativa del fitoplancton nel mare Mediterraneo. Atti della Soc. Ital. di Sci. Nat. 84:105-113.

Triposolenia bicornis

Rampi, L., 1942
Il Fitoplancton mediterraneo: Problemi ed affinita interoceaniche. Boll. di Pesca di Piscicoltura e di Idrobiologia, Anno 18, Fasc. 4:7-19.

Triposolenia depressa

Balech, Enrique, 1962
Tintinnoinea y Dinoflagellata del Pacifico segun material de las Expediciones NORPAC y DOWNWIND del Instituto Scripps de Oceanografia. Revista, Mus. Argentino Ciencias Nat. "Bernardino Rivadavia", Ciencias Zool., 7(1):1-253.

Triposolenia depressa, n. sp

Kofoid, C. A., 1906.
XIII. Dinoflagellata of the San Diego Region. II. On Triposolenia, a new genus of the Dinophysidae. Contributions from the Laboratory of the Marine Biological Association of San Diego. Univ. Calif. Publ. Zool. 3(6):93-116, Pls. 15-17.

Triposolenia depressa

Kofoid, C. A. and T. Skogsberg, 1928
XXXV. The Dinoflagellata: The Dinophysiodae. Reports on the scientific results of the expedition to the Eastern Tropical Pacific, in charge of Alexander Agassiz, by the U. S. Fish Commission Steamer "Albatross" from October 1904 to March 1905----. Mem. M. C. Z. 51:766 pp., 31 pls.

Triposolenia exilis n.sp.

Kofoid, C. A., 1906.
XIII. Dinoflagellata of the San Diego Region. II. On Triposolenia, a new genus of the Dinophysidae. Contributions from the Laboratory of the Marine Biological Association of San Diego. Univ. Calif. Publ. Zool. 3(6):93-116, Pls. 15-17.

Triposolenia intermedia n.sp.

Kofoid, C. A. and T. Skogsberg, 1928
XXXV. The Dinoflagellata: The Dinophysiodae. Reports on the scientific results of the expedition to the Eastern Tropical Pacific, in charge of Alexander Agassiz, by the U. S. Fish Commission Steamer "Albatross" from October 1904 to March 1905----. Mem. M. C. Z. 51:766 pp., 31 pls.

Triposolenia longicornis

Kofoid, C. A. and T. Skogsberg, 1928
XXXV. The Dinoflagellata: The Dinophysiodae. Reports on the scientific results of the expedition to the Eastern Tropical Pacific, in charge of Alexander Agassiz, by the U. S. Fish Commission Steamer "Albatross" from October 1904 to March 1905----. Mem. M. C. Z. 51:766 pp., 31 pls.

Triposolenia ramiciformis n.sp.

Kofoid, C. A., 1906.
XIII. Dinoflagellata of the San Diego Region. II On Triposolenia, a new genus of the Dinophysidae. Contributions from the Laboratory of the Marine Biological Association of San Diego. Univ. Calif Publ. Zool. 3(6):93-116, Pls. 15-17.

Triposolenia ramiciformis

Kofoid, C. A. and T. Skogsberg, 1928
XXXV. The Dinoflagellata: The Dinophysiodae. Reports on the scientific results of the expedition to the Eastern Tropical Pacific, in charge of Alexander Agassiz, by the U. S. Fish Commission Steamer "Albatross" from October 1904 to March 1905----. Mem. M. C. Z. 51:766 pp., 31 pls.

Triposolenia truncata

Jörgensen, E., 1923
Mediterranean Dinophysiaceae. Rept. Danish Oceanogr. Expeds. 1908-10, to the Mediterranean and adjacent seas, Vol.II, Biol. J 2, 48 pp., 64 text figs.

Triposolenia truncata n.sp.

Kofoid, C. A., 1906.
XIII. Dinoflagellata of the San Diego Region. II. On Triposolenia, a new genus of the Dinophysidae. Contributions from the Laboratory of the Marine Biological Association of San Diego. Univ. Calif. Publ. Zool. 3(6):93-116, Pls. 15-17.

Triposolenia truncata

Käsler, R., 1938
Die Verbreitung der Dinophysiales im Sudatlantischen Ozean. Wiss. Ergeb. Deutschen Atlantischen Expedition----"Meteor" 1925-1927, 12(2):162-237, text figs. 85-118.

Triposolenia truncata

Kofoid, C. A. and T. Skogsberg, 1928
XXXV. The Dinoflagellata: The Dinophysiodae. Reports on the scientific results of the expedition to the Eastern Tropical Pacific, in charge of Alexander Agassiz, by the U. S. Fish Commission Steamer "Albatross" from October 1904 to March 1905----. Mem. M. C. Z. 51:766 pp., 31 pls.

Triposolenia truncata

Rampi, L., 1945
Osservazioni sulla distribuzione qualitativa del fitoplancton nel mare Mediterraneo. Atti della Soc. Ital. di Sci. Nat. 84:105-113.

Triposolenia truncata

Rampi, L., 1942
Il Fitoplancton mediterraneo: Problemi ed affinita interoceaniche. Boll. di Pesca di Piscicoltura e di Idrobiologia, Anno 18, Fasc. 4:7-19.

Warnowia atra

Wood, E.J.F., 1963
Dinoflagellates in the Australian region.
II. Recent collections.
Comm. Sci. and Industr. Res. Org., Div.
Fish. and Oceanogr., Techn. Paper, No. 14:
55 pp.

Warnowia parva

Hulburt, E.M., 1957.
The taxonomy of unarmored Dinophyceae of shallow embayments on Cape Cod, Massachusetts.
Biol. Bull., 112(2):196-219.

Warnowia pulchra

Greuet, Claude, 1968.
Organisation ultrastructurale de l'ocelle de deux peridiniens warnowiidae, Erythropsis pavillardi Kofoid et Swezy et Warnowia pulchra Schiller.
Protistologica 4(2):209-228.

Warnowia rosea

Wood, E.J.F., 1963
Dinoflagellates in the Australian region.
II. Recent collections.
Comm. Sci. and Industr. Res. Org., Div.
Fish. and Oceanogr., Techn. Paper, No. 14:
55 pp.

Warnowia rubescens

Hada, Yoshina, 1970. (1970)
The protozoan plankton of the Antarctic and Subantarctic seas.
Scient. Repts, Japan. Antarct. Res. Exped., (E)31:51 pp.

Warnowia subnigra

Wood, E.J.F., 1963
Dinoflagellates in the Australian region.
II. Recent collections.
Comm. Sci. and Industr. Res. Org., Div.
Fish. and Oceanogr., Techn. Paper, No. 14:
55 pp.

Warnowia violescens

Wood, E.J.F., 1963
Dinoflagellates in the Australian region.
II. Recent collections.
Comm. Sci. and Industr. Res. Org., Div.
Fish. and Oceanogr., Techn. Paper, No. 14:
55 pp.

Warnowia voracis

Wood, E.J.F., 1963
Dinoflagellates in the Australian region.
II. Recent collections.
Comm. Sci. and Industr. Res. Org., Div.
Fish. and Oceanogr., Techn. Paper, No. 14: 55 pp.

Woloszynskia micra

Leadbeater, B., and J.D. Dodge, 1967.
Fine structure of the dinoflagellate transverse flagellum.
Nature, Lond., 213(5074):421-422.

Xanthidium brachiolatum

Pavillard, J., 1905
Recherches sur la flore pelagique (Phytoplankton) de l'Etang de Thau. Theses presentees a la Fac. Sci., Paris, 116 pp., 3 pls.

Xanthidium coronatum n.sp.

Pavillard, J., 1905
Recherches sur la flore pelagique (Phytoplankton) de l'Etang de Thau. Theses presentees a la Fac. Sci., Paris, 116 pp., 3 pls.

Xanthidium multispinosum

Pavillard, J., 1905
Recherches sur la flore pelagique (Phytoplankton) de l'Etang de Thau. Theses presentees a la Fac. Sci., Paris, 116 pp., 3 pls.

silicoflagellates

Brunel, J., 1962
Le phytoplancton de la Baie de Chaleurs.
Inst. Botan., Univ. Montréal, Contrib.
No. 77: 365 pp., 66 pls.

silicoflagellates

Coste, Bernard, Hans-Joachim Minas et Pierre Nival 1969.
Distribution superficielle des taux de production organique primaire et des silicoflagellés entre la Sardaigne et la Tunisie (fevrier 1968).
Tethys 1(3): 573-580.

silicoflagellates

Dumitrica, Paulian, 1973
Paleocene, Late Oligocene and Post-Oligocene silicoflagellates in southwestern Pacific sediments cored on DSDP leg 21. Initial Repts, Deep Sea Drilling Project, 21:837-883.

silicoflagellates

Frenguelli, J., 1960.
Diatomeas y silicoflagelados recogidas en Tierra Adélia durante las Expediciones Polares Francesas de Paul-Emile VICTOR (1950-1952).
Revue Algologique, (1):1-47.

silicoflagellates, lists of spp.

Frenguelli, Joaquin, and Hector Antonio Orlando, 1959.
Operacion MERLUZA. Diatomeas y silicoflagelados del plancton del "VI Crucero".
Servicio Hidrogr. Naval., Argentina, Publ. No. H. 619: 5-62.

silicoflagellates

Frenguelli, J. and H. A. Orlando, 1958.
Diatomeas y silicoflagelados del Sector Antartico Sudamericano.
Publ. Inst. Ant. Argentino, 5: 191 pp.

silicoflagellates

Gemeinhardt, K., 1934.
Die Silicoflagellaten des Südatlantischen Ozeans Wiss. Ergeb. Deutschen Atlantischen Exped. "Meteor", 1925-1927, 12(1):274-312.

silicoflagellates

Glezer Z.I. 1966.
Silicoflagellatophyceae. (In Russian)
Flora sporovykh rastenii, SSSR,
Akad. Nauk. Inst IMYA Komarova, 7.
Translation: Israel Program for Scientific Translations, Jerusalem 1970: 363 pp.

silicoflagelates

Herrera, Juan, y Ramon Margalef, 1963
Hidrografía y fitoplancton de la costa comprendida entre Castellón y la desembocadura del Ebro, de julio de 1960 a junio de 1961.
Inv. Pesq., Barcelona, 24:33-112.

silicoflagellates

Hovasse, R., 1946.
Flagellés a squelette silicieux: Silicoflagellés et Ebridies provenant du plancton recueilli au cours des campagnes scientifiques du Prince Albert ler de Monaco (1885-1912). Rés. Camp. Sci. Monaco, No. 107:19 pp., 1 pl.

silicoflagellates, list of spp

Kozlova, O.G., and V.V. Mukhina, 1967.
Diatoms and silicoflagellates in suspension and floor sediments of the Pacific Ocean.
Int. Geol. Rev., 9(10):1322-1342.

(Translated from: Geokhimiya Kremnezema, NK 65-12 (51):192-218 (1966).

silicoflagellates

Kozlova, O.G., and V.V. Mukhina, 1966.
Diatoms and silicoflagellates in suspension and in the bottom sediments of the Pacific Ocean. (In Russian).
In: Geochemistry of silica, N.M. Strakhov, editor, Isdatel. "Nauka", Moskva, 192-218.

silicoflagellates

Kruger, D., 1950.
Variations quantitatives des protistes marins au voisinage du Port d'Alger durant l'hiver 1949-1950. Bull. Inst. Océan., Monaco, No. 978:20 pp., 5 textfigs.

silicoflagellates

*Loeblich, Alfred R. III, Laurel A. Loeblich, Helen Tappan and Alfred R. Loeblich, Jr., 1968.
Annotated index of fossil and Recent silicoflagellates and ebridians with descriptions and illustrations of validly proposed taxa.
Mem. geol. Soc. Am., 106:319 pp.

silicoflagellata

Marshall, S. M., 1934
The silicoflagellata and tintinnoinea.
Great Barrier Reef Exped., 1928-29. Sci. Repts. 4(15):623-664, 43 text figs.

silicoflagellates

Martini, Erlend, 1971.
Neogene silicoflagellates from the equatorial Pacific. Initial Repts Deep Sea Drill. Proj. 7(2): 1695-1708.

silicoflagellates

Mukina, V.V., 1971.
Problems of diatom and silicoflagellate Quaternary stratigraphy in the equatorial Pacific Ocean. In: Micropaleontology of oceans, B.M. Funnell and W.R. Riedel, editors, Cambridge Univ. Press. 423-431.

silicoflagellates

Navarro, F. de P., and L. Bellon Uriarte, 1945.
Catálogo de la flora del Mar de Baleares (con exclusion de las diatomeas), Notas y Res. Inst. Español Ocean., 2nd ser., No. 124:160-295.

silicoflagellates

Travers, A., et M. Travers 1968.
Les silicoflagellés du Golfe de Marseille.
Mar. Biol. 1(4):285-288.

silicoflagellates

Wood, E.J. Ferguson 1968.
Dinoflagellates of the Caribbean Sea and adjacent areas.
Univ. Miami Press, 143 pp.

silicoflagellate distribution

Kozlova, O.G., 1971.
The main features of diatom and silicoflagellate distribution in the Indian Ocean. In: Micropalaeontology of oceans, B.M. Funnell and W.R. Riedel, editors, Cambridge Univ. Press, 271-275.

silicoflagellates, lists of spp

Balech, Enrique, 1971.
Microplancton de la campaña oceanografica: Productividad III.
Revta Mus. argent. Cienc. Nat. Bernadina Rivadavia, Hydrobiol. 3(1):1-202, 39 pls.

silicoflagellates, lists of spp.

Falcão Paredes, Jorge 1969/70.
Subsídios para o conhecimento do plancton marinho de Cabo Verde. I. Diatomáceas, Silicoflagelados e Dinoflagelados.
Memórias. Inst. Invest. cient. Moçambique (A) 10: 3-107.

silicoflagellates, lists of spp.

Frenguelli, Joaquin, y Hector A. Orlando, 1958.
Diatomeas y silicoflagelados del sector Antartico Sudamericano.
Inst. Antartico Argentino, Publ., No. 5:191 pp.

silicoflagellates, lists of spp.

*Hada, Yoshina, 1970. (1970)
The protozoan plankton of the Antarctic and Subantarctic seas.
Scient. Repts, Japan. Antarct. Res. Exped., (E)31:51 pp.

silicoflagellates, lists of spp.

Lackey, James B., and Elsie W. Lackey, 1963
Microscopic algae and protozoa in the waters near Plymouth in August 1962.
J. Mar. Biol. Assoc., U.K., 43(3):797-805.

silicoflagellates, lists of spp.

Lecal, J. 1967.
Le nannoplancton des côtes d'Israel.
Hydrobiologia 29(3/4): 305-387.

Silicoflagellata, lists of species

Skolka, V.H., 1961
Données sur le phytoplancton des parages prébosphoriques de la Mer Noire.
Rapp. Proc. Verb., Réunions, Comm. Int. Expl. Sci. Mer. Méditerranée, Monaco, 16(2):129-132.

silicoflagellates, lists of sp.

Stroukuna, V.G., 1950.
[Phytoplankton of the Black Sea in the vicinity of Karadaga and its seasonal dynamics.]
Trudy Karadagsk Biol. Sta., 10:38-52.

silicoflagellates, lists of spp.

Magazzù, Giuseppe, e Carlo Andreoli 1971.
Trasferimenti fitoplanctonici attraverso lo Stretto di Messina in relazione alle condizioni idrologiche.
Boll. Pesca Piscic. Idrobiol. 26(1/2): 125-193

Cannopilus hemisphaericus

Lemmermann, E., 1908.
XXI. Flagellatae, Chlorophyceae, Coccosphaerales und Silicoflagellatae. Nordisches Plankton, Bot. Teil:1-40, 135 textfigs.

Dictyocha

Brunel, J., 1962
Le phytoplancton de la Baie de Chaleurs.
Inst. Botan., Univ. Montréal, Contrib. No. 77: 365 pp., 66 pls.

Dictyocha bioctonarius

Marshall, S. M., 1934
The silicoflagellata and tintinnoinea.
Great Barrier Reef Exped., 1928-29. Sci. Repts. 4(15):623-664, 43 text figs.

Dictiocha fibula

Ercegovic, A., 1936
Etudes qualitative et quantitatives du phytoplancton dans les eaux cotières de l'Adriatique oriental moyen au cours de l'année 1934. Acta Adriatica 1(9):1-126

Dictyocha fibula

Forti, A., 1922
Ricerche sulla flora pelagica (fitoplancton) di Quarto dei Mille. Mem. R. Com. Talass. Ital. 97:248 pp., 13 pls.

Dictyocha fibula

Gran, H.H., and T. Braarud, 1935
A quantitative study of the phytoplankton in the Bay of Fundy and the Gulf of Maine (including observations on hydrography, chemistry, and turbidity).
J. Biol. Bd., Canada, 1(5):279-467, 69 text figs.

Dictyocha fibula

Herrera, Juan, y Ramón Margalef, 1963
Hidrografía y fitoplancton de la costa comprendida entre Castellón y la desembocadura del Ebro, de julio de 1960 a junio de 1961.
Inv. Pesq., Barcelona, 24:33-112.

Dictyocha fibula

Hovasse, R., 1946.
Flagellés à squelette siliceux: Silicoflagellés et Ebriidés provenant du plancton recueilli au cours des campagnes scientifiques du Prince Albert 1er de Monaco (1885-1912). Rés. Camp. Sci. Monaco, No. 107:19 pp., 1 pl.

Dictyocha fibula

Jørgensen, E., 1905
B. Protistplankton and the diatoms in bottom samples. Hydrographical and biological investigations in Norwegian fjords. Bergens Mus. Skr. 7: 49-225.

Dictyocha fibula

Jorgensen, E., 1900
Protophyten und Protozoën im Plankton aus der Norwegischen Westkerste. Bergens Mus. Aarb. 1899(6): 95 pp., 5 pls., 83 tables.

Dictyocha fibula

Lillick, L.C., 1938
Preliminary report of the phytoplankton of the Gulf of Maine. Am. Mid. Nat. 20(3):624-640, 1 text figs 37 tables.

Dictyocha fabula

Schröder, B., 1900
Phytoplankton des Golfes von Neapel nebst vergleichenden Ausblicken auf das atlantischen Ozean. Mitt. Zool. Stat. Neapel, 14:1-38.

Dictyocha fibula

Travers, A., et M. Travers 1968.
Les silicoflagellés du Golfe de Marseille.
Mar. Biol. 1(4): 285-288.

Dictyocha fibula

Van Valkenburg, Shirley D. 1971.
Observations on the fine structure of Dictyocha fibula Ehrenberg. 1. The skeleton. 2. The protoplast.
J. Phycol. 7(2): 113-118; 118-132.

Dictyocha fibula

Van Valkenburg, Shirley D., and Richard E. Norris 1970.
The growth and morphology of the silicoflagellate Dictyocha fibula Ehrenberg in culture.
J. Phycol. 6(1): 48-54.

Dictyocha fibula var. messanensis

Lillick, L.C., 1938
Preliminary report of the phytoplankton of the Gulf of Maine. Am. Mid. Nat. 20(3):624-640, 1 text figs 37 tables.

Dictyocha Fibula

Lillick, L.C., 1937
Seasonal studies of the phytoplankton off Woods Hole, Massachusetts. Biol. Bull. LXXIII (3):488-503, 3 text figs.

Dictyocha fibula

Nival, Paul, 1965.
Sur le cycle de Dictyocha fibula Ehrenberg dans les eaux de surface de la rade de Villefranche-sur-Mer.
Cahiers, Biol. Mar., Roscoff, 6(1):67-85.

Dictyocha fibula messanensis

Pavillard, J., 1905
Recherches sur la flore pélagique (Phytoplankton) de l'Etang de Thau. Theses presentees a la Fac. Sci., Paris, 116 pp., 3 pls.

Dictyocha fibula

Rampi, L., 1948
Ricerche sul fitoplancton del Mar Ligure. (8)1 Silicoflagellati della acque di Sanremo. Atti della Società Italiani di Scienze Naturali, 87:4 pp., 9 text figs.

Dictyocha fibula

Gemeinhardt, K., 1934.
Die Silicoflagellaten des Südatlantischen Ozeans.
Wiss. Ergeb. Deutschen Atlantischen Exped. "Meteor", 1925-1927, 12(1):274-312.

forma aspera
forma rhombica
var. messensis
forma spinosa
var. stapedia
var. aculeata

Dictyocha fibula and forma

Lemmermann, E., 1908.
XXI. Flagellatae, Chlorophyceae, Coccosphaerales und Silicoflagellatae. Nordisches Plankton, Bot. Teil:1-40, 135 textfigs.

Dictyocha fibula var. aculeata

Rampi, L., 1948
Ricerche sul fitoplancton del Mar Ligure. (8)1 Silicoflagellati della acque di Sanremo. Atti della Società Italiani di Scienze Naturali, 87:4 pp., 9 text figs.

Dictyocha fibula hexagona n.var.

Marshall, S. M., 1934
The silicoflagellata and tintinnoinea.
Great Barrier Reef Exped., 1928-29. Sci. Repts. 4(15):623-664, 43 text figs.

Dictyocha fibula longispina

Meunier, A., 1919
Microplankton de la Mer Flamande. 4. Les Tintinnides et Coetera. Mem. Mus. Roy. Hist. Nat., Belgique, 8(2):59 pp., Pls. 22-23.

Dictyocha fibula var. pentagona

Rampi, L., 1948
Ricerche sul fitoplancton del Mar Ligure. (8)1 Silicoflagellati della acque di Sanremo. Atti della Società Italiani di Scienze Naturali, 87:4 pp., 9 text figs.

Dictyocha fibula var. messanensis

Rampi, L., 1948
Ricerche sul fitoplancton del Mar Ligure. (8)1 Silicoflagellati della acque di Sanremo. Atti della Società Italiani di Scienze Naturali, 87:4 pp., 9 text figs.

Dictyocha fibula spinosa

Forti, A., 1922
Ricerche sulla flora pelagica (fitoplancton) di Quarto dei Mille. Mem. R. Com. Talass. Ital. 97:248 pp., 13 pls.

Dictyocha fibula stapedia

Marshall, S. M., 1934
The silicoflagellata and tintinnoinea.
Great Barrier Reef Exped., 1928-29. Sci. Repts. 4(15):623-664, 43 text figs.

Dictyocha fibula var. stapedia

Rampi, L., 1948
Ricerche sul fitoplancton del Mar Ligure. (8)1 Silicoflagellati della acque di Sanremo. Atti della Società Italiani di Scienze Naturali, 87:4 pp., 9 text figs.

Dictyocha navicula
Lemmermann, E., 1908.
XXI. Flagellatae, Chlorophyceae, Coccosphaerales, und Silicoflagellatae. Nordisches Plankton, Bot. Teil:1-40, 135 textfigs.

Dictyocha octonaria
Balech, Enrique, 1964.
El plancton de Mar del Plata durante el periodo 1961-1962.
Bol. Inst. Biol. Mar., Buenos Aires, (4):56 pp.

Dictyocha octonarius
Rampi, L., 1948
Ricerche sul fitoplancton del Mar Ligure. (8)1 Silicoflagellati della acque di Sanremo.
Atti della Società Italiani di Scienze Naturali, 87:4 pp., 9 text figs.

Dictyocha polyactis
Rampi, L., 1948
Ricerche sul fitoplancton del Mar Ligure. (8)1 Silicoflagellati della acque di Sanremo.
Atti della Società Italiani di Scienze Naturali, 87:4 pp., 9 text figs.

Dictyocha quadrata
Lemmermann, E., 1908.
XXI. Flagellatae, Chlorophyceae, Coccosphaerales und Silicoflagellatae. Nordisches Plankton, Bot. Teil:1-40, 135 textfigs.

Dictyocha speculum
Avaria P., Sergio 1970.
Fitoplancton de la expedición del Dona Berta en la zona Puerto Montt-Aysen.
Rev. Biol. mar. Valparaiso 14(2): 1-17.

Dictyocha speculum
Boney, A.D., 1973
Observations on the silicoflagellate Dictyocha speculum Ehrenb. from the Firth of Clyde. J. mar. biol. Ass. U.K. 53(2):263-268.

Distephanus speculum
Braarud, T., 1945
A phytoplankton survey of the polluted waters of inner Oslo Fjord. Hvalrådets Skrifter, No.28, 142 pp., 19 text figs., 17 tables.

Dictyocha speculum
Gran, H.H., 1897
Protophyta: Diatomaceae, Silico-flagellata and Cilioflagellata. Den Norske Nordhavs Expedition 1876-1878, h. 24, 36 pp., 4 pls.

Dictyocha speculum
Hovasse, R., 1946.
Flagellés à squelette silicieux: Silicoflagellés et Ebriidés provenant du plancton recueilli au cours des campagnes scientifiques du Prince Albert ler de Monaco. Rés. Camp. Sci., Monaco, 107:19 pp., 1 pl. (1885-1912)

Distephanus speculum
Iselin, C., 1930
A report on the coastal waters of Labrador based on explorations of the "Chance" during the summer of 1926. Proc. Am. Acad. Arts Sci., 66(1):1-37, 14 text figs.

Distephanes speculum
Jorgensen, E., 1900
Protophyten und Protozoën im Plankton aus der Norwegischen Westkerste. Bergens Mus. Aarb. 1899(6): 95 pp., 5 pls., 83 tables.

Dictyocha speculum
Rampi, L., 1948
Ricerche sul fitoplancton del Mar Ligure. (8)1 Silicoflagellati della acque di Sanremo.
Atti della Società Italiani di Scienze Naturali, 87:4 pp., 9 text figs.

Dictyocha speculum
Schröder, B., 1900
Phytoplankton des Golfes von Neapel nebst vergleichenden Ausblicken auf das atlantischen Ozean. Mitt. Zool. Stat. Neapel, 14:1-38.

Dictiocha staurodon
Ercegovic, A., 1936
Etudes qualitative et quantitatives du phytoplancton dans les eaux cotières de l'Adriatique oriental moyen au cours de l'année 1934. Acta Adriatica 1(9):1-126

Dictyocha staurodon
Lemmermann, E., 1908.
XXI. Flagellatae, Chlorophyceae, Coccosphaerales und Silicoflagellatae. Nordisches Plankton, Bot. Teil:1-40, 135 textfigs.

Dictyocha Staurodon
Pavillard, J., 1905
Recherches sur la flore pelagique (Phytoplankton) de l'Etang de Thau. Theses presentees a la Fac. Sci., Paris, 116 pp., 3 pls.

Distephanus speculum
*Hada, Yoshine, 1970.   (1970)
The protozoan plankton of the Antarctic and Subantarctic seas.
Scient. Repts, Japan. Antarct. Res. Exped., (E)31:51 pp.

Distephanum sp.
Morse, D.C., 1947
Some observations on seasonal variations in plankton population Patuxant River, Maryland 1943-1945. Bd. Nat. Res., Publ. No.65, Chesapeake Biol. Lab., 31, 3 figs.

Distephanus crux
Ercegovic, A., 1936
Etudes qualitative et quantitatives du phytoplancton dans les eaux cotières de l'Adriatique oriental moyen au cours de l'année 1934. Acta Adriatica 1(9):1-126

Distephanes crux
Lemmermann, E., 1908.
XXI. Flagellatae, Chlorophyceae, Coccosphaerales und Silicoflagellatae. Nordisches Plankton, Bot. Teil:1-40, 135 textfigs.

Distephanus crus
Forti, A., 1922
Ricerche sulla flora pelagica (fitoplancton) di Quarto dei Mille. Mem. R. Com. Talass. Ital. 97:248 pp., 13 pls.

Distephanus speculum
Brunel, J., 1962
Le phytoplancton de la Baie de Chaleurs. Inst. Botan., Univ. Montréal, Contrib. No. 77: 365 pp., 66 pls.

Distephanus speculum
Ercegovic, A., 1936
Etudes qualitative et quantitatives du phytoplancton dans les eaux cotières de l'Adriatique oriental moyen au cours de l'année 1934. Acta Adriatica 1(9):1-126

Distephanes speculum
Gemeinhardt, K., 1934.
Die Silicoflagellaten des Südatlantischen Ozeans. Wiss. Ergeb. Deutschen Atlantischen Exped. "Meteor", 1925-1927, 12(1):274-312.
forma brevispinosa
forma robusta
var. regularis
var. pentagonus
var. septenarius

Distephanus speculum
Gran, H.H., 1897
Protophyta: Diatomaceae, Silico-flagellata and Cilioflagellata. Den Norske Nordhavs Expedition 1876-1878, h. 24, 36 pp., 4 pls.

Distephanus speculum
Gran, H.H., and T. Braarud, 1935
A quantitative study of the phytoplankton in the Bay of Fundy and the Gulf of Maine (including observations on hydrography, chemistry, and turbidity). J. Biol. Bd. Canada, 1(5):279-467, 69 text figs.

Distephanus speculum
Herrera, Juan, y Ramon Margalef, 1963
Hidrografía y fitoplancton de la costa comprendida entre Castellón y la desembocadura del Ebro, de julio de 1960 a junio de 1961. Inv. Pesq., Barcelona, 24:33-112.

Distephanus speculum
Hovasse, R., 1946.
Flagellés à squelette silicieux:Silicoflagellés et Ebriidés provenant du plancton recueilli au cours des campagnes scientifiques du Prince Albert ler de Monaco (1885-1912). Rés. Camp. Sci Monaco, 107:19 pp., 1 pl.

Distephanes speculum
Jørgensen, E., 1905
B. Protistplankton and the diatoms in bottom samples. Hydrographical and biological investigations in Norwegian fjords. Bergens Mus. Skr. 7: 49-225.

Distephanus speculum
Lillick, L.C., 1938
Preliminary report of the phytoplankton of the Gulf of Maine. Am. Mid. Nat. 20(3):624-640, 1 text figs, 37 tables.

Distephanus speculum
Lillick, L.C., 1937
Seasonal studies of the phytoplankton off Woods Hole, Massachusetts. Biol. Bull. LXXIII (3):488-503, 3 text figs.

Distephanus speculum
Meunier, A., 1919
Microplankton de la Mer Flamande. 4. Les Tintinnides et Coetera. Mem. Mus. Roy. Hist. Nat., Belgique, 8(2):59pp., Pls. 22-23.

Distephanes speculum
Stæmann-Nielsen, Einar, 1951
The marine vegetation of the Isefjord. A study on ecology and production. Medd. Komm. Danmarks Fiskeri-og Havundersøgelser. Ser. Plankton. 5(4); 114pp., 46 text figs.

Hannaites quadria n.gen. n.sp.
Mandra, York T. 1969.
A new genus of Silicoflagellata from an Eocene South Atlantic deep-sea core (Protozoa: Mastigophora).
Occ. Pap. Calif. Acad. Sci. 77:7pp.

ebridinians
*Loeblich, Alfred R. III, Laurel A. Loeblich, Helen Tappan and Alfred R. Loeblich,Jr. 1968.
Annotated index of fossil and Recent silicoflagellates and ebridinians with descriptions and illustrations of calidly proposed taxa. Mem.geol.Soc.Am., 106:319 pp.

Ebria sp.
Morse, D.C., 1947
Some observations on seasonal variations in plankton population Patuxant River, Maryland 1943-1945. Bd. Nat. Res., Publ. No.65, Chesapeake Biol. Lab., 31, 3 figs.

Oceanographic Index: Marine Organisms Cumulation, 1946-1973

Lillick, L.C., 1938 Ebria antiqua
Preliminary report of the phytoplankton of the Gulf of Maine. Am. Mid. Nat. 20(3):624-640, 1 text fig. 37 tables.

Ebria bipartita
Florin, M-B., 1948
9. Diatomeae in submarine cores from the Tyrrhenian Sea. Medd. Ocean. Inst., Göteborg, 15 (Göteborgs Kungl. Vetenskaps-och Viterrhets Samhälles Handlingar, Sjätte Foljden, Ser. B 5(13):80-88.

Ebria tripartita
Gran, H.H., and T. Braarud, 1935
A quantitative study of the phytoplankton in the Bay of Fundy and the Gulf of Maine (including observations on hydrography, chemistry, and turbidity). J. Biol. Bd., Canada, 1(5):279-467, 69 text figs.

Ebria tripartita
Halim, Y., 1965.
Microplancton des eaux Egyptiennes. II. Chrysomonedines; Ebriediens et dinoflagellés nouveaux on d'interet biogeographique. Rapp. Proc. Verb. Reunions, Comm. Int. Expl. Sci., Mer Mediterranee, Monaco, 18(2):373-379.

Ebria tripartita
Hovasse, R., 1946.
Flagellés à squelette silicieux:Silicoflagellés et Ebriides provenant du plancton recueilli au cours des campagnes scientifiques du Prince Albert ler de Monaco (1885-1912). Rés. Camp. Sci. Monaco, 107:19 pp., 1 pl.

Ebria tripartita
Levander, K.M., 1947
Plankton gesammelt in den Jahren 1899-1910 an den Küsten Finnlands. Finnländische Hydrographisch-Biologische Untersuchungen (aus dem Wasserbiologischen Laboratorin der Societas Scientiarum Fennica) No.11: 40 pp., 6 diagrams, 13 pls., tables.

Ebria tripartita
Lemmermann, E., 1908.
XXI. Flagellatae, Chlorphyceae, Coccosphaerales und Silicoflagellatae. Nordisches Plankton, Bot. Teil:1-40, 135 textfigs.

Ebria tripartita
Lillick, L.C., 1938
Preliminary report of the phytoplankton of the Gulf of Maine. Am. Mid. Nat. 20(3):624-640, 1 text fig. 37 tables.

Ebria tripartita
Lillick, L.C., 1937
Seasonal studies of the phytoplankton off Woods Hole, Massachusetts. Biol. Bull. LXXIII (3):488-503, 3 text figs.

Ebria tripartita
Marshall, S.M., 1947
An experiment in marine fish cultivation: III. The plankton of a fertilized loch. Proc. Roy. Soc., Edinburgh, Sect.B., 63, Pt.I(3):21-33, 7 text figs.

Ebria tripartita
Meunier, A., 1919
Microplankton de la Mer Flamande. 4. Les Tintinnides et Coetera. Mem. Mus. Roy. Hist. Nat., Belgique, 8(2):59pp., Pls. 22-23.

Ebria tripartita
Pettersson, H., 1934
Scattering and extinction of light in sea-water. Meddelanden från Göteborgs Högskolas Oceanografiska Institut 9 (Göteborgs Kungl. vetenskaps-och vitterhetssamhälles Handlingar. Femte Foljden, Ser.B, 4(4)):1-18.

Ebria tripartita
Rumkówna, A., 1948
List of the phytoplankton species occurring in the superficial water layers in the Gulf of Gdańsk. Bull. Lab. mar., Gdynia, No. 4: 139-141 with tables in back.

Ebria tripartita
Stemann-Nielsen, Einar, 1951
The marine vegetation of the Isefjord. A study on ecology and production. Medd. Komm. Danmarks Fiskeri-og Havundersøgelser. Ser. Plankton. 5(4); 114pp., 46 text figs.

Mesocena polymorpha
Lemmermann, E., 1908.
XXI. Flagellatae, Chlorophyceae, Coccosphaerales, und Silicoflagellatae. Nordisches Plankton, Bot. Teil:1-40, 135 textfigs.

Mesocena polymorpha bioctonaria
Ercegovic, A., 1936
Etudes qualitative et quantitatives du phytoplancton dans les eaux cotières de l'Adriatique oriental moyen au cours de l'année 1934. Acta Adriatica 1(9):1-126

Octactis octonaria
Hovasse, R., 1946.
Flagellés à squelette silicieux: Silicoflagellés et Ebriides provenant du plancton recueilli au cours des campagnes scientifiques du Prince Albert ler de Monaco (1885-1912). Rés. Camp. Sci. Monaco, 107:19 pp., 1 pl.

Protozoa (Rhizopoda)
Boltovskoy, Estaban, y Haydee Lena,1966.
Contribución al conocimiento de las tecamebas de Ushuaia (Tierra del Fuego, Argentina). Neotropica, 12(38):55-65.

protozoa
Bunt, John S., 1970.
Preliminary observations on the growth of a naked marine ameba. Bull. mar. Sci., 20(2): 315-330.

Protozoa (rhizopods)
Golemansky, Vassil, 1970.
Rhizopodes nouveaux du psammon littoral de la mer Noire (note préliminaire).
Protistologica 6 (4): 365-371.

Thecamoebians, lists of spp.
Grossman, Stuart, 1967.
Ecology of Rhizopoda and Ostracoda of southern Pamlico Sound region, North Carolina. I. Living and subfossil rhizpod and ostracode populations. Paleontol.Contrib..Univ.Kansas,44(1):7-82.

thecamoebids
Page, Frederick C., 1971.
A comparative study of five freshwater and marine species of Thecamoebidae.
Trans. Am. microscop. Soc., 90(2):157-173

Entamoeba
Kott Hanna and Yehuda Kott,1970.
Viability of Entamoeba histolytica cysts exposed in sea water.
Rev. int. Océanogr. méd. 18-19:85-96

Centropyxis constricta
Boltovskoy, Estaban, y Haydee Lena,1966.
Contribución al conocimiento de las tecamebas de Ushuaia (Tierra del Fuego, Argentina). Neotropica, 12(38):55-65.

Cyclopxis arenata
Boltovskoy, Estaban, y Haydee Lena,1966.
Contribución al conocimiento de las tecamebas de Ushuaia (Tierra del Fuego, Argentina). Neotropica, 12(38):55-65.

Difflugia acuminata
Boltovskoy, Estaban, y Haydee Lena,1966.
Contribución al conocimiento de las tecamebas de Ushuaia (Tierra del Fuego, Argentina). Neotropica, 12(38):55-65.

Difflugia avellana
Boltovskoy, Estaban, y Haydee Lena,1966.
Contribución al conocimiento de las tecamebas de Ushuaia (Tierra del Fuego, Argentina). Neotropica,12(38):55-65.

Difflugia globularis
Boltovskoy, Estaban, y Haydee Lena,1966.
Contribución al conocimiento de las tecamebas de Ushuaia (Tierra del Fuego, Argentina). Neotropica, 12(38):55-65.

Difflugia globulosa
Boltovskoy, Estaban, y Haydee Lena,1966.
Contribución al conocimiento de las tecamebas de Ushuaia (Tierra del Fuego, Argentina). Neotropica, 12(38):55-65.

Difflugia mitriformis
Boltovskoy, Estaban, y Haydee Lena,1966.
Contribución al conocimiento de las tecamebas de Ushuaia (Tierra del Fuego, Argentina). Neotropica, 12(38):55-65.

Difflugia pyriformis
Boltovskoy, Estaban, y Haydee Lena, 1966.
Contribución al conocimiento de las tecamebas de Ushuaia (Tierra del Fuego, Argentina). Neotropica, 12(38):55-65.

Difflugia urceolata
Boltovskoy, Estaban, y Haydee Lena,1966.
Contribución al conocimiento de las tecamebas de Ushuaia (Tierra del Fuego, Argentina). Neotropica, 12(38):55-65.

lesquereusia modesta
Boltovskoy, Estaban, y Haydee Lena, 1966.
Contribución al conocimiento de las tecamebas de Ushuaia (Tierra del Fuego, Argentina). Neotropica, 12(38):55-65.

Fontigulasia compressa
Boltovskoy, Estaban, y Haydee Lena,1966.
Contribución al conocimiento de las tecamebas de Ushuaia (Tierra del Fuego, Argentina). Neotropica, 12(38):55-65.

Trigonopyxis arcula
Boltovskoy, Estaban, y Haydee Lena, 1966.
Contribución al conocimiento de las tecamebas de Ushuaia (Tierra del Fuego, Argentina). Neotropica, 12(38):55-65.

foraminifera
Acosta, J.T., 1940.
Algunos foraminiferos nuevos de las costas cubanas. Torreia, No. 5:3-6, 1 pl.

foraminifera
Akers, W.H., 1972.
Larger foraminifera from hole 98. Initial Reports, Deep Sea Drilling Project, Glomar Challenger 11: 545-546.

Foraminifera
Albanic, A.D., 1965.
The Foraminifera in a sample dredged from the vicinity of Salisbury Island, Durban Bay, South Africa.
Contrib. Cushman Found., Foram. Res., 16(2):60-66.

foraminifera
Alfirević, Slobodan, 1969.
Quelques observations sur les relations écologiques des foraminifères adriatiques.
Rapp. P.-v. Reun. Commn int. Explor. scient. Mer. Mediterr., 19(4): 655-657.

Foraminifera, lists of spp.
Alfirević, Slobodan, 1964.
Contribution a la connaissance de l'appartenance systematique des foraminiferes Adriatiques. (In Jugoslavian; French resume).
Acta Adriatica, 11(1):19-28.

Oceanographic Index: Marine Organisms Cumulation, 1946-1973

Foraminifera
Anderson, G.J., 1963.
Distribution patterns of Recent Foraminifera of the Bering Sea.
Micropaleontology, 9(3):305-317.

Foraminifera
Antony, A., 1968.
Studies on the shelf water Foraminifera of the Kerala Coast. Bull. Dept. mar. Biol. Oceanogr. Univ. Kerala, 4: 11-154.

---

foraminifera
Arrhenius, G., 1952.
Sediment cores from the East Pacific.
Repts. Swedish Deep-sea Exped., 1947-1948, 5(1): 227 pp., textfigs., with appendix of plates.

---

foraminifera
Arrhenius, G., 1950.
Foraminifera and deep sea stratigraphy. Science 111:288.

---

foraminifera
Ascoli, Piero, 1964.
Preliminary ecological study on Ostracoda from bottom cores of the Adriatic Sea.
Pubbl., Staz. Zool., Napoli, 33(Suppl.):213-246.

Foraminifera
Avnimelch, M., 1959.
Report on foraminiferal sands from the Red Sea coast. Contribution to the Knowledge of the Red Sea, Bull., Haifa, Israel, No. 20:4.

---

foraminifera
Ayala-Castañares, A. 1966.
Investigaciones sobre foraminiferos recientes de Mexico.
Revista Soc. Mexicana Hist. Nat. 27:7-21.

Foraminifera
Bandy, Orville L., 1967.
Foraminiferal definition of the boundaries of the Pleistocene in southern California, U.S.A.
Progress in Oceanography, 4:27-49.

Foraminifera
Bandy, Orville L., 1963
Aquitanian planktonic Foraminifera from Erben Guyot.
Science, 140(3574):1402.

---

foraminifera
Bandy, O.L., 1956.
Ecology of foraminifera in northeastern Gulf of Mexico. U.S.G.S. Prof. Paper 274-G:179-202, Pls. 29-31.

---

foraminifera
Bandy, O.L., and R.E. Arnal, 1960
Concepts of foraminiferal paleoecology. Bull. Amer. Assoc. Petr. Geol., 44(12): 1921-1932.

Foraminifera
Bandy, O.L., and R.E. Arnal, 1957.
Distribution of Recent Foraminifera of west coast of Central America.
Bull. Amer. Assoc. Petr. Geol., 41(9):2037-2053.

Foraminifera
Bandy, Orville L., James C. Ingle, Jr., and William E. Frerichs, 1967.
Isomorphism in "Sphaeroidinella" and "Sphaercidinellopsis".
Micropaleontology, 13(4):483-488.

---

foraminifera
Bandy, Orville L., James C. Ingle, Jr., and Johanna M. Resig, 1965.
Foraminiferal trends, Hyperion outfall, Calif.
Limnol. Oceanogr., 10(3):314-332.

---

foraminifera
Bandy, Orville L., James C. Ingle, Jr., and Johanna M. Resig, 1964.
Foraminifera, Los Angeles County outfall area, California.
Limnology and Oceanography, 9(1):124-137.

---

foraminifera
Bandy, Orville L., James C. Ingle, Jr., and Johanna M. Resig, 1964.
Foraminiferal trends, Laguna Beach outfall area, California.
Limnology and Oceanography, 9(1):112-123.

---

foraminifera
Bandy, Orville L., and Kelvin S. Rodolfo, 1964.
Distribution of Foraminifera and sediments, Peru-Chile Trench area.
Deep-Sea Res., 11(5):817-837.

---

foraminifera
Barbieri, F., and F. Medioli, 1972.
Upper Oligocene-lower Miocene microfacies from a Caribbean seamount (Aves swell) with infiltrated Pleistocene foraminifera.
Revista esp. Micropaleontol. 6(1):97-103

---

foraminifera
Bartlett, G.A., 1966.
The significance of foraminiferal distribution in waters influenced by the Gulf of St. Lawrence. (Abstract only).
Second Int. Oceanogr. Congr., 30 May-9 June 1966. Abstracts, Moscow:22-23.

Foraminifera
Bartlett, Grant A., and Linda Molinsky, 1972.
Foraminifera and the Holocene history of the Gulf of St. Lawrence.
Can. J. Earth Sci. 9(9):1204-1215.

Foraminifera
Bé, Allan W.H., 1960
Ecology of Recent planktonic Foraminifera. 2 Bathymetric and seasonal distributions in the Sargasso Sea off Bermuda. Micropaleontology 6(4): 373-392.

Foraminifera
Bé, Allan W.H., 1960
209. Some observations on Arctic planktonic Foraminifera; Contr. Cushman Found., Foraminiferal Res. 11(2): 64-68.

Foraminifera
Bé, Allan W.H., and William H. Hamlin, 1964.
North Atlantic planktonic Foraminifera during the summer of 1962. (Abstract).
Geol. Soc. Amer., Special Paper, No. 76:11.

Foraminifera
Beigbeder, Yvonne, et Marie Moulinier, 1966.
Fonds sedimentaires et foraminiferes dans la baie de Saint-Brieuc.
C.R. hebd. Séanc., Acad. Sci., Paris (D)263(4): 324-327.

---

foraminifera
Belyaeva, N.V., 1970.
Distribution of planktonic foraminiferal tests in sediments from the rift zone of the Indian Ocean. (In Russian; English abstract).
Okeanologiia, 10(4): 681-685.

Foraminifera
Belyaeva, N.V., 1964.
Distribution of planktonic Foraminifera in the water and on the floor in the Indian Ocean. (In Russian; English abstract).
Investigation of marine bottom sediments and suspended matter.
Trudy Inst. Okeanol., Akad. Nauk, SSSR, 68:12-83.

---

foraminifera
Belyaeva, N.B., and Kh. M. Saidova, 1965.
The correlation between benthic and planktonic Foraminifera in the surface layer of the sediments of the Pacific Ocean. (In Russian).
Okeanologiia, Akad. Nauk, SSSR, 5(6):1010-1014

Foraminifera
Berger, Wolfgang H., 1970.
Planktonic Foraminifera: selective solution and the lysocline.
Mar. Geol. 8(2):111-138.

Foraminifera
Bermudez, Pedro J. 1969.
Cuaternario y Recente en Venezuela.
Memoria, Soc. Cienc. nat. La Salle, 29(82): 43-59.

---

foraminifera
Bermudez, Pedro J., 1963
Estudio sistematico de los foraminiferos del Golfo de Cariaco.
Bol. Inst. Oceanogr., Univ. de Oriente, Cumana, Venezuela, 2(2):3-267, 29 pls.

---

foraminifera
Bermudez, P.J., 1961
Los foraminiferos planctonicos.
Mem. Soc. Cienc. Nat. La Salle, 21(59): 111-138.

---

foraminifera
Bermudez, Pedro J., 1960.
Foraminiferos planctónicos del Golfo de Venezuela.
Memoria Soc. Ciencias Naturales la Salle, 20(55):58-76.

---

foraminifera
Bermudez, Pedro J., 1960.
Foraminiferos planctonicos del Golfo de Venezuela.
Memoria Tercer Congreso Geologico Venezolano, 2: 905-927.

---

foraminifera
Bermúdez, P.J., 1939.
Resultados de la Primera Expedicion en las Antillas del Ketch Atlantis bajo los auspicios de las Universidades de Harvard y Habana. Nuevo Género y Especies nuevas de Foraminiferos. Mem. Soc. Hist. Nat. 13(4):247-251.

---

foraminifera
Bermúdez, P.J., 1939.
Resultados de la Primera Expedicion en las Antillas del Ketch Atlantis bajo los auspicios de las Universidades de Harvard y Habana. Foraminiferos del Género Recurvoides, descripción de una Especie nueva. Mem. Soc. Cub. Hist. Nat. 13(2):57-61, 1 pl.

---

foraminifera
Bermúdez, P.J., 1939.
Resultados de la Primera Expedicion en las Antillas del Ketch Atlantis bajo los auspicios de las Universidades de Harvard y Habana. Nuevo Género y Especies nuevas de Foraminiferos. Mem. Soc. Cub. Hist. Nat. 13(1):9-12, 2 pls.

---

foraminifera
Bermudez, Pedro J. y Hector A. Gamez, 1966.
Estudio paleontologico de una seccion del Eoceno grupo Punto Carrero de la Isla Margarita Venezuela. Memoria, Soc. Cienc. nat. La Salle, 26(75)205-209.

Foraminifera
Blanc-Vernet, L., 1969.
Contribution à l'Etude des foraminifères de Méditerranée: relations entre la microfaune et le sédiment; biocoenoses actuelles thanatocoenoses pliocènes et quaternaires.
Rec. Trav. Sta. mar. Endoume 48(64): 315pp

Foraminifera
Blanc-Vernet, Laure, 1965.
Nouvelles données sur les foraminifères des Gris sous-marins du canyon de la Cassidaigne la Ciotat (Bouches-du-Rhône)
Rec. Trav. Sta. Mar. Endoume, Bull. 39(55):303-304

697

Foraminifera

Boltovskoy, E., 1970.
Distribution of the marine littoral Foraminifera in Argentina, Uruguay and Southern Brazil. Marine Biol., 6(4): 335-344.

Foraminifera

Blanc-Vernet, Laure, 1965.
Note sur la repartition des Foraminiferes au voisinage des cotes de Terre Adelie.
Rec. Trav. Sta Mar., Endoume, 52(36):191-205.
Also:
Exped. Polaires Francaises, Missions Paul-Emile Victor, No. 276.

foraminifera

Blanc-Vernhet, L., 1963.
Note préliminaire sur les Foraminifères des fonds détritiques côtiers et de la vase terrigène côtière dans la baie de Marseille.
Rec. Trav. Sta. Mar., Endoume, Bull., 30(45):83-93.

Foraminifera

Boltovstoy, Esteban, 1965.
Datos nuevos con respecto a la ublicacion de la zona de convergencia subtropical/subantarctic en base al esutdio de los foraminiforos planctonicos.
Anais Acad. bras. Cienc., 37(Supl.):146-155.

Foraminifera

Boltovskoy, Esteban, 1965.
Los foraminiferos recientes: biologie, metodos de estudio, aplicacion oceanografica.
Eudeba, Editorial Universitaria de Buenos Aires, 510 pp.

Foraminifera (living)

Boltovskoy, Esteban, 1964.
Seasonal occurrences of some living Foraminifera in Puerto Deseado (Patagonia, Argentina).
J. du Cons., 29(2):136-145.

foraminifera

Boltovskoy, Esteban, 1964.
Provincias zoogeograficas de America del Sur y su sector Antartico segun los foraminiferos bentonicos.
Bol. Inst. Biol. Mar., Mar del Plata, Argentina, No. 7:93-99.

Foraminifera

Boltovskoy, Estaban, 1963
Foraminiferos y sus relaciones con el medio.
Rev. Mus. Argentina Ciencias Nat. "Bernardino Rivadavia" e Inst. Nacional, Invest. Ciencias Nat. Hidrobiol., 1(2):21-104.

foraminifera

Boltovskoy, Esteban, 1963
The littoral foraminiferal biocenoses of Puerto Deseado (Patagonia, Argentina).
Contrib. Cushman Found. Foram. Res., 14(2):58-70.
Also in:
Centro Invest. Biol. Mar., Estacion Puerto Deseardo, Contrib. Cient., N$_o$. 3:

foraminifera

Boltovskoy, E., 1962.
Plankton foraminifera as indicators of different water masses in the South Atlantic.
Micropaleontology, 8(3):403-408.

foraminifera

Boltovskoy, Estaban, 1961
Algunos foraminiferos nuevos de las aguas brasilenas (Protozoa).
Neotropica, 7:73-79.

foraminifera

Boltovskoy, Esteban, 1961
Foraminiferos de la plataforma continental entre el Cabo Santo Tome y la desembocadura del Rio de la Plata.
Revista. Mus. Argentino Ciencias Nat. "Bernardino Rivadavia", Ciencias Zool., 6(6):249-346.

Foraminifera

Boltovskoy, Estaban, 1959
Foraminifera as biological indicators in the study of ocean currents. Micropaleontology, 5(4): 473-481.

foraminifera

Boltovskoy, Estaban, 1959.
La Corriente de Malvinas (un estudio en base a la investigacion de Foraminifera).
Servicion de Hidrografia Naval, Argentina, Publ. H. 1015:96 pp.

foraminifera

Boltovskoy, E., 1957.
Los foraminiferos del estuario del rio de La Plata y su zona de influencia.
Ciencias Zool., 6(1): 1-77.
Rev. in Bol. Inst. Ocean., Sao Paulo, 8(1/.2): 255-256.

foraminifera

Boltovskoy, E., 1956.
Diccionario foraminiferologoico plurilingüe, en cinco idiomas: Inglés, Español, Alemán, Francés y Ruso con indices alphabéticos.
S.H. Pub., Misc., Ministerio de Marina, Direccion General de Navegacion e Hidrografia, Republica Argentina, No. 1001:196 pp.

foraminifera

Boltovskoy, E., y A. Boltovskoy, 1968.
Foraminiferos y tecamebas de la parte inferior del Rio Quequen Grande.
Revta Mus. argent. Cienc. nat. Bernardino Rivadavia Inst. nac. Hidrobiol., 2(4):127-164.

foraminifera

Boltovskoy, E., and F. Theyer, 1970.
Foraminiferos recientes de Chile central.
Revta Mus. argent. Cienc. nat. Bernardino Rivadavia, (Hidrobiol) 2(9): 279-378

Foraminifera

Bradshaw, John S., 1968.
Environmental parameters and marsh Foraminifera
Limnol. Oceanogr., 13(1):26-38.

foraminifera

Bradshaw, J.S., 1955.
Preliminary laboratory experiments on ecology of foraminiferal populations. Micropaleontology 1(4):351

foraminifera

Buzas, Martin A. 1972.
Foraminifera of the Chesapeake Bay.
Chesapeake Sci. 13 (Suppl.): 597-599

Foraminifera

Buzas, Martin A., 1968.
On the spatial distribution of Foraminifera.
Contrib. Cushman fdn foramin. Res., 19(1):1-11.

Foraminifera

Buzas, Martin A., 1968.
Foraminifera from the Hadley Harbor complex, Massachusetts.
Smithsonian Misc.Coll., 152(8)(Publ.4727):1-26.

Foraminifera

Buzas, Martin A., 1965.
The distribution and abundance of Foraminifera in Long Island Sound.
Smithsonian Misc. Coll. (Publ. 4604), 149(1):1-99.

Foraminfera

Church, Clifford C., 1968.
Shallow water Foraminifera from Cape San Lucas, Lower California.
Proc. Calif. Acad. Sci., (4)30(17):357-380.

foraminifera

Colom, G., 1970.
Estudio de los foraminiferos de muestras de fondo de la costa de Barcelona.
Inv. pesq., Barcelona, 34(2): 355-384

foraminifera

Colom, G., 1963.
Los foraminiferos de la Ria de Vigo.
Inv. Pesq., Barcelona, 23:71-89.

Foraminifera

Cooper, Susan C., 1964.
Benthonic Foraminifera of the Chukchi Sea.
Contrib. Cushman Found. Foraminiferal Res., 15(3):79-104.

Foraminifera

Cooper, Susan C., and Joe S. Creager, 1961.
Benthic Foraminifera of the Chukchi Sea. (Abstract).
Tenth Pacific Sci. Congr., Honolulu, 21 Aug.-6 Sept., 1961, Abstracts of Symposium Papers, 368-369.

Foraminifera

Duplessy, Jean Claude, Claude Lalou and Annie Claude Vinot, 1970.
Differential isotopic fractionation in benthic Foraminifera and paleotemperatures reassessed.
Science, 168(3928): 250-251.

foraminifera

Ehrenberg, C.G., 1873.
Mikrogeologische Studien über das kleinste Leben der Meeres-Tiefgründe aller Zonen und dessen geologischen Einfluss. Abhandl. Königl. Akad. Wissensch., Berlin, 1872,:131-397, 12 pls., 1 map

Foraminifera

Ellison, R., M. Nichols and J. Hughes, 1965.
Distribution of Recent Foraminifera in the Rappahannock River Estuary.
Virginia Inst. Mar. Sci., Spec. Sci. Rept., No. 47:35 pp. (unpublished manuscript).

foraminifera

Emiliani, C., 1954.
Depth habitats of some species of foraminifera as indicated by oxygen isotope ratios.
Am. J. Sci. 252:149-158, 4 textfigs.
(pelagic)

foraminifera

Enbysk, Betty Joyce, 1961.
Foraminifera from northeast Pacific cores. (Abstract).
Tenth Pacific Sci. Congr., Honolulu, 21 Aug.-6 Sept., 1961, Abstracts of Symposium Papers, 369-370.

foraminifera

Fayose, Emmanuel A., 1970.
Preliminary account of the distribution of Recent Foraminifera off the Bight of Benin, Tarkwa Bay.
Bull. Inst. fondament. Afrique Noire (A) 32(3):594-606.

Foraminifera

Fowler, Gerald A., 1966.
Notes on Late Tertiary Foraminifera from off the central coast of Oregon.
Ore Bin, 28(3):53-60.
Also in: Coll. Repr., Dep. Oceanogr., Oregon State Univ., 5.

Foraminifera

Frankel, J.J., 1964.
Recent Foraminifera filled and encrusted with pyrite from Durban Bay.
S. African J. Sci., 60(10):299.

Foraminifera

*Funnell, B.M., 1967.
Foraminifera and Radiolaria as depth indicators in the marine environment.
Marine Geol., 5(5/6):333-347.

foraminifera

Funnell, B.M., 1964.
Studies in North Atlantic geology and paleontology. 1. Upper Cretaceous.
Geol. Mag., 101(5):421-434.

Foraminifera

Gabel, B., 1971.
Die Foraminiferen der Nordsee. Helgoländer wiss. Meeresunters 22(1): 1-65.

Foraminifera

Ganapati, P.N., and P. Satyavati, 1958
Report on the Foraminifera in bottom sediments in the Bay of Bengal off the east coast of India. Andhra Univ. Mem., Ocean., 2: 100-127.

foraminifera

Gibson, Thomas G., 1965.
Eocene and Miocene rocks off the northeast coast of the United States.
Deep-Sea Res., 12(6):975-981.

foraminifera

Giunta, M., 1955.
Studio delle microfauna contenuta in cinque saggi di fondo prelevati presso S. Margherita Ligure e Chiavari (Genova).
Arch. Ocean. e Limnol. 10(1/2):67-108, 2 pls.

foraminifera

Green, Keith, E., 1959.
Ecology of some Arctic Foraminifera.
Sci. Studies Fletcher's Ice Island, T-3, Vol. 1,
Air Force Cambridge Res. Center, Geophys., Res. Pap., No. 63:59-81.

Foraminifera

*Haake, Friedrich-Wilhelm, 1967.
Zum Jahresgang von Populationen einer Foraminiferen-Art in der westlichen Ostsee.
Meyniana, 17:13-27.

Foraminifera

Hansen, Hans Jørgen, 1965.
On the sedimentology and the quantitative distribution of living Foraminifera in the northern part of the Øresund.
Ophelia. 2(2):323-331.

Foraminifera

Hay, W.W., and D.S. Marszalek, 1962.
Fossil Foraminifera in Adriatic beach sands.
Contributions, Cushman Found. Foram. Res., 14(1):16.

foraminifera

Haynes, John, 1964.
Live and dead Foraminifera between the Sarns, Cardigan Bay.
Nature, 204(4960):774.

foraminifera

T.D. Adams, K. Atkinson, E.A. Fayose, D. Haman
Haynes J.R. 1973. K.H. James, J.A. Johnson + J. Scott
Cardigan Bay Recent foraminifera
(cruises of the R.V. Antur, 1962-1964).
Bull. Brit. Mus. (N.H.) Zool., Suppl. 4: 245 pp., 33 pls

foraminifera

Hedley, R.H., C.M. Hurdle and I.D.J. Burdett, 1967.
The marine fauna of new Zealand: intertidal Foraminifera of the Corallina officinalis zone.
Bull. N.Z. Dept. scient. indust. Res. 180:1-86.
(Mem. N.Z. Oceanogr. Inst. 38).

Foraminifera

Hofker, J., Dr., 1968.
Foraminifera from the Bay of Jakarta, Java.
Bijdragen tot de Dierkunde, 37:11-59.

Foraminifera

Holtedahl, Hans, 1965.
Recent turbidites in the Hardangerfjord, Norway.
In: Submarine geology and geophysics, Colston Papers, W.F. Whittard and R. Bradshaw, editors, Butterworth's, London, 107-140.

Foraminifera

Hornibrook, N. de B., 1965.
A preliminary statement on the types of the New Zealand Tertiary Foraminifera described in the reports of the Novara Expedition.
New Zealand J. Geol. Geophys., 8(3):830-836.

foraminifera (Campanian)

Hottinger, L., 1972.
Campanian larger foraminifera from site 98, leg 11 of the deep sea drilling project (Northwest Providence Channel, Bahama Islands).
Initial Reports, Deep Sea Drilling Project, Glomar Challenger 11: 595-605.

foraminifera

Huang, Tun-Yow, 1961.
Smaller Foraminifera from beach sands at Tanmenkang, Paohao-tao, Penghu.
Proc. Geol. Soc., China, (4):83-90.

foraminifera

Hulme, S.G., 1964.
Recent Foraminifera from Manukau Harbour, Auckland, New Zealand.
New Zealand J. Sci., 7(3):305-340.

Foremifera

Iaccarino, Silvia, 1965.
Ricerche preliminari sui Foraminiferi contenuti in 3 carote prelevate nel Mare Ligure (Ia Spezia)
Boll. Soc. Geol. Ital., 83(1):1-17.

foraminifera

Ikeya, Noriyuki 1970.
Populational ecology of benthonic Foraminifera in Ishikari Bay, Hokkaido, Japan.
Rec. oceanogr. Wks, Japan 10(2):173-191.

foraminifera

Jarke, Joachim, 1960.
Beitrag zur Kenntnis der Foraminiferenfauna der mittleren und westlichen Barents-See.
Int. Revue ges. Hydrobiol., 45(4):581-654.
Also in:
Deutsches Hydrogr. Inst., Ozeanogr., 1960.

foraminifera, anatomy

Jepps, M.W., 1946
Are there Triflagellate Gametes in the Foraminifera. Nature 157:374

Foraminifera (Tertiary)

Johnson, David A., and Frances L. Parker 1972.
Tertiary Radiolaria and Foraminifera from the equatorial Pacific.
Micropaleontology 18(2):129-143

Foraminifera

Jones, James I., and Wayne D. Bock, 1964.
Trace element distribution in some living and fossil Foraminifera from South Florida, Bahaman and Caribbean waters. (Abstract).
Geol. Soc., Amer., Special Paper, No. 76:88-89.

foraminifera

Kane, Julian 1968.
Foraminifera.
Sea Frontiers 14(3):176-186 (popular)

foraminifera

Kennett, James P. 1968.
6. Ecology and distribution of Foraminifera.
Fauna of the Ross Sea.
Bull. N.Z. Dept. sci. Ind. Res. (Mem. N.Z. Oceanogr. Inst.) 46) 186: 1-46.

Foraminifera

Kennett, James P., 1966.
Foraminiferal evidence of a shallow calcium carbonate solution boundary, Ross Sea, Antarctica.
Science, 153(3732):191-193.

foraminifera

Khusid, T.A., 1971.
Distribution of foraminiferal taxocoenoses and genocoenoses on the South American borderland in the Pacific Ocean. (In Russian; English abstract). Okeanologiia 11(2): 266-269.

Foraminifera

Kure, G., 1961
Foraminifères et Ostracodes de l'étang de Thau.
Rev. Trav. Inst. Pêches Marit., 25(2):133-217.

Foraminifera

Kustanowich, S., 1962.
A foraminiferal fauna from Capricorn Seamount, Southwest Equatorial Pacific.
New Zealand J. Geol. and Geophys., 5(3):427-434.

foraminifera

Kuwano, Y., 1956.
Invertebrate fauna of the intertidal zone of the Tokara Islands. XII. Foraminifera.
Publ. Seto Mar. Biol. Lab., 5(2)(16):273-282.

foraminifera

Lankford, R.R., 1959.
Distribution and ecology of Foraminifera from East Mississippi delta margin.
Bull. Amer. Assoc. Petr. Geol., 43(9):2068-2099.

foraminifera

Le Calvez, Jean et Yolande, 1958.
Repartition des Foraminifères dans la Baie de Villefranche.
Ann. Inst. Océan., Monaco, 35:159-234.
Also in Trav. Sta. Zool., Villefranche-sur-Mer, 17(5):

foraminifera

Le Calvez, Yolande 1972
Etude écologique de quelques foraminifères de la côte saharienne de l'Atlantique.
Rev. Trav. Inst. Pêches marit. 36(3): 245-254.

Foraminifera

Le Calvez, Yolande, and Gilbert Boillot, 1967.
Etudes des foraminifères contenus dans les sédiments actuels de la Manche occidentale.
Revue Géogr. phys. Géol. dyn., (2)9(5):391-408.

Foraminifera

Lee, J.J., W.A. Muller, R.J. Stone, M.E. McEnery and W. Zucker, 1969.
Standing crop of Foraminifera in sublittoral epiphytic communities of a Long Island salt marsh. Mar. Biol., 4(1): 44-61.

foraminifera

Levy, Alain 1967
Contribution à l'étude des foraminifères des rechs du Roussillon et du plateau continental de bordure.
Vie Milieu (B) 18 (1B): 63-102

Foraminifera

Lidz, Louis, 1965.
Sedimentary environment and foraminiferal parameters: Nantucket Bay, Massachusetts.
Limnol. Oceanogr., 10(3):392-402.

Foraminifera

Lukina, T.G., 1969.
Distribution of foraminifers in the central part of the Pacific. (In Russian; English abstract).
Zool. Zh. 48(10): 1445-1450.

Foraminifera

Lukina, T.G., 1967.
The distribution of calcaerous Foraminifera in the central Pacific. (In Russian).
Dokl. Akad. Nauk, SSSR, 177(5):1205-1207.

foraminifera

Lukina T.G. 1967.
Distribution of deep-water fauna of agglutinized Foraminifera in the central Pacific. (In Russian).
Dokl. Akad. Nauk, SSSR, 173(6):1431-1433

Foraminifera

Luterbacher, Hanspeter, 1972.
Foraminifera from the Lower Cretaceous and Upper Jurassic of the Northwestern Atlantic.
Initial Reports, Deep Sea Drilling Project. Glomar Challenger 11: 561-593.

foraminifera

Lynts, George W., 1965.
Observations on Some Recent Florida Bay Foraminifera.
Contrib. Cushman Found., Foram. Res., 16(2):67-69.

Foraminifera

Lynts, G.W., 1962.
Distribution of Recent Foraminifera in Upper Florida Bay and associated sounds.
Contr. Cushman Found. Foram. Res., 13(4):127-144.

foraminifera

Mateu Mateu, Guillermo 1970.
Estudio sistemático y bioecológico de los foraminíferos vivientes de los litorales de Cataluña y Baleares.
Trab. Inst. Esp. Oceanogr. 38:84pp., 28 pls.

Foraminifera

Mateu, Guillermo, 1965.
Datos para el estudio de los Foraminíferos algícolas del litoral de Blanes (gerona).
Publ. Inst. Biol. Aplic. 39:129-135.

Foraminifera

McCrone, Alistair W., and Charles Schafer, 1966.
Geochemical and sedimentary environments of Foraminifera in the Hudson River estuary, New York.
Micropaleontology, 12(4):505-509.

foraminifera

McKenzie, K.G., 1962.
13. A record of Foraminifera from Oyster Harbour, near Albany, Western Australia.
J. R. Soc., Western Australia, 45(4):117-132.

Foraminifera

Moncharmont-zei, Maria, 1964.
Studio ecologico sui Foraminiferi del Golfo di Pozzuoli (Napoli).
Pubbl. Staz. Zool., Napoli, 34:160-184.

foraminifera

Moncharmont-Zei, Maria, 1962.
I Foraminiferi del Banco delle Vedove (Golfo di Napoli).
Pubbl. Staz. Zool., Napoli, 32, (Suppl.):442-482.

foraminifera

Murray, John W., 1971.
Living foraminiferids of tidal marshes: a review.
J. foram. Res. 1(4):153-161

Foraminifera

Murray, J.W., 1966.
The Foraminiferida of the Persian Gulf. 5. The shelf off the Trucial Coast.
Palaeogr., Palaeoclimatol., Palaeoecol., 2(3):267-278.

Foraminifera

Murray, J.W., 1966.
The Foraminiferida of the Persian Gulf. 3. The Halat Al Bahrani region.
Palaeogr., Palaeoclimatol., Palaeoecol., 2(1):59-68.

foraminifera, ecology of

Murray, J.W., 1963.
Ecological experiments on Foraminiferida.
J. Mar. Biol. Assoc., U.K., 43(3):621-642.

foraminifera

Myers, E. H. 1943
Life activities of foraminifera in relation to marine ecology. Proc. Am. Phil. Soc., 86(3):439-458, 1 pl.

foraminifera

Myers, E. H., 1942.
Biological evidence as to the rate at which tests of Foraminifera are contributed to marine sediments. J. Palaeont. 16(3):397-398.

foraminifera

Myers, E.H., 1942
A quantitative study of the productivity of the foraminifera in the sea. Proc. Am. Phil. Soc. 85(4):325-342, 1 pl.

foraminifera

Myers, E. H., 1942.
Rate at which Foraminifera are contributed to marine sediments. J. Sed. Petr. 12(2):92-95.

foraminifera

Myers, E.H., 1938
The present state of our knowledge of the life history of the foraminifera. Nat. Acad. Sci., Proc. 24:10-17

Foraminifera

Narchi, Walter, 1962.
Sobre Nonionidae, Globorotalidae e Orbulinidae recentes do Brasil.
Bol. Inst. Oceanogr., Sao Paulo, Brasil, 12(3):23-46.

Foraminifera

Natland, M.L., 1938.
New species of Foraminifera from the west coast of North America and from later Tertiary of the Los Angeles Basin.
S.I.O. Bull., Techn. Ser., 4:137-164.

Foraminifera

Nørvang, Aksel, 1961.
Schizamminida, a new family of Foraminifera.
Atlantide Rept., Sci. Res., Danish Exped., Coasts of Tropical West Africa, 6:169-201. (1945-1946).

Foraminifera

Parker, Frances L., 1964.
Foraminifera from the experimental Mohole drilling near Guadalupe Island, Mexico.
J. Paleontology, 38(4):617-636.

Foraminifera

Parker, F.L., 1962.
Planktonic foraminiferal species in Pacific sediments.
Micropaleontology, 8(2):219-254.

foraminifera

Parker, F.L., 1955.
Distribution of planktonic foraminifera in some Mediterranean.
Pap. Mar. Biol. and Oceanogr., Deep-Sea Res., Suppl. to Vol. 3:204-211.

foraminifera

Parker, F.L., 1952.
Foraminiferal distribution in the Long Island Sound-Buzzards Bay area. Bull. M.C.Z. 106(10):427-473, 5 pls., 6 tables, 2 textfigs.

Foraminifera

Parker, F.L., 1952.
No. 9. Foraminifera species off Portsmouth, New Hampshire. Bull. M.C.Z. 106:391-423, 6 pls.

Foraminifera

Parker, F.L., and W.D. Athern, 1959.
Ecology of marsh Foraminifera in Popponesset Bay, Massachusetts.
J. Paleontology, 33(2):333-343.

foraminifera

Parker, F.L., F.B. Phleger and J.F. Pierson, 1954.
Ecology of Foraminifera from San Antonio Bay and environs, southwest Texas.
Cushman Found. Foram. Res., Spec. Publ. No. 2:1-72, 4 pls., 48 textfigs.

foraminifera

Phleger, Fred B 1970.
Foraminiferal populations and marine marsh processes. Limnol. Oceanogr., 15(4): 522-534.

Foraminifera

Phleger, Fred B., 1965.
Patterns of marsh Foraminifera, Galveston, Texas.
Limnol. and Oceanogr., Redfield Vol., Suppl. to 10:R169-R184.

foraminifera

Phleger, Fred B, 1964.
Patterns of living benthonic Foraminifer, Gulf of California.
In: Marine geology of the Gulf of California, a symposium, Amer. Assoc. Petr. Geol., Memoir, T. van Andel and G.G. Shor, Jr., editors, 3:377-394.

Foraminifera

Phleger, Fred B, 1964.
Foraminiferal ecology and marine geology.
Marine Geology, 1(1):16-43.

Foraminifera

Phleger, Fred B., 1961.
Sedimentary patterns of benthonic Foraminifera in the eastern Pacific. (Abstract).
Tenth Pacific Sci. Congr., Honolulu, 21 Aug.-6 Sept., 1961, Abstracts of Symposium Papers, 383-384.

Foraminifera

Phleger, Fred B, 1960
Ecology and distribution of Recent Foraminifera. The Johns Hopkins University Press, 297 pp. $7.50.

Reviewed by W. Schott in Deep-Sea Res., 7(4):301-302.

foraminifera

Phleger, F.B, 1955.
Ecology of Foraminifera in southeastern Mississippi delta area.
Bull. Amer. Assoc. Petr. Geol. 39(5):712-752, 40 textfigs.

Foraminifera

Phleger, F.B, 1955.
Foraminiferal faunas in cores offshore from the Mississippi Delta.
Pap. Mar. Biol. and Oceanogr., Deep-Sea Res., Suppl. to Vol. 3:45-57.

foraminifera

Phleger, F. B, 1954.
Ecology of Foraminifera and associated microorganisms from Mississippi Sound and environs.
Bull. Amer. Assoc. Petr. Geol. 38(4):584-647, 3 pls., 28 textfigs.

foraminifera

Phleger, F. B, 1954.
Foraminifera and deep-sea research.
Deep-Sea Res. 2(1):1-23.

foraminifera

Phleger, F.B, 1952.
No. 8. Foraminifera ecology off Portsmouth, New Hampshire. Bull. Mus. Com. Zool. 106:315-390, tables, 25 textfigs.

foraminifera

Phleger, F. B, 1951.
Ecology of Foraminifera, northwest Gulf of Mexico
Pt. 1. Foraminifera distribution. Mem. G.S.A.
No. 46:88 pp., 44 textfigs., 37 loose tables.

foraminifera

Phleger, F. B, and F. L. Parker, 1951.
Ecology of Foraminifera, northwest Gulf of Mexico
Pt. II. Foraminifera species. Mem. G.S.A., No.
46:64 pp., 19 pls.

Foraminifera

Phleger, F. b, 1949.
The Foraminifera. In: The Boylston Street Fish-
weir II. Papers Robert S. Peabody Found. Archeol.
4(1):99-108.

foraminifera

Phleger, Fred B, jr., 1948.
Foraminifera of a submarine core from the
Caribbean Sea. Medd.Ocean. Inst., Göteborg, 16
(Göteborgs Kungl. Vetenskaps- och Vitterhets-
Samhälles Handlingar, Sjätte Följden, Ser. B,
5(14):9 pp., 1 pl., 2 tables.

foraminifera

Phleger, F. B, jr., 1947.
Foraminifera of three submarine cores from the
Tyrrhenian Sea. Göteborgs Kungl. Vetenskaps-
och Vitterhets- samhälles Handl., Sjätte Följden
Ser. B, 5(5):1-19, 1 textfig. (Medd. från
Oceanografiska Institutet i Göteborg, 13). With
a foreword by H. Pettersson.

foraminifera

Phleger, F.B., jr. 1945
Vertical distribution of pelagic
foraminifera. Am. J. Sci., 243:
377-383, 1 textfigs.

Foraminifera

Phleger, F.B., jr., and W.A. Hamilton, 1946
Foraminifera of two submarine cores from
the North Atlantic Basin. Bull. G.S.A. 57:
951-966, 1 pl., 3 figs.

foraminifera

Phleger, F.B, F.L. Parker and J.F. Peirson, 1953.
Sediment cores from the North Atlantic Ocean.
Repts. Swedish Deep-Sea Exped., 1947-1948, 7(1):
1-122, 12 pls., 26 textfigs.

Foraminifera

Phleger, F. B, F.L. Parker, and J.F. Pierson, 1951.
Foraminifera in North Atlantic deep sea cores.
(Abstr.). Bull. G.S.A. 62:1470.

foraminifera

Phleger, Fred B, and W.R. Walton, 1950.
Ecology of marsh and bay Foraminifera,
Barnstable, Mass. Am. J. Sci. 248(4):274-294,
5 tables, 2 pls., 2 textfigs.

Foraminifera

Polski, W., 1959
Foraminiferal biofacies off the North Asiatic
coast.
J. Paleont., 33(4): 569-587.

Foraminifera

Rao, K. Kameswara 1969.
Foraminifera of the Gulf of Cambay.
J. Bombay nat. hist. Soc. 66(3): 584-596

foraminifera

Reiss, Z., 1971.
Progress and problems of foraminiferal systema-
tics. In: Micropaleontology of oceans, B.M.
Funnell and W.R. Riedel, editors, Cambridge
Univ. Press, 633-638.

Foraminifera

Reiss, Z., K. Klug and P. Merling, 1961.
II. Notes on Foraminifera from Israel. 10.
Recent Foraminifera from the Mediterranean and
Red Sea coasts of Israel.
Ministry of Development, Geol. Survey, Bull., No
32:27-28.

Foraminifera

Richter, Gotthard, 1966.
Typische Foraminiferen-Gemeinschaften rezenter und
subfossiler Wattensedimente.
Veroff. Inst. Meeresforsch., Bremerh., Sonderband
II:301-302.

foraminifera

Ruddiman, W.F., D.S. Tolderlund and A.W.H. Be
1970,
Foraminiferal evidence of a modern warming of
the North Atlantic Ocean. Deep-Sea Res., 17(1):
141-155.

Foraminifera

Said, R., 1953.
Foraminifera of Great Pond, Falmouth, Massachusetts. Contr. Cushman Found. Foram. Res. 4(1):
7-14, 1 table, 3 textfigs.

Foraminifera

Saidova, Kh. M, 1967.
The biomass and quantitative distribution
of live Foraminifera in the region of the
Kuril-Kamchatka Trench (In Russian).
Dokl. Akad. Nauk. SSSR, 174(1): 207-209

foraminifera

Saidova, Ch. M., 1960. Stratigraphy of sediments
and paleogeography of the northeast sector of the Pacific
Ocean according to sea-bottom foraminifers. Morsk.
Geol., Doklady Sovetsk. Geol., 21 Sess., Inst. Geol.
Kongress, 59-68.
(English abstract)

Foraminifera (bottom)

Saidova, Kh. M., 1965.
The distribution of bottom Foraminifera in the
Pacific Ocean. (In Russian).
Okeanologiia, Akad. Nauk, SSSR, 5(1):99-110.

Foraminifera

Saidova, H.M., 1964.
Distribution of bottom Foraminifera and strati-
graphy of sediment in the north-eastern Pacific.
Investigations of marine bottom sediments and
suspended matter. (In Russian; English abstract).
Trudy Inst. Okeanol., Akad. Nauk, SSSR, 68:84-119

foraminifera

Saidova, Kh. M., 1961.
The quantitative distribution of bottom Foramin-
ifera in Antarctica.
Doklady, Akad. Nauk, SSSR, 139(4):967-969.

foraminifera

Saidova, Kh. M., 1957.
Quantitative distribution of Foraminifera in the
Okhotsk Sea. Doklady Akad. Nauk, SSSR, 114(6):
1302-1305.

foraminifera

Saidova, Kh. M., 1957.
The distribution of Foraminifera in the deposits
of the Okhotsk Sea. Doklady Akad. Nauk, SSSR, 115
(6):1213-1216.

foraminifera

Saidova, Kh. M., 1956.
Methods for the isolation of Foraminifera from
bottom sediments. Trudy Inst. Okeanol., 19:294-
296.

foraminifera

Schjedrina, Z.G., 1958.
Die Foraminiferenfauna des Kurilen-Kamtschatka
Grabens. Trudy Inst. Okeanol., 27:161-179.

Foraminifera

Schlenz, Erika, 1965.
Outline of the ecological distribution of the
Foraminifera of the Brazilian continental shelf.
(abstract).
Anais Acad. bras. Cienc., 37(Supl.):325.

Foraminifera

Schnitker Detmar 1969.
Distribution of Foraminifera on a
portion of the continental shelf of the
Golfe de Gascogne (Bay of Biscay).
Bull. Centre Rech. Pau- SNPA 3(1): 33-64.

Foraminifera

Seibold, Eugen, and Ilse Seibold, 1960.
Foraminifera in sponge bioherms and bedded
limestones of the Manm, South Germany.
Micropaleontology, 6(3):301-306.

foraminifera

Seiglie, George A., 1971.
Distribution of foraminifers in the
Cabo Rojo platform and their paleoecological
significance.
Revista esp. Micropaleontol. 3(1):5-34.

foraminifera

Seiglie, George A. 1970.
The distribution of the foraminifers
in the Yabucoa Bay, southeastern Puerto
Rico and its paleoecological significance.
Revista esp. Micropaleont. 2(2):183-208.

foraminifera

Seiglie, George A., 1966.
Distribution of foraminifers in the sediments of
Araya - Los Testigos shelf and upper slope.
Carib.J.Sci., 6(3/4):93-117.

Foraminifera

Seiglie, A., George A., 1965.
Notas sobre las familias Pegiliidae y
Siphoninidae (Foraminiferida) Genero y especies
nuevos.
Carib. J. Sci., 5(1/2):9-13.

Foraminifera

Seiglie, George A., 1965.
Some observations on Recent Foraminifera from
Venezuela.
Contrib. Cushman Found., Foram. Res., 16(2):70-
73.

Foraminifera

Seiglie, George A., 1964.
New and rare foraminifers from Los Testigos
reefs, Venezuela.
Caribbean J. Sci., 4(4):497-512.

Foraminifera

Seiglie, George A., 1964.
Algunas foraminiferos arenaceos recientes de
Venezuela.
Bol. Inst. Oceanogr., Univ. Oriente, Venezuela,
3(1/2):5-14.

foraminifera

Seiglie, George A. y Pedro J. Bermudez, 1965.
Observaciones sobre Foraminifero Rotal; for
mer con cameras suplementarias o estructuras
semejantes.
Bol. Inst. Oceanogr., Univ. Oriente, 4(1):155-
171.

Foraminifera

Sen Gupta, Barun K., 1967.
Distribution of Foraminifera in the sediments
of the Grand Banks: A preliminary report.
Marit. Sed., Halifax, N.S. 3(2/3):61-63.

Foraminifera

Sliter William V., 1970
Inner-neritic Bolivinitidae from
the eastern Pacific margin.
Micropaleontology 16(2): 155-170

Foraminifera

Smith, A. Barrett, 1963.
Distribution of living planktonic Foraminifera
in the northeastern Pacific.
Contrib. Cushman Found, Foram. Res., 14(1):1-15.

foraminifera
Smith, P.B. and C. Emiliani 1968.
Oxygen-isotope analysis of recent tropical Pacific benthonic Foraminifera.
Science 160 (3834):1335-1336.

Foraminifera
Souaya, Fernand Joseph, 1966.
Miocene Foraminifera of the Gulf of Suez, region, U.A.R.
Micropaleontology, 12(4):493-504.

foraminifera
Stainforth, R.M., 1960.
Estado actual de las correlaciones transatlanticas del Oligo-Mioceno por medio de foraminiferos planctonicos.
Memoria Tercer Congreso Geologico Venezolano, 1:382-406.

foraminifera
Stehli, F.G., and C.E. Helsley, 1963.
Paleontologic technique for defining ancient pole positions.
Science, 142(3595):1057-1059.

Foraminifera
Stschedrina, Z. G., 1958.
Foraminifera fauna of the East Antarctic.
Inform. Biull. Sovetsk. Antarkt. Eksped., (3):51-54.

foraminifera
Stubbings, H.G., 1939.
Stratification of biological remains in marine deposits. John Murray Exped., 1933-34, Sci. Repts. 3(3):159-192, 4 textfigs.

foraminifera
Takayanagi, Y., 1953.
Distribution of recent foraminifera from adjacent seas of Japan.
Rec. Ocean. Wks., Japan, n.s., 1(2):78-85.

Foraminifera
*Tinoco, Ivan M., 1965/1966.
Foraminiferos do atol das Rocas.
Trabhs Int.Oceanogr., Univ.Fed.Pernambuco, Recife, (7/8):91-106.

foraminifera
Todd, Ruth, 1965.
The Foraminifera of the Tropical Pacific collections of the "Albatross", 1899-1900. 4. Rataliform families and planktonic families.
U.S. Nat. Mus., Bull., 161:1-139.

Foraminifera (taxonomy)
Todd, Ruth and Doris Low, 1964
Cenomanian (Cretaceous) Foraminifera from the Puerto Rico Trench.
Deep-Sea Research, 11(3):395-414.

foraminifera
Todd, R., & D. Low, 1961
Near-shore foraminifera of Martha's Vineyard Island, Massachusetts.
Contr. Cushman Found. Foram. Res., 12(1): 5-21.

Foraminifera
Uchio, Takayasu, 1964.
Influence of the River Shinano on Foraminifera and sediment grain-size distributions.
In: Papers in Marine Geology, R.L. Miller, Editor, Macmillan Co., N.Y., 411-428.

Foraminifera
Uchio, Takayasu, 1962
Influence of the River Shinano on Foraminifera and sediment grain size distributions.
Publ. Seto Mar. Biol. Lab., 10(2) (Art. 18): 363-392.

Foraminifera
Uchio, Takayasu, 1962
Influence of the River Shinano on Foraminifera and sediment grain size distributions. (In Japanese; English abstract).
J. Oceanogr. Soc., Japan. 20th Ann. Vol., 15-24.

Foraminifera
Uchio, T., 1959.
Ecology of shallow-water Foraminifera off the coast of Noribetsu, southwestern Hokkaido, Japan.
Publ. Seto Mar. Biol. Lab., 7(3):295-302.

Foraminifera
Vilks G., E.N. Anthony and W.T. Williams, 1970.
Application of association-analysis to distribution studies of Recent Foraminifera.
Can. J. Earth Sci., 7(6): 1462-1469.

foraminifera
Waller, H.O., 1960
Foraminiferal biofacies of the South China coast. J. Paleont., 34(6): 1164-1182.

foraminifera
Waller, H.O., & W. Polski, 1959
Planktonic foraminifera of the Asiatic shelf.
Contr. Cushman Found. Foram. Res., 10(4): 123-126.

Foraminifera
Wilcoxon, James A., 1964.
275. Distribution of Foraminifera off the Southern Atlantic coast of the United States.
Contrib. Cushman Found. Foram. Res., 15(1):1-24.

Foraminifera, abundance of
Cifelli, R., and K.N., Sachs, Jr., 1966.
Abundance relationships of planktonic Foraminifera and Radiolaria.
Deep-Sea Res., 13(4):751-753.

Foraminifera, anat.-physiol.
Bé, Allan W.H., 1964.
Influence of depth on shell growth in planktonic Foraminifera. (Abstract).
Geol. Soc., Amer., Special Paper, No. 76:10-11.

foraminifera, anat.
Febvre-Chevalier, Colette et M. Pierre Lecher, 1971.
Etude ultrastructurale des lamelles annelées intracytoplasmiques chez les Foraminifères et les Radiolaires Phaeodariés. C.R. Acad. Sci. Paris 272: 1264-1267.

foraminifera, anatomy
Hofker J., Sr., 1971.
Wall-structure of Globigerines and Globorotaliid Foraminifera.
Revista esp. Micropaleont. 3 (1): 35-60.

foraminifera, anat. physiol.
Lipps, Jere H., and Malcolm G. Erskian, 1969.
Plastogamy in Foraminifera: Glabratella ornatissima (Cushman).
J. Protozool. 16(3): 422-425.

foraminifera, anat. physiol.
Muller, William A., and John J. Lee 1969.
Apparent indispensability of bacteria in foraminiferan nutrition.
J. Protozool. 16(3): 471-478.

Foraminifera, anatomy
Pessagno, Emile A., Jr. and Kei Miyano, 1968.
Notes on the wall structure of Globigerinacea.
Micropaleontology, 14(1):38-50.

foraminifera, anat. physiol.
Hansen, H.J. and Z. Reiss 1972
Scanning electron microscopy of wall structures in some benthonic and planktonic Foraminiferida.
Revista esp. Micropaleont. 4(2): 169-179.

Foraminifera, anat, physiol.
Lengsfeld, A.M., 1969.
Zum Feinbau der Foraminifere Allogromia laticollaris. 1. Mitteilung: Zellen mit ausgestreckten und eingezogenen Rhizopodien.
Helgoländer wiss. Meeresunters, 19(2): 230-261.

Foraminifera, anat., physiol.
Lengsfeld, A.M., 1969.
Nahrungsaufnahme und Verdauung bei der Foraminifera Allogromia laticollaris.
Helgoländer wiss. Meeresunters, 19(3): 385-400.

Foraminifera, anatl, physiol.
Lengsfeld, A.M., 1969.
Zum Feinbau der Foraminifere Allogromia laticollaris. II. Mitteilung: Ausgestreckte und durch Abreissen isolierte Rhizopodien. Helgoländer wiss. Meeresunters, 19(2): 262-283.

Foraminafera, anat.physiol.
Marszalek, Donald S., Ramil C. Wright and William W. Hay, 1969.
Function of the test in Foraminifera. Trans. Gulf Coast Ass. geol. Socs, 19: 341-352.

foraminifera, anat.
Nyholm Karl-Georg. 1973.
The ultrastructure of the test in the Foraminiferan Glandulina.
Zoon 1(1): 11-15

foraminifera, anat.-physiol.
Wiles, William W., 1967.
Pleistocene changes in the pore concentration of a planktonic foraminiferal species from the Pacific Ocean.
Progress in Oceanography, 4:153-160.

foraminiferal associations
Tufescu, Mircea 1973.
Les associations de foraminifères du nord-ouest de la mer Noire.
Revista esp. Micropaleontol. 5(1):15-32.

foraminifera (attached)
Delaca, Ted E., and Jere H. Lipps 1972
The mechanism and adaptive significance of attachment and substrate pitting in the foraminiferan Rosalina globularis D'Orbigny.
J. foram. Res. 2(2): 68-72.

foraminifera, benthic
Allen, George P., et Annick Pujos-Lamy 1970.
Application de l'analyse factorielle (Mode-Q) à l'étude micropaléontologique d'une carotte marine.
Bull. Soc. géol. France (7) 12 (5): 916-925

foraminifera, benthic

Bandy, Orville L., Hans G. Lindenberg and Edith Vincent 1971.
History of research, Indian Ocean foraminifera.
J. mar. biol. Ass. India 13(1): 86-104

foraminifera, benthonic

Bock, Wayne D. 1971.
Paleoecology of a section cored on the Nicaragua Rise, Caribbean Sea.
Micropaleontology 17(2), 181-196.

foraminifera, benthonic

Bock, Wayne D., 1969.
Thalassia testudinum, a habitat and means of dispersal for shallow water benthonic Foraminifera. Trans. Gulf Coast Ass. geol. Socs, 19: 337-340.

foraminifera, benthic

Chekhovskaya, M.P., 1973
Distribution of benthonic foraminifers in the north-eastern part of the Bering Sea. (In Russian; English abstract).
Okeanologiia 13(4):691-696.

Foraminifera, benthonic

Devdariani, A.S. and Kh.M. Saidova, 1972
Algorithmic reconstructions of the latest tectonic movements of the ocean floor (based on benthonic foraminifera). (In Russian; English abstract). Okeanologiia 12(6): 1113-1117.

Foraminifera, benthic

Dupeuble, Pierre Alain, Robert Mathieu, Iradj Momeni, Armelle Poignant, Marie Rosset-Moulinier, Armelle Rouvillois et Maria Ubaldo, 1972. Travaux récents sur les foraminifères actuels des côtes Françaises de la Manche. Mém. Bur. Rech. géol. min. CNEXO 79: 97-100.

foraminifera, benthonic

El-Wakeel, S.K., H.F. Abdou and S.D. Wahby, 1970
Foraminifera from bottom sediments of Lake Maryut and Lake Manzalah, Egypt. Bull. Inst. Oceanogr. Fish. U.A.R. 1: 427-448.

foraminifera species

Emiliani, Cesare, 1969
A new paleontology.
Micropaleontology 15(3): 265-300

foraminifera benthonic, lists of spp.

Frerichs, W.E. 1970.
Distribution and ecology of benthonic Foraminifera in the sediments of the Andaman Sea.
Contrib. Cushman Fdn foramin. Res. 21(4): 123-147

Foraminifera, benthonic

Greiner, Gary O.G., 1970.
Distribution of major benthic foraminiferal groups on the Gulf of Mexico continental shelf
Micropaleontology 16(1): 83-101

Foramifera, benthic

Hadley R.H., C.M. Hurdle, and I.D.J. Burdett, 1965.
A foraminiferal fauna from the western continental shelf, North Island, New Zealand.
N.Z., Dept. Sci. Ind. Res., Bull. 163(N.Z. Oceanogr. Inst. Mem. No. 25):46 pp.

foraminifera, benthonic

Hooper, Kenneth, 1970.
The distribution of modern benthonic foraminifera in the northwest Gulf of St. Lawrence.
Marit. Sediments 6(2):74-78

foraminifera, benthic

Ikeya, Noriyuki, 1970.
Populational ecology of benthonic Foraminifera in Ishikari Bay, Hokkaido, Japan. Rec. oceanogr Wks. Japan, n.s., 10(2): 173-191.

foraminifera, benthic (lists of spp.)

Moulinier, Marie, 1967.
Répartition des foraminifères benthiques dans les sédiments de la Baie de Seine entre le Contentin et le méridien de Ouistreham.
Cah. océanogr., 19(6):477-494.

foraminifera, benthic

Murray, John W. 1973.
Distribution and ecology of living benthic foraminifera.
Heinemann, London xii + 274 pp + 12 pls.
£7.50

foraminifera, benthic

Murray, John W., 1970.
Foraminifera of the western approaches to the English Channel.
Micropaleontology 16(4):471-485.

Foraminifera, benthic

*Murray, John W., 1968.
The living Foraminiferida of Christ Church Harbour, England.
Micropaleontology. 14(1):83-96.

foraminifera, benthic

Murray, J.W., 1965.
Significance of benthic foraminiferids in plankton samples.
J. Palaeont., 39(1):156-157.
abstract JMBA, UK 45(3):794.

foraminifera, benthic

Murray, J.W., 1967.
Production in benthic foraminiferids.
J. nat. Hist 1(6): 61-68.

foraminifera, benthic

Pastouret, L. 1970.
Distribution des foraminifères benthiques dans une carotte de Méditerranée orientale.
Bull. Mus. Anthropol. préhist. Monaco, 16: 155-171

Foraminifera, benthonic

Phleger, Fred B., 1965.
Depth patterns of benthonic Foraminifera in the eastern Pacific.
Progress in Oceanography, 3:273-287.

foraminifera, benthic

Phleger, Fred B and Andrew Soutar 1973.
Production of benthic foraminifera in three east Pacific oxygen minima.
Micropaleontology 19(1): 110-115

foraminifera, benthic

Pujos, Michel, 1971.
Quelques exemples de distribution des foraminifères benthiques sur le plateau continental du Golfe de Gascogne. Relation entre microfaune et environnement.
Cah. océanogr. 23(5): 445-453.

foraminifera, benthic

Pujos-Lamy, Annick 1972.
Les foraminifères agglutinants du Quaternaire récent dans le domaine profond du Golfe de Gascogne
Boreas, Oslo, 1: 185-198.

foraminifera, benthic

Pujos, Michel 1970.
Influence des eaux de type méditerranéen sur la répartition de certains foraminifères benthiques dans le Golfe de Gascogne.
Cah. océanogr. 22(8):827-831

foraminifera, benthic

Rao, K. Kameswara, 1970.
Foraminifera of the Gulf of Cambay.
J. Bombay nat. hist. Soc. 67(2): 259-273

Foraminifera, benthic

Rouvillois, Armelle, 1966.
Contribution à l'étude micropaléontologique de la baie du Roi, au Spitzberg.
Revue Micropaléont., 9 (3):169-176.

Foraminifera, benthonic

Saidova, Kh. M., 1970.
Bottom of the Pacific divided into areas according to the benthonic foraminifera present. (In Russian).
Dokl. Akad. Nauk SSSR, 192(5): 1145-1148.

foraminifera benthonic

Saidova, Kh. M., 1967.
Distribution of benthonic Foraminifera and depths of the Pacific Ocean during Holocene-Wisconsin. (In Russian; English abstract).
Okeanologiia, Akad. Nauk, SSSR, 7(3):483-489.

foraminifera benthonic

Saidova, H.M., 1967.
Sediment stratigraphy and Paleogeography of the Pacific Ocean by benthonic Foraminifera during the Quaternary.
Progress in Oceanography, 4:143-151.

Foraminifera, benthic

Saidova, Kh. M., 1966.
Foraminiferal bottom fauna of the Pacific. (In Russian; English abstract).
Okeanologiia, Akad. Nauk, SSSR, 6(2):276-284.

Foraminifera, benthic

Saidova, Kh. M., 1966.
The distribution of benthic agglutinating foraminiferal species in the Pacific Ocean. (in Russian)
Okeanologiia, Akad. Nauk, SSSR, 6(1):144-147.

foraminifera, benthic

Schafer, Charles T., 1972.
Sampling and spatial distribution of benthonic foraminifera. Limnol. Oceanogr. 16(6): 944-951.

foraminifera, benthic

Schafer, C.T. 1970.
Studies of benthonic foraminifera in the Restigouche Estuary: 1. Faunal distribution patterns near pollution sources.
Marit. Sed, 6(3): 121-134.
Halifax

foraminifera, benthic

Schafer, C.T. 1967.
Preliminary survey of the distribution of living benthonic Foraminifera in Northumberland Strait.
Marit. Sed., Halifax, 3(4): 105-108

foraminifera, benthic

Sellier de Civrieux, J.M. y Jaime Bonilla Ruíz, 1971
La influencia de los parametros fisico-quimicos del fondo en las facies de foraminiferos bentonicos. (In Spanish; English abstract). Bol. Inst. Oceanogr. Univ. Oriente, Venezuela 10(2):15-34.

foraminifera, benthic (lists of spp.)

Sellier de Civrieux, J.M. y Pedro J. Bermudez 1973.
Ecologia y distribución de foraminiferos bentonicos del Golfo de Santa Fe (Venezuela)
Revista esp. Micropaleont. 5(1): 33-80.

foraminifera, benthonic

Slesser, D.H. 1970.
Benthonic foraminiferal ecology in Covehead Bay, Prince Edward Island - a preliminary study.
Marit. Sediments 6(2): 48-64.

foraminifera, benthic

Sliter, William V., 1971.
Predation on benthic foraminifers.
J. Foram. Res. 1(6): 20-29.

foraminifera, benthic

Sliter, William B., and Robert A. Baker 1972.
Cretaceous bathymetric distribution of benthic foraminifers.
J. foram. Res. 2(4): 167-183

foraminifera, benthonic

Streeter, S. Stephen, 1972.
Living benthonic foraminifera of the Gulf of California, a factor analysis of Phleger's (1964) data. Micropaleontology 18(1): 64-73.

foraminifera, benthic

Tapley, Sandra 1969.
Foraminiferal analysis of the Miramichi Estuary.
Marit. Sediments 5(1): 30-39.

foraminifera, benthic

Theyer, Fritz, 1971.
Benthic foraminiferal trends, Pacific-Antarctic Basin. Deep-Sea Res. 18(7): 723-738.

foraminifera, benthic

Theyer, Fritz, 1971.
Benthic foraminifera: bathymetric patterns of calcareous and arenaceous assemblages.
Nature, Lond., 229(7): 207-209

Foraminifera, benthonic

Uchio, Takayasu, 1960
Benthonic Foraminifera of the Antarctic Ocean. Biological results of the Japanese Antarctic Research Expedition, 12. Spec. Publ., Seto Mar. Biol. Lab., 19 pp.

foraminifera biomass

Saidova, Kh.M., 1971.
On Foraminifera distribution near the Pacific Coast of South America. (In Russian; English abstract). Okeanologiia 11(2): 256-265.

foraminifera, calcification of

Orr, W.N., 1967.
Secondary calcification in the foraminiferal genus Globorotalia.
Science, 157(3796):1554-1555.

foraminifera, chemistry of

Belyaeva, N.V., 1973
Peculiarities of the chemical composition of planktonic foraminiferal tests. (In Russian; English abstract). Okeanologiia 13(2):303-306.

Foraminifera, chemistry

Hecht, Alan D. and Samuel M. Savin, 1970.
Oxygen-18 studies of recent planktonic Foraminifera: comparisons of phenotypes and of test parts. Science, 170 (3953): 69-71.

foraminifera, chemistry

King, Jr., Kenneth and P.E. Hare, 1972.
Amino acid composition of planktonic Foraminifera: a paleobiochemical approach to evolution. Science 175 (4029): 1461-1463.

foraminifera, coiling of

Bandy, Orville L., 1967.
Foraminiferal definition of the boundaries of the Pleistocene in southern California, U.S.A.
Progress in Oceanography, 4:27-49.

foraminifera, coiling ratio

Echols, Ronald J., 1973
Foraminifera, leg 19, deep sea drilling project. Initial Repts Deep Sea Drilling Project 19:721-735.

Foraminifera, coiling direction

Robinson, E., 1969.
Coiling directions in planktonic Foraminifera from the coastal group of Jamaica. Trans. Gulf Coast Ass. geol. Socs, 19: 555-558.

foraminifera, coiling ratios

Thiede, Jörn, 1971.
Variations in coiling ratios of Holocene planktonic foraminifera. Deep-Sea Res., 18(8): 823-831.

Foraminifera, cultures of

Adshead, Patricia C., 1967.
Collection and laboratory maintenance of living planktonic Foraminifera.
Micropaleontology, 13(1):32-40.

foraminifera, species diversity

Gibson, Thomas G., and Martin A. Buzas 1973
Species diversity: patterns in modern and Miocene Foraminifera of the eastern margin of North America.
Bull. geol. Soc. Am. 84(6): 217-238.

foraminifera, species diversity

Ikeya, Noriyuki, 1971.
Species diversity of benthonic Foraminifera, off the Shimokita Peninsula, Pacific Coast of north Japan. Rec. oceanogr. Wks, Japan n.s. 11(1): 27-37.

Foraminifera, ecology of

Berger, Wolfgang H., 1969.
Ecologic patterns of living planktonic Foraminifera.
Deep-Sea Res., 16(1):1-24.

Foraminifera, ecology of

Hillman, Norman S., 1966.
Ecology of skeletal plankton.
Antarctic J., United States, 1(5):214-215.

foraminifera, ecology

Lipps, J.H. and J.W. Valentine 1970
The role of Foraminifera in the trophic structure of marine communities.
Lethaia 3(3): 279-286

foraminifera, growth of

Röttger, R., 1972
Analyse von Wachstumskurven von Heterostegina depressa. Mar. Biol. 17(3): 228-242.

Foraminifera, life cycle

Boltovskoy, Esteban, 1965.
Beitrage zur Kenntnis der Jahreszyklen der Foraminiferen.
Int. Rev. ges. Hydrobiol., 50(2):293-296.

Foraminifera, lists of spp.

Aljinović, Slobodan 1969
Sur la microfaune des foraminifères adriatiques et les espèces nouvelles pour cette mer.
Rapp. P.-v. Réun. Comm int. Explor. scient. Mer Mediterr., 19(4): 651-654.

Foraminifera, lists of spp.

Antony, A., 1968.
Studies on the shelf water Foraminifera of the Kerala coast.
Bull. Dept. Mar. Biol. Oceanogr. Univ. Kerala 4:11-154

Foraminifera, lists of spp

Beljaeva, N.V., 1960
Distribution of Foraminifera in the western part of the Bering Sea. Trudy Inst. Okeanol., 32: 158-170.

Foraminifera, lists of spp.

Benda, W.K., and H.S. Puri, 1962.
The distribution of Foraminifera and Ostracoda off the Gulf Coast of the Cape Romano area, Florida.
Trans. Gulf Coast Assoc. geol. Soc., 12:303-342.

Foraminifera, lists spp

Bermúdez, Pedro J., 1965.
Resultado preliminar del estudio de los foraminíferos encontrados en muestras de sedimentos recogidas por la Estación de Investigaciones Marinas de Margarita en la fosa de Cariaco.
Informe de Progresso del Estudio Hidrografico de la Fosa de Cariaco, Fundación La Salle de Ciencias Naturales Estación de Investigaciones Marinas de Margarita, Caracas, Sept. 1965, (mimeographed):25-26.

foraminifera, lists of spp. (some new)

Bermudez, Pedro J., 1960.
Contribucion al estudio de las Globigerinidea de la region Caribe-Antillana (Paleoceno-Reciente).
Memoria Tercer Congreso Geologico Venezolano, 3: 1119-1393.

foraminifera, lists of spp.

Betjeman, K.J. 1969.
Recent Foraminifera from the western Continental shelf of Western Australia.
Contrib. Cushman Fdn foramin. Res 20(4): 119-136.

foraminifera, lists of spp.

Bhalla, S.N., 1970
Foraminifera from Marine beach sands, Madras, and faunal provinces of the Indian Ocean.
Contrib. Cushman Fdn foramin. Res. 21(6): 156-163

Oceanographic Index: Marine Organisms Cumulation, 1946-1973

---

foraminifera, lists of spp.

Blow, W.H., 1970.
Deep sea drilling project, leg 2 Foraminifera from selected samples. Initial Reports of the Deep Sea Drilling Project, Glomar Challenger 2: 357-365.

---

foraminifera, lists of spp.

Boltovskoy, Esteban, 1959.
Foraminiferos recientes del sur de Brasil y sus relaciones con los de Argentina e India del Oeste.
Servicio de Hidrografia Naval, Argentina, Publ., H. 1005:124 pp.

---

foraminifera, lists of spp.

Boltovskoy, Esteban, and Haydée Lena 1970.
Additional note on unrecorded Foraminifera from littoral of Puerto Deseado (Patagonia, Argentina).
Contrib. Cushman Fdn foramin. Res. 21(4): 148-155.

---

foraminifera, lists of spp.

Boltovskoy, Esteban, and Haydée Lena, 1969.
Seasonal occurrences, standing crop and production in benthic Foraminifera of Puerto Deseado.
Contrib. Cushman Fdn foramin. Res. 20(3): 87-95.

Foraminifera, lists of spp.

Boltovskoy, E., y H. Lena, 1966.
Foraminiferos recientes de la zona litoral de Pernambuco (Brasil).
Revta Mus. argent. Cienc. nat. Bernadino Rivadavia Inst. nac. Invest. Cienc. nat., Hidrobiol. 1(8):269-367.

Foraminifera, lists of spp.

Boltovskoy, Esteban, and Haydee Lena, 1966.
Unrecorded Foraminifera from the littoral of Puerto Deseado.
Contrib. Cushman Fdn foramin. Res., 17(4):144-149.

---

foraminifera, lists of species

Brenner, G.J., 1962
Results of the Puritan-American Museum of Natural History Expedition to Western Mexico. 14. A zoogeographical analysis of some shallow water Foraminifera in the Gulf of California.
Bull. Amer. Mus. Nat. Hist., 123(5):255-297.

---

foraminifera, lists of spp.

Burckle, Lloyd H., Tsunemasa Saito and Maurice Ewing, 1967.
A Cretaceous (Turonian) core from the Naturaliste Plateau, southeast Indian Ocean. Deep-Sea Research, 14(4): 421-426.

Foraminifera, lists of spp.

Chen, Chin, 1966.
Calcareous zooplankton in the Scotia Sea and Drake Passage.
Nature, 212(5063):678-681.

Foraminifera, lists of spp.

Chiji, Manzo, and Silvio M. Lopez, 1968.
Regional foraminiferal assemblages in Tanabe Bay, Kii Peninsula, Central Japan.
Publ. Seto mar. biol. Lab., 16(2): 85-125.

Foraminifera, lists of spp

Cifelli, Richard, Vaughan T. Bowen and Raymond Siever, 1966.
Cemented foraminiferal oozes from the Mid-Atlantic Ridge.
Nature, 209 (5018):32-34.

Foraminifera, lists of spp.

Cita, M.B., and M.A. Chierici, 1963
Crociera Talassografica Adriatica 1955. V. Ricerche sui Foraminiferi contenuti in 18 carote prelavate sul fondo del mare Adriatico.
Arch. Oceanogr. & Limnol., 12(3):297-360.

Foraminifera, lists of spp.

Cita, Maria Bianca, and Sara d'Onofrio, 1967.
Climatic fluctuations in submarine cores from the Adriatic Sea.
Progress in Oceanography, 4:161-178.

---

foraminifera, lists of spp.

Cita, Maria B. (Catherine Nigrini) and Stefan Gartner, 1970.
Biostratigraphy. Initial Reports of the Deep Sea Drilling Project, Glomar Challenger, 2: 391-411.

Foraminifera, lists of spp.

Cole, W.S., 1957.
Larger Foraminifera from Eniwetok Atoll drill holes. Bikini and nearby atolls, Marshall Is.
Geol. Sur. Prof. Pap., 260-V:743-784.

---

foraminifera, lists of spp.

De Medeiros Tinoco, Ivan, 1959.
Classificao sistematica dos foraminiferos dos testemunhos de sondagens submarinas efetuadas pelo Navio Escola "Almirante Saldanha" na embocadura do Rio Amazonas (I)
Trabalhos, Inst. Ocean. e Oceanogr., 1(1):107-112.

Foraminifera, lists of spp.

Eade, J.V., 1967.
New Zealand Recent Foraminifera of the families Islandiellidae and Cassidulinidae.
N.Z. Jl mar. Freshwat. Res. 1(4):421-454.

Foraminifera, lists of spp.

Fierro, G., 1961.
Foraminiferi di sedimenti del Mar Ligure.
Rapp. Proc. Verb. Réunions, Comm. Int. Expl. Sci., Mer Méditerranée, Monaco, 16(3):737-744.

Foraminifera, lists of spp.

Forti, Ieda R.S., and Erica Roettger, 1964
Further observations on the seasonal variations of mixohaline Foraminifera from the Patos Lagoon, southern Brazil.
Arch. Oceanogr. Limnol., 15(1):55-61

Foraminifera, lists of spp.

Frerichs, William E. 1971.
Paleobathymetric trends of Neogene foraminiferal assemblages and sea floor tectonism in the Andaman Sea area.
Mar. Geol. 11(3): 159-173

Foraminifera, lists of spp.

Grossman, Stuart, 1967.
Ecology of Rhizopoda and Ostracoda of southern Pamlico Sound Region, North Carolina. I. Living and subfossil rhizopod and ostracode populations.
Paleontol. Contrib. Univ. Kansas, 44(1):7-82.

Foraminifera, lists of spp.

Hofker, J., Sr. 1971.
The Foraminifera of Piscadera Bay, Curaçao.
Studies Fauna Curaçao and other Carib. Is. 35(127):1-62.

Foraminifera, lists of spp.

Hofker, J. Sr., 1969.
Recent Foraminifera from Barbados
Studies Fauna Curaçao and other Carib. Is. 31:1-158
(115)

Foraminifera, lists of spp.

Hofker, J. Sr. 1964.
Foraminifera from the tidal zone in the Netherlands Antilles and the other West Indian islands.
Studies Fauna Curaçao and other Carib. Is. 21(83):1-119

Foraminifera, lists of spp.

Hooper, Kenneth, 1969.
A re-evaluation of eastern Mediterranean Foraminifera using factor-vector analyses.
Contrib. Cushman Fdn foramin. Res. 20(4): 147-151

Foraminifera, lists of spp.

Huang, Tunyow 1970.
New Foraminiferida from the Taiwan Strait, Taiwan, China.
Proc. Geol. Soc. China 13:108-114.

Foraminifera, lists of spp

Iaccarino, Silvia, 1967.
Ricerche sui foraminiferi dell'alto Adriatico estme di 3a campioni di fondo raccolti nella Crociera Adriatica invernale 1966 della N/O Bannock
Arch. Oceanogr. Limnol., 15(1):1-54

Foraminifera, lista of spp.

Kaesler, Roger L., 1966.
Quantitative re-evaluation of ecology and distribution of Recent Foraminifera and Ostracoda of Todos Santos Bay, Baja California, Mexico.
Univ. Kansas, Paleontol. Contrib. Paper 10: 50pp.

Foraminifera, lists of spp.

Kustanowich, S., 1964.
Foraminifera of Milford Sound.
New Zealand Dept. Sci. Ind. Res., Bull., No. 157; New Zealand Oceanogr. Inst., Memoir, No. 17:49-63

---

foraminifera, lists of spp.

Lankford, Robert R., and Fred B Phleger 1973
Foraminifera from the nearshore turbulent zone, western North America.
J. foram. Res. 3(3): 101-132.

Foraminifera, lists of spp.

Le Campion, J., 1968
Foraminifères des principaux biotopes du Bassin d'Arcachon et du proche océan (inventaire faunistique).
Bull. Cent. Etud. Rech. sci. Biarritz. 7(2): 207-391

---

foraminifera, lists of spp.

Lena, Haydee, 1966.
Foraminiferos recientes de Ushuaia (Tierra del Fuego, Argentina).
Ameghiniana, Buenos Aires, 4(9):311-336.

Foraminifera, lists of spp.

Lidz, Louis, 1966.
Deep-sea Pleistocene biostratigraphy.
Science, 154(3755):1448-1451.

---

foraminifera, lists of spp

Macarovici, N., and Bica Cehan Ionesi, 1962.
Distributions des Foraminifères sur la plate-forme continental du nord-ouest de la Mer Noire.
Trav. Mus. Hist. Nat. "Gr. Antipa", Bucarest, 3: 45-60.

Resumes in Roumainian and Russian

Foraminifera, lists of spp.

Murray, J.W., 1965.
On the Foraminiferida of the Plymouth region. Jour. Mar. Biol. Assoc., U.K. 45(2):481-505.

Foraminifera, lists of spp.

Murray, J.W., 1965.
The Foraminiferida of the Persian Gulf. 2. The Abu Dhabi region.
Palaeogr. Palaeoclimatol. Palaeoecol., 1(4): 307-332.

Foraminifera, lists of spp.
Olson, R.K. and R. Goll, 1970.
Biostratigraphy. Initial Repts. Deep Sea Drilling Project, Glomar Challenger 5: 557-567.

foraminifera, lists of spp.
Pessagno, Emile A., Jr. and Jose F. Longoria, 1973
Shore laboratory report on Mesozoic foraminifera, leg 17. Initial Repts Deep Sea Drilling Proj. 17:891-894.

foraminifera, lists of spp.
Rao, K. Kameswara, 1971.
On some foraminifera from the northeastern part of the Arabian Sea.
Proc. Indian Acad. Sci. (B) 73(4):155-178.

Foraminifera, lists of spp
Saidova, Kh. M., 1960
[Distribution of Foraminifera in the bottom sediments of the Okhotsk Sea.] Trudy Inst. Okeanol., 32: 96-157.

Foraminifera, lists of spp.
Schafer, C.T., 1969.
Distribution of Foraminifera along the west coasts of Hudson and James bays, a preliminary report.
Maritime Sediments 5(3):90-94

foraminifera, lists of spp.
Schnitker, Detmar 1971.
Distribution of foraminifera on the North Carolina continental shelf.
Tulane Studies Geol. Paleontol. 8(4):169-215

foraminifera, lists of spp.
Seiglie, George A. 1971.
A preliminary note on the relationships between foraminifera and pollution in two Puerto Rican bays.
Carib. J. Sci. 11(1/2): 93-98.

foraminifera, lists of spp.
Seiglie, George A., 1971.
Foraminiferos de las bahias de Mayagüez y Añasco, y sus alrededores, oeste de Puerto Rico. II. Foraminiferos aglutinados. Distribución general y especies de la plataforma exterior.
Revista esp. Micropaleont. 3(3): 255-276

Foraminifera, lists of spp.
Seiglie, George A., 1969
Notes on species of the genera Buliminella and Bulimina (Foraminiferida). Carib. J. Sci., 9(3/4): 93-116.

Foraminefera, lists of spp.
Seiglie, George A., 1967.
Systematics of the foraminifera from Araya - Los Testigos Shelf and Upper Slope, Venezuela, with special reference to suborder Rotaliina and its distribution.
Carib. J. Sci. 7(3/4):95-133

foraminifera, lists of spp.
Seiglie, George A., and Pedro J. Bermudez, 1963.
Distribucion de los foraminiferos del Golfo de Cariaco.
Bol. Inst. Oceanogr., Univ. de Oriente, Venezuela, 2(1):5-87.

Foraminifera, lists of spp.
Smith, Patsy B. 1973.
Foraminifera of the North Pacific Ocean: a systematic study of Foraminifera from lat 25° to 55°N.
Prof. Pap. U.S. Geol. Surv. 766: 27pp., pls.

Foraminifera, lists of spp.
Smith, Patsy B., 1964.
Recent Foraminifera off Central America. Ecology of benthonic species.
Geol. Survey Prof. Paper, 429-B:55 pp.

Foraminifera, lists of spp.
Tamanova, S.V., and Z.T. Shchedrina,1965.
The Foraminifera fauna in the bottom sediments of the Norwegian Sea. (In Russian).
Trudy vses. nauchno-issled. Inst. morsk. ryb. Khoz. Okeanogr. (VNIRO). 57:285-296.

Foraminifera, lists of spp
Uchio, Takayasu, 1968.
Foraminiferal assemblages in the vicinity of the Seto Marine Biological Laboratory, Shirahama-cho, Wakayama-ken, Japan.
Publs Seto mar. biol. Lab., 15(5):399-417.

Foraminifera, lists of spp.
Uchio, Takayasu, 1962
Recent Foraminifera thanatocoenoses of beach and nearshore sediments along the coast of Wakayama-Ken, Japan.
Publ. Seto Mar. Biol. Lab., 10(1) (Article 8): 133-144.

foraminifera (lists of spp.)
University of Southern California, Allan Hancock Foundation, 1965.
An oceanographic and biological survey of the southern California mainland shelf.
State of California, Resources Agency, State Water Quality Control Board, Publ. No. 27:232 pp. Appendix, 445 pp.

Foraminifera, lists of spp.
Vilks, Gustavs, 1969.
Recent Foraminifera in the Canadian Arctic.
Micropaleontology, 15(1):35-60.

Foraminifera, lists of spp
Waldron, R.P., 1963.
A seasonal ecological study of Foraminifera from Timbalier Bay, Louisiana.
Gulf Res. Repts., 1(4):132-188.

foraminifera, lists of spp.
Walton, William R., 1964.
Ecology of benthonic Foraminifera in the Tampa-Sarasota Bay area, Florida.
In: Papers in Marine Geology, R.L. Miller, Editor, Macmillan Co., N.Y., 429-454.

foraminifera, lists of spp.
Wright, Ramil, 1968.
Miliolidae (foraminiferos) recientes del estuario del Rio Quequeh Grande.
Revta Mus. argent. Cienc. nat. Bernardino Rivadavia, Hidrobiol., 2(7):225-256.

Foraminifera (methods)
Boltovskoy, Esteban, 1966.
Methods for sorting of Foraminifera from plankton samples.
J. Paleontol., 40 (5):1244-1246.

foraminifera, preservation
Saidova, Kh.M., 1968.
Preservation of Foraminifera in water, sediment and intestine of sediment feeders. (In Russian).
Dokl. Akad. Nauk, SSSR, 182(2):453-456.

foraminifera, pyritization of
Seiglie, George A. 1973.
Pyritization in living foraminifera
J. foram. Res. 3(1):1-6.

Foraminifera, solution of
*Ruddiman, William F., and Bruce C. Heezen, 1967.
Differential solution of planktonic Foraminifera.
Deep-Sea Res., 14(6):801-808.

Foraminifera oxygen isotopes
van Donk, Jan and Guy Mathieu, 1969.
Oxygen isotope compositions of Foraminifera and water samples from the Arctic Ocean. J. geophys. Res., 74(13): 3396-3407.

foraminifera, standing crop
Matera, N.J. and J.J. Lee, 1972.
Environmental factors affecting the standing crop of foraminifera in sublittoral and psammolittoral communities of a Long Island salt marsh. Mar. Biol. 14(2): 89-103.

Foraminifera, planktonic
Bandy, Orville L., 1966.
Restrictions of the "Orbalina" datum.
Micropaleontology. 12(1):79-86.

foraminifera, planktonic
Barash, M.S., 1971.
The vertical and horizontal distribution of planktonic Foraminifera in Quaternary sediments of the Atlantic Ocean. In: Micropaleontology of oceans, B.M. Funnell and W.R. Riedel, editors, Cambridge Univ. Press, 433-442.

foraminifera, planktonic
Barash, M.S., 1971.
Paleoclimatic reconstructions based on Quaternary planktonic Foraminifera of the Atlantic Ocean. (In Russian; English abstract).
Okeanologiia 11(6): 1049-1056.

Foraminifera, planktonic
Barash, M.S., 1965.
Distribution of planktonic Foraminifera in sediments of the northern Atlantic Ocean. (in Russian; English abstract).
Okeanolog. Issled. Rezult. Issled. Programme Mezhd. Geofiz. Goda, Mezhd. Geofiz. Komitet Presidiume, Akad. Nauk, SSSR. No. 13:225-235.

foraminifera, planktonic
Bartlett, Grant A. 1967.
Planktonic Foraminifera - new dimensions with the scanning electron microscope.
Can. J. Earth Sci. 5(2): 231-233.

Foraminifera, planktonic (lists of spp).
Bé, Allan W.H., 1968.
Shell porosity of Recent planktonic Foraminifera as a climatic index.
Science, 161(3844):881-884.

foraminifera, planktonic
Bé, A.W.H. 1967.
Foraminifera families Globigerinidae and Globorotaliidae.
Fiches d'Ident. Cons. perm. int. Expl. Mer, Zooplankton 108: 9pp.

Foraminifera, planktonic
Bé, Allan W.H.,1967.
Zoogeography of Antarctic and subantarctic planktonic Foraminifera in the Atlantic and Pacific Ocean sectors.
Antarctic Jl, U.S.A. 2(5):188-189.

Foraminifera, planktonic
Bé, Allan W.H., and David B. Ericson, 1963.
Aspects of calcification in planktonic Foraminifera (Sarcodina).
Annals, New York Acad. Sci., 109(1):65-81.

Foraminifera, planktonic
Be, Allan W.H., and Leroy Lott, 1964.
Shell growth and structure of planktonic Foraminifera.
Science, 145(3634):823-824.

foraminifera, planktonic

Bé, A.W.H. and D.S. Tolderlund, 1971. Distribution and ecology of living planktonic Foraminifera in surface waters of the Atlantic and Indian Oceans. In: Micropalaeontology of oceans, B.M. Funnell and W.R. Riedel, editors, Cambridge Univ. Press, 105-149.

foraminifera, planktonic

Bé, Allan W.H., Gustavs Vilks and Leroy Lott, 1971. Winter distribution of planktonic foraminifera between the Grand Banks and the Caribbean. Micropaleontology 17(1): 31-42

Foraminifera, plankton

Beard, John H., 1969. Pleistocene paleotemperature record based on planktonic foraminifers, Gulf of Mexico. Trans. Gulf Coast Ass. geol. Socs, 19: 535-553.

Foraminifera, planktonic

Belyaeva, N.V. 1969. The distribution of planktonic foraminifers in the water and sediment of the Antarctic Ocean. (In Russian; English abstract). Okeanologiia, 9(6): 1063-1070.

foraminifera, planktonic

Belyaeva, N.V., 1969. Thanatocoenoses of planktonic foraminifers on the Pacific floor. Okeanologiia 9(3): 500-504. (In Russian, English abstract)

FORAMINIFERA, planktonic

Belyaeva, N.V., 1968. Quantitative distribution of planktonic foraminiferal tests in Recent sediments of the Pacific Ocean. (In Russian; English abstract). Okeanologiia, Akad. Nauk, SSSR, 8(1): 111-115.

foraminifera, planktonic

Belyaeva, N.V., 1967. Distribution of planktonic foraminiferal tests over the bottom of the Bay of Bengal and some questions concerning methods of foraminiferal analysis. (In Russian; English abstract). Okeanologiia, Akad. Nauk. SSSR, 7(4): 645-654.

Foraminifera, planktonic

Belyaeva, N.V., 1965. Distribution of planktonic Foraminifera in the Indian Ocean. (In Russian; English abstract). Okeanolog. Issled. Rezult. Issled. Programme Mezhd. Geofiz. Goda, Mezhd. Geofiz. Komitet Presidiume, Akad. Nauk, SSSR, No. 13; 205-211.

Foraminifera, planktonic

Belieeva, N.V., and Kh. M. Saidova, 1965. Relations between the benthic and planktonic Foraminifera in the uppermost layers of Pacific sediments. (In Russian). Okeanologiia, Akad. Nauk, SSSR, 5(6): 1010-1014.
Translation: Scripta Tecnica, Inc. for AGU, 5(6): 56-59.

Foraminifera, planktonic

Berger, Wolfgang H., 1968. Planktonic Foraminifera: selective solution and paleoclimatic interpretation. Deep-Sea Res. 15(1): 31-43.

foraminifera, planktonic

Berger, Wolfgang H. and David J.W. Piper, 1972. Planktonic foraminifera: differential settling, dissolution and redeposition. Limnol. Oceanogr. 17(2): 275-287.

foraminifera, planktonic

Berggren, W.A. 1971. Paleogene planktonic foraminiferal faunas on Legs I-IV (Atlantic Ocean), JOIDES Deep-Sea Drilling Program - a Synthesis. Proc. II Plankton. Conf., Roma 1970, A. Farinacci, editor, 57-77.

foraminifera, planktonic

Berggren, W.A. 1971. Multiple phylogenetic zonations of the Cenozoic based on planktonic Foraminifera. Proc. II Plankton. Conf. Roma 1970, A. Farinacci, editor, 41-56

foraminifera, planktonic

Berggren, William A. 1969. Rates of evolution in some Cenozoic planktonic Foraminifera. Micropaleontology 15(3): 351-365.

Foraminifera, planktonic

Berggren, W.A., 1965. Further comments on planktonic Foraminifera in the Type Thanetian. Contrib. Cushman Found. Foram. Res., 16(3): 125-127.

foraminifera, planktonic

Boltovskoy, Esteban 1973. Daily vertical migration and absolute abundance of living planktonic foraminifera. J. foram. Res. 3(2): 89-94.

foraminifera, planktonic

Boltovskoy Esteban, 1971. Ecology of the planktonic foraminifera living in the surface layer of Drake Passage. Micropaleontology, 17(6): 53-68.

foraminifera, planktonic

Boltovskoy, E., 1971. Planktonic foraminiferal assemblages of the epipelagic zone and their thanatocoenoses. In: Micropaleontology of oceans, B.M. Funnell and W.R. Riedel, editors, Cambridge Univ. Press. 277-288.

foraminifera, planktonic

Boltovskoy, Esteban, 1966. Zonación en las latitudes altas del Pacifico sur segun los foraminiferos planctonicos vivos. Revta Mus. argent. Cienc. nat. Bernardino Rivadavia Inst. nac. Invest. Cienc. nat. Hidrobiol., 2(1): 1-56.

Foraminifera, planktonic

Blow, W.H., 1970. Validity of biostratigraphic correlations based on the Globigerinacea. Micropaleontology, 16(3): 257-268.

foraminifera, planktonic

Boltovskoy, Esteban, 1969. Thanatocenosis de foraminiferos planctonicos en el Estrecho de Mozambique. Revista esp. Micropaleont., 1(2): 117-129

Foraminifera, planktonic

Boltovskoy, Esteban, 1969. Distribution of planktonic Foraminifera as indicators of water masses in the western part of the tropical Atlantic. Actes Symp. Oceanogr. Ressources halieut. Atlant. trop., Abidjan, 20-28 Oct. 1966, UNESCO 45-55.

foraminifera, planktonic

#Boltovskoy, Esteban, 1967. Campaña oceanografica "Corrientes Drake VI" (distribucion de masas de aguas superficiales segun el plancton). Bol. Servicio Hidrografia Naval, Armada Argentina, 4(1): 5-15.

Foraminifera, planktonic

Boltovskoy, Esteban, 1966. Resultados oceanograficos sobre la base del estudio del plancton durante la campaña "Cosetri V". Bol. Servicio Hidrogr. Naval, Publ., Argentina. H. 106: 105-114.

foraminifera (planktonic)

Boltovskoy, Esteban, 1964. Distribucion de los foraminiferos planctonicos vivos en el Atlantico ecuatorial, parte oeste. Rep. Argentina, Sec. Marina, Serv. Hidrogr. Naval, Publ H. 639: 54 pp.
English summary

Foraminifera planktonic

Boltovskoy, Esteban, y Demetrio Boltovskoy 1970. Foraminiferos planctonicos vivos del Mar de la Flota (Antártica). Revista Española Micropaleontologia 2(1): 27-44.

foraminifera, planktonic

Boltovskoy, Esteban and Haydée Lena, 1970. On the decomposition of the protoplasm and the sinking velocity of the planktonic foraminifera. Int. Revue ges. Hydrobiol. 55(5): 797-804.

FORAMINIFERA, PLANKTONIC

Caron, Michele, 1972. Planktonic foraminifera from the Upper Cretaceous of site 98, leg 11, DSDP. Initial Reports, Deep Sea Drilling Project, Glomar Challenger 11, 551-559.

Foraminifera, planktonic

Cifelli, Richard, 1972. The holotypes of Pulvinulina crassata var. densa Cushman and Globigerina spinuloinflata Brady. J. foram. Res. 2(3): 157-159.

foraminifera, planktonic

Cifelli, Richard 1971. On the temperature relationships of planktonic Foraminifera. J. foram. Res. 1(4): 170-177.

foraminifera, planktonic

Cifelli, Richard, 1969. Radiation of Cenozoic planktonic Foraminifera. Syst. Zool. 18(2): 154-168

Foraminifera, planktonic

Cifelli, Richard, 1965. Late Tertiary planktonic Foraminifera associated with a basaltic boulder from the Mid-Atlantic Ridge. J. Mar. Res., 22(3): 73-87.

Foraminifera, planktonic

Cifelli, Richard, 1965. Planktonic Foraminifera from the western North Atlantic. Smithsonian Misc. Coll., (Publ. 4599), 148(4): 1-35.

Foraminifera, planktonic

Cifelli, Richard, 1962. Some dynamic aspects of the distribution of planktonic Foraminifera in the western North Atlantic. J. Mar. Res., 20(3): 201-213.

Oceanographic Index: Marine Organisms Cumulation, 1946-1973

Foraminifera, planktonic

Cifelli, Richard and Roberta K. Smith, 1970.
Distribution of planktonic Foraminifera in
the vicinity of The North Atlantic Current.
Smithson Contrib. Paleobiol. 4: 52 pp.

Foraminifera, planktonic

Cita, Maria Bianca, and Sara d'Onofrio, 1967.
Climatic fluctuations in submarine cores from the
Adriatic Sea.
Progress in Oceanography, 4:161-178.

Foraminifera, planktonic

Collen, J.D. and P. Velle 1973.
Pliocene planktonic Foraminifera, southern
North Island, New Zealand.
J. foram. Res. 3(1): 13-29.

---

foraminifera, pelagic

Dasch, E. Julius, and Pierre E. Biscaye 1971
Isotopic composition of strontium in
Cretaceous-to-Recent pelagic Foraminifera.
Earth planet. Sci. Lett. 11(3): 201-204

Foraminifera, planktonic

Douglas, Robert G., and Clay Rankin, 1969
Cretaceous planktonic Foraminifera from Bornholm
and their zoogeographic significance.
Lethaia, 2(3): 185-217.

---

foraminifera, planktonic

Emiliani, C., 1971.
Depth habitats of growth stages of pelagic
Foraminifera. Science 173(4002): 1122-1124.

Foraminifera, planktonic

Fierro, Giuliano, e Paolo Ceretti, 1966
I foraminiferi planctonici in alcuni
campioni di fondo del Mare Tirreno.
Atti Acad. Ligure, 22(1): 134-150.

---

foraminifera, planktonic

Frerichs, William E., 1971
Planktonic foraminifera in the sediments
of the Andaman Sea.
J. Foram. Res. 1(1): 1-14.

---

foraminifera, planktonic

Frerichs, William E., Mary E. Heiman,
Leon E. Borgman and Allan W.H. Bé, 1972
Latitudinal variations in planktonic
foraminiferal test porosity. I. Optical studies.
J. foram. Res. 2(1): 6-13.

Foraminifera, planktonic

Hadley, W.H., C.M. Hurdle, and I.D.J. Burdett, 1965.
A foraminiferal fauna from the western
continental shelf, North Island, New Zealand.
N.Z. Dept. Sci. Ind. Res., Bull. 163(N.Z.
Oceanogr. Inst. Mem. No. 25):46 pp.

---

foraminifera, planktonic

Hecht, Alan D., and Samuel M. Savin 1972
Phenotypic variation and oxygen
isotope ratios in Recent planktonic
Foraminifera.
J. foram. Res. 2(2): 55-67.

Foraminifera, planktonic

Hemleben Christoph 1969
Zur Morphogenese planktonischer
Foraminiferen.
Zitteliana 1: 91-132 pp.

Foraminifera, planktonic

Herb, René, 1968.
Recent planktonic Foraminifera from sediments
of the Drake Passage, Southern Ocean.
Eclogae geol. Helv. 61(2): 467-480

Foraminifera, planktonic

Herman, Yvonne 1969.
Arctic Ocean Quaternary microfauna
and its relation to paleoclimatology.
Palaeogeogr. Palaeoclimatol. Palaeoecol. 6(4): 251-276.

Foraminifera, planktonic

Honjo, Susumo, and W.A. Berggren, 1967.
Scanning electron microscope studies of
planktonic Foraminifera.
Micropaleontology 13 (4):393-406.

---

foraminifera, planktonic

Jenkins, D. Graham, 1965.
Planktonic Foraminifera and Tertiary intercontinental correlations.
Micropaleontology, 11(3):265-277.

Foraminifera, planktonic

Jones, James I., 1969.
Planktonic Foraminifera as indicator organisms
in the eastern Atlantic Equatorial Current System.
Actes Symp. Oceanogr. Ressources halieut.
Atlant. trop., Abidjan, 20-28 Oct. 1966, UNESCO 213-230.

Foraminifera, planktonic

Jones, James I., 1968.
The relationship of planktonic foraminiferal
populations to water masses in the western
Caribbean and lower Gulf of Mexico.
Bull. mar. Sci., 18(4):946-982.

Foraminifera, planktonic

Jones, James I., 1967.
Significance of distribution of planktonic
Foraminifera in the Equatorial Atlantic Undercurrent.
Micropaleontology, 13(4):489-501.

---

foraminifera, planktonic

Kennett, James P., 1973
Middle and Late Cenozoic planktonic
foraminiferal biostratigraphy of the
southwest Pacific - DSDP leg 21.
Initial Repts, Deep Sea Drilling Project
21:575-639.

Foraminifera, pelagic

Lidz, Barbara, Alexis Kehm and Hendrick Miller, 1968.
Depth habitats of pelagic Foraminifera during
the Pleistocene.
Nature, Lond., 217(5125):245-247.

Foraminifera, planktonic

Lidz, Louis, 1966.
Planktonic Foraminifera in the water column of
the mainland shelf off Newport Beach, California.
Limnol. Oceanogr., 11(2):257-263.

Foraminifera, planktonic

Lipps, Jere H., and Paul H. Ribbe, 1967.
Electron-probe microanalysis of planktonic
Foraminifera.
J. Paleont., 41(2):492-496.

---

foraminifera, planktonic

Luterbacher, Hanspeter, 1972.
Paleocene and Eocene plankton foraminifera,
leg 11, DSDP. Initial Reports, Deep Sea Drilling
Project, Glomar Challenger 11, 547-550.

---

foraminifera, planktonic

Lynts, Georges W., and Charles F. Stehman 1971
Factor-vector models of Middle Eocene
planktonic foraminiferal fauna of Core
6282 Northeast Providence Channel.
Revta esp. Micropaleontol. 3(2): 205-213.

Foraminifera, planktonic

McTavish, R.A., 1966.
Planktonic Foraminifera from the Malaita Group,
British Solomon Islands.
Micropaleontology, 12(1):1-36.

---

foraminifera, planktonic

Parker, Frances L. 1973.
Correlations by planktonic foraminifera
of some Tertiary localities in northern
Italy.
Micropaleontology 19(2): 235-238.

---

foraminifera, planktonic

Parker, Frances L. 1973.
Living planktonic foraminifera
from the Gulf of California.
J. foram. Res. 3(2):70-77.

---

foraminifera, planktonic

Parker, Frances L., 1971.
Distribution of planktonic Foraminifera in
recent deep-sea sediments. In: Micropaleontology of oceans, B.M. Funnell and W.R. Riedel,
editors, Cambridge Univ. Press, 289-307.

Foraminifera, Planktonic

Parker, Frances L., 1965.
Irregular distributions of planktonic Foraminifera and stratigraphic correlation.
Progress in Oceanography, 3:267-272.

Foraminifera, planktonic

Parker, Frances L. and Wolfgang H. Berger, 1971.
Faunal and solution patterns of planktonic
Foraminifera in surface sediments of the
South Pacific. Deep-Sea Res., 18(1): 73-107.

Foraminifera, planktonic

Postuma, J.A. 1971.
Manual of planktonic Foraminifera.
Elsevier Publ. Co. 420 pp., 162 pls. $29.00

Foraminifera, planktonic

Reiss, Z., and B. Luz, 1970.
Test formation pattern in planktonic
foraminiferids.
Revista Española Micropaleontología 2(1): 85-96

---

foraminifera, planktonic

Rosenberg-Herman, Yvonne, 1965.
Étude des sédiments quaternaires de la Mer Rouge
Ann. Inst. Océanogr., Monaco, 42(3):339-415, 12 pls.

---

foraminifera, planktonic

Rotschy, Francoise; et Hervé Chamley 1971.
Comparaison des données des foraminifères planctoniques et des minéraux
argileux dans une carotte nord-atlantique.
Eclogae Geol. Helv. 64(2): 279-289.

Foraminifera, planktonic

Saito, Tsunemase, and Allan W.H. Bé, 1964.
Planktonic Foraminifera from the American
Oligocene.
Science, 145(3633):702-705.

---

foraminifera, planktonic

Seiglie, George A., y Oscar Cucurullo, Jr. 1971.
Foraminiferos planctónicos de las localidades
tipo de la "Caliza Mas Adentro" y de la
"Arcilla Mas", Mioceno y Plioceno, Santo
Domingo.
Carib. J. Sci. 11 (3/4):101-122.

foraminifera, planktonic

Sen Gupta, Barun K., 1971.
The benthonic foraminifera of the
Tail of the Grand Banks.
Micropaleontology. 17(1): 69-98.

foraminifera, planktonic

Sliter, William V., 1972.
Upper Cretaceous planktonic foraminiferal zoogeography and ecology - eastern Pacific margin.
Palaeogeogr. Palaeoclimatol. Palaeoecol. 12(6/2):15-31.

Foraminifera, planktonic

Srinivasan, M.S., 1968.
Late Eocene and Early Oligocene planktonic Foraminifera from Port Elizabeth and Cape Foulwind, New Zealand.
Contrib. Cushman Fdn. foramin. Res.19(4):142-159.

foraminifera, planktonic

Stehman, Charles F. 1972.
Planktonic foraminifera in Baffin Bay, Davis Strait and the Labrador Sea.
Marit. Sediments 8 (1): 13-19

foraminifera, planktonic

Thiede, Jörn, 1973.
Planktonic foraminifera in hemipelagic sediments: shell preservation off Portugal and Morocco.
Bull. Geol. Soc. Am. 84 (8): 2749-2754.

Foraminifera, planktonic

Uchio, Takayasu, 1960.
Planktonic Foraminifera of the Antarctic Ocean. Biological Results of the Japanese Antarctic Research Expedition. Spec. Publ., Seto Mar. Biol. Lab., 9 pp.

Foraminifera, planktonic

Uchio, T., 1959.
Planktonic Foraminifera off the coast of Boso Peninsula and Kinkazan, Japan.
J. Oceanogr. Soc., Japan, 15(3):137-142.

foraminifera, planktonic coiling direction

Bolli, H. M., 1971.
The direction of coiling in planktonic Foraminifera. In: Micropalaeontology of oceans, B.M. Funnell and W.R. Riedel, editors, Cambridge Univ. Press, 639-648.

foraminifera, planktonic, coiling of

Hofker, J., 1972.
Is the direction of coiling in the early stages of an evolution of planktonic foraminifera at random? (50% right and 50% left). Revista esp. Micropaleont. 6 (1): 11-17.

Foraminifera, planktonic, distribution

Bé, Allen W.H., and William H. Hamlin, 1967.
Ecology of Recent planktonic Foraminiferae. 3. Distribution in the North Atlantic during the summer of 1962.
Micropaleontology, 13(1):87-106.

Foraminifera, planktonic, diversity of

Berger, Wolfgang H. and Frances L. Parker 1970.
Diversity of planktonic Foraminifera in deep-sea sediments.
Science 168 (3937): 1345-1347.

foraminifera, planktonic, lists of spp.

Bandy, Orville L., 1967.
Cretaceous planktonic foraminiferal zonation.
Micropaleontology, 13(1):1-31.

Foraminifera, lists of spp.

Barash, M.S., 1964.
Ecology of planktonic Foraminifera in the North Atlantic and their significance for stratigraphic investigations. Regularity of the distribution of oceanic plankton. (In Russian; English abstract). Trudy Inst. Okeanol., Akad. Nauk, SSSR, 65:229-258.

Foraminifera, planktonic, lists of spp

Bé, Allen W.H., and William H. Hamlin, 1967.
Ecology of Recent planktonic Foraminifera. 3. Distribution in the North Atlantic during the summer of 1962.
Micropaleontology, 13(1):87-106.

foraminifera, planktonic, lists of spp.

Berggren, W.A., 1973
Cenozoic biostratigraphy and Paleobiogeography of the North Atlantic.
Initial Repts Deep-Sea Drill. Proj. 12(NSFSP-12): 965-1001.

foraminifera, lists of spps

Bermudez, Pedro J., 1960
Foraminiferos planctonicos del Golfo de Venezuela. Mem. Soc. Cien. Nat. la Salle 20(55): 58-76.

Foraminifera, planktonic (lists of spp.)

Bermudez, Pedro J., y Hans M. Bolli 1969.
Consideraciones sobre los sedimentos del Mioceno medio al Reciente de las costas central y oriental de Venezuela. 3. Los Foraminiferos planctonicos.
Bol. Geol. Ministerio Minas Hidrocarburos, Venezuela 10(20):137-223.

Foraminifera, planktonic, lists of spp.

Berthois, L., A. Crosnier et Y. Le Calvez, 1968.
Contribution a l'étude sédimentologique du plateau continental dans la Baie de Biafra.
Cah. ORSTOM, sér. Océanogr., 6(3/4): 55-86.

Foraminifera, planktonic, lists of spp.

Bhatt, D.K., 1969.
Planktonic Foraminifera from sediments off the Vishakhapatnam Coast, India.
Contrib. Cushman Fdn foramin. Res.20(1):30-36.

Foraminifera, lists of spp. (planktonic)

Blow, W.H., 1970.
Deep sea drilling project, leg 4 Foraminifera from selected samples. Initial Reports of the Deep Sea Drilling Project, Glomar Challenger, 4: 383-400.

Foraminifera (planktonic), lists of spp.

Blow, W.H., 1970.
Deep sea drilling project, leg 3 Foraminifera from selected samples. Initial Reports of the Deep Sea Drilling Project, Glomar Challenger 3: 629-661.

Foraminifera (planktonic), lists of spp.

Bolli, Hans M., 1970.
The Foraminifera of sites 23-31, leg 4. Initial Reports of the Deep Sea Drilling Project, Glomar Challenger, 4:577-643.

foraminifera, planktonic (lists of spp)

Boltovskoy, Esteban, 1969.
Living planktonic Foraminifera at the 90°E meridian from the equator to the Antarctic.
Micropaleontology 15(0):237-255.

Foraminifera, lists of spp.

Boltovskoy, Esteban, 1969.
Tanatocenosis de foraminiferos planctonicos en el Estrecho de Mozambique.
Revta esp. Micropaleontol. 1 (2): 117-129

Foraminifera, planktonic (lists of spp

Boltovskoy, Esteban, 1968.
Living planktonic Foraminifera of the eastern part of the tropical Atlantic.
Revue. Micropaleontol. 11(2):85-98.

foraminifera (planktonic) lists of spp.

Cifelli, Richard, 1967.
Distributional analysis of North Atlantic foraminifera collected in 1961 during Cruises 17 and 21 of the R/V chain.
Contrib. Cushman Found., Foram. Res. 18(3): 118-127

foraminifera, lists of spp.

Miró, Montserrat D. de 1973
Foraminiferos planctónicos vivos de las aguas superficiales de la región de afloramiento del noroeste africano.
Result. Exped. cient. Cornide de Saavedra, Madrid 2:95-108.

foraminifera, planktonic (lists of spp.)

de Miro, Montserrat, D., 1965 (1967).
Comparación de la fauna foraminiferos de los sedimentos de la Fosa de Cariaco con la del area oceanica adyacente.
Memoria Soc. Cienc. nat. La Salle, 25(70/71/72):225-260.

foraminifera, planktonic (lists of spp.

DeMiro, Montserrat, and José Ana Marval, 1967.
Foraminiferos planctonicos vivos de la fosa de Cariaco y del talud continental de Venezuela.
Memoria Soc. Cienc. nat. LaSalle, 27(76):11-34.

foraminifera, planktonic (lists of spp.)

Jenkins, D. Graham, and William N. Orr, 1971.
Cenozoic planktonic foraminiferal zonation and the problem of test solutions.
Revista esp. Micropaleont. 3(3):301-304.

foraminifera, planktonic (lists of spp.)

Kustanowich, S., 1963.
Distribution of planktonic foraminifera in surface sediments of the south-west Pacific Ocean
New Zealand J. Geol. Geophys., 6(4):534-565.

Foraminifera, planktonic, lists of spp

Ruddiman, William F., 1969.
Recent planktonic Foraminifera: dominance and diversity in North Atlantic surface sediments.
Science, 164(3884):1164-1167.

foraminifera (planktonic biostratigraphy)

Poag, C. Wylie, 1972.
Neogene planktonic foraminiferal biostratigraphy of the western North Atlantic: DSDP Leg 11. Initial Reports, Deep Sea Drilling Project, Glomar Challenger 11: 483-543.

foraminifera, planktonic (zonation of)

Berggren, W.A. 1971
Neogene chronostratigraphy, planktonic foraminiferal zonation and the radiometric time scale.
Földtani Közlöny [Bull. Hungarian Geol. Soc.] 101 (2/3): 162-169.

Foraminifera, taxonomy

Nyholm, Karl-Georg, 1973.
To the study of vertical and horizontal taxonomy in Foraminifera.
Zoon 1 (1): 3-9.

foraminifera, Allogromia marina n.sp.
Nyholm, Karl-Georg and Ineg Pertz, 1973.
To the biology of the monothalamous foraminifer Allogromia marina n.sp.
Zoon 1(2): 89-93

Ammonia beccarii
Brooks, Albert L., 1967.
Standing crop, vertical distribution and morphometrics of Ammonia beccarii (Linne).
Limnol. Oceanogr., 12(4):667-684.

"Anomalina" elegans
Hansen, Hans Jørgen, 1967.
Description of seven type specimens of Foraminifera designated by D'Orbigny, 1826.
Biol. Meddelser K. Danske Vidensk. Selskab. 23(16):1-12.

Asterigerina rosacea
Hansen, Hans Jørgen, 1967.
Description of seven type specimens of Foraminifera designated by D'Orbigny, 1826.
Biol. Meddelser, K. Danske Vidensk. Selskab. 23(16):1-12.

Foraminifera, seasonal variations
Reiter, M., 1959
Seasonal variations in intertidal foraminifera of Santa Monica Bay, California.
J. Paleont., 33(4): 606-630.

foraminifera, solubility of
Pytkowicz, R.M., and G.A. Fowler, 1967.
Solubility of Foraminifera in seawater at high pressures.
Geochem. J. Japan, 1(4):169-182

foraminifera species diversity
Huang Tunyow, 1972.
Species diversity of benthonic foraminifers in the Taiwan Strait, Taiwan, China.
Proc. Geol. Soc. China 15:99-110.

foraminifera tests
Deuser, W.G., 1968.
Postdepositional changes in the oxygen isotope ratios of Pleistocene foraminifera tests in the Red Sea. J. geophys. Res., 73(10): 3311-3314.

foraminifera tests
Petelin, V.P., 1970.
The composition of agglutinated material of some recent foraminiferal tests. (In Russian; English abstract). Okeanologiia, 10(1): 63-75.

foraminifera, vertical migration of
Boltovskoy, Estaban, 1966.
La zona de convergencia subtropical/subantartica en el Oceano Atlantico (partie occidental). (Un estudio en base a la investigación de Foraminiferos-indicadores).
Republico Argentia, Sec. Marina, Serv. Hidrogr. Naval, Publ., H. 640:69 pp.

Foraminifera, zonation of
Bandy, Orville L., and Mary E. Wade, 1967.
Miocene-Pliocene boundaries in deep-water environments.
Progress in Oceanography, 4:51-66.

foraminiferal zonation
Bronnimann, Paul and Johanna Resig, 1971.
A Neogene globigerinacean biochronologic timescale of the southwestern Pacific. Initial Repts Deep Sea Drill. Proj. 7(2): 1235-1469.

Bolivina arctica n.sp.
Herman, Yvonne, 1973.
Bolivina arctica, a new benthonic foraminifera from Arctic Ocean sediments.
J. foram. Res. 3(3): 137-141

Bolivina argentea
Lutze, Gerhard F., 1964.
Statistical investigations on the variability of Bolivina argentea Cushman.
Contrib. Cushman Foundation, Foraminiferal Res., 15(3):105-116.

Bolivina daggarius
Parker, F.L., 1955.
Bolivina daggarius nom. nov.
Contr. Cushman Found., Foram. Res. 6(1):52.

Bolivina doniezi
Sliter, William V., 1970
Bolivina doniezi Cushman and Wickenden in clone culture.
Contrib. Cushman Fdn foramin Res. 21(3):87-99

Bolivina vaughani
Lidz, Louis, 1966.
Planktonic Foraminifera in the water column of the mainland shelf off Newport Beach, California.
Limnol. Oceanogr., 11(2):257-263.

Bulimena marginata
Hay, William W., and Philip A. Sandberg, 1967.
The scanning electron microscope, a major break-through for micropaleontology.
Micropaleontology, 13(4):407-418.

Buliminoides, spp.
Seiglie, George A., 1970.
Additional observations on the foraminiferal genus Buliminoides Cushman.
Contrib. Cushman Fdn foramin. Res. 21(3):112-115.

Buliminoides
Seiglie, George A. 1969.
Observaciones sobre el genero de foraminiferos Buliminoides Cushman.
Revista esp. Micropaleontol. 1(3): 327-333.

Buliminoides stainforthi nsp.
Seiglie, George A., 1965.
Un genero nuevo y dos especies nuevas de foraminiferos de los Testigos, Venezuela.
Bol. Inst. Oceanogr., Univ. Oriente, 4(1):51-59.

Candeina nitida
Bé, Allan W.H., and William H. Hamlin, 1967.
Ecology of Recent planktonic Foraminifera. 3. Distribution in the North Atlantic during the summer of 1962.
Micropaleontology, 13(1):87-106.

Candeina nitida
Boltovskoy, Estaban, 1964.
Distribucio de los foraminiferos planctonicos vivos en el Atlantico ecuatorial, parte oeste. (Expedicion "Equalant").
Rep. Argentina, Sec. de Marina, Servicio Hidrogr. Naval, Publ., H. 639:54 pp.

Cenchridium spp.
Loeblich, Alfred R. III,1968.
Cenchridium: foraminiferan not dinoflagellate.
Botanica mar., 11(¼):127-128.

Clavihedbergella, phylogeny of
Bandy, Orville L., 1967.
Cretaceous planktonic foraminiferal zonation.
Micropaleontology, 13(1):1-31.

Cribononion excavatum
Haeke, Friedrich-Wilhelm, 1967.
Zum Jahresgang von Populationen einer Foraminiferen-Art in der westlichen Ostsee.
Meyniana, 17:13-27.

Discobotellina biperforata
Stephenson, W., and May Rees, 1965.
Ecological and life history studies upon a large foraminiferan (Discobotellina biperforata Collins 1958) from Moreton Bay, Queensland. II. Aquarium observations.
Univ. Queensland Pap. (Zool.)2(12):239-255.

Discobotellina biperforata
Stephenson, W., and May Rees, 1965.
Ecological and life history studies upon a large foraminiferan (Discobotellina biperforata Collins 1958) from Moreton Bay, Queensland. I. The life cycle and nature of the test.
Univ. Queensland Pap. (Zool.) 2(10):207-223.

Eggerella advena
Resig, Johanna M., 1963.
Size relationships of Eggerella advena to sediment and depth substrate.
In: Essays in Marine Geology in honor of K.O. Emery, Thomas Clements, Editor, Univ. Southern California Press, 121-126.

Ehrenbergina scamnicola
Seiglie, George A., and Pedro J. Bermudez, 1971.
Foraminiferos de las Bahias de Mayaguez y Añasco y de sus alrededores. I. Ehrenbergina scamnicola sp. nov.
Revista esp. Micropaleontol. 3(1): 67-70.

Elphidium crispum
Moulinier, Marie, 1966.
Variabilité d'une population d'Elphidium de la rade de Brest (N.Finistère) apparentés à Elphidium crispum (Linné).
Revue Micropaléont. 9(3):194-200.

Elphidium alvarezzianum
Lévy, Alain, 1966.
Contribution à l'étude écologique et micropaleontologique de quelques Elphidium (Foraminifères) du Roussellon. Description d'une nouvelle espèce: E. Cuvillieri n. sp.
Vie et Milieu, (A), 17(1):1-8.

Elphidium cuvillieri n. sp.
Lévy, Alain, 1966.
Contribution a l'étude écologique et micropaleontologique de quelques Elphidium (Foraminifères) du Roussellon. Description d'une nouvelle espèce: E. Cuvillieri n. sp.
Vie et Milieu, (A), 17(1):1-8.

Elphidium excavatum
Lévy, Alain, 1966.
Contribution à l'étude écologique et micropaléontologique de quelques Elphidium (Foraminifères) du Roussellon. Description d'une nouvelle espèce: E. Cuvillieri n. sp.
Vie et Milieu, (A), 17(1):1-8.

Elphidium gunteri
Lévy, Alain, 1966.
Contribution a l'étude écologique et micropaleontologique de quelques Elphidium (Foraminifères) du Roussellon. Description d'une nouvelle espèce: E. Cuvillieri n. sp.
Vie et Milieu, (A), 17(1):1-8.

Elphidium lidoense
Lévy, Alain, 1966.
Contribution à l'étude écologique et micropaleontologique de quelques Elphidium (Foraminifères) du Roussellon. Description d'une nouvelle espèce: E. Cuvillieri n. sp.
Vie et Milieu, (A), 17(1):1-8.

Elphidium Macellum aculeatum
Moulinier, Marie, 1966.
Variabilité d'une population d'Elphidium de la rade de Brest (N. Finistère), apparentés à Elphidium crispum (Linné).
Revue Micropaléont. 9(3):194-200.

Elphidium oceanicum
Lévy, Alain, 1966.
Contribution a l'etude écologique et micropaleontologique de quelques Elphidium (Foresminiferes) du Roussellon. Description d'une nouvelle espèce: E. Cuvillieri n. sp.
Vie et Milieu, (A), 17(1):1-8.

Fasciolites oblonga
Hansen, Hans Jørgen 1967.
Description of seven type specimens of Foraminifera designated by D'Orbigny, 1826.
Biol. Meddeler K. Danske Videnskab. Selskab. 23(16):1-12.

Fissurina pellucida n.sp.
Bock, Wayne D., 1968.
Two new species of Foraminifera from the Florida Keys.
Contr. Cushman fdn foramin. Res., 19(1):27-29.

Globaquadrina dehiscens
Cifelli, Richard, 1965.
Late Tertiary planktonic Foraminifera associated with a basaltic boulder from the Mid-Atlantic Ridge.
J. Mar. Res., 22(3): 73-87.

abstract.

Globigeraspis indica n.sp.
Tewari, B.S. and M.P. Singh, 1967(1968).
Two new planktonic Foraminifera from Kutch.
Res. Bull. Panjab Univ., N.S., 18(3/4):425-427.

Globigerina, phylogeny of
Bandy, Orville L., 1967.
Cretaceous planktonic foraminiferal zonation.
Micropaleontology, 13(1):1-31.

Globigerina sp. A
Boltovskoy, Estaban, 1964.
Distribucio de los foraminiferos planctonicos vivos en el Atlantico ecuatorial, parte oeste. (Expedicion "Equalant").
Rep. Argentina, Sec. de Marina, Servicio Hidrogr. Naval, Publ., H. 639:54 pp.

Globigerina
Hofker, J., 1959.
On the splitting of Globigerina.
Contrib. Cushman Found. Foram. Res. No. 191, Vol. 10, pt. 1, pps. 1-9.

Globigerina
Seiglie, George A., 1963.
Una nueva especie del genero Globigerina del Reciente de Venezuela.
Bol. Inst. Oceanogr., Univ. de Oriente, Venezuela, 2(1):89-93.

?Globigerina cf. G. aequilateralis
Cifelli, Richard, 1965.
Late Tertiary planktonic Foraminifera associated with a basaltic boulder from the Mid-Atlantic Ridge.
J. Mar. Res., 22(3): 73-87.

abstract.

Globigerina altispira altispira
Cifelli, Richard, 1965.
Late Tertiary planktonic Foraminifera associated with a basaltic boulder from the Mid-Atlantic Ridge.
J. Mar. Res., 22(3): 73-87.

abstract.

Globigerina apertura
Cifelli, Richard, 1965.
Late Tertiary planktonic Foraminifera associated with a basaltic boulder from the Mid-Atlantic Ridge.
J. Mar. Res., 22(3): 73-87.

abstract.

Globigerina bradyi
Smith, A. Barrett, 1963.
Distribution of living planktonic Foraminifera in the northeastern Pacific.
Contrib., Cushman Found. Foram. Res., 14(1):1-15.

Globigerina bulloides (photo
Adshead, Patricia C., 1967.
Collection and laboratory maintenance of living planktonic Foraminifera.
Micropaleontology, 13(1):32-40.

Globigerina bulloides
Bartlett, Grant A., 1967.
Scanning electron microscope: potentials in the morphology of microorganisms.
Science, 158(3806):1318-1319.

Globigerina bulloides
Bé, Allen W.H., and William H. Hamlin, 1967.
Ecology of Recent planktonic Foraminifera. 3. Distribution in the North Atlantic during the summer of 1962.
Micropaleontology, 13(1):87-106.

Globigerina bulloides
Boltovskoy, Estaban, 1966.
La zona de convergencia subtropical/subantartica en el Oceano Atlantico (partie occidental), (Un estudio en base a la investigación de Foraminiferos- indicadores).
Republico Argentina, Sec. Marina, Serv. Hidrogr. Naval, Publ. H. 640:69 pp.

Globigerina bulloides
Boltovskoy, Estaban, 1966.
Zonación en las latitudes altas del Pacifico sur segun los foraminiferos planctonicos vivos.
Revta Mus. argent. Cienc. nat. Bernardino Rivadavia Inst. nac. Invest. Cienc. nat. Hidrobiol., 2(1):1-56.

Globigerina bulloides
Boltovskoy, Estaban, 1964.
Distribucio de los foraminiferos planctonicos vivos en el Atlantico ecuatorial, parte oeste. (Expedicion "Equalant").
Rep. Argentina, Sec. de Marina, Servicio Hidrogr. Naval, Publ., H.639:54 pp.

Globigerina cf. G. bulloides
Cifelli, Richard, 1965.
Late Tertiary planktonic Foraminifera associated with a basaltic boulder from the Mid-Atlantic Ridge.
J. Mar. Res., 22(3): 73-87.

abstract.

Globigerina bulloides
Cita, M.B., and M.A. Chierici, 1963
Crociera Talassografica Adriatica 1955. V. Ricerche sui Foraminiferi contenuti in 18 carote prelavate sul fondo del mare Adriatico.
Arch. Oceanogr. & Limnol., 12(3):297-360.

planktonic

Globigerina bulloides
Febvre-Chevalier, Colette 1971.
Constitution ultrastructurale de Globigerina bulloides D'Orbigny, 1826 (Rhizopoda-Foraminifera).
Protistologica 7(3): 311-324

Globigerina bulloides
Hada, Yoshina, 1970. (1970)
The protozoan plankton of the Antarctic and Subantarctic seas.
Scient. Repts, Japan. Antarct. Res. Exped., (E)31:51 pp.

Globigerina bulloides
Lidz, Louis, 1966.
Planktonic Foraminifera in the water column of the mainland shelf off Newport Beach, California
Limnol. Oceanogr., 11(2):257-263.

Globigerina bulloides
Lipps, J.H., and J.E. Warme, 1966.
Planktonic foraminiferal biofacies in the Okhotsk.
Contrib. Cushman Found., Foram. Res., 17(4):125-134.

Globigerina bulloides
Parker, Frances L., 1962.
Planktonic foraminiferal species in Pacific sediments.
Micropaleontology, 8(2):219-254.

Globigerina bulloides
Schulz, B., and A. Wulff, 1929
Hydrographie und Oberflächenplankton des westlichen Barentsmeeres im Sommer 1927.
Ber. deutschen wissensch. Komm. F. Meeresforsch. n.s. 4(5):232-372, 13 tables, 25 text figs.

Globigerina bulloides
Smith, A. Barrett, 1963.
Distribution of living planktonic Foraminifera in the northeastern Pacific.
Contrib., Cushman Found., Foram. Res., 14(1):1-15

Globigerina calida n.sp.
Parker, Frances L., 1962.
Planktonic foraminiferal species in Pacific sediments.
Micropaleontology, 8(2):219-254.

Globigerina concinna
Cita, M.B., and M.A. Chierici, 1963
Crociera Talassografica Adriatica 1955. V. Ricerche sui Foraminiferi contenuti in 18 carote prelavate sul fondo del mare Adriatico.
Arch. Oceanogr. & Limnol., 12(3):297-360.

planktonic

Globigerina digitata
Parker, Frances L., 1962.
Planktonic foraminiferal species in Pacific sediments.
Micropaleontology, 8(2):219-254.

Globigerina dutertrei
Boltovskoy, Estaban, 1966.
La zona de convergencia subtropical/subantartica en el Oceano Atlantico (partie occidental), (Un estudio en base a la investigacion de Foraminiferos-indicadores).
Republico Argentina, Sec. Marina, Serv. Hidrogr. Naval, Publ. H. 640:68 pp.

Globigerina dutertri
Boltovskoy, Estaban, 1966.
Zonación en las latitudes altas del Pacifico sur segun los foraminiferos planctonicos vivos.
Revta Mus. argent. Cienc. nat. Bernardino Rivadavia Inst. nac. Invest. Cienc. nat. Hidrobiol., 2(1):1-56.

Globigerina dutertrei
Jones, James I., 1967.
Significance of distribution of planktonic Foraminifera in the Equatorial Atlantic Undercurrent.
Micropaleontology, 13(4):489-501.

Globigerina dutertrei
Boltovskoy, Estaban, 1964.
Distribucio de los foraminiferos planctonicos vivos en el Atlantico ecuatorial, parte oeste. (Expedicion "Equalant").
Rep. Argentina, Sec. de Marina, Servicio Hidrogr. Naval, Publ., H. 639:54 pp.

Globigerina dutertrei
Cita, M.B., and M.A. Chierici, 1963
Crociera Talassografica Adriatica 1955. V. Ricerche sui Foraminiferi contenuti in 18 carote prelavate sul fondo del mare Adriatico.
Arch. Oceanogr. & Limnol., 12(3):297-360.

Globigerina eggeri

Cita, M.B., and M.A. Chierici, 1963
Crociera Talassografica Adriatica 1955.
V. Ricerche sui Foraminiferi contenuti
in 18 carote prelavate sul fondo del
mare Adriatico.
Arch. Oceanogr. & Limnol., 12(3):297-360.

Globigerina eggeri

Lipps, J.H., and J.E. Warme. 1966.
Planktonic foraminiferal biofacies in the
Okhotsk.
Contrib. Cushman Found., Foram. Res., 17(4):125-134.

Globigerina eggeri

Smith, A. Barrett, 1963.
Distribution of living planktonic Foraminifera
in the northeastern Pacific.
Contrib. Cushman Found. Foram. Res., 14(1):1-15.

Globigerina eggeri

Wiles, William W., 1967.
Pleistocene changes in the pore concentration of
a planktonic foraminiferal species from the
Pacific Ocean.
Progress in Oceanography, 4:153-160.

Globigerina falconensis

Boltovskoy, Estaban, 1966.
Zonación en las latitudes altas del Pacifico
sur segun los foraminiferos planctonicos vivos
Revta Mus. argent. Cienc. nat. Bernardino
Rivadavia Inst. nac. Invest. Cienc. nat.
Hidrobiol., 2(1):1-56.

Globigerina falconensis

Parker, Frances L., 1962.
Planktonic foraminiferal species in Pacific
sediments.
Micropaleontology, 8(2):219-254.

Globigerina glutinata

Cifelli, Richard, 1965.
Late Tertiary planktonic Foraminifera
associated with a basaltic boulder from
the Mid-Atlantic Ridge.
J. Mar. Res., 22(3): 73-87.

Globigerina glutinata

Cita, M.B., and M.A. Chierici, 1963
Crociera Talassografica Adriatica 1955.
V. Ricerche sui Foraminiferi contenuti
in 18 carote prelavate sul fondo del mare
Adriatico.
Arch. Oceanogr. & Limnol., 12(3):297-360.

Globigerina glutinata

Smith, A. Barrett, 1963.
Distribution of living planktonic Foraminifera
in the northeastern Pacific.
Contrib. Cushman Found. Foram. Res., 14(1):1-15.

Globigerina incompta n.sp.

Cifelli, Richard, 1961.
Globigerina incompta, a new species of pelagic
Foraminifera from the North Atlantic.
Contrib. Cushman Found., Foram. Res., 12(3):83-86.

Globigerina mexicana

Blow, W.H., and Tsunemasa Saito, 1968.
The morphology and taxonomy of Globigerina
mexicana Cushman, 1925.
Micropaleontology, 14(3):357-360.

Globigerina nepenthes

Bandy, Orville L., and Mary E. Wade, 1967.
Miocene-Pliocene boundaries in deep-water
environments.
Progress in Oceanography, 4:51-66.

Globigerina nepenthes

Cifelli, Richard, 1965.
Late Tertiary planktonic Foraminifera
associated with a basaltic boulder from
the Mid-Atlantic Ridge.
J. Mar. Res., 22(3): 73-87.

Globigerina pachyderma

Bandy, Orville L. 1968.
Cycles in Neogene paleoceanography
and eustatic changes.
Palaeogr. Palaeoclimatol. Palaeoecol. 5(1):
63-75.

Globigerina pachyderma

Bandy, Orville L., 1967.
Foraminiferal definition of the boundaries of
the Pleistocene in southern California, U.S.A.
Progress in Oceanography, 4:27-49.

Globigerina pachyderma

Bé, Allan W.H., and William H. Hamlin, 1967.
Ecology of Recent planktonic Foraminifera. 3.
Distribution in the North Atlantic during the
summer of 1962.
Micropaleontology, 13(1):87-106.

Globigerina pachyderma

Boltovskoy, Estaban, 1966.
Zonacion en las latitudes altas del Pacifico
sur segun los foraminiferos planctonicos vivos.
Revta Mus. argent. Cienc. nat. Bernardino Rivadavia Inst. nac. Invest. Cienc. nat. Hidrobiol.,
2(1): 1-56.

Globigerina pachyderma

Boltovskoy, Estaban, 1966.
La zona de convergencia subtropical/
subantartica en el Oceano Atlantico (partie
occidental). (Un estudio en base a la investifación de Foraminiferos-indicadores).
Republico Argentina, Sec. Marina, Serv. Hidrogr.
Naval, Publ., H. 640:69 pp.

Globigerina pachyderma

Kennett, James P., 1970
Comparison of Globigerina pachyderma
(Ehrenberg) in Arctic and Antarctic areas
Contrib. Cushman Fdn foramin. Res. 21(2):
47-49.

Globigerina pachyderma

Cita, M.B., and M.A. Chierici, 1963
Crociera Talassografica Adriatica 1955.
V. Ricerche sui Foraminiferi contenuti
in 18 carote prelavate sul fondo del
mare Adriatico.
Arch. Oceanogr. & Limnol., 12(3):297-360.

Globigerina pachyderma

Kennett, James P., 1968.
Latitudinal variation in Globigerina pachyderma
(Ehrenberg) in surface sediments of the southwest Pacific Ocean.
Micropaleontology 14(3):305-318.

Globigerina pachyderma

Lidz, Louis, 1966.
Planktonic Foraminifera in the water column of
the mainland shelf off Newport Beach, California
Limnol. Oceanogr., 11(2):257-263.

Globigerina pachyderma

Lipps, J.H., and J.E. Warme, 1966.
Planktonic foraminiferal biofacies in the
Okhotsk.
Contrib. Cushman Found., Foram. Res., 17(4):125-134.

Globigerina pachyderma

Malmgren, Björn, and James P. Kennett
1972.
Biometric analysis of phenotypic variation:
Globigerina pachyderma (Ehrenberg) in
the South Pacific Ocean.
Micropaleontology 18(2):241-248.

Globigerina pachyderma

Parker, Frances L., 1962.
Planktonic foraminiferal species in Pacific
sediments.
Micropaleontology, 8(2):219-254.

Globigerina pachyderma

Smith, A. Barrett, 1963.
Distribution of living planktonic Foraminifera
in the northeastern Pacific.
Contrib. Cushman Found. Foram. Res., 14(1):1-15.

"Globigerina pseudoiota"

Berggren, William A., Richard K. Olsson
and Richard A. Reyment 1967.
Origin and development of the foraminiferal genus Pseudohastigerina Banner
and Blow, 1959. I. Taxonomy and phylogeny
II. Biometric analysis.
Micropaleontology 13(3):265-288.

Globigerina quinqueloba

Bé, Allen W.H., and H. Hamlin, 1967.
Ecology of Recent planktonic Foraminifera. 3.
Distribution in the North Atlantic during the
summer of 1962.
Micropaleontology, 13(1):87-106.

Globigerina quinqueloba

Cita, M.B., and M.A. Chierici, 1963
Crociera Talassografica Adriatica 1955.
V. Ricerche sui Foraminiferi contenuti
in 18 carote prelavate sul fondo del
mare Adriatico.
Arch. Oceanogr. & Limnol., 12(3):297-360.

Globigerina quinqueloba

Hay, William W., and Philip A. Sandberg, 1967.
The scanning electron microscope, a major
breek-through for micropaleontology.
Micropaleontology, 13(4):407-418.

Globigerina quinqueloba

Lidz, Louis, 1966.
Planktonic Foraminifera in the water column of
the mainland shelf off Newport Beach, California
Limnol. Oceanogr., 11(2):257-263.

Globigerina quinqueloba

Lipps, J.H., and J.E. Warme, 1966.
Planktonic foraminiferal biofacies in the
Okhotsk.
Contrib. Cushman Found., Foram. Res., 17(4):125-134.

Globigerina quinqueloba

Parker, Frances L., 1962.
Planktonic foraminiferal species in Pacific
sediments.
Micropaleontology, 8(2):219-254.

Globigerina quinquelobata

Smith, A. Barrett, 1963.
Distribution of living planktonic Foraminifera
in the northeastern Pacific.
Contrib. Cushman Found. Foram. Res., 14(1):1-15.

Globigerina rubescens

Bé, Allan W.H., and William H. Hamlin, 1967.
Ecology of Recent planktonic Foraminifera. 3.
Distribution in the North Atlantic during the
summer of 1962.
Micropaleontology, 13(1):87-106.

globigerina rubescens

Boltovskoy, Estaban, 1966.
La zona de convergencia subtropical/
subantartica en el Oceano Atlantico (partie
occidental), Un estudio en base a la
investigación de Foraminiferos-indicadores).
Republico Argentina, Sec. Marina, Serv. Hidrogr.
Naval, Publ., H. 640:69 pp.

Globigerina rubescens

Boltovskoy, Estaban, 1964.
Distribucio de los foraminiferos planctonicos
vivos en el Atlantico ecuatorial, parte oeste.
(Expedicion "Equalant").
Rep. Argentina, Sec. de Marina, Servicio Hidrogr.
Naval, Publ., H. 639:54 pp.

Globigerina rubescens
Frerichs, William E., 1968.
Pleistocene-Recent boundary and Wisconsin glacial stratigraphy in the northern Indian Ocean.
Science, 159(3822):1456-1458.

Globigerina rubescens
Parker, Frances L., 1962.
Planktonic foraminiferal species in Pacific sediments.
Micropaleontology, 8(2):219-254.

Globigerina rubra
Emiliani, Cesar 1971.
Isotopic paleotemperatures and shell morphology of Globigerinoides rubra in the type section for the Pliocene-Pleistocene boundary.
Micropaleontology 17(2):233-238.

Globigerina utilisendex n.sp.
Jenkins, D. Graham, and W.N. Orr, 1973.
Globigerina utilisindex n.sp. from the upper Eocene-Oligocene of the eastern equatorial Pacific.
J. foram. Res. 3(3):133-136.

Globigerinatheka Kutchensis n.sp
Tewari, B.S., and M.P. Singh, 1967(1968)
Two new planktonic Foraminifera from Kutch.
Res. Bull. Panjab Univ., N.S., 18(3/4):425-427.

Globigerinella adamsi
Parker, Frances L., 1962.
Planktonic foraminiferal species in Pacific sediments.
Micropaleontology, 8(2):219-254.

Globigerinella aequilateralis
Bé, Allen W.H., and William H. Hamlin, 1967.
Ecology of Recent planktonic Foraminifera. 3. Distribution in the North Atlantic during the summer of 1962.
Micropaleontology, 13(1):87-106.

globigerinella aequilateralis
Boltovskoy, Estaban, 1966.
La zona de convergencia subtropical/subantartica en el Oceano Atlantico (partie occidental). (Un estudio en base a la investigación de Foraminiferos-indicadores).
Republico Argentina, Sec. Marina, Serv. Hidrogr. Naval, Publ., H. 640:69 pp.

Globigerinella aequilateralis
Boltovskoy, Estabab, 1964.
Distribucio de los foraminiferos planctonicos vivos en el Atlantico ecuatorial, parte oeste. (Expedicion "Equalant").
Rep. Argentina, Sec. de Marina, Servicio Hidrogr. Naval, Publ., H. 639:54 pp.

Globigerinella siphonifera
Honjo, Susumo, and W.A. Berggren, 1967.
Scanning electron microscope studies of planktonic Foraminifera.
Micropaleontology 13(4):393-406.

Globigerinella siphonophora
Parker, Frances L., 1962.
Planktonic foraminiferal species in Pacific sediments.
Micropaleontology, 8(2):219-254.

Globigerinelloides, phylogeny
Bandy, Orville L., 1967.
Cretaceous planktonic foraminiferal zonation.
Micropaleontology, 13(1):1-31.

globigerinita cretacea
Boltovskoy, Estaban, 1966.
La zona de convergencia subtropical/subantartica en el Oceano Atlantico (partie occidental). (Un estudio en base a la investigación de Foraminiferos-indicadores).
Republico Argentina, Sec. Marina, Serv. Hidrogr. Naval, Publ., H. 640:69 pp.

Globigerinita glutinata
Bé, Allan W.H., and William H. Hamlin, 1967.
Ecology of Recent planktonic Foraminifera. 3. Distribution in the North Atlantic during the summer of 1962.
Micropaleontology, 13(1):87-106.

Globigerinita glutinata
Boltovskoy, Estaban, 1966.
Zonación en las latitudes altas del Pacifico sur segun los foraminifeos planctonicos vivos.
Revta Mus. argent. Cienc. nat. Bernardino Rivadavia Inst. nac. Invest. Cienc. nat. Hidrobiol., 2(1): 1-56.

globigerinita glutinata
Boltovskoy, Estaban, 1966.
La zona de convergencia subtropical/subantartica en el Oceano Atlantico (partie occidental). (Un estudio en base a la investigación de Foraminiferos-indicadores).
Republico Argentina, Sec. Marina, Serv. Hidrogr. Naval, Publ., H. 640:69 pp.

Globigerinita glutinata
Boltovskoy, Estaban, 1964.
Distribucio de los foraminiferos planctonicos vivos en el Atlantico ecuatorial, parte oeste. (Expedicion "Equalant".)
Rep. Argentina, Sec. De Marina, Servicio Hidrogr. Naval, Publ., H. 639:54 pp.

Globigerinita glutinata
Lipps, J.H., and J.E. Warme, 1966.
Planktonic foraminiferal biofacies in the Okhotsk.
Contrib. Cushman Found., Foram. Res., 17(4):125-134.

globigerinita humilis
Boltovskoy, Estaban, 1966.
La zona de convergencia subtropical/subantartica en el Oceano Atlantico (partie occidental). (Un estudio en base a la investigación de Foraminiferos-indicadores).
Republico Argentina, Sec. Marina, Serv. Hidrogr. Naval, Publ., H. 640:69 pp.

Globigerinita cf. humilis
Boltovskoy, Estaban, 1964.
Distribucio de los foraminiferos planctonicos vivos en el Atlantico ecuatorial, parte oeste. (Expedicion "Equalant").
Rep. Argentina, Sec. de Marina, Servicio Hidrogr. Naval, Publ., H. 639:54 pp.

Globigerinita uvula
Boltovskoy, Estaban, 1966.
Zonación en las latitudes altas del Pacifico sur segun los foraminifeos planctonicos vivos.
Revta Mus. argent. Cienc. nat. Bernardino Rivadavia Inst. nac. Invest. Cienc. nat. Hidrobiol., 2(1): 1-56.

Globigerinita uvula
Boltovskoy, Estaban, 1966.
La zona de convergencia subtropical/subantartica en el Oceano Atlantico (partie occidental). (Un estudio en base a la investigación de Foraminiferos-indicadores).
Republico Argentina, Sec. Marina, Serv. Hidrogr. Naval, Publ., H. 640:69 pp.

Globigerinita uvula
Lipps, J.H., and J.E. Warme, 1966.
Planktonic foraminiferal biofacies in the Okhotsk.
Contrib. Cushman Found., Forem. Res., 17(4):125-134.

Globigerinoides
Scott, G.H. 1968.
Comparison of lower Miocene Globigerinoides from the Caribbean and from New Zealand.
N.Z. Jl. Geol. Geophys. 11(2): 376-390.

Globigerinoides
Scott, G.H. 1968.
Comparison of the primary aperture of Globigerinoides from the lower Miocene of Trinidad and New Zealand.
N.Z. Jl Geol. Geophys. 11(2): 356-375.

Globigerinoides adriatica
Cita, M.B., and M.A. Chierici, 1963
Crociera Talassografica Adriatica 1955. V. Ricerche sui Foraminiferi contenuti in 18 carote prelavate sul fondo del mare Adriatico.
Arch. Oceanogr. & Limnol., 12(3):297-360.

Globigerinoides conglotatus
Bandy, Orville L., James C. Ingle, Jr., and William E. Frerichs, 1967.
Isomorphism in "Sphaeroidinella" and "Sphaeroidinellopsis".
Micropaleontology, 13(4):483-488.

Globigerinoides conglobatus
Bé, Allan W.H., and William H. Hamlin, 1967.
Ecology of Recent planktonic Foraminifera. 3. Distribution in the North Atlantic during the summer of 1962.
Micropaleontology, 13(1):87-106.

Globigerinoides conglobatus
Boltovskoy, Estaban, 1966.
La zona de convergencia subtropical/subantartica en el Oceano Atlantico (partie occidental). Un estudio en base a la investigación de Foraminiferos-indicadores).
Republico Argentina, Sec. Marina, Serv. Hidrogr. Naval, Publ., H. 640:69 pp.

Globigerinoides conglobatus
Boltovskoy, Estaban, 1964.
Distribucio de los foraminiferos planctonicos vivos en el Atlantico ecuatorial, parte oeste. (Expedicion "Equalant").
Rep. Argentina, Sec. de Marina, Servicio Hidrogr. Naval, Publ., H. 639:54 pp.

Globigerinoides conglobata
Cita, M.B., and M.A. Chierici, 1963
Crociera Talassografica Adriatica 1955. V. Ricerche sui Foraminiferi contenuti in 18 carote prelavate sul fondo del mare Adriatico.
Arch. Oceanogr. & Limnol., 12(3):297-360.

Globigerinoides conglobatus
Parker, Frances L., 1962.
Planktonic foraminiferal species in Pacific sediments.
Micropaleontology, 8(2):219-254.

Globigerinoides conglobatus gomitulus
Cifelli, Richard, 1965.
Late Tertiary planktonic Foraminifera associated with a basaltic boulder from the Mid-Atlantic Ridge.
J. Mar. Res., 22(3): 73-87.

Globigerinoides "elongatus"
Boltovskoy, Estaban, 1964.
Distribucio de los foraminiferos planctonicos vivos en el Atlantico ecuatorial, parte oeste. (Expedicion "Equalant").
Rep. Argentina, Sec. de Marina, Servicio Hidrogr. Naval, Publ., H. 639:54 pp.

Globigerinoides elongata
Cita, M.B., and M.A. Chierici, 1963
Crociera Talassografica Adriatica 1955. V. Ricerche sui Foraminiferi contenuti in 18 carote prelavate sul fondo del mare Adriatico.
Arch. Oceanogr. & Limnol., 12(3):297-360.

Globigerinoides haitiensis
Bermudez, Pedro J., y Julio R. Farias 1971.
Globigerinoides haitiensis (Coryell y Rivero), un foraminifero planctonico del Terciario superior de la region Caribe Antillana
Mem. Soc. Cienc. nat. La Salle 31 (90): 299-308

Globigerinoides gomitulus
Cita, M.B., and M.A. Chierici, 1963
Crociera Talassografica Adriatica 1955.
V. Ricerche sui Foraminiferi contenuti
in 18 carote prelavate sul fondo del mare
Adriatico.
Arch. Oceanogr. & Limnol., 12(3):297-360.

Globigerinoides helicina
Cita, M.B., and M.A. Chierici, 1963
Crociera Talassografica Adriatica 1955.
V. Ricerche sui Foraminiferi contenuti
in 18 carote prelavate sul fondo del mare
Adriatico.
Arch. Oceanogr. & Limnol., 12(3):297-360.

Globigerinoides obliquus
Cifelli, Richard, 1965.
Late Tertiary planktonic Foraminifera
associated with a basaltic boulder from
the Mid-Atlantic Ridge.
J. Mar. Res., 22(3): 73-87.

Globigerinoides quadrilobatus sacculifer
Parker, Frances L., 1962.
Planktonic foraminiferal species in Pacific
sediments.
Micropaleontology, 8(2):219-254.

Globigerinoides ruber
Bé, Allan W.H., and William H. Hamlin, 1967.
Ecology of Recent planktonic Foraminifera. 3.
Distribution in the North Atlantic during the
summer of 1962.
Micropaleontology, 13(1):87-106.

Globigerinoides ruber
Boltovskoy, Estaban, 1966.
La zona de convergencia subtropical/
subantartica en el Oceano Atlantico (partie
occidental). (Un estudio en base a la
investigación de Foraminiferos-indicadores).
Republico Argentina, Sec. Marina, Serv. Hidrogr.
Naval, Publ., H. 640:69 pp.

Globigerinoides ruber
Boltovskoy, Estaban, 1964.
Distribucio de los foraminiferos planctonicos
vivos en el Atlantico ecuatorial, parte oeste.
(Expedicion "Equalant").
Rep. Argentina, Sec. de Marina, Servicio Hidrogr.
Naval, Publ., H. 639:54 pp.

Globigerinoides rubra
Christiansen, Bengt O., 1965.
A bottom form of the planktonic foraminifer
Globigerinoides rubra (d'Orbigny, 1839).
Pubbl. Staz. Zool., Napoli, 34:197-202.

Globigerinoides ruber
Cifelli, Richard, 1965.
Late Tertiary planktonic Foraminifera
associated with a basaltic boulder from
the Mid-Atlantic Ridge.
J. Mar. Res., 22(3): 73-87.

Globigerinoides rubra
Cita, M.B., and M.A. Chierici, 1963
Crociera Talassografica Adriatica 1955.
V. Ricerche sui Foraminiferi contenuti
in 18 carote prelavate sul fondo del mare
Adriatico.
Arch. Oceanogr. & Limnol., 12(3):297-360.

Globigeriroides rubra
Emiliani, Cesare, 1971.
Isotopic paleotemperatures and shell morphology of Globigerinoides rubra in the type section for the Pliocene-Pleistocene boundary.
Micropaleontology 17(2): 233-238.

GLOBIGERINOIDES RUBER
Glaçon, Gengett, et Jacques Sigal, 1969.
Précisions morphologiques sur la paroi du test de Globorotalia truncatulinoides (d'Orbigny), Globigerinoides ruber (d'Orbigny) et Globigerinoides trilobus (Reuss). Réflexions sur la valeur taxinomique des détails observés.
C. r. hebd. Séanc. Acad. Sci., Paris, (D) 269(11): 987-989

Globigerinoides ruber
Hemleben Christoph 1969
Zur Morphogenese planktonischer Foraminiferen.
Zitteliana 1: 91-132 pp.

Globigerinoides rubra
Honjo, Susumo, and W.A. Berggren, 1967.
Scanning electron microscope studies of
planktonic Foraminifera.
Micropaleontology 13(4):393-406.

Globigerinoides ruber
Jones, James I., 1967.
Significance of distribution of planktonic
Foraminifera in the Equatorial Atlantic Undercurrent.
Micropaleontology, 13(4):489-501.

Globigerinoides ruber
Orr, William N., 1969.
Variation and distribution of Globigerinoides ruber in the Gulf of Mexico.
Micropaleontology, 15(3): 373-379.

Globigerinoides ruber
Parker, Frances L., 1962.
Planktonic foraminiferal species in Pacific
sediments.
Micropaleontology, 8(2):219-254.

Globigerinoides sacculifer
Bé, Allan W.H., and William H. Hamlin, 1967.
Ecology of Recent planktonic Foraminifera. 3.
Distribution in the North Atlantic during the
summer of 1962.
Micropaleontology, 13(1):87-106.

Globigerinoides sacculifer
Bé, Allan W.H., 1965.
The influence of depth on shell growth in
Globigerinoides sacculifer (Brady).
Micropaleontology, 11(1):81-97.

Globigerinoides sacculifera
Cita, M.B., and M.A. Chierici, 1963
Crociera Talassografica Adriatica 1955.
V. Ricerche sui Foraminiferi contenuti
in 18 carote prelavate sul fondo del mare
Adriatico.
Arch. Oceanogr. & Limnol., 12(3):297-360.

Globigerinoides sacculifer
Hemleben Christoph 1969
Zur Morphogenese planktonischer Foraminiferen.
Zitteliana 1: 91-132 pp.

Globigerinoides tenellus
Boltovskoy, Estaban, 1966.
La zona de convergencia subtropical/
subantartica en el Oceano Atlantico (partie
occidental). (Un estudio en base a la
investigación de Foraminiferos-indicadores).
Republico Argentina, Sec. Marina, Serv. Hidrogr.
Naval, Publ., H. 640:69 pp.

Globigerinoides trilobus
Boltovskoy, Estaban, 1966.
La zona de convergencia subtropical/
subantartica en el Oceano Atlantico (partie
occidental). (Un estudio en base a la
investigación de Foraminiferos-indicadores).
Republico Argentia, Sec. Marina, Serv. Hidrogr.
Naval, Publ., H. 640:69 pp.

Globigerinoides trilobus + var.
Boltovskoy, Estaban, 1964.
Distribucio de los foraminiferos planctonicos
vivos en el Atlantico ecuatorial, parte oeste.
(Ep edicion "Equalant").
Rep. Argentina, Sec. de Marina, Servicio Hidrogr.
Naval, Publ. H. 639: 54 pp.

Globigerinoides cf. triloba
Cita, M.B., and M.A. Chierici, 1963
Crociera Talassografica Adriatica 1955.
V. Ricerche sui Foraminiferi contenuti
in 18 carote prelavate sul fondo del mare
Adriatico.
Arch. Oceanogr. & Limnol., 12(3):297-360.

GLOBIGERINOIDES TRILOBUS
Glaçon, Gengett, et Jacques Sigal, 1969.
Précisions morphologiques sur la paroi du test de Globorotalia truncatulinoides (d'Orbigny), Globigerinoides ruber (d'Orbigny) et Globigerinoides trilobus (Reuss). Réflexions sur la valeur taxinomique des détails observés.
C. r. hebd. Séanc. Acad. Sci., Paris, (D) 269(11): 987-989

Globigerinoides trilobus
Jones, James I., 1967.
Significance of distribution of planktonic
Foraminifera in the Equatoriel Atlantic Undercurrent.
Micropaleontology, 13(4):489-501.

Globigerinoides trilobus
Scott G.H. 1971.
Phyletic trees for trans-Atlantic lower Neogene Globigerinoides.
Revista esp. Micropaleont. 3(3): 283-292.

Globigerinoides trilobus f. dehiscens
Blanc-Vernet, Laure et Hervé Chamley, 1971.
Sédimentation à attapulgite et Globigerinoides trilobus f. dehiscens (P. et J.) dans une carotte profonde de Méditerranée orientale.
Deep-Sea Res, 18(6): 631-637.

Globigerinoides trilobus sacculifer
Bandy, Orville L., James C. Ingle, Jr., and William E. Frerichs, 1967.
Isomorphism in "Sphaeroidinells" and
"Sphaeroidinellopsis."
Micropaleontology, 13(4):483-488.

Globigerinoides trilobus trilobus
Cifelli, Richard, 1965.
Late Tertiary planktonic Foraminifera
associated with a basaltic boulder from
the Mid-Atlantic Ridge.
J. Mar. Res., 22(3): 73-87.

Globoquadrina dutertrei
Bé, Allan W.H., and William H. Hamlin, 1967.
Ecology of Recent planktonic Foraminifera. 3.
Distribution in the North Atlantic during the
summer of 1962.
Micropaleontology, 13(1):87-106.

Globorotalia sp. ?
Boltovskoy, Estaban, 1966.
La zona de convergencia subtropical/
subantartica en el Oceano Atlantico (partie
occidental). (Un estudio en base a la
investigación de Foraminiferos-indicadores).
Republico Argentia, Sec. Marina, Serv. Hidrogr.
Naval, Publ., H. 640:69 pp.

Globorotalia
Orr, W.N., 1967.
Secondary calcification in the foraminiferal
genus Globorotalia.
Science, 157(3796):1554-1555.

Globorotalia cf. G. acostaensis
Cifelli, Richard, 1965.
Late Tertiary planktonic Foraminifera
associated with a basaltic boulder from
the Mid-Atlantic Ridge.
J. Mar. Res., 22(3): 73-87.

Globorotalia cavernula n.sp.

Bé, Allan W.H. 1967.
Globorotalia cavernula, a new species of planktonic Foraminifera from the Subantarctic Pacific Ocean.
Contrib. Cushman Found. Foram. Res., 18(3):128-132.

Globorotalia chapmani

Berggren, William A., Richard K. Olsson and Richard A. Reyment 1967.
Origin and development of the foraminiferal genus Pseudohastigerina Banner and Blow, 1959. I. Taxonomy and phylogeny. II. Biometric analysis.
Micropaleontology, 13(3): 265-288.

Globorotalia crassaformis

Bé, Allan W.H., and William H. Hamlin, 1967.
Ecology of Recent planktonic Foraminifera. 3. Distribution in the North Atlantic during the summer of 1962.
Micropaleontology, 13(1):87-106.

Globorotalia crassaformis

Boltovskoy, Esteban, 1964.
Distribucio de los foraminiferos planctonicos vivos en el Atlantico ecuatorial, parte oeste. (Expedicion "Equalant").
Rep. Argentina, Sec. de Marina, Servicion Hidrogr. Naval, Publ., H. 639:54 pp.

globorotalia crassaformis

Jones, James I., 1967.
Significance of distribution of planktonic Foraminifera in the Equatorial Atlantic Undercurrent.
Micropaleontology, 13(4):489-501.

Globorotalia crassaformis

Lidz, Barbara 1972.
Globorotalia crassaformis morphotype variations in Atlantic and Caribbean deep-sea cores.
Micropaleontology 18(2):194-211

Globorotalia cultrata

Jones, James I., 1967.
Significance of distribution of planktonic Foraminifera in the Equatorial Atlantic Undercurrent.
Micropaleontology, 13(4):489-501.

Globorotalia fohsi

Bandy, Orville L., Edith Vincent and Ramil C. Wright 1969.
Chronologic relationships of orbulines to the Globorotalia fohsi lineage.
Revista esp. Micropaleontol. 1(2): 131-145.

Globorotalia fohsi

Bolli, Hans M., 1967.
The subspecies of Globorotalia fohsi Cushman and Ellisor and the zones based on them.
Micropaleontology, 13(4):502-512.

Globorotalia fohsi

Cifelli, Richard, 1968.
A note on the holotype of Globorotalia fohsi Cushman and Ellisor.
Micropaleontology, 14(3):369-370.

Globorotalia fohsi robusta

Cifelli, Richard, 1965.
Late Tertiary planktonic Foraminifera associated with a basaltic boulder from the Mid-Atlantic Ridge.
J. Mar. Res., 22(3): 73-87.

Globorotalia hirsuta

Bé, Allan W.H., and William H. Hamlin, 1967.
Ecology of Recent planktonic Foraminifera. 3. Distribution in the North Atlantic during the summer of 1962.
Micropaleontology, 13(1):87-106.

Globorotalia hirsuta

Boltovskoy, Esteban, 1966.
La zona de convergencia subtropical/subantartica en el Oceano Atlantico (partie occidental). (Un estudio en base a la investigación de Foraminiferos-indicadores).
Republico Argentia, Sec. Marina, Serv. Hidrogr. Naval, Publ. H. 640:69 pp.

Globorotalia hirsuta

Boltovskoy, Esteban, 1964.
Distribucio de los foraminiferos planctonicos vivos en el Atlantico ecuatorial, parte oeste. (Expedicion "Equalant").
Rep. Argentina, Sec. de Marina, Servicio Hidrogr. Naval, Publ., H. 639:54 pp.

Globorotalia inflata

Bé, Allan W.H., and William H. Hamlin, 1967.
Ecology of Recent planktonic Foraminifera. 3. Distribution in the North Atlantic during the summer of 1962.
Micropaleontology, 13(1):87-106.

Globorotalia inflata

Blanc-Vernet, L., et L. Pastouret, 1969.
Précisions sur la valeur du foraminifère Globorotalia inflata (D'Orb.) comme critère climatique en Méditerranée.
Tethys 1(2): 535-538.

Glorborotalia inflata

Boltovskoy, Esteban, 1966.
Zonación en las latitudes altas del Pacifico sur segun los foraminiferos planctonicos vivos.
Revta Mus. argent. Cienc. nat. Bernardino Rivadavia Inst. nac. Invest. Cienc. nat. Hidrobiol., 2(1): 1-56.

Globorotalia inflata

Boltovskoy, Esteban, 1966.
La zona de convergencia subtropical/subantartica en el Oceano Atlantico (partie occidental). (Un estudio en base a la investigación de Foraminiferos-indicadores).
Republico Argentia, Sec. Marina, Serv. Hidrogr. Naval, Publ., H. 640:69 pp.

Globorotalia inflata

Cita, M.B., and M.A. Chierici, 1963
Crociera Talassografica Adriatica 1955.
V. Ricerche sui Foraminiferi contenuti in 18 carote prelavate sul fondo del mare Adriatico.
Arch. Oceanogr. & Limnol., 12(3):297-360.

Globorotalia menardii

Bé, Allan W.H., and William H. Hamlin, 1967.
Ecology of Recent planktonic Foraminifera. 3. Distribution in the North Atlantic during the summer of 1962.
Micropaleontology, 13(1):87-106.

Globorotalia menardii

Bé, Allen W.H., Andrew McIntyre and Dee L. Breger 1966.
Shell microstructure of a planktonic foraminifer, Globorotalia menardii (d'Orbigny).
Eclogae geol. Helvetiae 59(2): 885-896, 7pls.

Globorotalia menardii

Boltovskoy, Esteban, 1966.
La zona de convergencia subtropical/subantartica en el Oceano Atlantico (partie occidental). (Un estudio en base a la investigación de Foraminiferos-indicadores).
Republico Argentia, Sec. Marina, Serv. Hidrogr. Naval, Publ. H. 640:69 pp.

Globorotalia menardii

Frerichs, William E., 1968.
Pleistocene-Recent boundary and Wisconsin glacial stratigraphy in the northern Indian Ocean.
Science, 159(3822):1456-1458.

Globorotalia menardii

Hay, William W., and Philip A. Sandberg, 1967.
The scanning electron microscope, a major break-through for micropaleontology.
Micropaleontology, 13(4):407-418.

Globorotalia menardii

Scott, G.H. 1973.
Ontogeny and shape in Globorotalia menardii.
J. foram. Res. 3(3):142-146.

Globorotalia menardii & var.

Boltovskoy, Esteban, 1964.
Distribucio de los foraminiferos planctonicos vivos en el Atlantico ecuatorial, parte oeste. (Expedicion "Equalant").
Rep. Argentina, Sec. De Marina, Servicio Hidrogr. Naval, Publ., H. 639:54 pp.

Globorotalia menardii flexuosa

Bé, Allan W.H. and Andrew McIntyre, 1970.
Globorotalia menardii flexuosa (Koch): An 'extinct' foraminiferal subspecies living in the northern Indian Ocean. Deep-Sea Res., 17(3): 595-601.

Globorotalia pachyderma

Jenkins, D. Graham 1967.
Recent distribution, origin and coiling ratio changes in Globorotalia pachyderma.
Micropaleontology 13(2): 195-203.

Globorotalia punctaculata

Cita, M.B., and M.A. Chierici, 1963
Crociera Talassografica Adriatica 1955.
V. Ricerche sui Foraminiferi contenuti in 18 carote prelavate sul fondo del mare Adriatico.
Arch. Oceanogr. & Limnol., 12(3):297-360.

Globorotalia scitula

Bé, Allan W.H., and William H. Hamlin, 1967.
Ecology of Recent planktonic Foraminifera. 3. Distribution in the North Atlantic during the summer of 1962.
Micropaleontology, 13(1):87-106.

Globorotalia scitula

Boltovskoy, Esteban, 1966.
Zonación en las latitudes altas del Pacifico sur segun los foraminiferos planctonicos vivos.
Revta Mus. argent. Cienc. nat. Bernardino Rivadavia Inst. nac. Invest. Cienc. nat. Hidrobiol., 2(1):1-56.

Globorotalia scitula

Boltovskoy, Esteban, 1966.
La zona de convergencia subtropical/subantartica en el Oceano Atlantico (partie occidental). (Un estudio en base a la investigación de Foraminiferos-indicadores).
Republico Argentia, Sec. Marina, Serv. Hidrogr. Naval, Publ., H. 640:69 pp.

Globorotalia scitula

Boltovskoy, Esteban, 1964.
Distribucio de los foraminiferos planctonicos vivos en el Atlantico ecuatorial, parte oeste. (Expedicion "Equalant").
Rep. Argentina, Sec. de Marina, Servicio Hidrogr. Naval, Publ., H. 639:54 pp.

Globorotalia scitula gigantea

Cifelli, Richard, 1965.
Late Tertiary planktonic Foraminifera associated with a basaltic boulder from the Mid-Atlantic Ridge.
J. Mar. Res., 22(3): 73-87.

Globorotalia tosaensis

Berggren, W.A., J.D. Phillips, A. Bertels and D. Wall, 1967.
Late Pliocene-Pleistocene stratigraphy in deep sea cores from the south-central North Atlantic.
Nature, 216(5112):253-255.

Globorotalia scitula

Cita, M.B., and M.A. Chierici, 1963
Crociera Talassografica Adriatica 1955.
V. Ricerche sui Foraminiferi contenuti in 18 carote prelavate sul fondo del mare Adriatico.
Arch. Oceanogr. & Limnol., 12(3):297-360.

Globorotalia scitula

Smith, A. Barrett, 1963.
Distribution of living planktonic Foraminifera in the northeastern Pacific.
Contrib., Cushman Found. Foram. Res., 14(1):1-15.

Globorotalia truncatulnoides

Babcock, Laurel C. 1970
Micropaleontologic correlation of a group of cores from a submarine plain Southeast of Puerto Rico.
Naval Underwater Systems Center NUSC Rept TD93: 12 pp. (Unpublished manuscript)

Globorotalia truncatuloides

Bé, Allan W.H., and William H. Hamlin, 1967.
Ecology of Recent planktonic Foraminifera. 3. Distribution in the North Atlantic during the summer of 1962.
Micropaleontology, 13(1):87-106.

Globorotalia truncatulinoides

Belderson, R.H., and A.S. Laughton, 1966.
Correlation of some Atlantic turbidites.
Sedimentology 7(2):103-116.

Globorotalia truncatulinoides

Berggren, W.A., J.D. Phillips, A. Bertels and D. Wall, 1967.
Late Pliocene-Pleistocene stratigraphy in deep sea cores from the south-central North Atlantic.
Nature, 216(5112):253-255.

Globorotalia truncatuloides

Blanc, François, Laure Blanc-Vernet et Joël Le Campion 1972.
Application paléoécologique de la méthode d'analyse factorielle en composantes principales: interprétation des microfaunes de foraminifères planctoniques Quaternaires en Méditerranée. I. Étude des espèces de Méditerranée occidentale
Tethys 4(5):761-768

Globorotalia truncatulinoides

Boltovskoy, Estaban, 1966.
Zonación en las latitudes altas del Pacifico sur segun les foraminiferos planctonicos vivos. Revta Mus. argent. Cienc. nat. Bernardino Rivadavia Inst. nac. Invest. Cienc. nat. Hidrobiol., 2(1): 1-56.

Globorotalia truncatulinoides

Boltovskoy, Estaban, 1966.
La zona de convergencia subtropical/subantartica en el Oceano Atlantico (partie occidental), (Un estudio en base a la investigación de Foraminiferos-indicadores).
Republico Argentia, Sec. Marina, Serv. Hidrogr. Naval, Publ., H. 640:69 pp.

Globorotalia truncatulinoides

Cita, M.B., and M.A. Chierici, 1963
Crociera Talassografica Adriatica 1955. V. Ricerche sui Foraminiferi contenuti in 18 carote prelavate sul fondo del mare Adriatico.
Arch. Oceanogr. & Limnol., 12(3):297-360.

Globorotalia truncatulinoides

Ericson, D.B., G. Wollin and J. Wollin, 1955.
Coiling direction of Globorotalia truncatulinoides in deep-sea cores. Deep-Sea Res. 2(2):152-158, 4 textfigs.

Globorotalia truncatulinoides

Glaçon, Georgette, et Jacques Sigal, 1969.
Précisions morphologiques sur la paroi du test de Globorotalia truncatulinoides (d'Orbigny), Globigerinoides ruber (d'Orbigny) et Globigerinoides trilobus (Reuss). Réflexions sur la valeur taxonomique des détails observés.
C. r. hebd. Séanc. Acad. Sci., Paris, (D) 269 (11): 987-989

Globorotalia truncatulinoides

Kennett, James P., 1968.
Globorotalia truncatulinoides as a paleo-oceanographic index.
Science, 159(3822):1461-1463.

Globorotalia truncatuloides

Theyer, F., 1973
Reply
Nature (phys. Sci.) Lond. 244 (133): 47-45.

Globorotalia truncatuloides

Watkins, N.D., J.P. Kennett and P. Vella 1973.
Palaeomagnetism and the Globorotalia truncatuloides datum in the Tasman Sea and Southern Ocean.
Nature (phys. sci.) Lond. 244 (133): 45-47.

Globotruncana, phylogeny of

Bandy, Orville L., 1967.
Cretaceous planktonic foraminiferal zonation.
Micropaleontology, 13(1):1-31.

Gublerina, phylogeny of

Bandy, Orville L., 1967.
Cretaceous planktonic foraminiferal zonation.
Micropaleontology, 13(1):1-31.

Gyrodina orbicularis

Hansen Hans Jørgen 1967.
Description of seven type specimens of Foraminifera designated by D'Orbigny 1826.
Biol. Meddeleser K. Danske Videnskab. Selskab. 23 (16): 1-12.

Hantkennia alabamensis

Honjo, Susumo, and W.A. Berggren, 1967.
Scanning electron microscope studies of planktonic Foraminifera.
Micropaleontology 13(4):393-406.

Hastigerina aequilateralis

Cita, M.B., and M.A. Chierici, 1963
Crociera Talassografica Adriatica 1955. V. Ricerche sui Foraminiferi contenuti in 18 carote prelavate sul fondo del mare Adriatico.
Arch. Oceanogr. & Limnol., 12(3):297-360.

Hastigerina eocenica

Berggren, William A., Richard K. Olsson and Richard A. Reyment 1967.
Origin and development of the foraminiferal genus Pseudohastigerina Banner and Blow, 1959. I. Taxonomy and phylogeny. II. Biometric analysis.
Micropaleontology 13(3): 265-288.

Hastigerina pelagica

Bé, Allan W.H., and William H. Hamlin, 1967.
Ecology of Recent planktonic Foraminifera. 3. Distribution in the North Atlantic during the summer of 1962.
Micropaleontology, 13(1):87-106.

Hastigerina pelagica

Boltovskoy, Estaban, 1966.
La zona de convergencia subtropical/subantartica en el Oceano Atlantico (partie occidental), (Un estudio en base a la investigación de Foraminiferos-indicadores).
Republico Argentia, Sec. Marina, Serv. Hidrogr. Naval, Publ., H. 640:69 pp.

Hastigerina pelagica

Boltovskoy, Estaban, 1964.
Distribucio de los foraminiferos planctonicos vivos en el Atlantico ecuatorial, parte oeste. (Expedicion "Equalant").
Rep. Argentina, Sec. de Marina, Servicio Hidrogr. Naval, Publ., H. 639:54 pp.

Hastigerina pelagica

Hemleben Christoph 1969
Zur Morphogenese planktonischer Foraminiferen.
Zitteliana 1: 91-132 pp.

Hastigerina pelagica

Parker, Frances L., 1962.
Planktonic foraminiferal species in Pacific sediments.
Micropaleontology, 8(2):219-254.

Hastigerinoides, phylogeny of

Bandy, Orville L., 1967.
Cretaceous planktonic foraminiferal zonation.
Micropaleontology, 13(1):1-31.

Hedbergella, phylogeny of

Bandy, Orville L., 1967.
Cretaceous planktonic foraminiferal zonation.
Micropaleontology, 13(1):1-31.

Hemidiscella palabunda n. sp.

Bock, Wayne D., 1968.
Two new species of Foraminifera from the Florida Keys.
Contr. Cushman fdn foramin. Res., 19(1):27-29.

Heterohelix, phylogeny of

Bandy, Orville L., 1967.
Cretaceous planktonic foraminiferal zonation.
Micropaleontology, 13(1):1-31.

Heterostegina depressa

Dietz-Elbrächter, G., 1971.
Untersuchungen über die Zooxanthellen der Foraminifere Heterostegina depressa Orbigny 1826.
Meteor-Forsch.-Ergebnisse (C) 6: 41-47

Heterostegina depressa

Lutze, G.F., B. Grabert und E. Seibold 1971.
Lebendbeobachtungen an Gross-Foraminiferen (Heterostegina) aus dem Persischen Golf.
Meteor-Forsch.-Ergebnisse (C) 6: 21-40

Heterostegina depressa

Röttger, R., 1972.
Die Kultur von Heterostegina depressa (Foraminifera: Nummulitidae). Mar. Biol. 15(2): 150-159.

Heterostegina depressa

Röttger, R. and W.H. Berger, 1972.
Benthic Foraminifera: morphology and growth in clone cultures of Heterostegina depressa.
Mar. Biol, 15(1): 89-94.

Homotrema rubrum

MacKenzie, F.T., L.D. Kulm, R.L. Cooley and J.T. Barnhart, 1965.
Homotrema rubrum (Lamarck), a sediment transport indicator.
J. Sed. Petr., 35(1):265-272.

Marginopora

Smith, Roberta K., 1968.
An intertidal Marginopora colony in Suva Harbor, Fiji.
Contr. Cushman fdn foramin. Res., 19(1):12-17.

Neogloboquadrina n. gen.

Bandy, Orville L., William E. Frerichs, and Edith Vincent, 1967.
Origin, development and geologic significance of Neogloboquadrina Frerichs and Vincent, gen. nov.
Contrib. Cushman Fdn. foramin. Res., 18(4):152-157.

"Nummulina" discoidalis
Hansen, Hans Jørgen 1967.
Description of seven type specimens of Foraminifera designated by D'Orbigny, 1826.
Biol. Meddeleser, K. Danske Videnskab. Selskab. 23(10):1-12.

Orbulina
Jenkins, D. Graham, 1968.
Acceleration of the evolutionary rate in the Orbulina lineage.
Contrib. Cushman Fdn. foramin. Res., 19(4):133-139.

Orbulina bilobata
Cita, M.B., and M.A. Chierici, 1963
Crociera Talassografica Adriatica 1955.
V. Ricerche sui Foraminiferi contenuti in 18 carote prelavate sul fondo del mare Adriatico.
Arch. Oceanogr. & Limnol., 12(3):297-360.

Orbulina datum
Soediono, H., 1967.
In defense of the Orbulina datum: a review of current opinion.
Geol. en Mijnbouw, 46(10):363-368.

Orbulina suturalis
Cita, M.B., and M.A. Chierici, 1963
Crociera Talassografica Adriatica 1955.
V. Ricerche sui Foraminiferi contenuti in 18 carote prelavate sul fondo del mare Adriatico.
Arch. Oceanogr. & Limnol., 12(3):297-360.

Orbulina universa
Bé, Allan W.H., and William H. Hamlin, 1967.
Ecology of Recent planktonic Foraminifera. 3. Distribution in the North Atlantic during the summer of 1962.
Micropaleontology, 13(1):87-106.

Orbulina universa
Bé, Allan W.H., Stanley M. Harrison and Leroy Lott, 1973.
Orbulina universa d'Orbigny in the Indian Ocean.
Micropaleontology 19(2): 150-192.

Orbulina universa
Boltovskoy, Esteban, 1966.
La zone de convergencia subtropical/subantertica en el Oceano Atlantico (partie occidental). (Un estudio en base a la investigación de Foraminiferos-indicadores).
Republico Argentia, Sec. Marine, Serv. Hidrogr. Naval, Publ., H. 640:69 pp.

Orbulina universa
Boltovskoy, Esteban, 1964.
Distribucio de los foraminiferos planctonicos vivos en el Atlantico ecuatorial, parte oeste. (Expedicion "Equalant").
Rep. Argentina, Sec. de Marina, Servicion Hidrogr. Naval, Publ., H. 639:54 pp.

Orbulina universa
Cita, M.B., and M.A. Chierici, 1963
Crociera Talassografica Adriatica 1955.
V. Ricerche sui Foraminiferi contenuti in 18 carote prelavate sul fondo del mare Adriatico.
Arch. Oceanogr. & Limnol., 12(3):297-360.

Orbulina universa
Smith, A. Barrett, 1963.
Distribution of living planktonic Foraminifera in the northeastern Pacific.
Contrib., Cushman Found. Foram. Res., 14(1):1-15.

Orbulina universa suturalis
Cifelli, Richard, 1965.
Late Tertiary planktonic Foraminifera associated with a basaltic boulder from the Mid-Atlantic Ridge.
J. Mar. Res., 22(3): 73-87.

Pararotalia bisaculeata
Hansen, Hans Jørgen 1967.
Description of seven type specimens of Foraminifera designated by D'Orbigny, 1826.
Biol. Meddeleser, K. Danske Videnskab. Selskab. 23(16):1-12.

Parvigenerina spiculsta nsp.
Seiglie, George A., 1965.
Un genero nuevo y dos especies nuevas de foraminiferos de los Testigos, Venezuela.
Bol. Inst. Oceanogr., Univ. Oriente, 4(1):51-59.

Planctostoma bermudezi n. sp.
Seiglie, George A. 1972.
Foraminifers of the Mayaguez and Añasco bays and their surroundings, western Puerto Rico. 3. Planctostoma bermudezi, sp. nov.
Revista esp. Micropaleontol. 6(1): 5-9

Planoglobulina, phylogeny of
Bandy, Orville L., 1967.
Cretaceous planktonic foraminiferal zonation.
Micropaleontology, 13(1):1-31.

Planomalvina, phylogeny
Bandy, Orville L., 1967.
Cretaceous planktonic foraminiferal zonation.
Micropaleontology, 13(1):1-31.

Polystomammina ngen.
Seiglie, George A., 1965.
Un genero nuevo y dos especies nuevas de foraminiferos de los Testigos, Venezuela.
Bol. Inst. Oceanogr., Univ. Oriente, 4(1):51-59.

Praeglobotruncana, phylogeny of
Bandy, Orville L., 1967.
Cretaceous planktonic foraminiferal zonation.
Micropaleontology, 13(1):1-31.

Praeorbulina glomerosa curva
Honjo, Susumo, and W.A. Berggren, 1967.
Scanning electron microscope studies of planktonic Foraminifera.
Micropaleontology 13(4):393-406.

Proxifrons
Vella, Paul, 1964.
The type species of the foraminiferal genus Proxifrons, a correction.
New Zealand Jour. Geol. & Geophys., 7(2):402.

Pseudohastigerina micra
Berggren, William A., Richard K. Olsson and Richard A. Reyment 1967.
Origin and developement of the foraminiferal genus Pseudohastigerina Banner and Blow, 1959. I. Taxonomy and phylogeny. II. Biometric analysis.
Micropaleontology, 13(3): 265-288.

Pseudohastigerina sherkiverensis n.sp.
Berggren, William A., Richard K. Olsson and Richard A. Reyment 1967.
Origin and development of the foraminiferal genus Pseudohastigerina Banner and Blow, 1959. I. Taxonomy and phylogeny. II. Biometric analysis.
Micropaleontology 13(3): 265-288.

Pseudohastigerina wilcoxensis
Berggren, William A., Richard K. Olsson and Richard A. Reyment 1967.
Origin and development of the foraminiferal genus Pseudohastigerina Banner and Blow 1959. I. Taxonomy and phylogeny. II. Biometric analysis.
Micropaleontology 13(3): 265-288.

Pulleniatina
Banner, F.T., and W.H. Blow 1967.
The origin, evolution and taxonomy of the foraminiferal genus Pulleniatina Cushman 1927.
Micropaleontology 13(2): 133-162.

Pulleniatina obliquiloculata
Bé, Allan W.H., and William H. Hamlin, 1967.
Ecology of Recent planktonic Foraminifera. 3. Distribution in the North Atlantic during the summer of 1962.
Micropaleontology, 13(1):87-106.

Pulleniatina obliquiloculata
Boltovskoy, Estaban, 1964.
Distribucio de los foraminiferos planctonicos vivos en el Atlantico ecuatorial, parte oeste. (Expedicion "Equalant").
Rep. Argentina, Sec. de Marina, Servicio Hidrogr. Naval, Publ., H. 639:54 pp.

Pulleniatina obliquiloculata
Jones, James I., 1967.
Significance of distribution of planktonic Foraminifera in the Equatorial Atlantic Undercurrent.
Micropaleontology, 13(4):489-501.

Pulleniatina spectabilis n.sp.
Parker, Frances L., 1965.
A new planktonic species (Foraminiferida) from the Pliocene of Pacific deep-sea cores.
Contrib. Cushman Found., Foram. Res., 16(4): 151-152.

Quinqueloculina suberiana
Hay, William W., and Philip A. Sandberg, 1967.
The scanning electron microscope, a major break-through for micropaleontology.
Micropaleontology, 13(4):407-418.

Rosalina
Todd, Ruth 1965.
A new Rosalina (Foraminifera) parasitic on a bivalve.
Deep-Sea Research 12(6): 831-837.

Rosalina adhaerens
Murray, J.W., 1965.
The Foraminiferida of the Persian Gulg.
I. Rosalina adhaerens Sp. Nov.
Ann. Mag. Nat. Hist., (13)8(85):77-79.

Rosalina floridana
Angell, R.W., 1967.
The process of chamber formation in the foraminifer Rosalina floridana (Cushman).
J. Protozool., 14(4):566-574.

Rosalina floridana
Angell, Robert W., 1967.
Test recalcification in Rosalina floridana (Cushman).
Contrib. Cushman Fdn. foramin. Res., 18(4):176-177.

Rosalina floridana
Hedley, R.H., and J. St. J. Wakefield, 1967.
Clone culture studies of a new rodalinid foraminifer from Plymouth, England and Wellington, New Zealand.
J. mar. biol. Ass., U.K., 47(1):121-128.

Rosalina globularis
Hedley, R.H., and J. St. J. Wakefield, 1967.
Clone culture studies of a new rosalinid foraminifer from Plymouth, England and Wellington, New Zealand.
J. mar. biol. Ass., U.K., 47(1):121-128.

Rosalina leoi n.sp.
Hedley, R.H., and J. St. J. Wakefield, 1967.
Clone culture studies of a new rodalinid foraminifer from Plymouth, England and Wellington, New Zealand.
J. mar. biol. Ass., U.K., 47(1):212-128.

Rotalipora, phylogeny of
Bandy, Orville L., 1967.
Cretaceous planktonic foraminiferal zonation.
Micropaleontology, 13(1):1-31.

Ruboglobigerina, phylogeny of
Bandy, Orville L., 1967.
Cretaceous planktonic foraminiferal zonation.
Micropaleontology, 13(1):1-31.

Schackoina, phylogeny of
Bandy, Orville L., 1967.
Cretaceous planktonic foraminiferal zonation.
Micropaleontology, 13(1):1-31.

Sepidocyclinas
Seiglie, George A., 1965.
Notas sobre tres Sepidocyclinas de Cuba.
Bol. Inst. Oceanogr., Univ Oriente, 4(1):191-213.

Sphaeroidinella dehiscens
Bandy, Orville L., James C. Ingle, Jr., and William E. Frerichs, 1967.
Isomorphism in "Sphaeroidinella" and "Sphaeroidinellopsis".
Micropaleontology, 13(4):483-488.

Sphaeroidinella kochi
Cifelli, Richard, 1965.
Late Tertiary planktonic Foraminifera associated with a basaltic boulder from the Mid-Atlantic Ridge.
J. Mar. Res., 22(3): 73-87.

Sphaeroidinella seminulina
Cifelli, Richard, 1965.
Late Tertiary planktonic Foraminifera associated with a basaltic boulder from the Mid-Atlantic Ridge.
J. Mar. Res., 22(3): 73-87.

Sphaeroidinellopsis
Bandy, Orville, L., James C. Ingle Jr., and William E. Frerichs, 1967.
Isomorphism in "Sphaeroidinella" and "Sphaeroidinellopsis."
Micropaleontology, 13(4):483-488.

Spiculosiphon radiata
Christiansen, Bengt O., 1964.
Spiculosiphon radiata, a new Foraminifera from northern Norway.
Astarte, Tromsø, No. 25:8 pp.

Spiroloculina hyalina
Arnold, Zach M., 1964.
Biological observations on the foraminifer, Spiroloculina hyalina Schulze.
Univ. California, Publ. Zool., 72:78 pp.

Syringammina tasmanensis nsp
Lewis, K. B., 1966.
A giant foraminifer: a new species of Syringammina from the New Zealand region.
New Zealand J. Sci., 9(1):114-123.

Tetromphalus
Todd, Ruth 1971.
Tetromphalus (Foraminifera) from Midway.
J. foram. Res. 1(4): 162-169.

Textularia gibbosa
Hansen, Hans Jørgen 1967.
Description of seven type specimens of Foraminifera designated by D'Orbigny 1826.
Biol. Meddeleser, K. Danske Videnskab. Selskab. 23(16):1-12.

Tosaia hanzawa
Seiglie, George A., and Pedro J. Bermudez 1966.
Notes on genus Tosaia takayanagi in America and description of a new species.
Carib. J. Sci. 6(1/2): 65-69.

Tosaia lowmani n. sp.
Seiglie, George A., and Pedro J. Bermudez 1969.
Tosaia lowmani, a new species from off the Pacific coast of Panama.
Contrib. Cushman Fdn foramin. Res. 20(3), 96-98.

Tosaia weaveri n.sp.
Seiglie George A., and Pedro J. Bermudez 1966.
Notes on genus Tosaia takayanagi in America and description of a new species
Carib. J. Sci 6(1/2):65-69.

Triloculina echinata
Rao, K. Kameswara 1970
On a little known miliolid foraminifer from north-eastern part of The Arabian Sea.
Current Sci. 39 (4): 87-88.

Trochammina squamata
Hedley, R.H., C.M. Hurdle and I.D.J. Burdett, 1964.
Trochammina squamata Jones and Parker (Foraminifera) with observations on some closely related species.
New Zealand J. Sci., 7(3):417-426.

Turborotalia pachyderma
Bandy, Orville L., E. Ann Butler, and Ramil C. Wright, 1969.
Alaskan upper Miocene marine glacial deposits and the Turborotalia pachyderma datum plane.
Science 166 (3905): 607-609.

Turborotalia scitula
Lipps, J.H., and J.E. Warme, 1966.
Planktonic foraminiferal biofacies in the Okhotsk.
Contrib. Cushman Found., Foram. Res., 17(4):125-134.

Raphidiophrys marina n.sp.
Ostenfeld, C.H., 1904.
On two new marine species of Heliozoa occurring in the plankton of the North Sea and Skager Rak. Medd. Komm. Havundersøgelser, Ser. Plankton, 1(2): 5pp.

radiolaria
Arrhenius, G., 1952.
Sediment cores from the East Pacific.
Repts. Swedish Deep-sea Exped., 1947-1948, 5(1): 227 pp., textfigs., with appendix of plates.

radiolaria
Benson, R.N., 1973
Radiolaria, leg 12, deep sea drilling project. Initial Repts Deep Sea Drill. Proj. 12 (NSFSP-12): 1085-1113.

Radiolarians
Bigelow, H.B., and M. Leslie, 1930
Reconnaissance of the waters and plankton of Monterey Bay, July 1928.
Bull. M.C.Z., 70(5):429-481, 43 text figs.

radiolarians
Cachon-Enjumet, Monique, 1964.
L'évolution sporogenetique des Phaeodaries (radiolaires).
Comptes Rendus, Acad. Sci., Paris, 259:2677-2679.
Also in:
Trav. Sta. Zool., Villefranche-sur-Mer, 24.1964

Radiolarians
Cooper, L.H.N. 1965.
Radiolarians as possible chronometers of continental drift.
Progress in Oceanography, 3:71-82.

Radiolaria
Deflandre-Rigaud, Martha 1969.
Remarques sur la nomenclature des radiolaires. 1. Haeckel 1887 et le Challenger.
Bull. Mus. natn. Hist. Nat. Paris (2) 40 (5): 1071-1092.

radiolaria
Dumitrica, Paulian, 1973
Paleocene radiolaria, DSDP leg 21.
Initial Repts, Deep Sea Drilling Project, 21:787-815.

Radiolaria
Dumitrica, Paulian, 1973
Phaeodarian radiolaria in southwest Pacific sediments cored during leg 21 of the deep sea drilling project.
Initial Repts, Deep Sea Drilling Project, 21:751-785.

radiolaria
Foreman, Helen P., 1973
Radiolaria of leg 10 with systematics and ranges for the families Amphipyndacidae, Artostrobiidae, and Theoperidae. Initial Repts Deep Sea Drill. Proj. 10(NSFSP-10): 407-474.

radiolaria
Foreman, Helen P., 1971.
Cretaceous radiolaria, Leg 7, DSDP. Initial Repts Deep Sea Drill. Proj. 7(2): 1673-1693.

Radiolaria
Friend, Jennifer K. and William R. Riedel 1967
Cenozoic orosphaerid radiolarians from tropical Pacific sediments.
Micropaleontology 13(2): 217-232.

Radiolaria
*Funnell, B.M., 1967.
Foraminifera and Radiolaria as depth indicators in the marine environment.
Marine Geol., 5(5/6):333-347.

Radiolaria, fossil
Goll, Robert M. 1972.
Systematics of eight Tholospyris taxa (Trissocyclidae, Radiolaria)
Micropaleontology 18(4): 443-475

Radiolaria
Goll, R.M., 1968.
Classification and phylogeny of Cenozoic Trissoayclidae (Radiolaria) in the Pacific and Caribbean basins.
J. Paleont., 42(6):1409-1432.

radiolaria, fossil
Hays, James D., 1965.
Radiolaria and Late Tertiary and Quaternary history of Antarctic seas.
In: Biology of Antarctic seas, II.
Antarctic Res. Ser., Amer. Geophys. Union, 5:125-184.

Radiolaria
*Hays, James D., and Neil D. Opdyke, 1967.
Antarctic Radiolaria, magnetic reversals and climatic change.
Science, 158(3804):1001-1011.

radiolarians

Kamelava, N.N. 1967.
The role of Radiolaria in the evaluation of the primary production in the Red Sea and the Gulf of Aden.
Dokl. Akad. Nauk SSSR 172 (6):1430-1433.

Radiolaria

*Koslova, G.E., 1967.
On phylogenetic relation between the suborders Discoidea and Larcoidea (Radiolaria, Spumellaria). (In Russian; English abstract).
Zool. Zh., 46(9):1311-1320.

radiolarians

Kuzmina, A.I. 1972.
On finding of Radiolaria in the Pacific tropical waters. (In Russian)
Izv. Tichookean. nauchno-issled. Inst. rybn. Choz. Okean. (TINRO) 81: 234-241.

Radiolaria (Tertiary)

Johnson, David A., and Frances L. Parker 1972.
Tertiary Radiolaria and Foraminifera from the equatorial Pacific.
Micropaleontology 18(2):129-143.

radiolarians

Meyer, K., 1934.
Die geographische Verbreitung der Tripyleen Radiolarien des Südatlantischen Ozeans.
Wiss. Ergeb. Deutschen Atlantischen Exped. "Meteor", 1925-1927, 12(1):133-198.

Radiolaria

Moore, T.C. 1972.
Mid-Tertiary evolution of the radiolarian genus Calocycletta.
Micropaleontology 18(2): 144-152.

Radiolaria

Nakaseko, Kojiro, 1959
On superfamily Liosphaericae (Radiolaria) from sediments in the sea near Antarctica.
On Radiolaria from sediments in the sea near Antarctica. Biol. Res., Japan. Antarctic Res. Exped., Spec. Publ. Seto Mar. Biol. Lab., 13 pp.

Radiolaria

Nigrini, Catherine 1967.
Radiolaria in pelagic sediments from the Indian and Atlantic oceans.
Bull. Scripps Inst. Oceanogr. 11:106pp.

radiolaria

Petrushevskaia, M.G., 1966.
Radiolaria in the plankton and in the bottom sediments. (In Russian).
In: Geochemistry of silica, N.M. Strakhov, editor, Isdatel."Nauka", Moskva, 219-245.

radiolaria

Riedel, W.R., 1971.
Cenozoic radiolaria from the western tropical Pacific, Leg 7. Initial Repts. Deep Sea Drill. Proj. 7(2): 1529-1672.

Radiolaria, lists of spp.

Riedel, William R., 1958
Radiolaria in Antarctic sediments.
Repts. B.A.N.Z. Antarctic Res. Exped. 1929-31 B, 6(10):219 - 255.

radiolaria

Riedel, W.R., 1957.
Radiolaria: a preliminary stratigraphy.
Repts. Swedish Deep-Sea Exped., 6(Sediment Cores from the West Pacific, No. 3):61-96, 4 pls.

radiolaria

Riedel, W.R., 1952.
Tertiary Radiolaria in western Pacific sediments.
Medd. Ocean. Inst., Göteborg, 19: 18 pp., 2 pls. [Göteborgs K. Vetenskaps- och Vitterhets-Handl., Sjätte Följden, B, 6(3):]

Radiolaria

Riedel, W.R., 1951.
Number of Radiolaria in sediments. Nature 167 (No. 4237):75.

Radiolaria

Riedel, William R., and Helen P. Foreman, 1961
Type specimens of North American Paleozoic Radiolaria.
J. Paleontology, 35(3):628-632.

radiolaria

Riedel, W.R., and J. Schlocker, 1956.
Radiolaria from the Franciscan group, Belmont, California. Micropaleontology 2(4):357-360.

radiolaria

Sanfilippo, Annika and W.R. Riedel, 1973
Cenozoic Radiolaria (exclusive of Theoperids, Artostrobiids and Amphipynidacids) from the Gulf of Mexico, deep sea drilling project leg 10.
Initial Repts Deep Sea Drill. Proj. 10(NSFSP-10): 475-611.

Radiolaria, abundance of

Cifelli, R., and K.N. Sachs, Jr., 1966.
Abundance relationships of planktonic Foraminifera and Radiolaria.
Deep-Sea Res., 13(4):751-753.

radiolarians, anat. physiol

Cachon, Jean, et Monique Cachon 1972.
Le système axopodial des radiolaires sphaeroidés.
Arch. Protistenk. 114 (3): 291-307.

radiolarians, anat. physiol.

Cachon, J., M. Cachon and M. Petrushevskaya 1972.
Irregular meshworks in radiolarian skeletons. (In Russian. English abstract).
Zool. Zh. 51(6):904-969.

radiolarians, anat. physiol.

Cachon, Jean et Monique Cachon, 1971
Recherches sur le métabolisme de la silice chez les Radiolaires. Absorption et excrétion. C.r. acad. Sci. Paris 272:1652-1654.

Radiolaria, anat-physiol

Cachon, Jean, et Monique Cachon-Enjumet, 1965.
Etude cytologique et caryologique d'un Phaeodérié bathy pélagique Planktonetta atlantica Borgert.
Bull. Inst. Océanogr., Monaco, 64(1330):23pp.

radiolarians, anat.

Febvre-Chevalier, Colette et M. Pierre Lecher, 1971.
Etude ultrastructurale des lamelles annelées intracytoplasmiques chez les Foraminifères et les Radiolaires Phaeodariens. C.R. Acad. Sci. Paris 272: 1264-1267.

radiolarians, anat.

Hollande, André, Jean Cachon et Monique Cachon-Enjumet, 1965.
L'infrastructure des axopodes chez les radiolaires Sphaerellaires Periaxoplastidies.
C. R. Séanc. Hebd., Acad. Sci., Paris, 261: 1388-1391.
Also in:
Trav. Sta. Zool., Villefranche-sur-Mer, 25.

radiolarians, anat. physiol

*Kozlova, G.E., 1967.
The structure patterns of the skeleton of radiolarians from the family Porodiscidae. (In Russian; English abstract).
Zool. Zh., 46(8):1163-1173.

radiolaria, anat.

Lecal, J., 1963.
A propos de la structure des spicules de Sticholonche zanclea Hertwig (Radiolaires)
Rapp. Proc. Verb. Réunions, Comm. Int. Expl. Sci. Mer Méditerranée, Monaco, 17(2):505-506.

radiolaria, dissolution of

*Berger, Wolfgang H., 1968.
Radiolarian skeletons: solution at depth.
Science, 159(3820):1237-1239.

Radiolaria, dissolution of

Moore, Theodore C. Jr. 1969.
Radiolaria: change in skeletal weight and resistance to solution.
Bull. geol. Soc. Am. 80(10): 2103-2108.

radiolaria, distribution of

Casey, R.E., 1971.
Distribution of polycystine radiolaria in the oceans in relation to physical and chemical conditions. In: Micropaleontology of oceans, B.M. Funnell and W.R. Riedel, editors, Cambridge Univ. Press, 151-159.

Radiolaria, extinction of

*Black, D.I., 1967.
Cosmic ray effects and faunal extinctions at geomagnetic field reversals.
Earth Planet. Sci. Letters, 3(3):225-236.

radiolaria (fossil), lists of spp.

Benson, R.N., 1973
Radiolaria, leg 12, deep sea drilling project. Initial Repts Deep Sea Drill. Proj. 12 (NSFSP-12): 1085-1113.

radiolaria, lists of spp.

Björklund, Kjell R. 1973.
Radiolarians from the surface sediment in Lindåspollene, western Norway.
Sarsia 53: 71-75.

Radiolaria, lists of spp.

Cita, Maria B., and Stefan Gartner, 1970.
Biostratigraphy. Initial Reports of the Deep Sea Drilling Project, Glomar Challenger, 2: 391-411.

Radiolarians, lists of spp

Falcão Paredes, Jorge, 1969/70
Subsídios para o conhecimento do plâncton marinho de Cabo Verde. II Tintinídeos, Foraminíferos e Radiolários.
Mems Inst. Invest. cient. Moçambique (A) 10:109-143

radiolaria (fossil) lists of spp.

Foreman, Helen P., 1973
Radiolaria of leg 10 with systematics and ranges for the families Amphipyndacidae, Artostrobiidae, and Theoperidae. Initial Repts Deep Sea Drill. Proj. 10(NSFSP-10): 407-474.

Oceanographic Index: Marine Organisms Cumulation, 1946-1973

Kimor, Baruch, 1971. Radiolaria — lists of spp.
Some considerations on the distribution of Acantharia and Radiolaria in the eastern Mediterranean.
Rapp. P.-v. Comm. int. Explor. scient. mer Medit. 20(3):349-351.

radiolaria, lists of spp.
Kling, Stanley A., 1973
Radiolaria from the eastern North Pacific, deep sea drilling project, leg 18. Initial Repts Deep Sea Drilling Proj. 18:617-671.

Radiolaria, lists of spp.
Moore, T.C., 1971.
Radiolaria. Initial Repts Deep Sea Drilling Project, 8: 727-775.

Radiolaria, lists of spp.
Nigrini, Catherine A. 1968.
Radiolaria from eastern tropical Pacific sediments.
Micropaleontology 14(1): 51-63.

Radiolaria, lists of spp.
Resig, Johanna M., Hsin-Yi Ling and Carol J. Stadum 1970.
Micropaleontology of a Miocene core from the western tropical Pacific.
Pacific Science 24(4): 421-432.

Radiolaria, lists of spp.
Riedel, W.R. and Annika Sanfilippo, 1970.
Radiolaria, leg 4, deep sea drilling project. Initial Reports of the Deep Sea Drilling Project Glomar Challenger, 4: 503-575.

radiolaria (fossil) lists of spp.
Sanfilippo, Annika and W.R. Riedel, 1973
Cenozoic Radiolaria (exclusive of Theoperids, Artostrobiids and Amphipynidacids) from the Gulf of Mexico, deep sea drilling project leg 10. Initial Repts Deep Sea Drill. Proj. 10(NSFSP-10): 475-611.

radiolaria, phylogeny
Sanfilippo, Annika and W.R. Riedel, 1973
Cenozoic Radiolaria (exclusive of Theoperids, Artostrobiids and Amphipynidacids) from the Gulf of Mexico, deep sea drilling project leg 10. Initial Repts Deep Sea Drill. Proj. 10(NSFSP-10): 475-611.

radiolaria, skeletons
*Berger, Wolfgang H., 1968.
Radiolarian skeletons: solution at depth. Science, 159(3820):1237-1239.

Acanthosphaera dodecaspinosa
Riedel, William R., and Helen F. Foreman, 1961
Type specimens of North American Paleozoic Radiolaria.
J. Paleonotology, 35(3):628-632.

Acanthosphaera grandispinosa
Riedel, William R., and Helen F. Foreman, 1961
Type specimens of North American Paleozoic Radiolaria.
J. Paleonotology, 35(3):628-632.

Acanthosphaera hirsuta
Riedel, William R., and Helen F. Foreman, 1961
Type specimens of North American Paleozoic Radiolaria.
J. Paleonotology, 35(3):628-632.

Acanthosphaera microspinosa
Riedel, William R., and Helen F. Foreman, 1961
Type specimens of North American Paleozoic Radiolaria.
J. Paleonotology, 35(3):628-632.

Actinomma (Actinommura) antarctica n.sp.
Nakaseko, Kojiro, 1959
On superfamily Liosphaericae (Radiolaria) from sediments in the sea near Antarctica. On Radiolaria from sediments in the sea near Antarctica. Biol. Res., Japan, Antarctic Res. Exped., Spec. Publ. Seto Mar. Biol. Lab., 13 pp.

Actinomma (Actinommilla) capillaceum n.sp.
Nakaseko, Kojiro, 1959
On superfamily Liosphaericae (Radiolaria) from sediments in the sea near Antarctica. On Radiolaria from sediments in the sea near Antarctica. Biol. Res., Japan, Antarctic Res. Exped., Spec. Publ. Seto Mar. Biol. Lab., 13 pp.

Actinomma (Actinommilla) erinaceum n.sp.
Nakaseko, Kojiro, 1959
On superfamily Liosphaericae (Radiolaria) from sediments in the sea near Antarctica. On Radiolaria from sediments in the sea near Antarctica. Biol. Res., Japan, Antarctic Res. Exped., Spec. Publ. Seto Mar. Biol. Lab., 13 pp.

Actinomma (Actinomma) yosii n.sp.
Nakaseko, Kojiro, 1959
On superfamily Liosphaericae (Radiolaria) from sediments in the sea near Antarctica. On Radiolaria from sediments in the sea near Antarctica. Biol. Res., Japan, Antarctic Res. Exped., Spec. Publ. Seto Mar. Biol. Lab., 13 pp.

Anthocyrtidium cineraria
Riedel, W.R., 1957.
Radiolaria: a preliminary stratigraphy.
Repts., Swedish Deep-Sea Exped., 6(Sediment Cores from the West Pacific, No. 3):61-96, 4 pls.

Astrophacus cingulatus
Riedel, William R., and Helen F. Foreman, 1961
Type specimens of North American Paleozoic Radiolaria.
J. Paleonotology, 35(3):628-632.

Calocyclas turris
Riedel, W.R., 1957.
Radiolaria: a preliminary stratigraphy.
Repts., Swedish Deep-Sea Exped., 6(Sediment Cores from the West Pacific, No. 3):61-96, 4 pls.

Calocyclas virginis
Riedel, W.R., 1957.
Radiolaria: a preliminary stratigraphy.
Repts., Swedish Deep-Sea Exped., 6(Sediment Cores from the West Pacific, No. 3):61-96, 4 pls.

Carposphaera equalis
Riedel, William R., and Helen F. Foreman, 1961
Type specimens of North American Paleozoic Radiolaria.
J. Paleonotology, 35(3):628-632.

Carposphaera magna
Riedel, William R., and Helen F. Foreman, 1961
Type specimens of North American Paleozoic Radiolaria.
J. Paleonotology, 35(3):628-632.

Cenosphaera (Cyrtidosphaera) a sp.
Nakaseko, Kojiro, 1959
On superfamily Liosphaericae (Radiolaria) from sediments in the sea near Antarctica. On Radiolaria from sediments in the sea near Antarctica. Biol. Res., Japan, Antarctic Res. Exped., Spec. Publ. Seto Mar. Biol. Lab., 13 pp.

Cenosphaera (Cyrtidosphaera) antarctica, n.sp.
Nakaseko, Kojiro, 1959
On superfamily Liosphaericae (Radiolaria) from sediments in the sea near Antarctica. On Radiolaria from sediments in the sea near Antarctica. Biol. Res., Japan, Antarctic Res. Exped., Spec. Publ. Seto Mar. Biol. Lab., 13 pp.

Cenosphaera (Phormosphaera) b sp.
Nakaseko, Kojiro, 1959
On superfamily Liosphaericae (Radiolaria) from sediments in the sea near Antarctica. On Radiolaria from sediments in the sea near Antarctica. Biol. Res., Japan, Antarctic Res. Exped., Spec. Publ. Seto Mar. Biol. Lab., 13 pp.

Cenosphaera hexagonalis
Riedel, William R., and Helen F. Foreman, 1961
Type specimens of North American Paleozoic Radiolaria.
J. Paleonotology, 35(3):628-632.

Cenosphaera (Phormosphaera) nagatai n.sp.
Nakaseko, Kojiro, 1959
On superfamily Liosphaericae (Radiolaria) from sediments in the sea near Antarctica. On Radiolaria from sediments in the sea near Antarctica. Biol. Res., Japan, Antarctic Res. Exped., Spec. Publ. Seto Mar. Biol. Lab., 13 pp.

Cenosphaera semiequalis
Riedel, William R., and Helen F. Foreman, 1961
Type specimens of North American Paleozoic Radiolaria.
J. Paleonotology, 35(3):628-632.

Cenosphaera variabilis
Riedel, William R., and Helen F. Foreman, 1961
Type specimens of North American Paleozoic Radiolaria.
J. Paleonotology, 35(3):628-632.

Cenosphaera (Cyrtidosphaera) yosii n. sp.
Nakaseko, Kojiro, 1959
On superfamily Liosphaericae (Radiolaria) from sediments in the sea near Antarctica. On Radiolaria from sediments in the sea near Antarctica. Biol. Res., Japan, Antarctic Res. Exped., Spec. Publ. Seto Mar. Biol. Lab., 13 pp.

Cromyesphaera nipponica n.sp.
Nakaseko, Kojiro, 1959
On superfamily Liosphaericae (Radiolaria) from sediments in the sea near Antarctica. On Radiolaria from sediments in the sea near Antarctica. Biol. Res., Japan, Antarctic Res. Exped., Spec. Publ. Seto Mar. Biol. Lab., 13 pp.

Cytocladus dozieli n.sp.
Resletniak, V.V., 1956.
A new species of giant radiolarian of the genus Cytocladus Schr. (Radiolaria) from the Barents Sea. Trudy Inst. Okeanol., 18:10-12.

Eucyrtidium delmontense
Riedel, W.R., 1957.
Radiolaria: a preliminary stratigraphy.
Repts., Swedish Deep-Sea Exped., 6(Sediment Cores from the West Pacific, No. 3):61-96, 4 pls.

Eusyringium fistuligerum
Riedel, W.R., 1957.
Radiolaria: a preliminary stratigraphy.
Repts. Swedish Deep-Sea Exped., 6(Sediment Cores from the West Pacific, No. 3):61-96, 4 pls.

Haliomma perfecta
Riedel, William R., and Helen F. Foreman, 1961
Type specimens of North American Paleozoic Radiolaria.
J. Paleonotology, 35(3):628-632.

Heliosphaera alternata
Riedel, William R., and Helen F. Foreman, 1961
Type specimens of North American Paleozoic Radiolaria.
J. Paleonotology, 35(3):628-632.

Heliosphaera macrospinosa
Riedel, William R., and Helen F. Foreman, 1961
Type specimens of North American Paleozoic Radiolaria.
J. Paleonotology, 35(3):628-632.

Hexalonche sp.
Riedel, William R., and Helen F. Foreman, 1961
Type specimens of North American Paleozoic Radiolaria.
J. Paleonotology, 35(3):628-632.

Hexastylidium variatum
Riedel, William R., and Helen F. Foreman, 1961
Type specimens of North American Paleozoic Radiolaria.
J. Paleonotology, 35(3):628-632.

Hexastylus basiporosus
Riedel, William R., and Helen F. Foreman, 1961
Type specimens of North American Paleozoic Radiolaria.
J. Paleonotology, 35(3):628-632.

Liosphaera (Craspedomma) antarctica, n.sp.
Nakaseko, Kojiro, 1959
On superfamily Liosphaericae (Radiolaria) from sediments in the sea near Antarctica. On Radiolaria from sediments in the sea near Antarctica. Biol. Res., Japan, Antarctic Res. Exped., Spec. Publ. Seto Mar. Biol. Lab., 13 pp.

Panarium penultimum n.sp.
Riedel, W.R., 1957.
Radiolaria: a preliminary stratigraphy.
Repts., Swedish Deep-Sea Exped., 6(Sediment Cores from the West Pacific, No. 3):61-96, 4 pls.

Phormocyrtis embolum
Riedel, W.R., 1957.
Radiolaria: a preliminary stratigraphy.
Repts., Swedish Deep-Sea Exped., 6(Sediment Cores from the West Pacific, No. 3):61-96, 4 pls.

Plegmosphaera spiculata
Riedel, William R., and Helen F. Foreman, 1961
Type specimens of North American Paleozoic Radiolaria.
J. Paleonotology, 35(3):628-632.

Pterocanium praetextum
Riedel, W.R., 1957.
Radiolaria: a preliminary stratigraphy.
Repts., Swedish Deep-Sea Exped., 6(Sediment Cores from the West Pacific, No. 3):61-96, 4 pls.

Pterocanium prismatium n.sp.
Riedel, W.R., 1957.
Radiolaria: a preliminary stratigraphy.
Repts., Swedish Deep-Sea Exped., 6(Sediment Cores from the West Pacific, No. 3):61-96, 4 pls.

Rhodosphaera sp.
Riedel, William R., and Helen F. Foreman, 1961
Type specimens of North American Paleozoic Radiolaria.
J. Paleonotology, 35(3):628-632.

Spaeropyle robusta n.sp.
Kling, Stanley A., 1973
Radiolaria from the eastern North Pacific, deep sea drilling project, leg 18. Initial Repts Deep Sea Drilling Proj, 18:617-671.

Spongoplegma brevisphaera
Riedel, William R., and Helen F. Foreman, 1961
Type specimens of North American Paleozoic Radiolaria.
J. Paleonotology, 35(3):628-632.

Spongoplegma longispinosa
Riedel, William R., and Helen F. Foreman, 1961
Type specimens of North American Paleozoic Radiolaria.
J. Paleonotology, 35(3):628-632.

Staurodoras brevispinosa
Riedel, William R., and Helen F. Foreman, 1961
Type specimens of North American Paleozoic Radiolaria.
J. Paleonotology, 35(3):628-632.

Stichocorys wolffii
Riedel, W.R., 1957.
Radiolaria: a preliminary stratigraphy.
Repts. Swedish Deep-Sea Exped., 6(Sediment Cores from the West Pacific, No. 3):61-96, 4 pls.

Sticholonone zanclea
Balecn, Enrique, 1964.
El plancton de Mar del Plata durante el periodo 1961-1962.
Bol. Inst. Biol. Mar., Buenos Aires, (4):56 pp. 49 pp + plates.

Stylosphaera quasiobtusa
Riedel, William R., and Helen F. Foreman, 1961
Type specimens of North American Paleozoic Radiolaria.
J. Paleonotology, 35(3):628-632.

Staurosphaera rotunda
Riedel, William R., and Helen F. Foreman, 1961
Type specimens of North American Paleozoic Radiolaria.
J. Paleonotology, 35(3):628-632.

Staurostylus varispinatus
Riedel, William R., and Helen F. Foreman, 1961
Type specimens of North American Paleozoic Radiolaria.
J. Paleonotology, 35(3):628-632.

Thecosphaera (Thecosphaeromma) antarctica n.sp.
Nakaseko, Kojiro, 1959
On superfamily Liosphaericae (Radiolaria) from sediments in the sea near Antarctica. On Radiolaria from sediments in the sea near Antarctica. Biol. Res., Japan, Antarctic Res. Exped., Spec. Publ. Seto Mar. Biol. Lab., 13 pp.

Thecosphaera hexpenetrata
Riedel, William R., and Helen F. Foreman, 1961
Type specimens of North American Paleozoic Radiolaria.
J. Paleonotology, 35(3):628-632.

Thecosphaera (Thecosphaera) miocenica
Nakaseko, Kojiro, 1959
On superfamily Liosphaericae (Radiolaria) from sediments in the sea near Antarctica. On Radiolaria from sediments in the sea near Antarctica. Biol. Res., Japan, Antarctic Res. Exped., Spec. Publ. Seto Mar. Biol. Lab., 13 pp.

Tympanidium binoctonum
Riedel, W.R., 1957.
Radiolaria: a preliminary stratigraphy.
Repts., Swedish Deep-Sea Exped., 6(Sediment Cores from the West Pacific, No. 3):61-96, 4 pls.

Xiphostylus inclinatus
Riedel, William R., and Helen F. Foreman, 1961
Type specimens of North American Paleozoic Radiolaria.
J. Paleonotology, 35(3):628-632.

protozoa (ciliates), lists of spp.
Berrer, Arthur Charles, 1963.
Morphology and ecology of the benthic ciliated Protozoa of Alligator Harbor, Florida.
Arch. Protistenkunde, 106(4):465-534.

Protozoa (Ciliata)
Borror, Arthur C., 1963.
Morphology and ecology of some uncommon ciliates from Alligator Harbor, Florida.
Trans. Amer. Microsc. Soc., 82(2):125-131.

Protozoa, ciliates
*Fenchel, Tom, 1969
The ecology of marine microbenthos. IV Structure and function of the benthic ecosystem, its chemical and physical factors and the microfauna communities with special reference to the ciliated Protozoa.
Ophelia, 6:1-182.
benthos
Protozoa, ciliates

ciliates
Fenchel, Tom., 1966.
On the ciliated Protozoa inhabiting the mantle cavity of lamellibranchs.
Malacologia, 5(1):35-36.

ciliates
Fenchel, Tom, 1965.
On the ciliate fauna associated with the marine species of the amphipod genus Gammarus J.G. Fabricius.
Ophelia, 2(2): 281-303.

ciliates
Fenchel, Tom., 1965.
Ciliates from Scandinavian molluscs.
Ophelia, 2(1):71-74.

protozoa, ciliates
Hamilton, R.D. and Janet E. Preslan, 1970.
Observations on the continuous culture of a planktonic phagotrophic protozoan. J. exp. mar Biol. Ecol., 5(1): 94-104.

ciliates
Hamilton, R.D. and Janet E. Preslan, 1969.
Cultural characteristics of a pelagic marine hymenostome ciliate, Uronema sp. J. exp. mar. Biol. Ecol., 4(1): 90-99.

ciliates, distribution of
Scheltema, R.S. 1973.
On an unusual means by which the sessile marine ciliate Folliculina simplex maintains its widespread geographical distribution.
Netherl. J. Sea. Res. 7:122-125.

Ciliates, effect of

Purdom, C.E. and A.E. Howard, 1971.
Ciliate infestations: a problem in marine fish farming. J. Cons. int. Explor. Mer, 33(3): 511-514.

ciliates, lists of spp.

Lackey, James B., 1967.
The microbiota of estuaries and their roles.
In: Estuaries, G.H. Lauff, editor, Publs Am. Ass. Advmt Sci., 83:291-302.

ciliates, lists of spp.

Lackey, James B., and Elsie W. Lackey, 1963
Microscopic algae and protozoa in the waters near Plymouth in August 1962.
J. Mar. Biol. Assoc., U.K., 43(3):797-805.

ciliates, lists of spp. a

Margalef, Ramón, 1973
Distribución de los ciliados planctónicos en la región de afloramiento del noroeste de África (Campaña Sahara II del Cornide de Saavedra). Result. Exped. cient. Cornide de Saavedra, Madrid 2: 109-124.

ciliates, lists of spp. a

Santhakumari V., and N. Balakrishnan Nair 1973
Ciliates from marine wood-boring molluscs.
Treubia 28(2): 41-58

ciliates

Vacelet, E., 1960.
Note preliminaire sur la faune infusorienne des "sables à Amphioxus" de la Baie de Marseille.
Rec. Trav. Sta. Mar. d'Endoume, 33(20):53-57.

ciliates

Leegaard, C., 1915.
Untersuchungen über einige Planktonciliaten des Meeres. Nyt Mag. f. Naturvid. 53:1-37, 24 text-figs.

ciliates, lists of spp.

Bock, Karl Jürgen, 1960
Biologische Untersuchungen, inbesondere der Ciliatenfaun, in durch Abwässer belasteten Schlei (westliche Ostsee). Kieler Meeresf., 16(1): 57-69.

ciliates, lists of spp.

Borrer, A.C., 1962.
Ciliate protozoa of the Gulf of Mexico.
Bull. Mar. Sci., Gulf and Caribbean, 12(3): 333-349.

ciliates, lists of species

Ganapati, P. N., and M. V. N. Rao, 1958.
Systematic survey of marine ciliates from Visakhapatnam
Andhra U., Mem. Ocean. 2:75-90.

ciliates

Margalef, R., 1963.
Rôle des ciliés dans le cycle de la vie pélagique en Méditerranée.
Rapp. Proc. Verb. Réunions, Comm. Int. Expl. Sci., Mer Méditerranée, Monaco, 17(2):511-512.

ciliates, sand-dwelling, lists of spp.

Petran, Adriana, 1965.
Cercetări asupra faunei de ciliate psamobionte la plajele din sudul litoralului Românesc al Mării Negre.
In: Ecologie marina, M. Băcescu, redactor, Edit. Acad. Republ. Pop. Romîne, Bucaresti, 2:169-191.

Protozoa (Infusoria)

Zaika, V.E. and T. Yu. Averina, 1968.
Numbers of Infusoria in plankton of the Sevastopol Bay of the Black Sea. (In Russian; English abstract).
Okeanologiia, Akad. Nauk, SSSR, 8(6):1071-1073.

ciliates, lists of spp.

Burkovsky, I.V. 1971.
A comparative study of the ecology of free-living ciliates in the Rugozersky Inlet (Kandalaksha Bay, White Sea). (In Russian, English abstract).
Zool. Zh. 50(12): 1773-1779.

ciliates, lists of spp.

Petran, Adriana 1971.
Sur la faune des ciliés des sédiments sablonneux du littoral roumain de la Mer Noire.
Cercetări marine, Constant., (1): 149-166

Ciliospina norvegica n.gen., n.sp.

Leegaard, L., 1915.
Untersuchungen über einige Planktonciliaten des Meeres. Nyt Mag. f. Naturvid. 53:1-37, 24 text-figs.

Cothurnia aplatita flexa

Felinska, Maria, 1965.
Marine Ciliata from Plymouth: Peritricha, Vaginicolidae.
Jour. Mar. Biol. Assoc., U.K., 45(1):229-239.

Cothurnia auriculata

Felinska, Maria, 1965.
Marine Ciliata from Plymouth: Peritricha, Vaginicolidae.
Jour. Mar. Biol. Assoc., U.K., 45(1):229-239.

Cothurnia ceramicola

Felinska, Maria, 1965.
Marine Ciliata from Plymouth: Peritricha, Vaginicolidae.
Jour. Mar. Biol. Assoc., U.K., 45(1):229-239.

Cothurnia coarctata

Felinska, Maria, 1965.
Marine Ciliata from Plymouth: Peritricha, Vaginicolidae.
Jour. Mar. Biol. Assoc., U.K., 45(1):229-239.

Cothurnia collaris incisa

Felinska, Maria, 1965.
Marine Ciliata from Plymouth: Peritricha, Vaginicolidae.
Jour. Mar. Biol. Assoc., U.K., 45(1):229-239.

Cothurnia compressa

Felinska, Maria, 1965.
Marine Ciliata from Plymouth: Peritricha, Vaginicolidae.
Jour. Mar. Biol. Assoc., U.K., 45(1):229-239.

Cothurnia elegans

Felinska, Maria, 1965.
Marine Ciliata from Plymouth: Peritricha, Vaginicolidae.
Jour. Mar. Biol. Assoc., U.K., 45(1):229-239.

Cothurnia elongata n.sp.

Felinska, Maria, 1965.
Marine Ciliata from Plymouth: Peritricha, Vaginicolidae.
Jour. Mar. Biol. Assoc., U.K., 45(1):229-239.

Cothurnia inclinans n.sp.

Felinska, Maria, 1965.
Marine Ciliata from Plymouth: Peritricha, Vaginicolidae.
Jour. Mar. Biol. Assoc., U.K., 45(1):229-239.

Cothurnia maritima

Felinska, Maria, 1965.
Marine Ciliata from Plymouth: Peritricha, Vaginicolidae.
Jour. Mar. Biol. Assoc., U.K., 45(1):229-239.

Cothurnia parvula n.sp.

Felinska, Maria, 1965.
Marine Ciliata from Plymouth: Peritricha, Vaginicolidae.
Jour. Mar. Biol. Assoc., U.K., 45(1):229-239.

Cothurnia recurva

Felinska, Maria, 1965.
Marine Ciliata from Plymouth: Peritricha, Vaginicolidae.
Jour. Mar. Biol. Assoc., U.K., 45(1):229-239.

Cothurnia simplex

Felinska, Maria, 1965.
Marine Ciliata from Plymouth: Peritricha, Vaginicolidae.
Jour. Mar. Biol. Assoc., U.K., 45(1):229-239.

Cyclotrichium meunieri

Avaria P., Sergio, 1970.
Observación de un fenomeno de marea roja en la Bahia de Valparaiso.
Rev. Biol. mar., Valparaiso 14(1): 1-5.

Cyclotrichium meunieri

Bary, B.M., and R.G. Stuckey, 1950.
An occurrence in Wellington Harbour of Cyclotrichium meunieri Powers, a ciliate causing red water, with some additions to its morphology.
Trans. Roy. Soc., New Zealand, 78(1):86-92, Pl. 13.

Cyclotrichium Meunieri

McAlice, Bernard J., 1968.
An occurrence of ciliate red water in the Gulf of Maine.
J. Fish. Res. Bd., Can. 25(8): 1749-1751.

Cyclotrichium meunieri sp. nov.,

Powers, P., 1932.
Cyclotrichium meunieri sp. nov., (Protozoa Ciliata); cause of red water in the Gulf of Maine
Biol. Bull. 63:74-80.

Cyclotrichium meunieri

Ryther, John H., 1967.
Occurrence of red water off Peru.
Nature, Lond., 214(5095):1318-1319.

Euplotes neapolitanus n.sp.

Wichterman, Ralph, 1964.
Description and life cycle of Euplotes neapolitanus sp. nov. (Protozoa, Ciliophora, Hypotrichida) from the Gulf of Naples.
Trans. Amer. Microsc. Soc., 83(3):362-370.

ciliates Folliculina simplex

Scheltema, R.S. 1973.
On an unusual means by which the sessile marine ciliate Folliculina simplex maintains its widespread geographical distribution
Netherl. J. Sea Res. 7: 122-125.

Laboea sp.

Gran, H.H., and T. Braarud, 1935
A quantitative study of the phytoplankton in the Bay of Fundy and the Gulf of Maine (including observations on hydrography, chemistry, and turbidity).
J. Biol. Bd., Canada, 1(5):279-467, 69 text figs.

Laboea acuminata n.sp.

Leegaard, C., 1915.
Untersuchungen über einige Planktonciliaten des Meeres. Nyt Mag. f. Naturvid. 53:1-37, 24 text-figs.

Laboea capitata n.sp.

Leegaard, C., 1915.
Untersuchungen über einige Planktonciliaten des Meeres. Nyt Mag. f. Naturvid. 53:1-37, 24 text-figs.

Laboea compressa n.sp.

Leegaard, C., 1915.
Untersuchunge über einige Planktonciliaten des Meeres. Nyt Mag. f. Naturvid. 53:1-37, 24 text-figs.

Laboea conica
Braarud, T., 1945
A phytoplankton survey of the polluted waters of inner Oslo Fjord. Hvalrådets Skrifter, No.28, 142 pp., 19 text figs., 17 tables.

Laboea conica
Braarud, T., 1934
A note on the phytoplankton of the Gulf of Maine in the summer of 1933. Biol. Bull. 67(1):76-82. (Contribution No.46 of the Woods Hole Oceanographic Institution)

Laboea conica
Braarud, T., and Adam Bursa, 1939
On the phytoplankton of the Oslo Fjord, 1933-1934. Hvaldrådets Skr. No.19:1-63; 9 text figs. Reviewed. J. du. Cons. 14(3):418-420. A.C. Gardiner.

Laboea conica
Grøntved, J., 1949
Investigations on the phytoplankton in the Danish Waddensea in July 1941. Medd. Komm. Danmarks Fiskeri og Havundersøgelser, ser. Plankton, 5(2):55 pp., 2 pls., 38 text figs.

Laboea conica
Leegaard, C., 1915.
Untersuchungen über einige Planktonciliaten des Meeres. Nyt Mag. f. Naturvid. 53:1-37, 24 text-figs.

Laboea constricta
Leegaard, C., 1915.
Untersuchungen über einige Planktonciliaten des Meeres. Nyt Mag. f. Naturvid. 53:1-37, 24 text-figs.

Laboea coronata n.sp.
Leegaard, C., 1915.
Untersuchungen über einige Planktonciliaten des Meeres. Nyt Mag. f. Naturvid. 53:1-37, 24 text-figs.

Laboea cornuta n.sp.
Leegaard, C., 1915.
Untersuchungen über einige Planktonciliaten des Meeres. Nyt Mag. f. Naturvid. 53:1-37, 24 text-figs.

Laboea crassula
Braarud, T., 1945
A phytoplankton survey of the polluted waters of inner Oslo Fjord. Hvalrådets Skrifter, No.28, 142 pp., 19 text figs., 17 tables.

Laboea crassula n.sp.
Leegaard, C., 1915.
Untersuchungen über einige Planktonciliaten des Meeres. Nyt Mag. f. Naturvid. 53:1-37, 24 text-figs.

Laboea delicatissima
Braarud, T., 1945
A phytoplankton survey of the polluted waters of inner Oslo Fjord. Hvalrådets Skrifter, No.28, 142 pp., 19 text figs., 17 tables.

Laboea delicatissima
Braarud, T., and Adam Bursa, 1939
On the phytoplankton of the Oslo Fjord, 1933-1934. Hvalrådets Skr. No.19:1-63; 9 text figs. Reviewed. J. du. Cons. 14(3):418-420. A.C. Gardiner.

Laboea delicatissima n.sp.
Leegaard, C., 1915.
Untersuchungen über einige Planktonciliaten des Meeres. Nyt Mag. f. Naturvid. 53:1-37, 24 text-figs.

Laboea emergens
Braarud, T., 1945
A phytoplankton survey of the polluted waters of inner Oslo Fjord. Hvalrådets Skrifter, No.28, 142 pp., 19 text figs., 17 tables.

Laboea emergens
Braarud, T., 1934
A note on the phytoplankton of the Gulf of Maine in the summer of 1933. Biol. Bull. 67(1):76-82. (Contribution No.46 of the Woods Hole Oceanographic Institution)

Laboea emergens
Braarud, T., and Adam Bursa, 1939
On the phytoplankton of the Oslo Fjord, 1933-1934. Hvaldrådets Skr. No.19:1-63; 9 text figs. Reviewed. J. du. Cons. 14(3):418-420. A.C. Gardiner.

Laboea emergens
Leegaard, C., 1915.
Untersuchungen über einige Planktonciliaten des Meeres. Nyt Mag. f. Naturvid. 53:1-37, 24 text-figs.

Laboea pulchra n.sp.
Leegaard, C., 1915.
Untersuchungen über einige Planktonciliaten des Meeres. Nyt Mag. f. Naturvid. 53:1-37, 24 text-figs.

Laboea reticulata
Braarud, T., 1934
A note on the phytoplankton of the Gulf of Maine in the summer of 1933. Biol. Bull. 67(1):76-82.

Laboea reticulata n.sp.
Leegaard, C., 1915.
Untersuchungen über einige Planktonciliaten des Meeres. Nyt Mag.f. Naturvid. 53:1-37, 24 text-figs.

Laboea strobila
Braarud, T., 1945
A phytoplankton survey of the polluted waters of inner Oslo Fjord. Hvalrådets Skrifter, No.28, 142 pp., 19 text figs., 17 tables.

Laboea strobila
Braarud, T., 1934
A note on the phytoplankton of the Gulf of Maine in the summer of 1933. Biol. Bull. 67(1):76-82.

Laboea strobila
Braarud, T., and Adam Bursa, 1939
On the phytoplankton of the Oslo Fjord, 1933-1934. Hvaldrådets Skr. No.19:1-63; 9 text figs. Reviewed. J. du. Cons. 14(3):418-420. A.C. Gardiner.

Laboea strobila
Leegaard, C., 1915.
Untersuchungen über einige Planktonciliaten des Meeres. Nyt Mag. f. Naturvid. 53:1-37, 24 text-figs.

Laboea vestita
Braarud, T., 1945
A phytoplankton survey of the polluted waters of inner Oslo Fjord. Hvalrådets Skrifter, No.28, 142 pp., 19 text figs., 17 tables.

Laboea vestita
Grøntved, J., 1949
Investigations on the phytoplankton in the Danish Waddensea in July 1941. Medd. Komm. Danmarks Fiskeri og Havundersøgelser, ser. Plankton, 5(2):55 pp., 2 pls., 38 text figs.

Laboea vestita n.sp.
Leegaard, C., 1915.
Untersuchungen über einige Planktonciliaten des Meeres. Nyt Mag. f. Naturvid. 53:1-37, 24 text-figs.

Lionotus lamellus
*Hada, Yoshina, 1970.   (1970)
The protozoan plankton of the Antarctic and Subantarctic seas.
Scient. Repts, Japan. Antarct. Res. Exped., (E)31:51 pp.

Lohmaniella n.gen.
Leegaard, C., 1915.
Untersuchungen über einige Planktonciliaten des Meeres. Nyt Mag. f. Naturvid. 53:1-37, 24 text-figs.

Lohmaniella oviformis
Braarud, T., 1945
A phytoplankton survey of the polluted waters of inner Oslo Fjord. Hvalrådets Skrifter, No.28, 142 pp., 19 text figs., 17 tables.

Lohmanniella oviformis
Braarud, T., 1934
A note on the phytoplankton of the Gulf of Maine in the summer of 1933. Biol. Bull. 67(1):76-82.

Lohmanniella oviformis
Braarud, T., and Adam Bursa, 1939
On the phytoplankton of the Oslo Fjord, 1933-1934. Hvaldrådets Skr. No.19:1-63; 9 text figs. Reviewed. J. du. Cons. 14(3):418-420. A.C. Gardiner.

Lohmanniella oviformis
Gran, H.H., and T. Braarud, 1935
A quantitative study of the phytoplankton in the Bay of Fundy and the Gulf of Maine (including observations on hydrography, chemistry, and turbidity). J. Biol. Bd., Canada, 1(5):279-467, 69 text figs.

Lohmanniella oviformis
Grøntved, J., 1949
Investigations on the phytoplankton in the Danish Waddensea in July 1941. Medd. Komm. Danmarks Fiskeri og Havundersøgelser, ser. Plankton, 5(2):55 pp., 2 pls., 38 text figs.

Lohmanniella oviformis
Leegaard, C., 1915.
Untersuchungen über einige Planktonciliaten des Meeres. Nyt Mag. f. Naturvid. 53:1-37, 24 text-figs.

Lohmaniella oviformis
Pettersson, H., 1934
Scattering and extinction of light in sea-water. Meddelanden från Göteborgs Högskolas Oceanografiska Institut: 9 (Göteborgs Kungl. vetenskaps-och vitterhets-samhälles Handlingar. Femte Foljden, Ser.B, 4(4)):1-16.

Lohmanniella spiralis
Braarud, T., 1945
A phytoplankton survey of the polluted waters of inner Oslo Fjord. Hvalrådets Skrifter, No.28, 142 pp., 19 text figs., 17 tables.

Lohmanniella spiralis
Braarud, T., 1934
A note on the phytoplankton of the Gulf of Maine in the summer of 1933. Biol. Bull. 67(1):76-82.

Lohmanniella spiralis
Braarud, T., and Adam Bursa, 1939
On the phytoplankton of the Oslo Fjord, 1933-1934. Hvaldrådets Skr. No.19:1-63; 9 text figs. Reviewed. J. du. Cons. 14(3):418-420. A.C. Gardiner.

Lohmanniella spiralis
Gran, H.H., and T. Braarud, 1935
A quantitative study of the phytoplankton in the Bay of Fundy and the Gulf of Maine (including observations on hydrography, chemistry, and turbidity). J. Biol. Bd., Canada, 1(5):279-467, 69 text figs.

Lohmanniella spiralis
Leegaard, C., 1915.
Untersuchungen über einige Planktonciliaten des Meeres. Nyt Mag. f. Naturvid. 53:1-37, 24 text-figs.

Mesodinium sp.
Braarud, T., 1945
A phytoplankton survey of the polluted waters of inner Oslo Fjord. Hvalrådets Skrifter, No.28, 142 pp., 19 text figs., 17 tables.

Mesodinium sp.

Quayle, D.B.
Paralytic shellfish poisoning in British Colombia.
Bull. Fish Res. Bd. Can., 168: 68 pp.

Mesodinium pulex "rubrum"

Fonds M., and D. Eisma, 1967.
Upwelling water as a possible cause of red plankton bloom along the Dutch coast.
Neth. J. Sea Res. 3(5): 458-463.

Mesodinium rubrum

Braarud, T., 1945
A phytoplankton survey of the polluted waters of inner Oslo Fjord. Hvalrådets Skrifter, No.28, 142 pp., 19 text figs., 17 tables.

Mesodinium rubrum

Braarud, T., 1934
A note on the phytoplankton of the Gulf of Maine in the summer of 1933. Biol. Bull. 67(1):76-82.

Mesodinium rubrum

Braarud, T., and Adam Bursa, 1939
On the phytoplankton of the Oslo Fjord, 1933-1934. Hvalrådets Skr. No.19:1-63; 9 text figs. Reviewed. J. du. Cons. 14(3): 418-420. A.C. Gardiner.

Mesodinium rubrum

Fenchel, Tom, 1968.
On "red water" in the Isefjord (inner Danish waters) caused by the ciliate Mesodinium rubrum. Ophelia, 5(2):245-253.

Mesodinium rubrum

Gran, H.H., and T. Braarud, 1935
A quantitative study of the phytoplankton in the Bay of Fundy and the Gulf of Maine (including observations on hydrography, chemistry, and turbidity). J. Biol. Bd. Canada, 1(5):279-467, 69 text figs.

Mesodinium rubrum

Perano, T.R., and D.J. Blackbourn, 1968
Pigments of the ciliate Mesodinium rubrum (Lohmann).
Netherlands J. Sea Res. 4(1): 27-31.

Mesodinium rubrum

Rothe, F., 1941
Quantitative Untersuchunger über die Plankton verteilung in der östlichen Ostsee. Ber. Deut. Wiss. Komm. fur Meeresforschung. n.f. X(3):291-368, 33 text figs.

Mesodinium rubrum

Taylor, F.J.R., D.J. Blackbourn and Janice Blackbourn, 1971.
The red-water ciliate Mesodinium rubrum and its "incomplete symbionts": a review including new ultrastructural observations.
J. Fish. Res. Bd. Can. 28(3): 391-407.

Mesodinium rubrum

Taylor, F.J.R., D.J. Blackbourn and Janice Blackbourn, 1969
Ultrastructure of the chloroplasts and associated structures within the marine ciliate Mesodinium rubrum (Lohman)
Nature, Lond., 224 (5221): 819-821.

ciliates

Metanyctotherus rancurelli n.sp.

Laval, Michèle, et Michel Tuffrau 1973.
Les ciliés endocommensaux d'un taret de Côte d'Ivoire, Teredo adami, mollusque Teredinidae. I. Infraciliature et polymorphisme de Metanyctotherus rancurelli sp. nov. (Hétérotriche).
Protistologica 9(1): 149-157.

Planktonetta atlantica

Cachon, Jean, et Monique Cachon-Enjumet, 1965.
Etude cytologique et caryologique d'un Phaeodarie bathy pélagique Planktonetta atlantica Borgert.
Bull. Inst. Océanogr., Monaco. 64(1330):23pp.

Pleuronema marinum

Hausmann, K., 1973
Cytologische Studien an Trichocysten VI. Feinstruktur und Funktionsmodus der Trichocysten des Flagellaten Oxyrrhis marina und des Ciliaten Pleuronema marinum. Helgoländer wiss. Meeresunters. 25(1):39-62.

Pyxicola socialis

Felinska, Maria, 1965.
Marine Ciliata from Plymouth: Peritricha, Vaginicolidae.
Jour. Mar. Biol. Assoc., U.K. 45(1):229-239.

Strombidium sp.

Braarud, T., and Adam Bursa, 1939
On the phytoplankton of the Oslo Fjord, 1933-1934. Hvalrådets Skr. No.19:1-63; 9 text figs. Reviewed. J. du. Cons. 14(3): 418-420. A.C. Gardiner.

Strombidium sp.

Gran, H.H., and T. Braarud, 1935
A quantitative study of the phytoplankton in the Bay of Fundy and the Gulf of Maine (including observations on hydrography, chemistry, and turbidity). J. Biol. Bd. Canada, 1(5):279-467, 69 text figs.

Strombidium acutum n.sp.

Leegaard, C., 1915.
Untersuchungen über einige Planktonciliaten des Meeres. Nyt Mag. f. Naturvid. 53:1-37, 24 text-figs.

Strombidium oblongum

Braarud, T., 1945
A phytoplankton survey of the polluted waters of inner Oslo Fjord. Hvalrådets Skrifter, No.28, 142 pp., 19 text figs., 17 tables.

Strombidium oblongum n.sp.

Leegaard, C., 1915.
Untersuchungen über einige Planktonciliaten des Meeres. Nyt Mag. f. Naturvid. 53:1-37, 24 text-figs.

Strombidium spiniferum n.sp.

Leegaard, C., 1915.
Untersuchungen über einige Planktonciliaten des Meeres. Nyt Mag. f. Naturvid. 53:1-37, 24 text-figs.

Strobilidium antarcticum

*Hada, Yoshina, 1970. (1970)
The protozoan plankton of the Antarctic and Subantarctic seas.
Scient. Repts, Japan. Antarct. Res. Exped., (E)31:51 pp.

Strobilidium conicum

*Hada, Yoshina, 1970. (1970)
The protozoan plankton of the Antarctic and Subantarctic seas.
Scient. Repts, Japan. Antarct. Res. Exped., (E)31:51 pp.

Strobilidium diversum

*Hada, Yoshina, 1970. (1970)
The protozoan plankton of the Antarctic and Subantarctic seas.
Scient. Repts, Japan. Antarct. Res. Exped., (E)31:51 pp.

Strobilidium eligans n. sp.

*Hada, Yoshina, 1970. (1970)
The protozoan plankton of the Antarctic and Subantarctic seas.
Scient. Repts, Japan. Antarct. Res. Exped., (E)31:51 pp.

Strobilidium elongatum

*Hada, Yoshina, 1970. (1970)
The protozoan plankton of the Antarctic and Subantarctic seas.
Scient. Repts, Japan. Antarct. Res. Exped., (E)31:51 pp.

Strobilidium striatum

*Hada, Yoshina, 1970. (1970)
The protozoan plankton of the Antarctic and Subantarctic seas.
Scient. Repts, Japan. Antarct. Res. Exped., (E)31:51 pp.

Strobilidium sulcatum

*Hada, Yoshina, 1970. (1970)
The protozoan plankton of the Antarctic and Subantarctic seas.
Scient. Repts, Japan. Antarct. Res. Exped., (E)31:51 pp.

Strobilidium syowaensis n. sp.

*Hada, Yoshina, 1970. (1970)
The protozoan plankton of the Antarctic and Subantarctic seas.
Scient. Repts, Japan. Antarct. Res. Exped., (E)31:51 pp.

Tiarina antarctica n. sp.

*Hada, Yoshina, 1970. (1970)
The protozoan plankton of the Antarctic and Subantarctic seas.
Scient. Repts, Japan. Antarct. Res. Exped., (E)31:51 pp.

Tiarina fusca

*Hada, Yoshina, 1970. (1970)
The protozoan plankton of the Antarctic and Subantarctic seas.
Scient. Repts, Japan. Antarct. Res. Exped., (E)31:51 pp.

Thuricola valvata

Felinska, Maria, 1965.
Marine Ciliata from Plymouth: Peritricha, Vaginicolidae.
Jour. Mar. Biol. Assoc., U.K. 45(1):229-239.

Vorticella marina

de Sousa e Silva, E., 1956.
Contribution à l'étude du microplancton de Dakar et des regions maritimes voisines.
Bull. I.F.A.N. 8(2):335-371, 7 pls.

Vorticella microstoma

de Sousa e Silva, E., 1956.
Contribution à l'étude du microplancton de Dakar et des regions maritimes voisines.
Bull. I.F.A.N. 8(2):335-371, 7 pls.

Woodania conicoides

Braarud, T., 1934
A note on the phytoplankton of the Gulf of Maine in the summer of 1933. Biol. Bull. 67(1):76-82.

Woodania conicoides n.gen., n.sp.

Leegaard, C., 1915.
Untersuchungen über einige Planktonciliaten des Meeres. Nyt Mag. f. Naturvid. 53:1-37, 24 text-figs.

Zoothamnium

Herman, Sidney S., and Joseph R. Mihursky, 1964
Infestation of the copepod Acartia tonsa with the stalked ciliate Zoothamnium.
Science, 146(3643):543-544.

Zoothamnium pelagicum

*Laval, Michèle, 1968.
Zoothamnium pelagicum Du Plessis, cilié péritriche planctonique: morphologie, croissance et comportement.
Protistologica, 4(3):333-363.

Oceanographic Index: Marine Organisms Cumulation, 1946-1973

tintinnids

Balech, Enrique, 1968.
Algunas especies nuevas o interesantes de tintinnidos del Golfo de Mexico y Caribe.
Revta Mus. argent. Cienc. nat. Bernardino Rivadavia Inst. nac. Hidrobiol. 2(5):165-197.

tintinnids

Balech, Enrique, 1962
Tintinnoinea y Dinoflagellata del Pacifico segun material de las Expediciones NORPAC y DOWNWIND del Instituto Scripps de Oceanografia.
Revista. Mus. Argentino Ciencias Nat. "Bernardino Rivadavia". Ciencias Zool., 7(1):1-253.

tintinnids

Balech, E., 1959.
Tintinnoinea del Mediterraneo.
Trav. Sta. Zool., Villefranche-sur-Mer, 18(17): Reprinted from:
Trab. Inst. Español Ocean., Madrid, 19 Sept. 1959(28):88 pp., 22 pls.

tintinnids

Balech, E., 1951.
Nuevos datos sobre Tintinnoinea de Argentina y Uraguay. Physis 20(58):291-302, 16 textfigs.

tintinnids

Balech, Enrique, 1948
Tintinnoinea de Atlantida (R.O. Del Uruguay) (Protozoa, Ciliata Oligotr.). Comm. Mus. Argentino Ciencias Nat. "Bernardino Rivadavia", Ser.Cien.Zool. No.7, 23 pp., 8 pls. of 107 figs.

tintinnids

Balech, E., 1945.
Tintinnoinea de Quequén. Physis 20(55):1-15, 3 pls., of 30 figs.

tintinnids

Balech, E., 1941.
Tintinnoineos del estrecho Le Maire. Physis XIX (52):245-252, 10 textfigs.

Tintinnoinea

Biernacka, I., 1948
(Tintinnoinea in the Gulf of Gdansk and adjoining waters) Builetyn Morskiego Laboratorium Rybackiego w Gdyni dawniej-Stacji Morskiej w. Helu. No. 4:73-91, 4 text figs., 1 pl. with 21 figs. (Bull. Lab. Mar., Gdynia, formerly Bull. Sta. Hel.)

tintinnids

Brandes, C. -H., 1939(1951).
Über die räumlichen und zeitlichen Unterschiede in der Zusammensetzung des Ostseeplanktons. Mitt. Hamburg Zool. Mus. u. Inst. 48:1-47, 23 textfigs.

Tintinnoina

Campbell, A. S., 1942
The oceanic Tintinnoina of the plankton gathered during the last cruise of the CARNEGIE. Sci. Res. Cruise VII of the Carnegie, 1928-1929, Biol. 2:163 pp., 128 figs.

tintinnids

Duran, M., 1957.
Nota sobre algunos tintinnoineas del plancton de Puerto Rico. Invest. Pesquera, Barcelona, 8:97-120.

tintinnids

Duran, M., 1953.
Contribución al estudio de los Tintinnidos del plancton de las costas de Castellón (Mediterraneo occidental). Nota II. Publ. Inst. Biol. Aplic. 12:79-95, 23 textfigs.

tintinnids

Duran, M., 1951.
Contribución al estudio de los tintinnidos del plancton de las costas de Castellón (Mediterraneo occidental). Publ. Inst. Biol. Aplic. 8:101-120, 29 textfigs.

tintinnids

Gaarder, K.R., 1946
Tintinnoidea. Rep.Sci.Res. "Michael Sars" N.Atlantic Deep-Sea Exped., 1910, 2(1): 37 pp., 23 text figs.

tintinnids, lists of spp.

Genovese, S., G. Gangemi e F. De Domenio 1972.
Campagna estiva 1970 della n/o Bannock nel Mar Tirreno - misure di produzione primaria lungo la trasversale Palermo-Cagliari.
Boll. Pesca Piscic. Idrobiol. 27(1):139-157.

tintinnids

Hada, Yoshine, 1957
[The Tintinnoinea, useful microplankton for judging oceanographic conditions.]
Info. Bull. Plankton, Japan, (5):10-12.

Tintinnoidea

Hofker, J., 1931
Studien uber Tintinnoidea. Arch. f. Protistenk 75(3):315-402, 89 text figs.

tintinnids

Johnson, M. W., 1949.
Relation of plankton to hydrographic conditions in Sweetwater Lake. J. Am Water Works Assoc. 41(4):347-356, 12 textfigs.

Tintinnidae

Jørgensen, E., 1924
Mediterranean Tintinnidae. Rept. Danish Oceanogr. Exped. 1908-10 to the Mediterranean and adjacent seas, Vol.II, Biol. J 3, 110 pp., 114 text figs.

tintinnids

Jørgensen, E., 1900.
Ueber die Tintinnodeen der norwegischen Westküste Bergens Mus. Aarb., 1899(2):48 pp., 3 pls.

tintinnids

Kofoid, C.A. and A.S. Campbell, 1939
Reports on the scientific results of the expedition to the eastern tropical Pacific in charge of Alexander Agassiz, by the U.S. Fish Commission steamer "Albatross" from October 1904 to March 1905-----XXXVII. The Ciliata: The Tintinnoinea. Bull. Mus. Comp. Zool. 84: 473 pp., 36 pls.

Tintinnoinea

Kofoid, C. A. and A. S. Campbell, 1929
A conspectus of the marine and freshwater Ciliata belonging to the suborder Tintinnoinea, with descriptions of new species principally from the Agassiz expedition to the eastern tropical Pacific, 1904-1905. Univ. Calif. Publ. Zool. 34:1-403, 697 text figs.

tintinnids

Komarovsky, B., 1962
Tintinnina from the vicinity of the Straits of Tiran and Massawa Region. Contributions to the knowledge of the Red Sea, No. 25. Sea Fish. Res. Sta., Haifa, Israel, Bull. No. 30: 48-56.

tintinnids

Komarovsky, B., 1959
The Tintinnina of the Gulf of Eylath (Aqaba).
Contrib. Knowl. Red Sea, Bull. Sea Fish. Res. Sta.Haifa, Israel 21 (14):1-40.

tintinnids

Kruger, D., 1950.
Variations quantitatives des protistes marins au voisinage du Port d'Alger durant l'hiver 1949-1950. Bull. Inst. Océan., Monaco, No. 978:20 pp., 5 textfigs.

Tintinnids

Laackmann, H., 1910
Die Tintinnodeen der Deutschen Sudpolar Expedition 1901-1903. Deutsch.Sudpolar-Exped., 1901-1903, 11 Zool. 3: 343-496. Pls. 33-51.

tintinnoinea

Marshall, S. M., 1934
The silicoflagellata and tintinnoinea. Great Barrier Reef Exped., 1928-29. Sci. Repts. 4(15):623-664, 43 text figs.

tintinnids

Massuti Alzamora, M., 1929.
Contribución al estudio de los infusorios de la Bahia de Palma de Mallorca. Nota secunda. Notas y Res., Inst. Español Ocean., 2nd ser., No. 32:10 pp., 38 textfigs.

tintinnids

Massuti, M., and F. de P. Navarro, 1950.
Tintinídos y Copépodos planctónicos del Mar de Alboran. (Campana del "Xauen" en Agosto y Septiembre de 1948). Bol. Inst. Español Ocean., No. 37:28 pp., 7 textfigs.

tintinnids

Meunier, A., 1919.
Microplankton de la Mer Flamande. 4. Les Tintinnides et Coetera. Mem. Mus. Roy. Hist. Nat., Belgique, 8(2):59pp., Pls. 22-23.

tintinnids

Posta, Annette, 1963.
Relation entre l'évolution de quelques tintinnides de la rade de Villefranche et la température de l'eau.
Cahiers, Biol. Mar., Roscoff, 4:201-210.
Also in:
Trav. Sta. Zool., Villefranche-sur-Mer, 22(5).

tintinntids

Rampi, L., 1950
I Tintinnoidi delle acque di Monaco raccolti d'all'Eider nell'anno 1913. Bull. Inst. Ocean. Monaco, No. 965: 7 pp., 1 textfig.

tintinnids

Rampi, L., 1948
Sur quelques Tintinnides (Infusoires louques) du Pacifique subtropical (Recoltes Alain Gerbault) Bull. l'Inst. Ocean., Monaco, No.938, 4 pp.

tintinnids

Rampi, L., 1939
Primo contributo alla conoscenza dei tintinnoidi do Maro Ligure. Atti della Soc. Ital. di Sci. Nat. 78:67-81, 58 text figs.

tintinnids

Roxas, H.A., 1941.
Marine protozoa of the Philippines.
Philippine J. Sci. 74:91-136, 17 pls., 2 textfigs.

tintinnids

Sousa e Silva, E. de, 1950.
Les tintinnides de la Baie de Cascais. Bull. Inst. Océan., Monaco, No. 979:28 pp., 4 pls.

tintinnids

Tappan, Helen, and Alfred P. Loeblich, Jr., 1968.
Lorica composition of modern and fossil Tintinnida (ciliate Protozoa), systematics, geologic distribution, and some new Tertiary taxa.
J. Paleont., 42(6):1378-1394.

tintinnids

Travers, Anne, et Marc Travers 1973.
Présence en Méditerranée du genre Salpingacantha Kofoid et Campbell (Ciliés Oligotriches Tintinnides). Rapp. Proc.-V. Réun. Commn int. Explor. scient. Mer Medit. Monaco 21(8):429-432.

tintinnids

Travers, Marc, 1969.
Contribution à l'étude du phytoplancton et des tintinnides de la région de Tuléar (Madagascar). II. Les pigments planctoniques.
Rec. Trav. Sta. mar. Endoume, hors sér. Suppl. 9:49-57.

tintinnids

Tregouboff, G., 1956.
Rapport sur les travaux concernant le plancton Mediterranéen publiés entre Novembre 1952 et Novembre 1954.
Rapp. Proc. Verb., Comm. Int. Expl. Sci., Mer Mediterranee, 13:65-100

Tintinnids, anat.-physiol.

Biernacka, Izabela, 1965.
Ausscheidung gehäusebildender Substanzen durch reife Formen gewisser Arten der Gattung Tintinnopsis Stein.
Acta Protozoologica, Warszawa, 3(23):265-268.

tintinnids, life history

Biernacka, I., 1952.
[Studies on the reproduction of some species of the genus Tintinnopsis Stein.] (English summary).
Ann. Univ. Marie Curie Sklodowska, Sect. C, 6(6):211-247.

Tintinnids, anatomy

Entz, G., Jr. 1909
Studien über organisation und biologie der Tintinniden. Arch. f. Protistenkunde 15:93-226, Pls. 8-21, text figs.

tintinnids, anat. physiol.

Gold, Kenneth 1968.
Some observations on the biology of Tintinnopsis sp.
J. Protozool. 15(1): 193-194.

tintinnids, anat.

Laval, Michele, 1971.
Mise en évidence par la microscopie électronique d'un organite d'un type nouveau chez les Ciliés Tintinnides.
C.r. Acad. Sci. Paris 273:1383-1386.

tintinnids, distribution of

Zeitzschel, Bernt, 1966.
Die Verbreitung der Tintinnen im Nordatlantik.
Veröff. Inst. Meeresforsch., Bremerh., Sonderband II:293-300.

tintinnids, ecology of

Zeitzschel, Bernt 1967.
Die Bedeutung der Tintinnen als Glied der Nahrungskette.
Helgoländer Wiss. Meeresunters. 15(1/4): 589-600.

tintinnids, lists of spp.

Balech, Enrique, 1971
Dinoflagelados y tintinnidos del Golfo de México y Caribe: sus relaciones con el Atlántico Ecuatorial. Symp. Investigations and resources of the Caribbean Sea and adjacent regions, UNESCO, 18-26 Nov. 1968, Curaçao: 297-301.

tintinnids, lists of spp.

Balech, Enrique, 1967.
Dinoflagellates and tintinnids in the northeastern Gulf of Mexico.
Bull. mar. Sci., Miami, 17(2):280-298.

tintinnids, lists of spp.

Hada, Yoshine, 1970. (1970)
The protozoan plankton of the Antarctic and Subantarctic seas.
Scient. Repts, Japan. Antarct. Res. Exped., (E)31:51 pp.

Tintinnids, lists of spp.

Lecal, J., 1967.
Le nannoplancton des côtes d'Israel.
Hydrobiologia, 29(3/4):305-387

tintinnids, lists of spp.

Loeblich, Alfred R., Jr., and Helen Tappan 1968.
Annotated index to genera, subgenera and suprageneric taxa of the ciliate order Tintinnida.
J. Protozool. 15(1): 185-192.

tintinnids, lists of spp.

Portugal, Instituto Hidrográfico 1973
CAPEC-II Janeiro/Fevereiro-1971: resultados preliminares 9: 149 pp.

tintinnids, lists of spp.

Portugal, Instituto Hidrográfico 1973
Companha oceanografica para Apoio as Pescas do continente. CAPECI (12 Outubro a 15 de Novembro de 1970. Resultados Preliminares, 6:123 pp.

tintinnids, list of species

Rampi, L., 1948.
I. Tintinnoidi delle acque di San Remo. Boll. Pesca, Piscicolt., Idrobiol., n.s., 3(1):50-56.

tintinnids, lists of spp.

Travers, Anne, et Marc Travers 1970 (1971)
Catalogue des tintinnides (ciliés oligotriches) récoltés dans le Golfe de Marseille de 1962 à 1964.
Téthys 2(3): 639-646.

tintinnids lists of spp.

Travers, A. et M., 1965.
Introduction a l'étude du phytoplancton et des tintinnides de la region de Tuléar (Madagascar).
Rec. Trav. Sta. Mar. Endoume, hors sér., Suppl. 125-162.

tintinnids, lists of spp.

Vitiello, Pierre, 1964.
Contribution a l'etude des tintinnides de la baie d'Alger.
Pelagos, Bull. Inst. Oceanogr., Alger, 2(2):5-41
(pp numbers of this article duplicate those of previous one by Francis Bernard!)

Acanthostomella conicoides

Kofoid, C.A. and A.S. Campbell,1939
Reports on the scientific results of the expedition to the eastern tropical Pacific in charge of Alexander Agassiz, by the U.S. Fish Commission steamer "Albatross" from October 1904 to March 1905-----XXXVII. The Ciliata: The Tintinnoimea. Bull. Mus. Comp. Zool. 84: 473 pp., 36 pls.

Acanthostomella conicoides n.sp.

Kofoid, C. A. and A. S. Campbell, 1929
A conspectus of the marine and freshwater Ciliata belonging to the suborder Tintinnoinea, with descriptions of new species principally from the Agassiz expedition to the eastern tropical Pacific, 1904-1905. Univ. Calif. Publ. Zool. 34:1-403, 697 text figs.

Acanthostomella elongata

Campbell, A. S., 1942
The oceanic Tintinnoina of the plankton gathered during the last cruise of the CARNEGIE. Sci. Res. Cruise VII of the Carnegie, 1928-1929, Biol. 2:163 pp., 128 figs.

Acanthostomella elongata n.sp.

Kofoid, C. A. and A. S. Campbell, 1929
A conspectus of the marine and freshwater Ciliata belonging to the suborder Tintinnoinea, with descriptions of new species principally from the Agassiz expedition to the eastern tropical Pacific, 1904-1905. Univ. Calif. Publ. Zool. 34:1-403, 697 text figs.

Acanthostomella gracilis

Campbell, A. S., 1942
The oceanic Tintinnoina of the plankton gathered during the last cruise of the CARNEGIE. Sci. Res. Cruise VII of the Carnegie, 1928-1929, Biol. 2:163 pp., 128 figs.

Acanthostomella gracilis

Kofoid, C. A. and A. S. Campbell, 1929
A conspectus of the marine and freshwater Ciliata belonging to the suborder Tintinnoinea, with descriptions of new species principally from the Agassiz expedition to the eastern tropical Pacific, 1904-1905. Univ. Calif. Publ. Zool. 34:1-403, 697 text figs.

Acanthostomella lata

Balech, Enrique, 1968.
Algunas especies nuevas o interesantes do tintinnidos del Golfo de Mexico y Caribe.
Revta Mus. argent. Cienc. nat. Bernardino Rivadavia Inst. nac. Hidrobiol. 2(5):165-197.

Acanthostomella lata

Campbell, A. S., 1942
The oceanic Tintinnoina of the plankton gathered during the last cruise of the CARNEGIE. Sci. Res. Cruise VII of the Carnegie, 1928-1929, Biol. 2:163 pp., 128 figs.

Acanthostomella lata

Gaarder, K.R., 1946
Tintinnoidea. Rep.Sci.Res. "Michael Sars" N.Atlantic Deep-Sea Exped., 1910, 2(1): 37 pp., 23 text figs.

Acanthostomella lata

Kofoid, C.A. and A.S. Campbell,1939
Reports on the scientific results of the expedition to the eastern tropical Pacific in charge of Alexander Agassiz, by the U.S. Fish Commission steamer "Albatross" from October 1904 to March 1905-----XXXVII. The Ciliata: The Tintinnoimea. Bull. Mus. Comp. Zool. 84: 473 pp., 36 pls.

Acanthostomella lata n.sp.

Kofoid, C. A. and A. S. Campbell, 1929
A conspectus of the marine and freshwater Ciliata belonging to the suborder Tintinnoinea, with descriptions of new species principally from the Agassiz expedition to the eastern tropical Pacific, 1904-1905. Univ. Calif. Publ. Zool. 34:1-403, 697 text figs.

Acanthostomella minutissima

Balech, Enrique, 1968.
Algunas especies nuevas o interesantes do tintinnidos del Golfo de Mexico y Caribe.
Revta Mus. argent. Cienc. nat. Bernardino Rivadavia Inst. nac. Hidrobiol. 2(5):165-197.

Acanthostomella minutissima

Campbell, A. S., 1942
The oceanic Tintinnoina of the plankton gathered during the last cruise of the CARNEGIE. Sci. Res. Cruise VII of the Carnegie, 1928-1929, Biol. 2:163 pp., 128 figs.

Oceanographic Index: Marine Organisms Cumulation, 1946-1973

Acanthostomella minutissima

Gaarder, K.R., 1946
Tintinnoidea. Rep.Sci.Res."Michael Sars" N.Atlantic Deep-Sea Exped., 1910, 2(1): 37 pp., 23 text figs.

Acanthostomella minutissima

Hasle, Grethe Rytter, 1960
Phytoplankton and ciliate species from the Tropical Pacific. Skr. Norske Videnskaps-Akad., Oslo, 1. Mat.-Nat. Kl., 1960(2): 1-50.

Acanthostomella minutissima

Kofoid, C.A. and A.S. Campbell, 1939
Reports on the scientific results of the expedition to the eastern tropical Pacific in charge of Alexander Agassiz, by the U.S. Fish Commission steamer "Albatross" from October 1904 to March 1905-----XXXVII. The Ciliata: The Tintinnoimea. Bull.Mus. Comp.Zool. 84: 473 pp., 36 pls.

Acanthostomella minutissima n.sp.

Kofoid, C. A. and A. S. Campbell, 1929
A conspectus of the marine and fresh-water Ciliata belonging to the suborder Tintinnoinea, with descriptions of new species principally from the Agassiz expedition to the eastern tropical Pacific, 1904-1905. Univ. Calif. Publ. Zool. 34:1-403, 697 text figs.

Acanthostomella norvegica

Braarud, T., 1934
A note on the phytoplankton of the Gulf of Maine in the summer of 1933. Biol. Bull. 67(1):76-82.

Acanthostomella norvegica

Campbell, A. S., 1942
The oceanic Tintinnoina of the plankton gathered during the last cruise of the CARNEGIE. Sci. Res. Cruise VII of the Carnegie, 1928-1929, Biol. 2:163 pp., 128 figs.

Acanthostomella norvegica

Gaarder, K.R., 1946
Tintinnoidea. Rep.Sci.Res."Michael Sars" N.Atlantic Deep-Sea Exped., 1910, 2(1): 37 pp., 23 text figs.

Acanthostomella norvegica

Gran, H.H., and T. Braarud, 1935
A quantitative study of the phytoplankton in the Bay of Fundy and the Gulf of Maine (including observations on hydrography, chemistry, and turbidity). J. Biol. Bd., Canada, 1(5):279-467, 69 text figs.

Acanthostomella norvegica

Kofoid, C. A. and A. S. Campbell, 1929
A conspectus of the marine and fresh-water Ciliata belonging to the suborder Tintinnoinea, with descriptions of new species principally from the Agassiz expedition to the eastern tropical Pacific, 1904-1905. Univ. Calif. Publ. Zool. 34:1-403, 697 text figs.

Acanthostomella obtusa

Balech, Enrique, 1968.
Algunas especies nuevas o interesantes do tintinnidos del Golfo de Mexico y Caribe. Revta Mus. argent. Cienc. nat. Bernardino Rivadavia Inst. nac. Hidrobiol. 2(5):165-197.

Acanthostomella obtusa

Campbell, A. S., 1942
The oceanic Tintinnoina of the plankton gathered during the last cruise of the CARNEGIE. Sci. Res. Cruise VII of the Carnegie, 1928-1929, Biol. 2:163 pp., 128 figs.

Acanthostomella obtusa

Kofoid, C.A. and A.S. Campbell,1939
Reports on the scientific results of the expedition to the eastern tropical Pacific in charge of Alexander Agassiz, by the U.S. Fish Commission steamer "Albatross" from October 1904 to March 1905-----XXXVII. The Ciliata: The Tintinnoimea. Bull.Mus. Comp.Zool. 84: 473 pp., 36 pls.

Albatrossiella n. gen.

Kofoid, C. A. and A. S. Campbell, 1929
A conspectus of the marine and fresh-water Ciliata belonging to the suborder Tintinnoinea, with descriptions of new species principally from the Agassiz expedition to the eastern tropical Pacific, 1904-1905. Univ. Calif. Publ. Zool. 34:1-403, 697 text figs.

Albatrossiella agassizi

Kofoid, C.A. and A.S. Campbell,1939
Reports on the scientific results of the expedition to the eastern tropical Pacific in charge of Alexander Agassiz, by the U.S. Fish Commission steamer "Albatross" from October 1904 to March 1905-----XXXVII. The Ciliata: The Tintinnoimea. Bull.Mus. Comp.Zool. 84: 473 pp., 36 pls.

Albatrossiella agassizi n.sp.

Kofoid, C. A. and A. S. Campbell, 1929
A conspectus of the marine and fresh-water Ciliata belonging to the suborder Tintinnoinea, with descriptions of new species principally from the Agassiz expedition to the eastern tropical Pacific, 1904-1905. Univ. Calif. Publ. Zool. 34:1-403, 697 text figs.

Albatrossiella filigera n.sp.

Kofoid, C. A. and A. S. Campbell, 1929
A conspectus of the marine and fresh-water Ciliata belonging to the suborder Tintinnoinea, with descriptions of new species principally from the Agassiz expedition to the eastern tropical Pacific, 1904-1905. Univ. Calif. Publ. Zool. 34:1-403, 697 text figs.

Albatrossiella minutissima

Kofoid, C. A. and A. S. Campbell, 1929
A conspectus of the marine and fresh-water Ciliata belonging to the suborder Tintinnoinea, with descriptions of new species principally from the Agassiz expedition to the eastern tropical Pacific, 1904-1905. Univ. Calif. Publ. Zool. 34:1-403, 697 text figs.

Amphorella sp.

Massutí Algamora, M., 1949
Estudio de diez y seis muestras de plancton del Golfo de Nápoles. Publ. Inst. Biol. Appl. 5:85-94, 1 fold-in table.

Amphorella amphora

Balech, E., 1959.
Tintinnoinea del Mediterraneo. Trav. Sta. Zool., Villefranche-sur-Mer, 18(17): Reprinted from: Trab. Inst. Español Oceanogr., Madrid, 19 Sept. 1959(28):88 pp., 22 pls.

Amphorella amphora

Balech, E., 1945.
Tintinnoinea de Quequén. Physis 20(55):1-15, 3 pls.

Amphorella amphora

Campbell, A. S., 1942
The oceanic Tintinnoina of the plankton gathered during the last cruise of the CARNEGIE. Sci. Res. Cruise VII of the Carnegie, 1928-1929, Biol. 2:163 pp., 128 figs.

Amphorella amphora

Kofoid, C.A. and A.S. Campbell,1939
Reports on the scientific results of the expedition to the eastern tropical Pacific in charge of Alexander Agassiz, by the U.S. Fish Commission steamer "Albatross" from October 1904 to March 1905-----XXXVII. The Ciliata: The Tintinnoimea. Bull.Mus. Comp.Zool. 84: 473 pp., 36 pls.

Amphorella amphora

Kofoid, C. A. and A. S. Campbell, 1929
A conspectus of the marine and fresh-water Ciliata belonging to the suborder Tintinnoinea, with descriptions of new species principally from the Agassiz expedition to the eastern tropical Pacific, 1904-1905. Univ. Calif. Publ. Zool. 34:1-403, 697 text figs.

Amphorella ampla n.sp.

Jørgensen, E., 1900.
Ueber die Tintinnodeen der norwegischen Westküste. Bergens Mus. Aarb. 1899(2):48 pp., 3 pls.

Amphorella brandti

Kofoid, C. A. and A. S. Campbell, 1929
A conspectus of the marine and fresh-water Ciliata belonging to the suborder Tintinnoinea, with descriptions of new species principally from the Agassiz expedition to the eastern tropical Pacific, 1904-1905. Univ. Calif. Publ. Zool. 34:1-403, 697 text figs.

Amphorella brandti

Marshall, S. M., 1934
The silicoflagellata and tintinnoinea. Great Barrier Reef Exped., 1928-29. Sci. Repts. 4(15):623-664, 43 text figs.

Amphorella brandti

Rampi, L., 1939
Primo contributo alla conoscenza dei tintinnoidi do Maro Ligure. Atti della Soc. Ital. di Sci. Nat. 78:67-81, 58 text figs.

Amphorella calida n.sp.

Kofoid, C. A. and A. S. Campbell, 1929
A conspectus of the marine and fresh-water Ciliata belonging to the suborder Tintinnoinea, with descriptions of new species principally from the Agassiz expedition to the eastern tropical Pacific, 1904-1905. Univ. Calif. Publ. Zool. 34:1-403, 697 text figs.

Amphorella fusiformis

Meunier, A., 1919
Microplankton de la Mer Flamanda. 4. Les Tintinnides et Coetera. Mem. Mus. Roy. Hist. Nat., Belgique, 8(2):59pp., Pls. 22-23.

Amphorella ganymedes

Duran, M., 1951
Contribucion al estudio de los tintinidos del plancton de los costas de Castellon. (Mediterraneo occidental). Publ. Inst. Biol. Aplic., Barcelone, 8: 101-120, 29 text figs.

Amphorella ganymedes

Hofker, J., 1931
Studien uber Tintinnoidea. Arch. f. Protistenk 75(3):315-402, 89 text figs.

Amphorella ganymedes

Jörgensen, E., 1924
Mediterranean Tintinnidae. Rept. Danish Oceanogr. Exped. 1908-10 to the Mediterranean and adjacent seas, Vol.II, Biol. J 3, 110 pp., 114 text figs.

Amphorella gracilis

Duran, M., 1951
Contribucion al estudio de los tintinidos del plancton de los costas de Castellon. (Mediterraneo occidental). Publ. Inst. Biol. Aplic., Barcelone, 8: 101-120, 29 text figs.

Amphorella gracilis n.sp.

Jörgensen, E., 1924
Mediterranean Tintinnidae. Rept. Danish Oceanogr. Exped. 1908-10 to the Mediterranean and adjacent seas, Vol.II, Biol. J 3, 110 pp., 114 text figs.

Amphorella infundibulum n.sp.

Kofoid, C. A. and A. S. Campbell, 1929
A conspectus of the marine and fresh-water Ciliata belonging to the suborder Tintinnoinea, with descriptions of new species principally from the Agassiz expedition to the eastern tropical Pacific, 1904-1905. Univ. Calif. Publ. Zool. 34:1-403, 697 text figs.

Amphorella intumescens n.sp.

Jörgensen, E., 1924
Mediterranean Tintinnidae. Rept. Danish Oceanogr. Exped. 1908-10 to the Mediterranean and adjacent seas, Vol.II, Biol. J 3, 110 pp., 114 text figs.

Amphorella laackmanni n.nom.

Jörgensen, E., 1924
Mediterranean Tintinnidae. Rept. Danish Oceanogr. Exped. 1908-10 to the Mediterranean and adjacent seas, Vol.II, Biol. J 3, 110 pp., 114 text figs.

Amphorella laackmanni

Kofoid, C. A. and A. S. Campbell, 1929
A conspectus of the marine and fresh-water Ciliata belonging to the suborder Tintinnoinea, with descriptions of new species principally from the Agassiz expedition to the eastern tropical Pacific, 1904-1905. Univ. Calif. Publ. Zool. 34:1-403, 697 text figs.

Amphorella laackmanni

Marshall, S. M., 1934
The silicoflagellata and tintinnoinea. Great Barrier Reef Exped., 1928-29. Sci. Repts. 4(15):623-664, 43 text figs.

Amphorella minor

Balech, E., 1945.
Tintinnoinea de Quequén. Physis 20(55):1-15, 3 pls.

Amphorella minor

Kofoid, C.A. and A.S. Campbell, 1939
Reports on the scientific results of the expedition to the eastern tropical Pacific in charge of Alexander Agassiz, by the U.S. Fish Commission steamer"Albatross" from October 1904 to March 1905------XXXVII. The Ciliata: The Tintinnoimea. Bull.Mus. Comp.Zool. 84: 473 pp., 36 pls.

Amphorella minor

Kofoid, C. A. and A. S. Campbell, 1929
A conspectus of the marine and fresh-water Ciliata belonging to the suborder Tintinnoinea, with descriptions of new species principally from the Agassiz expedition to the eastern tropical Pacific, 1904-1905. Univ. Calif. Publ. Zool. 34:1-403, 697 text figs.

Amphorella minor

Komarovsky, B., 1959
The Tintinnina of the Gulf of Eylath (Aqaba). Contrib. Knowl. Red Sea, Bull. Sea Fish. Res. Sta.Haifa, Israel 21 (14):1-40.

Amphorella minor

Marshall, S. M., 1934
The silicoflagellata and tintinnoinea. Great Barrier Reef Exped., 1928-29. Sci. Repts. 4(15):623-664, 43 text figs.

Amphorella minor

Rampi, L., 1939
Primo contributo alla conoscenza dei tintinnoidi do Mare Ligure. Atti della Soc. Ital. di Sci. Nat. 78:67-81, 58 text figs.

Amphorella oxyura n.sp.

Jörgensen, E., 1924
Mediterranean Tintinnidae. Rept. Danish Oceanogr. Exped. 1908-10 to the Mediterranean and adjacent seas, Vol.II, Biol. J 3, 110 pp., 114 text figs.

Amphorella pachytoecus n.sp.

Jörgensen, E., 1924
Mediterranean Tintinnidae. Rept. Danish Oceanogr. Exped. 1908-10 to the Mediterranean and adjacent seas, Vol.II, Biol. J 3, 110 pp., 114 text figs.

Amphorella pyramidata n.sp.

Jörgensen, E., 1924
Mediterranean Tintinnidae. Rept. Danish Oceanogr. Exped. 1908-10 to the Mediterranean and adjacent seas, Vol.II, Biol. J 3, 110 pp., 114 text figs.

Amphorella quadrilineata / Amphorella quadrilineata var. minor

Balech, E., 1959.
Tintinnoinea del Mediterraneo. Trav. Sta. Zool., Villefranche-sur-Mer, 18(17); Reprinted from:
Trab. Inst. Español Oceanogr., Madrid, 19 Sept. 1959(28):88 pp., 22 pls.

Amphorella quadrilineata

Balech, Enrique, 1944
Contribucion al conocimiento del Plancton de Lennox y Cabo de Hornos. Physis XIX:423-446, 6 pls. with 67 figs.

Amphorella quadrilineata

Campbell, A. S., 1942
The oceanic Tintinnoina of the plankton gathered during the last cruise of the CARNEGIE. Sci. Res. Cruise VII of the Carnegie, 1928-1929, Biol. 2:163 pp., 128 figs.

Amphorella quadrilineata

Duran, M., 1951
Contribucion al estudio de los tintinidos del plancton de los costas de Castellon. (Mediterranes occidental). Publ. Inst. Biol. Aplic., Barcelone, 8: 101-120, 29 text figs.

Amphorella quadrilineata

Gaarder, K.R., 1946
Tintinnoidea. Rep.Sci.Res."Michael Sars" N.Atlantic Deep-Sea Exped., 1910, 2(1): 37 pp., 23 text figs.

Amphorella quadrilineata

Hofker, J., 1931
Studien uber Tintinnoidea. Arch. f. Protistenk 75(3):315-402, 89 text figs.

Amphorella quadrilineata

Jörgensen, E., 1924
Mediterranean Tintinnidae. Rept. Danish Oceanogr. Exped. 1908-10 to the Mediterranean and adjacent seas, Vol.II, Biol. J 3, 110 pp., 114 text figs.

Amphorella quadrilineata

Jørgensen, E., 1905
B.Protistplankton and the diatoms in bottom samples. Hydrographical and biological investigations in Norwegian fiords. Bergens Mus. Skr. 7: 49-225.

Amphorella quadrilineata

Jørgensen, E., 1900.
Ueber die Tintinnodeen der norwegischen Westküste Bergens Mus. Aarb., 1899(2):48 pp., 3 pls.

Amphorella quadrilineata

Kofoid, C.A. and A.S. Campbell,1939
Reports on the scientific results of the expedition to the eastern tropical Pacific in charge of Alexander Agassiz, by the U.S. Fish Commission steamer"Albatross" from October 1904 to March 1905------XXXVII. The Ciliata: The Tintinnoimea. Bull.Mus. Comp.Zool. 84: 473 pp., 36 pls.

Amphorella quadrilineata

Kofoid, C. A. and A. S. Campbell, 1929
A conspectus of the marine and fresh-water Ciliata belonging to the suborder Tintinnoinea, with descriptions of new species principally from the Agassiz expedition to the eastern tropical Pacific, 1904-1905. Univ. Calif. Publ. Zool. 34:1-403, 697 text figs.

Amphorella quadrilineata

Komarovsky, B., 1959
The Tintinnina of the Gulf of Eylath (Aqaba). Contrib. Knowl. Red Sea, Bull. Sea Fish. Res. Sta.Haifa, Israel 21 (14):1-40.

Amphorella quadrilineata

Marshall, S. M., 1934
The silicoflagellata and tintinnoinea. Great Barrier Reef Exped., 1928-29. Sci. Repts. 4(15):623-664, 43 text figs.

Amphorella quadrilineata

Marshall, S. M., 1933
The production of microplankton in the Great Barrier Reef Region. Brit. Mus. (N.H.) Great Barrier Reef Exped. 1928-29, Sci. Repts. II(5):111-157, 14 text figs.

Amphorella quadrilineata

Rampi, L., 1939
Primo contributo alla conoscenza dei tintinnoidi do Mare Ligure. Atti della Soc. Ital. di Sci. Nat. 78:67-81, 58 text figs.

Amphorella quadrilineata

Sousa e Silva, E. de, 1950.
Les tintinnides de la Baie de Cascais. Bull. Inst. Océan., Monaco, No. 979:28 pp., 4 pls.

Amphorella steenstrupii

Delegazione Italiana della Commissione Internazionale per l'Esplorazione Scientifica del Mediterraneo, 1941
Note sul plancton della Laguna veneta. [Memoria CCLXXXIX] , Arch. di Ocean. e Limn. Anno I, Fasc. I, 1941 XIX: 31-57 pp.

Amphorella steenstrupii

Duran, M., 1951
Contribucion al estudio de los tintinidos del plancton de los costas de Castellon. (Mediterranes occidental). Publ. Inst. Biol. Aplic., Barcelone, 8: 101-120, 29 text figs.

Amphorella steenstrupi

Jörgensen, E., 1924
Mediterranean Tintinnidae. Rept. Danish Oceanogr. Exped. 1908-10 to the Mediterranean and adjacent seas, Vol.II, Biol. J 3, 110 pp., 114 text figs.

Amphorella steenstrupii

Jørgensen, E., 1905
B.Protistplankton and the diatoms in bottom samples. Hydrographical and biological investigations in Norwegian fiords. Bergens Mus. Skr. 7: 49-225.

Amphorella steenstrupi

Jørgensen, E., 1900.
Ueber die Tintinnodeen der norwegischen Westküste. Bergens Mus. Aarb., 1899(2):48 pp., 3 pls.

Amphorella subulata

Jørgensen, E., 1900.
Ueber die Tintinnodeen der norwegischen Westküste Bergens Mus. Aarb., 1899(2):48 pp., 3 pls.

Amphorella tetragona

Duran, M., 1951
Contribucion al estudio de los tintinidos del plancton de los costas de Castellon. (Mediterranes occidental). Publ. Inst. Biol. Aplic., Barcelone, 8: 101-120, 29 text figs.

Amphorella tetragona n.sp.

Jörgensen, E., 1924
Mediterranean Tintinnidae. Rept. Danish Oceanogr. Exped. 1908-10 to the Mediterranean and adjacent seas, Vol.II, Biol. J 3, 110 pp., 114 text figs.

Amphorella torulata n.sp.

Jörgensen, E., 1924
Mediterranean Tintinnidae. Rept. Danish Oceanogr. Exped. 1908-10 to the Mediterranean and adjacent seas, Vol.II, Biol. J 3, 110 pp., 114 text figs.

Amphorella trachelium n.sp.

Jörgensen, E., 1924
Mediterranean Tintinnidae. Rept. Danish Oceanogr. Exped. 1908-10 to the Mediterranean and adjacent seas, Vol.II, Biol. J 3, 110 pp., 114 text figs.

Amphorellopsis n. gen.

Kofoid, C. A. and A. S. Campbell, 1929
A conspectus of the marine and freshwater Ciliata belonging to the suborder Tintinnoinea, with descriptions of new species principally from the Agassiz expedition to the eastern tropical Pacific, 1904-1905. Univ. Calif. Publ. Zool. 34:1-403, 697 text figs.

Amphorellopsis acanthurus

Kofoid, C.A. and A.S. Campbell, 1939
Reports on the scientific results of the expedition to the eastern tropical Pacific in charge of Alexander Agassiz, by the U.S. Fish Commission steamer"Albatross" from October 1904 to March 1905------XXXVII. The Ciliata: The Tintinnoimea. Bull.Mus. Comp.Zool. 84: 473 pp., 36 pls.

Amphorellopsis acantharus n.sp.

Kofoid, C. A. and A. S. Campbell, 1929
A conspectus of the marine and freshwater Ciliata belonging to the suborder Tintinnoinea, with descriptions of new species principally from the Agassiz expedition to the eastern tropical Pacific, 1904-1905. Univ. Calif. Publ. Zool. 34:1-403, 697 text figs.

Amphorellopsis acuta

Duran, M., 1957.
Nota sobre algunos tintinnoineas del plancton de Puerto Rico. Invest. Pesqueria, Barcelona, 8: 97-120.

Amphorellopsis acuta

Gaarder, K.R., 1946
Tintinnoidea. Rep.Sci.Res."Michael Sars" N.Atlantic Deep-Sea Exped., 1910, 2(1): 37 pp., 23 text figs.

Amphorellopsis acuta

Kofoid, C.A. and A.S. Campbell, 1939
Reports on the scientific results of the expedition to the eastern tropical Pacific in charge of Alexander Agassiz, by the U.S. Fish Commission steamer"Albatross" from October 1904 to March 1905------XXXVII. The Ciliata: The Tintinnoimea. Bull.Mus. Comp.Zool. 84: 473 pp., 36 pls.

Amphorellopsis acuta

Kofoid, C. A. and A. S. Campbell, 1929
A conspectus of the marine and freshwater Ciliata belonging to the suborder Tintinnoinea, with descriptions of new species principally from the Agassiz expedition to the eastern tropical Pacific, 1904-1905. Univ. Calif. Publ. Zool. 34:1-403, 697 text figs.

Amphorellopsis acuta

Marshall, S. M., 1934
The silicoflagellata and tintinnoinea. Great Barrier Reef Exped., 1928-29. Sci. Repts. 4(15):623-664, 43 text figs.

Amphorellopsis laevis

Kofoid, C.A. and A.S. Campbell, 1939
Reports on the scientific results of the expedition to the eastern tropical Pacific in charge of Alexander Agassiz, by the U.S. Fish Commission steamer"Albatross" from October 1904 to March 1905------XXXVII. The Ciliata: The Tintinnoimea. Bull.Mus. Comp.Zool. 84: 473 pp., 36 pls.

Amphorellopsis laevis n.sp.

Kofoid, C. A. and A. S. Campbell, 1929
A conspectus of the marine and freshwater Ciliata belonging to the suborder Tintinnoinea, with descriptions of new species principally from the Agassiz expedition to the eastern tropical Pacific, 1904-1905. Univ. Calif. Publ. Zool. 34:1-403, 697 text figs.

Amphorellopsis quadrangula

Kofoid, C.A. and A.S. Campbell, 1939
Reports on the scientific results of the expedition to the eastern tropical Pacific in charge of Alexander Agassiz, by the U.S. Fish Commission steamer"Albatross" from October 1904 to March 1905------XXXVII. The Ciliata: The Tintinnoimea. Bull.Mus. Comp.Zool. 84: 473 pp., 36 pls.

Amphorellopsis quadrangula n.sp.

Kofoid, C. A. and A. S. Campbell, 1929
A conspectus of the marine and freshwater Ciliata belonging to the suborder Tintinnoinea, with descriptions of new species principally from the Agassiz expedition to the eastern tropical Pacific, 1904-1905. Univ. Calif. Publ. Zool. 34:1-403, 697 text figs.

Amphorellopsis tenuis

Cosper, T.C., 1972.
The identification of tintinnids (Protozoa: Ciliata: Tintinnida) of the St. Andrew Bay system, Florida. Bull. mar. Sci. Miami 22(2): 391-418.

Amphorellopsis tetragona

Kofoid, C. A. and A. S. Campbell, 1929
A conspectus of the marine and freshwater Ciliata belonging to the suborder Tintinnoinea, with descriptions of new species principally from the Agassiz expedition to the eastern tropical Pacific, 1904-1905. Univ. Calif. Publ. Zool. 34:1-403, 697 text figs.

Amphorellopsis tropica

Kofoid, C.A. and A.S. Campbell, 1939
Reports on the scientific results of the expedition to the eastern tropical Pacific in charge of Alexander Agassiz, by the U.S. Fish Commission steamer"Albatross" from October 1904 to March 1905------XXXVII. The Ciliata: The Tintinnoimea. Bull.Mus. Comp.Zool. 84: 473 pp., 36 pls.

Amphorellopsis tropica n.sp.

Kofoid, C. A. and A. S. Campbell, 1929
A conspectus of the marine and freshwater Ciliata belonging to the suborder Tintinnoinea, with descriptions of new species principally from the Agassiz expedition to the eastern tropical Pacific, 1904-1905. Univ. Calif. Publ. Zool. 34:1-403, 697 text figs.

Amphorellopsis turbinata

Kofoid, C.A. and A.S. Campbell, 1939
Reports on the scientific results of the expedition to the eastern tropical Pacific in charge of Alexander Agassiz, by the U.S. Fish Commission steamer"Albatross" from October 1904 to March 1905------XXXVII. The Ciliata: The Tintinnoimea. Bull.Mus. Comp.Zool. 84: 473 pp., 36 pls.

Amphorellopsis turbinea n.sp.

Kofoid, C. A. and A. S. Campbell, 1929
A conspectus of the marine and freshwater Ciliata belonging to the suborder Tintinnoinea, with descriptions of new species principally from the Agassiz expedition to the eastern tropical Pacific, 1904-1905. Univ. Calif. Publ. Zool. 34:1-403, 697 text figs.

Amphorides brandti

Cosper, T.C., 1972.
The identification of tintinnids (Protozoa: Ciliata: Tintinnida) of the St. Andrew Bay system, Florida. Bull. mar. Sci. Miami 22(2): 391-418.

Amplectella n. gen.

Kofoid, C. A. and A. S. Campbell, 1929
A conspectus of the marine and freshwater Ciliata belonging to the suborder Tintinnoinea, with descriptions of new species principally from the Agassiz expedition to the eastern tropical Pacific, 1904-1905. Univ. Calif. Publ. Zool. 34:1-403, 697 text figs.

Amplectella ampla

Kofoid, C.A. and A.S. Campbell, 1939
Reports on the scientific results of the expedition to the eastern tropical Pacific in charge of Alexander Agassiz, by the U.S. Fish Commission steamer"Albatross" from October 1904 to March 1905------XXXVII. The Ciliata: The Tintinnoimea. Bull.Mus. Comp.Zool. 84: 473 pp., 36 pls.

Amplectella ampla n.sp.

Kofoid, C. A. and A. S. Campbell, 1929
A conspectus of the marine and freshwater Ciliata belonging to the suborder Tintinnoinea, with descriptions of new species principally from the Agassiz expedition to the eastern tropical Pacific, 1904-1905. Univ. Calif. Publ. Zool. 34:1-403, 697 text figs.

Amplectella bulbosa n.sp.

Kofoid, C.A. and A.S. Campbell, 1939
Reports on the scientific results of the expedition to the eastern tropical Pacific in charge of Alexander Agassiz, by the U.S. Fish Commission steamer"Albatross" from October 1904 to March 1905------XXXVII. The Ciliata: The Tintinnoimea. Bull.Mus. Comp.Zool. 84: 473 pp., 36 pls.

Amplectella collaria

Campbell, A. S., 1942
The oceanic Tintinnoina of the plankton gathered during the last cruise of the CARNEGIE. Sci. Res. Cruise VII of the Carnegie, 1928-1929, Biol. 2:163 pp., 128 figs.

Amplectella collaria

Gaarder, K.R., 1946
Tintinnoidea. Rep.Sci.Res."Michael Sars" N.Atlantic Deep-Sea Exped., 1910, 2(1): 37 pp., 23 text figs.

Amplectella collaria

Kofoid, C.A. and A.S. Campbell, 1939
Reports on the scientific results of the expedition to the eastern tropical Pacific in charge of Alexander Agassiz, by the U.S. Fish Commission steamer"Albatross" from October 1904 to March 1905------XXXVII. The Ciliata: The Tintinnoimea. Bull.Mus. Comp.Zool. 84: 473 pp., 36 pls.

Amplectella collaria

Kofoid, C. A. and A. S. Campbell, 1929
A conspectus of the marine and freshwater Ciliata belonging to the suborder Tintinnoinea, with descriptions of new species principally from the Agassiz expedition to the eastern tropical Pacific, 1904-1905. Univ. Calif. Publ. Zool. 34:1-403, 697 text figs.

Amplectella insignis

Kofoid, C. A. and A. S. Campbell, 1929
A conspectus of the marine and freshwater Ciliata belonging to the suborder Tintinnoinea, with descriptions of new species principally from the Agassiz expedition to the eastern tropical Pacific, 1904-1905. Univ. Calif. Publ. Zool. 34:1-403, 697 text figs.

Amplectella monocollaria

Campbell, A. S., 1942
The oceanic Tintinnoina of the plankton gathered during the last cruise of the CARNEGIE. Sci. Res. Cruise VII of the Carnegie, 1928-1929, Biol. 2:163 pp., 128 figs.

Amplectella monocollaria

Kofoid, C.A. and A.S. Campbell, 1939
Reports on the scientific results of the expedition to the eastern tropical Pacific in charge of Alexander Agassiz, by the U.S. Fish Commission steamer "Albatross" from October 1904 to March 1905------XXXVII. The Ciliata: The Tintinnoinea. Bull. Mus. Comp. Zool. 84: 473 pp., 36 pls.

Amplectella monocollaria

Kofoid, C. A. and A. S. Campbell, 1929
A conspectus of the marine and freshwater Ciliata belonging to the suborder Tintinnoinea, with descriptions of new species principally from the Agassiz expedition to the eastern tropical Pacific, 1904-1905. Univ. Calif. Publ. Zool. 34:1-403, 697 text figs.

Amplectella occidentalis

Campbell, A. S., 1942
The oceanic Tintinnoina of the plankton gathered during the last cruise of the CARNEGIE. Sci. Res. Cruise VII of the Carnegie, 1928-1929, Biol. 2:163 pp., 128 figs.

Amplectella occidentalis

Gaarder, K.R., 1946
Tintinnoidea. Rep.Sci.Res. "Michael Sars" N.Atlantic Deep-Sea Exped., 1910, 2(1): 37 pp., 23 text figs.

Amplectella occidentalis

Kofoid, C.A. and A.S. Campbell, 1939
Reports on the scientific results of the expedition to the eastern tropical Pacific in charge of Alexander Agassiz, by the U.S. Fish Commission steamer "Albatross" from October 1904 to March 1905------XXXVII. The Ciliata: The Tintinnoinea. Bull. Mus. Comp. Zool. 84: 473 pp., 36 pls.

Amplectella occidentalis n.sp.

Kofoid, C. A. and A. S. Campbell, 1929
A conspectus of the marine and freshwater Ciliata belonging to the suborder Tintinnoinea, with descriptions of new species principally from the Agassiz expedition to the eastern tropical Pacific, 1904-1905. Univ. Calif. Publ. Zool. 34:1-403, 697 text figs.

Amplectella praeacuta

Campbell, A. S., 1942
The oceanic Tintinnoina of the plankton gathered during the last cruise of the CARNEGIE. Sci. Res. Cruise VII of the Carnegie, 1928-1929, Biol. 2:163 pp., 128 figs.

Amplectella praeacuta

Kofoid, C.A. and A.S. Campbell, 1939
Reports on the scientific results of the expedition to the eastern tropical Pacific in charge of Alexander Agassiz, by the U.S. Fish Commission steamer "Albatross" from October 1904 to March 1905------XXXVII. The Ciliata: The Tintinnoinea. Bull. Mus. Comp. Zool. 84: 473 pp., 36 pls.

Amplectella praeacuta n.sp.

Kofoid, C. A. and A. S. Campbell, 1929
A conspectus of the marine and freshwater Ciliata belonging to the suborder Tintinnoinea, with descriptions of new species principally from the Agassiz expedition to the eastern tropical Pacific, 1904-1905. Univ. Calif. Publ. Zool. 34:1-403, 697 text figs.

Amplectella quadricollaria n.sp.

Kofoid, C. A. and A. S. Campbell, 1929
A conspectus of the marine and freshwater Ciliata belonging to the suborder Tintinnoinea, with descriptions of new species principally from the Agassiz expedition to the eastern tropical Pacific, 1904-1905. Univ. Calif. Publ. Zool. 34:1-403, 697 text figs.

Amplectellopsis n. gen.

Kofoid, C. A. and A. S. Campbell, 1929
A conspectus of the marine and freshwater Ciliata belonging to the suborder Tintinnoinea, with descriptions of new species principally from the Agassiz expedition to the eastern tropical Pacific, 1904-1905. Univ. Calif. Publ. Zool. 34:1-403, 697 text figs.

Amplectellopsis angularis

Balech, Enrique, 1962
Tintinnoinea y Dinoflagellata del Pacifico segun material de las Expediciones NORPAC y DOWNWIND del Instituto Scripps de Oceanografia. Revista, Mus. Argentino Ciencias Nat. "Bernardino Rivadavia", Ciencias Zool., 7(1):1-253.

Amplectellopsis angularis

Campbell, A. S., 1942
The oceanic Tintinnoina of the plankton gathered during the last cruise of the CARNEGIE. Sci. Res. Cruise VII of the Carnegie, 1928-1929, Biol. 2:163 pp., 128 figs.

Amplectellopsis angularis

Kofoid, C.A. and A.S. Campbell, 1939
Reports on the scientific results of the expedition to the eastern tropical Pacific in charge of Alexander Agassiz, by the U.S. Fish Commission steamer "Albatross" from October 1904 to March 1905------XXXVII. The Ciliata: The Tintinnoinea. Bull. Mus. Comp. Zool. 84: 473 pp., 36 pls.

Amplectellopsis angularis n.sp.

Kofoid, C. A. and A. S. Campbell, 1929
A conspectus of the marine and freshwater Ciliata belonging to the suborder Tintinnoinea, with descriptions of new species principally from the Agassiz expedition to the eastern tropical Pacific, 1904-1905. Univ. Calif. Publ. Zool. 34:1-403, 697 text figs.

Amplectellopsis biedermanni

Kofoid, C.A. and A.S. Campbell, 1939
Reports on the scientific results of the expedition to the eastern tropical Pacific in charge of Alexander Agassiz, by the U.S. Fish Commission steamer "Albatross" from October 1904 to March 1905------XXXVII. The Ciliata: The Tintinnoinea. Bull. Mus. Comp. Zool. 84: 473 pp., 36 pls.

Amplectellopsis bredermanni n.sp.

Kofoid, C. A. and A. S. Campbell, 1929
A conspectus of the marine and freshwater Ciliata belonging to the suborder Tintinnoinea, with descriptions of new species principally from the Agassiz expedition to the eastern tropical Pacific, 1904-1905. Univ. Calif. Publ. Zool. 34:1-403, 697 text figs.

Brandtiella palliata

Gaarder, K.R., 1946
Tintinnoidea. Rep.Sci.Res. "Michael Sars" N.Atlantic Deep-Sea Exped., 1910, 2(1): 37 pp., 23 text figs.

Brandtiella palliata

Campbell, A. S., 1942
The oceanic Tintinnoina of the plankton gathered during the last cruise of the CARNEGIE. Sci. Res. Cruise VII of the Carnegie, 1928-1929, Biol. 2:163 pp., 128 figs.

Brandtiella palliata

Kofoid, C.A. and A.S. Campbell, 1939
Reports on the scientific results of the expedition to the eastern tropical Pacific in charge of Alexander Agassiz, by the U.S. Fish Commission steamer "Albatross" from October 1904 to March 1905------XXXVII. The Ciliata: The Tintinnoinea. Bull. Mus. Comp. Zool. 84: 473 pp., 36 pls.

Brandtiella palliata n. gen.

Kofoid, C. A. and A. S. Campbell, 1929
A conspectus of the marine and freshwater Ciliata belonging to the suborder Tintinnoinea, with descriptions of new species principally from the Agassiz expedition to the eastern tropical Pacific, 1904-1905. Univ. Calif. Publ. Zool. 34:1-403, 697 text figs.

Bursaopsis n. gen.

Kofoid, C. A. and A. S. Campbell, 1929
A conspectus of the marine and freshwater Ciliata belonging to the suborder Tintinnoinea, with descriptions of new species principally from the Agassiz expedition to the eastern tropical Pacific, 1904-1905. Univ. Calif. Publ. Zool. 34:1-403, 697 text figs.

Bursaopsis bursa

Gaarder, K.R., 1946
Tintinnoidea. Rep.Sci.Res. "Michael Sars" N.Atlantic Deep-Sea Exped., 1910, 2(1): 37 pp., 23 text figs.

Bursaopsis bursa

Kofoid, C. A. and A. S. Campbell, 1929
A conspectus of the marine and freshwater Ciliata belonging to the suborder Tintinnoinea, with descriptions of new species principally from the Agassiz expedition to the eastern tropical Pacific, 1904-1905. Univ. Calif. Publ. Zool. 34:1-403, 697 text figs.

Bursaopsis conicoides n. sp.

*Hada, Yoshina, 1970.   (1970)
The protozoan plankton of the Antarctic and Subantarctic seas.
Scient. Repts, Japan. Antarct. Res. Exped., (E)31:51 pp.

Bursaopsis fergusonii

Kofoid, C. A. and A. S. Campbell, 1929
A conspectus of the marine and freshwater Ciliata belonging to the suborder Tintinnoinea, with descriptions of new species principally from the Agassiz expedition to the eastern tropical Pacific, 1904-1905. Univ. Calif. Publ. Zool. 34:1-403, 697 text figs.

Bursaopsis obliqua

Kofoid, C. A. and A. S. Campbell, 1929
A conspectus of the marine and freshwater Ciliata belonging to the suborder Tintinnoinea, with descriptions of new species principally from the Agassiz expedition to the eastern tropical Pacific, 1904-1905. Univ. Calif. Publ. Zool. 34:1-403, 697 text figs.

Bursaopsis ollula n. sp.

*Hada, Yoshina, 1970.   (1970)
The protozoan plankton of the Antarctic and Subantarctic seas.
Scient. Repts, Japan. Antarct. Res. Exped., (E)31:51 pp.

Bursaopsis punctostriata

Kofoid, C. A. and A. S. Campbell, 1929
A conspectus of the marine and freshwater Ciliata belonging to the suborder Tintinnoinea, with descriptions of new species principally from the Agassiz expedition to the eastern tropical Pacific, 1904-1905. Univ. Calif. Publ. Zool. 34:1-403, 697 text figs.

Bursaopsis quinquealata

Kofoid, C. A. and A. S. Campbell, 1929
A conspectus of the marine and freshwater Ciliata belonging to the suborder Tintinnoinea, with descriptions of new species principally from the Agassiz expedition to the eastern tropical Pacific, 1904-1905. Univ. Calif. Publ. Zool. 34:1-403, 697 text figs.

Bursaopsis striata

Kofoid, C. A. and A. S. Campbell, 1929
A conspectus of the marine and freshwater Ciliata belonging to the suborder Tintinnoinea, with descriptions of new species principally from the Agassiz expedition to the eastern tropical Pacific, 1904-1905. Univ. Calif. Publ. Zool. 34:1-403, 697 text figs.

Bursaopsis vitrea

Kofoid, C. A. and A. S. Campbell, 1929
A conspectus of the marine and freshwater Ciliata belonging to the suborder Tintinnoinea, with descriptions of new species principally from the Agassiz expedition to the eastern tropical Pacific, 1904-1905. Univ. Calif. Publ. Zool. 34:1-403, 697 text figs.

Canthariella sp.

Hasle, Grethe Rytter, 1960
Phytoplankton and ciliate species from the Tropical Pacific.
Skr. Norske Vidensk-Akad., Oslo, 1. Mat.-Nat. Kl., 1960(2): 1-50.

Canthariella n. gen.

Kofoid, C. A. and A. S. Campbell, 1929
A conspectus of the marine and freshwater Ciliata belonging to the suborder Tintinnoinea, with descriptions of new species principally from the Agassiz expedition to the eastern tropical Pacific, 1904-1905. Univ. Calif. Publ. Zool. 34:1-403, 697 text figs.

Canthariella brevis

Kofoid, C.A. and A.S. Campbell, 1939
Reports on the scientific results of the expedition to the eastern tropical Pacific in charge of Alexander Agassiz, by the U.S. Fish Commission steamer "Albatross" from October 1904 to March 1905——XXXVII. The Ciliata: The Tintinnoinea. Bull. Mus. Comp. Zool. 84: 473 pp., 36 pls.

Canthariella brevis n.sp.

Kofoid, C. A. and A. S. Campbell, 1929
A conspectus of the marine and freshwater Ciliata belonging to the suborder Tintinnoinea, with descriptions of new species principally from the Agassiz expedition to the eastern tropical Pacific, 1904-1905. Univ. Calif. Publ. Zool. 34:1-403, 697 text figs.

Canthariella pyramidata

Kofoid, C. A. and A. S. Campbell, 1929
A conspectus of the marine and freshwater Ciliata belonging to the suborder Tintinnoinea, with descriptions of new species principally from the Agassiz expedition to the eastern tropical Pacific, 1904-1905. Univ. Calif. Publ. Zool. 34:1-403, 697 text figs.

Canthariella septinaria

Campbell, A. S., 1942
The oceanic Tintinnoina of the plankton gathered during the last cruise of the CARNEGIE. Sci. Res. Cruise VII of the Carnegie, 1928-1929, Biol. 2:163 pp., 128 figs.

Canthariella septinaria

Kofoid, C.A. and A.S. Campbell, 1939
Reports on the scientific results of the expedition to the eastern tropical Pacific in charge of Alexander Agassiz, by the U.S. Fish Commission steamer "Albatross" from October 1904 to March 1905——XXXVII. The Ciliata: The Tintinnoinea. Bull. Mus. Comp. Zool. 84: 473 pp., 36 pls.

Canthariella septinaria n.sp.

Kofoid, C. A. and A. S. Campbell, 1929
A conspectus of the marine and freshwater Ciliata belonging to the suborder Tintinnoinea, with descriptions of new species principally from the Agassiz expedition to the eastern tropical Pacific, 1904-1905. Univ. Calif. Publ. Zool. 34:1-403, 697 text figs.

Canthariella truncata

Campbell, A. S., 1942
The oceanic Tintinnoina of the plankton gathered during the last cruise of the CARNEGIE. Sci. Res. Cruise VII of the Carnegie, 1928-1929, Biol. 2:163 pp., 128 figs.

Canthariella truncata n.sp.

Kofoid, C. A. and A. S. Campbell, 1929
A conspectus of the marine and freshwater Ciliata belonging to the suborder Tintinnoinea, with descriptions of new species principally from the Agassiz expedition to the eastern tropical Pacific, 1904-1905. Univ. Calif. Publ. Zool. 34:1-403, 697 text figs.

Clevea melchersi n.gen., n.sp.

Balech, Enrique, 1948
Tintinnoinea de Atlantida (R.O. Del Uruguay) (Protozoa, Ciliata Oligotr.). Comm. Mus. Argentino Ciencias Nat. "Bernardino Rivadavia", Ser. Cien. Zool. No.7, 23 pp., 8 pls. of 107 figs.

Climacocylis sp.

Gaarder, K.R., 1946
Tintinnoidea. Rep.Sci.Res. "Michael Sars" N.Atlantic Deep-Sea Exped., 1910, 2(1): 37 pp., 23 text figs.

Climacocylis n.gen.

Jörgensen, E., 1924
Mediterranean Tintinnidae. Rept. Danish Oceanogr. Exped. 1908-10 to the Mediterranean and adjacent seas, Vol.II, Biol. J 3, 110 pp., 114 text figs.

Climacocylis digitula

Balech, Enrique, 1971
Microplancton del Atlantico ecuatorial oeste (Equalant 1)
Publ. Serv. Hidrograf. Naval, Argentina H. 654: 103 pp., 122 figs.

Climacocylis digitula

Kofoid, C.A. and A.S. Campbell, 1939
Reports on the scientific results of the expedition to the eastern tropical Pacific in charge of Alexander Agassiz, by the U.S. Fish Commission steamer "Albatross" from October 1904 to March 1905——XXXVII. The Ciliata: The Tintinnoinea. Bull. Mus. Comp. Zool. 84: 473 pp., 36 pls.

Climacocylis digitula n.sp.

Kofoid, C. A. and A. S. Campbell, 1929
A conspectus of the marine and freshwater Ciliata belonging to the suborder Tintinnoinea, with descriptions of new species principally from the Agassiz expedition to the eastern tropical Pacific, 1904-1905. Univ. Calif. Publ. Zool. 34:1-403, 697 text figs.

Climacocylis elongata

Kofoid, C.A. and A.S. Campbell, 1939
Reports on the scientific results of the expedition to the eastern tropical Pacific in charge of Alexander Agassiz, by the U.S. Fish Commission steamer "Albatross" from October 1904 to March 1905——XXXVII. The Ciliata: The Tintinnoimea. Bull. Mus. Comp. Zool. 84: 473 pp., 36 pls.

Climacocylis elongata n.sp.

Kofoid, C. A. and A. S. Campbell, 1929
A conspectus of the marine and freshwater Ciliata belonging to the suborder Tintinnoinea, with descriptions of new species principally from the Agassiz expedition to the eastern tropical Pacific, 1904-1905. Univ. Calif. Publ. Zool. 34:1-403, 697 text figs.

Climacocylis leiospiralis n.sp.

Kofoid, C.A. and A.S. Campbell, 1939
Reports on the scientific results of the expedition to the eastern tropical Pacific in charge of Alexander Agassiz, by the U.S. Fish Commission steamer "Albatross" from October 1904 to March 1905——XXXVII. The Ciliata: The Tintinnoimea. Bull. Mus. Comp. Zool. 84: 473 pp., 36 pls.

Climacocylis scalaria

Balech, Enrique, 1962
Tintinnoinea y Dinoflagellata del Pacifico segun material de las Expediciones NORPAC & DOWNWIND del Instituto Scripps de Oceanografia.
Revista, Mus. Argentino Ciencias Nat. "Bernardino Rivadavia", Ciencias Zool., 7(1):1-253.

Climacocylis scalaria

Campbell, A. S., 1942
The oceanic Tintinnoina of the plankton gathered during the last cruise of the CARNEGIE. Sci. Res. Cruise VII of the Carnegie, 1928-1929, Biol. 2:163 pp., 128 figs.

Climacocylis scalaria

Jörgensen, E., 1924
Mediterranean Tintinnidae. Rept. Danish Oceanogr. Exped. 1908-10 to the Mediterranean and adjacent seas, Vol.II, Biol. J 3, 110 pp., 114 text figs.

Climacocylis scalaria

Kofoid, C.A. and A.S. Campbell, 1939
Reports on the scientific results of the expedition to the eastern tropical Pacific in charge of Alexander Agassiz, by the U.S. Fish Commission steamer "Albatross" from October 1904 to March 1905——XXXVII. The Ciliata: The Tintinnoinea. Bull. Mus. Comp. Zool. 84: 473 pp., 36 pls.

Climacocylis scalaria

Kofoid, C. A. and A. S. Campbell, 1929
A conspectus of the marine and freshwater Ciliata belonging to the suborder Tintinnoinea, with descriptions of new species principally from the Agassiz expedition to the eastern tropical Pacific, 1904-1905. Univ. Calif. Publ. Zool. 34:1-403, 697 text figs.

Climacocylis scalaria

Komarovsky, B., 1959
The Tintinnina of the Gulf of Eylath (Aqaba). Contrib. Knowl. Red Sea, Bull. Sea Fish. Res. Sta. Haifa, Israel 21 (14):1-40.

Climacocylis scalaria

Marshall, S. M., 1934
The silicoflagellata and tintinnoinea. Great Barrier Reef Exped., 1928-29. Sci. Repts. 4(15):623-664, 43 text figs.

Climacocylis scalaria

Rampi, L., 1948
Sur quelques Tintinnides (Infusoires louques) du Pacifique subtropical (Recoltes Alain Gerbault) Bull. l'Inst. Ocean., Monaco, No.938, 4 pp.

Climacocylis scalaria

Rampi, L., 1939
Primo contributo alla conoscenza dei tintinnoidi do Maro Ligure. Atti della Soc. Ital. di Sci. Nat. 78:67-81, 58 text figs.

Climacocylis scalaroides

Campbell, A. S., 1942
The oceanic Tintinnoina of the plankton gathered during the last cruise of the CARNEGIE. Sci. Res. Cruise VII of the Carnegie, 1928-1929, Biol. 2:163 pp., 128 figs.

Climacocylis scalaroides

Kofoid, C.A. and A.S. Campbell, 1939
Reports on the scientific results of the expedition to the eastern tropical Pacific in charge of Alexander Agassiz, by the U.S. Fish Commission steamer "Albatross" from October 1904 to March 1905——XXXVII. The Ciliata: The Tintinnoimea. Bull. Mus. Comp. Zool. 84: 473 pp., 36 pls.

Climacocylis scalaroides n.sp.

Kofoid, C. A. and A. S. Campbell, 1929
A conspectus of the marine and freshwater Ciliata belonging to the suborder Tintinnoinea, with descriptions of new species principally from the Agassiz expedition to the eastern tropical Pacific, 1904-1905. Univ. Calif. Publ. Zool. 34:1-403, 697 text figs.

Climacocylis scalaroides

Marshall, S. M., 1934
The silicoflagellata and tintinnoinea. Great Barrier Reef Exped., 1928-29. Sci. Repts. 4(15):623-664, 43 text figs.

Climacocylis scalaroides marshallae

Balech, Enrique, 1964
El plancton de Mar del Plata durante el periodo 1961-1962.
Bol. Inst. Biol. Mar., Buenos Aires, (4):56 pp.

Climacocylis scalaroides marshallae

Balech, Enrique, 1964
El plancton de Mar del Plata durante el periodo 1961-1962 (Buenos Aires, Argentina). Bol. Inst. Biol. Mar., Mar del Plata, Argentina No. 4: 49 pp.

Climacocylis sipho

Campbell, A. S., 1942
The oceanic Tintinnoina of the plankton gathered during the last cruise of the CARNEGIE. Sci. Res. Cruise VII of the Carnegie, 1928-1929, Biol. 2:163 pp., 128 figs.

Oceanographic Index: Marine Organisms Cumulation, 1946-1973

Climacocylis sipho

Kofoid, C.A. and A.S. Campbell, 1939
Reports on the scientific results of the expedition to the eastern tropical Pacific in charge of Alexander Agassiz, by the U.S. Fish Commission steamer "Albatross" from October 1904 to March 1905-----XXXVII. The Ciliata: The Tintinnoimea. Bull. Mus. Comp. Zool. 84: 473 pp., 36 pls.

Climacocylis sipho

Kofoid, C. A. and A. S. Campbell, 1929
A conspectus of the marine and fresh-water Ciliata belonging to the suborder Tintinnoinea, with descriptions of new species principally from the Agassiz expedition to the eastern tropical Pacific, 1904-1905. Univ. Calif. Publ. Zool. 34:1-403, 697 text figs.

Codonaria n.gen.

Kofoid, C.A. and A.S. Campbell, 1939
Reports on the scientific results of the expedition to the eastern tropical Pacific in charge of Alexander Agassiz, by the U.S. Fish Commission steamer "Albatross" from October 1904 to March 1905-----XXXVII. The Ciliata: The Tintinnoimea. Bull. Mus. Comp. Zool. 84: 473 pp., 36 pls.

Codonaria angusta

Campbell, A. S., 1942
The oceanic Tintinnoina of the plankton gathered during the last cruise of the CARNEGIE. Sci. Res. Cruise VII of the Carnegie, 1928-1929, Biol. 2:163 pp., 128 figs.

Codonaria australis

Kofoid, C.A. and A.S. Campbell, 1939
Reports on the scientific results of the expedition to the eastern tropical Pacific in charge of Alexander Agassiz, by the U.S. Fish Commission steamer "Albatross" from October 1904 to March 1905-----XXXVII. The Ciliata: The Tintinnoimea. Bull. Mus. Comp. Zool. 84: 473 pp., 36 pls.

Codonaria benguelensis

Campbell, A. S., 1942
The oceanic Tintinnoina of the plankton gathered during the last cruise of the CARNEGIE. Sci. Res. Cruise VII of the Carnegie, 1928-1929, Biol. 2:163 pp., 128 figs.

Codonaria benguelensis

Kofoid, C.A. and A.S. Campbell, 1939
Reports on the scientific results of the expedition to the eastern tropical Pacific in charge of Alexander Agassiz, by the U.S. Fish Commission steamer "Albatross" from October 1904 to March 1905-----XXXVII. The Ciliata: The Tintinnoimea. Bull. Mus. Comp. Zool. 84: 473 pp., 36 pls.

Codonaria cistellula

Balech, Enrique, 1962
Tintinnoinea y Dinoflagellata del Pacifico segun material de las Expediciones NORPAC y DOWNWIND del Instituto Scripps de Oceanografia. Revista, Mus. Argentino Ciencias Nat. "Bernardino Rivadavia", Ciencias Zool., 7(1):1-253.

Codonaria cistellula

Balech, E., 1959.
Tintinnoinea del Mediterraneo. Trav. Sta. Zool., Villefranche-sur-Mer, 18(17): Reprinted from: Trab. Inst. Español Oceanogr., Madrid, 19 Sept. 1959(28):88 pp., 22 pls.

Codonaria cistellula

Kofoid, C.A. and A.S. Campbell, 1939
Reports on the scientific results of the expedition to the eastern tropical Pacific in charge of Alexander Agassiz, by the U.S. Fish Commission steamer "Albatross" from October 1904 to March 1905-----XXXV The Ciliata: The Tintinnoimea. Bull. Mus. Comp. Zool. 84: 473 pp., 36 pls.

Codonaria fimbriata

Balech, Enrique, 1948
Tintinnoinea de Atlantida (R.O. Del Uruguay) (Protozoa, Ciliata Oligotr.). Comm. Mus. Argentino Ciencias Nat. "Bernardino Rivadavia", Ser. Cien. Zool. No. 7, 23 pp., 8 pls. of 107 figs.

Codonaria lata

Campbell, A. S., 1942
The oceanic Tintinnoina of the plankton gathered during the last cruise of the CARNEGIE. Sci. Res. Cruise VII of the Carnegie, 1928-1929, Biol. 2:163 pp., 128 figs.

Codonaria lata

Kofoid, C.A. and A.S. Campbell, 1939
Reports on the scientific results of the expedition to the eastern tropical Pacific in charge of Alexander Agassiz, by the U.S. Fish Commission steamer "Albatross" from October 1904 to March 1905-----XXXVII. The Ciliata: The Tintinnoimea. Bull. Mus. Comp. Zool. 84: 473 pp., 36 pls.

Codonaria mucronata

Campbell, A. S., 1942
The oceanic Tintinnoina of the plankton gathered during the last cruise of the CARNEGIE. Sci. Res. Cruise VII of the Carnegie, 1928-1929, Biol. 2:163 pp., 128 figs.

Codonaria mucronata

Kofoid, C.A. and A.S. Campbell, 1939
Reports on the scientific results of the expedition to the eastern tropical Pacific in charge of Alexander Agassiz, by the U.S. Fish Commission steamer "Albatross" from October 1904 to March 1905-----XXXVII. The Ciliata: The Tintinnoimea. Bull. Mus. Comp. Zool. 84: 473 pp., 36 pls.

Codonaria oceanica

Campbell, A. S., 1942
The oceanic Tintinnoina of the plankton gathered during the last cruise of the CARNEGIE. Sci. Res. Cruise VII of the Carnegie, 1928-1929, Biol. 2:163 pp., 128 figs.

Codonaria oceanica

Kofoid, C.A. and A.S. Campbell, 1939
Reports on the scientific results of the expedition to the eastern tropical Pacific in charge of Alexander Agassiz, by the U.S. Fish Commission steamer "Albatross" from October 1904 to March 1905-----XXXVII. The Ciliata: The Tintinnoimea. Bull. Mus. Comp. Zool. 84: 473 pp., 36 pls.

Codonella sp.

Braarud, T., 1945
A phytoplankton survey of the polluted waters of inner Oslo Fjord. Hvalrådets Skrifter, No. 28, 142 pp., 19 text figs., 17 tables.

Codonella sp.

Delegazione Italiana della Commissione Internazionale per l'Esplorazione Scientifica del Mediterraneo, 1941
Note sul plancton della Laguna veneta. [Memoria CCLXXIX], Arch. di Ocean. e Limn. Anno I, Fasc. I, 1941 XIX: 31-57 pp.

Codonella sp.

Gran, H.H., and T. Braarud, 1935
A quantitative study of the phytoplankton in the Bay of Fundy and the Gulf of Maine (including observations on hydrography, chemistry, and turbidity). J. Biol. Bd., Canada, 1(5):279-467, 69 text figs.

Codonella sp.

Hasle, Grethe Rytter, 1960
Phytoplankton and ciliate species from the Tropical Pacific. Skr. Norske Videnskaps-Akad., Oslo, 1. Mat.-Nat. Kl., 1960(2): 1-50.

Codonella sp.

Morse, D.C., 1947
Some observations on seasonal variations in plankton population Patuxent River, Maryland 1943-1945. Bd. Nat. Res., Publ. No. 65, Chesapeake Biol. Lab., 31, 3 figs.

Codonella acerca

Campbell, A. S., 1942
The oceanic Tintinnoina of the plankton gathered during the last cruise of the CARNEGIE. Sci. Res. Cruise VII of the Carnegie, 1928-1929, Biol. 2:163 pp., 128 figs.

Codonella acerca n.sp.

Jörgensen, E., 1924
Mediterranean Tintinnidae. Rept. Danish Oceanogr. Exped. 1908-10 to the Mediterranean and adjacent seas, Vol. II, Biol. J 3, 110 pp., 114 text figs.

Codonella acerca

Kofoid, C. A. and A. S. Campbell, 1929
A conspectus of the marine and fresh-water Ciliata belonging to the suborder Tintinnoinea, with descriptions of new species principally from the Agassiz expedition to the eastern tropical Pacific, 1904-1905. Univ. Calif. Publ. Zool. 34:1-403, 697 text figs.

Codonella acerca

Komarovsky, B., 1959
The Tintinnina of the Gulf of Eylath (Aqaba). Contrib. Knowl. 3rd Sea, Bull. Sea Fish. Res. Sta. Haifa, Israel 21 (14):1-40.

Codonella acerca

Sousa e Silva, E. de, 1950.
Les tintinnides de la Baie de Cascais. Bull. Inst. Océan., Monaco, No. 979:28 pp., 4 pls.

Codonella acuta

Balech, Enrique, 1962
Tintinnoinea y Dinoflagellata del Pacifico segun material de las Expediciones NORPAC y DOWNWIND del Instituto Scripps de Oceanografia. Revista, Mus. Argentino Ciencias Nat. "Bernardino Rivadavia", Ciencias Zool., 7(1):1-253.

Codonella acuta

Campbell, A. S., 1942
The oceanic Tintinnoina of the plankton gathered during the last cruise of the CARNEGIE. Sci. Res. Cruise VII of the Carnegie, 1928-1929, Biol. 2:163 pp., 128 figs.

Codonella acuta

Kofoid, C.A. and A.S. Campbell, 1939
Reports on the scientific results of the expedition to the eastern tropical Pacific in charge of Alexander Agassiz, by the U.S. Fish Commission steamer "Albatross" from October 1904 to March 1905-----XXXVII. The Ciliata: The Tintinnoimea. Bull. Mus. Comp. Zool. 84: 473 pp., 36 pls.

Codonella acuta n. sp.

Kofoid, C. A. and A. S. Campbell, 1929
A conspectus of the marine and fresh-water Ciliata belonging to the suborder Tintinnoinea, with descriptions of new species principally from the Agassiz expedition to the eastern tropical Pacific, 1904-1905. Univ. Calif. Publ. Zool. 34:1-403, 697 text figs.

Codonella acutula

Kofoid, C.A. and A.S. Campbell, 1939
Reports on the scientific results of the expedition to the eastern tropical Pacific in charge of Alexander Agassiz, by the U.S. Fish Commission steamer "Albatross" from October 1904 to March 1905-----XXXVII. The Ciliata: The Tintinnoimea. Bull. Mus. Comp. Zool. 84: 473 pp., 36 pls.

Oceanographic Index: Marine Organisms Cumulation, 1946-1973

Codonella acutula n. sp.

Kofoid, C. A. and A. S. Campbell, 1929
A conspectus of the marine and fresh-water Ciliata belonging to the suborder Tintinnoinea, with descriptions of new species principally from the Agassiz expedition to the eastern tropical Pacific, 1904-1905. Univ. Calif. Publ. Zool. 34:1-403, 697 text figs.

Codonella amphorella

Balech, Enrique, 1962
Tintinnoinea y Dinoflagellata del Pacifico segun material de las Expediciones NORPAC y DOWNWIND del Instituto Scripps de Oceanografia. Revista, Mus. Argentino Ciencias Nat. "Bernardino Rivadavia". Ciencias Zool., 7(1):1-253.

Codonella amphorella

Campbell, A. S., 1942
The oceanic Tintinnoina of the plankton gathered during the last cruise of the CARNEGIE. Sci. Res. Cruise VII of the Carnegie, 1928-1929, Biol. 2:163 pp., 128 figs.

Codonella amphorella

Entz, G., Jr., 1909
Studien über organisation und biologie der Tintinniden. Arch. f. Protistenkunde 15:93-226, Pls. 8-21, text figs.

Codonella amphorella

Gaarder, K.R., 1946
Tintinnoidea. Rep.Sci.Res."Michael Sars" N.Atlantic Deep-Sea Exped., 1910, 2(1): 37 pp., 23 text figs.

Codonella amphorella

Jörgensen, E., 1924
Mediterranean Tintinnidae. Rept. Danish Oceanogr. Exped. 1908-10 to the Mediterranean and adjacent seas, Vol.II, Biol. J 3, 110 pp., 114 text figs.

Codonella amphorella

Kofoid, C.A. and A.S. Campbell, 1939
Reports on the scientific results of the expedition to the eastern tropical Pacific in charge of Alexander Agassiz, by the U.S. Fish Commission steamer"Albatross" from October 1904 to March 1905-----XXXVII. The Ciliata: The Tintinnoinea. Bull.Mus. Comp.Zool. 84: 473 pp., 36 pls.

Codonella amphorella

Kofoid, C. A. and A. S. Campbell, 1929
A conspectus of the marine and fresh-water Ciliata belonging to the suborder Tintinnoinea, with descriptions of new species principally from the Agassiz expedition to the eastern tropical Pacific, 1904-1905. Univ. Calif. Publ. Zool. 34:1-403, 697 text figs.

Codonella amphorella

Laackmann, H., 1910
Die Tintinnodeen der Deutschen Südpolar Expedition 1901-1903. Deutsch. Südpolar Exped. 11, Zool. 3(4):341-496, pls. 33-51.

Codonella angusta n. sp.

Kofoid, C. A. and A. S. Campbell, 1929
A conspectus of the marine and fresh-water Ciliata belonging to the suborder Tintinnoinea, with descriptions of new species principally from the Agassiz expedition to the eastern tropical Pacific, 1904-1905. Univ. Calif. Publ. Zool. 34:1-403, 697 text figs.

Codonella apicata

Balech, Enrique, 1962
Tintinnoinea y Dinoflagellata del Pacifico segun material de las Expediciones NORPAC y DOWNWIND del Instituto Scripps de Oceanografia. Revista, Mus. Argentino Ciencias Nat. "Bernardino Rivadavia". Ciencias Zool., 7(1):1-253.

Codonella apicata

Kofoid, C.A. and A.S. Campbell, 1939
Reports on the scientific results of the expedition to the eastern tropical Pacific in charge of Alexander Agassiz, by the U.S. Fish Commission steamer"Albatross" from October 1904 to March 1905-----XXXVII. The Ciliata: The Tintinnoinea. Bull.Mus. Comp.Zool. 84: 473 pp., 36 pls.

Codonella apicata n. sp.

Kofoid, C. A. and A. S. Campbell, 1929
A conspectus of the marine and fresh-water Ciliata belonging to the suborder Tintinnoinea, with descriptions of new species principally from the Agassiz expedition to the eastern tropical Pacific, 1904-1905. Univ. Calif. Publ. Zool. 34:1-403, 697 text figs.

Codonella apicata

Rampi, L., 1939
Primo contributo alla conoscenza dei tintinnoidi do Maro Ligure. Atti della Soc. Ital. di Sci. Nat. 78:67-81, 58 text figs.

Codonella aspera

Balech, Enrique, 1962
Tintinnoinea y Dinoflagellata del Pacifico segun material de las Expediciones NORPAC y DOWNWIND del Instituto Scripps de Oceanografia. Revista, Mus. Argentino Ciencias Nat. "Bernardino Rivadavia". Ciencias Zool., 7(1):1-253.

Codonella aspera

Balech, E., 1959.
Tintinnoinea del Mediterraneo. Trav. Sta. Zool., Villefranche-sur-Mer, 18(17): Reprinted from: Trab. Inst. Espanol Oceanogr., Madrid, 19Sept. 1959(28):88 pp., 22 pls.

Codonella aspera

Kofoid, C.A. and A.S. Campbell, 1939
Reports on the scientific results of the expedition to the eastern tropical Pacific in charge of Alexander Agassiz, by the U.S. Fish Commission steamer"Albatross" from October 1904 to March 1905-----XXXVII. The Ciliata: The Tintinnoinea. Bull.Mus. Comp.Zool. 84: 473 pp., 36 pls.

Codonella aspera n. sp.

Kofoid, C. A. and A. S. Campbell, 1929
A conspectus of the marine and fresh-water Ciliata belonging to the suborder Tintinnoinea, with descriptions of new species principally from the Agassiz expedition to the eastern tropical Pacific, 1904-1905. Univ. Calif. Publ. Zool. 34:1-403, 697 text figs.

Codonella aspera

Posta, Annette, 1963.
Relation entre l'évolution de quelques tintinnides de la rade de Villefranche et la température de l'eau. Cahiers, Biol. Mar. Roscoff, 4:201-210.

Also in:
Trav. Sta. Zool., Villefranche-sur-Mer, 22(5):

Codonella australis n. sp.

Kofoid, C. A. and A. S. Campbell, 1929
A conspectus of the marine and fresh-water Ciliata belonging to the suborder Tintinnoinea, with descriptions of new species principally from the Agassiz expedition to the eastern tropical Pacific, 1904-1905. Univ. Calif. Publ. Zool. 34:1-403, 697 text figs.

Codonella benguelensis n. sp.

Kofoid, C. A. and A. S. Campbell, 1929
A conspectus of the marine and fresh-water Ciliata belonging to the suborder Tintinnoinea, with descriptions of new species principally from the Agassiz expedition to the eastern tropical Pacific, 1904-1905. Univ. Calif. Publ. Zool. 34:1-403, 697 text figs.

Codonella brevicollis

Kofoid, C.A. and A.S. Campbell, 1939
Reports on the scientific results of the expedition to the eastern tropical Pacific in charge of Alexander Agassiz, by the U.S. Fish Commission steamer"Albatross" from October 1904 to March 1905-----XXXVII. The Ciliata: The Tintinnoinea. Bull.Mus. Comp.Zool. 84: 473 pp., 36 pls.

Codonella brevicollis

Kofoid, C. A. and A. S. Campbell, 1929
A conspectus of the marine and fresh-water Ciliata belonging to the suborder Tintinnoinea, with descriptions of new species principally from the Agassiz expedition to the eastern tropical Pacific, 1904-1905. Univ. Calif. Publ. Zool. 34:1-403, 697 text figs.

Codonella campanula

Fol, H., 1884.
Sur la famille des Tintinnodea. Recueil Zool. Suisse 1:27-64, Pls. 4-5.

Codonella campanula

Schodduyn, M., 1926
Observations faites dans la baie d'Ambleteuse (Pas de Calais). Bull. Inst. Ocean., Monaco, No. 482: 64 pp.

Codonella cistellula

Duran, M., 1951
Contribucion al estudio de los tintinidos del plancton de los costas de Castellon. (Mediterraneo occidental). Publ. Inst. Biol. Aplic., Barcelone, 8: 101-120, 29 text figs.

Codonella cistellula

Entz, G., Jr., 1909
Studien über organisation und biologie der Tintinniden. Arch. f. Protistenkunde 15:93-226, Pls. 8-21, text figs.

Codonella cistellula

Hofker, J., 1931
Studien uber Tintinnoidea. Arch. f. Protistenk 75(3):315-402, 89 text figs.

Codonella cistellula

Jörgensen, E., 1924
Mediterranean Tintinnidae. Rept. Danish Oceanogr. Exped. 1908-10 to the Mediterranean and adjacent seas, Vol.II, Biol. J 3, 110 pp., 114 text figs.

Codonella cistellula

Laackmann, H., 1910
Die Tintinnodeen der Deutschen Südpolar Expedition 1901-1903. Deutsch. Südpolar Exped. 11, Zool. 3(4):341-496, pls. 33-51.

Codonella cistellula

Sousa e Silva, E. de, 1950.
Les tintinnides de la Baie de Cascais. Bull. Inst. Océan., Monaco, No. 979:28 pp., 4 pls.

Codonella cordata n.sp.

Kofoid, C. A. and A. S. Campbell, 1929
A conspectus of the marine and fresh-water Ciliata belonging to the suborder Tintinnoinea, with descriptions of new species principally from the Agassiz expedition to the eastern tropical Pacific, 1904-1905. Univ. Calif. Publ. Zool. 34:1-403, 697 text figs.

Codonella cratera

Kofoid, C. A. and A. S. Campbell, 1929
A conspectus of the marine and fresh-water Ciliata belonging to the suborder Tintinnoinea, with descriptions of new species principally from the Agassiz expedition to the eastern tropical Pacific, 1904-1905. Univ. Calif. Publ. Zool. 34:1-403, 697 text figs.

Codonella cratera

Schulz, H., 1965.
Die Tintinnoinea des Elbe Aestuars. Arch. Fischereiwiss., 15(3):216-225.

Codonella cuspidata

Balech, Enrique, 1962
Tintinnoinea y Dinoflagellata del Pacifico segun material de las Expediciones NORPAC y DOWNWIND del Instituto Scripps de Oceanografia. Revista, Mus. Argentino Ciencias Nat. "Bernardino Rivadavia", Ciencias Zool., 7(1):1-253.

Codonella cuspidata

Kofoid, C.A. and A.S. Campbell, 1939
Reports on the scientific results of the expedition to the eastern tropical Pacific in charge of Alexander Agassiz, by the U.S. Fish Commission steamer "Albatross" from October 1904 to March 1905-----XXXVII. The Ciliata: The Tintinnoimea. Bull. Mus. Comp. Zool. 84: 473 pp., 36 pls.

Codonella cuspidata n. sp.

Kofoid, C. A. and A. S. Campbell, 1929
A conspectus of the marine and freshwater Ciliata belonging to the suborder Tintinnoinea, with descriptions of new species principally from the Agassiz expedition to the eastern tropical Pacific, 1904-1905. Univ. Calif. Publ. Zool. 34:1-403, 697 text figs.

Codonella dadayi n. sp.

Kofoid, C. A. and A. S. Campbell, 1929
A conspectus of the marine and freshwater Ciliata belonging to the suborder Tintinnoinea, with descriptions of new species principally from the Agassiz expedition to the eastern tropical Pacific, 1904-1905. Univ. Calif. Publ. Zool. 34:1-403, 697 text figs.

Codonella diomedae

Kofoid, C.A. and A.S. Campbell, 1939
Reports on the scientific results of the expedition to the eastern tropical Pacific in charge of Alexander Agassiz, by the U.S. Fish Commission steamer "Albatross" from October 1904 to March 1905-----XXXVII. The Ciliata: The Tintinnoimea. Bull. Mus. Comp. Zool. 84: 473 pp., 36 pls.

Codonella diomedae n.sp.

Kofoid, C. A. and A. S. Campbell, 1929
A conspectus of the marine and freshwater Ciliata belonging to the suborder Tintinnoinea, with descriptions of new species principally from the Agassiz expedition to the eastern tropical Pacific, 1904-1905. Univ. Calif. Publ. Zool. 34:1-403, 697 text figs.

Codonella ecaudata

Kofoid, C. A. and A. S. Campbell, 1929
A conspectus of the marine and freshwater Ciliata belonging to the suborder Tintinnoinea, with descriptions of new species principally from the Agassiz expedition to the eastern tropical Pacific, 1904-1905. Univ. Calif. Publ. Zool. 34:1-403, 697 text figs.

Codonella elongata

Gaarder, K.R., 1946
Tintinnoidea. Rep. Sci. Res. "Michael Sars" N. Atlantic Deep-Sea Exped., 1910, 2(1): 37 pp., 23 text figs.

Codonella elongata

Kofoid, C.A. and A.S. Campbell, 1939
Reports on the scientific results of the expedition to the eastern tropical Pacific in charge of Alexander Agassiz, by the U.S. Fish Commission steamer "Albatross" from October 1904 to March 1905-----XXXVII. The Ciliata: The Tintinnoimea. Bull. Mus. Comp. Zool. 84: 473 pp., 36 pls.

Codonella elongata

Kofoid, C. A. and A. S. Campbell, 1929
A conspectus of the marine and freshwater Ciliata belonging to the suborder Tintinnoinea, with descriptions of new species principally from the Agassiz expedition to the eastern tropical Pacific, 1904-1905. Univ. Calif. Publ. Zool. 34:1-403, 697 text figs.

Codonella elongata

Massuti Algamora, M., 1949
Estudio de diez y seis muestras de plancton del Golfo de Nápoles. Publ. Inst. Biol. Appl. 5:85-94, 1 fold-in table.

Codonella elongata

Sousa e Silva, E. de, 1950.
Les tintinnides de la Baie de Cascais. Bull. Inst. Océan., Monaco, No. 979:28 pp., 4 pls.

Codonella erythraeensis

Kofoid, C. A. and A. S. Campbell, 1929
A conspectus of the marine and freshwater Ciliata belonging to the suborder Tintinnoinea, with descriptions of new species principally from the Agassiz expedition to the eastern tropical Pacific, 1904-1905. Univ. Calif. Publ. Zool. 34:1-403, 697 text figs.

Codonella galea

Balech, Enrique, 1962
Tintinnoinea y Dinoflagellata del Pacifico segun material de las Expediciones NORPAC y DOWNWIND del Instituto Scripps de Oceanografia. Revista, Mus. Argentino Ciencias Nat. "Bernardino Rivadavia", Ciencias Zool., 7(1):1-253.

Codonella galea

Balech, E., 1959.
Tintinnoinea del Mediterraneo. Trav. Sta. Zool., Villefranche-sur-Mer, 18(17): Reprinted from: Trab. Inst. Español Oceanogr., Madrid, 19 Sept. 1959(28):88 pp., 22 pls.

Codonella galea

Campbell, A. S., 1942
The oceanic Tintinnoina of the plankton gathered during the last cruise of the CARNEGIE. Sci. Res. Cruise VII of the Carnegie, 1928-1929, Biol. 2:163 pp., 128 figs.

Codonella galea

Duran, M., 1951
Contribucion al estudio de los tintinidos del plancton de los costas de Castellon. (Mediterraneo occidental). Publ. Inst. Biol. Aplic., Barcelone, 8: 101-120, 29 text figs.

Codonella galea

Entz, G., Jr., 1909
Studien über organisation und biologie der Tintinniden. Arch. f. Protistenkunde 15:93-226, Pls. 8-21, text figs.

Codonella galea

Fol, H., 1884.
Sur la famille des Tintinnodea. Recueil Zool. Suisse 1:27-64, Pls. 4-5.

Codonella galea

Jörgensen, E., 1924
Mediterranean Tintinnidae. Rept. Danish Oceanogr. Exped. 1908-10 to the Mediterranean and adjacent seas, Vol. II, Biol. J 3, 110 pp., 114 text figs.

Codonella galea

Kofoid, C.A. and A.S. Campbell, 1939
Reports on the scientific results of the expedition to the eastern tropical Pacific in charge of Alexander Agassiz, by the U.S. Fish Commission steamer "Albatross" from October 1904 to March 1905-----XXXVII. The Ciliata: The Tintinnoimea. Bull. Mus. Comp. Zool. 84: 473 pp., 36 pls.

Codonella galea

Kofoid, C. A. and A. S. Campbell, 1929
A conspectus of the marine and freshwater Ciliata belonging to the suborder Tintinnoinea, with descriptions of new species principally from the Agassiz expedition to the eastern tropical Pacific, 1904-1905. Univ. Calif. Publ. Zool. 34:1-403, 697 text figs.

Codonella galea

Komarovsky, B., 1959
The Tintinnina of the Gulf of Eylath (Aqaba). Contrib. Knowl. Red Sea, Bull. Sea Fish. Res. Sta. Haifa, Israel 21 (14):1-40.

Codonella galea

Laackmann, H., 1910
Die Tintinnodeen der Deutschen Südpolar Expedition 1901-1903. Deutsch. Südpolar Exped. 11, Zool. 3(4):341-496, pls. 33-51.

Codonella galea

Rampi, L., 1948
Sur quelques Tintinnides (Infusoires louques) du Pacifique subtropical (Recoltes Alain Gerbault) Bull. l'Inst. Ocean., Monaco, No. 938, 4 pp.

Codonella galea

Rampi, L., 1939
Primo contributo alla conoscenza dei tintinnoidi do Mare Ligure. Atti della Soc. Ital. di Sci. Nat. 78:67-81, 58 text figs.

Codonella galea

Sousa e Silva, E. de, 1950.
Les tintinnides de la Baie de Cascais. Bull. Inst. Océan., Monaco, No. 979:28 pp., 4 pls.

Codonella grahami n.sp.

Campbell, A. S., 1942
The oceanic Tintinnoina of the plankton gathered during the last cruise of the CARNEGIE. Sci. Res. Cruise VII of the Carnegie, 1928-1929, Biol. 2:163 pp., 128 figs.

Codonella inflata n. sp.

Kofoid, C. A. and A. S. Campbell, 1929
A conspectus of the marine and freshwater Ciliata belonging to the suborder Tintinnoinea, with descriptions of new species principally from the Agassiz expedition to the eastern tropical Pacific, 1904-1905. Univ. Calif. Publ. Zool. 34:1-403, 697 text figs.

Codonella lagenula

Jørgensen, E., 1900.
Ueber die Tintinnodeen der norwegischen Westküste Bergens Mus. Aarb., 1899(2):48 pp., 3 pls.

Codonella lagenula

Kofoid, C. A. and A. S. Campbell, 1929
A conspectus of the marine and freshwater Ciliata belonging to the suborder Tintinnoinea, with descriptions of new species principally from the Agassiz expedition to the eastern tropical Pacific, 1904-1905. Univ. Calif. Publ. Zool. 34:1-403, 697 text figs.

Codonella lagenula ovata

Jørgensen, E., 1905
B. Protistplankton and the diatoms in bottom samples. Hydrographical and biological investigations in Norwegian fjords. Bergens Mus. Skr. 7: 49-225.

Codonella lariana

Kofoid, C. A. and A. S. Campbell, 1929
A conspectus of the marine and freshwater Ciliata belonging to the suborder Tintinnoinea, with descriptions of new species principally from the Agassiz expedition to the eastern tropical Pacific, 1904-1905. Univ. Calif. Publ. Zool. 34:1-403, 697 text figs.

Codonella lata n. sp.

Kofoid, C. A. and A. S. Campbell, 1929.
A conspectus of the marine and freshwater Ciliata belonging to the suborder Tintinnoinea, with descriptions of new species principally from the Agassiz expedition to the eastern tropical Pacific, 1904-1905. Univ. Calif. Publ. Zool. 34:1-403, 697 text figs.

Codonella lata

Sousa e Silva, E. de, 1950.
Les tintinnides de la Baie de Cascais. Bull. Inst. Océan., Monaco, No. 979:28 pp., 4 pls.

Codonella laticollis

Kofoid, C. A. and A. S. Campbell, 1929
A conspectus of the marine and freshwater Ciliata belonging to the suborder Tintinnoinea, with descriptions of new species principally from the Agassiz expedition to the eastern tropical Pacific, 1904-1905. Univ. Calif. Publ. Zool. 34:1-403, 697 text figs.

Codonella? morchella

Laackmann, H., 1910
Die Tintinnodeen der Deutschen Südpolar Expedition 1901-1903. Deutsch. Südpolar Exped. 11, Zool. 3(4):341-496, pls. 33-51.

Codonella mucronata n. sp.

Kofoid, C. A. and A. S. Campbell, 1929
A conspectus of the marine and freshwater Ciliata belonging to the suborder Tintinnoinea, with descriptions of new species principally from the Agassiz expedition to the eastern tropical Pacific, 1904-1905. Univ. Calif. Publ. Zool. 34:1-403, 697 text figs.

Codonella nationalis

Campbell, A. S., 1942
The oceanic Tintinnoina of the plankton gathered during the last cruise of the CARNEGIE. Sci. Res. Cruise VII of the Carnegie, 1928-1929, Biol. 2:163 pp., 128 figs.

Codonella nationalis

Duran, M., 1951
Contribucion al estudio de los tintinidos del plancton de los costas de Castellon. (Mediterraneo occidental). Publ. Inst. Biol. Aplic., Barcelone, 8: 101-120, 29 text figs.

Codonella nationalis

Entz, G., Jr. 1909
Studien über organisation und biologie der Tintinniden. Arch. f. Protistenkunde 15:93-226, Pls. 8-21, text figs.

Codonella nationalis

Gaarder, K.R., 1946
Tintinnoidea. Rep.Sci.Res."Michael Sars" N.Atlantic Deep-Sea Exped., 1910, 2(1): 37 pp., 23 text figs.

Codonella nationalis

Kofoid, C.A. and A.S. Campbell, 1939
Reports on the scientific results of the expedition to the eastern tropical Pacific in charge of Alexander Agassiz, by the U.S. Fish Commission steamer "Albatross" from October 1904 to March 1905-----XXXVII. The Ciliata: The Tintinnoimea. Bull.Mus. Comp.Zool. 84: 473 pp., 36 pls.

Codonella nationalis

Kofoid, C. A. and A. S. Campbell, 1929
A conspectus of the marine and freshwater Ciliata belonging to the suborder Tintinnoinea, with descriptions of new species principally from the Agassiz expedition to the eastern tropical Pacific, 1904-1905. Univ. Calif. Publ. Zool. 34:1-403, 697 text figs.

Codonella nationalis

Hofker, J., 1931
Studien uber Tintinnoidea. Arch. f. Protistenk 75(3):315-402, 89 text figs.

Codonella nationalis

Jörgensen, E., 1924
Mediterranean Tintinnidae. Rept. Danish Oceanogr. Exped. 1908-10 to the Mediterranean and adjacent seas, Vol.II, Biol. J 3, 110 pp., 114 text figs.

Codonella nationalis

Sousa e Silva, E. de, 1950.
Les tintinnides de la Baie de Cascais. Bull. Inst. Océan., Monaco, No. 979:28 pp., 4 pls.

Codonella nationales

Laackmann, H., 1910
Die Tintinnodeen der Deutschen Südpolar Expedition 1901-1903. Deutsch. Südpolar Exped. 11, Zool. 3(4):341-496, pls. 33-51.

Codonella nucula

Fol, H., 1884
Sur la famille des Tintinnodea. Recueil Zool. Suisse 1:27-64, Pls. 4-5.

Codonella oceanica

Gaarder, K.R., 1946
Tintinnoidea. Rep.Sci.Res."Michael Sars" N.Atlantic Deep-Sea Exped., 1910, 2(1): 37 pp., 23 text figs.

Codonella oceanica

Kofoid, C. A. and A. S. Campbell, 1929
A conspectus of the marine and freshwater Ciliata belonging to the suborder Tintinnoinea, with descriptions of new species principally from the Agassiz expedition to the eastern tropical Pacific, 1904-1905. Univ. Calif. Publ. Zool. 34:1-403, 697 text figs.

Codonella oceanica

Rampi, L., 1948
Sur quelques Tintinnides (Infusoires louques) du Pacifique subtropical (Recoltes Alain Gerbault) Bull. l'Inst. Ocean., Monaco, No.938, 4 pp.

Codonella oceanica

Rampi, L., 1939
Primo contributo alla conoscenza dei tintinnoidi do Maro Ligure. Atti della Soc. Ital. di Sci. Nat. 78:67-81, 58 text figs.

Codonella oceanica

Sousa e Silva, E. de, 1950.
Les tintinnides de la Baie de Cascais. Bull. Inst. Océan., Monaco, No. 979:28 pp., 4 pls.

Codonella aff. olla

Balech, Enrique, 1962
Tintinnoinea y Dinoflagellata del Pacifico segun material de las Expediciones NORPAC y DOWNWIND del Instituto Scripps de Oceanografia. Revista. Mus. Argentino Ciencias Nat. "Bernardino Rivadavia", Ciencias Zool., 7(1):1-253.

Codonella olla

Campbell, A. S., 1942
The oceanic Tintinnoina of the plankton gathered during the last cruise of the CARNEGIE. Sci. Res. Cruise VII of the Carnegie, 1928-1929, Biol. 2:163 pp., 128 figs.

Codonella olla n. sp.

Kofoid, C. A. and A. S. Campbell, 1929
A conspectus of the marine and freshwater Ciliata belonging to the suborder Tintinnoinea, with descriptions of new species principally from the Agassiz expedition to the eastern tropical Pacific, 1904-1905. Univ. Calif. Publ. Zool. 34:1-403, 697 text figs.

Codonella olla minor n.var.

Komarovsky, B., 1959
The Tintinnina of the Gulf of Eylath (Aqaba). Contrib. Knowl. Red Sea, Bull. Sea Fish. Res. Sta.Haifa, Israel 21 (14):1-40.

Codonella orthoceras

Entz, G., Jr. 1909
Studien über organisation und biologie der Tintinniden. Arch. f. Protistenkunde 15:93-226, Pls. 8-21, text figs.

Codonella orthoceros

Jørgensen, E., 1900.
Ueber die Tintinnodeen der norwegischen Westküste. Bergens Mus. Aarb., 1899(2):48 pp., 3 pls.

Codonella orthocercus

Laackmann, H., 1910
Die Tintinnodeen der Deutschen Südpolar Expedition 1901-1903. Deutsch. Südpolar Exped. 11, Zool. 3(4):341-496, pls. 33-51.

Codonella orthoceros

Schodduyn, M., 1926
Observations faites dans la baie d'Ambleteuse (Pas de Calais). Bull. Inst. Ocean., Monaco, No. 482: 64 pp.

Codonella pacifica

Kofoid, C.A. and A.S. Campbell, 1939
Reports on the scientific results of the expedition to the eastern tropical Pacific in charge of Alexander Agassiz, by the U.S. Fish Commission steamer "Albatross" from October 1904 to March 1905-----XXXVII. The Ciliata: The Tintinnoimea. Bull.Mus. Comp.Zool. 84: 473 pp., 36 pls.

Codonella pacifica n. sp.

Kofoid, C. A. and A. S. Campbell, 1929
A conspectus of the marine and freshwater Ciliata belonging to the suborder Tintinnoinea, with descriptions of new species principally from the Agassiz expedition to the eastern tropical Pacific, 1904-1905. Univ. Calif. Publ. Zool. 34:1-403, 697 text figs.

Codonella perforata

Balech, Enrique, 1962
Tintinnoinea y Dinoflagellata del Pacifico segun material de las Expediciones NORPAC y DOWNWIND del Instituto Scripps de Oceanografia. Revista. Mus. Argentino Ciencias Nat. "Bernardino Rivadavia", Ciencias Zool., 7(1):1-253.

Codonella perforata

Entz, G., Jr. 1909
Studien über organisation und biologie der Tintinniden. Arch. f. Protistenkunde 15:93-226, Pls. 8-21, text figs.

Codonella perforata

Jörgensen, E., 1924
Mediterranean Tintinnidae. Rept. Danish Oceanogr. Exped. 1908-10 to the Mediterranean and adjacent seas, Vol.II, Biol. J 3, 110 pp., 114 text figs.

Codonella perforata

Kofoid, C.A. and A.S. Campbell, 1939
Reports on the scientific results of the expedition to the eastern tropical Pacific in charge of Alexander Agassiz, by the U.S. Fish Commission steamer "Albatross" from October 1904 to March 1905-----XXXVII. The Ciliata: The Tintinnoimea. Bull.Mus. Comp.Zool. 84: 473 pp., 36 pls.

Codonella perforata

Kofoid, C. A. and A. S. Campbell, 1929
A conspectus of the marine and freshwater Ciliata belonging to the suborder Tintinnoinea, with descriptions of new species principally from the Agassiz expedition to the eastern tropical Pacific, 1904-1905. Univ. Calif. Publ. Zool. 34:1-403, 697 text figs.

Codonella perforata

Komarovsky, B., 1959
The Tintinnina of the Gulf of Eylath (Aqaba). Contrib. Knowl. Red Sea, Bull. Sea Fish. Res. Sta.Haifa, Israel 21 (14):1-40.

Codonella perforata
Laackmann, H., 1910
Die Tintinnodeen der Deutschen Südpolar Expedition 1901-1903. Deutsch. Südpolar Exped. 11, Zool. 3(4):341-496, pls. 33-51.

Codonella perforata
Sousa e Silva, E. de, 1950.
Les tintinnides de la Baie de Cascais. Bull. Inst. Océan., Monaco, No. 979;28 pp., 4 pls.

Codonella poculum
Campbell, A. S., 1942
The oceanic Tintinnoina of the plankton gathered during the last cruise of the CARNEGIE. Sci. Res. Cruise VII of the Carnegie, 1928-1929, Biol. 2:163 pp., 128 figs.

Codonella poculum n. sp.
Kofoid, C. A. and A. S. Campbell, 1929
A conspectus of the marine and freshwater Ciliata belonging to the suborder Tintinnoinea, with descriptions of new species principally from the Agassiz expedition to the eastern tropical Pacific, 1904-1905. Univ. Calif. Publ. Zool. 34:1-403, 697 text figs.

Codonella rapa
Campbell, A. S., 1942
The oceanic Tintinnoina of the plankton gathered during the last cruise of the CARNEGIE. Sci. Res. Cruise VII of the Carnegie, 1928-1929, Biol. 2:163 pp., 128 figs.

Codonella rapa
Kofoid, C.A. and A.S. Campbell, 1939
Reports on the scientific results of the expedition to the eastern tropical Pacific in charge of Alexander Agassiz, by the U.S. Fish Commission steamer "Albatross" from October 1904 to March 1905-----XXXVJI. The Ciliata: The Tintinnoimea. Bull.Mus. Comp.Zool. 84: 473 pp., 36 pls.

Codonella rapa n.sp.
Kofoid, C. A. and A. S. Campbell, 1929
A conspectus of the marine and freshwater Ciliata belonging to the suborder Tintinnoinea, with descriptions of new species principally from the Agassiz expedition to the eastern tropical Pacific, 1904-1905. Univ. Calif. Publ. Zool. 34:1-403, 697 text figs.

Codonella recta
Campbell, A. S., 1942
The oceanic Tintinnoina of the plankton gathered during the last cruise of the CARNEGIE. Sci. Res. Cruise VII of the Carnegie, 1928-1929, Biol. 2:163 pp., 128 figs.

Codonella recta n. sp.
Kofoid, C. A. and A. S. Campbell, 1929
A conspectus of the marine and freshwater Ciliata belonging to the suborder Tintinnoinea, with descriptions of new species principally from the Agassiz expedition to the eastern tropical Pacific, 1904-1905. Univ. Calif. Publ. Zool. 34:1-403, 697 text figs.

Codonella relicta
Kofoid, C. A. and A. S. Campbell, 1929
A conspectus of the marine and freshwater Ciliata belonging to the suborder Tintinnoinea, with descriptions of new species principally from the Agassiz expedition to the eastern tropical Pacific, 1904-1905. Univ. Calif. Publ. Zool. 34:1-403, 697 text figs.

Codonella robusta n. sp.
Kofoid, C. A. and A. S. Campbell, 1929
A conspectus of the marine and freshwater Ciliata belonging to the suborder Tintinnoinea, with descriptions of new species principally from the Agassiz expedition to the eastern tropical Pacific, 1904-1905. Univ. Calif. Publ. Zool. 34:1-403, 697 text figs.

Codonella saccus n. sp.
Kofoid, C. A. and A. S. Campbell, 1929
A conspectus of the marine and freshwater Ciliata belonging to the suborder Tintinnoinea, with descriptions of new species principally from the Agassiz expedition to the eastern tropical Pacific, 1904-1905. Univ. Calif. Publ. Zool. 34:1-403, 697 text figs.

Codonella sphaerica
Kofoid, C. A. and A. S. Campbell, 1929
A conspectus of the marine and freshwater Ciliata belonging to the suborder Tintinnoinea, with descriptions of new species principally from the Agassiz expedition to the eastern tropical Pacific, 1904-1905. Univ. Calif. Publ. Zool. 34:1-403, 697 text figs.

Codonella tropica
Campbell, A. S., 1942
The oceanic Tintinnoina of the plankton gathered during the last cruise of the CARNEGIE. Sci. Res. Cruise VII of the Carnegie, 1928-1929, Biol. 2:163 pp., 128 figs.

Codonella tropica
Kofoid, C.A. and A.S. Campbell, 1939
Reports on the scientific results of the expedition to the eastern tropical Pacific in charge of Alexander Agassiz, by the U.S. Fish Commission steamer "Albatross" from October 1904 to March 1905-----XXXVJI. The Ciliata: The Tintinnoimea. Bull.Mus. Comp.Zool. 84: 473 pp., 36 pls.

Codonella tropica n. sp.
Kofoid, C. A. and A. S. Campbell, 1929
A conspectus of the marine and freshwater Ciliata belonging to the suborder Tintinnoinea, with descriptions of new species principally from the Agassiz expedition to the eastern tropical Pacific, 1904-1905. Univ. Calif. Publ. Zool. 34:1-403, 697 text figs.

Codonella ventricosa
Fol, H., 1884.
Sur la famille des Tintinnodea. Recueil Zool. Suisse 1:27-64, Pls. 4-5.

Codonella ventricosa
Jørgensen, E., 1905
B.Protistplankton and the diatoms in bottom samples. Hydrographical and biological investigations in Norwegian fjords. Bergens Mus. Skr. 7: 49-225.

Codonella ventricosa
Jørgensen, E., 1900.
Ueber die Tintinnodeen der norwegischen Westküste Bergens Mus. Aarb., 1899(2):48 pp., 3 pls.

Codonelloides n.subgen.
Kofoid, C.A. and A.S. Campbell, 1939
Reports on the scientific results of the expedition to the eastern tropical Pacific in charge of Alexander Agassiz, by the U.S. Fish Commission steamer "Albatross" from October 1904 to March 1905-----XXXVJI. The Ciliata: The Tintinnoimea. Bull.Mus. Comp.Zool. 84: 473 pp., 36 pls.

Codonellopsis n.gen.
Jörgensen, E., 1924
Mediterranean Tintinnidae. Rept. Danish Oceanogr. Exped. 1908-10 to the Mediterranean and adjacent seas, Vol.II, Biol. J 3, 110 pp., 114 text figs.

Codonellopsis n.subgen.
Kofoid, C.A. and A.S. Campbell, 1939
Reports on the scientific results of the expedition to the eastern tropical Pacific in charge of Alexander Agassiz, by the U.S. Fish Commission steamer "Albatross" from October 1904 to March 1905-----XXXVJI. The Ciliata: The Tintinnoimea. Bull.Mus. Comp.Zool. 84: 473 pp., 36 pls.

Codonellopsis sp.
Schulz, H., 1965.
Die Tintinnoinea des Elbe-Aestuars. Arch. Fischereiwiss., 15(3):216-225.

Codonellopsis sp.
Sousa e Silva, E. de, 1950.
Les tintinnides de la Baie de Cascais. Bull. Inst. Océan., Monaco, No. 979;28 pp., 4 pls.

Codonellopsis aleutiensis n.sp.
Campbell, A. S., 1942
The oceanic Tintinnoina of the plankton gathered during the last cruise of the CARNEGIE. Sci. Res. Cruise VII of the Carnegie, 1928-1929, Biol. 2:163 pp., 128 figs.

Codonellopsis americana
Campbell, A. S., 1942
The oceanic Tintinnoina of the plankton gathered during the last cruise of the CARNEGIE. Sci. Res. Cruise VII of the Carnegie, 1928-1929, Biol. 2:163 pp., 128 figs.

Codonellopsis americana
Duran, M., 1957.
Nota sobre algunos tintinnoineos del plancton de Puerto Rico. Invest. Pesqueria, Barcelona, 8: 97-120.

Codonellopsis americana
Kofoid, C.A. and A.S. Campbell, 1939
Reports on the scientific results of the expedition to the eastern tropical Pacific in charge of Alexander Agassiz, by the U.S. Fish Commission steamer "Albatross" from October 1904 to March 1905-----XXXVJI. The Ciliata: The Tintinnoimea. Bull.Mus. Comp.Zool. 84: 473 pp., 36 pls.

Codonellopsis americana n.sp.
Kofoid, C. A. and A. S. Campbell, 1929
A conspectus of the marine and freshwater Ciliata belonging to the suborder Tintinnoinea, with descriptions of new species principally from the Agassiz expedition to the eastern tropical Pacific, 1904-1905. Univ. Calif. Publ. Zool. 34:1-403, 697 text figs.

Codonellopsis americana
Rampi, L., 1939
Primo contributo alla conoscenza dei tintinnoidi do Maro Ligure. Atti della Soc. Ital. di Sci. Nat. 78:67-81, 58 text figs.

Codonellopsis biedermanni
Campbell, A. S., 1942
The oceanic Tintinnoina of the plankton gathered during the last cruise of the CARNEGIE. Sci. Res. Cruise VII of the Carnegie, 1928-1929, Biol. 2:163 pp., 128 figs.

Codonellopsis biedermanni
Kofoid, C.A. and A.S. Campbell, 1939
Reports on the scientific results of the expedition to the eastern tropical Pacific in charge of Alexander Agassiz, by the U.S. Fish Commission steamer "Albatross" from October 1904 to March 1905-----XXXVJI. The Ciliata: The Tintinnoimea. Bull.Mus. Comp.Zool. 84: 473 pp., 36 pls.

Codonellopsis biedermanni
Kofoid, C. A. and A. S. Campbell, 1929
A conspectus of the marine and freshwater Ciliata belonging to the suborder Tintinnoinea, with descriptions of new species principally from the Agassiz expedition to the eastern tropical Pacific, 1904-1905. Univ. Calif. Publ. Zool. 34:1-403, 697 text figs.

Codonellopsis brasiliensis
Balech, Enrique, 1971
Microplancton del Atlantico ecuatorial oeste (Equalant 1)
Publ. Serv. Hidrograf. Naval, Argentina H. 654: 103 pp., 122 figs.

Codonellopsis brasiliensis

Campbell, A. S., 1942
The oceanic Tintinnoina of the plankton gathered during the last cruise of the CARNEGIE. Sci. Res. Cruise VII of the Carnegie, 1928-1929, Biol. 2:163 pp., 128 figs.

Codonellopsis brasiliensis

Kofoid, C. A. and A. S. Campbell, 1929
A conspectus of the marine and fresh-water Ciliata belonging to the suborder Tintinnoinea, with descriptions of new species principally from the Agassiz expedition to the eastern tropical Pacific, 1904-1905. Univ. Calif. Publ. Zool. 34:1-403, 697 text figs.

Codonellopsis brevicaudata

Kofoid, C. A. and A. S. Campbell, 1929
A conspectus of the marine and fresh-water Ciliata belonging to the suborder Tintinnoinea, with descriptions of new species principally from the Agassiz expedition to the eastern tropical Pacific, 1904-1905. Univ. Calif. Publ. Zool. 34:1-403, 697 text figs.

Codonellopsis brevicaudata

Marshall, S. M., 1934
The silicoflagellata and tintinnoinea. Great Barrier Reef Exped., 1928-29. Sci. Repts. 4(15):623-664, 43 text figs.

Codonellopsis bulbulus

Kofoid, C. A. and A. S. Campbell, 1929
A conspectus of the marine and fresh-water Ciliata belonging to the suborder Tintinnoinea, with descriptions of new species principally from the Agassiz expedition to the eastern tropical Pacific, 1904-1905. Univ. Calif. Publ. Zool. 34:1-403, 697 text figs.

Codonellopsis californiensis

Kofoid, C.A. and A.S. Campbell, 1939
Reports on the scientific results of the expedition to the eastern tropical Pacific in charge of Alexander Agassiz, by the U.S. Fish Commission steamer "Albatross" from October 1904 to March 1905-----XXXVII. The Ciliata: The Tintinnoimea. Bull. Mus. Comp. Zool. 84: 473 pp., 36 pls.

Codonellopsis californiensis n. sp.

Kofoid, C. A. and A. S. Campbell, 1929
A conspectus of the marine and fresh-water Ciliata belonging to the suborder Tintinnoinea, with descriptions of new species principally from the Agassiz expedition to the eastern tropical Pacific, 1904-1905. Univ. Calif. Publ. Zool. 34:1-403, 697 text figs.

Codonellopsis contracta

Balech, Enrique, 1962
Tintinnoinea y Dinoflagellata del Pacifico segun material de las Expediciones NORPAC y DOWNWIND del Instituto Scripps de Oceanografia. Revista, Mus. Argentino Ciencias Nat. "Bernardino Rivadavia", Ciencias Zool., 7(1):1-253.

Codonellopsis contracta

Balech, E., 1945.
Tintinnoinea de Quequén. Physis 20(55):1-15, 3 pls.

Codonellopsis contracta

Balech, Enrique, 1944
Contribucion al conocimiento del Plancton de Lennox y Cabo de Hornos. Physis XIX:423-446, 6 pls. with 67 figs.

Codonellopsis contracta

Campbell, A. S., 1942
The oceanic Tintinnoina of the plankton gathered during the last cruise of the CARNEGIE. Sci. Res. Cruise VII of the Carnegie, 1928-1929, Biol. 2:163 pp., 128 figs.

Codonellopsis contracta

Kofoid, C.A. and A.S. Campbell, 1939
Reports on the scientific results of the expedition to the eastern tropical Pacific in charge of Alexander Agassiz, by the U.S. Fish Commission steamer "Albatross" from October 1904 to March 1905-----XXXVII. The Ciliata: The Tintinnoimea. Bull. Mus. Comp. Zool. 84: 473 pp., 36 pls.

Codonellopsis contracta n.sp.

Kofoid, C. A. and A. S. Campbell, 1929
A conspectus of the marine and fresh-water Ciliata belonging to the suborder Tintinnoinea, with descriptions of new species principally from the Agassiz expedition to the eastern tropical Pacific, 1904-1905. Univ. Calif. Publ. Zool. 34:1-403, 697 text figs.

Codonellopsis cordata

Kofoid, C.A. and A.S. Campbell, 1939
Reports on the scientific results of the expedition to the eastern tropical Pacific in charge of Alexander Agassiz, by the U.S. Fish Commission steamer "Albatross" from October 1904 to March 1905-----XXXVII. The Ciliata: The Tintinnoimea. Bull. Mus. Comp. Zool. 84: 473 pp., 36 pls.

Codonellopsis ecaudata

Balech, Enrique, 1962
Tintinnoinea y Dinoflagellata del Pacifico segun material de las Expediciones NORPAC y DOWNWIND del Instituto Scripps de Oceanografia. Revista, Mus. Argentino Ciencias Nat. "Bernardino Rivadavia", Ciencias Zool., 7(1):1-253.

Codonellopsis ecaudata

Campbell, A. S., 1942
The oceanic Tintinnoina of the plankton gathered during the last cruise of the CARNEGIE. Sci. Res. Cruise VII of the Carnegie, 1928-1929, Biol. 2:163 pp., 128 figs.

Codonellopsis ecaudata

Kofoid, C.A. and A.S. Campbell, 1939
Reports on the scientific results of the expedition to the eastern tropical Pacific in charge of Alexander Agassiz, by the U.S. Fish Commission steamer "Albatross" from October 1904 to March 1905-----XXXVII. The Ciliata: The Tintinnoimea. Bull. Mus. Comp. Zool. 84: 473 pp., 36 pls.

Codonellopsis eylathensis n. sp.

Komarovsky, B., 1959
The Tintinnina of the Gulf of Eylath (Aqaba). Contrib. Knowl. Red Sea, Bull. Sea Fish. Res. Sta. Haifa, Israel 21 (14):1-40.

Codonellopsis frigida n.sp.

Balech, Enrique, 1958.
Plancton de la Campaña Antartica Argentina, 1954-1955. Physis, 21(60):75-108.

Codonellopsis gaussi

Balech, Enrique, 1958.
Plancton de la Campaña Antartica Argentina, 1954-1955. Physis, 21(60):75-108.

Codonellopsis gaussi

Balech, E., 1947
Contribution al conocimiento del plancton antartico. Plankton del Mar de Bellinghausen. Physis 20:75-91, 76 figs. on 8 pls.

Codonellopsis gaussi

Balech, Enrique, 1947
Contribucion al conocimiento del Plancton antartico Plancton del Mar de Bellingshausen. Physis 22(56):75-91, 7 pls. with 76 figs.

Codonellopsis gaussi

*Hada, Yoshina, 1970. (1970)
The protozoan plankton of the Antarctic and Subantarctic seas. Scient. Repts, Japan. Antarct. Res. Exped., (E)31:51 pp.

Codonellopsis gaussi

Kofoid, C. A. and A. S. Campbell, 1929
A conspectus of the marine and fresh-water Ciliata belonging to the suborder Tintinnoinea, with descriptions of new species principally from the Agassiz expedition to the eastern tropical Pacific, 1904-1905. Univ. Calif. Publ. Zool. 34:1-403, 697 text figs.

Codonellopsis glacilis

*Hada, Yoshina, 1970. (1970)
The protozoan plankton of the Antarctic and Subantarctic seas. Scient. Repts, Japan. Antarct. Res. Exped., (E)31:51 pp.

Codonellopsis glacialis

Kofoid, C. A. and A. S. Campbell, 1929
A conspectus of the marine and fresh-water Ciliata belonging to the suborder Tintinnoinea, with descriptions of new species principally from the Agassiz expedition to the eastern tropical Pacific, 1904-1905. Univ. Calif. Publ. Zool. 34:1-403, 697 text figs.

Codonellopsis globosa n.sp.

Kofoid, C. A. and A. S. Campbell, 1929
A conspectus of the marine and fresh-water Ciliata belonging to the suborder Tintinnoinea, with descriptions of new species principally from the Agassiz expedition to the eastern tropical Pacific, 1904-1905. Univ. Calif. Publ. Zool. 34:1-403, 697 text figs.

Codonellopsis indica n.sp.

Kofoid, C. A. and A. S. Campbell, 1929
A conspectus of the marine and fresh-water Ciliata belonging to the suborder Tintinnoinea, with descriptions of new species principally from the Agassiz expedition to the eastern tropical Pacific, 1904-1905. Univ. Calif. Publ. Zool. 34:1-403, 697 text figs.

Codonellopsis indica

Marshall, S. M., 1934
The silicoflagellata and tintinnoinea. Great Barrier Reef Exped., 1928-29. Sci. Repts. 4(15):623-664, 43 text figs.

Codonellopsis inflata

Campbell, A. S., 1942
The oceanic Tintinnoina of the plankton gathered during the last cruise of the CARNEGIE. Sci. Res. Cruise VII of the Carnegie, 1928-1929, Biol. 2:163 pp., 128 figs.

Codonellopsis inflata

Kofoid, C.A. and A.S. Campbell, 1939
Reports on the scientific results of the expedition to the eastern tropical Pacific in charge of Alexander Agassiz, by the U.S. Fish Commission steamer "Albatross" from October 1904 to March 1905-----XXXVII. The Ciliata: The Tintinnoimea. Bull. Mus. Comp. Zool. 84: 473 pp., 36 pls.

Codonellopsis inflata n.sp

Kofoid, C. A. and A. S. Campbell, 1929
A conspectus of the marine and fresh-water Ciliata belonging to the suborder Tintinnoinea, with descriptions of new species principally from the Agassiz expedition to the eastern tropical Pacific, 1904-1905. Univ. Calif. Publ. Zool. 34:1-403, 697 text figs.

Codonellopsis inornata

Kofoid, C. A. and A. S. Campbell, 1929
A conspectus of the marine and fresh-water Ciliata belonging to the suborder Tintinnoinea, with descriptions of new species principally from the Agassiz expedition to the eastern tropical Pacific, 1904-1905. Univ. Calif. Publ. Zool. 34:1-403, 697 text figs.

Codonellopsis lata n.sp.
Kofoid, C. A. and A. S. Campbell, 1929
A conspectus of the marine and fresh-water Ciliata belonging to the suborder Tintinnoinea, with descriptions of new species principally from the Agassiz expedition to the eastern tropical Pacific, 1904-1905. Univ. Calif. Publ. Zool. 34:1-403, 697 text figs.

Codonellopsis lagenula
Gaarder, K.R., 1946
Tintinnoidea. Rep.Sci.Res."Michael Sars" N.Atlantic Deep-Sea Exped., 1910, 2(1): 37 pp., 23 text figs.

Codonellopsis lagenula
Jörgensen, E., 1924
Mediterranean Tintinnidae. Rept. Danish Oceanogr. Exped. 1908-10 to the Mediterranean and adjacent seas, Vol.II, Biol. J 3, 110 pp., 114 text figs.

Codonellopsis longa
Campbell, A. S., 1942
The oceanic Tintinnoina of the plankton gathered during the last cruise of the CARNEGIE. Sci. Res. Cruise VII of the Carnegie, 1928-1929, Biol. 2:163 pp., 128 figs.

Codonellopsis longa
Kofoid, C.A. and A.S. Campbell, 1939
Reports on the scientific results of the expedition to the eastern tropical Pacific in charge of Alexander Agassiz, by the U.S. Fish Commission steamer "Albatross" from October 1904 to March 1905-----XXXVII. The Ciliata: The Tintinnoimea. Bull.Mus. Comp.Zool. 84: 473 pp., 36 pls.

Codonellopsis longa n.sp.
Kofoid, C. A. and A. S. Campbell, 1929
A conspectus of the marine and fresh-water Ciliata belonging to the suborder Tintinnoinea, with descriptions of new species principally from the Agassiz expedition to the eastern tropical Pacific, 1904-1905. Univ. Calif. Publ. Zool. 34:1-403, 697 text figs.

Codonellopsis longa
Komarovsky, B., 1959
The Tintinnina of the Gulf of Eylath (Aqaba). Contrib. Knowl. Red Sea, Bull. Sea Fish. Res. Sta.Haifa, Israel 21 (14):1-40.

Codonellopsis lusitanica
Balech, Enrique, 1948
Tintinnoinea de Atlantida (R.O. Del Uruguay) (Protozoa, Ciliata Oligotr.). Comm. Mus. Argentino Ciencias Nat. "Bernardino Rivadavia", Ser.Cien.Zool. No.7, 23 pp., 8 pls. of 107 figs.

Codonellopsis lusitanica
Kofoid, C. A. and A. S. Campbell, 1929
A conspectus of the marine and fresh-water Ciliata belonging to the suborder Tintinnoinea, with descriptions of new species principally from the Agassiz expedition to the eastern tropical Pacific, 1904-1905. Univ. Calif. Publ. Zool. 34:1-403, 697 text figs.

Codonellopsis lusitanica n.sp.
Jörgensen, E., 1924
Mediterranean Tintinnidae. Rept. Danish Oceanogr. Exped. 1908-10 to the Mediterranean and adjacent seas, Vol.II, Biol. J 3, 110 pp., 114 text figs.

Codonellopsis meridionalis
Campbell, A. S., 1942
The oceanic Tintinnoina of the plankton gathered during the last cruise of the CARNEGIE. Sci. Res. Cruise VII of the Carnegie, 1928-1929, Biol. 2:163 pp., 128 figs.

Codonellopsis meridionalis
Kofoid, C.A. and A.S. Campbell, 1939
Reports on the scientific results of the expedition to the eastern tropical Pacific in charge of Alexander Agassiz, by the U.S. Fish Commission steamer "Albatross" from October 1904 to March 1905-----XXXVII. The Ciliata: The Tintinnoimea. Bull.Mus. Comp.Zool. 84: 473 pp., 36 pls.

Codonellopsis meridionalis n.sp.
Kofoid, C. A. and A. S. Campbell, 1929
A conspectus of the marine and fresh-water Ciliata belonging to the suborder Tintinnoinea, with descriptions of new species principally from the Agassiz expedition to the eastern tropical Pacific, 1904-1905. Univ. Calif. Publ. Zool. 34:1-403, 697 text figs.

Codonellopsis minor
Campbell, A. S., 1942
The oceanic Tintinnoina of the plankton gathered during the last cruise of the CARNEGIE. Sci. Res. Cruise VII of the Carnegie, 1928-1929, Biol. 2:163 pp., 128 figs.

Codonellopsis minor
Kofoid, C.A. and A.S. Campbell, 1939
Reports on the scientific results of the expedition to the eastern tropical Pacific in charge of Alexander Agassiz, by the U.S. Fish Commission steamer "Albatross" from October 1904 to March 1905-----XXXVII. The Ciliata: The Tintinnoimea. Bull.Mus. Comp.Zool. 84: 473 pp., 36 pls.

Codonellopsis minor
Kofoid, C. A. and A. S. Campbell, 1929
A conspectus of the marine and fresh-water Ciliata belonging to the suborder Tintinnoinea, with descriptions of new species principally from the Agassiz expedition to the eastern tropical Pacific, 1904-1905. Univ. Calif. Publ. Zool. 34:1-403, 697 text figs.

Codonellopsis morchella
Duran, M., 1957.
Nota sobre algunos tintinnoineas del plancton de Puerto Rico. Invest. Pesqueria, Barcelona, 8: 97-120.

Codonellopsis morchella
Duran, M., 1951
Contribucion al estudio de los tintinidos del plancton de los costas de Castellon. (Mediterraneo occidental). Publ. Inst. Biol. Aplic., Barcelone, 8: 101-120, 29 text figs.

Codonellopsis morchella
Hofker, J., 1931
Studien uber Tintinnoidea. Arch. f. Protistenk 75(3):315-402, 89 text figs.

Codonellopsis morchella
Jörgensen, E., 1924
Mediterranean Tintinnidae. Rept. Danish Oceanogr. Exped. 1908-10 to the Mediterranean and adjacent seas, Vol.II, Biol. J 3, 110 pp., 114 text figs.

Codonellopsis morchella
Kofoid, C. A. and A. S. Campbell, 1929
A conspectus of the marine and fresh-water Ciliata belonging to the suborder Tintinnoinea, with descriptions of new species principally from the Agassiz expedition to the eastern tropical Pacific, 1904-1905. Univ. Calif. Publ. Zool. 34:1-403, 697 text figs.

Codonellopsis morchella
Komarovsky, B., 1959
The Tintinnina of the Gulf of Eylath (Aqaba). Contrib. Knowl. Red Sea, Bull. Sea Fish. Res. Sta.Haifa, Israel 21 (14):1-40.

Codonellopsis morchela
Massutí Algamora, M., 1949
Estudio de diez y seis muestras de plancton del Golfo de Nápoles. Publ. Inst. Biol. Appl. 5:85-94, 1 fold-in table.

Codonellopsis morchella
Rampi, L., 1948
Sur quelques Tintinnides (Infusoires louques) du Pacifique subtropical (Recoltes Alain Gerbault) Bull. l'Inst. Ocean., Monaco, No.938, 4 pp.

Codonellopsis obconica n.sp.
Kofoid, C. A. and A. S. Campbell, 1929
A conspectus of the marine and fresh-water Ciliata belonging to the suborder Tintinnoinea, with descriptions of new species principally from the Agassiz expedition to the eastern tropical Pacific, 1904-1905. Univ. Calif. Publ. Zool. 34:1-403, 697 text figs.

Codonellopsis obesa
Balech, E., 1951.
Nuevos datos sobre Tintinnoinea de Argentina y Uruguay. Physis 20(58):291-302, 16 textfigs.

Codonellopsis obesa n.sp
Balech, Enrique, 1948
Tintinnoinea de Atlantida (R.O. Del Uruguay) (Protozoa, Ciliata Oligotr.). Comm. Mus. Argentino Ciencias Nat. "Bernardino Rivadavia", Ser.Cien.Zool. No.7, 23 pp., 8 pls. of 107 figs.

Codonellopsis orientalis
Campbell, A. S., 1942
The oceanic Tintinnoina of the plankton gathered during the last cruise of the CARNEGIE. Sci. Res. Cruise VII of the Carnegie, 1928-1929, Biol. 2:163 pp., 128 figs.

Codonellopsis orthoceras
Balech, Enrique, 1962
Tintinnoinea y Dinoflagellata del Pacifico segun material de las Expediciones NORPAC y DOWNWIND del Instituto Scripps de Oceanografia. Revista, Mus. Argentino Ciencias Nat. "Bernardino Rivadavia", Ciencias Zool., 7(1):1-253.

Codonellopsis orthoceras
Balech, E., 1959.
Tintinnoinea del Mediterraneo. Trav. Sta. Zool., Villefranche-sur-Mer, 18(17): Reprinted from: Trab. Inst. Español Oceanogr., Madrid, 19Sept. 1959(28):88 pp., 22 pls.

Codonellopsis orthoceras
Duran, M., 1951
Contribucion al estudio de los tintinidos del plancton de los costas de Castellon. (Mediterraneo occidental). Publ. Inst. Biol. Aplic., Barcelone, 8: 101-120, 29 text figs.

Codonellopsis orthoceras
Gaarder, K.R., 1946
Tintinnoidea. Rep.Sci.Res."Michael Sars" N.Atlantic Deep-Sea Exped., 1910, 2(1): 37 pp., 23 text figs.

Codonellopsis orthoceras
Hofker, J., 1931
Studien uber Tintinnoidea. Arch. f. Protistenk 75(3):315-402, 89 text figs.

Codonellopsis orthoceras
Jörgensen, E., 1924
Mediterranean Tintinnidae. Rept. Danish Oceanogr. Exped. 1908-10 to the Mediterranean and adjacent seas, Vol.II, Biol. J 3, 110 pp., 114 text figs.

Codonellopsis orthoceras
Kofoid, C. A. and A. S. Campbell, 1929
A conspectus of the marine and fresh-water Ciliata belonging to the suborder Tintinnoinea, with descriptions of new species principally from the Agassiz expedition to the eastern tropical Pacific, 1904-1905. Univ. Calif. Publ. Zool. 34:1-403, 697 text figs.

Codonellopsis orthocercus

Komarovsky, B., 1959
The Tintinnina of the Gulf of Eylath (Aqaba). Contrib. Knowl. Red Sea, Bull. Sea Fish. Res. Sta. Haifa, Israel 21 (14):1-40.

Codonellopsis orthoceras

Rampi, L., 1948
Sur quelques Tintinnides (Infusoires louques) du Pacifique subtropical (Recoltes Alain Gerbault) Bull. l'Inst. Ocean., Monaco, No. 938, 4 pp.

Codonellopsis orthoceros

Rampi, L., 1939
Primo contributo alla conoscenza dei tintinnoidi de Maro Ligure. Atti della Soc. Ital. di Sci. Nat. 78:67-81, 58 text figs.

Codonellopsis orthocercas

Sousa e Silva E. de, 1950.
Les tintinnides de la Baie de Cascais. Bull. Inst. Océan., Monaco, No. 979:28 pp., 4 pls.

Codonellopsis ostenfeldi

Kofoid, C. A. and A. S. Campbell, 1929
A conspectus of the marine and freshwater Ciliata belonging to the suborder Tintinnoinea, with descriptions of new species principally from the Agassiz expedition to the eastern tropical Pacific, 1904-1905. Univ. Calif. Publ. Zool. 34:1-403, 697 text figs.

Codonellopsis ostenfeldi

Komarovsky, B., 1962
Tintinnina from the vicinity of the Straits of Tiran and Massawa Region. Contributions to the knowledge of the Red Sea, No. 25. Sea Fish. Res. Sta. Haifa, Israel, Bull. No. 30:48-56.

Codonellopsis ostenfeldii

Marshall, S. M., 1934
The silicoflagellata and tintinnoinea. Great Barrier Reef Exped., 1928-29. Sci. Repts. 4(15):623-664, 43 text figs.

Codonellopsis ostenfeldii

Marshall, S. M., 1933
The production of microplankton in the Great Barrier Reef Region. Brit. Mus. (N.H.) Great Barrier Reef Exped. 1928-29, Sci. Repts. II(5):111-157, 14 text figs.

Codonellopsis ostenfeldi

Wang, C. C., and D. Nie, 1932
A survey of the marine protozoa of Amoy. Contr. Biol. Lab. Sci. Soc., China, Zool. Ser., 8:285-385, 89 text figs.

Codonellopsis ovata

Kofoid, C. A. and A. S. Campbell, 1929
A conspectus of the marine and freshwater Ciliata belonging to the suborder Tintinnoinea, with descriptions of new species principally from the Agassiz expedition to the eastern tropical Pacific, 1904-1905. Univ. Calif. Publ. Zool. 34:1-403, 697 text figs.

Codonellopsis pacifica

Campbell, A. S., 1942
The oceanic Tintinnoina of the plankton gathered during the last cruise of the CARNEGIE. Sci. Res. Cruise VII of the Carnegie, 1928-1929, Biol. 2:163 pp., 128 figs.

Codonellopsis pacifica

Kofoid, C.A. and A.S. Campbell, 1939
Reports on the scientific results of the expedition to the eastern tropical Pacific in charge of Alexander Agassiz, by the U.S. Fish Commission steamer "Albatross" from October 1904 to March 1905-----XXXVII. The Ciliata: The Tintinnoimea. Bull. Mus. Comp. Zool. 84: 473 pp., 36 pls.

Codonellopsis pacifica

Kofoid, C. A. and A. S. Campbell, 1929
A conspectus of the marine and freshwater Ciliata belonging to the suborder Tintinnoinea, with descriptions of new species principally from the Agassiz expedition to the eastern tropical Pacific, 1904-1905. Univ. Calif. Publ. Zool. 34:1-403, 697 text figs.

Codonellopsis parva

Campbell, A. S., 1942
The oceanic Tintinnoina of the plankton gathered during the last cruise of the CARNEGIE. Sci. Res. Cruise VII of the Carnegie, 1928-1929, Biol. 2:163 pp., 128 figs.

Codonellopsis parva

Kofoid, C.A. and A.S. Campbell, 1939
Reports on the scientific results of the expedition to the eastern tropical Pacific in charge of Alexander Agassiz, by the U.S. Fish Commission steamer "Albatross" from October 1904 to March 1905-----XXXVII. The Ciliata: The Tintinnoimea. Bull. Mus. Comp. Zool. 84: 473 pp., 36 pls.

Codonellopsis parva n.sp.

Kofoid, C. A. and A. S. Campbell, 1929
A conspectus of the marine and freshwater Ciliata belonging to the suborder Tintinnoinea, with descriptions of new species principally from the Agassiz expedition to the eastern tropical Pacific, 1904-1905. Univ. Calif. Publ. Zool. 34:1-403, 697 text figs.

Codonellopsis parvicollis n.sp.

Marshall, S. M., 1934
The silicoflagellata and tintinnoinea. Great Barrier Reef Exped., 1928-29. Sci. Repts. 4(15):623-664, 43 text figs.

Codonellopsis pura

Campbell, A. S., 1942
The oceanic Tintinnoina of the plankton gathered during the last cruise of the CARNEGIE. Sci. Res. Cruise VII of the Carnegie, 1928-1929, Biol. 2:163 pp., 128 figs.

Codonellopsis pura

Kofoid, C.A. and A.S. Campbell, 1939
Reports on the scientific results of the expedition to the eastern tropical Pacific in charge of Alexander Agassiz, by the U.S. Fish Commission steamer "Albatross" from October 1904 to March 1905-----XXXVII. The Ciliata: The Tintinnoimea. Bull. Mus. Comp. Zool. 84: 473 pp., 36 pls.

Codonellopsis pura

Kofoid, C. A. and A. S. Campbell, 1929
A conspectus of the marine and freshwater Ciliata belonging to the suborder Tintinnoinea, with descriptions of new species principally from the Agassiz expedition to the eastern tropical Pacific, 1904-1905. Univ. Calif. Publ. Zool. 34:1-403, 697 text figs.

Codonellopsis pusilla

Balech, E., 1959.
Tintinnoinea del Mediterraneo. Trav. Sta. Zool., Villefranche-sur-Mer, 18(17): Reprinted from: Trab. Inst. Español Oceanogr., Madrid, 19 Sept. 1959 (28):88 pp., 22 pls.

Codonellopsis pusilla

Campbell, A. S., 1942
The oceanic Tintinnoina of the plankton gathered during the last cruise of the CARNEGIE. Sci. Res. Cruise VII of the Carnegie, 1928-1929, Biol. 2:163 pp., 128 figs.

Codonellopsis pusilla

Kofoid, C. A. and A. S. Campbell, 1929
A conspectus of the marine and freshwater Ciliata belonging to the suborder Tintinnoinea, with descriptions of new species principally from the Agassiz expedition to the eastern tropical Pacific, 1904-1905. Univ. Calif. Publ. Zool. 34:1-403, 697 text figs.

Codonellopsis robusta

Kofoid, C.A. and A.S. Campbell, 1939
Reports on the scientific results of the expedition to the eastern tropical Pacific in charge of Alexander Agassiz, by the U.S. Fish Commission steamer "Albatross" from October 1904 to March 1905-----XXXVII. The Ciliata: The Tintinnoimea. Bull. Mus. Comp. Zool. 84: 473 pp., 36 pls.

Codonellopsis robusta n.sp.

Kofoid, C. A. and A. S. Campbell, 1929
A conspectus of the marine and freshwater Ciliata belonging to the suborder Tintinnoinea, with descriptions of new species principally from the Agassiz expedition to the eastern tropical Pacific, 1904-1905. Univ. Calif. Publ. Zool. 34:1-403, 697 text figs.

Codonellopsis rotunda n.sp.

Wang, C. C., and D. Nie, 1932
A survey of the marine protozoa of Amoy. Contr. Biol. Lab. Sci. Soc., China, Zool. Ser., 8:285-385, 89 text figs.

Codonellopsis schabi

Balech, E., 1959.
Tintinnoinea del Mediterraneo. Trav. Sta. Zool., Villefranche-sur-Mer, 18(17): Reprinted from: Trab. Inst. Español Oceanogr., Madrid, 19 Sept. 1959, (28):88 pp., 22 pls.

Codonellopsis schabi

Duran, M., 1957.
Nota sobre algunos tintinnoineas del plancton de Puerto Rico. Invest. Pesqueria, Barcelona, 8: 97-120.

Codonellopsis schabi

Kofoid, C. A. and A. S. Campbell, 1929
A conspectus of the marine and freshwater Ciliata belonging to the suborder Tintinnoinea, with descriptions of new species principally from the Agassiz expedition to the eastern tropical Pacific, 1904-1905. Univ. Calif. Publ. Zool. 34:1-403, 697 text figs.

Codonellopsis schabi

Komarovsky, B., 1962
Tintinnina from the vicinity of the Straits of Tiran and Massawa Region. Contributions to the knowledge of the Red Sea, No. 25. Sea Fish. Res. Sta. Haifa, Israel, Bull. No. 30:48-56.

Codonellopsis schabi

Posta, Annette, 1963.
Relation entre l'évolution de quelques tintinnides de la rade de Villefranche et la température de l'eau.
Cahiers, Biol. Mar., Roscoff, 4:201-210.

Also in:
Trav. Sta. Zool., Villefranche-sur-Mer, 22(5);

Codonellopsis soyai n. sp.

Hada, Yoshina, 1970.   (1970)
The protozoan plankton of the Antarctic and Subantarctic seas.
Scient. Repts, Japan. Antarct. Res. Exped., (E)31:51 pp.

Codonellopsis speciosa

Campbell, A. S., 1942
The oceanic Tintinnoina of the plankton gathered during the last cruise of the CARNEGIE. Sci. Res. Cruise VII of the Carnegie, 1928-1929, Biol. 2:163 pp., 128 figs.

Codonellopsis speciosa

Kofoid, C.A. and A.S. Campbell, 1939
Reports on the scientific results of the expedition to the eastern tropical Pacific in charge of Alexander Agassiz, by the U.S. Fish Commission steamer "Albatross" from October 1904 to March 1905-----XXXVII. The Ciliata: The Tintinnoimea. Bull. Mus. Comp. Zool. 84: 473 pp., 36 pls.

Codonellopsis speciosa n.sp.

Kofoid, C. A. and A. S. Campbell, 1929
A conspectus of the marine and freshwater Ciliata belonging to the suborder Tintinnoinea, with descriptions of new species principally from the Agassiz expedition to the eastern tropical Pacific, 1904-1905. Univ. Calif. Publ. Zool. 34:1-403, 697 text figs.

Codonellopsis tessellata

Kofoid, C. A. and A. S. Campbell, 1929
A conspectus of the marine and fresh-water Ciliata belonging to the suborder Tintinnoinea, with descriptions of new species principally from the Agassiz expedition to the eastern tropical Pacific, 1904-1905. Univ. Calif. Publ. Zool. 34:1-403, 697 text figs.

Codonellopsis tropica

Kofoid, C.A. and A.S. Campbell, 1939
Reports on the scientific results of the expedition to the eastern tropical Pacific in charge of Alexander Agassiz, by the U.S. Fish Commission steamer "Albatross" from October 1904 to March 1905-----XXXVII. The Ciliata: The Tintinnoimea. Bull. Mus. Comp. Zool. 84: 473 pp., 36 pls.

Codonellopsis tropica n.sp.

Kofoid, C. A. and A. S. Campbell, 1929
A conspectus of the marine and fresh-water Ciliata belonging to the suborder Tintinnoinea, with descriptions of new species principally from the Agassiz expedition to the eastern tropical Pacific, 1904-1905. Univ. Calif. Publ. Zool. 34:1-403, 697 text figs.

Codonellopsis tuberculata

Gran, H.H., and T. Braarud, 1935
A quantitative study of the phytoplankton in the Bay of Fundy and the Gulf of Maine (including observations on hydrography, chemistry, and turbidity). J. Biol. Bd., Canada, 1(5):279-467, 69 text figs.

Codonellopsis (?) tuberculata

Jörgensen, E., 1924
Mediterranean Tintinnidae. Rept. Danish Oceanogr. Exped. 1908-10 to the Mediterranean and adjacent seas, Vol.II, Biol. J 3, 110 pp., 114 text figs.

Codonellopsis tuberculata

Kofoid, C. A. and A. S. Campbell, 1929
A conspectus of the marine and fresh-water Ciliata belonging to the suborder Tintinnoinea, with descriptions of new species principally from the Agassiz expedition to the eastern tropical Pacific, 1904-1905. Univ. Calif. Publ. Zool. 34:1-403, 697 text figs.

Codonellopsis turbinella n.sp.

Kofoid, C. A. and A. S. Campbell, 1929
A conspectus of the marine and fresh-water Ciliata belonging to the suborder Tintinnoinea, with descriptions of new species principally from the Agassiz expedition to the eastern tropical Pacific, 1904-1905. Univ. Calif. Publ. Zool. 34:1-403, 697 text figs.

Codonellopsis turgescens

Campbell, A. S., 1942
The oceanic Tintinnoina of the plankton gathered during the last cruise of the CARNEGIE. Sci. Res. Cruise VII of the Carnegie, 1928-1929, Biol. 2:163 pp., 128 figs.

Codonellopsis turgescens

Kofoid, C.A. and A.S. Campbell, 1939
Reports on the scientific results of the expedition to the eastern tropical Pacific in charge of Alexander Agassiz, by the U.S. Fish Commission steamer "Albatross" from October 1904 to March 1905-----XXXVII. The Ciliata: The Tintinnoimea. Bull. Mus. Comp. Zool. 84: 473 pp., 36 pls.

Codonellopsis turgescens n.sp.

Kofoid, C. A. and A. S. Campbell, 1929
A conspectus of the marine and fresh-water Ciliata belonging to the suborder Tintinnoinea, with descriptions of new species principally from the Agassiz expedition to the eastern tropical Pacific, 1904-1905. Univ. Calif. Publ. Zool. 34:1-403, 697 text figs.

Codonellopsis turgida

Campbell, A. S., 1942
The oceanic Tintinnoina of the plankton gathered during the last cruise of the CARNEGIE. Sci. Res. Cruise VII of the Carnegie, 1928-1929, Biol. 2:163 pp., 128 figs.

Codonellopsis turgida

Kofoid, C.A. and A.S. Campbell, 1939
Reports on the scientific results of the expedition to the eastern tropical Pacific in charge of Alexander Agassiz, by the U.S. Fish Commission steamer "Albatross" from October 1904 to March 1905-----XXXVII. The Ciliata: The Tintinnoimea. Bull. Mus. Comp. Zool. 84: 473 pp., 36 pls.

Codonellopsis turgida n.sp.

Kofoid, C. A. and A. S. Campbell, 1929
A conspectus of the marine and fresh-water Ciliata belonging to the suborder Tintinnoinea, with descriptions of new species principally from the Agassiz expedition to the eastern tropical Pacific, 1904-1905. Univ. Calif. Publ. Zool. 34:1-403, 697 text figs.

Codonopsis n.gen.

Kofoid, C.A. and A.S. Campbell, 1939
Reports on the scientific results of the expedition to the eastern tropical Pacific in charge of Alexander Agassiz, by the U.S. Fish Commission steamer "Albatross" from October 1904 to March 1905-----XXXVII. The Ciliata: The Tintinnoimea. Bull. Mus. Comp. Zool. 84: 473 pp., 36 pls.

Codonopsis ollula

Balech, Enrique, 1962
Tintinnoinea y Dinoflagellata del Pacifico segun material de las Expediciones NORPAC y DOWNWIND del Instituto Scripps de Oceanografia. Revista, Mus. Argentino Ciencias Nat. "Bernardino Rivadavia", Ciencias Zool., 7(1):1-253.

Codonopsis ollula

Campbell, A. S., 1942
The oceanic Tintinnoina of the plankton gathered during the last cruise of the CARNEGIE. Sci. Res. Cruise VII of the Carnegie, 1928-1929, Biol. 2:163 pp., 128 figs.

Codonopsis ollula

Kofoid, C.A. and A.S. Campbell, 1939
Reports on the scientific results of the expedition to the eastern tropical Pacific in charge of Alexander Agassiz, by the U.S. Fish Commission steamer "Albatross" from October 1904 to March 1905-----XXXVII. The Ciliata: The Tintinnoimea. Bull. Mus. Comp. Zool. 84: 473 pp., 36 pls.

Coniocylis

Fol, H., 1884.
Sur la famille des Tintinnodea. Recueil Zool. Suisse 1:27-64, Pls. 4-5.

Cordonellopsis sp.

de Sousa e Silva, E., 1956.
Contribution à l'étude du microplancton de Dakar et des regions maritimes voisines. Bull. I.F.A.N., 8(2):335-371, 7 pls.

Coxliella sp.

Balech, Enrique, 1962
Tintinnoinea y Dinoflagellata del Pacifico segun material de las Expediciones NORPAC y DOWNWIND del Instituto Scripps de Oceanografia. Revista, Mus. Argentino Ciencias Nat. "Bernardino Rivadavia", Ciencias Zool., 7(1):1-253.

Coxliella sp.

Balech, E., 1959.
Tintinnoinea del Mediterraneo. Trav. Sta. Zool., Villefranche-sur-Mer, 18(17): Reprinted from: Trab. Inst. Español Oceanogr., Madrid, 19 Sept. 1959(28):88 pp., 22 pls.

Coxliella (Protocochliella) ampla

Duran, M., 1957.
Nota sobre algunos tintinoineas del plancton de Puerto Rico. Invest. Pesqueria, Barcelona, 8: 97-120.

Coxliella ampla

Gaarder, K.R., 1946
Tintinnoidea. Rept.Sci.Res."Michael Sars" N.Atlantic Deep-Sea Exped., 1910, 2(1): 37 pp., 23 text figs.

Coxliella ampla

Jörgensen, E., 1924
Mediterranean Tintinnidae. Rept. Danish Oceanogr. Exped. 1908-10 to the Mediterranean and adjacent seas, Vol.II, Biol. J 3, 110 pp., 114 text figs.

Coxliella ampla

Kofoid, C. A. and A. S. Campbell, 1929
A conspectus of the marine and fresh-water Ciliata belonging to the suborder Tintinnoinea, with descriptions of new species principally from the Agassiz expedition to the eastern tropical Pacific, 1904-1905. Univ. Calif. Publ. Zool. 34:1-403, 697 text figs.

Coxliella annulata

Balech, E., 1959.
Tintinnoinea del Mediterraneo. Trav. Sta. Zool., Villefranche-sur-Mer, 18(17): Reprinted from: Trab. Inst. Español Oceanogr., Madrid, 19 Sept. 1959(28):88 pp., 22 pls.

Coxliella annulata

Delegazione Italiana della Commissione Internazionale per l'Esplorazione Scientifica del Mediterraneo, 1941
Note sul plancton della Laguna veneta. [Memoria CCLXXXIX], Arch. di Ocean. e Limn. Anno I, Fasc. I, 1941 XIX: 31-57 pp.

Coxliella annulata

Jörgensen, E., 1924
Mediterranean Tintinnidae. Rept. Danish Oceanogr. Exped. 1908-10 to the Mediterranean and adjacent seas, Vol.II, Biol. J 3, 110 pp., 114 text figs.

Coxliella annulata

Kofoid, C. A. and A. S. Campbell, 1929
A conspectus of the marine and fresh-water Ciliata belonging to the suborder Tintinnoinea, with descriptions of new species principally from the Agassiz expedition to the eastern tropical Pacific, 1904-1905. Univ. Calif. Publ. Zool. 34:1-403, 697 text figs.

Coxliella annulata

Rampi, L., 1939
Primo contributo alla conoscenza dei tintinnoidi do Mare Ligure. Atti della Soc. Ital. di Sci. Nat. 78:67-81, 58 text figs.

Coxliella bolivari

Duran, M., 1957.
Nota sobre algunos tintinoineas del plancton de Puerto Rico. Invest. Pesqueria, Barcelona, 8: 97-120.

Coxliella sp. (C. Calyptra?)

Balech, Enrique, 1971
Microplancton del Atlantico ecuatorial oeste (Equalant 1) Publ. Serv. Hidrograf. Naval, Argentina H. 654: 103 pp., 122 figs.

Coxliella calyptra

Kofoid, C. A. and A. S. Campbell, 1929
A conspectus of the marine and fresh-water Ciliata belonging to the suborder Tintinnoinea, with descriptions of new species principally from the Agassiz expedition to the eastern tropical Pacific, 1904-1905. Univ. Calif. Publ. Zool. 34:1-403, 697 text figs.

Coxliella cymatiocoides n.sp.

Kofoid, C. A. and A. S. Campbell, 1929
A conspectus of the marine and fresh-water Ciliata belonging to the suborder Tintinnoinea, with descriptions of new species principally from the Agassiz expedition to the eastern tropical Pacific, 1904-1905. Univ. Calif. Publ. Zool. 34:1-403, 697 text figs.

Coxliella decipiens n.sp.

Jörgensen, E., 1924
Mediterranean Tintinnidae. Rept. Danish Oceanogr. Exped. 1908-10 to the Mediterranean and adjacent seas, Vol.II, Biol. J 3, 110 pp., 114 text figs.

Coxliella decipiens

Kofoid, C. A. and A. S. Campbell, 1929
A conspectus of the marine and fresh-water Ciliata belonging to the suborder Tintinnoinea, with descriptions of new species principally from the Agassiz expedition to the eastern tropical Pacific, 1904-1905. Univ. Calif. Publ. Zool. 34:1-403, 697 text figs.

Coxliella decipiens

Komarovsky, B., 1959
The Tintinnina of the Gulf of Eylath (Aqaba). Contrib. Knowl. Red Sea, Bull. Sea Fish. Res. Sta.Haifa, Israel 21 (14):1-40.

Coxliella decipiens

Massutí Algamora, M., 1949
Estudio de diez y seis muestras de plancton del Golfo de Nápoles. Publ. Inst. Biol. Appl. 5:85-94, 1 fold-in table.

Coxliella declivis

Balech, Enrique, 1962
Tintinnoinea y Dinoflagellata del Pacifico segun material de las Expediciones NORPAC y DOWNWIND del Instituto Scripps de Oceanografia. Revista, Mus. Argentino Ciencias Nat. "Bernardino Rivadavia", Ciencias Zool., 7(1):1-253.

Coxliella declivis

Campbell, A. S., 1942
The oceanic Tintinnoina of the plankton gathered during the last cruise of the CARNEGIE. Sci. Res. Cruise VII of the Carnegie, 1928-1929, Biol. 2:163 pp., 128 figs.

Coxliella declivis

Kofoid, C.A. and A.S. Campbell, 1939
Reports on the scientific results of the expedition to the eastern tropical Pacific in charge of Alexander Agassiz, by the U.S. Fish Commission steamer "Albatross" from October 1904 to March 1905-----XXXVII. The Ciliata: The Tintinnoimea. Bull.Mus. Comp.Zool. 84: 473 pp., 36 pls.

Coxliella declivis n.sp.

Kofoid, C. A. and A. S. Campbell, 1929
A conspectus of the marine and fresh-water Ciliata belonging to the suborder Tintinnoinea, with descriptions of new species principally from the Agassiz expedition to the eastern tropical Pacific, 1904-1905. Univ. Calif. Publ. Zool. 34:1-403, 697 text figs.

Coxliella fabricatrix n.sp.

Kofoid, C.A. and A.S. Campbell, 1939
Reports on the scientific results of the expedition to the eastern tropical Pacific in charge of Alexander Agassiz, by the U.S. Fish Commission steamer "Albatross" from October 1904 to March 1905-----XXXVII. The Ciliata: The Tintinnoimea. Bull.Mus. Comp.Zool. 84: 473 pp., 36 pls.

Coxliella fasciata

Balech, Enrique, 1971
Microplancton del Atlantico ecuatorial oeste (Equalant 1)
Publ. Serv. Hidrograf. Naval, Argentina H. 654: 103 pp., 122 figs.

Coxliella fasciata

Balech, Enrique, 1962
Tintinnoinea y Dinoflagellata del Pacifico segun material de las Expediciones NORPAC y DOWNWIND del Instituto Scripps de Oceanografia. Revista, Mus. Argentino Ciencias Nat. "Bernardino Rivadavia", Ciencias Zool., 7(1):1-253.

Coxliella fasciata

Campbell, A. S., 1942
The oceanic Tintinnoina of the plankton gathered during the last cruise of the CARNEGIE. Sci. Res. Cruise VII of the Carnegie, 1928-1929, Biol. 2:163 pp., 128 figs.

Coxliella fasciata

Jörgensen, E., 1924
Mediterranean Tintinnidae. Rept. Danish Oceanogr. Exped. 1908-10 to the Mediterranean and adjacent seas, Vol.II, Biol. J 3, 110 pp., 114 text figs.

Coxliella fasciata

Kofoid, C.A. and A.S. Campbell, 1939
Reports on the scientific results of the expedition to the eastern tropical Pacific in charge of Alexander Agassiz, by the U.S. Fish Commission steamer "Albatross" from October 1904 to March 1905-----XXXVII. The Ciliata: The Tintinnoimea. Bull.Mus. Comp.Zool. 84: 473 pp., 36 pls.

Coxliella fasciata

Kofoid, C. A. and A. S. Campbell, 1929
A conspectus of the marine and fresh-water Ciliata belonging to the suborder Tintinnoinea, with descriptions of new species principally from the Agassiz expedition to the eastern tropical Pacific, 1904-1905. Univ. Calif. Publ. Zool. 34:1-403, 697 text figs.

Coxliella fasciata

Komarovsky, B., 1959
The Tintinnina of the Gulf of Eylath (Aqaba). Contrib. Knowl. Red Sea, Bull. Sea Fish. Res. Sta.Haifa, Israel 21 (14):1-40.

Coxliella fasciata

Laackmann, H., 1910
Die Tintinnodeen der Deutschen Südpolar Expedition 1901-1903. Deutsch. Südpolar Exped. 11, Zool. 3(4):341-496, pls. 33-51.

Coxliella fasciata

Rampi, L., 1939
Primo contributo alla conoscenza dei tintinnoidi do Maro Ligure. Atti della Soc. Ital. di Sci. Nat. 78:67-81, 58 text figs.

Coxliella frigida

Balech, E., 1947
Contribution al conocimiento del plancton antartico. Plankton del Mar de Bellinghausen. Physis 20:75-91, 76 figs. on 8 pls.

Coxliella frigida

Balech, Enrique, 1947
Contribucion al conocimiento del Plancton antartico Plancton del Mar de Bellinghausen. Physis 22(56):75-91, 7 pls. with 76 figs.

Coxliella frigida

Kofoid, C. A. and A. S. Campbell, 1929
A conspectus of the marine and fresh-water Ciliata belonging to the suborder Tintinnoinea, with descriptions of new species principally from the Agassiz expedition to the eastern tropical Pacific, 1904-1905. Univ. Calif. Publ. Zool. 34:1-403, 697 text figs.

Coxliella frigida

Laackmann, H., 1910
Die Tintinnodeen der Deutschen Südpolar Expedition 1901-1903. Deutsch. Südpolar Exped. 11, Zool. 3(4):341-496, pls. 33-51.

Coxliella helix

Campbell, A. S., 1942
The oceanic Tintinnoina of the plankton gathered during the last cruise of the CARNEGIE. Sci. Res. Cruise VII of the Carnegie, 1928-1929, Biol. 2:163 pp., 128 figs.

Coxliella helix

Duran, M., 1951
Contribucion al estudio de los tintinidos del plancton de los costas de Castellon. (Mediterranes occidental). Publ. Inst. Biol. Aplic., Barcelone, 8: 101-120, 29 text figs.

Coxliella helix

Jörgensen, E., 1924
Mediterranean Tintinnidae. Rept. Danish Oceanogr. Exped. 1908-10 to the Mediterranean and adjacent seas, Vol.II, Biol. J 3, 110 pp., 114 text figs.

Coxliella helix

Kofoid, C. A. and A. S. Campbell, 1929
A conspectus of the marine and fresh-water Ciliata belonging to the suborder Tintinnoinea, with descriptions of new species principally from the Agassiz expedition to the eastern tropical Pacific, 1904-1905. Univ. Calif. Publ. Zool. 34:1-403, 697 text figs.

Coxliella helix

Massutí Algamora, M., 1949
Estudio de diez y seis muestras de plancton del Golfo de Nápoles. Publ. Inst. Biol. Appl. 5:85-94, 1 fold-in table.

Coxliella helix

Schulz, H., 1965.
Die Tintinnoinea des Elbe-Aestuars. Arch. Fischereiwiss., 15(3):216-225.

Coxliella intermedia

Balech, Enrique, 1958.
Plancton de la Campana Antartica Argentina, 1954-1955. Physis, 21(60):75-108.

Coxliella intermedia

Kofoid, C. A. and A. S. Campbell, 1929
A conspectus of the marine and fresh-water Ciliata belonging to the suborder Tintinnoinea, with descriptions of new species principally from the Agassiz expedition to the eastern tropical Pacific, 1904-1905. Univ. Calif. Publ. Zool. 34:1-403, 697 text figs.

Coxliella intermedia

Laackmann, H., 1910
Die Tintinnodeen der Deutschen Südpolar Expedition 1901-1903. Deutsch. Südpolar Exped. 11, Zool. 3(4):341-496, pls. 33-51.

Coxliella laciniosa

Balech, Enrique, 1962
Tintinnoinea y Dinoflagellata del Pacifico segun material de las Expediciones NORPAC y DOWNWIND del Instituto Scripps de Oceanografia. Revista, Mus. Argentino Ciencias Nat. "Bernardino Rivadavia", Ciencias Zool., 7(1):1-253.

Coxliella laciniosa

Campbell, A. S., 1942
The oceanic Tintinnoina of the plankton gathered during the last cruise of the CARNEGIE. Sci. Res. Cruise VII of the Carnegie, 1928-1929, Biol. 2:163 pp., 128 figs.

Coxliella laciniosa

Duran, M., 1951
Contribucion al estudio de los tintinidos del plancton de los costas de Castellon. (Mediterranes occidental). Publ. Inst. Biol. Aplic., Barcelone, 8: 101-120, 29 text figs.

Coxliella laciniosa

Gaarder, K.R., 1946
Tintinnoidea. Rep.Sci.Res. "Michael Sars" N.Atlantic Deep-Sea Exped., 1910, 2(1): 37 pp., 23 text figs.

Coxliella laciniosa

Jörgensen, E., 1924
Mediterranean Tintinnidae. Rept. Danish Oceanogr. Exped. 1908-10 to the Mediterranean and adjacent seas, Vol.II, Biol. J 3, 110 pp., 114 text figs.

Coxliella laciniosa

Kofoid, C.A. and A.S. Campbell, 1939
Reports on the scientific results of the expedition to the eastern tropical Pacific in charge of Alexander Agassiz, by the U.S. Fish Commission steamer "Albatross" from October 1904 to March 1905-----XXXVII. The Ciliata: The Tintinnoimea. Bull.Mus. Comp.Zool. 84: 473 pp., 36 pls.

Coxliella laciniosa

Kofoid, C. A. and A. S. Campbell, 1929
A conspectus of the marine and fresh-water Ciliata belonging to the suborder Tintinnoinea, with descriptions of new species principally from the Agassiz expedition to the eastern tropical Pacific, 1904-1905. Univ. Calif. Publ. Zool. 34:1-403, 697 text figs.

Coxliella laciniosa
Laackmann, H., 1910
Die Tintinnodeen der Deutschen Südpolar Expedition 1901-1903. Deutsch. Südpolar Exped. 11, Zool. 3(4):341-496, pls. 33-51.

Coxliella laciniosa
Marshall, S. M., 1934
The silicoflagellata and tintinnoinea. Great Barrier Reef Exped., 1928-29. Sci. Repts. 4(15):623-664, 43 text figs.

Coxliella laciniosa
Marshall, S. M., 1933
The production of microplankton in the Great Barrier Reef Region. Brit. Mus. (N.H.) Great Barrier Reef Exped. 1928-29, Sci. Repts. II(5):111-157, 14 text figs.

Coxliella laciniosa
Rampi, L., 1948
Sur quelques Tintinnides (Infusoires louques) du Pacifique subtropical (Recoltes Alain Gerbault). Bull. l'Inst. Ocean., Monaco, No. 938, 4 pp.

Coxliella laciniosa
Rampi, L., 1939
Primo contributo alla conoscenza dei tintinnoidi do Maro Ligure. Atti della Soc. Ital. di Sci. Nat. 78:67-81, 58 text figs.

Coxliella laciniosa
Sousa e Silva, E. de, 1950.
Les tintinnides de la Baie de Cascais. Bull. Inst. Ocean., Monaco, No. 979:28 pp., 4 pls

Coxliella longa
Balech, Enrique, 1962
Tintinnoinea y Dinoflagellata del Pacifico segun material de las Expediciones NORPAC y DOWNWIND del Instituto Scripps de Oceanografia. Revista, Mus. Argentino Ciencias Nat. "Bernardino Rivadavia", Ciencias Zool., 7(1):1-253.

Coxliella longa
Campbell, A. S., 1942
The oceanic Tintinnoina of the plankton gathered during the last cruise of the CARNEGIE. Sci. Res. Cruise VII of the Carnegie, 1928-1929, Biol. 2:163 pp., 128 figs.

Coxliella longa
Cosper, T.C., 1972.
The identification of tintinnids (Protozoa: Ciliata: Tintinnida) of the St. Andrew Bay system, Florida. Bull. mar. Sci. Miami 22(2):391-418.

Coxliella longa
Kofoid, C.A. and A.S. Campbell, 1939
Reports on the scientific results of the expedition to the eastern tropical Pacific in charge of Alexander Agassiz, by the U.S. Fish Commission steamer "Albatross" from October 1904 to March 1905-----XXXVII. The Ciliata: The Tintinnoimea. Bull. Mus. Comp. Zool. 84: 473 pp., 36 pls.

Coxliella longa
Kofoid, C. A. and A. S. Campbell, 1929
A conspectus of the marine and freshwater Ciliata belonging to the suborder Tintinnoinea, with descriptions of new species principally from the Agassiz expedition to the eastern tropical Pacific, 1904-1905. Univ. Calif. Publ. Zool. 34:1-403, 697 text figs.

Coxliella mariana
Balech, Enrique, 1971
Microplancton del Atlantico ecuatorial oeste (Equalant 1) Publ. Serv. Hidrograf. Naval, Argentina H. 654: 103 pp., 122 figs.

Coxliella massuti
Balech, Enrique, 1971
Microplancton del Atlantico ecuatorial oeste (Equalant 1) Publ. Serv. Hidrograf. Naval, Argentina H. 654: 103 pp., 122 figs.

Coxliella massuti n.sp.
Duran, M., 1953.
Contribución al estudio de los Tintinnidos del plancton de las costas de Castellón (Mediterráneo occidental). Nota II. Publ. Inst. Biol. Aplic. 12:79-95, 23 textfigs.

Coxliella meunieri n.sp.
Kofoid, C. A. and A. S. Campbell, 1929
A conspectus of the marine and freshwater Ciliata belonging to the suborder Tintinnoinea, with descriptions of new species principally from the Agassiz expedition to the eastern tropical Pacific, 1904-1905. Univ. Calif. Publ. Zool. 34:1-403, 697 text figs.

Coxliella meunieri minor n.var
Komarovsky, B., 1959
The Tintinnina of the Gulf of Eylath (Aqaba). Contrib. Knowl. Red Sea, Bull. Sea Fish. Res. Sta.Haifa, Israel 21 (14):1-40.

Coxliella minor
Balech, Enrique, 1958.
Plancton de la Campaña Antartico Argentina, 1954-1955. Physis, 21(60):75-108.

Coxliella minor
Balech, E., 1947
Contribution al conocimiento del plancton antarctico. Plankton del Mar de Bellinghausen. Physis 20:75-91, 76 figs. on 8 pls.

Coxliella minor
Balech, Enrique, 1947
Contribucion al conocimiento del Plancton antartico Plancton del Mar de Bellingshausen. Physis 22(56):75-91, 7 pls. with 76 figs.

Coxliella minor
de Sousa e Silva, E., 1956.
Contribution à l'étude du microplancton de Dakar et des regions maritimes voisines. Bull. I.F.A.N., 8(2):335-371, 7 pls.

Coxliella minor
Kofoid, C. A. and A. S. Campbell, 1929
A conspectus of the marine and freshwater Ciliata belonging to the suborder Tintinnoinea, with descriptions of new species principally from the Agassiz expedition to the eastern tropical Pacific, 1904-1905. Univ. Calif. Publ. Zool. 34:1-403, 697 text figs.

Coxliella minor
Laackmann, H., 1910
Die Tintinnodeen der Deutschen Südpolar Expedition 1901-1903. Deutsch. Südpolar Exped. 11, Zool. 3(4):341-496, pls. 33-51.

Coxliella nana n. sp
Balech, Enrique, 1968.
Algunas especies nuevas o interesantes do tintinnidos del Golfo de Mexico y Caribe. Revta Mus. argent. Cienc. nat. Bernardino Rivadavia Inst. nac. Hidrobiol. 2(5):165-197.

Coxliella oviformis
Kofoid, C. A. and A. S. Campbell, 1929
A conspectus of the marine and freshwater Ciliata belonging to the suborder Tintinnoinea, with descriptions of new species principally from the Agassiz expedition to the eastern tropical Pacific, 1904-1905. Univ. Calif. Publ. Zool. 34:1-403, 697 text figs.

Coxliella pelagica
Balech, Enrique, 1971
Microplancton del Atlantico ecuatorial oeste (Equalant 1) Publ. Serv. Hidrograf. Naval, Argentina H. 654: 103 pp., 122 figs.

Coxliella pelagica
Kofoid, C.A. and A.S. Campbell, 1939
Reports on the scientific results of the expedition to the eastern tropical Pacific in charge of Alexander Agassiz, by the U.S. Fish Commission steamer "Albatross" from October 1904 to March 1905-----XXXVII. The Ciliata: The Tintinnoimea. Bull. Mus. Comp. Zool. 84: 473 pp., 36 pls.

Coxliella pelagica n.sp.
Kofoid, C. A. and A. S. Campbell, 1929
A conspectus of the marine and freshwater Ciliata belonging to the suborder Tintinnoinea, with descriptions of new species principally from the Agassiz expedition to the eastern tropical Pacific, 1904-1905. Univ. Calif. Publ. Zool. 34:1-403, 697 text figs.

Coxliella pseudannulata
Balech, Enrique, 1971
Microplancton del Atlantico ecuatorial oeste (Equalant 1) Publ. Serv. Hidrograf. Naval, Argentina H. 654: 103 pp., 122 figs.

Coxliella pseudannulata
Campbell, A. S., 1942
The oceanic Tintinnoina of the plankton gathered during the last cruise of the CARNEGIE. Sci. Res. Cruise VII of the Carnegie, 1928-1929, Biol. 2:163 pp., 128 figs.

Coxliella pseudannulata
Jörgensen, E., 1924
Mediterranean Tintinnidae. Rept. Danish Oceanogr. Exped. 1908-10 to the Mediterranean and adjacent seas, Vol.II, Biol. J 3, 110 pp., 114 text figs.

Coxliella pseudannulata
Kofoid, C. A. and A. S. Campbell, 1929
A conspectus of the marine and freshwater Ciliata belonging to the suborder Tintinnoinea, with descriptions of new species principally from the Agassiz expedition to the eastern tropical Pacific, 1904-1905. Univ. Calif. Publ. Zool. 34:1-403, 697 text figs.

Coxliella scalaria
Laackmann, H., 1910
Die Tintinnodeen der Deutschen Südpolar Expedition 1901-1903. Deutsch. Südpolar Exped. 11, Zool. 3(4):341-496, pls. 33-51.

Coxliella tubularis
Kofoid, C. A. and A. S. Campbell, 1929
A conspectus of the marine and freshwater Ciliata belonging to the suborder Tintinnoinea, with descriptions of new species principally from the Agassiz expedition to the eastern tropical Pacific, 1904-1905. Univ. Calif. Publ. Zool. 34:1-403, 697 text figs.

Craterella n.gen.
Kofoid, C. A. and A. S. Campbell, 1929
A conspectus of the marine and freshwater Ciliata belonging to the suborder Tintinnoinea, with descriptions of new species principally from the Agassiz expedition to the eastern tropical Pacific, 1904-1905. Univ. Calif. Publ. Zool. 34:1-403, 697 text figs.

Craterella acuta
Gaarder, K.R., 1946
Tintinnoidea. Rep.Sci.Res. "Michael Sars" N.Atlantic Deep-Sea Exped., 1910, 2(1): 37 pp., 23 text figs.

Craterella acuta n.sp.
Kofoid, C. A. and A. S. Campbell, 1929
A conspectus of the marine and freshwater Ciliata belonging to the suborder Tintinnoinea, with descriptions of new species principally from the Agassiz expedition to the eastern tropical Pacific, 1904-1905. Univ. Calif. Publ. Zool. 34:1-403, 697 text figs.

Craterella aperta n.sp.

Marshall, S. M., 1934
The silicoflagellata and tintinnoinea.
Great Barrier Reef Exped., 1928-29. Sci. Repts. 4(15):623-664, 43 text figs.

Craterella armilla

Campbell, A. S., 1942
The oceanic Tintinnoina of the plankton gathered during the last cruise of the CARNEGIE. Sci. Res. Cruise VII of the Carnegie, 1928-1929, Biol. 2:163 pp., 128 figs.

Craterella armilla

de Sousa e Silva, E., 1956.
Contribution à l'étude du microplancton de Dakar et des régions maritimes voisines.
Bull. I.F.A.N., 8(2):335-371, 7 pls.

Craterella armilla

Gaarder, K.R., 1946
Tintinnoidea. Rep.Sci.Res."Michael Sars" N.Atlantic Deep-Sea Exped., 1910, 2(1): 37 pp., 23 text figs.

Craterella armilla

Kofoid, C.A. and A.S. Campbell, 1939
Reports on the scientific results of the expedition to the eastern tropical Pacific in charge of Alexander Agassiz, by the U.S. Fish Commission steamer "Albatross" from October 1904 to March 1905-----XXXVII. The Ciliata: The Tintinnoimea. Bull.Mus. Comp.Zool. 84: 473 pp., 36 pls.

Craterella armilla

Kofoid, C. A. and A. S. Campbell, 1929
A conspectus of the marine and fresh-water Ciliata belonging to the suborder Tintinnoinea, with descriptions of new species principally from the Agassiz expedition to the eastern tropical Pacific, 1904-1905. Univ. Calif. Publ. Zool. 34:1-403, 697 text figs.

Craterella obscura

Duran, E., 1957.
Nota sobre algunos tintinnoineas del plancton de Puerto Rico. Invest. Pesqueria, Barcelona, 8: 97-120.

Craterella obscura

Gaarder, K.R., 1946
Tintinnoidea. Rep.Sci.Res."Michael Sars" N.Atlantic Deep-Sea Exped., 1910, 2(1): 37 pp., 23 text figs.

Craterella obscura

Kofoid, C. A. and A. S. Campbell, 1929
A conspectus of the marine and fresh-water Ciliata belonging to the suborder Tintinnoinea, with descriptions of new species principally from the Agassiz expedition to the eastern tropical Pacific, 1904-1905. Univ. Calif. Publ. Zool. 34:1-403, 697 text figs.

Cratella obscura

Komarovsky, B., 1959
The Tintinnina of the Gulf of Eylath (Aqaba). Contrib. Knowl. Red Sea, Bull. Sea Fish. Res. Sta.Haifa, Israel 21 (14):1-40.

Craterella oxyura

Kofoid, C. A. and A. S. Campbell, 1929
A conspectus of the marine and fresh-water Ciliata belonging to the suborder Tintinnoinea, with descriptions of new species principally from the Agassiz expedition to the eastern tropical Pacific, 1904-1905. Univ. Calif. Publ. Zool. 34:1-403, 697 text figs.

Craterella perminuta n. sp.

*Hada, Yoshina, 1970.   (1970)

The protozoan plankton of the Antarctic and Subantarctic seas.
Scient. Repts, Japan. Antarct. Res. Exped., (E)31.51 pp.

Craterella protuberans

Kofoid, C.A. and A.S. Campbell, 1939
Reports on the scientific results of the expedition to the eastern tropical Pacific in charge of Alexander Agassiz, by the U.S. Fish Commission steamer "Albatross" from October 1904 to March 1905-----XXXVII. The Ciliata: The Tintinnoimea. Bull.Mus. Comp.Zool. 84: 473 pp., 36 pls.

Craterella protuberans n. sp.

Kofoid, C. A. and A. S. Campbell, 1929
A conspectus of the marine and fresh-water Ciliata belonging to the suborder Tintinnoinea, with descriptions of new species principally from the Agassiz expedition to the eastern tropical Pacific, 1904-1905. Univ. Calif. Publ. Zool. 34:1-403, 697 text figs.

Craterella torulata

Kofoid, C. A. and A. S. Campbell, 1929
A conspectus of the marine and fresh-water Ciliata belonging to the suborder Tintinnoinea, with descriptions of new species principally from the Agassiz expedition to the eastern tropical Pacific, 1904-1905. Univ. Calif. Publ. Zool. 34:1-403, 697 text figs.

Craterella urceolata

Campbell, A. S., 1942
The oceanic Tintinnoina of the plankton gathered during the last cruise of the CARNEGIE. Sci. Res. Cruise VII of the Carnegie, 1928-1929, Biol. 2:163 pp., 128 figs.

Craterella urceolata

Gaarder, K.R., 1946
Tintinnoidea. Rep.Sci.Res."Michael Sars" N.Atlantic Deep-Sea Exped., 1910, 2(1): 37 pp., 23 text figs.

Craterella urceolata

Kofoid, C.A. and A.S. Campbell, 1939
Reports on the scientific results of the expedition to the eastern tropical Pacific in charge of Alexander Agassiz, by the U.S. Fish Commission steamer "Albatross" from October 1904 to March 1905-----XXXVII. The Ciliata: The Tintinnoimea. Bull.Mus. Comp.Zool. 84: 473 pp., 36 pls.

Craterella urceolata

Kofoid, C. A. and A. S. Campbell, 1929
A conspectus of the marine and fresh-water Ciliata belonging to the suborder Tintinnoinea, with descriptions of new species principally from the Agassiz expedition to the eastern tropical Pacific, 1904-1905. Univ. Calif. Publ. Zool. 34:1-403, 697 text figs.

Cratella urceolata

Komarovsky, B., 1959
The Tintinnina of the Gulf of Eylath (Aqaba). Contrib. Knowl. Red Sea, Bull. Sea Fish. Res. Sta.Haifa, Israel 21 (14):1-40.

Cricundella n. gen.

Kofoid, C. A. and A. S. Campbell, 1929
A conspectus of the marine and fresh-water Ciliata belonging to the suborder Tintinnoinea, with descriptions of new species principally from the Agassiz expedition to the eastern tropical Pacific, 1904-1905. Univ. Calif. Publ. Zool. 34:1-403, 697 text figs.

Cricundella quadridivisa

Campbell, A. S., 1942
The oceanic Tintinnoina of the plankton gathered during the last cruise of the CARNEGIE. Sci. Res. Cruise VII of the Carnegie, 1928-1929, Biol. 2:163 pp., 128 figs.

Cricundella quadricincta

Kofoid, C.A. and A.S. Campbell, 1939
Reports on the scientific results of the expedition to the eastern tropical Pacific in charge of Alexander Agassiz, by the U.S. Fish Commission steamer "Albatross" from October 1904 to March 1905-----XXXVII. The Ciliata: The Tintinnoimea. Bull.Mus. Comp.Zool. 84: 473 pp., 36 pls.

Cricundella quadricincta n.sp.

Kofoid, C. A. and A. S. Campbell, 1929
A conspectus of the marine and fresh-water Ciliata belonging to the suborder Tintinnoinea, with descriptions of new species principally from the Agassiz expedition to the eastern tropical Pacific, 1904-1905. Univ. Calif. Publ. Zool. 34:1-403, 697 text figs.

Cricundella quadridivisa

Kofoid, C.A. and A.S. Campbell, 1939
Reports on the scientific results of the expedition to the eastern tropical Pacific in charge of Alexander Agassiz, by the U.S. Fish Commission steamer "Albatross" from October 1904 to March 1905-----XXXVII. The Ciliata: The Tintinnoimea. Bull.Mus. Comp.Zool. 84: 473 pp., 36 pls.

Cricundella quadridivisa n.sp.

Kofoid, C. A. and A. S. Campbell, 1929
A conspectus of the marine and fresh-water Ciliata belonging to the suborder Tintinnoinea, with descriptions of new species principally from the Agassiz expedition to the eastern tropical Pacific, 1904-1905. Univ. Calif. Publ. Zool. 34:1-403, 697 text figs.

Cricundella tridivisa

Kofoid, C. A. and A. S. Campbell, 1929
A conspectus of the marine and fresh-water Ciliata belonging to the suborder Tintinnoinea, with descriptions of new species principally from the Agassiz expedition to the eastern tropical Pacific, 1904-1905. Univ. Calif. Publ. Zool. 34:1-403, 697 text figs.

Cymatocylis spp.

Argentina, Secretaria de Marina, Servicio de Hidrografia Naval, 1962.
Plancton de las campanas oceanograficas DRAKE I y II.
Publico. H. 627:57.

Cymatocylis spp.

Balech, Enrique, 1958.
Plancton de la Campaña Antartico Argentina, 1954-1955.
Physis, 21(60):75-108.

Cymatocylis n.gen.

Laackmann, H., 1910
Die Tintinnodeen der Deutschen Südpolar Expedition 1901-1903. Deutsch. Südpolar Exped. 11, Zool. 3(4):341-496, pls. 33-51.

Cymatocylis affinis

Kofoid, C. A. and A. S. Campbell, 1929
A conspectus of the marine and fresh-water Ciliata belonging to the suborder Tintinnoinea, with descriptions of new species principally from the Agassiz expedition to the eastern tropical Pacific, 1904-1905. Univ. Calif. Publ. Zool. 34:1-403, 697 text figs.

Cymatocylis affinis n.sp.

Laackmann, H., 1910
Die Tintinnodeen der Deutschen Südpolar Expedition 1901-1903. Deutsch. Südpolar Exped. 11, Zool. 3(4):341-496, pls. 33-51.

Cymatocylis affinoides nom. nov.

Kofoid, C. A. and A. S. Campbell, 1929
A conspectus of the marine and fresh-water Ciliata belonging to the suborder Tintinnoinea, with descriptions of new species principally from the Agassiz expedition to the eastern tropical Pacific, 1904-1905. Univ. Calif. Publ. Zool. 34:1-403, 697 text figs.

Cymatocylis antarctica

Kofoid, C. A. and A. S. Campbell, 1929
A conspectus of the marine and fresh-water Ciliata belonging to the suborder Tintinnoinea, with descriptions of new species principally from the Agassiz expedition to the eastern tropical Pacific, 1904-1905. Univ. Calif. Publ. Zool. 34:1-403, 697 text figs.

Cymatocylis brevicandata

*Hada, Yoshina, 1970.   (1970)

The protozoan plankton of the Antarctic and Subantarctic seas.
Scient. Repts, Japan. Antarct. Res. Exped., (E)31.51 pp.

Cymatocylis brevicaudata

Kofoid, C. A. and A. S. Campbell, 1929
A conspectus of the marine and freshwater Ciliata belonging to the suborder Tintinnoinea, with descriptions of new species principally from the Agassiz expedition to the eastern tropical Pacific, 1904-1905. Univ. Calif. Publ. Zool. 34:1-403, 697 text figs.

Cymatocylis calycifornis

*Hada, Yoshina, 1970. (1970)
The protozoan plankton of the Antarctic and Subantarctic seas.
Scient. Repts, Japan. Antarct. Res. Exped., (E) 31:51 pp.

Cymatocylis calyciformis

Kofoid, C. A. and A. S. Campbell, 1929
A conspectus of the marine and freshwater Ciliata belonging to the suborder Tintinnoinea, with descriptions of new species principally from the Agassiz expedition to the eastern tropical Pacific, 1904-1905. Univ. Calif. Publ. Zool. 34:1-403, 697 text figs.

Cymatocylis calyciformis

Laackmann, H., 1910
Die Tintinnodeen der Deutschen Südpolar Expedition 1901-1903. Deutsch. Südpolar Exped. 11, Zool. 3(4):341-496, pls. 33-51.

Cymatocylis calycina

Kofoid, C. A. and A. S. Campbell, 1929
A conspectus of the marine and freshwater Ciliata belonging to the suborder Tintinnoinea, with descriptions of new species principally from the Agassiz expedition to the eastern tropical Pacific, 1904-1905. Univ. Calif. Publ. Zool. 34:1-403, 697 text figs.

Cymatocylis calyx n.sp.

Kofoid, C. A. and A. S. Campbell, 1929
A conspectus of the marine and freshwater Ciliata belonging to the suborder Tintinnoinea, with descriptions of new species principally from the Agassiz expedition to the eastern tropical Pacific, 1904-1905. Univ. Calif. Publ. Zool. 34:1-403, 697 text figs.

Cymatocylis conica

Kofoid, C. A. and A. S. Campbell, 1929
A conspectus of the marine and freshwater Ciliata belonging to the suborder Tintinnoinea, with descriptions of new species principally from the Agassiz expedition to the eastern tropical Pacific, 1904-1905. Univ. Calif. Publ. Zool. 34:1-403, 697 text figs.

Cymatocylis contracta n.sp.

Kofoid, C. A. and A. S. Campbell, 1929
A conspectus of the marine and freshwater Ciliata belonging to the suborder Tintinnoinea, with descriptions of new species principally from the Agassiz expedition to the eastern tropical Pacific, 1904-1905. Univ. Calif. Publ. Zool. 34:1-403, 697 text figs.

Cymatocylis convallaria

Balech, E., 1947
Contribution al conocimiento del plancton antarctico. Plankton del Mar de Bellinghausen. Physis 20:75-91, 76 figs. on 8 pls.

Cymatocylis convallaria

Balech, Enrique, 1947
Contribucion al conocimiento del Plancton antartico Plancton del Mar de Bellinghausen. Physis 22(56):75-91, 7 pls. with 76 figs.

Cymatocylis convallaria

Kofoid, C. A. and A. S. Campbell, 1929
A conspectus of the marine and freshwater Ciliata belonging to the suborder Tintinnoinea, with descriptions of new species principally from the Agassiz expedition to the eastern tropical Pacific, 1904-1905. Univ. Calif. Publ. Zool. 34:1-403, 697 text figs.

Cymatocylis convallaria n.sp.

Laackmann, H., 1910
Die Tintinnodeen der Deutschen Südpolar Expedition 1901-1903. Deutsch. Südpolar Exped. 11, Zool. 3(4):341-496, pls. 33-51.

Cymatocylis crassa nom.nov.

Kofoid, C. A. and A. S. Campbell, 1929
A conspectus of the marine and freshwater Ciliata belonging to the suborder Tintinnoinea, with descriptions of new species principally from the Agassiz expedition to the eastern tropical Pacific, 1904-1905. Univ. Calif. Publ. Zool. 34:1-403, 697 text figs.

Cymatocylis cristallina

*Hada, Yoshina, 1970. (1970)
The protozoan plankton of the Antarctic and Subantarctic seas.
Scient. Repts, Japan. Antarct. Res. Exped., (E) 31:51 pp.

Cymatocylis cristallina

Kofoid, C. A. and A. S. Campbell, 1929
A conspectus of the marine and freshwater Ciliata belonging to the suborder Tintinnoinea, with descriptions of new species principally from the Agassiz expedition to the eastern tropical Pacific, 1904-1905. Univ. Calif. Publ. Zool. 34:1-403, 697 text figs.

Cymatocylis crystallina n.sp.

Laackmann, H., 1910
Die Tintinnodeen der Deutschen Südpolar Expedition 1901-1903. Deutsch. Südpolar Exped. 11, Zool. 3(4):341-496, pls. 33-51.

Cymatocylis cucullus nom.nov.

Kofoid, C. A. and A. S. Campbell, 1929
A conspectus of the marine and freshwater Ciliata belonging to the suborder Tintinnoinea, with descriptions of new species principally from the Agassiz expedition to the eastern tropical Pacific, 1904-1905. Univ. Calif. Publ. Zool. 34:1-403, 697 text figs.

Cymatocylis cylindrella nom.nov.

Kofoid, C. A. and A. S. Campbell, 1929
A conspectus of the marine and freshwater Ciliata belonging to the suborder Tintinnoinea, with descriptions of new species principally from the Agassiz expedition to the eastern tropical Pacific, 1904-1905. Univ. Calif. Publ. Zool. 34:1-403, 697 text figs.

Cymatocylis cylindroides nom. nov.

Kofoid, C. A. and A. S. Campbell, 1929
A conspectus of the marine and freshwater Ciliata belonging to the suborder Tintinnoinea, with descriptions of new species principally from the Agassiz expedition to the eastern tropical Pacific, 1904-1905. Univ. Calif. Publ. Zool. 34:1-403, 697 text figs.

Cymatocylis cylindrus nom. nov.

Kofoid, C. A. and A. S. Campbell, 1929
A conspectus of the marine and freshwater Ciliata belonging to the suborder Tintinnoinea, with descriptions of new species principally from the Agassiz expedition to the eastern tropical Pacific, 1904-1905. Univ. Calif. Publ. Zool. 34:1-403, 697 text figs.

Cymatocylis digitabulum n.sp.

Kofoid, C. A. and A. S. Campbell, 1929
A conspectus of the marine and freshwater Ciliata belonging to the suborder Tintinnoinea, with descriptions of new species principally from the Agassiz expedition to the eastern tropical Pacific, 1904-1905. Univ. Calif. Publ. Zool. 34:1-403, 697 text figs.

Cymatocylis digitulus

Balech, E., 1947
Contribution al conocimiento del plancton antarctico. Plankton del Mar de Bellinghausen. Physis 20:75-91, 76 figs. on 8 pls.

Cymatocylis digitulus

Balech, Enrique, 1947
Contribucion al conocimiento del Plancton antartico Plancton del Mar de Bellinghausen. Physis 22(56):75-91, 7 pls. with 76 figs.

Cymatocylis digitulus n.sp.

Kofoid, C. A. and A. S. Campbell, 1929
A conspectus of the marine and freshwater Ciliata belonging to the suborder Tintinnoinea, with descriptions of new species principally from the Agassiz expedition to the eastern tropical Pacific, 1904-1905. Univ. Calif. Publ. Zool. 34:1-403, 697 text figs.

Cymatocylis digitulus n.sp.

Kofoid, C. A. and A. S. Campbell, 1929
A conspectus of the marine and freshwater Ciliata belonging to the suborder Tintinnoinea, with descriptions of new species principally from the Agassiz expedition to the eastern tropical Pacific, 1904-1905. Univ. Calif. Publ. Zool. 34:1-403, 697 text figs.

Cymatocylis diminuta n.sp.

Kofoid, C. A. and A. S. Campbell, 1929
A conspectus of the marine and freshwater Ciliata belonging to the suborder Tintinnoinea, with descriptions of new species principally from the Agassiz expedition to the eastern tropical Pacific, 1904-1905. Univ. Calif. Publ. Zool. 34:1-403, 697 text figs.

Cymatocylis drygalskii

Balech, E., 1947
Contribution al conocimiento del plancton antarctico. Plankton del Mar de Bellinghausen. Physis 20:75-91, 76 figs. on 8 pls.

Cymatocylis drygalskii

Balech, Enrique, 1947
Contribucion al conocimiento del Plancton antartico Plancton del Mar de Bellinghausen. Physis 22(56):75-91, 7 pls. with 76 figs.

Cymatocylis drygalskii

Kofoid, C. A. and A. S. Campbell, 1929
A conspectus of the marine and freshwater Ciliata belonging to the suborder Tintinnoinea, with descriptions of new species principally from the Agassiz expedition to the eastern tropical Pacific, 1904-1905. Univ. Calif. Publ. Zool. 34:1-403, 697 text figs.

Cymatocylis drygalskii

Laackmann, H., 1910
Die Tintinnodeen der Deutschen Südpolar Expedition 1901-1903. Deutsch. Südpolar Exped. 11, Zool. 3(4):341-496, pls. 33-51.

Cymatocylis ecaudata n.sp.

Kofoid, C. A. and A. S. Campbell, 1929
A conspectus of the marine and freshwater Ciliata belonging to the suborder Tintinnoinea, with descriptions of new species principally from the Agassiz expedition to the eastern tropical Pacific, 1904-1905. Univ. Calif. Publ. Zool. 34:1-403, 697 text figs.

Cymatocylis everta n.sp.

Kofoid, C. A. and A. S. Campbell, 1929
A conspectus of the marine and freshwater Ciliata belonging to the suborder Tintinnoinea, with descriptions of new species principally from the Agassiz expedition to the eastern tropical Pacific, 1904-1905. Univ. Calif. Publ. Zool. 34:1-403, 697 text figs.

Cymatocylis flava

Kofoid, C. A. and A. S. Campbell, 1929
A conspectus of the marine and freshwater Ciliata belonging to the suborder Tintinnoinea, with descriptions of new species principally from the Agassiz expedition to the eastern tropical Pacific, 1904-1905. Univ. Calif. Publ. Zool. 34:1-403, 697 text figs.

Cymatocylis flava n.sp.

Laackmann, H., 1910
Die Tintinnodeen der Deutschen Südpolar Expedition 1901-1903. Deutsch. Südpolar Exped. 11, Zool. 3(4):341-496, pls. 33-51.

Cymatocylis folliculus nom. nov.

Kofoid, C. A. and A. S. Campbell, 1929
A conspectus of the marine and freshwater Ciliata belonging to the suborder Tintinnoinea, with descriptions of new species principally from the Agassiz expedition to the eastern tropical Pacific, 1904-1905. Univ. Calif. Publ. Zool. 34:1-403, 697 text figs.

Cymatocylis gaussi nom. nov.

Kofoid, C. A. and A. S. Campbell, 1929
A conspectus of the marine and freshwater Ciliata belonging to the suborder Tintinnoinea, with descriptions of new species principally from the Agassiz expedition to the eastern tropical Pacific, 1904-1905. Univ. Calif. Publ. Zool. 34:1-403, 697 text figs.

Cymatocylis glans n.sp.

Kofoid, C. A. and A. S. Campbell, 1929
A conspectus of the marine and freshwater Ciliata belonging to the suborder Tintinnoinea, with descriptions of new species principally from the Agassiz expedition to the eastern tropical Pacific, 1904-1905. Univ. Calif. Publ. Zool. 34:1-403, 697 text figs.

Cymatocylis incondita nom. nov.

Kofoid, C. A. and A. S. Campbell, 1929
A conspectus of the marine and freshwater Ciliata belonging to the suborder Tintinnoinea, with descriptions of new species principally from the Agassiz expedition to the eastern tropical Pacific, 1904-1905. Univ. Calif. Publ. Zool. 34:1-403, 697 text figs.

Cymatocylis kerguielensis

*Hada, Yoshina, 1970. (1970)
The protozoan plankton of the Antarctic and Subantarctic seas.
Scient. Repts, Japan. Antarct. Res. Exped., (E)31:51 pp.

Cymatocylis kerguelensis

Kofoid, C. A. and A. S. Campbell, 1929
A conspectus of the marine and freshwater Ciliata belonging to the suborder Tintinnoinea, with descriptions of new species principally from the Agassiz expedition to the eastern tropical Pacific, 1904-1905. Univ. Calif. Publ. Zool. 34:1-403, 697 text figs.

Cymatocylis kerguelensis n.sp.

Laackmann, H., 1910
Die Tintinnodeen der Deutschen Südpolar Expedition 1901-1903. Deutsch. Südpolar Exped. 11, Zool. 3(4):341-496, pls. 33-51.

Cymatocylis labiosa nom. nov.

Kofoid, C. A. and A. S. Campbell, 1929
A conspectus of the marine and freshwater Ciliata belonging to the suborder Tintinnoinea, with descriptions of new species principally from the Agassiz expedition to the eastern tropical Pacific, 1904-1905. Univ. Calif. Publ. Zool. 34:1-403, 697 text figs.

Cymatocylis meridiana nom. nov.

Kofoid, C. A. and A. S. Campbell, 1929
A conspectus of the marine and freshwater Ciliata belonging to the suborder Tintinnoinea, with descriptions of new species principally from the Agassiz expedition to the eastern tropical Pacific, 1904-1905. Univ. Calif. Publ. Zool. 34:1-403, 697 text figs.

Cymatocylis minor

Kofoid, C. A. and A. S. Campbell, 1929
A conspectus of the marine and freshwater Ciliata belonging to the suborder Tintinnoinea, with descriptions of new species principally from the Agassiz expedition to the eastern tropical Pacific, 1904-1905. Univ. Calif. Publ. Zool. 34:1-403, 697 text figs.

Cymatocylis nobilis

Kofoid, C. A. and A. S. Campbell, 1929
A conspectus of the marine and freshwater Ciliata belonging to the suborder Tintinnoinea, with descriptions of new species principally from the Agassiz expedition to the eastern tropical Pacific, 1904-1905. Univ. Calif. Publ. Zool. 34:1-403, 697 text figs.

Cymatocylis nobilis

Laackmann, H., 1910
Die Tintinnodeen der Deutschen Südpolar Expedition 1901-1903. Deutsch. Südpolar Exped. 11, Zool. 3(4):341-496, pls. 33-51.

Cymatocylis ovata

Balech, Enrique, 1947
Contribucion al conocimiento del Plancton antartico Plancton del Mar de Bellingshausen. Physis 22(56):75-91, 7 pls. with 76 figs.

Cymatocylis ovata

Balech, E., 1947
Contribution al conocimiento del plancton antartico. Plankton del Mar de Bellinghausen. Physis 20:75-91, 76 figs. on 8 pls.

Cymatocylis ovata

Kofoid, C. A. and A. S. Campbell, 1929
A conspectus of the marine and freshwater Ciliata belonging to the suborder Tintinnoinea, with descriptions of new species principally from the Agassiz expedition to the eastern tropical Pacific, 1904-1905. Univ. Calif. Publ. Zool. 34:1-403, 697 text figs.

Cymetocylis parva

*Hada, Yoshina, 1970. (1970)
The protozoan plankton of the Antarctic and Subantarctic seas.
Scient. Repts, Japan. Antarct. Res. Exped., (E)31:51 pp.

Cymatocylis parva

Kofoid, C. A. and A. S. Campbell, 1929
A conspectus of the marine and freshwater Ciliata belonging to the suborder Tintinnoinea, with descriptions of new species principally from the Agassiz expedition to the eastern tropical Pacific, 1904-1905. Univ. Calif. Publ. Zool. 34:1-403, 697 text figs.

Cymatocylis parva

Laackmann, H., 1910
Die Tintinnodeen der Deutschen Südpolar Expedition 1901-1903. Deutsch. Südpolar Exped. 11, Zool. 3(4):341-496, pls. 33-51.

Cymatocylis robusta

Kofoid, C. A. and A. S. Campbell, 1929
A conspectus of the marine and freshwater Ciliata belonging to the suborder Tintinnoinea, with descriptions of new species principally from the Agassiz expedition to the eastern tropical Pacific, 1904-1905. Univ. Calif. Publ. Zool. 34:1-403, 697 text figs.

Cymatocylis scyphus nom. nov.

Kofoid, C. A. and A. S. Campbell, 1929
A conspectus of the marine and freshwater Ciliata belonging to the suborder Tintinnoinea, with descriptions of new species principally from the Agassiz expedition to the eastern tropical Pacific, 1904-1905. Univ. Calif. Publ. Zool. 34:1-403, 697 text figs.

Cymatocylis simplex

Kofoid, C. A. and A. S. Campbell, 1929
A conspectus of the marine and freshwater Ciliata belonging to the suborder Tintinnoinea, with descriptions of new species principally from the Agassiz expedition to the eastern tropical Pacific, 1904-1905. Univ. Calif. Publ. Zool. 34:1-403, 697 text figs.

Cymatocylis situla nom. nov.

Kofoid, C. A. and A. S. Campbell, 1929
A conspectus of the marine and freshwater Ciliata belonging to the suborder Tintinnoinea, with descriptions of new species principally from the Agassiz expedition to the eastern tropical Pacific, 1904-1905. Univ. Calif. Publ. Zool. 34:1-403, 697 text figs.

Cymatocylis subconica n.sp.

Kofoid, C. A. and A. S. Campbell, 1929
A conspectus of the marine and freshwater Ciliata belonging to the suborder Tintinnoinea, with descriptions of new species principally from the Agassiz expedition to the eastern tropical Pacific, 1904-1905. Univ. Calif. Publ. Zool. 34:1-403, 697 text figs.

Cymatocylis tubulosa nom. nov.

Kofoid, C. A. and A. S. Campbell, 1929
A conspectus of the marine and freshwater Ciliata belonging to the suborder Tintinnoinea, with descriptions of new species principally from the Agassiz expedition to the eastern tropical Pacific, 1904-1905. Univ. Calif. Publ. Zool. 34:1-403, 697 text figs.

Cymatocylis typica

Kofoid, C. A. and A. S. Campbell, 1929
A conspectus of the marine and freshwater Ciliata belonging to the suborder Tintinnoinea, with descriptions of new species principally from the Agassiz expedition to the eastern tropical Pacific, 1904-1905. Univ. Calif. Publ. Zool. 34:1-403, 697 text figs.

Cymatocylis urnula

Kofoid, C. A. and A. S. Campbell, 1929
A conspectus of the marine and freshwater Ciliata belonging to the suborder Tintinnoinea, with descriptions of new species principally from the Agassiz expedition to the eastern tropical Pacific, 1904-1905. Univ. Calif. Publ. Zool. 34:1-403, 697 text figs.

Cymatocylis vanhöffeni

Kofoid, C. A. and A. S. Campbell, 1929
A conspectus of the marine and freshwater Ciliata belonging to the suborder Tintinnoinea, with descriptions of new species principally from the Agassiz expedition to the eastern tropical Pacific, 1904-1905. Univ. Calif. Publ. Zool. 34:1-403, 697 text figs.

Cymatocylis vanhoffeni

Laackmann, H., 1910
Die Tintinnodeen der Deutschen Südpolar Expedition 1901-1903. Deutsch. Südpolar Exped. 11, Zool. 3(4):341-496, pls. 33-51.

Cymatocylis ventricosoides n.sp.

Kofoid, C. A. and A. S. Campbell, 1929
A conspectus of the marine and freshwater Ciliata belonging to the suborder Tintinnoinea, with descriptions of new species principally from the Agassiz expedition to the eastern tropical Pacific, 1904-1905. Univ. Calif. Publ. Zool. 34:1-403, 697 text figs.

Cyttarocylis acuminata

Kofoid, C. A. and A. S. Campbell, 1929
A conspectus of the marine and freshwater Ciliata belonging to the suborder Tintinnoinea, with descriptions of new species principally from the Agassiz expedition to the eastern tropical Pacific, 1904-1905. Univ. Calif. Publ. Zool. 34:1-403, 697 text figs.

Cyttarocylis acutiformis

Balech, Enrique, 1962
Tintinnoinea y Dinoflagellata del Pacifico segun material de las Expediciones NORPAC y DOWNWIND del Instituto Scripps de Oceanografia.
Revista, Mus. Argentino Ciencias Nat. "Bernardino Rivadavia", Ciencias Zool., 7(1):1-253.

Cyttarocylis acutiformis

Campbell, A. S., 1942
The oceanic Tintinnoina of the plankton gathered during the last cruise of the CARNEGIE. Sci. Res. Cruise VII of the Carnegie, 1928-1929, Biol. 2:163 pp., 128 figs.

Cyttarocylis acutiformis

Kofoid, C.A. and A.S. Campbell, 1939
Reports on the scientific results of the expedition to the eastern tropical Pacific in charge of Alexander Agassiz, by the U.S. Fish Commission steamer "Albatross" from October 1904 to March 1905----XXXVII. The Ciliata: The Tintinnoinea. Bull. Mus. Comp. Zool. 84: 473 pp., 36 pls.

Cyttarocylis acutiformis n.sp.

Kofoid, C. A. and A. S. Campbell, 1929
A conspectus of the marine and freshwater Ciliata belonging to the suborder Tintinnoinea, with descriptions of new species principally from the Agassiz expedition to the eastern tropical Pacific, 1904-1905. Univ. Calif. Publ. Zool. 34:1-403, 697 text figs.

Cyttarocylis acutiformis

Rampi, L., 1948
Sur quelques Tintinnides (Infusoires louques) du Pacifique subtropical (Recoltes Alain Gerbault) Bull. l'Inst. Ocean., Monaco, No.938, 4 pp.

Cyttarocylis (Coxliella) annulata

Entz, G., Jr. 1909
Studien über organisation und biologie der Tintinniden. Arch. f. Protistenkunde 15:93-226, Pls. 8-21, text figs.

Cyttarocylis annulata

Jørgensen, E., 1900.
Ueber die Tintinnodeen der norwegischen Westküste Bergens Mus. Aarb., 1899(2):48 pp., 3 pls.

Cyttarocylis arcuata?

Entz, G., Jr., 1909
 Studien über organisation und biologie der Tintinniden. Arch. f. Protistenkunde 15:93-226, Pls. 8-21, text figs.

Cyttarocylis brandti

Campbell, A. S., 1942
 The oceanic Tintinnoina of the plankton gathered during the last cruise of the CARNEGIE. Sci. Res. Cruise VII of the Carnegie, 1928-1929, Biol. 2:163 pp., 128 figs.

Cyttarocylis brandti

Kofoid, C.A. and A.S. Campbell, 1939
 Reports on the scientific results of the expedition to the eastern tropical Pacific in charge of Alexander Agassiz, by the U.S. Fish Commission steamer "Albatross" from October 1904 to March 1905-----XXXVII. The Ciliata: The Tintinnoimea. Bull. Mus. Comp. Zool. 84: 473 pp., 36 pls.

Cyttarocylis brandti n.sp.

Kofoid, C. A. and A. S. Campbell, 1929
 A conspectus of the marine and freshwater Ciliata belonging to the suborder Tintinnoinea, with descriptions of new species principally from the Agassiz expedition to the eastern tropical Pacific, 1904-1905. Univ. Calif. Publ. Zool. 34:1-403, 697 text figs.

Cyttarocylis brandti

Rampi, L., 1939
 Primo contributo alla conoscenza dei tintinnoidi de Maro Ligure. Atti della Soc. Ital. di Sci. Nat. 78:67-81, 58 text figs.

Cyttarocylis cassis

Balech, Enrique, 1962
 Tintinnoinea y Dinoflagellata del Pacifico segun material de las Expediciones NORPAC y DOWNWIND del Instituto Scripps de Oceanografia. Revista. Mus. Argentino Ciencias Nat. "Bernardino Rivadavia", Ciencias Zool., 7(1):1-253.

Cyttarocylis cassis

Balech, E., 1959.
 Tintinnoinea del Mediterraneo. Trav. Sta. Zool., Villefranche-sur-Mer, 18(17): Reprinted from: Trab. Inst. Español Oceanogr., Madrid, 19 Sept. 1959(28):88 pp., 22 pls.

Cyttarocylis cassis

Campbell, A. S., 1942
 The oceanic Tintinnoina of the plankton gathered during the last cruise of the CARNEGIE. Sci. Res. Cruise VII of the Carnegie, 1928-1929, Biol. 2:163 pp., 128 figs.

Cyttarocylis cassis

Duran, M., 1951
 Contribucion al estudio de los tintinidos del plancton de los costas de Castellon. (Mediterranes occidental). Publ. Inst. Biol. Aplic., Barcelone, 8: 101-120, 29 text figs.

Cyttarocylis cassis

Entz, G., Jr., 1909
 Studien über organisation und biologie der Tintinniden. Arch. f. Protistenkunde 15:93-226, Pls. 8-21, text figs.

Cyttarocylis cassis

Fol, H., 1884.
 Sur la famille des Tintinnodea. Recueil Zool. Suisse 1:27-64, Pls. 4-5.

Cyttarocylis cassis

Gaarder, K.R., 1946
 Tintinnoidea. Rep.Sci.Res. "Michael Sars" N.Atlantic Deep-Sea Exped., 1910, 2(1): 37 pp., 23 text figs.

Cyttarocylis cassis

Hofker, J., 1931
 Studien uber Tintinnoidea. Arch. f. Protistenk 75(3):315-402, 89 text figs.

Cyttarocylis cassis

Jörgensen, E., 1924
 Mediterranean Tintinnidae. Rept. Danish Oceanogr. Exped. 1908-10 to the Mediterranean and adjacent seas, Vol.II, Biol. J 3, 110 pp., 114 text figs.

Cyttarocylis cassis

Kofoid, C.A. and A.S. Campbell, 1939
 Reports on the scientific results of the expedition to the eastern tropical Pacific in charge of Alexander Agassiz, by the U.S. Fish Commission steamer "Albatross" from October 1904 to March 1905-----XXXVII. The Ciliata: The Tintinnoimea. Bull. Mus. Comp. Zool. 84: 473 pp., 36 pls.

Cyttarocylis cassis

Kofoid, C. A. and A. S. Campbell, 1929
 A conspectus of the marine and freshwater Ciliata belonging to the suborder Tintinnoinea, with descriptions of new species principally from the Agassiz expedition to the eastern tropical Pacific, 1904-1905. Univ. Calif. Publ. Zool. 34:1-403, 697 text figs.

Cyttarocylis cassis

Laackmann, H., 1910
 Die Tintinnodeen der Deutschen Südpolar Expedition 1901-1903. Deutsch. Südpolar Exped. 11, Zool. 3(4):341-496, pls. 33-51.

Cyttarocylis cassis

Rampi, L., 1948
 Sur quelques Tintinnides (Infusoires louques) du Pacifique subtropical (Recoltes Alain Gerbault) Bull. l'Inst. Ocean., Monaco, No.938, 4 pp.

Cyttarocylis cistellula

Fol, H., 1884.
 Sur la famille des Tintinnodea. Recueil Zool. Suisse 1:27-64, Pls. 4-5.

Cyttarocylis conica

Campbell, A. S., 1942
 The oceanic Tintinnoina of the plankton gathered during the last cruise of the CARNEGIE. Sci. Res. Cruise VII of the Carnegie, 1928-1929, Biol. 2:163 pp., 128 figs.

Cyttarocylis conica

Gaarder, K.R., 1946
 Tintinnoidea. Rep.Sci.Res. "Michael Sars" N.Atlantic Deep-Sea Exped., 1910, 2(1): 37 pp., 23 text figs.

Cyttarocylis conica

Kofoid, C.A. and A.S. Campbell, 1939
 Reports on the scientific results of the expedition to the eastern tropical Pacific in charge of Alexander Agassiz, by the U.S. Fish Commission steamer "Albatross" from October 1904 to March 1905-----XXXVII. The Ciliata: The Tintinnoimea. Bull. Mus. Comp. Zool. 84: 473 pp., 36 pls.

Cyttarocylis conica

Kofoid, C. A. and A. S. Campbell, 1929
 A conspectus of the marine and freshwater Ciliata belonging to the suborder Tintinnoinea, with descriptions of new species principally from the Agassiz expedition to the eastern tropical Pacific, 1904-1905. Univ. Calif. Publ. Zool. 34:1-403, 697 text figs.

Cyttarocylis denticulata

Iselin, C., 1930
 A report on the coastal waters of Labrador based on explorations of the "Chance" during the summer of 1926. Proc. Am. Acad. Arts Sci., 66(1):1-37, 14 text figs.

Cyttarocylis denticulata

Jørgensen, E., 1905
 B.Protistplankton and the diatoms in bottom samples. Hydrographical and biological investigations in Norwegian fiords. Bergens Mus. Skr. 7: 49-225.

Cyttarocylis denticulata

Jørgensen, E., 1900.
 Ueber die Tintinnodeen der norwegischen Westküste Bergens Mus. Aarb., 1899(2):48 pp., 3 pls.

Cyttarocylis denticulatum

Marukawa, H., 1921
 Plankton lists and some new species of copepods, from the northern waters of Japan. Bull. Inst. Ocean., No.384, 15 pp., 3 pls., 1 chart. Monaco

Cyttarocylis Ehrenbergii

Entz, G., Jr., 1909
 Studien über organisation und biologie der Tintinniden. Arch. f. Protistenkunde 15:93-226, Pls. 8-21, text figs.

Cyttarocylis ehrenbergii adriatica

Entz, G., Jr., 1909
 Studien über organisation und biologie der Tintinniden. Arch. f. Protistenkunde 15:93-226, Pls. 8-21, text figs.

Cyttarocylis ehrenbergii

Jørgensen, E., 1900.
 Ueber die Tintinnodeen der norwegischen Westküste Bergens Mus. Aarb., 1899(2):48 pp., 3 pls.

Cyttarocylis Ehrenbergi

Marukawa, H., 1921
 Plankton lists and some new species of copepods, from the northern waters of Japan. Bull. Inst. Ocean., No.384, 15 pp., 3 pls., 1 chart. Monaco

Cyttarocylis Ehrenbergii

Meunier, A., 1919
 Microplankton de la Mer Flamande. 4. Les Tintinnides et Coetera. Mem. Mus. Roy. Hist. Nat., Belgique, 8(2):59pp., Pls. 22-23.

Cyttarocylis encecryphalus

Rampi, L., 1948
 Sur quelques Tintinnides (Infusoires louques) du Pacifique subtropical (Recoltes Alain Gerbault) Bull. l'Inst. Ocean., Monaco, No.938, 4 pp.

Cyttarocylis eucecryphalus

Balech, Enrique, 1962
 Tintinnoinea y Dinoflagellata del Pacifico segun material de las Expediciones NORPAC y DOWNWIND del Instituto Scripps de Oceanografia. Revista. Mus. Argentino Ciencias Nat. "Bernardino Rivadavia", Ciencias Zool., 7(1):1-253.

Cyttarocylis eucecryphalus

Balech, E., 1959.
 Tintinnoinea del Mediterraneo. Trav. Sta. Zool., Villefranche-sur-Mer, 18(17): Reprinted from: Trab. Inst. Español Oceanogr., Madrid, 19 Sept. 1959(28):88 pp., 22 pls.

Cyttarocylis eucecryphalus

Campbell, A. S., 1942
 The oceanic Tintinnoina of the plankton gathered during the last cruise of the CARNEGIE. Sci. Res. Cruise VII of the Carnegie, 1928-1929, Biol. 2:163 pp., 128 figs.

Cyttarocylis eucecryphalus

Duran, M., 1951
 Contribucion al estudio de los tintinidos del plancton de los costas de Castellon. (Mediterranes occidental). Publ. Inst. Biol. Aplic., Barcelone, 8: 101-120, 29 text figs.

Cyttarocylis eucecryphalus

Jörgensen, E., 1924
 Mediterranean Tintinnidae. Rept. Danish Oceanogr. Exped. 1908-10 to the Mediterranean and adjacent seas, Vol.II, Biol. J 3, 110 pp., 114 text figs.

Cyttarocylis eucecryphalus

Kofoid, C.A. and A.S. Campbell, 1939
 Reports on the scientific results of the expedition to the eastern tropical Pacific in charge of Alexander Agassiz, by the U.S. Fish Commission steamer "Albatross" from October 1904 to March 1905-----XXXVII. The Ciliata: The Tintinnoimea. Bull. Mus. Comp. Zool. 84: 473 pp., 36 pls.

Cyttarocylis eucecryphalus
Kofoid, C. A. and A. S. Campbell, 1929
A conspectus of the marine and fresh-water Ciliata belonging to the suborder Tintinnoinea, with descriptions of new species principally from the Agassiz expedition to the eastern tropical Pacific, 1904-1905. Univ. Calif. Publ. Zool. 34:1-403, 697 text figs.

Cyttarocylis eucercryphalus
Komarovsky, B., 1959
The Tintinnina of the Gulf of Eylath (Aqaba). Contrib. Knowl. Red Sea, Bull. Sea Fish. Res. Sta.Haifa, Israel 21 (14):1-40.

Cyttarocylis (Coxliella) helix
Entz, G., Jr. 1909
Studien über organisation und biologie der Tintinniden. Arch. f. Protistenkunde 15:93-226, Pls. 8-21, text figs.

Cyttarocylis helix
Jørgensen, E., 1900
Ueber die Tintinnodeen der norwegischen Westküste Bergens Mus. Aarb., 1899(2):48 pp., 3 pls.

Cyttarocylis longa
Campbell, A. S., 1942
The oceanic Tintinnoina of the plankton gathered during the last cruise of the CARNEGIE. Sci. Res. Cruise VII of the Carnegie, 1928-1929, Biol. 2:163 pp., 128 figs.

Cyttarocylis longa
Gaarder, K.R., 1946
Tintinnoidea. Rep.Sci.Res."Michael Sars" N.Atlantic Deep-Sea Exped., 1910, 2(1): 37 pp., 23 text figs.

Cyttarocylis longa
Kofoid, C.A. and A.S. Campbell, 1939
Reports on the scientific results of the expedition to the eastern tropical Pacific in charge of Alexander Agassiz, by the U.S. Fish Commission steamer"Albatross" from October 1904 to March 1905-----XXXVII. The Ciliata: The Tintinnoimea. Bull.Mus. Comp.Zool. 84: 473 pp., 36 pls.

Cyttarocylis longa n.sp.
Kofoid, C. A. and A. S. Campbell, 1929
A conspectus of the marine and fresh-water Ciliata belonging to the suborder Tintinnoinea, with descriptions of new species principally from the Agassiz expedition to the eastern tropical Pacific, 1904-1905. Univ. Calif. Publ. Zool. 34:1-403, 697 text figs.

Cyttarocylis magna
Balech, E., 1959.
Tintinnoinea del Mediterraneo. Trav. Sta. Zool., Villefranche-sur-Mer, 18(17): Reprinted from: Trab. Inst. Español Oceanogr., Madrid, 19 Sept. 1959(28):88 pp., 22 pls.

Cyttarocylis magna
Campbell, A. S., 1942
The oceanic Tintinnoina of the plankton gathered during the last cruise of the CARNEGIE. Sci. Res. Cruise VII of the Carnegie, 1928-1929, Biol. 2:163 pp., 128 figs.

Cyttarocylis magna
Gaarder, K.R., 1946
Tintinnoidea. Rep.Sci.Res."Michael Sars" N.Atlantic Deep-Sea Exped., 1910, 2(1): 37 pp., 23 text figs.

Cyttarocylis magna
Kofoid, C.A. and A.S. Campbell, 1939
Reports on the scientific results of the expedition to the eastern tropical Pacific in charge of Alexander Agassiz, by the U.S. Fish Commission steamer"Albatross" from October 1904 to March 1905-----XXXVII. The Ciliata: The Tintinnoimea. Bull.Mus. Comp.Zool. 84: 473 pp., 36 pls.

Cyttarocylis magna
Kofoid, C. A. and A. S. Campbell, 1929
A conspectus of the marine and fresh-water Ciliata belonging to the suborder Tintinnoinea, with descriptions of new species principally from the Agassiz expedition to the eastern tropical Pacific, 1904-1905. Univ. Calif. Publ. Zool. 34:1-403, 697 text figs.

Cyttarocylis mucronata
Campbell, A. S., 1942
The oceanic Tintinnoina of the plankton gathered during the last cruise of the CARNEGIE. Sci. Res. Cruise VII of the Carnegie, 1928-1929, Biol. 2:163 pp., 128 figs.

Cyttarocylis mucronata
Kofoid, C.A. and A.S. Campbell, 1939
Reports on the scientific results of the expedition to the eastern tropical Pacific in charge of Alexander Agassiz, by the U.S. Fish Commission steamer"Albatross" from October 1904 to March 1905-----XXXVII. The Ciliata: The Tintinnoimea. Bull.Mus. Comp.Zool. 84: 473 pp., 36 pls.

Cyttarocylis mucronata n.sp.
Kofoid, C. A. and A. S. Campbell, 1929
A conspectus of the marine and fresh-water Ciliata belonging to the suborder Tintinnoinea, with descriptions of new species principally from the Agassiz expedition to the eastern tropical Pacific, 1904-1905. Univ. Calif. Publ. Zool. 34:1-403, 697 text figs.

Cyttarocylis norvegica
Jørgensen, E., 1905
B.Protistplankton and the diatoms in bottom samples. Hydrographical and biological investigations in Norwegian fjords. Bergens Mus. Skr. 7: 49-225.

Cyttarocylis norvegica
Jørgensen, E., 1900.
Ueber die Tintinnodeen der norwegischen Westküste Bergens Mus. Aarb., 1899(2):48 pp., 3 pls.

Cyttarocylis obtusa
Campbell, A. S., 1942
The oceanic Tintinnoina of the plankton gathered during the last cruise of the CARNEGIE. Sci. Res. Cruise VII of the Carnegie, 1928-1929, Biol. 2:163 pp., 128 figs.

Cyttarocylis obtusa n.sp.
Kofoid, C. A. and A. S. Campbell, 1929
A conspectus of the marine and fresh-water Ciliata belonging to the suborder Tintinnoinea, with descriptions of new species principally from the Agassiz expedition to the eastern tropical Pacific, 1904-1905. Univ. Calif. Publ. Zool. 34:1-403, 697 text figs.

Cyttarocylis ollula
Kofoid, C. A. and A. S. Campbell, 1929
A conspectus of the marine and fresh-water Ciliata belonging to the suborder Tintinnoinea, with descriptions of new species principally from the Agassiz expedition to the eastern tropical Pacific, 1904-1905. Univ. Calif. Publ. Zool. 34:1-403, 697 text figs.

Cyttarocylis (Xystonella) paradoxa
Entz, G., Jr. 1909
Studien über organisation und biologie der Tintinniden. Arch. f. Protistenkunde 15:93-226, Pls. 8-21, text figs.

Cyttarocylis plagiostoma
Entz, G., Jr. 1909
Studien über organisation und biologie der Tintinniden. Arch. f. Protistenkunde 15:93-226, Pls. 8-21, text figs.

Cyttarocylis plagiostoma
Hofker, J., 1931
Studien uber Tintinnoidea. Arch. f. Protistenk 75(3):315-402, 89 text figs.

Cyttarocylis plagiostoma
Kofoid, C. A. and A. S. Campbell, 1929
A conspectus of the marine and fresh-water Ciliata belonging to the suborder Tintinnoinea, with descriptions of new species principally from the Agassiz expedition to the eastern tropical Pacific, 1904-1905. Univ. Calif. Publ. Zool. 34:1-403, 697 text figs.

Cyttarocylis plagiostoma
Laackmann, H., 1910
Die Tintinnodeen der Deutschen Südpolar-Expedition 1901-1903. Deutsch. Südpolar Exped. 11, Zool. 3(4):341-496, pls. 33-51.

Cyttarocylis ricta
Kofoid, C.A. and A.S. Campbell, 1939
Reports on the scientific results of the expedition to the eastern tropical Pacific in charge of Alexander Agassiz, by the U.S. Fish Commission steamer"Albatross" from October 1904 to March 1905-----XXXVII. The Ciliata: The Tintinnoimea. Bull.Mus. Comp.Zool. 84: 473 pp., 36 pls.

Cyttarocylis ricta n.sp.
Kofoid, C. A. and A. S. Campbell, 1929
A conspectus of the marine and fresh-water Ciliata belonging to the suborder Tintinnoinea, with descriptions of new species principally from the Agassiz expedition to the eastern tropical Pacific, 1904-1905. Univ. Calif. Publ. Zool. 34:1-403, 697 text figs.

Cyttarocylis rotundata
Kofoid, C. A. and A. S. Campbell, 1929
A conspectus of the marine and fresh-water Ciliata belonging to the suborder Tintinnoinea, with descriptions of new species principally from the Agassiz expedition to the eastern tropical Pacific, 1904-1905. Univ. Calif. Publ. Zool. 34:1-403, 697 text figs.

Cyttarocylis serrata
Entz, G., Jr. 1909
Studien über organisation und biologie der Tintinniden. Arch. f. Protistenkunde 15:93-226, Pls. 8-21, text figs.

Cyttarocylis serrata
Jørgensen, E., 1905
B.Protistplankton and the diatoms in bottom samples. Hydrographical and biological investigations in Norwegian fjords. Bergens Mus. Skr. 7: 49-225.

Cyttarocylis serrata
Jørgensen, E., 1900.
Ueber die Tintinnodeen der norwegischen Westküste Bergens Mus. Aarb., 1899(2):48 pp., 3 pls.

Cyttarocylis serrata
Meunier, A., 1919
Microplankton de la Mer Flamande. 4. Les Tintinnides et Coetera. Mem. Mus. Roy. Hist. Nat., Belgique, 8(2):59pp., Pls. 22-23.

Cyttarocylis spiralis
Meunier, A., 1919
Microplankton de la Mer Flamande. 4. Les Tintinnides et Coetera. Mem. Mus. Roy. Hist. Nat., Belgique, 8(2):59pp., Pls. 22-23.

Cyttarocyclis (Xystonella) treforti
Entz, G., Jr. 1909
Studien über organisation und biologie der Tintinniden. Arch. f. Protistenkunde 15:93-226, Pls. 8-21, text figs.

Dadayiella n. gen.
Kofoid, C. A. and A. S. Campbell, 1929
A conspectus of the marine and fresh-water Ciliata belonging to the suborder Tintinnoinea, with descriptions of new species principally from the Agassiz expedition to the eastern tropical Pacific, 1904-1905. Univ. Calif. Publ. Zool. 34:1-403, 697 text figs.

Dadayiella ganymedes
Balech, E., 1959.
Tintinnoinea del Mediterraneo. Trav. Sta. Zool., Villefranche-sur-Mer, 18(17): Reprinted from: Trab. Inst. Español Oceanogr., Madrid, 19 Sept. 1959 (28):88 pp., 22 pls.

Dadayiella acuta

Gaarder, K.R., 1946
Tintinnoidea. Rep. Sci. Res. "Michael Sars" N. Atlantic Deep-Sea Exped., 1910, 2(1): 37 pp., 23 text figs.

Dadayiella acuta

Kofoid, C. A. and A. S. Campbell, 1929
A conspectus of the marine and fresh-water Ciliata belonging to the suborder Tintinnoinea, with descriptions of new species principally from the Agassiz expedition to the eastern tropical Pacific, 1904-1905. Univ. Calif. Publ. Zool. 34:1-403, 697 text figs.

Dadayiella acutiformis

Campbell, A. S., 1942
The oceanic Tintinnoina of the plankton gathered during the last cruise of the CARNEGIE. Sci. Res. Cruise VII of the Carnegie, 1928-1929, Biol. 2:163 pp., 128 figs.

Dadayella acutiformis n.sp.

Kofoid, C.A. and A.S. Campbell, 1939
Reports on the scientific results of the expedition to the eastern tropical Pacific in charge of Alexander Agassiz, by the U.S. Fish Commission steamer "Albatross" from October 1904 to March 1905-----XXXVII. The Ciliata: The Tintinnoinea. Bull. Mus. Comp. Zool. 84: 473 pp., 36 pls.

Dadayiella bulbosa

Campbell, A. S., 1942
The oceanic Tintinnoina of the plankton gathered during the last cruise of the CARNEGIE. Sci. Res. Cruise VII of the Carnegie, 1928-1929, Biol. 2:163 pp., 128 figs.

Dadayiella bulbosa

Kofoid, C.A. and A.S. Campbell, 1939
Reports on the scientific results of the expedition to the eastern tropical Pacific in charge of Alexander Agassiz, by the U.S. Fish Commission steamer "Albatross" from October 1904 to March 1905-----XXXVII. The Ciliata: The Tintinnoimea. Bull. Mus. Comp. Zool. 84: 473 pp., 36 pls.

Dadayiella bulbosa

Kofoid, C. A. and A. S. Campbell, 1929
A conspectus of the marine and fresh-water Ciliata belonging to the suborder Tintinnoinea, with descriptions of new species principally from the Agassiz expedition to the eastern tropical Pacific, 1904-1905. Univ. Calif. Publ. Zool. 34:1-403, 697 text figs.

Dadayella bulbosa

Komarovsky, B., 1959
The Tintinnina of the Gulf of Eylath (Aqaba). Contrib. Knowl. Red Sea, Bull. Sea Fish. Res. Sta. Haifa, Israel 21 (14):1-40.

Dadayiella curta

Kofoid, C.A. and A.S. Campbell, 1939
Reports on the scientific results of the expedition to the eastern tropical Pacific in charge of Alexander Agassiz, by the U.S. Fish Commission steamer "Albatross" from October 1904 to March 1905-----XXXVII. The Ciliata: The Tintinnoimea. Bull. Mus. Comp. Zool. 84: 473 pp., 36 pls.

Dadayiella curta n.sp.

Kofoid, C. A. and A. S. Campbell, 1929
A conspectus of the marine and fresh-water Ciliata belonging to the suborder Tintinnoinea, with descriptions of new species principally from the Agassiz expedition to the eastern tropical Pacific, 1904-1905. Univ. Calif. Publ. Zool. 34:1-403, 697 text figs.

Dadayiella cuspis

Campbell, A. S., 1942
The oceanic Tintinnoina of the plankton gathered during the last cruise of the CARNEGIE. Sci. Res. Cruise VII of the Carnegie, 1928-1929, Biol. 2:163 pp., 128 figs.

Dadayiella cuspis

Kofoid, C.A. and A.S. Campbell, 1939
Reports on the scientific results of the expedition to the eastern tropical Pacific in charge of Alexander Agassiz, by the U.S. Fish Commission steamer "Albatross" from October 1904 to March 1905-----XXXVII. The Ciliata: The Tintinnoimea. Bull. Mus. Comp. Zool. 84: 473 pp., 36 pls.

Dadayiella cuspis n.sp.

Kofoid, C. A. and A. S. Campbell, 1929
A conspectus of the marine and fresh-water Ciliata belonging to the suborder Tintinnoinea, with descriptions of new species principally from the Agassiz expedition to the eastern tropical Pacific, 1904-1905. Univ. Calif. Publ. Zool. 34:1-403, 697 text figs.

Dadayiella ganymedes

Campbell, A. S., 1942
The oceanic Tintinnoina of the plankton gathered during the last cruise of the CARNEGIE. Sci. Res. Cruise VII of the Carnegie, 1928-1929, Biol. 2:163 pp., 128 figs.

Dadayiella ganymedes

Gaarder, K.R., 1946
Tintinnoidea. Rep. Sci. Res. "Michael Sars" N. Atlantic Deep-Sea Exped., 1910, 2(1): 37 pp., 23 text figs.

Dadayiella ganymedes

*Hada, Yoshina, 1970. (1970)
The protozoan plankton of the Antarctic and Subantarctic seas.
Scient. Repts, Japan. Antarct. Res. Exped., (E) 31:51 pp.

Dadayiella ganymedes

Kofoid, C.A. and A.S. Campbell, 1939
Reports on the scientific results of the expedition to the eastern tropical Pacific in charge of Alexander Agassiz, by the U.S. Fish Commission steamer "Albatross" from October 1904 to March 1905-----XXXVII. The Ciliata: The Tintinnoimea. Bull. Mus. Comp. Zool. 84: 473 pp., 36 pls.

Dadayiella ganymedes

Kofoid, C. A. and A. S. Campbell, 1929
A conspectus of the marine and fresh-water Ciliata belonging to the suborder Tintinnoinea, with descriptions of new species principally from the Agassiz expedition to the eastern tropical Pacific, 1904-1905. Univ. Calif. Publ. Zool. 34:1-403, 697 text figs.

Dadayella ganymedes

Komarovsky, B., 1959
The Tintinnina of the Gulf of Eylath (Aqaba). Contrib. Knowl. Red Sea, Bull. Sea Fish. Res. Sta. Haifa, Israel 21 (14):1-40.

Dadayiella ganymedes

Marshall, S. M., 1934
The silicoflagellata and tintinnoinea. Great Barrier Reef Exped., 1928-29. Sci. Repts. 4(15):623-664, 43 text figs.

Dadayiella ganymedes

Marshall, S. M., 1933
The production of microplankton in the Great Barrier Reef Region. Brit. Mus. (N.H.) Great Barrier Reef Exped. 1928-29, Sci. Repts. II(5):111-157, 14 text figs.

Dadayella ganymedes

Rampi, L., 1939
Primo contributo alla conoscenza dei tintinnoidi de Maro Ligure. Atti della Soc. Ital. di Sci. Nat. 78:67-81, 58 text figs.

Dadayiella jorgenseni n.sp.

Kofoid, C. A. and A. S. Campbell, 1929
A conspectus of the marine and fresh-water Ciliata belonging to the suborder Tintinnoinea, with descriptions of new species principally from the Agassiz expedition to the eastern tropical Pacific, 1904-1905. Univ. Calif. Publ. Zool. 34:1-403, 697 text figs.

Dadayiella pachytoecus

Balech, Enrique, 1968.
Algunas especies nuevas o interesantes do tintinnidos del Golfo de Mexico y Caribe. Revta Mus. argent. Cienc. nat. Bernardino Rivadavia Inst. nac. Hidrobiol. 2(5):165-197.

Dadayiella pachytoecus

Kofoid, C. A. and A. S. Campbell, 1929
A conspectus of the marine and fresh-water Ciliata belonging to the suborder Tintinnoinea, with descriptions of new species principally from the Agassiz expedition to the eastern tropical Pacific, 1904-1905. Univ. Calif. Publ. Zool. 34:1-403, 697 text figs.

Daturella sp.

Balech, E., 1959.
Tintinnoinea del Mediterranea.
Trav. Sta. Zool., Villefranche-sur-Mer, 18(17): Reprinted from:
Trab. Inst. Español Oceanogr., Madrid, 19 Sept. 1959(28):88 pp., 22 pls.

Daturella n. gen.

Kofoid, C. A. and A. S. Campbell, 1929
A conspectus of the marine and fresh-water Ciliata belonging to the suborder Tintinnoinea, with descriptions of new species principally from the Agassiz expedition to the eastern tropical Pacific, 1904-1905. Univ. Calif. Publ. Zool. 34:1-403, 697 text figs.

Daturella angusta n.sp.

Kofoid, C. A. and A. S. Campbell, 1929
A conspectus of the marine and fresh-water Ciliata belonging to the suborder Tintinnoinea, with descriptions of new species principally from the Agassiz expedition to the eastern tropical Pacific, 1904-1905. Univ. Calif. Publ. Zool. 34:1-403, 697 text figs.

Daturella balechei n.sp.

de Sousa e Silva, E., 1956.
Contribution à l'étude du microplancton de Dakar et des regions maritimes voisines.
Bull. I.F.A.N., 8(2):335-371, 7 pls.

Daturella datura

Kofoid, C. A. and A. S. Campbell, 1929
A conspectus of the marine and fresh-water Ciliata belonging to the suborder Tintinnoinea, with descriptions of new species principally from the Agassiz expedition to the eastern tropical Pacific, 1904-1905. Univ. Calif. Publ. Zool. 34:1-403, 697 text figs.

Daturella emarginata

Kofoid, C. A. and A. S. Campbell, 1929
A conspectus of the marine and fresh-water Ciliata belonging to the suborder Tintinnoinea, with descriptions of new species principally from the Agassiz expedition to the eastern tropical Pacific, 1904-1905. Univ. Calif. Publ. Zool. 34:1-403, 697 text figs.

Daturella frigida n. sp.

*Hada, Yoshina, 1970. (1970)
The protozoan plankton of the Antarctic and Subantarctic seas.
Scient. Repts, Japan. Antarct. Res. Exped., (E) 31:51 pp.

Daturella gaussi

Gaarder, K.R., 1946
Tintinnoidea. Rep. Sci. Res. "Michael Sars" N. Atlantic Deep-Sea Exped., 1910, 2(1): 37 pp., 23 text figs.

Daturella gaussi n.sp.

Kofoid, C. A. and A. S. Campbell, 1929
A conspectus of the marine and fresh-water Ciliata belonging to the suborder Tintinnoinea, with descriptions of new species principally from the Agassiz expedition to the eastern tropical Pacific, 1904-1905. Univ. Calif. Publ. Zool. 34:1-403, 697 text figs.

Daturella luanae n.sp.

Marshall, S. M., 1934
The silicoflagellata and tintinnoinea. Great Barrier Reef Exped., 1928-29. Sci. Repts. 4(15):623-664, 43 text figs.

Daturella magna

Gaarder, K.R., 1946
Tintinnoidea. Rep.Sci.Res."Michael Sars" N.Atlantic Deep-Sea Exped., 1910, 2(1): 37 pp., 23 text figs.

Daturella magna

Kofoid, C.A. and A.S. Campbell, 1939
Reports on the scientific results of the expedition to the eastern tropical Pacific in charge of Alexander Agassiz, by the U.S. Fish Commission steamer "Albatross" from October 1904 to March 1905-----XXXVII. The Ciliata: The Tintinnoimea. Bull.Mus. Comp.Zool. 84: 473 pp., 36 pls.

Daturella magna n.sp.

Kofoid, C. A. and A. S. Campbell, 1929
A conspectus of the marine and fresh-water Ciliata belonging to the suborder Tintinnoinea, with descriptions of new species principally from the Agassiz expedition to the eastern tropical Pacific, 1904-1905. Univ. Calif. Publ. Zool. 34:1-403, 697 text figs.

Daturella ora

Kofoid, C.A. and A.S. Campbell, 1939
Reports on the scientific results of the expedition to the eastern tropical Pacific in charge of Alexander Agassiz, by the U.S. Fish Commission steamer "Albatross" from October 1904 to March 1905-----XXXVII. The Ciliata: The Tintinnoimea. Bull.Mus. Comp.Zool. 84: 473 pp., 36 pls.

Daturella ora n.sp.

Kofoid, C. A. and A. S. Campbell, 1929
A conspectus of the marine and fresh-water Ciliata belonging to the suborder Tintinnoinea, with descriptions of new species principally from the Agassiz expedition to the eastern tropical Pacific, 1904-1905. Univ. Calif. Publ. Zool. 34:1-403, 697 text figs.

Daturella recta

Kofoid, C.A. and A.S. Campbell, 1939
Reports on the scientific results of the expedition to the eastern tropical Pacific in charge of Alexander Agassiz, by the U.S. Fish Commission steamer "Albatross" from October 1904 to March 1905-----XXXVII. The Ciliata: The Tintinnoimea. Bull.Mus. Comp.Zool. 84: 473 pp., 36 pls.

Daturella recta n.sp.

Kofoid, C. A. and A. S. Campbell, 1929
A conspectus of the marine and fresh-water Ciliata belonging to the suborder Tintinnoinea, with descriptions of new species principally from the Agassiz expedition to the eastern tropical Pacific, 1904-1905. Univ. Calif. Publ. Zool. 34:1-403, 697 text figs.

Daturella stramonium

Balech, Enrique, 1962
Tintinnoinea y Dinoflagellata del Pacifico segun material de las Expediciones NORPAC y DOWNWIND del Instituto Scripps de Oceanografia.
Revista, Mus. Argentino Ciencias Nat. "Bernardino Rivadavia", Ciencias Zool., 7(1):1-253.

Daturella stramonium

Campbell, A. S., 1942
The oceanic Tintinnoina of the plankton gathered during the last cruise of the CARNEGIE. Sci. Res. Cruise VII of the Carnegie, 1928-1929, Biol. 2:163 pp., 128 figs.

Daturella stramonium

Gaarder, K.R., 1946
Tintinnoidea. Rep.Sci.Res."Michael Sars" N.Atlantic Deep-Sea Exped., 1910, 2(1): 37 pp., 23 text figs.

Daturella stramonium

Kofoid, C.A. and A.S. Campbell, 1939
Reports on the scientific results of the expedition to the eastern tropical Pacific in charge of Alexander Agassiz, by the U.S. Fish Commission steamer "Albatross" from October 1904 to March 1905-----XXXVII. The Ciliata: The Tintinnoimea. Bull.Mus. Comp.Zool. 84: 473 pp., 36 pls.

Daturella stramonium n.sp.

Kofoid, C. A. and A. S. Campbell, 1929
A conspectus of the marine and fresh-water Ciliata belonging to the suborder Tintinnoinea, with descriptions of new species principally from the Agassiz expedition to the eastern tropical Pacific, 1904-1905. Univ. Calif. Publ. Zool. 34:1-403, 697 text figs.

Daturella striata

Kofoid, C.A. and A.S. Campbell, 1939
Reports on the scientific results of the expedition to the eastern tropical Pacific in charge of Alexander Agassiz, by the U.S. Fish Commission steamer "Albatross" from October 1904 to March 1905-----XXXVII. The Ciliata: The Tintinnoimea. Bull.Mus. Comp.Zool. 84: 473 pp., 36 pls.

Daturella striata n.sp.

Kofoid, C. A. and A. S. Campbell, 1929
A conspectus of the marine and fresh-water Ciliata belonging to the suborder Tintinnoinea, with descriptions of new species principally from the Agassiz expedition to the eastern tropical Pacific, 1904-1905. Univ. Calif. Publ. Zool. 34:1-403, 697 text figs.

Dictyocysta ampla

Kofoid, C.A. and A.S. Campbell, 1939
Reports on the scientific results of the expedition to the eastern tropical Pacific in charge of Alexander Agassiz, by the U.S. Fish Commission steamer "Albatross" from October 1904 to March 1905-----XXXVII. The Ciliata: The Tintinnoimea. Bull.Mus. Comp.Zool. 84: 473 pp., 36 pls.

Dictyocysta ampla n.sp.

Kofoid, C. A. and A. S. Campbell, 1929
A conspectus of the marine and fresh-water Ciliata belonging to the suborder Tintinnoinea, with descriptions of new species principally from the Agassiz expedition to the eastern tropical Pacific, 1904-1905. Univ. Calif. Publ. Zool. 34:1-403, 697 text figs.

Dictyocysta apiculata

Kofoid, C. A. and A. S. Campbell, 1929
A conspectus of the marine and fresh-water Ciliata belonging to the suborder Tintinnoinea, with descriptions of new species principally from the Agassiz expedition to the eastern tropical Pacific, 1904-1905. Univ. Calif. Publ. Zool. 34:1-403, 697 text figs.

Dictyocysta atlantica

Kofoid, C. A. and A. S. Campbell, 1929
A conspectus of the marine and fresh-water Ciliata belonging to the suborder Tintinnoinea, with descriptions of new species principally from the Agassiz expedition to the eastern tropical Pacific, 1904-1905. Univ. Calif. Publ. Zool. 34:1-403, 697 text figs.

Dictyocysta californiensis

Balech, Enrique, 1962
Tintinnoinea y Dinoflagellata del Pacifico segun material de las Expediciones NORPAC y DOWNWIND del Instituto Scripps de Oceanografia.
Revista, Mus. Argentino Ciencias Nat. "Bernardino Rivadavia", Ciencias Zool., 7(1):1-253.

Dictyocysta californiensis

Kofoid, C.A. and A.S. Campbell, 1939
Reports on the scientific results of the expedition to the eastern tropical Pacific in charge of Alexander Agassiz, by the U.S. Fish Commission steamer "Albatross" from October 1904 to March 1905-----XXXVII. The Ciliata: The Tintinnoimea. Bull.Mus. Comp.Zool. 84: 473 pp., 36 pls.

Dictyocysta californiensis n.sp.

Kofoid, C. A. and A. S. Campbell, 1929
A conspectus of the marine and fresh-water Ciliata belonging to the suborder Tintinnoinea, with descriptions of new species principally from the Agassiz expedition to the eastern tropical Pacific, 1904-1905. Univ. Calif. Publ. Zool. 34:1-403, 697 text figs.

Dictyocysta dilatata

Campbell, A. S., 1942
The oceanic Tintinnoina of the plankton gathered during the last cruise of the CARNEGIE. Sci. Res. Cruise VII of the Carnegie, 1928-1929, Biol. 2:163 pp., 128 figs.

Dictyocysta dilatata

Gaarder, K.R., 1946
Tintinnoidea. Rep.Sci.Res."Michael Sars" N.Atlantic Deep-Sea Exped., 1910, 2(1): 37 pp., 23 text figs.

Dictyocysta dilatata

Kofoid, C.A. and A.S. Campbell, 1939
Reports on the scientific results of the expedition to the eastern tropical Pacific in charge of Alexander Agassiz, by the U.S. Fish Commission steamer "Albatross" from October 1904 to March 1905-----XXXVII. The Ciliata: The Tintinnoimea. Bull.Mus. Comp.Zool. 84: 473 pp., 36 pls.

Dictyocysta dilatata

Kofoid, C. A. and A. S. Campbell, 1929
A conspectus of the marine and fresh-water Ciliata belonging to the suborder Tintinnoinea, with descriptions of new species principally from the Agassiz expedition to the eastern tropical Pacific, 1904-1905. Univ. Calif. Publ. Zool. 34:1-403, 697 text figs.

Dictyocysta duplex

Gaarder, K.R., 1946
Tintinnoidea. Rep.Sci.Res."Michael Sars" N.Atlantic Deep-Sea Exped., 1910, 2(1): 37 pp., 23 text figs.

Dictyocysta duplex

Kofoid, C.A. and A.S. Campbell, 1939
Reports on the scientific results of the expedition to the eastern tropical Pacific in charge of Alexander Agassiz, by the U.S. Fish Commission steamer "Albatross" from October 1904 to March 1905-----XXXVII. The Ciliata: The Tintinnoimea. Bull.Mus. Comp.Zool. 84: 473 pp., 36 pls.

Dictyocysta duplex

Kofoid, C. A. and A. S. Campbell, 1929
A conspectus of the marine and fresh-water Ciliata belonging to the suborder Tintinnoinea, with descriptions of new species principally from the Agassiz expedition to the eastern tropical Pacific, 1904-1905. Univ. Calif. Publ. Zool. 34:1-403, 697 text figs.

Dictyocysta duplex

Komarovsky, B., 1962
Tintinnina from the vicinity of the Straits of Tiran and Massawa Region. Contributions to the knowledge of the Red Sea, No. 25.
Sea Fish. Res. Sta. Haifa, Israel, Bull. No. 30:48-56.

Dictyocysta elegans

Argentina, Secretaria de Marina, Servicio de Hidrografia Naval, 1962.
Plancton de las campanas oceanograficas DRAKE I y II.
Publico, H. 627:57.

Dictyocysta elegans

Campbell, A. S., 1942
The oceanic Tintinnoina of the plankton gathered during the last cruise of the CARNEGIE. Sci. Res. Cruise VII of the Carnegie, 1928-1929, Biol. 2:163 pp., 128 figs.

Dictyocysta elegans

Delegazione Italiana della Commissione Internazionale per l'Esplorazione Scientifica del Mediterraneo, 1941
Note sul plancton della Laguna veneta. [Memoria CCLXXIX] , Arch. di Ocean. e Limn. Anno I, Fasc. I, 1941 XIX: 31-57 pp.

Dictyocysta elegans

Duran, M., 1951
Contribucion al estudio de los tintinidos del plancton de los costas de Castellon. (Mediterranes occidental). Publ. Inst. Biol. Aplic., Barcelone, 8: 101-120, 29 text figs.

Dictyocysta elegans
Entz, G., Jr., 1909
Studien über organisation und biologie der Tintinniden. Arch. f. Protistenkunde 15:93-226, Pls. 8-21, text figs.

Dictyocysta elegans
Gaarder, K.R., 1946
Tintinnoidea. Rep.Sci.Res."Michael Sars" N.Atlantic Deep-Sea Exped., 1910, 2(1): 37 pp., 23 text figs.

Dictyocysta elegans
Gran, H.H., and T. Braarud, 1935
A quantitative study of the phytoplankton in the Bay of Fundy and the Gulf of Maine (including observations on hydrography, chemistry, and turbidity). J. Biol. Bd., Canada, 1(5):279-467, 69 text figs.

Dictyocysta elegans
Jörgensen, E., 1924
Mediterranean Tintinnidae. Rept. Danish Oceanogr. Exped. 1908-10 to the Mediterranean and adjacent seas, Vol.II, Biol. J 3, 110 pp., 114 text figs.

Dictyocysta elegans
Jörgensen, E., 1900.
Ueber die Tintinnodeen der norwegischen Westküste Bergens Mus. Aarb., 1899(2):48 pp., 3 pls.

Dictyocysta elegans
Kofoid, C. A. and A. S. Campbell, 1929
A conspectus of the marine and freshwater Ciliata belonging to the suborder Tintinnoinea, with descriptions of new species principally from the Agassiz expedition to the eastern tropical Pacific, 1904-1905. Univ. Calif. Publ. Zool. 34:1-403, 697 text figs.

Dictyocysta elegans
Laackmann, H., 1910
Die Tintinnodeen der Deutschen Südpolar Expedition 1901-1903. Deutsch. Südpolar Exped. 11, Zool. 3(4):341-496, pls. 33-51.

Dictyocysta elegans
Posta, Annette, 1963.
Relations entre l'évolution de quelques tintinnides de la rade de Villefranche et la température de l'eau.
Cahiers, Biol. Mar., Roscoff, 4:201-210.
Also in:
Trav. Sta. Zool., Villefranche-sur-Mer, 22(5);

Dictyocysta elegans
Sousa e Silva, E. de, 1950.
Les tintinnides de la Baie de Cascais. Bull. Inst. Océan., Monaco, No. 979:28 pp., 4 pls.

Dictyocysta elegans lepida
Balech, Enrique, 1962
Tintinnoinea y Dinoflagellata del Pacifico segun material de las Expediciones NORPAC y DOWNWIND del Instituto Scripps de Oceanografia.
Revista, Mus. Argentino Ciencias Nat. "Bernardino Rivadavia", Ciencias Zool., 7(1):1-253.

Dictyocysta elegans lepida
Balech, E., 1959.
Tintinnoinea del Mediterraneo.
Trav. Sta. Zool., Villefranche-sur-Mer, 18(17):
Reprinted from:
Trab. Inst. Español Oceanogr., Madrid, 19 Sept. 1959(28):88 pp., 22 pls.

Dictyocysta elegans speciosa
Balech, Enrique, 1962
Tintinnoinea y Dinoflagellata del Pacifico segun material de las Expediciones NORPAC y DOWNWIND del Instituto Scripps de Oceanografia.
Revista, Mus. Argentino Ciencias Nat. "Bernardino Rivadavia", Ciencias Zool., 7(1):1-253.

Dictyocysta elegans speciosa
Balech, E., 1959.
Tintinnoinea del Mediterraneo.
Trav. Sta. Zool., Villefranche-sur-Mer, 18(17):
Reprinted from:
Trab. Inst. Español Oceanogr., Madrid, 19 Sept. 1959(28):88 pp., 22 pls.

Dictyocysta entzi
Balech, E., 1959.
Tintinnoinea del Mediterraneo.
Trav. Sta. Zool., Villefranche-sur-Mer, 18(17):
Reprinted from:
Trab. Inst. Español Oceanogr., Madrid, 19 Sept. 1959(28):88 pp., 22 pls.

Dictyocysta entzi n.sp.
Jörgensen, E., 1924
Mediterranean Tintinnidae. Rept. Danish Oceanogr. Exped. 1908-10 to the Mediterranean and adjacent seas, Vol.II, Biol. J 3, 110 pp., 114 text figs.

Dictyocysta entzi
Kofoid, C. A. and A. S. Campbell, 1929
A conspectus of the marine and freshwater Ciliata belonging to the suborder Tintinnoinea, with descriptions of new species principally from the Agassiz expedition to the eastern tropical Pacific, 1904-1905. Univ. Calif. Publ. Zool. 34:1-403, 697 text figs.

Dictyocysta extensa
Balech, Enrique, 1962
Tintinnoinea y Dinoflagellata del Pacifico segun material de las Expediciones NORPAC y DOWNWIND del Instituto Scripps de Oceanografia.
Revista, Mus. Argentino Ciencias Nat. "Bernardino Rivadavia", Ciencias Zool., 7(1):1-253.

Dictyocysta extensa
Balech, E., 1959.
Tintinnoinea del Mediterraneo.
Trav. Sta. Zool., Villefranche-sur-Mer, 18(17):
Reprinted from:
Trab. Inst. Español Oceanogr., Madrid, 19 Sept. 1959, (28):88 pp., 22 pls.

Dictyocysta extensa n.sp.
Kofoid, C. A. and A. S. Campbell, 1929
A conspectus of the marine and freshwater Ciliata belonging to the suborder Tintinnoinea, with descriptions of new species principally from the Agassiz expedition to the eastern tropical Pacific, 1904-1905. Univ. Calif. Publ. Zool. 34:1-403, 697 text figs.

Dictyocysta fenestrata
Kofoid, C.A. and A.S. Campbell, 1939
Reports on the scientific results of the expedition to the eastern tropical Pacific in charge of Alexander Agassiz, by the U.S. Fish Commission steamer"Albetross" from October 1904 to March 1905------XXXVII. The Ciliata: The Tintinnoimea. Bull.Mus. Comp.Zool. 84: 473 pp., 36 pls.

Dictyocysta fenestrata n.sp.
Kofoid, C. A. and A. S. Campbell, 1929
A conspectus of the marine and freshwater Ciliata belonging to the suborder Tintinnoinea, with descriptions of new species principally from the Agassiz expedition to the eastern tropical Pacific, 1904-1905. Univ. Calif. Publ. Zool. 34:1-403, 697 text figs.

Dictyocysta fundlandica
Kofoid, C. A. and A. S. Campbell, 1929
A conspectus of the marine and freshwater Ciliata belonging to the suborder Tintinnoinea, with descriptions of new species principally from the Agassiz expedition to the eastern tropical Pacific, 1904-1905. Univ. Calif. Publ. Zool. 34:1-403, 697 text figs.

Dictyocysta grandis
Kofoid, C. A. and A. S. Campbell, 1929
A conspectus of the marine and freshwater Ciliata belonging to the suborder Tintinnoinea, with descriptions of new species principally from the Agassiz expedition to the eastern tropical Pacific, 1904-1905. Univ. Calif. Publ. Zool. 34:1-403, 697 text figs.

Dictyocysta inaequalis
Campbell, A. S., 1942
The oceanic Tintinnoina of the plankton gathered during the last cruise of the CARNEGIE. Sci. Res. Cruise VII of the Carnegie, 1928-1929, Biol. 2:163 pp., 128 figs.

Dictyocysta inaequalis n.sp.
Kofoid, C. A. and A. S. Campbell, 1929
A conspectus of the marine and freshwater Ciliata belonging to the suborder Tintinnoinea, with descriptions of new species principally from the Agassiz expedition to the eastern tropical Pacific, 1904-1905. Univ. Calif. Publ. Zool. 34:1-403, 697 text figs.

Dictyocysta lata
Campbell, A. S., 1942
The oceanic Tintinnoina of the plankton gathered during the last cruise of the CARNEGIE. Sci. Res. Cruise VII of the Carnegie, 1928-1929, Biol. 2:163 pp., 128 figs.

Dictyocysta lata
Kofoid, C.A. and A.S. Campbell, 1939
Reports on the scientific results of the expedition to the eastern tropical Pacific in charge of Alexander Agassiz, by the U.S. Fish Commission steamer"Albetross" from October 1904 to March 1905------XXXVII. The Ciliata: The Tintinnoimea. Bull.Mus. Comp.Zool. 84: 473 pp., 36 pls.

Dictyocysta lata n.sp.
Kofoid, C. A. and A. S. Campbell, 1929
A conspectus of the marine and freshwater Ciliata belonging to the suborder Tintinnoinea, with descriptions of new species principally from the Agassiz expedition to the eastern tropical Pacific, 1904-1905. Univ. Calif. Publ. Zool. 34:1-403, 697 text figs.

Dictyocysta lata
Sousa e Silva, E. de, 1950.
Les tintinnides de la Baie de Cascais. Bull. Inst. Océan., Monaco, No. 979:28 pp., 4 pls.

Dictyocysta lepida
Balech, Enrique, 1944
Contribucion al conocimiento del Plancton de Lennox y Cabo de Hornos. Physis XIX:423-446, 6 pls. with 67 figs.

Dictyocysta lepida
Campbell, A. S., 1942
The oceanic Tintinnoina of the plankton gathered during the last cruise of the CARNEGIE. Sci. Res. Cruise VII of the Carnegie, 1928-1929, Biol. 2:163 pp., 128 figs.

Dictyocysta lepida
Delegazione Italiana della Commissione Internazionale per l'Esplorazione Scientifica del Mediterraneo, 1941
Note sul plancton della Laguna veneta. [Memoria CCLXXIX], Arch. di Ocean. e Limn. Anno I, Fasc. I, 1941 XIX: 31-57 pp.

Dictyocysta lepida
Duran, M., 1951
Contribucion al estudio de los tintinidos del plancton de los costas de Castellon. (Mediterranes occidental). Publ. Inst. Biol. Aplic., Barcelone, 8: 101-120, 29 text figs.

Dictyocysta lepida
Gran, H.H., and T. Braarud, 1935
A quantitative study of the phytoplankton in the Bay of Fundy and the Gulf of Maine (including observations on hydrography, chemistry, and turbidity). J. Biol. Bd., Canada, 1(5):279-467, 69 text figs.

Dictyocysta lepida
Hofker, J., 1931
Studien uber Tintinnoidea. Arch. f. Protistenk 75(3):315-402, 89 text figs.

Dictyocysta lepida
Jörgensen, E., 1924
Mediterranean Tintinnidae. Rept. Danish Oceanogr. Exped. 1908-10 to the Mediterranean and adjacent seas, Vol.II, Biol. J 3, 110 pp., 114 text figs.

Dictyocysta lepida
Kofoid, C.A. and A.S. Campbell, 1939
Reports on the scientific results of the expedition to the eastern tropical Pacific in charge of Alexander Agassiz, by the U.S. Fish Commission steamer"Albetross" from October 1904 to March 1905------XXXVII. The Ciliata: The Tintinnoimea. Bull.Mus. Comp.Zool. 84: 473 pp., 36 pls.

Dictyocysta lepida
Kofoid, C. A. and A. S. Campbell, 1929
A conspectus of the marine and fresh-water Ciliata belonging to the suborder Tintinnoinea, with descriptions of new species principally from the Agassiz expedition to the eastern tropical Pacific, 1904-1905. Univ. Calif. Publ. Zool. 34:1-403, 697 text figs.

Dictyocysta lepida
Sousa e Silva, E. de, 1950.
Les tintinnides de la Baie de Cascais. Bull. Inst. Océan., Monaco, No. 979:28 pp., 4 pls.

Dictyocysta lepida
Rampi, L., 1948
Sur quelques Tintinnides (Infusoires louques) du Pacifique subtropical (Recoltes Alain Gerbault) Bull. l'Inst. Ocean., Monaco, No.938, 4 pp.

Dictyocysta magna
Campbell, A. S., 1942
The oceanic Tintinnoina of the plankton gathered during the last cruise of the CARNEGIE. Sci. Res. Cruise VII of the Carnegie, 1928-1929, Biol. 2:163 pp., 128 figs.

Dictyocysta magna n.sp.
Kofoid, C. A. and A. S. Campbell, 1929
A conspectus of the marine and fresh-water Ciliata belonging to the suborder Tintinnoinea, with descriptions of new species principally from the Agassiz expedition to the eastern tropical Pacific, 1904-1905. Univ. Calif. Publ. Zool. 34:1-403, 697 text figs.

Dictyocysta mexicana
Kofoid, C.A. and A.S. Campbell, 1939
Reports on the scientific results of the expedition to the eastern tropical Pacific in charge of Alexander Agassiz, by the U.S. Fish Commission steamer "Albatross" from October 1904 to March 1905------XXXVII. The Ciliata: The Tintinnoinea. Bull.Mus. Comp.Zool. 84: 473 pp., 36 pls.

Dictyocysta mexicana n.sp.
Kofoid, C. A. and A. S. Campbell, 1929
A conspectus of the marine and fresh-water Ciliata belonging to the suborder Tintinnoinea, with descriptions of new species principally from the Agassiz expedition to the eastern tropical Pacific, 1904-1905. Univ. Calif. Publ. Zool. 34:1-403, 697 text figs.

Dictyocysta minor
Campbell, A. S., 1942
The oceanic Tintinnoina of the plankton gathered during the last cruise of the CARNEGIE. Sci. Res. Cruise VII of the Carnegie, 1928-1929, Biol. 2:163 pp., 128 figs.

Dictyocysta minor
Gaarder, K.R., 1946
Tintinnoidea. Rep.Sci.Res."Michael Sars" N.Atlantic Deep-Sea Exped., 1910, 2(1): 37 pp., 23 text figs.

Dictyocysta minor
Kofoid, C.A. and A.S. Campbell, 1939
Reports on the scientific results of the expedition to the eastern tropical Pacific in charge of Alexander Agassiz, by the U.S. Fish Commission steamer "Albatross" from October 1904 to March 1905------XXXVII. The Ciliata: The Tintinnoinea. Bull.Mus. Comp.Zool. 84: 473 pp., 36 pls.

Dictyocysta minor
Kofoid, C. A. and A. S. Campbell, 1929
A conspectus of the marine and fresh-water Ciliata belonging to the suborder Tintinnoinea, with descriptions of new species principally from the Agassiz expedition to the eastern tropical Pacific, 1904-1905. Univ. Calif. Publ. Zool. 34:1-403, 697 text figs.

Dictyocysta mitra
Balech, Enrique, 1962
Tintinnoinea y Dinoflagellata del Pacifico segun material de las Expediciones NORPAC y DOWNWIND del Instituto Scripps de Oceanografia. Revista. Mus. Argentino Ciencias Nat. "Bernardino Rivadavia", Ciencias Zool., 7(1):1-253.

Dictyocysta mitra
Balech, E., 1959.
Tintinnoinea del Mediterraneo. Trav. Sta. Zool. Villefranche-sur-Mer, 18(17); Reprinted from:
Trab. Inst. Español Oceanogr., Madrid, 19 Sept. 1959(28):88 pp., 22 pls.

Dictyocysta mitra
Campbell, A. S., 1942
The oceanic Tintinnoina of the plankton gathered during the last cruise of the CARNEGIE. Sci. Res. Cruise VII of the Carnegie, 1928-1929, Biol. 2:163 pp., 128 figs.

Dictyocysta mitra
Duran, M., 1951
Contribucion al estudio de los tintinidos del plancton de los costas de Castellon. (Mediterranes occidental). Publ. Inst. Biol. Aplic., Barcelone, 8: 101-120, 29 text figs.

Dictyocysta mitra
Entz, G., Jr. 1909
Studien über organisation und biologie der Tintinniden. Arch. f. Protistenkunde 15:93-226, Pls. 8-21, text figs.

Dictyocysta mitra
Hofker, J., 1931
Studien uber Tintinnoidea. Arch. f. Protistenk 75(3):315-402, 89 text figs.

Dictyocysta mitra
Jörgensen, E., 1924
Mediterranean Tintinnidae. Rept. Danish Oceanogr. Exped. 1908-10 to the Mediterranean and adjacent seas, Vol.II, Biol. J 3, 110 pp., 114 text figs.

Dictyocysta mitra
Kofoid, C.A. and A.S. Campbell, 1939
Reports on the scientific results of the expedition to the eastern tropical Pacific in charge of Alexander Agassiz, by the U.S. Fish Commission steamer "Albatross" from October 1904 to March 1905------XXXVII. The Ciliata: The Tintinnoinea. Bull.Mus. Comp.Zool. 84: 473 pp., 36 pls.

Dictyocysta mitra
Kofoid, C. A. and A. S. Campbell, 1929
A conspectus of the marine and fresh-water Ciliata belonging to the suborder Tintinnoinea, with descriptions of new species principally from the Agassiz expedition to the eastern tropical Pacific, 1904-1905. Univ. Calif. Publ. Zool. 34:1-403, 697 text figs.

Dictyocysta mitra
Laackmann, H., 1910
Die Tintinnodeen der Deutschen Südpolar Expedition 1901-1903. Deutsch. Südpolar Exped. 11, Zool. 3(4):341-496, pls. 33-51.

Dictyocysta mülleri
Campbell, A. S., 1942
The oceanic Tintinnoina of the plankton gathered during the last cruise of the CARNEGIE. Sci. Res. Cruise VII of the Carnegie, 1928-1929, Biol. 2:163 pp., 128 figs.

Dictyocysta mülleri n.sp.
Jörgensen, E., 1924
Mediterranean Tintinnidae. Rept. Danish Oceanogr. Exped. 1908-10 to the Mediterranean and adjacent seas, Vol.II, Biol. J 3, 110 pp., 114 text figs.

Dictyocysta mülleri
Gaarder, K.R., 1946
Tintinnoidea. Rep.Sci.Res."Michael Sars" N.Atlantic Deep-Sea Exped., 1910, 2(1): 37 pp., 23 text figs.

Dictyocysta mülleri
Kofoid, C.A. and A.S. Campbell, 1939
Reports on the scientific results of the expedition to the eastern tropical Pacific in charge of Alexander Agassiz, by the U.S. Fish Commission steamer "Albatross" from October 1904 to March 1905------XXXVII. The Ciliata: The Tintinnoinea. Bull.Mus. Comp.Zool. 84: 473 pp., 36 pls.

Dictyocysta mülleri
Kofoid, C. A. and A. S. Campbell, 1929
A conspectus of the marine and fresh-water Ciliata belonging to the suborder Tintinnoinea, with descriptions of new species principally from the Agassiz expedition to the eastern tropical Pacific, 1904-1905. Univ. Calif. Publ. Zool. 34:1-403, 697 text figs.

Dictyocysta mülleri
Sousa e Silva, E. de, 1950.
Les tintinnides de la Baie de Cascais. Bull. Inst. Océan., Monaco, No. 979:28 pp., 4 pls.

Dictyocysta nidulus
Gaarder, K.R., 1946
Tintinnoidea. Rep.Sci.Res."Michael Sars" N.Atlantic Deep-Sea Exped., 1910, 2(1): 37 pp., 23 text figs.

Dictyocysta nidulus n.sp.
Kofoid, C. A. and A. S. Campbell, 1929
A conspectus of the marine and fresh-water Ciliata belonging to the suborder Tintinnoinea, with descriptions of new species principally from the Agassiz expedition to the eastern tropical Pacific, 1904-1905. Univ. Calif. Publ. Zool. 34:1-403, 697 text figs.

Dictyocysta obtusa
Kofoid, C. A. and A. S. Campbell, 1929
A conspectus of the marine and fresh-water Ciliata belonging to the suborder Tintinnoinea, with descriptions of new species principally from the Agassiz expedition to the eastern tropical Pacific, 1904-1905. Univ. Calif. Publ. Zool. 34:1-403, 697 text figs.

Dictyocysta obtusa
Rampi, L., 1939
Primo contributo alla conoscenza dei tintinnoidi do Mare Ligure. Atti della Soc. Ital. di Sci. Nat. 78:67-81, 58 text figs.

Dictyocysta occidentalis
Campbell, A. S., 1942
The oceanic Tintinnoina of the plankton gathered during the last cruise of the CARNEGIE. Sci. Res. Cruise VII of the Carnegie, 1928-1929, Biol. 2:163 pp., 128 figs.

Dictyocysta occidentalis
Kofoid, C.A. and A.S. Campbell, 1939
Reports on the scientific results of the expedition to the eastern tropical Pacific in charge of Alexander Agassiz, by the U.S. Fish Commission steamer "Albatross" from October 1904 to March 1905------XXXVII. The Ciliata: The Tintinnoinea. Bull.Mus. Comp.Zool. 84: 473 pp., 36 pls.

Dictyocysta occidentalis n.sp.
Kofoid, C. A. and A. S. Campbell, 1929
A conspectus of the marine and fresh-water Ciliata belonging to the suborder Tintinnoinea, with descriptions of new species principally from the Agassiz expedition to the eastern tropical Pacific, 1904-1905. Univ. Calif. Publ. Zool. 34:1-403, 697 text figs.

Dictyocysta ovalis
Kofoid, C. A. and A. S. Campbell, 1929
A conspectus of the marine and fresh-water Ciliata belonging to the suborder Tintinnoinea, with descriptions of new species principally from the Agassiz expedition to the eastern tropical Pacific, 1904-1905. Univ. Calif. Publ. Zool. 34:1-403, 697 text figs.

Dictyocysta pacifica
Campbell, A. S., 1942
The oceanic Tintinnoina of the plankton gathered during the last cruise of the CARNEGIE. Sci. Res. Cruise VII of the Carnegie, 1928-1929, Biol. 2:163 pp., 128 figs.

Dictyocysta pacifica
Kofoid, C.A. and A.S. Campbell, 1939
Reports on the scientific results of the expedition to the eastern tropical Pacific in charge of Alexander Agassiz, by the U.S. Fish Commission steamer "Albatross" from October 1904 to March 1905------XXXVII. The Ciliata: The Tintinnoimea. Bull. Mus. Comp. Zool. 84: 473 pp., 36 pls.

Dictyocysta pacifica n.sp.
Kofoid, C. A. and A. S. Campbell, 1929
A conspectus of the marine and freshwater Ciliata belonging to the suborder Tintinnoinea, with descriptions of new species principally from the Agassiz expedition to the eastern tropical Pacific, 1904-1905. Univ. Calif. Publ. Zool. 34:1-403, 697 text figs.

Dictyocysta pacifica
Komarovsky, B., 1959
The Tintinnina of the Gulf of Eylath (Aqaba). Contrib. Knowl. Red Sea, Bull. Sea Fish. Res. Sta. Haifa, Israel 21 (14):1-40.

Dictyocysta polygonata
de Sousa e Silva, E., 1956.
Contribution à l'étude du microplancton de Dakar et des regions maritimes voisines. Bull. I.F.A.N., 8(2):335-371, 7 pls.

Dictyocysta polygonata
Kofoid, C.A. and A.S. Campbell, 1939
Reports on the scientific results of the expedition to the eastern tropical Pacific in charge of Alexander Agassiz, by the U.S. Fish Commission steamer "Albatross" from October 1904 to March 1905------XXXVII. The Ciliata: The Tintinnoimea. Bull. Mus. Comp. Zool. 84: 473 pp., 36 pls.

Dictyocysta polygonata n.sp.
Kofoid, C. A. and A. S. Campbell, 1929
A conspectus of the marine and freshwater Ciliata belonging to the suborder Tintinnoinea, with descriptions of new species principally from the Agassiz expedition to the eastern tropical Pacific, 1904-1905. Univ. Calif. Publ. Zool. 34:1-403, 697 text figs.

Dictyocysta poligonata
Rampi, L., 1939
Primo contributo alla conoscenza dei tintinnoidi de Mare Ligure. Atti della Soc. Ital. di Sci. Nat. 78:67-81, 58 text figs.

Dictyocysta reticulata
Campbell, A. S., 1942
The oceanic Tintinnoina of the plankton gathered during the last cruise of the CARNEGIE. Sci. Res. Cruise VII of the Carnegie, 1928-1929, Biol. 2:163 pp., 128 figs.

Dictyocysta reticulata
Gaarder, K.R., 1946
Tintinnoidea. Rep. Sci. Res. "Michael Sars" N. Atlantic Deep-Sea Exped., 1910, 2(1): 37 pp., 23 text figs.

Dictyocysta reticulata
Kofoid, C.A. and A.S. Campbell, 1939
Reports on the scientific results of the expedition to the eastern tropical Pacific in charge of Alexander Agassiz, by the U.S. Fish Commission steamer "Albatross" from October 1904 to March 1905------XXXVII. The Ciliata: The Tintinnoimea. Bull. Mus. Comp. Zool. 84: 473 pp., 36 pls.

Dictyocysta reticulata n.sp.
Kofoid, C. A. and A. S. Campbell, 1929
A conspectus of the marine and freshwater Ciliata belonging to the suborder Tintinnoinea, with descriptions of new species principally from the Agassiz expedition to the eastern tropical Pacific, 1904-1905. Univ. Calif. Publ. Zool. 34:1-403, 697 text figs.

Dictyocysta reticulata
Marshall, S. M., 1934
The silicoflagellata and tintinnoinea. Great Barrier Reef Exped., 1928-29. Sci. Repts. 4(15):623-664, 43 text figs.

Dictyocysta reticulata
Rampi, L., 1939
Primo contributo alla conoscenza dei tintinnoidi de Mare Ligure. Atti della Soc. Ital. di Sci. Nat. 78:67-81, 58 text figs.

Dictyocysta ridulus
Campbell, A. S., 1942
The oceanic Tintinnoina of the plankton gathered during the last cruise of the CARNEGIE. Sci. Res. Cruise VII of the Carnegie, 1928-1929, Biol. 2:163 pp., 128 figs.

Dictyocysta speciosa
Campbell, A. S., 1942
The oceanic Tintinnoina of the plankton gathered during the last cruise of the CARNEGIE. Sci. Res. Cruise VII of the Carnegie, 1928-1929, Biol. 2:163 pp., 128 figs.

Dictyocysta speciosa
Gaarder, K.R., 1946
Tintinnoidea. Rep. Sci. Res. "Michael Sars" N. Atlantic Deep-Sea Exped., 1910, 2(1): 37 pp., 23 text figs.

Dictyocysta speciosa
Kofoid, C. A. and A. S. Campbell, 1929
A conspectus of the marine and freshwater Ciliata belonging to the suborder Tintinnoinea, with descriptions of new species principally from the Agassiz expedition to the eastern tropical Pacific, 1904-1905. Univ. Calif. Publ. Zool. 34:1-403, 697 text figs.

Dictyocysta spinosa
Campbell, A. S., 1942
The oceanic Tintinnoina of the plankton gathered during the last cruise of the CARNEGIE. Sci. Res. Cruise VII of the Carnegie, 1928-1929, Biol. 2:163 pp., 128 figs.

Dictyocysta spinosa
Kofoid, C.A. and A.S. Campbell, 1939
Reports on the scientific results of the expedition to the eastern tropical Pacific in charge of Alexander Agassiz, by the U.S. Fish Commission steamer "Albatross" from October 1904 to March 1905------XXXVII. The Ciliata: The Tintinnoimea. Bull. Mus. Comp. Zool. 84: 473 pp., 36 pls.

Dictyocysta spinosa n.sp.
Kofoid, C. A. and A. S. Campbell, 1929
A conspectus of the marine and freshwater Ciliata belonging to the suborder Tintinnoinea, with descriptions of new species principally from the Agassiz expedition to the eastern tropical Pacific, 1904-1905. Univ. Calif. Publ. Zool. 34:1-403, 697 text figs.

Dictyocysta templum
Delegazione Italiana della Commissione Internazionale per l'Esplorazione Scientifica del Mediterraneo, 1941
Note sul plancton della Laguna veneta. [Memoria CCLXXXI], Arch. di Ocean. e Limn. Anno I, Fasc. I, 1941 XIX: 31-57 pp.

Dictyocysta templum
Entz, G., Jr., 1909
Studien über organisation und biologie der Tintinniden. Arch. f. Protistenkunde 15:93-226, Pls. 8-21, text figs.

Dictyocysta templum
Fol, H., 1884.
Sur la famille des Tintinnodea. Recueil Zool. Suisse 1:27-64, Pls. 4-5.

Dictyocysta templum
Jørgensen, E., 1905
B. Protistplankton and the diatoms in bottom samples. Hydrographical and biological investigations in Norwegian fjords. Bergens Mus. Skr. 7: 49-225.

Dictyocysta templum
Jørgensen, E., 1900
Ueber die Tintinnodeen der norwegischen Westküste. Bergens Mus. Aarb., 1899(2):48 pp., 3 pls.

Dictyocysta templum
Laackmann, H., 1910
Die Tintinnodeen der Deutschen Südpolar Expedition 1901-1903. Deutsch. Südpolar Exped. 11, Zool. 3(4):341-496, pls. 33-51.

Dictyocysta tiara
Campbell, A. S., 1942
The oceanic Tintinnoina of the plankton gathered during the last cruise of the CARNEGIE. Sci. Res. Cruise VII of the Carnegie, 1928-1929, Biol. 2:163 pp., 128 figs.

Dictyocysta tiara
Kofoid, C.A. and A.S. Campbell, 1939
Reports on the scientific results of the expedition to the eastern tropical Pacific in charge of Alexander Agassiz, by the U.S. Fish Commission steamer "Albatross" from October 1904 to March 1905------XXXVII. The Ciliata: The Tintinnoimea. Bull. Mus. Comp. Zool. 84: 473 pp., 36 pls.

Dictyocysta tiara
Kofoid, C. A. and A. S. Campbell, 1929
A conspectus of the marine and freshwater Ciliata belonging to the suborder Tintinnoinea, with descriptions of new species principally from the Agassiz expedition to the eastern tropical Pacific, 1904-1905. Univ. Calif. Publ. Zool. 34:1-403, 697 text figs.

Epicancella nervosa
Balech, Enrique, 1962
Tintinnoinea y Dinoflagellata del Pacifico segun material de las Expediciones NORPAC y DOWNWIND del Instituto Scripps de Oceanografia. Revista, Mus. Argentino Ciencias Nat. "Bernardino Rivadavia", Ciencias Zool., 7(1):1-253.

Epicancella nervosa
Campbell, A. S., 1942
The oceanic Tintinnoina of the plankton gathered during the last cruise of the CARNEGIE. Sci. Res. Cruise VII of the Carnegie, 1928-1929, Biol. 2:163 pp., 128 figs.

Epicancella nervosa
Kofoid, C.A. and A.S. Campbell, 1939
Reports on the scientific results of the expedition to the eastern tropical Pacific in charge of Alexander Agassiz, by the U.S. Fish Commission steamer "Albatross" from October 1904 to March 1905------XXXVII. The Ciliata: The Tintinnoimea. Bull. Mus. Comp. Zool. 84: 473 pp., 36 pls.

Epicancella nervosa
Komarovsky, B., 1959
The Tintinnina of the Gulf of Eylath (Aqaba). Contrib. Knowl. Red Sea, Bull. Sea Fish. Res. Sta. Haifa, Israel 21 (14):1-40.

Epicranella n.gen.
Kofoid, C. A. and A. S. Campbell, 1929
A conspectus of the marine and freshwater Ciliata belonging to the suborder Tintinnoinea, with descriptions of new species principally from the Agassiz expedition to the eastern tropical Pacific, 1904-1905. Univ. Calif. Publ. Zool. 34:1-403, 697 text figs.

Epicranella bella

Kofoid, C.A. and A.S. Campbell, 1939
Reports on the scientific results of the expedition to the eastern tropical Pacific in charge of Alexander Agassiz, by the U.S. Fish Commission steamer "Albatross" from October 1904 to March 1905-----XXXVII. The Ciliata: The Tintinnoimea. Bull. Mus. Comp. Zool. 84: 473 pp., 36 pls.

Epicranella bella n.sp.

Kofoid, C. A. and A. S. Campbell, 1929
A conspectus of the marine and fresh-water Ciliata belonging to the suborder Tintinnoinea, with descriptions of new species principally from the Agassiz expedition to the eastern tropical Pacific, 1904-1905. Univ. Calif. Publ. Zool. 34:1-403, 697 text figs.

Epicranella bellissima n.sp.

Kofoid, C.A. and A.S. Campbell, 1939
Reports on the scientific results of the expedition to the eastern tropical Pacific in charge of Alexander Agassiz, by the U.S. Fish Commission steamer "Albatross" from October 1904 to March 1905-----XXXVII. The Ciliata: The Tintinnoimea. Bull. Mus. Comp. Zool. 84: 473 pp., 36 pls.

Epicranella dextra n.sp.

Kofoid, C.A. and A.S. Campbell, 1939
Reports on the scientific results of the expedition to the eastern tropical Pacific in charge of Alexander Agassiz, by the U.S. Fish Commission steamer "Albatross" from October 1904 to March 1905-----XXXVII. The Ciliata: The Tintinnoimea. Bull. Mus. Comp. Zool. 84: 473 pp., 36 pls.

Epicranella magnifica n.sp.

Kofoid, C.A. and A.S. Campbell, 1939
Reports on the scientific results of the expedition to the eastern tropical Pacific in charge of Alexander Agassiz, by the U.S. Fish Commission steamer "Albatross" from October 1904 to March 1905-----XXXVII. The Ciliata: The Tintinnoimea. Bull. Mus. Comp. Zool. 84: 473 pp., 36 pls.

Epicranella prismatica

Kofoid, C.A. and A.S. Campbell, 1939
Reports on the scientific results of the expedition to the eastern tropical Pacific in charge of Alexander Agassiz, by the U.S. Fish Commission steamer "Albatross" from October 1904 to March 1905-----XXXVII. The Ciliata: The Tintinnoimea. Bull. Mus. Comp. Zool. 84: 473 pp., 36 pls.

Epicranella prismatida n. sp.

Kofoid, C. A. and A. S. Campbell, 1929
A conspectus of the marine and fresh-water Ciliata belonging to the suborder Tintinnoinea, with descriptions of new species principally from the Agassiz expedition to the eastern tropical Pacific, 1904-1905. Univ. Calif. Publ. Zool. 34:1-403, 697 text figs.

Epiorella n.gen.

Kofoid, C.A. and A.S. Campbell, 1939
Reports on the scientific results of the expedition to the eastern tropical Pacific in charge of Alexander Agassiz, by the U.S. Fish Commission steamer "Albatross" from October 1904 to March 1905-----XXXVII. The Ciliata: The Tintinnoimea. Bull. Mus. Comp. Zool. 84: 473 pp., 36 pls.

Epiorella acuta

Campbell, A. S., 1942
The oceanic Tintinnoina of the plankton gathered during the last cruise of the CARNEGIE. Sci. Res. Cruise VII of the Carnegie, 1928-1929, Biol. 2:163 pp., 128 figs.

Epiorella acuta

Kofoid, C.A. and A.S. Campbell, 1939
Reports on the scientific results of the expedition to the eastern tropical Pacific in charge of Alexander Agassiz, by the U.S. Fish Commission steamer "Albatross" from October 1904 to March 1905-----XXXVII. The Ciliata: The Tintinnoimea. Bull. Mus. Comp. Zool. 84: 473 pp., 36 pls.

Epiorella brandti

Campbell, A. S., 1942
The oceanic Tintinnoina of the plankton gathered during the last cruise of the CARNEGIE. Sci. Res. Cruise VII of the Carnegie, 1928-1929, Biol. 2:163 pp., 128 figs.

Epiorella curta

Campbell, A. S., 1942
The oceanic Tintinnoina of the plankton gathered during the last cruise of the CARNEGIE. Sci. Res. Cruise VII of the Carnegie, 1928-1929, Biol. 2:163 pp., 128 figs.

Epiorella curta

Kofoid, C.A. and A.S. Campbell, 1939
Reports on the scientific results of the expedition to the eastern tropical Pacific in charge of Alexander Agassiz, by the U.S. Fish Commission steamer "Albatross" from October 1904 to March 1905-----XXXVII. The Ciliata: The Tintinnoimea. Bull. Mus. Comp. Zool. 84: 473 pp., 36 pls.

Epiorella curta

Komarovsky, B., 1959
The Tintinnina of the Gulf of Eylath (Aqaba). Contrib. Knowl. Red Sea, Bull. Sea Fish. Res. Sta. Haifa, Israel 21 (14):1-40.

Epiorella healdi

Campbell, A. S., 1942
The oceanic Tintinnoina of the plankton gathered during the last cruise of the CARNEGIE. Sci. Res. Cruise VII of the Carnegie, 1928-1929, Biol. 2:163 pp., 128 figs.

Epiorella healdi

de Sousa e Silva, E., 1956.
Contribution à l'étude du microplancton de Dakar et des regions maritimes voisines. Bull. I.F.A.N., 8(2):335-371, 7 Pls.

Epiorella healdi

Kofoid, C.A. and A.S. Campbell, 1939
Reports on the scientific results of the expedition to the eastern tropical Pacific in charge of Alexander Agassiz, by the U.S. Fish Commission steamer "Albatross" from October 1904 to March 1905-----XXXVII. The Ciliata: The Tintinnoimea. Bull. Mus. Comp. Zool. 84: 473 pp., 36 pls.

Epiorella ralumensis

Kofoid, C.A. and A.S. Campbell, 1939
Reports on the scientific results of the expedition to the eastern tropical Pacific in charge of Alexander Agassiz, by the U.S. Fish Commission steamer "Albatross" from October 1904 to March 1905-----XXXVII. The Ciliata: The Tintinnoimea. Bull. Mus. Comp. Zool. 84: 473 pp., 36 pls.

Epiplocylis n.gen.

Jörgensen, E., 1924
Mediterranean Tintinnidae. Rept. Danish Oceanogr. Exped. 1908-10 to the Mediterranean and adjacent seas, Vol.II, Biol. J 3, 110 pp., 114 text figs.

Epiplocylis acuminata

Balech, Enrique, 1962
Tintinnoinea y Dinoflagellata del Pacifico segun material de las Expediciones NORPAC y DOWNWIND del Instituto Scripps de Oceanografia. Revista, Mus. Argentino Ciencias Nat. "Bernardino Rivadavia", Ciencias Zool., 7(1):1-253.

Epiplocylis acuminata

Balech, E., 1959.
Tintinnoinea del Mediterraneo. Trav. Sta. Zool., Villefranche-sur-Mer, 18(17): Reprinted from: Trab. Inst. Español Oceanogr., Madrid, 19 Sept. 1959(28):88 pp., 22 pls.

Epiplocylis acuminata

Duran, M., 1951
Contribucion al estudio de los tintinidos del plancton de las costas de Castellon. (Mediterraneo occidental). Publ. Inst. Biol. Aplic., Barcelona, 8: 101-120, 29 text figs.

Epiplocylis acuminata

Gaarder, K.R., 1946
Tintinnoidea. Rep.Sci.Res. "Michael Sars" N.Atlantic Deep-Sea Exped., 1910, 2(1): 37 pp., 23 text figs.

Epiplocylis acuminata

Jörgensen, E., 1924
Mediterranean Tintinnidae. Rept. Danish Oceanogr. Exped. 1908-10 to the Mediterranean and adjacent seas, Vol.II, Biol. J 3, 110 pp., 114 text figs.

Epiplocylis acuminata

Kofoid, C. A. and A. S. Campbell, 1929
A conspectus of the marine and fresh-water Ciliata belonging to the suborder Tintinnoinea, with descriptions of new species principally from the Agassiz expedition to the eastern tropical Pacific, 1904-1905. Univ. Calif. Publ. Zool. 34:1-403, 697 text figs.

Epiplocylis acuminata

Rampi, L., 1939
Primo contributo alla conoscenza dei tintinnoidi do Maro Ligure. Atti della Soc. Ital. di Sci. Nat. 78:67-81, 58 text figs.

Epiplocylis acuta

Gaarder, K.R., 1946
Tintinnoidea. Rep.Sci.Res. "Michael Sars" N.Atlantic Deep-Sea Exped., 1910, 2(1): 37 pp., 23 text figs.

Epiplocylis acuta n.sp.

Kofoid, C. A. and A. S. Campbell, 1929
A conspectus of the marine and fresh-water Ciliata belonging to the suborder Tintinnoinea, with descriptions of new species principally from the Agassiz expedition to the eastern tropical Pacific, 1904-1905. Univ. Calif. Publ. Zool. 34:1-403, 697 text figs.

Epiplocylis atlantica

Campbell, A. S., 1942
The oceanic Tintinnoina of the plankton gathered during the last cruise of the CARNEGIE. Sci. Res. Cruise VII of the Carnegie, 1928-1929, Biol. 2:163 pp., 128 figs.

Epiplocylis atlantica n.sp.

Kofoid, C. A. and A. S. Campbell, 1929
A conspectus of the marine and fresh-water Ciliata belonging to the suborder Tintinnoinea, with descriptions of new species principally from the Agassiz expedition to the eastern tropical Pacific, 1904-1905. Univ. Calif. Publ. Zool. 34:1-403, 697 text figs.

Epiplocylis atlantica

Komarovsky, B., 1959
The Tintinnina of the Gulf of Eylath (Aqaba). Contrib. Knowl. Red Sea, Bull. Sea Fish. Res. Sta. Haifa, Israel 21 (14):1-40.

Epiplocylis blanda

Campbell, A. S., 1942
The oceanic Tintinnoina of the plankton gathered during the last cruise of the CARNEGIE. Sci. Res. Cruise VII of the Carnegie, 1928-1929, Biol. 2:163 pp., 128 figs.

Epiplocylis blanda

Gaarder, K.R., 1946
Tintinnoidea. Rep.Sci.Res. "Michael Sars" N.Atlantic Deep-Sea Exped., 1910, 2(1): 37 pp., 23 text figs.

Epiplocylis blanda

Kofoid, C.A. and A.S. Campbell, 1939
Reports on the scientific results of the expedition to the eastern tropical Pacific in charge of Alexander Agassiz, by the U.S. Fish Commission steamer "Albatross" from October 1904 to March 1905-----XXXVII. The Ciliata: The Tintinnoimea. Bull. Mus. Comp. Zool. 84: 473 pp., 36 pls.

Epiplocylis blanda

Kofoid, C. A. and A. S. Campbell, 1929
A conspectus of the marine and fresh-water Ciliata belonging to the suborder Tintinnoinea, with descriptions of new species principally from the Agassiz expedition to the eastern tropical Pacific, 1904-1905. Univ. Calif. Publ. Zool. 34:1-403, 697 text figs.

Epiplocylis blanda

Komarovsky, B., 1959
The Tintinnina of the Gulf of Eylath (Aqaba). Contrib. Knowl. Red Sea, Bull. Sea Fish. Res. Sta. Haifa, Israel 21 (14):1-40.

Epiplocylis blanda

Marshall, S. M., 1934
The silicoflagellata and tintinnoinea. Great Barrier Reef Exped., 1928-29. Sci. Repts. 4(15):623-664, 43 text figs.

Epiplocylis brandti

Gaarder, K.R., 1946
Tintinnoidea. Rep.Sci.Res."Michael Sars" N.Atlantic Deep-Sea Exped., 1910, 2(1): 37 pp., 23 text figs.

Epiplocylis brandti n.sp.

Kofoid, C. A. and A. S. Campbell, 1929
A conspectus of the marine and fresh-water Ciliata belonging to the suborder Tintinnoinea, with descriptions of new species principally from the Agassiz expedition to the eastern tropical Pacific, 1904-1905. Univ. Calif. Publ. Zool. 34:1-403, 697 text figs.

Epiplocylis bruhni

Balech, Enrique, 1962
Tintinnoinea y Dinoflagellata del Pacifico segun material de las Expediciones NORPAC y DOWNWIND del Instituto Scripps de Oceanografia. Revista, Mus. Argentino Ciencias Nat. "Bernardino Rivadavia", Ciencias Zool., 7(1):1-253.

Epiplocylis bruhni

Kofoid, C. A. and A. S. Campbell, 1929
A conspectus of the marine and fresh-water Ciliata belonging to the suborder Tintinnoinea, with descriptions of new species principally from the Agassiz expedition to the eastern tropical Pacific, 1904-1905. Univ. Calif. Publ. Zool. 34:1-403, 697 text figs.

Epiplocylis calyx

Kofoid, C. A. and A. S. Campbell, 1929
A conspectus of the marine and fresh-water Ciliata belonging to the suborder Tintinnoinea, with descriptions of new species principally from the Agassiz expedition to the eastern tropical Pacific, 1904-1905. Univ. Calif. Publ. Zool. 34:1-403, 697 text figs.

Epiplocylis carnegiei n.sp.

Campbell, A. S., 1942
The oceanic Tintinnoina of the plankton gathered during the last cruise of the CARNEGIE. Sci. Res. Cruise VII of the Carnegie, 1928-1929, Biol. 2:163 pp., 128 figs.

Epiplocylis constricta

Campbell, A. S., 1942
The oceanic Tintinnoina of the plankton gathered during the last cruise of the CARNEGIE. Sci. Res. Cruise VII of the Carnegie, 1928-1929, Biol. 2:163 pp., 128 figs.

Epiplocylis constricta

Gaarder, K.R., 1946
Tintinnoidea. Rep.Sci.Res."Michael Sars" N.Atlantic Deep-Sea Exped., 1910, 2(1): 37 pp., 23 text figs.

Epiplocylis constricta

Kofoid, C.A. and A.S. Campbell, 1939
Reports on the scientific results of the expedition to the eastern tropical Pacific in charge of Alexander Agassiz, by the U.S. Fish Commission steamer "Albatross" from October 1904 to March 1905-----XXXVII. The Ciliata: The Tintinnoimea. Bull.Mus. Comp.Zool. 84: 473 pp., 36 pls.

Epiplocylis constricta n.sp.

Kofoid, C. A. and A. S. Campbell, 1929
A conspectus of the marine and fresh-water Ciliata belonging to the suborder Tintinnoinea, with descriptions of new species principally from the Agassiz expedition to the eastern tropical Pacific, 1904-1905. Univ. Calif. Publ. Zool. 34:1-403, 697 text figs.

Epiplocylis constricta

Marshall, S. M., 1934
The silicoflagellata and tintinnoinea. Great Barrier Reef Exped., 1928-29. Sci. Repts. 4(15):623-664, 43 text figs.

Epiplocyloides curta

Duran, M., 1957
Nota sobre algunos tintinnoineas del plancton de Puerto Rico. Invest. Pesquerias, Barcelona, 8: 97-120.

Epiplocylis curta n.sp.

Kofoid, C. A. and A. S. Campbell, 1929
A conspectus of the marine and fresh-water Ciliata belonging to the suborder Tintinnoinea, with descriptions of new species principally from the Agassiz expedition to the eastern tropical Pacific, 1904-1905. Univ. Calif. Publ. Zool. 34:1-403, 697 text figs.

Epiplocylis deflexa

Balech, Enrique, 1962
Tintinnoinea y Dinoflagellata del Pacifico segun material de las Expediciones NORPAC y DOWNWIND del Instituto Scripps de Oceanografia. Revista, Mus. Argentino Ciencias Nat. "Bernardino Rivadavia", Ciencias Zool., 7(1):1-253.

Epiplocylis deflexa

Campbell, A. S., 1942
The oceanic Tintinnoina of the plankton gathered during the last cruise of the CARNEGIE. Sci. Res. Cruise VII of the Carnegie, 1928-1929, Biol. 2:163 pp., 128 figs.

Epiplocylis deflexa

Kofoid, C.A. and A.S. Campbell, 1939
Reports on the scientific results of the expedition to the eastern tropical Pacific in charge of Alexander Agassiz, by the U.S. Fish Commission steamer "Albatross" from October 1904 to March 1905-----XXXVII. The Ciliata: The Tintinnoimea. Bull.Mus. Comp.Zool. 84: 473 pp., 36 pls.

Epiplocylis deflexa n.sp.

Kofoid, C. A. and A. S. Campbell, 1929
A conspectus of the marine and fresh-water Ciliata belonging to the suborder Tintinnoinea, with descriptions of new species principally from the Agassiz expedition to the eastern tropical Pacific, 1904-1905. Univ. Calif. Publ. Zool. 34:1-403, 697 text figs.

Epiplocylis deflexa

Marshall, S. M., 1934
The silicoflagellata and tintinnoinea. Great Barrier Reef Exped., 1928-29. Sci. Repts. 4(15):623-664, 43 text figs.

Epiplocylus deflexa

Rampi, L., 1948
Sur quelques Tintinnides (Infusoires louques) du Pacifique subtropical (Recoltes Alain Gerbault) Bull. l'Inst. Ocean., Monaco, No.938, 4 pp.

Epiplocylis exigua

Campbell, A. S., 1942
The oceanic Tintinnoina of the plankton gathered during the last cruise of the CARNEGIE. Sci. Res. Cruise VII of the Carnegie, 1928-1929, Biol. 2:163 pp., 128 figs.

Epiplocylis exigua

Kofoid, C.A. and A.S. Campbell, 1939
Reports on the scientific results of the expedition to the eastern tropical Pacific in charge of Alexander Agassiz, by the U.S. Fish Commission steamer "Albatross" from October 1904 to March 1905-----XXXVII. The Ciliata: The Tintinnoimea. Bull.Mus. Comp.Zool. 84: 473 pp., 36 pls.

Epiplocylis exigua n.sp.

Kofoid, C. A. and A. S. Campbell, 1929
A conspectus of the marine and fresh-water Ciliata belonging to the suborder Tintinnoinea, with descriptions of new species principally from the Agassiz expedition to the eastern tropical Pacific, 1904-1905. Univ. Calif. Publ. Zool. 34:1-403, 697 text figs.

Epiplocylis exigua

Marshall, S. M., 1934
The silicoflagellata and tintinnoinea. Great Barrier Reef Exped., 1928-29. Sci. Repts. 4(15):623-664, 43 text figs.

Epiplocylus exigua

Rampi, L., 1948
Sur quelques Tintinnides (Infusoires louques) du Pacifique subtropical (Recoltes Alain Gerbault) Bull. l'Inst. Ocean., Monaco, No.938, 4 pp.

Epiplocylis exquisita

Campbell, A. S., 1942
The oceanic Tintinnoina of the plankton gathered during the last cruise of the CARNEGIE. Sci. Res. Cruise VII of the Carnegie, 1928-1929, Biol. 2:163 pp., 128 figs.

Epiplocylis exquisita

Kofoid, C.A. and A.S. Campbell, 1939
Reports on the scientific results of the expedition to the eastern tropical Pacific in charge of Alexander Agassiz, by the U.S. Fish Commission steamer "Albatross" from October 1904 to March 1905-----XXXVII. The Ciliata: The Tintinnoimea. Bull.Mus. Comp.Zool. 84: 473 pp., 36 pls.

Epiplocylis exquisita n.sp.

Kofoid, C. A. and A. S. Campbell, 1929
A conspectus of the marine and fresh-water Ciliata belonging to the suborder Tintinnoinea, with descriptions of new species principally from the Agassiz expedition to the eastern tropical Pacific, 1904-1905. Univ. Calif. Publ. Zool. 34:1-403, 697 text figs.

Epiplocylis exquisita

Komarovsky, B., 1959
The Tintinnina of the Gulf of Eylath (Aqaba). Contrib. Knowl. Red Sea, Bull. Sea Fish. Res. Sta. Haifa, Israel 21 (14):1-40.

Epiplocylis freymadli

Kofoid, C. A. and A. S. Campbell, 1929
A conspectus of the marine and fresh-water Ciliata belonging to the suborder Tintinnoinea, with descriptions of new species principally from the Agassiz expedition to the eastern tropical Pacific, 1904-1905. Univ. Calif. Publ. Zool. 34:1-403, 697 text figs.

Epiplocylis healdi n.sp.

Kofoid, C. A. and A. S. Campbell, 1929
A conspectus of the marine and fresh-water Ciliata belonging to the suborder Tintinnoinea, with descriptions of new species principally from the Agassiz expedition to the eastern tropical Pacific, 1904-1905. Univ. Calif. Publ. Zool. 34:1-403, 697 text figs.

Epiplocylis healdi

Marshall, S. M., 1934
The silicoflagellata and tintinnoinea. Great Barrier Reef Exped., 1928-29. Sci. Repts. 4(15):623-664, 43 text figs.

Epiplocylis impensa

Campbell, A. S., 1942
The oceanic Tintinnoina of the plankton gathered during the last cruise of the CARNEGIE. Sci. Res. Cruise VII of the Carnegie, 1928-1929, Biol. 2:163 pp., 128 figs.

Epiplocylis impensa

Kofoid, C.A. and A.S. Campbell, 1939
Reports on the scientific results of the expedition to the eastern tropical Pacific in charge of Alexander Agassiz, by the U.S. Fish Commission steamer "Albatross" from October 1904 to March 1905-----XXXVII. The Ciliata: The Tintinnoimea. Bull. Mus. Comp. Zool. 84: 473 pp., 36 pls.

Epiplocylis impensa n.sp.

Kofoid, C. A. and A. S. Campbell, 1929
A conspectus of the marine and fresh-water Ciliata belonging to the suborder Tintinnoinea, with descriptions of new species principally from the Agassiz expedition to the eastern tropical Pacific, 1904-1905. Univ. Calif. Publ. Zool. 34:1-403, 697 text figs.

Epiplocylis inconspicuata

Campbell, A. S., 1942
The oceanic Tintinnoina of the plankton gathered during the last cruise of the CARNEGIE. Sci. Res. Cruise VII of the Carnegie, 1928-1929, Biol. 2:163 pp., 128 figs.

Epiplocylis inconspicuata n.sp.

Kofoid, C. A. and A. S. Campbell, 1929
A conspectus of the marine and fresh-water Ciliata belonging to the suborder Tintinnoinea, with descriptions of new species principally from the Agassiz expedition to the eastern tropical Pacific, 1904-1905. Univ. Calif. Publ. Zool. 34:1-403, 697 text figs.

Epiplocylis inflata

Campbell, A. S., 1942
The oceanic Tintinnoina of the plankton gathered during the last cruise of the CARNEGIE. Sci. Res. Cruise VII of the Carnegie, 1928-1929, Biol. 2:163 pp., 128 figs.

Epiplocylis inflata n.sp.

Kofoid, C. A. and A. S. Campbell, 1929
A conspectus of the marine and fresh-water Ciliata belonging to the suborder Tintinnoinea, with descriptions of new species principally from the Agassiz expedition to the eastern tropical Pacific, 1904-1905. Univ. Calif. Publ. Zool. 34:1-403, 697 text figs.

Epiplocylis laackmanni nom. nov.

Kofoid, C. A. and A. S. Campbell, 1929
A conspectus of the marine and fresh-water Ciliata belonging to the suborder Tintinnoinea, with descriptions of new species principally from the Agassiz expedition to the eastern tropical Pacific, 1904-1905. Univ. Calif. Publ. Zool. 34:1-403, 697 text figs.

Epiplocylis labiosa

Balech, Enrique, 1962
Tintinnoinea y Dinoflagellata del Pacifico segun material de las Expediciones NORPAC y DOWNWIND del Instituto Scripps de Oceanografia. Revista, Mus. Argentino Ciencias Nat. "Bernardino Rivadavia", Ciencias Zool., 7(1):1-253.

Epiplocylis labiosa

Campbell, A. S., 1942
The oceanic Tintinnoina of the plankton gathered during the last cruise of the CARNEGIE. Sci. Res. Cruise VII of the Carnegie, 1928-1929, Biol. 2:163 pp., 128 figs.

Epiplocylis labiosa n.sp.

Kofoid, C. A. and A. S. Campbell, 1929
A conspectus of the marine and fresh-water Ciliata belonging to the suborder Tintinnoinea, with descriptions of new species principally from the Agassiz expedition to the eastern tropical Pacific, 1904-1905. Univ. Calif. Publ. Zool. 34:1-403, 697 text figs.

Epiplocylis lata

Campbell, A. S., 1942
The oceanic Tintinnoina of the plankton gathered during the last cruise of the CARNEGIE. Sci. Res. Cruise VII of the Carnegie, 1928-1929, Biol. 2:163 pp., 128 figs.

Epiplocylis lata

Kofoid, C.A. and A.S. Campbell, 1939
Reports on the scientific results of the expedition to the eastern tropical Pacific in charge of Alexander Agassiz, by the U.S. Fish Commission steamer "Albatross" from October 1904 to March 1905-----XXXVII. The Ciliata: The Tintinnoimea. Bull. Mus. Comp. Zool. 84: 473 pp., 36 pls.

Epiplocylis lata n.sp.

Kofoid, C. A. and A. S. Campbell, 1929
A conspectus of the marine and fresh-water Ciliata belonging to the suborder Tintinnoinea, with descriptions of new species principally from the Agassiz expedition to the eastern tropical Pacific, 1904-1905. Univ. Calif. Publ. Zool. 34:1-403, 697 text figs.

Epiplocylis lineata n.sp.

Kofoid, C. A. and A. S. Campbell, 1929
A conspectus of the marine and fresh-water Ciliata belonging to the suborder Tintinnoinea, with descriptions of new species principally from the Agassiz expedition to the eastern tropical Pacific, 1904-1905. Univ. Calif. Publ. Zool. 34:1-403, 697 text figs.

Epiplocylis mira n.sp.

Balech, Enrique, 1958
Plancton de la Campana Antartico Argentina, 1954-1955. Physis, 21(60):75-108.

Epiplocylis mucronata

Balech, Enrique, 1962
Tintinnoinea y Dinoflagellata del Pacifico segun material de las Expediciones NORPAC y DOWNWIND del Instituto Scripps de Oceanografia. Revista, Mus. Argentino Ciencias Nat. "Bernardino Rivadavia", Ciencias Zool., 7(1):1-253.

Epiplocylis mucronata

Kofoid, C. A. and A. S. Campbell, 1929
A conspectus of the marine and fresh-water Ciliata belonging to the suborder Tintinnoinea, with descriptions of new species principally from the Agassiz expedition to the eastern tropical Pacific, 1904-1905. Univ. Calif. Publ. Zool. 34:1-403, 697 text figs.

Epiplocylus mucronata

Rampi, L., 1948
Sur quelques Tintinnides (Infusoires louques) du Pacifique subtropical (Recoltes Alain Gerbault) Bull. l'Inst. Ocean. Monaco, No.938, 4 pp.

Epiplocylis nervosa

Gaarder, K.R., 1946
Tintinnoidea. Rep.Sci.Res. "Michael Sars" N.Atlantic Deep-Sea Exped., 1910, 2(1): 37 pp., 23 text figs.

Epiplocylis nervosa

Kofoid, C. A. and A. S. Campbell, 1929
A conspectus of the marine and fresh-water Ciliata belonging to the suborder Tintinnoinea, with descriptions of new species principally from the Agassiz expedition to the eastern tropical Pacific, 1904-1905. Univ. Calif. Publ. Zool. 34:1-403, 697 text figs.

Epiplocylis obtusa n.sp.

Kofoid, C. A. and A. S. Campbell, 1929
A conspectus of the marine and fresh-water Ciliata belonging to the suborder Tintinnoinea, with descriptions of new species principally from the Agassiz expedition to the eastern tropical Pacific, 1904-1905. Univ. Calif. Publ. Zool. 34:1-403, 697 text figs.

Epiplocylis pacifica

Campbell, A. S., 1942
The oceanic Tintinnoina of the plankton gathered during the last cruise of the CARNEGIE. Sci. Res. Cruise VII of the Carnegie, 1928-1929, Biol. 2:163 pp., 128 figs.

Epiplocylis pacifica

Kofoid, C.A. and A.S. Campbell, 1939
Reports on the scientific results of the expedition to the eastern tropical Pacific in charge of Alexander Agassiz, by the U.S. Fish Commission steamer "Albatross" from October 1904 to March 1905-----XXXVII. The Ciliata: The Tintinnoimea. Bull. Mus. Comp. Zool. 84: 473 pp., 36 pls.

Epiplocylis pacifica n.sp.

Kofoid, C. A. and A. S. Campbell, 1929
A conspectus of the marine and fresh-water Ciliata belonging to the suborder Tintinnoinea, with descriptions of new species principally from the Agassiz expedition to the eastern tropical Pacific, 1904-1905. Univ. Calif. Publ. Zool. 34:1-403, 697 text figs.

Epiplocylis ralumensis

Kofoid, C. A. and A. S. Campbell, 1929
A conspectus of the marine and fresh-water Ciliata belonging to the suborder Tintinnoinea, with descriptions of new species principally from the Agassiz expedition to the eastern tropical Pacific, 1904-1905. Univ. Calif. Publ. Zool. 34:1-403, 697 text figs.

Epiplocylis ralumensis

Marshall, S. M., 1934
The silicoflagellata and tintinnoinea. Great Barrier Reef Exped., 1928-29. Sci. Repts. 4(15):623-664, 43 text figs.

Epiplocylis reticulata

Hofker, J., 1931
Studien uber Tintinnoidea. Arch. f. Protistenk 75(3):315-402, 89 text figs.

Epiplocylis reticulata

Kofoid, C. A. and A. S. Campbell, 1929
A conspectus of the marine and fresh-water Ciliata belonging to the suborder Tintinnoinea, with descriptions of new species principally from the Agassiz expedition to the eastern tropical Pacific, 1904-1905. Univ. Calif. Publ. Zool. 34:1-403, 697 text figs.

Epiplocylis sargassensis

Campbell, A. S., 1942
The oceanic Tintinnoina of the plankton gathered during the last cruise of the CARNEGIE. Sci. Res. Cruise VII of the Carnegie, 1928-1929, Biol. 2:163 pp., 128 figs.

Epiplocylis sargassensis

Kofoid, C.A. and A.S. Campbell, 1939
Reports on the scientific results of the expedition to the eastern tropical Pacific in charge of Alexander Agassiz, by the U.S. Fish Commission steamer "Albatross" from October 1904 to March 1905-----XXXVII. The Ciliata: The Tintinnoimea. Bull. Mus. Comp. Zool. 84: 473 pp., 36 pls.

Epiplocylis sargassensis

Kofoid, C. A. and A. S. Campbell, 1929
A conspectus of the marine and fresh-water Ciliata belonging to the suborder Tintinnoinea, with descriptions of new species principally from the Agassiz expedition to the eastern tropical Pacific, 1904-1905. Univ. Calif. Publ. Zool. 34:1-403, 697 text figs.

Epiplocylis symmetrica n.sp.

Kofoid, C.A. and A.S. Campbell, 1939
Reports on the scientific results of the expedition to the eastern tropical Pacific in charge of Alexander Agassiz, by the U.S. Fish Commission steamer "Albatross" from October 1904 to March 1905-----XXXVII. The Ciliata: The Tintinnoimea. Bull. Mus. Comp. Zool. 84: 473 pp., 36 pls.

Epiplocylis undella

Balech, Enrique, 1962
Tintinnoinea y Dinoflagellata del Pacifico segun material de las Expediciones NORPAC y DOWNWIND del Instituto Scripps de Oceanografia. Revista, Mus. Argentino Ciencias Nat. "Bernardino Rivadavia", Ciencias Zool., 7(1):1-253.

Epiplocylis undella

Campbell, A. S., 1942
The oceanic Tintinnoina of the plankton gathered during the last cruise of the CARNEGIE. Sci. Res. Cruise VII of the Carnegie, 1928-1929, Biol. 2:163 pp., 128 figs.

Epiplocylis undella

Jörgensen, E., 1924
Mediterranean Tintinnidae. Rept. Danish Oceanogr. Exped. 1908-10 to the Mediterranean and adjacent seas, Vol.II, Biol. J 3, 110 pp., 114 text figs.

Epiplocylis undella

Kofoid, C.A. and A.S. Campbell, 1939
Reports on the scientific results of the expedition to the eastern tropical Pacific in charge of Alexander Agassiz, by the U.S. Fish Commission steamer "Albatross" from October 1904 to March 1905-----XXXVII. The Ciliata: The Tintinnoimea. Bull.Mus. Comp.Zool. 84: 473 pp., 36 pls.

Epiplocylis undella

Kofoid, C. A. and A. S. Campbell, 1929
A conspectus of the marine and fresh-water Ciliata belonging to the suborder Tintinnoinea, with descriptions of new species principally from the Agassiz expedition to the eastern tropical Pacific, 1904-1905. Univ. Calif. Publ. Zool. 34:1-403, 697 text figs.

Epiplocylis undella

Komarovsky, B., 1959
The Tintinnina of the Gulf of Eylath (Aqaba). Contrib. Knowl. Red Sea, Bull. Sea Fish. Res. Sta.Haifa, Israel 21 (14):1-40.

Epiplocylis undella

Marshall, S. M., 1934
The silicoflagellata and tintinnoinea. Great Barrier Reef Exped., 1928-29. Sci. Repts. 4(15):623-664, 43 text figs.

Epiplocyloides antarctica n. sp.

Argentina, Secretaria de Marina, Servicio de Hidrografia Naval, 1962.
Plancton de las companas oceanograficas DRAKE I y II.
Publico. H. 627:57.

Epirhabdonella coronata n.sp.

Kofoid, C.A. and A.S. Campbell, 1939
Reports on the scientific results of the expedition to the eastern tropical Pacific in charge of Alexander Agassiz, by the U.S. Fish Commission steamer "Albatross" from October 1904 to March 1905-----XXXVII. The Ciliata: The Tintinnoimes. Bull.Mus. Comp.Zool. 84: 473 pp., 36 pls.

Epirhabdonella mucronata n.sp.

Kofoid, C.A. and A.S. Campbell, 1939
Reports on the scientific results of the expedition to the eastern tropical Pacific in charge of Alexander Agassiz, by the U.S. Fish Commission steamer "Albatross" from October 1904 to March 1905-----XXXVII. The Ciliata: The Tintinnoimes. Bull.Mus. Comp.Zool. 84: 473 pp., 36 pls.

Epirhabdosella n. gen.

Campbell, A. S., 1942
The oceanic Tintinnoina of the plankton gathered during the last cruise of the CARNEGIE. Sci. Res. Cruise VII of the Carnegie, 1928-1929, Biol. 2:163 pp., 128 figs.

Epirhabdosella cuneolata

Campbell, A. S., 1942
The oceanic Tintinnoina of the plankton gathered during the last cruise of the CARNEGIE. Sci. Res. Cruise VII of the Carnegie, 1928-1929, Biol. 2:163 pp., 128 figs.

Eutintinnus (Odontotintinnus) sp n.sp. ?

Balech, E., 1941.
Tintinnoineos del estrecho Le Maire. Physis XIX(52):245-252, 10 textfigs.

Eutintinnus sp.

Morse, D.C., 1947
Some observations on seasonal variations in plankton population Patuxant River, Maryland 1943-1945. Bd. Nat. Res., Publ. No.65, Chesapeake Biol. Lab., 31, 3 figs.

Eutintinnus apertus

Balech, Enrique, 1962
Tintinnoinea y Dinoflagellata del Pacifico segun material de las Expediciones NORPAC y DOWNWIND del Instituto Scripps de Oceanografia.
Revista, Mus. Argentino Ciencias Nat. "Bernardino Rivadavia", Ciencias Zool., 7(1):1-253.

Eutintinnus apertus

Balech, E., 1959.
Tintinnoinea del Mediterraneo.
Trav. Sta. Zool., Villefranche-sur-Mer, 18(17): Reprinted from:
Trab. Inst. Español Oceanogr., Madrid, 19 Sept. 1959(28):88 pp., 22 pls.

Eutintinnus apertus

Campbell, A. S., 1942
The oceanic Tintinnoina of the plankton gathered during the last cruise of the CARNEGIE. Sci. Res. Cruise VII of the Carnegie, 1928-1929, Biol. 2:163 pp., 128 figs.

Eutintinnus apertus

Hasle, Grethe Rytter, 1960
Phytoplankton and ciliate species from the Tropical Pacific.
Skr. Norske Videnskaps-Akad., Oslo, 1. Mat.-Nat. Kl., 1960(2): 1-50.

Eutintinnus apertus

Kofoid, C.A. and A.S. Campbell, 1939
Reports on the scientific results of the expedition to the eastern tropical Pacific in charge of Alexander Agassiz, b the U.S. Fish Commission steamer "Albatros from October 1904 to March 1905-----XXXVJ The Ciliata: The Tintinnoimea. Bull.Mus. Comp.Zool. 84: 473 pp., 36 pls.

Eutintinnus apertus

Komarovsky, B., 1959
The Tintinnina of the Gulf of Eylath (Aqaba). Contrib. Knowl. Red Sea, Bull. Sea Fish. Res. Sta.Haifa, Israel 21 (14):1-40.

Eutintinnus apertus curta n. var.

Komarovsky, B., 1959
The Tintinnina of the Gulf of Eylath (Aqaba). Contrib. Knowl. Red Sea, Bull. Sea Fish. Res. Sta.Haifa, Israel 21 (14):1-40.

Eutintinnus asymmetricus n.sp

Balech, Enrique, 1968.
Algunas especies nuevas o interesantes do tintinnidos del Golfo de Mexico y Caribe.
Revta Mus. argent. Cienc. nat. Bernardino Rivadavia Inst. nac. Hidrobiol. 2(5):165-197.

Eutintinnus australis

Balech, E., 1945.
Tintinnoinea de Quequen. Physis 20(55):1-15, 3 pls.

Eutintinnus australis nom.nov

Balech, Enrique, 1944
Contribucion al conocimiento del Plancton de Lennox y Cabo de Hornos. Physis XIX:423-446, 6 pls. with 67 figs.

Eutintinnus birictus

Balech, Enrique, 1962
Tintinnoinea y Dinoflagellata del Pacifico segun material de las Expediciones NORPAC y DOWNWIND del Instituto Scripps de Oceanografia.
Bd. Nat. Res., Mus. Argentino Ciencias Nat. "Bernardino Rivadavia", Ciencias Zool., 7(1):1-253.

Eutintinnus birictus

Campbell, A. S., 1942
The oceanic Tintinnoina of the plankton gathered during the last cruise of the CARNEGIE. Sci. Res. Cruise VII of the Carnegie, 1928-1929, Biol. 2:163 pp., 128 figs.

Eutintinnus birictus

Kofoid, C.A. and A.S. Campbell, 1939
Reports on the scientific results of the expedition to the eastern tropical Pacific in charge of Alexander Agassiz, by the U.S. Fish Commission steamer "Albatross" from October 1904 to March 1905-----XXXVII. The Ciliata: The Tintinnoimea. Bull. Mus. Comp.Zool. 84: 473 pp., 36 pls.

Eutintinnus brandti

Campbell, A. S., 1942
The oceanic Tintinnoina of the plankton gathered during the last cruise of the CARNEGIE. Sci. Res. Cruise VII of the Carnegie, 1928-1929, Biol. 2:163 pp., 128 figs.

Eutintinnus brandti

Kofoid, C.A. and A.S. Campbell, 1939
Reports on the scientific results of the expedition to the eastern tropical Pacific in charge of Alexander Agassiz, by the U.S. Fish Commission steamer "Albatross" from October 1904 to March 1905-----XXXVII. The Ciliata: The Tintinnoimea. Bull.Mus. Comp.Zool. 84: 473 pp., 36 pls.

Eutintinnus brandti

Komarovsky, B., 1959
The Tintinnina of the Gulf of Eylath (Aqaba). Contrib. Knowl. Red Sea, Bull. Sea Fish. Res. Sta.Haifa, Israel 21 (14):1-40.

Eutintinnus colligatus

Balech, Enrique, 1962
Tintinnoinea y Dinoflagellata del Pacifico segun material de las Expediciones NORPAC y DOWNWIND del Instituto Scripps de Oceanografia.
Revista, Mus. Argentino Ciencias Nat. "Bernardino Rivadavia", Ciencias Zool., 7(1):1-253.

Eutintinnus colligatus

Campbell, A. S., 1942
The oceanic Tintinnoina of the plankton gathered during the last cruise of the CARNEGIE. Sci. Res. Cruise VII of the Carnegie, 1928-1929, Biol. 2:163 pp., 128 figs.

Eutintinnus colligatus

Kofoid, C.A. and A.S. Campbell, 1939
Reports on the scientific results of the expedition to the eastern tropical Pacific in charge of Alexander Agassiz, by the U.S. Fish Commission steamer "Albatross" from October 1904 to March 1905-----XXXVII. The Ciliata: The Tintinnoimea. Bull.Mus. Comp.Zool. 84: 473 pp., 36 pls.

Eutintinnus elegans

Balech, E., 1959.
Tintinnoinea del Mediterraneo.
Trav. Sta. Zool., Villefranche-sur-Mer, 18(17): Reprinted from:
Trab. Inst. Español Oceanogr., Madrid, 19 Sept. 1959(28):88 pp., 22 pls.

Eutintinnus elegans mihi

Balech, Enrique, 1944
Contribucion al conocimiento del Plancton de Lennox y Cabo de Hornos. Physis XIX:423-446, 6 pls. with 67 figs.

Eutintinnus (Eutintinnus) elegans n.sp.

Balech, E., 1941.
Tintinnoineos del estrecho Le Maire. Physis XIX (52):245-252, 10 textfigs.

Eutintinnus elegans

Campbell, A. S., 1942
The oceanic Tintinnoina of the plankton gathered during the last cruise of the CARNEGIE. Sci. Res. Cruise VII of the Carnegie, 1928-1929, Biol. 2:163 pp., 128 figs.

Eutintinnus elegans

Komarovsky, B., 1959
The Tintinnina of the Gulf of Eylath (Aqaba). Contrib. Knowl. Red Sea, Bull. Sea Fish. Res. Sta. Haifa, Israel 21 (14):1-40.

Eutintinnus elongatus

Campbell, A. S., 1942
The oceanic Tintinnoina of the plankton gathered during the last cruise of the CARNEGIE. Sci. Res. Cruise VII of the Carnegie, 1928-1929, Biol. 2:163 pp., 128 figs.

Eutintinnus elongatus

Kofoid, C.A. and A.S. Campbell, 1939
Reports on the scientific results of the expedition to the eastern tropical Pacific in charge of Alexander Agassiz, by the U.S. Fish Commission steamer "Albatross" from October 1904 to March 1905------XXXVII. The Ciliata: The Tintinnoimea. Bull. Mus. Comp. Zool. 84: 473 pp., 36 pls.

Eutintinnus erythraensis, n.sp.

Komarovsky, B., 1962
Tintinnina from the vicinity of the Straits of Tiran and Massawa Region. Contributions to the knowledge of the Red Sea, No. 25. Sea Fish. Res. Sta., Haifa, Israel, Bull. No. 30:48-56.

Eutintinnus fraknoi

Balech, Enrique, 1962
Tintinnoinea y Dinoflagellata del Pacifico segun material de las Expediciones NORPAC y DOWNWIND del Instituto Scripps de Oceanografia. Revista, Mus. Argentino Ciencias Nat. "Bernardino Rivadavia", Ciencias Zool., 7(1):1-253.

Eutintinnus fraknoi

Balech, E., 1959.
Tintinnoinea del Mediterraneo. Trav. Sta. Zool., Villefranche-sur-Mer, 18(17): Reprinted from: Trab. Inst. Español Oceanogr., Madrid, 19 Sept. 1959(28):88 pp., 22 pls.

Eutintinnus fraknóii

Campbell, A. S., 1942
The oceanic Tintinnoina of the plankton gathered during the last cruise of the CARNEGIE. Sci. Res. Cruise VII of the Carnegie, 1928-1929, Biol. 2:163 pp., 128 figs.

Eutintinnus fraknoi

Duran, M., 1957.
Nota sobre algunos tintinnoineas del plancton de Puerto Rico. Invest. Pesqueria, Barcelona, 8: 97-120.

Eutintinnus fraknoii

Kofoid, C.A. and A.S. Campbell, 1939
Reports on the scientific results of the expedition to the eastern tropical Pacific in charge of Alexander Agassiz, by the U.S. Fish Commission steamer "Albatross" from October 1904 to March 1905------XXXVII. The Ciliata: The Tintinnoimea. Bull. Mus. Comp. Zool. 84: 473 pp., 36 pls.

Eutintinnus fraknoii

Komarovsky, B., 1959
The Tintinnina of the Gulf of Eylath (Aqaba). Contrib. Knowl. Red Sea, Bull. Sea Fish. Res. Sta. Haifa, Israel 21 (14):1-40.

Eutintinnus latus

Balech, Enrique, 1962
Tintinnoinea y Dinoflagellata del Pacifico segun material de las Expediciones NORPAC y DOWNWIND del Instituto Scripps de Oceanografia. Revista, Mus. Argentino Ciencias Nat. "Bernardino Rivadavia", Ciencias Zool., 7(1):1-253.

Eutintinnus latus

Campbell, A. S., 1942
The oceanic Tintinnoina of the plankton gathered during the last cruise of the CARNEGIE. Sci. Res. Cruise VII of the Carnegie, 1928-1929, Biol. 2:163 pp., 128 figs.

Eutintinnus latus

Komarovsky, B., 1959
The Tintinnina of the Gulf of Eylath (Aqaba). Contrib. Knowl. Red Sea, Bull. Sea Fish. Res. Sta. Haifa, Israel 21 (14):1-40.

Eutintinnus lususundae

Balech, Enrique, 1962
Tintinnoinea y Dinoflagellata del Pacifico segun material de las Expediciones NORPAC y DOWNWIND del Instituto Scripps de Oceanografia. Revista, Mus. Argentino Ciencias Nat. "Bernardino Rivadavia", Ciencias Zool., 7(1):1-253.

Eutintinnus lusus-undae

Balech, E., 1959.
Tintinnoinea del Mediterraneo. Trav. Sta. Zool., Villefranche-sur-Mer, 18(17): Reprinted from: Trab. Inst. Español Oceanogr., Madrid, 19 Sept. 1959(28):88 pp., 22 pls.

Eutintinnus lusus-undae

Balech, E., 1945.
Tintinnoinea de Quequén. Physis 20(55):1-15, 3 pls.

Eutintinnus lusus-undae

Campbell, A. S., 1942
The oceanic Tintinnoina of the plankton gathered during the last cruise of the CARNEGIE. Sci. Res. Cruise VII of the Carnegie, 1928-1929, Biol. 2:163 pp., 128 figs.

Eutintinnus lusus-undae

Duran, M., 1957.
Nota sobre algunos tintinnoineas del plancton de Puerto Rico. Invest. Pesqueria, Barcelona, 8: 97-120.

Eutintinnus lusus-undae

Kofoid, C.A. and A.S. Campbell, 1939
Reports on the scientific results of the expedition to the eastern tropical Pacific in charge of Alexander Agassiz, by the U.S. Fish Commission steamer "Albatross" from October 1904 to March 1905------XXXVII. The Ciliata: The Tintinnoimea. Bull. Mus. Comp. Zool. 84: 473 pp., 36 pls.

Eutintinnus lusus-undae

Komarovsky, B., 1959
The Tintinnina of the Gulf of Eylath (Aqaba). Contrib. Knowl. Red Sea, Bull. Sea Fish. Res. Sta. Haifa, Israel 21 (14):1-40.

Eutintinnus macilentus

Balech, Enrique, 1962
Tintinnoinea y Dinoflagellata del Pacifico segun material de las Expediciones NORPAC y DOWNWIND del Instituto Scripps de Oceanografia. Revista, Mus. Argentino Ciencias Nat. "Bernardino Rivadavia", Ciencias Zool., 7(1):1-253.

Eutintinnus macilentus

Balech, E., 1959.
Tintinnoinea del Mediterraneo. Trav. Sta. Zool., Villefranche-sur-Mer, 18(17): Reprinted from: Trab. Inst. Español Oceanogr., Madrid, 19 Sept. 1959(28):88 pp., 22 pls.

Eutintinnus macilentus

Campbell, A. S., 1942
The oceanic Tintinnoina of the plankton gathered during the last cruise of the CARNEGIE. Sci. Res. Cruise VII of the Carnegie, 1928-1929, Biol. 2:163 pp., 128 figs.

Eutintinnus macilentus

Kofoid, C.A. and A.S. Campbell, 1939
Reports on the scientific results of the expedition to the eastern tropical Pacific in charge of Alexander Agassiz, by the U.S. Fish Commission steamer "Albatross" from October 1904 to March 1905------XXXVII. The Ciliata: The Tintinnoimea. Bull. Mus. Comp. Zool. 84: 473 pp., 36 pls.

Eutintinnus macilentus

Komarovsky, B., 1959
The Tintinnina of the Gulf of Eylath (Aqaba). Contrib. Knowl. Red Sea, Bull. Sea Fish. Res. Sta. Haifa, Israel 21 (14):1-40.

Eutintinnus magnificus n. sp.

Campbell, A. S., 1942
The oceanic Tintinnoina of the plankton gathered during the last cruise of the CARNEGIE. Sci. Res. Cruise VII of the Carnegie, 1928-1929, Biol. 2:163 pp., 128 figs.

Eutintinnus medius

Balech, Enrique, 1962
Tintinnoinea y Dinoflagellata del Pacifico segun material de las Expediciones NORPAC y DOWNWIND del Instituto Scripps de Oceanografia. Revista, Mus. Argentino Ciencias Nat. "Bernardino Rivadavia", Ciencias Zool., 7(1):1-253.

Eutintinnus medius

Campbell, A. S., 1942
The oceanic Tintinnoina of the plankton gathered during the last cruise of the CARNEGIE. Sci. Res. Cruise VII of the Carnegie, 1928-1929, Biol. 2:163 pp., 128 figs.

Eutintinnus medius

Kofoid, C.A. and A.S. Campbell, 1939
Reports on the scientific results of the expedition to the eastern tropical Pacific in charge of Alexander Agassiz, by the U.S. Fish Commission steamer "Albatross" from October 1904 to March 1905------XXXVII. The Ciliata: The Tintinnoimea. Bull. Mus. Comp. Zool. 84: 473 pp., 36 pls.

Eutintinnus medius

Komarovsky, B., 1959
The Tintinnina of the Gulf of Eylath (Aqaba). Contrib. Knowl. Red Sea, Bull. Sea Fish. Res. Sta. Haifa, Israel 21 (14):1-40.

Eutintinnus pacificus

Campbell, A. S., 1942
The oceanic Tintinnoina of the plankton gathered during the last cruise of the CARNEGIE. Sci. Res. Cruise VII of the Carnegie, 1928-1929, Biol. 2:163 pp., 128 figs.

Eutintinnus pacificus

Kofoid, C.A. and A.S. Campbell, 1939
Reports on the scientific results of the expedition to the eastern tropical Pacific in charge of Alexander Agassiz, by the U.S. Fish Commission steamer "Albatross" from October 1904 to March 1905------XXXVII. The Ciliata: The Tintinnoimea. Bull. Mus. Comp. Zool. 84: 473 pp., 36 pls.

Eutintinnus perminutus

Balech, Enrique, 1962
Tintinnoinea y Dinoflagellata del Pacifico segun material de las Expediciones NORPAC y DOWNWIND del Instituto Scripps de Oceanografia. Revista, Mus. Argentino Ciencias Nat. "Bernardino Rivadavia", Ciencias Zool., 7(1):1-253.

Eutintinnus perminutus

Campbell, A. S., 1942
The oceanic Tintinnoina of the plankton gathered during the last cruise of the CARNEGIE. Sci. Res. Cruise VII of the Carnegie, 1928-1929, Biol. 2:163 pp., 128 figs.

Eutintinnus perminutus

Kofoid, C.A. and A.S. Campbell, 1939
Reports on the scientific results of the expedition to the eastern tropical Pacific in charge of Alexander Agassiz, by the U.S. Fish Commission steamer "Albatross" from October 1904 to March 1905-----XXXVII. The Ciliata: The Tintinnoimea. Bull.Mus. Comp.Zool. 84: 473 pp., 36 pls.

Eutintinnus pinguis

Balech, Enrique, 1962
Tintinnoinea y Dinoflagellata del Pacifico segun material de las Expediciones NORPAC y DOWNWIND del Instituto Scripps de Oceanografia. Revista, Mus. Argentino Ciencias Nat. "Bernardino Rivadavia", Ciencias Zool., 7(1):1-253.

Eutintinnus pinguis

Campbell, A. S., 1942
The oceanic Tintinnoina of the plankton gathered during the last cruise of the CARNEGIE. Sci. Res. Cruise VII of the Carnegie, 1928-1929, Biol. 2:163 pp., 128 figs.

Eutintinnus pinguis

Kofoid, C.A. and A.S. Campbell, 1939
Reports on the scientific results of the expedition to the eastern tropical Pacific in charge of Alexander Agassiz, by the U.S. Fish Commission steamer "Albatross" from October 1904 to March 1905-----XXXVII. The Ciliata: The Tintinnoimea. Bull.Mus. Comp.Zool. 84: 473 pp., 36 pls.

Eutintinnus procurrerens

Balech, Enrique, 1962
Tintinnoinea y Dinoflagellata del Pacifico segun material de las Expediciones NORPAC y DOWNWIND del Instituto Scripps de Oceanografia. Revista, Mus. Argentino Ciencias Nat. "Bernardino Rivadavia", Ciencias Zool., 7(1):1-253.

Eutintinnus procurrerens

Campbell, A. S., 1942
The oceanic Tintinnoina of the plankton gathered during the last cruise of the CARNEGIE. Sci. Res. Cruise VII of the Carnegie, 1928-1929, Biol. 2:163 pp., 128 figs.

Eutintinnus procurrerens

Kofoid, C.A. and A.S. Campbell, 1939
Reports on the scientific results of the expedition to the eastern tropical Pacific in charge of Alexander Agassiz, by the U.S. Fish Commission steamer "Albatross" from October 1904 to March 1905-----XXXVII. The Ciliata: The Tintinnoimea. Bull.Mus. Comp.Zool. 84: 473 pp., 36 pls.

Eutintinnus rugosus

Balech, Enrique, 1948
Tintinnoinea de Atlantida (R.O. Del Uruguay) (Protozoa, Ciliata Oligotr.). Comm. Mus. Argentino Ciencias Nat. "Bernardino Rivadavia", Ser.Cien.Zool. No.7, 23 pp., 8 pls. of 107 figs.

Eutintinnus rugosus

Balech, E., 1945.
Tintinnoinea de Quequén. Physis 20(55):1-15, 3 pls.

Eutintinnus (Odontotintinnus) rugosus

Balech, Enrique, 1944
Contribucion al conocimiento del Plancton de Lennox y Cabo de Hornos. Physis XIX:423-446, 6 pls. with 67 figs.

Eutintinnus rugosus

Kofoid, C.A. and A.S. Campbell, 1939
Reports on the scientific results of the expedition to the eastern tropical Pacific in charge of Alexander Agassiz, by the U.S. Fish Commission steamer "Albatross" from October 1904 to March 1905-----XXXVII. The Ciliata: The Tintinnoimea. Bull.Mus. Comp.Zool. 84: 473 pp., 36 pls.

Eutintinnus similis n.sp.

Balech, Enrique, 1962
Tintinnoinea y Dinoflagellata del Pacifico segun material de las Expediciones NORPAC y DOWNWIND del Instituto Scripps de Oceanografia. Revista, Mus. Argentino Ciencias Nat. "Bernardino Rivadavia", Ciencias Zool., 7(1):1-253.

Eutintinnus stramentus

Balech, Enrique, 1962
Tintinnoinea y Dinoflagellata del Pacifico segun material de las Expediciones NORPAC y DOWNWIND del Instituto Scripps de Oceanografia. Revista, Mus. Argentino Ciencias Nat. "Bernardino Rivadavia", Ciencias Zool., 7(1):1-253.

Eutintinnus stramentus

Campbell, A. S., 1942
The oceanic Tintinnoina of the plankton gathered during the last cruise of the CARNEGIE. Sci. Res. Cruise VII of the Carnegie, 1928-1929, Biol. 2:163 pp., 128 figs.

Eutintinnus stramentus

Kofoid, C.A. and A.S. Campbell, 1939
Reports on the scientific results of the expedition to the eastern tropical Pacific in charge of Alexander Agassiz, by the U.S. Fish Commission steamer "Albatross" from October 1904 to March 1905-----XXXVII. The Ciliata: The Tintinnoimea. Bull.Mus. Comp.Zool. 84: 473 pp., 36 pls.

Eutintinnus subrugosus

Balech, Enrique, 1944
Contribucion al conocimiento del Plancton de Lennox y Cabo de Hornos. Physis XIX:423-446, 6 pls. with 67 figs.

Eutintinnus (Odontotintinnus) subrugosus n.sp.

Balech, E., 1941.
Tintinnoineos del estrecho Le Maire. Physis XIX(52):245-252, 10 textfigs.

Eutintinnus tenuis

Balech, Enrique, 1962
Tintinnoinea y Dinoflagellata del Pacifico segun material de las Expediciones NORPAC y DOWNWIND del Instituto Scripps de Oceanografia. Revista, Mus. Argentino Ciencias Nat. "Bernardino Rivadavia", Ciencias Zool., 7(1):1-253.

Eutintinnus tenuis

Campbell, A. S., 1942
The oceanic Tintinnoina of the plankton gathered during the last cruise of the CARNEGIE. Sci. Res. Cruise VII of the Carnegie, 1928-1929, Biol. 2:163 pp., 128 figs.

Eutintinnus tenuis

Cosper, T.C., 1972.
The identification of tintinnids (Protozoa: Ciliata: Tintinnida) of the St. Andrew Bay system, Florida. Bull. mar. Sci. Miami 22(2): 391-418.

Eutintinnus tenuis

Kofoid, C.A. and A.S. Campbell, 1939
Reports on the scientific results of the expedition to the eastern tropical Pacific in charge of Alexander Agassiz, by the U.S. Fish Commission steamer "Albatross" from October 1904 to March 1905-----XXXVII. The Ciliata: The Tintinnoimea. Bull.Mus. Comp.Zool. 84: 473 pp., 36 pls.

Eutintinnus tubiformis

Campbell, A. S., 1942
The oceanic Tintinnoina of the plankton gathered during the last cruise of the CARNEGIE. Sci. Res. Cruise VII of the Carnegie, 1928-1929, Biol. 2:163 pp., 128 figs.

Eutintinnus tubiformis

Kofoid, C.A. and A.S. Campbell, 1939
Reports on the scientific results of the expedition to the eastern tropical Pacific in charge of Alexander Agassiz, by the U.S. Fish Commission steamer "Albatross" from October 1904 to March 1905-----XXXVII. The Ciliata: The Tintinnoimea. Bull.Mus. Comp.Zool. 84: 473 pp., 36 pls.

Eutintinnus tubulosus

Balech, Enrique, 1962
Tintinnoinea y Dinoflagellata del Pacifico segun material de las Expediciones NORPAC y DOWNWIND del Instituto Scripps de Oceanografia. Revista, Mus. Argentino Ciencias Nat. "Bernardino Rivadavia", Ciencias Zool., 7(1):1-253.

Eutintinnus tubulosus

Balech, E., 1959.
Tintinnoinea del Mediterraneo. Trav. Sta. Zool., Villefranche-sur-Mer, 18(17): Reprinted from: Trab. Inst. Español Oceanogr., Madrid, 19 Sept. 1959(28):88 pp., 22 pls.

Eutintinnus tubulosus

Campbell, A. S., 1942
The oceanic Tintinnoina of the plankton gathered during the last cruise of the CARNEGIE. Sci. Res. Cruise VII of the Carnegie, 1928-1929, Biol. 2:163 pp., 128 figs.

Eutintinnus tubulosus

Cosper, T.C., 1972.
The identification of tintinnids (Protozoa: Ciliata: Tintinnida) of the St. Andrew Bay system, Florida. Bull. mar. Sci. Miami 22(2): 391-418.

Eutinnus tubulosus

Kofoid, C.A. and A.S. Campbell, 1939
Reports on the scientific results of the expedition to the eastern tropical Pacific in charge of Alexander Agassiz, by the U.S. Fish Commission steamer "Albatross" from October 1904 to March 1905-----XXXVII. The Ciliata: The Tintinnoimea. Bull.Mus. Comp.Zool. 84: 473 pp., 36 pls.

Eutintinnus turgescens

Kofoid, C.A. and A.S. Campbell, 1939
Reports on the scientific results of the expedition to the eastern tropical Pacific in charge of Alexander Agassiz, by the U.S. Fish Commission steamer "Albatross" from October 1904 to March 1905-----XXXVII. The Ciliata: The Tintinnoimea. Bull.Mus. Comp.Zool. 84: 473 pp., 36 pls.

Favella n.gen.

Jörgensen, E., 1924
Mediterranean Tintinnidae. Rept. Danish Oceanogr. Exped. 1908-10 to the Mediterranean and adjacent seas, Vol.II, Biol. J 3, 110 pp., 114 text figs.

Favella sp.

Morse, D.C., 1947
Some observations on seasonal variations in plankton population Patuxant River, Maryland 1943-1945. Bd. Nat. Res., Publ. No.65, Chesapeake Biol. Lab., 31, 3 figs.

Favella aciculifera n.sp.

Jörgensen, E., 1924
Mediterranean Tintinnidae. Rept. Danish Oceanogr. Exped. 1908-10 to the Mediterranean and adjacent seas, Vol.II, Biol. J 3, 110 pp., 114 text figs.

Favella aciculifera

Kofoid, C. A. and A. S. Campbell, 1929
A conspectus of the marine and freshwater Ciliata belonging to the suborder Tintinnoinea, with descriptions of new species principally from the Agassiz expedition to the eastern tropical Pacific, 1904-1905. Univ. Calif. Publ. Zool. 34:1-403, 697 text figs.

Favella adriatica

Campbell, A. S., 1942
The oceanic Tintinnoina of the plankton gathered during the last cruise of the CARNEGIE. Sci. Res. Cruise VII of the Carnegie, 1928-1929, Biol. 2:163 pp., 128 figs.

Favella adriatica
Duran, M., 1951
Contribucion al estudio de los tintinidos del plancton de los costas de Castellon. (Mediterraneo occidental). Publ. Inst. Biol. Aplic., Barcelona, 8: 101-120, 29 text figs.

Favella adriatica
Jörgensen, E., 1924
Mediterranean Tintinnidae. Rept. Danish Oceanogr. Exped. 1908-10 to the Mediterranean and adjacent seas, Vol. II, Biol. J 3, 110 pp., 114 text figs.

Favella adriatica
Kofoid, C. A. and A. S. Campbell, 1929
A conspectus of the marine and freshwater Ciliata belonging to the suborder Tintinnoinea, with descriptions of new species principally from the Agassiz expedition to the eastern tropical Pacific, 1904-1905. Univ. Calif. Publ. Zool. 34:1-403, 697 text figs.

Favella adriatica
Rampi, L., 1939
Primo contributo alla conoscenza dei tintinnoidi de Maro Ligure. Atti della Soc. Ital. di Sci. Nat. 78:67-81, 58 text figs.

Favella amoyensis n. sp.
Wang, C. C., and D. Nie, 1932
A survey of the marine protozoa of Amoy. Contr. Biol. Lab. Sci. Soc., China, Zool. Ser., 8:285-385, 89 text figs.

Favella arcuata
Gaarder, K. R., 1946
Tintinnoidea. Rep. Sci. Res. "Michael Sars" N. Atlantic Deep-Sea Exped., 1910, 2(1): 37 pp., 23 text figs.

Favella arcuata
Kofoid, C. A. and A. S. Campbell, 1929
A conspectus of the marine and freshwater Ciliata belonging to the suborder Tintinnoinea, with descriptions of new species principally from the Agassiz expedition to the eastern tropical Pacific, 1904-1905. Univ. Calif. Publ. Zool. 34:1-403, 697 text figs.

Favella arcuata
Rampi, L., 1939
Primo contributo alla conoscenza dei tintinnoidi de Maro Ligure. Atti della Soc. Ital. di Sci. Nat. 78:67-81, 58 text figs.

Favella attingata
Gaarder, K. R., 1946
Tintinnoidea. Rep. Sci. Res. "Michael Sars" N. Atlantic Deep-Sea Exped., 1910, 2(1): 37 pp., 23 text figs.

Favella attingata n.sp.
Kofoid, C. A. and A. S. Campbell, 1929
A conspectus of the marine and freshwater Ciliata belonging to the suborder Tintinnoinea, with descriptions of new species principally from the Agassiz expedition to the eastern tropical Pacific, 1904-1905. Univ. Calif. Publ. Zool. 34:1-403, 697 text figs.

Favella azorica
Balech, E., 1959.
Tintinnoinea del Mediterraneo. Trav. Sta. Zool., Villefranche-sur-Mer, 18(17): Reprinted from: Trab. Inst. Español Oceanogr., Madrid, 19 Sept. 1959(28):88 pp., 22 pls.

Favella azorica
Campbell, A. S., 1942
The oceanic Tintinnoina of the plankton gathered during the last cruise of the CARNEGIE. Sci. Res. Cruise VII of the Carnegie, 1928-1929, Biol. 2:163 pp., 128 figs.

Favella azorica
Duran, M., 1951
Contribucion al estudio de los tintinidos del plancton de los costas de Castellon. (Mediterraneo occidental). Publ. Inst. Biol. Aplic., Barcelona, 8: 101-120, 29 text figs.

Favella azorica
Jörgensen, E., 1924
Mediterranean Tintinnidae. Rept. Danish Oceanogr. Exped. 1908-10 to the Mediterranean and adjacent seas, Vol. II, Biol. J 3, 110 pp., 114 text figs.

Favella azorica
Kofoid, C. A. and A. S. Campbell, 1939
Reports on the scientific results of the expedition to the eastern tropical Pacific in charge of Alexander Agassiz, by the U.S. Fish Commission steamer "Albatross" from October 1904 to March 1905-----XXXVII. The Ciliata: The Tintinnoinea. Bull. Mus. Comp. Zool. 84: 473 pp., 36 pls.

Favella azorica
Kofoid, C. A. and A. S. Campbell, 1929
A conspectus of the marine and freshwater Ciliata belonging to the suborder Tintinnoinea, with descriptions of new species principally from the Agassiz expedition to the eastern tropical Pacific, 1904-1905. Univ. Calif. Publ. Zool. 34:1-403, 697 text figs.

Favella azorica
Komarovsky, B., 1959
The Tintinnina of the Gulf of Eylath (Aqaba). Contrib. Knowl. Red Sea, Bull. Sea Fish. Res. Sta. Haifa, Israel 21 (14):1-40.

Favella azorica
Marshall, S. M., 1934
The silicoflagellata and tintinnoinea. Great Barrier Reef Exped., 1928-29. Sci. Repts. 4(15):623-664, 43 text figs.

Favella azorica
Rampi, L., 1948
Sur quelques Tintinnides (Infusoires louques) du Pacifique subtropical (Recoltes Alain Gerbault) Bull. l'Inst. Ocean., Monaco, No. 938, 4 pp.

Favella azorica
Rampi, L., 1939
Primo contributo alla conoscenza dei tintinnoidi de Maro Ligure. Atti della Soc. Ital. di Sci. Nat. 78:67-81, 58 text figs.

Favella azorica
Vitiello, Pierre, 1964.
Contribution a l'etude des tintinnides de la baie d'Alger. Pelagos, Bull. Inst. Oceanogr., Alger, 2(2):5-41

(pp numbers of this article duplicate those of previous one by Francis Bernard!)

Favella brevis nom. nov.
Kofoid, C. A. and A. S. Campbell, 1929
A conspectus of the marine and freshwater Ciliata belonging to the suborder Tintinnoinea, with descriptions of new species principally from the Agassiz expedition to the eastern tropical Pacific, 1904-1905. Univ. Calif. Publ. Zool. 34:1-403, 697 text figs.

Favella brevis.
Sousa e Silva, E. de, 1950.
Les tintinnides de la Baie de Cascais. Bull. Inst. Océan., Monaco, No. 979:28 pp., 4 pls.

Favella campanula
Balech, E., 1959.
Tintinnoinea del Mediterraneo. Trav. Sta. Zool., Villefranche-sur-Mer, 18(17): Reprinted from: Trab. Inst. Español Oceanogr., Madrid, 19 Sept. 1959(28):88 pp., 22 pls.

Favella campanula
Duran, M., 1957.
Nota sobre algunos tintinnoineas del plancton de Puerto Rico. Invest. Pesqueria, Barcelona, 8: 97-120.

Favella campanula
Kofoid, C. A. and A. S. Campbell, 1929
A conspectus of the marine and freshwater Ciliata belonging to the suborder Tintinnoinea, with descriptions of new species principally from the Agassiz expedition to the eastern tropical Pacific, 1904-1905. Univ. Calif. Publ. Zool. 34:1-403, 697 text figs.

Favella composita
Kofoid, C. A. and A. S. Campbell, 1929
A conspectus of the marine and freshwater Ciliata belonging to the suborder Tintinnoinea, with descriptions of new species principally from the Agassiz expedition to the eastern tropical Pacific, 1904-1905. Univ. Calif. Publ. Zool. 34:1-403, 697 text figs.

Favella composita
Rampi, L., 1939
Primo contributo alla conoscenza dei tintinnoidi de Maro Ligure. Atti della Soc. Ital. di Sci. Nat. 78:67-81, 58 text figs.

Favella confessa
Kofoid, C. A. and A. S. Campbell, 1929
A conspectus of the marine and freshwater Ciliata belonging to the suborder Tintinnoinea, with descriptions of new species principally from the Agassiz expedition to the eastern tropical Pacific, 1904-1905. Univ. Calif. Publ. Zool. 34:1-403, 697 text figs.

Favella denticulata
Schulz, B., and A. Wulff, 1929
Hydrographie und Oberflächen plankton des westlichen Barentsmeeres im Sommer 1927. Ber. deutschen wissensch. Komm. F. Meeresforsch. n.s. 4(5):232-372, 13 tables, 25 text figs.

Favella edentata
Schulz, B., and A. Wulff, 1929
Hydrographie und Oberflächen plankton des westlichen Barentsmeeres im Sommer 1927. Ber. deutschen wissensch. Komm. F. Meeresforsch. n.s. 4(5):232-372, 13 tables, 25 text figs.

Favella ehrenbergi
Balech, E., 1959.
Tintinnoinea del Mediterraneo. Trav. Sta. Zool., Villefranche-sur-Mer, 18(17): Reprinted from: Trab. Inst. Español Oceanogr., Madrid, 19 Sept. 1959(28):88 pp., 22 pls.

Favella ehrenbergi
Delegazione Italiana della Commissione Internazionale per l'Esplorazione Scientifica del Mediterraneo, 1941
Note sul plancton della Laguna veneta. [Memoria CCLXXXIX], Arch. di Ocean. e Limn. Anno I, Fasc. I, 1941 XIX: 31-57 pp.

Favella ehrenbergii
Hofker, J., 1931
Studien uber Tintinnoidea. Arch. f. Protistenk 75(3):315-402, 89 text figs.

Favella ehrenbergi
Jörgensen, E., 1924
Mediterranean Tintinnidae. Rept. Danish Oceanogr. Exped. 1908-10 to the Mediterranean and adjacent seas, Vol. II, Biol. J 3, 110 pp., 114 text figs.

Favella ehrenbergii
Kofoid, C. A. and A. S. Campbell, 1929
A conspectus of the marine and freshwater Ciliata belonging to the suborder Tintinnoinea, with descriptions of new species principally from the Agassiz expedition to the eastern tropical Pacific, 1904-1905. Univ. Calif. Publ. Zool. 34:1-403, 697 text figs.

Favella ehrenbergi
Massuti Algamora, M., 1949
Estudio de diez y seis muestras de plancton del Golfo de Nápoles. Publ. Inst. Biol. Appl. 5:85-94, 1 fold-in table.

Favella ehrenbergii
Needler, A. B., 1949.
Paralytic shellfish poisoning and Goniaulax tamarensis. J. Fish. Res. Bd., Canada, 7(8): 490-504, 3 textfigs.

Favella ehrenbergi
Sousa e Silva, E. de, 1950.
Les tintinnides de la Baie de Cascais. Bull. Inst. Océan., Monaco, No. 979:28 pp., 4 pls.

Favella elongata, n.sp.
Roxas, H.A., 1941.
Marine protozoa of the Philippines.
Philippine J. Sci. 74:91-136, 17 pls., 2 textfigs

Favella fistulicauda n.sp.
Jörgensen, E., 1924
Mediterranean Tintinnidae. Rept. Danish Oceanogr. Exped. 1908-10 to the Mediterranean and adjacent seas, Vol.II, Biol. J 3, 110 pp., 114 text figs.

Favella fistulicauda
Kofoid, C. A. and A. S. Campbell, 1929
A conspectus of the marine and freshwater Ciliata belonging to the suborder Tintinnoinea, with descriptions of new species principally from the Agassiz expedition to the eastern tropical Pacific, 1904-1905. Univ. Calif. Publ. Zool. 34:1-403, 697 text figs.

Favella Branciscana n. sp.
Kofoid, C. A. and A. S. Campbell, 1929
A conspectus of the marine and freshwater Ciliata belonging to the suborder Tintinnoinea, with descriptions of new species principally from the Agassiz expedition to the eastern tropical Pacific, 1904-1905. Univ. Calif. Publ. Zool. 34:1-403, 697 text figs.

Favella helgolandica
Kofoid, C. A. and A. S. Campbell, 1929
A conspectus of the marine and freshwater Ciliata belonging to the suborder Tintinnoinea, with descriptions of new species principally from the Agassiz expedition to the eastern tropical Pacific, 1904-1905. Univ. Calif. Publ. Zool. 34:1-403, 697 text figs.

Favella infundibulum n.sp.
Kofoid, C. A. and A. S. Campbell, 1929
A conspectus of the marine and freshwater Ciliata belonging to the suborder Tintinnoinea, with descriptions of new species principally from the Agassiz expedition to the eastern tropical Pacific, 1904-1905. Univ. Calif. Publ. Zool. 34:1-403, 697 text figs.

Favella markusovszkyi
Duran, M., 1957.
Nota sobre algunos tintinnoineas del plancton de Puerto Rico. Invest. Pesqueria, Barcelona, 8: 97-120.

Favella markusovzkyi
Jörgensen, E., 1924
Mediterranean Tintinnidae. Rept. Danish Oceanogr. Exped. 1908-10 to the Mediterranean and adjacent seas, Vol.II, Biol. J 3, 110 pp., 114 text figs.

Favella markusovszkyi
Kofoid, C. A. and A. S. Campbell, 1929
A conspectus of the marine and freshwater Ciliata belonging to the suborder Tintinnoinea, with descriptions of new species principally from the Agassiz expedition to the eastern tropical Pacific, 1904-1905. Univ. Calif. Publ. Zool. 34:1-403, 697 text figs.

Favella markusovszkyi
Massutí Algamora, M., 1949
Estudio de diez y seis muestras de plancton del Golfo de Nápoles. Publ. Inst. Biol. Appl. 5:85-94, 1 fold-in table.

Favella markusovszkyi
Rampi, L., 1939
Primo contributo alla conoscenza dei tintinnoidi do Maro Ligure. Atti della Soc. Ital. di Sci. Nat. 78:67-81, 58 text figs.

Favella markusovskyi
Sousa e Silva, E. de, 1950.
Les tintinnides de la Baie de Cascais. Bull. Inst Ocśan., Monaco, No. 979:28 pp., 4 pls.

Favella meunieri n.sp.
Kofoid, C. A. and A. S. Campbell, 1929
A conspectus of the marine and freshwater Ciliata belonging to the suborder Tintinnoinea, with descriptions of new species principally from the Agassiz expedition to the eastern tropical Pacific, 1904-1905. Univ. Calif. Publ. Zool. 34:1-403, 697 text figs.

Favella minutissima n.sp
Campbell, A. S., 1942
The oceanic Tintinnoina of the plankton gathered during the last cruise of the CARNEGIE. Sci. Res. Cruise VII of the Carnegie, 1928-1929, Biol. 2:163 pp., 128 figs.

Favella panamensis
Cosper, T.C., 1972.
The identification of tintinnids (Protozoa: Ciliata: Tintinnida) of the St. Andrew Bay system, Florida. Bull. mar. Sci. Miami 22(2): 391-418.

Favella panamensis
Duran, M., 1957.
Nota sobre algunos tintinnoineas del plancton de Puerto Rico. Invest. Pesqueria, Barcelona, 8: 97-120.

Favella panamensis
Kofoid, C.A. and A.S. Campbell, 1939
Reports on the scientific results of the expedition to the eastern tropical Pacific in charge of Alexander Agassiz, by the U.S. Fish Commission steamer "Albatross" from October 1904 to March 1905-----XXXVII. The Ciliata: The Tintinnoinea. Bull.Mus. Comp.Zool. 84: 473 pp., 36 pls.

Favella panamensis n.sp.
Kofoid, C. A. and A. S. Campbell, 1929
A conspectus of the marine and freshwater Ciliata belonging to the suborder Tintinnoinea, with descriptions of new species principally from the Agassiz expedition to the eastern tropical Pacific, 1904-1905. Univ. Calif. Publ. Zool. 34:1-403, 697 text figs.

Favella panamensis
Kokubo, S., and S. Sato., 1947
Plankters in JÛ-San Gata. Physiol. and Ecol. (Japan) 1(4):1-16, 3 text figs., tables.

Favella panamensis
Komarovsky, B., 1962
Tintinnina from the vicinity of the Straits of Tiran and Massawa Region. Contributions to the knowledge of the Red Sea, No. 25. Sea Fish. Res. Sta., Haifa, Israel, Bull. No. 30:48-56.

Favella panamensis
Wang, C. C., and D. Nie, 1932
A survey of the marine protozoa of Amoy. Contr. Biol. Lab. Sci. Soc., China, Zool. Ser., 8:285-385, 89 text figs.

Favella philippinensis n.sp.
Roxas, H.A., 1941.
Marine protozoa of the Philippines.
Philippine J. Sci. 74:91-136, 17 pls., 2 textfigs

Favella quequenensis
Balech, Enrique, 1948
Tintinnoinea de Atlantida (R.O. del Uruguay) (Protozoa, Ciliata Oligotr.). Comm. Mus. Argentino Ciencias Nat. "Bernardino Rivadavia", Ser.Cien.Zool. No.7, 23 pp., 8 pls. of 107 figs.

Favella quequenense n.sp.
Balech, E., 1945.
Tintinnoinea de Quequén. Physis 20(55):1-15, 3 pls.

Favella septentrionalis n.sp.
Campbell, A. S., 1942
The oceanic Tintinnoina of the plankton gathered during the last cruise of the CARNEGIE. Sci. Res. Cruise VII of the Carnegie, 1928-1929, Biol. 2:163 pp., 128 figs.

Favella serrata
Balech, E., 1959.
Tintinnoinea del Mediterraneo.
Trav. Sta. Zool., Villefranche-sur-Mer, 18(17):
Reprinted from:
Trab. Inst. Español Oceanogr., Madrid, 19 Sept. 1959(28):88 pp., 22 pls.

Favella serrata
Campbell, A. S., 1942
The oceanic Tintinnoina of the plankton gathered during the last cruise of the CARNEGIE. Sci. Res. Cruise VII of the Carnegie, 1928-1929, Biol. 2:163 pp., 128 figs.

Favella serrata
Gaarder, K.R., 1946
Tintinnoidea. Rep.Sci.Res."Michael Sars" N.Atlantic Deep-Sea Exped., 1910, 2(1): 37 pp., 23 text figs.

Favella serrata
Gran, H.H., and T. Braarud, 1935
A quantitative study of the phytoplankton in the Bay of Fundy and the Gulf of Maine (including observations on hydrography, chemistry, and turbidity). J. Biol. Bd., Canada, 1(5):279-467, 69 text figs.

Favella serrata
Jörgensen, E., 1924
Mediterranean Tintinnidae. Rept. Danish Oceanogr. Exped. 1908-10 to the Mediterranean and adjacent seas, Vol.II, Biol. J 3, 110 pp., 114 text figs.

Favella serrata
Massutí Algamora, M., 1949
Estudio de diez y seis muestras de plancton del Golfo de Nápoles. Publ. Inst. Biol. Appl. 5:85-94, 1 fold-in table.

Favella serrata
Kofoid, C. A. and A. S. Campbell, 1929
A conspectus of the marine and freshwater Ciliata belonging to the suborder Tintinnoinea, with descriptions of new species principally from the Agassiz expedition to the eastern tropical Pacific, 1904-1905. Univ. Calif. Publ. Zool. 34:1-403, 697 text figs.

Favella serrata
Sousa e Silva, E. de, 1950.
Les tintinnides de la Baie de Cascais. Bull. Inst. Ocśan., Monaco, No. 979:28 pp., 4 pls.

Favella simplex n.sp.
Roxas, H.A., 1941.
Marine protozoa of the Philippines.
Philippine J. Sci. 74:91-136, 17 pls., 2 textfigs

Favella taraikaensis
Balech, E., 1951.
Nuevos datos sobre Tintinnoinea de Argentina y Uruguay. Physis 20(58):291-302, 16 textfigs.

Favella undulata n.sp.
Wang, C. C., and D. Nie, 1932
A survey of the marine protozoa of Amoy. Contr. Biol. Lab. Sci. Soc., China, Zool. Ser., 8:285-385, 89 text figs.

Helicostoma notata
*Hada, Yoshina, 1970. (1970)
The protozoan plankton of the Antarctic and Subantarctic seas.
Scient. Repts, Japan. Antarct. Res. Exped., (E)31:51 pp.

Helicostomella sp. n.gen.
Jörgensen, E., 1924
Mediterranean Tintinnidae. Rept. Danish Oceanogr. Exped. 1908-10 to the Mediterranean and adjacent seas, Vol.II, Biol. J 3, 110 pp., 114 text figs.

Helicostomella sp.
Morse, D.C., 1947
Some observations on seasonal variations in plankton population Patuxant River, Maryland 1943-1945. Bd. Nat. Res., Publ. No.65, Chesapeake Biol. Lab. 31, 3 figs.

Helicostomella antarctica
*Hada, Yoshina, 1970. (1970)
The protozoan plankton of the Antarctic and Subantarctic seas.
Scient. Repts, Japan. Antarct. Res. Exped., (E)31:51 pp.

Helicostomella edentata
Kofoid, C. A. and A. S. Campbell, 1929
A conspectus of the marine and freshwater Ciliata belonging to the suborder Tintinnoinea, with descriptions of new species principally from the Agassiz expedition to the eastern tropical Pacific, 1904-1905. Univ. Calif. Publ. Zool. 34:1-403, 697 text figs.

Helicostomella fusiformis
Kofoid, C. A. and A. S. Campbell, 1929
A conspectus of the marine and freshwater Ciliata belonging to the suborder Tintinnoinea, with descriptions of new species principally from the Agassiz expedition to the eastern tropical Pacific, 1904-1905. Univ. Calif. Publ. Zool. 34:1-403, 697 text figs.

Helicostomella kiliensis
Kofoid, C. A. and A. S. Campbell, 1929
A conspectus of the marine and freshwater Ciliata belonging to the suborder Tintinnoinea, with descriptions of new species principally from the Agassiz expedition to the eastern tropical Pacific, 1904-1905. Univ. Calif. Publ. Zool. 34:1-403, 697 text figs.

Helicostomella Lemairei
Balech, Enrique, 1944
Contribucion al conocimiento del Plancton de Lennox y Cabo de Hornos. Physis XIX:423-446, 6 pls. with 67 figs.

Helicostomella lemairei n.sp.
Balech, E., 1941.
Tintinnoineos del estrecho Le Maire. Physis XIX(52):245-252, 10 textfigs.

Helicostomella longa
Campbell, A. S., 1942
The oceanic Tintinnoina of the plankton gathered during the last cruise of the CARNEGIE. Sci. Res. Cruise VII of the Carnegie, 1928-1929, Biol. 2:163 pp., 128 figs.

Helicostomella longa
Kofoid, C. A. and A. S. Campbell, 1929
A conspectus of the marine and freshwater Ciliata belonging to the suborder Tintinnoinea, with descriptions of new species principally from the Agassiz expedition to the eastern tropical Pacific, 1904-1905. Univ. Calif. Publ. Zool. 34:1-403, 697 text figs.

Helicostomella longa
Kofoid, C.A. and A.S. Campbell, 1939
Reports on the scientific results of the expedition to the eastern tropical Pacific in charge of Alexander Agassiz, by the U.S. Fish Commission steamer "Albatross" from October 1904 to March 1905-----XXXVII. The Ciliata: The Tintinnoimea. Bull.Mus. Comp.Zool. 84: 473 pp., 36 pls.

Helicostomella subulata
Balech, Enrique, 1962.
Notulas de la Estacion Hidrobiologica de Puerto Quequen. Revista, Mus. Argentino, Ciencias Nat., "Bernardino Rivadavia", Ciencias Zool., 8(6):81-87.

Helicostomella subulata
Balech, E., 1959.
Tintinnoinea del Mediterraneo. Trav. Sta. Zool., Villefranche-sur-Mer, 18(1):
Reprinted from:
Trab. Inst. Español Oceanogr., Madrid, 19 Sept. 1959(28):88 pp., 22 pls.

Helicostomella subulata
Braarud, T., 1945
A phytoplankton survey of the polluted waters of inner Oslo Fjord. Hvalrådets Skrifter, No.28, 142 pp., 19 text figs., 17 tables.

Helicostomella subulata
Braarud, T., and Adam Bursa, 1939
On the phytoplankton of the Oslo Fjord, 1933-1934. Hvalrådets Skr. No.19:1-63; 9 text figs. Reviewed. J. du. Cons. 14(3): 418-420. A.C. Gardiner.

Helicostomella subulata
Cosper, T.C., 1972.
The identification of tintinnids (Protozoa: Ciliata: Tintinnida) of the St. Andrew Bay system, Florida. Bull. mar. Sci. Miami 22(2): 391-418.

Helicostomella subulata
Gaarder, K.R., 1946
Tintinnoidea. Rep.Sci.Res. "Michael Sars" N.Atlantic Deep-Sea Exped., 1910, 2(1): 37 pp., 23 text figs.

Helicostomella subulata
Gran, H.H., and T. Braarud, 1935
A quantitative study of the phytoplankton in the Bay of Fundy and the Gulf of Maine (including observations on hydrography, chemistry, and turbidity). J. Biol. Bd., Canada, 1(5):279-467, 69 text figs.

Helicostomella subulata
Grøntved, J., 1949
Investigations on the phytoplankton in the Danish Waddensea in July 1941. Medd. Komm. Danmarks Fiskeri og Havundersøgelser, ser. Plankton, 5(2):55 pp., 2 pls., 38 text figs.

Helicostomella subulata
Hofker, J., 1931
Studien uber Tintinnoidea. Arch. f. Protistenk 75(3):315-402, 89 text figs.

Helicostomella subulata
Jörgensen, E., 1924
Mediterranean Tintinnidae. Rept. Danish Oceanogr. Exped. 1908-10 to the Mediterranean and adjacent seas, Vol.II, Biol. J 3, 110 pp., 114 text figs.

Helicostomella subulata
Kofoid, C. A. and A. S. Campbell, 1929
A conspectus of the marine and freshwater Ciliata belonging to the suborder Tintinnoinea, with descriptions of new species principally from the Agassiz expedition to the eastern tropical Pacific, 1904-1905. Univ. Calif. Publ. Zool. 34:1-403, 697 text figs.

Helicostomella subulata
Levander, K.M., 1947
Plankton gesammelt in den Jahren 1899-1910 an den Küsten Finnlands. Finnländische Hydrographisch-Biologische Untersuchunger (aus dem Wasserbiologischen Laboratorien der Societas Scientiarum Fennica) No.11:40 pp., 6 diagrams, 13 pls., tables.

Hystonella treforti
Hofker, J., 1931
Studien uber Tintinnoidea. Arch. f. Protistenk 75(3):315-402, 89 text figs.

Laackmanniella n. gen.
Kofoid, C. A. and A. S. Campbell, 1929
A conspectus of the marine and freshwater Ciliata belonging to the suborder Tintinnoinea, with descriptions of new species principally from the Agassiz expedition to the eastern tropical Pacific, 1904-1905. Univ. Calif. Publ. Zool. 34:1-403, 697 text figs.

Laackmanniella naviculaefera
*Hada, Yoshina, 1970. (1970)
The protozoan plankton of the Antarctic and Subantarctic seas. Scient. Repts, Japan. Antarct. Res. Exped., (E)31:51 pp.

Laackmaniella naviculaefera
Kofoid, C. A. and A. S. Campbell, 1929
A conspectus of the marine and freshwater Ciliata belonging to the suborder Tintinnoinea, with descriptions of new species principally from the Agassiz expedition to the eastern tropical Pacific, 1904-1905. Univ. Calif. Publ. Zool. 34:1-403, 697 text figs.

Laackmaniella prolongata
Balech, E., 1947
Contribution al conocimiento del plancton antarctico. Plankton del Mar de Bellinghausen. Physis 20:75-91, 76 figs. on 8 pls.

Laackmaniella prolongata
Balech, Enrique, 1947
Contribucion al conocimiento del Plancton antartico Plancton del Mar de Bellinghausen. Physis 22(56):75-91, 7 pls. with 76 figs.

Laackmanniella prolongata
Kofoid, C. A. and A. S. Campbell, 1929
A conspectus of the marine and freshwater Ciliata belonging to the suborder Tintinnoinea, with descriptions of new species principally from the Agassiz expedition to the eastern tropical Pacific, 1904-1905. Univ. Calif. Publ. Zool. 34:1-403, 697 text figs.

Leprotintinnopsis bottnicus
Rumkówna, A., 1948
List of the phytoplankton species occurring in the superficial water layers in the Gulf of Gdańsk. Bull. Lab. mar., Gdynia, No. 4: 139-141 with tables in back.

Leprotintinnus spp.
Braarud, T., 1945
A phytoplankton survey of the polluted waters of inner Oslo Fjord. Hvalrådets Skrifter, No.28, 142 pp., 19 text figs., 17 tables.

Leprotintinnus sp.
Morse, D.C., 1947
Some observations on seasonal variations in plankton population Patuxant River, Maryland 1943-1945. Bd. Nat. Res., Publ. No.65, Chesapeake Biol. Lab., 31, 3 figs.

Leprotintinnus bottnicus
Biernacka, I., 1948
(Tintinnoinea in the Gulf of Gdansk and adjoining waters) Builetyn Morskiego Laboratorium Rybackiego w Gdyni dawniej-Stacji Morskiej w. Helu. 4:73-91, 4 text figs., 1 pl. with 21 figs. (Bull. Lab. Mar. Gdynia, formerly Bull. Sta. Hel.)

Leprotintinnus bottnicus
Dolgopol'skaia, M.A., and V.L. Pauli, 1964.
Plankton of the Azov Sea. (In Russian). Trudy, Sevastopol Biol. Stants., Akad. Nauk, SSSR, 15:118-151.

Leprotintinnus bottnicus
Hofker, J., 1931
Studien uber Tintinnoidea. Arch. f. Protistenk 75(3):315-402, 89 text figs.

Leprotintinnus botticus n.gen.
Jørgensen, E., 1900.
Ueber die Tintinnodeen der norwegischen Westküste Bergens Mus. Aarb., 1899(2):48 pp., 3 pls.

Leprotintinnus bottnicus
Kofoid, C. A. and A. S. Campbell, 1929
A conspectus of the marine and freshwater Ciliata belonging to the suborder Tintinnoinea, with descriptions of new species principally from the Agassiz expedition to the eastern tropical Pacific, 1904-1905. Univ. Calif. Publ. Zool. 34:1-403, 697 text figs.

Leprotintinnus bottnicus
Levander, K.M., 1947
Plankton gesammelt in den Jahren 1899-1910 an den Küsten Finnlands. Finnländische Hydrographisch-Biologische Untersuchunger (aus dem Wasserbiologischen Laboratorin der Societas Scientiarum Fennica) No.11:40 pp., 6 diagrams, 13 pls., tables.

Leprotintinnus bottnicus
Schulz, H., 1965.
Die Tintinnoinea des Elbe-Aestuars. Arch. Fischereiwiss., 15(3):216-225.

Leprotintinnus gaussi
Laackmann, H., 1910
Die Tintinnodeen der Deutschen Südpolar Expedition 1901-1903. Deutsch. Südpolar Exped. 11, Zool. 3(4):341-496, pls. 33-51.

Leprotintinnus glacialis
Laackmann, H., 1910
Die Tintinnodeen der Deutschen Südpolar Expedition 1901-1903. Deutsch. Südpolar Exped. 11, Zool. 3(4):341-496, pls. 33-51.

Leprotintinnus naviculaeferus
Laackmann, H., 1910
Die Tintinnodeen der Deutschen Südpolar Expedition 1901-1903. Deutsch. Südpolar Exped. 11, Zool. 3(4):341-496, pls. 33-51.

Leprotintinnus neriticus
Kofoid, C. A. and A. S. Campbell, 1929
A conspectus of the marine and freshwater Ciliata belonging to the suborder Tintinnoinea, with descriptions of new species principally from the Agassiz expedition to the eastern tropical Pacific, 1904-1905. Univ. Calif. Publ. Zool. 34:1-403, 697 text figs.

Leprotintinnus nordqvisti
Duran, M., 1957.
Nota sobre algunos tintinnoineas del plancton de Puerto Rico. Invest. Pesqueria, Barcelona, 8:97-120.

Leprotintinnus nordqvisti
Kofoid, C. A. and A. S. Campbell, 1929
A conspectus of the marine and freshwater Ciliata belonging to the suborder Tintinnoinea, with descriptions of new species principally from the Agassiz expedition to the eastern tropical Pacific, 1904-1905. Univ. Calif. Publ. Zool. 34:1-403, 697 text figs.

Leprotintinnus nordquisti
Marshall, S. M., 1934
The silicoflagellata and tintinnoinea. Great Barrier Reef Exped., 1928-29. Sci. Repts. 4(15):623-664, 43 text figs.

Leprotintinnus nordqvisti
Marshall, S. M., 1933
The production of microplankton in the Great Barrier Reef Region. Brit. Mus. (N.H.) Great Barrier Reef Exped. 1928-29, Sci. Repts. II(5):111-157, 14 text figs.

Leprotintinnus nordqvisti
Wang, C. C., and D. Nie, 1932
A survey of the marine protozoa of Amoy. Contr. Biol. Lab., Sci. Soc., China, Zool. Ser., 8:285-385, 89 text figs.

Leprotintinnus pellucidus
Dolgopol'skaia, M.A., and V.L. Pauli, 1964.
Plankton of the Azov Sea. (In Russian). Trudy, Sevastopol Biol. Stants., Akad. Nauk. SSSR, 15:118-151.

Leprotintinnus pellucidus
Jørgensen, E., 1905
B.Protistplankton and the diatoms in bottom samples. Hydrographical and biological investigations in Norwegian fjords. Bergens Mus. Skr. 7: 49-225.

Leprotintinnus pellucidus
Kofoid, C. A. and A. S. Campbell, 1929
A conspectus of the marine and freshwater Ciliata belonging to the suborder Tintinnoinea, with descriptions of new species principally from the Agassiz expedition to the eastern tropical Pacific, 1904-1905. Univ. Calif. Publ. Zool. 34:1-403, 697 text figs.

Leprotintinnus pellucidus
Schulz, B., and A. Wulff, 1929
Hydrographie und Oberflächen plankton des westlichen Barentsmeeres im Sommer 1927. Ber. deutschen wissensch. Komm. F. Meeresforsch. n.s. 4(5):232-372, 13 tables, 25 text figs.

Leprotintinnus pellucidus
Schulz, H., 1965.
Die Tintinnoinea des Elbe-Aestuars. Arch. Fischereiwiss., 15(3):216-225.

Leptrotintinnus prolongatus
Laackmann, H., 1910
Die Tintinnodeen der Deutschen Südpolar Expedition 1901-1903. Deutsch. Südpolar Exped. 11, Zool. 3(4):341-496, pls. 33-51.

Leprotintinnus simplex
Kofoid, C. A. and A. S. Campbell, 1929
A conspectus of the marine and freshwater Ciliata belonging to the suborder Tintinnoinea, with descriptions of new species principally from the Agassiz expedition to the eastern tropical Pacific, 1904-1905. Univ. Calif. Publ. Zool. 34:1-403, 697 text figs.

Leprotintinnus tuberculosum n.sp.
Roxas, H.A., 1941.
Marine protozoa of the Philippines. Philippine J. Sci. 74:91-136, 17 pls., 2 textfigs

Metacylis spp.
Cosper, T.C., 1972.
The identification of tintinnids (Protozoa: Ciliata: Tintinnida) of the St. Andrew Bay system, Florida. Bull. mar. Sci. Miami 22(2): 391-418.

Metacylis sp.
de Sousa e Silva, E., 1956.
Contribution à l'étude du microplancton de Dakar et des regions maritimes voisines. Bull. I.F.A.N., 8(2):335-371, 7 pls.

Metacylis n.gen.
Jörgensen, E., 1924
Mediterranean Tintinnidae. Rept. Danish Oceanogr. Exped. 1908-10 to the Mediterranean and adjacent seas, Vol.II, Biol. J 3, 110 pp., 114 text figs.

Metacylis sp.
Morse, D.C., 1947
Some observations on seasonal variations in plankton population Patuxant River, Maryland 1943-1945. Bd. Nat. Res., Publ. No.65, Chesapeake Biol. Lab., 31, 3 figs.

Metacylis angulata n sp.
Lackey, James B., and Enrique Balech, 1966.
A new marine tintinnid.
Trans. Am. microsc. Soc., 85(4):575-578.

Metacylis annulata
Kofoid, C. A. and A. S. Campbell, 1929
A conspectus of the marine and freshwater Ciliata belonging to the suborder Tintinnoinea, with descriptions of new species principally from the Agassiz expedition to the eastern tropical Pacific, 1904-1905. Univ. Calif. Publ. Zool. 34:1-403, 697 text figs.

Metacylis annulifera
Balech, E., 1951.
Nuevos datos sobre Tintinnoinea de Argentina y Uruguay. Physis 20(58):291-302, 16 textfigs.

Metacylis annulifera
Kofoid, C. A. and A. S. Campbell, 1929
A conspectus of the marine and freshwater Ciliata belonging to the suborder Tintinnoinea, with descriptions of new species principally from the Agassiz expedition to the eastern tropical Pacific, 1904-1905. Univ. Calif. Publ. Zool. 34:1-403, 697 text figs.

Metacylis conica
Kofoid, C.A. and A.S. Campbell, 1939
Reports on the scientific results of the expedition to the eastern tropical Pacific in charge of Alexander Agassiz, by the U.S. Fish Commission steamer "Albatross" from October 1904 to March 1905-----XXXVII. The Ciliata: The Tintinnoimea. Bull.Mus. Comp.Zool. 84: 473 pp., 36 pls.

Metacylis conica n.sp.
Kofoid, C. A. and A. S. Campbell, 1929
A conspectus of the marine and freshwater Ciliata belonging to the suborder Tintinnoinea, with descriptions of new species principally from the Agassiz expedition to the eastern tropical Pacific, 1904-1905. Univ. Calif. Publ. Zool. 34:1-403, 697 text figs.

Metaclys corbula
Duran, M., 1957.
Nota sobre algunos tintinnoineas del plancton de Puerto Rico. Invest. Pesqueria, Barcelona, 8: 97-120.

Metacylis corbula
*Hada, Yoshina, 1970.   (1970)
The protozoan plankton of the Antarctic and Subantarctic seas.
Scient. Repts, Japan. Antarct. Res. Exped., (E)31:51 pp.

Metacylis corbula
Kofoid, C.A. and A.S. Campbell, 1939
Reports on the scientific results of the expedition to the eastern tropical Pacific in charge of Alexander Agassiz, by the U.S. Fish Commission steamer "Albatross" from October 1904 to March 1905-----XXXVII. The Ciliata: The Tintinnoimea. Bull.Mus. Comp.Zool. 84: 473 pp., 36 pls.

Metacylis corbula n.sp.
Kofoid, C. A. and A. S. Campbell, 1929
A conspectus of the marine and freshwater Ciliata belonging to the suborder Tintinnoinea, with descriptions of new species principally from the Agassiz expedition to the eastern tropical Pacific, 1904-1905. Univ. Calif. Publ. Zool. 34:1-403, 697 text figs.

Metacylis corbula
Marshall, S. M., 1934
The silicoflagellata and tintinnoinea. Great Barrier Reef Exped., 1928-29. Sci. Repts. 4(15):623-664, 43 text figs.

Metacylis hemisphaerica n.s
Roxas, H.A., 1941.
Marine protozoa of the Philippines. Philippine J. Sci. 74:91-136, 17 pls., 2 textfig

Metacylis jörgenseni
Balech, E., 1959.
Tintinnoinea del Mediterraneo. Trav. Sta. Zool., Villefranche-sur-Mer, 18(17): Reprinted from:
Trab. Inst. Español Oceanogr., Madrid, 19 Sept. 1959(28):88 pp., 22 pls.

Metacylis jörgensenii
Kofoid, C. A. and A. S. Campbell, 1929
A conspectus of the marine and freshwater Ciliata belonging to the suborder Tintinnoinea, with descriptions of new species principally from the Agassiz expedition to the eastern tropical Pacific, 1904-1905. Univ. Calif. Publ. Zool. 34:1-403, 697 text figs.

Metacylis joergensenii
Rampi, L., 1939
Primo contributo alla conoscenza dei tintinnoidi do Maro Ligure. Atti della Soc. Ital. di Sci. Nat. 78:67-81, 58 text figs.

Metacylis kofoidi n.sp.
Roxas, H.A., 1941.
Marine protozoa of the Philippines. Philippine J. Sci. 74:91-136, 17 pls., 2 textfigs

Metacylis lucasensis
de Sousa e Silva, E., 1956.
Contribution à l'étude du microplancton de Dakar et des regions maritimes voisines. Bull. I.F.A.N., 8(2):335-371, 7 pls.

Metacylis lucasensis
Gaarder, K.R., 1946
Tintinnoidea. Rep.Sci.Res. "Michael Sars" N.Atlantic Deep-Sea Exped.,1910, 2(1): 37 pp., 23 text figs.

Metacylis lucasensis
Kofoid, C.A. and A.S. Campbell, 1939
Reports on the scientific results of the expedition to the eastern tropical Pacific in charge of Alexander Agassiz, by the U.S. Fish Commission steamer "Albatross" from October 1904 to March 1905-----XXXVII. The Ciliata: The Tintinnoimea. Bull.Mus. Comp.Zool. 84: 473 pp., 36 pls.

Metacylis lucasensis n.sp.
Kofoid, C. A. and A. S. Campbell, 1929
A conspectus of the marine and freshwater Ciliata belonging to the suborder Tintinnoinea, with descriptions of new species principally from the Agassiz expedition to the eastern tropical Pacific, 1904-1905. Univ. Calif. Publ. Zool. 34:1-403, 697 text figs.

Metacylis mediterranea
Duran, M., 1951
Contribucion al estudio de los tintinidos del plancton de los costas de Castellon. (Mediterraneo occidental). Publ. Inst. Biol. Aplic., Barcelone, 8: 101-120, 29 text figs.

Metacylis mediterranea
Kofoid, C. A. and A. S. Campbell, 1929
A conspectus of the marine and freshwater Ciliata belonging to the suborder Tintinnoinea, with descriptions of new species principally from the Agassiz expedition to the eastern tropical Pacific, 1904-1905. Univ. Calif. Publ. Zool. 34:1-403, 697 text figs.

Metacylis mediterranea
Jörgensen, E., 1924
Mediterranean Tintinnidae. Rept. Danish Oceanogr. Exped. 1908-10 to the Mediterranean and adjacent seas, Vol.II, Biol. J 3, 110 pp., 114 text figs.

Metacylis mereschkowskyi
Balech, Enrique, 1968.
Algunas especies nuevas o interesantes do tintinnidos del Golfo de Mexico y Caribe. Revta Mus. argent. Cienc. nat. Bernardino Rivadavia Inst. nac. Hidrobiol. 2(5):165-197.

Metacylis mereschkowskii
Cosper, T.C., 1972.
The identification of tintinnids (Protozoa: Ciliata: Tintinnida) of the St. Andrew Bay system, Florida. Bull. mar. Sci. Miami 22(2): 391-418.

Metacylis mereschkowskii n.sp.
Kofoid, C. A. and A. S. Campbell, 1929
A conspectus of the marine and freshwater Tintinnoinea, with descriptions of new species principally from the Agassiz expedition to the eastern tropical Pacific, 1904-1905. Univ. Calif. Publ. Zool. 34:1-403, 697 text figs.

Metacylis pontica
Kofoid, C. A. and A. S. Campbell, 1929
A conspectus of the marine and freshwater Ciliata belonging to the suborder Tintinnoinea, with descriptions of new species principally from the Agassiz expedition to the eastern tropical Pacific, 1904-1905. Univ. Calif. Publ. Zool. 34:1-403, 697 text figs.

Metacylis rossica nom. nov.
Kofoid, C. A. and A. S. Campbell, 1929
A conspectus of the marine and freshwater Ciliata belonging to the suborder Tintinnoinea, with descriptions of new species principally from the Agassiz expedition to the eastern tropical Pacific, 1904-1905. Univ. Calif. Publ. Zool. 34:1-403, 697 text figs.

Metaclys tropica n.sp.
Duran, M., 1957.
Nota sobre algunos tintinnoineas del plancton de Puerto Rico. Invest. Pesqueria, Barcelona, 8: 97-120.

Metacylis vitreoides nom. nov.
Kofoid, C. A. and A. S. Campbell, 1929
A conspectus of the marine and freshwater Ciliata belonging to the suborder Tintinnoinea, with descriptions of new species principally from the Agassiz expedition to the eastern tropical Pacific, 1904-1905. Univ. Calif. Publ. Zool. 34:1-403, 697 text figs.

Odontophorella serrulata
Kofoid, C.A. and A.S. Campbell, 1939
Reports on the scientific results of the expedition to the eastern tropical Pacific in charge of Alexender Agassiz, by the U.S. Fish Commission steamer "Albatross" from October 1904 to March 1905-----XXXVII. The Ciliata: The Tintinnoimea. Bull.Mus. Comp.Zool. 84: 473 pp., 36 pls.

Odontophorella serrulata n.gen., n. sp.
Kofoid, C. A. and A. S. Campbell, 1929
A conspectus of the marine and freshwater Ciliata belonging to the suborder Tintinnoinea, with descriptions of new species principally from the Agassiz expedition to the eastern tropical Pacific, 1904-1905. Univ. Calif. Publ. Zool. 34:1-403, 697 text figs.

Ormosella sp
Balech, Enrique, 1968.
Algunas especies nuevas o interesantes do tintinnidos del Golfo de Mexico y Caribe. Revta Mus. argent. Cienc. nat. Bernardino Rivadavia Inst. nac. Hidrobiol. 2(5):165-197.

Ormosella n. gen.
Kofoid, C. A. and A. S. Campbell, 1929
A conspectus of the marine and freshwater Ciliata belonging to the suborder Tintinnoinea, with descriptions of new species principally from the Agassiz expedition to the eastern tropical Pacific, 1904-1905. Univ. Calif. Publ. Zool. 34:1-403, 697 text figs.

Ormosella apsteini
Campbell, A. S., 1942
The oceanic Tintinnoina of the plankton gathered during the last cruise of the CARNEGIE. Sci. Res. Cruise VII of the Carnegie, 1928-1929, Biol. 2:163 pp., 128 figs.

Ormosella apsteini
Kofoid, C.A. and A.S. Campbell, 1939
Reports on the scientific results of the expedition to the eastern tropical Pacific in charge of Alexender Agassiz, by the U.S. Fish Commission steamer "Albatross" from October 1904 to March 1905-----XXXVII. The Ciliata: The Tintinnoimea. Bull.Mus. Comp.Zool. 84: 473 pp., 36 pls.

Ormosella apsteini n.sp.
Kofoid, C. A. and A. S. Campbell, 1929
A conspectus of the marine and freshwater Ciliata belonging to the suborder Tintinnoinea, with descriptions of new species principally from the Agassiz expedition to the eastern tropical Pacific, 1904-1905. Univ. Calif. Publ. Zool. 34:1-403, 697 text figs.

Ormosella breslaui
Balech, Enrique, 1968.
Algunas especies nuevas o interesantes do tintinnidos del Golfo de Mexico y Caribe. Revta Mus. argent. Cienc. nat. Bernardino Rivadavia Inst. nac. Hidrobiol. 2(5):165-197.

Ormosella bresslaui
Gaarder, K.R., 1946
Tintinnoidea. Rep.Sci.Res. "Michael Sars" N.Atlantic Deep-Sea Exped., 1910, 2(1): 37 pp., 23 text figs.

Ormosella bresslaui
Kofoid, C.A. and A.S. Campbell, 1939
Reports on the scientific results of the expedition to the eastern tropical Pacific in charge of Alexender Agassiz, by the U.S. Fish Commission steamer "Albatross" from October 1904 to March 1905-----XXXVII. The Ciliata: The Tintinnoimea. Bull.Mus. Comp.Zool. 84: 473 pp., 36 pls.

Ormosella bresslaui n.sp.
Kofoid, C. A. and A. S. Campbell, 1929
A conspectus of the marine and freshwater Ciliata belonging to the suborder Tintinnoinea, with descriptions of new species principally from the Agassiz expedition to the eastern tropical Pacific, 1904-1905. Univ. Calif. Publ. Zool. 34:1-403, 697 text figs.

Ormodella cornucopia
Kofoid, C.A. and A.S. Campbell, 1939
Reports on the scientific results of the expedition to the eastern tropical Pacific in charge of Alexender Agassiz, by the U.S. Fish Commission steamer "Albatross" from October 1904 to March 1905-----XXXVII. The Ciliata: The Tintinnoimea. Bull.Mus. Comp.Zool. 84: 473 pp., 36 pls.

Ormosella cornucopia nom. nov.
Kofoid, C. A. and A. S. Campbell, 1929
A conspectus of the marine and freshwater Ciliata belonging to the suborder Tintinnoinea, with descriptions of new species principally from the Agassiz expedition to the eastern tropical Pacific, 1904-1905. Univ. Calif. Publ. Zool. 34:1-403, 697 text figs.

Ormosella haeckeli
Kofoid, C.A. and A.S. Campbell, 1939
Reports on the scientific results of the eastern tropical Pacific in charge of Alexender Agassiz, by the U.S. Fish Commission steamer "Albatross" from October 1904 to March 1905-----XXXVII. The Ciliata: The Tintinnoimea. Bull.Mus. Comp.Zool. 84: 473 pp., 36 pls.

Ormosella haeckeli n.sp.
Kofoid, C. A. and A. S. Campbell, 1929
A conspectus of the marine and freshwater Ciliata belonging to the suborder Tintinnoinea, with descriptions of new species principally from the Agassiz expedition to the eastern tropical Pacific, 1904-1905. Univ. Calif. Publ. Zool. 34:1-403, 697 text figs.

Ormosella schmidti
Kofoid, C.A. and A.S. Campbell, 1939
Reports on the scientific results of the expedition to the eastern tropical Pacific in charge of Alexender Agassiz, by the U.S. Fish Commission steamer "Albatross" from October 1904 to March 1905-----XXXVII. The Ciliata: The Tintinnoimea. Bull.Mus. Comp.Zool. 84: 473 pp., 36 pls.

Ormosella schmidti n.sp.
Kofoid, C. A. and A. S. Campbell, 1929
A conspectus of the marine and freshwater Ciliata belonging to the suborder Tintinnoinea, with descriptions of new species principally from the Agassiz expedition to the eastern tropical Pacific, 1904-1905. Univ. Calif. Publ. Zool. 34:1-403, 697 text figs.

Ormosella schweyeri
Balech, Enrique, 1968.
Algunas especies nuevas o interesantes do tintinnidos del Golfo de Mexico y Caribe. Revta Mus. argent. Cienc. nat. Bernardino Rivadavia Inst. nac. Hidrobiol. 2(5):165-197.

Ormosella schweyeri
Kofoid, C.A. and A.S. Campbell, 1939
Reports on the scientific results of the expedition to the eastern tropical Pacific in charge of Alexender Agassiz, by the U.S. Fish Commission steamer "Albatross" from October 1904 to March 1905-----XXXVII. The Ciliata: The Tintinnoimea. Bull.Mus. Comp.Zool. 84: 473 pp., 36 pls.

Ormosella schweyeri n.sp.
Kofoid, C. A. and A. S. Campbell, 1929
A conspectus of the marine and freshwater Ciliata belonging to the suborder Tintinnoinea, with descriptions of new species principally from the Agassiz expedition to the eastern tropical Pacific, 1904-1905. Univ. Calif. Publ. Zool. 34:1-403, 697 text figs.

Ormosella trachelium
Balech, Enrique, 1968.
Algunas especies nuevas o interesantes do tintinnidos del Golfo de Mexico y Caribe. Revta Mus. argent. Cienc. nat. Bernardino Rivadavia Inst. nac. Hidrobiol. 2(5):165-197.

Ormosella trachelium
Kofoid, C. A. and A. S. Campbell, 1929
A conspectus of the marine and freshwater Ciliata belonging to the suborder Tintinnoinea, with descriptions of new species principally from the Agassiz expedition to the eastern tropical Pacific, 1904-1905. Univ. Calif. Publ. Zool. 34:1-403, 697 text figs.

Parafavella sp.
Braarud, T., 1945
A phytoplankton survey of the polluted waters of inner Oslo Fjord. Hvalrådets Skrifter, No.28, 142 pp., 19 text figs., 17 tables.

Parafavella sp.
Braarud, T., and Adam Bursa, 1939
On the phytoplankton of the Oslo Fjord, 1933-1934. Hvalrådets Skr. No.19:1-63; 9 text figs. Reviewed. J. du. Cons. 14(3): 418-420. A.C. Gardiner.

Parafavella n. gen.
Kofoid, C. A. and A. S. Campbell, 1929
A conspectus of the marine and freshwater Ciliata belonging to the suborder Tintinnoinea, with descriptions of new species principally from the Agassiz expedition to the eastern tropical Pacific, 1904-1905. Univ. Calif. Publ. Zool. 34:1-403, 697 text figs.

Parafavella sp.
Gran, H.H., and T. Braarud, 1935
A quantitative study of the phytoplankton in the Bay of Fundy and the Gulf of Maine (including observations on hydrography, chemistry, and turbidity). J. Biol. Bd., Canada, 1(5):279-467, 69 text figs.

Parafavella acuta
Kofoid, C. A. and A. S. Campbell, 1929
A conspectus of the marine and freshwater Ciliata belonging to the suborder Tintinnoinea, with descriptions of new species principally from the Agassiz expedition to the eastern tropical Pacific, 1904-1905. Univ. Calif. Publ. Zool. 34:1-403, 697 text figs.

Parafavella affinis n.sp.
Campbell, A. S., 1942
The oceanic Tintinnoina of the plankton gathered during the last cruise of the CARNEGIE. Sci. Res. Cruise VII of the Carnegie, 1928-1929, Biol. 2:163 pp., 128 figs.

Parafavella calycina
Kofoid, C. A. and A. S. Campbell, 1929
A conspectus of the marine and freshwater Ciliata belonging to the suborder Tintinnoinea, with descriptions of new species principally from the Agassiz expedition to the eastern tropical Pacific, 1904-1905. Univ. Calif. Publ. Zool. 34:1-403, 697 text figs.

Parafavella curvata n.sp.
Kofoid, C. A. and A. S. Campbell, 1929
A conspectus of the marine and freshwater Ciliata belonging to the suborder Tintinnoinea, with descriptions of new species principally from the Agassiz expedition to the eastern tropical Pacific, 1904-1905. Univ. Calif. Publ. Zool. 34:1-403, 697 text figs.

Parafavella cylindrica
Campbell, A. S., 1942
The oceanic Tintinnoina of the plankton gathered during the last cruise of the CARNEGIE. Sci. Res. Cruise VII of the Carnegie, 1928-1929, Biol. 2:163 pp., 128 figs.

Parafavella cylindrica
Kofoid, C. A. and A. S. Campbell, 1929
A conspectus of the marine and freshwater Ciliata belonging to the suborder Tintinnoinea, with descriptions of new species principally from the Agassiz expedition to the eastern tropical Pacific, 1904-1905. Univ. Calif. Publ. Zool. 34:1-403, 697 text figs.

Parafavella denticulata
Braarud, T., 1945
A phytoplankton survey of the polluted waters of inner Oslo Fjord. Hvalrådets Skrifter, No.28, 142 pp., 19 text figs., 17 tables.

Parafavella denticulata
Braarud, T., and Adam Bursa, 1939
On the phytoplankton of the Oslo Fjord, 1933-1934. Hvalrådets Skr. No.19:1-63; 9 text figs. Reviewed. J. du. Cons. 14(3): 418-420. A.C. Gardiner.

Parafavella denticulata
Campbell, A. S., 1942
The oceanic Tintinnoina of the plankton gathered during the last cruise of the CARNEGIE. Sci. Res. Cruise VII of the Carnegie, 1928-1929, Biol. 2:163 pp., 128 figs.

Parafavella denticulata
Gaarder, K.R., 1946
Tintinnoidea. Rep.Sci.Res."Michael Sars" N.Atlantic Deep-Sea Exped., 1910, 2(1): 37 pp., 23 text figs.

Parafavella denticulata
Kofoid, C. A. and A. S. Campbell, 1929
A conspectus of the marine and freshwater Ciliata belonging to the suborder Tintinnoinea, with descriptions of new species principally from the Agassiz expedition to the eastern tropical Pacific, 1904-1905. Univ. Calif. Publ. Zool. 34:1-403, 697 text figs.

Parafavella digitalis nom. nov.
Kofoid, C. A. and A. S. Campbell, 1929
A conspectus of the marine and freshwater Ciliata belonging to the suborder Tintinnoinea, with descriptions of new species principally from the Agassiz expedition to the eastern tropical Pacific, 1904-1905. Univ. Calif. Publ. Zool. 34:1-403, 697 text figs.

Parafavella dilatata
Campbell, A. S., 1942
The oceanic Tintinnoina of the plankton gathered during the last cruise of the CARNEGIE. Sci. Res. Cruise VII of the Carnegie, 1928-1929, Biol. 2:163 pp., 128 figs.

Parafavella dilatata
Kofoid, C. A. and A. S. Campbell, 1929
A conspectus of the marine and freshwater Ciliata belonging to the suborder Tintinnoinea, with descriptions of new species principally from the Agassiz expedition to the eastern tropical Pacific, 1904-1905. Univ. Calif. Publ. Zool. 34:1-403, 697 text figs.

Parafavella edentata
Gaarder, K.R., 1946
Tintinnoidea. Rep.Sci.Res."Michael Sars" N.Atlantic Deep-Sea Exped., 1910, 2(1): 37 pp., 23 text figs.

Parafavella edentata
Kofoid, C. A. and A. S. Campbell, 1929
A conspectus of the marine and freshwater Ciliata belonging to the suborder Tintinnoinea, with descriptions of new species principally from the Agassiz expedition to the eastern tropical Pacific, 1904-1905. Univ. Calif. Publ. Zool. 34:1-403, 697 text figs.

Parafavella elegans
Kofoid, C. A. and A. S. Campbell, 1929
A conspectus of the marine and freshwater Ciliata belonging to the suborder Tintinnoinea, with descriptions of new species principally from the Agassiz expedition to the eastern tropical Pacific, 1904-1905. Univ. Calif. Publ. Zool. 34:1-403, 697 text figs.

Parafavella gigantea
Campbell, A. S., 1942
The oceanic Tintinnoina of the plankton gathered during the last cruise of the CARNEGIE. Sci. Res. Cruise VII of the Carnegie, 1928-1929, Biol. 2:163 pp., 128 figs.

Parafavella gigantea
Kofoid, C. A. and A. S. Campbell, 1929
A conspectus of the marine and freshwater Ciliata belonging to the suborder Tintinnoinea, with descriptions of new species principally from the Agassiz expedition to the eastern tropical Pacific, 1904-1905. Univ. Calif. Publ. Zool. 34:1-403, 697 text figs.

Parafavella gigantea
Ling, Hsin-Yi, 1965.
The tintinnid Parafavella gigantea (Brandt) Kofoid & Campbell, 1929, in the North Pacific Ocean.
J. Paleontol., 39(4):721-723.

Parafavella greenlandica n.sp.
Kofoid, C. A. and A. S. Campbell, 1929
A conspectus of the marine and freshwater Ciliata belonging to the suborder Tintinnoinea, with descriptions of new species principally from the Agassiz expedition to the eastern tropical Pacific, 1904-1905. Univ. Calif. Publ. Zool. 34:1-403, 697 text figs.

Parafavella hadai n. sp.
Campbell, A. S., 1942
The oceanic Tintinnoina of the plankton gathered during the last cruise of the CARNEGIE. Sci. Res. Cruise VII of the Carnegie, 1928-1929, Biol. 2:163 pp., 128 figs.

Parafavella hemifusus
Kofoid, C. A. and A. S. Campbell, 1929
A conspectus of the marine and freshwater Ciliata belonging to the suborder Tintinnoinea, with descriptions of new species principally from the Agassiz expedition to the eastern tropical Pacific, 1904-1905. Univ. Calif. Publ. Zool. 34:1-403, 697 text figs.

Parafavella inflata n.sp.
Kofoid, C. A. and A. S. Campbell, 1929
A conspectus of the marine and freshwater Ciliata belonging to the suborder Tintinnoinea, with descriptions of new species principally from the Agassiz expedition to the eastern tropical Pacific, 1904-1905. Univ. Calif. Publ. Zool. 34:1-403, 697 text figs.

Parafavella media
Kofoid, C. A. and A. S. Campbell, 1929
A conspectus of the marine and freshwater Ciliata belonging to the suborder Tintinnoinea, with descriptions of new species principally from the Agassiz expedition to the eastern tropical Pacific, 1904-1905. Univ. Calif. Publ. Zool. 34:1-403, 697 text figs.

Parafavella obtusa
Kofoid, C. A. and A. S. Campbell, 1929
A conspectus of the marine and freshwater Ciliata belonging to the suborder Tintinnoinea, with descriptions of new species principally from the Agassiz expedition to the eastern tropical Pacific, 1904-1905. Univ. Calif. Publ. Zool. 34:1-403, 697 text figs.

Parafavella obtusangula
Campbell, A. S., 1942
The oceanic Tintinnoina of the plankton gathered during the last cruise of the CARNEGIE. Sci. Res. Cruise VII of the Carnegie, 1928-1929, Biol. 2:163 pp., 128 figs.

Parafavella obtusangula
Kofoid, C. A. and A. S. Campbell, 1929
A conspectus of the marine and freshwater Ciliata belonging to the suborder Tintinnoinea, with descriptions of new species principally from the Agassiz expedition to the eastern tropical Pacific, 1904-1905. Univ. Calif. Publ. Zool. 34:1-403, 697 text figs.

Parafavella parumdentata
Campbell, A. S., 1942
The oceanic Tintinnoina of the plankton gathered during the last cruise of the CARNEGIE. Sci. Res. Cruise VII of the Carnegie, 1928-1929, Biol. 2:163 pp., 128 figs.

Parafavella parumdentata
Kofoid, C. A. and A. S. Campbell, 1929
A conspectus of the marine and freshwater Ciliata belonging to the suborder Tintinnoinea, with descriptions of new species principally from the Agassiz expedition to the eastern tropical Pacific, 1904-1905. Univ. Calif. Publ. Zool. 34:1-403, 697 text figs.

Parafavella promissa
Campbell, A. S., 1942
The oceanic Tintinnoina of the plankton gathered during the last cruise of the CARNEGIE. Sci. Res. Cruise VII of the Carnegie, 1928-1929, Biol. 2:163 pp., 128 figs.

Parafavella robusta
Kofoid, C. A. and A. S. Campbell, 1929
A conspectus of the marine and freshwater Ciliata belonging to the suborder Tintinnoinea, with descriptions of new species principally from the Agassiz expedition to the eastern tropical Pacific, 1904-1905. Univ. Calif. Publ. Zool. 34:1-403, 697 text figs.

Parafavella rotundata
Kofoid, C. A. and A. S. Campbell, 1929
A conspectus of the marine and freshwater Ciliata belonging to the suborder Tintinnoinea, with descriptions of new species principally from the Agassiz expedition to the eastern tropical Pacific, 1904-1905. Univ. Calif. Publ. Zool. 34:1-403, 697 text figs.

Parafavella subedentata

Kofoid, C. A. and A. S. Campbell, 1929
A conspectus of the marine and fresh-water Ciliata belonging to the suborder Tintinnoinea, with descriptions of new species principally from the Agassiz expedition to the eastern tropical Pacific, 1904-1905. Univ. Calif. Publ. Zool. 34:1-403, 697 text figs.

Parafavella subrotundata

Kofoid, C. A. and A. S. Campbell, 1929
A conspectus of the marine and fresh-water Ciliata belonging to the suborder Tintinnoinea, with descriptions of new species principally from the Agassiz expedition to the eastern tropical Pacific, 1904-1905. Univ. Calif. Publ. Zool. 34:1-403, 697 text figs.

Parafavella subula n.sp.

Kofoid, C. A. and A. S. Campbell, 1929
A conspectus of the marine and fresh-water Ciliata belonging to the suborder Tintinnoinea, with descriptions of new species principally from the Agassiz expedition to the eastern tropical Pacific, 1904-1905. Univ. Calif. Publ. Zool. 34:1-403, 697 text figs.

Parafavella ventricosa

Campbell, A. S., 1942
The oceanic Tintinnoina of the plankton gathered during the last cruise of the CARNEGIE. Sci. Res. Cruise VII of the Carnegie, 1928-1929, Biol. 2:163 pp., 128 figs.

Parafavella ventricosa

Kofoid, C. A. and A. S. Campbell, 1929
A conspectus of the marine and fresh-water Ciliata belonging to the suborder Tintinnoinea, with descriptions of new species principally from the Agassiz expedition to the eastern tropical Pacific, 1904-1905. Univ. Calif. Publ. Zool. 34:1-403, 697 text figs.

Parapetalotricha meridiana n.gen. n. sp.

*Hada, Yoshina, 1970.   (1970)
The protozoan plankton of the Antarctic and Subantarctic seas.
Scient. Repts. Japan. Antarct. Res. Exped., (E)31:51 pp.

Parundella aciculifera

Kofoid, C.A. and A.S. Campbell, 1939
Reports on the scientific results of the expedition to the eastern tropical Pacific in charge of Alexander Agassiz, by the U.S. Fish Commission steamer "Albatross" from October 1904 to March 1905-----XXXVII. The Ciliata: The Tintinnoimea. Bull.Mus. Comp.Zool. 84: 473 pp., 36 pls.

Parundella aculeata

Balech, Enrique, 1962
Tintinnoinea y Dinoflagellata del Pacifico segun material de las Expediciones NORPAC y DOWNWIND del Instituto Scripps de Oceanografia.
Revista, Mus. Argentino Ciencias Nat. "Bernardino Rivadavia", Ciencias Zool., 7(1):1-253.

Parundella aculeata

Balech, E., 1959.
Tintinnoinea del Mediterraneo.
Trav. Sta. Zool., Villefranche-sur-Mer, 18(17):
Reprinted from:
Trab. Inst. Español Oceanogr., Madrid, 19 Sept. 1959(28):88 pp., 22 pls.

Parundella aculeata

Campbell, A. S., 1942
The oceanic Tintinnoina of the plankton gathered during the last cruise of the CARNEGIE. Sci. Res. Cruise VII of the Carnegie, 1928-1929, Biol. 2:163 pp., 128 figs.

Parundella aculeata

Gaarder, K.R., 1946
Tintinnoidea. Rep.Sci.Res."Michael Sars" N.Atlantic Deep-Sea Exped., 1910, 2(1): 37 pp., 23 text figs.

Parundella aculeata

Kofoid, C.A. and A.S. Campbell, 1939
Reports on the scientific results of the expedition to the eastern tropical Pacific in charge of Alexander Agassiz, by the U.S. Fish Commission steamer "Albatross" from October 1904 to March 1905-----XXXVII. The Ciliata: The Tintinnoimea. Bull.Mus. Comp.Zool. 84: 473 pp., 36 pls.

Parundella aculeata

Kofoid, C. A. and A. S. Campbell, 1929
A conspectus of the marine and fresh-water Ciliata belonging to the suborder Tintinnoinea, with descriptions of new species principally from the Agassiz expedition to the eastern tropical Pacific, 1904-1905. Univ. Calif. Publ. Zool. 34:1-403, 697 text figs.

Parundella aculeata

Komarovsky, B., 1959
The Tintinnina of the Gulf of Eylath (Aqaba). Contrib. Knowl. Red Sea, Bull. Sea Fish. Res. Sta.Haifa, Israel 21 (14):1-40.

Parundella acuta

Campbell, A. S., 1942
The oceanic Tintinnoina of the plankton gathered during the last cruise of the CARNEGIE. Sci. Res. Cruise VII of the Carnegie, 1928-1929, Biol. 2:163 pp., 128 figs.

Parundella acuta

Kofoid, C. A. and A. S. Campbell, 1929
A conspectus of the marine and fresh-water Ciliata belonging to the suborder Tintinnoinea, with descriptions of new species principally from the Agassiz expedition to the eastern tropical Pacific, 1904-1905. Univ. Calif. Publ. Zool. 34:1-403, 697 text figs.

Parundella attenuata

Gaarder, K.R., 1946
Tintinnoidea. Rep.Sci.Res."Michael Sars" N.Atlantic Deep-Sea Exped., 1910, 2(1): 37 pp., 23 text figs.

Parundella attenuata

Kofoid, C.A. and A.S. Campbell, 1939
Reports on the scientific results of the expedition to the eastern tropical Pacific in charge of Alexander Agassiz, by the U.S. Fish Commission steamer "Albatross" from October 1904 to March 1905-----XXXVII. The Ciliata: The Tintinnoimea. Bull.Mus. Comp.Zool. 84: 473 pp., 36 pls.

Parundella attenuata n.sp.

Kofoid, C. A. and A. S. Campbell, 1929
A conspectus of the marine and fresh-water Ciliata belonging to the suborder Tintinnoinea, with descriptions of new species principally from the Agassiz expedition to the eastern tropical Pacific, 1904-1905. Univ. Calif. Publ. Zool. 34:1-403, 697 text figs.

Parundella caudata

Campbell, A. S., 1942
The oceanic Tintinnoina of the plankton gathered during the last cruise of the CARNEGIE. Sci. Res. Cruise VII of the Carnegie, 1928-1929, Biol. 2:163 pp., 128 figs.

Parundella caudata

Gaarder, K.R., 1946
Tintinnoidea. Rep.Sci.Res."Michael Sars" N.Atlantic Deep-Sea Exped., 1910, 2(1): 37 pp., 23 text figs.

Parundella caudata

Kofoid, C.A. and A.S. Campbell, 1939
Reports on the scientific results of the expedition to the eastern tropical Pacific in charge of Alexander Agassiz, by the U.S. Fish Commission steamer "Albatross" from October 1904 to March 1905-----XXXVII. The Ciliata: The Tintinnoimea. Bull.Mus. Comp.Zool. 84: 473 pp., 36 pls.

Parundella caudata

Kofoid, C. A. and A. S. Campbell, 1929
A conspectus of the marine and fresh-water Ciliata belonging to the suborder Tintinnoinea, with descriptions of new species principally from the Agassiz expedition to the eastern tropical Pacific, 1904-1905. Univ. Calif. Publ. Zool. 34:1-403, 697 text figs.

Parundella clavus

Kofoid, C.A. and A.S. Campbell, 1939
Reports on the scientific results of the expedition to the eastern tropical Pacific in charge of Alexander Agassiz, by the U.S. Fish Commission steamer "Albatross" from October 1904 to March 1905-----XXXVII. The Ciliata: The Tintinnoimea. Bull.Mus. Comp.Zool. 84: 473 pp., 36 pls.

Parundella clavus n.sp.

Kofoid, C. A. and A. S. Campbell, 1929
A conspectus of the marine and fresh-water Ciliata belonging to the suborder Tintinnoinea, with descriptions of new species principally from the Agassiz expedition to the eastern tropical Pacific, 1904-1905. Univ. Calif. Publ. Zool. 34:1-403, 697 text figs.

Parundella conica

Balech, Enrique, 1968.
Algunas especies nuevas o interesantes de tintinnidos del Golfo de Mexico y Caribe.
Revta Mus. argent. Cienc. nat. Bernardino Rivadavia Inst. nac. Hidrobiol. 2(5):165-197.

Parundella difficilis

Gaarder, K.R., 1946. Rep.Sci.Res."Michael Sars" N.Atlantic Deep-Sea Exped., 1910, 2(1): 37 pp., 23 text figs.

Parundella difficilis

Kofoid, C.A. and A.S. Campbell, 1939
Reports on the scientific results of the expedition to the eastern tropical Pacific in charge of Alexander Agassiz, by the U.S. Fish Commission steamer "Albatross" from October 1904 to March 1905-----XXXVII. The Ciliata: The Tintinnoimea. Bull.Mus. Comp.Zool. 84: 473 pp., 36 pls.

Parundella difficilis n.sp.

Kofoid, C. A. and A. S. Campbell, 1929
A conspectus of the marine and fresh-water Ciliata belonging to the suborder Tintinnoinea, with descriptions of new species principally from the Agassiz expedition to the eastern tropical Pacific, 1904-1905. Univ. Calif. Publ. Zool. 34:1-403, 697 text figs.

Parundella elongata n.sp.

Kofoid, C.A. and A.S. Campbell, 1939
Reports on the scientific results of the expedition to the eastern tropical Pacific in charge of Alexander Agassiz, by the U.S. Fish Commission steamer "Albatross" from October 1904 to March 1905-----XXXVII. The Ciliata: The Tintinnoimea. Bull.Mus. Comp.Zool. 84: 473 pp., 36 pls.

Parundella gigantea

Kofoid, C.A. and A.S. Campbell, 1939
Reports on the scientific results of the expedition to the eastern tropical Pacific in charge of Alexander Agassiz, by the U.S. Fish Commission steamer "Albatross" from October 1904 to March 1905-----XXXVII. The Ciliata: The Tintinnoimea. Bull.Mus. Comp.Zool. 84: 473 pp., 36 pls.

Parundella gigantea n.sp.

Kofoid, C. A. and A. S. Campbell, 1929
A conspectus of the marine and fresh-water Ciliata belonging to the suborder Tintinnoinea, with descriptions of new species principally from the Agassiz expedition to the eastern tropical Pacific, 1904-1905. Univ. Calif. Publ. Zool. 34:1-403, 697 text figs.

Parundella grandis

Gaarder, K.R., 1946
Tintinnoidea. Rep.Sci.Res."Michael Sars" N.Atlantic Deep-Sea Exped., 1910, 2(1): 37 pp., 23 text figs.

Parundella grandis n.sp.

Kofoid, C. A. and A. S. Campbell, 1929
A conspectus of the marine and freshwater Ciliata belonging to the suborder Tintinnoinea, with descriptions of new species principally from the Agassiz expedition to the eastern tropical Pacific, 1904-1905. Univ. Calif. Publ. Zool. 34:1-403, 697 text figs.

Parundella humerosa

Kofoid, C.A. and A.S. Campbell, 1939
Reports on the scientific results of the expedition to the eastern tropical Pacific in charge of Alexander Agassiz, by the U.S. Fish Commission steamer "Albatross" from October 1904 to March 1905-----XXXVII. The Ciliata: The Tintinnoimea. Bull. Mus. Comp. Zool. 84: 473 pp., 36 pls.

Parundella humerosa n.sp.

Kofoid, C. A. and A. S. Campbell, 1929
A conspectus of the marine and freshwater Ciliata belonging to the suborder Tintinnoinea, with descriptions of new species principally from the Agassiz expedition to the eastern tropical Pacific, 1904-1905. Univ. Calif. Publ. Zool. 34:1-403, 697 text figs.

Parundella inflata

Campbell, A. S., 1942
The oceanic Tintinnoina of the plankton gathered during the last cruise of the CARNEGIE. Sci. Res. Cruise VII of the Carnegie, 1928-1929, Biol. 2:163 pp., 128 figs.

Parundella inflata

Kofoid, C.A. and A.S. Campbell, 1939
Reports on the scientific results of the expedition to the eastern tropical Pacific in charge of Alexander Agassiz, by the U.S. Fish Commission steamer "Albatross" from October 1904 to March 1905-----XXXVII. The Ciliata: The Tintinnoimea. Bull. Mus. Comp. Zool. 84: 473 pp., 36 pls.

Parundella inflata n.sp.

Kofoid, C. A. and A. S. Campbell, 1929
A conspectus of the marine and freshwater Ciliata belonging to the suborder Tintinnoinea, with descriptions of new species principally from the Agassiz expedition to the eastern tropical Pacific, 1904-1905. Univ. Calif. Publ. Zool. 34:1-403, 697 text figs.

Parundella inflata

Komarovsky, B., 1959
The Tintinnina of the Gulf of Eylath (Aqaba). Contrib. Knowl. Red Sea, Bull. Sea Fish. Res. Sta. Haifa, Israel 21 (14):1-40.

Parundella invaginata

de Sousa e Silva, E., 1956.
Contribution à l'étude du microplancton de Dakar et des regions maritimes voisines. Bull. I.F.A.N., 8(2):335-371, 7 pl

Parundella invaginata

Kofoid, C.A. and A.S. Campbell, 1939
Reports on the scientific results of the expedition to the eastern tropical Pacific in charge of Alexander Agassiz, by the U.S. Fish Commission steamer "Albatross" from October 1904 to March 1905-----XXXVII. The Ciliata: The Tintinnoimea. Bull. Mus. Comp. Zool. 84: 473 pp., 36 pls.

Parundella invaginata

Kofoid, C. A. and A. S. Campbell, 1929
A conspectus of the marine and freshwater Ciliata belonging to the suborder Tintinnoinea, with descriptions of new species principally from the Agassiz expedition to the eastern tropical Pacific, 1904-1905. Univ. Calif. Publ. Zool. 34:1-403, 697 text figs.

Parundella lachmanni

Campbell, A. S., 1942
The oceanic Tintinnoina of the plankton gathered during the last cruise of the CARNEGIE. Sci. Res. Cruise VII of the Carnegie, 1928-1929, Biol. 2:163 pp., 128 figs.

Parundella lachmanni

Kofoid, C. A. and A. S. Campbell, 1929
A conspectus of the marine and freshwater Ciliata belonging to the suborder Tintinnoinea, with descriptions of new species principally from the Agassiz expedition to the eastern tropical Pacific, 1904-1905. Univ. Calif. Publ. Zool. 34:1-403, 697 text figs.

Parundella lagena n.sp.

Kofoid, C. A. and A. S. Campbell, 1929
A conspectus of the marine and freshwater Ciliata belonging to the suborder Tintinnoinea, with descriptions of new species principally from the Agassiz expedition to the eastern tropical Pacific, 1904-1905. Univ. Calif. Publ. Zool. 34:1-403, 697 text figs.

Parundella lata n.sp.

Gaarder, K.R., 1946
Tintinnoidea. Rep.Sci.Res. "Michael Sars" N.Atlantic Deep-Sea Exped., 1910, 2(1): 37 pp., 23 text figs.

Parundella lohmanni

Balech, E., 1959.
Tintinnoinea del Mediterraneo.
Trav. Sta. Zool., Villefranche-sur-Mer, 18(17):
Reprinted from:
Trab. Inst. Español Oceanogr., Madrid, 19 Sept. 1959(28):88 pp., 22 pls.

Parundella lohmanni

Gaarder, K.R., 1946
Tintinnoidea. Rep.Sci.Res. "Michael Sars" N.Atlantic Deep-Sea Exped., 1910, 2(1): 37 pp., 23 text figs.

Parundella lohmanni

Kofoid, C. A. and A. S. Campbell, 1929
A conspectus of the marine and freshwater Ciliata belonging to the suborder Tintinnoinea, with descriptions of new species principally from the Agassiz expedition to the eastern tropical Pacific, 1904-1905. Univ. Calif. Publ. Zool. 34:1-403, 697 text figs.

Parundella lohmanni

Sousa e Silva, E. de, 1950.
Les tintinnides de la Baie de Cascais. Bull. Inst. Océan., Monaco, No. 979:28 pp., 4 pls.

Parundella longa

Campbell, A. S., 1942
The oceanic Tintinnoina of the plankton gathered during the last cruise of the CARNEGIE. Sci. Res. Cruise VII of the Carnegie, 1928-1929, Biol. 2:163 pp., 128 figs.

Parundella longa

Gaarder, K.R., 1946
Tintinnoidea. Rep.Sci.Res. "Michael Sars" N.Atlantic Deep-Sea Exped., 1910, 2(1): 37 pp., 23 text figs.

Parundella longa

Kofoid, C. A. and A. S. Campbell, 1929
A conspectus of the marine and freshwater Ciliata belonging to the suborder Tintinnoinea, with descriptions of new species principally from the Agassiz expedition to the eastern tropical Pacific, 1904-1905. Univ. Calif. Publ. Zool. 34:1-403, 697 text figs.

Parundella major

Kofoid, C. A. and A. S. Campbell, 1929
A conspectus of the marine and freshwater Ciliata belonging to the suborder Tintinnoinea, with descriptions of new species principally from the Agassiz expedition to the eastern tropical Pacific, 1904-1905. Univ. Calif. Publ. Zool. 34:1-403, 697 text figs.

Parundella messinensis

Gaarder, K.R., 1946
Tintinnoidea. Rep.Sci.Res. "Michael Sars" N.Atlantic Deep-Sea Exped., 1910, 2(1): 37 pp., 23 text figs.

Parundella messinensis

Kofoid, C.A. and A.S. Campbell, 1939
Reports on the scientific results of the expedition to the eastern tropical Pacific in charge of Alexander Agassiz, by the U.S. Fish Commission steamer "Albatross" from October 1904 to March 1905-----XXXVII. The Ciliata: The Tintinnoimea. Bull. Mus. Comp. Zool. 84: 473 pp., 36 pls.

Parundella messinensis

Kofoid, C. A. and A. S. Campbell, 1929
A conspectus of the marine and freshwater Ciliata belonging to the suborder Tintinnoinea, with descriptions of new species principally from the Agassiz expedition to the eastern tropical Pacific, 1904-1905. Univ. Calif. Publ. Zool. 34:1-403, 697 text figs.

Parundella messanensis

Rampi, L., 1948
Sur quelques Tintinnides (Infusoires louques) du Pacifique subtropical (Recoltes Alain Gerbault) Bull. l'Inst. Ocean., Monaco, No.938, 4 pp.

Parundella minor

Kofoid, C. A. and A. S. Campbell, 1929
A conspectus of the marine and freshwater Ciliata belonging to the suborder Tintinnoinea, with descriptions of new species principally from the Agassiz expedition to the eastern tropical Pacific, 1904-1905. Univ. Calif. Publ. Zool. 34:1-403, 697 text figs.

Parundella pellucida

Campbell, A. S., 1942
The oceanic Tintinnoina of the plankton gathered during the last cruise of the CARNEGIE. Sci. Res. Cruise VII of the Carnegie, 1928-1929, Biol. 2:163 pp., 128 figs.

Parundella pellucida

Kofoid, C. A. and A. S. Campbell, 1929
A conspectus of the marine and freshwater Ciliata belonging to the suborder Tintinnoinea, with descriptions of new species principally from the Agassiz expedition to the eastern tropical Pacific, 1904-1905. Univ. Calif. Publ. Zool. 34:1-403, 697 text figs.

Parundella praetenuis

Balech, Enrique, 1968.
Algunas especies nuevas o interesantes do tintinnidos del Golfo de Mexico y Caribe. Revta Mus. argent. Cienc. nat. Bernardino Rivadavia Inst. nac. Hidrobiol. 2(5):165-197.

Parundella praetenuis

Campbell, A. S., 1942
The oceanic Tintinnoina of the plankton gathered during the last cruise of the CARNEGIE. Sci. Res. Cruise VII of the Carnegie, 1928-1929, Biol. 2:163 pp., 128 figs.

Parundella praetenuis

Kofoid, C.A. and A.S. Campbell, 1939
Reports on the scientific results of the expedition to the eastern tropical Pacific in charge of Alexander Agassiz, by the U.S. Fish Commission steamer "Albatross" from October 1904 to March 1905-----XXXVII. The Ciliata: The Tintinnoimea. Bull. Mus. Comp. Zool. 84: 473 pp., 36 pls.

Parundella praetenuis n.sp.

Kofoid, C. A. and A. S. Campbell, 1929
A conspectus of the marine and freshwater Ciliata belonging to the suborder Tintinnoinea, with descriptions of new species principally from the Agassiz expedition to the eastern tropical Pacific, 1904-1905. Univ. Calif. Publ. Zool. 34:1-403, 697 text figs.

Parundella spinosa n.sp.

Kofoid, C. A. and A. S. Campbell, 1929
A conspectus of the marine and freshwater Ciliata belonging to the suborder Tintinnoinea, with descriptions of new species principally from the Agassiz expedition to the eastern tropical Pacific, 1904-1905. Univ. Calif. Publ. Zool. 34:1-403, 697 text figs.

Parundella translucens

Kofoid, C. A. and A. S. Campbell, 1929
A conspectus of the marine and freshwater Ciliata belonging to the suborder Tintinnoinea, with descriptions of new species principally from the Agassiz expedition to the eastern tropical Pacific, 1904-1905. Univ. Calif. Publ. Zool. 34:1-403, 697 text figs.

Petalotricha ampulla

Balech, Enrique, 1968.
Algunas especies nuevas o interesantes do tintinnidos del Golfo de Mexico y Caribe. Revta Mus. argent. Cienc. nat. Bernardino Rivadavia Inst. nac. Hidrobiol. 2(5):165-197.

Petalotricha ampulla

Balech, Enrique, 1962
Tintinnoinea y Dinoflagellata del Pacifico segun material de las Expediciones NORPAC y DOWNWIND del Instituto Scripps de Oceanografia. Revista, Mus. Argentino Ciencias Nat. "Bernardino Rivadavia", Ciencias Zool., 7(1):1-253.

Petalotricha ampula

Balech, E., 1959.
Tintinnoinea del Mediterraneo. Trav. Sta. Zool., Villefranche-sur-Mer, 18(17):
Reprinted from:
Trab. Inst. Español Oceanogr., Madrid, 19 Sept. 1959, (28):88pp., 22 pls.

Petalotricha ampulla

Campbell, A. S., 1942
The oceanic Tintinnoina of the plankton gathered during the last cruise of the CARNEGIE. Sci. Res. Cruise VII of the Carnegie, 1928-1929, Biol. 2:163 pp., 128 figs.

Petalotricha ampulla

Duran, M., 1951
Contribucion al estudio de los tintinidos del plancton de los costas de Castellon. (Mediterraneo occidental). Publ. Inst. Biol. Aplic., Barcelone, 8: 101-120, 29 text figs.

Petalotricha ampulla

Entz, G., Jr., 1909
Studien über organisation und biologie der Tintinniden. Arch. f. Protistenkunde 15:93-226, Pls. 8-21, text figs.

Petalotricha ampulla

Gaarder, K.R., 1946
Tintinnoidea. Rep.Sci.Res."Michael Sars" N.Atlantic Deep-Sea Exped., 1910, 2(1): 37 pp., 23 text figs.

Petalotricha ampulla

Hofker, J., 1931
Studien uber Tintinnoidea. Arch. f. Protistenk 75(3):315-402, 89 text figs.

Petalotricha ampulla

Jörgensen, E., 1924
Mediterranean Tintinnidae. Rept. Danish Oceanogr. Exped. 1908-10 to the Mediterranean and adjacent seas, Vol.II, Biol. J 3, 110 pp., 114 text figs.

Petalotricha ampulla

Kofoid, C. A. and A. S. Campbell, 1929
A conspectus of the marine and freshwater Ciliata belonging to the suborder Tintinnoinea, with descriptions of new species principally from the Agassiz expedition to the eastern tropical Pacific, 1904-1905. Univ. Calif. Publ. Zool. 34:1-403, 697 text figs.

Petalotricha ampulla

Komarovsky, B., 1959
The Tintinnina of the Gulf of Eylath (Aqaba). Contrib. Knowl. Red Sea, Bull. Sea Fish. Res. Sta.Haifa, Israel 21 (14):1-40.

Petalotricha ampulla

Laackmann, H., 1910
Die Tintinnodeen der Deutschen Südpolar Expedition 1901-1903. Deutschen Südpolar Expedition 11, Zool.3 (4): 341-496, pls. 33-51.

Petalotricha ampulla

Rampi, L., 1948
Sur quelques Tintinnides (Infusoires louqes) du Pacifique subtropical (Recoltes Alain Gerbault) Bull. l'Inst. Ocean., Monaco, No.938, 4 pp.

Petalotricha ampulla

Rampi, L., 1939
Primo contributo alla conoscenza dei tintinnoidi do Maro Ligure. Atti della Soc. Ital. di Sci. Nat. 78:67-81, 58 text figs.

Petalotricha capsa

Campbell, A. S., 1942
The oceanic Tintinnoina of the plankton gathered during the last cruise of the CARNEGIE. Sci. Res. Cruise VII of the Carnegie, 1928-1929, Biol. 2:163 pp., 128 figs.

Petalotricha capsa

Kofoid, C.A. and A.S. Campbell,1939
Reports on the scientific results of the expedition to the eastern tropical Pacific in charge of Alexander Agassiz, by the U.S. Fish Commission steamer"Albatross" from October 1904 to March 1905-----XXXVII. The Ciliata: The Tintinnoimea. Bull.Mus. Comp.Zool. 84: 473 pp., 36 pls.

Petalotricha capsa

Kofoid, C. A. and A. S. Campbell, 1929
A conspectus of the marine and freshwater Ciliata belonging to the suborder Tintinnoinea, with descriptions of new species principally from the Agassiz expedition to the eastern tropical Pacific, 1904-1905. Univ. Calif. Publ. Zool. 34:1-403, 697 text figs.

Petalotricha entzi

Kofoid, C. A. and A. S. Campbell, 1929
A conspectus of the marine and freshwater Ciliata belonging to the suborder Tintinnoinea, with descriptions of new species principally from the Agassiz expedition to the eastern tropical Pacific, 1904-1905. Univ. Calif. Publ. Zool. 34:1-403, 697 text figs.

Petalotricha foli

Campbell, A. S., 1942
The oceanic Tintinnoina of the plankton gathered during the last cruise of the CARNEGIE. Sci. Res. Cruise VII of the Carnegie, 1928-1929, Biol. 2:163 pp., 128 figs.

Petalotricha foli

Kofoid, C.A. and A.S. Campbell,1939
Reports on the scientific results of the expedition to the eastern tropical Pacific in charge of Alexander Agassiz, by the U.S. Fish Commission steamer"Albatross" from October 1904 to March 1905-----XXXVII. The Ciliata: The Tintinnoimea. Bull.Mus. Comp.Zool. 84: 473 pp., 36 pls.

Petalotricha foli n.sp.

Kofoid, C. A. and A. S. Campbell, 1929
A conspectus of the marine and freshwater Ciliata belonging to the suborder Tintinnoinea, with descriptions of new species principally from the Agassiz expedition to the eastern tropical Pacific, 1904-1905. Univ. Calif. Publ. Zool. 34:1-403, 697 text figs.

Petalotricha indica

Kofoid, C. A. and A. S. Campbell, 1929
A conspectus of the marine and freshwater Ciliata belonging to the suborder Tintinnoinea, with descriptions of new species principally from the Agassiz expedition to the eastern tropical Pacific, 1904-1905. Univ. Calif. Publ. Zool. 34:1-403, 697 text figs.

Petalotricha major

Campbell, A. S., 1942
The oceanic Tintinnoina of the plankton gathered during the last cruise of the CARNEGIE. Sci. Res. Cruise VII of the Carnegie, 1928-1929, Biol. 2:163 pp., 128 figs.

Petalotricha major

Gaarder, K.R., 1946
Tintinnoidea. Rep.Sci.Res."Michael Sars" N.Atlantic Deep-Sea Exped., 1910, 2(1): 37 pp., 23 text figs.

Petalotricha major

Kofoid, C.A. and A.S. Campbell,1939
Reports on the scientific results of the expedition to the eastern tropical Pacific in charge of Alexander Agassiz, by the U.S. Fish Commission steamer"Albatross" from October 1904 to March 1905-----XXXVII The Ciliata: The Tintinnoimea. Bull.Mus. Comp.Zool. 84: 473 pp., 36 pls.

Petalotricha major

Kofoid, C. A. and A. S. Campbell, 1929
A conspectus of the marine and freshwater Ciliata belonging to the suborder Tintinnoinea, with descriptions of new species principally from the Agassiz expedition to the eastern tropical Pacific, 1904-1905. Univ. Calif. Publ. Zool. 34:1-403, 697 text figs.

Petalotricha major

Komarovsky, B., 1959
The Tintinnina of the Gulf of Eylath (Aqaba). Contrib. Knowl. Red Sea, Bull. Sea Fish. Res. Sta.Haifa, Israel 21 (14):1-40.

Petalotricha major

Rampi, L., 1939
Primo contributo alla conoscenza dei tintinnoidi do Maro Ligure. Atti della Soc. Ital. di Sci. Nat. 78:67-81, 58 text figs.

Petalotricha mira n.sp.

Kofoid, C. A. and A. S. Campbell, 1929
A conspectus of the marine and freshwater Ciliata belonging to the suborder Tintinnoinea, with descriptions of new species principally from the Agassiz expedition to the eastern tropical Pacific, 1904-1905. Univ. Calif. Publ. Zool. 34:1-403, 697 text figs.

Petalotricha pacifica

Kofoid, C.A. and A.S. Campbell,1939
Reports on the scientific results of the expedition to the eastern tropical Pacific in charge of Alexander Agassiz, by the U.S. Fish Commission steamer"Albatross" from October 1904 to March 1905-----XXXVII The Ciliata: The Tintinnoimea. Bull.Mus. Comp.Zool. 84: 473 pp., 36 pls.

Petalotricha pacifica n.sp.

Kofoid, C. A. and A. S. Campbell, 1929
A conspectus of the marine and freshwater Ciliata belonging to the suborder Tintinnoinea, with descriptions of new species principally from the Agassiz expedition to the eastern tropical Pacific, 1904-1905. Univ. Calif. Publ. Zool. 34:1-403, 697 text figs.

Petalotricha pacifica

Komarovsky, B., 1959
The Tintinnina of the Gulf of Eylath (Aqaba). Contrib. Knowl. Red Sea, Bull. Sea Fish. Res. Sta.Haifa, Israel 21 (14):1-40.

Petalotricha rotorhabdonella curta

Kofoid, C. A. and A. S. Campbell, 1929
A conspectus of the marine and freshwater Ciliata belonging to the suborder Tintinnoinea, with descriptions of new species principally from the Agassiz expedition to the eastern tropical Pacific, 1904-1905. Univ. Calif. Publ. Zool. 34:1-403, 697 text figs.

Petalotricha serrata

Campbell, A. S., 1942
The oceanic Tintinnoina of the plankton gathered during the last cruise of the CARNEGIE. Sci. Res. Cruise VII of the Carnegie, 1928-1929, Biol. 2:163 pp., 128 figs.

Petalotricha serrata n.sp.

Kofoid, C. A. and A. S. Campbell, 1929
A conspectus of the marine and freshwater Ciliata belonging to the suborder Tintinnoinea, with descriptions of new species principally from the Agassiz expedition to the eastern tropical Pacific, 1904-1905. Univ. Calif. Publ. Zool. 34:1-403, 697 text figs.

Poroecus annulatus

Kofoid, C.A. and A.S. Campbell, 1939
Reports on the scientific results of the expedition to the eastern tropical Pacific in charge of Alexander Agassiz, by the U.S. Fish Commission steamer "Albatross" from October 1904 to March 1905-----XXXVII. The Ciliata: The Tintinnoimea. Bull. Mus. Comp. Zool. 84: 473 pp., 36 pls.

Poroecus annulatus n.sp.

Kofoid, C. A. and A. S. Campbell, 1929
A conspectus of the marine and freshwater Ciliata belonging to the suborder Tintinnoinea, with descriptions of new species principally from the Agassiz expedition to the eastern tropical Pacific, 1904-1905. Univ. Calif. Publ. Zool. 34:1-403, 697 text figs.

Poroecus apicatus

Campbell, A. S., 1942
The oceanic Tintinnoina of the plankton gathered during the last cruise of the CARNEGIE. Sci. Res. Cruise VII of the Carnegie, 1928-1929, Biol. 2:163 pp., 128 figs.

Poroecus apicatus

Kofoid, C.A. and A.S. Campbell, 1939
Reports on the scientific results of the expedition to the eastern tropical Pacific in charge of Alexander Agassiz, by the U.S. Fish Commission steamer "Albatross" from October 1904 to March 1905-----XXXVII. The Ciliata: The Tintinnoimea. Bull. Mus. Comp. Zool. 84: 473 pp., 36 pls.

Poroecus apicatus n.sp.

Kofoid, C. A. and A. S. Campbell, 1929
A conspectus of the marine and freshwater Ciliata belonging to the suborder Tintinnoinea, with descriptions of new species principally from the Agassiz expedition to the eastern tropical Pacific, 1904-1905. Univ. Calif. Publ. Zool. 34:1-403, 697 text figs.

Poroecus apiculatus

Balech, Enrique, 1968.
Algunas especies nuevas o interesantes do tintinnidos del Golfo de Mexico y Caribe. Revta Mus. argent. Cienc. nat. Bernardino Rivadavia Inst. nac. Hidrobiol. 2(5):165-197.

Poroecus apiculatus

Gaarder, K.R., 1946
Tintinnoidea. Rep.Sci.Res. "Michael Sars" N.Atlantic Deep-Sea Exped., 1910, 2(1): 37 pp., 23 text figs.

Poroecus apiculatus

Jörgensen, E., 1924
Mediterranean Tintinnidae. Rept. Danish Oceanogr. Exped. 1908-10 to the Mediterranean and adjacent seas, Vol.II, Biol. J 3, 110 pp., 114 text figs.

Poroecus apiculatus

Kofoid, C.A. and A.S. Campbell, 1939
Reports on the scientific results of the expedition to the eastern tropical Pacific in charge of Alexander Agassiz, by the U.S. Fish Commission steamer "Albatross" from October 1904 to March 1905-----XXXVII. The Ciliata: The Tintinnoimea. Bull. Mus. Comp. Zool. 84: 473 pp., 36 pls.

Poroecus apiculatus

Kofoid, C. A. and A. S. Campbell, 1929
A conspectus of the marine and freshwater Ciliata belonging to the suborder Tintinnoinea, with descriptions of new species principally from the Agassiz expedition to the eastern tropical Pacific, 1904-1905. Univ. Calif. Publ. Zool. 34:1-403, 697 text figs.

Poroecus brandti nom.nov.

Kofoid, C. A. and A. S. Campbell, 1929
A conspectus of the marine and freshwater Ciliata belonging to the suborder Tintinnoinea, with descriptions of new species principally from the Agassiz expedition to the eastern tropical Pacific, 1904-1905. Univ. Calif. Publ. Zool. 34:1-403, 697 text figs.

Poroecus curtus

Balech, Enrique, 1968.
Algunas especies nuevas o interesantes do tintinnidos del Golfo de Mexico y Caribe. Revta Mus. argent. Cienc. nat. Bernardino Rivadavia Inst. nac. Hidrobiol. 2(5):165-197.

Poroecus curtus

Gaarder, K.R., 1946
Tintinnoidea. Rep.Sci.Res. "Michael Sars" N.Atlantic Deep-Sea Exped., 1910, 2(1): 37 pp., 23 text figs.

Poroecus curtus

Hasle, Grethe Rytter, 1960
Phytoplankton and ciliate species from the Tropical Pacific. Skr. Norske Videnskaps-Akad., Oslo, 1. Mat.-Nat. Kl., 1960(2): 1-50.

Poroecus curtus

Kofoid, C.A. and A.S. Campbell, 1939
Reports on the scientific results of the expedition to the eastern tropical Pacific in charge of Alexander Agassiz, by the U.S. Fish Commission steamer "Albatross" from October 1904 to March 1905-----XXXVII. The Ciliata: The Tintinnoimea. Bull. Mus. Comp. Zool. 84: 473 pp., 36 pls.

Poroecus curtus n.sp.

Kofoid, C. A. and A. S. Campbell, 1929
A conspectus of the marine and freshwater Ciliata belonging to the suborder Tintinnoinea, with descriptions of new species principally from the Agassiz expedition to the eastern tropical Pacific, 1904-1905. Univ. Calif. Publ. Zool. 34:1-403, 697 text figs.

Poroecus curtus

Komarovsky, B., 1959
The Tintinnina of the Gulf of Eylath (Aqaba). Contrib. Knowl. Red Sea, Bull. Sea Fish. Res. Sta.Haifa, Israel 21 (14):1-40.

Poroecus tubulosus n.sp

Balech, Enrique, 1968.
Algunas especies nuevas o interesantes do tintinnidos del Golfo de Mexico y Caribe. Revta Mus. argent. Cienc. nat. Bernardino Rivadavia Inst. nac. Hidrobiol. 2(5):165-197.

Proplectella sp.

Balech, Enrique, 1962
Tintinnoinea y Dinoflagellata del Pacifico segun material de las Expediciones NORPAC & DOWNWIND del Instituto Scripps de Oceanografia. Revista, Mus. Argentino Ciencias Nat. "Bernardino Rivadavia", Ciencias Zool., 7(1):1-253.

Proplectella

Kofoid, C. A. and A. S. Campbell, 1929
A conspectus of the marine and freshwater Ciliata belonging to the suborder Tintinnoinea, with descriptions of new species principally from the Agassiz expedition to the eastern tropical Pacific, 1904-1905. Univ. Calif. Publ. Zool. 34:1-403, 697 text figs.

Proplectella n. gen.

Kofoid, C. A. and A. S. Campbell, 1929
A conspectus of the marine and freshwater Ciliata belonging to the suborder Tintinnoinea, with descriptions of new species principally from the Agassiz expedition to the eastern tropical Pacific, 1904-1905. Univ. Calif. Publ. Zool. 34:1-403, 697 text figs.

Proplectella acuta

Campbell, A. S., 1942
The oceanic Tintinnoina of the plankton gathered during the last cruise of the CARNEGIE. Sci. Res. Cruise VII of the Carnegie, 1928-1929, Biol. 2:163 pp., 128 figs.

Proplectella acuta

Gaarder, K.R., 1946
Tintinnoidea. Rep.Sci.Res. "Michael Sars" N.Atlantic Deep-Sea Exped., 1910, 2(1): 37 pp., 23 text figs.

Proplectella acuta

Kofoid, C. A. and A. S. Campbell, 1929
A conspectus of the marine and freshwater Ciliata belonging to the suborder Tintinnoinea, with descriptions of new species principally from the Agassiz expedition to the eastern tropical Pacific, 1904-1905. Univ. Calif. Publ. Zool. 34:1-403, 697 text figs.

Proplectella acuta

Marshall, S. M., 1934
The silicoflagellata and tintinnoinea. Great Barrier Reef Exped., 1928-29. Sci. Repts. 4(15):623-664, 43 text figs.

Proplectella amphora

Campbell, A. S., 1942
The oceanic Tintinnoina of the plankton gathered during the last cruise of the CARNEGIE. Sci. Res. Cruise VII of the Carnegie, 1928-1929, Biol. 2:163 pp., 128 figs.

Proplectella amphora

Kofoid, C.A. and A.S. Campbell, 1939
Reports on the scientific results of the expedition to the eastern tropical Pacific in charge of Alexander Agassiz, by the U.S. Fish Commission steamer "Albatross" from October 1904 to March 1905-----XXXVII. The Ciliata: The Tintinnoimea. Bull. Mus. Comp. Zool. 84: 473 pp., 36 pls.

Proplectella amphora n.sp.

Kofoid, C. A. and A. S. Campbell, 1929
A conspectus of the marine and freshwater Ciliata belonging to the suborder Tintinnoinea, with descriptions of new species principally from the Agassiz expedition to the eastern tropical Pacific, 1904-1905. Univ. Calif. Publ. Zool. 34:1-403, 697 text figs.

Proplectella angustior

Campbell, A. S., 1942
The oceanic Tintinnoina of the plankton gathered during the last cruise of the CARNEGIE. Sci. Res. Cruise VII of the Carnegie, 1928-1929, Biol. 2:163 pp., 128 figs.

Proplectella angustior

Gaarder, K.R., 1946
Tintinnoidea. Rep.Sci.Res. "Michael Sars" N.Atlantic Deep-Sea Exped., 1910, 2(1): 37 pp., 23 text figs.

Proplectella angustior

Kofoid, C. A. and A. S. Campbell, 1929
A conspectus of the marine and freshwater Ciliata belonging to the suborder Tintinnoinea, with descriptions of new species principally from the Agassiz expedition to the eastern tropical Pacific, 1904-1905. Univ. Calif. Publ. Zool. 34:1-403, 697 text figs.

Proplectella angustior

Komarovsky, B., 1959
The Tintinnina of the Gulf of Eylath (Aqaba). Contrib. Knowl. Red Sea, Bull. Sea Fish. Res. Sta.Haifa, Israel 21 (14):1-40.

Proplectella angustior

Rampi, L., 1948
Sur quelques Tintinnides (Infusoires louques) du Pacifique subtropical (Recoltes Alain Gerbault) Bull. l'Inst. Ocean., Monaco, No.938, 4 pp.

Proplectella angustior

Rampi, L., 1939
Primo contributo alla conoscénza dei tintinnoidi do Maro Ligure. Atti della Soc. Ital. di Sci. Nat. 78:67-81, 58 text figs.

Proplectella aulti n. sp.

Campbell, A. S., 1942
The oceanic Tintinnoina of the plankton gathered during the last cruise of the CARNEGIE. Sci. Res. Cruise VII of the Carnegie, 1928-1929, Biol. 2:163 pp., 128 figs.

Proplectella biangulata

Campbell, A. S., 1942
The oceanic Tintinnoina of the plankton gathered during the last cruise of the CARNEGIE. Sci. Res. Cruise VII of the Carnegie, 1928-1929, Biol. 2:163 pp., 128 figs.

Proplectella biangulata

Kofoid, C.A. and A.S. Campbell, 1939
Reports on the scientific results of the expedition to the eastern tropical Pacific in charge of Alexander Agassiz, by the U.S. Fish Commission steamer "Albatross" from October 1904 to March 1905-----XXXVII. The Ciliata: The Tintinnoimea. Bull. Mus. Comp. Zool. 84: 473 pp., 36 pls.

Proplectella biangulata n.sp.

Kofoid, C. A. and A. S. Campbell, 1929
A conspectus of the marine and fresh-water Ciliata belonging to the suborder Tintinnoinea, with descriptions of new species principally from the Agassiz expedition to the eastern tropical Pacific, 1904-1905. Univ. Calif. Publ. Zool. 34:1-403, 697 text figs.

Proplectella claparedei

Balech, Enrique, 1968.
Algunas especies nuevas o interesantes do tintinnidos del Golfo de Mexico y Caribe. Revta Mus. argent. Cienc. nat. Bernardino Rivadavia Inst. nac. Hidrobiol. 2(5):165-197.

Proplectella claparedei

Balech, Enrique, 1962
Tintinnoinea y Dinoflagellata del Pacifico segun material de las Expediciones NORPAC y DOWNWIND del Instituto Scripps de Oceanografia. Revista, Mus. Argentino Ciencias Nat. "Bernardino Rivadavia", Ciencias Zool., 7(1):1-253.

Proplectella claparedei

Balech, E., 1959.
Tintinnoinea del Mediterraneo. Trav. Sta. Zool., Villefranche-sur-Mer, 18(17): Reprinted from: Trab. Inst. Español Oceanogr., Madrid, 19 Sept. 1959(28):88 pp., 22 pls.

Proplectella claparedei

Campbell, A. S., 1942
The oceanic Tintinnoina of the plankton gathered during the last cruise of the CARNEGIE. Sci. Res. Cruise VII of the Carnegie, 1928-1929, Biol. 2:163 pp., 128 figs.

Proplectella claparedei

Gaarder, K.R., 1946
Tintinnoidea. Rep.Sci.Res. "Michael Sars" N.Atlantic Deep-Sea Exped., 1910, 2(1): 37 pp., 23 text figs.

Proplectella claparedei

Kofoid, C.A. and A.S. Campbell, 1939
Reports on the scientific results of the expedition to the eastern tropical Pacific in charge of Alexander Agassiz, by the U.S. Fish Commission steamer "Albatross" from October 1904 to March 1905-----XXXVII. The Ciliata: The Tintinnoimea. Bull. Mus. Comp. Zool. 84: 473 pp., 36 pls.

Proplectella clarapedei

Kofoid, C. A. and A. S. Campbell, 1929
A conspectus of the marine and fresh-water Ciliata belonging to the suborder Tintinnoinea, with descriptions of new species principally from the Agassiz expedition to the eastern tropical Pacific, 1904-1905. Univ. Calif. Publ. Zool. 34:1-403, 697 text figs.

Proplectella claparedei

Komarovsky, B., 1959
The Tintinnina of the Gulf of Eylath (Aqaba). Contrib. Knowl. Red Sea, Bull. Sea Fish. Res. Sta.Haifa, Israel 21 (14):1-40.

Proplectella claparedei

Posta, Annette, 1963.
Relation entre l'évolution de quelques tintinnides de la rade de Villefranche et la température de l'eau. Cahiers, Biol. Mar., Roscoff, 4:201-210.
Also in:
Trav. Sta. Zool., Villefranche-sur-Mer, 22(5).

Proplectella claparedei

Sousa e Silva, E. de, 1950.
Les tintinnides de la Baie de Cascais. Bull. Inst. Océan., Monaco, No. 979:28 pp., 4 plas.

Proplectella columbiana

Kofoid, C. A. and A. S. Campbell, 1929
A conspectus of the marine and fresh-water Ciliata belonging to the suborder Tintinnoinea, with descriptions of new species principally from the Agassiz expedition to the eastern tropical Pacific, 1904-1905. Univ. Calif. Publ. Zool. 34:1-403, 697 text figs.

Proplectella cuspidata

Campbell, A. S., 1942
The oceanic Tintinnoina of the plankton gathered during the last cruise of the CARNEGIE. Sci. Res. Cruise VII of the Carnegie, 1928-1929, Biol. 2:163 pp., 128 figs.

Proplectella cuspidata

Kofoid, C.A. and A.S. Campbell, 1939
Reports on the scientific results of the expedition to the eastern tropical Pacific in charge of Alexander Agassiz, by the U.S. Fish Commission steamer "Albatross" from October 1904 to March 1905-----XXXVII. The Ciliata: The Tintinnoimea. Bull. Mus. Comp. Zool. 84: 473 pp., 36 pls.

Proplectella cuspidata n.sp.

Kofoid, C. A. and A. S. Campbell, 1929
A conspectus of the marine and fresh-water Ciliata belonging to the suborder Tintinnoinea, with descriptions of new species principally from the Agassiz expedition to the eastern tropical Pacific, 1904-1905. Univ. Calif. Publ. Zool. 34:1-403, 697 text figs.

Proplectella ellipsoida

Campbell, A. S., 1942
The oceanic Tintinnoina of the plankton gathered during the last cruise of the CARNEGIE. Sci. Res. Cruise VII of the Carnegie, 1928-1929, Biol. 2:163 pp., 128 figs.

Proplectella ellipsoida

Gaarder, K.R., 1946
Tintinnoidea. Rep.Sci.Res. "Michael Sars" N.Atlantic Deep-Sea Exped., 1910, 2(1): 37 pp., 23 text figs.

Proplectella ellipsoida

Kofoid, C.A. and A.S. Campbell, 1939
Reports on the scientific results of the expedition to the eastern tropical Pacific in charge of Alexander Agassiz, by the U.S. Fish Commission steamer "Albatross" from October 1904 to March 1905-----XXXVII. The Ciliata: The Tintinnoimea. Bull. Mus. Comp. Zool. 84: 473 pp., 36 pls.

Proplectella ellipsoida n.sp.

Kofoid, C. A. and A. S. Campbell, 1929
A conspectus of the marine and fresh-water Ciliata belonging to the suborder Tintinnoinea, with descriptions of new species principally from the Agassiz expedition to the eastern tropical Pacific, 1904-1905. Univ. Calif. Publ. Zool. 34:1-403, 697 text figs.

Proplectella expolita

Campbell, A. S., 1942
The oceanic Tintinnoina of the plankton gathered during the last cruise of the CARNEGIE. Sci. Res. Cruise VII of the Carnegie, 1928-1929, Biol. 2:163 pp., 128 figs.

Proplectella fastigata

Gaarder, K.R., 1946
Tintinnoidea. Rep.Sci.Res. "Michael Sars" N.Atlantic Deep-Sea Exped., 1910, 2(1): 37 pp., 23 text figs.

Proplectella fastigata

Hofker, J., 1931
Studien uber Tintinnoidea. Arch. f. Protistenk 75(3):315-402, 89 text figs.

Proplectella fastigata

Kofoid, C.A. and A.S. Campbell, 1939
Reports on the scientific results of the expedition to the eastern tropical Pacific in charge of Alexander Agassiz, by the U.S. Fish Commission steamer "Albatross" from October 1904 to March 1905-----XXXVII. The Ciliata: The Tintinnoimea. Bull. Mus. Comp. Zool. 84: 473 pp., 36 pls.

Proplectella fastigata

Kofoid, C. A. and A. S. Campbell, 1929
A conspectus of the marine and fresh-water Ciliata belonging to the suborder Tintinnoinea, with descriptions of new species principally from the Agassiz expedition to the eastern tropical Pacific, 1904-1905. Univ. Calif. Publ. Zool. 34:1-403, 697 text figs.

Proplectella fastigata

Komarovsky, B., 1959
The Tintinnina of the Gulf of Eylath (Aqaba). Contrib. Knowl. Red Sea, Bull. Sea Fish. Res. Sta.Haifa, Israel 21 (14):1-40.

Proplectella fastigata

Rampi, L., 1948
Sur quelques Tintinnides (Infusoires louques) du Pacifique subtropical (Recoltes Alain Gerbault) Bull. l'Inst. Ocean., Monaco, No.938, 4 pp.

Proplectella fastigata

Rampi, L., 1939
Primo contributo alla conoscenza dei tintinnoidi do Maro Ligure. Atti della Soc. Ital. di Sci. Nat. 78:67-81, 58 text figs.

Proplectella globosa

Campbell, A. S., 1942
The oceanic Tintinnoina of the plankton gathered during the last cruise of the CARNEGIE. Sci. Res. Cruise VII of the Carnegie, 1928-1929, Biol. 2:163 pp., 128 figs.

Proplectella globosa

Kofoid, C.A. and A.S. Campbell, 1939
Reports on the scientific results of the expedition to the eastern tropical Pacific in charge of Alexander Agassiz, by the U.S. Fish Commission steamer "Albatross" from October 1904 to March 1905-----XXXVII. The Ciliata: The Tintinnoimea. Bull. Mus. Comp. Zool. 84: 473 pp., 36 pls.

Proplectella globosa

Kofoid, C. A. and A. S. Campbell, 1929
A conspectus of the marine and fresh-water Ciliata belonging to the suborder Tintinnoinea, with descriptions of new species principally from the Agassiz expedition to the eastern tropical Pacific, 1904-1905. Univ. Calif. Publ. Zool. 34:1-403, 697 text figs.

Proplectella globosa

Rampi, L., 1948
Sur quelques Tintinnides (Infusoires louques) du Pacifique subtropical (Recoltes Alain Gerbault) Bull. l'Inst. Ocean., Monaco, No.938, 4 pp.

Proplectella grandis

Kofoid, C. A. and A. S. Campbell, 1929
A conspectus of the marine and freshwater Ciliata belonging to the suborder Tintinnoinea, with descriptions of new species principally from the Agassiz expedition to the eastern tropical Pacific, 1904-1905. Univ. Calif. Publ. Zool. 34:1-403, 697 text figs.

Proplectella merriami n. sp.

Campbell, A. S., 1942
The oceanic Tintinnoina of the plankton gathered during the last cruise of the CARNEGIE. Sci. Res. Cruise VII of the Carnegie, 1928-1929, Biol. 2:163 pp., 128 figs.

Proplectella ostenfeldi

Campbell, A. S., 1942
The oceanic Tintinnoina of the plankton gathered during the last cruise of the CARNEGIE. Sci. Res. Cruise VII of the Carnegie, 1928-1929, Biol. 2:163 pp., 128 figs.

Proplectella ostenfeldi

Kofoid, C.A. and A.S. Campbell, 1939
Reports on the scientific results of the expedition to the eastern tropical Pacific in charge of Alexander Agassiz, by the U.S. Fish Commission steamer "Albatross" from October 1904 to March 1905-----XXXVII. The Ciliata: The Tintinnoimea. Bull.Mus. Comp.Zool. 84: 473 pp., 36 pls.

Proplectella ostenfeldi n. sp.

Kofoid, C. A. and A. S. Campbell, 1929
A conspectus of the marine and freshwater Ciliata belonging to the suborder Tintinnoinea, with descriptions of new species principally from the Agassiz expedition to the eastern tropical Pacific, 1904-1905. Univ. Calif. Publ. Zool. 34:1-403, 697 text figs.

Proplectella ostendfeldi

Komarovsky, B., 1959
The Tintinnina of the Gulf of Eylath (Aqaba). Contrib. Knowl. Red Sea, Bull. Sea Fish. Res. Sta.Haifa, Israel 21 (14):1-40.

Proplectella ostenfeldi

Rampi, L., 1939
Primo contributo alla conoscenza dei tintinnoidi de Mare Ligure. Atti della Soc. Ital. di Sci. Nat. 78:67-81, 58 text figs.

Proplectella ovata

Campbell, A. S., 1942
The oceanic Tintinnoina of the plankton gathered during the last cruise of the CARNEGIE. Sci. Res. Cruise VII of the Carnegie, 1928-1929, Biol. 2:163 pp., 128 figs.

Proplectella ovata

Gaarder, K.R., 1946
Tintinnoidea. Rep.Sci.Res."Michael Sars" N.Atlantic Deep-Sea Exped., 1910, 2(1): 37 pp., 23 text figs.

Proplectella ovata

Kofoid, C.A. and A.S. Campbell, 1939
Reports on the scientific results of the expedition to the eastern tropical Pacific in charge of Alexander Agassiz, by the U.S. Fish Commission steamer "Albatross" from October 1904 to March 1905-----XXXVII. The Ciliata: The Tintinnoimea. Bull.Mus. Comp.Zool. 84: 473 pp., 36 pls.

Proplectella ovata

Kofoid, C. A. and A. S. Campbell, 1929
A conspectus of the marine and freshwater Ciliata belonging to the suborder Tintinnoinea, with descriptions of new species principally from the Agassiz expedition to the eastern tropical Pacific, 1904-1905. Univ. Calif. Publ. Zool. 34:1-403, 697 text figs.

Proplectella parva

Balech, Enrique, 1962
Tintinnoinea y Dinoflagellata del Pacifico segun material de las Expediciones NORPAC y DOWNWIND del Instituto Scripps de Oceanografia. Revista, Mus. Argentino Ciencias Nat. "Bernardino Rivadavia", Ciencias Zool., 7(1):1-253.

Proplectella parva

Campbell, A. S., 1942
The oceanic Tintinnoina of the plankton gathered during the last cruise of the CARNEGIE. Sci. Res. Cruise VII of the Carnegie, 1928-1929, Biol. 2:163 pp., 128 figs.

Proplectella parva

Kofoid, C.A. and A.S. Campbell, 1939
Reports on the scientific results of the expedition to the eastern tropical Pacific in charge of Alexander Agassiz, by the U.S. Fish Commission steamer "Albatross" from October 1904 to March 1905-----XXXVII. The Ciliata: The Tintinnoimea. Bull.Mus. Comp.Zool. 84: 473 pp., 36 pls.

Proplectella parva n.sp.

Kofoid, C. A. and A. S. Campbell, 1929
A conspectus of the marine and freshwater Ciliata belonging to the suborder Tintinnoinea, with descriptions of new species principally from the Agassiz expedition to the eastern tropical Pacific, 1904-1905. Univ. Calif. Publ. Zool. 34:1-403, 697 text figs.

Proplectella pentagona

Campbell, A. S., 1942
The oceanic Tintinnoina of the plankton gathered during the last cruise of the CARNEGIE. Sci. Res. Cruise VII of the Carnegie, 1928-1929, Biol. 2:163 pp., 128 figs.

Proplectella pentagona

Kofoid, C.A. and A.S. Campbell, 1939
Reports on the scientific results of the expedition to the eastern tropical Pacific in charge of Alexander Agassiz, by the U.S. Fish Commission steamer "Albatross" from October 1904 to March 1905-----XXXVII. The Ciliata: The Tintinnoimea. Bull.Mus. Comp.Zool. 84: 473 pp., 36 pls.

Proplectella pentagona

Kofoid, C. A. and A. S. Campbell, 1929
A conspectus of the marine and freshwater Ciliata belonging to the suborder Tintinnoinea, with descriptions of new species principally from the Agassiz expedition to the eastern tropical Pacific, 1904-1905. Univ. Calif. Publ. Zool. 34:1-403, 697 text figs.

Proplectella perpusilla

Balech, Enrique, 1971
Microplancton del Atlantico ecuatorial oeste (Equalant 1) Publ. Serv. Hidrograf. Naval, Argentina H. 654: 103 pp., 122 figs.

Proplectella perpusilla

Balech, Enrique, 1962
Tintinnoinea y Dinoflagellata del Pacifico segun material de las Expediciones NORPAC y DOWNWIND del Instituto Scripps de Oceanografia. Revista, Mus. Argentino Ciencias Nat. "Bernardino Rivadavia", Ciencias Zool., 7(1):1-253.

Proplectella perpusilla

Campbell, A. S., 1942
The oceanic Tintinnoina of the plankton gathered during the last cruise of the CARNEGIE. Sci. Res. Cruise VII of the Carnegie, 1928-1929, Biol. 2:163 pp., 128 figs.

Proplectella perpusilla

Kofoid, C.A. and A.S. Campbell, 1939
Reports on the scientific results of the expedition to the eastern tropical Pacific in charge of Alexander Agassiz, by the U.S. Fish Commission steamer "Albatross" from October 1904 to March 1905-----XXXVII. The Ciliata: The Tintinnoimea. Bull.Mus. Comp.Zool. 84: 473 pp., 36 pls.

Proplectella perpusilla n.sp.

Kofoid, C. A. and A. S. Campbell, 1929
A conspectus of the marine and freshwater Ciliata belonging to the suborder Tintinnoinea, with descriptions of new species principally from the Agassiz expedition to the eastern tropical Pacific, 1904-1905. Univ. Calif. Publ. Zool. 34:1-403, 697 text figs.

Proplectella perpusilla

Komarovsky, B., 1959
The Tintinnina of the Gulf of Eylath (Aqaba). Contrib. Knowl. Red Sea, Bull. Sea Fish. Res. Sta.Haifa, Israel 21 (14):1-40.

Proplectella perpusilla

Marshall, S. M., 1934
The silicoflagellata and tintinnoinea. Great Barrier Reef Exped., 1928-29. Sci. Repts. 4(15):623-664, 43 text figs.

Proplectella praelonga

Campbell, A. S., 1942
The oceanic Tintinnoina of the plankton gathered during the last cruise of the CARNEGIE. Sci. Res. Cruise VII of the Carnegie, 1928-1929, Biol. 2:163 pp., 128 figs.

Proplectella praelonga

Gaarder, K.R., 1946
Tintinnoidea. Rep.Sci.Res."Michael Sars" N.Atlantic Deep-Sea Exped., 1910, 2(1): 37 pp., 23 text figs.

Proplectella praelonga

Kofoid, C.A. and A.S. Campbell, 1939
Reports on the scientific results of the expedition to the eastern tropical Pacific in charge of Alexander Agassiz, by the U.S. Fish Commission steamer "Albatross" from October 1904 to March 1905-----XXXVII. The Ciliata: The Tintinnoimea. Bull.Mus. Comp.Zool. 84: 473 pp., 36 pls.

Proplectella praelonga n.sp.

Kofoid, C. A. and A. S. Campbell, 1929
A conspectus of the marine and freshwater Ciliata belonging to the suborder Tintinnoinea, with descriptions of new species principally from the Agassiz expedition to the eastern tropical Pacific, 1904-1905. Univ. Calif. Publ. Zool. 34:1-403, 697 text figs.

Proplectella praelonga

Komarovsky, B., 1959
The Tintinnina of the Gulf of Eylath (Aqaba). Contrib. Knowl. Red Sea, Bull. Sea Fish. Res. Sta.Haifa, Israel 21 (14):1-40.

Proplectella subacuta

Campbell, A. S., 1942
The oceanic Tintinnoina of the plankton gathered during the last cruise of the CARNEGIE. Sci. Res. Cruise VII of the Carnegie, 1928-1929, Biol. 2:163 pp., 128 figs.

Proplectella subacuta

Kofoid, C. A. and A. S. Campbell, 1929
A conspectus of the marine and freshwater Ciliata belonging to the suborder Tintinnoinea, with descriptions of new species principally from the Agassiz expedition to the eastern tropical Pacific, 1904-1905. Univ. Calif. Publ. Zool. 34:1-403, 697 text figs.

Proplectella subangulata

Kofoid, C.A. and A.S. Campbell, 1939
Reports on the scientific results of the expedition to the eastern tropical Pacific in charge of Alexander Agassiz, by the U.S. Fish Commission steamer "Albatross" from October 1904 to March 1905-----XXXVII. The Ciliata: The Tintinnoimea. Bull.Mus. Comp.Zool. 84: 473 pp., 36 pls.

Proplectella subangulata n.sp.

Kofoid, C. A. and A. S. Campbell, 1929
A conspectus of the marine and fresh-water Ciliata belonging to the suborder Tintinnoinea, with descriptions of new species principally from the Agassiz expedition to the eastern tropical Pacific, 1904-1905. Univ. Calif. Publ. Zool. 34:1-403, 697 text figs.

Proplectella subangulata

Komarovsky, B., 1959
The Tintinnina of the Gulf of Eylath (Aqaba). Contrib. Knowl. Red Sea, Bull. Sea Fish. Res. Sta. Haifa, Israel 21 (14):1-40.

Proplectella subcaudata

Balech, Enrique, 1968.
Algunas especies nuevas o interesantes do tintinnidos del Golfo de Mexico y Caribe. Revta Mus. argent. Cienc. nat. Bernardino Rivadavia Inst. nac. Hidrobiol. 2(5):165-197.

Proplectata subcaudata

Balech, Enrique, 1962
Tintinnoinea y Dinoflagellata del Pacifico segun material de las Expediciones NORPAC y DOWNWIND del Instituto Scripps de Oceanografia. Revista, Mus. Argentino Ciencias Nat. "Bernardino Rivadavia", Ciencias Zool., 7(1):1-253.

Proplectella subcaudata

Campbell, A. S., 1942
The oceanic Tintinnoina of the plankton gathered during the last cruise of the CARNEGIE. Sci. Res. Cruise VII of the Carnegie, 1928-1929, Biol. 2:163 pp., 128 figs.

Proplectella subcaudata

Gaarder, K.R., 1946
Tintinnoidea. Rep.Sci.Res."Michael Sars" N.Atlantic Deep-Sea Exped., 1910, 2(1): 37 pp., 23 text figs.

Proplectella subcaudata

Kofoid, C. A. and A. S. Campbell, 1929
A conspectus of the marine and fresh-water Ciliata belonging to the suborder Tintinnoinea, with descriptions of new species principally from the Agassiz expedition to the eastern tropical Pacific, 1904-1905. Univ. Calif. Publ. Zool. 34:1-403, 697 text figs.

Proplectella tenuis

Campbell, A. S., 1942
The oceanic Tintinnoina of the plankton gathered during the last cruise of the CARNEGIE. Sci. Res. Cruise VII of the Carnegie, 1928-1929, Biol. 2:163 pp., 128 figs.

Proplectella tenuis

Gaarder, K.R., 1946
Tintinnoidea. Rep.Sci.Res."Michael Sars" N.Atlantic Deep-Sea Exped., 1910, 2(1): 37 pp., 23 text figs.

Proplectella tenuis

Kofoid, C.A. and A.S. Campbell, 1939
Reports on the scientific results of the expedition to the eastern tropical Pacific in charge of Alexander Agassiz, by the U.S. Fish Commission steamer "Albetross" from October 1904 to March 1905-----XXXVII. The Ciliata: The Tintinnoimea. Bull.Mus. Comp.Zool. 84: 473 pp., 36 pls.

Proplectella tenuis n.sp.

Kofoid, C. A. and A. S. Campbell, 1929
A conspectus of the marine and fresh-water Ciliata belonging to the suborder Tintinnoinea, with descriptions of new species principally from the Agassiz expedition to the eastern tropical Pacific, 1904-1905. Univ. Calif. Publ. Zool. 34:1-403, 697 text figs.

Proplectella tenuis

Komarovsky, B., 1959
The Tintinnina of the Gulf of Eylath (Aqaba). Contrib. Knowl. Red Sea, Bull. Sea Fish. Res. Sta. Haifa, Israel 21 (14):1-40.

Proplectella tenuis

Marshall, S. M., 1934
The silicoflagellata and tintinnoinea. Great Barrier Reef Exped., 1928-29. Sci. Repts. 4(15):623-664, 43 text figs.

Proplectella tumida

Campbell, A. S., 1942
The oceanic Tintinnoina of the plankton gathered during the last cruise of the CARNEGIE. Sci. Res. Cruise VII of the Carnegie, 1928-1929, Biol. 2:163 pp., 128 figs.

Proplectella tumida

Kofoid, C.A. and A.S. Campbell, 1939
Reports on the scientific results of the expedition to the eastern tropical Pacific in charge of Alexander Agassiz, by the U.S. Fish Commission steamer "Albetross" from October 1904 to March 1905-----XXXVII. The Ciliata: The Tintinnoimea. Bull.Mus. Comp.Zool. 84: 473 pp., 36 pls.

Proplectella tumida n.sp.

Kofoid, C. A. and A. S. Campbell, 1929
A conspectus of the marine and fresh-water Ciliata belonging to the suborder Tintinnoinea, with descriptions of new species principally from the Agassiz expedition to the eastern tropical Pacific, 1904-1905. Univ. Calif. Publ. Zool. 34:1-403, 697 text figs.

Proplectella urna

Campbell, A. S., 1942
The oceanic Tintinnoina of the plankton gathered during the last cruise of the CARNEGIE. Sci. Res. Cruise VII of the Carnegie, 1928-1929, Biol. 2:163 pp., 128 figs.

Proplectella urna

Kofoid, C.A. and A.S. Campbell, 1939
Reports on the scientific results of the expedition to the eastern tropical Pacific in charge of Alexander Agassiz, by the U.S. Fish Commission steamer "Albetross" from October 1904 to March 1905-----XXXVII. The Ciliata: The Tintinnoimea. Bull.Mus. Comp.Zool. 84: 473 pp., 36 pls.

Proplectella urna n.sp.

Kofoid, C. A. and A. S. Campbell, 1929
A conspectus of the marine and fresh-water Ciliata belonging to the suborder Tintinnoinea, with descriptions of new species principally from the Agassiz expedition to the eastern tropical Pacific, 1904-1905. Univ. Calif. Publ. Zool. 34:1-403, 697 text figs.

Prostelidiella phialia

Kofoid, C.A. and A.S. Campbell, 1939
Reports on the scientific results of the expedition to the eastern tropical Pacific in charge of Alexander Agassiz, by the U.S. Fish Commission steamer "Albetross" from October 1904 to March 1905-----XXXVII. The Ciliata: The Tintinnoimea. Bull.Mus. Comp.Zool. 84: 473 pp., 36 pls.

Protocymatocylis conicoides nom nov.

Kofoid, C. A. and A. S. Campbell, 1929
A conspectus of the marine and fresh-water Ciliata belonging to the suborder Tintinnoinea, with descriptions of new species principally from the Agassiz expedition to the eastern tropical Pacific, 1904-1905. Univ. Calif. Publ. Zool. 34:1-403, 697 text figs.

Pseudometacylis ornata n.gen.,n.sp

Balech, Enrique, 1968.
Algunas especies nuevas o interesantes do tintinnidos del Golfo de Mexico y Caribe. Revta Mus. argent. Cienc. nat. Bernardino Rivadavia Inst. nac. Hidrobiol. 2(5):165-197.

Protocymatocylis pseudoconica

*Hada, Yoshina, 1970.   (1970)
The protozoan plankton of the Antarctic and Subantarctic seas. Scient. Repts, Japan. Antarct. Res. Exped., (E)31:51 pp.

Protocymatocylis subrotundata

*Hada, Yoshina, 1970.   (1970)
The protozoan plankton of the Antarctic and Subantarctic seas. Scient. Repts, Japan. Antarct. Res. Exped., (E)31:51 pp.

Protocymatocylis subrotundata

Kofoid, C. A. and A. S. Campbell, 1929
A conspectus of the marine and fresh-water Ciliata belonging to the suborder Tintinnoinea, with descriptions of new species principally from the Agassiz expedition to the eastern tropical Pacific, 1904-1905. Univ. Calif. Publ. Zool. 34:1-403, 697 text figs.

Protodymatocylis vas n.sp.

Kofoid, C. A. and A. S. Campbell, 1929
A conspectus of the marine and fresh-water Ciliata belonging to the suborder Tintinnoinea, with descriptions of new species principally from the Agassiz expedition to the eastern tropical Pacific, 1904-1905. Univ. Calif. Publ. Zool. 34:1-403, 697 text figs.

Protorhabdonella sp.

Argentina, Secretaria de Marina, Servicio de Hidrografia Naval, 1962.
Plancton de las companas oceanograficas DRAKE I y II. Publico, H. 627:57.

Protorhabdonella n.gen.

Jörgensen, E., 1924
Mediterranean Tintinnidae. Rept. Danish Oceanogr. Exped. 1908-10 to the Mediterranean and adjacent seas, Vol.II, Biol. J 3, 110 pp., 114 text figs.

Protorhabdonella amor simplex

Komarovsky, B., 1959
The Tintinnina of the Gulf of Eylath (Aqaba). Contrib. Knowl. Red Sea, Bull. Sea Fish. Res. Sta. Haifa, Israel 21 (14):1-40.

Protorhabodonella curta

Balech, E., 1951.
Nuevos datos sobre Tintinnoinea de Argentina y Uraguay. Physis 20(58):291-302, 16 textfigs.

Protorhabdonella curta

Balech, Enrique, 1944
Contribucion al conocimiento del Planeton de Lennox y Cabo de Hornos. Physis XIX:423-446, 6 pls. with 67 figs.

Protorhabdella curta

Campbell, A. S., 1942
The oceanic Tintinnoina of the plankton gathered during the last cruise of the CARNEGIE. Sci. Res. Cruise VII of the Carnegie, 1928-1929, Biol. 2:163 pp., 128 figs.

Protorhabdonella curta

Duran, M., 1951
Contribucion al estudio de los tintinidos del plancton de los costas de Castellon. (Mediterranes occidental). Publ. Inst. Biol. Aplic., Barcelone, 8: 101-120, 29 text figs.

Protorhabdonella curta

Jörgensen, E., 1924
Mediterranean Tintinnidae. Rept. Danish Oceanogr. Exped. 1908-10 to the Mediterranean and adjacent seas, Vol.II, Biol. J 3, 110 pp., 114 text figs.

Oceanographic Index: Marine Organisms Cumulation, 1946-1973

Protorhabdonella curta

Kofoid, C.A. and A.S. Campbell, 1939
Reports on the scientific results of the expedition to the eastern tropical Pacific in charge of Alexander Agassiz, by the U.S. Fish Commission steamer "Albatross" from October 1904 to March 1905-----XXXVII. The Ciliata: The Tintinnoinea. Bull. Mus. Comp. Zool. 84: 473 pp., 36 pls.

Protorhabdonella curta

Marshall, S. M., 1934
The silicoflagellata and tintinnoinea. Great Barrier Reef Exped., 1928-29. Sci. Repts. 4(15):623-664, 43 text figs.

Protorhabdonella mira

Balech, Enrique, 1962
Tintinnoinea y Dinoflagellata del Pacifico segun material de las Expediciones NORPAC y DOWNWIND del Instituto Scripps de Oceanografia. Revista. Mus. Argentino Ciencias Nat. "Bernardino Rivadavia", Ciencias Zool., 7(1):1-253.

Protorhabdonella mira

Kofoid, C.A. and A.S. Campbell, 1939
Reports on the scientific results of the expedition to the eastern tropical Pacific in charge of Alexander Agassiz, by the U.S. Fish Commission steamer "Albatross" from October 1904 to March 1905-----XXXVII. The Ciliata: The Tintinnoimea. Bull. Mus. Comp. Zool. 84: 473 pp., 36 pls.

Protorhabdonella simplex

Balech, Enrique, 1962
Tintinnoinea y Dinoflagellata del Pacifico segun material de las Expediciones NORPAC y DOWNWIND del Instituto Scripps de Oceanografia. Revista. Mus. Argentino Ciencias Nat. "Bernardino Rivadavia", Ciencias Zool., 7(1):1-253.

Protorhabdella simplex

Campbell, A. S., 1942
The oceanic Tintinnoina of the plankton gathered during the last cruise of the CARNEGIE. Sci. Res. Cruise VII of the Carnegie, 1928-1929, Biol. 2:163 pp., 128 figs.

Protonhabdonella simplex

Delegazione Italiana della Commissione Internazionale per l'Esplorazione Scientifica del Mediterraneo, 1941
Note sul plancton della Laguna veneta. [Memoria CCLXXXIX], Arch. di Ocean. e Limn. Anno I, Fasc. I, 1941 XIX: 31-57 pp.

Protorhabdonella simplex

Jörgensen, E., 1924
Mediterranean Tintinnidae. Rept. Danish Oceanogr. Exped. 1908-10 to the Mediterranean and adjacent seas, Vol. II, Biol. J 3, 110 pp., 114 text figs.

Protorhabdonella simplex

Kofoid, C.A. and A.S. Campbell, 1939
Reports on the scientific results of the expedition to the eastern tropical Pacific in charge of Alexander Agassiz, by the U.S. Fish Commission steamer "Albatross" from October 1904 to March 1905-----XXXVII. The Ciliata: The Tintinnoimea. Bull. Mus. Comp. Zool. 84: 473 pp., 36 pls.

Protorhabdonella simplex

Kofoid, C. A. and A. S. Campbell, 1929
A conspectus of the marine and fresh-water Ciliata belonging to the suborder Tintinnoinea, with descriptions of new species principally from the Agassiz expedition to the eastern tropical Pacific, 1904-1905. Univ. Calif. Publ. Zool. 34:1-403, 697 text figs.

Protorhabdonella simplex

Marshall, S. M., 1934
The silicoflagellata and tintinnoinea. Great Barrier Reef Exped., 1928-29. Sci. Repts. 4(15):623-664, 43 text figs.

Protorhabdonella simplex

Marshall, S. M., 1933
The production of microplankton in the Great Barrier Reef Region. Brit. Mus. (N.H.) Great Barrier Reef Exped. 1928-29, Sci. Repts. II(5):111-157, 14 text figs.

Protorhabdonella striatura

Balech, Enrique, 1971
Microplancton del Atlantico ecuatorial oeste (Equalant 1) Publ. Serv. Hidrograf. Naval, Argentina H. 654: 103 pp., 122 figs.

Protorhabdella striatura

Campbell, A. S., 1942
The oceanic Tintinnoina of the plankton gathered during the last cruise of the CARNEGIE. Sci. Res. Cruise VII of the Carnegie, 1928-1929, Biol. 2:163 pp., 128 figs.

Protorhabdonella striatura

Kofoid, C.A. and A.S. Campbell, 1939
Reports on the scientific results of the expedition to the eastern tropical Pacific in charge of Alexander Agassiz, by the U.S. Fish Commission steamer "Albatross" from October 1904 to March 1905-----XXXVII. The Ciliata: The Tintinnoimea. Bull. Mus. Comp. Zool. 84: 473 pp., 36 pls.

Protorhabdonella striatura nom. nov.

Kofoid, C. A. and A. S. Campbell, 1929
A conspectus of the marine and fresh-water Ciliata belonging to the suborder Tintinnoinea, with descriptions of new species principally from the Agassiz expedition to the eastern tropical Pacific, 1904-1905. Univ. Calif. Publ. Zool. 34:1-403, 697 text figs.

Protorhabdonella ventricosa

Kofoid, C. A. and A. S. Campbell, 1929
A conspectus of the marine and fresh-water Ciliata belonging to the suborder Tintinnoinea, with descriptions of new species principally from the Agassiz expedition to the eastern tropical Pacific, 1904-1905. Univ. Calif. Publ. Zool. 34:1-403, 697 text figs.

Ptychocylis acuminata

Laackmann, H., 1910
Die Tintinnodeen der Deutschen Südpolar Expedition 1901-1903. Deutsch. Südpolar Exped. 11, Zool. 3(4):341-496, pls. 33-51.

Ptychocylis acuta

Campbell, A. S., 1942
The oceanic Tintinnoina of the plankton gathered during the last cruise of the CARNEGIE. Sci. Res. Cruise VII of the Carnegie, 1928-1929, Biol. 2:163 pp., 128 figs.

Ptychocylis acuta

Kofoid, C. A. and A. S. Campbell, 1929
A conspectus of the marine and fresh-water Ciliata belonging to the suborder Tintinnoinea, with descriptions of new species principally from the Agassiz expedition to the eastern tropical Pacific, 1904-1905. Univ. Calif. Publ. Zool. 34:1-403, 697 text figs.

Ptychocylis (Rhabdonella) amor

Entz, G., Jr., 1909
Studien über organisation und biologie der Tintinniden. Arch. f. Protistenkunde 15:93-226, Pls. 8-21, text figs.

Ptychocylis Amphorella n.sp.

Meunier, A., 1919
Microplankton de la Mer Flamande. 4. Les Tintinnides et Coetera. Mem. Mus. Roy. Hist. Nat., Belgique, 8(2):59 pp., Pls. 22-23.

Ptychocylis arctica

Campbell, A. S., 1942
The oceanic Tintinnoina of the plankton gathered during the last cruise of the CARNEGIE. Sci. Res. Cruise VII of the Carnegie, 1928-1929, Biol. 2:163 pp., 128 figs.

Ptychocylis arctica

Kofoid, C. A. and A. S. Campbell, 1929
A conspectus of the marine and fresh-water Ciliata belonging to the suborder Tintinnoinea, with descriptions of new species principally from the Agassiz expedition to the eastern tropical Pacific, 1904-1905. Univ. Calif. Publ. Zool. 34:1-403, 697 text figs.

Ptychocylis arctica

Schulz, B., and A. Wulff, 1929
Hydrographie und Oberflächen plankton des westlichen Barentsmeeres im Sommer 1927. Ber. deutschen wissensch. Komm. F. Meeresforsch. n.s. 4(5):232-372, 13 tables, 25 text figs.

Ptychocylis basicurvata

Kofoid, C. A. and A. S. Campbell, 1929
A conspectus of the marine and fresh-water Ciliata belonging to the suborder Tintinnoinea, with descriptions of new species principally from the Agassiz expedition to the eastern tropical Pacific, 1904-1905. Univ. Calif. Publ. Zool. 34:1-403, 697 text figs.

Ptychocylis calyx

Laackmann, H., 1910
Die Tintinnodeen der Deutschen Südpolar Expedition 1901-1903. Deutschen Südpolar Expedition 11, Zool. 3 (4): 341-496, pls. 33-51.

Ptychocylis cylindrica

Kofoid, C. A. and A. S. Campbell, 1929
A conspectus of the marine and fresh-water Ciliata belonging to the suborder Tintinnoinea, with descriptions of new species principally from the Agassiz expedition to the eastern tropical Pacific, 1904-1905. Univ. Calif. Publ. Zool. 34:1-403, 697 text figs.

Ptychocylis drygalskii

Campbell, A. S., 1942
The oceanic Tintinnoina of the plankton gathered during the last cruise of the CARNEGIE. Sci. Res. Cruise VII of the Carnegie, 1928-1929, Biol. 2:163 pp., 128 figs.

Ptychocylis drygalskii

Kofoid, C. A. and A. S. Campbell, 1929
A conspectus of the marine and fresh-water Ciliata belonging to the suborder Tintinnoinea, with descriptions of new species principally from the Agassiz expedition to the eastern tropical Pacific, 1904-1905. Univ. Calif. Publ. Zool. 34:1-403, 697 text figs.

Ptychocylis glacialis

Kofoid, C. A. and A. S. Campbell, 1929
A conspectus of the marine and fresh-water Ciliata belonging to the suborder Tintinnoinea, with descriptions of new species principally from the Agassiz expedition to the eastern tropical Pacific, 1904-1905. Univ. Calif. Publ. Zool. 34:1-403, 697 text figs.

Ptychocylis minor

Kofoid, C. A. and A. S. Campbell, 1929
A conspectus of the marine and fresh-water Ciliata belonging to the suborder Tintinnoinea, with descriptions of new species principally from the Agassiz expedition to the eastern tropical Pacific, 1904-1905. Univ. Calif. Publ. Zool. 34:1-403, 697 text figs.

Ptychocylis minor

Campbell, A. S., 1942
The oceanic Tintinnoina of the plankton gathered during the last cruise of the CARNEGIE. Sci. Res. Cruise VII of the Carnegie, 1928-1929, Biol. 2:163 pp., 128 figs.

Ptychocylis nervosa

Laackmann, H., 1910
Die Tintinnodeen der Deutschen Südpolar Expedition 1901-1903. Deutschen Südpolar Expedition 11, Zool.3 (4): 341-496, pls. 33-51.

Ptychocylis obtusa

Campbell, A. S., 1942
The oceanic Tintinnoina of the plankton gathered during the last cruise of the CARNEGIE. Sci. Res. Cruise VII of the Carnegie, 1928-1929, Biol. 2:163 pp., 128 figs.

Ptychocylis obtusa

Gaarder, K.R., 1946
Tintinnoidea. Rep.Sci.Res."Michael Sars" N.Atlantic Deep-Sea Exped., 1910, 2(1): 37 pp., 23 text figs.

Ptychocylis obtusa

Gran, H.H., and T. Braarud, 1935
A quantitative study of the phytoplankton in the Bay of Fundy and the Gulf of Maine (including observations on hydrography, chemistry, and turbidity). J. Biol. Bd., Canada, 1(5):279-467, 69 text figs.

Ptychocylis obtusa

Kofoid, C. A. and A. S. Campbell, 1929
A conspectus of the marine and freshwater Ciliata belonging to the suborder Tintinnoinea, with descriptions of new species principally from the Agassiz expedition to the eastern tropical Pacific, 1904-1905. Univ. Calif. Publ. Zool. 34:1-403, 697 text figs.

Ptychocylis obtusa

Schulz, B., and A. Wulff, 1929
Hydrographie und Oberflächen plankton des westlichen Barentsmeeres im Sommer 1927. Ber. deutschen wissensch. Komm. F. Meeresforsch. n.s. 4(5):232-572, 13 tables, 25 text figs.

Ptychocylis orthoceros n.sp.

Entz, G., Jr. 1909
Studien über organisation und biologie der Tintinniden. Arch. f. Protistenkunde 15:93-226, Pls. 8-21, text figs.

Ptychocylis ostenfeldi n.sp.

Kofoid, C. A. and A. S. Campbell, 1929
A conspectus of the marine and freshwater Ciliata belonging to the suborder Tintinnoinea, with descriptions of new species principally from the Agassiz expedition to the eastern tropical Pacific, 1904-1905. Univ. Calif. Publ. Zool. 34:1-403, 697 text figs.

Ptychocylis repanda

Kofoid, C. A. and A. S. Campbell, 1929
A conspectus of the marine and freshwater Ciliata belonging to the suborder Tintinnoinea, with descriptions of new species principally from the Agassiz expedition to the eastern tropical Pacific, 1904-1905. Univ. Calif. Publ. Zool. 34:1-403, 697 text figs.

Ptychocylis reticulata

Laackmann, H., 1910
Die Tintinnodeen der Deutschen Südpolar Expedition 1901-1903. Deutsch. Südpolar Exped. 11, Zool. 3(4):341-496, pls. 33-51.

Ptychocylis undella

Laackmann, H., 1910
Die Tintinnodeen der Deutschen Südpolar Expedition 1901-1903. Deutschen Südpolar Expedition 11, Zool.3 (4): 341-496, pls. 33-51.

Ptychocylis urnula

Campbell, A. S., 1942
The oceanic Tintinnoina of the plankton gathered during the last cruise of the CARNEGIE. Sci. Res. Cruise VII of the Carnegie, 1928-1929, Biol. 2:163 pp., 128 figs.

Ptychocylis urnula

Gaarder, K.R., 1946
Tintinnoidea. Rep.Sci.Res."Michael Sars" N.Atlantic Deep-Sea Exped., 1910, 2(1): 37 pp., 23 text figs.

Ptychocylis urnula

Gran, H.H., and T. Braarud, 1935
A quantitative study of the phytoplankton in the Bay of Fundy and the Gulf of Maine (including observations on hydrography, chemistry, and turbidity). J. Biol. Bd., Canada, 1(5):279-467, 69 text figs.

Ptychocylis urnula

Jørgensen, E., 1905
B.Protistplankton and the diatoms in bottom samples. Hydrographical and biological investigations in Norwegian fjords. Bergens Mus. Skr. 7: 49-225.

Ptychocylis urnula

Iselin, C., 1930
A report on the coastal waters of Labrador based on explorations of the "Chance" during the summer of 1926. Proc. Am. Acad. Arts Sci., 66(1):1-37, 14 text figs.

Ptychocylis urnula

Jørgensen, E., 1900.
Ueber die Tintinnodeen der norwegischen Westküste. Bergens Mus. Aarb., 1899(2):48 pp., 3 pls.

Ptychocylis urnula

Kofoid, C. A. and A. S. Campbell, 1929
A conspectus of the marine and freshwater Ciliata belonging to the suborder Tintinnoinea, with descriptions of new species principally from the Agassiz expedition to the eastern tropical Pacific, 1904-1905. Univ. Calif. Publ. Zool. 34:1-403, 697 text figs.

Ptychocylis urnula

Schulz, B., and A. Wulff, 1929
Hydrographie und Oberflächen plankton des westlichen Barentsmeeres im Sommer 1927. Ber. deutschen wissensch. Komm. F. Meeresforsch. n.s. 4(5):232-572, 13 tables, 25 text figs.

Ptychocylus urnula

Marukawa, H., 1921
Plankton lists and some new species of copepods, from the northern waters of Japan. Bull. Inst. Ocean., No.384, 15 pp., 3 pls., 1 chart. Monaco

Ptychocylis urnula

Rampi, L., 1939
Primo contributo alla conoscenza dei tintinnoidi do Maro Ligure. Atti della Soc. Ital. di Sci. Nat. 78:67-81, 58 text figs.

Ptychocylis wailesi n.sp.

Kofoid, C. A. and A. S. Campbell, 1929
A conspectus of the marine and freshwater Ciliata belonging to the suborder Tintinnoinea, with descriptions of new species principally from the Agassiz expedition to the eastern tropical Pacific, 1904-1905. Univ. Calif. Publ. Zool. 34:1-403, 697 text figs.

Rhabdonella sp.

Duran, M., 1951
Contribucion al estudio de los tintinidos del plancton de los costas de Castellon. (Mediterraneo occidental). Publ. Inst. Biol. Aplic., Barcelone, 8: 101-120, 29 text figs.

Rhabdonella aberrans

Kofoid, C.A. and A.S. Campbell,1939
Reports on the scientific results of the expedition to the eastern tropical Pacific in charge of Alexander Agassiz, by the U.S. Fish Commission steamer "Albetross" from October 1904 to March 1905-----XXXVII. The Ciliata: The Tintinnoimea. Bull.Mus. Comp.Zool. 84: 473 pp., 36 pls.

Rhabdonella aberrans n.sp.

Kofoid, C. A. and A. S. Campbell, 1929
A conspectus of the marine and freshwater Ciliata belonging to the suborder Tintinnoinea, with descriptions of new species principally from the Agassiz expedition to the eastern tropical Pacific, 1904-1905. Univ. Calif. Publ. Zool. 34:1-403, 697 text figs.

Rhabdonella amor

Balech, Enrique, 1962
Tintinnoinea y Dinoflagellata del Pacifico segun material de las Expediciones NORPAC y DOWNWIND del Instituto Scripps de Oceanografia. Revista. Mus. Argentino Ciencias Nat. "Bernardino Rivadavia", Ciencias Zool., 7(1):1-253.

Rhabdonella amor

Campbell, A. S., 1942
The oceanic Tintinnoina of the plankton gathered during the last cruise of the CARNEGIE. Sci. Res. Cruise VII of the Carnegie, 1928-1929, Biol. 2:163 pp., 128 figs.

Rhabdonella armor

Gaarder, K.R., 1946
Tintinnoidea. Rep.Sci.Res."Michael Sars" N.Atlantic Deep-Sea Exped., 1910, 2(1): 37 pp., 23 text figs.

Rhabdonella amor

Jörgensen, E., 1924
Mediterranean Tintinnidae. Rept. Danish Oceanogr. Exped. 1908-10 to the Mediterranean and adjacent seas, Vol.II, Biol. J 3, 110 pp., 114 text figs.

Rhabdonella amor

Kofoid, C.A. and A.S. Campbell,1939
Reports on the scientific results of the expedition to the eastern tropical Pacific in charge of Alexander Agassiz, by the U.S. Fish Commission steamer "Albetross" from October 1904 to March 1905-----XXXVII. The Ciliata: The Tintinnoimea. Bull.Mus. Comp.Zool. 84: 473 pp., 36 pls.

Rhabdonella amor

Kofoid, C. A. and A. S. Campbell, 1929
A conspectus of the marine and freshwater Ciliata belonging to the suborder Tintinnoinea, with descriptions of new species principally from the Agassiz expedition to the eastern tropical Pacific, 1904-1905. Univ. Calif. Publ. Zool. 34:1-403, 697 text figs.

Rhabdonella amor

Laackmann, H., 1910
Die Tintinnodeen der Deutschen Südpolar Expedition 1901-1903. Deutschen Südpolar Expedition 11, Zool.3 (4): 341-496, pls. 33-51.

Rhabdonella amor

Marshall, S. M., 1934
The silicoflagellata and tintinnoinea. Great Barrier Reef Exped., 1928-29. Sci. Repts. 4(15):623-664, 43 text figs.

Rhabdonella amor

Rampi, L., 1948
Sur quelques Tintinnides (Infusoires louques) du Pacifique subtropical (Recoltes Alain Gerbault) Bull. l'Inst. Ocean., Monaco, No.938, 4 pp.

Rhabdonella amor simplex

Delegazione Italiana della Commissione Internazionale per l'Esplorazione Scientifica del Mediterraneo, 1941
Note sul plancton della Laguna veneta. [Memoria CCLXXXIX] , Arch. di Ocean. e Limn. Anno I, Fasc. I, 1941 XIX: 31-57 pp.

Rhabdonella anadyomene

Kofoid, C. A. and A. S. Campbell, 1929
A conspectus of the marine and freshwater Ciliata belonging to the suborder Tintinnoinea, with descriptions of new species principally from the Agassiz expedition to the eastern tropical Pacific, 1904-1905. Univ. Calif. Publ. Zool. 34:1-403, 697 text figs.

Rhabdonella apophysata

Jörgensen, E., 1924
Mediterranean Tintinnidae. Rept. Danish Oceanogr. Exped. 1908-10 to the Mediterranean and adjacent seas, Vol.II, Biol. J 3, 110 pp., 114 text figs.

Rhabdonella apophysata

Laackmann, H., 1910
Die Tintinnodeen der Deutschen Südpolar Expedition 1901-1903. Deutschen Südpolar Expedition 11, Zool.3 (4): 341-496, pls. 33-51.

Rhabdonella brandti

Campbell, A. S., 1942
The oceanic Tintinnoina of the plankton gathered during the last cruise of the CARNEGIE. Sci. Res. Cruise VII of the Carnegie, 1928-1929, Biol. 2:163 pp., 128 figs.

Rhabdonella brandti

Gaarder, K.R., 1946
Tintinnoidea. Rep.Sci.Res."Michael Sars" N.Atlantic Deep-Sea Exped., 1910, 2(1): 37 pp., 23 text figs.

Rhabdonella brandti nom. nov.

Kofoid, C. A. and A. S. Campbell, 1929
A conspectus of the marine and fresh-water Ciliata belonging to the suborder Tintinnoinea, with descriptions of new species principally from the Agassiz expedition to the eastern tropical Pacific, 1904-1905. Univ. Calif. Publ. Zool. 34:1-403, 697 text figs.

Rhabdonella brandti

Komarovsky, B., 1959
The Tintinnina of the Gulf of Eylath (Aqaba). Contrib. Knowl. Red Sea, Bull. Sea Fish. Res. Sta.Haifa, Israel 21 (14):1-40.

Rhabdonella brandti

Marshall, S. M., 1934
The silicoflagellata and tintinnoinea. Great Barrier Reef Exped., 1928-29. Sci. Repts. 4(15):623-664, 43 text figs.

Rhabdonella chavesi

Gaarder, K.R., 1946
Tintinnoidea. Rep.Sci.Res."Michael Sars" N.Atlantic Deep-Sea Exped., 1910, 2(1): 37 pp., 23 text figs.

Rhabdonella chavesi

Kofoid, C. A. and A. S. Campbell, 1929
A conspectus of the marine and fresh-water Ciliata belonging to the suborder Tintinnoinea, with descriptions of new species principally from the Agassiz expedition to the eastern tropical Pacific, 1904-1905. Univ. Calif. Publ. Zool. 34:1-403, 697 text figs.

Rhabdonella chilensis

Fofoid, C.A. and A.S. Campbell, 1939
Reports on the scientific results of the expedition to the eastern tropical Pacific in charge of Alexander Agassiz, by the U.S. Fish Commission steamer"Albatross" from October 1904 to March 1905-----XXXVII. The Ciliata: The Tintinnoimea. Bull.Mus. Comp.Zool. 84: 473 pp., 36 pls.

Rhabdonella chilensis n.sp.

Kofoid, C. A. and A. S. Campbell, 1929
A conspectus of the marine and fresh-water Ciliata belonging to the suborder Tintinnoinea, with descriptions of new species principally from the Agassiz expedition to the eastern tropical Pacific, 1904-1905. Univ. Calif. Publ. Zool. 34:1-403, 697 text figs.

Rhabdonella conica

Campbell, A. S., 1942
The oceanic Tintinnoina of the plankton gathered during the last cruise of the CARNEGIE. Sci. Res. Cruise VII of the Carnegie, 1928-1929, Biol. 2:163 pp., 128 figs.

Rhabdonella conica

Gaarder, K.R., 1946
Tintinnoidea. Rep.Sci.Res."Michael Sars" N.Atlantic Deep-Sea Exped., 1910, 2(1): 37 pp., 23 text figs.

Rhabdonella conica

Fofoid, C.A. and A.S. Campbell, 1939
Reports on the scientific results of the expedition to the eastern tropical Pacific in charge of Alexander Agassiz, by the U.S. Fish Commission steamer"Albatross" from October 1904 to March 1905-----XXXVII. The Ciliata: The Tintinnoimea. Bull.Mus. Comp.Zool. 84: 473 pp., 36 pls.

Rhabdonella conica n.sp.

Kofoid, C. A. and A. S. Campbell, 1929
A conspectus of the marine and fresh-water Ciliata belonging to the suborder Tintinnoinea, with descriptions of new species principally from the Agassiz expedition to the eastern tropical Pacific, 1904-1905. Univ. Calif. Publ. Zool. 34:1-403, 697 text figs.

Rhabdonella conica

Rampi, L., 1948
Sur quelques Tintinnides (Infusoires louques) du Pacifique subtropical (Recoltes Alain Gerbault) Bull. l'Inst. Ocean., Monaco, No.938, 4 pp.

Rhabdonella conica

Rampi, L., 1939
Primo contributo alla conoscenza dei tintinnoidi do Maro Ligure. Atti della Soc. Ital. di Sci. Nat. 78:67-81, 58 text figs.

Rhabdonella cornucopia

Balech, Enrique, 1962
Tintinnoinea y Dinoflagellata del Pacifico segun material de las Expediciones NORPAC y DOWNWIND del Instituto Scripps de Oceanografia. Revista, Mus. Argentino Ciencias Nat. "Bernardino Rivadavia", Ciencias Zool., 7(1):1-253.

Rhabdonella cornucopia

Campbell, A. S., 1942
The oceanic Tintinnoina of the plankton gathered during the last cruise of the CARNEGIE. Sci. Res. Cruise VII of the Carnegie, 1928-1929, Biol. 2:163 pp., 128 figs.

Rhabdonella cornucopia

Fofoid, C.A. and A.S. Campbell, 1939
Reports on the scientific results of the expedition to the eastern tropical Pacific in charge of Alexander Agassiz, by the U.S. Fish Commission steamer"Albatross" from October 1904 to March 1905-----XXXVII. The Ciliata: The Tintinnoimea. Bull.Mus. Comp.Zool. 84: 473 pp., 36 pls.

Rhabdonella cornucopia n.sp.

Kofoid, C. A. and A. S. Campbell, 1929
A conspectus of the marine and fresh-water Ciliata belonging to the suborder Tintinnoinea, with descriptions of new species principally from the Agassiz expedition to the eastern tropical Pacific, 1904-1905. Univ. Calif. Publ. Zool. 34:1-403, 697 text figs.

Rhabdonella cuspidata

Campbell, A. S., 1942
The oceanic Tintinnoina of the plankton gathered during the last cruise of the CARNEGIE. Sci. Res. Cruise VII of the Carnegie, 1928-1929, Biol. 2:163 pp., 128 figs.

Rhabdonella cuspidata

Fofoid, C.A. and A.S. Campbell, 1939
Reports on the scientific results of the expedition to the eastern tropical Pacific in charge of Alexander Agassiz, by the U.S. Fish Commission steamer"Albatross" from October 1904 to March 1905-----XXXVII. The Ciliata: The Tintinnoimea. Bull.Mus. Comp.Zool. 84: 473 pp., 36 pls.

Rhabdonella cuspidata

Kofoid, C. A. and A. S. Campbell, 1929
A conspectus of the marine and fresh-water Ciliata belonging to the suborder Tintinnoinea, with descriptions of new species principally from the Agassiz expedition to the eastern tropical Pacific, 1904-1905. Univ. Calif. Publ. Zool. 34:1-403, 697 text figs.

Rhabdonella elegans

Balech, Enrique, 1962
Tintinnoinea y Dinoflagellata del Pacifico segun material de las Expediciones NORPAC y DOWNWIND del Instituto Scripps de Oceanografia. Revista, Mus. Argentino Ciencias Nat. "Bernardino Rivadavia", Ciencias Zool., 7(1):1-253.

Rhabdonella elegans

Campbell, A. S., 1942
The oceanic Tintinnoina of the plankton gathered during the last cruise of the CARNEGIE. Sci. Res. Cruise VII of the Carnegie, 1928-1929, Biol. 2:163 pp., 128 figs.

Rhabdonella elegans

Gaarder, K.R., 1946
Tintinnoidea. Rep.Sci.Res."Michael Sars" N.Atlantic Deep-Sea Exped., 1910, 2(1): 37 pp., 23 text figs.

Rhabdonella elegans n.sp.

Jörgensen, E., 1924
Mediterranean Tintinnidae. Rept. Danish Oceanogr. Exped. 1908-10 to the Mediterranean and adjacent seas, Vol.II, Biol. J 3, 110 pp., 114 text figs.

Rhabdonella elegans

Fofoid, C.A. and A.S. Campbell, 1939
Reports on the scientific results of the expedition to the eastern tropical Pacific in charge of Alexander Agassiz, by the U.S. Fish Commission steamer"Albatross" from October 1904 to March 1905-----XXXVII. The Ciliata: The Tintinnoimea. Bull.Mus. Comp.Zool. 84: 473 pp., 36 pls.

Rhabdonella elegans

Kofoid, C. A. and A. S. Campbell, 1929
A conspectus of the marine and fresh-water Ciliata belonging to the suborder Tintinnoinea, with descriptions of new species principally from the Agassiz expedition to the eastern tropical Pacific, 1904-1905. Univ. Calif. Publ. Zool. 34:1-403, 697 text figs.

Rhabdonella exilis

Campbell, A. S., 1942
The oceanic Tintinnoina of the plankton gathered during the last cruise of the CARNEGIE. Sci. Res. Cruise VII of the Carnegie, 1928-1929, Biol. 2:163 pp., 128 figs.

Rhabdonella exilis

Fofoid, C.A. and A.S. Campbell, 1939
Reports on the scientific results of the expedition to the eastern tropical Pacific in charge of Alexander Agassiz, by the U.S. Fish Commission steamer"Albatross" from October 1904 to March 1905-----XXXVII. The Ciliata: The Tintinnoimea. Bull.Mus. Comp.Zool. 84: 473 pp., 36 pls.

Rhabdonella exilis n.sp.

Kofoid, C. A. and A. S. Campbell, 1929
A conspectus of the marine and fresh-water Ciliata belonging to the suborder Tintinnoinea, with descriptions of new species principally from the Agassiz expedition to the eastern tropical Pacific, 1904-1905. Univ. Calif. Publ. Zool. 34:1-403, 697 text figs.

Rhabdonella fenestrata n.sp.

Roxas, H.A., 1941.
Marine protozoa of the Philippines. Philippine J. Sci. 74:91-136, 17 pls., 2 textfigs

Rhabdonella hebe

Campbell, A. S., 1942
The oceanic Tintinnoina of the plankton gathered during the last cruise of the CARNEGIE. Sci. Res. Cruise VII of the Carnegie, 1928-1929, Biol. 2:163 pp., 128 figs.

Rhabdonella hebe

Gaarder, K.R., 1946
Tintinnoidea. Rep.Sci.Res."Michael Sars" N.Atlantic Deep-Sea Exped., 1910, 2(1): 37 pp., 23 text figs.

Rhabdonella hebe

Kofoid, C.A. and A.S. Campbell, 1939
Reports on the scientific results of the expedition to the eastern tropical Pacific in charge of Alexander Agassiz, by the U.S. Fish Commission steamer "Albatross" from October 1904 to March 1905-----XXXVII. The Ciliata: The Tintinnoimea. Bull. Mus. Comp. Zool. 84: 473 pp., 36 pls.

Rhabdonella hebe

Kofoid, C. A. and A. S. Campbell, 1929
A conspectus of the marine and fresh-water Ciliata belonging to the suborder Tintinnoinea, with descriptions of new species principally from the Agassiz expedition to the eastern tropical Pacific, 1904-1905. Univ. Calif. Publ. Zool. 34:1-403, 697 text figs.

Rhabdonella henseni

Balech, Enrique, 1962
Tintinnoinea y Dinoflagellata del Pacifico segun material de las Expediciones NORPAC y DOWNWIND del Instituto Scripps de Oceanografia. Revista, Mus. Argentino Ciencias Nat. "Bernardino Rivadavia", Ciencias Zool., 7(1):1-253.

Rhabdonella henseni

Campbell, A. S., 1942
The oceanic Tintinnoina of the plankton gathered during the last cruise of the CARNEGIE. Sci. Res. Cruise VII of the Carnegie, 1928-1929, Biol. 2:163 pp., 128 figs.

Rhabdonella henseni

Gaarder, K.R., 1946
Tintinnoidea. Rep.Sci.Res."Michael Sars" N.Atlantic Deep-Sea Exped., 1910, 2(1): 37 pp., 23 text figs.

Rhabdonella henseni

Kofoid, C.A. and A.S. Campbell, 1939
Reports on the scientific results of the expedition to the eastern tropical Pacific in charge of Alexander Agassiz, by the U.S. Fish Commission steamer "Albatross" from October 1904 to March 1905-----XXXVII. The Ciliata: The Tintinnoimea. Bull. Mus. Comp. Zool. 84: 473 pp., 36 pls.

Rhabdonella henseni

Kofoid, C. A. and A. S. Campbell, 1929
A conspectus of the marine and fresh-water Ciliata belonging to the suborder Tintinnoinea, with descriptions of new species principally from the Agassiz expedition to the eastern tropical Pacific, 1904-1905. Univ. Calif. Publ. Zool. 34:1-403, 697 text figs.

Rhabdonella hydria

Kofoid, C. A. and A. S. Campbell, 1929
A conspectus of the marine and fresh-water Ciliata belonging to the suborder Tintinnoinea, with descriptions of new species principally from the Agassiz expedition to the eastern tropical Pacific, 1904-1905. Univ. Calif. Publ. Zool. 34:1-403, 697 text figs.

Rhabdonella hydria

Rampi, L., 1939
Primo contributo alla conoscenza dei tintinnoidi do Maro Ligure. Atti della Soc. Ital. di Sci. Nat. 78:67-81, 58 text figs.

Rhabdonella hydria

Sousa e Silva, E. de, 1950.
Les tintinnides de la Baie de Cascais. Bull. Inst. Océan., Monaco, No. 979:28 pp., 4 pls.

Rhabdonella indica

Balech, Enrique, 1962
Tintinnoinea y Dinoflagellata del Pacifico segun material de las Expediciones NORPAC y DOWNWIND del Instituto Scripps de Oceanografia. Revista, Mus. Argentino Ciencias Nat. "Bernardino Rivadavia", Ciencias Zool., 7(1):1-253.

Rhabdonella indica

Campbell, A. S., 1942
The oceanic Tintinnoina of the plankton gathered during the last cruise of the CARNEGIE. Sci. Res. Cruise VII of the Carnegie, 1928-1929, Biol. 2:163 pp., 128 figs.

Rhabdonella indica

Kofoid, C.A. and A.S. Campbell, 1939
Reports on the scientific results of the expedition to the eastern tropical Pacific in charge of Alexander Agassiz, by the U.S. Fish Commission steamer "Albatross" from October 1904 to March 1905-----XXXVII. The Ciliata: The Tintinnoimea. Bull. Mus. Comp. Zool. 84: 473 pp., 36 pls.

Rhabdonella indica

Kofoid, C. A. and A. S. Campbell, 1929
A conspectus of the marine and fresh-water Ciliata belonging to the suborder Tintinnoinea, with descriptions of new species principally from the Agassiz expedition to the eastern tropical Pacific, 1904-1905. Univ. Calif. Publ. Zool. 34:1-403, 697 text figs.

Rhabdonella indica

Komarovsky, B., 1959
The Tintinnina of the Gulf of Eylath (Aqaba). Contrib. Knowl. Red Sea, Bull. Sea Fish. Res. Sta. Haifa, Israel 21 (14):1-40.

Rhabdonella inflata

Campbell, A. S., 1942
The oceanic Tintinnoina of the plankton gathered during the last cruise of the CARNEGIE. Sci. Res. Cruise VII of the Carnegie, 1928-1929, Biol. 2:163 pp., 128 figs.

Rhabdonella inflata

Kofoid, C.A. and A.S. Campbell, 1939
Reports on the scientific results of the expedition to the eastern tropical Pacific in charge of Alexander Agassiz, by the U.S. Fish Commission steamer "Albatross" from October 1904 to March 1905-----XXXVII. The Ciliata: The Tintinnoimea. Bull. Mus. Comp. Zool. 84: 473 pp., 36 pls.

Rhabdonella inflata n.sp.

Kofoid, C. A. and A. S. Campbell, 1929
A conspectus of the marine and fresh-water Ciliata belonging to the suborder Tintinnoinea, with descriptions of new species principally from the Agassiz expedition to the eastern tropical Pacific, 1904-1905. Univ. Calif. Publ. Zool. 34:1-403, 697 text figs.

Rhabdonella lohmanni

Balech, Enrique, 1962
Tintinnoinea y Dinoflagellata del Pacifico segun material de las Expediciones NORPAC y DOWNWIND del Instituto Scripps de Oceanografia. Revista, Mus. Argentino Ciencias Nat. "Bernardino Rivadavia", Ciencias Zool., 7(1):1-253.

Rhabdonella lohmanni

Campbell, A. S., 1942
The oceanic Tintinnoina of the plankton gathered during the last cruise of the CARNEGIE. Sci. Res. Cruise VII of the Carnegie, 1928-1929, Biol. 2:163 pp., 128 figs.

Rhabdonella lohmanni

Kofoid, C.A. and A.S. Campbell, 1939
Reports on the scientific results of the expedition to the eastern tropical Pacific in charge of Alexander Agassiz, by the U.S. Fish Commission steamer "Albatross" from October 1904 to March 1905-----XXXVII. The Ciliata: The Tintinnoimea. Bull. Mus. Comp. Zool. 84: 473 pp., 36 pls.

Rhabdonella lohmanni n.sp.

Kofoid, C. A. and A. S. Campbell, 1929
A conspectus of the marine and fresh-water Ciliata belonging to the suborder Tintinnoinea, with descriptions of new species principally from the Agassiz expedition to the eastern tropical Pacific, 1904-1905. Univ. Calif. Publ. Zool. 34:1-403, 697 text figs.

Rhabdonella poculum

Campbell, A. S., 1942
The oceanic Tintinnoina of the plankton gathered during the last cruise of the CARNEGIE. Sci. Res. Cruise VII of the Carnegie, 1928-1929, Biol. 2:163 pp., 128 figs.

Rhabdonella poculum

Kofoid, C. A. and A. S. Campbell, 1929
A conspectus of the marine and fresh-water Ciliata belonging to the suborder Tintinnoinea, with descriptions of new species principally from the Agassiz expedition to the eastern tropical Pacific, 1904-1905. Univ. Calif. Publ. Zool. 34:1-403, 697 text figs.

Rhabdonella poculum

Kofoid, C.A. and A.S. Campbell, 1939
Reports on the scientific results of the expedition to the eastern tropical Pacific in charge of Alexander Agassiz, by the U.S. Fish Commission steamer "Albatross" from October 1904 to March 1905-----XXXVII. The Ciliata: The Tintinnoimea. Bull. Mus. Comp. Zool. 84: 473 pp., 36 pls.

Rhabdonella poculum

Komarovsky, B., 1959
The Tintinnina of the Gulf of Eylath (Aqaba). Contrib. Knowl. Red Sea, Bull. Sea Fish. Res. Sta. Haifa, Israel 21 (14):1-40.

Rhabdonella poculum

Rampi, L., 1948
Sur quelques Tintinnides (Infusoires louques) du Pacifique subtropical (Recoltes Alain Gerbault) Bull. l'Inst. Ocean., Monaco, No.938, 4 pp.

Rhabdonella quantula

Campbell, A. S., 1942
The oceanic Tintinnoina of the plankton gathered during the last cruise of the CARNEGIE. Sci. Res. Cruise VII of the Carnegie, 1928-1929, Biol. 2:163 pp., 128 figs.

Rhabdonella quantula

Kofoid, C.A. and A.S. Campbell, 1939
Reports on the scientific results of the expedition to the eastern tropical Pacific in charge of Alexander Agassiz, by the U.S. Fish Commission steamer "Albatross" from October 1904 to March 1905-----XXXVII. The Ciliata: The Tintinnoimea. Bull. Mus. Comp. Zool. 84: 473 pp., 36 pls.

Rhabdonella quantula

Kofoid, C. A. and A. S. Campbell, 1929
A conspectus of the marine and fresh-water Ciliata belonging to the suborder Tintinnoinea, with descriptions of new species principally from the Agassiz expedition to the eastern tropical Pacific, 1904-1905. Univ. Calif. Publ. Zool. 34:1-403, 697 text figs.

Rhabdonella quantula

Marshall, S. M., 1934
The silicoflagellata and tintinnoinea. Great Barrier Reef Exped., 1928-29. Sci. Repts. 4(15):623-664, 43 text figs.

Rhabdonella spiralis

Balech, Enrique, 1962
Tintinnoinea y Dinoflagellata del Pacifico segun material de las Expediciones NORPAC y DOWNWIND del Instituto Scripps de Oceanografia. Revista, Mus. Argentino Ciencias Nat. "Bernardino Rivadavia", Ciencias Zool., 7(1):1-253.

Rhabdonella spiralis

Balech, E., 1959.
Tintinnoinea del Mediterraneo. Trav. Sta. Zool., Villefranche-sur-Mer, 18(17): Reprinted from: Trab. Inst. Español Oceanogr., Madrid, 19 Sept. 1959(28):88 pp., 22 pls.

Rhabdonella spiralis

Campbell, A. S., 1942
The oceanic Tintinnoina of the plankton gathered during the last cruise of the CARNEGIE. Sci. Res. Cruise VII of the Carnegie, 1928-1929, Biol. 2:163 pp., 128 figs.

Rhabdonella spiralis

Delegazione Italiana della Commissione Internazionale per l'Esplorazione Scientifica del Mediterraneo, 1941
Note sul plancton della Laguna veneta. Memoria CCLXXXI , Arch. di Ocean. e Limn. Anno I, Fasc. I, 1941 XIX: 31-57 pp.

Rhabdonella spiralis

Duran, M., 1951
Contribucion al estudio de los tintinidos del plancton de los costas de Castellon. (Mediterraneo occidental). Publ. Inst. Biol. Aplic., Barcelone, 8: 101-120, 29 text figs.

Rhabdonella spiralis

Entz, G., Jr., 1909
Studien über organisation und biologie der Tintinniden. Arch. f. Protistenkunde 15:93-226, Pls. 8-21, text figs.

Rhabdonella spiralis

Hofker, J., 1931
Studien uber Tintinnoidea. Arch. f. Protistenk 75(3):315-402, 89 text figs.

Rhabdonella spiralis

Jörgensen, E., 1924
Mediterranean Tintinnidae. Rept. Danish Oceanogr. Exped. 1908-10 to the Mediterranean and adjacent seas, Vol.II, Biol. J 3, 110 pp., 114 text figs.

Rhabdonella spiralis

Kofoid, C.A. and A.S. Campbell, 1939
Reports on the scientific results of the expedition to the eastern tropical Pacific in charge of Alexander Agassiz, by the U.S. Fish Commission steamer "Albatross" from October 1904 to March 1905-----XXXVII. The Ciliata: The Tintinnoinea. Bull.Mus. Comp.Zool. 84: 473 pp., 36 pls.

Rhabdonella spiralis

Kofoid, C. A. and A. S. Campbell, 1929
A conspectus of the marine and freshwater Ciliata belonging to the suborder Tintinnoinea, with descriptions of new species principally from the Agassiz expedition to the eastern tropical Pacific, 1904-1905. Univ. Calif. Publ. Zool. 34:1-403, 697 text figs.

Rhabdonella spiralis

Laackmann, H., 1910
Die Tintinnodeen der Deutschen Südpolar Expedition 1901-1903. Deutschen Südpolar Expedition 11, Zool.3 (4): 341-496, pls. 33-51.

Rhabdonella spiralis

Marshall, S. M., 1934
The silicoflagellata and tintinnoinea. Great Barrier Reef Exped., 1928-29. Sci. Repts. 4(15):623-664, 43 text figs.

Rhabdonella spiralis

Rampi, L., 1939
Primo contributo alla conoscenza dei tintinnoidi do Maro Ligure. Atti della Soc. Ital. di Sci. Nat. 78:67-81, 58 text figs.

Rhabdonella spiralis

Sousa e Silva, E. de, 1950.
Les tintinnides de la Baie de Cascais. Bull. Inst. Océan., Monaco, No. 979:28 pp., 4 pls.

Rhabdonella striata

Campbell, A. S., 1942
The oceanic Tintinnoina of the plankton gathered during the last cruise of the CARNEGIE. Sci. Res. Cruise VII of the Carnegie, 1928-1929, Biol. 2:163 pp., 128 figs.

Rhabdonella striata

Kofoid, C.A. and A.S. Campbell, 1939
Reports on the scientific results of the expedition to the eastern tropical Pacific in charge of Alexander Agassiz, by the U.S. Fish Commission steamer "Albatross" from October 1904 to March 1905-----XXXVII. The Ciliata: The Tintinnoimea. Bull.Mus. Comp.Zool. 84: 473 pp., 36 pls.

Rhabdonella striata

Kofoid, C. A. and A. S. Campbell, 1929
A conspectus of the marine and freshwater Ciliata belonging to the suborder Tintinnoinea, with descriptions of new species principally from the Agassiz expedition to the eastern tropical Pacific, 1904-1905. Univ. Calif. Publ. Zool. 34:1-403, 697 text figs.

Rhabdonella striata

Rampi, L., 1948
Sur quelques Tintinnides (Infusoires louques) du Pacifique subtropical (Recoltes Alain Gerbault) Bull. l'Inst. Ocean., Monaco, No.938, 4 pp.

Rhabdonella torta

Campbell, A. S., 1942
The oceanic Tintinnoina of the plankton gathered during the last cruise of the CARNEGIE. Sci. Res. Cruise VII of the Carnegie, 1928-1929, Biol. 2:163 pp., 128 figs.

Rhabdonella torta

Kofoid, C.A. and A.S. Campbell, 1939
Reports on the scientific results of the expedition to the eastern tropical Pacific in charge of Alexander Agassiz, by the U.S. Fish Commission steamer "Albatross" from October 1904 to March 1905-----XXXVII. The Ciliata: The Tintinnoimea. Bull.Mus. Comp.Zool. 84: 473 pp., 36 pls.

Rhabdonella torta n.sp.

Kofoid, C. A. and A. S. Campbell, 1929
A conspectus of the marine and freshwater Ciliata belonging to the suborder Tintinnoinea, with descriptions of new species principally from the Agassiz expedition to the eastern tropical Pacific, 1904-1905. Univ. Calif. Publ. Zool. 34:1-403, 697 text figs.

Rhabdonella valdestriata

Campbell, A. S., 1942
The oceanic Tintinnoina of the plankton gathered during the last cruise of the CARNEGIE. Sci. Res. Cruise VII of the Carnegie, 1928-1929, Biol. 2:163 pp., 128 figs.

Rhabdonella valdestriata

Kofoid, C.A. and A.S. Campbell, 1939
Reports on the scientific results of the expedition to the eastern tropical Pacific in charge of Alexander Agassiz, by the U.S. Fish Commission steamer "Albatross" from October 1904 to March 1905-----XXXVII. The Ciliata: The Tintinnoimea. Bull.Mus. Comp.Zool. 84: 473 pp., 36 pls.

Rhabdonella valdestriata

Kofoid, C. A. and A. S. Campbell, 1929
A conspectus of the marine and freshwater Ciliata belonging to the suborder Tintinnoinea, with descriptions of new species principally from the Agassiz expedition to the eastern tropical Pacific, 1904-1905. Univ. Calif. Publ. Zool. 34:1-403, 697 text figs.

Rhabdonella valdestriata

Komarovsky, B., 1959
The Tintinnina of the Gulf of Eylath (Aqaba). Contrib. Knowl. Red Sea, Bull. Sea Fish. Res. Sta.Haifa, Israel 21 (14):1-40.

Rhabdonellopsis n. gen.

Kofoid, C. A. and A. S. Campbell, 1929
A conspectus of the marine and freshwater Ciliata belonging to the suborder Tintinnoinea, with descriptions of new species principally from the Agassiz expedition to the eastern tropical Pacific, 1904-1905. Univ. Calif. Publ. Zool. 34:1-403, 697 text figs.

Rhabdonellopsis apophysata

Campbell, A. S., 1942
The oceanic Tintinnoina of the plankton gathered during the last cruise of the CARNEGIE. Sci. Res. Cruise VII of the Carnegie, 1928-1929, Biol. 2:163 pp., 128 figs.

Rhabdonellopsis apophysata

Balech, Enrique, 1962
Tintinnoinea y Dinoflagellata del Pacifico segun material de las Expediciones NORPAC y DOWNWIND del Instituto Scripps de Oceanografia. Revista, Mus. Argentino Ciencias Nat. "Bernardino Rivadavia", Ciencias Zool., 7(1):1-253.

Rhabdonellopsis apophysata

Gaarder, K.R., 1946
Tintinnoidea. Rep.Sci.Res. "Michael Sars" N.Atlantic Deep-Sea Exped., 1910, 2(1): 37 pp., 23 text figs.

Rhabdonellopsis apophysata

Kofoid, C. A. and A. S. Campbell, 1929
A conspectus of the marine and freshwater Ciliata belonging to the suborder Tintinnoinea, with descriptions of new species principally from the Agassiz expedition to the eastern tropical Pacific, 1904-1905. Univ. Calif. Publ. Zool. 34:1-403, 697 text figs.

Rhabdonellopsis composita

Campbell, A. S., 1942
The oceanic Tintinnoina of the plankton gathered during the last cruise of the CARNEGIE. Sci. Res. Cruise VII of the Carnegie, 1928-1929, Biol. 2:163 pp., 128 figs.

Rhabdonellopsis composita

Kofoid, C. A. and A. S. Campbell, 1929
A conspectus of the marine and freshwater Ciliata belonging to the suborder Tintinnoinea, with descriptions of new species principally from the Agassiz expedition to the eastern tropical Pacific, 1904-1905. Univ. Calif. Publ. Zool. 34:1-403, 697 text figs.

Rhodonellopsis constricta n.sp.

Kofoid, C.A. and A.S. Campbell, 1939
Reports on the scientific results of the expedition to the eastern tropical Pacific in charge of Alexander Agassiz, by the U.S. Fish Commission steamer "Albatross" from October 1904 to March 1905-----XXXVII. The Ciliata: The Tintinnoimea. Bull.Mus. Comp.Zool. 84: 473 pp., 36 pls.

Rhabdonellopsis intermedia

Campbell, A. S., 1942
The oceanic Tintinnoina of the plankton gathered during the last cruise of the CARNEGIE. Sci. Res. Cruise VII of the Carnegie, 1928-1929, Biol. 2:163 pp., 128 figs.

Rhodonellopsis intermedia

Kofoid, C.A. and A.S. Campbell, 1939
Reports on the scientific results of the expedition to the eastern tropical Pacific in charge of Alexander Agassiz, by the U.S. Fish Commission steamer "Albatross" from October 1904 to March 1905-----XXXVII. The Ciliata: The Tintinnoimea. Bull.Mus. Comp.Zool. 84: 473 pp., 36 pls.

Rhabdonellopsis intermedia n.sp.

Kofoid, C. A. and A. S. Campbell, 1929
A conspectus of the marine and freshwater Ciliata belonging to the suborder Tintinnoinea, with descriptions of new species principally from the Agassiz expedition to the eastern tropical Pacific, 1904-1905. Univ. Calif. Publ. Zool. 34:1-403, 697 text figs.

Rhapdonellopsis intermedia

Marshall, S. M., 1934
The silicoflagellata and tintinnoinea. Great Barrier Reef Exped., 1928-29. Sci. Repts. 4(15):623-664, 43 text figs.

Rhabdonellopsis longicaulis

Campbell, A. S., 1942
The oceanic Tintinnoina of the plankton gathered during the last cruise of the CARNEGIE. Sci. Res. Cruise VII of the Carnegie, 1928-1929, Biol. 2:163 pp., 128 figs.

Rhabdonellopsis longicaulis

Gaarder, K.R., 1946
Tintinnoidea. Rep.Sci.Res."Michael Sars" N.Atlantic Deep-Sea Exped., 1910, 2(1): 37 pp., 23 text figs.

Rhodonellopsis longicaulis

Kofoid, C.A. and A.S. Campbell,1939
Reports on the scientific results of the expedition to the eastern tropical Pacific in charge of Alexander Agassiz, by the U.S. Fish Commission steamer"Albatross" from October 1904 to March 1905-----XXXVII. The Ciliata: The Tintinnoimea. Bull.Mus. Comp.Zool. 84: 473 pp., 36 pls.

Rhabdonellopsis longicaulis n.sp.

Kofoid, C. A. and A. S. Campbell, 1929
A conspectus of the marine and freshwater Ciliata belonging to the suborder Tintinnoinea, with descriptions of new species principally from the Agassiz expedition to the eastern tropical Pacific, 1904-1905. Univ. Calif. Publ. Zool. 34:1-403, 697 text figs.

Rhabdonellopsis longicaulis

Rampi, L., 1948
Sur quelques Tintinnides (Infusoires louques) du Pacifique subtropical (Recoltes Alain Gerbault) Bull. l'Inst. Ocean., Monaco, No.938, 4 pp.

Rhabdonellopsis minima

Balech, Enrique, 1962
Tintinnoinea y Dinoflagellata del Pacifico segun material de las Expediciones NORPAC y DOWNWIND del Instituto Scripps de Oceanografia.
Revista, Mus. Argentino Ciencias Nat. "Bernardino Rivadavia", Ciencias Zool., 7(1):1-253.

Rhabdonellopsis minima

Campbell, A. S., 1942
The oceanic Tintinnoina of the plankton gathered during the last cruise of the CARNEGIE. Sci. Res. Cruise VII of the Carnegie, 1928-1929, Biol. 2:163 pp., 128 figs.

Rhodonellopsis minima

Kofoid, C.A. and A.S. Campbell,1939
Reports on the scientific results of the expedition to the eastern tropical Pacific in charge of Alexander Agassiz, by the U.S. Fish Commission steamer"Albatross" from October 1904 to March 1905-----XXXVII. The Ciliata: The Tintinnoimea. Bull.Mus. Comp.Zool. 84: 473 pp., 36 pls.

Rhabdonellopsis minima n.sp.

Kofoid, C. A. and A. S. Campbell, 1929
A conspectus of the marine and freshwater Ciliata belonging to the suborder Tintinnoinea, with descriptions of new species principally from the Agassiz expedition to the eastern tropical Pacific, 1904-1905. Univ. Calif. Publ. Zool. 34:1-403, 697 text figs.

Rhabdonellopsis triton

Balech, Enrique, 1962
Tintinnoinea y Dinoflagellata del Pacifico segun material de las Expediciones NORPAC y DOWNWIND del Instituto Scripps de Oceanografia.
Revista, Mus. Argentino Ciencias Nat. "Bernardino Rivadavia", Ciencias Zool., 7(1):1-253.

Rhabdonellopsis triton

Campbell, A. S., 1942
The oceanic Tintinnoina of the plankton gathered during the last cruise of the CARNEGIE. Sci. Res. Cruise VII of the Carnegie, 1928-1929, Biol. 2:163 pp., 128 figs.

Rhodonellopsis triton

Kofoid, C.A. and A.S. Campbell,1939
Reports on the scientific results of the expedition to the eastern tropical Pacific in charge of Alexander Agassiz, by the U.S. Fish Commission steamer"Albatross" from October 1904 to March 1905-----XXXVII. The Ciliata: The Tintinnoimea. Bull.Mus. Comp.Zool. 84: 473 pp., 36 pls.

Rhabdonellopsis triton

Kofoid, C. A. and A. S. Campbell, 1929
A conspectus of the marine and freshwater Ciliata belonging to the suborder Tintinnoinea, with descriptions of new species principally from the Agassiz expedition to the eastern tropical Pacific, 1904-1905. Univ. Calif. Publ. Zool. 34:1-403, 697 text figs.

Rhabdonellopsis triton

Komarovsky, B., 1962
Tintinnina from the vicinity of the Straits of Tiran and Massawa Region. Contributions to the knowledge of the Red Sea, No. 25. Sea Fish. Res. Sta., Haifa, Israel, Bull. No. 30:48-56.

Rhabdosella cuneolata

Balech, Enrique, 1968.
Algunas especies nuevas o interesantes do tintinnidos del Golfo de Mexico y Caribe.
Revta Mus. argent. Cienc. nat. Bernardino Rivadavia Inst. nac. Hidrobiol. 2(5):165-197.

Rhabdosella cuneolata

Kofoid, C.A. and A.S. Campbell,1939
Reports on the scientific results of the expedition to the eastern tropical Pacific in charge of Alexander Agassiz, by the U.S. Fish Commission steamer"Albatross" from October 1904 to March 1905-----XXXVII. The Ciliata: The Tintinnoimea. Bull.Mus. Comp.Zool. 84: 473 pp., 36 pls.

Rhabdosella octogenata

Kofoid, C.A. and A.S. Campbell,1939
Reports on the scientific results of the expedition to the eastern tropical Pacific in charge of Alexander Agassiz, by the U.S. Fish Commission steamer"Albatross" from October 1904 to March 1905-----XXXVII. The Ciliata: The Tintinnoimea. Bull.Mus. Comp.Zool. 84: 473 pp., 36 pls.

Salpingacantha n. gen.

Kofoid, C. A. and A. S. Campbell, 1929
A conspectus of the marine and freshwater Ciliata belonging to the suborder Tintinnoinea, with descriptions of new species principally from the Agassiz expedition to the eastern tropical Pacific, 1904-1905. Univ. Calif. Publ. Zool. 34:1-403, 697 text figs.

Salpingacantha

Travers, Anne, et Marc Travers 1973.
Présence en Méditerranée du genre Salpingacantha Kofoid et Campbell (Ciliés Oligotriches Tintinnides. Rapp. Proc.-V. Réun. Commn int. Explor. scient. Mer Medit. Monaco 21(8):429-432.

Salpingacantha ampla

Gaarder, K.R., 1946
Tintinnoidea. Rep.Sci.Res."Michael Sars" N.Atlantic Deep-Sea Exped., 1910, 2(1): 37 pp., 23 text figs.

Salpingacantha ampla

Kofoid, C.A. and A.S. Campbell,1939
Reports on the scientific results of the expedition to the eastern tropical Pacific in charge of Alexander Agassiz, by the U.S. Fish Commission steamer"Albatross" from October 1904 to March 1905-----XXXVII. The Ciliata: The Tintinnoimea. Bull.Mus. Comp.Zool. 84: 473 pp., 36 pls.

Salpingacantha ampla n.sp.

Kofoid, C. A. and A. S. Campbell, 1929
A conspectus of the marine and freshwater Ciliata belonging to the suborder Tintinnoinea, with descriptions of new species principally from the Agassiz expedition to the eastern tropical Pacific, 1904-1905. Univ. Calif. Publ. Zool. 34:1-403, 697 text figs.

Salpingacantha crenulata

Campbell, A. S., 1942
The oceanic Tintinnoina of the plankton gathered during the last cruise of the CARNEGIE. Sci. Res. Cruise VII of the Carnegie, 1928-1929, Biol. 2:163 pp., 128 figs.

Salpingacantha crenulata

Kofoid, C.A. and A.S. Campbell,1939
Reports on the scientific results of the expedition to the eastern tropical Pacific in charge of Alexander Agassiz, by the U.S. Fish Commission steamer"Albatross" from October 1904 to March 1905-----XXXVII. The Ciliata: The Tintinnoimea. Bull.Mus. Comp.Zool. 84: 473 pp., 36 pls.

Salpingacantha crenulata

Kofoid, C. A. and A. S. Campbell, 1929
A conspectus of the marine and freshwater Ciliata belonging to the suborder Tintinnoinea, with descriptions of new species principally from the Agassiz expedition to the eastern tropical Pacific, 1904-1905. Univ. Calif. Publ. Zool. 34:1-403, 697 text figs.

Salpingacantha exilis

Kofoid, C.A. and A.S. Campbell,1939
Reports on the scientific results of the expedition to the eastern tropical Pacific in charge of Alexander Agassiz, by the U.S. Fish Commission steamer"Albatross" from October 1904 to March 1905-----XXXVII. The Ciliata: The Tintinnoimea. Bull.Mus. Comp.Zool. 84: 473 pp., 36 pls.

Salpingacantha exilis n.sp.

Kofoid, C. A. and A. S. Campbell, 1929
A conspectus of the marine and freshwater Ciliata belonging to the suborder Tintinnoinea, with descriptions of new species principally from the Agassiz expedition to the eastern tropical Pacific, 1904-1905. Univ. Calif. Publ. Zool. 34:1-403, 697 text figs.

Salpingacantha perca

Campbell, A. S., 1942
The oceanic Tintinnoina of the plankton gathered during the last cruise of the CARNEGIE. Sci. Res. Cruise VII of the Carnegie, 1928-1929, Biol. 2:163 pp., 128 figs.

Salpingacantha perca

Kofoid, C.A. and A.S. Campbell,1939
Reports on the scientific results of the expedition to the eastern tropical Pacific in charge of Alexander Agassiz, by the U.S. Fish Commission steamer"Albatross" from October 1904 to March 1905-----XXXVII. The Ciliata: The Tintinnoimea. Bull.Mus. Comp.Zool. 84: 473 pp., 36 pls.

Salpingacantha perca n.sp.

Kofoid, C. A. and A. S. Campbell, 1929
A conspectus of the marine and freshwater Ciliata belonging to the suborder Tintinnoinea, with descriptions of new species principally from the Agassiz expedition to the eastern tropical Pacific, 1904-1905. Univ. Calif. Publ. Zool. 34:1-403, 697 text figs.

Salpingacantha simplex nom. nov.

Kofoid, C. A. and A. S. Campbell, 1929
A conspectus of the marine and freshwater Ciliata belonging to the suborder Tintinnoinea, with descriptions of new species principally from the Agassiz expedition to the eastern tropical Pacific, 1904-1905. Univ. Calif. Publ. Zool. 34:1-403, 697 text figs.

Salpingacantha undata

Campbell, A. S., 1942
The oceanic Tintinnoina of the plankton gathered during the last cruise of the CARNEGIE. Sci. Res. Cruise VII of the Carnegie, 1928-1929, Biol. 2:163 pp., 128 figs.

Salpingacantha undata

Gaarder, K.R., 1946
Tintinnoidea. Rep.Sci.Res."Michael Sars" N.Atlantic Deep-Sea Exped., 1910, 2(1): 37 pp., 23 text figs.

Salpingacantha undata

Kofoid, C.A. and A.S. Campbell, 1939
Reports on the scientific results of the expedition to the eastern tropical Pacific in charge of Alexander Agassiz, by the U.S. Fish Commission steamer "Albatross" from October 1904 to March 1905-----XXXVII. The Ciliata: The Tintinnoinea. Bull. Mus. Comp. Zool. 84: 473 pp., 36 pls.

Salpingacantha undata

Kofoid, C. A. and A. S. Campbell, 1929
A conspectus of the marine and fresh-water Ciliata belonging to the suborder Tintinnoinea, with descriptions of new species principally from the Agassiz expedition to the eastern tropical Pacific, 1904-1905. Univ. Calif. Publ. Zool. 34:1-403, 697 text figs.

Salpingacantha unguiculata

Kofoid, C.A. and A.S. Campbell, 1939
Reports on the scientific results of the expedition to the eastern tropical Pacific in charge of Alexander Agassiz, by the U.S. Fish Commission steamer "Albatross" from October 1904 to March 1905-----XXXVII. The Ciliata: The Tintinnoinea. Bull. Mus. Comp. Zool. 84: 473 pp., 36 pls.

Salpingacantha unguiculata

Kofoid, C. A. and A. S. Campbell, 1929
A conspectus of the marine and fresh-water Ciliata belonging to the suborder Tintinnoinea, with descriptions of new species principally from the Agassiz expedition to the eastern tropical Pacific, 1904-1905. Univ. Calif. Publ. Zool. 34:1-403, 697 text figs.

Salpingella sp.

Balech, Enrique, 1962
Tintinnoinea y Dinoflagellata del Pacifico segun material de las Expediciones NORPAC y DOWNWIND del Instituto Scripps de Oceanografia. Revista, Mus. Argentino Ciencias Nat. "Bernardino Rivadavia", Ciencias Zool., 7(1):1-253.

Salpingella n.gen.

Jörgensen, E., 1924
Mediterranean Tintinnidae. Rept. Danish Oceanogr. Exped. 1908-10 to the Mediterranean and adjacent seas, Vol.II, Biol. J 3, 110 pp., 114 text figs.

Salpingella sp.

Sousa e Silva, E. de, 1950.
Les tintinnides de la Baie de Cascais. Bull. Inst. Océan., Monaco, No. 979:28 pp., 4 pls.

Salpingella acuminata

Balech, E., 1959.
Tintinnoinea del Mediterraneo. Trav. Sta. Zool., Villefranche-sur-Mer, 18(17):
Reprinted from:
Trab. Inst. Español Oceanogr., Madrid, 19 Sept., 1959(28):88 pp., 22 pls.

Salpingella acuminata

Braarud, T., and Adam Bursa, 1939
On the phytoplankton of the Oslo Fjord, 1933-1934. Hvalrådets Skr. No.19:1-63; 9 text figs. Reviewed. J. du. Cons. 14(3): 418-420. A.C. Gardiner.

Salpingella acuminata

Campbell, A. S., 1942
The oceanic Tintinnoina of the plankton gathered during the last cruise of the CARNEGIE. Sci. Res. Cruise VII of the Carnegie, 1928-1929, Biol. 2:163 pp., 128 figs.

Salpingella acuminata

Duran, M., 1951
Contribucion al estudio de los tintinidos del plancton de los costas de Castellon. (Mediterranes occidental). Publ. Inst. Biol. Aplic., Barcelone, 8: 101-120, 29 text figs.

Salpingella acuminata

Gaarder, K.R., 1946
Tintinnoidea. Rep.Sci.Res."Michael Sars" N.Atlantic Deep-Sea Exped., 1910, 2(1): 37 pp., 23 text figs.

Salpingella acuminata

Gran, H.H., and T. Braarud, 1935
A quantitative study of the phytoplankton in the Bay of Fundy and the Gulf of Maine (including observations on hydrography, chemistry, and turbidity). J. Biol. Bd. Canada, 1(5):279-467, 69 text figs.

Salpingella acuminata

Hofker, J., 1931
Studien uber Tintinnoidea. Arch. f. Protistenk 75(3):315-402, 89 text figs.

Salpingella acuminata

Jörgensen, E., 1924
Mediterranean Tintinnidae. Rept. Danish Oceanogr. Exped. 1908-10 to the Mediterranean and adjacent seas, Vol.II, Biol. J 3, 110 pp., 114 text figs.

Salpingella acuminata

Kofoid, C.A. and A.S. Campbell, 1939
Reports on the scientific results of the expedition to the eastern tropical Pacific in charge of Alexander Agassiz, by the U.S. Fish Commission steamer "Albatross" from October 1904 to March 1905-----XXXVII. The Ciliata: The Tintinnoinea. Bull. Mus. Comp. Zool. 84: 473 pp., 36 pls.

Salpingella acuminata

Kofoid, C. A. and A. S. Campbell, 1929
A conspectus of the marine and fresh-water Ciliata belonging to the suborder Tintinnoinea, with descriptions of new species principally from the Agassiz expedition to the eastern tropical Pacific, 1904-1905. Univ. Calif. Publ. Zool. 34:1-403, 697 text figs.

Salpingella acuminata

Komarovsky, B., 1959
The Tintinnina of the Gulf of Eylath (Aqaba). Contrib. Knowl. Red Sea, Bull. Sea Fish. Res. Sta.Haifa, Israel 21 (14):1-40.

Salpingella acuminata

Sousa e Silva, E. de, 1950.
Les tintinnides de la Baie de Cascais. Bull. Inst. Océan., Monaco, No. 979:28 pp., 4 pls.

Salpingella acuminatoides

Kofoid, C. A. and A. S. Campbell, 1929
A conspectus of the marine and fresh-water Ciliata belonging to the suborder Tintinnoinea, with descriptions of new species principally from the Agassiz expedition to the eastern tropical Pacific, 1904-1905. Univ. Calif. Publ. Zool. 34:1-403, 697 text figs.

Salpingella alata n.sp.

Kofoid, C. A. and A. S. Campbell, 1929
A conspectus of the marine and fresh-water Ciliata belonging to the suborder Tintinnoinea, with descriptions of new species principally from the Agassiz expedition to the eastern tropical Pacific, 1904-1905. Univ. Calif. Publ. Zool. 34:1-403, 697 text figs.

Salpingella altiplicata

Kofoid, C. A. and A. S. Campbell, 1929
A conspectus of the marine and fresh-water Ciliata belonging to the suborder Tintinnoinea, with descriptions of new species principally from the Agassiz expedition to the eastern tropical Pacific, 1904-1905. Univ. Calif. Publ. Zool. 34:1-403, 697 text figs.

Salpingella attenuata

Campbell, A. S., 1942
The oceanic Tintinnoina of the plankton gathered during the last cruise of the CARNEGIE. Sci. Res. Cruise VII of the Carnegie, 1928-1929, Biol. 2:163 pp., 128 figs.

Salpingella attenuata

Kofoid, C.A. and A.S. Campbell, 1939
Reports on the scientific results of the expedition to the eastern tropical Pacific in charge of Alexander Agassiz, by the U.S. Fish Commission steamer "Albatross" from October 1904 to March 1905-----XXXVII. The Ciliata: The Tintinnoinea. Bull. Mus. Comp. Zool. 84: 473 pp., 36 pls.

Salpingella attenuata

Kofoid, C. A. and A. S. Campbell, 1929
A conspectus of the marine and fresh-water Ciliata belonging to the suborder Tintinnoinea, with descriptions of new species principally from the Agassiz expedition to the eastern tropical Pacific, 1904-1905. Univ. Calif. Publ. Zool. 34:1-403, 697 text figs.

Salpingella attenuata

Komarovsky, B., 1959
The Tintinnina of the Gulf of Eylath (Aqaba). Contrib. Knowl. Red Sea, Bull. Sea Fish. Res. Sta.Haifa, Israel 21 (14):1-40.

Salpingella attenuata

Rampi, L., 1948
Sur quelques Tintinnides (Infusoires louques) du Pacifique subtropical (Recoltes Alain Gerbault) Bull. l'Inst. Ocean., Monaco, No.938, 4 pp.

Salpingella costata

Kofoid, C. A. and A. S. Campbell, 1929
A conspectus of the marine and fresh-water Ciliata belonging to the suborder Tintinnoinea, with descriptions of new species principally from the Agassiz expedition to the eastern tropical Pacific, 1904-1905. Univ. Calif. Publ. Zool. 34:1-403, 697 text figs.

Salpingella cuneolata n.sp.

Kofoid, C. A. and A. S. Campbell, 1929
A conspectus of the marine and fresh-water Ciliata belonging to the suborder Tintinnoinea, with descriptions of new species principally from the Agassiz expedition to the eastern tropical Pacific, 1904-1905. Univ. Calif. Publ. Zool. 34:1-403, 697 text figs.

Salpingella curta

Balech, E., 1959.
Tintinnoinea del Mediterraneo. Trav. Sta. Zool., Villefranche-sur-Mer, 18(17):
Reprinted from:
Trab. Inst. Español Oceanogr., Madrid, 19 Sept., 1959(28):88 pp., 22 pls.

Salpingella curta

Campbell, A. S., 1942
The oceanic Tintinnoina of the plankton gathered during the last cruise of the CARNEGIE. Sci. Res. Cruise VII of the Carnegie, 1928-1929, Biol. 2:163 pp., 128 figs.

Salpingella curta

Gaarder, K.R., 1946
Tintinnoidea. Rep.Sci.Res."Michael Sars" N.Atlantic Deep-Sea Exped., 1910, 2(1): 37 pp., 23 text figs.

Salpingella curta

Kofoid, C.A. and A.S. Campbell, 1939
Reports on the scientific results of the expedition to the eastern tropical Pacific in charge of Alexander Agassiz, by the U.S. Fish Commission steamer "Albatross" from October 1904 to March 1905-----XXXVII. The Ciliata: The Tintinnoinea. Bull. Mus. Comp. Zool. 84: 473 pp., 36 pls.

Salpingella curta n.sp.

Kofoid, C. A. and A. S. Campbell, 1929
A conspectus of the marine and fresh-water Ciliata belonging to the suborder Tintinnoinea, with descriptions of new species principally from the Agassiz expedition to the eastern tropical Pacific, 1904-1905. Univ. Calif. Publ. Zool. 34:1-403, 697 text figs.

Oceanographic Index: Marine Organisms Cumulation, 1946-1973

Salpingella decurtata

Balech, E., 1959.
Tintinnoinea del Mediterraneo.
Trav. Sta. Zool., Villefranche-sur-Mer, 18(17):
Reprinted from:
Trab. Inst. Español Oceanogr., Madrid, 19 Sept. 1959(28):88 pp., 22 pls.

Salpingella decurtata

Campbell, A. S., 1942
The oceanic Tintinnoina of the plankton gathered during the last cruise of the CARNEGIE. Sci. Res. Cruise VII of the Carnegie, 1928-1929, Biol. 2:163 pp., 128 figs.

Salpingella decurtata

Duran, M., 1951
Contribucion al estudio de los tintinidos del plancton de los costas de Castellon. (Mediterraneo occidental). Publ. Inst. Biol. Aplic., Barcelone, 8: 101-120, 29 text figs.

Salpingella decurtata

Gaarder, K.R., 1946
Tintinnoidea. Rep.Sci.Res."Michael Sars" N.Atlantic Deep-Sea Exped., 1910, 2(1): 37 pp., 23 text figs.

Salpingella decurtata n.sp.

Jörgensen, E., 1924
Mediterranean Tintinnidae. Rept. Danish Oceanogr. Exped. 1908-10 to the Mediterranean and adjacent seas, Vol.II, Biol. J 3, 110 pp., 114 text figs.

Salpingella decurtata

Kofoid, C. A. and A. S. Campbell, 1929
A conspectus of the marine and freshwater Ciliata belonging to the suborder Tintinnoinea, with descriptions of new species principally from the Agassiz expedition to the eastern tropical Pacific, 1904-1905. Univ. Calif. Publ. Zool. 34:1-403, 697 text figs.

Salpingella decurtata

Komarovsky, B., 1959
The Tintinnina of the Gulf of Eylath (Aqaba). Contrib. Knowl. Red Sea, Bull. Sea Fish. Res. Sta.Haifa, Israel 21 (14):1-40.

Salpiginella decurtata

Rampi, L., 1939
Primo contributo alla conoscenza dei tintinnoidi de Mare Ligure. Atti della Soc. Ital. di Sci. Nat. 78:67-81, 58 text figs.

Salpingella decurtata

Sousa e Silva, E. de, 1950.
Les tintinnides de la Baie de Cascais. Bull. Inst. Océan., Monaco, No. 979:28 pp., 4 pls.

Salpingella decurtata var. joergenseni n.var.

Duran, M., 1953.
Contribución al estudio de los Tintinnidos del plancton de las costas de Castellón (Mediterráneo occidental). Nota II. Publ. Inst. Biol. Aplic. 12:79-95, 23 textfigs.

Salpingella expansa

Fofoid, C.A. and A.S. Campbell,1939
Reports on the scientific results of the expedition to the eastern tropical Pacific in charge of Alexander Agassiz, by the U.S. Fish Commission steamer "Albatross" from October 1904 to March 1905-----XXXVII. The Ciliata: The Tintinnoimea. Bull.Mus. Comp.Zool. 84: 473 pp., 36 pls.

Salpingella expansa n.sp.

Kofoid, C. A. and A. S. Campbell, 1929
A conspectus of the marine and freshwater Ciliata belonging to the suborder Tintinnoinea, with descriptions of new species principally from the Agassiz expedition to the eastern tropical Pacific, 1904-1905. Univ. Calif. Publ. Zool. 34:1-403, 697 text figs.

Salpingella faurei

Campbell, A. S., 1942
The oceanic Tintinnoina of the plankton gathered during the last cruise of the CARNEGIE. Sci. Res. Cruise VII of the Carnegie, 1928-1929, Biol. 2:163 pp., 128 figs.

Salpingella faurei

Fofoid, C.A. and A.S. Campbell,1939
Reports on the scientific results of the expedition to the eastern tropical Pacific in charge of Alexander Agassiz, by the U.S. Fish Commission steamer "Albatross" from October 1904 to March 1905-----XXXVII. The Ciliata: The Tintinnoimea. Bull.Mus. Comp.Zool. 84: 473 pp., 36 pls.

Salpingella faurei n.sp.

Kofoid, C. A. and A. S. Campbell, 1929
A conspectus of the marine and freshwater Ciliata belonging to the suborder Tintinnoinea, with descriptions of new species principally from the Agassiz expedition to the eastern tropical Pacific, 1904-1905. Univ. Calif. Publ. Zool. 34:1-403, 697 text figs.

Salpingella glockentägeri

Balech, E., 1959.
Tintinnoinea del Mediterraneo.
Trav. Sta. Zool., Villefranche-sur-Mer, 18(17):
Reprinted from:
Trab. Inst. Español Oceanogr., Madrid, 19 Sept. 1959(28):88 pp., 22 pls.

Salpingella glockentögeri

Campbell, A. S., 1942
The oceanic Tintinnoina of the plankton gathered during the last cruise of the CARNEGIE. Sci. Res. Cruise VII of the Carnegie, 1928-1929, Biol. 2:163 pp., 128 figs.

Salpingella glockentogeri

Fofoid, C.A. and A.S. Campbell,1939
Reports on the scientific results of the expedition to the eastern tropical Pacific in charge of Alexander Agassiz, by the U.S. Fish Commission steamer "Albatross" from October 1904 to March 1905-----XXXVII. The Ciliata: The Tintinnoimea. Bull.Mus. Comp.Zool. 84: 473 pp., 36 pls.

Salpingella glockentögeri

Kofoid, C. A. and A. S. Campbell, 1929
A conspectus of the marine and freshwater Ciliata belonging to the suborder Tintinnoinea, with descriptions of new species principally from the Agassiz expedition to the eastern tropical Pacific, 1904-1905. Univ. Calif. Publ. Zool. 34:1-403, 697 text figs.

Salpingella gracilis

Balech, Enrique, 1962
Tintinnoinea y Dinoflagellata del Pacifico segun material de las Expediciones NORPAC y DOWNWIND del Instituto Scripps de Oceanografia.
Revista, Mus. Argentino Ciencias Nat. "Bernardino Rivadavia", Ciencias Zool., 7(1):1-253.

Salpingella gracilis

Campbell, A. S., 1942
The oceanic Tintinnoina of the plankton gathered during the last cruise of the CARNEGIE. Sci. Res. Cruise VII of the Carnegie, 1928-1929, Biol. 2:163 pp., 128 figs.

Salpingella gracilis

Gaarder, K.R., 1946
Tintinnoidea. Rep.Sci.Res."Michael Sars" N.Atlantic Deep-Sea Exped., 1910, 2(1): 37 pp., 23 text figs.

Salpingella gracilis

Fofoid, C.A. and A.S. Campbell,1939
Reports on the scientific results of the expedition to the eastern tropical Pacific in charge of Alexander Agassiz, by the U.S. Fish Commission steamer "Albatross" from October 1904 to March 1905-----XXXVII. The Ciliata: The Tintinnoimea. Bull.Mus. Comp.Zool. 84: 473 pp., 36 pls.

Salpingella gracilis n.sp.

Kofoid, C. A. and A. S. Campbell, 1929
A conspectus of the marine and freshwater Ciliata belonging to the suborder Tintinnoinea, with descriptions of new species principally from the Agassiz expedition to the eastern tropical Pacific, 1904-1905. Univ. Calif. Publ. Zool. 34:1-403, 697 text figs.

Salpingella incurva

Campbell, A. S., 1942
The oceanic Tintinnoina of the plankton gathered during the last cruise of the CARNEGIE. Sci. Res. Cruise VII of the Carnegie, 1928-1929, Biol. 2:163 pp., 128 figs.

Salpingella incurva n.sp.

Fofoid, C.A. and A.S. Campbell,1939
Reports on the scientific results of the expedition to the eastern tropical Pacific in charge of Alexander Agassiz, by the U.S. Fish Commission steamer "Albatross" from October 1904 to March 1905-----XXXVII. The Ciliata: The Tintinnoimea. Bull.Mus. Comp.Zool. 84: 473 pp., 36 pls.

Salpingella aff. jugosa

Balech, Enrique, 1962
Tintinnoinea y Dinoflagellata del Pacifico segun material de las Expediciones NORPAC y DOWNWIND del Instituto Scripps de Oceanografia.
Revista, Mus. Argentino Ciencias Nat. "Bernardino Rivadavia", Ciencias Zool., 7(1):1-253.

Salpingella jugosa

Campbell, A. S., 1942
The oceanic Tintinnoina of the plankton gathered during the last cruise of the CARNEGIE. Sci. Res. Cruise VII of the Carnegie, 1928-1929, Biol. 2:163 pp., 128 figs.

Salpingella jugosa

Fofoid, C.A. and A.S. Campbell,1939
Reports on the scientific results of the expedition to the eastern tropical Pacific in charge of Alexander Agassiz, by the U.S. Fish Commission steamer "Albatross" from October 1904 to March 1905-----XXXVII. The Ciliata: The Tintinnoimea. Bull.Mus. Comp.Zool. 84: 473 pp., 36 pls.

Salpingella jugosa n. sp.

Kofoid, C. A. and A. S. Campbell, 1929
A conspectus of the marine and freshwater Ciliata belonging to the suborder Tintinnoinea, with descriptions of new species principally from the Agassiz expedition to the eastern tropical Pacific, 1904-1905. Univ. Calif. Publ. Zool. 34:1-403, 697 text figs.

Salpingella laackmanni

Balech, Enrique, and Sayed Z. El-Sayed, 1965.
The microplankton of the Weddell Sea.
In: Biology of Antarctic seas. II.
Antarctic Res. Ser. Amer. Geophys. Union, 5:107-124.

Salpingella laackmanni nom. nov.

Kofoid, C. A. and A. S. Campbell, 1929
A conspectus of the marine and freshwater Ciliata belonging to the suborder Tintinnoinea, with descriptions of new species principally from the Agassiz expedition to the eastern tropical Pacific, 1904-1905. Univ. Calif. Publ. Zool. 34:1-403, 697 text figs.

Salpingella laminata n.sp.

Fofoid, C.A. and A.S. Campbell,1939
Reports on the scientific results of the expedition to the eastern tropical Pacific in charge of Alexander Agassiz, by the U.S. Fish Commission steamer "Albatross" from October 1904 to March 1905-----XXXVII. The Ciliata: The Tintinnoimea. Bull.Mus. Comp.Zool. 84: 473 pp., 36 pls.

Salpingella lineata

Kofoid, C. A. and A. S. Campbell, 1929
A conspectus of the marine and freshwater Ciliata belonging to the suborder Tintinnoinea, with descriptions of new species principally from the Agassiz expedition to the eastern tropical Pacific, 1904-1905. Univ. Calif. Publ. Zool. 34:1-403, 697 text figs.

Salpingella minutissima

Gaarder, K.R., 1946
Tintinnoidea. Rep.Sci.Res."Michael Sars" N.Atlantic Deep-Sea Exped., 1910, 2(1): 37 pp., 23 text figs.

Oceanographic Index: Marine Organisms Cumulation, 1946-1973

Salpingella minutissima

Kofoid, C.A. and A.S. Campbell, 1939
Reports on the scientific results of the expedition to the eastern tropical Pacific in charge of Alexander Agassiz, by the U.S. Fish Commission steamer "Albatross" from October 1904 to March 1905----- XXXVII. The Ciliata: The Tintinnoimea. Bull. Mus. Comp. Zool. 84: 473 pp., 36 pls.

Salpingella minutissima n.sp.

Kofoid, C. A. and A. S. Campbell, 1929
A conspectus of the marine and freshwater Ciliata belonging to the suborder Tintinnoinea, with descriptions of new species principally from the Agassiz expedition to the eastern tropical Pacific, 1904-1905. Univ. Calif. Publ. Zool. 34:1-403, 697 text figs.

Salpingella octogenata n.sp.

Kofoid, C. A. and A. S. Campbell, 1929
A conspectus of the marine and freshwater Ciliata belonging to the suborder Tintinnoinea, with descriptions of new species principally from the Agassiz expedition to the eastern tropical Pacific, 1904-1905. Univ. Calif. Publ. Zool. 34:1-403, 697 text figs.

Salpingella regulata

Kofoid, C. A. and A. S. Campbell, 1929
A conspectus of the marine and freshwater Ciliata belonging to the suborder Tintinnoinea, with descriptions of new species principally from the Agassiz expedition to the eastern tropical Pacific, 1904-1905. Univ. Calif. Publ. Zool. 34:1-403, 697 text figs.

Salpingella ricta

Campbell, A. S., 1942
The oceanic Tintinnoina of the plankton gathered during the last cruise of the CARNEGIE. Sci. Res. Cruise VII of the Carnegie, 1928-1929, Biol. 2:163 pp., 128 figs.

Salpingella ricta

Kofoid, C.A. and A.S. Campbell, 1939
Reports on the scientific results of the expedition to the eastern tropical Pacific in charge of Alexander Agassiz, by the U.S. Fish Commission steamer "Albatross" from October 1904 to March 1905----- XXXVII. The Ciliata: The Tintinnoimea. Bull. Mus. Comp. Zool. 84: 473 pp., 36 pls.

Salpingella ricta n.sp.

Kofoid, C. A. and A. S. Campbell, 1929
A conspectus of the marine and freshwater Ciliata belonging to the suborder Tintinnoinea, with descriptions of new species principally from the Agassiz expedition to the eastern tropical Pacific, 1904-1905. Univ. Calif. Publ. Zool. 34:1-403, 697 text figs.

Salpingella rotundata

Balech, E., 1959.
Tintinnoinea del Mediterraneo.
Trav. Sta. Zool., Villefranche-sur-Mer, 18(17):
Reprinted from:
Trab. Inst. Español Oceanogr., Madrid, 19 Sept. 1959(28):88 pp., 22 pls.

Salpingella rotundata

Campbell, A. S., 1942
The oceanic Tintinnoina of the plankton gathered during the last cruise of the CARNEGIE. Sci. Res. Cruise VII of the Carnegie, 1928-1929, Biol. 2:163 pp., 128 figs.

Salpingella rotundata

Kofoid, C.A. and A.S. Campbell, 1939
Reports on the scientific results of the expedition to the eastern tropical Pacific in charge of Alexander Agassiz, by the U.S. Fish Commission steamer "Albatross" from October 1904 to March 1905----- XXXVII. The Ciliata: The Tintinnoimea. Bull. Mus. Comp. Zool. 84: 473 pp., 36 pls.

Salpingella rotundata n.sp.

Kofoid, C. A. and A. S. Campbell, 1929
A conspectus of the marine and freshwater Ciliata belonging to the suborder Tintinnoinea, with descriptions of new species principally from the Agassiz expedition to the eastern tropical Pacific, 1904-1905. Univ. Calif. Publ. Zool. 34:1-403, 697 text figs.

Salpingella secata

Campbell, A. S., 1942
The oceanic Tintinnoina of the plankton gathered during the last cruise of the CARNEGIE. Sci. Res. Cruise VII of the Carnegie, 1928-1929, Biol. 2:163 pp., 128 figs.

Salpingella secata

Kofoid, C.A. and A.S. Campbell, 1939
Reports on the scientific results of the expedition to the eastern tropical Pacific in charge of Alexander Agassiz, by the U.S. Fish Commission steamer "Albatross" from October 1904 to March 1905----- XXXVII. The Ciliata: The Tintinnoimea. Bull. Mus. Comp. Zool. 84: 473 pp., 36 pls.

Salpingella secata

Kofoid, C. A. and A. S. Campbell, 1929
A conspectus of the marine and freshwater Ciliata belonging to the suborder Tintinnoinea, with descriptions of new species principally from the Agassiz expedition to the eastern tropical Pacific, 1904-1905. Univ. Calif. Publ. Zool. 34:1-403, 697 text figs.

Salpingella sinistra n.sp.

Kofoid, C.A. and A.S. Campbell, 1939
Reports on the scientific results of the expedition to the eastern tropical Pacific in charge of Alexander Agassiz, by the U.S. Fish Commission steamer "Albatross" from October 1904 to March 1905----- XXXVII. The Ciliata: The Tintinnoimea. Bull. Mus. Comp. Zool. 84: 473 pp., 36 pls.

Salpingella subconica

Balech, Enrique, 1962
Tintinnoinea y Dinoflagellata del Pacifico segun material de las Expediciones NORPAC y DOWNWIND del Instituto Scripps de Oceanografia.
Revista, Mus. Argentino Ciencias Nat. "Bernardino Rivadavia", Ciencias Zool., 7(1):1-253.

Salpingella subconica

Balech, E., 1959.
Tintinnoinea del Mediterraneo.
Trav. Sta. Zool., Villefranche-sur-Mer, 18(17):
Reprinted from:
Trab. Inst. Español Oceanogr., Madrid, 19 Sept. 1959(28):88 pp., 22 pls.

Salpingella subconica

Campbell, A. S., 1942
The oceanic Tintinnoina of the plankton gathered during the last cruise of the CARNEGIE. Sci. Res. Cruise VII of the Carnegie, 1928-1929, Biol. 2:163 pp., 128 figs.

Salpingella subconica

Kofoid, C.A. and A.S. Campbell, 1939
Reports on the scientific results of the expedition to the eastern tropical Pacific in charge of Alexander Agassiz, by the U.S. Fish Commission steamer "Albatross" from October 1904 to March 1905----- XXXVII. The Ciliata: The Tintinnoimea. Bull. Mus. Comp. Zool. 84: 473 pp., 36 pls.

Salpingella subconica n.sp.

Kofoid, C. A. and A. S. Campbell, 1929
A conspectus of the marine and freshwater Ciliata belonging to the suborder Tintinnoinea, with descriptions of new species principally from the Agassiz expedition to the eastern tropical Pacific, 1904-1905. Univ. Calif. Publ. Zool. 34:1-403, 697 text figs.

Salpingella subconica

Marshall, S. M., 1934
The silicoflagellata and tintinnoinea.
Great Barrier Reef Exped., 1928-29. Sci. Repts. 4(15):623-664, 43 text figs.

Salpingella tuba n.sp.

Kofoid, C.A. and A.S. Campbell, 1939
Reports on the scientific results of the expedition to the eastern tropical Pacific in charge of Alexander Agassiz, by the U.S. Fish Commission steamer "Albatross" from October 1904 to March 1905----- XXXVII. The Ciliata: The Tintinnoimea. Bull. Mus. Comp. Zool. 84: 473 pp., 36 pls.

Salpingelloides n. gen.

Campbell, A. S., 1942
The oceanic Tintinnoina of the plankton gathered during the last cruise of the CARNEGIE. Sci. Res. Cruise VII of the Carnegie, 1928-1929, Biol. 2:163 pp., 128 figs.

Steenstrupiella n. gen.

Kofoid, C. A. and A. S. Campbell, 1929
A conspectus of the marine and freshwater Ciliata belonging to the suborder Tintinnoinea, with descriptions of new species principally from the Agassiz expedition to the eastern tropical Pacific, 1904-1905. Univ. Calif. Publ. Zool. 34:1-403, 697 text figs.

Steenstrupiella entzi nom. nov.

Kofoid, C. A. and A. S. Campbell, 1929
A conspectus of the marine and freshwater Ciliata belonging to the suborder Tintinnoinea, with descriptions of new species principally from the Agassiz expedition to the eastern tropical Pacific, 1904-1905. Univ. Calif. Publ. Zool. 34:1-403, 697 text figs.

Steenstrupiella gracilis

Balech, Enrique, 1971
Microplancton del Atlantico ecuatorial oeste (Equalant 1)
Publ. Serv. Hidrograf. Naval, Argentina H. 654: 103 pp., 122 figs.

Steenstrupiella gracilis

Campbell, A. S., 1942
The oceanic Tintinnoina of the plankton gathered during the last cruise of the CARNEGIE. Sci. Res. Cruise VII of the Carnegie, 1928-1929, Biol. 2:163 pp., 128 figs.

Steenstrupiella gracilis

de Sousa e Silva, E., 1956.
Contribution à l'étude du microplancton de Dakar et des regions maritimes voisines.
Bull. I.F.A.N., 8(2):335-371, 7 pls.

Steenstrupiella gracilis

Kofoid, C.A. and A.S. Campbell, 1939
Reports on the scientific results of the expedition to the eastern tropical Pacific in charge of Alexander Agassiz, by the U.S. Fish Commission steamer "Albatross" from October 1904 to March 1905----- XXXVII. The Ciliata: The Tintinnoimea. Bull. Mus. Comp. Zool. 84: 473 pp., 36 pls.

Steenstrupiella gracilis

Kofoid, C. A. and A. S. Campbell, 1929
A conspectus of the marine and freshwater Ciliata belonging to the suborder Tintinnoinea, with descriptions of new species principally from the Agassiz expedition to the eastern tropical Pacific, 1904-1905. Univ. Calif. Publ. Zool. 34:1-403, 697 text figs.

Steenstrupiella gracilis

Komarovsky, B., 1959
The Tintinnina of the Gulf of Eylath (Aqaba).
Contrib. Knowl. Red Sea, Bull. Sea Fish. Res. Sta. Haifa, Israel 21 (14):1-40.

Steenstrupiella intumescens

Balech, Enrique, 1962
Tintinnoinea y Dinoflagellata del Pacifico segun material de las Expediciones NORPAC y DOWNWIND del Instituto Scripps de Oceanografia.
Revista, Mus. Argentino Ciencias Nat. "Bernardino Rivadavia", Ciencias Zool., 7(1):1-253.

Steenstrupiella intumescens

Kofoid, C. A. and A. S. Campbell, 1929
A conspectus of the marine and freshwater Ciliata belonging to the suborder Tintinnoinea, with descriptions of new species principally from the Agassiz expedition to the eastern tropical Pacific, 1904-1905. Univ. Calif. Publ. Zool. 34:1-403, 697 text figs.

Steenstrupiella intumescens

Komarovsky, B., 1959
The Tintinnina of the Gulf of Eylath (Aqaba). Contrib. Knowl. Red Sea, Bull. Sea Fish. Res. Sta. Haifa, Israel 21 (14):1-40.

Steenstrupiella intumescens

Marshall, S. M., 1934
The silicoflagellata and tintinnoinea. Great Barrier Reef Exped., 1928-29. Sci. Repts. 4(15):623-664, 43 text figs.

Steenstrupiella pozzii

Balech, Enrique, 1962
Tintinnoinea y Dinoflagellata del Pacifico segun material de las Expediciones NORPAC y DOWNWIND del Instituto Scripps de Oceanografia. Revista, Mus. Argentino Ciencias Nat. "Bernardino Rivadavia", Ciencias Zool., 7(1):1-253.

Steenstrupiella pozzii

Balech, Enrique, 1944
Contribucion al conocimiento del Plancton de Lennox y Cabo de Hornos. Physis XIX:423-446, 6 pls. with 67 figs.

Steenstrupiella pozzii n.sp.

Balech, E., 1941.
Tintinnoineos del estrecho Le Maire. Physis XIX(52):245-252, 10 textfigs.

Steenstrupiella robusta

Campbell, A. S., 1942
The oceanic Tintinnoina of the plankton gathered during the last cruise of the CARNEGIE. Sci. Res. Cruise VII of the Carnegie, 1928-1929, Biol. 2:163 pp., 128 figs.

Steenstrupiella robusta

Kofoid, C.A. and A.S. Campbell, 1939
Reports on the scientific results of the expedition to the eastern tropical Pacific in charge of Alexander Agassiz, by the U.S. Fish Commission steamer "Albetross" from October 1904 to March 1905------XXXVII. The Ciliata: The Tintinnoinea. Bull. Mus. Comp. Zool. 84: 473 pp., 36 pls.

Steenstrupiella robusta n.sp.

Kofoid, C. A. and A. S. Campbell, 1929
A conspectus of the marine and fresh-water Ciliata belonging to the suborder Tintinnoinea, with descriptions of new species principally from the Agassiz expedition to the eastern tropical Pacific, 1904-1905. Univ. Calif. Publ. Zool. 34:1-403, 697 text figs.

Steenstrupiella steenstrupii

Balech, Enrique, 1962
Tintinnoinea y Dinoflagellata del Pacifico segun material de las Expediciones NORPAC y DOWNWIND del Instituto Scripps de Oceanografia. Revista, Mus. Argentino Ciencias Nat. "Bernardino Rivadavia", Ciencias Zool., 7(1):1-253.

Steentrupiella steenstrupii

Balech, E., 1959.
Tintinnoinea del Mediterraneo. Trav. Sta. Zool., Villefranche-sur-Mer, 18(17): Reprinted from: Trab. Inst. Español Oceanogr., Madrid, 19 Sept., 1959(28):88 pp., 22 pls.

Steenstrupella steenstrupi

Delegazione Italiana della Commissione Internazionale per l'Esplorazione Scientifica del Mediterraneo, 1941
Note sul plancton della Laguna veneta. [Memoria CCLXXXI] , Arch. di Ocean. e Limn. Anno I, Fasc. I, 1941 XIX: 31-57 pp.

Steenstrupiella steenstrupi

Gaarder, K.R., 1946
Tintinnoidea. Rep.Sci.Res. "Michael Sars" N.Atlantic Deep-Sea Exped., 1910, 2(1): 37 pp., 23 text figs.

Steenstrupiella steenstrupia

Kofoid, C.A. and A.S. Campbell,1939
Reports on the scientific results of the expedition to the eastern tropical Pacific in charge of Alexander Agassiz, by the U.S. Fish Commission steamer "Albetross" from October 1904 to March 1905------XXXVII. The Ciliata: The Tintinnoinea. Bull. Mus. Comp. Zool. 84: 473 pp., 36 pls.

Steenstrupiella steenstrupii

Kofoid, C. A. and A. S. Campbell, 1929
A conspectus of the marine and fresh-water Ciliata belonging to the suborder Tintinnoinea, with descriptions of new species principally from the Agassiz expedition to the eastern tropical Pacific, 1904-1905. Univ. Calif. Publ. Zool. 34:1-403, 697 text figs.

Steenstrupiella steenstrupii

Komarovsky, B., 1959
The Tintinnina of the Gulf of Eylath (Aqaba). Contrib. Knowl. Red Sea, Bull. Sea Fish. Res. Sta. Haifa, Israel 21 (14):1-40.

Steenstrupiella steenstrupii

Marshall, S. M., 1934
The silicoflagellata and tintinnoinea. Great Barrier Reef Exped., 1928-29. Sci. Repts. 4(15):623-664, 43 text figs.

Steenstrupiella steenstrupii

Rampi, L., 1939
Primo contributo alla conoscenza dei tintinnoidi de Mare Ligure. Atti della Soc. Ital. di Sci. Nat. 78:67-81, 58 text figs.

Stelidiella n. gen.

Kofoid, C. A. and A. S. Campbell, 1929
A conspectus of the marine and fresh-water Ciliata belonging to the suborder Tintinnoinea, with descriptions of new species principally from the Agassiz expedition to the eastern tropical Pacific, 1904-1905. Univ. Calif. Publ. Zool. 34:1-403, 697 text figs.

Stelidiella fenestrata

Kofoid, C.A. and A.S. Campbell,1939
Reports on the scientific results of the expedition to the eastern tropical Pacific in charge of Alexander Agassiz, by the U.S. Fish Commission steamer "Albetross" from October 1904 to March 1905------XXXVII. The Ciliata: The Tintinnoinea. Bull. Mus. Comp. Zool. 84: 473 pp., 36 pls.

Stelidiella fenestrata n.sp.

Kofoid, C. A. and A. S. Campbell, 1929
A conspectus of the marine and fresh-water Ciliata belonging to the suborder Tintinnoinea, with descriptions of new species principally from the Agassiz expedition to the eastern tropical Pacific, 1904-1905. Univ. Calif. Publ. Zool. 34:1-403, 697 text figs.

Stelidiella phialia n.sp.

Kofoid, C. A. and A. S. Campbell, 1929
A conspectus of the marine and fresh-water Ciliata belonging to the suborder Tintinnoinea, with descriptions of new species principally from the Agassiz expedition to the eastern tropical Pacific, 1904-1905. Univ. Calif. Publ. Zool. 34:1-403, 697 text figs.

Stelidiella simplex

Kofoid, C.A. and A.S. Campbell,1939
Reports on the scientific results of the expedition to the eastern tropical Pacific in charge of Alexander Agassiz, by the U.S. Fish Commission steamer "Albetross" from October 1904 to March 1905------XXXVII. The Ciliata: The Tintinnoinea. Bull. Mus. Comp. Zool. 84: 473 pp., 36 pls.

Stelidiella simplex n.sp.

Kofoid, C. A. and A. S. Campbell, 1929
A conspectus of the marine and fresh-water Ciliata belonging to the suborder Tintinnoinea, with descriptions of new species principally from the Agassiz expedition to the eastern tropical Pacific, 1904-1905. Univ. Calif. Publ. Zool. 34:1-403, 697 text figs.

Stelidiella stalidium

Gaarder, K.R., 1946
Tintinnoidea. Rep.Sci.Res. "Michael Sars" N.Atlantic Deep-Sea Exped., 1910, 2(1): 37 pp., 23 text figs.

Stelidiella stelidium

Kofoid, C.A. and A.S. Campbell,1939
Reports on the scientific results of the expedition to the eastern tropical Pacific in charge of Alexander Agassiz, by the U.S. Fish Commission steamer "Albetross" from October 1904 to March 1905------XXXVII. The Ciliata: The Tintinnoinea. Bull. Mus. Comp. Zool. 84: 473 pp., 36 pls.

Stelidiella stelidium

Kofoid, C. A. and A. S. Campbell, 1929
A conspectus of the marine and fresh-water Ciliata belonging to the suborder Tintinnoinea, with descriptions of new species principally from the Agassiz expedition to the eastern tropical Pacific, 1904-1905. Univ. Calif. Publ. Zool. 34:1-403, 697 text figs.

Stenosemella sp.

Argentina, Secretaria de Marina, Servicio de Hidrografia Naval, 1962.
Plancton de las campanas oceanograficas DRAKE I y II. Publico, H. 627:57.

Stenosemella sp.

Braarud, T., 1945
A phytoplankton survey of the polluted waters of inner Oslo Fjord. Hvalrådets Skrifter, No.28, 142 pp., 19 text figs., 17 tables.

Stenosemella sp.

Braarud, T., and Adam Bursa, 1939
On the phytoplankton of the Oslo Fjord, 1933-1934. Hvalrådets Skr. No.19:1-63; 9 text figs. Reviewed. J. du. Cons. 14(3): 418-420. A.C. Gardiner.

Stenosemella sp.

Duran, M., 1957.
Nota sobre algunos tintinnoineas del plancton de Puerto Rico. Invest. Pesqueria, Barcelona, 8: 97-120.

Stenosemella sp.

Gran, H.H., and T. Braarud, 1935
A quantitative study of the phytoplankton in the Bay of Fundy and the Gulf of Maine (including observations on hydrography, chemistry, and turbidity). J. Biol. Bd., Canada, 1(5):279-467, 69 text figs.

Stenosemella n.gen.

Jörgensen, E., 1924
Mediterranean Tintinnidae. Rept. Danish Oceanogr. Exped. 1908-10 to the Mediterranean and adjacent seas, Vol.II, Biol. J 3, 110 pp., 114 text figs.

Stenosemella avellana

Balech, Enrique, 1948
Tintinnoinea de Atlantida (R.O. Del Uruguay) (Protozoa, Ciliata Oligotr.). Comm. Mus. Argentino Ciencias Nat. "Bernardino Rivadavia". Ser.Cien.Zool. No.7, 23 pp., 8 pls. of 107 figs.

Stenosemella avellana

*Hada, Yoshina, 1970. (1970)
The protozoan plankton of the Antarctic and Subantarctic seas. Scient. Repts, Japan. Antarct. Res. Exped., (E)31:51 pp.

Stenosemella avellana

Kofoid, C. A. and A. S. Campbell, 1929
A conspectus of the marine and fresh-water Ciliata belonging to the suborder Tintinnoinea, with descriptions of new species principally from the Agassiz expedition to the eastern tropical Pacific, 1904-1905. Univ. Calif. Publ. Zool. 34:1-403, 697 text figs.

Stenosemella expansa

Kofoid, C. A. and A. S. Campbell, 1929
A conspectus of the marine and fresh-water Ciliata belonging to the suborder Tintinnoinea, with descriptions of new species principally from the Agassiz expedition to the eastern tropical Pacific, 1904-1905. Univ. Calif. Publ. Zool. 34:1-403, 697 text figs.

Stenosemella inflata n. sp.

Kofoid, C. A. and A. S. Campbell, 1929
A conspectus of the marine and freshwater Ciliata belonging to the suborder Tintinnoinea, with descriptions of new species principally from the Agassiz expedition to the eastern tropical Pacific, 1904-1905. Univ. Calif. Publ. Zool. 34:1-403, 697 text figs.

Stenosemella monacense n. sp.

Rampi, L., 1950
I Tintinnoidi delle acque di Monaco raccolti d'all'Eider nell'anno 1913. Bull. Inst. Ocean. Monaco, No. 965: 7 pp., 1 textfig.

Stenosemella nivalis

Balech, E., 1959.
Tintinnoinea del Mediterraneo.
Trav. Sta. Zool., Villefranche-sur-Mer, 18(17):
Reprinted from:
Trab. Inst. Español Oceanogr., Madrid, 19 Sept. 1959(28):88 pp., 22 pls.

Stenosemella nivalis

Campbell, A. S., 1942
The oceanic Tintinnoina of the plankton gathered during the last cruise of the CARNEGIE. Sci. Res. Cruise VII of the Carnegie, 1928-1929, Biol. 2:163 pp., 128 figs.

Stenosemella nivalis

Campbell, A. S., 1931.
The membranelles of Stenosemella nivalis. Anat. Rec. 47:347-348 (absbr. 151).

Stenosemella nivalis

Gaarder, K.R., 1946
Tintinnoidea. Rep.Sci.Res."Michael Sars" N.Atlantic Deep-Sea Exped., 1910, 2(1): 37 pp., 23 text figs.

Stenosemella nivalis

*Hada, Yoshina, 1970.   (1970)
The protozoan plankton of the Antarctic and Subantarctic seas.
Scient. Repts, Japan. Antarct. Res. Exped., (E)31:51 pp.

Stenosemella nivalis

Kofoid, C.A. and A.S. Campbell,1939
Reports on the scientific results of the expedition to the eastern tropical Pacific in charge of Alexander Agassiz, by the U.S. Fish Commission steamer"Albatross" from October 1904 to March 1905-----XXXVII. The Ciliata: The Tintinnoinea. Bull. Mus. Comp.Zool. 84: 473 pp., 36 pls.

Stenosemella nivalis

Kofoid, C. A. and A. S. Campbell, 1929
A conspectus of the marine and freshwater Ciliata belonging to the suborder Tintinnoinea, with descriptions of new species principally from the Agassiz expedition to the eastern tropical Pacific, 1904-1905. Univ. Calif. Publ. Zool. 34:1-403, 697 text figs.

Stenosomella nivalis

Marshall, S. M., 1934
The silicoflagellata and tintinnoinea. Great Barrier Reef Exped., 1928-29. Sci. Repts. 4(15):623-664, 43 text figs.

Stenosemella nivalis

Schulz, H., 1965.
Die Tintinnoinea des Elbe-Aestuars.
Arch. Fischereiwiss., 15(3):216-225.

Stenosemella nucula

Duran, M., 1951
Contribucion al estudio de los tintinidos del plancton de los costas de Castellon. (Mediterranes occidental). Publ. Inst. Biol. Aplic., Barcelone, 8: 101-120, 29 text figs.

Stenosomella nucula

Hofker, J., 1931
Studien uber Tintinnoidea. Arch. f. Protistenk 75(3):315-402, 89 text figs.

Stenosemella nucula

Jörgensen, E., 1924
Mediterranean Tintinnidae. Rept. Danish Oceanogr. Exped. 1908-10 to the Mediterranean and adjacent seas, Vol.II, Biol. J 3, 110 pp., 114 text figs.

Stenosemella nucula

Sousa e Silva, E. de, 1950.
Les tintinnides de la Baie de Cascais. Bull. Inst. Océan., Monaco, No. 979:28 pp., 4 pls.

Stenosemella oliva

Gaarder, K.R., 1946
Tintinnoidea. Rep.Sci.Res."Michael Sars" N.Atlantic Deep-Sea Exped., 1910, 2(1): 37 pp., 23 text figs.

Stenosemella oliva

Kofoid, C. A. and A. S. Campbell, 1929
A conspectus of the marine and freshwater Ciliata belonging to the suborder Tintinnoinea, with descriptions of new species principally from the Agassiz expedition to the eastern tropical Pacific, 1904-1905. Univ. Calif. Publ. Zool. 34:1-403, 697 text figs.

Stenosemella oliva

Schulz, H., 1965.
Die Tintinnoinea des Elbe-Aestuars.
Arch. Fischereiwiss., 15(3):216-225.

Stenosemella pacifica

Balech, Enrique, 1968.
Algunas especies nuevas o interesantes do tintinnidos del Golfo de Mexico y Caribe.
Revta Mus. argent. Cienc. nat. Bernardino Rivadavia Inst. nac. Hidrobiol. 2(5):165-197.

Stenosemella pacifica

Kofoid, C. A. and A. S. Campbell, 1929
A conspectus of the marine and freshwater Ciliata belonging to the suborder Tintinnoinea, with descriptions of new species principally from the Agassiz expedition to the eastern tropical Pacific, 1904-1905. Univ. Calif. Publ. Zool. 34:1-403, 697 text figs.

Stenosemella perpusilla n. sp.

*Hada, Yoshina, 1970.   (1970)
The protozoan plankton of the Antarctic and Subantarctic seas.
Scient. Repts, Japan. Antarct. Res. Exped., (E)31:51 pp.

Stenosemella producta

Kofoid, C. A. and A. S. Campbell, 1929
A conspectus of the marine and freshwater Ciliata belonging to the suborder Tintinnoinea, with descriptions of new species principally from the Agassiz expedition to the eastern tropical Pacific, 1904-1905. Univ. Calif. Publ. Zool. 34:1-403, 697 text figs.

Stenosemella punctata

Sousa e Silva, E. de, 1950.
Les tintinnides de la Baie de Cascais. Bull. Inst. Océan., Monaco, No. 979:28 pp., 4 pls.

Stenosemella steini

Kofoid, C. A. and A. S. Campbell, 1929
A conspectus of the marine and freshwater Ciliata belonging to the suborder Tintinnoinea, with descriptions of new species principally from the Agassiz expedition to the eastern tropical Pacific, 1904-1905. Univ. Calif. Publ. Zool. 34:1-403, 697 text figs.

Stenosomella steini

Levander, K.M., 1947
Plankton gesammelt in den Jahren 1899-1910 an den Küsten Finnlands. Finnländische Hydrographisch-Biologische Untersuchunger (aus dem Wasserbiologischen Laboratorin der Societas Scientiarum Fennica) No.11:40 pp., 6 diagrams, 13 pls., tables.

Stenosemella ventricosa

Balech, E., 1959.
Tintinnoinea del Mediterranea.
Trav. Sta. Zool., Villefranche-sur-Mer, 18(17):
Reprinted from:
Trab. Inst. Español Oceanogr., Madrid, 19 Sept. 1959(28):88 pp., 22 pls.

Stenosemmella ventricosa

Biernacka, I., 1948
(Tintinnoinea in the Gulf of Gdansk and adjoining waters) Builetyn Morskiego Laboratorium Rybackiego w Gdyni dawniej-Stacji Morskiej w. Helu. No. 4:73-91, 4 text figs., 1 pl. with 21 figs. (Bull. Lab. Mar., Gdynia, formerly Bull. Sta. Hel.)

Stenosemella ventricosa

Duran, Miguel, 1957.
Nota sobre algunos tintinoineos del plancton de Puerto Rico.
Inv. Pesq., Barcelona, 8:97-120.

Stenosemella ventricosa

Duran, M., 1951
Contribucion al estudio de los tintinidos del plancton de los costas de Castellon. (Mediterranes occidental). Publ. Inst. Biol. Aplic., Barcelone, 8: 101-120, 29 text figs.

Stenosemella ventricosa

Gaarder, K.R., 1946
Tintinnoidea. Rep.Sci.Res."Michael Sars" N.Atlantic Deep-Sea Exped., 1910, 2(1): 37 pp., 23 text figs.

Stenosomella ventricosa

Hofker, J., 1931
Studien uber Tintinnoidea. Arch. f. Protistenk 75(3):315-402, 89 text figs.

Stenosemella ventricosa

Jörgensen, E., 1924
Mediterranean Tintinnidae. Rept. Danish Oceanogr. Exped. 1908-10 to the Mediterranean and adjacent seas, Vol.II, Biol. J 3, 110 pp., 114 text figs.

Stenosemella ventricosa

Kofoid, C. A. and A. S. Campbell, 1929
A conspectus of the marine and freshwater Ciliata belonging to the suborder Tintinnoinea, with descriptions of new species principally from the Agassiz expedition to the eastern tropical Pacific, 1904-1905. Univ. Calif. Publ. Zool. 34:1-403, 697 text figs.

Stenosemella ventricosa

Posta, Annette, 1963.
Relation entre l'évolution de quelques tintinnides de la rade de Villefranche et de la température de l'eau.
Cahiers, Biol. Mar., Roscoff, 4:201-210.
Also in:
Trav. Sta. Zool., Villefranche-sur-Mer, 22(5).

Stenosemella ventricosa

Rampi, L., 1939
Primo contributo alla conoscenza dei tintinnoidi del Mare Ligure. Atti della Soc. Ital. di Sci. Nat. 78:67-81, 58 text figs.

Stenosemella ventricosa

Rumkówna, A., 1948
[List of the phytoplankton species occurring in the superficial water layers in the Gulf of Gdansk] Bull. Lab. mar., Gdynia, No. 4: 139-141 with tables in back.

Stenosemella ventricosa

Sousa e Silva E. de, 1950.
Les tintinnides de la Baie de Cascais. Bull. Inst. Océan., Monaco, No. 979:28 pp., 4 pls.

Stenosemella ventricosa

Vitiello, Pierre, 1964.
Contribution a l'etude des tintinnides de la baie d'Alger.
Pelagos, Bull. Inst. Oceanogr., Alger, 2(2):5-41

(pp numbers of this article duplicate those of previous one by Francis Bernard!)

Stylicauda platensis n.gen.

Balech, E., 1951.
Nuevos datos sobre Tintinnoinea de Argentina y Uruguay. Physis 20(58):291-302, 16 textfigs.

Styicauda platensis

Cosper, T.C., 1972.
The identification of tintinnids (Protozoa: Ciliata: Tintinnida) of the St. Andrew Bay system, Florida. Bull. mar. Sci. Miami 22(2): 391-418.

Tintinnidium ampullarum n.sp.
Roxas, H.A., 1941.
Marine protozoa of the Philippines.
Philippine J. Sci. 74:91-136, 17 pls., 2 textfigs

Tintinnidium cylindricum n.sp.
Roxas, H.A., 1941.
Marine protozoa of the Philippines.
Philippine J. Sci. 78:91-136, 17 pls., 2 textfigs

Tintinnidium fluviatile
Kofoid, C. A. and A. S. Campbell, 1929
A conspectus of the marine and fresh-water Ciliata belonging to the suborder Tintinnoinea, with descriptions of new species principally from the Agassiz expedition to the eastern tropical Pacific, 1904-1905. Univ. Calif. Publ. Zool. 34:1-403, 697 text figs.

Tintinnidium fluviatile
Schulz, H., 1965.
Die Tintinnoinea des Elbe-Aestuars.
Arch. Fischereiwiss., 15(3):216-225.

Tintinnidium incertum
Aurich, H. J., 1949.
Die Verbreitung des Nannoplanktons im Oberflächenwasser vor der Nordfroesischen Küste. Ber. Deutschen Wiss. Komm. f. Meeresf., n.f., 11(4): 403-405, 2 figs.

Tintinnidium incertum
Hofker, J., 1931
Studien uber Tintinnoidea. Arch. f. Protistenk 75(3):315-402, 89 text figs.

Tintinnidium incertum
Kofoid, C. A. and A. S. Campbell, 1929
A conspectus of the marine and fresh-water Ciliata belonging to the suborder Tintinnoinea, with descriptions of new species principally from the Agassiz expedition to the eastern tropical Pacific, 1904-1905. Univ. Calif. Publ. Zool. 34:1-403, 697 text figs.

Tintinnidium incertum
Schulz, H., 1965.
Die Tintinnoinea des Elbe-Aestuars.
Arch. Fischereiwiss., 15(3):216-225.

Tintinnidium inquilinum
Kofoid, C. A. and A. S. Campbell, 1929
A conspectus of the marine and fresh-water Ciliata belonging to the suborder Tintinnoinea, with descriptions of new species principally from the Agassiz expedition to the eastern tropical Pacific, 1904-1905. Univ. Calif. Publ. Zool. 34:1-403, 697 text figs.

Tintinnidium mucicola
Hofker, J., 1931
Studien uber Tintinnoidea. Arch. f. Protistenk 75(3):315-402, 89 text figs.

Tintinnidium muciola
Kofoid, C. A. and A. S. Campbell, 1929
A conspectus of the marine and fresh-water Ciliata belonging to the suborder Tintinnoinea, with descriptions of new species principally from the Agassiz expedition to the eastern tropical Pacific, 1904-1905. Univ. Calif. Publ. Zool. 34:1-403, 697 text figs.

Tintinnidium neapolitanum
Kofoid, C. A. and A. S. Campbell, 1929
A conspectus of the marine and fresh-water Ciliata belonging to the suborder Tintinnoinea, with descriptions of new species principally from the Agassiz expedition to the eastern tropical Pacific, 1904-1905. Univ. Calif. Publ. Zool. 34:1-403, 697 text figs.

Tintinnidium primitvum
Kofoid, C. A. and A. S. Campbell, 1929
A conspectus of the marine and fresh-water Ciliata belonging to the suborder Tintinnoinea, with descriptions of new species principally from the Agassiz expedition to the eastern tropical Pacific, 1904-1905. Univ. Calif. Publ. Zool. 34:1-403, 697 text figs.

Tintinnidium pusillum
Kofoid, C. A. and A. S. Campbell, 1929
A conspectus of the marine and fresh-water Ciliata belonging to the suborder Tintinnoinea, with descriptions of new species principally from the Agassiz expedition to the eastern tropical Pacific, 1904-1905. Univ. Calif. Publ. Zool. 34:1-403, 697 text figs.

Tintinnidium ranunculi
Kofoid, C. A. and A. S. Campbell, 1929
A conspectus of the marine and fresh-water Ciliata belonging to the suborder Tintinnoinea, with descriptions of new species principally from the Agassiz expedition to the eastern tropical Pacific, 1904-1905. Univ. Calif. Publ. Zool. 34:1-403, 697 text figs.

Tintinnidium semiciliatum
Kofoid, C. A. and A. S. Campbell, 1929
A conspectus of the marine and fresh-water Ciliata belonging to the suborder Tintinnoinea, with descriptions of new species principally from the Agassiz expedition to the eastern tropical Pacific, 1904-1905. Univ. Calif. Publ. Zool. 34:1-403, 697 text figs.

Tintinnopsis spp.
Balech, E., 1959.
Tintinnoinea del Mediterraneo.
Trav. Sta. Zool., Villefranche-sur-Mer, 18(17):
Reprinted from:
Trab. Inst. Español Oceanogr., Madrid, 19 Sept. 1959(28):88 pp., 22 pls.

Tintinnopsis sp.
Duran, Miguel, 1957.
Nota sobre algunos tintinnoineos del plancton de Puerto Rico.
Inv. Pesq., Barcelona, 8:97-120.

Tintinnopsis sp.
Gold, Kenneth 1968.
Some observations on the biology of Tintinnopsis sp.
J. Protozool. 15(1): 193-194.

Tintinnopsis spp.
Gran, H.H., and T. Braarud, 1935
A quantitative study of the phytoplankton in the Bay of Fundy and the Gulf of Maine (including observations on hydrography, chemistry, and turbidity).
J. Biol. Bd., Canada, 1(5):279-467, 69 text figs.

Tintinopsis sp.
Kokubo, S., and S. Sato, 1947
Plankterz in Jū-San Gata. Physiol. and Ecol. (Japan) 1(4):1-16, 3 text figs., tables.

Tintinnopsis sp.
Marshall, S.M., 1947
An experiment in marine fish cultivation: III. The plankton of a fertilized loch. Proc. Roy. Soc., Edinburgh, Sect.B., 63, Pt.1(3):21-33, 7 text figs.

Tintinnopsis sp.
Morse, D.C., 1947
Some observations on seasonal variations in plankton population Patuxant River, Maryland 1943-1945. Bd. Nat. Res., Publ. No.65, Chesapeake Biol. Lab., 31, 3 figs.

Tintinnopsis Brandt
Levander, K.M., 1947
Plankton gesammelt in den Jahren 1899-1910 an den Küsten Finnlands.
Finnländische Hydrographisch-Biologische Untersuchunger (aus dem Wasserbiologischen Laboratorin der Societas Scientiarum Fennica) No.11:40 pp., 6 diagrams, 13 pls., tables.

Tintinnopsis sp.
Posta, Annette, 1963.
Relation entre l'évolution de quelques tintinnides de la rade de Villefranche et la température de l'eau.
Cahiers, Biol. Mar., Roscoff, 4:201-210.
Also in:
Trav. Sta. Zool., Villefranche-sur-Mer, 22(5).

Tintinnopsis acuminata
Kofoid, C. A. and A. S. Campbell, 1929
A conspectus of the marine and fresh-water Ciliata belonging to the suborder Tintinnoinea, with descriptions of new species principally from the Agassiz expedition to the eastern tropical Pacific, 1904-1905. Univ. Calif. Publ. Zool. 34:1-403, 697 text figs.

Tintinnopsis acuminatus
Marukawa, H., 1921
Plankton lists and some new species of copepods, from the northern waters of Japan.
Bull. Inst. Ocean., No.384, 15 pp., 3 pls., 1 chart. Monaco

Tintinnopsis acuminata
Meunier, A., 1919
Microplankton de la Mer Flamande. 4. Les Tintinnides et Coetera. Mem. Mus. Roy. Hist. Nat., Belgique, 8(2):59pp., Pls. 22-23.

Tintinnopsis acuta
Meunier, A., 1919
Microplankton de la Mer Flamande. 4. Les Tintinnides et Coetera. Mem. Mus. Roy. Hist. Nat., Belgique, 8(2):59pp., Pls. 22-23.

Tintinnopsis amphora nom. nov.
Kofoid, C. A. and A. S. Campbell, 1929
A conspectus of the marine and fresh-water Ciliata belonging to the suborder Tintinnoinea, with descriptions of new species principally from the Agassiz expedition to the eastern tropical Pacific, 1904-1905. Univ. Calif. Publ. Zool. 34:1-403, 697 text figs.

Tintinnopsis angulata
Delegazione Italiana della Commissione Internazionale per l'Esplorazione Scientifica del Mediterraneo, 1941
Note sul plancton della Laguna veneta.
[Memoria CCLXXXIX], Arch. di Ocean. e Limn. Anno I, Fasc. I, 1941 XIX: 31-57 pp.

Tintinnopsis angulata
Jörgensen, E., 1924
Mediterranean Tintinnidae. Rept. Danish Oceanogr. Exped. 1908-10 to the Mediterranean and adjacent seas, Vol.II, Biol. J 3, 110 pp., 114 text figs.

Tintinnopsis angulata
Kofoid, C. A. and A. S. Campbell, 1929
A conspectus of the marine and fresh-water Ciliata belonging to the suborder Tintinnoinea, with descriptions of new species principally from the Agassiz expedition to the eastern tropical Pacific, 1904-1905. Univ. Calif. Publ. Zool. 34:1-403, 697 text figs.

Tintinnopsis angusta
Kofoid, C. A. and A. S. Campbell, 1929
A conspectus of the marine and fresh-water Ciliata belonging to the suborder Tintinnoinea, with descriptions of new species principally from the Agassiz expedition to the eastern tropical Pacific, 1904-1905. Univ. Calif. Publ. Zool. 34:1-403, 697 text figs.

Tintinnopsis annulata
Entz, G., Jr. 1909
Studien über organisation und biologie der Tintinniden. Arch. f. Protistenkunde 15:93-226, Pls. 8-21, figs.

Tintinnopsis annulata
Kofoid, C. A. and A. S. Campbell, 1929
A conspectus of the marine and fresh-water Ciliata belonging to the suborder Tintinnoinea, with descriptions of new species principally from the Agassiz expedition to the eastern tropical Pacific, 1904-1905. Univ. Calif. Publ. Zool. 34:1-403, 697 text figs.

Tintinnopsis aperta
Balech, Enrique, 1948
Tintinnoinea de Atlantida (R.O. Del Uruguay) (Protozoa, Ciliata Oligotr.). Comm. Mus. Argentino Ciencias Nat. "Bernardino Rivadavia", Ser.Cien.Zool. No.7, 23 pp., 8 pls. of 107 figs.

Tintinnopsis aperta
Kofoid, C. A. and A. S. Campbell, 1929
A conspectus of the marine and fresh-water Ciliata belonging to the suborder Tintinnoinea, with descriptions of new species principally from the Agassiz expedition to the eastern tropical Pacific, 1904-1905. Univ. Calif. Publ. Zool. 34:1-403, 697 text figs.

Tintinnopsis amphistoma n.sp.
Balech, E., 1951.
Nuevos datos sobre Tintinnoinea de Argentina y Uruguay. Physis 20(58):291-302, 16 textfigs.

Tintinnopsis avellana n.sp.
Meunier, A., 1919
Microplankton de la Mer Flamande. 4. Les Tintinnides et Coetera. Mem. Mus. Roy. Hist. Nat., Belgique, 8(2):59pp., Pls. 22-23.

Tintinnopsis bacillaria n. sp.
*Hada, Yoshina, 1970. (1970)
The protozoan plankton of the Antarctic and Subantarctic seas.
Scient. Repts, Japan. Antarct. Res. Exped., (E)31:51 pp.

Tintinnopsis bacoorensis n.sp.
Roxas, H.A., 1941.
Marine protozoa of the Philippines.
Philippine J. Sci. 74:91-136, 17 pls., 2 textfigs.

Tintinnopsis baltica
Balech, E., 1959.
Tintinnoinea del Mediterraneo.
Trav. Sta. Zool., Villefranche-sur-Mer, 18(17):
Reprinted from:
Trab. Inst. Español Oceanogr., Madrid, 19 Sept. 1959(28):88 pp., 22 pls.

Tintinnopsis baltica
Balech, Enrique, 1948
Tintinnoinea de Atlantida (R.O. Del Uruguay) (Protozoa, Ciliata Oligotr.). Comm. Mus. Argentino Ciencias Nat. "Bernardino Rivadavia", Ser.Cien.Zool. No.7, 23 pp., 8 pls. of 107 figs.

Tintinnopsis baltica
Balech, E., 1945.
Tintinnoinea de Quequén. Physis 20(55):1-15, 3 pls.

Tintinnopsis baltica
Biernacka, I., 1948
(Tintinnoinea in the Gulf of Gdansk and adjoining waters) Builetyn Morskiego Laboratorium Rybackiego w Gdyni dawniej-Stacji Morskiej w. Helu. No. 4:73-91, 4 text figs., 1 pl. with 21 figs. (Bull. Lab. Mar., Gdynia, formerly Bull. Sta. Hel.)

Tintinnopsis baltica
Kofoid, C. A. and A. S. Campbell, 1929
A conspectus of the marine and freshwater Ciliata belonging to the suborder Tintinnoinea, with descriptions of new species principally from the Agassiz expedition to the eastern tropical Pacific, 1904-1905. Univ. Calif. Publ. Zool. 34:1-403, 697 text figs.

Tintinnopsis baltica
Levander, K.M., 1947
Plankton gesammelt in den Jahren 1899-1910 an den Küsten Finnlands. Finnländische Hydrographisch-Biologische Untersuchunger (aus dem Wasserbiologischen Laboratorin der Societas Scientiarum Fennica) No.11:40 pp., 6 diagrams, 13 pls., tables.

Tintinnopsis baltica
Rumkówna, A., 1948
[List of the phytoplankton species occurring in the superficial water layers in the Gulf of Gdańsk] Bull. Lab. mar., Gdynia, No. 4: 139-141 with tables in back.

Tintinnopsis baltica
Sousa e Silva E. de, 1950.
Les tintinnides de la Baie de Cascais. Bull. Inst. Océan., Monaco, No. 979:28 pp., 4 pls.

Tintinnopsis bermudensis
Kofoid, C. A. and A. S. Campbell, 1929
A conspectus of the marine and freshwater Ciliata belonging to the suborder Tintinnoinea, with descriptions of new species principally from the Agassiz expedition to the eastern tropical Pacific, 1904-1905. Univ. Calif. Publ. Zool. 34:1-403, 697 text figs.

Tintinnopsis beroidea
Balech, E., 1959.
Tintinnoinea del Mediterraneo.
Trav. Sta. Zool., Villefranche-sur-Mer, 18(17):
Reprinted from:
Trab. Inst. Espanol Oceanogr., Madrid, 19 Sept. 1959(28):88 pp. 22 pls.

Tintinnopsis beroidea
Biernacka, I., 1948
(Tintinnoinea in the Gulf of Gdansk and adjoining waters) Builetyn Morskiego Laboratorium Rybackiego w Gdyni dawniej-Staoji Morskiej w. Helu. No. 4:73-91, 4 text figs., 1 pl. with 21 figs. (Bull. Lab. Mar., Gdynia, formerly Bull. Sta. Hel.)

Tintinnopsis beroidea
Braarud, T., and Adam Bursa, 1939
On the phytoplankton of the Oslo Fjord, 1933-1934. Hvalrådets Skr. No.19:1-63; 9 text figs. Reviewed. J. du Cons. 14(3): 418-420. A.C. Gardiner.

Tintinnopsis beroidea
Cosper, T.C., 1972.
The identification of tintinnids (Protozoa: Ciliata: Tintinnida) of the St. Andrew Bay system, Florida. Bull. mar. Sci. Miami 22(2): 391-418.

Tintinnopsis beroidea
Delegazione Italiana della Commissione Internazionale per l'Esplorazione Scientifica del Mediterraneo, 1941
Note sul plancton della Laguna veneta. [Memoria CCLXXIX], Arch. di Ocean. e Limn. Anno I, Fasc. I, 1941 XIX: 31-57 pp.

Tintinnopsis beroidea
Duran, Miguel, 1957.
Nota sobre algunos tintinnoineos del plancton de Puerto Rico.
Inv. Pesc., Barcelona, 8:97-120.

Tintinnopsis beroidea
Entz, G., Jr., 1909
Studien über organisation und biologie der Tintinniden. Arch. f. Protistenkunde 15:93-226, Pls. 8-21, text figs.

Tintinnopsis beroida
Gaarder, K.R., 1946
Tintinnoidea. Rep.Sci.Res."Michael Sars" N.Atlantic Deep-Sea Exped., 1910, 2(1): 37 pp., 23 text figs.

Tintinnopsis beroidea
Gillbricht, M., 1954.
Das Verhalten von Zooplankton - Vorzugsweise von Tintinnopsis beroidea Entz - gegenüber Thermohalinen Sprungschicht. Kurze Mitt., Inst. Fischereibiol., Univ. Hamburg, No. 5:32-44, 5 textfigs.

Tintinnopsis beroidea
Gold, K., 1971.
Growth characteristics of the mass-reared tintinnid Tintinnopsis beroidea. Marine Biol., 8(2): 105-108.

Tintinnopsis beroidea
Gold, K. and U. Pollingher, 1971.
Microgamete formation and the growth rate of Tintinnopsis beroidea. Mar. Biol. 11(4): 324-329.

Tintinnopsis beroidea
Grøntved, J., 1949
Investigations on the phytoplankton in the Danish Waddensea in July 1941. Medd. Komm. Danmarks Fiskeri og Havundersøgelser, ser. Plankton, 5(2):55 pp., 2 pls., 38 text figs.

Tintinnopsis beroidea
Hofker, J., 1931
Studien uber Tintinnoidea. Arch. f. Protistenk 75(3):315-402, 89 text figs.

Tintinnopsis beroidea
Jörgensen, E., 1924
Mediterranean Tintinnidae. Rept. Danish Oceanogr. Exped. 1908-10 to the Mediterranean and adjacent seas, Vol.II, Biol. J 3, 110 pp., 114 text figs.

Tintinnopsis beroidea
Jørgensen, E., 1900.
Ueber die Tintinnodeen der norwegischen Westküste Bergens Mus. Aarb., 1899(2):48 pp., 3 pls.

Tintinnopsis beroidea
Kofoid, C.A. and A.S. Campbell, 1939
Reports on the scientific results of the expedition to the eastern tropical Pacific in charge of Alexander Agassiz, by the U.S. Fish Commission steamer "Albatross" from October 1904 to March 1905----- XXXVII. The Ciliata: The Tintinnoimea. Bull. Mus. Comp.Zool. 84: 473 pp., 36 pls.

Tintinnopsis beroidea
Kofoid, C. A. and A. S. Campbell, 1929
A conspectus of the marine and freshwater Ciliata belonging to the suborder Tintinnoinea, with descriptions of new species principally from the Agassiz expedition to the eastern tropical Pacific, 1904-1905. Univ. Calif. Publ. Zool. 34:1-403, 697 text figs.

Tintinnopsis beroidea
Meunier, A., 1919
Microplankton de la Mer Flamande. 4. Les Tintinnides et Coetera. Mem. Mus. Roy. Hist. Nat., Belgique, 8(2):59pp., Pls. 22-23.

Tintinnopsis beroidea
Rampi, L., 1939
Primo contributo alla conoscenza dei tintinnoidi de Maro Ligure. Atti della Soc. Ital. di Sci. Nat. 78:67-81, 58 text figs.

Tintinnopsis beroidea
Rumkówna, A., 1948
[List of the phytoplankton species occurring in the superficial water layers in the Gulf of Gdańsk.] Bull. Lab. mar., Gdynia, No. 4: 139-141 with tables in back.

Tintinnopsis beroides
Vitiello, Pierre, 1964.
Contribution a l'etude des tintinnides de la baie d'Alger.
Pelagos, Bull. Inst. Oceanogr., Alger, 2(2):5-41
(pp numbers of this article duplicate those of previous one by Francis Bernard!)

Tintinnopsis Bornandi?
Entz, G., Jr., 1909
Studien über organisation und biologie der Tintinniden. Arch. f. Protistenkunde 15:93-226, Pls. 8-21, text figs.

Tintinnopsis bornandi
Kofoid, C. A. and A. S. Campbell, 1929
A conspectus of the marine and freshwater Ciliata belonging to the suborder Tintinnoinea, with descriptions of new species principally from the Agassiz expedition to the eastern tropical Pacific, 1904-1905. Univ. Calif. Publ. Zool. 34:1-403, 697 text figs.

Tintinnopsis bottnica
Levander, K.M., 1947
Plankton gesammelt in den Jahren 1899-1910 an den Küsten Finnlands. Finnländische Hydrographisch-Biologische Untersuchunger (aus dem Wasserbiologischen Laboratorin der Societas Scientiarum Fennica) No.11:40 pp., 6 diagrams, 13 pls., tables.

Tintinnopsis brandti
Balech, Enrique, 1968.
Algunas especies nuevas o interesantes do tintinnidos del Golfo de Mexico y Caribe.
Revta Mus. argent. Cienc. nat. Bernardino Rivadavia Inst. nac. Hidrobiol. 2(5):165-197.

Tintinnopsis brandti
Cosper, T.C., 1972.
The identification of tintinnids (Protozoa: Ciliata: Tintinnida) of the St. Andrew Bay system, Florida. Bull. mar. Sci. Miami 22(2): 391-418.

Tintinnopsis brandti
Kofoid, C. A. and A. S. Campbell, 1929
A conspectus of the marine and freshwater Ciliata belonging to the suborder Tintinnoinea, with descriptions of new species principally from the Agassiz expedition to the eastern tropical Pacific, 1904-1905. Univ. Calif. Publ. Zool. 34:1-403, 697 text figs.

Tintinnopsis brandti

Lindquist, Armin, 1959
Studien über das Zooplankton der Bottensee II. Zur Verbreitung und Zusammensetzung des Zooplanktons.
Inst. Mar. Res., Lysekil, Ser. Biol., Rept., No. 11:136 pp.

Tintinnopsis brasiliensis n. sp.

Kofoid, C. A. and A. S. Campbell, 1929
A conspectus of the marine and freshwater Ciliata belonging to the suborder Tintinnoinea, with descriptions of new species principally from the Agassiz expedition to the eastern tropical Pacific, 1904-1905. Univ. Calif. Publ. Zool. 34:1-403, 697 text figs.

Tintinnopsis bulbulus n.sp.

Meunier, A., 1919
Microplankton de la Mer Flamande. 4. Les Tintinnides et Coetera. Mem. Mus. Roy. Hist. Nat., Belgique, 8(2):59pp., Pls. 22-23.

Tintinnopsis bütschlii

Balech, E., 1959.
Tintinnoinea del Mediterraneo.
Trav. Sta. Zool., Villefranche-sur-Mer, 18(17):
Reprinted from:
Trab. Inst. Español Oceanogr., Madrid, 19 Sept. 1959(28):88pp., 22 pls.

Tintinnopsis bütschlii

Balech, Enrique, 1948
Tintinnoinea de Atlantida (R.O. Del Uruguay) (Protozoa, Ciliata Oligotr.). Comm. Mus. Argentino Ciencias Nat. "Bernardino Rivadavia", Ser.Cien.Zool. No.7, 23 pp., 8 pls. of 107 figs.

Tintinnopsis bütschlii var. mortensenii

Balech, Enrique, 1948
Tintinnoinea de Atlantida (R.O. Del Uruguay) (Protozoa, Ciliata Oligotr.). Comm. Mus. Argentino Ciencias Nat. "Bernardino Rivadavia", Ser.Cien.Zool. No.7, 23 pp., 8 pls. of 107 figs.

Tintinnopsis bütschlii

Biernacka, I., 1948
(Tintinnoinea in the Gulf of Gdansk and adjoining waters) Builetyn Morskiego Laboratorium Rybackiego w Gdyni dawniej-Stacji Morskiej w. Helu. No. 4:73-91, 4 text figs., 1 pl. with 21 figs. (Bull. Lab. Mar., Gdynia, formerly Bull. Sta. Hel.)

Tintinnopsis Butschlii

Entz, G., Jr., 1909
Studien über organisation und biologie der Tintinniden. Arch. f. Protistenkunde 15:93-226, Pls. 8-21, text figs.

Tintinnopsis bütschlii

Kofoid, C. A. and A. S. Campbell, 1929
A conspectus of the marine and freshwater Ciliata belonging to the suborder Tintinnoinea, with descriptions of new species principally from the Agassiz expedition to the eastern tropical Pacific, 1904-1905. Univ. Calif. Publ. Zool. 34:1-403, 697 text figs.

Tintinnopsis bütschlii

Rumkówna, A., 1948
[List of the phytoplankton species occurring in the superficial water layers in the Gulf of Gdańsk] Bull. Lab. mar., Gdynia, No. 4: 139-141 with tables in back.

Tintinnopsis bütschlii

Sousa e Silva E. de, 1950.
Les tintinnides de la Baie de Cascais. Bull. Inst. Océan., Monaco, No. 979:28 pp., 4 pls.

Tintinnopsis bütschlii var. mortenseni

Komarovsky, B., 1962
Tintinnina from the vicinity of the Straits of Tiran and Massawa Region. Contributions to the knowledge of the Red Sea, No. 25.
Sea Fish. Res. Sta., Haifa, Israel, Bull. No. 30:48-56.

Tintinnopsis bütschlii

Schulz, H., 1965.
Die Tintinnoinea des Elbe-Aestuars.
Arch. Fischereiwiss., 15(3):216-225.

Tintinnopsis campanula

Balech, E., 1959.
Tintinnoinea del Mediterraneo.
Trav. Sta. Zool., Villefranche-sur-Mer, 18(17):
Reprinted from:
Trab. Inst. Español Oceanogr., Madrid, 19 Sept. 1959(28):88 pp., 22 pls.

Tintinnopsis campanula

Biernacka, Izabela, 1965.
Ausscheidung gehäusebildender Substanzen durch reife Formen gewisser Arten der Gattung Tintinnopsis Stein.
Acta Protozoologica, Warszawa, 3(23):265-268.

Tintinnopsis campanula

Biernacka, I., 1952.
[Studies on the reproduction of some species of the genus Tintinnopsis Stein.](English summary).
Ann. Univ. Marie Curie Sklodowska, Sect. C, 6(6):211-247.

Tintinnopsis campanula

Biernacka, I., 1948
(Tintinnoinea in the Gulf of Gdansk and adjoining waters) Builetyn Morskiego Laboratorium Rybackiego w Gdyni dawniej-Stacji Morskiej w. Helu. No. 4:73-91, 4 text figs., 1 pl. with 21 figs. (Bull. Lab. Mar., Gdynia, formerly Bull. Sta. Hel.)

Tintinnopsis campanula

Braarud, T., 1945
A phytoplankton survey of the polluted waters of inner Oslo Fjord. Hvalrådets Skrifter, No.28, 142 pp., 19 text figs., 17 tables.

Tintinnopsis campanula

Delegazione Italiana della Commissione Internazionale per l'Esplorazione Scientifica del Mediterraneo, 1941
Note sul plancton della Laguna veneta.
[Memoria CCLXXIX], Arch. di Ocean. e Limn. Anno I, Fasc. I, 1941 XIX: 31-57 pp.

Tintinnopsis campanula

Duran, M., 1951
Contribucion al estudio de los tintinidos del plancton de los costas de Castellon. (Mediterraneo occidental). Publ. Inst. Biol. Aplic., Barcelone, 8: 101-120, 29 text figs.

Tintinnopsis campanula

Entz, G., Jr., 1909
Studien über organisation und biologie der Tintinniden. Arch. f. Protistenkunde 15:93-226, Pls. 8-21, text figs.

Tintinnopsis campanula

Gaarder, K.R., 1946
Tintinnoidea. Rep.Sci.Res. "Michael Sars" N.Atlantic Deep-Sea Exped., 1910, 2(1): 37 pp., 23 text figs.

Tintinnopsis campanula

Hofker, J., 1931
Studien uber Tintinnoidea. Arch. f. Protistenk 75(3):315-402, 89 text figs.

Tintinnopsis campanula

Jørgensen, E., 1924
Mediterranean Tintinnidae. Rept. Danish Oceanogr. Exped. 1908-10 to the Mediterranean and adjacent seas, Vol.II, Biol. J 3, 110 pp., 114 text figs.

Tintinnopsis campanula

Jørgensen, E., 1905
B. Protistplankton and the diatoms in bottom samples. Hydrographical and biological investigations in Norwegian fjords. Bergens Mus. Skr. 7: 49-225.

Tintinnopsis campanula

Jørgensen, 1900.
Ueber die Tintinnodeen der norwegischen Westküste. Bergens Mus. Aarb., 1899(2):48 pp., 3 pls.

Tintinnopsis campanula

Kofoid, C. A. and A. S. Campbell, 1929
A conspectus of the marine and freshwater Ciliata belonging to the suborder Tintinnoinea, with descriptions of new species principally from the Agassiz expedition to the eastern tropical Pacific, 1904-1905. Univ. Calif. Publ. Zool. 34:1-403, 697 text figs.

Tintinnopsis campanula

Levander, K.M., 1947
Plankton gesammelt in den Jahren 1899-1910 an den Küsten Finnlands.
Finnländische Hydrographisch-Biologische Untersuchunger (aus dem Wasserbiologischen Laboratorin der Societas Scientiarum Fennica) No.11:40 pp., 6 diagrams, 13 pls., tables.

Tintinnopsis campanula

Meunier, A., 1919
Microplankton de la Mer Flamande. 4. Les Tintinnides et Coetera. Mem. Mus. Roy. Hist. Nat., Belgique, 8(2):59pp., Pls. 22-23.

Tintinnopsis campanula

Rampi, L., 1939
Primo contributo alla conoscenza dei tintinnoidi do Mare Ligure. Atti della Soc. Ital. di Sci. Nat. 78:67-81, 58 text figs.

Tintinnopsis campanula

Rumkówna, A., 1948
[List of the phytoplankton species occurring in the superficial water layers in the Gulf of Gdańsk] Bull. Lab. mar., Gdynia, No. 4: 139-141 with tables in back.

Tintinnopsis campanula

Schulz, H., 1965.
Die Tintinnoinea des Elbe-Aestuars.
Arch. Fischereiwiss., 15(3):216-225.

Tintinnopsis campanula

Sousa e Silva E. de, 1950.
Les tintinnides de la Baie de Cascais. Bull. Inst. Océan., Monaco, No. 979:28 pp., 4 pls.

Tintinnopsis capitonis n.sp

Balech, Enrique, 1968.
Algunas especies nuevas o interesantes do tintinnidos del Golfo de Mexico y Caribe.
Revta Mus. argent. Cienc. nat. Bernardino Rivadavia Inst. nac. Hidrobiol. 2(5):165-197.

Tintinnopsis capulus

Kofoid, C. A. and A. S. Campbell, 1929
A conspectus of the marine and freshwater Ciliata belonging to the suborder Tintinnoinea, with descriptions of new species principally from the Agassiz expedition to the eastern tropical Pacific, 1904-1905. Univ. Calif. Publ. Zool. 34:1-403, 697 text figs.

Tintinnopsis chyzeri

Kofoid, C. A. and A. S. Campbell, 1929
A conspectus of the marine and freshwater Ciliata belonging to the suborder Tintinnoinea, with descriptions of new species principally from the Agassiz expedition to the eastern tropical Pacific, 1904-1905. Univ. Calif. Publ. Zool. 34:1-403, 697 text figs.

Tintinnopsis cincta

Entz, G., Jr., 1909
Studien über organisation und biologie der Tintinniden. Arch. f. Protistenkunde 15:93-226, Pls. 8-21, text figs.

Tintinnopsis cincta

Kofoid, C. A. and A. S. Campbell, 1929
A conspectus of the marine and freshwater Ciliata belonging to the suborder Tintinnoinea, with descriptions of new species principally from the Agassiz expedition to the eastern tropical Pacific, 1904-1905. Univ. Calif. Publ. Zool. 34:1-403, 697 text figs.

Tintinnopsis cincta

Rumkówna, A., 1948
[List of the phytoplankton species occurring in the superficial water layers in the Gulf of Gdańsk] Bull. Lab. mar., Gdynia, No. 4: 139-141 with tables in back.

Tintinnopsis cochleata

Kofoid, C. A. and A. S. Campbell, 1929
A conspectus of the marine and freshwater Ciliata belonging to the suborder Tintinnoinea, with descriptions of new species principally from the Agassiz expedition to the eastern tropical Pacific, 1904-1905. Univ. Calif. Publ. Zool. 34:1-403, 697 text figs.

Tintinnopsis compressa

Balech, E., 1959.
Tintinnoinea del Mediterraneo.
Trav. Sta. Zool., Villefranche-sur-Mer, 18(17):
Reprinted from:
Trab. Inst. Español Oceanogr., Madrid, 19 Sept. 1959(28):88 pp., 22 pls.

Tintinnopsis compressa

Jörgensen, E., 1924
Mediterranean Tintinnidae. Rept. Danish Oceanogr. Exped. 1908-10 to the Mediterranean and adjacent seas, Vol.II, Biol. J 3, 110 pp., 114 text figs.

Tintinnopsis compressa

Kofoid, C. A. and A. S. Campbell, 1929
A conspectus of the marine and freshwater Ciliata belonging to the suborder Tintinnoinea, with descriptions of new species principally from the Agassiz expedition to the eastern tropical Pacific, 1904-1905. Univ. Calif. Publ. Zool. 34:1-403, 697 text figs.

Tintinnopsis compressa

Marshall, S. M., 1934
The silicoflagellata and tintinnoinea. Great Barrier Reef Exped., 1928-29. Sci. Repts. 4(15):623-664, 43 text figs.

Tintinnopsis compressa

Rampi, L., 1939
Primo contributo alla conoscenza dei tintinnoidi do Maro Ligure. Atti della Soc. Ital. di Sci. Nat. 78:67-81, 58 text figs.

Tintinnopsis compressa

Sousa e Silva E. de, 1950.
Les tintinnides de la Baie de Cascais. Bull. Inst. Océan., Monaco, No. 979:28 pp., 4 pls.

Tintinnopsis (Paratintinnopsis) corniger

Balech, Enrique, 1968.
Algunas especies nuevas o interesantes do tintinnidos del Golfo de Mexico y Caribe.
Revta Mus. argent. Cienc. nat. Bernardino Rivadavia Inst. nac. Hidrobiol. 2(5):165-197.

Tintinnopsis coronata

Gaarder, K.R., 1946
Tintinnoidea. Rep.Sci.Res."Michael Sars" N.Atlantic Deep-Sea Exped., 1910, 2(1): 37 pp., 23 text figs.

Tintinnopsis coronata nom. nov.

Kofoid, C. A. and A. S. Campbell, 1929
A conspectus of the marine and freshwater Ciliata belonging to the suborder Tintinnoinea, with descriptions of new species principally from the Agassiz expedition to the eastern tropical Pacific, 1904-1905. Univ. Calif. Publ. Zool. 34:1-403, 697 text figs.

Tintinnopsis cyathus

Kofoid, C. A. and A. S. Campbell, 1929
A conspectus of the marine and freshwater Ciliata belonging to the suborder Tintinnoinea, with descriptions of new species principally from the Agassiz expedition to the eastern tropical Pacific, 1904-1905. Univ. Calif. Publ. Zool. 34:1-403, 697 text figs.

Tintinnopsis cyathus

Schulz, H., 1965.
Die Tintinnoinea des Elbe Aestuars.
Arch. Fischereiwiss., 15(3):216-225.

Tintinnopsis cyathus

Sousa e Silva E. de, 1950.
Les tintinnides de la Baie de Cascais. Bull. Inst. Océan., Monaco, No. 979:28 pp., 4 pls.

Tintinnopsis cylindrata nom. nov.

Kofoid, C. A. and A. S. Campbell, 1929
A conspectus of the marine and freshwater Ciliata belonging to the suborder Tintinnoinea, with descriptions of new species principally from the Agassiz expedition to the eastern tropical Pacific, 1904-1905. Univ. Calif. Publ. Zool. 34:1-403, 697 text figs.

Tintinnopsis cylindrica

Biernacka, I., 1948
(Tintinnoinea in the Gulf of Gdansk and adjoining waters) Builetyn Morskiego Laboratorium Rybackiego w Gdyni dawniej-Stacji Morskiej w. Helu. No. 4:73-91, 4 text figs., 1 pl. with 21 figs. (Bull. Lab. Mar., Gdynia, formerly Bull. Sta. Hel.)

Tintinnopsis cylindrica

Cosper, T.C., 1972.
The identification of tintinnids (Protozoa: Ciliata: Tintinnida) of the St. Andrew Bay system, Florida. Bull. mar. Sci. Miami 22(2): 391-418.

Tintinnopsis cylindrica

Kofoid, C. A. and A. S. Campbell, 1929
A conspectus of the marine and freshwater Ciliata belonging to the suborder Tintinnoinea, with descriptions of new species principally from the Agassiz expedition to the eastern tropical Pacific, 1904-1905. Univ. Calif. Publ. Zool. 34:1-403, 697 text figs.

Tintinnopsis cylindrica

Marshall, S. M., 1934
The silicoflagellata and tintinnoinea. Great Barrier Reef Exped., 1928-29. Sci. Repts. 4(15):623-664, 43 text figs.

Tintinnopsis cylindrica

Rumkówna, A., 1948
[List of the phytoplankton species occurring in the superficial water layers in the Gulf of Gdansk] Bull. Lab. mar., Gdynia, No. 4: 139-141 with tables in back.

Tintinnopsis cylindrica

Wang, C. C., and D. Nie, 1932
A survey of the marine protozoa of Amoy. Contr. Biol. Lab. Sci. Soc., China, Zool. Ser., 8:285-385, 89 text figs.

Tintinnopsis dadayi

Duran, M., 1957.
Nota sobre algunos tintinoineas del plancton de Puerto Rico. Invest. Pesqueria, Barcelona, 8:97-120.

Tintinnopsis dadayi

Kofoid, C. A. and A. S. Campbell, 1929
A conspectus of the marine and freshwater Ciliata belonging to the suborder Tintinnoinea, with descriptions of new species principally from the Agassiz expedition to the eastern tropical Pacific, 1904-1905. Univ. Calif. Publ. Zool. 34:1-403, 697 text figs.

Tintinnopsis Davidoffi

Marukawa, H., 1921
Plankton lists and some new species of copepods, from the northern waters of Japan. Bull. Inst. Ocean. No.384, 15 pp., 3 pls., 1 chart. Monaco

Tintinnopsis denticulata n. sp.

Kofoid, C. A. and A. S. Campbell, 1929
A conspectus of the marine and freshwater Ciliata belonging to the suborder Tintinnoinea, with descriptions of new species principally from the Agassiz expedition to the eastern tropical Pacific, 1904-1905. Univ. Calif. Publ. Zool. 34:1-403, 697 text figs.

Tintinnopsis sp cf T directa

Balech, Enrique, 1968.
Algunas especies nuevas o interesantes do tintinnidos del Golfo de Mexico y Caribe.
Revta Mus. argent. Cienc. nat. Bernardino Rivadavia Inst. nac. Hidrobiol. 2(5):165-197.

Tintinnopsis ecaudata n. sp.

Kofoid, C. A. and A. S. Campbell, 1929
A conspectus of the marine and freshwater Ciliata belonging to the suborder Tintinnoinea, with descriptions of new species principally from the Agassiz expedition to the eastern tropical Pacific, 1904-1905. Univ. Calif. Publ. Zool. 34:1-403, 697 text figs.

Tintinnopsis elongata

Kofoid, C. A. and A. S. Campbell, 1929
A conspectus of the marine and freshwater Ciliata belonging to the suborder Tintinnoinea, with descriptions of new species principally from the Agassiz expedition to the eastern tropical Pacific, 1904-1905. Univ. Calif. Publ. Zool. 34:1-403, 697 text figs.

Tintinnopsis elongata

Kofoid, C. A. and A. S. Campbell, 1929
A conspectus of the marine and freshwater Ciliata belonging to the suborder Tintinnoinea, with descriptions of new species principally from the Agassiz expedition to the eastern tropical Pacific, 1904-1905. Univ. Calif. Publ. Zool. 34:1-403, 697 text figs.

Tintinnopsis entzii

Kofoid, C. A. and A. S. Campbell, 1929
A conspectus of the marine and freshwater Ciliata belonging to the suborder Tintinnoinea, with descriptions of new species principally from the Agassiz expedition to the eastern tropical Pacific, 1904-1905. Univ. Calif. Publ. Zool. 34:1-403, 697 text figs.

Tintinnopsis everta nom. nov.

Kofoid, C. A. and A. S. Campbell, 1929
A conspectus of the marine and freshwater Ciliata belonging to the suborder Tintinnoinea, with descriptions of new species principally from the Agassiz expedition to the eastern tropical Pacific, 1904-1905. Univ. Calif. Publ. Zool. 34:1-403, 697 text figs.

Tintinnopsis fennica

Kofoid, C.A. and A.S. Campbell, 1939
Reports on the scientific results of the expedition to the eastern tropical Pacific in charge of Alexander Agassiz, by the U.S. Fish Commission steamer "Albatross" from October 1904 to March 1905----- XXXVII. The Ciliata: The Tintinnoimea. Bull.Mus. Comp.Zool. 84: 473 pp., 36 pls.

Tintinnopsis fennica nom. nov.

Kofoid, C. A. and A. S. Campbell, 1929
A conspectus of the marine and freshwater Ciliata belonging to the suborder Tintinnoinea, with descriptions of new species principally from the Agassiz expedition to the eastern tropical Pacific, 1904-1905. Univ. Calif. Publ. Zool. 34:1-403, 697 text figs.

Tintinnopsis fimbriata

*Hada, Yoshina, 1970.   (1970)

The protozoan plankton of the Antarctic and Subantarctic seas.
Scient. Repts, Japan. Antarct. Res. Exped., (E)31:51 pp.

Tintinnopsis fimbriata

Hofker, J., 1931
Studien uber Tintinnoidea. Arch. f. Protistenk 75(3):315-402, 89 text figs.

Tintinnopsis fimbriata

Kofoid, C. A. and A. S. Campbell, 1929
A conspectus of the marine and freshwater Ciliata belonging to the suborder Tintinnoinea, with descriptions of new species principally from the Agassiz expedition to the eastern tropical Pacific, 1904-1905. Univ. Calif. Publ. Zool. 34:1-403, 697 text figs.

Tintinnopsis fimbriata n.sp.

Meunier, A., 1919
Microplankton de la Mer Flamande. 4. Les Tintinnides et Coetera. Mem. Mus. Roy. Hist. Nat., Belgique, 8(2):59pp., Pls. 22-23.

Tintinnopsis fistularis n.sp.

Meunier, A., 1919
Microplankton de la Mer Flamande. 4. Les Tintinnides et Coetera. Mem. Mus. Roy. Hist. Nat., Belgique, 8(2):59pp., Pls. 22-23.

Tintinnopsis Frokubi

Marukawa, H., 1921
Plankton lists and some new species of copepods, from the northern waters of Japan. Bull. Inst. Ocean. No.384, 15 pp., 3 pls., 1 chart. Monaco

Tintinnopsis fusiformis

Kofoid, C. A. and A. S. Campbell, 1929
A conspectus of the marine and fresh-water Ciliata belonging to the suborder Tintinnoinea, with descriptions of new species principally from the Agassiz expedition to the eastern tropical Pacific, 1904-1905. Univ. Calif. Publ. Zool. 34:1-403, 697 text figs.

Tintinnopsis glans

Balech, Enrique, 1948
Tintinnoinea de Atlantida (R.O. Del Uruguay) (Protozoa, Ciliata Oligotr.). Comm. Mus. Argentino Ciencias Nat. "Bernardino Rivadavia", Ser.Cien.Zool. No.7, 23 pp., 8 pls. of 107 figs.

Tintinnopsis glans

Balech, E., 1945.
Tintinnoinea de Quequén. Physis 20(55):1-15, 3 pls.

Tintinnopsis glans

*Hada, Yoshina, 1970.   (1970)
The protozoan plankton of the Antarctic and Subantarctic seas.
Scient. Repts, Japan. Antarct. Res. Exped., (E)31:51 pp.

Tintinnopsis glans n.sp.

Meunier, A., 1919
Microplankton de la Mer Flamanda. 4. Les Tintinnides et Coetera. Mem. Mus. Roy. Hist. Nat., Belgique, 8(2):59pp., Pls. 22-23.

Tintinnopsis gracilis

Balech, Enrique, 1948
Tintinnoinea de Atlantida (R.O. Del Uruguay) (Protozoa, Ciliata Oligotr.). Comm. Mus. Argentino Ciencias Nat. "Bernardino Rivadavia", Ser.Cien.Zool. No.7, 23 pp., 8 pls. of 107 figs.

Tintinnopsis gracilis

Balech, E., 1945.
Tintinnoinea de Quequén. Physis 20(55):1-15, 3 pls.

Tintinnopsis gracilis n. sp.

Kofoid, C. A. and A. S. Campbell, 1929
A conspectus of the marine and fresh-water Ciliata belonging to the suborder Tintinnoinea, with descriptions of new species principally from the Agassiz expedition to the eastern tropical Pacific, 1904-1905. Univ. Calif. Publ. Zool. 34:1-403, 697 text figs.

Tintinnopsis gracilis

Marshall, S. M., 1934
The silicoflagellata and tintinnoinea.
Great Barrier Reef Exped., 1928-29. Sci. Repts. 4(15):623-664, 43 text figs.

Tintinnopsis gracilis

Wang, C. C., and D. Nie, 1932
A survey of the marine protozoa of Amoy.
Contr. Biol. Lab. Sci. Soc., China, Zool. Ser., 8:285-385, 89 text figs.

Tintinnopsis hemispiralis n.sp.

Kiang-teh, Y., 1956.
[Three new species of Tintinnopsis collected from Kiaochow Bay, Shantung, China.]
Acta Sci. Nat., 2(4):64-69.

Tintinnopsis heroidea

Levander, K.M., 1947
Plankton gesammelt in den Jahren 1899-1910 an den Küsten Finnlands.
Finnländische Hydrographisch-Biologische Untersuchungen (aus dem Wasserbiologischen Laboratorium der Societas Scientiarum Fennica) No.11:40 pp., 6 diagrams, 13 pls., tables.

Tintinnopsis illinoisensis

Kofoid, C. A. and A. S. Campbell, 1929
A conspectus of the marine and fresh-water Ciliata belonging to the suborder Tintinnoinea, with descriptions of new species principally from the Agassiz expedition to the eastern tropical Pacific, 1904-1905. Univ. Calif. Publ. Zool. 34:1-403, 697 text figs.

Tintinnopsis incurvata

Kofoid, C. A. and A. S. Campbell, 1929
A conspectus of the marine and fresh-water Ciliata belonging to the suborder Tintinnoinea, with descriptions of new species principally from the Agassiz expedition to the eastern tropical Pacific, 1904-1905. Univ. Calif. Publ. Zool. 34:1-403, 697 text figs.

Tintinnopsis infundibulum

Kofoid, C. A. and A. S. Campbell, 1929
A conspectus of the marine and fresh-water Ciliata belonging to the suborder Tintinnoinea, with descriptions of new species principally from the Agassiz expedition to the eastern tropical Pacific, 1904-1905. Univ. Calif. Publ. Zool. 34:1-403, 697 text figs.

Tintinnopsis karjacensis

Biernacka, I., 1948
(Tintinnoinea in the Gulf of Gdansk and adjoining waters) Builetyn Morskiego Laboratorium Rybackiego w Gdyni dawniej-Stacji Morskiej w. Helu. No. 4:73-91, 4 text figs., 1 pl. with 21 figs. (Bull. Lab. Mar., Gdynia, formerly Bull. Sta. Hel.)

Tintinnopsis karajacensis

Kofoid, C.A. and A.S. Campbell, 1939
Reports on the scientific results of the expedition to the eastern tropical Pacific in charge of Alexander Agassiz, by the U.S. Fish Commission steamer "Albatross" from October 1904 to March 1905----XXXVII. The Ciliata: The Tintinnoinea. Bull. Mus. Comp.Zool. 84: 473 pp., 36 pls.

Tintinnopsis karajacensis

Kofoid, C. A. and A. S. Campbell, 1929
A conspectus of the marine and fresh-water Ciliata belonging to the suborder Tintinnoinea, with descriptions of new species principally from the Agassiz expedition to the eastern tropical Pacific, 1904-1905. Univ. Calif. Publ. Zool. 34:1-403, 697 text figs.

Tintinnopsis Karajacensis

Rumkówna, A., 1948
[List of the phytoplankton species occurring in the superficial water layers in the Gulf of Gdańsk] Bull. Lab. mar., Gdynia, No. 4: 139-141 with tables in back.

Tintinnopsis karajacensis

Schulz, H., 1965.
Die Tintinnoinea des Elbe-Aestuars.
Arch. Fischereiwiss., 15(3):216-225.

Tintinnopsis kiaochouensis n.sp.

Kiang-teh, Y., 1956.
[Three new species of Tintinnopsis collected from Kiaochow Bay, Shantung, China.]
Acta Sci. Nat., 2(4):64-69.

Tintinnopsis kofoidi

Balech, E., 1951.
Nuevos datos sobre Tintinnoinea de Argentina y Uraguay. Physis 20(58):291-302, 16 textfigs.

Tintinnopsis kofoidi

Balech, Enrique, 1948
Tintinnoinea de Atlantida (R.O. Del Uruguay) (Protozoa, Ciliata Oligotr.). Comm. Mus. Argentino Ciencias Nat. "Bernardino Rivadavia", Ser.Cien.Zool. No.7, 23 pp., 8 pls. of 107 figs.

Tintinnopsis kofoidi

Cosper, T.C., 1972.
The identification of tintinnids (Protozoa: Ciliata: Tintinnida) of the St. Andrew Bay system, Florida. Bull. mar. Sci. Miami 22(2): 391-418.

Tintinnopsis lacustris

Biernacka, I., 1948
(Tintinnoinea in the Gulf of Gdansk and adjoining waters) Builetyn Morskiego Laboratorium Rybackiego w Gdyni dawniej-Stacji Morskiej w. Helu. No. 4:73-91, 4 text figs., 1 pl. with 21 figs. (Bull. Lab. Mar., Gdynia, formerly Bull. Sta. Hel.)

Tintinnopsis lacustris

Rumkówna, A., 1948
[List of the phytoplankton species occurring in the superficial water layers in the Gulf of Gdańsk] Bull. Lab. mar., Gdynia, No. 4: 139-141 with tables in back.

Tintinnopsis lata

Balech, E., 1945.
Tintinnoinea de Quequén. Physis 20(55):1-15, 3 pls.

Tintinnopsis lata

Kofoid, C. A. and A. S. Campbell, 1929
A conspectus of the marine and fresh-water Ciliata belonging to the suborder Tintinnoinea, with descriptions of new species principally from the Agassiz expedition to the eastern tropical Pacific, 1904-1905. Univ. Calif. Publ. Zool. 34:1-403, 697 text figs.

Tintinnopsis lata

Meunier, A., 1919
Microplankton de la Mer Flamanda. 4. Les Tintinnides et Coetera. Mem. Mus. Roy. Hist. Nat., Belgique, 8(2):59pp., Pls. 22-23.

Tintinnopsis levigata

Balech, Enrique, 1964
El plancton de Mar del Plata durante el periodo 1961-1962 (Buenos Aires, Argentina).
Bol. Inst. Biol. Mar., Mar del Plata, Argentina No. 4: 56 pp.

Tintinnopsis levigata

Balech, E., 1959.
Tintinnoinea del Mediterraneo.
Trav. Sta. Zool., Villefranche-sur-Mer, 18(17): Reprinted from:
Trab. Inst. Español Oceanogr., Madrid, 19 Sept. 1959(28):88 pp., 22 pls.

Tintinnopsis levigata

Cosper, T.C., 1972.
The identification of tintinnids (Protozoa: Ciliata: Tintinnida) of the St. Andrew Bay system, Florida. Bull. mar. Sci. Miami 22(2): 391-418.

Tintinnopsis levigata nom. nov.

Kofoid, C. A. and A. S. Campbell, 1929
A conspectus of the marine and fresh-water Ciliata belonging to the suborder Tintinnoinea, with descriptions of new species principally from the Agassiz expedition to the eastern tropical Pacific, 1904-1905. Univ. Calif. Publ. Zool. 34:1-403, 697 text figs.

Tintinnopsis levigata

Sousa e Silva E. de, 1950.
Les tintinnides de la Baie de Cascais. Bull. Inst. Océan., Monaco, No. 979:28 pp., 4 pls.

Tintinnopsis lindeni (?)

Jörgensen, E., 1924
Mediterranean Tintinnidae. Rept. Danish Oceanogr. Exped. 1908-10 to the Mediterranean and adjacent seas, Vol.II, Biol. J 3, 110 pp., 114 text figs.

Tintinnopsis lindeni

Kofoid, C. A. and A. S. Campbell, 1929
A conspectus of the marine and fresh-water Ciliata belonging to the suborder Tintinnoinea, with descriptions of new species principally from the Agassiz expedition to the eastern tropical Pacific, 1904-1905. Univ. Calif. Publ. Zool. 34:1-403, 697 text figs.

Tintinnopsis lobiancoi

Balech, E., 1945.
Tintinnoinea de Quequén. Physis 20(55):1-15, 3 pls.

Tintinnopsis lobiaco

Biernacka, I., 1948
(Tintinnoinea in the Gulf of Gdansk and adjoining waters) Builetyn Morskiego Laboratorium Rybackiego w Gdyni dawniej-Stacji Morskiej w. Helu. No. 4:73-91, 4 text figs., 1 pl. with 21 figs. (Bull. Lab. Mar., Gdynia, formerly Bull. Sta. Hel.)

Tintinnopsis Lobiancoi

Entz, G., Jr., 1909
Studien über organisation und biologie der Tintinniden. Arch. f. Protistenkunde 15:93-226, Pls. 8-21, text figs.

Tintinnopsis lobiancoi

Gaarder, K.R., 1946
Tintinnoidea. Rep.Sci.Res."Michael Sars" N.Atlantic Deep-Sea Exped., 1910, 2(1): 37 pp., 23 text figs.

Tintinnopsis lobiancoi

Kofoid, C. A. and A. S. Campbell, 1929
A conspectus of the marine and freshwater Ciliata belonging to the suborder Tintinnoinea, with descriptions of new species principally from the Agassiz expedition to the eastern tropical Pacific, 1904-1905. Univ. Calif. Publ. Zool. 34:1-403, 697 text figs.

Tintinnopsis lobiancoi

Komarovsky, B., 1962
Tintinnina from the vicinity of the Straits of Tiran and Massawa Region. Contributions to the knowledge of the Red Sea, No. 25. Sea Fish. Res. Sta., Haifa, Israel, Bull. No. 30:48-56.

Tintinnopsis Lobiancoi

Meunier, A., 1919
Microplankton de la Mer Flamande. 4. Les Tintinnides et Coetera. Mem. Mus. Roy. Hist. Nat., Belgique, 8(2):59pp., Pls. 22-23.

Tintinnopsis lobiancoi

Rampi, L., 1939
Primo contributo alla conoscenza dei tintinnoidi de Maro Ligure. Atti della Soc. Ital. di Sci. Nat. 78:67-81, 58 text figs.

Tintinnopsis lobiancoi

Rumkówna, A., 1948
[List of the phytoplankton species occurring in the superficial water layers in the Gulf of Gdansk] Bull. Lab. mar., Gdynia, No. 4: 139-141 with tables in back.

Tintinnopsis lobiancoi

Schulz, H., 1965.
Die Tintinnoinea des Elbe-Aestuars. Arch. Fischereiwiss., 15(3):216-225.

Tintinnopsis lobiancoi

Sousa e Silva E. de, 1950.
Les tintinnides de la Baie de Cascais. Bull. Inst. Océan. Monaco, No. 979:28 pp., 4 pls.

Tintinnopsis lohmanni

Biernacka, Izabela, 1965.
Ausscheidung gehäusebildender Substanzen durch reife Formen gewisser Arten der Gattung Tintinnopsis Stein. Acta Protozoologica, Warszawa, 3(23):265-268.

Tintinnopsis lohmanni

Biernacka, I., 1952.
[Studies on the reproduction of some species of the genus Tintinnopsis Stein.](English summary). Ann. Univ. Marie Curie Sklodowska, Sect. C, 6(6):211-247.

Tintinnopsis lohmanni

Biernacka, I., 1948
(Tintinnoinea in the Gulf of Gdansk and adjoining waters) Builetyn Morskiego Laboratorium Rybackiego w Gdyni dawniej-Stacji Morskiej w. Helu. No. 4:73-91, 4 text figs., 1 pl. with 21 figs. (Bull. Lab. Mar., Gdynia, formerly Bull. Sta. Hel.)

Tintinnopsis lohmanni

Hofker, J., 1931
Studien uber Tintinnoidea. Arch. f. Protistenk 75(3):315-402, 89 text figs.

Tintinnopsis lohmanni

Rumkówna, A., 1948
[List of the phytoplankton species occurring in the superficial water layers in the Gulf of Gdansk] Bull. Lab. mar., Gdynia, No. 4: 139-141 with tables in back.

Tintinnopsis loricata

Wang, C. C., and D. Nie, 1932
A survey of the marine protozoa of Amoy. Contr. Biol. Lab. Sci. Soc., China, Zool. Ser., 8:285-385, 89 text figs.

Tintinnopsis maculosa

Kofoid, C. A. and A. S. Campbell, 1929
A conspectus of the marine and freshwater Ciliata belonging to the suborder Tintinnoinea, with descriptions of new species principally from the Agassiz expedition to the eastern tropical Pacific, 1904-1905. Univ. Calif. Publ. Zool. 34:1-403, 697 text figs.

Tintinnopsis magna

Kofoid, C. A. and A. S. Campbell, 1929
A conspectus of the marine and freshwater Ciliata belonging to the suborder Tintinnoinea, with descriptions of new species principally from the Agassiz expedition to the eastern tropical Pacific, 1904-1905. Univ. Calif. Publ. Zool. 34:1-403, 697 text figs.

Tintinnopsis major

Biernacka, I., 1948
(Tintinnoinea in the Gulf of Gdansk and adjoining waters) Builetyn Morskiego Laboratorium Rybackiego w Gdyni dawniej-Stacji Morskiej w. Helu. No. 4:73-91, 4 text figs., 1 pl. with 21 figs. (Bull. Lab. Mar., Gdynia, formerly Bull. Sta. Hel.)

Tintinnopsis major

Kofoid, C. A. and A. S. Campbell, 1929
A conspectus of the marine and freshwater Ciliata belonging to the suborder Tintinnoinea, with descriptions of new species principally from the Agassiz expedition to the eastern tropical Pacific, 1904-1905. Univ. Calif. Publ. Zool. 34:1-403, 697 text figs.

Tintinnopsis major

Rumkówna, A., 1948
[List of the phytoplankton species occurring in the superficial water layers in the Gulf of Gdansk] Bull. Lab. mar., Gdynia, No. 4: 139-141 with tables in back.

Tintinnopsis manilensis n.sp.

Roxas, H.A., 1941.
Marine protozoa of the Philippines. Philippine J. Sci. 74:91-136, 17 pls., 2 textfigs.

Tintinnopsis mayeri

Kofoid, C. A. and A. S. Campbell, 1929
A conspectus of the marine and freshwater Ciliata belonging to the suborder Tintinnoinea, with descriptions of new species principally from the Agassiz expedition to the eastern tropical Pacific, 1904-1905. Univ. Calif. Publ. Zool. 34:1-403, 697 text figs.

Tintinnopsis meunieri

Biernacka, I., 1952.
[Studies on the reproduction of some species of the genus Tintinnopsis Stein.](English summary). Ann. Univ. Marie Curie Sklodowska, Sect. C, 6(6):211-247.

Tintinnopsis meunieri

Biernacka, I., 1948
(Tintinnoinea in the Gulf of Gdansk and adjoining waters) Builetyn Morskiego Laboratorium Rybackiego w Gdyni dawniej-Stacji Morskiej w. Helu. No. 4:73-91, 4 text figs., 1 pl. with 21 figs. (Bull. Lab. Mar., Gdynia, formerly Bull. Sta. Hel.)

Tintinnopsis meunieri

Dolgopol'skaia, M.A., and V.L. Pauli, 1964.
Plankton of the Azov Sea. (In Russian). Trudy. Sevastopol Biol. Stants., Akad. Nauk. SSSR, 15:118-151.

Tintinnopsis meunieri nom. nov.

Kofoid, C. A. and A. S. Campbell, 1929
A conspectus of the marine and freshwater Ciliata belonging to the suborder Tintinnoinea, with descriptions of new species principally from the Agassiz expedition to the eastern tropical Pacific, 1904-1905. Univ. Calif. Publ. Zool. 34:1-403, 697 text figs.

Tintinnopsis meunieri

Rumkówna, A., 1948
[List of the phytoplankton species occurring in the superficial water layers in the Gulf of Gdansk] Bull. Lab. mar., Gdynia, No. 4: 139-141 with tables in back.

Tintinnopsis meunieri

Schulz, H., 1965.
Die Tintinnoinea des Elbe-Aestuars. Arch. Fischereiwiss., 15(3):216-225.

Tintinnopsis minima n.sp.

Wang, C. C., and D. Nie, 1932
A survey of the marine protozoa of Amoy. Contr. Biol. Lab. Sci. Soc., China, Zool. Ser., 8:285-385, 89 text figs.

Tintinnopsis minuta

Kofoid, C. A. and A. S. Campbell, 1929
A conspectus of the marine and freshwater Ciliata belonging to the suborder Tintinnoinea, with descriptions of new species principally from the Agassiz expedition to the eastern tropical Pacific, 1904-1905. Univ. Calif. Publ. Zool. 34:1-403, 697 text figs.

Tintinnopsis minuta

Rampi, L., 1939
Primo contributo alla conoscenza dei tintinnoidi de Maro Ligure. Atti della Soc. Ital. di Sci. Nat. 78:67-81, 58 text figs.

Tintinnopsis mortenseni

Cosper, T.C., 1972.
The identification of tintinnids (Protozoa: Ciliata: Tintinnida) of the St. Andrew Bay system, Florida. Bull. mar. Sci. Miami 22(2): 391-418.

Tintinnopsis mortensenii

Duran, M., 1957.
Nota sobre algunos tintinnoineas del plancton de Puerto Rico. Invest. Persqueria, Barcelona, 8: 97-120.

Tintinnopsis mortensenii

Kofoid, C.A. and A.S. Campbell, 1939
Reports on the scientific results of the expedition to the eastern tropical Pacific in charge of Alexander Agassiz, by the U.S. Fish Commission steamer "Albatross from October 1904 to March 1905------XXXVII The Ciliata: The Tintinnoinea. Bull. Mus. Comp. Zool. 84: 473 pp., 36 pls.

Tintinnopsis mortensii

Kofoid, C. A. and A. S. Campbell, 1929
A conspectus of the marine and freshwater Ciliata belonging to the suborder Tintinnoinea, with descriptions of new species principally from the Agassiz expedition to the eastern tropical Pacific, 1904-1905. Univ. Calif. Publ. Zool. 34:1-403, 697 text figs.

Tintinnopsis mortensenii

Marshall, S. M., 1934
The silicoflagellata and tintinnoinea. Great Barrier Reef Exped., 1928-29. Sci. Repts. 4(15):623-664, 43 text figs.

Tintinnopsis Mortenseni

Marukawa, H., 1921
Plankton lists and some new species of copepods, from the northern waters of Japan. Bull. Inst. Ocean. No.384, 15 pp., 3 pls., 1 chart. Monaco

Tintinnopsis mulctrella n. sp.

Kofoid, C. A. and A. S. Campbell, 1929
A conspectus of the marine and freshwater Ciliata belonging to the suborder Tintinnoinea, with descriptions of new species principally from the Agassiz expedition to the eastern tropical Pacific, 1904-1905. Univ. Calif. Publ. Zool. 34:1-403, 697 text figs.

Tintinnopsis nana

Balech, E., 1959.
Tintinnoinea del Mediterraneo. Trav. Sta. Zool., Villefranche-sur-Mer, 18(17): Reprinted from: Trab. Inst. Espanol Oceanogr., Madrid, 19 Sept. 1959(28):88 pp., 22 pls.

Tintinnopsis nana

Hofker, J., 1931
Studien über Tintinnoidea. Arch. f. Protistenk 75(3):315-402, 89 text figs.

Tintinnopsis nana

Kofoid, C. A. and A. S. Campbell, 1929
A conspectus of the marine and fresh-water Ciliata belonging to the suborder Tintinnoinea, with descriptions of new species principally from the Agassiz expedition to the eastern tropical Pacific, 1904-1905. Univ. Calif. Publ. Zool. 34:1-403, 697 text figs.

Tintinnopsis nitida

Jørgensen, E., 1905
B. Protistplankton and the diatoms in bottom samples. Hydrographical and biological investigations in Norwegian fjords. Bergens Mus. Skr. 7: 49-225.

Tintinnopsis nitida

Kofoid, C. A. and A. S. Campbell, 1929
A conspectus of the marine and fresh-water Ciliata belonging to the suborder Tintinnoinea, with descriptions of new species principally from the Agassiz expedition to the eastern tropical Pacific, 1904-1905. Univ. Calif. Publ. Zool. 34:1-403, 697 text figs.

Tintinnopsis nitida

Rumkówna, A., 1948
[List of the phytoplankton species occurring in the superficial water layers in the Gulf of Gdańsk] Bull. Lab. mar., Gdynia, No. 4: 139-141 with tables in back.

Tintinnopsis nucula

Biernacka, I., 1948
(Tintinnoinea in the Gulf of Gdansk and adjoining waters) Builetyn Morskiego Laboratorium Rybackiego w Gdyni dawniej-Stacji Morskiej w. Helu. No. 4:73-91, 4 text figs., 1 pl. with 21 figs. (Bull. Lab. Mar., Gdynia, formerly Bull. Sta. Hel.)

Tintinnopsis nucula

Entz, G., Jr. 1909
Studien über organisation und biologie der Tintinniden. Arch. f. Protistenkunde 15:93-226, Pls. 8-21, text figs.

Tintinnopsis nucula

Kofoid, C. A. and A. S. Campbell, 1929
A conspectus of the marine and fresh-water Ciliata belonging to the suborder Tintinnoinea, with descriptions of new species principally from the Agassiz expedition to the eastern tropical Pacific, 1904-1905. Univ. Calif. Publ. Zool. 34:1-403, 697 text figs.

Tintinnopsis nucula

Rumkówna, A., 1948
[List of the phytoplankton species occurring in the superficial water layers in the Gulf of Gdańsk] Bull. Lab. mar., Gdynia, No. 4: 139-141 with tables in back.

Tintinnopsis orientalis n. sp

Kofoid, C. A. and A. S. Campbell, 1929
A conspectus of the marine and fresh-water Ciliata belonging to the suborder Tintinnoinea, with descriptions of new species principally from the Agassiz expedition to the eastern tropical Pacific, 1904-1905. Univ. Calif. Publ. Zool. 34:1-403, 697 text figs.

Tintinnopsis ornata n.sp.

Kofoid, C.A. and A.S. Campbell, 1939
Reports on the scientific results of the expedition to the eastern tropical Pacific in charge of Alexander Agassiz, by the U.S. Fish Commission steamer "Albatross" from October 1904 to March 1905-----XXXVII. The Ciliata: The Tintinnoimea. Bull. Mus. Comp. Zool. 84: 473 pp., 36 pls.

Tintinnopsis ovalis

Kofoid, C. A. and A. S. Campbell, 1929
A conspectus of the marine and fresh-water Ciliata belonging to the suborder Tintinnoinea, with descriptions of new species principally from the Agassiz expedition to the eastern tropical Pacific, 1904-1905. Univ. Calif. Publ. Zool. 34:1-403, 697 text figs.

Tintinnopsis pallida

Kofoid, C. A. and A. S. Campbell, 1929
A conspectus of the marine and fresh-water Ciliata belonging to the suborder Tintinnoinea, with descriptions of new species principally from the Agassiz expedition to the eastern tropical Pacific, 1904-1905. Univ. Calif. Publ. Zool. 34:1-403, 697 text figs.

Tintinnopsis pamamensis

Kofoid, C.A. and A.S. Campbell, 1939
Reports on the scientific results of the expedition to the eastern tropical Pacific in charge of Alexander Agassiz, by the U.S. Fish Commission steamer "Albatross" from October 1904 to March 1905-----XXXVII. The Ciliata: The Tintinnoimea. Bull. Mus. Comp. Zool. 84: 473 pp., 36 pls.

Tintinnopsis panamaensis n. sp.

Kofoid, C. A. and A. S. Campbell, 1929
A conspectus of the marine and fresh-water Ciliata belonging to the suborder Tintinnoinea, with descriptions of new species principally from the Agassiz expedition to the eastern tropical Pacific, 1904-1905. Univ. Calif. Publ. Zool. 34:1-403, 697 text figs.

Tintinnopsis parva

Balech, E., 1951.
Nuevos datos sobre Tintinnoinea de Argentina y Uraguay. Physis 20(58):291-302, 16 textfigs.

Tintinnopsis parva?

Balech, Enrique, 1948
Tintinnoinea de Atlantida (R.O. Del Uruguay) (Protozoa, Ciliata Oligotr.). Comm. Mus. Argentino Ciencias Nat. "Bernardino Rivadavia", Ser.Cien.Zool. No.7, 23 pp., 8 pls. of 107 figs.

Tintinnopsis parva

Biernacka, I., 1948
(Tintinnoinea in the Gulf of Gdansk and adjoining waters) Builetyn Morskiego Laboratorium Rybackiego w Gdyni dawniej-Stacji Morskiej w. Helu. No. 4:73-91, 4 text figs., 1 pl. with 21 figs. (Bull. Lab. Mar., Gdynia, formerly Bull. Sta. Hel.)

Tintinnopsis parva

Kofoid, C. A. and A. S. Campbell, 1929
A conspectus of the marine and fresh-water Ciliata belonging to the suborder Tintinnoinea, with descriptions of new species principally from the Agassiz expedition to the eastern tropical Pacific, 1904-1905. Univ. Calif. Publ. Zool. 34:1-403, 697 text figs.

Tintinnopsis parva

Rumkówna, A., 1948
[List of the phytoplankton species occurring in the superficial water layers in the Gulf of Gdańsk] Bull. Lab. mar., Gdynia, No. 4: 139-141 with tables in back.

Tintinnopsis parvula

Aurich, H. J., 1949.
Die Verbreitung des Nannoplanktons im Oberflächenwasser vor der Nordfriesischen Kuste. Ber. Deutschen Wiss. Komm. f. Meeresf., n.f., II(4): 403-405, 2 figs.

Tintinnopsis pavula

Balech, E., 1959.
Tintinnoinea del Mediterraneo. Trav. Sta. Zool., Villefranche-sur-Mer, 18(17): Reprinted from: Trab. Inst. Español Oceanogr., Madrid, 19 Sept. 1959(28):88 pp., 22 pls.

Tintinnopsis parvula

Balech, Enrique, 1948
Tintinnoinea de Atlantida (R.O. Del Uruguay) (Protozoa, Ciliata Oligotr.). Comm. Mus. Argentino Ciencias Nat. "Bernardino Rivadavia", Ser.Cien.Zool. No.7, 23 pp., 8 pls. of 107 figs.

Tintinnopsis parvula

Balech, E., 1945.
Tintinnoinea de Quequén. Physis 20(55):1-15, 3 pls.

Tintinnopsis parvula

Biernacka, I., 1948
(Tintinnoinea in the Gulf of Gdansk and adjoining waters) Builetyn Morskiego Laboratorium Rybackiego w Gdyni dawniej-Stacji Morskiej w. Helu. No. 4:73-91, 4 text figs., 1 pl. with 21 figs. (Bull. Lab. Mar., Gdynia, formerly Bull. Sta. Hel.)

Tintinnopsis parvula

Cosper, T.C., 1972.
The identification of tintinnids (Protozoa: Ciliata: Tintinnida) of the St. Andrew Bay system, Florida. Bull. mar. Sci. Miami 22(2): 391-418.

Tintinnopsis parvula

Gaarder, K.R., 1946
Tintinnoidea. Rep.Sci.Res. "Michael Sars" N.Atlantic Deep-Sea Exped., 1910, 2(1): 37 pp., 23 text figs.

Tintinnopsis parvula

Levander, K.M., 1947
Plankton gesammelt in den Jahren 1899-1910 an den Küsten Finnlands. Finnländische Hydrographisch-Biologische Untersuchunger (aus dem Wasserbiologischen Laboratorin der Societas Scientiarum Fennica) No.11:40 pp., 6 diagrams, 13 pls., tables.

Tintinnopsis parvula

Rumkówna, A., 1948
[List of the phytoplankton species occurring in the superficial water layers in the Gulf of Gdańsk] Bull. Lab. mar., Gdynia, No. 4: 139-141 with tables in back.

Tintinnopsis parvula

Schulz, H., 1965.
Die Tintinnoinea des Elbe-Aestuars. Arch. Fischereiwiss., 15(3):216-225.

Tintinnopsis patens n.sp.

Hasle, Grethe Rytter, 1960
Phytoplankton and ciliate species from the Tropical Pacific. Skr. Norske Videnskaps-Akad., Oslo, 1. Mat.-Nat. Kl., 1960(2): 1-50.

Tintinnopsis patula

Kofoid, C. A. and A. S. Campbell, 1929
A conspectus of the marine and fresh-water Ciliata belonging to the suborder Tintinnoinea, with descriptions of new species principally from the Agassiz expedition to the eastern tropical Pacific, 1904-1905. Univ. Calif. Publ. Zool. 34:1-403, 697 text figs.

Tintinnopsis penrhyrensis n.sp.

Campbell, A. S., 1942
The oceanic Tintinnoina of the plankton gathered during the last cruise of the CARNEGIE. Sci. Res. Cruise VII of the Carnegie, 1928-1929, Biol. 2:163 pp., 128 figs.

Tintinnopsis petapa n. sp. (1970)

*Hada, Yoshina, 1970.
The protozoan plankton of the Antarctic and Subantarctic seas. Scient. Repts. Japan. Antarct. Res. Exped., (E)31:51 pp.

Tintinnopsis pistillum

Kofoid, C. A. and A. S. Campbell, 1929
A conspectus of the marine and fresh-water Ciliata belonging to the suborder Tintinnoinea, with descriptions of new species principally from the Agassiz expedition to the eastern tropical Pacific, 1904-1905. Univ. Calif. Publ. Zool. 34:1-403, 697 text figs.

Tintinnopsis plagiostoma

Kofoid, C.A. and A.S. Campbell, 1939
Reports on the scientific results of the expedition to the eastern tropical Pacific in charge of Alexander Agassiz, by the U.S. Fish Commission steamer "Albatross" from October 1904 to March 1905-----XXXVII. The Ciliata: The Tintinnoimees. Bull. Mus. Comp. Zool. 84: 473 pp., 36 pls.

Tintinnopsis plagiostoma

Kofoid, C. A. and A. S. Campbell, 1929
A conspectus of the marine and fresh-water Ciliata belonging to the suborder Tintinnoinea, with descriptions of new species principally from the Agassiz expedition to the eastern tropical Pacific, 1904-1905. Univ. Calif. Publ. Zool. 34:1-403, 697 text figs.

Tintinnopsis plagiostoma
Sousa e Silva E. de, 1950.
Les tintinnides de la Baie de Cascais. Bull. Inst. Ocean., Monaco, No. 979:28 pp., 4 pls.

Tintinnopsis platensis
Balech, E., 1945.
Tintinnoinea de Quequén. Physis 20(55):1-15, 3 pls.

Tintinnopsis platensis
Kofoid, C. A. and A. S. Campbell, 1929
A conspectus of the marine and freshwater Ciliata belonging to the suborder Tintinnoinea, with descriptions of new species principally from the Agassiz expedition to the eastern tropical Pacific, 1904-1905. Univ. Calif. Publ. Zool. 34:1-403, 697 text figs.

Tintinnopsis producta n.sp.
Meunier, A., 1919
Microplankton de la Mer Flamande. 4. Les Tintinnides et Coetera. Mem. Mus. Roy. Hist. Nat., Belgique, 8(2):59pp., Pls. 22-23.

Tintinnopsis prowazeki
Kofoid, C. A. and A. S. Campbell, 1929
A conspectus of the marine and freshwater Ciliata belonging to the suborder Tintinnoinea, with descriptions of new species principally from the Agassiz expedition to the eastern tropical Pacific, 1904-1905. Univ. Calif. Publ. Zool. 34:1-403, 697 text figs.

Tintinnopsis radix
Balech, E., 1959.
Tintinnoinea del Mediterraneo. Trav. Sta. Zool., Villefranche-sur-Mer, 18(17): Reprinted from: Trab. Inst. Español Oceanogr., Madrid, 19 Sept. 1959(28):88 pp., 22 pls.

Tintinnopsis radix
Balech, E., 1951.
Nuevos datos sobre Tintinnoinea de Argentina y Uruguay. Physis 20(58):291-302, 16 textfigs.

Tintinnopsis radix
Balech, E., 1945.
Tintinnoinea de Quequén. Physis 20(55):1-15, 3 pls.

Tintinnopsis radix
Cosper, T.C., 1972.
The identification of tintinnids (Protozoa: Ciliata: Tintinnida) of the St. Andrew Bay system, Florida. Bull. mar. Sci. Miami 22(2): 391-418.

Tintinnopsis radix
Delegazione Italiana della Commissione Internazionale per l'Esplorazione Scientifica del Mediterraneo, 1941
Note sul plancton della Laguna veneta. [Memoria CCLXXXIX], Arch. di Ocean. e Limn. Anno I, Fasc. I, 1941 XIX: 31-57 pp.

Tintinnopsis radix
Duran, M., 1951
Contribucion al estudio de los tintinidos del plancton de los costas de Castellon. (Mediterraneo occidental). Publ. Inst. Biol. Aplic., Barcelone, 8: 101-120, 29 text figs.

Tintinnopsis radix
*Hada, Yoshina, 1970.   (1970)
The protozoan plankton of the Antarctic and Subantarctic seas. Scient. Repts, Japan. Antarct. Res. Exped., (E)31:51 pp.

Tintinnopsis radix
Jörgensen, E., 1924
Mediterranean Tintinnidae. Rept. Danish Oceanogr. Exped. 1908-10 to the Mediterranean and adjacent seas, Vol.II, Biol. J 3, 110 pp., 114 text figs.

Tintinnopsis radix
Kofoid, C.A. and A.S. Campbell, 1939
Reports on the scientific results of the expedition to the eastern tropical Pacific in charge of Alexander Agassiz, by the U.S. Fish Commission steamer"Albatross" from October 1904 to March 1905-----XXXVII. The Ciliata: The Tintinnoimea. Bull.Mus. Comp.Zool. 84: 473 pp., 36 pls.

Tintinnopsis radix
Kofoid, C. A. and A. S. Campbell, 1929
A conspectus of the marine and freshwater Ciliata belonging to the suborder Tintinnoinea, with descriptions of new species principally from the Agassiz expedition to the eastern tropical Pacific, 1904-1905. Univ. Calif. Publ. Zool. 34:1-403, 697 text figs.

Tintinnopsis radix
Komarovsky, B., 1962
Tintinnina from the vicinity of the Straits of Tiran and Massawa Region. Contributions to the knowledge of the Red Sea, No. 25. Sea Fish. Res. Sta., Haifa, Israel, Bull. No. 30:48-56.

Tintinnopsis radix
Macdonald, R., 1933
An examination of plankton hauls made in the Suez Canal during the year 1928. Fish. Res. Dis., Notes & Mem. No.3, 11 pp., 1 chart.

Tintinnopsis radix
Marshall, S. M., 1934
The silicoflagellata and tintinnoinea. Great Barrier Reef Exped., 1928-29. Sci. Repts. 4(15):623-664, 43 text figs.

Tintinnopsis radix
Rampi, L., 1939
Primo contributo alla conoscenza dei tintinnoidi de Mare Ligure. Atti della Soc. Ital. di Sci. Nat. 78:67-81, 58 text figs.

Tintinnopsis rapa
Gaarder, K.R., 1946
Tintinnoidea. Rep.Sci.Res."Michael Sars" N.Atlantic Deep-Sea Exped., 1910, 2(1): 37 pp., 23 text figs.

Tintinnopsis rapa
Kofoid, C. A. and A. S. Campbell, 1929
A conspectus of the marine and freshwater Ciliata belonging to the suborder Tintinnoinea, with descriptions of new species principally from the Agassiz expedition to the eastern tropical Pacific, 1904-1905. Univ. Calif. Publ. Zool. 34:1-403, 697 text figs.

Tintinnopsis rara
Campbell, A. S., 1942
The oceanic Tintinnoina of the plankton gathered during the last cruise of the CARNEGIE. Sci. Res. Cruise VII of the Carnegie, 1928-1929, Biol. 2:163 pp., 128 figs.

Tintinnopsis rara n.sp.
Kofoid, C.A. and A.S. Campbell, 1939
Reports on the scientific results of the expedition to the eastern tropical Pacific in charge of Alexander Agassiz, by the U.S. Fish Commission steamer"Albatross" from October 1904 to March 1905-----XXXVII. The Ciliata: The Tintinnoimea. Bull.Mus. Comp.Zool. 84: 473 pp., 36 pls.

Tintinnopsis reflexa
Kofoid, C. A. and A. S. Campbell, 1929
A conspectus of the marine and freshwater Ciliata belonging to the suborder Tintinnoinea, with descriptions of new species principally from the Agassiz expedition to the eastern tropical Pacific, 1904-1905. Univ. Calif. Publ. Zool. 34:1-403, 697 text figs.

Tintinnopsis rotundata
Kofoid, C. A. and A. S. Campbell, 1929
A conspectus of the marine and freshwater Ciliata belonging to the suborder Tintinnoinea, with descriptions of new species principally from the Agassiz expedition to the eastern tropical Pacific, 1904-1905. Univ. Calif. Publ. Zool. 34:1-403, 697 text figs.

Tintinnopsis rotundata
Marshall, S. M., 1934
The silicoflagellata and tintinnoinea. Great Barrier Reef Exped., 1928-29. Sci. Repts. 4(15):623-664, 43 text figs.

Tintinnopsis rotundata
Wang, C. C., and D. Nie, 1932
A survey of the marine protozoa of Amoy. Contr. Biol. Lab. Sci. Soc., China, Zool. Ser., 8:285-385, 89 text figs.

Tintinnopsis sacculus
Kofoid, C.A. and A.S. Campbell, 1939
Reports on the scientific results of the expedition to the eastern tropical Pacific in charge of Alexander Agassiz, by the U.S. Fish Commission steamer"Albatross" from October 1904 to March 1905-----XXXVII. The Ciliata: The Tintinnoimea. Bull.Mus. Comp.Zool. 84: 473 pp., 36 pls.

Tintinnopsis sacculus
Kofoid, C. A. and A. S. Campbell, 1929
A conspectus of the marine and freshwater Ciliata belonging to the suborder Tintinnoinea, with descriptions of new species principally from the Agassiz expedition to the eastern tropical Pacific, 1904-1905. Univ. Calif. Publ. Zool. 34:1-403, 697 text figs.

Tintinnopsis schotti
Kofoid, C.A. and A.S. Campbell, 1939
Reports on the scientific results of the expedition to the eastern tropical Pacific in charge of Alexander Agassiz, by the U.S. Fish Commission steamer"Albatross" from October 1904 to March 1905-----XXXVII. The Ciliata: The Tintinnoimea. Bull.Mus. Comp.Zool. 84: 473 pp., 36 pls.

Tintinnopsis schotti
Kofoid, C. A. and A. S. Campbell, 1929
A conspectus of the marine and freshwater Ciliata belonging to the suborder Tintinnoinea, with descriptions of new species principally from the Agassiz expedition to the eastern tropical Pacific, 1904-1905. Univ. Calif. Publ. Zool. 34:1-403, 697 text figs.

Tintinnopsis sinuata
Kofoid, C. A. and A. S. Campbell, 1929
A conspectus of the marine and freshwater Ciliata belonging to the suborder Tintinnoinea, with descriptions of new species principally from the Agassiz expedition to the eastern tropical Pacific, 1904-1905. Univ. Calif. Publ. Zool. 34:1-403, 697 text figs.

Tintinnopsis spiralis nom. nov.
Kofoid, C. A. and A. S. Campbell, 1929
A conspectus of the marine and freshwater Ciliata belonging to the suborder Tintinnoinea, with descriptions of new species principally from the Agassiz expedition to the eastern tropical Pacific, 1904-1905. Univ. Calif. Publ. Zool. 34:1-403, 697 text figs.

Tintinnopsis spiralis
Wang, C. C., and D. Nie, 1932
A survey of the marine protozoa of Amoy. Contr. Biol. Lab. Sci. Soc., China, Zool. Ser., 8:285-385, 89 text figs.

Tintinnopsis strigosa
Kofoid, C. A. and A. S. Campbell, 1929
A conspectus of the marine and freshwater Ciliata belonging to the suborder Tintinnoinea, with descriptions of new species principally from the Agassiz expedition to the eastern tropical Pacific, 1904-1905. Univ. Calif. Publ. Zool. 34:1-403, 697 text figs.

Tintinnopsis strigosa n.sp.
Meunier, A., 1919
Microplankton de la Mer Flamande. 4. Les Tintinnides et Coetera. Mem. Mus. Roy. Hist. Nat., Belgique, 8(2):59pp., Pls. 22-23.

Tintinnopsis subacuta
Biernacka, I., 1952.
[Studies on the reproduction of some species of the genus Tintinnopsis Stein.](English summary). Ann. Univ. Marie Curie Sklodowska, Sect. C, 6(6):211-247.

Tintinnopsis subacuta
Biernacka, I., 1948
(Tintinnoinea in the Gulf of Gdansk and adjoining waters) Builetyn Morskiego Laboratorium Rybackiego w Gdyni dawniej Stacji Morskiej w. Helu. No. 4:73-91, 4 text figs., 1 pl. with 21 figs. (Bull. Lab. Mar., Gdynia, formerly Bull. Sta. Hel.)

Tintinnopsis subacuta
Dolgopol'skaia, M.A., and V.L. Pauli, 1964.
Plankton of the Azov Sea. (In Russian).
Trudy. Sevastopol Biol. Stants., Akad. Nauk.
SSSR, 15:118-151.

Tintinnopsis subacuta n.sp.
Jørgensen, E., 1900.
Ueber die Tintinnodeen der norwegischen Westküste
Bergens Mus. Aarb., 1899(2):48 pp., 3 pls.

Tintinnopsis subacuta
Kofoid, C. A. and A. S. Campbell, 1929
A conspectus of the marine and fresh-water Ciliata belonging to the suborder Tintinnoinea, with descriptions of new species principally from the Agassiz expedition to the eastern tropical Pacific, 1904-1905. Univ. Calif. Publ. Zool. 34:1-403, 697 text figs.

Tintinnopsis subacuta
Rumkówna, A., 1948
[List of the phytoplankton species occurring in the superficial water layers in the Gulf of Gdańsk] Bull. Lab. mar., Gdynia, No. 4: 139-141 with tables in back.

Tintinnopsis subacuta
Schulz, H., 1965.
Die Tintinnoinea des Elbe-Aestuars.
Arch. Fischereiwiss., 15(3):216-225.

Tintinnopsis tenuis
Balech, Enrique, 1968.
Algunas especies nuevas o interesantes do tintinnidos del Golfo de Mexico y Caribe.
Revta Mus. argent. Cienc. nat. Bernardino Rivadavia Inst. nac. Hidrobiol. 2(5):165-197.

Tintinnopsis tocantinensis?
Balech, Enrique, 1948
Tintinnoinea de Atlantida (R.O. Del Uruguay) (Protozoa, Ciliata Oligotr.). Comm. Mus. Argentino Ciencias Nat. "Bernardino Rivadavia", Ser.Cien.Zool. No.7, 23 pp., 8 pls. of 107 figs.

Tintinnopsis tocantinensis
Cosper, T.C., 1972.
The identification of tintinnids (Protozoa: Ciliata: Tintinnida) of the St. Andrew Bay system, Florida. Bull. mar. Sci. Miami 22(2): 391-418.

Tintinnopsis tocantinensis n. sp.
Kofoid, C. A. and A. S. Campbell, 1929
A conspectus of the marine and fresh-water Ciliata belonging to the suborder Tintinnoinea, with descriptions of new species principally from the Agassiz expedition to the eastern tropical Pacific, 1904-1905. Univ. Calif. Publ. Zool. 34:1-403, 697 text figs.

Tintinnopsis tocantinensis
Marshall, S. M., 1934
The silicoflagellata and tintinnoinea.
Great Barrier Reef Exped., 1928-29. Sci. Repts. 4(15):623-664, 43 text figs.

Tintinnopsis tocantinensis
Wang, C. C., and D. Nie, 1932
A survey of the marine protozoa of Amoy.
Contr. Biol. Lab. Sci. Soc., China, Zool. Ser., 8:285-385, 89 text figs.

Tintinnopsis tregouboffi n.sp.
Balech, Enrique, 1964
El plancton de Mar del Plata durante el periodo 1961-1962 (Buenos Aires, Argentina).
Bol. Inst. Biol. Mar., Mar del Plata, Argentina No. 4: 49 pp.
56

Tintinnopsis tsingtaoensis n.sp.
Kiang-teh, Y., 1956.
[Three new species of Tintinnopsis collected from Kiaochow Bay, Shantung, China.]
Acta Sci. Nat., 2(4):64-69.

Tintinnopsis tubulosoides
Balech, Enrique, 1944
Contribucion al conocimiento del Plancton de Lennox y Cabo de Hornos. Physis XIX:423-446, 6 pls. with 67 figs.

Tintinnopsis tubulosoides
Kofoid, C. A. and A. S. Campbell, 1929
A conspectus of the marine and fresh-water Ciliata belonging to the suborder Tintinnoinea, with descriptions of new species principally from the Agassiz expedition to the eastern tropical Pacific, 1904-1905. Univ. Calif. Publ. Zool. 34:1-403, 697 text figs.

Tintinnopsis tubulosa
Biernacka, I., 1948
(Tintinnoinea in the Gulf of Gdansk and adjoining waters) Builetyn Morskiego Laboratorium Rybackiego w Gdyni dawniej-Stacji Morskiej w. Helu. No. 4:73-91, 4 text figs., 1 pl. with 21 figs. (Bull. Lab. Mar., Gdynia, formerly Bull. Sta. Hel.)

Tintinnopsis tubulosa
Cosper, T.C., 1972.
The identification of tintinnids (Protozoa: Ciliata: Tintinnida) of the St. Andrew Bay system, Florida. Bull. mar. Sci. Miami 22(2): 391-418.

Tintinnopsis tubulosa
Grøntved, J., 1949
Investigations on the phytoplankton in the Danish Waddensea in July 1941. Medd. Komm. Danmarks Fiskeri og Havundersøgelser, ser. Plankton, 5(2):55 pp., 2 pls., 38 text figs.

Tintinnopsis tubulosa
Hofker, J., 1931
Studien uber Tintinnoidea. Arch. f. Protistenk 75(3):315-402, 89 text figs.

Tintinnopsis tubulosa
Kofoid, C. A. and A. S. Campbell, 1929
A conspectus of the marine and fresh-water Ciliata belonging to the suborder Tintinnoinea, with descriptions of new species principally from the Agassiz expedition to the eastern tropical Pacific, 1904-1905. Univ. Calif. Publ. Zool. 34:1-403, 697 text figs.

Tintinnopsis tubulosa
Levander, K.M., 1947
Plankton gesammelt in den Jahren 1899-1910 an den Küsten Finnlands.
Finnländische Hydrographisch-Biologische Untersuchunger (aus dem Wasserbiologischen Laboratorin der Societas Scientiarum Fennica) No.11:40 pp., 6 diagrams, 13 pls., tables.

Tintinnopsis tubulosa
Lindquist, Armin, 1959
Studien über das Zooplankton der Bottensee II. Zur Verbreitung und Zusammensetzung des Zooplanktons.
Inst. Mar. Res., Lysekil. Ser. Biol., Rept. No. 11:136 pp.

Tintinnopsis tubulosa
Rumkówna, A., 1948
[List of the phytoplankton species occurring in the superficial water layers in the Gulf of Gdańsk] Bull. Lab. mar., Gdynia, No. 4: 139-141 with tables in back.

Tintinnopsis turbinata
Balech, E., 1951.
Nuevos datos sobre Tintinnoinea de Argentina y Uruguay. Physis 20(58):291-302, 16 textfigs.

Tintinnopsis turbinata n.sp.
Balech, Enrique, 1948
Tintinnoinea de Atlantida (R.O. Del Uruguay) (Protozoa, Ciliata Oligotr.). Comm. Mus. Argentino Ciencias Nat. "Bernardino Rivadavia", Ser.Cien.Zool. No.7, 23 pp., 8 pls. of 107 figs.

Tintinnopsis turbo
Biernacka, I., 1948
(Tintinnoinea in the Gulf of Gdansk and adjoining waters) Builetyn Morskiego Laboratorium Rybackiego w Gdyni dawniej-Stacji Morskiej w. Helu. No. 4:73-91, 4 text figs., 1 pl. with 21 figs. (Bull. Lab. Mar., Gdynia, formerly Bull. Sta. Hel.)

Tintinnopsis turbo n.sp.
Meunier, A., 1919
Microplankton de la Mer Flamande. 4. Les Tintinnides et Coetera. Mem. Mus. Roy. Hist. Nat., Belgique, 8(2):59pp., Pls. 22-23.

Tintinnopsis turbo
Kofoid, C. A. and A. S. Campbell, 1929
A conspectus of the marine and fresh-water Ciliata belonging to the suborder Tintinnoinea, with descriptions of new species principally from the Agassiz expedition to the eastern tropical Pacific, 1904-1905. Univ. Calif. Publ. Zool. 34:1-403, 697 text figs.

Tintinnopsis turbo
Rumkówna, A., 1948
[List of the phytoplankton species occurring in the superficial water layers in the Gulf of Gdańsk] Bull. Lab. mar., Gdynia, No. 4: 139-141 with tables in back.

Tintinnopsis turbo
Schulz, H., 1965.
Die Tintinnoinea des Elbe-Aestuars.
Arch. Fischereiwiss., 15(3):216-225.

Tintinnopsis turgida n. sp.
Kofoid, C. A. and A. S. Campbell, 1929
A conspectus of the marine and fresh-water Ciliata belonging to the suborder Tintinnoinea, with descriptions of new species principally from the Agassiz expedition to the eastern tropical Pacific, 1904-1905. Univ. Calif. Publ. Zool. 34:1-403, 697 text figs.

Tintinnopsis undella
Gaarder, K.R., 1946
Tintinnoidea. Rep.Sci.Res. "Michael Sars" N.Atlantic Deep-Sea Exped., 1910, 2(1): 37 pp., 23 text figs.

Tintinnopsis undella
Kofoid, C. A. and A. S. Campbell, 1929
A conspectus of the marine and fresh-water Ciliata belonging to the suborder Tintinnoinea, with descriptions of new species principally from the Agassiz expedition to the eastern tropical Pacific, 1904-1905. Univ. Calif. Publ. Zool. 34:1-403, 697 text figs.

Tintinnopsis undella
Sousa e Silva, E. de, 1950.
Les tintinnides de la Baie de Cascais. Bull. Inst. Océan., Monaco, No. 979:28 pp., 4 pls.

Tintinnopsis urniger
Kofoid, C. A. and A. S. Campbell, 1929
A conspectus of the marine and fresh-water Ciliata belonging to the suborder Tintinnoinea, with descriptions of new species principally from the Agassiz expedition to the eastern tropical Pacific, 1904-1905. Univ. Calif. Publ. Zool. 34:1-403, 697 text figs.

Tintinnopsis urnula ?
Iselin, C., 1930
A report on the coastal waters of Labrador based on explorations of the "Chance" during the summer of 1926. Proc. Am. Acad. Arts Sci., 66(1):1-37, 14 text figs.

Tintinnopsis urnula
Kofoid, C. A. and A. S. Campbell, 1929
A conspectus of the marine and fresh-water Ciliata belonging to the suborder Tintinnoinea, with descriptions of new species principally from the Agassiz expedition to the eastern tropical Pacific, 1904-1905. Univ. Calif. Publ. Zool. 34:1-403, 697 text figs.

Tintinnopsis uruguayensis n.sp.
Balech, Enrique, 1948
Tintinnoinea de Atlantida (R.O. Del Uruguay) (Protozoa, Ciliata Oligotr.). Comm. Mus. Argentino Ciencias Nat. "Bernardino Rivadavia", Ser.Cien.Zool. No.7, 23 pp., 8 pls. of 107 figs.

Tintinnopsis vasculum n.sp.
Meunier, A., 1919
Microplankton de la Mer Flamande. 4. Les Tintinnides et Coetera. Mem. Mus. Roy. Hist. Nat., Belgique, 8(2):59pp., Pls. 22-23.

Tintinnopsis vasculum
Sousa e Silva E. de, 1950.
Les tintinnides de la Baie de Cascais. Bull. Inst. Océan., Monaco, No. 979:28 pp., 4 pls.

Tintinnopsis ventricosa
Entz, G., Jr. 1909
Studien über organisation und biologie der Tintinniden. Arch. f. Protistenkunde 15:93-226, Pls. 8-21, text figs.

Tintinnopsis ventricosoides
Gaarder, K.R., 1946
Tintinnoidea. Rep.Sci.Res."Michael Sars" N.Atlantic Deep-Sea Exped., 1910, 2(1): 37 pp., 23 text figs.

Tintinnopsis ventricosa
Meunier, A., 1919
Microplankton de la Mer Flamande. 4. Les Tintinnides et Coetera. Mem. Mus. Roy. Hist. Nat., Belgique, 8(2):59pp., Pls. 22-23.

Tintinnopsis ventricoides
Braarud, T., and Adam Bursa, 1939
On the phytoplankton of the Oslo Fjord, 1933-1934. Hvalrådets Skr. No.19:1-63; 9 text figs. Reviewed. J. du. Cons. 14(3): 418-420. A.C. Gardiner.

Tintinnopsis ventricosa
Levander, K.M., 1947
Plankton gesammelt in den Jahren 1899-1910 an den Küsten Finnlands. Finnländische Hydrographisch-Biologische Untersuchunger (aus dem Wasserbiologischen Laboratorin der Societas Scientiarum Fennica) No.11:40 pp., 6 diagrams, 13 pls., tables.

Tintinnopsis vosmaeri
Kofoid, C. A. and A. S. Campbell, 1929
A conspectus of the marine and fresh-water Ciliata belonging to the suborder Tintinnoinea, with descriptions of new species principally from the Agassiz expedition to the eastern tropical Pacific, 1904-1905. Univ. Calif. Publ. Zool. 34:1-403, 697 text figs.

Tintinnopsis wailesi n. sp.
Kofoid, C. A. and A. S. Campbell, 1929
A conspectus of the marine and fresh-water Ciliata belonging to the suborder Tintinnoinea, with descriptions of new species principally from the Agassiz expedition to the eastern tropical Pacific, 1904-1905. Univ. Calif. Publ. Zool. 34:1-403, 697 text figs.

Tintinnus sp.
Duran, M., 1951
Contribucion al estudio de los tintinidos del plancton de los costas de Castellon. (Mediterranes occidental). Publ. Inst. Biol. Aplic., Barcelone, 8: 101-120, 29 text figs.

Tintinnus sp.
Gran, H.H. and T. Braarud, 1935
A quantitative study of the phytoplankton in the Bay of Fundy and the Gulf of Maine (including observations on hydrography, chemistry, and turbidity). J. Biol. Bd., Canada, 1(5):279-467, 69 text figs.

Tintinnus acuminatus
Jørgensen, E., 1905
Tintinnoidea. B.Protistplankton and the diatoms in bottom samples. Hydrographical and biological investigations in Norwegian fjords. Bergens Mus. Skr. 7: 49-225.

Tintinnus acuminatoides secata. n.sp.n.var.
Laackmann, H., 1910
Die Tintinnodeen der Deutschen Südpolar Expedition 1901-1903. Deutsch. Südpolar Exped. 11, Zool. 3(4):341-496, pls. 33-51.

Tintinnus acuminatus
Entz, G., Jr. 1909
Studien über organisation und biologie der Tintinniden. Arch. f. Protistenkunde 15:93-226, Pls. 8-21, text figs.

Tintinnus acuminatus
Jørgensen, E., 1900.
Ueber die Tintinnodeen der norwegischen Westküste Bergens Mus. Aarb., 1899(2):48 pp., 3 pls.

Tintinnus acuminoides n.sp.
Laackmann, H., 1910
Die Tintinnodeen der Deutschen Südpolar Expedition 1901-1903. Deutschen Südpolar Expedition 11, Zool.3 (4): 341-496, pls. 33-51.

Tintinnus amphora
Laackmann, H., 1910
Die Tintinnodeen der Deutschen Südpolar Expedition 1901-1903. Deutschen Südpolar Expedition 11, Zool.3 (4): 341-496, pls. 33-51.

Tintinnus ampulla
Fol, H., 1884.
Sur la famille des Tintinnodea. Recueil Zool. Suisse 1:27-64, Pls. 4-5.

Tintinnus angustatus
Kofoid, C. A. and A. S. Campbell, 1929
A conspectus of the marine and fresh-water Ciliata belonging to the suborder Tintinnoinea, with descriptions of new species principally from the Agassiz expedition to the eastern tropical Pacific, 1904-1905. Univ. Calif. Publ. Zool. 34:1-403, 697 text figs.

Tintinnus apertus
Gaarder, K.R., 1946
Tintinnoidea. Rep.Sci.Res."Michael Sars" N.Atlantic Deep-Sea Exped., 1910, 2(1): 37 pp., 23 text figs.

Tintinnus apertus n.sp.
Kofoid, C. A. and A. S. Campbell, 1929
A conspectus of the marine and fresh-water Ciliata belonging to the suborder Tintinnoinea, with descriptions of new species principally from the Agassiz expedition to the eastern tropical Pacific, 1904-1905. Univ. Calif. Publ. Zool. 34:1-403, 697 text figs.

Tintinnus apertus
Marshall, S. M., 1934
The silicoflagellata and tintinnoinea. Great Barrier Reef Exped., 1928-29. Sci. Repts. 4(15):623-664, 43 text figs.

Tintinnus apertus
Rampi, L., 1939
Primo contributo alla conoscenza dei tintinnoidi do Mare Ligure. Atti della Soc. Ital. di Sci. Nat. 78:67-81, 58 text figs.

Tintinnus attenuatus n.sp.
Kofoid, C. A. and A. S. Campbell, 1929
A conspectus of the marine and fresh-water Ciliata belonging to the suborder Tintinnoinea, with descriptions of new species principally from the Agassiz expedition to the eastern tropical Pacific, 1904-1905. Univ. Calif. Publ. Zool. 34:1-403, 697 text figs.

Tintinnus attenuatus
Marshall, S. M., 1934
The silicoflagellata and tintinnoinea. Great Barrier Reef Exped., 1928-29. Sci. Repts. 4(15):623-664, 43 text figs.

Tintinnus birictus n.sp.
Kofoid, C. A. and A. S. Campbell, 1929
A conspectus of the marine and fresh-water Ciliata belonging to the suborder Tintinnoinea, with descriptions of new species principally from the Agassiz expedition to the eastern tropical Pacific, 1904-1905. Univ. Calif. Publ. Zool. 34:1-403, 697 text figs.

Tintinnus brandti n.sp.
Kofoid, C. A. and A. S. Campbell, 1929
A conspectus of the marine and fresh-water Ciliata belonging to the suborder Tintinnoinea, with descriptions of new species principally from the Agassiz expedition to the eastern tropical Pacific, 1904-1905. Univ. Calif. Publ. Zool. 34:1-403, 697 text figs.

Tintinnus brandti
Sousa e Silva, E. de, 1950.
Les tintinnides de la Baie de Cascais. Bull. Inst. Océan., Monaco, No. 979:28 pp., 4 pls.

Tintinnus bulbosus
Entz, G., Jr. 1909
Studien über organisation und biologie der Tintinniden. Arch. f. Protistenkunde 15:93-226, Pls. 8-21, text figs.

Tintinnus bulbosus
Laackmann, H., 1910
Die Tintinnodeen der Deutschen Südpolar Expedition 1901-1903. Deutschen Südpolar Expedition 11, Zool.3 (4): 341-496, pls. 33-51.

Tintinnus colligatus n.sp.
Kofoid, C. A. and A. S. Campbell, 1929
A conspectus of the marine and fresh-water Ciliata belonging to the suborder Tintinnoinea, with descriptions of new species principally from the Agassiz expedition to the eastern tropical Pacific, 1904-1905. Univ. Calif. Publ. Zool. 34:1-403, 697 text figs.

Tintinnus costatus
Laackmann, H., 1910
Die Tintinnodeen der Deutschen Südpolar Expedition 1901-1903. Deutsch. Südpolar Exped. 11, Zool. 3(4):341-496, pls. 33-51.

Tintinnus datura
Laackmann, H., 1910
Die Tintinnodeen der Deutschen Südpolar Expedition 1901-1903. Deutschen Südpolar Expedition 11, Zool.3 (4): 341-496, pls. 33-51.

Tintinnus elegans
Kofoid, C. A. and A. S. Campbell, 1929
A conspectus of the marine and fresh-water Ciliata belonging to the suborder Tintinnoinea, with descriptions of new species principally from the Agassiz expedition to the eastern tropical Pacific, 1904-1905. Univ. Calif. Publ. Zool. 34:1-403, 697 text figs.

Tintinnus elongatus
Gaarder, K.R., 1946
Tintinnoidea. Rep.Sci.Res."Michael Sars" N.Atlantic Deep-Sea Exped., 1910, 2(1): 37 pp., 23 text figs.

Tintinnus elongatus
Kofoid, C. A. and A. S. Campbell, 1929
A conspectus of the marine and fresh-water Ciliata belonging to the suborder Tintinnoinea, with descriptions of new species principally from the Agassiz expedition to the eastern tropical Pacific, 1904-1905. Univ. Calif. Publ. Zool. 34:1-403, 697 text figs.

Tintinnus elongatus
Rampi, L., 1939
Primo contributo alla conoscenza dei tintinnoidi do Mare Ligure. Atti della Soc. Ital. di Sci. Nat. 78:67-81, 58 text figs.

Tintinnus elongatus
Sousa e Silva, E. de, 1950.
Les tintinnides de la Baie de Cascais. Bull. Inst Océan., Monaco, No. 979:28 pp., 4 pls.

Tintinnus emarginatus
Entz, G., Jr. 1909
Studien über organisation und biologie der Tintinniden. Arch. f. Protistenkunde 15:93-226, Pls. 8-21, text figs.

Tintinnus fraknoi
Delegazione Italiana della Commissione Internazionale per l'Esplorazione Scientifica del Mediterraneo, 1941
Note sul plancton della Laguna veneta. [Memoria CCLXXXI], Arch. di Ocean. e Limn. Anno I, Fasc. I, 1941 XIX: 31-57 pp.

Tintinnus fraknoi
Duran, M., 1951
Contribucion al estudio de los tintinidos del plancton de los costas de Castellon. (Mediterraneo occidental). Publ. Inst. Biol. Aplic., Barcelone, 8: 101-120, 29 text figs.

Tintinnus fraknoi
Entz, G., Jr. 1909
Studien über organisation und biologie der Tintinniden. Arch. f. Protistenkunde 15:93-226, Pls. 8-21, text figs.

Tintinnus fraknoi
Gaarder, K.R., 1946
Tintinnoidea. Rep.Sci.Res."Michael Sars" N.Atlantic Deep-Sea Exped., 1910, 2(1): 37 pp., 23 text figs.

Tintinnus frakenoii
Hofker, J., 1931
Studien uber Tintinnoidea. Arch. f. Protistenk 75(3):315-402, 89 text figs.

Tintinnus fraknoii
Jörgensen, E., 1924
Mediterranean Tintinnidae. Rept. Danish Oceanogr. Exped. 1908-10 to the Mediterranean and adjacent seas, Vol.II, Biol. J 3, 110 pp., 114 text figs.

Tintinnus fraknoii
Kofoid, C. A. and A. S. Campbell, 1929
A conspectus of the marine and freshwater Ciliata belonging to the suborder Tintinnoinea, with descriptions of new species principally from the Agassiz expedition to the eastern tropical Pacific, 1904-1905. Univ. Calif. Publ. Zool. 34:1-403, 697 text figs.

Tintinnus fraknoi
Laackmann, H., 1910
" Die Tintinnodeen der Deutschen Sudpolar Expedition 1901-1903. Deutschen Südpolar Expedition 11, Zool.3 (4): 341-496, pls. 33-51.

Tintinnus fraknoii
Rampi, L., 1939
Primo contributo alla conoscenza dei tintinnoidi do Maro Ligure. Atti della Soc. Ital. di Sci. Nat. 78:67-81, 58 text figs.

Tintinnus fraknoii
Sousa e Silva, E. de, 1950.
Les tintinnides de la Baie de Cascais. Bull. Inst. Océan., Monaco, No. 979:28 pp., 4 pls.

Tintinnus glockentögeri
Laackmann, H., 1910
" Die Tintinnodeen der Deutschen Sudpolar Expedition 1901-1903. Deutschen Südpolar Expedition 11, Zool.3 (4): 341-496, pls. 33-51.

Tintinnus inquilinus
Duran, M., 1951
Contribucion al estudio de los tintinidos del plancton de los costas de Castellon. (Mediterraneo occidental). Publ. Inst. Biol. Aplic., Barcelone, 8: 101-120, 29 text figs.

Tintinnus inquilinus
Entz, G., Jr. 1909
Studien über organisation und biologie der Tintinniden. Arch. f. Protistenkunde 15:93-226, Pls. 8-21, text figs.

Tintinnus inquilinus
Hofker, J., 1931
Studien uber Tintinnoidea. Arch. f. Protistenk 75(3):315-402, 89 text figs.

Tintinnus lusus-undae
Hofker, J., 1931
Studien uber Tintinnoidea. Arch. f. Protistenk 75(3):315-402, 89 text figs.

Tintinnus inquilinus
Jörgensen, E., 1924
Mediterranean Tintinnidae. Rept. Danish Oceanogr. Exped. 1908-10 to the Mediterranean and adjacent seas, Vol.II, Biol. J 3, 110 pp., 114 text figs.

Tintinnus inquilinus
Laackmann, H., 1910
" Die Tintinnodeen der Deutschen Sudpolar Expedition 1901-1903. Deutschen Südpolar Expedition 11, Zool.3 (4): 341-496, pls. 33-51.

Tintinnus latus
Gaarder, K.R., 1946
Tintinnoidea. Rep.Sci.Res."Michael Sars" N.Atlantic Deep-Sea Exped., 1910, 2(1): 37 pp., 23 text figs.

Tintinnus latus
Kofoid, C. A. and A. S. Campbell, 1929
A conspectus of the marine and freshwater Ciliata belonging to the suborder Tintinnoinea, with descriptions of new species principally from the Agassiz expedition to the eastern tropical Pacific, 1904-1905. Univ. Calif. Publ. Zool. 34:1-403, 697 text figs.

Tintinnus latus
Sousa e Silva, E. de, 1950.
Les tintinnides de la Baie de Cascais. Bull. Inst. Océan., Monaco, No. 979:28 pp., 4 pls.

Tintinnus lusus-undae
Duran, M., 1951
Contribucion al estudio de los tintinidos del plancton de los costas de Castellon. (Mediterraneo occidental). Publ. Inst. Biol. Aplic., Barcelone, 8: 101-120, 29 text figs.

Tintinnus lususundae
Entz, G., Jr., 1909
Studien über organisation und biologie der Tintinniden. Arch. f. Protistenkunde 15:93-226, Pls. 8-21, text figs.

Tintinnus lusus-undae
Gaarder, K.R., 1946
Tintinnoidea. Rep.Sci.Res."Michael Sars" N.Atlantic Deep-Sea Exped., 1910, 2(1): 37 pp., 23 text figs.

Tintinnus lusus-undae
Jörgensen, E., 1924
Mediterranean Tintinnidae. Rept. Danish Oceanogr. Exped. 1908-10 to the Mediterranean and adjacent seas, Vol.II, Biol. J 3, 110 pp., 114 text figs.

Tintinnus lusus undae
Jørgensen, E., 1900.
Ueber die Tintinnodeen der norwegischen Westküste Bergens Mus. Aarb., 1899(2):48 pp., 3 pls.

Tintinnus lusus-undae
Kofoid, C. A. and A. S. Campbell, 1929
A conspectus of the marine and freshwater Ciliata belonging to the suborder Tintinnoinea, with descriptions of new species principally from the Agassiz expedition to the eastern tropical Pacific, 1904-1905. Univ. Calif. Publ. Zool. 34:1-403, 697 text figs.

Tintinnus lusus-undae
Laackmann, H., 1910
" Die Tintinnodeen der Deutschen Sudpolar Expedition 1901-1903. Deutschen Südpolar Expedition 11, Zool.3 (4): 341-496, pls. 33-51.

Tintinnus lusus-undae
Marshall, S. M., 1934
The silicoflagellata and tintinnoinea. Great Barrier Reef Exped., 1928-29. Sci. Repts. 4(15):623-664, 43 text figs.

Tintinnus lusus-undae
Massuti Algamora, M., 1949
Estudio de diez y seis muestras de plancton del Golfo de Nápoles. Publ. Inst. Biol. Appl. 5:85-94, 1 fold-in table.

Tintinnus lusus undae
Rampi, L., 1948
Sur quelques Tintinnides (Infusoires louques) du Pacifique subtropical (Recoltes Alain Gerbault) Bull. l'Inst. Ocean., Monaco, No.938, 4 pp.

Tintinnus lusus-undae
Sousa e Silva, E. de, 1950.
Les tintinnides de la Baie de Cascais. Bull. Inst. Océan., Monaco, No. 979:28 pp., 4 pls.

Tintinnus macilentus
Gaarder, K.R., 1946
Tintinnoidea. Rep.Sci.Res."Michael Sars" N.Atlantic Deep-Sea Exped., 1910, 2(1): 37 pp., 23 text figs.

Tintinnus macilentus
Kofoid, C. A. and A. S. Campbell, 1929
A conspectus of the marine and freshwater Ciliata belonging to the suborder Tintinnoinea, with descriptions of new species principally from the Agassiz expedition to the eastern tropical Pacific, 1904-1905. Univ. Calif. Publ. Zool. 34:1-403, 697 text figs.

Tintinnus macilentus
Rampi, L., 1939
Primo contributo alla conoscenza dei tintinnoidi do Maro Ligure. Atti della Soc. Ital. di Sci. Nat. 78:67-81, 58 text figs.

Tintinnus maculatus
Kofoid, C. A. and A. S. Campbell, 1929
A conspectus of the marine and freshwater Ciliata belonging to the suborder Tintinnoinea, with descriptions of new species principally from the Agassiz expedition to the eastern tropical Pacific, 1904-1905. Univ. Calif. Publ. Zool. 34:1-403, 697 text figs.

Tintinnus medius n.sp.
Kofoid, C. A. and A. S. Campbell, 1929
A conspectus of the marine and freshwater Ciliata belonging to the suborder Tintinnoinea, with descriptions of new species principally from the Agassiz expedition to the eastern tropical Pacific, 1904-1905. Univ. Calif. Publ. Zool. 34:1-403, 697 text figs.

Tintinnus minimus n.sp.
Entz, G., Jr., 1909
Studien über organisation und biologie der Tintinniden. Arch. f. Protistenkunde 15:93-226, Pls. 8-21, text figs.

Tintinnus mirabilis n.sp.
Kofoid, C. A. and A. S. Campbell, 1929
A conspectus of the marine and fresh-water Ciliata belonging to the suborder Tintinnoinea, with descriptions of new species principally from the Agassiz expedition to the eastern tropical Pacific, 1904-1905. Univ. Calif. Publ. Zool. 34:1-403, 697 text figs.

Tintinnus pacificus n.sp.
Kofoid, C. A. and A. S. Campbell, 1929
A conspectus of the marine and fresh-water Ciliata belonging to the suborder Tintinnoinea, with descriptions of new species principally from the Agassiz expedition to the eastern tropical Pacific, 1904-1905. Univ. Calif. Publ. Zool. 34:1-403, 697 text figs.

Tintinnus pacificus
Marshall, S. M., 1934
The silicoflagellata and tintinnoinea. Great Barrier Reef Exped., 1928-29. Sci. Repts. 4(15):623-664, 43 text figs.

Tintinnus palliatus
Laackmann, H., 1910
Die Tintinnodeen der Deutschen Südpolar Expedition 1901-1903. Deutschen Südpolar Expedition 11, Zool.3 (4): 341-496, pls. 33-51.

Tintinnus pectinis nom. nov.
Kofoid, C. A. and A. S. Campbell, 1929
A conspectus of the marine and fresh-water Ciliata belonging to the suborder Tintinnoinea, with descriptions of new species principally from the Agassiz expedition to the eastern tropical Pacific, 1904-1905. Univ. Calif. Publ. Zool. 34:1-403, 697 text figs.

Tintinnus perminutus n.sp.
Kofoid, C. A. and A. S. Campbell, 1929
A conspectus of the marine and fresh-water Ciliata belonging to the suborder Tintinnoinea, with descriptions of new species principally from the Agassiz expedition to the eastern tropical Pacific, 1904-1905. Univ. Calif. Publ. Zool. 34:1-403, 697 text figs.

Tintinnus perminutus
Rampi, L., 1939
Primo contributo alla conoscenza dei tintinnoidi do Mare Ligure. Atti della Soc. Ital. di Sci. Nat. 78:67-81, 58 text figs.

Tintinnus pinguis
Gaarder, K.R., 1946
Tintinnoidea. Rep.Sci.Res."Michael Sars" N.Atlantic Deep-Sea Exped., 1910, 2(1): 37 pp., 23 text figs.

Tintinnus pinquis n.sp.
Kofoid, C. A. and A. S. Campbell, 1929
A conspectus of the marine and fresh-water Ciliata belonging to the suborder Tintinnoinea, with descriptions of new species principally from the Agassiz expedition to the eastern tropical Pacific, 1904-1905. Univ. Calif. Publ. Zool. 34:1-403, 697 text figs.

Tintinnus procurrerens n.sp.
Kofoid, C. A. and A. S. Campbell, 1929
A conspectus of the marine and fresh-water Ciliata belonging to the suborder Tintinnoinea, with descriptions of new species principally from the Agassiz expedition to the eastern tropical Pacific, 1904-1905. Univ. Calif. Publ. Zool. 34:1-403, 697 text figs.

Tintinnus quinqueelatos
Laackmann, H., 1910
Die Tintinnodeen der Deutschen Südpolar Expedition 1901-1903. Deutsch. Südpolar Exped. 11, Zool. 3(4):341-496, pls. 33-51.

Tintinnus rectus
Kofoid, C. A. and A. S. Campbell, 1929
A conspectus of the marine and fresh-water Ciliata belonging to the suborder Tintinnoinea, with descriptions of new species principally from the Agassiz expedition to the eastern tropical Pacific, 1904-1905. Univ. Calif. Publ. Zool. 34:1-403, 697 text figs.

Tintinnus rugosus n.sp.
Kofoid, C. A. and A. S. Campbell, 1929
A conspectus of the marine and fresh-water Ciliata belonging to the suborder Tintinnoinea, with descriptions of new species principally from the Agassiz expedition to the eastern tropical Pacific, 1904-1905. Univ. Calif. Publ. Zool. 34:1-403, 697 text figs.

Tintinnus spiralis
Fol, H., 1884.
Sur la famille des Tintinnodea. Recueil Zool. Suisse 1:27-64, Pls. 4-5.

Tintinnus steenstrupi
Laackmann, H., 1910
Die Tintinnodeen der Deutschen Südpolar Expedition 1901-1903. Deutschen Südpolar Expedition 11, Zool.3 (4): 341-496, pls. 33-51.

Tintinnus stramentus n.sp.
Kofoid, C. A. and A. S. Campbell, 1929
A conspectus of the marine and fresh-water Ciliata belonging to the suborder Tintinnoinea, with descriptions of new species principally from the Agassiz expedition to the eastern tropical Pacific, 1904-1905. Univ. Calif. Publ. Zool. 34:1-403, 697 text figs.

Tintinnus stramentus
Marshall, S. M., 1934
The silicoflagellata and tintinnoinea. Great Barrier Reef Exped., 1928-29. Sci. Repts. 4(15):623-664, 43 text figs.

Tintinnus subulatus
Entz, G., Jr., 1909
Studien über organisation und biologie der Tintinniden. Arch. f. Protistenkunde 15:93-226, Pls. 8-21, text figs.

Tintinnus subulatus
Levander, K.M., 1947
Plankton gesammelt in den Jahren 1899-1910 an den Küsten Finnlands. Finnländische Hydrographisch-Biologische Untersuchungen (aus dem Wasserbiologischen Laboratorin der Societas Scientiarum Fennica) No.11:40 pp., 6 diagrams, 13 pls., tables.

Tintinnus tenua n.sp.
Kofoid, C. A. and A. S. Campbell, 1929
A conspectus of the marine and fresh-water Ciliata belonging to the suborder Tintinnoinea, with descriptions of new species principally from the Agassiz expedition to the eastern tropical Pacific, 1904-1905. Univ. Calif. Publ. Zool. 34:1-403, 697 text figs.

Tintinnus tenua
Rampi, L., 1948
Sur quelques Tintinnides (Infusoires louques) du Pacifique subtropical (Recoltes Alain Gerbault) Bull. l'Inst. Ocean., Monaco, No.938, 4 pp.

Tintinnus tubiformis n.sp.
Kofoid, C. A. and A. S. Campbell, 1929
A conspectus of the marine and fresh-water Ciliata belonging to the suborder Tintinnoinea, with descriptions of new species principally from the Agassiz expedition to the eastern tropical Pacific, 1904-1905. Univ. Calif. Publ. Zool. 34:1-403, 697 text figs.

Tintinnus tubulosus
Gaarder, K.R., 1946
Tintinnoidea. Rep.Sci.Res."Michael Sars" N.Atlantic Deep-Sea Exped., 1910, 2(1): 37 pp., 23 text figs.

Tintinnus tubus
Kofoid, C. A. and A. S. Campbell, 1929
A conspectus of the marine and fresh-water Ciliata belonging to the suborder Tintinnoinea, with descriptions of new species principally from the Agassiz expedition to the eastern tropical Pacific, 1904-1905. Univ. Calif. Publ. Zool. 34:1-403, 697 text figs.

Tintinnus turgescens n.sp.
Kofoid, C. A. and A. S. Campbell, 1929
A conspectus of the marine and fresh-water Ciliata belonging to the suborder Tintinnoinea, with descriptions of new species principally from the Agassiz expedition to the eastern tropical Pacific, 1904-1905. Univ. Calif. Publ. Zool. 34:1-403, 697 text figs.

Tintinnus turris n.sp.
Kofoid, C. A. and A. S. Campbell, 1929
A conspectus of the marine and fresh-water Ciliata belonging to the suborder Tintinnoinea, with descriptions of new species principally from the Agassiz expedition to the eastern tropical Pacific, 1904-1905. Univ. Calif. Publ. Zool. 34:1-403, 697 text figs.

Tintinnus undatus
Laackmann, H., 1910
Die Tintinnodeen der Deutschen Südpolar Expedition 1901-1903. Deutschen Südpolar Expedition 11, Zool.3 (4): 341-496, pls. 33-51.

Tintinnus ventricosus
Fol, H., 1884.
Sur la famille des Tintinnodea. Recueil Zool. Suisse 1:27-64, Pls. 4-5.

Undella aculeata n.nom.
Jörgensen, E., 1924
Mediterranean Tintinnidae. Rept. Danish Oceanogr. Exped. 1908-10 to the Mediterranean and adjacent seas, Vol.II, Biol. J 3, 110 pp., 114 text figs.

Undella antarctica n. sp.
*Hada, Yoshina, 1970. (1970)
The protozoan plankton of the Antarctic and Subantarctic seas.
Scient. Repts, Japan. Antarct. Res. Exped., (E) 31:51 pp.

Undella armata
Laackmann, H., 1910
Die Tintinnodeen der Deutschen Südpolar Expedition 1901-1903. Deutschen Südpolar Expedition 11, Zool.3 (4): 341-496, pls. 33-51.

Undella attenuata
Campbell, A. S., 1942
The oceanic Tintinnoina of the plankton gathered during the last cruise of the CARNEGIE. Sci. Res. Cruise VII of the Carnegie, 1928-1929, Biol. 2:163 pp., 128 figs.

Undella attenuata
Kofoid, C. A. and A. S. Campbell, 1929
A conspectus of the marine and fresh-water Ciliata belonging to the suborder Tintinnoinea, with descriptions of new species principally from the Agassiz expedition to the eastern tropical Pacific, 1904-1905. Univ. Calif. Publ. Zool. 34:1-403, 697 text figs.

Undella attenuata
Rampi, L., 1939
Primo contributo alla conoscenza dei tintinnoidi do Mare Ligure. Atti della Soc. Ital. di Sci. Nat. 78:67-81, 58 text figs.

Undella azorica?
Entz, G., Jr., 1909
Studien über organisation und biologie der Tintinniden. Arch. f. Protistenkunde 15:93-226, Pls. 8-21, text figs.

Undella bulla

Kofoid, C.A. and A.S. Campbell, 1939
Reports on the scientific results of the expedition to the eastern tropical Pacific in charge of Alexander Agassiz, by the U.S. Fish Commission steamer "Albatross" from October 1904 to March 1905-----XXXVII. The Ciliata: The Tintinnoinea. Bull.Mus. Comp.Zool. 84: 473 pp., 36 pls.

Undella bulla n.sp.

Kofoid, C. A. and A. S. Campbell, 1929
A conspectus of the marine and freshwater Ciliata belonging to the suborder Tintinnoinea, with descriptions of new species principally from the Agassiz expedition to the eastern tropical Pacific, 1904-1905. Univ. Calif. Publ. Zool. 34:1-403, 697 text figs.

Undella californiensis

Campbell, A. S., 1942
The oceanic Tintinnoina of the plankton gathered during the last cruise of the CARNEGIE. Sci. Res. Cruise VII of the Carnegie, 1928-1929, Biol. 2:163 pp., 128 figs.

Undella californiensis

Kofoid, C.A. and A.S. Campbell, 1939
Reports on the scientific results of the expedition to the eastern tropical Pacific in charge of Alexander Agassiz, by the U.S. Fish Commission steamer "Albatross" from October 1904 to March 1905-----XXXVII. The Ciliata: The Tintinnoinea. Bull.Mus. Comp.Zool. 84: 473 pp., 36 pls.

Undella californiensis n.sp.

Kofoid, C. A. and A. S. Campbell, 1929
A conspectus of the marine and freshwater Ciliata belonging to the suborder Tintinnoinea, with descriptions of new species principally from the Agassiz expedition to the eastern tropical Pacific, 1904-1905. Univ. Calif. Publ. Zool. 34:1-403, 697 text figs.

Undella carnegiei n. sp.

Campbell, A. S., 1942
The oceanic Tintinnoina of the plankton gathered during the last cruise of the CARNEGIE. Sci. Res. Cruise VII of the Carnegie, 1928-1929, Biol. 2:163 pp., 128 figs.

Undella caudata

Jörgensen, E., 1924
Mediterranean Tintinnidae. Rept. Danish Oceanogr. Exped. 1908-10 to the Mediterranean and adjacent seas, Vol.II, Biol. J 3, 110 pp., 114 text figs.

Undella caudata

Jörgensen, E., 1905
B. Protistplankton and the diatoms in bottom samples. Hydrographical and biological investigations in Norwegian fjords. Bergens Mus. Skr. 7: 49-225.

Undella claparedei

Duran, M., 1951
Contribucion al estudio de los tintinidos del plancton de los costas de Castellon. (Mediterraneo occidental). Publ. Inst. Biol. Aplic., Barcelone, 8: 101-120, 29 text figs.

Undella claparedi

Entz, G., Jr., 1909
Studien über organisation und biologie der Tintinniden. Arch. f. Protistenkunde 15:93-226, Pls. 8-21, text figs.

Undella claparedei

Jörgensen, E., 1924
Mediterranean Tintinnidae. Rept. Danish Oceanogr. Exped. 1908-10 to the Mediterranean and adjacent seas, Vol.II, Biol. J 3, 110 pp., 114 text figs.

Undella claparedei

Laackmann, H., 1910
Die Tintinnodeen der Deutschen Südpolar Expedition 1901-1903. Deutschen Südpolar Expedition 11, Zool.3 (4): 341-496, pls. 33-51.

Undella clevei

Balech, E., 1959
Tintinnoinea del Mediterraneo. Trav. Sta. Zool., Villefranche-sur-Mer, 18(17): Reprinted from: Trab. Inst. Español Oceanogr., Madrid, 19 Sept. 1959(28):88 pp., 22 pls.

Undella clevei

Campbell, A. S., 1942
The oceanic Tintinnoina of the plankton gathered during the last cruise of the CARNEGIE. Sci. Res. Cruise VII of the Carnegie, 1928-1929, Biol. 2:163 pp., 128 figs.

Undella clevei

Duran, M., 1951
Contribucion al estudio de los tintinidos del plancton de los costas de Castellon. (Mediterraneo occidental). Publ. Inst. Biol. Aplic., Barcelone, 8: 101-120, 29 text figs.

Undella clevei n.sp.

Jörgensen, E., 1924
Mediterranean Tintinnidae. Rept. Danish Oceanogr. Exped. 1908-10 to the Mediterranean and adjacent seas, Vol.II, Biol. J 3, 110 pp., 114 text figs.

Undella clevei

Kofoid, C. A. and A. S. Campbell, 1929
A conspectus of the marine and freshwater Ciliata belonging to the suborder Tintinnoinea, with descriptions of new species principally from the Agassiz expedition to the eastern tropical Pacific, 1904-1905. Univ. Calif. Publ. Zool. 34:1-403, 697 text figs.

Undella clevei

Rampi, L., 1939
Primo contributo alla conoscenza dei tintinnoidi de Mare Ligure. Atti della Soc. Ital. di Sci. Nat. 78:67-81, 58 text figs.

Undella collaria

Jörgensen, E., 1924
Mediterranean Tintinnidae. Rept. Danish Oceanogr. Exped. 1908-10 to the Mediterranean and adjacent seas, Vol.II, Biol. J 3, 110 pp., 114 text figs.

Undella collaria

Laackmann, H., 1910
Die Tintinnodeen der Deutschen Südpolar Expedition 1901-1903. Deutschen Südpolar Expedition 11, Zool.3 (4): 341-496, pls. 33-51.

Undella declivis

Campbell, A. S., 1942
The oceanic Tintinnoina of the plankton gathered during the last cruise of the CARNEGIE. Sci. Res. Cruise VII of the Carnegie, 1928-1929, Biol. 2:163 pp., 128 figs.

Undella declivis

Kofoid, C.A. and A.S. Campbell, 1939
Reports on the scientific results of the expedition to the eastern tropical Pacific in charge of Alexander Agassiz, by the U.S. Fish Commission steamer "Albatross from October 1904 to March 1905-----XXXVII The Ciliata: The Tintinnoinea. Bull.Mus. Comp.Zool. 84: 473 pp., 36 pls.

Undella declives n.sp.

Kofoid, C. A. and A. S. Campbell, 1929
A conspectus of the marine and freshwater Ciliata belonging to the suborder Tintinnoinea, with descriptions of new species principally from the Agassiz expedition to the eastern tropical Pacific, 1904-1905. Univ. Calif. Publ. Zool. 34:1-403, 697 text figs.

Undella dilatata

Campbell, A. S., 1942
The oceanic Tintinnoina of the plankton gathered during the last cruise of the CARNEGIE. Sci. Res. Cruise VII of the Carnegie, 1928-1929, Biol. 2:163 pp., 128 figs.

Undella dilatata

Gaarder, K.R., 1946
Tintinnoidea. Rep.Sci.Res. "Michael Sars" N.Atlantic Deep-Sea Exped., 1910, 2(1): 37 pp., 23 text figs.

Undella dilatata

Kofoid, C.A. and A.S. Campbell, 1939
Reports on the scientific results of the expedition to the eastern tropical Pacific in charge of Alexander Agassiz, by the U.S. Fish Commission steamer "Albatross" from October 1904 to March 1905-----XXXVII. The Ciliata: The Tintinnoinea. Bull.Mus. Comp.Zool. 84: 473 pp., 36 pls.

Undella dilatata n.sp.

Kofoid, C. A. and A. S. Campbell, 1929
A conspectus of the marine and freshwater Ciliata belonging to the suborder Tintinnoinea, with descriptions of new species principally from the Agassiz expedition to the eastern tropical Pacific, 1904-1905. Univ. Calif. Publ. Zool. 34:1-403, 697 text figs.

Undella dohrni

Jörgensen, E., 1924
Mediterranean Tintinnidae. Rept. Danish Oceanogr. Exped. 1908-10 to the Mediterranean and adjacent seas, Vol.II, Biol. J 3, 110 pp., 114 text figs.

Undella dohrnii

Kofoid, C. A. and A. S. Campbell, 1929
A conspectus of the marine and freshwater Ciliata belonging to the suborder Tintinnoinea, with descriptions of new species principally from the Agassiz expedition to the eastern tropical Pacific, 1904-1905. Univ. Calif. Publ. Zool. 34:1-403, 697 text figs.

Undella filigera n.sp.

Laackmann, H., 1910
Die Tintinnodeen der Deutschen Südpolar Expedition 1901-1903. Deutschen Südpolar Expedition 11, Zool.3 (4): 341-496, pls. 33-51.

Undella hadai n.sp.

Balech, Enrique, 1962
Tintinnoinea y Dinoflagellata del Pacifico segun material de las Expediciones NORPAC y DOWNWIND del Instituto Scripps de Oceanografia. Revista, Mus. Argentino Ciencias Nat. "Bernardino Rivadavia", Ciencias Zool., 7(1):1-253.

Undella hawaiensis n. sp.

Campbell, A. S., 1942
The oceanic Tintinnoina of the plankton gathered during the last cruise of the CARNEGIE. Sci. Res. Cruise VII of the Carnegie, 1928-1929, Biol. 2:163 pp., 128 figs.

Undella hemispherica

Campbell, A. S., 1942
The oceanic Tintinnoina of the plankton gathered during the last cruise of the CARNEGIE. Sci. Res. Cruise VII of the Carnegie, 1928-1929, Biol. 2:163 pp., 128 figs.

Undella hemispherica

Kofoid, C.A. and A.S. Campbell, 1939
Reports on the scientific results of the expedition to the eastern tropical Pacific in charge of Alexander Agassiz, by the U.S. Fish Commission steamer "Albatross" from October 1904 to March 1905-----XXXVII. The Ciliata: The Tintinnoinea. Bull.Mus. Comp.Zool. 84: 473 pp., 36 pls.

Undella hemisphaerica

Kofoid, C. A. and A. S. Campbell, 1929
A conspectus of the marine and freshwater Ciliata belonging to the suborder Tintinnoinea, with descriptions of new species principally from the Agassiz expedition to the eastern tropical Pacific, 1904-1905. Univ. Calif. Publ. Zool. 34:1-403, 697 text figs.

Undella hemisphaerica
Laackmann, H., 1910
" Die Tintinnodeen der Deutschen Südpolar Expedition 1901-1903. Deutschen Südpolar Expedition 11, Zool.3 (4): 341-496, pls. 33-51.

Undella hemispherica
Marshall, S. M., 1934
The silicoflagellata and tintinnoinea. Great Barrier Reef Exped., 1928-29. Sci. Repts. 4(15):623-664, 43 text figs.

Undella heros
Entz, G., Jr., 1909
Studien über organisation und biologie der Tintinniden. Arch. f. Protistenkunde 15:93-226, Pls. 8-21, text figs.

Undella heros
Laackmann, H., 1910
" Die Tintinnodeen der Deutschen Südpolar Expedition 1901-1903. Deutschen Südpolar Expedition 11, Zool.3 (4): 341-496, pls. 33-51.

Undella hyalina
Balech, Enrique, 1962
Tintinnoinea y Dinoflagellata del Pacifico segun material de las Expediciones NORPAC y DOWNWIND del Instituto Scripps de Oceanografia. Revista, Mus. Argentino Ciencias Nat. "Bernardino Rivadavia", Ciencias Zool., 7(1):1-253.

Undella hyalina
Balech, E., 1959.
Tintinnoinea del Mediterraneo.
Trav. Sta. Zool., Villefranche-sur-Mer, 18(17):
Reprinted from:
Trab. Inst. Español Oceanogr., Madrid, 19 Sept. 1959, (28):88 pp., 22 pls.

Undella hyalina
Campbell, A. S., 1942
The oceanic Tintinnoina of the plankton gathered during the last cruise of the CARNEGIE. Sci. Res. Cruise VII of the Carnegie, 1928-1929, Biol. 2:163 pp., 128 figs.

Undella hyalina
Duran, M., 1951
Contribucion al estudio de los tintinidos del plancton de los costas de Castellon. (Mediterranes occidental). Publ. Inst. Biol. Aplic., Barcelone, 8: 101-120, 29 text figs.

Undella hyalina
Gaarder, K.R., 1946
Tintinnoidea. Rep.Sci.Res."Michael Sars" N.Atlantic Deep-Sea Exped., 1910, 2(1): 37 pp., 23 text figs.

Undella hyalina
Jörgensen, E., 1924
Mediterranean Tintinnidae. Rept. Danish Oceanogr. Exped. 1908-10 to the Mediterranean and adjacent seas, Vol.II, Biol. J 3, 110 pp., 114 text figs.

Undella hyalina
Kofoid, C.A. and A.S. Campbell, 1939
Reports on the scientific results of the expedition to the eastern tropical Pacific in charge of Alexander Agassiz, by the U.S. Fish Commission steamer"Albatross" from October 1904 to March 1905-----XXXVII. The Ciliata: The Tintinnoimea. Bull.Mus. Comp.Zool. 84: 473 pp., 36 pls.

Undella hyalina
Kofoid, C. A. and A. S. Campbell, 1929
A conspectus of the marine and freshwater Ciliata belonging to the suborder Tintinnoinea, with descriptions of new species principally from the Agassiz expedition to the eastern tropical Pacific, 1904-1905. Univ. Calif. Publ. Zool. 34:1-403, 697 text figs.

Undella hyalina
Laackmann, H., 1910
" Die Tintinnodeen der Deutschen Südpolar Expedition 1901-1903. Deutschen Südpolar Expedition 11, Zool.3 (4): 341-496, pls. 33-51.

Undella hyalinella
Campbell, A. S., 1942
The oceanic Tintinnoina of the plankton gathered during the last cruise of the CARNEGIE. Sci. Res. Cruise VII of the Carnegie, 1928-1929, Biol. 2:163 pp., 128 figs.

Undella hyalinella n.sp.
Kofoid, C. A. and A. S. Campbell, 1929
A conspectus of the marine and freshwater Ciliata belonging to the suborder Tintinnoinea, with descriptions of new species principally from the Agassiz expedition to the eastern tropical Pacific, 1904-1905. Univ. Calif. Publ. Zool. 34:1-403, 697 text figs.

Undella hyalinella
Kofoid, C.A. and A.S. Campbell, 1939
Reports on the scientific results of the expedition to the eastern tropical Pacific in charge of Alexander Agassiz, by the U.S. Fish Commission steamer"Albatross" from October 1904 to March 1905-----XXXVII. The Ciliata: The Tintinnoimea. Bull.Mus. Comp.Zool. 84: 473 pp., 36 pls.

Undella junei n. comb.
Balech, Enrique, 1971
Microplancton del Atlantico ecuatorial oeste (Equalant 1)
Publ. Serv. Hidrograf. Naval, Argentina H. 654: 103 pp., 122 figs.

Undella lachmanni
Laackmann, H., 1910
" Die Tintinnodeen der Deutschen Südpolar Expedition 1901-1903. Deutschen Südpolar Expedition 11, Zool.3 (4): 341-496, pls. 33-51.

Undella lohmanni n.sp.
Jörgensen, E., 1924
Mediterranean Tintinnidae. Rept. Danish Oceanogr. Exped. 1908-10 to the Mediterranean and adjacent seas, Vol.II, Biol. J 3, 110 pp., 114 text figs.

Undella mammilata n.sp.
Kofoid, C.A. and A.S. Campbell, 1939
Reports on the scientific results of the expedition to the eastern tropical Pacific in charge of Alexander Agassiz, by the U.S. Fish Commission steamer"Albatross" from October 1904 to March 1905-----XXXVII The Ciliata: The Tintinnoimea. Bull.Mus. Comp.Zool. 84: 473 pp., 36 pls.

Undella marsupialis
Duran, M., 1951
Contribucion al estudio de los tintinidos del plancton de los costas de Castellon. (Mediterranes occidental). Publ. Inst. Biol. Aplic., Barcelone, 8: 101-120, 29 text figs.

Undella marsupialis
Entz, G., Jr., 1909
Studien über organisation und biologie der Tintinniden. Arch. f. Protistenkunde 15:93-226, Pls. 8-21, text figs.

Undella marsupialis
Jörgensen, E., 1924
Mediterranean Tintinnidae. Rept. Danish Oceanogr. Exped. 1908-10 to the Mediterranean and adjacent seas, Vol.II, Biol. J 3, 110 pp., 114 text figs.

Undella marsupialis
Laackmann, H., 1910
" Die Tintinnodeen der Deutschen Südpolar Expedition 1901-1903. Deutschen Südpolar Expedition 11, Zool.3 (4): 341-496, pls. 33-51.

Undella media n.sp.
Kofoid, C.A. and A.S. Campbell, 1939
Reports on the scientific results of the expedition to the eastern tropical Pacific in charge of Alexander Agassiz, by the U.S. Fish Commission steamer"Albatross" from October 1904 to March 1905-----XXXVII. The Ciliata: The Tintinnoimea. Bull.Mus. Comp.Zool. 84: 473 pp., 36 pls.

Undella messinensis
Jörgensen, E., 1924
Mediterranean Tintinnidae. Rept. Danish Oceanogr. Exped. 1908-10 to the Mediterranean and adjacent seas, Vol.II, Biol. J 3, 110 pp., 114 text figs.

Undella messinensis
Laackmann, H., 1910
" Die Tintinnodeen der Deutschen Südpolar Expedition 1901-1903. Deutschen Südpolar Expedition 11, Zool.3 (4): 341-496, pls. 33-51.

Undella monocollaria n.sp.
Laackmann, H., 1910
" Die Tintinnodeen der Deutschen Südpolar Expedition 1901-1903. Deutschen Südpolar Expedition 11, Zool.3 (4): 341-496, pls. 33-51.

Undella ostenfeldi n.sp.
Kofoid, C. A. and A. S. Campbell, 1929
A conspectus of the marine and freshwater Ciliata belonging to the suborder Tintinnoinea, with descriptions of new species principally from the Agassiz expedition to the eastern tropical Pacific, 1904-1905. Univ. Calif. Publ. Zool. 34:1-403, 697 text figs.

Undella parva
Campbell, A. S., 1942
The oceanic Tintinnoina of the plankton gathered during the last cruise of the CARNEGIE. Sci. Res. Cruise VII of the Carnegie, 1928-1929, Biol. 2:163 pp., 128 figs.

Undella parva
Kofoid, C.A. and A.S. Campbell, 1939
Reports on the scientific results of the expedition to the eastern tropical Pacific in charge of Alexander Agassiz, by the U.S. Fish Commission steamer"Albatross" from October 1904 to March 1905-----XXXVII. The Ciliata: The Tintinnoimea. Bull.Mus. Comp.Zool. 84: 473 pp., 36 pls.

Undella parva n.sp.
Kofoid, C. A. and A. S. Campbell, 1929
A conspectus of the marine and freshwater Ciliata belonging to the suborder Tintinnoinea, with descriptions of new species principally from the Agassiz expedition to the eastern tropical Pacific, 1904-1905. Univ. Calif. Publ. Zool. 34:1-403, 697 text figs.

Undella pellucida n.sp.
Jörgensen, E., 1900.
Ueber die Tintinnodeen der norwegischen Westküste Bergens Mus. Aarb., 1899(2):48 pp., 3 pls.

Undella peruana
Campbell, A. S., 1942
The oceanic Tintinnoina of the plankton gathered during the last cruise of the CARNEGIE. Sci. Res. Cruise VII of the Carnegie, 1928-1929, Biol. 2:163 pp., 128 figs.

Undella peruana
Kofoid, C.A. and A.S. Campbell, 1939
Reports on the scientific results of the expedition to the eastern tropical Pacific in charge of Alexander Agassiz, by the U.S. Fish Commission steamer"Albatross" from October 1904 to March 1905-----XXXVII. The Ciliata: The Tintinnoimea. Bull.Mus. Comp.Zool. 84: 473 pp., 36 pls.

Undella peruana n.sp.

Kofoid, C. A. and A. S. Campbell, 1929
A conspectus of the marine and freshwater Ciliata belonging to the suborder Tintinnoinea, with descriptions of new species principally from the Agassiz expedition to the eastern tropical Pacific, 1904-1905. Univ. Calif. Publ. Zool. 34:1-403, 697 text figs.

Undella pistillum

Campbell, A. S., 1942
The oceanic Tintinnoina of the plankton gathered during the last cruise of the CARNEGIE. Sci. Res. Cruise VII of the Carnegie, 1928-1929, Biol. 2:163 pp., 128 figs.

Undella pistillum

Kofoid, C.A. and A.S. Campbell, 1939
Reports on the scientific results of the expedition to the eastern tropical Pacific in charge of Alexander Agassiz, by the U.S. Fish Commission steamer "Albetross" from October 1904 to March 1905-----XXXVII. The Ciliata: The Tintinnoimea. Bull. Mus. Comp. Zool. 84: 473 pp., 36 pls.

Undella pistillum n.sp.

Kofoid, C. A. and A. S. Campbell, 1929
A conspectus of the marine and freshwater Ciliata belonging to the suborder Tintinnoinea, with descriptions of new species principally from the Agassiz expedition to the eastern tropical Pacific, 1904-1905. Univ. Calif. Publ. Zool. 34:1-403, 697 text figs.

Undella pusilla

Kofoid, C. A. and A. S. Campbell, 1929
A conspectus of the marine and freshwater Ciliata belonging to the suborder Tintinnoinea, with descriptions of new species principally from the Agassiz expedition to the eastern tropical Pacific, 1904-1905. Univ. Calif. Publ. Zool. 34:1-403, 697 text figs.

Undella stenfeldi

Campbell, A. S., 1942
The oceanic Tintinnoina of the plankton gathered during the last cruise of the CARNEGIE. Sci. Res. Cruise VII of the Carnegie, 1928-1929, Biol. 2:163 pp., 128 figs.

Undella stenfeldi

Kofoid, C.A. and A.S. Campbell, 1939
Reports on the scientific results of the expedition to the eastern tropical Pacific in charge of Alexander Agassiz, by the U.S. Fish Commission steamer "Albetross" from October 1904 to March 1905-----XXXVII. The Ciliata: The Tintinnoimea. Bull. Mus. Comp. Zool. 84: 473 pp., 36 pls.

Undella subacuta

Duran, M., 1951
Contribucion al estudio de los tintinidos del plancton de los costas de Castellon. (Mediterraneo occidental). Publ. Inst. Biol. Aplic., Barcelone, 8: 101-120, 29 text figs.

Undella subacuta

Jörgensen, E., 1924
Mediterranean Tintinnidae. Rept. Danish Oceanogr. Exped. 1908-10 to the Mediterranean and adjacent seas, Vol. II, Biol. J 3, 110 pp., 114 text figs.

Undella tenuirostris

Laackmann, H., 1910
Die Tintinnodeen der Deutschen Südpolar Expedition 1901-1903. Deutschen Südpolar Expedition 11, Zool. 3 (4): 341-496, pls. 33-51.

Undella tricollaria

Jörgensen, E., 1924
Mediterranean Tintinnidae. Rept. Danish Oceanogr. Exped. 1908-10 to the Mediterranean and adjacent seas, Vol. II, Biol. J 3, 110 pp., 114 text figs.

Undella tricollaria n.sp.

Laackmann, H., 1910
Die Tintinnodeen der Deutschen Südpolar Expedition 1901-1903. Deutschen Südpolar Expedition 11, Zool. 3 (4): 341-496, pls. 33-51.

Undella turgida

Balech, Enrique, 1971
Microplancton del Atlantico ecuatorial oeste (Equalant 1)
Publ. Serv. Hidrograf. Naval, Argentina H. 654: 103 pp., 122 figs.

Undella turgida

Balech, Enrique, 1968.
Algunas especies nuevas o interesantes do tintinnidos del Golfo de Mexico y Caribe. Revta Mus. argent. Cienc. nat. Bernardino Rivadavia Inst. nac. Hidrobiol. 2(5):165-197.

Undella turgida

Balech, Enrique, 1962
Tintinnoinea y Dinoflagellata del Pacifico segun material de las Expediciones NORPAC y DOWNWIND del Instituto Scripps de Oceanografia. Revista, Mus. Argentino Ciencias Nat. "Bernardino Rivadavia", Ciencias Zool., 7(1):1-253.

Undella turgida

Campbell, A. S., 1942
The oceanic Tintinnoina of the plankton gathered during the last cruise of the CARNEGIE. Sci. Res. Cruise VII of the Carnegie, 1928-1929, Biol. 2:163 pp., 128 figs.

Undella turgida

Kofoid, C.A. and A.S. Campbell, 1939
Reports on the scientific results of the expedition to the eastern tropical Pacific in charge of Alexander Agassiz, by the U.S. Fish Commission steamer "Albetross" from October 1904 to March 1905-----XXXVII. The Ciliata: The Tintinnoimea. Bull. Mus. Comp. Zool. 84: 473 pp., 36 pls.

Undella turgida n.sp.

Kofoid, C. A. and A. S. Campbell, 1929
A conspectus of the marine and freshwater Ciliata belonging to the suborder Tintinnoinea, with descriptions of new species principally from the Agassiz expedition to the eastern tropical Pacific, 1904-1905. Univ. Calif. Publ. Zool. 34:1-403, 697 text figs.

Undella turgida

Marshall, S. M., 1934
The silicoflagellata and tintinnoinea. Great Barrier Reef Exped., 1928-29. Sci. Repts. 4(15):623-664, 43 text figs.

Undellapsis n. gen.

Kofoid, C. A. and A. S. Campbell, 1929
A conspectus of the marine and freshwater Ciliata belonging to the suborder Tintinnoinea, with descriptions of new species principally from the Agassiz expedition to the eastern tropical Pacific, 1904-1905. Univ. Calif. Publ. Zool. 34:1-403, 697 text figs.

Undellopsis angularis

Kofoid, C.A. and A.S. Campbell, 1939
Reports on the scientific results of the expedition to the eastern tropical Pacific in charge of Alexander Agassiz, by the U.S. Fish Commission steamer "Albetross" from October 1904 to March 1905-----XXXVII. The Ciliata: The Tintinnoimea. Bull. Mus. Comp. Zool. 84: 473 pp., 36 pls.

Undellopsis angulata n.sp.

Kofoid, C.A. and A.S. Campbell, 1939
Reports on the scientific results of the expedition to the eastern tropical Pacific in charge of Alexander Agassiz, by the U.S. Fish Commission steamer "Albetross" from October 1904 to March 1905-----XXXVII. The Ciliata: The Tintinnoimea. Bull. Mus. Comp. Zool. 84: 473 pp., 36 pls.

Undellopsis anularis n.sp.

Kofoid, C. A. and A. S. Campbell, 1929
A conspectus of the marine and freshwater Ciliata belonging to the suborder Tintinnoinea, with descriptions of new species principally from the Agassiz expedition to the eastern tropical Pacific, 1904-1905. Univ. Calif. Publ. Zool. 34:1-403, 697 text figs.

Undellopsis bicollaria n.sp.

Kofoid, C. A. and A. S. Campbell, 1929
A conspectus of the marine and freshwater Ciliata belonging to the suborder Tintinnoinea, with descriptions of new species principally from the Agassiz expedition to the eastern tropical Pacific, 1904-1905. Univ. Calif. Publ. Zool. 34:1-403, 697 text figs.

Undellopsis cubitum

Campbell, A. S., 1942
The oceanic Tintinnoina of the plankton gathered during the last cruise of the CARNEGIE. Sci. Res. Cruise VII of the Carnegie, 1928-1929, Biol. 2:163 pp., 128 figs.

Undellopsis cubitum

Kofoid, C.A. and A.S. Campbell, 1939
Reports on the scientific results of the expedition to the eastern tropical Pacific in charge of Alexander Agassiz, by the U.S. Fish Commission steamer "Albetross" from October 1904 to March 1905-----XXXVII. The Ciliata: The Tintinnoimea. Bull. Mus. Comp. Zool. 84: 473 pp., 36 pls.

Undellopsis cubitum n.sp.

Kofoid, C. A. and A. S. Campbell, 1929
A conspectus of the marine and freshwater Ciliata belonging to the suborder Tintinnoinea, with descriptions of new species principally from the Agassiz expedition to the eastern tropical Pacific, 1904-1905. Univ. Calif. Publ. Zool. 34:1-403, 697 text figs.

Undellopsis entzi

Campbell, A. S., 1942
The oceanic Tintinnoina of the plankton gathered during the last cruise of the CARNEGIE. Sci. Res. Cruise VII of the Carnegie, 1928-1929, Biol. 2:163 pp., 128 figs.

Undellopsis entzi

Kofoid, C.A. and A.S. Campbell, 1939
Reports on the scientific results of the expedition to the eastern tropical Pacific in charge of Alexander Agassiz, by the U.S. Fish Commission steamer "Albetross" from October 1904 to March 1905-----XXXVII. The Ciliata: The Tintinnoimea. Bull. Mus. Comp. Zool. 84: 473 pp., 36 pls.

Undellopsis entzi n.sp.

Kofoid, C. A. and A. S. Campbell, 1929
A conspectus of the marine and freshwater Ciliata belonging to the suborder Tintinnoinea, with descriptions of new species principally from the Agassiz expedition to the eastern tropical Pacific, 1904-1905. Univ. Calif. Publ. Zool. 34:1-403, 697 text figs.

Undellopsis insignata

Campbell, A. S., 1942
The oceanic Tintinnoina of the plankton gathered during the last cruise of the CARNEGIE. Sci. Res. Cruise VII of the Carnegie, 1928-1929, Biol. 2:163 pp., 128 figs.

Undellopsis insignata

Kofoid, C.A. and A.S. Campbell, 1939
Reports on the scientific results of the expedition to the eastern tropical Pacific in charge of Alexander Agassiz, by the U.S. Fish Commission steamer "Albetross" from October 1904 to March 1905-----XXXVII. The Ciliata: The Tintinnoimea. Bull. Mus. Comp. Zool. 84: 473 pp., 36 pls.

Undellopsis insignata n.sp.

Kofoid, C. A. and A. S. Campbell, 1929
A conspectus of the marine and freshwater Ciliata belonging to the suborder Tintinnoinea, with descriptions of new species principally from the Agassiz expedition to the eastern tropical Pacific, 1904-1905. Univ. Calif. Publ. Zool. 34:1-403, 697 text figs.

Undellopsis lineata

Campbell, A. S., 1942
The oceanic Tintinnoina of the plankton gathered during the last cruise of the CARNEGIE. Sci. Res. Cruise VII of the Carnegie, 1928-1929, Biol. 2:163 pp., 128 figs.

Undellopsis lineata n.sp.

Kofoid, C. A. and A. S. Campbell, 1929
A conspectus of the marine and freshwater Ciliata belonging to the suborder Tintinnoinea, with descriptions of new species principally from the Agassiz expedition to the eastern tropical Pacific, 1904-1905. Univ. Calif. Publ. Zool. 34:1-403, 697 text figs.

Undellopsis marsupialis

Balech, E., 1959.
Tintinnoinea del Mediterranea.
Trav. Sta. Zool., Villefranche-sur-Mer, 18(17):
Reprinted from:
Trab. Inst. Español Oceanogr., Madrid, 19 Sept. 1959(28):88 pp., 22 pls.

Undellopsis marsupialis

Campbell, A. S., 1942
The oceanic Tintinnoina of the plankton gathered during the last cruise of the CARNEGIE. Sci. Res. Cruise VII of the Carnegie, 1928-1929, Biol. 2:163 pp., 128 figs.

Undellopsis marsupialis

Gaarder, K.R., 1946
Tintinnoidea. Rep.Sci.Res. "Michael Sars" N.Atlantic Deep-Sea Exped., 1910, 2(1): 37 pp., 23 text figs.

Undellopsis marsupialis

Kofoid, C. A. and A. S. Campbell, 1929
A conspectus of the marine and freshwater Ciliata belonging to the suborder Tintinnoinea, with descriptions of new species principally from the Agassiz expedition to the eastern tropical Pacific, 1904-1905. Univ. Calif. Publ. Zool. 34:1-403, 697 text figs.

Undellopsis marsupialis

Rampi, L., 1939
Primo contributo alla conoscenza dei tintinnoidi do Mare Ligure. Atti della Soc. Ital. di Sci. Nat. 78:67-81, 58 text figs.

Undellopsis pacifica

Campbell, A. S., 1942
The oceanic Tintinnoina of the plankton gathered during the last cruise of the CARNEGIE. Sci. Res. Cruise VII of the Carnegie, 1928-1929, Biol. 2:163 pp., 128 figs.

Undellopsis pacifica n.sp.

Kofoid, C. A. and A. S. Campbell, 1929
A conspectus of the marine and freshwater Ciliata belonging to the suborder Tintinnoinea, with descriptions of new species principally from the Agassiz expedition to the eastern tropical Pacific, 1904-1905. Univ. Calif. Publ. Zool. 34:1-403, 697 text figs.

Undellopsis pacifica

Kofoid, C.A. and A.S. Campbell, 1939
Reports on the scientific results of the expedition to the eastern tropical Pacific in charge of Alexander Agassiz, by the U.S. Fish Commission steamer "Albatross" from October 1904 to March 1905-----XXXVII. The Ciliata: The Tintinnoimea. Bull.Mus. Comp.Zool. 84: 473 pp., 36 pls.

Undellopsis subangulata

Kofoid, C. A. and A. S. Campbell, 1929
A conspectus of the marine and freshwater Ciliata belonging to the suborder Tintinnoinea, with descriptions of new species principally from the Agassiz expedition to the eastern tropical Pacific, 1904-1905. Univ. Calif. Publ. Zool. 34:1-403, 697 text figs.

Undellopsis tricollaria

Campbell, A. S., 1942
The oceanic Tintinnoina of the plankton gathered during the last cruise of the CARNEGIE. Sci. Res. Cruise VII of the Carnegie, 1928-1929, Biol. 2:163 pp., 128 figs.

Undellopsis tricollaria

Gaarder, K.R., 1946
Tintinnoidea. Rep.Sci.Res. "Michael Sars" N.Atlantic Deep-Sea Exped., 1910, 2(1): 37 pp., 23 text figs.

Undellopsis tricollaria

Kofoid, C.A. and A.S. Campbell, 1939
Reports on the scientific results of the expedition to the eastern tropical Pacific in charge of Alexander Agassiz, by the U.S. Fish Commission steamer "Albatross" from October 1904 to March 1905-----XXXVII. The Ciliata: The Tintinnoimea. Bull.Mus. Comp.Zool. 84: 473 pp., 36 pls.

Undellopsis tricollaria

Kofoid, C. A. and A. S. Campbell, 1929
A conspectus of the marine and freshwater Ciliata belonging to the suborder Tintinnoinea, with descriptions of new species principally from the Agassiz expedition to the eastern tropical Pacific, 1904-1905. Univ. Calif. Publ. Zool. 34:1-403, 697 text figs.

Undellopsis truncata n.sp.

Kofoid, C.A. and A.S. Campbell, 1939
Reports on the scientific results of the expedition to the eastern tropical Pacific in charge of Alexander Agassiz, by the U.S. Fish Commission steamer "Albatross" from October 1904 to March 1905-----XXXVII. The Ciliata: The Tintinnoimea. Bull.Mus. Comp.Zool. 84: 473 pp., 36 pls.

Undellopsis umbilicata

Kofoid, C.A. and A.S. Campbell, 1939
Reports on the scientific results of the expedition to the eastern tropical Pacific in charge of Alexander Agassiz, by the U.S. Fish Commission steamer "Albatross" from October 1904 to March 1905-----XXXVII. The Ciliata: The Tintinnoimea. Bull.Mus. Comp.Zool. 84: 473 pp., 36 pls.

Undellopsis umbilicata n.sp.

Kofoid, C. A. and A. S. Campbell, 1929
A conspectus of the marine and freshwater Ciliata belonging to the suborder Tintinnoinea, with descriptions of new species principally from the Agassiz expedition to the eastern tropical Pacific, 1904-1905. Univ. Calif. Publ. Zool. 34:1-403, 697 text figs.

Xystonella sp.

Balech, Enrique, 1962
Tintinnoinea y Dinoflagellata del Pacifico segun material de las Expediciones NORPAC y DOWNWIND del Instituto Scripps de Oceanografia.
Revista, Mus. Argentino Ciencias Nat. "Bernardino Rivadavia", Ciencias Zool., 7(1):1-253.

Xystonella acus

Balech, Enrique, 1962
Tintinnoinea y Dinoflagellata del Pacifico segun material de las Expediciones NORPAC y DOWNWIND del Instituto Scripps de Oceanografia.
Revista, Mus. Argentino Ciencias Nat. "Bernardino Rivadavia", Ciencias Zool., 7(1):1-253.

Xystonella acus

Campbell, A. S., 1942
The oceanic Tintinnoina of the plankton gathered during the last cruise of the CARNEGIE. Sci. Res. Cruise VII of the Carnegie, 1928-1929, Biol. 2:163 pp., 128 figs.

Xystonella acus

Jörgensen, E., 1924
Mediterranean Tintinnidae. Rept. Danish Oceanogr. Exped. 1908-10 to the Mediterranean and adjacent seas, Vol.II, Biol. J 3, 110 pp., 114 text figs.

Xystonella acus

Kofoid, C.A. and A.S. Campbell,1939
Reports on the scientific results of the expedition to the eastern tropical Pacific in charge of Alexander Agassiz, by the U.S. Fish Commission steamer "Albatross" from October 1904 to March 1905-----XXXVII. The Ciliata: The Tintinnoimea. Bull.Mus. Comp.Zool. 84: 473 pp., 36 pls.

Xystonella acus

Kofoid, C. A. and A. S. Campbell, 1929
A conspectus of the marine and freshwater Ciliata belonging to the suborder Tintinnoinea, with descriptions of new species principally from the Agassiz expedition to the eastern tropical Pacific, 1904-1905. Univ. Calif. Publ. Zool. 34:1-403, 697 text figs.

Xystonella acus

Laackmann, H., 1910
Die Tintinnodeen der Deutschen Südpolar Expedition 1901-1903. Deutsch. Südpolar Exped. 11, Zool. 3(4):341-496, pls. 33-51.

Xystonella clavata

Campbell, A. S., 1942
The oceanic Tintinnoina of the plankton gathered during the last cruise of the CARNEGIE. Sci. Res. Cruise VII of the Carnegie, 1928-1929, Biol. 2:163 pp., 128 figs.

Xystonella clavata

Kofoid, C.A. and A.S. Campbell,1939
Reports on the scientific results of the expedition to the eastern tropical Pacific in charge of Alexander Agassiz, by the U.S. Fish Commission steamer "Albatross" from October 1904 to March 1905-----XXXVII. The Ciliata: The Tintinnoimea. Bull.Mus. Comp.Zool. 84: 473 pp., 36 pls.

Xystonella clavata

Kofoid, C. A. and A. S. Campbell, 1929
A conspectus of the marine and freshwater Ciliata belonging to the suborder Tintinnoinea, with descriptions of new species principally from the Agassiz expedition to the eastern tropical Pacific, 1904-1905. Univ. Calif. Publ. Zool. 34:1-403, 697 text figs.

Xystonella clavata

Komarovsky, B., 1959
The Tintinnina of the Gulf of Eylath (Aqaba). Contrib. Knowl. Red Sea, Bull. Sea Fish. Res. Sta.Haifa, Israel 21 (14):1-40.

Xystonella coronata

Kofoid, C. A. and A. S. Campbell, 1929
A conspectus of the marine and freshwater Ciliata belonging to the suborder Tintinnoinea, with descriptions of new species principally from the Agassiz expedition to the eastern tropical Pacific, 1904-1905. Univ. Calif. Publ. Zool. 34:1-403, 697 text figs.

Xystonella curticauda n.sp

Campbell, A. S., 1942
The oceanic Tintinnoina of the plankton gathered during the last cruise of the CARNEGIE. Sci. Res. Cruise VII of the Carnegie, 1928-1929, Biol. 2:163 pp., 128 figs.

Xystonella cymatica

Laackmann, H., 1910
Die Tintinnodeen der Deutschen Südpolar Expedition 1901-1903. Deutsch. Südpolar Exped. 11, Zool. 3(4):341-496, pls. 33-51.

Xystonella dicynatica

Laackmann, H., 1910
Die Tintinnodeen der Deutschen Südpolar Expedition 1901-1903. Deutsch. Südpolar Exped. 11, Zool. 3(4):341-496, pls. 33-51.

Xystonella flemingi n. sp.

Campbell, A. S., 1942
The oceanic Tintinnoina of the plankton gathered during the last cruise of the CARNEGIE. Sci. Res. Cruise VII of the Carnegie, 1928-1929, Biol. 2:163 pp., 128 figs.

Xystonella hastata

Laackmann, H., 1910
Die Tintinnodeen der Deutschen Südpolar Expedition 1901-1903. Deutsch. Südpolar Exped. 11, Zool. 3(4):341-496, pls. 33-51.

Xystonella lanceolata

Balech, Enrique, 1962
Tintinnoinea y Dinoflagellata del Pacifico segun material de las Expediciones NORPAC y DOWNWIND del Instituto Scripps de Oceanografia. Revista. Mus. Argentino Ciencias Nat. "Bernardino Rivadavia". Ciencias Zool., 7(1):1-253.

Xystonella lanceolata

Campbell, A. S., 1942
The oceanic Tintinnoina of the plankton gathered during the last cruise of the CARNEGIE. Sci. Res. Cruise VII of the Carnegie, 1928-1929, Biol. 2:163 pp., 128 figs.

Xystonella lanceolata

Jörgensen, E., 1924
Mediterranean Tintinnidae. Rept. Danish Oceanogr. Exped. 1908-10 to the Mediterranean and adjacent seas, Vol.II, Biol. J 3, 110 pp., 114 text figs.

Xystonella lanceolata

Kofoid, C.A. and A.S. Campbell, 1939
Reports on the scientific results of the expedition to the eastern tropical Pacific in charge of Alexander Agassiz, by the U.S. Fish Commission steamer "Albatross" from October 1904 to March 1905-----XXXVII. The Ciliata: The Tintinnoimea. Bull.Mus. Comp.Zool. 84: 473 pp., 36 pls.

Xystonella lanceolata

Kofoid, C.A. and A.S. Campbell, 1939
Reports on the scientific results of the expedition to the eastern tropical Pacific in charge of Alexander Agassiz, by the U.S. Fish Commission steamer "Albatross" from October 1904 to March 1905-----XXXVII. The Ciliata: The Tintinnoimea. Bull.Mus. Comp.Zool. 84: 473 pp., 36 pls.

Xystonella lanceolata

Laackmann, H., 1910
Die Tintinnodeen der Deutschen Südpolar Expedition 1901-1903. Deutsch. Südpolar Exped. 11, Zool. 3(4):341-496, pls. 33-51.

Xystonella lanceolata

Marshall, S. M., 1934
The silicoflagellata and tintinnoinea. Great Barrier Reef Exped., 1928-29. Sci. Repts. 4(15):623-664, 43 text figs.

Xystonella lohmanni

Campbell, A. S., 1942
The oceanic Tintinnoina of the plankton gathered during the last cruise of the CARNEGIE. Sci. Res. Cruise VII of the Carnegie, 1928-1929, Biol. 2:163 pp., 128 figs.

Xystonella lohmani

Duran, M., 1951
Contribucion al estudio de los tintinidos del plancton de los costas de Castellon. (Mediterranes occidental). Publ. Inst. Biol. Aplic., Barcelone, 8: 101-120, 29 text figs.

Xystonella lohmanni

Kofoid, C. A. and A. S. Campbell, 1929
A conspectus of the marine and freshwater Ciliata belonging to the suborder Tintinnoinea, with descriptions of new species principally from the Agassiz expedition to the eastern tropical Pacific, 1904-1905. Univ. Calif. Publ. Zool. 34:1-403, 697 text figs.

Xystonella lohmanni

Gaarder, K.R., 1946
Tintinnoidea. Rep.Sci.Res."Michael Sars" N.Atlantic Deep-Sea Exped., 1910, 2(1): 37 pp., 23 text figs.

Xystonella lohmanni

Rampi, L., 1939
Primo contributo alla conoscenza dei tintinnoidi do Maro Ligure. Atti della Soc. Ital. di Sci. Nat. 78:67-81, 58 text figs.

Xystonella longicauda

Balech, Enrique, 1962
Tintinnoinea y Dinoflagellata del Pacifico segun material de las Expediciones NORPAC y DOWNWIND del Instituto Scripps de Oceanografia. Revista. Mus. Argentino Ciencias Nat. "Bernardino Rivadavia". Ciencias Zool., 7(1):1-253.

Xystonella longicauda

Balech, E., 1959.
Tintinnoinea del Mediterraneo. Trav. Sta. Zool., Villefranche-sur-Mer, 18(17): Reprinted from: Trab. Inst. Español Oceanogr., Madrid, 19 Sept. 1959(28):88 pp., 22 pls.

Xystonella longicauda

Jörgensen, E., 1924
Mediterranean Tintinnidae. Rept. Danish Oceanogr. Exped. 1908-10 to the Mediterranean and adjacent seas, Vol.II, Biol. J 3, 110 pp., 114 text figs.

Xystonella longicauda

Kofoid, C.A. and A.S. Campbell, 1939
Reports on the scientific results of the expedition to the eastern tropical Pacific in charge of Alexander Agassiz, by the U.S. Fish Commission steamer "Albatross" from October 1904 to March 1905-----XXXVII. The Ciliata: The Tintinnoimea. Bull.Mus. Comp.Zool. 84: 473 pp., 36 pls.

Xystonella longicauda

Laackmann, H., 1910
Die Tintinnodeen der Deutschen Südpolar Expedition 1901-1903. Deutsch. Südpolar Exped. 11, Zool. 3(4):341-496, pls. 33-51.

Xystonella longicauda

Rampi, L., 1939
Primo contributo alla conoscenza dei tintinnoidi do Maro Ligure. Atti della Soc. Ital. di Sci. Nat. 78:67-81, 58 text figs.

Xystonella longicaudata

Kofoid, C. A. and A. S. Campbell, 1929
A conspectus of the marine and freshwater Ciliata belonging to the suborder Tintinnoinea, with descriptions of new species principally from the Agassiz expedition to the eastern tropical Pacific, 1904-1905. Univ. Calif. Publ. Zool. 34:1-403, 697 text figs.

Xystonella minuscula

Campbell, A. S., 1942
The oceanic Tintinnoina of the plankton gathered during the last cruise of the CARNEGIE. Sci. Res. Cruise VII of the Carnegie, 1928-1929, Biol. 2:163 pp., 128 figs.

Xystonella minuscula

Kofoid, C.A. and A.S. Campbell, 1939
Reports on the scientific results of the expedition to the eastern tropical Pacific in charge of Alexander Agassiz, by the U.S. Fish Commission steamer "Albatross" from October 1904 to March 1905-----XXXVII. The Ciliata: The Tintinnoimea. Bull.Mus. Comp.Zool. 84: 473 pp., 36 pls.

Xystonella minuscula n.sp.

Kofoid, C. A. and A. S. Campbell, 1929
A conspectus of the marine and freshwater Ciliata belonging to the suborder Tintinnoinea, with descriptions of new species principally from the Agassiz expedition to the eastern tropical Pacific, 1904-1905. Univ. Calif. Publ. Zool. 34:1-403, 697 text figs.

Xystonella minuscula

Rampi, L., 1948
Sur quelques Tintinnides (Infusoires louques) du Pacifique subtropical (Recoltes Alain Gerbault) Bull. l'Inst. Ocean., Monaco, No.938, 4 pp.

Xystonella paradoxa

Laackmann, H., 1910
Die Tintinnodeen der Deutschen Südpolar Expedition 1901-1903. Deutsch. Südpolar Exped. 11, Zool. 3(4):341-496, pls. 33-51.

Xystonella scandens

Kofoid, C. A. and A. S. Campbell, 1929
A conspectus of the marine and freshwater Ciliata belonging to the suborder Tintinnoinea, with descriptions of new species principally from the Agassiz expedition to the eastern tropical Pacific, 1904-1905. Univ. Calif. Publ. Zool. 34:1-403, 697 text figs.

Xystonella scandens

Komarovsky, B., 1962
Tintinnina from the vicinity of the Straits of Tiran and Massawa Region. Contributions to the knowledge of the Red Sea, No. 25. Sea Fish. Res. Sta., Haifa, Israel. Bull. No. 30:48-56.

Xystonella treforti

Balech, Enrique, 1962
Tintinnoinea y Dinoflagellata del Pacifico segun material de las Expediciones NORPAC y DOWNWIND del Instituto Scripps de Oceanografia. Revista. Mus. Argentino Ciencias Nat. "Bernardino Rivadavia". Ciencias Zool., 7(1):1-253.

Xystonella treforti

Balech, E., 1959.
Tintinnoinea del Mediterraneo. Trav. Sta. Zool., Villefranche-sur-Mer, 18(17): Reprinted from: Trab. Inst. Español Oceanogr., Madrid, 19 Sept. 1959(28):88 pp., 22 pls.

Xystonella treforti

Campbell, A. S., 1942
The oceanic Tintinnoina of the plankton gathered during the last cruise of the CARNEGIE. Sci. Res. Cruise VII of the Carnegie, 1928-1929, Biol. 2:163 pp., 128 figs.

Xystonella treforti

Duran, M., 1951
Contribucion al estudio de los tintinidos del plancton de los costas de Castellon. (Mediterranes occidental). Publ. Inst. Biol. Aplic., Barcelone, 8: 101-120, 29 text figs.

Xystonella treforti

Gaarder, K.R., 1946
Tintinnoidea. Rep.Sci.Res."Michael Sars" N.Atlantic Deep-Sea Exped., 1910, 2(1): 37 pp., 23 text figs.

Xystonella treforti

Jörgensen, E., 1924
Mediterranean Tintinnidae. Rept. Danish Oceanogr. Exped. 1908-10 to the Mediterranean and adjacent seas, Vol.II, Biol. J 3, 110 pp., 114 text figs.

Xystonella treforti

Kofoid, C.A. and A.S. Campbell, 1939
Reports on the scientific results of the expedition to the eastern tropical Pacific in charge of Alexander Agassiz, by the U.S. Fish Commission steamer "Albatross" from October 1904 to March 1905-----XXXVII. The Ciliata: The Tintinnoimea. Bull.Mus. Comp.Zool. 84: 473 pp., 36 pls.

Xystonella treforti

Kofoid, C. A. and A. S. Campbell, 1929
A conspectus of the marine and freshwater Ciliata belonging to the suborder Tintinnoinea, with descriptions of new species principally from the Agassiz expedition to the eastern tropical Pacific, 1904-1905. Univ. Calif. Publ. Zool. 34:1-403, 697 text figs.

Xystonella treforti

Komarovsky, B., 1959
The Tintinnina of the Gulf of Eylath (Aqaba). Contrib. Knowl. Red Sea, Bull. Sea Fish. Res. Sta.Haifa, Israel 21 (14):1-40.

Xystonella treforti

Laackmann, H., 1910
Die Tintinnodeen der Deutschen Südpolar Expedition 1901-1903. Deutsch. Südpolar Exped. 11, Zool. 3(4):341-496, pls. 33-51.

Xystonella treforti

Marshall, S. M., 1934
The silicoflagellata and tintinnoinea. Great Barrier Reef Exped., 1928-29. Sci. Repts. 4(15):623-664, 43 text figs.

Xystonella treforti

Rampi, L., 1939
Primo contributo alla conoscenza dei tintinnoidi do Maro Ligure. Atti della Soc. Ital. di Sci. Nat. 78:67-81, 58 text figs.

Xystonella treforti

Sousa e Silva, E. de, 1950.
Les tintinnides de la Baie de Cascais. Bull. Inst. Océan., Monaco, No. 979:28 pp., 4 pls.

Xystonellopsis n.gen.

Jörgensen, E., 1924
Mediterranean Tintinnidae. Rept. Danish Oceanogr. Exped. 1908-10 to the Mediterranean and adjacent seas, Vol.II, Biol. J 3, 110 pp., 114 text figs.

Xystonellopsis abbreviata

Campbell, A. S., 1942
The oceanic Tintinnoina of the plankton gathered during the last cruise of the CARNEGIE. Sci. Res. Cruise VII of the Carnegie, 1928-1929, Biol. 2:163 pp., 128 figs.

Xystonellopsis abbreviata n.sp.

Kofoid, C. A. and A. S. Campbell, 1929
A conspectus of the marine and freshwater Ciliata belonging to the suborder Tintinnoinea, with descriptions of new species principally from the Agassiz expedition to the eastern tropical Pacific, 1904-1905. Univ. Calif. Publ. Zool. 34:1-403, 697 text figs.

Xystonellopsis abbreviata

Kofoid, C.A. and A.S. Campbell, 1939
Reports on the scientific results of the expedition to the eastern tropical Pacific in charge of Alexander Agassiz, by the U.S. Fish Commission steamer "Albatross" from October 1904 to March 1905-----XXXVII. The Ciliata: The Tintinnoimea. Bull.Mus. Comp.Zool. 84: 473 pp., 36 pls.

Xystonellopsis aciculifera

Balech, Enrique, 1968.
Algunas especies nuevas o interesantes do tintinnidos del Golfo de Mexico y Caribe. Revta Mus. argent. Cienc. nat. Bernardino Rivadavia Inst. nac. Hidrobiol. 2(5):165-197.

Xystonellopsis acuminata

Campbell, A. S., 1942
The oceanic Tintinnoina of the plankton gathered during the last cruise of the CARNEGIE. Sci. Res. Cruise VII of the Carnegie, 1928-1929, Biol. 2:163 pp., 128 figs.

Xystonellopsis acuminata

Kofoid, C.A. and A.S. Campbell, 1939
Reports on the scientific results of the expedition to the eastern tropical Pacific in charge of Alexander Agassiz, by the U.S. Fish Commission steamer "Albatross" from October 1904 to March 1905-----XXXVII. The Ciliata: The Tintinnoimea. Bull.Mus. Comp.Zool. 84: 473 pp., 36 pls.

Xystonellopsis acuminata n.sp.

Kofoid, C. A. and A. S. Campbell, 1929
A conspectus of the marine and freshwater Ciliata belonging to the suborder Tintinnoinea, with descriptions of new species principally from the Agassiz expedition to the eastern tropical Pacific, 1904-1905. Univ. Calif. Publ. Zool. 34:1-403, 697 text figs.

Xystonellopsis apicata

Kofoid, C. A. and A. S. Campbell, 1929
A conspectus of the marine and freshwater Ciliata belonging to the suborder Tintinnoinea, with descriptions of new species principally from the Agassiz expedition to the eastern tropical Pacific, 1904-1905. Univ. Calif. Publ. Zool. 34:1-403, 697 text figs.

Xystonellopsis armata

Campbell, A. S., 1942
The oceanic Tintinnoina of the plankton gathered during the last cruise of the CARNEGIE. Sci. Res. Cruise VII of the Carnegie, 1928-1929, Biol. 2:163 pp., 128 figs.

Xystonellopsis armata

Gaarder, K.R., 1946
Tintinnoidea. Rep.Sci.Res. "Michael Sars" N.Atlantic Deep-Sea Exped., 1910, 2(1): 37 pp., 23 text figs.

Xystonellopsis armata

Kofoid, C.A. and A.S. Campbell, 1939
Reports on the scientific results of the expedition to the eastern tropical Pacific in charge of Alexander Agassiz, by the U.S. Fish Commission steamer "Albatross" from October 1904 to March 1905-----XXXVII. The Ciliata: The Tintinnoimea. Bull.Mus. Comp.Zool. 84: 473 pp., 36 pls.

Xystonellopsis armata

Kofoid, C. A. and A. S. Campbell, 1929
A conspectus of the marine and freshwater Ciliata belonging to the suborder Tintinnoinea, with descriptions of new species principally from the Agassiz expedition to the eastern tropical Pacific, 1904-1905. Univ. Calif. Publ. Zool. 34:1-403, 697 text figs.

Xystonellopsis brandti

Balech, Enrique, 1962
Tintinnoinea y Dinoflagellata del Pacifico segun material de las Expediciones NORPAC y DOWNWIND del Instituto Scripps de Oceanografia. Revista, Mus. Argentino Ciencias Nat. "Bernardino Rivadavia", Ciencias Zool., 7(1):1-253.

Xystonellopsis brandti

Campbell, A. S., 1942
The oceanic Tintinnoina of the plankton gathered during the last cruise of the CARNEGIE. Sci. Res. Cruise VII of the Carnegie, 1928-1929, Biol. 2:163 pp., 128 figs.

Xystonellopsis brandti

Jörgensen, E., 1924
Mediterranean Tintinnidae. Rept. Danish Oceanogr. Exped. 1908-10 to the Mediterranean and adjacent seas, Vol.II, Biol. J 3, 110 pp., 114 text figs.

Xystonellopsis brandti

Kofoid, C.A. and A.S. Campbell, 1939
Reports on the scientific results of the expedition to the eastern tropical Pacific in charge of Alexander Agassiz, by the U.S. Fish Commission steamer "Albatross" from October 1904 to March 1905-----XXXVII. The Ciliata: The Tintinnoimea. Bull.Mus. Comp.Zool. 84: 473 pp., 36 pls.

Xystonellopsis brandti

Kofoid, C. A. and A. S. Campbell, 1929
A conspectus of the marine and freshwater Ciliata belonging to the suborder Tintinnoinea, with descriptions of new species principally from the Agassiz expedition to the eastern tropical Pacific, 1904-1905. Univ. Calif. Publ. Zool. 34:1-403, 697 text figs.

Xystonellopsis clevei

Kofoid, C.A. and A.S. Campbell, 1939
Reports on the scientific results of the expedition to the eastern tropical Pacific in charge of Alexander Agassiz, by the U.S. Fish Commission steamer "Albatross" from October 1904 to March 1905-----XXXVII. The Ciliata: The Tintinnoimea. Bull.Mus. Comp.Zool. 84: 473 pp., 36 pls.

Xystonellopsis clevei n.sp

Kofoid, C. A. and A. S. Campbell, 1929
A conspectus of the marine and freshwater Ciliata belonging to the suborder Tintinnoinea, with descriptions of new species principally from the Agassiz expedition to the eastern tropical Pacific, 1904-1905. Univ. Calif. Publ. Zool. 34:1-403, 697 text figs.

Xystonellopsis conicacauda

Campbell, A. S., 1942
The oceanic Tintinnoina of the plankton gathered during the last cruise of the CARNEGIE. Sci. Res. Cruise VII of the Carnegie, 1928-1929, Biol. 2:163 pp., 128 figs.

Xystonellopsis conicacauda

Kofoid, C.A. and A.S. Campbell, 1939
Reports on the scientific results of the expedition to the eastern tropical Pacific in charge of Alexander Agassiz, by the U.S. Fish Commission steamer "Albatross" from October 1904 to March 1905-----XXXVII. The Ciliata: The Tintinnoimea. Bull.Mus. Comp.Zool. 84: 473 pp., 36 pls.

Xystonellopsis conicacauda n.sp.

Kofoid, C. A. and A. S. Campbell, 1929
A conspectus of the marine and freshwater Ciliata belonging to the suborder Tintinnoinea, with descriptions of new species principally from the Agassiz expedition to the eastern tropical Pacific, 1904-1905. Univ. Calif. Publ. Zool. 34:1-403, 697 text figs.

Xystonellopsis constricta

Balech, Enrique, 1962
Tintinnoinea y Dinoflagellata del Pacifico segun material de las Expediciones NORPAC y DOWNWIND del Instituto Scripps de Oceanografia. Revista, Mus. Argentino Ciencias Nat. "Bernardino Rivadavia", Ciencias Zool., 7(1):1-253.

Xystonellopsis constricta

Kofoid, C.A. and A.S. Campbell, 1939
Reports on the scientific results of the expedition to the eastern tropical Pacific in charge of Alexander Agassiz, by the U.S. Fish Commission steamer "Albatross" from October 1904 to March 1905-----XXXVII. The Ciliata: The Tintinnoimea. Bull.Mus. Comp.Zool. 84: 473 pp., 36 pls.

Xystonellopsis constricta n.sp.

Kofoid, C. A. and A. S. Campbell, 1929
A conspectus of the marine and freshwater Ciliata belonging to the suborder Tintinnoinea, with descriptions of new species principally from the Agassiz expedition to the eastern tropical Pacific, 1904-1905. Univ. Calif. Publ. Zool. 34:1-403, 697 text figs.

Xystonellopsis crassispinosa

Campbell, A. S., 1942
The oceanic Tintinnoina of the plankton gathered during the last cruise of the CARNEGIE. Sci. Res. Cruise VII of the Carnegie, 1928-1929, Biol. 2:163 pp., 128 figs.

Xystonellopsis crassispinosa

Gaarder, K.R., 1946
Tintinnoidea. Rep.Sci.Res. "Michael Sars" N.Atlantic Deep-Sea Exped., 1910, 2(1): 37 pp., 23 text figs.

Xystonellopsis crassispinosa

Kofoid, C.A. and A.S. Campbell, 1939
Reports on the scientific results of the expedition to the eastern tropical Pacific in charge of Alexander Agassiz, by the U.S. Fish Commission steamer "Albatross" from October 1904 to March 1905-----XXXVII. The Ciliata: The Tintinnoimea. Bull.Mus. Comp.Zool. 84: 473 pp., 36 pls.

Xystonellopsis crassispinosa n.sp

Kofoid, C. A. and A. S. Campbell, 1929
A conspectus of the marine and freshwater Ciliata belonging to the suborder Tintinnoinea, with descriptions of new species principally from the Agassiz expedition to the eastern tropical Pacific, 1904-1905. Univ. Calif. Publ. Zool. 34:1-403, 697 text figs.

Xystonellopsis cyclas

Campbell, A. S., 1942
The oceanic Tintinnoina of the plankton gathered during the last cruise of the CARNEGIE. Sci. Res. Cruise VII of the Carnegie, 1928-1929, Biol. 2:163 pp., 128 figs.

Xystonellopsis cyclas

Kofoid, C.A. and A.S. Campbell, 1939
Reports on the scientific results of the expedition to the eastern tropical Pacific in charge of Alexander Agassiz, by the U.S. Fish Commission steamer "Albatross" from October 1904 to March 1905-----XXXVII. The Ciliata: The Tintinnoimea. Bull. Mus. Comp. Zool. 84: 473 pp., 36 pls.

Xystonellopsis cyclas n.sp.

Kofoid, C. A. and A. S. Campbell, 1929
A conspectus of the marine and fresh-water Ciliata belonging to the suborder Tintinnoinea, with descriptions of new species principally from the Agassiz expedition to the eastern tropical Pacific, 1904-1905. Univ. Calif. Publ. Zool. 34:1-403, 697 text figs.

Xystonellopsis cymatica

*Balech, Enrique, 1968.
Algunas especies nuevas o interesantes do tintinnidos del Golfo de Mexico y Caribe. Revta Mus. argent. Cienc. nat. Bernardino Rivadavia Inst. nac. Hidrobiol. 2(5):165-197.

Xystonellopsis cymatica

Campbell, A. S., 1942
The oceanic Tintinnoina of the plankton gathered during the last cruise of the CARNEGIE. Sci. Res. Cruise VII of the Carnegie, 1928-1929, Biol. 2:163 pp., 128 figs.

Xystonellopsis cymatica

Gaarder, K.R., 1946
Tintinnoidea. Rep.Sci.Res. "Michael Sars" N.Atlantic Deep-Sea Exped., 1910, 2(1): 37 pp., 23 text figs.

Xystonellopsis cymatica

Jörgensen, E., 1924
Mediterranean Tintinnidae. Rept. Danish Oceanogr. Exped. 1908-10 to the Mediterranean and adjacent seas, Vol.II, Biol. J 3, 110 pp., 114 text figs.

Xystonellopsis cymatica

Kofoid, C.A. and A.S. Campbell, 1939
Reports on the scientific results of the expedition to the eastern tropical Pacific in charge of Alexander Agassiz, by the U.S. Fish Commission steamer "Albatross" from October 1904 to March 1905-----XXXVII. The Ciliata: The Tintinnoimea. Bull. Mus. Comp. Zool. 84: 473 pp., 36 pls.

Xystonellopsis cymatica

Kofoid, C. A. and A. S. Campbell, 1929
A conspectus of the marine and fresh-water Ciliata belonging to the suborder Tintinnoinea, with descriptions of new species principally from the Agassiz expedition to the eastern tropical Pacific, 1904-1905. Univ. Calif. Publ. Zool. 34:1-403, 697 text figs.

Xystonellopsis dahli

Balech, Enrique, 1962
Tintinnoinea y Dinoflagellata del Pacifico segun material de las Expediciones NORPAC y DOWNWIND del Instituto Scripps de Oceanografia. Revista, Mus. Argentino Ciencias Nat. "Bernardino Rivadavia", Ciencias Zool., 7(1):1-253.

Xystonellopsis dahli

Kofoid, C.A. and A.S. Campbell, 1939
Reports on the scientific results of the expedition to the eastern tropical Pacific in charge of Alexander Agassiz, by the U.S. Fish Commission steamer "Albatross" from October 1904 to March 1905-----XXXVII. The Ciliata: The Tintinnoimea. Bull. Mus. Comp. Zool. 84: 473 pp., 36 pls.

Xystonellopsis dahli

Kofoid, C. A. and A. S. Campbell, 1929
A conspectus of the marine and fresh-water Ciliata belonging to the suborder Tintinnoinea, with descriptions of new species principally from the Agassiz expedition to the eastern tropical Pacific, 1904-1905. Univ. Calif. Publ. Zool. 34:1-403, 697 text figs.

Xystonellopsis dicymatica

Balech, Enrique, 1971
Microplancton del Atlantico ecuatorial oeste (Equalant 1)
Publ. Serv. Hidrograf. Naval, Argentina H. 654: 103 pp., 122 figs.

Xystonellopsis dicymatica

Campbell, A. S., 1942
The oceanic Tintinnoina of the plankton gathered during the last cruise of the CARNEGIE. Sci. Res. Cruise VII of the Carnegie, 1928-1929, Biol. 2:163 pp., 128 figs.

Xystonellopsis dicymatica

Gaarder, K.R., 1946
Tintinnoidea. Rep.Sci.Res. "Michael Sars" N.Atlantic Deep-Sea Exped., 1910, 2(1): 37 pp., 23 text figs.

Xystonellopsis dicymatica

Kofoid, C.A. and A.S. Campbell, 1939
Reports on the scientific results of the expedition to the eastern tropical Pacific in charge of Alexander Agassiz, by the U.S. Fish Commission steamer "Albatross" from October 1904 to March 1905-----XXXVII. The Ciliata: The Tintinnoimea. Bull. Mus. Comp. Zool. 84: 473 pp., 36 pls.

Xystonellopsis dicymatica

Kofoid, C. A. and A. S. Campbell, 1929
A conspectus of the marine and fresh-water Ciliata belonging to the suborder Tintinnoinea, with descriptions of new species principally from the Agassiz expedition to the eastern tropical Pacific, 1904-1905. Univ. Calif. Publ. Zool. 34:1-403, 697 text figs.

Xystonellopsis dilatata

Kofoid, C. A. and A. S. Campbell, 1929
A conspectus of the marine and fresh-water Ciliata belonging to the suborder Tintinnoinea, with descriptions of new species principally from the Agassiz expedition to the eastern tropical Pacific, 1904-1905. Univ. Calif. Publ. Zool. 34:1-403, 697 text figs.

Xystonellopsis epigrus

Gaarder, K.R., 1946
Tintinnoidea. Rep.Sci.Res. "Michael Sars" N.Atlantic Deep-Sea Exped., 1910, 2(1): 37 pp., 23 text figs.

Xystonellopsis epigrus

Kofoid, C.A. and A.S. Campbell, 1939
Reports on the scientific results of the expedition to the eastern tropical Pacific in charge of Alexander Agassiz, by the U.S. Fish Commission steamer "Albatross" from October 1904 to March 1905-----XXXVII. The Ciliata: The Tintinnoimea. Bull. Mus. Comp. Zool. 84: 473 pp., 36 pls.

Xystonellopsis epigrus n.sp.

Kofoid, C. A. and A. S. Campbell, 1929
A conspectus of the marine and fresh-water Ciliata belonging to the suborder Tintinnoinea, with descriptions of new species principally from the Agassiz expedition to the eastern tropical Pacific, 1904-1905. Univ. Calif. Publ. Zool. 34:1-403, 697 text figs.

Xystonellopsis favata

Campbell, A. S., 1942
The oceanic Tintinnoina of the plankton gathered during the last cruise of the CARNEGIE. Sci. Res. Cruise VII of the Carnegie, 1928-1929, Biol. 2:163 pp., 128 figs.

Xystonellopsis favata

Kofoid, C.A. and A.S. Campbell, 1939
Reports on the scientific results of the expedition to the eastern tropical Pacific in charge of Alexander Agassiz, by the U.S. Fish Commission steamer "Albatross" from October 1904 to March 1905-----XXXVII. The Ciliata: The Tintinnoimea. Bull. Mus. Comp. Zool. 84: 473 pp., 36 pls.

Xystonellopsis favata

Kofoid, C. A. and A. S. Campbell, 1929
A conspectus of the marine and fresh-water Ciliata belonging to the suborder Tintinnoinea, with descriptions of new species principally from the Agassiz expedition to the eastern tropical Pacific, 1904-1905. Univ. Calif. Publ. Zool. 34:1-403, 697 text figs.

Xystonellopsis gaussi

Campbell, A. S., 1942
The oceanic Tintinnoina of the plankton gathered during the last cruise of the CARNEGIE. Sci. Res. Cruise VII of the Carnegie, 1928-1929, Biol. 2:163 pp., 128 figs.

Xystonellopsis gaussi

Gaarder, K.R., 1946
Tintinnoidea. Rep.Sci.Res. "Michael Sars" N.Atlantic Deep-Sea Exped., 1910, 2(1): 37 pp., 23 text figs.

Xystonellopsis gaussi

Kofoid, C.A. and A.S. Campbell, 1939
Reports on the scientific results of the expedition to the eastern tropical Pacific in charge of Alexander Agassiz, by the U.S. Fish Commission steamer "Albatross" from October 1904 to March 1905-----XXXVII. The Ciliata: The Tintinnoimea. Bull. Mus. Comp. Zool. 84: 473 pp., 36 pls.

Xystonellopsis gaussi

Kofoid, C. A. and A. S. Campbell, 1929
A conspectus of the marine and fresh-water Ciliata belonging to the suborder Tintinnoinea, with descriptions of new species principally from the Agassiz expedition to the eastern tropical Pacific, 1904-1905. Univ. Calif. Publ. Zool. 34:1-403, 697 text figs.

Xystonellopsis hastata

Balech, Enrique, 1971
Microplancton del Atlantico ecuatorial oeste (Equalant 1)
Publ. Serv. Hidrograf. Naval, Argentina H. 654: 103 pp., 122 figs.

Xystonellopsis hastata

Campbell, A. S., 1942
The oceanic Tintinnoina of the plankton gathered during the last cruise of the CARNEGIE. Sci. Res. Cruise VII of the Carnegie, 1928-1929, Biol. 2:163 pp., 128 figs.

Xystonellopsis hastata

Gaarder, K.R., 1946
Tintinnoidea. Rep.Sci.Res. "Michael Sars" N.Atlantic Deep-Sea Exped., 1910, 2(1): 37 pp., 23 text figs.

Xystonellopsis hastata

Kofoid, C.A. and A.S. Campbell, 1939
Reports on the scientific results of the expedition to the eastern tropical Pacific in charge of Alexander Agassiz, by the U.S. Fish Commission steamer "Albatross" from October 1904 to March 1905-----XXXVII. The Ciliata: The Tintinnoimea. Bull. Mus. Comp. Zool. 84: 473 pp., 36 pls.

Xystonellopsis hastata

Kofoid, C. A. and A. S. Campbell, 1929
A conspectus of the marine and fresh-water Ciliata belonging to the suborder Tintinnoinea, with descriptions of new species principally from the Agassiz expedition to the eastern tropical Pacific, 1904-1905. Univ. Calif. Publ. Zool. 34:1-403, 697 text figs.

Xystonellopsis heroica

Campbell, A. S., 1942
The oceanic Tintinnoina of the plankton gathered during the last cruise of the CARNEGIE. Sci. Res. Cruise VII of the Carnegie, 1928-1929, Biol. 2:163 pp., 128 figs.

Xystonellopsis heroida

Kofoid, C.A. and A.S. Campbell, 1939
Reports on the scientific results of the expedition to the eastern tropical Pacific in charge of Alexander Agassiz, by the U.S. Fish Commission steamer "Albatross" from October 1904 to March 1905-----XXXVII. The Ciliata: The Tintinnoimea. Bull. Mus. Comp. Zool. 84: 473 pp., 36 pls.

Xystonellopsis heroica n.sp.

Kofoid, C. A. and A. S. Campbell, 1929
A conspectus of the marine and fresh-water Ciliata belonging to the suborder Tintinnoinea, with descriptions of new species principally from the Agassiz expedition to the eastern tropical Pacific, 1904-1905. Univ. Calif. Publ. Zool. 34:1-403, 697 text figs.

Xystonellopsis heros

Balech, Enrique, 1962
Tintinnoinea y Dinoflagellata del Pacifico segun material de las Expediciones NORPAC y DOWNWIND del Instituto Scripps de Oceanografia. Revista, Mus. Argentino Ciencias Nat. "Bernardino Rivadavia", Ciencias Zool., 7(1):1-253.

Xystonellopsis heros

Campbell, A. S., 1942
The oceanic Tintinnoina of the plankton gathered during the last cruise of the CARNEGIE. Sci. Res. Cruise VII of the Carnegie, 1928-1929, Biol. 2:163 pp., 128 figs.

Xystonellopsis heros

Kofoid, C.A. and A.S. Campbell,1939
Reports on the scientific results of the expedition to the eastern tropical Pacific in charge of Alexander Agassiz, by the U.S. Fish Commission steamer "Albatross" from October 1904 to March 1905-----XXXVII. The Ciliata: The Tintinnoimea. Bull. Mus. Comp. Zool. 84: 473 pp., 36 pls.

Xystonellopsis heros

Kofoid, C. A. and A. S. Campbell, 1929
A conspectus of the marine and fresh-water Ciliata belonging to the suborder Tintinnoinea, with descriptions of new species principally from the Agassiz expedition to the eastern tropical Pacific, 1904-1905. Univ. Calif. Publ. Zool. 34:1-403, 697 text figs.

Xystonellopsis inaequalis

Campbell, A. S., 1942
The oceanic Tintinnoina of the plankton gathered during the last cruise of the CARNEGIE. Sci. Res. Cruise VII of the Carnegie, 1928-1929, Biol. 2:163 pp., 128 figs.

Xystonellopsis inaequalis

Gaarder, K.R., 1946
Tintinnoidea, Rep.Sci.Res. "Michael Sars" N.Atlantic Deep-Sea Exped., 1910, 2(1): 37 pp., 23 text figs.

Xystonellopsis inaequalis

Kofoid, C.A. and A.S. Campbell,1939
Reports on the scientific results of the expedition to the eastern tropical Pacific in charge of Alexander Agassiz, by the U.S. Fish Commission steamer "Albatross" from October 1904 to March 1905-----XXXVII. The Ciliata: The Tintinnoimea. Bull. Mus. Comp. Zool. 84: 473 pp., 36 pls.

Xystonellopsis inaequalis n.sp.

Kofoid, C. A. and A. S. Campbell, 1929
A conspectus of the marine and fresh-water Ciliata belonging to the suborder Tintinnoinea, with descriptions of new species principally from the Agassiz expedition to the eastern tropical Pacific, 1904-1905. Univ. Calif. Publ. Zool. 34:1-403, 697 text figs.

Xystonellopsis Krämmeri

Campbell, A. S., 1942
The oceanic Tintinnoina of the plankton gathered during the last cruise of the CARNEGIE. Sci. Res. Cruise VII of the Carnegie, 1928-1929, Biol. 2:163 pp., 128 figs.

Xystonellopsis krämeri

Kofoid, C.A. and A.S. Campbell,1939
Reports on the scientific results of the expedition to the eastern tropical Pacific in charge of Alexander Agassiz, by the U.S. Fish Commission steamer "Albatross" from October 1904 to March 1905-----XXXVII. The Ciliata: The Tintinnoimea. Bull. Mus. Comp. Zool. 84: 473 pp., 36 pls.

Xystonellopsis krämeri

Kofoid, C. A. and A. S. Campbell, 1929
A conspectus of the marine and fresh-water Ciliata belonging to the suborder Tintinnoinea, with descriptions of new species principally from the Agassiz expedition to the eastern tropical Pacific, 1904-1905. Univ. Calif. Publ. Zool. 34:1-403, 697 text figs.

Xystonellopsis laticincta

Campbell, A. S., 1942
The oceanic Tintinnoina of the plankton gathered during the last cruise of the CARNEGIE. Sci. Res. Cruise VII of the Carnegie, 1928-1929, Biol. 2:163 pp., 128 figs.

Xystonellopsis laticincta

Kofoid, C.A. and A.S. Campbell,1939
Reports on the scientific results of the expedition to the eastern tropical Pacific in charge of Alexander Agassiz, by the U.S. Fish Commission steamer "Albatross" from October 1904 to March 1905-----XXXVII. The Ciliata: The Tintinnoimea. Bull. Mus. Comp. Zool. 84: 473 pp., 36 pls.

Xystonellopsis laticincta n.sp.

Kofoid, C. A. and A. S. Campbell, 1929
A conspectus of the marine and fresh-water Ciliata belonging to the suborder Tintinnoinea, with descriptions of new species principally from the Agassiz expedition to the eastern tropical Pacific, 1904-1905. Univ. Calif. Publ. Zool. 34:1-403, 697 text figs.

Xystonellopsis mascarensis n.sp.

Kofoid, C. A. and A. S. Campbell, 1929
A conspectus of the marine and fresh-water Ciliata belonging to the suborder Tintinnoinea, with descriptions of new species principally from the Agassiz expedition to the eastern tropical Pacific, 1904-1905. Univ. Calif. Publ. Zool. 34:1-403, 697 text figs.

Xystonellopsis ornata

Campbell, A. S., 1942
The oceanic Tintinnoina of the plankton gathered during the last cruise of the CARNEGIE. Sci. Res. Cruise VII of the Carnegie, 1928-1929, Biol. 2:163 pp., 128 figs.

Xystonellopsis ornata

Kofoid, C.A. and A.S. Campbell,1939
Reports on the scientific results of the expedition to the eastern tropical Pacific in charge of Alexander Agassiz, by the U.S. Fish Commission steamer "Albatross" from October 1904 to March 1905-----XXXVII. The Ciliata: The Tintinnoimea. Bull. Mus. Comp. Zool. 84: 473 pp., 36 pls.

Xystonellopsis ornata

Kofoid, C. A. and A. S. Campbell, 1929
A conspectus of the marine and fresh-water Ciliata belonging to the suborder Tintinnoinea, with descriptions of new species principally from the Agassiz expedition to the eastern tropical Pacific, 1904-1905. Univ. Calif. Publ. Zool. 34:1-403, 697 text figs.

Xystonellopsis paradoxa

Balech, Enrique, 1962
Tintinnoinea y Dinoflagellata del Pacifico segun material de las Expediciones NORPAC y DOWNWIND del Instituto Scripps de Oceanografia. Revista, Mus. Argentino Ciencias Nat. "Bernardino Rivadavia", Ciencias Zool., 7(1):1-253.

Xystonellopsis paradoxa

Balech, E., 1959.
Tintinnoina del Mediterraneo. Trav. Sta. Zool., Villefranche-sur-Mer, 18(17): Reprinted from:
Trab. Inst. Español Oceanogr., Madrid, 19 Sept. 1959(28):88 pp., 22 pls.

Xystonellopsis paradoxa

Campbell, A. S., 1942
The oceanic Tintinnoina of the plankton gathered during the last cruise of the CARNEGIE. Sci. Res. Cruise VII of the Carnegie, 1928-1929, Biol. 2:163 pp., 128 figs.

Xystonellopsis paradoxa

Duran, M., 1951
Contribucion al estudio de los tintinidos del plancton de los costas de Castellon. (Mediterranes occidental). Publ. Inst. Biol. Aplic., Barcelona, 8: 101-120, 29 text figs.

Xystonellopsis paradoxa

Gaarder, K.R., 1946
Tintinnoidea. Rep.Sci.Res. "Michael Sars" N.Atlantic Deep-Sea Exped., 1910, 2(1): 37 pp., 23 text figs.

Xystonellopsis paradoxa

Jörgensen, E., 1924
Mediterranean Tintinnidae. Rept. Danish Oceanogr. Exped. 1908-10 to the Mediterranean and adjacent seas, Vol.II, Biol. J 3, 110 pp., 114 text figs.

Xystonellopsis paradoxa

Kofoid, C.A. and A.S. Campbell,1939
Reports on the scientific results of the expedition to the eastern tropical Pacific in charge of Alexander Agassiz, by the U.S. Fish Commission steamer "Albatross" from October 1904 to March 1905-----XXXVII. The Ciliata: The Tintinnoimea. Bull. Mus. Comp. Zool. 84: 473 pp., 36 pls.

Xystonellopsis paradoxa

Kofoid, C. A. and A. S. Campbell, 1929
A conspectus of the marine and fresh-water Ciliata belonging to the suborder Tintinnoinea, with descriptions of new species principally from the Agassiz expedition to the eastern tropical Pacific, 1904-1905. Univ. Calif. Publ. Zool. 34:1-403, 697 text figs.

Xystonellopsis parva n.sp.

Kofoid, C.A. and A.S. Campbell,1939
Reports on the scientific results of the expedition to the eastern tropical Pacific in charge of Alexander Agassiz, by the U.S. Fish Commission steamer "Albatross" from October 1904 to March 1905-----XXXVII. The Ciliata: The Tintinnoimea. Bull. Mus. Comp. Zool. 84: 473 pp., 36 pls.

Xystonellopsis pinnata

Kofoid, C.A. and A.S. Campbell,1939
Reports on the scientific results of the expedition to the eastern tropical Pacific in charge of Alexander Agassiz, by the U.S. Fish Commission steamer "Albatross" from October 1904 to March 1905-----XXXVII. The Ciliata: The Tintinnoimea. Bull. Mus. Comp. Zool. 84: 473 pp., 36 pls.

Xystonellopsis pinnata n.sp.

Kofoid, C. A. and A. S. Campbell, 1929
A conspectus of the marine and fresh-water Ciliata belonging to the suborder Tintinnoinea, with descriptions of new species principally from the Agassiz expedition to the eastern tropical Pacific, 1904-1905. Univ. Calif. Publ. Zool. 34:1-403, 697 text figs.

Xystonellopsis pulchra

Balech, Enrique, 1962
Tintinnoinea y Dinoflagellata del Pacifico segun material de las Expediciones NORPAC y DOWNWIND del Instituto Scripps de Oceanografia. Revista, Mus. Argentino Ciencias Nat. "Bernardino Rivadavia", Ciencias Zool., 7(1):1-253.

Xystonellopsis pulchra

Campbell, A. S., 1942
The oceanic Tintinnoina of the plankton gathered during the last cruise of the CARNEGIE. Sci. Res. Cruise VII of the Carnegie, 1928-1929, Biol. 2:163 pp., 128 figs.

Oceanographic Index: Marine Organisms Cumulation, 1946-1973

Xystonellopsis pulchra
Kofoid, C.A. and A.S. Campbell, 1939
Reports on the scientific results of the expedition to the eastern tropical Pacific in charge of Alexander Agassiz, by the U.S. Fish Commission steamer "Albetross" from October 1904 to March 1905-----XXXVII. The Ciliata: The Tintinnoimea. Bull. Mus. Comp. Zool. 84: 473 pp., 36 pls.

Xystonellopsis pulchra
Kofoid, C.A. and A.S. Campbell, 1929
A conspectus of the marine and fresh-water Ciliata belonging to the suborder Tintinnoinea, with descriptions of new species principally from the Agassiz expedition to the eastern tropical Pacific, 1904-1905. Univ. Calif. Publ. Zool. 34:1-403, 697 text figs.

Xystonellopsis scyphium n.sp.
Jörgensen, E., 1924
Mediterranean Tintinnidae. Rept. Danish Oceanogr. Exped. 1908-10 to the Mediterranean and adjacent seas, Vol.II, Biol. J 3, 110 pp., 114 text figs.

Xystonellopsis scyphium
Kofoid, C.A. and A.S. Campbell, 1929
A conspectus of the marine and fresh-water Ciliata belonging to the suborder Tintinnoinea, with descriptions of new species principally from the Agassiz expedition to the eastern tropical Pacific, 1904-1905. Univ. Calif. Publ. Zool. 34:1-403, 697 text figs.

Xystonellopsis spicata
Gaarder, K.R., 1946
Tintinnoidea. Rep.Sci.Res."Michael Sars" N.Atlantic Deep-Sea Exped., 1910, 2(1): 37 pp., 23 text figs.

Xystonellopsis spicata
Kofoid, C.A. and A.S. Campbell, 1939
Reports on the scientific results of the expedition to the eastern tropical Pacific in charge of Alexander Agassiz, by the U.S. Fish Commission steamer "Albetross" from October 1904 to March 1905-----XXXVII. The Ciliata: The Tintinnoimea. Bull. Mus. Comp. Zool. 84: 473 pp., 36 pls.

Xystonellopsis tenuirostris
Balech, Enrique, 1962
Tintinnoinea y Dinoflagellata del Pacifico segun material de las Expediciones NORPAC y DOWNWIND del Instituto Scripps de Oceanografia. Revista, Mus. Argentino Ciencias Nat. "Bernardino Rivadavia", Ciencias Zool., 7(1):1-253.

Xystonellopsis tenuirostris
Campbell, A.S., 1942
The oceanic Tintinnoina of the plankton gathered during the last cruise of the CARNEGIE. Sci. Res. Cruise VII of the Carnegie, 1928-1929, Biol. 2:163 pp., 128 figs.

Xystonellopsis tenuirostris
Kofoid, C.A. and A.S. Campbell, 1929
A conspectus of the marine and fresh-water Ciliata belonging to the suborder Tintinnoinea, with descriptions of new species principally from the Agassiz expedition to the eastern tropical Pacific, 1904-1905. Univ. Calif. Publ. Zool. 34:1-403, 697 text figs.

Xystonellopsis torta
Kofoid, C.A. and A.S. Campbell, 1939
Reports on the scientific results of the expedition to the eastern tropical Pacific in charge of Alexander Agassiz, by the U.S. Fish Commission steamer "Albetross" from October 1904 to March 1905-----XXXVII. The Ciliata: The Tintinnoimea. Bull. Mus. Comp. Zool. 84: 473 pp., 36 pls.

Xystonellopsis torta
Kofoid, C.A. and A.S. Campbell, 1929
A conspectus of the marine and freshwater Ciliata belonging to the suborder Tintinnoinea, with descriptions of new species principally from the Agassiz expedition to the eastern tropical Pacific, 1904-1905. Univ. Calif. Publ. Zool. 34:1-403, 697 text figs.

Xystonellopsis tropica n.sp.
Kofoid, C.A. and A.S. Campbell, 1939
Reports on the scientific results of the expedition to the eastern tropical Pacific in charge of Alexander Agassiz, by the U.S. Fish Commission steamer "Albetross" from October 1904 to March 1905-----XXXVII. The Ciliata: The Tintinnoimea. Bull. Mus. Comp. Zool. 84: 473 pp., 36 pls.

Xystonellopsis turgida n.sp.
Kofoid, C.A. and A.S. Campbell, 1939
Reports on the scientific results of the expedition to the eastern tropical Pacific in charge of Alexander Agassiz, by the U.S. Fish Commission steamer "Albetross" from October 1904 to March 1905-----XXXVII. The Ciliata: The Tintinnoimea. Bull. Mus. Comp. Zool. 84: 473 pp., 36 pls.

coccolithophorids
Bernard, F., 1953.
Rôle des Flagellés calcaires dans la fertilité et la sédimentation en mer profonde. Deep-sea Res. 1(1):34-46, 3 textfigs.

coccolithophorids
Bernard, F., 1939.
Coccolithophorides nouveaux ou peu connus observés à Monaco en 1938. Arch. Zool. Exp. Gen. 81(Notes et Revue No. 1): 33-44, 2 textfigs.

coccolithophorids
Black, M., and B. Barnes, 1961.
Coccoliths and discoasters from the floor of the South Atlantic Ocean. J.R. Microsc. Soc., (3), 80(2):137-147.

coccoliths
Boganov, Iu. A., and M.G. Vshakova, 1966.
Coccoliths of discoaster group Tan Syn Hok in aqueous suspensions from the Pacific. (In Russian). Dokl., Akad. Nauk, SSSR. 171(2):465-467.

coccolithophorids
Braarud, T., 1962.
Electron microscope studies of coccoliths in ocean deposits. Nature, 193(4820):1035-1036.

coccolithophorids
Braarud, T., 1954.
Studiet av planktonalger i elektronmikroskop. Blyttia 2:102-108, 4 pls.

coccolithophores
Braarud, T., G. Deflandre, P. Halldal and E. Kamptner, 1955.
Terminology, nomenclature and systematics of the Coccolithophoridae. Micropaleontology, 1(2):157-159.

cocolithophorids
Bramlett, M.N., 1958.
Significance of coccolithophorids in calcium-carbonate deposition. Bull. G.S.A., 69(1):121-126

coccolithophorids
Bucalossi, G., 1960.
Étude quantitative des variations du phytoplancton dans la baie d'Alger en fonction du milieu (novembre 1959 à mai 1960). Bull. Inst. Océanogr., Monaco, 57(1189):1-40.

coccolithophorids
Cohen, C.L.D. 1965.
Coccoliths and discoasters from the bottom sediments of the Adriatic Rapp. P.-v. Réun, Comm. int. Explor. scient. Mer Médit. 18(3):959.

coccolithophores
Cohen, C.L.D. 1965.
Coccoliths and discoasters. Geol. en Mijnbouw, 44(10):337-344.

Coccolithophorids, new species, (fossils)
Gartner, Stefan, Jr., 1967.
Calcareous nannofossils from Neogene of Trinidad, Jamaica, and Gulf of Mexico. Paleontol. Contrib., Univ. Kansas, (29):1-6.

coccolithophorids
Halldal, P., and J. Markali, 1958.
Electron microscope studies on coccolithophorids from the Norwegian Sea, the Gulf Stream and Mediterranean. Avhandl. Norske Videnskaps-Akad., Oslo, 1. Math. Naturvidensk. Kl., 1955(1):1-30, 27 pls.

coccolithophorids, anat.
Halldal, P., and J. Markali, 1954.
Morphology and microstructure of coccoliths studied in the electron microscope. Observations on Anthosphaera robusta and Calyptrosphaera papillifera. Nytt Mag. Bot. 2:117-119, 2 pls.

coccolithophorids
Halldal, P., and J. Markali, 1954.
Observations on coccoliths of Syracosphaera mediterranea Lohm., S. pulchra Lohm., and S. molischi Schill. in the electron microscope. J. du Cons. 19(3):329-336, 6 textfigs.

coccolithophorids
Hay, W.W., and Kenneth M. Towe, 1962.
Electron-microscope studies of Braarudosphaera bigelowi and some related coccolithophorids. Science, 137(3528):426-428.

coccoliths
Kamptner, Erwin, 1941
Die Coccolithineen der Südwestküste von Istrien. A .. Naturhist. Mus., Wien, 51: 54-149, pls 1-15.

Coccolithophorids
Kruger, D., 1950.
Variations quantitatives des protistes marins au voisinage du Port d'Alger durant l'hiver 1949-1950. Bull. Inst. Océan, Monaco, No. 978:20 pp., 5 textfigs.

Coccolithophorids
Lecal, J. 1967.
Le nannoplancton des côtes d'Israel. Hydrobiologia 29 (3/4): 305-387.

Coccolithophorids, anat. phys.
Lecal, J. 1965.
A propos des modalités d'élaboration des formations épineuses des Coccolithophorides. Protistologica 1(2): 63-70.

coccolithophorides
Lecal, J., 1952.
Répartition en profondeur des coccolithophorides en quelques stations méditerranéenes occidentales Bull. Inst. Océan., Monaco, No. 1018:13 pp., 4 textfigs.

calcareous flagellates
Lecal-Schlauder, J., 1947
Repartition des Flagellés calcaires dans la baie d'Alger. Comp. Rend. Soc. Biol. 141: 403-405.

coccolithophorids
Lecal, J. and Mlle. Bernheim, 1960
Microstructure du squelette de quelques Coccolithophorides. Bull. Soc. Hist. Nat. Afrique du Nord, 51(4/6):273-297, 22 pls.

Coccolithophorids
Marshall, Harold C. 1968.
Coccolithophores in the northwest Sargasso Sea. Limnol. Oceanogr. 13(2): 370-376.

coccolithophorids
Martini, E., 1958.
Discoasteriden und verwandte Formen im NW-deutschen Eozän (Coccolithophorida). 1. Taxionomische Untersuchungen.
Senckenbergiana Lethaea, Frankfurt-am-Main, 39(5/6):353-388.

coccoliths
McIntyre, Andrew, 1967.
Coccoliths as paleoclimatic indicators of Pleistocene glaciation.
Science, 158(3806):1314-1317.

coccolithophorids
McIntyre, Andrew, and Allan W.H. Bé, 1967.
Modern Coccolithophoridae of the Atlantic Ocean. I. Placolith and cyrtoliths.
Deep-Sea Res., 14(5):561-597.

coccolithophores
McIntyre, Andrew, Allan W.H. Bé and Resa Preikstas, 1967.
Coccoliths and the Pliocene-Pleistocene boundary.
Progress in Oceanography, 4:3-25.

Coccolithophorids
Trégouboff, G., 1956.
Rapport sur les travaux concernant le plancton Méditerranéen publiés entre Novembre 1952 et Novembre 1954.
Rapp. Proc. Verb., Comm. Int. Expl. Sci., Mer Mediterranee, 13:65-100

coccolithophorids
Wauthy, B., R. Desrosières et J. Le Bourhis, 1967.
Importance présumée de l'ultraplancton dans les eaux tropicales oligotrophes du Pacifique central sud.
Cah. ORSTOM, Sér.Océanogr., 5(2):109-116.

coccolithophorids, anat. physiol. [a]
Burns, D.A. 1973.
Structural analysis of flanged coccoliths in sediments from the south west Pacific Ocean.
Revista esp. Micropaleontol. 5(1):147-160

coccolithophorids, anat.
Kamptner, E., 1952.
Das mikroscopische Studium des Skelettes der Coccolithineen. 1. Mikroscopie 7:232-243.

coccolithophorids, anatomy
Lecal-Schlauder, J., 1951
Recherches morphologiques et biologiques sur les Coccolithophorides Nord-Africains. Ann. Inst. Ocean 26(7): 255-362, Pls. 9-13, 45 text figs.

coccolith formation
Paasche E. 1968.
The effect of temperature, light intensity, and photoperiod of coccolith formation.
Limnol. Oceanogr. 13(1):178-181.

coccoliths, anat-physiol
Paasche, E., 1965.
The effect of 3-(p-chlorophenyl)-1, 1-dimethylurea (CMU) on photosynthesis and light-dependent coccolith formation in Coccolithus huxleyi. Physiol. Plant., 18(1): 138-145.

coccolithophorids, anat.-physiol.
Paasche, E., 1962.
Coccolith formation.
Nature, 193(4820):1094-1095.

coccolithophorids
Wilbur, Karl M., and Norimitsu Watabe, 1967.
Mechanisms of calcium carbonate deposition in coccolithophorids and molluscs.
Studies Trop. Oceanogr., Miami, 5:133-154.

coccolithophorids, distribution of
Gaarder, K.R., 1971.
Comments on the distribution of coccolithophorids in the oceans. In: Micropalaeontology of oceans, B.M. Funnell and W.R. Riedel, editors. Cambridge Univ. Press, 97-103.

coccolithophorid distribution [a]
Okada, Hisatake and Susumu Honjo, 1973.
The distribution of oceanic coccolithophorids in the Pacific. Deep-Sea Res. 20(4): 355-374.

coccolithophorids, ecology of
Hillman, Norman S., 1966.
Ecology of skeletal plankton.
Antarctic J., United States, 1(5):214-215.

coccolithophorids, life history
Rayns, D.G., 1962
Alternation of generations in a coccolithophorid, Cricosphaera carterae (Braarud & Fagerl.) Braarud.
J. Mar. Biol. Assoc., U.K., 42(3):481-484.

coccolithophorids, lists of spp.
Bernard, Francis, 1967.
Contribution à l'étude du nannoplancton, de 0 à 3000 m, dans les zones atlantiques lusitanienne et mauritanienne (Campagnes de la Calypso, 1960, et du Coriolus, 1964).
Pelagos, Alger, 7:1-81.

coccolithophorids, lists of spp.
Bernhard, M., L. Rampi e A. Zattera 1969.
La distribuzione del fitoplancton nel mar Ligure.
Pubbl. Staz. Zool. Napoli 37 (2 Suppl.):73-114

coccolithophorids (fossil) lists of spp. [a]
Bukry, David, 1973
Further comments on coccolith stratigraphy, leg 12, deep sea drilling project. Initial Repts Deep Sea Drill. Proj. 12 (NSFSP-12): 1071-1083.

coccolithophorids (fossil) lists of spp. [a]
Bukry, David, 1973
Coccolith stratigraphy, leg 10, deep sea drilling project. Initial Repts Deep Sea Drill. Proj. 10(NSFSP-10): 385-406.

coccolithophorids, lists of spp. [a]
Gaarder, Karen R. and Grethe R. Hasle, 1971.
Coccolithophorids of the Gulf of Mexico.
Bull. mar. Sci. Miami 21(2): 519-544.

coccolithophorids, lists of spp.
Hasle, Grethe Rytter, 1960
Plankton coccolithophorids from the Subantarctic and equatorial Pacific.
Nytt Mag. Botanikk, 8:77-88.

coccoliths, lists of spp.
Herrera, Juan, y Ramón Margalef, 1963
Hidrografía y fitoplancton de la costa comprendida entre Castellón y la desembocadura del Ebro, de julio de 1960 a junio de 1961.
Inv. Pesq., Barcelona, 24:33-112.

coccolithophores, lists of spp.
Hulburt, Edward M., 1964.
Succession and diversity in the plankton flora of the western North Atlantic.
Bull. Mar. Sci., Gulf and Caribbean, 14(1):33-44.

coccolithophorids, lists of spp.
Lackey, James B., and Elsie W. Lackey, 1963
Microscopic algae and protozoa in the waters near Plymouth in August 1962.
J. Mar. Biol. Assoc., U.K., 43(3):797-805.

coccolithophorids, lists of spp. [a]
Magazzù, Giuseppe, e Carlo Andreoli 1971.
Trasferimenti fitoplanctonici attraverso lo Stretto di Messina in relazione alle condizioni idrologiche.
Boll. Pesca Piscic. Idrobiol. 26(1/2): 125-193

coccolithophorids, lists of spp. [a]
Margalef, Ramón, 1973
Fitoplancton marino de la región de afloramiento del NW de África. II. Composición y distribución del fitoplancton (Campaña Sahara II del Cornide de Saavedra. Result. Exped. cient. Cornide de Saavedra, Madrid 2:65-94.

coccolithophorids, lists of spp.
Marshall, Harold G., 1966.
Observations on the vertical distribution of coccolithophores in the northwestern Sargasso Sea.
Limnol. Oceanogr., 11(3):432-435.

coccolithophorids, lists of spp.
Morozova-Vodianitskaia, H.V., and E.V. Belozorskaia, 1957.
The significance of the coccolithophorids and a single Pontosphaera in the plankton of the Black Sea. Trudy Sevastopol. Biol. Stan., 9:14-21.

coccolithophorids, lists of spp.
Norris, R.E., 1961.
Observations on phytoplankton organisms collected on the N.Z.O.I. Pacific Cruise, September 1958.
N.Z. J. Sci., 4(1):162-188.

coccolithophorids (fossil), lists of spp.
Perch-Nielsen, K., 1973
Remarks on Late Cretaceous to Pleistocene coccoliths from the North Atlantic.
Initial Repts Deep Sea Drill. Proj. 12 (NSFSP-12): 1003-1069.

Coccolithinae, lists of species
Skolka, V.H., 1961
Données sur le phytoplancton des parages prébosphoriques de la Mer Noire.
Rapp. Proc. Verb., Réunions, Comm. Int. Expl. Sci. Mer. Mediterranee, Monaco, 16(2):129-132

Coccoliths, lists of spp
Smayda, Theodore J., 1966.
A quantitative analysis of the phytoplankton of the Gulf of Panama. III General ecological conditions and the phytoplankton dynamics at 8o 45'N, 79o 23'W from November 1954 to May 1957.
Inter-Amer. Trop. Tuna Comm., Bull., 11(5): 355-612.

Coccolithophorids, lists of spp. [a]
Throndsen J. 1972.
Coccolithophorids from the Caribbean Sea.
Norw. J. Bot. 19(1):51-60

coccolithophorids, lists of spp.

Tratet, Gérard, 1964.
Variations du phytoplancton à Tanger.
Trav. Inst. Sci., Cherifien, Rabat, Ser. Botan., (29):204 pp.

coccolithophorides, lists of spp.

Travers, Anne, 1969.
Quelques observations sur les coccolithophorides du Golfe de Marseille.
Tethys 1(2): 241-248.

coccolithophorids, lists of spp.

Wilcoxon, James A., 1972.
Calcareous nannoplankton ranges, deep sea drilling project. Initial reports, Deep Sea Drilling Project, Glomar Challenger 11: 459-473.

Coccolithophorids, lists of spp.

Wood, E.J. Ferguson, 1965.
Protoplankton of the Benguela-Guinea current region.
Bull. Mar. Sci.,15(2):475-479.

coccolith ooze

Thompson, Geoffrey and V.T. Bowen, 1969.
Analyses of coccolith ooze from the deep tropical Atlantic.
J. mar. Res., 27(1):32-38.

coccolithophorid stratigraphy

Bukry, David, 1972.
Coccolith stratigraphy leg 11, deep sea drilling project. Initial Reports, Deep Sea Drilling Project, Glomar Challenger 11: 475-482.

coccoliths

Bukry, David, Stanley A. King, Michael K. Horn, Frank K. Manheim, 1970.
Geological significance of coccoliths in fine-grained carbonate bands of postglacial Black Sea sediments.
Nature, Lond. 226(5241):156-158

Coccolithophorids, taxonomy of

Black, Maurice, 1968.
Taxonomic problems in the study of coccoliths.
Palaeontology,11(5):793-813.

coccolithophorids, vertical distribution of

Marshall, Harold G., 1966.
Observations on the vertical distribution of coccolithophores in the northwestern Sargasso Sea.
Limnol. Oceanogr., 11(3):432-435.

Acanthoica acanthifera

Gran, H.H., and T. Braarud, 1935
A quantitative study of the phytoplankton in the Bay of Fundy and the Gulf of Maine (including observations on hydrography, chemistry, and turbidity).
J. Biol. Bd., Canada, 1(5):279-467, 69 text figs.

Acanthoica acanthifera

Halldal, P., and J. Markali, 1955.
Electron microscope studies on coccolithophorids from the Norwegian Sea, the Gulf Stream and Mediterranean.
Avhandl. Norske Videnskaps-Akad., Oslo, 1. Mat.-Naturvidensk. Kl., 1955(1):1-30, 27 pls.

Acanthoica acanthos

Lecal-Schlauder, J., 1951
Recherches morphologiques et biologiques sur les Coccolithophorides Nord-Arficanis. Ann. Inst. Ocean 26(3): 255-362, Pls. 9-13, 45 text figs.

Acanthoica acanthos

Lillick, L.C., 1938
Preliminary report of the phytoplankton of the Gulf of Maine. Am. Mid. Nat. 20(3):624-640, 1 text figs 37 tables.

Acanthoica acanthifera

Lillick, L.C., 1938
Preliminary report of the phytoplankton of the Gulf of Maine. Am. Mid. Nat. 20(3):624-640, 1 text figs 37 tables.

Acanthoica aculeata

Kamptner, Erwin, 1941
Die Coccolithineen der Südwestküste von Istrien. Ann. Naturhist. Mus., Wien,51: 54-149, pls 1-15.

Acanthoica bidentula n.sp.

Lecal-Schlauder, J., 1951
Recherches morphologiques et biologiques sur les Coccolithophorides Nord-Arficanis. Ann. Inst. Ocean 26(3): 255-362, Pls. 9-13, 45 text figs.

Acanthoica coronata

Lillick, L.C., 1938
Preliminary report of the phytoplankton of the Gulf of Maine. Am. Mid. Nat. 20(3):624-640, 1 text figs 37 tables.

Acanthoica cucullata n.sp.

Lecal-Schlauder, J., 1951
Recherches morphologiques et biologiques sur les Coccolithophorides Nord-Arficanis. Ann. Inst. Ocean 26(3): 255-362, Pls. 9-13, 45 text figs.

Acanthoica monospina

Lillick, L.C., 1938
Preliminary report of the phytoplankton of the Gulf of Maine. Am. Mid. Nat. 20(3):624-640, 1 text figs 37 tables.

Acanthoica quatrospina

Halldal, P., and J. Markali, 1955.
Electron microscope studies on coccolithophorids from the Norwegian Sea, the Gulf Stream and Mediterranean.
Avhandl. Norske Videnskaps-Akad., Oslo, 1. Mat.-Naturvidensk. Kl., 1955(1):1-30, 27 pls.

Acanthoica quattrospina mediterranea n.var.

Lecal-Schlauder, J., 1951
Recherches morphologiques et biologiques sur les Coccolithophorides Nord-Arficanis. Ann. Inst. Ocean 26(3): 255-362, Pls. 9-13, 45 text figs.

Acanthoica ordinata n.sp.

Kamptner, Erwin, 1941
Die Coccolithineen der Südwestküste von Istrien. Ann. Naturhist. Mus., Wien,51: 54-149, pls 1-15.

Acanthoica quattrospina

Kamptner, Erwin, 1941
Die Coccolithineen der Südwestküste von Istrien. Ann. Naturhist. Mus., Wien,51: 54-149, pls 1-15.

Acanthoica rubrus n.sp.

Kamptner, Erwin, 1941
Die Coccolithineen der Südwestküste von Istrien. Ann. Naturhist. Mus., Wien,51: 54-149, pls 1-15.

Acanthosolenia mediterranea

Lecal-Schlauder, J., 1951
Recherches morphologiques et biologiques sur les Coccolithophorides Nord-Arficanis. Ann. Inst. Ocean 26(3): 255-362, Pls. 9-13, 45 text figs.

Algirosphaera campanula

Lecal-Schlauder, J., 1951
Recherches morphologiques et biologiques sur les Coccolithophorides Nord-Arficanis. Ann. Inst. Ocean 26(3): 255-362, Pls. 9-13, 45 text figs.

Algerosphaera campanula

Lecal-Schlauder, J., 1947
Repartition des Flagellés calcaires dans la baie d'Alger. Comp. Rend. Soc. Biol. 141: 403-405.

Algirosphaera oriza

Lecal-Schlauder, J., 1951
Recherches morphologiques et biologiques sur les Coccolithophorides Nord-Arficanis. Ann. Inst. Ocean 26(3): 255-362, Pls. 9-13, 45 text figs.

Algirosphaera spinulosa

Lecal-Schlauder, J., 1951
Recherches morphologiques et biologiques sur les Coccolithophorides Nord-Arficanis. Ann. Inst. Ocean 26(3): 255-362, Pls. 9-13, 45 text figs.

Anacanthoica caballicanus n.sp.

Lecal, J. 1967.
Le nannoplancton des côtes d'Israel.
Hydrobiologia 29 (3/4):305-387.

Anoplosolenia sp.

Balech, Enrique, 1964
El plancton de Mar del Plata durante el periodo 1961-1962 (Buenos Aires, Argentina).
Bol. Inst. Biol. Mar., Mar del Plata, Argentina, No. 4: 49 pp.

Anoplosolenia brasiliensis

Balech, Enrique, 1964
El plancton de Mar del Plata durante el periodo 1961-1962 (Buenos Aires, Argentina).
Bol. Inst. Biol. Mar., Mar del Plata, Argentina, No. 4: 49 pp.

Anoplosolenia brasiliensis

Halldal, P., and J. Markali, 1955.
Electron microscope studies on coccolithophorids from the Norwegian Sea, the Gulf Stream and Mediterranean.
Avhandl. Norske Videnskaps-Akad., Oslo, 1. Mat.-Naturvidensk. Kl., 1955(1):1-30, 27 pls.

Anthosphaera aurea n.sp.

Bernard, F., and J. Lecal, 1960.
Plancton unicellulaire récolté dans l'océan Indien par le Charcot (1950) et le Norsel (1955-56).
Bull. Inst. Oceanogr., Monaco, No. 1166:59 pp.

Anthosphaera quadricornu

Halldal, P., and J. Markali, 1955.
Electron microscope studies on coccolithophorids from the Norwegian Sea, the Gulf Stream and Mediterranean.
Avhandl. Norske Videnskaps-Akad., Oslo, 1. Mat.-Naturvidensk. Kl., 1955(1):1-30, 27 pls.

Anthosphaera bicornu

Lecal-Schlauder, J., 1951
Recherches morphologiques et biologiques sur les Coccolithophorides Nord-Arficanis. Ann. Inst. Ocean 26(3): 255-362, Pls. 9-13, 45 text figs.

Anthosphaera bicornu

Lecal, J. and Mlle. Bernheim, 1960
Microstructure du squelette de quelques Coccolithophorides.
Bull. Soc. Hist. Nat. Afrique du Nord, 51(4/6):273-297, 22 pls.

Anthosphaera fragaria

Kamptner, Erwin, 1941
Die Coccolithineen der Südwestküste von Istrien. Ann. Naturhist. Mus., Wien,51: 54-149, pls 1-15.

Anthosphaera robusta

Braarud, T., 1954.
Studiet av planktonalger i elektronmikroskop.
Blyttia 2:102-108, 4 pls.

Anthosphaera robusta

Halldal, P., and J. Markali, 1954.
Morphology and microstructure of coccoliths studied in the electron microscope. Observations on Anthosphaera robusta and Calyptrosphaera papillifera. Nytt Mag. Bot. 2:117-119, 2 pls.

Anthosphaera robusta
Kamptner, Erwin, 1941
Die Coccolithineen der Südwestküste von Istrien. Ann. Naturhist. Mus., Wien, 51: 54-149, pls 1-15.

Anthosphaera robusta
Braarud, T., K.R. Gaarder and J. Grøntved, 1953.
The phytoplankton of the North Sea and adjacent waters in May 1948. Rapp. Proc. Verb., Cons. Perm. Int. Expl. Mer, 133:1-87, 29 tables, Pls. A-B, 18 textfigs.

Bernardosphaera stellata n.gen., n.sp.
Lecal-Schlauder, J., 1951
Recherches morphologiques et biologiques sur les Coccolithophorides Nord-Arficanis. Ann. Inst. Ocean 26(3): 255-362, Pls. 9-13, 45 text figs.

Braarudosphaera bigelowi
Braarud, T., 1954.
Studiet av planktonalger i elektronmikroskop. Blyttia 2:102-108, 4 pls.

Braarudosphaera bigelowi
Cohen, C.L.D., 1965.
Coccoliths and discoasters from Adriatic bottom sediments.
Leidse Geol. Mededelingen. 35: 1-44, 25 pl.

Braarudosphaera bigelowi
Hay, W.W., and Kenneth M. Towe, 1962.
Electron-microscope studies of Braarudosphaera bigelowi and some related coccolithophorids. Science, 137(3528):426-428.

Braarudosphaera Deflandrei n.sp.
Lecal-Schlauder, J., 1949.
Notes préliminaires sur les Coccolithophorides d'Afrique du Nord. Bull. Soc. Hist. Nat., Afrique du Nord, 40:160-167, Pl. 6, 6 textfigs.

Braarudosphaera deflandrei maroccana n.var.
Lecal-Schlauder, J., 1951
Recherches morphologiques et biologiques sur les Coccolithophorides Nord-Arficanis. Ann. Inst. Ocean 26(3): 255-362, Pls. 9-13, 45 text figs.

Braarudosphaera undata
Hay, W.W., Kenneth M. Towe, 1962.
Electron-microscope studies of Braarudosphaera bigelowi and some related coccolithophorids. Science, 137(3528):426-428.

Calciopappus caudatus
Gaarder, K.R., J. Markali and E. Ramsfjell, 1954.
Further observations on the coccolithophorid, Calciopappus caudatus. Norske Videnskaps-Akad., Oslo, 1954(1):1-9, 2 textfigs., 4 pls.

Calciosolenia Grani
Braarud, T., and Adam Bursa, 1939
On the phytoplankton of the Oslo Fjord, 1933-1934. Hvalrådets Skr. No.19:1-63; 9 text figs. Reviewed. J. du. Cons. 14(3): 418-420. A.C. Gardiner.

Calciosolenia grani var. cylindro-thecaeformis
Kamptner, Erwin, 1941
Die Coccolithineen der Südwestküste von Istrien. Ann. Naturhist. Mus., Wien, 51: 54-149, pls 1-15.

Calciosolenia granii cylindrothecaeformis
Michaelova, N.F., 1964.
New Black Sea form of the coccolithophorid Calciosolenia granii var. cylindrothecaeformis Schiller, (In Russian).
Trudy Sevastopol. Biol. Stants., Akad. Nauk,15: 50-52.

Calciosolenia granii cylindrothecaeformis

Calciosolenia sinuosa
Balech, Enrique, 1964.
El plancton de Mar del Plata durante el periodo 1961-1962.
Bol. Inst. Biol. Mar., Buenos Aires, (4):56 pp.
49 pg. + plates

Calciosolenia sinuosa
Halldal, P., and J. Markali, 1955.
Electron microscope studies on coccolithophorids from the Norwegia Sea, the Gulf Stream and Mediterranean.
Avhandl. Norske Videnskaps-Akad., Oslo, 1. Mat. Naturvidensk. Kl., 1955(1):1-30, 27 pls.

Calciosolenia sinuosa
Lecal-Schlauder, J., 1951
Recherches morphologiques et biologiques sur les Coccolithophorides Nord-Arficanis. Ann. Inst. Ocean 26(3): 255-362, Pls. 9-13, 45 text figs.

Calciosolenia sinuosa
Lecal-Schlauder, J., 1947
Repartition des Flagellés calcaires dans la baie d'Alger. Comp. Rend. Soc. Biol. 141: 403-405.

Calciosolenia tenuis n.sp.
Bernard, F., and J. Lecal, 1960.
Plancton unicellulaire récolté dans l'océan Indien par le Charcot (1950) et le Norsel (1955-56).
Bull. Inst. Oceanogr., Monaco, No. 1166:59 pp.

Calyptrosphaera acuta n.sp.
Lecal-Schlauder, J., 1951
Recherches morphologiques et biologiques sur les Coccolithophorides Nord-Arficanis. Ann. Inst. Ocean 26(3): 255-362, Pls. 9-13, 45 text figs.

Calyptrosphaera catillifera n.comb.
Gaarder, Karen Ringdal, 1962.
Electron microscope studies on Holococcolithophorids.
Nytt Mag. for Botanikk, 10:35-51.

Calyptrosphaera galea n.sp.
Lecal-Schlauder, J., 1951
Recherches morphologiques et biologiques sur les Coccolithophorides Nord-Arficanis. Ann. Inst. Ocean 26(3): 255-362, Pls. 9-13, 45 text figs.

Calyptrosphaera globosa
Ercegovic, A., 1936
Etudes qualitative et quantitatives du phytoplancton dans les eaux cotières de l'Adriatique oriental moyen au cours de l'année 1934. Acta Adriatica 1(9):1-126

Calyptrosphaera globosa
Lecal, J. and Mlle. Bernheim, 1960
Microstructure du squelette de quelques Coccolithophorides.
Bull. Soc. Hist. Nat., Afrique du Nord, 51(4/6):273-297, 22 pls.

Calyptrosphaera gracillima n.sp.
Kamptner, Erwin, 1941
Die Coccolithineen der Südwestküste von Istrien. Ann. Naturhist. Mus., Wien, 51: 54-149, pls 1-15.

Calyptrosphaera oblonga
Ercegovic, A., 1936
Etudes qualitative et quantitatives du phytoplancton dans les eaux cotières de l'Adriatique oriental moyen au cours de l'année 1934. Acta Adriatica 1(9):1-126

Calyptrosphaera oblonga
Gran, H.H., and T. Braarud, 1935
A quantitative study of the phytoplankton in the Bay of Fundy and the Gulf of Maine (including observations on hydrography, chemistry, and turbidity).
J. Biol. Bd., Canada, 1(5):279-467, 69 text figs.

Calyptrosphaera oblonga
Halldal, P., and J. Markali, 1955.
Electron microscope studies on coccolithophorids from the Norwegian Sea, Gulf Stream and Mediterranean.
Avhandl. Norske Videnskaps-Akad., Oslo, 1. Mat.-Naturvidensk. Kl., 1955(1):1-30, 27 pls.

Calyptrosphaera oblonga
Lecal J. 1967
Le nannoplancton des côtes d'Israel
Hydrobiologia 29(3/4): 305-387.

Calyptrosphaera oblonga
Lecal-Schlauder, J., 1947
Repartition des Flagellés calcaires dans la baie d'Alger. Comp. Rend. Soc. Biol. 141: 403-405.

Calyptrosphaera oblonga
Marshall, S.M., 1933
The production of microplankton in the Great Barrier Reef Region. Brit. Mus. (N.H.) Great Barrier Reef Exped. 1928-29, Sci. Repts. II(5):111-157, 14 text figs.

Calyptrosphaera papillifera n.sp.
cocco Halldal, P., 1953.
Phytoplankton investigations from Weather Ship M in the Norwegian Sea, 1948-49 (including observations during the "Armauer Hansen" cruise, July 1949). Hvalrådets Skrifter No. 38:91 pp., 20 tables, 21 textfigs.

Calyptrosphaera papillifera
Halldal, P., and J. Markali, 1954.
Morphology and microstructure of coccoliths studied in the electron microscope. Observations on Anthosphaera robusta and Calyptrosphaera papillifera. Nytt Mag. Bot. 2:117-119, 2 pls.

Calyptrosphaera pirus
Kamptner, Erwin, 1941
Die Coccolithineen der Südwestküste von Istrien. Ann. Naturhist. Mus., Wien, 51: 54-149, pls 1-15.

Calyptrosphaera quadridentata
Kamptner, Erwin, 1941
Die Coccolithineen der Südwestküste von Istrien. Ann. Naturhist. Mus., Wien, 51: 54-149, pls 1-15.

Calyptrosphaera sphaeroidea
Gaarder, Karen Ringdal, 1962.
Electron microscope studies on Holococcolithophorids.
Nytt Mag. for Botanikk, 10:35-51.

Calyptrosphaera sphaeroides
Hope, B., 1954.
Floristic and taxonomic observations on marine phytoplankton from Nordavåtn, near Bergen.
Nytt Mag. f. Botanikk 2:149-153, 1 fig.

Calyptrosphaera sphaeroidea
Klaveness, Dag, 1973
The microanatomy of Calyptrosphaera sphaeroidea, with some supplementary observations on the motile stage of Coccolithus pelagicus. Norwegian J. Bot. 20(2/3):151-162.

Calyptrosphaera superba n.sp.
Lecal-Schlauder, J., 1951
Recherches morphologiques et biologiques sur les Coccolithophorides Nord-Arficanis. Ann. Inst. Ocean 26(3): 255-362, Pls. 9-13, 45 text figs.

Calyptrosphaera superba
Lecal, J. and Mlle. Bernheim, 1960
Microstructure du squelette de quelques Coccolithophorides.
Bull. Soc. Hist. Nat., Afrique du Nord, 51(4/6):273-297, 22 pls.

Calyptrosphaera tholifer n.sp.
Kamptner, Erwin, 1941
Die Coccolithineen der Südwestküste von Istrien. Ann. Naturhist. Mus., Wien, 51: 54-149, pls 1-15.

Calyptrosphaera tuberigera n.sp.
Lecal-Schlauder, J., 1951
Recherches morphologiques et biologiques sur les Coccolithophorides Nord-Arficanis. Ann. Inst. Ocean 26(3): 255-362, Pls. 9-13, 45 text figs.

Calytrosphaera uvella
Lillick, L.C., 1938
Preliminary report of the phytoplankton of the Gulf of Maine. Am. Mid. Nat. 20(3):624-640, 1 text figs, 37 tables.

Uavosphaera nov. comb.
Lecal, J. and Mlle. Bernheim, 1960
Microstructure du squelette de quelques Coccolithophorides.
Bull. Soc. Hist. Nat., Afrique du Nord, 51(4/6):273-297, 22 pls.

Clavosphaera furonculata n.gen., n.sp.
Lecal-Schlauder, J., 1951
Recherches morphologiques et biologiques sur les Coccolithophorides Nord-Africanis. Ann. Inst. Ocean 26(3): 255-362, Pls. 9-13, 45 text figs.

Coccolithites dissitus n. sp.
Cohen, C.L.D., 1965.
Coccoliths and discoasters from Adriatic bottom sediments.
Leidse Geol. Mededelingen, 35: 1-44, 25 pls.

Coccolithites kueneni n. sp.
Cohen, C.L.D., 1965.
Coccoliths and discoasters from Adriatic bottom sediments.
Leidse Geol. Mededelingen, 35: 1-44, 25 pls.

Coccolithites mendicus n. sp.
Cohen, C.L.D., 1965.
Coccoliths and discoasters from Adriatic bottom sediments.
Leidse Geol. Mededelingen, 35: 1-44, 25 pls.

Coccolithites repletus n. sp.
Cohen, C.L.D., 1965.
Coccoliths and discoasters from Adriatic bottom sediments.
Leidse Geol. Mededelingen, 35: 1-44, 25 pls.

Coccolithites visiosus n. sp.
Cohen, C.L.D., 1965.
Coccoliths and discoasters from Adriatic bottom sediments.
Leidse Geol. Mededelingen, 35: 1-44, 25 pls.

Coccolithus
Bernard, F., 1963
Chute de Coccolithus en mer. Son influence sur les cycles vitaux.
Rapp. Proc. Verb. Reunions, Comm. Int. Expl. Sci. Mer Méditerranée, Monaco, 17(2):503.

Coccolithus sp.
Bernard, F., 1957.
Présence des flagellés marins Coccolithus et Exuviella dans le plancton de la Mer Morte.
C.R. Acad. Sci., Paris, 245(20):1754-1756.

Coccolithus ? sp.
Cohen, C.L.D., 1965.
Coccoliths and discoasters from Adriatic bottom sediments.
Leidse Geol. Mededelingen, 35: 1-44, 25 pls.

Coccolithus
Gartner, Stefan, Jr., 1967.
Calcareous nannofossils from Neogene of Trinidad, Jamaica, and Gulf of Maine.
Paleontol. Contrib. Univ. Kansas, (29):1-6.

Lecal, J. 1967.
Le nannoplancton des côtes d'Israel. *Cycolithus anulus n.sp.*
Hydrobiologia 29 (3/4): 305-387.

Coccolithus atlanticus
Cohen, C.L.D., 1965.
Coccoliths and discoasters from Adriatic bottom sediments.
Leidse Geol. Mededelingen, 35: 1-44, 25 pls.

Coccolithus brambatii n. sp.
Cohen, C.L.D., 1965.
Coccoliths and discoasters from Adriatic bottom sediments.
Leidse Geol. Mededelingen, 35: 1-44, 25 pls.

Coccolithus carteri
Kamptner, Erwin, 1941
Die Coccolithineen der Südwestküste von Istrien. Ann. Naturhist. Mus., Wien, 51: 54-149, pls 1-15.

Coccolithus cucullus n. sp.
Lecal, J. and Mlle. Bernheim, 1960
Microstructure du squelette de quelques Coccolithophorides.
Bull. Soc. Hist. Nat., Afrique du Nord, 51(4/6):273-297, 22 pls.

Coccolithus doronicoides n. sp.
Black, M., and B. Barnes, 1961.
Coccoliths and discoasters from the floor of the South Atlantic Ocean.
J. R. Microsc. Soc., (3) 80(2):137-147.

Coccolithus erythreus n. sp.
Bernard, F., and J. Lecal, 1960.
Plancton unicellulaire récolté dans l'océan Indien par le Charcot (1950) et le Norsel (1955-56).
Bull. Inst. Océanogr., Monaco, No. 1166:59 pp.

Coccolithus formosus n. comb.
Wise, Sherwood W., Jr., 1973
Calcareous nannofossils from cores recovered during leg 18, deep sea drilling project: biostratigraphy and observations of diagenesis. Initial Repts Deep Sea Drilling Proj. 18:569-615.

Coccolithus fragilis
Bernard, Francis, 1964.
Vitesses de chute chez Cyclococcolithus fragilis (Lohm.). Consequences pour le cycle vital des mers chaudes.
Pelagos, Bull. Inst. Océanogr., Alger, 1(2):1-34.

Coccolithus fragilus
Bernard, F., 1957.
Présence dans la Mer Morte (Israël) d'un plancton unicellulaire de type méditerranéen.
Bull. Soc. Hist. Nat., Afrique du Nord, 48(56):378-384.

Coccolithus fragilis, life history
Bernard, F., 1956.
Eaux atlantiques et méditerranéennes au large de l'Algérie. II. Courants et nannoplancton de 1951 à 1953. Ann. Inst. Océan., 31(4):231-334.

Coccolithus fragilis
Bernard, F., 1953.
Rôle des flagellés calcaires dans la fertilité et la sédimentation en mer profonde.
Deep-sea Res. 1(1):34-46, 3 textfigs.

Coccolithus fragilis
Bernard, F., 1949
Remarques sur la biologie du Coccolithus fragilis Lohm., flagellé calcaire dominant du plankton méditerranéen. Trav. Bot. Soc. d'Hist. Nat., L'Afrique du Nors, Mem. Hors-série, 2:21-28, 2 text figs.

Coccolithus fragilis
Bernard, F., 1948
Rôle des Flagellés dans la transparence marine en Méditerranée. Bull. Soc. Hist. Nat., Afrique du Nord, 38(1/9):74-79, 2 text figs.

Coccolithus fragilis
Bernard, F., 1948
Recherches préliminaires sur la fertilité marine au large d'Alger. J. du Cons. 15(3):260-267, 5 text figs.

Coccolithus fragilis
Bernard, F., 1948
Recherches sur le cycle du Coccolithus fragilis Lohm., Flagellé dominant des mers chaudes. J. du Cons 15(2):177-188, 2 text figs.

Coccolithus fragilis
Bucalossi, G., 1960.
Etude quantitative des variations du phytoplancton dans la baie d'Alger en fonction du milieu (novembre 1959 à mai 1960).
Bull. Inst. Océanogr., Monaco, 57(1189):1-40.

Coccolithus fragilis
Lecal, J., 1952.
Variabilité de la teneur en coccolithophorides de différentes stations de la Baie d'Alger en mars 1952. Bull. Soc. Hist. Nat., Afrique du Nord, 43(4/6):69-80, 5 textfigs.

Coccolithus fragilis
Lecal, J., 1952.
Influence de l'éloignement des côtes sur la végétation des flagellés calcaires.
Bull. Soc. Hist. Nat., Afrique du Nord, 43(4/6):81-85, 2 textfigs.

Coccolithus fragilis
Lecal, J., 1952.
Répartition en profondeur des coccolithophoirdes en quelques stations méditerranéennes occidentales.
Bull. Inst. Océan., Monaco, No. 1018:13 pp., 4 textfigs.

Coccolithus fragilis
Lecal-Schlauder, J., 1947
Repartition des Flagellés calcaires dans la baie d'Alger. Comp. Rend. Soc. Biol. 141:403-405.

Coccolithus fragilis
Tratet, Gérard, 1964.
Variations du phytoplancton à Tanger.
Trans. Inst. Sci., Cherifien, Rabat, Ser. Botan. (29):204 pp.

Coccolithus huxleyi
Berge, G., 1962
Discoloration of the sea due to Coccolithus huxleyi bloom.
Sarsia, (6):27-40.

Coccolithus huxleyi
Bernhard, M., preparator, 1965.
Studies on the radioactive contamination of the sea, annual report 1964.
Com. Naz. Energ. Nucleare, La Spezia, Rept., No. RT/BIO (65) 18:35 pp.

Coccolithus huxleyi
Black, M., and B. Barnes, 1961.
Coccoliths and discoasters from the floor of the South Atlantic Ocean.
J.R. Microsc. Soc., (3) 80(2):137-147.

Coccolithus huxleyi
Braarud, Trygve, 1963.
Reproduction in the marine coccolithophorid Coccolithus huxleyi in culture.
Publ. Staz. Zool., Napoli, 33:110-116.

Coccolithus huxleyi
Braarud, T., 1962
Species distribution in marine phytoplankton.
J. Oceanogr. Soc., Japan, 20th Ann. Vol., 628-649.

Coccolithus (Pontosphaera) huxleyi
Braarud, T., 1954.
Studiet av planktonalger i elektronmikroskop.
Blyttia 2:102-108, 4 pls.

Coccolithus (Pontosphaera) huxleyi
Braarud, T., K.R. Gaarder and J. Grøntved, 1953.
The phytoplankton of the North Sea and adjacent waters in May 1948. Rapp. Proc. Verb., Cons. Perm. Int. Expl. Mer, 133:1-87, 29 Tables, Pls. A-B, 18 textfigs.

Coccolithes Huxleyi
Braarud, T., K.R. Gaarder, J. Markeli and E. Nordli, 1952.
Coccolithophorids studied in the electron microscope. Observations on Coccolithus Huxleyi and Syracosphaera carterae. Nytt Mag. Bot. 1:129-133.

Coccolithus huxleyi
Carlucci, A.F., and Peggy M. Bowes, 1970.
Production of vitamin B$_{12}$, thiamine and biotin by phytoplankton.
J. Phycol. 6(4): 351-357.

Coccolithus huxleyi
Cohen, C.L.D., 1965.
Coccoliths and discoasters from Adriatic bottom sediments.
Leidse Geol. Mededelingen. 35: 1-44, 25 pls.

Coccolithus huxleyi
Eppley, Richard W., Jane N. Rogers, James J. McCarthy and Alain Sournia, 1971.
Light/dark periodicity in nitrogen assimilation of the marine phytoplankters Skeletonema costatum and Coccolithus huxleyi in N-limited chemostat cultures.
J. Phycol. 7(2): 150-154.

Coccolithus huxleyi
Gaarder, Karen Ringdal, and Grethe Rytter Hasle, 1961(1962).
On the assumed symbiosis between diatoms and coccolithophorids in Brenneckella.
Nytt Mag. Botanikk, 9:145-149.

Coccolithus huxleyi (electron micrograph)
Hasle, Grethe Rytter, 1960
Plankton coccolithophorids from the Subantarctic and equatorial Pacific.
Nytt Mag. Botanikk, 8:77-88.

Coccolithus huxleyi
Hulburt, Edward M., 1967.
A note on regional differences in phytoplankton during a crossing of the southern North Atlantic Ocean in January, 1967.
Deep-Sea Res., 14(6):685-690.

Coccolithus huxleyi
Klaveness, Dag 1972.
Coccolithus huxleyi (Lohmann) Kamptner 1. Morphological investigations on the vegetative cell and the process of coccolith formation.
Protistologica 8(3): 335-346.

Coccolithus huxleyi
Lecal, J. 1966.
Diversification des populations de Coccolithus huxleyi (Lohm.) Kamptner sous l'influence des facteurs écologiques.
Protistologica 2(4): 57-70.

Coccolithus huxleyi
Lecal, J. and Mlle. Bernheim, 1960
Microstructure du squelette de quelques Coccolithophorides.
Bull. Soc. Hist. Nat. Afrique du Nord, 51(4/6):273-297, 22 pls.

Coccolithus huxleyi
McIntyre, Andrew, and Allan W.H. Be, 1967.
Modern Coccolithophoridae of the Atlantic Ocean. I. Placoliths and cyrtoliths.
Deep-Sea Res., 14(5):561-597.

Coccolithus huxleyi
Mjaaland, Gunnar, 1956
Laboratory experiments with Coccolithus huxleyi
Oikos, 7(2): 251-255.

Coccolithus huxleyi
Paasche, E. 1966.
Action spectrum of coccolith formation.
Physiologia Plantarum 19(3): 770-779.

Coccolithus huxleyi
Paasche, E., 1965.
The effect of 3-(p-chlorophenyl)-1, 1-dimethylurea (CMU) on photosynthesis and light-dependent coccolith formation in Coccolithus huxleyi. Physiol. Plant., 18(1): 138-145.

Coccolithus huxleyi
Paasche, E., 1964.
A tracer study of the inorganic carbon uptake during coccolith formation and photosynthesis in the coccolithophorid Coccolithus huxleyi.
Physiol. Plant. Suppl. 3:82 pp.

Coccolithus huxleyi
Paasche, E., 1962.
Coccolith formation.
Nature, 193(4820):1094-1095.

Coccolithus huxleyi
Parsons, T.R., 1961
On the pigment composition of eleven species of marine phytoplankters.
J. Fish. Res. Bd., Canada 18(6): 1017-1025.

Coccolithus huxleyi
Ryther, John H., and Dana D. Kramer, 1961
Relative iron requirement of some coastal and offshore algae.
Ecology, 42(2): 444-446.

Coccolithus huxleyi
Steemann-Nielsen, E., 1966.
The uptake of free CO_2 and HCO_3^- during photosynthesis of plankton algae with special reference to the coccolithophorid Coccolithus huxleyi.
Physiologia Pl., 19:232-240.

Coccolithus huxleyi
Watabe, Norimitsu, and Karl M. Wilbur, 1966.
Effects of temperature on growth, calcification, and coccolith form in Coccolithus huxleyi (Coccolithineae).
Limnol. Oceanogr., 11(4):567-575.

Coccolithus leptoporus
Black, M., and B. Barnes, 1961.
Coccoliths and discoasters from the floor of the South Atlantic Ocean.
J.R. Microsc Soc (3) 80(2):137-147.

Coccolithus leptoporus
Kamptner, Erwin, 1941
Die Coccolithineen der Südwestküste von Istrien. Ann. Naturhist. Wien, 51: 54-149, pls 1-15.

Coccolithus miopelagicus emend.
Wise, Sherwood W., Jr., 1973
Calcareous nannofossils from cores recovered during leg 18, deep sea drilling project: biostratigraphy and observations of diagenesis. Initial Repts Deep Sea Drilling Proj. 18:569-615.

Coccolithus multiporus n.sp.
Lecal-Schlauder, J., 1951
Recherches morphologiques et biologiques sur les Coccolithophorides Nord-Africains. Ann. Inst. Ocean 26(3): 255-362, Pls. 9-13, 45 text figs.

Coccolithus pelagicus
Braarud, T., 1962
Species distribution in marine phytoplankton.
J. Oceanogr. Soc. Japan, 20th Ann. Vol., 628-649.

Coccolithus pelagicus
Braarud, T., 1934
A note on the phytoplankton of the Gulf of Maine in the summer of 1933. Biol. Bull. 67(1):76-82. (Contribution No.46 of the Woods Hole Oceanographic Institution)

Coccolithus pelagicus
Cohen, C.L.D., 1965.
Coccoliths and discoasters from Adriatic bottom sediments.
Leidse Geol. Mededelingen. 35: 1-44, 25 pls.

Coccolithus pelagicus
Gran, H.H., and T. Braarud, 1935
A quantitative study of the phytoplankton in the Bay of Fundy and the Gulf of Maine (including observations on hydrography, chemistry, and turbidity).
J. Biol. Bd., Canada, 1(5):279-467, 69 text figs.

Coccolithus pelagicus
Klaveness, Dag, 1973
The microanatomy of Calyptrosphaera sphaeroidea, with some supplementary observations on the motile stage of Coccolithus pelagicus. Norwegian J. Bot. 20(2/3):151-162.

Coccolithus pelagicus
Lecal, J. and Mlle. Bernheim, 1960
Microstructure du squelette de quelques Coccolithophorides.
Bull. Soc. Hist. Nat. Afrique du Nord, 51(4/6):273-297, 22 pls.

Coccolithus pelagicus
Lillick, L.C., 1938
Preliminary report of the phytoplankton of the Gulf of Maine. Am. Mid. Nat. 20(3):624-640, 1 text figs 37 tables.

Coccolithus pelagicus
Manton, I. and G.F. Leedale, 1969.
Observations on the microanatomy of Coccolithus pelagicus and Cricosphaera carterae, with special reference to the origin and nature of coccoliths and scales.
J.mar.biol.Ass., U.K. 49(1):1-16.

Coccolithus pelagicus
McIntyre, Andrew, and Allen W.H. Be, 1967.
Modern Coccolithophoridae of the Atlantic Ocean. I. Placoliths and cyrtoliths.
Deep-Sea Res., 14(5):561-597.

Coccolithus pelagicus
Parke, Mary, and Irene Adams, 1960.
The motile (Crystallolithus hyalinus Gaarder & Markali) and non motile phases in the life history of Coccolithus pelagicus (Wallich) Schiller.
J.M.B.A., U.K., 39(2):263-274.

Coccolithus pliopelagicus n.sp.
Wise, Sherwood W., Jr., 1973
Calcareous nannofossils from cores recovered during leg 18, deep sea drilling project: biostratigraphy and observations of diagenesis. Initial Repts Deep Sea Drilling Proj. 18:569-615.

Coccolithus oblonga
Lecal-Schlauder, J., 1947
Repartition des Flagellés calcaires dans la baie d'Alger. Comp. Rend. Soc. Biol. 141: 403-405.

Coccolithus aff. sarsiae
Cohen, C.L.D., 1965.
Coccoliths and discoasters from Adriatic bottom sediments.
Leidse Geol. Mededelingen. 35: 1-44, 25 pls.

Coccolithus sessilis
Lecal-Schlauder, J., 1951
Recherches morphologiques et biologiques sur les Coccolithophorides Nord-Africains. Ann. Inst. Ocean 26(3): 255-362, Pls. 9-13, 45 text figs.

Coccolithus tessallatus n.sp.
Lecal-Schlauder, J., 1951
Recherches morphologiques et biologiques sur les Coccolithophorides Nord-Africains. Ann. Inst. Ocean 26(3): 255-362, Pls. 9-13, 45 text figs.

Coccolithus wallichii
Bernard, F., 1953.
Rôle des flagellés calcaires dans la fertilité et la sédimentation en mer profonde.
Deep-sea Res. 1(1):34-46, 3 textfigs.

Coccolithophora leptopora
Forti, A., 1922
Ricerche sulla flora pelagica (fitoplancton) di Quarto dei Mille. Mem. R. Com. Talass. Ital. 97:248 pp., 13 pls.

Coriosphaera amplior n.sp.
Lecal-Schlauder, J., 1951
Recherches morphologiques et biologiques sur les Coccolithophorides Nord-Arficanis. Ann. Inst. Ocean 26(3): 255-362, Pls. 9-13, 45 text figs.

Corisphaera arethusae
Gaarder, Karen Ringdal, 1962.
Electron microscope studies on Holococcolithophorids.
Nytt Mag. for Botanikk, 10:35-51.

Corisphaera arethusa n.sp.
Kamptner, Erwin, 1941
Die Coccolithineen der Südwestküste von Istrien. Ann. Naturhist. Mus., Wien, 51: 54-149, pls 1-15.

Corisphaera coriolis n.sp.
#Bernard, Francis, 1967.
Contribution à l'étude du nannoplancton, de 0 à 3000 m, dans les zones atlantiques lusitanienne et mauritanienne (Campagnes de la Calypso, 1960, et du Coriolus, 1964).
Pelagos, Alger, 7:1-81.

Corisphaera corona n.sp.
Kamptner, Erwin, 1941
Die Coccolithineen der Südwestküste von Istrien. Ann. Naturhist. Mus., Wien, 51: 54-149, pls 1-15.

Corisphaera fagei
Bernard, Francis 1973.
Coccolithophorides géants, de 100 à 620μ, pris en décembre 1966 depuis la Bouée Laboratoire du C.N.E.X.O.
Rapp. Proc.-v. Reun. Commn int. Explor. scient. Mer Medit. Monaco 21(8):419-421.

Corisphaera fagei
Bernard, F., 1953.
Rôle des flagellés calcaires dans la fertilité et la sédimentation en mer profonde.
Deep-sea Res. 1(1):34-46, 3 textfigs.

Corisphaera fagei n.sp.
Bernard, F., 1939.
Coccolithophorides nouveaux ou peu connus observés à Monaco en 1938.
Arch. Zool. Exp. Gen. 81 (Notes et Revue No. 1): 33-44, 2 textfigs.

Corisphaera fagei
Lecal-Schlauder, J., 1947
Repartition des Flagellés calcaires dans la baie d'Alger. Comp. Rend. Soc. Biol. 141: 403-405.

Coriosphaera fibula n.sp.
Lecal-Schlauder, J., 1951
Recherches morphologiques et biologiques sur les Coccolithophorides Nord-Arficanis. Ann. Inst. Ocean 26(3): 255-362, Pls. 9-13, 45 text figs.

Corisphaera fournieri n.sp.
Bernard, Francis, 1967.
Contribution à l'étude du nannoplancton, de 0 à 3000 m, dans les zones atlantiques lusitanienne et mauritanienne (Campagnes de la Calypso, 1960, et du Coriolus, 1964).
Pelagos, Alger, 7:1-81.

Corisphaera gracilis
Gaarder, Karen Ringdal, 1962.
Electron microscope studies on Holococcolithophorids.
Nytt Mag. for Botanikk, 10:35-51.

Corisphaera gracilis
Kamptner, Erwin, 1941
Die Coccolithineen der Südwestküste von Istrien. Ann. Naturhist. Mus., Wien, 51: 54-149, pls 1-15.

Corisphaera hasleana, n.sp.
Gaarder, Karen Ringdal, 1962.
Electron microscope studies on Holococcolithophorids.
Nytt Mag. for Botanikk, 10:35-51.

Coriosphaera margaritacea n.sp.
Lecal-Schlauder, J., 1951
Recherches morphologiques et biologiques sur les Coccolithophorides Nord-Arficanis. Ann. Inst. Ocean 26(3): 255-362, Pls. 9-13, 45 text figs.

Corisphaera perennis
Lecal, J., 1952.
Variabilité de la teneur en coccolithophorides de différentes stations de la Baie d'Alger en mars 1952. Bull. Soc. Hist. Nat., Afrique du Nord, 43(4/6):69-80, 5 textfigs.

Corisphaera perennis
Lecal, J., 1952.
Influence de l'éloignement des côtes sur la végétation des flagellés calcaires.
Bull. Soc. Hist. Nat., Afrique du Nord, 43(4/6): 81-85, 2 textfigs.

Corisphaera perennis
Lecal-Schlauder, J., 1951
Recherches morphologiques et biologiques sur les Coccolithophorides Nord-Arficanis. Ann. Inst. Ocean 26(3): 255-362, Pls. 9-13, 45 text figs.

Corisphaera perennis
Lecal-Schlauder, J., 1947
Repartition des Flagellés calcaires dans la baie d'Alger. Comp. Rend. Soc. Biol. 141: 403-405.

Corisphaera perennis and var. alpha
Lecal, J. and Mlle. Bernheim, 1960
Microstructure du squelette de quelques Coccolithophorides.
Bull. Soc. Hist. Nat., Afrique du Nord, 51(4/6):273-297, 22 pls.

Corisphaera ponticulifera n.sp.
Kamptner, Erwin, 1941
Die Coccolithineen der Südwestküste von Istrien. Ann. Naturhist. Mus., Wien, 51: 54-149, pls 1-15.

Coriosphaera stellulata n.sp.
Lecal-Schlauder, J., 1951
Recherches morphologiques et biologiques sur les Coccolithophorides Nord-Arficanis. Ann. Inst. Ocean 26(3): 255-362, Pls. 9-13, 45 text figs.

Corisphaera strigilis n.sp
Gaarder, Karen Ringdal, 1962.
Electron microscope studies on Holococcolithophorids.
Nytt Mag for Botanikk, 10:35-51.

Corisphaera strigilis
Lecal, J. 1967.
Le nannoplancton des côtes d'Israel.
Hydrobiologia 29 (3/4): 305-387.

Corisphaera tenax n.sp.
Bernard, F., and J. Lecal, 1960.
Plancton unicellulaire récolté dans l'océan Indien par le Charcot (1950) et le Norsel (1955-56).
Bull. Inst. Océanogr., Monaco, No. 1166:1-59.

Corisphaera vivesi n.sp.
Bernard, Francis, 1967.
Contribution à l'étude du nannoplancton, de 0 à 3000 m, dans les zones atlantiques lusitanienne et mauritanienne (Campagnes de la Calypso, 1960, et du Coriolus, 1964).
Pelagos, Alger, 7:1-81.

Cribrosphaerella ehrenbergi
Cohen, C.L.D. 1965.
Coccoliths and discoasters from Adriatic bottom sediments.
Leidse Geol. Mededelingen. 35: 1-44, 25 pls.

Cricolithus adriaticus n.sp
Cohen, C.L.D., 1965.
Coccoliths and discoasters from Adriatic bottom sediments.
Leidse Geol. Mededelingen. 35: 1-44, 25 pls.

Cricolithus jonesi n.sp.
Cohen, C.L.D., 1965.
Coccoliths and discoasters f. Adriatic bottom sediments.
Leidse Geol. Mededelingen. 35: 1-44, 25 pls.

Cricolithus scabrosus n.sp
Cohen, C.L.D., 1965.
Coccoliths and discoasters f. Adriatic bottom sediments.
Leidse Geol. Mededelingen. 35: 1-44, 25 pls.

Cricosphaera n. gen.
Braarud, T., 1960
On the coccolithophorid genus Cricosphaera n. gen.
Nytt Magasin for Botanikk, 8: 211-212.

Cricosphaera carterae
Braarud, T., 1962
Species distribution in marine phytoplankton.
J. Oceanogr. Soc., Japan, 20th Ann. Vol. 628-649.

Cricosphaera (Syracosphaera) carterae
Braarud, T., 1960
On the coccolithophorid genus Cricosphaera n. gen.
Nytt Magasin for Botanikk, 8: 211-212.

Cricosphaera carterae
Manton, I. and G.F. Leedale, 1969.
Observations on the microanatomy of Coccolithus pelagicus and Cricosphaera carterae, with special reference to the origin and nature of coccoliths and scales.
J.mar.biol. Ass., U.K., 49(1):1-16.

Cricosphaera carterae
Rayns, D.G., 1962
Alternation of generations in a coccolithophorid, Cricosphaera carterae (Braarud & Fagerl.) Braarud.
J. Mar. Biol. Assoc., U.K., 42(3):481-484.

Cricosphaera elongata
Taylor, W. Rowland, 1964.
Inorganic nutrient requirements for marine phytoplankton organisms.
Narragansett Mar. Lab., Univ. Rhode Island, Occ. Publ., No. 2:17-24.

Crystallolithus braarudi n.sp.
Gaarder, Karen Ringdal, 1962.
Electron microscope studies on Holococcolithophorids.
Nytt Mag. for Botanikk, 10:35-51.

Crystallolithus hyalinus n.gen.,
Gaarder, K.R., and J. Markali, 1956. n.sp.
On the coccolithophorid Crystallolithus hyalinus n.gen., n.sp. Nytt Mag. Botanikk, 5:1-5.

Crystallolithus hyalinus
Parke, Mary and Irene Adams, 1960.
The motile (Crystallolithus hyalinus Gaarder & Markali) and non-motile phases in the life history of Coccolithus pelagicus (Wallich) Schiller.
J.M.B.A., U.K., 39(2):263-274.

Cycloccolithus
Gartner, Stefan, Jr., 1967.
Calcareous nannofossils from Neogene of Trinidad, Jamaica, and Gulf of Mexico.
Paleontol. Contrib., Univ. Kansas, (29):1-6.

Cyclococcolithus bolli n.sp.
Roth, Peter H. 1948.
Calcareous nannoplankton zonation of Oligocene sections in Alabama (U.S.A.) on the islands of Trinidad and Barbados (W.I.) and the Blake Plateau (E coast of Florida, U.S.A.)
Eclogae geol. Helv. 61(2): 459-465.

Cyclococcolithus cf. formosa
Cohen, C.L.D., 1965.
Coccoliths and discoasters from Adriatic bottom sediments.
Leidse Geol. Mededelingen. 35: 1-44, 25 pls.

Cyclococcolithus fragilis
Bernard, Francis, 1967.
Contribution à l'étude du nannoplancton, de 0 à 3000 m, dans les zones atlantiques lusitanienne et mauritanienne (Campagnes de la Calypso, 1960, et du Coriolus, 1964).
Pelagos, Alger, 7:1-81.

Cyclococcolithus fragilis
Bernard, Francis, 1963.
Vitesses de chute en mer des amas palmelloïdes de Cyclococcolithis. Ses conséquences pour le cycle vital des mers chaudes.
Pelagos, Bull. Inst. Océanogr., Alger, 1(2):5-34.

Cyclococcolithus fragilis
Bernard, F., and J. Lecal, 1960.
Plancton récolté dans l'océan Indien par le Charcot (1950) et le Norsel (1955-56).
Bull. Inst. Océanogr., Monaco, No. 1166:59 pp.

unicellulaire

Cyclococcolithus fragilis
Fournier, Robert O., 1968.
Observations of particulate organic carbon in the Mediterranean Sea and their relevance to the deep-living Coccolithophorid Cyclococcolithus fragilis.
Limnol. Oceanogr., 13(4):693-696.

Cyclococcolithus fragilis
Hasle, Grethe Rytter, 1960
Plankton coccolithophorids from the Subantarctic and equatorial Pacific.
Nytt Mag. Botanikk, 8:77-88.

Cyclococcolithus fragilis
McIntyre, Andrew, and Allen W.H. Be, 1967.
Modern Coccolithophoridae of the Atlantic Ocean. I. Placoliths and cyrtoliths.
Deep-Sea Res., 14(5):561-597.

Cyclococcolithus leptoporus
Cohen, C.L.D., 1965.
Coccoliths and discoasters from Adriatic bottom sediments.
Leidse Geol. Mededelingen. 35: 1-44, 25 pls.

Cyclococcolithus leptoporus
Gartner, Stefan, Jr., 1967.
Nannofossil species related to Cyclococcolithus leptoporus (Murray & Blackman).
Paleontol. Contrib., Univ. Kansas, (28):1-4.

Cyclococcolithus leptoporus (electron micrograph)
Hasle, Grethe Rytter, 1960
Plankton coccolithophorids from the Subantarctic and equatorial Pacific.
Nytt Mag. Botanikk, 8:77-88.

Cyclococcolithus leptoporus
McIntyre, Andrew, and Allen W.H. Be, 1967.
Modern Coccolithophoridae of the Atlantic Ocean. I. Placoliths and cyrtoliths.
Deep-Sea Res., 14(5):561-597.

Cyclolithella annulus n. comb.
McIntyre, Andrew, and Allen W.H. Be, 1967.
Modern Coccolithophoridae of the Atlantic Ocean. I. Placoliths and cyrtoliths.
Deep-Sea Res., 14(5):561-597.

?Cyclolithus rotundus
Cohen, C.L.D., 1965.
Coccoliths and discoasters from Adriatic bottom sediments.
Leidse Geol. Mededelingen. 35: 1-44, 25 pls.

Deflandrius intercisus
Cohen, C.L.D., 1965.
Coccoliths and discoasters from Adriatic bottom sediments.
Leidse Geol. Mededelingen. 35: 1-44, 25 pls.

Deutschlandia cinaria n.sp.
Lecal-Schlauder, J., 1951
Recherches morphologiques et biologiques sur les Coccolithophorides Nord-Africains. Ann. Inst. Ocean 26(3): 255-362, Pls. 9-13, 45 text figs.

Discoaster murrayi n.sp.
Black, M., and B. Barnes, 1961.
Coccoliths and discoasters from the floor of the South Atlantic Ocean.
J.R. Microsc. Soc., (3) 80(2):137-147.

Discoaster perplexus
Black, M., and B. Barnes, 1961.
Coccoliths and discoasters from the floor of the South Atlantic Ocean.
J.R. Microsc. Soc., (3) 80(2):137-147.

Discolithus cf. distinctus
Cohen, C.L.D., 1965.
Coccoliths and discoasters from Adriatic bottom sediments.
Leidse Geol. Mededelingen. 35: 1-44, 25 pls.

Discolithus aff. histricus
Cohen, C.L.D., 1965.
Coccoliths and discoasters from Adriatic bottom sediments.
Leidse Geol. Mededelingen. 35: 1-44, 25 pls.

Discolithus macroporus
Cohen, C.L.D., 1965.
Coccoliths and discoasters from Adriatic bottom sediments.
Leidse Geol. Mededelingen. 35: 1-44, 25 pls.

Discolithus phaseolus n.sp.
Black, M., and B. Barnes, 1961.
Coccoliths and discoasters from the floor of the South Atlantic Ocean.
J.R. Microsc. Soc., (3) 80(2):137-147.

Discolithus planus
Cohen, C.L.D., 1965.
Coccoliths and discoasters from Adriatic bottom sediments.
Leidse Geol. Mededelingen. 35: 1-44, 25 pls.

Discolithus aff. ponticulus
Cohen, C.L.D., 1965.
Coccoliths and discoasters from Adriatic bottom sediments.
Leidse Geol. Mededelingen. 35: 1-44, 25 pls.

Discosphaera tubifer
Cohen, C.L.D., 1965.
Coccoliths and discoasters from Adriatic bottom sediments.
Leidse Geol. Mededelingen. 35: 1-44, 25 pls.

?Discosphaera thomsonii
Cohen, C.L.D., 1965.
Coccoliths and discoasters from Adriatic bottom sediments.
Leidse Geol. Mededelingen. 35: 1-44, 25 pls.

Discosphaera tubifer
Halldal, P., and J. Markali, 1955.
Electron microscope studies on coccolithophorids from the Norwegian Sea, the Gulf Stream, the Mediterranean.
Avhandl. Norske Videnskaps-Akad., Oslo, 1. Mat.-Naturvidensk. Kl., 1955(1):1-30, 27 pls.

Discosphaera tubigera
Lecal-Schlauder, J., 1951
Recherches morphologiques et biologiques sur les Coccolithophorides Nord-Africains. Ann. Inst. Ocean 26(3): 255-362, Pls. 9-13, 45 text figs.

Discosphaera tubifer
Lecal-Schlauder, J., 1947
Repartition des Flagellés calcaires dans la baie d'Alger. Comp. Rend. Soc. Biol. 141: 403-405.

Discosphaera tubifer
Lillick, L.C., 1938
Preliminary report of the phytoplankton of the Gulf of Maine. Am. Mid. Nat. 20(3):624-640, 1 text figs 37 tables.

Discosphaera tubifer
Marshall, S.M., 1933
The production of microplankton in the Great Barrier Reef Region. Brit. Mus. (N.H.) Great Barrier Reef Exped. 1928-29, Sci. Repts. II(5):111-157, 14 text figs.

Discosphaera tubifer
Rumkówna, A., 1948
[List of the phytoplankton species occurring in the superficial water layers in the Gulf of Gdańsk] Bull. Lab. mar. Gdynia, No. 4: 139-141 with tables in back.

Discosphaera tubifera
McIntyre, Andrew, and Allen W.H. Be, 1967.
Modern Coccolithophoridae of the Atlantic Ocean. I. Placoliths and cyrtoliths.
Deep-Sea Res., 14(5):561-597.

Gephyrocapsa dentata n.sp.
Halldal, P., and J. Markali, 1955.
Electron microscope studies on coccolithophorids from the Norwegian Sea, the Gulf Stream and Mediterranean.
Avhandl. Norske Videnskaps-Akad., Oslo, 1. Mat.-Naturvidensk. Kl., 1955(1):1-30, 27 pls.

Gephyrocapsa ericsonii n.sp.
McIntyre, Andrew, and Allen W.H. Be, 1967.
Modern Coccolithophoridae of the Atlantic Ocean. I. Placoliths and cyrtoliths.
Deep-Sea Res., 14(5):561-597.

Gephyrocapsa gracillima n.sp.
Lecal, J. and Mlle. Bernheim, 1960
Microstructure du squelette de quelques Coccolithophorides.
Bull. Soc. Hist. Nat., Afrique du Nord, 51(4/6):273-297, 22 pls.

Gephyrocapsa oceanica
Black, M., and B. Barnes, 1961.
Coccoliths and discoasters from the floor of the South Atlantic Ocean.
J.R. Microsc. Soc., (3) 80(2):137-147.

Gephyrocapsa oceanica

Grindley, J.R., and F.J.R. Taylor, 1970.
Factors affecting plankton blooms in False Bay. Trans. roy. Soc. SAfr. 39(2): 201-210.

Gephyrocapsa oceanica

Halldal, P., and J. Markali, 1955.
Electron microscope studies on coccolithophorids from the Norwegian Sea, the Gulf Stream and Mediterranean.
Avhandl. Norske Videnskaps-Akad., Oslo, 1. Mat.-Naturvidensk. Kl., 1955(1):1-30, 27 pls.

Gephyrocapsa oceanica (electron micrograph)

Hasle, Grethe Rytter, 1960
Plankton coccolithophorids from the Subantarctic and equatorial Pacific.
Nytt Mag. Botanikk, 8:77-88.

Gyphyrocapsa oceanica

Lecal-Schlauder, J., 1951
Recherches morphologiques et biologiques sur les Coccolithophorides Nord-Africanis. Ann. Inst. Ocean 26(?): 255-362, Pls. 9-13, 45 text figs.

Gephyrocapsa oceanica

McIntyre, Andrew, and Allen W.H. Be, 1967.
Modern Coccolithophoridae of the Atlantic Ocean. I. Placoliths and cyrtoliths.
Deep-Sea Res., 14(5):561-597.

Gephyrocapsa protohuxleyi n.sp.

McIntyre, Andrew, 1970.
Gephyrocapsa protohuxleyi sp.n. a possible phyletic link and index fossil for the Pleistocene. Deep-Sea Res., 17(1): 187-190.

Gephyrocapsa undulatus n.sp.

Lecal, J. 1967.
Le nannoplancton des côtes d'Israel.
Hydrobiologia 29 (3/4): 305-387.

Gigantoshaera elephas n.gen.n.sp.

Bernard, Francis 1973.
Coccolithophorides géants, de 100 à 620μ, pris en décembre 1966 depuis la Bouée Laboratoire du C.N.E.X.O.
Rapp. Proc.-v. Reun. Commn int. Explor. scient. Mer Medit. Monaco 21(8):419-421.

Halopappus adriaticus

Ercegovic, A., 1936
Etudes qualitative et quantitatives du phytoplancton dans les eaux cotières de l'Adriatique oriental moyen au cours de l'année 1934. Acta Adriatica 1(9):1-126.

Helicopontosphaera

Haq, Bilal ul 1973.
Evolutionary trends in the Cenozoic coccolithophore genus Helicopontosphaera.
Micropaleontology 19 (1): 32-52.

Helicosphaera carteri

Black, M., and B. Barnes, 1961.
Coccoliths and discoasters from the floor of the South Atlantic Ocean.
J. R. Microsc. Soc., (3) 80(2):137-147.

Helicosphaera carteri

Cohen, C.L.D., 1965.
Coccoliths and discoasters from Adriatic bottom sediments.
Leidse Geol. Mededelingen. 35: 1-44, 25 pls.

Helicosphaera carteri

McIntyre, Andrew, and Allen W.H. Be, 1967.
Modern Coccolithophoridae of the Atlantic Ocean. I. Placoliths and cyrtoliths.
Deep-Sea Res., 14(5):561-597.

Helicosphaera hyalina n.sp.

Gaarder, Karen R., 1970.
Three new taxa of Coccolithineae.
Nytt Mag. Bot. 17(2): 113-126.

Helladosphaera aurisinae

Gaarder, Karen Ringdal, 1962.
Electron microscope studies on Holococcolithophorids.
Nytt Mag. for Botanikk, 10:35-51.

Helladosphaera aurisinae n.sp.

Kamptner, Erwin, 1941
Die Coccolithineen der Südwestküste von Istrien. Ann. Naturhist. Mus., Wien, 51: 54-149, pls 1-15.

Helladosphaera aurisinae

Lecal J. 1967.
Le nannoplancton des côtes d'Israel.
Hydrobiologia 29 (3/4): 305-387.

Helladosphaera cornifera

Gaarder, Karen Ringdal, 1962.
Electron microscope studies on Holococcolithophorids.
Nytt Mag. for Botanikk, 10:35-51.

Helladosphaera cornifera

Kamptner, Erwin, 1941
Die Coccolithineen der Südwestküste von Istrien. Ann. Naturhist. Mus., Wien, 51: 54-149, pls 1-15.

Helladosphaera fragaria n.comb.

Gaarder, Karen Ringdal, 1962.
Electron microscope studies on Holococcolithophorids.
Nytt Mag. for Botanikk, 10:35-51.

Helladosphaera lafourcadii n.sp.

Lecal, J. 1967.
Le nannoplancton des côtes d'Israel.
Hydrobiologia 29 (3/4): 305-387.

Helladosphaera richardi n.sp.

Bernard, F., 1939.
Coccolithophorides nouveaux ou peu connus observés à Monaco en 1938.
Arch. Zool. Exp. Gen. 81(Notes et Revue No. 1):

Helladosphaera strigillata n.sp.

Lecal-Schlauder, J., 1951
Recherches morphologiques et biologiques sur les Coccolithophorides Nord-Africanis. Ann. Inst. Ocean 26(?): 255-362, Pls. 9-13, 45 text figs.

Hololaminotus flosculus

Lecal, J. 1967.
Le nannoplancton des côtes d'Israel.
Hydrobiologia 29 (3/4): 305-387.

Homozygosphaera quadriperforata n.comb.

Gaarder, Karen Ringdal, 1962.
Electron microscope studies on Holococcolithophorids.
Nytt Mag. for Botanikk, 10:35-51.

Homozygosphaera tholifera

Halldal, P., and J. Markali, 1955.
Electron microscope studies on coccolithophorids from the Norwegian Sea, Gulf Stream and Mediterranean.
Avhandl. Norske Videnskaps-Akad., Oslo, 1. Mat.-Naturvidensk. Kl., 1955(1):1-30, 27 pls.

Homozygosphaera triarcha n.sp.

Halldal, P., and J. Markali, 1955.
Electron microscope studies on coccolithophorids from the Norwegian Sea, Gulf Stream and Mediterranean.
Avhandl. Norske Videnskaps-Akad., Oslo, 1. Mat.-Naturvidensk. Kl., 1955(1):1-30, 27 pls.

Homozygosphaera weltsteini

Halldal, P., and J. Markali, 1955.
Electron microscope studies on coccolithophorids from the Norwegian Sea, Gulf Stream and Mediterranean.
Avhandl. Norske Videnskaps-Akad., Oslo, 1.Mat.-Naturvidensk.Kl., 1955(1):1-30, 27 pls.

Homozygosphaera weltsteini

Lecal, J. 1967.
Le nannoplancton des côtes d'Israel.
Hydrobiologia 29 (3/4): 305-387.

Hymenomonas (Syracosphaera) carterae

Braarud, T., 1954.
Studiet av planktonalger i elektronmikroskop.
Blyttia 2:102-108, 4 pls.

Hymenomonas roseola

Braarud, T., 1954.
Coccolith morphology and taxonomic position of Hymenomonas roseola Stein and Syracosphaera carterae Braarud & Fagerland. Nytt Mag. Bot. 3:1-4, 1 fig., 2 pls.

Hymenomonas roseola

Braarud, T., 1954.
Studiet av planktonalger i elektronmikroskop.
Blyttia 2:102-108, 4 pls.

Kamptnerius magnificus

Cohen, C.L.D., 1965.
Coccoliths and discoasters from Adriatic bottom sediments.
Leidse Geol. Mededelingen. 35: 1-44, 25 pls.

Lacrymasphaera angelieri n.sp.

Lecal, J. 1967.
Le nannoplancton des côtes d'Israel.
Hydrobiologia 29 (3/4): 305-387.

Lithotoenia spp.

Bernard F., 1968.
Lettre à l'editeur: "A propos du genre Litnotoenia, Coccolithophorida Mediterraneene.
Protistologica, 4(2):283.

Lithotaenia n.g.

Bernard, Francis, 1965.
Le genre Lithotaenia n.g., type inedit de Coccolithophorides pelagiques en forme de rubans.
C. r. hebd. Seanc., Acad. Sci., Paris, 261(5):1420-1422.

Lithotaenia biconvexa n. sp.

Bernard, Francis, 1965
Monographie du genre Lithotaenia nov. gen., coccolithophorides pelagiques enforme de ruban.
Pelagos, Bull. Inst. Oceanogr., Alger, 2(4):7-43

Lithotaenia coriolis n. sp.

Bernard, Francis, 1965.
Monographie du genre Lithotaenia nov. gen., coccolithophorides pelagiques en forme de ruban.
Pelagos, Bull. Inst. Oceanogr., Alger, 2(4):7-43

Lithotaenia vivesi n. sp.
Bernard, Francis, 1965
Monographie du genre Lithotaenia nov. gen., coccolithophorides pelagiques en forme de ruban.
Pelagos, Bull. Inst. Oceanogr., Alger, 2(4):7-43

Lithotaenia vulgaris n. sp.
Bernard, Francis, 1965
Monographie du genre Lithotaenia nov. gen., coccolithophorides pelagiques en forme de ruban.
Pelagos, Bull. Inst. Oceanogr., Alger, 2(4):7-43

Lohmanosphaera adriatica
Lillick, L.C., 1938
Preliminary report of the phytoplankton of the Gulf of Maine. Am. Mid. Nat. 20(3):624-640, 1 text figs 37 tables.

Lohmannosphaera cedrus n.sp.
Lecal-Schlauder, J., 1951
Recherches morphologiques et biologiques sur les Coccolithophorides Nord-Arficanis. Ann. Inst. Ocean 26(3): 255-362, Pls. 9-13, 45 text figs.

Lohmannosphaera inclicus n.sp.
Lecal, J. and Mlle. Bernheim, 1960
Microstructure du squelette de quelques Coccolithophorides.
Bull. Soc. Hist. Nat., Afrique du Nord, 51(4/6):273-297, 22 pls.

Lohmannosphaera subclausa n. sp.
Gran, H.H., and T. Braarud, 1935
A quantitative study of the phytoplankton in the Bay of Fundy and the Gulf of Maine (including observations on hydrography, chemistry, and turbidity). J. Biol. Bd., Canada, 1(5):279-467, 69 text figs.

Lohmannosphaera tholica n.sp.
Lecal-Schlauder, J., 1951
Recherches morphologiques et biologiques sur les Coccolithophorides Nord-Arficanis. Ann. Inst. Ocean 26(3): 255-362, Pls. 9-13, 45 text figs.

Michaelsarsia aranea n.sp.
Lecal-Schlauder, J., 1951
Recherches morphologiques et biologiques sur les Coccolithophorides Nord-Arficanis. Ann. Inst. Ocean 26(3): 255-362, Pls. 9-13, 45 text figs.

Micrantholithus flos
Hay, W.W., and Kenneth M. Towe, 1962.
Electron-microscope studies of Braarudosphaera bigelowi and some related coccolithophorids. Science, 137(3528):426-428.

Neosphaera n. gen.
Lecal-Schlauder, J., 1949
Notes préliminaires sur les Coccolithophorides d'Afrique du Nord. Bull. Soc. Hist. Nat., Afrique du Nord, 40:160-167, Pl. 6, 6 textfigs.

Neosphaera coccolithomorpha
Lecal-Schlauder, J., 1951
Recherches morphologiques et biologiques sur les Coccolithophorides Nord-Arficanis. Ann. Inst. Ocean 26(3): 255-362, Pls. 9-13, 45 text figs.

Neosphaera coccolithomorpha n.sp.
Lecal-Schlauder, J., 1949
Notes préliminaires sur les Coccolithophorides d'Afrique du Nord. Bull. Soc. Hist. Nat., Afrique du Nord, 40:160-167, Pl. 6, 6 textfigs.

Neosphaera coccolithomorpha var. striata
Bernard, F., and J. Lecal, 1960.
Plancton unicellulaire récolté dans l'océan Indien par le Charcot (1950) et le Norsel (1955-56).
Bull. Inst. Océanogr., Monaco, No. 1166: 59 pp.

Ophiaster formosus
Gaarder, Karen Ringdal, 1967.
Observations on the genus Ophiaster Gran (Coccofithineae).
Sarsia, 29:183-192.

Ophiaster hydroides
Gaarder, Karen Ringdal, 1967,
Observations on the genus Ophiaster Gran (Coccolithineae).
Sarsia, 29:183-192.

Ophiaster hydroideus
Halldal, P., and J. Markali, 1955.
Electron microscope studies on coccolithophorids from the Norwegian Sea, the Gulf Stream and Mediterranean.
Avhandl Norske Videnskaps-Akad., Oslo, 1. Mat.-Naturvidensk. Kl., 1955(1):1-30, 27 pls.

Palusphaera vandeli n.sp.
Lecal, J. 1967.
Le nannoplancton des côtes d'Israel.
Hydrobiologia 29 (3/4):305-387.

Papposphaera lepida, n.gen., n.sp.
Tangen, Karl 1972.
Papposphaera lepida gen. nov. n. sp., a new marine coccolithophorid from Norwegian coastal waters.
Norw. J. Bot. 19 (3/4): 171-178

?Pemma rotundum
Cohen, C.L.D., 1965.
Coccoliths and discoasters from Adriatic bottom sediments.
Leidse Geol. Mededelingen 35: 1-44, 25 pls.

Pemma rotundum
Hay, W.W., and Kenneth M. Towe, 1962.
Electron-microscope studies of Braarudosphaera bigelowi and some related coccolithophorids.
Science, 137(3528):426-428.

Periphyllophora
Lecal-Schlauder, J., 1951
Recherches morphologiques et biologiques sur les Coccolithophorides Nord-Arficanis. Ann. Inst. Ocean 26(3): 255-362, Pls. 9-13, 45 text figs.

Periphyllophora mirabilis
Halldal, P., and J. Markali, 1955.
Electron microscope studies on coccolithophorids from the Norwegian Sea, the Gulf Stream and Mediterranean.
Avhandl. Norske Videnskaps-Akad., Oslo, 1.Mat.-Naturvidensk. Kl., 1955(1):1-30, 27 pls.

Periphyllophora mirabilis
Kamptner, Erwin, 1941
Die Coccolithineen der Südwestküste von Istrien. Ann. Naturhist. Mus., Wien, 51: 54-149, pls 1-15.

Pontosphaera Bigelowi n.sp.
Gran, H.H., and T. Braarud, 1935
A quantitative study of the phytoplankton in the Bay of Fundy and the Gulf of Maine (including observations on hydrography, chemistry, and turbidity). J. Biol. Bd., Canada, 1(5):279-467, 69 text figs.

Pontosphaera Bigelowi
Lillick, L.C., 1938
Preliminary report of the phytoplankton of the Gulf of Maine. Am. Mid. Nat. 20(3):624-640, 1 text figs 37 tables.

Pontosphaera caelamenisa n.sp.
Lecal-Schlauder, J., 1951
Recherches morphologiques et biologiques sur les Coccolithophorides Nord-Arficanis. Ann. Inst. Ocean 26(3): 255-362, Pls. 9-13, 45 text figs.

Pontosphaera discopora
Halldal, P., and J. Markali, 1955.
Electron microscope studies on coccolithophorids from the Norwegian Sea, the Gulf Stream and Mediterranean.
Avhandl. Norske Videnskaps-Akad., Oslo, 1. Mat.-Naturvidensk. Kl., 1955(1):1-30, 27 pls.

Pontosphaera ditrematolitha nsp.
Bursa, Adam S., 1961
The annual oceanographic cycle at Igloolik in the Canadian Arctic. II. The phytoplankton J. Fish. Res. Bd., Canada, 18(4):563-615.

Pontosphaera elleipsis n.sp.
Lecal, J. and Mlle. Bernheim, 1960
Microstructure du squelette de quelques Coccolithophorides.
Bull. Soc. Hist. Nat., Afrique du Nord, 51(4/6):273-297, 22 pls.

Pontosphaera haeckeli
Marshall, S. M., 1933
The production of microplankton in the Great Barrier Reef Region. Brit. Mus. (N.H.) Great Barrier Reef Exped. 1928-29, Sci. Repts. II(5):111-157, 14 text figs.

Pontosphaera Huxleyi
Braarud, T., 1945
A phytoplankton survey of the polluted waters of inner Oslo Fjord. Hvalrådets Skrifter, No.28, 142 pp., 19 text figs., 17 tables.

Pontosphaera huxleyi
Braarud, T., and Adam Bursa, 1939
On the phytoplankton of the Oslo Fjord, 1933-1934. Hvalrådets Skr. No.19:1-63; 9 text figs. Reviewed. J. du. Cons. 14(3): 418-420. A.C. Gardiner.

Pontosphaera Huxley
Ercegovic, A., 1936
Etudes qualitative et quantitatives du phytoplancton dans les eaux cotières de l'Adriatique oriental moyen au cours de l'année 1934. Acta Adriatica 1(9):1-126

Pontosphaera Huxleyi
Gran, H.H., and T. Braarud, 1935
A quantitative study of the phytoplankton in the Bay of Fundy and the Gulf of Maine (including observations on hydrography, chemistry, and turbidity). J. Biol. Bd., Canada, 1(5):279-467, 69 text figs.

Pontosphaera huxleyi
Kamptner, Erwin, 1941
Die Coccolithineen der Südwestküste von Istrien. Ann. Naturhist. Mus., Wien, 51: 54-149, pls 1-15.

Pontosphaera huxleyi
Lecal, J., 1952.
Variabilité de la teneur en coccolithophorides de différentes stations de la Baie d'Alger en mars 1952. Bull. Soc. Hist. Nat., Afrique du Nord, 43(4/6):69-80, 5 textfigs.

Pontosphaera huxleyi
Lecal, J., 1952.
Influence de l'éloignement des côtes sur la végétation des flagellés calcaires.
Bull. Soc. Hist. Nat., Afrique du Nord, 43(4/6): 81-85, 2 textfigs.

Pontosphaera huxleyi
Lecal, J., 1952.
Répartition en profondeur des Coccolithophorides en quelques stations méditerranéenes occidentales Bull. Inst. Océan., Monaco, No. 1018:13 pp., 4 textfigs.

Pontosphaera huxleyi
Lecal-Schlauder, J., 1951
Recherches morphologiques et biologiques sur les Coccolithophorides Nord-Arficanis. Ann. Inst. Ocean 26(3): 255-362, Pls. 9-13, 45 text figs.

Pontosphaera huxleyi
Lecal-Schlauder, J., 1947
Repartition des Flagellés calcaires dans la baie d'Alger. Comp. Rend. Soc. Biol. 141: 403-405.

Pontosphaera Huxleyi
Lillick, L.C., 1938
Preliminary report of the phytoplankton of the Gulf of Maine. Am. Mid. Nat. 20(3):624-640, 1 text fig, 37 tables.

Pontosphaera huxleyi
Morozova-Vodianitskaia, H.V., and E.V. Belozorskaia, 1957.
The significance of coccolithophorids and a single Pontosphaera in the plankton of the Black Sea. Trudy Sevastopol. Biol. Stan., 9:14-21

Pontosphaera huxleyi
Steemann-Nielsen, Einar, 1951
The marine vegetation of the Isefjord. A study on ecology and production. Medd. Komm. Danmarks Fiskeri-og Havundersøgelser. Ser. Plankton. 5(4); 114pp., 46 text figs.

Pontosphaera huxleyi
Tratet, Gerard, 1964.
Variations du phytoplancton à Tanger. Trav. Inst. Sci., Cherifien, Rabat, Ser. Botan., (29):204 pp.

Pontosphaera impliera n.sp.
Lecal, J. and Mlle. Bernheim, 1960
Microstructure du squelette de quelques Coccolithophorides. Bull. Soc. Hist. Nat., Afrique du Nord, 51(4/6):273-297, 22 pls.

Pontosphaera inermis
Ercegovic, A., 1936
Etudes qualitative et quantitatives du phytoplancton dans les eaux cotières de l'Adriatique oriental moyen au cours de l'année 1934. Acta Adriatica 1(9):1-126

Pontosphaera inermis
Lecal, J. and Mlle. Bernheim, 1960
Microstructure du squelette de quelques Coccolithophorides. Bull. Soc. Hist. Nat., Afrique du Nord, 51(4/6):273-297, 22 pls.

Pontosphaera margarita n.sp.
Lecal, J. and Mlle. Bernheim, 1960
Microstructure du squelette de quelques Coccolithophorides. Bull. Soc. Hist. Nat., Afrique du Nord, 51(4/6):273-297, 22 pls.

Pontosphaera nana
Halldal, P., and J. Markali, 1955.
Electron microscope studies on coccolithophorids from the Norwegian Sea, the Gulf Stream and Mediterranean. Avhandl. Norske Videnskaps-Akad., Oslo, 1. Mat.-Naturvidensk. Kl., 1955(1):1-30, 27 pls.

Pontosphaera nana n.sp.
Kamptner, Erwin, 1941
Die Coccolithineen der Südwestküste von Istrien. Ann. Naturhist. Mus., Wien, 51: 54-149, pls 1-15.

Pontosphaera nigra
Ercegovic, A., 1936
Etudes qualitative et quantitatives du phytoplancton dans les eaux cotières de l'Adriatique oriental moyen au cours de l'année 1934. Acta Adriatica 1(9):1-126

Pontosphaera ovalis
Ercegovic, A., 1936
Etudes qualitative et quantitatives du phytoplancton dans les eaux cotières de l'Adriatique oriental moyen au cours de l'année 1934. Acta Adriatica 1(9):1-126

Pontosphaera ovalis
Lillick, L.C., 1938
Preliminary report of the phytoplankton of the Gulf of Maine. Am. Mid. Nat. 20(3):624-640, 1 text fig, 37 tables.

Pontosphaera pietschmanni
Kamptner, Erwin, 1941
Die Coccolithineen der Südwestküste von Istrien. Ann. Naturhist. Mus., Wien, 51: 54-149, pls 1-15.

Pontosphaera roscoffensis
Dragesco, Jean, 1965.
Étude cytologique de quelques flagellés mésopsammiques. Cahiers Biol. Mar., Roscoff, 6(1):83-115.

Pontosphaera scutellum
Cohen, C.L.D., 1965.
Coccoliths and discoasters from Adriatic bottom sediments. Leidse Geol. Mededelingen. 35: 1-44, 25 pls.

Pontosphaera steueri
Kamptner, Erwin 1941
Die Coccolithineen der Südwestküste von Istrien. Ann. Naturhist. Mus. Wien 51: 54-149, pls 1-15.

Pontosphaera steueri
Lecal, J. and Mlle. Bernheim, 1960
Microstructure du squelette de quelques Coccolithophorides. Bull. Soc. Hist. Nat., Afrique du Nord, 51(4/6):273-297, 22 pls.

Pontosphaera syracusana
Lecal-Schlauder, J., 1951
Recherches morphologiques et biologiques sur les Coccolithophorides Nord-Africanis. Ann. Inst. Ocean 26(3): 255-362, Pls. 9-13, 45 text figs.

Pontosphaera triangularis
Lecal, J. and Mlle. Bernheim, 1960
Microstructure du squelette de quelques Coccolithophorides. Bull. Soc. Hist. Nat., Afrique du Nord, 51(4/6):273-297, 22 pls.

Pontosphaera variabilis n.sp.
Halldal, P., and J. Markali, 1955.
Electron microscope studies on coccolithophorids from the Norwegian Sea, the Gulf Stream and Mediterranean. Avhandl. Norske Videnskaps-Akad., Oslo, 1. Mat.-Naturvidenskap. Kl., 1955(1):1-30, 27 pls.

Pontosphaera verruca n.sp.
Lecal-Schlauder, J., 1951
Recherches morphologiques et biologiques sur les Coccolithophorides Nord-Africanis. Ann. Inst. Ocean 26(3): 255-362, Pls. 9-13, 45 text figs.

Pseudorhabdosphaera soliqua nov. comb.
Lecal, J. and Mlle. Bernheim, 1960
Microstructure du squelette de quelques Coccolithophorides. Bull. Soc. Hist. Nat., Afrique du Nord, 51(4/6):273-297, 22 pls.

Rhabdocyclus simplex n.sp.
Bernard, F., and J. Lecal, 1960.
Plancton unicellulaire récolté dans l'océan Indien par le Charcot (1950) et le Norsel (1955-56). Bull. Inst. Océanogr., Monaco, No. 1166:59 pp.

? Rhabdoderma claviger
Kufferath, H., 1957.
Examen microscopique d'ultraplancton recueilli au large de Monaco. Bull. Inst. Océan., Monaco, No. 1089:12 pp.

?Rhabdosphaera sp.
Cohen, C.L.D., 1965.
Coccoliths and discoasters from Adriatic bottom sediments. Leidse Geol. Mededelingen. 35: 1-44, 25 pls.

Rhabdosphaera n.sp.
Lecal-Schlauder, J., 1951
Recherches morphologiques et biologiques sur les Coccolithophorides Nord-Africanis. Ann. Inst. Ocean 26(3): 255-362, Pls. 9-13, 45 text figs.

Rhabdosphaera ampullacea n.sp.
Lecal-Schlauder, J., 1951
Recherches morphologiques et biologiques sur les Coccolithophorides Nord-Africanis. Ann. Inst. Ocean 26(3): 255-362, Pls. 9-13, 45 text figs.

Rhabdosphaera claviger
Cohen, C.L.D., 1965.
Coccoliths and discoasters from Adriatic bottom sediments. Leidse Geol. Mededelingen. 35: 1-44, 25 pls.

Rhabdosphaera erinaceus
Kamptner, Erwin, 1941
Die Coccolithineen der Südwestküste von Istrien. Ann. Naturhist. Mus., Wien, 51: 54-149, pls 1-15.

Rhabdosphaera hispida
Ercegovic, A., 1936
Etudes qualitative et quantitatives du phytoplancton dans les eaux cotières de l'Adriatique oriental moyen au cours de l'année 1934. Acta Adriatica 1(9):1-126

Rhabdosphaera longistylis
Ercegovic, A., 1936
Etudes qualitative et quantitatives du phytoplancton dans les eaux cotières de l'Adriatique oriental moyen au cours de l'année 1934. Acta Adriatica 1(9):1-126

Rhabdosphaera nigra
Kamptner, Erwin, 1941
Die Coccolithineen der Südwestküste von Istrien. Ann. Naturhist. Mus., Wien, 51: 54-149, pls 1-15.

Rhabdosphaera siliqua n.sp.
Lecal-Schlauder, J., 1951
Recherches morphologiques et biologiques sur les Coccolithophorides Nord-Africanis. Ann. Inst. Ocean 26(3): 255-362, Pls. 9-13, 45 text figs.

Rhabdosphaera stylifer
Cohen, C.L.D., 1965.
Coccoliths and discoasters from Adriatic bottom sediments. Leidse Geol. Mededelingen. 35: 1-44, 25 pls.

Rhabdosphaera stylifer
Ercegovic, A., 1936
Etudes qualitative et quantitatives du phytoplancton dans les eaux cotières de l'Adriatique oriental moyen au cours de l'année 1934. Acta Adriatica 1(9):1-126

Rhabdosphaera stylifera
Gran, H.H., and T. Braarud, 1935
A quantitative study of the phytoplankton in the Bay of Fundy and the Gulf of Maine (including observations on hydrography, chemistry, and turbidity). J. Biol. Bd. Canada, 1(5):279-467, 69 text figs.

Rhabdosphaera stylifera
Halldal, P., and J. Markali, 1955.
Electron microscope studies on coccolithophorids from the Norwegian Sea, the Gulf Stream and Mediterranean. Avhandl. Norske Videnskaps-Akad., Oslo, 1. Mat.-Naturvidensk. Kl., 1955(1):1-30, 27 pls.

Rhabdosphaera stylifera
Kamptner, Erwin, 1941
Die Coccolithineen der Südwestküste von Istrien. Ann. Naturhist. Mus., Wien, 51: 54-149, pls 1-15.

Rhabdosphaera stylifer
Lecal-Schlauder, J., 1947
Repartition des Flagellés calcaires dans la baie d'Alger. Comp. Rend. Soc. Biol. 141: 403-405.

Rhabdosphaera stylifer
Lillick, L.C., 1938
Preliminary report of the phytoplankton of the Gulf of Maine. Am. Mid. Nat. 20(3):624-640, 1 text figs 37 tables.

Rhabdosphaera stylifer
Marshall, S. M., 1933
The production of microplankton in the Great Barrier Reef Region. Brit. Mus. (N.H.) Great Barrier Reef Exped. 1928-29, Sci. Repts. II(5):111-157, 14 text figs.

Rhabdosphaera stylifera
McIntyre, Andrew, and Allen W.H. Be, 1967.
Modern Coccolithophoridae of the Atlantic Ocean. I. Placoliths and cyrtoliths. Deep-Sea Res., 14(5):561-597.

Rhabdosphaera subopaca n.sp.
Bernard, F., 1939.
Coccolithophorides nouveaux ou peu connus observés à Monaco en 1938. Arch. Zool. Exp. Gen. 81(Notes et Revue No. 1): 33-44, 2 textfigs.

Rhabdosphaera tignifer
Ercegovic, A., 1936
Etudes qualitative et quantitatives du phytoplancton dans les eaux cotières de l'Adriatique oriental moyen au cours de l'année 1934. Acta Adriatica 1(9):1-126

Rhabdosphaera tubulosa
Lillick, L.C., 1938
Preliminary report of the phytoplankton of the Gulf of Maine. Am. Mid. Nat. 20(3):624-640, 1 text figs 37 tables.

Rhabdosphaera tubulosa
Lillick, L.C., 1937
Seasonal studies of the phytoplankton off Woods Hole, Massachusetts. Biol. Bull. LXXIII (3):488-503, 3 text figs.

Rhabdosphaera vatricosa n. sp.
Cohen, C.L.D., 1965.
Coccoliths and discoasters from Adriatic bottom sediments. Leidse Geol. Mededelingen 35: 1-44, 25 pls.

Rhabdothorax regalis n.comb.
Gaarder, Karen R. and Berit R. Heimdal 1973
Light and scanning electron microscope observations on Rhabdothorax regale (Gaarder) Gaarder nov. comb. Norwegian J. Bot. 20(2/3):89-97.

Scalpholithus sp.
Cohen, C.L.D., 1965.
Coccoliths and discoasters from Adriatic bottom sediments. Leidse Geol. Mededelingen. 35: 1-44, 25 pls.

Scyphosphaera adriatica
Ercegovic, A., 1936
Etudes qualitative et quantitatives du phytoplancton dans les eaux cotières de l'Adriatique oriental moyen au cours de l'année 1934. Acta Adriatica 1(9):1-126

Scyphosphaera apsteini
Bernard, F., 1939.
Coccolithophorides nouveaux ou peu connus observés à Monaco en 1938. Arch. Zool. Exp. Gen. 81(Notes et Revue No. 1):33-44, 2 textfigs.

Scyphosphaera Apsteini
Ercegovic, A., 1936
Etudes qualitative et quantitatives du phytoplancton dans les eaux cotières de l'Adriatique oriental moyen au cours de l'année 1934. Acta Adriatica 1(9):1-126

Scyphosphaera Apsteini
Lillick, L.C., 1938
Preliminary report of the phytoplankton of the Gulf of Maine. Am. Mid. Nat. 20(3):624-640, 1 text figs 37 tables.

Scyphosphaera apsteini
Nival, P. 1969.
Données écologiques sur quelques protozoaires planctoniques rares en Méditerranée. Protistologica 5(2): 215-225.

Scyphosphaera apsteinii dilatata n.f.
Gaarder, Karen R., 1970.
Three new taxa of Coccolithineae. Nytt Mag. Bot. 17(2): 113-126

Scyphosphaera cordiformis
Ercegovic, A., 1936
Etudes qualitative et quantitatives du phytoplancton dans les eaux cotières de l'Adriatique oriental moyen au cours de l'année 1934. Acta Adriatica 1(9):1-126

Scyphosphaera mediterranea
Ercegovic, A., 1936
Etudes qualitative et quantitatives du phytoplancton dans les eaux cotières de l'Adriatique oriental moyen au cours de l'année 1934. Acta Adriatica 1(9):1-126

Scyphosphaera Molischii
Ercegovic, A., 1936
Etudes qualitative et quantitatives du phytoplancton dans les eaux cotières de l'Adriatique oriental moyen au cours de l'année 1934. Acta Adriatica 1(9):1-126

Scyphosphaera pulchra
Ercegovic, A., 1936
Etudes qualitative et quantitatives du phytoplancton dans les eaux cotières de l'Adriatique oriental moyen au cours de l'année 1934. Acta Adriatica 1(9):1-126

Scyphosphaera ovata
Ercegovic, A., 1936
Etudes qualitative et quantitatives du phytoplancton dans les eaux cotières de l'Adriatique oriental moyen au cours de l'année 1934. Acta Adriatica 1(9):1-126

Sphaerocalypta papillifera
Braarud, T., 1954.
Studiet av planktonalger i elektronmikroskop. Blyttia 2:102-108, 4 pls.

Syracolithus antarcticus n.sp.
Lecal, J. and Mlle. Bernheim, 1960
Microstructure du squeletto de quelques Coccolithophorides. Bull. Soc. Hist. Nat., Afrique du Nord, 51(4/6):273-297, 22 pls.

Syracolithus aperta nov. comb.
Lecal, J. and Mlle. Bernheim, 1960
Microstructure du squeletto de quelques Coccolithophorides. Bull. Soc. Hist. Nat., Afrique du Nord, 51(4/6):273-297, 22 pls.

Syracolithus corii nov. comb.
Lecal, J. and Mlle. Bernheim, 1960
Microstructure du squeletto de quelques Coccolithophorides. Bull. Soc. Hist. Nat., Afrique du Nord, 51(4/6):273-297, 22 pls.

Syracolithus corolla nsp
Lecal, J., 1965.
Coccolithophorides littoreaux de Banyuls. Vie et Milieu, Bull. Lab. Arago,(B)16(1):251-260.

Syracolithus dorisa nov. comb.
Lecal, J. and Mlle. Bernheim, 1960
Microstructure du squeletto de quelques Coccolithophorides. Bull. Soc. Hist. Nat., Afrique du Nord, 51(4/6):273-297, 22 pls.

Syracolithus globula n.sp.
Lecal, J. 1967.
Le nannoplancton des côtes d'Israel. Hydrobiologia 29(3/4): 305-387.

Syracolithus oculata nov. comb.
Lecal, J. and Mlle. Bernheim, 1960
Microstructure du squeletto de quelques Coccolithophorides. Bull. Soc. Hist. Nat., Afrique du Nord, 51(4/6):273-297, 22 pls.

Syracolithus orientalis n.sp.
Bernard, F., and J. Lecal, 1960.
Plancton unicellulaire récolté dans l'océan Indien par le Charcot (1950) et le Norsel (m955-56). Bull. Inst. Océanogr., Monaco, No. 1166:59 pp.

Syracolithus ossa n.sp.
Lecal, J. 1967.
Le nannoplancton des côtes d'Israel. Hydrobiologia 29(3/4): 305-387.

Syracolithus pastillusa nov. comb.
Lecal, J. and Mlle. Bernheim, 1960
Microstructure du squeletto de quelques Coccolithophorides. Bull. Soc. Hist. Nat., Afrique du Nord, 51(4/6):273-297, 22 pls.

Syracolithus scutata
Lecal, J. 1967.
Le nannoplancton des côtes d'Israel. Hydrobiologia 29(3/4): 305-387.

Syracolithus uninodata (nov. comb.
Lecal, J. and Mlle. Bernheim, 1960
Microstructure du squeletto de quelques Coccolithophorides. Bull. Soc. Hist. Nat., Afrique du Nord, 51(4/6):273-297, 22 pls.

Syracolithus verruca
Lecal, J. 1967.
Le nannoplancton des côtes d'Israel. Hydrobiologia 29(3/4): 305-387.

Syracolithus verruca nov. Comb.
Lecal, J. and Mlle. Bernheim, 1960
Microstructure du squeletto de quelques Coccolithophorides. Bull. Soc. Hist. Nat., Afrique du Nord, 51(4/6):273-297, 22 pls.

Syracorhabdus n. subgen.
Lecal, J. and Mlle. Bernheim, 1960
Microstructure du squelette de quelques Coccolithophorides.
Bull. Soc. Hist. Nat., Afrique du Nord, 51(4/6):273-297, 22 pls.

Syracorhabdus confusa
Lecal, J. 1967
Le nannoplancton des côtes d'Israel.
Hydrobiologia 29(3/4): 305-387.

Syracorhabdus fertii n.sp.
Lecal, J. 1967
Le nannoplancton des côtes d'Israel.
Hydrobiologia 29(3/4): 305-387.

Syracorhabdus lactarie nsp.
Lecal, J., 1965.
Coccolithophorides littoraux de Banyuls.
Vie et Milieu, Bull. Lab. Arago, (B)16(1):251-260.

Syracorhabdus molischi
Lecal, J., 1965.
Coccolithophorides littoraux de Banyuls.
Vie et Milieu, Bull. Lab. Arago, (B)16(1):251-260.

Syracorhabdus molischii nov. comb.
Lecal, J. and Mlle. Bernheim, 1960
Microstructure du squelette de quelques Coccolithophorides.
Bull. Soc. Hist. Nat., Afrique du Nord, 51(4/6):273-297, 22 pls.

Syracorhabdus nodosa nov. comb.
Lecal, J. and Mlle. Bernheim, 1960
Microstructure du squelette de quelques Coccolithophorides.
Bull. Soc. Hist. Nat., Afrique du Nord, 51(4/6):273-297, 22 pls.

Syracorhabdus ossa
Lecal, J. 1967
Le nannoplancton des côtes d'Israel.
Hydrobiologia 29(3/4): 305-387.

Syracorhabdus ossa nsp.
Lecal, J., 1965.
Coccolithophorides littoraux de Banyuls.
Vie et Milieu, Bull. Lab. Arago, (B)16(1):251-260

Syracorhabdus pulchra
Lecal, J. 1967
Le nannoplancton des côtes d'Israel.
Hydrobiologia 29(3/4): 305-387.

Syracorhabdus pulchra
Lecal, J., 1965.
Coccolithophorides littoraux de Banyuls.
Vie et Milieu, Bull. Lab. Arago, (B)16(1):251-260.

Syracorhabdus pulchra
Lecal, J. and Mlle. Bernheim, 1960
Microstructure du squelette de quelques Coccolithophorides.
Bull. Soc. Hist. Nat., Afrique du Nord, 51(4/6):273-297, 22 pls.

Syracorhabdus revisera
Lecal, J. 1967
Le nannoplancton des côtes d'Israel.
Hydrobiologia 29(3/4): 305-387.

Syracorhabdus revisera nsp.
Lecal, J., 1965.
Coccolithophorides littoraux de Banyuls.
Vie et Milieu, Bull. Lab. Arago, (B)16(1):251-260.

Syracosphaera sp.
Cohen, C.L.D., 1965.
Coccoliths and discoasters from Adriatic bottom sediments.
Leidse Geol. Mededelingen. 35: 1-44, 25 pls.

Syracosphaera spp.
Lillick, L.C., 1937
Seasonal studies of the phytoplankton off Woods Hole, Massachusetts. Biol. Bull. LXXIII (3):488-503, 3 text figs.

Scyphosphaera apsteini
Lecal-Schlauder, J., 1951
Recherches morphologiques et biologiques sur les Coccolithophorides Nord-Arfícanis. Ann. Inst. Ocean 26(3): 255-362, Pls. 9-13, 45 text figs.

Syracosphaera aperta
Lecal-Schlauder, J., 1947
Repartition des Flagellés calcaires dans la baie d'Alger. Comp. Rend. Soc. Biol. 141: 403-405.

Syracosphaera binodata
Kamptner, Erwin, 1941
Die Coccolithineen der Südwestküste von Istrien. Ann. Naturhist. Mus., Wien, 51: 54-149, pls 1-15.

Syracosphaera canariensis n.sp
Bernard, Francis, 1967.
Contribution à l'étude du nannoplancton, de 0 à 3000 m, dans les zones atlantiques lusitanienne et mauritanienne (campagnes de la Calypso, 1960, et du Coriolus, 1964). Pelagos, Alger, 7:1-81.

Syracosphaera carterae
Braarud, T., 1954.
Coccolith morphology and taxonomic position of Hymenomonas roseola and Syracosphaera carterae Braarud & Fagerland. Nytt Mag. Bot. 3:1-4, 1 fig., 2 pls.

Syracosphaera carterae
Braarud, T., K.R. Gaarder, J. Markeli and E. Nordli, 1952.
Coccolithophorids studied in the electron microscope. Observations on Coccolithas Huxleyi and Syracosphaera carterae. Nytt Mag. Bot. 1:129-133, 2 pls., 4 figs.

Syracosphaera carterae
Braarud, T., 1951.
Salinity as an ecological factor in marine phytoplankton. Physiol. Plant. 4:28-34, 3 textfigs.

Syracosphaera carterae n.sp.
Braarud, T., and E. Fagerland, 1946.
A Coccolithophoride in laboratory culture: Syracosphaera carterae n.sp. Avhandl. Norske Videnskaps-Akademi, Oslo, I. Mat-Naturv. Kl. 1946(2): 1-10, 1 pl.

Syracosphaera carterae
Parsons, T.R., 1961
On the pigment composition of eleven species of marine phytoplankters.
J. Fish. Res. Bd., Canada. 18(6):1017-1025.

Syracosphaera carterae
Parsons, T.R., K. Stephens and J.D.H. Strickland 1961
On the chemical composition of eleven species of marine phytoplankters.
J. Fish. Res. Bd., Canada. 18(6):1001-1016.

Syracosphaera catillifera
Kamptner, Erwin, 1941
Die Coccolithineen der Südwestküste von Istrien. Ann. Naturhist. Mus., Wien, 51: 54-149, pls 1-15.

Syracosphaera clypeata n.sp.
Lecal-Schlauder, J., 1951
Recherches morphologiques et biologiques sur les Coccolithophorides Nord-Arfícanis. Ann. Inst. Ocean 26(3): 255-362, Pls. 9-13, 45 text figs.

Syracosphaera confusa n.sp.
Halldal, P., and J. Markali, 1955.
Electron microscope studies on coccolithophorids from the Norwegian Sea, the Gulf Stream and Mediterranean.
Avhandl. Norske Videnskaps-Akad., Oslo, I. Mat.-Naturvidensk. Kl., 1955(1):1-30, 27 pls.

Syracosphaera corii
Kamptner, Erwin, 1941.
Die Coccolithineen der Südwestküste von Istrien.
Ann. Naturhist. Mus. Wien 51: 54-149, pls. 1-15.

Syracosphaera corii
Lecal-Schlauder, J., 1947
Repartition des Flagellés calcaires dans la baie d'Alger. Comp. Rend. Soc. Biol. 141: 403-405.

Syracosphaera cornus n.sp.
Kamptner, Erwin, 1941
Die Coccolithineen der Südwestküste von Istrien. Ann. Naturhist. Mus., Wien, 51: 54-149, pls 1-15.

Syracosphaera coronifera n.sp.
Lecal-Schlauder, J., 1951
Recherches morphologiques et biologiques sur les Coccolithophorides Nord-Arfícanis. Ann. Inst. Ocean 26(3): 255-362, Pls. 9-13, 45 text figs.

Syracosphaera dalmatica
Cohen, C.L.D., 1965.
Coccoliths and discoasters from Adriatic bottom sediments.
Leidse Geol. Mededelingen. 35: 1-44, 25 pls.

Syracosphaera dalmatica
Kamptner, Erwin, 1941
Die Coccolithineen der Südwestküste von Istrien. Ann. Naturhist. Mus., Wien, 51: 54-149, pls 1-15.

Syracosphaera dentata
Braarud, T., 1934
A note on the phytoplankton of the Gulf of Maine in the summer of 1933. Biol. Bull. 67(1):76-82. (Contribution No.46 of the Woods Hole Oceanographic Institution)

Syracosphaera echinus n.sp.
Bernard, Francis 1973.
Coccolithophorides géants, de 100 à 620μ, pris en décembre 1966 depuis la Bouée Laboratoire du C.N.E.X.O.
Rapp. Proc.-v. Reun. Commn int. Explor. scient. Mer Medit. Monaco 21(8):419-421.

Syracosphaera gigas n.sp.
Bernard, Francis 1973.
Coccolithophorides géants, de 100 à 620μ, pris en décembre 1966 depuis la Bouée Laboratoire du C.N.E.X.O.
Rapp. Proc.-v. Reun. Commn int. Explor. scient. Mer Medit. Monaco 21(8):419-421.

Syracosphaera heimii
Lecal-Schlauder, J., 1947
Repartition des Flagellés calcaires dans la baie d'Alger. Comp. Rend. Soc. Biol. 141: 403-405.

Syracosphaera Heimi
Tratet, Gérard, 1964.
Variations du phytoplancton à Tanger.
Trav. Inst. Sci., Cherifien, Rabat, Ser. Botan., (29):204 pp.

Syracosphaera histrica n.sp.
Kamptner, Erwin, 1941
Die Coccolithineen der Südwestküste von Istrien. Ann. Naturhist. Mus., Wien, 51: 54-149, pls 1-15.

Syracosphaera lamina n.sp.
Lecal-Schlauder, J., 1951
Recherches morphologiques et biologiques sur les Coccolithophorides Nord-Arficanis. Ann. Inst. Ocean 26(3): 255-362, Pls. 9-13, 45 text figs.

Syracosphaera maxima n.sp.
Halldal, P., and J. Markali, 1955.
Electron microscope studies on coccolithophorids from the Norwegian Sea, the Gulf Stream and Mediterranean.
Avhandl. Norske Videnskaps-Akad., Oslo, 1. Mat.-Naturvidensk. Kl., 1955(1):1-30, 27 pls.

Syracosphaera mediterrannea
Braarud, T., 1954.
Studiet av planktonalger i elektronmikroskop.
Blyttia 2:102-108, 4 pls.

Syracosphaera mediterranea
Halldal, P., and J. Markali, 1954.
Observations on coccoliths of Syracosphaera mediterranea Lohm., S. pulchra Lohm. and S. molischi Schill. in the electron microscope.
J. du Cons. 19(3):329-336, 6 textfigs.

Lillick, L.C., 1938 Syracosphaera mediteranea
Preliminary report of the phytoplankton of the Gulf of Maine. Am. Mid. Nat. 20(3):624-640, 1 text figs, 37 tables.

Syracosphaera molischi
Halldal, P., and J. Markali, 1954.
Observations on coccoliths of Syracosphaera mediterranea Lohm., S. pulchra Lohm., and S. molischi Schill. in the electron microscope.
J. du Cons. 19(3):329-336, 6 textfigs.

Syracosphaera molischi
Kamptner, Erwin, 1941
Die Coccolithineen der Südwestküste von Istrien. Ann. Naturhist. Mus., Wien, 51: 54-149, pls 1-15.

Syracosphaera nodosa
Halldal, P., and Markali, J., 1955.
Electron microscope studies on coccolithophorids from the Norwegian Sea, the Gulf Stream and Mediterranean.
Avhandl. Norske Videnskaps-Akad., Oslo, 1. Mat.-Naturvidensk. Kl., 1955(1):1-30, 27 pls.

Syracosphaera nodosa n.sp.
Kamptner, Erwin, 1941
Die Coccolithineen der Südwestküste von Istrien. Ann. Naturhist. Mus., Wien, 51: 54-149, pls 1-15.

Syracosphaera oculata n.sp.
Lecal-Schlauder, J., 1951
Recherches morphologiques et biologiques sur les Coccolithophorides Nord-Arficanis. Ann. Inst. Ocean 26(3): 255-362, Pls. 9-13, 45 text figs.

Syracosphaera pastillusa n.sp.
Lecal-Schlauder, J., 1951
Recherches morphologiques et biologiques sur les Coccolithophorides Nord-Arficanis. Ann. Inst. Ocean 26(3): 255-362, Pls. 9-13, 45 text figs.

Syracosphaera pila n.sp.
Lecal-Schlauder, J., 1951
Recherches morphologiques et biologiques sur les Coccolithophorides Nord-Arficanis. Ann. Inst. Ocean 26(3): 255-362, Pls. 9-13, 45 text figs.

Syracosphaera pirus n.sp.
Halldal, P., and J. Markali, 1955.
Electron microscope studies on coccolithophorids from the Norwegian Sea, the Gulf Stream and Mediterranean.
Avhandl. Norsek Videnskaps-Akad., Oslo, 1. Mat.-Naturvidensk. Kl., 1955(1):1-30, 27 pls.

Syracosphaera profunda
Bernard, F., 1953.
Rôle des flagellés calcaires dans la fertilité et la sédimentation en mer profonde.
Deep-sea Res. 1(1):34-46, 3 textfigs.

Syracosphaera profunda n.sp.
Bernard, F., 1939.
Coccolithophorides nouveaux eu peu connus observés à Monaco en 1938.
Arch. Zool. Exp. Gen. 81(Notes et Revue No. 1): 33-44, 2 textfigs.

Syracosphaera pulchra
Black, M., and B. Barnes, 1961.
Coccoliths and discoasters from the floor of the South Atlantic Ocean.
J.R. Microsc. Soc., (3). 80(2):137-147.

Syracosphaera pulchra
Cohen, C.L.D., 1965.
Coccoliths and discoasters from Adriatic bottom sediments.
Leidse Geol. Mededelingen, 35: 1-44, 25 pls.

Syracosphaera pulchra
Gran, H.H., and T. Braarud, 1935
A quantitative study of the phytoplankton in the Bay of Fundy and the Gulf of Maine (including observations on hydrography, chemistry, and turbidity).
J. Biol. Bd., Canada, 1(5):279-467, 69 text figs.

Syracosphaera pulchra n.sp.
Halldal, P., and J. Markali, 1955.
Electron microscope studies on coccolithophorids from the Norwegian Sea, the Gulf Stream and Mediterranean.
Avhandl. Norske Videnskaps-Akad., Oslo, 1. Mat.-Naturvidensk. Kl., 1955(1):1-30, 27 pls.

Syracosphaera pulchra
Halldal, P., and J. Markali, 1954.
Observations on coccoliths of Syracosphaera mediterranea Lohm., S. pulchra Lohm. and S. molischi Scil. in the electron microscope.
J. du Cons. 19(3):329-336, 6 textfigs.

Syracosphaera pulchra
Kamptner, Erwin, 1941
Die Coccolithineen der Südwestküste von Istrien. Ann. Naturhist. Mus., Wien, 51: 54-149, pls 1-15.

Syracosphaera pulchra
Lecal-Schlauder, J., 1951
Recherches morphologiques et biologiques sur les Coccolithophorides Nord-Arficanis. Ann. Inst. Ocean 26(3): 255-362, Pls. 9-13, 45 text figs.

Syracosphaera pulchra
Lecal-Schlauder, J., 1947
Repartition des Flagellés calcaires dans la baie d'Alger. Comp. Rend. Soc. Biol. 141: 403-405.

Syracosphaera pulchra
Lillick, L.C., 1938
Preliminary report of the phytoplankton of the Gulf of Maine. Am. Mid. Nat. 20(3):624-640, 1 text figs, 37 tables.

Syracosphaera pulchra
Marshall, S.M., 1933
The production of microplankton in the Great Barrier Reef Region. Brit. Mus. (N.H.) Great Barrier Reef Exped. 1928-29, Sci. Repts. II(5):111-157, 14 text figs.

Syracosphaera pulchroides n.sp.
Halldal, P., and J. Markali, 1955.
Electron microscope studies on coccolithophorids from the Norwegia Sea, the Gulf Stream and Mediterranean.
Avhandl. Norske Videnskaps-Akad., Oslo, 1. Mat.-Naturvidensk. Kl., 1955(1):1-30, 27 pls.

Syracosphaera quadriperforata
Kamptner, Erwin, 1941
Die Coccolithineen der Südwestküste von Istrien. Ann. Naturhist. Mus., Wien, 51: 54-149, pls 1-15.

Syracosphaera schilleri
Kamptner, Erwin, 1941
Die Coccolithineen der Südwestküste von Istrien. Ann. Naturhist. Mus., Wien, 51: 54-149, pls 1-15.

Syracosphaera scutata n.sp.
Lecal-Schlauder, J., 1951
Recherches morphologiques et biologiques sur les Coccolithophorides Nord-Arficanis. Ann. Inst. Ocean 26(3): 255-362, Pls. 9-13, 45 text figs.

Syracosphaera tuberculata
Halldal, P., and J. Markali, 1955.
Electron microscope studies on coccolithophorids from the Norwegian Sea, the Gulf Stream and Mediterranean.
Avhandl. Norske Videnskaps-Akad., Oslo, 1. Mat.-Naturvidensk. Kl., 1955(1):1-30, 27 pls.

Syracosphaera tuberculata
Kamptner, Erwin, 1941
Die Coccolithineen der Südwestküste von Istrien. Ann. Naturhist. Mus., Wien, 51: 54-149, pls 1-15.

Syracosphaera uninodata
Lecal-Schlauder, J., 1947
Repartition des Flagellés calcaires dans la baie d'Alger. Comp. Rend. Soc. Biol. 141: 403-405.

Syracosphaera uniperforata n.sp.
Bernard, Francis, 1967.
Contribution à l'étude du nannoplancton de 0 à 3000 m, dans les zones atlantiques lusitanienne et mauritanienne (Campagnes de la Calypso, 1960, et du Coriolus, 1964).
Pelagos, Alger, 7:1-81.

Tergestiella columnia n.sp.
Bernard, F., and J. Lecal, 1960.
Plancton unicellulaire récolté dans l'ocean Indien par le Charcot (1950) et le Norsel (1955-1956).
Bull. Inst. Océanogr., Monaco, No. 1166:59 pp.

Tergestriella adriatica n.gen., n.sp.
Kamptner, Erwin, 1941
Die Coccolithineen der Südwestküste von Istrien. Ann. Naturhist. Mus., Wien, 51: 54-149, pls 1-15.

Tergestiella gemma n.sp.
Lecal-Schlauder, J., 1951
Recherches morphologiques et biologiques sur les Coccolithophorides Nord-Arficanis. Ann. Inst. Ocean 26(3): 255-362, Pls. 9-13, 45 text figs.

Thoracosphaera heimi
Bernard, F., 1953.
Rôle des flagellés calcaires dans la fertilité et la sédimentation en mer profonde.
Deep-sea Res. 1(1):34-46, 3 textfigs.

Thoracosphaera heimi
Cohen, C.L.D., 1965.
Coccoliths and discoasters from Adriatic bottom sediments.
Leidse Geol. Mededelingen. 35: 1-44, 25 pls.

?Thoracosphaera imperforata
Cohen, C.L.D., 1965.
Coccoliths and discoasters from Adriatic bottom sediments.
Leidse Geol. Mededelingen. 35: 1-44, 25 pls.

Thoracosphaera pelagica
Lecal-Schlauder, J., 1951
Recherches morphologiques et biologiques sur les Coccolithophorides Nord-Africanis. Ann. Inst. Ocean 26(3): 255-362, Pls. 9-13, 45 text figs.

Thorosphaera elegans
Bernard, F., 1939.
Coccolithophorides nouveaux ou peu connus observés à Monaco en 1938.
Arch. Zool. Exp. Gen. 81(Notes et Revue No. 1): 33-44, 2 textfigs.

(Thorosphaera elegans)
Lecal-Schlauder, J., 1951
Recherches morphologiques et biologiques sur les Coccolithophorides Nord-Africanis. Ann. Inst. Ocean 26(3): 255-362, Pls. 9-13, 45 text figs.

Thorosphaera flabellata n.sp.
Halldal, P., and J. Markali, 1955.
Electron microscope studies on coccolithophorids from the Norwegian Sea, the Gulf Stream and Mediterranean.
Avhandl. Norske Videnskaps-Akad., Oslo, 1. Mat. Naturvidensk. Kl., 1955(1):1-30, 27 pls.

Tremalithus sertus n.sp.
Bernard, F., and J. Lecal, 1960.
Plancton unicellulaire récolté dans l'océan Indien par le Charcot (1950) et le Norsel (1955-1956).
Bull. Inst. Océanogr., Monaco, No. 1166:59 pp.

Trochischia sp.
Braarud, T., 1945
A phytoplankton survey of the polluted waters of inner Oslo Fjord. Hvalrådets Skrifter, No.28, 142 pp., 19 text figs., 17 tables.

Trochiscia clevei
Kokubo, S., and S. Sato., 1947
Plankters in Jū-San Gata. Physiol. and Ecol. (Japan) 1(4):1-16, 3 text figs., tables.

Trochischia coronata
Forti, A., 1922
Ricerche sulla flora pelagica (fitoplancton) di Quarto dei Mille. Mem. R. Com. Talass. Ital. 97:248 pp., 13 pls.

Trochischia Clevei
Forti, A., 1922
Ricerche sulla flora pelagica (fitoplancton) di Quarto dei Mille. Mem. R. Com. Talass. Ital. 97:248 pp., 13 pls.

Trochischia paucispinosa
Forti, A., 1922
Ricerche sulla flora pelagica (fitoplancton) di Quarto dei Mille. Mem. R. Com. Talass. Ital. 97:248 pp., 13 pls.

Umbellosphaera irregularis n.sp.
Markali, J., and E. Paasche, 1955.
On two new species of Umbellosphaera, a new marine coccolithophorid genus. Nytt Mag., Bot., 4:95-100, 6 pls.

Umbellosphaera irregularis
McIntyre, Andrew, and Allen W.H. Be, 1967.
Modern Coccolithophoridae of the Atlantic Ocean. I. Placoliths and cyrtoliths.
Deep-Sea Res., 14(5):561-597.

Umbellosphaera tenuis
Cohen, C.L.D., 1965.
Coccoliths and discoasters from Adriatic bottom sediments.
Leidse Geol. Mededelingen. 35: 1-44, 25 pls.

Zygolithus dubius
Cohen, C.L.D., 1965.
Coccoliths and discoasters from Adriatic bottom sediments.
Leidse Geol. Mededelingen. 35: 1-44, 25 pls.

Umbellosphaera tenuis n.sp.
Markali, J., and E. Paasche, 1955.
On two new species of Umbellosphaera, a new marine coccolithophorid genus. Nytt Mag., Bot., 4: 95-100, 6 pls.

Umbellosphaera tenuis
McIntyre, Andrew, and Allen W.H. Be, 1967.
Modern Coccolithophoridae of the Atlantic Ocean. I. Placoliths and cyrtoliths.
Deep-Sea Res., 14(5):561-597.

Umbilicosphaera hulburtiana n.sp.
Gaarder, Karen R. 1970.
Three new taxa of Coccolithineae.
Nytt Mag. Bot. 17(2):113-126.

Umbilicosphaera mirabilis
Black, M., and B. Barnes, 1961.
Coccoliths and discoasters from the floor of the South Atlantic Ocean.
J. R. Microsc. Soc., (3) 80(2):137-147.

Umbilicosphaera mirabilis
McIntyre, Andrew, and Allen W.H. Be, 1967.
Modern Coccolithophoridae of the Atlantic Ocean. I. Placoliths and cyrtoliths.
Deep-Sea Res., 14(5):561-597.

Umbilicosphaera rosaceus
Lecal, J. 1967.
Le nannoplancton des côtes d'Israel.
Hydrobiologia 29(3/4):305-387.

Zygrahablithus bijugatus
Cohen, C.L.D., 1965.
Coccoliths and discoasters from Adriatic bottom sediments.
Leidse Geol. Mededelingen. 35: 1-44, 25 pls.

Zygrhablithus ? turris liffeli
Cohen, C.L.D., 1965.
Coccoliths and discoasters from Adriatic bottom sediments.
Leidse Geol. Mededelingen. 35: 1-44, 25 pls.

Zygosphaera acuta n. sp.
Bernard, Francis, 1967.
Contribution à l'étude du nannoplancton, de 0 à 3000 m, dans les zones atlantiques lusitanienne et mauritanienne (Campagnes de la Calypso, 1960, et du Coriolus, 1964).
Pelagos, Alger, 7:1-81.

Zygosphaera debilis n.sp.
Kamptner, Erwin, 1941
Die Coccolithineen der Südwestküste von Istrien. Ann. Naturhist. Mus., Wien, 51: 54-149, pls 1-15.

Zygosphaera divergens n.sp.
Halldal, P., J. Markali, 1955.
Electron microscope studies on coccolithophorids from the Norwegian Sea, the Gulf Stream and Mediterranean.
Avhandl. Norske Videnskaps-Akad., Oslo, 1. Mat.-Naturvidensk. Kl., 1955(1):1-30, 27 pls.

Zygosphaera hellenica
Kamptner, Erwin, 1941
Die Coccolithineen der Südwestküste von Istrien. Ann. Naturhist. Mus., Wien, 51: 54-149, pls 1-15.

Zygosphaera minor
Lecal-Schlauder, J., 1947
Repartition des Flagellés calcaires dans la baie d'Alger. Comp. Rend. Soc. Biol. 141: 403-405.

Zygosphaera regalis n.sp.
Lecal-Schlauder, J., 1951
Recherches morphologiques et biologiques sur les Coccolithophorides Nord-Africanis. Ann. Inst. Ocean 26(3): 255-362, Pls. 9-13, 45 text figs.

Zygosphaera wettsteini
Kamptner, Erwin, 1941
Die Coccolithineen der Südwestküste von Istrien. Ann. Naturhist. Mus., Wien, 51: 54-149, pls 1-15.

fungi
Alderman, D.J., and E.B. Gareth Jones 1971.
Physiological requirements of two marine phycomycetes, Althornia crouchii and Ostracoblabe implexa.
Trans. Br. mycol. Soc. 57(2): 213-225.

fungi
Aleem, A.A., 1955.
Marine fungi from the west coast of Sweden.
Ark. Botan. 3(1/2):1-33.

fungi
Anasaki, Seibin, Kiichi Akino, and Tetsuo Tomiyama, 1968.
A comparison of some physiological aspects in a marine Pythium on the host and on the artificial medium.
Bull. Misaki mar. biol. Inst., Kyoto Univ., 12: 203-206.

fungi
Artemtchuk, N.J. 1972.
The fungi of the White Sea. III. New Phycomycetes discovered in the Great Salma Strait of the Kandalaksha Bay.
Veröff. Inst. Meeresforsch. Bremerh. 13: (2):231-237.

fungi
Barghoorn, Elso S., and David H. Linder, 1944.
Marine fungi: their taxonomy and biology.
Farlowia, 1:395-467.

fungi
Brooks, R.D., R.D. Goos and J.McN. Sieburth, 1972.
Fungal infestation of the surface and interior vessels of freshly collected driftwood. Mar. Biol. 16(4): 274-278.

fungi
Buck, John D., and Leonard J. Greenfield, 1964.
Calcification in marine-occurring yeasts.
Bull. Mar. Sci., Gulf and Caribbean, 14(2):239-245.

fungi
Buck, J.D., S.P. Meyers and K.M. Kamp, 1962
Marine bacteria with antiyeast activity.
Science, 138(3547):1339-1340.

fungi
Cavaliere, A.R. 1968.
Marine fungi of Iceland: a preliminary account of the Ascomycetes.
Mycologia 60(3): 475-479.

fungi

Cavaliere, A.R., and Randall S. Alberte 1970.
Fungi in animal shell fragments.
J. Elisha Mitchell scient. Soc. 86(4):203-206.

Fungi

Chakravarty, Dilip K., 1970.
Production of pure culture of Lagenisma coscinodisci Drebes parasitising the marine diatom Coscinodiscus.
Veröff. Inst. Meeresforsch., Bremerh. 12(3): 305-312.

fungi

Chakravarty, Dilip K., 1969.
Zum Kulturverhalten des marinen parasitischen Pilzes Lagenisma coscinodisci.
Veröff. Inst. Meeresforsch. Bremerhaven, 11(2), 304-312.

fungi

Clokie, Julian 1970.
Some substrate relationships of the family Thraustochytriaceae.
Veröff. Inst. Meeresforsch. Bremerh. 12(3): 329-351.

fungi

Colwell, Rita 1972.
Bacteria, yeasts, viruses and related microorganisms of the Chesapeake Bay.
Chesapeake Sci. 13 (Suppl.): 569-570.

fungi

Cribb, A.B., and Joan W. Cribb, 1960.
Marine fungi from Queensland, III.
Univ. Queensland Papers, Dept. Botany, 4(2/4): 39-44.

fungi

Cribb, A.B., and J.W. Cribb, 1960.
Some marine fungi on algae in European herbaria.
Univ. Queensland Papers, Dept. Botany, 4(2/4): 45-48.

fungi

Desai, A.J., and S.M. Betrabet 1971.
Cellulytic activity of marine fungi.
Curr. Sci. 40(16): 423-426.

fungi

Fell, J.W., and Christopher Martin, 1967.
Distribution of Antarctic marine fungi.
Antarctic Jl., U.S.A., 2(5):193-194.

fungi

Fell, Jack W., and Christopher Martin, 1966.
Distribution of Antarctic marine fungi.
Antarctic J., United States, 1(5):216.

fungi

Fell, Jack W., Christopher Martin and John J. Walsh, 1966.
Distribution of marine fungi.
Antarctic J., U.S., 1(4):127w

fungi

Freudenthal, Hugo D., and George von Baumgarten, 1965.
The effect of pressure on the growth of marine yeasts. (abstract).
Ocean Sci. and Ocean Eng., Mar. Techn. Soc.,- Amer. Soc. Limnol. Oceanogr., 2:1110.

Fungi

Gaertner, Alwin, 1970.
Beobachtungen über die Sporulation der dickwandigen Sporangien von Thraustochytrium kinnei Gaertner.
Veröff. Inst. Meeresforsch. Bremerh. 12(3) 321-327.

fungi

Gaertner, A., 1970.
Einiges zur Kultur mariner niederer Pilze.
Helgoländer wiss. Meeresunters, 20(1/4): 29-38.

fungi

Gaertner, Alwin 1968.
Die Fluktuationen mariner niederer Pilze in der Deutschen Bucht 1965 und 1966.
Mar. Mykologie, Veröff. Inst. Meeresforsch. Bremerh. Sonderband 3:105-117.

fungi

Gaertner, Alwin 1968.
Eine Methode des quantitativen Nachweises niederer mit Pollen köderbarer Pilze im Meerwasser und im Sediment.
Mar. Mykologie, Veröff. Inst. Meeresforsch. Bremerh. Sonderband 3:75-89.

fungi

Gaertner, Alwin, 1967.
Niedere mit Pollen köderbare Pilze in der südlichen Nordsee.
Veröff. Inst. Meeresforsch. Bremerh., 10(3): 159-165.

fungi

Gaertner, Alwin, 1966.
Vorkommen, Physiologie und Verteilung "Mariner Niederer Pilze" (Aquatic Phycomycetes).
Veröff. Inst. Meeresforsch., Bremerh., Sonderband II:221-234.

fungi

Gessner, R.V., R.D. Goos and J. McN. Sieburth, 1972.
The fungal microcosm of the internodes of Spartina alterniflora. Mar. Biol. 16(4): 269-273.

fungi

Gold, Harvey S., 1959.
Distribution of some lignicolous Ascomycetes and Fungi Imperfecti in an estuary.
J. Elisha Mitchell Sci. Soc., 75:25-26.

fungi

Goldstein, Solomon, and Louis Moriber, 1966.
Biology of a problematic marine fungus, Dermocystidium sp. 1. Development and cytology.
Arch. Mikrobiol., 53(1):1-11.

fungi

Grein, A., and S.P. Meyers, 1958.
Growth characteristics and antibiotic production of Actinomycetes isolated from littoral sediments and materials suspended in sea water.
J. Bact., 76(5):457-463.

fungi

Höhnk, Willy 1969.
Über den pilzlichen Befall kalkiger Hartteile von Meerestieren.
Ber. dt. wiss. Komm. Meeresforsch. n.f. 20(2):129-140.

fungi

Höhnk, Willy, 1967
Über die submersen Pilze an der rumänischen Schwarzmeerküste nahe Constanza.
Veröff. Inst. Meeresforsch. Bremerh. 10(3):149-158.

fungi

Höhnk, Willy, 1965.
Weitere Daten zur Verbreitung der marinen Pilze.
Veröff. Inst. Meeresforsch., Bremerh., Sonderband II:209-219.

fungi, role of

Höhnk, Willy, 1962
Hinweise auf die Rolle der marinen Pilze im Stoffhaushalt des Meeres.
Kieler Meeresf., 18(3) (Sonderheft): 145-150.

fungi

Hoppe, Hans-Georg 1972.
Untersuchungen zur Ökologie der Hefen im Bereich der westlichen Ostsee.
Kieler Meeresf. 28(1):54-77.

fungi

Johnson, T.W., Jr., 1967.
The estuarine mycoflora.
In: Estuaries, G.H. Lauff, editor, Publs Am. Ass. Advmt Sci., 83:303-305.

fungi

Johnson, T.W. Jr. 1966.
Trichomycetes in species of Hemigrapsus.
J. Elisha Mitchell scient. Soc. 82(1): 5-6.

fungi

Johnson, T.W., Jr., 1966.
A Lagenidium in the marine diatom Coscinodiscus centralis.
Mycologia, 58(1):131-135.

fungi

Johnson, T.W., Jr. 1966.
Chytridiomycetes and Oomycetes in marine phytoplankton.
Nova Hedwigia 10(3/4): 579-588.

fungi

Johnson, T.W., Jr., 1966.
Fungi in planktonic Synedra from brackish waters.
Mycologia, 58(3):373-382.

fungi

Johnson, T.W., Jr., 1966.
Fungi in planktonic Synedra from brackish waters.
Mycologia, 58(3):373-382.

fungi

Johnson, T.W., Jr., 1958.
Some lignicolous fungi from the North Carolina coast.
J. Elisha Mitchell Sci. Soc., 74(1):42-48.

fungi

Johnson, T.W., and S.P. Meyers, 1957.
Literature on halophilus and halolimnic fungi.
Bull. Mar. Sci. Gulf and Caribbean, 7(4):330-359.

FUNGI

Johnson, T. W., Jr. and Frederick K. Sparrow, Jr., 1961
Fungi in oceans and estuaries.
J. Cramer, Publ., Weinheim, 668 pp. $30.00

fungi

Jones, E.B. Gareth 1968
The distribution of marine fungi on wood submerged in the sea.
Proc. 1st Int. Biodeterioration Symp. (Biodeterioration of Materials: Microbiological and Allied Aspects, Elsevier Publ. Ltd.) 460-485.

fungi

Jones, E.B. Gareth, 1968.
Marine fungi.
Current Sci. 37(13): 378-379

Jones, E.B. Gareth, 1963. **fungi**
Observations on the fungal succession on wood
test blocks submerged in the sea.
J. Inst. Wood Sci., (11):14-23.

Jones, E.B.G., 1963. **fungi**
Marine fungi. II. Ascomycetes and Deuteromycetes
from submerged wood and drift Spartina.
Trans. British Mycol. Soc., 46:135-144.
Abstract in:
J. Mar. Biol. Assoc., U.K., 44(1):274.

Jones, E.B.G., 1962. **fungi**
Marine fungi. I.
Trans. British Mycol. Soc., 45:93-114.
Abstract in:
J. Mar. Biol. Assoc., U.K., 44(1):273.

Jones, E.B. Gareth, 1962. **fungi**
The Taxonomy and biology of the Pyrenomycetes.
Trans. British Mycol. Soc., 45(4):587-588.

Jones, E. B. Gareth, and D. H Jennings, 1964 **FUNGI**
The effect of salinity on the growth of marine
fungi in comparison with non-marine species.
Trans. Brit. Mycol. Soc., 47(4): 619-625.

Karling, John S., 1968. **fungi**
Zoosporic fungi of Oceania. 5.
Cladochytriaceae, Catenariaceae and
Blastocladiaceae.
Nova Hedwigia, 15(6):191-201

Kohlmeyer, Jan 1973. **fungi** a
Spathulosporales, a new order and
possible missing link between
Laboulbeniales and Pyrenomycetes.
Mycologia 65(3):614-647.

Kohlmeyer, Jan 1972. **fungi** a
A revision of Halosphaeriaceae.
Can. J. Bot. 50(9): 1951-1963.

Kohlmeyer, Jan 1972. **fungi**
Parasitic Hologuignardia oceanica
(Ascomycetes) and hyperparasitic
Sphaceloma cecidii sp. nov.
(Deuteromycetes) in drift Sargassum
in North Carolina.
J. Elisha Mitchell scient. Soc. 88(4):255-259.

Kohlmeyer, J., 1971. **fungi** a
Fungi from the Sargasso Sea. Mar. Biol., 8(4):
344-350.

Kohlmeyer, Jan 1970. **fungi**
Ein neuer Ascomycet auf Hydrozoen
im Südatlantik.
Ber. dtsch. Bot. Ges. 83(9/10):505-509.

Kohlmeyer, Jan, 1969. **fungi**
Deterioration of wood by marine fungi in the
deep sea.
Materials performance and the deep sea, Spec.
Techn. Publ., Am. Soc. Test. Materials, 445:
20-30.

Kohlmeyer, Jan., 1968. **fungi**
The first ascomycete from the deep sea.
J. Elisha Mitchell Scient. Soc., 84(1):239-241.

Kohlmeyer, J., 1968. **fungi**
Revisions and descriptions of algicolous marine
fungi.
Phytopathol. Z., 63: 341-363.

Kohlmeyer, J., 1967. **fungi**
Intertidal and phycophilous fungi from Tenerife
(Canary Islands).
Trans. Br. mycol. Soc., 50(1):137-147.

Kohlmeyer, Jan, 1966. **fungi**
Ecological observations on arenicolous marine
fungi.
Zeits. Allg. Mikrobiol., 6(2):95-106.

Kohlmeyer, J., 1964. **fungi**
Pilzfunde am Meer.
Zeits. fur Pilzkunde, 30(2):43-51.

Kohlmeyer, J., 1963. **fungi**
Parasitische und epiphytische Pilze auf
Meeresalgen.
Nova Hedwigia, 5:127-146.

Kohlmeyer, J., 1963. **fungi**
Fungi marini novi vel critici.
Nova Hedwigia, 6:297-329.

Kohlmeyer, Jan, 1963. **fungi**
Répartition de champignons marins (Ascomycetes et
Fungi imperfecti) dans la Mediterranée.
Rapp. Proc. Verb., Réunions, Comm. Int., Expl.
Sci., Mer Méditerranée, Monaco, 17(3):723-730.

Kohlmeyer, Jan, 1962. **fungi**
Heophile Pilze von den Ufern Frankreichs.
Nova Hedwigia, 4(3/4):389-420.

Kohlmeyer, J., 1961. **fungi**
Synoptic plates for quick determination of
marine Deuteromycetes and Ascomycetes.
Nova Hedwigia, 3(2 + 3): 383-398.

Kohlmeyer, J., 1960. **fungi**
Wood-inhabiting marine fungi from the Pacific
Northwest and California.
Nova Hedwigia, 2(1/2 & 3):293-344.

Kohlmeyer, J., 1959. **fungi**
Neufunde holzbesiedelnder Meerespilze.
Nova Hedwigia, 1(1):77-100.

Kohlmeyer, Jan, and Erika Kohlmeyer 1971 **fungi**
Marine fungi from tropical America and
Africa.
Mycologia 63(4): 831-861.

Kohlmeyer, J., and E. Kohlmeyer, 1966. **fungi**
On the life history of marine ascomycetes:
Halosphaeria mediosetigera and H. circumvestita.
Nova Hedwigia, 12(1/2):189-202.

Kriss, A.E., and I.N. Mitskevich, 1957. **fungi**
On a new class of microorganism living in the
deep seas and oceans. Uspechi Sovremennoi Biol.,
44(2(5)):269-280.

Kriss, A.E., E.A. Rukina and A.S. Tikhonenko, **fungi**
1952.
Distribution of yeast organisms in the sea.
(In Russian)
Zhurn. Obsh. Biol., 13(3):232.

Krumbein, W.E., 1971. **fungi** a
Manganese-oxidizing fungi and bacteria.
Naturwissenschaften 58(1):56-57.

Lundström-Eriksson, A., and B. Norkrans, 1965 **Fungi**
Studies on marine occurring yeasts,
relations to inorganic nitrogen compounds,
especially hydroxylamine.
Arch. Mikrobiol., 62(4): 373-383

Madri, Peter P., George Claus and **fungi**
Elisabeth E. Moss 1966.
Infectivity of pathogenic fungi in a
simulated marine environment.
Revista Biol. Brasileira e Portuguesa
Biol. Geral. 5(3/4): 341-381.

Meyers, Samuel P. 1968. **fungi**
Observations on the physiological ecology
of marine fungi.
Bull. Misaki mar. biol. Inst. Kyoto Univ.
12: 207-225.

Schneider, Joachim 1968. **fungi**
Über niedere Phycomyceten der
westlichen Ostsee.
Mar. Mykologie, Veröff. Inst. Meeresforsch.
Bremerh. Sonderband 3: 93-103.

Meyers, Samuel P., William A. Feder, and **fungi**
King Mon Tsue, 1963
Nutritional relationships among certain fila-
mentous fungi and a marine nematode.
Science, 141(3580):520-522.

Meyers, S.P., D.G. Ahearn, W. Gunkel and F.J. **fungi**
Roth, Jr., 1967.
Yeasts from the North Sea.
Marine Biol., 1(2):118-123.

Meyers, Samuel P., Bryce Prindle, Katherine **fungi**
Kamp, and Joseph U. Levi, 1963
Degradation of manila cordage by marine fungi
analysis of breaking strength tests.
Tappi, Tech. Assoc. Pulp and Paper Industry,
46(2):164A-167A.

Meyers, Samuel P., and Ernest S. Reynolds*, **fungi**
1963
Degradation of lignocellulose material by
marine fungi.
In: Symposium on Marine Microbiology, Carl H.
Oppenheimer, Compiler and Editor, Charles C.
Thomas, Publisher, Springfield, Illinois, pp.
315-328.

*deceased 31 May 1961.

Meyers, S.P., and E.S. Reynolds, 1960. **fungi**
Occurrence of Lignicolous fungi in northern
Atlantic and Pacific marine localities.
Canadian J. Bot., 38:217-226.

Meyers, Samuel P., and Ernest S. Reynolds, 1959. **fungi**
Growth and cellulolytic activity of lignicolous
Deuteromycetes from Marine localities.
Canadian J. Microbiology, 5:493-503.

fungi

Meyers, Samuel P., and Ernest S. Reynolds, 1958.
A wood incubation method for the study of lignicolous marine fungi.
Bull. Mar. Sci., Gulf & Caribbean, 8:342-347.

fungi

Meyers, S.P., and E.S. Reynolds, 1957.
Incidence of marine fungi in relation to marine borers. Science 126(3280):969.

fungi

Mitchell, R., and C. Wirsen, 1968.
Lysis of non-marine fungi by marine micro-organisms.
J. gen. Microbiol. 52(3): 335-345.

fungi

Moore, Royal, and Samuel P. Meyers, 1959.
Thalassiomycetes. 1. Principles of delimitation of the marine Mycota with the description of a new aquatically adapted Deuteromycete genus.
Mycologia, 51(6):871-876.

fungi

Morii, Hideaki 1973.
Yeasts predominating in the stomach of marine little toothed whales.
Bull. Jap. Soc. scient. Fish. 39(3): 333.

fungi

Morris, E.O., 1969.
Yeasts of marine origin. In: Oceanography and marine biology: an annual review, H. Barnes, editor, George Allen & Unwin Ltd., 6: 201-230.

fungi

Novozhilova, M.I., 1955.
[The quantitative characteristics, species composition and distribution of the yeast-like organisms in the Black Sea, the Sea of Okhotsk and the Pacific Ocean.]
Trudy Inst. Mikrobiol., 1955(4):155-195.

cited in Zhurn Bot., 1956, No. 52820.

fungi

Paterson, Robert A. and Hugh Rooney 1972.
Aquatic fungi: Their occurrence in the McMurdo oasis.
Antarctic J. U.S.A. 7(4):85-86.

fungi

Pugh, C.J.F., 1961
Fungal colonization of a developing salt marsh.
Nature, 190(4780): 1032-1033.

fungi

Ray, D.L., and D.E. Stuntz, 1959.
Possible relation between marine fungi and Limnoria attack on submerged wood.
Science 129(3341):93-94.

fungi

Rheinheimer Gerhard und Karl-Heinz Kullmann 1972
Untersuchungen über den Bakterien- und Hefegehalt von Wasser und Sand an einem Badestrand der Ostseeküste.
Kieler Meeresf. 28(2): 204-212.

fungi

Ritchie, Don, 1957.
Salinity optima for marine fungi affected by temperature.
Amer. J. Botany, 44:870-874.

fungi

Ritchie, D., 1954.
A fungous flora of the sea. Science 120(3119): 578-579, 1 textfig.

fungi

Roach, A.W., & J.K.G. Silvey, 1959
The occurrence of marine actinomycetes in Texas Gulf Coast substrates.
Amer. Midland Naturalist, 62(2):482-499.

fungi

Roth, Frank J., Donald G. Ahearn, Jack W. Fell, Samuel P. Meyers and Sally A. Meyer 1962
Ecology and taxonomy of yeasts isolated from various marine substrates.
Limnology and Oceanography, 7(2):178-185.

fungi

Schneider, J., 1972
Niedere pilze als Testorganismen für Schadstoffe in Meer- und Brackwasser die Wirkung von Schwermetallverbindungen und Phenol auf Thraustochytrium striatum. Mar. Biol. 16(3): 214-225.

fungus

Schneider, Joachim 1970.
Labyrinthula-Befall an niederen Pilzen (Thraustochytrium spec.) aus der Flensburger Förde.
Kieler Meeresforsch. 25(2): 314-315.

Lulworthia floridana

Meyers, Samuel P. 1966.
Variability in growth and reproduction of the marine fungus, Lulworthia floridana.
Helgoländer wiss. Meeresunters. 13(4): 436-443

Fungi

Schneider, Joachim, 1970.
Zur Taxonomie Verbreitung und Ökologie einiger mariner Phycomyceten.
Kieler Meeresforsch. 25(2):316-327

schizomycetes

Schreiber, Franca, e Giandomenico Rottini 1970.
Schizomiceti marini nel Golfo di Trieste carica batterica e fattori idrologici.
Boll. Pesca Piscic. Idrobiol. 25(1): 36-50

fungi

Seshadri, R., K. Krishnamurthy, and V.P. Ramamurthy, 1966.
Bacteria and yeasts in marine and estuarine waters of Portonovo (S. India).
Bull. Dept. Mar. Biol. Oceanog., Annam. Univ., 2: 5-11.

fungi

Sguros, Peter L., Samuel P. Meyers and Jacqueline Simms, 1962.
Role of marine fungi in the biochemistry of the oceans. 1. Establishment of quantitative technique for cultivation, growth measurement and production of inocula.
Mycologia, 54(5):521-535.

fungi

Sguros, Peter L., and Jacqueline Simms, 1964.
Role of marine fungi in the biochemistry of the oceans. IV. Growth responses to seawater inorganic macroconstituents.
J. Bacteriol., 88(12):346-355.

fungi

Sguros, Peter L., and Jacqueline Simms, 1963
Role of marine fungi in the biochemistry of the oceans. III. Growth factor requirements of the Ascomycete Halosphaeria mediosetigera.
Canadian J. Microbiology, 9:585-591.

fungi

Sguros, Peter L., and Jacqueline Simms, 1963.
Role of marine fungi in the biochemistry of the oceans. II. Effect of glucose, inorganic nitrogen and tris(hydroxymethyl)aminomethane on growth and pH changes in synthetic media.
Mycologia, 55(6):728-741.

fungi

Shearer, C.A. 1972
Fungi of the Chesapeake Bay.
Chesapeake Sci. 13 (Suppl.): 571-572.

fungi

Shinano, Haruo 1973.
Studies of marine bacteria taking part in the precipitation of calcium carbonate VI. A taxonomic study of marine bacteria taking part in the precipitation of calcium carbonate (4). VII A taxonomic study of the yeasts of marine origin taking part in the precipitation of calcium carbonate. (In Japanese; English abstract)
Bull. Jap. Soc. scient. Fish. 39(1):85-90; 91-95. p 95 en Jpn

fungi

Simard, R.E., and A.C. Blackwood, 1971.
Ecological studies on yeasts in the St. Lawrence river.
Can. J. Microbiol. 17:353-357.

fungi

Siegenthaler, Paul Andre, Melvin M. Belsky and Solomon Goldstein, 1967.
Phosphate uptake in an obligately marine fungus: a specific requirement for sodium.
Science, 155(3758):93-94.

fungi

Sparrow, Frederick Kroeber Jr. 1968.
Remarks on the Thraustochytriaceae.
Mar. Mykologie, Veröff. Inst. Meeresforsch. Bremerh. Sonderband 3: 7-17.

fungi

Sparrow, Frederick K., Jr., 1960 (2nd edit.)
Aquatic Phycomycetes.
Univ. Michigan Press, 1187 pp.

fungi

Suehiro S., 1963
Studies on the marine yeasts. III. Yeasts isolated from the mud of tideland. (In Japanese; English abstract).
Sci. Bull. Fac. Agric. Kyushu Univ., 20(2):223-227.

Also in:
Contrib. Dept. Fish. and Fish. Res. Lab., Kyushu Univ., No. 9(1963)

fungi

Suehiro, Sumio, 1962
Studies on the marine yeasts. II. Yeasts isolated from Thalassiosira subtilis (marine diatom) decayed in flasks. (In Japanese, English resume).
Sci. Bull., Fac. Agric., Kyushu Univ., 20(1): 101-105.

Also in: Contrib., Dept. Fish. and Fish. Res. Lab., Kyushu Univ., No. 8.

fungi

Suehiro, Sumio, and Yukio Tomiyasu, 1962
Studies on the marine yeasts. V. Yeasts isolated from seaweeds.
J. Fac. Agric. Kyushu Univ., 12(3):163-169.

Also in:
Contrib., Dept. Fish. and Fish. Res. Lab., Kyushu Univ., No. 8.

fungi

Suehiro, Sumio, Yukio Tomiyasu and Otohiko Tanaka 1962
Studies on the marine yeasts. IV. Yeasts isolated from marine plankton.
J. Fac. Agric., Kyushu Univ., 12(3):155-162.

Also in: Contrib. Dept. Fish. and Fish. Res. Lab., Kyushu Univ., No. 8.

fungi

Tubaki, Keisuke, 1968.
Studies on the Japanese marine fungi: lignicolous group II.
Publs Seto mar. biol. Lab., 15(5):357-372.

fungi

Ulken, Annemarie 1970.
Phycomyceten aus der Mangrove bei Cananéia (São Paulo, Brasilien).
Veröff. Inst. Meeresforsch. Bremerh. 12(3): 313-319.

fungi

Ulken, Annemarie 1969.
Über das Vorkommen niederer saprophytischer Phycomyceten [Chytridiales] im Bassin d'Arcachon [Frankreich].
Veröff. Inst. Meeresforsch. Bremerhaven 11(2):303-308.

fungi

Ulken, Annemarie, 1968.
Über zwei marine niedere Pilze vom Meeresboden der Nordsee.
Mar. Mykologie, Veröff. Inst. Meeresforsch. Bremerh. Sonderband 3: 71-73

fungi

Ulken, Annemarie, 1967.
Einige Beobachtungen über das Vorkommen von Phycomyceten aus der Reihe der Chytridiales im brackigen und marinen Wasser.
Veröff. Inst. Meeresforsch. Bremerh. 10(3):167-172.

fungi

Ulken, Annemarie, 1966.
Untersuchungen über marine Pilze im äquatorialen Atlantik vor der Küste Brasiliens.
Veröff. Inst. Meeresforsch, Bremerh., 10(2):107-115.

fungi

Van Uden, N., and J.W. Fell, 1968.
Marine yeasts. In: Advances in microbiology of the sea, M.R. Droop and E.J.F. Wood, editors, Academic Press, 167-201.

fungi, anat.-physiol.

Vishniac, Helen S., 1960
Salt requirements of marine phycomycetes. Limnol. & Oceanogr., 5(4):362-365.

fungi

Vishniac, H.S., 1956.
On the ecology of the lower marine fungi.
Biol. Bull. 111(3):410-414.

fungi

Vishniac, Helen S., 1955.
Marine mycology.
Trans. N.Y. Acad. Sci., II, 17:353.

fungi

Westheide, W., 1968.
Zur quantitativen Verteilung von Bakterien und Hefen in einem Gezeitenstrand der Nordseeküste.
Marine Biol., 1(4):336-347.

fungi

Weyland, H., 1969.
Actinomycetes in North Sea and Atlantic Ocean sediments.
Nature, Lond., 223(5208): 858.

fungi

Willoughby, L.G. 1968.
Aquatic Actinomycetales with particular reference to the Actinoplanaceae.
Mar. Mykologie Veröff. Inst. Meeresforsch. Bremerh. Sonderband 3:19-25.

fungi

Zelezinskaya, L.M. 1972.
On studying of fungi infections of sea Cladocera
Gidrol. Zh. 1972(4): 85-87

fungi, anat. physiol.

Kirk, Paul W. Jr. 1966.
Morphogenesis and microscopic cytochemistry of marine pyrenomycete ascospores.
Nova Hedwigia, Beihefte 22:128 pp.

fungi, lignicolous

Tubaki, K. 1968.
On Japanese lignicolous marine fungi.
Bull. Misaki mar. biol. Inst. Kyoto Univ. 12:195-202.

fungi, anat.-physiol.

Meyers, Samuel P., and J. Simms, 1967.
Thalassiomycetes IX. Comparative studies of reproduction in marine Ascomycetes.
Bull. mar. Sci., Miami, 17(1):132-148.

fungi, deep-sea

Kohlmeyer, Jan, 1968.
The first Ascomycete from the deep sea.
J. Elisha Mitchell sci. Soc., 84(1):239-241.

fungi

Andrews, J.D., 1965.
Infection experiments in nature with Dermocystidium marinum in Chesapeake Bay.
Chesapeake Science, 6(1):60-67.

fungi, effect of

Chandramohan, D., S. Ramu and R. Natarajan 1972.
Cellulolytic activity of marine Streptomycetes.
Current Sci. 41(7): 245-246.

fungi, effect of

de Simone, E., E. Campanile, V. Ferro, R. de Fusco et S. Grosso 1972.
Champignons producteurs d'antibiotiques dans l'eau de mer.
Rev. int. Océanogr. méd. 15-19:125-124

fungi, effect of

Jones, E.B. Gareth, and T. Le Campion-Alsumard 1970.
The biodeterioration of polyurethane by marine fungi.
Int. Biotetn. Bull. 6(3): 119-124.

fungi, effect of

Kim, Juhee, and Claude E. ZoBell 1972.
Agarase, amylase, cellulase and chitinase activity at deep-sea pressures.
J. oceanogr. Soc. Japan 28(4): 131-137

fungi, effect of

Kohlmeyer, J., 1972.
Marine fungi deteriorating chitin of hydrozoa and keratin-like annelid tubes. Mar. Biol. 12(4): 277-284.

fungi, effect of

Meyers, S.P., and E. Scott, 1968.
Cellulose degradation by Lulworthia floridana and other lignicolous marine fungi.
Marine Biol., 2(1):41-46.

fungi, effect of

Sato, T., and Y. Okuda, 1964.
Studies on species of fungi isolated from marine products. II. On fungi from marine products of all kinds. (In Japanese; English abstract).
Sci. Repts. Hokkaido Fish. Exper. Sta., (2):79-86.

fungi, effect of

Willingham, C.A., A.W. Roach and J.K.G. Silvey, 1966.
Comparative studies of substrate degradation by marine-occurring Actinomycetes.
Am. Midland Nat., 75(1):232-241.

fungi, lists of spp.

Ahearn, D.G., F.J. Roth, Jr., and S.P. Meyers 1968.
Ecology and characterization of yeasts from aquatic regions of South Florida.
Mar. Biol. 1(4): 291-308.

fungi, lists of spp.

Borut, S.Y., and T.W. Johnson, Jr., 1962.
Some biological observations on fungi in estuarine sediments.
Mycologia, 54(2):181-193.

fungi, lists of spp.

Fell, J.W., 1967.
Distribution of yeasts in the Indian Ocean.
Bull. mar. Sci., Miami, 17(2):454-470.

fungi (yeasts), lists of spp.

Hoppe, Hans-Georg, 1972.
Taxonomische Untersuchungen an Hefen aus der westlichen Ostsee.
Kieler Meeresf. 28(2): 219-226

fungi, lists of spp.

Kohlmeyer, J. 1971
Annotated check list of New England marine fungi.
Trans. Br. mycol. Soc. 57(3): 473-492

fungi, lists of spp.

Kohlmeyer, Jan 1968.
Dänische Meerespilze (Ascomycetes).
Ber. dt. bot. Ges. 81(1/2): 53-61.

fungi, lists of spp

Kohlmeyer, Jan, 1968.
Marine fungi from the tropics.
Mycologia, 60(2):252-270.

fungi, lists of spp.

Kohlmeyer, Jan & Erica, 1964.
Synoptic plates of higher marine fungi.
J. Cramer, Weinheim, 64 pp.

fungi, lists of spp.

Meyers, Samuel P., Donald G. Ahearn and Frank J. Roth, Jr., 1967.
Mycological investigations of the Black Sea.
Bull. mar. Sci., Miami, 17(3):576-596.

fungi, lists of spp.

Scott, W.W., 1962.
The aquatic phycomycetous flora of marine and brackish waters in the vicinity of Gloucester Point, Virginia.
Virginia Inst. Mar. Sci., Spec. Sci. Rept., 36: 16 pp.

fungi, lists of spp.

Siepmann, Rolf, and Willy Höhnk, 1962.
Über Hefen und einiger Pilze (Fungi imp., Hyphales) aus dem Nordatlantik.
Veröffentlichungen Inst. Meeresf., Bremerhaven, 8(1):79-97.

fungi (yeasts), lists of spp.

Taysi, I., and N. van Uden, 1964.
Occurrence and population densities of yeast species in an estuarine-marine area.
Limnology and Oceanography, 9(1):42-45.

fungi, lower

Schneider, Joachim 1970.
3.7 Niedere Pilze. Chemische, mikrobiologische und planktologische Untersuchungen in der Schlei im Hinblick auf deren Abwasserbelastung.
Kieler Meeresforsch 26(2): 173-178.

Yeasts

Ahearn, Donald G., Frank J. Roth, Jr., and Samuel P. Meyers 1962.
A comparative study of marine and terrestrial strains of Rhodotorula.
Canadian J. Microbiol. 8:121-132.

yeasts

Fell, J.W., D.G. Ahearn, S.P. Meyers and F.J. Roth, Jr., 1960.
Isolation of yeasts from Biscayne Bay, Florida, and adjacent benthic areas.
Limnol. and Oceanogr., 5(4):366-371.

yeasts

Fell, Jack W., Ingrid L. Hunter and Adele S. Tallman 1973
Marine basidiomycetous yeasts (Rhodosporidium spp.n.) with tetrapolar and multiple allelic bipolar mating systems.
Can. J. Microbiol. 19 (5): 643-657.

fungi (yeasts)

Norkrans, Birgitta 1968.
Studies on marine occurring yeasts: respiration, fermentation and salt tolerance.
Arch. Mikrobiol. 62(4): 358-372.

fungi (yeasts)

Rheinheimer, Gerhard, 1970.
Mikrobiologische und chemische Untersuchungen in der Flensburger Förde.
Ber. dt. wiss. Komm. Meeresforsch. 21(3/4): 420-429.

fungi (yeasts)

Shirano, H., 1962.
Studies on yeasts isolated from various areas of the North Pacific. (In Japanese; English summary).
Bull. Jap. Soc. Sci. Fish., 28(11):1113-1122.

fungi (yeasts)

Van Uden, N., 1967.
Occurrence and origin of yeasts in estuaries.
In: Estuaries, G.H. Lauff, editor, Publs Am. Ass. Advmt Sci., 83:306-310.

yeasts

van Uden, N., and R. Castelo Branco, 1963
Distribution and population densities of yeast species in Pacific water, air, animals and kelp off Southern California.
Limnology and Oceanography, 8(3):323-329.

yeasts, lists of spp.

Fell, J.W., and N. Van Uden, 1963
Yeasts in marine environments.
In: Symposium on Marine Microbiology, Carl H. Oppenheimer, Compiler and Editor, Charles C. Thomas, Publisher, Springfield, Illinois, 329-340.

Ascomycetes

Meyers, Samuel P. and Ernest S. Reynolds, 1960
Cellulolytic activity of Lignicolous marine Ascomycetes and Deuteromycetes.
Development in Industrial Microbiology, Plenum Press, N.Y. 1: 157-168.

Deuteromycetes

Meyers, Samuel P. and Ernest S. Reynolds, 1960
Cellulolytic activity of Lignicolous marine Ascomycetes and Deuteromycetes.
Development in Industrial Microbiology, Plenum Press, N.Y. 1: 157-168.

Alternaria maritima

Jones, E.B. Gareth, 1962
Marine fungi.
Trans. British Mycolog. Soc., 45(1): 93-114.

Alternaria maritima (imperfecti)

Kohlmeyer, Jan, 1962.
Haophile Pilze von den Ufern Frankreichs.
Nova Hedwigia, 4(3/4):389-420.

Antennospora caribbean n. gen. n. sp.

Meyers, Samuel P., 1957.
Taxonomy of marine pyrenomycetes.
Mycologia 49(4):475-528.

Amphisphaeria biturbinata

Meyers, Samuel P., 1957.
Taxonomy of marine pyrenomycetes.
Mycologia 49(4):475-528.

Amphisphaeria maritima

Meyers, Samuel P., 1957.
Taxonomy of marine pyrenomycetes.
Mycologia 49(4):475-528.

Amphisphaeria posidoniae (ascomycetes)

Kohlmeyer, Jan, 1962.
Haophile Pilze von den Ufern Frankreichs.
Nova Hedwigia, 4(3/4):389-420.

Amphisphaeria posidoniae

Meyers, Samuel P., 1957.
Taxonomy of marine pyrenomycetes.
Mycologia 49(4):475-528.

Arenariomyces quadri-remis comb nov.

Meyers, Samuel P., 1957.
Taxonomy of marine pyrenomycetes.
Mycologia 49(4):475-528.

Arenariomyces salina sp.nov

Meyers, Samuel P., 1957.
Taxonomy of marine pyrenomycetes.
Mycologia 49(4):475-528.

Arenariomyces trifurcatus

Meyers, Samuel P., 1957.
Taxonomy of marine pyrenomycetes.
Mycologia 49(4):475-528.

Brachysporium helgolandicium n.sp.

Schaumann, K., 1973
Brachysporium helgolandicum nov. sp., ein neuer Deutermycet auf Treibborke im Meer.
Helgoländer wiss. Meeresunters. 25(1):26-34.

Buellia haliotrepha n.sp.

Kohlmeyer, Jan, and Erika Kohlmeyer, 1965
New marine fungi from mangroves and trees along eroding shorelines.
Nova Hedwigia, 9:89-104.

Cerabosphaeria sp.

Meyers, Samuel P., 1957.
Taxonomy of marine pyrenomycetes.
Mycologia 49(4):475-528.

Cercospora salina

Jones, E.B. Gareth, 1962
Marine fungi.
Trans. British Mycolog. Soc., 45(1): 93-114.

Ceriosporopsis calyptrata

Kohlmeyer, Jan, 1962.
Haophile Pilze von den Ufern Frankreichs.
Nova Hedwigia, 4(3/4):389-420.

Ceriosporopsis cambrensis

Meyers, Samuel P., 1957.
Taxonomy of marine pyrenomycetes.
Mycologia 49(4):475-528.

Cerisporopsis halima

Jones, E.B. Gareth, 1963
Marine Fungi. II. Ascomycetes and Deuteromycetes from submerged wood and drift Spartina.
Trans. Brit. Mycol. Soc., 46(1):135-144.

Ceriosporopsis halima

Jones, E.B. Gareth, 1962
Marine fungi.
Trans. British Mycolog. Soc., 45(1): 93-114.

Ceriosporopsis halima

Kohlmeyer, Jan, 1962.
Haophile Pilze von den Ufern Frankreichs.
Nova Hedwigia, 4 (3/4):389-420.

Ceriosporopsis halima

Meyers, Samuel P., 1957.
Taxonomy of marine pyrenomycetes.
Mycologia 49(4):475-528.

Ceriosporopsis hamata

Jones, E.B. Gareth, 1962
Marine fungi.
Trans. British Mycolog. Soc., 45(1): 93-114.

Ceriosporopsis hamata

Meyers, Samuel P., 1957.
Taxonomy of marine pyrenomycetes.
Mycologia 49(4):475-528.

Ceriosporopsis longissima

Kohlmeyer, J., 1963.
Fungi marini novi vel critici.
Nova Hedwigia, 6:297-329.

Ceriosporopsis longissima nsp.

Kohlmeyer, Jan, 1962.
Haophile Pilze von den Ufern Frankreichs.
Nova Hedwigia, 4(3/4):389-420.

Chadefaudia balliae n.sp.

Kohlmeyer Jan 1973.
Chadefaudia balliae, a new species of Ascomycetes on Ballia in Australia.
Mycologia 65(1): 244-248

Chaetomium erectum
Jones, E.B. Gareth, 1962
Marine fungi.
Trans. British Mycolog. Soc., 45(1): 93-114.

Chaetomium globosum
Jones, E.B. Gareth, 1962
Marine fungi.
Trans. British Mycolog. Soc., 45(1): 93-114.

Chaetosphaeria chaetosa nsp.
Kohlmeyer, J., 1963.
Fungi marini novi vel critici.
Nova Hedwigia, 6:297-329.

Cirrenalia macrocephala
Jones, E.B. Gareth, 1962
Marine fungi.
Trans. British Mycolog. Soc., 45(1): 93-114.

Cirrenalis macrocephala
Kohlmeyer, J., 1963.
Fungi marini novi vel critici.
Nova Hedwigia, 6:297-329.

Cirrenalis macrocephala
Kohlmeyer, Jan, 1962.
Heophile Pilze von den Ufern Frankreichs.
Nova Hedwigia, 4(3/4):389-420.

Cirrenalia pygmea n. sp.
Kohlmeyer, Jan, 1966.
Neue Meerespilze an Mangroven.
Ber. Deutsch. Botan. Gesellschaft, 79(1):27-37.

Corollospora
Kohlmeyer, Jan, 1966.
Ascospore morphology in Corollospora.
Mycologia, 18(2):281-288.

Corollospora cristata
Kohlmeyer, J., 1963.
Fungi marini novi vel critici.
Nova Hedwigia, 6:297-329.

Corollospora maritima
Kohlmeyer, J., 1963.
Fungi marini novi vel critici.
Nova Hedwigia, 6:297-329.

Corollospora puchella n. sp.
Kohlmeyer, Jan, Ingeborg Scmidt and N. Balakrishnan Nair, 1967.
Corollospora (Ascomycetes-) aus dem Indischen Ozean und der Ostsee.
Ber.dt.bot.Ges.80(2):98-102.

Corollospora trifurcata
Kohlmeyer, J., 1963.
Fungi marini novi vel critici.
Nova Hedwigia, 6:297-329.

Cryptococcus albidus
Siepmann, Rolf, and Willy Höhnk, 1962.
Über Hefen und einige Pilze (Fungi imp., Hyphales) aus den Nordatlantik.
Veröff. Inst. Meeresf., Bremerhaven, 8(1):79-97.

Cryptococcus laurenti
Siepmann, Rolf, and Willy Höhnk, 1962.
Über Hefen und einige Pilze (Fungi imp., Hyphales) aus den Nordatlantik.
Veröff. Inst. Meeresf., Bremerhaven, 8(1):79-97.

Culcitalna achraspora
Jones, E.B. Gareth, 1963
Marine Fungi. II. Ascomycetes and Deuteromycetes from submerged wood and drift Spartina.
Trans. Brit. Mycol. Soc., 46(1):135-144.

Culcitalna achraspora
Kohlmeyer, J., 1963.
Fungi marini novi vel critici.
Nova Hedwigia, 6:297-329.

Debaryomyces subglobosus
Siepmann, Rolf, and Willy Höhnk, 1962.
Über Hefen und einige Pilze (Fungi imp., Hyphales) aus den Nordatlantik.
Veröff. Inst. Meeresf., Bremerhaven, 8(1):79-97.

Debaryomyces kloeckeri
Siepmann, Rolf, and Willy Höhnk, 1962.
Über Hefen und einige Pilze (Fungi imp., Hyphales) aus den Nordatlantik.
Veröff. Inst. Meeresf., Bremerhaven, 8(1):79-97.

Dermocystidium marinum
Andrews, J.D., 1965.
Infection experiments in nature with Dermocystidium marinum in Chesapeake Bay.
Chesapeake Science, 6(1):60-67.

Dermocystidium marinum
Andrews, J.D., and W. Hewatt, 1957.
Oyster mortality studies in Virginia. II. The fungous disease caused by Dermocystidium marinum in oysters in Chesapeake Bay.
Ecological Monographs, 27(1):1-26.

Dermocystidium marinum
Beckett, Robert L., 1967.
Occurrence of the fungus Dermocystidium marinum in the American oyster in Chincotuaque Bay.
Chesapeake Sci., 8(4):261-262.

Dermocystidium marinum
Ray, Sammy M., 1964?
Notes on the occurrence of Dermocystidium marinum on the Gulf of Mexico coast during 1961 and 1962.
1963 Proc. Nat. Shellfish. Assoc., 54:45-54.

Dermocystidium marinum
Ray, Sammy M., 1964?
A review of the culture method for detecting Dermocystidium marinum with suggested modifications and precautions.
1963 Proc. Nat. Shellfish. Assoc., 55-59.

Dictyosporium pelagica
Jones, E.B. Gareth, 1963
Marine Fungi. II. Ascomycetes and Deuteromycetes from submerged wood and drift Spartina.
Trans. Brit. Mycol. Soc., 46(1):135-144.

Dictyosporium toruloides
Jones, E.B Gareth, 1963
Marine Fungi. II. Ascomycetes and Deuteromycetes from submerged wood and drift Spartina.
Trans. Brit. Mycol. Soc., 46(1):135-144.

Didymodamarospora sp.
Johnson, Terry W., Jr., and Harvey S. Gold, 1958.
Didymodamarospora a new genus of fungi from fresh and marine waters.
J. Elisha Mitchell Sci. Soc., 73:103-108.

Didymodamarospora n gen
Johnson, Terry W., Jr., and Harvey S. Gold, 1957
Didymodamarospora a new genus of fungi from fresh and marine waters.
J. Elisha Mitchell Sci. Soc., 73:103-108.

Didymosphaeria enalis n. sp.
Kohlmeyer, Jan, 1966.
Neue Meerespilze an Mangroven.
Ber. Deutsch. Botan. Gesellschaft, 79(1):27-37.

Didymosphaeria fucicola
Meyers, Samuel P., 1957.
Taxonomy of marine pyrenomycetes.
Mycologia 49(4):475-528.

Didymosphaeria maritima
Meyers, Samuel P., 1957.
Taxonomy of marine pyrenomycetes.
Mycologia 49(4):475-528.

Didymosphaeria pelvetiana
Meyers, Samuel P., 1957.
Taxonomy of marine pyrenomycetes.
Mycologia 49(4):475-528.

Digitatispora marina
Kohlmeyer, J., 1963.
Fungi marini novi vel critici.
Nova Hedwigia, 6:297-329.

Ectogella
Johnson, T.W. Jr., 1966
Ectogella in marine species of Licmophora.
J. Elisha Mitchell scient. Soc. 82(1):25-29

Fusarium conglutinans
Siepmann, Rolf, and Willy Höhnk, 1962.
Über Hefen und einige Pilze (Fungi imp., Hyphales) aus den Nordatlantik.
Veröff. Inst. Meeresf., Bremerhaven, 8(1):79-97.

Fusarium solani
Siepmann, Rolf, and Willy Höhnk, 1962.
Über Hefen und einige Pilze (Fungi imp., Hyphales) aus den Nordatlantik.
Veröff. Inst. Meeresf., Bremerhaven, 8(1):79-97.

Gnomonia salina n.sp.
Jones, E.B. Gareth, 1962
Marine fungi.
Trans. British Mycolog. Soc., 45(1): 93-114.

Guignardia alaskane
Meyers, Samuel P., 1957.
Taxonomy of marine pyrenomycetes.
Mycologia 49(4):475-528.

Guignardia irritans
Meyers, Samuel P., 1957.
Taxonomy of marine pyrenomycetes.
Mycologia 49(4):475-528.

Guignardia prasiolae
Meyers, Samuel P., 1957.
Taxonomy of marine pyrenomycetes.
Mycologia 49(4):475-528.

Guignardia tumefaciens
Meyers, Samuel P., 1957.
Taxonomy of marine pyrenomycetes.
Mycologia 49(4):475-528.

Guignardia ulvae
Meyers, Samuel P., 1957.
Taxonomy of marine pyrenomycetes.
Mycologia 49(4):475-528.

Haligena elaterophora
Kohlmeyer, J., 1963.
Fungi marini novi vel critici.
Nova Hedwigia, 6:297-329.

Haligena spartinae
Jones, E.B. Gareth, 1962
Haligena spartinae sp. nov., a Pyrenomycete on Spartina townsendii.
Trans. Brit. Mycol. Soc., 45(2):245-248.

Haligena viscidula n. sp.
Kohlmeyer, Jan, and Erika Kohlmeyer, 1965
New marine fungi from mangroves and trees along eroding shorelines.
Nova Hedwigia, 9:89-104.

Haloguignardia oceanica
Kohlmeyer Jan 1972.
Parasitic Haloguignardia oceanica (Ascomycetes) and hyperparasitic Sphaceloma cecidii sp. nov. (Deuteromycetes) in drift Sargassum in North Carolina.
J. Elisha Mitchell scient. Soc. 88(4):255-259.

Halonectria milfordensis n. gen., n.sp.
Jones, E.B. Gareth, 1965.
Halonectria milfordensis gen. et sp. nov., a marine pyrenomycete on submerged wood.
Trans. Brit. mycol. Soc., 48(2):287-290.

Halosphaera appendiculata
Jones, E.B. Gareth, 1962
Marine fungi.
Trans. British Mycolog. Soc., 45(1): 93-114.

Halisphaeria appendiculata
Kohlmeyer, J., 1963.
Fungi marini novi vel critici.
Nova Hedwigia, 6:297-329.

Halisphaeria appendiculata
Kohlmeyer, Jan, 1962.
Heophile Pilze von den Ufern Frankreichs.
Nova Hedwigia, 4(3/4):389-420.

Halosphaeria appendiculata
Meyers, Samuel P., 1957.
Taxonomy of marine pyrenomycetes.
Mycologia 49(4):475-528.

Halisphaeria circumvestita
Kohlmeyer, J., 1963.
Fungi marini novi vel critici.
Nova Hedwigia, 6:297-329.

Halisphaeria mediosetigera
Kohlmeyer, J., 1963.
Fungi marini novi vel critici.
Nova Hedwigia, 6:297-329.

Halosphaeria circumvestita
Kohlmeyer, Jan, 1962.
Heophile Pilze von den Ufern Frankreichs.
Nova Hedwigia, 4(3/4):389-420.

Halosphaeria mediosetigera
Kohlmeyer, Jan, 1962.
Heophile Pilze von den Ufern Frankreichs.
Nova Hedwigia, 4(3/4):389-420.

Halosphaeria circumvestita
Jones, E.B. Gareth, 1963
Marine Fungi. II. Ascomycetes and Deuteromycetes from submerged wood and drift Spartina.
Trans. Brit. Mycol. Soc., 46(1):135-144.

Halosphaeria torquata
Jones, E.B. Gareth, 1963
Marine Fungi. II. Ascomycetes and Deuteromycetes from submerged wood and drift Spartina.
Trans. Brit. Mycol. Soc., 46(1):135-144.

Halisphaeria tubulifera
Kohlmeyer, J., 1963.
Fungi marini novi vel critici.
Nova Hedwigia, 6:297-329.

Halotthia n.gen.
Kohlmeyer, J., 1963.
Zwei neue Ascomyceten-Gattungen auf Posidonia-Rhizomen.
Nova Hedwigia, 6:7-13.

Halotthia posidoniae
Kohlmeyer, J., 1963.
Zwei neue Ascomyceten-Gattungen auf Posidonia-Rhizomen.
Nova Hedwigia, 6:7-13.

Hansenula carlifornica
Siepmann, Rolf, and Willy Höhnk, 1962.
Über Hefen und einige Pilze (Fungi imp.), Hyphales aus den Nordatlantik.
Veröff. Inst. Meeresf., Bremerhaven, 8(1):79-97.

Helicoma maritimum
Jones, E.B. Gareth, 1962
Marine fungi.
Trans. British Mycolog. Soc., 45(1): 93-114.

Helicoma maritimum
Kohlmeyer, J., 1963.
Fungi marini novi vel critici.
Nova Hedwigia, 6:297-329.

Helicoma maritimum
Kohlmeyer, Jan, 1962.
Heophile Pilze von den Ufern Frankreichs.
Nova Hedwigia, 4(3/4):389-420.

Helicoma salinum
Jones, E.B. Gareth, 1962
Marine fungi.
Trans. British Mycolog. Soc., 45(1): 93-114.

Herpotrichiella ciliomaris
Kohlmeyer, J., 1963.
Fungi marini novi vel critici.
Nova Hedwigia, 6:297-329.

Humicola alopallonella
Jones, E.B. Gareth, 1963
Marine Fungi. II. Ascomycetes and Deuteromycetes from submerged wood and drift Spartina
Trans. Brit. Mycol. Soc., 46(1):135-144.

Humicola alopallonella
Kohlmeyer, J., 1963.
Fungi marini novi vel critici.
Nova Hedwigia, 6:297-329.

Hypoderma laminariae
Meyers, Samuel P., 1957.
Taxonomy of marine pyrenomycetes.
Mycologia 49(4):475-528.

Hydronectria tethys n. sp.
Kohlmeyer, Jan, and Erika Kohlmeyer, 1965
New marine fungi from mangroves and trees along eroding shorelines.
Nova Hedwigia, 9:89-104.

Keissleriella blepharospora n.sp.
Kohlmeyer, Jan, and Erika Kohlmeyer, 1965
New marine fungi from mangroves and trees along eroding shorelines.
Nova Hedwigia, 9:89-104.

Ichthyophonus
Ruggieri, George D., Ross F. Nigrelli, P.M. Powles and D.G. Garnett, 1970
Epizootics in yellowtail flounder, Limanda ferruginea Storer, in the western North Atlantic caused by Ichthyophonus.
Zoologica, N.Y., 55(3):57-62.

an ubiquitous parasitic fungus

Labyrinthula (myxomyceta)
Schneider, Joachim, 1970.
Labyrinthula-Befall an niederen Pilzen (Thraustochytrium spec) aus der Flensburger Förde.
Kieler Meeresforsch. 25(2):314-315

Lagenisma coscinodisci gen. nov., spec. nov.
Drebes, Gerhard 1968.
Lagenisma coscinodisci gen. nov., sp. nov., ein Vertreter der Lagenidiales in der marinen Diatomee Coscinodiscus.
Mar. Mykologie. Veröff. Inst. Meeresforsch. Bremerh. Sonderband 3:67-69.

Lentescospora submarina
Kohlmeyer, J., 1963.
Fungi marini novi vel critici.
Nova Hedwigia, 6:297-329.

Lentescospora submarina
Kohlmeyer, Jan, 1962.
Heophile Pilze von den Ufern Frankreichs.
Nova Hedwigia, 4(3/4):389-420.

Lentescospora submarina
Meyers, Samuel P., 1957.
Taxonomy of marine pyrenomycetes.
Mycologia 49(4):475-528.

Leptosphaeria albopunctata
Jones, E.B. Gareth, 1962
Marine fungi.
Trans. British Mycolog. Soc., 45(1): 93-114.

Leptosphaeria albopunctata
Meyers, Samuel P., 1957.
Taxonomy of marine pyrenomycetes.
Mycologia 49(4):475-528.

Leptosphaeria avicenniae
Kohlmeyer, Jan, and Erika Kohlmeyer, 1965
New marine fungi from mangroves and trees along eroding shorelines.
Nova Hedwigia, 9:89-104.

Leptosphaeria chondri
Meyers, Samuel P., 1957.
Taxonomy of marine pyrenomycetes.
Mycologia 49(4):475-528.

Leptosphaeria discors
Johnson, T.W., Jr., 1956.
Ascus development and spore discharge in Leptosphaeria discors, a marine and brackish-water fungus.
Bull. Mar. Sci., Gulf and Caribbean, 6(4):349-358.

Leptosphaeria discors
Jones, E.B. Gareth, 1962
Marine fungi.
Trans. British Mycolog. Soc., 45(1): 93-114.

Leptosphaeria discors

Meyers, Samuel P., 1957.
Taxonomy of marin pyrenomycetes.
Mycologia 49(4):475-528.

Leptosphaeria halima

Meyers, Samuel P., 1957.
Taxonomy of marine pyrenomycetes.
Mycologia 49(4):475-528.

Leptosphaeria macrosporidium n.sp.

Jones, E.B. Gareth, 1962
Marine fungi.
Trans. British Mycolog. Soc., 45(1):
93-114.

Leptosphaeria marina

Meyers, Samuel P., 1957.
Taxonomy of marine pyrenomycetes.
Mycologia 49(4):475-528.

Leptosphaeria maritima

Meyers, Samuel P., 1957.
Taxonomy of marine pyrenomycetes.
Mycologia 49(4):475-528.

Leptosphaeria orae-maris

Jones, E.B. Gareth, 1962
Marine fungi.
Trans. British Mycolog. Soc., 45(1):
93-114.

Leptosphaeria orae-maris

Kohlmeyer, Jan, 1962.
Haophile Pilze von den Ufern Frankreichs.
Nova Hedwigia, 4(3/4):389-420.

Leptosphaeria orae-maris

Meyers, Samuel P., 1957.
Taxonomy of marine pyrenomycetes.
Mycologia 49(4):475-528.

Leptosphaeria pelagica n.sp.

Jones, E.B. Gareth, 1962
Marine fungi.
Trans. British Mycolog. Soc., 45(1):
93-114.

Leptosphaeria typharum

Jones, E.B. Gareth, 1962
Marine fungi.
Trans. British Mycolog. Soc., 45(1):
93-114.

Leucothrix mucor

Brock, Thomas D., 1966.
The habitat of Leucothrix mucor, a widespread marine microorganism.
Limnol. Oceanogr., 11(2):303-307.

Ligninicola laevis

Meyers, Samuel P., 1957.
Taxonomy of marine pyrenomycetes.
Mycologia 49(4):475-528.

Lulworthia sp.

Kohlmeyer, J., 1963.
Fungi marini novi vel critici.
Nova Hedwigia, 6:297-329.

Lulworthia floridana

Jones, E.B. Gareth, 1962
Marine fungi.
Trans. British Mycolog. Soc., 45(1):
93-114.

Lulworthia floridana

Kohlmeyer, Jan, 1962.
Haophile Pilze von den ufern Frankreichs.
Nova Hedwigia, 4(3/4):389-420.

Lulworthia floridana n. sp.

Meyers, Samuel P., 1957.
Taxonomy of marine pyrenomycetes.
Mycologia 49(4):475-528.

Lulworthia fucicola

Jones, E.B. Gareth, 1963
Marine Fungi. II. Ascomycetes and Deuteromycetes from submerged wood and drift Spartina.
Trans. Brit. Mycol. Soc., 46(1):135-144.

Lulworthia cf. fucucola

Kohlmeyer, J., 1963.
Fungi marini novi vel critici.
Nova Hedwigia, 6:297-329.

Lulworthia grandispora n.sp.

Meyers, Samuel P., 1957.
Taxonomy of marine pyrenomycetes.
Mycologia 49(4):475-528.

Lulworthia halima

Kohlmeyer, J., 1963.
Fungi marini novi vel critici.
Nova Hedwigia, 6:297-329.

Lulworthia medusa

Jones, E.B. Gareth, 1963
Marine Fungi. II. Ascomycetes and Deuteromycetes from submerged wood and drift Spartina.
Trans. Brit. Mycol. Soc., 46(1):135-144.

Lulworthia medusa

Kohlmeyer, J., 1963.
Fungi marini novi vel critici.
Nova Hedwigia, 6:297-329.

Lulworthia medusa biscaynia n. var.

Meyers, Samuel P., 1957.
Taxonomy of marine pyrenomycetes.
Mycologia 49(4):475-528.

Lulworthia opaca

Jones, E.B. Gareth, 1963
Marine Fungi. II. Ascomycetes and Deuteromycetes from submerged wood and drift Spartina.
Trans. Brit. Mycol. Soc., 46(1):135-144.

Lulworthia purpurea

Jones, E.B. Gareth, 1962
Marine fungi.
Trans. British Mycolog. Soc., 45(1):
93-114.

Lulworthia rufa

Jones, E.B. Gareth, 1962
Marine fungi.
Trans. British Mycolog. Soc., 45(1):
93-114.

Lulworthia submersa

Jones, E.B. Gareth, 1962
Marine fungi.
Trans. British Mycolog. Soc., 45(1):
93-114.

Macrophoma gymnogongri

Jones, E.B. Gareth, 1962
Marine fungi.
Trans. British Mycolog. Soc., 45(1):
93-114.

Massariella maritima

Meyers, Samuel P., 1957.
Taxonomy of marine pyrenomycetes.
Mycologia 49(4):475-528.

Melanopsamma balani

Meyers, Samuel P., 1957.
Taxonomy of marine pyrenomycetes.
Mycologia 49(4):475-528.

Melanopsamma oystophorae

Meyers, Samuel P., 1957.
Taxonomy of marine pyrenomycetes.
Mycologia 49(4):475-528.

Melanopsamma tregouboiri

Meyers, Samuel P., 1957
Taxonomy of marine pyrenomycetes.
Mycologia 49(4):475-528.

Metasphaeria australiensis

Meyers, Samuel P., 1957.
Taxonomy of marine pyrenomycetes.
Mycologia 49(4):475-528.

Microthalia maritima

Kohlmeyer, Jan, 1962.
Haophile Pilze von den Ufern Frankreichs.
Nova Hedwigia, 4(3/4):389-420.

Monodictys putredinis

Jones, E.B. Gareth, 1963
Marine Fungi. II. Ascomycetes and Deuteromycetes from submerged wood and drift Spartina.
Trans. Brit. Mycol. Soc., 46(1):135-144.

Mycaureola dilseae

Meyers, Samuel P., 1957.
Taxonomy of marine pyrenomycetes.
Mycologia 49(4):475-528.

Mycosphaerella ascophylli

Meyers, Samuel P., 1957.
Taxonomy of marine pyrenomycetes.
Mycologia 49(4):475-528.

Mycosphaerella pneumatophorae n. sp.

Kohlmeyer, Jan, 1966.
Neue Meerespilze an Mangroven.
Ber. Deutsch. Botan. Gesellschaft, 79(1):27-37.

Mycosphaerella pelvetiae
Meyers, Samuel P., 1957.
Taxonomy of marine pyrenomycetes.
Mycologia 49(4):475-528.

Nautosphaeria cristaminuta n.gen., n.sp.
Jones, E.B. Gareth, 1964.
Nautosphaeria cristaminuta gen. et sp. nov., a marine pyrenomycete on submerged wood.
Trans. Brit. Mycol. Soc., 47(1):97-101.

Nais enornata n.gen. n.sp.
Kohlmeyer, Jan, 1962.
Haophile Pilze von den Ufern Frankreichs.
Nova Hedwigia, 4(3/4):389-420.

Nia vibrissa
Kohlmeyer, J., 1963.
Fungi marini novi vel critici.
Nova Hedwigia, 6:207-329.

Nia vibrissa, n.gen., n.sp.
Moore, Royal, and Samuel P. Meyers, 1959.
Thalassiomycetes 1. Principles of delimitation of the marine Mycota with the description of a new aquatically adapted Deuteromycete genus.
Mycologia, 51(6):871-876.

Olpidiopsis feldmanni
Dixon, P.S., 1963
Studies on marine Fungi. I. The occurrence of Olpidiopsis feldmannii in the British Isles.
Nova Hedwigia, 5(1/2):341-346.

Oneadia ascophylli
Meyers, Samuel P., 1957.
Taxonomy of marine pyrenomycetes.
Mycologia 49(4):475-528.

Oneadia pelvetiana
Meyers, Samuel P., 1957.
Taxonomy of marine pyrenomycetes.
Mycologia 49(4):475-528.

Ophiobolus laminariae
Meyers, Samuel P., 1957.
Taxonomy of marine pyrenomycetes.
Mycologia 49(4):475-528.

Ophiobolus maritimus
Meyers, Samuel P., 1957.
Taxonomy of marine pyrenomycetes.
Mycologia 49(4):475-528.

Ophiobolus salina nom. nov.
Meyers, Samuel P., 1957.
Taxonomy of marine pyrenomycetes.
Mycologia 49(4):475-528.

Peritrichospora Crustata
Kohlmeyer, Jan, 1962.
Haophile Pilze von den Ufern Frankreichs.
Nova Hedwigia, 4 (3/4):389-420.

Peritrichospora intergra
Jones, E.B. Gareth, 1962
Marine fungi.
Trans. British Mycolog. Soc., 45(1):93-114.

Peritrichospora integra
Kohlmeyer, Jan, 1962.
Haophile Pilze von den Ufern Frankreichs.
Nova Hedwigia, 4 (3/4):389-420.

Peritrichospora integra
Meyers, Samuel P., 1957.
Taxonomy of marine pyrenomycetes.
Mycologia 49(4):475-528.

Peritrichospora lacera
Meyers, Samuel P., 1957.
Taxonomy of marine pyrenomycetes.
Mycologia 49(4):475-528.

Peritrichospora trifurcata
Kohlmeyer, Jan, 1962.
Haophile Pilze von den Ufern Frankreichs.
Nova Hedwigia, 4(3/4):389-420.

Phoma sp.
Jones, E.B. Gareth, 1962
Marine fungi.
Trans. British Mycolog. Soc., 45(1):93-114.

Phyllachorella oceanica
Meyers, Samuel P., 1957.
Taxonomy of marine pyrenomycetes.
Mycologia 49(4):475-528.

Physalospora sp.
Meyers, Samuel P., 1957.
Taxonomy of marine Pyrenomycetes.
Mycologia 49(4):475-528.

Piricauda pelagica
Kohlmeyer, J., 1963.
Fungi marini novi vel critici.
Nova Hedwigia, 6:207-329.

Piricauda pelagica
Kohlmeyer, Jan, 1962.
Haophile Pilze von den Ufern Frankreichs.
Nova Hedwigia, 4(3/4):389-420.

Placostroma laminariae
Meyers, Samuel P., 1957.
Taxonomy of marine pyrenomycetes.
Mycologia 49(4):475-528.

Placostroma pelvetiae
Meyers, Samuel P., 1957.
Taxonomy of marine pyrenomycetes.
Mycologia 49(4):475-528.

Pleospora gaudefroyi
Kohlmeyer, Jan, 1962.
Über Pleospore gaudefroyi Patouillard.
Willdenowia, Mitt. Bot. Garten u. Mus., Berlin-Dahlem, 3(2):315-324.

Pleospora herbarum
Jones, E.B. Gareth, 1962
Marine fungi.
Trans. British Mycolog. Soc., 45(1):93-114.

Pleospora laminariana
Meyers, Samuel P., 1957.
Taxonomy of marine pyrenomycetes.
Mycologia 49(4):475-528.

Pleospora maritima
Meyers, Samuel P., 1957.
Taxonomy of marine pyrenomycetes.
Mycologia 49(4):475-528.

Pleospora cf. pelagica
Kohlmeyer, Jan, 1962.
Haophile Pilze von den Ufern Frankreichs.
Nova Hedwigia, 4 (3/4):389-420.

Pleospora pelagica
Meyers, Samuel P., 1957.
Taxonomy of marine pyrenomycetes.
Mycologia 49(4):475-528.

Pleospora pelvetia
Meyers, Samuel P., 1957.
Taxonomy of marine pyrenomycetes.
Mycologia 49(4):475-528.

Pontoporeia n.gen.
Kohlmeyer, J., 1963.
Zwei neue Ascomyceten-Gattungen auf Posidonia-Rhizomen.
Nova Hedwigia, 6;7-13.

Pontoporeia biturbinata
Kohlmeyer, J., 1963.
Zwei neue Ascomyceten-Gattungen auf Posidonia-Rhizomen.
Nova Hedwigia, 6:7-13.

Pullularia pullulans
Siepmann, Rolf, and Willy Höhnk, 1962.
Über Hefen und einige Pilze (Fungi imp., Hyphales) aus den Nordatlantik.
Veröff. Inst. Meeresf., Bremerhaven, 8(1):79-97

Prymnesium parvum
Doig, M.T. III, and D.F. Martin 1973.
Anticoagulant properties of a red tide toxin.
Toxicon 11 (4): 351-355.

Prymnesium parvum
Valkanov, Alexander, 1964
Untersuchungen uber Prymnesium parvum Carter und seine toxische Einwirkung auf die Wasserorganismen.
Kieler Meeresforsch., 20(1):65-81.

Pythium sp.
Fuller, Melvin S., Billy Lewis and Philip Cook, 1966.
Occurence of Pythium sp. on the marine alga Porphyia.
Mycologia, 58(2):313-318.

Remispora cucullata n.sp.
Kohlmeyer, Jan, 1964.
A new marine ascomycete from wood.
Mycologia, 56(5):770-774.

Remispora hamata
Kohlmeyer, J., 1963.
Fungi marini novi vel critici.
Nova Hedwigia, 6:297-329.

Remispora hamata
Kohlmeyer, Jan, 1962.
Haophile Pilze von den Ufern Frankreichs.
Nova Hedwigia, 4 (3/4):389-420.

Remispora maritima
Kohlmeyer, J., 1963.
Fungi marini novi vel critici.
Nova Hedwigia, 6:297-329.

Remispora maritima
Kohlmeyer, Jan, 1962.
Haophile Pilze von den Ufern Frankreichs.
Nova Hedwigia, 4 (3/4):389-420.

Remispora maritima
Meyers, Samuel P., 1957.
Taxonomy of marine pyrenomycetes.
Mycologia 49(4):475-528.

Remispora pilleata n.sp.
Kohlmeyer, J., 1963.
Fungi marini novi vel critici.
Nova Hedwigia, 6:297-329.

Remispora quadri-remis
Kohlmeyer, J., 1963.
Fungi marini novi vel critici.
Nova Hedwigia, 6:297-329.

Remispora quadriremis
Kohlmeyer, Jan, 1962.
Haophile Pilze von den Ufern Frankreichs.
Nova Hedwigia, 4 (3/4):389-420.

Remispora trullifer n.sp.
Kohlmeyer, J., 1963. Spelling?
Fungi marini novi vel critici.
Nova Hedwigia, 6:297-329.

Rhodotorula glutinis
Siepmann, Rolf, and Willy Höhnk, 1962.
Über Hefen und einige Pilze (Fungi imp., Hyphales) aus den Nordatlantik.
Veröff. Inst. Meeresf., Bremerhaven, 8(1):79-97.

Rhodotorula mucilaginosa
Siepmann, Rolf, and Willy Höhnk, 1962.
Über Hefen und einige Pilze (Fungi imp., Hyphales) aus den Nordatlantik.
Veröff. Inst. Meeresf., Bremerhaven, 8(1):79-97.

Rhodotorula rubra
Siepmann, Rolf, and Willy Höhnk, 1962.
Über Hefen und einige Pilze (Fungi imp., Hyphales) aus den Nordatlantik.
Veröff. Inst. Meeresf., Bremerhaven, 8(1):79-97.

Rhodotorula texensis
Siepmann, Rolf, and Willy Höhnk, 1962.
Über Hefen und einige Pilze (Fungi imp., Hyphales) aus den Nordatlantik.
Veröff. Inst. Meeresf., Bremerhaven, 8(1):79-97.

Rosellinia laminariana
Meyers, Samuel P., 1957.
Taxonomy of marine pyrenomycetes.
Mycologia 49(4):475-528.

Saccharomyces aestuarii n.sp.
Fell, Jack W., 1961
A new species of Saccharomyces isolated from a subtropical estuary.
Antonie Van Leeuwenhoek, 27:27-30.

Samarosporella pelagica
Meyers, Samuel P., 1957.
Taxonomy of marine pyrenomycetes.
Mycologia 49(4):475-528.

Saprolegnia parasitica
Harrison, J.L., and E.B. Gareth Jones 1971.
Salinity tolerance of Saprolegnia parasitica Coker.
Mycopathologia Mycol. appl. 43(3/4):297-307.

Sphaerulina amicta n.sp.
Kohlmeyer, Jan, 1962.
Haophile Pilze von den Ufern Frankreichs.
Nova Hedwigia, 4 (3/4):389-420.

Sphaerulina orae-maris
Jones, E.B. Gareth, 1963
Marine Fungi. II. Ascomycetes and Deuteromycetes from submerged wood and drift Spartina.
Trans. Brit. Mycol. Soc., 46(1):135-144.

Sphaerulina orae-maris
Meyers, Samuel P., 1957.
Taxonomy of marine pyrenomycetes.
Mycologia 49(4):475-528.

Sphaerulina pedicellata
Jones, E.B. Gareth, 1962
Marine fungi.
Trans. British Mycolog. Soc., 45(1):93-114.

Sphaerulina pedicellata
Meyers, Samuel P., 1957.
Taxonomy of marine pyrenomycetes.
Mycologia 49(4):475-528.

Spathulospora antarctica n.sp. [a]
Kohlmeyer, Jan 1973.
Spathulosporales, a new order and possible missing link between Laboulbeniales and Pyrenomycetes.
Mycologia 65(3):614-647.

Spathulospora calva n.sp. [a]
Kohlmeyer, Jan 1973.
Spathulosporales, a new order and possible missing link between Laboulbeniales and Pyrenomycetes.
Mycologia 65(3):614-647.

Spathulosporia lanata n.sp. [a]
Kohlmeyer, Jan 1973.
Spathulosporales, a new order and possible missing link between Laboulbeniales and Pyrenomycetes.
Mycologia 65(3):614-647.

Spathulospora phycophila [a]
Kohlmeyer, Jan 1973.
Spathulosporales, a new order and possible missing link between Laboulbeniales and Pyrenomycetes.
Mycologia 65(3):614-647.

Sphaceloma cecidii n.sp. [a]
Kohlmeyer, Jan 1972.
Parasitic Haloguignardia oceanica (Ascomycetes) and hyperparasitic Sphaceloma cecidii sp. nov. (Deuteromycetes) in drift Sargassum in North Carolina.
J. Elisha Mitchell scient. Soc. 88(4):255-259.

Speira pelagica
Kohlmeyer, J., 1963.
Fungi marini novi vel critici.
Nova Hedwigia, 6:297-329.

Speira pelagica
Kohlmeyer, Jan, 1962.
Haophile Pilze von den Ufern Frankreichs.
Nova Hedwigai, 4(3/4):389-420.

Sphaerulina amicta
Kohlmeyer, J., 1963.
Fungi marini novi vel critici.
Nova Hedwigia, 6:297-329.

Sporidesmium salinum n.sp.
Jones, E.B. Gareth, 1963
Marine Fungi. II. Ascomycetes and Deuteromycetes from submerged wood and drift Spartina.
Trans. Brit. Mycol. Soc., 46(1):135-144.

Stemphylium maritimum
Jones, E.B. Gareth, 1962
Marine fungi.
Trans. British Mycolog. Soc., 45(1):93-114.

Sterigmatomyces n.gen.
Fell, J.W., 1966.
Sterigmatomyces, a new fungal genus from marine areas.
Antonie Van Leeuwenhoek, 32(1):99-104. (not seen)

Thraustochytrium roseum
Siegenthaler, Paul Andre, Melvin M. Belsky and Solomon Goldstein, 1967.
Phosphate uptake in an obligately marine fungus: a specific requirement for sodium.
Science, 155(3758):93-94.

Thraustochytrium striatum
Schneider, Joachim, 1967.
Ein neuer mariner Phycomycet auf der Kieler Bucht (Thraustochytrium striatum spec. nov.).
Kieler Meeresforsch., 23(1):16-20.

Torpedospora radiata
Kohlmeyer, J., 1963.
Fungi marini novi vel critici.
Nova Hedwigia, 6:297-329.

Torpedospora radiata n. gen. n. sp.
Meyers, Samuel P., 1957.
Taxonomy of marine pyrenomycetes.
Mycologia 49(4):475-528.

Torulopsis candida
Siepmann, Rolf, and Willy Höhnk, 1962.
Über Hefen und einige Pilze (Fungi imp., Hyphales) aus den Nordatlantik.
Veröff. Inst. Meeresf., Bremerhaven, 8(1):79-97.

Trailia ascophylli
Meyers, Samuel P., 1957.
Taxonomy of marine pyrenomycetes.
Mycologia 49(4):475-528.

Trichosporon atlanticum,
Siepmann, Rolf, and Willy Höhnk, 1962. n.sp.
Über Hefen und einige Pilze (Fungi imp., Hyphales) aus den Nordatlantik.
Veröff. Inst. Meeresf., Bremerhaven, 8(1):79-97.

Trichosporon cutaneum
Siepmann, Rolf, and Willy Höhnk, 1962.
Über Hefen und einige Pilze (Fungi imp., Hyphales) aus den Nordatlantik.
Veröff. Inst. Meeresf., Bremerhaven, 8(1):79-97.

Trichosporon maritimum
Siepmann, Rolf, and Willy Höhnk, 1962. n.sp.
Über Hefen und einige Pilze (Fungi imp., Hyphales) aus den Nordatlantik.
Veröff. Inst. Meeresf., Bremerhaven, 8(1):79-97.

Turgidosculum n.gen. [a]
Kohlmeyer, Jan, and Erika Kohlmeyer 1972.
A new genus of marine Ascomycetes on Ulva vexata Setch. et Gard.
Bot. Jahrb. Syst. 92 (2/3): 429-432.

Varicosporina ramulosa
Meyers, Samuel P., and Jan J. Kohlmeyer, 1965.
Varicosporina ramulosa fen. nov. sp. nov., an aquatic hyphomycete from marine areas.
Canadian J. Bot., 43:915-921.

Zignoella calospora
Meyers, Samuel P., 1957.
Taxonomy of marine pyrenomycetes.
Mycologia 49(4):475-528.

Zignoella cubensis
Meyers, Samuel P., 1957.
Taxonomy of marine pyrenomycetes.
Mycologia 49(4):475-528.

Zignoella enormis
Meyers, Samuel P., 1957.
Taxonomy of marine pyrenomycetes.
Mycologia 49(4):475-528.